# PLATELETS

## Second Edition

# PLATELETS
## Second Edition

*Editor*

**Alan D. Michelson, M.D.**

*Director, Center for Platelet Function Studies*
*Professor of Pediatrics, Medicine, and Pathology*
*University of Massachusetts Medical School*
*Worcester, Massachusetts*

 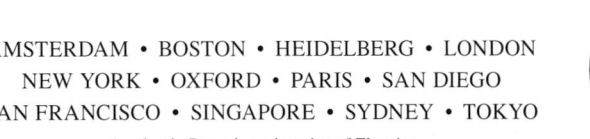 

ELSEVIER

AMSTERDAM • BOSTON • HEIDELBERG • LONDON
NEW YORK • OXFORD • PARIS • SAN DIEGO
SAN FRANCISCO • SINGAPORE • SYDNEY • TOKYO
Academic Press is an imprint of Elsevier

**Cover image:** *Platelet Shape Change and Aggregation.* Scanning electron micrographs of resting (lower left), partially activated (lower middle), and fully activated platelets (upper right), showing the accompanying shape changes, formation of filopodia and lamellipodia, and platelet aggregation. Image generously provided by John W. Weisel, Chandrasekaran Nagaswami, and Rustem I. Litvinov, Department of Cell and Developmental Biology, University of Pennsylvania School of Medicine, Philadelphia, PA, U.S.A.

**Academic Press is an imprint of Elsevier**
30 Corporate Drive, Suite 400, Burlington, MA 01803, USA
525 B Street, Suite 1900, San Diego, California 92101-4495, USA
84 Theobald's Road, London WC1X 8RR, UK

**This book is printed on acid-free paper.**

*Library of Congress Cataloging-in-Publication Data*

Application Submitted

*British Library Cataloguing-in-Publication Data*
A catalogue record for this book is available from the British Library.

ISBN 13: 978-0-12-369367-9
ISBN 10: 0-12-369367-5

For information on all Academic Press publications
visit our Web site at www.books.elsevier.com

Printed in Canada
06  07  08  09  10  9  8  7  6  5  4  3  2  1

Working together to grow
libraries in developing countries

www.elsevier.com  |  www.bookaid.org  |  www.sabre.org

ELSEVIER     BOOK AID International     Sabre Foundation

To Lee Ann, Daniel, and Sarah

# Contents

## Part One

## Platelet Biology

*Part Two*

# Tests of Platelet Function

*Part Three*

# The Role of Platelets in Disease

*Part Four*

# Disorders of Platelet Number and Function

# Contributors

**Vahid Afshar-Kharghan, MD**
Assistant Professor of Medicine
Thrombosis Research Section
Baylor College of Medicine
Houston, Texas

**Ramtin Agah, MD**
Assistant Professor of Medicine
Division of Cardiology
Program in Human Molecular Biology & Genetics
University of Utah
Salt Lake City, Utah

**Robert K. Andrews, PhD**
Senior Research Fellow
Monash University
Melbourne, Australia

**Richard H. Aster, MD**
Senior Investigator, Blood Research Institute, BloodCenter
   of Wisconsin
Professor of Medicine, Medical College of Wisconsin
Milwaukee, Wisconsin

**Ben Atkinson, PhD**
Fellow in Medicine
Harvard Medical School
Boston, Massachusetts

**Eric H. Awtry, MD, FACC**
Director of Education, Division of Cardiology
Boston Medical Center
Assistant Professor of Medicine
Boston University School of Medicine
Boston, Massachusetts

**Wadie F. Bahou, MD**
Professor of Medicine and Program in Genetics
Vice Dean for Scientific Affairs
Stony Brook University School of Medicine
Stony Brook, New York

**Marc R. Barnard, MS**
Laboratory Supervisor, Center for Platelet Function
   Studies
Department of Pediatrics
University of Massachusetts Medical School
Worcester, Massachusetts

**Anthony A. Bavry, MD**
Interventional Cardiology Fellow
Department of Cardiovascular Medicine
Cleveland Clinic Foundation
Cleveland, Ohio

**Arnold S. Bayer, MD**
Professor of Medicine
David Geffen School of Medicine at UCLA
Associate Chief, Division of Infectious Diseases
LAC-Harbor UCLA Medical Center
Los Angeles Biomedical Research Institute
Torrance, California

**Richard C. Becker, Jr., MD**
Professor of Medicine
Divisions of Cardiology and Hematology
Duke University School of Medicine
Durham, North Carolina

**Wolfgang Bergmeier, PhD**
Junior Investigator, CBR Institute for Biomedical
   Research
Instructor in Pathology, Harvard Medical School
Boston, Massachusetts

**Michael C. Berndt, PhD**
Head and Professor of Immunology
Deputy Dean of Research
Faculty of Medicine, Nursing and Health Sciences
Monash University
Clayton, Victoria, Australia

**Deepak L. Bhatt, MD, FACC, FSCAI, FESC, FACP**
Associate Director, Cleveland Clinic Cardiovascular
  Coordinating Center
Staff, Cardiac, Peripheral, and Carotid Intervention
Associate Professor of Medicine
Department of Cardiovascular Medicine
Cleveland Clinic Foundation
Cleveland, Ohio

**Nicola Bizzaro, MD**
Director, Department of Clinical Pathology
St. Antonio Hospital
Tolmezzo, Italy

**Morris A. Blajchman, MD, FRCP(C)**
Head, Transfusion Medicine Services, McMaster/
  Hamilton Regional Laboratory Medicine Program
Professor of Medicine and Pathology, McMaster
  University
Medical Director, Hamilton Centre, Canadian Blood
  Services
Hamilton, Ontario, Canada

**Beth A. Bouchard, PhD**
Research Assistant Professor of Biochemistry
University of Vermont College of Medicine
Burlington, Vermont

**Lawrence F. Brass, MD, PhD**
Professor of Medicine, Pathology and Pharmacology
Vice Chair for Research, Department of Medicine
Associate Dean and Director, Combined Degree and
  Physician Scholars Program
University of Pennsylvania School of Medicine
Philadelphia, Pennsylvania

**Paul F. Bray, MD**
Professor of Medicine
Chief, Thrombosis Research Section
Baylor College of Medicine
Houston, Texas

**Carol Briggs, BSc, FIBMS**
Department of Haematology Evaluations
University College London Hospital
London, United Kingdom

**Alexander Brill, MD, PhD**
Coagulation Unit
Hadassah-University Medical Center
Jerusalem, Israel

**James B. Bussel, MD**
Professor of Pediatrics, Obstetrics and Gynecology
Director, Platelet Disorders Center
Weill Medical College of Cornell University
New York Presbyterian Hospital
New York, New York

**Saulius Butenas, PhD**
Research Associate Professor of Biochemistry
University of Vermont College of Medicine
Burlington, Vermont

**Marco Cattaneo, MD**
Director, Unità di Ematologia e Trombosi
Professor of Internal Medicine
Ospedale San Paolo
Università di Milano
Milan, Italy

**Beng H. Chong, MBBS, PhD**
Head, Department of Medicine
Professor of Medicine
St. George Clinical School
University of New South Wales
Sydney, Australia

**Kenneth J. Clemetson, PhD, CChem, FRSC**
Professor of Biochemistry
Theodor Kocher Institute
University of Berne
Berne, Switzerland

**Jeannine M. Clemetson, PhD**
Theodor Kocher Institute
University of Berne
Berne, Switzerland

**Barry S. Coller, MD**
David Rockefeller Professor of Medicine
Head, Laboratory of Blood and Vascular Biology
Physician-in-Chief, Rockefeller University Hospital
Vice President for Medical Affairs
Rockefeller University
New York, New York

**Lawrence E. Crawford, MD**
Assistant Professor of Medicine
Division of Cardiology
Duke University Medical Center
Durham, North Carolina

**Philip G. de Groot, PhD**
Department of Clinical Chemistry and Haematology
Professor of Biochemistry
University Medical Centre
Utrecht, The Netherlands

**Gregory J. del Zoppo, MD**
Associate Professor, Department of Molecular and
  Experimental Medicine
The Scripps Research Institute
Member, Division of Hematology/Medical Oncology
Scripps Clinic
La Jolla, California

**Christophe Dubois, PhD**
Fellow in Medicine
Harvard Medical School
Boston, Massachusetts

**Wolfgang G. Eisert, MD, PhD**
Boehringer Ingelheim GmbH
Ingelheim, Germany
Professor of Biophysics
Director, Center for Thrombosis and Atherosclerosis
  Research
University of Hannover
Hannover, Germany

**Garret A. FitzGerald, MD**
Director, Institute for Translational Medicine and
  Therapeutics
Chair, Department of Pharmacology
Professor of Medicine and Pharmacology
University of Pennsylvania School of Medicine
Philadelphia, Pennsylvania

**John L. Francis, PhD**
Director, Institute of Translational Research
Director, Center for Hemostasis and Thrombosis
Florida Hospital
Orlando, Florida

**Jane E. Freedman, MD**
Associate Professor of Medicine and Pharmacology
Boston University School of Medicine
Boston, Massachusetts

**John Freedman, MD**
Director, Transfusion Medicine
St Michael's Hospital
Professor of Medicine, Laboratory Medicine, and
  Pathobiology
University of Toronto
Toronto, Canada

**A. L. Frelinger, III, PhD**
Associate Director, Center for Platelet Function Studies
Associate Professor of Pediatrics and Molecular Genetics
  & Microbiology
University of Massachusetts Medical School
Worcester, Massachusetts

**Susanne Fries, MD**
Research Associate
Institute for Translational Medicine and Therapeutics
University of Pennsylvania School of Medicine
Philadelphia, Pennsylvania

**Barbara C. Furie, PhD**
Professor of Medicine
Harvard Medical School
Boston, Massachusetts

**Bruce Furie MD**
Professor of Medicine
Harvard Medical School
Boston, Massachusetts

**Mark I. Furman, MD**
Director, Interventional Cardiology
UMass Memorial Health Care
Associate Director, Center for Platelet Function Studies
Associate Professor of Medicine, Pediatrics, and Cell
  Biology
University of Massachusetts Medical School
Worcester, Massachusetts

**Ángel García-Alonso, PhD**
RIAIDT — Edificio CACTUS
Universidade de Santiago de Compostela
Campus Universitario Sur
Santiago de Compostela, Spain

**Pascal J. Goldschmidt, MD**
Senior Vice President for Medical Affairs and Dean
Miller School of Medicine
University of Miami
Miami, Florida

**Tilo Grosser, MD**
Research Assistant Professor
Institute for Translational Medicine and Therapeutics
University of Pennsylvania School of Medicine
Philadelphia, Pennsylvania

**George N. M. Gurguis, MB, BCh**
Staff Physician, VA North Texas Health Care Systems
Associate Professor of Psychiatry
University of Texas Southwestern Medical School
Dallas, Texas

**Paul Harrison, PhD**
Clinical Scientist
Oxford Haemophilia Centre and Thrombosis Unit
Churchill Hospital
Oxford, United Kingdom

**John H. Hartwig, PhD**
Professor of Medicine
Division of Hematology
Brigham and Women's Hospital
Harvard Medical School
Boston, Massachusetts

**Yasuo Ikeda, MD**
Dean Professor,
Division of Hematology, Department of Internal Medicine
Keio University School of Medicine
Tokyo, Japan

**Sara J. Israels, MD, FRCPC**
Professor
Department of Pediatrics and Child Health
University of Manitoba
Winnipeg, Manitoba, Canada

**Joseph E. Italiano Jr., PhD**
Assistant Professor of Medicine
Hematology Division
Brigham and Women's Hospital
Harvard Medical School
Boston, Massachusetts

**Lisa K. Jennings, PhD**
Professor, Department of Medicine
Director, Vascular Biology Center of Excellence
Director, TN-AR-MS Clinical Research Consortium
University of Tennessee Health Science Center
Memphis, Tennessee

**Cécile Kaplan, MD**
Director, Platelet Immunology Laboratory
National Institute of Blood Transfusion
Paris, France

**Simon Karpatkin, MD**
Professor of Medicine
Director of Hematology
New York University School of Medicine
New York, New York

**David M. Keeling, MD**
Consultant Haematologist
Oxford Haemophilia Centre and Thrombosis Unit
Churchill Hospital
Oxford, United Kingdom

**Yukio Kimura, PhD**
Senior Clinical Researcher
Clinical Planning & Control Department
Pharmaceutical Development Department
Pharmaceutical Division
Kowa Co. Ltd.
Tokyo, Japan

**Carla D. Kurkjian, MD**
Fellow, Hematology-Oncology Section
University of Oklahoma Health Sciences Center
Oklahoma City, Oklahoma

**David J. Kuter, MD, DPhil**
Chief of Hematology, Massachusetts General Hospital
Professor of Medicine, Harvard Medical School
Boston, Massachusetts

**Michele P. Lambert, MD**
Clinical Associate in Pediatrics
Division of Pediatric Hematology
University of Pennsylvania School of Medicine
Philadelphia, Pennsylvania

**David H. Lee, MD**
Associate Professor, Departments of Medicine, Pathology
  and Molecular Medicine
Queen's University
Kingston, Ontario, Canada

**Jack Levin, MD**
Professor of Laboratory Medicine
Professor of Medicine
University of California School of Medicine
San Francisco, California

**Qiao-Xin Li, PhD**
Senior Research Fellow
Department of Pathology
University of Melbourne
Parkville, Victoria, Australia

**Zongdong Li, PhD**
Assistant Professor of Medicine
New York University School of Medicine
New York, New York

**Stuart E. Lind, MD**
Professor of Medicine
University of Colorado Health Sciences Center
Aurora, Colorado

**Matthew D. Linden, PhD**
Instructor in Pediatrics
Center for Platelet Function Studies
University of Massachusetts Medical School
Worcester, Massachusetts

**Neuza H. M. Lopes, MD**
Research Associate in Medicine
Division of Cardiology
Duke University Medical Center
Durham, North Carolina

**José A. López, MD**
Executive Vice-President for Research
Puget Sound Blood Center
Professor of Medicine and Biochemistry
University of Washington
Seattle, Washington

**Joseph Loscalzo, MD, PhD**
Chairman, Department of Medicine, Brigham and
    Women's Hospital
Hersey Professor of the Theory and Practice of Medicine,
    Harvard Medical School
Boston, Massachusetts

**Yan-qing Ma, PhD**
Postdoctoral Fellow
Department of Molecular Cardiology
Lerner Research Institute
Cleveland Clinic
Cleveland, Ohio

**Samuel J. Machin, FRCPath, FRCP**
Professor of Haematology
University College London
London, United Kingdom

**Kenneth G. Mann, PhD**
Professor of Biochemistry
University of Vermont College of Medicine
Burlington, Vermont

**Pier Mannuccio Mannucci, MD**
Professor and Chairman of Medicine
Director, Angelo Bianchi Bonomi Hemophilia and
    Thrombosis Center
University of Milan
Milan, Italy

**Bradley A. Maron, MD**
Clinical Fellow in Medicine
Harvard Medical School
Boston, Massachusetts

**Colin L. Masters, MD**
Professor of Pathology
University of Melbourne
Mental Health Research Institute of Victoria
Parkville, Victoria, Australia

**Keith R. McCrae, MD**
Professor of Medicine and Pathology
Case Western Reserve University School of Medicine
Cleveland, Ohio

**Rodger P. McEver, MD**
Vice President of Research
Eli Lilly Distinguished Chair in Biomedical Research
Oklahoma Medical Research Foundation
Oklahoma City, Oklahoma

**Barbara Menart, PhD**
German Diabetes Clinic
German Diabetes Center
Leibniz Institute at the Heinrich Heine University
Düsseldorf, Germany

**Alan D. Michelson, MD**
Director, Center for Platelet Function Studies
Professor of Pediatrics, Medicine, and Pathology
University of Massachusetts Medical School
Worcester, Massachusetts

**Joel Moake, MD**
Professor of Medicine, Baylor College of Medicine
Associate Director, Biomedical Engineering Laboratory,
    Rice University
Houston, Texas

**Neil Murray, MD**
Senior Lecturer and Honorary Consultant in Neonatal
    Medicine
Hammersmith Hospital
Imperial College London
London, United Kingdom

**Michael A. Nardi, BS, MS**
Associate Professor of Pediatrics
New York University School of Medicine
New York, New York

**Debra K. Newman, PhD**
Associate Investigator
Blood Research Institute
BloodCenter of Wisconsin
Associate Professor, Department of Microbiology
Medical College of Wisconsin
Milwaukee, Wisconsin

**Peter J. Newman, PhD**
Vice President for Research
Walter A. Schroeder Associate Director, Blood Research
    Institute
BloodCenter of Wisconsin
Professor, Department of Pharmacology and Cellular
    Biology
Medical College of Wisconsin
Milwaukee, Wisconsin

**Mary Lynn Nierodzik, MD**
Associate Director, Bellevue Hospital Cancer Center
Assistant Professor of Medicine
New York University School of Medicine
New York, New York

**Rienk Nieuwland, PhD**
Head, Laboratory of Experimental Clinical Chemistry
Academic Medical Center
University of Amsterdam
Amsterdam, The Netherlands

**Melanie Novinska, PhD**
Research Fellow
Blood Research Institute
BloodCenter of Wisconsin
Milwaukee, Wisconsin

**Alan T. Nurden, PhD**
Research Director, CNRS
Centre de Référence des Pathologies Plaquettaires
Plateforme Technologique et d'Innovation Biomédicale
Hôpital Xavier Arnozan
Pessac, France

**Paquita Nurden, MD, PhD**
Coordinateur
Centre de Référence des Pathologies Plaquettaires
Hôpital Cardiologique
Pessac, France

**Peter L. Perrotta, MD**
Associate Professor of Pathology
West Virginia University
Morgantown, West Virginia

**Michelle M. Pesho, PhD**
Postdoctoral Fellow
Department of Molecular Cardiology
Lerner Research Institute
Cleveland Clinic
Cleveland, Ohio

**Edward F. Plow, PhD**
Chair, Department of Molecular Cardiology
Head of Research, Joseph J. Jacobs Center for Thrombosis
    and Vascular Biology
Cleveland Clinic
Cleveland, Ohio

**Mortimer Poncz, MD**
Professor of Pediatrics
University of Pennsylvania School of Medicine
Philadelphia, Pennsylvania

**Man-Chiu Poon, MD, MSc, FRCP(C), FACP**
Professor of Medicine, Pediatrics and Oncology
Director, Southern Alberta Hemophilia Clinic
University of Calgary and Calgary Health Region
Calgary, Alberta, Canada

**Nicholas Prévost, PhD**
Post-Doctoral Fellow
Division of Hematology-Oncology
Department of Medicine
University of California San Diego
La Jolla, California

**A. Koneti Rao, MBBS**
Professor of Medicine, Thrombosis Research and
    Pharmacology
Director, Sol Sherry Thrombosis Research Center
Chief, Hematology Division
Temple University School of Medicine
Philadelphia, Pennsylvania

**Vipul Rathore, PhD**
Research Scientist
Blood Research Institute
BloodCenter of Wisconsin
Milwaukee, Wisconsin

**Guy L. Reed, MD**
GRA Eminent Scholar
Kupperman Professor & Chair
Cardiovascular Medicine
Medical College of Georgia
Augusta, Georgia

**Sybille Rex, PhD**
Research Associate
Department of Medicine
Boston University School of Medicine
Boston, Massachusetts

**Christine S. Rinder, MD**
Associate Professor of Anesthesiology and Laboratory
    Medicine
Yale University School of Medicine
New Haven, Connecticut

**Henry M. Rinder, MD**
Associate Professor of Laboratory Medicine and Internal
    Medicine (Hematology)
Yale University School of Medicine
Director, Hematology Laboratories
Yale-New Haven Hospital
New Haven, Connecticut

**Irene Roberts, MD**
Professor of Paediatric Haematology
Hammersmith and St. Mary's Hospitals
Imperial College London
London, United Kingdom

**Zaverio M. Ruggeri, MD**
Professor, Department of Molecular and Experimental
    Medicine
Head, Division of Experimental Hemostasis and
    Thrombosis
Director, Roon Research Center for Arteriosclerosis and
    Thrombosis
The Scripps Research Institute
La Jolla, California

**Brian Savage, PhD**
Staff Scientist, Department of Molecular and
    Experimental Medicine
Division of Experimental Hemostasis and Thrombosis
Roon Research Center for Arteriosclerosis and
    Thrombosis
The Scripps Research Institute
La Jolla, California

**Naphtali Savion, PhD**
Professor of Clinical Biochemistry
Goldschleger Eye Research Institute
Sackler Faculty of Medicine, Tel Aviv University
Sheba Medical Center
Tel Hashomer, Israel

**Yotis Senis, PhD**
Centre for Cardiovascular Sciences
Division of Medical Sciences
Institute of Biomedical Research
The Medical School
University of Birmingham
Edgbaston, Birmingham, United Kingdom

**Sanford J. Shattil, MD**
Professor and Chief
Division of Hematology-Oncology
Department of Medicine
University of California San Diego
La Jolla, California

**Jan J. Sixma, MD, PhD**
Department of Clinical Chemistry and Haematology
Professor Emeritus Hematology
University Medical Centre
Utrecht, The Netherlands

**Brian R. Smith, MD**
Professor and Chair of Laboratory Medicine
Professor of Internal Medicine and Pediatrics
Yale University School of Medicine
New Haven, Connecticut

**Edward L. Snyder, MD**
Professor, Laboratory Medicine
Director, Blood Bank/Apheresis Services
Yale University School of Medicine
Yale-New Haven Hospital
New Haven, Connecticut

**Michael Sobel, MD**
Chief of Surgery, VA Puget Sound Health Care System
Professor and Vice-Chair, Department of Surgery
University of Washington School of Medicine
Seattle, Washington

**Timothy J. Stalker, PhD**
Department of Medicine
University of Pennsylvania
Philadelphia, Pennsylvania

**Steven R. Steinhubl, MD**
Director of Cardiovascular Education and Clinical
    Research
Associate Professor of Medicine
Gill Heart Institute
University of Kentucky
Lexington, Kentucky

**Bernd Stratmann, PhD**
Research Director, Diabetes Center
Heart and Diabetes Center NRW
Ruhr-University Bochum
Bad Oeynhausen, Germany

**Augeste Sturk, PhD**
Head, Department of Clinical Chemistry
Professor of Clinical Chemistry
Academic Medical Center
University of Amsterdam
Amsterdam, The Netherlands

**Toshiki Sudo, PhD**
Researcher
First Institute of New Drug Discovery
Otsuka Pharmaceutical Co. Ltd.
Tokushima, Japan

**Ayalew Tefferi, MD**
Consultant, Mayo Clinic
Professor of Medicine and Hematology
Mayo College of Medicine
Rochester, Minnesota

**Michael G. Tomlinson, PhD**
Centre for Cardiovascular Sciences
Division of Medical Sciences
Institute of Biomedical Research
The Medical School
University of Birmingham
Edgbaston, Birmingham, United Kingdom

**Eric J. Topol, MD**
Professor of Genetics
Case Western Reserve University
Cleveland, Ohio

**Paula B. Tracy, PhD**
Professor of Biochemistry
University of Vermont College of Medicine
Burlington, Vermont

**Diethelm Tschoepe, MD**
Director, Diabetes Center
Professor of Medicine
Heart and Diabetes Center NRW
Ruhr-University Bochum
Bad Oeynhausen, Germany

**David Varon, MD**
Director, Coagulation Unit
Hadassah-University Medical Center
Jerusalem, Israel

**K. Vinod Vijayan, PhD**
Assistant Professor of Medicine
Thrombosis Research Section
Baylor College of Medicine
Houston, Texas

**Denisa D. Wagner, PhD**
Senior Investigator, CBR Institute for Biomedical
    Research
Professor of Pathology, Harvard Medical School
Boston, Massachusetts

**Steve P. Watson, PhD**
British Heart Foundation Professor in Cardiovascular
    Sciences
Centre for Cardiovascular Sciences
Division of Medical Sciences
Institute of Biomedical Research
The Medical School
University of Birmingham
Edgbaston, Birmingham, United Kingdom

**Gilbert C. White, II, MD**
Executive Vice President, BloodCenter of Wisconsin
Director, Blood Research Institute
Associate Dean for Research, Medical College of
    Wisconsin
Milwaukee, Wisconsin

**James G. White, MD**
Regents Professor
Laboratory Medicine—Pathology and Pediatrics
University of Minnesota School of Medicine
Minneapolis, Minnesota

**Melanie McCabe White, BA**
Staff Scientist
Vascular Biology Center of Excellence
University of Tennessee Health Science Center
Memphis, Tennessee

**David A. Wilcox, PhD**
Associate Professor of Pediatrics
Medical College of Wisconsin
Associate Investigator, Blood Research Institute
BloodCenter of Wisconsin
Milwaukee, Wisconsin

**Donna S. Woulfe, PhD**
Assistant Professor of Medicine
Center for Translational Medicine
Thomas Jefferson University
Philadelphia, Pennsylvania

**Michael R. Yeaman, PhD**
Professor of Medicine
David Geffen School of Medicine at UCLA
Los Angeles Biomedical Research Institute
Torrance, California

**Li Zhu, MD, PhD**
Department of Medicine
University of Pennsylvania
Philadelphia, Pennsylvania

# Preface

The goal of this book is to be a comprehensive and definitive source of knowledge about platelets. *Platelets* integrates the entire field of platelet biology, pathophysiology, and clinical medicine.

Platelets are small cells of great importance in many pathophysiological processes, including thrombosis, hemorrhage, inflammation, and cancer. In addition to primary disorders of platelet number and function, platelets have a critical role in coronary artery disease (the most common cause of human death in developed countries) and in other common diseases including stroke, peripheral vascular disease, and diabetes mellitus. The intended audience for *Platelets* includes hematologists, cardiologists, stroke physicians, blood bankers, pathologists, and researchers in thrombosis and hemostasis, as well as students and fellows in these fields.

The first edition of this book was the winner of the **2002 Best Book in Medical Science Award** from the *Association of American Publishers* — a tribute to the 108 contributing authors from 13 countries. In this second edition of *Platelets,* each chapter has been completely updated and 14 new chapters have been added. The authors of each of the 71 chapters are world leaders in their fields. As in the first edition, the book is organized into eight parts:

Part One: Platelet Biology

Part Two: Tests of Platelet Function

Part Three: The Role of Platelets in Disease

Part Four: Disorders of Platelet Number and Function

Part Five: Pharmacology: Antiplatelet Therapy

Part Six: Pharmacology: Therapy to Increase Platelet Numbers and/or Function

Part Seven: Platelet Transfusion Medicine

Part Eight: Gene Therapy for Platelet Disorders

---

I thank Peter Newburger and Marianne Felice for their endorsement of a platelet-friendly work environment. I also thank my wonderful colleagues at the Center for Platelet Function Studies — Larry Frelinger, Mark Furman, Marc Barnard, Matt Linden, Marsha Fox, and Youfu Li — without whom this book could not have been created.

*Alan D. Michelson*

A companion web site for this book is available at:
http://books.elsevier.com/companions/0123693675

# Foreword

# A Brief History of Ideas about Platelets in Health and Disease

**Barry S. Coller**

*Laboratory of Blood and Vascular Biology, Rockefeller University, New York, New York*

Credible claims have been made by or on behalf of a number of different scientists to be recognized as the first to describe the blood platelet.[1] For example, Osler[2] undoubtedly described platelets in his 1873 and 1874 papers (Fig. 1),[1,2] recognizing their disklike structure, and that while they circulated singly in the blood, they rapidly formed aggregates when removed from the blood vessel. Osler was not certain, however, whether the platelets were normal blood elements or exogenous "organisms."[1,2] Seven years later, the elegant and compelling studies of Bizzozero[3,4] in 1881 and 1882 not only identified the platelet anatomically, but also assigned it roles in both hemostasis and experimental thrombosis. Thus, using intravital microscopy of mesenteric venules in guinea pigs, Bizzozero also observed that platelets are disk shaped and circulate in isolation, but he went on to demonstrate that they adhere to the blood vessel wall at sites of injury and then form aggregates. He also observed that leukocytes were recruited into the platelet aggregates, thus also offering the first description of platelet–leukocyte interactions (Fig. 2). Using human biopsy materials, Osler later built on Bizzozero's observations, and in studies conducted from 1881 to 1886 clearly established the role of blood platelets (which he termed *blood plaques*) in human thrombotic disease, identifying them in white thrombi forming on "atheromatous ulcers" and vegetations on heart valves and in aortic aneurysms (Fig. 3).[1,5]

Ironically, 12 years before his publications on platelets, Bizzozero[6] was also the first to identify bone marrow megakaryocytes (Fig. 4),[7] but he never recognized them as the precursors of platelets. That discovery was made by Wright[8] in 1906, who noted the similarities in shape and color of the red-to-violet granules in platelets and megakaryocytes using his new polychrome staining solution (Wright's stain).

The intervening 125 years since Bizzozero's pioneering studies have seen a remarkable increase in our understanding of hemostasis and thrombosis, and the important roles that platelets play in both these phenomena. Thus, it is timely that Dr. Alan Michelson, who himself has made many important discoveries regarding platelet function and pioneered the innovative application of flow cytometry to the study of platelets, has once again brought together contributions from the leaders in this discipline to provide a comprehensive review of the topic in this second edition of *Platelets*. In this Foreword, I offer a brief and highly selective historical analysis of platelet topics in the hope of providing both a sense of the sweep of intellectual development in the field and a context for understanding our current concepts and likely future directions. Those interested in a more systematic history of the discovery of the platelet and its functions can consult a number of excellent references.[1,9–16] In addition, the regular Historical Sketch section of the *Journal of Thrombosis and Haemostasis* provides extremely valuable first-hand accounts of discoveries related to platelets by prominent members of the platelet research community.[17–24]

## I. The Role of Platelets in Human Hemostasis, Platelet Transfusion, and Bleeding Time

Among the first, and still most compelling, evidence that platelets are crucial for human hemostasis, and that platelet transfusions can restore hemostatic competence to individuals with low platelet counts, are the data from a single patient reported by Duke, a student of Wright, in 1910 (Fig. 5).[14,25,26] A 20-year-old Armenian man was admitted to Massachusetts General Hospital on May 8, 1909, with a platelet count of approximately 6000 platelets/μL and profound mucocutaneous bleeding. On May 11 he developed uncontrollable epistaxis that left him nearly moribund. In desperation, a donor was selected from among his young Armenian friends and a direct blood transfusion was arranged. A "large" amount of blood was transfused, as judged by the increase in the pulse of the donor as well as

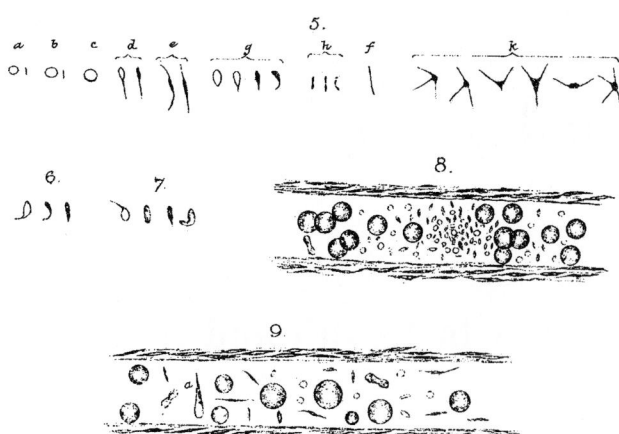

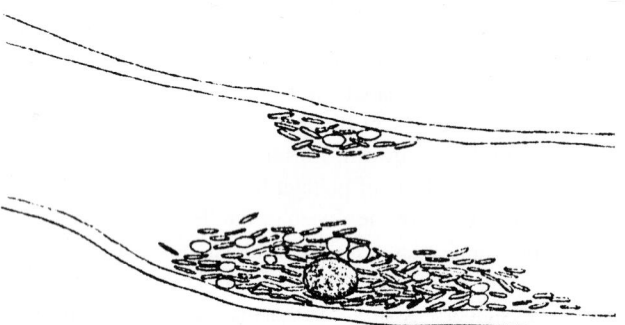

**Figure 1.** Drawings by Osler (1874)[1,2] of "corpuscles" from rat blood (5–7) and within a vein in the loose connective tissue of a rat (8 and 9). (From Osler,[2] with permission.)

**Figure 2.** Drawing by Bizzozero (1882)[4] of an intravital microscopic image of a platelet and leukocyte deposition on a damaged blood vessel in guinea pig mesentery. (From Bizzozero,[4] with permission.)

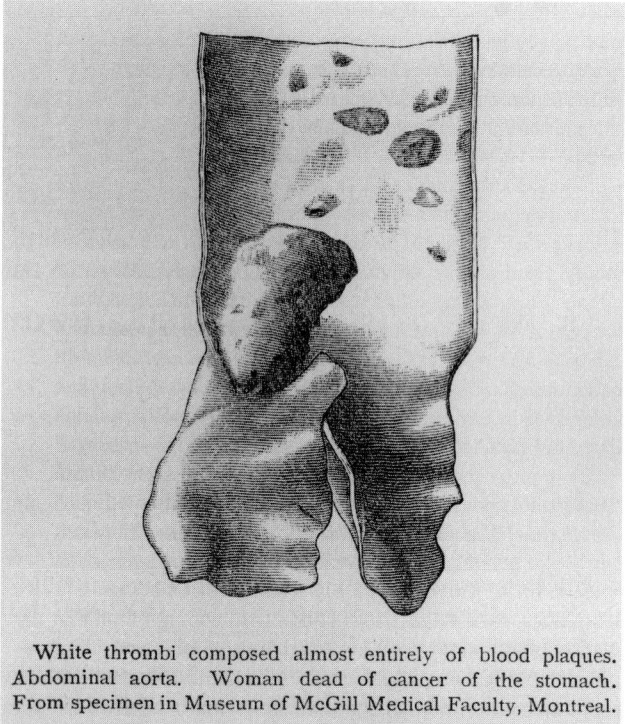

White thrombi composed almost entirely of blood plaques. Abdominal aorta. Woman dead of cancer of the stomach. From specimen in Museum of McGill Medical Faculty, Montreal.

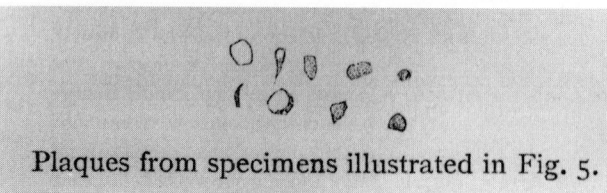

Plaques from specimens illustrated in Fig. 5.

**Figure 3.** Figures 5 and 6 from Osler (1886)[5] demonstrating "white thrombi" in the aorta of a patient who died with cancer of the stomach and the "plaques" (platelets) that could be retrieved from the thrombi. (From Osler,[5] with permission.)

the increase in the platelet count in the recipient to 123,000 platelets/μL! All overt signs of bleeding improved immediately after the transfusion. To monitor the response, Duke used the bleeding time assay he had just developed, analyzing the time for bleeding to stop from a wound in the ear lobe. The improvement in the bleeding time correlated with the improvement in clinical bleeding (Fig. 6). Because the patient's coagulation time was normal throughout, Duke's data provided strong evidence that the low platelet count was responsible for both the clinical hemorrhage and the prolonged bleeding time, and that normal coagulation does not ensure a normal bleeding time or adequate hemostasis in the face of thrombocytopenia.

Despite this remarkable success, many obstacles to routine platelet transfusion remained. These began to be systematically addressed 35 years later when U.S. government support of research in this area was spurred by the observation that thrombocytopenic purpura was a major cause of death after radiation exposure from atomic weapons

(Fig. 7).[27,28] Further impetus for improving platelet transfusion support came soon thereafter, when it became apparent that thrombocytopenia was a major cause of death from the then-new chemotherapeutic agents. In fact, successful platelet transfusion therapy was absolutely crucial for the development of modern chemotherapy.[28,29]

Many dramatic improvements in platelet transfusion therapy were made during the next decades, including improved anticoagulants, development of a differential centrifugation technique to produce platelet concentrates, recognition of the benefits of room temperature storage, improved plastics and conditions of bag storage to facilitate gas exchange and the maintenance of pH, cryopreservation, human leukocyte antigen (HLA) matching, or partial matching to improve platelet responses in patients refractory to random donor platelets, platelet "cross-matching," single-donor platelet collection via continuous-flow centrifugation, blood screening assays to prevent viral transmission, hepa-

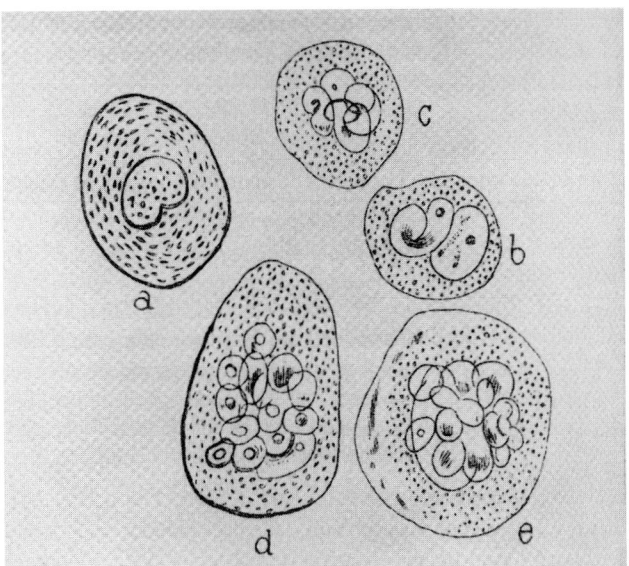

**Figure 4.** Drawings by Bizzozero (1869)[6] of giant bone marrow cells with multilobular nuclei, now recognizable as megakaryocytes. (From Baerg,[7] with permission.)

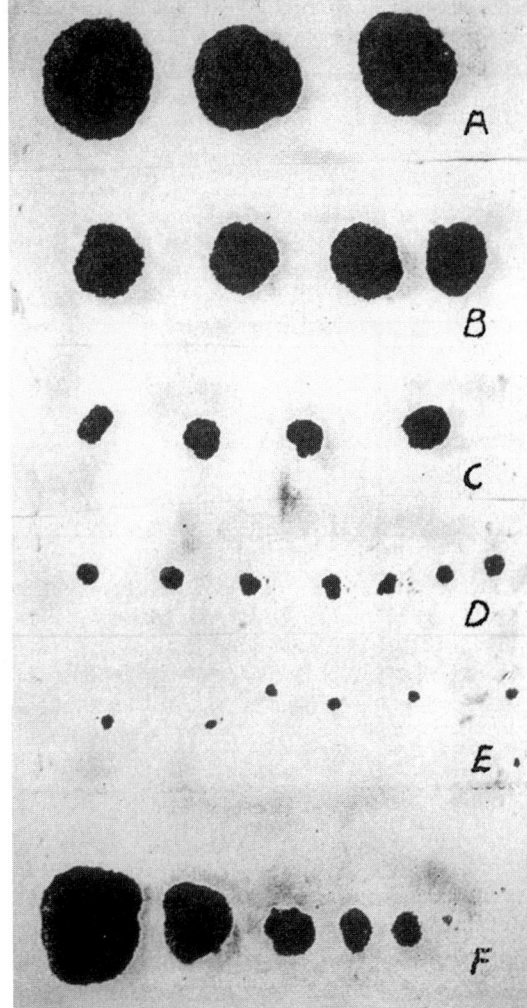

**Figure 5.** Bleeding time results from patient described by Duke[25] in 1910. The series from rows A to D are blots from the ear wound made when the patient's platelet count was 3000 platelets/µL. The blots in series A were taken immediately after the ear was pricked, series B began at 20 minutes, series C began at 40 minutes, series D began at 60 minutes, and series E began at 80 minutes. The bleeding time was 90 minutes. In row F are the blots obtained after transfusion (platelet count, 110,000 platelets/µL), demonstrating a bleeding time of 3 minutes. (From Duke,[25] with permission.)

titis B vaccination, prestorage leukoreduction to decrease febrile transfusion reactions, ultraviolet irradiation and leukoreduction to decrease immunogenicity, development of synthetic storage solutions, new methods to eliminate viable pathogens by damaging their DNA, and attempts to define better the most appropriate platelet count at which to transfuse platelets (see Chapter 69).[30–40] Ironically, one of the most important advances in platelet transfusion therapy had nothing to do with the platelet product, but rather the recognition that aspirin inhibits platelet function and thus aspirin should not be used as an antipyretic or analgesic in patients who are thrombocytopenic.[41]

Although attempts to develop substitutes for fresh platelets that could secure hemostasis in thrombocytopenic patients have been ongoing for more than 40 years,[12,42] this goal has not yet been achieved (see Chapter 70) despite the advances in our understanding of platelet physiology. The need for such agents grew more apparent in the 1980s and 1990s as a result of the dramatic increase in the demand for platelet transfusions, especially with the growth of bone marrow and stem cell transplantation and autologous stem cell reconstitution, reaching approximately 8,330,000 units of platelets transfused in the United States in 1992.[43] Since then, growth has slowed substantially, with just over 9,000,000 units transfused in both 1997 and 1999, the last years for which data are available.[44] There is also an increased appreciation of the seriousness of platelet transfusion-transmitted bacterial infections, a consequence of room temperature platelet storage, and now estimated to cause life-threatening sepsis in 1 in 100,000 recipients.[45] As a result, new methods to detect bacterial contamination before releasing platelets for transfusion have been implemented, but their efficacy is not absolute.[45] Storage of platelets at 4°C would dramatically reduce this risk, but it has been known for more than 50 years, since the studies of Zucker and Borelli,[46] that exposure to cold temperature affects platelet morphology. Subsequent studies demonstrated that exposure to cold results in platelet activation,[47,48] seriously compromising the *in vivo* survival and hemostatic effectiveness of the platelets.[49] Basic studies led by Hoffmeister[50,51] in Stossel's group have provided exciting new insights into the

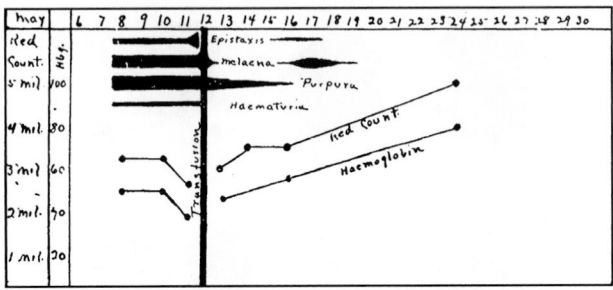

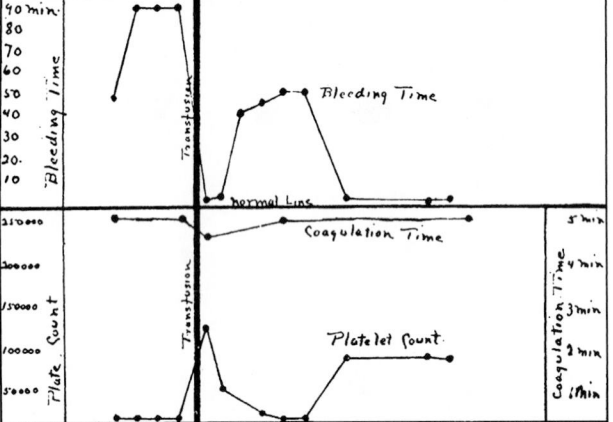

**Figure 6.** The clinical course of the patient reported by Duke.[25] The transfusion was performed at 2 AM on May 12. (From Duke,[25] with permission.)

## TABLE I

### INCIDENCE OF HEMORRHAGIC MANIFESTATIONS IN JAPANESE RADIATION CASUALTIES AT HIROSHIMA AND NAGASAKI

| Exposure Group* | Hiroshima | | Nagasaki | |
|---|---|---|---|---|
| | No. | Per cent | No. | Per cent |
| A | 550 | 96 | 248 | 66 |
| B | 485 | 43 | 610 | 42 |
| C | 227 | 12.5 | 385 | 24 |
| D | 137 | 8.5 | 93 | 13 |
| E | 52 | 7.0 | 36 | 6 |

* A to E represents decreasing exposure to the nuclear radiation.

**Figure 7.** Risk of developing hemorrhagic purpura after exposure to the atomic bomb explosions at Nagasaki and Hiroshima as a function of distance from the explosion. (From Cronkite et al.[27] with permission.)

mechanism responsible for the rapid clearance of platelets incubated in the cold, identifying the lectinlike interaction between integrin αMβ2 on hepatic macrophages and β-N-acetylglucosamine residues on clustered platelet glycoprotein (GP) Ib molecules as playing a central role. Most important, they developed a simple method to add a galactose residue to the glycan (by adding uridine diphosphate galactose and taking advantage of the platelet's own galactosyl transferase) and thus restore near-normal platelet survival. If this method can be applied to human platelets collected for transfusion, it might allow safe storage at 4°C and thus potentially both reduce the risk of platelet transfusion-associated bacterial transmission and prolong the allowable time for platelet storage.

Currently, a number of different factors are influencing the field of platelet transfusion therapy, including the tendency to transfuse platelets only at lower platelet counts and with smaller numbers of platelets than in the past (see Chapter 69)[36,41]; the decrease in bone marrow and stem cell transplantation and autologous stem cell reconstitution for breast cancer[52]; the emergence of cord blood as a source of stem cells, because megakaryocyte reconstitution is often delayed with this source of stem cells[53–55]; the development of new conditioning regimens for stem cell reconstitution that produce much less myelosuppression than in the past[56]; the availability of interleukin (IL)-11, the first agent approved by the U.S. Food and Drug Administration to increase the platelet count[57]; the ingenious approaches that resulted in the identification and subsequent production of thrombopoietin[58]; ongoing clinical trials of thrombopoietin and development of related molecules that can activate the thrombopoietin receptor (mpl) (see Chapter 66)[59–63]; the identification of additional chemokines and growth factors that can stimulate thrombopoiesis, such as stromal-derived factor-1 and fibroblast growth factor-4[64]; and multiple attempts to develop platelet substitutes (see Chapter 70).[65–73]

Advances in stem cell biology, including the potential for patient-specific somatic nuclear transfer into enucleate oocytes, coupled with advances in *ex vivo* expansion of megakaryocytes from cord blood hematopoietic precursors,[74–77] now suggest the future possibility of producing tissue-type-specific platelets *ex vivo* for patients refractory to platelet transfusion as a result of alloimmunization. The major obstacle to this approach is our limited understanding of the signals that initiate the terminal differentiation of megakaryocytes into platelets (see Chapter 2).

## II. Immune Thrombocytopenia

Clinical descriptions of purple skin discoloration due to hemorrhage ("purpura") extend back to Hippocrates, but Werlhof, who was physician to King George II of England in the German states, and Behrens independently described cases of what was most likely immune thrombocytopenia in 1735.[9,13,78] Almost 150 years later, the association between purpura and thrombocytopenia was made by Krauss in 1883 and Denys in 1887, and confirmed by Hayem in 1890 to 1891, who also noted that the remaining platelets (which he termed *hematoblasts,* believing they were primitive erythrocytes) were large and did not support clot retraction normally.[9,10] In 1907, Cole[79] produced an antiserum to platelets that did not react with erythrocytes, thus demonstrating fundamental immunological differences (and presumably different lineages) between platelets and erythrocytes. Later, similar antisera were used to produce animal models of immune thrombocytopenia.[80] In 1916, Kaznelson,[81] a medical student in Prague, suggested splenectomy as a therapy for thrombocytopenia. Professor Schloffer followed his suggestion and the first patient so treated enjoyed an increase in her platelet count from 200 to 500,000 platelets/μL![9,81]

The pioneering and courageous studies of Harrington and colleagues,[82] in which blood components from patients with immune thrombocytopenia were infused into volunteers, including Harrington himself, demonstrated that the agent producing thrombocytopenia in idiopathic thrombocytopenia (ITP) could be passively transferred with plasma (Fig. 8). Harrington's group[83] then proceeded to define many important aspects of immune thrombocytopenia, including the crucial role of antibodies, and the beneficial effects of splenectomy and glucocorticoid therapy on platelet clearance.

Immune mechanisms also underlie HIV-associated thrombocytopenia[84,85] (see Chapter 47) and many of the drug-induced thrombocytopenias (see Chapter 49),[86,87] including the uniquely dangerous heparin-induced thrombocytopenia when it is associated with life-threatening thrombotic complications (see Chapter 48).[87,88] The association of immune thrombocytopenia with *Helicobacter pylori* infection, and the clinical improvement enjoyed by some patients when the *H. pylori* is eradicated, stimulated new thinking about the pathophysiology and therapy of immune thrombocytopenia, but unfortunately, therapeutic results have not been consistent (see Chapter 46).[88–90] Although steroids and splenectomy remain among the most important treatments for immune thrombocytopenia, the introduction of intravenous immunoglobulin G (a major scientific and technical advance, given the many challenges to manufacturing a product that would not produce severe systemic reactions)[91,92] and later anti-Rh D therapy (for patients fortunate enough to be Rh D positive)[93] provided additional effective agents (see Chapter 46).[94] Knowledge of the role of antiplatelet antibodies in the pathophysiology of the disorder provided the rationale for the use of the anti-B cell (anti-CD20) monoclonal antibody derivative rituximab, thus opening a new therapeutic chapter by producing sustained responses in a sizable percentage of patients refractory to other treatments (see Chapter 46).[88] As we continue to learn more about the immune response, additional rational targets will

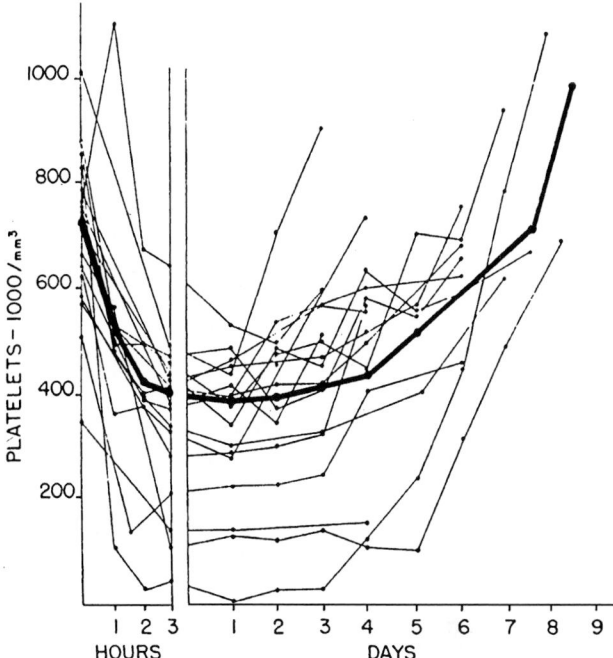

**Figure 8.** Data from Harrington, Minnich, and Hollingsworth[82] demonstrating that infusion of plasma from patients with idiopathic thrombocytopenic purpura could decrease the platelet count in normal individuals. (From Harrington et al.,[82] with permission.)

undoubtedly be identified, offering the hope for even better therapy.[95–97]

Patients with severe, refractory thrombocytopenia still, however, present a profound clinical challenge.[98] There is a rationale for considering treatment with thrombopoietin or other similar agents based on the finding that patients with immune thrombocytopenia do not have very high levels of thrombopoietin,[99–101] and the fascinating clinical observation of an improvement in the platelet count of a patient with immune thrombocytopenia in association with an elevation in thrombopoietin in response to exposure to high altitude.[102] Although, a pilot study of recombinant IL-11 in patients with refractory ITP gave disappointing results, there is great optimism about the positive data using agents that can activate the thrombopoietin receptor to increase the platelet count.[103] Nonmyeloablative bone marrow reconstitution offers yet another approach to patients with refractory immune thrombocytopenia that may prove valuable as regimens continue to improve in both safety and efficacy.[56]

## III. Alloantigens, Single Nucleotide Polymorphisms, and Alloimmune Thrombocytopenias

There is currently considerable interest in single nucleotide polymorphisms (SNPs) because of the remarkable progress made in sequencing the human genome and the recognition that identifying such polymorphisms may be useful in assessing disease predisposition and response to medications.[105–107] Studies of platelets have made major contributions to this field. The clinical syndrome of neonatal alloimmune thrombocytopenia, in which maternal antibody to fetal platelet antigens inherited from the father cross the placenta and induce immune-mediated thrombocytopenia in the fetus, was recognized as early as 1959, and posttransfusion purpura, also resulting from alloantigens, was first described soon thereafter (see Chapter 53).[108–111] The pioneering studies of Shulman and colleagues[109,110] produced a systematic characterization of important platelet alloantigens at the serological level. Later, in landmark studies, Newman and his colleagues[112] applied the polymerase chain reaction to the trace amounts of messenger RNA present in platelets. This new technology very rapidly led to the identification of the SNPs that result in the alloantigens on platelet glycoproteins, thus establishing the molecular biological basis of these disorders (see Chapter 14).[113] Furthermore, recognition that only women with certain HLA-DR histocompatibility antigens are at risk of making alloimmune antibodies to certain polymorphisms has provided important molecular insights into the immune response (see Chapter 53).[114–117] Most recently, in fulfillment of some of the hoped-for benefits in defining SNPs, reports of associations between platelet SNPs and differences in platelet function, as well as predispositions to hemorrhagic and thrombotic diseases, have begun to emerge, although differences in results from study to study indicate considerable complexity (see Chapter 14).[118–126] Furthermore, preliminary data suggest that platelet SNPs may have pharmacogenetic significance, because at least one SNP appears to affect the response to aspirin as an antiplatelet agent.[127,128]

## IV. Platelet Physiology, Assays of Platelet "Adhesiveness," Adenosine Diphosphate, Platelet Aggregation, and Myocardial Infarction

The fundamental features of platelet physiology began to be characterized at the molecular level by a small group of investigators from the late 1940s to the mid-1960s.[11–13,17–24] These included studies of platelet serotonin and vasospasm, the release reaction and viscous metamorphosis (the loss of definable boundaries between platelets some time after platelets form aggregates), clot retraction, and the effects of thrombin and collagen on platelet clumping. Attempts to quantify platelet "adhesiveness" resulted in the development of a large number of techniques, but the method of Hellem[129] deserves singling out, because his glass bead platelet retention assay led him to discover an association between hema-

tocrit and platelet adhesiveness. Hellem[129] therefore hypothesized that erythrocytes contain a factor that can activate platelets. He then went on to isolate such a factor from erythrocytes and found that its activity was stable even when boiling water was added to the red cells! In 1961, Gaarder and colleagues[130] identified this substance as adenosine diphosphate (ADP). Modified versions of the glass bead platelet retention test provided other crucial insights into platelet physiology, including the definition of the abnormalities in platelet function in uremia, afibrinogenemia, Glanzmann thrombasthenia, Bernard–Soulier syndrome, and von Willebrand disease. Moreover, the fundamental discovery that high shear conditions accentuate the abnormality in von Willebrand disease was first made using this assay (described later).[131–133] The test also was extremely important in early attempts to purify von Willebrand factor, because at that time no other *in vitro* assay of von Willebrand factor activity was available.[134] Standardization of the platelet retention assay, however, proved very difficult[135] and thus it now is primarily of historical significance (see Chapter 23).

The need for more quantitative and robust methods of assessing platelet function led O'Brien[136] and Born[137] to develop independently turbidometric platelet aggregometry in 1962. In this technique, the transmission of light through a cuvette containing stirred platelets at 37°C is continuously recorded. When aggregates form after the addition of an aggregating agent, there is an increase in light transmission. When large aggregates form and move into and out of the light path, the increase in light transmission is accompanied by wide excursions of the recording (see Chapter 26). This technique revealed that different agonists produce different patterns of aggregation. ADP, for example, produces a double wave of platelet aggregation at certain doses. The second wave depends on the induction of the release reaction, during which platelets release the contents of their granules, including the platelets' own storage pool of ADP (see Chapter 26).

*In vivo* animal studies with ADP later provided a crucial link between platelet aggregate formation and myocardial infarction. Thus, building on clinical observations of platelet microemboli traversing the retinal vasculature during transient ischemic attacks and autopsy data demonstrating platelet aggregates in the distal coronary artery circulation in patients suffering myocardial infarctions (reviewed by Jorgensen[24]), Jørgensen and other members of the group led by Fraser Mustard demonstrated that infusing ADP into the aorta and/or coronary arteries of pigs resulted in the development of platelet microthrombi, hypotension, and transient arrest of blood flow in the microcirculation, culminating in gross infarction, death from arrhythmia, or electrocardiographic ST-segment changes indicative of ischemia in almost 90% of animals.[138] The causal role of platelets in the phenomenon was established by showing protection from cardiac dysfunction by inducing thrombocytopenia in the pigs using $^{32}P$ before infusing ADP.

## V. Aspirin, Arachidonic Acid, Prostaglandins, Eicosanoids, Nitric Oxide, and Endothelial Ecto-ADPase

Gastrointestinal bleeding was linked to aspirin ingestion as early as 1938,[139] and prolongation of the bleeding time by aspirin ingestion, especially in patients with von Willebrand disease or telangiectasias, was demonstrated in the 1950s and early to mid-1960s.[16] In fact, excessive prolongation of the bleeding time by aspirin ("aspirin tolerance time") was proposed by Quick[140] as a method to enhance the sensitivity of diagnosing von Willebrand disease. Nonetheless, there was no widespread appreciation that aspirin specifically interfered with platelet function until 1967, when Weiss (personal communication, 2001), noting his own increase in bleeding from razor nicks when on aspirin, and building on his previous studies of patients with abnormalities in platelet ADP release and the exciting new observations by Hovig,[22,141] and Spaet and Zucker[142] that exposure of platelets to connective tissue results in platelet release of ADP,[17] demonstrated with Aledort and Kochwa that aspirin inhibits platelet aggregation initiated by connective tissue.[143] Very soon thereafter, the inhibitory effects of aspirin on the second wave of platelet aggregation induced by ADP and other platelet functions were reported.[144–147] These observations ushered in a new era in platelet research that was followed by exciting findings by investigators in many fields related to the important role of arachidonic acid in cell biology, and the discovery of prostaglandins, thromboxanes, and the cyclooxygenase enzymes.[16,148–152] These observations also led, after some painful false starts,[153] to the use of aspirin as the first clinical antiplatelet therapy, which we now recognize reduces the mortality of myocardial infarction by almost one-quarter, the risk of subsequent vascular events after a first event by about one-third, and the ischemic complications of percutaneous coronary interventions by at least one-half (see Chapter 60).[154,155] In addition to the medical benefits of aspirin treatment, the economic benefits of this discovery probably exceed many billions of dollars each year in both lower health care costs and increased economic productivity.

Marcus,[156] who was the first to describe the extraordinarily high content of arachidonic acid in platelet phospholipids, also described the complex transcellular eicosanoid metabolism that occurs between platelets, endothelial cells, and leukocytes.[157] Studies by Moncada and colleagues[158] and Weksler and colleagues[159] demonstrated that endothelial cells synthesize prostaglandin $I_2$ ($PGI_2$, prostacyclin), providing the first biochemical evidence that the lack of platelet

reactivity of normal endothelium is an active process, not as had previously been proposed simply a function of the inert nature of the surface of the endothelium (see Chapter 13). This theme was reinforced by the subsequent discovery that the nitric oxide (NO) synthesized by endothelial cells (originally termed *endothelial-derived relaxation factor,* or EDRF, because of its effects on smooth muscle cells) is a potent inhibitor of platelet function and acts synergistically with prostacyclin.[160–162] Finally, endothelial cells were found by Marcus and colleagues[163,164] to have an ecto-ADPase (CD39) activity capable of rapidly metabolizing the platelet activator ADP to the platelet inhibitor adenosine (see Chapter 13). Thus, endothelial prostacyclin, NO, and CD39 are part of a highly coordinated, active system designed to inhibit platelet activation that works in concert with endothelial thrombomodulin, which regulates the level and proteolytic specificity of the potent platelet activator thrombin. The precise contributions of each of these elements to basal platelet function and platelet function after vascular injury in different vascular beds are currently being defined by experiments in transgenic animals.[165]

## VI. Clot Retraction, Glanzmann Thrombasthenia, GPIIb-IIIa Receptor, and GPIIb-IIIa Antagonists

The observation that clots formed by shed blood undergo retraction within minutes to hours probably extends back to antiquity. Hewson, who discovered fibrinogen in 1770, understood the importance of fibrin in clot retraction,[166] but Hayem is credited with ascribing a central role to platelets in clot retraction in the 19th century.[11] Ultimately, studies of clot retraction resulted in the discovery by Bettex–Galland and Lüscher[167,168] in 1959 that platelets contain large amounts of the contractile proteins actin and myosin, which they termed *thrombosthenin* ("the strength of the clot"). This was the first time these "muscle" proteins were isolated from a nonmuscle cell, a discovery with profound implications for understanding the role of these proteins in cell motility in many other nonmuscle cells. Ultimately, the identification of nonmuscle myosin type IIA in platelets paved the way for the discovery of mutations in the gene for this protein (*MHY9*) as contributing to a group of autosomal dominant giant platelet disorders, including May–Hegglin anomaly and the Fechtner, Sebastian, Epstein, and Alport–like syndromes (see Chapter 57). More immediately, however, the early recognition of the contribution of platelets to clot retraction provided a platelet function assay that could be used to diagnose both qualitative and quantitative disorders of platelets and to monitor platelet transfusion therapy.[28]

Thus in 1918, when the Swiss pediatrician Glanzmann studied a group of patients with a bleeding diathesis and found them to have normal platelet counts but poor clot retraction, he termed the disorder *thrombasthenia* ("weak platelet").[169] Subsequent studies by groups led by Zucker and colleagues[12,170] and Caen and colleagues[171] defined the platelet defect in Glanzmann thrombasthenia as an inability to aggregate in response to the usual platelet agonists such as ADP and epinephrine. The deficiency in platelet fibrinogen found in these patients ultimately led to the recognition that platelets aggregate *in vitro* by binding fibrinogen to their surface, with the fibrinogen acting as a bridging molecule.[172–176] The molecular basis of Glanzmann thrombasthenia was revealed in pioneering studies by groups led by Nurden and Caen[177] and by Phillips and colleagues[178] that demonstrated abnormalities in two surface glycoproteins termed *GPIIb* and *GPIIIa* based on their electrophoretic mobility. Additional studies performed by many outstanding laboratories demonstrated that these two glycoproteins form a complex that acts as a receptor for fibrinogen and for a number of other adhesive glycoproteins, including von Willebrand factor, that contain arginine–glycine–aspartic acid (RGD) sequences (see Chapter 8).[172–175] Moreover, as the cloning and sequencing of many different receptors progressed, it became apparent that the GPIIb-IIIa receptor is a member of a large family of receptors termed *integrins* that extend back in evolution to *Drosophila* and are involved in cell adhesion and aggregation, as well as protein trafficking and bidirectional signaling (see Chapter 17).[179–181] Several other integrin receptors also bind ligands containing the RGD sequence. Molecular biological analysis of the defects in GPIIb-IIIa (renamed αIIbβ3 according to integrin nomenclature) causing Glanzmann thrombasthenia have provided important information linking structure to biogenesis and function (reviewed by Coller et al.[182]; and in Chapter 57). Mice deficient in β3, and thus lacking both αIIbβ3 and αVβ3 receptors, have many of the clinical and laboratory features characteristic of Glanzmann thrombasthenia.[183] They are also protected from developing thrombosis.[184] These mice are providing important new insights into the roles of αVβ3 and/or αIIbβ3 in a variety of different phenomena, including tumor angiogenesis, wound healing, osteoclast bone resorption, and αIIbβ3-mediated signal transduction.[185–188] They also provide an excellent model for testing gene therapy of Glanzmann thrombasthenia (see Chapter 71).[189]

The development of monoclonal antibodies to αIIbβ3 and the ability to analyze patient DNA via PCR translated into direct patient benefit in the form of new methods for carrier detection and prenatal diagnosis in families with Glanzmann thrombasthenia.[176,190–193] Moreover, improved understanding of ligand binding to αIIbβ3 led to the development of drugs that inhibit the αIIbβ3 receptor (see Chapter 62). The latter, which includes a chimeric monoclonal antibody fragment and low-molecular weight molecules patterned after RGD and related sequences, have proved

efficacious and safe in preventing ischemic complications of percutaneous coronary interventions and unstable angina.[194–198] These drugs represent the first rationally designed antiplatelet therapies and thus mark an important milestone in moving from serendipity to purposeful drug development based on a molecular understanding of platelet function. Another one of the monoclonal antibodies to αIIbβ3 proved useful in studies of the crystal structure of αIIbβ3, because it stabilized the αIIbβ3 headpiece complex during purification and crystallization.[199] The resulting high-resolution structure has provided detailed information on the ligand binding pocket, the structural basis of the specificity of the low-molecular weight drugs for αIIbβ3, and the likely conformational changes associated with receptor activation.[199] Mapping the epitope on β3 of the chimeric monoclonal antibody drug has also provided valuable insights regarding how it prevents ligand binding and how it differs from the other αIIbβ3 antagonist drugs.[200]

## VII. Bernard–Soulier Syndrome, the GPIb-IX-V Complex, von Willebrand Factor, Shear, ADAMTS-13, and Thrombotic Thrombocytopenic Purpura

In 1948, Bernard and Soulier[15,201] reported on two children of consanguineous parents with a mucocutaneous bleeding disorder, giant-size platelets, and thrombocytopenia. Subsequent studies conducted in many laboratories identified abnormalities in platelet adhesion resulting from an inability of patient platelets to interact with von Willebrand factor.[202–206] In 1971, Howard and Firkin,[207,208] while trying to develop an animal model of thrombocytopenia, discovered that the antibiotic ristocetin, which is akin to vancomycin but was removed from clinical use because it caused thrombocytopenia in humans, could induce platelet aggregation, but only in the presence of von Willebrand factor. Later Brinkhous and colleagues[209] made parallel findings about the snake venom botrocetin. These extremely important discoveries provided crucial tools that facilitated the *in vitro* study of the interaction between platelets and von Willebrand factor. They also permitted the practical development, first by Weiss,[210] of functional assays for plasma von Willebrand factor that have been valuable in the diagnosis and therapy of von Willebrand disease.

Building on the pioneering studies by Nurden and Caen[211] that identified a defect in platelet GPIb as a cause of Bernard–Soulier syndrome, many others provided important insights into the structure and function of the receptor complex made up of GPIbα, GPIbβ, GPIX, and GPV, all of which are members of the leucine-rich repeat family (see Chapter 7).[212–216] Patients have been identified with defects in GPIbα, GPIbβ, and GPIX, but not GPV (see Chapter

57).[182,216] Thus, the identification of the molecular defect in Bernard–Soulier syndrome made an important contribution to defining the central role of the binding of von Willebrand factor to GPIbα in platelet adhesion. Studies with monoclonal antibodies to GPIbα that inhibited both von Willebrand factor binding and ristocetin-induced platelet agglutination confirmed the functional importance of this interaction and provided tools for rapidly diagnosing Bernard–Soulier syndrome.[217,218] They also provided insights into the region of GPIbα involved in von Willebrand factor binding,[219] and these inferences have been elegantly confirmed by atomic resolution structural data derived from X-ray crystallographic studies of isolated regions of GPIbα and von Willebrand factor.[220–222] In turn, the structural data provide new opportunities to develop novel antithrombotic therapeutics that interfere with these interactions. In addition, the availability of mouse models of von Willebrand disease, GPIb deficiency, and GPV deficiency[223–225] is opening exciting new paths for understanding the biological roles of the molecular interactions between GPIb and von Willebrand factor, the pathophysiology of the Bernard–Soulier syndrome, and the interactions between thrombin and platelets (discussed later).

The discovery of the shear dependence of the interaction between von Willebrand factor and platelet glycoprotein receptors GPIb and GPIIb-IIIa can be traced back to early observations using the glass bead platelet retention test, in which the defect in von Willebrand disease was found to be accentuated by increasing the speed of blood flow.[226] The importance of shear in understanding platelet function was defined by many laboratories[215,227–233] and helped to explain the importance of the very large size of von Willebrand factor and its multimeric composition.[228,234–240] Ultimately, shear was also found to function as a cofactor for the cleavage of ultrahigh-molecular weight multimers of von Willebrand factor by the enzyme ADAMTS-13. The discovery of ADAMTS-13 reads like a detective story, starting with the original clinical descriptions of thrombotic thrombocytopenic purpura by Moschcowitz[241] in 1924 and inherited relapsing hemolytic–uremic syndrome by Schulman and colleagues[242] in 1960 and Upshaw[243] in 1978, with the groups led by Moake, Furlan, Tsai, Sadler, and Ginsburg all making vital contributions.[18–20,244] Building on careful clinical observations, the key clues included (a) finding increased amounts of unusually large von Willebrand multimers in the plasma of patients with thrombotic thrombocytopenic purpura; (b) identifying and purifying a plasma enzyme that could cleave von Willebrand factor multimers, especially under high shear; (c) recognizing that most patients with thrombotic thrombocytopenic purpura have autoantibodies to the enzyme that cleaves von Willebrand factor multimers; and (d) identifying mutations in ADAMTS-13 that cause the inherited form of the disorder.[18–20] Still to be defined, however, are the precise triggers that initiate acute episodes of thrombotic thrombocytopenic purpura, but endothelial activation is likely to play an important role (see Chapter 50).

## VIII. Thrombin-Induced Platelet Activation

Thrombin was shown to aggregate platelets 50 years ago,[12,245] but the precise mechanisms involved remained mysterious for a long time because, despite being able to identify platelet proteins that could bind thrombin with variable affinities, deciphering which interactions actually led to signal transduction was a challenging task.[246] Thrombin appeared to act both as an enzyme (because enzymatic activity was required to initiate aggregation) and as a ligand (because catalytically inactive thrombin could enhance the response to catalytically active thrombin).[247] Phillips and Agin[248] established that thrombin cleaved platelet GPV and platelets from patients with Bernard–Soulier syndrome, which lack functional GPIb and GPV molecules, had decreased platelet aggregation in response to thrombin.[249] The role of GPV remained mysterious, however, because the time course of GPV cleavage did not correlate well with platelet activation.[249,250,251] The landmark studies of Coughlin and colleagues using a functional expression cloning strategy opened a new era by identifying a novel group of seven transmembrane receptors that could be activated by cleavage of the amino-terminal region and creation of a new "tethered ligand" that could then bind to another site on the receptor to initiate activation (protease-activated receptors or PARs; see Chapter 9).[252] Compelling support for the tethered ligand hypothesis came from studies demonstrating that short peptides from the tethered ligand region could themselves activate platelets.[252] Subsequent studies identified the presence of at least two such receptors on human platelets that have thrombin cleavage sites (PAR-1 and PAR-4), whereas parallel studies in mice failed to identify a PAR-1 homologue but did identify a receptor homologous to PAR-4 and another receptor from the same family, PAR-3.[253–255] Phillips' group established that the platelets of mice deficient in GPV, but not normal mice, could aggregate in response to proteolytically inactive thrombin and that this response depended on binding of thrombin to GPIb.[256] Similar responses could be obtained after thrombin cleaved platelet GPV from normal platelets. Thus, the dual nature of thrombin-induced platelet aggregation may be the result of the presence of at least two separate mechanisms: cleavage of the PAR receptors and thrombin cleavage of GPV followed by thrombin-initiated signaling through binding to GPIb. There may, however, be links between these mechanisms, because binding of thrombin to GPIb may accelerate hydrolysis of PAR-1.[257] Therapeutic agents designed to inhibit thrombin-induced platelet activation through the PAR system are already in clinical development.

# IX. ADP Receptors

After the discovery of the platelet aggregating ability of ADP, a number of separate ADP-initiated functions were defined, including increasing the intracytoplasmic calcium concentration,[258,259] changing platelet shape from disk to spiny sphere,[260,261] and reducing cyclic adenosine monophosphate (AMP); and preventing the increase in cyclic AMP in response to agents known to activate adenylyl cyclase.[262,263] ADP binding studies failed to identify unequivocally a specific receptor, and pharmacological studies using congeners of ADP, as well as the related compounds ATP and adenosine, demonstrated a complex set of agonist and antagonist interactions that were difficult to account for on the basis of a single receptor mechanism.[263] Ultimately, an ADP-activated seven transmembrane receptor in the P2Y class, $P2Y_1$, linked to $Gq$ was cloned and found to play an important role in calcium mobilization, shape change, and platelet aggregation.[264,265] In addition, a ligand-gated $Ca^{++}$ channel receptor of the P2X family, $P2X_1$, was cloned from platelets and thought to act as a second ADP receptor, although it was not clear that it had a role in platelet aggregation.[263,266] Inferential observations, some obtained using inhibitors of ADP-induced aggregation, suggested that the $P2Y_1$ receptor mediated shape change and contributed to platelet aggregation, but this receptor could not account for the cyclic AMP lowering effect of ADP, nor did it appear to mediate the effects of ticlopidine and clopidogrel, two clinically important antiplatelet drugs that selectively inhibit ADP-induced platelet aggregation.[263,267] Thus, another ADP receptor was postulated to exist,[268] and using an ingenious strategy based on the presumptive linkage of the unknown receptor to $Gi$ and adenylyl cyclase (because of the effect on lowering cyclic AMP), Hollopeter and colleagues[269] cloned another seven transmembrane ADP receptor, named $P2Y_{12}$, with all the requisite characteristics of the postulated receptor (see Chapter 10). Moreover, they demonstrated a mutation in the receptor in a patient with a bleeding disorder associated with a selective defect in ADP-induced platelet aggregation. The $P2X_1$ receptor also appears to contribute to platelet activation, but it preferentially responds to adenosine triphosphate rather than ADP.[270,271] The remarkable success of clopidogrel as an antithrombotic agent has encouraged the development of a new generation of $P2Y_{12}$ antagonists, several of which are currently in advanced stages of clinical assessment (see Chapter 61).[272–275]

# X. Platelets and Metastases

Almost 40 years ago, Gasic and colleagues,[276,277] working on the hypothesis that sialic acid residues on the glycoproteins of tumor cells and/or endothelial cells were important in metastasis formation, injected neuraminidase into mice and demonstrated a marked reduction in metastasis formation when $TA_3$ ascites tumor cells were subsequently injected. Further studies, however, demonstrated that the neuraminidase injection caused thrombocytopenia (presumably by decreasing the surface change on platelets[278]) and that thrombocytopenia produced by alternative methods also protected against metastasis.[276] Additional studies demonstrated that platelets can interact directly with some tumors and that some tumors can induce or enhance platelet aggregation.[279–282] In some experimental models, inhibiting the GPIIb-IIIa receptor could also decrease metastasis formation.[279,283] Moreover, because both platelets and some tumors can facilitate thrombin formation, and because thrombin, either directly or indirectly, may have both growth-promoting and growth-inhibiting effects, the potential mechanisms involved are complex.[279,284,285] The discovery that platelets contain high concentrations of vascular endothelial growth factor, a factor implicated in tumor angiogenesis, and can release it when activated by agonists or tumor cells, adds yet another dimension to this evolving story (see Chapter 42).[286–288] Whether antiplatelet therapy can be exploited to decrease metastasis formation in humans remains an unanswered, but intriguing, question.[282]

# XI. Reflections on the Past and Thoughts about the Future

Staggering progress has been made in understanding platelet physiology and the pathophysiology of disorders of platelet number and function in the past 125 years, but most particularly in the past 40 years. Advances in the broad disciplines of biochemistry, molecular biology, and, most recently, structural biology have made possible an increasingly detailed understanding of the fundamental cellular and molecular interactions, now reaching atomic resolution! It is predictable that genomics, microarray-based gene expression analysis, proteomics, and systems biology will accelerate future progress in complex areas of platelet physiology, especially signal transduction, granule secretion, and cytoskeletal organization, and excellent examples of the application of such approaches have already been published.[289–296] Gene therapy of platelet disorders is no longer science fiction, because important proof-of-concept experiments have already been performed successfully.[189,297] Similarly, additional advances in structural biology, especially ones that make it more tractable to determine the structures of membrane proteins, will predictably provide exciting new information that will add new dimensions to our understanding of platelet function and open new opportunities for rational drug design. When combined with data regarding SNPs, these observations may allow antiplatelet therapy to lead the way in pharmacogenetics, in which the choice and dose of a drug is tailored to the individual patient's genetic makeup. New advances in imaging, especially dynamic imaging, are providing extraordinary insights into phenomena such as platelet formation from megakaryocytes,[298] and permitting the integration of the biochemical and cell biological data related to signal transduction and cytoskeletal dynamics.

Transgenic mice and mice in which genes have been "knocked out" have revolutionized platelet research, providing the opportunity to assess the role of single genes, or combinations of genes, on platelet function.[299–302] The identification of similarities between human platelets and zebrafish thrombocytes opens up even greater opportunities for identifying the role of individual genes in platelet function via mutagenesis screens,[303] and this model system has already provided provocative information on the greater efficacy of "young" versus "old" thrombocytes.[304] These approaches have the added benefit of permitting assessment of platelet function in intact animals, thus providing confirmation of *in vitro* and *ex vivo* findings. As with all new technologies, cautions are advisable because, for example, mouse platelets are known to differ from human platelets in a number of fundamental ways, including platelet size, platelet number, and the PAR receptor repertoire.[299] Nonetheless, studies of the fundamental aspects of platelet function are quite reassuring about the similarities between human and mouse platelets, and so there is reason to be optimistic that many, but perhaps not all, observations made in mice can be extrapolated to humans. The development of well-defined and robust mouse models of thrombosis and hemostasis, as well as complex biological phenomena related to platelet function, such as vascular injury, atherosclerosis, metastasis formation, and inflammation, are especially important.[305] It is both fitting and ironic, therefore, that intravital microscopy,[306–308] the same technique that Bizzozero[4] used to make his landmark discoveries about platelets after a long period of dormancy, is once again cherished as a most powerful method to reveal the platelet's contributions to hemostasis and disease.

# References

1. Robb–Smith, A. H. (1967). Why the platelets were discovered. *Br J Haematol, 13,* 618–637.

2. Osler, W. (1874). An account of certain organisms occurring in the liquor sanguinis. *Proc R Soc Lond, 22,* 391–398.

3. Bizzozero, G. (1881). Su di un nuovo elemento morfologico del sangue dei mammiferi e della sua importanza nella trombosi e nella coagulazione. *L'Osservatore, 17,* 785–787.

4. Bizzozero, J. (1882). Uber einen neuen formbestandteil des blutes und dessen rolle bei der thrombose und blutgerinnung. *Virchows Archiv, 90,* 261–332.

5. Osler, W. (1886). On certain problems in the physiology of the blood corpuscles. *The Medical News, 48,* 421–425.

6. Bizzozero, G. (1869). Sul midollo delle ossa. Il Morgagni.

7. Baerg, A. (1958). The hematologic work of Giulio Bizzozero. *Sci Med Hal, 7,* 45–63.

8. Wright, J. H. (1906). The origin and nature of blood plates. *Boston Med Surg J, 154,* 643–645.

9. Jones, H. W., & Tocantins, L. M. (1933). The history of purpura hemorrhagica. *Annals Medical History, 5,* 349–364.

10. Tocantins, L. M. (1948). Historical notes on blood platelets. *Blood, 3,* 1073–1082.

11. Budtz–Olsen, O. E. (Ed.). (1951). *Clot retraction.* Springfield: Charles Thomas.

12. Marcus, A. J., & Zucker, M. B. (Eds.) (1965). *The physiology of blood platelets.* New York: Gurne and Stratton.

13. Spaet, T. H. (1980). Platelets: The blood dust. In M. Wintrobe, (Ed.), *Blood, pure and eloquent* (pp. 549–571). New York: McGraw-Hill.

14. Brinkhous, K. M. (1983). W. W. Duke and his bleeding time test. A commentary on platelet function. *JAMA, 250,* 1210–1214.

15. Bernard, J. (1983). History of congenital hemorrhagic thrombocytopathic dystrophy. *Blood Cells, 9,* 179.

16. de Gaetano, G. (2001). Historical overview of the role of platelets in hemostasis and thrombosis. *Haematologica, 86,* 349–356.

17. Weiss, H. J. (2003). The discovery of the antiplatelet effect of aspirin: A personal reminiscence. *J Thromb Haemost, 1,* 1869–1875.

18. Moake, J. L. (2004). Defective processing of unusually large von Willebrand factor multimers and thrombotic thrombocytopenic purpura. *J Thromb Haemost, 2,* 1515–1521.

19. Furlan, M. (2004). Proteolytic cleavage of von Willebrand factor by ADAMTS-13 prevents uninvited clumping of blood platelets. *J Thromb Haemost, 2,* 1505–1509.

20. Tsai, H. M. (2004). A journey from sickle cell anemia to ADAMTS13. *J Thromb Haemost, 2,* 1510–1514.

21. Sakariassen, K. S., Turitto, V. T., & Baumgartner, H. R. (2004). Recollections of the development of flow devices for studying mechanisms of hemostasis and thrombosis in flowing whole blood. *J Thromb Haemost, 2,* 1681–1690.

22. Hovig, T. (2005). The early discoveries of collagen-platelet interaction and studies on its role in hemostatic plug formation. *J Thromb Haemost, 3,* 1–6.

23. Solum, N. O., & Clemetson, K. J. (2005). The discovery and characterization of platelet GPIb. *J Thromb Haemost, 3,* 1125–1132.

24. Jorgensen, L. (2005). ADP-induced platelet aggregation in the microcirculation of pig myocardium and rabbit kidneys. *J Thromb Haemost, 3,* 1119–1124.

25. Duke, W. W. (1910). The relation of blood platelets to hemorrhagic disease. Description of a method for determining the bleeding time and the coagulation time and report of three cases of hemorrhagic disease relieved by transfusion. *JAMA, 55,* 1185–1192.

26. Duke, W. W. (1912). The pathogenesis of purpura hemorrhagica with especial reference to the part played by blood-platelets. *Arch Intern Med, 10,* 445–469.

27. Cronkite, E. P., Jacobs, G. J., Brecher, G., et al. (1952). The hemorrhagic phase of the acute radiation syndrome due to exposure of the whole body to penetrating ionizing radiation. *Am J Roentgenol Radium Ther Nuc Med, 67,* 796–803.

28. Gardner, F. H., & Cohen, P. (1960). The value of platelet transfusions. *Med Clin N Am, 44,* 1425–1439.

29. Cronkite, E. P., & Jackson, D. P. (1959). The use of platelet transfusions in hemorrhagic disease. In: L. M. Toscautine (Ed.), *Prog Hematol* (Vol. 2, p. 239). New York: Grune and Stratton.

30. (1987). Platelet transfusion therapy. National Institutes of Health Consensus Conference. *Transfus Med Rev, 1,* 195–200.

31. Schiffer, C. A. (1987). Management of patients refractory to platelet transfusion — An evaluation of methods of donor selection. *Prog Hematol, 15,* 91–113.

32. Heyman, M. R., & Schiffer, C. A. (1990). Platelet transfusion therapy for the cancer patient. *Semin Oncol, 17,* 198–209.

33. Slichter, S. J. (1990). Platelet transfusion therapy. *Hematol Oncol Clin N Am, 4,* 291–311.

34. Carmen, R. (1993). The selection of plastic materials for blood bags. *Transfus Med Rev, 7,* 1–10.

35. Murphy, S. (1994). Metabolic patterns of platelets — Impact on storage for transfusion. *Vox Sanguinis, 67*(Suppl. 3), 271–273.

36. Rinder, H. M., Arbini, A. A., & Snyder, E. L. (1999). Optimal dosing and triggers for prophylactic use of platelet transfusions. *Curr Opin Hematol, 6,* 437–441.

37. Pamphilon, D. H. (1999). The rationale and use of platelet concentrates irradiated with ultraviolet-B light. *Transfus Med Rev, 13,* 323–333.

38. Murphy, S. (1999). The efficacy of synthetic media in the storage of human platelets for transfusion. *Transfus Med Rev, 13,* 153–163.

39. Roddie, P. H., Turner, M. L., & Williamson, L. M. (2000). Leucocyte depletion of blood components. *Blood Rev, 14,* 145–156.

40. Corash, L. (2000). New technologies for the inactivation of infectious pathogens in cellular blood components and the development of platelet substitutes. *Baillieres Best Pract Res Clin Haematol, 13,* 549–563.

41. Beutler, E. (1993). Platelet transfusions: The 20,000/microL trigger. *Blood, 81,* 1411–1413.

42. Sorensen, D. K., Cronkite, E. P., & Bond, V. P. (1959). Effectiveness of transfusions of fresh and lyophilized platelets in controlling bleeding due to thrombocytopenia. *J Clin Invest, 38,* 1689.

43. Wallace, E. L., Churchill, W. H., Surgenor, D. M., et al. (1998). Collection and transfusion of blood and blood components in the United States, 1994. *Transfusion, 38,* 625–636.

44. Sullivan, M. T., & Wallace, E. L. (2005). Blood collection and transfusion in the United States in 1999. *Transfusion, 45,* 141–148.

45. (2005). Fatal bacterial infections associated with platelet transfusions — United States, 2004. *MMWR Morbidity and Mortality Weekly Report, 54,* 168–170.

46. Zucker, M. B., & Borelli, J. (1954). Reversible alterations in platelet morphology produced by anticoagulants and by cold. *Blood, 9,* 602–608.

47. Kattlove, H. E., & Alexander, B. (1971). The effect of cold on platelets. I. Cold-induced platelet aggregation. *Blood, 38,* 39–48.

48. Peerschke, E. I., & Zucker, M. B. (1981). Fibrinogen receptor exposure and aggregation of human blood platelets produced by ADP and chilling. *Blood, 57,* 663–670.

49. Filip, D. J., & Aster, R. H. (1978). Relative hemostatic effectiveness of human platelets stored at 4 degrees and 22 degrees C. *J Lab Clin Med, 91,* 618–624.

50. Hoffmeister, K. M., Felbinger, T. W., Falet, H., et al. (2003). The clearance mechanism of chilled blood platelets. *Cell, 112,* 87–97.

51. Hoffmeister, K. M., Josefsson, E. C., Isaac, N. A., et al. (2003). Glycosylation restores survival of chilled blood platelets. *Science, 301,* 1531–1534.

52. Stadtmauer, E. A., O'Neill, A., Goldstein, L. J., et al. (2000). Conventional-dose chemotherapy compared with high-dose chemotherapy plus autologous hematopoietic stem-cell transplantation for metastatic breast cancer. Philadelphia Bone Marrow Transplant Group. *N Engl J Med, 342,* 1069–1076.

53. Rubinstein, P., Carrier, C., Scaradavou, A., et al. (1998). Outcomes among 562 recipients of placental-blood transplants from unrelated donors. *N Engl J Med, 339,* 1565–1577.

54. Chao, N. J., Emerson, S. G., & Weinberg, K. I. (2004). Stem cell transplantation (cord blood transplants). *Hematology (Am Soc Hematol Educ Program),* 354–371.

55. Rocha, V., Sanz, G., & Gluckman, E. (2004). Umbilical cord blood transplantation. *Curr Opin Hematol, 11,* 375–385.

56. Shimoni, A., & Nagler, A. (2004). Nonmyeloablative stem cell transplantation: Lessons from the first decade of clinical experience. *Curr Hematol Rep, 3,* 242–248.

57. Kaye, J. A. (1998). FDA licensure of NEUMEGA to prevent severe chemotherapy-induced thrombocytopenia. *Stem Cells, 16*(Suppl. 2), 207–223.

58. Kaushansky, K. (1995). Thrombopoietin: The primary regulator of platelet production. *Blood, 86,* 419–431.

59. Kuter, D. J. (2000). Future directions with platelet growth factors. *Semin Hematol, 37,* 41–49.

60. Orita, T., Tsunoda, H., Yabuta, N., et al. (2005). A novel therapeutic approach for thrombocytopenia by minibody agonist of the thrombopoietin receptor. *Blood, 105,* 562–566.

61. Inagaki, K., Oda, T., Naka, Y., et al. (2004). Induction of megakaryocytopoiesis and thrombocytopoiesis by JTZ-132, a novel small molecule with thrombopoietin mimetic activities. *Blood, 104,* 58–64.

62. Geissler, K., Yin, J. A., Ganser, A., et al. (2003). Prior and concurrent administration of recombinant human megakaryocyte growth and development factor in patients receiving consolidation chemotherapy for de novo acute myeloid leukemia — A randomized, placebo-controlled, double-blind safety and efficacy study. *Ann Hematol, 82,* 677–683.

63. Vadhan–Raj, S., Patel, S., Bueso–Ramos, C., et al. (2003). Importance of predosing of recombinant human thrombopoietin to reduce chemotherapy-induced early thrombocytopenia. *J Clin Oncol, 21,* 3158–3167.

64. Avecilla, S. T., Hattori, K., Heissig, B., et al. (2004). Chemokine-mediated interaction of hematopoietic progenitors with the bone marrow vascular niche is required for thrombopoiesis. *Nat Med, 10,* 64–71.

65. Coller, B. S., Springer, K. T., Beer, J. H., et al. (1992). Thromboerythrocytes. In vitro studies of a potential autologous, semi-artificial alternative to platelet transfusions. *J Clin Invest, 89,* 546–555.

66. Adam, G., & Livne, A. A. (1992). Erythrocytes with covalently bound fibrinogen as a cellular replacement for the treatment of thrombocytopenia. *Eur J Clin Invest, 22,* 105–112.

67. Chao, F. C., Kim, B. K., Houranieh, A. M., et al. (1996). Infusible platelet membrane microvesicles: A potential transfusion substitute for platelets. *Transfusion, 36,* 536–542.

68. Alving, B. M., Reid, T. J., Fratantoni, J. C., et al. (1997). Frozen platelets and platelet substitutes in transfusion medicine. *Transfusion, 37,* 866–876.

69. Alving, B. (1998). Potential for synthetic phospholipids as partial platelet substitutes. *Transfusion, 38,* 997–998.

70. Kitaguchi, T., Murata, M., Iijima, K., et al. (1999). Characterization of liposomes carrying von Willebrand factor-binding domain of platelet glycoprotein Ibalpha: A potential substitute for platelet transfusion. *Biochem Biophys Res Commun, 261,* 784–789.

71. Levi, M., Friederich, P. W., Middleton, S., et al. (1999). Fibrinogen-coated albumin microcapsules reduce bleeding in severely thrombocytopenic rabbits. *Nat Med, 5,* 107–111.

72. Blajchman, M. A. (2003). Substitutes and alternatives to platelet transfusions in thrombocytopenic patients. *J Thromb Haemost, 1,* 1637–1641.

73. Alving, B. M. (2002). Platelet substitutes: The reality and the potential. *Vox Sang, 83*(Suppl. 1), 287–288.

74. Pick, M., Eldor, A., Grisaru, D., et al. (2002). Ex vivo expansion of megakaryocyte progenitors from cryopreserved umbilical cord blood. A potential source of megakaryocytes for transplantation. *Exp Hematol, 30,* 1079–1087.

75. Sasayama, N., Kashiwakura, I., Tokushima, Y., et al. (2001). Expansion of megakaryocyte progenitors from cryopreserved leukocyte concentrates of human placental and umbilical cord blood in short-term liquid culture. *Cytotherapy, 3,* 117–126.

76. Lam, A. C., Li, K., Zhang, X. B., et al. (2001). Preclinical ex vivo expansion of cord blood hematopoietic stem and progenitor cells: Duration of culture; the media, serum supplements, and growth factors used; and engraftment in NOD/SCID mice. *Transfusion, 41,* 1567–1576.

77. Decaudin, D., Vautelon, J. M., Bourhis, J. H., et al. (2004). Ex vivo expansion of megaKaryocyle precurser cells in autologous stem cell transplantation for relapsed malignant lymphoma. *Bone Marrow Transplant, 34,* 1089–1093.

78. Werlhof, P. G., & Wichmann, J. E. (1775). Hannoverae imp fratorem Helwingiorium. *Opera Medica, 830,* 748.

79. Cole, R. I. (1907). Note on the production of an agglutinating serum for blood platelets. *Bull Johns Hopkins Hosp, 18,* 261–262.

80. Ledingham, J. C. G., & Bedson, S. P. (1916). Experimental purpura. *Lancet, 1,* 311–316.

81. Kaznelson, P. (1916). Verschwinden den haemorrhagischen Diathese bei einem Fallen von essentieller Thrombopenie (Frank) nach Milz Extirpation. Splenogene thrombolitische Purpura. *Wien Klin Wochenschr, 29,* 1451–1454.

82. Harrington, J., Minnich, V., & Hollingsworth, J. W. (1951). Demonstration of a thrombocytopenic factor in the blood of patients with thrombocytopenic purpura. *Lab Clin Med, 38,* 1–10.

83. Harrington, W. J., Sprague, C. C., & Minnich, V. (1953). Immunologic mechanisms in idiopathic and neonatal thrombocytopenic purpura. *Ann Intern Med, 38,* 433.

84. Louache, F., & Vainchenker, W. (1994). Thrombocytopenia in HIV infection. *Curr Opin Hematol, 1,* 369–372.

85. Karpatkin, S., Nardi, M., Lennette, E. T., et al. (1988). Anti-human immunodeficiency virus type 1 antibody complexes on platelets of seropositive thrombocytopenic homosexuals and narcotic addicts. *Proc Nat Acad Sci USA, 85,* 9763–9767.

86. George, J. N., Raskob, G. E., Shah, S. R., et al. (1998). Drug-induced thrombocytopenia: A systematic review of published case reports. *Ann Intern Med, 129,* 886–890.

87. Warkentin, T. E., & Kelton, J. G. (1990). Heparin and platelets. *Hematol Oncol Clin N Am, 4,* 243–264.

88. Cines, D. B., Bussel, J. B., McMillan, R. B., et al. (2004). Congenital and acquired thrombocytopenia. *Hematology (Am Soc Hematol Educ Program),* 390–406.

89. Fujimura, K. (2005). *Helicobacter pylori* infection and idiopathic thrombocytopenic purpura. *Int J Hematol, 81,* 113–118.

90. Jackson, S., Beck, P. L., Pineo, G. F., et al. (2005). *Helicobacter pylori* eradication: Novel therapy for immune thrombocytopenic purpura? A review of the literature. *Am J Hematol, 78,* 142–150.

91. Lusher, J. M., & Warrier, I. (1987). Use of intravenous gamma globulin in children and adolescents with idiopathic thrombocytopenic purpura and other immune thrombocytopenias. *Am J Med, 83,* 10–16.

92. Bussel, J. B. (1989). The use of intravenous gamma-globulin in idiopathic thrombocytopenic purpura. *Clin Immunol Immunopathol, 53,* S147–S155.

93. Scaradavou, A., & Bussel, J. B. (1998). Clinical experience with anti-D in the treatment of idiopathic thrombocytopenic purpura. *Semin Hematol, 35,* 52–57.

94. George, J. N., Woolf, S. H., Raskob, G. E., et al. (1996). Idiopathic thrombocytopenic purpura: A practice guideline developed by explicit methods for the American Society of Hematology. *Blood, 88,* 3–40.

95. Saleh, M. N., Gutheil, J., Moore, M., et al. (2000). A pilot study of the anti-CD20 monoclonal antibody rituximab in patients with refractory immune thrombocytopenia. *Semin Oncol, 27,* 99–103.

96. Bussel, J. B. (2000). Overview of idiopathic thrombocytopenic purpura: New approach to refractory patients. *Semin Oncol, 27,* 91–98.

97. George, J. N., Kojouri, K., Perdue, J. J., et al. (2000). Management of patients with chronic, refractory idiopathic thrombocytopenic purpura. *Semin Hematol, 37,* 290–298.

98. Vesely, S. K., Perdue, J. J., Rizvi, M. A., et al. (2004). Management of adult patients with persistent idiopathic thrombocytopenic purpura following splenectomy: A systematic review. *Ann Intern Med, 140,* 112–120.

99. Kosugi, S., Kurata, Y., Tomiyama, Y., et al. (1996). Circulating thrombopoietin level in chronic immune thrombocytopenic purpura. *Br J Haematol, 93,* 704–706.

100. Porcelijn, L., Folman, C. C., Bossers, B., et al. (1998). The diagnostic value of thrombopoietin level measurements in thrombocytopenia. *Thromb Haemost, 79,* 1101–1105.

101. Aledort, L. M., Hayward, C. P., Chen, M. G., et al. (2004). Prospective screening of 205 patients with ITP, including diagnosis, serological markers, and the relationship between platelet counts, endogenous thrombopoietin, and circulating antithrombopoietin antibodies. *Am J Hematol, 76,* 205–213.

102. Krafft, A., Huch, R., Hartmann, S., et al. (2004). Combined thrombopoietin and platelet response to altitude in a patient with autoimmune thrombocytopenia. *Thromb Haemost, 91,* 626–627.

103. Bussel, J. B., Mukherjee, R., & Stone, A. J. (2001). A pilot study of rhuIL-11 treatment of refractory ITP. *Am J Hematol, 66,* 172–177.

104. Bussel, J. B., Kuter, D. J., George, J. N., et al. (2006). Effect of a normal thrombopoiesis-stimulating protein (AMG 531) in chronic immune thrombocytopenic purpura. *N Engl J Med* (in press).

105. Newton–Cheh, C., & Hirschhorn, J. N. (2005). Genetic association studies of complex traits: Design and analysis issues. *Mutat Res, 573,* 54–69.

106. Carlson, C. S., Eberle, M. A., Kruglyak, L., et al. (2004). Mapping complex disease loci in whole-genome association studies. *Nature, 429,* 446–452.

107. Guttmacher, A. E., & Collins, F. S. (2002). Genomic medicine — a primer. *N Engl J Med, 347,* 1512–1520.

108. van Loghen, J. J., Dormeijer, H., & van der Hart, M. (1959). Serological and genetical studies on a platelet antigen. *Vox Sanguinis, 4,* 161.

109. Shulman, N. R., Marder, V. J., & Hiler, M. C. (1964). Platelet and leukocyte isoantigens and their antibodies: Serologic, physiologic, and clinical studies. *Prog Hematol, 4,* 222.

110. Shulman, N. R., Aster, R. H., & Leitner, A. (1961). Immunoreactions involving platelets. V. Post-transfusion purpura due to a complement-fixing antibody against a genetically-controlled platelet antigen: A proposed mechanism for thrombocytopenia and its relevance in "autoimmunity." *J Clin Invest, 40,* 1597.

111. Morrison, F. S., & Mollison, P. L. (1966). Post-transfusion purpura. *N Engl J Med, 275,* 243–248.

112. Newman, P. J., Gorski, J., White, G. C., et al. (1988). Enzymatic amplification of platelet-specific messenger RNA using the polymerase chain reaction. *J Clin Invest, 82,* 739–743.

113. Valentin, N., & Newman, P. J. (1994). Human platelet alloantigens. *Curr Opin Hematol, 1,* 381–387.

114. Mueller–Eckhardt, C., Mueller–Eckhardt, G., Willen–Ohff, H., et al. (1985). Immunogenicity of and immune response to the human platelet antigen Zwa is strongly associated with HLA-B8 and DR3. *Tissue Antigens, 26,* 71–76.

115. de Waal, L. P., van Dalen, C. M., Engelfriet, C. P., et al. (1986). Alloimmunization against the platelet-specific Zwa antigen, resulting in neonatal alloimmune thrombocytopenia or posttransfusion purpura, is associated with the supertypic DRw52 antigen including DR3 and DRw6. *Hum Immuno, 17,* 45–53.

116. L'Abbe, D., Tremblay, L., Filion, M., et al. (1992). Alloimmunization to platelet antigen HPA-1a (PIA1) is strongly associated with both HLA-DRB3*0101 and HLA-DQB1*0201. *Hum Immuno, 34,* 107–114.

117. Williamson, L. M., Hackett, G., Rennie, J., et al. (1998). The natural history of fetomaternal alloimmunization to the platelet-specific antigen HPA-1a (PlA1, Zwa) as determined by antenatal screening. *Blood, 92,* 2280–2287.

118. Di Paola, J., Federici, A. B., Mannucci, P. M., et al. (1999). Low platelet alpha2beta1 levels in type I von Willebrand disease correlate with impaired platelet function in a high shear stress system. *Blood, 93,* 3578–3582.

119. Vijayan, K. V., Goldschmidt–Clermont, P. J., Roos, C., et al. (2000). The Pl(A2) polymorphism of integrin beta(3) enhances outside-in signaling and adhesive functions. *J Clin Invest, 105,* 793–802.

120. Michelson, A. D., Furman, M. I., Goldschmidt–Clermont, P., et al. (2000). Platelet GP IIIa Pl(A) polymorphisms display different sensitivities to agonists. *Circulation, 101,* 1013–1018.

121. Beer, J. H., Pederiva, S., & Pontiggia, L. (2000). Genetics of platelet receptor single-nucleotide polymorphisms: Clinical implications in thrombosis. *Ann Med, 32*(Suppl. 1), 10–14.

122. Bray, P. F. (2000). Platelet glycoprotein polymorphisms as risk factors for thrombosis. *Curr Opin Hematol, 7,* 284–289.

123. Kritzik, M., Savage, B., Nugent, D. J., et al. (1998). Nucleotide polymorphisms in the alpha2 gene define multiple alleles that are associated with differences in platelet α2 β1 density. *Blood, 92,* 2382–2388.

124. Vijayan, K. V., Liu, Y., Sun, W., et al. (2005). The Pro33 isoform of integrin beta3 enhances outside-in signaling in human platelets by regulating the activation of serine/threonine phosphatases. *J Biol Chem, 280,* 21756–21762.

125. Feng, D., Lindpaintner, K., Larson, M. G., et al. (2001). Platelet glycoprotein IIIa Pl(a) polymorphism, fibrinogen, and platelet aggregability: The Framingham Heart Study. *Circulation, 104,* 140–144.

126. Vijayan, K. V., Huang, T. C., Liu, Y., et al. (2003). Shear stress augments the enhanced adhesive phenotype of cells expressing the Pro33 isoform of integrin beta3. *FEBS Lett, 540,* 41–46.

127. Cooke, G. E., Bray, P. F., Hamlington, J. D., et al. (1998). PlA2 polymorphism and efficacy of aspirin. *Lancet, 351,* 1253.

128. Szczeklik, A., Undas, A., Sanak, M., et al. (2000). Relationship between bleeding time, aspirin and the PlA1/A2 poly-

morphism of platelet glycoprotein IIIa. *Br J Haematol, 110,* 965–967.

129. Hellem, A. J. (1960). The adhesiveness of human blood platelets in vitro. *Scand J Clin Lab Invest, 12,* 117.

130. Gaarder, A., Jonsen, J., Laland, S., et al. (1961). Adenosine diphosphate in red cells as a factor in the adhesiveness of human blood platelets. *Nature, 192,* 531–532.

131. Salzman, E. W., & Neri, L. L. (1966). Adhesiveness of blood platelets in uremia. *Thromb Diath Haemorr, 15,* 84–92.

132. Bowie, E. J., Owen, C. A., Jr., Thompson, J. H., et al. (1969). Platelet adhesiveness in von Willebrand's disease. *Am J Clin Pathol, 52,* 69–77.

133. Zucker, M. B., Brownlea, S., & McPherson, J. (1987). Insights into the mechanism of platelet retention in glass bead columns. *Ann N Y Acad Sci, 516,* 398–406.

134. Bouma, B. N., Wiegerinck, Y., Sixma, J. J., et al. (1972). Immunological characterization of purified anti-haemophilic factor A (factor VIII) which corrects abnormal platelet retention in Von Willebrand's disease. *Nat New Biol, 236,* 104–106.

135. Coller, B. S., & Zucker, M. B. (1971). Reversible decrease in platelet retention by glass bead columns (adhesiveness) induced by disturbing the blood. *Proc Soc Exp Biol Med, 136,* 769–771.

136. O'Brien, J. (1962). Platelet aggregation. II. Some results from a new method of study. *J Clin Pathol, 15,* 452–481.

137. Born, G. V. (1962). Aggregation of blood platelets by adenosine diphosphate and its reversal. *Nature, 194,* 927–929.

138. Jorgensen, L., Rowsell, H. C., Hovig, T., et al. (1967). Adenosine diphosphate-induced platelet aggregation and myocardial infarction in swine. *Lab Invest, 17,* 616–644.

139. Douthwaite, A. H., & Lintott, G. A. M. (1938). Gastroscopic observation of the effect of aspirin and certain other substances on the stomach. *Lancet, 2,* 1222.

140. Quick, A. J. (1970). *Bleeding problems in clinical medicine.* Philadelphia: WB Saunders.

141. Hovig, T. (1963). Release of a platelet-aggregatiing substance (adenosine diphosphate) from rabbit blood platelets induced by saline "extract" of tendons. *Thromb Diath Haemorr, 143,* 264–278.

142. Spaet, T. H., & Zucker, M. B. (1964). Mechanism of platelet plug formation and role of adenosine diphosphate. *Am J Physiol, 206,* 1267–1274.

143. Weiss, H. J., & Aledort, L. M. (1967). Impaired platelet–connective-tissue reaction in man after aspirin ingestion. *Lancet, 2,* 495–497.

144. Zucker, M. B., & Peterson, J. (1968). Inhibition of adenosine diphosphate-induced secondary aggregation and other platelet functions by acetylsalicylic acid ingestion. *Proc Soc Exper Biol Med, 127,* 547–551.

145. Weiss, H. J., Aledort, L. M., & Kochwa, S. (1968). The effect of salicylates on the hemostatic properties of platelets in man. *J Clin Invest, 47,* 2169–2180.

146. O'Brien, J. R. (1968). Effect of salicylates on human platelets. *Lancet, 1,* 1431.

147. Evans, G., Packham, M. A., Nishizawa, E. E., et al. (1968). The effect of acetylsalicyclic acid on platelet function. *J Exp Med, 128,* 877–894.

148. Smith, J. B., & Willis, A. L. (1971). Aspirin selectively inhibits prostaglandin production in human platelets. *Nat New Biol, 231,* 235–237.

149. Ferreira, S. H., Moncada, S., & Vane, J. R. (1971). Indomethacin and aspirin abolish prostaglandin release from the spleen. *Nat New Biol, 231,* 237–239.

150. Vane, J. R. (1971). Inhibition of prostaglandin synthesis as a mechanism of action for aspirin-like drugs. *Nat New Biol, 231,* 232–235.

151. Marcus, A. J. (1979). The role of prostaglandins in platelet function. *Prog Hematol, 11,* 147–171.

152. Samuelsson, B. (1983). From studies of biochemical mechanism to novel biological mediators: Prostaglandin endoperoxides, thromboxanes, and leukotrienes. Nobel lecture, 8 December 1982. *Biosci Rep, 3,* 791–813.

153. Mann, C. C. (Ed.) (1971). The aspirin wars: Money, medicine, and 100 years of rampant competition. Knopf.

154. Barnathan, E. S., Schwartz, J. S., Taylor, L., et al. (1987). Aspirin and dipyridamole in the prevention of acute coronary thrombosis complicating coronary angioplasty. *Circulation, 76,* 125–134.

155. (1994). Collaborative overview of randomised trials of antiplatelet therapy — I: Prevention of death, myocardial infarction, and stroke by prolonged antiplatelet therapy in various categories of patients. Antiplatelet Trialists' Collaboration. *Br Med J, 308,* 81–106.

156. Marcus, A. J. (1978). The role of lipids in platelet function: With particular reference to the arachidonic acid pathway. *J Lip Res, 19,* 793–826.

157. Marcus, A. J., Safier, L. B., Broekman, M. J., et al. (1995). Thrombosis and inflammation as multicellular processes: Significance of cell–cell interactions. *Thromb Haemost, 74,* 213–217.

158. Moncada, S., Gryglewski, R., Bunting, S., et al. (1976). An enzyme isolated from arteries transforms prostaglandin endoperoxides to an unstable substance that inhibits platelet aggregation. *Nature, 263,* 663–665.

159. Weksler, B. B., Marcus, A. J., & Jaffe, E. A. (1977). Synthesis of prostaglandin I2 (prostacyclin) by cultured human and bovine endothelial cells. *Proc Nat Acad Sci USA, 74,* 3922–3926.

160. Radomski, M. W., Palmer, R. M. J., & Moncada, S. (1987). The anti-aggregating properties of vascular endothelium: Interactions between prostacyclin and nitric oxide. *Br J Pharmacol, 92,* 639–646.

161. Radomski, M. W., Palmer, R. M., & Moncada, S. (1987). Endogenous nitric oxide inhibits human platelet adhesion to vascular endothelium. *Lancet, 2,* 1057–1058.

162. Loscalzo, J. (2001). Nitric oxide insufficiency, platelet activation, and arterial thrombosis. *Circ Res, 88,* 756–762.

163. Marcus, A. J., Broekman, M. J., Drosopoulos, J. H., et al. (1997). The endothelial cell ecto-ADPase responsible for inhibition of platelet function is CD39. *J Clin Invest, 99,* 1351–1360.

164. Marcus, A. J., Broekman, M. J., Drosopoulos, J. H., et al. (2005). Role of CD39 (NTPDase-1) in thromboregulation, cerebroprotection, and cardioprotection. *Sem Thromb Hemost, 31,* 234–246.

165. Rosenberg, R. D., & Aird, W. C. (1999). Vascular-bed — specific hemostasis and hypercoagulable states. *N Engl J Med, 340,* 1555–1564.

166. Gulliver, G. (Ed.). (1846). *The works of William Hewson.* London: F.R.S. Syndenham Society.

167. Bettex–Galland, M., & Luscher, E. F. (1959). Extraction of an actomyosin-like protein from human thrombocytes. *Lancet, 184,* 276–277.

168. Bettex–Galland, M., & Clemetson, K. J. (2005). First isolation of actomyosin from a non-muscle cell: First isolated platelet protein. *J Thromb Haemost, 3,* 834–839.

169. Glanzmann, E. (1918). Hereditäre hämmorhagische thrombasthenie. *Beitr Pathologie Bluplätchen J Kinderkt, 88,* 113.

170. Zucker, M. B., Pert, J. H., & Hilgartner, M. W. (1966). Platelet function in a patient with thrombasthenia. *Blood, 28,* 524–534.

171. Caen, J. P., Castaldi, P. A., Leclerc, J. C., et al. (1966). Congenital bleeding disorders with long bleeding time and normal platelet count. I. Glanzmann's thrombasthenia. *Am J Med, 41,* 4.

172. Peerschke, E. I. (1985). The platelet fibrinogen receptor. *Sem Hematol, 22,* 241–259.

173. Bennett, J. S. (1985). The platelet–fibrinogen interaction. In J. N. George, A. T. Nurden, & D. R. Phillips (Eds.), *Platelet membrane glycoproteins* (pp. 193). New York: Plenum.

174. Phillips, D. R., Charo, I. F., Parise, L. V., et al. (1988). The platelet membrane glycoprotein IIb-IIIa complex. *Blood, 71,* 831–843.

175. Plow, E. F., & Ginsberg, M. H. (1989). Cellular adhesion: GPIIb-IIIa as a prototypic adhesion receptor. *Prog Hem Thromb, 9,* 117–156.

176. Coller, B. S., Seligsohn, U., Peretz, H., et al. (1994). Glanzmann thrombasthenia: New insights from an historical perspective. *Sem Hematol, 31,* 301–311.

177. Nurden, A. T., & Caen, J. P. (1974). An abnormal platelet glycoprotein pattern in three cases of Glanzmann's thrombasthenia. *Br J Haematol, 28,* 253–260.

178. Phillips, D. R., Jenkins, C. S., Luscher, E. F., et al. (1975). Molecular differences of exposed surface proteins on thrombasthenic platelet plasma membranes. *Nature, 257,* 599–600.

179. Poncz, M., Eisman, R., Heidenreich, R., et al. (1987). Structure of the platelet membrane glycoprotein IIb. Homology to the alpha subunits of the vitronectin and fibronectin membrane receptors. *J Biol Chem, 262,* 8476–8482.

180. Fitzgerald, L. A., Steiner, B., Rall, S. C., et al. (1987). Protein sequence of endothelial glycoprotein IIIa derived from a cDNA clone. Identity with platelet glycoprotein IIIa and similarity to "integrin." *J Biol Chem, 262,* 3936–3939.

181. Hynes, R. O. (1987). Integrins: A family of cell surface receptors. *Cell, 48,* 549–554.

182. Coller, B. S., French, D. L., & Mitchell, B. (2005). Hereditary qualitative platelet disorders. In M. A. Lichtman, B. Beutler, T. J. Kipps, U. Seligsohn, K. Kaushansky, & J. T. Prchal (Eds.), *Williams hematology.* New York: McGraw-Hill.

183. Hodivala–Dilke, K. M., Tsakiris, D. A., Rayburn, H., et al. (1999). Beta3-integrin-deficient mice are a model for Glanzmann thrombasthenia showing placental defects and reduced survival. *J Clin Invest, 103,* 229–238.

184. Smyth, S. S., Reis, E. D., Vaananen, H., et al. (2001). Variable protection of β3-integrin-deficient mice from thrombosis initiated by different mechanisms. *Blood, 98,* 1055–1062.

185. Law, D. A., DeGuzmann, F. R., Heiser, P., et al. (1999). Integrin cytoplasmic tyrosine motif is required for outside-in alphaIIbbeta3 signalling and platelet function. *Nature, 401,* 808–811.

186. Reynolds, L. E., Conti, F. J., Lucas, M., et al. (2005). Accelerated re-epithelialization in beta3-integrin-deficient mice is associated with enhanced TGF-beta1 signaling. *Nat Med, 11,* 167–174.

187. Reynolds, L. E., Wyder, L., Lively, J. C., et al. (2002). Enhanced pathological angiogenesis in mice lacking beta3 integrin or beta3 and beta5 integrins. *Nat Med, 8,* 27–34.

188. McHugh, K. P., Hodivala–Dilke, K., Zheng, M. H., et al. (2000). Mice lacking beta3 integrins are osteosclerotic because of dysfunctional osteoclasts. *J Clin Invest, 105,* 433–440.

189. Fang, J., Hodivala–Dilke, K., Johnson, B. D., et al. (2005). Therapeutic expression of the platelet-specific integrin, αIIbβ3, in a murine model for Glanzmann thrombasthenia. *Blood, 106,* 2671–2679.

190. Wautier, J. L., & Gruel, Y. (1989). Prenatal diagnosis of platelet disorders. *Baillieres Clinical Haematology, 2,* 569–583.

191. Champeix, P., Forestier, F., Daffos, F., et al. (1988). Prenatal diagnosis of a molecular variant of Glanzmann's thrombasthenia. *Curr Stud Hematol Blood Transfu, 55,* 180–183.

192. Seligsohn, U., Mibashan, R. S., Rodeck, C. H., et al. (1985). Prenatal diagnosis of Glanzmann's thrombasthenia. *Lancet, 2,* 1419.

193. Peretz, H., Seligsohn, U., Zwang, E., et al. (1991). Detection of the Glanzmann's thrombasthenia mutations in Arab and Iraqi–Jewish patients by polymerase chain reaction and restriction analysis of blood or urine samples. *Thromb Haemost, 66,* 500–504.

194. Coller, B. S. (2001). Anti-GPIIb/IIIa drugs: Current strategies and future directions. *Thromb Haemost, 86,* 437–443.

195. Phillips, D. R., & Scarborough, R. M. (1997). Clinical pharmacology of eptifibatide. *Am J Cardiol, 80,* 11B–20B.

196. Scarborough, R. M. (1999). Development of eptifibatide. *Am Heart J, 138,* 1093–1104.

197. McClellan, K. J., & Goa, K. L. (1998). Tirofiban. A review of its use in acute coronary syndromes. *Drugs, 56,* 1067–1080.

198. Chew, D. P., & Moliterno, D. J. (2000). A critical appraisal of platelet glycoprotein IIb/IIIa inhibition. *J Am Coll Cardiol, 36,* 2028–2035.

199. Xiao, T., Takagi, J., Coller, B. S., et al. (2004). Structural basis for allostery in integrins and binding to fibrinogen-mimetic therapeutics. *Nature, 432,* 59–67.

200. Artoni, A., Li, J., Mitchell, B., et al. (2004). Integrin β3 regions controlling binding of murine mAb 7E3: Implications for the mechanism of integrin αIIbβ3 activation. *Proc Nat Acad Sci USA, 101,* 13114–13120.

201. Bernard, J., & Soulier, J.-P. (1948). Sur une nouvelle variete de dystrophie thrombocytaire-hemorragipare congenitale. *Semin Hop Paris, 24,* 3217.

202. Weiss, H. J., Tschopp, T. B., Baumgartner, H. R., et al. (1974). Decreased adhesion of giant (Bernard–Soulier) platelets to subendothelium. Further implications on the role of the von Willebrand factor in hemostasis. *Am J Med, 57,* 920–925.

203. Howard, M. A., Hutton, R. A., & Hardisty, R. M. (1973). Hereditary giant platelet syndrome: A disorder of a new aspect of platelet function. *Br Med J, 2,* 586–588.

204. Bithell, T. C., Parekh, S. J., & Strong, R. R. (1972). Platelet-function studies in the Bernard–Soulier syndrome. *Ann N Y Acad Sci, 201,* 145–160.

205. Grottum, K. A., & Solum, N. O. (1969). Congenital thrombocytopenia with giant platelets: A defect in the platelet membrane. *Br J Haematol, 16,* 277–290.

206. Grottum, K. A., & Solum, N. O. (1969). Congenital thrombocytopenia with giant platelets: A defect in the platelet membrane. *Br J Haematol, 16,* 277–290.

207. Howard, M. A., & Firkin, B. G. (1971). Ristocetin — a new tool in the investigation of platelet aggregation. *Thromb Diath Haemorrhag, 26,* 362–369.

208. Howard, M. A., Sawers, R. J., & Firkin, B. G. (1973). Ristocetin: A means of differentiating von Willebrand's disease into two groups. *Blood, 41,* 687–690.

209. Brinkhous, K. M., Barnes, D. S., Potter, J. Y., et al. (1981). Von Willebrand syndrome induced by a Bothrops venom factor: Bioassay for venom coagglutinin. *Proc Nat Acad Sci USA, 78,* 3230–3234.

210. Weiss, H. J., Hoyer, L. W., Rickles, F. R., et al. (1973). Quantitative assay of a plasma factor deficient in von Willebrand's disease that is necessary for platelet aggregation. Relationship to factor VIII procoagulant activity and antigen content. *J Clin Invest, 52,* 2708–2716.

211. Nurden, A. T., & Caen, J. P. (1975). Specific roles for platelet surface glycoproteins in platelet function. *Nature, 255,* 720–722.

212. Clemetson, K. J., McGregor, J. L., James, E., et al. (1982). Characterization of the platelet membrane glycoprotein abnormalities in Bernard–Soulier syndrome and comparison with normal by surface-labeling techniques and high-resolution two-dimensional gel electrophoresis. *J Clin Invest, 70,* 304–311.

213. Berndt, M. C., Gregory, C., Chong, B. H., et al. (1983). Additional glycoprotein defects in Bernard–Soulier's syndrome: Confirmation of genetic basis by parental analysis. *Blood, 62,* 800–807.

214. Ruggeri, Z. (1991). The platelet glycoprotein Ib-IX complex. *Prog Hem Thromb, 10,* 35–68.

215. Roth, G. J. (1991). Developing relationships: Arterial platelet adhesion, glycoprotein Ib, and leucine-rich glycoproteins. *Blood, 77,* 5–19.

216. Lopez, J. A., Andrews, R. K., Afshar–Kharghan, V., et al. (1998). Bernard–Soulier syndrome. *Blood, 91,* 4397–4418.

217. Coller, B. S., Peerschke, E. I., Scudder, L. E., et al. (1983). Studies with a murine monoclonal antibody that abolishes ristocetin-induced binding of von Willebrand factor to platelets: Additional evidence in support of GPIb as a platelet receptor for von Willebrand factor. *Blood, 61,* 99–110.

218. Montgomery, R. R., Kunicki, T. J., Taves, C., et al. (1983). Diagnosis of Bernard–Soulier syndrome and Glanzmann's thrombasthenia with a monoclonal assay on whole blood. *J Clin Invest, 71,* 385.

219. Cauwenberghs, N., Vanhoorelbeke, K., Vauterin, S., et al. (2001). Epitope mapping of inhibitory antibodies against platelet glycoprotein Ibalpha reveals interaction between the leucine-rich repeat N-terminal and C-terminal flanking domains of glycoprotein Ibalpha. *Blood, 98,* 652–660.

220. Emsley, J., Cruz, M., Handin, R., et al. (1998). Crystal structure of the von Willebrand factor A1 domain and implications for the binding of platelet glycoprotein Ib. *J Biol Chem, 273,* 10396–10401.

221. Vasudevan, S., Roberts, J. R., McClintock, R. A., et al. (2000). Modeling and functional analysis of the interaction between von Willebrand factor A1 domain and glycoprotein Ibalpha. *J Biol Chem, 275,* 12763–12768.

222. Cruz, M. A., Diacovo, T. G., Emsley, J., et al. (2000). Mapping the glycoprotein Ib-binding site in the von Willebrand factor A1 domain. *J Biol Chem, 275,* 19098–19105.

223. Ware, J., Russell, S., & Ruggeri, Z. M. (2000). Generation and rescue of a murine model of platelet dysfunction: The Bernard–Soulier syndrome. *Proc Nat Acad Sci USA, 97,* 2803–2808.

224. Kahn, M. L., Diacovo, T. G., Bainton, D. F., et al. (1999). Glycoprotein V-deficient platelets have undiminished thrombin responsiveness and do not exhibit a Bernard–Soulier phenotype. *Blood, 94,* 4112–4121.

225. Ramakrishnan, V., Reeves, P. S., DeGuzman, F., et al. (1999). Increased thrombin responsiveness in platelets from mice lacking glycoprotein V. *Proc Nat Acad Sci USA, 96,* 13336–13341.

226. O'Brien, J. R., & Heywood, J. B. (1967). Some interactions between human platelets and glass: Von Willebrand's disease compared with normal. *J Clin Pathol, 20,* 56–64.

227. Sixma, J. J., Sakariassen, K. S., Beeser–Visser, N. H., et al. (1984). Adhesion of platelets to human artery subendothelium: Effect of factor VIII-von Willebrand factor of various multimeric composition. *Blood, 63,* 128–139.

228. Peterson, D. M., Stathopoulos, N. A., Giorgio, T. D., et al. (1987). Shear-induced platelet aggregation requires von Willebrand factor and platelet membrane glycoproteins Ib and IIb-IIIa. *Blood, 69,* 625–628.

229. Ikeda, Y., Handa, M., Kawano, K., et al. (1991). The role of von Willebrand factor and fibrinogen in platelet aggregation under varying shear stress. *J Clin Invest, 87,* 1234–1240.

230. Kroll, M. H., Hellums, J. D., McIntire, L. V., et al. (1996). Platelets and shear stress. *Blood, 88,* 1525–1541.

231. Ruggeri, Z. M. (1997). Mechanisms initiating platelet thrombus formation. *Thromb Haemost, 78,* 611–616.

232. Wu, Y. P., Vink, T., Schiphorst, M., et al. (2000). Platelet thrombus formation on collagen at high shear rates is mediated by von Willebrand factor–glycoprotein Ib interaction and inhibited by von Willebrand factor–glycoprotein IIb/IIIa interaction. *Arterioscl Thromb Vasc Biol, 20,* 1661–1667.

233. Reininger, A. J., Heijnen, H. F., Shumann, H., et al. (2006). Mechanism of platelet adhesion to von Willebrand factor and microparticle formation under high shear stress. *Blood, 107,* 3537–3545.

234. Ginsburg, D., Handin, R. I., Bonthron, D. T., et al. (1985). Human von Willebrand factor (vWF): Isolation of complementary DNA (cDNA) clones and chromosomal localization. *Science, 228,* 1401–1406.

235. Loscalzo, J., Fisch, M., & Handin, R. I. (1985). Solution studies of the quaternary structure and assembly of human von Willebrand factor. *Biochemistry, 24,* 4468–4475.

236. Weiss, H. J., Hawiger, J., Ruggeri, Z. M., et al. (1989). Fibrinogen-independent platelet adhesion and thrombus formation on subendothelium mediated by glycoprotein IIb-IIIa complex at high shear rate. *J Clin Invest, 83,* 288–297.

237. Pannekoek, H., & Voorberg, J. (1989). Molecular cloning, expression and assembly of multimeric von Willebrand factor. *Baillieres Clin Haematol, 2,* 879–896.

238. Meyer, D., Pietu, G., Fressinaud, E., et al. (1991). von Willebrand factor: Structure and function. *Mayo Clin Proc, 66,* 516–523.

239. Ginsburg, D. (1991). The von Willebrand factor gene and genetics of von Willebrand's disease. *Mayo Clin Proc, 66,* 506–515.

240. Goto, S., Salomon, D. R., Ikeda, Y., et al. (1995). Characterization of the unique mechanism mediating the shear-dependent binding of soluble von Willebrand factor to platelets. *J Biol Chem, 270,* 23352–23361.

241. Moschcowitz, E. (1924). Hyaline thrombosis of the terminal arterioles and capillaries: A hitherto undescribed disease. *Proc N Y Pathol Soc, 24,* 21–24.

242. Schulman, I., Pierce, M., Lukens, A., et al. (1960). Studies on thrombopoiesis. I. A factor in normal human plasma required for platelet production; chronic thrombocytopenia due to its deficiency. *Blood, 16,* 943–957.

243. Upshaw, J. D., Jr. (1978). Congenital deficiency of a factor in normal plasma that reverses microangiopathic hemolysis and thrombocytopenia. *N Engl J Med, 298,* 1350–1352.

244. Levy, G. G., Motto, D. G., & Ginsburg, D. (2005). ADAMTS13 turns 3. *Blood, 106,* 11–17.

245. Bounameaux, H. (1956). Recherches sur le mécanisme de la rétraction du caillot et de la métamorphose visqueuse des plaquettes. *Experientia, 12,* 355.

246. Jamieson, G. A. (1997). Pathophysiology of platelet thrombin receptors. *Thromb Haemost, 78,* 242–246.

247. Phillips, D. R. (1974). Thrombin interaction with human platelets. Potentiation of thrombin-induced aggregation and release by inactivated thrombin. *Thromb Diath Haemorr, 32,* 207–215.

248. Phillips, D. R., & Agin, P. P. (1977). Platelet plasma membrane glycoproteins. Identification of a proteolytic substrate for thrombin. *Biochem Biophys Res Commun, 75,* 940–947.

249. Jamieson, G. A., & Okumura, T. (1978). Reduced thrombin binding and aggregation in Bernard–Soulier platelets. *J Clin Invest, 61,* 861–864.

250. McGowan, E. B., Ding, A., & Detwiler, T. C. (1983). Correlation of thrombin-induced glycoprotein V hydrolysis and platelet activation. *J Biol Chem, 258,* 11243–11248.

251. Bienz, D., Schnippering, W., & Clemetson, K. J. (1986). Glycoprotein V is not the thrombin activation receptor on human blood platelets. *Blood, 68,* 720–725.

252. Vu, T. K., Hung, D. T., Wheaton, V. I., et al. (1991). Molecular cloning of a functional thrombin receptor reveals a novel proteolytic mechanism of receptor activation. *Cell, 64,* 1057–1068.

253. Kahn, M. L., Zheng, Y. W., Huang, W., et al. (1998). A dual thrombin receptor system for platelet activation. *Nature, 394,* 690–694.

254. Nakanishi–Matsui, M., Zheng, Y. W., Sulciner, D. J., et al. (2000). PAR3 is a cofactor for PAR4 activation by thrombin. *Nature, 404,* 609–613.

255. Faruqi, T. R., Weiss, E. J., Shapiro, M. J., et al. (2000). Structure-function analysis of protease-activated receptor 4 tethered ligand peptides. Determinants of specificity and utility in assays of receptor function. *J Biol Chem, 275,* 19728–19734.

256. Ramakrishnan, V., DeGuzman, F., Bao, M., et al. (2001). A thrombin receptor function for platelet glycoprotein Ib-IX unmasked by cleavage of glycoprotein V. *Proc Nat Acad Sci USA, 98,* 1823–1828.

257. De Candia, E., Hall, S. W., Rutella, S., et al. (2001). Binding of thrombin to glycoprotein Ib accelerates the hydrolysis of Par-1 on intact platelets. *J Biol Chem, 276,* 4692–4698.

258. Hallam, T. J., & Rink, T. J. (1985). Responses to adenosine diphosphate in human platelets loaded with the fluorescent calcium indicator quin2. *J Physiol, 368,* 131–146.

259. Ware, J. A., Johnson, P. C., Smith, M., et al. (1986). Effect of common agonists on cytoplasmic ionized calcium concentration in platelets. Measurement with 2-methyl-6-methoxy 8-nitroquinoline (quin2) and aequorin. *J Clin Invest, 77,* 878–886.

260. Caen, J. P., & Michel, H. (1972). Platelet shape change and aggregation. *Nature, 240,* 148–149.

261. Peerschke, E. I., & Zucker, M. B. (1980). Relationship of ADP-induced fibrinogen binding to platelet shape change and aggregation elucidated by use of colchicine and cytochalasin B. *Thromb Haemost, 43,* 58–60.

262. Salzman, E. W. (1972). Cyclic AMP and platelet function. *N Engl J Med, 286,* 358–363.

263. Hourani, S. M., & Hall, D. A. (1994). Receptors for ADP on human blood platelets. *Trends Pharmacol Sci, 15,* 103–108.

264. Jin, J., Daniel, J. L., & Kunapuli, S. P. (1998). Molecular basis for ADP-induced platelet activation. II. The P2Y1 receptor mediates ADP-induced intracellular calcium mobilization and shape change in platelets. *J Biol Chem, 273,* 2030–2034.

265. Leon, C., Hechler, B., Freund, M., et al. (1999). Defective platelet aggregation and increased resistance to thrombosis

in purinergic P2Y(1) receptor-null mice. *J Clin Invest, 104,* 1731–1737.

266. Sun, B., Li, J., Okahara, K., et al. (1998). P2X1 purinoceptor in human platelets. Molecular cloning and functional characterization after heterologous expression. *J Biol Chem, 273,* 11544–11547.

267. Geiger, J., Brich, J., Honig–Liedl, P., et al. (1999). Specific impairment of human platelet P2Y(AC) ADP receptor-mediated signaling by the antiplatelet drug clopidogrel. *Arterioscler Thromb Vasc Biol, 19,* 2007–2011.

268. Cattaneo, M., & Gachet, C. (1999). ADP receptors and clinical bleeding disorders. *Arterioscler Thromb Vasc Biol, 19,* 2281–2285.

269. Hollopeter, G., Jantzen, H. M., Vincent, D., et al. (2001). Identification of the platelet ADP receptor targeted by antithrombotic drugs. *Nature, 409,* 202–207.

270. Mahaut–Smith, M. P., Ennion, S. J., Rolf, M. G., et al. (2000). ADP is not an agonist at P2X(1) receptors: Evidence for separate receptors stimulated by ATP and ADP on human platelets. *Br J Pharmacol, 131,* 108–114.

271. Hechler, B., Cattaneo, M., & Gachet, C. (2005). The P2 receptors in platelet function. *Sem Thromb Hemost, 31,* 150–161.

272. Wiviott, S. D., Antman, E. M., Winters, K. J., et al. (2005). Randomized comparison of prasugrel (CS-747, LY640315), a novel thienopyridine P2Y12 antagonist, with clopidogrel in percutaneous coronary intervention: Results of the Joint Utilization of Medications to Block Platelets Optimally (JUMBO)-TIMI 26 trial. *Circulation, 111,* 3366–3373.

273. van Giezen, J. J., & Humphries, R. G. (2005). Preclinical and clinical studies with selective reversible direct P2Y12 antagonists. *Sem Thromb Hemost, 31,* 195–204.

274. Boeynaems, J. M., van Giezen, H., Savi, P., et al. (2005). P2Y receptor antagonists in thrombosis. *Curr Opin Investig Drugs, 6,* 275–282.

275. Gachet, C., & Hechler, B. (2005). The platelet P2 receptors in thrombosis. *Sem Thromb Hemost, 31,* 162–167.

276. Gasic, G. J., Gasic, T. B., & Stewart, C. C. (1968). Antimetastatic effects associated with platelet reduction. *Proc Nat Acad Sci USA, 61,* 46–52.

277. Gasic, G. J. (1984). Role of plasma, platelets, and endothelial cells in tumor metastasis. *Cancer Metastasis Rev, 3,* 99–114.

278. Greenberg, J., Packham, M. A., Cazenave, J. P., et al. (1975). Effects on platelet function of removal of platelet sialic acid by neuraminidase. *Lab Invest, 32,* 476–484.

279. Nierodzik, M. L., Klepfish, A., & Karpatkin, S. (1995). Role of platelets, thrombin, integrin IIb-IIIa, fibronectin and von Willebrand factor on tumor adhesion in vitro and metastasis in vivo. *Thromb Haemost, 74,* 282–290.

280. Zacharski, L. R., Rickles, F. R., Henderson, W. G., et al. (1982). Platelets and malignancy. Rationale and experimental design for the VA Cooperative Study of RA-233 in the treatment of cancer. *Am J Clin Oncol, 5,* 593–609.

281. Mehta, P. (1984). Potential role of platelets in the pathogenesis of tumor metastasis. *Blood, 63,* 55–63.

282. Jurasz, P., Alonso–Escolano, D., & Radomski, M. W. (2004). Platelet–cancer interactions: Mechanisms and pharmacology of tumour cell-induced platelet aggregation. *Br J Pharmacol, 143,* 819–826.

283. Karpatkin, S., Pearlstein, E., Ambrogio, C., et al. (1988). Role of adhesive proteins in platelet tumor interaction in vitro and metastasis formation in vivo. *J Clin Invest, 81,* 1012–1019.

284. Huang, Y. Q., Li, J. J., & Karpatkin, S. (2000). Thrombin inhibits tumor cell growth in association with up-regulation of p21(waf/cip1) and caspases via a p53-independent, STAT-1-dependent pathway. *J Biol Chem, 275,* 6462–6468.

285. Hejna, M., Raderer, M., & Zielinski, C. C. (1999). Inhibition of metastases by anticoagulants. *J Nat Cancer Inst, 91,* 22–36.

286. Amirkhosravi, A., Amaya, M., Siddiqui, F., et al. (1999). Blockade of GPIIb/IIIa inhibits the release of vascular endothelial growth factor (VEGF) from tumor cell-activated platelets and experimental metastasis. *Platelets, 10,* 285–292.

287. Mohle, R., Green, D., Moore, M. A., et al. (1997). Constitutive production and thrombin-induced release of vascular endothelial growth factor by human megakaryocytes and platelets. *Proc Nat Acad Sci USA, 94,* 663–668.

288. Verheul, H. M., Hoekman, K., Luykx–de Bakker, S., et al. (1997). Platelet: Transporter of vascular endothelial growth factor. *Clin Cancer Res, 3,* 2187–2190.

289. Gevaert, K., Eggermont, L., Demol, H., et al. (2000). A fast and convenient MALDI-MS based proteomic approach: Identification of components scaffolded by the actin cytoskeleton of activated human thrombocytes. *J Biotechnol, 78,* 259–269.

290. Marcus, K., Immler, D., Sternberger, J., et al. (2000). Identification of platelet proteins separated by two-dimensional gel electrophoresis and analyzed by matrix assisted laser desorption/ionization-time of flight-mass spectrometry and detection of tyrosine-phosphorylated proteins. *Electrophoresis, 21,* 2622–2636.

291. Gravel, P., Sanchez, J. C., Walzer, C., et al. (1995). Human blood platelet protein map established by two-dimensional polyacrylamide gel electrophoresis. *Electrophoresis, 16,* 1152–1159.

292. Asakawa, J., Neel, J. V., Takahashi, N., et al. (1988). Heterozygosity and ethnic variation in Japanese platelet proteins. *Human Genet, 78,* 1–8.

293. Maguire, P. B., Foy, M., & Fitzgerald, D. J. (2005). Using proteomics to identify potential therapeutic targets in platelets. *Biochem Soc Trans, 33,* 409–412.

294. Garcia, A. (2006). Proteome analysis of signaling cascades in human platelets. *Blood Cells, Mol, and Diseases, 36,* 152–156.

295. Perrotta, P. L., & Bahou, W. F. (2004). Proteomics in platelet science. *Curr Hematol Rep, 3,* 462–469.

296. Marcus, K., & Meyer, H. E. (2004). Two-dimensional polyacrylamide gel electrophoresis for platelet proteomics. *Methods Mol Biol, 273,* 421–434.

297. Wilcox, D. A., Olsen, J. C., Ishizawa, L., et al. (2000). Megakaryocyte-targeted synthesis of the integrin beta(3)-subunit results in the phenotypic correction of Glanzmann thrombasthenia. *Blood, 95,* 3645–3651.

298. Italiano, J. E., Jr. Lecine, P., Shivdasani, R. A., et al. (1999). Blood platelets are assembled principally at the ends of proplatelet processes produced by differentiated megakaryocytes. *J Cell Biol, 147,* 1299–1312.

299. Tsakiris, D. A., Scudder, L., Hodivala–Dilke, K. M., et al. (1999). Hemostasis in the mouse (*Mus musculus*): A review. *Thromb Haemost, 81,* 177–188.

300. Reilly, M. P., & McKenzie, S. E. (2002). Insights from mouse models of heparin-induced thrombocytopenia and thrombosis. *Curr Opin Hematol, 9,* 395–400.

301. McKenzie, S. E. (2002). Humanized mouse models of FcR clearance in immune platelet disorders. *Blood Rev, 16,* 3–5.

302. Degen, J. L. (2001). Genetic interactions between the coagulation and fibrinolytic systems. *Thromb Haemost, 86,* 130–137.

303. Jagadeeswaran, P., Sheehan, J. P., Craig, F. E., et al. (1999). Identification and characterization of zebrafish thrombocytes. *Br J Haematol, 107,* 731–738.

304. Thattaliyath, B., Cykowski, M., & Jagadeeswaran, P. (2005). Young thrombocytes initiate the formation of arterial thrombi in zebrafish. *Blood, 106,* 118–124.

305. Carmeliet, P., Moons, L., & Collen, D. (1998). Mouse models of angiogenesis, arterial stenosis, atherosclerosis and hemostasis. *Cardiovasc Res, 39,* 8–33.

306. Ni, H., Denis, C. V., Subbarao, S., et al. (2000). Persistence of platelet thrombus formation in arterioles of mice lacking both von Willebrand factor and fibrinogen. *J Clin Invest, 106,* 385–392.

307. Minamitani, H., Tsukada, K., Sekizuka, E., et al. (2003). Optical bioimaging: From living tissue to a single molecule: Imaging and functional analysis of blood flow in organic microcirculation. *J Pharmacol Sci, 93,* 227–233.

308. Celi, A., Merrill–Skoloff, G., Gross, P., et al. (2003). Thrombus formation: Direct real-time observation and digital analysis of thrombus assembly in a living mouse by confocal and widefield intravital microscopy. *J Thromb Haemost, 1,* 60–68.

# Platelet Biology

# The Evolution of Mammalian Platelets

**Jack Levin**

*Departments of Laboratory Medicine and Medicine, University of California School of Medicine,
San Francisco, California*

## I. Introduction

The mammalian platelet is derived from the cytoplasm of megakaryocytes, the only polyploid hematopoietic cell. Polyploid megakaryocytes and their progeny, nonnucleated platelets, are found only in mammals. In all other animal species, cells involved in hemostasis and blood coagulation are nucleated. The nucleated cells primarily involved in nonmammalian, vertebrate hemostasis are designated *thrombocytes* to distinguish them from nonnucleated platelets.

In many invertebrates, only one type of cell circulates in the blood (or hemolymph), and this single type of cell is typically involved in multiple defense mechanisms of the animal, including hemostasis. These cells are capable of aggregating and sealing wounds. This process is probably the earliest cell-based hemostatic function. A comparison of amebocytes (the only type of circulating cell in the hemolymph of the horseshoe crab, *Limulus polyphemus*) with human platelets provides a basis for understanding the many nonhemostatic functions of platelets.[1-4] Platelets appear to possess many of the multiple capabilities that characterize "primitive" circulating amebocytes, only one of whose functions is hemostasis. For example, platelets possess rudimentary bactericidal and phagocytic activity (see Chapter 40). They have been shown to interact with bacteria, endotoxins, viruses, parasites, and fungi. Platelets are not only important for the maintenance of hemostasis but are also inflammatory cells (see Chapter 39). Despite these overall similarities in function, there remains no proof that invertebrate blood cells evolved into platelets. Furthermore, the fact that nonnucleated platelets and their polyploid megakaryocyte progenitors in the bone marrow are present only in mammals suggests that some important feature of mammalian physiology benefits from this unique mechanism for the production of anucleate cells from the cytoplasm of a larger cell, for the apparently major purpose of supporting hemostasis. However, because both monotremes, which are egg-laying mammals, and marsupials, which have a non-placental pregnancy, possess megakaryocytes and platelets, it is apparent that neither live birth nor the presence of a placenta accounts for the evolution of platelets in mammals. Therefore, the biological advantage gained from the presence of polyploid megakaryocytes as the origin of nonnucleated platelet progeny remains unidentified.

## II. Invertebrates

In many marine invertebrates, only one type of cell circulates in the blood or is present in the coelomic fluid. This single cell type plays multiple roles in the defense mechanisms of the animal, including hemostasis. Such cells are capable of aggregating and sealing wounds. Although the biochemical basis for adhesion of these cells is not understood, the participation of cell aggregation in invertebrate hemostasis suggests that this process is perhaps the earliest cell-based hemostatic function (Figs. 1-1 and 1-2).[1-4] The hemocytes of the ascidian *Halocynthia roretzi,* for example, aggregate after removal from the hemolymph (i.e., the circulating body fluid, equivalent to blood, in animals with open circulatory systems).[3] Aggregation depends on divalent cations and is inhibited by ethylene diamine tetraacetic acid (EDTA). Similar to the response of mammalian platelets to vascular trauma, repeated sampling of hemolymph from the same area of an individual *H. roretzi* results in hemocytes that are increasingly activated, as measured by aggregometry (Fig. 1-3). Furthermore, the aggregated hemocytes release a factor into the plasma that induces additional aggregation. The hemocytes of the California mussel *Mytilus californianus* rapidly aggregate and adhere to foreign surfaces after the removal of blood from the animal.[5] Aggregation *in vitro* was shown to be a two-step process and distinct from adhesion. Adhesion is $Ca^{++}$- or $Mg^{++}$-dependent.[5] The parallels with mammalian platelet function are evident.

The enormous range of types of cell-based coagulation in insects has been described by Grégoire.[6] The hemolymph of many insects contains coagulocytes, which on contact

PLATELETS

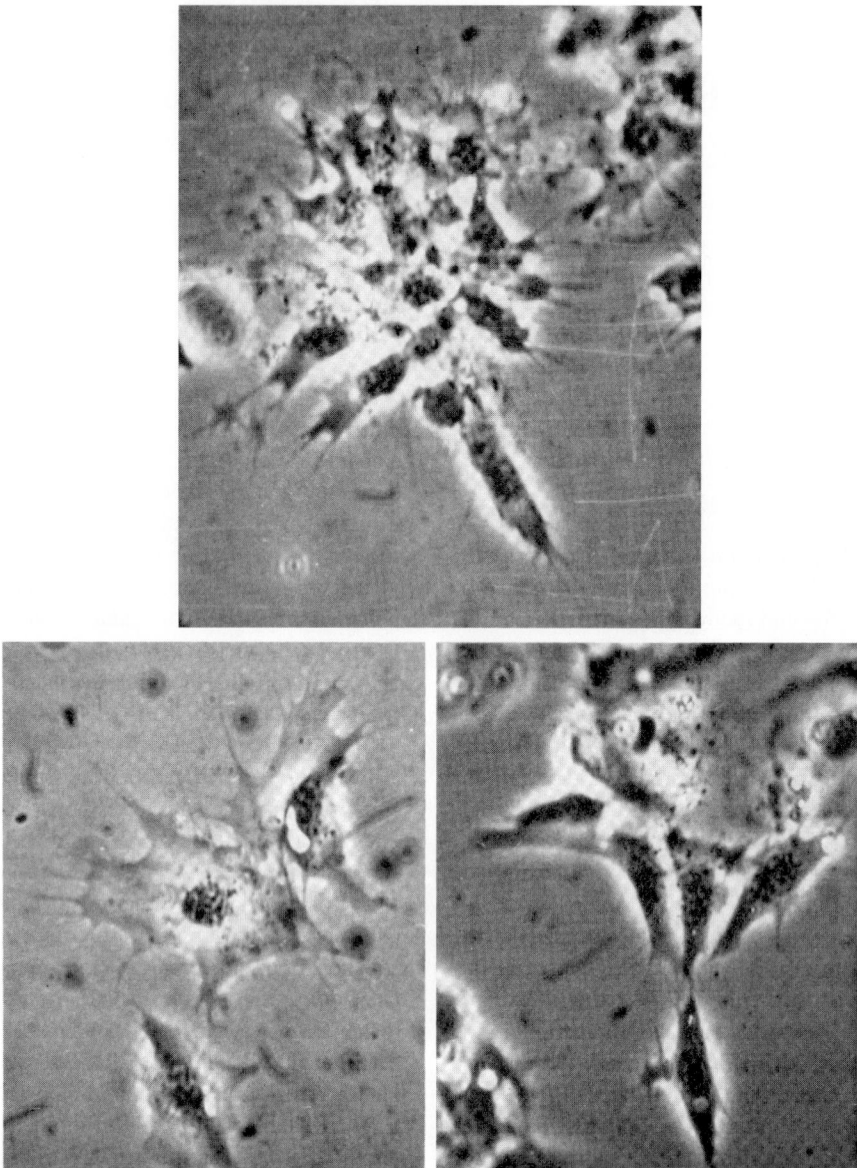

**Figure 1-1.** Aggregation and alteration in the shape of *Limulus* amebocytes during cellular clotting on a glass surface. Note the prominent nucleus and the variety of shapes that occur after activation. The cells have also become degranulated. (See also Figures 1-11, 1-13, and 1-14.) Magnification ×200 (upper) and ×320 (lower). (From Levin and Bang,[1] with permission.)

with a foreign surface extrude long, straight, threadlike processes that may contain cytoplasmic granules. These cytoplasmic extensions mesh with similar processes from other coagulocytes, creating a hemostatic plug. Examples of coagulocyte aggregation and cytoplasmic expansions are shown in Figs. 1-4 and 1-5.

In some invertebrates, hemostasis is provided entirely by cell aggregation at the site of a wound.[7–9] In others, the hemocytes contain coagulation factors or clottable protein that are released after activation and/or aggregation of the cells.[10–13] In the Arthropoda, as in other invertebrate classes,

mechanisms of hemostasis vary widely. In decapod crustaceans (e.g., *Homarus americanus,* the American lobster), hyaline cells, a type of hemocyte, initiate coagulation when they lyse.[14] In some arthropods, circulating amebocytes release a coagulase that activates a clottable protein already in the plasma.[15] Plasma from the hemolymph of *L. polyphemus,* the American horseshoe crab, normally lacks coagulation factors.[11–12] However, after activation, amebocytes, the only type of circulating blood cell in *Limulus,* release a cascade of coagulation factors that result in coagulation of the plasma.[16,17]

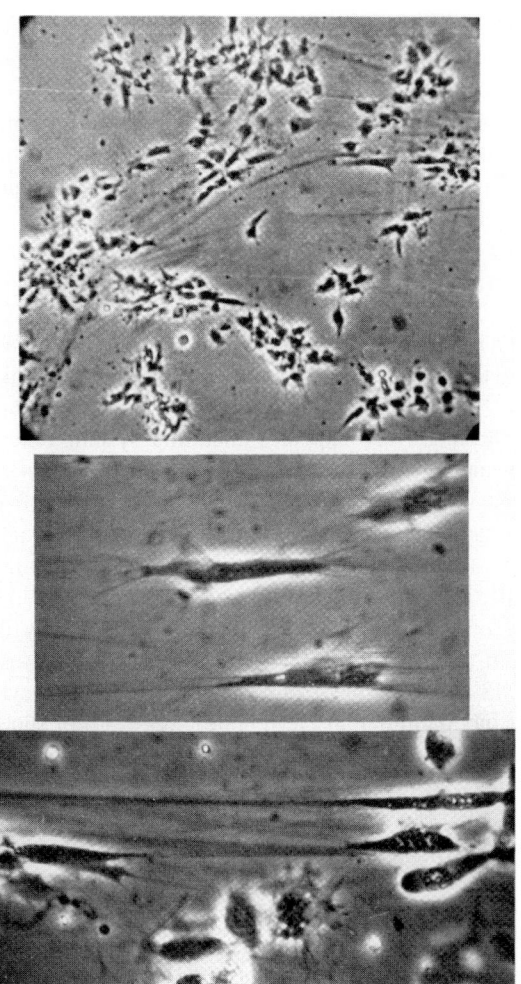

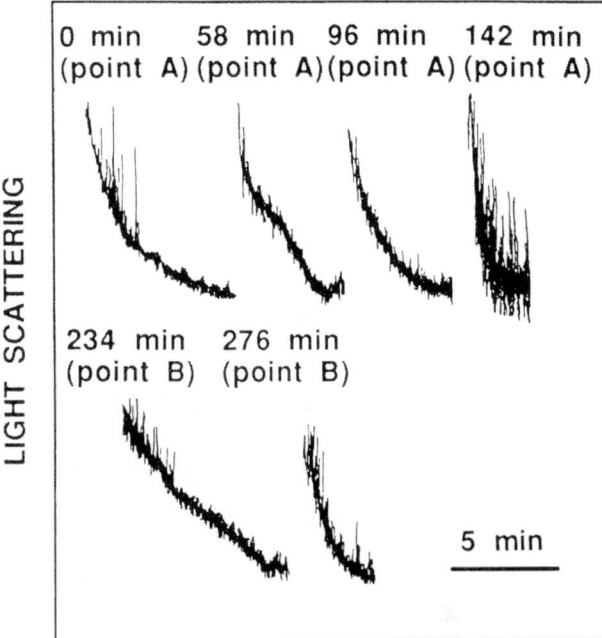

**Figure 1-3.** Activation of hemocyte aggregation in *Halocynthia roretzi*. Point A represents the point on the tunic of the animal through which the hemolymph was repeatedly taken; point B represents a different point on the tunic. The time between the initial collection of hemolymph (0 minute) and subsequent collections is indicated above each tracing. The bar represents 5 minutes. The extent of light scattering is shown on the ordinate in arbitrary units. (From Takahashi et al.,[3] with permission.)

**Figure 1-2.** Appearance of long filamentous processes after the activation of *Limulus* amebocytes. These processes are often connected with those of other amebocytes. Magnification ×128 (upper) and ×320 (middle and lower). (From Levin and Bang,[1] with permission.)

## III. Nonmammalian Vertebrates

Nonmammalian vertebrates have nucleated, often spindle-shaped thrombocytes, the first cells to evolve that specialize in hemostasis.[7,10,18] Thrombocytes are found in fish, and in some species multiple types of thrombocytes have been described.[19,20] However, some reports of multiple types of thrombocytes may have inadvertently described technical artifacts of the methods of blood collection, which have resulted in cell activation and morphological alterations of some of the thrombocytes. Thrombocytes have been variously described as spindle shaped, spiked, spherical, oval, or teardrop in appearance, with a few but variable number of cytoplasmic granules. In addition, size differences may exist between immature and mature thrombocytes.[21] Peri-

odic acid–Schiff (PAS)-positive cytoplasmic granules are sometimes described. Quantification of fish thrombocytes has been made difficult by the often-described resemblance between fish thrombocytes and lymphocytes, and the tendency of thrombocytes to clump.[19] The most extensive quantitative study, based on Wright's-stained blood smears, has reported that 52% (mean; range, 46–61%) of white blood cells in 121 species of marine fishes are thrombocytes.[22] In that study, thrombocytes were the most common type of white blood cell. Examples of thrombocytes in cartilagenous fish are shown in Fig. 1-6.[23] Some species of estuarine cyprinodontiform fishes have been described as having a seasonal variation in the ratio of mature to immature circulating thrombocytes.[21] Immature thrombocytes reached their highest levels during July and August, which the authors interpreted as indicating an increased rate of thrombopoiesis. Fish thrombocytes contain bands of microtubules and in at least some species are described as phagocytic.[24,25] In contrast to platelets, with the exception of fowl, thrombocytes are not typically aggregated by adenosine diphosphate (ADP), nor by epinephrine.[7,26] Zebrafish *(Danio rerio)* thrombocytes have recently been demonstrated *in vivo* to play a role in the development of arterial thrombi.[27]

Avian thrombocytes are similar in appearance to the thrombocytes in fish and are believed to be produced by mononuclear precursors in the bone marrow.[28] Four developmental stages of thrombocyte precursors have been described in the chick embryo and collectively account for 0.6 to 2.4% of nucleated cells in the bone marrow.[29] PAS-positive cytoplasmic granules were present in all stages of the thrombocytic lineage. An extensive ultrastructural study of six species of domestic birds concluded that although thrombocytes are similar in size to lymphocytes, thrombocytes have a denser nucleus and a very highly vacuolated cytoplasm.[30] The developmental stages of the thrombocyte lineage in chicken are shown in Fig. 1-7.[31] Phagocytosis has been demonstrated.[32] Similar to fish, thrombocytes are the most common white blood cell in the chicken. Similar to platelets, avian thrombocytes contain 5-hydroxytryptamine[33] and release what appears to be β-thromboglobulin during the release reaction.[34]

The thrombocytes of at least some birds, amphibians, reptiles, and fish have a membrane system referred to as the surface-connected canalicular system (SCCS).[35–37] This system is also a feature of mammalian platelets and has been linked to their derivation from the cytoplasm of megakaryocytes.[38] However, the presence of the SCCS in nonmammalian thrombocytes indicates that this system reflects an important function of blood cells that play a major role in hemostasis, and that the SCCS need not be derived from the demarcation membrane system (DMS) of megakaryocyte cytoplasm. Thrombocytes from an alligator (a reptile) and a bullfrog (an amphibian) are shown in Fig. 1-8. Bovine platelets are apparently an exception in that they do not contain an SCCS, although bovine megakaryocytes have a DMS.[38,39] Multiple examples of the thrombocytes of other nonmammalian vertebrate species are shown in Fig. 1-9.[40–44]

## IV. Comparative Hemostasis

An overview of comparative hemostasis is provided in Table 1-1.[45] Because of the great heterogeneity in types of blood cells and coagulation mechanisms in invertebrates, any attempt to summarize the characteristics of hemostasis in these animals cannot avoid oversimplification and inaccuracy. Overall, however, it is clear that when cells are present in the invertebrate circulation or coelomic fluid, they always play a role in hemostasis.[7,46]

Among the extensive original studies of the blood by William Hewson (1739–1774) is a highly instructive plate that illustrates the "red particles of the blood" in a wide variety of animals (Fig. 1-10).[47] The following statements in Hewson's text[47] accompany this plate:

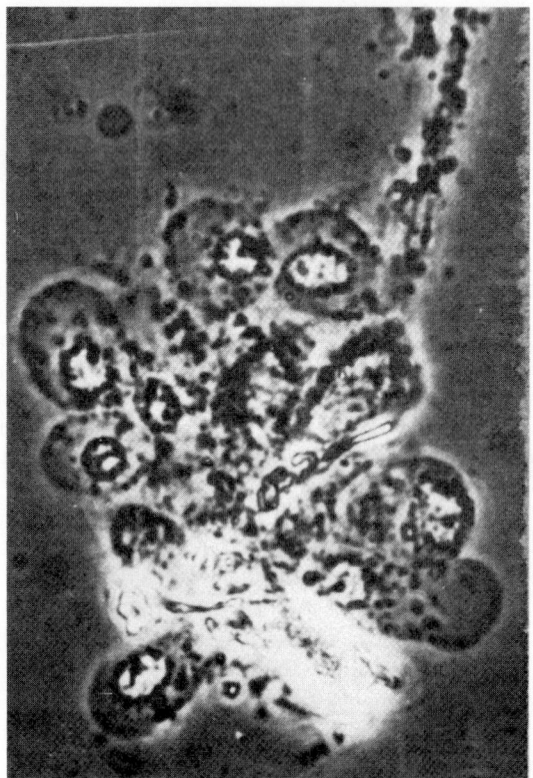

**Figure 1-4.** *Dytiscus marginalis* L. (Coleoptera). Clustering of coagulocytes around a fragment of cuticle stimulates the reaction of hemolymph at wound sites. Magnification ×800. (From Grégoire,[6] with permission.)

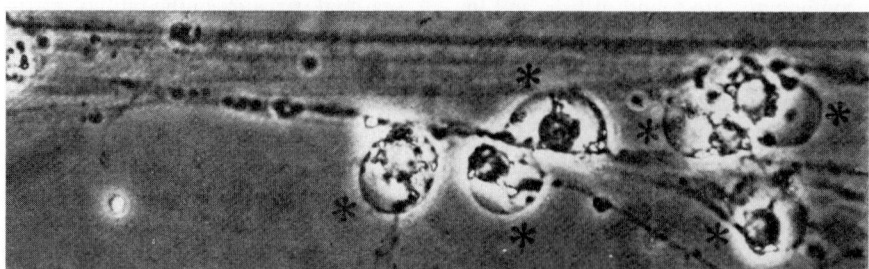

**Figure 1-5.** *Erodius tibialis* (Coleoptera). The threadlike cytoplasmic processes are shown, carrying along the granules produced by six coagulocytes (asterisks). Magnification ×960. (From Grégoire,[6] with permission.)

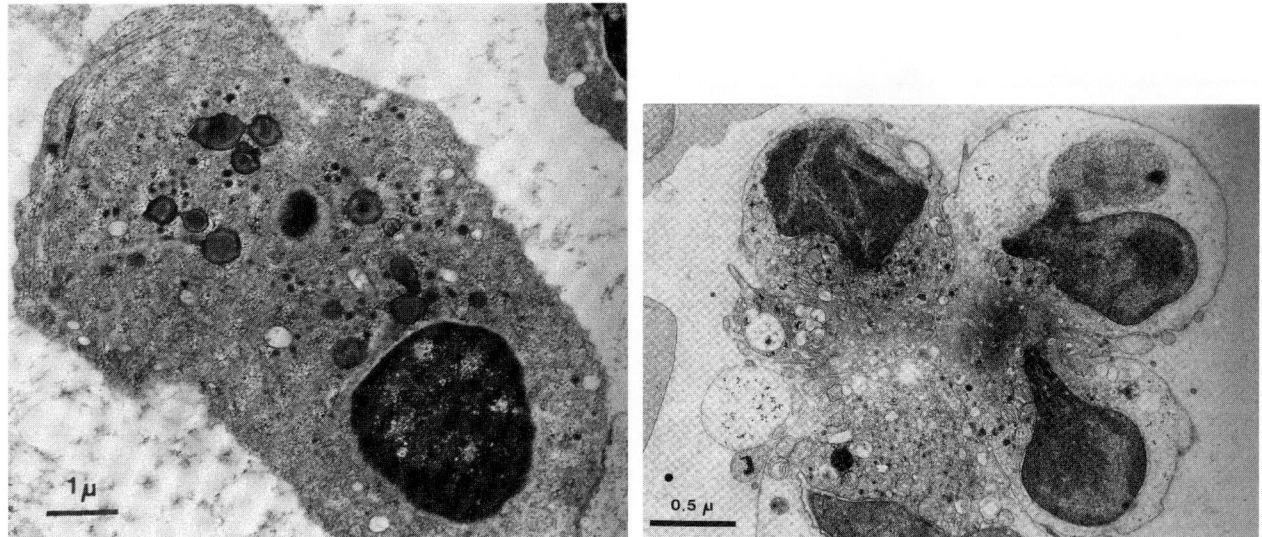

**Figure 1-6.** Transmission electron micrographs of thrombocytes in cartilaginous fish (Chondrichthyes). Left: Dogfish *(Squalus acanthias)* thrombocyte. Note the single nucleus in the lower region of the cell. A group of peripheral microtubules, sectioned lengthwise, are present in the upper left part of the cell. Right: Skate *(Raja eglanteria)* thrombocytes after their incubation with skate thrombin. Four thrombocytes have aggregated, with a loss of distinct cytoplasmic boundaries. The aggregation and fusion of nucleated thrombocytes after exposure to thrombin resembles the response of human platelets to thrombin. In contrast to platelets, adenosine diphosphate does not cause aggregation of thrombocytes. (From Lewis,[23] with permission.)

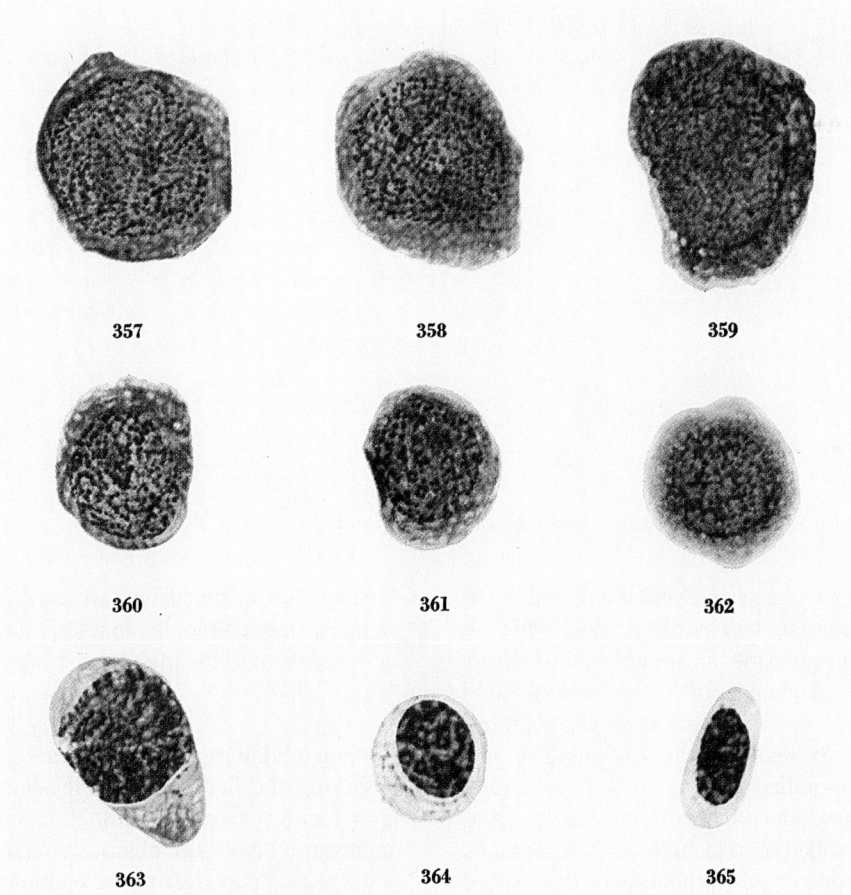

**Figure 1-7.** Camera lucida drawings of the stages of maturation of the thrombocyte lineage in the bone marrow of the white leghorn chicken. Thromboblasts (357, 358); early immature thrombocytes (359–362); midimmature thrombocyte (363); late immature thrombocyte (364); mature thrombocyte (365). (From Lucas and Jamroz,[31] with permission.)

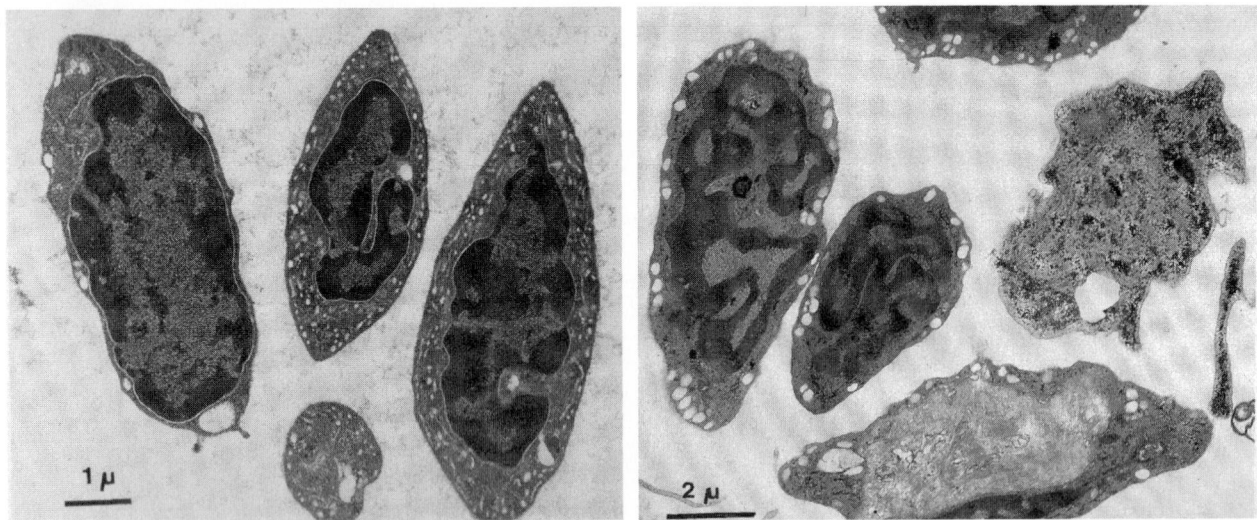

**Figure 1-8.** Transmission electron micrographs of thrombocytes. Left: Alligator *(Alligator mississippiensis)* thrombocytes. The three thrombocytes demonstrate a spindle shape, large nuclei, and a fine open canalicular system. Right: Bullfrog *(Rana catesbeiana)* thrombocytes. Two of the thrombocytes contain large nuclei. The inclusion in the lowest cell may be a phagocytosed red blood cell. (From Lewis,[23] with permission.)

**Table 1-1: Summary of Hemostatic Mechanisms So Far Identified in Various Groups of Animals**

|  | Inverte-brates | Cyclo-stomes | Elasmo-branchs | Bony fish | Amphi-bians | Reptiles | Birds | Mammals |
|---|---|---|---|---|---|---|---|---|
| Thrombocytes/platelets | "+" | + | + | + | + | + | + | + |
|   Adhesion | + |  |  |  | + |  |  | + |
|   Aggregation | + | + |  |  | + | + | + | + |
|   Retraction | + | + | + | + | + |  | + | + |
|   Viscous metamorphosis | "+" |  |  |  | + |  | + | + |
|   Coagulation factor | + | + | + | + | + |  |  | + |
| Vessel contraction | + |  |  |  | + |  | + | + |
| Plasma coagulation | + | + | + | + | + | + | + | + |
|   Fibrinogen → fibrin[a] | "+" | + | + | + | + | + | + | + |
|   Prothrombin → thrombin |  | + | + | + | + | + | + | + |
| Spontaneous fibrinolysis | 0 | 0 | + | 0 | + | 0 | + | + |

Adapted from Hawkey.[45] See also Needham.[8]

[a]In some examples, the term *clottable protein* would be more appropriate.

In the blood of some insects the vesicles [blood cells] are not red, but white, as may easily be observed in a lobster (which Linnaeus calls an insect), one of whose legs being cut off, a quantity of a clear sanies flows from it; this after being some time exposed to the air jellies, but less firmly than the blood of more perfect animals. When it is jellied it is found to have several white filaments; these are principally the vesicles concreted, as I am persuaded from the following experiment. . . . There is a curious change produced in their shape by being exposed to the air, for soon after they are received on the glass they are corrugated, or from a flat shape are changed into irregular spheres. This change takes place so rapidly, that it requires great expedition to apply them to the microscope soon enough to observe it. (pp. 233–234)

Hewson (and his colleague Magnus Falconar) were actually observing the hemocytes of lobsters becoming activated after removal from the animal, changing shape, and then aggregating. Note the obviously altered appearance of the hemocytes, in contrast to the multiple examples of genuine "red particles of the blood" (in Fig. 1-10, an original Hewson plate; compare his Figs. 11 and 12).

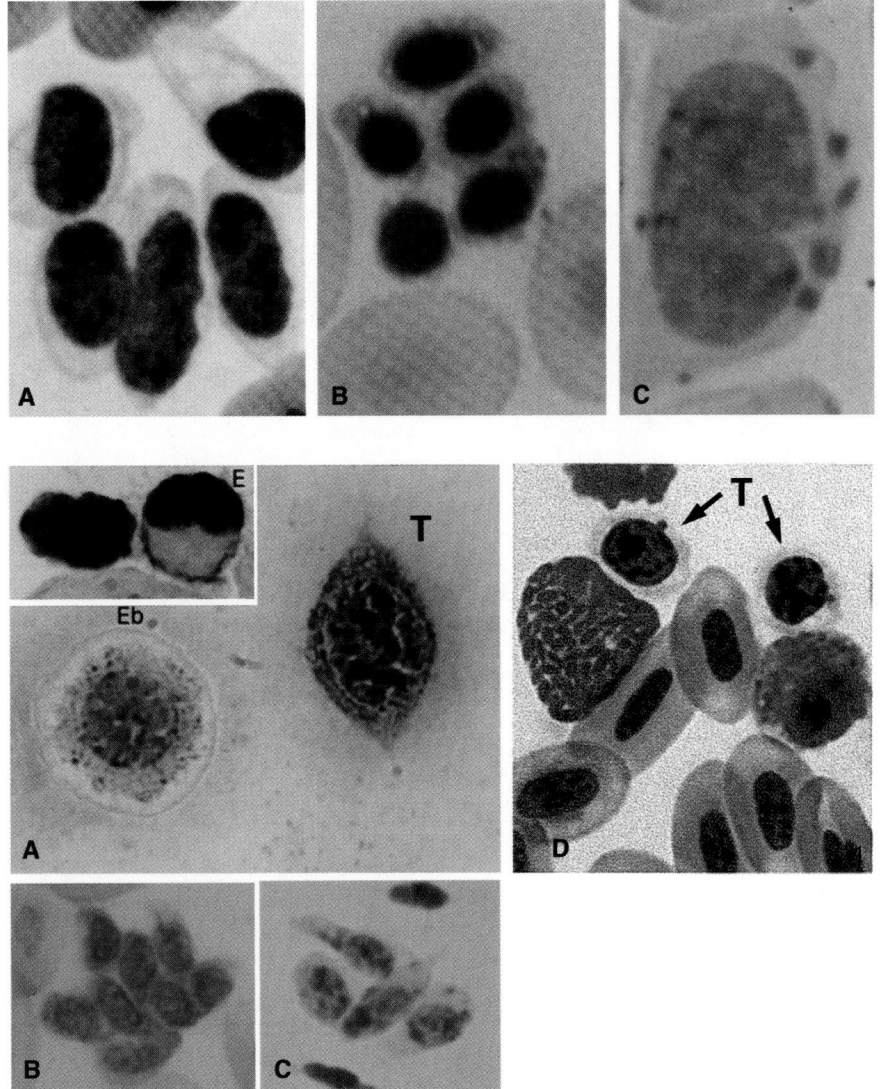

**Figure 1-9.** Thrombocytes of the common lizard (*Podarcis s. sicula* Raf.), dogfish *(Scyliorhynus stellaris L.),* electric ray *(Torpedo marmorata),* red-legged partridge *(Alectoris rufa rufa L.),* and eagle *(Aquila rapax).* Top panel: A. Group of lizard thrombocytes (May Grünwald Giemsa). B. Lizard thrombocytes positively stained for acid phosphatase. C. Dogfish thrombocyte with granules stained for β-glucuronidase. Lower left panel, A–C: A. Torpedo thrombocyte (T) positively stained and basophilic erythroblast (Eb) negatively stained for platelet factor 4 (PF4). [Insert: Thrombocyte (left) positive and eosinophilic granulocyte (E) negative for PF4]. B. Red-legged partridge thrombocytes positively stained for α-naphthylacetate esterase. C. Red-legged partridge thrombocytes positively stained for acid phosphatase. Lower right panel, D: Eagle thrombocytes (T), in a field that contains a heterophile (right), basophile (left), and many nucleated red blood cells. In this figure, the relative sizes of the thrombocytes in the species shown are not to scale. Note the tendency of thrombocytes to clump (Top panel: A and B; Lower left panel: B and C). The cleft that sometimes appears in the nucleus of thrombocytes has been described in many reports (Top panel: C). (From Pica et al.,[40–42] D'Ippolito et al.,[43] and Jain,[44] with permission.)

## V. A Comparison of Human Platelets and *Limulus* Amebocytes

*L. polyphemus,* the horseshoe crab, is the last surviving member of the class merostomata, which included marine spiders. The *Limulus* amebocyte, which is the only type of circulating blood cell in the hemolymph, has probably been the most intensely studied of the invertebrate blood cells involved in hemostasis and blood coagulation.[1,48] The concentration of amebocytes in the blood of adult limuli is approximately $15 \times 10^9$/L. Amebocytes are nucleated cells that are approximately the size of mammalian monocytes. Their cytoplasm is packed with granules (Fig. 1-11).[49] After activation on a foreign surface (or by bacterial endotoxins), amebocytes spread and degranulate (Fig. 1-11, right panel). Degranulation is associated with exocytosis.[11,50–52] The amebocytes, which are normally discoid (like platelets),

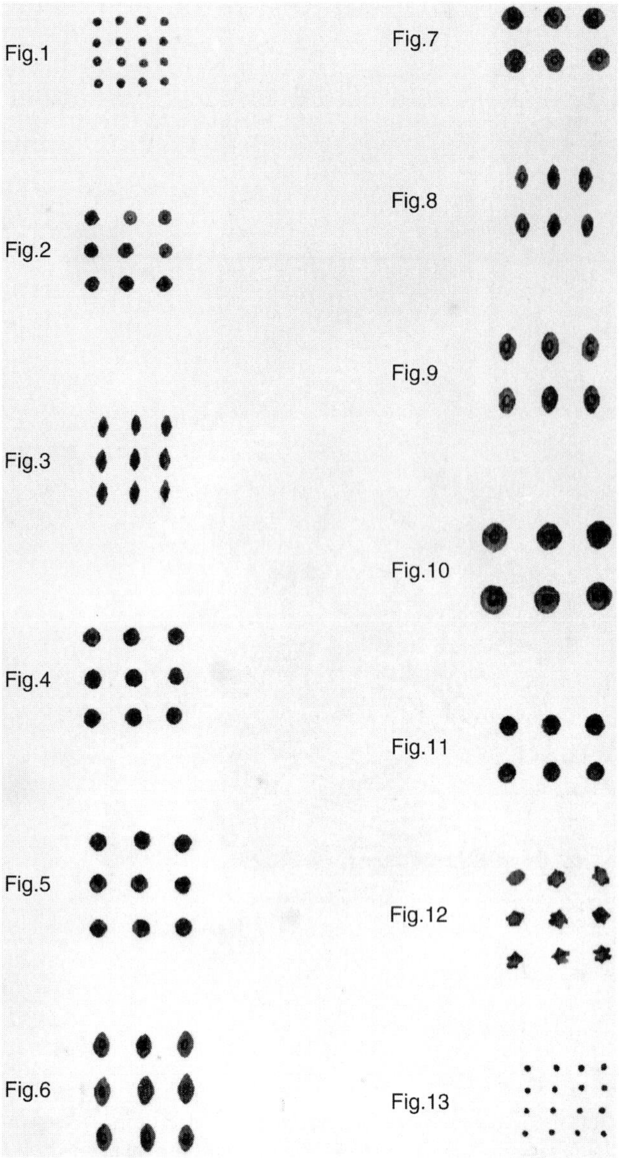


**Figure 1-10.** "A comparative view of the flat vesicles of the blood in different animals, exhibiting their size and shape" (ref. 47, p. 312). The size of the red blood cell in man (Fig. 2) is compared with that in 23 animals grouped according to the similar size of their red blood cells, thus yielding the 12 figures (Figs. 1–12) in this plate. Figs. 11 and 12 show the hemocytes of the lobster before and after the shape change produced by a foreign surface. Note that these are the only cells observed among the 24 animals studied that changed shape and aggregated after removal from the animal (see text). Fig. 13 demonstrates "milk globules." (From Hewson[47] (1846), Plate V from Section III (originally published in 1777); from the author's collection.)

**Table 1-2: Comparison of *Limulus* Amebocytes and Mammalian Platelets**

| Characteristic or Function | *Limulus* Amebocyte | Mammalian Platelet |
|---|---|---|
| Hemostasis | Essential | Essential |
| Blood coagulation | Essential | Ancillary |
| Clottable protein | Yes | Yes |
| Nucleus | Yes | No |
| Viscous metamorphosis | Yes (?) | Yes |
| Cellular processes | Yes | Yes |
| Granules | Yes | Yes |
| Response to endotoxin | Yes | Yes |
| Phagocytosis | Yes | Yes (?) |
| Antibacterial function | Yes | Yes |
| Motility | Yes | Yes |

develop pseudopodia and microspikes.[53] These cells are also capable of retraction (Fig. 1-12),[54] apparently similar to the process of clot retraction induced by mammalian platelets.

Fig. 1-13[55] shows the ultrastructure of a *Limulus* amebocyte. The cytoplasm contains at least two types of granules (Fig. 1-14)[56] that are biochemically different and contain all the components of the blood coagulation mechanism.[17,48,56] Marginal microtubule bands are present (Fig. 1-15).

Mammalian platelets are appropriately considered as functioning primarily to support a range of hemostatic mechanisms, both by maintaining the integrity of blood vessels and by contributing to the process of blood coagulation. However, platelets have many other capabilities, although often rudimentary, that appear to be unrelated to their hemostatic function. Despite approximately 400 million years of evolution, mammalian platelets retain many of the functions of *Limulus* amebocytes (and of many other comparable invertebrate blood cells). This phenomenon may have resulted from the retention of multiple functions previously found in a single, "all-purpose" circulating cell type, such as the *Limulus* amebocyte, only one of whose functions was hemostasis.

The multiple characteristics shared by mammalian platelets and *Limulus* amebocytes are summarized in Table 1-2. Platelets have rudimentary bactericidal and some phagocytic or phagocytic–like activity.[57–60] However, a recent ultrastructural study of interactions between human platelets and various-size particles concluded that platelets were not true phagocytes, and that the uptake of bacteria involved the channels of the open canalicular system (OCS).[61]

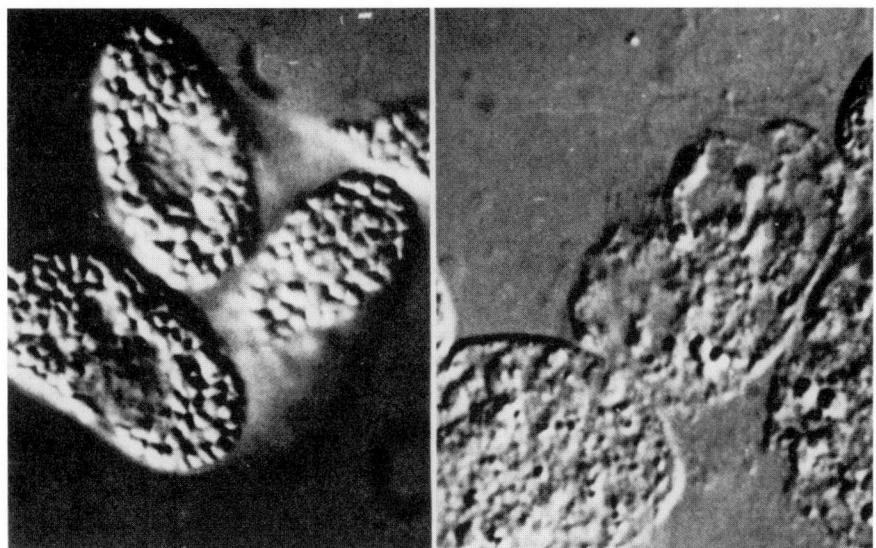

**Figure 1-11.** *Limulus* amebocytes. Three intact normal cells are shown on the left. The cytoplasm is packed with granules. Flattened, spread, degranulated amebocytes, after exposure to a foreign surface or a bacterial endotoxin, are shown on the right. Differential interference phase microscopy does not reveal the single large nucleus, which is located in the apparently depressed area in the middle of two of the cells in the left panel. Original magnification ×1,000. (From Levin,[49] with permission.)

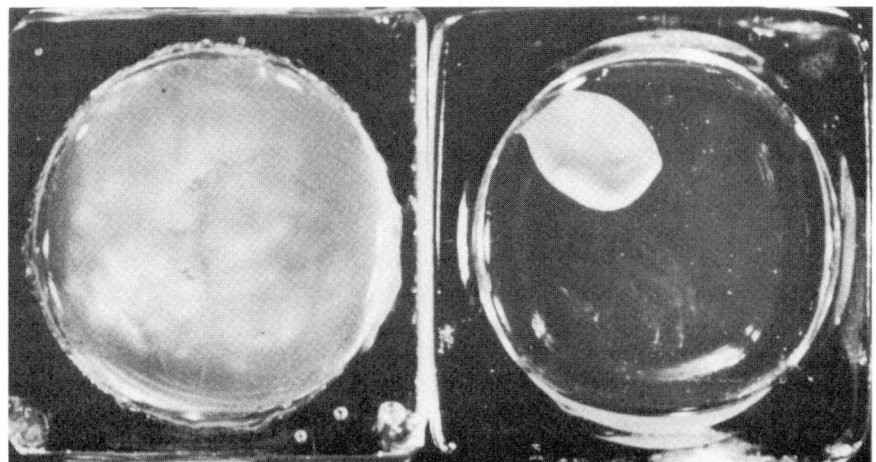

**Figure 1-12.** Amebocyte tissue can be prepared for study by collecting blood under aseptic and endotoxin-free conditions in embryo watchglasses. After removal from *Limulus,* the blood cells settle and aggregate into a tissuelike mass that, after an extended period *in vitro,* undergoes contraction. In the right-hand watchglass, the mass is 1 day old. In the upper left of this watchglass, the amebocyte tissue mass has contracted into the compact, white, buttonlike mass. The fluid medium was *Limulus* plasma. (From Söderhäll et al.,[54] with permission.)

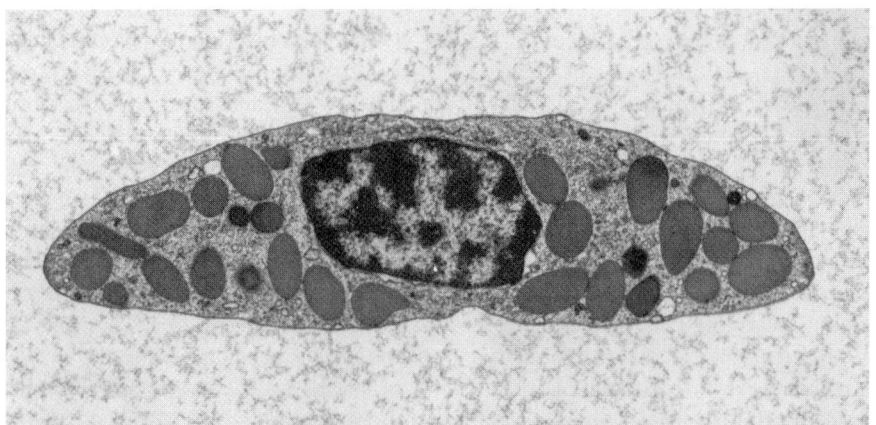

**Figure 1-13.** Longitudinal section of a *Limulus* amebocyte. In this transmission electron micrograph, the cell is spindle shaped. A longitudinal section cut at right angles to this one would reveal a more oval shape. Large, homogeneous secretory granules are present. Magnification ×7,000. (From Copeland and Levin,[55] with permission.)

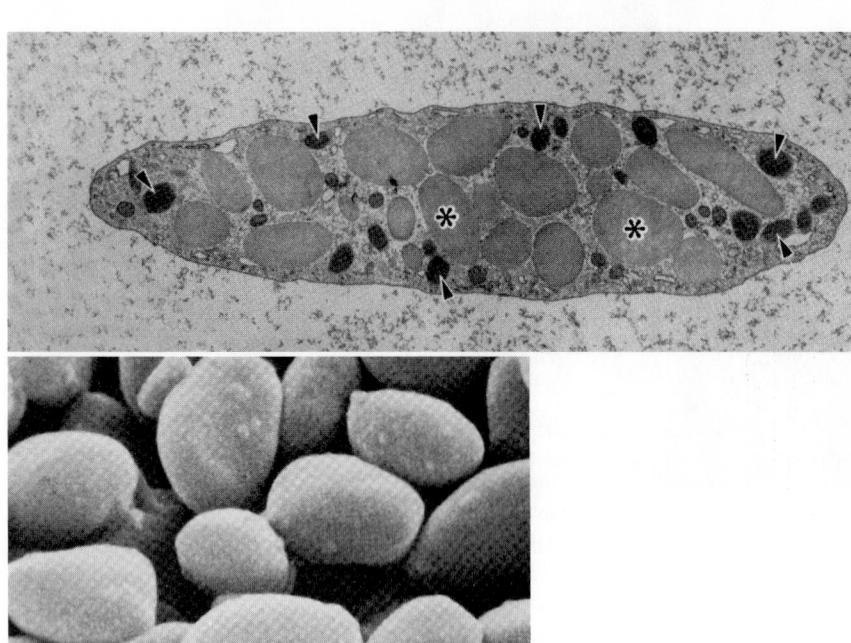

**Figure 1-14.** Top: *Limulus* amebocyte showing both major (asterisks) and minor (arrows) granules. Note the marked density of the smaller class of granules. In this section, the large nucleus is not present. Magnification ×6,140. (From Copeland and Levin,[55] with permission.) Bottom: Scanning electron micrograph of a preparation of intact cytoplasmic granules obtained from *Limulus* amebocytes. Magnification ×10,000. (From Mürer et al.,[56] with permission.)

Platelets contain endotoxin-binding substances[62] and have been shown to interact with bacteria, endotoxins, viruses, and fungi.[63–67] Murine platelets have been shown to express a functional toll-like receptor-4 (TLR4), the receptor for bacterial endotoxin (LPS).[68] The extensive role of platelets in antimicrobial host defense has been reviewed (see Chapter 40).[67,69] *Staphylococcus* can stimulate human platelets to undergo the release reaction.[70] Microbicidal proteins polymorphic protein-1 (PMP-1) and PMP-2 are small cationic peptides released by rabbit platelets, which disrupt the membrane of *S. aureus* and cause cell death.[71] It has been demonstrated that staphylococci have a receptor for the $F_c$ fragment of immunoglobulin G (IgG) that provides a mechanism for aggregation of human platelets by the formation of a complex composed of bacteria, IgG, and platelets.[72] Furthermore, it has been shown that platelet binding by *S.*

*aureus* is mediated, at least in part, by direct binding of clumping factor A (ClfA) to a novel 118-kDa platelet membrane receptor.[73] Certain surface proteins of *Streptococcus gordonii* bind to the extracellular portion of the platelet membrane glycoprotein (GP) Ibα.[74] Pneumococcus *(S. pneumoniae)* and vaccinia virus can induce the release of serotonin from human platelets.[75,76] Other investigations have demonstrated that two endocarditis-producing strains of *S. viridans* can activate human platelets *in vitro,*[77] apparently by direct platelet–bacterial interaction. Fibrinogen in conjunction with other unidentified plasma factors was required for platelet aggregation and secretion. The aggregation of human platelets by *S. viridans, S. pyogenes,* and *S. sanguis* has been demonstrated.[78–80] *S. mitis* has been shown to possess at least two genomic regions that contribute to platelet binding.[81]

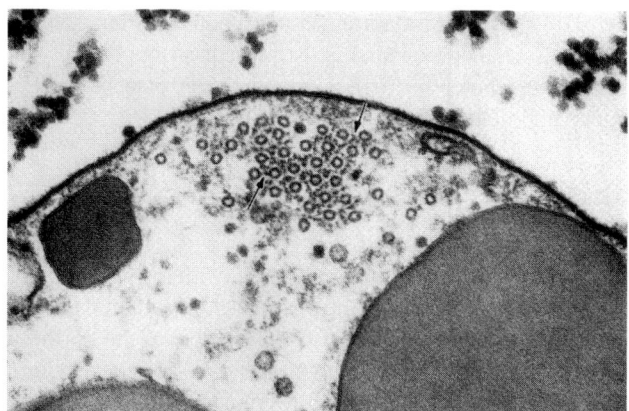

**Figure 1-15.** A marginal microtubule band is demonstrated in a cross-section of a *Limulus* amebocyte. Projections can be seen leading from one microtubule to another (arrows). These projections may serve to stabilize the microtubule band and are likely related to microtubule-associated proteins present in other systems. Magnification×104,000. (From Tablin and Levin,[53] with permission.)

Human platelets have been reported to be cytotoxic for parasites by a mechanism involving an IgE receptor on the platelet surface.[82,83] Platelets also have been shown to play a role in the excretion of *Schistosoma mansoni* in the stool of mice.[84] Other studies have demonstrated that platelets enhance the adherence of schistosome eggs to endothelial cells, and that interleukin (IL)-6, produced by activated monocytes, markedly increases the cytotoxicity of platelets against schistosomula.[85] Under certain circumstances, platelets demonstrate a chemotactic response and migrate[86,87] as do amebocytes.[88] Collectively, these and other observations indicate that bacteria, viruses, fungi, and parasites are capable of interacting with platelets. Depending upon the nature of the infectious particles and other poorly understood variables, this interaction can result in platelet aggregation, release of platelet constituents, phagocytosis of the infectious agent, and ultimately a shortened platelet life span (see Chapter 40 for further details).

## VI. The Evolution of Hemostasis and Blood Coagulation

None of the previously described functions seem necessary for the current role of platelets in mammalian hemostasis. Therefore, it is likely that some of the capabilities of mammalian platelets are vestiges of functions originally present in the more "primitive" yet multicompetent cells from which mammalian blood cells have evolved. The roles of the amebocyte in providing hemostasis and controlling infection, and the reaction of the amebocyte to endotoxin, suggest that in various mammals the response of platelets and the blood

coagulation system to Gram-negative infections or endotoxins is an evolutionary remnant of an ancient mechanism. The limited ability of mammalian platelets to phagocytose (or internalize) particles and kill bacteria may be another remnant of functions that are more important in amebocytes (and the hemostatic cells of other invertebrates). Thus, these two cell types, amebocytes from an ancient marine invertebrate and platelets from mammals, have remarkably similar characteristics. The relative importance of their functions has changed with evolution, but after millions of years, coagulation and antibacterial mechanisms remain at least partly linked. An example of the association between hemostatic and antibacterial functions in platelets is demonstrated by the observation that *S. aureus* induces platelet aggregation through a fibrinogen-dependent mechanism[89] (see Chapter 40).

Quick[90] wrote: "Even in man one does not only see residua of the more elementary hemostatic reactions, but actually these phylogenetically older mechanisms still function effectively, but in a restricted manner" (p. 2). And Laki,[91] referring to the evolution of blood coagulation and hemostasis, stated: "... man, who must realize that the imprints of a very distant past are still with him" (p. 305). Insightfully, Quick[90] also considered "the possibility that the clotting mechanism may not have had as its primary purpose, hemostasis, but rather defense against bacterial invasion and repair of tissue" (p. 5). Needham[8] has articulated the same hypothesis. In summary, platelets are not only important for the maintenance of hemostasis, but are also inflammatory cells (see Chapter 39) that have important roles in antimicrobial host defense mechanisms[67,92] (see Chapter 40).

Although the similarities in functions of platelets and amebocytes are consistent with the evolution of plasmatic blood coagulation in mammals from initially cell-based mechanisms in invertebrates, there is no proof of such an evolutionary trail, despite the parallel cellular functions and the similarity of the enzymatic components — for example, the serine proteases upon which *Limulus* blood coagulation is based and some mammalian blood coagulation factors that are also serine proteases. Nevertheless, as eloquently stated by Quick,[90]

The appearance in blood of a soluble protein that could be gelated appears *a priori* as another abrupt imposition of a new hemostatic process, but it is likely that such is not the case. It is probable that this new material in the plasma represents a transfer of an intracellular constituent to an extracellular state. If one looks upon the platelet as the descendant of a primitive coalescing cell, then one can understand why it has the power to aggregate and why it still contains fibrinogen. (p. 2)

Some evolutionary aspects of blood coagulation have been reviewed.[7,15,16,25] In addition, the apparently sudden evolutionary appearance of nonnucleated platelets and megakaryocytes in mammals must be considered. This seems to be a marked departure from all other groups of animals, even taking into account the evolutionary concept of punctuated equilibria.[93,94] However, as thoughtfully stated by Ratcliffe and Millar,[95] "attempts to trace the origin of vertebrate blood cells within the invertebrates will at best be speculative and based on the assumption that comparisons with living species are valid since the enigmatic ancestral forms are no longer available" (p. 2). These authors also wrote: "It is most likely that these cells (i.e., invertebrate thrombocyte-like cells) are analogous rather than homologous to vertebrate platelets. They (i.e., platelets) may have arisen by convergent evolution due to similar evolutionary/environmental pressures" (p. 12). Intriguingly, it has been shown that the surface of avian thrombocytes presents analogues of human platelet GPIIb and GPIIIa, which are recognized by both polyclonal antibodies specific for the human GPIIb and GPIIIa subunits, and monoclonal anti-GPIIb-IIIa complex-specific antibodies.[96–98] Zebrafish thrombocytes also have been demonstrated to have platelet GPIb and GPIIb-IIIa-like complexes on their surfaces.[99] In addition, Zebrafish thrombocyte precursors are c-mpl positive, and injection of an anti-sense c-mpl morpholino into embryos eliminated the production of thrombocytes.[100]

## VII. Megakaryocytes and Mammals

Nonnucleated platelets, and presumably their polyploid megakaryocyte progenitors in the bone marrow, are present only in mammals. This suggests that some important feature of mammalian physiology benefits from this unique mechanism for producing an unprecedented anucleate cell from the cytoplasm of a larger cell, for the apparently major purpose of supporting hemostasis. However, because platelets have the rudimentary capacity to perform some of the functions that are carried out by other blood cell types in mammals, and because there is increased recognition that they also play a role in nonhemostatic defense mechanisms, we must be cautious about assuming that hemostasis is the only major platelet function. Also, all nonmammalian animal species require a mechanism to prevent hemorrhage or uncontrolled loss of body fluids, and most (nonmammalian vertebrates being particularly relevant) have effective hemostatic mechanisms that do not involve the megakaryocyte/platelet axis. We must ask, therefore, what is the biological advantage that led to the establishment and persistence of this cell lineage in mammals?

Members of the vertebrate class Mammalia are characterized by body hair, mammary glands, and viviparous birth, except in the egg-laying monotremes. The presence of a placenta is also characteristic of most, but not all, pregnant female mammals. The two variations in birth mechanism,

oviparous birth in monotremes and viviparous but nonplacental pregnancy in marsupials, present an opportunity to explore the potential association between placental pregnancy and platelet formation in mammals.

### A. Marine Mammals

Marine mammals are unique because, with the exception of the duck-billed platypus, these are the only mammals that live in a marine environment. The platelets of the northern elephant seal *(Mirounga angustirostris)*, collected in sodium citrate, demonstrate decreased responsiveness to thrombin, ADP, and ristocetin; shape change and aggregation were not produced by collagen or epinephrine in the absence of divalent cations.[101] However, their platelets are morphologically similar to other mammalian platelets. Similarly decreased responsiveness of the platelets of the killer whale *(Orcinus orca)* to agonists was noted in experiments that utilized platelet-rich plasma prepared in sodium citrate.[102] Only limited data are available for platelet counts in marine mammals.[103–108] Platelet counts in selected marine mammals are shown in Table 1-3. Megakaryocytes are present in the bone marrow of the sea lion *(Zalophus californianus)* and are detectable with a polyclonal antihuman factor VIII-related antigen/von Willebrand factor antibody (Levin, J., unpublished observations).

### B. Order Monotremata

Monotremes, considered the most primitive form of mammals, have birdlike and reptilian features. The females lay eggs. Monotremes are represented by the aquatic duck-billed platypus and insectivorous echidna (spiny anteater). One report on echidna platelets described two possible types: "elongated, spindle-shaped structures with a tendency to intertwine together or as normal platelets with spreading and aggregating activity" (ref. 109, p. 218). Hawkey[109] suggested that the presence of two types of hemostatic cells in the echidna might indicate a link between the spindle-shaped thrombocytes of nonmammalian vertebrates and typical mammalian platelets. Platelet counts in echidna were 200 to $250 \times 10^9$/L. Another study[110] of the echidna reported platelet levels of approximately 500 to $650 \times 10^9$/L, and that "giant multinucleated cells with a few platelet-like bodies within the cytoplasm" were present in the bone marrow (p. 1133). These giant cells were believed to be megakaryocytes. Other studies of the echidna *Tachyglossus aculeatus* reported platelet counts of 300 to $320 \times 10^9$/L[111] and 205 to $682 \times 10^9$/L (18 animals).[112]

Platelet levels of approximately 400 to $450 \times 10^9$/L were reported for the duck-billed platypus.[113] The same investigators[114] described platypus platelets as "anucleate, circular

### Table 1-3: Platelet Counts in Selected Marine Mammals

| Name | Order[a] | Platelet Count ($\times 10^9$/L)[b] |
|---|---|---|
| Commerson's dolphin | Cetacea | 185 (10) |
| Common dolphin | Cetacea | 80 (2) |
| Beluga whale | Cetacea | 106 (16) |
| Pilot whale | Cetacea | 85 (2) |
| White-sided dolphin | Cetacea | 120 (10) |
| Killer whale | Cetacea | 163 (18) |
| Bottlenose dolphin | Cetacea | 165 (96) |
| Northern fur seal | Pinnipedia | 428 (17) |
| Steller's sea lion | Pinnipedia | 243 (5) |
| Northern elephant seal | Pinnipedia | 437 (149) |
| Harp seal | Pinnipedia | 500 (12) |
| Harbor seal | Pinnipedia | 314 (24) |
| Gray seal | Pinnipedia | 378 (9) |
| California sea lion | Pinnipedia | 280 (26) |
| Florida manatee | Sirenia | 283 (n.a.) |
| Polar bear | Carnivora | 317 (n.a.) |

Data from Reidarson et al.[108] (Almost all the published platelet counts of marine mammals have been provided by the clinical laboratory of Sea World.)
[a]The order (subclass) within the class Mammalia is shown.
[b]Mean platelet count. The number of animals studied is shown in parentheses.
n.a., not available.

### Table 1-4: Platelet Counts in Selected Mammals

| Name | Order[a] | Platelet Count ($\times 10^9$/L)[b] |
|---|---|---|
| Echidna | Monotremata | 549 (3) |
| Quokka | Marsupialia | 1180 (1) |
| Wallaby | Marsupialia | 390 (1) |
| Opossum | Marsupialia | 498 (9) |
| Hedgehog | Insectivora | 113 (2) |
| Bat | Chiroptera | 819 (5) |
| Armadillo | Edentata | 357 (6) |
| Porpoise | Cetacea | 136 (12) |
| Seal | Pinnipedia | 651 (5) |
| Elephant | Proboscidea | 620 (6) |
| Manatee | Sirenia | 347 (5) |
| Baboon | Primate | 299 (7) |
| Chimpanzee | Primate | 406 (5) |
| Monkey | Primate | 510 (9) |

Data were derived from Lewis.[23]
[a]The order (subclass) within the class Mammalia is shown.
[b]Mean platelet count. The number of animals studied is shown in parentheses.

and 2 to 5 μm in diameter, with occasional large platelets (up to 8 μm) seen" (p. 423). Their electron microscopy studies demonstrated a homogeneous population of cells with typical platelet organelles and ultrastructure, including parallel bundles of microtubules.[114] This detailed report emphasized that platypus platelets were similar in appearance and size to those of other mammals, including marsupials. The two types of platelets that Hawkey[109] suggested were present in the echidna were not observed. Another study, based on 56 platypuses, reported that platelet levels ranged from 315 to $2144 \times 10^9$/L.[112] Previous attempts by this author (J. L.) to obtain specimens of platypus bone marrow to study megakaryocytes have failed because of Australian and U.S. regulations restricting export and import of specimens from endangered wildlife species.

### C. Order Marsupialia

Marsupials give birth to live but very immature young, after a nonplacental pregnancy. This order is represented by the opossum, kangaroo, wombat, and bandicoot. An extensive histochemical study of the blood cells of the marsupial *Trichosurus vulpecula,* the Australian brush-tailed possum, provided multiple photographs of typical mammalian platelets with essentially the same histochemical characteristics as human platelets.[115] Overall platelet size was approximately the same as that of human platelets, but a greater proportion of large platelets was reported to be present. No quantitative data were provided. Opossum platelets are shown in Fig. 1-16. For comparison, electron micrographs of platelets in other mammalian orders are shown in Fig. 1-17. In other investigations, platelet levels in five different species of marsupials ranged from approximately 200 to $500 \times 10^9$/L[109,110,116,117] and in a sixth species (*Setonix brachyurus,* the quokka) from 425 to $1180 \times 10^9$/L (Table 1-4). The mean platelet level was reported to be $153 \times 10^9$/L in the common wombat (23 animals) and to range from 136 to $485 \times 10^9$/L in the red-necked wallaby[118] (Table 1-4). Release of serotonin, a characteristic of mammalian platelets, was also demonstrated for marsupial platelets.[116] In the two marsupial species whose bone marrow has been studied, megakaryocytes were described, but not illustrated.[110] Fig. 1-18 presents micrographs of the bone marrow of the South American opossum (*Monodelphis domestica*), which provide multiple examples of megakaryocytes in this marsupial.

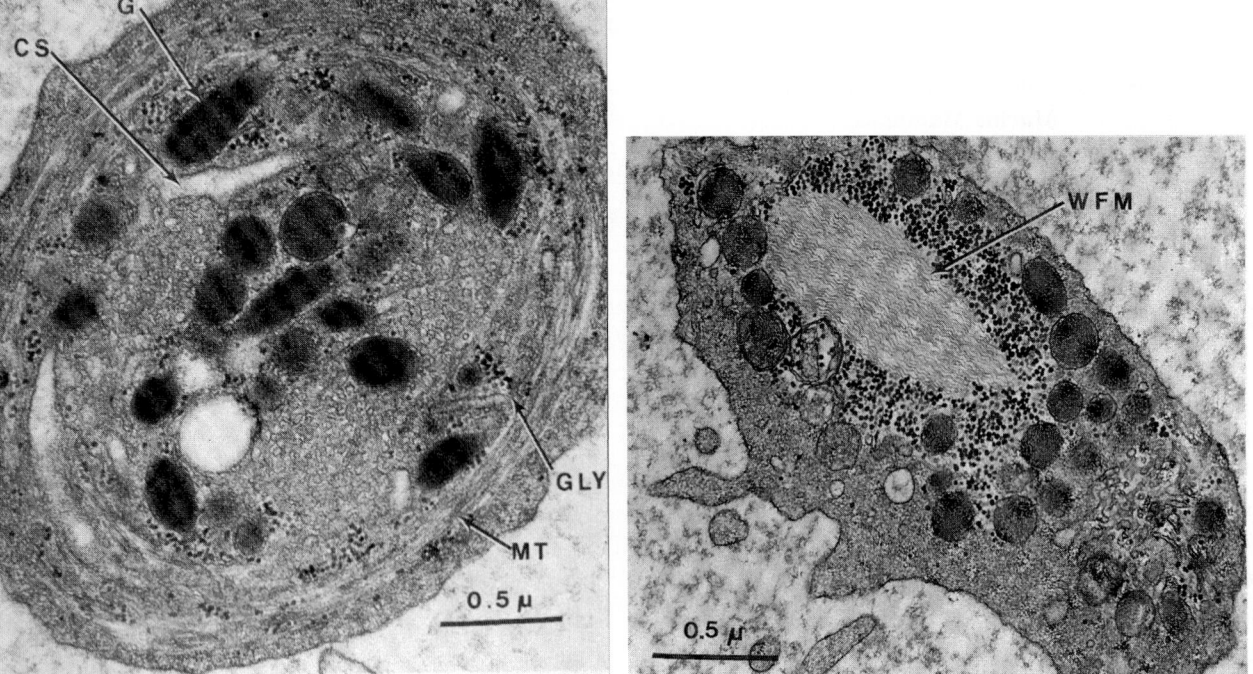

**Figure 1-16.** Transmission electron micrographs of opossum *(Didelphis marsupialis virginiana)* platelets. Nonplacental mammals have platelets, as do all mammals. Left: A circumferential band of microtubules (MT) is clearly shown, and an open canalicular system (CS) is present. A large granule (G) and scattered glycogen particles (GLY) are present. Right: A large mass of wavy fibrillar material (WFM) is present. This fibrillar material was observed in approximately 10% of all platelets; its composition is unknown. (From Lewis,[23] with permission.)

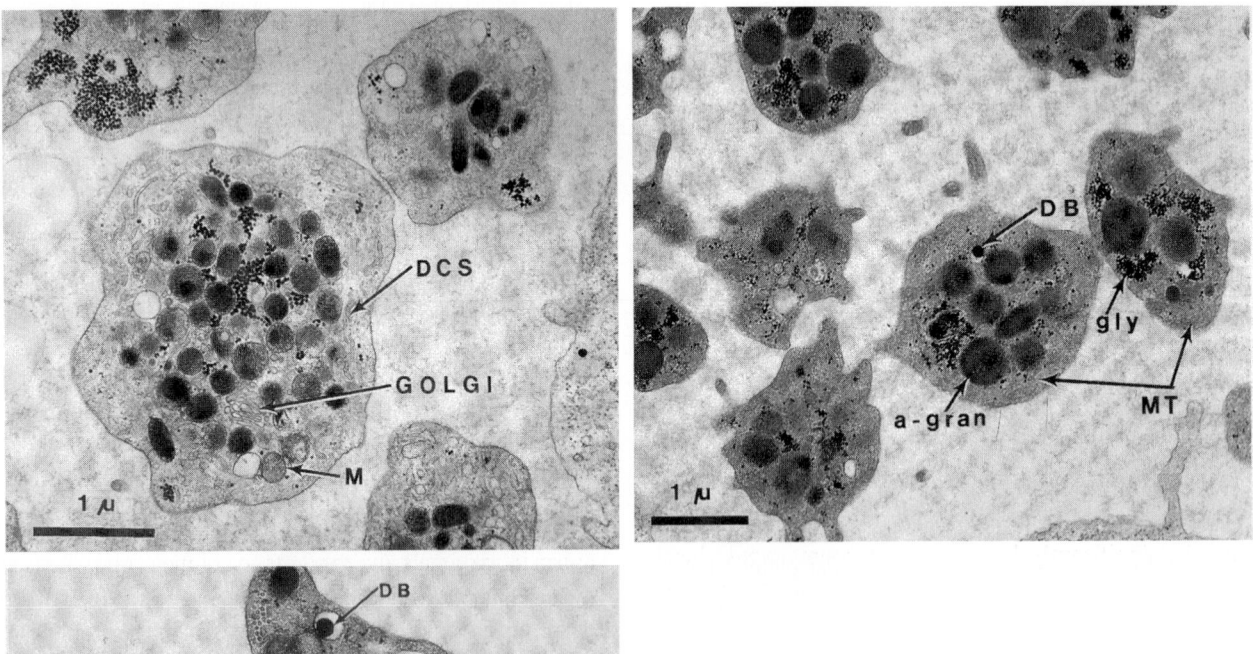

**Figure 1-17.** Transmission electron micrographs of platelets. Left: Hedgehog *(Erinaceus europaeus)* platelets. These platelets contain many granules. A dense canalicular system (DCS), a Golgi apparatus (GOLGI), and mitochondria (M) are present. Bottom: Two monkey *(Macaca mulatta)* platelets. These contain large clumps of glycogen (GLY) and an open canalicular system (OCS). Microtubules (MT), mitochondria (M), dense bodies (DB), and α-granules (a) are also present. Right: Elephant *(Elephas maximus)* platelets. α-Granules (a-gran), dense bodies (DB), glycogen particles (GLY), and microtubules (MT) are shown. An open canalicular system is also present. (From Lewis,[23] with permission.)

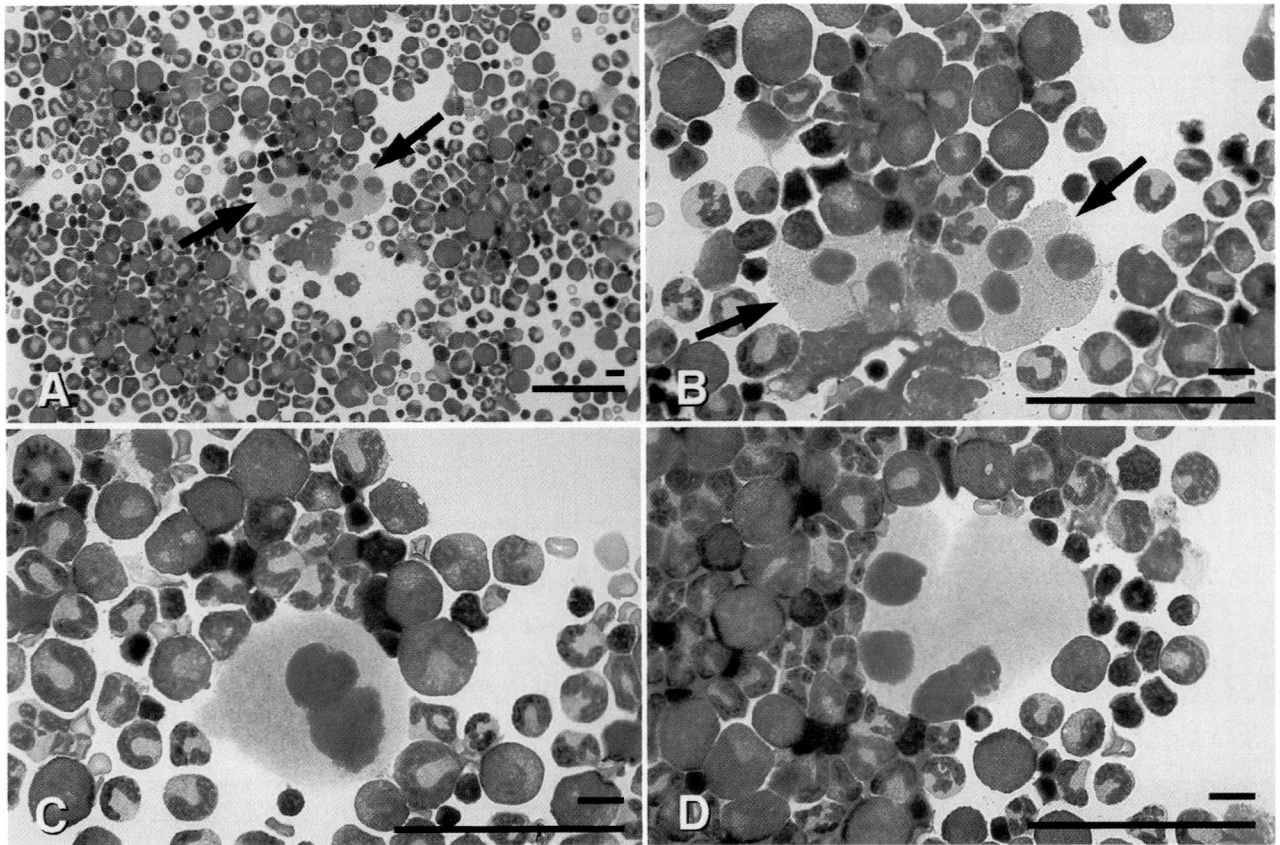

**Figure 1-18.** Specimens of bone marrow from the South American opossum *(Monodelphis domestica)* stained with Wright-Giemsa. The megakaryocytes are the large, multinucleated cells, with light blue or pinkish cytoplasm. A. Two adjacent megakaryocytes (indicated by the arrows; original magnification ×100). B. The same two megakaryocytes are shown at a higher power (original magnification ×250). C. A single megakaryocyte is detectable (original magnification ×250). D. A very large megakaryocyte is shown (original magnification ×250). The black bars in the right lower corner of each panel indicate 10 μm (shorter bar) and 50 μm (longer bar). A previous report also described marsupial megakaryocytes as multinucleated cells.[110] This is in important contrast to the megakaryocytes of other mammals, such as human and rodent, which contain a single, multilobed, polyploid nucleus.

### D. Platelet Levels

Interestingly, in most of the marsupials and echidnas studied, the previously described inverse relationship between mammalian body size and platelet level in a limited group of animals was not evident.[119] On the basis of the previously published nomogram and platelet levels in some additional mammals, significantly higher platelet counts would have been expected in the smaller monotremes and marsupials. Perhaps the failure of this relationship to hold reflects the primitive nature of these two mammalian orders. Platelet counts in selected mammals are shown in Table 1-4, and summarized elsewhere.[120–122] Mean platelet volume is not related to body mass.[121–123]

### E. A Comparison of Nonplacental and Placental Mammals

Blood coagulation has been described in both monotremes and marsupials as resembling human blood coagulation; clot retraction, which is consistent with normal platelet function, also has been observed in both orders.[110,116] Based on the previous comparisons of the hemostatic systems of the aplacental and placental mammals, it appears that there is no specific association between either the development of a placenta during pregnancy (or the occurrence of live birth) or the appearance of a markedly different hemostatic mechanism in mammals and the presence of megakaryocytes and platelets. Therefore, the suggestion that the evolution of platelets resulted from unique hemostatic requirements imposed by placental birth cannot be supported.[124] Nevertheless, a potential role for platelets during pregnancy in mammals has been described.[125] The preimplantation embryo produces platelet activating factor (PAF), which has been described as then activating platelets in the microvascular bed of the oviduct in mice.[125] These same investigators suggested that platelets may contribute to the establishment of early pregnancy, because administration of inhibitors of PAF to mice reduced the number of implantation sites. Furthermore, they concluded that activated platelets may produce a range of molecules that support the attachment of the embryo to the uterine surface.

## VIII. Conclusions

There is, presumably, a biological advantage gained from the presence of polyploid megakaryocytes as the source for nonnucleated platelet progeny in mammals, but this advantage has not yet been identified. It has been pointed out that the abolition of mitosis provides a basis for increasing RNA synthesis and therefore an increased potential for protein synthesis in the cell.[126] In addition, cell growth and differentiation can continue, uninterrupted by nuclear and cell divisions. Nagl[126] has also suggested that regulation of specific gene activity in a polyploid cell is more efficient than in a group of diploid cells. An obvious benefit is the ability of a single megakaryocyte to produce many hundreds or thousands of platelets. However, augmented production can be achieved by other means: The bone marrow can markedly and adequately increase the production of red and white blood cells without resorting to a mechanism based on polyploid progenitors and cytoplasmic fragmentation. Furthermore, the many animal species with nucleated thrombocytes have seemingly adequate hemostasis, and, as yet, platelets have not been found to be mandatory for any special feature of mammalian blood coagulation that is not present in nonmammals. However, the mechanism by which platelets are produced by megakaryocytes does allow for the rapid release of larger than normal platelets.[127,128] These cells are more biologically active than platelets produced under steady-state conditions, and therefore may constitute an attempt to provide a maximally effective response to a pathophysiological emergency.

Another unexplained element is the presence of high concentrations of acetyl cholinesterase (AChE) in the megakaryocytes of only some mammalian species, such as the mouse, rat, and cat, but not humans.[129] Why should only select species have megakaryocytes that produce high concentrations of AChE? Mechanisms for the regulation of megakaryocytopoiesis and of platelet function appear identical regardless of the presence or absence of AChE in megakaryocytes. In addition, inhibitors of AChE do not have any detectable effect on platelet aggregation, platelet factor 3 availability, or plasma clot retraction.[130]

The elucidation of the evolutionary event or events that resulted in the appearance of mammalian megakaryocytes and platelets, as well as the potential biological advantage of this system, remain elusive. Because both monotremes, which are egg-laying mammals, and marsupials, which have a nonplacental pregnancy, possess megakaryocytes and platelets, it is apparent that neither live birth nor the presence of a placenta accounts for the evolution of platelets in

mammals. Comparative molecular genetic studies are likely to provide further insights.

## Acknowledgment

I express my personal and professional gratitude to the late Dr. Frederik B. Bang for introducing me to *Limulus* more than 40 years ago, and for providing the intellectual guidance that made it possible for me to begin the studies that constitute a significant component of this chapter. I also wish to acknowledge the superb and supportive environment of the Marine Biological Laboratory, Woods Hole, Massachusetts, which nurtured my interests in comparative hemostasis, and the many staff members and scientists who provided me with the information and techniques that I required to pursue my experimental goals.

I thank the staff of the Medical Library at the San Francisco Veterans Administration Medical Center for their consistent and effective support of my efforts to access a widely scattered literature, much of which was published decades before the advent of databases. The unfailing ability of Ms. Nadine Walas to obtain copies of papers in obscure journals or in volumes published at the beginning of the 20th century was especially appreciated.

This chapter is dedicated to the late Dr. Jessica H. Lewis, whose instructive studies provided many of the figures that appear in this analysis.

Supported in part by the Veterans Administration.

## References

1. Levin, J., & Bang, F. B. (1964). A description of cellular coagulation in the *Limulus*. *Bull Johns Hopkins Hosp, 115,* 337–345.
2. Kenny, D. M., Belamarich, F. A., & Shepro, D. (1972). Aggregation of horseshoe crab *(Limulus polyphemus)* amebocytes and reversible inhibition of aggregation by EDTA. *Biol Bull, 143,* 548–567.
3. Takahashi, H., Azumi, K., & Yokosawa, H. (1994). Hemocyte aggregation in the solitary ascidian *Halocynthia roretzi:* Plasma factors, magnesium ion, and met-lys-bradykinin induce the aggregation. *Biol Bull, 186,* 247–253.
4. Goffinet, G., & Grégoire, C. (1975). Coagulocyte alterations in clotting hemolymph of *Carausius morosus* L. *Arch Int de Physiol Biochim, 83,* 707–722.
5. Chen, J. H., & Bayne, C. J. (1995). Bivalve mollusc hemocyte behaviors: Characterization of hemocyte aggregation and adhesion and their inhibition in the California mussel *(Mytilus californianus)*. *Biol Bull, 188,* 255–266.
6. Grégoire, C. (1984). Haemolymph coagulation in insects and taxonomy. *Bull K Belg Inst Nat Wet, 55,* 3–48.
7. Ratnoff, O. D. (1987). The evolution of hemostatic mechanisms. *Perspect Biol Med, 31,* 4–33.

8. Needham, A. E. (1970). Haemostatic mechanisms in the invertebrata. *Symp Zool Soc Lond, 27,* 19–44.

9. Dales, R. P., & Dixon, L. R. J. (1981). Polychaetes. In N. A. Ratcliffe & A. F. Rowley (Eds.), *Invertebrate blood cells* (Vol. 1, pp. 35–74). London: Academic Press.

10. Ravindranath, M. H. (1980). Haemocytes in haemolymph coagulation of arthropods. *Biol Rev, 55,* 139–170.

11. Levin, J., & Bang, F. B. (1964). The role of endotoxin in the extracellular coagulation of *Limulus* blood. *Bull Johns Hopkins Hosp, 115,* 265–274.

12. Levin, J., & Bang, F. B. (1968). Clottable protein in *Limulus:* Its localization and kinetics of its coagulation by endotoxin. *Thromb Diath Haemorr, 19,* 186–197.

13. Madaras, F., Parkin, J. D., & Castaldi, P. A. (1979). Coagulation in the sand crab *(Ovalipes bipustulatus). Thromb Haemost, 42,* 734–742.

14. Hose, J. E., Martin, G. G., & Gerard, A. S. (1990). A decapod hemocyte classification scheme integrating morphology, cytochemistry, and function. *Biol Bull, 178,* 33–45.

15. Spurling, N. W. (1981). Comparative physiology of blood clotting. *Comp Biochem Physiol, 68A,* 541–548.

16. Levin, J. (1985). The role of amebocytes in the blood coagulation mechanism of the horseshoe crab *Limulus polyphemus.* In W. Cohen (Ed.), *Blood cells of marine invertebrates: Experimental systems in cell biology and comparative physiology* (pp. 145–163). New York: Alan R. Liss.

17. Young, N. S., Levin, J., & Prendergast, R. A. (1972). An invertebrate coagulation system activated by endotoxin: Evidence for enzymatic mediation. *J Clin Invest, 51,* 1790–1797.

18. Schneider, W., & Gattermann, N. (1994). Megakaryocytes: Origin of bleeding and thrombotic disorders. *Europ J Clin Invest, 24*(Suppl. 1), 16–20.

19. Ellis, A. E. (1976). Leukocytes and related cells in the plaice. *Pleuronectes platessa. J Fish Biol, 8,* 143–156.

20. Ellis, A. E. (1977). The leukocytes of fish: A review. *J Fish Biol, 11,* 453–491.

21. Gardner, G. R., & Yevich, P. P. (1969). Studies on the blood morphology of three estuarine cyprinodontiform fishes. *J Fish Res Board Can, 26,* 433–447.

22. Saunders, D. C. (1966). Differential blood cell counts of 121 species of marine fishes in Puerto Rico. *Trans Amer Microsc Soc, 85,* 427–449.

23. Lewis, J. H. (1996). *Comparative hemostasis in vertebrates* (pp. 3–322). New York: Plenum Press.

24. Ferguson, H. W. (1976). The ultrastructure of plaice *(Pleuronectes platessa)* leucocytes. *J Fish Biol, 8,* 139–142.

25. Rowley, A. F., Hill, D. J., Ray, C. E., & Munro, R. (1997). Haemostasis in fish — an evolutionary perspective. *Thromb Haemost, 77,* 227–233.

26. Belamarich, F. A., Fusari, M. H., Shepro, D., & Kien, M. (1966). *In vitro* studies of aggregation of non-mammalian thrombocytes. *Nature, 212,* 1579–1580.

27. Thattaliyath, B., Cykowski, M., & Jagadeeswaran, P. (2005). Young thrombocytes initiate the formation of arterial thrombi in zebrafish. *Blood, 106,* 118–124.

28. Campbell, T. W. (2000). Normal hematology of psittacines. In B. F. Feldman, J. G. Zinkl, & N. C. Jain (Eds.), *Schalm's veterinary hematology* (5th ed, pp. 1155–1163). Philadelphia: Lippincott Williams & Wilkins.

29. Stalsberg, H., & Prydz, H. (1963). Studies on chick embryo thrombocytes. I. Morphology and development. *Thromb Diath Haemorrh, 9,* 279–290.

30. Maxwell, M. H. (1974). An ultrastructural comparison of the mononuclear leucocytes and thrombocytes in six species of domestic bird. *J Anat, 117,* 69–80.

31. Lucas, A. M., & Jamroz, C. (1961). *Atlas of avian hematology* (Agriculture monograph 25). Washington, DC: US Department of Agriculture.

32. Carlson, H. C., Sweeny, P. R., & Tokaryk, J. M. (1968). Demonstration of phagocytic and trephocytic activities of chicken thrombocytes by microscopy and vital staining techniques. *Avian Dis, 12,* 700–715.

33. Kuruma, I., Okada, T., Kataoka, K., & Sorimachi, M. (1970). Ultrastructural observation of 5-hydroxytryptamine-storing granules in the domestic fowl thrombocytes. *Z Zellforsch, 108,* 268–281.

34. Wachowicz, B., & Krajewski, T. (1979). The released proteins from avian thrombocytes. *Thromb Haemost, 42,* 1289–1295.

35. Daimon, T., Mizuhira, V., Takahashi, I., & Uchida, K. (1979). The surface connected canalicular system of carp *(Cyprinus carpio)* thrombocytes: Its fine structure and three-dimensional architecture. *Cell Tissue Res, 203,* 355–365.

36. Daimon, T., Mizuhira, V., & Uchida, K. (1979). Fine structural distribution of the surface-connected canalicular system in frog thrombocytes. *Cell Tissue Res, 201,* 431–439.

37. Daimon, T., & Uchida, K. (1978). Electron microscopic and cytochemical observations on the membrane systems of the chicken thrombocyte. *J Anat, 125,* 11–21.

38. Stenberg, P. E., & Levin, J. (1989). Mechanisms of platelet production. *Blood Cells, 15,* 23–47.

39. Zucker–Franklin, D., Benson, K. A., & Myers, K. M. (1985). Absence of a surface-connected canalicular system in bovine platelets. *Blood, 65,* 241–244.

40. Pica, A., Della Corte, F., Grimaldi, M. C., & D'Ippolito, S. (1986). The blood cells of the common lizard *(Podarcis s. sicula* Raf.). *Arch Ital Anat Embriol, 91,* 301–320.

41. Pica, A., Lodato, A., Grimaldi, M. C., & Della Corte, F. (1990). Morphology, origin and functions of the thrombocytes of *Elasmobranchs. Arch Ital Anat Embriol, 95,* 187–207.

42. Pica, A., Lodato, A., Grimaldi, M. C., Della Corte, F., & Galderisi, U. (1993). A study of the bone marrow precursors and hemoglobin of the blood cells of the red-legged partridge *(Alectoris rufa rufa* L.). *Ital J Anat Embryol, 98,* 277–292.

43. D'Ippolito, S., Pica, A., Grimaldi, M. C., & Della Corte, F. (1985). The blood cells and their precursors in the hemopoietic organs of the dogfish *Scyliorhynus stellaris* L. *Arch Ital Anat Embriol, 90,* 31–46.

44. Jain, N. C. (1986). Blood cells and parasites of birds. In N. C. Jain (Ed.), *Schalm's veterinary hematology* (4th ed., plate XXV). Philadelphia: Lea & Febiger.

45. Hawkey, C. M. (1970). General summary and conclusions. *Symp Zool Soc Lond, 27,* 217–229.

46. Belamarich, F. A. (1976). Hemostasis in animals other than mammals: The role of cells. In T. H. Spaet (Ed.), *Progress in hemostasis and thrombosis* (Vol. 3, pp. 191–209). New York: Grune and Stratton.

47. Hewson, W. (1846). *The works of William Hewson, F.R.S.,* G. Gulliver, Ed. (pp. 211–244). London: C. and J. Adlard.

48. Iwanaga, S., Kawabata, S. I., & Muta, T. (1998). New types of clotting factors and defense molecules found in horseshoe crab hemolymph: Their structures and functions. *J Biochem, 123,* 1–15.

49. Levin, J. (1988). The horseshoe crab: A model for Gram-negative sepsis in marine organisms and humans. In J. Levin, H. R. Buller, J. W. ten Cate, S. J. H. van Deventer, & A. Sturk (Eds.), *Bacterial endotoxins. Pathophysiological effects, clinical significance, and pharmacological control* (pp. 3–15). New York: Alan R. Liss.

50. Armstrong, P. B., & Rickles, F. R. (1982). Endotoxin-induced degranulation of the *Limulus* amebocyte. *Exp Cell Res, 140,* 15–24.

51. Dumont, J. N., Anderson, E., & Winmer, G. (1966). Some cytologic characteristics of the hemocytes of *Limulus* during clotting. *J Morphol, 119,* 181–208.

52. Ornberg, R. L., & Reese, T. S. (1981). Beginning of exocytosis captured by rapid-freezing of *Limulus* amebocytes. *J Cell Biol, 909,* 40–54.

53. Tablin, F., & Levin, J. (1988). The fine structure of the amebocyte in the blood of *Limulus polyphemus.* II. The amebocyte cytoskeleton: A morphological analysis of native, activated, and endotoxin-stimulated amebocytes. *Biol Bull, 175,* 417–429.

54. Söderhäll, K., Levin, J., & Armstrong, P. B. (1985). The effects of β1,3-glucans on blood coagulation and amebocyte release in the horseshoe crab, *Limulus polyphemus. Biol Bull, 169,* 661–674.

55. Copeland, D. E., & Levin, J. (1985). The fine structure of the amebocyte in the blood of *Limulus polyphemus.* I. Morphology of the normal cell. *Biol Bull, 169,* 449–457.

56. Mürer, E. H., Levin, J., & Holme, R. (1975). Isolation and studies of the granules of the amebocytes of *Limulus polyphemus,* the horseshoe crab. *J Cell Physiol, 86,* 533–542.

57. Glynn, M. F., Movat, H. Z., Murphy, E. A., & Mustard, J. F. (1965). Study of platelet adhesiveness and aggregation, with latex particles. *J Lab Clin Med, 65,* 179–201.

58. Levin, J. (1976). Blood coagulation in the horseshoe crab *(Limulus polyphemus):* A model for mammalian coagulation and hemostasis. In *Animal models of thrombosis and hemorrhagic diseases* (pp. 87–96). Washington, DC: US Department of Health, Education and Welfare. Publication No. (NIH) 76–982.

59. Lewis, J. C., Maldonado, J. E., & Mann, K. G. (1976). Phagocytosis in human platelets: Localization of acid phosphatase-positive phagosomes following latex uptake. *Blood, 47,* 833–840.

60. Youssefian, T., Drouin, A., Massé, J. M., Guichard, J., & Cramer, E. M. (2002). Host defense role of platelets: Engulfment of HIV and *Staphylococcus aureus* occurs in a specific subcellular compartment and is enhanced by platelet activation. *Blood, 99,* 4021–4029.

61. White, J. G. (2005). Platelets are covercytes, not phagocytes: Uptake of bacteria involves channels of the open canalicular system. *Platelets, 16,* 121–131.

62. Springer, G. F., & Adye, J. C. (1975). Endotoxin-binding substances from human leukocytes and platelets. *Infect Immun, 12,* 978–986.

63. Clawson, C. C. (1973). Platelet interaction with bacteria. III. Ultrastructure. *Am J Pathol, 70,* 449–472.

64. MacIntyre, D. E., Allen, A. P., Thorne, K. J. I., Glauert, A. M., & Gordon, J. L. (1977). Endotoxin-induced platelet aggregation and secretion. I. Morphological changes and pharmacological effects. *J Cell Sci, 28,* 211–223.

65. Levin, J. (1996). Bleeding with infectious diseases. In O. D. Ratnoff & C. D. Forbes (Eds.), *Disorders of hemostasis* (3rd ed., pp. 339–356). Philadelphia: WB Saunders.

66. Clawson, C. C. (1995). Platelets in bacterial infections. In M. Joseph (Ed.), *Immunopharmacology of platelets* (pp. 83–124). London: Harcourt Brace.

67. Yeaman, M. R. (1997). The role of platelets in antimicrobial host defense. *Clinical Infections Diseases, 25,* 951–970.

68. Andonegui, G., Kerfoot, S. M., McNagny, K., Ebbert, K. V. J., Patel, K. D., & Kubes, P. (2005). Platelets express functional toll-like receptor-4. *Blood, 106,* 2417–2423.

69. Klinger, M. H. F. (1997). Platelets and inflammation. *Anat Embryol, 196,* 1–11.

70. Clawson, C. C., Rao, G. H. R., & White, J. G. (1975). Platelet interaction with bacteria. IV. Stimulation of the release reaction. *Am J Pathol, 81,* 411–419.

71. Yeaman, M. R., Bayer, A. S., Koo, S. P., Foss, W., & Sullam, P. M. (1998). Platelet microbicidal proteins and neutrophil defensin disrupt the *Staphylococcus aureus* cytoplasmic membrane by distinct mechanisms of action. *J Clin Invest, 101,* 178–187.

72. Hawiger, J., Steckley, S., Hammond, D., Cheng, C., Timmons, S., Glick, A. D., & Des Prez, R. M. (1979). Staphylococci-induced human platelet injury mediated by protein A and immunoglobulin G Fc fragment receptor. *J Clin Invest, 64,* 931–937.

73. Siboo, I. R., Cheung, A. L., Bayer, A. S., & Sullam, P. M. (2001). Clumping factor A mediates binding of *Staphylococcus aureus* to human platelets. *Infect Immun, 69,* 3120–3127.

74. Bensing, B. A., López, J. A., & Sullam, P. M. (2004). The *Streptococcus gordonii* surface proteins GspB and Hsa mediate binding to sialylated carbohydrate epitopes on the platelet membrane glycoprotein Ibα. *Infect Immun, 72,* 6528–6537.

75. Zimmerman, T. S., & Spiegelberg, H. L. (1975). Pneumococcus-induced serotonin release from human platelets. Identification of the participating plasma/serum factor as immunoglobulin. *J Clin Invest, 56,* 828–834.

76. Bik, T., Sarov, I., & Livne, A. (1982). Interaction between vaccinia virus and human blood platelets. *Blood, 59,* 482–487.

77. Sullam, P. M., Valone, F. H., & Mills, J. (1987). Mechanisms of platelet aggregation by viridans group streptococci. *Infect Immun, 55,* 1743–1750.

78. Sullam, P. M., Jarvis, G. A., & Valone, F. H. (1988). Role of immunoglobulin G in platelet aggregation by viridans group streptococci. *Infect Immun, 56,* 2907–2911.

79. Herzberg, M. C., Brintzenhofe, K. L., & Clawson, C. C. (1983). Aggregation of human platelets and adhesion of *Streptococcus sanguis. Infect Immun, 39,* 1457–1469.

80. Kurpiewski, G. E., Forrester, L. J., Campbell, B. J., & Barrett, J. T. (1983). Platelet aggregation by *Streptococcus pyogenes. Infect Immun, 39,* 704–708.

81. Bensing, B. A., Rubens, C. E., & Sullam, P. M. (2001). Genetic loci of *Streptococcus mitis* that mediate binding to human platelets. *Infect Immun, 69,* 1373–1380.

82. Capron, A., Ameisen, J. C., Joseph, M., Auriault, C., Tonnel, A. B., & Caen, J. (1985). New functions for platelets and their pathological implications. *Int Arch Allergy Appl Immunol, 77,* 107–114.

83. Pancré, V., & Auriault, C. (1995). Platelets in parasitic diseases. In M. Joseph (Ed.), *Immunopharmacology of platelets* (pp. 125–135). London: Harcourt Brace.

84. Ngaiza, J. R., & Doenhoff, M. J. (1990). Blood platelets and schistosome egg excretion. *Proc Soc Exp Biol Med, 193,* 73–79.

85. Pancré, V., Monté, D., Delanoye, A., Capron, A., & Auriault, C. (1990). Interleukin-6 is the main mediator of the interaction between monocytes and platelets in the killing of *Schistosoma mansoni. Eur Cytokine Net, 1,* 15–19.

86. Lowenhaupt, R. W., Miller, M. A., & Glueck, H. I. (1973). Platelet migration and chemotaxis demonstrated *in vitro. Thromb Res, 3,* 477–487.

87. Lowenhaupt, R. W., Glueck, H. I., Miller, M. A., & Kline, D. L. (1977). Factors which influence blood platelet migration. *J Lab Clin Med, 90,* 37–45.

88. Armstrong, P. B. (1979). Motility of the *Limulus* blood cell. *J Cell Sci, 37,* 169–180.

89. Bayer, A. S., Sullam, P. M., Ramos, M., Li, C., Cheung, A. L., & Yeaman, M. R. (1995). *Staphylococcus aureus* induces platelet aggregation via a fibrinogen-dependent mechanism which is independent of principal platelet glycoprotein IIb/IIIa fibrinogen-binding domains. *Infect Immun, 63,* 3634–3641.

90. Quick, A. J. (1967). Hemostasis as an evolutionary development. *Thromb Diath Haemorrh, 18,* 1–11.

91. Laki, K. (1972). Our ancient heritage in blood clotting and some of its consequences. *Ann NY Acad Sci, 202,* 297–307.

92. Page, C. P. (1989). Platelets as inflammatory cells. *Immunopharmacol, 17,* 51–59.

93. Gould, S. J. (1977). Bushes and ladders in human evolution. In *Ever since Darwin. Reflections in natural history* (pp. 56–62). New York: WW Norton.

94. Gould, S. J. (1980). The episodic nature of evolutionary change. In *The panda's thumb. More reflections in natural history* (pp. 179–185). New York: WW Norton.

95. Ratcliffe, N. A., & Millar, D. A. (1988). Comparative aspects and possible phylogenetic affinities of vertebrate and invertebrate blood cells. In A. F. Rowley & N. A. Ratcliffe (Eds.), *Vertebrate blood cells* (pp. 1–17). Cambridge, UK: Cambridge University Press.

96. Kunicki, T. J., & Newman, P. J. (1985). Synthesis of analogs of human platelet membrane glycoprotein IIb-IIIa complex by chicken peripheral blood thrombocytes. *Proc Nat Acad Sci USA, 82,* 7319–7323.

97. Lacoste–Eleaume, A. S., Bleux, C., Quéré, P., Coudert, F., Corbel, C., & Kanellopoulos–Langevin, C. (1994). Biochemical and functional characterization of an avian homolog of the integrin GPIIb-IIIa present on chicken thrombocytes. *Exp Cell Res, 213,* 198–209.

98. Ody, C., Vaigot, P., Quéré, P., Imhof, B. A., & Corbel, C. (1999). Glycoprotein IIb-IIIa is expressed on avian multilineage hematopoietic progenitor cells. *Blood, 93,* 2898–2906.

99. Jagadeeswaran, P., Sheehan, J. P., Craig, F. E., & Troyer, D. (1999). Identification and characterization of zebrafish thrombocytes. *Br J Haematol, 107,* 731–738.

100. Lin, H. F., Traver, D., Zhu, H., Dooley, K., Paw, B. H., Zon, L. I., & Handin, R. I. (2005). Analysis of thrombocyte development in CD41-GFP transgenic zebrafish. *Blood, 106,* 3803–3810.

101. Field, C. L., Walker, N. J., & Tablin, F. (2001). Northern elephant seal platelets: Analysis of shape change and response to platelet agonists. *Thromb Res, 101,* 267–277.

102. Patterson, W. R., Dalton, L. M., McGlasson, D. L., & Cissik, J. H. (1993). Aggregation of killer whale platelets. *Thromb Res, 70,* 225–231.

103. Bossart, G. D., Reidarson, T. H., Dierauf, L. A., & Duffield, D. A. (2001). Clinical pathology. In L. A. Dierauf & F. M. D. Gulland (Eds.), *CRC handbook of marine mammal medicine* (2nd ed., pp. 383–436). Boca Raton, FL: CRC Press.

104. Reidarson, T. H. (2003). Cetacea (whales, dolphins, porpoises). In M. E. Fowler & R. E. Miller (Eds.), *Zoo and wild animal medicine* (5th ed., pp. 442–458). St. Louis: Saunders.

105. Gage, L. J. (2003). Pinnipedia (seals, sea lions, walruses). In M. E. Fowler & R. E. Miller (Eds.), *Zoo and wild animal medicine* (5th ed., pp. 459–475). St. Louis: Saunders.

106. Murphy, D. (2003). Sirenia. In M. E. Fowler & R. E. Miller (Eds.), *Zoo and wild animal medicine* (5th ed., pp. 476–481). St. Louis: Saunders.

107. Walsh, M. T., & Bossart, G. D. (1999). Manatee medicine. In M. E. Fowler & R. E. Miller (Eds.), *Zoo and wild animal medicine* (4th ed., pp. 507–516). St. Louis: WB Saunders.

108. Reidarson, T. H., Duffield, D., & McBain, J. (2000). Normal hematology of marine mammals. In B. F. Feldman, J. G. Zinkl, & N. C. Jain (Eds.), *Schalm's veterinary hematology* (5th ed., pp. 1164–1173). Philadelphia: Lippincott Williams & Wilkins.

109. Hawkey, C. M. (1975). *Comparative mammalian haematology. Cellular components and blood coagulation of captive wild animals* (pp. 218–227). London: William Heinemann Medical Books.

110. Lewis, J. H., Phillips, L. L., & Hann, C. (1968). Coagulation and hematological studies in primitive Australian mammals. *Comp Biochem Physiol, 25,* 1129–1135.

111. Bolliger, A., & Backhouse, T. C. (1960). Blood studies on the echidna *Tachyglossus aculeatus. Proc Zool Soc Lond, 135*(Pt 1), 91–97.

112. Booth, R. J. (2003). Monotremata (echidna, platypus). In
     M. E. Fowler & R. E. Miller (Eds.), *Zoo and wild animal
     medicine* (5th ed., pp. 278–287). St. Louis: WB Saunders.
113. Whittington, R. J., & Grant, T. R. (1983). Haematology and
     blood chemistry of the free-living platypus, *Ornithorhyn-
     chus anatinus* (Shaw) (Monotremata: Ornithorhynchidae).
     *Aust J Zool, 31,* 475–482.
114. Canfield, P. J., & Whittington, R. J. (1983). Morphological
     observations on the erythrocytes, leukocytes and platelets of
     free-living platypuses, *Ornithorhynchus anatinus* (Shaw)
     (Monotremata: Ornithorhynchidae). *Aust J Zool, 31,* 421–
     432.
115. Barbour, R. A. (1972). The leukocytes and platelets of a
     marsupial, *Trichosurus vulpecula.* A comparative morpho-
     logical, metrical, and cytochemical study. *Arch Histol Jpn,
     34,* 311–360.
116. Fantl, P., & Ward, H. A. (1957). Comparison of blood
     clotting in marsupials and man. *Aust J Exp Biol, 35,*
     209–224.
117. Lewis, J. H. (1975). Comparative hematology: Studies on
     opossums *Didelphis marsupialis (Virginianus). Comp
     Biochem Physiol, 51A,* 275–280.
118. Holz, P. (2003). Marsupialia (marsupials). In M. E. Fowler
     & R. E. Miller (Eds.), *Zoo and wild animal medicine* (5th
     ed., pp. 288–303). St. Louis: Saunders.
119. Nakeff, A., & Ingram, M. (1970). Platelet count: Volume
     relationships in four mammalian species. *J Appl Physiol, 28,*
     530–533.
120. Jain, N. C. (1993). *Essentials of veterinary hematology*
     (pp. 54–71). Philadephia: Lea & Febiger.
121. Boudreaux, M. K., & Ebbe, S. (1998). Comparison of platelet
     number, mean platelet volume and platelet mass in five mam-
     malian species. *Comp Haematol Intern, 8,* 16–20.
122. Ebbe, S., & Boudreaux, M. K. (1998). Relationship of mega-
     karyocyte ploidy with platelet number and size in cats, dog,
     rabbits and mice. *Comp Haematol Intern, 8,* 21–25.
123. Jain, N. C. (1993). *Essentials of veterinary hematology*
     (pp. 105–132). Philadephia: Lea & Febiger.
124. Martin, J. F. (1994). Punctuated equilibrium in evolution and
     the platelet. *Europ J Clin Invest, 24,* 291. (letter).
125. Burton, G., O'Neill, C., & Saunders, D. M. (1991). Platelets
     and pregnancy. In C. P. Page (Ed.), *The platelet in health
     and disease* (pp. 191–209). London: Blackwell Scientific
     Publications.
126. Nagl, W. (1978). Functional significance of endo-cycles. In
     *Endopolyploidy and polyteny in differentiation and evolu-
     tion* (pp. 154–157). Amsterdam: North-Holland Publishing.
127. Stenberg, P. E., & Levin, J. (1989). Ultrastructural analysis
     of acute immune thrombocytopenia in mice: Dissociation
     between alterations in megakaryocytes and platelets. *J Cell
     Physiol, 141,* 160–169.
128. Stenberg, P. E., Levin, J., Baker, G., Mok, Y., & Corash, L.
     (1991). Neuraminidase-induced thrombocytopenia in mice:
     Effects on thrombopoiesis. *J Cell Physiol, 147,* 7–16.
129. Jackson, C. W. (1973). Cholinesterase as a possible marker
     for early cells of the megakaryocytic series. *Blood, 42,*
     413–421.
130. Smith, M. J., Braem, B., & Davis, K. D. (1980). Human pla-
     telet acetylcholinesterase: The effects of anticholinesterases
     on platelet function. *Thromb Haemost, 42,* 1615–1619.

# CHAPTER 2

# Megakaryocyte Development and Platelet Formation

**Joseph E. Italiano, Jr. and John H. Hartwig**

*Hematology Division, Brigham and Women's Hospital, Harvard Medical School, Boston, Massachusetts*

## I. Introduction

Megakaryocytes are highly specialized precursor cells that function solely to produce and release platelets into the circulation. Understanding mechanisms by which megakaryocytes develop and give rise to platelets has fascinated hematologists for more than a century. Megakaryocytes are descended from pluripotent stem cells and undergo multiple DNA replications without cell divisions by the unique process of endomitosis. Upon completion of endomitosis, polyploid megakaryocytes begin a rapid cytoplasmic expansion phase characterized by the formation of an elaborate demarcation membrane system (DMS) and the accumulation of cytoplasmic proteins and granules essential for platelet function. During the final stages of development, the megakaryocyte cytoplasm undergoes a massive reorganization into beaded cytoplasmic extensions called *proplatelets*. This chapter focuses on the development of megakaryocytes and evaluates the proposed mechanisms and sites of platelet formation. In addition, we review the proplatelet theory of platelet biogenesis and discuss the cytoskeletal mechanics of platelet formation. Finally, we consider how insights gained from knockout animal models and human diseases have increased our understanding of megakaryocyte development and platelet formation.

## II. Megakaryocyte Development

### A. The Hematopoietic Program

Megakaryocytes, like all other cells in the blood, develop from a master stem cell. In adults, these hematopoietic stem cells reside primarily in the bone marrow.[1,2] During mammalian development, stem cells also successively populate the embryonic yolk sac, fetal liver, and spleen. A detailed model of hematopoiesis has emerged from experiments analyzing the effects of hematopoietic growth factors on marrow cells contained in a semisolid media. In the current model of hematopoiesis, stem cells either self-renew or commit to a specific cellular lineage that ultimately gives rise to mature blood cells.[1,2] Stem cells appear to lose developmental potential gradually and ultimately become restricted to a specific blood cell lineage.

### B. Committed Megakaryocyte Progenitor Cells

Committed megakaryocyte precursor cells develop from pluripotential hematopoietic progenitors (Fig. 2-1). All hematopoietic progenitors express surface CD34 and CD41, and commitment to the megakaryocyte lineage is indicated by expression of CD61 (integrin β3, GPIIIa) and elevated CD41 (integrin αIIb, GPIIb) levels. From the committed myeloid progenitor cell (colony-forming unit–granulocyte-erythroid-macrophage-megakaryocyte [CFU-GEMM]), there is strong evidence for a bipotential progenitor intermediate between the pluripotential stem cell and the committed precursor that can give rise to biclonal colonies composed of megakaryocytic and erythroid cells.[3–5] Diploid precursors that are committed to the megakaryocyte lineage have traditionally been divided into two colonies based on their functional capacities.[6–9] The megakaryocyte burst-forming cell is a primitive progenitor that develops from a mixed lineage, bipotential erythroid/megakaryocytic cell. The morphology of the megakaryocyte burst-forming cell does not resemble a mature megakaryocyte, but rather resembles a small lymphocyte. Its high proliferation capacity gives rise to large megakaryocyte colonies. Under appropriate culture conditions, the megakaryocyte burst-forming cell can develop into 40 to 500 megakaryocytes within 1 week. The colony-forming cell is a more mature megakaryocyte progenitor that gives rise to a colony containing from 3 to 50 mature megakaryocytes that vary in their proliferation potential. Megakaryocyte progenitors can be readily identified in bone marrow by immunoperoxidase and AChE labeling.[10–12] Although both human megakaryocyte colony-forming and burst-forming cells express

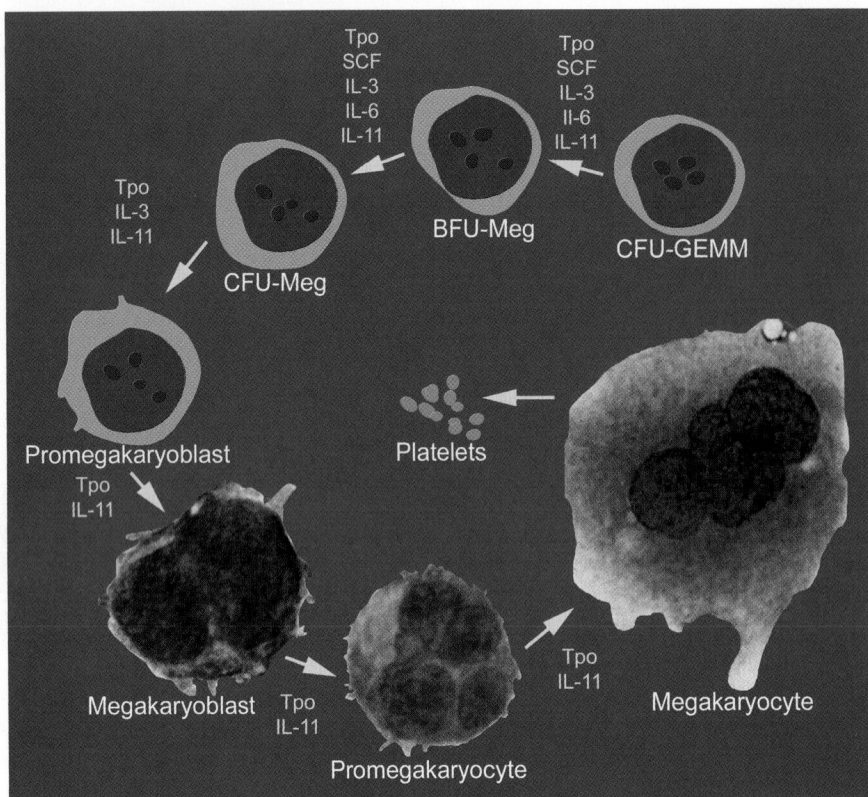

**Figure 2-1.** Megakaryocyte and platelet development. From a committed myeloid progenitor cell, colony-forming unit–granulocyte–erythroid–macrophage–megakaryocyte (CFU-GEMM), there is evidence for a common intermediate cell (not depicted) that differentiates into the megakaryocytic, basophilic, and erythroid lineages. The burst-forming unit–megakaryocyte (BFU-Meg) is committed to mega-karyocyte differentiation. Both CFU-Meg and BFU-Meg express CD34, CD33, and CD41. The CD41 (GPIIb) cell surface antigen is a megakaryocyte lineage marker. The promegakaryoblast is the first morphologically recognizable megakaryocyte precursor in bone marrow. Megakaryoblasts are 15 to 50 μm in diameter, with large oval nuclei, and a basophilic cytoplasm lacking granules. They have two sets of chromosomes (4N). Promegakaryocytes are 20 to 80 μm in diameter and have a polychromatic staining cytoplasm. Megakaryocytes are the largest hematopoietic cells in the bone marrow, with diameters as large as 150 μm, and they have a highly lobulated multilobed nucleus. The cytoplasm stains basophilic. The functions of specific cytokines in megakaryocyte development have been studied in detail. Interleukin (IL)-3, by itself, supports the early stages of megakaryocyte development up to the promegakaryoblast stage before polyploidization. Thrombopoietin (Tpo) is the principal regulator of thrombopoiesis and affects all stages of megakaryocyte development. IL-6, IL-11, and stem cell factor (SCF; kit ligand) also stimulate specific stages of megakaryocyte development, but function only in concert with Tpo or IL-3.

the CD34 antigen, only colony-forming cells express the HLA-DR antigen.[13]

### C. Immediate Megakaryocyte Precursors

Various classification schemes based on morphological features, histochemical staining, and biochemical markers have been used to categorize different stages of megakaryocyte development. In general, three types of morphologies can be identified in bone marrow. The promegakaryoblast is the first recognizable megakaryocyte precursor (Fig. 2-1). The megakaryoblast, or stage I megakaryocyte, is a more

mature cell that has a distinct morphology.[14] The mega-karyoblast has a kidney-shaped nucleus with two sets of chromosomes (4N). It is 10 to 50 μm in diameter and appears intensely basophilic in Romanovsky-stained marrow preparations as a result of the large quantity of ribosomes, although the cytoplasm at this stage lacks granules. The megakaryo-blast displays blebbing of the plasma membrane, a high nuclear-to-cytoplasmic ratio and, in rodents, is AChE positive. The promegakaryocyte, or stage II megakaryocyte, is 20 to 80 μm in diameter with a polychromatic cytoplasm. The cytoplasm of the promegakaryocyte is less basophilic than the megakaryoblast and now contains developing granules.

## D. *Endomitosis and Polyploid Formation*

Megakaryocytes undergo endomitosis and become polyploid through repeated cycles of DNA replication without cell division.[15–18] At the end of the proliferation phase, mononuclear megakaryocyte precursors exit the diploid state to differentiate and undergo endomitosis, resulting in a cell that contains multiples of a normal diploid chromosome content (i.e., 4N, 16N, 32N, 64N).[19] Although the number of endomitotic cycles can range from two to six, the majority of megakaryocytes undergo three endomitotic cycles to attain a DNA content of 16N. Megakaryocyte polyploidization results in a functional gene amplification, the likely purpose of which is an increase in protein synthesis in parallel with cell enlargement.[20] It was initially postulated that polyploidization may result from an absence of mitosis after each round of DNA replication. However, recent studies of primary megakaryocytes in culture indicate endomitosis does not result from a complete absence of mitosis, but rather a prematurely terminated mitosis.[20–22] Megakaryocyte progenitors initiate the cycle and undergo a short G1 phase, a typical 6- to 7-hour S phase for DNA synthesis, a short G2 phase, followed by endomitosis.[15] Megakaryocytes begin the mitotic cycle and proceed from prophase to anaphase A, but do not enter anaphase B, telophase, or undergo cytokinesis. During polyploidization of megakaryocytes, the nuclear envelope breaks down and an abnormal spherical mitotic spindle forms (Fig. 2-2). Each spindle attaches chromosomes that align to a position equidistant from the spindle poles (metaphase). Sister chromatids segregate and begin to move toward their respective poles (anaphase A). However, the spindle poles fail to move apart and do not undergo the separation typically observed during anaphase B. Individual chromatids are not moved to the poles, and subsequently a nuclear envelope reassembles around the entire set of sister chromatids, forming a single, enlarged, but lobed, nucleus with multiple chromosome copies. The cell then skips telophase and cytokinesis to enter G1. This failure to separate sets of daughter chromosomes fully may prevent the formation of a nuclear envelope around each individual set of chromosomes.[21,22]

In most cell types, checkpoints and feedback controls ensure that DNA replication and cell division are tightly coupled. Megakaryocytes appear to be the exception to this rule, indicating that they have managed to deregulate this process. Recent work by a number of laboratories has focused on identifying the signals that regulate polyploidization in megakaryocytes.[23] It has been postulated that endomitosis is the consequence of a reduction in mitosis promoting factor (MPF) activity, a multiprotein complex consisting of Cdc2 and cyclin B.[24,25] MPF has a kinase activity that is necessary for entry of cells into mitosis. In most cell types, newly synthesized cyclin B binds to Cdc2 and

**Figure 2-2.** Multiple mitotic spindle pole formation during the endomitotic process of thrombopoietin-treated megakaryocytes. Interconnected mitotic spindle poles were observed by antitubulin immunofluorescence confocal microscopy. Multiple mitotic spindle poles linked by microtubules have formed within this megakaryocyte, which is in the metaphase stage of mitosis. Spindle microtubules radiate from each aster to form an abnormal conformation. Megakaryocytes become polyploid through repeated cycles of DNA replication without concomitant cell division resulting from a block in anaphase B of mitosis. (Courtesy of W. Vainchencker and L. Roy.)

produces active MPF, whereas cyclin degradation at the end of mitosis inactivates MPF. Conditional mutations in strains of budding and fission yeast that inhibit either cyclin B or cdc2 cause them to go through an additional round of DNA replication without mitosis.[26,27] In addition, studies using a human erythroleukemia cell line have demonstrated that these cells contain inactive cdc2 during polyploidization and investigations with phorbol ester-induced Meg T cells have demonstrated that cyclin B is absent in this cell line during endomitosis.[28,29] However, it has been difficult to deduce the role of MPF activity in promoting endomitosis, because these cell lines have a curtailed ability to undergo endomitosis. Furthermore, experiments using normal megakaryocytes in culture have demonstrated normal levels of cyclin B and cdc2 with functional mitotic kinase activity in megakaryocytes undergoing mitosis, suggesting that endomitosis can be regulated by signaling pathways other than MPF.[21,22] In addition, it has recently been demonstrated that the molecular programming involved in endomitosis is characterized by the mislocalization or absence of at least two critical regulators of mitosis: Aurora-B/AIM-1 and sur-vivin.[30] An alternative hypothesis that endomitosis is driven by the inhibition of microtubule-based forces during anaphase B is based on the observation that mitosis proceeds normally up to anaphase B, but is then blocked in spindle pole elongation. Spindle pole separation during anaphase B is believed to be a consequence of the sliding of antiparallel and interdigitating nonkinetochore (polar) microtubules past one another.[31] The mitotic kinesinlike protein 1 localizes at regions of overlapping microtubules during anaphase B and can slide microtubules past one another *in vitro*.[32] Therefore, the lack of spindle pole separation during endomitosis may result from a failure to organize antiparallel overlapping microtubules into the correct configuration and/or the absence of signals that localize or activate the kinesin motor molecule that provides force for sliding.

### E. Megakaryocyte Cytoplasmic Maturation

After the process of endomitosis is completed, the megakaryocyte begins a maturation stage in which the cytoplasm rapidly fills with platelet-specific proteins, organelles, and membrane systems that will ultimately be subdivided and packaged into platelets. During this stage of maturation, the megakaryocyte cytoplasm acquires its distinct ultrastructural features, including the development of a DMS, the assembly of a dense tubular system, and the formation of granules.

**1. Demarcation Membrane System.** One of the most striking features of a mature megakaryocyte is its elaborate DMS (Fig. 2-3)[33] — an extensive network of membrane channels composed of flattened cisternae and tubules. The suborganization of the megakaryocyte cytoplasm into membrane-delineated "platelet territories" was first reported by Kautz and De Marsh,[34] and a detailed description of these membranes by Yamada[35] soon followed. The DMS is detectable in promegakaryocytes, but becomes most evident in mature megakaryocytes in which it permeates the megakaryocyte cytoplasm, except for a rim of cortical cytoplasm from which it is excluded (Fig. 2-3). It has been proposed that the DMS derives from megakaryocyte plasma membrane in the form of tubular invaginations.[36] The DMS is in contact with the external milieu and can be labeled with extracellular tracers, such as ruthenium red, lanthanum salts, and tannic acid.[37–39] The exact function of this elaborate smooth membrane system has been hotly debated for many years. Initially, it was postulated to play a central role in platelet formation by defining preformed "platelet territories" within the megakaryocyte cytoplasm (discussed later). However, recent studies more strongly suggest that the DMS functions primarily as a membrane reserve for proplatelet formation and extension. The DMS has also been

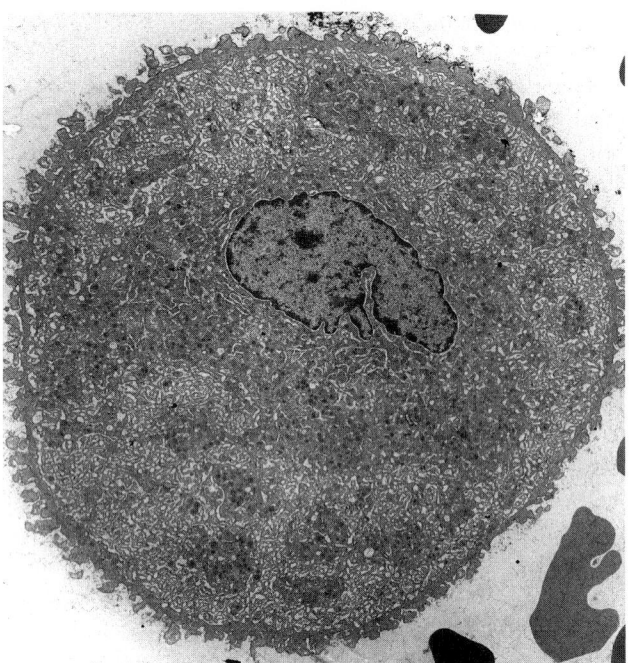

**Figure 2-3.** Electron micrograph of a thin section through a mature megakaryocyte having a well-defined demarcation membrane system (DMS). The DMS is a smooth membrane system organized into a network of narrow channels homogeneously distributed throughout the cytoplasm (×5000). The DMS has been proposed to originate from the invagination of plasma membrane and to function as a membrane reservoir for proplatelet formation or as a mechanism to subdivide the megakaryocyte cytoplasm into "platelet fields." (From Zucker–Franklin,[33] with permission.)

proposed to mature into the OCS of the mature platelet, which functions as a channel for the secretion of granule contents. However, bovine megakaryocytes, which have a well-defined DMS, produce platelets that do not develop an OCS, suggesting that the OCS is not necessarily a remnant of the DMS.[39,40]

**2. Dense Tubular System.** Megakaryocytes contain a dense tubular system.[41] The dense tubular system is believed to be the site of prostaglandin synthesis in platelets.[42] The dense tubular system does not stain with extracellular membrane tracers, indicating it is not in contact with the external environment.

**3. Granules.** Megakaryocyte maturation is characterized by the progressive formation and appearance of a variety of secretory granules. The most abundant are α granules, which contain proteins essential for platelet adhesion during vascular repair (see Chapter 15). These granules are typically 200 to 500 nm in diameter and have spherical shapes with a dark central core. They are present in early-stage megakaryocytes and originate from the trans-Golgi network, where their characteristic dark nucleoid core becomes visible within the budding vesicles.[43] α Granules acquire their molecular contents from both endogenous protein synthesis and by uptake and packaging of plasma proteins by receptor-mediated endocytosis and pinocytosis.[44] Endogenously synthesized proteins such as platelet factor 4, β-thromboglobulin, and von Willebrand factor are detected in megakaryocytes before endocytosed proteins such as fibrinogen. In addition, synthesized proteins predominate in the juxtanuclear Golgi area, whereas endocytosed proteins are localized in the peripheral regions of the cell.[45] It has been well documented that uptake and delivery of fibrinogen to α granules is mediated by integrin αIIbβ3.[46–48] Several membrane proteins critical to platelet function are also packaged into α granules, including integrin αIIbβ3, P-selectin (CD62P), and CD36. Although little is known about the intracellular tracking of proteins in megakaryocytes and platelets, experiments using ultrathin cryosectioning and immunoelectron microscopy suggest multivesicular bodies are a crucial intermediate stage in the formation of platelet α granules.[49] Multivesicular bodies are prominent in cultured megakaryocytes, but less numerous in bone marrow megakaryocytes. During megakaryocyte development, these large (up to 0.5 μm) multivesicular bodies undergo a gradual transition from granules containing 30- to 70-nm internal vesicles to granules containing predominantly dense material. Internalization kinetics of exogenous bovine–serum albumin–gold particles and of fibrinogen position the multivesicular bodies and α granules sequentially in the endocytic pathway. Multivesicular bodies contain the secretory proteins von Willebrand factor and β-thromboglobulin, the platelet-specific membrane protein P-selectin, and the lysosomal membrane protein CD63, suggesting they are a precursor organelle for α granules.[49] Dense granules (or dense bodies), approximately 250 nm in size, identified in electron micrographs by virtue of their electron-dense cores, contain a variety of hemostatically active substances that are released upon platelet activation, including serotonin, catecholamines, ADP, adenosine 5′-triphosphate (ATP), and calcium. Immunoelectron microscopy studies have also indicated that multivesicular bodies are an intermediary stage of dense granule maturation and constitute a sorting compartment between α granules and dense granules.[50]

## III. Platelet Formation

### A. Mechanisms of Platelet Production

Although it has been universally accepted that platelets derive from megakaryocytes, the mechanisms by which platelets form and release from these precursor cells remain controversial. Throughout the years, several models of platelet production have been proposed (Fig. 2-4). These include (a) platelet budding, (b) cytoplasmic fragmentation via the DMS, and (c) proplatelet formation. Past studies attempting to discriminate between these mechanisms of platelet biogenesis have been hampered by the requirement of sampling bone marrow to obtain megakaryocytes, the relative infrequency of megakaryocytes in the marrow, and the lack of *in vitro* systems that faithfully reconstitute platelet formation. However, the discovery of thrombopoietin (TPO), a cytokine that binds to the megakaryocyte-specific receptor c-MPL and promotes the growth and development of megakaryocyte precursors (see Chapter 66), has led to the emergence of culture systems that recapitulate platelet biogenesis and resulted in a new understanding of the terminal differentiation phase of thrombopoiesis. Several models of platelet biogenesis are discussed next.

**1. Budding from the Megakaryocyte Surface.** Based on scanning electron micrographs of megakaryocytes with apparent platelet-size blebs on their surface, it was proposed that platelets shed from the periphery of the megakaryocyte cytoplasm.[51,52] Examination of these structures by thin-section electron microscopy, however, revealed that these blebs did not contain platelet organelles, an observation inconsistent with the concept of platelet budding as a mechanism for platelet release. In addition, the platelet buds were probably confused with the pseudopods that extend from mature megakaryocytes during the initial stages of proplatelet formation.

**2. Cytoplasmic Fragmentation via the Demarcation Membrane System.** The DMS, described in detail by Yamada[35] in 1957, has been proposed to define preformed "platelet territories" within the cytoplasm of the megakaryocyte. Microscopists recognized that maturing megakaryocytes became filled with membranes and platelet-specific organelles (Fig. 2-3), and postulated that these membranes formed a system that defined territories or fields for developing platelets.[53] Release of individual platelets was proposed to occur by a massive fragmentation of the megakaryocyte cytoplasm along DMS fracture lines residing between these fields. The DMS model predicts that platelets form through an extensive internal membrane reorganization process.[54] Tubular membranes, which may originate from invagination of the megakaryocyte plasma membrane, are predicted to interconnect and branch, forming a continuous network throughout the cytoplasm. The fusion of adjacent tubules has been proposed as a mechanism to generate a flat membrane that ultimately surrounds the cytoplasm of a putative platelet.[53] Models attempting to use the DMS to explain how the megakaryocyte cytoplasm becomes subdivided into platelet volumes and enveloped by its own membrane have lost support because of several inconsistent observations. For example, if platelets are delineated within the megakaryocyte cytoplasm by the DMS, then platelet fields should exhibit structural characteristics of platelets, which is not

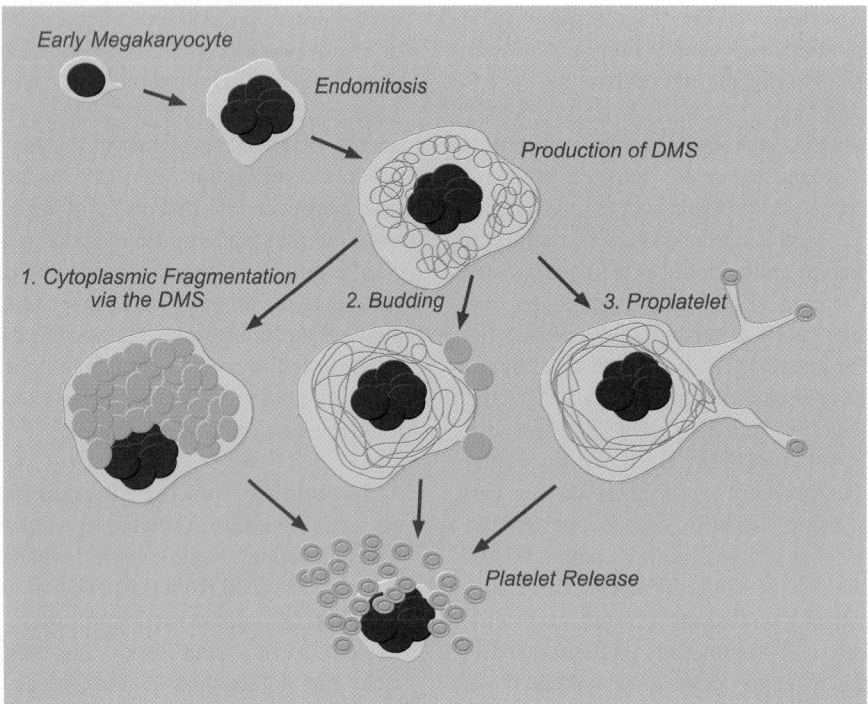

**Figure 2-4.** Mechanisms proposed for platelet production. Three models have been proposed to explain the mechanics of platelet production. The first is cytoplasmic fragmentation via the DMS. In this model, the DMS defines predetermined platelet-size fields within the megakaryocyte cytoplasm. Platelets form when the cytoplasm fragments along these DMS fracture lines. The second is platelet budding. In this model, platelets pinch off from blebs protruded at the megakaryocyte periphery. The third is proplatelet formation. The flow model of proplatelet formation requires platelets to form through proplatelet intermediate structures, which are long cytoplasmic extensions that appear as platelet-size beads linked together by thin cytoplasmic strands. In this model, the DMS functions primarily as a membrane reservoir for extension of proplatelets.

the case.[55] Platelet territories within the megakaryocyte cytoplasm lack marginal microtubule coils, the most characteristic feature of resting platelet structure (see Chapters 3 and 4). In addition, there are no studies directly demonstrating that platelet fields shatter into mature, functional platelets. In contrast, studies that focused on the DMS of megakaryocytes before and after proplatelet retraction induced by microtubule depolymerizing agents suggest this specialized membrane system may function primarily as a membrane reservoir that evaginates to provide plasma membrane for the growth of proplatelets.[56] Radley and Haller[56] considered the name *DMS* to be a misnomer, and suggested *invagination membrane system* as a more suitable name to describe this membranous network.

### 3. Proplatelet Formation

*a. The Proplatelet Theory.* The term *proplatelet* is generally used to describe long (up to millimeters in length), thin cytoplasmic processes emanating from megakaryocytes.[57] These extensions are typically characterized by multiple platelet-size swellings linked together by thin cytoplasmic bridges and are thought to represent intermediate

structures in the megakaryocyte-to-platelet transition. The concept of platelets arising from these pseudopodialike structures dates originally to 1906, when Wright[58] recognized that platelets originate from megakaryocytes and described "the detachment of plate-like fragments or segments from pseudopods" of megakaryocytes. Thiery and Bessis,[59] and Behnke,[60] later described in more detail the structure of these cytoplasmic processes extending from megakaryocytes during platelet formation. The classic "proplatelet theory" was introduced by Becker and DeBruyn,[57] who proposed that megakaryocytes form long pseudopodlike processes that subsequently fragment to generate individual platelets. In this early model, the DMS was still proposed to subdivide the megakaryocyte cytoplasm into platelet areas. Radley and Haller[56] later developed the "flow model," which postulated that platelets derived exclusively from the interconnected platelet-size beads connected along the shaft of proplatelets, and suggested that the DMS did not function to define platelet fields, but as a reservoir of surface membrane to be evaginated during proplatelet formation. Developing platelets were assumed to become encased by plasma membrane only as proplatelets were formed.

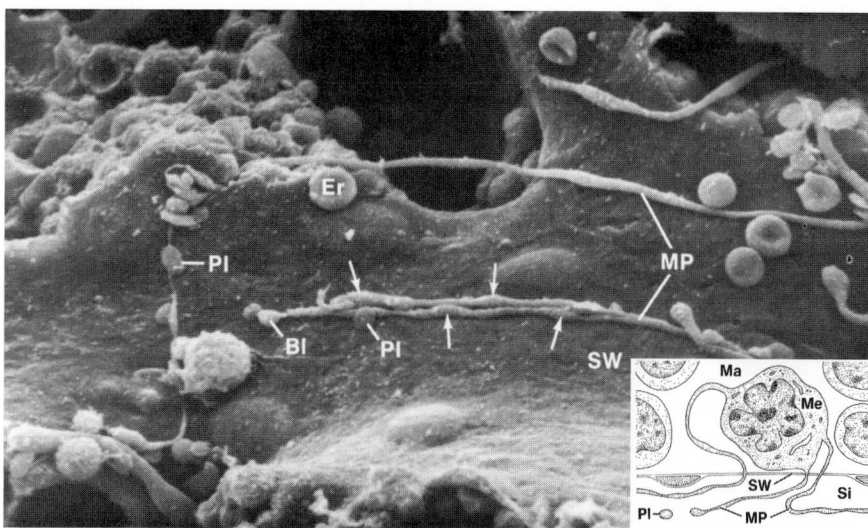

**Figure 2-5.**    Scanning electron micrograph showing proplatelet extensions in the interior of a blood sinus in the bone marrow. Proplatelet processes (MP), some more than 100 μm long, extend from the extravascular compartment and sinusoidal wall (SW) into the sinusoids. The beaded contour of these extensions resembles chains of putative platelets joined by narrow cytoplasmic links. The megakaryocyte extensions frequently terminate as bulbous tips (Bl) that are similar in size and morphology to circulating platelets (Pl). Platelets (Pl) can be compared in size with erythrocytes (Er). Inset. Schematic drawing of a megakaryocyte (Me) body with proplatelet processes (MP). The cell body of the megakaryocyte is located in the bone marrow (Ma). Proplatelet processes (MP) are extended into the lumen of myeloid sinusoids (Si). (From Kessel and Kardon,[69] with permission.)

The bulk of experimental evidence now supports a modified proplatelet model of platelet formation. Proplatelets have been observed (a) both *in vivo*[61] and *in vitro,* and maturation of proplatelets yields platelets that are structurally and functionally similar to blood platelets[62,63]; (b) in a wide range of mammalian species, including mice, rats, guinea pigs, dogs, cows, and humans[62,64–68]; (c) extending from megakaryocytes in the bone marrow through junctions in the endothelial lining of blood sinuses, where they have been hypothesized to be released into the circulation and undergo further fragmentation into individual platelets (Fig. 2-5)[69–72]; and (d) to be absent in mice lacking two distinct hematopoietic transcription factors. Such mice fail to produce proplatelets in culture and exhibit severe thrombocytopenia.[73–75] Taken together, these findings establish an important role for proplatelet formation in thrombopoiesis.

*b. Proplatelet Morphogenesis.* The development of megakaryocyte cultures that faithfully reconstitute platelet formation has provided systems to study megakaryocytes in the act of forming proplatelets *in vitro.* Video microscopy reveals both temporal and spatial change leading to the formation of proplatelets[76] (Fig. 2-6). Transformation of the megakaryocyte cytoplasm concentrates virtually all the intracellular contents into proplatelet extensions and their platelet-size particles, which, in the final stages, appear as beads linked by thin cytoplasmic bridges. The transformation unfolds over 4 to 10 hours and commences with the erosion of one pole (Fig. 2-6, asterisk at 0 hr) of the mega-karyocyte cytoplasm. Thick pseudopodia initially form and subsequently elongate into thin tubules of uniform diameter of 2 to 4 μm. These slender tubules, in turn, undergo a dynamic bending and branching process, and develop periodic densities along their length. Eventually, the megakaryocyte is transformed into a residual naked nuclei surrounded by an anastomosing network of proplatelet processes. Megakaryocyte maturation ends when a rapid retraction separates the proplatelet fragments from the cell body releasing them into culture (Fig. 2-6, asterisk at 10 hr). The subsequent rupture of the cytoplasmic bridges between platelet-size segments is thought to release individual platelets into the circulation.

*c. The Cytoskeletal Mechanics of Proplatelet Formation.* The cytoskeleton of the mature platelet plays a crucial role in maintaining the discoid shape of the resting platelet and is responsible for the shape change associated with platelet activation (see Chapter 4). This same set of cytoskeletal proteins provides the force to bring about the shape changes associated with megakaryocyte maturation.[77] Two cytoskeletal polymer systems exist in megakaryocytes: actin and tubulin. Both of these proteins reversibly polymerize into cytoskeletal filaments. Accumulating evidence supports a model of platelet formation in which microtubules and actin filaments play a crucial role (Fig. 2-7).

i. Microtubules Power Proplatelet Elongation. Proplatelet formation is dependent on microtubule function, because

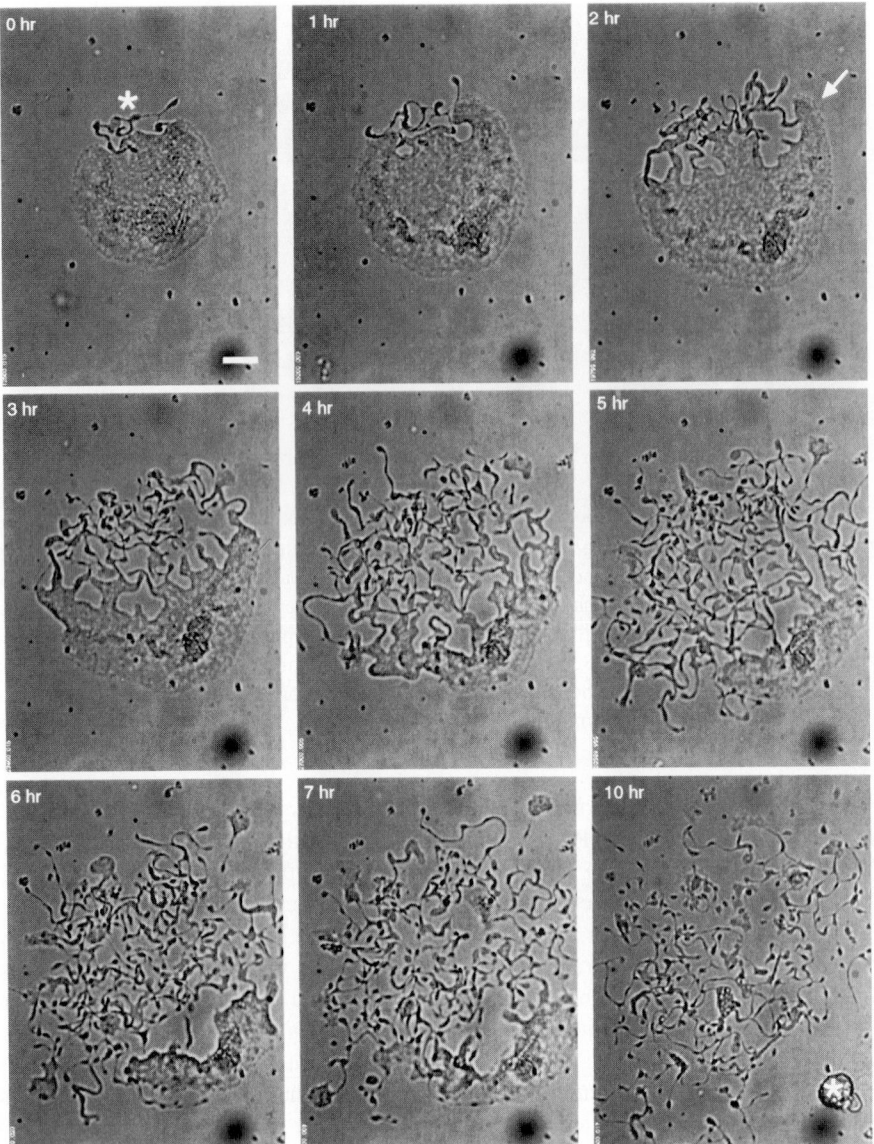

**Figure 2-6.** Video-enhanced light microscopy of a mouse megakaryocyte forming proplatelets *in vitro*. During the initial stages of proplatelet formation, the megakaryocyte spreads and its cortical cytoplasm begins to unravel at one pole. (This zone of erosion is labeled with a white asterisk in the first panel.) As the cell spreads, the cytoplasm at the erosion site is remodeled into large pseudopodia (white arrow at 2 hr) that lengthen and thin over time, forming narrow tubes of 2 to 4 μm in diameter. Proplatelet extensions frequently bend (white arrow at 2 hr), and bending sites subsequently bifurcate to generate new proplatelet processes. In this manner, the entire cytoplasmic volume of the megakaryocyte is converted into branched proplatelet extensions, and proplatelet ends are dramatically increased. Proplatelets also develop segmented constrictions along their length that impart a beaded appearance. The process of proplatelet elaboration ends in a rapid retraction that separates strands of proplatelets from the residual naked nucleus (asterisk in lower right at 10 hr). The scale bar is 20 μm. (From Italiano, Lecine, Shivdasani, and Hartwig,[76] with permission from the Rockefeller University Press.)

treatment of megakaryocytes with drugs that depolymerize microtubules, such as nocodazole or vincristine, blocks proplatelet formation.[55,56,65,66,78] Microtubules, hollow polymers assembled from αβ-tubulin dimers, are the major structural components of the engine that powers proplatelet elongation. Examination of the microtubule cytoskeletons of proplatelet-producing megakaryocytes provides clues regarding

how microtubules mediate platelet development.[76,79] The microtubule cytoskeleton in megakaryocytes undergoes a dramatic reorganization during proplatelet production. In immature megakaryocytes without proplatelets, microtubules spiral out from the cell center to the cortex. As blunt pseudopodia form during the initial stage of proplatelet formation, cortical microtubules consolidate into thick bundles

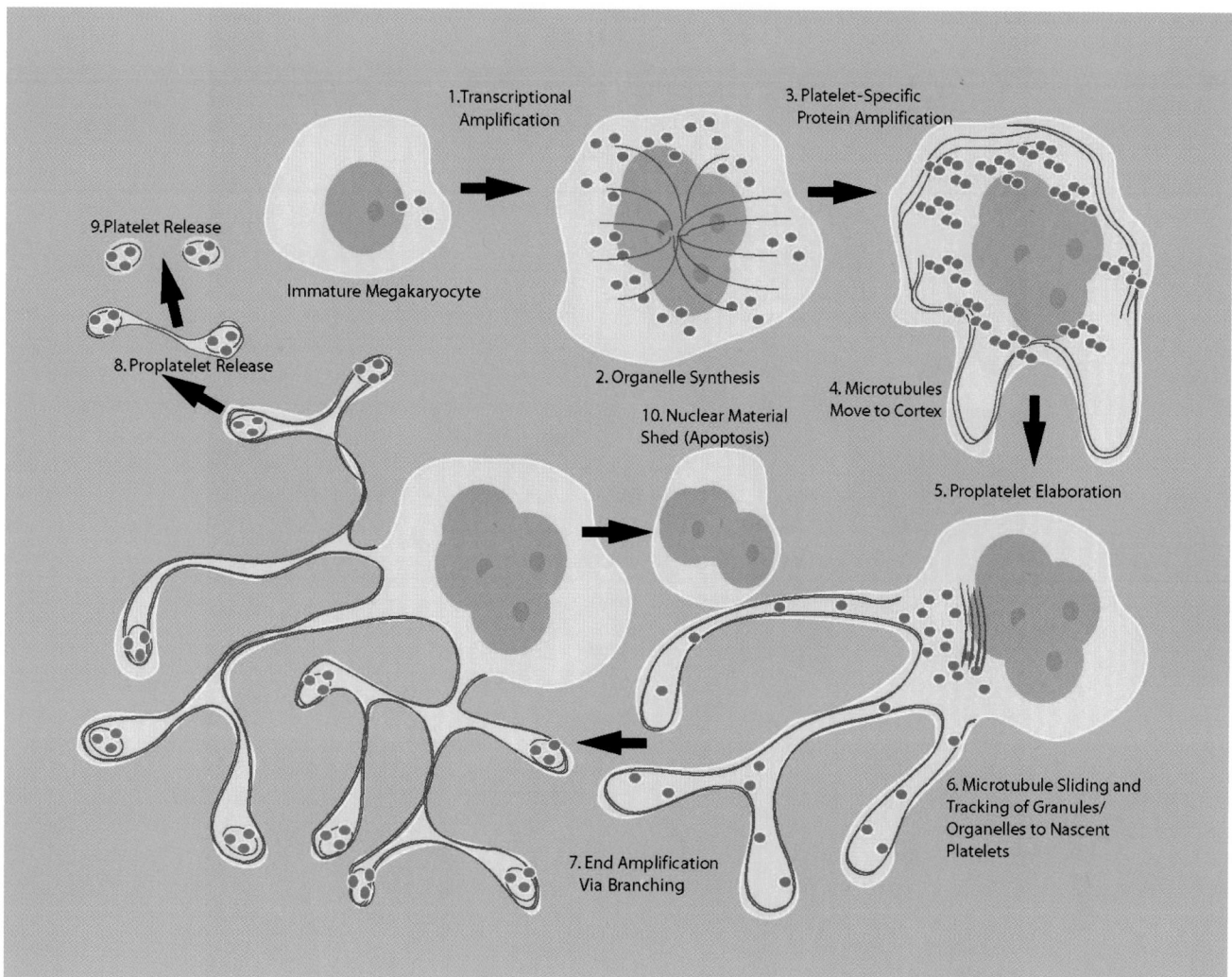

**Figure 2-7.** Proplatelet model detailing some of the cytoskeletal mechanics of platelet biogenesis. As megakaryocytes transition from immature cells to release platelets, a systematic series of events occurs. (1) After commitment to the megakaryocyte lineage, cells undergo endomitosis and transcriptional activation. (2) During cytoplasmic maturation and expansion, the megakaryocyte undergoes organelle synthesis and (3) platelet-specific protein amplification. Microtubule arrays emanating from a centrosome are clearly established. (4) Prior to the onset of proplatelet formation, centrosomes disassemble and microtubules translocate to the cell cortex. (5) Proplatelet production initiates with the formation of large pseudopodia that elongate and form thin proplatelet processes with bulbous ends; these ends contain a peripheral bundle of microtubules that loops upon itself and forms a teardrop-shape structure. (6) Sliding of overlapping microtubules drives proplatelet elongation as organelles are moved individually over microtubules into proplatelet ends, where nascent platelets assemble. (7) Growth and extension of proplatelet processes is associated with repeated actin-dependent bending and bifurcation, which amplifies free proplatelet ends. Proplatelets form constrictions along their length, giving them a beaded appearance. (8) The entire megakaryocyte cytoplasm is converted into a mass of proplatelets, which release from the megakaryocyte body after a rapid retraction. A barbell of two platelet-size particles linked by a small cytoplasmic bridge may provide an intermediate stage. (9) Proplatelets undergo further fragmentation into individual platelets. (10) The extruded naked nucleus remaining after near-complete shedding of the megakaryocyte cytoplasm undergoes apoptosis.

situated just beneath the plasma membrane of these structures (Fig. 2-8A, C). When pseudopodia begin to elongate (at an average rate of 0.85 μm/minute), microtubules form thick linear arrays that line the entire length of the proplatelet extensions (Fig. 2-8B). The microtubule bundles are thickest in the portion of the proplatelet near the body of the megakaryocyte, but thin to bundles of 5 to 10 microtubules near proplatelet ends. The distal end of each proplatelet always has a platelet-size enlargement containing a microtubule bundle that loops just beneath the plasma membrane and reenters the shaft to form a teardrop-shape structure (Fig. 2-8D). Because microtubule coils similar to those

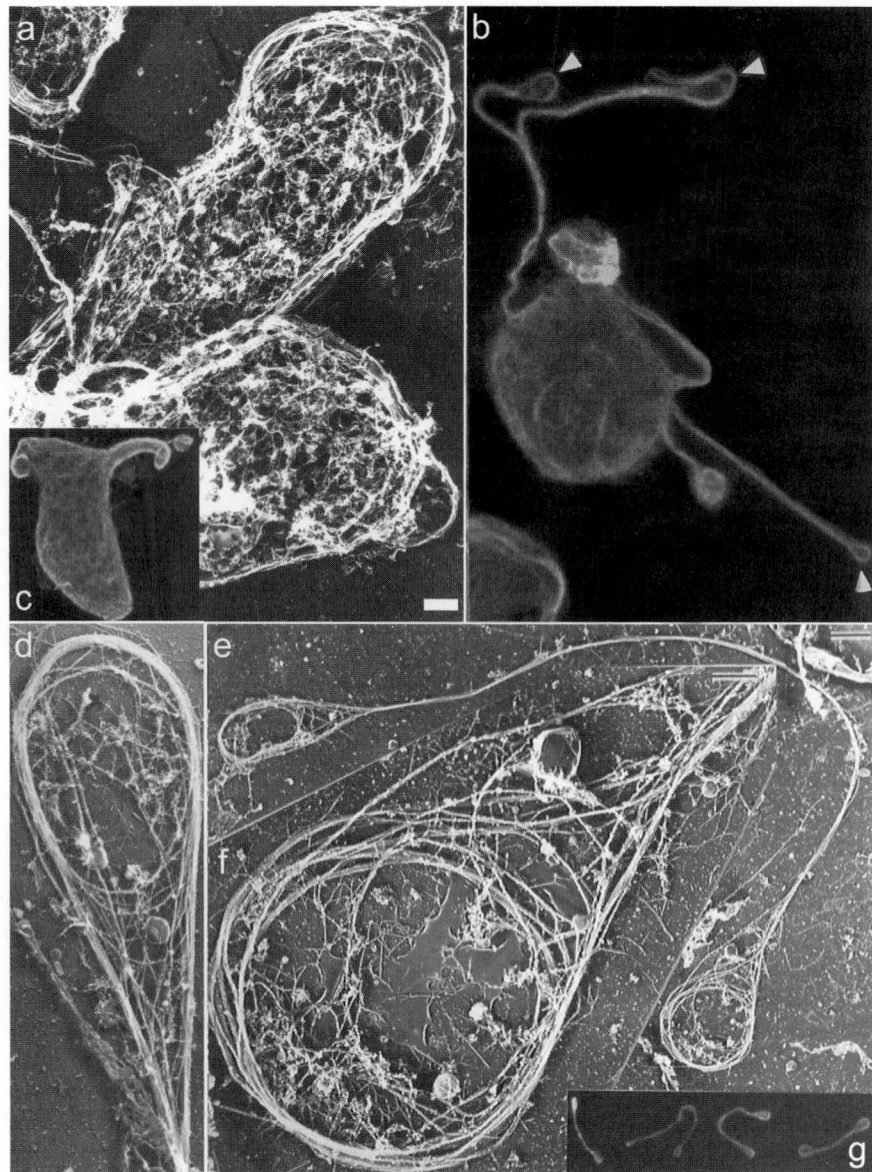

**Figure 2-8.** Organization of microtubules in megakaryocyte pseudopodia and proplatelets. A. Electron micrograph showing the structure of the megakaryocyte cytoskeleton during early proplatelet formation. At the initial stage of proplatelet formation, blunt pseudopodia form in the area of cytoplasmic erosion. Microtubules consolidate into thick bundles situated just beneath the plasma membrane of these pseudopodia. The bar is 1 μm. B. Antitubulin staining of an early megakaryocyte and its proplatelet extensions demonstrating that microtubules form loops at the ends of proplatelets. As the pseudopodia elongate and thin, microtubules form linear arrays along the length of the cytoplasmic extensions. The distal end of each proplatelet has a teardrop-shape enlargement that contains a microtubule loop (white arrowheads). C. Antitubulin immunofluorescent staining reveals a cortical concentration of microtubules in the pseudopodia. D. Electron micrograph showing the organization of microtubules within the bulbous tips of proplatelets. Each end of the proplatelet contains a microtubule bundle that loops beneath the plasma membrane and reenters the shaft to form a teardrop-shape structure. E. Low-magnification electron micrograph showing the microtubule-based cytoskeleton of a representative released proplatelet form. A microtubule bundle lines the shaft of the proplatelet. The bar is 0.5 μm. F. Proplatelet ends have microtubule bundles arranged into teardrop-shape loops, similar to those at the ends of proplatelets extending from megakaryocytes. Microtubule coils similar in size and structure to those observed in mature resting platelets are formed within these loops. The scale bar is 0.5 μm. G. Gallery of released proplatelet forms stained for tubulin by immunofluorescence confocal microscopy. (From Italiano et al.,[76] with permission of The Rockefeller University Press.)

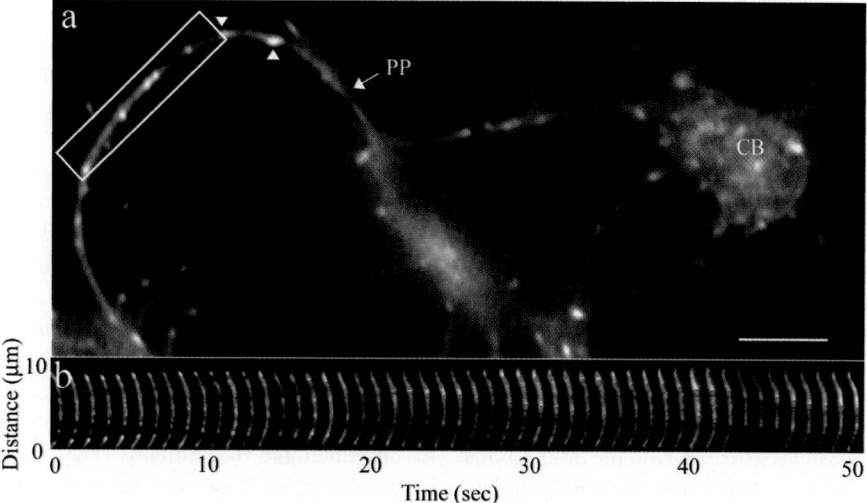

**Figure 2-9.** Visualization of microtubule assembly in living megakaryocytes expressing end-binding protein three (EB3)–green fluorescent protein (GFP). A. First frame from a time-lapse movie of a living megakaryocyte that was retrovirally directed to express EB3-GFP. The cell body (CB) is at the right of the micrograph and proplatelets (PP) extend to the left. EB3-GFP labels growing microtubule plus ends in a characteristic "comet" staining pattern (arrowheads) that has a bright front and dim tail. Moving comets are found along the proplatelets as well as in the megakaryocyte cell body. The scale bar is 5 μm. B. Kymograph (movement over time) of the boxed region in panel A. Images are every 1 second. EB3-GFP comets undergo bidirectional movements in proplatelets, demonstrating that microtubules are organized as bipolar arrays. Some EB3-GFP comets move tipward and are highlighted in green; others that move toward the cell body are highlighted in red.

observed in blood platelets are detected only at the ends of proplatelets and not within the platelet-size beads found along the length of proplatelets, mature platelets are formed only at the ends of proplatelets.

In recent studies, direct visualization of microtubule dynamics in living megakaryocytes using green fluorescent protein (GFP) technology has provided clues regarding how microtubules power proplatelet elongation.[79] End-binding protein three (EB3), a microtubule plus end-binding protein associated only with growing microtubules, fused to GFP was expressed in murine megakaryocytes and used as a marker to follow microtubule plus end dynamics. Immature megakaryocytes without proplatelets use a centrosomal-coupled microtubule nucleation/assembly reaction, which appears as a prominent starburst pattern when visualized with EB3-GFP. Microtubules assemble only from the centrosomes and grow outward into the cell cortex, where they turn and run in parallel with the cell edges.[79] However, just before proplatelet production commences, centrosomal assembly ceases and microtubules begin to consolidate into the cortex. Fluorescence time-lapse microscopy of proplatelet-producing megakaryocytes expressing EB3-GFP reveals that as proplatelets elongate, microtubule assembly occurs continuously throughout the entire proplatelet, including the shaft, swellings, and tip (Fig. 2-9A). The rates of microtubule polymerization (average, 10.2 μm/min) are approximately 10-fold faster than the proplatelet growth rate, suggesting polymerization and proplatelet elongation are not

tightly coupled. The EB3-GFP studies also revealed that microtubules polymerize in both directions in proplatelets (e.g., both toward the tips and cell body (Fig. 2-9B).[79] This demonstrates that the microtubules composing the bundles have a mixed polarity.

Although microtubules are continuously polymerizing in proplatelets, polymerization *per se* does not provide the force for proplatelet elongation. Proplatelets continue to elongate at normal rates even when microtubule polymerization is temporarily inhibited by drugs that block net microtubule assembly, suggesting another mechanism for proplatelet elongation.[79] Consistent with this idea, proplatelets have an inherent microtubule sliding mechanism. Cytoplasmic dynein, a minus-end microtubule molecular motor protein, localizes along the microtubules of the proplatelet and appears to contribute directly to microtubule sliding, because inhibition of dynein, through disassembly of the dynactin complex, prevents proplatelet formation.[79] Microtubule sliding can also be reactivated in detergent permeabilized proplatelets. ATP, known to support the enzymatic activity of microtubule-based molecular motors, activates proplatelet elongation in permeabilized proplatelets that contain both dynein and its regulatory complex, dynactin.[79] Thus, dynein-facilitated microtubule sliding appears to be the key event in driving proplatelet elongation.

ii. Actin-Dependent Proplatelet Branching. Each megakaryocyte has been estimated to generate and release thousands of platelets.[80–82] Analysis of time-lapsed video microscopy of proplatelet development from megakaryocytes grown *in vitro* has revealed that ends of proplatelets are amplified in a dynamic process that repeatedly bends and bifurcates the proplatelet shaft.[76] End amplification initiates when a proplatelet shaft is bent into a sharp kink, which then folds back on itself, forming a loop in the microtubule bundle. The new loop eventually elongates, forming a new proplatelet shaft branching from the side of the original proplatelet. Loops lead the proplatelet tip, and define the site where nascent platelets will assemble and where platelet-specific contents are trafficked. In marked difference to the microtubule-based motor that elongates proplatelets, actin-based force is used to bend the proplatelet in end amplification. Megakaryocytes treated with the actin toxins cytochalasin or latrunculin can only extend long, unbranched proplatelets that are decorated with few swellings along their length.[76] Despite extensive characterization of actin filament dynamics during platelet activation (see Chapter 4), how actin participates in this reaction and the cytoplasmic signals that regulate bending have yet to be determined. Electron microscopy and phalloidin staining of megakaryocytes undergoing proplatelet formation indicate that actin filaments are distributed throughout the proplatelet and are particularly abundant within swellings and at proplatelet branch points.[65,76,83] One likely possibility is that proplatelet bending and branching is powered by the actin-based molecular motor myosin. Interestingly, a mutation in the nonmuscle myosin heavy chain-A gene in humans results in a disease called May–Hegglin anomaly,[84,85] characterized by thrombocytopenia with giant platelets (see Chapter 54). Studies also indicate that protein kinase Cα associates with aggregated actin filaments in megakaryocytes undergoing proplatelet formation, and inhibition of protein kinase Cα or integrin signaling pathways prevent actin filament aggregation and proplatelet formation in megakaryocytes.[83] However, the role of actin filament dynamics in platelet biogenesis remains unclear.

iii. Organelle Transport in Proplatelets. In addition to playing a crucial role in proplatelet elongation, the microtubules lining the shafts of proplatelets serve a secondary function — tracks for the transport of membrane, organelles, and granules into proplatelets and assembling platelets at proplatelet ends. Individual organelles are sent from the cell body into the proplatelets, where they move bidirectionally until they are captured at proplatelet ends[86] (Fig. 2-10). Immunofluorescence and electron microscopy studies indicate that organelles are in direct contact with microtubules, and actin poisons do not diminish organelle motion. Therefore, movement appears to involve microtubule-based forces. Bidirectional organelle movement is conveyed in part by the bipolar organization of microtubules within the proplatelet, as kinesin-coated beads move bidirectionally over the microtubule arrays of permeabilized proplatelets. Of the two major microtubule motors — kinesin and dynein — only the plus-end-directed kinesin is situated in a pattern similar to organelles and granules, and is likely responsible for transporting these elements along microtubules.[86] It appears that a twofold mechanism of organelle and granule movement occurs in platelet assembly. First, organelles and granules travel along microtubules and, second, the microtubules themselves can slide bidirectionally in relation to other motile filaments to indirectly move organelles along proplatelets in a "piggyback" fashion.

iv. Proplatelet Release. *In vivo,* proplatelets extend into bone marrow vascular sinusoids, where they may be released and enter the bloodstream. The actual events surrounding platelet release *in vivo* have not been identified because of the rarity of megakaryocytes within the bone marrow. The events leading up to platelet release within cultured murine megakaryocytes have been documented. After complete conversion of the megakaryocyte cytoplasm into a network of proplatelets, a retraction event occurs that releases individual proplatelets from the proplatelet mass.[76] Proplatelets are released as chains of platelet-size particles. Maturation of platelets occurs at the ends of proplatelets. Microtubules filling the shaft of proplatelets are reorganized into microtubule coils as platelets release from the end of each proplatelet. Many of the proplatelets released into megakaryocyte cultures remain connected by thin cytoplasmic strands (Fig. 2-8E–G). The most abundant forms release as barbell shapes composed of two plateletlike swellings, each with a microtubule coil, that are connected by a thin cytoplasmic strand containing a microtubule bundle. Proplatelet termini are the only regions of proplatelets in which a single microtubule can roll into a coil, having dimensions similar to the microtubule coil of the platelet in circulation. The mechanism of microtubule coiling remains to be elucidated, but is likely to involve microtubule motor proteins such as dynein or kinesin. Because platelet maturation is limited to these sites, efficient platelet production requires the generation of a large number of proplatelet ends during megakaryocyte development. Although the actual release event has yet to be visually captured, the platelet-size particle must be liberated as the proplatelet shaft narrows and fragments.

### B. The Sites of Platelet Formation In Vivo

Although megakaryocytes arise in the bone marrow, they can migrate into the bloodstream and, as a consequence, platelet formation may also occur at nonmarrow sites. Platelet biogenesis has been proposed to take place in many different tissues, including the bone marrow, lungs, and blood. Specific stages of platelet development have been observed in all three locations.

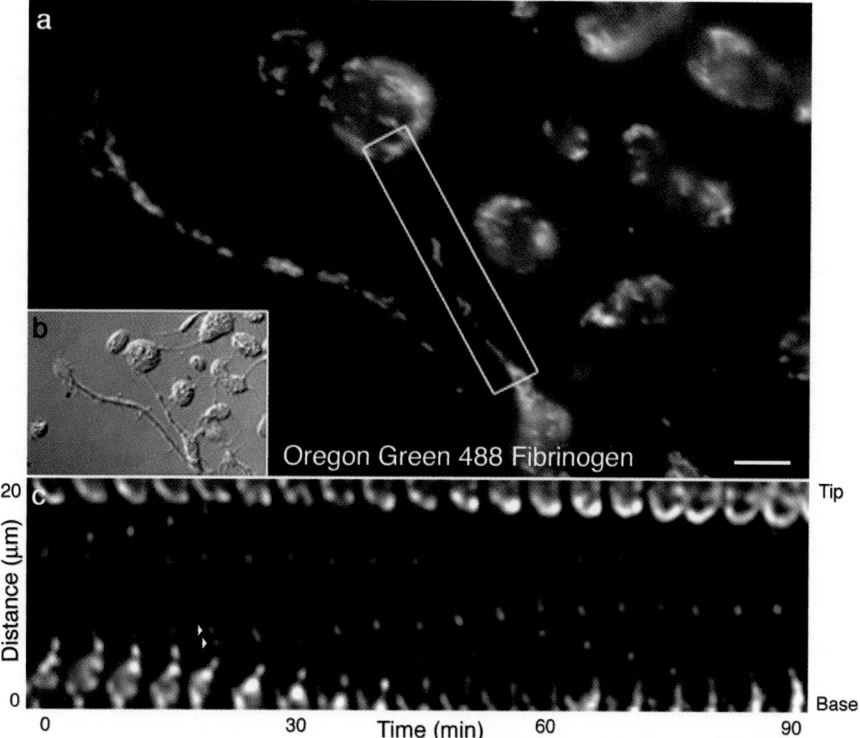

**Figure 2-10.** Organelles move bidirectionally in proplatelets. α Granules in proplatelets translocate bidirectionally. The α granules were labeled by incubating megakaryocytes with Oregon Green 488 fibrinogen conjugate, which is taken up and stored in α granules. The distribution and dynamics of the labeled α granules were followed by time-lapse fluorescence microscopy. A. Micrograph of a proplatelet field labeled with Oregon Green 488 fibrinogen conjugate. Stained α granules appear as punctate spots. Scale bar, 5 μm. B. Imaged differential interference contrast micrograph. C. Kymograph (movement over time) showing time-lapse from the boxed region in panel A. Fluorescent images of the labeled α granules were taken every 5 minutes. Two α granules highlighted in green (white arrowheads at 20 min) come together and move toward the right until one separates (60 min) and then moves towards the left (60–75 min). An α granule, highlighted in blue, remains stationary during the recording period. (From Richardson, Shivdasani, Boers, Hartwig, and Italiano,[86] with permission.)

**1. Platelet Formation in the Bone Marrow.** Megakaryocytes cultured *in vitro* outside the confines of the bone marrow can form highly developed proplatelets in suspension, suggesting that direct interaction with the bone marrow environment is not a requirement for platelet production. Nevertheless, the efficiency of platelet production in culture appears to be diminished relative to that observed *in vivo,* and the bone marrow environment composed of a complex adherent cell population could play a role in platelet formation by direct cell contact or secretion of cytokines.[87] Scanning electron micrographs of bone marrow megakaryocytes extending proplatelets through junctions in the endothelial lining into the sinusoidal lumen have been published, suggesting platelet production occurs in the bone marrow (Fig. 2-5).[60,70,71,88,89] Bone marrow megakaryocytes are strategically located in the extravascular space on the abluminal side of sinus endothelial cells and appear to send beaded proplatelet projections into the lumen of sinusoids.[70,72] Electron micrographs show that these cells are anchored to the endothelium by organelle-free projections extended by the

megakaryoctes.[72] Several observations suggest that thrombopoiesis is dependent on the direct cellular interaction of megakaryocytes with bone marrow endothelial cell-specific adhesion molecules.[90] It has been demonstrated that the translocation of megakaryocyte progenitors to the vicinity of bone marrow vascular sinusoids is sufficient to induce megakaryocyte maturation.[91] Implicated in this process are the chemokines stromal cell-derived factor 1 (SDF-1) and FGF-4, which are known to induce expression of adhesion molecules, including very late antigen (VLA)-4 on megakaryocytes and VCAM-1 on bone marrow endothelial cells.[92,93] Disruption of bone marrow endothelial cell VE-cadherin-mediated homotypic intercellular adhesion interactions results in a profound inability of the vascular niche to support megakaryocyte differentiation and to act as a conduit to the bloodstream.[91]

**2. Platelet Formation in the Bloodstream.** It is unclear whether individual platelets are released from proplatelets into the sinus lumen or whether megakaryocytes preferentially release large cytoplasmic processes into the sinus lumen that later fragment into individual platelets within the circulation. Behnke and Forer[94] have proposed that the final stages of platelet maturation occur exclusively in the circulation. In this model of thrombopoiesis, megakaryocyte fragments released into the blood become transformed into platelets while in the circulation. This theory is supported by several observations. First, the presence of megakaryocytes,[95–98] and megakaryocyte processes[66,94,99] that are sometimes beaded in blood, has been amply documented. Megakaryocyte fragments can represent up to 5 to 20% of the platelet mass in plasma. Second, these megakaryocyte fragments, when isolated from platelet-rich plasma, have been reported to elongate, undergo curving and bending motions, and eventually fragment to form disk-shape structures resembling chains of platelets.[94] Third, because both cultured human and mouse megakaryocytes can form functional platelets *in vitro,* neither the bone marrow environment nor the pulmonary circulation is essential for platelet formation and release.[62,76,100] Lastly, many of the platelet-size particles generated in these *in vitro* systems still remain attached by small cytoplasmic bridges. It is possible that the shear forces encountered in the circulation or an unidentified fragmentation factor in blood may play a crucial role in separating proplatelets into individual platelets.

**3. Platelet Formation in the Lung.** Megakaryocytes have been identified in intravascular sites within the lung, leading to a theory that platelets are formed from their parent cell predominantly in the pulmonary circulation.[82,98,101–106] Aschoff[107] first described pulmonary megakaryocytes in 1893 and suggested they originated in the marrow, migrated into the bloodstream, and, because of their massive size, were lodged in the capillary bed of the lung, where they released platelets. This mechanism requires the migration of megakaryocytes from the bone marrow into the circulation. Although the size of megakaryocytes would seem limiting, the transmigration of entire megakaryocytes through endothelial apertures of approximately 3 to 6 μm in diameter into the circulation has been recorded in electron micrographs and by living microscopy of rabbit bone marrow.[108,109] Megakaryocytes express the chemokine receptor CXCR4 and can respond to the CXCR4 ligand SDF-1 in chemotaxis assays.[110] However, both mature megakaryocytes and platelets are nonresponsive to SDF-1, suggesting the CXCR4 signaling pathway may be turned off during late stages of megakaryocyte development. This may provide a simple mechanism for retaining immature megakaryocytes in the marrow and permitting mature megakaryocytes to enter the circulation, where they can liberate platelets.[111,112] Megakaryocytes are remarkably abundant in the lung and the pulmonary circulation, and some have estimated that 250,000 megakaryocytes reach the lung every hour. In addition, platelet counts are higher in the pulmo-

nary vein than in the pulmonary artery, providing further evidence that the pulmonary bed contributes to platelet formation.[98,103,105] In humans, megakaryocytes are 10 times more concentrated in pulmonary arterial blood than in blood obtained from the aorta.[113] Kaufman and colleagues[81] investigated the possibility of platelet formation in the lung by rearranging the pulmonary vessels in a dog such that the blood from the right heart perfused first the right lung and then was directed to the left lung. The majority of megakaryocytes were found in the right lung, suggesting filtration of megakaryocytes by the pulmonary circulation. Despite these observations, the estimated contribution of pulmonary megakaryocytes to total platelet production remains unclear, as values have been estimated from 7 to 100%.[102,114,115] Furthermore, experimental results using accelerated models of thrombopoiesis in mice suggest that the fraction of platelet production occurring in the murine lung is insignificant. Davis and colleagues[116] reported that megakaryocytes and their naked nuclei were rarely observed in lung tissue even after strong stimulation of thrombopoiesis. In theory, proplatelets as well as megakaryocytes may also reach the pulmonary circulation and complete their development into platelets in lung capillaries.

### C. Apoptosis and Platelet Production

The process of platelet formation in megakaryocytes exhibits some characteristics related to apoptosis, including cytoskeletal reorganization, membrane condensation, and ruffling. These similarities have led to further investigations aimed at determining whether apoptosis is a major force driving proplatelet formation and platelet release. Apoptosis (programmed cell death) is responsible for destruction of the nucleus in senescent megakaryocytes.[117] However, it is thought that a specialized apoptotic process may lead to platelet assembly and release. Apoptosis has been described in megakaryocytes[118] and was found to be more prominent in mature megakaryocytes as opposed to immature cells.[119,120] A number of apoptotic factors, both proapoptotic and antiapoptotic, have been identified in megakaryocytes (reviewed by Kaluzhny and Ravid[121]). Apoptosis inhibitory proteins, such as Bcl-2 and Bcl-x$_L$, are expressed in early megakaryocytes. When overexpressed in megakaryocytes, both factors inhibit proplatelet formation.[122,123] Bcl-2 is absent in mature blood platelets, and Bcl-x$_L$ is absent from senescent megakaryocytes,[124] consistent with a role for apoptosis in mature megakaryocytes. Proapoptotic factors, including caspases and nitric oxide, are also expressed in megakaryocytes. Evidence indicating a role for caspases in platelet assembly is strong. Caspase activation has been established as a requirement for proplatelet formation. Caspases 3 and 9 are active in mature megakaryocytes, and inhibition of these caspases blocks proplatelet formation.[122] Nitric oxide has been implicated in the release of platelet-size particles from the mega-karyocytic cell line Meg-01 and may work in conjunction with TPO to augment platelet release.[125,126] Other proapo-ptotic factors expressed in megakaryocytes and thought to be involved in platelet production include transforming growth factor-β1 and SMAD proteins.[127] Of interest is the distinct accumulation of apoptotic factors in mature mega-karyocytes and mature platelets.[128] For example, caspases 3 and 9 are active in terminally differentiated megakaryo-cytes. However, only caspase 3 is abundant in platelets,[129] whereas caspase 9 is absent.[128] Similarly, caspase 12, found in megakaryocytes, is absent in platelets.[130] These data support differential mechanisms for programmed cell death in platelets and megakaryocytes, and suggest selective deliv-ery and restriction of apoptotic factors to nascent platelets during proplatelet-based platelet assembly.

## IV. Regulation of Megakaryocyte Development and Platelet Formation

Megakaryocyte development and platelet formation are regulated at multiple levels by many different cytokines (Fig. 2-1).[131] IL-3, a cytokine produced by both mast cells and T lymphocytes, can independently stimulate the early stages of megakaryocyte development up to the endomito-tic phase.[132,133] The full development of megakaryocytes requires at least a second cytokine, such as TPO. TPO, a cytokine that was purified and cloned by five separate groups in 1995, is the principal regulator of thrombopoiesis.[134] The biological effects of TPO and other platelet growth factors are described in detail in Chapter 66. TPO regulates all stages of megakaryocyte development — from the hema-topoietic stem cell stage through cytoplasmic maturation.[134] Kit ligand, also known as *stem cell factor, steel factor,* or *mast cell growth factor,* is a cytokine that exists in both soluble and membrane-bound forms and influences primi-tive hematopoietic cells.[135–137] Cytokines such as IL-6, IL-11, and kit ligand also regulate stages of megakaryocyte devel-opment at multiple levels, but appear to function only in concert with TPO or IL-3. Interestingly, TPO and the other cytokines mentioned earlier are not essential for the final stages of thrombopoiesis (proplatelet and platelet produc-tion) *in vitro.*[62] In fact, TPO appears to inhibit proplatelet formation by mature human megakaryocytes *in vitro.*[138]

## V. Murine Model Systems and Human Diseases as Tools to Study Platelet Biogenesis

### A. Transcription Factors

Genetic studies indicate that a transcriptional program con-trols the differentiation of megakaryocyte development.

**1. GATA-1.** The zinc finger protein GATA-1 is a tran-scription factor that plays a critical role in driving the expression of genes essential for megakaryocyte maturation. GATA proteins were initially thought to regulate red blood cell maturation because genetic disruption of the GATA-1 gene in mice results in embryonic lethality resulting from a block in erythropoiesis.[139] However, several observations also implicate GATA-1 as an important regulator in both early and late megakaryocyte differentiation.[140] First, forced expression of GATA-1 in the early myeloid cell line 416b induced megakaryocyte differentiation of these cells.[141] Second, Shivdasani and colleagues[74] used targeted mutagen-esis of regulatory elements within the GATA-1 locus to generate mice with a selective loss of GATA-1 in the mega-karyocyte lineage. These knockdown mice express suffi-cient levels of GATA-1 in erythroid cells to circumvent the embryonic lethality caused by anemia. GATA-1 deficiency in megakaryocytes leads to severe thrombocytopenia. Plate-let counts are reduced to approximately 15% of normal, and the small number of circulating platelets are typically round and significantly larger than usual. These mice have an increased number of small megakaryocytes that exhibit an increased rate of proliferation. The small cytoplasmic volume of GATA-1-deficient megakaryocytes typically con-tains an excess of rough endoplasmic reticulum, very few platelet-specific granules, and an underdeveloped or disor-ganized DMS, suggesting that maturation of megakaryo-cytes is arrested in GATA-1-deficient megakaryocytes.[142] The later role for GATA-1 in platelet biogenesis probably signifies its control over the expression of the nuclear factor-erythroid 2 (NF-E2) p45 subunit.

A family with X-linked dyserythropoietic anemia and thrombocytopenia resulting from a mutation in GATA-1 has been described.[143] A single nucleotide substitution in the amino-terminal zinc finger of GATA-1 inhibits the interac-tion of GATA-1 with its essential cofactor, friend of GATA-1 (FOG-1).[144] Although the megakaryocytes in affected family members are abundant, they are unusually small and exhi-bit several abnormal features, including an abundance of smooth endoplasmic reticulum, an underdeveloped DMS, and a lack of granules. These observations suggest an essen-tial role for the FOG-1–GATA-1 interaction in thrombopoi-esis. Genetic elimination of FOG in mice unexpectedly resulted in specific ablation of the megakaryocyte lineage, suggesting a GATA-1-independent role for FOG in early stages of megakaryocyte development.[145]

**2. Nuclear Factor-Erythroid 2.** NF-E2 is a basic leucine zipper transcription factor that appears to function as the major regulator of megakaryocyte maturation and platelet biogenesis. NF-E2 is a heterodimeric protein com-posed of a p45 subunit found only in the erythroid and megakaryocytic lineages and a smaller (p18) subunit from the more ubiquitously expressed Maf family.[146] NF-E2 was initially thought to be a transcription factor that specifically drove the expression of genes essential for erythropoiesis, but mice lacking p45 NF-E2 do not exhibit defects in erythropoiesis. Instead, p45 NF-E2-deficient mice die from hemorrhage shortly after birth as a result of a complete lack of circulating platelets.[73] These megakaryocytes undergo normal endomitosis and proliferate in response to TPO. NF-E2-deficient mice produce increased numbers of megakaryocytes that are larger than normal, contain fewer granules, exhibit a highly disorganized DMS, and fail to produce proplatelets *in vitro*,[75] a phenotype indicative of a very late block in megakaryocyte maturation. Mice lacking the p18 subunit of NF-E2, called mafG, also exhibit abnormal megakaryocyte development with accompanying thrombocytopenia.[147] NF-E2 appears to control the transcription of a limited number of genes involved in cytoplasmic maturation and platelet formation, and most likely functions directly downstream of GATA-1.[142] Shivdasani and colleagues[75] have generated a subtracted complementary DNA (cDNA) library that is enriched in transcripts downregulated in NF-E2 knockout megakaryocytes, such as β1 tubulin (see Section V.B), and this approach has begun to define the downstream targets of NF-E2 and allowed analysis of their precise role in the terminal stages of megakaryocyte differentiation. Megakaryocytes from NF-E2-deficient mice also lack the Rab27b protein, and expression of a dominant negative construct of Rab27 clearly inhibits proplatelet formation.[148] The functional loss of Rab27 may explain the similarities between NF-E2-deficient megakaryocytes and those from gunmetal mice, which exhibit deficient Rab isoprenylation and macrothrombocytopenia with reduced granules.[149]

### B. Cytoskeletal Proteins

**1. β1 Tubulin.** The importance of β1 tubulin in platelet biogenesis has been established by several observations. Immunofluorescence studies show β1 tubulin to be the major component of the proplatelet microtubule cytoskeleton.[150,151] β1 tubulin is expressed exclusively in platelets and megakaryocytes during late stages of megakaryocyte development. Reorganization of the megakaryocyte microtubule cytoskeleton and assembly of the marginal microtubule coil are important steps in platelet formation. Messenger RNA (mRNA) subtraction between wild-type and NF-E2-deficient megakaryocytes demonstrates that β1 tubulin is a downstream effector of the transcription factor NF-E2. β1 tubulin protein and mRNA are virtually absent from NF-E2-deficient megakaryocytes.[150] Genetic elimination of β1 tubulin in mice results in thrombocytopenia. β1 tubulin-deficient mice have circulating platelet counts less than 50% of normal. This reduction in platelet number appears to be the result of a defect in generating proplatelets, because megakaryocytes from β1 tubulin knockout mice fail to form

proplatelets *in vitro*. β1 tubulin-deficient platelets are spherical in shape, and this appears to be the result of defective marginal bands with fewer microtubule coilings (see Fig. 4-8). Although normal platelets have a marginal band that consists of 8 to 12 coils, β1 tubulin knockout platelets contain only two to three coils.[152] A human β1 tubulin functional substitution (AG → CC) inducing both structural and functional platelet alterations has been described.[153] Interestingly, this Q43P β1-tubulin variant was found in 10.6% of the general population and in 24.2% of 33 unrelated patients with undefined congenital macrothrombocytopenia. Electron microscopy revealed enlarged spherocytic platelets with a disrupted marginal band and structural alterations. Platelets with the Q43P β1-tubulin variant showed a mild platelet dysfunction, with reduced ATP secretion, thrombin–receptor activating peptide (TRAP)-induced aggregation, and impaired adhesion to collagen under flow conditions. A more than doubled prevalence of the β1-tubulin variant was observed in healthy subjects not undergoing ischemic events, suggesting it could confer an evolutionary advantage and protective cardiovascular role.

**2. Glycoprotein Ib-IX-V Complex.** Resting blood platelets express approximately 25,000 to 30,000 copies of the $(GPIb_{\alpha\beta}-IX)_2$-V complex (also called the von Willebrand factor receptor, see Chapter 7) that, by binding to activated von Willebrand factor, initiates hemostasis by causing rolling of platelets over the vascular surface.[154] In addition to this crucial role, the GPIb-IX-V complex is also an important structural component of the resting platelet cytoskeleton that functions as the major membrane–actin filament linkage in platelets (see Chapter 4). The importance of this linkage is demonstrated by the abnormal morphology and extreme fragility of platelets observed in Bernard–Soulier syndrome,[155] which lack the GPIb-IX-V complex.[156] Bernard–Soulier syndrome is characterized by severe bleeding, giant spherical platelets, and thrombocytopenia (see Chapter 57). The platelet count ranges from approximately 30 to 200 × $10^9$/L and the platelet life span is shortened to approximately 50% of normal. Morphological studies of megakaryocytes from patients with Bernard–Soulier syndrome show increased volume and ploidy, an abnormal distribution of the DMS, a nonhomogeneous distribution of granules, and randomly distributed microtubules.[157] These observations are consistent with both a defect in platelet formation and dysfunction of platelets.[158,159] Recently, a GPIbα-deficient mouse has been generated that recapitulates the human phenotype of Bernard–Soulier syndrome, providing further evidence that the GPIb-IX-V-filamin–actin linkage may play a crucial role in platelet morphogenesis.[160]

**3. Nonmuscle Myosin Heavy Chain A.** May–Hegglin anomaly (MHA), the most common form of inherited giant platelet disorders, was first described by May in 1909[161] and later by Hegglin[162] in 1945. This rare autosomal dominant platelet disorder is characterized by giant platelets, thrombocytopenia, leukocyte inclusions, and mild bleeding tendency (see Chapter 54). Giant platelets have a dispersed organization of microtubules. The disease appears to be the result of a mutation in the gene encoding nonmuscle heavy chain 9 (MYH9), which encodes a 224-kD polypeptide that makes up 2 to 5% of the total platelet protein.[84,85,163] Myosin II is an ATPase motor molecule that binds to actin filaments and generates force for contraction. Each myosin has two heads and a long, rodlike tail. The major function of the rodlike tail of myosin II is to permit the molecules to assemble into bipolar filaments. This assembly is crucial for the function of myosin II, and the hematological phenotype of MHA may be the result of a block in the polymerization of myosin II into filaments during megakaryocyte development and platelet formation. The most common mutations in MYH9 — lesions in the rod — cause defects in nonmuscle myosin IIA assembly *in vitro*.[164] Mutations in MYH9 are also responsible for Fechtner and Sebastian syndromes, which are also autosomal dominant macrothrombocytopenias characterized by thrombocytopenia, leukocyte inclusions, and giant platelets (see Chapter 54).[85,163] In contrast to MHA, Fechtner syndrome is associated with cataracts, nephritis, and hearing disability.[165] Sebastian syndrome can be differentiated from MHA by ultrastructural leukocyte inclusion properties.[166]

### C. Rab Geranylgeranyl Transferase

Gunmetal mice have a coat color mutation along with prolonged bleeding, macrothrombocytopenia, and a deficiency in α and dense granule contents (see Chapter 15). Megakaryocytes are increased in number in gunmetal mice and have an abnormal intracellular membrane system. Platelet synthesis is decreased.[167,168] The molecular basis for these defects is a mutation in the α subunit of the Rab geranylgeranyl transferase, an enzyme that transfers geranylgeranyl groups to Rab GTPase proteins.[169] This modification is essential for membrane localization. Rab proteins are small Ras-related GTPases that are typically involved in aspects of vesicle transport in the secretory and endocytic pathways. Impaired prenylation of Rab geranylgeranyl substrates in gunmetal mice may prevent specific Rab proteins from associating with membranes and may possibly inhibit membrane remodeling and granule packaging during megakaryocyte development.

# References

1. Golde, D. (1991). The stem cell. *Sci Am, 265,* 86–93.
2. Ogawa, M. (1993). Differentiation and proliferation of hematopoietic stem cells. *Blood, 81,* 2844–2853.
3. Debili, N., Coulombel, L., & Croisille, L. (1996). Characterization of a bipotent erythro-megakaryocytic progenitor in human bone marrow. *Blood, 88,* 1284–1296.
4. Hunt, P. (1995). A bipotential megakaryocyte/erythrocyte progenitor cell: The link between erythropoiesis and megakaryopoiesis becomes stronger. *J Lab Clin Med, 125,* 303–304.
5. McDonald, T., & Sullivan, P. (1993). Megakaryocytic and erythrocytic cell lines share a common precursor cell. *Exp Hematol, 21,* 1316–1320.
6. Nakeff, A., & Daniels–McQueen, S. (1976). In vitro colony assay for a new class of megakaryocyte precursor: Colony-forming unit megakaryocyte (CFU-M). *Proc Soc Exp Biol Med, 151,* 587–590.
7. Levin, J. (1983). Murine megakaryocytopoiesis in vitro: An analysis of culture systems used for the study of megakaryocyte colony-forming cells and of the characteristics of megakaryocyte colonies. *Blood, 61,* 617–623.
8. Long, M., Gragowski, L., & Heffner, C. (1985). Phorbol diesters stimulate the development of an early murine progenitor cell. The burst-forming unit — megakaryocyte. *J Clin Invest, 76,* 431–438.
9. Williams, N., Eger, R., & Jackson, H. (1982). Two-factor requirement for murine megakaryocyte colony formation. *J Cell Physiol, 110,* 101–104.
10. Lev–Lehman, E., Deutsh, V., Eldor, A., & Soreq, H. (1997). Immature human megakaryocytes produce nuclear associated acetylcholinesterase. *Blood, 89,* 3644–3653.
11. Breton–Gorius, J., & Guichard, J. (1972). Ultrastructural localization of peroxidase activity in human platelets and megakaryocytes. *Am J Pathol, 66,* 227–293.
12. Jackson, C. (1973). Cholinesterase as a possible marker for early cells of themegakaryocytic series. *Blood, 42,* 413–421.
13. Briddell, R., Brandt, J., Stravena, J., Srour, E., & Hoffman, R. (1989). Characterization of the human burst-forming unit–megakaryocyte. *Blood, 74,* 145–151.
14. Long, M., Williams, N., & Ebbe, S. (1982). Immature megakaryocytes in the mouse: Physical characteristics, cell cycle status, and in vitro responsiveness to thrombopoietic stimulatory factor. *Blood, 59,* 569–575.
15. Odell, T., Jackson, C. J., & Reiter, R. (1968). Generation cycle of rat megakaryocytes. *Exp Cell Res, 53,* 321.
16. Ebbe, S., & Stohlman, F. (1965). Megakaryocytopoiesis in the rat. *Blood, 26,* 20–34.
17. Ebbe, S. (1976). Biology of megakaryocytes. *Prog Hemost Thromb, 3,* 211–229.
18. Therman, E., Sarto, G., & Stubblefiels, P. (1983). Endomitosis: A reappraisal. *Hum Genet, 63,* 13–18.
19. Odell, T., Jackson, C., & Friday, T. (1970). Megakaryocytopoiesis in rats with special reference to polyploidy. *Blood, 35,* 775–782.
20. Raslova, H., Roy, L., Vourc'h, C., et al. (2003). Megakaryocyte polyploidization is associated with a functional gene amplification. *Blood, 101,* 541–544.
21. Nagata, Y., Muro, Y., & Todokoro, K. (1997). Thrombopoietin-induced polyploidization of bone marrow megakaryocytes is due to a unique regulatory mechanism in late mitosis. *J Cell Biol, 139,* 449–457.

22. Vitrat, N., Cohen–Solal, K., Pique, C., et al. (1998). Endomitosis of human megakaryocytes are due to abortive mitosis. *Blood, 91,* 3711–3723.

23. Ravid, K., Lu, J., Zimmet, J. M., & Jones, M. R. (2002). Roads to polyploidy: The megakaryocyte example. *J Cell Physiol, 190,* 7–20.

24. Wang, Z., Zhang, Y., Kamen, D., Lee, E., & Ravid, K. (1995). Cyclin D3 is essential for megakaryocytopoiesis. *Blood, 86,* 3783–3788.

25. Gu, X. F., Allain, A., Li, L., et al. (1993). Expression of cyclin B in megakaryocytes and cells of other hematopoietic lineages. *C R Acad Sci III, 316,* 1438–1445.

26. Hayles, J., Fisher, D., & Woodlard, A. (1994). Temporal order of S phase and mitosis in fission yeast is determined by the state of the p34cdc22–mitotic B cyclin complex. *Cell, 78,* 813–822.

27. Broek, D., Bartlett, R., & Crawford, K. (1991). Involvement of p34cdc2 in establishing the dependency of S phase on mitosis. *Nature, 349,* 388–393.

28. Zhang, Y., Wang, Z., & Ravid, K. (1996). The cell cycle in polyploid megakaryocytes is associated with reduced activity of cyclin B1-dependent cdc2 kinase. *J Biol Chem, 271,* 4266–4272.

29. Datta, N. S., Williams, J. L., Caldwell, J., Curry, A. M., Ashcraft, E. K., & Long, M. W. (1996). Novel alterations in CDK1/cyclin B1 kinase complex formation occur during the acquisition of a polyploid DNA content. *Mol Biol Cell, 7,* 209–223.

30. Zhang, Y., Nagata, Y., Yu, G., et al. (2004). Aberrant quantity and localization of Aurora-B/AIM-1 and survivin during megakaryocyte polyploidization and the consequences of Aurora-B/AIM-1-deregulated expression. *Blood, 103,* 3717–3726.

31. Masuda, H., McDonald, K. L., & Cande, W. Z. (1988). The mechanism of anaphase spindle elongation: Uncoupling of tubulin incorporation and microtubule sliding during in vitro spindle reactivation. *J Cell Biol, 107,* 623–633.

32. Nislow, C., Lombillo, V. A., Kuriyama, R., & McIntosh, J. (1992). A plus-end-directed motor enzyme that moves antiparallel microtubules in vitro localizes to the interzone of mitotic spindles. *Nature (Lond), 359,* 543–547.

33. Zucker–Franklin, D. (1988). Chapter 10. In D. Zucker–Franklin et al. (Eds.), *Atlas of blood cells, function and pathology* (Vol. 2). Philadelphia: Lea & Febiger.

34. Kautz, J., & De Marsh, Q. B. (1955). Electron microscopy of sectioned blood and bone marrow elements. *Rev Hematol, 10,* 314–323; discussion, 324–344.

35. Yamada, F. (1957). The fine structure of the megakaryocyte in the mouse spleen. *Acta Anat, 29,* 267–290.

36. Behnke, O. (1968). An electron microscope study of megakaryocytes of rat bone marrow. I. The development of the demarcation membrane system and the platelet surface coat. *J Ultrastruct Res, 24,* 412–428.

37. Bentfield–Barker, M. E., & Bainton, D. (1977). Ultrastructure of rat megakaryocytes after prolonged thrombocytopenia. *J Ultrastruct Res, 61,* 201–214.

38. Behnke, O. (1968). An electron microscope study of megakaryocytes of rat bone marrow. I. The development of the demarcation membrane system and the platelet surface coat. *J Ultrastruct Res, 24,* 412–433.

39. Nakao, K., & Angrist, A. (1968). Membrane surface specialization of blood platelet and megakaryocyte. *Nature, 217,* 960–961.

40. Zucker–Franklin, D., Benson, K., & Myers, K. (1985). Absence of a surface-connected canalicular system in bovine platelets. *Blood, 65,* 241–244.

41. Daimon, T., & Gotoh, Y. (1982). Cytochemical evidence of the origin of the dense tubular system in the mouse platelet. *Histochem, 76,* 189–196.

42. Gerrard, J. M., White, J. G., Rao, G. H., & Townsend, D. (1976). Localization of platelet prostaglandin production in the platelet dense tubular system. *Am J Pathol, 101,* 283–298.

43. Jones, O. P. (1960). Origin of megakaryocyte granules from Golgi vesicles. *Anat Rec, 138,* 105–114.

44. Handagama, P., George, J., Shuman, M., McEver, R., & Bainton D. F. (1987). Incorporation of circulating protein into megakaryocyte and platelet granules. *PNAS, 84,* 861–865.

45. de Larouziere, V., Brouland, J. P., Souni, F., Drouet, L., & Cramer, E. (1998). Inverse immunostaining pattern for synthesized versus endocytosed alpha-granule proteins in human bone marrow megakaryocytes. *Br J Haematol, 101,* 618–625.

46. Coller, B. S., Seligsohn, U., West, S. M., Scudder, L. E., & Norton, K. J. (1991). Platelet fibrinogen and vitronectin in Glanzmann thrombasthenia: Evidence consistent with specific roles for glycoprotein IIb/IIIA and alpha V beta 3 integrins in platelet protein trafficking. *Blood, 78,* 2603–2610.

47. Handagama, P., Bainton, D. F., Jacques, Y., Conn, M. T., Lazarus, R. A., & Shuman, M. A. (1993). Kistrin, an integrin antagonist, blocks endocytosis of fibrinogen into guinea pig megakaryocyte and platelet alpha-granules. *J Clini Invest, 91,* 193–200.

48. Handagama, P., Scarborough, R. M., Shuman, M. A., & Bainton, D. F. (1993). Endocytosis of fibrinogen into megakaryocyte and platelet alpha-granules is mediated by alpha IIb beta 3 (glycoprotein IIb-IIIa). *Blood, 82,* 135–138.

49. Heijnen, H. F., Debili, N., Vainchencker, W., Breton–Gorius, J., Geuze, H. J., & Sixma, J. J. (1998). Multivesicular bodies are an intermediate stage in the formation of platelet alpha-granules. *Blood, 91,* 2313–2325.

50. Youssefian, T., & Cramer, E. M. (2000). Megakaryocyte dense granule components are sorted in multivesicular bodies. *Blood, 95,* 4004–4007.

51. Djaldetti, M., Fishman, P., Bessler, H., & Notti, I. (1979). SEM observations on the mechanism of platelet release from megakaryocytes. *Thromb Haemost, 42,* 611–620.

52. Ihzumi, T., Hattori, A., Sanada, M., & Muto, M. (1977). Megakaryocyte and platelet formation: A scanning electron microscope study in mouse spleen. *Arch Histol Jpon, 40,* 305–320.

53. Shaklai, M., & Tavassoli, M. (1978). Demarcation membrane system in rat megakaryocyte and the mechanism of platelet formation: A membrane reorganization process. *J Ultrastruct Res, 62,* 270–285.

54. Kosaki, G. (2005). In vivo platelet production from mature megakaryocytes: Does platelet release occur via proplatelets? *Int J Hematol, 81*, 208–219.

55. Radley, J., & Hatshorm, M. (1987). Megakaryocyte fragments and the microtubule coil. *Blood Cells, 12*, 603–608.

56. Radley, J. M., & Haller, C. J. (1982). The demarcation membrane system of the megakaryocyte: A misnomer? *Blood, 60*, 213–219.

57. Becker, R. P., & DeBruyn, P. P. (1976). The transmural passage of blood cells into myeloid sinusoids and the entry of platelets into the sinusoidal circulation: A scanning electron microscopic investigation. *Am J Anat, 145*, 1046–1052.

58. Wright, J. (1906). The origin and nature of blood platelets. *Boston Med Surg J, 154*, 643–645.

59. Thiery, J. B., & Bessis, M. (1956). Platelet genesis from megakaryocytes observed in live cells. *Acad Sci, 242*, 290.

60. Behnke, O. (1969). An electron microscope study of the rat megakaryocyte. II. Some aspects of platelet release and microtubules. *J Ultrastruct Res, 26*, 111–129.

61. Schmitt, A., Guichard, J., Masse, J., Debili, N., & Cramer, E. M. (2001). Of mice and men: Comparison of the ultrastructure of megakaryocytes and platelets. *Exp Hematol, 29*, 1295–1302.

62. Choi, E. S., Nichol, J. L., Hokom, M. M., Hornkohl, A. C., & Hunt, P. (1995). Platelets generated in vitro from proplatelet-displaying human megakaryocytes are functional. *Blood, 85*, 402–413.

63. Cramer, E. M., Norol, F., Guichard, J., et al. (1997). Ultrastructure of platelet formation by human megakaryocytes cultured with the Mpl ligand. *Blood, 89*, 2336–2346.

64. Leven, R. M. (1987). Megakaryocyte motility and platelet formation. *Scanning Micros, 1*, 1701–1709.

65. Tablin, F., Castro, M., & Leven, R. M. (1990). Blood platelet formation in vitro: The role of the cytoskeleton in megakaryocyte fragmentation. *J Cell Sci, 97*, 59–70.

66. Handagama, P. J., Feldman, B. F., Jain, N. C., Farver, T. B., & Kono, C. S. (1987). In vitro platelet release by rat megakaryocytes: Effect of metabolic inhibitors and cytoskeletal disrupting agents. *Am J Vet Res, 48*, 1142–1146.

67. Miyazaki, H., Inoue, H., Yanagida, M., et al. (1992). Purification of rat megakaryocyte colony-forming cells using monoclonal antibody against rat platelet glycoprotein IIb/IIIa. *Exp Hematol, 20*, 855–861.

68. Choi, E. (1997). Regulation of proplatelet and platelet formation in vitro. *Thrombopoiesis and Thrombopoietins: Molecular, Cellular, Pre Clinical, and Clinical Biology, ••,* 271–284.

69. Kessal, K. G., & Kardon, R. H. (1979). *Tissues and organs: A text: Atlas of scanning electron microscopy.* San Francisco: Freeman.

70. Lichtman, M. A., Chamberlain, J. K., Simon, W., & Santillo, P. A. (1978). Parasinusoidal location of megakaryocytes in marrow: A determinant of platelet release. *Am J Hematol, 4*, 303–312.

71. Scurfield, G., & Radley, J. M. (1981). Aspects of platelet formation and release. *Am J Hematol, 10*, 285–296.

72. Tavassoli, M., & Aoki, M. (1989). Localization of megakaryocytes in the bone marrow. *Blood Cells, 15*, 3–14.

73. Shivdasani, R. A., Rosenblatt, M. F., Zucker–Franklin, D., et al. (1995). Transcription factor NF-E2 is required for platelet formation independent of the actions of thrombopoietin/MGDF in megakaryocyte development. *Cell, 81*, 695–704.

74. Shivdasani, R. A., Fujiwara, Y., McDevitt, M. A., & Orkin, S. H. (1997). A lineage-selective knockout establishes the critical role of transcription factor GATA-1 in megakaryocyte growth and platelet development. *EMBO J, 16*, 3965–3973.

75. Lecine, P., Villeval, J., Vyas, P., Swencki, B., Yuhui, X., & Shivdasani, R. A. (1998). Mice lacking transcription factor NF-E2 provide in vivo validation of the proplatelet model of thrombocytopoiesis and show a platelet production defect that is intrinsic to megakaryocytes. *Blood, 92*, 1608–1616.

76. Italiano, J. E. J., Lecine, P., Shivdasani, R. A., & Hartwig, J. H. (1999). Blood platelets are assembled principally at the ends of proplatelet processes produced by differentiated megakaryocytes. *J Cell Biol, 147*, 1299–1312.

77. Patel, S. R., Hartwig, J. H., & Italiano, Jr., J. E. (2005). The biogenesis of platelets from megakaryocyte proplatelets. *J Clin Invest, 115*, 3348–3354.

78. Hunt, P., Hokom, M., Wiemann, B., Leven, R. M., & Arakawa, T. (1993). Megakaryocyte proplatelet-like process formation in vitro is inhibited by serum prothrombin, a process which is blocked by matrix-bound glycosaminoglycans. *Exp Hematol, 21*, 372–381.

79. Patel, S. R., Richardson, J., Schulze, H., et al. (2005). Differential roles of microtubule assembly and sliding in proplatelet formation by megakaryocytes. *Blood, 15*, 4076–4085.

80. Harker, L. A., & Finch, C. A. (1969). Thrombokinetics in man. *J Clin Invest, 48*, 963–974.

81. Kaufman, R., Airo, R., & Pollack, S. (1965). Origin of pulmonary megakaryocytes. *Blood, 25*, 767–775.

82. Trowbridge, E. A., Martin, J. F., Slater, D. N., Kishk, Y. T., Warren, C. W., Harley, P. J., & Woodcock, B. (1984). The origin of platelet count and volume. *Clin Phys Physiol Meas, 5*, 145–156.

83. Rojnuckarin, P., & Kaushansky, K. (2001). Actin reorganization and proplatelet formation in murine megakaryocytes: The role of protein kinase calpha. *Blood, 97*, 154–161.

84. Kelley, M. J., Jawien, W., Ortel, T. L., & Korczak, J. F. (2000). Mutation of MYH9, encoding non-muscle myosin heavy chain A, in May–Hegglin anomaly. *Nat Gene, 26*, 106–108.

85. Kunishima, S., Kojima, T., Matsushita, T., et al. (2001). Mutations in the NMMHC-A gene cause autosomal dominant macrothrombocytopenia with leukocyte inclusions (May–Hegglin anomaly/Sebastian syndrome). *Blood, 97*, 1147–1149.

86. Richardson, J., Shivdasani, R., Boers, C., Hartwig, J., & Italiano, Jr., J. E. (2005). Mechanisms of organelle transport and capture along proplatelets during platelet production. *Blood, 106*, 4066–4075.

87. Mohandas, N., & Prenant, M. (1978). Three-dimensional model of bone marrow. *Blood, 51,* 633–643.

88. Radley, J. M. (1986). Ultastructural aspects of platelet formation. *Pro Clin Biol Res, 215,* 387–398.

89. Radley, J. M., & Scurfield, G. (1980). The mechanism of platelet release. *Blood, 56,* 996–999.

90. Kopp, H. G., Avecilla, S. T., Hooper, A. T., & Rafii, S. (2005). The bone marrow vascular niche: Home of HSC differentiation and mobilization. *Physiol, 20,* 349–356.

91. Avecilla, S. T., Hattori, K., Heissig, B., et al. (2004). Chemokine-mediated interaction of hematopoietic progenitors with the bone marrow vascular niche is required for thrombopoiesis. *Nat Med, 10,* 64–71.

92. Avraham, H., Banu, N., Scadden, D. T., Abraham, J., & Groopman, J. E. (1994). Modulation of megakaryocytopoiesis by human basic fibroblast growth factor. *Blood, 83,* 2126–2132.

93. Avraham, H., Cowley, S., Chi, S. Y., Jiang, S., & Groopman, J. E. (1993). Characterization of adhesive interactions between human endothelial cells and megakaryocytes. *J Clin Invest, 91,* 2378–2384.

94. Behnke, O., & Forer, A. (1998). From megakaryocytes to platelets: Platelet morphogenesis takes place in the bloodstream. *Eur J Haematol Suppl, 60,* 3–23.

95. Minot, G. (1922). Megakaryocytes in the peripheral circulation. *J Exp Med, 36,* 1–4.

96. Hansen, M., & Pedersen, N. T. (1978). Circulating megakaryocytes in blood from the antecubital vein in healthy, adult human. *Scan J Haematol, 20,* 371–376.

97. Efrati, P., & Rozenszajn, L. (1960). The morphology of buffy coat in normal human adults. *Blood, 16,* 1012–1019.

98. Melamed, M. R., Cliffen, E. E., Mercer, C., & Koss, G. (1966). The megakaryocyte blood count. *Am J Med Sci, 252,* 301–309.

99. Tong, M., Seth, P., & Pennington, D. G. (1987). Proplatelets and stress platelets. *Blood, 69,* 522–528.

100. Cramer, L. (1997). Molecular mechanism of actin-dependent retrograde flow in lamellipodia of motile cells. *Front Bio, 2,* 260–270.

101. Howell, W., & Donahue, D. (1937). The production of blood platelets in the lungs. *J Exp Med, 65,* 171–203.

102. Kaufman, R. M., Airo, R., Pollack, S., & Crosby, W. H. (1965). Circulating megakaryocytes and platelet release in the lung. *Blood, 26,* 720–728.

103. Pederson, N. T. (1974). The pulmonary vessels as a filter for circulating megakaryocytes in rats. *Scan J Haematol, 13,* 225–231.

104. Trowbridge, E. A., Martin, J. F., & Slater, D. N. (1982). Evidence for a theory of physical fragmentation of megakaryocytes, implying that all platelets are produced in the pulmonary circulation. *Thromb Res, 28,* 461–475.

105. Scheinin, T., & Koivuneimi, A. (1963). Megakaryocytes in the pulmonary circulation. *Blood, 22,* 82–87.

106. Jordan, H. (1940). Origin and significance of megakaryocytes of lungs. *Anat Rec, 77,* 91–101.

107. Aschoff, L. (1893). Ueber capillare embolie von riesenkernhaltigen zellen. *Arch Pathol Anat Physiol, 134,* 11–14.

108. Kinosita, R., & Ohno, S. (1958). Biodynamics of thrombopoiesis. In S. Johnson, J. Rebuck, & R. Horn (Eds.), *Blood platelets.* Boston: Little, Brown.

109. Tavassoli, M., & Aoki, M. (1981). Migration of entire megakaryocytes through the marrow–blood barrier. *Br J Haematol, 48,* 25–29.

110. Hamada, T., Mohle, R., Hesselgesser, J., et al. (1998). Transendothelial migration of megakaryocytes in response to stromal cell-derived factor 1 (SDF-1) enhances platelet formation. *J Exp Med, 188,* 539–548.

111. Riviere, C., Subra, F., Cohen–Solal, K., Cordette–Lagarde, Letestu, R., & Auclair, C. (1999). Phenotypic and functional evidence for the expression of CXCR receptor during megakaryocytopoiesis. *Blood, 93,* 1511–1523.

112. Kowalska, M. A., Ratajczak, J., Hoxie, J., et al. (1999). Megakaryocyte precursors, megakaryocytes and platelets express the HIV co-receptor CXCR4 on their surface: Determination of response to stromal-derived factor-1 by megakaryocytes and platelets. *Br J Haematol, 104,* 220–229.

113. Levine, R. F., Eldor, A., Shoff, P. K., Kirwin, S., Tenza, D., & Cramer, E. M. (1993). Circulating megakaryocytes. Delivery of large numbers of intact, mature megakaryocytes to the lungs. *Eur J Haematol, 51,* 233–246.

114. Crosby, W. (1976). Delivery of megakaryocytes by the marrow. The derivation of platelets. In S. Seno, F. Takaku, & S. Irino (Eds.), *Topics in hematology* (pp. 416–419). Amsterdam: Excerpta Medica.

115. Pedersen, N. (1978). Occurrence of megakaryocytes in various vessels and their retention in the pulmonary capillaries in man. *Scan J Haematol, 21,* 369–375.

116. Davis R. E., Stenberg, P. E., Levin, J., & Beckstead, J. A. (1997). Localization of megakaryocytes in normal mice and following administration of platelet antiserum, 5-fluorouracil, or radiostrontium: Evidence for the site of platelet production. *Exp Hematol, 25,* 638–648.

117. Gordge, M. P. (2005). Megakaryocyte apoptosis: Sorting out the signals. *Br J Pharmacol, 145,* 271–273.

118. Radley, J. M., & Haller, C. J. (1983). Fate of senescent megakaryocytes in the bone marrow. *Br J Haematol, 53,* 277–287.

119. Falcieri, E., Bassini, A., Pierpaoli, S., et al. (2000). Ultrastructural characterization of maturation, platelet release, and senescence of human cultured megakaryocytes. *Anat Rec, 258,* 90–99.

120. Zauli, G., Vitale, M., Falcieri, E., et al. (1997). In vitro senescence and apoptotic cell death of human megakaryocytes. *Blood, 90,* 2234–2243.

121. Kaluzhny, Y., & Ravid, K. (2004). Role of apoptotic processes in platelet biogenesis. *Acta Haematol, 111,* 67–77.

122. de Botton, S., Sabri, S., Daugas, E., et al. (2002). Platelet formation is the consequence of caspase activation within megakaryocytes. *Blood, 100,* 1310–1317.

123. Kaluzhny, Y., Yu, G., Sun, S., et al. (2002). BclxL overexpression in megakaryocytes leads to impaired platelet fragmentation. *Blood, 100,* 1670–1678.

124. Sanz, C., Benet, I., Richard, C., et al. (2001). Antiapoptotic protein Bcl-x(L) is up-regulated during megakaryocytic

differentiation of CD34(+) progenitors but is absent from senescent megakaryocytes. *Exp Hematol, 29,* 728–735.

125. Battinelli, E., & Loscalzo, J. (2000). Nitric oxide induces apoptosis in megakaryocytic cell lines. *Blood, 95,* 3451–3459.

126. Battinelli, E., Willoughby, S. R., Foxall, T., Valeri, C. R., & Loscalzo, J. (2001). Induction of platelet formation from megakaryocytoid cells by nitric oxide. *Proc Nat Acad Sci USA, 98,* 14458–14463.

127. Kim, J. A., Jung, Y. J., Seoh, J. Y., Woo, S. Y., Seo, J. S., & Kim, H. L. (2002). Gene expression profile of megakaryocytes from human cord blood CD34(+) cells ex vivo expanded by thrombopoietin. *Stem Cells, 20,* 402–416.

128. Clarke, M. C., Savill, J., Jones, D. B., Noble, B. S., & Brown, S. B. (2003). Compartmentalized megakaryocyte death generates functional platelets committed to caspase-independent death. *J Cell Biol, 160,* 577–587.

129. Brown, S. B., Clarke, M. C., Magowan, L., Sanderson, H., & Savill, J. (2000). Constitutive death of platelets leading to scavenger receptor-mediated phagocytosis. A caspase-independent cell clearance program. *J Biol Chem, 275,* 5987–5996.

130. Kerrigan, S. W., Gaur, M., Murphy, R. P., Shattil, S. J., & Leavitt, A. D. (2004). Caspase-12: A developmental link between G-protein-coupled receptors and integrin alphaIIb-beta3 activation. *Blood, 104,* 1327–1334.

131. Kaushansky, K. Regulation of megakaryocytopoiesis.

132. Segal, G., Stueve, T., & Adamson, J. (1988). Analysis of murine megakaryocyte ploidy and size: Effects of interleukin-3. *J Cell Physiol, 137,* 537–544.

133. Yang, Y., Ciarletta, A., & Temple, P. (1986). Human IL-3 (multi-CSF): Identification by expression cloning of a novl hematopoietic growth factor related to murine IL-3. *Cell, 47,* 3–10.

134. Kaushansky, K. (1995). Thrombopoietin: The primary regulator of platelet production. *Blood, 85,* 419–431.

135. Martin, F., Suggs, S., & Langley, K. (1990). Primary structure and functional expression of rat and human stem cell factor. *Cell, 63,* 203–210.

136. Flanagan, J., & Leder, P. (1990). The kit ligand: A cell surface molecule altered in steel mutant fibroblasts. *Cell, 63,* 185–194.

137. Briddell, R., Bruno, E., & Cooper, R. (1991). Effect of c-kit ligand on in vitro human megakaryocytopoiesis. *Blood, 78,* 2854–2859.

138. Ito, T., Ishida, Y., Kashiwagi, R., & Kuriya, S. (1996). Recombinant human c-Mpl ligand is not a direct stimulator of proplatelet formation of human megakaryocytes. *Br J Haematol, 94,* 387–390.

139. Pevny, L., Simon, M., Robertson, E., et al. (1991). Erythroid differentiation in chimeric mice blocked by a targeted mutation in the gene for transcription factor GATA-1. *Nature,* 349.

140. Centurione, L., Di Baldassarre, A., Zingariello, M., et al. (2004). Increased and pathologic emperipolesis of neutrophils within megakaryocytes associated with marrow fibrosis in GATA-1(low) mice. *Blood, 104,* 3573–3580.

141. Visvader, J., Elefanty, A., Strasser, A., & Adams, J. (1992). GATA-1 but not SCL induces megakaryocytic differentiation in an early myeloid line. *EMBO, 11,* 4557–4564.

142. Vyas, P., Ault, K., Jackson, C., Orkin, S. H., & Shivdasani, R. A. (1999). Consequences of GATA-1 deficiency in megakaryocytes and platelets. *Blood, 93,* 2867–2875.

143. Nichols, K. E., Crispino, J. D., Poncz, M., et al. (2000). Familial dyserythropoietic anaemia and thrombocytopenia due to an inherited mutation in GATA1. *Nat Gene,* 266–270.

144. Mehaffey, M. G., Newton, A. L., Gandhi, M. J., Crossley, M., & Drachman, J. G. (2001). X-linked thrombocytopenia caused by a novel mutation of GATA-1. *Blood, 98,* 2681–2688.

145. Tsang, A. P., Fujiwara, Y., Hom, D. B., & Orkin, S. H. (1998). Failure of megakaryopoiesis and arrested erythropoiesis in mice lacking the GATA-1 transcriptional cofactor FOG. *Genes Dev, 12,* 1176–1188.

146. Andrews, N., Erdjument–Bromage, H., Davidson, M., Tempst, P., & Orkin, S. (1993). Erythroid transcription factor NF-E2 is a hematopoietic-specific basic-leucine zipper protein. *Nature, 362,* 722–728.

147. Shavit, J., Motohashi, H., Onodera, K., Akasaka, J., Yamamoto, M., & Engel, J. (1998). Impaired megakaryopoiesis and behavioral defects in mafG-null mutant mice. *Genes Dev, 12,* 2164–2174.

148. Tiwari, S., Italiano, J. E., Jr., Barral, D. C., et al. (2003). A role for Rab27b in NF-E2-dependent pathways of platelet formation. *Blood, 102,* 3970–3979.

149. Novak, E. K., Reddington, M., Zhen, L., et al. (1995). Inherited thrombocytopenia caused by reduced platelet production in mice with the gunmetal pigment gene mutation. *Blood, 85,* 1781–1789.

150. Lecine, P., Italiano, J. E. J., Kim, S., Villeval, J., & Shivdasani, R. A. (2000). Hematopoietic-specific B1 tubulin participates in a pathway of platelet biogenesis dependent on the transcription factor NF-E2. *Blood, 96,* 1366–1373.

151. Wang, D., Villasante, A., Lewis, S. A., & Cowan, N. J. (1985). The mammalian B-tubulin repertoire: Hematopoietic expression of a novel B-tubulin isotype. *J Cell Biol, 103,* 1903–1910.

152. Schwer, H. D., Lecine, P., Tiwari, S., Italiano, J. E., Jr., Hartwig, J. H., & Shivdasani, R. A. (2001). A lineage-restricted and divergent B-tubulin isoform is essential for the biogenesis, structure and function of blood platelets. *Curr Biol, 11,* 579–586.

153. Freson, K., De Vos, R., Wittevrongel, C., et al. (2005). The TUBB1 Q43P functional polymorphism reduces the risk of cardiovascular disease in men by modulating platelet function and structure. *Blood, 106,* 2356–2362.

154. Andrews, R., Shen, Y., Gardiner, E., Dong, J., Lopez, J., & Berndt, M. (1999). The glycoprotein Ib-IX-V complex in platelet adhesion and signaling. *Thromb Haemost, 82,* 357–364.

155. Lopez, J. A., Andrews, R. K., Afshar–Kharghan, V., & Berndt, M. (1998). Bernard–Soulier Syndrome. *Blood, 91,* 4397–4418.

156. Clemetson, K. J., McGregor, J. L., James, E., Dechavanne, M., & Luscher, E. F. (1982). Characterization of the platelet

membrane glycoprotein abnormalities in Bernard–Soulier syndrome and comparison with normal by surface-labeling techniques and high-resolution two-dimensional gel electrophoresis. *J Clin Invest, 70,* 304–311.

157. Tomer, A., Scharf, R., McMillan, R., Ruggeri, Z., & Harker, L. (1994). Bernard–Soulier syndrome: Quantitative characterization of megakaryocytes and platelets by flow cytometric and platelet kinetic measurements. *Eur J Haematol, 52,* 193–200.

158. Nurden, P., & Nurden, A. (1996). Giant platelets, megakaryocytes and the expression of glycoprotein Ib-IX complexes. *CR Acad Sci Paris, 319,* 717–726.

159. Kass, L., Leichtman, D. A., & Beals, T. F. (1977). Megakaryocytes in the giant platelet syndrome. A cytochemical and ultrastructural study. *Thromb Haemost, 38,* 652–659.

160. Ware, J., Russell, S., & Ruggeri, Z. M. (2000). Generation and rescue of a murine model of platelet dysfunction: The Bernard–Soulier syndrome. *Blood, 97,* 2803–2808.

161. May, R. (1909). Leokozyteneinschlusse. *Deutsch Arch Klin Med, 96,* 1–6.

162. Hegglin, R. (1945). Gleichzeitige konstitutionelle veranderungen an neurtophilen und thrombocyten. *Helv Med Acta, 12,* 439–440.

163. Seri, M., Cusano, R., Gangarossa, S., et al. (2000). Mutations in MYH9 result in the May–Hegglin anomaly, and Fechtner and Sebastian syndromes. The May–Hegglin/Fechtner Syndrome Consortium. *Nat Gene, 26,* 103–105.

164. Franke, J. D., Dong, F., Rickoll, W. L., Kelley, M. J., & Kiehart, D. P. (2005). Rod mutations associated with MYH9-related disorders disrupt nonmuscle myosin-IIA assembly. *Blood, 105,* 161–169.

165. Peterson, L., Rao, K., Crosson, J., & White, J. (1985). Fechtner syndrome — A variant of Alport's syndrome with leukocyte inclusions and macrothrombocytopenia. *Blood, 65,* 397–406.

166. Greinacher, A., Nieuwenhuis, H., & White, J. (1990). Sebastian platelet syndrome: A new variant of the hereditary macrothrombocytopenia with leukocyte inclusions. *Blut, 61,* 282–288.

167. Novak, E., Reddington, M., Zhen, L., et al. (1995). *Blood, 85,* 1781–1789.

168. Swank, R., Jiang, S., Reddington, M., et al. (1993). *Blood, 81,* 2626–2635.

169. Detter, J., Zhang, Q., Mules, E., et al. (2000). Rab geranylgeranyl transferase alpha mutation in the gunmetal mouse reduces Rab prenylation and platelet synthesis. *Blood, 97,* 4144–4149.

# CHAPTER 3

# Platelet Structure

**James G. White**

*Department of Laboratory Medicine and Pathology, University of Minnesota, Minneapolis, Minnesota*

## I. Introduction

Platelets are the smallest of the many types of cells in circulating blood, averaging only 2.0 to 5.0 μm in diameter, 0.5 μm in thickness, and having a mean cell volume of 6 to 10 femtoliters.[1-4] Inconsequential size does not bother the platelet. It prefers to remain obscure throughout its 7- to 10-day life span. If it can reach its final destination in the reticuloendothelial system without having exercised a function in hemostasis or been involved in thrombotic events, its life would be considered a complete success. Yet, the life of the host in whom the platelet resides is not easy. Sooner or later, a generation of platelets will be called upon to exercise functions useful or essential for host survival. Unfortunately, platelet function important to host preservation may also contribute to host destruction. Thus, it is important to understand platelet structure, biochemistry, physiology, and pathology to foster normal function and block involvement in occlusive vascular and thrombotic disease. The purpose of this chapter is to contribute to these objectives by presenting current knowledge of platelet structure and structural physiology. Occasionally, examples of abnormal platelet structure will be included to assist in the understanding of normal platelet morphology and function.

## II. Peripheral Zone

### A. General Features

The platelet plasma membrane is relatively smooth compared with that of leukocytes in circulating blood, but low-voltage, high-resolution scanning electron microscopy suggested it has a fine, rugose appearance, resembling the gyri and sulci on the surface of the brain (Fig. 3-1).[5] The tiny folds may provide additional membrane needed when platelets spread on surfaces. Small openings of the surface connected open canalicular system (SCCS) are randomly dispersed on the otherwise featureless platelet exterior (Fig.

3-2). Thin sections and whole-mount preparations reveal that the platelet plasma membrane has a thicker exterior coat, or glycocalyx, than other blood cells. The use of cytochemical agents, including horse radish peroxidase, ruthenium red, lanthanum nitrate, phosphotungstic acid, and tannic acid have confirmed this impression.[6]

The lipid bilayer of the peripheral zone on which the glycocalyx rests is a typical unit membrane and does not differ in appearance from the membrane covering other cells.[7] Yet, it serves an extremely important role in the acceleration of clotting (see Chapter 19), a function not shared by other cells in circulating blood. The lipid bilayer is incompressible and cannot stretch. Therefore any contribution to the increased surface area of spread platelets must come from tiny folds on the exposed surface and internalized membrane provided by channels of the open canalicular system (OCS) (Figs. 3-3–3-6).[8]

The submembrane area is also a vital component of the peripheral zone. Under certain conditions, a relatively regular system of thin filaments resembling actin filaments can be identified.[9] The submembrane filaments have an important role in the shape change and translocation of receptors and particles over the exterior surface of the cell. A complete description of this filament system and its interaction with other filamentous elements of the platelet cytoplasm is in Chapter 4.

### B. Specific Features

**1. Glycocalyx.** The exterior surface of the platelet, together with other constituents of the peripheral zone, does not serve merely as a barrier to separate internal contents of platelets from the external milieu. Rather, it is a very dynamic structure that serves as the site of first contact, sensing changes in the vascular compartment requiring the hemostatic response of platelets at sites of vessel injury. The glycocalyx is covered with major and minor glycoprotein receptors necessary to facilitate platelet adhesion to a

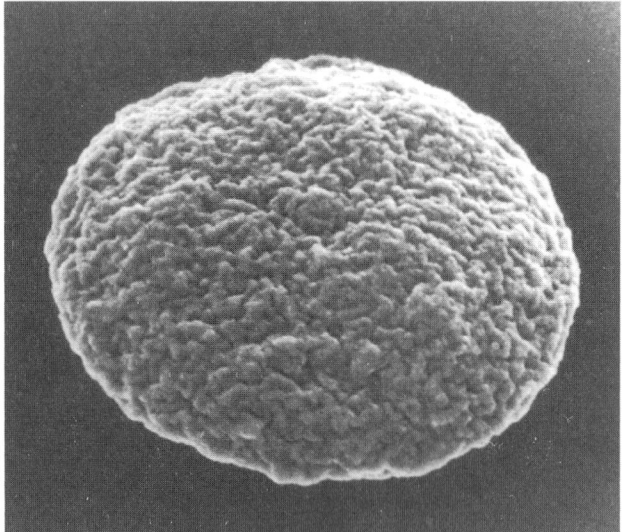

**Figure 3-1.** Discoid platelet photographed in the low-voltage, high-resolution scanning electron microscope (LVHR-SEM). The outside of the cell resembles the surface of the brain. Gyri and sulci alternate in a convoluted fashion resulting in the wrinkled appearance. Magnification ×30,000.

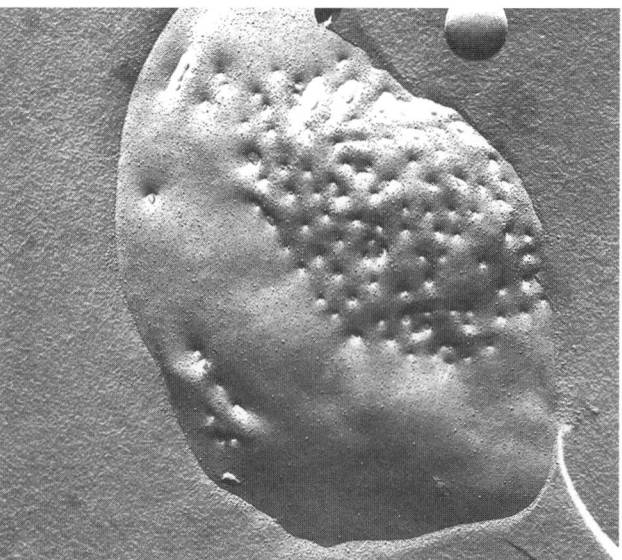

**Figure 3-2.** Replica of freeze-fractured platelet reveals communications between channels of the surface connected canalicular system (SCCS) and the surface membrane. The fracture plane has cleaved the lipid bilayer, exposing the outside of the inner portion. Numerous opening pores for channels of the open canalieular system (OCS) are clustered in one area, and others are apparent along the outer edge. Magnification ×26,000.

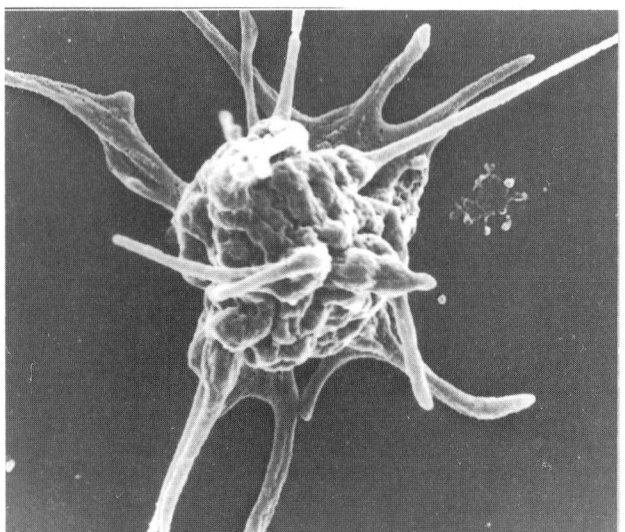

**Figure 3-3.** Early dendritic platelet viewed by LVHR-SEM. Fine processes extend in all directions. Surfaces of the pseudopods are smooth compared with the central body from which they extend. Convolutions similar to those on the discoid cell in Figure 3-1 are present on the platelet body. Magnification ×13,000.

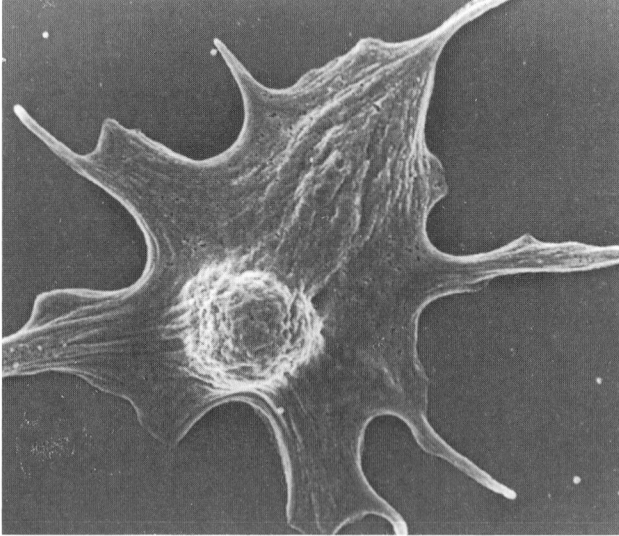

**Figure 3-4.** Early spread platelet. The central body of the cell remains convoluted, but is gradually disappearing as the cytoplasm spreads and fills the spaces between pseudopods. Magnification ×11,000.

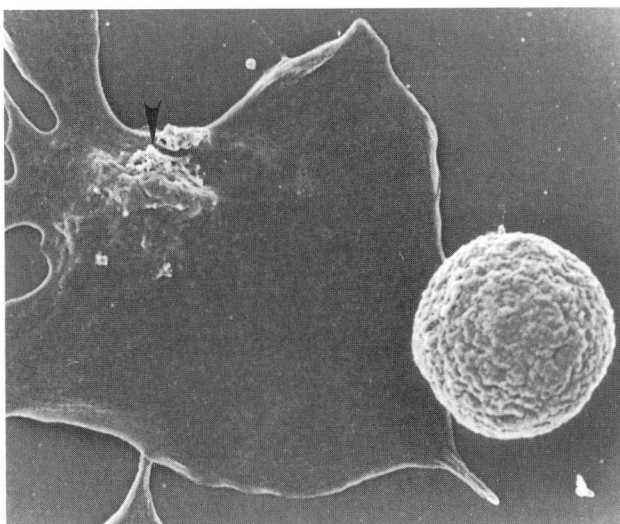

**Figure 3-5.** Spread platelet and discoid cell. The central body has almost disappeared on the spread cell, but reveals several pores of the OCS (↑). The marked difference in surface areas of the smooth spread cell and the convoluted discoid platelet is apparent. Magnification ×15,000.

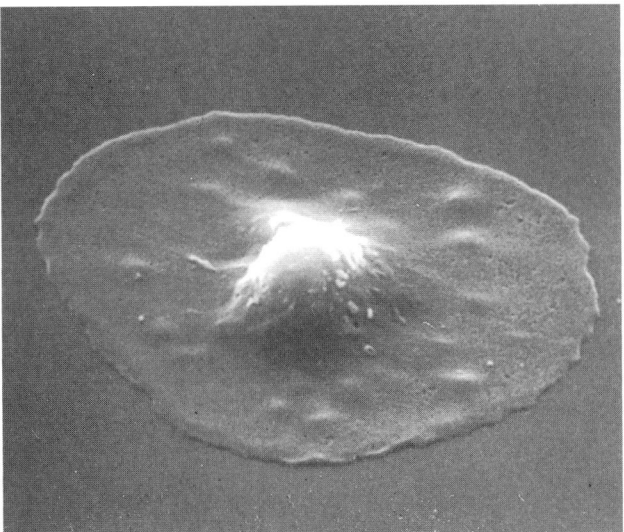

**Figure 3-6.** Spread platelet viewed by conventional SEM resembles the spread cell shown in the Figure 3-5. Magnification ×9000.

damaged surface, to trigger full activation of the platelet, to promote platelet aggregation and interaction with other cellular elements, and to accelerate the process of clot reaction.[10–12] Chapters 6 to 12 are devoted to a full description of platelet receptors. Here we restrict our discussion to receptor mobility and its role in the platelet response.

The principle platelet glycoprotein receptors involved in hemostasis are the glycoprotein (GP) Ib-IX-V complex (Chapter 7), and integrin αIIbβ3 (the GPIIb-IIIa complex,

Chapter 8). There are about 25,000 GPIb-IX receptors and about 80,000 GPIIb-IIIa receptors covering the outside surface and lining channels of the OCS on and in resting platelets. The GPIb-IX complex is coupled to the submembrane cytoskeleton by actin binding protein (filamin), and GPIIb-IIIa to the same cytoskeleton system via the cytoplasmic tails of αIIb and β3. Exposure of vascular subendothelium under high shear results in the immediate attachment of GPIb-IX to von Willebrand factor covering collagen fibers in the wound.[13] Two collagen receptors, GPVI and integrin α2β1, stabilize the attachment, and GPVI together with GPIb-IX activates the GPIIb-IIIa complex that binds fibrinogen and fibronectin at the damaged site. GPIb-IX binding to VWF also triggers actin filament formation and cytoskeletal assembly in newly adherent platelets, together with secretion of products maintained in platelet storage organelles. Additional platelets adhere to spread cells on the damaged surface, binding fibrinogen via newly expressed GPIIb-IIIa complexes and resulting in the formation of platelet aggregates.

Both GPIIb-IIIa and GPIb-IX are mobile receptors, and their ability to move is extremely important for their function in hemostasis (Chapters 7 and 8). Considerable emphasis has been placed on the ability of both receptor complexes to move spontaneously when platelets are activated in suspension or after spreading on surfaces. Unfortunately, the emphasis may have obscured the major role of mobility. One need only to fix platelets activated in suspension or on surfaces, label them with specific ligands or antibodies and electron-dense markers, then observe the cells in the scanning or transmission electron microscope.[14–19] For this purpose we have used fibrinogen-coated gold (Fgn/Au) particles or monoclonal antibodies to GPIIb-IIIa to identify the fibrinogen receptor (Fig. 3-7), and bovine VWF with a polyclonal antibody to VWF followed by protein A gold or specific antibodies to GPIb-IX detected by antiimmunoglobulins coupled to gold particles (Fig. 3-8). The studies have shown that the receptors remain randomly distributed and unmoved on platelets fixed following activation in suspension or on surfaces (Figs. 3-9 and 3-10). The same finding is obtained on platelets activated in suspension with thrombin, spread on surfaces, and fixed only after exposure to high shear force in the Baumgartner apparatus.[16] These studies have shown that spontaneous receptor translocation on platelets activated under any conditions does not take place.

Yet, the GPIIb-IIIa and GPIb-IX-V complexes can move.[17,19] If fibrinogen–gold (Fgn/Au) particles to bind GPIIb-IIIa are combined with platelets activated in suspension and are fixed at various intervals afterward, the receptor–ligand complexes are translocated across the surface membrane and moved into the channels of the OCS. A similar process takes place when spread platelets are incubated with Fgn/Au, allowed to rest on drops of Hank's balanced salt solution, and fixed at intervals. The Fgn/Au and

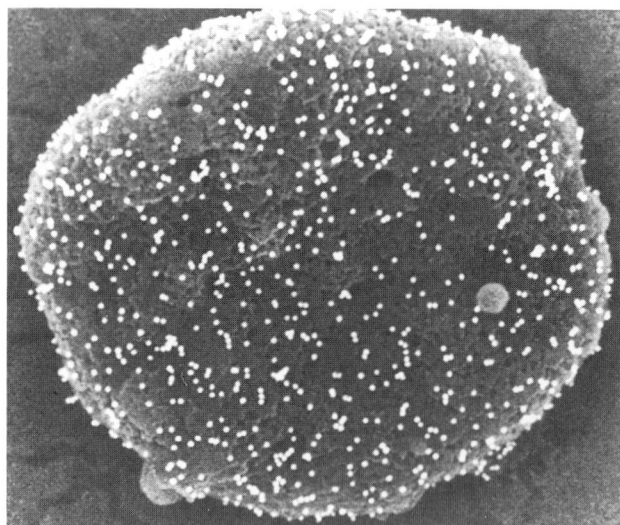

**Figure 3-7.** Cytochalasin E-treated discoid platelet exposed to fibrinogen-coated 20-nm gold particles after mounting and fixation on a glass chip. The gold particles are dispersed evenly over the cell surface from edge to edge. Magnification ×40,000.

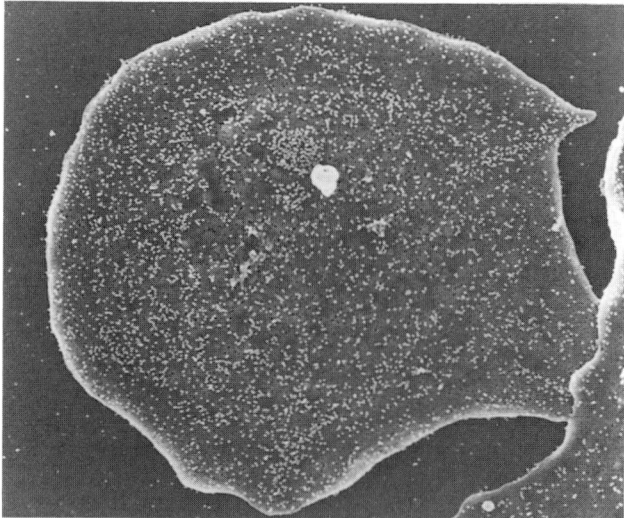

**Figure 3-9.** Platelet allowed to spread on glass for 10 minutes, fixed in 0.01% glutaraldehyde, then exposed to fibrinogen-coated gold (Fgn/Au). The Fgn/Au particles cover the spread cell surface from edge to edge. Magnification ×9000.

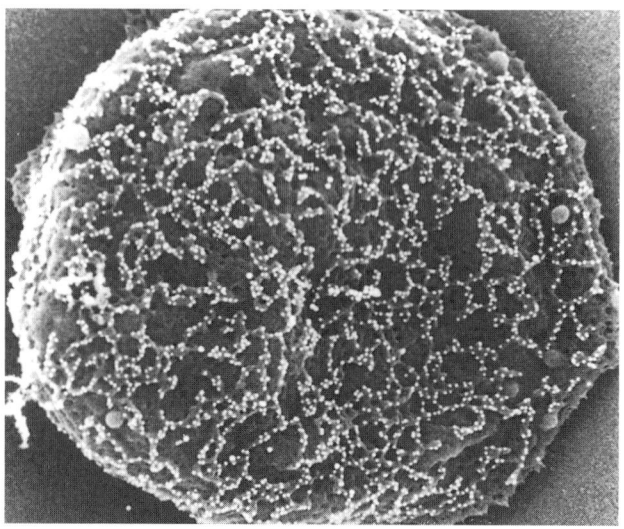

**Figure 3-8.** Cytochalasin E-treated discoid platelet fixed in 0.01% glutaraldehyde, covered with bovine plasma as the source of VWF, exposed to anti-VWF antibody, and stained with staph protein A gold (PAG 10), critical point dried, and viewed in the LVHR-SEM. Multimers of von Willebrand factor attached to GPIb-IX receptors and delineated by anti-von Willebrand factor antibody and PAG 10 are randomly dispersed like a cargo net over the platelet surface. Magnification ×30,000.

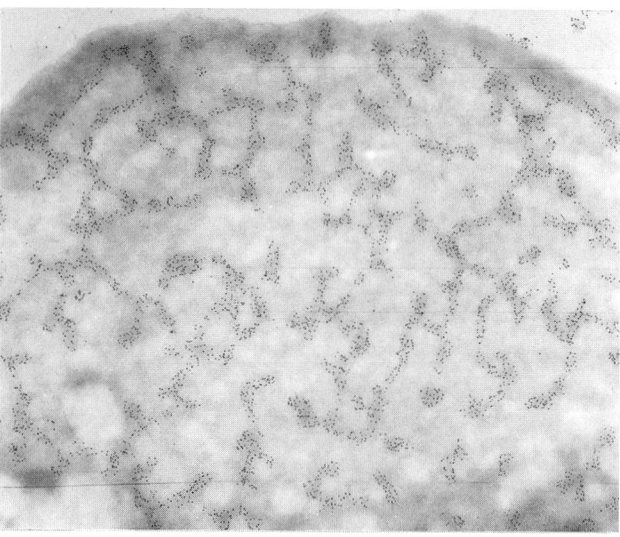

**Figure 3-10.** Platelet allowed to spread on formvar for 10 minutes, fixed in 0.01% glutaraldehyde, covered with bovine plasma, washed, exposed to anti-VWF antibody, then PAG 10 to cover the entire surface of the spread cell. Magnification ×10,000.

attached GPIIb-IIIa receptors are translocated from pseudopodia and bodies of dendritic cells into OCS channels (Fig. 3-11) and from peripheral margins of fully spread platelets to caps on cell centers (Fig. 3-12). The reason for the difference is that most OCS channels have been extruded from the interior to become part of the exposed surface of the spread cell or are compressed by spreading and are not available to take up GPIIb-IIIa–Fgn/Au complexes.

However, movement of GPIIb-IIIa–Fgn/Au complexes into the OCS of dendritic cells or the central zone of spread cells does not completely clear the fibrinogen receptors from peripheral zones.[5,20] Fixation of platelets that have transferred the receptor–ligand complexes to the OCS or platelet centers, then reexposure to Fgn/Au, reveals GPIIb-IIIa

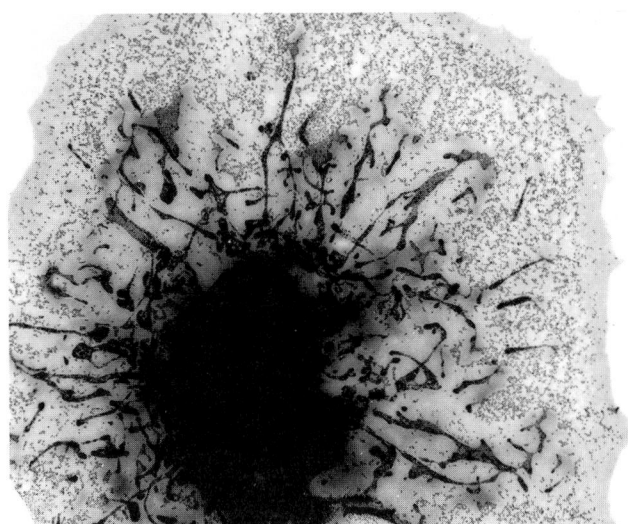

**Figure 3-11.** Fully spread platelet exposed to Fgn/Au for 5 minutes before fixation. Fgn/Au spherules have moved away from the peripheral margin and concentrated in channels of the OCS. Magnification ×6500.

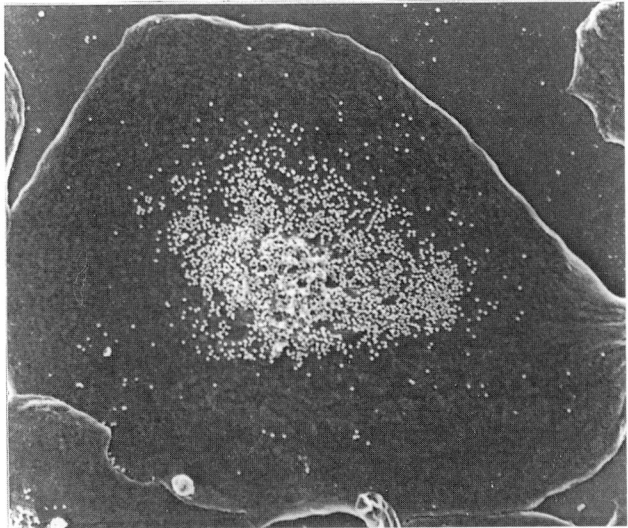

**Figure 3-12.** Fully spread platelet exposed to Fgn/Au for 5 minutes, rinsed, and rested on a drop of Hank's balanced salt solution (HBSS) for 15 minutes. Fgn/Au particles have moved from the peripheral areas of the cell to form a cap in the spread platelet center. Magnification ×11,000.

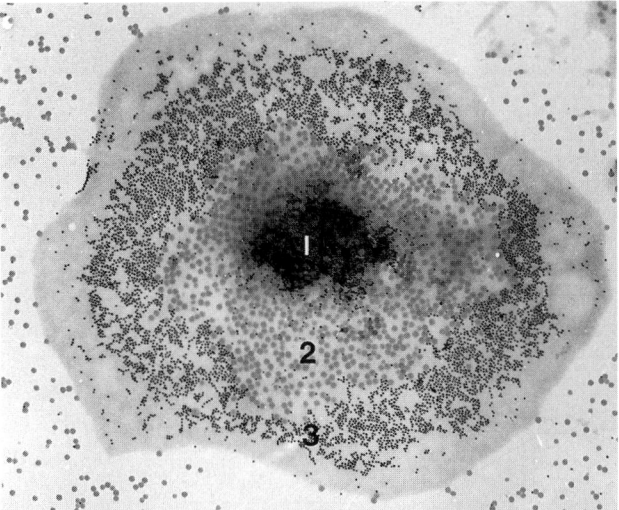

**Figure 3-13.** Fully spread platelet exposed to Fgn/Au for 5 minutes, rested on HBSS for 15 minutes, rinsed, then exposed in the same manner to small latex spheres, rinsed, then exposed in spheres, rinsed, exposed again to Fgn/Au, and then fixed. Fgn/Au from the first exposure is concentrated in the cell center (1). Latex particles form a ring around the cap of Fgn/Au (2), which is encircled by the second wave of Fgn/Au (3). Magnification ×10,000.

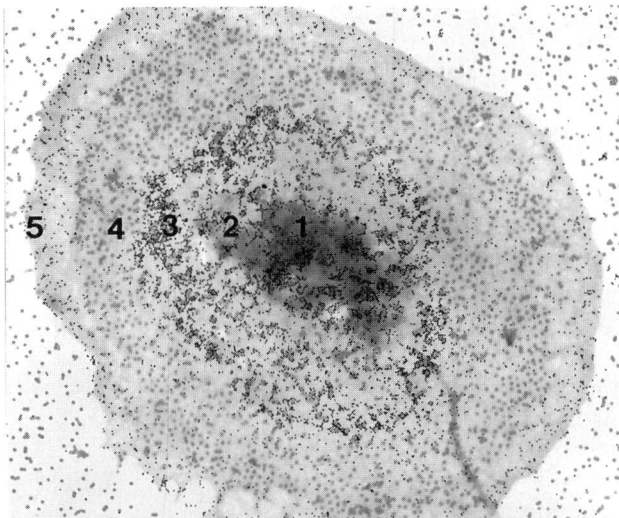

**Figure 3-14.** Spread platelet exposed in the same manner as the cell in Figure 3-13 to Fgn/Au and small latex particles. Fgn/Au has formed a cap in the cell center (1), surrounded by a ring of latex (2), in turn encircled by Fgn/Au (3), then latex spherules (4), and, at the margin, Fgn/Au (5). Magnification ×9000.

receptors are still present edge to edge on surface-activated platelets (Fig. 3-13). The experiment can be done in such a way as to show that at least three separate waves of GPIIb-IIIa–Fgn/Au can move into rings around the first cap of receptor–ligand complexes on the spread platelet center and, after fixation, residual GPIIb-IIIa receptors remain on the peripheral margin (Fig. 3-14). It appears that the glycocalyx has an almost inexhaustible supply of GPIIb-IIIa receptors that can become available as platelets spread on foreign surfaces or on other platelets.

GPIb-IX can also move.[12,18] Addition of anti-VWF antibody and protein A gold to an incubation medium containing washed human platelets and bovine plasma as a source of VWF, followed by fixation at intervals, revealed clear-

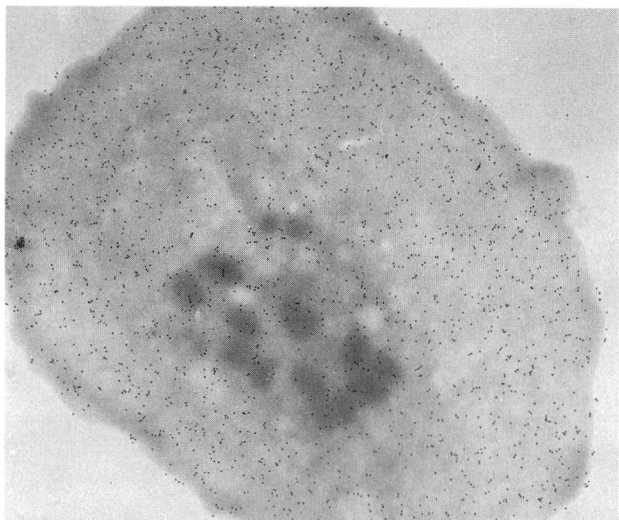

**Figure 3-15.** Surface-activated platelet exposed to thrombin at 37°C and then maintained at 4°C for 20 minutes before exposure to a cocktail of monoclonal antibodies against GPIb-IX, including AP1 and 6D1, before fixation in 0.01% glutaraldehyde, followed by goat antimouse IgG coupled to 10-nm gold particles. GPIb-IX receptors marked by gold particles remain randomly dispersed from edge to edge on the spread cell. Magnification ×15,000.

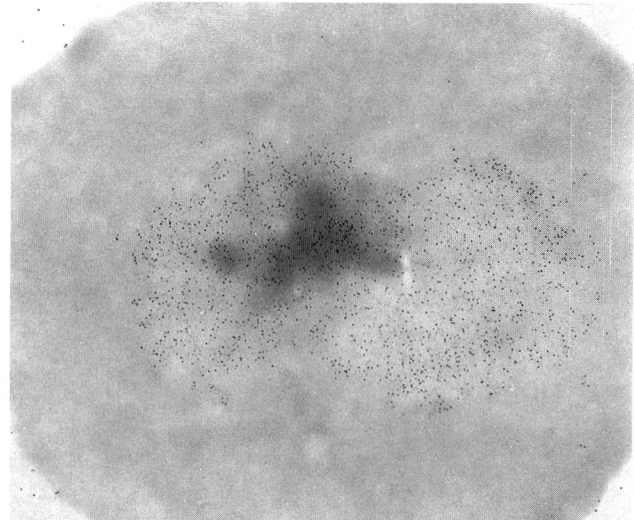

**Figure 3-16.** Surface-activated platelet incubated at 4°C for 20 minutes and exposed to anti-GPIb antibody and PAG 10 before fixation. Gold particles marking sites of GPIb-IX are concentrated in a cap over the central region of the cell. Magnification ×15,000.

ance of GPIb-IX-VWF–anti-VWF complexes to channels of the OCS.[21] If the antibody to GPIb-IX was not included, movement to the OCS did not occur. Translocation of GPIb-IX could also be induced on spread platelets.[17,19] Spreading of the platelets alone or exposure of spread platelets to VWF did not cause movement of GPIb-IX complexes (Fig. 3-15). However, when VWF, anti-VWF, and protein A gold were added to spread cells, the GPIb-IX–VWF–anti-VWF– protein A gold complexes moved from peripheral margins and concentrated in caps over the platelet centers (Fig. 3-16). Fixation of spread platelets with caps of GPIb-IX–VWF and exposure a second time to VWF revealed GPIb-IX receptors were still present on the spread cell from the margins of the cap to the edges of the peripheral zone. Thus, just as is the case with the GPIIb-IIIa receptor, the GPIb-IX–VWF receptor complexes can move, but they do not move spontaneously on platelets activated in suspension or on surfaces. What, then, is important about the ability of GPIb-IX to move?

Platelets driven to an area of denuded vascular surface under high shear force have only nano- to microseconds during which to interact. Mobile but resilient GPIb-IX receptors can soften the impact by moving upon attachment to VWF. Moving upon contact would slow the cell sufficiently to permit it to bind more VWF multimers and enhance adhesion. If the GPIb-IX receptors were rigid, interaction time would be reduced and the platelet would be more likely to bounce off the damaged surface. On the other

hand, if the GPIb-IX receptors were as mobile as other studies have suggested, they might not hold the platelet firmly enough or long enough to facilitate adhesion. The loosely attached cell might tether under high shear force and break loose, possibly leaving only a vesicular remnant on the damaged surface.

The ability of GPIb-IX–VWF complexes to move in the lipid bilayer of the exposed surface of spread platelets has an even more important relation to events following attachment than to the process of adhesion itself.[22] Following the binding of discoid platelets to the damaged vascular surface, the cells must spread to establish a hemostatic plug. Multimers of VWF on the injured surface are unlikely to move. As a result, the GPIb-IX receptors form immobile associations with VWF. The receptor–ligand complexes cannot move in the plane of the downside platelet surface to channels of the OCS. However, the downside surface and membranes lining the channels of the OCS can move through the immobilized GPIb-IX–VWF complexes, thus permitting the platelet to spread during assembly of new actin filaments. If the GPIb-IX receptors were not mobile, then the platelet surface membrane could not move through them, and spreading would be limited or impossible.

The ability of GPIIb-IIIa to move is important for the same reason. During the platelet–vessel wall interaction, GPIIb-IIIa receptors form relatively immobile complexes with fibrinogen/fibrin and fibronectin in the wound site, and cannot move toward and into the OCS on the downside of the platelet membrane. Instead, as GPIb-IX bound to von Willebrand factor stimulates rapid assembly of cytoplasmic actin during the process of spreading, the downside membrane and channels of the OCS move through the receptor ligand complexes to secure maximum coverage of the denuded vascular surface. If this were not the mechanism, the only way platelets could spread would be if the GPIIb-IIIa receptor–ligand complexes on the downside securing platelets to the damaged surface could tear loose as the cells spread. It is unlikely that such an event could occur. Therefore, the importance of GPIIb-IIIa mobility, like the mobility of GPIb-IX, is not to be cleared to the OCS following surface activation, but to secure the movement of the exposed surface and channels of the OCS to cover the site of injury as rapidly as possible.

**2. Unit Membrane.** The lipid bilayer of the surface membrane and the linings of the OCS are morphologically similar, if not identical, to the appearance of unit membranes covering other cells. Yet, it is significantly different because of the vital role the platelet unit membrane serves in the acceleration of blood coagulation (Chapter 19). Clotting is initiated at the site of vascular injury by tissue factor, which serves as a cofactor for the coagulant protein, factor VIIa, in converting factor X to Xa. Factor Xa, in association with factor Va, converts prothrombin to thrombin on the surface of the anionic phospholipid, phosphatidylserine, provided by the surface of activated platelets.

The origin of tissue factor was believed to be extravascular, becoming available from damaged tissue after vascular injury (Chapter 19). However, it is now known that tissue factor also circulates in blood under normal conditions, and may be increased during some disease states.[23,24] Most of the circulating tissue factor has been shown to be present in cell-derived microparticles, or microvesicles, believed to originate from monocyte/macrophages that may be taken up by platelets (Fig. 3-17).[25] The microparticles have been shown to bind to activated platelets through a molecular bridge between P-selectin glycoprotein ligand 1 (PSGL-1) on the microparticles and P-selectin on activated platelets.[25]

Platelets also contain tissue factor encrypted with the cholesterol-rich lipid islands in the resting platelet surface membrane (Fig. 3-18).[26] Following platelet activation, the lipid islands become associated with the outer layer of the unit membrane, bringing anionic phosphatidylserine to the exposed surface. Tissue factor associated with the lipid islands is also exposed on the platelet surface in an inactive form. However, when microparticles are released from the activated platelet surface, the tissue factor associated with them becomes decrypted. It can then interact with coagulation factors VIIa, Va, and Xa, binding to phosphatidylserine exposed on the microparticles surface. Thus platelets can foster the generation of thrombin on their activated surface as well as on microparticles shed at the site of vascular injury. The actual formation of microparticles on the activated platelet surface has not been well visualized. However, Heijnen and colleagues[27] used immunogold ultracytochemistry to identify microparticles released from platelets, and others have identified them by flow cytometry or following the exposure of platelets to conditions of high shear (see Chapter 20). A clearer definition of the formation of the microparticles and their cleavage from the surface of activated platelets will undoubtedly be forthcoming.

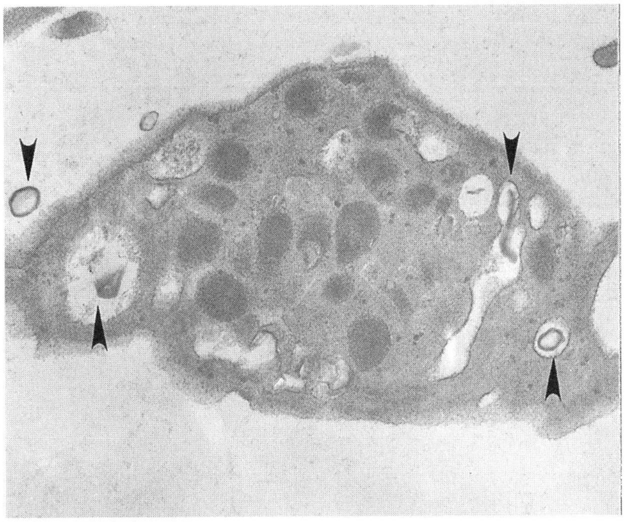

**Figure 3-17.** Human platelet fixed 2 minutes after incubation with placental tissue factor. Vesicles of placental tissue factor (arrowheads) are present in surrounding plasma and in channels of the OCS. (arrowheads) Magnification ×28,000.

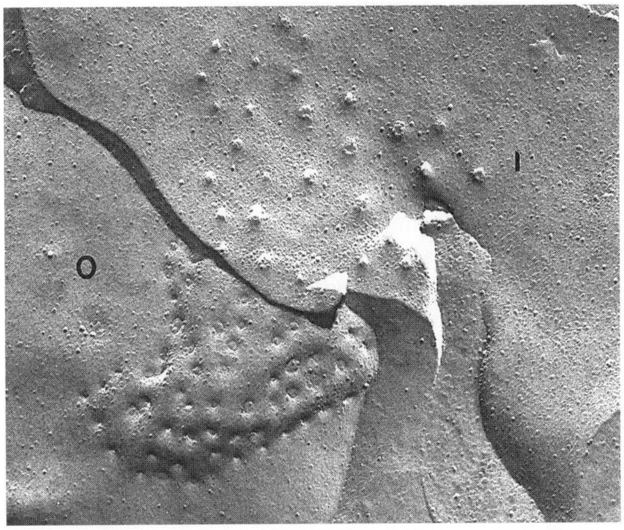


**Figure 3-18.** The freeze–fracture technique splits the lipid bilayer of the platelet unit membrane revealing the outside (O) of the inner layer of one platelet and the inside (I) of the outer layer of another. Both surfaces are covered by intramembranous particles. There are no structural differences evident in replicas of resting and activated platelets prepared by this method. As a result, this technique reveals the central area of the lipid bilayer, but not the location of cholesterol-rich lipid islands. Magnification ×50,000.

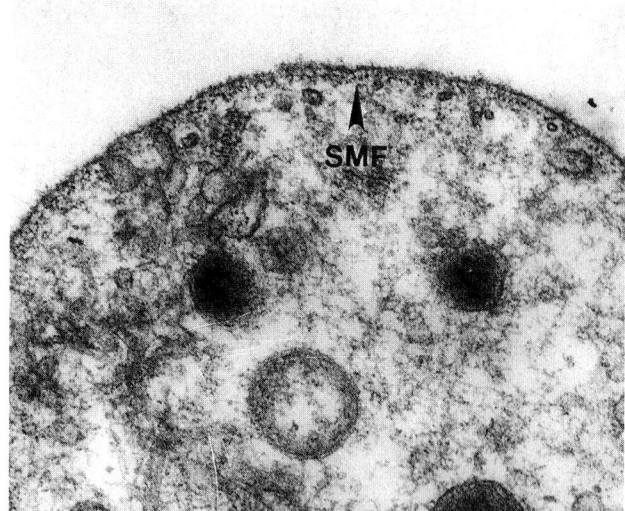

**Figure 3-19.** Thin section of a platelet exposed to mild osmotic shock. A fine layer of submembrane filaments (SMF) is revealed lying just under the surface membrane. Magnification ×45,000.

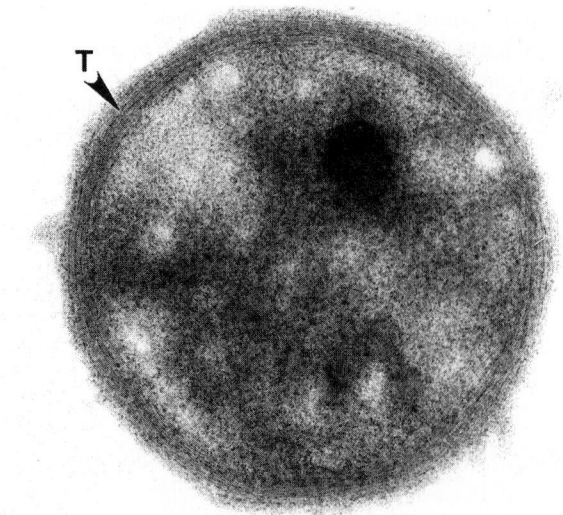

**Figure 3-20.** Whole-mount preparation of discoid platelet negatively stained after simultaneous fixation and detergent extraction. Removal of the surface membrane permits visualization of the microtubule (T) coil. Residual granule substance appears relatively electron dense. Negatively stained microfilaments are evident in the cytoplasm. Magnification ×22,000.

**3. The Submembrane Area.** The area lying just under the unit membrane of the peripheral zone is critically important to platelet function. Organelles in the cytoplasm of resting platelets never appear to contact the submembrane zone. It remains distinct and separated so that it can serve its many functions. The cytoplasmic domains of all transmembrane receptors interact in the submembrane area, with numerous protein constituents regulating the signaling processes of platelet activation (see Chapter 16). A significant number of these proteins, including the actin-binding protein (filamin) linked to the GPIb-IX-V complex and the cytoplasmic tails of GPIIb and GPIIIa, are associated with calmodulin, myosin, and the short actin filaments making up the membrane contractile cytoskeleton (see Chapter 4).[6,28] The filaments are difficult to define in the resting cell, but modest forms of shock cause a relatively regular system of filaments resembling actin to appear in the submembrane area (Fig. 3-19).[29] The contractile system of this zone is involved in the translocation of receptor complexes, including GPIb-IX-V and GPIIb-IIIa, on the platelet exterior surface. The importance of transmembrane associations of the receptors with the submembrane cytoskeleton in adhesion and spreading at sites of vascular injury are discussed earlier and in Chapter 4.

## III. The Sol–Gel Zone

### A. General Aspects

Light and phase contrast microscopic studies revealed the presence of organelles inside blood platelets, but the transparency of the matrix prevented recognition of other structural components. The formed bodies were thought to be dispersed in fluid suspension or salt solution, the clarity of which suggested the term *hyaloplasm*. Although it is now apparent that the internal environment of platelets is made up of many structural elements, the idea that these components were in solution or suspension remained prevalent for many years.

When thin sections of platelets were examined using the electron microscope, the internal matrix appeared to consist of an irregular meshwork of fibrous material in which formed organelles were imbedded. It was suggested that the matrix was an artifact created by precipitation of protein from solution during fixation. To some extent, this concept was correct. However, the asymmetrical shape of platelets and separation of granules from one another visible in the light microscope should have suggested that the hyaloplasm must be viscous. Because high internal viscosity results, to a large extent, from molecular interaction, it was predictable that protein polymers would be present in the platelet hyaloplasm.

The introduction of glutaraldehyde fixation at 37°C prior to treatment with osmic acid at low temperature revealed that the matrix of the platelet was a dense mat of fibrous elements (Figs. 3-20–3-22). Investigations have shown that changes in the state of polymerization and movement of the fibrous components of the matrix are intimately related to support of the platelet discoid shape and to internal contraction. Because it is clear that the matrix inside platelets resembles a liquid gel, it seemed appropriate to focus on this characteristic by renaming the hyaloplasm the *sol–gel zone*. The biochemistry of contractile elements is discussed in Chapter 4.

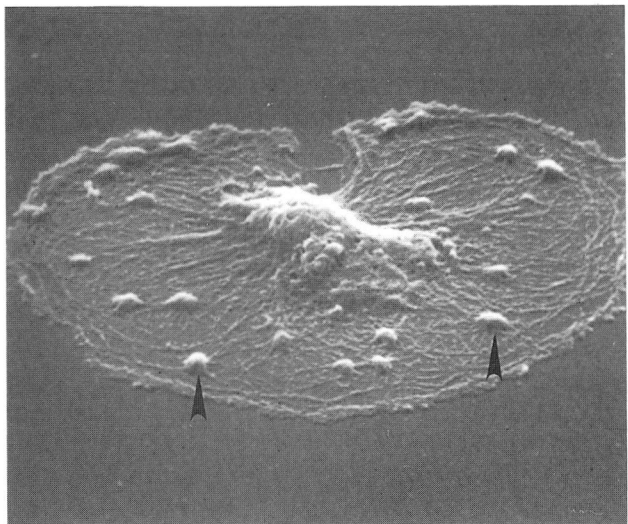

**Figure 3-21.** Spread platelet extracted with Triton X-100 and examined in a conventional SEM. The cytoplasm is filled with detergent-resistant actin filaments. Small raised areas in the cytoplasm of the extracted cell (arrowheads) are adhesion plaques. Magnification ×10,000.

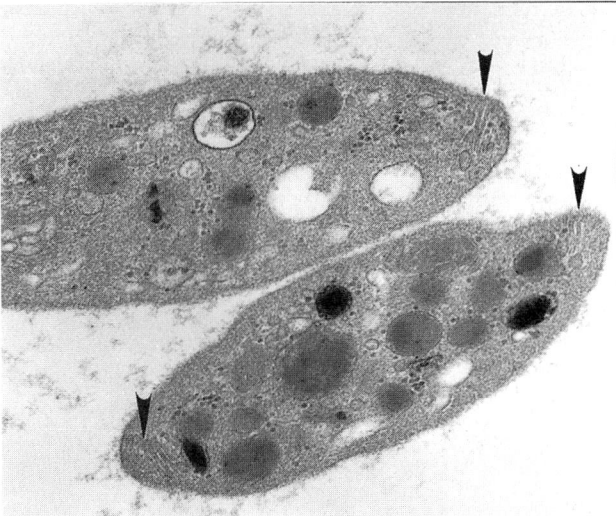

**Figure 3-23.** Cross-sections of two discoid platelets revealing the circumferential coils of microtubules (arrowheads). Magnification ×22,000.

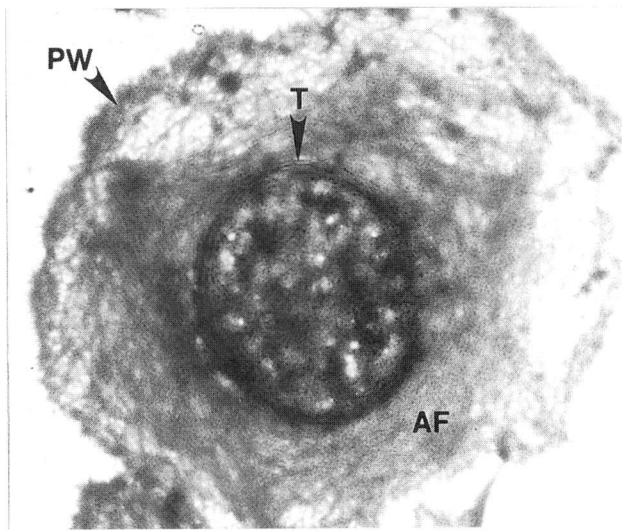

**Figure 3-22.** Spread platelet cytoskeleton prepared from citrated platelet-rich plasma. Coils of the circumferential microtubule (T) are constricted in the cell center. Actin microfilaments (AF) form a peripheral weave (PW) at the cell border and are organized in concentric layers around the MT coil. Magnification ×8000.

## B. Specific Features

**1. Microtubules.** In addition to the subsurface membrane contractile filament system discussed earlier, there are two other filament systems in the platelet cytoplasm. One is the circumferential coil of microtubules, which is a cytoskeletal support system. The other, the actomyosin filament system, is involved in shape change, internal transformation,

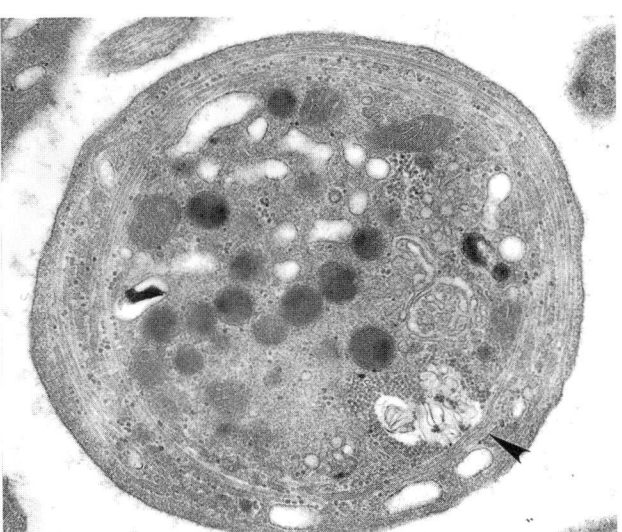

**Figure 3-24.** Discoid platelet sectioned in the equatorial plane revealing several loops of the microtubular coil (arrowheads). Magnification ×22,000.

and, ultimately, contraction of the hemostatic plug and retraction of clots.[30]

Cross-sections of fixed platelets revealed microtubules as a group of 3 to 24 circular profiles, each approximately 25 nm in diameter, at the polar ends of the lentiform cell (Figs. 3-23 and 3-24).[31–33] When platelets are sectioned in the equatorial plane, the bundle of tubules is apparent just under the cell wall along its greatest circumference. Circumferential tubules are always slightly separated from each other, although small bridges between tubules can occasionally be identified (Figs. 3-25 and 3-26). Although the bundle lies close to the cell wall, it never appears to contact it. The

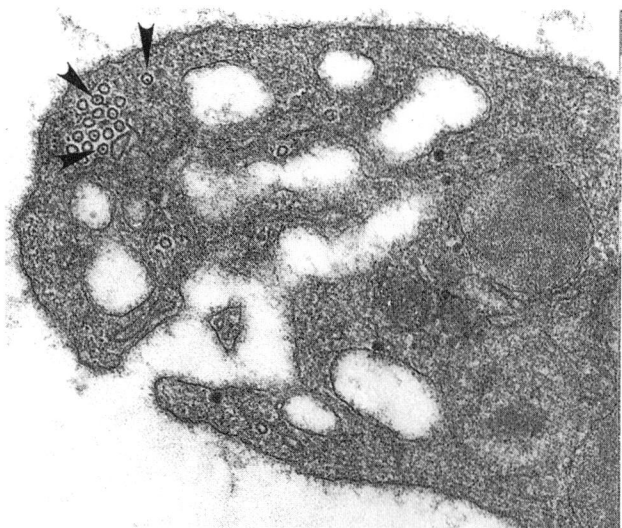

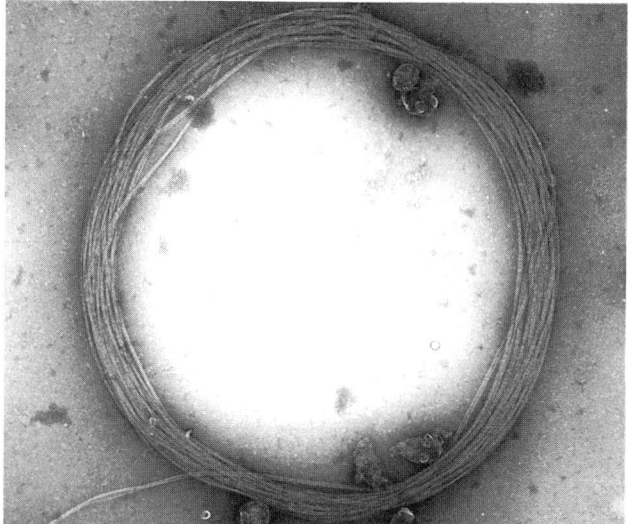

**Figure 3-25.** Cross-sections of a discoid platelet showing a circumferential coil of microtubules at higher magnification. Filaments (arrowheads) running down the core can be found in some tubules. Magnification ×55,000.

**Figure 3-27.** Microtubule coil isolated from human platelet after simultaneous fixation and detergent extraction in suspension. The coil has many loops, but appears to consist of one microtubule. Magnification ×13,000.

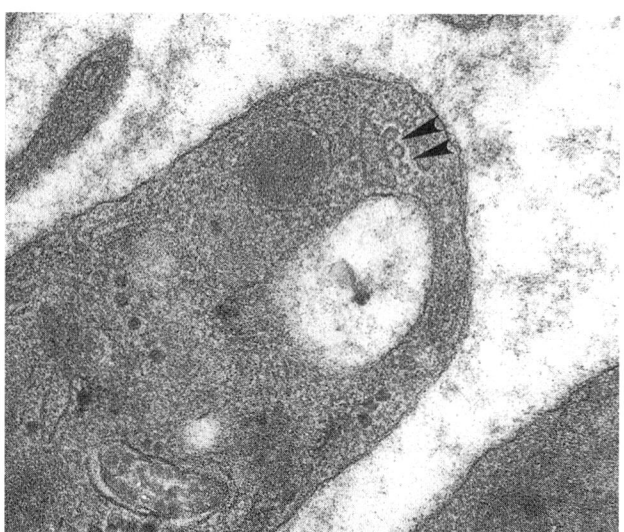

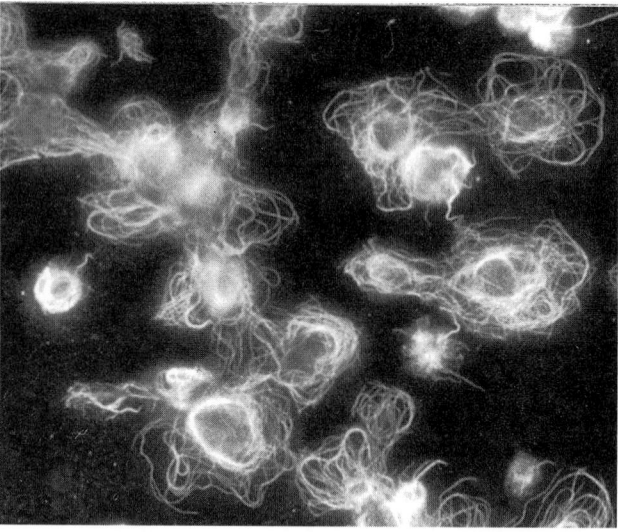

**Figure 3-26.** Cross-section of a discoid platelet. Filaments (arrowheads) connecting microtubules are evident. Magnification ×60,000.

**Figure 3-28.** Giant platelets from a patient with Epstein syndrome, settled on glass and stained with fluorescein labeled anti-tubulin antibody. The microtubule coil in these cells is twisted like a ball of yarn. Magnification ×1000.

space between the tubules and the surface is often occupied by submembrane filaments cut in cross-section. In the unactivated, circulating platelet, evidence suggests that tubulin, the protein constituent of microtubules, exists almost exclusively in the polymerized (microtubule) form.[34]

Circumferential coils of microtubules have been separated from platelets before and after fixation.[35,36] Examination by the whole-mount technique or by scanning electron microscopy revealed that the coil is not a bundle of microtubules bound together. Instead, it is a single microtubule

coiled on itself many times, much like the lariat the cowboy carries on his saddle to rope animals (Fig. 3-27). The peripherally oriented microtubule is present even in large platelets, but in giant platelet syndromes the coil is replaced by an irregular mass of tubules resembling a ball of yarn (Fig. 3-28).[37] It is uncertain whether the unorganized mass of microtubules is the result of being in a giant platelet, or the cause of the spherical form assumed by the very large cells.

The location of the circumferential band of microtubules in the equatorial plane just under the cell wall in discoid platelets suggested its participation in cytoskeletal support. Further proof for this role has come from several experimental approaches. Platelets were long known to lose their discoid form and become irregular and relatively spherical when exposed to low temperatures for brief periods. Discoid shape was restored if the chilled cells were warmed to 37°C. The loss of discoid form was associated with the disappearance of the circumferential band of microtubules, and recovery was associated with reformation of the bundle in its usual position under the cell surface.[38] Colchicine, vincristine, and vinblastine, agents that inhibit mitosis by disassembling or preventing formation of microtubules, were also found to dissolve platelet microtubules.[39] The disappearance of the circumferential band of microtubules in the platelets treated with these agents resulted in loss of the platelet discoid shape. These studies strongly supported the concept that circumferential microtubules were involved in maintaining platelet lentiform appearance.

Other experiments, however, were not as easily explained. For example, when platelets were exposed to EDTA, they lost their discoid shape and became "spiny spheres."[40] Yet, the band of microtubules remained intact and organelles were randomly dispersed within it. Thus, shape change could take place in platelets despite persistence of the peripherally located circumferential coil of microtubules. However, if the blood was collected in citrate anticoagulant first, then the platelets were washed and resuspended in buffer containing EDTA, cell discoid shape and circumferential microtubule coils were retained, although the effects of EDTA on membranes were the same as in platelets anticoagulated with EDTA alone.[41] Therefore, the effect of EDTA on platelet shape change was the result of the interaction with exposed membranes of the surface and the OCS, and not on the microtubule coils.

The bovine platelet also raised a question regarding the central role of circumferential microtubules in cytoskeletal support of discoid shape. Bovine platelets must be exposed to colchicine or vinca alkaloids for at least 1 hour longer than human cells to disassemble microtubule rings. Chilling requires up to 2 hours to dissolve bovine platelet microtubules, and their removal does not cause loss of discoid shape. Even after disappearance of the rings in bovine platelets, the cells remain discoid and just as resistant to aspiration into a micropipette as untreated, control platelets.[42] The basis for these differences in human and bovine platelets appears to be the result of variations in their intrinsic anatomy and functional expression.[43] A peripheral arrangement of assembled actin filaments plays a role in supporting the discoid shape of bovine platelets. As a result, both disassembly of microtubules by colchicine and prevention or reversal of actin filament assembly by cytochalasin B was required to cause bovine platelets to lose their discoid form.

Perhaps the strongest arguments for actin filaments, rather than microtubules, supporting the discoid form of human platelets were raised by the work of Winokur and Hartwig.[44] Their study indicated that the spectrin-reinforced surface membrane cytoskeleton and the actin-rich cytoplasmic cytoskeleton are responsible for maintaining discoid form, for shape change caused by surface or suspension activation, and for recovery when stimulation is incomplete. This concept has led to a new proposal to explain the mechanism involved in shape changes induced in platelets by chilling and rewarming. Exposure to low temperature caused severing of actin filaments associated with the surface membrane cytoskeleton. Cold also caused an elevation of cytoplasmic calcium, activation of gelsolin, uncapping and severing of established cytoplasmic actin filaments, formation of multiple nucleation sites, and rapid assembly of new actin filaments. The cooled platelets lost their discoid form, became swollen in appearance, and extended filopodia and lamellipodia. Rewarming to 24 to 37°C did not restore the irregular, cold platelets to their resting lenslike form. Rather, the new actin filament bundles formed in the cold underwent reorganization, resulting in elimination of pseudopodia and formation of irreversible platelet spheres.

This challenge to the concept placing circumferential microtubules as the major cytoskeletal system for maintaining platelet discoid shape was retested using the chilling–rewarming model together with microtubule stabilizing (Taxol) and destabilizing agents.[45] Washed platelet samples were rested at 37°C and chilled to 4°C; chilled and rewarmed to 37°C for 60 minutes; or chilled, rewarmed, and exposed to the same cycle in the presence or absence of vincristine or Taxol and fixed for study by disseminated interference phase contrast microscopy and electron microscopy. Rhodamine phalloidin and flow cytometry were used to measure changes in actin filament assembly. Chilling caused loss of disc shape, pseudopod extension, disassembly of microtubule coils, and assembly of new actin filaments. Rewarming resulted in restoration of platelet discoid shape, pseudopod retraction, disassembly of new actin filaments, and reassembly of circumferential microtubule coils. Vincristine converted discoid platelets to rounded cells that extended pseudopods when chilled, and retracted them when rewarmed, leaving spheres that could undergo the same sequence of changes when chilled and rewarmed again. Taxol prevented cold-induced disassembly of microtubules and limited pseudopod formation. Rewarming caused retraction of pseudopods on Taxol-treated discoid cells. Cytochalasin B, an agent that blocks new actin filament assembly, alone or in combination with Taxol, inhibited the cold-induced shape change but not dilation of the OCS. Rewarming eliminated OCS dilation and restored lentiform appearance. The results indicate that microtubule coils are the major structural elements responsible for platelet discoid shape and for its restoration

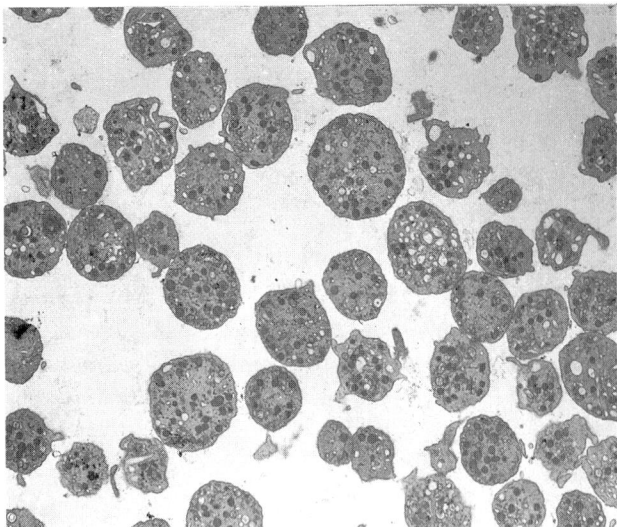

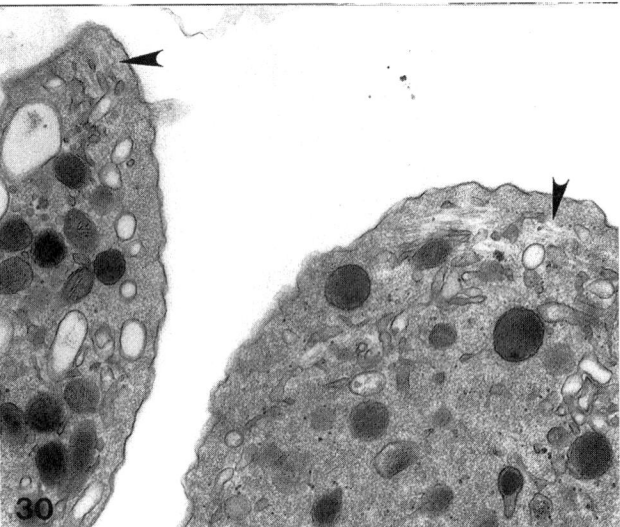

**Figure 3-29.** Thin section of platelets from a patient with platelet spherocytosis. Approximately 98% of the circulating platelets are spherical and lack circumferential coils of microtubules. Magnification ×1500.

**Figure 3-30.** Thin section of two cells from the patient with platelet spherocytosis (see Fig. 3-29) after exposure to Taxol, an agent that stimulates microtubule assembly. The platelets have assembled microtubule coils (arrowheads) and recovered their discoid shape. Magnification ×22,000.

after submaximal stimulation or rewarming of chilled platelets.

Questions may have remained after these investigations, but they have been largely resolved by the study of a child with a new bleeding syndrome.[46] The patient has a prolonged bleeding time, a history of easy bruising since infancy, and abnormal platelet aggregation despite normal levels of GPIb-IX and GPIIb-IIIa receptors. Platelet dense body frequency was normal, indicating the absence of storage pool deficiency. Light and electron microscopy revealed that the platelets were spherical in form. Thin sections of the platelet spherocytes revealed an absence of microtubules and microtubule coils in 98% of the cells (Fig. 3-29). The 2% of platelets that contained microtubule coils were discoid. Immunofluorescence microscopy with an antibody against tubulin, the protein precursor of microtubules, revealed amounts of tubulin similar to normal platelets, but virtually none was assembled into tubules or coils. Addition of Taxol, an agent that stabilizes microtubules or stimulates their assembly, caused formation of microtubules and microtubule coils in the platelets and their conversion to disks (Fig. 3-30). Thus, results of the studies in this patient demonstrate that the circumferential coil of microtubules is the major cytoskeletal support system responsible for maintaining the discoid shape of human platelets.

**2. Microfilaments.** Platelets are, in essence, muscle cells. Systems of contractile microfilaments serve major roles in platelet physiology (see Chapter 4). In resting cells, only half the actin molecules are assembled into filaments.[47] A portion of these are associated with the inside of the platelet surface membrane and constitute the submembrane

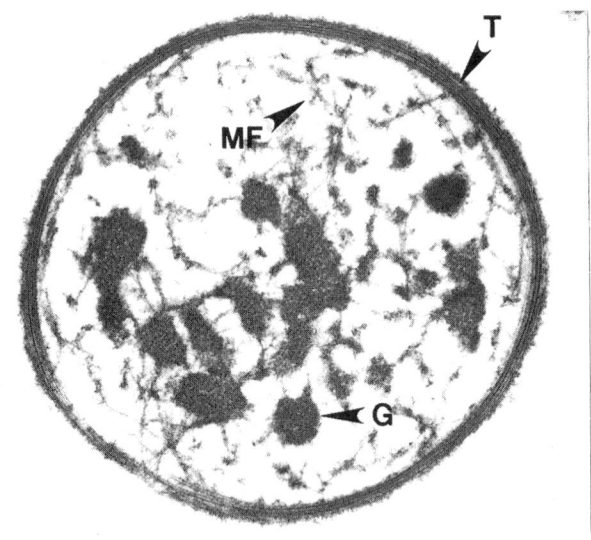

**Figure 3-31.** Thin section of a discoid platelet cytoskeleton fixed after detergent extraction in the presence of lysine and phalloidin. The microtubule (T) is well preserved. A rough amorphous coat, most likely submembrane filaments, is evident on its outer surface. Remnants of α granules (G) are suspended in a matrix of microfilaments (MF) resistant to detergent extraction. Magnification ×22,000.

contractile cytoskeleton discussed earlier. The larger proportion of assembled actin filaments make up the cytoplasmic actin filament cytoskeleton (Figs. 3-31 and 3-32). The cytoplasmic actin filament cytoskeleton has a function in platelet physiology separate from that of the submembrane actin cytoskeleton. In the resting cell, it serves as the matrix

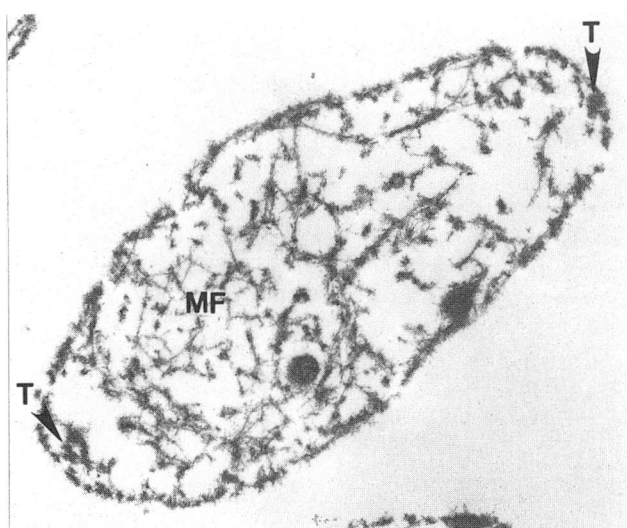

**Figure 3-32.** Cross-section of a discoid platelet cytoskeleton prepared as in Figure 3-31. The cell is from a sample of C-PRP treated with Taxol ($10^{-4}$ M) before extraction. Microtubule (T) profiles are evident at the polar ends of the cell. A meshwork of microfilaments (MF) replaces the cytoplasmic matrix. Although the surface membrane is no longer present, a fine amorphous layer remains in its place, probably submembrane filaments. Magnification ×22,000.

on which all organelles and other structural components are suspended and maintained separate from each other and the cell wall.[30] Other molecular species may help in this function, but the cytoplasmic actin filament cytoskeleton serves the major role.

Following platelet activation in suspension or on surfaces, the cytoplasmic actomyosin cytoskeleton has a unique role in contractile physiology. It constricts the circumferential microtubule coils and drives the α granules and dense bodies into close association in platelet centers (Figs. 3-33 and 3-34).[48] If stimulation is strong enough, granule and dense body contents are secreted to the exterior via channels of the OCS (Figs. 3-35 and 3-36),[30,49] leaving behind a dense, central mass of actomyosin (Fig. 3-34). The centripetal movement of the cytoplasmic actomyosin cytoskeleton in activated platelets is not blocked by cytochalasins B or E.[50] However, these agents, which inhibit new actin filament assembly, disconnect the cytoplasmic actomyosin cytoskeleton from the circumferential coil of microtubules and the secretory organelles.[51,52] As a result, the circumferential coils remain at the periphery of activated cytochalasin B- or E-treated cells, and organelles retain their random distribution while cytoplasmic actomyosin becomes concentrated in a central mass. Thus, platelet contractile physiology is highly specialized to serve multiple functions in this tiny cell.

**3. Glycogen.** The cytoplasm of platelets is rich in glycogen. Individual particles are randomly distributed in the

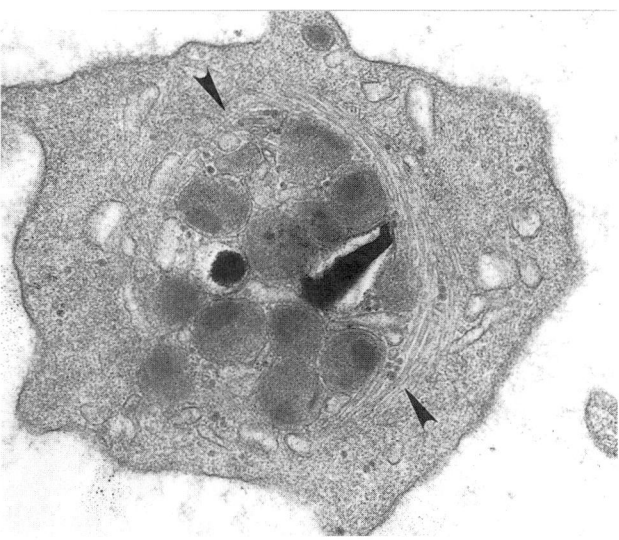

**Figure 3-33.** Internal transformation. The platelet in this example has been stimulated by thrombin. Organelles are concentrated in the cell center within tight rings of constricted microtubule coils (arrowheads). Magnification ×30,000.

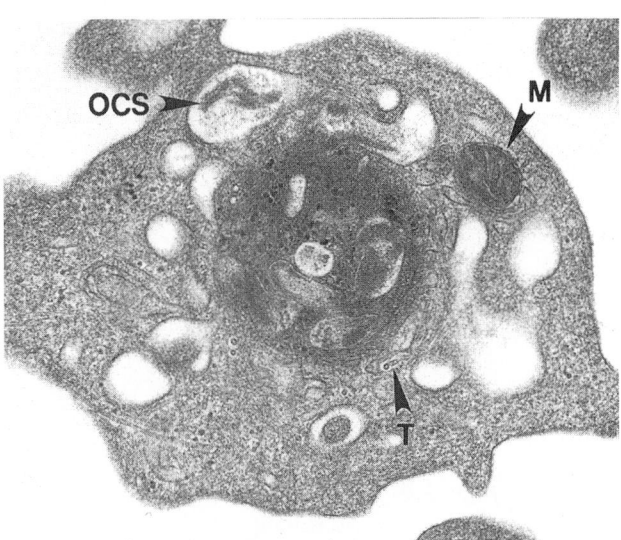

**Figure 3-34.** Internal transformation. Platelet from sample of washed platelets exposed to thrombin 5 U/mL in the presence of EDTA. Granules are concentrated in a dense mass of contractile gel and constricted microtubules (T). Mitochondria (M) are seldom incorporated into the central mass. Fibrinlike material is apparent in channels of the OCS. Magnification ×30,000.

cytoplasm, as are masses of glycogen particles (Fig. 3-37). It is unknown why glycogen is distributed in this manner. However, we have observed that membrane segments often appear associated with masses of glycogen particles. Occasionally the membranes are in the form of circles within the mass of particles, appearing to isolate the enclosed glycogen (Figs. 3-37 and 3-38). Although the evidence is limited, it

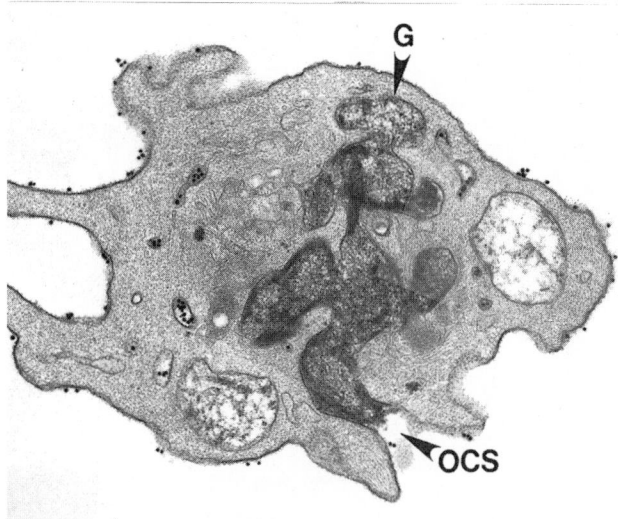

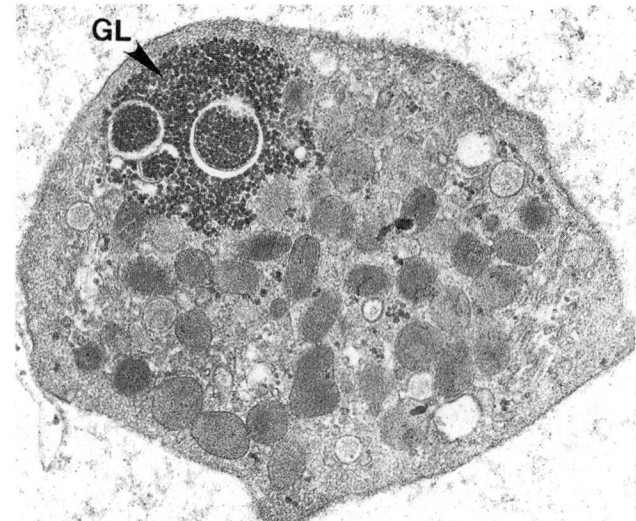

**Figure 3-35.** Platelet from a sample of washed cells exposed to thrombin 5 U/mL in the presence of Fgn/Au particles, fixed at 30 seconds, and stained with tannic acid. The reaction product formed by the interaction of osmic acid and tannic acid, osmium black, selectively stains fibrinogen and fibrin. Osmium black has stained fibrinogen/fibrin in α granules (G) fused to channels of the OCS delivering their contents to the cell exterior. Magnification ×25,000.

**Figure 3-37.** A mass of glycogen (GL) in this platelet contains two areas encircled by membranes. This may be a mechanism for glycosome formation. Magnification ×28,000.

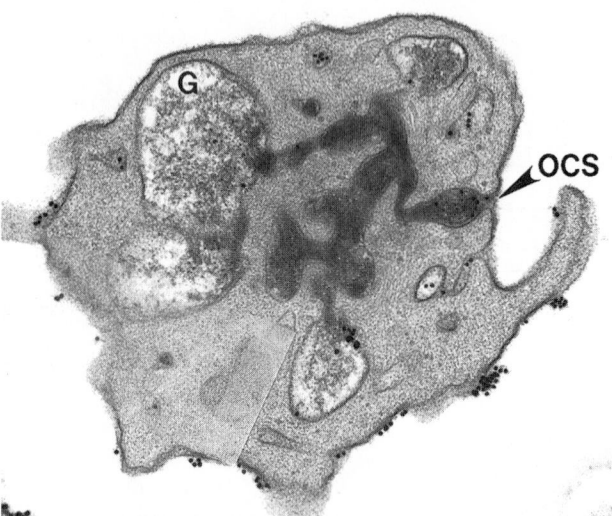

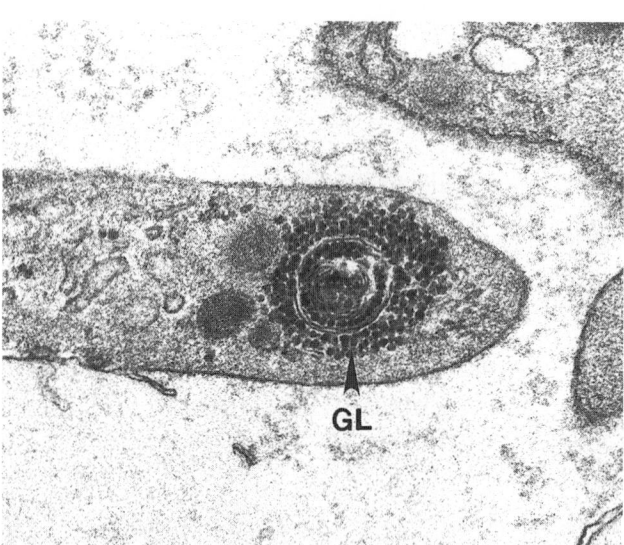

**Figure 3-36.** Platelet from a sample treated in the same manner as the cell in Figure 3-35 and stained with tannic acid during fixation. Osmium black reaction product stains fibrinogen/fibrin being extruded from α granules (G) to channels of the OCS and to the platelet exterior. Magnification ×25,000.

**Figure 3-38.** A mass of glycogen (GL) in this platelet contains a single area encircled by a membrane. This may be the genesis of a glycosome. Magnification ×43,000.

suggests the possibility that platelet glycosomes[53] are formed in this manner. Other suggestive evidence is the strong association of glycogen particles with the tubular membrane inclusions found in the Medich giant platelet disorder (Fig. 3-39).[54] It has been suggested that the membranes with which glycogen associates are derived from residual smooth

endoplasmic reticulum following the loss of ribosomes. However, channels of the dense tubular system (DTS) are known to originate from the smooth endoplasmic reticulum, but reveal no affinity for glycogen.

**4. Smooth and Coated Vesicles.** Small, smooth and coated vesicles the sizes of those budding from the trans-Golgi face of the Golgi zone in the parent megakaryocyte are difficult to identify in normal circulating platelets. Smooth vesicles could easily be mistaken for cross-sectional channels of the OCS or dense tubular system. Coated

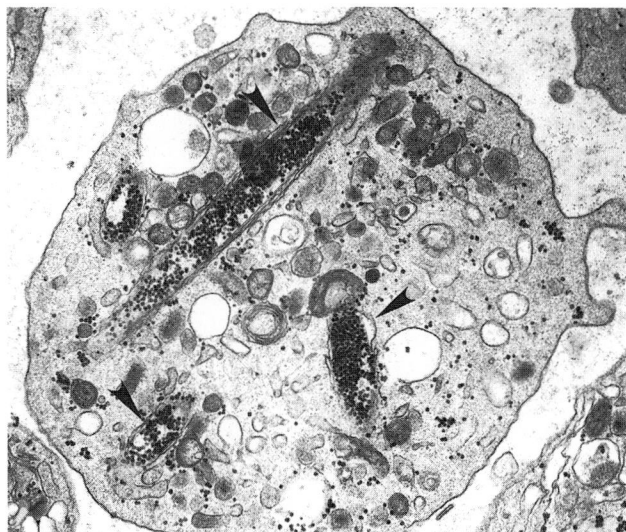

**Figure 3-39.** Platelet from a patient with the Medich giant plate-let syndrome. Three tubular inclusions (arrowheads) cut in longi-tudinal section are present. Glycogen particles adhere to the inner membrane of each. Magnification ×18,000.

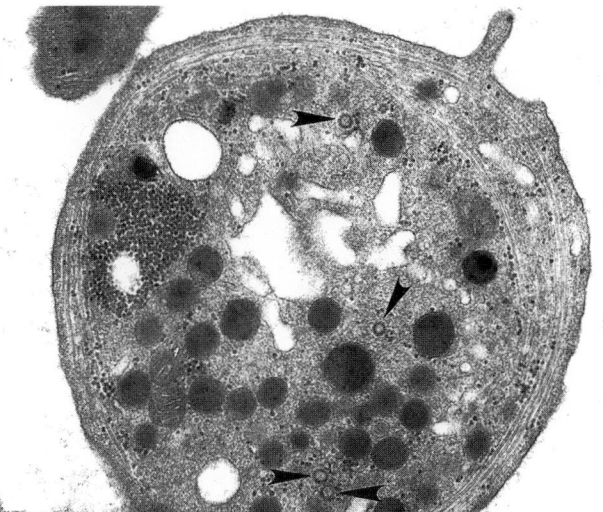

**Figure 3-40.** Thin section of a platelet containing four clathrin-coated vesicles (arrowheads). Examples like this are rare. Magni-fication ×25,000.

vesicles covered with knobs of clathrin are more easily rec-ognized than smooth vesicles.[55,56] However, they are hard to find because there are so few of them. The most observed in any single section of a normal platelet in my experience is four (Fig. 3-40).

Platelets from patients with the white platelet syndrome present a different appearance.[57] Many of these have carried forward Golgi complexes from the parent megakaryocyte, and innumerable smooth vesicles and some coated vesicles fill the cytoplasm close to the Golgi complexes. These find-ings in white platelet syndrome support the observation that there are very few vesicles in most normal platelets.

## IV. Organelle Zone

### A. General Aspects

Platelets contain three major types of secretory organelles, including α granules, dense bodies (δ granules), and lyso-somes. Occasional multivesicular bodies are also present. Multivesicular bodies develop in the megakaryocyte by fusion of the small vesicles budding from the trans-Golgi zone of the Golgi complex, and may serve as sorting stations in the development of α granules, dense bodies, and lyso-somes.[27] Small numbers of relatively simple mitochondria are present in the platelet cytoplasm. They serve an impor-tant role in energy metabolism. Other membrane-enclosed organelles or structures are also present in the cell, such as glycosomes,[53] electron-dense chains and clusters,[58] and tubular inclusions.[54]

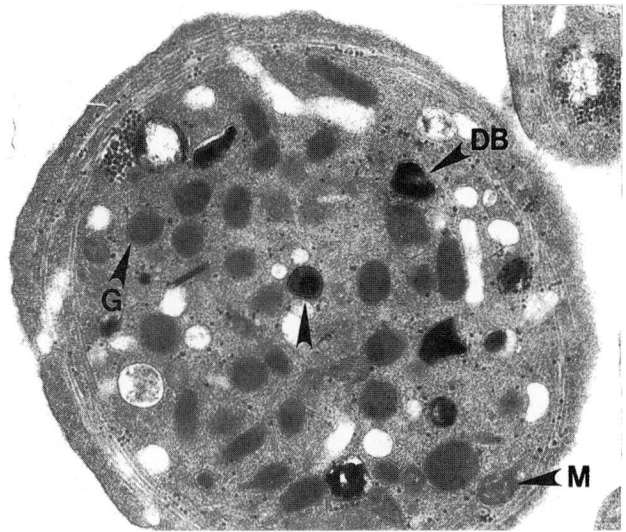

**Figure 3-41.** Thin section of a discoid platelet containing many α granules (G), a few mitochondria (M), and several dense bodies (DB). The α granules and dense bodies demonstrate considerable variation in shape and internal appearance. One α granule (arrow-head) contains a nucleoid with the same opacity as a dense body. Magnification ×25,000.

### B. Specific Features

**1. α Granules.** α Granules are the most numerous of the platelet organelles (Fig. 3-41).[59–61] The number depends on the size of the platelet and the presence of other space-occupying structures, such as masses of glycogen. There are usually 40 to 80 α granules per platelet, but large and giant cells may have well over 100. α Granules in resting platelets

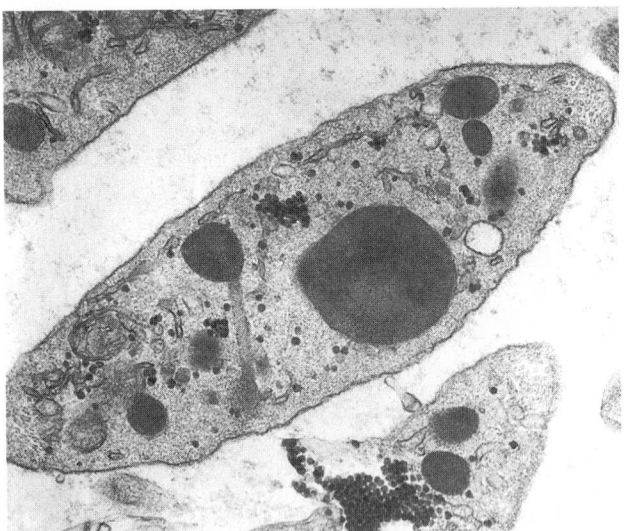

**Figure 3-42.** Normal discoid platelet containing a giant α granule. Magnification ×32,000.

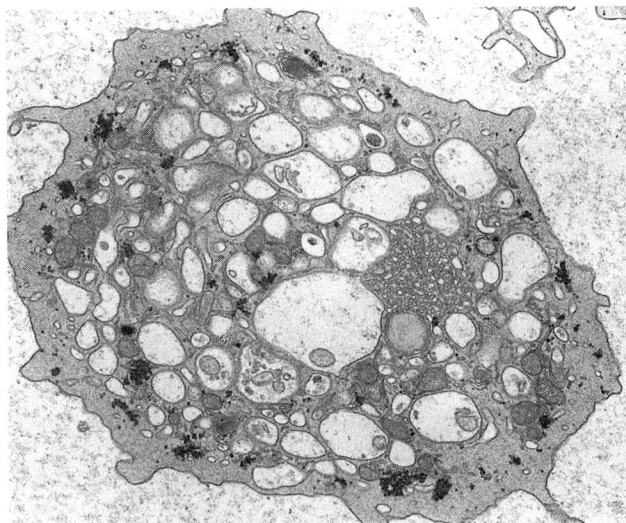

**Figure 3-43.** Large platelet from a patient with gray platelet syndrome. The megakaryocytes form α granules, but the organelles cannot retain their contents. As a result, the circulating platelets contain many empty vacuoles that were α granules. Magnification ×11,000.

remain separated from each other, indicating that the cytoplasmic matrix has an organized substructure. However, giant α granules are present in platelets from normal individuals (Fig. 3-42), and during long-term storage *in vitro,* α granules fuse with each other. Fusion of granules is the first physical sign of injury in the stored cell.[62] Giant α granules, also developing by fusion, are common in platelets from patient with the Paris–Trousseau–Jacobsen syndrome[63] and hypogranular disorders such as the white platelet syndrome[57] and the Medich giant platelet disorder.[54] α Granules are virtually absent in the giant platelets from individuals with the gray platelet syndrome.[64,65] Their megakaryocytes can make α granules, but the enclosing membranes fail to develop structure-linked latency. As a result, granule contents leak from most of the organelles while still in the megakaryocyte, leaving behind empty α granule vacuoles that fill the cytoplasm of circulating platelets in patients with the gray platelet syndrome (Fig. 3-43).

α Granules are round to oval in shape and 200 to 500 nm in diameter. The interior substructure is divided into zones. The submembrane zone contains von Willebrand factor, organized into tubelike structures,[66] similar to those found in Weibel–Palade bodies in endothelial cells (Fig. 3-44). The peripheral zone is less electron dense than the central zone. Yet, in human platelets it contains many proteins, including those synthesized by the megakaryocytes; platelet-selective proteins including coagulation factor V, thrombospondin, P-selectin, and VWF; and proteins synthesized in other cells and taken up by platelets, such as fibrinogen. Except for the zones of varying density, the proteins appear unorganized in human α granules. However, in equine platelet α granules, fibrinogen appears separated and sandwiched between other proteins of the α granule matrix.

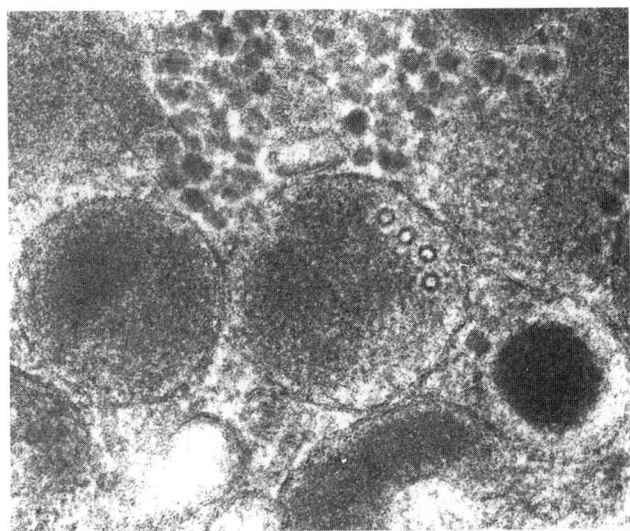

**Figure 3-44.** Platelet α granule containing four small tubular inclusions in its matrix just under the enclosing membrane. This is the form assumed by von Willebrand factor and is identical to the polymers of von Willebrand factor in endothelial cell Weibel–Palade bodies. Magnification ×100,000.

The central zone of the human platelet α granule is more dense than the peripheral zone. In some α granules, the central zone compares in density with nearby dense bodies (Figs. 3-41 and 3-45). The basis for the electron opacity is not known, but it suggests that proteins with binding sites for heavy metals may be present in this zone. Proteins in the central zone may be the earliest to be synthesized by the megakaryocyte and delivered to the developing α granules.

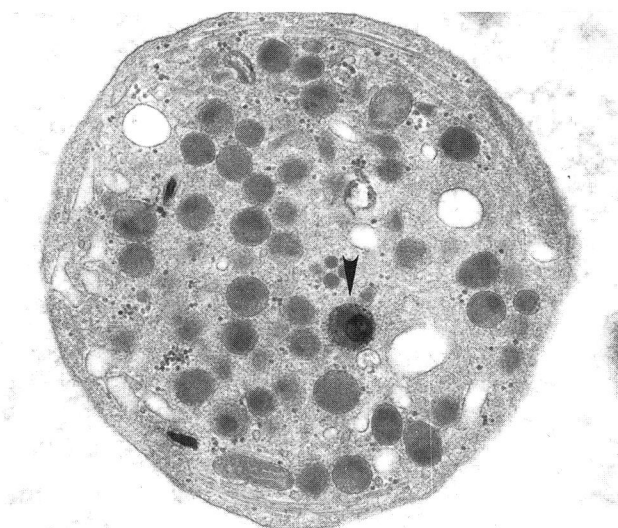

**Figure 3-45.** Discoid platelet with one α granule (arrowhead) containing a nucleoid with the same electron opacity as a dense body. Magnification ×22,000.

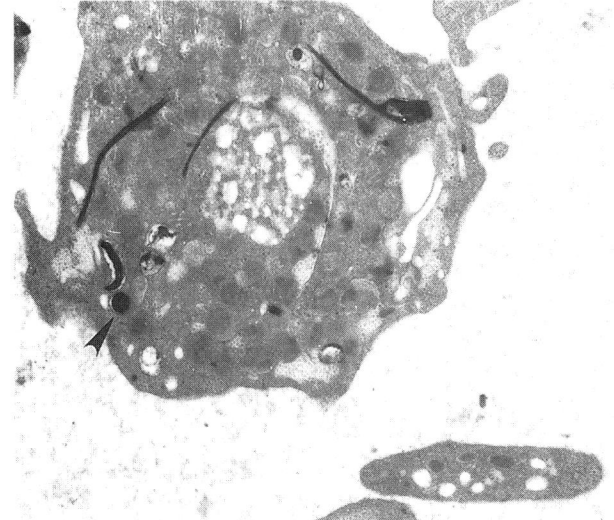

**Figure 3-47.** Large platelet containing four whiplike extensions from dense bodies. Dense bodies without whips (arrowhead) are also present in this cell. Magnification ×15,000.

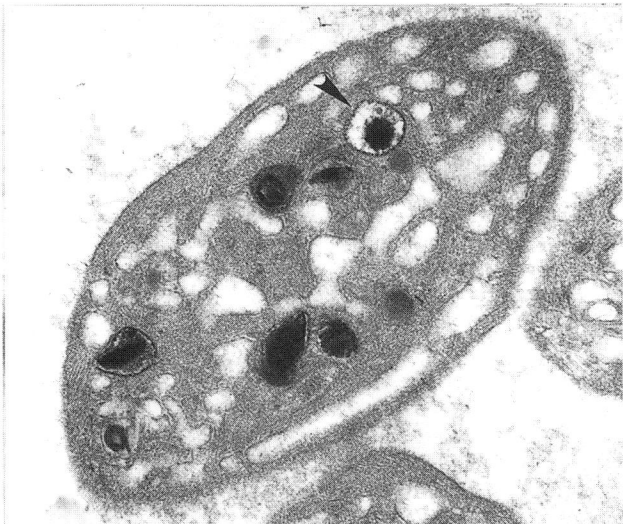

**Figure 3-46.** Discoid platelet showing the marked variation in morphology of dense bodies. One resembles a "bull's eye." Several others contain a granular substance around the dense core resembling the matrix material of the α granule. The appearance tends to support the concept that dense bodies and α granules may originate from the same multivesicular bodies. Magnification ×22,000.

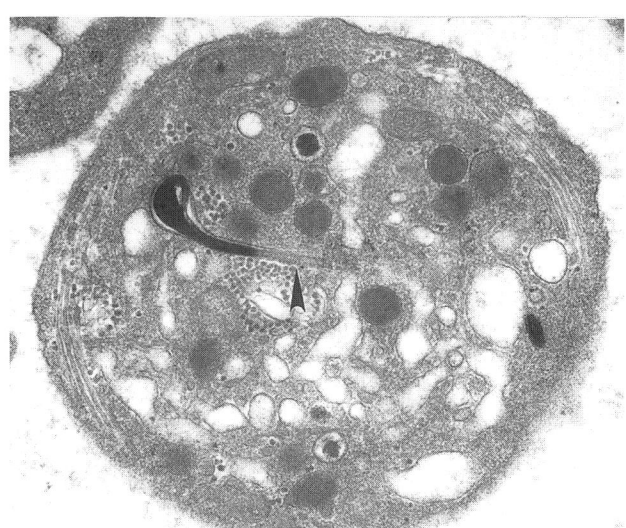

**Figure 3-48.** Discoid platelet containing several dense bodies, one with a whiplike extension. The tail of the whip (arrowhead) lacks electron-dense substance. Magnification ×22,000.

A more complete description of α granule chemical contents is given in Chapter 15.

**2. Dense Bodies.** Human platelet dense bodies are smaller than the α granules, fewer in number, and have high morphological variability (Figs. 3-41, 3-46, and 3-47).[67] Thin sections average approximately 1 to 1.4 dense bodies per platelet. Their most distinguishing feature is the

intensely, electron-opaque spherical body within the organelle, but separated from the enclosing membrane by a seemingly empty space (Figs. 3-46 and 3-48). This feature led to the term *bull's eye* to describe dense bodies. However, some dense bodies are irregular in form, have filaments extending from the dense inner core to the enclosing membrane, or contain a granulelike substance filling the usually empty space (Fig. 3-46). Whether the latter structures are dense bodies or α granules with electron-dense central zones is uncertain (Figs. 3-41 and 3-45). It is possible that if both α granules and dense bodies are processed and

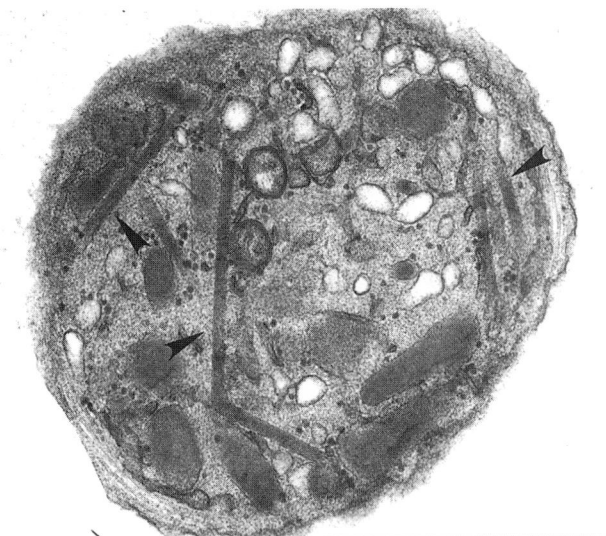

**Figure 3-49.** Discoid human platelet containing α granules resembling rods, or rodlike extensions of α granules (arrowheads). Magnification ×28,000.

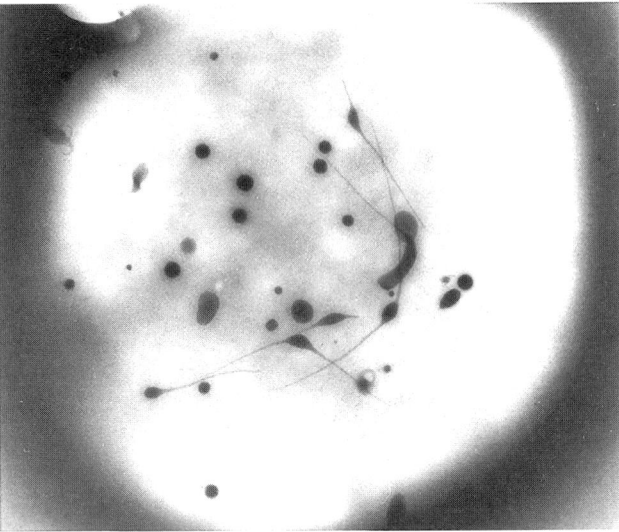

**Figure 3-50.** Whole-mount preparation of an unfixed, unstained human platelet. Dense bodies are inherently electron opaque, whereas the platelet is more electron transparent than the surrounding formvar surface. As a result, the dense bodies and other electron opaque structures are easily identified. Magnification ×13,000.

formed in multivesicular bodies of the megakaryocytes, some products specific for one organelle might find their way into another.[27] The only other evidence supporting this possibility is the presence of some identical receptors in the enclosing membranes of dense bodies, α granules, and the platelet surface membrane.[68] Some dense bodies have long, whiplike extensions (Figs. 3-47 and 3-48), which might suggest an origin different from α granules. However, α granules with rodlike extensions have been observed (Fig. 3-49). Compared with the whiplike extensions from dense bodies, however, rodlike extensions from α granules are rare.

The whole-mount technique is of particular value for the study of human platelet dense bodies.[68] A small drop of platelet-rich plasma is placed on a plastic coated (Formvar) 300 mesh copper electron microscope grid for 10 to 15 seconds. The excess plasma is washed from the grid with one to two drops of distilled water, and excess fluid is removed by touching the grid edge with filter paper. Care must be taken not to wet the bottom side of the grid. Platelets retain their discoid form. Their phospholipid-rich membranes have a lenslike effect on the electron beam, causing the platelet cytoplasm to appear relatively transparent. Dense bodies are inherently electron opaque and their spherical margins are sharply delineated (Fig. 3-50). As a result, dense bodies can be counted easily and their volume determined by special imaging techniques. The marked variation in dense body shape is more clearly apparent in whole mounts than in thin sections. There are, on average, four to eight dense bodies per platelet (range, 0–24). Long tails can extend from two or more sides on the same dense body (Figs. 3-51 and 3-52). It is difficult to envision how such a

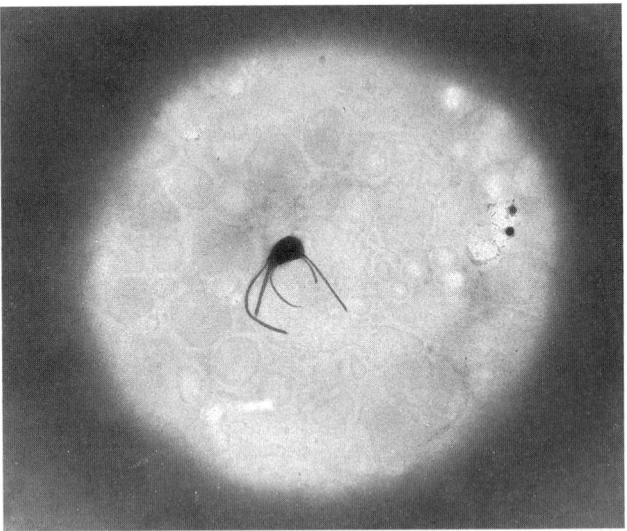

**Figure 3-51.** Whole-mount platelet containing a dense body with five whiplike extensions. Magnification ×13,000.

structure could be formed by multivesicular bodies in the megakaryocyte. Dense bodies are rich in adenine nucleotides, including ATP and ADP, serotonin, pyrophosphate, calcium, and magnesium (see Chapter 15). Analytical electron microscopy has shown that human dense bodies are particularly rich in calcium, and the complexes that calcium forms with pyrophosphate and serotonin are responsible for the electron opacity of platelet dense bodies.

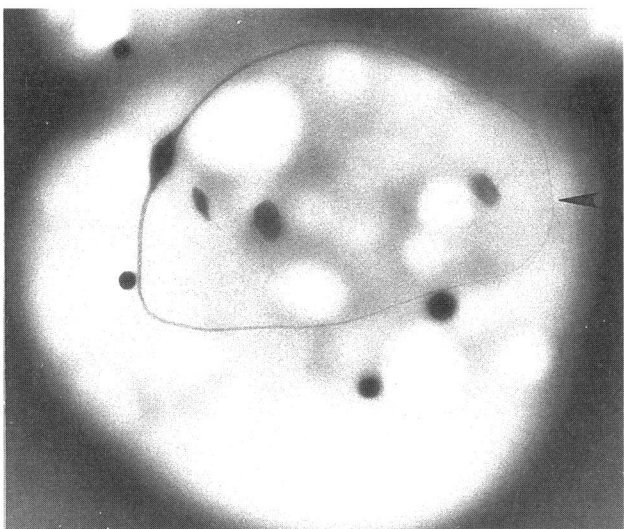

**Figure 3-52.** Whole-mount platelet containing a number of dense bodies, one with whiplike processes extending from opposite sides. The extensions appear to be fused (arrowhead) or continuous, resulting in a complete circle. Magnification ×20,000.

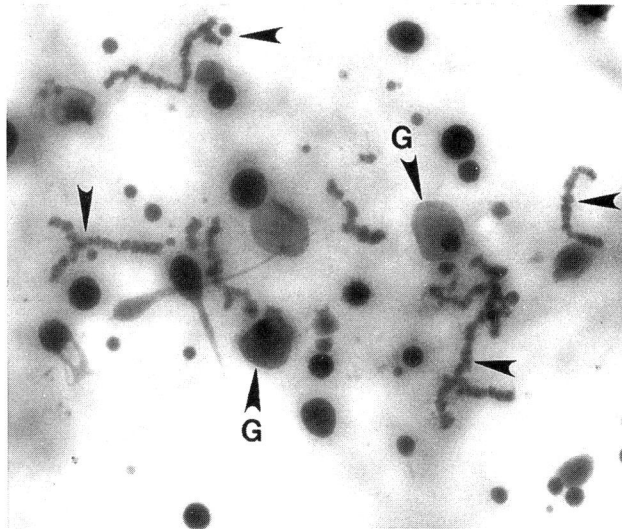

**Figure 3-54.** Whole mount of a human platelet showing a variety of electron-dense structures in addition to several dense bodies. Several beadlike chains are present (arrowheads). Also, two α granules (G) with dense nucleoids are visible. Magnification ×22,000.

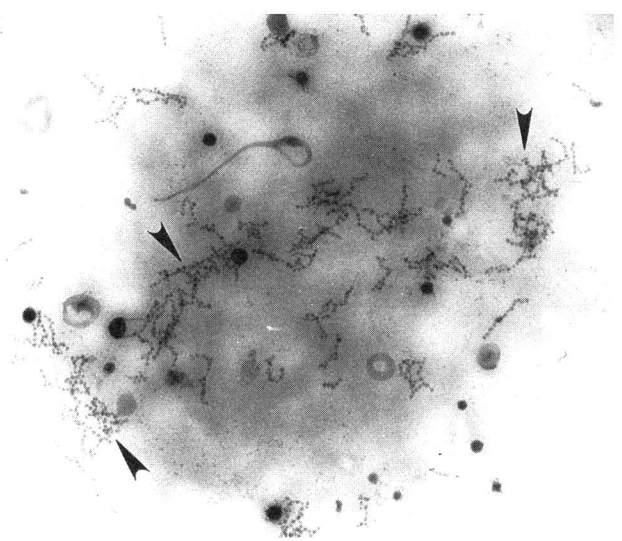

**Figure 3-53.** Low-magnification micrograph of a whole-mount platelet containing numerous electron-dense beadlike chains (arrowheads) as well as dense bodies. Magnification ×13,000.

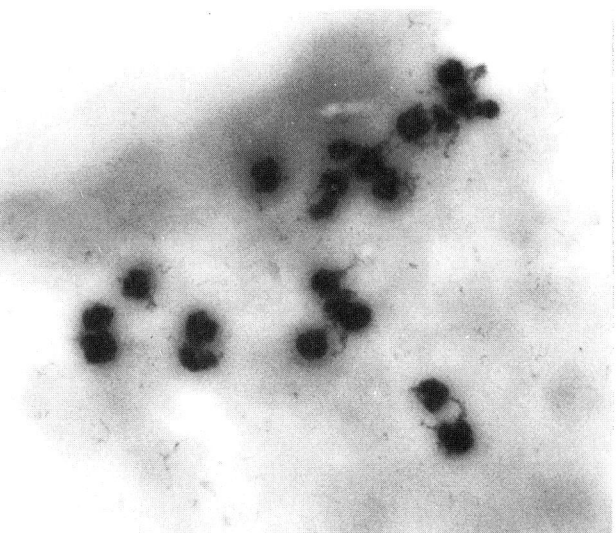

**Figure 3-55.** Whole mount of human platelet that contains numerous electron-dense clusters that could be confused with dense bodies. However, they lack the "bull's eye" appearance of dense bodies and have uneven, rather than sharp, edges. Magnification ×25,000.

**3. Electron-Dense Chains and Clusters.** Dense bodies are not the only inherently opaque structures in the platelet cytoplasm. Small, electron-dense hexagonal beads usually gathered into chains (Figs. 3-53 and 3-54) or assembled into clusters (Fig. 3-55) resembling dense bodies, but lacking their sharp margins and smooth appearance, are best observed in whole-mount preparations.[69] The bead chains and clusters are not as apparent in thin sections. However, study of their appearance in whole-mount platelets facilitated their recognition in sections of plastic-embedded cells

(Fig. 3-56). Individual beads were difficult to identify, but chains of beads were enclosed within unit membranes (Fig. 3-57). The clusters appeared to be groups of beads about the size of dense bodies but with irregular, fuzzy borders and without the "bull's eye" appearance.

The frequency of chains and clusters in normal and abnormal platelets was examined by the whole-mount method. Normal adults have chains and/or clusters in 2 to

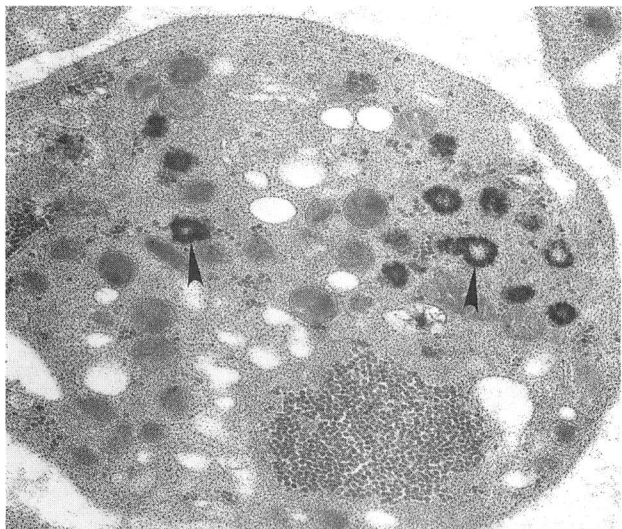

**Figure 3-56.** Thin section of discoid platelet containing several membrane-enclosed clusters (arrowheads). Magnification ×17,000.

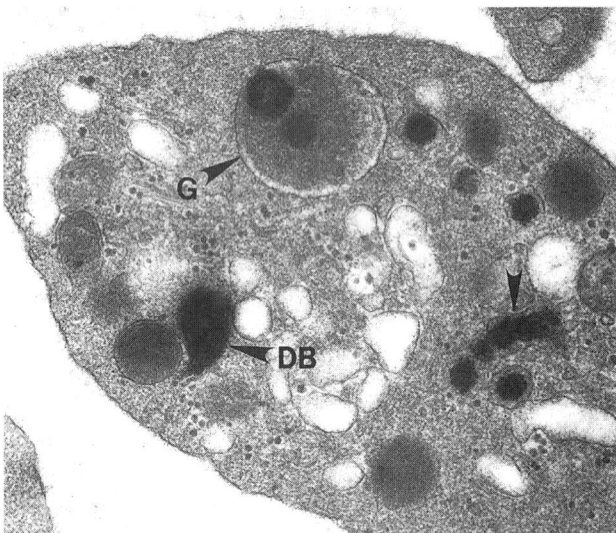

**Figure 3-57.** Thin section of a platelet containing a membrane-enclosed chain of beads and clusters (arrowheads). Several dense bodies (DB) are present, and a large α granule (G) with a normal nucleoid and an eccentric, electron-dense nucleoid. Magnification ×32,000.

22% of their platelets. Newborn and infant platelets contained chains or clusters in 2% of their cells. Chains and clusters were more common in platelets from individuals older than 40 years of age, and random individuals have been studied who contained chains and clusters in up to 58% of their platelets.

The possibility that chains and clusters might be stages in the development of dense bodies required consideration. Therefore, platelets from patients with storage pool deficiency disorders, including Hermansky–Pudlak syndrome, Chediak–Higashi syndrome, and platelet storage pool disease were studied — all of whose platelets were markedly deficient in or lacked dense bodies.[70] Platelets from these individuals contained the same frequency of chains and clusters in their platelets as observed in platelets from normal control subjects. Thus, chains and clusters do not appear to be stages in the development of dense bodies.

The origin of chains and clusters remains unknown. They could arise from the trans-Golgi face of the Golgi apparatus as vesicles, in the same manner as the precursor vesicles for α granules, dense bodies, and lysosomes. However, it is more likely that the membranes enclosing bead chains and clusters arise as segments of tubules making up the trans-Golgi face of Golgi zones in the megakaryocyte. Another possible source are the membranes of the dense tubular system of the circulating platelet, which serves as the reservoir for calcium involved in platelet activation.[71] Some elements of the DTS may take up an excessive amount of calcium in keeping the cytoplasmic content very low. The excess amount of calcium may precipitate, forming bead, chains and clusters.

Chains and clusters in platelets have been recognized for several years, but they have not been discussed in chapters on platelet structure because these organelles have no known function. Now, because of their similarity to dense bodies, their presence needs to be recognized. Newer imaging techniques are being used to assess the number and volume of dense bodies in platelets that depend only on the electron opacity of structures in platelet whole mounts.[72,73] Because this procedure may be important for identifying patients with Hermansky–Pudlak syndrome or mild forms of platelet storage pool deficiency, it is vital to recognize the contribution of chains and clusters to the images.

**4. Lysosomes.** Human platelets contain few lysosomes. Ultrastructural cytochemistry using cerium phosphate to capture the product of acid phosphatase activity revealed no more than three and usually zero to one lysosome per platelet (Fig. 3-58). The organelles showing acid phosphatase reaction product are spherical in form and slightly smaller than α granules. They contain at least 13 acid hydrolases together with cathepsin D and E, WAMP-1, LAMP-2, and CD63 (see also Chapter 15).

The function of platelet lysosomes in the physiology of hemostasis is unknown. Their contents, together with products stored in α granules and dense bodies, can be released when platelets are exposed to maximum stimulation *in vitro,* but acid hydrolases may not be secreted at sites of vascular injury *in vivo.* Particles such as latex spherules, taken up into the OCS of platelets, are often transferred to α granules, but may also react with lysosomes.[74] Thus, platelet lysosomes may serve as an endosomal digestion compartment in platelets. However, platelets are covercytes, not

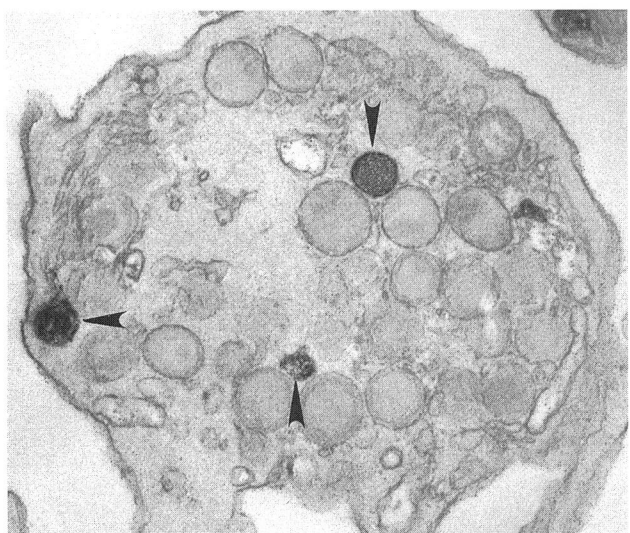

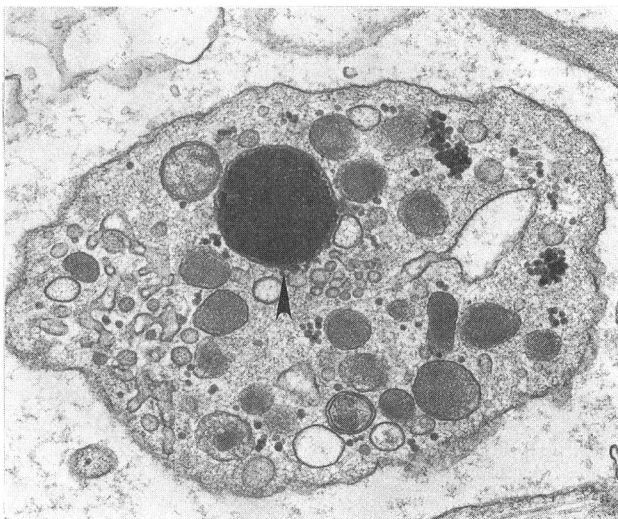

**Figure 3-58.** Platelet from a sample stained cytochemically for acid phosphatase using cerium as the capture agent. Three small lysosomes containing reaction product (arrowheads) are present in the cell. Magnification ×28,000.

**Figure 3-60.** White platelet syndrome platelet containing a giant glycosome (arrowhead). Magnification ×25,000.

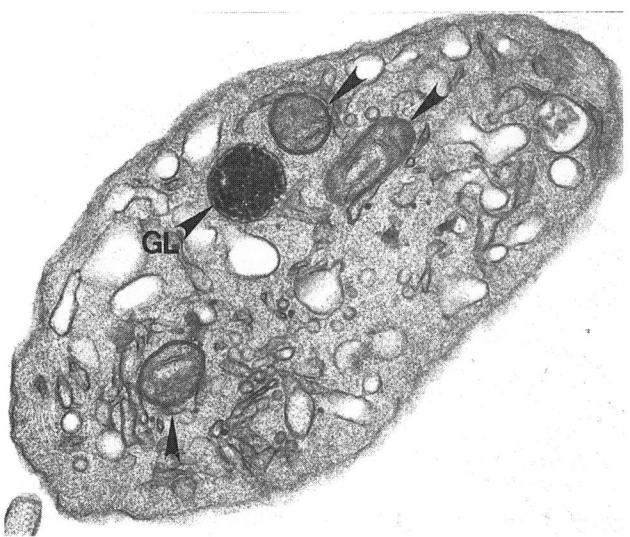

**Figure 3-59.** Discoid platelet containing mitochondria (arrowheads) and a glycosome (GL). Magnification ×30,000.

phagocytes, and do not isolate engulfed bacteria into sealed phagosomes, where they could be killed by lysosomal enzymes.[75] Therefore, the lysosomes in platelets appear to be vestigal remnants with no known significant role in platelet function.

**5. Platelet Glycosomes.** Another organelle in platelets, like chains and clusters, that has received little attention is the platelet glycosome.[53] The glycosome is round to oval in shape and similar in size to α granules or other organelles (Figs. 3-59 and 3-60). Glycosomes are enclosed within a

typical unit membrane and their content of glycogen particles is identical in appearance to particles of glycogen lying free in the cytoplasm or condensed into glycogen masses (Figs. 3-37 and 3-38).

There are other structures in platelets that resemble glycosomes. α Granules may contain glycogen and may be confused with glycosomes. Particles of glycogen are commonly observed in α granules of bovine platelets and, less often, in α granules of human cells. The basis for their presence in these organelles is unknown.

**6. Tubular Inclusions.** There is another structure seen in human platelets that may resemble the glycosome, depending on the plane of section. Tubular inclusions with open ends were frequently observed in platelets of one patient with the Medich giant platelet disorder (Fig. 3-39)[54] and are common in platelets from the Wistar–Furth rat. Longitudinal sections reveal cytoplasmic constituents within the multilayer membranes of the tubes, and glycogen particles are prominent. Cross-sections of the tubular inclusions are often filled with glycogen (Fig. 3-61). The multilamellar membrane distinguishes this structure from the glycosome.

**7. Mitochondria.** Platelet mitochondria are few in number and structurally simple. Yet, they contribute significantly to the energy metabolism of the cell. Blockade of anaerobic glycolysis does not cause a drop in the level of platelet ATP, nor does it inhibit normal platelet function. Thus, mitochondria alone can support platelet energy requirements. During platelet hemostatic reactions the mitochondria become more electron dense, suggesting a change in energy state. Mitochondria also contain calcium and are considered by some workers to be the source of calcium

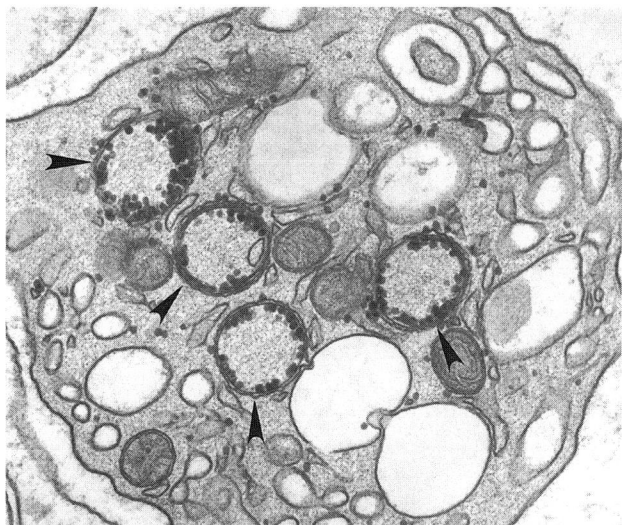

**Figure 3-61.** Medich giant platelet syndrome platelet containing cross-sections of four tubular inclusions (arrowheads). The matrix has a different consistency than surrounding cytoplasm. Glycogen particles are clustered on the inner membranes. Magnification ×33,000.

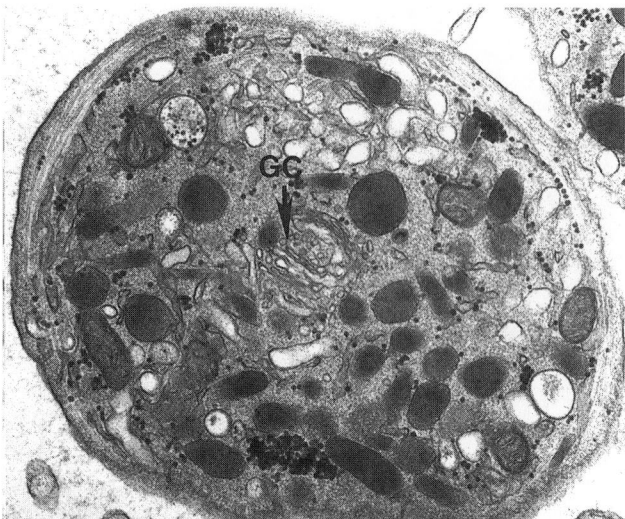

**Figure 3-62.** Discoid human platelet containing a Golgi complex (GC) in its cytoplasm. Magnification ×22,000.

important in platelet activation. However, more evidence supports the DTS as the sarcoplasmic reticulum of platelets important in calcium flux[71] and extracellular calcium as the second source.

## V. Platelet Membrane Systems

### A. General Aspects

Golgi zones are membrane systems ordinarily confined to the megakaryocyte and are therefore discussed in Chapter 2. However, residual membrane elements of the Golgi apparatus are found in thin sections of 1 of every 100 to 500 platelets. (Fig. 3-62). The reason for discussing them here is the discovery that Golgi zones are common in platelets from patients with hypogranular platelet syndromes.[76] The SCCS is part of the platelet surface membrane[77–79] and could have been discussed under the section dealing with the peripheral zone. However, some workers believe the OCS is a unique membrane system that is not involved in activities of the exposed surface membrane. Therefore, it is discussed here as one of the internal platelet membrane systems. The DTS is formed by residual channels of the rough and smooth endoplasmic reticulum. Together with the channels of the OCS, elements of the DTS form membrane complexes.[80,81] They closely resemble the sarcoplasmic reticulum of embryonic muscle cells, but the precise function of platelet membrane complexes is not known.

### B. Specific Features

**1. Golgi Complexes.** Golgi complexes consisting of parallel-associated, flattened sacules are prominent in the perinuclear cytoplasm of the megakaryocyte during granulopoiesis. When organelle formation is completed, the extensively developed, highly complex Golgi zones move to the periphery of the megakaryocyte cytoplasm and almost completely disappear before proplatelets develop. Only residual elements consisting of a few parallel-associated, flattened sacules with no budding vesicles are found in less than 1% of circulating platelets (Fig. 3-62). Its mission was accomplished before the circulating platelet was born.

But not always. Platelets from patients with some of the hypogranular platelet syndromes carry significant numbers of Golgi complexes to the peripheral blood.[76] This is particularly true for the white platelet syndrome.[57] Patients with this disorder have mild thrombocytopenia, increased mean platelet volume, decreased sensitivity to aggregating agents, and prolonged bleeding times. Four to 13 percent of their platelets contain large, fully developed Golgi complexes actively budding smooth and coated vesicles and frequently associated with centrioles (Figs. 3-63 and 3-64). As many as seven Golgi complexes and five centrioles have been observed in single platelets. The findings indicate that platelets from patients with some hypogranular platelet disorders are continuing the process of granulopoiesis into circulating platelets. Recognition that this phenomenon can occur and is characteristic of the white platelet syndrome will prevent patients from being diagnosed incorrectly and possibly considered to have a leukemic disorder.

**2. The Surface-Connected Open Canalicular System.** The OCS is not only connected to the platelet surface

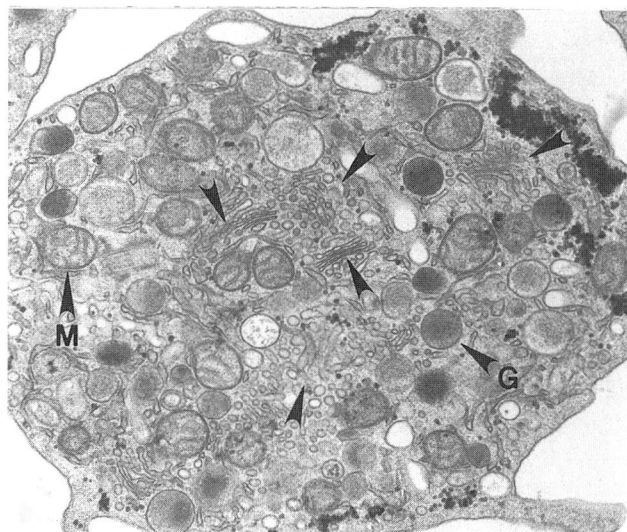

**Figure 3-63.** Platelet from a patient with white platelet syndrome. There are five well-developed Golgi complexes (arrowheads) shedding numerous smooth and coated vesicles in the cytoplasm. Mitochondria (M) are more numerous than α granules (G). Magnification ×22,000.

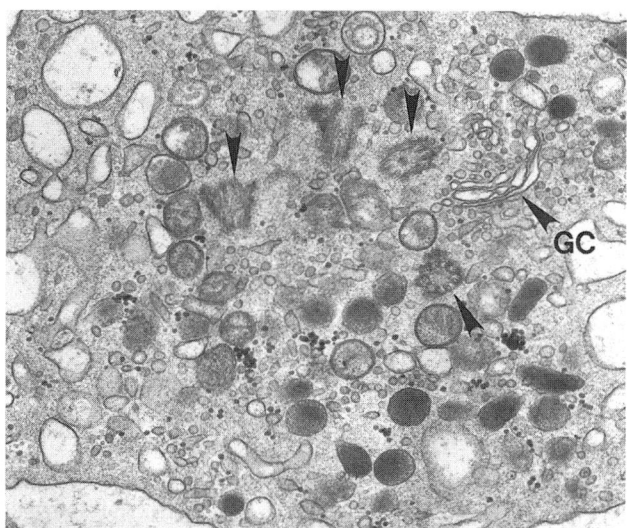

**Figure 3-64.** White platelet syndrome platelet containing a large Golgi complex (GC) and four centrioles (arrowheads). Magnification ×22,000.

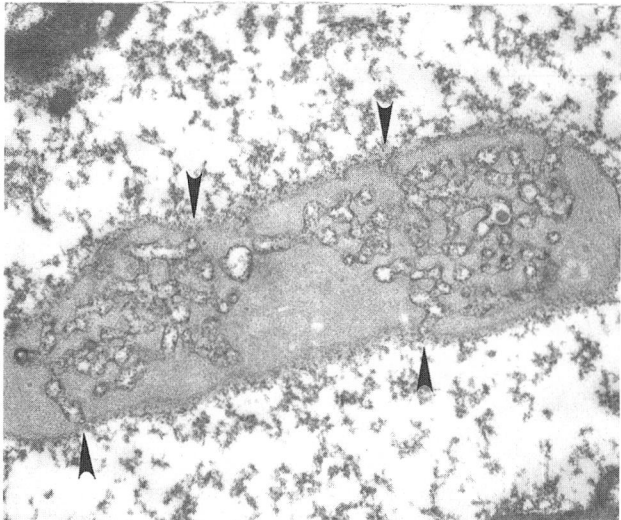

**Figure 3-65.** Discoid platelet stained with tannic acid during fixation. Osmium black, the reaction product of tannic acid and osmic acid, stains the glycocalyx covering the surface and lining channels of the OCS. The OCS channels form an interconnecting lattice (arrowheads) stretching from one side of the platelet to the other. Magnification ×22,000.

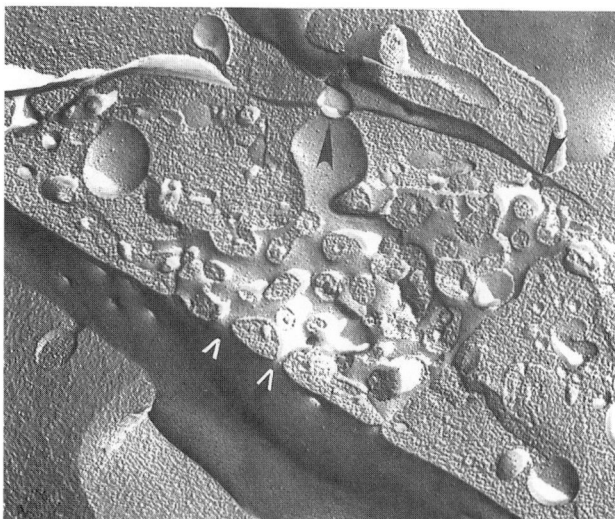

**Figure 3-66.** Replica of a freeze-fractured platelet. Channels of the OCS form an interconnecting lattice (arrowheads) that stretches from one side of the cell to the other. Magnification ×28,000.

membrane; it is surface membrane.[77–79] OCS channels are tortuous invaginations of surface membrane tunneling through the cytoplasm in a serpentine manner (Figs. 3-65 and 3-66). Studies using electron-dense tracers and tannic acid have shown that channels of the OCS are patent in activated and aggregated platelets, as well as in unstimulated cells. Channels of the OCS greatly expand the total surface area of the platelet exposed to circulating plasma[82] and provide a means for chemical and particulate substances

to reach the deepest recesses of the cell.[83] As a result, the OCS may be the major route for uptake and transfer of products, such as fibrinogen, from plasma to platelet α granules (Fig. 3-67).[84,85] In addition, channels of the OCS serve as conduits for the discharge of products stored in secretory organelles during the platelet release reaction (Figs. 3-35, 3-36, and 3-68).[49] Bovine and equine platelets do not have the OCS found in human platelets. After activation, the

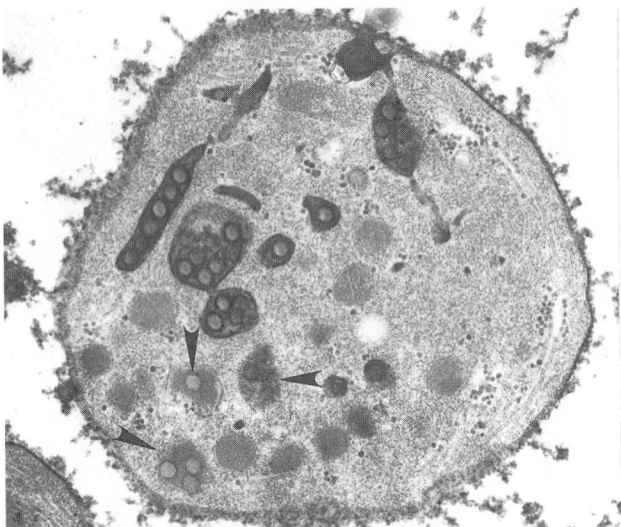

**Figure 3-67.** Platelet from a sample of C-PRP incubated with small latex particles for 30 minutes and stained with tannic acid during fixation. The cell has retained its discoid shape during uptake of the latex spherules into channels of the OCS stained by osmium black. Some latex spherules have been transferred to α granules (arrowheads) during this process. Magnification ×25,000.

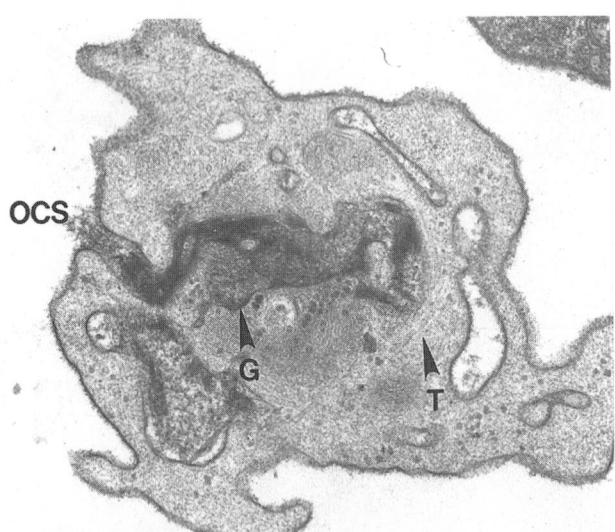

**Figure 3-68.** Washed platelet exposed to thrombin 5 U/mL for 1 minute in the presence of EDTA. Granules (G) stained by osmium black have secreted their fibrinogen/fibrin content to the OCS and through the channels to the cell exterior. Internal transformation bringing the microtubule coil (T) toward the cell center drives the release process. Magnification ×••.

granules in bovine and equine platelets fuse directly with the cell surface and with each other, and thereby extrude their contents to the exterior.[86]

Channels of the OCS also have major roles in the platelet's hemostatic reaction. After adhesion to a damaged vascular surface, the platelet extends filopodia to bind firmly to

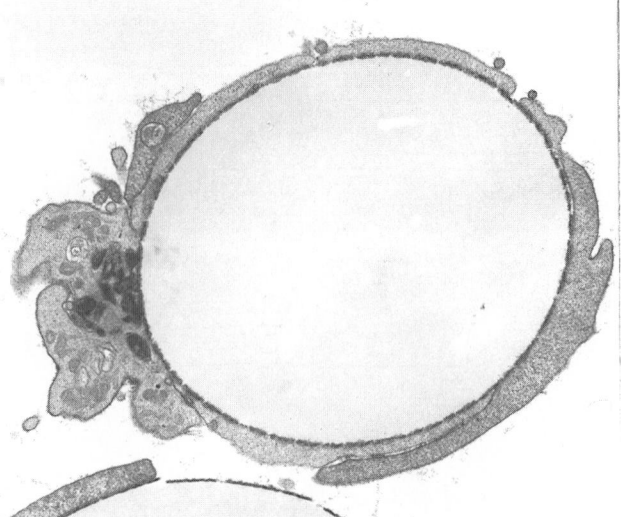

**Figure 3-69.** Incubation of platelets with small latex particles brings the particles to the OCS, as shown in Figure 3-67. Incubation with large latex spheres brings the OCS to the particle. Most of the platelets have spread fully on this spherule. One cell is in process and its OCS, stained by osmium black, is spreading on the latex surface. Magnification ×13,000.

the injured area. This is followed rapidly by the assembly of cytoplasmic actin and spreading of the platelet to cover as much area as possible (see Chapter 4). The result is an increase of up to 420% of the exposed surface area during conversion from a discoid platelet to a fully spread cell. Evaginated channels of the OCS are the major source of membrane for the expanded surface area of the spread platelet.[87] Bovine and equine platelets that lack the OCS of human platelets do not spread; they unfold.

The spreading reaction of human platelets is not limited to their hemostatic response on damaged vascular surfaces. They respond to almost any foreign surface in a similar manner.[84] Platelets can ingest small particles such as thorium dioxide, colloidal carbon, Fgn/Au, and small and medium-size (0.312 μm) latex spherules without undergoing shape change or internal transformation.[6,83] Some of the particles will be moved to α granules (Fig. 3-67) without causing platelet activation, supporting a role for the OCS in the uptake and transfer of products from plasma to organelles.[88] When particles are too large to move in to the OCS, the platelet moves the OCS to the particle, just as it does to flat surfaces (Fig. 3-69). This has been demonstrated recently with regard to platelet–bacterial interaction, using tannic acid as an electron-dense tracer. Bacteria ingested by platelets appear to be in phagocytic vacuoles just as they do when taken up by neutrophils.[89] However, tannic acid staining reveals that the engulfment vacuoles containing bacteria are lined on their interior surface with the reaction product of tannic acid and osmic acid, osmium black (Figs. 3-70–3-72). Channels of the OCS also stained by the reaction product

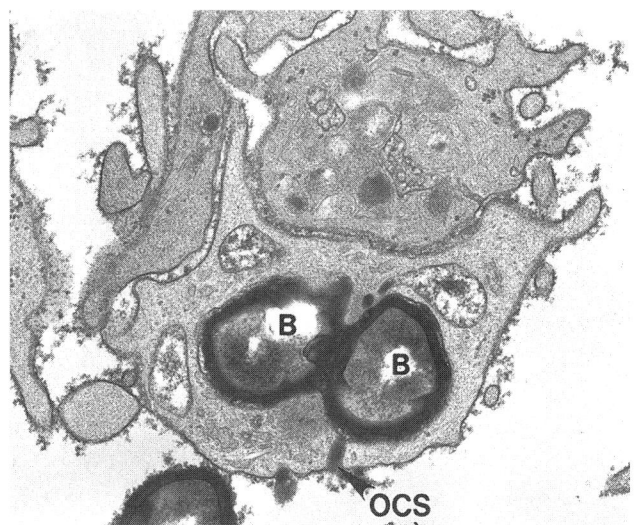

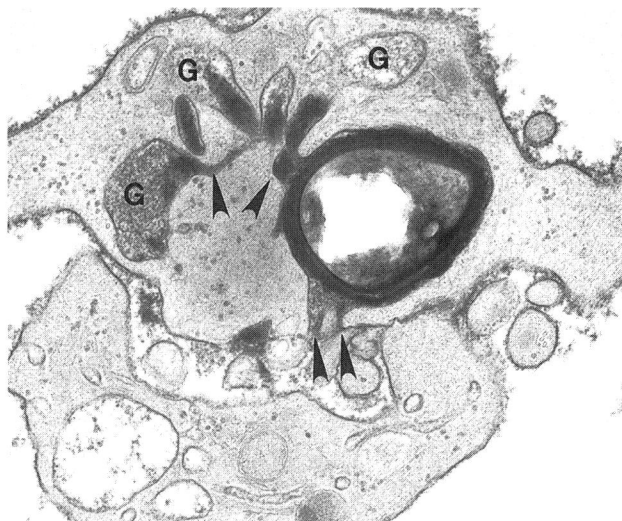

**Figure 3-70.** Platelets incubated with *Staphylococcus aureus* 502A for 60 minutes and stained with tannic acid during fixation. Osmium black stains the glycocalyx covering the platelet outside surface and lining channels of the OCS. It also stains the inside of membranes enclosing the two bacteria (B). Magnification ×27,000.

**Figure 3-72.** Platelet incubated with *S. aureus* for 60 minutes and fixed in solution containing tannic acid. Osmium black stains the glycocalyx of the outside surface, lining channels of the OCS (arrowheads), the interior of the engulfment vacuole containing an organism, and α granules (G) secreting fibrinogen into the OCS and into the engulfment vacuole. Magnification ×27,000.

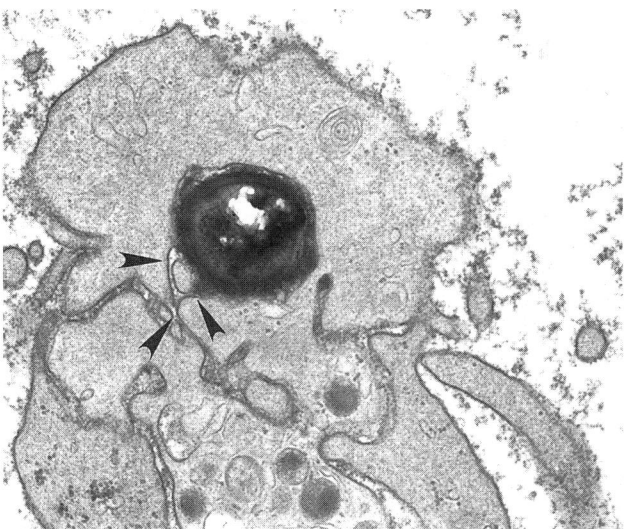

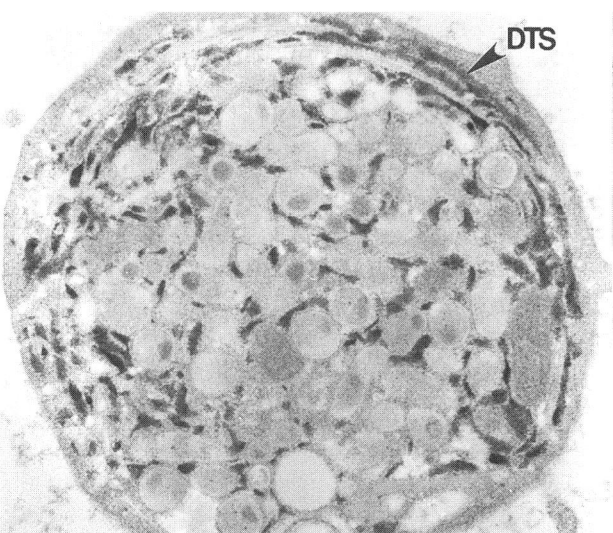

**Figure 3-71.** Platelets incubated with *S. aureus* 502A for 60 minutes then fixed in solutions containing tannic acid. Osmium black covers the exposed surface membrane and linings of OCS (arrowheads) channels. Its also stains the inside of the membrane enclosing the organism. The OCS is clearly involved in the process and responds in the same manner as it does to large latex spheres or flat surfaces (Fig. 3-69). Magnification ×27,000.

**Figure 3-73.** Platelet from a sample of PRP fixed in glutaraldehyde–paraformaldehyde before incubation with diaminobenzidine to detect endogenous peroxidase. Reaction product is localized to channels of the DTS. Magnification ×30,000.

communicate directly with the exterior surface also covered with osmium black. Neutrophils fixed in the same manner after taking up bacteria are covered by the reaction product, but it never appears in the phagocytic vacuoles. Thus, platelets are covercytes, not phagocytes, and the OCS is critically involved in spreading on any surface.[75]

**3. Dense Tubular System.** Channels of the DTS are distinguished from the clear canaliculi of the OCS by an amorphous material similar in opacity to surrounding cytoplasm concentrated within them[77] that stains for peroxidase[80] and binds lead,[90] indicating calcium binding sites and enzymes involved in prostaglandin synthesis (Figs. 3-73 and 3-74).[91] Like the OCS, channels of the DTS are randomly

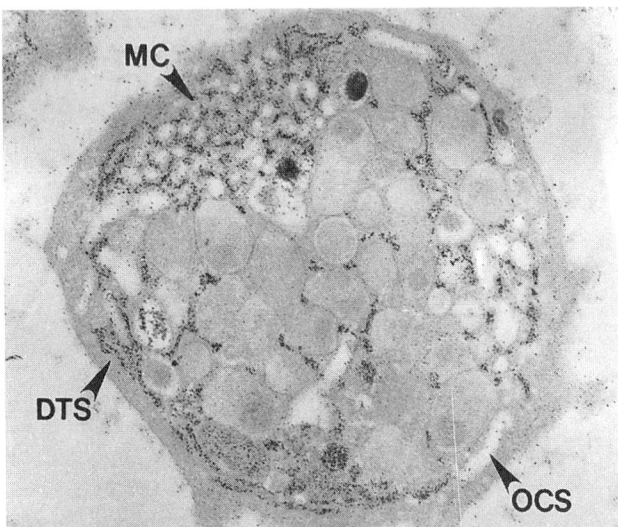

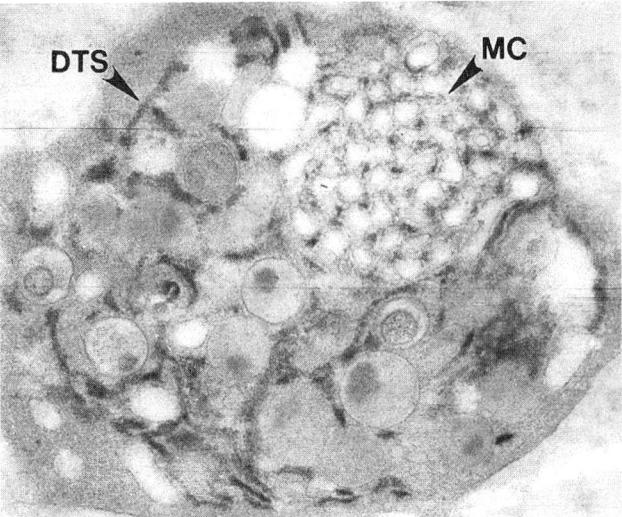

**Figure 3-74.** Platelet from sample of PRP fixed in glutaraldehyde and incubated in lead-containing solution for 60 minutes. Lead taken up by the cell is specifically localized to channels of the DTS, including those forming a membrane complex (MC) with channels of the OCS. Magnification ×30,000.

**Figure 3-75.** Platelet from a sample of PRP incubated for peroxidase activity. Reaction product is localized to elements of the DTS, including those associated with channels of the OCS in a membrane complex (MC). Magnification ×40,000.

dispersed in the platelet cytoplasm. In addition, a channel or two of the DTS can be identified in close association with the circumferential band of microtubules in most thin sections of platelets. The DTS originates from rough endoplasmic reticulum in the parent megakaryocyte and is, therefore, residual smooth endoplasmic reticulum. There is no physical communication between channels of the OCS and DTS. Thus, platelets have two discrete membrane systems not found in other blood cells: the OCS derived from the plasma membrane and demarcation membrane system of the megakaryocyte, and the DTS representing residual smooth endoplasmic reticulum of the parent cell.

**4. Membrane Complexes.** However, the OCS and DTS are not completely isolated membrane systems. Close inspection of thin sections of well-preserved platelets revealed that canaliculi of the OCS and DTS form intimate physical relationships in nearly every cell.[80] The association of the two channel systems was usually restricted to one or two areas of the cytoplasm, and in most examples these areas were eccentrically located. Elements of the OCS in such areas were gathered in clusters or groups. Even though the open channels were closely approximated, small canaliculi of the DTS could be identified interspersed between them. The relationship was particularly prominent in platelets stained for peroxidase activity in which dense reaction product delineated channels of the DTS and outlined their extremely close relationship to clusters of open canaliculi (Fig. 3-75). Examination of the membrane complexes at higher magnification revealed that elements of the DTS

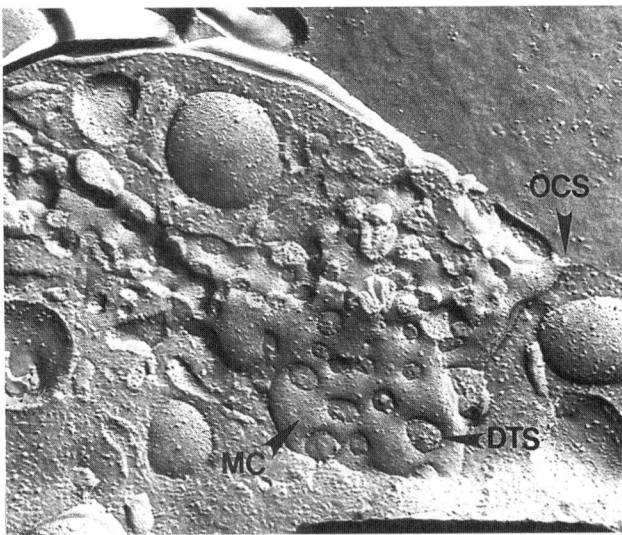

**Figure 3-76.** Replica of a freeze-fractured platelet demonstrating the fenestrated membrane complex (MC) formed by elements of the OCS and channels of the DTS. A channel of the DTS is present in each window formed by the interconnected elements of the OCS. Magnification ×50,000.

were the only structures interspersed between open canaliculi, and that membranes of the two channel systems were practically in apposition. Replicas of freeze-fractured platelets revealed an identical arrangement of the two channel systems in complexes (Fig. 3-76).[92] When studied by this method, the OCS is clearly a fenestrated membrane system, and channels of the DTS are located in each window or fenestra. The membrane complexes in platelets are very

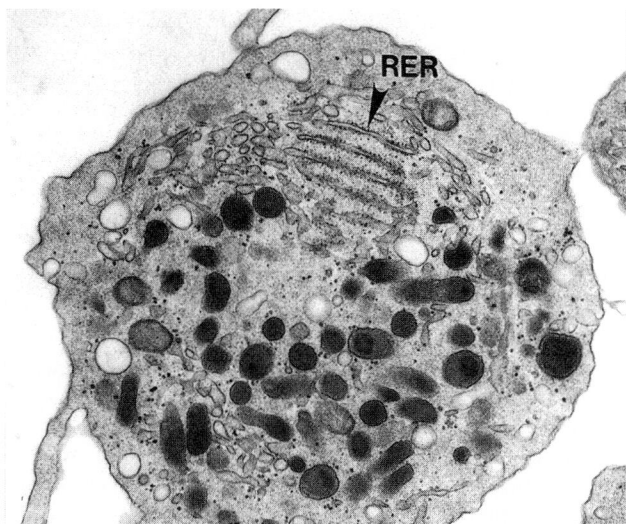

**Figure 3-77.** Platelet from a patient with the white platelet syndrome containing residual elements of rough endoplasmic reticulum (RER). Magnification ×21,000.

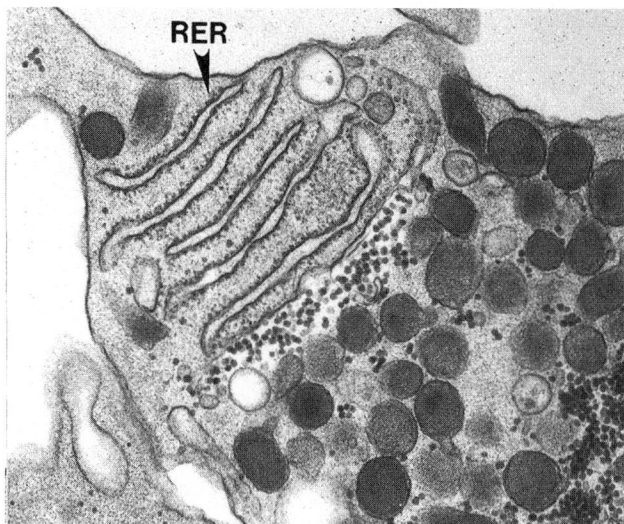

**Figure 3-78.** Platelet from a patient with immune thrombocytopenic purpura. The cell is normal except for retention of elements of rough endoplasmic reticulum (RER). Magnification ×30,000.

similar to arrangements of the sarcoplasmic reticulum in embryonic muscle cells.

**5. Rough Endoplasmic Reticulum.** The rarest of the membrane systems found in circulating platelets are channels of rough endoplasmic reticulum. The rough endoplasmic reticulum in the parent megakaryocyte has disappeared by the time granulopoiesis has been completed and before proplatelets are formed. However, in patients with rapid platelet turnover resulting from immune thrombocytopenia,

a few cells will contain channels of rough endoplasmic reticulum. The channels are usually in parallel association and are studded with ribosomes (Figs. 3-77 and 3-78). Ribosomal complexes are often present in platelet cytoplasm adjacent to the rough endoplasmic reticulum. Aside from such examples, ribosomal complexes and rough endoplasmic reticulum are not found in normal human platelets. Smooth endoplasmic reticulum making up the DTS is the final stage in the life of the rough endoplasmic reticulum.

## References

1. Donné, A. (1842). De porigine des globules du sang, de leur mode de formation et du leur fin. *Comptes Render Seances de L'Academia de Sciences, 14,* 366–368.
2. Bizzozero, J. (1882). Ueber einen neuen formbestandheil des blutes und dessen rolle bei der thrombose und der blutgerinnung. *Arch Pathol Anat Physiol, 90,* 261–332.
3. Tocantins, L. M. (1938). The mammalian blood platelet in health and disease. *Medicine, 17,* 155–260.
4. Bessis, M. (1973). *Living blood cells and their ultrastructure* (pp. 767). New York: Springer-Verlag.
5. White, J. G., & Escolar, G. (1993). Current concepts of platelet membrane response to surface activation. *Platelets, 4,* 175–198.
6. White, J. G. (1971). Platelet morphology. In S. E. Johnson (Ed.), *The circulating platelet* (pp. 45–121). New York: Academic Press.
7. White, J. G., & Conrad, W. J. (1973). The fine structure of freeze-fractured blood platelets. *Am J Pathol, 70,* 45–56.
8. Behnke, O. (1970). The morphology of blood platelet membrane systems. *Seminarsin Haematology, 3,* 3–16.
9. White, J. G. (1969). The submembrane filaments of blood platelets. *Am J Pathol, 56,* 267–277.
10. Clemetson, K. J. (1985). Glycoproteins of the platelet plasma membrane. In J. N. George, A. T. Nurden, & D. R. Phillips (Eds.), *Platelet membrane glycoproteins* (pp. 145–169). New York: Plenum Press.
11. Phillips, D. R. (1985). Receptors for platelet agonists. In J. N. George, A. T. Nurden, & D. R. Phillips (Eds.), *Platelet membrane glycoproteins* (pp. 51–85). New York: Plenum Press.
12. Kunicki, T. J. (1988). Platelet glycoproteins antigens and immune receptors. In G. A. Jamieson (Ed.), Platelet membrane receptors: Molecular biology, immunology, biochemistry and pathology. *Prog Clin Biol Res. 283,* 87–125.
13. White, J. G. (1987). Platelet structural physiology: The ultrastructure of adhesion, secretion and aggregation in arterial thrombosis. *Cardiovasc Clin, 18,* 13–33.
14. White, J. G., & Escolar, G. (1990). Fibrinogen receptors do not undergo spontaneous redistribution on surface-activated platelets. *Arteriosclerosis, 10,* 738–744.
15. Escolar, G., Clemetson, K., & White, J. G. (1994). Persistence of mobile receptors on surface- and suspension-activated platelets. *J Lab Clin Med, 123,* 536–546.
16. White, J. G., Krumwiede, M. D., Cocking–Johnson, D., & Escolar, G. (1994). Influence of combined thrombin stimulation, surface activation, and receptor occupancy on organiza-

tion of GPIb/IX receptors on human platelets. *Br J Haematol, 88,* 137–148.

17. White, J. G., Krumwiede, M. D., Cocking–Johnson, D. J., & Escolar, G. (1995). Dynamic redistribution of glycoprotein Ib/IX on surface-activated platelets. A second look. *Am J Pathol, 147,* 1057–1067.

18. White, J. G., Krumwiede, M. D., Cocking–Johnson, D., Rao, G. H., & Escolar, G. (1995). Retention of glycoprotein Ib/IX receptors on external surfaces of thrombin-activated platelets in suspension. *Blood, 86,* 3468–3478.

19. White, J. G., Krumwiede, M. D., Cocking–Johnson, D. K., & Escolar, G. (1995). Redistribution of GPIb/IX and GPIIb/IIIa during spreading of discoid platelets. *Br J Haematol, 90,* 633–644.

20. White, J. G. (1990). Separate and combined interactions of fibrinogen-gold and latex with surface activated platelets. *Am J Pathol, 137,* 989–998.

21. White, J. G., Krumwiede, M. D., Cocking–Johnson, D., & Escolar, G. (1996). Uptake of vWF-anti-vWF complexes by platelets in suspension. *Arterioscler Thromb Vasc Biol, 16,* 868–877.

22. White, J. G. (1993). Functional significance of mobile receptors on human platelets. *Arterioscler Thromb, 13,* 1236–1243.

23. Koyama, T., Nishida, K., Ohdama, S., et al. (1994). Determination of plasma tissue factor antigen and its clinical significance. *Br J Haematol, 87,* 343–347.

24. Giesen, P. L., Rauch, U., Bohrmann, B., et al. (1999). Blood-borne tissue factor: Another view of thrombosis. *Proc Nat Acad Sci USA, 96,* 2311–2315.

25. Falati, S., Liu, Q., Gross, P., et al. (2003). Accumulation of tissue factor into developing thrombi in vivo is dependent upon microparticle P-selectin glycoprotein ligand 1 and platelet P-selectin. *J Exp Med, 197,* 1585–1598.

26. del Conde, I., Shrimpton, C. N., Thiagarajan, P., & López, J. A. (2005). Tissue factor-bearing microvesicles arise from lipid rafts and fuse with activated platelets to initiate coagulation. *Blood, 106,* 1604–1611.

27. Heijnen, H. F. J., Debili, N., Vainchenker, W., Breton–Gorius, J., Geuze, H. J., & Sixma, J. (1998). Multivesicular bodies are an intermediate stage in the formation of platelet α granules. *Blood, 91,* 2313–2325.

28. Hartwig, J. H., & DeSisto, M. (1991). The cytoskeleton of resting human blood platelet: Structure of the membrane skeleton and its attachment to actin filaments. *J Cell Biol, 112,* 407–425.

29. White, J. G. (1969). The submembrane filaments of blood platelets. *Am J Pathol, 56,* 267–277.

30. Escolar, G., Krumwiede, M., & White, J. G. (1986). Organization of the actin cytoskeleton of resting and activated platelets in suspension. *Am J Pathol, 123,* 86–94.

31. Haydon, G. B., & Taylor, A. B. (1956). Microtubules in hamster platelets. *J Cell Biol, 26,* 673–675.

32. Behnke, O. (1965). Further studies on microtubules: A marginal bundle in human and rat thrombocytes. *J Ultrastruct Res, 13,* 469–477.

33. Bessis, M., & Breton–Gorius, J. (1965). Les microtubules et les fibrilles dan le plaquettes etalies. *Nouv Rev Fr Hematol, 5,* 657–662.

34. Pipeleers, D. G., Pipeleers–Marichal, M. A., & Kipnis, D. M. (1977). Physiological regulation of total tubulin and polymerized tubulin synthesis. *J Cell Biol, 74,* 351–357.

35. White, J. G., & Krumwiede, M. (1985). Isolation of microtubule coils from normal human platelets. *Blood, 65,* 1028–1032.

36. White, J. G., Radha, E., & Krumwiede, M. (1986). Isolation of circumferential microtubules from platelets without simultaneous fixation. *Blood, 67,* 873–877.

37. White, J. G., & Sauk, J. J. (1984). The organization of microtubules and microtubule coils in giant platelet disorders. *Am J Pathol, 116,* 514–522.

38. White, J. G., & Krivit, W. (1967). An ultrastructural basis for the shape change induced in platelets by chilling. *Blood, 30,* 625–635.

39. White, J. G. (1968). Effects of colchicine and vinca alkaloids on human platelets: I. Influence on platelet microtubules and contractile function. *Am J Pathol, 53,* 281–291.

40. Zucker, M. B., & Borrelli, J. (1954). Reversible alterations in platelet morphology produced by anticoagulants and by cold. *Blood, 9,* 602–608.

41. White, J. G., Krumwiede, M., & Escolar, G. (1999). EDTA induced changes in platelet structure and function: Influence on particle uptake. *Platelets, 10,* 327–337.

42. Smith, C. M., II, Burris, S. M., Weiss, D. J., & White, J. G. (1989). Comparison of bovine and human platelet deformability, using micropipette elastimetry. *Am J Veter Res, 50,* 34–38.

43. Takeuchi, K., Kuroda, K., Ishigami, M., & Nakamura, T. (1990). Actin cytoskeleton of resting bovine platelets. *Exp Cell Res, 186,* 374–830.

44. Winokur, R., & Hartwig, J. (1995). Mechanism of shape change in chilled human platelets. *Blood, 85,* 1796–1804.

45. White, J. G., & Rao, G. H. R. (1998). Microtubule coils versus the surface membrane cytoskeleton in maintenance and restoration of platelet discoid shape. *Am J Pathol, 165,* 597–610.

46. White, J. G., & de Alarcon, P. (2002). Platelet spherocytosis: A new bleeding disorder. *Am J Hematol, 70,* 158–166.

47. Fox, J. E. B., & Phillips, D. R. (1981). Inhibition of actin polymerization in blood platelets by cytochalasins. *Nature, 292,* 650–651.

48. White, J. G. (1968). Fine structural alterations induced in platelets by adenosine diphosphate. *Blood, 31,* 604–622.

49. White, J. G., & Krumwiede, M. (1987). Further studies of the secretory pathway in thrombin-stimulated human platelets. *Blood, 69,* 1196–1203.

50. White, J. G. (1968). Effects of colchicine and vinca alkaloids on human platelets. I. Influence on platelet microtubules and contractile function. *Am J Pathol, 53,* 281–291.

51. White, J. G. (1971). Platelet microtubules and microfilaments: Effects of cytochalasin B on structure and function. In J. Caen (Ed.), *Platelet aggregation* (pp. 15–52). Paris: Masson & Cie.

52. Lefebvre, P., White, J. G., Krumwiede, M. D., & Cohen, I. (1993). Role of actin on platelet function. *Eur J Cell Biol, 62,* 194–204.

53. White, J. G. (1999). Platelet glycosomes. *Platelets, 10,* 242–246.

54. White, J. G. (2004). Medich giant platelets disorder: A unique α granule deficiency. I. Structural abnormalities. *Platelets, 15,* 345–354.

55. Morgenstern, E. (1982). Coated membranes in blood platelets. *Euro J Cell Biol, 26,* 315–318.

56. Behnke, O. (1989). Coated pits and vesicles transfer plasma components to platelet granules. *Thromb Haemost, 62,* 718–722.

57. White, J. G., Key, N. S., King, R. A., & Vercellotti, G. M. (2004). The white platelet syndrome: A new autosomal dominant platelet disorder. 1. Structural abnormalities. *Platelets, 15,* 173–184.

58. White, J. G. (2002). Electron dense chains and clusters in human platelets. *Platelets, 13,* 317–325.

59. White, J. G., & Gerrard, J. M. (1980). The cell biology of platelets. In G. Weissman (Ed.), *Handbook of inflammation: The cell biology of inflammation* (pp. 83–143). New York: Elsevier/North Holland.

60. King, S. M., & Reed, G. L. (2002). Development of platelet secretory organelles. *Sem in Cell and Deve Biol, 13,* 293–302.

61. Reed, G. L. (2004). Platelet secretory mechanisms. *Sem Thromb Hemost, 30,* 441–450.

62. White, J. G., & Clawson, C. C. (1980). Development of giant granules in platelets during prolonged storage. *Am J Pathol, 101,* 635–645.

63. Krishnamurti, L., Neglia, J. P., Nagarajan, R., Berray, S. A., Lohr, J., Hirsh, B., & White, J. G. (2001). Paris–Trousseau syndrome platelets in a child with Jacobsen's syndrome. *Am J Hematol, 66,* 295–299.

64. Raccuglia, G. (1971). Gray platelet syndrome: A variety of qualitative platelet disorder. *Am J Med, 95,* 455–462.

65. White, J. G. (1979). Ultrastructural studies of the gray platelet syndrome. *Am J Pathol, 95,* 455–462.

66. White, J. G. (1968). Tubular elements in platelet granules. *Blood, 32,* 148–156.

67. Berger, G., Masse, J. M., & Cramer, E. M. (1996). Alpha-granule membrane mirrors the platelet plasma membrane and contains the glycoproteins Ib, IX and V. *Blood, 87,* 1385–1395.

68. White, J. G. (1992). The dense bodies of human platelets. In K. M. Myers & C. D. Barnes (Eds.), *The platelet amine storage granule* (pp. 31–50). Boca Raton, FL: CRC Press.

69. White, J. G. (2002). Electron dense chains and clusters in human platelets. *Platelets, 13,* 317–325.

70. White, J. G. (2003). Electron dense chains and clusters in platelets from patients with storage pool deficiency disorders. *J Thromb Haemost, 1,* 1–6.

71. Statland, B., Heagan, B., & White, J. G. (1969). Uptake of calcium by platelet relaxing factor. *Nature, 223,* 521–522.

72. Athota, K. P., McKenzie, M. M., Colomeni, E. P., Smith, M. R., & Gunning, W. T. (1998). Decreased adenine nucleotide content in platelet dense granule volume deficiency: Confirmation of a new bleeding defect. *Blood, 92,* 34a.

73. Lachant, N. A., & Gunning, W. T. (2001). Platelet dense granule distribution in normal subjects and individuals with storage pool disease. *Blood, 98,* 252a.

74. Lewis, J. C., Maldonado, J. E., & Mann, K. G. (1976). Phagocytosis in human platelets: Localization of acid phosphatase-positive phagosomes following latex uptake. *Blood, 47,* 833–840.

75. White, J. G. (2005). Platelets are covercytes, not phagocytes: Uptake of bacteria involves channels of the open canalicular system. *Platelets, 16,* 121–131.

76. White, J. G. (2005). Golgi complexes in hypogranular platelet disorders. *Platelets, 16,* 51–60.

77. Behnke, O. (1967). Electron microscopic observations on the membrane systems of the rat blood platelet. *Anat Rec, 158,* 121–137.

78. Behnke, O. (1968). An electron microscope study of the megakaryocyte of the rat bone marrow: I. The development of the demarcation membrane system and the platelet surface coat. *J Ultrastruc Res, 158,* 121–137.

79. Breton–Gorius, J. (1975). Development of two distinct membrane systems associated in giant complexes in pathological megakaryocytes. *Sem Haematol, 8,* 49–67.

80. White, J. G. (1972). Interaction of membrane systems in blood platelets. *Am J Pathol, 66,* 295–312.

81. White, J. G. (1999). Platelet membrane interactions. *Platelets, 10,* 368–381.

82. Frojmovic, M. M., Wong, T., & White, J. G. (1992). Platelet plasma membrane is equally distributed between surface and osmotically-evaginable surface-connecting membrane, independent of size, subpopulation or species. *Nouv Rev Fr Hematol, 34,* 99–110.

83. White, J. G. (1972). Uptake of latex particles by blood platelets: Phagocytosis or sequestration? *Am J Pathol, 69,* 439–458.

84. White, J. G., & Escolar, G. (1991). The blood platelet open canalicular system: A two-way street. *Euro J Cell Biol, 56,* 233–242.

85. Escolar, G., & White, J. G. (1991). The platelet open canalicular system: A final common pathway. *Blood Cells, 17,* 467–485.

86. White, J. G. (1987). The secretory pathway of bovine platelets. *Blood, 69,* 878–885.

87. Escolar, G., Leistikow, E., & White, J. G. (1989). The fate of the open canalicular system in surface and suspension-activated platelets. *Blood, 74,* 1983–1988.

88. White, J. G., & Clawson, C. C. (1982). Effects of small latex particle uptake on the surface connected canalicular system of blood platelets: a freeze-fracture and cytochemical study. *Diagno Histopathol, 5,* 3–10.

89. Youssefian, Y., Drouin, A., Masse, J. M., Guichard, J., & Cramer, E. M. (2002). Host defense role of platelets: Engulfment of HIV and *Staphylococcus aureus* occurs in a specific subcellular component and is enhanced by platelet activation. *Blood, 99,* 4021–4029.

90. White, J. G., & Gerrard, J. M. (1976). Ultrastructural features of abnormal blood platelets. A review. *Am J Pathol, 83,* 589–632.

91. Gerrard, J. M., White, J. G., Rao, G. H., & Townsend, D. (1976). Localization of platelet prostaglandin production in the platelet dense tubular system. *Am J Pathol, 83,* 283–298.

92. White, J. G., & Clawson, C. C. (1980). The surface-connected canalicular system of blood platelets — A fenestrated membrane system. *Am J Pathol, 101,* 353–364.

# CHAPTER 4

# The Platelet Cytoskeleton

## John H. Hartwig

*Hematology Division, Brigham and Women's Hospital, Boston, Massachusetts*

## I. Introduction

Platelets are subcellular fragments released from megakaryocytes that circulate in blood as small discs. Their small size and discoid shape result in their being pushed to the vessel edge by blood flow, which posits them near the apical surface of the endothelium, allowing them to be in the right place to detect and respond to vascular damage rapidly. When encountering such surface-bound activating factors as linearized von Willebrand factor (VWF) bound to collagen and/or soluble factors released into blood, platelets avidly react, bind, spread, secrete, and interact with one another and with fibrin to form a plug that seals the damaged surface. Plug formation requires platelets to undergo rapid morphological changes as they convert from their resting discoid forms to their active shapes. Spreading allows platelets to flatten over the damaged surface, whereas the elaboration of long filopods facilitates the recruitment of additional platelets into the wound site. Recruitment of additional cells is accomplished by both the delivery of P-selectin receptors to the platelet surface and the release of attractive molecules such as ADP and serotonin from platelet granules during secretion, by the production and release of thromboxane, and by a conformational change that activates the major surface platelet integrin, αIIbβ3, the receptor for fibrinogen/fibrin.

The unique discoid shape of the circulating platelet is maintained by a sturdy internal cytoskeleton composed of the polymers of actin and tubulin and their associated proteins. Shape change requires the remodeling of the resting cytoskeleton and the assembly of new cytoplasmic actin filaments. This chapter discusses the cytoskeletal fibers that support the resting platelet shape and reorganize to transform the cell into its active form.

## II. The Structure of the Resting Platelet

Platelets release from the ends of megakaryocyte proplatelets (see Chapter 2) as discs with dimensions of approxi-

mately 3.0 by 0.5 μm (Fig. 4-1A) and normally circulate in humans for around 7 days. Although platelets vary somewhat in size and in their granule contents, there is remarkable consistency in the internal cytoskeleton that both supports and provides the cell with its discoid shape. The surface of each disc is featureless, lacking protrusions, except for membrane infoldings, which are the entrances into an extensive conduit system of internal membranes called the open canalicular system (OCS). These entrances, which appear as pits in metal replicas of the platelet surface (Fig. 4-1C), are semiselective, allowing small molecules to enter, but restricting the entry of larger proteins such as antibodies. Whether the composition of the OCS lipid membrane or its glycoprotein (GP) content is identical to that of the plasma membrane is unknown. The OCS itself is an anastomosing network of membrane channels or tubules that run throughout the platelet. In the active platelet, the OCS serves as a conduit into which granules fuse and release their contents, and as a source of surface membrane for cell spreading.

Packaged internally in the cytoplasm are specific platelet granules (Fig. 4-1A) and normal cellular organelles such as mitochondria, lysosomes, and residual packages of endoplasmic reticulum membrane classically called the dense membrane system (see Chapter 3). Granules are of two types: α and dense (see Chapters 3 and 15). α Granules are the larger of the two (0.2–0.4 μm in diameter), store matrix adhesive proteins, and have glycoprotein receptors embedded in their limiting membranes that promote adhesion between platelets and the matrix. In particular, P-selectin, not expressed on the surface of the resting platelet, is stored in the membranes of the α granules (see Chapter 12) as well as a portion of the major platelet adherence receptors, GPIb-IX-V (a receptor for VWF, see Chapter 7) and the integrin αIIbβ3 (the receptor for fibrinogen, see Chapter 8). Adhesive components within the granules include the matrix adhesive proteins, fibrinogen, fibronectin, thrombospondin, vitronectin, and VWF. The second and smaller type of platelet granule is the dense granule. Each platelet contains a

75

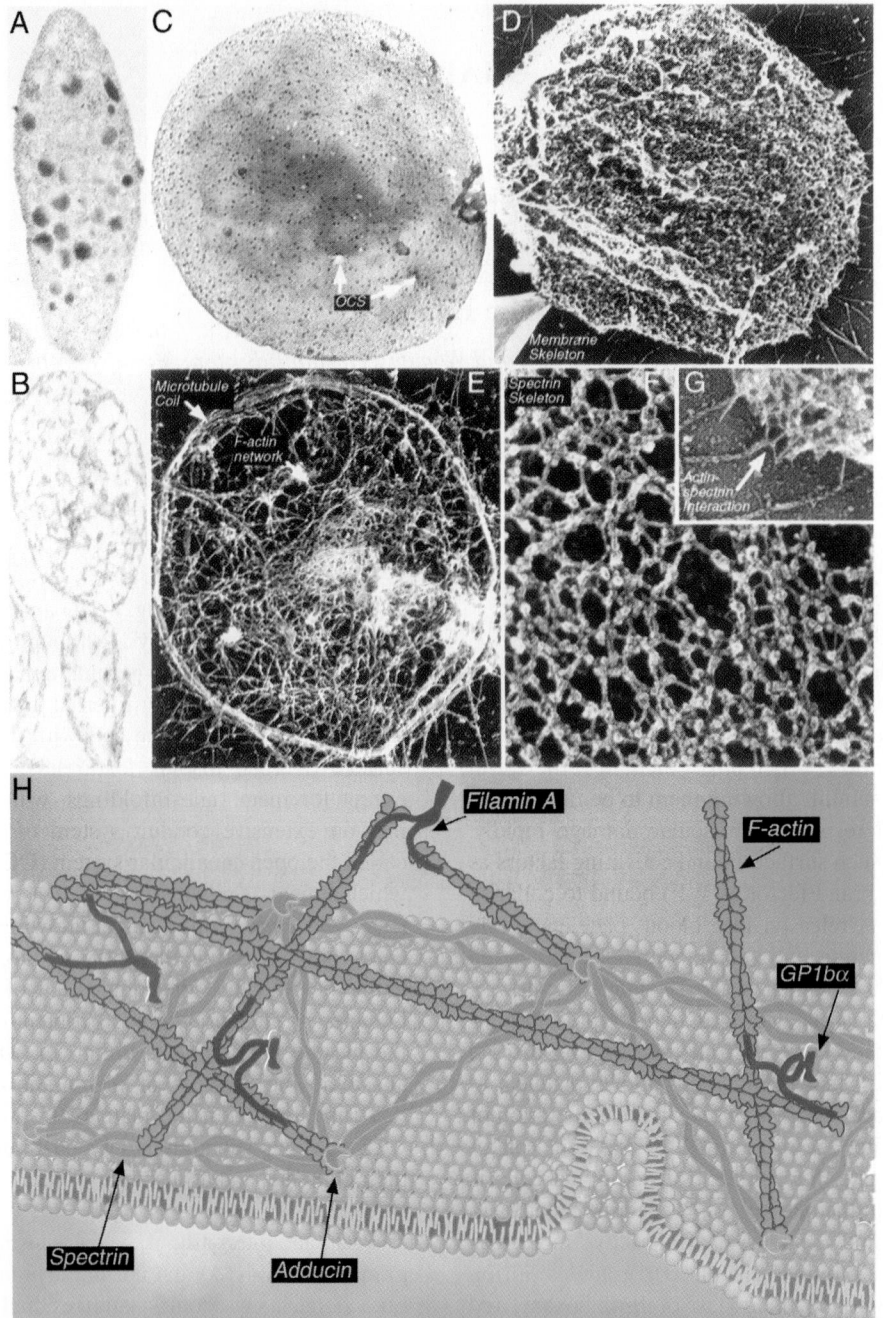

**Figure 4-1.** Structure of the resting platelet. A. Transection of a resting platelet. The resting cell contains numerous α granules, a few dense granules, mitochondria, and membrane-lined tunnels that correspond to the open canalicular system (OCS). The microtubule coil is cut in cross-section at each pole of the cell. B. Structure of the cytoskeleton of the resting cell after embedding and thin sectioning. Platelets were extracted with Triton X-100 in buffers that preserve the actin cytoskeleton. In thin section, the membrane skeleton appears as dense fuzz that lines the cell surface. The actin cytoskeleton within the platelet cytoplasm appears as a sparse fibrous network. C. Surface of the resting platelet. Rapid freezing, freeze-drying, and metal coating of platelets reveal that their disc surface is smooth and lacks membrane protrusions. The pitlike openings into the OCS (white arrows) are apparent. The plasma membrane has a granular surface containing small particles that correspond to extracellular domains of membrane glycoproteins. D. The membrane skeleton of the resting platelet. Treatment of platelets with nonionic detergents such as Triton X-100 reveals the spectrin-based membrane skeleton. Seen from the surface, the membrane skeleton is an extremely dense network that prohibits visualization of the underlying actin filaments. E. Microtubule coil and actin cytoskeleton. The underlying actin and tubulin cytoskeleton can be separated from the membrane skeleton (D and F) by mechanical shearing forces. A single microtubule is coiled 8 to 12 times. The actin cytoskeleton is a three-dimensional network of approximately 1-μm-long filaments. F. The fine structure of the spectrin-based membrane skeleton. Removal of some of the actin filaments and the bulk of the membrane glycoproteins reveals details of the spectrin lattice. This lattice is composed of thin, elongated strands decorated by globular particles. These may correspond to the membrane anchors for the lattice. G. Attachment of the spectrin-based membrane skeleton to actin filaments. Multiple spectrin molecules bind to the barbed ends of actin filaments. The polarity of actin filaments was determined using myosin S1, which decorates filaments, allowing a pointed and barbed end to be defined. H. Diagram summarizing the organization of the platelet spectrin-based membrane skeleton. Spectrin strands laminate the cytoplasmic surface of the plasma membrane and OCS. Strands interconnect using the barbed ends of actin filaments. Adducin binding to the barbed actin filament end helps target them to spectrin. Actin filaments are crosslinked by filamin, which provides the major membrane–cytoskeletal connection linking actin to the cytoplasmic tail of the GPIbα chain of the von Willebrand factor receptor.

small number of these granules, which are approximately 0.15 μm in diameter. Dense granules have electron opaque cores and carry the soluble activating agents ADP and serotonin, as well as divalent cations (see Chapters 3 and 15). A small amount of the total platelet P-selectin is stored in the membrane of the dense granules.

## III. The Cytoskeleton of the Resting Platelet

The cytoskeleton (Figs. 4-1B–H) serves as a system of molecular struts and girders that defines the discoid shape of the resting platelet and maintains cell integrity as platelets encounter high fluid shear forces generated by blood flow over the endothelium. Critical components of this system are, from the plasma membrane inward, a spectrin-based skeleton[1,2] (Figs. 4-1B and D) that is adherent to the cytoplasmic side of the plasma membrane, a microtubule coil that runs along the perimeter of the disc[3] and hence lines the thin axis of the cell (Fig. 4-1E), and a rigid network of crosslinked actin filaments that fills the cytoplasmic space of the cell (Fig. 4-1E). Actin filaments are polarized structures. The polarity can be defined using the stereospecific binding of myosin heads to the filament.[4] Myosin heads decorate the filament (Fig. 4-1G) allowing a pointed and barbed end to be defined. Functionally, this definition is of importance. The barbed filament end is the only end in cells onto which actin assembles and has the higher affinity for actin monomers. This is the end of the actin filament to which the cytochalasin family of actin inhibitors binds. The pointed end has the lower affinity for monomers and is the end from which actin disassembles in cells. Platelets express high levels of cytoskeletal proteins, and the amount and function of these proteins are listed in Table 4-1.[5–58]

### A. The Spectrin-Based Membrane Skeleton

Platelets are one of only two cells with a membrane skeleton that has been visualized at high resolution, the other being the erythrocyte. Both the erythroid and the platelet membrane skeleton are two-dimensional assemblies of spectrin strands (Fig. 4-1F) that interconnect to one another at their ends by binding to actin filaments. Each molecular end of the spectrin molecule has an actin binding site.

In the erythrocyte, the 200-nm-long spectrin strands observed in the electron microscope are tetramers composed of head-to-head aggregates of αβ chains.[59–63] The individual subunit chains self-associate first into heterodimers in a side-by-side configuration and an antiparallel fashion with respect to their amino and carboxyl termini. Heterodimers then further associate to form bipolar heterotetramers. The amino terminus of the α chain[64] and the carboxyl terminus of the β chain interact to form a self-association site, whereas the amino terminus of the β subunit

of each heterodimer contains the actin binding site.[65] Binding sites for membrane glycoprotein attachments reside along the side of each heterodimer and near the actin binding domain of the β chain. A binding site for ankyrin sits on the β chain,[66] near its carboxyl terminus, which connects spectrin to the integral membrane glycoprotein, band 3, and anchors the spectrin strands on the cytoplasmic surface of the plasma membrane. Protein 4.1 binds to the β chain near the actin binding site. This interaction helps both to stabilize the actin–spectrin interaction and to link it to glycophorin C. The actin binding sites are used to propagate the spectrin network. In the erythrocyte membrane skeleton, the actin joints between spectrin strands are short filaments made from 14 monomers of actin, each 37 nm in length.[67,68] Their short size is regulated by limiting the amount of actin protein expressed through the binding of two tropomyosin molecules to the sides of the oligomer, one in each helix groove of the filament, and by the capping of both their pointed and barbed ends by tropomodulin 1 and 3[45,69,70] and αβ-adducin[71–73] respectively.

Platelets express approximately 2000 spectrin molecules, which are composed of the nonerythroid subunits.[2,74,75] Although considerably less is known about how the spectrin–actin network forms and is connected to the plasma membrane in the platelet relative to the erythrocyte, certain differences between the two membrane skeletons have been defined. First, the spectrin strands composing the platelet membrane skeleton interconnect using the ends of long actin filaments instead of short actin oligomers, as in the erythrocyte (Figs. 4-1G and H).[75] The actin filament ends arrive at the plasma membrane originating from filaments in the cytoplasm (Fig. 4-1G). Hence, the spectrin lattice and the actin network in reality form a single, continuous ultrastructure. Second, tropomodulin 1, a protein that caps in conjunction with tropomyosin the pointed end of erythroid actin and stabilizes the short filaments in erythrocytes, is not expressed in platelets. In platelets, a portion of the pointed ends are capped by the Arp2/3 complex[4] instead, and biochemical experiments have revealed that a substantial number (approximately 2000) of these ends are free in the resting platelet.[76] Third, although little tropomodulin protein is expressed in platelets, αγ-adducins are abundantly expressed, and this protein caps many of the barbed ends of the filaments composing the resting actin cytoskeleton.[9] Capping of barbed filament ends by adducin also serves the function of targeting them to the spectrin-based membrane skeleton, because the affinity of spectrin for adducing–actin complexes is greater than for either actin or adducin alone.[72,77–79] Platelet glycoproteins involved in attaching spectrin to the membrane remain to be defined. Nonerythroid β-spectrin subunits contain a pleckstrin homology domain of approximately 100 amino acids inserted near their C-terminus that will associate spectrin molecules with polyphosphoinositides that densely coat the cytoplasmic side of the plasma

JOHN H. HARTWIG

## Table 4-1: Platelet Cytoskeletal Components

| Protein | Molecules/ Platelet | Molecules in Resting Cytoskeleton | Molecules in Active Cytoskeleton | Regulation of Cytoskeletal Activity[a] | Function | References |
|---|---|---|---|---|---|---|
| Actin<br>F-actin | $2 \times 10^6$ | $8 \times 10^5$<br>25,000 | $1.6 \times 10^6$<br>10,000–20,000 | | Cellular and intracellular movements | 5, 6 |
| GPIbαβ-IX– V | 25,000 | 12,000–25,000 | 12,000–25,000 | 14–3-3ξ phosphorylation | VWF receptor, attach filamin to membrane | 7, 8 |
| Nonerythroid spectrin | 2,000 | 2,000 | 2,000 | Proteolysis | Assemble 2-D membrane skeleton; bind F-actin ends and adducin | 9 |
| Adducin<br>α<br>γ | 8,000 | 6,000 | 2,500 | −Calcium<br>−Phosphorylation<br>−Phosphoinositides | Target barbed actin filament ends to spectrin, cap barbed ends | 9 |
| Filamin<br>FLNa<br>FLNb<br>FLNc | 13,000<br>12,000<br>500<br>? | 6,500 | 10,000 | | Attach F-actin to GPIbα, crosslink F-actin, target partner proteins to the membrane–cytoskeletal interface | 10–13 |
| CapZ | 20,000 | 5,000 | 14,000 | −Phosphoinositides | Terminate actin filament assembly | 14 |
| Arp2/3 complex | 8,000 | 2,500 | 5–6,000 | +WASp family members, cortactin<br>+Phosphoinositides | Actin nucleation, pointed end capping, filament crosslinking | 15–17 |
| Gelsolin | 20,000 | >100 | 2,000, 10,000 G-A complexes | + Calcium<br>−Phosphoinositides | Sever actin filaments, increase barbed ends | 18–23 |
| Profilin | 150,000 | | | −Phosphoinositides | Transfer of actin monomer to barbed filament ends | 24, 25 |
| Cofilin | 100,000 | >100 | 10,000 | −Phosphorylation, ppIs | Increase filament dynamics by promoting filament disassembly | 26–28 |
| β4-thymosin | $2 \times 10^6$ | | | | Monomer sequestration | 29–32 |
| α-Actinin | 12,000 | 1,200 | 8,000 | −Calcium | | 33 |
| Myosin II | 12,000 | 5,000 | | + Calcium (+myosin light-chain kinase) +/−Phosphorylation | Filament sliding in the direction of the pointed ends | 34, 35 |
| Talin | | | | +Phosphorylation | Integrin binding protein involved in adhesion site function | 36–38 |
| Vinculin | | | | + Phosphorylation + Phosphoinositides | Actin filament binding, targeting of Arp2/3 complex | 39 |
| Paxillin | | | | | | 40, 41 |
| Zyxin | | | | | | 42–44 |

**Table 4-1: Platelet Cytoskeletal Components—*Continued***

| Protein | Molecules/ Platelet | Molecules in Resting Cytoskeleton | Molecules in Active Cytoskeleton | Regulation of Cytoskeletal Activity[a] | Function | References |
|---|---|---|---|---|---|---|
| Tropomodulin 3 | | | | | | 45 |
| VASP | | | | + Phosphorylation | Increase actin monomer concentration locally, prevent capZ binding to filament barbed ends, associate with adhesion site partners using EVH domain | 43, 46–50 |
| Cortactin | | | | ?Phosphorylation | Activate the Arp2/3 complex | 51, 52 |
| WASp WAVE | | | | +Cdc42 binding +Phosphorylation +ppIs | Activate the Arp2/3 complex | 53–57 |
| Wip | | | | | WASp partner | 58 |

[a]+, positive; −, negative modulation.

G-A, gelsolin–actin; ppIs, polyphosphoinositides; 2-D, two-dimensional; VASP, vasodilator-stimulated phosphoprotein; VWF, von Willebrand factor; WASp, Wiskott–Aldrich syndrome protein; WAVE, WASP family verprolin–homologous protein.

membrane of the resting platelet,[80] and thus may help to bring spectrin to the membrane.

## B. The Cytoplasmic Actin Network

Actin is the most abundant of all the platelet proteins. The concentration of actin in a platelet is 0.55 mM, which translates into approximately two million copies per platelet.[5,6] Of these molecules, 800,000 assemble to form the 2000 to 5000 linear actin polymers that form the cytoskeleton of the resting cell.[75] All evidence indicates that the filaments of the resting platelet are interconnected at various points to form a mechanically rigid cytoplasmic network, because platelets express high concentrations of actin crosslinking proteins including filamin (FLNa and FLNb)[10,81–83] and α-actinin[33] (each is present at a molar ratio to actin of 1 : 100). Both FLN and α-actinin are homodimers in solution.

FLN homodimers are elongated strands in solution (Figs. 4-2A and B).[84–89] Subunits are composed of a flexible backbone of 24 repeats, each approximately 100 amino acids in length, that fold into IgG-like β-barrels[84,90] connected to an amino terminus actin binding site that shares homology with other actin binding proteins of the spectrin family. The repetitive nature of the backbone is interrupted by insertions between repeats 15 and 16, and 23 and 24. The C-terminal repeat (24) of subunits self-associates to form homodimers (Fig. 4-2). Molecules tend to assume V shapes — the self-association site is the vertex of the V, whereas the actin binding sites are at the free ends. Inclusion of the first hinge insertion in FLN depends on alternative RNA splicing. There are three FLN genes (X, 3, and 7) corresponding to FLNa, FLNb, and FLNc, and all three filamin isoforms have been reported to be expressed in platelets.[11] FLNa is the most abundant form expressed in platelets (12,000 molecules/ platelet) and is in approximately 20-fold excess of FLNb. Only the FLNb isoform lacking the first hinge is expressed.[11] Filamin is now recognized to be a prototype "scaffolding molecule" that collects binding partners (Fig. 4-2C) and localizes them adjacent to the plasma membrane.[10,83] Partners bound by filamin members include the small GTPase, ralA, rac, rho, and cdc42, with ralA binding in a GTP-dependent manner;[91] the exchange factors Trio and Toll; a rhoGAP called FILGAP (Ohta and colleagues, manuscript submitted); kinases and phosphatases; and transmembrane proteins. The majority of these partner proteins interact within the carboxyl terminal third of FLNa. Of particular importance for the structural stability of the resting platelet is an interaction that occurs between FLNa/b and the cytoplasmic tail of the GPIbα subunit of the VWF receptor (VWFR). There is a binding site on FLNa repeat 17 for residues 556–575 of the cytoplasmic tail of GPIbα.[87,88] FLNa repeats are β-barrel structures composed of eight short β-sheet strands interconnected by turn regions. The corresponding binding site on GPIbα for FLN is also a β-strand and posits in a hydrophobic pocket that forms between the C and D β-strands of FLNa repeat 17 (Fig. 4-2D).[92] The GPIbα binding

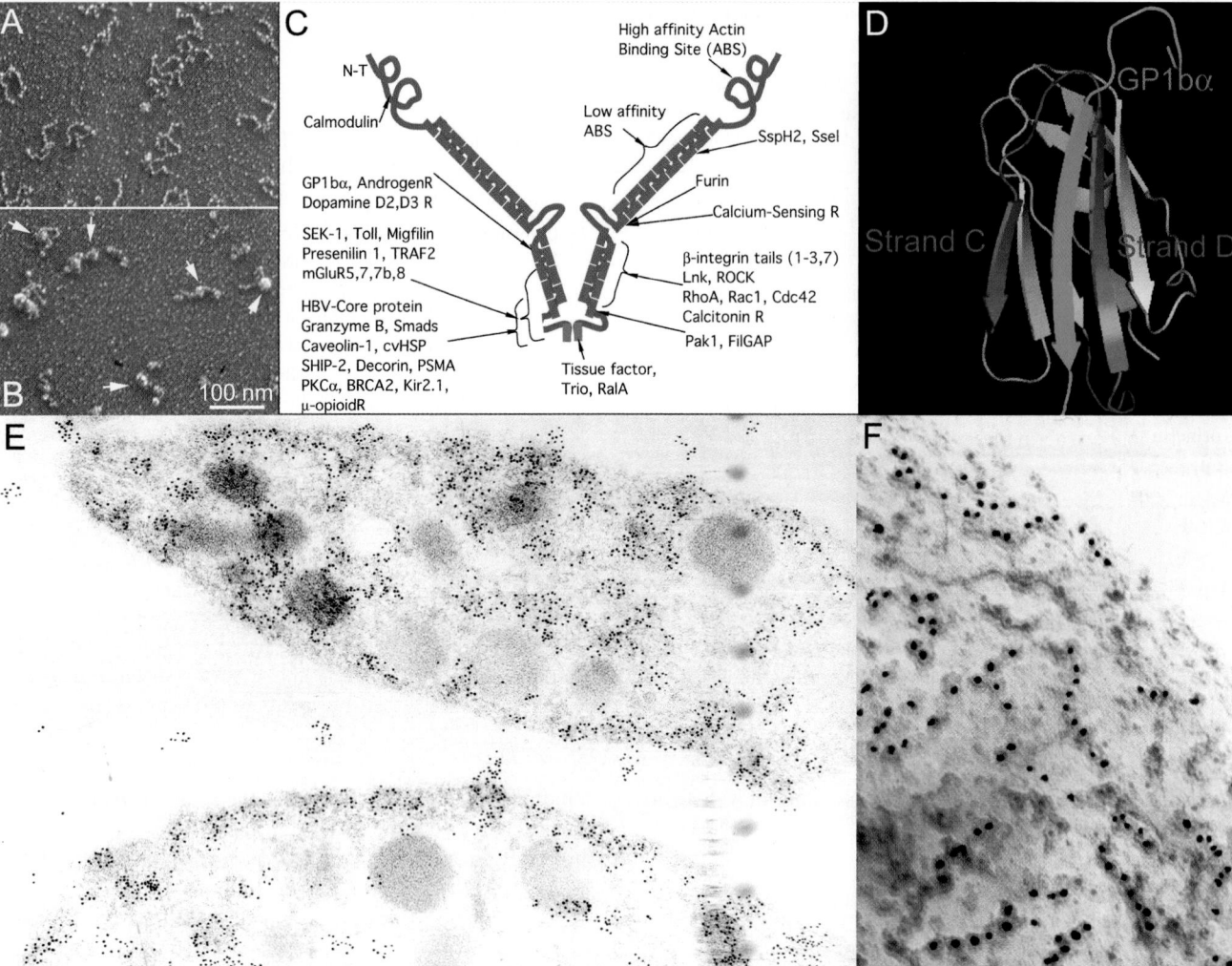

**Figure 4-2.** Role of filamin A (FLNa) in controlling the distribution of the von Willebrand factor receptor (VWFR, GPIb-IX-V complex) on the platelet surface. FLNa regulates the movement of VWFR in the plane of the plasma membrane by attaching it to underlying actin filaments. A. FLNa molecules are homodimers with contour lengths of 160 nm. B. When isolated platelet FLNa–VWFR complexes are visualized in the electron microscope, the VWFR complex sits near the molecular middle of the FLNa strands (white arrowheads). C. Structural organization of FLNa molecules and known binding partners. Each subunit has an N-terminal actin binding site joined to 24 repeat motifs, each approximately 100 residues in length.[84] The repeats are β-barrel structures that are believed to interconnect like beads on a string (see D). The subunits self-associate into molecules using the most C-T repeat motif. With many binding partners (>40) now described, FLNa participates in signaling cascades by spatially collecting and concentrating signaling proteins at the plasma membrane–cytoskeletal interface, and may possibly function as an organizing center for actin network rearrangements. The relative position of known binding partner proteins along the FLNa subunit is indicated. Important partner interactions that may be dependent on filamin include GTPase targeting, charging, and linkage of the actin cytoskeleton to membrane glycoproteins such as GPIbα and β-integrins. D. FLNa was first shown to bind to GPIbα in 1985 independently by Fox[85] and Okita and Colleagues.[86] Andrews and Fox[87,88] later delimited the binding region for FLNa on the cytoplasmic tail of GPIbα to approximately 20 residues (556–577) whereas Meyer and colleagues[89] identified FLNa repeats 17–19 to be the corresponding binding site on FLNa for the GPIbα tail. The FLNa binding site for GPIbα has been restricted to repeat 17, and the crystal structure of the repeat 17–GPIbα interfaces determined. FLNa repeat 17 has eight strands of β-structure that are separated by loops or turns. The GPIbα domain that interacts with FLNa is also a β-sheet, and it binds between the C and D strands of FLNa repeat 17 in a groove made by these strands. E, F. Consequences of the FLN–GPIbα interaction in the platelet. E. Immunogold electron micrographs showing the membranous localization of FLNa in the resting platelet. F. Organization of GPIb on the surface of the resting platelet. Staining of platelets with anti-GPIb antibodies and 10-nm immunogold. Linear arrays of gold label on the platelet surface reflect the underlying connections to actin filaments.

site on FLNb, also reported to bind GPIbα, is likely to be identical, because the critical binding sequences of the two repeat 17 differ by one conservative amino acid substitution. The platelet content of FLNc, and whether FLNc can bind GPIbα, remains to be determined. Biochemical experiments have shown that most of the platelet FLN (≥90%) is in complex with GPIbα.[93] This GP1bα–FLN interaction has three consequences. First, it posits FLN's self-association domain (Fig. 4-2C) and its associated partner proteins at the plasma membrane interface while dangling FLN's actin binding sites into the cytoplasm. Second, because a large fraction of FLN is bound to actin, it aligns much of the VWFR on the surface of the platelet over the underlying cytoplasmic filaments (Fig. 4-2F). Third, because the FLN linkages between actin filaments and the VWFR pass through the pores of the spectrin lattice, they restrain the molecular movement of the spectrin strands in this lattice and hold the lattice in compression (Fig. 4-1H).

Besides binding to FLN, the cytoplasmic tail of GP1bα as well as its associated GPIbβ subunit, binds to the adaptor protein 14-3-3ζ.[94] This interaction is positively regulated by phosphorylation. Release of 14-3-3ζ after engagement of the VWFR to the A1 domain of VWF may be critical for signals to pass from the VWFR to αIIbβ3.

The roles of the other actin crosslinking proteins in the actin organization of the resting cell are less well defined. As mentioned earlier, α-actinin is abundant in the platelet, and a portion of this protein is bound to the resting cytoskeleton.[33] The α-actinin molecule is a small rod-shaped homodimer composed of antiparallel and overlapped subunits. Like filamin, α-actinin has an amino terminal actin binding domain followed by a rod domain composed of six end-to-end linked repeats, each having approximately 100 amino acids. Each repeat is believed to be composed of three α-helical domains. High-resolution localization of α-actinin in the resting or active cytoskeleton has not been determined.

### C. The Microtubule Coil

Platelets, once released into blood by megakaryocytes (see Chapter 2), are thought to contain a single microtubule that is approximately 100 μm in length. To fit this length of microtubule inside the resting platelet, it is wound 8 to 12 times into a coil. The coil sits in the cytoplasm, just beneath the plasma membrane, along the thin edge of each disc (Fig. 4-1E) and is solely responsible for distorting the cell into a discoid shape. Microtubules are rigid, hollow polymers assembled from tubulin heterodimeric subunits. Each microtubule is composed of 13 stacks of αβ-tubulin subunits, each arranged in linear head-to-tail aggregates called *protofilaments*. Like actin, microtubules are dynamic, polarized structures with definable plus and minus ends. The plus end

has the higher affinity for tubulin dimmers, and growth at this end is blocked by colchicines and vinblastine. Platelets express four different β-tubulins[95] (β1, β2, β4, β5) at a total concentration of 70 μM. This yields $3 \times 10^5$ tubulin subunits per platelet, of which 1.6 to $2 \times 10^5$ are required to generate a 100-μm-long microtubule, in good agreement with estimates of 50 to 60% of the total platelet tubulin as polymer. β1 is the dominant isoform expressed, and gene knockout experiments have demonstrated that platelets lacking β1 are spherical, failing to develop the characteristic discoid shape of normal platelets because they have aberrant microtubule coils.[95,96] The microtubule forming the coil can be disassembled in platelets by chilling them — a phenomenon that is also associated with a rounding up of platelets.[97] A human polymorphism that changes Q43 to P has recently been discovered in β1-tubulin and may be carried by as many as 10% of the population.[98] Heterozygous carriers show substantially reduced levels of β1-tubulin expression and a large fraction of platelets that are spherocytic. Mouse platelets lacking β1-tubulin circulate only as spherocytes.

Microtubule dynamics and microtubule-based motors can be involved in certain aspects of cytoplasmic motility. Platelets have been shown to express both types of microtubule motors: cytoplasmic dynein[99] and kinesin. Dynein is a minus-end-directed motor[100] that can transport cargo vesicles toward the minus end,[101] or slide microtubules past one another. Kinesin is a plus-end-directed motor expressed in many different isoforms and is believed to be involved in vesicular trafficking in cells.[101] Whether the assembly–disassembly of microtubules or microtubule sliding, postulated to generate certain cytoplasmic movement such as chromosome separation,[102] is involved in aspects of platelet spreading remains to be established. Resting platelets maintain approximately 50% of their total tubulin as polymer, and changes in polymer content during activation have not been described. However, the rapid reorganization of the microtubule coil that occurs in the activated platelet suggests either that fragmentation of the coil into numerous smaller microtubules follows stimulation or that a portion of the coil disassembles while new individual microtubules assemble.

## IV. The Structure of the Activated Platelet

The major physiological function of circulating platelets is to detect damage to the walls of the blood vessels. This feat is accomplished by the expression of surface receptors that recognize exposed connective tissue components, normally covered by endothelial cells, and/or by the release of soluble factors by endothelial and other connective tissue cells that attract platelets. When damage is detected, platelets respond rapidly. They attach, change shape, and spread over the damaged area. Activation also initiates certain platelet responses including (a) secretion (Chapter 15), which moves

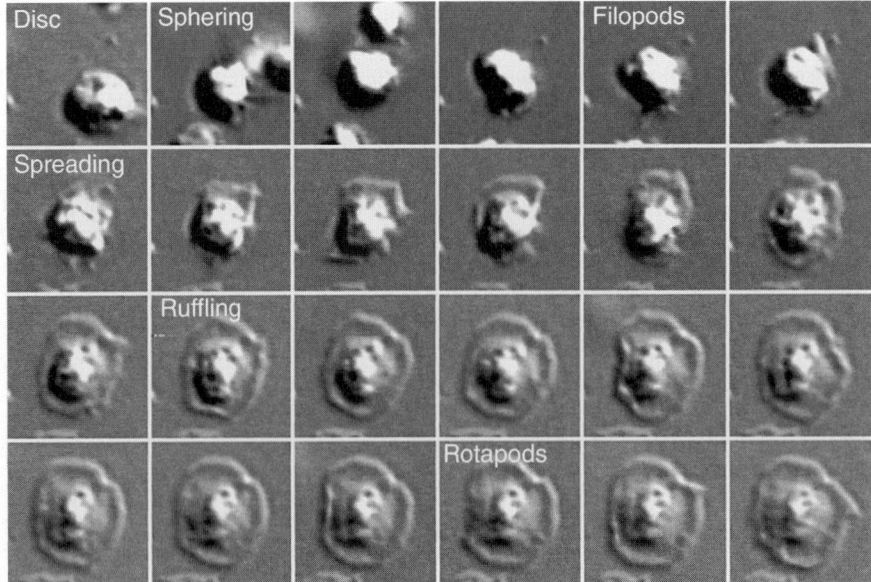

**Figure 4-3.** Structural changes that occur after surface activation, observed using differential contrast optics in a light microscope. The platelet was tethered to a glass surface coated with the A1 domain of VWF and then activated by ligation of the protease-activated receptor 1 receptor (addition of 25 μM TRAP peptide). The first change that can be recognized is a transformation from the discoid to a more spherical shape. Filopods are then elaborated from the surface of the spheres. The spreading of the platelet onto the surface follows filopodial growth. This occurs through the growth of many lamellipods, which then coalesce to form single, large circumferential lamellae. Spreading greatly increases the surface area of the cell. With continued incubation, ruffling activity is observed along the cell edges, and novel filopods are extended from the central dense aggregate and rotated around the cell. This figure shows video frames taken over a 15-minute time period.

adhesive receptors onto the platelet surface and releases agonist compounds to attract and activate additional platelets and leukocytes from the blood; (b) activation of the biochemical synthetic pathway that produces and releases thromboxane, another potent platelet agonist (Chapter 31); and (c) activation of the platelet surface integrins,[103] particularly the major platelet integrin, αIIbβ3,[104,105] the receptor for fibrinogen and VWF (Chapters 8 and 17). The αIIbβ3 receptor is the most abundant of all the platelet surface receptors, but is maintained in an inactive state in the circulating cell. After activation, however, signals are generated inside platelets that alter the conformation of αIIbβ3, permitting ligand binding (see Chapters 8 and 17).

### A. Mechanism of Shape Change

Platelet shape change is a complex process that depends on the cytoplasmic dynamics of the actin polymer.[106] To spread, the resting platelet must reorganize its cytoskeleton and assemble new actin filaments. Hence, proteins that regulate actin architecture and dynamics control shape change. Table 4-1 lists proteins important for this actin transformation in the blood platelet. Actin regulatory proteins that control filament dynamics can be classified by how they influence actin

filament assembly–disassembly.[107,108] Key proteins control the availability of filament ends, preventing monomer addition or subtraction, or influence the kinetics of monomer association, dissociation, and nucleotide content. Only monomers having bound ATP can efficiently assemble onto the barbed filament ends.[109,110] The function of these important proteins is discussed within the context of the shape transformation.

Platelet shape change follows a reproducible temporal sequence.[74] Figure 4-3 follows a platelet as it activates and spreads on a surface. The first event that is observed as the platelet makes contact with the surface is that the discoid shape is lost and it becomes rounded or spheroid in form. Next, fingerlike projections grow from the cell periphery. Then the platelet flattens over surfaces and broad lamellae are extended (Figs. 4-3 and 4-4). As the platelet flattens, granules and organelles are squeezed into the center of the cell, resulting in a fried egg appearance. Finally, a dynamic phase of membrane motility begins at various points along the lamellae; membrane ruffles form and retract inward. Unique blunt filopods extend from the cell center and are rotated around the cell periphery. In the electron microscope, a dense network of approximately 0.5-μm-long actin filaments is found to fill the lamellipodial spaces. These cortical networks derive from the assembly of new actin

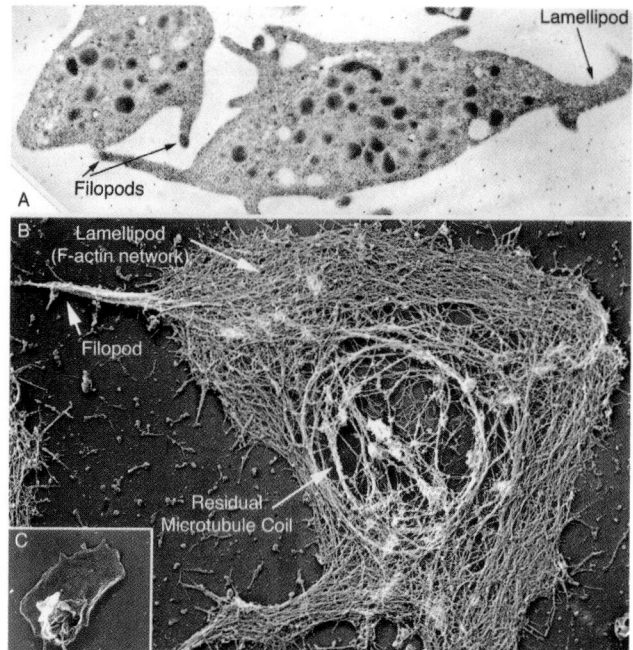

**Figure 4-4.** Structure of the activated platelet in the electron microscope. A. Thin section through platelets activated with thrombin for 15 seconds. At this time point, the cells have elaborated primarily filopods. One larger protrusion has also been extended. B. Cytoskeleton of the active platelet. Platelet spreading is accompanied by a striking reorganization of the actin cytoskeleton. Regions that correspond to lamellae are found densely filled with a three-dimensional network of short actin filaments. The filaments in these regions have been assembled after platelet activation and are not found if actin assembly is prevented using cytochalasin. Filopods that form at the cell edge are filled with long filaments with roots that are coalesced from filaments in the cytoskeletal center. In this specimen, the microtubule coil has been partially compressed into the cell center and some individual microtubule fibers can be observed to radiate outward from this residual coil. C. Surface of the activated platelet.

filaments, which doubles the filament content of the cell.[74] The filament assembly reaction is so robust that platelets have a modified secretory process. Exocytosis of granules occurs primarily in the OCS as the granules concentrate in the cell center, in part as a consequence of the assembly of this dense cortical actin network that prevents the close approach of granules to the plasma membrane. Filopods that grow from the cell edges are cored by bundles of long F-actin that coalesce from filaments originating in the cell center. Activation also reorganizes the microtubule coil of the platelet.[5] In some cells, as shown in Figure 4-4B, a residual microtubule coil remains partially intact and is compressed into the cell center. In other cells, the microtubule coil fragments and/or new microtubules are generated, as disparate microtubules are now found running from the

cell center into the newly elaborated cellular protrusions, particularly into the filopods.

As discussed earlier, the first change that can be recognized as the platelets assume new shapes is a disc-to-sphere transformation. Both the cytoplasmic signal that triggers this rounding process and the protein that affects the transformation are well understood. Platelet rounding is induced by a transient rise in cytosolic calcium that follows receptor ligation. Calcium mobilization into the cytosol is mediated by the activation of phospholipase C.[111] This enzyme hydrolyzes the membrane-bound polyphosphoinositide $(PI)_{4,5}P_2$ at the cytoplasmic surface of the plasma membrane to form the second messengers, diacylglycerol and inositol triphosphate ($IP_3$). Because $IP_3$ is soluble, it freely diffuses in the cytoplasm and moves to receptors on the dense membrane system of the platelet where it binds, initiating the release of stored calcium.[112] Intracellular calcium increases to near-micromolar levels after activation of this pathway, and further increases to near 10 μM may be reached by opening the calcium channels in the plasma membrane. Phospholipase Cβ is activated by the βγ-subunit of trimeric G-proteins, which couples it to the serpentine receptors on the platelet surface.[113] These receptors include the protease-activated receptor (PAR) family (PAR-1, PAR-4 in humans;[114–117] see Chapter 9), the ADP receptors ($P2Y_1$ and $P2Y_{12}$; see Chapter 10), and the serotonin receptor (see Chapter 6).

Calcium can also be mobilized through other receptors. Phospholipase Cγ activation is coupled to the immunoreceptor tyrosine-based activation motif (ITAM) domains of the γ-subunit associated with the collagen receptor, GPVI, and the platelet FcγRIIA receptor.[118] Phospholipase Cγ is activated through phosphorylation, but is recruited to the membrane only by the production of D3-containing phosphoinositides.

Blunting of the normal calcium transient in platelets by loading them with cytoplasmic calcium chelator (EGTA-AM) inhibits platelet rounding, and platelets treated in this fashion develop elongated, aberrant filopods.[76] The underlying basis of the aberrant protrusive behavior is a markedly diminished capacity to generate barbed ends for actin assembly, a consequence of the lack of the normal filament fragmentation mechanism used to generate new ends and to remodel the filaments composing the resting actin cytoskeleton.[119,120] Residual filament assembly from the ends of preexisting long filaments leads to the growth of the aberrant filopodia. Filaments that assemble under these conditions elongate in a direction parallel to the plasma membrane, which distorts the discoid shapes into barbell conformations, or forms abnormal filopods and blebs at the cell surface.

The protein activated by calcium to remodel the cytoskeleton is gelsolin. Gelsolin, a globular molecule of 81-kDa,[18,19,121] is an abundant platelet protein (5 μM, 20,000 copies per platelet).[122,123] First discovered in 1979 as a result of its ability to solate F-actin gels in a calcium-dependent fashion,[19] gelsolin has been extremely well characterized biochemically, its atomic structure solved,[124] and its physiological function in platelets and other cells established by gene knockout experiments. Gelsolin is composed of six repeat motifs and contains multiple calcium binding domains, two actin binding sites, one of which overlaps with a phospholipid binding domain.[125–134] In the absence of calcium, the repeats of gelsolin fold upon themselves to generate a globular structure that has its actin binding sites sequestered and, hence, a molecule that cannot bind to actin. Calcium binding opens the molecule, allowing actin access to its actin binding sites. Both of gelsolin's actin binding sites interact to generate the noncovalent actin filament-severing process. Gelsolin first binds along the filament side using one of its actin binding sites. Actin binding activates the second site, causing it to intercalate into the filament, mediating the severing process. After fragmenting a filament, gelsolin remains bound to the barbed end of the filament. The only agents that have been shown to disassociate gelsolin from these barbed ends are the phospholipids, the phosphoinositide family,[126] and lysophosphatidic acid.[135] More important, platelets from mice that lack gelsolin fail to undergo the normal shape change reaction.[120] Gelsolin-deficient platelets respond weakly to all agonists, spread poorly, and have a reduced capacity to nucleate and assemble actin filaments compared with normal wild-type platelets.[120] The normal filament-severing reaction that mediates the disc-to-sphere transition is all but absent in gelsolin-deficient platelets. The cortex of gelsolin[-/-] platelets remains replete with long actin filaments after stimulation with agonists, and these platelets have difficulty in generating actin networks and lamellae.

In the resting platelet, more than 98% of the cytoplasmic gelsolin is soluble in the cytoplasm and not associated with actin. Treatment of platelets with thrombin leads to the complexing of gelsolin to actin within seconds.[123,136] The kinetics of this reaction parallels calcium release, filament fragmentation, and cell rounding (Fig. 4-5). This interaction of gelsolin with actin is partially reversible and approximately 50% of the total gelsolin dissociates from actin, exposing barbed filament ends. Dissociation of gelsolin from actin correlates with the times of maximal actin filament assembly in the platelet.

### B. Actin Assembly and Cell Spreading

Actin filament assembly provides the force for protrusive activity in platelets. The assembly reaction, which doubles the filament content of cells, is driven by the generation of barbed-end nucleation sites after receptor ligation.[76] These nucleation sites are formed either by exposure of the barbed ends of preexisting actin filaments[76,137] or new sites formed

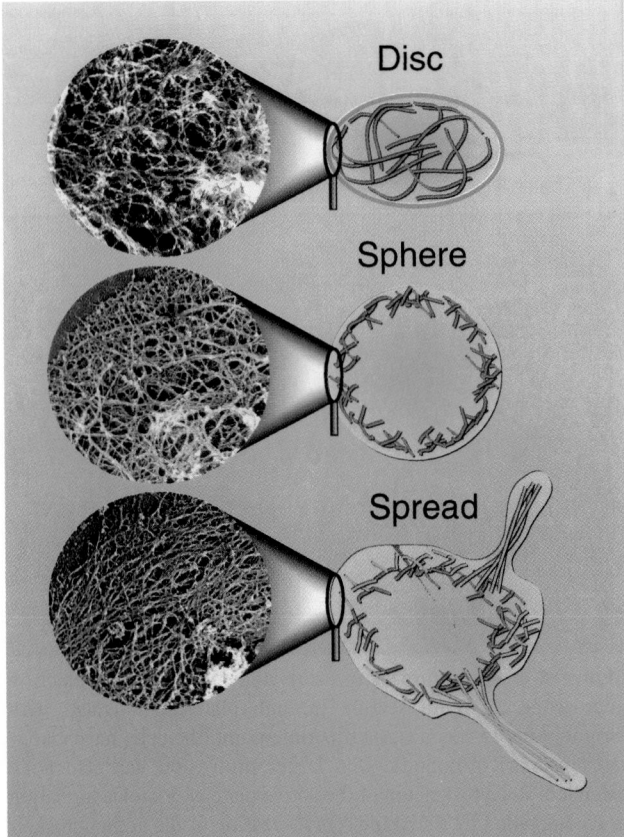

**Figure 4-5.** Cytoskeletal changes of the activated platelet. As platelets convert from discs into their activated forms, the internal actin cytoskeleton is remodeled. In the first step, which corresponds to the rounding phase, the long filaments that exist in the resting cell are converted into many short filaments. This process is mediated by the actin-severing property of gelsolin (see Figure 4-6) and requires intracellular calcium to increase to near-micromolar concentrations. Filament severing can be visualized by preventing the growth of new filaments with cytochalasin. To spread, the rounded platelet must assemble new filaments. It does this at the cell cortex by elongating the barbed end of the short filament fragments and by generating new barbed ends for monomer addition.

by the activation of the Arp2/3 complex, which are equivalent in function to the barbed ends of the preformed filaments.[16,17] Barbed ends alone have the capacity to initiate actin assembly because they have a higher affinity for actin subunits than do the proteins used by the cell to sequester them.

Two proteins are expressed in platelets with a primary function of interacting with and sequestering monomers of actin (Fig. 4-6). The first, and most abundant of the two, is β4-thymosin. Platelets contain 0.55 mM β4-thymosin, a concentration that ensures equilimolarity to actin.[29–32] β4-thymosin binds actin monomers with an affinity greater than the pointed filament end has for monomers, allowing it to

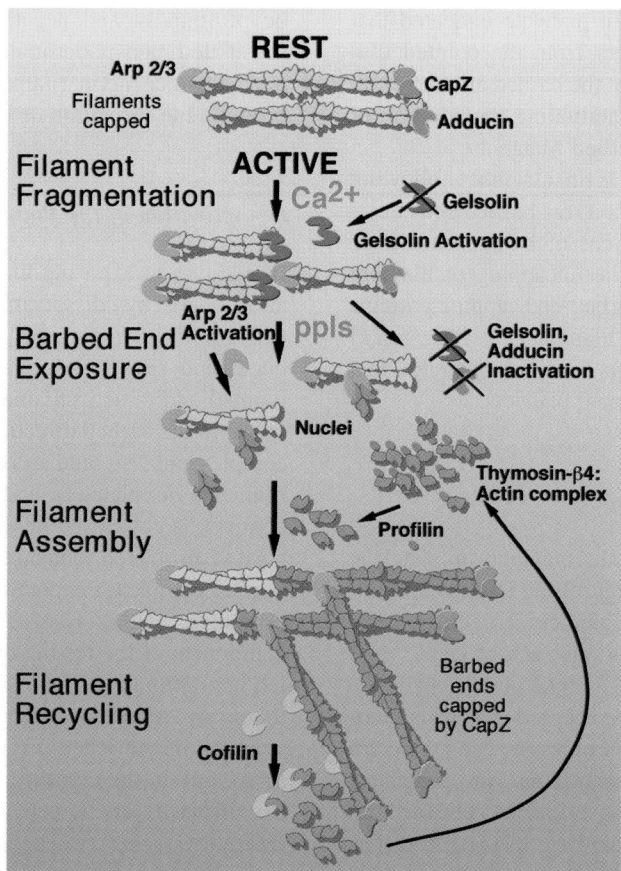

**Figure 4-6.** Regulation of the platelet actin assembly by actin regulatory proteins. (REST) Resting platelets contain approximately 2000 to 4000 actin filaments. More than 98% of these filaments have their barbed ends capped. Two proteins known to participate in capping the barbed ends of the resting platelet are adducin and capZ. Because a large fraction of the Arp2/3 complex is a component of the resting actin cytoskeleton (approximately 25%), it is possible that approximately half the filaments are also capped on their pointed ends by this complex. Approximately 2000 free pointed ends have been measured by biochemical assay. (ACTIVE) The first physiological event in cell activation is the release of internally stored calcium into the cytoplasm. One important target of calcium is the protein gelsolin. Calcium binding to gelsolin opens the molecule, allowing it to bind to F-actin, interdigitate into the filament, and sever it. This process fragments the preexisting filaments. After severing F-actin, gelsolin remains bound to the barbed filament end, preventing new assembly even after intracellular calcium returns to nanomolar levels. Net filament assembly begins when barbed ends are formed and exposed. Critical cofactors in this process are membrane-bound polyphosphoinositides — in particular, $PI_{4,5}P_2$, $PI_{3,4}P_2$, and $PI_{3,4,5}P_3$, as well as the WASp family of proteins. Polyphosphoinositides (ppIs) target filament polymerization, bring together regulatory components at the cytoplasmic surface of the plasma membrane, and release barbed-end capping proteins from filament ends. Targets of ppIs include gelsolin, capZ, adducin, and the Arp2/3 complex, where ppIs are a necessary cofactor for activation along with WASp family members or cortactin. The exposure of barbed ends is sufficient to initiate actin filament assembly. The driving force for assembly is a large pool of unpolymerized actin that is stored in complex with β4-thymosin. Actin, complexed to β4-thymosin, cannot add to the pointed end of actin filaments but, because the barbed end has a higher affinity for actin than β4-thymosin, will add to the barbed end. Addition at the barbed end is facilitated by profiling, which acts to shuttle actin subunits to actin filament barbed ends. Recapping of the barbed filament ends primarily by capZ terminates filament assembly.

drain monomers from this filament end, but of lower affinity for monomer than the barbed end, allowing filament assembly when barbed ends are exposed.[138] Because resting platelets have very few exposed barbed ends, β4-thymosin can maintain a large pool of unpolymerized actin, and 60% of the total actin in the platelet is bound to β4-thymosin. The affinity of β4-thymosin for monomer is strongly influenced by the nucleotide bound to actin. The affinity of β4-thymo-

sin is highest for actin with a bound ATP compared with actin with a bound ADP.[138] This suggests, based on the pool size, that β4-thymosin sequesters primarily ATP-actin in the platelet, the nucleotide-containing form of actin that is assembly competent for the barbed end. A second protein involved in sequestration of monomer and in stimulating the assembly of actin at the barbed ends of actin nuclei is profilin. Platelets contain a 50-μM concentration of profilin.[139–141] Early experiments in platelets indicated that profilin could sequester monomers from the pointed filament end like β4-thymosin, but not the barbed ends. Profilin also serves as a transfer factor, facilitating the addition of ATP-actin monomers onto the barbed filament end.

As discussed earlier, platelet actin assembly following the addition of agonists begins when free barbed end equivalents are generated.[137] Two sources of barbed ends have been identified in platelets: (a) the uncapping of filament ends previously sequestered by barbed-end capping proteins and (b) *de novo* nucleation of filaments by the Arp2/3 complex. Biochemical experiments have demonstrated that both pathways function in stimulated platelets.

## C. Barbed-End Capping/Uncapping

Barbed-end capping proteins are abundant in platelets, attesting to their importance in controlling the accessibility of these ends to regulate the dynamics of cellular actin. Platelets contain 5-μM concentrations each of capZ[142] and gelsolin,[123,143] and 3 μM adducin.[9] CapZ and adducin are constitutively active and bind to free barbed ends with nanomolar affinity.[144] Gelsolin is inactive unless cytosolic calcium increases to micromolar concentrations in platelets.[74] However, once active, filament severing by gelsolin both creates new barbed ends and caps them in marked difference to the activities of capZ or adducin, which can only bind to existing ends. In the resting platelet, all the gelsolin is inactive (i.e., free and not bound to the actin cytoskeleton). Because more than 98% of the barbed ends are capped in the resting cell,[75] adducin and capZ must account for this capping activity, and 80% and 35% of the total adducin and capZ are bound to the resting actin cytoskeleton respectively (Fig. 4-6, REST). Studies comparing the kinetics of association–dissociation of these capping molecules with actin in activated platelets revealed that capZ functions to terminate the actin filament assembly, and adducin helps to start this reaction by dissociating from the actin cytoskeleton. In support of this notion, the content of capZ in the cytoskeleton of the active platelet increases steadily as new actin filaments form until a steady state occurs when approximately 60% of the capZ is bound as the actin filament content plateaus at 80% of the total. Adducin, on the other hand, dissociates from the actin cytoskeleton as cells activate, decreasing to a minimum of approximately 25% bound after filament assembly is complete. Gelsolin also plays a major role in generating new filament ends for the actin assembly reaction in the active platelet. The calcium-induced severing increases the number of filaments (and barbed ends) by 5- to 10-fold. A large fraction of the gelsolin, all bound to actin in the first seconds after cell activation, partially dissociates. Dissociation temporally correlates with the time of maximal actin filament assembly. Filament capping by capZ can also

be profoundly and negatively influenced by vasodilator-stimulated phosphoprotein (VASP), and activated VASP may enhance actin filament length in cells and thereby promote the formation of filopods.[145]

## D. Activation of the Platelet Arp2/3 Complex

In addition to exposing filament ends as a source of nucleation sites by dissociating capping proteins, platelets activate the cytoplasmic Arp2/3 complex to create new nucleation sites.[16,17] Active Arp2/3 complexes mimic the pointed end of actin filaments to induce the growth of filaments in the barbed direction.[146] Because of this, the Arp2/3 complex can also bind and cap the pointed end of preformed filaments in platelets and in other cells.[147] The Arp2/3 complex is composed of seven polypeptides, two of which have actin-related sequences, named Arp2 and Arp3.[148] The Arp2/3 complex is expressed at high levels in platelets, with estimates ranging from 2 to 10 μM, and is a prominent component of the resting cytoskeleton. In the resting platelet, approximately 25 to 30% of the total Arp2/3 is bound to the cytoskeleton. After platelet activation, the Arp2/3 complex is enriched in the platelet actin cytoskeleton, increasing in the cytoskeleton by threefold and concentrating primarily at the cell periphery in zones of new actin assembly.

Different signaling pathways intersect in the platelet cytoplasm to regulate the activity of the Arp2/3 complex. A bacterial protein, ActA, was the first agent discovered to activate the Arp2/3 complex.[149] ActA plays a critical role in generating the actin tails that move the intracellular pathogen *Listera monocytogenes* through the cytoplasm of cells. This bacterial protein is expressed asymmetrically on the surface, predominantly at one pole, of this bacteria. Binding to ActA stimulates the ability of Arp2/3 to nucleate actin filaments. ActA-related sequences have now been identified in the proteome of mammalian cells. Both vinculin[150] and zyxin,[36,151] components of cellular adhesion sites, contain such domains, suggesting roles in activating the Arp2/3 complex at points of cellular adhesion. More recently, it has been recognized that Wiskott–Aldrich syndrome protein (WASp) family members activate the Arp2/3 complex.[53,152–157] The family of proteins includes WASp, N-WASp, and WAVE/SCAR. Mutations in the WASp gene result in Wiskott–Aldrich syndrome,[158–161] a severe, inherited X-linked recessive hematopoietic disorder characterized by T-cell immunodeficiency and thrombocytopenia (see Chapter 54). WASp is expressed specifically in hematopoietic cells;[162,163] N-WASp and WAVE/SCAR are ubiquitously expressed. Human platelets, however, express little, if any, N-WASp or WAVE-1.

WASp family members share functional domains.[53,153–155,164] The C-terminus of each is composed of a conserved VCA domain, which has homology to verprolin (V), and cofilin (C), and a unique acidic region. This C-T region from N-WASp binds specifically to the Arp2/3 complex and activates its nucleation activity.[53] In the resting cell, this domain is cryptic, sequestered by interactions between it and domains in the amino terminus of the protein. The amino termini of WASp and N-WASp each contain a pleckstrin homology domain used for lipid binding, a WIP binding site, and a CRIB domain that binds GTP-cdc42.[53,165,166] Binding of cdc42 and polyphosphoinositides (ppIs) to WASp and N-WASp are believed to affect a conformational change in each protein, opening them to permit binding to the Arp2/3 complex. However, WASp family members cannot be the only proteins that can activate the Arp2/3 complex, because platelets from Wiskott–Aldrich syndrome patients have platelets that are completely normal in their actin assembly-driven responses to agonists, despite lacking all three of these proteins. Activation of the Arp2/3 complex *in vitro* has also been achieved with cortactin,[51,52,167–170] a protein abundantly expressed in platelets known to be phosphorylated after platelet activation,[168] and by yeast Abp1p,[171] a member of the drebrins protein family.

## V. Signals Activating Actin Assembly

Barbed-end capping proteins, as a group, are inactive when bound to membrane phospholipids of the phosphoinositide family. Both D3- ($PI_{3,4}P_2$ and $PI_{3,4,5}P_3$) and D4- ($PI_4P$ and $PI_{4,5}P_2$) containing phosphoinositides are potent inhibitors of this group of proteins.[172,173] Resting platelets contain little, if any, D3-containing phosphoinositides, whereas the D4-containing phosphoinositides $PI_4P$ and $PI_{4,5}P_2$ represent approximately 1 to 2% of the total membrane lipid. $PI_4P$ and $PI_{4,5}P_2$ are each present in the resting platelet at cellular concentrations of approximately 200 μM. This corresponds to approximately $10^6$ copies expressed on the cytoplasmic face of the plasma membrane of the resting platelet. After platelet stimulation, $PI_4P$ and $PI_{4,5}P_2$ contents initially diminish for 1 to 10 seconds, when phospholipase C activity is maximal, then subsequently increase to as much as 40% over resting levels. Measurements of the content of $PI_{4,5}P_2$, however, underestimate the amount that is produced by the activated cell, and studies measuring the amount of $IP_3$ produced suggest that the entire population is turned over five- to sixfold in the activated cell.

Platelet activation also stimulates the activity of PI-3 kinase activity, and the levels of D3-containing phosphoinositides increase dramatically after receptor ligation.[174] $PI_{3,4,5}P_3$ is formed first from the phosphorylation of $PI_{4,5}P_2$ in the D3 position. While the production of this lipid fades within 1 to 2 minutes after cell activation, $PI_{3,4}P_2$ content increases and stays elevated for 2 to 10 minutes.[174] The total content of D3-containing phosphoinositides never increases

to more than 2 to 5% of the total phosphoinositides, or approximately 50,000 per platelet.

Phosphoinositides have been linked to the actin assembly reaction of the platelet. First, their synthesis correlates with the times of maximal actin assembly.[137,175] Second, the addition of reagents to permeabilized platelets and neutrophils that sequester ppIs, or overexpression of pleckstrin homology domains that bind to ppIs, block either barbed-end production and/or actin assembly and cell movement *per se*.[176] Third, the addition of PI kinases to permeabilized cells or the overexpression of PI-3 and PI-5 kinases leads to barbed-end exposure, membrane ruffling, and enhanced actin assembly.[175] Coupling of PI kinases to receptor ligation is through trimeric G-proteins and small GTPases, rac, rho, cdc42,[177] and arf6.[178,179] There are two forms of activity of PI-3 kinases. One couples directly to the βγ-subunit of the trimeric G-protein, the other p85/p110 is coupled to receptor tyrosine phosphorylation and has been reported to use both rac and rho. Type Iα PI-5 kinase has been reported to be activated by both rac and arf6. The small GTPases in turn are coupled to receptors by upstream guanine exchange factors, in particular VAV1,[180,181] VAV2,[182] and SOS.

Polyphosphoinositides serve not only to initiate actin filament assembly directly, but also to target this assembly to the membrane–cytoskeletal interface. This targeting is critical, because all filament assembly begins just beneath the plasma membrane.

## VI. Membrane Dynamics and Actin Filament Turnover in Platelets

In motile cells, actin filaments are dynamic. As cells crawl, actin filaments assemble in the leading edge, a process balanced by filament disassembly in other regions of the cell. This assembly–disassembly process, termed *filament turnover,* is related to the speed of cell movement; the faster the cell crawls, the faster filaments turn over. Filament turnover is controlled by a number of factors. Barbed-end bias filament assembly is driven by ATP hydrolysis and, as filaments age, ATP is hydrolyzed to release inorganic phosphate (Pi) and leave subunits with a bound ADP molecule. ADP subunits are recognized by a number of proteins involved in accelerating the actin filament disassembly process. The primary proteins involved in the process are the filament-severing proteins gelsolin[119,183] and cofilin.[184] Gelsolin plays a major role in the turnover process, generating pointed filament ends as it severs filaments, as evidenced by the finding that cells lacking gelsolin turn over filaments much more slowly than gelsolin-expressing cells[119] and greatly increase their content of actin filaments.[183] The binding of cofilin to actin subunits in filaments destabilizes them, releasing them from the filament. Cofilin is present in the platelet at high concentrations of approximately 30 μM. The activity of cofilin is regulated by phosphorylation. Phosphorylation on serine 3 prevents binding to actin. In the resting platelet, a large fraction of the cofilin is phosphorylated and hence is inactive.[184,185] Both gelsolin and cofilin have higher affinities for actin bound to ADP relative to ATP, allowing them to target old filaments.

Although filament turnover rates have not been measured in platelets, it seems likely that such turnover occurs, principally after platelet activation. Cofilin is rapidly dephosphorylated after cell stimulation[184] and can remain in this active state for a considerable time if integrin receptors become ligated.

# VII. Platelet Contraction

Platelets are the force-generating components of clot retraction. After activation, the major platelet integrin, αIIbβ3 (Chapter 8), becomes tethered to underlying actin filaments by binding interactions between it and a still unresolved number of adhesion site proteins.[186,187] In platelets, important adhesion site proteins include talin,[36,37,188] filamin,[189] paxillin,[190] zyxin,[42,43] α-actinin,[191] tensin, moesin,[192] and vinculin.[193] These proteins are brought into large macromolecular complexes in processes positively regulated by kinases. Critical kinases include focal adhesion kinase (FAK),[194–197] src,[198–201] and phosphoinositol 3-kinase,[200,202] among others. Once tethered to actin, cytoplasmic myosin is the molecular motor that applies the contractile force on actin filaments.

Platelets express two forms of myosin II (nonmuscle myosin IIA and B) and very likely other types of nonfilamentous myosin. Myosin II is distinguished from its counterparts by its ability to self-aggregate into small bipolar filaments approximately 300 nm in length, each containing 28 molecules.[203] Myosin II is a 500-kDa hexamer composed of two 220-kDa elongated heavy chains, two 20-kDa light chains, and two 15-kDa light chains. Its activity is regulated by phosphorylation of the 20-kDa light chains, which promotes a conformational change that favors filament assembly. Two signaling pathways converge at this phosphorylation site. The major pathway leading to activity of myosin light-chain kinase is through calcium/calmodulin, and myosin II is rapidly phosphorylated after platelet activation at the times of maximal calcium mobilization.[34] This indicates that contractile force is being applied to actin filaments early in the activation process, even before actin filament assembly is complete.[204] The second pathway leading to myosin is from the small GTPase rhoA, which binds and activates rho kinase.[205,206] Rho kinase promotes myosin light-chain phosphorylation by phosphorylating and inhibiting the myosin light-chain phosphatase.

The early functions of myosin are likely to be secretory and to involve the modulation of membrane structure receptors. Myosin II activity is required to downregulate the

GPIb-IX-V complex from the surface of platelets in suspension, to aggregate granules into a dense zone in the center of the active cell, and to contract clots. Inhibition of myosin II, however, has little effect on the actin filament assembly reaction or shape change *per se*.

# VIII. Diseases of the Platelet Cytoskeleton

## A. Bernard-Soulier Syndrome

Bernard-Soulier syndrome (BSS) is the best understood disease of the platelet cytoskeleton[207] (see Chapter 57). This syndrome is characterized by thrombocytopenia with abnormally large and fragile platelets in the circulation.[208] Platelets from BSS fail to express stable GPIb-IX-V complex on their surface, a consequence of mutations in GPIbα, GPIbβ, and GPIX[209–217] that terminate expression or alter conformation of this receptor, thereby prohibiting surface expression. GPV is not required for stable expression of a functional complex.[218] Although a loss of the GPIb-IX-V complex, the VWFR, has serious physiological consequences for normal platelet activation at wound sites, the effects of its loss on cytoskeletal structure are also grave. The GPIb-IX-V complex is the major link between the plasma membrane and the underlying actin cytoskeleton, accounting for 85% of the linkages to actin. Its linkage to actin is arbitrated by the two platelet filamins FLNa and FLNb, each of which can bind two GPIb-IX-V complexes. Loss of this linkage swells the spectrin-based membrane skeleton, because the GPIb-IX-V-filamin–actin interaction restricts the mobility of the spectrin lattice. This explains the large size of BSS platelets and their fragility, which results from having a membrane skeleton that is more loosely associated with its underlying actin cytoskeleton. Figure 4-7 shows this loss of actin filament attachment to the membrane skeleton in BSS platelets compared with normal platelets. Thrombocytopenia is most likely the result of increased clearance of BSS platelets, although defects in production are possible.

## B. May–Hegglin Anomaly

Diseases correlated with large platelets have long been recognized and include May–Hegglin,[219–220] Sebastian, and Fechtner syndromes characterized by macrothrombocytopenia (see Chapter 54). In addition, giant platelets from these individuals are known to contain abnormally large inclusions/granules. The biochemical defects in these three related platelet syndromes are mutations in the myosin II gene. May–Hegglin anomaly results from mutations in the myosin heavy chain, leading to failure of myosin II to assemble into filaments.[221–223] As discussed in Chapter 2, myosin filaments are very likely involved in platelet

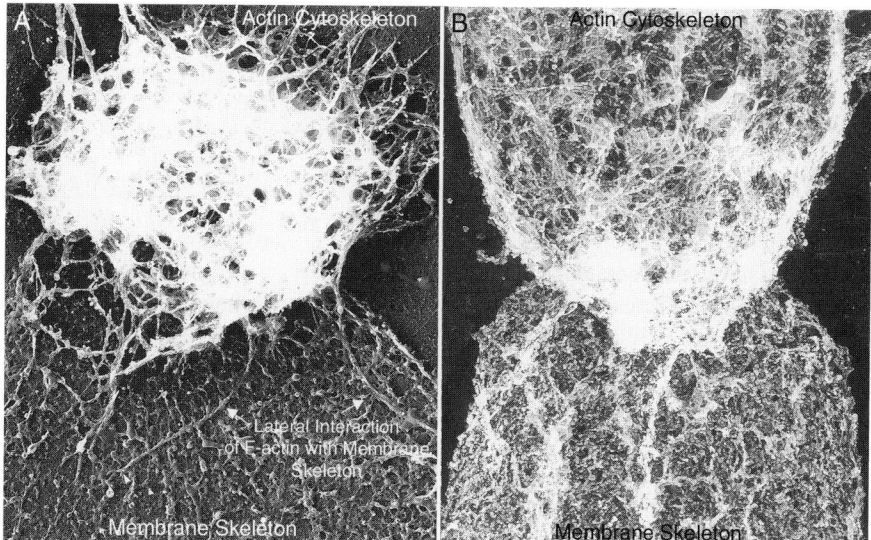

**Figure 4-7.** Altered attachment of the spectrin-based membrane skeleton to the actin cytoskeleton in platelets from a patient with Bernard-Soulier syndrome (BSS) lacking surface expression of the GPIb-IX-V complex. A. Actin attachments to membrane skeleton in the normal platelet. Actin filaments, coming from the cytoskeleton, run along the cytoplasmic surface of the membrane skeleton and end in their spectrin attachments. The lateral associations of actin filaments with the membrane skeleton are maintained by filamin, which links the GPIb-IX-V receptor to actin. This interaction is visualized by centrifuging platelet cytoskeletons onto glass coverslips at high shear forces, which fractures them, revealing both the cytoplasmic surface of the membrane skeleton and the internal cytoskeletal actin filling. Actin filaments can be observed to radiate out from the cytoskeletal mass to the membrane surface. B. Loss of lateral actin filament associations in BSS platelet cytoskeleton. When the fracturing procedure is repeated on the cytoskeleton of platelets from BSS patients, the cytoskeleton is found to be weakly associated with the membrane skeleton, and few actin filament interactions with the membrane skeleton are observed.

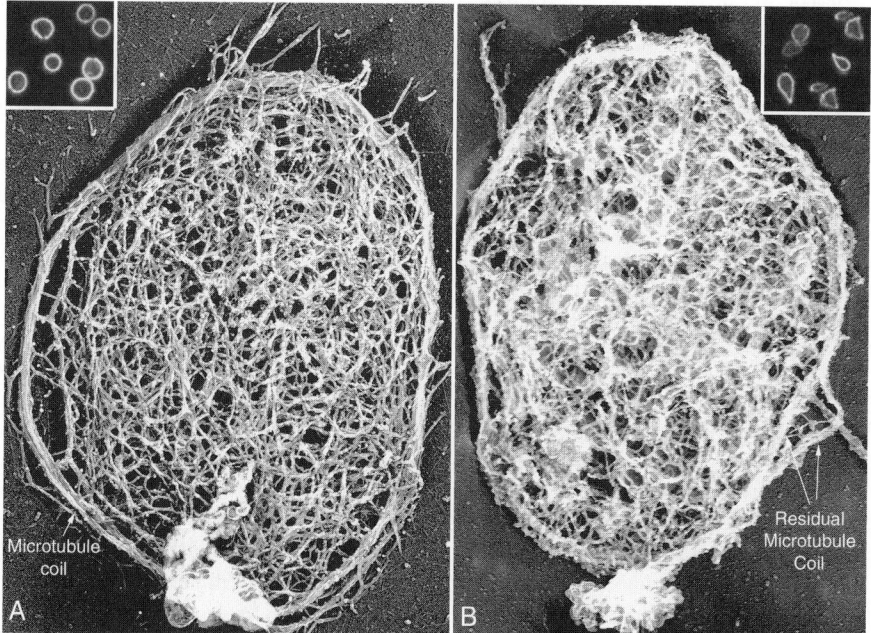

**Figure 4-8.** Altered structure of the resting platelet microtubule coil in mouse platelets lacking β1-tubulin. A. Cytoskeleton of a normal resting platelet. The microtubule coil is maintained in these preparations and encircles the actin cytoskeleton. This coil is believed to be composed of a single microtubule that is curled 8 to 12 times. Inset. Immunofluorescent labeling of tubulin in the resting cell. B. β1-Tubulin-deficient platelets, which polymerize considerably less tubulin than their wild-type counterparts, assemble an abnormal microtubule coil. These platelets are not discoid, and, in cytoskeletal preparations, the microtubule coil is thin, composed of fewer loops than in normal cells, and is loosely wrapped. Inset. Immunofluorescent labeling of tubulin in the β1-tubulin-deficient platelets.

biogenesis, particularly in the proplatelet end amplification. Large platelets, diminished in number, are a likely consequence of this defect.

### C. β1-Tubulin Deficient Platelets

Transgenic mice lacking β1-tubulin, the major β-tubulin isoform expressed in platelets, have been created.[95] These animals have mild thrombocytopenia, suggesting that their ability to produce platelets is only slightly compromised. Examination of the platelets from these animals reveal spherical forms and few, if any, discoid forms. In these platelets, aberrant microtubule coils form that have greatly diminished coil mass and the integrity of the microtubule coil itself is lost, with breaks and bends apparent (Fig. 4-8). These studies therefore directly demonstrate a role for the microtubule coil in stabilizing the discoid shapes of resting platelets.

## References

1. Boyles, J., Fox, J. E. B., Phillips, D. R., & Stenberg, P. E. (1985). Organization of the cytoskeleton in resting, discoid platelets: Preservation of actin filaments by a modified fixation that prevents osmium damage. *J Cell Biol, 101,* 1463–1472.
2. Fox, J., Boyles, J., Berndt, M., Steffen, P., & Anderson, L. (1988). Identification of a membrane skeleton in platelets. *J Cell Biol, 106,* 1525–1538.
3. Kenney, D., & Linck, R. (1985). The cytoskeleton of unstimulated blood platelets: Structure and composition of the isolated marginal microtubular band. *J Cell Sci, 78,* 1–22.
4. Huxley, H. (1963). Electron microscopic studies on the structure of natural and synthetic protein filaments from striated muscle. *J Mol Biol, 3,* 281–308.
5. Nachmias, V. (1980). Cytoskeleton of human platelets at rest and after spreading. *J Cell Biol, 86,* 795–802.
6. Nachmias, V. T., & Yoshida, K.-I. (1988). The cytoskeleton of the blood platelet: A dynamic structure. *Adv Cell Biol, 2,* 181–211.
7. Fox, J., Aggerbeck, L., & Berndt, M. (1988). Structure of the glycoprotein Ib-IX complex from platelet membranes. *J Biol Chem, 263,* 4882–4890.
8. Du, X., Harris, S., Tetaz, T., Ginsberg, M., & Berndt, M. (1994). Association of a phospholipase A2 (14-3-3 protein) with the platelet glycoprotein Ib-IX complex. *J Biol Chem, 269,* 18287–18294.
9. Barkalow, K., Italiano, Jr., J., Matsuoka, Y., Bennett, V., & Hartwig, J. (2003). a-Adducin dissociates from F-actin filaments and spectrin during platelet activation. *J Cell Biol, 161,* 557–570.
10. Feng, Y., & Walsh, C. (2004). The many faces of filamin: A versatile molecular scaffold for cell motility and signalling. *Nature Cell Biol, 6,* 1034–1038.
11. Takafuta, T., Wu, G., Murphy, G., & Shapiro, S. (1998). Human β-filamin is a new protein that interacts with the cytoplasmic tail of glycoprotein Ibα. *J Biol Chem, 273,* 17531–17538.
12. Thompson, T., Chan, Y.-M., Hack, A., Brosius, M., Rajala, M., Lidov, H., McNally, E., Watkins, S., & Kunkel, L. (2000). Filamin 2 (FLN2): A muscle-specific sacroglycan interacting protein. *J Cell Biol, 148,* 115–126.
13. Bröcker, F., et al. (1999). Assignment of human filamin gene FLNB to human chromosome band 3p14.3 and identification of YACs containing the complete FLNB transcribed region. *Cytogenet Cell Genet, 85,* 267–268.
14. Barkalow, K., Witke, W., Kwiatkowski, D., & Hartwig, J. (1996). Coordinated regulation of platelet actin filament barbed ends by gelsolin and capping protein. *J Cell Biol, 134,* 389–399.
15. Welch, M., & Mitchison, T. (1998). Purification and assay of the platelet Arp2/3 complex. *Methods Enzymol, 298,* 52–61.
16. Falet, H., Hoffmeister, K., Neujahr, R., & Hartwig, J. (2002). Normal Arp2/3 complex activation in Wiskott–Aldrich syndrome platelets. *Blood, 100,* 2113–2122.
17. Falet, H., Hoffmeister, K., Neujahr, R., Italiano, Jr., J., Stossel, T., Southwick, F., & Hartwig, J. (2002). Importance of free barbed ends for Arp2/3 complex function in platelets and fibroblasts. *Proc Natl Acad Sci USA, 99,* 16782–16788.
18. Kwiatkowski, D. J., Stossel, T. P., Orkin, S. H., Mole, J. E., Colten, H. R., & Yin, H. L. (1986). Plasma and cytoplasmic gelsolins are encoded by a single gene and contain a duplicated actin-binding domain. *Nature, 323,* 455–458.
19. Yin, H. L., & Stossel, T. P. (1979). Control of cytoplasmic actin gel-sol transformation by gelsolin, a calcium-dependent regulatory protein. *Nature, 281,* 583–586.
20. Janmey, P. A., Chaponnier, C., Lind, S. E., Zaner, K. S., Stossel, T. P., & Yin, H. L. (1985). Interactions of gelsolin and gelsolin actin complexes with actin. Effects of calcium on actin nucleation, filament severing and end blocking. *Biochem, 24,* 3714–3723.
21. Janmey, P. A., Iida, K., Yin, H. L., & Stossel, T. P. (1987). Polyphosphoinositide micelles and polyphosphoinositide-containing vesicles dissociate endogenous gelsolin–actin complexes and promote actin assembly from the fast-growing end of actin filaments blocked by gelsolin. *J Biol Chem, 262,* 12228–12236.
22. Kwiatkowski, D. J., Janmey, P. A., Mole, J. E., & Yin, H. L. (1985). Isolation and properties of two actin-binding domains in gelsolin. *J Biol Chem, 260,* 15232–15238.
23. Lind, S. E., Janmey, P. A., Herbert, T., Chaponnier, C., Yin, H., & Stossel, T. (1986). Interaction of actin with profilin and gelsolin during platelet activation. *Blood, 68,* 321a.
24. Lassing, I., & Lindberg, U. (1988). Specificity of the interaction between phosphatidylinositol 4,5-bisphosphate and the profiling : actin complex. *J Cell Biochem, 37,* 255–267.
25. Markey, F., Lindberg, U., & Eriksson, L. (1978). Human platelets contain profilin, a potential regulator of actin polymerizability. *FEBS Lett, 88,* 75–79.

26. Bamburg, J. (1999). Proteins of the ADF/cofilin family: Essential regulators of actin dynamics. *Ann Rev Cell Dev, 15,* 185–230.

27. Ghosh, M., Song, X., Mouneimne, G., Sidani, M., Lawrence, D., & Condeelis, J. (2004). Cofilin promotes actin polymerization and defines the direction of cell motility. *Science, 304,* 743–746.

28. Yamaguchi, H., Lorenz, M., Kempiak, S., Sarmiento, C., Coniglio, S., Symons, M., Segall, J., Eddy, R., Miki, H., Takenawa, T., & Condeelis, J. (2005). Molecular mechanisms of invadopodium formation: The role of the N-WASP-Arp2/3 complex pathway and cofilin. *J Cell Biol, 168,* 441–452.

29. Nachmias, V. (1993). Small actin-binding proteins: The β-thymosin family. *Curr Opin Cell Biol, 5,* 56–62.

30. Nachmias, V., Cassimeris, L., Golla, R., & Safer, D. (1993). Thymosin beta 4 (Tβ4) in activated platelets. *Eur J Cell Biol, 61,* 314–320.

31. Safer, D., Elzinga, M., & Nachmias, V. T. (1991). Thymosin B4 and Fx, an actin-sequestering peptide, are indistinguishable. *J Biol Chem, 266,* 4029–4032.

32. Safer, D., & Nachmias, V. (1994). Beta thymosins as actin binding peptides. *BioEssays, 16,* 473–479.

33. Rosenberg, S., Stracher, A., & Burridge, K. (1981). Isolation and characterization of a calcium-sensitive α-actinin-like protein from human platelet cytoskeletons. *J Biol Chem, 256,* 12986–12991.

34. Adelstein, R. S., & Conti, M. A. (1975). Phosphorylation of platelet myosin increases actin activated myosin ATPase activity. *Nature, 256,* 597–598.

35. Takashima, T., Matsumura, S., Kariya, T., Sunaga, T., & Kumon, A. (1988). Studies on the physical states of human platelet myosin in crude extracts. *J Biochem, 104,* 1027–1035.

36. Beckerle, M., Miller, D., Bertagnolli, M., & Locke, S. (1989). Activation-dependent redistribution of the adhesion plaque protein, talin, in intact human platelets. *J Cell Biol, 109,* 3333–3346.

37. Knezevic, I., Leisner, T., & Lam, S.-T. (1996). Direct binding of the platelet integrin αIIbβ3 (GPIIb-IIIa) to talin. *J Biol Chem, 271,* 16416–16421.

38. O'Halloran, T., Beckerle, M. C., & Burridge, K. (1985). Identification of talin as a major cytoplasmic protein implicated in platelet activation. *Nature, 317,* 449–451.

39. Asyee, G. M., Sturk, A., & Muszbek, L. (1987). Association of vinculin to the platelet cytoskeleton during thrombin-induced aggregation. *Exp Cell Res, 168,* 358–364.

40. Turner, C. E., Glenney, J. J., & Burridge, K. (1990). Paxillin: A new vinculin-binding protein present in focal adhesions. *J Cell Biol, 111*(3), 1059–1068.

41. Salgia, R., Li, J.-L., Lo, S., Brunkhorst, B., Kansas, G., Sobhany, E., Sun, Y., Pisick, E., Hallek, M., Ernst, T., Tantravahi, R., Chen, L., & Griffin, J. (1995). Molecular cloning of human paxillin, a focal adhesion protein phosphorylated by P210[BCR/ABL]. *J Biol Chem, 270,* 5039–5047.

42. Crawford, A., Michelsen, J., & Beckerle, M. (1992). An interaction between zyxin and alpha-actinin. *J Cell Biol, 116,* 1381–1393.

43. Reinhard, M., Tripier, D., & Walter, U. (1995). Identification, purification and characterization of a zyxin-related protein that binds the focal adhesion and microfilament protein VASP (vasodilator-stimulated phosphoprotein). *Proc Natl Acad Sci USA, 92,* 7956–7960.

44. Reinhard, M., Zumbrunn, J., Jaquemar, D., Kuhn, M., Walter, U., & Trueb, B. (1999). An α-actinin binding site of zyxin is essential for subcellular zyxin localization and α-actinin recruitment. *J Biol Chem, 274,* 13410–13418.

45. Fischer, R., Fritz–Six, K., & Fowler, V. (2003). Pointed-end capping by tropomodulin 3 negative regulates endothelial cell motility. *J Cell Biol, 161,* 371–380.

46. Reinhard, M., Giehl, K., Abel, K., Haffner, C., Jarchau, T., Hoppe, V., Jockusch, B., & Walter, U. (1995). The proline-rich focal adhesion and microfilament protein VASP is a ligand for profilins. *EMBO J, 14,* 1583–1589.

47. Reinhard, M., Rudiger, M., Jockusch, B., & Walter, U. (1996). VASP interaction with vinculin: A recurring theme of interactions with proline-rich motifs. *FEBS Letters, 399,* 103–107.

48. Reinhard, M., Halbrügge, M., Scheer, U., Wiegand, C., Jockusch, B. M., & Walter, U. (1992). The 46/50 kDa phosphoprotein VASP purified from human platelets is a novel protein associated with actin filaments and focal contacts. *EMBO J, 11,* 2063–2070.

49. Rottner, K., Behrendt, B., Small, J., & Wehland, J. (1999). VASP dynamics during lamellipodia protrusion. *Nature Cell Biol, 5,* 321–322.

50. Massberg, S., Gruner, S., Konrad, I., Garcia Arguinzonis, G., Eigenthaler, M., Hemler, K., Kersting, J., Schulz, C., Muller, I., Besta, F., Niewsandt, B., Heinzmann, U., Walter, U., & Gawaz, M. (2004). Enhanced in vivo platelet adhesion in vasodilator-stimulated phosphoprotein (VASP)-deficient mice. *Blood, 103,* 136–142.

51. Wu, H., & Parsons, J. (1993). Cortactin, an 80/85-kilodalton pp60src substrate, is a filamentous actin-binding protein enriched in the cell cortex. *J Cell Biol, 120,* 1417–1426.

52. Weaver, A., Heuser, J., Karginov, A., Lee, W., Parsons, J., & Cooper, J. (2002). Interaction of cortactin and N-WASp with Arp2/3 complex. *Curr Biol, 6,* 1270–1278.

53. Rohatgi, R., Ma, L., Miki, H., Lopez, M., Kirchhausen, T., Takenawa, T., & Kirschner, M. (1999). The interaction between N-WASP and the Arp2/3 complex links Cdc42-dependent signals to actin assembly. *Cell, 97,* 221–231.

54. Snapper, S. B., & Rosen, F. S. (1999). The Wiskott–Aldrich syndrome protein (WASP): Roles in signaling and cytoskeletal organization. *Annu Rev Immunol, 17,* 905–929.

55. Suetsugu, S., Miki, H., & Takenawa, T. (1999). Identification of two human WAVE/SCAR homologues as general actin regulatory molecules which associate with the Arp2/3 complex. *Biochem Biophys Res Commun, 260,* 296–302.

56. Dahl, J., Wang–Dunlop, J., Gonzales, C., Goad, M., Mark, R., & Kwak, S. (2003). Characterization of the WAVE1 knock-out mouse: Implications for CNS development. *J Neurosci, 23,* 3343–3352.

57. Oda, A., Miki, H., Wada, I., Yamaguchi, H., Yamazaki, D., Suetsugu, S., Nakajima, M., Nakayama, A., Okawa, K., Miyazaki, H., Matsuno, K., Ochs, H., Machesky, L., Fujita,

H., & Takenawa, T. (2005). WAVE/scars in platelets. *Blood, 105,* 3141–3148.

58. Vaduva, G., Martinez–Quiles, N., Anton, I., Martin, N., Geha, R., Hopper, A., & Ramesh, N. (1999). The human WASP-interacting protein, WIP, activates the cell polarity pathway in yeast. *J Biol Chem, 274,* 17103–17108.

59. Glenney, J., Glenney, P., & Weber, K. (1982). Erythroid spectrin, brain fodrin, and intestinal brush border proteins (TW-260/240) are related molecules containing a common calmodulin-binding subunit bound to a variant cell type-specific subunit. *Proc Natl Acad Sci USA, 79,* 4002–4005.

60. Shotton, D., Burke, B., & Branton, D. (1978). The molecular structure of human erythrocyte spectrin. Biophysical and electron microscopic studies. *J Mol Biol, 131,* 303–329.

61. Tyler, J., Hargreaves, W., & Branton, D. (1979). Purification of two spectrin-binding proteins: Biochemical and electron microscopic evidence for site-specific reassociation between spectrin and bands 2.1 and 4.1. *Proc Natl Acad Sci USA, 76,* 5192–5196.

62. Tyler, J. M., Anderson, J. M., & Branton, D. (1980). Structural comparison of several actin-binding molecules. *J Cell Biol, 85,* 489–495.

63. Ungewickell, E., & Gratzer, W. (1978). Self-association of human spectrin. A thermodynamic and kinetic study. *Eur J Biochem, 88,* 379–385.

64. Wasenius, V.-M., Saraste, M., Salven, P., Eramaa, M., Holm, L., & Lehto, V.-P. (1989). Primary structure of the brain alpha-spectrin. *J Cell Biol, 108,* 79–93.

65. Winkelmann, J., Chang, J.-G., Tse, W., Scarpa, A., Marchesi, V., & Forget, B. (1990). Full-length sequence of the cDNA for human erythroid β-spectrin. *J Biol Chem, 264,* 11827–11832.

66. Bennett, V. (1979). Immunoreactive forms of human erythrocyte ankyrin are present in diverse cells and tissues. *Nature, 281,* 597–599.

67. Liu, S.-C., Derick, L., & Palek, J. (1987). Visualization of the hexogonal lattice in the erythrocyte membrane skeleton. *J Cell Biol, 104,* 527–536.

68. Liu, S.-C., Windisch, P., Kim, S., & Palek, J. (1984). Oligomeric states of spectrin in normal erythrocyte membranes: Biochemical and electron microscopic studies. *Cell, 37,* 587–594.

69. Fowler, V., Sussmann, M., Miller, P., Flucher, B., & Daniels, M. (1993). Tropomodulin is associated with the free (pointed) ends of the thin filaments in rat skeletal muscle. *J Cell Biol, 120,* 411–420.

70. Fowler, V. M. (1990). Tropomodulin: A cytoskeletal protein that binds to the end of erythrocyte tropomyosin and inhibits tropomyosin binding to actin. *J Cell Biol, 111,* 471–482.

71. Nehls, V., Drenckhahn, D., Joshi, R., & Bennett, V. (1991). Adducin in erythrocyte precursor cells of rats and humans: Expression and compartmentalization. *Blood, 78,* 1692–1696.

72. Kuhlman, P., Hughes, C., Bennett, V., & Fowler, V. (1996). A new function for adducin. Calcium/calmodulin-regulated capping of the barbed ends of actin filaments. *J Biol Chem, 271,* 7986–7991.

73. Mische, S., Mooseker, M. S., & Morrow, J. S. (1987). Erythrocyte adducin: A calmodulin-regulated actin-bundling protein that stimulates spectrin-actin binding. *J Cell Biol, 105,* 2837–2845.

74. Fox, J., Reynolds, C., Morrow, J., & Phillips, D. (1987). Spectrin is associated with membrane-bound actin filaments in platelets and is hydrolyzed by the $Ca^{2+}$-dependent protease during platelet activation. *Blood, 69,* 537–545.

75. Hartwig, J., & DeSisto, M. (1991). The cytoskeleton of the resting human blood platelet: Structure of the membrane skeleton and its attachment to actin filaments. *J Cell Biol, 112,* 407–425.

76. Hartwig, J. (1992). Mechanism of actin rearrangements mediating platelet activation. *J Cell Biol, 118,* 1421–1442.

77. Kaiser, H., O'Keefe, E., & Bennett, V. (1989). Adducin: Ca++-dependent association with sites of cell–cell contact. *J Cell Biol, 109,* 557–569.

78. Matsuoka, Y., Li, X., & Bennett, V. (1998). Adducin is an in vivo substrate for protein kinase C: Phosphorylation in the MARCKS-related domain inhibits activity in promoting spectrin–actin complexes and occurs in many cells, including dendritic spines of neurons. *J Cell Biol, 142,* 485–497.

79. Matsuoka, Y., Li, X., & Bennett, V. (2000). Adducin: Structure, function, and regulation. *Cell Mot Life Sci, 57,* 884–895.

80. Wang, D., & Shaw, G. (1995). The association of the C-terminal region of betaIepsilon2 spectrin to brain membranes is mediated by a pH domain, does not require membrane proteins, and coincides with an inositol-1,4,5 triphosphate binding site. *Biochem Biophys Res Commun, 217,* 608–615.

81. Rosenberg, S., & Stracher, A. (1982). Effect of actin-binding protein on the sedimentation properties of actin. *J Cell Biol, 94,* 51–55.

82. Rosenberg, S., Stracher, A., & Lucas, R. (1981). Isolation and characterization of actin and actin-binding protein from human platelets. *J Cell Biol, 91,* 201–211.

83. Stossel, T., Condeelis, J., Cooley, L., Hartwig, J., Noegel, A., Schleicher, M., & Shapiro, S. (2001). Filamins as integrators of cell mechanics and signalling. *Nature Reviews, 2,* 138–145.

84. Gorlin, J., Yamin, R., Egan, S., Stewart, M., Stossel, T., Kwiatkowski, D., & Hartwig, J. (1990). Human endothelial actin-binding protein (ABP-280, non-muscle filamin): A molecular leaf spring. *J Cell Biol, 111,* 1089–1105.

85. Fox, J. (1985). Identification of actin-binding protein as the protein linking the membrane skeleton to glycoproteins on platelet plasma membrane. *J Biol Chem, 260,* 11970–11977.

86. Okita, L., Pidard, D., Newman, P., Montogomery, R., & Kunicki, T. (1985). On the association of glycoprotein Ib and actin-binding protein in human platelets. *J Cell Biol, 100,* 317–321.

87. Andrews, R., & Fox, J. (1991). Interaction of purified actin-binding protein with the platelet membrane glycoprotein Ib-IX complex. *J Biol Chem, 266,* 7144–7147.

88. Andrews, R., & Fox, J. (1992). Identification of a region in the cytoplasmic domain of the platelet membrane glycopro-

tein Ib-IX complex that binds to purified actin-binding protein. *J Cell Biol, 267,* 18605–18611.

89. Meyer, S., Zuerbig, S., Cunninghan, C., Hartwig, J., Bissel, T., Gardner, K., & Fox, J. (1997). Identification of the region in actin-binding protein that binds to the cytoplasmic domain of glycoprotein Ibα. *J Biol Chem, 272,* 2914–2919.

90. Fucini, P., & Al, E. (1999). Molecular architecture of the rod domain of the dicytostelium gelation factor (ABP-120). *J Mol Biol, 291,* 1017–1023.

91. Ohta, Y., Suxuki, N., Nakamura, S., Hartwig, J., & Stossel, T. (1999). The small GTPase RalA targets filamin to induce filopodia. *Proc Natl Acad Sci USA, 96,* 2122–2128.

92. Nakamura, F., Pudas, R., Heikkinen, O., Permi, P., Kilpeläinen, I., Munday, A., Hartwig, J., Stossel, T., & Ylänne, J. (2005). The structure of the GP1b-filamin A complex. *Blood,* Submitted.

93. Kovacsovics, T., & Hartwig, J. (1996). Thrombin-induced GPIb-IX centralization on the platelet surface requires actin assembly and myosin II activation. *Blood, 87,* 618–629.

94. Du, X., Fox, J., & Pei, S. (1996). Identification of a binding sequence for the 14-3-3 protein within the cytoplasmic domain of the adhesion receptor, platelet glycoprotein Ibα. *J Biol Chem, 267,* 18605.

95. Schwer, H., Lecine, P., Tiwari, S., Italiano, Jr., J., Hartwig, J., & Shivdasani, R. (2001). A lineage-restricted and divergent β tubulin isoform is essential for the biogenesis, structure and function of mammalian blood platelets. *Curr Biol, 11,* 579–586.

96. Italiano, Jr., J., Bergmeier, W., Tiwari, S., Falet, H., Hartwig, J., Hoffmeister, K., Andre, P., Wagner, D., & Shivdasani, R. (2003). Mechanisms and implications of platelet discoid shape. *Blood, 101,* 4789–4796.

97. White, J. (1982). Influence of taxol on the response of platelets to chilling. *Am J Pathol, 108,* 184.

98. Freson, K., De Vos, R., Wittevrognel, C., Thys, C., Defoor, J., Vanhees, L., Vermylen, J., Peerlinck, K., & Van Geet, C. (2005). The β1-tubulin Q43P functional polymorphism reduces the risk of cardiovascular disease in men by modulating platelet function and structure. *Blood, 106,* 2356–2362.

99. Rothwell, S., & Calvert, V. (1997). Activation of human platelets causes post-translational modifications to cytoplasmic dynein. *Thromb Haemost, 78,* 910–918.

100. Gibbons, I. (1996). The role of dynein in microtubule-based motility. *Cell Struct Funct, 21,* 343–349.

101. Schroer, T. A., Steuer, E. R., & Sheetz, M. P. (1989). Cytoplasmic dynein is a minus end-directed motor for membranous organelles. *Cell, 56(6),* 937–946.

102. Steuer, E. R., Wordeman, L., Schroer, T. A., & Sheetz, M. P. (1990). Localization of cytoplasmic dynein to mitotic spindles and kinetochores [see comments]. *Nature, 345(6272),* 266–268.

103. Ginsberg, M. H., Du, X., & Plow, E. (1992). Inside-out integrin signaling. *Curr Opin Cell Biol, 4,* 766–771.

104. Shattil, S., Hoxie, J., Cunningham, M., & Brass, L. (1985). Changes in platelet membrane glycoprotein IIb-IIIa complex during platelet activition. *J Biol Chem, 260,* 11107–11114.

105. Shattil, S., Kashiwagi, H., & Pampori, N. (1998). Integrin signaling: The platelet paradigm. *Blood, 91,* 2645–2657.

106. Fox, J. E. B., & Phillips, D. R. (1981). Inhibition of actin polymerization in blood platelets by cytochalasins. *Nature, 292,* 650–652.

107. Carlier, M. (1998). Control of actin dynamics. *Curr Opin Cell Biol, 10,* 45–51.

108. Carlier, M.-F. (1993). Dynamic actin. *Curr Biol, 3,* 321–323.

109. Pollard, T. D., & Mooseker, M. S. (1981). Direct measurement of actin polymerization rate constants by electron microscopy of actin filaments nucleated by isolated microvillus cores. *J Cell Biol, 88,* 654–659.

110. Pollard, T. D., & Weeds, A. G. (1984). The rate constant for ATP hydrolysis by polymerized actin. *FEBS Lett, 170,* 94–98.

111. Berridge, M. J. (1984). Inositol trisphosphate and diacylglycerol as second messengers. *Biochem, 220,* 345–360.

112. Sage, S., & Rink, T. (1987). The kinetics of changes in intracellular calcium concentration in fura-2-loaded human platelets. *J Biol Chem, 262,* 16364–16369.

113. Brass, L. (1999). More pieces of the platelet activation puzzle slide into place. *J Clin Invest, 104,* 1663–16654.

114. Vu, T., Hung, D., Wheaton, V., & Coughlin, S. (1991). Molecular cloning of a functional thrombin receptor reveals a novel proteolytic mechanism for receptor activation. *Cell, 64,* 1057–1068.

115. Molino, M., Bainton, D., Hoxie, J., Coughlin, S., & Brass, L. (1997). Thrombin receptors on human platelets. *J Biol Chem, 272,* 6011–6017.

116. Kahn, M., Nakanishi–Matsui, M., Shapiro, M., Ishihara, H., & Coughlin, S. (1999). Protease-activated receptors 1 and 4 mediate activation of human platelets by thrombin. *J Clin Invest, 103,* 879–887.

117. Shapiro, M., Weiss, E., Faruqi, T., & Coughlin, S. (2000). Protease-activated receptors 1 and 4 are shut off with distinct kinetics after activation by thrombin. *J Biol Chem, 275,* 25216–25221.

118. Watson, S. P., & Gibbins, J. (1998). Collagen receptor signalling in platelets: Extending the role of the ITAM. *Immunol Today, 19,* 260–264.

119. McGrath, J., Osborn, E., Tardy, Y., Dewey, Jr., C., & Hartwig, J. (2000). Regulation of the actin cycle in vivo by actin filament severing. *Proc Natl Acad Sci USA, 97,* 6532–6537.

120. Witke, W., Sharpe, A., Hartwig, J., Azuma, T., Stossel, T., & Kwiatkowski, D. (1995). Hemostatic, inflammatory, and fibroblast responses are blunted in mice lacking gelsolin. *Cell, 81,* 41–51.

121. Yin, H. L., & Stossel, T. P. (1980). Purification and structural properties of gelsolin, a Ca2+-activated regulatory protein of macrophages. *J Biol Chem, 255,* 9490–9493.

122. Hartwig, J., Chambers, K., & Stossel, T. (1989). Association of gelsolin with actin filaments and cell membranes of macrophages and platelets. *J Cell Biol, 108,* 467–479.

123. Lind, S., Yin, H. L., & Stossel, T. P. (1982). Human platelets contain gelsolin, a regulator of actin filament length. *J Clin Invest, 69,* 1384–1387.

124. McLaughlin, P., Gooch, J., Mannherz, H.-G., & Weeds, A. (1993). Structure of gelsolin segment 1-actin complex

and the mechanism of filament severing. *Nature, 364,* 685–692.

125. Janmey, P., Lamb, J., Allen, P., & Matsudaira, P. (1992). Phosphoinositide-binding peptides derived from the sequence of gelsolin and villin. *J Biol Chem, 267,* 11818–11823.

126. Janmey, P., & Stossel, T. (1987). Modulation of gelsolin function by phosphatidylinositol 4,5-bisphosphate. *Nature, 325,* 362–364.

127. Janmey, P., & Stossel, T. (1989). Gelsolin–polyphosphoinositide interaction. Full expression of gelsolin-inhibiting function by polyphosphoinositides in vesicular form and inactivation by dilution, aggregation, or masking of the inositol head group. *J Biol Chem, 264,* 4825–4831.

128. Selve, N., & Wegner, A. (1986). Rate of treadmilling of actin filaments in vitro. *J Mol Biol, 187,* 627–631.

129. Selve, N., & Wegner, A. (1987). pH-dependent rate of formation of the gelsolin–actin complex from gelsolin and monomeric actin. *Eur J Biochem, 168,* 111–115.

130. Weeds, A., & Maciver, S. (1993). F-actin capping proteins. *Curr Opin Cell Biol, 5,* 63–69.

131. Weeds, A. G., Harris, H., Gratzer, W., & Gooch, J. (1986). Interactions of pig plasma gelsolin with G-actin. *Eur J Biochem, 161,* 77–84.

132. Yin, H. L., Albrecht, J., & Fattoum, A. (1981). Identification of gelsolin, a Ca2+-dependent regulatory protein of actin gel sol transformation. Its intracellular distribution in a variety of cells and tissues. *J Cell Biol, 91,* 901–906.

133. Yin, H. L., Iida, K., & Janmey, P. A. (1988). Identification of a polyphosphoinositide-modulated domain in gelsolin which binds to the sides of actin filaments. *J Cell Biol, 106,* 805–812.

134. Yin, H. L., Zaner, K. S., & Stossel, T. P. (1980). Ca2+ control of actin gelation. *J Biol Chem, 255,* 9494–9500.

135. Meerschaert, K., De Corte, V., De Ville, Y., Vandekerckhove, J., & Gettemans, J. (1998). Gelsolin and functionally similar actin-binding proteins are regulated by lysophosphatidic acid. *EMBO J, 15,* 5923–5932.

136. Kurth, M., & Bryan, J. (1984). Platelet activation induces the formation of a stable gelsolin–actin complex from monomeric gelsolin. *J Biol Chem, 259,* 7473–7479.

137. Hartwig, J., Bokoch, G., Carpenter, C., Janmey, P., Taylor, L., Toker, A., & Stossel, T. (1995). Thrombin receptor ligation and activated Rac uncap actin filament barbed ends through phosphoinositide synthesis in permeabilized human platelets. *Cell, 82,* 643–653.

138. Carlier, M.-F., Didry, D., Erk, I., Lepault, J., Van Troys, M., Vanderkerckhove, J., Perelroizen, I., Yin, H., Doi, Y., & Pantaloni, D. (1996). Tβ4 is not a simple G-actin sequestering protein and interacts with F-actin at high concentration. *J Biol Chem, 271,* 9231–9239.

139. Carlsson, L., Markey, F., Blikstad, I., Persson, T., & Lindberg, U. (1979). Reorganization of actin in platelets stimulated by thrombin as measured by the DNase I inhibition assay. *Proc Natl Acad Sci USA, 76,* 6376–6380.

140. Carlsson, L., Nystrom, L. E., Sundkvist, I., Markey, F., & Lindberg, U. (1976). Profilin, a low molecular weight protein controlling actin polymerisability. In S. V. Perry, A. Margreth, & R. S. Adelstein (Eds.), *Contractile systems in non-muscle tissues.* Amsterdam: Elsevier/North-Holland Biomedical Press.

141. Carlsson, L., Nystrom, L. E., Sundkvist, I., Markey, F., & Lindberg, U. (1977). Actin polymerizability is influenced by profilin, a low molecular weight protein in non-muscle cells. *J Mol Biol, 115,* 465–483.

142. Barkalow, K., & Hartwig, J. (1995). The role of actin filament barbed-end exposure in cytoskeletal dynamics and cell motility. *Biochem Soc Trans, 23,* 451–456.

143. Lind, S. E., Janmey, P. A., Chaponnier, C., Herbert, T.-J., & Stossel, T. P. (1987). Reversible binding of actin to gelsolin and profilin in human platelet extracts. *J Cell Biol, 105,* 833–842.

144. Schafer, D., Jennings, P., & Cooper, J. (1996). Dynamics of capping protein and actin assembly in vitro: Uncapping barbed ends by polyphosphoinositides. *J Cell Biol, 135,* 169–179.

145. Mejillano, M., Kojima, S., Applewhite, D., Gertler, F., Svitkina, T., & Borisy, G. (2004). Lamellipodial versus filopodial mode of the actin nanomachinery: Pivotal role of the filament barbed end. *Cell, 118,* 363–373.

146. Machesky, L., & Gould, K. (1999). The Arp2/3 complex: A multifunctional actin organizer. *Curr Opin Cell Biol, 11,* 117–121.

147. Mullins, R., Heuser, J., & Pollard, T. (1998). The interaction of Arp2/3 complex with actin: Nucleation, high affinity pointed end capping, and formation of branched networks of filaments. *Proc Natl Acad Sci USA, 95,* 6181–6186.

148. Machesky, L., Atkinson, S., Ampe, C., Vandekerckhoe, J., & Pollard, T. (1994). Purification of a cortical actin complex containing two unconventional actins from Acanthamoeba by affinity chromatography on profilin–agarose. *J Cell Biol, 127,* 107–115.

149. Welch, M., Rosenblatt, J., Skoble, J., Portnoy, D., & Mitchison, T. (1998). Interaction of human Arp2/3 complex and the *Listeria monocytogenes* ActA protein in actin filament nucleation. *Science, 281,* 105–108.

150. Laine, R., Zeile, W., Kang, F., Purich, D., & Southwick, F. (1997). Vinculin proteolysis unmasks an ActA homolog for actin-based *Shigella* motility. *J Cell Biol, 138,* 1255–1264.

151. Beckerle, M. (1998). Spatial control of actin filament assembly: Lessons from *Listeria. Cell, 95,* 741–748.

152. Bear, J., Rawls, J., & Saxe III, C. (1998). SCAR, a WASP-related protein, isolated as a suppressor of receptor defects in late *Dictyostelium* development. *J Cell Biol, 142,* 1325–1335.

153. Machesky, L., Mullins, R., Higgs, H., Kaiser, D., Blanchoin, L., May, R., Hall, M., & Pollard, T. (1999). SCAR, a WASp-related protein, activates dendritic nucleation of actin filaments by the ARP2/3 complex. *Proc Natl Acad Sci USA, 96,* 3739–3744.

154. Miki, H., Miura, K., & Takenawa, T. (1996). N-WASP, a novel actin-depolymerizing protein, regulates the cortical cytoskeletal rearrangement in a PIP2-dependent manner downstream of tyrosine kinases. *EMBO J, 15,* 5326–5335.

155. Miki, H., Sasaki, T., Takai, Y., & Takenawa, T. (1998). Induction of filopodium formation by a WASP-related actin-depolymerizing protein N-WASP. *Nature, 391,* 93–96.

156. Miki, H., Suetsugu, S., & Takenawa, T. (1998). WAVE, a novel WASP-family protein involved in actin reorganization induced by Rac. *EMBO J, 17,* 6932–6941.

157. Winter, D., Lechler, T., & Li, R. (1999). Activation of the yeast Arp2/3 complex by Bee1p, a WASP-family protein. *Curr Biol, 9,* 501–506.

158. Aspenstrom, P., Lindberg, U., & Hall, A. (1996). The two GTPases, Cdc42 and Rac, bind directly to a protein implicated in the immunodeficiency disorder Wiskott–Aldrich syndrome. *Curr Biol, 6,* 70–75.

159. Derry, J., Ochs, H., & Francke, U. (1994). Identification of a novel gene mutated in Wiskott–Aldrich syndrome. *Cell, 78,* 635–644.

160. Kolluri, R., Tolias, K., Carpenter, C., Rosen, F., & Kirchhausen, T. (1996). Direct interaction of the Wiskott–Aldrich syndrome protein with the GTPase, Cdc42. *Proc Natl Acad Sci USA, 93,* 5615–5618.

161. Symons, M., Derry, J., Karlak, B., Jiang, S., Lemahieu, V., McCormick, F., Francke, U., & Abo, A. (1996). Wiskott–Aldrich syndrome protein, a novel effector for the GTPase CDC42Hs, is implicated in actin polymerization. *Cell, 84,* 723–734.

162. Snapper, S., Rosen, F., Mizoguchi, E., Cohen, P., Khan, W., Liu, C., Hagemann, T., Kwan, S., Ferrini, R., Davidson, L., Bhan, A., & Alt, F. (1998). Wiskott–Aldrich syndrome protein-deficient mice reveal a role for WASP in T but not B cell activation. *Immunity, 9,* 81–91.

163. Snapper, S., Takeshima, F., Anton, I., Liu, C.-H., Thomas, S., Nguyen, D., Dudley, D., Fraser, H., Purich, D., Klein, C., Bronson, R., Mulligan, R., Southwick, F., Geha, R., Goldberg, M., Rosen, F., Hartwig, J., & Alt, F. (2001). N-WASP deficiency reveals distinct pathways for cell surface projections and the actin-based motility of intracellular microorganisms. *Nature Cell Biol, 3,* 897–904.

164. Ramesh, N., Antón, I., Martínez–Quiles, N., & Geha, R. (1999). Waltzing with WASP. *Trends Cell Biol, 9,* 15–19.

165. Egile, C., Loisel, T., Laurent, V., Li, R., Pantaloni, D., Sansonetti, S., & Carlier, M.-F. (1999). Activation of the CDC42 effector N-WASP by the *Shigella flexneri* IcsA protein promotes actin nucleation by Arp2/3 complex and bacterial actin-based motility. *J Cell Biol, 146,* 1319–1332.

166. Higgs, H., & Pollard, T. (2000). Activation by Cdc42 and PIP2 of Wiskott–Aldrich syndrome protein (WASp) stimulates actin nucleation by Arp2/3 complex. *J Cell Biol, 150,* 1311–1320.

167. Uruno, T. J., Zhang, P., Fan, Y.-X., Egile, C., Li, R., Mueller, S., & Zhan, X. (2001). Activation of Arp2/3 complex-mediated actin polymerization by cortactin. *Nature Cell Biol, 3,* 259–266.

168. Ozawa, K., Kashiwada, K., Takahashi, M., & Sobue, K. (1995). Translocation of cortactin (p80/85) to the actin-based cytoskeleton during thrombin receptor-mediated platelet activation. *Exp Cell Res, 221,* 197–204.

169. Weed, S., Karginov, V., Schafer, D., Weaver, A., Kinley, A., Cooper, J., & Parsons, J. (2000). Cortactin localization to sites of actin assembly in lamellipodia requires interactions with F-actin and the Arp2/3 complex. *J Cell Biol, 151,* 29–40.

170. Head, J., Jiang, D., Li, M., Zorn, L., Schaefer, E., Parsons, J., & Weed, S. (2003). Cortactin tyrosine phosphorylation requires rac1 activity and association with the cortical actin cytoskeleton. *Mol Biol Cell, 14,* 3216–3229.

171. Goode, B., Rodal, A., Barnes, G., & Drubin, D. (2001). Activation of the Arp2/3 complex by the actin filament binding protein Abp1p. *J Cell Biol, 153,* 627–634.

172. Janmey, P. (1994). Phosphoinositides and calcium as regulators of cellular actin assembly and disassembly. *Annu Rev Physiol, 56,* 169–191.

173. Janmey, P. (1998). The cytoskeleton and cell signaling component localization and mechanical coupling. *Physiol Rev, 78,* 763–781.

174. Kovacsovics, T., Bachelot, C., Toker, A., Vlahos, C., Duckworth, B., Cantley, L., & Hartwig, J. (1995). Phosphoinositide 3-kinase inhibition spares actin assembly in activating platelets, but reverses platelet aggregation. *J Biol Chem, 270,* 11358–11366.

175. Tolias, K., Hartwig, J., Ishihara, H., Shibisaki, Y., Erickson, J., Cantley, L., & Carpenter, C. (2000). Type Iα phosphatidylinositol-4-phosphate 5-kinase mediates Rac-dependent actin assembly. *Curr Biol, 10,* 153–156.

176. Abrams, C., Zhang, J., Downes, C., Tang, X., Zhao, W., & Rittenhouse, S. (1996). Phosphopleckstrin inhibits Gβγ-activable platelet phosphatidylinositol-4,5-bisphosphate 3-kinase. *J Biol Chem, 271,* 25192–25197.

177. Tolias, K., Cantley, L., & Carpenter, C. (1995). Rho family GTPases bind to phosphoinositide kinases. *J Biol Chem, 270,* 17656–17659.

178. Honda, A., Nogami, M., Yokozeki, T., Yamazaki, M., Nakamura, H., Watanage, H., Kawamoto, K., Nakayama, K., Morris, A., Frohman, M., & Kanaho, Y. (1999). Phosphatidylinositol 4-phosphate 5-kinase a is α downstream effector of the small G protein ARF6 in membrane ruffle formation. *Cell, 99,* 521–532.

179. Zhang, Q., Calafat, J., Janssen, H., & Greenberg, S. (1999). ARF6 is required for growth factor- and rac-mediated membrane ruffling in macrophages at a stage distal to rac membrane ruffling. *Mol Cell Biol, 19,* 8158–8168.

180. Cichowski, K., Brugge, J., & Brass, L. (1996). Thrombin receptor activation and integrin engagement stimulate tyrosine phosphorylation of the proto-oncogene product, p95[vav], in platelets. *J Biol Chem, 271,* 7544–7550.

181. Fischer, K.-D., Kong, Y.-Y., Nishina, H., Tedford, K., Marengere, L., Kozieradzki, I., Sasaki, T., Starr, M., Chan, G., Garderer, S., Nghiem, M., Bouchard, D., Barbacid, M., Bernstein, A., & Penninger, J. (1998). Vav is a regulator of cytoskeletal reorganization mediated by the T-cell receptor. *Curr Biol, 8,* 554–562.

182. Abe, K., Rossman, K., Liu, B., Ritola, K., Chiang, D., Campbell, S., Burridge, K., & Der, C. (2000). Vav2 is an activator of cdc42, rac1, and rhoA. *J Biol Chem, 275,* 10141–10149.

183. Azuma, T., Witke, W., Stossel, T., Hartwig, J., & Kwiatkowski, D. (1998). Gelsolin is a downstream effector of rac for fibroblast motility. *EMBO J, 17,* 1362–1370.

184. Falet, H., Chang, G., Brohard–Bohn, B., Rendu, F., & Hartwig, J. (2005). Integrin αIIbb3 signals lead cofilin to

accelerate platelet actin dynamics. *Am J Physiol Cell Physiol, 289,* C819–C825.

185. Davidson, M., & Haslam, R. (1994). Dephosphorylation of cofilin in stimulated platelets roles for GTP-binding protein and Ca2+. *Biochem J, 301,* 41–47.

186. Burridge, K., & Fath, K. (1989). Focal contacts: Transmembrane links between the extracellular matrix and the cytoskeleton. *Bioessays, 10*(4), 104–108.

187. Burridge, K., Nuckolls, G., Otey, C., Pavalko, F., Simon, K., & Turner, C. (1990). Actin-membrane interaction in focal adhesions. *Cell Differ Dev, 32*(3), 337–342.

188. Burridge, K., & Mangeat, P. (1984). An interaction between vinculin and talin. *308,* 744–746.

189. Pavalko, F., Otey, C., & Burridge, K. (1989). Identification of a filamin isoform enriched at the ends of stress fibers in chicken embryo fibroblasts. *J Cell Sci, 94,* 109–118.

190. Turner, C. E., Kramarcy, N., Sealock, R., & Burridge, K. (1991). Localization of paxillin, a focal adhesion protein, to smooth muscle dense plaques, and the myotendinous and neuromuscular junctions of skeletal muscle. *Exp Cell Res, 192*(2), 651–655.

191. Pavalko, F. M., Otey, C. A., Simon, K. O., & Burridge, K. (1991). Alpha-actinin: A direct link between actin and integrins. *Biochem Soc Trans, 19*(4), 1065–1069.

192. Nakamura, F., Huang, L., Pestonjamasp, K., Luna, E., & Furthmayr, H. (1999). Regulation of F-actin binding to platelet moesin in vitro by both phosphorylation of threonine 558 and polyphosphatidylinositides. *Mol Biol Cell, 10,* 2669–2685.

193. Nachmias, V., & Golla, R. (1991). Vinculin in relation to stress fibers in spread platelets. *Cell Motil Cytoskel, 20,* 190–202.

194. Haimovich, B., Regan, C., DiFazio, L., Ginalis, E., Ji, P., Purohit, U., Rowley, R., Bolen, J., & Greco, R. (1996). The FcγRII receptor triggers pp125FAK phosphorylation in platelets. *J Biol Chem, 271,* 16332–16337.

195. Huang, M.-M., Lipfert, L., Cunningham, M., Brugge, J., Ginsberg, M., & Shattil, S. (1993). Adhesive ligand binding to integrin αIIbβ3 stimulates tyrosine phosphorylation of novel protein substrates before phosphorylation of pp125FAK. *J Biol Chem, 122,* 473–483.

196. Lipfert, L., Haimovich, B., Schaller, M., Cobb, B., Parsons, J., & Brugge, J. (1992). Integrin-dependent phosphorylation and activation of the protein tyrosine kinase pp125FAK in platelets. *J Cell Biol, 119,* 905–912.

197. Rankin, S., & Rozengurt, E. (1994). Platelet-derived growth factor modulation of focal adhesion kinase (p125$^{FAK}$) and paxillin tyrosine phosphorylation in Swiss 3T3 cells. *J Biol Chem, 269,* 704–710.

198. Grondin, P., Plantavid, M., Sultan, C., Breton, M., Mauco, G., & Chap, H. (1991). Interaction of pp60c-src, phospholipase C, inositol-lipid, and diacylglycerol kinases with the cytoskeleton of thrombin-stimulated platelets. *J Biol Chem, 266,* 15705–15709.

199. Horvath, A., Muszbek, L., & Kellie, S. (1992). Translocation of pp60$^{c\text{-}src}$ to the cytoskeleton during platelet aggregation. *EMBO J, 11,* 855–861.

200. Meng, F., & Lowell, C. (1998). A β$_1$ integrin signaling pathway involving src-family kinases, Cbl and PI-3 kinase is required for macrophage spreading and migration. *EMBO J, 17,* 4391–4403.

201. Wu, H., Reynolds, A., Kanner, S., Vines, R., & Parsons, J. (1991). Identification and characterization of a novel cytoskeleton-associated pp60$^{src}$ substrate. *Mol Cell Biol, 11,* 5113–5124.

202. Jackson, S., Schoenwaelder, S., Yuan, Y., Rabinowitz, I., Salem, H., & Mitchell, C. (1994). Adhesion receptor activation of phosphatidylinositol 3-kinase. Von Willebrand factor stimulates the cytoskeletal association and activation of phosphatidylinositol 3-kinase and pp60$^{c\text{-}src}$ in human platelets. *J Biol Chem, 269,* 27093–27099.

203. Niederman, R., & Pollard, T. (1975). Human platelet myosin II. In vitro assembly and structure of myosin filaments. *J Cell Biol, 67,* 72–92.

204. Nachmias, V. T. (1985). Reversible association of myosin with the platelet cytoskeleton. *Nature, 313,* 70–72.

205. Essler, M., Amano, M., Kruse, H.-J., Kaibuchi, K., Weber, P., & Aepfelbacher, M. (1998). Thrombin inactivates myosin light chain phosphatase via rho and its target rho kinase in human endothelial cells. *J Biol Chem, 273,* 21867–21874.

206. Klages, B., Brandt, U., Simon, M., Schultz, G., & Offermanns, S. (1999). Activation of G12/G13 results in shape change and rho/rho kinase-mediated myosin light chain phosphorylation in mouse platelets. *J Cell Biol, 144,* 745–754.

207. Bernard, J., & Soulier, J. (1948). Sur une nouvelle variété de dystrophie thrombocytaire hemorragique congenitale. *Semaine des Hopitaux de Paris, 24,* 3217–3223.

208. McGill, M., Jamieson, G., Drouin, J., Cho, M., & Rock, G. (1984). Morphometric analysis of platelets in Bernard–Soulier syndrome: Size and configuration in patients and carriers. *Thromb Haemost, 52,* 37–41.

209. Berndt, M., Gregory, C., Chong, B., Zola, H., & Castaldi, P. (1983). Additional glycoprotein defects in Bernard–Soulier's syndrome: Confirmation of genetic basis by parental analysis. *Blood, 62,* 800–807.

210. Clemetson, K., McGregor, J., James, E., Dechavanne, M., & Luscher, E. (1982). Characterization of the platelet membrane glycoprotein abnormalities in Bernard–Soulier syndrome and comparison with normal by surface-labeling techniques and high-resolution two-dimensional gel electrophoresis. *J Clin Invest, 70,* 304–311.

211. Devine, D., Currie, M., Rosse, W., & Greenberg, C. (1987). Pseudo-Bernard–Soulier syndrome: Thrombocytopenia caused by autoantibody to platelet glycoprotein Ib. *Blood, 70,* 428–431.

212. Drouin, J., McGregor, J., Parmentier, S., Izaguirre, C., & Clemetson, K. (1988). Residual amounts of glycoprotein Ib concomitant with near-absence of glycoprotein IX in platelets of Bernard–Soulier patients. *Blood, 72,* 1086–1088.

213. Ludlow, L., Schick, B., Budarf, M., Driscoll, D., Zackai, E., Cohen, A., & Konkle, B. (1996). Indentification of a mutation in the GATA binding site of the platelet glycoprotein Ibβ promoter resulting in the Bernard–Soulier syndrome. *J Biol Chem, 271,* 22076–22080.

214. Marco, L., Mazzucato, M., Fabris, F., Roia, D., Coser, P., Girolami, A., Vicente, V., & Ruggeri, Z. (1990). Variant

Bernard–Soulier syndrome type bolzano. *Blood, 86,* 25–31.

215. Ware, J., Russell, S., & Ruggeri, Z. (2000). Generation and rescue of a murine model of platelet dysfunction: The Bernard–Soulier syndrome. *Proc Natl Acad Sci USA, 97,* 2803–2808.

216. Ware, J., Russell, S., Vicente, V., Scharf, R., Tomer, A., McMillan, R., & Ruggeri, Z. (1990). Nonsense mutation in the glycoprotein Ibα coding sequence associated with Bernard–Soulier syndrome. *Proc Natl Acad Sci USA, 87,* 2026–2030.

217. Kunishima, S., Miura, H., Fukutani, H., et al. (1994). Bernard–Soulier Kagoshima: Ser444 → stop mutation of the glycoprotein (gp) Ib alpha and surface expression of gpIb beta and gpIX. *Blood, 84,* 3356–3362.

218. Lanza, F., Morales, M., de la Salle, C., Cazenave, J.-P., Clemetson, K., Shimomura, T., & Phillips, D. (1993). Cloning and characterization of the gene encoding the human platelet glycoprotein V. A member of the leucine-rich glycoprotein family cleaved during thrombin-induced platelet activation. *J Biol Chem, 268,* 20801–20807.

219. Hegglin, R. (1945). Gleichzeitige Konstilutionelle veranderungen an neutrophilen and thrombocyten. *Helv Med Acta, 12,* 439–440.

220. May, R. (1909). Leukozyteneinschlusse. *Deusch Arch Klin Med, 96,* 439–440.

221. Consortium, T. M.-H. F. S. (2000). Mutations in MYH9 result in the May–Hegglin anomaly, and Fechtner and Sebastian syndromes. *Nat Genet, 26,* 103–105.

222. Kelley, M., Jawien, W., Ortel, T., & Korczak, J. (2000). Mutation of MYH9, encoding non-muscle myosin heavy chain A, in May–Hegglin anomaly. *Nat Genet, 26,* 106–108.

223. Kunishima, S., Kojima, T., Matsushita, T., Tanaka, T., Tsunusawa, M., Furukawa, Y., Nakamura, Y., Okamura, T., Amemiya, N., Nakayama, T., & Siato, H. (2001). Mutations in the NMMHC-A gene cause autosomal dominant macrothrombocytopenia with leukocyte inclusions (May–Hegglin anomaly/Sebastian syndrome). *Blood, 97,* 1147–1149.

CHAPTER 5

# Platelet Genomics and Proteomics

**Ángel García,[1] Yotis Senis,[2] Michael G. Tomlinson,[2] and Steve P. Watson[2]**

[1]*Oxford Glycobiology Institute, University of Oxford, Oxford, United Kingdom*
*Universidade de Santiago de Compostela, Santiago de Compostela, Spain*
[2]*Institute of Biomedical Research, University of Birmingham, Birmingham, United Kingdom*

## I. Introduction

The sequencing of the human genome has provided the blueprint of the cell, but it is the complexity of the protein makeup, which is determined by alternative splicing and posttranslational modifications, that governs the complexity of a cell. Thus, although molecular biology-based methodology is able to provide information on the proteins that are predicted to be expressed in the cell (the field of genomics), it is the protein studies, or proteomics, that generate the essential information required for a full understanding of cellular function. The fields of genomics and proteomics, however, are at very different stages of development. The ability to amplify RNA or DNA means that, in nucleated cells that can be purified to homogeneity using mild procedures, it is relatively easy to get a "snapshot" of expressed genes using microarrays or serial analysis of gene expression (SAGE). On the other hand, highly specialized equipment and reagents are required to map the full proteome of a cell, and currently the field is a long way from achieving this goal.

In this chapter we describe the use of genomic and proteomic approaches to provide information on proteins that are present in platelets, including discussion of the strengths and limitations of available methodologies. We then compare the two approaches and their application in facilitating our understanding of platelet function.

## II. Platelet Genomics

In this chapter we define the term *genomics* as the measurement of gene expression on a large scale and the use of the resulting information to understand how genes function together in the whole organism. The human genome is currently estimated to contain between 20,000 and 25,000 genes.[1] Two major goals in the current postgenomic era are to determine which genes are expressed in individual cell types (the "transcriptome") and to elucidate their functional roles. This

has been facilitated by two techniques for large-scale genomic profiling, namely DNA microarrays[2,3] and SAGE.[4] Public sequence databases now contain an extensive collection of DNA microarray and SAGE data from a wide range of cell and tissue types in many different species.[5–8] Bioinformatic cluster analyses have revealed a striking property of such large data sets — namely, that genes with similar expression patterns over multiple experiments tend to have similar functions.[9] This can be useful in identifying groups of genes that work together on a particular signaling pathway or cellular process, and in assigning potential functions to novel genes.

The application of genomics to the anucleate platelet is technically challenging because of relatively low messenger RNA (mRNA) levels and is of questionable value, as discussed later in this chapter. The use of genomics to study megakaryocytes may be more useful, but *in vitro* culture methods for their expansion to sufficient number and purity levels have only recently been developed. Despite these issues, several groups have recently published genomic analyses of the transcriptomes of platelets and megakaryocytes. In this section we review these studies, after a brief overview of the DNA microarray and SAGE methodologies.

### A. DNA Microarrays

DNA microarrays represent a relatively quick and cost-effective way to study gene expression on a genomic scale. Expression data can be generated for thousands of genes in many different cell types and tissues exposed to a variety of stimuli. DNA microarray technology is based on the principle that each DNA strand in a complex sample has the capacity to recognize and hybridize to its complementary sequence immobilized onto a specific region of the DNA microarray. Such DNA microarrays are generally made on glass slides in one of two ways: by printing DNA (often polymerase chain reaction [PCR] products or oligonucleotides) using a robotic arrayer, as developed in the laboratory of Patrick Brown at Stanford University;[2] or by *in situ* synthesis of oligonucle-

99

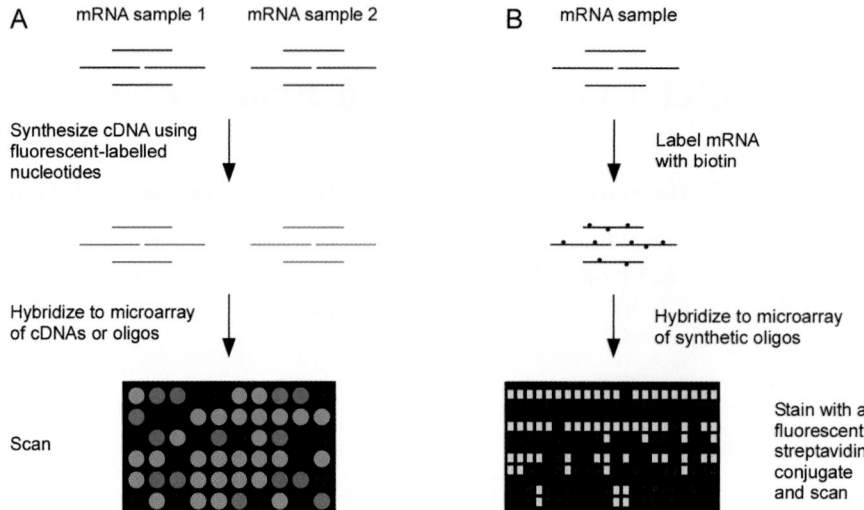

**Figure 5-1.** Schematic of the two major types of DNA microarray. A. The printed complementary DNA (cDNA) or oligonucleotide microarray. In this example, 60 data spots show the hybridization efficiency to 60 different printed cDNAs or oligonucleotides representing 60 genes. A green signal indicates relatively high gene expression in sample 1, a red signal indicates high gene expression in sample 2, a yellow signal indicates equivalent gene expression in both samples, and gray indicates lack of gene expression in both. B. The synthetic oligonucleotide microarray. In this example of the Affymetrix type of microarray, each gene on the microarray is represented by a row of 22 different 25-mer oligonucleotides, each with its own mismatch control in the row below. The data for four different genes are shown, with yellow indicating detectable hybridization and gray indicating no hybridization. Computational analyses determine whether individual genes are "present" in the mRNA sample (e.g., the gene at the top of the microarray), "equivocal," or "absent" (e.g., the gene at the bottom of the microarray).

otides using a process termed *photolithography,* as developed by Affymetrix.[3] These methods can produce microarrays with thousands of discrete DNA sequences. For example, Affymetrix GeneChips contain 600,000 different oligonucleotides per square centimeter (www.affymetrix.com).

The major advantage of the microarray technology developed by the Brown Laboratory is the relative ease with which the entire system can be set up in an academic laboratory (for instructions see www.cmgm.stanford.edu/pbrown), allowing investigators to produce their own custom microarrays in a cost-effective manner.[9] For such microarrays, each experiment is generally designed to compare the relative abundance of gene sequences between two different DNA or RNA samples (Fig. 5-1A). The two samples are initially labeled with a different fluorescent dye, such as Cy3 (red) or Cy5 (green), and are then mixed and hybridized to the microarray. A microscope is used to measure the fluorescence of the two dyes for each spot, and the red-to-green ratio is a measure of the relative abundance of the gene in the two nucleic acid samples. Multiple experimental samples can be compared indirectly through the use of multiple microarrays, each hybridized with one common reference sample, to allow normalization, and one experimental sample.[10]

The main advantage of the synthetic oligonucleotide array method is the greater reliability of the data as a result of the use of multiple "oligos" per gene, including mismatch controls for each.[11] The disadvantage is that high-tech microarray manufacture cannot be readily established in the academic setting. Instead, DNA microarrays are purchased from companies such as Affymetrix. As mentioned, the Affymetrix GeneChips typically contain 22 different 25-mer oligonucleotides (termed *probes*) for each gene. For each of these 22 probes, a mismatch probe with a single, central nucleotide substitution is also synthesized as a control for nonspecific hybridization. Experiments are designed to analyze a single DNA or RNA sample that is fluorescently labeled, hybridized to the GeneChip, and the signals detected by laser confocal fluorescence scanning (Fig. 5-1B). The relatively high reproducibility and accuracy of synthetic oligonucleotide arrays allows for direct comparison of data between experiments, after normalization of the overall median or mean intensity of each GeneChip to a common standard.[10]

### B. Serial Analysis of Gene Expression

SAGE provides a quantitative measure of mRNA levels for the vast majority of expressed genes.[4] This is an advantage over DNA microarrays that are dependent on hybridization efficiencies and are therefore not quantitative. Moreover, DNA microarrays can only detect those genes that are

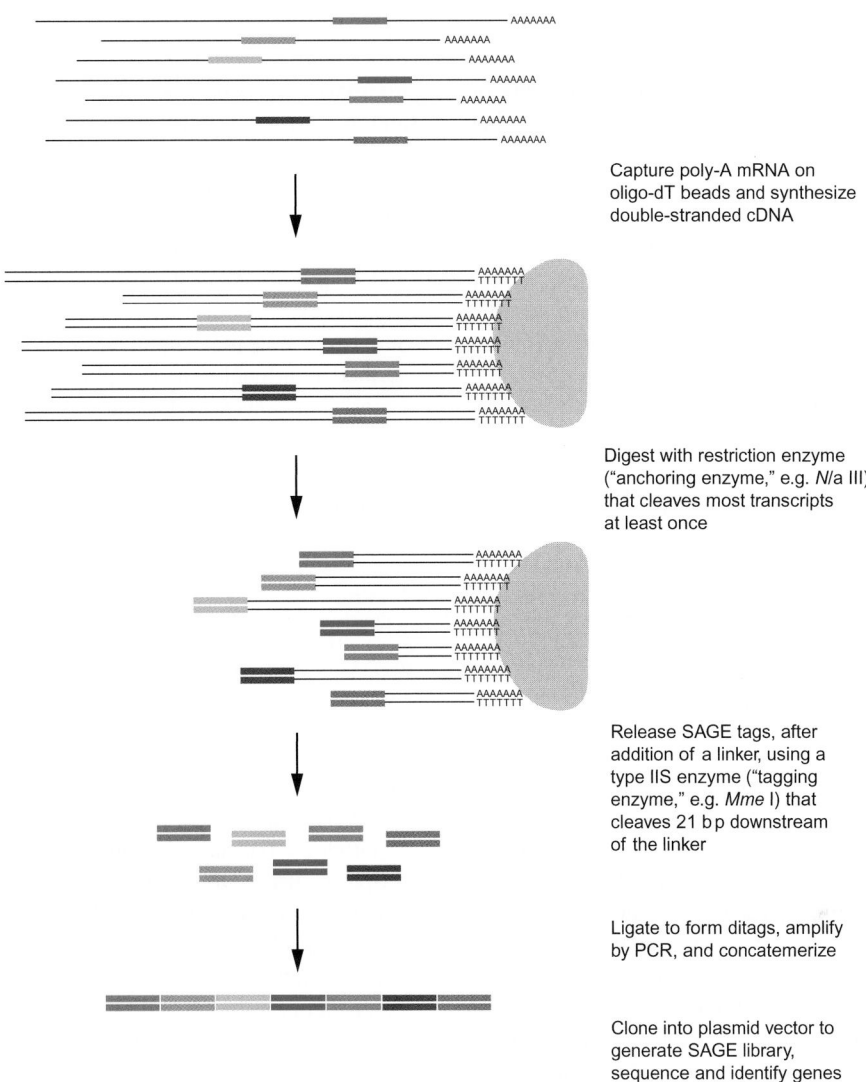

Capture poly-A mRNA on
oligo-dT beads and synthesize
double-stranded cDNA

Digest with restriction enzyme
("anchoring enzyme," e.g. *N*/a III)
that cleaves most transcripts
at least once

Release SAGE tags, after
addition of a linker, using a
type IIS enzyme ("tagging
enzyme," e.g. *Mme* I) that
cleaves 21 bp downstream
of the linker

Ligate to form ditags, amplify
by PCR, and concatemerize

Clone into plasmid vector to
generate SAGE library,
sequence and identify genes

**Figure 5-2.** Schematic of Serial Analysis of Gene Expansion (SAGE). This modification of the original SAGE method is termed *long-SAGE*. It uses *Nla III* as the anchoring enzyme and *Mme I* as the tagging enzyme to generate a 21-bp tag that is specific to each mRNA species.

present on the array. However, the generation of SAGE data is time-consuming and labor intensive in comparison with the relatively high-throughput microarrays.

The SAGE methodology is based on two principles. The first is that a short sequence tag (typically 14 bp), with a defined position on an mRNA molecule, is sufficient to identify the mRNA. The second is that a single sequencing reaction can serially identify multiple sequence tags, if they are concatenized within a single DNA molecule.[4] The basic method is shown in Figure 5-2. Briefly, mRNA is used as a template to synthesize cDNA, from which a single specific sequence tag is excised for each cDNA molecule. Single tags are joined together to form concatemers, which are cloned into an appropriate vector, to produce what is termed a *SAGE library,* and then sequenced. Each tag sequence can be matched

to an mRNA species by comparison with the "SAGEmap" reference sequence database, a public resource run by the National Center for Biotechnology Information (NCBI).[12]

SAGEmap was initially established as a repository for SAGE data generated by the scientific community, and now holds sequence data from more than 300 human and 200 mouse SAGE libraries, each containing an average of 60,000 and 78,000 sequence tags respectively (www.ncbi.nlm.nih. gov/SAGE). Because of the quantitative nature of SAGE, the sequence data from individual SAGE libraries can be directly compared to determine expression levels of individual genes (relative to the total mRNA pool) across a range of cell and tissue types. However, the identification of novel genes from SAGE data is not straightforward. For example, a recent analysis of 29 SAGE libraries showed that approximately 60% of unassigned tags that had been considered to represent novel genes were actually the result of nonrandom sequencing errors in highly abundant tags or linker sequences. Indeed, Cobbold et al.[13] concluded that after a SAGE library size exceeds 30,000 tags, artifactual tags accumulate more rapidly than new gene tags. This problem may be overcome by recent modifications to the basic SAGE method that produce longer sequence tags of 21 bp (LongSAGE)[14] or 26 bp (SuperSAGE)[15]. These longer SAGE tags are an advantage because they more reliably match to specific mRNAs. Moreover, they are of sufficient length to search the entire genome, and therefore have the potential to confirm the identity of novel genes.

## C. The Platelet and Megakaryocyte Transcriptomes

Platelets have extremely low levels of mRNA, which have been estimated at 12,500 times lower than in leukocytes, for example.[16] This is not surprising, given that platelets do not possess a nucleus and therefore cannot transcribe mRNA to replenish their residual megakaryocyte-derived mRNA. Nevertheless, upon activation, platelets can translate mRNA into protein.[17] One example of this is the inflammatory cytokine IL-1β,[18] thus suggesting how platelets might affect inflammatory processes.[19] However, the importance of protein synthesis in platelets is not clear, and it is possible that the translational machinery is simply carried over from the megakaryocyte, which has a high capacity for protein synthesis.

The relatively low level of mRNA in platelets presents two problems for the application of genomic techniques to characterize the platelet transcriptome. First, on a technical level, the platelet starting material must be highly pure, because even a low level contamination with nucleated cells or platelet-size cell fragments can make a substantial contribution to the observed transcriptome. Second, on a philosophical level, it becomes questionable whether knowledge of the platelet transcriptome is actually useful. Indeed, the fact that the mRNA level does not necessarily equate to protein level is particularly pertinent to the platelet. For example, an important platelet protein, encoded by a relatively unstable mRNA species, might not be detected in the transcriptome, whereas the opposite would be true for a megakaryocyte-derived protein with no function in a platelet, encoded by a long-lived mRNA. Thus, platelet mRNA levels may be misleading for platelet researchers.

Despite these difficulties in the application of genomics to platelets, such studies have been performed. Lindemann et al.[17,18] were the first to use array technology to identify platelet mRNA species by using a relatively small, arrayed library of 588 cDNAs on nylon membranes probed with [32]P-labeled human platelet cDNA. The first large-scale genomic experiment was done by Gnatenko et al.[20] using SAGE to profile human platelet mRNA (Table 5-1). As discussed previously, SAGE is an appropriate method for the unequivocal determination of the platelet transcriptome, but microarrays are not, because they do not detect all genes and are not quantitative. More important, Gnatenko et al.[20] used a rigorous two-step platelet purification method involving filtration and CD45[+] depletion to remove leukocytes, and verified its efficacy by an inability to "PCR-amplify" leukocyte genes. A LongSAGE library was generated of 2033 tags and, strikingly, 89% of these tags represented mitochondrial transcripts.[20] The mitochondrial genome contains 13 genes and two ribosomal subunits, which are presumably transcribed continually in the platelet, in contrast to nuclear genes. The platelet transcriptome is thus dominated by mitochondria-derived mRNA. Only 126 nuclear-derived genes were identified in the relatively small SAGE library of Gnatenko et al.,[20] and because 120 of these were detected three times or less, the data do not provide any substantial insight into the true diversity of the platelet transcriptome. It is unlikely that future researchers will consider it worthwhile to repeat the study of Gnatenko et al.[20] on a larger scale, because to generate a useful library of 30,000 nuclear tags, approximately 300,000 total sequence tags would be required, which in our experience equates to about 13,000 individual sequencing reactions. Moreover, the level of artifacts resulting from nonrandom sequence errors will be relatively high in such a large SAGE library.[13]

Other studies have attempted to analyze the platelet transcriptome using DNA microarrays (Table 5-1) that, although not truly quantitative, are able to identify nuclear-derived sequences. Gnatenko et al.[20] used an Affymetrix GeneChip in their study in addition to LongSAGE. Three independent experiments were carried out, each using purified platelets from a different healthy donor to minimize individual differences in gene expression, and average signal intensities were used by the Affymetrix software to determine whether genes were present. The percentage of platelet-expressed genes was 15 to 17% compared with typically 30 to 50% for other cell types, which is consistent with the relatively low level of nuclear-derived mRNA. It is also possible that a substantial proportion of platelet mRNA was not represented on the microarray. Of those that were detected, 32% were classed as "unknown function," 11% as "metabolism," and 11% as "receptor/signaling" genes. An analysis of the top 50 platelet-expressed genes found that more than 75% were not expressed by leukocytes, suggesting that platelet and leukocyte transcriptomes have distinct profiles.[20]

McRedmond et al.[21] used an identical Affymetrix GeneChip to that used in the study by Gnatenko et al.[20] Two GeneChip experiments were performed using pooled platelets from multiple donors and, despite the fact that filtration and CD45[+] depletion were not used, the data were in good agreement with that of Gnatenko et al.[20] in that 45 genes were common to both top-50 lists of identified genes.[21] Further analyses of the top 50 platelet-expressed genes iden-

**Table 5-1: DNA Microarray and SAGE Analyses of the Platelet Transcriptome**

| Study | Gene Profiling Method(s) | mRNA Source | Number of Genes Identified | Key Findings |
|---|---|---|---|---|
| Gnatenko et al.[20] | a) Affymetrix HG-U95Av2 human array (12,600 genes) <br> b) LongSAGE | Human platelets collected by apheresis, filtered through a 5 μm filter, depleted of CD45$^+$ cells | a) 2,147 (17% of all genes on the array) <br> b) 2,033 total tags, including 233 non-mitochondrial tags representing 126 genes | a) Major classes of gene function were "unknown function" (32%), "metabolism" (11%), "receptor/signalling" (11%) <br> b) 89% of tags represented mitochondrial transcripts |
| Bugert et al.[22] | MWG Biotech 10K oligonucleotide (50-mer) human array (9,850 genes) | Human platelets from pooled platelet concentrates, depleted 3× with leukocyte depletion filters | 1,526 (15% of all genes on the array) – data available online: www.ma.uni-heidelberg.de/inst/iti/plt_array.xls | The major classes of gene function that were overrepresented in the identified genes were "glycoproteins/integrins" and "receptors" |
| McRedmond et al.[21] | Affymetrix HG-U95Av2 human array (12,600 genes) | Human platelets from whole blood collected by centrifugation | 2,928 (23% of all genes on the array) | The platelet transcriptome correlated well with the proteome, analyzed in the same study |

tified 46% of these as *bona fide* platelet proteins, as detected using proteomics,[21] suggesting that the platelet transcriptome mirrors the proteome.

A third microarray study was reported by Bugert et al.[22] using printed 50-mer oligonucleotides. Compared with Affymetrix GeneChips, such printed microarrays are even less suited to a comprehensive transcriptome analysis of a single cell type because the data are dependent on hybridization efficiency to a single oligonucleotide, rather than multiple oligonucleotides that include mismatch negative controls. As discussed earlier, the printed microarrays are most useful for the high-throughput analyses of multiple cell types to identify, for example, genes that are specific to platelets. Nevertheless, the study by Bugert et al.[22] was well conducted; platelets were from multiple donors and were depleted of leukocytes by filtration, which was confirmed by an inability to PCR-amplify genomic DNA.[22] Six individual microarray experiments were performed and the data, presented as mean signal intensity, can be accessed at the University of Heidelberg website (Table 5-1). To summarize the data briefly, 15% of the genes on the microarray were identified as platelet expressed, a value consistent with the Affymetrix studies,[20,21] and glycoproteins, integrins, and receptors were overrepresented.[22]

Confirmation of SAGE and DNA microarray data is often desirable for a gene of interest, but standard reverse transcriptase PCR is not straightforward for platelet mRNA because of the potential for contamination with leukocyte mRNA. Fink et al.[16] have recently addressed this problem through the development of a real-time PCR method following laser-assisted microdissection. In brief, ultraviolet laser photolysis was used to select a leukocyte-free area corresponding to about 50,000 platelets from a cytospin preparation. Microdissection was used to extract these platelets to a reaction tube for subsequent real-time PCR analyses to confirm mRNA expression.[16]

The megakaryocyte is more amenable to genomic analysis than the platelet because it is nucleated and therefore transcriptionally active. One potential problem is the isolation of megakaryocytes in sufficient number and purity, but this obstacle has been overcome in recent years by the development of *ex vivo* culture systems to expand megakaryocytes from bone marrow progenitors.[23] One can now envisage the identification of genes that are important for megakaryocytopoiesis, through DNA microarray analyses of megakaryocytes at different stages of differentiation. Furthermore, it is reasonable to assume that the transcriptome of a mature megakaryocyte would resemble the nuclear-derived transcriptome of a platelet, and SAGE would be a useful tool for this type of gene profiling.

Three groups have reported genomic analyses of human megakaryocytes (Table 5-2). Kim et al.[24,25] have used both SAGE[24] and a printed 50-mer oligonucleotide microarray[25] to compare megakaryocyte and nonmegakaryocyte fractions of bone marrow cells that had been expanded *ex vivo* with thrombopoietin. The two different profiling methods provided data that correlated well, and although the expression levels of most genes were similar between the megakaryocyte and nonmegakaryocyte fractions, substantial differences were still found for many genes.[24,25] Approximately half of the top 50 most abundant SAGE tags were housekeeping ribosomal genes involved in protein synthesis.[24] We have

**Table 5-2: DNA Microarray and SAGE Analyses of the Megakaryocyte Transcriptome**

| Study | Gene Profiling Methods | mRNA Source | No. of Genes Identified | Key Findings |
|---|---|---|---|---|
| Kim et al.[24] | SAGE | Human meg (CD41$^+$) and nonmeg (CD41$^-$) fractions from cord blood CD34$^+$ cells expanded *ex vivo* with TPO for 10 days | 20,580 tags from meg 18,329 tags from nonmeg, representing a total of 8,976 genes | The major transcripts in meg versus nonmeg samples |
| Kim[25] | Macrogen II-10 K oligonucleotide (50-mer) array (10,108 human genes) | Same mRNA as above | Not reported | A correlation between SAGE and DNA microarray data |
| Shim et al.[27] | Affymetrix HG-U95Av2 human array (12,600 genes) | Human megs from bone marrow CD34$^+$CD38$^{lo}$ cells expanded *ex vivo* with TPO, SCF, IL-3, and IL-6 for 0 or 10 days | 229 genes upregulated in megs, 74 genes downregulated in megs; data available online at ArrayExpress[7] | Eight of the top 25 most highly upregulated genes were hemostasis related. Of the 304 differentially expressed genes, 85% of adhesion/receptor and 68% of signal transduction genes were upregulated; 70% of transcription factor genes were downregulated. |
| Tenedini et al.[28] | Affymetrix HG-U133A human array (14,500 genes) | Human megs (CD41$^+$) from bone marrow CD34$^+$ cells, either controls or ET patients, expanded *ex vivo* with TPO for 14 days | 4,655 genes from normal megs, 4,637 genes from ET megs; data available online at the NCBI GEO | In ET megs, several proapoptotic genes were downregulated and several antiapoptotic genes were upregulated. |

ET, essential thrombocythemia; IL, interleukin; megs, megakaryocytes; SCF, stem cell factor; TPO, thrombopoietin.

recently generated a LongSAGE library of 53,046 sequence tags to analyze the transcriptome of mouse megakaryocytes, cultured *ex vivo* from bone marrow progenitors.[26] Consistent with Kim et al.,[24] our most abundant tags were dominated by ribosomal genes. To identify megakaryocyte-specific genes, we discarded ribosomal genes and compared our library with 30 other SAGE libraries, from a variety of mouse cell types, deposited in the NCBI SAGEmap database.[12] Interestingly, of the 28 most megakaryocyte-specific genes, 18 encoded transmembrane proteins. These included integrin $\alpha_{IIb}$, glycoprotein (GP) Ibα, GPIbβ, GPV, GPIX, P-selectin, TLT-1, PAR3, P2X$_1$, the thrombopoietin receptor, and several novel receptors.[26] This type of analysis highlights both the utility of SAGE and the information that can be gleaned through genomic analyses of large data sets.

Two recent studies have used Affymetrix GeneChips to profile human megakaryocytes. Shim et al.[27] have compared *ex vivo*-expanded megakaryocytes with their CD34$^+$CD38$^{lo}$ bone marrow progenitors. Tenedini et al.[28] have compared megakaryocytes expanded *ex vivo* from bone marrow CD34$^+$ progenitor cells of either healthy control subjects or essential thrombocythemia patients. The results of these studies are summarized in Table 5-2. Such comparative types of experi-

ments, although not exhaustive because of the limitations of DNA microarrays, are powerful methods for the identification of genes involved in megakaryocyte differentiation, effector function, and tumorigenesis.

The DNA microarray and SAGE studies described here have provided new insights into the platelet and megakaryocyte transcriptomes. However, a new genomic profiling technique, termed *massively parallel signature sequencing* (MPSS),[29] will enable a complete characterization of these transcriptomes and may be the only way to achieve this for the platelet. In common with SAGE, MPSS generates short (20-bp) sequence tags that can be used to identify the parent mRNA species. The advantage of MPSS is its capacity to sequence one million tags simultaneously. The sequencing is performed on microbeads using multiple cycles of a ligation-based method that sequences 4 bp per cycle.[29] MPSS is potentially less prone to sequencing errors and is certainly more cost-effective than LongSAGE. In our experience, three million MPSS tags can be generated at a similar cost to 150,000 LongSAGE tags. MPSS is currently available through Lynx Therapeutics (www.lynxgen.com/wt/home.php3), where the entire process is carried out upon mRNA provided by the researcher. Although the technique is still in its infancy, several groups have published MPSS data, including a recent identification of two to three million tags from each of 32 normal human tissues and two cancer cell lines.[30]

In summary, several recent genomic studies have profiled mRNA expression in platelets and megakaryocytes (Tables 5-1 and 5-2). Further SAGE and, more ideally, MPSS studies are required to provide the complete platelet and mega-karyocyte transcriptomes. Such data will be useful for the identification of new platelet/megakaryocyte-specific genes and thus new potential drug targets for anti- or proplatelet therapy. Furthermore, the future generation of a "platelet chip" (a synthetic oligonucleotide microarray representing all platelet-expressed genes and their polymorphisms) would be useful in the clinic for high-throughput characterization of platelet-based bleeding disorders.

## III. Platelet Proteomics

The complexity of an organism is primarily within the makeup of its proteins — the proteome. This can be defined as the whole set of proteins present in a cell, tissue, body fluid, or organism at one time, including all the isoforms and posttranslational variants.[31] As a result of differential splicing and translation, each human gene may encode many different proteins. Taking into account the numerous posttranslational modifications (PTMs), the estimated 20,000 to 25,000 human genes may generate approximately one million distinguishable functional entities at the protein level. The analysis of the complete protein set is an analytical challenge that is made more difficult by the broad dynamic range of protein quantities expressed and the fact that proteins, unlike DNA sequences, cannot be amplified.

### A. Mass Spectrometry-Based Proteomics

Mass spectrometry (MS) is currently the most important proteomic tool. During the last decade, changes in MS instrumentation and techniques have revolutionized the analysis of proteins. Two ionization techniques developed in the late 1980s were the reason for this step forward and contributed to the development of proteomics research: electrospray ionization (ESI) and matrix-assisted laser desorption ionization (MALDI).[32,33] These methods solved the problem of generating ions from large, nonvolatile analytes, such as proteins and peptides, without significant analyte fragmentation, which is the reason they are called "soft" ionization methods.[34,35] MS-based proteomics, which routinely achieves femtomole sensitivity, has a growing role in biomedical research, where limited sample material is available.

Proteomics technology allows a comprehensive and efficient analysis of the proteome. Proteome analysis involves separating up to several thousand proteins at a time, most commonly by two-dimensional gel electrophoresis (2-DE).[36] 2-DE separates proteins according to their isoelectric point (pI) and size (molecular weight). Recent technical improvements in 2-DE-based proteomics include a better solubilization of membrane proteins; enrichment of low-abundance proteins by sample prefractionation; novel, highly sensitive water-soluble fluorescent dyes; and narrow pI range 2-D gels (*zoom* gels) for a detailed analysis of the region of interest.[37–39] After staining of the gels, information on the presence of proteins, their up- or downregulation, and changes in the pI resulting from potential PTMs, is provided by extensive image analysis. Proteins of interest can be excised, in-gel trypsinized, and identified by application of powerful liquid chromatography–electrospray ionization–tandem mass spectrometry (LC-EI-MS/MS), which provides data on the protein amino acid sequence (Fig. 5-3). The use of subcellular prefractionation techniques in combination with one-dimensional sodium dodecyl sulfate–polyacrylamide gel electrophoresis (1-D SDS-PAGE) and LC-MS/MS or multidimensional nanoscale capillary LC-MS/MS overcomes problems that arise from the fact that 2-DE cannot properly resolve many high-molecular weight proteins or very basic and hydrophobic ones.[40,41] Powerful databases, such as SWISS-PROT and TrEMBL, can be used for protein identification. This approach, however, is not always absolutely accurate, and possible explanations for a failure in protein identification are (a) protein digestion, peptide recovery, and peptide mass spectrometric analysis become unmatched because of low protein quantity in conjunction with peptide hydrophobicity; (b) tryptic cleavage does not result in peptides of the appropriate size for detection in the experimental mass range; (c) corresponding gene sequences are not yet entered into the genome databases; and (d) ambiguous patterns of fragmentation that result in incorrect sequence predictions. The last of these possibilities can be kept to a minimum through the use of stringent criteria on which to accept a sequence prediction.[42]

### B. Methods of Analysis of the Platelet Proteome

Platelets are a good target for a proteomics approach because they are anucleate, can be obtained in high yield, and are relatively easily separated from other blood cells. In our experience, 100 mL of blood yield approximately $2 \times 10^{10}$ platelets, from which it is possible to obtain 16 to 24 mg of protein, depending on the method of protein extraction. As an example, a high-resolution *zoom* 2-D gel (18 × 18 cm) may need on the order of 700 µg of protein for a full analysis; therefore, a large amount of blood is not required.

**1. Sample Preparation.** The method used for platelet isolation is crucial. The goal is to obtain a highly purified

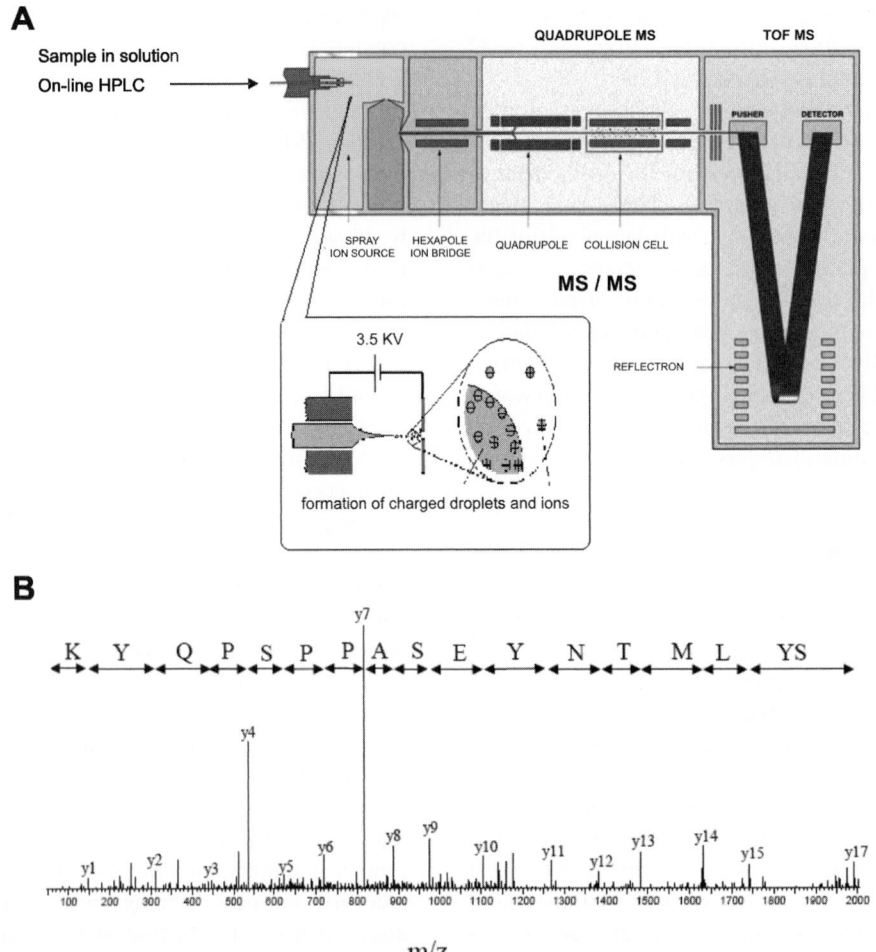

**Figure 5-3.** Peptide sequencing using MS/MS. A. Representation of a typical quadrupole-time-of-flight (TOF) mass spectrometer (modified with permission from Waters Corporation). Peptides are separated by high-pressure liquid chromatography (HPLC) in fine microcapillaries and are eluted into an electrospray ion source, where they are nebulized in small, highly charged droplets. After evaporation, multiply protonated peptides, $(M + nH)^{n+}$ and enter the mass spectrometer. A mass spectrum of all the peptides eluted at this time point is generated. A given peptide can be selected by the computer for MS/MS analysis. The peptide is first isolated at the quadrupole section of the mass spectrometer and is fragmented by collision with gas. The corresponding peptide fragments enter the TOF section and are pulsed onto the detector, where they are recorded to produce the MS/MS spectrum. B. Sequence analysis of peptides by collision-induced dissociation (CID). CID causes cleavage at the various amide bonds of the peptide to generate a series of fragments that differ by a single amino acid residue. The main ions generated are ions of type y and b, which contain the C terminus or the N terminus respectively. As an example, an MS/MS profile that corresponds to a peptide that allowed the identification of *protein C7orf24* (Swiss-Prot accession number: O75223) in human platelets is shown. The y ions that resulted from the peptide fragmentation are highlighted. (From García et al.,[77] with permission.)

preparation, while minimizing platelet activation. It is important to minimize contamination from other blood cells, such as red cells and leukocytes, and from plasma as far as possible; otherwise, the most abundant proteins in these sources will be detected. Contamination can be minimized by taking the upper third of the platelet-rich plasma (PRP) and using leukocyte removal filters.[43] Platelets resuspended from PRP contain a low concentration of plasma proteins and should therefore undergo an additional centrifugation to minimize this source of contamination. Purity

of isolated platelets can be estimated by microscopic inspection and by the absence of proteins specific for other blood cells, such as globins.

It is essential that extreme care is taken to avoid activation of platelets during their preparation. There is, however, no standard accepted method of platelet preparation and it is likely that the choice of anticoagulant, buffer, centrifugation speeds, and general methodology will influence the nature of the proteome. For example, the use of cAMP-elevating agents will lead to phosphorylation of a range of substrates for protein kinase A, thereby inducing formation of new protein forms.

A key issue is to improve and standardize the conditions for protein extraction. In particular, it is essential to inhibit proteases and quantitatively enriching proteins, while leaving behind substances such as salts, lipids, and nucleic acids, which would interfere with any further proteomic analysis. The method of choice for protein extraction, including the sample buffer in which proteins have to be solubilized prior to their separation/analysis, varies depending on the proteomics approach. In the case of 2-DE, it is extremely important to avoid the presence of salts in the sample, because this would interfere with the isoelectric focusing (IEF) step. To obtain a highly purified protein sample and improve the quality of the 2-D gels, especially in the basic region, it is advisable to carry out protein precipitation and delipidation. Protein precipitation, and subsequently solubilizing the pellet in IEF-compatible sample solution, is generally used to concentrate and selectively separate proteins in the sample from interfering substances. Our preferred option is to lyse platelets in liquid nitrogen, followed by TCA/acetone protein precipitation in the presence of protease inhibitors. The protein pellet is then washed in cold acetone to remove lipids and is finally resuspended in a sample buffer that retains the native charge.[44] Sample preparation and sample buffer vary between research groups, thereby influencing interreproducibility of platelet proteomics analyses.

**2. 2-D Gel Electrophoresis.** 2-DE has been used to analyze platelet biology for more than 30 years. The very early studies separated proteins using nonreducing and reducing SDS-PAGE, and led to the identification and naming of many of the platelet glycoproteins.[45] Subsequently, SDS-PAGE was combined with IEF to obtain a greater separation of proteins. Prior to the application of MS, however, researchers were reliant on the use of immunoblotting and N-terminal sequencing approaches for mapping of the proteome. For example, in 1995, Gravel and coworkers[46] first attempted to map cytosolic and membrane platelet proteins by 2-DE. They identified approximately 25 spots using three different methods: matching of the platelet gels with other 2-DE reference maps, immunoblotting with chemiluminescence detection, and N-terminal sequencing. The proteins identified corresponded to various G-protein subunits, cytoskeletal proteins, and proteins common to liver, blood cells, and plasma.[46]

However, as discussed earlier, a major leap forward in proteomics came about almost 15 years ago with the discovery of electrospray and MALDI, methods that allow ionization of large biomolecules. By taking advantage of these ionization techniques, MS-based proteomics has become the key tool in protein analysis. In recent years, a small number of research groups have focused on a detailed analysis of the platelet proteome, many by taking advantage of the powerful combination of 2-DE for protein separation and MS for protein analysis (Table 5-3).

In 2000, Marcus and collaborators[47] identified 186 protein features analyzed by MALDI–time-of-flight (TOF)–MS following their separation by 2-DE. Many of these proteins were cytoskeletal.[47] This study represented the first pI 3- to 10-platelet proteome 2-DE map. Two reports from the Glycobiology Institute at Oxford University make up the broadest investigation so far on the 2-DE human platelet proteome.[43,44] Together, the two reports provide a high-resolution 2-DE platelet proteome map comprising more than 2000 protein features. Proteins were separated by 2-DE using narrow pH gradients during IEF (4–5, 5–6, 4–7, and 6–11), and 9 to 16% SDS-PAGE gradient gels in the second dimension (18 × 18 cm; Fig. 5-4). Gels were stained with a highly sensitive fluorescent dye, and the corresponding protein spots were excised, in-gel trypsinized, and analyzed by LC-MS/MS. Overall, more than 1000 proteins were identified, corresponding to 411 open reading frames. The list was rich in proteins involved in signaling (24%) and protein synthesis (22%). The large number of signaling proteins, which includes kinases, G proteins, adapters, and protein substrates, was to be expected in view of the ability of platelets to undergo powerful and rapid activation after damage to the vasculature, although it should be added that this is also a very broad group, with several proteins being allocated because of their regulation by phosphorylation. Nevertheless, it is a striking feature that many well-known signaling proteins were not detected using this technology, such as, for example, the major components of the collagen receptor GPVI-based signaling cascade, including Syk and PLCγ2, as presumably many are present at levels that fall below the detection limits of this approach. On the face of it, the presence of a large number of proteins involved in protein translation, transcription, and regulation of the cell cycle is a surprise in view of the limited degree of protein synthesis that takes place in platelets, which is of uncertain significance (as discussed earlier). However, the presence of these proteins may simply represent an unavoidable carryover of protein synthesis machinery from the megakaryocyte. Strikingly, approximately 45% of the identified proteins had never previously been reported in platelets, including 15 hypothetical proteins that had not been described in any other cell type, emphasizing the power of the approach.

One class of proteins that is particularly underrepresented in the 2-DE proteome analysis is that of membrane proteins, because many of them come out of solution during the IEF step or are not solubilized by the IEF sample buffer. An estimated 30% of all proteins can be described as membrane proteins, and are made up of both transmembrane proteins and proteins that associate with the membrane, for example, as a consequence of palmitoylation or myris-

**Table 5-3: Major Contributions to Platelet Proteomics Research**

| Study | Separation Method | MS Approach | Key Achievements |
|---|---|---|---|
| Marcus et al.[47] | 2-DE (*pI* 3–10) | MALDI-TOF | First platelet 2-DE proteome map; 186 protein features identified |
| O'Neill et al.[43] García et al.[44] | 2-DE (*zoom* gels) | LC-MS/MS (Q-TOF) | Broadest investigation so far on the human 2-DE platelet proteome. More than 1000 protein features identified, corresponding to 411 different ORFs. |
| Maguire et al.[73] | IP plus 2-DE (pI 3–10) | MALDI-TOF | Immunoprecipitation-based approach to analyze the phosphotyrosine proteome from thrombin-activated platelets. Ten proteins identified by a combination of MS and Western blotting. |
| Gevaert et al.[55] Gevaert et al.[56] | COFRADIC™ | LC-MS/MS (Q-TOF) | Novel gel-free method applied to platelet proteomics research. Identification of more than 300 proteins, with special success on identifying membrane proteins. |
| Marcus et al.[74] | 2-DE (*zoom* gels) IP plus 1D-SDS-PAGE | MALDI-TOF LC-MS/MS (Ion trap) | Differential analysis of phosphorylated proteins in resting and thrombin-stimulated platelets. Seventy-seven differentially regulated protein features detected; 55 identified. Several Ser– and Thr–phosphorylation sites mapped on five proteins. |
| García et al.[52] | 1-D SDS-PAGE | LC-MS/MS (Q-TOF) | Identification of more than 20 proteins present in GEMs fractions of platelets in a basal state, including stomatin, Yes, Lyn, flotillins, and GPIV. |
| Coppinger et al.[54] | MudPIT | LC-MS/MS (Ion trap) | First proteomics analysis of the platelet releasate, focusing on the analysis of those proteins released by human platelets after thrombin activation. Eighty-one proteins consistently identified in at least two independent experiments. |
| García et al.[75] | 2-DE (*zoom* gels) | LC-MS/MS (Q-TOF) | Differential proteome analysis of TRAP-activated platelets. Sixty-two differentially regulated protein features detected; 41 identified. Identifications included novel platelet signaling proteins, such as Dok-2, and first mapping of phosphorylation sites on RGS18. |

2-DE, two-dimensional gel electrophoresis; COFRADIC, combined fractional diagonal chromatography; Dok-2, downstream of tyrosine kinase-2; GEM, glycolipid-enriched membrane domains; IP, immunoprecipitation; LC-MS/MS, liquid chromatography–tandem mass spectrometry; MALDI, matrix-assisted laser desorption ionization; MudPIT, multidimensional protein identification technology; ORFs, open reading frames; Q, quadrupole; RGS18, regulator of G-protein signaling 18; SDS-PAGE, sodium dodecyl sulfate–polyacrylamide gel electrophoresis; TOF, time-of-flight; TRAP, thrombin-receptor activating peptide.

toylation. However, just more than 3% of the proteins identified in the 2-DE studies performed by the Oxford Glycobiology Institute were transmembrane proteins, which falls a long way short of the expected number.[44]

**3. Subcellular Prefractionation Combined with 1-D Gel Electrophoresis.** Strategies for prefractionation of samples prior to electrophoresis and LC-MS/MS represent the most promising approach for increasing the number of protein components that can be identified. Prefractionation methods include sequential extraction with increasingly stronger solubilization solutions, selective precipitation, affinity purification of individual proteins or protein complexes, chromatography separation methods, removal of the most abundant proteins to detect the less abundant ones, and subcellular fractionation.[48–50] Furthermore, combining these

strategies with 1-D SDS-PAGE can have a significant impact on the identification of many of the proteins that are notoriously difficult to resolve using 2-DE, including membrane proteins.

One example of the successful application of this approach is the study of specialized regions of the plasma membrane that are rich in signaling proteins, known as *glycolipid-enriched membrane domains* (GEMs) or *lipid rafts*. These regions are rich in glycosphingolipids and cholesterol, along with many key membrane receptors and intracellular signaling proteins. Highly enriched raft fractions can be isolated by sucrose gradient chromatography, using specific detergents, such as Brji58 or Triton X-100, prior to analysis by SDS-PAGE and MS. The application of this approach by two groups has led to the identification of a number of proteins not previously found using 2-DE,

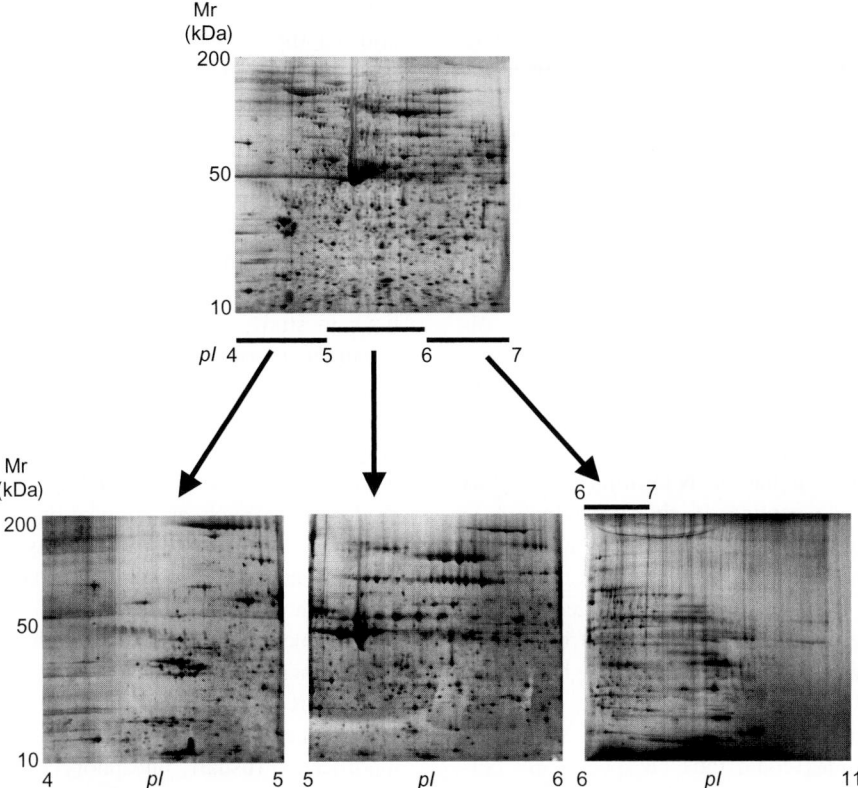

**Figure 5-4.** Two-dimensional electrophoresis proteome map of the human platelet. The narrow *pI* range gels 4–5, 5–6, 4–7, and 6–11 are shown. Protein identifications are available at www.bioch.ox.ac.uk/glycob/ogp. (From García et al.,[77] with permisson.)

including the membrane proteins flotillin-1, flotillin-2, and GPIV, and the tyrosine-protein kinases Yes and Lyn.[51,52] Thus, the use of a combination of different separation and analytical techniques is required for the analysis of the full platelet proteome.

**4. Multidimensional Protein Identification Technology and Platelet Proteomics Research.** "Shotgun proteomics" refers to the direct analysis of complex protein mixtures to generate rapidly a global profile of the protein complement within the mixture. This approach has been facilitated by the use of multidimensional protein identification technology (MudPIT), which incorporates multidimensional high-pressure liquid chromatography (LC/LC), tandem mass spectromety (MS/MS), and database-searching algorithms. This technology, initially applied to the investigation of the yeast proteome, is based on the combination of strong cation exchange and reverse-phase (RP) chromatography to achieve two-dimensional separation prior to MS/MS.[41,53]

MudPIT has been recently applied to platelet proteomics research to characterize the platelet releasate.[54] First, platelets were stimulated with 0.5 U/mL thrombin for 3 minutes to achieve maximum release of all granule contents. The releasate fraction was harvested by centrifugation, digested with trypsin, and the resulting peptides loaded under pressure onto a nanocapillary column containing both strong cation exchange and RP materials. Seven successive salt elutions and high-pressure liquid chromatography (HPLC) cycles were used to separate the peptides that were then analyzed by MS/MS, leading to protein identification. In this way, 81 proteins were identified in three experiments. Reproducibility of identification for a repeat analysis of a single sample was about 80%. This suggested that about 20% of the variation in the results arises from missed identifications resulting from saturation of the MudPIT analysis. Interestingly, some of the proteins identified were not previously attributed to platelets, and three of them — secretogranin III, cyclophilin A, and calumenin — were confirmed to localize in platelets and to be released upon activation. Furthermore, these three proteins were also specifically identified in human atherosclerotic lesions, with evident clinical implications.

**5. Shotgun Proteomics: The COFRADIC™ Approach.** In 2002, Gevaert and colleagues[55] reported a novel gel-free shotgun proteomics technology called combined fractional diagonal chromatography (COFRADIC). Central to this method is a modification reaction that alters the retention behavior of specific peptides on RP columns prior to MS/MS analysis. Initially, COFRADIC was developed for the isolation of methionine-containing peptides from a tryptic peptide mixture, prior to protein identification by MS, and applied to investigate the *Escherichia coli* proteome.[55] The procedure is about 100 times more sensitive than 2-DE analysis and can be carried out in a fully automated manner.

COFRADIC technologies were recently applied to platelet proteomics research. In a first paper, Gevaert and collaborators[55] applied this method to isolate N-terminal peptides from cytosolic and membrane skeleton protein fractions of human platelets. The main advantage of this method is that it reduces the complexity of the peptide sample because each protein has one N terminus and is thus represented by only one peptide. In this procedure, free amino groups in proteins were first blocked by acetylation and then digested with trypsin. After an RP chromatographic fractionation of the generated peptide mixture, internal peptides were blocked using 2,4,6-trinitrobenzenesulfonic acid; in this way, they displayed a strong hydrophobic shift and therefore segregated from the unaltered N-terminal peptides during a second identical separation step. N-terminal peptides could thus be specifically collected for LC-MS/MS. By using this technique, 264 proteins were identified in total, many of them highly abundant cytoskeletal proteins.

In a more recent paper, Gevaert and collaborators[56] presented a COFRADIC approach in which cysteine-containing peptides were isolated from a complete platelet proteome digest and used to identify their precursor proteins after LC-MS/MS analysis. First, cysteines were converted to hydrophobic residues by mixed disulfide formation with Ellman's reagent. Proteins were subsequently digested with trypsin and the generated peptide mixture fractionated by RP-HPLC. Cysteinyl peptides were isolated out of each primary fraction by a reduction step followed by a secondary peptide separation on the same column, performed under identical conditions as for the primary separation. The reducing agent removed the covalently attached group from the cysteine side chain, making cysteine peptides more hydrophilic so they could be specifically collected during the secondary separation and finally used to identify their precursor proteins using automated LC-MS/MS. This procedure efficiently isolated cysteine peptides, making the sample mixture less complex for further analysis. The approach led to the identification of 163 different platelet proteins with a broad range of functions and abundance.

The main disadvantage of the COFRADIC technology compared with the 2-DE approach is the lack of information regarding the pI of the proteins identified, which would miss relevant information about potential PTMs. However, both methods complement each other, and thus COFRADIC allowed the identification of several membrane-spanning proteins not previously identified by 2-DE, demonstrating that gel-free methods are able to overcome some of the limitations of the 2-DE approach.

## 6. Mapping the Platelet Surface Proteome as an Example of How to Study a Platelet Subcompartment.

One platelet compartment of special interest from a physiological and therapeutic perspective that has so far been difficult to study using proteomics is the platelet surface or plasma membrane. Membrane proteins can be classified as either "integral" (intrinsic) or "peripheral" (extrinsic). Integral membrane proteins are anchored to the phospholipid bilayer directly through either a membrane-spanning domain, which consists of α-helices or a β-barrel, or lipid modification, such as prenylation, myristoylation, palmitoylation, and glycosylphosphatidylinositol anchoring. In contrast, peripheral membrane proteins are attached to the plasma membrane through noncovalent interactions with integral membrane proteins or with the lipid itself. Many of the interactions of the group of "peripheral" proteins take place in response to cell activation, either as a result of a change in composition of the membrane or through covalent modification (usually phosphorylation) of an integral membrane protein. For example, tyrosine phosphorylation of many integral membrane proteins leads to recruitment of cytosolic proteins. This is the case for the membrane adapter LAT, which recruits cytosolic proteins such as phospholipase Cγ2 and the adapters Grb2 and Gads following its tyrosine phosphorylation downstream of activation of the collagen receptor GPVI.[57] Similarly, elevation of the minor membrane phospholipid phosphatidylinositol 3,4,5-trisphosphate leads to recruitment of a number of cytosolic proteins containing PH domains,[58] whereas expression of phosphatidylserine on the outside of stimulated platelets leads to binding of components of the coagulation cascade.[59]

Identifying the complete repertoire of proteins expressed on the surface of resting and activated platelets has important implications for the understanding of how platelets function under normal physiological conditions as well as under pathological conditions. However, as illustrated by the previous discussion, establishing the criteria to classify a protein as a membrane protein, or indeed demonstrate its association with the membrane, is subject to debate and experimental conditions. It has been estimated that as many as 70% of all known drug targets are plasma membrane proteins.[60] Therefore, mapping the platelet surface proteome may lead to identification of novel drug targets and a better understanding of signaling complexes that form on its inner surface.

As discussed earlier, there is a dramatic underrepresentation of membrane proteins in 2-D gels, such that their analysis requires use of one or more alternative experimental approaches. Of 311 proteins identified on human platelets in the pI range 5–11 by García et al.[44] using a classical 2-DE/MS approach, only 3% or nine membrane proteins were identified. Gevaert et al.[56] identified 11 proteins that contain at least one membrane-spanning domain out of 163 different proteins identified using a nongel COFRADIC proteomic approach. The reason for the underrepresentation of membrane proteins by the traditional proteomics approach that involves 2-DE is that the physicochemical properties of many membrane proteins, particularly those with large hydrophobic domains (multiple transmembrane domains), are such that they do not solubilize well in the nondetergent IEF sample buffer, and those that do solubilize are prone to precipitate at their pI.[61,62] The limitation in the dynamic range of detection of proteomics is also an issue because many membrane proteins are typically present at low levels, as is the hydrophobicity of their transmembrane regions and the lack of hypoin sites.[61] These factors, therefore, render it essential to enrich surface proteins prior to resolving them by a chromatographic technique that does not involve IEF.

Cell surface proteins can be enriched by affinity chromatography either using a specific ligand or a more global approach based, for example, on the use of plant lectins or by a combination of surface biotinylation and purification with immobilized avidin. The former approach takes advantage of the fact that lectins interact specifically and reversibly with sugar residues, and that the majority of cell surface proteins are glycosylated to varying degrees. Bound proteins can be competed off with the sugar residue or an ionic gradient. Furthermore, different lectins target distinct classes of proteins. For example, wheat germ agglutinin binds molecules containing N-acetyl-β-D-glucosamine residues, whereas concanavalin A binds molecules containing α-D-mannosyl residues, α-D-glucosyl residues, and branched mannoses.[63] One drawback of this approach, however, is that not all surface proteins are glycosylated or have an appropriate sugar residue to support binding. An additional consideration is that, in a detergent-solubilized cell lysate, the lectin will bind proteins present in intracellular secretory granules, because these also tend to be highly glycosylated.

An alternative approach to lectin affinity chromatography is biotin–avidin affinity chromatography. This technique involves biotin-labeling surface proteins with a water-soluble, membrane-impermeable reagent, followed by disrupting the cells either with a detergent or by sonication and affinity isolating the biotinylated proteins with immobilized avidin[64] (Fig. 5-5). Sulfosuccinimidyl-2-[biotinamido]ethyl-1,3-dithiopropionate (sulfo-NHS-SS-biotin) is a commercially available biotinylation reagent. The labeling is based on the reaction of primary amines, primarily on lysine residues on the extracellular region of membrane proteins with N-hydroxysulfosuccinimde (NHS) esters on the biotinylation reagent. This approach has the advantages that the majority of surface proteins are biotinylated and that the interaction with avidin is of very high affinity ($k_d$ approximately $10^{-15}$ M), enabling use of stringent washing conditions and therefore removal of many associated proteins. Nevertheless, a limited degree of contamination with intracellular proteins is expected, because the biotin label is not fully membrane impermeable. The use of modified forms of biotinylation reagents and avidin make it less likely that intracellular proteins will be labeled and easier to release bound proteins prior to analysis, although with the caveat that this may increase contamination with cytosolic proteins.[65] There is also a possibility that the biotin molecule may interfere with the ability of trypsin to cleave the proteins. Despite these limitations, this approach has been successfully used to identify surface proteins on both prokaryotic and eukaryotic cells.[66,67] Zhao and coworkers[68] used a biotin-directed affinity purification method to prepare membrane proteins from a human lung cancer cell line.[68] This involved biotinylating cell surface membrane proteins in viable cells followed by affinity enrichment using streptavidin conjugated to beads and removal of associated cytosolic proteins by harsh washes with high-salt and high-pH buffers. Isolated membrane proteins were subsequently resolved by 1-D SDS-PAGE and identified by nano-LC/MS/MS analysis.[68] Among 898 unique proteins identified in this study, 781 were annotated and 117 were unclassified. Of the classified proteins, 526 were integral plasma membrane proteins and 118 were cytosolic.[68] Another cell surface biotinylation/streptavidin affinity chromatography approach described by Pierce and coworkers[69] identified 42 plasma membrane proteins from a murine T-cell hybridoma and 46 from unfractionated primary murine splenocytes. In their study, surface proteins isolated by biotinylation/streptavidin chromatography were resolved by solution-phase IEF followed by 1-D SDS-PAGE prior to in-gel trypsinization and identification by LC/MS/MS analysis. We have recently used a combination of lectin and biotinylation/avidin affinity chromatography, followed by 1-D SDS-PAGE and HPLC/MS/MS analysis to isolate and identify novel platelet plasma membrane receptors. To date we have identified more than 42 known platelet surface proteins, including integrins $\alpha_{IIb}$, $\beta_3$, $\alpha_2$, $\beta_1$, all of the components of the GPIb-IX-V complex, GPVI, CD9, and P2Y$_{12}$, plus approximately 14 novel platelet plasma membrane proteins, using these two affinity chromatography approaches (Senis, Martin, Garcia, Wakelam, and Watson, unpublished results). One of the limitations of this approach is its limited ability to identify multiple-spanning transmembrane proteins, such as seven transmembrane receptors and tetraspanins.

A further approach to providing samples for analysis is to separate plasma and intracellular membranes using free-flow electrophoresis. This involves separating the plasma membrane from intracellular membranes by electrophoretic means after treatment of the extracellular membrane with

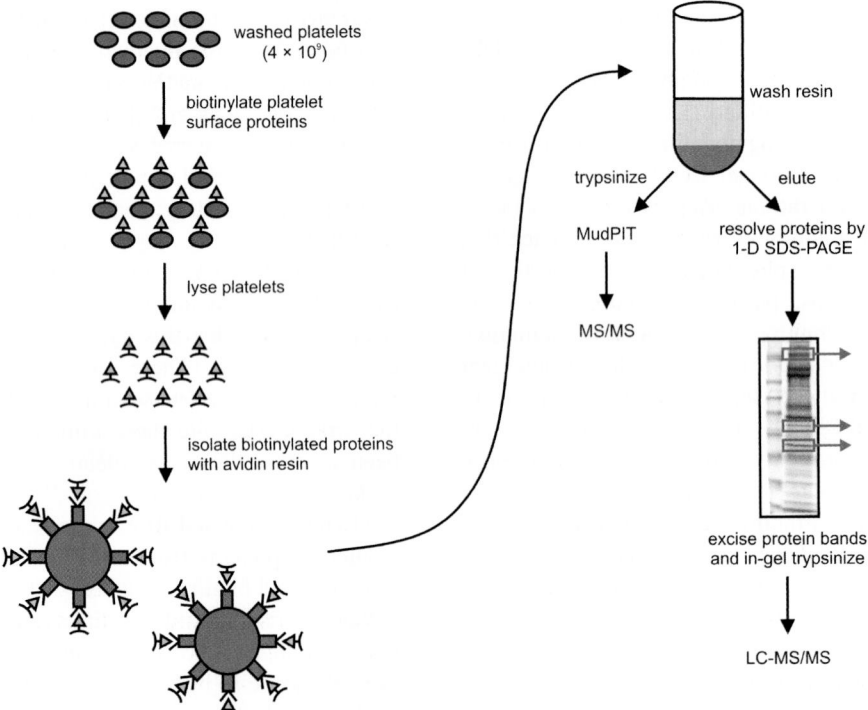

**Figure 5-5.** Enrichment of platelet surface receptors by biotinylation and avidin affinity chromatography. Platelet surface proteins are biotinylated with a water-soluble, membrane-impermeable biotinylation reagent such as sulfo-NHS-SS-biotin. Platelets are then lysed either with a detergent or by sonication. Surface membranes are affinity isolated with avidin coupled to a resin. The resin is washed with high-salt and high-pH buffers to remove associated proteins, such as actin and signaling molecules, that may mask low-abundance receptors. Proteins bound to the beads can then be trypsinized off the beads and resolved by 2-D HPLC (cation exchange chromatography followed by RP chromatography), also referred to as MudPIT. MudPIT is performed in line with tandem mass spectrometry. An alternative approach for identifying the platelet surface proteins is to elute the bound proteins from the resin by boiling the resin in 2 × Laemmli buffer containing a reducing agent such as DTT, resolving the proteins by 1-D-SDS-PAGE, staining the gel with *Colloidal Coomassie,* excising protein bands, and performing in-gel trypsinization. Tryptic fragments are then eluted from the gel slices and identified by LC-MS/MS analysis. Mass spectra obtained by either of these approaches are subsequently compared with those found in databases, such as Swiss-PROT/TrEMBL using search software such as SEQUEST or MASCOT.

neuraminidase, which reduces its negative charge by virtue of removal of sialic acid residues from glycosidic side chains on platelet surface glycoproteins.[70] The mixture of plasma and intracellular membranes are separated from the granules by sorbitol density gradient centrifugation, and the plasma and intracellular membranes are subsequently separated by free-flow electrophoresis. Although this approach works well, it is labor intensive and requires specialized equipment. A further limitation of this approach is that membrane-associated cytoskeletal proteins, such as actin and filamin, are separated in the plasma membrane fraction and can mask low-abundance receptors. Recent work using free-flow electrophoresis to prepare purified platelet plasma membranes followed by 1-D SDS-PAGE and HPLC/MS/MS analysis has identified many known and novel platelet surface proteins (Senis, Martin, Garcia, Authi, and Watson, unpublished results). These findings correlate well with the results obtained with the affinity chromatography approaches described earlier.

**7. Proteome Analysis of Resting and Activated Platelets.** The proteome of a cell can undergo considerable changes with time and in response to stimulation. In the case of the platelet proteome, the major changes that take place upon activation are predicted to involve either degradation or PTMs, because platelets have a very limited capacity for *de novo* protein synthesis. The major PTM is protein phosphorylation on serine, threonine, or tyrosine residues, because this plays a major role in modulating signaling cascades in platelets. Identifying which proteins undergo phosphorylation will likely provide important information about the molecular mechanisms regulating platelet activation.

Several research groups have used MS to study the proteome of activated human platelets. The majority of these studies have focused on the thrombin receptor signaling cascade, because thrombin is one of the most powerful platelet agonists, mediating potent activation of serine/threonine kinases and relatively weak activation of tyrosine kinases. Immler and coworkers[71] first attempted to characterize tyrosine phosphorylation signaling events induced by thrombin on human platelets by 2-DE and LC-MS/MS. They identified several phosphorylated proteins, some of which corresponded to different phosphorylation states of myosin light chain. In 2000, Gevaert and coworkers[72] compared cytoskeletal preparations of basal and thrombin-activated platelets by a combination of 2-DE and MALDI-MS. They identified 27 proteins, most of which were F-actin-binding proteins that translocate to the platelet cytoskeleton after thrombin stimulation. A subsequent study in 2002 by Maguire and coworkers[73] used an immunoprecipitation-based approach to analyze the phosphotyrosine proteome of thrombin-activated platelets. Tyrosine-phosphorylated proteins isolated using the monoclonal antibody 4G10 were separated by 2-DE. Sixty-seven proteins were detected as unique in the thrombin-activated platelet proteome compared with the resting platelet proteome. Ten of these proteins were identified by a combination of Western blotting and MALDI-TOF MS and included FAK, Syk, Alk-4, P2X6, and MAPK kinase kinase.[73] One factor that should be considered with respect to this approach is that several of the proteins may have been pulled down as part of protein complexes and thus independent verification is required of regulation through protein tyrosine phosphorylation.

More recently, different protein separation and analytical methods have been used to obtain a more comprehensive analysis of signaling cascades activated in platelets. Marcus and coworkers[74] used various techniques to identify phosphotyrosine-containing proteins in thrombin-activated platelets. In one approach, platelet proteins were radiolabeled with $^{32}$P and separated by 2-DE using different pI ranges, and phosphorylated proteins were detected by autoradiography. This study also separated phosphorylated proteins by immunoprecipitation with the 4G10 antiphosphotyrosine antibody. Phosphoproteins were resolved by 1-D SDS-PAGE or 2-DE, protein bands and spots of interest were excised from the gels, trypsinized within the gel, and analyzed by MALDI-TOF-MS and nano-LC-ESI-MS/MS. The authors identified 55 proteins in the pH 4 to 7 and 6 to 11 regions of 2-DE gels, most of which corresponded to different isoforms of pleckstrin and cytoskeletal proteins. Twenty-one of the proteins had previously been reported to be phosphorylated.

One of the approaches that has been recently used with success to analyze signaling cascades in platelets involves the separation of the whole proteome, without any prefractionation, by high-resolution 2-DE to detect differentially regulated features that can be analyzed by MS. García and colleagues[75] have used this approach to investigate intracellular signaling cascades in platelets stimulated with thrombin-receptor activating peptide, which is able to activate the main thrombin receptor PAR-1. Forty-one differentially regulated protein features were identified by LC-MS/MS and were found to derive from 31 different genes, most of them corresponding to signaling proteins. Several of the proteins identified had not previously been reported in platelets, including the adapter downstream of tyrosine kinase 2 (Dok-2). Further studies revealed that the change in mobility of Dok-2 was caused by tyrosine phosphorylation. García et al.[75] also provided the first demonstration of phosphorylation of the regulator of G-protein signaling, RGS18, and mapped one of the phosphorylation sites by MS/MS. This report has set the stage for future studies[76] and illustrates how the application of proteomics to the analysis of signaling cascades in platelets can lead to the identification of novel signaling molecules that play important roles in platelet activation.

## IV. Overview and Future Directions

The previous text illustrates how powerful insights into platelet biology have been made in recent years through the application of genomic and proteomic technologies, with each having its strengths and limitations. Genomics has made a dramatic contribution to our understanding of the function of a large and diverse number of cell types, but has special problems in platelets because of the absence of a nucleus. As such, the overall value of global genomic data with respect to platelets would appear to be limited, especially in the absence of a strong correlation between gene and protein expression. The application of genomics to platelets may therefore be restricted to a few specialist applications, such as analysis of polymorphisms in platelet-specific proteins (see Chapter 14) or development of platelet-specific microarrays that measure platelet-specific genes. On the other hand, a list of mRNAs detected in platelets is of limited use in the absence of supporting information, because only a fraction of the genes are likely to encode proteins that have functional relevance.

The application of proteomics to platelets has, in theory, several advantages over the use of genomics, because platelets can be easily isolated and purified, and the absence of a nucleus facilitates detection of important regulatory proteins. Although contamination with other cell types is also less of an issue, it remains a concern because of the wide variation in protein expression in all cells. There are, however, major limitations in the application of proteomics at the present time, including difficulties in detecting particular groups of proteins and mapping of PTMs. Concentration of platelet organelles and specific proteins can be used to address many of these concerns, but requires specialist expertise and reagents and is time-consuming. In light of this, it is difficult to envision a way in which proteomics can be automated such that it will find utility in the clinic for processing large numbers of samples. The development of platelet-based antibody arrays may offer a more advantageous approach, although a lack of antibodies of sufficient affinity and selectivity is a major issue.

An area that is important for future development is that of mapping of sites of PTM, such as determination of sites of protein phosphorylation or lipid modification. Although this information can be predicted to some extent from bioinformatics and studies in other cell types, it is important that each modification is determined in the platelet itself, because there may be cell-specific routes of protein regulation. The powerful set of MS-based approaches available today means that, in theory, it is now possible to identify almost all sites of PTM. However, in reality, few proteins are present at high enough levels, or the stoichiometry of modification is too low, to enable these to be mapped in an intact cell. This is an area that will benefit from important technical breakthroughs during the course of the next few years. Quantitation is another important area in proteomics that suffers from methodological limitations, most notably protein losses during extraction. Determination of the levels of protein expression and stoichiometry of PTM has important implications for the mechanisms underlying platelet regulation and, in particular, for investigation of platelet-based disorders, many of which will be the result of quantitative rather than qualitative changes. Such information could further guide development of platelet-targeted antibody arrays.

A key question that emerges from these considerations is the number of proteins that play "important" roles in regulating platelet activity. The answer to this is unclear, because we are a long way from understanding the function of many of the expressed proteins. To find a way to address this, it will be necessary to compile the mounting volume of genomic, proteomic, and functional data, so that they are both readily accessible and helpful. A recent report from the International Society on Thrombosis and Haemostasis Scientific Subcommittee on Platelet Physiology has recommended the setting up of a database describing proteins that are expressed in platelets and the way they change upon activation and in disease.[42] Ideally, the database should include information from genomic and proteomic analyses, and functional studies. It will be important for the database to be maintained with widely accessible and established databases, such as those maintained by SwissProt and HuPo. Such a database would serve as a powerful resource to further our understanding of platelet biology.

## Acknowledgment

The authors acknowledge support from the BBSRC, British Heart Foundation, Medical Research Council, The Oxford Glycobiology Institute and Wellcome Trust. S.P.W. holds a British Heart Foundation Chair and M.G.T. is supported by an MRC New Investigator Award.

## References

1. Lander, E. S., Linton, L. M., & Birren, B. (2004). Finishing the euchromatic sequence of the human genome. *Nature, 431,* 931–945.
2. Schena, M., Shalon, D., Davis, R. W., & Brown, P. O. (1995). Quantitative monitoring of gene expression patterns with a complementary DNA microarray. *Science, 270,* 467–470.
3. Lockhart, D. J., Dong, H., Byrne, M. C., et al. (1996). Expression monitoring by hybridization to high-density oligonucleotide arrays. *Nat Biotechnol, 14,* 1675–1680.
4. Velculescu, V. E., Zhang, L., Vogelstein, B., & Kinzler, K. W. (1995). Serial analysis of gene expression. *Science, 270,* 484–487.
5. Barrett, T., Suzek, T. O., Troup, D. B., et al. (2005). NCBI GEO: Mining millions of expression profiles — database and tools. *Nucleic Acids Res, 33,* D562–D566.
6. Edgar, R., Domrachev, M., & Lash, A. E. (2002). Gene expression omnibus: NCBI gene expression and hybridization array data repository. *Nucleic Acids Res, 30,* 207–210.
7. Parkinson, H., Sarkans, U., Shojatalab, M., et al. (2005). ArrayExpress — a public repository for microarray gene expression data at the EBI. *Nucleic Acids Res, 33,* D553–D555.
8. Su, A. I., Wiltshire, T., Batalov, S., et al. (2004). A gene atlas of the mouse and human protein-encoding transcriptomes. *Proc Natl Acad Sci USA, 101,* 6062–6067.
9. Brown, P. O., & Botstein, D. (1999). Exploring the new world of the genome with DNA microarrays. *Nat Genet, 21,* 33–37.
10. Holloway, A. J., van Laar, R. K., Tothill, R. W., & Bowtell, D. D. (2002). Options available — from start to finish — for obtaining data from DNA microarrays II. *Nat Genet, 32*(Suppl),481–489.
11. Lipshutz, R. J., Fodor, S. P., Gingeras, T. R., & Lockhart, D. J. (1999). High density synthetic oligonucleotide arrays. *Nat Genet, 21,* 20–24.
12. Lash, A. E., Tolstoshev, C. M., Wagner, L., et al. (2000). SAGEmap: A public gene expression resource. *Genome Res, 10,* 1051–1060.
13. Cobbold, S. P., Nolan, K. F., Graca, L., et al. (2005). Regulatory T cells and dendritic cells in transplantation tolerance: Molecular markers and mechanisms. *Immunol Rev, 196,* 109–124.
14. Saha, S., Sparks, A. B., Rago, C., et al. (2005). Using the transcriptome to annotate the genome. *Nat Biotechnol, 20,* 508–512.
15. Matsumura, H., Ito, A., Saitoh, H., et al. (2005). SuperSAGE. *Cell Microbiol, 7,* 11–18.
16. Fink, L., Holschermann, H., Kwapiszewska, G., et al. (2003). Characterization of platelet-specific mRNA by real-time PCR after laser-assisted microdissection. *Thromb Haemost, 90,* 749–756.

17. Lindemann, S., Tolley, N. D., Eyre, J. R., Kraiss, L. W., Mahoney, T. M., & Weyrich, A. S. (2001). Integrins regulate the intracellular distribution of eukaryotic initiation factor 4E in platelets. A checkpoint for translational control. *J Biol Chem, 276,* 33947–33951.

18. Lindemann, S., Tolley, N. D., Dixon, D. A., et al. (2001). Activated platelets mediate inflammatory signaling by regulated interleukin 1beta synthesis. *J Cell Biol, 154,* 485–490.

19. Weyrich, A. S., Lindemann, S., Tolley, N. D., et al. (2004). Change in protein phenotype without a nucleus: Translational control in platelets. *Semin Thromb Hemost, 30,* 491–498.

20. Gnatenko, D. V., Dunn, J. J., McCorkle, S. R., Weissmann, D., Perrotta, P. L., & Bahou, W. F. (2003). Transcript profiling of human platelets using microarray and serial analysis of gene expression. *Blood, 101,* 2285–2293.

21. McRedmond, J. P., Park, S. D., Reilly, D. F., et al. (2004). Integration of proteomics and genomics in platelets: A profile of platelet proteins and platelet-specific genes. *Mol Cell Proteomics, 3,* 133–144.

22. Bugert, P., Dugrillon, A., Gunaydin, A., Eichler, H., & Kluter, H. (2003). Messenger RNA profiling of human platelets by microarray hybridization. *Thromb Haemost, 90,* 738–748.

23. Guerriero, R., Testa, U., Gabbianelli, M., et al. (1995). Unilineage megakaryocytic proliferation and differentiation of purified hematopoietic progenitors in serum-free liquid culture. *Blood, 86,* 3725–3736.

24. Kim, J. A., Jung, Y. J., Seoh, J. Y., Woo, S. Y., Seo, J. S., & Kim, H. L. (2002). Gene expression profile of megakaryocytes from human cord blood CD34(+) cells ex vivo expanded by thrombopoietin. *Stem Cells, 20,* 402–416.

25. Kim, H. L. (2003). Comparison of oligonucleotide-microarray and serial analysis of gene expression (SAGE) in transcript profiling analysis of megakaryocytes derived from CD34+ cells. *Exp Mol Med, 35,* 460–466.

26. Tomlinson, M. G., Heath, V. L., Dumon, S., Cobbold, S. P., Frampton, J., & Watson, S. P. *Long SAGE analysis of the mouse megakaryocyte transcriptome.* Unpublished manuscript.

27. Shim, M. H., Hoover, A., Blake, N., Drachman, J. G., & Reems, J. A. (2004). Gene expression profile of primary human CD34+CD38lo cells differentiating along the megakaryocyte lineage. *Exp Hematol, 32,* 638–648.

28. Tenedini, E., Fagioli, M. E., Vianelli, N., et al. (2004). Gene expression profiling of normal and malignant CD34-derived megakaryocytic cells. *Blood, 104,* 3126–3135.

29. Brenner, S., Johnson, M., Bridgham, J., et al. (2000). Gene expression analysis by massively parallel signature sequencing (MPSS) on microbead arrays. *Nat Biotechnol, 18,* 630–634.

30. Chen, Y. T., Scanlan, M. J., Venditti, C. A., et al. (2005). Identification of cancer/testis-antigen genes by massively parallel signature sequencing. *Proc Natl Acad Sci USA, 102,* 7940–7945.

31. Tyers, M., & Mann, M. (2003). From genomics to proteomics. *Nature, 422,* 193–197.

32. Fenn, J. B., Mann, M., Meng, C. K., Wong, S. F., & Whitehouse, C. M. (1989). Electrospray ionization for mass spectrometry of large biomolecules. *Science, 246,* 64–71.

33. Karas, M., & Hillenkamp, F. (1988). Laser desorption ionization of proteins with molecular masses exceeding 10,000 daltons. *Anal Chem, 60,* 2299–2301.

34. Aebersold, R., & Goodlett, D. R. (2001). Mass spectrometry in proteomics. *Chem Rev, 101,* 269–295.

35. Aebersold, R., & Mann, M. (2003). Mass spectrometry-based proteomics. *Nature, 422,* 198–207.

36. Gorg, A., Obermaier, C., Boguth, G., et al. (2000). The current state of two-dimensional electrophoresis with immobilized pH gradients. *Electrophoresis, 21,* 1037–1053.

37. Herbert, B. (1999). Advances in protein solubilisation for two-dimensional electrophoresis. *Electrophoresis, 20,* 660–663.

38. Hoving, S., Voshol. H., & van Oostrum, J. (2000). Towards high performance two-dimensional gel electrophoresis using ultrazoom gels. *Electrophoresis, 21,* 2617–2621.

39. Lopez, M. F., Kristal, B. S., Chernokalskaya, E., et al. (2000). High-throughput profiling of the mitochondrial proteome using affinity fractionation and automation. *Electrophoresis, 21,* 3427–3440.

40. Link, A. J., Eng, J., Schieltz, D. M., et al. (1999). Direct analysis of protein complexes using mass spectrometry. *Nat Biotechnol, 17,* 676–682.

41. Washburn, M. P., Wolters, D., & Yates, J. R., 3rd. (2001). Large-scale analysis of the yeast proteome by multidimensional protein identification technology. *Nat Biotechnol, 19,* 242–247.

42. Watson, S. P., Bahou, W. F., Fitzgerald, D., Ouweuhand, W., Rao, K., & Leavitt, A. D. (2005). Mapping the platelet proteome: A report from the ISTH Platelet Physiology Subcommittee. *J Thromb Haemost,* 2098–2101.

43. O'Neill, E. E., Brock, C. J., von Kriegsheim, A. F., et al. (2002). Towards complete analysis of the platelet proteome. *Proteomics, 2,* 288–305.

44. García, A., Prabhakar, S., Brock, C. J., et al. (2004). Extensive analysis of the human platelet proteome by two-dimensional gel electrophoresis and mass spectrometry. *Proteomics, 4,* 656–668.

45. Jenkins, C. S., Phillips, D. R., Clemetson, K. J., Meyer, D., Larrieu, M. J., & Luscher, E. F. (1976). Platelet membrane glycoproteins implicated in ristocetin-induced aggregation. Studies of the proteins on platelets from patients with Bernard–Soulier syndrome and von Willebrand's disease. *J Clin Invest, 57,* 112–124.

46. Gravel, P., Sanchez, J. C., Walzer, C., et al. (1995). Human blood platelet protein map established by two-dimensional polyacrylamide gel electrophoresis. *Electrophoresis, 16,* 1152–1159.

47. Marcus, K., Immler, D., Sternberger, J., & Meyer, H. E. (2000). Identification of platelet proteins separated by two-dimensional gel electrophoresis and analyzed by matrix assisted laser desorption/ionization-time of flight-mass spectrometry and detection of tyrosine-phosphorylated proteins. *Electrophoresis, 21,* 2622–2636.

48. Fey, S. J., & Larsen, P. M. (2001). 2D or not 2D. Two-dimensional gel electrophoresis. *Curr Opin Chem Biol, 5,* 26–33.

49. Gorg, A., Boguth, G., Kopf, A., Reil, G., Parlar, H., & Weiss, W. (2002). Sample prefractionation with Sephadex isoelectric

focusing prior to narrow pH range two-dimensional gels. *Proteomics, 2,* 1652–1657.

50. Righetti, P. G., Castagna, A., Herbert, B., Reymond, F., & Rossier, J. S. (2003). Prefractionation techniques in proteome analysis. *Proteomics, 3,* 1397–1407.

51. Mairhofer, M., Steiner, M., Mosgoeller, W., Prohaska, R., & Salzer, U. (2002). Stomatin is a major lipid-raft component of platelet alpha granules. *Blood, 100,* 897–904.

52. García, A., Zitzmann, N., & Watson, S. P. (2004). Analyzing the platelet proteome. *Semin Thromb Hemost, 30,* 485–489.

53. Peng, J., Elias, J. E., Thoreen, C. C., Licklider, L. J., & Gygi, S. P. (2003). Evaluation of multidimensional chromatography coupled with tandem mass spectrometry (LC/LC-MS/MS) for large-scale protein analysis: The yeast proteome. *J Proteome Res, 2,* 43–50.

54. Coppinger, J. A., Cagney, G., Toomey, S., et al. (2004). Characterization of the proteins released from activated platelets leads to localization of novel platelet proteins in human atherosclerotic lesions. *Blood, 103,* 2096–2104.

55. Gevaert, K., Goethals, M., Martens, L., et al. (2003). Exploring proteomes and analyzing protein processing by mass spectrometric identification of sorted N-terminal peptides. *Nat Biotechnol, 21,* 566–569.

56. Gevaert, K., Ghesquiere, B., Staes, A., et al. (2004). Reversible labeling of cysteine-containing peptides allows their specific chromatographic isolation for non-gel proteome studies. *Proteomics, 4,* 897–908.

57. Asazuma, N., Wilde, J. I., Berlanga, O., et al. (2000). Interaction of linker for activation of T cells with multiple adapter proteins in platelets activated by the glycoprotein VI-selective ligand, convulxin. *J Biol Chem, 275,* 33427–33434.

58. Vanhaesebroeck, B., Leevers, S. J., Ahmadi, K., et al. (2001). Synthesis and function of 3-phosphorylated inositol lipids. *Annu Rev Biochem, 70,* 535–602.

59. Monroe, D. M., Hoffman, M., & Roberts, H. R. (2002). Platelets and thrombin generation. *Arterioscler Thromb Vasc Biol, 22,* 1381–1389.

60. Hopkins, A. L., & Groom, C. R. (2002). The druggable genome. *Nat Rev Drug Discov, 1,* 727–730.

61. Eichacker, L. A., Granvogl, B., Mirus, O., Muller, B. C., Miess, C., & Schleiff, E. (2004). Hiding behind hydrophobicity. Transmembrane segments in mass spectrometry. *J Biol Chem, 279,* 50915–50922.

62. Hartinger, J., Stenius, K., Hogemann, D., & Jahn, R. (1996). 16-BAC/SDS-PAGE: A two-dimensional gel electrophoresis system suitable for the separation of integral membrane proteins. *Anal Biochem, 240,* 126–133.

63. Robertson, E. R., & Kennedy, J. F. (1996). Glycoproteins: A consideration of the potential problems and their solutions with respect to purification and characterisation. *Bioseparation, 6,* 1–15.

64. Jang, J. H., & Hanash, S. (2003). Profiling of the cell surface proteome. *Proteomics, 3,* 1947–1954.

65. Gauthier, D. J., Gibbs, B. F., Rabah, N., & Lazure, C. (2004). Utilization of a new biotinylation reagent in the development of a nondiscriminatory investigative approach for the study of cell surface proteins. *Proteomics, 4,* 3783–3790.

66. Sabarth, N., Lamer, S., Zimny–Arndt, U., Jungblut, P. R., Meyer, T. F., & Bumann, D. (2002). Identification of surface proteins of *Helicobacter pylori* by selective biotinylation, affinity purification, and two-dimensional gel electrophoresis. *J Biol Chem, 277,* 27896–27902.

67. Shin, B. K., Wang, H., Yim, A. M., et al. (2003). Global profiling of the cell surface proteome of cancer cells uncovers an abundance of proteins with chaperone function. *J Biol Chem, 278,* 7607–7616.

68. Zhao, Y., Zhang, W., & Kho, Y. (2004). Proteomic analysis of integral plasma membrane proteins. *Anal Chem, 76,* 1817–1823.

69. Pierce, M. J., Wait, R., Begum, S., Saklatvala, J., & Cope, A. P. (2004). Expression profiling of lymphocyte plasma membrane proteins. *Mol Cell Proteomics, 3,* 56–65.

70. Authi, K. S. (1996). Preparation of highly purified human platelet plasma and intracellular membranes using high voltage free flow electrophoresis and methods to study Ca2+ regulation. In S. P. Watson & K. S. Authi (Eds.), *Platelets: A practial approach* (Vol. 167, pp. 91–111). Oxford University Press.

71. Immler, D., Gremm, D., Kirsch, D., Spengler, B., Presek, P., & Meyer, H. E. (1998). Identification of phosphorylated proteins from thrombin-activated human platelets isolated by two-dimensional gel electrophoresis by electrospray ionization–tandem mass spectrometry (ESI-MS/MS) and liquid chromatography–electrospray ionization–mass spectrometry (LC-ESI-MS). *Electrophoresis, 19,* 1015–1023.

72. Gavaert, K., Eggermont, L., Demol, H., & Vanderkerckhove, J. (2000). A fast and convenient MALDI-MS based proteomic approach: Identification of components scaffolded by the actin cytoskeleton of activated human thrombocytes. *J Biotech, 78,* 259–269.

73. Maguire, P. B., Wynne, K. J., Harney, D. F., O'Donoghue, N. M., Stephens, G., & Fitzgerald, D. J. (2002). Identification of the phosphotyrosine proteome from thrombin activated platelets. *Proteomics, 2,* 642–648.

74. Marcus, K., Moebius, J., & Meyer, H. E. (2003). Differential analysis of phosphorylated proteins in resting and thrombin-stimulated human platelets. *Anal Bioanal Chem, 376,* 973–993.

75. García, A., Prabhakar, S., Hughan, S., et al. (2004). Differential proteome analysis of TRAP-activated platelets: Involvement of DOK-2 and phosphorylation of RGS proteins. *Blood, 103,* 2088–2095.

76. Weyrich, A. S., & Zimmerman, G. A. (2004). Propelling the platelet proteome. *Blood, 103,* 1979.

77. García, A., Watson, S. P., Dwek, R. A., & Zitzmann, N. (2005). Applying proteomics technology to platelet research. *Mass Spectrometry Rev, 24,* 918–930.

# CHAPTER 6

# Platelet Receptors

**Kenneth J. Clemetson and Jeannine M. Clemetson**

*Theodor Kocher Institute, University of Berne, Berne, Switzerland*

## I. Introduction

Platelet receptors, as the contacts between platelets and their external world, determine the reactivity of platelets with a wide range of agonists and adhesive proteins. To a great extent it is the surface receptors on platelets that, together with their granules, determine the specific cellular identity of platelets. Many of the other biological mechanisms present in platelets are shared with other cells, including cytoplasmic enzymes and signal transduction molecules, cytoskeletal components, and housekeeping enzymes. Unlike the great majority of cells, platelets lack a nucleus and hence cannot adapt to different situations by *de novo* protein synthesis, although there is some evidence for residual protein synthetic capacity from messenger RNA (mRNA) carried over from megakaryocytes.[1,2] Thus, platelets need to be equipped with a wide range of presynthesized molecules ready to carry out various physiological functions and deal with pathological events. One way that platelets can alter their "phenotype" to adapt to different situations is through the many types of receptors that are present in platelet storage granule membranes and only expressed on the platelet surface after activation (e.g., P-selectin [see Chapter 12]). Furthermore, because of the relatively small size of platelets, their membrane and membrane proteins represent a larger proportion of cellular mass compared with other cells. As a result of improved technology, the last few years have seen an increase in the number of receptors, with demonstrable physiological functions, detected on platelets.

Because the main function of platelets is hemostasis, it is not surprising that their major receptors have a direct role in this process, either in activating platelets or as adhesive receptors interacting with damaged cell walls or with other platelets to contribute to thrombus formation. The variety of different receptors implicated in this process, as well as in both positive and negative feedback loops, emphasizes their importance in platelet functions and in the ways that platelets adapt their function to different situations. The difference between physiological hemostasis and pathological thrombosis is initially very small, and constant regulation of responses is critical to the preservation of this difference. In addition, it is increasingly recognized that platelets are not only involved in hemostasis, but have a range of other less well-understood functions, such as inflammation (see Chapter 39), antimicrobial host defense (see Chapter 40), tumor growth and metastasis (see Chapter 42), and angiogenesis (see Chapter 41). Platelets express a range of receptors with no obvious direct role in hemostasis that could be implicated in other activities, such as immunological defense against viruses, bacteria, and parasites, among other pathogens (see Chapter 40). The fact that the normal human platelet count is approximately $250 \times 10^9$/L, whereas only approximately $10 \times 10^9$ platelets/L are adequate to prevent bleeding (see Chapter 69), may also argue for additional platelet functions. Many of the receptors present in low copy numbers, which do not seem to have major direct effects in platelet activation, act synergistically with major agonists and have critical roles in regulating overall platelet responsiveness, so that they are nevertheless physiologically important. Examples of these are the Gas6, leptin, and insulin receptors, and no doubt there will be others identified in this category in the future. A new category of platelet receptors enhances aggregation downstream of platelet–platelet contacts via integrin $\alpha$IIb$\beta$3, including PEAR1 and the Eph/ephrin system, and therefore affects the stability of thrombi.

It has been nearly 90 years since the first disorder involving platelet receptors was described[3] and the second was nearly 60 years ago.[4] More detailed analysis of platelets from these patients in the mid-1970s and early 1980s identified the affected receptors and provided evidence for their function.[5–10] Since then there has been a steady acceleration in knowledge about platelet receptors and their function. The number of known receptors has expanded dramatically and there have been breakthroughs in the identification and characterization of major receptors, such as those for collagen,[11,12] adenosine diphosphate (ADP), and adenosine triphosphate (ATP).[13–15] Platelets share many receptors with other cell types, although some receptors are only expressed on platelets or shared with a few other cells. Undoubtedly, there will still be surprises in store in terms of platelet receptors and their function.

This chapter surveys and summarizes the current state of knowledge about platelet receptors and their structure and function. The receptors have been listed by approximate importance of their role in platelet function and by structural class. A number of major platelet receptors have become research disciplines in their own right and are discussed in detail in the following chapters: the glycoprotein (GP) Ib-IX-V complex (Chapter 7), integrin αIIbβ3 (Chapter 8), thrombin receptors (Chapter 9), P2 receptors (Chapter 10), platelet–endothelial cell adhesion molecule 1 (PECAM-1; Chapter 11), P-selectin (Chapter 12), and receptors for coagulation factors (Chapter 19). Platelet receptor polymorphisms are discussed in Chapter 14.

## II. Integrins

The integrins are a major class of adhesive and signaling molecules present on most cell types.[16] They consist of noncovalently associated heterodimers of α- and β-subunits and are generally involved in linking adhesive molecules to the cellular cytoskeleton. Integrins usually exist in two affinity states, low and high, that are altered by cytoplasmic signaling and phosphorylation of their cytoplasmic domains. Platelets have members of three families of integrins (β1, β2, and β3) and, in total, six different integrins: α2β1, α5β1, α6β1, αLβ2, αIIbβ3, and αvβ3.

### A. β3 Family

**1. αIIbβ3.** αIIbβ3 (the GPIIb-IIIa complex) is the only integrin expressed uniquely on platelets. αIIbβ3 is the major platelet integrin (and receptor) with 50,000 to 80,000 copies per platelet. Its absence or deficiency leads to Glanzmann thrombasthenia, the most common bleeding disorder caused by a platelet receptor defect (see Chapter 57). Integrin αIIbβ3, essential for platelet aggregation, is discussed in detail in Chapter 8.

**2. αvβ3.** Integrin αvβ3 is present on platelets in rather small amounts (several hundred copies per platelet) and its function remains unclear. On other cells, αvβ3 is a vitronectin receptor.[17] Because Glanzmann thrombasthenia can be caused by defects in either the αIIb or the β3 gene, αvβ3 is either expressed or not. When β3 is affected, αvβ3 as well as αIIbβ3 are not expressed. Although Glanzmann thrombasthenia patients who have this type of defect do not appear to be more seriously affected than those with an αIIb defect, and hence intact (and often higher expression of)

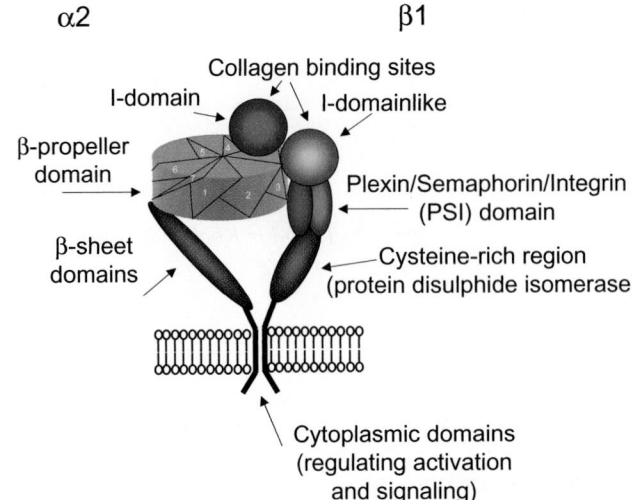

**Figure 6-1.** Model of α2β1 integrin structure based on sequence similarities between the domains and other molecules. The collagen-binding outer I-domain of the α2 subunit is inserted between repeats 2 and 3 of the seven β-propeller structure, which in turn contacts a putative MIDAS (metal ion-dependent adhesion site) or I-domainlike structure in the β1-subunit. The N-terminus of the subunit folds into a plexin–semaphorin–integrinlike domain under the putative MIDAS domain near a domain with no known sequence similarities. The I-domainlike structure of β1 may also participate in collagen binding. The domain of β1 above the membrane is very rich in disulfide bridges and has endogenous protein disulfide isomerase activity responsible for regulating conformational changes of the integrin in response to signaling via the cytoplasmic domains. These changes alter the conformation of the α2 I-domain and its avidity for collagen.

αvβ3, some differences in vitronectin transport have been observed.[18]

### B. β1 Family

**1. α2β1.** The second most important platelet integrin after αIIbβ3, α2β1, is also known on platelets as GPIa-IIa and on lymphocytes as VLA2.[19] There are 2000 to 4000 copies per platelet. α2β1 is well established as a major collagen adhesion receptor on platelets and a wide range of other cells. On platelets it is not a laminin receptor. The expression level is controlled by silent polymorphisms in the promoter and has been related to the incidence of cardiovascular disease (see Chapter 14).

Since the crystal structure of the extracellular domains of αvβ3 was determined,[20] computer modeling of other integrins has allowed the preparation of models that are useful in assessing structure/function relationships (Fig. 6-1). In all α-subunits of integrins, the N-terminal contains seven tandem repeats with internal homology that can be folded into a seven-bladed β-propeller structure.[21] This β-propeller

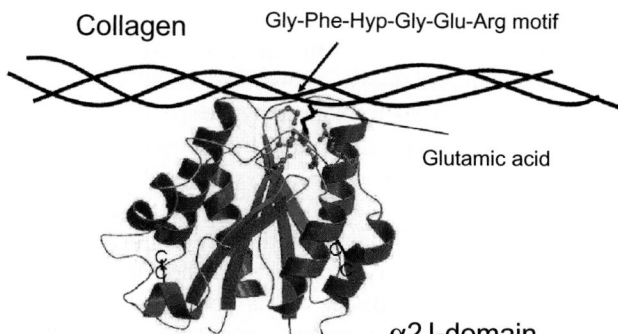

**Figure 6-2.** Model of the α2-subunit I-domain complexed with a triple-helical collagen peptide containing the sequence Gly-Phe-Hyp-Gly-Glu-Arg. The I-domain folds into a Rossmann-type structure with five parallel and one antiparallel β-strands in the center and seven α-helices round the outside. The metal ion ($Mg^{2+}$ or $Mn^{2+}$) is coordinated at the top of the I-domain at the center of an octahedron with three residues from the I-domain, two water molecules, and the glutamate side chain from a collagen peptide. (Based upon the results of Emsley et al.[22,23])

also has three to four $Ca^{2+}$ coordination sites. Some α-subunits, including α2, contain an inserted (or I-) domain of about 200 amino acids, between the second and third repeats, which forms a protrusion on the upper surface of the propeller (Fig. 6-1). The I-domain folds into five parallel and one antiparallel β-strands surrounded by seven α-helices including a metal coordination site. The I-domain on the α2-subunit clearly prefers $Mg^{2+}/Mn^{2+}$ in this site.

The delineation of the X-ray crystallographic structure of both the collagen binding I-domain of α2 and its complex with collagen-related peptides[22,23] was a major breakthrough. The $Mg^{2+}$ ion complexed by the I-domain is critical for interacting with the collagen, completing the six coordination sites with a glutamate residue on the collagen (Fig. 6-2). Peptides containing an aspartic acid residue in place of glutamic acid did not bind to the I-domain,[24] showing that the longer sidechain is necessary to reach and coordinate with the metal ion.

The N-terminal domain on the β-subunit has sequence similarities to MIDAS (metal ion-dependent adhesion site) and I-domains, and it is thought that it can also be folded into a related structure. The β-subunit preference as cation for this site is probably $Ca^{2+}$. Most, if not all, integrins undergo conformational changes to upregulate binding to their ligands. One theory of collagen interactions with platelets was the so-called "two-step, two-site" model in which platelets first bind to collagen by α2β1 and are then activated via a second receptor,[25] thought to be GPVI. This has been a highly controversial area during the past few years, with the involvement of each receptor in collagen-induced responses being estimated differently at various stages, to return currently to somewhere near the initial estimate.

Collagen-related peptides (CRPs) containing the GFOGER sequence bind to the I-domain without apparently needing platelet activation, but there is recent evidence for changes in the affinity of α2β1 for collagen after platelet activation[26] supporting a model for platelet adhesion to collagen in which the platelet is first activated by interactions between specific sites on collagen and GPVI on the platelet. Under high shear stress conditions, neither α2β1 nor GPVI is adequate to initiate adhesion, and both GPIb on the platelet and von Willebrand factor (VWF) in the plasma are essential for platelet interactions with collagen (see Chapter 18). Binding between platelet GPIb and VWF–collagen on the injured vessel wall causes the platelets to roll on the subendothelium until they adhere firmly or return to the circulation. Integrins responsible for the firm adhesion become activated on the platelet and can attach strongly to collagen in the case of α2β1 or fibrinogen/fibrin in the case of αIIbβ3 (see Chapters 17 and 18). Additional interactions involving α5β1 or α6β1 and fibronectin or laminin respectively may also be involved. Because Glanzmann thrombasthenia platelets still adhere to subendothelium, it can be deduced that αIIbβ3 is not the main or only integrin involved. Both α2 and β1 null mice have been produced recently,[27-30] and both show a relatively normal phenotype with some defective platelet responses to collagen. The mice had normal bleeding times and their platelets showed essentially normal adhesion to collagen and aggregation to equine tendon collagen. A noticeable difference was with "soluble" collagen, to which wild-type platelets gave a clear aggregation response but $β1^{-/-}$ platelets did not.[27] The platelets from α2 null mice showed more pronounced differences from the wild type than those from the β1 null mice.

In humans, information on α2 functional defects comes from a few patients. In 1995, a patient with a myeloproliferative disorder was shown to have a specific deficiency in platelet α2β1, and lacked aggregation and adhesion responses to collagen.[31] This patient had a prolonged bleeding time and marked thrombocytosis. Two female patients were described earlier with mild bleeding disorders.[32,33] These patients had a defective response to collagen but not to other agonists, and both surface labeling as well as two-dimensional gel electrophoresis studies demonstrated low levels or absent α2β1. One of these patients also had apparent defects in thrombospondin 1 content and, soon after the initial set of studies, passed through menopause, after which collagen responsiveness recovered. The other was younger at the time of original testing, but also recently passed through menopause with concomitant recovery of platelet functions. In this patient, the interaction of patient blood with rabbit subendothelium was examined in a perfusion chamber.[32] Although normal human platelets showed the expected behavior with a thick coating of shape-changed, degranulated platelets on the subendothelial surface, in this patient the rare platelets that adhered did not change shape or degranulate. Higher magnification electron microscopy of the adherent platelets showed very weak interaction with the subendothelial surface at only a few points of contact, and no surface spreading like normal platelets. GPIb–VWF interaction was adequate for some platelet adhesion, perhaps helped by minor integrins (α5β1, α6β1), but in the absence of α2β1 there was no firm adhesion or general platelet activation. Despite the presumed presence of GPVI and interaction with collagen, the platelets were not activated. However, it cannot be excluded that these platelets showed other defects.

When platelets are activated by collagen in a stirred suspension, antagonists of α2β1, such as antibodies[34,35] or snake venom proteins,[36,37] prevent the aggregation response despite the presence of GPVI on the platelets and GPVI binding sites on the collagen. Cleavage of β1 on platelets by a specific snake venom metalloprotease also prevents collagen-induced signaling.[38] Therefore, the question has been raised whether α2β1 is involved, not only in platelet adhesion to collagen, but also in the platelet signaling response. Some snake venom proteins block α2β1[36,37] whereas others bind to α2 and cleave β1.[39,40] Other snake venom proteins block the α2β1 binding site on collagen.[41] Snake venom proteins have also been reported that activate platelets via an α2β1 binding mechanism[42,43] independently of GPVI.[44] In all these cases there are still questions of specificity that have not been satisfactorily resolved, such as whether GPVI is cleaved or whether other receptors are also implicated.

Whether α2β1 shows various activation states was difficult to test for a long time because of the lack of appropriate reagents. However, using soluble collagen fragments Jung and Moroi[45] were able to show that collagen binding to α2β1 is increased by platelet activation by a number of agonists. Although released ADP had a major role at low agonist concentrations, it was less critical at higher concentrations, implying the involvement of direct signaling pathways.[26] Various collagens also contain binding sites that are recognized by different activation states of α2β1.[46] Integrins are well known to signal in both directions across cell membranes and interact with a wide range of cytoplasmic proteins.[47] Integrin cytoplasmic domains are critical for cellular functions, such as regulating their affinity state. Results with platelets from FcRγ[−/−] mice also support a role for this subunit of GPVI/FcRγ in activation of integrins before platelet adhesion to collagen (see Section V.A).

A crystal structure of the α2 I-domain in complex with the EMS16 C-type lectin from *Echis multisquamatus,* that blocks collagen-induced platelet and endothelial functions, clearly showed that EMS16 does so by binding to the α2 I-domain over the collagen binding site, thus preventing access by collagen.[48]

Several groups have shown that endogenous protein disulfide isomerase (PDI) activity is critical for integrin activation.[49–51] Integrins, in particular the β-subunits, are rich in cysteine residues and contain CGXC sequences within the cysteine-rich repeats, characteristic for PDI. It was shown earlier that integrins are activated by treatment with low amounts of thiol reagents such as dithiothreitol.[52–54] Integrin activation can be blocked by the specific PDI inhibitor bacitracin.[51,55] The α2β1 integrin also has these properties[50] and rearrangement of the cysteine-rich domains of β1 by endogenous PDI activity probably regulates the high affinity state of the I-domain on the α2-subunit.[56]

A wide distribution of α2β1 expression levels among normal donors was linked to silent polymorphisms in the gene for α2.[57,58] Various studies suggested possible effects of these polymorphisms on the incidence of cardiovascular disease (see Chapter 14).[59–64] The C807T polymorphism in α2β1 was linked to a bleeding tendency in patients with reduced levels of plasma or platelet VWF,[65] as well as to reduced adhesion to collagen in blood from healthy donors.[66]

**2. α5β1.** α5β1 is thought to be the fibronectin receptor on platelets and to have a supplementary role in platelet adhesion at injury sites.[67]

**3. α6β1.** α6β1 is thought to be the laminin receptor on platelets and to have a supplementary role in platelet adhesion at injury sites.[68]

### C.  β2 Family

**1. αLβ2.** For a long time it was thought that platelets did not express β2 integrins, but then the αLβ2 integrin was detected on the surface of activated, but not resting, human platelets, suggesting that it is expressed on granule membranes.[69] Later studies in mice indicated that platelet β2, but not β1 or β3, expression decreases when thrombocytopoiesis is stimulated.[70] Ablation of the β2 gene in mice results in platelets with a significantly shorter life span and that do not localize at a local inflammation site, unlike wild-type platelets.[71] Caspase activities were also raised in β2[−/−] mice, suggesting that this integrin regulates caspase activition.[71]

## III. Leucine-Rich Repeat (LRR) Family

### A.  GPIb-IX-V Complex

The LRR family is represented in platelets by the GPIb-IX-V complex with approximately 50,000 copies per platelet — the second most common platelet receptor (after integrin αIIbβ3). Its absence or deficiency leads to Bernard–Soulier syndrome, the second most common (but still very rare) bleeding disorder linked to a platelet receptor (see Chapter 57). The GPIb-IX-V complex, essential for platelet adhesion under high shear, is discussed in detail in Chapter 7.

## B. Toll-like Receptors

Toll-like receptors (TLRs) are the second class of LRR proteins to be detected on platelets, although in much smaller amounts than the GPIb-IV-V complex. The LRRs within TLRs are of two types, either with the tight α-helical coils (such as found in ribonuclease inhibitor) or the more open, less tightly curved coils (found in the GPIb-IX-V family). This allows various types of interaction. Several recent studies have provided evidence for either the presence or associated function of several members of this class of receptor on platelets. TLRs are type I transmembrane receptors with an extracellular domain containing many LRRs, a transmembrane domain, and an intracellular toll/interleukin (IL)-1 receptor domain (Fig. 6-3). In typical immune cells, stimulation of TLRs by their ligands, generally related to innate immunity requirements such as bacterial, viral, or parasite cell wall components, activates a signaling cascade leading to production of proinflammatory cytokines and immune responses. Shiraki et al.[72] detected TLR1 and TLR6 mRNA in platelets and Meg-01 cells by reverse transcriptase–polymerase chain reaction (RT-PCR) and on Meg-01 cells by flow cytometry. Cognasse et al.[73] found evidence by flow cytometry for several TLRs on platelets: weak expression of platelet surface TLR2, TLR4 (approximately 60% of platelets positive), and TLR9, and somewhat higher levels (TLR2, approximately 50%; TLR4, approximately 80%; and TLR9, approximately 35%) after permeabilization. TLR4 is widely distributed in many cell types and even low expression levels are adequate for responses to lipopolysaccharide (LPS) or endotoxin.[74] It requires an additional small molecule MD-2 belonging to the ML family of lipoprotein binding proteins and that associates with TLR4 near to its N-terminal domain (Fig. 6-3). The availability of TRL4$^{-/-}$ mice has allowed rigorous testing for a role in platelets.[75] In wild-type mice about half were positive for TLR4 by flow cytometry. This was confirmed by $^{125}$I-anti-TLR4 monoclonal antibody binding; wild-type platelets were positive whereas TLR4$^{-/-}$ platelets were not. An important criterion was the necessity for fixing before staining, implying downregulation of TLR4 after antibody binding, which might explain earlier difficulties in detecting this receptor on platelets. In this study,[75] TLR4 was present on about 40% of human platelets. TLR4 was also detected on megakaryocytes, with expression levels increasing with maturation (assessed by CD41 levels). Treatment with LPS, the TLR4 ligand, caused wild-type platelets to adhere significantly more to fibrinogen, at about 50% of levels induced by thrombin. LPS did not affect the behavior of TLR4$^{-/-}$ platelets, although they responded normally to thrombin. LPS did not induce P-selectin expression on wild-type platelets, unlike thrombin, and even when LPS was injected *in vivo,* the platelets sequestered in the lungs did not show enhanced P-selectin. TLR4$^{-/-}$ mice did not show thrombocytopenia

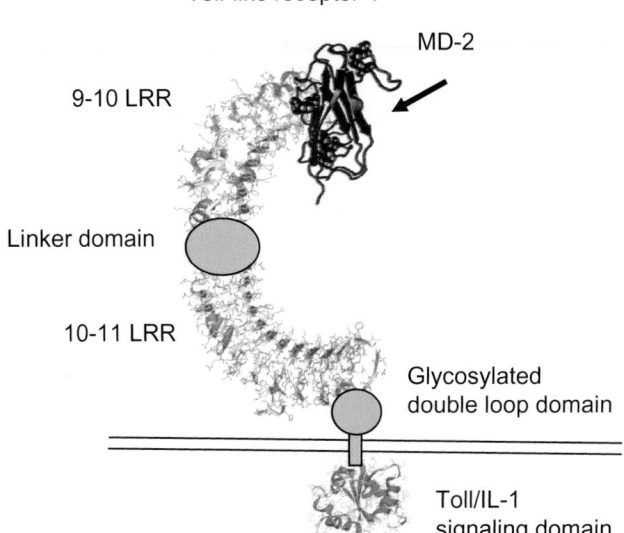

Toll-like receptor 4

**Figure 6-3.** Model of toll-like receptor 4/MD-2 complex. Two domains formed by about 9 to 11 leucine-rich repeats (LRR) are separated by a less well-defined region that contains two cysteines likely to form a disulfide bridge-linked loop and to be a flexible linking domain. The region between the LRRs and the membrane is predicted to contain the classic double-loop domain found in most LRR molecules such as GPIb-IX-V. The cytoplasmic region folds into a toll/interleukin (IL)-1 signaling domain as in most toll-like receptors. The MD-2 subunit is shown associated with the N-terminal domain.

and platelet sequestration in the lungs. Wild-type platelets do not accumulate in lungs of TLR4$^{-/-}$ mice, showing that the presence of TLR4 in the lung is also critical. This was shown to be a neutrophil-dependent phenomenon.[75]

The structure of TLR4 (modeled on other LRR-containing molecules, particularly GPIb and TLR3) is shown in Figure 6-3. There are two domains formed by about 9 to 11 LRRs that are separated by a less well-defined region that is unlikely to fold in this manner, contains two cysteines that might form a disulfide bridge, and seems to be a flexible linking domain. The region between the LRRs and the membrane seems to contain the classic double-loop domain found in most LRR molecules. The cytoplasmic region folds into a toll/IL-1 signaling domain as in most TLRs.

## IV. Seven Transmembrane Receptors

The seven transmembrane receptor family is the major agonist receptor family in cells generally and is also very well represented on platelets. Members of this family are still being identified and characterized.

## A. Thrombin Receptors

Thrombin receptors are major representatives of the seven transmembrane receptor group on platelets, because thrombin is a critical platelet agonist. The first receptor to be identified and characterized was protease activation receptor 1 (PAR1).[76] Unlike other members of the seven transmembrane family (discussed later), PAR class receptors have a distinctive mechanism of activation, involving specific cleavage of the N-terminal extracellular domain, which exposes a new N-terminus that, by refolding, acts as ligand to the receptor. The cleaved N-terminal peptide can also activate platelets, but the receptor involved has not yet been established.[77] Human platelets have about 2500 copies of the PAR1 receptor, which responds to levels of thrombin of about 1 nM. Other PAR class receptors have also been characterized on platelets.[78] PAR2 receptor is cleaved and activated by trypsin and related proteases, but not by thrombin. Mouse (but not human) platelets have PAR3, which delivers thrombin to PAR4 and increases sensitivity, whereas mouse PAR1 seems unimportant. Both human and mouse platelets have PAR4,[78] which is sensitive to 10 times higher concentrations of thrombin than PAR1, possibly to handle situations in which platelets are exposed to high doses and PAR1 is downregulated. Platelet thrombin receptors are discussed in detail in Chapter 9.

## B. ADP Receptors

ADP is an important primary platelet agonist and is also a critical autocrine via secretion from platelet-dense granules. Although ADP was one of the earliest identified platelet agonists, ADP receptors have only recently been definitively identified.[13,14,79–82] The current model for ADP and ATP receptors on platelets involves three separate components. $P2Y_1$ and $P2Y_{12}$ are ADP receptors and belong to the seven transmembrane family; $P2X_1$ is an ATP receptor, which is a calcium channel that belongs to a different structural family. ADP and ATP receptors are discussed in detail in Chapter 10.

## C. Prostaglandin Family Receptors

**1. Thromboxane (TX) Receptors.** Platelets only have the A-type TXA2/prostaglandin (PG) H2 receptor[83] with a mass of 57 kDa and two N-glycosylation sites. The TXA2/PGH2 receptor is coupled to signal transduction via several G proteins including Gq, $Gi_2$, and $G_{12/13}$ to activate phospholipase A2 and phospholipase C. TXA2 receptor agonists induce tyrosine phosphorylation of several signaling proteins, including p72[SYK].[84] The A-type TXA2/PGH2 receptor is an important receptor for autocrine amplification of plate-

let activation following stimulation by primary agonists (see Chapter 31).

**2. PGI2 Receptor.** The PGI2 (prostacyclin) receptor,[85] the major inhibitory prostaglandin receptor on platelets, binds prostacyclin released from endothelial cells to maintain platelets in a resting state. It is coupled to Gs to activate adenylate cyclase. The PGI2 receptor is discussed in detail in Chapter 13.

**3. PGD2 Receptor.** It is generally accepted that platelets have PGD2 receptors that are similar to the vascular receptors, but distinct from other platelet prostaglandin receptors.[86]

**4. PGE2 Receptor.** PGE2 potentiates platelet responses to ADP and collagen at low concentrations and inhibits aggregation at higher concentrations. It was recently shown that activation of the mouse EP3 receptor for PGE2 inhibits cyclic adenosine monophosphate (cAMP) production and promotes platelet aggregation.[87] The α-type of the PGE2 receptor has been cloned from HEL cells and, when expressed in cells, showed ligand-binding characteristics like the platelet receptor.[88]

## D. Lipid Receptors

**1. Platelet Activating Factor (PAF) Receptor.** PAF is a lipid, 1-alkyl-2-acetyl-sn-glycero-3-phosphocholine, which generates biological responses at levels as low as 10 fM and is a major mediator of inflammation. Human platelets have 300 PAF receptors with a $K_d$ of about 0.2 nM.[89] PAF receptors have been cloned from various species, and the expressed glycoproteins have masses in the 50- to 60-kDa range.[89] The platelet receptor is thought to be coupled via Gq and Gi proteins. PAF receptors are important pharmacological targets for antagonist research.

**2. Lysophosphatidic Acid Receptor.** Lysophosphatidic acid added to platelet suspensions causes shape change, release of storage granules, and aggregation. It is also produced and secreted by activated platelets and thus acts as an autocrine agonist. Platelets are thought to have lysophosphatidic acid receptors of the seven transmembrane domain class with a mass in the 38- to 40-kDa range, implying low levels of glycosylation.[90] Platelets may also have sphingosine-1-phosphate receptors.[91]

## E. Chemokine Receptors

Platelets were the source for the first members of the chemokine family to be described: platelet factor 4 and β-thromboglobulin. More recently, platelets were shown to contain other chemokines including ENA-78 (epithelial neutrophil activating peptide) and RANTES (regulated upon activation, normal T cell expressed and presumably secreted). The presence of chemokine receptors has been more controversial, but good evidence has now accumulated for platelet receptors CXCR4 (receptor 4 for chemokines containing the Cys-X-Cys motif) and CCR4 (receptor 4 for chemokines containing the Cys-Cys motif), and for lower but still functional amounts of CCR1 and CCR3.[92–95] A few more recently identified chemokine receptors have still not been tested for in platelets. CXCR4 is the unique receptor for stromal cell-derived factor 1 and is also found on megakaryocytes. CXCR4 may have a role in trafficking between the liver and the bone marrow during megakaryocytopoiesis, and, together with thrombopoietin, may also have a role in maturation of megakaryocytes (see Chapters 2 and 66). Less is known about possible roles of CCR4, but because this is the receptor for macrophage-derived chemokines, and thymus and activation-regulated chemokine (TARC) produced by monocytes, it may be involved in platelet–monocyte interactions, which are common after vessel damage. TARC was also shown to be present in platelet α-granules and secreted upon activation. CCR1 and CCR3 are both receptors for RANTES, released during platelet activation, and thus could participate in autocrine feedback mechanisms, as well as in activation by other RANTES-producing cells. The fractalkine receptor (CX3CR1), which has an unusual structure, including a stem domain holding it out from the platelet surface, is also present on platelets and plays a role in atherosclerosis.[96]

### F. Other Seven Transmembrane Receptors

**1. $V_{1a}$ Vasopressin Receptor.** Vasopressin is an agonist capable of inducing a rapid activation of platelets, but the level of this activation is dependent on plasma adenosine levels and, normally reversible, it is strongly increased in the presence of theophylline, which antagonizes the adenosine receptor.[97] In platelets, the $V_{1a}$ receptor is coupled via $Gq_{11}$.

**2. $A2_a$-Adenosine Receptor.** Adenosine was first identified as a platelet inhibitor in 1962.[98] On platelets it acts via the $A2_a$-adenosine receptor coupled to Gs protein to stimulate adenylate cyclase. This receptor has been cloned.[99] Plasma adenosine is a strong inhibitor of platelet activation by vasopressin or PAF.

**3. β2-Adrenergic Receptor.** Epinephrine only induces platelet aggregation in the presence of other agonists, and its cloned β2-adrenergic receptor is thought to couple only to the Gs class of G proteins.[100] Epinephrine activates plate-

let nitric oxide synthase by increasing cAMP in a $Ca^{2+}$-independent manner.

**4. Serotonin Receptor.** The major platelet serotonin (5-hydroxytryptamin 5-HT) receptor is $5\text{-HT}_{2A}$. Serotonin $5\text{-HT}_{2A}$ receptors are essential for a large number of physiological functions in the central nervous system and the periphery. In platelets, $5\text{-HT}_{2A}$ is coupled to G proteins, and its occupation by serotonin leads to calcium signaling.[101] Serotonin is thus a further autocrine agonist, because it is a major component of dense granules and is released upon platelet activation. The T102C polymorphism of the $5\text{-HT}_{2A}$ receptor gene is thought to lead to an upregulation of signaling and has been associated with nonfatal acute myocardial infarction.[102] Clinical depression is an independent risk factor for increased mortality in patients following acute coronary events. Increased platelet activity has been suggested as the mechanism responsible. A number of studies have shown that there is an increase in platelet $5\text{-HT}_{2A}$ expression in acute depression,[103] but this is still a controversial area (see Chapter 44). The serotonin reuptake receptor does not belong to the seven transmembrane class and is discussed in Section XI.D.

**5. Dopamine Receptors.** Recent studies have provided evidence for the presence of D3 and D5 dopamine receptors on platelets.[104] A 42-kDa glycoprotein was identified earlier.

## V. Immunoglobulin Superfamily

### A. GPVI

GPVI is one of the important members of the immunoglobulin (Ig) superfamily on platelets and, in addition to α2β1, is the other major established platelet receptor for collagen. Although GPVI was recognized earlier on platelets,[9] its function was then obscure. The detection of patients lacking or deficient in GPVI was an important step in identifying this glycoprotein as a critical collagen receptor. The first patient to be identified had platelets with a specific collagen-response defect.[105] Her plasma contained antibodies that recognized GPVI in normal platelets. The antibodies from this patient were used to characterize GPVI, as well as to identify other Japanese patients with a defect in this receptor. At least three other patients were identified, including two whose platelets completely lacked surface expression of GPVI, whereas the platelets of the third had 10% of the normal amounts.[106,107] The human gene sequence,[11,12,108] the mouse gene sequence,[12] and the genomic structure of GPVI[109] are now established. The molecular reason for the absence of GPVI in these patients is likely to be the presence of autoantibodies to GPVI, because treatment of either mouse platelets *in vivo*[110] or human platelets *ex vivo*[111] with anti-GPVI monoclonal antibodies leads to loss of GPVI from the platelet surface by induction of metalloproteases. Expression of GPVI requires its signaling subunit FcRγ,[112] whereas expression of FcRγ does not require GPVI. In mouse platelets, ablation of the FcRγ gene led to the absence of GPVI,[113] whereas FcRγ was unaffected by ablation of GPVI.[114] Although most patients with GPVI-deficient platelets have been detected in Japan, implying that the deficiency is more common in Japanese, it is also possible that other ethnic groups with more thrombogenic diets and lifestyles do not show the mild bleeding tendency that has permitted its detection in Japan. A similar type of case was recently described from the United States.[115]

GPVI belongs to the Ig superfamily[11] and is predominantly a signaling molecule. It complexes with the common FcRγ chain, which is critical for the signaling process[116] to give the structure shown in Fig. 6-4. GPVI consists of two Ig C2 loops that contain the collagen-binding domain. The single N-glycosylation site is on the outer Ig-C2-like domain. GPVI shows a higher degree of similarity to FcαR and to some of the natural killer receptor family than to other Fc receptors such as FcRI or FcRIII. Two studies to define the collagen-binding site, using antibodies and amino acid substitution, point to two regions on the N-terminal first loop.[117,118] The binding site for convulxin is still not known. Although experiments with collagen and convulxin subunits[119] or CRP and convulxin[120] have shown partial cross-inhibition of binding, the only reason why convulxin might need to use exactly the collagen-binding site on GPVI would be if it originally started as a GPVI inhibitory molecule and then, by multimerization, progressed to an activating molecule. As an activating molecule it only needs to be able to bind with high-avidity and cluster GPVI molecules. Unlike collagen- and CRP-binding to α2β1, where the conformation of the collagen appears less critical, binding to GPVI has a strict requirement for the triple-helical conformation.[121]

To make the receptor region readily accessible to the platelet environment, the two Ig C2 domains are held out from the platelet membrane by a mucinlike domain with a high content of serine and threonine residues (Fig. 6-4), somewhat like GPIbα (see Chapter 7), although not as long. GPVI has long been known to be an acidic, strongly glycosylated and sialylated molecule, and O-glycosylation of this region (Fig. 6-4) is responsible for the bulk of the molecular mass difference between the theoretical amino acid mass of 35 kDa and the observed mass of about 62 kDa.[11] The transmembrane region has an arginine in the third amino acid position, also like FcαR[122] and some members of the activating natural killer receptor family,[123,124] forming a salt bridge to an aspartic acid residue in the transmembrane region of the FcRγ subunit of the complex (Fig. 6-4). The cytoplasmic domain has some similarity to other receptors of this class, but this is rapidly lost in the direction of the C-terminus. GPVI, in which the transmembrane arginine has been mutated or the

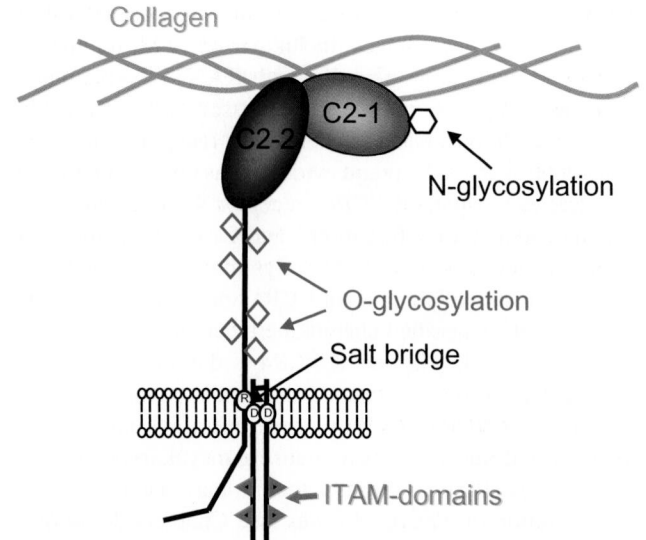

**Figure 6-4.**   Model of GPVI complexed with triple-helical collagen peptide containing the Gly-Pro-Hyp sequences. The two loops fold into globular structures made up of β-strands, forming hydrogen bonds to the collagen triple helix. These domains are held out from the platelet surface by an O-glycosylated mucinlike stalk. The third residue of the transmembrane domain is arginine, which forms a salt bridge to an aspartic acid residue in the transmembrane domain of the Fcγ subunit, helping to stabilize the complex. The immunoreceptor tyrosine-based activatory motifs (ITAMs) are phosphorylated by src family tyrosine kinases following receptor clustering. The p72[SYK] kinase binds to the phosphorylated ITAM domains and is activated, leading to phospholipase Cγ2 stimulation, the major effector pathway in response to collagen.

cytoplasmic region deleted, and to which FcRγ does not bind, does not give a Ca²⁺ signal in response to convulxin when expressed in RBL-3H3 cells, whereas the wild type does.[125] This emphasizes the importance in GPVI-mediated signal transduction of the interaction with the FcRγ subunit and with cytoplasmic components, including src family kinases like Fyn and Lyn. Signaling via GPVI/FcRγ has many similarities with that via immune cell receptors, in particular T-cell receptors, and is still under active investigation.

The genomic structure of the GPVI gene has been available since early 2000 via the Human Genome project,[109] together with the chromosomal localization on 19q13.4. The gene consists of eight exons. Exons 1 and 2 encode the signal sequence of 20 amino acids; exons 3 and 4, the two Ig-C2-like domains respectively; and exons 5 to 7, the mucinlike region. Exon 8 encodes the transmembrane region and the cytoplasmic domain, as well as the 3′ untranslated region and a poly-A signal. In mRNA from CMK megakaryoblastic leukemia cells stimulated with phorbol esters, Ezumi et al.[109] could demonstrate three splice variants for GPVI mRNA: GPVI-1, containing the full sequence; GPVI-2, lacking exon 5; and GPVI-3, with a four-nucleotide insertion leading to a frame shift and an amino acid sequence that did not include a transmembrane sequence. When expressed in COS-7 cells, GPVI-1 and -2 bound biotinylated CRP, whereas GPVI-3-expressing cells did not (presumably it was not in the membrane). It is unclear whether platelet GPVI has more than one splicing variant, which might be an alternative explanation to partial proteolysis for the double band found in gel electrophoresis separations of GPVI since its earliest detection.[9] No difference in function of GPVI-1 or GPVI-2 was found. GPVI-2 protrudes less from the cell membrane, like the smaller size polymorphisms of GPIbα, and might possibly be less active *in vivo*.

Although some studies have indicated that GPVI may also have a role in platelet adhesion to collagen,[126] this might be via activation of α2β1. Other studies on direct binding suggest that, at most, this is a very minor role.[125] It cannot be completely excluded yet that, like α2β1, the affinity of GPVI for collagen can be regulated on either the platelet side or the collagen side. Many of the studies into platelet activation via GPVI and signaling pathways have used CRP consisting of $(GPO)_n$ sequences organized in triple helices[127] or the snake C-type lectin, convulxin,[119,128] as agonists. Both of these ligands are thought to work by clustering GPVI molecules on the platelet surface and, for example, bring together kinases and exclude phosphatases to start a local signal chain of events that, among other consequences, activates phospholipase Cγ2,[129] releasing the second messengers 1,4,5-inositol trisphosphates and diacylglycerol that are responsible for raising $Ca^{2+}$ levels and activating protein kinase C. Although much of the signaling observed in platelets using GPVI-specific reagents resembles very closely that produced by collagen, it should not be forgotten that this refers predominantly to tyrosine phosphorylation and not to threonine and serine phosphorylation, nor to other second messenger systems, for the simple reason that tyrosine phosphorylation is technically easier to observe. Studies with GPVI-negative platelets[121,130] showed that, despite the lack of platelet aggregation, signaling nevertheless does occur, most likely via α2β1. In experiments with normal platelets, blocking GPVI by $F_{ab}$ fragments of anti-GPVI antibodies or via convulxin subunits resulted in inhibition of platelet aggregation.[11,119,128] Thus, a complete response to collagen requires interactions with (at least) both GPVI and α2β1. Purified GPVI or recombinant GPVI was also able to prevent collagen activation of platelets, but only if allowed to bind to the collagen and saturate binding sites first.[11,12] Jandrot–Perrus et al.[12] noted that the recombinant GPVI inhibited convulxin more readily than collagen, and suggested that the GPVI binding site on collagen might have restricted access. This could also be the result of the refolding of the recombinant protein, if collagen and convulxin do not use (completely) the same binding site on GPVI.[120]

More recent studies expressing wild-type and modified GPVI in cells in which FcRγ, but not α2β1, is present have confirmed these findings.[125] In the absence of α2β1, collagen does not induce $Ca^{2+}$ signaling, whereas convulxin gives a strong signal and CRP gives a weak signal compared with wild-type platelets. This reduced sensitivity to CRP could imply either that CRP is not as specific for GPVI as previously supposed and/or that GPVI may have more than one affinity state for the GPO recognition sequence. Convulxin does not necessarily need to have the same binding site on GPVI as collagen or CRP,[120] as long as its binding constant is sufficiently high. Why can collagen not activate GPVI-expressing cells, whereas convulxin can? There are obvious potential answers to this question, but it is difficult to demonstrate that they are the correct ones. The difference in distribution of GPVI binding sites on convulxin and on collagen is one clear possibility. Convulxin is a covalent tetramer of a heterodimer and therefore has at least four binding sites for GPVI. However, under nondenaturing conditions, convulxin migrates on gel filtration as a much larger complex (300 kDa), and electron microscopic studies also showed large hexagonal-shaped structures,[131] indicating that convulxin is able to form large noncovalently linked multimers. Thus, convulxin is able to cluster large numbers of GPVI molecules and their associated signaling systems, and this is why it is highly active even at low concentrations. On the other hand, collagen contains a large number of binding sites for different molecules, including α2β1, VWF, and GPVI. Each triple-helical collagen molecule has these binding sites distributed along its length. The collagen molecules are organized together to form macromolecular fibrils, which are the structures that platelets generally encounter. The GPVI binding sites are likely to be, like those for α2β1,[132] sufficiently far apart that, when α2β1 is absent or is blocked, the collagen is unable to cluster GPVI, or the GPVI molecules bound to the collagen are not close enough together to be able to initiate the signal cascades. In contrast to the "two-step, two-site" model of collagen activation, this is the "matrix" model, because it assumes that collagen has to impose a change in receptor distribution on the platelet surface. The necessity for a matrix model does not preclude changes in the activity state of receptors, such as proposed in "two-step, two-site" models.

The requirement for both α2β1 and GPVI for platelet aggregation to collagen, as well as the observation that cells expressing only GPVI but not α2β1 do not show a $Ca^{2+}$ response to collagen but do to convulxin,[125] might be thought to exclude a model in which binding to GPVI activates platelets and hence upregulates α2β1. However, such models do not take into account the distribution of receptor binding sites on collagen. In addition, at low GPVI expression levels, convulxin may be a much more efficient GPVI-binding and -clustering molecule than either collagen or CRP, and less susceptible to minor folding irregularities.

During the last few years, several studies have looked at the role of collagen receptors *in vivo* in various thrombosis models. Many of these studies have shown that GPVI has a major role in thrombus formation, although not all pointed in this direction — perhaps related to the thrombosis model used. Although some models are based on direct injection of fibrillar collagen plus epinephrine into an artery,[133] others use injuries to the vessel wall as the trigger for thrombus formation. One type of study uses laser-induced injury or mechanical injury,[134,135] whereas others use chemically induced injury by FeCl₃ to induce loss of endothelial cells.[136] Because the injuries caused by these procedures have not been examined histologically, it is not certain how they differ in detail. Massberg et al.[136] perfused recombinant soluble GPVI into mice, with an FeCl₃-induced vascular injury type of thrombosis model, and virtually completely inhibited thrombosis. On the other hand, Gruner et al.,[135] using a mechanical-induced aorta injury type of thrombosis model, found little inhibition by recombinant soluble GPVI, but strong inhibition by monoclonal anti-GPVI antibodies. This apparent contradiction has drawn attention to possible differences between various kinds of injury, as well as to other aspects of the models used. Thus, the effectiveness of recombinant soluble GPVI in a thrombosis model may be related to the quantity and type of collagen exposed, which is known to vary with depth in the subendothelium. Some thrombosis models, such as laser injury, may reflect activation of the endothelium rather than any major exposure of subendothelium. On the other hand, the effectiveness of monoclonal anti-GPVI antibodies may reflect their ability to activate platelets and downregulate receptors, including GPVI, but possibly others. For example, it is well-known that the GPIb complex is downregulated and therefore less effective after platelet activation.[137,138]

Several recent reports have examined the role of GPVI in pathological problems *in vivo.* Gawaz et al.[139] showed that radiolabeled GPVI could be used in scintillographic imaging of vascular lesions in mice because it binds specifically to the injured region, again indicating that collagen is exposed at these sites. Penz et al.[140] showed that human atheromatous plaque from patients with carotid stenosis contained collagen type I and type III structures that were able to activate platelets. Blockage or absence of GPVI was able to prevent thrombus formation, whereas blockage of α2β1 had little effect. Similarly, blockage of collagen with anticollagen antibodies, or its degradation with collagenase, prevented thrombus formation. All these studies reinforce the idea that GPVI is a potential target for antithrombotic therapy.

### B. FcγRIIA

Another major member of the Ig family present on human platelets is FcγRIIA (CD32), which is a low-affinity receptor for the IgG Fc domain. Like GPVI, FcγRIIA has two C-2-like Ig loops, but it signals directly via its cytoplasmic domain, which contains two immunoreceptor tyrosine-based activatory motif domains. It has an Mr of approximately 40 to 42 kDa under both reducing and nonreducing conditions. FcγRIIA has a role in immunological defense against bacteria, viruses, and parasites (see Chapter 40). The significance of FcγRIIA also lies in the problems caused by a variety of autoimmune and alloimmune disorders involving antigen–antibody clusters that cause platelet activation by clustering FcγRIIA. Thus, nearly all polyclonal antibodies against platelet receptors, and many monoclonal antibodies, activate platelets by clustering this receptor. Why some monoclonals have this effect whereas others do not has not yet been explained satisfactorily, but is probably related to the way in which the Fc part of the IgG is presented from the binding epitope. There has been a lot of interest in the FcγRIIA receptor because of its role in heparin-induced thrombocytopenia.[141] Antibodies develop against complexes of heparin and the platelet α-granule chemokine (platelet factor 4), and the complexes cause platelet activation via FcγRIIA, leading to thrombocytopenia. Heparin-induced thrombocytopenia is discussed in detail in Chapter 48. FcγRIIA has an R/H polymorphism at residue 131, which affects the affinity of Fc binding (see Chapter 14). Although it was thought that this could be related to susceptibility to heparin-induced thrombocytopenia, this appears not to be the case. Mouse platelets lack both the FcγRIIA receptor and immune thrombocytopenia-type disorders. The latter can be mimicked by transfection of the gene introducing expression of FcγRIIA on platelets.[142] As in many other aspects of platelet activation, lipid rafts may be involved in signaling via FcγRIIA.[143]

### C. FcεRI

Platelets can be activated by IgE, indicating that they are involved in defense mechanisms against parasites as well as in allergies (see Chapter 40). Several authors have shown that platelets express FcεRI, the high-affinity receptor for IgE.[144,145] This was detected as protein by flow cytometry and as mRNA by PCR. There was a wide heterogeneity in expression levels among normal donors, with a low proportion of platelets also positive for the low-affinity receptor, FcεRII (CD23). In megakaryocytes, FcεRI was only detected in the cytoplasm. Activation of platelets via FcεRI caused platelets to release serotonin and RANTES, again supporting a role for platelets in allergic inflammation.[145] Cross-linking of FcεRI triggered platelet cytotoxicity for *Schistosoma mansoni* larvae (see Chapter 40).

### D. Platelet and T-Cell Antigen 1 (PTA-1 [TLiSA1, DNAM-1, CD226])

T-lineage-specific activation antigen (TLiSA1) was originally described as being involved in the differentiation of human cytotoxic T cells. Because this antigen was subsequently identified on platelets and was shown to be involved in platelet activation, it was renamed PTA-1.[146] It is an unusual member of the Ig superfamily with two V domains only in the extracellular region. More recently, this molecule has been classified as CD226, and a wide range of monoclonal antibodies were prepared.[147] The polio virus receptor (CD155) and the adherens junction protein, nectin-2 (CD112), were identified as two potential ligands for CD226.[148] There is evidence that CD226 mediates platelet and megakaryocyte adhesion to endothelial cells.[149]

### E. Junction Adhesion Molecules (JAMs)

**1. JAM-1 (F11).** A monoclonal antibody against platelets was found to recognize two membrane proteins of 32 kDa and 35 kDa, and to activate platelets by cross-linking these molecules to FcγRIIA.[150] This molecule (F11) belongs to the Ig superfamily with two Ig domains in the extracellular region. F11 was subsequently shown to be identical to the previously identified JAM-1, which has been shown to localize in cell–cell contact areas by homophilic interactions when expressed in CHO cells.[151] Although its function in platelets is not yet clear, JAM-1 is known to aid in the formation of tight junctions in endothelial cells. One possible function might therefore be to seal the edges of a thrombus to surrounding endothelial cells. A role in platelet–platelet aggregate stabilization also cannot be excluded.

**2. JAM-3.** JAM-3 was also identified on platelets as a 43-kDa glycoprotein.[152] It was shown to be a type I transmembrane receptor, containing two Ig-like domains, that is a counterreceptor for the leukocyte integrin Mac-1, thereby mediating leukocyte–platelet interactions. It may thus have a role in inflammatory vascular pathologies such as atherothrombosis.

### F. Intercellular Adhesion Molecule 2 (ICAM-2)

ICAM-2 (CD102) is the only member of the ICAM subfamily detected on platelets, with $3000 \pm 230$ copies per platelet.[153] The platelet molecule migrates as a 59-kDa glycoprotein under reducing conditions. It is composed of two C2 Ig domains, a transmembrane region, and a cytoplasmic tail. Megakaryocytes are also strongly positive to specific staining. ICAM-2 is the only known β2 integrin ligand on platelets and is important for platelet adhesion to neutrophils.[154]

### G. PECAM-1

PECAM-1 is a 130-kDa glycoprotein composed of six C2 Ig domains, a transmembrane region, and a short cytoplasmic tail, which contains two immunoreceptor tyrosine-based inhibitory domains (ITIM).[155,156] PECAM-1 is expressed by platelets, endothelial cells, and most subsets of leukocytes. PECAM-1 is discussed in detail in Chapter 11.

### H. CD47

Integrin-associated protein (CD47) is a receptor for the cell-binding domain of thrombospondin (TSP) family members, a ligand for the transmembrane signaling protein signal-regulatory protein α, and a component of a supramolecular complex containing specific integrins, heterotrimeric G proteins, and cholesterol.[157] CD47 occurs in all cells, including platelets. It is an unusual member of the Ig superfamily with a V-like extracellular domain, five membrane-spanning domains, and a short cytoplasmic domain.[158] Peptides containing a VVM motif in the C-terminal domain of TSPs are agonists for CD47, initiating heterotrimeric Gi protein signaling that augments the functions of integrins of the β1, β2, and β3 families, thus modulating a range of cell activities including platelet activation, cell motility and adhesion, and leukocyte adhesion, migration, and phagocytosis.[157] In platelets, CD47 associates with and regulates the function of integrin αIIbβ3[159] and modulates the function of platelet integrin α2β1, a collagen receptor. The CD47 agonist peptide, 4N1K (KRFYVVMWKK), derived from the cell-binding domain, synergizes with soluble collagen in aggregating platelet-rich plasma. 4N1K and intact TSP-1 also induce the aggregation of washed, unstirred platelets on immobilized collagen with a rapid increase in tyrosine phosphorylation. The effects of TSP-1 and 4N1K on platelet aggregation are absolutely dependent on CD47, based on platelets from CD47$^{-/-}$ mice. PGE$_1$ prevents 4N1K-dependent aggregation on immobilized collagen, but does not inhibit the 4N1K peptide stimulation of α2β1-dependent platelet spreading.[160]

### I. Endothelial Cell-Selective Adhesion Molecule (ESAM)

ESAM is a member of the Ig superfamily structurally similar to the JAMs,[161] with an extracellular domain containing two Ig domains, one C2 and one V type. The cytoplasmic domain contains an SH3 binding domain and a C-terminal PDZ target domain. Although first characterized in endothelial cells,[161] subsequent studies showed that ESAM is also expressed in platelets.[162] It belongs to the tight junction adhesion receptors and colocalizes with other members of this family in the endothelium. No bleeding problems were reported when ESAM was deleted in mice.[163]

### J. TREM-Like Transcript 1 (TLT-1)

TLT-1, like other TREM (triggering receptor expressed on myeloid cells) family receptors, is an Ig superfamily member with a single V-type extracellular domain but, unlike the others, does not interact with the activating subunit DAP12. TLT-1 has a longer cytoplasmic domain than the others with two splice variants, one of 199 amino acids and the other of 126 amino acids, differing in their cytoplasmic domains. The shorter splice variant contains a canonical C-terminal ITIM domain and a membrane proximal nonclassical sequence resembling one known to recruit SHP-2. TLT-1 has been detected exclusively in megakaryocyte and platelet α-granules and is redistributed to the platelet surface upon activation.[164,165] The physiological ligand for TLT-1 is not yet known. Surprisingly, cross-linking TLT-1 to FcεRI increased, rather than decreased, calcium signaling.

## VI. C-Type Lectin Receptor Family

### A. P-Selectin (CD62P)

The selectins are an important group of adhesive receptors present on platelets (P-selectin), endothelial cells (E- and P-selectin), and lymphocytes (L-selectin). Their main function involves multiple, transient weak interactions with carbohydrate ligands expressed on other cells, thereby allowing the development of stronger, more stable binding via other ligands and receptors. P-selectin is discussed in detail in Chapter 12.

### B. CD72

CD72, a member of the C-type lectin family (also called Lyb-2), is a 45-kDa type II transmembrane protein normally found as a homodimer.[166] The cytoplasmic domain contains an ITIM motif which, when phosphorylated, can recruit the SHP-1 tyrosine phosphatase. CD72 was detected on human platelets and may be a receptor for CD100, also recently found on platelets.[167]

### C. CD93 (C1q-Rp)

CD93, or C1q-Rp, is a 120-kDa O-sialoglycoprotein, the equivalent of the AA4 antigen in mice, and is classed as a defense collagen receptor for C1q. CD93 is a member of the C-type lectin family and the N-terminal domain has this fold. The domain structure composition and organization closely resembles that of thrombomodulin.[168] It is strongly expressed in platelets and megakaryocytes (as well as endothelial cells, natural killer [NK] cells, and monocytes) both on the cell surface and in intracytoplasmic vesicles. CD93 ectodomain was also detected in plasma, but the cellular source is not clearly established. Monocytes, neutrophils, and endothelial cells are possible sources, but platelets should also be considered.[169]

### D. Other C-Type Lectin Receptors

Small C-type lectin receptors are widespread, particularly in immune system cells, and are generally thought to recognize carbohydrate structures. Because platelets are increasingly considered to have defensive functions (see Chapters 39 and 40), and as more receptors associated with innate immunity (e.g., TLR, discussed in Section III.B) are recognized on platelets, it seems likely that the diversity of small C-type lectin receptors of the NK class identified on platelets will also increase. Indeed, one of these, CLEC-2, was recently detected on platelets and was shown to have a role in platelet activation by the snake C-type lectin, aggretin or rhodocytin.[170]

## VII. Tetraspanins

Tetraspanins are a group of membrane proteins that, as the name implies, contain four membrane-spanning domains. They are thought to have important functions in signal transduction across the cell membrane, in complexes with other membrane receptors, particularly including integrins. Platelets contain several members of this group of molecules, but the role of some of these is still poorly understood. There is still some controversy about the role of Fc receptors in tetraspanin function, with some authors claiming that activation of platelets via antitetraspanin antibodies is independent of Fc receptors,[171] whereas others support a model in which tetraspanins modulate Fc receptor functions.[172]

### A. CD9

CD9 is the main member of the tetraspanin group present on platelets.[173] It is a 24-kDa protein with a single N-glycosylation site and four transmembrane domains. Although many data have been accumulated implicating CD9 in several platelet functions, particularly in connection with the integrins αIIbβ3 and α6β1, the way in which CD9 acts is still obscure.

### B. CD63

CD63 was first detected as a marker of platelet activation, which is increased on the surface of platelets after granule release. It was subsequently found to be a 53-kDa lysosomal membrane protein with four transmembrane domains and three putative N-glycosylation sites.[174] CD63 has been found to be associated with integrins[175] and is used as a platelet activation marker (see Chapter 30). Recent studies have shown that CD63 does not affect adhesion, but modulates platelet spreading and platelet tyrosine phosphorylation on immobilized fibrinogen.[176] CD63 also coprecipitated with a lipid kinase thought to be PI4 kinase type II.

## C. CD82

CD82 remains one of the more obscure members of the tetraspanin family present on platelets. It has a mass of 52 to 53 kDa and is glycosylated with three putative N-glycosylation sites.[177]

### D. Platelet and Endothelial Cell Tetraspan Antigen 3 (PETA-3, CD151)

CD151 is highly expressed in megakaryocytes and, to a lesser extent, in platelets. It was recognized as a 27-kDa platelet surface receptor of the tetraspanin family, with a single N-glycosylation site.[178] As with other members of this family, the function is not yet well established. There is evidence that CD151 associates preferentially with integrins of the β1 family and has a role in signal transduction.[179] Recent studies indicate that it may have a role in linking activated protein kinase C with integrins. Platelets from CD151-deficient mice have impaired "outside-in" αIIbβ3 signaling responses to a variety of agonists,[180] whereas inside-out responses were normal.

## VIII. Glycosyl Phosphatidylinositol (GPI)-Anchored Proteins

Platelets contain at least five glycoproteins that are linked by GPI anchors. GPI-linked receptors have a poorly understood role associated with lipid rafts and signal transduction. The GPI-linked receptors identified in platelets include the classic CD55 (decay accelerating factor [DAF]) and CD59, which are also present on a wide range of blood cells to protect them against complement when this is activated (e.g., during defense against bacteria).[181] In addition, activated but not resting platelets express a 170-kDa glycoprotein, now classified as CD109, which carries the *Gov* alloantigen, as well as ABO oligosaccharides, but which has no recognized function.[182] CD109 was recently cloned and identified as a member of the α2-macroglobulin/complement thioester-containing gene family.[183,184] It may be activated by proteolytic cleavage and could then react covalently with carbohydrate or protein targets. There is also a heavily

glycosylated 500-kDa GPI-anchored glycoprotein that has not yet been characterized. Because there is good evidence that platelets contain glycosaminoglycan-carrying receptors, and glypican is a GPI-linked member of this family, it might be a likely candidate. These GPI-linked proteins are all affected by the clonal absence of the GPI anchor in paroxysmal nocturnal hemoglobinuria.[181]

The normal prion protein (PrP^C) has been recently reported to be present on the platelet surface, linked by a GPI anchor, has a mass of 27 to 30 kDa, and sheds when platelets are activated or during storage of platelet concentrates. PrP^C is associated with granules and its surface expression is increased about twofold (from 1800–4300 copies) after activation.[185] Platelets are calculated to contain more than 96% of the PrP^C present in adult blood and may thus have a role in the development of novel Creuzfeld–Jakob disease. PrP^C on mononuclear cells and platelets is sensitive to enzymatic treatment of cells by proteinase K and phosphatidylinositol-specific phospholipase C. PrP^C was detected by immunofluorescent confocal microscopy in the surface-connected canalicular system and in CD63-negative granules — probably α-granules. In megakaryocytes, staining for PrP^C increased with ploidy and appeared to be evenly distributed throughout the cytoplasm. Thus, megakaryocytes are likely to be the source of PrP^C in platelets.

## IX. Glycosaminoglycan-Carrying Receptors

In addition to the GPI-linked glypican (see Section VIII), there is some evidence for the presence on platelets of other members of the glycosaminoglycan-carrying receptor family, such as syndecan and perlican. Recently, evidence was reported for the expression and function of syndecan 4 in human platelets.[186] When antithrombin binds to syndecan 4, ADP- and ATP-dependent platelet activation is reduced, and syndecan 4 shedding from activated platelets is inhibited. Glycosaminoglycans of heparin and chondroitin sulfate families associated with the platelet surface have been implicated in both heparin-induced thrombocytopenia as well as in the platelet responses to chemokines.[95,187]

## X. Tyrosine Kinase Receptors

These receptors belong to a family that is activated by dimerization through ligand binding to the extracellular domains, leading to tyrosine phosphorylation of each individual protein by the kinase activity of the other.

### A. Thrombopoietin Receptor (c-mpl, CD110)

The classic representative of the tyrosine kinase receptor family on platelets is the thrombopoietin receptor:

molecular mass, 80 to 84 kDa; $56 \pm 17$ copies per platelet; high-affinity binding sites ($K_d$, $163 \pm 31$ pM). After adding thrombopoietin to platelets, 80% of the binding sites are internalized within an hour and are not recycled. Although the main function of this receptor is at the megakaryocyte level, its presence on platelets is critical for modulating the total cytokine mass and thus regulating platelet production. In addition, it may have a role in regulating the sensitivity of platelet responses to other agonists. This receptor and its ligand are discussed in detail in Chapter 66.

### B. Leptin Receptor

Leptin, the 167 amino acid protein product of the *ob* gene, regulates energy storage by fat in mammals. Absence of leptin, or mutations leading to defective function, cause hyperphagia and obesity that can be corrected by administration of leptin.[188] Higher leptin levels increase weight loss by decreasing appetite and increasing metabolism, linking nutritional status with eating and energy consumption. In addition to adipocytes, leptin is expressed in a range of tissues and has other physiological functions. The presence of the leptin receptor, the product of the *db* (diabetes) gene, on platelets was first described in 1999, when the long form of the receptor, which had been detected in megakaryocytes by PCR, was found using Western blotting methods as an approximately 130-kDa protein.[189] This study also showed that leptin 30 to 100 ng/mL enhanced platelet aggregation responses to threshold amounts of ADP (1 μM), but had no effect alone. However, leptin 50 ng/mL did cause enhanced tyrosine phosphorylation of platelet proteins. The leptin receptor is a class I cytokine receptor with four conserved cysteine residues and the WSXWS five amino acid motif, which lacks intrinsic tyrosine kinase activity but is directly coupled to Janus kinase and STAT proteins and is activated by homo- or heterodimerization. A recent study has implicated phosphodiesterase 3A in signaling downstream of the leptin receptor.[190]

Both leptin-negative and leptin receptor-negative mice had delayed thrombotic occlusion, unstable thrombi, and a higher tendency to embolize than wild-type mice.[191] In leptin-deficient but not receptor-deficient mice, this defect could be corrected by administration of leptin. Because obese people often have high leptin levels and also respond poorly to leptin administration, it has been suggested that this leptin resistance may be the result of desensitization of receptors to the leptin signal.[192] It has also been proposed that a secreted form of the receptor from alternative splicing, or cleavage of an extracellular domain, may be involved in regulating leptin activity.[193] Platelet activation might be the source of such soluble leptin receptor, although so far there is no evidence. Although several plausible mechanisms for leptin resistance have been proposed, there is as yet no hard evidence for any of these.[194] Blockage of endogenous circulating leptin by a specific antibody protected mice against arterial and venous thrombosis.[195] Platelet responses to leptin vary widely between individuals, with approximately 40% responding and 60% not responding. All individuals express the signaling form of the receptor, but the non-responders have lower levels and somewhat lower avidity.[196] Differences in N-glycosylation of the leptin receptor have been reported and it was postulated that this might affect folding and hence function, but no experimental evidence is yet available.

### C. Tie-1 (Tyrosine Kinase with Immunoglobulin and Epidermal Growth Factor Homology-1) Receptor

Tie-1 is expressed predominantly on endothelial cells and has a critical role in angiogenesis. However, it is also present on platelets as a 110-kDa molecule.[197] Surface expression is increased after platelet activation. Tie-2/Tek is not present on platelets. Tie-1 and -2 have been recently identified as receptors for angiopoietin 1 and 2, but the function of Tie-1 on platelets is not yet established.

### D. Insulin Receptor

Like many other cells, platelets express insulin receptors in low numbers.[198] There is some evidence that the platelet insulin receptor is functional and that platelets show metabolic changes in response to insulin. Binding of insulin in physiological amounts to platelets increased adenylate cyclase-linked prostacyclin receptor numbers on the cell surface, which was shown to be directly related to the ADP ribosylation of Giα.[199]

### E. Platelet-Derived Growth Factor (PDGF) Receptor

PDGF is a dimeric, disulfide-bridged protein released from α-granules upon platelet activation. Platelets only have the PDGF α-receptor.[200] PDGF binding dimerizes this receptor, initiating a tyrosine phosphorylation-based signaling cascade.

### F. Gas6 Receptors (Axl, Sky, and Mer)

This group of receptors belongs to the Tyro3 receptor subfamily of single transmembrane tyrosine kinase receptors. They have similar domain structures, containing two extracellular N-terminal Ig-like domains, as well as two fibronectin-III-like domains followed by a tyrosine kinase domain at the C-terminal cytoplasmic end. All three were shown to be present in both mouse and human platelets by flow cytometry.[201] An earlier study using RT-PCR detected Mer but not Axl.[202] Mer-deficient mice generated by targeted ablation of the Mer gene had impaired platelet aggregation to collagen, U46619, and PAR4 agonist peptide at low concentrations *in vitro* and inhibition of acute arterial thrombosis *in vivo;* however, responses to ADP were normal. Blocking antibodies to Sky or Mer inhibited platelet aggregation to ADP or TRAP by 80%, whereas a stimulatory anti-Axl antibody increased platelet responses to these agonists. Mice lacking Gas6 receptors, like Gas6-deficient mice, are protected against thrombosis in model systems, but do not show any increased bleeding in standard assays.

### G. Platelet-Endothelial Aggregation Receptor-1 (PEAR1)

PEAR1 was detected during a search for platelet components that are tyrosine phosphorylated downstream of platelet activation and aggregation.[203] PEAR1 is a type 1 membrane protein, with 15 extracellular EGF-type repeats and numerous tyrosines in the cytoplasmic domain, that is highly expressed in platelets and endothelial cells. Induction of platelet aggregation by physiological agonists leads to phosphorylation of residues Y925, S953, and S1029. Inhibition of integrin αIIbβ3 prevents phosphorylation of Y925. The ShcB adaptor protein associates with PEAR1 after tyrosine phosphorylation. Centrifugation of platelets, bringing them into close association in the pellet, also leads to phosphorylation of Y925 that is not inhibited by blockage of αIIbβ3. Overall, the results point to homodimeric binding of PEAR1 stabilizing platelet–platelet interactions downstream of aggregation via αIIbβ3 and its ligands.

### H. Ephrins and Eph Kinases

The Eph kinases are a large family of receptor tyrosine kinases with an extracellular ligand-binding domain and a cytoplasmic tyrosine kinase domain. Two Eph kinases, EphA4 and EphB1, and at least one ligand, ephrinB1, have been reported to be expressed on human platelets.[204] In activated platelets, EphA4 associates with src kinases Fyn and Lyn. EphA4 is also constitutively associated with integrin αIIbβ3. Experimental evidence supports a model in which platelet aggregation via fibrinogen and αIIbβ3 (or possibly sustained, close platelet–platelet contact) leads to Eph/ephrin interactions supporting thrombus growth by enhanced outside-in signaling via αIIbβ3 (see Chapters 16 and 17).

## XI. Miscellaneous Platelet Membrane Glycoproteins

### A. CD36 (GPIV, GPIIIb)

CD36 is the general name for a molecule that was also named GPIV or GPIIIb in platelets. Many early studies of this glycoprotein were carried out on platelets. There was considerable interest when CD36 was found to be absent in a small population (4–7%) of healthy donors.[205] This deficiency was at first thought to be confined to Japanese, but then other deficient individuals were found in different East Asian populations.[206] More recent studies showed that CD36 is also absent in a minority (7–10%) of the sub-Saharan populations of Africa[207] and in a very small proportion (approximately 0.3%) of populations in other parts of the world. Is CD36 an adhesive receptor? This has been a controversial area and is even now not completely resolved. Two major adhesive proteins, collagen[208] and TSP,[209] have been suggested as ligands for CD36, but it is clear that other proteins also bind to this receptor, including plasmodium-infected erythrocytes, and that its main function may be as a scavenger for oxidized lipoproteins[210] or as a transporter of long-chain fatty acids.[211] Other investigators found little evidence for a role of CD36 as a collagen receptor.[212,213] The molecular basis of CD36 deficiency has been identified as a polymorphism in codon 90, which, if expressed, leads to a Pro→Ser shift.[214] CD36 has been reported to be a good specific marker for lipid rafts in platelets.[215]

### B. C1q Receptor

C1q has been shown to modulate platelet interactions with collagen and immune complexes, and has been identified at sites of vascular injury and inflammation, as well as in atherosclerotic lesions. C1qR/p33 is a single-chain, multiligand binding protein with an apparent molecular mass of 33 kDa by sodium dodecyl sulfate–polyacrylamide gel electrophoresis (SDS-PAGE), but a multimer of 97.2 kDa by gel filtration under nondissociating conditions.[216] Crystallographic evidence suggests that C1qR may associate to form a doughnut-shaped ternary complex.[217] C1qR was originally isolated from a membrane preparation of a lymphoblastoid cell line (Raji), but has been shown to have a wide cellular distribution, including platelets and endothelial cells.[218,219] On both resting and activated platelets, C1qR surface expression is poor, but is greatly enhanced after platelet adhesion to immobilized fibrinogen or fibronectin.[220] *Staphylococcus aureus* protein A recognizes platelet C1qR/p33, providing a mechanism for staphylococci to interact with platelets.[221] C1q and the outer envelope protein of HIV, gp120, have several structural and functional similarities. The recombinant form of the 33-kDa protein, which binds to the globular "heads" of C1q (gC1q-R/p33), inhibited the growth of different HIV-1 strains in cell cultures.[222] Another class of receptor for C1q, CD93 (C1q-Rp), was discussed under C-type lectins in Section VI.C.

### C. C3-Specific Binding Protein (Membrane Cofactor Protein, CD46)

A C3-specific binding protein isolated from [125]I surface-labeled human platelets by affinity chromatography gave two bands with masses of 53 kDa and 64 kDa on SDS-PAGE, similar to human leukocyte iC3- and C3b-binding glycoprotein (CD46). This membrane glycoprotein is structurally and antigenically distinct from CD55 (DAF), a complement regulatory protein on human platelet membranes (see Section VIII). DAF and CD46 have complementary activity profiles because DAF can prevent assembly of and dissociate the C3 convertases but has no cofactor activity, whereas C3-specific binding protein has cofactor activity but no decay accelerating activity. These two proteins were proposed to function together to prevent autologous complement activation.[223] More recently, CD46 was found to be highly enriched in plateletlike bodies released from differentiating CMK myeloid cells and is thought to have a role in microparticle function.[224]

### D. Serotonin Reuptake Receptor

The serotonin transporter in platelets is the same protein that is present in other cells storing 5-HT, such as the presynaptic uptake mechanism in neurons. It is a 74-kDa glycoprotein, with two N-glycosylation sites, that has 12 transmembrane domains with both N- and C-termini on the cytoplasmic side of the membrane.[225] Numerous studies have associated its expression levels with a variety of disorders ranging from depression to alcoholism to romantic love (see Chapter 44). Hyperserotonemia in autism appears to be the result of enhanced 5-HT uptake, because free 5-HT levels are normal, and the reported excess of the long/long 5-HTTLPR genotype in autism could provide a partial molecular explanation.[226] Serotonin-specific reuptake inhibitors are known to influence susceptibility to cardiovascular diseases and, *in vitro,* directly inhibit platelet function.[227]

### E. Lysosomal-Associated Membrane Proteins 1 and 2 (LAMP-1, CD107a; LAMP-2, CD107b)

LAMP-1 and -2 are both membrane-associated proteins found in platelet lysosomes and dense granules.[228,229] They have been used as markers of platelet activation (see Chapter 30), but their functions remain unknown.

### F. CD40 Ligand (CD40L, CD154)

CD154 is the ligand for CD40 and is reported to be present in platelet granules.[230] It is a 39-kDa type II membrane glycoprotein. Platelets are capable of expressing CD154 on their surface within minutes of stimulation. Platelet CD154 can interact with CD40 on endothelial cells and cause an inflammatory response.[230] Mutations in the CD154 gene are associated with a rare immunodeficiency state, X-linked

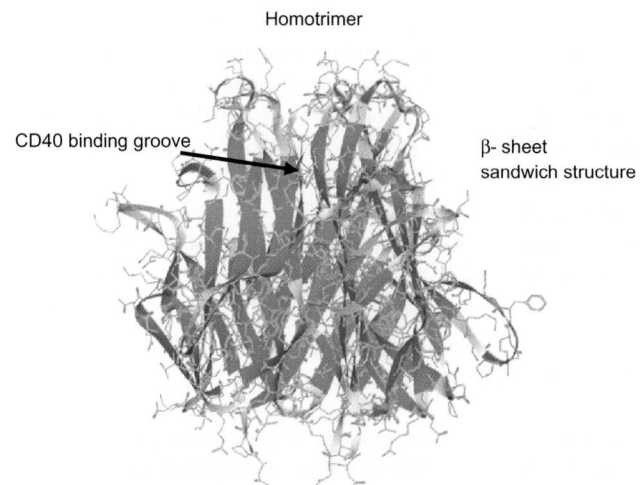

Figure 6-5. Model of the trimeric complex of CD154 (CD40L). CD154 is a type II membrane glycoprotein that mediates cell activation via interaction with CD40. It folds into a tissue necrosis factor (TNF)–homologous globular domain (shown here), a long extracellular stalk, a short transmembrane segment, and a small cytoplasmic domain. The TNF-like domain is characterized by a β-sheet sandwich fold. The binding site for CD40 coincides with a shallow cleft along the interface between adjacent monomers (facing the viewer), including residues from both monomers. Signaling in the CD40-expressing cell is thought to be induced by trimeric clustering being imposed on CD40 by the CD154 trimer, whether present on the cell surface or as a soluble extracellular complex.

hyper-IgM syndrome. The absence of CD154 on platelets is diagnostic for this disease.[231] The duration of CD154 expression after release from the granules is closely regulated by expression of CD40 on the platelet surface, and binding of these two molecules leads to loss of the extracellular domain of CD154 as a soluble molecule.[232] The CD154/CD40 system is considered to have an important role in maintaining inflammatory responses and to be involved in atherosclerosis.[233,234] Figure 6-5 shows a model of the trimeric structure of the extracellular domain of CD154 based on its crystal structure. It is thought that the soluble form of CD154 also maintains this trimeric structure, allowing it to interact with and activate cells expressing CD40. CD154 (CD40L) is also discussed in Chapter 39.

### G. P-Selectin Glycoprotein Ligand 1 (PSGL-1, CD162)

PSGL-1 (discussed in detail in Chapter 12) has been known since 1993 as the counterreceptor for P-selectin on monocytes, neutrophils, and a variety of lymphocytes.[235] It is a highly O-glycosylated type I transmembrane protein, with sulfated tyrosines in its N-terminal region forming an important part of the binding site, and it exists as a disulfide-linked dimer. PSGL-1 has also been detected on platelets, particularly on young ones, and anti-PSGL-1 antibodies reduced platelet rolling on mesenteric venules, suggesting that PSGL-1 participates in platelet–vessel interactions.[236]

### H. P2X₁

P2X₁ is the receptor responsible for rapid $Ca^{2+}$ entry into platelets in response to ATP.[237] P2X₁ belongs to the ADP- or ATP-driven calcium channel family of purinergic receptors, which have three transmembrane domains and form a trimer to create the calcium channel in the membrane.[15] A patient was described with a heterozygous-specific defect in platelet response to ADP, with a mutation in the P2X₁ gene leading to deletion of a single leucine in a series of four in the second transmembrane domain.[238] Because the other allele was normal, apparently even one defective molecule in the trimer is sufficient to prevent the channel from operating normally. P2X₁ is discussed in detail in Chapter 10.

### I. Tight Junction Receptors

In addition to JAM-1 and JAM-3 (discussed in Section V.E), two other tight junction receptors have been identified in platelets. These are occludin and zonula occludens protein 1. Occludin, with a molecular mass of approximately 65 kDa, was identified as the first component of the tight junction strand itself.[239,240] Occludin is an integral membrane protein comprised of four transmembrane domains, a long COOH-terminal cytoplasmic domain, a short NH2-terminal cytoplasmic domain, two extracellular loops, and one intracellular turn. Zonula occludens protein 1 is a 220-kDa peripheral membrane protein expressed in tight junctions of both epithelial and endothelial cells.[241,242] Occludin and zonula occludens protein 1 were shown to be present in tight junctionlike structures between aggregated platelets.[243] The junctional protein E-cadherin has also been reported to be present in platelets.[244]

### J. Tumor Necrosis Factor (TNF) Receptor

TNF was identified as a product of lymphocytes and macrophages that causes the lysis of certain cell types, particularly tumor cells. TNF and similar molecules were later found to form a TNF superfamily. The receptors for these ligands also form a TNF receptor (TNFR) superfamily (reviewed by Locksley et al.[245]). TNFR-like receptors are type 1 transmembrane proteins that use a scaffold of disulfide bridges to form elongated structures. These form "cysteine-rich" domains containing 40 amino acid repeats with three intrachain disulfides made up of six highly conserved cysteines. The functional receptor is a trimer and these elongated structures fit into the grooves between individual chains of the ligand trimer forming a 3 : 3 complex. The cytoplasmic domains of TNFRs are fairly short and provide docking sites for two major classes of signaling molecules. These are TNFR-associated factors (TRAFs) and "death domain" molecules. Binding of TNF to TNFR1 recruits TNFR1-associated death domain protein followed by several other signaling molecules (FAS-associated death domain protein, TRAF-2, and the death domain kinase, receptor interacting protein), generally leading to caspase activation and other aspect of apoptosis.

Platelets did not show direct responses to physiological amounts of TNF-α (20–60 pg/mL), but these amounts enhanced platelet responses to collagen.[246] This effect was prevented by the addition of TNFR inhibitor. However, TNF-α did not enhance thrombin- or ADP-induced platelet aggregation. Flow cytometric analysis provided evidence for the presence of TNFR1 and TNFR2 on platelets in a ratio of about 2 : 1. Because TNF-α is upregulated in heart failure patients and related to severity, TNFR1 and TNFR2 might contribute to the increased platelet activity observed in such cardiovascular diseases.[246]

### K. Semaphorin 3A Receptors: Neuropilin 1/Plexin A

Semaphorin 3A is a secreted disulfide-bound homodimer involved in the nervous system in growth cone collapse and axon growth repulsion. Recently, recombinant semaphorin 3A was used to study effects on platelets, which have some characteristics similar to neurons. Functional semaphorin 3A receptors were detected on platelets, and dose-dependent saturable binding of semaphorin 3A was observed.[247] Semaphorin 3A is produced by endothelial cells and inhibits integrin function. It inhibited platelet integrin αIIbβ3 activation by a wide range of agonists, as well as adhesion and spreading on fibrinogen-coated surfaces. The receptor for semaphorin 3A is a complex of neutropilin 1 and plexin A. The first contains the binding site and the second transmits the signals via its cytoplasmic domain. Rac1, a Rho family small G protein has been identified as a regulator of the actin cytoskeleton downstream of plexin A.

### L. CD100 (Sema4D)

CD100 is a 150-kDa member of the class IV semaphorin family consisting of a sema domain, an Ig-like domain, a lysine-rich stretch, a transmembrane domain, and a cytoplasmic domain with tyrosine and serine phosphorylation sites. On cell surfaces, CD100 forms a 300-kDa disulfide-linked dimer. The extracellular domain contains N-glycosylation sites and a metalloprotease cleavage site on the membrane side of the disulfide linkage. Known cell surface receptors for CD100 are plexin B1 and CD72. CD100 was recently reported to be present on the surface of both human and mouse platelets.[167]

## M. Peroxisome Proliferator-Activated Receptor γ (PPARγ): An Intracellular Receptor for Lysophosphatidic Acid

PPARγ is a transcription factor that regulates genes involved in energy metabolism. It is activated by several lipid ligands and, recently, the physiologically important lysophosphatidic acid was shown to activate it.[248] Thrombin-activated platelets are a major source of lysophosphatidic acid in plasma. PPARγ is present in platelets, and activation by lysophosphatidic acid may provide an additional feedback mechanism.

## Acknowledgment

Research performed at the Theodor Kocher Institute was supported by the Swiss National Science Foundation, grants 31-063868.00 and 31-107754.04.

## References

1. Wicki, A. N., Walz, A., Gerber–Huber, S. N., Wenger, R. H., Vornhagen, R., & Clemetson, K. J. (1989). Isolation and characterization of human blood platelet mRNA and construction of a cDNA library in lambda gt11. Confirmation of the platelet derivation by identification of GPIb coding mRNA and cloning of a GPIb coding cDNA insert. *Thromb Haemost, 61,* 448–453.
2. Lindemann, S., Tolley, N. D., Eyre, J. I., Kraiss, L. W., Mahoney, T. M., & Weyrich, A. S. (2001). Integrins regulate the intracellular distribution of eukaryotic initiation factor 4E in platelets: A checkpoint for translational control. *J Biol Chem, 276,* 33947–33951.
3. Glanzmann, E. (1918). Hereditäre hämorrhagische Thrombasthenie: Ein Beitrage zur Pathologie der Blutplättchen. *J Kinderkr, 88,* 113–141.
4. Bernard, J., & Soulier, J. P. (1948). Sur une nouvelle variété de dystrophie thrombocytaire-hémorrhagipare congénitale. *Semin Hôp Paris, 24,* 3217–3223.
5. Nurden, A. T., & Caen, J. P. (1974). An abnormal platelet glycoprotein pattern in three cases of Glanzmann's thrombasthenia. *Br J Haematol, 28,* 253–260.
6. Nurden, A. T., & Caen, J. P. (1975). Specific roles for platelet surface glycoproteins in platelet function. *Nature, 255,* 720–722.
7. Jenkins, C. S., Phillips, D. R., Clemetson, K. J., Meyer, D., Larrieu, M. J., & Luscher, E. F. (1976). Platelet membrane glycoproteins implicated in ristocetin-induced aggregation. Studies of the proteins on platelets from patients with Bernard–Soulier syndrome and von Willebrand's disease. *J Clin Invest, 57,* 112–124.
8. Phillips, D. R., & Agin, P. P. (1977). Platelet membrane defects in Glanzmann's thrombasthenia. Evidence for decreased amounts of two major glycoproteins. *J Clin Invest, 60,* 535–545.
9. Clemetson, K. J., McGregor, J. L., James, E., Dechavanne, M., & Luscher, E. F. (1982). Characterization of the platelet membrane glycoprotein abnormalities in Bernard–Soulier syndrome and comparison with normal by surface-labeling techniques and high-resolution two-dimensional gel electrophoresis. *J Clin Invest, 70,* 304–311.
10. Berndt, M. C., Gregory, C., Chong, B. H., Zola, H., & Castaldi, P. A. (1983). Additional glycoprotein defects in Bernard–Soulier's syndrome: Confirmation of genetic basis by parental analysis. *Blood, 62,* 800–807.
11. Clemetson, J. M., Polgar, J., Magnenat, E., Wells, T. N., & Clemetson, K. J. (1999). The platelet collagen receptor glycoprotein VI is a member of the immunoglobulin superfamily closely related to FcαR and the natural killer receptors. *J Biol Chem, 274,* 29019–29024.
12. Jandrot–Perrus, M., Busfield, S., Lagrue, A. H., Xiong, X., Debili, N., Chickering, T., Couedic, J. P., Goodearl, A., Dussault, B., Fraser, C., Vainchenker, W., & Villeval, J. L. (2000). Cloning, characterization, and functional studies of human and mouse glycoprotein VI: A platelet-specific collagen receptor from the immunoglobulin superfamily. *Blood, 96,* 1798–1807.
13. Ayyanathan, K., Webbs, T. E., Sandhu, A. K., Athwal, R. S., Barnard, E. A., & Kunapuli, S. P. (1996). Cloning and chromosomal localization of the human $P_2Y_1$ purinoceptor. *Biochem Biophys Res Commun, 218,* 783–788.
14. Hollopeter, G., Jantzen, H. M., Vincent, D., Li, G., England, L., Ramakrishnan, V., Yang, R. B., Nurden, P., Nurden, A., Julius, D., & Conley, P. B. (2001). Identification of the platelet ADP receptor targeted by antithrombotic drugs. *Nature, 409,* 202–207.
15. Sun, B., Li, J., Okahara, K., & Kambayashi, J. (1998). $P2X_1$ purinoceptor in human platelets. Molecular cloning and functional characterization after heterologous expression. *J Biol Chem, 273,* 11544–11547.
16. Bouvard, D., Brakebusch, C., Gustafsson, E., Aszodi, A., Bengtsson, T., Berna, A., & Fassler, R. (2001). Functional consequences of integrin gene mutations in mice. *Circ Res, 89,* 211–223.
17. Pytela, R., Pierschbacher, M. D., & Ruoslahti, E. (1985). A 125/115-kDa cell surface receptor specific for vitronectin interacts with the arginine-glycine-aspartic acid adhesion sequence derived from fibronectin. *Proc Natl Acad Sci USA, 82,* 5766–5770.
18. Coller, B. S., Seligsohn, U., West, S. M., Scudder, L. E., & Norton, K. J. (1991). Platelet fibrinogen and vitronectin in Glanzmann thrombasthenia: Evidence consistent with specific roles for glycoprotein IIb/IIIa and αvβ3 integrins in platelet protein trafficking. *Blood, 78,* 2603–2610.
19. Staatz, W. D., Rajpara, S. M., Wayner, E. A., Carter, W. G., & Santoro, S. A. (1989). The membrane glycoprotein Ia-IIa

(VLA-2) complex mediates the Mg$^{2+}$-dependent adhesion of platelets to collagen. *J Cell Biol, 108,* 1917–1924.

20. Xiong, J. P., Stehle, T., Diefenbach, B., Zhang, R., Dunker, R., Scott, D. L., Joachimiak, A., Goodman, S. L., & Arnaout, M. A. (2001). Crystal structure of the extracellular segment of integrin αvβ3. *Science, 294,* 339–345.

21. Springer, T. A. (1997). Folding of the N-terminal, ligand-binding region of integrin α-subunits into a β-propeller domain. *Proc Natl Acad Sci USA, 94,* 65–72.

22. Emsley, J., King, S. L., Bergelson, J. M., & Liddington, R. C. (1997). Crystal structure of the I domain from integrin α2β1. *J Biol Chem, 272,* 28512–28517.

23. Emsley, J., Knight, C. G., Farndale, R. W., Barnes, M. J., & Liddington, R. C. (2000). Structural basis of collagen recognition by integrin α2β1. *Cell, 101,* 47–56.

24. Knight, C. G., Morton, L. F., Peachey, A. R., Tuckwell, D. S., Farndale, R. W., & Barnes, M. J. (2000). The collagen-binding A-domains of integrins α1β1 and α2β1 recognize the same specific amino acid sequence, GFOGER, in native (triple-helical) collagens. *J Biol Chem, 275,* 35–40.

25. Santoro, S. A., Walsh, J. J., Staatz, W. D., & Baranski, K. J. (1991). Distinct determinants on collagen support α2β1 integrin-mediated platelet adhesion and platelet activation. *Cell Regul, 2,* 905–913.

26. Jung, S. M., & Moroi, M. (2000). Signal-transducing mechanisms involved in activation of the platelet collagen receptor integrin α2β1. *J Biol Chem, 275,* 8016–8026.

27. Nieswandt, B., Brakebusch, C., Bergmeier, W., Schulte, V., Bouvard, D., Mokhtari–Nejad, R., Lindhout, T., Heemskerk, J. W., Zirngibl, H., & Fassler, R. (2001). Glycoprotein VI but not α2β1 integrin is essential for platelet interaction with collagen. *EMBO J, 20,* 2120–2130.

28. Kuijpers, M. J., Schulte, V., Bergmeier, W., Lindhout, T., Brakebusch, C., Offermanns, S., Fassler, R., Heemskerk, J. W., & Nieswandt, B. (2003). Complementary roles of glycoprotein VI and α2β1 integrin in collagen-induced thrombus formation in flowing whole blood *ex vivo. FASEB J, 17,* 685–687.

29. Chen, J., Diacovo, T. G., Grenache, D. G., Santoro, S. A., & Zutter, M. M. (2002). The α2 integrin subunit-deficient mouse: A multifaceted phenotype including defects of branching morphogenesis and hemostasis. *Am J Pathol, 161,* 337–344.

30. Grenache, D. G., Coleman, T., Semenkovich, C. F., Santoro, S. A., & Zutter, M. M. (2003). α2β1 integrin and development of atherosclerosis in a mouse model: Assessment of risk. *Arterioscler Thromb Vasc Biol, 23,* 2104–2109.

31. Handa, M., Watanabe, K., Kawai, Y., Kamata, T., Koyama, T., Nagai, H., & Ikeda, Y. (1995). Platelet unresponsiveness to collagen: Involvement of glycoprotein Ia-IIa (α2β1 integrin) deficiency associated with a myeloproliferative disorder. *Thromb Haemost, 73,* 521–528.

32. Nieuwenhuis, H. K., Akkerman, J. W., Houdijk, W. P., & Sixma, J. J. (1985). Human blood platelets showing no response to collagen fail to express surface glycoprotein Ia. *Nature, 318,* 470–472.

33. Kehrel, B., Balleisen, L., Kokott, R., Mesters, R., Stenzinger, W., Clemetson, K. J., & van de Loo, J. (1988). Deficiency of intact thrombospondin and membrane glycoprotein Ia in platelets with defective collagen-induced aggregation and spontaneous loss of disorder. *Blood, 71,* 1074–1078.

34. Coller, B. S., Beer, J. H., Scudder, L. E., & Steinberg, M. H. (1989). Collagen-platelet interactions: Evidence for a direct interaction of collagen with platelet GPIa/IIa and an indirect interaction with platelet GPIIb/IIIa mediated by adhesive proteins. *Blood, 74,* 182–192.

35. Keely, P. J., & Parise, L. V. (1996). The α2β1 integrin is a necessary co-receptor for collagen-induced activation of Syk and the subsequent phosphorylation of phospholipase Cγ2 in platelets. *J Biol Chem, 271,* 26668–26676.

36. Wang, R., Kini, R. M., & Chung, M. C. (1999). Rhodocetin, a novel platelet aggregation inhibitor from the venom of *Calloselasma rhodostoma* (Malayan pit viper): Synergistic and noncovalent interaction between its subunits. *Biochemistry, 38,* 7584–7593.

37. Marcinkiewicz, C., Lobb, R. R., Marcinkiewicz, M. M., Daniel, J. L., Smith, J. B., Dangelmaier, C., Weinreb, P. H., Beacham, D. A., & Niewiarowski, S. (2000). Isolation and characterization of EMS16, a C-lectin type protein from *Echis multisquamatus* venom, a potent and selective inhibitor of the α2β1 integrin. *Biochemistry, 39,* 59–9867.

38. Kamiguti, A. S., Theakston, R. D., Watson, S. P., Bon, C., Laing, G. D., & Zuzel, M. (2000). Distinct contributions of glycoprotein VI and α2β1 integrin to the induction of platelet protein tyrosine phosphorylation and aggregation. *Arch Biochem Biophys, 374,* 356–362.

39. De Luca, M., Ward, C. M., Ohmori, K., Andrews, R. K., & Berndt, M. C. (1995). Jararhagin and jaracetin: Novel snake venom inhibitors of the integrin collagen receptor, α2β1. *Biochem Biophys Res Commun, 206,* 570–576.

40. Kamiguti, A. S., Markland, F. S., Zhou, Q., Laing, G. D., Theakston, R. D., & Zuzel, M. (1997). Proteolytic cleavage of the β1 subunit of platelet α2β1 integrin by the metallo-proteinase jararhagin compromises collagen-stimulated phosphorylation of pp72. *J Biol Chem, 272,* 32599–32605.

41. Zhou, Q., Dangelmaier, C., & Smith, J. B. (1996). The hemorrhagin catrocollastatin inhibits collagen-induced platelet aggregation by binding to collagen via its disintegrin-like domain. *Biochem Biophys Res Commun, 219,* 720–726.

42. Huang, T. F., Liu, C. Z., & Yang, S. H. (1995). Aggretin, a novel platelet-aggregation inducer from snake (*Calloselasma rhodostoma*) venom, activates phospholipase C by acting as a glycoprotein Ia/IIa agonist. *Biochem J, 309,* 1021–1027.

43. Chung, C. H., Au, L. C., & Huang, T. F. (1999). Molecular cloning and sequence analysis of aggretin, a collagen-like platelet aggregation inducer. *Biochem Biophys Res Commun, 263,* 723–727.

44. Navdaev, A., Clemetson, J. M., Polgar, J., Kehrel, B. E., Glauner, M., Magnenat, E., Wells, T. N., & Clemetson, K. J. (2001). Aggretin, a heterodimeric C-type lectin from *Calloselasma rhodostoma* (Malayan pit viper) stimulates platelets by binding to α2β1 integrin and GPIb, activating Syk and PLCγ2, but does not involve the GPVI/Fcγ collagen receptor. *J Biol Chem, 276,* 20882–20889.

45. Jung, S. M., & Moroi, M. (1998). Platelets interact with soluble and insoluble collagens through characteristically different reactions. *J Biol Chem, 273,* 14827–14837.

46. Siljander, P. R., Hamaia, S., Peachey, A. R., Slatter, D. A., Smethurst, P. A., Ouwehand, W. H., Knight, C. G., & Farndale, R. W. (2004). Integrin activation state determines selectivity for novel recognition sites in fibrillar collagens. *J Biol Chem, 279,* 47763–47772.

47. Liu, S., Calderwood, D. A., & Ginsberg, M. H. (2000). Integrin cytoplasmic domain-binding proteins. *J Cell Sci, 113,* 3563–3571.

48. Horii, K., Okuda, D., Morita, T., & Mizuno, H. (2004). Crystal structure of EMS16 in complex with the integrin α2-I domain. *J Mol Biol, 341,* 519–527.

49. Chen, K., Lin, Y., & Detwiler, T. C. (1992). Protein disulfide isomerase activity is released by activated platelets. *Blood, 79,* 2226–2228.

50. Lahav, J., Gofer–Dadosh, N., Luboshitz, J., Hess, O., & Shaklai, M. (2000). Protein disulfide isomerase mediates integrin-dependent adhesion. *FEBS Lett, 475,* 89–92.

51. O'Neill, S., Robinson, A., Deering, A., Ryan, M., Fitzgerald, D. J., & Moran, N. (2000). The platelet integrin αIIbβ3 has an endogenous thiol isomerase activity. *J Biol Chem, 275,* 36984–36990.

52. Zucker, M. B., & Masiello, N. C. (1984). Platelet aggregation caused by dithiothreitol. *Thromb Haemost, 51,* 119–124.

53. Peerschke, E. I. (1995). Regulation of platelet aggregation by post-fibrinogen binding events. Insights provided by dithiothreitol-treated platelets. *Thromb Haemost, 73,* 862–867.

54. Gofer–Dadosh, N., Klepfish, A., Schmilowitz, H., Shaklai, M., & Lahav, J. (1997). Affinity modulation in platelet α2β1 following ligand binding. *Biochem Biophys Res Commun, 232,* 724–727.

55. Essex, D. W., & Li, M. (1999). Protein disulphide isomerase mediates platelet aggregation and secretion. *Br J Haematol, 104,* 448–454.

56. Lahav, J., Wijnen, E. M., Hess, O., Hamaia, S. W., Griffiths, D., Makris, M., Knight, C. G., Essex, D. W., & Farndale, R. W. (2003). Enzymatically catalyzed disulfide exchange is required for platelet adhesion to collagen via integrin α2β1. *Blood, 102,* 2085–2092.

57. Kunicki, T. J., Kritzik, M., Annis, D. S., & Nugent, D. J. (1997). Hereditary variation in platelet integrin α2β1 density is associated with two silent polymorphisms in the α2 gene coding sequence. *Blood, 89,* 1939–1943.

58. Kritzik, M., Savage, B., Nugent, D. J., Santoso, S., Ruggeri, Z. M., & Kunicki, T. J. (1998). Nucleotide polymorphisms in the α2 gene define multiple alleles that are associated with differences in platelet α2β1 density. *Blood, 92,* 2382–2388.

59. Moshfegh, K., Wuillemin, W. A., Redondo, M., Lammle, B., Beer, J. H., Liechti–Gallati, S., & Meyer, B. J. (1999). Association of two silent polymorphisms of platelet glycoprotein Ia/IIa receptor with risk of myocardial infarction: A case–control study. *Lancet, 353,* 351–354.

60. Carlsson, L. E., Santoso, S., Spitzer, C., Kessler, C., & Greinacher, A. (1999). The α2 gene coding sequence T807/A873 of the platelet collagen receptor integrin α2β1 might

61. Reiner, A. P., Kumar, P. N., Schwartz, S. M., Longstreth, Jr., W. T., Pearce, R. M., Rosendaal, F. R., Psaty, B. M., & Siscovick, D. S. (2000). Genetic variants of platelet glycoprotein receptors and risk of stroke in young women. *Stroke, 31,* 1628–1633.

62. von Beckerath, N., Koch, W., Mehilli, J., Bottiger, C., Schomig, A., & Kastrati, A. (2000). Glycoprotein Ia gene C807T polymorphism and risk for major adverse cardiac events within the first 30 days after coronary artery stenting. *Blood, 95,* 3297–3301.

63. Matsubara, Y., Murata, M., Maruyama, T., Handa, M., Yamagata, N., Watanabe, G., Saruta, T., & Ikeda, Y. (2000). Association between diabetic retinopathy and genetic variations in α2β1 integrin, a platelet receptor for collagen. *Blood, 95,* 1560–1564.

64. Roest, M., Banga, J. D., Grobbee, D. E., de Groot, P. G., Sixma, J. J., Tempelman, M. J., & van Der Schouw, Y. T. (2000). Homozygosity for 807 T polymorphism in α2 subunit of platelet α2β1 is associated with increased risk of cardiovascular mortality in high-risk women. *Circulation, 102,* 1645–1650.

65. Di Paola, J., Federici, A. B., Mannucci, P. M., Canciani, M. T., Kritzik, M., Kunicki, T. J., & Nugent, D. (1999). Low platelet α2β1 levels in type I von Willebrand disease correlate with impaired platelet function in a high shear stress system. *Blood, 93,* 3578–3582.

66. Roest, M., Sixma, J. J., Wu, Y. P., Ijsseldijk, M. J., Tempelman, M., Slootweg, P. J., de Groot, P. G., & van Zanten, G. H. (2000). Platelet adhesion to collagen in healthy volunteers is influenced by variation of both α2β1 density and von Willebrand factor. *Blood, 96,* 1433–1437.

67. Piotrowicz, R. S., Orchekowski, R. P., Nugent, D. J., Yamada, K. Y., & Kunicki, T. J. (1988). Glycoprotein Ic-IIa functions as an activation-independent fibronectin receptor on human platelets. *J Cell Biol, 106,* 1359–1364.

68. Sonnenberg, A., Modderman, P. W., & Hogervorst, F. (1988). Laminin receptor on platelets is the integrin VLA-6. *Nature, 336,* 487–489.

69. Philippeaux, M. M., Vesin, C., Tacchini–Cottier, F., & Piguet, P. F. (1996). Activated human platelets express β2 integrin. *Eur J Haematol, 56,* 130–137.

70. Guo, J., & Piguet, P. F. (1998). Stimulation of thrombocytopoiesis decreases platelet β2 but not β1 or β3 integrins. *Br J Haematol, 100,* 712–719.

71. Piguet, P. F., Vesin, C., & Rochat, A. (2001). β2 Integrin modulates platelet caspase activation and life span in mice. *Eur J Cell Biol, 80,* 171–177.

72. Shiraki, R., Inoue, N., Kawasaki, S., Takei, A., Kadotani, M., Ohnishi, Y., Ejiri, J., Kobayashi, S., Hirata, K., Kawashima, S., & Yokoyama, M. (2004). Expression of toll-like receptors on human platelets. *Thromb Res, 113,* 379–385.

73. Cognasse, F., Hamzeh, H., Chavarin, P., Acquart, S., Genin, C., & Garraud. O. (2005). Evidence of toll-like receptor molecules on human platelets. *Immunol Cell Biol, 83,* 196–198.

74. Visintin, A., Mazzoni, A., Spitzer, J. H., Wyllie, D. H., Dower, S. K., & Segal, D. M. (2001). Regulation of toll-like

receptors in human monocytes and dendritic cells. *J Immunol, 166,* 249–255.

75. Andonegui, G., Kerfoot, S. M., McNagny, K., Ebbert, K. V., Patel, K. D., & Kubes, P. (2005). Platelets express functional toll-like receptor-4 (TLR4). *Blood, 106,* 2417–2423.

76. Vu, T. K., Hung, D. T., Wheaton, V. I., & Coughlin, S. R. (1991). Molecular cloning of a functional thrombin receptor reveals a novel proteolytic mechanism of receptor activation. *Cell, 64,* 1057–1068.

77. Furman, M. I., Liu, L., Benoit, S. E., Becker, R. C., Barnard, M. R., & Michelson, A. D. (1998). The cleaved peptide of the thrombin receptor is a strong platelet agonist. *Proc Natl Acad Sci USA, 95,* 3082–3087.

78. Xu, W. F., Andersen, H., Whitmore, T. E., Presnell, S. R., Yee, D. P., Ching, A., Gilbert, T., Davie, E. W., & Foster, D. C. (1998). Cloning and characterization of human protease-activated receptor 4. *Proc Natl Acad Sci USA, 95,* 6642–6646.

79. Leon, C., Vial, C., Cazenave, J. P., & Gachet, C. (1996). Cloning and sequencing of a human cDNA encoding endothelial P2Y1 purinoceptor. *Gene, 171,* 295–297.

80. Janssens, R., Communi, D., Pirotton, S., Samson, M., Parmentier, M., & Boeynaems, J. M. (1996). Cloning and tissue distribution of the human P2Y1 receptor. *Biochem Biophys Res Commun, 221,* 588–593.

81. Zhang, F. L., Luo, L., Gustafson, E., Lachowicz, J., Smith, M., Qiao, X., Liu, Y. H., Chen, G., Pramanik, B., Laz, T. M., Palmer, K., Bayne, M., & Monsma, Jr., F. J. (2001). ADP is the cognate ligand for the orphan G protein-coupled receptor SP1999. *J Biol Chem, 276,* 8608–8615.

82. MacKenzie, A. B., Mahaut–Smith, M. P., & Sage, S. O. (1996). Activation of receptor-operated cation channels via P2X1 not P2T purinoceptors in human platelets. *J Biol Chem, 271,* 2879–2881.

83. Hirata, M., Hayashi, Y., Ushikubi, F., Yokota, Y., Kageyama, R., Nakanishi, S., & Narumiya, S. (1991). Cloning and expression of cDNA for a human thromboxane A2 receptor. *Nature, 349,* 617–620.

84. Maeda, H., Inazu, T., Nagai, K., Maruyama, S., Nakagawara, G., & Yamamura, H. (1995). Possible involvement of protein-tyrosine kinases such as p72$^{SYK}$ in the disc–sphere change response of porcine platelets. *J Biochem, 117,* 1201–1208.

85. Katsuyama, M., Sugimoto, Y., Namba, T., Irie, A., Negishi, M., Narumiya, S., & Ichikawa, A. (1994). Cloning and expression of a cDNA for the human prostacyclin receptor. *FEBS Lett, 344,* 74–78.

86. Boie, Y., Sawyer, N., Slipetz, D. M., Metters, K. M., & Abramovitz, M. (1995). Molecular cloning and characterization of the human prostanoid DP receptor. *J Biol Chem, 270,* 18910–18916.

87. Fabre, J. E., Nguyen, M., Athirakul, K., Coggins, K., McNeish, J. D., Austin, S., Parise, L. K., FitzGerald, G. A., Coffman, T. M., & Koller, B. H. (2001). Activation of the murine EP3 receptor for PGE2 inhibits cAMP production and promotes platelet aggregation. *J Clin Invest, 107,* 603–610.

88. Kunapuli, S. P., Fen Mao, G., Bastepe, M., Liu–Chen, L. Y., Li, S., Cheung, P. P., DeRiel, J. K., & Ashby, B. (1994). Cloning and expression of a prostaglandin E receptor EP3 subtype from human erythroleukaemia cells. *Biochem J, 298,* 263–267.

89. Burgers, J. A., & Akkerman, J. W. (1993). Regulation of the receptor for platelet-activating factor on human platelets. *Biochem J, 291,* 157–161.

90. Bandoh, K., Aoki, J., Hosono, H., Kobayashi, S., Kobayashi, T., Murakami–Murofushi, K., Tsujimoto, M., Arai, H., & Inoue, K. (1999). Molecular cloning and characterization of a novel human G-protein-coupled receptor, EDG7, for lysophosphatidic acid. *J Biol Chem, 274,* 27776–27785.

91. Motohashi, K., Shibata, S., Ozaki, Y., Yatomi, Y., & Igarashi, Y. (2000). Identification of lysophospholipid receptors in human platelets: The relation of two agonists, lysophosphatidic acid and sphingosine 1-phosphate. *FEBS Lett, 468,* 189–193.

92. Kowalska, M. A., Ratajczak, J., Hoxie, J., Brass, L. F., Gewirtz, A., Poncz, M., & Ratajczak, M. Z. (1999). Megakaryocyte precursors, megakaryocytes and platelets express the HIV co-receptor CXCR4 on their surface: Determination of response to stromal-derived factor-1 by megakaryocytes and platelets. *Br J Haematol, 104,* 220–229.

93. Abi–Younes, S., Sauty, A., Mach, F., Sukhova, G. K., Libby, P., & Luster, A. D. (2000). The stromal cell-derived factor-1 chemokine is a potent platelet agonist highly expressed in atherosclerotic plaques. *Circ Res, 86,* 131–138.

94. Kowalska, M. A., Ratajczak, M. Z., Majka, M., Jin, J., Kunapuli, S., Brass, L., & Poncz, M. (2000). Stromal cell-derived factor-1 and macrophage-derived chemokine: two chemokines that activate platelets. *Blood, 96,* 50–57.

95. Clemetson, K. J., Clemetson, J. M., Proudfoot, A. E., Power, C. A., Baggiolini, M., & Wells, T. N. (2000). Functional expression of CCR1, CCR3, CCR4, and CXCR4 chemokine receptors on human platelets. *Blood, 96,* 4046–4054.

96. Schafer, A., Schulz, C., Eigenthaler, M., Fraccarollo, D., Kobsar, A., Gawaz, M., Ertl, G., Walter, U., & Bauersachs, J. (2004). Novel role of the membrane-bound chemokine fractalkine in platelet activation and adhesion. *Blood, 103,* 407–412.

97. Thibonnier, M., Auzan, C., Madhun, Z., Wilkins, P., Berti–Mattera, L., & Clauser, E. (1994). Molecular cloning, sequencing, and functional expression of a cDNA encoding the human V$_{1a}$ vasopressin receptor. *J Biol Chem, 269,* 3304–3310.

98. Born, G. V. R. (1962). Aggregation of blood platelets by adenosine diphosphate and its reversal. *Nature, 194,* 927–929.

99. Le, F., Townsend–Nicholson, A., Baker, E., Sutherland, G. R., & Schofield, P. R. (1996). Characterization and chromosomal localization of the human A2a adenosine receptor gene: ADORA2A. *Biochem Biophys Res Commun, 223,* 461–467.

100. Caron, M. G., Kobilka, B. K., Frielle, T., Bolanowski, M. A., Benovic, J. L., & Lefkowitz, R. J. (1988). Cloning of the cDNA and genes for the hamster and human β2-adrenergic receptors. *J Recept Res, 8,* 7–21.

101. Kagaya, A., Mikuni, M., Yamamoto, H., Muraoka, S., Yamawaki, S., & Takahashi, K. (1992). Heterologous

supersensitization between serotonin2 and α2-adrenergic receptor-mediated intracellular calcium mobilization in human platelets. *J Neural Transm Gen Sect, 88,* 25–36.

102. Yamada, S., Akita, H., Kanazawa, K., Ishida, T., Hirata, K., Ito, K., Kawashima, S., & Yokoyama, M. (2000). T102C polymorphism of the serotonin 5-HT$_{2A}$ receptor gene in patients with non-fatal acute myocardial infarction. *Atherosclerosis, 150,* 143–148.

103. Serres, F., Azorin, J. M., Valli, M., & Jeanningros, R. (1999). Evidence for an increase in functional platelet 5-HT$_{2A}$ receptors in depressed patients using the new ligand [$^{125}$I]-DOI. *Eur J Psychiatry, 14,* 451–457.

104. Ricci, A., Bronzetti, E., Mannino, F., Mignini, F., Morosetti, C., Tayebati, S. K., & Amenta, F. (2001). Dopamine receptors in human platelets. *Naunyn Schmiedebergs Arch Pharmacol, 363,* 376–382.

105. Sugiyama, T., Okuma, M., Ushikubi, F., Sensaki, S., Kanaji, K., & Uchino, H. (1987). A novel platelet aggregating factor found in a patient with defective collagen-induced platelet aggregation and autoimmune thrombocytopenia. *Blood, 69,* 1712–1720.

106. Moroi, M., Jung, S. M., Okuma, M., & Shinmyozu, K. (1989). A patient with platelets deficient in glycoprotein VI that lack both collagen-induced aggregation and adhesion. *J Clin Invest, 84,* 1440–1445.

107. Arai, M., Yamamoto, N., Moroi, M., Akamatsu, N., Fukutake, K., & Tanoue, K. (1995). Platelets with 10% of the normal amount of glycoprotein VI have an impaired response to collagen that results in a mild bleeding tendency. *Br J Haematol, 89,* 124–130.

108. Miura, Y., Ohnuma, M., Jung, S. M., & Moroi, M. (2000). Cloning and expression of the platelet-specific collagen receptor glycoprotein VI. *Thromb Res, 98,* 301–309.

109. Ezumi, Y., Uchiyama, T., & Takayama, H. (2000). Molecular cloning, genomic structure, chromosomal localization, and alternative splice forms of the platelet collagen receptor glycoprotein VI. *Biochem Biophys Res Commun, 277,* 27–36.

110. Nieswandt, B., Schulte, V., Bergmeier, W., Mokhtari–Nejad, R., Rackebrandt, K., Cazenave, J. P., Ohlmann, P., Gachet, C., & Zirngibl, H. (2001). Long-term antithrombotic protection by *in vivo* depletion of platelet glycoprotein VI in mice. *J Exp Med, 193,* 459–469.

111. Stephens, G., Yan, Y., Jandrot–Perrus, M., Villeval, J. L., Clemetson, K. J., & Phillips, D. R. (2005). Platelet activation induces metalloproteinase-dependent GP VI cleavage to down-regulate platelet reactivity to collagen. *Blood, 105,* 186–191.

112. Tsuji, M., Ezumi, Y., Arai, M., & Takayama, H. (1997). A novel association of Fc receptor γ-chain with glycoprotein VI and their co-expression as a collagen receptor in human platelets. *J Biol Chem, 272,* 23528–23531.

113. Nieswandt, B., Bergmeier, W., Schulte, V., Rackebrandt, K., Gessner, J. E., & Zirngibl, H. (2000). Expression and function of the mouse collagen receptor glycoprotein VI is strictly dependent on its association with the FcRγ chain. *J Biol Chem, 275,* 23998–24002.

114. Kato, K., Kanaji, T., Russell, S., Kunicki, T. J., Furihata, K., Kanaji, S., Marchese, P., Reininger, A., Ruggeri, Z. M., &

Ware, J. (2003). The contribution of glycoprotein VI to stable platelet adhesion and thrombus formation illustrated by targeted gene deletion. *Blood, 102,* 1701–1707.

115. Boylan, B., Chen, H., Rathore, V., Paddock, C., Salacz, M., Friedman, K. D., Curtis, B. R., Stapleton, M., Newman, D. K., Kahn, M. L., & Newman, P. J. (2004). Anti-GPVI-associated ITP: An acquired platelet disorder caused by autoantibody-mediated clearance of the GPVI/FcRgamma-chain complex from the human platelet surface. *Blood, 104,* 1350–1355.

116. Gibbins, J. M., Okuma, M., Farndale, R., Barnes, M., & Watson, S. P. (1997). Glycoprotein VI is the collagen receptor in platelets which underlies tyrosine phosphorylation of the Fc receptor γ-chain. *FEBS Lett, 413,* 255–259.

117. Smethurst, P. A., Joutsi–Korhonen, L., MN, O. C., Wilson, E., Jennings, N. S., Garner, S. F., Zhang, Y., Knight, C. G., Dafforn, T. R., Buckle, A., MJ, I. J., De Groot, P. G., Watkins, N. A., Farndale, R. W., & Ouwehand, W. H. (2004). Identification of the primary collagen-binding surface on human glycoprotein VI by site-directed mutagenesis and by a blocking phage antibody. *Blood, 103,* 903–911.

118. Lecut, C., Arocas, V., Ulrichts, H., Elbaz, A., Villeval, J. L., Lacapere, J. J., Deckmyn, H., & Jandrot–Perrus, M. (2004). Identification of residues within human glycoprotein VI involved in the binding to collagen: Evidence for the existence of distinct binding sites. *J Biol Chem, 279,* 52293–52299.

119. Polgar, J., Clemetson, J. M., Kehrel, B. E., Wiedemann, M., Magnenat, E. M., Wells, T. N. C., & Clemetson, K. J. (1997). Platelet activation and signal transduction by convulxin, a C-type lectin from *Crotalus durissus terrificus* (tropical rattlesnake) venom via the p62/GPVI collagen receptor. *J Biol Chem, 272,* 13576–13583.

120. Niedergang, F., Alcover, A., Knight, C. G., Farndale, R. W., Barnes, M. J., Francischetti, I. M., Bon, C., & Leduc, M. (2000). Convulxin binding to platelet receptor GPVI: Competition with collagen related peptides. *Biochem Biophys Res Commun, 273,* 246–250.

121. Kehrel, B., Wierwille, S., Clemetson, K. J., Anders, O., Steiner, M., Knight, C. G., Farndale, R. W., Okuma, M., & Barnes, M. J. (1998). Glycoprotein VI is a major collagen receptor for platelet activation: It recognizes the platelet-activating quaternary structure of collagen, whereas CD36, glycoprotein IIb/IIIa, and von Willebrand factor do not. *Blood, 91,* 491–499.

122. Morton, H. C., van den Herik–Oudijk, I. E., Vossebeld, P., Snijders, A., Verhoeven, A. J., Capel, P. J., & van de Winkel, J. G. (1995). Functional association between the human myeloid immunoglobulin A Fc receptor (CD89) and FcR γ chain. Molecular basis for CD89/FcR γ chain association. *J Biol Chem, 270,* 29781–29787.

123. Maeda, A., Kurosaki, M., & Kurosaki, T. (1998). Paired immunoglobulin-like receptor (PIR)-A is involved in activating mast cells through its association with Fc receptor γ chain. *J Exp Med, 188,* 991–995.

124. Taylor, L. S., & McVicar, D. W. (1999). Functional association of FcεRIγ with arginine (632) of paired immunoglobulin-like receptor (PIR)-A3 in murine macrophages. *Blood, 94,* 1790–1796.

125. Zheng, Y.-M., Liu, C., Chen, H., Locke, D., Ryan, J. C., & Kahn, M. L. (2001). Expression of the platelet receptor GPVI confers signaling via the Fc receptor γ chain in response to the snake venom convulxin but not to collagen. *J Biol Chem, 276,* 12999–13006.

126. Moroi, M., Jung, S. M., Shinmyozu, K., Tomiyama, Y., Ordinas, A., & Diaz–Ricart, M. (1996). Analysis of platelet adhesion to a collagen-coated surface under flow conditions: The involvement of glycoprotein VI in the platelet adhesion. *Blood, 88,* 2081–2092.

127. Morton, L. F., Hargreaves, P. G., Farndale, R. W., Young, R. D., & Barnes, M. J. (1995). Integrin α2β1-independent activation of platelets by simple collagen-like peptides: Collagen tertiary (triple-helical) and quaternary (polymeric) structures are sufficient alone for α2β1-independent platelet reactivity. *Biochem J, 306,* 337–344.

128. Jandrot–Perrus, M., Lagrue, A. H., Okuma, M., & Bon, C. (1997). Adhesion and activation of human platelets induced by convulxin involve glycoprotein VI and integrin α2β1. *J Biol Chem, 272,* 27035–27041.

129. Asselin, J., Gibbins, J. M., Achison, M., Lee, Y. H., Morton, L. F., Farndale, R. W., Barnes, M. J., & Watson, S. P. (1997). A collagen-like peptide stimulates tyrosine phosphorylation of syk and phospholipase Cγ2 in platelets independent of the integrin α2β1. *Blood, 89,* 1235–1242.

130. Ichinohe, T., Takayama, H., Ezumi, Y., Arai, M., Yamamoto, N., Takahashi, H., & Okuma, M. (1997). Collagen-stimulated activation of Syk but not c-Src is severely compromised in human platelets lacking membrane glycoprotein VI. *J Biol Chem, 272,* 63–68.

131. Marlas, G., Joseph, D., & Huet, C. (1983). Isolation and electron microscope studies of a potent platelet-aggregating glycoprotein from the venom of *Crotalus durissus cascavella. Biochimie, 65,* 405–416.

132. Xu, Y., Gurusiddappa, S., Rich, R. L., Owens, R. T., Keene, D. R., Mayne, R., Hook, A., & Hook, M. (2000). Multiple binding sites in collagen type I for the integrins α1β1 and α2β1. *J Biol Chem, 275,* 38981–38989.

133. He, L., Pappan, L. K., Grenache, D. G., Li, Z., Tollefsen, D. M., Santoro, S. A., & Zutter, M. M. (2003). The contributions of the α2β1 integrin to vascular thrombosis in vivo. *Blood, 102,* 3652–3657.

134. Nonne, C., Lenain, N., Hechler, B., Mangin, P., Cazenave, J. P., Gachet, C., & Lanza, F. (2005). Importance of platelet phospholipase Cγ2 signaling in arterial thrombosis as a function of lesion severity. *Arterioscler Thromb Vasc Biol, 25,* 1293–1298.

135. Gruner, S., Prostredna, M., Koch, M., Miura, Y., Schulte, V., Jung, S. M., Moroi, M., & Nieswandt, B. (2005). Relative antithrombotic effect of soluble GPVI dimer compared with anti-GPVI antibodies in mice. *Blood, 105,* 1492–1499.

136. Massberg, S., Konrad, I., Bultmann, A., Schulz, C., Munch, G., Peluso, M., Lorenz, M., Schneider, S., Besta, F., Muller, I., Hu, B., Langer, H., Kremmer, E., Rudelius, M., Heinzmann, U., Ungerer, M., & Gawaz, M. (2004). Soluble glycoprotein VI dimer inhibits platelet adhesion and aggregation to the injured vessel wall in vivo. *FASEB J, 18,* 397–399.

137. Hourdille, P., Heilmann, E., Combrie, R., Winckler, J., Clemetson, K. J., & Nurden, A. T. (1990). Thrombin induces a rapid redistribution of glycoprotein Ib-IX complexes within the membrane systems of activated human platelets. *Blood, 76,* 1503–1513.

138. Michelson, A. D., Benoit, S. E., Furman, M. I., Barnard, M. R., Nurden, P., & Nurden, A. T. (1996). The platelet surface expression of glycoprotein V is regulated by two independent mechanisms: Proteolysis and a reversible cytoskeletal-mediated redistribution to the surface-connected canalicular system. *Blood, 87,* 1396–1408.

139. Gawaz, M., Konrad, I., Hauser, A. I., Sauer, S., Li, Z., Wester, H. J., Bengel, F. M., Schwaiger, M., Schomig, A., Massberg, S., & Haubner, R. (2005). Non-invasive imaging of glycoprotein VI binding to injured arterial lesions. *Thromb Haemost, 93,* 910–913.

140. Penz, S., Reininger, A. J., Brandl, R., Goyal, P., Rabie, T., Bernlochner, I., Rother, E., Goetz, C., Engelmann, B., Smethurst, P. A., Ouwehand, W. H., Farndale, R., Nieswandt, B., & Siess, W. (2005). Human atheromatous plaques stimulate thrombus formation by activating platelet glycoprotein VI. *FASEB J, 19,* 898–909.

141. Carlsson, L. E., Santoso, S., Baurichter, G., Kroll, H., Papenberg, S., Eichler, P., Westerdaal, N. A., Kiefel, V., van de Winkel, J. G., & Greinacher, A. (1998). Heparin-induced thrombocytopenia: New insights into the impact of the FcγRIIa-R-H131 polymorphism. *Blood, 92,* 1526–1531.

142. McKenzie, S. E., Taylor, S. M., Malladi, P., Yuhan, H., Cassel, D. L., Chien, P., Schwartz, E., Schreiber, A. D., Surrey, S., & Reilly, M. P. (1999). The role of the human Fc receptor FcγRIIA in the immune clearance of platelets: A transgenic mouse model. *J Immunol, 162,* 4311–4318.

143. Bodin, S., Viala, C., Ragab, A., & Payrastre, B. (2003). A critical role of lipid rafts in the organization of a key FcγRIIa-mediated signaling pathway in human platelets. *Thromb Haemost, 89,* 318–330.

144. Joseph, M., Gounni, A. S., Kusnierz, J. P., Vorng, H., Sarfati, M., Kinet, J. P., Tonnel, A. B., Capron, A., & Capron, M. (1997). Expression and functions of the high-affinity IgE receptor on human platelets and megakaryocyte precursors. *Eur J Immunol, 27,* 2212–2218.

145. Hasegawa, S., Pawankar, R., Suzuki, K., Nakahata, T., Furukawa, S., Okumura, K., & Ra, C. (1999). Functional expression of the high affinity receptor for IgE (FcεRI) in human platelets and its intracellular expression in human megakaryocytes. *Blood, 93,* 2543–2551.

146. Sherrington, P. D., Scott, J. L., Jin, B., Simmons, D., Dorahy, D. J., Lloyd, J., Brien, J. H., Aebersold, R. H., Adamson, J., Zuzel, M., & Burns, G. F. (1997). TLiSA1 (PTA1) activation antigen implicated in T cell differentiation and platelet activation is a member of the immunoglobulin superfamily exhibiting distinctive regulation of expression. *J Biol Chem, 272,* 21735–21744.

147. Jia, W., Liu, X. S., Zhu, Y., Li, Q., Han, W. N., Zhang, Y., Zhang, J. S., Yang, K., Zhang, X. H., & Jin, B. Q. (2000). Preparation and characterization of mabs against different epitopes of CD226 (PTA1). *Hybridoma, 19,* 489–494.

148. Bottino, C., Castriconi, R., Pende, D., Rivera, P., Nanni, M., Carnemolla, B., Cantoni, C., Grassi, J., Marcenaro, S., Reymond, N., Vitale, M., Moretta, L., Lopez, M., & Moretta, A. (2003). Identification of PVR (CD155) and Nectin-2

(CD112) as cell surface ligands for the human DNAM-1 (CD226) activating molecule. *J Exp Med, 198,* 557–567.

149. Kojima, H., Kanada, H., Shimizu, S., Kasama, E., Shibuya, K., Nakauchi, H., Nagasawa, T., & Shibuya, A. (2003). CD226 mediates platelet and megakaryocytic cell adhesion to vascular endothelial cells. *J Biol Chem, 278,* 36748–36753.

150. Gupta, S. K., Pillarisetti, K., & Ohlstein, E. H. (2000). Platelet agonist F11 receptor is a member of the immunoglobulin superfamily and identical with junctional adhesion molecule (JAM): Regulation of expression in human endothelial cells and macrophages. *IUBMB Life, 50,* 51–56.

151. Naik, U. P., Naik, M. U., Eckfeld, K., Martin–DeLeon, P., & Spychala, J. (2001). Characterization and chromosomal localization of JAM-1, a platelet receptor for a stimulatory monoclonal antibody. *J Cell Sci, 114,* 539–547.

152. Santoso, S., Sachs, U. J., Kroll, H., Linder, M., Ruf, A., Preissner, K. T., & Chavakis, T. (2002). The junctional adhesion molecule 3 (JAM-3) on human platelets is a counterreceptor for the leukocyte integrin Mac-1. *J Exp Med, 196,* 679–691.

153. Diacovo, T. G., DeFougerolles, A. R., Bainton, D. F., & Springer, T. A. (1994). A functional integrin ligand on the surface of platelets: Intercellular adhesion molecule-2. *J Clin Invest, 94,* 1243–1251.

154. Weber, C., & Springer, T. A. (1997). Neutrophil accumulation on activated, surface-adherent platelets in flow is mediated by interaction of Mac-1 with fibrinogen bound to $\alpha IIb\beta 3$ and stimulated by platelet-activating factor. *J Clin Invest, 100,* 2085–2093.

155. Newman, P. J., Berndt, M. C., Gorski, J., White II, G. C., Lyman, S., Paddock, C., & Muller, W. A. (1990). PECAM-1 (CD31) cloning and relation to adhesion molecules of the immunoglobulin gene superfamily. *Science, 247,* 1219–1222.

156. Newman, D. K., Hamilton, C., & Newman, P. J. (2001). Inhibition of antigen-receptor signaling by platelet endothelial cell adhesion molecule-1 (CD31) requires functional ITIMs, SHP-2, and p56[lck]. *Blood, 97,* 2351–2357.

157. Brown, E. J., & Frazier, W. A. (2001). Integrin-associated protein (CD47) and its ligands. *Trends Cell Biol, 11,* 130–135.

158. Lindberg, F. P., Gresham, H. D., Schwarz, E., & Brown, E. J. (1993). Molecular cloning of integrin-associated protein: An immunoglobulin family member with multiple membrane-spanning domains implicated in $\alpha v\beta 3$-dependent ligand binding. *J Cell Biol, 123,* 485–496.

159. Chung, J., Gao, A. G., & Frazier, W. A. (1997). Thrombspondin acts via integrin-associated protein to activate the platelet integrin $\alpha IIb\beta 3$. *J Biol Chem, 272,* 14740–14746.

160. Chung, J., Wang, X. Q., Lindberg, F. P., & Frazier, W. A. (1999). Thrombospondin-1 acts via IAP/CD47 to synergize with collagen in $\alpha 2\beta 1$-mediated platelet activation. *Blood, 94,* 642–648.

161. Hirata, K., Ishida, T., Penta, K., Rezaee, M., Yang, E., Wohlgemuth, J., & Quertermous, T. (2001). Cloning of an immunoglobulin family adhesion molecule selectively expressed by endothelial cells. *J Biol Chem, 276,* 16223–16231.

162. Nasdala, I., Wolburg–Buchholz, K., Wolburg, H., Kuhn, A., Ebnet, K., Brachtendorf, G., Samulowitz, U., Kuster, B., Engelhardt, B., Vestweber, D., & Butz, S. (2002). A transmembrane tight junction protein selectively expressed on endothelial cells and platelets. *J Biol Chem, 277,* 16294–16303.

163. Ishida, T., Kundu, R. K., Yang, E., Hirata, K., Ho, Y. D., & Quertermous, T. (2003). Targeted disruption of endothelial cell-selective adhesion molecule inhibits angiogenic processes in vitro and in vivo. *J Biol Chem, 278,* 34598–34604.

164. Barrow, A. D., Astoul, E., Floto, A., Brooke, G., Relou, I. A., Jennings, N. S., Smith, K. G., Ouwehand, W., Farndale, R. W., Alexander, D. R., & Trowsdale, J. (2004). Cutting edge: TREM-like transcript-1, a platelet immunoreceptor tyrosine-based inhibition motif encoding costimulatory immunoreceptor that enhances, rather than inhibits, calcium signaling via SHP-2. *J Immunol, 172,* 5838–5842.

165. Washington, A. V., Schubert, R. L., Quigley, L., Disipio, T., Feltz, R., Cho, E. H., & McVicar, D. W. (2004). A TREM family member, TLT-1, is found exclusively in the $\alpha$-granules of megakaryocytes and platelets. *Blood, 104,* 1042–1047.

166. Nakayama, E., von Hoegen, I., & Parnes, J. R. (1989). Sequence of the Lyb-2 B-cell differentiation antigen defines a gene superfamily of receptors with inverted membrane orientation. *Proc Natl Acad Sci USA, 86,* 1352–1356.

167. Brass, L. F., Stalker, T. J., Zhu, L., Lu, B., Woulfe, D. S., & Prevost, N. (2004). Boundary events: Contact-dependent and contact-facilitated signaling between platelets. *Semin Thromb Hemost, 30,* 399–410.

168. Dean, Y. D., McGreal, E. P., Akatsu, H., & Gasque, P. (2000). Molecular and cellular properties of the rat AA4 antigen, a C-type lectin-like receptor with structural homology to thrombomodulin. *J Biol Chem, 275,* 34382–34392.

169. Bohlson, S. S., Silva, R., Fonseca, M. I., & Tenner, A. J. (2005). CD93 is rapidly shed from the surface of human myeloid cells and the soluble form is detected in human plasma. *J Immunol, 175,* 1239–1247.

170. Suzuki–Inoue, K., Fuller, G. L., Garcia, A., Eble, J. A., Pohlmann, S., Inoue, O., Gartner, T. K., Hughan, S. C., Pearce, A. C., Laing, G. D., Theakston, R. D., Schweighoffer, E., Zitzmann, N., Morita, T., Tybulewicz, V. L., Ozaki, Y., & Watson, S. P. (2006). A novel Syk-dependent mechanism of platelet activation by the C-type lectin receptor CLEC-2. *Blood, 107,* 542–549.

171. Wu, H., Li, J., Peng, L., Liu, H., Wu, W., Zhou, Y., Hou, Q., & Ke, D. (2000). Anti-human platelet tetraspanin (CD9) monoclonal antibodies induce platelet integrin $\alpha IIb\beta 3$ activation in an Fc receptor-independent fashion. *Chin Med Sci J, 15,* 145–149.

172. Moseley, G. W. (2005). Tetraspanin-Fc receptor interactions. *Platelets, 16,* 3–12.

173. Boucheix, C., Benoit, P., Frachet, P., Billard, M., Worthington, R. E., Gagnon, J., & Uzan, G. (1991). Molecular cloning of the CD9 antigen. A new family of cell surface proteins. *J Biol Chem, 266,* 117–122.

174. Metzelaar, M. J., Wijngaard, P. L., Peters, P. J., Sixma, J. J., Nieuwenhuis, H. K., & Clevers, H. C. (1991). CD63 antigen. A novel lysosomal membrane glycoprotein, cloned by a screening procedure for intracellular antigens in eukaryotic cells. *J Biol Chem, 266,* 3239–3245.

175. Israels, S. J., McMillan–Ward, E. M., Easton, J., Robertson, C., & McNicol, A. (2001). CD63 associates with the αIIbβ3 integrin-CD9 complex on the surface of activated platelets. *Thromb Haemost, 85,* 134–141.

176. Israels, S. J., & McMillan–Ward, E. M. (2005). CD63 modulates spreading and tyrosine phosphorylation of platelets on immobilized fibrinogen. *Thromb Haemost, 93,* 311–318.

177. Martens, L., Van Damme, P., Van Damme, J., Staes, A., Timmerman, E., Ghesquiere, B., Thomas, G. R., Vandekerckhove, J., & Gevaert, K. (2005). The human platelet proteome mapped by peptide-centric proteomics: A functional protein profile. *Proteomics, 5,* 3193–3204.

178. Fitter, S., Tetaz, T. J., Berndt, M. C., & Ashman, L. K. (1995). Molecular cloning of cDNA encoding a novel platelet–endothelial cell tetra-span antigen, PETA-3. *Blood, 86,* 1348–1355.

179. Fitter, S., Sincock, P. M., Jolliffe, C. N., & Ashman, L. K. (1999). Transmembrane 4 superfamily protein CD151 (PETA-3) associates with β1 and αIIbβ3 integrins in haemopoietic cell lines and modulates cell-cell adhesion. *Biochem J, 338,* 61–70.

180. Lau, L. M., Wee, J. L., Wright, M. D., Moseley, G. W., Hogarth, P. M., Ashman, L. K., & Jackson, D. E. (2004). The tetraspanin superfamily member CD151 regulates outside-in integrin αIIbβ3 signaling and platelet function. *Blood, 104,* 2368–2375.

181. Polgar, J., Clemetson, J. M., Gengenbacher, D., & Clemetson, K. J. (1993). Additional GPI-anchored glycoproteins on human platelets that are absent or deficient in paroxysmal nocturnal haemoglobinuria. *FEBS Lett, 327,* 49–53.

182. Kelton, J. G., Smith, J. W., Horsewood, P., Warner, M. N., Warkentin, T. E., Finberg, R. W., & Hayward, C. P. (1998). ABH antigens on human platelets: Expression on the glycosyl phosphatidylinositol-anchored protein CD109. *J Lab Clin Med, 132,* 142–148.

183. Lin, M., Sutherland, D. R., Horsfall, W., Totty, N., Yeo, E., Nayar, R., Wu, X. F., & Schuh, A. C. (2002). Cell surface antigen CD109 is a novel member of the α2-macroglobulin/C3, C4, C5 family of thioester-containing proteins. *Blood, 99,* 1683–1691.

184. Solomon, K. R., Sharma, P., Chan, M., Morrison, P. T., & Finberg, R. W. (2004). CD109 represents a novel branch of the α2-macroglobulin/complement gene family. *Gene, 327,* 171–183.

185. Starke, R., Harrison, P., Mackie, I., Wang, G., Erusalimsky, J. D., Gale, R., Masse, J. M., Cramer, E., Pizzey, A., Biggerstaff, J., & Machin, S. (2005). The expression of prion protein (PrP(C)) in the megakaryocyte lineage. *J Thromb Haemost, 3,* 1266–1273.

186. Kaneider, N. C., Feistritzer, C., Gritti, D., Mosheimer, B. A., Ricevuti, G., Patsch, J. R., & Wiedermann, C. J. (2005). Expression and function of syndecan-4 in human platelets. *Thromb Haemost, 93,* 1120–1127.

187. Polgar, J., Eichler, P., Greinacher, A., & Clemetson, K. J. (1998). Adenosine diphosphate (ADP) and ADP receptor play a major role in platelet activation/aggregation induced by sera from heparin-induced thrombocytopenia patients. *Blood, 91,* 549–554.

188. Pelleymounter, M. A., Cullen, M. J., Baker, M. B., Hecht, R., Winters, D., Boone, T., & Collins, F. (1995). Effects of the obese gene product on body weight regulation in ob/ob mice. *Science, 269,* 540–543.

189. Nakata, M., Yada, T., Soejima, N., & Maruyama, I. (1999). Leptin promotes aggregation of human platelets via the long form of its receptor. *Diabetes, 48,* 426–429.

190. Elbatarny, H. S., & Maurice, D. H. (2005). Leptin-mediated activation of human platelets: Involvement of a leptin receptor and phosphodiesterase 3A containing cellular signaling complex. *Am J Physiol Endocrinol Metab, 289,* E695–E702.

191. Konstantinides, S., Schafer, K., Koschnick, S., & Loskutoff, D. J. (2001). Leptin-dependent platelet aggregation and arterial thrombosis suggests a mechanism for atherothrombotic disease in obesity. *J Clin Invest, 108,* 1533–1540.

192. El-Haschimi, K., Pierroz, D. D., Hileman, S. M., Bjorbaek, C., & Flier, J. S. (2000). Two defects contribute to hypothalamic leptin resistance in mice with diet-induced obesity. *J Clin Invest, 105,* 1827–1832.

193. Ge, H., Huang, L., Pourbahrami, T., & Li, C. (2002). Generation of soluble leptin receptor by ectodomain shedding of membrane-spanning receptors in vitro and in vivo. *J Biol Chem, 277,* 45898–45903.

194. Zabeau, L., Lavens, D., Peelman, F., Eyckerman, S., Vandekerckhove, J., & Tavernier, J. (2003). The ins and outs of leptin receptor activation. *FEBS Lett, 546,* 45–50.

195. Konstantinides, S., Schafer, K., Neels, J. G., Dellas, C., Loskutoff, D. J., & Koschnick, S. (2004). Inhibition of endogenous leptin protects mice from arterial and venous thrombosis. *Arterioscler Thromb Vasc Biol, 24,* 2196–2201.

196. Giandomenico, G., Dellas, C., Czekay, R. P., Koschnick, S., & Loskutoff, D. J. (2005). The leptin receptor system of human platelets. *J Thromb Haemost, 3,* 1042–1049.

197. Tsiamis, A. C., Hayes, P., Box, H., Goodall, A. H., Bell, P. R., & Brindle, N. P. (2000). Characterization and regulation of the receptor tyrosine kinase Tie-1 in platelets. *J Vasc Res, 37,* 437–442.

198. Hajek, A. S., & Joist, J. H. (1992). Platelet insulin receptor. *Methods Enzymol, 215,* 398–403.

199. Kahn, N. N. (1998). Insulin-induced expression of prostacyclin receptors on platelets is mediated through ADP-ribosylation of Giα protein. *Life Sci, 63,* 2031–2038.

200. Vassbotn, F. S., Havnen, O. K., Heldin, C. H., & Holmsen, H. (1994). Negative feedback regulation of human platelets via autocrine activation of the platelet-derived growth factor α-receptor. *J Biol Chem, 269,* 13874–13879.

201. Gould, W. R., Baxi, S. M., Schroeder, R., Peng, Y. W., Leadley, R. J., Peterson, J. T., & Perrin, L. A. (2005). Gas6 receptors Axl, Sky and Mer enhance platelet activation and regulate thrombotic responses. *J Thromb Haemost, 3,* 733–741.

202. Chen, C., Li, Q., Darrow, A. L., Wang, Y., Derian, C. K., Yang, J., de Garavilla, L., Andrade–Gordon, P., & Damiano, B. P. (2004). Mer receptor tyrosine kinase signaling participates in platelet function. *Arterioscler Thromb Vasc Biol, 24,* 1118–1123.

203. Nanda, N., Bao, M., Lin, H., Clauser, K., Komuves, L., Quertermous, T., Conley, P. B., Phillips, D. R., & Hart, M. J. (2005). Platelet endothelial aggregation receptor 1 (PEAR1), a novel epidermal growth factor repeat-containing transmembrane receptor, participates in platelet contact-induced activation. *J Biol Chem, 280,* 24680–24689.

204. Prevost, N., Woulfe, D. S., Jiang, H., Stalker, T. J., Marchese, P., Ruggeri, Z. M., & Brass, L. F. (2005). Eph kinases and ephrins support thrombus growth and stability by regulating integrin outside-in signaling in platelets. *Proc Natl Acad Sci USA, 102,* 9820–9825.

205. Yamamoto, N., Ikeda, H., Tandon, N. N., Herman, J., Tomiyama, Y., Mitani, T., Sekiguchi, S., Lipsky, R., Kralisz, U., & Jamieson, G. A. (1990). A platelet membrane glycoprotein (GP) deficiency in healthy blood donors: Naka– platelets lack detectable GPIV (CD36). *Blood, 76,* 1698–1703.

206. Kashiwagi, H., Tomiyama, Y., Kosugi, S., Shiraga, M., Lipsky, R. H., Nagao, N., Kanakura, Y., Kurata, Y., & Matsuzawa, Y. (1995). Family studies of type II CD36 deficient subjects: Linkage of a CD36 allele to a platelet-specific mRNA expression defect(s) causing type II CD36 deficiency. *Thromb Haemost, 74,* 758–763.

207. Lee, K., Godeau, B., Fromont, P., Plonquet, A., Debili, N., Bachir, D., Reviron, D., Gourin, J., Fernandez, E., Galacteros, F., & Bierling, P. (1999). CD36 deficiency is frequent and can cause platelet immunization in Africans. *Transfusion, 39,* 873–879.

208. Diaz–Ricart, M., Tandon, N. N., Carretero, M., Ordinas, A., Bastida, E., & Jamieson, G. A. (1993). Platelets lacking functional CD36 (glycoprotein IV) show reduced adhesion to collagen in flowing whole blood. *Blood, 82,* 491–496.

209. Asch, A. S., Liu, I., Briccetti, F. M., Barnwell, J. W., Kwakye–Berko, F., Dokun, A., Goldberger, J., & Pernambuco, M. (1993). Analysis of CD36 binding domains: Ligand specificity controlled by dephosphorylation of an ectodomain. *Science, 262,* 1436–1440.

210. Endemann, G., Stanton, L. W., Madden, K. S., Bryant, C. M., White, R. T., & Protter, A. A. (1993). CD36 is a receptor for oxidized low density lipoprotein. *J Biol Chem, 268,* 11811–11816.

211. Coburn, C. T., Knapp, Jr., F. F., Febbraio, M., Beets, A. L., Silverstein, R. L., & Abumrad, N. A. (2000). Defective uptake and utilization of long chain fatty acids in muscle and adipose tissues of CD36 knockout mice. *J Biol Chem, 275,* 32523–32529.

212. Kehrel, B., Kronenberg, A., Rauterberg, J., Niesing–Bresch, D., Niehues, U., Kardoeus, J., Schwippert, B., Tschöpe, D., van de Loo, J., & Clemetson, K. J. (1993). Platelets deficient in glycoprotein IIIb aggregate normally to collagens type I and III but not to collagen type V. *Blood, 82,* 3364–3370.

213. Daniel, J. L., Dangelmaier, C., Strouse, R., & Smith, J. B. (1994). Collagen induces normal signal transduction in platelets deficient in CD36 (platelet glycoprotein IV). *Thromb Haemost, 71,* 353–356.

214. Kashiwagi, H., Tomiyama, Y., Honda, S., Kosugi, S., Shiraga, M., Nagao, N., Sekiguchi, S., Kanayama, Y., Kurata, Y., & Matsuzawa, Y. (1995). Molecular basis of CD36 deficiency. Evidence that a 478C→T substitution (proline90→serine) in CD36 cDNA accounts for CD36 deficiency. *J Clin Invest, 95,* 1040–1046.

215. Gousset, K., Tsvetkova, N. M., Crowe, J. H., & Tablin, F. (2004). Important role of raft aggregation in the signaling events of cold-induced platelet activation. *Biochim Biophys Acta, 1660,* 7–15.

216. Peerschke, E. I. B., & Ghebrehiwet, B. (1994). Platelet membrane receptors for the complement component C1q. *Semin Hematol, 31,* 320–328.

217. Jiang, J., Zhang, Y., Krainer, A. R., & Xu, R. M. (1999). Crystal structure of human p32, a doughnut-shaped acidic mitochondrial matrix protein. *Proc Natl Acad Sci USA, 96,* 3572–3577.

218. Ghebrehiwet, B., & Peerschke, E. I. (1998). Structure and function of gC1q-R: A multiligand binding cellular protein. *Immunobiology, 199,* 225–238.

219. Peerschke, E. I., & Ghebrehiwet, B. (1998). Platelet receptors for the complement component C1q: Implications for hemostasis and thrombosis. *Immunobiology, 199,* 239–249.

220. Peerschke, E. I. B., Reid, K. B. M., & Ghebrehiwet, B. (1994). Identification of a novel 33-kDa C1q-binding site on human blood platelets. *J Immunol, 152,* 5896–5901.

221. Nguyen, T., Ghebrehiwet, B., & Peerschke, E. I. (2000). *Staphylococcus aureus* protein A recognizes platelet gC1qR/p33: A novel mechanism for staphylococcal interactions with platelets. *Infect Immunol, 68,* 2061–2068.

222. Szabo, J., Cervenak, L., Toth, F. D., Prohaszka, Z., Horvath, L., Kerekes, K., Beck, Z., Bacsi, A., Erdei, A., Peerschke, E. I., Fust, G., & Ghebrehiwet, B. (2001). Soluble gC1q-R/p33, a cell protein that binds to the globular "heads" of C1q, effectively inhibits the growth of HIV-1 strains in cell cultures. *Clin Immunol, 99,* 222–231.

223. Yu, G. H., Holers, V. M., Seya, T., Ballard, L., & Atkinson, J. P. (1986). Identification of a third component of complement-binding glycoprotein of human platelets. *J Clin Invest, 78,* 494–501.

224. Elward, K., Griffiths, M., Mizuno, M., Harris, C. L., Neal, J. W., Morgan, B. P., & Gasque, P. (2005). CD46 plays a key role in tailoring innate immune recognition of apoptotic and necrotic cells. *J Biol Chem, 280,* 36342–36354.

225. Lesch, K. P., Wolozin, B. L., Murphy, D. L., & Reiderer, P. (1993). Primary structure of the human platelet serotonin uptake site: Identity with the brain serotonin transporter. *J Neurochem, 60,* 2319–2322.

226. Yirmiya, N., Pilowsky, T., Nemanov, L., Arbelle, S., Feinsilver, T., Fried, I., & Ebstein, R. P. (2001). Evidence for an association with the serotonin transporter promoter region polymorphism and autism. *Am J Med Genet, 105,* 381–386.

227. Maurer–Spurej, E. (2005). Serotonin reuptake inhibitors and cardiovascular diseases: A platelet connection. *Cell Mol Life Sci, 62,* 159–170.

228. Febbraio, M., & Silverstein, R. L. (1990). Identification and characterization of LAMP-1 as an activation-dependent platelet surface glycoprotein. *J Biol Chem, 265,* 18531–18537.

229. Silverstein, R. L., & Febbraio, M. (1992). Identification of lysosome-associated membrane protein-2 as an activation-dependent platelet surface glycoprotein. *Blood, 80,* 1470–1475.

230. Henn, V., Slupsky, J. R., Grafe, M., Anagnostopoulos, I., Forster, R., Muller–Berghaus, G., & Kroczek, R. A. (1998). CD40 ligand on activated platelets triggers an inflammatory reaction of endothelial cells. *Nature, 391,* 591–594.

231. Inwald, D. P., Peters, M. J., Walshe, D., Jones, A., Davies, E. G., & J. Klein, N. (2000). Absence of platelet CD40L identifies patients with X-linked hyper IgM syndrome. *Clin Exp Immunol, 120,* 499–502.

232. Henn, V., Steinbach, S., Buchner, K., Presek, P., & Kroczek, R. A. (2001). The inflammatory action of CD40 ligand (CD154) expressed on activated human platelets is temporally limited by coexpressed CD40. *Blood, 98,* 1047–1054.

233. Danese, S., & Fiocchi, C. (2005). Platelet activation and the CD40/CD40 ligand pathway: Mechanisms and implications for human disease. *Crit Rev Immunol, 25,* 103–122.

234. Langer, F., Ingersoll, S. B., Amirkhosravi, A., Meyer, T., Siddiqui, F. A., Ahmad, S., Walker, J. M., Amaya, M., Desai, H., & Francis, J. L. (2005). The role of CD40 in CD40L- and antibody-mediated platelet activation. *Thromb Haemost, 93,* 1137–1146.

235. Sako, D., Chang, X.-J., Barone, K. M., Vachino, G., White, H. M., Shaw, G., Veldman, G. M., Bean, K. M., Ahern, T. J., Furie, B., Cumming, D. A., & Larsen, G. R. (1993). Expression cloning of a functional glycoprotein ligand for P-selectin. *Cell, 75,* 1179–1186.

236. Frenette, P. S., Denis, C. V., Weiss, L., Jurk, K., Subbarao, S., Kehrel, B., Hartwig, J. H., Vestweber, D., & Wagner, D. D. (2000). P-selectin glycoprotein ligand 1 (PSGL-1) is expressed on platelets and can mediate platelet–endothelial interactions in vivo. *J Exp Med, 191,* 1413–1422.

237. Clifford, E. E., Parker, K., Humphreys, B. D., Kertesy, S. B., & Dubyak, G. R. (1998). The P2X1 receptor, an adenosine triphosphate-gated cation channel, is expressed in human platelets but not in human blood leukocytes. *Blood, 91,* 3172–3181.

238. Oury, C., Toth–Zsamboki, E., van Geet, C., Thys, C., Wei, L., Nilius, B., Vermylen, J., & Hoylaerts, M. F. (2000). A natural dominant negative P2X$_1$ receptor due to deletion of a single amino acid residue. *J Biol Chem, 275,* 22611–22614.

239. Furuse, M., Hirase, T., Itoh, M., Nagafuchi, A., Yonemura, S., Tsukita, S., & Tsukita, S. (1993). Occludin: A novel integral membrane protein localizing at tight junctions. *J Cell Biol, 123,* 1777–1788.

240. Saitou, M., Ando–Akatsuka, Y., Itoh, M., Furuse, M., Inazawa, J., Fujimoto, K., & Tsukita, S. (1997). Mammalian occludin in epithelial cells: Its expression and subcellular distribution. *Eur J Cell Biol, 73,* 222–231.

241. Stevenson, B. R., Siliciano, J. D., Mooseker, M. S., & Goodenough, D. A. (1986). Identification of ZO-1: A high molecular weight polypeptide associated with the tight junction (zonula occludens) in a variety of epithelia. *J Cell Biol, 103,* 755–766.

242. Willott, E., Balda, M. S., Fanning, A. S., Jameson, B., van Itallie, C., & Anderson, J. M. (1993). The tight junction protein ZO-1 is homologous to the *Drosophila* discs–large tumor suppressor protein of septate junctions. *Proc Natl Acad Sci USA, 90,* 7834–7838.

243. Boeckel, N., Brodde, M., Kehrel, B., & Morgenstern, E. (2001). Tight junction proteins in platelet contacts. *Thrombosis and Haemostasis, 86,* 1206.

244. Elrod, J. W., Park, J. H., Oshima, T., Sharp, C. D., Minagar, A., & Alexander, J. S. (2003). Expression of junctional proteins in human platelets. *Platelets, 14,* 247–251.

245. Locksley, R. M., Killeen, N., & Lenardo, M. J. (2001). The TNF and TNF receptor superfamilies: Integrating mammalian biology. *Cell, 104,* 487–501.

246. Pignatelli, P., De Biase, L., Lenti, L., Tocci, G., Brunelli, A., Cangemi, R., Riondino, S., Grego, S., Volpe, M., & Violi, F. (2005). Tumor necrosis factor-α as trigger of platelet activation in patients with heart failure. *Blood, 106,* 1992–1994.

247. Kashiwagi, H., Shiraga, M., Kato, H., Kamae, T., Yamamoto, N., Tadokoro, S., Kurata, Y., Tomiyama, Y., & Kanakura, Y. (2005). Negative regulation of platelet function by a secreted cell repulsive protein, semaphorin 3A. *Blood, 106,* 913–921.

248. McIntyre, T. M., Pontsler, A. V., Silva, A. R., St Hilaire, A., Xu, Y., Hinshaw, J. C., Zimmerman, G. A., Hama, K., Aoki, J., Arai, H., & Prestwich, G. D. (2003). Identification of an intracellular receptor for lysophosphatidic acid (LPA): LPA is a transcellular PPARgamma agonist. *Proc Natl Acad Sci USA, 100,* 131–136.

# CHAPTER 7

# The Glycoprotein Ib-IX-V Complex

**Robert K. Andrews,[1] Michael C. Berndt,[1] and José A. López[2]**

[1]*Department of Immunology, Monash University, Clayton, Australia*
[2]*Puget Sound Blood Center, Research Division, University of Washington, Seattle, Washington*

## I. Introduction/Structure

One of the most fundamental homeostatic mechanisms is hemostasis, the ability to arrest blood loss after traumatic injury. In mammals, this is initiated by the adhesion of circulating blood platelets to the damaged vessel wall, culminating in platelet plug formation. Ironically, however, when triggered under pathological conditions, this normally protective cascade of events results in arterial thrombosis and is responsible for major clinical sequelae such as acute myocardial infarction and ischemic stroke. The glycoprotein (GP) Ib-IX-V complex is a pivotal platelet receptor in initiating and propagating both hemostasis and thrombosis.

A history of the GPIb-IX-V complex could begin with the discovery of the congenital bleeding disorder Bernard–Soulier syndrome in 1948, later attributed to a deficiency of functional GPIb-IX-V expression; followed by the purification of GPIb-IX from human platelets in 1985; and cloning of the GPIbα, GPIbβ, GPIX, and GPV subunits between 1987 and 1993.[1–5] The first role uncovered for GPIb-IX-V was as an adhesion receptor for the multimeric adhesive ligand von Willebrand factor (VWF) in the arterial circulation. In ensuing years, the number of ligands, counterreceptors, and functions for GPIb-IX-V has rapidly proliferated. It is now known that GPIb-IX-V (a) is a major platelet receptor for VWF, thrombospondin-1 (TSP), the leukocyte integrin αMβ2, and P-selectin expressed on activated endothelial cells or platelets; (b) acts as a classic signaling receptor to propagate downstream events that activate the platelet integrins, αIIbβ3 and α2β1, leading to stable platelet adhesion and aggregation; (c) promotes procoagulant activity on platelets by binding α-thrombin, factors XI and XII, and kininogen; (d) controls platelet size and shape by attachment to the cytoskeleton *via* filamin; and (e) regulates clearance of platelets from the circulation by mediating recognition of platelets by αMβ2 on liver macrophages, a process underlying rapid clearance from the circulation of refrigerated platelets if used for transfusion. This chapter examines these functions of GPIb-IX-V in cell adhesion and assembly of procoagulant activity. In addition, we also discuss the important contribution to GPIb-IX-V signaling of associated receptors such as GPVI/FcRγ-chain, FcγRIIa, and integrin α2β1, and the localization of subpopulations of GPIb-IX-V in cholesterol-rich lipid rafts. With the multiplicity and diversity of these functions, it is, at times, difficult to know where to begin in explaining or understanding GPIb-IX-V. One way to do this is to resort to chicken genomics.

Genomic analysis of the common red junglefowl, *Gallus gallus*, reveals a homologue of human GPIbα (the major ligand-binding subunit of GPIb-IX-V), with conserved structural features critical for the function of human GPIbα (Fig. 7-1). Chickens contain nucleated thrombocytes — circulating immune cells that also perform the hemostatic role of anucleate human platelets. Despite the intervening eons since evolutionary divergence of chickens from the mammalian lineage leading to humans, GPIbα is remarkably well preserved. In humans, GPIbα is a type I, membrane-spanning subunit containing an N-terminal, ligand-binding domain (residues 1–282), a sialomucin core, a transmembrane region, and a cytoplasmic tail (Fig. 7-2A). The major ligand-binding domain at the N-terminus of GPIbα (His1-Glu282), separated from the membrane by the sialomucin macroglycopeptide, consists of seven tandem leucine-rich repeats (LRR), each of approximately 24 amino acids together spanning residues 36 to 200, an N-terminal capping sequence (residues 1–35) representing a partial LRR stabilized by a disulfide bond between Cys4 and Cys17, a C-terminal flanking sequence (residues 201–268) containing two disulfide bonds at Cys209-Cys248 and Cys211-Cys264 forming a disulfide knot, and an anionic sequence (Asp269-Glu282) that includes sulfation sites at Tyr276, Tyr278, and Tyr279.[1,3–5]

The following points can be made about human and *G. gallus* GPIbα. First, within the N-terminal ligand-binding domain, 145 of 300 residues are identical or conservative substitutions, with minimal gapping required for alignment (Fig. 7-1A). Thus, the number and arrangement of LRRs is identical. This is significant because in human GPIbα, VWF

## A. GPIbα Ligand-binding N-terminal Domain:

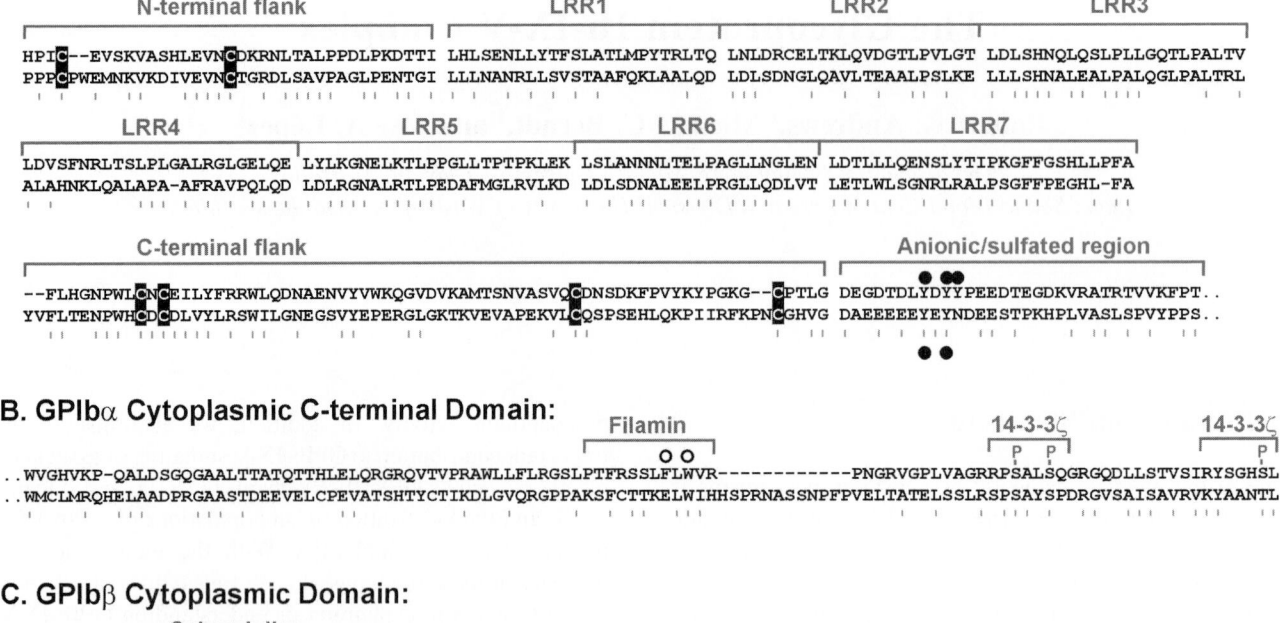

## B. GPIbα Cytoplasmic C-terminal Domain:

## C. GPIbβ Cytoplasmic Domain:

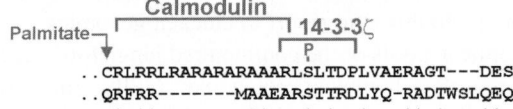

## D. GPIbα-binding domain of VWF A1 (478-718):

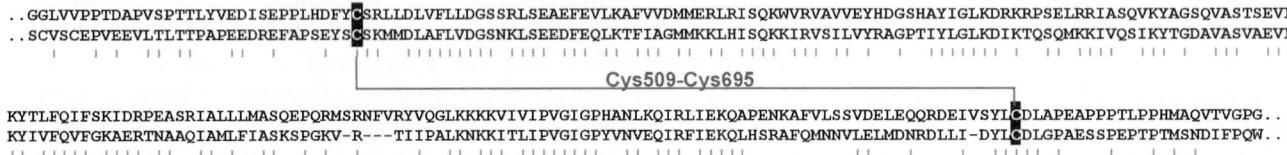

**Figure 7-1.** Primary sequences of human and chicken (*Gallus gallus*) GPIb. A. N-terminal ligand-binding domain of human (upper sequence, residues 1–300) and chicken (lower sequence) GPIbα, consisting of the N-terminal disulfide-looped flanking sequence, the leucine-rich repeat (LRR) domains, the C-terminal disulfide-looped flanking sequence, and the anionic/sulfated region. Cys residues forming Cys4-Cys17, Cys209-Cys248, and Cys211-Cys264 disulfide bonds in the human sequence are highlighted. Sulfated tyrosine residues at Tyr276, Tyr278, and Tyr279 in human GPIbα are indicated by the solid circles. B. The C-terminal cytoplasmic tail of human (upper sequence, residues 512–610) and chicken (lower sequence) GPIbα. The proposed binding site for filamin (Pro561-Arg572) including critical residues, Trp570, and Phe568 (open circles); and phosphorylation sites at Ser587, Ser590, and Ser609 within consensus binding sites for 14-3-3ζ on human GPIbα are indicated. C. The C-terminal cytoplasmic tail of human (upper sequence, residues 148–181) and chicken (lower sequence) GPIbβ. The palmitoylation site at Cys148, calmodulin-binding sequence at Arg149-Arg164, and the 14-3-3ζ binding site encompassing the protein kinase A-dependent phosphorylation site at Ser166 on human GPIbβ are indicated. D. Comparison of human and *Gallus gallus* VWF A1 domain sequences spanning the Cys509-Cys695 disulfide bond. Short gray lines under the *Gallus gallus* sequences indicate residues identical to humans or conservative substitutions (A/L/I/V; Q/E/D/N; R/K; S/T).

binding to the ligand-binding domain (His1-Glu282) is conformation dependent, and precise orientation of sequences flanking the LRRs is both critical for VWF binding and sensitive to either hydrodynamic shear stress or point mutations disrupting conformation (see Section II. A.1). The two Cys residues in the N-terminal flank and the four Cys residues in the C-terminal flank are conserved in

chicken GPIbα. Two of three sulfated tyrosines are conserved, and in the chicken sequence these also meet the consensus for sulfation. N-linked glycosylation sites (NXS/T) at Asn21 and Asn159 of human GPIbα, however, are not conserved.

Second, the cytoplasmic domain of chicken GPIbα contains 38 of approximately 100 residues that are identical or

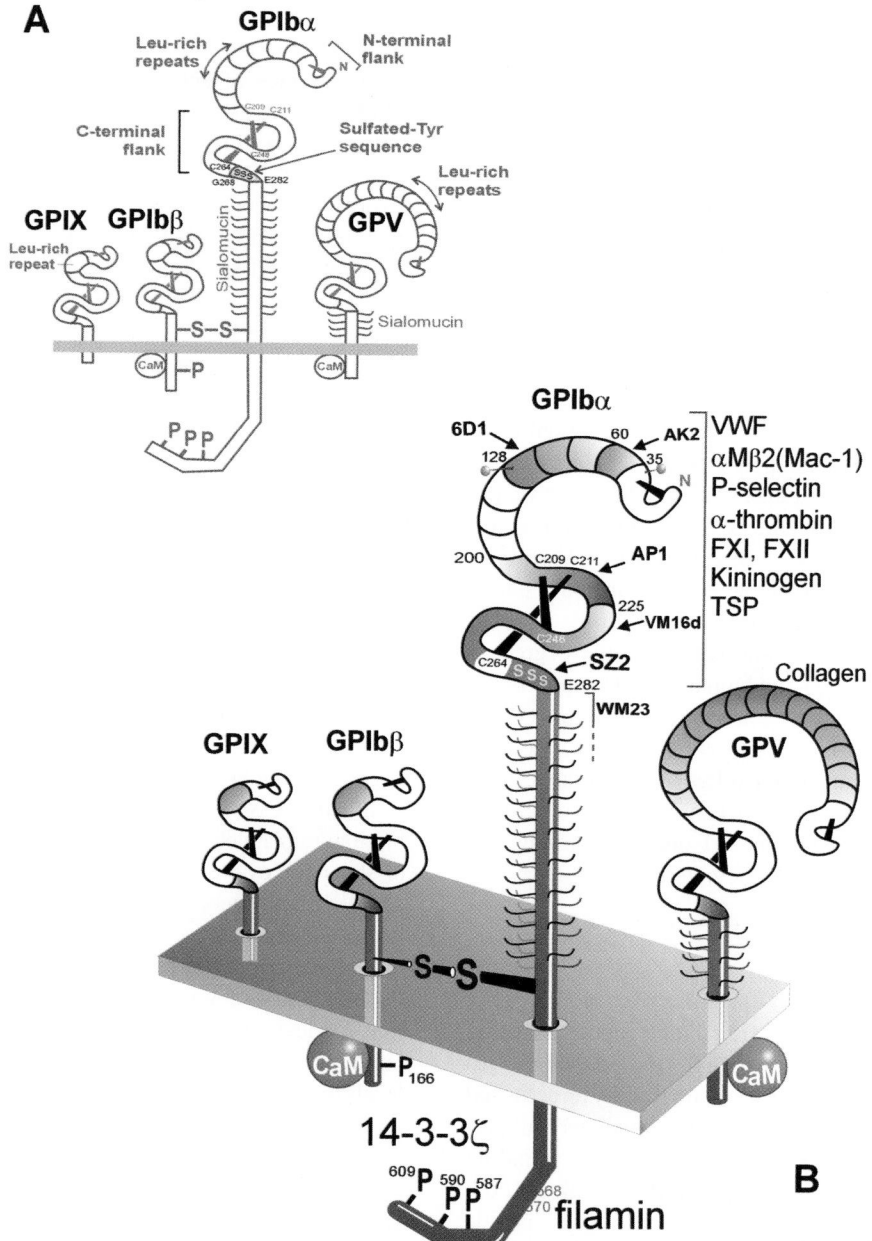

**Figure 7-2.** Structure of the GPIb-IX-V complex. A. Structural domains of GPIb-IX-V, showing the extracellular leucine-rich repeats and disulfide-looped N- and C-terminal flanking sequences of the GPIbα, GPIbβ, GPIX, and GPV subunits. The location of the anionic/sulfated Tyr sequence is indicated. The location of phosphorylation sites within the cytoplasmic tails of GPIbβ (Ser166) and GPIbα (Ser587/Ser590 and Ser 609) are also indicated. Juxtamembrane cytoplasmic sequences of GPIbβ and GPV contain calmodulin binding sites. B. Functional binding sites of GPIbα include the N-terminal 282 residues of GPIbα, which contains noncontiguous binding sites for the listed ligands as well as epitopes for a number of monoclonal anti-GPIbα antibodies (arrows) that inhibit binding of VWF (AK2, 6D1, AP1), P-selectin (AK2, SZ2), αMβ2 (AP1, VM16d), thrombospondin (AP1, SZ2), and/or α-thrombin (VM16d, SZ2). Collagen binds to the extracellular domain of GPV. Filamin binds to the central region of the GPIbα cytoplasmic domain, whereas more C-terminal phosphorylation sites on GPIbα (Ser587/Ser590 and Ser 609), and GPIbβ (Ser166), regulate interaction of 14-3-3ζ with GPIb-IX-V in resting platelets. Circles represent N-linked glycosylation sites (NXS/T) at Asn21 and Asn159. Calmodulin association with GPIbβ and GPV may regulate surface expression of GPIb-IX-V, and possibly other aspects of GPIb-IX-V-dependent platelet activation. CaM, calmodulin; FXI, factor XI; FXII, factor XII; TSP, thrombospondin-1; VWF, von Willebrand factor.

with conservative substitutions, but diverging within the filamin-binding region of human GPIbα (Fig. 7-1B). A sequence required for filamin binding in human GPIbα ([561]PTFRSSLFLWVR[572]), spanning critical residues Phe568/Trp570,[6] is replaced by the sequence KSFCTTKELWIH in chicken GPIbα, and this is followed by a 12-residue insert sequence not present in human GPIbα. However, two phosphorylation sites at Ser587 and Ser590 in human GPIbα[7] are conserved, whereas human Ser609 (which is constitutively phosphorylated),[8] is replaced at the homologous position by threonine in *G. gallus*, and thus may also be phosphorylated.

Third, chicken GPIbα also contains a Cys-Cys motif near the N-terminal side of the predicted transmembrane domain. In human GPIbα, one of these Cys residues forms a disulfide bond to GPIbβ.[4] Functional sequences within the cytoplasmic tail of human GPIbβ, for binding calmodulin and 14-3-3ζ (incorporating a protein kinase A [PKA]-dependent phosphorylation site at Ser166), are only partly conserved in chicken GPIbβ (Fig. 7-1C). With the possible exception of the filamin binding sequence, the degree of similarity between chicken and human GPIbα supports the functional importance of both extracellular and intracellular regions of the receptor. The long evolutionary history of this platelet receptor presumably underlies its remarkable functional diversity. Human and chicken VWF show similar homology in the GPIbα-binding A1 domain spanning the Cys509-Cys695 disulfide bond (Fig. 7-1D).

## II. Function

Functions of GPIb-IX-V include regulating platelet adhesion to subendothelial matrix, endothelial cells, or leukocytes (Section II.A); assembly of procoagulant activity on activated platelets (Section II.B); and signaling, mediated in part by the association of GPIb-IX-V with accessory signaling receptors and localization in lipid rafts (Section II.C). Signaling responses to engagement of GPIb-IX-V include dissociation of signaling proteins such as 14-3-3ζ or calmodulin, and activation of αIIbβ3 and α2β1 (Section III). Many, if not all, of these interactions involving GPIb-IX-V, both outside and inside the membrane, are intricately interrelated and coregulated. The network of adhesive interactions supports crosstalk between platelets, endothelial cells, and leukocytes (Fig. 7-3A), relevant to thrombosis, inflammation, and other vascular pathophysiologies.

### A. Adhesive Functions

In this section, ligands of GPIb-IX-V are discussed in terms of their physiological role, binding sites, and regulation, focusing mainly on recent developments in these areas.

**1. VWF.** VWF, a multimeric adhesive glycoprotein present in plasma and subendothelial matrix, is expressed on the surface of activated endothelial cells. The mature VWF subunit of 2050 residues (approximately 275 kDa) consists of modular domains D'-D3-A1-A2-A3-D4-B1-B2-B3-C1-C2. VWF forms disulfide-linked multimers up to 20,000 kDa. Weibel–Palade bodies in endothelial cells contain ultralarge VWF (ULVWF) that, upon stimulation, is secreted and tethered to the endothelial cell surface *via* an association with P-selectin.[9,10] Release of VWF into plasma is controlled by proteolysis of ULVWF. ULVWF is hyperadhesive and requires processing to smaller, less adhesive forms *via* proteolysis by ADAMTS-13 (a disintegrin and metalloproteinase with thrombospondin motif), which cleaves at Tyr1605/Met1606 within the A2 domain.[11] In von Willebrand disease, dysfunction or deficiency of VWF results in bleeding. On the other hand, increased plasma levels of ULVWF, resulting from ADAMTS-13 deficiency, for example, produce a prothrombotic phenotype often associated with thrombotic thrombocytopenia purpura (Chapter 50).[12] VWF is also secreted from the α-granules of activated platelets, where, as in endothelial cell Weibel–Palade bodies, it is colocalized with P-selectin. When released upon platelet activation, it promotes platelet aggregation *in vivo* (see also Chapter 18).[13]

Shear stress plays a paramount role in regulating VWF binding to platelet GPIb-IX-V. A balance between the effect of shear forces on the ADAMTS-13-dependent breakdown of VWF multimers and the activation of VWF binding to its platelet receptor is likely to control the initiation and development of thrombus at arterial shear rates. Normally, the interaction of VWF with GPIb-IX-V requires a stimulus that induces an active conformation of either the ligand and/or the receptor (Fig. 7-3B). This may involve pathological levels of shear stress, or immobilization of VWF on collagen or other components of subendothelial matrix, allowing VWF to bind GPIbα with increased affinity.[1,10,14] Such regulation prevents unwanted binding of GPIbα on circulating platelets to plasma VWF. ULVWF, by contrast, is able to bind GPIb-IX-V spontaneously.[15] Similarly, congenital gain-of-function mutations within the GPIbα-binding VWF A1 domain, or in the first C-terminal disulfide loop of GPIbα (Gly233/Val or Met239/Val), also induce VWF binding to GPIb-IX-V.[16,17] *In vitro*, VWF binding to platelet GPIb-IX-V is induced by nonphysiological modulators such as the bacterial glycopeptide ristocetin or the snake toxin botrocetin, as described further later.

The dynamics of GPIbα/VWF-dependent platelet adhesion have been studied using endothelial matrix, collagen-coated surfaces in flow chambers, and intravital microscopy in live mice or rats at low-physiological (<600 s$^{-1}$), high-physiological (600–900 s$^{-1}$), and pathological (>1500 s$^{-1}$) shear rates. At high or pathological shear, the GPIbα–VWF interaction is necessary to slow platelet velocity sufficiently

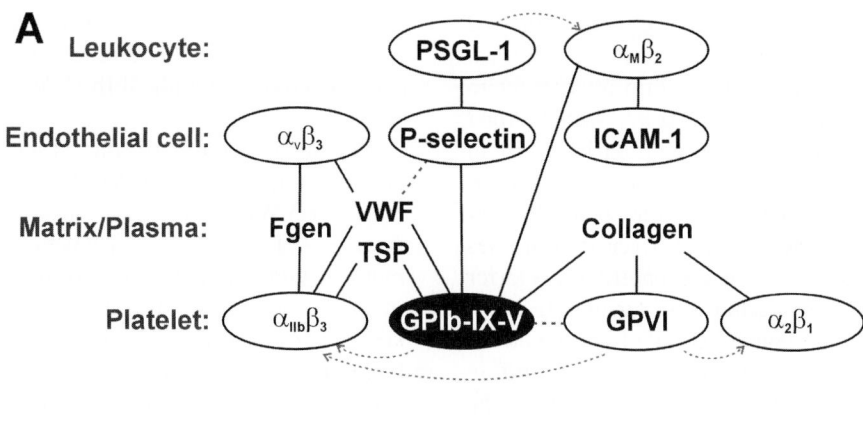

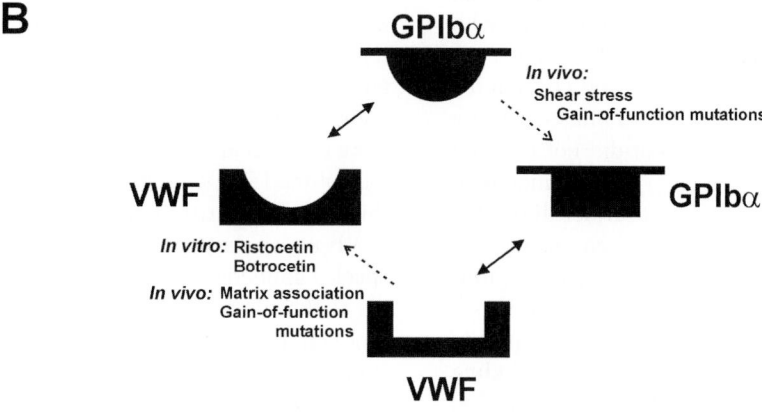

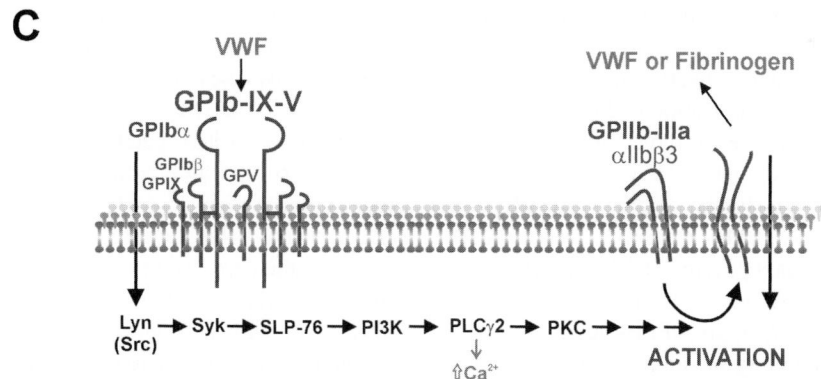

**Figure 7-3.** Adhesive interactions of GPIb-IX-V. A. Interactions of GPIb-IX-V with ligands in matrix or plasma, or counterreceptors on endothelial cells or leukocytes (solid lines), allowing GPIb-IX-V to play a central role in regulating platelet–endothelial cell–leukocyte networking in vascular processes. The association of VWF strings with endothelial P-selectin and the coassociation of GPIb-IX-V and GPVI on platelets are indicated by the dashed line. Dashed arrows show how integrin αIIbβ3 on platelets and αMβ2 on leukocytes may be activated by signals from GPIb-IX-V/GPVI or P-selectin glycoprotein ligand 1 respectively. P-selectin is expressed on both activated endothelial cells and activated platelets. B. A simple scheme showing how interaction between GPIbα and VWF may occur after activation of receptor and/or ligand, resulting in a compatible "conformation." (This does not, however, exclude the possibility of an interaction between activated forms of both ligand and receptor.) C. A pathway leading from engagement of GPIb-IX-V to activation of αIIbβ3 on human platelets. PI3K, phosphoinositide 3 kinase; PKC, protein kinase C; PLCγ2, phospholipase γ2; VWF, von Willebrand factor.

to enable GPVI-collagen-mediated platelet activation, and activation of αIIbβ3 or other integrins.[13,14,18] In contrast, activated αIIbβ3 only supports direct firm attachment to fibrinogen at shear stresses below 900 s$^{-1}$, with GPIbα/VWF being essential at high shear, even if αIIbβ3 is in an activated form.[18,19] The GPIbα–VWF interaction is fast on-rate/fast off-rate, and allows rapid translocation or rolling of platelets in continuous contact with the surface at shear rates up to 6000 s$^{-1}$. Additional evidence supporting a role for GPIbα/VWF in the initial attachment of circulating platelets to sites of vessel injury comes not only from the bleeding abnormality associated with dysfunctional GPIbα in Bernard–Soulier syndrome[3] (Chapter 57), but also from the effect of disrupting the GPIbα–VWF interaction on arterial thrombus formation in primates. Pretreating baboons with inhibitory anti-GPIbα Fab fragments reduced platelet deposition onto collagen at shear rates of 700 to 1000 s$^{-1}$, but the treatment was ineffective if administered 6 minutes after a thrombus had formed.[20,21] This treatment did not increase the bleeding time. The importance of VWF deposition onto collagen in the subendothelium *in vivo* is further illustrated by blocking the VWF A3 domain–collagen interaction. When VWF is unable to bind to collagen, there is a significant inhibition of thrombus under high shear conditions.[22] Similarly, C1q-TNF-related protein 1, a normal component of the vessel wall, is capable of blocking the binding of VWF to collagen, and prevents thrombus formation in a Folts model of arterial thrombosis when injected intravenously.[23]

In endothelial cells, a mechanosensory complex upstream of shear responses and integrin activation pathways is made up of platelet–endothelial cell adhesion molecule 1 (PECAM-1; Chapter 11), vascular/endothelial (VE)-cadherin, and vascular endothelial growth factor receptor 2 (VEGF-R2).[24,25] In this system, PECAM-1 transmits shear force, VE-cadherin acts as an adaptor, and VEGF-R2 activates phosphoinositide (PI) 3-kinase, leading to integrin activation. Activation of the complex leads to NFκB-mediated gene transcription of proinflammatory mediators and expression of E-selectin, intercellular adhesion molecule (ICAM)-1 and vascular cell adhesion molecule 1.[24] Platelets express PECAM-1, and recent studies suggest a role for this protein in inhibiting GPIb-IX-V/VWF-dependent platelet aggregation[26,27] (Section II.C.3). It is possible that PECAM-1 serves a similar role as a mechanosensory receptor on human platelets, although GPIbα remains a strong candidate for mediating shear-dependent platelet activation. VWF binds to GPIb-IX-V under the influence of high shear forces by induction of conformational changes in either the receptor, or VWF, or both.[3,28,29] Anti-GPIbα antibodies that inhibit ristocetin-dependent binding of VWF to GPIbα under static conditions also inhibit shear-dependent platelet aggregation.[30] GPIbα/VWF-dependent adhesion under high

shear conditions and subsequent activation of αIIbβ3 can be induced in CHO cells and other cell lines. These observations are consistent with GPIb-IX-V being sensitive to shear stress.

Crystal structures of the N-terminal ligand-binding domain of GPIbα, the VWF A1 domain, and the GPIbα fragment–VWF A1 complex (nonglycosylated and/or mutant fragments of each, with limited detail on the anionic sulfated motif of GPIbα) have provided useful information on the interaction between VWF and GPIbα, and identify some key residues involved.[31-33] Simultaneously, these structures highlight the gaps in our understanding of how the interaction is regulated between the native proteins *in vivo*. Residues of GPIbα in direct contact with VWF-A1 are clustered toward the N-terminal and C-terminal ends of the ligand-binding region. In particular, a β-finger structure within the C-terminal flank domain of unliganded GPIbα forms part of a stabilized β-sheet in the complex with VWF A1.[33] Although there is minimal direct contact with VWF A1 at the concave face of the LRR sequence, 60 to 128 (shaded in Fig. 7-2B), this region is critical for GPIbα-dependent adhesion to VWF under shear conditions, based on analysis of cross-species chimeras of GPIbα. A chimeric mutant GPIbα in which the human sequence spanning residues 60 to 128 was replaced by the corresponding canine sequence did not support cell rolling on immobilized human VWF, even though insertion of the canine sequence maintains GPIbα in a functional form recognized by conformation-sensitive antibodies and capable of binding VWF in the presence of botrocetin (species indiscriminate).[34] Residues 60 to 128 constitute a negatively charged patch on GPIbα, complementary to a positive patch on human VWF A1, implying that electrostatic interactions are important for platelet adhesion under flow.[1,35]

Another notable feature of the N-terminal domain of GPIbα is the capacity to bind structurally different ligands through overlapping, but not identical, binding sites. Anti-GPIbα monoclonal antibodies mapping to different epitopes within the N-terminal region of GPIbα[34,36] differentially inhibit binding of VWF and other ligands (Fig. 7-2B). Inhibition of VWF binding by these antibodies depends on whether the VWF-GPIbα interaction is induced by ristocetin, botrocetin, or shear stress,[34] suggesting a more complicated mechanism for VWF recognition by GPIbα. For the most part, ristocetin and shear stress induce an interaction indistinguishable by a panel of inhibitory antibodies against either the N-terminal domain of GPIbα or the A1 domain of VWF.[30] Botrocetin, however, is particularly interesting, in that it binds and activates VWF A1 by a unique mechanism.[37] This C-type lectinlike snake toxin initially makes no contact with GPIbα, but after binding of botrocetin/VWF, "slides around the A1 surface to form a new interface."[37] Botrocetin contains an anionic sequence analogous to the anionic/sulfated sequence of GPIbα, Asp269-Glu282,[16] and extends the negative patch on GPIbα to facilitate VWF binding. Botrocetin, like other snake toxins, acts indiscriminately across species[34] and may reflect the aforementioned conservation of GPIbα from chicken to human (Fig. 7-1A).

The length of the sialomucin core of GPIbα, separating the ligand-binding domain from the membrane, also appears to be an important factor in regulating platelet adhesion to VWF.[38] The length of this region in human GPIbα depends upon which of several polymorphic variants of the GPIbα gene encodes it. This polymorphism is based on a variable number of tandem repeats, with four variants each specified by distinct alleles and varying in the number (1–4) of tandem repeats of a 13-amino acid sequence. Each repeat would be expected to add approximately 2.5 nm to the overall length of GPIbα. An in-frame deletion of 27 bp within the macro-glycopeptide region of GPIbα does not affect expression levels, but results in a gain-of-VWF-binding function in a patient with this defect, or if the corresponding mutant is expressed in CHO cells.[39] These studies point to the potential importance of regions outside the N-terminal domain (His1-Glu282) in regulating ligand binding by GPIbα.[40]

**2. TSP.** TSP is a recently identified ligand for GPIb-IX-V on platelets. The interaction between these two molecules is capable of supporting platelet adhesion at high shear rates (up to 4000 s$^{-1}$) in the absence of VWF.[41] The TSP binding site on GPIbα partially overlaps the VWF site and is inhibited by the anti-GPIbα antibodies AP1 and SZ2. The relative contribution of TSP-GPIbα *versus* VWF-GPIbα in shear-induced thrombus formation is not yet established under different pathophysiological conditions; however, the purported prevalence of TSP in atherosclerotic plaque possibly points to a potential pathological role for the TSP–GPIbα interaction. TSP is also likely to affect platelet thrombus formation by inhibiting VWF cleavage by ADAMTS-13.[42]

**3. Collagen.** The major collagen receptors on platelets seem to be GPVI and the integrin α2β1, with evidence suggesting both play significant hemostatic roles *in vivo*[43–47] (see also Chapter 18). Because the A1 domain of VWF binds GPIbα and the A3 domain binds collagen, GPIb-IX-V is also a major, but indirect, collagen receptor on platelets through VWF bridging interactions.[46] GPVI, however, is the major signaling collagen receptor on platelets, because engagement of GPVI by collagen is a potent trigger of platelet activation, and GPVI deficiency impairs thrombus formation at arterial shear rates.[48,49] Sarratt et al.[50] recently suggested that collagen-induced "signaling and adhesion are intimately and inextricably linked," and their studies of human platelets, or mouse platelets lacking GPVI and/or α2β1, further support this view. As discussed in Section

II.C, GPIb-IX-V and GPVI are physically and functionally associated on platelets, potentially acting as mutual accessory signaling proteins.[51] In this respect, Moog and colleagues[52] reported in 2001 that platelet GPV also binds collagen and supports platelet–collagen interactions. Whether this is directly related to the GPIb-IX-V/GPVI association is unclear.

**4. P-Selectin.** In 1999, Romo and colleagues[53] reported that platelet GPIbα recognized P-selectin. This was the first example of a cell surface counterreceptor for GPIb-IX-V.[53] Binding of P-selectin to GPIbα is inhibited by the anti-GPIbα antibodies AK2 and SZ2, the latter antibody result suggesting involvement of the anionic/sulfated tyrosine sequence of GPIbα in binding (Fig. 7-2B). Significantly, the P-selectin/GPIbα interaction provides a mechanism for platelet adhesion to activated endothelial cells or activated (mural) platelets that express P-selectin. Both wild-type and P-selectin-deficient platelets roll on stimulated endothelium *in vivo* by virtue of endothelial P-selectin and independent of the activation state of the platelets.[54,55] Activation of endothelium results in a fourfold increase in the number of interacting platelets, but there is no interaction of platelets with activated or unactivated endothelial cells deficient in P-selectin, suggesting the interaction between platelets and endothelium requires endothelial P-selectin and a platelet counterreceptor likely to be GPIbα. In this regard, anti-GPIbα antibodies attenuate endotoxin-induced interactions of platelets and leukocytes with rat venular endothelium *in vivo*.[56]

In both the α-granules of platelets and Weibel–Palade bodies of endothelial cells, P-selectin is colocalized with VWF, and is rapidly surface expressed after activation. As described in Section II.A.1, P-selectin anchors ULVWF to the endothelium, providing an adhesive surface and another mechanism for GPIbα-dependent platelet adhesion at sites of injury or infection.[9,10] This is important because GPIbα interacting with P-selectin or P-selectin/VWF would allow *unactivated* circulating platelets to adhere to activated endothelium at high shear stresses, mimicking E- or P-selectin-mediated adhesion of leukocytes.[55] Other mechanisms for platelets to interact with the vessel wall require platelet activation (e.g., adhesion mediated by activated platelet αIIbβ3 and endothelial αvβ3, bridged by VWF or fibrinogen).[57] As also discussed in Chapter 12, there are structural and functional similarities between GPIbα and the leukocyte receptor for P-selectin (P-selectin glycoprotein ligand 1 [PSGL-1]).[16,55] Both receptors are sialomucins, and both possess an anionic sequence within their ligand binding domains ([269]DEGTDLsYDsYsYPEED[283] in GPIbα *cf.* [1]QATEsYEsYLDsYDFLPETE[17] in PSGL-1; sY, sulfotyrosine). In addition to supporting adhesion of leukocytes *via* PSGL-1 to P-selectin on activated platelets or endothelial cells, soluble P-selectin binding to PSGL-1 also stimulates release of tissue factor-bearing microparticles from monocytes. Circulating microparticles are highly prothrombotic, become incorporated into a developing thrombus, and promote activation of factor VII.[55,58] A major outcome of PSGL-1 engagement on leukocytes is inside-out activation of integrin αMβ2, which binds to ICAM-1 on endothelial cells (Fig. 7-3A). Surprisingly, αMβ2 is yet another counterreceptor for GPIb-IX-V.

**5. αMβ2 (Mac-1).** The leukocyte integrin αMβ2 was identified as a GPIb-IX-V counterreceptor in 2000, when it was proposed that this interaction mediates adhesion and migration of leukocytes on mural vascular thrombi.[59] Both monocytic cells and αMβ2-transfected cell lines specifically adhered to purified GPIbα. This binding was regulated by the αM I-domain, which is homologous to the A1 domain of VWF, which also binds GPIbα. αMβ2-dependent leukocyte adhesion is inhibited by pretreating platelets with mocarhagin, a cobra metalloproteinase that cleaves the N-terminal ligand-binding domain of GPIbα (Glu282/Asp283), or by anti-GPIbα monoclonal antibodies AP1 and VM16d. (AP1, but not VM16d, inhibits VWF A1 binding to GPIbα, suggesting nonidentical binding sites [Fig. 7-2B]). These studies support a role for the αMβ2/GPIb-IX-V interaction in forming leukocyte–platelet aggregates in, for example, vascular inflammation associated with thrombosis or atherosclerosis. Other evidence supports a role for adherent platelets in early stages of atherosclerotic plaque formation.[60] A second important function of the αMβ2/GPIb-IX-V interaction is in the clearance of cold-stored platelets. The short lifetime of platelets that have been refrigerated (<15°C) for even a short period prior to transfusion depends on cold-induced clustering of GPIbα, accelerating αMβ2-dependent clearance by phagocytic liver macrophages or Kupffer cells.[61] The clearance mechanism involves lectinlike activity of αMβ2 recognizing β-N-acetylglucosamine residues of N-linked glycans on GPIbα,[62] which may have further implications for GPIb-IX-V functions (Section II.C.5).

**6. Nonphysiological Ligands.** There are two main categories of nonphysiological GPIb-IX-V ligands: snake venom toxins and bacteria. Snake venom proteins targeting GPIbα, which either inhibit VWF binding or induce GPIbα-dependent platelet activation (mimic VWF) by cross-linking GPIbα and/or GPIbα/GPVI, are described elsewhere.[63–65] The role of bacteria binding to platelets through GPIbα may be of great importance in pathology or in increasing thrombotic risk (see Chapter 40). GPIbα is a high-abundance sialomucin on platelets, and GPIbα-linked sialic acid is recognized by cell wall proteins of *Streptococcus gordonii* as one of several mechanisms for bacteria–platelet interactions. *Staphylococcus aureus* uses a different

mechanism that involves a bridging interaction with GPIbα mediated by VWF. Yet another mechanism is direct binding of *S. sanguis* to the N-terminal domain of GPIbα.[66–68]

### B. Procoagulant Function

**1. α-Thrombin.** It has long been known that GPIb-IX-V is a high-affinity binding site for α-thrombin on platelets, but it has been only recently that the full implications of this interaction have emerged. Thrombin can activate platelets by either proteolytic or nonproteolytic mechanisms, and intact GPIbα facilitates the platelet response to low, but not high, doses of thrombin. For example, Bernard–Soulier syndrome platelets lacking functional GPIbα show an impaired response to thrombin and a prolonged lag time before aggregation.[3] Other α-thrombin receptors on platelets include the G protein-coupled, protease-activated receptors PAR-1 and PAR-4 (Chapter 9). Thrombin's interaction with GPIbα has two potential roles. First, thrombin is involved in the regulation of coagulation by localizing thrombin and factor XI (see Section II.B.2) on GPIbα. Second, thrombin's interaction with GPIbα can activate platelets in two ways: either by acting as a cofactor for the thrombin-dependent activation of PAR1 or by signaling directly *via* GPIbα in the absence of GPV.[69] The understanding of thrombin–GPIbα interactions has been advanced by the solution of crystal structures of the complex, or more precisely, complexes, because there are two distinct ways in which thrombin can bind GPIbα These different interactions may have profound consequences on both thrombin activity and platelet activation.[70–74]

Thrombin binding to the N-terminal region of GPIbα (His1-Glu282) is inhibited by the anti-GPIbα antibodies SZ2, which maps to the anionic/sulfated tyrosine sequence, Asp269-Glu282, and VM16d, which maps to the C-terminal flanking sequence 226 to 268 (Fig. 7-2B). Crystal structures used either a GPIbα fragment (N-terminal residues 1–290) and Phe-Pro-Arg-chloromethylketone-inhibited thrombin,[70] or GPIbα 1-279 (with N-linked glycosylation sites at Asn21 and Asn159 substituted by Glu residues) and DFP-thrombin.[71] One GPIbα–thrombin complex involves mainly the fibrinogen recognition exosite I of thrombin contacting residues 151 to 284 of GPIbα, particularly residue 279. The other complex involves mostly heparin-binding exosite II of thrombin in contact with residues 274, 275, and 277 of GPIbα. The complexes reveal a "striking charge complementarity between interacting surfaces of GPIbα and thrombin."[71] Thrombin bound to GPIbα (*via* exosite II) accelerates PAR1 cleavage and platelet activation, with a six- to sevenfold increase in $k_{cat}/K_m$ for PAR1 proteolysis on a platelet surface where there is intact GPIbα.[73,75]

By an analogous mechanism, thrombin associated with GPIbα *via* exosite II is also ideally oriented to bind (*via* exosite I) and activate factor XI, which may bind at a neighboring GPIbα subunit[76] The two GPIbα–thrombin interactions would also allow thrombin to cross-link GPIbα, leading to thrombin-dependent platelet activation (as for other multivalent ligands). In this case, intact GPV in the GPIb-IX-V complex acts as a brake on platelet activation in response to thrombin, whereas removing GPV facilitates GPIbα/thrombin-dependent platelet activation.[69,72] In this respect, catalytically inactive thrombin induces thrombosis in GPV-null mice. Indeed, platelet GPV (approximately 85 kDa) is a thrombin substrate, and cleavage releases a soluble GPV ectodomain fragment, GPVf1 (approximately 69 kDa), which then enables GPIbα-dependent platelet activation by thrombin. GPIbα also mediates thrombin-dependent fibrin(ogen) cross-linking at the surface of activated platelets,[77,78] as well as regulating thrombin's effects on coagulation factor XI.

**2. Factor XI.** Binding of factor XI to GPIbα was first demonstrated in 2002, providing an unexpected mechanism for localizing factor XI to the activated platelet surface, leading to enhanced procoagulant activity.[79] The soluble ectodomain fragment of GPIbα (glycocalicin) markedly enhances activation of factor XI and, with the addition of kininogen (see Section II.B.3), can approximate the rate of activation at the activated platelet surface.[79] The factor XI binding site is not identical to that of thrombin, although there is some overlap. The anti-GPIbα antibody SZ2 against the anionic/sulfated sequence (Asp269-Glu282) inhibits both interactions. Factor XI binding also involves upstream LRR sequences of GPIbα, and it is conceivable that localization of α-thrombin and factor XI on GPIbα accelerates generation of factor XIa.[76,79,80] Formation of this procoagulant complex requires platelet activation and appears to involve a subset of the GPIb-IX-V complex located within lipid rafts. Disrupting rafts by cholesterol depletion inhibits the process.[81]

**3. Kininogen.** In 1999, it was shown that high-molecular weight kininogen exhibits zinc-dependent binding to platelet GPIbα,[82] an interaction that in turn may regulate thrombin-dependent activation of factor XI, as discussed in Section II.B.2. More recent studies suggest that the presence of high-molecular weight kininogen may also enhance leukocyte–platelet adhesion by promoting the interaction between GPIbα and αMβ2.[83] The anti-GPIbα monoclonal antibodies SZ2 (against the anionic sequence Asp269-Glu282) and AP1 (against the C-terminal flanking sequence 201–225) differentially inhibit binding of kininogen and αMβ2 respectively, but the mechanism underlying the regulation of αMβ2 binding by kininogen is still not defined.

**4. Factor XII.** Human factor XIIa binding to GPIbα may regulate assembly of the procoagulant complex involving kininogen/thrombin/factor XI.[84] Factor XII competes with kininogen binding and, more important, inhibits thrombin-dependent platelet aggregation associated with thrombin binding to GPIbα. Factor XII does not affect activation of PAR1 by agonist peptides, which act independently of GPIbα.[84] Furthermore, factor XII deficiency may increase thrombotic risk, which may be consistent with a role in controlling procoagulant activity and thrombosis associated with GPIbα.

### C. Topographical Associations

**1. GPVI/FcRγ.** In the wake of abundant circumstantial evidence for a direct link between the platelet receptors GPIb-IX-V and GPVI, recent direct evidence that these receptors are in fact physically and functionally associated suggests they may form a unique adhesion/signaling complex in resting or activated platelets.[51,85] GPVI is coimmunoprecipitated with GPIbα from platelet lysates, independent of either intact lipid rafts or attachment of GPIb-IX-V to the platelet cytoskeleton. Furthermore, proteolytic removal of the GPIbα N-terminus by mocarhagin not only inhibits binding of VWF to GPIbα, but also inhibits platelet aggregation induced by the GPVI-selective agonist, cross-linked collagen-related peptide (CRP). The GPIbα antibody SZ2, against the anionic/sulfated sequence (Fig. 7-2B), but not other antibodies (AK2, VM16d), also blocks CRP-dependent platelet aggregation. GPIb-IX-V and GPVI, therefore, appear to act as mutual accessory signaling proteins, with ligand binding to GPVI inducing signals that in part require GPIb-IX-V, and *vice versa*, particularly at threshold agonist concentrations. Baker et al.[86] also reported a synergy of signaling responses to VWF/collagen acting at GPIb-IX-V/GPVI. Consistent with this supposition, GPVI signaling contributes to platelet–VWF adhesion under shear, because GPVI-deficient platelets show impaired thrombus formation at arterial shear rates.[48,49] The interaction of GPIbα and GPVI involves their ectodomains, because a soluble extracellular fragment of GPIbα (glycocalicin) coimmunoprecipitates a soluble GPVI ectodomain fragment.[51]

GPVI contains two immunoglobulin domains in the extracellular region, a mucinlike domain, a transmembrane domain, and a cytoplasmic tail (see Chapter 6, Fig. 6-4). In addition to the physiological ligand, collagen, GPVI also binds CRP and snake venom proteins, such as convulxin of the C-type lectinlike family and alborhagin of the metalloproteinase–disintegrin family.[46,63,64] GPVI signals in three ways: by a functional coassociation with FcR γ-chain (FcRγ; required for GPVI surface expression); by a direct association with Src family kinases, Fyn or Lyn, *via* a proline-rich cytoplasmic sequence that binds Src homology-3 domains; and by activation of $Ca^{2+}$ signals involving a juxtamembrane calmodulin-binding sequence.[46,47,87–89] GPVI is associated with FcRγ *via* a salt bridge in the transmembrane domain. Cross-linking of GPVI/FcRγ by multivalent GPVI ligands results in concomitant cross-linking of immunoreceptor tyrosine-based activatory motif (ITAM) domains in the cytoplasmic tail of FcRγ, which in turn activates Syk kinase. GPIb-IX-V also utilizes FcRγ-dependent signaling pathways (see Section III), consistent with GPIb-IX-V/GPVI association, and coimmunoprecipitation of GPIb-IX and FcRγ from detergent (Brij 35) lysates of human platelets.[90] GPIbα and FcRγ mutually translocate to the cytoskeleton in platelets stimulated by a GPIbα-selective agonist (mucetin) that cross-links GPIbα.[65]

**2. FcγRIIa.** In addition to FcRγ, the Fc receptor FcγRIIa is also involved in GPIb-IX-V signaling. A topographical association and functional interplay between these receptors was shown on platelets by fluorescence energy transfer,[91] and an interaction between GPIbα and FcγRIIa was also revealed by yeast two-hybrid analysis.[92] In the latter study, a cytoplasmic sequence — [542]Arg-Gly-Arg[544] — was identified as being important for FcγRIIa binding. (Incidentally, this motif is not conserved in chicken GPIbα [Fig. 7-1B]). FcγRIIa is also coisolated with GPIb-IX in lipid raft fractions on sucrose density gradients[93,94] (Section II.C.5).

It is interesting to compare FcγRIIa with GPVI/FcRγ, because all these receptors associate with GPIb-IX-V. Like GPVI, FcγRIIa has two extracellular immunoglobulin domains. Like FcRγ, FcγRIIa contains an ITAM domain within the cytoplasmic tail, which upon receptor cross-linking activates Syk kinase. FcγRIIa therefore resembles a hybrid of GPVI (extracellular region) and FcRγ (cytoplasmic domain). It would be intriguing to define how the interaction of GPIb-IX-V/FcγRIIa compares with GPIb-IX-V/GPVI/FcRγ with respect to signaling output.

**3. PECAM-1.** Platelets express PECAM-1 (Chapter 11). Recent studies suggest a role for this receptor in inhibiting GPIb-IX-V-dependent aggregation of human or mouse platelets mediated by αIIbβ3, and thrombus formation in mouse models.[26,27,95] This inhibition involves an immunoreceptor tyrosine-based inhibitory motif (ITIM) domain in the PECAM-1 cytoplasmic domain, which upon receptor clustering becomes phosphorylated and recruits tyrosine phosphatases (such as SHP-1 and SHP-2) through binding to phosphotyrosine *via* Src-homology 2 domains. There is no definitive evidence for a direct topographical link between PECAM-1 and GPIb-IX-V on platelets. However, like GPIb-IX-V (Section II.C.2), PECAM-1 physically and functionally associates with FcγRIIa on human platelets, and also negatively regulates collagen-dependent stimulation of GPVI/FcRγ.[96] This network of associations between GPIb-IX-V, GPVI, ITAM-containing (activating) FcRγ and

FcγRIIa, and ITIM-containing (inhibitory) PECAM-1, suggests an intricate adhesion signaling complex regulating activation of αIIbβ3 after platelets adhere to a surface. Like GPIb-IX-V and GPVI (Section III.A.2), platelet PECAM-1 is regulated by calmodulin associated with a juxtamembrane sequence in the cytoplasmic domain. Dissociation of calmodulin leads to proteolysis of PECAM-1, providing a potential mechanism for regulation of PECAM-1 functions.[97]

**4. α2β1.** A possible role for collagen binding to integrin α2β1 facilitating signaling *via* GPVI has been evaluated (Section II.A.3), although, like other integrins, α2β1 shows increased ligand-binding affinity *after* platelets are activated. Collagen binding to α2β1 is enhanced by signals from GPVI and GPIb-IX-V, reinforcing platelet–collagen adhesion. The link between α2β1 and GPVI (Section II.A.3) also implies a connection with GPIb-IX-V, because GPV binds collagen[52] (Section II.A.3), and GPIb-IX-V is associated with GPVI[51] (Section II.C.1). Using activation-dependent antibodies against the ligand-binding I domain of α2, Cruz et al.[98] showed a direct link between VWF binding to GPIbα and activation of α2β1, thereby increasing its affinity for collagen.

**5. Lipid Rafts.** The topographical/functional association of GPIb-IX-V and other receptors needs to be considered in light of recent evidence that a subpopulation of GPIb-IX-V is located in cholesterol-rich lipid raft domains, and has enhanced VWF-binding affinity.[93,94] GPIX and GPIbβ contain cysteinyl palmitoylation sites at the cytoplasmic side of the transmembrane domain, and the extent of palmitoylation correlates closely with raft localization. Although GPIb-IX-V and GPVI association can occur outside of rafts, and is not affected by disrupting rafts with methyl-β-cyclodextrin (MβCD),[51] raft association may be critical for GPIb-IX-V signaling, as also shown for GPVI/FcRγ.[99–101] In fact, disrupting rafts with MβCD impairs VWF binding or VWF-dependent platelet aggregation under static or high shear conditions.[93] FcγRIIa is also raft associated in activated platelets, consistent with an association with GPIb-IX-V.[93,94,101] One reason for receptor signaling in rafts is the presence in these domains of signaling/adapter proteins, such as LAT and SLP-76. Lipid rafts associate with the cytoskeleton in activated platelets, a process requiring engagement of αIIbβ3, which itself is not raft associated.[102] This raft–cytoskeleton linkage may also depend on GPIbα/filamin attachment[94] (discussed later).

Lipid rafts are also implicated in the initiation and maintenance of procoagulant activity on platelets (Section II.B). GPIb-IX-V localized in rafts appears to play a key role in assembling thrombin, coagulation factor XI, and kininogen on activated platelets.[81,94] According to a model in which thrombin bound to GPIbα *via* exosite II is oriented to activate factor XI on a nearby GPIbα (Section II.B), clustering of GPIb-IX-V within lipid rafts would be required for efficient thrombin-dependent activation of factor XI within a proposed quaternary procoagulant complex ([GPIbα]₂/thrombin/factor XI). In this regard, low-temperature-induced clustering of GPIbα, which together with glycosylation promotes αMβ2-mediated platelet clearance[61] (Section II.A.5), also enhances binding of multivalent VWF,[94] observations that could in part be explained by a coalescence of GPIb-IX-V-bearing lipid rafts. This role of rafts in controlling the topography of GPIb-IX-V (and associated receptors) on resting and activated platelets, and hence platelet adhesiveness, procoagulant activity, and reactivity, warrants further investigation.

## III. Signaling

GPIb-IX-V signaling is reviewed in this section by presenting an overall picture of interactions occurring at the cytoplasmic face of the receptor (Section III.A), and signaling molecules in the pathway leading from GPIb-IX-V to activation of αIIbβ3 (Section III.B). It is important to note, however, that the model of how GPIb-IX-V signals in human platelets is currently incomplete.

### A. Interactions at the Cytoplasmic Domain

**1. 14-3-3ζ.** The signaling protein 14-3-3ζ (the predominant 14-3-3 isoform in platelets) forms a homodimer, and interacts with multiple sites in the cytoplasmic domain of GPIb-IX-V.[1,7,103] One site in the central region of the GPIbβ tail is within a motif containing the PKA-dependent phosphorylation site at Ser166 ([163]RLpSLTDP[171]; pS, phosphoserine), and a motif in the C-terminal region of GPIbα contains phosphorylation sites at Ser587 and Ser590 ([585]RPpSALpSQG[592]). These sites resemble, but are not identical to, consensus 14-3-3ζ-binding motifs in other proteins (RSXSXP or RXSXS/TXP). However, a unique site for binding 14-3-3ζ occurs at the extreme C-terminus of GPIbα and contains a constitutive phosphorylation site at Ser609 ([606]SGHpSL[610] stop). The arrangement of these sites is shown in Figures 7-1B and 7-1C. A toggle switch mechanism has been proposed,[104] whereby a dimer of 14-3-3ζ anchored at the constitutively phosphorylated Ser609 motif on GPIbα can also interact with either (a) the upstream GPIbα sequence surrounding Ser587/590 or (b) the Ser166-containing motif on GPIbβ, and that these interactions regulate the affinity of GPIb-IX-V for VWF. This (a) *versus* (b) configuration is proposed to be regulated by PKA-dependent phosphorylation of GPIbβ. Activation of PKA

inhibits platelet activation and controls the ability of GPIbβ to regulate actin polymerization, and might also explain why Ser609 is constitutively phosphorylated to maintain 14-3-3ζ attachment.

But how does 14-3-3ζ regulate VWF binding to GPIb-IX-V? The answer is not simple. With respect to 14-3-3ζ binding to GPIbβ, mutation of Ser166, preventing phosphorylation and attenuating 14-3-3ζ association,[105] differentially regulates VWF binding: Replacing Ser166 with Ala or Gly respectively up- or downregulates VWF binding.[106,107] Several mechanisms can be inferred from these observations, but none is compelling. For example, the mutations may alter the conformation of the cytoplasmic tail, which may be critical, or they may interfere with or augment the interactions of the tail with other proteins such as calmodulin or filamin.

The role for 14-3-3ζ association with GPIbα may be clearer. Studies in which wild-type or C-terminally truncated GPIbα (as GPIb-IX) was cotransfected with αIIbβ3 in CHO cells suggest that an intact 14-3-3ζ binding site encompassing the C-terminal five residues GPIbα is required for VWF/GPIbα-dependent activation of αIIbβ3 and, further, that activation of ERK-1/2 kinase is a likely downstream effector of this pathway.[103] Dai et al.[104] showed that a membrane-permeable peptide inhibitor of the 14-3-3ζ–GPIbα interaction, MPαC, inhibits VWF binding to platelets and VWF-mediated platelet adhesion under flow conditions. Furthermore, disrupting 14-3-3ζ binding at the GPIbα C-terminus prevented the enhanced VWF binding to GPIb-IX otherwise seen upon dephosphorylation of GPIbβ. Finally, a recent study suggests that 14-3-3ζ associated with the C-terminus of GPIbα regulates PI 3-kinase-dependent signaling and proliferation in megakaryocytes.[108]

**2. Calmodulin.** Calmodulin binds to two sites in the cytoplasmic domain of GPIb-IX-V, a juxtamembrane positive-charge/hydrophobic sequence in GPIbβ immediately upstream of the 14-3-3ζ-binding site (Fig. 7-1C), and within the 15-residue tail of GPV (Fig. 7-2B).[109] The calmodulin binding site in these subunits is in a similar juxtamembrane position as the calmodulin binding site in GPVI, a receptor that is associated with GPIb-IX-V[51,87,89,109,110] (Section II.C.1). In GPVI, calmodulin binding regulates GPVI-dependent Ca²⁺ signals independently of FcRγ- and Src-related pathways.[89] A similar role is conceivable for calmodulin associated with GPIb-IX-V. Another important role for calmodulin associated with GPVI is to regulate metalloproteinase-mediated shedding of the GPVI ectodomain, releasing a soluble approximately 55-kDa fragment.[111] Ligand binding to GPVI induces activation-dependent dissociation of calmodulin,[87] whereas calmodulin inhibitors bypass this requirement for activation and directly induce metalloproteinase-dependent GPVI shedding.[111] The soluble GPVI fragment is detectable in normal human plasma, attesting to the likely physiological significance of this process.[112] Expression levels of GPVI correlate in a graded fashion with the thrombotic responsiveness of platelets to collagen.[50,113] In circulating thrombocytes in zebrafish, changes in expression of GPVI (and other receptors) with age correlate with decreased participation of the thrombocytes in thrombus formation, and young thrombocytes are more prevalent in microthrombi.[114]

Calmodulin inhibitors also induce metalloproteinase-dependent ectodomain shedding of platelet GPV. In this case, the sheddase has been identified as ADAM-17 (tumor necrosis factor-α-converting enzyme).[115] A soluble ectodomain fragment of GPIbα (glycocalicin) is also released from the platelet membrane by metalloproteinase activity.[116] GPIbα shedding can be induced by treating platelets with the mitochondrial poison CCCP, which mimics platelet aging and, as for GPV, shedding involves ADAM-17.[117] Glycocalicin circulates in normal plasma at 1 to 3 μg/mL, with levels rising in thrombocytopenia caused by platelet destruction or disease.[118] Like GPV and GPVI, surface expression of GPIbα might be regulated by the association and dissociation of calmodulin with the cytoplasmic tail of GPIbβ. Together, these data favor a central role for calmodulin in regulating functional surface expression of GPIb-IX-V and GPVI, the key receptor complexes initiating platelet thrombus formation, and point to a potential antithrombotic role for calmodulin-targeting pharmaceuticals.[110]

**3. PI 3-Kinase.** PI 3-kinase is associated with GPIb-IX-V and can be coimmunoprecipitated with the receptor from resting platelet lysates.[119] The interaction with GPIb-IX-V involves the PI 3-kinase regulatory p85 subunit, which is associated with the catalytic p110 subunit. Together with 14-3-3ζ and calmodulin, PI 3-kinase translocates to the Triton-insoluble actin cytoskeleton in activated platelets, but the association of these proteins is not dependent on the GPIb-IX-V–cytoskeletal interaction, because the interactions are not significantly affected by dissociation of filamin from GPIbα (using N-ethylmaleimide) or depolymerization of actin filaments (using DNAse I). It is still unknown whether PI 3-kinase is directly linked to GPIb-IX-V or is associated as part of a signaling complex involving 14-3-3ζ. PI 3-kinase certainly potentiates activation of αIIbβ3 in response to some agonists, although its precise role in GPIb-IX-V signaling is undecided.[120–122] PI 3-kinase inhibitors block GPIb-IX-V/VWF-dependent platelet aggregation, although this may involve PI 3-kinase activated downstream of other GPIb-IX-V-associated receptors (Section III.B). Nevertheless, PI 3-kinase is a potential antithrombotic target, and a new class of isoform-selective inhibitors attenuate formation of occlusive thrombi *in vivo*.[123] In this

study, the type Ia PI 3-kinase (p110b isoform) was found to regulate the formation and stability of platelet thrombi, and isoform-specific inhibitors prevented formation of αIIbβ3-dependent adhesion contacts.

**4. Filamin and the Platelet Cytoskeleton.** Filamin-1, also known as actin binding protein, is a homodimeric multifunctional protein that associates with actin filaments in the platelet cytoskeleton, and links this structure to other structural/signaling proteins such as α-actinin and to membrane receptors such as GPIb-IX-V (see also Chapter 4). In platelets, a cytoskeleton is made up of long cytoplasmic actin filaments, and a membrane skeleton is composed of short submembranous actin filaments. Although the interaction between GPIb-IX-V and filamin/membrane skeleton was identified by Fox[124] more than 20 years ago, many unanswered questions remain regarding exactly how this interaction is regulated; the nature of the binding sites; and the functional consequences of the interaction on GPIbα-dependent adhesion, platelet size and shape, and raft localization of GPIb-IX-V and other receptors.

A number of studies suggest filamin binds to the central portion of the GPIbα cytoplasmic tail (Figs. 7-1B and 7-2B). Most recently, Trp570 and Phe568 within the sequence Pro561-Arg572 were identified as important for filamin association[6] (Section I). This site, upstream of the 14-3-3ζ binding motifs centered on the phosphorylation sites at Ser587/Ser590 and Ser609, is important for anchoring GPIb-IX to the membrane skeleton, and for cell adhesion to VWF under high shear conditions. The effect of filamin on VWF binding to GPIbα, however, is obscured by conflicting data in the literature. Truncation, deletion, or point mutations of the GPIbα cytoplasmic domain that disrupt filamin binding have been shown either to increase or decrease VWF binding,[125–128] probably depending upon differences in the experimental systems used for analysis and parameters such as shear rate, expression levels, and/or other methodological variables.[127] Regardless of the discrepancies, it appears that cytoskeletal association affects the adhesive function of GPIb-IX-V, although how this may be relevant to different pathophysiologal circumstances is yet to be established.

Another line of enquiry has shown that the cytoplasmic tail of GPIbα plays a role in maintaining normal platelet size. In this case, the large-platelet phenotype observed in a mouse model of Bernard–Soulier syndrome is rescued by reintroducing the GPIbα tail in the form of a fusion protein with the extracellular domain of interleukin 4 receptor.[129] This is in contrast to the giant platelets seen in patients with Bernard–Soulier syndrome, type Bolzano, which carry a mutation of GPIbα that abrogates VWF binding, but with substantial residual expression of GPIb-IX-V on the platelet surface.[130] Filamin also regulates normal membrane insertion of GPIb-IX. In GPIb-IX-transfected cells, filamin binds

GPIbα within the endoplasmic reticulum and regulates trafficking and surface expression.[131,132] Filamin associated with GPIb-IX-V also regulates actin rearrangement during spreading and aggregation by associating with Src homology 2 domain-containing inositol polyphosphate 5-phosphatase-2 (SHIP-2). SHIP-2 hydrolyzes phosphatidylinositol 3,4 triphosphate (PtdIns[3,4,5]P3), forming phosphatidylinositol 3,4 bisphosphate (PtdIns[3,4]P2) and regulates membrane ruffling *via* an interaction with filamin in motile cells. Dyson et al.[133] showed that SHIP-2 forms an active complex with filamin, actin, and GPIb-IX-V, and that this complex mediates localized hydrolysis of PtdIns(3,4,5)P3, thereby regulating submembranous actin polymerization.

Recent evidence also suggests that the platelet membrane skeleton localizes receptors to lipid raft domains. Disruption of the GPIb-IX-V-filamin interaction, using the reagent N-ethylmaleimide, alters the amount of the receptor found in lipid raft fractions, and thus the adhesive function of GPIb-IX-V.[94] Given the association of GPIb-IX-V with other platelet receptors such as GPVI, FcRγ, and FcγRIIa (Section II.C), the GPIb-IX-V–cytoskeleton interaction may be critical in targeting receptors to rafts, and thus platelet adhesiveness to VWF and reactivity to other activating stimuli.

### B. Signaling from GPIb-IX-V to αIIbβ3

The arrow in the equation GPIb-IX-V → activated αIIbβ3 (and other integrins) represents the essence of understanding GPIb-IX-V signaling, because in pathophysiology this is the major consequence of engagement of GPIb-IX-V and the transition of circulating platelets to an adhered, activated, and aggregated state. GPIb-IX-V does not have built-in tyrosine kinase activity, is not directly coupled to G proteins, and does not contain phosphorylatable tyrosine residues that recruit signaling molecules. Nevertheless, it utilizes all these avenues to transmit signals by associating with other receptors. Evidence for pathways related to GPIb-IX-V signaling in human platelets is rather scant. Nevertheless, some signaling molecules involved in GPIb-IX-V-related pathways have been identified and are summarized in Figure 7-3C.[120,122,134–136]

Key elements in the signaling pathway downstream of GPIb-IX-V signaling include Src family kinases that activate phospholipase γ2 (PLCγ2); PLCγ2, which leads to elevation of cytosolic $Ca^{2+}$ by generating $IP_3$ that acts on $Ca^{2+}$ stores; and PI 3-kinase potentiating signals leading to activation of αIIbβ3 (Section III.A.3). Fc receptor cross-linking and activation of ITAM domains also activates Syk kinase, which in turn activates PLCγ2 (Sections II.C.1 and II.C.2). Activation of ERK-1/2 kinase occurs downstream of Src activation, and this is implicated in the pathway leading from VWF binding to GPIb-IX-V to activation of

αIIbβ3 in both human platelets and GPIb-IX/αIIbβ3-transfected cells[103,137] (Section III.A.1).

More contentious is the role of cGMP-dependent protein kinase (PKG) as an activator of VWF-dependent signals leading to activation of αIIbβ3. Li et al.[138] originally described this novel pathway, based on defective responsiveness of platelets to VWF in PKG-deficient mice, and the blockade of aggregation of human platelets by PKG inhibitors (and, conversely, enhanced aggregation by elevation of cGMP). Contradictory studies by others have led to ongoing debate over the relevance of cGMP/PKG and the experimental/physiological conditions used to demonstrate its role.[139–141] Such qualification is common in much of the discussion of signaling in human platelets, reflecting analytical challenges (see also Sections III.A.1 and III.A.4).

Another role for PKG may be to regulate platelet secretion of agonists such as adenosine diphosphate (ADP) and thromboxane $A_2$ (TXA$_2$) in response to engagement of GPIb-IX-V or other receptors.[142] VWF binding to GPIb-IX-V leads to secretion of ADP and TXA$_2$, which, acting in an autocrine fashion, stimulate platelet G protein-coupled receptors. The purinergic ADP receptors on platelets are P2Y$_1$ (G$_{αq}$ linked and leading to elevation of cytosolic $Ca^{2+}$) and P2Y$_{12}$ (G$_{αi}$ linked and diminishing cAMP/cAMP-dependent PKA phosphorylation) (Chapter 10). As discussed earlier, GPIb-IX-V signaling is inextricably linked to signals from GPVI/FcRγ and FcγRIIa, and α2β1, as well as αIIbβ3.[120,122,134–136] Signaling is initiated by cross-linking of GPIb-IX-V by VWF or other multivalent ligands.[85,143,144] A receptor cross-linking mechanism for activation and signaling *via* accessory receptors such as GPVI and Fc receptors is familiar from studies of T-cell family receptors and accessory receptors that these proteins resemble.

## IV. The End of the Beginning

Current evidence evokes an adhesion signaling complex on platelets involving GPIb-IX-V. In rafts, this could consist of GPIb-IX-V associated with GPVI, FcRγ, and FcγRIIa linked to the cytoskeleton *via* GPIbα-filamin, linked directly to 14-3-3ζ and calmodulin, and linked directly or indirectly to PI 3-kinase and Src kinases. Engagement of GPIbα/GPVI by VWF/collagen at high shear rates, or activating platelets by other agonists at low shear, could induce translocation of 14-3-3ζ, calmodulin, and PI 3-kinase to the cytoskeleton; elevation of cytosolic $Ca^{2+}$; disruption of filamin attachment to GPIb-IX-V; and initiation of signals leading to activation of αIIbβ3. Complexes involving GPIb-IX-V could exist outside rafts, with a lower capacity for signaling, but a capacity to regulate activation-independent adhesion/agglutination or submembranous distribution of 14-3-3ζ/

calmodulin. Calmodulin may control GPIb-IX-V/GPVI expression by regulating metalloproteinase-mediated ectodomain shedding of GPVI, GPV, and/or GPIbα. Further investigation of this receptor–signaling–cytoskeleton–raft complex on platelets should lead to a greater understanding of how it forms, how it is regulated, and how it contributes to the functional role of platelets in controlling thrombosis, coagulation, and vascular cell adhesion.

# References

1. Andrews, R. K., Gardiner, E. E., Shen, Y., Whisstock, J. C., & Berndt, M. C. (2003). Molecules in focus: Glycoprotein Ib-IX-V. *Int J Biochem Cell Biol, 35,* 1170–1174.
2. Berndt, M. C., Gregory, C., Kabral, A., Zola, H., Fournier, D., & Castaldi, P. A. (1985). Purification and preliminary characterization of the glycoprotein Ib complex in the human platelet membrane. *Eur J Biochem, 151,* 637–649.
3. López, J. A., Andrews, R. K., Afshar–Kharghan, V., & Berndt, M. C. (1998). Bernard–Soulier syndrome. *Blood, 91,* 4397–4418.
4. López, J. A. (1994). The platelet glycoprotein Ib-IX complex. *Blood Coag Fibrinol, 5,* 97–119.
5. López, J. A., Chung, D. W., Fujikawa, K., Hagen, F. S., Papayannopoulou, T., & Roth, G. J. (1987). Cloning of the α chain of human platelet glycoprotein Ib: A transmembrane protein with homology to leucine-rich α 2-glycoprotein. *Proc Natl Acad Sci USA, 84,* 5615–5619.
6. Cranmer, S. L., Pikovski, I., Mangin, P., Thompson, P. E., Domagala, T., Frazzetto, M., Salem, H. H., & Jackson, S. P. (2005). Identification of a unique filamin A binding region within the cytoplasmic domain of glycoprotein Ibα. *Biochem J, 387,* 849–858.
7. Mangin, P., David, T., Lavaud, V., Cranmer, S. L., Pikovski, I., Jackson, S. P., Berndt, M. C., Cazenave, J. P., Gachet, C., & Lanza, F. (2004). Identification of a novel 14-3-3ζ binding site within the cytoplasmic tail of platelet glycoprotein Ibα. *Blood, 104,* 420–427.
8. Bodnar, R. J., Gu, M., Li, Z., Englund, G. D., & Du, X. (1999). The cytoplasmic domain of the platelet glycoprotein Ibα is phosphorylated at serine[609]. *J Biol Chem, 274,* 33474–33479.
9. Padilla, A., Moake, J. L., Bernardo, A., Ball, C., Wang, Y., Arya, M., Nolasco, L., Turner, N., Berndt, M. C., Anvari, B., López, J. A., & Dong, J. F. (2004). P-selectin anchors newly released ultra-large von Willebrand factor multimers to the endothelial cell surface. *Blood, 103,* 2150–2156.
10. Dong, J. F. (2005). Cleavage of ultra-large von Willebrand factor by ADAMTS-13 under flow conditions. *J Thromb Haemost, 3,* 1710–1716.
11. Dong, J. F., Moake, J. L., Nolasco, L., Bernardo, A., Arceneaux, W., Shrimpton, C. N., Schade, A. J., McIntire, L. V., Fujikawa, K., & López, J. A. (2002). ADAMTS-13 rapidly cleaves newly secreted ultra-large von Willebrand factor multimers on the endothelial surface under flowing conditions. *Blood, 100,* 4033–4039.
12. Lammle, B., Kremer Hovinga, J. A., & Alberio, L. (2005). Thrombotic thrombocytopenic purpura. *J Thromb Haemost, 3,* 1663–1675.
13. Kulkarni, S., Dopheide, S. M., Yap, C. L., Ravanat, C., Freund, M., Mangin, P., Heel, K. A., Street, A., Harper, I. S., Lanza F., & Jackson, S. P. (2000). A revised model of platelet aggregation. *J Clin Invest, 105,* 783–791.
14. Kroll, M. H., Hellums, J. D., McIntire, L. V., Schafer, A. I., & Moake, J. L. (1996). Platelets and shear stress. *Blood, 88,* 1525–1541.
15. Arya, M., Anvari, B., Romo, G. M., Cruz, M. A., Dong, J. F., McIntire, L. V., Moake, J. L., & López, J. A. (2002). Ultra-large multimers of von Willebrand factor form spontaneous high-strength bonds with the platelet glycoprotein Ib-IX complex: Studies using optical tweezers. *Blood, 99,* 3971–3977.
16. Andrews, R. K., López, J. A., & Berndt, M. C. (1997). Molecular mechanisms of platelet adhesion and activation. *Int J Biochem Cell Biol, 29,* 91–105.
17. Nurden, A. T. (2005). Qualitative disorders of platelets and megakaryocytes. *J Thromb Haemost, 3,* 1773–1782.
18. Savage, B., Saldivar, E., & Ruggeri, Z. M. (1996). Initiation of platelet adhesion by arrest onto fibrinogen or translocation on von Willebrand factor. *Cell, 84,* 289–297.
19. Goto, S., Ikeda, Y., Saldivar, E., & Ruggeri, Z. M. (1998). Distinct mechanisms of platelet aggregation as a consequence of different shearing flow conditions. *J Clin Invest, 101,* 479–486.
20. Cauwenberghs, N., Meiring, M., Vauterin, S., van Wyk, V., Lamprecht, S., Roodt, J. P., Novak, L., Harsfalvi, J., Deckmyn, H., & Kotze, H. F. (2000). Antithrombotic effect of platelet glycoprotein Ib-blocking monoclonal antibody Fab fragments in nonhuman primates. *Arterioscler Thromb Vasc Biol, 20,* 1347–1353.
21. Wu, D., Meiring, M., Kotze, H. F., Deckmyn, H., & Cauwenberghs, N. (2002). Inhibition of platelet glycoprotein Ib, glycoprotein IIb/IIIa, or both by monoclonal antibodies prevents arterial thrombosis in baboons. *Arterioscler Thromb Vasc Biol, 22,* 323–328.
22. Wu, D., Vanhoorelbeke, K., Cauwenberghs, N., Meiring, M., Depraetere, H., Kotze, H. F., & Deckmyn, H. (2002). Inhibition of the von Willebrand (VWF)-collagen interaction by an antihuman VWF monoclonal antibody results in abolition of *in vivo* arterial platelet thrombus formation in baboons. *Blood, 99,* 3623–3628.
23. Lasser, G., Guchhait, P., Ellsworth, J., Sheppard, P., Lewis, K., Bishop, P., Cruz, M. A., López, J. A., & Fruebis, J. (2006). C1q-TNF related protein-1 (CTRP-1), a vascular wall protein that inhibits collagen-induced platelet aggregation by blocking VWF binding to collagen. *Blood. 107,* 423–430.
24. Tzima, E., Irani-Tehrani, M., Kiosses, W. B., Dejana, E., Schultz, D. A., Engelhardt, B., Cao, G., DeLisser, H., & Schwartz, M. A. (2005). A mechanosensory complex that mediates the endothelial cell response to fluid shear stress. *Nature, 437,* 426–431.
25. Fleming, I., Fisslthaler, B., Dixit, M., & Busse, R. (2005). Role of PECAM-1 in the shear-stress-induced activation of

Akt and the endothelial nitric oxide synthase (eNOS) in endothelial cells. *J Cell Sci, 118,* 4103–4111.

26. Rathore, V., Stapleton, M. A., Hillery, C. A., Montgomery, R. R., Nichols, T. C., Merricks, E. P., Newman, D. K., & Newman, P. J. (2003). PECAM-1 negatively regulates GPIb/V/IX signaling in murine platelets. *Blood, 102,* 3658–3664.

27. Falati, S., Patil, S., Gross, P. L., Stapleton, M., Merrill–Skoloff, G., Barrett, N. E., Pixton, K. L., Weiler, H., Cooley, B., Newman, D. K., Newman, P. J., Furie, B. C., Furie, B., & Gibbins, J. M. (2006). Platelet PECAM-1 inhibits thrombus formation in vivo. *Blood, 107,* 535–541.

28. Peterson, D. M., Stathopoulos, N. A., Giorgio, T. D., Hellums, J. D., & Moake, J. L. (1987). Shear-induced platelet aggregation requires von Willebrand factor and platelet membrane glycoproteins Ib and IIb-IIIa. *Blood, 69,* 625–628.

29. Siedlecki, C. A., Lestini, B. J., Kottke–Marchant, K. K., Eppell, S. J., Wilson, D. L., & Marchant, R. E. (1996). Shear-dependent changes in the three dimensional structure of human von Willebrand factor. *Blood, 88,* 2939–2950.

30. Dong, J. F., Berndt, M. C., Schade, A., McIntire, L. V., Andrews, R. K., & López, J. A. (2001). Ristocetin-dependent, but not botrocetin-dependent, binding of von Willebrand factor to the platelet glycoprotein Ib-IX-V complex correlates with shear-dependent interactions. *Blood, 97,* 162–168.

31. Uff, S., Clemetson, J. M., Harrison, T., Clemetson, K. J., & Emsley, J. (2002). Crystal structure of the platelet glycoprotein Ibα N-terminal domain reveals an unmasking mechanism for receptor activation. *J Biol Chem, 277,* 35657–35663.

32. Huizinga, E. G., Tsuji, S., Romijn, R. A., Schiphorst, M. E., de Groot, P. G., Sixma, J. J., & Gros, P. (2002). Structures of glycoprotein Ibα and its complex with von Willebrand factor A1 domain. *Science, 297,* 1176–1179.

33. Dumas, J. J., Kumar, R., McDonagh, T., Sullivan, F., Stahl, M. L., Somers, W. S., & Mosyak, L. (2004). Crystal structure of the wild-type von Willebrand factor A1-glycoprotein Ibα complex reveals conformation differences with a complex bearing von Willebrand disease mutations. *J Biol Chem, 279,* 23327–23334.

34. Shen, Y., Romo, G. M., Dong, J. F., Schade, A., McIntire, L. V., Kenny, D., Whisstock, J. C., Berndt, M. C., López, J. A., & Andrews, R. K. (2000). Requirement of leucine-rich repeats of glycoprotein (GP) Ibα for shear-dependent and static binding of von Willebrand factor to the platelet membrane GPIb-IX-V complex. *Blood, 95,* 903–910.

35. Whisstock, J. C., Shen, Y., López, J. A., Andrews, R. K., & Berndt, M. C. (2002). Molecular modeling of the seven tandem leucine-rich repeats within the ligand-binding region of platelet glycoprotein Ibα. *Thromb Haemostas, 87,* 329–333.

36. Shen, Y., Dong, J. F., Romo, G. M., Arceneaux, W., Aprico, A., Gardiner, E. E., López, J. A., Berndt, M. C., & Andrews, R. K. (2002). Functional analysis of the C-terminal flanking sequence of platelet glycoprotein Ibα using canine-human chimeras. *Blood, 99,* 145–150.

37. Fukuda, K., Doggett, T., Laurenzi, I. J., Liddington, R. C., & Diacovo, T. G. (2005). The snake venom protein botroce-tin acts as a biological brace to promote dysfunctional platelet aggregation. *Nat Struct Mol Biol, 12,* 152–159.

38. Li, C. Q., Dong, J. F., & López, J. A. (2002). The mucin-like macroglycopeptide region of glycoprotein Ibα is required for cell adhesion to immobilized von Willebrand factor (VWF) under flow but not for static VWF binding. *Thromb Haemost, 88,* 673–677.

39. Othman, M., Notley, C., Lavender, F. L., White, H., Byrne, C. D., Lillicrap, D., & O'Shaughnessy, D. F. (2005). Identification and functional characterization of a novel 27-bp deletion in the macroglycopeptide-coding region of the *GPIBA* gene resulting in platelet-type von Willebrand disease. *Blood, 105,* 4330–4336.

40. Chen, J., & López, J. A. (2005). The mysteries of a platelet adhesion receptor. *Blood, 105,* 4154–4155.

41. Jurk, K., Clemetson, K. J., de Groot, P. G., Brodde, M. F., Steiner, M., Savion, N., Varon, D., Sixma, J. J., Van Aken, H., & Kehrel, B. E. (2003). Thrombospondin-1 mediates platelet adhesion at high shear via glycoprotein Ib (GPIb): An alternative/backup mechanism to von Willebrand factor. *FASEB J, 17,* 1490–1492.

42. Bonnefoy, A., Daenens, K., Feys, H. B., De Vos, R., Vandervoort, P., Vermylen, J., Lawler, J., & Hoylaerts, M. F. (2006). Thrombospondin-1 controls vascular platelet recruitment and thrombus adherence in mice by protecting (sub)endothelial VWF from cleavage by ADAMTS-13. *Blood, 107,* 955–964.

43. Savage, B., Ginsberg, M. H., & Ruggeri, Z. M. (1999). Influence of fibrillar collagen structure on the mechanisms of platelet thrombus formation under flow. *Blood, 94,* 2704–2715.

44. Polanowska–Grabowska, R., Gibbins, J. M., & Gear, A. R. (2003). Platelet adhesion to collagen and collagen-related peptide under flow. Roles of the α2β1 integrin, GPVI, and Src tyrosine kinases. *Arterioscler Thromb Vasc Biol, 23,* 1934–1940.

45. Siljander, P. R., Munnix, I. C., Smethurst, P. A., Deckmyn, H., Lindhout, T., Ouwehand, W. H., Farndale, R. W., & Heemskerk, J. W. (2004). Platelet receptor interplay regulates collagen-induced thrombus formation in flowing human blood. *Blood, 103,* 1333–1341.

46. Nieswandt, B., & Watson, S. P. (2003). Platelet-collagen interaction: Is GPVI the central receptor? *Blood, 102,* 449–461.

47. Farndale, R. W., Sixma, J. J., Barnes, M. J., & de Groot, P. G. (2004). The role of collagen in thrombosis and hemostasis. *J Thromb Haemost, 2,* 561–573.

48. Goto, S., Tamura, N., Handa, S., Arai, M., Kodama, K., & Takayama, H. (2002). Involvement of glycoprotein VI in platelet thrombus formation on both collagen and von Willebrand factor surfaces under flow conditions. *Circulation, 106,* 266–272.

49. Massberg, S., Gawaz, M., Gruner, S., Schulte, V., Konrad, I., Zohlnhofer, D., Heinzmann, U., & Nieswandt, B. (2003). A crucial role of glycoprotein VI for platelet recruitment to the injured arterial wall *in vivo. J Exp Med, 197,* 41–49.

50. Sarratt, K. L., Chen, H., Zutter, M. M., Santoro, S. A., Hammer, D. A., & Kahn, M. L. (2005). GPVI and α2β1 play independent critical roles during platelet adhesion and

aggregate formation to collagen under flow. *Blood, 106,* 1268–1277.

51. Arthur, J. F., Gardiner, E. E., Matzaris, M., Taylor, S. G., Wijeyewickrema, L., Ozaki, Y., Kahn, M. L., Andrews, R. K., & Berndt, M. C. (2005). Glycoprotein VI is associated with GPIb-IX-V on the membrane of resting and activated platelets. *Thromb Haemost, 93,* 716–723.

52. Moog, S., Mangin, P., Lenain, N., Strassel, C., Ravanat, C., Schuhler, S., Freund, M., Santer, M., Kahn, M., Nieswandt, B., Gachet, C., Cazenave, J. P., & Lanza, F. (2001). Platelet glycoprotein V binds to collagen and participates in platelet adhesion and aggregation. *Blood, 98,* 1038–1046.

53. Romo, G. M., Dong, J. F., Schade, A. J., Gardiner, E. E., Kansas, G. S., Li, C. Q., McIntire, L. V., Berndt, M. C., & López, J. A. (1999). The glycoprotein Ib-IX-V complex is a platelet counterreceptor for P-selectin. *J Exp Med, 190,* 803–814.

54. Frenette, P. S., Johnson, R. C., Hynes, R. O., & Wagner, D. D. (1995). Platelets roll on stimulated endothelium *in vivo:* An interaction mediated by endothelial P-selectin. *Proc Natl Acad Sci USA, 92,* 7450–7454.

55. Polgar, J., Matuskova, J., & Wagner, D. D. (2005). The P-selectin, tissue factor, coagulation triad. *J Thromb Haemost,* 1590–1596.

56. Katayama, T., Ikeda, Y., Handa, M., Tamatani, T., Sakamoto, S., Ito, M., Ishimura, Y., & Suematsu, M. (2000). Immunoneutralization of glycoprotein Ibα attenuates endotoxin-induced interactions of platelets and leukocytes with rat venular endothelium *in vivo. Circ Res, 86,* 1031–1037.

57. Bombeli, T., Schwartz, B. R., & Harlan, J. M. (1998). Adhesion of activated platelets to endothelial cells: Evidence for a GPIIb-IIIa-dependent bridging mechanism and novel roles for endothelial intercellular adhesion molecule 1 (ICAM-1), $\alpha_v\beta_3$ integrin, and GPIbα. *J Exp Med, 187,* 329–339.

58. del Conde, I., Shrimpton, C. N., Thiagarajan, P., & López, J. A. (2005). Tissue-factor-bearing microvesicles arise from lipid rafts and fuse with activated platelets to initiate coagulation. *Blood, 106,* 1604–1611.

59. Simon, D. I., Chen, Z., Xu, H., Li, C. Q., Dong, J. F., McIntire, L. V., Ballantyne, C. M., Zhang, L., Furman, M. I., Berndt, M. C., & López, J. A. (2000). Platelet glycoprotein Ibα is a counterreceptor for the leukocyte integrin Mac-1 (CD11b/CD18). *J Exp Med, 192,* 193–204.

60. Massberg, S., Brand, K., Gruner, S., Page, S., Muller, E., Muller, I., Bergmeier, W., Richter, T., Lorenz, M., Konrad, I., Nieswandt, B., & Gawaz, M. (2002). A critical role of platelet adhesion in the initiation of atherosclerotic lesion formation. *J Exp Med, 196,* 887–896.

61. Hoffmeister, K. M., Felbinger, T. W., Falet, H., Denis, C. V., Bergmeier, W., Mayadas, T. N., von Andrian, U. H., Wagner, D. D., Stossel, T. P., & Hartwig, J. H. (2003). The clearance mechanism of chilled blood platelets. *Cell, 112,* 87–97.

62. Hoffmeister, K. M., Josefsson, E. C., Isaac, N. A., Clausen, H., Hartwig, J. H., & Stossel, T. P. (2003). Glycosylation restores survival of chilled blood platelets. *Science, 301,* 1531–1534.

63. Wijeyewickrema, L. C., Berndt, M. C., & Andrews, R. K. (2005). Snake venom probes of platelet adhesion receptors and their ligands. *Toxicon, 45,* 1051–1061.

64. Lu, Q., Clemetson, J. M., & Clemetson, K. J. (2005). Snake venoms and hemostasis. *J Thromb Haemost, 3,* 1791–1799.

65. Lu, Q., Clemetson, J. M., & Clemetson, K. J. (2005). Translocation of GPIb and Fc receptor γ-chain to cytoskeleton in mucetin-activated platelets. *J Thromb Haemost, 3,* 2065–2076.

66. Kerrigan, S. W., Douglas, I., Wray, A., Heath, J., Byrne, M. F., Fitzgerald, D., & Cox, D. (2002). A role for glycoprotein Ib in *Streptococcus sanguis*-induced platelet aggregation. *Blood, 100,* 509–516.

67. Bensing, B. A., López, J. A., & Sullam, P. M. (2004). The *Streptococcus gordonii* surface proteins GspB and Hsa mediate binding to sialylated carbohydrate epitopes on the platelet membrane glycoprotein Ibα. *Infect Immun, 72,* 6528–6537.

68. Pawar, P., Shin, P. K., Mousa, S. A., Ross, J. M., & Konstantopoulos, K. (2004). Fluid shear regulates the kinetics and receptor specificity of *Staphylococcus aureus* binding to activated platelets. *J Immunol, 173,* 1258–1265.

69. Ramakrishnan, V., DeGuzman, F., Bao, M., Hall, S. W., Leung, L. L., & Phillips, D. R. (2001). A thrombin receptor function for platelet glycoprotein Ib-IX unmasked by cleavage of glycoprotein V. *Proc Natl Acad Sci USA, 98,* 1823–1828.

70. Celikel, R., McClintock, R. A., Roberts, J. R., Mendolicchio, G. L., Ware, J., Varughese, K. I., & Ruggeri, Z. M. (2003). Modulation of α-thrombin function by distinct interactions with platelet glycoprotein Ibα. *Science, 301,* 218–221.

71. Dumas, J. J., Kumar, R., Seehra, J., Somers, W. S., & Mosyak, L. (2003). Crystal structure of the GPIbα-thrombin complex essential for platelet aggregation. *Science, 301,* 222–226.

72. Adam, F., Bouton, M. C., Huisse, M. G., & Jandrot–Perrus, M. (2003). Thrombin interaction with platelet membrane glycoprotein Ibα. *Trends Mol Med, 9,* 461–464.

73. De Cristofaro, R., & De Candia, E. (2003). Thrombin domains: Structure, function and interaction with platelet receptors. *J Thromb Thrombolysis, 15,* 151–163.

74. Vanhoorelbeke, K., Ulrichts, H., Romijn, R. A., Huizinga, E. G., & Deckmyn, H. (2004). The GPIbα-thrombin interaction: Far from crystal clear. *Trends Mol Med, 10,* 33–39.

75. De Candia, E., Hall, S. W., Rutella, S., Landolfi, R., Andrews, R. K., & De Cristofaro, R. (2001). Binding of thrombin to glycoprotein Ib accelerates the hydrolysis of PAR-1 on intact platelets. *J Biol Chem, 276,* 4692–4698.

76. Yun, T. H., Baglia, F. A., Myles, T., Navaneetham, D., López, J. A., Walsh, P. N., & Leung, L. L. (2003). Thrombin activation of factor XI on activated platelets requires the interaction of factor XI and platelet glycoprotein Ibα with thrombin anion-binding exosites I and II, respectively. *J Biol Chem, 278,* 48112–48119.

77. Beguin, S., Keularts, I., Al Dieri, R., Bellucci, S., Caen, J., & Hemker, H. C. (2004). Fibrin polymerization is crucial for thrombin generation in platelet-rich plasma in a VWF-GPIb-dependent process, defective in Bernard–Soulier syndrome. *J Thromb Haemost, 2,* 170–176.

78. Soslau, G., & Favero, M. (2004). The GPIb-thrombin pathway: Evidence for a novel role of fibrin in platelet aggregation. *J Thromb Haemost, 2,* 522–524.

79. Baglia, F. A., Badellino, K. O., Li, C. Q., López, J. A., & Walsh, P. N. (2002). Factor XI binding to the platelet glycoprotein Ib-IX-V complex promotes factor XI activation by thrombin. *J Biol Chem, 277*, 1662–1668.

80. Baglia, F. A., Shrimpton, C. N., Emsley, J., Kitagawa, K., Ruggeri, Z. M., López, J. A., & Walsh, P. N. (2004). Factor XI interacts with the leucine-rich repeats of glycoprotein Ibα on the activated platelet. *J Biol Chem, 279*, 49323–49329.

81. Baglia, F. A., Shrimpton, C. N., López, J. A., & Walsh, P. N. (2003). The glycoprotein Ib-IX-V complex mediates localization of factor XI to lipid rafts on the platelet membrane. *J Biol Chem, 278*, 21744–21750.

82. Joseph, K., Nakazawa, Y., Bahou, W. F., Ghebrehiwet, B., & Kaplan, A. P. (1999). Platelet glycoprotein Ib: A zinc-dependent binding protein for the heavy chain of high-molecular-weight kininogen. *Mol Med, 5*, 555–563.

83. Chavakis, T., Santoso, S., Clemetson, K. J., Sachs, U. J., Isordia–Salas, I., Pixley, R. A., Nawroth, P. P., Colman, R. W., & Preissner, K. T. (2003). High molecular weight kininogen regulates platelet–leukocyte interactions by bridging Mac-1 and glycoprotein Ib. *J Biol Chem, 278*, 45375–45381.

84. Bradford, H. N., Pixley, R. A., & Colman, R. W. (2000). Human factor XII binding to the glycoprotein Ib-IX-V complex inhibits thrombin-induced platelet aggregation. *J Biol Chem, 275*, 22756–22763.

85. Andrews, R. K., Gardiner, E. E., Shen, Y., & Berndt, M. C. (2004). Platelet interactions in thrombosis. *IUBMB Life, 56*, 13–18.

86. Baker, J., Griggs, R. K., Falati, S., & Poole, A. W. (2004). GPIb potentiates GPVI-induced responses in human platelets. *Platelets, 15*, 207–214.

87. Andrews, R. K., Suzuki–Inoue, K., Shen, Y., Tulasne, D., Watson, S. P., & Berndt, M. C. (2002). Interaction of calmodulin with the cytoplasmic domain of platelet glycoprotein VI. *Blood, 99*, 4219–4221.

88. Suzuki–Inoue, K., Tulasne, D., Bori–Sanz, T., Inoue, O., Jung, S. M., Moroi, M., Shen, Y., Andrews, R. K., Berndt, M. C., & Watson, S. P. (2002). Association of fyn and lyn with the proline rich domain of GPVI regulates intracellular signalling. *J Biol Chem, 277*, 21561–21566.

89. Locke, D., Liu, C., Peng, X., Chen, H., & Kahn, M. L. (2003). Fc Rγ-independent signaling by the platelet collagen receptor glycoprotein VI. *J Biol Chem, 278*, 15441–15448.

90. Wu, Y., Suzuki–Inoue, K., Satoh, K., Asazuma, N., Yatomi, Y., Berndt, M. C., & Ozaki, Y. (2001). Role of Fc receptor γ-chain in platelet glycoprotein Ib-mediated signaling. *Blood, 97*, 3836–3845.

91. Sullam, P. M., Hyun, W. C., Szollosi, J., Dong, J. F., Foss, W. M., & López, J. A. (1998). Physical proximity and functional interplay of the glycoprotein Ib-IX-V complex and the Fc receptor FcγRIIA on the platelet plasma membrane. *J Biol Chem, 273*, 5331–5336.

92. Sun, B., Li, J., & Kambayashi, J. (1999). Interaction between GPIbα and FcγIIA receptor in human platelets. *Biochem Biophys Res Commun, 266*, 24–27.

93. Shrimpton, C. N., Borthakur, G., Larrucea, S., Cruz, M. A., Dong, J. F., & López, J. A. (2002). Localization of the adhesion receptor glycoprotein Ib-IX-V complex to lipid rafts is

94. required for platelet adhesion and activation. *J Exp Med, 196*, 1057–1066.

94. López, J. A., del Conde, I., & Shrimpton, C. N. (2005). Receptors, rafts, and microvesicles in thrombosis and inflammation. *J Thromb Haemost, 3*, 1737–1744.

95. Wee, J. L., & Jackson, D. E. (2005). The Ig-ITIM superfamily member, PECAM-1 regulates the "outside-in" signalling properties of integrin αIIbβ3 in platelets. *Blood, 106*, 3816–3823.

96. Thai le, M., Ashman, L. K., Harbour, S. N., Hogarth, P. M., & Jackson, D. E. (2003). Physical proximity and functional interplay of PECAM-1 with the Fc receptor FcγRIIa on the platelet plasma membrane. *Blood, 102*, 3637–3645.

97. Wong, M. X., Harbour, S. N., Wee, J. L., Lau, L. M., Andrews, R. K., & Jackson, D. E. (2004). Proteolytic cleavage of platelet endothelial cell adhesion molecule-1 (PECAM-1/CD31) is regulated by a calmodulin-binding motif. *FEBS Lett, 568*, 70–78.

98. Cruz, M. A., Chen, J., Whitelock, J. L., Morales, L. D., & López, J. A. (2005). The platelet glycoprotein Ib–von Willebrand factor interaction activates the collagen receptor α2β1 to bind collagen: Activation-dependent conformational change of the α2-I domain. *Blood, 105*, 1986–1991.

99. Locke, D., Chen, H., Liu, Y., Liu, C., & Kahn, M. L. (2002). Lipid rafts orchestrate signaling by the platelet receptor glycoprotein VI. *J Biol Chem, 277*, 18801–18809.

100. Ezumi, Y., Kodama, K., Uchiyama, T., & Takayama, H. (2002). Constitutive and functional association of the platelet collagen receptor glycoprotein VI-Fc receptor γ-chain complex with membrane rafts. *Blood, 99*, 3250–3255.

101. Bodin, S., Tronchere, H., & Payrastre, B. (2003). Lipid rafts are critical membrane domains in blood platelet activation processes. *Biochim Biophys Acta, 1610*, 247–257.

102. Bodin, S., Soulet, C., Tronchere, H., Sie, P., Gachet, C., Plantavid, M., & Payrastre, B. (2005). Integrin-dependent interaction of lipid rafts with the actin cytoskeleton in activated human platelets. *J Cell Sci, 118*, 759–769.

103. Gu, M., Xi, X., Englund, G. D., Berndt, M. C., & Du, X. (1999). Analysis of the roles of 14-3-3 in the platelet glycoprotein Ib-IX-mediated activation of integrin αIIbβ3 using a reconstituted mammalian cell expression model. *J Cell Biol, 147*, 1085–1096.

104. Dai, K., Bodnar, R., Berndt, M. C., & Du, X. (2005). A critical role for 14-3-3ζ protein in regulating the VWF binding function of platelet glycoprotein Ib-IX and its therapeutic implications. *Blood, 106*, 1975–1981.

105. Feng, S., Christodoulides, N., Resendiz, J. C., Berndt, M. C., & Kroll, M. H. (2000). Cytoplasmic domains of GPIbα and GPIbβ regulate 14-3-3ζ binding to GPIb/IX/V. *Blood, 95*, 551–557.

106. Bodnar, R. J., Xi, X., Li, Z., Berndt, M. C., & Du, X. (2002). Regulation of glycoprotein Ib-IX-von Willebrand factor interaction by cAMP-dependent protein kinase-mediated phosphorylation at Ser166 of glycoprotein Ibβ. *J Biol Chem, 277*, 47080–47087.

107. Perrault, C., Mangin, P., Santer, M., Baas, M. J., Moog, S., Cranmer, S. L., Pikovski, I., Williamson, D., Jackson, S. P., Cazenave, J. P., & Lanza, F. (2003). Role of the intracellular domains of GPIb in controlling the adhesive

properties of the platelet GPIb/V/IX complex. *Blood, 101,* 3477–3484.

108. Kanaji, T., Russell, S., Cunningham, J., Izuhara, K., Fox, J. E., & Ware, J. (2004). Megakaryocyte proliferation and ploidy regulated by the cytoplasmic tail of glycoprotein Ibα. *Blood, 104,* 3161–3168.

109. Andrews, R. K., Munday, A. D., Mitchell, C. A., & Berndt, M. C. (2001). Interaction of calmodulin with the cytoplasmic domain of the glycoprotein Ib-IX-V complex. *Blood, 98,* 681–687.

110. Gardiner, E. E., Arthur, J. F., Berndt, M. C., & Andrews, R. K. (2005). Role of calmodulin in platelet receptor function. *Curr Med Chem Cardiovasc Hematol Agents, 3,* 173–179.

111. Gardiner, E. E., Arthur, J. F., Kahn, M. L., Berndt, M. C., & Andrews, R. K. (2004). Regulation of platelet membrane levels of glycoprotein VI by a platelet-derived metalloproteinase. *Blood, 104,* 3611–3617.

112. Boylan, B., Chen, H., Rathore, V., Paddock, C., Salacz, M., Friedman, K. D., Curtis, B. R., Stapleton, M., Newman, D. K., Kahn, M. L., & Newman, P. J. (2004). Anti-GPVI-associated ITP: An acquired platelet disorder caused by autoantibody-mediated clearance of the GPVI/FcRγ-chain complex from the human platelet surface. *Blood, 104,* 1350–1355.

113. Chen, H., Locke, D., Liu, Y., Liu, C., & Kahn, M. L. (2002). The platelet receptor GPVI mediates both adhesion and signaling responses to collagen in a receptor density-dependent fashion. *J Biol Chem, 277,* 3011–3019.

114. Thattaliyath, B., Cykowski, M., & Jagadeeswaran, P. (2005). Young thrombocytes initiate the formation of arterial thrombi in zebrafish. *Blood, 106,* 118–124.

115. Rabie, T., Strehl, A., Ludwig, A., & Nieswandt, B. (2005). Evidence for a role of ADAM17 (TACE) in the regulation of platelet glycoprotein V. *J Biol Chem, 280,* 14462–14468.

116. Fox, J. E. (1994). Shedding of adhesion receptors from the surface of activated platelets. *Blood Coagul Fibrinolysis, 5,* 291–304.

117. Bergmeier, W., Piffath, C. L., Cheng, G., Dole, V. S., Zhang, Y., von Andrian, U. H., & Wagner, D. D. (2004). Tumor necrosis factor-α-converting enzyme (ADAM17) mediates GPIbα shedding from platelets *in vitro* and *in vivo. Circ Res, 95,* 677–683.

118. Coller, B. S., Kalomiris, E., Steinberg, M., & Scudder, L. E. (1984). Evidence that glycocalicin circulates in normal plasma. *J Clin Invest, 73,* 794–799.

119. Munday, A. D., Berndt, M. C., & Mitchell, C. A. (2000). Phosphoinositide 3-kinase forms a complex with platelet membrane glycoprotein Ib-IX-V complex and 14-3-3ζ. *Blood, 96,* 577–584.

120. Jackson, S. P., Nesbitt, W. S., & Kulkarni, S. (2003). Signaling events underlying thrombus formation. *J Thromb Haemost, 1,* 1602–1612.

121. Jackson, S. P., Yap, C. L., & Anderson, K. E. (2004). Phosphoinositide 3-kinases and the regulation of platelet function. *Biochem Soc Trans, 32,* 387–392.

122. Ozaki, Y., Asazuma, N., Suziki–Inoue, K., & Berndt, M. C. (2005). Platelet GPIb-IX-V-dependent signaling. *J Thromb Haemost, 3,* 1745–1751.

123. Jackson, S. P., Schoenwaelder, S. M., Goncalves, I., Nesbitt, W. S., Yap, C. L., Wright, C. E., Kenche, V., Anderson, K. E., Dopheide, S. M., Yuan, Y., Sturgeon, S. A., Prabaharan, H., Thompson, P. E., Smith, G. D., Shepherd, P. R., Daniele, N., Kulkarni, S., Abbott, B., Saylik, D., Jones, C., Lu, L., Giuliano, S., Hughan, S. C., Angus, J. A., Robertson, A. D., & Salem, H. H. (2005). PI 3-kinase p110β: A new target for antithrombotic therapy. *Nat Med, 11,* 507–514.

124. Fox, J. E. (1993). The platelet cytoskeleton. *Thromb Haemost, 70,* 884–893.

125. Englund, G. D., Bodnar, R. J., Li, Z., Ruggeri, Z. M., & Du, X. (2001). Regulation of von Willebrand factor binding to the platelet glycoprotein Ib-IX by a membrane skeleton-dependent inside-out signal. *J Biol Chem, 276,* 16952–16959.

126. Williamson, D., Pikovski, I., Cranmer, S. L., Mangin, P., Mistry, N., Domagala, T., Chehab, S., Lanza, F., Salem, H. H., & Jackson, S. P. (2002). Interaction between platelet glycoprotein Ibα and filamin-1 is essential for glycoprotein Ib/IX receptor anchorage at high shear. *J Biol Chem, 277,* 2151–2159.

127. Schade, A. J., Arya, M., Gao, S., Diz–Kucukkaya, R., Anvari, B., McIntire, L. V., López, J. A., & Dong, J. F. (2003). Cytoplasmic truncation of glycoprotein Ibα weakens its interaction with von Willebrand factor and impairs cell adhesion. *Biochemistry, 42,* 2245–2251.

128. Feng, S., Resendiz, J. C., Lu, X., & Kroll, M. H. (2003). Filamin A binding to the cytoplasmic tail of glycoprotein Ibα regulates von Willebrand factor-induced platelet activation. *Blood, 102,* 2122–2129.

129. Kanaji, T., Russell, S., & Ware, J. (2002). Amelioration of the macrothrombocytopenia associated with the murine Bernard–Soulier syndrome. *Blood, 100,* 2102–2107.

130. De Marco, L., Mazzucato, M., Fabris, F., De Roia, D., Coser, P., Girolami, A., Vicente, V., & Ruggeri, Z. M. (1990). Variant Bernard–Soulier syndrome type Bolzano. A congenital bleeding disorder due to a structural and functional abnormality of the platelet glycoprotein Ib-IX complex. *J Clin Invest, 86,* 25–31.

131. Meyer, S. C., Sanan, D. A., & Fox, J. E. (1998). Role of actin-binding protein in insertion of adhesion receptors into the membrane. *J Biol Chem, 273,* 3013–3020.

132. Feng, S., Lu, X., & Kroll, M. H. (2005). Filamin A binding stabilizes nascent glycoprotein Ibα trafficking and thereby enhances its surface expression. *J Biol Chem, 280,* 6709–6715.

133. Dyson, J. M., Munday, A. D., Kong, A. M., Huysmans, R. D., Matzaris, M., Layton, M. J., Nandurkar, H. H., Berndt, M. C., & Mitchell, C. A. (2003). SHIP-2 forms a tetrameric complex with filamin, actin, and GPIb-IX-V: Localization of SHIP-2 to the activated platelet actin cytoskeleton. *Blood, 102,* 940–948.

134. Canobbio, I., Balduini, C., & Torti, M. (2004). Signalling through the platelet glycoprotein Ib-IX-V complex. *Cell Signalling, 16,* 1329–1344.

135. Gibbins, J. M. (2004). Platelet adhesion signalling and the regulation of thrombus formation. *J Cell Sci, 117,* 3415–3425.

136. Liu, J., Pestina, T., Berndt, M. C., Jackson, C. W., & Gartner, T. K. (2005). Botrocetin/vWf-induced signaling through GPIb-IX-V that produces TXA$_2$ in an $\alpha_{IIb}\beta_3$ and aggregation-independent manner. *Blood, 106,* 2750–2756.

137. Garcia, A., Quinton, T. M., Dorsam, R. T., & Kunapuli, S. P. (2005). Src family kinase and Erk-mediated thromboxane A$_2$ generation is essential for VWF/GPIb-induced fibrinogen receptor activation in human platelets. *Blood, 106,* 3410–3414.

138. Li, Z., Xi, X., Gu, M., Feil, R., Ye, R. D., Eigenthaler, M., Hofmann, F., & Du, X. (2003). A stimulatory role for cGMP-dependent protein kinase in platelet activation. *Cell, 112,* 77–86.

139. Marshall, S. J., Senis, Y. A., Auger, J. M., Feil, R., Hofmann, F., Salmon, G., Peterson, J. T., Burslem, F., & Watson, S. P. (2004). GPIb-dependent platelet activation is dependent on Src kinases but not MAP kinase or cGMP-dependent kinase. *Blood, 103,* 2601–2609.

140. Walter, U., & Gambaryan S. (2004). Roles of cGMP/cGMP-dependent protein kinase in platelet activation. *Blood, 104,* 2609.

141. Du, X., Marjanovic, J. A., & Li, Z. (2004). On the roles of cGMP and glycoprotein Ib in platelet activation. *Blood, 103,* 4371–4372.

142. Li, Z., Zhang, G., Marjanovic, J. A., Ruan, C., & Du, X. (2004). A platelet secretion pathway mediated by cGMP-dependent protein kinase. *J Biol Chem, 279,* 42469–42475.

143. Schulte am Esch, II, J., Cruz, M. A., Siegel, J. B., Anrather, J., & Robson, S. C. (1997). Activation of human platelets by the membrane-expressed A1 domain of von Willebrand factor. *Blood, 90,* 4425–4437.

144. Kasirer–Friede, A., Cozzi, M. R., Mazzucato, M., De Marco, L., Ruggeri, Z. M., & Shattil, S. J. (2004). Signaling through GP Ib-IX-V activates $\alpha_{IIb}\beta_3$ independently of other receptors. *Blood, 103,* 3403–3411.

# CHAPTER 8

# Integrin αIIbβ3

**Edward F. Plow, Michelle M. Pesho, and Yan-Qing Ma**

*Department of Molecular Cardiology, Cleveland Clinic Foundation, Cleveland, Ohio*

## I. Introduction

Platelets contribute to the normal circulation of blood through the preservation of vascular integrity and prevention of hemorrhage after injury. Although the role of platelets in thrombus formation is a necessary defense mechanism, excessive thrombus formation on unstable atherosclerotic plaques can obstruct blood flow and lead to acute myocardial infarction or stroke.[1-3] Therefore, platelet-induced thrombus formation is both a pivotal physiological and pathological response. The capacity of platelets to form a thrombus depends on their ability to aggregate. At a molecular level, platelet aggregation is mediated by a specific receptor on the platelet surface, αIIbβ3 (also known as glycoprotein [GP] IIb-IIIa), a member of the integrin family.[4,5] Central to the function of integrin αIIbβ3 is its capacity to undergo activation, a transition from a low-affinity state (resting state) to a high-affinity state (active state) for its extracellular ligands. This transformation of αIIbβ3 allows it to bind divalent fibrinogen or multivalent von Willebrand factor (VWF),[6,7] which can act as bridging molecules between platelets to form aggregates. Recognition of other ligands, notably vitronectin, fibronectin, and thrombospondin,[8-10] by αIIbβ3 may help to mediate platelet adhesion to the subendothelial matrix and to regulate platelet aggregation. The key role of αIIbβ3 in mediating platelet aggregation, and the demonstration many years ago that blockade of this function inhibits this response,[11,12] defined this receptor as a potential target for antithrombotic therapy. Indeed, this therapeutic strategy has come to realization. Inhibition of platelet aggregation using intravenously administered antibody, peptide, and nonpeptide derivatives as αIIbβ3 (GPIIb-IIIa) antagonists has been accepted clinically and is used broadly to treat and prevent thrombosis in the settings of percutaneous coronary intervention and acute coronary syndromes[13] (see Chapter 62). Considerable information on αIIbβ3 has evolved from numerous basic and clinical investigations, but a full understanding of its functions is far from resolved. αIIbβ3 is the focus of this chapter, with a primary emphasis on its structure and activation. The molecular events that regulate activation of αIIbβ3 have been termed collectively *inside-out signaling*. The *outside-in signaling* that results from ligand binding to αIIbβ3 is considered separately in Chapter 17.

## II. αIIbβ3 as an Integrin and a Platelet Protein

Integrins are a family of adhesion molecules that mediate cell–cell and cell–matrix interactions. The integrin family is the most diverse of the cell adhesion receptors: there are 18 different α-subunits and eight different β-subunits that associate noncovalently to form 24 different integrins.[14] However, there are only two β3 integrins: αIIbβ3 and αVβ3. They share the same β-subunit, and the α-subunit sequences are 36% identical.[15] Both αIIbβ3 and αVβ3 are expressed on platelets.[16,17] Although αIIbβ3 is restricted to platelets, megakaryocytes, and some tumor cells,[18-20] αVβ3 is expressed more widely, including endothelial cells, certain leukocytes, smooth muscle cells, platelets, and many tumor cells.[21]

On platelets, αIIbβ3 is the major integral plasma membrane protein. It represents 3% of the total protein of the cells and 17% of the platelet membrane protein mass.[22] Altogether, there are 80,0000 to 100,000 copies of αIIbβ3 per platelet.[23-25] Platelet α-granule membranes also contain αIIbβ3, and this pool can become accessible and is functional upon platelet stimulation.[26,27]

## III. Structure of αIIbβ3

Like all integrins, αIIbβ3 is an α/β heterodimer in which the two subunits associate noncovalently to form a 1 : 1 complex.[28] Each subunit is the product of a separate gene on chromosome 17, where the two genes are adjacent at q21-23.[29] Although both subunits are synthesized as single glycosylated polypeptide chains, αIIb undergoes proteolytic cleavage into a heavy and light chain, which remain disulfide linked.[30] Such processing occurs in the α-subunits of several integrins but not in the αV subunit, and processing is not necessary for the function of αIIbβ3.[31] As expressed on the platelet surface, αIIb, also known as GPIIb, is composed of 1008 amino acids.[32] The mature β3 subunit, also known as GPIIIa, contains 762 amino acids.[33,34] Each subunit consists of a large extracellular region, transmembrane region, and a short cytoplasmic tail,[33–36] and is in a type I orientation with the N-terminus in the extracellular domain and the C-terminus in the cytoplasmic domain. In αIIb, it is the light chain that contains the transmembrane and cytoplasmic tail.

## A. Extracellular Domain

The following description is based on the crystal structures of αIIbβ3 and αVβ3.[35,36] The individual domains that compose each subunit are shown in Figure 8-1. The β-propeller domain of the α-subunit contains the N-terminus and is composed of a series of amino acid repeats that are distributed into seven "blades."[35] There are four divalent ion-binding sites within these blades. The cation binding sites are formed from oxygens within aspartic acids and asparagines as well as backbone oxygens, and the bound cations help maintain rigidity in the interface between the β-propeller and the thigh domain. The thigh, calf-1, and calf-2 domains, which constitute the remaining portion of the extracellular domain of the α-subunit, are arranged primarily into a β-sheet. There is a "genu" located in a small area between the thigh and the calf-1 domains. The genu is highly acidic and also coordinates a calcium ion. The large hydrophobic interface between calf-1 and calf-2, which spans hundreds of angstroms, aids in structural rigidity.

The N-terminal portion of the β-subunit is comprised of the β3 A-domain that contains six β-sheets encircled by eight α-helices. The β3 A-domain contains three metal sites for binding ligands, as shown in Figure 8-2.[37,38] The metal ion adhesion site, MIDAS, is located at the top of the central β-strand and is formed by Asp119, Ser121, Ser123, Glu220, and Asp251. Adjacent to the MIDAS is the ADMIDAS, which coordinates a metal ion through Ser123, Asp126, Asp127, and Asp251. The ligand-induced metal binding site, LIMBS, is formed by Asp158, Asn215, Asp217, Pro219, and Glu220. The β A-domain and the hybrid domain have several contact points with each other. The interface is a large area with both hydrophilic and hydrophobic residues. The hybrid domain structure resembles the I-set immunoglobulin domains.[39] The PSI domain connects the hybrid to the EGF-1 domain. There are four EGF domains rich in cysteines and three disulfide bonds. These domains are rod shaped and connected by disulfide bridges. There are a few

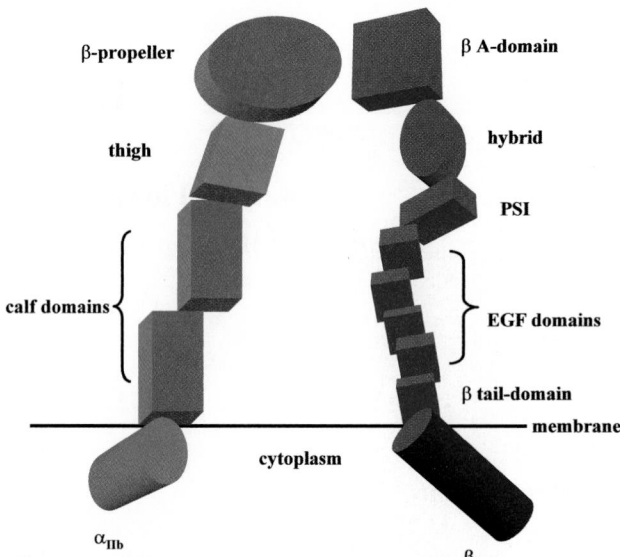

**Figure 8-1.** Cartoon depicting the domains of the αIIb (green) and β3 (blue) subunits in an extended conformation.

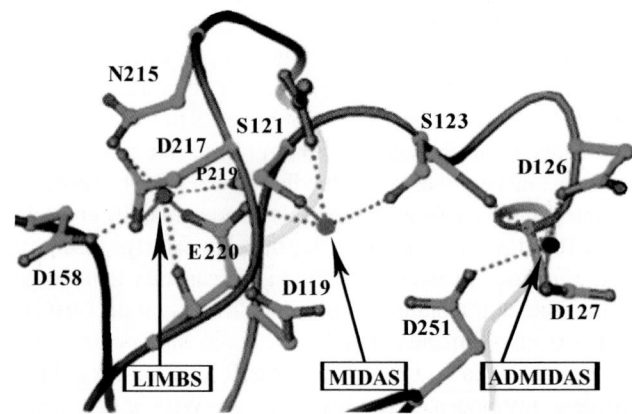

**Figure 8-2.** The β3 A-domain metal sites. The amino acid residues involved in binding divalent cations are labeled by the single letter code. Carbon atoms are shown in green, nitrogens in blue, and oxygens in red. From left to right, the metal sites are the LIMBS, MIDAS, and ADMIDAS, and their bound cations are shown as gray, green, and purple spheres respectively. (Modified from Xiong et al.,[38] with permission.)

weak interactions linking the EGF-4 and the last domain of the β3 subunit, the β-tail domain. The β-tail domain is membrane proximal and contains a four-stranded β-sheet. A genu also exists in the β3 subunit between the hybrid, 2 EGF, and PSI domains. The presence of a genu within both subunits allows the receptor to transit between a bent and an extended conformation, a transition that may be important in the activation of the receptor.

The main area of contact between the α- and β-subunits involves the propeller and β A-domain, respectively. Arg261 of the β A-domain extends into the propeller's core and is

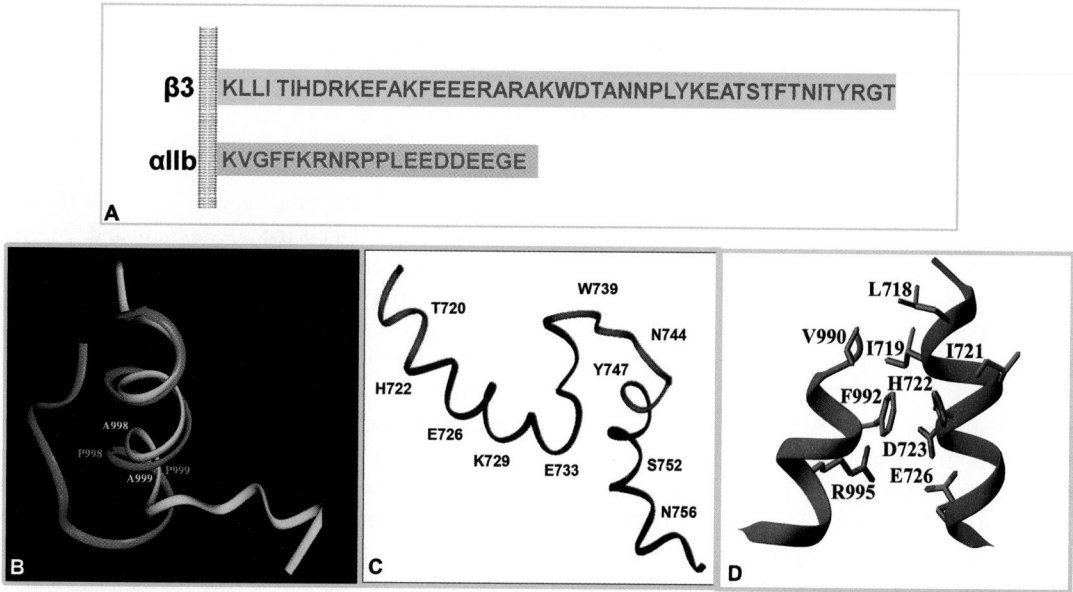

**Figure 8-3.** The amino acid sequences and structures of the individual αIIb and β3 cytoplasmic tails and their complex. A. The amino acid sequences of αIIb and β3 tails. B, C. The structures of αIIb and β3 tails respectively in DPC micelles. The C-terminal turn structure of the αIIb tail is perturbed by PP/AA mutation. D. Multiple contacts formed between the membrane-proximal helices of αIIb and β3 tails in aqueous solution. (Panels B and C from Vinogradova et al.,[42,44] with permission. Panel D from Vinogradova et al.,[43] with permission.)

confined by two groupings of aromatic amino acids. The upper grouping consists of a serine, a phenylalanine, and three tyrosine residues. The bottom grouping contains three phenylalanine and two tyrosine residues that contact Arg261 directly. The propeller–β A-domain interface has several additional hydrophobic and ionic contacts that maintain the large surface of interaction. The α- and β-subunits also have additional contacts between the propeller, thigh, and calf domains of the α-subunit, with several of the EGF domains in the β-subunit. However, in the crystal structure these contacts appear minimal and sporadic, and may not exist in the native receptor.

Electron microscopy of αIIbβ3 revealed the overall structure is assembled into a globular "head" with two protruding "stalks."[40,41] The "head" consists of the β-propeller domain from the α-subunit and the A-domain and hybrid domain from the β-subunit.[35] The α-"stalk" consists of the thigh and two calf domains, whereas the β-"stalk" contains the PSI, the four EGF domains, and the β-tail domain. In addition to this extended conformation, electron microscopy also revealed more compact organizations for the extracellular domain, which is discussed later.

## B. Cytoplasmic Domain

The structures of the cytoplasmic tails (CT) of the αIIb and β3 subunits were determined by nuclear magnetic resonance

(NMR).[42–45] The αIIb tail contains 20 amino acid residues with an α-helix in the positively charged N-terminal region that is highly conserved among integrin α-subunits (Fig. 8-3A). The peptide forms a turn at its Asn-Arg-Pro-Pro (NRPP) sequence that allows the negatively charged C-terminal loop to contact the N-terminal region (Fig. 8-3B). Electrostatic interactions via salt bridges occur between the N- and C-termini, providing structural stability. A calcium binding site in the C-terminal created by Arg997, Glu1001, Asp1003, and Asp1004 also helps to stabilize the structure. The β3 tail consists of 47 amino acids with a long α-helical structure emanating from the membrane, a brief interrupting loop, and a second shorter α-helix in the N-terminal (Fig. 8-3C). The Asn-Pro-Leu-Tyr (NPLY) sequence, residues 744 to 747, within the turn is found broadly in transmembrane proteins,[46] where it forms a similar structure. The extreme C-terminus contains a sequence, NITY, that is also capable of forming a turn, but this structure has not been detected.

The CT of αIIb and β3 interact with each other through their membrane-proximal helical regions. The contacts between the α-helices are both electrostatic and hydrophobic. The hydrophobic residues involved are valines, leucines, isoleucines, and phenylalanines. The electrostatic interactions are salt bridges involving arginines and aspartic acids or glutamic acids. Perturbation of the CT complex maintained through these interactions is central to the mechanism of αIIbβ3 activation and is discussed later in Section IV.C.

The structures of the transmembrane domains of the αIIb and β3 subunits have not been solved, but the sequences of these 23 amino acid stretches are compatible with the formation of typical transmembrane helices. Furthermore, the helices are likely to interact. Mutational analyses and cross-linking approaches support the occurrence of such interhelical interactions.[47] Thus, cobbling together direct structural information and predictive analyses, a full picture of the structure of αIIbβ3 has emerged.

## C. Ligand Binding

αIIbβ3 can bind multiple ligands, including the extracellular matrix proteins fibrinogen, fibronectin, vitronectin, thrombospondin-1, and VWF.[9] αIIbβ3 recognizes a simple peptide sequence, Arg-Gly-Asp (RGD), which is present in many αIIbβ3 ligands and is also recognized by several other integrins.[48,49] Both fibrinogen and VWF, the ligands that directly support platelet aggregation, contain RGD sequences, and their binding to the receptor is inhibited by small RGD-containing peptides. Other integrins also recognize fibrinogen by binding to its RGD sequences.[50–52] However, these observations do not indicate that αIIbβ3 recognizes the RGD sequences in both ligands and that both ligands bind to the same site in αIIbβ3. Indeed, the RGD sequences are not the sites that αIIbβ3 uses to engage soluble fibrinogen, and some studies suggest that fibrinogen and VWF interact with different sites in the receptor.[53–55] The basis for what appears to be contradictory statements is the presence of an alternative and preferred binding site within fibrinogen for αIIbβ3. Fibrinogen contains a Lys-Gln-Ala-Gly-Asp-Val (KQAGDV) sequence at the C-terminal of its γ-chain. It is this sequence, rather than either of the two RGD sequences in its Aα-chain, that is the primary mediator of its binding to αIIbβ3. However, it is the RGD sequence in the C-terminal of its Aα-chain that is recognized by αvβ3, establishing a noteworthy distinction in the specificity of the two β3 integrins.[51,56,57]

Based on the crystal structure of αvβ3, the main site of ligand interaction lies between the seven-bladed β-propeller of the αv subunit and the A-like domain of the β3 subunit.[35] The crystal structure of αvβ3 in complex with an RGD ligand,[37] as well as subsequent structures of various antagonists bound to αIIbβ3,[36] support the involvement of these same two domains as the major contact sites for ligand. Metal ions are required for the binding of fibrinogen and VWF to the β3 integrins, for the binding of most ligands to integrins, and for the aggregation response of platelets.[58–60] The MIDAS bound cation is involved in coordinating to an aspartic acid in the bound ligand, the ADMIDAS cation plays a role in regulating ligand binding, and the LIMBS cation helps to stabilize the ligand-receptor complex.[38]

Based upon differences in the crystal structures of the unoccupied, ligand occupied, and receptor constrained within a resting state, it is surmised that residues within the cation binding sites undergo extensive movement during these transitions. These movements include changes in the position of ligand and metal binding β1-α1 loop, the α1-helix and several other key loops,[36] and the α7-helix. The latter helix connects the β A-domain and the hybrid domain, and movements of the domains are deduced from differences in the crystal structures of the unoccupied, which remodel the interface between these two domains. Such changes are believed to create a pathway for the transition of conformational changes that lead to outside-in signaling associated with ligand binding.

The genu within the αIIb and β3 subunits allows the receptor to exist in two extreme conformations: a bent state or an extended state.[61] Most electron micrographs of αIIbβ3 with bound fibrinogen are in an extended conformation with the ligand bound to the headpiece.[41] Crystal structures of αvβ3 showed a bent conformation.[35] These difference led to the hypothesis that activation of the integrins occurs through a "switchblade" motion in which the headpiece moves away from the membrane and allows for access of large ligands to their binding sites within the headpiece.[62,63] This extension is driven by a downward movement of the β A-domain α7-helix that prompts the hybrid and PSI domains to move away from the α-subunit, resulting in a separation between the α- and β-"stalks."[36,62] These two extreme states, as well as an intermediate state that has been observed by electron microscopy, are believed to correspond to different activation states of αIIbβ3. In the bent conformation, αIIbβ3 is in a low-affinity resting state; in an extended form, but with a closed headpiece, αIIbβ3 is in an intermediate-affinity state; and in an extended form with open headpiece, αIIbβ3 is in a high-affinity state for ligand.[36,61] However, electron microscopic studies of the extracellular domain of αvβ3 in complex with a large ligand was observed in a bent, rather than an extended, conformation,[64] a challenge to the notion that the switchblade extension is necessary for ligand engagement.

# IV. "Inside-Out" Signaling and αIIbβ3 Activation

The changes in conformation that occur in the extracellular domain of αIIbβ3 enhance its affinity for soluble ligand, affinity modulation. Affinity modulation is distinguished from avidity modulation, in which clustering of integrins enhances and stabilizes interactions with ligand. Both affinity and avidity modulation occur during αIIbβ3 activation,[65,66] and these events are initiated when platelets are stimulated with any one or more of several agonists. The changes in integrin receptors, including affinity/avidity

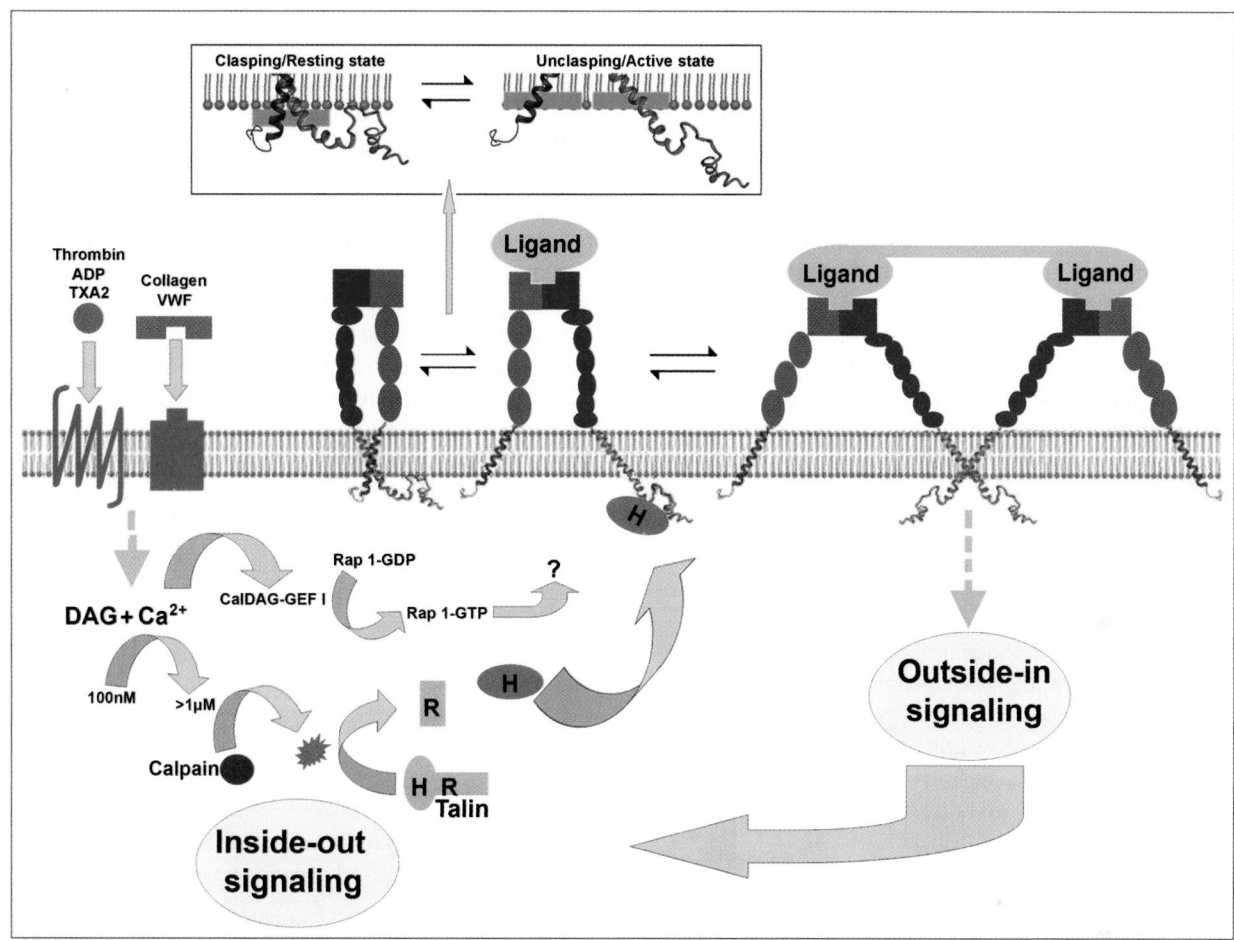

**Figure 8-4.** Inside-out signaling and clasping/unclasping model. Inside-out signaling can be induced by different agonists, which initiate different pathways that lead to the cytoplasmic tails (CTs) of αIIbβ3. Talin-H (H) is one of the potential candidates to bind directly to the β3-tail and unclasp the membrane-proximal complex of αIIb and β3 CT. This unclasping triggers a conformational switch in the integrin extracellular domain, resulting in its conversion from a resting to an activated state, in which it is competent to bind soluble ligand. The subsequent outside-in signaling initiated by ligand binding further propagates and enhances the inside-out signaling. The clasping/unclasping model and potential membrane insertion of membrane-proximal regions of αIIb and β3 CT during inside-out signaling are shown in greater detail in the insert. (Portions of this figure adapted from Vinogradova et al.[44] and Qin,[75] with permission.)

modulation of the receptors *per se* and the events that begin with agonist stimulation of the cell are referred to collectively as *inside-out signaling.*[67-71] After ligand binds to αIIbβ3, "outside-in" signaling is initiated. This process is initiated by the conformational changes that are induced by ligand binding,[36,72,73] which, in turn, change the CTs such that they acquire sites for interaction with cytoskeletal and signaling proteins, such as tyrosine and serine/threonine kinases[69,70,74,75] (Fig. 8-4).

## A. Initiation of "Inside-Out" Signaling

Inside-out signaling can be initiated by the engagement of various platelet adhesion or G protein-coupled receptors, or by model agonists such as phorbol myristate acetate or calcium ionophore. Collagen and VWF, exposed in subendothelial matrix after vascular injury, are clearly important initiators of αIIbβ3 activation. These stimuli engage GPVI and the GPIb-IX-V complex to trigger inside-out signaling. Collagen is able to activate platelets and induces shape change, aggregation, and secretion.[76] Polymeric collagen binds to GPVI on platelets and causes clustering of GPVI and its associated Fc receptor γ-chain, subsequently activating phospholipase C (PLCγ2) through Syk.[77-82] The active PLCγ hydrolyzes phosphatidylinositol-4,5-bisphosphate (PIP$_2$) to form IP$_3$ and diacylglycerol (DAG). IP$_3$ ultimately mobilizes Ca$^{2+}$, and DAG activates protein kinase C (PKC), which initiates protein phosphorylation events. Both mediators participate in αIIbβ3 activation. This process is supported by collagen binding to α2β1 simultaneously.[83-85] αIIbβ3 activation can also be initiated by shear stress-induced VWF binding to the GPIb-IX-V complex.[86–91] This pathway can be reconstituted in Chinese hamster ovary (CHO) cells cotransfected with GPIb/αIIbβ3,[89,90,92] a system that has allowed for the identification of some of the key steps in this inside-out signaling pathway.

Other major agonists, thrombin, adenosine diphosphate (ADP), and thromboxane $A_2$ ($TXA_2$), are generated rapidly within the microenvironment of a vascular injury and developing thrombus. Thrombin activates platelets mainly through protease-activated receptor (PAR) family members (Chapter 9). PAR1 and PAR4 are expressed on human platelets.[93,94] Activation occurs when thrombin binds and cleaves the PARs, and the new N-terminus serves as a tethered ligand.[95] Three ADP receptors, $P2Y_1$, $P2Y_{12}$, and $P2X_1$, have been identified on platelets (Chapter 10), and all are required for optimal platelet activation by ADP.[96–99] Of these, $P2Y_1$ is responsible for platelet shape change and calcium mobilization, and $P2Y_{12}$ is critically involved in αIIbβ3 activation, inhibition of adenylyl cyclase, and stabilization of platelet aggregation.[100] Two splice variants of the $TXA_2$ receptors, designated TPα and TPβ, have been identified (Chapter 31).[101] All of the membrane proteins that recognize these agonists are G protein-coupled receptors. Their engagement leads to activation of $PLC_β$, another PLC isoform in platelets. Once activated, $PLC_β$ hydrolyzes $PIP_2$ to $IP_3$ and DAG, which raises cytosolic $Ca^{2+}$ and activates PKC respectively (see also Chapter 16).

Although each of the stimuli identified here is capable of initiating the inside-out signaling events leading to αIIbβ3 activation, it is likely that they act cooperatively in physiological settings. Many stimuli can act synergistically to activate αIIbβ3.[102] As only one of many potential scenarios, damage to the endothelium will expose collagen and results in turbulent shear that allows engagement of collagen receptors and supports VWF binding to GPIb-IX-V on platelets. The signaling induced through these receptors leads to activation of αIIbβ3 and initiation of thrombus formation. As platelets aggregate, they produce/release contents that include $TXA_2$ and ADP. These agonists activate αIIbβ3 on circulating platelets, leading to their recruitment into the developing thrombus. These events are all occurring in a microenvironment of vascular injury in which tissue factor is being expressed and initiating the intrinsic coagulation pathway leading to thrombin generation (Chapter 19). Thrombin activates additional platelets and their αIIbβ3, potentially leading to formation of a large thrombus.

One of the most complex and as yet unresolved issues regarding inside-out signaling in platelets is how the signals induced by different agonists converge to *touch* the cytoplasmic domain of αIIbβ3. The common event induced by the agonists is the activation of one or more PLC isoforms, thereby elevating cytosolic $Ca^{2+}$ concentration and activating PKC. In resting platelets, cytosolic $Ca^{2+}$ is maintained at approximately 100 nM, whereas it can exceed 1 μM upon platelet activation. $Ca^{2+}$ at micromolar levels in platelets substantially regulates multiple events, including μ-calpain activation.[103–106] Activated PKC regulates serine/threonine phosphorylation events that are required for the activities of αIIbβ3. In response to calcium and DAG, CalDAG-GEFI, a guanine nucleotide exchange factor, activates Rap1,[107,108] which are small GTPases of the Ras family. Murine knockouts have demonstrated that CalDAG-GEF1 or Rap1b deficiency results in compromised αIIbβ3 activation and defective platelet function.[109,110] In addition, PI 3-kinases are also involved in regulation of αIIbβ3 activation[111] through $G_i$-dependent activation of Rap1b.[112] The mechanisms by which these signals are integrated to activate αIIbβ3 are still unclear, but are likely to depend on the binding to or modification of the inner leaflet of the platelet membrane or proteins that bind to the CT of αIIbβ3.[113]

### B. Integrin αIIbβ3 Tail-Binding Proteins

More than 20 cytoplasmic proteins have been shown to interact with the αIIb or β3 CT (Table 8-1). The list includes cytoskeletal and signaling proteins as well as molecules with undefined functions. The roles of these binding part-

**Table 8-1:  Cytoplasmic Tail-Binding Proteins**

| Protein | Binds To | Reference |
|---|---|---|
| *Cytoskeletal proteins* | | |
| Talin | αIIb and β3 | 114–117 |
| Myosin | β3 | 118 |
| Skelemin | β3 | 119 |
| Filamin | αIIbβ3 | 120 |
| α-actinin | β3 | 121 |
| *Adaptor and signaling proteins* | | |
| Paxillin | β3 | 122 |
| Shc | β3 | 123, 124 |
| Grb2 | β3 | 123 |
| *Protein kinases and phosphatase* | | |
| Src | β3 | 125–127 |
| Csk | β3 | 125 |
| Syk | β3 | 125, 126 |
| ILK | β3 | 128 |
| FAK | β3 | 122 |
| PP1c | αIIb | 129 |
| *Others* | | |
| BiP | αIIbβ3 | 130 |
| Calreticulin | α | 131, 132 |
| CIB | αIIb | 133–135 |
| ICln | αIIb | 136 |
| β3-endotoxin | β3 | 137, 138 |
| Aup1 | αIIb | 139 |

ners in "inside-out" signaling in platelets remain uncertain. Many of the characterized interactions depend upon clustering of the occupied integrin and are, therefore, influential on outside-in signaling. At this time, the one interacting molecule that has been most firmly linked to αIIbβ3 activation is the cytoskeleton protein talin. Talin is an abundant cytosolic protein that is capable of linking integrins to the actin cytoskeleton (Chapter 4) either directly or indirectly via its interactions with vinculin and α-actinin.[113,121,140] Undifferentiated cells deficient in talin fail to form vinculin or paxillin-containing focal adhesions,[141] and inactivation of the talin gene in mice is embryonically lethal.[142] Talin is composed of a 47-kDa head domain (talin-H) and a 190-kDa rod domain (talin-R), which align in an antiparallel arrangement to form a homodimer. Each talin domain contains at least one binding site for the β3 CT. Talin-H is composed of a FERM (band 4.1, ezrin, radinxin, and moesin) domain,[113] which consists of three subdomains: $F_1$, $F_2$, and $F_3$.[143] Talin-H contains a high-affinity site that interacts at the NPLY motif of the β3 CT.[114,144–146] Talin-H also interacts with a second less well-defined motif within the membrane-proximal region of the β3 CT.[43,147] Fragments composed of the $F_2$–$F_3$ subdomains or $F_3$ subdomain alone bind to the β3 CT with a similar affinity to talin-H.[145] A lower affinity binding site for the β3 CT is localized in the C-terminal region of rod domain.[144,147,148] Cooperative binding of the talin-H and talin-R domains to the β3 tail has been suggested.[149] In addition, it has been suggested that talin also associates with the αIIb CT.[115]

Talin-H, but not talin-R, domain induces integrin αIIbβ3 activation in transfected CHO cells.[114,116,146] The β3 CT binding site in talin-H is masked in intact talin, and an exposing event is required for talin-H to become an activator of αIIbβ3. Calpain cleavage[149] or phosphoinositide binding[150] are such exposing events. Calpain itself becomes activated as a consequence of increases in cytosolic $Ca^{2+}$, which occur upon agonist stimulation of platelets, and cleavage of talin is detected within activated platelets.[151,152] Alternatively, the integrin binding sites in talin can also be exposed by conformation change induced by its binding of phosphoinositide PI4,5P$_2$. The kinase, PIP4,5 kinase, which generates the product, binds directly to talin.[150] These events permit the talin-H domain to bind to the NPLY[747] motif and the membrane-proximal region of the β3 tail to induce integrin activation.[43,144,147] NMR studies show that the talin-H binding site in the membrane-proximal region of the β3 CT overlaps extensively with the αIIb binding site and effectively displaces the αIIb CT from the β3 CT. This perturbation of the CT complex initiates integrin activation[43] (Fig. 8-4).

Besides talin, β3-endonexin, a 111-amino acid protein present in platelets,[137] can also induce αIIbβ3 activation. This activity was demonstrated by overexpression of β3-endonexin in CHO cells expressing αIIbβ3. β3-endonexin binds to the C-terminal aspect of β3 CT, involving the NITY motif.[137,138,153]

In addition, several αIIb tail-binding proteins have been identified that could play a role in integrin activation. They include calcium integrin-binding protein,[133,134,154] a chloride channel regulatory protein Icln,[136] ancient ubiquitous protein 1,[139] and protein phosphatase 1.[129] All these proteins are present in platelets and appear to bind to the membrane-proximal region of the αIIb CT. However, the roles of these interactions in αIIbβ3 activation are unclear at this time.

In addition to cytoplasmic proteins, several membrane proteins also associate with αIIbβ3. These include CD47/IAP,[155] CD98,[156] CD36,[157] CD9,[158] CD63,[159] and CD151.[160] CD47 is a thrombospondin 1 receptor and may regulate αIIbβ3 activation through a $G_i$ pathway.[161–163] The importance of CD98 in integrin activation was demonstrated by an expression-cloning screen for proteins that reversed the inhibitory effect of the $β_{1A}$ CT on αIIbβ3 activation.[156,164] CD36, known to associate with nonreceptor tyrosine kinases pp[60Fyn], pp[54/58Lyn], and pp[62Yes], associates with αIIbβ3 on the surface of resting platelets.[157,165] CD36 antibodies can activate αIIbβ3 through an Fc receptor-dependent pathway, suggesting that CD36, αIIbβ3, and the Fc receptor may form a ternary complex during platelet activation.[166,167] Antibodies to CD9 can activate platelets and induce an association between CD9 and αIIbβ3.[158] The CD9–αIIbβ3 complex can also recruit CD63 into a trimolecular complex on activated platelets.[159] Although CD151 potentially interacts with αIIbβ3 and its associated proteins, such as CD9 and CD63, studies of CD151$^{-/-}$ platelets suggest that it may be involved in outside-in rather than inside-out signaling.[160,168]

Interestingly, the platelet membrane itself may be involved in regulating inside-out signaling via regulated contacts with CTs.[44] The five amino acids in αIIb and β3 CT proximal to the membrane are hydrophobic in character and their positioning in or out of the membrane can vary.[169] Movement of these proximal sequences into the membrane may occur during integrin activation. Besides the membrane-proximal region, the membrane-distal region of the β3 CT containing the NPLY motif also binds to DPC, providing an additional site at which the platelet membrane may regulate integrin activation.

## C. Initiation of Inside-Out Signaling at the CT of αIIbβ3: A Clasping/Unclasping Model

The CTs of integrin are very small compared with other signaling receptors, and they lack intrinsic catalytic activity[69,170] (Fig. 8-3A). Although serine/threonine and tyrosine phosphorylation events occur on the CT, these modifications seem more involved in outside-in rather than inside-out signaling. Hence, activation of αIIbβ3 must depend on changes in the spatial organization induced by their interactions with specific cytoplasmic components[75] (discussed earlier). The spatial organization that appears to be altered by such regulators seems to be the interaction between the αIIb and β3 CT.

This noncovalent complex is formed by hydrophobic and electrostatic contacts between the membrane-proximal helices of the CT[43] (Fig. 8-3D). The hydrophobic contacts are αIIb(V990)-β3(L718), αIIb(V990)-β3(I719), αIIb (F992)-β3(I721), and αIIb(F992)-β3(H722). The electrostatic interface involves side chains of the following pairs: αIIb(R995 guanidyl group)-β3(H722 imidazole group), αIIb(R995 guanidyl group)-β3(D723 carboxyl group), and αIIb(R995 guanidyl group)-β3(E726 carboxyl group). Point mutations in these two helices or deletions of either CT that destroy specific contacts enhance αIIbβ3 activation.[68,171,172] The residues that mediate binding between the CTs are highly conserved in all integrin α- and β-subunits, suggesting that a CT complex is present in most if not all integrins. Several lines of evidence supported the existence of a CT complex prior to the NMR solution of its structure. Among the most compelling observations was a study showing that mutation of either αIIb(R995) or β3(D723) led to integrin activation whereas the double mutation of αIIb(R995D)-β3(D723R) sustained the resting state.[173] This *charge reversal* strategy is strong circumstantial evidence for formation of a salt bridge between αIIb(R995D)-β3(D723R). As noted earlier, overexpression of talin can lead to integrin activation, and binding of the talin head domain to the β3 CT dissociates the CT complex. Hence, it is now broadly accepted that the "clasp" between the CT maintains the integrin in the resting state and the "unclasping" induced by talin, other activators, or mutations is a trigger of inside-out signaling (Fig. 8-4). This model is supported by the suppression of activation by mutations that prevent unclasping of the CT complex and by fluorescence resonance energy transfer (FRET) experiments. In these FRET experiments, fluorescent probes engineered onto the CT quench each other when in close proximity, and the fluorescence signal increases as the probes separate. In the resting state, fluorescence quenching is extensive, and mutations that disrupt the CT complex and activate αIIbβ3 increase the fluorescence.[174]

How the signal of CT separation is transmitted to the extracellular domain is less certain. Mutations within the predicted transmembrane helices can lead to αIIbβ3 activation.[47,175,176] These mutations are likely to disrupt the αIIb and β3 transmembrane helices and lead to their dissociation. This dissociation may then continue into the extracellular stalk regions. Hence, introduction of disulfide bonds that lock α/β-helices or membrane-proximal α-/β-stalks together abrogate αIIbβ3 activation.[177,178] A separation mechanism may even regulate activation of the ligand binding site. There are data to suggest that the shape of the headpiece formed by the αIIb β-propeller and the β3 A-domain for ligand binding is also altered during integrin αIIbβ3 activation.[179] Alternately, as discussed earlier, the transition may involve the change from the bent to the extended conformation. These processes are not mutually exclusive.

A potential mechanism by which unclasping of the CT transmits a signal into the transmembrane region is insertion of the unclasped CT.[44] The membrane-proximal helices are hydrophobic in character. As they separate and their hydrophobic residues become exposed to an unfavorable aqueous environment, they may rapidly translocate into the more hydrophobic milieu of the membrane (Fig. 8-4). This hypothesis is based on NMR experiments in aqueous and DPC micelles.[43,44] The CT complex is observed in the aqueous environment but is dissociated in the presence of DPC as the micelles bind to the membrane-proximal helices and unclasp the CT complex. The individual CT helices insert their first five amino acids into the micelles. Such insertion could provide the energy to disrupt associations of the transmembrane helices, thereby propagating the inside-out signal initiated by unclasping.

Although the membrane-distal regions of αIIb and β3 CTs are not directly involved in formation of the clasp, they may play a role in regulating integrin activation. The acidic C-terminal resides of αIIb CT fold back and interact with the positively charged membrane-proximal region, which may stabilize the resting state of αIIbβ3[42] (Fig. 8-3B). Double mutations of P989$^A$/P990$^A$ in the αIIb CT to destabilize the folded structure can lead to activation of αIIbβ3.[180] The membrane-distal region of β3 plays a more direct role in αIIbβ3 activation. A point mutation at Y$^{747}$A (which perturbs the NPLY$^{747}$ turn motif), an S$^{752}$P change (a naturally occurring point mutant that destroys the membrane-distal helix), or truncations at F$^{754}$, Y$^{747}$, or T$^{741}$ sites, all inhibit integrin activation.[92,181–183] These mutations are all well beyond the membrane-proximal helical domain of β3 CT.

The molecular events that occur during inside-out signaling and αIIbβ3 activation in platelets are undoubtedly much more complicated than depicted in this clasping/unclasping model. The pathways that lead from the multiple agonists and their receptors on the platelet surface and the cross-talk between cytosolic signals have yet to be integrated into a cohesive hypothesis. The roles of the numerous CT binding proteins in inside-out signaling remain particularly vague at this time. For example, it is unclear whether proteins that bind to sites in the β3 CT that overlap with the talin binding site also can serve as activators. Even in the case of talin, direct evidence for its participation in αIIbβ3 activation in platelets needs to be solidified.[151,184] Thus, an in-depth understanding of the molecular details involved in inside-out signaling and αIIbβ3 activation remains to be resolved.

## V. Conclusion

Much progress has been made in providing high-resolution structures of integrin αIIbβ3. These structures provide detailed insights into the molecular architecture of the receptor and the precise contacts that are involved in its engagement of ligands. The snapshots provided by these structures also allow for construction of detailed hypotheses to explain how the functions of αIIbβ3 and other integrins are regulated. A next step in our understanding will come as these hypotheses are tested and the dynamic changes that control the functional responses of the receptor in real time are unraveled. αIIbβ3 (GPIIb-IIIa) antagonists are in clinical use and are effective in preventing acute thrombotic events (Chapter 62), but more "gentle" platelet antagonists, such as aspirin (Chapter 60) and clopidogrel (Chapter 61), are more effective as palliative drugs. Drugs that target specific events leading to activation of αIIbβ3 could offer a specificity and safety advantage over aspirin and clopidogrel, as well as over the current αIIbβ3 inhibitors. Because thrombotic diseases remain the leading cause of mortality in the Western world, identifying new and different approaches to alter the function of αIIbβ3 represents an interesting target for drug development, and one that will depend upon a fuller understanding of the activation and functions of αIIbβ3.

# References

1. Adams, P. C., Badimon, J. J., Badimon, L., Chesebro, J. H., & Fuster, V. (1987). Role of platelets in atherogenesis: Relevance to coronary arterial restenosis after angioplasty. *Cardiovasc Clin, 18,* 49–71.
2. Ross, R. (1993). Atherosclerosis: A defense mechanism gone awry. *Am J Pathol, 143,* 987–1002.
3. Libby, P. (1995). Molecular bases of the acute coronary syndromes. *Circulation, 91,* 2844–2850.
4. Bennett, J. S. (1990). The molecular biology of platelet membrane proteins. *Semin Hematol, 27,* 186–204.
5. Plow, E. F., & Byzova, T. (1999). The biology of glycoprotein IIb-IIIa. *Coron Artery Dis, 10,* 547–551.
6. Plow, E. F., D'Souza, S. E., & Ginsberg, M. H. (1992). Ligand binding to GPIIb-IIIa: A status report. *Semin Thromb Hemost, 18,* 324–332.
7. Savage, B., Cattaneo, M., & Ruggeri, Z. M. (2001). Mechanisms of platelet aggregation. *Curr Opin Hematol, 8,* 270–276.
8. Hynes, R. O. (1992). Integrins: Versatility, modulation, and signaling in cell adhesion. *Cell, 69,* 11–25.
9. Plow, E. F., Haas, T. A., Zhang, L., Loftus, J., & Smith, J. W. (2000). Ligand binding to integrins. *J Biol Chem, 275,* 21785–21788.
10. Plow, E. F., & Shattil, S. J. (2001). Integrin αIIbβ3 and platelet aggregation. In R. W. Colman, J. Hirsh, V. J. Marder, A. W. Clowes, & J. N. George (Eds.), *Hemostasis and thrombosis: Basic principles and clinical practice* (pp. 479–491). Philadelphia: Lippincott Williams & Wilkins.
11. Plow, E. F., & Marguerie, G. A. (1982). Inhibition of fibrinogen binding to human platelets by the tetrapeptide glycyl-L-prolyl-L-arginyl-L-proline. *Proc Natl Acad Sci USA, 79,* 3711–3715.
12. Coller, B. S., Peerschke, E. I., Scudder, L. E., & Sullivan, C. A. (1983). A murine monoclonal antibody that completely blocks the binding of fibrinogen to platelets produces a thrombasthenic-like state in normal platelets and binds to glycoproteins IIb and/or IIIa. *J Clin Invest, 72,* 325–338.
13. Bhatt, D. L., & Topol, E. J. (2000). Current role of platelet glycoprotein IIb/IIIa inhibitors in acute coronary syndromes. *J Am Med Assoc, 284,* 1549–1558.
14. Hynes, R. O. (2002). Integrins: Bidirectional, allosteric signaling machines. *Cell, 110,* 673–687.
15. Smith, J. W., Piotrowicz, R. S., & Mathis, D. (1994). A mechanism for divalent cation regulation of β3-integrins. *J Biol Chem, 269,* 960–967.
16. Lam, S. C., Plow, E. F., D'Souza, S. E., et al. (1989). Isolation and characterization of a platelet membrane protein related to the vitronectin receptor. *J Biol Chem, 264,* 3742–3749.
17. Lawler, J., & Hynes, R. O. (1989). An integrin receptor on normal and thrombasthenic platelets that bind thrombospondin. *Blood, 74,* 2022–2027.
18. Uzan, G., Prenant, M., Prandini, M.-H., Martin, F., & Marguerie, G. (1991). Tissue-specific expression of the platelet GPIIb gene. *J Biol Chem, 266(14),* 8932–8939.
19. Trikha, M., Timar, J., Lundy, S. K., et al. (1996). Human prostate carcinoma cells express functional αIIbβ3 integrin. *Cancer Res, 56,* 5071–5078.
20. Butler–Zimrin, A. E., Bennett, J. S., Poncz, M., et al. (1987). Isolation and characterization of cDNA clones for the platelet membrane glycoproteins IIb and IIIa. *Thromb Haemost, 58,* 319a.
21. Cheresh, D. A. (1992). Structural and biologic properties of integrin-mediated cell adhesion. *Clin Lab Med, 12,* 217.
22. Phillips, D. R., Charo, I. F., Parise, L. V., & Fitzgerald, L. A. (1988). The platelet membrane glycoprotein IIb-IIIa complex. *Blood, 71,* 831–843.
23. Niiya, K., Hodson, E., Bader, R., et al. (1987). Increased surface expression of the membrane glycoprotein IIb/IIIa complex induced by platelet activation. Relationship to the binding of fibrinogen and platelet aggregation. *Blood, 70,* 475–483.
24. Wagner, C. L., Mascelli, M. A., Neblock, D. S., et al. (1996). Analysis of GPIIb/IIIa receptor number by quanitification of 7E3 binding to human platelets. *Blood, 88,* 907–914.
25. Stouffer, G. A., & Smyth, S. S. (2003). Effects of thrombin on interactions between beta3-integrins and extracellular matrix in platelets and vascular cells. *Arterioscler Thromb Vasc Biol, 23,* 1971–1978.
26. Wencel–Drake, J. D., Plow, E. F., Kunicki, T. J., et al. (1986). Localization of internal pools of membrane glycoproteins involved in platelet adhesive responses. *Am J Pathol, 124,* 324–334.
27. Woods, V. L., Wolff, L. E., & Keller, D. M. (1986). Resting platelets contain a substantial centrally located pool of glycoprotein IIb-IIIa complex which may be accessible to some but not other extracellular proteins. *J Biol Chem, 261,* 15242–15251.
28. Fujimura, K., & Phillips, D. R. (1983). Calcium cation regulation of glycoprotein IIb-IIIa complex formation in

platelet plasma membranes. *J Biol Chem, 258,* 10247–10252.

29. Sosnoski, D. M., Emanuel, B. S., Hawkins, A. L., et al. (1988). Chromosomal localization of the genes for the vitronectin and fibronectin receptors α subunits and for platelet glycoproteins IIb and IIIa. *J Clin Invest, 81,* 1993–1998.

30. Duperray, A., Berthier, R., Chagnon, E., et al. (1987). Biosynthesis and processing of platelet GPIIb-IIIa in human megakaryocytes. *J Cell Biol, 104,* 1665–1673.

31. Loftus, J. C., Plow, E. F., Jennings, L. K., & Ginsberg, M. H. (1988). Alternative proteolytic processing of platelet membrane GPIIb. *J Biol Chem, 263,* 11025–11028.

32. Poncz, M., Eisman, R., Heidenreich, R., et al. (1987). Structure of the platelet membrane glycoprotein IIb. Homology to the alpha subunits of the vitronectin and fibronectin membrane receptors. *J Biol Chem, 262,* 8476–8482.

33. Zimrin, A. B., Gidwitz, S., Lord, S., et al. (1990). The genomic organization of platelet glycoprotein IIIa. *J Biol Chem, 265,* 8590–8595.

34. Fitzgerald, L. A., Steiner, B., Rall, S. C. J., Lo, S. S., & Phillips, D. R. (1987). Protein sequence of endothelial glycoprotein IIIa derived from a cDNA clone. Identity with platelet glycoprotein IIIa and similarity to "integrin." *J Biol Chem, 262,* 3936–3939.

35. Xiong, J. P., Stehle, T., Diefenbach, B., et al. (2001). Crystal structure of the extracellular segment of integrin alpha Vbeta3. *Science, 294,* 339–345.

36. Xiao, T., Takagi, J., Coller, B. S., Wang, J. H., & Springer, T. A. (2004). Structural basis for allostery in integrins and binding to fibrinogen-mimetic therapeutics. *Nature, 432,* 59–67.

37. Xiong, J. P., Stehle, T., Zhang, R., et al. (2002). Crystal structure of the extracellular segment of integrin alpha Vbeta3 in complex with an Arg-Gly-Asp ligand. *Science, 296,* 151–155.

38. Xiong, J. P., Stehle, T., Goodman, S. L., & Arnaout, M. A. (2003). Integrins, cations and ligands: Making the connection. *J Thromb Haemost, 7,* 1642–1654.

39. Chothia, C., & Jones, E. Y. (1997). The molecular structure of cell adhesion molecules. *Annu Rev Biochem, 66,* 823–862.

40. Carrell, N. A., Fitzgerald, L. A., Steiner, B., Erickson, H. P., & Phillips, D. R. (1985). Structure of human platelet membrane glycoproteins IIb and IIIa as determined by electron microscopy. *J Biol Chem, 260,* 1743–1749.

41. Weisel, J. W., Nagaswami, C., Vilaire, G., & Bennett, J. S. (1992). Examination of the platelet membrane GPIIb-IIIa complex and its interaction with fibrinogen and other ligands by electron microscopy. *J Biol Chem, 267,* 16637–16643.

42. Vinogradova, O., Haas, T., Plow, E. F., & Qin, J. (2000). A structural basis for integrin activation by the cytoplasmic tail of the alpha IIb-subunit. *Proc Natl Acad Sci USA, 97,* 1450–1455.

43. Vinogradova, O., Velyvis, A., Velyviene, A., et al. (2002). A structural mechanism of integrin αIIbβ3 "inside-out" activation as regulated by its cytoplasmic face. *Cell, 110,* 587–597.

44. Vinogradova, O., Vaynberg, J., Kong, X., et al. (2004). Membrane-mediated structural transitions at the cytoplas-

mic face during integrin activation. *Proc Natl Acad Sci USA, 101,* 4094–4099.

45. Ulmer, T. S., Yaspan, B., Ginsberg, M. H., & Campbell, I. D. (2001). NMR analysis of structure and dynamics of the cytosolic tails of integrin alphaIIbbeta3 in aqueous solution. *Biochemistry, 40,* 7498–7508.

46. Beglova, N., & Blacklow, S. C. (2005). The LDL receptor: How acid pulls the trigger. *Trends Biochem Sci, 30,* 309–317.

47. Partridge, A. W., Liu, S., Kim, S., Bowie, J. U., & Ginsberg, M. H. (2005). Transmembrane domain helix packing stabilizes integrin alphaIIbbeta3 in the low affinity state. *J Biol Chem, 280,* 7294–7300.

48. Ruoslahti, E. (1991). Integrins. *J Clin Invest, 87,* 1–5.

49. Ruoslahti, E. (1996). RGD and other recognition sequences for integrins. *Annu Rev Cell Biol, 12,* 697–715.

50. Felding–Habermann, B., Ruggeri, Z. M., & Cheresh, D. A. (1992). Distinct biological consequences of integrin $\alpha_v\beta_3$-mediated melanoma cell adhesion to fibrinogen and its plasmic fragments. *J Biol Chem, 267,* 5070–5077.

51. Cheresh, D. A., Berliner, S. A., Vicente, V., & Ruggeri, Z. M. (1989). Recognition of distinct adhesive sites on fibrinogen by related integrins on platelets and endothelial cells. *Cell, 58,* 945–953.

52. Suehiro, K., Gailit, J., & Plow, E. F. (1997). Fibrinogen is a ligand for integrin $\alpha_5\beta_1$ on endothelial cells. *J Biol Chem, 272,* 5360–5366.

53. Farrell, D. H., Thiagarajan, P., Chung, D. W., & Davie, E. W. (1992). Role of fibrinogen alpha and gamma chain sites in platelet aggregation. *Proc Natl Acad Sci USA, 89,* 10729–10732.

54. Beacham, D. A., Wise, R. J., Turci, S. M., & Handin, R. I. (1992). Selective inactivation of the arg-gly-asp-ser (RGDS) binding site in von Willebrand factor by site-directed mutagenesis. *J Biol Chem, 267,* 3409–3415.

55. Cierniewski, C. S., Byzova, T., Papierak, M., et al. (1999). Peptide ligands can bind to distinct sites in integrin $\alpha_{IIb}\beta_3$ and elicit different functional responses. *J Biol Chem, 274,* 16923–16932.

56. Farrell, D. H., & Thiagarajan, P. (1994). Binding of recombinant fibrinogen mutants to platelets. *J Biol Chem, 269,* 226–231.

57. Smith, R. A., Mosesson, M. W., Rooney, M. M., et al. (1997). The role of putative fibrinogen Aα-, Bβ-, and gammaA-chain integrin binding sites in endothelial cell-mediated clot retraction. *J Biol Chem, 272,* 22080–22085.

58. Mustard, J. F., Packham, M. A., Kinlough–Rathbone, R. L., Perry, D. W., & Regoeczi, E. (1978). Fibrinogen and ADP-induced platelet aggregation. *Blood, 52,* 453–465.

59. Marguerie, G. A., Edgington, T. S., & Plow, E. F. (1980). Interaction of fibrinogen with its platelet receptor as part of a multistep reaction in ADP-induced platelet aggregation. *J Biol Chem, 255,* 154–161.

60. Bennett, J. S., & Vilaire, G. (1979). Exposure of platelet fibrinogen receptors by ADP and epinephrine. *J Clin Invest, 64,* 1393–1401.

61. Takagi, J., Petre, B. M., Walz, T., & Springer, T. A. (2002). Global conformational rearrangements in integrin extracel-

lular domains in outside-in and inside-out signaling. *Cell, 110,* 599–611.

62. Luo, B. H., Takagi, J., & Springer, T. A. (2004). Locking the beta3 integrin I-like domain into high and low affinity conformations with disulfides. *J Biol Chem, 279,* 10215–10221.

63. Takagi, J., DeBottis, D. P., Erickson, H. P., & Springer, T. A. (2002). The role of the specificity-determining loop of the integrin beta subunit I-like domain in autonomous expression, association with the alpha subunit, and ligand binding. *Biochemistry, 41,* 4339–4347.

64. Adair, B. D., Xiong, J. P., Maddock, C., et al. (2005). Three-dimensional EM structure of the ectodomain of integrin αVβ3 in a complex with fibronectin. *J Cell Biol, 168,* 1109–1118.

65. Hato, T., Pampori, N., & Shattil, S. J. (1998). Complementary roles for receptor clustering and conformational change in the adhesive and signaling functions of integrin αIIbβ3. *J Cell Biol, 141,* 1685–1695.

66. Buensucesco, C., de Virgilio, M., & Shattil, S. J. (2003). Detection of integrin alpha IIbbeta 3 clustering in living cells. *J Biol Chem, 278,* 15217–15224.

67. Ginsberg, M. H., Du, X., & Plow, E. F. (1992). Inside-out integrin signaling. *Curr Opin Cell Biol, 4,* 766.

68. O'Toole, T. E., Katagiri, Y., Faull, R. J., et al. (1994). Integrin cytoplasmic domains mediate inside-out signal transduction. *J Cell Biol, 124,* 1047–1059.

69. Schwartz, M. A., Schaller, M. D., & Ginsberg, M. H. (1995). Integrins: Emerging paradigms of signal transduction. *Annu Rev Cell Biol, 11,* 549–599.

70. Shattil, S. J., Ginsberg, M. H., & Brugge, J. S. (1994). Adhesive signaling in platelets. *Curr Opin Cell Biol, 6,* 695–704.

71. Clark, E. A., & Brugge, J. S. (1995). Integrins and signal transduction pathways: The road taken. *Science, 268,* 233–239.

72. Frelinger, A. L., III, Lam, S. C. T., Plow, E. F., et al. (1988). Occupancy of an adhesive glycoprotein receptor modulates expression of an antigenic site involved in cell adhesion. *J Biol Chem, 263,* 12397–12402.

73. Parise, L. V., Helgerson, S. L., Steiner, B., Nannizzi, L., & Phillips, D. R. (1987). Synthetic peptides derived from fibrinogen and fibronectin change the conformation of purified platelet glycoprotein IIb-IIIa. *J Biol Chem, 262,* 12597–12602.

74. Dedhar, S., & Hannigan, G. E. (1996). Integrin cytoplasmic interactions and bidirectional transmembrane signalling. *Curr Opin Cell Biol, 8,* 657–669.

75. Qin, J., Vinogradova, O., & Plow, E. F. (2004). Integrin bidirectional signaling: A molecular view. *PLoS Biol, 2,* e169.

76. Packham, M. A., Guccione, M. A., Greenberg, J. P., Kinlough–Rathbone, R. L., & Mustard, J. F. (1977). Release of 14C-serotonin during initial platelet changes induced by thrombin, collagen, or A23187. *Blood, 50,* 915–926.

77. Moroi, M., & Jung, S. M. (2004). Platelet glycoprotein VI: Its structure and function. *Thromb Res, 114,* 221–233.

78. Nieswandt, B., & Watson, S. P. (2003). Platelet–collagen interaction: Is GPVI the central receptor? *Blood, 102,* 449–461.

79. Poole, A., Gibbins, J. M., Turner, M., et al. (1997). The Fc receptor gamma-chain and the tyrosine kinase Syk are essential for activation of mouse platelets by collagen. *EMBO J, 16,* 2333–2341.

80. Shattil, S. J., & Brass, L. F. (1987). Induction of the fibrinogen receptor on human platelets by intracellular mediators. *J Biol Chem, 262,* 992–1000.

81. Asazuma, N., Wilde, J. I., Berlanga, O., et al. (2000). Interaction of linker for activation of T cells with multiple adapter proteins in platelets activated by the glycoprotein VI-selective ligand, convulxin. *J Biol Chem, 275,* 33427–33434.

82. Falet, H., Barkalow, K. L., Pivniouk, V. I., et al. (2000). Roles of SLP-76, phosphoinositide 3-kinase, and gelsolin in the platelet shape changes initiated by the collagen receptor GPVI/FcR gamma-chain complex. *Blood, 96,* 3786–3792.

83. Zheng, Y. M., Liu, C., Chen, H., et al. (2001). Expression of the platelet receptor GPVI confers signaling via the Fc receptor gamma-chain in response to the snake venom convulxin but not to collagen. *J Biol Chem, 276,* 12999–13006.

84. Nieuwenhuis, H. K., Akkerman, J. W. N., Houdijk, W. P. M., & Sixma, J. J. (1985). Human blood platelets showing no response to collagen fail to express surface glycoprotein Ia. *Nature, 318,* 470–472.

85. Holtkotter, O., Nieswandt, B., Smyth, N., et al. (2002). Integrin alpha 2-deficient mice develop normally, are fertile, but display partially defective platelet interaction with collagen. *J Biol Chem, 277,* 10789–10794.

86. Kroll, M. H., Hellums, J. D., McIntire, L. V., Schafer, A. I., & Moake, J. L. (1996). Platelets and shear stress. *Blood, 88,* 1525–1541.

87. Andrews, R. K., Kroll, M. H., Ward, C. M., et al. (1996). Binding of a novel 50-kilodalton alboaggregin from *Trimeresurus albolabris* and related viper venom proteins to the platelet membrane glycoprotein Ib-IX-V complex. Effect on platelet aggregation and glycoprotein Ib-mediated platelet activation. *Biochemistry, 35,* 12629–12639.

88. Feng, S., Christodoulides, N., Resendiz, J. C., Berndt, M. C., & Kroll, M. H. (2000). Cytoplasmic domains of GpIbalpha and GpIbbeta regulate 14-3-3zeta binding to GpIb/IX/V. *Blood, 95,* 551–557.

89. Gu, M., Xi, X., Englund, G. D., Berndt, M. C., & Du, X. (1999). Analysis of the roles of 14-3-3 in the platelet glycoprotein Ib-IX-mediated activation of integrin alpha(IIb) beta(3) using a reconstituted mammalian cell expression model. *J Cell Biol, 147,* 1085–1096.

90. Zaffran, Y., Meyer, S. C., Negrescu, E., Reddy, K. B., & Fox, J. E. (2000). Signaling across the platelet adhesion receptor glycoprotein Ib-IX induces alpha IIbbeta 3 activation both in platelets and a transfected Chinese hamster ovary cell system. *J Biol Chem, 275,* 16779–16787.

91. Chen, J., & Lopez, J. A. (2005). Interactions of platelets with subendothelium and endothelium. *Microcirculation, 12,* 235–246.

92. Xi, X., Bodnar, R. J., Li, Z., Lam, S. C., & Du, X. (2003). Critical roles for the COOH-terminal NITY and RGT sequences of the integrin beta3 cytoplasmic domain in inside-out and outside-in signaling. *J Cell Biol, 162,* 329–339.

93. Kahn, M. L., Zheng, Y. W., Huang, W., et al. (1998). A dual thrombin receptor system for platelet activation. *Nature, 394,* 690–694.

94. Xu, W. F., Andersen, H., Whitmore, T. E., et al. (1998). Cloning and characterization of human protease-activated receptor 4. *Proc Natl Acad Sci USA, 95,* 6642–6646.

95. Vu, T. K., Wheaton, V. I., Hung, D. T., Charo, I., & Coughlin, S. R. (1991). Domains specifying thrombin–receptor interaction. *Nature, 353,* 674–677.

96. Hollopeter, G., Jantzen, H. M., Vincent, D., et al. (2001). Identification of the platelet ADP receptor targeted by antithrombotic drugs. *Nature, 409,* 202–207.

97. Fabre, J. E., Nguyen, M., Latour, A., et al. (1999). Decreased platelet aggregation, increased bleeding time and resistance to thromboembolism in P2Y1-deficient mice. *Nat Med, 5,* 1199–1202.

98. MacKenzie, A. B., Mahaut–Smith, M. P., & Sage, S. O. (1996). Activation of receptor-operated cation channels via P2X1 not P2T purinoceptors in human platelets. *J Biol Chem, 271,* 2879–2881.

99. Kunapuli, S. P. (1998). Molecular physiology of platelet ADP receptors. *Drug Dev Res, 45,* 135–139.

100. Dorsam, R. T., & Kunapuli, S. P. (2004). Central role of the P2Y12 receptor in platelet activation. *J Clin Invest, 113,* 340–345.

101. Hirata, T., Ushikubi, F., Kakizuka, A., Okuma, M., & Narumiya, S. (1996). Two thromboxane A2 receptor isoforms in human platelets. Opposite coupling to adenylyl cyclase with different sensitivity to Arg60 to Leu mutation. *J Clin Invest, 97,* 949–956.

102. Abrams, C. S. (2005). Intracellular signaling in platelets. *Curr Opin Hematol, 12,* 401–405.

103. Fox, J. E. B., Reynolds, C. C., & Phillips, D. R. (1983). Calcium-dependent proteolysis occurs during platelet aggregation. *J Biol Chem, 258,* 9973–9981.

104. Saido, T. C., Shibata, M., Takenawa, T., Murofushi, H., & Suzuki, K. (1992). Positive regulation of μ-calpain action by polyphosphoinositides. *J Biol Chem, 267,* 24585–24590.

105. Tsujinaka, T., Sakon, M., Kambayashi, J., & Kosaki, G. (1982). Cleavage of cytoskeletal proteins by two forms of Ca2+ activated neutral proteases in human platelets. *Thromb Res, 28,* 149–156.

106. Suzuki, K., Imajoh, S., Emori, Y., et al. (1987). Calcium-activated neutral protease and its endogenous inhibitor. Activation at the cell membrane and biological function. *FEBS Lett, 220,* 271–277.

107. Kawasaki, H., Springett, G. M., Toki, S., et al. (1998). A Rap guanine nucleotide exchange factor enriched highly in the basal ganglia. *Proc Natl Acad Sci U S A, 95,* 13278–13283.

108. Dupuy, A. J., Morgan, K., von Lintig, F. C., et al. (2001). Activation of the Rap1 guanine nucleotide exchange gene, CalDAG-GEF I, in BXH-2 murine myeloid leukemia. *J Biol Chem, 276,* 11804–11811.

109. Crittenden, J. R., Bergmeier, W., Zhang, Y., et al. (2004). CalDAG-GEFI integrates signaling for platelet aggregation and thrombus formation. *Nat Med, 10,* 982–986.

110. Chrzanowska–Wodnicka, M., Smyth, S. S., Schoenwaelder, S. M., Fischer, T. H., & White, G. C. (2005). Rap1b is required for normal platelet function and hemostasis in mice. *J Clin Invest, 115,* 680–687.

111. Chen, J., De, S., Damron, D. S., et al. (2004). Impaired platelet responses to thrombin and collagen in AKT-1-deficient mice. *Blood, 104,* 1703–1710.

112. Woulfe, D., Jiang, H., Mortensen, R., Yang, J., & Brass, L. F. (2002). Activation of Rap1B by G(i) family members in platelets. *J Biol Chem, 277,* 23382–23390.

113. Rees, D. J., Ades, S. E., Singer, S. J., & Hynes, R. O. (1990). Sequence and domain structure of talin. *Nature, 347,* 685–689.

114. Calderwood, D. A., Zent, R., Grant, R., et al. (1999). The talin head domain binds to integrin beta subunit cytoplasmic tails and regulates integrin activation. *J Biol Chem, 274,* 28071–28074.

115. Knezevic, I., Leisner, T. M., & Lam, S. C. T. (1996). Direct binding of the platelet integrin $\alpha_{IIb}\beta_3$ (GPIIb-IIIa) to talin: Evidence that interaction is mediated through the cytoplasmic domains of both $\alpha_{IIb}$ and $\beta_3$. *J Biol Chem, 271,* 16416–16421.

116. Patil, S., Jedsadayanmata, A., Wencel–Drake, J. D., et al. (1999). Identification of a talin-binding site in the integrin beta(3) subunit distinct from the NPLY regulatory motif of post-ligand binding functions. The talin n-terminal head domain interacts with the membrane-proximal region of the beta(3) cytoplasmic tail. *J Biol Chem, 274,* 28575–28583.

117. Pfaff, M., Liu, S., Erle, D. J., & Ginsberg, M. H. (1998). Integrin beta cytoplasmic domains differentially bind to cytoskeletal proteins. *J Biol Chem, 273,* 6104–6109.

118. Jenkins, A. L., Nannizzi–Alaimo, L., Silver, D., et al. (1998). Tyrosine phosphorylation of the $\beta_3$ cytoplasmic domain mediates integrin-cytoskeletal interactions. *J Biol Chem, 273,* 13878–13885.

119. Reddy, K. B., Gascard, P., Price, M. G., Negrescu, E. V., & Fox, J. E. (1998). Identification of an interaction between the m-band protein skelemin and beta-integrin subunits. Colocalization of a skelemin-like protein with beta1- and beta3-integrins in non-muscle cells. *J Biol Chem, 273,* 35039–35047.

120. Goldmann, W. H. (2000). Kinetic determination of focal adhesion protein formation. *Biochem Biophys Res Commun, 271,* 553–557.

121. Otey, C. A., Pavalko, F. M., & Burridge, K. (1990). An interaction between alpha-actinin and the $beta_1$ integrin subunit *in vitro*. *J Cell Biol, 11,* 721–729.

122. Schaller, M. D., Otey, C. A., Hildebrand, J. D., & Parsons, J. T. (1995). Focal adhesion kinase and paxillin bind to peptides mimicking beta integrin cytoplasmic domains. *J Cell Biol, 130,* 1181–1187.

123. Law, D. A., Nannizzi–Alaimo, L., & Phillips, D. R. (1996). Outside-in integrin signal transduction. $\alpha_{IIb}\beta_3$-(GPIIb-IIIa) tyrosine phosphorylation induced by platelet aggregation. *J Biol Chem, 271,* 10811–10815.

124. Cowan, K. J., Law, D. A., & Phillips, D. R. (2000). Identification of Shc as the primary protein binding to the tyrosine-phosphorylated $\beta_3$ subunit of $\alpha_{IIb}\beta_3$ during outside-in integrin platelet signaling. *J Biol Chem, 275,* 36423–36429.

125. Obergfell, A., Eto, K., Mocsai, A., et al. (2002). Coordinate interactions of Csk, Src, and Syk kinases with [alpha]IIb [beta]3 initiate integrin signaling to the cytoskeleton. *J Cell Biol, 157,* 265–275.

126. de, V. M., Kiosses, W. B., & Shattil, S. J. (2004). Proximal, selective, and dynamic interactions between integrin alpha-IIbbeta3 and protein tyrosine kinases in living cells. *J Cell Biol, 165,* 305–311.

127. Arias–Salgado, E. G., Lizano, S., Sarkar, S., et al. (2003). Src kinase activation by direct interaction with the integrin beta cytoplasmic domain. *Proc Natl Acad Sci USA, 100,* 13298–13302.

128. Hannigan, G. E., Leung–Hagesteijn, C., Fitz–Gibbon, L., et al. (1996). Regulation of cell adhesion and anchorage-dependent growth by a new $\beta_1$-integrin-linked protein kinase. *Nature, 379,* 91–96.

129. Vijayan, K. V., Liu, Y., Li, T. T., & Bray, P. F. (2004). Protein phosphatase 1 associates with the integrin alphaIIb subunit and regulates signaling. *J Biol Chem, 279,* 33039–33042.

130. Kahn, M. J., Kieber–Emmons, T., Vilaire, G., et al. (1996). Effect of mutagenesis of GPIIb amino acid 273 on the expression and conformation of the platelet integrin GPIIb-IIIa. *Biochemistry, 35,* 14304–14311.

131. Leung–Hagesteijn, C. Y., Milankov, K., Michalak, M., Wilkins, J., & Dedhar, S. (1994). Cell attachment to extracellular matrix substrates is inhibited upon downregulation of expression of calreticulin, an intracellular integrin alpha-subunit-binding protein. *J Cell Sci, 107,* 589–600.

132. Rojiani, M. V., Finlay, B. B., Gray, V., & Dedhar, S. (1991). In vitro interaction of a polypeptide homologous to human Ro/SS-A antigen (calreticulin) with a highly conserved amino acid sequence in the cytoplasmic domain of integrin α subunits. *Biochemistry, 30,* 9859–9866.

133. Shock, D. D., Naik, U. P., Brittain, J. E., et al. (1999). Calcium-dependent properties of CIB binding to the integrin αIIb cytoplasmic domain and translocation to the platelet cytoskeleton. *Biochem J, 342,* 729–735.

134. Naik, U. P., Patel, P. M., & Parise, L. V. (1997). Identification of a novel calcium-binding protein that interacts with the integrin $\alpha_{IIb}$ cytoplasmic domain. *J Biol Chem, 272,* 4651–4654.

135. Vallar, L., Melchior, C., Plancon, S., et al. (1999). Divalent cations differentially regulate integrin $\alpha_{IIb}$ cytoplasmic tail binding to $\beta_3$ and to calcium- and integrin-binding protein. *J Biol Chem, 274,* 17257–17266.

136. Larkin, D., Murphy, D., Reilly, D. F., et al. (2004). ICln, a novel integrin alphaIIbbeta3-associated protein, functionally regulates platelet activation. *J Biol Chem, 279,* 27286–27293.

137. Shattil, S. J., O'Toole, T., Eigenthaler, M., et al. (1995). Beta 3-endonexin, a novel polypeptide that interacts specifically with the cytoplasmic tail of the integrin beta 3 subunit. *J Cell Biol, 131,* 807–816.

138. Eigenthaler, M., Hofferer, L., Shattil, S. J., & Ginsberg, M. H. (1997). A conserved sequence motif in the integrin beta3 cytoplasmic domain is required for its specific interaction with beta3-endonexin. *J Biol Chem, 272,* 7693–7698.

139. Kato, A., Kawamata, N., Tamayose, K., et al. (2002). Ancient ubiquitous protein 1 binds to the conserved membrane-proximal sequence of the cytoplasmic tail of the integrin alpha subunits that plays a crucial role in the inside-out signaling of alpha IIbbeta 3. *J Biol Chem, 277,* 28934–28941.

140. Burridge, K., Chrzanowska–Wodnicka, M. (1996). Focal adhesions, contractility, and signaling. *Annu Rev Cell Dev Biol, 12,* 463–518.

141. Priddle, H., Hemmings, L., Monkley, S., et al. (1998). Disruption of the talin gene compromises focal adhesion assembly in undifferentiated but not differentiated embryonic stem cells. *J Cell Biol, 142,* 1121–1133.

142. Monkley, S. J., Zhou, X. H., Kinston, S. J., et al. (2000). Disruption of the talin gene arrests mouse development at the gastrulation stage. *Dev Dyn, 219,* 560–574.

143. Pearson, M. A., Reczek, D., Bretscher, A., & Karplus, P. A. (2000). Structure of the ERM protein moesin reveals the FERM domain fold masked by an extended actin binding tail domain. *Cell, 101,* 259–270.

144. Garcia–Alvarez, B., de Pereda, J. M., Calderwood, D. A., et al. (2003). Structural determinants of integrin recognition by talin. *Mol Cell, 11,* 49–58.

145. Calderwood, D. A., Yan, B., de Pereda, J. M., et al. (2002). The phosphotyrosine binding (PTB)-like domain of talin activates integrins. *J Biol Chem, 277,* 21749–21758.

146. Tadokoro, S., Shattil, S. J., Eto, K., et al. (2003). Talin binding to integrin β tails: A final common step in integrin activation. *Science, 302,* 103–106.

147. Ulmer, T. S., Calderwood, D. A., Ginsberg, M. H., & Campbell, I. D. (2003). Domain-specific interactions of talin with the membrane-proximal region of the integrin beta3 subunit. *Biochemistry, 42,* 8307–8312.

148. Xing, B., Jedsadayanmata, A., & Lam, S. C. (2001). Localization of an integrin binding site to the C terminus of talin. *J Biol Chem, 276,* 44373–44378.

149. Yan, B., Calderwood, D. A., Yaspan, B., & Ginsberg, M. H. (2001). Calpain cleavage promotes talin binding to the beta 3 integrin cytoplasmic domain. *J Biol Chem, 276,* 28164–28170.

150. Martel, V., Racaud–Sultan, C., Dupe, S., et al. (2001). Conformation, localization, and integrin binding of talin depend on its interaction with phosphoinositides. *J Biol Chem, 276,* 21217–21227.

151. Inomata, M., Hayashi, M., Ohno–Iwashita, Y., et al. (1996). Involvement of calpain in integrin-mediated signal transduction. *Arch Biochem Biophys, 328,* 129–134.

152. Hayashi, M., Suzuki, H., Kawashima, S., Saido, T. C., & Inomata, M. (1999). The behavior of calpain-generated N- and C-terminal fragments of talin in integrin-mediated signaling pathways. *Arch Biochem Biophys, 371,* 133–141.

153. Kashiwagi, H., Schwartz, M. A., Eigenthaler, M., et al. (1997). Affinity modulation of platelet integrin $\alpha_{IIb}\beta_3$ by $\beta_3$-endonexin, a selective binding partner of the $\beta_3$ integrin cytoplasmic tail. *J Cell Biol, 137,* 1433–1443.

154. Barry, W. T., Boudignon–Proudhon, C., Shock, D. D., et al. (2002). Molecular basis of CIB binding to the integrin alpha IIb cytoplasmic domain. *J Biol Chem, 277,* 28877–28883.

155. Brown, E., Hooper, L., Ho, T., & Gresham, H. (1990). Integrin-associated protein: A 50-kD plasma membrane antigen physically and functionally associated with integrins. *J Cell Biol, 111,* 2785–2794.

156. Zent, R., Fenczik, C. A., Calderwood, D. A., et al. (2000). Class- and splice variant-specific association of CD98 with integrin beta cytoplasmic domains. *J Biol Chem, 275,* 5059–5064.

157. Dorahy, D. J., Berndt, M. C., Shafren, D. R., & Burns, G. F. (1996). CD36 is spatially associated with glycoprotein IIb-IIIa (alpha IIb beta 3) on the surface of resting platelets. *Biochem Biophys Res Commun, 218,* 575–581.

158. Slupsky, J. R., Seehafer, J. G., Tang, S. C., Masellis–Smith, A., & Shaw, A. R. (1989). Evidence that monoclonal antibodies against CD9 antigen induce specific association between CD9 and the platelet glycoprotein IIb-IIIa complex. *J Biol Chem, 264,* 12289–12293.

159. Israels, S. J., Millan–Ward, E. M., Easton, J., Robertson, C., & McNicol, A. (2001). CD63 associates with the alphaIIb beta3 integrin-CD9 complex on the surface of activated platelets. *Thromb Haemost, 85,* 134–141.

160. Fitter, S., Sincock, P. M., Jolliffe, C. N., & Ashman, L. K. (1999). Transmembrane 4 superfamily protein CD151 (PETA-3) associates with $\beta_1$ and $\alpha_{IIb}\beta_3$ integrins in haemopoietic cell lines and modulates cell-cell adhesion. *Biochem J, 338,* 61–70.

161. Chung, J., Gao, A. G., & Frazier, W. A. (1997). Thrombospondin acts via integrin-associated protein to activate the platelet integrin $\alpha_{IIb}\beta_3$. *J Biol Chem, 272,* 14740–14746.

162. Dorahy, D. J., Thorne, R. F., Fecondo, J. V., & Burns, G. F. (1997). Stimulation of platelet activation and aggregation by a carboxyl-terminal peptide from thrombospondin binding to the integrin-associated protein receptor. *J Biol Chem, 272,* 1323–1330.

163. Frazier, W. A., Gao, A. G., Dimitry, J., et al. (1999). The thrombospondin receptor integrin-associated protein (CD47) functionally couples to heterotrimeric Gi. *J Biol Chem, 274,* 8554–8560.

164. Fenczik, C. A., Sethi, T., Ramos, J. W., Hughes, P. E., & Ginsberg, M. H. (1997). Complementation of dominant suppression implicates CD98 in integrin activation. *Nature, 390,* 81–85.

165. Huang, M. M., Bolen, J. B., Barnwell, J. W., Shattil, S. J., & Brugge, J. S. (1991). Membrane glycoprotein IV (CD36) is physically associated with the Fyn, Lyn, and Yes proteintyrosine kinases in human platelets. *Proc Natl Acad Sci USA, 88,* 7844–7848.

166. Aiken, M. L., Ginsberg, M. H., Byers–Ward, V., & Plow, E. F. (1990). Effects of OKM5, a monoclonal antibody to glycoprotein IV, on platelet aggregation and thrombospondin surface expression. *Blood, 76,* 2501–2509.

167. Leung, L. L. K. (1984). Role of thrombospondin in platelet aggregation. *J Clin Invest, 74,* 1764–1772.

168. Lau, L. M., Wee, J. L., Wright, M. D., et al. (2004). The tetraspanin superfamily member CD151 regulates outside-in integrin alphaIIbbeta3 signaling and platelet function. *Blood, 104,* 2368–2375.

169. Armulik, A., Nilsson, I., von Heijne, G., & Johansson, S. (1999). Determination of the border between the transmembrane and cytoplasmic domains of human integrin subunits. *J Biol Chem, 274,* 37030–37034.

170. Sastry, S. K., & Horwitz, A. F. (1993). Integrin cytoplasmic domains: Mediators of cytoskeletal linkages and extra- and intracellular initiated transmembrane signaling. *Curr Opin Cell Biol, 5,* 819–831.

171. O'Toole, T. E., Mandelman, D., Forsyth, J., et al. (1991). Modulation of the affinity of integrin alpha$_{IIb}$beta$_3$ (GPIIb-IIIa) by the cytoplasmic domain of alpha$_{IIb}$. *Science, 254,* 845–847.

172. Hughes, P. E., O'Toole, T. E., Ylanne, J., Shattil, S. J., & Ginsberg, M. H. (1995). The conserved membrane-proximal region of an integrin cytoplasmic domain specifies ligand binding affinity. *J Biol Chem, 270,* 12411–12417.

173. Hughes, P. E., Diaz–Gonzalez, F., Leong, L., et al. (1996). Breaking the integrin hinge. A defined structural constraint regulates integrin signaling. *J Biol Chem, 271,* 6571–6574.

174. Kim, M., Carman, C. V., & Springer, T. A. (2003). Bidirectional transmembrane signaling by cytoplasmic domain separation in integrins. *Science, 301,* 1720–1725.

175. Li, R., Mitra, N., Gratkowski, H., et al. (2003). Activation of integrin αIIbβ3 by modulation of transmembrane helix associations. *Science, 300,* 795–798.

176. Luo, B. H., Carman, C. V., Takagi, J., & Springer, T. A. (2005). Disrupting integrin transmembrane domain heterodimerization increases ligand binding affinity, not valency or clustering. *Proc Natl Acad Sci USA, 102,* 3679–3684.

177. Kamata, T., Handa, M., Sato, Y., Ikeda, Y., & Aiso, S. (2005). Membrane-proximal α/β stalk interactions differentially regulate integrin activation. *J Biol Chem, 280,* 24775–24783.

178. Luo, B. H., Springer, T. A., & Takagi, J. (2004). A specific interface between integrin transmembrane helices and affinity for ligand. *PLoS Biol, 2,* 776–786.

179. Litvinov, R. I., Nagaswami, C., Vilaire, G., et al. (2004). Functional and structural correlations of individual alphaI-Ibbeta3 molecules. *Blood, 104,* 3979–3985.

180. Leisner, T. M., Wencel–Drake, J. D., Wang, W., & Lam, S. C. (1999). Bidirectional transmembrane modulation of integrin αIIbβ3 conformations. *J Biol Chem, 274,* 12945–12949.

181. O'Toole, T. E., Ylanne, J., & Culley, B. M. (1995). Regulation of integrin affinity states through an NPXY motif in the β subunit cytoplasmic domain. *J Biol Chem, 270,* 8553–8558.

182. Chen, Y.-P., Djaffar, I., Pidard, D., et al. (1992). Ser-752 → Pro mutation in the cytoplasmic domain of integrin β₃ subunit and defective activation of platelet integrin αIIbβ₃ (GPIIb-IIIa) in a variant of Glanzmann's thrombasthenia. *Proc Natl Acad Sci USA, 89,* 10169–10173.

183. Chen, Y. P., O'Toole, T. E., Shipley, T., et al. (1994). "Insideout" signal transduction initiated by isolated integrin cytoplasmic domains. *J Biol Chem, 269,* 18307–18310.

184. Fox, J. E. B., Goll, D. E., Reynolds, C. C., & Phillips, D. R. (1993). Evidence that activation of platelet calpain is induced as a consequence of binding of adhesive ligand to the integrin, glycoprotein IIb-IIIa. *J Cell Biol, 120,* 1501–1507.

# CHAPTER 9

# Thrombin Receptors

## Wadie F. Bahou

*Department of Medicine and Program in Genetics, State University of New York,
Stony Brook, New York*

## I. Introduction

Thrombin is a multifunctional serine protease that is distinctly unique among coagulation proteins, possessing both procoagulant and anticoagulant properties.[1] These properties bestow upon thrombin a critical role in the regulation of normal hemostasis and exaggerated thrombosis. The precursor protein prothrombin is synthesized exclusively in the liver as an inactive zymogen with a tightly regulated plasmatic concentration of 100 μg/mL and a circulating half-life of about 72 hours.[2] The conversion of prothrombin to its 38-kD active protease (α-thrombin) requires the assembly of a prothrombinase complex comprised of prothrombin, coagulation factor Xa, and the active cofactor Va on the surface of a cellular phospholipid membrane requiring calcium ions (see Chapter 19).[3,4] α-Thrombin contains a typical active site catalytic triad composed of histidine 365, aspartic acid 419, and serine 527 that are in close proximity and responsible for the charge relay system evident in all serine proteases.[5] An anion binding exosite confers binding specificity to various thrombin substrates.[6] Although thrombin substrate specificity is similar to that of trypsin for small peptides, thrombin's action on larger proteins is more highly selective.[7-9] Virtually all thrombin substrates contain an arginine or lysine adjacent to the sessile bond, a representative cleavage site containing the consensus X-Pro-Arg-↓-X.[7] Known thrombin substrates include coagulation factor VIII, factor V, and factor XIII[10-13]; critical cleavages of fibrinogen at Arg[16]-Gly[17] within the Aα-chain and Arg[14]-Gly[15] within the Bβ-chain regulate fibrin polymerization and stabilization of the fibrin clot.[7]

Activated thrombin circulates briefly in the circulation, rapidly neutralized by antithrombin III,[14,15] with less physiologically relevant inhibition by α2-macroglobulin[16] and α1-antiprotease.[17] Thrombin binding with endothelial cell thrombomodulin results in a functional transformation, converting thrombin's procoagulant properties to those of an anticoagulant, by cleaving and activating protein C.[18,19] Acti-vated protein C inactivates factors Va and VIIIa, serving to attenuate thrombin generation.[20] Despite this rapid clearance of soluble thrombin, evidence exists that clot-bound thrombin may remain active for up to 2 weeks, providing a continual source for thrombus propagation and cellular activation.[21]

## II. Cellular Actions of Thrombin

In addition to its key role in regulating the coagulation cascade, thrombin demonstrates profound effects on diverse cells involved in regulation of the thrombotic response (Fig. 9-1). Thrombin is a potent platelet activator *in vivo*[22-26] and *in vitro*.[27,28] In addition to its well-characterized effects on intracellular signaling cascades, resulting in platelet degranulation and aggregation (discussed later), thrombin also induces cell-surface expression of the adhesion molecule P-selectin and CD40 ligand,[29,30] and activation of the integrin αIIbβ3, which binds fibrinogen and von Willebrand factor (VWF) to mediate aggregation.[31,32] Thrombin activation of endothelial cells stimulates the production of prostaglandin I2,[33] platelet activating factor,[34,35] plasminogen activator inhibitor-1 (PAI-1),[36] and platelet-derived growth factor.[37] *In vivo*, thrombin causes release of VWF[38] and expression of P-selectin[39] at the cytoplasmic membrane on endothelial cells. These actions result in platelet and leukocyte adhesion on the endothelial surface. Thrombin also stimulates endothelial cell release of the profibrinolytic protein tissue-type plasminogen activator (t-PA),[40-42] thereby serving to modulate fibrin degradation on the cell surface. Thrombin-stimulated endothelial cells also produce increased amounts of nitric oxide, which is a potent vasodilator and platelet inhibitor (Chapter 13).[43] Activated endothelial cells also undergo change in shape and increase their permeability in response to thrombin,[44] resulting in local transudation of proteins and edema.[45] Direct effects on monocyte chemotaxis,[46]

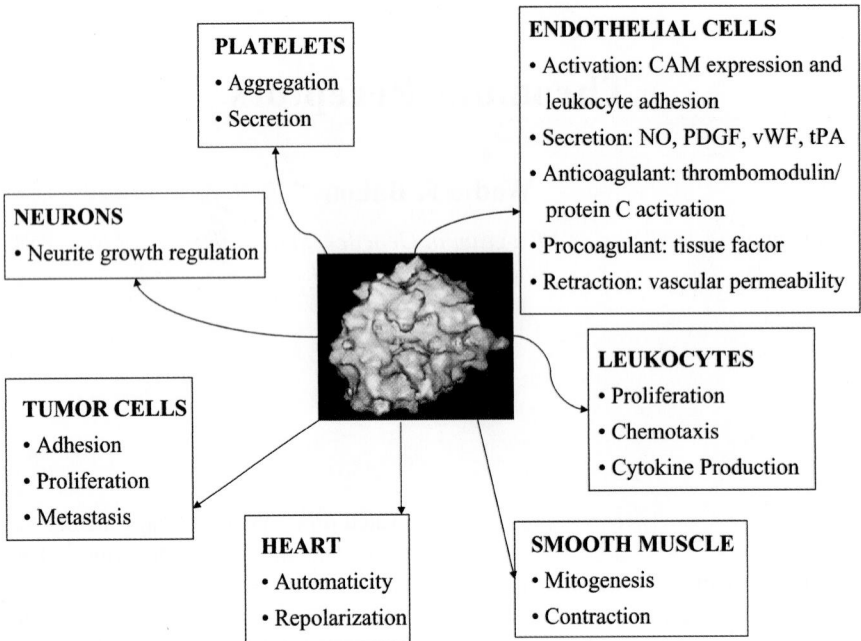

**Figure 9-1.** Cellular effects of thrombin (center). Refer to text for details. CAM, cellular adhesion molecule; NO, nitric oxide; PDGF, platelet-derived growth factor; tPA, tissue plasminogen activator; VWF, von Willebrand factor.

neutrophil activation,[47] and both monocyte and neutrophil adhesion[48–51] link thrombin activity with inflammatory responses. Known thrombin cellular effects involving other diverse tissues not directly involved in hemostatic regulation such as neurons,[52] periodontitis,[53] tumor metastases,[54–56] and cardiac effects[57,58] have been described and are outlined in Fig. 9-1.

Thrombin communicates with cells through a unique class of cell-surface protease activated receptors (PARs)[59,60] that are members of a larger family of G protein-coupled seven transmembrane domain receptors that include α- and β-adrenergic receptors, muscarinic, and serotonin receptors, among others.[61] Unlike the latter group of receptors, which signal through standard receptor/ligand interactions, PARs are activated by a unique proteolytic cleavage within the first extracellular loop, or by synthetic peptidomimetics corresponding to the new N-termini generated after receptor cleavage.[59,62] To date, four PARs have been identified and characterized (enumerated chronologically), three of which (PAR1[59,60], PAR3[63], and PAR4[64]) are substrates for thrombin (Table 9-1). In contrast, PAR2 is activated by mast cell tryptase, trypsin, and the tissue factor–factor VIIa–factor Xa complex, but not by thrombin.[65–67] At the molecular level, the PARs share a common structural organization, suggesting evolution from a common ancestral gene. Evolutionary cross-species differences between murine and human receptor systems suggest functional divergences yet to be fully elucidated (discussed later).

## A. Thrombin-Cleaved PARs

**1. PAR1.** Initially isolated and characterized in 1991, human PAR1 is the predominant receptor for thrombin-mediated platelet activation, and the prototypic and best-characterized cell-surface functional thrombin receptor.[59,60] PAR1 messenger RNA (mRNA) is differentially and widely expressed in human tissues, although it is readily detected in cells intimately involved in hemostatic regulation (i.e., platelets, vascular endothelial cells, and vascular smooth muscle cells).[59,68,69] The PAR1-predicted translation product encodes a 425-amino acid backbone that is heavily glycosylated, resulting in a molecular weight approximating 70 kD by sodium dodecyl sulfate–polyacrylamide gel electrophoresis.[59,60,68,70] PAR1 has a relatively long aminoterminal exodomain consisting of 99 amino acid residues uniquely evolved to facilitate interaction with its specific protease, and to enhance receptor recognition and cleavage (Fig. 9-2).[70–88] PAR1 is efficiently cleaved and activated by α-thrombin with an $EC_{50}$ for thrombin-stimulated phosphoinositide hydrolysis approximating 50 to 200 pM.[59,63] At concentrations approximating 10 nM, α-thrombin cleaves almost all of the cell-surface PAR1 within 1 minute.[81] Other receptor-activating proteases include granzyme A,[89] meizothrombin,[90] trypsin,[91] and factor Xa.[92,93] Inactivating proteases (i.e., those that cleave elsewhere within the receptor) include cathepsin G,[94] plasmin,[95] elastase,[96] and proteinase 3.[96] A schema outlining the current model for PAR1 cleavage and activation is outlined in Fig. 9-2.

**Table 9-1: Classification of PARs**

|  | PAR1 | PAR2 | PAR3 | PAR4 |
|---|---|---|---|---|
| Chromosome | 5q13.3 | 5q13.3 | 5q13.3 | 19p12 |
| Number of amino acids | 425 | 397 | 374 | 385 |
| Gene structure | 2 exons | 2 exons | 2 exons | 2 exons |
| Cleavage site | $Arg^{41} \downarrow Ser^{42}$ | $Arg^{42} \downarrow Asp^{43}$ | $Lys^{38} \downarrow Thr^{39}$ | $Arg^{47} \downarrow Gly^{48}$ |
| Agonist | Thrombin ($EC_{50}$ ~100 pM) Granzyme A Mezothrombin Trypsin Plasmin Factor Xa | Trypsin Tryptase Factor Xa Factor VIIa MT-SP1 | Thrombin ($EC_{50}$ ~0.2 nM) | Thrombin ($EC_{50}$ ~5 nM) Trypsin Cathepsin G |
| Peptide agonist[a] | SFLLR[b] | SLIGK | — | GYPGQV |
| Hirudinlike sequence | DKEYPF | — | FEEFP | — |
| Platelets/megakaryocytes |  |  |  |  |
| Human | Yes | No | Minimal | Yes |
| Murine | No | No | Yes | Yes |
| Vascular expression |  |  |  |  |
| Endothelium | Yes | Yes | Yes | Yes[c] |
| Smooth muscle cells | Yes | No | No | ? |
| Leukocytes | Yes | Yes | No | Yes[c] |
| Vascular functions | Platelet activation Thrombosis Embryonic development | Inflammation | Platelet activation Thrombosis[c] | Platelet activation Thrombosis Inflammation |

[a]Refers to endogenous agonist.

[b]Activates both PAR1 and PAR2.

[c]Best characterized in mice.

**2. PAR3.** In humans, PAR3 mRNA is expressed in the bone marrow and vascular endothelial cells, but minimally in platelets with no more than 150 to 200 PAR3 receptors *per* platelet.[97] *In situ* hybridization revealed that PAR3 is highly expressed in murine splenic and bone marrow megakaryocytes — cells that do not express murine PAR1.[63] Like PAR1, PAR3 is also a thrombin substrate, with 20 nM α-thrombin cleaving up to 80% of PAR3-transfected COS cells within 5 minutes.[63] The $EC_{50}$ for thrombin-induced phosphoinositide hydrolysis is comparable with that of PAR1 (0.2 nM). Thrombin inactivation by its proteolytic inhibitor D-phenylalanyl-L-prolyl-L-arginyl-chloromethyl ketone (PPACK) abrogates signaling in PAR3-transfected cells, even at concentrations as high as 1 μM, confirming that proteolytic cleavage is necessary for receptor activation.[63] Human PAR1 and PAR3 have about 27% amino acid sequence similarity.[63] The specific thrombin cleavage recognition sequence found in the amino-terminal exodomain of human PAR3 is located at the $^{35}LPIK \downarrow TFRGAP^{44}$ junction (between amino acid residues $Lys^{38}$ and $Thr^{39}$).[63,98] Similar to the structure of PAR1 is the presence within PAR3 of a hirudinlike sequence

($^{48}FEEFP^{52}$), which is also identified (carboxyl to the cleavage site) and facilitates the protease–receptor interaction.[71–73,77] Like PAR1, PAR3 is postulated to utilize the hirudinlike domain for thrombin interaction, supported by mutagenesis studies within this sequence that shift the dose–response curve for thrombin activation toward the right by about 10-fold.[63]

The human PAR3 cleavage site $^{35}LPIK \downarrow TFRGAP^{44}$ is specific and responsive to thrombin alone,[63,98] and PAR3 cleavage is prevented by substitution of proline for threonine at amino acid residue 39.[63] Other arginine/lysine-specific serine proteases (factor Xa, trypsin, factor VIIa, t-PA, or plasmin) demonstrate little or no ability to cleave PAR3.[63] Unlike PAR1, synthetic peptidomimetics based on the N-terminal sequence of the newly generated tethered ligand (TFRGAP and TFRGAPPNS) resulted in little or no activation of PAR3 even at concentrations as high as 100 μM.[63] Despite this observation, evidence suggests that PAR3's molecular mechanism of activation is comparable with that of PAR1 (i.e., signaling mediated *via* a tethered ligand mechanism). Thus, substitution of $Ala^{40}$ for $Phe^{40}$ generated a receptor that failed to signal upon thrombin

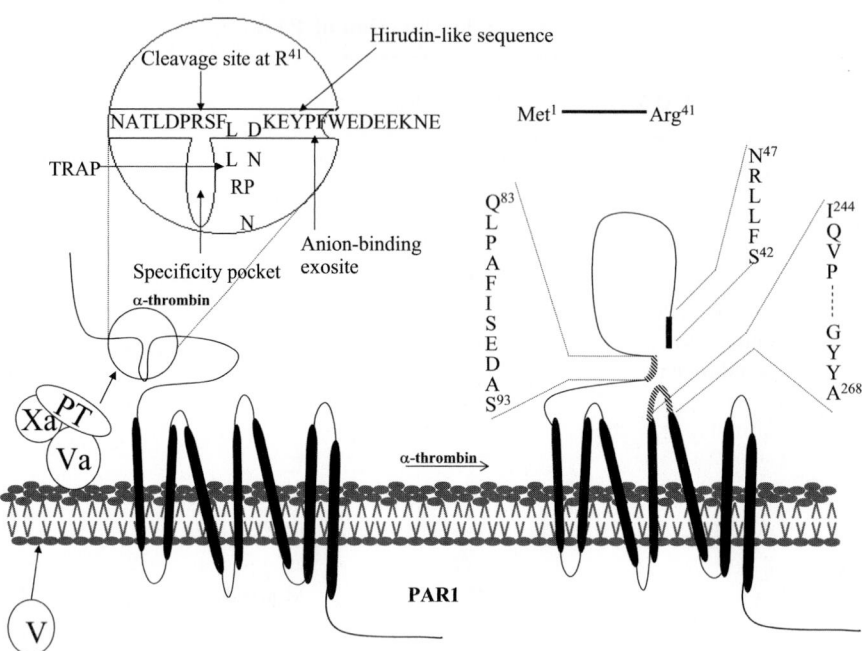

**Figure 9-2.** Molecular mechanism of PAR1 activation. α-Thrombin is generated from the assembly of the prothrombinase complex upon a platelet membrane. The PAR1 N-terminus interacts with thrombin's anion exosite through an acidic hirudinlike binding domain sequence [50]DKYEPF[55] (found carboxyl to the cleavage site),[60,73,74] thereby facilitating cleavage at the LDPR[41]/S[42]FLLR sessile bond.[60,73] A comparable [38]LDPR↓S[42] sequence is also utilized as the thrombin cleavage site in protein C when thrombin is specifically bound to thrombomodulin.[20,74] Free thrombin unattached to thrombomodulin is unable to cleave protein C because of the inhibitory effect of the P3 aspartate residue in [39]DPR[41].[75,76] Formation of the thrombin–thrombomodulin complex results in a conformational change in thrombin's active center that can accommodate the [39]DPR[41] sequence with subsequent cleavage of protein C.[71] By extrapolation from the thrombin–thrombomodulin model, thrombin binding to the hirudinlike domain would predictably cause a similar conformational change, with subsequent receptor cleavage.[71–73,77] This irreversible cleavage generates a new amino terminus, leading to "self-activation" *via* a tethered ligand mechanism. Critical intramolecular interactions involving the second extracellular loop (hatched),[78] the tethered ligand (solid rectangle), and an 11-mer peptide within the long extracellular domain (hatched)[79,80] may serve as the "binding pocket" for downstream receptor-coupling events,[62,81] although intermolecular activating mechanisms may exist with much less efficiency.[62] Synthetic peptides containing at least the first five amino acids (SFLLR) of the tethered ligand (referred to as thrombin receptor activating peptides, or TRAP) are able to effect receptor activation independently of receptor proteolysis[61,72,73,82] with evidence that the protonated Ser[42] amino group and the Phe[43] side chain appear to be especially critical for peptidomimetic function.[83–86] After thrombin cleavage, the Met[1]-Arg[41] peptide fragment is released from the platelet surface,[87] with evidence that the cleavage product may directly activate platelets through a poorly characterized mechanism.[88]

cleavage.[63] Phe[40] would be postulated to be critical in PAR3's intramolecular interaction and signaling, analogous to Phe[43] in PAR1.[83]

**3. PAR4.** PAR4 is the most recently identified member of the PAR family and a third thrombin receptor in humans.[64] Alignment of the human PAR4's 397-amino acid sequence with other known PARs indicates that it has about 33% amino acid homology with PAR1, PAR2, and PAR3.[64,98] However, PAR4's amino-terminal exodomain and intracellular cytoplasmic domain have very little or no amino acid sequence similarity to the corresponding regions in the three other PARs.[64] Northern blot analysis demonstrated that the PAR4 gene is widely expressed in human tissues, with especially high expression patterns in lung, pancreas, thyroid, testis, and small intestine.[64] No PAR4 expression is detected in the brain, kidney, or spinal cord,[64] although studies in murine models suggest that PAR4 may be

expressed on endothelial cells and leukocytes.[99] Although less abundant than PAR1, PAR4 mRNA is readily detected in human platelets by reverse transcription–polymerase chain reaction (RT-PCR).[64]

Like PAR1 and PAR3, thrombin is a primary activator of PAR4, although it is 1 to 2 logs less responsive to thrombin when compared with PAR1 and PAR3.[64,100] The EC$_{50}$ for thrombin- and trypsin-induced phosphoinositide hydrolysis mediated by PAR4 activation is about 5 nM.[64] The likely explanation for this higher thrombin requirement for PAR4 activation is the absence of the thrombin-interactive hirudin-like exodomain, which is present within PAR1 and PAR3.[64,100] Other arginine–lysine-specific serine proteases, including factors VIIa, IXa, and XIa, plasmin, and urokinase, demonstrate little or no ability to activate PAR4.[64] Minimal activation responses were evident using high (100 nM) concentrations of factor Xa.[64] Embedded within its first exodomain is the thrombin cleavage recognition site,

[44]PAPR↓GYPGQV[53], specifically cleaved at the Arg[47] and Gly[48] bond.[64] Mutagenesis of Arg[47] to Ala[47] generates a receptor unable to respond to either thrombin or trypsin.[64] More recent evidence exists that PAR4 is activatable by neutrophil-derived cathepsin G, with cleavage at the identical Arg[47]↓Gly[48] sessile bond.[101]

Comparable with human PAR1 and PAR2 (but not with PAR3), human PAR4 may be activated by peptidomimetics that are homologous to the N-terminal amino acid sequences of the tethered ligand.[64,102] The Tyr[49] in the second position of the tethered ligand is important in the function of this domain.[102] Its peptidomimetic GYPGQV is able to activate PAR4, although at a higher $EC_{50}$ (100 μM)[64] compared with that of SFLLRN for PAR1 ($EC_{50} \approx 5$ μM).[59] Another peptidomimetic AYPGKF is 10-fold more potent than GYPGQV and elicited PAR4 responses comparable with that of thrombin.[102]

### B. Thrombin Noncleavable PAR

**1. PAR2.** Human PAR2 was identified and characterized during a search for substance K receptor homologues.[66] Unlike the other PARs, which are cleaved by thrombin, PAR2 is unique in that it is cleaved by trypsin or by mast cell-derived tryptases, but not by thrombin.[103,104] PAR2 is therefore not a thrombin receptor, although its structural and functional mechanisms of activation are comparable with that of other PARs.

Northern blot analysis of mRNA demonstrated that the PAR2 gene is widely expressed in human tissues, with especially high expression patterns in the liver, kidney, pancreas, small intestine, and colon.[102,103] It is also detected in human endothelial cells, keratinocytes, and smooth muscles of the aorta and coronary artery, but not in the brain or skeletal muscles.[103,105] PAR2 is not expressed in human platelets.[98,106]

Similar to other PARs, human PAR2 is cleaved by its proteases at a specific [33]SKGR↓SLIGK[41] cleavage site.[103] Noncleavable PAR2 mutants cannot be activated by trypsin, although activation by synthetic peptidomimetics is unaffected.[66] PAR2, although not activatable by thrombin, may be activated directly by coagulation factor Xa, and possibly by the tissue factor–factor VIIa complex.[67] Like PAR1, PAR2 appears to mediate proliferative responses as evaluated in primary cultures of vascular endothelial cells.[105,106] Additional functional roles for PAR2 in embryonic development,[107] inflammation,[108–110] vascular hemodynamic responses,[111,112] and thrombosis have been suggested[67] (discussed later).

## III. Role of PARs in Disease

### A. Thrombosis

Although thrombin is clearly one of the most potent activators of human platelets *in vitro*, distinguishing thrombin's procoagulant effects (mediated through the coagulation cascade) from its diverse cellular effects (e.g., platelet, endothelial, vascular smooth muscle cell activation, etc.) in the regulation of thrombosis remains a concerted challenge.[113] This issue is relevant for strategies designed to inhibit PAR cellular activation (discussed later). Local concentrations of thrombin generation *in vivo* are diluted by normal blood flow and by endogenous plasma inhibitors. Nonetheless, thrombin concentrations within the vicinity of a thrombus may be as high as 140 nM, with persistence for up to 10 days,[114] and blood coagulation monitoring suggests that the effective thrombin concentration involved in fibrinogen cleavage approximates 2.5 nM.[115] These concentrations are greater than the $EC_{50}$ for human platelet activation.

*In vivo*, a critical role for thrombin-associated platelet activation in the generation and propagation of arterial thromboses has been suggested, based on the efficacy of heparin to reduce platelet and fibrinogen deposition in models of arterial injury.[116] That this represents an antithrombin effect — as opposed to an antifactor Xa effect — has been suggested by the use of low-molecular weight heparins.[117] More selective antithrombin therapy using the synthetic competitive thrombin inhibitor argatroban[26] or hirudin, which does not directly block other mediators of platelet aggregation such as thromboxane $A_2$, serotonin, adenosine diphosphate (ADP), or collagen, abolishes the development of mural thrombosis and markedly limits platelet deposition to a single layer *in vivo* after arterial injury.[117] As important, evidence exists that thrombin inhibition is associated with diminished platelet activation markers *in vivo*,[118] and that direct PAR1 inhibition can abrogate arterial thrombosis in platelet-dependent nonhuman primate animal models.[119,120] Finally, a PAR1 polymorphic sequence from the 5'-regulatory region of the gene has been implicated as exerting a protective effect in men for development of venous thrombosis.[121]

### B. Atherosclerotic Coronary Artery Disease

PARs may be involved in atherosclerotic pathogenesis, specifically coronary artery disease, and to some extent restenosis postangioplasty. PAR1 expression and message are upregulated in advanced human atherosclerotic vessels,[122] after experimental injury in animal models,[123] and are induced by balloon angioplasty in the rat and baboon.[124] *In situ* hybridization of atherosclerotic vessels demonstrated high levels of PAR1 expression in areas rich in macrophages, and in areas of proliferating smooth muscle cells and mesenchymallike intimal cells.[123] Earlier lesions consisting of fatty streaks demonstrated PAR1 expression in the intima alone, without significant expression in the underlying media.[122] In addition to its rapid upregulation during mechanical vascular injury, PAR1 expression is transiently upregulated by low — but not high — shear stress.[125]

Although PAR1 antisense oligodeoxynucleotides can inhibit vascular smooth muscle cell thrombin responsiveness *in vitro,*[126] they appear ineffective in blocking neointimal hyperplasia following intimal injury *in vivo.*[127] In contrast, neointimal smooth muscle cell accumulation can be attenuated using anti-PAR1 antibodies in a rat angioplasty model, confirming that PAR1 activation is involved in the proliferation and accumulation of neointimal smooth muscle cells induced by balloon injury, and that strategies directed at PAR1 blockade may have clinical relevance for postangioplasty restenosis.[128] Similar beneficial results using neointimal hyperplasia as the end point have been suggested using a carotid injury model in PAR1[-/-] mice, although the results did not reach statistical significance.[129]

## C. Inflammation

PARs provide an important link between hemostatic and inflammatory pathways,[130] best exemplified by thrombin's well-recognized proinflammatory effects, and reinforced by the regulatory functions of kininogens and bradykinin in PAR1 activation (discussed later). Thrombin activation of endothelial cells facilitates leukocyte adhesion by stimulation of cell-surface adhesion molecule expression,[38,131,132] and enhances platelet activating factor formation, a potent neutrophil stimulant.[35] Thrombin also stimulates interleukins 6 and 8 release from endothelial cells,[133] and increases endothelial cell permeability, in part by gap junction regulation.[44,134] Thrombin is also a mitogen for lymphocytes[135] and is chemotactic for monocytes.[46,136,137] Linking coagulant protein generation with proinflammatory stimuli is not limited to thrombin effects, however. It is known that factor XII,[138] factor VIIa,[139,140] and factor Xa[141] also activate various cellular types, with evidence that some of factor X's effects are mediated by binding to effector protease receptor 1.[142] More recently, coagulation factor Xa and possibly the tissue factor–VIIa complex have been shown to activate PAR2,[67] known to be expressed on vascular endothelial cells,[106] and upregulated during endothelial cell stimulation using various inflammatory stimuli.[143]

Disseminated intravascular coagulation with small-vessel thrombosis can develop in the setting of strong systemic inflammatory stimuli such as bacteremia, and/or with congenital deficiencies that modulate thrombin production, such as antithrombin III deficiency or factor V Leiden.[144] Furthermore, proinflammatory roles for coagulant proteins *in vivo* have been suggested by studies demonstrating amelioration of bacterial-induced endotoxic shock after treatment with activated protein C[145] or antitissue factor antibodies.[146] Similarly, in a murine model of injury-induced crescentic glomerulonephritis, hirudin was shown to attenuate renal injury and glomerular crescent formation by reducing T-cell and macrophage infiltration, and fibrin deposition.[147]

Comparable results were evident in PAR1[-/-] mice, suggesting that this receptor was involved in inflammatory cell-mediated renal injury.[147] Similar supportive observations implicate PARs and/or the coagulant pathways in the pathogenesis of bronchial inflammation,[110,148] mast cell-induced neurogenic inflammation,[108] periodontal disease,[53,149] antigen-induced arthritis,[150] and myocardial infarction.[151] Although a majority of reports suggest that PAR2 functions in the proinflammatory pathway, some evidence exists that PAR2 agonists may be protective in murine models of inflammation or ischemia.[110,152,153] Given their pleotropic signaling pathways, coupled with evidence for PAR cross-talk,[106] more research will be required to dissect specifically the biochemical functions of PARs in specific human diseases.

# IV. Molecular and Developmental Genetics of PARs

## A. PAR Molecular Genetics

Discordancy analysis of PCR products from a human–rodent hybrid cell mapping panel decisively established the PAR1 gene to be located on human chromosome 5.[154] In addition, cytogenetic localization using fluorescence *in situ* hybridization (FISH) detected a unique fluorescent signal localizing the PAR1 gene to the region 5q13.[154] Subsequent studies confirmed that the genes for PAR2 and PAR3 were clustered within the same region of the human genome, colocalizing with the gene for PAR1.[97,103,155–157] Furthermore, the human PAR3 gene is uniquely located within another gene (IQGAP2) identified within the PAR gene cluster.[158] In contrast, the PAR4 gene is located in the p12 region of chromosome 19.[65] Genomic characterization of PARs 1 through 4 has now been completed, with evidence for further similarities.[154–157,159] The gene structures share a common characteristic in that the coding regions of all these genes are contained within two exons. The first of the two exons is smaller, encompassing approximately 30 amino acids. The majority of the coding sequence is located in the larger second exon, which contains the protease cleavage site. Exon 1 and exon 2 are separated by introns of variable size (Fig. 9-3).[160–162] These similarities among the PAR genes suggest a relatively recent gene duplication event, and evolution from a common ancestral gene.

## B. PAR Developmental Genetics

Murine model systems have provided powerful tools for studies designed to elucidate the role of the coagulation system in cellular and embryonic development. Although the studies remain ongoing, there appear to be some parallels that clearly identify a key role for thrombin-generating

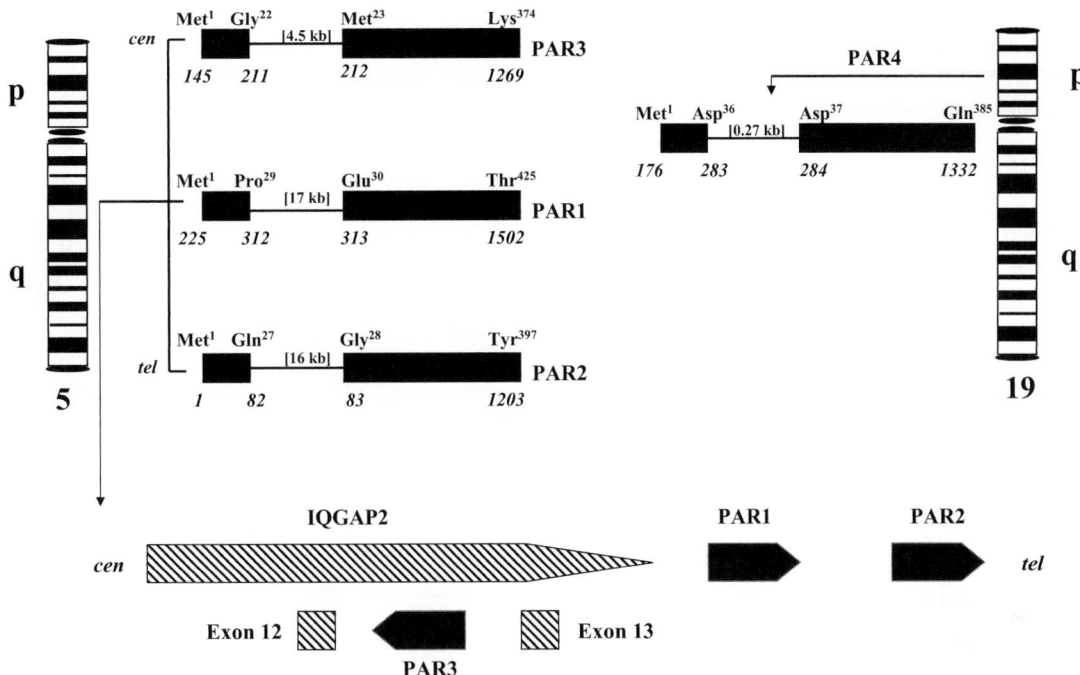

**Figure 9-3.** Molecular genetics of PARs. The genes for PAR1, PAR2, and PAR3 are located at chromosome 5q13, whereas the PAR4 gene is located on chromosome 19p12. The interorder chromosomal assignments for PARs 1 through 3 are updated from previously published reports to conform with more recently generated consensus data from the human genome databases.[97,159,160] Human PAR3 is situated within the 12th intron of IQGAP2, transcribed off the negative strand,[158] in contrast to the murine PAR homologue, which is similarly organized, but transcribed off the forward strand[161] (not shown). The nucleotide (italics) and amino acid designations are from published complementary DNA sequences for PAR1,[59] PAR2,[103] PAR3,[63,103] and PAR4,[64] aligned with genomic sequences using the BLAST homology program.[162] For each PAR gene, the smaller box represents exon 1 and the larger box represents exon 2. The approximate size of intronic sequences is delineated in brackets in kilobases (kb). Cen, centromere; p, short arm; q, long arm; tel, telomere.

pathways (and/or their cellular receptors) in functions beyond normal hemostasis, with relevance to normal embryogenesis. For example, a fundamental role for PAR1 in embryonic development has been suggested by generating thrombin receptor-deficient (PAR1⁻/⁻) knockout mice. Disruption of the murine gene resulted in approximately 50% lethality of homozygous (PAR1⁻/⁻) mice at embryonic days 9 to 10, although the other half developed normally, apparently with no hemorrhagic diatheses.[163,164] Although PAR1⁻/⁻ fibroblasts lost thrombin responsiveness, PAR1⁻/⁻ platelets continued to respond normally to thrombin, now known to be related to PAR3 signaling as the predominant murine thrombin receptor.[165-167] It is intriguing that these embryonic deaths corresponded to the development of the circulatory system, and that the growth deficits of PAR1⁻/⁻ mice displayed delayed vascularization of the yolk sac, possibly intimating a role for PAR1 in vasculogenesis.[163,164] Furthermore, these initial observations provided strong presumptive evidence that thrombin-induced platelet activation was not causally implicated in the defective hemostasis evident in these mice.

Examination of other transgenic knockout studies relevant to thrombin generation and PAR signaling are outlined

in Figure 9-4. It is intriguing that the timing and phenotype of embryonic lethality in PAR1⁻/⁻ embryos corresponded to that of mice deficient in factor V[168] and prothrombin,[169,170] consistent with a role for thrombin generation or cellular signaling in vascular remodeling or its integrity. More pronounced degrees of embryonic lethality were evident in tissue factor-deficient knockout mice resulting from fatal hemorrhage, presumably associated with defects in vascular integrity.[171-173] In contrast, targeted disruption of the factor VII gene (the protease cofactor for tissue factor) had no effect on embryonic development, although factor VII⁻/⁻ embryos did display postpartum hemorrhage.[174] One explanation for the normal development of factor VII⁻/⁻ mice is possible trace amounts of transplacental (maternal) factor VII transmission from heterozygote mothers. More recently, similar results have been seen with targeted disruption of the factor X gene, with factor X⁻/⁻ mice displaying partial embryonic lethality.[175] Nearly one third of factor X⁻/⁻ pups died by embryonic day 11.5 to 12.5, although no histological defects in the vasculature of affected embryos or their yolk sacs were observed.[175] Targeted disruption of the thrombomodulin gene results in an embryonic lethal mutation by embryonic day 9.5 not clearly related to defective

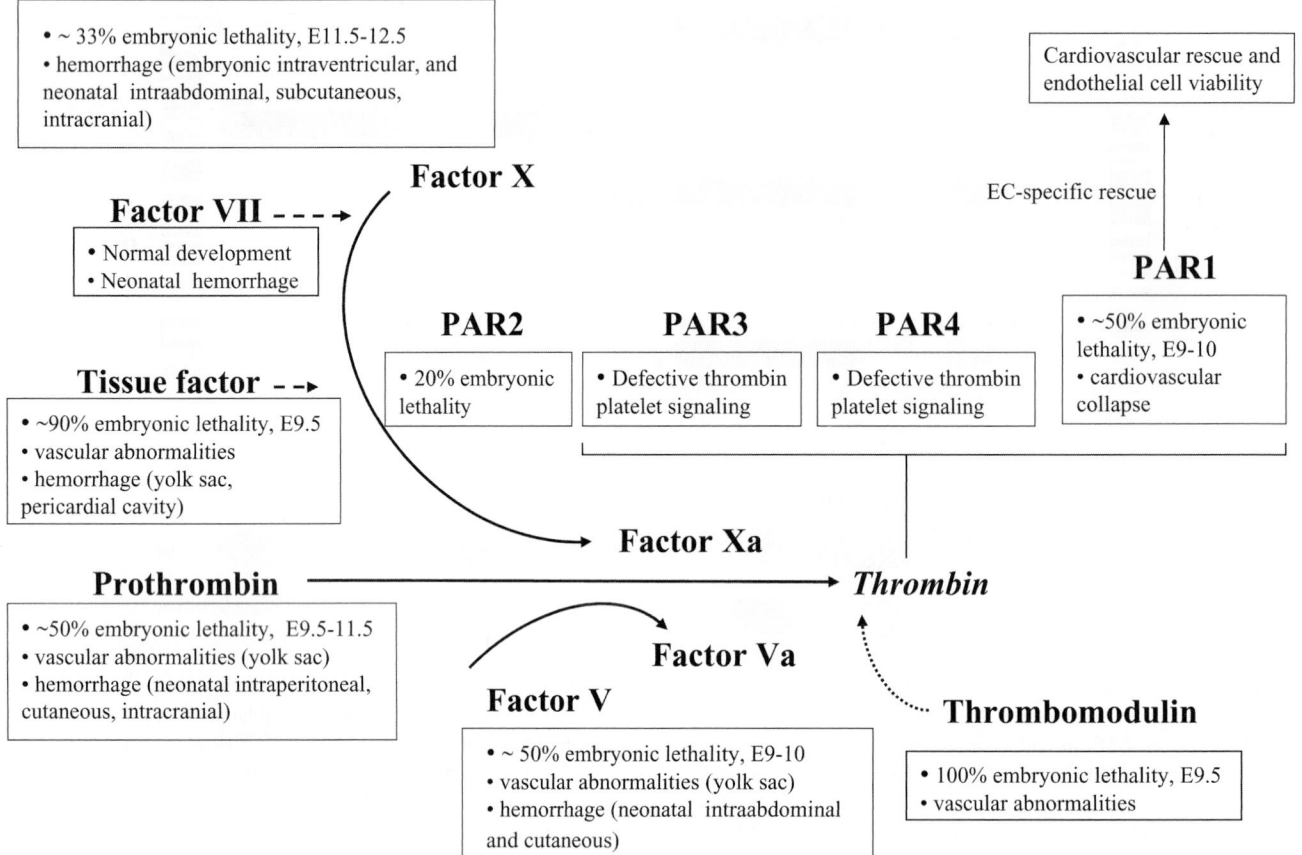

**Figure 9-4.** Phenotypes of transgenic knockout mice targeting coagulation proteins involved in thrombin-generating pathways. Solid arrows refer to those coagulant proteins directly involved in thrombin generation, dashed arrows are those leading indirectly to thrombin generation by way of factor Xa generation, and the dotted arrow refers to thrombomodulin, a natural endothelial cell anticoagulant. E, embryonic day; EC, endothelial cell. Refer to text for details.

vasculogenesis, but apparently resulting from dysfunctional maternal–embryonic interactions in the parietal yolk sac.[176] These collective observations clearly demonstrate that the serine proteases involved in thrombin-generating pathways play key roles in embryonic development, presumably related to defects in vascular integrity or development, distinct from platelet activation. Interestingly, there appears to be minimal coordinate expression of prothrombin and PAR1 during embryonic development, suggesting that other endogenous PAR1 activators may exist during development.[177–179] Targeted expression of PAR1 to the vascular endothelial compartment of PAR1-null mice rescued the lethal phenotype, with evidence for normal vascular development and resolution of the abnormal bleeding causally implicated in intrauterine death. Thus, loss of PAR1 signaling in endothelial cells probably explains the death at midgestation seen with knockout of tissue factor, factor V, prothrombin, and possibly factor X. In the latter cases, intrauterine death is the result of defective thrombin generation, whereas in

PAR1-null mice, the intrauterine collapse is related to loss of the endothelial cellular receptor mediating the thrombin response.[180]

A unifying explanation for these collective defects strongly implicated coagulation proteases in the maintenance of (endothelial cell) vascular wall integrity.[113] Although the vascular fragility seen in PAR1$^{-/-}$ embryos is likely related to defective thrombin signaling, ancillary roles of the tissue factor–factor VIIa–factor Xa complex remain possible. Thrombin's effects are protean: (a) it generates a provisional fibrin scaffold that attracts migrating endothelial cells, (b) it increases expression of vascular endothelial growth factor (VEGF) receptors, (c) it promotes production and extravasation of matrix proteins, (d) it loosens endothelial cells from the extracellular matrix by upregulating and activating metalloproteinases,[181] (e) it induces endothelial cell proliferation and migration, and (f) it activates the hypoxia-inducible transcription factor HIF-1α, thereby upregulating expression and production of numerous angiogenic molecules.[182] Similarly, tissue factor is known to stabilize newly forming fragile vessels by recruitment of supporting pericytes,[172] and the tissue factor–factor VIIa complex could stimulate angiogenesis by downregulating the angiogenic inhibitor thrombospondin 1, while upregulating VEGF, fibroblast growth factor 5, collagenases, and their receptors (u-PAR).[182] The proteolytic activity of factor VIIa (independent of thrombin formation) is also involved in angiogenic signaling. Recent evidence suggests that PAR2 is likely activated by factor VIIa in the vessel wall. This activation is most evident when sufficient tissue factor and factor Xa are present, the anticipated scenario in the angiogenic endothelium. Interestingly, the tissue factor–factor VIIa–factor Xa complex has been demonstrated to activate PAR1.[93] Thus, although thrombin is likely to be the primary PAR1 activator for vessel stabilization, upstream coagulation proteases may also be involved. At this point, it remains unknown if angiogenesis in pathological conditions is impaired in PAR1$^{-/-}$ mice.

## V. Thrombin Signaling in Platelets

The fate of PAR1 differs depending on the cellular type, although general mechanisms of regulation typically seen with noncleavable G protein-coupled receptors (GPCRs) are evident. These mechanisms include receptor desensitization and internalization, although uniquely adapted for the self-activating nature of a cleavable receptor that contains its own "ligand."[144]

### A. Cell-Surface Consequences of PAR Activation

Based on antibody binding studies, there are about 1500 to 2000 PAR1 receptors *per* human platelet, and on a resting platelet approximately two thirds of the receptors are located on the plasma membrane.[70,84] Prior to activation, the remainder are present in the membranes of the intracellular surface connecting system, a structure that is contiguous with the platelet plasma membrane, and that is exposed upon platelet activation.[183] PAR1 blocking antibodies directed at the cleavage site attenuate human platelet activation by thrombin alone,[68,70,84] whereas antibodies directed to a discrete region of the N-terminal extension block both thrombin- and peptidomimetic-induced activation.[79] Platelet activation by nonthrombin agonists exposes those PAR1 receptors initially present in the canalicular system, thereby increasing the number of cell-surface receptors available for cleavage. Thrombin activation efficiently cleaves the majority of the receptors, with subsequent internalization or receptor shedding into platelet microparticles during platelet aggregation.[183] Unlike the situation seen in vascular endothelial cells (discussed later), platelets do not appear to have a storage pool of

receptors and have minimal capacity for protein synthesis. Platelets are thus essentially capable of responding once to thrombin, not an unexpected evolutionary adaptation given their short circulatory half-lives and functionally terminal roles in hemostasis. Although it is evident that other proteases such as cathepsin G,[94,184] plasmin,[95] and chymotrypsin[185] are capable of disabling PAR1 by cleavage at alternative sites within the receptor, their physiological role as regulators of thrombin-induced platelet activation remain unclear.

The fate of PARs on generally quiescent and longer surviving vascular endothelial cells is different than that of platelets. PAR1 is readily detectable on endothelial cells,[79,92] with estimates as high as 10$^6$ receptors/cell.[186] Stimulation of endothelial cells with thrombin results in receptor internalization and cellular desensitization. Activated PAR1 is rapidly uncoupled from signaling and subsequent receptor activation, although nearly half of the receptors remain present on the cell surface. The uncoupling is most likely related to intracytoplasmic phosphorylation events, possibly involving one or more members of the GRK family of receptor kinases.[187] Unlike other GPCRs, the receptor does not recycle, but rather is delivered to the lysosomal compartment for degradation. Like platelets, endothelial cell PAR1 is used once, but unlike platelets, the endothelial cell-surface compartment is readily restored within 2 hours from a preformed intracellular pool.[92,188,189] Subsequent protein synthesis replenishes both the cell-surface and intracellular storage pool of receptors. The availability of a rapidly mobilized intracellular storage pool of receptors ensures that stationary and long-lived endothelial cells are able to respond to subsequent thrombin stimuli within a relatively short period, a situation of less relevance for an activated platelet with a terminal function. Comparable mechanisms regulating PAR2 activation and desensitization appear evident in endothelial cells, with suggestions that heterologous cross-desensitization of PAR1 and PAR2 may represent an additional method for regulating PAR function.[106,183]

### B. Molecular Mechanisms of Intracellular Signaling

Upon PAR activation, thrombin-induced intermediate signaling pathways involve phosphoinositide hydrolysis, protein phosphorylation, an increase in intracytosolic free calcium, and suppression of cAMP synthesis, signals that are generally identifiable at thrombin concentrations approximating 100 pM (Fig. 9-5).[190–199] These distinct but converging pathways ultimately lead to cytoskeletal actin reorganization and integrin activation, the critical common pathways mediating platelet adhesion and aggregation. The heterotrimeric αβγ G proteins mediate these diverse signaling cascades, and platelets are known to contain G$_{\alpha s}$, G$_{\alpha i}$, G$_{\alpha q}$, and G$_{\alpha 12/13}$, implicated in platelet thrombin signaling events (see Chapter 16).[200]

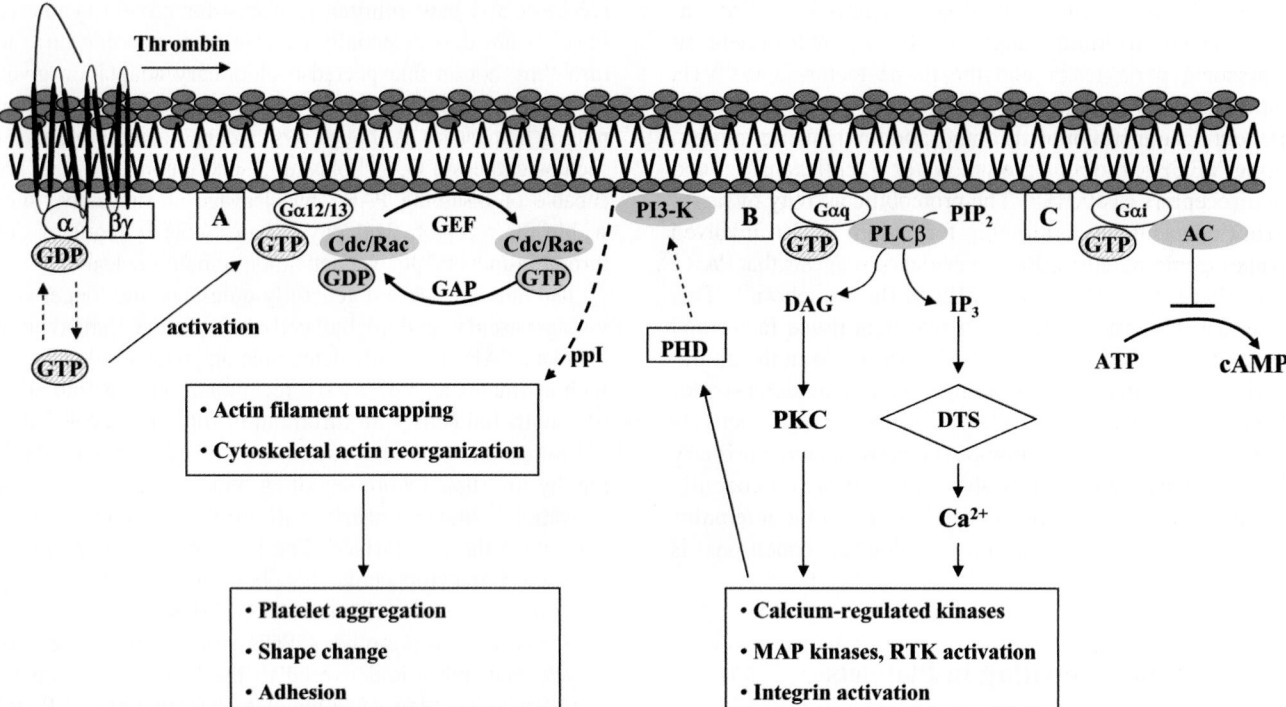

**Figure 9-5.** Thrombin-mediated intracellular signaling cascades. In quiescent platelets, PARs are associated with heterotrimeric $\alpha\beta\gamma$ G proteins, with the $\alpha$-subunit maintained in the inactive (GDP-bound) state. G protein activation is regulated by the binding and hydrolysis of GTP. Agonist stimulation by thrombin promotes the release of GDP with replacement by cytosolic GTP. A conformational change of the $\alpha$-subunit results in its dissociation from $G_{\beta\gamma}$, leaving both in their active states. The subsequent hydrolysis of the GTP-bound form to GDP is mediated by GTPase activating proteins (GAPs), which result in reassociation with $G_{\beta\gamma}$, pending subsequent cycles of receptor-mediated signaling. Upon thrombin stimulation, PARs couple to members of the $G_{\alpha12/13}$ (A), $G_{\alpha q}$ (B), and $G_{\alpha i}$ (C) G protein families, with downstream activation to a host of intracellular effectors, associated with $G_\alpha$ charging of GTP. $G_{\alpha i}$ inhibits adenylyl cyclase (AC), resulting in the diminished levels of cAMP and enhanced platelet responsiveness. The $G_{\alpha q}$-subunit activates phospholipase C$\beta$ (PLC$\beta$), resulting in the generation of 1,4,5-inositol trisphosphates (IP$_3$) from phosphatidylinositol 4,5-bisphosphate (PIP$_2$), and the release of intracytosolic calcium from the platelet dense tubular system (DTS). Both of these signaling pathways are partially (if not fully) mediated through PAR1 activation.[190] Concomitant activation of protein kinase C (PKC) from diacylglycerol (DAG) provides for a pathway linked to activation of calcium-regulated kinases, MAP kinases,[191] receptor tyrosine kinases (RTK), and integrins, among others. The $G_{\alpha12/13}$-subunits are coupled to guanine nucleotide exchange factors (GEFs) such as p115RhoGEF,[192–194] although the upstream signals leading to thrombin-induced GTP-charging of Rac1 and Cdc42 remain poorly characterized. Activated phosphoinositide kinases (such as PI3-K)[195] facilitate recruitment and attachment of various signaling proteins (including those containing pleckstrin homology domains [PHD]) to the platelet inner membrane.[196–198] The generation of D3-, D4-, or D5-phosphorylated phosphoinositides (ppI) appears to be terminal events for PAR1-mediated actin filament uncapping and assembly.[199]

Platelets minimally contain at least two of the known phospholipase C isoforms: PLC$\beta$ and PLC$\gamma$.[201] PLC$\beta$ is primarily responsible for the immediate burst of phosphoinositide hydrolysis that occurs during platelet activation by thrombin, with resultant generation of diacylglycerol and 1,4,5-inositol trisphosphate.[202–205] The release of intracellular calcium and subsequent rise in cytosolic Ca$^{2+}$ concentration results in tyrosine phosphorylation of other downstream effector proteins, and activation of other signaling cascades including those linked to integrin $\alpha_{IIb}\beta_3$ activation, MAP kinase pathways, and so forth.[200,202] In general, PLC$\beta$1 and PLC$\beta$3 respond optimally to $G_\alpha$ (especially members of the $G_{\alpha q}$ family), whereas PLC$\beta$2 may respond better to $G_{\beta\gamma}$.[200]

Thrombin results in the release of arachidonic acid from human platelets[206] and endothelial cells.[207] Phospholipase A$_2$ releases arachidonate from the C2 position of membrane phospholipids such as phosphatidylcholine, believed to occur primarily in the membranes of the dense tubular system.[208] Both GTP and GTP$\gamma$S can cause arachidonate release in permeabilized platelets, and pertussis toxin can inhibit thrombin-induced release, suggesting that the response to thrombin is mediated by $G_{\beta\gamma}$ derived from $G_i$. Although PLA$_2$ has been shown to be a substrate for phosphorylation by MAP kinases, it would appear that phosphorylation does not regulate arachidonate release in platelets.[209]

The molecular mechanisms that link thrombin activation to the profound shape change that develops during platelet adhesion and aggregation remain poorly characterized, although they retain as their penultimate end point the process of actin polymerization. Platelets from $G_{\alpha q}$-deficient mice are globally defective in platelet activation, demonstrating unresponsiveness to a variety of physiological platelet agonists,[201] whereas $G_{\alpha 12/13}$-subunits appear to be involved in platelet shape change, with a more selective impairment of thrombin-induced fibroblast migration change.[211–213] Although the effector proteins linking these proximal membrane events to cytoskeletal actin polymerization remain largely uncharacterized, evidence exists that the guanine nucleotide exchanger p115RhoGEF functions as an intermediary for $G_{\alpha 12/13}$.[192,193] Rho guanosine triphosphates (GTPases) form a subgroup of the Ras superfamily that are largely involved in the regulation of cytoskeletal organization in response to extracellular stimuli. Like all members of the superfamily, the activity of Rho GTPases is determined by the ratio of their GTP/GDP-bound forms, which are regulated by the opposing effects of guanine nucleotide exchange factors (GEFs) — known to enhance the exchange of bound GDP for GTP — and the GTPase-activating proteins (GAPs), which increase the intrinsic rate of hydrolysis of bound GTP (Fig. 9-5). In addition, Rho-like GTPases are regulated further by guanine nucleotide dissociation inhibitors, which can inhibit both the exchange of GTP and the hydrolysis of bound GDP.[213]

Remodeling of actin filaments transforms the normally quiescent discoid platelet to a contractile sphere extruding lamellae and filopodia (see Chapter 4). The latter are presumably involved in mediating fibrin–platelet or platelet–platelet interaction, whereas the former structures may be involved in platelet adhesion for plugging vascular leaks.[199] $\alpha$-Thrombin stimulation of platelets specifically leads to activation of the Rho GTPases rac1 and cdc42, although their different postactivation subcellular distributions suggest divergent roles in platelet actin assembly.[214] Although the precise role of cdc42 in platelet activation remains unknown, a more refined role for rac1 in actin filament uncapping and actin polymerization has been postulated.[199] GTP-bound rac1 can directly bind and stimulate phosphatidylinositol-3-kinase (PI3-K), which is also coactivated in thrombin-stimulated platelets.[215] Activated PI3-K results in the generation of polyphosphoinositides, which can lead to cortical actin assembly and the formation of lamellipodia.[216,217] Although D3-containing phosphoinositides appear to be necessary for actin assembly mediated through some platelet receptors, their generation does not appear to be required for actin assembly and platelet shape change initiated through PAR1 stimulation, suggesting the presence of other pathways for thrombin-stimulated actin reorganization.[218]

## C. Dual-Receptor Signaling in Platelets

A dual-receptor platelet signaling system exists in both humans and mice, initially elucidated by transgenic knockout studies. In mice, PAR3 and PAR4 are the primary thrombin receptors, whereas in humans PAR1 and PAR4 are the primary receptor systems.[100] PAR3-deficient mice developed normally and had no spontaneous bleeding but displayed an abnormal response to thrombin, with loss of platelet responses to low-dose (1 nM), but not to high-dose (30 nM), thrombin. GYPGKF, a mouse PAR4 peptidomimetic, was found to activate both wild-type and PAR3-deficient mice.[100] It is now known that PAR4 is responsible for the delayed residual signaling seen at high concentrations of thrombin in mouse platelets. PAR3 is the primary mediator of thrombin signaling whereas PAR4 subserves a role as "secondary" receptor in mouse platelets. PAR3 may actually function as a cofactor for PAR4 cleavage and activation at low thrombin concentrations, because PAR3 activation in the absence of PAR4 does not result in signaling.[219] It remains unclear whether this phenomenon occurs *via* receptor heterodimerization,[220] or whether a comparable mechanism is present in thrombin-mediated human activation systems.

A dual-receptor system for thrombin-induced platelet activation is also present in humans.[221] Thus, PAR4-blocking immunoglobulin (Ig)G had no effect on platelet aggregation in response to low-dose (1 nM) thrombin. In contrast, PAR1-inhibitory IgG completely attenuated platelet activation at 1 nM thrombin, but only modestly at 30 nM thrombin. Blocking both PAR1 and PAR4 resulted in complete abrogation of platelet response to concentrations as high as 30 nM thrombin. Further supportive evidence for this dual-receptor model system has been presented by studying intracellular cytosolic calcium patterns. Separation of a biphasic $Ca^{2+}$ response in platelets using PAR antagonists and agonists reveals two discrete components — a rapid spike response mediated through PAR1 activation and a slower, more prolonged response *via* PAR4 activation.[82,222] These data suggest the presence of the same dual-receptor system evident in mouse platelets, but in the case of human platelets, PAR1 is the primary mediator of thrombin-induced platelet activation whereas PAR4 assumes the same "backup" receptor role. These data would suggest that complete inhibition of thrombin-mediated platelet activation requires strategies targeting both PAR1 and PAR4.

## D. Non-PAR Thrombin Receptors

The platelet GPIb-IX-V complex, a receptor for both VWF and thrombin, is one of the most abundant glycoprotein receptors on the platelet surface with nearly 25,000 receptors *per* platelet (see Chapter 7).[223] GPIb is a heterodimer

composed of a larger 143-kD α-polypeptide (GPIbα) covalently linked *via* a disulfide bond to a smaller 22-kD β-
subunit (GPIbβ). The GPIbα-GPIbβ complex forms a
noncovalent complex with platelet GPIX and GPV, in
approximate stoichiometric ratios (GPIb to IX to V) of 2 :
2 : 1.[224] Although GPIb clearly functions as a primary platelet cytoadhesive receptor for VWF interaction during situations of high shear, there is abundant evidence that it also
binds to α-thrombin.[225–227] The binding site for thrombin has
been localized to a sequence spanning amino acid residues
271 to 284 in the extracytoplasmic domain of GPIbα.[228]
Peptidomimetics of this site inhibit α-thrombin binding to
GPIb as well as platelet activation and aggregation induced
by subnanomolar thrombin concentration. The presence
within this region of negatively charged amino acid residues
as well as sulfated tyrosine residues presumably function as
thrombin-binding sequences comparable with those identifiable in hirudin.[229,230] The binding affinities for this region
display a $k_d$ approximating $10^{-8}$ M, consistent with a previously characterized "high-affinity" binding site for thrombin.[225,228,230,231] The presence of alternative high-affinity
binding sites on platelets discrete from the GPIb-IX-V
complex has also been suggested.[232]

The relationship between these high-affinity binding
sites and thrombin-mediated platelet activation remains
somewhat controversial. Thus, the interaction has been variably suggested to be functionally relevant,[227,230,233,234] irrelevant,[235] or serves as a negative modulator by functioning as
a thrombin-sequestering system.[236] Nonetheless, based on
various antibody inhibition studies and incomplete thrombin responses evident in platelets from patients with
Bernard–Soulier syndrome that lack GPIb,[230,234,237,238] it is
reasonable to conclude that optimal thrombin responsiveness requires an intact GPIb complex, although not directly
functioning in the signal transduction mechanism. This
ancillary effect of GPIb does not appear to be mediated *via*
a direct cell-surface interaction between GPIb and platelet
PARs,[144,230,234,237,238] but rather may be relevant for localizing
α-thrombin to the platelet membrane for facilitation of its
action on cell-surface PAR substrates. This dichotomy may
be explained by recent works suggesting a role for GPV in
thrombin-induced platelet signaling. It is known that α-
thrombin cleaves GPV, an event that was thought to be
unrelated to platelet activation.[239–242] However, a putative
negative modulatory function of GPV in platelet activation
was recently described in GPV-null mice.[243] GPV$^{-/-}$ platelets
demonstrate exaggerated response to α-thrombin.[243] According to the proposed model, thrombin-induced GPV cleavage
unmasks a GPIbα thrombin binding site within the GPIb-
IX-V complex, allowing both proteolytically active and
inactive thrombin to initiate platelet activation, which is
apparently mediated by thrombin-induced ADP secretion
and P2Y$_{12}$ binding.[244] These data suggest that GPV inhibits
the ability of thrombin to function as a receptor ligand, a

novel observation that may help elucidate the role of this
complex in thrombin functions.

## VI. Development of PAR Inhibitors

Agents that block thrombin-induced platelet responsiveness
could potentially be useful therapeutic agents for coronary
artery and cerebrovascular syndromes, although the evidence
for this remains largely inferential and based on efficacy of
thrombin inhibition. There are several strategies by which
thrombin-induced platelet activation may be blocked: (a) by
inhibiting thrombin production, (b) by inactivating circulating thrombin, (c) by blocking interaction of thrombin with
its receptors and/or receptor cleavage, (d) by inhibiting intramolecular binding of the tethered ligand with the body of the
receptor, or (e) by targeted intracellular signaling inhibition.
With regard to strategies a and b, total thrombin blockade
may be problematic, because it blocks thrombin's important
fibrinogen-cleaving functions. The ideal agent, therefore, is
one that will selectively inhibit platelet activation at the
receptor level, without interfering with thrombin's normal
hemostatic functions. Monoclonal or polyclonal antibodies
directed against the PAR1 and/or PAR4 cleavage sites inhibit
*in vitro* platelet thrombin responsiveness at variable concentrations of thrombin.[79,84] Likewise, antisense oligonucleotides
directed against PAR1 can effectively modulate cellular
thrombin responses, although such strategies are not applicable for anucleate human platelets.[106,126] Finally, the presence of dual thrombin receptor systems on human platelets
creates a unique problem for thrombin blockade. Because
inhibition of PAR1 attenuates platelet thrombin responses at
low — but not high — thrombin concentrations,[221] isolated
use of PAR1 inhibitors would expectedly result in incomplete
inhibition. Whether such inhibition would be effective as an
antithrombotic in situations of "low thrombin generation"
remains to be established.

Selective approaches to inhibit PAR-specific activation
have recently been described, although they remain in the
developmental stages. With relevance to thrombin–receptor
"interactions," previous studies had indicated that bradykinin,[221,245] and both high-[246] and low-molecular weight kininogens[247] inhibit thrombin-induced platelet activation. A core
pentapeptide derived from these sequences (RPPGF) was
subsequently shown to inhibit platelet activation and thrombin cleavage of PAR1 by binding to the PAR1 tethered
ligand sequence NATLDPRSFLLR (Fig. 9-2). This appeared
to function as a selective platelet thrombin inhibitor because
it failed to interfere with ADP, collagen, or U46619-induced
platelet activation.[248] *In vivo* confirmation of its potential
efficacy has been established in an electrically induced
canine coronary thrombosis model, an effect apparently
brought about by inhibition of PAR-mediated platelet
activation.[240]

Inhibitory peptide antagonists have also been designed by applying iterations of the PAR1 peptide sequence.[249] For example, based on analysis of key functional and spatial groups of the PAR1 agonist peptide SFLLRN, a prototype PAR1 antagonist containing a rigid indole template has been developed.[250] This compound (RWJ-58259) was found to bind selectively and inhibit PAR1 activation without displaying PAR1 agonist or thrombin inhibitory activity. The peptide inhibits intracellular calcium mobilization and platelet aggregation, and displays no activity against PAR2, PAR3, or PAR4. Flow cytometry confirmed that this compound directly inhibited PAR1 activation and internalization, without affecting exodomain cleavage. At high concentration of α-thrombin and SFLLRN, RWJ-58259 failed to completely block platelet activation, most likely related to PAR4 saturation. Subsequent *in vivo* studies in a nonhuman primate thrombosis model demonstrated reduced platelet deposition in developing mural thrombi, with significantly extended time to occlusion in electrolytically injured carotid arteries.[119] Although this compound remains in the early phases of development, it suggests that PAR1-specific inhibition may ameliorate thrombus formation in large-vessel thrombosis models.

Finally, an approach to target peptide residues corresponding to human PAR1 and PAR4 intracellular 3 (i3) loops has recently been described.[251] This group used palmitoylation to partition peptides across the platelet lipid bilayer, while simultaneously serving as hydrophobic anchors designed to embed the peptide chemically into the intracellular lipid interface. PAR1-based palmitoylated peptides ("pepducins") were subsequently shown to block platelet aggregation to low-dose, but not to high-dose, thrombin, as predicted because of persistent PAR4 activation. In contrast, PAR4-based pepducins partially inhibited platelet aggregation to fully saturating concentrations of thrombin — capable of activating both PAR1 and PAR4 — providing evidence for cross-inhibition of PAR1. These initial *in vitro* results[251] were subsequently confirmed using an *in vivo* murine model in which preinfusion of the PAR4 pepducin effectively prolonged the bleeding time (as a measure of abnormal hemostasis), while partially inhibiting systemic platelet activation (a surrogate marker for an antithrombotic effect).[252] On the most parsimonious level, pepducins essentially function as dominant negative PAR inhibitors by blocking the interaction of PAR1/PAR4 i3 loops with their cognate G proteins. This implies a functional interaction of the pepducin with its homologous segment within the intact i3 loop. Alternatively, the micromolar excess of intracellular pepducins may simply function as "sponges" that efficiently bind effector G proteins, obviating a need for direct receptor interaction and blockade. Such an effect would likely occur on the platelet inner membrane, given the structure of pepducins and the submembranous localization of the activated G proteins. Although both models imply a direct interaction

of pepducins with G proteins, the cross-inhibitory effects of PAR4 pepducins on PAR1 activation may provide additional plausible clues. One explanation is the presence of a short stretch of basic residues within both PAR1 and PAR4 i3 loops that may function as key sequences for G protein receptor coupling. This explanation, however, is not entirely consistent with prior evidence that other intracellular segments contain key recognition sequences for G protein coupling.[253] Alternative, more speculative mechanisms could include functional receptor cooperativity as previously suggested for endothelial cell PAR1 and PAR2,[106] receptor clustering (physically inhibited by membrane-tethered pepducins), or requirements for physical interactions such as heterodimerization.[254] The latter explanation is especially intriguing given evidence for PAR3/PAR4 cooperativity,[219] although confirmatory support for heterodimerization is lacking.[255] Furthermore, although C-terminal domains have been implicated in such GPCR heterodimers, no prior evidence exists that the i3 loop may participate in comparable functions. A number of issues and challenges persist for successful progress in this area[256]: (a) the optimal therapeutic intervention should effectively inhibit both PAR1 and PAR4, suggesting that PAR4-optimized pepducins may be superior targets; (b) although the PAR4 pepducin inhibited systemic platelet activation, its efficacy as an antithrombotic awaits testing in a more formalized model of arterial thrombosis; (c) the PAR3/PAR4 thrombin receptor system in murine platelets appears to have a mechanism of activation distinct from that of human platelets, in which PAR1 is the predominant thrombin receptor, with no evidence for PAR4 cooperativity; thus, *in vivo* efficacy of PAR4-based pepducins as antiplatelet agents in humans remains unestablished; and (d) PAR1 displays a wide distribution in cells and tissues (including vascular endothelial cells), and its inhibition in these cells may be associated with untoward effects, such as inhibition of nitric oxide release — a potent inhibitor of platelet activation.

# References

1. Leung, L. L., Lawrence, L., & Gibbs, C. (1997). Modulation of thrombin's procoagulant and anticoagulant properties. *Thromb Haemost, 78,* 577–580.
2. Blanchard, R. A., Furie, B. C., Jorgensen, J. J., et al. (1981). Acquired vitamin K-dependent carboxylation deficiency in liver disease. *Proc Natl Acad Sci USA, 76,* 491.
3. Mann, K. G., Jenny, R. J., & Krishnaswamy, S. (1984). Cofactor proteins in the assembly and expression of blood clotting enzyme couples. *Ann Intern Med, 57,* 915.
4. Rosing, J., Tans, G., Govers–Riemslag, J. W. P., et al. (1980). The role of phospholipids and factor Va in the prothrombinase complex. *J Biol Chem, 255,* 274.
5. Neurath, H. (1984). Evolution of proteolytic enzymes. *Science, 224,* 350–357.

6. Maraganore, J. M., Bourdon, P., Jablonski, J., Ramachandran, K., & Fenton, J. W. (1990). Design and characterization of Hirulogs: A novel class of bivalent peptide inhibitors of thrombin. *Biochemistry, 29,* 7095–7101.

7. Blomback, B., Blomback, M., Hessel, B., et al. (1967). Structure of N-terminal fragments of fibrinogen and specificity of thrombin. *Nature, 215,* 1445.

8. Morita, T., Kato, H., Iwanaga, S., et al. (1977). New fluorogenic substrates for α-thrombin, factor Xa, kallikreins and urokinase. *J Biol Chem, 82,* 1495.

9. Sherry, S., & Troll, W. (1954). The action of thrombin on synthetic substrates. *J Biol Chem, 208,* 95.

10. Pittman, D., & Kaufman, R. (1988). Proteolytic requirements for thrombin activation of anti-hemophilic factor (factor VIII). *Proc Natl Acad Sci USA, 85,* 2429.

11. Nesheim, M. E., & Mann, K. G. (1979). Thrombin-catalyzed activation of single-chain bovine factor V. *J Biol Chem, 254,* 10952.

12. Esmon, C. T. (1979). The subunit structure of thrombin-activated factor V: Isolation of activated factor V, separation of subunits and reconstitution of biological activity. *J Biol Chem, 1979, 254.*

13. Schwartz, M. L., Pizzo, S. V., Hill, R. L., et al. (1973). Human factor XIII from plasma and platelets. *J Biol Chem, 248,* 1395.

14. Abildgaard, U. (1969). Binding of thrombin to antithrombin III. *Scand J Clin Lab Invest, 24,* 89.

15. Damus, P. S., Hicks, M., & Rosenberg, R. D. (1973). Anticoagulant action of heparin. *Nature, 246,* 355.

16. Downing, M. R., Bloom, J. W., & Mann, K. G. (1978). Comparison of the inhibition of thrombin by three plasma protease inhibitors. *Biochemistry, 17,* 2649.

17. Matheson, N. R., & Travis, J. (1976). Inactivation of human thrombin in the presence of human α₁-proteinase inhibitor. *Biochem J, 159,* 495.

18. Esmon, C. T. (1995). Thrombomodulin as a model of molecular mechanisms that modulate protease specificity and function at the vessel surface. *FASEB J, 9,* 946–955.

19. Dang, Q., Vindigni, A., & DiCera, E. (1995). An allosteric switch controls the procoagulant and anticoagulant activities of thrombin. *Proc Natl Acad Sci USA, 92,* 5977–5981.

20. Shuman, M. A. (1976). The measurement of thrombin in clotting blood by radioimmunoassay. *J Clin Invest, 58,* 1249–1258.

21. Bar-Shavit, R., Eldor, A., & Vlodavsky, I. (1989). Binding of thrombin to subendothelial extracellular matrix: Protection and expression of functional properties. *J Clin Invest, 84,* 1096–1104.

22. Eidt, J., Allison, P., Nobel, S., et al. (1989). Thrombin is an important mediator of platelet aggregation in stenosed canine coronary arteries with endothelial injury. *J Clin Invest, 84,* 18–27.

23. Fitzgerald, D., & Fitzgerald, G. (1989). Role of thrombin and thromboxane A2 in reocclusion following coronary thrombolysis. *Proc Natl Acad Sci USA, 86,* 7585–7589.

24. Hansen, S., & Harker, L. A. (1988). Interruption of acute platelet-dependent thrombosis by the synthetic antithrombin PPACK. *Proc Natl Acad Sci USA, 85,* 3184–3188.

25. Heras, M., Chesebro, J., Penny, W., Bailey, K., Badimon, L., & Fuster, V. (1989). Effects of thrombin inhibition on the development of acute platelet–thrombus deposition during angioplasty in pigs: Heparin versus hirudin, a specific thrombin inhibitor. *Circulation, 79,* 657–665.

26. Jang, I.-K., Gold, H. K., Ziskind, A. A., et al. (1989). Prevention of platelet-rich arterial thrombosis by selective thrombin inhibition. *Circulation, 81,* 219–225.

27. Davey, M., & Luscher, E. (1967). Actions of thrombin and other coagulant and proteolytic enzymes on blood platelets. *Nature, 216,* 857–858.

28. Berndt, M., & Phillips, D. (1981). Platelet membrane proteins: Composition and receptor function. In J. L. Gordon (Ed.), *Platelets in biology and pathology* (pp. 43–74). New York: Elsevier/North Holland Biomedical Press.

29. Stenberg, P. E., McEver, R. P., Shuman, M. A., Jacques, Y. V., & Bainton, D. F. (1985). A platelet alpha-granule membrane protein (GMP-140) is expressed on the plasma membrane after activation. *J Cell Biol, 101,* 880–886.

30. Henn, V., et al. (1998). CD40 ligand on activated platelets triggers an inflammatory reaction of endothelial cells. *Nature, 391,* 591–594.

31. Hughes, P. E., & Pfaff, M. (1998). Integrin affinity modulation. *Trends Cell Biol, 8,* 359–364.

32. Savage, B., Shattil, S., & Ruggeri, Z. M. (1992). Modulation of platelet function through adhesion receptors. A dual role for glycoprotein IIb-IIIa (integrin alpha IIb beta 3) mediated by fibrinogen and glycoprotein Ib-von Willebrand factor. *J Biol Chem, 267,* 11300–11306.

33. Weksler, B., Ley, C., & Jaffe, E. A. (1986). Stimulation of endothelial cell prostacyclin production by thrombin trypsin, and the ionophore A23187. *Ann NY Acad Sci, 485,* 349–368.

34. Prescott, S., Zimmerman, G., & McIntyre, T. (1979). Human endothelial cells in culture produce platelet-activating factor when stimulated by thrombin. *Proc Natl Acad Sci USA, 63,* 580–587.

35. Carveth, H., Shaddy, R., Whatley, R., McIntyre, T., Prescott, S., & Zimmerman, G. (1992). Regulation of platelet-activating factor (PAF) synthesis and PAF-mediated neutrophil adhesion to endothelial cells activated by thrombin. *Semin Thromb Hemost, 18,* 126–134.

36. Yamamoto, C., Kaji, T., Sakamoto, M., & Koizumi, F. (1992). Effect of endothelin on the release of tissue plasminogen activator and plasminogen activator inhibitor-1 from cultured human endothelial cells and interaction with thrombin. *Thromb Res, 67,* 619–624.

37. Daniel, T., Gibbs, V. C., Milfay, D., et al. (1986). Thrombin stimulates *c-sis* gene expression in microvascular endothelial cells. *J Biol Chem, 261,* 9579–9582.

38. Hattori, R., Hamilton, K. K., Fugate, R. D., McEver, R. P., & Sims, P. J. (1989). Stimulated secretion of endothelial von Willebrand factor is accompanied by rapid redistribution to the cell surface of the intracellular granule membrane protein GMP-140. *J Biol Chem, 264,* 7768–7771.

39. Zimmerman, G., McIntyre, T., & Prescott, S. M. (1986). Thrombin stimulates neutrophil adherence by an endothelial

cell-dependent mechanism. *Ann NY Acad Sci, 485,* 349–368.

40. Levin, E. G., Marzec, U., Anderson, J., & Harker, L. A. (1984). Thrombin stimulates tissue plasminogen activator release from cultured human endothelial cells. *J Clin Invest, 74,* 1988–1995.

41. Levin, E. G., Stern, D. M., Nawroth, P. P., Marlar, R. A., Fair, D. S., Fenton, J. W., II, & Harker, L. A. (1986). Specificity of the thrombin-induced release of tissue plasminogen activator from cultured human endothelial cells. *Thromb Haemost, 56,* 115–119.

42. Dichek, D., & Quertermous, T. (1989). Thrombin regulation of mRNA levels of tissue plasminogen activator and plasminogen activator inhibitor-1 in cultured human umbilical vein endothelial cells. *Blood, 74,* 222–228.

43. Tesfamariam, B., Allen, G., Normandin, K., & Antonaccio, M. (1993). Involvement of the tethered ligand receptor in thrombin-induced entothelium-mediated relaxations. *Am Physiol Cell Physiol, 265,* 1744–1749.

44. Lum, H., & Malik, A. B. (1994). Regulation of vascular endothelial barrier function. *Am Physiol Cell Physiol, 267,* 1223–1241.

45. DeMichele, M., & Minnear, F. (1992). Modulation of vascular endothelial permeability by thrombin. *Semin Thromb Hemost, 18,* 287–295.

46. Naldini, A., Sower, L., Bocci, V., Meyers, B., & Carney, D. (1998). Thrombin receptor expression and responsiveness of human monocytic cells to thrombin is linked to interferon-induced cellular differentiation. *J Cell Pathol, 177,* 76–84.

47. Drake, W., & Issekut, A. (1992). A role for α-thrombin in polymorphonuclear leukocyte recruitment during inflammation. *Thromb Haemost, 18,* 333–340.

48. Kaplanski, G., Fabrigoule, M., Boulay, V., Dinarello, C., Bongrand, P., & Farnarier, C. (1997). Thrombin induces endothelial-type II activation in vitro: IL-1 and TNF-x-independent IL-8 secretion and E-selectin expression. *J Immunol, 158,* 5435–5441.

49. Bizios, R., Lai, L., Fenton, J., & Malik, A. B. (1986). Thrombin-induced chemotaxis and aggregation of neutrophils. *J Cell Physiol, 128,* 485–490.

50. Zimmerman, G., McIntyre, T., & Prescott, S. (1985). Thrombin stimulates the adherence of neutrophils to human endothelial cells in vitro. *J Clin Invest, 76,* 2235–2246.

51. Kaplanski, G., Marin, V., Fabrigoule, M., Boulay, V., Benoliel, A., Bongrand, P., Kaplanski, S., & Farnarier, C. (1998). Thrombin-activated human endothelial cells support monocyte adhesion in vitro following expression of intercellular adhesion molecule-1 (ICAM-1; CD54) and vascular cell adhesion molecule-1 (VCAM-1; CD106). *Blood, 92,* 1259–1267.

52. Turgeon, V., Salman, N., & Houenou, L. (2001). Thrombin: A neuronal cell modulator. *Thromb Res, 99,* 417–427.

53. Lourbakos, A., Chinni, C., Thompson, P., Potempa, J., Travis, J., Mackie, E., & Pike, R. (1998). Cleavage and activation of proteinase-activated receptor-2 on human neutrophils by gingipain-R from *Porphyromonas gingivalis. FEBS Lett, 435,* 45–48.

54. Nierodzik, M., Kajumo, F., & Karpatkin, S. (1992). Effect of thrombin treatment of tumor cells on adhesion of tumor cells to platelets *in vitro* and tumor metastasis *in vivo. Cancer Res, 52,* 3267–3272.

55. Wojtukiewicz, M. Z., Tang, D. G., Ciarelli, J. J., Nelson, K. K., Walz, D. A., Diglio, C. A., Mammen, E. F., & Honn, K. V. (1993). Thrombin increases the metastatic potential of tumor cells. *Int J Cancer, 54,* 793–806.

56. Nierodzik, M. L., Plotkin, A., Kajumo, F., & Kartpatkin, S. (1991). Thrombin stimulates tumor-platelet adhesion *in vitro* and metastasis *in vivo. J Clin Invest, 87,* 299–336.

57. Steinberg, S., Robinson, R., Lieberman, H., Stern, D., & Rosen, M. (1991). Thrombin modulates phosphoinositide metabolism, cytosolic calcium, and impulse initiation in the heart. *Cir Res, 68,* 1216–1229.

58. Jacobsen, A., Du, X., Lambert, K., Dart, A., & Woodcock, E. (1996). Arrythmogenic action of thrombin during myocardial reperfusion via release of inositol 1,4,5-triphosphate. *Circulation, 93,* 23–26.

59. Vu, T., Hung, D., Wheaton, V., & Coughlin, S. (1991). Molecular cloning of a functional thrombin receptor reveals a novel proteolytic mechanism of receptor activation. *Cell, 64,* 1057–1068.

60. Rasmussen, U. B., et al. (1991). cDNA cloning and expression of a hamster alpha-thrombin receptor coupled to Ca mobilization. *FEBS Lett, 288,* 123–128.

61. Dohlman, H., Thorner, J., Caron, M., et al. (1992). Model systems for the study of seven-transmembrane segment receptors. *Annu Rev Biochem, 60,* 651–688.

62. Chen, J., Ishii, M., Wang, L., Ishii, K., & Coughlin, S. R. (1994). Thrombin receptor activation. Confirmation of the intramolecular tethered liganding hypothesis and discovery of an alternative intermolecular liganding mode. *J Biol Chem, 269,* 16041–16045.

63. Ishihara, H., Connolly, A., Zeng, D., Kahn, M. L., Zheng, W., Timmons, C., Tram, T., & Coughlin, S. R. (1997). Protease-activated receptor 3 is a second thrombin receptor in humans. *Nature, 386,* 502–506.

64. Xu, W., Andersen, H., Whitmore, T. E., Presnell, S. R., Yee, D. P., Shing, A., Gilbert, T., Davie, E. W., & Foster, D. C. (1998). Cloning and characterization of human protease-activated receptor 4. *Proc Natl Acad Sci USA, 95,* 6642–6646.

65. Molino, M., Barnathan, E. S., Numerof, R., Clark, J., Dreyer, M., Cumashi, A., Hoxie, J. A., Schecter, N., Woolkalis, M., & Brass, L. F. (1997). Interactions of mast cell tryptase with thrombin receptors and PAR-2. *J Biol Chem, 272,* 4043–4049.

66. Nystedt, S., Emilsson, K., Wahlestedt, C., & Sundelin, J. (1994). Molecular cloning of a potential proteinase activated receptor. *Proc Natl Acad Sci USA, 91,* 9208–9212.

67. Camerer, E., Huang, W., & Coughlin, S. R. (2000). Tissue factor- and factor X-dependent activation of PAR2 by factor VIIa. *Proc Natl Acad Sci USA, 97,* 5255–5260.

68. Hung, D. T., Vu, T.-K., Wheaton, V. I., Ishii, K., & Coughlin, S. R. (1992). Cloned platelet thrombin receptor is necessary for thrombin-induced platelet activation. *J Clin Invest, 89,* 1350–1353.

69. Ngaiza, J., & Jaffe, E. (1991). A 14 amino acid peptide derived from the amino terminus of the cleaved thrombin receptor elevates intracellular calcium and stimulates prostacyclin production in human endothelial cells. *Biochem Biophys Res Commun, 179,* 1656–1661.

70. Brass, L., Vassallo, R., Belmonte, E., Ahuja, M., Cichowski, K., & Hoxie, J. (1992). Structure and function of the human platelet thrombin receptor. Studies using monoclonal antibodies directed against a defined domain within the receptor N terminus. *J Biol Chem, 267,* 13795–13798.

71. Ishii, K., Gertstzen, R. E., Zheng, Y. W., Welsh, J., Turck, C. W., & Coughlin, S. R. (1995). Determinants of thrombin receptor cleavage receptor domains involved, specificity, and role of the P3 aspartate. *J Biol Chem, 270,* 16435–16440.

72. Vu, T., Wheaton, V., Hung, D., Charo, I., & Coughlin, S. (1991). Domains specifying thrombin-receptor interaction. *Nature (Lond), 353,* 674–677.

73. Liu, L.-W., Vu, T.-K., Esmon, C., & Coughlin, S. (1991). The region of the thrombin receptor resembling hirudin binds to thrombin and alters enzyme specificity. *J Biol Chem, 266,* 16977–16980.

74. Esmon, C. T. (1987). The regulation of natural anticoagulant pathways. *Science, 235,* 1348–1352.

75. Ehrlich, H., Grinnell, B. W., Jaskunas, S., Esmon, C. T., Yan, S., & Bang, N. (1990). Recombinant human protein C derivatives: Altered response to calcium resulting in enhanced activation by thrombin. *EMBO J, 9,* 2367–2373.

76. LeBonniec, B., & Esmon, C. T. (1991). Glu-192-Gln substitution in thrombin mimics the catalytic switch induced by thrombomodulin. *Proc Natl Acad Sci USA, 88,* 7371–7375.

77. Mathews, I. I., Padmanabhan, K. P., Ganesh, V., Tulinsky, A., Ishii, M., Chen, J., Turck, C. W., Coughlin, S. R., & Fenton, J. W., II. (1994). Crystallographic structures of thrombin complexed with thrombin receptor peptides: Existence of expected and novel binding modes. *Biochemistry, 33,* 3266–3279.

78. Gertstzen, R. E., Chen, J. I., Ishii, M., Ishii, K., Wang, L., Nanevicz, T., Turck, C. W., Vu, T.-K., & Coughlin, S. R. (1994). Specificity of the thrombin receptor for agonist peptide is defined by its extracellular surface. *Nature, 368,* 648.

79. Bahou, W., Coller, B., Potter, C., Norton, K., Kutok, J., & Goligorsky, M. (1993). The thrombin receptor extracellular domain contains sites crucial for peptide ligand-induced activation. *J Clin Invest, 91,* 1405–1413.

80. Nanevicz, T., Wang, L., Chen, M., Ishii, M., & Coughlin, S. R. (1996). Thrombin receptor activating mutations. Alteration of an extracellular agonist recognition domain causes constitutive signaling. *J Biol Chem, 271,* 702–706.

81. O'Brien, P. J., et al. (2000). Thrombin responses in human endothelial cells. Contributions from receptors other than PAR1 include the transactivation of PAR2 thrombin-cleaved PAR1. *J Biol Chem, 275,* 13502–13509.

82. Bahou, W. F., Coller, B. S., Potter, C. L., Norton, K. J., Kutok, J. L., & Goligorsky, M. S. (1993). The thrombin receptor extracellular domain contains sites crucial for peptide ligand-induced activation. *J Clin Invest, 91,* 1405–1413.

83. Scarborough, R., Naughton, M., Teng, W., Hung, D., Rose, J., Vu, T., Wheaton, V., Turck, C., & Coughlin, S. (1992). Tethered ligand agonist peptides: Structural requirements for thrombin receptor activation reveal mechanism of proteolytic unmasking of agonist function. *J Biol Chem, 267,* 13146–14149.

84. Norton, K. J., Scarborough, R. M., Kutok, J. L., Escobedo, M. A., Nannizzi, L., & Coller, B. S. (1993). Immunologic analysis of the cloned platelet thrombin receptor activation mechanism: Evidence supporting receptor cleavage, release of the N-terminal peptide, and insertion of the tethered ligand into a protected environment. *Blood, 82,* 2125–2136.

85. Coller, B. S., Springer, K. T., Scudder, L. E., Kutok, J. L., Ceruso, M., & Prestwich, G. D. (1993). Substituting isoserine for serine in the thrombin receptor activation peptide SFLLRN confers resistance to aminopeptidase M-induced cleavage and inactivation. *J Biol Chem, 268,* 20741–20743.

86. Vassallo, J., Kieber–Emmons, T., Cichowski, K., & Brass, L. F. (1992). Structure–function relationships in the activation of platelet thrombin receptors by receptor-derived peptides. *J Biol Chem, 267,* 6081–6085.

87. Ramachandran, B., Klufas, A., Molino, M., Ahuja, M., Hoxie, J. A., & Brass, L. F. (1997). Release of the thrombin receptor (PAR-1) N-terminus from the surface of human platelets activated by thrombin. *Thromb Haemost, 78,* 1119–1124.

88. Furman, M. I., Lu, L., Benoit, S. E., Becker, R. C., Barnard, M. R., & Michelson, A. D. (1998). The cleaved peptide of the thrombin receptor is a strong platelet agonist. *Proc Natl Acad Sci USA, 95,* 3082–3087.

89. Suidan, H. S., Bouvier, J., Schaerer, E., Stone, S. R., Monard, D., & Tschopp, J. (1994). Granzyme A released upon stimulation of cytotoxic T lymphocytes activates the thrombin receptor on neuronal cells and astrocytes. *Proc Natl Acad Sci USA, 91,* 8112–8116.

90. Kaufmann, R., Zieger, M., Tausch, S., Henklein, P., & Nowak, G. (2000). Meizothrombin, an intermediate of prothrombin activation stimulates human glioblastoma cells by interaction with PAR-1 type thrombin receptors. *J Neurosci Res, 59,* 643–648.

91. Brass, L. F., Manning, D. R., Williams, A. G., Woolkalis, M. J., & Poncz, M. (1991). Receptor and G protein-mediated responses to thrombin in HEL cells. *J Biol Chem, 266,* 958–965.

92. Molino, M., Woolkalis, M. J., Reavey–Cantwell, J., Practico, D., Andrade–Gordon, P., & Brass, L. (1997). Endothelial cell thrombin receptors and PAR-2: Two protease-activated receptors located in a single cellular environment. *J Biol Chem, 272,* 11133–11141.

93. Riewald, M., Kravchenko, V., Petrovan, R., O'Brien, P., Brass, L., Ulevitch, R., & Ruf, W. (2001). Gene induction by coagulation factor Xa is mediated by activation of protease-activated receptor 1. *Blood, 97,* 3109–3116.

94. Molino, M., Blanchard, N., Belmonte, E., Tarver, A. P., Abrams, C., Hoxie, J. A., Cerletti, C., & Brass, L. F. (1995). Proteolysis of the human platelet and endothelial cell throm-

bin receptor by neutrophil-derived cathepsin. *J Biol Chem, 270,* 11168–11175.

95. Kimura, M., Andersen, T. T., Fenton, J. W., II, Bahou, W. F., & Aviv, A. (1996). Plasmin–platelet interaction involves cleavage of functional thrombin receptor. *Am J Physiol, 271,* C54–C60.

96. Renesto, P., Si-Tahar, M., Moniatte, M., Balloy, V., Van Dorsselaer, A., Pidard, D., & Chignard, M. (1997). Specific inhibition of thrombin-induced cell activation by the neutrophil proteinases elastase, cathepsin G, and proteinase 3: Evidence for distinct cleavage sites within the aminoterminal domain of the thrombin receptor. *Blood, 89,* 1944–1953.

97. Schmidt, V. A., Nierman, W. C., Maglott, D. R., Cupit, L. D., Moskowitz, K. A., Wainer, J. A., & Bahou, W. F. (1998). The human proteinase-activated receptor-3 (PAR-3) gene. Identification within a PAR gene cluster and characterization in vascular endothelial cells and platelets. *J Biol Chem, 273,* 15061–15068.

98. Cupit, L. D., Schmidt, V. A., & Bahou, W. F. (1999). Proteolytically activated receptor-3. A member of an emerging gene family of protease receptors expressed on vascular endothelial cells and platelets. *Trends Cardiovasc Med, 9,* 42–48.

99. Major, C. D., Santulli, R. J., Derian, C. K., & Andrade–Gordon, P. (2003). Extracellular mediators in atherosclerosis and thrombosis: Lessons from thrombin receptor knockout mice. *Arterioscler Thromb Vasc Biol, 23,* 931–939.

100. Kahn, M. L., Zheng, Y. W., Huang, W., Bigornia, V., Zeng, D., Moff, S., Farese, J., Tam, C., & Coughlin, S. R. (1998). A dual thrombin receptor system for platelet activation. *Nature, 394,* 690–694.

101. Sambrano, G. R., et al. (2000). Cathepsin G activates protease-activated receptor-4 in human platelets. *J Biol Chem, 275,* 6819–6823.

102. Faruqi, T., Weiss, E., Shapiro, M. J., Huang, W., & Coughlin, S. R. (2000). Structure-function analysis of protease-activated receptor 4 tethered ligand peptides. Determinants of specificity and utility in assays of receptor function. *J Biol Chem,* 19728–19734.

103. Nystedt, S., Emilsson, K., Larsson, A. K., Strombeck, B., & Sundelin, J. (1995). Molecular cloning and functional expression of the gene encoding the human proteinase-activated receptor 2. *Eur J Biochem, 232,* 84–89.

104. Mirza, H., Schmidt, V., Derian, C., Jesty, J., & Bahou, W. (1997). Mitogenic responses mediated through the proteinase activated receptor-2 are induced by mast cell α- and β-tryptases. *Blood, 90,* 3914–3922.

105. Molino, M., Raghunath, P., KuO, A., Ahuja, M., Hoxie, J., Brass, L., & Barnathan, E. (1998). Differential expression of functional protease-activated receptor-2 (PAR-2) in human vascular smooth muscle cells. *Arterioscl, Thromb Vasc Biol, 18,* 825–832.

106. Mirza, H., Yatsula, V., & Bahou, W. F. (1996). The proteinase activated receptor-2 (PAR-2) mediates mitogenic responses in human vascular endothelial cells. Molecular characterization and evidence for functional coupling to the thrombin receptor. *J Clin Invest, 97,* 1705–1714.

107. Jenkins, A. L., Chinni, C., DeNiesse, M., & Blackhart, B. D. (2000). Expression of protease-activated receptor-2 during development. *Dev Dyn, 218,* 465–471.

108. Steinhoff, M., Vergnolle, N., Young, S., Tognetto, M., Amadesi, S., Ennes, H., Trevisani, M., Hollenberg, M., Wallace, J., Caughey, G. H., Mitchell, S., Williams, L., Geppetti, P., Mayer, E., & Bunnett, N. (2000). Agonists of proteinase-activated receptor 2 induce inflammation by a neurogenic mechanism. *Nat Med, 6,* 151–158.

109. Vergnolle, N., Hollenberg, M. D., Sharkey, K., & Wallace, J. (1999). Characterization of the inflammatory response to proteinase-activated receptor-2 (PAR-2)-activating peptides in the rat paw. *Br J Pharm, 127,* 1053–1090.

110. Cocks, T., Fong, B., Chow, J., Anderson, G., Frauman, A., Goldie, R., Henry, P., Carr, M., Hamilton, R., & Moffatt, J. (1999). A protective role for protease-activated receptors in the airways. *Nature, 398,* 156–160.

111. Damiano, B. P., D'Andrea, M. R., de Gavavilla, L., Cheung, W. M., & Andrade–Gordon, P. (1999). Increased expression of protease activated receptor-2 (PAR-2) in balloon-injured rat carotid artery. *Thromb Haemost, 81,* 808–814.

112. Sobey, C., Moffatt, J., & Cocks, T. (1999). Evidence for selective effects of chronic hypertension on cerebral artery vasodilation to protease-activated receptor-2 activation. *Stroke, 30,* 1933–1940.

113. Bahou, W. F. (2003). Thrombin's faces revealed: Cellular effects include induction of neoangiogenesis. *J Thromb Haemost, 1,* 2078–2080.

114. Walz, D. A., Anderson, G. F., Ciaglowski, R. E., Aiken, M., & Fenton, J. W. (1985). Thrombin-elicited contractile responses of aortic smooth muscle. *Proc Soc Exp Biol Med, 180,* 518–526.

115. Brummel, K., Butenas, S., & Mann, K. G. (1999). An integrated study of fibrinogen during blood coagulation. *J Biol Chem, 274,* 22862–22870.

116. Chesebro, J., Zoldhhelyi, P., & Fuster, V. (1991). Pathogenesis of thrombosis in unstable angina. *Am J Cardiol, 68,* 2B-10B.

117. Heras, M., Chesebro, J., Webster, M., Mruk, J., Grill, D., Penny, W., Bowie, E. J. W., Badimon, L., & Fuster, V. (1990). Hirudin, heparin and placebo during deep arterial injury in the pig: The *in vivo* role of thrombin in platelet-mediated thrombosis. *Circulation, 82,* 1476–1484.

118. Practico, D., Murphy, N., & Fitzgerald, D. (1997). Interaction of a thrombin inhibitor and a platelet GP IIb/IIIa antagonist in vivo: Evidence that thrombin mediates platelet aggregation and subsequent thromboxane A2 formation during coronary thrombolysis. *J Pharmacol Exp Ther, 281,* 1178–1185.

119. Derian, C. K., Damiano, B. P., Addo, M. F., Darrow, A. L., D'Andrea, M. R., Nedelman, M., Zhang, H. C., Maryanoff, B. E., & Andrade–Gordon, P. (2003). Blockade of the thrombin receptor protease-activated receptor-1 with a small-molecule antagonist prevents thrombus formation and vascular occlusion in nonhuman primates. *J Pharmacol Exp Ther, 304,* 855–861.

120. Cook, J., Sitko, G., Bednar, B., Condra, C., Mellott, M., Feng, D., Nutt, R., Shafer, J., Gould, R., & Connolly, T. (1995). An

antibody against the exosite of the cloned thrombin receptor inhibits experimental arterial thrombosis in the African green monkey. *Circulation, 91,* 2961–2971.

121. Arnaud, E., Nicaud, V., Poirier, O., Rendu, F., Alhenc–Gelas, M., Fiessinger, J., Emmerich, J., & Aiach, M. (2000). Protective effect of a thrombin receptor (protease-activated receptor 1) gene polymorphism toward venous thromboembolism. *Arterioscl Thromb Vasc Biol, 20,* 585–592.

122. Nelken, N. A., Soifer, S. J., O'Keefe, J., Vu, T.-K., H., Charo, I. F., & Coughlin, S. R. (1992). Thrombin receptor expression in normal and atherosclerotic human arteries. *J Clin Invest, 90,* 1614–1621.

123. Baykal, D., Schmedtje, J., & Runge, M. (1995). Role of the thrombin receptor in restenosis and atherosclerosis. *Am J Cardiol, 75,* 82B–87B.

124. Wilcox, J., Rodriguez, J., Subramanian, R., Ollerenshaw, J., Zhong, C., Hayzer, D., Horaist, C., Hanson, S., Lumdsen, A., Salam, T., et al. (1994). Characterization of thrombin receptor expression during vascular lesion formation. *Circ Res, 75,* 1029–1038.

125. Papadaki, M., Ruef, J., Nguyen, K., Li, F., Patterson, C., Eskin, S., McIntire, L., & Runge, M. (1998). Differential regulation of protease activated receptor-1 and tissue plasminogen activator expression by shear stress in vascular smooth muscle cells. *Circ Res, 83,* 1027–1034.

126. Chaikoff, E. L., Caban, R., Yan, C., Rao, G., & Runge, M. (1995). Growth-related responses in arterial smooth muscle cells are arrested by thrombin receptor antisense sequences. *J Biol Chem, 270,* 7431–7436.

127. Herbert, J., Guy, A., Lamarche, I., Mares, A., Savi, P., & Dol, F. (1997). Intimal hyperplasia following vascular injury is not inhibited by an antisense thrombin receptor oligodeoxynucleotide. *J Cell Physiol, 170,* 106–114.

128. Takada, M., Tanaka, H., Yamada, T., Ito, O., Kogushi, M., Yanagimachi, M., Kawamura, T., Musha, T., Yoshida, F., Ito, M., Kobayashi, H., Yoshitake, S., & Saito, I. (1998). Antibody to thrombin receptor inhibits neointimal smooth muscle cell accumulation without causing inhibition of platelet aggregation or altering hemostatic parameters after angioplasty in rat. *Circ Res, 82,* 980–987.

129. Cheung, W. M., D'Andrea, M. R., Andrade–Gordon, P., & Damiano, B. P. (1999). Altered vascular injury responses in mice deficient in protease-activated receptor-1. *Arterioscler Thromb Vasc Biol, 19,* 3014–3024.

130. Esmon, C. T. (2000). Introduction: Are natural anticoagulants candidates for modulating the inflammatory response to endotoxin? *Blood, 95,* 1113–1116.

131. Fenton, J. (1988). Regulation of thrombin generation and functions. *Semin Thromb Hemost, 14,* 234–240.

132. Shuman, M. (1986). Thrombin–cellular interactions. *Ann NY Acad Sci, 485,* 229–238.

133. Johnson, K., Choi, Y., DeGroot, E., Samuels, I., Creasey, A., & Aarden, L. (1998). Potential mechanisms for a proinflammatory vascular cytokine response to coagulation activation. *J Immunol, 160,* 5130–5135.

134. Cirino, G., et al. (1996). Thrombin functions as an inflammatory mediator through activation of its receptor. *J Exp Med, 183,* 821–827.

135. Naldini, A., Carney, D. H., Bocci, V., Klimpel, K., Assuncion, M., Soares, L., & Kimpel, G. (1993). T cell proliferative responses and cytokine production. *Cell Immunol, 147,* 367–377.

136. Bar-Shavit, R., Kahn, A., Wilner, G. D., et al. (1983). Monocyte chemotaxis: Stimulation by specific exosite region in thrombin. *Science, 220,* 728–731.

137. Bar-Shavit, R., Kahn, A., Fenton, J. W., II, & Wilner, G. D. (1983). Chemotactic response of monocytes to thrombin. *J Cell Biol, 96,* 282–285.

138. Schmeidler–Sapiro, K. T., Ratnoff, O. D., & Gordon, E. M. (1991). Mitogenic effects of coagulation factor XII and factor XIIa on Hep $G_2$ cells. *Proc Natl Acad Sci USA, 88,* 4382–4385.

139. Cunningham, M., Romas, P., Hutchinson, P., Holdsworth, S., & Tipping, P. (1999). Tissue factor and factor VIIa receptor/ligand interactions induce proinflammatory effects in macrophages. *Blood, 94,* 3413–3420.

140. Rottingen, J., Enden, T., Camerer, E., Iversen, J., & Prydz, H. (1995). Binding of human factor VIIa to tissue factor induces cytosolic $Ca^{2+}$ signals in J82 cells, transfected GDS-1 cells, Madin–Darby canine kidney cells and in human endothelial cells induced to synthesize tissue factor. *J Biol Chem, 270,* 4650–4660.

141. Camerer, E., Rottingen, J., Iversen, J., & Prydz, H. (1996). Coagulation factors VII and X induce $Ca^{2+}$ oscillations in Madin–Darby canine kidney cells only when proteolytically active. *J Biol Chem, 271,* 29042.

142. Nicholson, A. C., Nachman, R. L., Altieri, D. C., Summers, B. D., Ruf, W., Edgington, T. S., & Hajjar, D. P. (1996). Effector cell protease receptor-1 is a vascular receptor for coagulation factor Xa*. *J Biol Chem, 271,* 28407–28413.

143. Nystedt, S., Ramakrishnan, S., & Sundelin, J. (1996). The proteinase-activated receptor 2 is induced by inflammatory mediators in human endothelial cells. *J Biol Chem, 271,* 14910–14915.

144. Coughlin, S. R. (2000). Thrombin signalling and protease-activated receptors. *Nature, 407,* 258–264.

145. Taylor, J., Chang, A., Esmon, C. T., D'Angelo, A., Vigano–D'angelo, S., & Blick, K. (1987). Protein C prevents the coagulopathic and lethal effects of *E. coli* infusion in the baboon. *J Clin Invest, 97,* 918–925.

146. Taylor, J., Chang, A., Ruf, W., Morrissey, J., Hinshaw, L., Catlett, R., Blick, K., & Edginton, T. (1991). Lethal, *E. coli* septic shock is prevented by blocking tissue factor with monoclonal antibody. *Circ Shock, 1991:33*–127134.

147. Cunningham, M., Rondeau, E., Chen, X., Coughlin, S. R., Holdsworth, S., & Tipping, P. (2000). Protease-activated receptor 1 mediates thrombin-dependent, cell-mediated renal inflammation in crescentic glomemlonephritis. *J Exp Med, 191,* 455–462.

148. Hauck, R., Schulz, C., Schomig, A., Hoffman, R., & Panettieri, J. (1999). alpha-Thrombin stimulates contraction of human bronchial rings by activation of protease-activated receptors. *Am J Phys, 277,* L22–L29.

149. Hou, L., Ravenall, S., Macey, M., Harriott, P., Kapas, S., & Howells, G. L. (1998). Protease-activated receptors and their

role in IL-6 and NF-IL-6 expression in human gingival fibroblasts. *J Periodontal Res, 33,* 205–211.

150. Ferrell, W. R., Lockhart, J. C., Kelso, E. B., Dunning, L., Plevin, R., Meek, S. E., Smith, A. J., Hunter, G. D., McLean, J. S., McGarry, F., Ramage, R., Jiang, L., Kanke, T., & Kawagoe, J. (2003). Essential role for proteinase-activated receptor-2 in arthritis. *J Clin Invest, 111,* 35–41.

151. Ehrlich, J., Boyle, E., Labriola, J., Kovacich, J., Santucci, R., Fearns, C., Morgan, E., Yun, W., Luther, T., Kojikawa, O., Martin, T., Pohlman, T., Vernier, E., & Mackman, N. (2000). Inhibition of the tissue factor-thrombin pathway limits infarct size after myocardial ischemia–reperfusion injury by reducing inflammation. *Am J Pathol, 157,* 1849–1862.

152. Milia, A. F., Salis, M. B., Stacca, T., Pinna, A., Madeddu, P., Trevisani, M., Geppetti, P., & Emanueli, C. (2002). Protease-activated receptor-2 stimulates angiogenesis and accelerates hemodynamic recovery in a mouse model of hindlimb ischemia. *Circ Res, 91,* 346–352.

153. Fiorucci, S., Mencarelli, A., Palazzetti, B., Distrutti, E., Vergnolle, N., Hollenberg, M. D., Wallace, J. L., Morelli, A., & Cirino, G. (2001). Proteinase-activated receptor 2 is an anti-inflammatory signal for colonic lamina propria lymphocytes in a mouse model of colitis. *Proc Natl Acad Sci USA, 98,* 13936–13941.

154. Bahou, W. F., Nierman, W. C., Durkin, A. S., Potter, C. L., & Demetrick, D. J. (1993). Chromosomal assignment of the human thrombin receptor gene: Localization to region q13 of chromosome 5. *Blood, 82,* 1532–1537.

155. Schmidt, V., Vitale, E., & Bahou, W. (1996). Genomic cloning and characterization of the human thrombin receptor gene: Evidence for a novel gene family that includes PAR-2. *J Biol Chem, 271,* 9307–9312.

156. Kahn, M. L., Hammes, S. R., Botka, C., & Coughlin, S. R. (1998). Gene and locus structure and chromosomal localization of the protease-activated receptor gene family. *J Biol Chem, 273,* 23290–23296.

157. Guyonett, D., Boissear, J., Arvelier, B., & Nurden, A. (1998). Protease-activated receptor genes are clustered on 5q13. *Blood, 92,* 25.

158. Schmidt, V. A., Scudder, L., Devoe, C. E., Bernards, A., Cupit, L. D., & Bahou, W. F. (2003). IQGAP2 functions as a GTP-dependent effector protein in thrombin-induced platelet cytoskeletal reorganization. *Blood, 101,* 3021–3028.

159. Venter, J., Adams, M., Sutton, G., Kerlavage, A., Smith, H., & Hunkapiller, M. (1998). Shotgun sequencing of the human genome. *Science, 280,* 1540–1542.

160. Schmidt, V. A., Nierman, W. C., Feldblyum, T. V., Maglott, D. R., & Bahou, W. F. (1997). The human thrombin receptor and proteinase activated receptor-2 genes are tightly linked on chromosome 5q13. *Br J Haematol, 97,* 523–529.

161. Cupit, L. D., Schmidt, V. A., Miller, F., & Bahou, W. F. (2004). Distinct PAR/IQGAP expression patterns during murine development: Implications for thrombin-associated cytoskeletal reorganization. *Mamm Genome, 15,* 618–629.

162. Altschul, S. F., Gish, W., Miller, W., Myers, E. W., & Lipman, D. J. (1990). Basic local alignment search tool. *J Mol Biol, 215,* 403–410.

163. Connolly, A., Ishihara, H., Kahn, M. L., Farese, R. V., & Coughlin, S. R. (1996). Role of the thrombin receptor in development and evidence for a second receptor. *Nature, 38,* 516–519.

164. Darrow, A. L., Fung–Leung, W., Ye, R. D., Santulli, R. J., Cheung, W. M., Derian, C. K., Burns, C. L., Damiano, B. P., Zhou, L., Keena, C. M., Peterson, P., & Andrade–Gordon, P. (1996). Biological consequences of thrombin receptor deficiency in mice. *Thromb Haemost, 76,* 860–866.

165. Connolly, T. M., Condra, C., Feng, D.-M., Cook, J. J., Stranieri, M. T., Reilly, C. F., Nutt, R. F., & Gould, R. J. (1994). Species variability in platelet and other cellular responsiveness to thrombin receptor-derived peptides. *Thromb Haemost, 72,* 627–633.

166. Derian, C. K., Santulli, R. J., Tomko, K. A., Haertlein, B., & Andrade–Gordon, P. (1995). Species differences in platelet responses to thrombin and SFLLKN. Receptor-mediated calcium mobilization and aggregation and regulation by protein kinases. *Thromb Res, 6,* 505–519.

167. Kinlough–Rathbone, R. L., Rand, M., & Packham, M. A. (1993). Rabbit and rat platelets do not respond to thrombin receptor peptides that activate human platelets. *Blood, 82,* 103–106.

168. Cui, J., O'Shea, K. S., Purkayastha, A., Saunders, T. L., & Ginsburg, D. (1996). Fatal haemorrhage and incomplete block to embryogenesis in mice lacking coagulation factor V. *Nature (Lond), 384,* 66–68.

169. Sun, W. Y., Witte, D. P., Degen, J. L., Colbert, M. C., Burkart, M. C., Holmbäck, K., Xiao, Q., Bugge, T., H., & Degen, S. J. (1998). Prothrombin deficiency results in embryonic and neonatal lethality in mice. *Proc Natl Acad Sci USA, 95,* 7597–7602.

170. Xue, J., Wu, Q., Westfield, L. A., Tuley, E. A., Lu, D., Zhang, Q., Shim, K., Zheng, X., & Sadler, J. E. (1998). Incomplete embryonic lethality and fatal neonatal hemorrhage caused by prothrombin deficiency in mice. *Proc Natl Acad Sci USA, 95,* 7603–7607.

171. Bugge, T., H., Xiao, Q., Kombrinck, K., Flock, M., Holmbäck, K., Danton, M., Colbert, M. C., et al. (1996). Fatal embryonic bleeding events in mice lacking tissue factor, the cell-associated initiator of blood coagulation. *Proc Natl Acad Sci USA, 93,* 6258–6263.

172. Carmeliet, P., Mackman, N., Moons, Luther, T., Gressens, P., Van Vlaenderen, I., Demunck, Kasper, M., Breier, G., Evrard, P., Muller, M., Risau, W., Edgington, T., & Coleen, D. (1996). Role of tissue factor in embryonic blood vessel development. *Nature, 383,* 73–75.

173. Toomey, J., Kratzer, K., Lasky, N., Stanton, J., & Broze, J. (1996). Targeted disruption of the murine tissue factor gene results in embryonic lethality. *Blood, 88,* 1583–1587.

174. Rosen, E., Chan, J., Idusogie, E., et al. (1997). Mice lacking factor VII develop normally but suffer fatal perinatal bleeding. *Nature, 390,* 290–294.

175. Dewerchin, M., Liang, Z., Moons, L., Carmeliet, P., Castellino, F., Collen, D., & Rosen, E. (2000). Blood coagulation factor X deficiency causes partial embryonic lethality and fatal neonatal bleeding in mice. *Thromb Haemost, 83,* 185–190.

176. Rosenberg, R. D. (1996). The absence of the blood clotting regulator thrombomodulin causes embryonic lethality in mice before development of a functional cardiovascular system. *Thromb Haemost, 74,* 52–57.

177. Suidan, H. S., Niclou, S. P., Dreessen, J., Beltraminelli, N., & Monard, D. (1996). The thrombin receptor is present in myoblasts and its expression is repressed upon fusion. *J Biol Chem, 271,* 29162–29169.

178. Brass, L. F., & Molino, M. (1997). Protease-activated G protein-coupled receptors on human platelets and endothelial cells. *Thromb Haemost, 78,* 234–241.

179. Soifer, S., Peters, K., O'Keefe, J., & Coughlin, S. R. (1991). Disparate temporal expression of the prothrombin and thrombin receptor genes during mouse development. *Am J Pathol, 144,* 60–69.

180. Griffin, C., Srinivasan, Y., Zheng, Y.-W., Huang, W., & Coughlin, S. R. (2001). A role for thrombin receptor signaling in endothelial cells during embryonic development. *Science, 293,* 1666–1669.

181. Zucker, S., Conner, C., DiMassimo, B., Ende, H., & Bahou, W. (1995). Thrombin induces cell surface activation of gelatinase A in vascular endothelial cells: Physiological regulation of angiogenesis. *J Biol Chem, 270,* 23730–23738.

182. Rao, L., & Pendurthi, U. (2001). Factor VIIIa-induced gene expression: Potential implications in pathophysiology. *Trends Cardiovasc Med, 11,* 14–21.

183. Molino, M., Bainton, D. F., Hoxie, J., Coughlin, S. R., & Brass, L. F. (1997). Thrombin receptors on human platelets. Initial localization and subsequent redistribution during platelet activation. *J Biol Chem, 272,* 6011–6017.

184. Parry, M., Myles, T., Tschopp, J., & Stone, S. R. (1996). Cleavage of the thrombin receptor: Identification of potential activators and inactivators. *Biochem J, 320,* 335–341.

185. Kinlough–Rathbone, R. L., Perry, D. W., & Packham, M. A. (1995). Contrasting effects of thrombin and the thrombin receptor peptide, SFLLRN, on aggregation and release of 14C-serotinin by human platelets pretreated with chymotrypsin or serratia marcescens protease. *Thromb Haemost, 73,* 122–125.

186. Tiruppathi, C., Lum, H., Andersen, T. T., Fenton, J. W., II, & Malik, A. B. (1992). Thrombin receptor 14-amino acid peptide binds to endothelial cells and stimulates calcium transients. *Am J Physiol, 263,* L595–L601.

187. Ishii, K., Chen, J., Ishii, M., Koch, W. J., Freedman, N. J., Lefkowitz, R. J., & Coughlin, S. R. (1994). Inhibition of thrombin receptor signaling by a G-protein coupled receptor kinase. Functional specificity among G-protein coupled receptor kinases. *J Biol Chem, 269,* 1125–1130.

188. Woolkalis, M. J., DeMelfi, J., Blanchard, N., Hoxie, J. A., & Brass, L. F. (1995). Regulation of thrombin receptors on human umbilical vein endothelial cells. *J Biol Chem, 270,* 9868–9875.

189. Horvat, R., & Palade, G. (1995). The functional thrombin receptor is associated with the plasma lemma and a large endosomal network in cultured human umbilical vein endothelial cells. *J Cell Sci, 108,* 1155–1164.

190. Hung, D. T., Wong, Y. H., Vu, T. K. H., & Coughlin, S. R. (1992). The cloned platelet thrombin receptor couples to at least two distinct effectors to stimulate both phosphoinositide hydrolysis and inhibit adenylyl cyclase. *J Biol Chem, 353,* 20831–20834.

191. Garrington, T., & Johnson, G. (1999). Organization and regulation of mitogen-activated protein kinase signaling pathways. *Current Opinion in Cell Biology, 11,* 211–218.

192. Hart, M. J., et al. (1998). Direct stimulation of the guanine nucleotide exchange activity of p115 RhoGEF by Gα. *Science, 280,* 2112–2114.

193. Kozasa, T., et al. (1998). RhoGEF, a GTPase-activating protein for Galpha2 and Galpha3. *Science, 280,* 2109–2111.

194. Klages, B., Brandt, U., Simon, M. I., Schultz, G., & Offermanns, S. (1999). Activation of G12/G13 results in shape change and rho/rho kinase-mediated myosin light chain phosphorylation in mouse platelets. *J Cell Biol, 144,* 745–754.

195. Stoyanov, B., et al. (1995). Cloning and characterization of a G protein-activated human phosphoinositide-3 kinase. *Science, 269,* 690–693.

196. Ferguson, K., Lemmon, M., Schlessinger, J., & Sigler, P. (1994). Crystal structure at 2.2. A resolution of the pleckstrin homology domain from human dynamin. *Cell, 79,* 199–209.

197. Yoon, H., Hajduk, P., Petros, A., Olejniczak, E., Meadows, R., & Fesik, S. (1994). Solution structure of a pleckstrin-homology domain. *Nature, 369,* 672–675.

198. Harlan, J., Hajduk, P., Yoon, H., & Fesik, S. (1994). Pleckstrin homology domains bind to phosphatidylinositol-4,5-bisphosphate. *Nature, 371,* 168–170.

199. Hartwig, J., Bokoch, G., Carpenter, C., Janmey, P., Taylor, L., Toker, A., & Stossel, T. (1995). Thrombin receptor ligation and activated Rac uncap actin filament barbed ends through phosphoinositide synthesis in permeabilized human platelets. *Cell, 82,* 643–653.

200. Brass, L. F., Manning, D. R., Cichowski, K., & Abrams, C. (1997). Signaling through G proteins in platelets to the integrins and beyond. *Thromb Haemost, 78,* 581–589.

201. Banno, Y., Yada, Y., & Nozawa, Y. (1988). Purification and characterization of membrane-bound phospholipase C specific for phosphoinositides from human platelets. *J Biol Chem, 263,* 11459–11465.

202. Berridge, M. J. (1993). Inositol trisphosphate and calcium signalling. *Nature, 361,* 315–325.

203. Cockcroft, S., & Thomas, G. M. (1992). Inositol-lipid-specific phospholipase C isoenzymes and their differential regulation by receptors. *Biochem J, 288,* 1–14.

204. Rhee, S. G., Suh, P. G., Ryu, S. H., & Lee, S. Y. (1989). Studies of inositol phospholipid-specific phospholipase C. *Science, 244,* 546–550.

205. Rhee, S. G., Kim, H., Suh, P. G., & Choi, W. C. (1991). Multiple forms of phosphoinositide-specific phospholipase C and different modes of activation. *Biochem Soc Trans, 19,* 337–341.

206. Rittenhous–Simmons, S. (1981). Differential activation of platelet phospholipases by thrombin and ionophore A23187. *J Biol Chem, 256,* 4153–4155.

207. Lollar, D., & Owen, W. (1980). Evidence that the effects of thrombin on arachidonate metabolism in cultured human

endothelial cells are not mediated by a high affinity receptor. *J Biol Chem, 255,* 8031–8034.

208. Laposata, M., Krueger, C., & Saffitz, J. (1987). Selective uptake of [3H] arachidonic acid into the dense tubular system of human platelet. *Blood, 70,* 832–837.

209. Kramer, R., Roberts, E., Um, S., Borsch–Haubold, A., & Watson, S. (1996). p38 Mitogen-activated protein kinase phosphorylates cytosolic phospholipase A₂ (cPLA₂) in thrombin-stimulated platelets. Evidence that proline-directed phosphorylation is not required for mobilization of arachidonic acid by cPLA₂. *J Biol Chem, 271,* 27723–27729.

210. Offermanns, S., Toombs, C. F., Hu, Y. H., & Simon, M. I. (1997). Defective platelet activation in G alpha(q)-deficient mice. *Nature, 389,* 183–186.

211. Offermanns, S., Laugwitz, K., Spicher, K., & Schultz, G. (1994). G proteins of the G12 family are activated via thromboxane A2 and thrombin receptors in human platelets. *Proc Natl Acad Sci USA, 91,* 504–508.

212. Offermanns, S., Mancino, V., Revel, J.-P., & Simon, M. I. (1997). Vascular system defects and impaired cell chemokinesis as a result of G α₁₃ deficiency. *Science, 275,* 533–536.

213. Van Aelst, L., & D'Souza–Schorey, C. (1997). Rho GTPases and signaling networks. *Genes Dev, 11,* 2295–2322.

214. Azim, A., Barkalow, K., Chou, J., & Hartwig, J. (2000). Activation of the small GTPases, rac and cdc42, after ligation of the platelet PAR-1 receptor. *Blood, 95,* 959–964.

215. Hartwig, J., Barkalow, K., Azim, A., & Italiano, J. (1999). The elegant platelet: Signals controlling actin assembly. *Thromb Haemost, 82,* 392.

216. Kotani, K., Yonezawa, K., Hara, K., Ueda, H., Kitamura, Y., Sakaue, H., Ando, A., Chavanieu, A., Calas, B., & Grigorescu, F. (1994). Involvement of phosphoinositide 3-kinase in insulin- or IGF-1-induced membrane ruffling. *EMBO J, 13,* 2313–2321.

217. Yoshikazu, S., Ishihara, H., Kizuki, N., Asano, T., Oka, Y., & Yazaki, Y. (1997). Massive actin polymerization induced by phosphatidylinositol-4-phosphate 5-kinase *in vivo. J Biol Chem, 272,* 7578–7581.

218. Kovacsovics, T., Bachelot, C., Toker, A., Vlahos, C., Duckworth, B., Cantley, L., & Hartwig, J. (1995). Phosphoinositide 3-kinase inhibition spares actin assembly in activating platelets but reverses platelet aggregation. *J Biol Chem, 270,* 11358–11366.

219. Nakanishi–Matsui, M., Zheng, Y. W., Sulciner, D., Weiss, E., Ludeman, M., & Coughlin, S. (2000). PAR3 is a cofactor for PAR4 activation by thrombin. *Nature, 404,* 609–613.

220. AbdAlla, S., Lother, H., & Quitterer, U. (2000). AT1-receptor heterodimers show enhanced G-protein activation and altered receptor sequestration. *Nature, 407,* 94–98.

221. Kahn, M. L., Nakanishi–Matsui, M., Shapiro, M. J., Ishihara, H., & Coughlin, S. R. (1999). Protease-activated receptors 1 and 4 mediate activation of human platelets by thrombin. *J Clin Invest, 103,* 879–887.

222. Covic, L., Gresser, A., & Kuliopulos, A. (2000). Biphasic kinetics of activation and signaling for PAR1 and PAR4 thrombin receptors in platelets. *Biochemistry, 39,* 5458–5467.

223. Coller, B. S., Peerschke, E., Scudder, L. E., & Sullivan, C. (1983). Studies with a murine monoclonal antibody that abolishes ristocetin-induced binding of von Willebrand factor to platelets: Additional evidence in support of GPIb as a platelet receptor for von Willebrand factor. *Blood, 61,* 99–110.

224. Clemetson, K. J. (2001). Platelet GPIb-V-1X complex. *Thromb Haemost, 78,* 266–270.

225. Harmon, J., & Jamieson, G. (1985). Thrombin binds to a high affinity ~900,000 Dalton site in human platelets. *Biochemistry, 24,* 58–64.

226. Harmon, J., & Jamieson, G. (1986). The glycocalicin portion of platelet glycoprotein Ib expresses both high and moderate affinity receptor sites for thrombin. A soluble radioreceptor assay for the interaction of thrombin with platelets. *J Biol Chem, 261,* 13224–13229.

227. DeMarco, L., Mazzucato, M., Masotti, A., Fenton, J., & Ruggeri, Z. (1991). Function of glycoprotein Ibα in platelet activation induced by α-thrombin. *J Biol Chem, 265,* 23776.

228. DeMarco, L., Mazzucato, M., Masotti, A., & Ruggeri, Z. M. (1994). Localization and characterization of an alpha-thrombin-binding site on platelet glycoprotein Ib alpha. *J Biol Chem, 269,* 6478–6484.

229. Ward, C. M., Andrews, R. K., Smith, A. I., & Berndt, M. C. (1996). Mocarhagin, a novel cobra venom metalloproteinase, cleaves the platelet von Willebrand factor receptor glycoprotein Ibα. Identification of the sulfated tyrosine/anionic sequence Tyr-276-Glu-282 of glycoprotein Ibα as a binding site for von Willebrand factor and α-thrombin. *Biochemistry, 35,* 4929–4938.

230. Mazzucato, M., DeMarco, L., Masotti, A., Pradella, P., Bahou, W. F., & Ruggeri, Z. M. (1998). Characterization of the initial α-thrombin interaction with glycoprotein Ibα in relation to platelet activation. *J Biol Chem, 273,* 1880–1887.

231. Marchese, P., Murata, M., Mazzucato, M., Pradella, P., DeMarco, L., Ware, J. A., & Ruggeri, Z. M. (1995). Identification of three tyrosine residues of glycoprotein Ibα with distinct roles in von Willebrand factor and α-thrombin binding. *J Biol Chem, 270,* 9571–9578.

232. Hayes, K., & Tracy, P. B. (1999). The platelet high affinity binding site for thrombin mimics hirudin, modulates thrombin-induced platelet activation, and is distinct from the glycoprotein Ib-IX-V complex. *J Biol Chem, 274,* 972–980.

233. Okumura, T., Hasitz, M., & Jamieson, G. A. (1978). Platelet glycocalicin interaction with thrombin and role as thrombin receptor of the platelet surface. *J Biol Chem, 253,* 3435–3443.

234. Greco, N. J., Tandon, N. N., Jones, G. D., Kornhauser, R., Jackson, B., Yammamoton, N., Tanoue, K., & Jamieson, G. A. (1996). Contributions of glycoprotein Ib and the seven transmembrane domain receptor to increases in platelet cytoplasmic [CA2+] induced by alpha-thrombin. *Biochemistry, 35,* 906–914.

235. Liu, L., Freedman, J., Hornstein, A., Fenton, J. W., II, Song, Y., & Ofosu, F. A. (1997). Binding of thrombin to the G-protein-linked receptor, and not to glycoprotein Ib, precedes

thrombin-mediated platelet activation. *J Biol Chem, 272,* 1997–2004.

236. Leong, L., Henriksen, R., Kermode, J., Rittenhouse, S. E., & Tracy, P. B. (1992). The thrombin high-affinity binding site on platelets is a negative regulator of thrombin-induced platelet acivation. Structure–function studies using two mutant thrombins, Quick I and Quick II. *Biochemistry, 31,* 2567–2576.

237. McNicol, A., Sutherland, M., Zou, R., & Dronin, J. (1996). Defective thrombin-induced calcium changes and aggregation of Bernard–Soulier platelets are not associated with deficient moderate-affinity receptors. *Arterioscl Thromb Vasc Biol, 16,* 628–632.

238. Greco, N. J., Jones, G., Tandon, N. N., Kornhauser, R., Jackson, B., & Jamieson, G. A. (1996). Differentiation of the two forms of GPIb functioning as receptors for α-thrombin and von Willebrand factor: $Ca^{2+}$ responses of protease-treated human platelets activated with α-thrombin and the tethered ligand peptide. *Biochemistry, 35,* 915–921.

239. Phillips, D. R., & Agin, P. P. (1977). Platelet plasma membrane glycoproteins. Identification of a proteolytic substrate for thrombin. *Biochem Biophys Res Commun, 75,* 940–947.

240. Mosher, D., Vaheri, A., Choate, J., & Gahmberg, C. (1979). Action of thrombin on surface glycoproteins of human platelets. *Blood, 53,* 437–445.

241. White, G. C. I., & Krupp, C. (1985). Glycoprotein V hydrolysis by thrombin. Lack of correlation with secretion. *Thromb Res, 38,* 641–648.

242. Bienz, D., Schnippering, W., & Clemetson, K. J. (1986). Glycoprotein V is not the thrombin activation receptor on human blood platelets. *Blood, 68,* 720–725.

243. Ramakrishnan, V., Reeves, P., DeGuzman, F., Deshpande, U., Madrid–Ministri, K., DuBridge, R., & Philipps, D. (1999). Increased thrombin responsiveness in platelets from mice lacking glycoprotein. *Proc Natl Acad Sci USA, 96,* 13336–13341.

244. Ramakrishnan, V., DeGuzman, F., Bao, M., Hall, S., Leung, L. L., & Philipps, D. (2001). A thrombin receptor function for platelet glycoprotein Ib-IX unmasked by cleavage of glycoprotein V. *Proc Natl Acad Sci USA, 98,* 1823–1828.

245. Shima, C., Majima, M., & Katori, M. (1992). A stable metabolite, Arg-Pro-Pro-Gly-Phe of bradykinin in the degradation pathway in human plasma. *Japan J Pharmacol, 60,* 111–119.

246. Puri, R., Zhou, F., Hu, C., Colman, R., & Colman, R. (1991). High molecular weight kininogen inhibits thrombin-induced platelet aggregation and cleavage of aggregin by inhibiting binding of thrombin to platelets. *Blood, 77,* 500–507.

247. Meloni, F., & Schmaier, A. (1991). Low molecular weight kininogen binds to platelets to modulate thrombin-induced platelet activzation. *J Biol Chem, 266,* 6786–6794.

248. Hasan, A., Amenta, S., & Schmaier, A. (1996). Bradykinin and its metabolite, Arg-Pro-Pro-Gly-Phe, and selective inhibitors of alpha-thrombin induced platelet activation. *Circulation, 94,* 517–528.

249. Bernatowicz, M., Klimas, C., Hartl, K., Peluso, M., Alegretto, N., & Seiler, S. M. (1996). Development of potent thrombin receptor antagonist peptides. *J Med Chem, 39,* 4879–4887.

250. Andrade–Gordon, P., Maryanoff, B. E., Derian, C. K., Zhang, H. C., Addo, M. F., Darrow, A. L., Eckardt, A. J., Hoekstra, W. J., McComsey, D. F., Oksenberg, D., Reynolds, E. E., Santulli, R. J., Scarborough, R., Smit, E. M., & White, K. B. (1999). Design, synthesis, and biological characterization of a peptide-mimetic antagonist for a tethered-ligand receptor. *Proc Natl Acad Sci USA, 96,* 12257–12262.

251. Covic, L., Gresser, A. L., Talavera, J., Swift, S., & Kuliopulos, A. (2002). Activation and inhibition of G protein-coupled receptors by cell-penetrating membrane-tethered peptides. *Proc Natl Acad Sci USA, 99,* 643–648.

252. Covic, L., Misra, M., Badar, J., Singh, C., & Kuliopulos, A. (2002). Pepducin-based intervention of thrombin-receptor signaling and systemic platelet activation. *Nat Med, 8,* 1161–1165.

253. Cotecchia, S., Ostrowski, J., Kjelsberg, M. A., Caron, M. G., & Lefkowitz, R. J. (1992). Discrete amino acid sequences of the alpha 1-adrenergic receptor determine the selectivity of coupling to phosphatidylinositol hydrolysis. *J Biol Chem, 267,* 1633–1639.

254. Verall, S., Ishii, M., Wang, M., Tram, T., & Coughlin, S. (1997). The thrombin receptor second cytoplasmic loop confers coupling to Gq-like G proteins in chimeric receptors. Additional evidence for a common transmembrane signaling and G protein coupling mechanism in G protein-coupled receptors. *J Biol Chem, 272,* 6898–6902.

255. Rocheville, M., Lange, D. C., Kumar, U., Patel, S. C., Patel, R. C., & Patel, Y. C. (2000). Receptors for dopamine and somatostatin: Formation of hetero-oligomers with enhanced functional activity. *Science, 288,* 154–157.

256. Bahou, W. F. (2002). Attacked from within, blood thins. *Nat Med, 8,* 1082–1083.

# CHAPTER 10

# The Platelet P2 Receptors

**Marco Cattaneo**

*Unità di Ematologia e Trombosi–Ospedale San Paolo, Università degli Studi di Milano, Milano, Italy*

## I. Introduction

Purine and pyrimidine nucleotides are extracellular signaling molecules that regulate the function of virtually every cell in the body. They are released from damaged cells or are secreted via nonlytic mechanisms and interact with specific receptors on the cell plasma membranes. These specific plasma membrane receptors are called *P2 receptors* and, according to their molecular structure, are divided into two subfamilies: G protein-linked or "metabotropic," termed *P2Y*, and ligand-gated ion channels or "ionotropic," termed *P2X*.[1,2]

P2Y receptors are seven-membrane-spanning proteins with a molecular mass of 41 to 53 kDa after glycosylation. The carboxyl terminal domain is on the cytoplasmic side, whereas the amino terminal domain is exposed to the extracellular environment. Common mechanisms of signal transduction are shared by most seven-membrane-spanning receptors, including activation of phospholipase C (PLC) and/or regulation of adenylyl cyclase activity. Eight P2Y receptors have been identified so far: $P2Y_1$, $P2Y_2$, $P2Y_4$, $P2Y_6$, $P2Y_{11}$, $P2Y_{12}$, $P2Y_{13}$, and $P2Y_{14}$. The missing numbers represent species orthologs that have been cloned from nonmammalian animals or receptors that have some sequence homology to P2Y receptors, but for which convincing evidence of responsiveness to nucleotides is missing. Additional P2Y receptors are expected to be identified among the several existing, orphan G protein-coupled receptors (GPCRs) for endogenous ligands. The recently deorphanized GPR80 (also called *GPR99*) that was proposed to be the $P2Y_{15}$ receptor was later shown not to be a genuine P2Y receptor.[3]

Pharmacologically, P2Y receptors can be subdivided into the adenine–nucleotide-preferring receptors mainly responding to adenosine diphosphate (ADP) and adenosine triphosphate (ATP) ($P2Y_1$, $P2Y_{11}$, $P2Y_{12}$, and $P2Y_{13}$), the uracil–nucleotide-preferring receptors ($P2Y_4$ and $P2Y_6$), responding to either uridine triphosphate or uridine diphosphate (UDP), receptors of mixed selectivity ($P2Y_2$ and rodent $P2Y_4$), and the UDP–glucose receptor ($P2Y_{14}$).[3] From a phylogenetic and structural point of view, two distinct P2Y receptor subgroups with a relatively high level of structural divergence have been identified: the first subgroup includes the $G_q$-coupled subtypes ($P2Y_1$, $P2Y_2$, $P2Y_4$, $P2Y_6$, and $P2Y_{11}$) and the second subgroup includes the $G_i$-coupled subtypes ($P2Y_{12}$, $P2Y_{13}$, and $P2Y_{14}$).[3]

P2X receptors are ligand-gated ion channels that mediate rapid changes in the membrane permeability of monovalent and divalent cations, including $Na^+$, $K^+$, and $Ca^{2+}$. P2X receptors range from 384 to 595 amino acids and have two transmembrane domains separated by a large extracellular region. Unlike the P2Y receptors, the amino and carboxyl terminal domains are both on the cytoplasmic side of the plasma membrane.[2] Seven P2X receptors have been identified so far ($P2X_1$–$P2X_7$).[2] They exist as homo- or hetero-oligomers (trimers or hexamers). At variance with P2Y receptors, which can bind different nucleotides, all P2X receptors are primarily ATP receptors.

Human platelets express at least three distinct receptors that interact with ADP or ATP: $P2Y_1$, $P2Y_{12}$, and $P2X_1$. They have the following order of expression: $P2Y_{12} \gg P2X_1 > P2Y_1$.[4]

## II. Roles of Adenine Nucleotides in Platelet Function

ADP, the first known low-molecular-weight platelet aggregating agent, is a weak platelet agonist. As such, it only induces shape change and reversible aggregation in human platelets, whereas the secretion of platelet-dense granule constituents and the ensuing secondary aggregation that are sometimes observed after stimulation with ADP of normal platelet-rich plasma (PRP) are triggered by thromboxane (TX) $A_2$, the synthesis of which is stimulated by platelet aggregation.[5] This phenomenon can be observed with the PRP of only a minority of normal individuals when the concentration of $Ca^{2+}$ in plasma is maintained at physiological levels,[6] but it is greatly enhanced and can be observed in most individuals when the concentration of plasma $Ca^{2+}$ is artifactually decreased to the micromolar level, such as in citrated PRP.[6–9] Although itself a weak agonist, ADP plays a key role in platelet function because, when it is secreted from the platelet-dense granules where it is stored, it amplifies the platelet responses induced by other platelet agonists[8,10] and stabilizes platelet aggregates.[11–13] Transduction of the ADP-induced signal involves inhibition of adenylyl cyclase and a concomitant transient increase in the concentration of cytoplasmic $Ca^{2+}$, resulting from both $Ca^{2+}$ influx and mobilization of internal stores (see Chapter 16).

The search for the platelet receptor(s) for ADP was hindered for many years by technical difficulties, such as, for example, the limited variety of labeled compounds available for binding studies and their breakdown by nucleotidases on the platelet membrane.[14] Proteins such as aggregin[15] and integrin αIIb (glycoprotein [GP] IIb),[16] which have ADP binding sites, were erroneously proposed as the elusive platelet ADP receptor.[14] Early studies using [β$^{32}$P]-ADP as a ligand indicated that the number of binding sites for ADP on platelets is approximately 1000 per cell.[14] This figure was confirmed by more recent studies. The problem of whether one or more receptors are involved in platelet activation by ADP has long been debated.[14,17,18] Some observations, based on different patterns of modulation of platelet responses to ADP by inhibitors, such as p-chloromercuribenzene sulfonic acid (pCMBS), suggested the possibility that two receptors are involved: one responsible for platelet shape change and aggregation, and one responsible for inhibition of adenylyl cyclase.[19,20] These early suggestions were later confirmed by the demonstration that the ADP receptor $P2Y_1$ is expressed in human platelets[21] and that specific antagonists, such as adenosine-2′,5′-diphosphate (A2P5P) or adenosine-3′,5′-diphosphate (A3P5P), inhibit platelet shape change and aggregation, but have no effect on ADP-induced inhibition of adenylyl cyclase,[22–25] whereas other inhibitors, such as ticlopidine, clopidogrel, or the ATP analogue 2-propylthio-β,γ-difluoromethylene ATP (AR-C66096MX) have the opposite effects.[23–29] Additional evidence came from the observation that the $P2Y_1$ receptor was normal in a patient with congenital impairment of platelet responses to ADP,[30] and that in $P2Y_1$ knockout mice the ability of ADP to induce platelet shape change and aggregation was abolished, whereas that of inhibiting adenylyl cyclase was maintained.[31,32] The second platelet ADP receptor was eventually cloned from human and rat platelet complementary DNA (cDNA) libraries, and designated $P2Y_{12}$.[33] $P2Y_{12}$ displayed pharmacological characteristics of a $G_i$-coupled ADP receptor,[33] was shown to be deficient in a patient with congenital impairment of platelet responses to ADP,[33] and was the target of the active metabolite of clopidogrel.[34]

It is now established that platelets express two receptors for ADP: the $G_q$-coupled $P2Y_1$ receptor, which mediates a transient increase in cytoplasmic $Ca^{2+}$, platelet shape change, and rapidly reversible aggregation, and the $G_i$-coupled $P2Y_{12}$ receptor, which mediates inhibition of adenylyl cyclase and amplifies the platelet aggregation response.[10] Concomitant activation of both the $G_q$ and $G_i$ pathways by ADP is necessary to elicit normal aggregation (Fig. 10-1).[10,35] The importance of concurrent activation of the $G_q$ and $G_i$ pathways for full platelet aggregation is highlighted by the observations that normal aggregation responses to ADP can be restored by epinephrine, which is coupled to an inhibitory G protein, $G_z$, in $P2Y_{12}$-deficient platelets,[22] and by serotonin, which is coupled to $G_q$, in $P2Y_1$ knockout platelets.[31,32]

ATP, being an antagonist of both $P2Y_1$ and $P2Y_{12}$, inhibits platelet activation by ADP.[36,37] However, through its interaction with $P2X_1$, it can also activate platelets by inducing a very rapid influx of $Ca^{2+}$ from the extracellular medium, which is associated with a transient platelet shape change.[38] Platelet activation by ATP amplifies the platelet responses to other agonists, especially during flow conditions that are characterized by high shear stress.[39–42]

The effects of ADP and ATP on platelet function are modulated *in vivo* by CD39, a nucleoside triphosphate diphosphohydrolase (NTPDase-1), and another ectonucleotidase, CD39L1/NTPDase-2.[43–46] NTPDase-1, which is mainly expressed by endothelial cells and vascular smooth muscle cells, hydrolyes both triphosphonucleotides and diphosphonucleotides, and inhibits ADP-induced platelet aggregation. NTPDase-2, which is associated with vascular adventitial cells, preferentially hydrolyes triphosphonucleotides. The important role of CD39 in modulating platelet function and thrombus formation *in vivo* has been demonstrated in experiments with CD39 knockout mice.[44] (CD39 is discussed in detail in Chapter 13.)

## III. $P2Y_1$

### A. $P2Y_1$ is Expressed on Platelets

The human $P2Y_1$ receptor is a 42-kDa protein that contains 373 amino acid residues, has the classic seven-transmembrane domain structure of GPCRs, and is expressed by virtually all tissues of the human body.[47–51] The $P2Y_1$ gene maps to chromosome 3q21-q25. In 1997, Léon et al. detected $P2Y_1$ messenger RNA (mRNA) in both megakaryoblastic cells and platelets, and showed that ADP and its analogs were agonists, whereas ATP and its analogs were antagonists of $Ca^{2+}$ movements.[21] The partial agonist behavior of ATP that was observed in some studies was the result of receptor activation by contaminating ADP with concomitant receptor blockade by ATP.[23,28]

Binding studies using the radioligand [$^{33}$P]MRS 2179, a $P2Y_1$ receptor antagonist, showed that washed human platelets display only 134 ± 8 binding sites per platelet with an

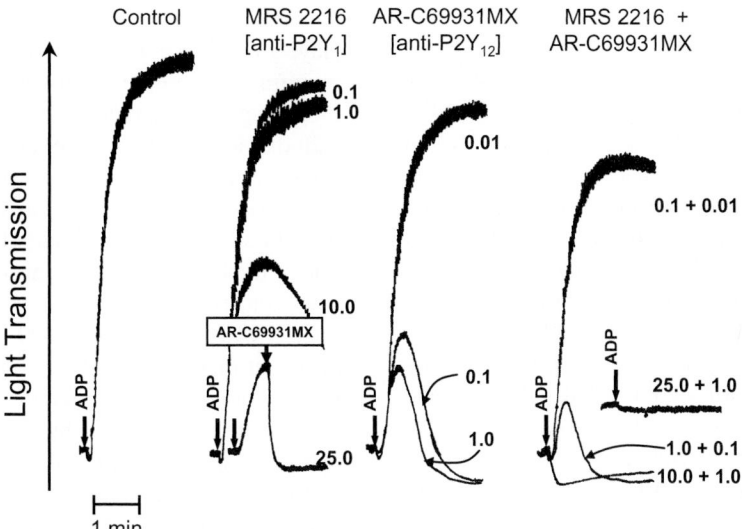

**Figure 10-1.** Effects of specific antagonists of P2Y₁ (MRS2216) and P2Y₁₂ (AR-C69931MX) added alone or in combination to normal platelet-rich plasma anticoagulated with the thrombin inhibitor D-phenylalanyl-L-prolyl-L-arginine chloromethyl ketone dihydrochloride (6 µM) on platelet aggregation induced by 5 µM ADP. Numbers next to each tracing refer to the antagonist concentration in micromoles per liter. In the presence of 25 µM MRS 2216, ADP did not induce platelet shape change, indicating that the antagonist completely abolished the function of P2Y₁. Yet, ADP still induced platelet aggregation, which was readily and completely reversed by the addition of AR-C69931MX (arrow), indicating that it was mediated by the P2Y₁₂ receptor. Representative of several experiments.

affinity (K$_d$) of 109 ± 18 nM.[52] The lower number of P2Y₁ receptors (5–35 receptors/platelet) that was calculated using another radioligand, [³H]MRS2279, a selective high-affinity, nonnucleotide P2Y₁ receptor antagonist,[53] is probably less accurate, because the experiments were performed using outdated human platelets,[54] which may have internalized P2Y₁ receptors[55,56] (discussed later). Overall, P2Y₁ accounts for about 20 to 30% of the total ADP binding sites on the platelet surface.[25,30] Immunogold labeling with a monoclonal antibody to the amino-terminal domain of P2Y₁ revealed that, although present at the platelet surface, P2Y₁ is also abundantly represented inside the platelet. Specifically, P2Y₁ was found in membranes of α-granules and elements of the open canalicular system.[57] Injection of Mpl ligand into mice upregulated P2Y₁ receptor mRNAs in megakaryocytes; however, platelets isolated from these mice did not exhibit a higher P2Y₁ receptor density or increased reactivity to ADP. Thus, the enhancement of P2Y₁ receptor expression induced by Mpl ligand in megakaryocytes may be an integral feature of their differentiation.[58]

## B. Role of P2Y₁ in ADP-Induced Platelet Activation

P2Y₁ mediates platelet shape change and transient, rapidly reversible aggregation induced by ADP, as shown in experiments with normal platelets in the presence of specific P2Y₁₂ antagonists (Fig. 10-1), or with platelets from patients who

are congenitally deficient in P2Y₁₂ or from P2Y₁₂ knockout mice.[23–25,59–63] P2Y₁ stimulation with ADP is also associated with secretion of α-granules from aspirin-treated platelets.[64] Overexpression of P2Y₁ in transgenic mice was associated with a shortened bleeding time and increased platelet reactivity to ADP or low concentrations of collagen, and enabled ADP to induce dense granule secretion.[65] Whether this last effect is directly related to the higher level of P2Y₁ expression or secondary to increased platelet aggregation is unclear.

It is well established that activation of the P2Y₁ receptor by ADP leads to the G$_q$-mediated activation of β-isoforms of phospholipase C (PLC) and to a transient increase in the concentration of intracellular Ca²⁺, mainly through the release of Ca²⁺ into the cytoplasm from intracellular stores, but partially by influx of Ca²⁺ from the external medium.[66] P2Y₁ and its downstream effector G$_q$ are also responsible for the activation of Rac.[67] The lack of ADP-induced platelet shape change in normal platelets in the presence of specific P2Y₁ antagonists or in platelets from P2Y₁ knockout mice is compatible with the hypothesis that it is mediated by the G$_q$ pathway only, as suggested also by the findings that G$_q$ knockout platelets do not undergo shape change upon stimulation by ADP.[68,69] However, the demonstration that the P2Y₁ receptor is also coupled to G₁₂/G₁₃-linked Rho-kinase activation,[70] which can mediate platelet shape change,[70,71] still leaves open the question of whether P2Y₁ is coupled to both pathways or to G$_q$ only.[69,70,72,73]

## C. Role of P2Y₁ in Platelet Responses to Agonists Other Than ADP

The $P2Y_1$ receptor plays an important role in mediating the potentiating effect of released ADP on platelet aggregation induced by agonists that trigger platelet secretion. Platelets from $P2Y_1$ knockout mice display delayed and reduced aggregation induced by collagen, compared with wild-type platelets.[31,32] In addition, the $P2Y_1$ receptor plays an essential role in platelet shape change induced by collagen when $TXA_2$ formation is prevented.[74] $P2Y_1$ also plays a critical role in the initial phases of platelet activation induced by chemokines interacting with CCR1, CCR3, CCR4, and CXCR4 chemokine receptors, including monocyte chemotactic protein 1, macrophage inflammatory peptide 1α, RANTES (regulated upon activation, normal T cell expressed and presumably secreted), TARC (thymus and activation regulated chemokine), macrophage-derived chemokine, and stromal cell-derived factor 1.[75,76] Although thrombin signals independently of released ADP, the relative role of ADP interaction with its two P2Y receptors in the reinforcement of platelet responses to protease-activated receptors (PARs), PAR1 and PAR4, is still unclear.[77–79]

When normal human platelets are exposed to very high shear stress, they undergo von Willebrand factor- and GPIb-dependent platelet aggregation, which is reinforced by released ADP interacting with both $P2Y_1$ and $P2Y_{12}$ receptors[73,80,81] (discussed later).

## D. Role of P2Y₁ in Thrombin Generation

Both the $P2Y_1$ and $P2Y_{12}$ ADP receptors were found to be implicated in collagen-induced exposure of tissue factor[82] and platelet microparticle formation[83] in whole blood, and to contribute to the interactions between platelets and leukocytes mediated by platelet P-selectin exposure, which results in tissue factor exposure at the surface of leukocytes.[84] However, the $P2Y_1$ receptor is not responsible for the potentiation of collagen-, thrombin-, PAR1-, or tissue factor-induced platelet procoagulant activity.[84–86] In vivo studies showed that thrombin generation following intravenous injection of tissue thromboplastin in $P2Y_1$ knockout mice is lower than in wild-type mice, indicating that $P2Y_1$ is indeed important for normal thrombin generation in whole blood.[87]

## E. Role of P2Y₁ in Platelet Thrombus Formation in Vitro and in Vivo

The $P2Y_1$ receptor was found to have a role in thrombus formation in vitro on collagen-coated surfaces, especially under flow conditions characterized by high shear.[80,81]

Studies in vivo showed that $P2Y_1$ knockout mice displayed resistance to systemic thromboembolism induced by infusion of a mixture of collagen and epinephrine[31,32] or tissue factor[87] with reduced mortality, platelet consumption, and the formation of thrombin–antithrombin complexes. These results could be reproduced by intravenous injection of the selective $P2Y_1$ antagonist MRS2179[87] or the more potent and stable compound MRS2500.[88,89] The important role of the $P2Y_1$ receptor in thrombus formation in vivo was later confirmed using $P2Y_1$ knockout mice or wild-type mice treated with MRS2179 or MRS2500 in models of experimental thrombosis involving injury in mouse mesenteric arteries with ferric chloride or laser injury.[89–91] The inhibition of thrombus formation accomplished by MRS2179 was found to be equivalent to that of clopidogrel-treated animals at the maximal effective dose and was potentiated by the concomitant administration of the two inhibitors.[90]

Altogether, these results demonstrate that the $P2Y_1$ receptor should be considered as an interesting target for antiplatelet compounds.

## F. Inherited Abnormalities of the P2Y₁ Receptor

A description of a patient with a history of bleeding after surgery and occasional weak ADP-induced platelet aggregation was published in abstract form by Oury et al.[92] in 1999. The defect was associated with normal $P2Y_1$ encoding regions in the patient's DNA, but reduced platelet levels of $P2Y_1$ mRNA (75% of normal), suggestive of deficient $P2Y_1$ gene transcription. Low $P2Y_1$ mRNA platelet levels were also found in a sister, son, and grandson of the index patient, but not in six other members of the family. Consistent with a deficiency of $P2Y_1$, ADP did not elicit intracellular $Ca^{2+}$ mobilization in the index patient and his family members with low $P2Y_1$ mRNA platelet levels. However, no further details of this family have been published in a full article since this 1999 abstract.[92]

A $P2Y_1$ gene dimorphism, 1622AG, was associated with a significant effect on platelet ADP response, with a greater response in carriers of the G allele (frequency 0.15). The response to all tested concentrations of ADP in GG homozygotes was higher than in AA homozygotes, but was greatest with 0.1 μM ADP (on average, 130% higher).[93]

## IV. P2Y₁₂

### A. P2Y₁₂ is Expressed on Platelets

In 2001, the $P2Y_{12}$ receptor was cloned from human and rat platelet cDNA libraries, using an expression cloning strategy in Xenopus oocytes designed to detect $G_i$-linked receptors through their coupling to cotransfected inward-rectifying K$^+$ channels,[33] and by screening an orphan receptor library as well.[94] Like the P2Y$_1$ gene, the P2Y$_{12}$ maps to chromosome 3q21-q25.[95] The human P2Y$_{12}$ receptor contains 342 amino acid residues, has a classic structure of a GPCR, and appears to have a more selective tissue distribution than P2Y$_1$, being restricted to platelets and subregions of the brain.[33] However, more recently, P2Y$_{12}$ mRNA has also been detected in other tissues, including vascular smooth muscle cells.[96]

P2Y$_{12}$ contains two potential N-linked glycosylation sites at its extracellular aminoterminus, which may modulate its activity. Indeed, studies of both tunicamycin treatment and site-directed mutagenesis have revealed a vital role of N-linked glycans in the receptor's signal transducing step but not in surface expression or ligand binding.[97] Studies of a patient with dysfunctional P2Y$_{12}$ revealed that the integrity of the highly conserved H-X-X-R/K motif in TM6 and of EL3 is important for receptor function[98] (discussed later). P2Y$_{12}$ has four Cys residues at positions 17, 97, 175, and 270. Studies by site-directed mutagenesis indicated the essential role of a disulfide bridge between Cys97 and Cys175 for receptor expression and suggested that Cys17 and Cys270 are targets of thiol reagents, including pCMBS and the active metabolites of thienopyridines.[19,99,100]

## B. Role of P2Y$_{12}$ in ADP-Induced Platelet Activation

ADP and its analogs stimulate P2Y$_{12}$-mediated inhibition of adenylyl cyclase, whereas ATP and its triphosphate analogs are antagonists at the P2Y$_{12}$ receptor, provided care is taken to remove diphosphate contaminants and to prevent the generation of diphosphate nucleotide derivatives by cell ectonucleotidases.[101] The P2Y$_{12}$ receptor is coupled to inhibition of adenylyl cyclase activity through activation of a G$_{\alpha i2}$ G protein subtype. This coupling was actually established before the molecular identification of the P2Y$_{12}$ receptor.[102] Studies of G$_{\alpha i2}$-deficient mice further indicated that the P2Y$_{12}$ receptor mainly activates G$_{\alpha i2}$, although it can also couple to other G$_{\alpha i}$ subtypes.[103] The G protein selectivity of the purified P2Y$_{12}$ receptor reconstituted into liposomes was more recently examined. The most robust coupling of P2Y$_{12}$ was to G$_{\alpha i2}$, but effective coupling also occurred to G$_{\alpha i1}$ and G$_{\alpha i3}$, whereas little or no coupling occurred to G$_{\alpha o}$ or G$_{\alpha q}$.[104] Activation of G$_{\alpha i2}$ by ADP is critical for integrin αIIbβ3 activation and platelet aggregation[103] and has a critical requirement for lipid rafts.[105] It must be noted, however, that, although inhibition of adenylyl cyclase via G$_{\alpha i2}$ is a key feature of platelet activation by ADP, it bears no causal relationship to platelet aggregation.[106–108] Therefore, other signaling events downstream of G$_{\alpha i2}$ are required for activation of integrin αIIbβ3 and subsequent platelet aggregation.[109]

Several studies suggested a crucial role for phosphoinositide 3-kinase (PI 3-K) in ADP-dependent, P2Y$_{12}$ receptor-mediated potentiation of platelet activation.[13,110,111] Platelets express three PI 3-K isoforms: PI 3-Kα, PI 3-Kβ, and PI 3-Kγ. Aggregation of, and fibrinogen binding to, PI 3-Kγ-deficient platelets were reduced in response to ADP, indicating that this PI 3-K subtype is required for ADP-induced full aggregation,[112] although other subtypes may play a role downstream of G$_i$.[113] In addition, studies of platelets from P2Y$_1$ knockout mice and of normal platelets in the presence of specific P2Y$_1$ antagonists showed that ADP, at higher concentrations than those commonly used to activate platelets, can induce slow and sustained platelet aggregation, which is not preceded by platelet shape change (Fig. 10-1).[31,32,114,115] This aggregation was independent of protein kinase C but dependent on PI 3-K.[115]

Two potential targets downstream of PI 3-K activation are the serine–threonine protein kinase B/Akt (PKB/Akt) and the small GTPase Rap1. The interaction of ADP with P2Y$_{12}$ leads to the phosphorylation of PKB/Akt in normal platelets,[112,116] but not in platelets that lack PI 3-Kγ.[112] More recently, G protein-gated, inwardly rectifying potassium channels have been shown to have a role in PKB/Akt phosphorylation and were postulated to be important functional effectors of the P2Y$_{12}$ receptor in human platelets.[117] Rap1 has been shown to contribute to integrin activation in several cell lines and to be activated by a calcium-dependent mechanism in platelets. Rap1 is also activated by G$_{\alpha i}$ family members in platelets.[118,119] In fact, platelets from mice lacking G$_{\alpha i}$ exhibited markedly reduced Rap1 activation in response to ADP, suggesting that Rap1 activation in platelets is dependent on the G$_{\alpha i}$-coupled P2Y$_{12}$ receptor.[118] This response to ADP was only partially impaired in mice lacking G$_{\alpha q}$,[118] suggesting the partial involvement of the G$_{\alpha q}$-coupled P2Y$_1$ receptor, which, in contrast, was apparently ruled out by other studies.[120] The activation of Rap1 by P2Y$_{12}$ does not require inhibition of adenylyl cyclase, but depends on PI 3-K activation.[118,121] The lack of expression of PI 3-Kγ does not compromise the ability of epinephrine or ADP to induce activation of Rap1B, which can still be inhibited by wortmannin, indicating that a different isoform of PI 3-K might be involved in this G$_{\alpha i}$-dependent activation of Rap1.[121] More recent studies indicated that PI 3-Kβ is involved in P2Y$_{12}$-mediated activation of Rap1B by ADP.[122]

P2Y$_{12}$ amplifies the mobilization of cytoplasmic Ca$^{2+}$ mediated by P2Y$_1$[60,123–126] and other receptors.[86,196] Whether this effect is mediated by inhibition of adenylyl cyclase, PI 3-K, or other, as-yet-unidentified effectors is currently unclear. Interestingly, G$_{\alpha i2}$ seems not to be essential for the described P2Y$_{12}$ effect, because the Ca$^{2+}$ response to ADP was not found to be diminished in platelets from the G$_{\alpha i2}$ knockout mouse.[118] It has also been suggested that P2Y$_{12}$, in addition to P2Y$_1$, is involved in ADP-induced platelet shape change[128]; however, this role of P2Y$_{12}$ was demonstrated using concentrations of the P2Y$_{12}$ antagonist AR-C69931MX that were two orders of magnitude higher than those required to inhibit P2Y$_{12}$ function fully, and was not confirmed in other studies.[72,129]

### C. Role of P2Y$_{12}$ in Platelet Responses to Agonists Other Than ADP

Although ADP by itself is unable to cause the release of platelet-dense granules, it can greatly amplify platelet secretion induced by agonists such as TXA$_2$[61,130,131] and thrombin receptor-activating peptide.[132] This effect, which is probably mediated by PI 3-K,[111,133] was observed both at physiological and micromolar concentrations of extracellular Ca$^{2+}$, in the presence of aspirin, and independent of the formation of large platelet aggregates, demonstrating that it is a direct effect of P2Y$_{12}$, rather than being secondary to P2Y$_{12}$ mediated potentiation of aggregation.[61,130–132]

P2Y$_{12}$ plays an essential role in the stabilization of platelet aggregates induced by thrombin[110–113] or TXA$_2$,[134] which is mediated by PI 3-K.[13] Interestingly, although in many cases epinephrine is able to rescue P2Y$_{12}$-deficient function, its ability to restore the stabilization of a thrombus was minimal in human platelets[11] or only partial in P2Y$_{12}$ knockout mice.[135] These observations reinforce the unique role of the P2Y$_{12}$ receptor in stabilization of platelet aggregates.

Studies of human platelets congenitally lacking P2Y$_{12}$ and of P2Y$_{12}$ knockout mice demonstrated that platelet aggregation and secretion induced by a range of platelet agonists, including the TXA$_2$ mimetic U46619, epinephrine, and low concentrations of collagen and thrombin, were impaired.[60,62,98,131,135,136] The central role of P2Y$_{12}$ in platelet responses to other agonists was confirmed by using specific receptor antagonists.[25,126,137–139] P2Y$_{12}$ is also implicated as an important cofactor of platelet aggregation and secretion induced by cross-linking FcγRIIa receptor with specific antibodies or by sera from patients with heparin-induced thrombocytopenia,[110,140–143] or when platelets are activated by collagen through the GPVI/tyrosine kinase/PLCγ2 pathway.[144] The P2Y$_{12}$ receptor is also responsible for the ability of ADP to restore collagen-induced aggregation in G$_{\alpha q}$-deficient mouse platelets.[69] In addition, P2Y$_{12}$ receptor stimulation by released ADP contributes to inhibition of adenylyl cyclase,[77,145] activation of serine–threonine kinase Akt in platelets,[116] tyrosine phosphorylation,[146,147] extracellular signal-regulated kinase 2 (ERK2) activation,[148,149] Rap1B activation,[78,120] Rac activation,[123] and Ca$^{2+}$ mobilization[86] induced by other agonists.

Early studies demonstrated the important role of P2Y$_{12}$ in shear-induced platelet aggregation long before its molecular identification, by using platelets from individuals treated with the anti-thrombotic drug ticlopidine[150] or from a patient with congenital impairment of platelet responses

to ADP.[150] This effect of P2Y$_{12}$, which was later confirmed using specific, direct P2Y$_{12}$ antagonists,[80,81,152,153] depends on PI 3-K activation.[154] Antagonism of both P2Y$_1$ and P2Y$_{12}$ receptors resulted in greater inhibition of shear-induced platelet aggregation than did the inhibition of either receptor alone.[80,81]

### D. Role of P2Y$_{12}$ in Thrombin Generation

As mentioned earlier, P2Y$_1$ and P2Y$_{12}$ have different roles in thrombin generation. P2Y$_{12}$ shares with P2Y$_1$ the ability to contribute to collagen-induced exposure of tissue factor[82] and platelet microparticle formation[83] in whole blood, and to contribute to the formation of platelet–leukocyte conjugates mediated by platelet surface P-selectin exposure, which result in tissue factor exposure at the surface of leukocytes.[84,155] However, only the P2Y$_{12}$ receptor was found to be involved in the exposure of phosphatidylserine by thrombin or other platelet agonists[84–86] and in tissue factor-induced thrombin formation in PRP.[84]

### E. Role of P2Y$_{12}$ in Platelet Thrombus Formation in Vitro and in Vivo

Several studies reported the important role of the P2Y$_{12}$ receptor in platelet thrombus formation and stabilization on collagen-coated surfaces under flow conditions.[80,81,152,153] Inhibition of both P2Y$_1$ and P2Y$_{12}$ receptors resulted in better inhibition of thrombus formation than inhibition of either receptor alone.

Studies of P2Y$_{12}$ knockout mice and of wild-type animals treated with P2Y$_{12}$ antagonists or inhibitors, using different models of experimental arterial and venous thrombosis, have clearly demonstrated the important role of this receptor in thrombogenesis in vivo.[63,134,156–169] In addition, the thienopyridines ticlopidine and clopidogrel, which irreversibly inhibit the platelet P2Y$_{12}$ receptor, are antithrombotic drugs of proven efficacy in patients with coronary artery, peripheral artery, or cerebrovascular diseases (see Chapter 61).

### F. Inherited Abnormalities of the P2Y$_{12}$ Receptor

**1. Severe P2Y$_{12}$ Deficiency.** The first patient with severe P2Y$_{12}$ deficiency (V. R.) was described in 1992 by Cattaneo et al.[60] V. R. is a man from southern Italy, age 49 years at the time of diagnosis, who had a lifelong history of excessive bleeding, a prolonged bleeding time (15–20 minutes) and abnormalities of platelet aggregation that are similar to those observed in patients with defects of platelet secretion (reversible aggregation in response to weak agonists and impaired aggregation in response to low concentrations of collagen or thrombin), except that the aggregation response to ADP was severely impaired, even at very high ADP concentrations (>10 μM). Other abnormalities of platelet function found in this patient were in common with those induced by thienopyridine compounds, including (a) no inhibition by ADP of prostaglandin $E_1$-stimulated platelet adenylyl cyclase, but normal inhibition by epinephrine; (b) normal shape change and borderline to normal (or mildly reduced) mobilization of cytoplasmic $Ca^{2+}$ induced by ADP; and (c) the presence of approximately 30% of the normal number of binding sites for [$^{33}$P]2MeS-ADP on fresh platelets[30] or [$^{3}$H]ADP on formalin-fixed platelets.[60] After the identification and cloning of P2Y$_{12}$, it was possible to characterize this defect at a molecular level. The patient's P2Y$_{12}$ gene displayed a homozygous 2-bp deletion in the open reading frame, located at base pair 294 from the start methionine (near the N-terminal end of the third transmembrane domain), thus shifting the reading frame for 33 residues before introducing a stop codon, causing a premature truncation of the protein.[169]

Four additional patients, one French man (M. L.),[136] two Italian sisters (I. G. and M. G.),[131] and a Japanese woman (OSP-1)[171] with very similar characteristics were described in the years 1995, 2000, and 2005. Similar to V. R., patients I. G. and M. G. displayed a homozygous single base pair deletion in the P2Y$_{12}$ gene occurring just beyond the third transmembrane domain, thus shifting the reading frame for 38 residues before introducing a stop codon, causing a premature truncation of the protein.[170] Patient OSP-1 was found to be homozygous for a single nucleotide substitution in the transduction initiation codon; transfection of the mutant P2Y$_{12}$ construct into human embryonic kidney 293 cells failed to express the P2Y$_{12}$ protein, demonstrating that the mutation was responsible for P2Y$_{12}$ deficiency in this patient.[171] In contrast, the molecular defect that is responsible for the abnormal phenotype of patient M. L. is less well defined. The patient has one mutant and one wild-type allele. The mutant allele contains a deletion of 2 bp within the coding region, at amino acid 240 (near the N-terminal end of the sixth transmembrane domain), thus shifting the reading frame for 28 residues before introducing a stop codon, causing a premature truncation of the protein.[33] Because biochemical studies of M. L.'s platelets indicated that he was completely defective for the $G_i$-linked receptor, this suggested that either his mutant allele functions as a dominant negative or that he has a second mutation that silences his wild-type allele.[33] The finding that his daughter, who had a heterozygous phenotype, inherited the mutant allele from her father and a normal allele from her mother, rules out the possibility that the mutant allele functions as a dominant negative, indicating that patient M. L. might have an additional, as-yet-unknown mutation that silences his normal allele.

**2. Heterozygous P2Y$_{12}$ Deficiency.** The study of the son of patient M. G. allowed the characterization of a heterozygous P2Y$_{12}$ defect.[131] His platelets bound intermediate levels of [$^{33}$P]2MeS-ADP and underwent a normal first wave of aggregation after stimulation with ADP, but did not secrete normal amounts of ATP after stimulation with different agonists. This secretion defect was not caused by impaired production of TXA$_2$ or low concentrations of platelet granule contents, and is therefore very similar to that described in patients with an ill-defined and probably heterogeneous group of congenital defects of platelet secretion, sometimes referred to by the general term *primary secretion defect* (PSD).[172] This defect, which is the most common congenital abnormality of platelet secretion, is characterized by abnormal/borderline-low platelet secretion induced by different agonists, a normal primary wave of aggregation induced by ADP, normal granule stores, and normal arachidonate metabolism. The results of this study,[131] therefore, confirmed a previous hypothesis that (some) patients with PSD are heterozygous for the severe defect of P2Y$_{12}$.[130] The important role of ADP's interaction with its P2Y$_{12}$ receptor in primary hemostasis is emphasized by the finding that the patient, like others with PSD, despite the mild defect of P2Y$_{12}$, has a prolongation of his bleeding time (13 minutes).

**3. Congenital Dysfunction of P2Y$_{12}$.** The patients with defective P2Y$_{12}$ function who have been described so far in this chapter have defective binding of ADP or its analogues to platelets, resulting from defective synthesis of the molecule. Another patient with a congenital bleeding disorder associated with abnormal P2Y$_{12}$-mediated platelet responses to ADP has been described whose platelets display the normal number of dysfunctional P2Y$_{12}$ receptors.[98] Platelets from this patient had normal shape change, but reduced and reversible aggregation in response to 4 μM ADP, similar to normal platelets with a blocked P2Y$_{12}$ receptor. The response to 20 μM ADP, albeit still decreased, was more pronounced and was further decreased by a P2Y$_{12}$ antagonist, indicating residual receptor function. ADP failed to lower the adenylyl cyclase activity stimulated by prostaglandin $E_1$ in the patient's platelets, even though the number and affinity of [$^{33}$P]-2MeS-ADP binding sites were normal. Analysis of the patient's P2Y$_{12}$ gene revealed, in one allele, a G-to-A transition changing the codon for Arg256 in TM6 to Gln and, in the other, a C-to-T transition changing the codon for Arg265 in EL3 to Trp. Neither mutation interfered with receptor surface expression, but both altered receptor function, because ADP inhibited the forskolin-induced increase of cyclic adenosine monophosphate (cAMP) markedly less in cells transfected with either mutant P2Y$_{12}$ than in wild-type cells. The results of these studies contribute to the development of concepts that can explain the sequence of events involved in the agonist-dependent activation of a GPCR. In accordance with previous studies of the P2Y₁ receptor,[173,174] they identify regions corresponding to the extracytoplasmic end of TM6 and EL3, with a structural integrity that is necessary for normal function.

**4. Diagnosis and Treatment of Congenital P2Y₁₂ Defects.** P2Y₁₂ defects should be suspected when ADP, even at relatively high concentrations (10 μM or higher), induces a slight and rapidly reversible aggregation that is preceded by normal shape change. Of the available confirmatory diagnostic tests, measurement of the platelet binding sites for radiolabeled 2MeS-ADP and inhibition of stimulated adenylyl cyclase by ADP (which can be tested by measuring the platelet levels of cAMP or of vasodilator-stimulated phosphoprotein phosphorylation[175]), the latter is preferred because it is easier to perform, cheaper, more specific, and sensitive not only to quantitative abnormalities of the receptor, but also to functional defects.

The intravenous infusion of the vasopressin analogue DDAVP (0.3 μg/kg) shortened the prolonged bleeding time of patient V. R. from 20 minutes to 8.5 minutes.[151] After the infusion of DDAVP, which was repeated twice at 24-hour intervals, the patient underwent a surgical intervention for disk hernia repair, which was not complicated by excessive bleeding. Although the efficacy of DDAVP in reducing bleeding complications of patients with defects of primary hemostasis is anecdotal, its administration is generally without serious side effects. Therefore, DDAVP can be recommended for the prophylaxis and treatment of bleeding episodes in these patients (see Chapter 67).

**5. Polymorphisms of the P2Y₁₂ Gene.** Four polymorphisms of the P2Y₁₂ gene were identified, which were in total linkage disequilibrium, determining haplotypes H1 and H2, with respective allelic frequencies of 0.86 and 0.14. H2 haplotype is a gain-of-function haplotype associated with increased ADP-induced platelet aggregation.[176] The H2 haplotype was more frequent among 184 patients with peripheral artery disease than in 330 age-matched control subjects (odds ratio, 2.3; confidence interval, 1.4–3.9; $p = 0.002$ after adjustment for diabetes, smoking, hypertension, hypercholesterolemia, and other selected platelet receptor gene polymorphisms).[177] The P2Y₁₂ H2 haplotype does not seem to affect the platelet response to a 600-mg loading dose of clopidogrel in coronary artery disease patients prior to stenting.[178]

# V. P2X₁

## A. *P2X₁ is Expressed on Platelets*

P2X₁ is a widely distributed ligand-gated ion channel. Its gene, which maps to chromosome 17p13.2, contains 12 exons. In 1996, MacKenzie et al.[179] obtained functional evidence of the existence of P2X₁ on platelets, which is responsible for a rapid phase of adenine nucleotide-evoked $Ca^{2+}$ entry via a cation channel. P2X₁ gene mRNA was subsequently found in platelets by several investigators.[4,180–183]

## B. *Role of P2X₁ in Platelet Activation and Thrombus Formation* in Vitro *and* in Vivo

For some time, it was believed that P2X₁ had no role in platelet function.[24,26,180,184,185] This was mainly for three reasons: (a) P2X₁ was considered an ADP receptor, and its role was therefore tested mostly on ADP-induced platelet aggregation, whereas it is now clear that P2X₁ is activated by ATP only[186] (the suggestion that a splice variant, P2X₁del, is expressed in platelets and, in contrast to the full-length P2X₁ wild-type receptor, is activated by ADP[187] was not confirmed by later studies[188]); (b) P2X₁ undergoes rapid desensitization during the preparation of platelet suspensions to study platelet function[38]; and (c) the lack of potent and selective P2X₁ receptor antagonists hindered the search for its functional role. In the year 2000, the seminal work by Rolf et al.[38] showed that when desensitization of the P2X₁ receptor is prevented by addition of a high concentration of apyrase (ATP-diphosphohydrolase E.C.3.6.1.5.), the selective P2X₁ receptor agonist α,β-methylene-ATP (α,β-Me-ATP) induces a rapid $Ca^{2+}$ influx associated with a transient shape change in human platelets. The filopodia that are formed by α,β-Me-ATP-activated platelets are much shorter than those observed in ADP-stimulated platelets.[189] This morphological difference of shape-changed platelets could at least partly account for the lack of formation of large aggregates observed after platelet stimulation with α,β-Me-ATP. P2X₁ receptor-mediated platelet shape change involves myosin light-chain phosphorylation, but not Rho kinase,[190] and is associated with centralization of secretory granules, although dense granule release has not been detected.[190,191] Overexpression of human P2X₁ receptor in transgenic mice did not lead to aggregation or secretion in response to α,β-Me-ATP, whereas $Ca^{2+}$ influx and shape change were enhanced.[192]

Despite the fact that P2X₁ receptor activation alone cannot lead to platelet aggregation, this receptor has been shown to contribute to the aggregation of human platelets in response to collagen[41,190,193] and to synergize with epinephrine and thrombopoietin in inducing platelet aggregation, mostly by accelerating the initial phase of platelet aggregation, rather than by increasing its extent.[194,195] The contribution of P2X₁ in the platelet response to collagen can be observed at low concentrations of collagen only, and probably involves the activation of the ERK2 mitogen-activated protein kinase, which enhances platelet secretion initiated by collagen.[190,191] A role for P2X$_1$-mediated ATP-gated Ca$^{2+}$ influx in the early collagen-evoked Ca$^{2+}$ signals was demonstrated more recently.[196]

The P2X$_1$ receptor has also been reported to synergize the ADP-induced Ca$^{2+}$ response of the P2Y$_1$ receptor.[124,197] The possibility that P2X$_1$ receptors synergize with the platelet P2Y receptors in platelet aggregation remains to be determined, although Rolf et al.[189] did not detect such synergy with P2Y$_{12}$.

Using a cone-and-plate viscometer that allows monitoring of shear-induced platelet aggregation in real time, Cattaneo et al.[39,40] showed that P2X$_1$ stimulation by ATP released from platelet granules accelerates and amplifies von Willebrand factor-mediated platelet aggregation induced by high shear stress (108 dynes/cm$^2$; Fig. 10-2). The shear dependence of the contribution of P2X$_1$ on platelet aggregation was also shown in experiments of thrombus formation on collagen-coated surfaces under flow conditions, which showed that the relative contribution of P2X$_1$ on thrombus growth was a function of the shear rate conditions.[39,40] The important role of P2X$_1$ in shear-induced platelet aggregation was confirmed by Oury et al.[42]

P2X$_1$ knockout mice have been shown to exhibit male infertility resulting from the loss of vas deferens contraction, indicating that the P2X$_1$ receptor is essential for normal male reproductive function.[198] The bleeding time is mildly prolonged in this mouse strain compared with the wild type. Study of P2X$_1$ knockout mice further indicated that the P2X$_1$ receptor contributes to thrombus formation on collagen-coated surfaces *in vitro* and to thrombosis of small arteries, in which the blood flow conditions are characterized by high shear.[193] In addition, these mice displayed resistance to systemic thromboembolism induced by infusion of a mixture of collagen and adrenaline.[193]

The shear dependence of the role of P2X$_1$ in thrombus formation makes it an attractive potential target for antithrombotic drugs, because its inhibition could theoretically affect thrombus formation at sites of very high shear, such as at sites of severe arterial stenosis caused by atheromas, without significantly affecting hemostatic platelet plug formation. Although the synthesis of several P2X$_1$ receptor antagonists has been reported, none of them appeared to display good affinity and selectivity for the P2X$_1$ receptor compared with the other P2 receptor subtypes. Recently, the synthesis of a new potent antagonist of the P2X$_1$ receptor, NF449, which is an analog of suramin, has been reported. NF449 displays good selectivity for the P2X$_1$ receptor versus the P2X$_3$, P2X$_7$,[198] and P2Y$_1$, P2Y$_2$, and P2Y$_{11}$ receptor subtypes.[199,200] *In vivo* studies showed that intravenous injection of 10 mg/kg NF449 into mice resulted in selective inhibition of the P2X$_1$ receptor and decreased intravascular platelet aggregation in a model of systemic thromboembolism, without prolongation of the bleeding time. At a higher dose

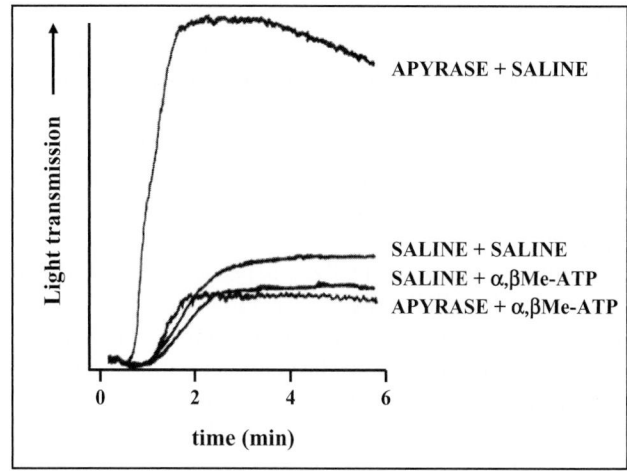

**Figure 10-2.** Platelet aggregation of normal PRP exposed to high shear stress (108 dynes/cm$^2$) in a cone-and-plate viscometer that allows monitoring of shear-induced platelet aggregation in real time. PRP was anticoagulated with 76 μM, PPACK to maintain physiological concentrations of plasma Ca$^{2+}$. Apyrase 1.5 U/mL, which was added to PRP immediately after its preparation in order to rescue P2X$_1$ (which is desensitized during the preparation of the PRP sample because of exposure of platelets to ATP released by blood cells), shortened the lag phase and increased the extent of shear-induced platelet aggregation compared with control (to which saline, instead of apyrase, had been added). The addition of α,β-Me-ATP (10 μM), a nonhydrolyzable P2X$_1$ agonist, immediately before shearing, inhibited shear-induced platelet aggregation in apyrase-PRP but not in saline-PRP. In the presence of α,β-Me-ATP, shear-induced platelet aggregations of apyrase-PRP and saline-PRP were very similar. The effect of α,β-Me-ATP in shear-induced platelet aggregation could be mimicked by NF264, a P2X$_1$ antagonist (not shown). Representative of eight experiments. Interpretation: Apyrase potentiates shear-induced platelet aggregation by rescuing the platelet P2X$_1$, whereas pretreatment of PRP with α,β-Me-ATP inhibits shear-induced platelet aggregation by causing rapid desensitization of P2X$_1$. The results of this study suggest that P2X$_1$ plays an important role in shear-induced platelet aggregation.[40]

(50 mg/kg), NF449 inhibited the three platelet P2 receptors, which led to a further reduction in platelet consumption.[201] NF449 also dose-dependently reduced the size of thrombi formed after laser-induced injury of mesenteric arterioles.[201] Another suramin analog, NF864, which was recently reported to be a very potent and selective antagonist of the platelet P2X$_1$ receptor,[202] is currently under further evaluation.

### C. Inherited Abnormalities of the P2X$_1$ Receptor

Oury et al.[203] described a patient with a severe bleeding diathesis associated with a naturally occurring dominant negative P2X$_1$ mutant, lacking one leucine within a stretch of four leucine residues in the second transmembrane domain (amino acids 351–354). However, the patient also displayed a severe defect of ADP-induced platelet aggregation that cannot be explained by the defect of $P2X_1$, and could by itself account for the bleeding diathesis of the patient. Therefore, the relationship between genotype and phenotype in the patient described by Oury et al.[203] remains unclear.

Screening for defective $P2X_1$-dependent platelet activation by α,β-Me-ATP in more than 100 consecutive patients with bleeding diatheses referred to a single institution for platelet function studies has been unsuccessful so far, indicating that defects in $P2X_1$ are either very rare or do not associate with a bleeding diathesis (Cattaneo, unpublished observations).

## VI. Interplay between the Platelet P2 Receptors

As discussed earlier, coactivation of both the $G_q$-coupled $P2Y_1$ and the $G_i$-coupled $P2Y_{12}$ receptors is necessary for normal ADP-induced platelet aggregation, because inhibition of either receptor is sufficient to inhibit it (Fig. 10-1).[35] This mechanism of ADP-induced platelet aggregation can be mimicked by coactivation of two non-ADP receptors coupled to $G_i$ and $G_q$, neither of which can cause platelet aggregation by themselves. When the $P2Y_{12}$ receptor is blocked by a selective antagonist, full platelet responses to ADP can be restored by simultaneous activation of the $α_{2A}$ adrenergic receptor, which is coupled to the inhibitory $G_z$.[35,59] Conversely, inhibition of $P2Y_1$ receptor-mediated responses by selective antagonists can be bypassed by stimulating the $G_q$ pathway with serotonin.[25,108]

A cross-talk among the three platelet P2 receptors exists at the level of mechanisms modulating the increase in cytoplasmic $Ca^{2+}$ levels, which still need to be completely understood.[123–125,196] Tolhurst et al.[204] recently showed that direct stimulation of both $P2Y_1$ and $P2Y_{12}$ receptors is required for complete activation of a nonselective cation channel that results in $Ca^{2+}$ and $Na^+$ entry in mouse megakaryocytes, which are a *bona fide* model to study P2 receptors signaling in platelets. Furthermore, $P2X_1$ receptors, after secretion of ATP, can contribute repetitively to the ADP-evoked currents and can act to accelerate the P2Y receptor currents.[204] The interplay between the three platelet P2 receptor may be very important in the pathogenesis of platelet thrombi, as suggested by the finding that concomitant inhibition of the three platelet P2 receptors completely abolished shear-induced platelet aggregation.[39]

A schematic representation of the main interactions between the three platelet P2 receptors in platelet function is shown in Fig. 10-3.

## VII. Desensitization of the Platelet P2 Receptors

Like other members of the GPCR family, the platelet P2 receptors readily undergo rapid agonist-induced desensitization. Thus, it has long been known that after being exposed to ADP, human platelets rapidly become unresponsive to a second stimulation with ADP.[8] There is no consensus on the extent of desensitization of each of the two platelet P2Y receptors and the relevant underlying mechanisms. Baurand et al.[55] initially showed that platelet desensitization to ADP is associated with no significant changes in the number of $P2Y_{12}$ sites, but with a decrease in the number of $P2Y_1$ platelet binding sites for [$^{33}$P]2MeS-ADP, which is caused by $P2Y_1$ receptor internalization through a clathrin-dependent pathway.[56] The same group later showed that a substantial fraction of $P2Y_{12}$ receptors was also rapidly internalized through a clathrin-independent pathway, but that this effect was very short-lived.[56] In contrast to the findings of Baurand et al.,[55] Hardy et al.[205] recently reported that both $P2Y_{12}$ and $P2Y_1$ undergo desensitization, and that $P2Y_{12}$ desensitization is mediated by G protein-coupled receptor kinases, whereas that of $P2Y_1$ is largely dependent on protein kinase C activity.[205]

The very rapid desensitization of the $P2X_1$ receptor,[38] which occurs within 20 msec of exposure to its agonist, and accounts for the difficulties that were encountered in the study of its role in platelet function, was mentioned in Section V.B.

## VIII. Conclusions

Several lines of evidence indicate that adenine nucleotides play a key role in the formation of the hemostatic plug and the pathogenesis of arterial thrombi: (a) they are contained at high concentrations in platelet-dense granules and are released when platelets are stimulated by other agents such as thrombin or collagen, thus modulating platelet aggregation; (b) thienopyridines, which inhibit ADP-induced platelet aggregation, are very effective antithrombotic drugs (see Chapter 61); and (c) patients with defects of ADP receptors or lacking adenine nucleotides in platelet granules have a bleeding diathesis. In the last few years, P2 receptors for ADP and ATP have been identified and characterized. Selective P2 receptor agonists and antagonists are under development, which will help in further defining the role of P2 receptors in the pathophysiology of hemostasis and thrombosis.[168,169,206–210] The results of clinical trials will tell us whether some of these antagonists will be useful in the treatment of patients with vascular disease (see Chapter 61).

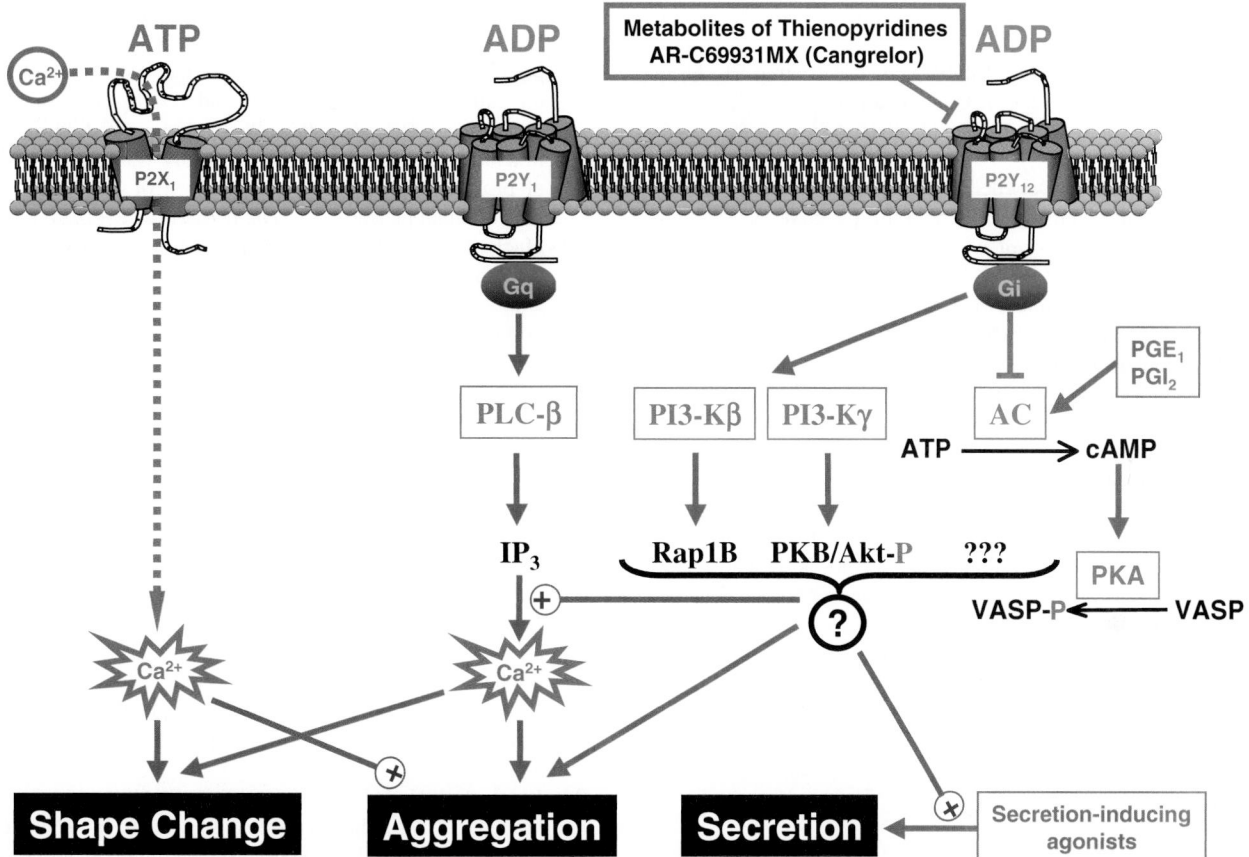

**Figure 10-3.** Simplified schematic representation of the effects of the interaction of adenine triphosphate (ATP) and adenosine diphosphate (ADP) with platelet P2 receptors (P2X$_1$, P2Y$_1$, P2Y$_{12}$) in platelet function. ATP activates the P2X$_1$ receptor, a ligand-gated ion channel, which is responsible for a rapid influx of Ca$^{2+}$ from the external medium and for platelet shape change, and amplifies platelet aggregation induced by other agonists, in particular under flow conditions that are characterized by high shear stress. ATP may also modulate platelet function by antagonizing (not shown) the effects of ADP on its P2 receptors: P2Y$_1$ and P2Y$_{12}$. Both P2Y$_1$ and P2Y$_{12}$, which are seven transmembrane G protein-coupled receptors for ADP, are essential for normal platelet response to ADP. P2Y$_1$, which is coupled to G$_q$, is responsible for intracellular Ca$^{2+}$ mobilization, platelet shape change, and initiation of aggregation. P2Y$_{12}$, which is coupled to G$_i$, is responsible for platelet aggregation (especially its stabilization). In addition, it amplifies the P2Y$_1$-mediated intracellular Ca$^{2+}$ mobilization and the secretion of platelet-dense granules stimulated by secretion-inducing agonists, such as thromboxane A$_2$ or thrombin. The signal transduction pathways responsible for platelet activation through P2Y$_{12}$ are only partially understood. Although P2Y$_{12}$ is coupled to inhibition of adenylyl cyclase (AC), this function is not causally related to P2Y$_{12}$-mediated platelet activation. However, it could have important implications *in vivo*, when platelets are exposed to the inhibitory prostaglandins (PG) E$_1$ or PGI$_2$, which increase platelet cyclic adenosine monophosphate (cAMP) levels by activating AC. Phosphoinositide 3-phosphate(PI3-K)-dependent Rap1B activation and serine–threonine protein kinase B/Akt (PKB/Akt) phosphorylation have a role in P2Y$_{12}$ receptor-mediated platelet activation (see text for further details). Many inhibitors of the P2X$_1$, P2Y$_1$, and P2Y$_{12}$ receptors have been developed. Only the P2Y$_{12}$ receptor inhibitor thienopyridines (which need to be transformed to active metabolites *in vivo* to inactivate P2Y$_{12}$) and AR-C69931MX (Cangrelor) are shown, because they are already used in clinical practice, or are in development as antithrombotic drugs (see Chapter 61). PLC-β, phospholipase C-β; IP$_3$, inosythol triphosphate; PKA, protein kinase A; VASP, vasodilator-stimulated phosphoprotein. Green arrow, activation; truncated red line, inhibition; blue line ending with +, amplification.

## References

1. Abbracchio, M. P., Boeynaems, J. M., Barnard, E. A., Boyer, J. L., & Kennedy, C. (2003). Characterization of the UDP-glucose receptor (re-named the P2Y14 receptor) adds diversity to the P2Y receptor family. *Trends Pharmacol Sci, 24,* 52–55.

2. Boeynaems, J. M., Communi, D., Suarez Gonzales, N., & Robaye, B. (2005). Overview of the P2 receptors. *Semin Thromb Hemost, 31,* 139–149.

3. Abbracchio, M. P., Burnstock, G., Boeynaems, J. M., Barnard, E. A., Boyer, J. M., Kennedy, C., et al. (2005). The recently deophanized GPR80 (GPR99) proposed to be the P2Y15

receptor is not a genuine P2Y receptor. *Trend Pharmacol Sci, 26,* 8–9.

4. Wang, L., Ostberg, O., Wihlborg, A. K., Brogren, H., Jern, S., & Erlinge, D. (2003). Quantification of ADP and ATP receptor expression in human platelets. *J Thromb Haemost, 1,* 330–336.

5. Huang, E. M., & Detwiler, T. C. (1986). Stimulus-response coupling mechanisms. In D. R. Phillips & M. A. Shuman (Eds.), *Biochemistry of platelets* (pp. 1–68). Orlando, FL: Academic Press.

6. Falcon, C. R., Cattaneo, M., Ghidoni, A., & Mannucci, P. M. (1993). The in vitro production of thromboxane B2 by platelets of diabetic patients is normal at physiological concentrations of ionised calcium. *Thromb Haemost, 70,* 389–392.

7. Mustard, J. F., Perry, D. W., Kinlough–Rathbone, R. L., & Packham, M. A. (1975). Factors responsible for ADP-induced release reaction of human platelets. *Am J Physiol, 228,* 1757–1765.

8. Packham, M. A., & Mustard, J. F. (2005). Platelet aggregation and adenosine diphosphate/adenosine triphosphate receptors: A historical perspective. *Semin Thromb Hemost, 31,* 129–138.

9. Cattaneo, M., Gachet, C., Cazenave, J. P., & Packham, M. A. (2002). Adenosine diphosphate (ADP) does not induce thromboxane A2 generation in human platelets. *Blood, 99,* 3868–3869.

10. Cattaneo, M., & Gachet, C. (1999). ADP receptors and clinical bleeding disorders. *Arterioscler Thromb Vasc Biol, 19,* 2281–2285.

11. Cattaneo, M., Canciani, M. T., Lecchi, A., Kinlough–Rathbone, R. L., Packham, M. A., Mannucci, P. M., & Mustard, J. F. (1990). Released adenosine diphosphate stabilizes thrombin-induced human platelet aggregates. *Blood, 75,* 1081–1086.

12. Cattaneo, M., Akkawat, B., Kinlough–Rathbone, R. L., Packham, M. A., Cimminiello, C., & Mannucci, P. M. (1994). Ticlopidine facilitates the deaggregation of human platelets aggregated by thrombin. *Thromb Haemost, 71,* 91–94.

13. Trumel, C., Payrastre, B., Plantavid, M., Hechler, B., Viala, C., Presek, P., Martinson, E. A., Cazenave, J. P., Chap, H., & Gachet, C. (1999). A key role of adenosine diphosphate in the irreversible platelet aggregation induced by the PAR1-activating peptide through the late activation of phosphoinositide 3-kinase. *Blood, 94,* 4156–4165.

14. Mills, D. C. (1996). ADP receptors on platelets. *Thromb Haemost, 76,* 835–856.

15. Colman, R. W. (1990). Aggregin: A platelet ADP receptor that mediates aggregation. *FASEB J, 3,* 1425–1435.

16. Greco, N. J., Yamamoto, N. N., Jackson, B., Tandon, N. N., Moos, M. J., & Jamieson, G. A. (1991). Identification of a nucleotide binding site on glycoprotein IIb. Relationship to platelet activation. *J Biol Chem, 266,* 13627–13633.

17. Cusack, N. J., & Hourani, S. M. O. (1982). Adenosine diphosphate antagonists and human platelets: No evidence that aggregation and inhibition of adenylate cyclase are mediated by different receptors. *Br J Pharmacol, 76,* 221–227.

18. Hourani, S. M. O., Welford, L. A., & Cusack, N. J. (1996). Effects of 2-methylthioadenosine 5′β, γ-methylenetriphosphonate and 2-ethylthioadenosine 5′-monophosphate on human platelet aggregation induced by adenosine 5′-diphosphate. *Drug Dev Res, 38,* 12–23.

19. Macfarlane, D. E., & Mills, D. C. (1981). Inhibition by ADP of prostaglandin induced accumulation of cyclic AMP in intact human platelets. *J Cyclic Nucleotide Res, 7*(1), 1–11.

20. MacFarlane, D. E., Srivastava, P. C., & Mills, D. C. (1983). 2-Methylthio-adenosine[beta32P]diphosphate. An agonist and radioligand for the receptor that inhibits the accumulation of cyclic AMP in intact blood platelets. *J Clin Invest, 71,* 420–428.

21. Léon, C., Hechler, B., Vial, C., Leray, C., Cazenave, J. P., & Gachet, C. (1997). The P2Y1 receptor is an ADP receptor antagonized by ATP and expressed in platelets and megakaryoblastic cells. *FEBS Lett, 403,* 26–30.

22. Leon, C., Vial, C., Gachet, C., Ohlmann, P., Hechler, B., Cazenave, J. P., Lecchi, A., & Cattaneo, M. (1999). The P2Y1 receptor is normal in a patient presenting a severe deficiency of ADP-induced platelet aggregation. *Thromb Haemost, 81,* 775–781.

23. Hechler, B., Eckly, A., Ohlmann, P., Cazenave, J. P., & Gachet, C. (1998). The P2Y1 receptor, necessary but not sufficient to support full ADP-induced platelet aggregation, is not the target of the drug clopidogrel. *Br J Haematol, 103,* 858–866.

24. Jin, J., Daniel, J. L., & Kunapuli, S. P. (1998). Molecular basis for ADP-induced platelet activation. II. The P2Y1 receptor mediates ADP-induced intracellular calcium mobilization and shape change in platelets. *J Biol Chem, 273,* 2030–2034.

25. Savi, P., Beauverger, P., Labouret, C., Delfaud, M., Salel, V., Kaghad, M., & Herbert, J. M. (1998). Role of P2Y1 purinoceptor in ADP-induced platelet activation. *FEBS Lett, 422,* 291–295.

26. Daniel, J. L., Dangelmaier, C., Jin, J., Ashby, B., Smith, J. B., & Kunapuli, S. P. (1998). Molecular basis for ADP-induced platelet activation. I. Evidence for three distinct ADP receptors on human platelets. *J Biol Chem, 273,* 2024–2029.

27. Geiger, J., Honig–Liedl, P., Schanzenbacher, P., & Walter, U. (1998). Ligand specificity and ticlopidine effects distinguish three human platelet ADP receptors. *Eur J Pharmacol, 351,* 235–246.

28. Fagura, M. S., Dainty, I. A., McKay, G. D., Kirk, I. P., Humphries, R. G., Robertson, M. J., Dougall, I. G., & Leff, P. P2Y1-receptors in human platelets which are pharmacologically distinct from P2Y(ADP) receptors. *Br J Pharmacol, 124,* 157–164.

29. Jantzen, H. M., Gousset, L., Bhaskar, V., Vincent, D., Tai, A., Reynolds, E. E., & Conley, P. B. (1999). Evidence for two distinct G-protein-coupled ADP receptors mediating platelet activation. *Thromb Haemost, 81,* 111–117.

30. Gachet, C., Cattaneo, M., Ohlmann, P., Hechler, B., Lecchi, A., Chevalier, J., Cassel, D., Mannucci, P. M., & Cazenave, J. P. (1995). Purinoceptors on blood platelets: Further pharmacological and clinical evidence to suggest the

presence of two ADP receptors. *Br J Haematol, 91,* 434–444.

31. Leon, C., Hechler, B., Freund, M., Eckly, A., Vial, C., Ohlmann, P., Dierich, A., LeMeur, M., Cazenave, J. P., & Gachet, C. (1999). Defective platelet aggregation and increased resistance to thrombosis in purinergic P2Y(1) receptor-null mice. *J Clin Invest, 104,* 1731–1737.

32. Fabre, J. E., Nguyen, M., Latour, A., Keifer, J. A., Audoly, L. P., Coffman, T. M., & Koller, B. H. (1999). Decreased platelet aggregation, increased bleeding time and resistance to thromboembolism in P2Y1-deficient mice. *Nat Med, 5,* 1199–1202.

33. Hollopeter, G., Jantzen, H. M., Vincent, D., Li, G., England, L., Ramakrishnan, V., Yang, R. B., Nurden, P., Nurden, A., Julius, D., & Conley, P. B. (2001). Identification of the platelet ADP receptor targeted by antithrombotic drugs. *Nature, 409,* 619–693.

34. Savi, P., Labouret, C., Delesque, N., Guette, F., Lupker, J., & Herbert, J. M. (2001). P2Y(12), a new platelet ADP receptor, target of clopidogrel. *Biochem Biophys Res Commun, 283,* 379–383.

35. Jin, J., & Kunapuli, S. P. (1998). Coactivation of two different G protein-coupled receptors is essential for ADP-induced platelet aggregation. *Proc Natl Acad Sci USA, 95,* 8070–8074.

36. Mustard, J. F., & Packham, M. A. (1970). Factors influencing platelet function: Adhesion, release and aggregation. *Pharmacol Rev, 22,* 97–187.

37. MacFarlane, D. E., & Mills, D. C. (1975). The effects of ATP on platelets: Evidence against the central role of released ADP in platelet aggregation. *Blood, 46,* 309–320.

38. Rolf, M. G., Brearley, C. A., & Mahaut–Smith, M. P. (2001). Platelet shape change evoked by selective activation of P2X1 purinoceptors with alpha,beta-methylene ATP. *Thromb Haemost, 85,* 303–308.

39. Cattaneo, M., Marchese, P., Jacobson, K. A., & Ruggeri, Z. (2002). New insights into the role of P2X1 in platelet function [abstract]. *Haematologica, 87,* 13–14.

40. Cattaneo, M., Marchese, P., Jacobson, K. A., & Ruggeri, Z. M. (2003). The P2X1 receptor for ATP plays an essential role in platelet aggregation at high shear [abstract] *Haematology J, 4*(Suppl 2), 150.

41. Oury, C., Toth–Zsamboki, E., Thys, C., Tytgat, J., Vermylen, J., & Hoylaerts, M. F. (2001). The ATP-gated P2X1 ion channel acts as a positive regulator of platelet responses to collagen. *Thromb Haemost, 86,* 1264–1271.

42. Oury, C., Sticker, E., Cornelissen, H., De Vos, R., Vermylen, J., & Hoylaerts, M. F. (2004). ATP augments von Willebrand factor-dependent shear-induced platelet aggregation through Ca2+-calmodulin and myosin light chain kinase activation. *J Biol Chem, 279,* 26266–26273.

43. Beukers, M. W., Pirovano, I. M., van Weert, A., Kerkhof, C. J., IJzerman, A. P., & Soudijn, W. (1993). Characterization of ecto-ATPase on human blood cells. A physiological role in platelet aggregation? *Biochem Pharmacol, 46,* 1959–1966.

44. Enjyoji, K., Sevigny, J., Lin, Y., Frenette, P. S., Christie, P. D., Esch, J. S., II, Imai, M., Edelberg, J. M., Rayburn, H.,

Lech, M., Beeler, D. L., Csizmadia, E., Wagner, D. D., Robson, S. C., & Rosenberg, R. D. (1999). Targeted disruption of cd39/ATP diphosphohydrolase results in disordered hemostasis and thromboregulation. *Nat Med, 5,* 1010–1017.

45. Marcus, A. J., Broekman, M. J., Drosopoulos, J. H. F., Olson, K. E., Islam, N., Pinsky, D. J., & Levi, R. (2005). Role of CD39 (NTPDase-1) in thromboregulation, cerebroprotection, and cardioprotection. *Semin Thromb Hemost, 31,* 234–246.

46. Robson, S. C., Wu, Y., Sun, X., Knosalla, C., Dwyer, K., & Eniyoji, K. (2005). Ectonucleotidases of CD39 family modulate vascular inflammation and thrombosis an transplantation. *Semin Thromb Hemost, 31,* 217–233.

47. Léon, C., Vial, C., Cazenave, J. P., & Gachet, C. (1996). Cloning and sequencing of a human cDNA encoding endothelial P2Y1 purinoceptor. *Gene, 171,* 295–297.

48. Hoffman, C., Moro, S., Nicholas, R. A., Harden, T. K., & Jacobson, K. A. (1999). The role of aminoacids in extracellular loops of the human P2Y1 receptor in surface expression and activation processes. *J Biol Chem, 274,* 14639–14647.

49. Communi, D., Janssens, R., Suarez–Huerta, N., Robaye, B., & Boeynaems, J. M. (2000). Advances in signalling by extracellular nucleotides. The role and transduction mechanisms of P2Y receptors. *Cell Signal, 12,* 351–360.

50. Ralevic, V., & Burnstock, G. (1998). Receptors for purines and pyrimidines. *Pharmacol Rev, 50,* 413–492.

51. Webb, T. E., Simon, J., Krishek, B. J., Bateson, A. N., Smart, T. G., King, B. F., Burnstock, G., & Barnard, E. A. (1993). Cloning and functional expression of a brain G-protein-coupled ATP receptor. *FEBS Lett, 324,* 219–225.

52. Baurand, A., Raboisson, P., Freund, M., Leon, C., Cazenave, J. P., Bourguignon, J. J., & Gachet, C. (2001). Inhibition of platelet function by administration of MRS2179, a P2Y1 receptor antagonist. *Eur J Pharmacol, 412,* 213–221.

53. Boyer, J. L., Adams, M., Ravi, R. G., Jacobson, K. A., & Harden, T. K. (2002). 2-Chloro N(6)-methyl-(N)-methano-carba-2′-deoxyadenosine-3′,5′-bisphosphate is a selective high affinity P2Y(1) receptor antagonist. *Br J Pharmacol, 135,* 1839–1840.

54. Waldo, G. L., Corbitt, J., Boyer, J. L., Ravi, G., Kim, H. S., Ji, X. D., Lacy, J., Jacobson, K. A., & Harden, T. K. (2002). Quantitation of the P2Y(1) receptor with a high affinity radiolabeled antagonist. *Mol Pharmacol, 62,* 1249–1257.

55. Baurand, A., Eckly, A., Bari, N., Leon, C., Hechler, B., Cazenave, J. P., & Gachet, C. (2000). Desensitization of the platelet aggregation response to ADP: Differential downregulation of the P2Y1 and P2cyc receptors. *Thromb Haemost, 84,* 484–491.

56. Baurand, A., Eckly, A., Hechler, B., Kauffenstein, G., Galzi, J. L., Cazenave, J. P., Leon, C., & Gachet, C. (2005). Differential regulation and relocalization of the platelet P2Y receptors after activation: A way to avoid loss of hemostatic properties? *Mol Pharmacol, 67,* 721–733.

57. Nurden, P., Poujol, C., Winckler, J., Combrie, R., Pousseau, N., Conley, P. B., Levy–Toledano, S., Habib, A., & Nurden, A. T. (2003). Immunolocalization of P2Y1 and TPalpha receptors in platelets showed a major pool associated with

the membranes of alpha-granules and the open canalicular system. *Blood, 101,* 1400–1408.

58. Hechler, B., Toselli, P., Ravanat, C., Gachet, C., & Ravid, K. (2001). Mpl ligand increases P2Y1 receptor gene expression in megakaryocytes with no concomitant change in platelet response to ADP. *Mol Pharmacol, 60,* 1112–1120.

59. Hechler, B., Léon, C., Vial, C., Vigne, P., Frelin, C., Cazenave, J. P., & Gachet, C. (1998). The P2Y1 receptor is necessary for adenosine 5′-diphosphate-induced platelet aggregation. *Blood, 92,* 152–159.

60. Cattaneo, M., Lecchi, A., Randi, A. M., McGregor, J. L., & Mannucci, P. M. (1992). Identification of a new congenital defect of platelet function characterized by severe impairment of platelet responses to adenosine diphosphate. *Blood, 80,* 2787–2796.

61. Cattaneo, M. (2005). The P2 receptors and congenital platelet function defects. *Semin Thromb Hemost, 31,* 168–173.

62. Foster, C. J., Prosser, D. M., Agans, J. M., Zhai, Y., Smith, M. D., Lachowicz, J. E., Zhang, F. L., Gustafson, E., Monsma, F. J., Jr., Wiekowski, M. T., Abbondanzo, S. J., Cook, D. N., Bayne, M. L., Lira, S. A., & Chintala, M. S. Molecular identification and characterization of the platelet ADP receptor targeted by thienopyridine antithrombotic drugs. *J Clin Invest, 107,* 1591–1598.

63. Andre, P., LaRocca, T., Delaney, S. M., Lin, P. H., Vincent, D., Sinha, U., Conley, P. B., & Phillips, D. R. (2003). Anticoagulants (thrombin inhibitors) and aspirin synergize with P2Y12 receptor antagonism in thrombosis. *Circulation, 108,* 2697–2703.

64. Quinton, T. M., Murugappan, S., Kim, S., Jin, J., & Kunapuli, S. P. (2004). Different, G. protein-coupled signaling pathways are involved in alpha granule release from human platelets. *J Thromb Haemost, 2,* 978–984.

65. Hechler, B., Zhang, Y., Eckly, A., Cazenave, J. P., Gachet, C., & Ravid, K. (2003). Lineage-specific overexpression of the P2Y1 receptor induces platelet hyper-reactivity in transgenic mice. *J Thromb Haemost, 1,* 155–163.

66. Sage, S. O., Reast, R., & Rink, T. J. (1990). ADP evokes biphasic Ca2+ influx in fura-2-loaded human platelets. Evidence for Ca2+ entry regulated by the intracellular Ca2+ store. *Biochem J, 265,* 675–680.

67. Soulet, C., Hechler, B., Gratacap, M. P., Plantavid, M., Offermans, S., Gachet, C., & Payrastre, B. (2005). A differential role of the platelet ADP receptors P2Y1 and P2Y12 in Rac activation. *J Thromb Haemost, 3,* 2296–2306.

68. Offermans, S., Toombs, C. F., Hu, Y. H., & Simon, M. I. (1997). Defective platelet activation in Galpha(q)-deficient mice. *Nature, 389,* 183–186.

69. Ohlmann, P., Eckly, A., Freund, M., Cazenave, J. P., Offermanns, S., & Gachet, C. (2000). ADP induces partial platelet aggregation without shape change and potentiates collagen-induced aggregation in the absence of Galphaq. *Blood, 96,* 2134–2139.

70. Paul, B. Z., Daniel, J. L., & Kunapuli, S. P. (1999). Platelet shape change is mediated by both calcium-dependent and -independent signalling pathways. Role of P160 Rho-associated coiled-coil-containing protein kinase in platelet shape change. *J Biol Chem, 274,* 28293–28300.

71. Bauer, M., Retzer, M., Wilde, J. I., Maschberger, P., Essler, M., Aepfelbacher, M., Watson, S. P., & Siess, W. (1999). Dichotomous regulation of myosin phosphorylation and shape change by Rho-kinase and calcium in intact human platelets. *Blood, 94,* 1665–1672.

72. Wilde, J. I., Retzer, M., Siess, W., & Watson, S. P. (2000). ADP-induced platelet shape change: An investigation of the signalling pathways involved and their dependence on the method of platelet preparation. *Platelets, 11,* 286–295.

73. Hechler, B., Cattaneo, M., & Gachet, C. (2005). The P2 receptors in platelet function. *Semin Thromb Hemos, 31,* 150–161.

74. Mangin, P., Ohlmann, P., Eckly, A., Cazenave, J. P., Lanza, F., & Gachet, C. (2004). The P2Y1 receptor plays an essential role in the platelet shape change induced by collagen when TxA2 formation is prevented. *J Thromb Haemost, 2,* 969–977.

75. Clemetson, K. J., Clemetson, J. M., Proudfoot, A. E., Power, C. A., Baggiolini, M., & Wells, T. N. (2000). Functional expression of CCR1, CCR3, CCR4, and CXCR4 chemokine receptors on human platelets. *Blood, 96,* 4046–4054.

76. Suttitanamongkol, S., & Gear, A. R. (2001). ADP receptor antagonists inhibit platelet aggregation induced by the chemokines SDF-1, MDC and TARC. *FEBS Lett, 490,* 84–87.

77. Kim, S., Foster, C., Lecchi, A., Quinton, T. M., Prosser, D. M., Jin, J., Cattaneo, M., & Kunapuli, S. P. (2002). Protease-activated receptors 1 and 4 do not stimulate G(i) signaling pathways in the absence of secreted ADP and cause human platelet aggregation independently of G(i) signaling. *Blood, 99,* 3629–3636.

78. Lova, P., Campus, F., Lombardi, R., Cattaneo, M., Sinigaglia, F., Balduini, C., & Torti, M. (2004). Contribution of protease-activated receptors 1 and 4 and glycoprotein Ib-IX-V in the G(i)-independent activation of platelet Rap1B by thrombin. *J Biol Chem, 279,* 25299–25306.

79. Adam, F., Verbeuren, T. J., Fauchere, J. L., Guillin, M. C., & Jandrot–Perrus, M. (2003). Thrombin-induced platelet PAR4 activation: Role of glycoprotein Ib and ADP. *J Thromb Haemost, 1,* 798–804.

80. Cattaneo, M., Savage, B., & Ruggeri, Z. M. (2001). Effects of pharmacological inhibition of the P2Y1 and P2Y12 ADP receptors on shear-induced platelet aggregation and platelet thrombus formation on a collagen-coated surface under flow conditions. *Blood, 98,* 239a.

81. Turner, N. A., Moake, J. L., & McIntire, L. V. (2001). Blockade of adenosine diphosphate receptors P2Y(12) and P2Y(1) is required to inhibit platelet aggregation in whole blood under flow. *Blood, 98,* 3340–3345.

82. Leon, C., Alex, M., Klocke, A., Morgenstern, E., Moosbauer, C., Eckly, A., Spannagl, M., Gachet, C., & Engelmann, B. (2004). Platelet ADP receptors contribute to the initiation of intravascular coagulation. *Blood, 103,* 594–600.

83. Takano, K., Asazuma, N., Satoh, K., Yatomi, Y., & Ozaki, Y. (2004). Collagen-induced generation of platelet-derived microparticles in whole blood is dependent on ADP released from red blood cells and calcium ions. *Platelets, 15,* 223–229.

84. Leon, C., Ravanat, C., Freund, M., Cazenave, J. P., & Gachet, C. (2003). Differential involvement of the P2Y1 and P2Y12 receptors in platelet procoagulant activity. *Arterioscler Thromb Vasc Biol, 23,* 1941–1947.

85. Dorsam, R. T., Tuluc, M., & Kunapuli, S. P. (2004). Role of protease-activated and ADP receptor subtypes in thrombin generation on human platelets. *J Thromb Haemost, 2,* 804–812.

86. van der Meijden, P. E., Feijge, M. A., Giesen, P. L., Huijberts, M., van Raak, L. P., & Heemskerk, J. W. (2005). Platelet P2Y12 receptors enhance signalling towards procoagulant activity and thrombin generation. A study with healthy subjects and patients at thrombotic risk. *Thromb Haemost, 93,* 1128–1136.

87. Leon, C., Freund, M., Ravanat, C., Baurand, A., Cazenave, J. P., & Gachet, C. (2001). Key role of the P2Y(1) receptor in tissue factor-induced thrombin-dependent acute thromboembolism: Studies in P2Y(1)-knockout mice and mice treated with a P2Y(1) antagonist. *Circulation, 103,* 718–723.

88. Cattaneo, M., Lecchi, A., Ohno, M., Joshi, B. V., Besada, P., Tchilibon, S., Lombardi, R., Bischofberger, N., Harden, T. K., & Jacobson, K. A. (2004). Antiaggregatory activity in human platelets of potent antagonists of the P2Y 1 receptor. *Biochem Pharmacol, 68,* 1995–2002.

89. Hechler, B., Nonne, C., Roh, E. J., Cattaneo, M., Cazenave, J. P., Lanza, F., Jacobson, K., & Gachet, C. (2005). MRS2500, a potent, selective and stable antagonist of the P2Y1 receptor, with strong antithrombotic activity in mice. *J Pharmacol Exp Ther,* ••

90. Lenain, N., Freund, M., Leon, C., Cazenave, J. P., & Gachet, C. (2003). Inhibition of localized thrombosis in P2Y1-deficient mice and rodents treated with MRS2179, a P2Y1 receptor antagonist. *J Thromb Haemost, 1,* 1144–1149.

91. Lenain, N., Freund, M., Hechler, B., Léon, C., Evans, R., Cazenave, J. P., & Gachet, C. (2003). The role of the P2Y1, P2Y12, and P2X1 platelet receptors in a laser induced model of arterial thrombosis in vivo [abstract]. *J Thromb Haemost,* ••.

92. Oury, C., Lenaerts, T., Peerlinck, K., Vermylen, J., & Hoylaerts, M. F. (1999). Congenital deficiency of the phospholipase C coupled platelet P2Y1 receptor leads to a mild bleeding disorder [abstract]. *Thromb Haemost, 85*(Suppl), 20.

93. Hetherington, S. L., Singh, R. K., Lodwick, D., Thompson, J. R., Goodall, A. H., & Samani, N. J. (2005). Dimorphism in the P2Y1 ADP receptor gene is associated with increased platelet activation response to ADP. *Arterioscler Thromb Vasc Biol, 25,* 252–257.

94. Zhang, F. L., Luo, L., Gustafson, E., Lachowicz, J., Smith, M., Qiao, X., Liu, Y. H., Chen, G., Pramanik, B., Laz, T. M., Palmer, K., Bayne, M., & Monsma, F. J., Jr. (2001). ADP is the cognate ligand for the orphan G protein-coupled receptor SP1999. *J Biol Chem, 276,* 8608–8615.

95. Takasaki, J., Kamohara, M., Saito, T., Matsumoto, M., Matsumoto, S., Ohishi, T., Soga, T., Matsushime, H., & Furuichi, K. (2001). Molecular cloning of the platelet P2T(AC) ADP receptor: Pharmacological comparison with another ADP receptor, the P2Y(1) receptor. *Mol Pharmacol, 60,* 432–439.

96. Wihlborg, A. K., Wang, L., Braun, O. O., Eyjolfsson, A., Gustafsson, R., Gudbjartsson, T., & Erlinge, D. (2004). ADP receptor P2Y12 is expressed in vascular smooth muscle cells and stimulates contraction in human blood vessels. *Arterioscler Thromb Vasc Biol, 24,* 1810–1815.

97. Zhong, X., Kriz, R., Seehra, J., & Kumar, R. (2004). N-linked glycosylation of platelet P2Y12 ADP receptor is essential for signal transduction but not for ligand binding or cell surface expression. *FEBS Lett, 562,* 111–117.

98. Cattaneo, M., Zighetti, M. L., Lombardi, R., Martinez, C., Lecchi, A., Conley, P. B., Ware, J., & Ruggeri, Z. M. (2003). Molecular bases of defective signal transduction in the platelet P2Y12 receptor of a patient with congenital bleeding. *Proc Natl Acad Sci USA 100,* 1978–1983.

99. Savi, P., Pereillo, J. M., Uzabiaga, M. F., Combalbert, J., Picard, C., Maffrand, J. P., Pascal, M., & Herbert, J. M. (2000). Identification and biological activity of the active metabolite of clopidogrel. *Thromb Haemost, 84,* 891–896.

100. Ding, Z., Kim, S., Dorsam, R. T., Jin, J., & Kunapuli, S. P. (2003). Inactivation of the human P2Y12 receptor by thiol reagents requires interaction with both extracellular cysteine residues, Cys17 and Cys270. *Blood, 101,* 3908–3914.

101. Kauffenstein, G., Hechler, B., Cazenave, J. P., & Gachet, C. (2004). Adenine triphosphate nucleotides are antagonists at the P2Y receptor. *J Thromb Haemost, 2,* 1980–1988.

102. Ohlmann, P., Laugwitz, K. L., Nurnberg, B., Spicher, K., Schultz, G., Cazenave, J. P., & Gachet, C. (1995). The human platelet ADP receptor activates Gi2 proteins. *Biochem J, 312,* 775–779.

103. Jantzen, H. M., Milstone, D. S., Gousset, L., Conley, P. B., & Mortensen, R. M. (2001). Impaired activation of murine platelets lacking G alpha(i2). *J Clin Invest, 108,* 477–483.

104. Bodor, E. T., Waldo, G. L., Hooks, S. B., Corbitt, J., Boyer, J. L., & Harden, T. K. (2003). Purification and functional reconstitution of the human P2Y12 receptor. *Mol Pharmacol, 64,* 1210–1216.

105. Quinton, T. M., Kim, S., Jin, J., & Kunapuli, S. P. (2005). Lipid rafts are required in Galpha(i) signaling downstream of the P2Y12 receptor during ADP-mediated platelet activation. *J Thromb Haemost, 3,* 1036–1041.

106. Haslam, R. J. (1973). Interactions of the pharmacological receptors of blood platelets with adenylate cyclase. *Ser Haematol, 6,* 333–350.

107. Savi, P., Pflieger, A. M., & Herbert, J. M. (1996). cAMP is not an important messenger for ADP-induced platelet aggregation. *Blood Coagul Fibrinolysis, 7,* 249–252.

108. Daniel, J. L., Dangelmaier, C., Jin, J., Kim, Y. B., & Kunapuli, S. P. (1999). Role of intracellular signaling events in ADP-induced platelet aggregation. *Thromb Haemost, 82,* 1322–1326.

109. Yang, J., Wu, J., Jiang, H., Mortensen, R., Austin, S., Manning, D. R., Woulfe, D., & Brass, L. F. (2002). Signaling through Gi family members in platelets. Redundancy and specificity in the regulation of adenylyl cyclase and other effectors. *J Biol Chem, 277,* 46035–46042.

110. Gratacap, M. P., Herault, J. P., Viala, C., Ragab, A., Savi, P., Herbert, J. M., Chap, H., Plantavid, M., & Payrastre, B. (2000). FcgammaRIIA requires a Gi-dependent pathway for an efficient stimulation of phosphoinositide 3-kinase, calcium mobilization, and platelet aggregation. *Blood, 96,* 3439–3446.

111. Dangelmaier, C., Jin, J., Smith, J. B., & Kunapuli, S. P. (2001). Potentiation of thromboxane A2-induced platelet secretion by Gi signaling through the phosphoinositide-3 kinase pathway. *Thromb Haemost, 85,* 341–348.

112. Hirsch, E., Bosco, O., Tropel, P., Laffargue, M., Calvez, R., Altruda, F., Wymann, M., & Montrucchio, G. (2001). Resistance to thromboembolism in PI3Kgamma-deficient mice. *FASEB J, 15,* 2019–2021.

113. Jackson, S. P., Yap, C. L., & Anderson, K. E. (2004). Phosphoinositide 3-kinases and the regulation of platelet function. *Biochem Soc Trans, 32,* 387–392.

114. Jarvis, G. E., Humphries, R. G., Robertson, M. J., & Leff, P. (2000). ADP can induce aggregation of human platelets via both P2Y(1) and P(2T) receptors. *Br J Pharmacol, 129,* 275–282.

115. Kauffenstein, G., Bergmeier, W., Eckly, A., Ohlmann, P., Leon, C., Cazenave, J. P., Nieswandt, B., & Gachet, C. (2001). The P2Y(12) receptor induces platelet aggregation through weak activation of the alpha(IIb)beta(3) integrin — a phosphoinositide 3-kinase-dependent mechanism. *FEBS Lett, 505,* 281–290.

116. Kim, S., Jin, J., & Kunapuli, S. P. (2004). Akt activation in platelets depends on Gi signaling pathways. *J Biol Chem, 279,* 4186–4195.

117. Shankar, H., Murugappan, S., Kim, S., Jin, J., Ding, Z., Wickman, K., & Kunapuli, S. P. (2004). Role of G protein-gated inwardly rectifying potassium channels in P2Y12 receptor-mediated platelet functional responses. *Blood, 104,* 1335–1343.

118. Woulfe, D., Jiang, H., Mortensen, R., Yang, J., & Brass, L. F. (2002). Activation of Rap1B by G(i) family members in platelets. *J Biol Chem, 277,* 23382–23390.

119. Lova, P., Paganini, S., Sinigaglia, F., Balduini, C., & Torti, M. (2002). A Gi-dependent pathway is required for activation of the small GTPase Rap1B in human platelets. *J Biol Chem, 277,* 12009–12015.

120. Larson, M. K., Chen, H., Kahn, M. L., Taylor, A. M., Fabre, J. E., Mortensen, R. M., Conley, P. B., & Parise, L. V. (2003). Identification of P2Y12-dependent and -independent mechanisms of glycoprotein VI-mediated Rap1 activation in platelets. *Blood, 101,* 1409–1415.

121. Lova, P., Paganini, S., Hirsch, E., Barberis, L., Wymann, M., Sinigaglia, F., Balduini, C., & Torti, M. (2003). A selective role for phosphatidylinositol 3,4,5-trisphosphate in the Gi-dependent activation of platelet Rap1B. *J Biol Chem, 278,* 131–138.

122. Jackson, S. P., Schoenwaelder, S. M., Goncalves, I., Nesbitt, W. S., Yap, C. L., Wright, C. E., Kenche, V., Anderson, K. E., Dopheide, S. M., Yuan, Y., Sturgeon, S. A., Prabaharan, H., Thompson, P. E., Smith, G. D., Shepherd, P. R., Daniele, N., Kulkarni, S., Abbott, B., Saylik, D., Jones, C., Lu, L., Giuliano, S., Hughan, S. C., Angus, J. A., Robertson, A. D.,

& Salem, H. H. (2005). PI 3-kinase p110beta: A new target for antithrombotic therapy. *Nat Med, 11,* 507–514.

123. Hardy, A. R., Jones, M. L., Mundell, S. J., & Poole, A. W. (2004). Reciprocal cross-talk between P2Y1 and P2Y12 receptors at the level of calcium signaling in human platelets. *Blood, 104,* 1745–1752.

124. Sage, S. O., Yamoah, E. H., & Heemskerk, J. W. (2008). The roles of P(2X1)and P(2T AC)receptors in ADP-evoked calcium signalling in human platelets. *Cell Calcium, 28,* 119–126.

125. Cattaneo, M., Lecchi, A., & Lombardi, R. (2003). Concomitant activation of both the P2Y1-driven Gq and the P2Y12-driven Gi pathways is necessary for normal ADP-induced mobilization of cytoplasmic Ca2+. *Hematol J, 4,* 132.

126. Fox, S. C., Behan, M. W., & Heptinstall, S. (2004). Inhibition of ADP-induced intracellular Ca2+ responses and platelet aggregation by the P2Y12 receptor antagonists AR-C69931MX and clopidogrel is enhanced by prostaglandin E1. *Cell Calcium, 35,* 39–46.

127. Storey, R. F., Sanderson, H. M., White, A. E., May, J. A., Cameron, K. E., & Heptinstall, S. (2000). The central role of the P(2T) receptor in amplification of human platelet activation, aggregation, secretion and procoagulant activity. *Br J Haematol, 110,* 925–934.

128. Jagroop, I. A., Burnstock, G., & Mikhailidis, D. P. (2003). Both the ADP receptors P2Y1 and P2Y12, play a role in controlling shape change in human platelets. *Platelets, 14,* 15–20.

129. Mateos–Trigos, G., Evans, R. J., & Heath, M. F. (2002). Effects of P2Y(1) and P2Y(12) receptor antagonists on ADP-induced shape change of equine platelets: Comparison with human platelets. *Platelets, 13,* 285–292.

130. Cattaneo, M., Lombardi, R., Zighetti, M. L., Gachet, C., Ohlmann, P., Cazenave, J. P., & Mannucci, P. M. (1997). Deficiency of [33P]2MeS-ADP binding sites on platelets with secretion defect, normal granule stores and normal thromboxane A2 production. Evidence that ADP potentiates platelet secretion independently of the formation of large platelet aggregates and thromboxane A2 production. *Thromb Haemost, 77,* 986–990.

131. Cattaneo, M., Lecchi, A., Lombardi, R., Gachet, C., & Zighetti, M. L. (2000). Platelets from a patient heterozygous for the defect of P2CYC receptors for ADP have a secretion defect despite normal thromboxane A2 production and normal granule stores: Further evidence that some cases of platelet "primary secretion defect" are heterozygous for a defect of P2CYC receptors. *Arterioscler Thromb Vasc Biol, 20,* E101–E106.

132. Lecchi, A., Cattaneo, M., & Mannucci, P. M. (2000). ADP potentiates platelet dense granule secretion induced by U46619 or TRAP through its interaction with the P2cyc receptor. *Haematologica, 85*(Suppl5), 83.

133. Li, Z., Zhang, G., Le Breton, G. C., Gao, X., Malik, A. B., & Du, X. (2003). Two waves of platelet secretion induced by thromboxane A2 receptor and a critical role for phosphoinositide 3-kinases. *J Biol Chem, 278,* 30725–30731.

134. Eckly, A., Gendrault, J. L., Hechler, B., Cazenave, J. P., & Gachet, C. (2001). Differential involvement of the P2Y1 and

P2YT receptors in the morphological changes of platelet aggregation. *Thromb Haemost, 85,* 694–701.

135. Andre, P., Delaney, S. M., LaRocca, T., Vincent, D., DeGuzman, F., Jurek, M., Koller, B., Phillips, D. R., & Conley, P. B. (2003). P2Y12 regulates platelet adhesion/activation, thrombus growth, and thrombus stability in injured arteries. *J Clin Invest, 112,* 398–406.

136. Nurden, P., Savi, P., Heilmann, E., Bihour, C., Herbert, J. M., Maffrand, J. P., & Nurden, A. (1995). An inherited bleeding disorder linked to a defective interaction between ADP and its receptor on platelets. Its influence on glycoprotein IIb-IIIa complex function. *J Clin Invest, 95,* 1612–1622.

137. Dorsam, R. T., Kim, S., Jin, J., & Kunapuli, S. P. (2002). Coordinated signaling through both G12/13 and G(i) pathways is sufficient to activate GPIIb/IIIa in human platelets. *J Biol Chem, 277,* 47588–47595.

138. Nieswandt, B., Schulte, V., Zywietz, A., Gratacap, M. P., & Offermanns, S. (2002). Costimulation of Gi- and G12/G13-mediated signaling pathways induces integrin alpha IIbbeta 3 activation in platelets. *J Biol Chem, 277,* 39493–39498.

139. Haseruck, N., Erl, W., Pandey, D., Tigyi, G., Ohlmann, P., Ravanat, C., Gachet, C., & Siess, W. (2004). The plaque lipid lysophosphatidic acid stimulates platelet activation and platelet–monocyte aggregate formation in whole blood: Involvement of P2Y1 and P2Y12 receptors. *Blood, 103,* 2585–2592.

140. Chacko, G. W., Brandt, J. T., Coggeshall, K. M., & Anderson, C. L. (1996). Phosphoinositide 3-kinase and p72syk non-covalently associate with the low affinity Fc gamma receptor on human platelets through an immunoreceptor tyrosine-based activation motif. Reconstitution with synthetic phosphopeptides. *J Biol Chem, 271,* 10775–10781.

141. Gratacap, M. P., Payrastre, B., Viala, C., Mauco, G., Plantavid, M., & Chap, H. (1998). Phosphatidylinositol 3,4,5-trisphosphate-dependent stimulation of phospholipase C-gamma2 is an early key event in FcgammaRIIA-mediated activation of human platelets. *J Biol Chem, 273,* 24314–24321.

142. Polgar, J., Eichler, P., Greinacher, A., & Clemetson, K. J. (1998). Adenosine diphosphate (ADP) and ADP receptor play a major role in platelet activation/aggregation induced by sera from heparin-induced thrombocytopenia patients. *Blood, 91,* 549–554.

143. Saci, A., Pain, S., Rendu, F., & Bachelot–Loza, C. (1999). Fc receptor-mediated platelet activation is dependent on phosphatidylinositol 3-kinase activation and involves p120(Cbl). *J Biol Chem, 274,* 1898–1904.

144. Nieswandt, B., Bergmeier, W., Eckly, A., Schulte, V., Ohlmann, P., Cazenave, J. P., Zirngibl, H., Offermanns, S., & Gachet, C. (2001). Evidence for cross-talk between glycoprotein VI and Gi-coupled receptors during collagen-induced platelet aggregation. *Blood, 97,* 3829–3835.

145. Paul, B. Z., Jin, J., & Kunapuli, S. P. (1999). Molecular mechanism of thromboxane A(2)-induced platelet aggregation. Essential role for p2t(ac) and alpha(2a) receptors. *J Biol Chem, 274,* 29108–29114.

146. Levy–Toledano, S., Maclouf, J., Rosa, J. P., Gallet, C., Valles, G., Nurden, P., & Nurden, A. T. (1998). Abnormal tyrosine phosphorylation linked to a defective interaction between ADP and its receptor on platelets. *Thromb Haemost, 80,* 463–468.

147. Falker, K., Lange, D., & Presek, P. (2005). P2Y12 ADP receptor-dependent tyrosine phosphorylation of proteins of 27 and 31 kDa in thrombin-stimulated human platelets. *Thromb Haemost, 93,* 880–888.

148. Falker, K., Lange, D., & Presek, P. (2004). ADP secretion and subsequent P2Y12 receptor signalling play a crucial role in thrombin-induced ERK2 activation in human platelets. *Thromb Haemost, 92,* 114–123.

149. Roger, S., Pawlowski, M., Habib, A., Jandrot–Perrus, M., Rosa, J. P., & Bryckaert, M. (2004). Costimulation of the Gi-coupled ADP receptor and the Gq-coupled TXA2 receptor is required for ERK2 activation in collagen-induced platelet aggregation. *FEBS Lett, 556,* 227–235.

150. Cattaneo, M., Lombardi, R., Bettega, D., Lecchi, A., & Mannucci, P. M. (1993). Shear-induced platelet aggregation is potentiated by desmopressin (DDAVP) and inhibited by ticlopidine. *Arterioscl Thromb, 13,* 393.

151. Cattaneo, M., Zighetti, M. L., Lombardi, R., & Mannucci, P. M. (1994). Role of ADP in platelet aggregation at high shear: Studies in a patient with congenital defect of platelet responses to ADP. *Br J Haematol, 88,* 826–829.

152. Goto, S., Tamura, N., Eto, K., Ikeda, Y., & Handa, S. (2002). Functional significance of adenosine 5'-diphosphate receptor (P2Y(12)) in platelet activation initiated by binding of von Willebrand factor to platelet GP Ibalpha induced by conditions of high shear rate. *Circulation, 105,* 2531–2536.

153. Goto, S., Tamura, N., Sakakibara, M., Ikeda, Y., & Handa, S. (2001). Effects of ticlopidine on von Willebrand factor-mediated shear-induced platelet activation and aggregation. *Platelets, 12,* 406–414.

154. Resendiz, J. C., Feng, S., Ji, G., Francis, K. A., Berndt, M. C., & Kroll, M. H. (2003). Purinergic P2Y12 receptor blockade inhibits shear-induced platelet phosphatidylinositol 3-kinase activation. *Mol Pharmacol, 63,* 639–645.

155. Storey, R. F., Judge, H. M., Wilcox, R. G., & Heptinstall, S. (2002). Inhibition of ADP-induced P-selectin expression and platelet–leukocyte conjugate formation by clopidogrel and the P2Y12 receptor antagonist AR-C69931MX but not aspirin. *Thromb Haemost, 88,* 488–494.

156. Freund, M., Mantz, F., Nicolini, P., Gachet, C., Mulvihill, J., Meyer, L., Beretz, A., & Cazenave, J. P. (1993). Experimental thrombosis on a collage coated arterioarterial shunt in rats: A pharmacological model to study antithrombotic agents inhibiting thrombin formation and platelet deposition. *Thromb Haemost, 69,* 515–521.

157. Yao, S. K., McNatt, J., Cui, K., Anderson, H. V., Maffrand, J. P., Buja, L. M., & Willerson, J. T. (1993). Combined ADP and thromboxane A2 antagonism prevents cyclic flow variations in stenosed and endothelium-injured arteries in non-human primates. *Circulation, 88,* 2888–2893.

158. Yao, S. K., Ober, J. C., Ferguson, J. J., Maffrand, J. P., Anderson, H. V., Buja, L. M., & Willerson, J. T. (1994). Clopidogrel is more effective than aspirin as adjuvant treatment to prevent reocclusion after thrombolysis. *Am J Physiol, 267,* H488–H493.

159. Yao, S. K., Ober, J. C., McNatt, J., Benedict, C. R., Rosolowsky, M., Anderson, H. V., Cui, K., Maffrand, J. P., Campbell, W. B., & Buja, L. M., et al. (1992). ADP plays an important role in mediating platelet aggregation and cyclic flow variations in vivo in stenosed and endothelium-injured canine coronary arteries. *Circ Res, 70,* 39–48.

160. Herbert, J. M., Bernat, A., & Maffrand, J. P. (1992). Importance of platelets in experimental venous thrombosis in the rat. *Blood, 80,* 2281–2286.

161. Herbert, J. M., Bernat, A., Sainte–Marie, M., Dol, F., & Rinaldi, M. (1993). Potentiating effect of clopidogrel and SR 46349, a novel 5-HT2 antagonist, on streptokinase-induced thrombolysis in the rabbit. *Thromb Haemost, 69,* 268–271.

162. Remijn, J. A., Wu, Y. P., Jeninga, E. H., MJ, I. J., van Willigen, G., de Groot, P. G., Sixma, J. J., Nurden, A. T., & Nurden, P. (2002). Role of ADP receptor P2Y(12) in platelet adhesion and thrombus formation in flowing blood. *Arterioscler Thromb Vasc Biol, 22,* 686–691.

163. Huang, J., Driscoll, E. M., Gonzales, M. L., Park, A. M., & Lucchesi, B. R. (2000). Prevention of arterial thrombosis by intravenously administered platelet P2T receptor antagonist AR-C69931MX in a canine model. *J Pharmacol Exp Ther, 295,* 492–499.

164. Humphries, R. G., Tomlinson, W., Clegg, J. A., Ingall, A. H., Kindon, N. D., & Leff, P. (1995). Pharmacological profile of the novel P2T-purinoceptor antagonist, FPL 67085 in vitro and in the anaesthetized rat in vivo. *Br J Pharmacol, 115,* 1110–1116.

165. van Gestel, M. A., Heemskerk, J. W., Slaaf, D. W., Heijnen, V. V., Reneman, R. S., & oude Egbrink, M. G. (2003). In vivo blockade of platelet ADP receptor P2Y12 reduces embolus and thrombus formation but not thrombus stability. *Arterioscler Thromb Vasc Biol, 23,* 518–523.

166. Wang, K., Zhou, X., Zhou, Z., Tarakji, K., Carneiro, M., Penn, M. S., Murray, D., Klein, A., Humphries, R. G., Turner, J., Thomas, J. D., Topol, E. J., & Lincoff, A. M. (2003). Blockade of the platelet P2Y12 receptor by AR C69931MX sustains coronary artery recanalization and improves the myocardial tissue perfusion in a canine thrombosis model. *Arterioscler Thromb Vasc Biol, 23,* 357–362.

167. Bauer, S. M. (2003). ADP receptor antagonists as antiplatelet therapeutics. *Expert Opin Emerg Drugs, 8,* 93–101.

168. Gachet, C. (2005). The platelet P2 receptors as molecular targets for old and new antiplatelet drugs. *Pharmacol Therapeut, 108,* 180–192.

169. Gachet, C., & Hechler, B. (2005). The platelet P2 receptors in thrombosis. *Semin Thromb Hemost, 31,* 162–167.

170. Conley, P. B., Jurek, M. M., Vincent, D., Lecchi, A., & Cattaneo, M. (2001). Unique mutations in the P2Y12 locus of patients with previously described defects in ADP-dependent aggregation. *Blood, 98,* 43b.

171. Shiraga, M., Miyata, S., Kato, H., Kashiwagi, H., Honda, S., Kurata, Y., Tomiyama, Y., & Kanakura, Y. (2005). Impaired platelet function in a patient with P2Y12 deficiency caused by a mutation in the translation initiation codon. *J Thromb Haemost, 3,* 2315–2323.

172. Cattaneo, M. (2003). Inherited platelet-based bleeding disorders. *J Thromb Haemost, 1,* 1628–1636.

173. Moro, S., Hoffmann, C., & Jacobson, K. A. (1998). Role of the extracellular loops of G protein-coupled receptors in ligand recognition: A molecular modeling study of the human P2Y1 receptor. *Biochemistry, 38,* 3498–3507.

174. Moro, S., Guo, D., Camaioni, E., Boyer, J. L., Harden, T. K., & Jacobson, K. A. (1998). Human P2Y1 receptor: Molecular modeling and site-directed mutagenesis as tools to identify agonist and antagonist recognition sites. *J Med Chem, 41,* 1456–1466.

175. Schwarz, U. R., Geiger, J., Walter, U., & Eigenthaler, M. (1999). Flow cytometry analysis of intracellular VASP phosphorylation for the assessment of activating and inhibitory signal transduction pathways in human platelets — definition and detection of ticlopidine/clopidogrel effects. *Thromb Haemost, 82,* 1145–1152.

176. Fontana, P., Dupont, A., Gandrille, S., Bachelot–Loza, C., Reny, J. L., Aiach, M., & Gaussem, P. (2003). Adenosine diphosphate-induced platelet aggregation is associated with P2Y12 gene sequence variations in healthy subjects. *Circulation, 108,* 989–995.

177. Fontana, P., Gaussem, P., Aiach, M., Fiessinger, J. N., Emmerich, J., & Reny, J. L. (2003). P2Y12 H2 haplotype is associated with peripheral arterial disease: A case–control study. *Circulation, 108,* 2971–2973.

178. von Beckerath, N., von Beckerath, O., Koch, W., Eichinger, M., Schomig, A., & Kastrati, A. (2005). P2Y12 gene H2 haplotype is not associated with increased adenosine diphosphate-induced platelet aggregation after initiation of clopidogrel therapy with a high loading dose. *Blood Coagul Fibrinolysis, 16,* 199–204.

179. MacKenzie, A. B., Mahaut–Smith, M. P., & Sage, S. O. (1996). Activation of receptor-operated cation channels via P2X1 not P2T purinoceptors in human platelets. *J Biol Chem, 271,* 2879–2881.

180. Vial, C., Hechler, B., Leon, C., Cazenave, J. P., & Gachet, C. (1997). Presence of P2X1 purinoceptors in human platelets and megakaryoblastic cell lines. *Thromb Haemost, 78,* 1500–1584.

181. Clifford, E. E., Parker, K., Humphreys, B. D., Kertesy, S. B., & Dubyak, G. R. (1998). The P2X1 receptor, an adenosine triphosphate-gated cation channel, is expressed in human platelets but not in human blood leukocytes. *Blood, 91,* 3172–3181.

182. Scase, T. J., Heath, M. F., Allen, J. M., Sage, S. O., & Evans, R. J. (1998). Identification of a P2X1 purinoceptor expressed on human platelets. *Biochem Biophys Res Commun, 242,* 525–528.

183. Sun, B., Li, J., Okahara, K., & Kambayashi, J. (1998). P2X1 purinoceptor in human platelets. Molecular cloning and functional characterization after heterologous expression. *J Biol Chem, 273,* 11544–11547.

184. Takano, S., Kimura, J., Matsuoka, I., & Ono, T. (1999). No requirement of P2X1 purinoceptors for platelet aggregation. *Eur J Pharmacol, 372,* 305–309.

185. Savi, P., Bornia, J., Salel, V., Delfaud, M., & Herbert, J. M. (1997). Characterization of P2X1 purinoreceptors on rat platelets: Effect of clopidogrel. *Br J Haematol, 98,* 880–886.

186. Mahaut–Smith, M. P., Ennion, S. J., Rolf, M. G., & Evans, R. J. (2000). ADP is not an agonist at P2X(1) receptors: Evidence for separate receptors stimulated by ATP and ADP on human platelets. *Br J Pharmacol, 131,* 108–114.

187. Greco, N. J., Tonon, G., Chen, W., Luo, X., Dalal, R., & Jamieson, G. A. (2001). Novel structurally altered P(2X1) receptor is preferentially activated by adenosine diphosphate in platelets and megakaryocytic cells. *Blood, 98,* 100–107.

188. Vial, C., Pitt, S. J., Roberts, J., Rolf, M. G., Mahaut–Smith, M. P., & Evans, R. J. (2003). Lack of evidence for functional ADP-activated human P2X1 receptors supports a role for ATP during hemostasis and thrombosis. *Blood, 102,* 3646–3651.

189. Rolf, M. G., & Mahaut–Smith, M. P. (2002). Effects of enhanced P2X1 receptor Ca2+ influx on functional responses in human platelets. *Thromb Haemost, 88,* 495–502.

190. Toth–Zsamboki, E., Oury, C., Cornelissen, H., De Vos, R., Vermylen, J., & Hoylaerts, M. F. (2003). P2X1-mediated ERK2 activation amplifies the collagen-induced platelet secretion by enhancing myosin light chain kinase activation. *J Biol Chem, 278,* 46661–46667.

191. Oury, C., Toth–Zsamboki, E., Vermylen, J., & Hoylaerts, M. F. (2002). P2X(1)-mediated activation of extracellular signal-regulated kinase 2 contributes to platelet secretion and aggregation induced by collagen. *Blood, 100,* 2499–2505.

192. Oury, C., Kuijpers, M. J., Toth–Zsamboki, E., Bonnefoy, A., Danloy, S., Vreys, I., Feijge, M. A., De Vos, R., Vermylen, J., Heemskerk, J. W., & Hoylaerts, M. F. (2003). Overexpression of the platelet P2X1 ion channel in transgenic mice generates a novel prothrombotic phenotype. *Blood, 101,* 3969–3976.

193. Hechler, B., Lenain, N., Marchese, P., Vial, C., Heim, V., Freund, M., Cazenave, J. P., Cattaneo, M., Ruggeri, Z. M., Evans, R., & Gachet, C. (2003). A role of the fast ATP-gated P2X1 cation channel in thrombosis of small arteries in vivo. *J Exp Med, 198,* 661–667.

194. Erhardt, J. A., Pillarisetti, K., & Toomey, J. R. (2003). Potentiation of platelet activation through the stimulation of P2X1 receptors. *J Thromb Haemost, 1,* 2626–2635.

195. Kawa, K. (2003). Thrombopoietin enhances rapid current responses mediated by P2X1 receptors on megakaryocytic cells in culture. *Jpn J Physiol, 53,* 287–299.

196. Fung, C. Y., Brearley, C. A., Farndale, R. W., & Mahaut–Smith, M. P. (2005). A major role for P2X1 receptors in the early collagen-evoked intracellular Ca2+ responses of human platelets. *Thromb Haemost, 94,* 37–40.

197. Vial, C., Rolf, M. G., Mahaut–Smith, M. P., & Evans, R. J. (2002). A study of P2X(1) receptor function in murine megakaryocytes and human platelets reveals synergy with P2Y receptors. *Br J Pharmacol, 135,* 363–372.

198. Mulryan, K., Gitterman, D. P., Lewis, C. J., Vial, C., Leckie, B. J., Cobb, A. L., Brown, J. E., Conley, E. C., Buell, G., Pritchard, C. A., & Evans, R. J. (2000). Reduced vas deferens contraction and male infertility in mice lacking P2X1 receptors. *Nature, 403,* 86–89.

199. Braun, K., Rettinger, J., Ganso, M., Kassack, M., Hildebrandt, C., Ullmann, H., Nickel, P., Schmalzing, G., & Lambrecht, G. (2001). NF449: A subnanomolar potency antagonist at recombinant rat P2X1 receptors. *Naunyn Schmiedebergs Arch Pharmacol, 364,* 285–290.

200. Kassack, M. U., Braun, K., Ganso, M., Ullmann, H., Nickel, P., Boing, B., Muller, G., & Lambrecht, G. (2004). Structure–activity relationships of analogues of NF449 confirm NF449 as the most potent and selective known P2X1 receptor antagonist. *Eur J Med Chem, 39,* 345–357.

201. Hechler, B., Magnenat, S., Zighetti, M. L., Kassack, M. U., Ullmann, H., Cazenave, J. P., Evans, R., Cattaneo, M., & Gachet, C. (2005). Inhibition of platelet functions and thrombosis through selective or nonselective inhibition of the platelet P2 receptors with increasing doses of NF449 [4,4′,4″,4″′-(carbonylbis(imino-5,1,3-benzenetriylbis (carbonylimino)))tetraki s-benzene-1,3-disulfonic acid octa-sodium salt]. *J Pharmacol Exp Ther, 314,* 232–243.

202. Horner, S., Menke, K., Hildebrandt, C., Kassack, M. U., Nickel, P., Ullmann, E., Mahaut–Smith, M. P., & Lambrecht, G. (2005). The novel suramin analogue NF864 selectively blocks P2X1 receptors in human platelets with potency in the nanomolar range. *Naunyn-Schmiedebergs Arch Pharmacol, 372,* 1–13.

203. Oury, C., Toth–Zsamboki, E., Van Geet, C., Thys, C., Wei, L., & Nilius, B., et al. (2000). A natural dominant negative P2X1 receptor due to deletion of a single amino acid residue. *J Biol Chem, 275,* 22611–22614.

204. Tolhurst, G., Vial, C., Leon, C., Gachet, C., Evans, R. J., & Mahaut–Smith, M. P. (2005). Interplay between P2Y, P2Y, and P2X(1) receptors in the activation of megakaryocyte cation influx currents by ADP: Evidence that the primary megakaryocyte represents a fully functional model of platelet P2 receptor signaling. *Blood, 106,* 1644–1651.

205. Hardy, A. R., Conley, P. B., Luo, J., Benovic, J. L., Poole, A. W., & Mundell, S. J. (2005). P2Y1 and P2Y12 receptors for ADP desensitize by distinct kinase-dependent mechanisms. *Blood, 105,* 3552–3560.

206. Jacobson, K. A., Mamedova, L., Joshi, B. V., Besada, P., & Costanzi, S. (2005). Molecular recognition at adenine nucleotide (P2) receptors in platelets. *Semin Thromb Hemost, 31,* 205–217.

207. Van Giezen, J. J. J., & Humphries, R. G. (2005). Preclinical and clinical studies with selective reversible direct P2Y12 antagonists. *Semin Thromb Hemost, 31,* 195–204.

208. Chhatriwala, M., Ravi, R. G., Patel, R. I., Boyer, J. L., Jacobson, K. A., & Harden, TK, ••. Induction of novel agonist selectivity for the ADP-activated P2Y1 receptor versus the ADP-activated P2Y12 and P2Y13 receptors by conformational constraint of an ADP analog. *J Pharmacol Exp Ther, 311,* 1038–1043.

209. Niitsu, Y., Jakubowski, J. A., Sugidachi, A., & Asai, F. (2005). Pharmacology of CS-747 (Prasugrel, LY640315), a novel, potent antiplatelet agent with in vivo P2Y12 receptor antagonist activity. *Semin Thromb Hemost, 31,* 184–194.

210. Xu, B., Stephens, A., Kirschenheuter, G., Greslin, A. F., Cheng, X., Sennelo, J., Cattaneo, M., Zighetti, M. L., Chen, A., Kim, S. A., Kim, H. S., Bischofberger, N., Cook, G., & Jacobson, K. A. (2002). Acyclic analogues of adenosine bisphosphates as P2Y receptor antagonists: Phosphate substitution leads to multiple pathways of inhibition of platelet aggregation. *J Med Chem, 45,* 5694–5709.

# CHAPTER 11

# PECAM-1

**Melanie S. Novinska, Vipul Rathore, Debra K. Newman, and Peter J. Newman**

*Blood Research Institute, BloodCenter of Wisconsin, Medical College of Wisconsin, Milwaukee, Wisconsin*

## I. Introduction

Platelet/endothelial cell adhesion molecule-1 (PECAM-1) was originally described in the mid-1980s as the CD31 differentiation antigen expressed on the surface of human granulocytes, monocytes, and platelets.[1-4] During the same period of time, several groups independently described the presence of an endothelial cell surface antigen — variously known as glycoprotein (GP) IIa[5] and then GPIIa′,[6] the hec7 antigen,[7] and EndoCAM[8] — that became highly enriched at cell–cell junctions. Screening of an endothelial cell complementary DNA expression library with a polyclonal antibody specific for platelet integral membrane proteins led to the cloning of a 130-kDa protein having homology with other recently cloned cell adhesion molecules (CAMs) containing extracellular immunoglobulin homology domains (so-called Ig-CAMs), and the protein was named PECAM-1 to denote its cloning origins, its family membership, and its likely function.[9] Antibody cross-reactivity analysis, together with its subsequent cloning from two leukocyte libraries,[10,11] established that the endothelial cell junctional protein, the CD31 differentiation antigen, and platelet PECAM-1 were identical entities, and permitted these previously disparate fields of research to be consolidated, thereby facilitating investigation of the role that this cell adhesion and signaling molecule plays in the biology of blood and vascular cells. A number of reviews exist on the function of PECAM-1 in endothelial cells[12] and leukocytes,[13] and on its signaling properties.[14] In contrast, this chapter focuses specifically on the role that PECAM-1 plays in platelet activation, thrombosis, and cardiovascular disease.

## II. PECAM-1 Genomic Organization and Protein Domain Structure

Human PECAM-1 is encoded by a 62-kb gene located at position 17q23 near the end of the long arm of chromosome 17,[15] whereas murine PECAM-1 is located at position F3-G1 of syntenic chromosome 6.[16] The human PECAM-1 gene (Fig. 11-1) is comprised of 16 exons, with exons 2 through 8 encoding its 574-amino acid extracellular domain, exon 9 encoding its 19-amino acid transmembrane domain, and exons 10 through 16 encoding a highly complex cytoplasmic domain that, because of the nature of its widely spaced exons, is subject to extensive alternative splicing.[17]

## III. Expression and Adhesive Properties of the Extracellular Domain

The extracellular domain of PECAM-1 is comprised of six Ig homology domains of the C2 subclass[9] that are highly glycosylated (red circles in Fig. 11-1), containing abundant amounts of ABO blood group antigens.[18,19] Monoclonal antibodies (mAbs) raised against the extracellular domain have been used to explore both PECAM-1 receptor density and function in human platelets. Scatchard analyses performed by four different groups using different anti-PECAM-1 mAbs found approximately 5000,[20] 5600,[21] 7760,[22] and 8800[23] molecules per resting platelet, with two of the four groups reporting a significant increase to 14,000 to 17,000 receptors on the platelet surface upon thrombin stimulation.[21,22] Whether this reflects the 30 to 40% increase in plasma membrane that is normally exposed as a result of platelet rounding or a separate internal pool of PECAM-1 that becomes externalized as a result of α-granule fusion is controversial, as one immunoelectron microscopic study found no PECAM-1 in α-granules,[22] whereas another reported significant α-granule-localized PECAM-1.[24] The level of expression of PECAM-1 on the platelet surface varies somewhat from individual to individual, as Mazurov et al.[23] evaluated PECAM-1 expression on platelets from a small number (n = 9) of donors and found a threefold variation, ranging from 5200 to 15,600 antibody binding sites/platelet.[23] More recent analysis of PECAM-1 expression performed on platelets from 125 normal blood donors (Fig. 11-2) shows that PECAM-1 levels vary approximately fourfold,

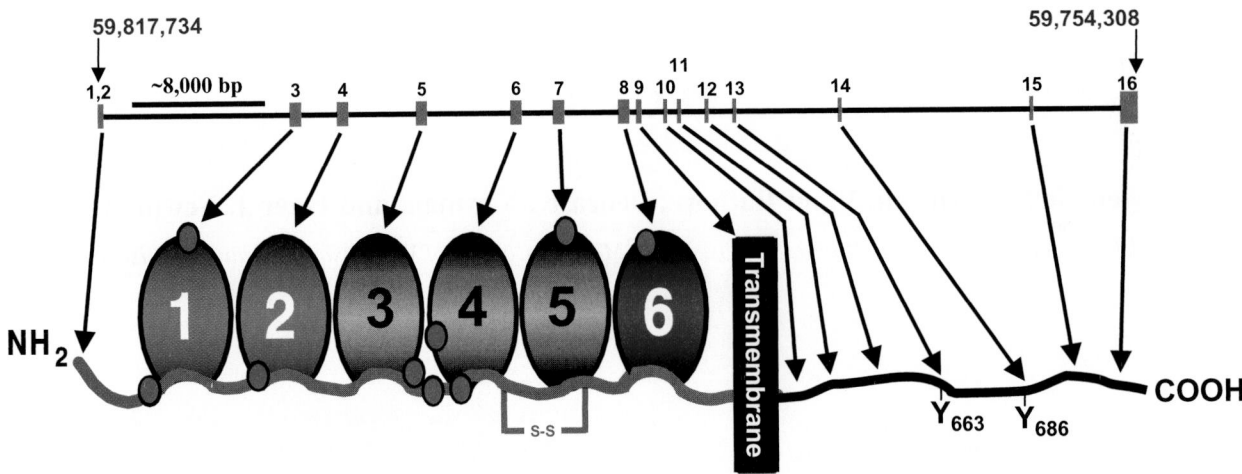

**Figure 11-1.** PECAM-1 genomic organization and protein domain structure. PECAM-1 is a 130-kDa type I transmembrane receptor glycoprotein encoded by a 62-kb gene located on the reverse strand of the long arm of chromosome 17, spanning nucleotides 59,754,308 to 59,817,734 of the human genome sequence. Exon 1 encodes the 5′-untranslated region and most of the signal peptide, whereas exon 2 encodes the remainder of the signal peptide and the mature N-terminus of the protein. Extracellular Ig type C2 domains 1 through 6 are encoded by exons 3 through 8 respectively, followed by a single-pass transmembrane domain encoded by exon 9. The 118-amino acid cytoplasmic domain is genetically and structurally complex, being encoded by the remainder of exon 9 and exons 10 through 16, all of which can be alternatively spliced to yield PECAM-1 isoforms with potentially distinct signaling properties. The molecular mass of the mature protein contributed by its 711 amino acids is only approximately 80 kDa, with the remaining 50 kDa accounted for by N-linked glycosylation at the nine sites denoted by small red circles. Immunoreceptor tyrosine-based inhibitory motifs encompassing tyrosine residues 663 and 686 are encoded by exons 13 and 14, and contribute importantly to PECAM-1-mediated signal transduction.

with PECAM-1 expression greater than 20,000 molecules/platelet in approximately 8% of normal individuals (M. Novinska, D. K. Newman, and P. J. Newman, unpublished observations). The functional consequences, if any, of variable PECAM-1 expression are not known.

Despite the fact that PECAM-1 (1) has the ability to bind homophilically with PECAM-1 molecules on adjacent cells,[25–27] (2) shares homology within its extracellular domain with cell–cell adhesion receptors like vascular cell adhesion molecule 1 (VCAM-1) and intercellular adhesion molecule 1 (ICAM),[9] and (3) has even been observed at sites of platelet–platelet contact (Fig. 11-3),[28] PECAM-1–PECAM-1 *adhesive* interactions do not appear to contribute importantly to platelet aggregation. That PECAM-1 is *not sufficient* to support platelet aggregation is shown by the failure of integrin αIIbβ3 (GPIIb-IIIa)-deficient Glanzmann thrombasthenic platelets, which contain a full complement of PECAM-1,[23,29] to aggregate in response to any physiological agonist. That PECAM-1 is *not necessary* for platelet–platelet interactions was demonstrated by the ability of PECAM-1-deficient platelets to aggregate normally in response to adenosine diphosphate (ADP).[30] (In fact, PECAM-1-deficient platelets are *hyper*responsive to certain agonists, as discussed later.)

Although not a normal physiological function of the extracellular domain, certain oral pharmaceuticals, includ-

ing the antithyroid drug carbimazole and the antiarrhythmic compound quinidine, have been shown, in rare instances, to react with PECAM-1 to produce neoepitopes that are immunogenic in humans and that result in the generation of antibodies that cause immune thrombocytopenia.[31]

## IV. Phosphorylation, Cytoskeletal Association, and Protein Interactions of the PECAM-1 Cytoplasmic Domain

### A. Cytoskeletal Association

Although largely uncoupled to the cytoskeleton in resting platelets, more than 50% of PECAM-1 receptors partition into the Triton-insoluble cytoskeleton when platelets are aggregated with thrombin.[28] Activation-dependent association of PECAM-1 with the underlying cytoskeleton appears to permit it to move large distances within the plane of the plasma membrane in spreading platelets, with a small proportion of receptors becoming localized at sites of platelet–platelet contact (Fig. 11-3), somewhat reminiscent of its enriched distribution at endothelial cell junctions. Neither the region of the PECAM-1 cytoplasmic domain responsible for mediating cytoskeletal attachment nor the influence of

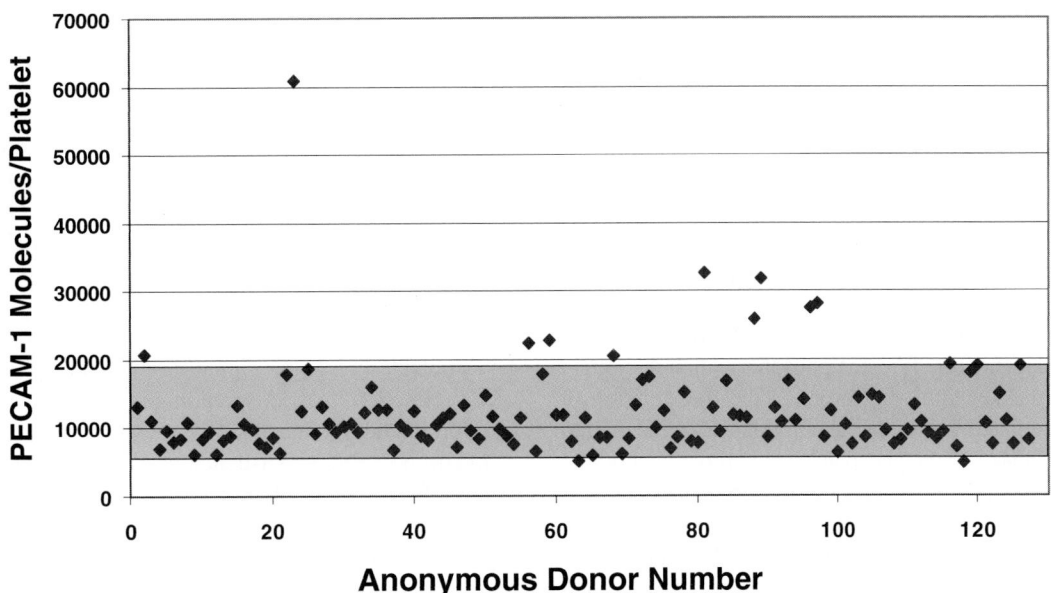

**Figure 11-2.** Variability in the expression of PECAM-1 on human platelets. Expression levels were determined by quantitative flow cytometry and were found to exhibit approximately fourfold variation within a normal range of approximately 5000 and 20,000 molecules per platelet. Note the presence of several outliers, including one individual who expressed more than 60,000 PECAM-1 molecules/platelet. The shaded box denotes expression values that lie within one standard deviation of the mean.

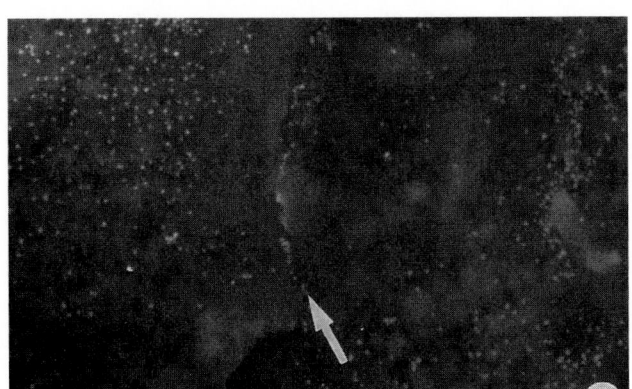

**Figure 11-3.** Enrichment of PECAM-1 at sites of platelet–platelet contact. Platelets were spread on a Formvar grid for 15 minutes at 37°C, fixed, and then stained with 18 nm gold particles that had been conjugated with the anti-PECAM-1 mAb, PECAM-1.3. Note the relatively even distribution of gold particles, except at sites of intercellular contact (arrow). Magnification, ×30,000. (From Newman et al.,[28] with permission.)

serine or tyrosine phosphorylation on its association with cytoskeletal elements has been characterized. Moesin, a member of the ezrin/radixin/moesin family of proteins that functions as cross-linkers between the actin cytoskeleton and the plasma membrane, has been reported to colocalize with PECAM-1 at the periphery of activated platelets, and can coimmunoprecipitate with PECAM-1 from thrombin-stimulated platelets.[32] More recently, calmodulin has also

been found to associate in platelets with the membrane-proximal, positively charged region of the PECAM-1 cytoplasmic domain.[33] The physiological significance of these interactions remains to be determined.

### B. Serine and Tyrosine Phosphorylation

The PECAM-1 cytoplasmic domain contains 12 serine residues, five tyrosine residues, and five threonine residues, all of which represent potential phosphorylation sites. In the early 1990s, three different research groups reported that PECAM-1, which is only minimally phosphorylated in resting platelets, becomes heavily phosphorylated within 3 minutes after addition of thrombin.[31,34,35] Phosphoaminoacid analysis showed that, under conditions that prevent cell–cell contact, phosphorylation is restricted to serine residues,[31,35] and pharmacological manipulation of serine kinase and phosphatase activity demonstrated that the phosphorylation state of PECAM-1 could be increased by phorbol 12-myristate 13-acetate (PMA) in a staurosporine-inhibitable manner,[34] implicating one or more isoforms of protein kinase C (PKC) as the enzyme responsible for phosphorylating PECAM-1 serine residues. Of the 12 potential serine phosphorylation sites, recent studies (C. M. Paddock, D. K. Newman, and P. J. Newman, unpublished) have shown that one of these, $Ser_{707}$, is constitutively phosphorylated in resting platelets, whereas phosphorylation of $Ser_{702}$ increases significantly when platelets are activated by a variety of physiological agonists. The identity of the phosphatases that dephosphorylate PECAM-1 serine residues remains unknown.

One of the more prominent and well-characterized features of the PECAM-1 cytoplasmic domain is the presence of two immunoreceptor tyrosine-based inhibitory motifs (ITIMs), encompassing tyrosine residues 663 and 686.[36] Phosphorylation of these two ITIM tyrosines, encoded by exons 13 and 14 respectively, has been shown to occur when platelets are stimulated through PAR1,[37–39] GPIb,[40] or GPVI,[39,41] and with one exception,[39] their phosphorylation has been observed to require integrin engagement (i.e., is platelet aggregation dependent). Tyrosine phosphorylation is thought to represent the key initiating event in the transmission of PECAM-1-mediated inhibitory signals, because the paired PECAM-1 ITIMs, when phosphorylated, function both to recruit and activate the cytosolic SH2 domain-containing protein–tyrosine phosphatase (PTPase), SHP-2,[37,42] and with 10-fold less affinity, the related PTPase, SHP-1.[43] PECAM-1 ITIMs have also been found to become hyperphosphorylated when PECAM-1 on the cell surface is cross-linked with bivalent anti-PECAM-1 antibodies,[38,44] or when the phosphatases that normally keep tyrosine phosphorylation in check are inhibited by pervanadate.[35] The possible physiological consequences of PECAM-1 tyrosine phosphorylation, although still largely unexplored, are discussed later.

# V. PECAM-1 as a Bidirectional Regulator of Platelet Reactivity and Thrombosis

## A. PECAM-1 as a Positive Modulator of Integrin Function

Like members of the integrin family of adhesion receptors, PECAM-1 participates in bidirectional signal transduction, both initiating and responding to changes in the cellular environment. The first experimental evidence that PECAM-1 might be involved in regulating the adhesive properties of cell surface integrins derived from a 1992 study by Tanaka et al.,[45] who showed that antibody-induced dimerization of PECAM-1 on the surface of T cells resulted in their increased adherence to the β1 integrin substrates VCAM-1 (via α4β1) and fibronectin (via α5β1). Monovalent Fab fragments of PECAM-1 mAbs were ineffective, suggesting that dimerization or oligomerization of PECAM-1 on the cell surface might be required for it to modulate integrin activation — a notion that was confirmed in a study in which oligomerized PECAM-1 was found to activate integrins in an antibody-independent manner.[46] Antibody-mediated cross-linking of PECAM-1 appears to be able to augment the adhesive properties of a broad range of integrin pairs in a variety of different cell types, as shown by the ability of anti-PECAM-1 mAbs to modulate β1 integrins in CD34+ hematopoietic progenitor cells,[47] neutrophils,[48] and eosinophils[49]; β2 inte-

grins in lymphokine-activated killer cells,[50] monocytes and neutrophils,[51] natural killer cells[52]; and the β3 integrins in endothelial cells.[49]

In human platelets, Varon and colleagues[38] have shown that certain anti-PECAM-1 mAbs can potentiate the aggregation response to G protein-coupled receptors for ADP or platelet-activating factor (PAF), and promote integrin-mediated deposition on extracellular matrix under conditions of shear. The physiological relevance of PECAM-1-mediated integrin activation in platelets has recently been demonstrated by Wee and Jackson,[53] who found that several key outside-in integrin signaling-mediated events, including clot retraction, platelet spreading and filipodia extension on immobilized fibrinogen, and FAK phosphorylation, are impaired in PECAM-1-deficient murine platelets. Precisely how PECAM-1 is able to augment integrin-mediated adhesion and/or outside-in signaling in platelets is not known, but there is evidence from other cellular systems that cross-linking PECAM-1 can result in activation of the small GTPase, Rap1,[54] which itself contributes to integrin activation.[55] Whether this is the case in platelets and whether PECAM-1 ITIMs and/or serine phosphorylation play a role in this process remain to be determined. However, it is clear that PECAM-1 is able to send "adhesion-strengthening" signals downstream of initial integrin engagement into a broad range of cell types.

## B. Effect of Anti-PECAM-1 Antibodies on Platelet Activation and Thrombus Formation

Numerous anti-PECAM-1 mAbs have been examined for their ability to affect either the rate or the extent of platelet adhesion and aggregation. von dem Borne and Modderman[56] reported that of the nine different anti-CD31 mAbs examined in the 1989 CD Workshop on Leukocyte Differentiation Antigens, none had functional effects on platelets. Other well-studied anti-PECAM-1 mAbs with no apparent effect on platelet aggregation include TM2 and TM3,[1] PECAM-1.3,[38] VM64,[23] 5.6E,[21] and RUU-PL 7E8 EC.[22] The latter mAb also had no observable effect on platelet adherence to collagen fibrils under flow conditions.[22] A few exceptions exist, however, as mAbs AAP2 and WM59 have been reported to inhibit platelet aggregation induced by weak platelet agonists like low-dose ADP or epinephrine.[21] Further cross-linking of bound anti-PECAM-1 mAbs, which results in rapid phosphorylation of PECAM-1 cytoplasmic ITIMs, has also been shown to render platelets refractory to activation, as demonstrated by reduced tyrosine phosphorylation of numerous platelet proteins, dampened platelet aggregation, and diminished secretion in response to collagen or thrombin.[44] Thus although the vast majority of anti-PECAM-1 mAbs are without effect on platelet function,[1,21–23,38,56] some are able to activate integrins

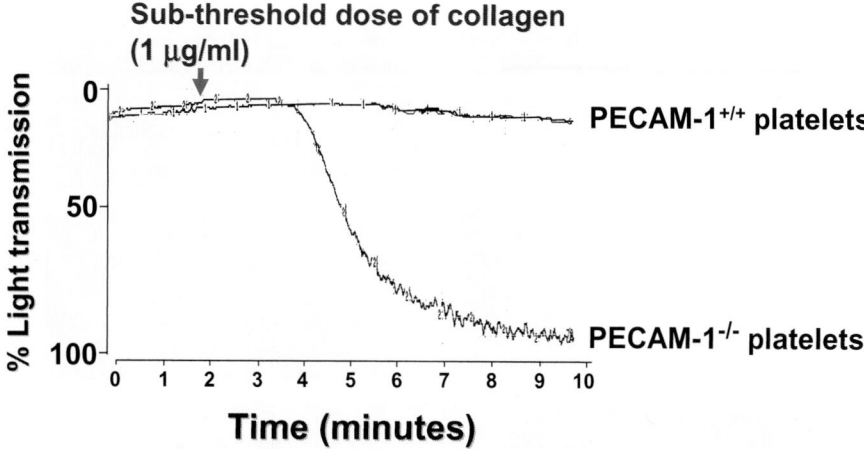

**Figure 11-4.** PECAM-1 regulates platelet activation by collagen. Platelets derived from PECAM-1-deficient mice exhibit exaggerated GPVI-mediated aggregation responses to low-dose collagen. Similar tracings are seen when platelets are stimulated with low-dose collagen-related peptide (CRP). The inhibitory function of PECAM-1 is thus most obvious at subthreshold levels of agonist, because a stronger stimulus can overcome PECAM-1-mediated inhibition. These observations suggest that one of the functions of PECAM-1 is to set a threshold for activation that prevents unwarranted and uncontrolled platelet responses under physiological conditions. (Adapted from Patil et al.,[66] with permission.)

and promote platelet aggregation,[38] whereas others,[21] especially when further cross-linked on the cell surface,[44] are able to send PECAM-1-dependent inhibitory signals into the cell. Similar inhibitory signals have been observed after incubation of platelets with a PECAM-1/IgG chimeric construct that binds homophilically to cell-surface PECAM-1.[41] PECAM-1-mediated delivery of inhibitory signals may be, at least in part, responsible for the reported ability of anti-PECAM-1 mAbs to delay platelet adhesion/aggregation responses at sites of endothelial injury.[57,58] As discussed later, loss of PECAM-1-mediated inhibitory signaling might be expected, at least in response to certain agonists, to result in a hyperaggregable platelet phenotype.

### C. PECAM-1 as an Inhibitory Receptor for ITAM-Bearing Platelet Agonist Receptors

Although adhesion strengthening by PECAM-1 can be a dominant phenotype downstream of initial integrin engagement,[38,53] or when PECAM-1 is cross-linked on the cell surface,[45-52] PECAM-1-mediated inhibitory signals appear to dominate when cells are activated via immunoreceptor tyrosine-based activation motif (ITAM)-bearing agonist receptors. Thus, PECAM-1 has been shown to dampen cellular responses that are triggered by activating the T-cell receptor,[59] the B-cell receptor,[60-63] FcεRI,[64] and FcγRIIa.[65] In addition, in several instances, its ability to inhibit ITAM-mediated activation signals has been shown to depend on the presence of SHP-2 (or SHP-1) and intact PECAM-1 ITIMs.[60,61] Likewise, although PECAM-1-deficient murine platelets aggregate and undergo granule

secretion normally when activated via G protein-coupled ADP[29] and thrombin[66] receptors, their adhesion, aggregation, and secretion responses to collagen, which binds the ITAM-bearing GPVI/FcRγ-chain complex, are markedly enhanced[41,66] (Fig. 11-4), and they form larger thrombi *in vitro* on immobilized collagen under conditions of flow.[41] Similar hyperactivation of PECAM-1-deficient platelets has been observed in response to signaling via the GPIb-IX-V complex[40] (Fig. 11-5,[66] upper panels); their engagement has been shown to activate the ITAM-bearing FcRγ chain.[68] Informatively, the kinetics of FcRγ-chain ITAM phosphorylation are enhanced in PECAM-1-deficient platelets,[40] providing biochemical evidence for the role of PECAM-1 in regulating the action of ITAM-bearing receptors. Finally, using intravital microscopy and laser-induced injury of cremaster muscle arterioles in mice (Chapter 34), Falati et al.[67] have recently demonstrated that PECAM-1 functions to inhibit thrombus formation *in vivo*. These authors found that that PECAM-1[-/-] mice formed thrombi more rapidly compared with control wild-type mice. The thrombi were larger in size and exhibited greater stability as well (Fig. 11-5, lower panels). This effect was also seen in PECAM-1-positive bone marrow chimeric mice with PECAM-1-deficient circulating blood cells, suggesting that the increased tendency toward thrombosis is a consequence of platelet, and not endothelial cell, PECAM-1 deficiency. Taken together, these data support the notion that PECAM-1 functions as an inhibitory receptor that limits *in vivo* platelet responses initiated by ITAM-bearing agonist receptors. As discussed in Section VI.B, natural variants of PECAM-1, like PECAM-1 deficiency, might predispose certain individuals to cardiovascular disease.

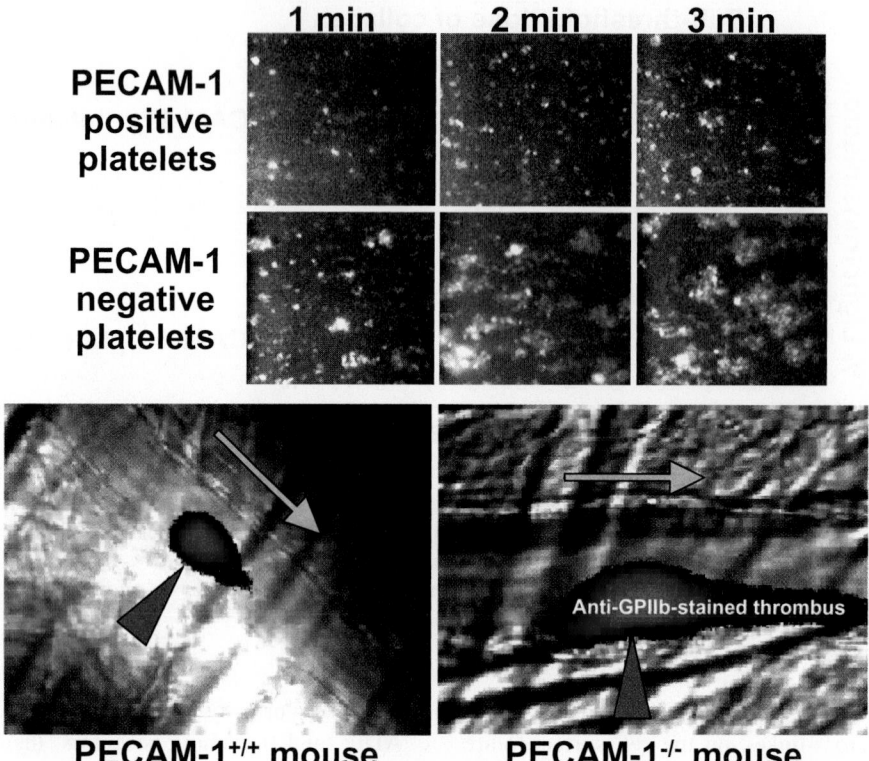

**Figure 11-5.** PECAM-1 negatively regulates thrombus formation *in vitro* and *in vivo*. Upper. Murine platelets from PECAM-1-deficient mice form thrombi more robustly than do wild-type murine platelets when flowed over immobilized von Willebrand factor at arterial shear rates. Lower. *In vivo* thrombus formation following laser-induced injury of wild-type or PECAM-1-deficient mice. Note that the size of the (red-stained) thrombus (indicated by blue arrow) is significantly inhibited when PECAM-1 is expressed on the platelet surface, demonstrating the role of this molecule as an inhibitory receptor *in vivo*. Yellow arrow indicates the direction of blood flow. (Adapted from Rathore et al.,[40] and Falati et al.,[67] with permission.)

## VI. Allelic and Soluble Isoforms of PECAM-1 and Their Association with Cardiovascular Disease

### A. PECAM-1 Shedding During Coronary Events

Like P-selectin,[69] L-selectin,[70] and ICAM-1,[71] a truncated, soluble form of PECAM-1 (sPECAM-1) has been found in normal human plasma in concentrations of nanograms per milliliter.[72] Although total sPECAM-1 is likely derived from several cellular sources, PECAM-1 can be specifically shed from human platelets when they are exposed to high sheer stress[73] in a process that may be regulated by activation-dependent dissociation of calmodulin from a positively charged region within the membrane-proximal region of the PECAM-1 cytoplasmic domain.[33] Although it probably does not reach concentrations sufficient to affect cell adhesion, increased levels of sPECAM-1 may serve as a sensitive reporter of platelet activation and/or vascular injury in coronary artery disease,[74,75] and a number of groups have reported

a rapid increase in sPECAM-1 levels after myocardial infarction.[76,77]

### B. PECAM Polymorphisms and Disease

Like most genes, PECAM-1 has variations within its nucleotide sequence, with polymorphisms in humans having been identified within the 5' untranslated region (UTR), the extracellular and cytoplasmic domains, and the 3' UTR. Four of these polymorphic sites are characterized by dimorphism, three of which encode amino acid substitutions within Ig domain 1 (exon 3, L98V), Ig domain 6 (exon 8, S536N), or the cytoplasmic domain (exon 12, R643G), and one of which occurs in the 3' UTR. Partial linkage disequilibrium among these four sites results in expression of four major PECAM-1 isoforms within the human population: LSRa, LSRg, VNGa, and VNGg (Fig. 11-6). Although the consequences of inheriting any one of these PECAM-1 variants on cellular physiology have not yet been determined, numerous studies have

| Allele | 5'UTR | | | | | | | | | | 3'UTR |
|--------|-------|--------|--------|--------|---------|---------|---------|---------|---------|---------|--------|
| LSRa | $g_{53}$ | $V_{53}$ | $N_{88}$ | $L_{98}$ | $N_{124}$ | $I_{348}$ | $D_{364}$ | $H_{457}$ | $S_{536}$ | $R_{643}$ | $a_{2479}$ |
| LSRg | $g_{53}$ | $V_{53}$ | $N_{88}$ | $L_{98}$ | $N_{124}$ | $I_{348}$ | $D_{364}$ | $H_{457}$ | $S_{536}$ | $R_{643}$ | $g_{2479}$ |
| VNGa | $g_{53}$ | $V_{53}$ | $N_{88}$ | $V_{98}$ | $N_{124}$ | $I_{348}$ | $D_{364}$ | $H_{457}$ | $N_{536}$ | $G_{643}$ | $a_{2479}$ |
| VNGg | $g_{53}$ | $V_{53}$ | $N_{88}$ | $V_{98}$ | $N_{124}$ | $I_{348}$ | $D_{364}$ | $H_{457}$ | $N_{536}$ | $G_{643}$ | $g_{2479}$ |

**Figure 11-6.** The four major allelic isoforms of human PECAM-1. 5' UTR and 3' UTR polymorphisms are in lowercase letters and show the nucleotide number based on the transcription start site of the PECAM-1 gene. Uppercase letters denote amino acids within the mature PECAM-1 protein that have been found to be polymorphic in humans. Sites at which nucleotide and amino acid variations are rare (frequency, <1%) are in black, whereas those at which differences are common are in color. Note that polymorphisms at positions 98, 536, and 643 often travel together, with an additional a/g polymorphism within the 3' UTR split among both major protein variants.

**Table 11-1: Diseases Associated with PECAM-1 SNPs**

| Amino Acid* | Disease Association | Reference |
|-------------|---------------------|-----------|
| $a_{53}$ (5' UTR) | Protective against | 78 |
| $V_{98}$ | CAD | 74, 75, 79 |
| $N_{536}$ | CAD | 80 |
| $S_{536}$ | MI | 79 |
| $R_{643}$ | MI | 79, 81, 82 |
| $G_{643}$ | CAS | 82 |

*Numbering based on mature amino acid sequence.
CAD, coronary artery disease; CAS, coronary artery stenosis; MI, myocardial infarction; SNPs, single nucleotide polymorphisms.

reported individual single nucleotide polymorphisms (SNPs) within the PECAM-1 gene that are associated with one or more forms of cardiovascular disease (see Chapter 14). As shown in Table 11-1,[78–82] these include a rare (gene frequency, <0.02) $a_{53}$ variant within the 5' UTR thought to be protective against atherosclerosis,[78] $V_{98}$ and $N_{536}$ PECAM-1 variants associated with coronary artery disease,[74,75,79,80] and $S_{536}$ and $R_{643}$ variants associated with a significantly increased risk of developing myocardial infarction.[79,81,82] Lastly, the $G_{643}$ variant of PECAM-1 has been linked to the development of coronary artery stenosis.[82] It would be of interest in the future to determine whether differences exist in either the signaling or adhesive properties of these four PECAM-1 allelic isoforms that might account for their association with cardiovascular disease, and whether differences in platelet versus endothelial PECAM-1 function contributes to disease.

## VII. Conclusions

The key points discussed in this chapter are the following:

1. PECAM-1 is a 130-kDa type I transmembrane glycoprotein expressed on the surface of platelets, leukocytes, and endothelial cells. Resting platelets express approximately 7500 PECAM-1 receptors/platelet, although fourfold variation exists within the human population.

2. Although PECAM-1 is not associated with the underlying cytoskeleton in resting platelets, more than 50% of PECAM-1 receptors partition into the Triton-insoluble fraction when platelets are stimulated with thrombin.

3. Although its extracellular domain has the ability to bind homophilically with PECAM-1 molecules on adjacent cells, PECAM-1 appears to exert its biological function in platelets mainly via signal transduction rather than adhesion.

4. The PECAM-1 cytoplasmic domain contains two ITIMs that become tyrosine phosphorylated downstream of integrin-mediated platelet aggregation, resulting in the recruitment of the protein-tyrosine phosphatase SHP-2.

5. In platelets, PECAM-1 functions as a bidirectional regulator of platelet reactivity and thrombosis, serving both as a positive modulator of integrin-mediated adhesion strengthening, and as a negative regulator of ITAM-1-bearing agonist receptors for collagen and immune complexes. Whether PECAM-1, on balance, serves to augment or suppress platelet reactivity and thrombus formation *in vivo* likely depends on the nature and extent of the stimulus.

6. Certain amino acid polymorphisms within the PECAM-1 gene have recently been associated with cardiovascular disease. How these affect the adhesive and/or signaling properties of PECAM-1 in platelets is a topic of current investigation.

## References

1. Ohto, H., Maeda, H., Shibata, Y., et al. (1985). A novel leukocyte differentiation antigen: Two monoclonal antibodies TM2 and TM3 define a 120-kd molecule present on neutrophils, monocytes, platelets, and activated lymphoblasts. *Blood, 66,* 873–881.

2. Goyert, S. M., Ferrero, E. M., Seremetis, S. V., et al. (1986). Biochemistry and expression of myelomonocytic antigens. *J Immunol, 137,* 3909–3914.

3. Lyons, A. B., Cooper, S. J., Cole, S. R., & Ashman, L. K. (1988). Human myeloid differentiation antigens identified by monoclonal antibodies to the myelomonocytic leukemia cell line RC-2A. *Pathology, 20,* 137–146.

4. Cabanas, C., Sanchez–Madrid, F., Bellon T., et al. (1989). Characterization of a novel myeloid antigen regulated during differentiation of monocytic cells. *Eur J Immunol, 19,* 1373–1378.

5. van Mourik, J. A., Leeksma, O. C., Reinders, J. H., de Groot, P. G., & Zandbergen–Spaargaren, J. (1985). Vascular endothelial cells synthesize a plasma membrane protein indistinguishable from platelet membrane glycoprotein IIa. *J Biol Chem, 260,* 11300–11306.

6. Giltay, J. C., Brinkman, H. J., Modderman, P. W., von dem Borne, A. E., & van Mourik, J. A. (1989). Human vascular endothelial cells express a membrane protein complex immunochemically indistinguishable from the platelet VLA-2 (glycoprotein Ia-IIa) complex. *Blood, 73,* 1235–1241.

7. Muller, W. A., Ratti, C. M., McDonnell, S. L., & Cohn, Z. A. (1989). A human endothelial cell-restricted externally disposed plasmalemmal protein enriched in intercellular junctions. *J Exp Med, 170,* 399–414.

8. Albelda, S. M., Oliver, P. D., Romer, L. H., & Buck, C. A. (1990). EndoCAM: A novel endothelial cell–cell adhesion molecule. *J Cell Biol, 110,* 1227–1237.

9. Newman, P. J., Berndt, M. C., Gorski, J., et al. (1990). PECAM-1 (CD31) cloning and relation to adhesion molecules of the immunoglobulin gene superfamily. *Science, 247,* 1219–1222.

10. Stockinger, H., Gadd, S. J., Eher, R., et al. (1990). Molecular characterization and functional analysis of the leukocyte surface protein CD31. *J Immunol, 145,* 3889–3897.

11. Simmons, D. L., Walker, C., Power, C., & Pigott, R. (1990). Molecular cloning of CD31, a putative intercellular adhesion molecule closely related to carcinoembryonic antigen. *J Exp Med, 171,* 2147–2152.

12. Newman, P. J., & Newman, D. K. (2006). Functions of PECAM-1 in the vascular endothelium. In W. C. Aird (Ed.), *Endothelial biomedicine.* Cambridge University Press.

13. Newman, P. J. (1997). The biology of PECAM-1. *J Clin Invest, 99,* 3–8.

14. Newman, P. J., & Newman, D. K. (2003). Signal transduction pathways mediated by PECAM-1. New roles for an old molecule in platelet and vascular cell biology. *Arterioscler Thromb Vasc Biol, 23,* 953–964.

15. Gumina, R. J., Kirschbaum, N., Rao, P. N., vanTuinen, P., & Newman, P. J. (1996). The human PECAM1 gene maps to 17q23. *Genomics, 34,* 229–232.

16. Xie, Y., & Muller, W. A. (1996). Fluorescence in situ hybridization mapping of the mouse platelet endothelial cell adhesion molecule-1 (PECAM-1) to mouse chromosome 6, region F3-G1. *Genomics, 37,* 226–228.

17. Kirschbaum, N. E., Gumina, R. J., & Newman, P. J. (1994). Organization of the gene for human platelet/endothelial cell adhesion molecule-1 (PECAM-1) reveals alternatively spliced isoforms and a functionally complex cytoplasmic domain. *Blood, 84,* 4028–4037.

18. Santoso, S., & Mueller–Eckhardt, C. (1993). PECAM is the major glycoprotein on platelets carrying blood group antigens [abstract]. *Thromb Haemost, 69,* 1191.

19. Curtis, B. R., Edwards, J. T., Hessner, M. J., Klein, J. P., & Aster, R. H. (2000). Blood group A and B antigens are strongly expressed on platelets of some individuals. *Blood, 96,* 1574–1581.

20. Newman, P. J. (1994). The role of PECAM-1 in vascular cell biology. In G. A. Fitzgerald, L. K. Jennings, & C. Patrono (Eds.). *Platelet-dependent vascular occlusion* (Vol. 714, pp. 165–174). New York: The New York Academy of Sciences.

21. Wu, X. W., & Lian, E. C. (1997). Binding properties and inhibition of platelet aggregation by a monoclonal antibody to CD31 (PECAM-1). *Arterioscler Thromb Vasc Biol, 17,* 3154–3158.

22. Metzelaar, M. J., Korteweg, J., Sixma, J. J., & Nieuwenhuis, H. K. (1991). Biochemical characterization of PECAM-1 (CD31 antigen) on human platelets. *Thromb Haemost, 66,* 700–707.

23. Mazurov, A. V., Vinogradov, D. V., Kabaeva, N. V., et al. (1991). A monoclonal antibody, VM64, reacts with a 130-kDa glycoprotein common to platelets and endothelial cells: Heterogeneity in antibody binding to human aortic endothelial cells. *Thromb Haemost, 66,* 494–499.

24. Cramer, E. M., Berger, G., & Berndt, M. C. (1994). Platelet alpha-granule and plasma membrane share two new components: CD9 and PECAM-1. *Blood, 84,* 1722–1730.

25. Sun, Q. H., DeLisser, H. M., Zukowski, M. M., et al. (1996). Individually distinct Ig homology domains in PECAM-1 regulate homophilic binding and modulate receptor affinity. *J Biol Chem, 271,* 11090–11098.

26. Sun, J., Williams, J., Yan, H. C., et al. (1996). Platelet endothelial cell adhesion molecule-1 (PECAM-1) homophilic adhesion is mediated by immunoglobulin-like domains 1 and 2 and depends on the cytoplasmic domain and the level of surface expression. *J Biol Chem, 271,* 18561–18570.

27. Newton, J. P., Buckley, C. D., Jones, E. Y., & Simmons, D. L. (1997). Residues on both faces of the first immunoglobulin fold contribute to homophilic binding sites on PECAM-1/CD31. *J Biol Chem, 272,* 20555–20563.

28. Newman, P. J., Hillery, C. A., Albrecht, R., et al. (1992). Activation-dependent changes in human platelet PECAM-1: Phosphorylation, cytoskeletal association, and surface membrane redistribution. *J Cell Biol, 119,* 239–246.

29. Burk, C. D., Newman, P. J., Lyman, S., et al. (1991). A deletion in the gene for glycoprotein IIb associated with Glanzmann's thrombasthenia. *J Clin Invest, 87,* 270–276.

30. Duncan, G. S., Andrew, D. P., Takimoto, H., et al. (1999). Genetic evidence for functional redundancy of platelet/endothelial cell adhesion molecule-1 (PECAM-1): CD31-deficient mice reveal PECAM-1-dependent and PECAM-1-independent functions. *J Immunol, 162,* 3022–3030.

31. Kroll, H., Sun, Q. H., & Santoso, S. (2000). Platelet endothelial cell adhesion molecule-1 (PECAM-1) is a target

glycoprotein in drug-induced thrombocytopenia. *Blood, 96,* 1409–1414.

32. Gamulescu, M. A., Seifert, K., Tingart, M., Falet, H., & Hoffmeister, K. M. (2003). Platelet moesin interacts with PECAM-1 (CD31). *Platelets, 14,* 211–217.

33. Wong, M. X., Harbour, S. N., Wee, J. L., et al. (2004). Proteolytic cleavage of platelet endothelial cell adhesion molecule-1 (PECAM-1/CD31) is regulated by a calmodulin-binding motif. *Febs Lett, 568,* 70–78.

34. Zehnder, J. L., Hirai, K., Shatsky, M., et al. (1992). The cell adhesion molecule CD31 is phosphorylated after cell activation. Down-regulation of CD31 in activated T lymphocytes. *J Biol Chem, 267,* 5243–5249.

35. Modderman, P. W., von dem Borne, A. E. G. K. R., & Sonnenberg, A. (1994). Tyrosine phosphorylation of P-selectin in intact platelets and in a disulfide-linked complex with immunoprecipitated pp60$^{c\text{-}src}$. *Biochem J, 299,* 613–621.

36. Newman, P. J. (1999). Switched at birth: A new family for PECAM-1. *J Clin Invest, 103,* 5–9.

37. Jackson, D. E., Ward, C. M., Wang, R., & Newman, P. J. (1997). The protein–tyrosine phosphatase SHP-2 binds PECAM-1 and forms a distinct signaling complex during platelet aggregation. Evidence for a mechanistic link between PECAM-1 and integrin-mediated cellular signaling. *J Biol Chem, 272,* 6986–6993.

38. Varon, D., Jackson, D. E., Shenkman, B., et al. (1998). Platelet/endothelial cell adhesion molecule-1 serves as a co-stimulatory agonist receptor that modulates integrin-dependent adhesion and aggregation of human platelets. *Blood, 91,* 500–507.

39. Cicmil, M., Thomas, J. M., Sage, T., et al. (2000). Collagen, convulxin, and thrombin stimulate aggregation-independent tyrosine phosphorylation of CD31 in platelets. Evidence for the involvement of Src family kinases. *J Biol Chem, 275,* 27339–27347.

40. Rathore, V., Stapleton, M. A., Hillery, C. A., et al. (2003). PECAM-1 negatively regulates GPIb/V/IX signaling in murine platelets. *Blood, 102,* 3658–3664.

41. Jones, K. L., Hughan, S. C., Dopheide, S. M., et al. (2001). Platelet endothelial cell adhesion molecule-1 is a negative regulator of platelet-collagen interactions. *Blood, 98,* 1456–1463.

42. Jackson, D. E., Kupcho, K. R., & Newman, P. J. (1997). Characterization of phosphotyrosine binding motifs in the cytoplasmic domain of platelet/endothelial cell adhesion molecule-1 (PECAM-1) that are required for the cellular association and activation of the protein-tyrosine phosphatase, SHP-2. *J Biol Chem, 272,* 24868–24875.

43. Hua, C. T., Gamble, J. R., Vadas, M. A., & Jackson, D. E. (1998). Recruitment and activation of SHP-1 protein-tyrosine phosphatase by human platelet endothelial cell adhesion molecule-1 (PECAM-1). *J Biol Chem, 273,* 28332–28340.

44. Cicmil, M., Thomas, J. M., Leduc, M., Bon, C., & Gibbins, J. M. (2002). Platelet endothelial cell adhesion molecule-1 signaling inhibits the activation of human platelets. *Blood, 99,* 137–144.

45. Tanaka, Y., Albelda, S. M., Horgan, K. J., et al. (1992). CD31 expressed on distinctive T cell subsets is a preferential ampli-fier of β1 integrin-mediated adhesion. *J Exp Med, 176,* 245–253.

46. Zhao, T., & Newman, P. J. (2001). Integrin activation by regulated dimerization and oligomerization of platelet endothelial cell adhesion molecule (PECAM)-1 from within the cell. *J Cell Biol, 152,* 65–73.

47. Leavesley, D. I., Oliver, J. M., Swart, B. W., et al. (1994). Signals from platelet/endothelial cell adhesion molecule enhance the adhesive activity of the very late antigen-4 integrin of human CD34$^+$ hemopoietic progenitor cells. *J Immunol, 153,* 4673–4683.

48. Pellegatta, F., Chierchia, S. L., & Zocchi, M. R. (1998). Functional association of platelet endothelial cell adhesion molecule-1 and phosphoinositide 3-kinase in human neutrophils. *J Biol Chem, 273,* 27768–27771.

49. Chiba, R., Nakagawa, N., Kurasawa, K., et al. (1999). Ligation of CD31 (PECAM-1) on endothelial cells increases adhesive function of αvβ3 integrin and enhances β1 integrin-mediated adhesion of eosinophils to endothelial cells. *Blood, 94,* 1319–1329.

50. Piali, L., Albelda, S. M., Baldwin, H. S., et al. (1993). Murine platelet endothelial cell adhesion molecule (PECAM-1/CD31) modulates β2 integrins on lymphokine-activated killer cells. *Eur J Immunol, 23,* 2464–2471.

51. Berman, M. E., & Muller, W. A. (1995). Ligation of platelet/endothelial cell adhesion molecule 1 (PECAM-1/CD31) on monocytes and neutrophils increases binding capacity of leukocyte CR3 (CD11b/CD18). *J Immunol, 154,* 299–307.

52. Berman, M. E., Xie, Y., & Muller, W. A. (1996). Roles of platelet endothelial cell adhesion molecule-1 (PECAM-1, CD31) in natural killer cell transendothelial migration and β2 integrin activation. *J Immunol, 156,* 1515–1524.

53. Wee, J. L., & Jackson, D. E. (2005). The Ig-ITIM superfamily member, PECAM-1 regulates the "outside-in" signaling properties of integrin αIIbβ3 in platelets. *Blood, ••.*

54. Reedquist, K. A., Ross, E., Koop, E. A., et al. (2000). The small GTPase, rap1, mediates CD31-induced integrin adhesion. *J Cell Biol, 148,* 1151–1158.

55. Bertoni, A., Tadokoro, S., Eto, K., et al. (2002). Relationships between Rap1b, affinity modulation of integrin αIIbβ3, and the actin cytoskeleton. *J Biol Chem, 277,* 25715–25721.

56. von dem Borne, A. E. G. K. R., & Modderman, P. W. (1989). Cluster report: CD31. In W. Knapp, B. Dorken, W. R. Gilks, et al. (Eds.), *Leukocyte typing IV. White cell differentiation antigens* (p. 995). Oxford: Oxford University Press.

57. Rosenblum, W. I., Murata, S., Nelson, G. H., et al. (1994). Anti-CD31 delays platelet adhesion/aggregation at sites of endothelial injury in mouse cerebral arterioles. *Am J Clin Pathol, 145,* 33–36.

58. Rosenblum, W. I., Nelson, G. H., Wormley, B., et al. (1996). Role of platelet–endothelial cell adhesion molecule (PECAM) in platelet adhesion/aggregation over injured but not denuded endothelium in vivo and ex vivo. *Stroke, 27,* 709–711.

59. Newton–Nash, D. K., & Newman, P. J. (1999). A new role for platelet–endothelial cell adhesion molecule-1 (CD31): Inhibition of TCR-mediated signal transduction. *J Immunol, 163,* 682–688.

60. Newman, D. K., Hamilton, C., & Newman, P. J. (2001). Inhibition of antigen-receptor signaling by platelet endothelial cell adhesion molecule-1 (CD31) requires functional ITIMs, SHP-2, and p56$^{lck}$. *Blood, 97,* 2351–2357.

61. Henshall, T. L., Jones, K. L., Wilkinson, R., & Jackson, D. E. (2001). Src homology 2 domain-containing protein-tyrosine phosphatases, SHP-1 and SHP-2, are required for platelet endothelial cell adhesion molecule-1/CD31-mediated inhibitory signaling. *J Immunol, 166,* 3098–3106.

62. Wilkinson, R., Lyons, A. B., Roberts, D., et al. (2002). Platelet endothelial cell adhesion molecule-1 (PECAM-1/CD31) acts as a regulator of B-cell development, B-cell antigen receptor (BCR)-mediated activation, and autoimmune disease. *Blood, 100,* 184–193.

63. Wong, M. X., & Jackson, D. E. (2004). Regulation of B cell activation by PECAM-1: Implications for the development of autoimmune disorders. *Curr Pharm Des, 10,* 155–161.

64. Wong, M. X., Roberts, D., Bartley, P. A., & Jackson, D. E. (2002). Absence of platelet endothelial cell adhesion molecule-1 (CD31) leads to increased severity of local and systemic IgE-mediated anaphylaxis and modulation of mast cell activation. *J Immunol, 168,* 6455–6462.

65. Thai, L. M., Ashman, L. K., Harbour, S. N., Hogarth, P. M., & Jackson, D. E. (2003). Physical proximity and functional interplay of PECAM-1 with the Fc receptor Fc gamma RIIa on the platelet plasma membrane. *Blood, 102,* 3637–3645.

66. Patil, S., Newman, D. K., & Newman, P. J. (2001). Platelet endothelial cell adhesion molecule-1 serves as an inhibitory receptor that modulates platelet responses to collagen. *Blood, 97,* 1727–1732.

67. Falati, S., Patil, S., Gross, P. L., et al. (2006). Platelet PECAM-1 inhibits thrombus formation in vivo. *Blood, 107,* 535–541.

68. Wu, Y., Suzuki–Inoue, K., Satoh, K., et al. (2001). Role of Fc receptor γ-chain in platelet glycoprotein Ib-mediated signaling. *Blood, 97,* 3836–3845.

69. Dunlop, L. C., Skinner, M. P., Bendall, L. J., et al. (1992). Characterization of GMP-140 (P-selectin) as a circulating plasma protein. *J Exp Med, 175,* 1147–1150.

70. Schleiffenbaum, B., Spertini, O., & Tedder, T. F. (1992). Soluble L-selectin is present in human plasma at high levels and retains functional activity. *J Cell Biol, 119,* 229–238.

71. Rothlein, R., Mainolfi, E. A., Czajkowski, M., & Marlin, S. D. (1991). A form of circulating ICAM-1 in human serum. *J Immunol, 147,* 3788–3793.

72. Goldberger, A., Middleton, K. A., Oliver, J. A., et al. (1994). Biosynthesis and processing of the cell adhesion molecule PECAM-1 includes production of a soluble form. *J Biol Chem, 269,* 17183–17191.

73. Naganuma, Y., Satoh, K., Yi, Q., et al. (2004). Cleavage of platelet endothelial cell adhesion molecule-1 (PECAM-1) in platelets exposed to high shear stress. *J Thromb Haemost, 2,* 1998–2008.

74. Wei, H., Fang, L., Chowdhury, S. H., et al. (2004). Platelet–endothelial cell adhesion molecule-1 gene polymorphism and its soluble level are associated with severe coronary artery stenosis in Chinese Singaporean. *Clin Biochem, 37,* 1091–1097.

75. Fang, L., Wei, H., Chowdhury, S. H., et al. (2005). Association of Leu125Val polymorphism of platelet endothelial cell adhesion molecule-1 (PECAM-1) gene & soluble level of PECAM-1 with coronary artery disease in Asian Indians. *Indian J Med Res, 121,* 92–99.

76. Serebruany, V. L., & Gurbel, P. A. (1999). Effect of thrombolytic therapy on platelet expression and plasma concentration of PECAM-1 (CD31) in patients with acute myocardial infarction. *Arterioscler Thromb Vasc Biol, 19,* 153–158.

77. Soeki, T., Tamura, Y., Shinohara, H., et al. (2003). Increased soluble platelet/endothelial cell adhesion molecule-1 in the early stages of acute coronary syndromes. *Int J Cardiol, 90,* 261–268.

78. Elrayess, M. A., Webb, K. E., Flavell, D. M., et al. (2003). A novel functional polymorphism in the PECAM-1 gene (53G > A) is associated with progression of atherosclerosis in the LOCAT and REGRESS studies. *Atherosclerosis, 168,* 131–138.

79. Sasaoka, T., Kimura, A., Hohta, S. A., et al. (2001). Polymorphisms in the platelet-endothelial cell adhesion molecule-1 (PECAM-1) gene, Asn563Ser and Gly670Arg, associated with myocardial infarction in the Japanese. *Ann NY Acad Sci, 947,* 259–269.

80. Song, F. C., Chen, A. H., Tang, X. M., et al. (2003). Association of platelet endothelial cell adhesion molecule-1 gene polymorphism with coronary heart disease. *Di Yi Jun Yi Da Xue Xue Bao, 23,* 156–158.

81. Listi, F., Candore, G., Lio, D., et al. (2004). Association between platelet endothelial cellular adhesion molecule-1 (PECAM-1/CD31) polymorphisms and acute myocardial infarction: A study in patients from Sicily. *Eur J Immunogenet, 31,* 175–178.

82. Elrayess, M. A., Webb, K. E., Bellingan, G. J., et al. (2004). R643G polymorphism in PECAM-1 influences transendothelial migration of monocytes and is associated with progression of CHD and CHD events. *Atherosclerosis, 177,* 127–135.

# CHAPTER 12

# P-Selectin/PSGL-1 and Other Interactions between Platelets, Leukocytes, and Endothelium

Rodger P. McEver

*Oklahoma Medical Research Foundation, University of Oklahoma Health Sciences Center, Oklahoma City, Oklahoma*

## I. Introduction

In response to hemorrhage, circulating platelets adhere to exposed subendothelial tissues and then recruit additional platelets into aggregates that function as procoagulant surfaces. In response to inflammatory stimuli, circulating leukocytes roll and then arrest on endothelial cells, and finally migrate into the surrounding tissues where they combat pathogens and repair tissue injury. These traditionally separate cellular adhesive contributions to coagulation and inflammation have become progressively understood to be linked. We now know that leukocytes roll and then arrest on activated platelets and that platelets roll on activated endothelial cells. Cell adhesion molecules that were first thought to contribute uniquely to either coagulation or inflammation are now known to contribute to both responses. This chapter focuses on multicellular adhesive interactions in the vasculature, with particular attention to the dual functions in coagulation and inflammation of the adhesion molecules P-selectin (CD62P), P-selectin glycoprotein ligand 1 (PSGL-1), and glycoprotein (GP) Ibα. (GPIbα is also discussed in Chapter 7. Inhibition of platelet function by the endothelium is discussed in Chapter 13.)

## II. Platelet Thrombus Formation on Subendothelial Surfaces under Flow

At regions of vascular injury under high shear stress, as in arterioles or stenotic arteries, platelets move rapidly on von Willebrand factor (VWF) that is bound to collagen on subendothelial surfaces (Fig. 12-1A; see also Chapter 18). This rapid movement is usually termed *translocation*, because rotation of the platelets on the surface is not readily observed.[1] However, end-over-end rolling of platelets has been noted.[2] As the translocating or rolling platelets become activated, they arrest on the vascular wall and form platelet plugs through interactions of β1 and β3 integrins with adhe-

sive ligands such as fibrinogen, fibronectin, and VWF itself.[3] The initial rolling on VWF is mediated through interactions with the platelet glycoprotein GPIb-IX-V complex, which consists of GPIbα, GPIbβ, GPIX, and GPV[4,5] (see also Chapter 7). The N-terminal 300 residues of the GPIbα subunit include a series of leucine-rich repeats with conserved N- and C-terminal flanking disulfide loops and an anionic peptide sequence with three tyrosine sulfates (Fig. 12-2). This region is projected above the platelet membrane by an extended mucin domain containing multiple O-glycans. The N-terminal region of GPIbα binds to the A1 domain of VWF. VWF is a multimeric protein composed of identical 250-kDa subunits. Each subunit has a single 24-kDa A1 domain that comprises residues 509 to 695. Endothelial cells initially secrete ultralarge multimers of VWF that are cleaved by the plasma protease ADAMTS-13 (see Chapter 50) to yield smaller multimers, which still typically exceed 1000 kDa.[6,7]

Microspheres coated with a recombinant fragment containing GPIbα residues 1 to 302 roll on immobilized VWF A1 domain.[8] This confirms that these regions are sufficient to confer adhesion under flow, although other cellular and molecular features modulate the interaction. These features include the concentration of GPIbα in lipid raft domains[9] and the projection of the ligand binding site above the membrane.[10] Gain-of-function mutations in either GPIbα or VWF increase binding affinity and slow rolling velocities.[11–13] A crystal structure has been determined for a complex of high-affinity mutants of VWF A1 domain and the GPIbα N-terminal region.[14] Regions of GPIbα at each end of the leucine-rich repeats bind, respectively, to the top and bottom of the A1 domain. The tyrosine-sulfated region does not contact VWF, even though mutants of GPIbα lacking tyrosine sulfates interact much less well with VWF under flow.[13] Mutations of GPIbα in other regions also reduce binding affinity to VWF, and transfected cells expressing these mutants roll more rapidly on VWF.[11,12,15] Conversely, patients with platelet-type and type 2B von Willebrand disease have respective mutations of GPIbα and

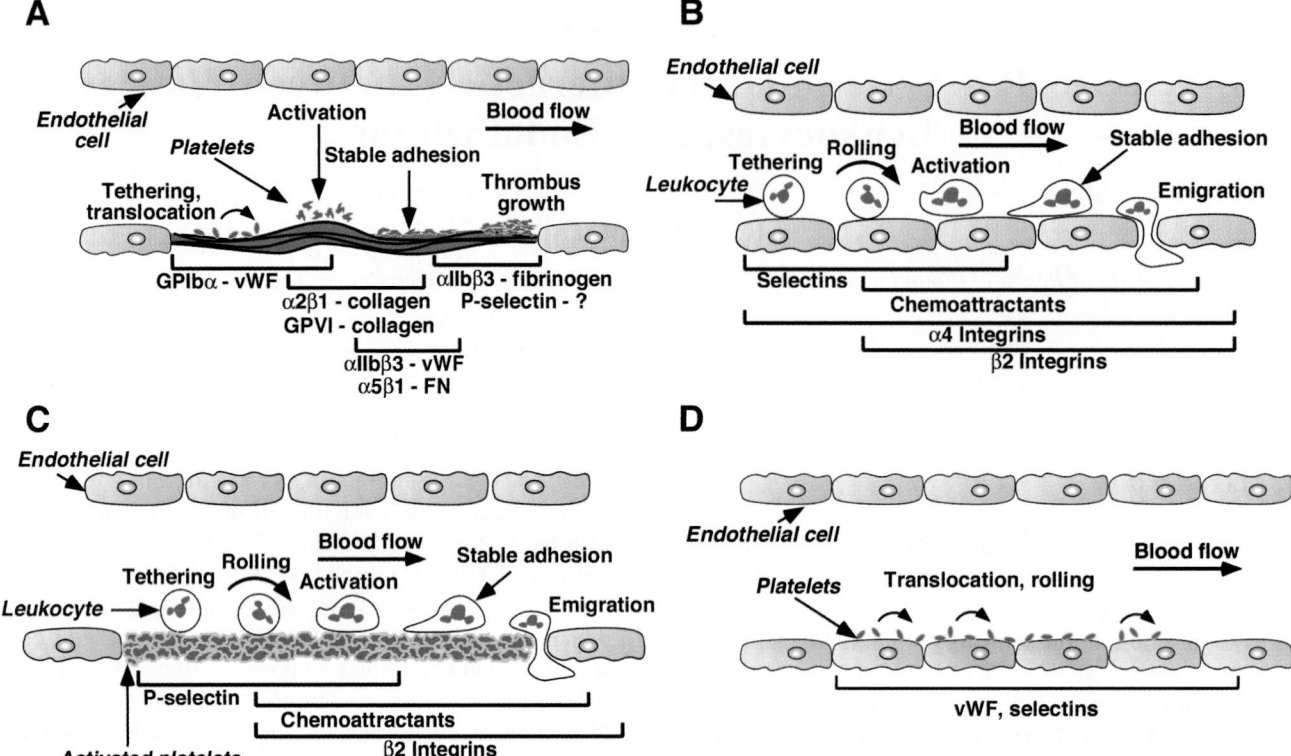

**Figure 12-1.** Multistep adhesive and signaling interactions of leukocytes and platelets with vascular surfaces in flow. A. Adhesion of platelets to subendothelial surfaces. B. Adhesion of leukocytes to activated endothelial cells. C. Adhesion of leukocytes to activated platelets. D. Adhesion of platelets to activated endothelial cells. vWF, von Willebrand factor; FN, fibronectin. See text for details.

VWF A1 that enhance binding, which results in spontaneous platelet agglutination.[16]

The fast rolling of platelets on VWF implies that adhesive bonds form and break rapidly and that the bonds have sufficient tensile strength to resist the forces applied to the adherent cells. Recent flow chamber experiments support this concept,[12,15,17] but studies with higher temporal resolution are required to address this issue definitively. Surprisingly, radioimmunoassay measurements in the fluid phase yielded orders-of-magnitude slower kinetic rates for interactions between GPIb and VWF.[18] This discrepancy has not yet been resolved. A cardinal feature of GPIbα-mediated platelet rolling on VWF is its flow dependency.[1] Arterial flow rates are required for platelets to tether to and roll on VWF. Platelets usually do not adhere at lower flow rates, and rolling platelets detach if arterial flow rates are reduced. Flow-enhanced rolling may be biologically important. It potentially ensures that platelets adhere to disrupted arterial surfaces where the requirement for platelets in hemostasis is most critical. Conversely, it is possible that disorders such as platelet-type and type 2B von Willebrand disease no longer exhibit flow-enhanced platelet adhesion, resulting in pathological platelet agglutination. The

mechanisms underlying the flow requirement for platelet rolling on VWF remain unclear.

Platelets translocating on subendothelial VWF then arrest and recruit additional platelets into growing thrombi (Fig. 12-1A). Stable adhesion to subendothelial tissues requires binding of platelet GPVI and integrin α2β1 to collagen, augmented by binding of platelet integrin αIIbβ3 to immobilized VWF, fibrinogen, and other ligands, and by binding of integrin α5β1 to fibronectin[19] (see Chapter 18 for further details). Binding of αIIbβ3 to the symmetrical binding sites on plasma and platelet-secreted fibrinogen leads to platelet cohesion and thrombus growth (see also Chapters 8 and 18). Platelet activation is required for integrin function. In flow, adhesion to VWF transduces signals through the GPIb-IX-V complex that are sufficient to activate integrins.[2,20–24] These signals may intersect with those from other mediators such as thrombin, secreted adenosine diphosphate (ADP), and collagen. Integrin-dependent platelet adhesion and cohesion occur only under very low shear stresses. Under high shear, platelet translocation through GPIbα-VWF interactions is essential for integrin-dependent adhesion to proceed.[25]

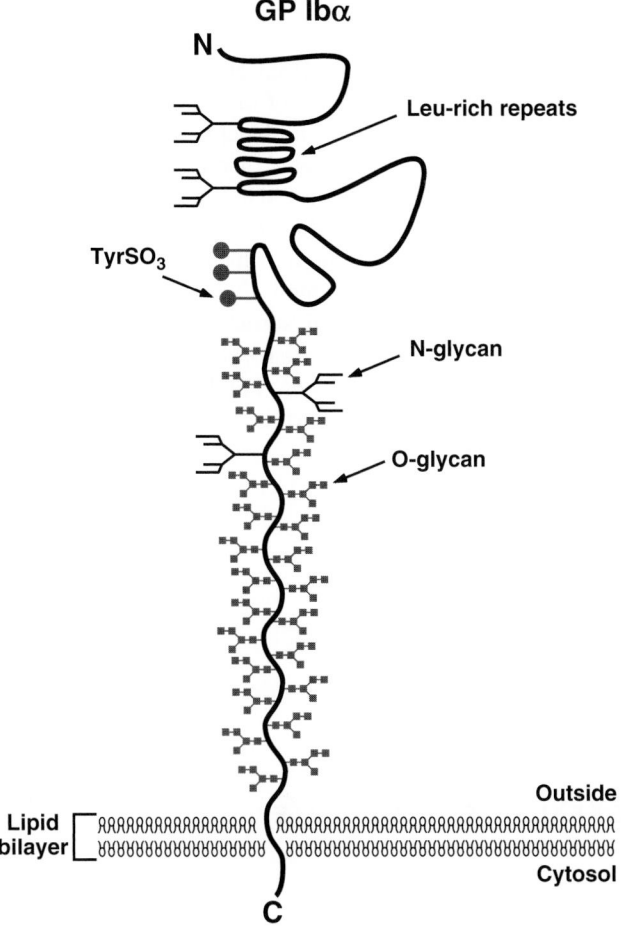

**Figure 12-2.** Schematic diagram of human platelet GPIbα. Not shown are the associated proteins GPIbβ, GPIX, and GPV.

## III. Leukocyte Adhesion to Endothelial Cells under Flow

At sites of inflammation, leukocytes tether, or form initial attachments, to vascular endothelial cells. The leukocytes then roll on the vessel wall. They finally arrest, spread, and emigrate between endothelial cells into the surrounding tissues (Fig. 12-1B).[26] Most leukocyte recruitment occurs in postcapillary venules, where shear stresses are low and the expression of some endothelial cell adhesion molecules is favored. However, leukocyte adhesion also occurs in inflamed arterioles and in large arteries, particularly at sites of bifurcation and large curvature, where shear stresses are not only low but also oscillatory and where atherosclerosis is most likely to develop.

In flow, interactions of selectins with cell-surface glycoconjugates mediate tethering and rolling adhesion of leukocytes on the vessel wall.[27,28] The selectins are type I membrane proteins, each with an N-terminal C-type lectin

domain, followed by an epidermal growth factor (EGF)-like motif, a series of short consensus repeats, a transmembrane domain, and a cytoplasmic tail. L-selectin on leukocytes binds to ligands on high endothelial venules of lymph nodes and on activated endothelial cells at sites of inflammation. P- and E-selectin, which are expressed on activated endothelial cells, bind to ligands on leukocytes. In response to thrombin, histamine, or other secretagogues, P-selectin is rapidly redistributed from the membranes of Weibel–Palade bodies to the endothelial cell surface. Inflammatory mediators such as tumor necrosis factor-α (TNF-α) transiently activate the E-selectin gene, resulting in slightly delayed delivery of E-selectin to the endothelial cell surface. P- and L-selectin are primarily responsible for tethering of flowing leukocytes to the endothelium, although they also support rolling. E-selectin is particularly important for slowing rolling velocities of leukocytes after they have tethered through P- or L-selectin.[29]

Slower rolling increases transit time through inflamed vessels, enhancing the probability of encountering chemokines or lipid autacoids presented on the endothelial cell surface.[30] These mediators activate leukocyte β2 integrins by transducing signals that intersect with those produced by engagement of selectin ligands or L-selectin.[27,31] Activation increases the binding affinity or avidity of β2 integrins to endothelial cell counterreceptors such as intercellular adhesion molecule 1 (ICAM-1) or to fibrinogen, which can bind simultaneously to ICAM-1 and to the β2 integrin Mac-1.[32] β2 integrins cooperate with selectins to slow rolling velocities further, which leads to stable adhesion.[33–35] Some leukocyte subsets express integrins α9β1, α4β1, or α4β7. These integrins, depending on their activation status, support both leukocyte rolling and firm adhesion by binding to endothelial cell counterreceptors such as vascular cell adhesion molecule 1 and mucosal addressin cell adhesion molecule 1.[36–38] However, they are less efficient in tethering leukocytes to the endothelium. Expression of particular combinations of adhesion molecules, chemokines, and chemokine receptors determines which classes of leukocytes accumulate at a specific inflammatory site.[26]

All three selectins bind with low affinity to sialylated and fucosylated oligosaccharides such as sialyl Lewis x (sLe$^x$). The sLe$^x$ determinant, NeuAcα2,3Galβ1,4(Fucα1,3)GlcNAcβ1-R, is a terminal component of some oligosaccharides on leukocytes and activated endothelial cells.[39] Binding requires fucose and, to a lesser degree, sialic acid. Targeted disruption of the genes encoding FTVII and FTIV, the α1,3-fucosyltransferases expressed in leukocytes and endothelial cells, eliminates selectin-dependent leukocyte rolling in the vasculature of mice.[40,41] Congenital defects in fucose metabolism in humans cause a similar phenotype.[42] A major question that is not yet fully answered is whether specific glycoproteins or glycolipids bearing sLe$^x$ are more important for mediating leukocyte interactions with selectins under flow. P- and L-selectin also bind to some sulfated carbohydrates, such as heparan sulfate and fucoidan, which lack sialic acid and fucose.[39] Furthermore, they bind preferentially to particular mucins (i.e., glycoproteins with multiple Ser/Thr-linked oligosaccharides and repeating peptide motifs). These mucins must be modified not only with sialic acid and fucose, but also with sulfate for optimal binding.

## A. Interactions of P-selectin with PSGL-1

The mucin that is most clearly defined as a biologically important selectin ligand is PSGL-1.[43] PSGL-1 is a homodi-meric mucin expressed on almost all leukocytes (Fig. 12-3). In affinity chromatography, blotting, or precipitation assays, it is the dominant ligand for P-selectin in cell lysates, even though it carries less than 1% of the total sLe$^x$ on the leukocyte surface.[44,45] Studies with blocking monoclonal antibodies (mAbs) and with targeted disruption of the PSGL-1 gene in mice have shown that PSGL-1 is the critical ligand for mediating leukocyte rolling on P-selectin in flow.[46–50] PSGL-1 also binds to L-selectin, which initiates leukocyte–leukocyte interactions that amplify leukocyte accumulation on inflamed endothelial cell surfaces in vitro[51,52] and in vivo.[53] Leukocytes may also use L-selectin to roll on leukocyte microparticles expressing PSGL-1 that have been deposited on the vascular surface.[54] Lectins usually bind

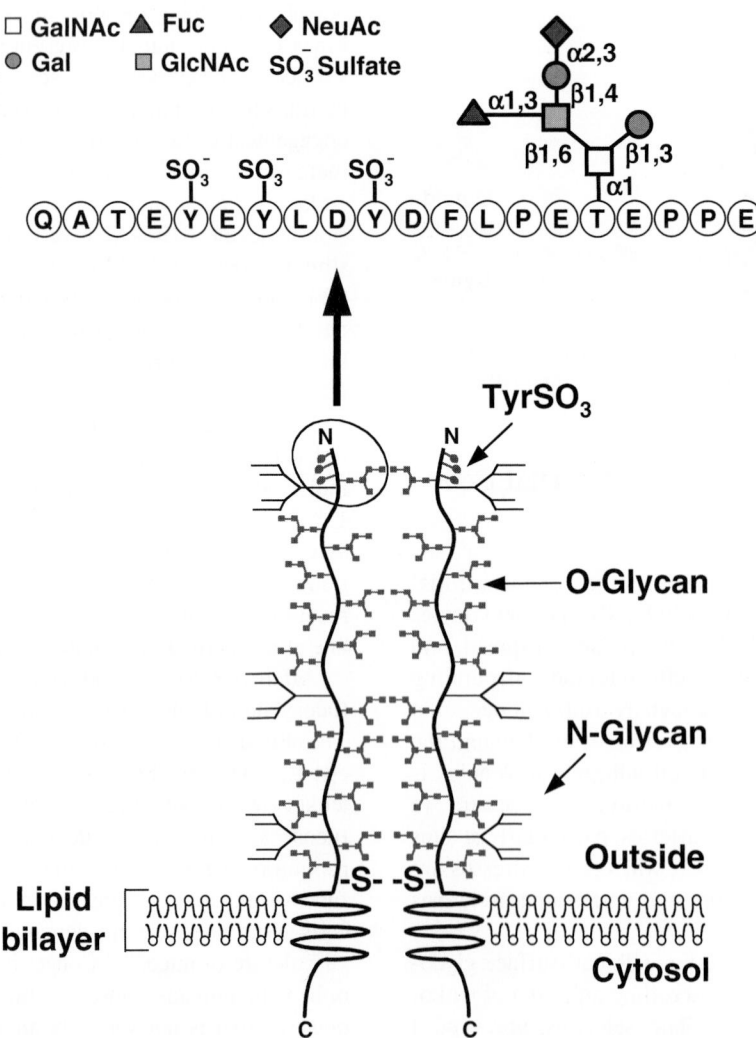

**Figure 12-3.** Schematic diagram of human leukocyte P-selectin glycoprotein ligand 1 (PSGL-1) and of the N-terminal glycosulfopeptide region that binds to P-selectin and L-selectin.

with low affinity but high avidity to multiple O-glycans clustered on the polypeptide backbone of mucins. However, only a minority of the O-glycans on PSGL-1 are fucosylated,[55] and most of the O-glycans appear to function only to extend the polypeptide backbone.[56] P- and L-selectin bind to a specific N-terminal region of PSGL-1 that includes the epitopes for the blocking mAbs.[46,51,56] In human PSGL-1, the binding site comprises a peptide sequence containing three tyrosine sulfate residues near a threonine to which a specific O-glycan is attached (Fig. 12-3). Biochemical assays and expression of recombinant forms of PSGL-1 have demonstrated that optimal binding to P-selectin requires sulfation of the tyrosines and addition of a branched, core 2 O-glycan capped with sLe$^x$ to the threonine.[57-62] The structural requirements for binding have been more definitively mapped by synthesis of glycosulfopeptides modeled after this region.[63,64] Each sulfate contributes to binding affinity, and the peptide itself confers weak binding. The position of the O-glycan in relation to the tyrosine sulfates is critical. Even the core backbone of the O-glycan is important: A short core 2 O-glycan capped with sLe$^x$ binds, whereas an isomeric extended core 1 O-glycan does not bind. The fucose moiety is essential for binding, whereas the sialic acid plays a lesser role. These studies reveal specific stereochemical requirements for optimal binding of PSGL-1 to P-selectin.

X-ray crystallography was used to solve the structure of the lectin and EGF domains of P-selectin complexed to a recombinant N-terminal fragment of PSGL-1.[65] The PSGL-1 fragment is sulfated on all three tyrosines, and the sequence of the core 2 O-glycan is nearly identical to the one in the synthetic glycosulfopeptides. The PSGL-1 fragment binds to a large but shallow surface on the lectin domain, opposite to where the EGF domain is attached. The fucose has multiple binding interactions, some of which participate in coordinating the single Ca$^{2+}$ ion in the lectin domain. The galactose and sialic acid residues make fewer contacts. Other regions of the lectin domain contact certain amino acids plus the sulfates of the middle and C-terminal tyrosines of PSGL-1. Monomeric P-selectin binds with equivalent affinity to native or recombinant PSGL-1 and to the synthetic glycosulfopeptides, with dissociation constants estimated at approximately 80 to 800 nM.[63-67] In contrast, P-selectin binds to carbohydrates containing only sLe$^x$ with dissociation constants of 1 to 10 mM.[68-70] The multiple binding interactions of carbohydrate, amino acids, and sulfate residues with the lectin domain explain why PSGL-1 binds to P-selectin with much higher affinity than oligosaccharides that contain only sLe$^x$. The shallow binding site on P-selectin may contribute to the rapid binding kinetics. The structural data do not offer immediate insight into how the P-selectin–PSGL-1 bond responds to force. The N-terminal tyrosine sulfate is not visualized in the cocrystal structure, which is surprising because a recombinant form of PSGL-1

or a glycosulfopeptide with sulfation restricted to this tyrosine still binds to P-selectin with appreciable affinity.[61,62,64] Interestingly, the N-terminal region of murine PSGL-1, which can bind to both human and murine P-selectin, has only two tyrosines that might be sulfated.[71] Compared with the human sequence, these tyrosines are positioned much closer to the two threonines that are the best candidates for modification by a fucosylated O-glycan. Mutagenesis studies suggest that only one tyrosine and one threonine contribute to binding.[72] Thus, there may be additional ways in which P-selectin can bind to both sLe$^x$ and a sulfated peptide segment. How P-selectin binds to heparin, fucoidan, sulfatides, or other sulfated carbohydrates also requires further study.

Leukocytes must adhere to vascular surfaces within the hydrodynamic environment of the circulation. The relative motion between the flowing cell and the vascular surface limits the time for a cell adhesion molecule to bind its ligand. This imposes kinetic constraints on selectin–ligand interactions. Therefore, fast association and dissociation rates for bonds between PSGL-1 and P-selectin have been suggested as requirements for leukocytes to roll on P-selectin in flow.[73] Surface plasmon resonance studies have confirmed rapid binding kinetics for unstressed bonds. The measured dissociation rate is approximately 1 per second, and the estimated association rate is a remarkably high $4.4 \times 10^6$ M$^{-1}$s$^{-1}$.[66] The rapid binding kinetics, particularly the fast association rate, may be important for leukocyte–endothelial cell interactions, because the surface densities of PSGL-1 and P-selectin appear to be much lower than those of GPIb$\alpha$ and VWF. There are approximately 25,000 PSGL-1 molecules on the surface of each leukocyte and approximately 25,000 GPIb$\alpha$ molecules on the surface of each platelet.[74,75] Because leukocytes are much larger than platelets, the density of PSGL-1 is much lower than the density of GPIb$\alpha$. Leukocytes roll on purified, immobilized P-selectin at densities as low as 10 to 25 sites/$\mu$m$^2$.[46,76] Cultured human endothelial cells stimulated with thrombin or histamine express approximately 25 to 50 P-selectin molecules/$\mu$m$^2$.[77] The density of P-selectin on activated endothelial cells *in vivo* is unknown, but immunohistochemical studies suggest that it is much lower than that of E-selectin.

The hydrodynamic environment also imposes mechanical constraints on selectin–ligand interactions. Adhesive bonds are subjected to force, which affects their dissociation rates. The most intuitively obvious response to force is an increase in dissociation rate that shortens bond lifetime. This phenomenon has been termed a *slip bond*. The first experimental demonstration of slip bonds was for interactions between P-selectin and PSGL-1. The lifetimes of transient tethers of neutrophils or PSGL-1-transfected cells on limiting densities of P-selectin were measured as a function of wall shear stress, or force on the tether.[62,78] The lifetimes appear to obey first-order kinetics, consistent with the tethers representing "quantal units" that may represent single bonds. The lifetimes shorten only gradually as force on the tether is increased. If the tether represents only one bond, the bond must have high mechanical strength. However, some transient tethers have more than one bond, even though they have lifetimes that appear to obey first-order dissociation kinetics.[79] It is possible that tethers at higher wall shear stresses require more bonds, which would distribute the force and thus lower the average force per bond. In this scenario, the strength of individual bonds need not be high.

A less intuitive response to force is a decrease in dissociation rate that lengthens bond lifetime. This phenomenon has been termed a *catch bond*. Although catch bonds were hypothesized many years ago,[80] they were experimentally demonstrated only recently, again for interactions between P-selectin and PSGL-1.[81] The catch bonds were observed at lower levels of force than those previously measured, explaining why they were not initially detected. As force initially increased, bond lifetimes lengthened — a characteristic of catch bonds — until the lifetimes reached an optimal level. As force increased further, bond lifetimes shortened — a characteristic of slip bonds. Thus, depending on the force level, interactions between P-selectin and PSGL-1 behaved as either catch bonds or slip bonds. This biphasic response to force introduces a new dimension into how cell adhesion molecules might function in mechanically stressful environments such as the circulation. The function of catch bonds between P-selectin and PSGL-1 is not clear, because they occur only at very low forces that would represent very low shear stresses in the circulation. In other molecular systems, however, catch bonds might have important biological functions. As mentioned earlier, interactions between L-selectin and PSGL-1 enable leukocytes to roll on already adherent leukocytes and on leukocyte fragments that have deposited on vascular surfaces. L-selectin-dependent rolling develops only above a certain shear threshold; below this shear threshold, leukocytes roll unstably and detach.[82,83] Such flow-enhanced rolling may prevent aggregation of free-flowing leukocytes, which express both L-selectin and PSGL-1. Transitions between catch bonds and slip bonds have been observed between L-selectin and PSGL-1,[84] and catch bonds have been shown to govern L-selectin-dependent cell rolling at threshold shear.[85] Catch bonds enable increasing force to convert short-lived tethers into longer-lived tethers, which decrease rolling velocities and increase the regularity of rolling steps as shear rises from the threshold to an optimal value. As shear increases above the optimum, transitions to slip bonds shorten tether lifetimes, which increase rolling velocities and decrease rolling regularity. Thus, force-dependent alterations of bond lifetimes govern L-selectin-dependent cell adhesion below and above the shear optimum. These findings establish the first biological function for catch bonds as a mechanism for flow-enhanced cell adhesion. Although

they have not been experimentally demonstrated, catch bonds are a potentially important mechanism to explain the flow-enhanced rolling of platelets through interactions between GPIbα and VWF. The structural basis for catch bonds is not understood. It is possible that certain levels of force cause the interacting molecules to slide across the binding interface, causing new interactions to form that compensate for the breakage of other interactions.

Microspheres decorated with PSGL-1 roll on purified P-selectin that is immobilized on the floor of a flow chamber.[86] This demonstrates that the intrinsic molecular interactions are sufficient to confer rolling adhesion under flow. However, the cell-surface organization of P-selectin and PSGL-1 further contributes to the efficiency of leukocyte tethering and rolling in flow. PSGL-1 is concentrated on the tips of microvilli, which enhances tethering rates.[46,87,88] Both P-selectin and PSGL-1 are long molecules that extend their binding sites well above the membrane.[56,75] P-selectin must have a minimal length to support leukocyte rolling *in vitro*,[89] and this may also be true for PSGL-1.[86] Molecular extension enhances bond formation and may reduce repulsion between the glycocalyces of apposing cells.[90] Adhesive forces applied through P-selectin/PSGL-1 bonds cause rolling leukocytes to extrude thin membrane tethers at the trailing edge of the cell. These tethers reduce the force required to balance the torque applied to the cell by fluid shear and prolong the time required for bond dissociation.[91,92] Tether extension and retraction is dynamic, rapidly increasing as wall shear stress increases and then rapidly decreasing as wall shear stress decreases.[93] Clustering of PSGL-1 in microvilli and of P-selectin in clathrin-coated pits may also delay bond dissociation.[94] Indeed, differential signaling of endothelial cells regulates the degree of clustering of P-selectin in clathrin-coated pits.[95] Dimerization through molecular self-association contributes to rolling efficiency. PSGL-1 forms noncovalent dimers through interactions of the transmembrane domain; dimerization may be further stabilized by formation of a single disulfide bond.[96] Mutation of the cysteine responsible for disulfide bond formation was reported to eliminate PSGL-1-dependent cell rolling on P-selectin,[97] but this has not been confirmed.[96] However, cells expressing a chimeric, monomeric form of PSGL-1 containing the transmembrane domain of CD43 roll less stably on P-selectin than cells expressing wild-type dimeric PSGL-1.[79] P-selectin forms dimers and oligomers, probably through interactions with the transmembrane domains.[75,98] All these changes delay dissociation of the last bonds at the trailing edge of the cell, where force is applied. This allows more time for new bonds to form at the leading edge of the cell. The result is slower and more regular rolling motions. As wall shear stress is increased, the velocities of leukocytes rolling on selectins tend to plateau. It has been suggested that intrinsic features of the selectin–ligand bond contribute to this "automatic braking system."[99] However, fixed cells or rigid microspheres displaying PSGL-1 do not stabilize their rolling velocities as wall shear stress increases.[86] This indicates that fixation-sensitive cellular features are critical for this phenomenon. Extension of thin membrane tethers will reduce the force on bonds, and enlargement of the contact area through shear-induced cellular deformation may increase the probability of forming bonds.

## IV. Leukocyte Adhesion to Platelets under Flow

P-selectin was originally described as a glycoprotein that is localized in the membranes of platelet α-granules.[100,101] Upon platelet activation, it is rapidly redistributed to the platelet surface, where it initiates adhesion to leukocytes.[102,103] In flow, neutrophils tether to and roll on adherent activated platelets *in vitro* and *in vivo* (Fig. 12-1C).[104–107] Rolling is completely blocked by mAbs to either P-selectin or PSGL-1. There are approximately 10,000 P-selectin molecules on each activated platelet, which translates into a density of approximately 350 sites/$\mu m^2$.[105,108,109] This is a much higher density than on activated endothelial cells *in vitro* and probably *in vivo,* which may explain why platelets do not require E-selectin to slow rolling velocities of leukocytes after they have tethered. The multistep process of leukocyte tethering and rolling, followed by leukocyte activation and firm adhesion, also occurs on activated platelets. Leukocyte activation occurs in part through signaling via PSGL-1, which may be particularly efficient because of the high density of P-selectin on platelets.[110–114] Platelet-secreted chemokines and lipid mediators also contribute to leukocyte signaling.[107,115–117]

Activated leukocytes use the β2 integrin Mac-1 (αMβ2, CD11b/CD18) and, to a lesser extent, LFA-1, to adhere firmly to platelets (Fig. 12-1C). In flow, these interactions do not develop if P-selectin or PSGL-1 is blocked. Platelets express ICAM-2, an Ig-like counterreceptor for LFA-1,[118] but ICAM-2 does not appear to play an important role in integrin-dependent adhesion to platelets. It has been suggested that fibrinogen stabilizes leukocyte adhesion by simultaneously binding to Mac-1 on leukocytes and αIIbβ3 on platelets.[107] However, neither αIIbβ3 antagonists nor β3-deficient platelets prevent leukocyte accumulation on activated platelets.[116] A more attractive candidate for a Mac-1 ligand on platelets is GPIbα, which has been shown to bind directly to Mac-1.[119] Binding requires the I-domain on the α-subunit of Mac-1. This is appealing because the I-domain is structurally related to the A1 domain of VWF, which also binds to GPIbα. However, studies with blocking mAbs suggest that the Mac-1 binding region on GPIbα, which includes the C-terminal region flanking the leucine-rich repeats, is not identical to the VWF binding site. Flowing monocytic cells roll on an immobilized proteolytic fragment of GPIbα; this resembles the GPIbα-mediated translocation

of platelets on VWF under flow.[119] It will be important to determine whether rolling of leukocytes on activated platelets through P-selectin–PSGL-1 interactions can be converted to firm adhesion through Mac-1–GPIbα interactions.

## V. Platelet Adhesion to Endothelial Cells under Flow

Unstimulated platelets roll on endothelial cells of postcapillary venules that are stimulated with calcium ionophore or TNF-α (Fig. 12-1D).[120] Platelets roll six- to ninefold faster than leukocytes. Studies with selectin-deficient mice indicate that platelets roll on endothelial cell P- and/or E-selectin, but P-selectin on platelets is not required. Platelets from mice deficient in both FTVII and FTIV roll normally, suggesting that selectin ligands on platelets do not require α1,3-fucosylation to function.[121] Two platelet ligands for P-selectin have been proposed: GPIbα and PSGL-1.[122,123] Selectin binding to fucosylated carbohydrates requires $Ca^{2+}$. In contrast, transfected Chinese hamster ovary cells expressing the GPIb-IX complex roll on P-selectin in a $Ca^{2+}$-independent manner, and do not require modification with either core 2-branched O-glycans or α1,3-fucosylation.[122] An mAb to the anionic tyrosine-sulfated region of GPIbα partially blocks rolling of human platelets on P-selectin *in vitro,* but an mAb to a more N-terminal epitope blocks rolling more effectively. An mAb to the N-terminal region of murine PSGL-1, which includes the tyrosine-sulfated region, partially blocks platelet rolling on P-selectin *in vivo.*[123] It has not been demonstrated that GPIbα or PSGL-1 on platelets must be tyrosine sulfated to bind P-selectin, although this is an attractive possibility. As discussed earlier, optimal binding of P-selectin to PSGL-1 requires cooperative interactions with both a core 2 O-glycan capped with sLe$^x$ and with a tyrosine-sulfated peptide component.[63,64] The lack of an obvious requirement for glycosylation suggests that platelet GPIbα and PSGL-1 bind to P-selectin with much lower affinity than leukocyte PSGL-1. The very high density of GPIbα on platelets may allow rapid, shear-dependent translocation on P-selectin, despite lower binding affinity. The smaller size of platelets relative to leukocytes also reduces the effective force on bonds formed with P-selectin in flow. The density of PSGL-1 on platelets appears to be very low,[123,124] making it difficult to explain how it could contribute significantly to platelet rolling on P-selectin.

Endothelial cells stimulated with secretagogues mobilize both P-selectin and ultralarge multimers of VWF from Weibel–Palade bodies. In larger venules, which have lower shear rates than smaller venules, unstimulated platelets translocate on VWF that is transiently bound to endothelial cells after stimulation with calcium ionophore or histamine.[125] Translocation is blocked by mAbs to either GPIbα or VWF. Platelet translocation on VWF is more rapid and irregular than the relatively smooth platelet rolling on endothelial cell P-selectin. Thus, GPIbα on platelets can support adhesive interactions with both P-selectin and VWF on stimulated endothelial cells. *In vitro,* the ultralarge multimers of VWF are anchored to activated human umbilical vein endothelial cells through interactions with P-selectin on the endothelial cell surface.[126] The domain of VWF that binds to P-selectin has not been characterized, and the role of P-selectin in anchoring VWF to activated endothelial cells *in vivo* has not been confirmed.

The function of platelet rolling on stimulated endothelial cells is not known. Platelet rolling may constitute a surveillance mechanism to accumulate platelets near a site of injury, making them available for immediate response.[120,127] If rolling platelets become activated, they may develop procoagulant surfaces. They may also secrete proinflammatory cytokines and chemokines that affect leukocyte responses or growth factors that stimulate wound repair. In some situations, activated platelets use P-selectin to roll on endothelial cells. This has been demonstrated for rolling of activated platelets through P-selectin on high endothelial venules of lymph nodes.[128] In lymph nodes, P-selectin appears to bind to endothelial cell mucins that primarily serve as ligands for L-selectin on circulating lymphocytes. These mucins create binding sites for both L- and P-selectin through presentation of O-glycans that are sulfated as well as sialylated and fucosylated.[129] Cytokines may induce expression of similar ligands on endothelial cells at some sites of inflammation.[129] At low shear rates, activated platelets also roll on vascular surfaces through interactions of platelet P-selectin with PSGL-1 on adherent leukocytes or leukocyte fragments.[130,131] *In vitro,* activation of platelets stabilizes their adhesion to endothelial cells, in part through bridging interactions of fibrinogen with αIIbβ3 on platelets and with ICAM-1 and αvβ3 on endothelial cells.[132] Whether these interactions contribute to platelet adhesion *in vivo* has not been determined.

## VI. Role of P-Selectin in Stabilizing Platelet Aggregates

The ability of unstimulated platelets to roll on endothelial cell P-selectin suggests that P-selectin may also promote platelet–platelet cohesion. With one exception,[133] earlier studies found no evidence for a role for P-selectin in platelet aggregation. More recent reports strongly support such a function (Fig. 12-1A). mAbs to the lectin domain of P-selectin, which block binding to leukocytes, have no effect on the initial phase of platelet aggregation induced by ADP, thrombin, or other agonists. However, they destabilize newly formed aggregates, reducing both their size and number.[134] Anti-P-selectin mAbs also significantly inhibit shear-induced platelet aggregation in a sensitive assay that detects

small groups of associated platelets.[135] P-selectin-deficient mice form less compact platelet layers on denuded arterial surfaces *in vivo.*[136] Kinetic analysis and blocking studies using anti-αIIbβ3 mAbs suggest that P-selectin stabilizes platelet aggregates only after they have formed through binding of fibrinogen to αIIbβ3.[134] This contrasts with the function of P-selectin in *initiating* platelet or leukocyte rolling in flow. Furthermore, mAbs to GPIbα or PSGL-1 do not destabilize platelet aggregates.[134] This suggests that platelets use another P-selectin ligand for this function. Close proximity of the apposing cell surfaces may be required for this P-selectin–ligand interaction, which is consistent with an early observation that P-selectin, but not αIIbβ3 or fibrinogen, is enriched at cell–cell contacts of stable platelet aggregates.[137] The high density of P-selectin may be important for multivalent binding to high-density, but low-affinity, ligands on apposing platelets. These ligands might be small glycoproteins, glycolipids, or sulfatides, which can bind to P-selectin on a closely apposed platelet but not on a free-flowing cell. A role for sulfatides is suggested by the observation that a sulfatide-binding protein destabilizes newly formed platelet aggregates.[138]

## VII. Multicellular Interactions of Platelets, Leukocytes, and Endothelial Cells under Flow

Fig. 12-4 summarizes the best-characterized molecular interactions that mediate adhesion of leukocytes to endothelium, leukocytes to platelets, platelets to endothelium, and platelets to platelets. These processes may cooperate to amplify platelet or leukocyte accumulation on vascular surfaces under flow. Activated platelets adherent to endothelium may recruit flowing leukocytes through P-selectin–PSGL-1 interactions, and then transfer these leukocytes to the endothelial cell surface, where they can interact with endothelial selectins.[139–142] Conversely, leukocytes adherent to inflamed venular endothelial cells may recruit circulating activated platelets through P-selectin–PSGL-1 interactions.[130,131,143] Leukocytes also accumulate on platelet thrombi that have deposited on subendothelial tissues at sites of vascular damage.[144] Platelet microparticles (see Chapter 20) generated at sites of injury express P-selectin, which may bridge adjacent leukocytes and increase accumulation of rolling leukocytes on endothelial selectins at sites of inflammation.[145] Similarly, leukocyte adhesion to P-selectin may stimulate production of leukocyte microparticles, which adhere through PSGL-1 to P-selectin on activated platelets or endothelial cells.[146–149] In addition, selectins and integrins cooperate to create stable multicellular interactions at sites of injury.[150]

Adhesive interactions enhance opportunities for signaling in leukocytes, platelets, and endothelial cells in the affected region. Signals can be transmitted directly through

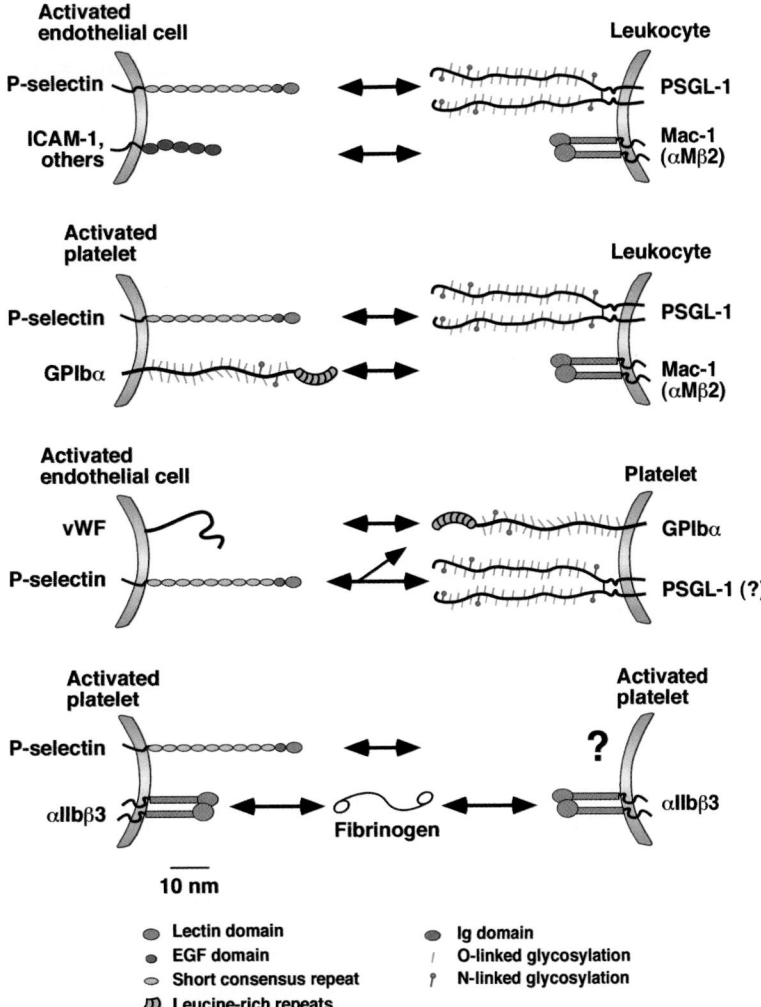

**Figure 12-4.** Multicellular adhesive interactions among leukocytes, platelets, and endothelial cells. Only the adhesion molecules emphasized in this chapter are shown. Molecular lengths were estimated from hydrodynamic data and electron microscopy.

adhesion receptors,[110,150] and these signals may intersect with those generated by chemokines, cytokines, proteases, or other mediators.[151] For example, monocytes that adhere to activated platelets secrete inflammatory chemokines; this response requires both engagement of PSGL-1 by P-selectin and binding of platelet-expressed RANTES (regulated upon activation, normal T cell expressed and presumably secreted) to its receptor on monocytes.[117] Activated platelets secrete a variety of inflammatory mediators, including chemokines such as neutrophil-activating peptide 2, which activate neutrophils[152] (see Chapter 39). Conversely, activated neutrophils release cathepsin G and other proteases that activate platelets[153,154] (see Chapter 39). Activated platelets express CD40 ligand and interleukin 1, which trigger expression of several adhesion molecules and chemokines.[155,156] Binding of platelet CD40 ligand to platelet integrin αIIbβ3 also stabilizes arterial thrombi.[157] Furthermore, adhesion of activated platelets to neutrophils facilitates transcellular

metabolism of inflammatory leukotrienes that cannot be synthesized by either cell alone.[158,159] Arachidonic acid in platelet-derived microparticles increases expression of ICAM-1 on endothelial cells.[160] *In vivo,* activated platelets induce endothelial cells to secrete VWF and mobilize P-selectin, which enhance platelet and leukocyte rolling on the vessel surface.[161] The effects require P-selectin on platelets and may be the result of mediators released from platelet–leukocyte interactions.[161] These signaling events among adherent cells serve to amplify the inflammatory and hemostatic responses, especially at sites of tissue injury.

In the absence of injury, endothelial cells secrete mediators that dampen platelet reactivity. The major products are nitric oxide (NO), prostacyclin (prostaglandin $I_2$), and the ectoADPase CD39, all of which inhibit platelet activation.[162] NO was originally described as a labile compound that dilates blood vessels. However, NO also inhibits platelet activation by stimulating guanylyl cyclase (thereby increasing cyclic guanosine 5′-monophosphate), inhibiting phosphoinositide 3-kinase, impairing capacitative calcium influx, and inhibiting cyclooxygenase-1.[163] The platelet inhibitory effects of NO synergize with those of prostacyclin, an arachidonic acid metabolite released by endothelial cells.[164] Prostacyclin increases cyclic adenosine monophosphate in platelets, which inhibits platelet aggregation and promotes platelet disaggregation as well.[165] Perhaps the most important endogenous inhibitor of platelet function is the endothelial-derived ecto-ADPase (CD39).[162,166] Ecto-ADPase on the surface of endothelial cells metabolizes ADP released from activated platelets, which potently inhibits platelet aggregation. Ecto-ADPase has major "thromboregulatory" effects in the absence of NO and prostacyclin, suggesting that it has a dominant function in limiting platelet reactivity. Vascular diseases such as diabetes and hypertension impair the endothelial production of NO, prostacyclin, and probably ecto-ADPase.[165] Such impairment, coupled with acute vascular injury that damages endothelial cells, reduces the major endothelial cell products that prevent platelet activation. The inhibition of platelet function by endothelial cell-derived NO, prostacyclin, and ecto-ADPase is discussed in more detail in Chapter 13.

## VIII. Consequences of Leukocyte–Platelet–Endothelial Cell Interactions *in Vivo*

Many models support a role for P-selectin in amplifying coagulation and thrombosis *in vivo*. Platelets, fibrin, and leukocytes accumulate on a graft implanted within an arteriovenous shunt in baboons.[167] An anti-P-selectin mAb blocks both leukocyte accumulation and fibrin formation, suggesting that leukocyte adhesion to activated platelets promotes fibrin deposition. P-selectin-deficient mice have a modest but consistent increase in bleeding time after amputation of the tip of the tail, and bleed more in the Shwartzman reaction, a model that involves close interaction between the inflammatory and hemostatic pathways.[168] mAbs to P-selectin reduce both leukocyte accumulation and thrombus formation in rat and baboon models of venous thrombosis.[169,170] Recombinant soluble PSGL-1 inhibits thrombus formation in feline and baboon models of venous thrombosis.[171,172] An mAb to P-selectin or recombinant soluble PSGL-1 accelerates pharmacological thrombolysis in primate and porcine models of arterial thrombosis.[173,174] Selectin inhibitors also reduce platelet and leukocyte accumulation on damaged canine arteries. This correlates with reduction in cyclic flow variations, a hallmark of recurrent platelet-dependent thrombus formation.[175,176]

How might P-selectin amplify coagulation or thrombosis? An emerging principle is that it promotes the recruitment of leukocytes or leukocyte microparticles that express tissue factor. Blood contains low levels of circulating microparticles with tissue factor on their surfaces.[177] These microparticles are derived from both neutrophils and monocytes, but it has not been established whether the parental cells synthesize the tissue factor or whether they ingest or adsorb tissue factor from another source. The tissue factor-rich microparticles accumulate on adherent activated platelets through P-selectin–PSGL-1 interactions in sufficient quantities to amplify fibrin formation.[147–149,178] Leukocyte microparticles bearing tissue factor and PSGL-1 arise primarily from cholesterol-rich lipid rafts.[179] After binding to activated platelets, the microparticles fuse with the platelet membrane in a process that is probably accelerated by the cholesterol-rich environment of the rafts. Fusion of leukocyte microparticles to platelets provides a mechanism to add tissue factor to the other procoagulant complexes on platelet thrombi at sites of injury. P-selectin on cell surfaces or isolated from platelets is reportedly sufficient to induce low-level expression of tissue factor messenger RNA and protein in monocytes,[180] but this observation has not been confirmed.[117] However, it is possible that P-selectin signaling through PSGL-1 may cooperate with signals from chemokines or other mediators to induce tissue factor synthesis. Platelets activated in whole blood may also use P-selectin to convert cryptic tissue factor on the surface of monocytes to an active form.[181,182]

*In vivo,* low-level inflammation might generate small quantities of tissue factor-rich leukocyte microparticles through P-selectin–PSGL-1 interactions. This could represent a priming event for a subsequent response to tissue injury, which could in turn amplify coagulation or thrombosis.[183] Supporting this notion, mice expressing a form of P-selectin without the cytoplasmic domain (ΔCT) constitutively express three- to fourfold higher levels of P-selectin on the endothelial cell surface, because the truncated molecule is not sorted in Weibel–Palade bodies or internalized efficiently.[184] The constitutive expression of P-selectin supports basal leukocyte rolling in postcapillary venules. In an *ex vivo* perfusion chamber, more fibrin is deposited at the site of platelet thrombus formation in ΔCT mice than in wild-type mice, whereas P-selectin-deficient mice deposit much less fibrin.[146] There is also less hemorrhage in the Shwartzman reaction in ΔCT mice than in wild-type mice. Strikingly, plasma from ΔCT mice has higher levels of leukocyte microparticles expressing tissue factor and clots faster than plasma from wild-type mice. Removal of microparticles by ultracentrifugation prolongs the clotting time.

The ΔCT mice have three fold higher levels of soluble P-selectin in plasma, which circulates as an approximately 100-kDa form that probably results from proteolytic cleavage of endothelial cell P-selectin.[184] The procoagulant phenotype of the ΔCT mice was ascribed to the elevated levels of soluble P-selectin, because intravenous infusion of a soluble P-selectin–immunoglobulin (Ig) chimera in wild-type mice reproduces the procoagulant features.[146] However, the level of soluble P-selectin in ΔCT mice is only 1 μg/mL, or approximately 10 nM. Assuming that plasma P-selectin remains monomeric, this is well below the estimated $K_d$ for binding of soluble, monomeric human P-selectin to leukocytes or to purified PSGL-1 ($K_d$ approximately 70–800 nM).[65–67,75] The plasma concentration of infused P-selectin Ig, although not measured, may have been 50 to 100 nM, and the higher avidity of the bivalent chimera may result in significantly greater binding to leukocytes.[75] Furthermore, signaling through PSGL-1, like through most receptors, appears to require divalent or multivalent interactions.[110–112,117,183] Thus, the increased soluble P-selectin in the $\Delta$CT mice may not directly activate leukocytes if it is monomeric and binds to only a small fraction of the PSGL-1 molecules on leukocytes. Instead, signaling through P-selectin-dependent leukocyte rolling might produce tissue factor-rich leukocyte microparticles that are procoagulant. Infusion of bivalent P-selectin Ig may mimic signaling during P-selectin-dependent leukocyte adhesion.[146,147] Normal human plasma also has low levels of soluble P-selectin, which may be derived from both proteolytic cleavage and from secretion of an alternatively spliced protein that lacks the transmembrane domain.[75,185,186] Plasma P-selectin levels increase three- to fourfold in some inflammatory and thrombotic disorders, but the levels remain well below the $K_d$ for binding to leukocytes.[187] Thus, the functional significance of plasma P-selectin remains unclear.[187]

The surfaces of activated platelets assemble the protease/cofactor complexes that generate thrombin[188] (see Chapter 19). Leukocyte adhesion to P-selectin may generate signals that amplify platelet activation, thereby enhancing the procoagulant properties of the platelet surface.[154] Circulating platelet microparticles are elevated in some disorders[189] (see Chapter 20). One of the procoagulant features of these microparticles may be signaling through multivalent binding of P-selectin on the microparticles to PSGL-1 on leukocytes.

Atherosclerosis (see Chapter 35) is now widely recognized to be an inflammatory disease, and accumulation of monocytes into the arterial wall is an early event.[190] Emigrated monocytes transform into lipid-laden foam cells, which secrete cytokines, growth factors, and other mediators that promote lesion development. Mice deficient in P-selectin, E-selectin, or ICAM-1 are variably protected from atherosclerosis in murine models.[191–193] P-selectin is particularly important in the pathogenesis of atherosclerosis in the apolipoprotein E-deficient mouse. Both platelet and endothelial cell P-selectin contribute to lesion formation.[194,195] The inflammatory response contributes to wound healing, but when dysregulated leads to fibrosis and inappropriate tissue remodeling. In models of arterial injury in normal or atherosclerotic backgrounds, mice deficient in P-selectin or Mac-1 accumulate fewer leukocytes into the vascular wall and are substantially protected from neointimal formation.[136,196,197] P-selectin on platelets is the major contributor to neointimal proliferation,[198,199] and a single injection of anti-P-selectin mAb is sufficient to block the proliferation.[200]

In wild-type mice, a carpet of platelets develops on the damaged vascular surface immediately after injury, and leukocytes adhere to the platelets.[136] Leukocytes do not adhere to the platelet carpet in P-selectin-deficient mice. Mice deficient in $\beta$3 integrins develop only a single layer of platelets on the damaged vessel, to which leukocytes still adhere. Strikingly, $\beta$3 integrin deficiency does not protect mice from neointimal formation.[136] These data suggest that leukocyte recruitment into damaged arterial walls is a major contributor to neointimal formation. Leukocyte recruitment is facilitated by release of platelet-derived chemokines such as RANTES as well as by adhesive interactions with P-selectin.[201] Emigrated leukocytes may release growth factors, proteases, and other mediators that enhance smooth muscle cell migration and proliferation.

Excessive accumulation of leukocytes contributes to the pathogenesis of many inflammatory disorders, including ischemia–reperfusion syndrome during unstable angina, myocardial infarction, stroke, and organ transplantation. Interactions of P-selectin with PSGL-1 make major contributions to these disorders. Activated complement and oxygen radicals, which are characteristically elaborated, mobilize P-selectin to the surface of endothelial cells *in vitro*.[202–204] Endothelial dysfunction decreases formation of NO, an oxygen-radical scavenger that may normally dampen the expression of P-selectin.[205] Hypoxia also translocates P-selectin to the surface of endothelial cells.[206,207] Ischemia–reperfusion induces expression of P-selectin on endothelial cells. Oxidized lipoproteins, cigarette smoke, and other insults activate both platelets and endothelial cells, promoting P-selectin-dependent platelet–leukocyte aggregates and leukocyte adhesion to vascular surfaces *in vitro* and *in vivo*.[139] Consistent with all these observations, mAbs and other P-selectin inhibitors significantly reduce neutrophil accumulation in many models of ischemia–reperfusion injury *in vivo*.[208–214]

Circulating P-selectin–PSGL-1-dependent leukocyte–platelet aggregates are increased in stable coronary artery disease,[215] unstable angina,[216] acute myocardial infarction,[217] venous insufficiency,[218] and during cardiopulmonary bypass.[219] Circulating leukocyte–platelet aggregates also increase after coronary angioplasty, especially in patients experiencing late ischemic events.[220] The biological and clinical significance of these circulating aggregates remains to be determined, but one consequence may be the release of mediators that activate vascular endothelial cells.[161]

## IX. Conclusions

The shear stresses in the circulation place kinetic and mechanical constraints on the adhesion of platelets and leukocytes to vascular surfaces. Both cell types use a multistep process of tethering and reversible adhesion (Figs. 12-1 and 12-4), which makes possible signaling events that facilitate stable adhesion. Platelets use interactions of GPIbα with VWF to tether to and translocate on subendothelial surfaces, whereas leukocytes use interactions of selectins with glycoconjugates to tether to and roll on activated endothelial cells or to adherent leukocytes or activated platelets. Activation of platelets or leukocytes allows shear-resistant adhesion through engagement of integrins with soluble or membrane-anchored ligands. Platelets also roll on activated endothelial cells through interactions with selectins or VWF. Conversion of these rolling events into stable adhesion may require platelet activation and integrin-dependent binding to leukocytes or to platelet thrombi at sites of vascular injury.

GPIbα and PSGL-1 have interesting structural and functional similarities, but important differences as well. Both molecules have long mucin stalks that serve to extend an N-terminal ligand-binding site well above the plasma membrane. This may be important for the tethering functions of these molecules in flow. Both ligand-binding sites include regions of tyrosine sulfation, which are important for GPIbα to bind to VWF and for PSGL-1 to bind to P-selectin. PSGL-1 uses a spatially precise combination of sulfated peptide and a specifically modified core 2 O-glycan to bind to P-selectin, which enhances binding affinity that may optimize rolling interactions despite the relatively low density of PSGL-1 on leukocytes. The structural features required for GPIbα to bind to VWF are partially understood, but the available data do not explain how tyrosine sulfation of GPIbα contributes to binding. The structural basis for binding of GPIbα to P-selectin and the biological significance of this binding are less well understood. High surface densities of GPIbα and its ligands may help compensate for lower affinity and suboptimal kinetic and mechanical properties of individual bonds. The unstressed binding kinetics of PSGL-1 to P-selectin and of GPIbα to VWF appear to be very fast. Depending on the degree of applied force, interactions of P-selectin with PSGL-1 undergo transitions from catch bonds to slip bonds. Whether interactions of GPIbα with VWF exhibit transitions between catch bonds and slip bonds remains to be determined.

The ability of GPIbα and P-selectin to bind to additional ligands broadens their biological functions. GPIbα mediates platelet translocation or rolling on P-selectin as well as on VWF. Interactions with Mac-1 promote monocyte rolling on purified GPIbα, but it is not yet known whether GPIbα–Mac-1 interactions can support firm adhesion of leukocytes to activated platelets. P-selectin stabilizes platelet aggregates by binding to a ligand that does not appear to be either PSGL-1 or GPIbα. Thus, P-selectin, a molecule known to initiate tethering and rolling under flow, may also stabilize cell adhesion after an integrin-dependent event.

The interactions of leukocytes with both activated platelets and endothelial cells add to the growing evidence that the hemostatic and inflammatory responses to tissue injury are intimately linked. Interactions of P-selectin with PSGL-1

and GPIbα with Mac-1 may concentrate leukocytes or leukocyte microparticles at sites of vascular injury and signal expression of tissue factor, chemokines, cytokines, growth factors, and oxygen-derived radicals that augment coagulation and promote wound repair. Conversely, excessive adhesion interactions may contribute to atheromas, thrombosis, and neointimal formation in the vascular system.

# References

1. Savage, B., Saldívar, E., & Ruggeri, Z. M. (1996). Initiation of platelet adhesion by arrest onto fibrinogen or translocation on von Willebrand factor. *Cell, 84*, 289–297.
2. Yuan, Y., Kulkarni, S., Ulsemer, P., Cranmer, *et al.* (1999). The von Willebrand factor-glycoprotein Ib/V/IX interaction induces actin polymerization and cytoskeletal reorganization in rolling platelets and glycoprotein Ib/V/IX-transfected cells. *J Biol Chem, 274*, 36241–36251.
3. Sadler, J. E. (2005). New concepts in von Willebrand disease. *Annu Rev Med, 56*, 173–191.
4. Berndt, M. C., Shen, Y., Dopheide, S. M., Gardiner, E. E., & Andrews, R. K. (2001). The vascular biology of the glycoprotein Ib-IX-V complex. *Thromb Haemost, 86*, 178–188.
5. Chen, J., & Lopez, J. A. (2005). Interactions of platelets with subendothelium and endothelium. *Microcirculation, 12*, 235–246.
6. Dong, J. F., Moake, J. L., Bernardo, A., *et al.* (2003). ADAMTS-13 metalloprotease interacts with the endothelial cell-derived ultra-large von Willebrand factor. *J Biol Chem, 278*, 29633–29639.
7. Dong, J. F., Moake, J. L., Nolasco, *et al.* (2002). ADAMTS-13 rapidly cleaves newly secreted ultralarge von Willebrand factor multimers on the endothelial surface under flowing conditions. *Blood, 100*, 4033–4039.
8. Marchese, P., Saldívar, E., Ware, J., & Ruggeri, Z. M. (1999). Adhesive properties of the isolated amino-terminal domain of platelet glycoprotein Ibα in a flow field. *Proc Natl Acad Sci USA, 96*, 7837–7842.
9. Shrimpton, C. N., Borthakur, G., Larrucea, S., Cruz, M. A., Dong, J. F., & Lopez, J. A. (2002). Localization of the adhesion receptor glycoprotein Ib-IX-V complex to lipid rafts is required for platelet adhesion and activation. *J Exp Med, 196*, 1057–1066.
10. Li, C. Q., Dong, J. F., & Lopez, J. A. (2002). The mucin-like macroglycopeptide region of glycoprotein Ibα is required for cell adhesion to immobilized von Willebrand factor (VWF) under flow but not for static VWF binding. *Thromb Haemost, 88*, 673–677.
11. Dong, J. F., Schade, A. J., Romo, G. M., *et al.* (2000). Novel gain-of-function mutations of platelet glycoprotein Ibα by valine mutagenesis in the Cys209-Cys248 disulfide loop. *J Biol Chem, 275*, 27663–27670.
12. Kumar, R. A., Dong, J. F., Thaggard, J. A., *et al.* (2003). Kinetics of GPIbα-vWF-A1 tether bond under flow: Effect of GPIbα mutations on the association and dissociation rates. *Biophys J, 85*, 4099–4109.

13. Dong, J. F., Ye, P., Schade, A. J., Gao, S., Romo, G. M., *et al.* (2001). Tyrosine sulfation of glycoprotein Ibα. *J Biol Chem, 276,* 16690–16694.

14. Huizinga, E. G., Tsuji, S., Romijn, R. A., *et al.* (2002). Structures of glycoprotein Ibα and its complex with von Willebrand factor A1 domain. *Science, 297,* 1176–1179.

15. Doggett, T. A., Girdhar, G., Lawshe, A., *et al.* (2003). Alterations in the intrinsic properties of the GPIbα-VWF tether bond define the kinetics of the platelet-type von Willebrand disease mutation, Gly233Val. *Blood, 102,* 152–160.

16. Miller, J. L. (1996). Platelet-type von Willebrand disease. *Thromb Haemost, 75,* 865–869.

17. Doggett, T. A., Girdhar, G., Lawshe, A., *et al.* (2002). Selectin-like kinetics and biomechanics promote rapid platelet adhesion in flow: The GPIbα-vWF tether bond. *Biophys J, 83,* 194–205.

18. Miura, S., Li, C. Q., Cao, Z., *et al.* (2000). Interaction of von Willebrand factor domain A1 with platelet glycoprotein Ibα-(1–289). Slow intrinsic binding kinetics mediate rapid platelet adhesion. *J Biol Chem, 275,* 7539–7546.

19. Savage, B., Almus–Jacobs, F., & Ruggeri, Z. M. (1998). Specific synergy of multiple substrate–receptor interactions in platelet thrombus formation under flow. *Cell, 94,* 657–666.

20. Gu, M., Xi, X., Englund, G. D., Berndt, M. C., & Du, X. (1999). Analysis of the roles of 14-3-3 in the platelet glycoprotein Ib-IX-mediated activation of integrin αIIbβ3 using a reconstituted mammalian cell expression model. *J Cell Biol, 147,* 1085–1096.

21. Yap, C. L., Hughan, S. C., Cranmer, S. L., *et al.* (2000). Synergistic adhesive interactions and signaling mechanisms operating between platelet glycoprotein Ib/IX and integrin αIIbβ3. Studies in human platelets and transfected Chinese hamster ovary cells. *J Biol Chem, 275,* 41377–41388.

22. Zaffran, Y., Meyer, S. C., Negrescu, E., Reddy, K. B., & Fox, J. E. (2000). Signaling across the platelet adhesion receptor glycoprotein Ib-IX induces αIIbβ3 activation both in platelets and a transfected Chinese hamster ovary cell system. *J Biol Chem, 275,* 16779–16787.

23. Garcia, A., Quinton, T. M., Dorsam, R. T., & Kunapuli, S. P. (2005). Src family kinase-mediated and Erk-mediated thromboxane A2 generation are essential for VWF/GPIb-induced fibrinogen receptor activation in human platelets. *Blood, 106,* 3410–3414.

24. Cruz, M. A., Chen, J., Whitelock, J. L., Morales, L. D., & Lopez, J. A. (2005). The platelet glycoprotein Ib-von Willebrand factor interaction activates the collagen receptor α2β1 to bind collagen: Activation-dependent conformational change of the α2-I domain. *Blood, 105,* 1986–1991.

25. Ruggeri, Z. M., Dent, J. A., & Saldivar, E. (1999). Contribution of distinct adhesive interactions to platelet aggregation in flowing blood. *Blood, 94,* 172–178.

26. Springer, T. A. (1995). Traffic signals on endothelium for lymphocyte recirculation and leukocyte emigration. *Annu Rev Physiol, 57,* 827–872.

27. Vestweber, D., & Blanks, J. E. (1999). Mechanisms that regulate the function of the selectins and their ligands. *Physiol Rev, 79,* 181–213.

28. McEver, R. P. (2002). Selectins: Lectins that initiate cell adhesion under flow. *Curr Opin Cell Biol, 14,* 581–586.

29. Kunkel, E. J., & Ley, K. (1996). Distinct phenotype of E-selectin-deficient mice — E-selectin is required for slow leukocyte rolling in vivo. *Circ Res, 79,* 1196–1204.

30. Jung, U., Norman, K. E., Scharffetter–Kochanek, K., Beaudet., A. L., & Ley, K. (1998). Transit time of leukocytes rolling through venules controls cytokine-induced inflammatory cell recruitment in vivo. *J Clin Invest, 102,* 1526–1533.

31. McIntyre, T. M., Prescott, S. M., Weyrich, A. S., & Zimmerman, G. A. (2003). Cell–cell interactions: Leukocyte–endothelial interactions. *Curr Opin Hematol, 10,* 150–158.

32. Harris, E. S., McIntyre, T. M., Prescott, S. M., & Zimmerman, G. A. (2000). The leukocyte integrins. *J Biol Chem, 275,* 23409–23412.

33. Kunkel, E. J., Dunne, J. L., & Ley, K. (2000). Leukocyte arrest during cytokine-dependent inflammation in vivo. *J Immunol, 164,* 3301–3308.

34. Forlow, S. B., White, E. J., Barlow, S. C., *et al.* (2000). Severe inflammatory defect and reduced viability in CD18 and E-selectin double-mutant mice. *J Clin Invest, 106,* 1457–1466.

35. Sigal, A., Bleijs, D. A., Grabovsky, V., *et al.* (2000). The LFA-1 integrin supports rolling adhesions on ICAM-1 under physiological shear flow in a permissive cellular environment. *J Immunol, 165,* 442–452.

36. Alon, R., Kassner, P. D., Carr, M. W., Finger, *et al.* (1995). The integrin VLA-4 supports tethering and rolling in flow on VCAM-1. *J Cell Biol, 128,* 1243–1253.

37. Berlin, C., Bargatze, R. F., Campbell, J. J., Von Andrian, *et al.* (1995). α4 Integrins mediate lymphocyte attachment and rolling under physiologic flow. *Cell, 80,* 413–422.

38. Taooka, Y., Chen, J., Yednock, T., & Sheppard, D. (1999). The integrin α9β1 mediates adhesion to activated endothelial cells and transendothelial neutrophil migration through interaction with vascular cell adhesion molecule-1. *J Cell Biol, 145,* 413–420.

39. Varki, A. (1997). Selectin ligands: Will the real ones please stand up? *J Clin Invest, 99,* 158–162.

40. Maly, P., Thall, A. D., Petryniak, B., Rogers, *et al.* (1996). The α(1,3)fucosyltransferase Fuc-TVII controls leukocyte trafficking through an essential role in L-, E-, and P-selectin ligand biosynthesis. *Cell, 86,* 643–653.

41. Homeister, J. W., Thall, A. D., Petryniak, *et al.* (2001). The α(1,3)fucosyltransferases FucT-IV and FucT-VII exert collaborative control over selectin-dependent leukocyte recruitment and lymphocyte homing. *Immunity, 15,* 115–126.

42. Etzioni, A., Frydman, M., Pollack, S., *et al.* (1992). Brief report: Recurrent severe infections caused by a novel leukocyte adhesion deficiency. *N Engl J Med, 327,* 1789–1792.

43. McEver, R. P., Cummings, R. D. (1997). Role of PSGL-1 binding to selectins in leukocyte recruitment. *J Clin Invest, 100,* 485–492.

44. Moore, K. L., Stults, N. L., Diaz, S., *et al.* (1992). Identification of a specific glycoprotein ligand for P-selectin (CD62) on myeloid cells. *J Cell Biol, 118,* 445–456.

45. Norgard, K. E., Moore, K. L., Diaz, S., *et al.* (1993). Characterization of a specific ligand for P-selectin on myeloid

cells. A minor glycoprotein with sialylated O-linked oligo-saccharides. *J Biol Chem, 268,* 12764–12774.

46. Moore, K. L., Patel, K. D., Bruehl, R. E., *et al.* (1995). P-selectin glycoprotein ligand-1 mediates rolling of human neutrophils on P-selectin. *J Cell Biol, 128,* 661–671.

47. Norman, K. E., Moore, K. L., McEver, R. P., & Ley, K. (1995). Leukocyte rolling in vivo is mediated by P-selectin glycoprotein ligand-1. *Blood, 86,* 4417–4421.

48. Borges, E., Eytner, R., Moll, T., *et al.* (1997). The P-selectin glycoprotein ligand-1 is important for recruitment of neutrophils into inflamed mouse peritoneum. *Blood, 90,* 1934–1942.

49. Yang, J., Hirata, T., Croce K., *et al.* (1999). Targeted gene disruption demonstrates that P-selectin glycoprotein ligand 1 (PSGL-1) is required for P-selectin-mediated but not E-selectin-mediated neutrophil rolling and migration. *J Exp Med, 190,* 1769–1782.

50. Xia, L., Sperandio, M., Yago, T., *et al.* (2002). P-selectin glycoprotein ligand-1-deficient mice have impaired leukocyte tethering to E-selectin under flow. *J Clin Invest, 109,* 939–950.

51. Walcheck, B., Moore, K. L., McEver, R. P., & Kishimoto, T. K. (1996). Neutrophil–neutrophil interactions under hydrodynamic shear stress involve L-selectin and PSGL-1: A mechanism that amplifies initial leukocyte accumulation on P-selectin in vitro. *J Clin Invest, 98,* 1081–1087.

52. Alon, R., Fuhlbrigge, R. C., Finger, E. B., & Springer, T. A. (1996). Interactions through L-selectin between leukocytes and adherent leukocytes nucleate rolling adhesions on selectins and VCAM-1 in shear flow. *J Cell Biol, 135,* 849–865.

53. Eriksson, E. E., Xie, X., Werr, J., Thoren, P., & Lindbom, L. (2001). Importance of primary capture and L-selectin-dependent secondary capture in leukocyte accumulation in inflammation and atherosclerosis in vivo. *J Exp Med, 194,* 205–218.

54. Sperandio, M., Smith, M. L., Forlow, S. B., *et al.* (2003). P-selectin glycoprotein ligand-1 mediates L-selectin-dependent leukocyte rolling in venules. *J Exp Med, 197,* 1355–1363.

55. Wilkins, P. P., McEver, R. P., & Cummings, R. D. (1996). Structures of the O-glycans on P-selectin glycoprotein ligand-1 from HL-60 cells. *J Biol Chem, 271,* 18732–18742.

56. Li, F., Erickson, H. P., James, J. A., *et al.* (1996). Visualization of P-selectin glycoprotein ligand-1 as a highly extended molecule and mapping of protein epitopes for monoclonal antibodies. *J Biol Chem, 271,* 6342–6348.

57. Wilkins, P. P., Moore, K. L., McEver, R. P., & Cummings, R. D. (1995). Tyrosine sulfation of P-selectin glycoprotein ligand-1 is required for high affinity binding to P-selectin. *J Biol Chem, 270,* 22677–22680.

58. Sako, D., Comess, K. M., Barone, K. M., *et al.* (1995). A sulfated peptide segment at the amino terminus of PSGL-1 is critical for P-selectin binding. *Cell, 83,* 323–331.

59. Pouyani, T., Seed, B. (1995). PSGL-1 recognition of P-selectin is controlled by a tyrosine sulfation consensus at the PSGL-1 amino terminus. *Cell, 83,* 333–343.

60. Li, F., Wilkins, P. P., Crawley, S., *et al.* (1996). Post-translational modifications of recombinant P-selectin glycoprotein ligand-1 required for binding to P- and E-selectin. *J Biol Chem, 271,* 3255–3264.

61. Liu, W. J., Ramachandran, V., Kang, J., *et al.* (1998). Identification of N-terminal residues on P-selectin glycoprotein ligand-1 required for binding to P-selectin. *J Biol Chem, 273,* 7078–7087.

62. Ramachandran, V., Nollert, M. U., Qiu, H., *et al.* (1999). Tyrosine replacement in P-selectin glycoprotein ligand-1 affects distinct kinetic and mechanical properties of bonds with P- and L-selectin. *Proc Natl Acad Sci USA, 96,* 13771–13776.

63. Leppänen, A., Mehta, P., Ouyangl, Y. B., *et al.* (1999). A novel glycosulfopeptide binds to P-selectin and inhibits leukocyte adhesion to P-selectin. *J Biol Chem, 274,* 24838–24848.

64. Leppänen, A., White, S. P., Helin, J., *et al.* (2000). Binding of glycosulfopeptides to P-selectin requires stereospecific contributions of individual tyrosine sulfate and sugar residues. *J Biol Chem, 275,* 39569–39578.

65. Somers, W. S., Tang, J., Shaw, G. D., & Camphausen, R. T. (2000). Insights into the molecular basis of leukocyte tethering and rolling revealed by structures of P- and E-selectin bound to SLe(X) and PSGL-1. *Cell, 103,* 467–479.

66. Mehta, P., Cummings, R. D., & McEver, R. P. (1998). Affinity and kinetic analysis of P-selectin binding to P-selectin glycoprotein ligand-1. *J Biol Chem, 273,* 32506–32513.

67. Croce, K., Freedman, S. J., Furie, B. C., & Furie, B. (1998). Interaction between soluble P-selectin and soluble P-selectin glycoprotein ligand 1: Equilibrium binding analysis. *Biochemistry, 37,* 16472–16480.

68. Poppe, L., Brown, G. S., Philo, J. S., Nikrad, P. V., & Shah, B. H. (1997). Conformation of sLex tetrasaccharide, free in solution and bound to E-, P-, and L-selectin. *J Am Chem Soc, 119,* 1727–1736.

69. Brandley, B. K., Kiso, M., Abbas, S., *et al.* (1993). Structure–function studies on selectin carbohydrate ligands. Modifications to fucose, sialic acid and sulphate as a sialic acid replacement. *Glycobiology, 3,* 633–639.

70. Koenig, A., Jain, R., Vig, R., *et al.* (1997). Selectin inhibition: Synthesis and evaluation of novel sialylated, sulfated and fucosylated oligosaccharides, including the major capping group of GlyCAM-1. *Glycobiology, 7,* 79–93.

71. Yang, J., Galipeau, J., Kozak, C. A., Furie, B. C., & Furie, B. (1996). Mouse P-selectin glycoprotein ligand-1: Molecular cloning, chromosomal localization, and expression of a functional P-selectin receptor. *Blood, 87,* 4176–4186.

72. Xia, L., Ramachandran, V., McDaniel, J. M., *et al.* (2003). N-terminal residues in murine P-selectin glycoprotein ligand-1 required for binding to murine P-selectin. *Blood, 101,* 552–559.

73. Lawrence, M. B., & Springer, T. A. (1991). Leukocytes roll on a selectin at physiologic flow rates: Distinction from and prerequisite for adhesion through integrins. *Cell, 65,* 859–873.

74. Coller, B. S., Peerschke, E. I., Scudder, L. E., & Sullivan, C. A. (1983). Studies with a murine monoclonal antibody

that abolishes ristocetin-induced binding of von Willebrand factor to platelets: Additional evidence in support of GPIb as a platelet receptor for von Willebrand factor. *Blood, 61,* 99–110.

75. Ushiyama, S., Laue, T. M., Moore, K. L., Erickson, H. P., & McEver, R. P. (1993). Structural and functional characterization of monomeric soluble P-selectin and comparison with membrane P-selectin. *J Biol Chem, 268,* 15229–15237.

76. Patel, K. D., Moore, K. L., Nollert, M. U., & McEver, R. P. (1995). Neutrophils use both shared and distinct mechanisms to adhere to selectins under static and flow conditions. *J Clin Invest, 96,* 1887–1896.

77. Hattori, R., Hamilton, K. K., Fugate, R. D., McEver, R. P., & Sims, P. J. (1989). Stimulated secretion of endothelial von Willebrand factor is accompanied by rapid redistribution to the cell surface of the intracellular granule membrane protein GMP-140. *J Biol Chem, 264,* 7768–7771.

78. Alon, R., Hammer, D. A., & Springer, T. A. (1995). Lifetime of the P-selectin: Carbohydrate bond and its response to tensile force in hydrodynamic flow. *Nature, 374,* 539–542.

79. Ramachandran, V., Yago, T., Epperson, T. K., et al. (2001). Dimerization of a selectin and its ligand stabilizes cell rolling and enhances tether strength in shear flow. *Proc Natl Acad Sci USA, 98,* 10166–10171.

80. Dembo, M., Torney, D. C., Saxman, K., & Hammer, D. (1988). The reaction-limited kinetics of membrane-to-surface adhesion and detachment. *Proc R Soc Lond B Biol Sci, 234,* 55–83.

81. Marshall, B. T., Long, M., Piper, J. W., et al. (2003). Direct observation of catch bonds involving cell-adhesion molecules. *Nature, 423,* 190–193.

82. Finger, E. B., Puri, K. D., Alon, R., et al. (1996). Adhesion through L-selectin requires a threshold hydrodynamic shear. *Nature, 379,* 266–269.

83. Lawrence, M. B., Kansas, G. S., Kunkel, E. J., & Ley, K. (1997). Threshold levels of fluid shear promote leukocyte adhesion through selectins (CD62L,P,E). *J Cell Biol, 136,* 717–727.

84. Sarangapani, K. K., Yago, T., Klopocki, A. G., et al. (2004). Low force decelerates L-selectin dissociation from P-selectin glycoprotein ligand-1 and endoglycan. *J Biol Chem, 279,* 2291–2298.

85. Yago, T., Wu, J., Wey, C. D., et al. (2004). Catch bonds govern adhesion through L-selectin at threshold shear. *J Cell Biol, 166,* 913–923.

86. Yago, T., Leppänen, A., Qiu, H., et al. (2002). Distinct molecular and cellular contributions to stabilizing selectin-mediated rolling under flow. *J Cell Biol, 158,* 787–799.

87. Von Andrian, U. H., Hasslen, S. R., Nelson, R. D., Erlandsen, S. L., & Butcher, E. C. (1995). A central role for microvillous receptor presentation in leukocyte adhesion under flow. *Cell, 82,* 989–999.

88. Bruehl, R. E., Moore, K. L., Lorant, D. E., et al. (1997). Leukocyte activation induces surface redistribution of P-selectin glycoprotein ligand-1. *J Leukoc Biol, 61,* 489–499.

89. Patel, K. D., Nollert, M. U., McEver, R. P. (1995). P-selectin must extend a sufficient length from the plasma membrane to mediate rolling of neutrophils. *J Cell Biol, 131,* 1893–1902.

90. Huang, J., Chen, J., Chesla, S. E., et al. (2004). Quantifying the effects of molecular orientation and length on two-dimensional receptor-ligand binding kinetics. *J Biol Chem, 279,* 44915–22923.

91. Shao, J. Y., Ting-Beall, H. P., & Hochmuth, R. M. (1998). Static and dynamic lengths of neutrophil microvilli. *Proc Natl Acad Sci USA, 95,* 6797–6802.

92. Schmidtke, D. W., & Diamond, S. L. (2000). Direct observation of membrane tethers formed during neutrophil attachment to platelets or P-selectin under physiological flow. *J Cell Biol, 149,* 719–729.

93. Ramachandran, V., Williams, M., Yago, T., et al. (2004). Dynamic alterations of membrane tethers stabilize leukocyte rolling on P-selectin. *Proc Natl Acad Sci USA, 101,* 13519–13524.

94. Setiadi, H., Sedgewick, G., Erlandsen, S. L., & McEver, R. P. (1998). Interactions of the cytoplasmic domain of P-selectin with clathrin-coated pits enhance leukocyte adhesion under flow. *J Cell Biol, 142,* 859–871.

95. Setiadi, H., & McEver, R. P. (2003). Signal-dependent distribution of cell surface P-selectin in clathrin-coated pits affects leukocyte rolling under flow. *J Cell Biol, 163,* 1385–1395.

96. Epperson, T. K., Patel, K. D., McEver, R. P., & Cummings, R. D. (2000). Noncovalent association of P-selectin glycoprotein ligand-1 and minimal determinants for binding to P-selectin. *J Biol Chem, 275,* 7839–7853.

97. Snapp, K. R., Craig, R., Herron, M., et al. (1998). Dimerization of P-selectin glycoprotein ligand-1 (PSGL-1) required for optimal recognition of P-selectin. *J Cell Biol, 142,* 263–270.

98. Barkalow, F. J., Barkalow, K. L., & Mayadas, T. N. (2000). Dimerization of P-selectin in platelets and endothelial cells. *Blood, 96,* 3070–3077.

99. Chen, S. Q., & Springer, T. A. (1999). An automatic braking system that stabilizes leukocyte rolling by an increase in selectin bond number with shear. *J Cell Biol, 144,* 185–200.

100. Stenberg, P. E., McEver, R. P., Shuman, M. A., Jacques, Y. V., & Bainton, D. F. (1985). A platelet alpha-granule membrane protein (GMP-140) is expressed on the plasma membrane after activation. *J Cell Biol, 101,* 880–886.

101. Berman, C. L., Yeo, E. L., Wencel-Drake, J. D., et al. (1986). A platelet alpha granule membrane protein that is associated with the plasma membrane after activation. *J Clin Invest, 78,* 130–137.

102. Larsen, E., Celi, A., Gilbertk, G. E., et al. (1989). PADGEM protein: A receptor that mediates the interaction of activated platelets with neutrophils and monocytes. *Cell, 59,* 305–312.

103. Hamburger, S. A., & McEver, R. P. (1990). GMP-140 mediates adhesion of stimulated platelets to neutrophils. *Blood, 75,* 550–554.

104. Buttrum, S. M., Hatton, R., & Nash, G. B. (1993). Selectin-mediated rolling of neutrophils on immobilized platelets. *Blood, 82,* 1165–1174.

105. Yeo, E. L., Sheppard, J.-A. I., & Feuerstein, I. A. (1994). Role of P-selectin and leukocyte activation in polymorphonuclear cell adhesion to surface adherent activated platelets under physiologic shear conditions (an injury vessel wall model). *Blood, 83*, 2498–2507.

106. Diacovo, T. G., Roth, S. J., Buccola, J. M., Bainton, D. F., & Springer, T. A. (1996). Neutrophil rolling, arrest, and transmigration across activated, surface-adherent platelets via sequential action of P-selectin and the β2-integrin CD11b/CD18. *Blood, 88*, 146–157.

107. Weber, C., & Springer, T. A. (1997). Neutrophil accumulation on activated, surface-adherent platelets in flow is mediated by interaction of Mac-1 with fibrinogen bound to αIIbβ3 and stimulated by platelet-activating factor. *J Clin Invest, 100*, 2085–2093.

108. McEver, R. P., Martin, M. N. (1984). A monoclonal antibody to a membrane glycoprotein binds only to activated platelets. *J Biol Chem, 259*, 9799–9804.

109. Hsu–Lin, S.-C., Berman, C. L., Furie, B. C., August, D., & Furie, B. (1984). A platelet membrane protein expressed during platelet activation and secretion. Studies using a monoclonal antibody specific for thrombin-activated platelets. *J Biol Chem, 259*, 9121–9126.

110. Hidari, K.I.-P.J., Weyrich, A. S., Zimmerman, G. A., & McEver, R. P. (1997). Engagement of P-selectin glycoprotein ligand-1 enhances tyrosine phosphorylation and activates mitogen-activated protein kinases in human neutrophils. *J Biol Chem, 272*, 28750–28756.

111. Blanks, J. E., Moll, T., Eytner, R., & Vestweber, D. (1998). Stimulation of P-selectin glycoprotein ligand-1 on mouse neutrophils activates β2-integrin mediated cell attachment to ICAM-1. *Eur J Immunol, 28*, 433–443.

112. Evangelista, V., Manarini, S., Sideri, R., et al. (1999). Platelet/polymorphonuclear leukocyte interaction: P-selectin triggers protein-tyrosine phosphorylation-dependent CD11b/CD18 adhesion: Role of PSGL-1 as a signaling molecule. *Blood, 93*, 876–885.

113. Yago, T., Tsukuda, M., & Minami, M. (1999). P-selectin binding promotes the adhesion of monocytes to VCAM-1 under flow conditions. *J Immunol, 163*, 367–373.

114. Ma, Y. Q., Plow, E. F., & Geng, J. G. (2004). P-selectin binding to P-selectin glycoprotein ligand-1 induces an intermediate state of αMβ2 activation and acts cooperatively with extracellular stimuli to support maximal adhesion of human neutrophils. *Blood, 104*, 2549–2556.

115. Weyrich, A. S., McIntyre, T. M., McEver, R. P., et al. (1995). Monocyte tethering by P-selectin regulates monocyte chemotactic protein-1 and tumor necrosis factor-α secretion. *J Clin Invest, 95*, 2297–2303.

116. Ostrovsky, L., King, A. J., Bond, S., et al. (1998). A juxtacrine mechanism for neutrophil adhesion on platelets involves platelet-activating factor and a selectin-dependent activation process. *Blood, 91*, 3028–3036.

117. Weyrich, A. S., Elstad, M. R., McEver, R. P., et al. (1996). Activated platelets signal chemokine synthesis by human monocytes. *J Clin Invest, 97*, 1525–1534.

118. Diacovo, T. G., DeFougerolles, A. R., Bainton, D. F., & Springer, T. A. (1994). A functional integrin ligand on the surface of platelets: Intercellular adhesion molecule-2. *J Clin Invest, 94*, 1243–1251.

119. Simon, D. I., Chen, Z. P., Xu, H., et al. (2000). Platelet glycoprotein Ibα is a counterreceptor for the leukocyte integrin Mac-1 (CD11b/CD18). *J Exp Med, 192*, 193–204.

120. Frenette, P. S., Johnson, R. C., Hynes, R. O., & Wagner, D. D. (1995). Platelets roll on stimulated endothelium in vivo: An interaction mediated by endothelial P-selectin. *Proc Natl Acad Sci USA, 92*, 7450–7454.

121. Frenette, P. S., Moyna, C., Hartwell, D. W., et al. (1998). Platelet–endothelial interactions in inflamed mesenteric venules. *Blood, 91*, 1318–1324.

122. Romo, G. M., Dong, J. F., Schade, A. J., et al. (1999). The glycoprotein Ib-IX-V complex is a platelet counterreceptor for P-selectin. *J Exp Med, 190*, 803–813.

123. Frenette, P. S., Denis, C. V., Weiss, L., et al. (2000). P-selectin glycoprotein ligand 1 (PSGL-1) is expressed on platelets and can mediate platelet–endothelial interactions in vivo. *J Exp Med, 191*, 1413–1422.

124. Laszik, Z., Jansen, P. J., Cummings, R. D., et al. (1996). P-selectin glycoprotein ligand-1 is broadly expressed in cells of myeloid, lymphoid, and dendritic lineage and in some nonhematopoietic cells. *Blood, 88*, 3010–3021.

125. André, P., Denis, C. V., Ware, J., et al. (2000). Platelets adhere to and translocate on von Willebrand factor presented by endothelium in simulated veins. *Blood, 96*, 3322–3328.

126. Padilla, A., Moake, J. L., Bernardo, A., et al. (2004). P-selectin anchors newly released ultralarge von Willebrand factor multimers to the endothelial cell surface. *Blood, 103*, 2150–2156.

127. Massberg, S., Enders, G., Leiderer, R., et al. (1998). Platelet–endothelial cell interactions during ischemia/reperfusion: The role of P-selectin. *Blood, 92*, 507–515.

128. Diacovo, T. G., Puri, K. D., Warnock, R. A., Springer, T. A., & Von Andrian, U. H. (1996). Platelet-mediated lymphocyte delivery to high endothelial venules. *Science, 273*, 252–255.

129. Rosen, S. D. (2004). Ligands for L-selectin: Homing, inflammation, and beyond. *Annu Rev Immunol, 22*, 129–156.

130. Russell, J., Cooper, D., Tailor, A., Stokes, K. Y., & Granger, D. N. (2003). Low venular shear rates promote leukocyte-dependent recruitment of adherent platelets. *Am J Physiol Gastrointest Liver Physiol, 284*, G123–G129.

131. Cooper, D., Russell, J., Chitman, K. D., Williams, M. C., Wolf, R. E., & Granger, D. N. (2004). Leukocyte dependence of platelet adhesion in postcapillary venules. *Am J Physiol Heart Circ Physiol, 286*, H1895–H1900.

132. Bombeli, T., Schwartz, B. R., & Harlan, J. M. (1998). Adhesion of activated platelets to endothelial cells: Evidence for a GPIIbIIIa-dependent bridging mechanism and novel roles for endothelial intercellular adhesion molecule 1 (ICAM-1), αvβ3 integrin, and GPIbα. *J Exp Med, 187*, 329–339.

133. Parmentier, S., McGregor, L., Catimel, B., Leung, L. L. K., & McGregor, J. L. (1991). Inhibition of platelet functions by a monoclonal antibody (LYP20) directed against a granule

membrane glycoprotein (GMP-140/PADGEM). *Blood, 77,* 1734–1739.

134. Merten, M., & Thiagarajan, P. (2000). P-selectin expression on platelets determines size and stability of platelet aggregates. *Circulation, 102,* 1931–1936.

135. Merten, M., Chow, T., Hellums, J. D., & Thiagarajan, P. (2000). A new role for P-selectin in shear-induced platelet aggregation. *Circulation, 102,* 2045–2050.

136. Smyth, S. S., Reis, E. D., Zhang, W., Fallon, J. T., Gordon, R. E., & Coller B. S. (2001). β3-integrin-deficient mice but not P-selectin-deficient mice develop intimal hyperplasia after vascular injury — Correlation with leukocyte recruitment to adherent platelets 1 hour after injury. *Circulation, 103,* 2501–2507.

137. Isenberg, W. M., McEver, R. P., Shuman, M. A., & Bainton, D. F. (1986). Topographic distribution of a granule membrane protein (GMP-140) that is expressed on the platelet surface after activation: An immunogold-surface replica study. *Blood Cells, 12,* 191–204.

138. Merten, M., & Thiagarajan, P. (2001). Role for sulfatides in platelet aggregation. *Circulation, 104,* 2955–2960.

139. Lehr, H.-A., Olofsson, A. M., Carew, T. E., *et al.* (1994). P-selectin mediates the interaction of circulating leukocytes with platelets and microvascular endothelium in response to oxidized lipoprotein in vivo. *Lab Invest, 71,* 380–386.

140. Theilmeier, G., Lenaerts, T., Remacle, C., *et al.* (1999). Circulating activated platelets assist THP-1 monocytoid/endothelial cell interaction under shear stress. *Blood, 94,* 2725–2734.

141. Kirton, C. M., & Nash, G. B. (2000). Activated platelets adherent to an intact endothelial cell monolayer bind flowing neutrophils and enable them to transfer to the endothelial surface. *J Lab Clin Med, 136,* 303–313.

142. Bernardo, A., Ball, C., Nolasco, L., Choi, H., Moake, J. L., & Dong, J. F. (2005). Platelets adhered to endothelial cell-bound ultra-large von Willebrand factor strings support leukocyte tethering and rolling under high shear stress. *J Thromb Haemost, 3,* 562–570.

143. Ludwig, R. J., Schultz, J. E., Boehncke, W. H., *et al.* (2004). Activated, not resting, platelets increase leukocyte rolling in murine skin utilizing a distinct set of adhesion molecules. *J Invest Dermatol, 122,* 830–836.

144. Merhi, Y., Provost, P., Chauvet, P., Théorêt, J. F., Phillips, M. L., & Latour, J. G. (1999). Selectin blockade reduces neutrophil interaction with platelets at the site of deep arterial injury by angioplasty in pigs. *Arterioscler Thromb Vasc Biol, 19,* 372–377.

145. Forlow, S. B., McEver, R. P., & Nollert, M. U. (2000). Leukocyte-leukocyte interactions mediated by platelet microparticles under flow. *Blood, 95,* 1317–1323.

146. André, P., Hartwell, D., Hrachovinová, I., Saffaripour, S., & Wagner, D. D. (2000). Pro-coagulant state resulting from high levels of soluble P-selectin in blood. *Proc Natl Acad Sci USA, 97,* 13835–13840.

147. Hrachovinova, I., Cambien, B., Hafezi–Moghadam, A., *et al.* (2003). Interaction of P-selectin and PSGL-1 generates microparticles that correct hemostasis in a mouse model of hemophilia A. *Nat Med, 9,* 1020–1025.

148. Falati, S., Liu, Q., Gross, P., *et al.* (2003). Accumulation of tissue factor into developing thrombi in vivo is dependent upon microparticle P-selectin glycoprotein ligand 1 and platelet P-selectin. *J Exp Med, 197,* 1585–1598.

149. Gross, P. L., Furie, B. C., Merrill-Skoloff, G., Chou, J., & Furie, B. (2005). Leukocyte versus microparticle-mediated tissue factor transfer during arteriolar thrombus development. *J Leukoc Biol,* ••.

150. Evangelista, V., Manarini, S., Rotondo, S., *et al.* (1996). Platelet/polymorphonuclear leukocyte interaction in dynamic conditions: Evidence of adhesion cascade and cross talk between P-selectin and the β2 integrin CD11b/CD18. *Blood, 88,* 4183–4194.

151. Zimmerman, G. A., McIntyre, T. M., & Prescott, S. M. (1996). Adhesion and signaling in vascular cell–cell interactions. *J Clin Invest, 98,* 1699–1702.

152. Yan, Z., Zhang, J., Holt, J. C., Stewart, G. J., Niewiarowski, S., & Poncz, M. (1994). Structural requirements of platelet chemokines for neutrophil activation. *Blood, 84,* 2329–2339.

153. Selak, M. A. (1992). Neutrophil elastase potentiates cathepsin G-induced platelet activation. *Thromb Haemost, 68,* 570–576.

154. LaRosa, C. A., Rohrer, M. J., Benoit, S. E., Rodino, L. J., Barnard, M. R., & Michelson, A. D. (1994). Human neutrophil cathepsin G is a potent platelet activator. *J Vasc Surg, 19,* 306–318.

155. Hawrylowicz, C. M., Howells, G. L., & Feldmann, M. (1991). Platelet-derived interleukin 1 induces human endothelial adhesion molecule expression and cytokine production. *J Exp Med, 174,* 785–790.

156. Henn, V., Slupsky, J. R., Gräfe, M., *et al.* (1998). CD40 ligand on activated platelets triggers an inflammatory reaction of endothelial cells. *Nature, 391,* 591–594.

157. Andre, P., Prasad, K. S., Denis, C. V., *et al.* (2002). CD40L stabilizes arterial thrombi by a β3 integrin-dependent mechanism. *Nat Med, 8,* 247–252.

158. Marcus, A. J. (1990). Eicosanoid interactions between platelets, endothelial cells, and neutrophils. *Methods Enzymol, 187,* 585–598.

159. Maugeri, N., Evangelista, V., Celardo, A., *et al.* (1994). Polymorphonuclear leukocyte-platelet interaction: Role of P-selectin in thromboxane B2 and leukotriene C4 cooperative synthesis. *Thromb Haemost, 72,* 450–456.

160. Barry, O. P., Pratico, D., Savani, R. C., & FitzGerald, G. A. (1998). Modulation of monocyte–endothelial cell interactions by platelet microparticles. *J Clin Invest, 102,* 136–144.

161. Dole, V. S., Bergmeier, W., Mitchell, H. A., Eichenberger, S. C., & Wagner, D. D. (2005). Activated platelets induce Weibel–Palade-body secretion and leukocyte rolling in vivo: Role of P-selectin. *Blood, 106,* 2334–2339.

162. Marcus, A. J., Safier, L. B., Broekman, M. J., *et al.* (1995). Thrombosis and inflammation as multicellular processes: Significance of cell-cell interactions. *Thromb Haemost, 74,* 213–217.

163. Loscalzo, J. (2001). Nitric oxide insufficiency, platelet activation, and arterial thrombosis. *Circ Res, 88,* 756–762.

164. Radomski, M. W., Palmer, R. M., & Moncada, S. (1987). The anti-aggregating properties of vasclar endothelium: Interactions between prostacyclin and nitric oxide. *Br J Pharmacol, 92*, 639–646.

165. Ware, J. A., & Heistad, D. D. (1993). Seminars in medicine of the Beth Israel Hospital, Boston. Platelet–endothelium interactions. *N Engl J Med, 328*, 628–635.

166. Marcus, A. J., Broekman, M. J., Drosopoulos, J. H., *et al.* (1997). The endothelial cell ecto-ADPase responsible for inhibition of platelet function is CD39. *J Clin Invest, 99*, 1351–1360.

167. Palabrica, T., Lobb, R., Furie, B. C., *et al.* (1992). Leukocyte accumulation promoting fibrin deposition is mediated in vivo by P-selectin on adherent platelets. *Nature, 359*, 848–851.

168. Subramaniam, M., Frenette, P. S., Saffaripour, S., *et al.* (1996). Defects in hemostasis in P-selectin-deficient mice. *Blood, 87*, 1238–1242.

169. Wakefield, T. W., Strieter, R. M., Downing, L. J., *et al.* (1996). P-selectin and TNF inhibition reduce venous thrombosis inflammation. *J Surg Res, 64*, 26–31.

170. Downing, L. J., Wakefield, T. W., Strieter, R. M., *et al.* (1997). Anti-P-selectin antibody decreases inflammation and thrombus formation in venous thrombosis. *J Vasc Surg, 25*, 816–827.

171. Eppihimer, M. J., & Schaub, R. G. (2000). P-selectin-dependent inhibition of thrombosis during venous stasis. *Arterioscler Thromb Vasc Biol, 20*, 2483–2488.

172. Myers, D., Wrobleski, S., Londy, F., *et al.* (2002). New and effective treatment of experimentally induced venous thrombosis with anti-inflammatory rPSGL-Ig. *Thromb Haemost, 87*, 374–382.

173. Toombs, C. F., DeGraaf, C. L., Martin, J. P., *et al.* (1995). Pretreatment with a blocking monoclonal antibody to P-selectin accelerates pharmacological thrombolysis in a primate model of arterial thrombosis. *J Pharmacol Exp Ther, 275*, 941–949.

174. Kumar, A., Villani, M. P., Patel, U. K., *et al.* (1999). Recombinant soluble form of PSGL-1 accelerates thrombolysis and prevents reocclusion in a porcine model. *Circulation, 99*, 1363–1369.

175. Ikeda, H., Ueyama, T., Murahara, T., *et al.* (1999). Adhesive interaction between P-selectin and sialyl Lewis x plays an important role in recurrent coronary arterial thrombosis in dogs. *Arterioscler Thromb Vasc Biol, 19*, 1083–1090.

176. Zoldhelyi, P., Beck, P. J., Bjercke, R. J., *et al.* (2000). Inhibition of coronary thrombosis and local inflammation by a noncarbohydrate selectin inhibitor. *Am J Physiol Heart Circ Physiol, 279*, H3065–H3075.

177. Giesen, P. L., Rauch, U., Bohrmann, B., *et al.* (1999). Blood-borne tissue factor: Another view of thrombosis. *Proc Natl Acad Sci USA, 96*, 2311–2315.

178. Rauch, U., Bonderman, D., Bohrmann, B., *et al.* (2000). Transfer of tissue factor from leukocytes to platelets is mediated by CD15 and tissue factor. *Blood, 96*, 170–175.

179. Del Conde, I., Shrimpton, C. N., Thiagarajan, P., & Lopez, J. A. (2005). Tissue-factor-bearing microvesicles arise from lipid rafts and fuse with activated platelets to initiate coagulation. *Blood, 106*, 1604–1611.

180. Celi, A., Pellegrini, G., Lorenzet, R., *et al.* (1994). P-selectin induces the expression of tissue factor on monocytes. *Proc Natl Acad Sci USA, 91*, 8767–8771.

181. Bach, R., & Rifkin, D. B. (1990). Expression of tissue factor procoagulant activity: Regulation by cytosolic calcium. *Proc Natl Acad Sci USA, 87*, 6995–6999.

182. Lindmark, E., Tenno, T., & Siegbahn, A. (2000). Role of platelet P-selectin and CD40 ligand in the induction of monocytic tissue factor expression. *Arterioscler Thromb Vasc Biol, 20*, 2322–2328.

183. Lorant, D. E., Patel, K. D., McIntyre, T. M., *et al.* (1991). Coexpression of GMP-140 and PAF by endothelium stimulated by histamine or thrombin: A juxtacrine system for adhesion and activation of neutrophils. *J Cell Biol, 115*, 223–234.

184. Hartwell, D. M., Mayadas, T. N., Berger, G., *et al.* (1998). Role of P-selectin cytoplasmic domain in granular targeting in vivo and in early inflammatory responses. *J Cell Biol, 143*, 1129–1141.

185. Ishiwata, N., Takio, K., Katayama, M., *et al.* (1994). Alternatively spliced isoform of P-selectin is present in vivo as a soluble molecule. *J Biol Chem, 269*, 23708–23715.

186. Michelson, A. D., Barnard, M. R., Hechtman, H. B., *et al.* (1996). In vivo tracking of platelets: Circulating degranulated platelets rapidly lose surface P-selectin but continue to circulate and function. *Proc Natl Acad Sci USA, 93*, 11877–11882.

187. Blann, A. D., Nadar, S. K., & Lip, G. Y. (2003). The adhesion molecule P-selectin and cardiovascular disease. *Eur Heart J, 24*, 2166–2179.

188. Bouchard, B. A., & Tracy, P. B. (2001). Platelets, leukocytes, and coagulation. *Curr Opin Hematol, 8*, 263–269.

189. VanWijk, M. J., VanBavel, E., Sturk, A., & Nieuwland, R. (2003). Microparticles in cardiovascular diseases. *Cardiovasc Res, 59*, 277–287.

190. Libby, P. (2002). Inflammation in atherosclerosis. *Nature, 420*, 868–874.

191. Dong, Z. M., Chapman, S. M., Brown, A. A., *et al.* (1998). Combined role of P- and E-selectins in atherosclerosis. *J Clin Invest, 102*, 145–152.

192. Dong, Z. M., Brown, A. A., Wagner, D. D. (2000). Prominent role of P-selectin in the development of advanced atherosclerosis in apoE-deficient mice. *Circulation, 101*, 2290–2295.

193. Collins, R. G., Velji, R., Guevara, N. V., Hicks, M. J., Chan, L., & Beaudet, A. L. (2000). P-selectin or intercellular adhesion molecule (ICAM)-1 deficiency substantially protects against atherosclerosis in apolipoprotein E-deficient mice. *J Exp Med, 191*, 189–194.

194. Burger, P. C., & Wagner, D. D. (2003). Platelet P-selectin facilitates atherosclerotic lesion development. *Blood, 101*, 2661–2666.

195. Huo, Y. Q., Schober, A., Forlow, S. B., *et al.* (2003). Circulating activated platelets exacerbate atherosclerosis in mice deficient in apolipoprotein E. *Nat Med, 9*, 61–67.

196. Manka, D., Collins, R., Ley, K., Beaudet, A. L., & Sarembock, I. J. (2000). Absence of P-selectin, but not ICAM-1, attenuates neointimal growth after arterial injury in apolipoprotein-E-deficient mice. *Circulation, 103*, 1000–1005.

197. Simon, D. I., Dhen, Z., Seifert, P., *et al.* (2000). Decreased neointimal formation in Mac-1(−/−) mice reveals a role for inflammation in vascular repair after angioplasty. *J Clin Invest, 105,* 293–300.

198. Wang, K., Zhou, X., Zhou, Z., *et al.* (2005). Platelet, not endothelial, P-selectin is required for neointimal formation after vascular injury. *Arterioscler Thromb Vasc Biol, 25,* 1584–1589.

199. Manka, D., Forlow, S. B., Sanders, J. M., *et al.* (2004). Critical role of platelet P-selectin in the response to arterial injury in apolipoprotein-E-deficient mice. *Arterioscler Thromb Vasc Biol, 24,* 1124–1129.

200. Phillips, J. W., Barringhaus, K. G., Sanders, J. M., *et al.* (2003). Single injection of P-selectin or P-selectin glycoprotein ligand-1 monoclonal antibody blocks neointima formation after arterial injury in apolipoprotein E-deficient mice. *Circulation, 107,* 2244–2249.

201. Schober, A., Manka, D., von Hundelshausen, P., *et al.* (2002). Deposition of platelet RANTES triggering monocyte recruitment requires P-selectin and is involved in neointima formation after arterial injury. *Circulation, 106,* 1523–1529.

202. Hattori, R., Hamilton, K. K., McEver, R. P., & Sims, P. J. (1989). Complement proteins C5b-9 induce secretion of high molecular weight multimers of endothelial von Willebrand factor and translocation of granule membrane protein GMP-140 to the cell surface. *J Biol Chem, 264,* 9053–9060.

203. Patel, K. D., Zimmerman, G. A., Prescott, S. M., McEver, R. P., & McIntyre, T. M. (1991). Oxygen radicals induce human endothelial cells to express GMP-140 and bind neutrophils. *J Cell Biol, 112,* 749–759.

204. Foreman, K. E., Vaporciyan, A. A., Bonish, B. K., *et al.* (1994). C5a-induced expression of P-selectin in endothelial cells. *J Clin Invest, 94,* 1147–1155.

205. Murohara, T., Parkinson, S. J., Waldman, S. A., & Lefer, A. M. (1995). Inhibition of nitric oxide biosynthesis promotes P-selectin expression in platelets — Role of protein kinase C. *Arterioscler Thromb Vasc Biol, 15,* 2068–2075.

206. Rainger, G. E., Fisher, A., Shearman, C., & Nash, G. B. (1995). Adhesion of flowing neutrophils to cultured endothelial cells after hypoxia and reoxygenation in vitro. *Am J Physiol Heart Circ Physiol, 269,* H1398–H1406.

207. Pinsky, D. J., Naka, Y., Liao, H., *et al.* (1996). Hypoxia-induced exocytosis of endothelial cell Weibel–Palade bodies. A mechanism for rapid neutrophil recruitment after cardiac preservation. *J Clin Invest, 97,* 493–500.

208. Winn, R. K., Liggitt, D., Vedder, N. B., Paulson, J. C., & Harlan, J. M. (1993). Anti-P-selectin monoclonal antibody attenuates reperfusion injury to the rabbit ear. *J Clin Invest, 92,* 2042–2047.

209. Weyrich, A. S., Ma, X., Lefer, D. J., Albertine, K. H., & Lefer, A. M. (1993). In vivo neutralization of P-selectin protects feline heart and endothelium in myocardial ischemia and reperfusion injury. *J Clin Invest, 91,* 2620–2629.

210. Davenpeck, K. L., Gauthier, T. W., Albertine, K. H., & Lefer, A. M. (1994). Role of P-selectin in microvascular leukocyte–endothelial interaction in splanchnic ischemia–reperfusion. *Am J Physiol Heart Circ Physiol, 267,* H622–H630.

211. Mulligan, M. S., Paulson, J. C., De Frees, S., *et al.* (1993). Protective effects of oligosaccharides in P-selectin-dependent lung injury. *Nature, 364,* 149–151.

212. Gauthier, T. W., Davenpeck, K. L., & Lefer, A. M. (1994). Nitric oxide attenuates leukocyte–endothelial interaction via P-selectin in splanchnic ischemia–reperfusion. *Am J Physiol Gastrointest Liver Physiol, 267,* G562–G568.

213. Lefer, D. J., Flynn, D. M., & Buda, A. J. (1996). Effects of a monoclonal antibody directed against P-selectin after myocardial ischemia and reperfusion. *Am J Physiol, 39,* H88–H98.

214. Singbartl, K., Green, S. A., & Ley, K. (2000). Blocking P-selectin protects from ischemia/reperfusion-induced acute renal failure. *FASEB J, 14,* 48–54.

215. Furman, M. I., Benoit, S. E., Barnard, M. R., *et al.* (1998). Increased platelet reactivity and circulating monocyte–platelet aggregates in patients with stable coronary artery disease. *J Am Coll Cardiol, 31,* 352–358.

216. Ott, I., Neumann, F. J., Gawaz, M., Schmitt, M., & Schomig, A. (1996). Increased neutrophil–platelet adhesion in patients with unstable angina. *Circulation, 94,* 1239–1246.

217. Neumann, F. J., Marx, N., Gawaz, M., *et al.* (1997). Induction of cytokine expression in leukocytes by binding of thrombin–stimulated platelets. *Circulation, 95,* 2387–2394.

218. Powell, C. C., Rohrer, M. J., Barnard, M. R., *et al.* (1999). Chronic venous insufficiency is associated with increased platelet and monocyte activation and aggregation. *J Vasc Surg, 30,* 844–851.

219. Rinder, C. S., Bonan, J. L., Rinder, H. M., *et al.* (1992). Cardiopulmonary bypass induces leukocyte–platelet adhesion. *Blood, 79,* 1201–1205.

220. Mickelson, J. K., Lakkis, N. M., Villarreal–Levy, G., Hughes, B. J., & Smith, C. W. (1996). Leukocyte activation with platelet adhesion after coronary angioplasty: A mechanism for recurrent disease? *J Am Coll Cardiol, 28,* 345–353.

# CHAPTER 13

# Inhibition of Platelet Function by the Endothelium

**Sybille Rex and Jane E. Freedman**

*Whitaker Cardiovascular Institute, Boston University School of Medicine,
Boston, Massachusetts*

## I. Introduction

Platelets circulate in a quiescent state as long as the endothelium remains biochemically and physically intact. Injury of a diseased or healthy vessel wall causes platelet adhesion initiated by the exposure of subendothelial components including von Willebrand factor (VWF), collagen, and fibronectin. Platelets bind to the exposed VWF via the glycoprotein (GP) Ib-IX-V receptor complex (see Chapter 7) that not only anchors the platelets to the injured vessel site but also leads to platelet activation via signaling events that involve the GPIIb-IIIa receptor (integrin αIIbβ3) (see Chapter 8). In addition, several collagen-binding proteins expressed on the platelet surface, such as GPVI, integrin α2β1, and CD36 (see Chapter 6), regulate collagen-induced platelet adhesion, specifically under flow conditions (see Chapter 18). These receptors interact in a synergistic way dependent on flow conditions, the extent of damage, and relative exposure of distinct types of collagen or extracellular matrix proteins. Subsequent platelet activation causes secretion of additional agonists such as adenosine diphosphate (ADP), thromboxane $A_2$ ($TXA_2$), and serotonin (see Chapter 15), which then leads to further platelet aggregation by recruitment of additional platelets to the growing thrombus. The formation of thrombin induces contraction and the formation of fibrin leads to the consolidation of the platelet plug.[1-3]

Arterial thrombosis in an atherosclerotic vessel is a pathological consequence of normal hemostasis.[4,5] Morphologically, thrombi and hemostatic plugs are similar but differing in their development. Arterial thrombi typically form at sites of pathologic vascular damage where atherosclerotic plaque has formed containing lipid-rich necrotic tissue and inflammatory cells.[6,7] These diseased lesions promote platelet adhesion, activation, secretion, recruitment, and consolidation to a greater extent as compared to an injured, but otherwise healthy blood vessel. Cell–cell and transcellular interactions play a major role in regulating and modulating the process of hemostasis and thrombus formation after vessel injury. A fine balance between prothrombotic and antithrombotic mediators and signaling molecules must be achieved to prevent an untoward pathophysiological outcome. During the adhesion, activation, and aggregation of platelets at an injured site, the endothelium responds by limiting the size and growth of the hemostatic plug or thrombus, or even reversing platelet reactivity. These responses are defined as endothelial thromboregulation[4,6,8] and require several forms of communication between endothelial cells and platelets. There are three primary (and functionally independent) pathways during the early stages of thromboregulation by which the endothelium controls platelet reactivity (Fig. 13-1 and Table 13-1):[4,6,8,9] (1) nitric oxide (NO); (2) the eicosanoid prostacyclin; and (3) the ecto-nucleotidase CD39. In this chapter, these three thromboregulatory mechanisms will be discussed in detail. In particular, this chapter will highlight the enzymes, receptors, and signaling pathways involved and discuss animal models, clinical investigations, and the role of these pathways in disease development, including relevant genetic variants.

## II. Nitric Oxide

### A. Biosynthesis of NO and Characteristics of Endothelial NO Synthase

NO, a well-characterized platelet inhibitor and vasodilator, was first discovered in endothelial cells by Furchgott and Zawadzki in 1980.[10] NO as a free radical of limited reactivity with an approximate diffusion rate of 50 μm/s in aqueous solution[11] and an *in vivo* half-life of about 10 seconds[12] diffuses readily across cellular compartments.[13] NO can exist in three interrelated redox forms which expand the biological spectrum of NO: the free radical NO, the nitrosonium anion ($NO^+$), and the nitrosyl anion ($NO^-$).[14,15] NO is inactivated by superoxide,[16] but stabilized by superoxide dismutase.[17,18] It can also bind to heme-containing proteins such as hemoglobin, to activate or inhibit their enzymatic activity.[15]

The biosynthesis of NO is carried out by a family of enzymes called nitric oxide synthases (NOS). The family

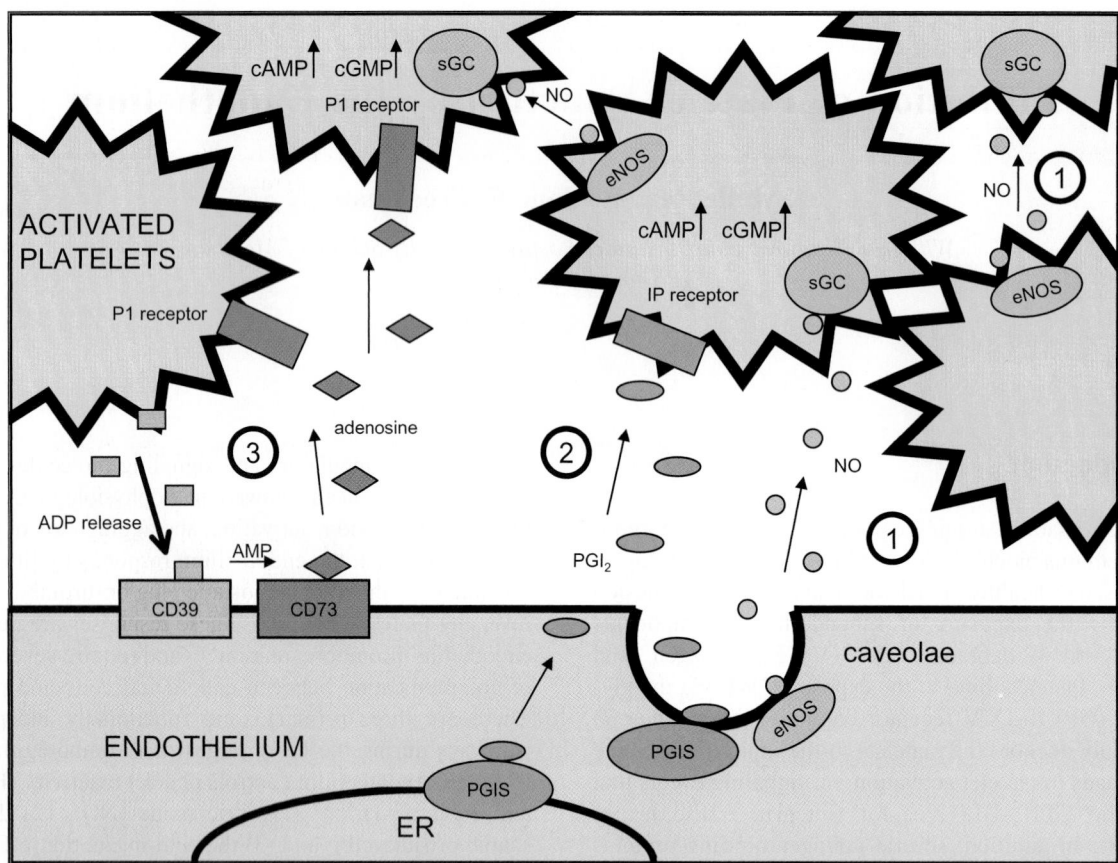

**Figure 13-1.** Overview of the three endothelial thromboregulatory mechanisms inhibiting platelet adhesion, activation, and aggregation: nitric oxide (NO), prostacyclin (PGI$_2$), and CD39. (1) Endothelial nitric oxide synthase (eNOS) located in caveolae of the plasma membrane releases NO which, upon diffusion across the platelet plasma membrane, stimulates soluble guanylyl cyclase (sGC) and leads to the upregulation of cyclic guanosine monophosphate (cGMP) production, thus resulting in inhibition of platelet adhesion to the endothelium. Activated platelets themselves produce NO to prevent an excessive recruitment of additional platelets. (2) Prostacyclin synthase (PGIS) located in the membrane of the endoplasmic reticulum (ER) and in caveolae of endothelial cells releases prostacyclin (PGI$_2$) into the bloodstream. After binding of prostacyclin to the prostacyclin receptor (IP receptor) on the platelet surface, G protein G$\alpha_s$ couples to adenylyl cyclase which subsequently results in an increase in cyclic adenosine monophosphate (cAMP) production that inhibits platelet activation and aggregation. (3) Adenosine diphosphate (ADP) released from activated platelets is hydrolyzed to adenosine monophosphate (AMP) by endothelial membrane-anchored CD39 to prevent uncontrolled platelet activation. In addition, AMP is converted into adenosine by endothelial membrane-anchored CD73. Adenosine binds to the platelet adenosine receptor (P1 receptor) causing an increase in cAMP production and thus further preventing platelet activation and aggregation.

consists of three members, which produce NO in either a constitutive or inducible manner: endothelial NOS (eNOS), inducible NOS (iNOS), and neuronal NOS (nNOS). Several recent reports also describe the existence of a distinct form of mitochondrial NOS (mtNOS).[19–22] eNOS is expressed in endothelial cells,[23,24] cardiomyocytes,[25] and bronchiolar epithelial cells.[26,27] Megakaryocytes and platelets also express eNOS in a constitutive manner.[28,29] The basal level of NO released from platelets has been reported to be similar to that from endothelial cells.[30]

eNOS catalyzes the multi-electron oxidation reaction of L-arginine with oxygen in a Ca$^{2+}$-dependent manner, forming L-citrulline and releasing NO.[31] Only active as homodimer, eNOS requires a number of essential cofactors: calmodulin/

Ca$^{2+}$, NADPH, tetrahydrobiopterin (BH$_4$), a Zn$^{2+}$-binding heme, flavin adenine dinucleotide (FAD), and flavin mononucleotide (FMN).[32–35] As a cytosolic enzyme of approximately 133 kDa, eNOS is composed of an N-terminal oxygenase domain and a C-terminal reductase domain. The oxygenase domain, which contains binding sites for heme, BH$_4$, and the substrate L-arginine, is linked by a calmodulin-recognition site to the reductase domain, which has binding sites for FAD, FMN, and NADPH.[36] The dimerization interface includes binding sites for BH$_4$ and the Zn$^{2+}$-binding heme,[37–39] of which the latter appears to be mandatory for dimer formation. An "N-terminal hook" domain,[39] as well as the presence of L-arginine, further stabilizes the dimerization process.[36]

**Table 13-1: Endothelial-Derived Mediators of Platelet-Dependent Thrombosis**

| Class of Thromboregulator | Nitrovasodilators | Eicosanoids | Ecto-nucleotidases |
|---|---|---|---|
| Type of Thromboregulator | NO | Prostacyclin | Enzymes CD39 and CD73, anchored to endothelial cell membrane |
| Site of Action | Released into blood-stream | Released into blood-stream | ADP/ATP removed from bloodstream adenosine released into bloodstream |
| Levels | Basal levels Increased upon stimulation | Induced upon stimulation | Basal levels |
| Endothelial Thromboregulating Enzyme | eNOS | Prostacyclin synthase | CD39, CD73 |
| Platelet Receptor for Endothelial Thromboregulator | NO diffusion, sGC | Prostacyclin receptor | Adenosine receptor (P1) |
| Thromboregulatory Mechanism to Inhibit Platelet Reactivity | Upregulation of platelet cGMP production | Upregulation of platelet cAMP production | Enzymatic removal of ADP secreted from activated platelets, adenosine-induced upregulation of platelet cAMP production |
| Platelet "Auto-Regulation" | Platelet eNOS (NO production), superoxide, ROS | $TXA_2$ synthesis (no $PGI_2$ synthesis) | ADP/ATP purinergic receptors (P2) CD39 (low expression) |
| Sensitivity to Aspirin | No | Yes | No |

ADP, adenosine diphosphate; ATP, adenosine triphosphate; cAMP, cyclic adenosine monophosphate; cGMP, cyclic guanosine 5′-monophosphate; eNOS, endothelial nitric oxide synthase; NO, nitric oxide; $PGI_2$, prostaglandin $I_2$; ROS, reactive oxygen species; sGC, soluble guanylyl cyclase; $TXA_2$, thromboxane $A_2$.

Intracellular localization, posttranslational modifications, and enzymatic activity of eNOS are tightly regulated. In the endoplasmic reticulum, eNOS becomes irreversibly myristoylated at the N-terminal residue Gly2, whereas reversible palmitoylation at residues Cys15 and Cys26 takes place in the Golgi apparatus. Both acylation processes are crucial for the optimal targeting of eNOS to the plasma membrane.[40–42] In the plasma membrane, eNOS is found to be highly enriched within caveolae,[43] which are specialized membrane invaginations rich in signaling molecules and coated with the protein caveolin.[44,45] The enzymatic activity is sevenfold higher in membrane-associated eNOS than in cytosolic eNOS, whereas in the plasma membrane its activity is about 10-fold greater in the caveolae fraction as compared to the whole plasma membrane.[45] It has been reported that eNOS binds to the scaffolding domain of caveolin, which has an inhibitory effect on eNOS activity *in vivo*.[46]

The activity of eNOS is regulated by several other mechanisms, such as changes in intracellular calcium concentration, phosphorylation and dephosphorylation at various tyrosine, serine, and threonine residues, as well as the association or dissociation of eNOS-interacting proteins (reviewed in [36,42,47]). For instance, mechanical stimuli such as shear flow in the blood vessel activate eNOS[48] by Akt-mediated phosphorylation at Ser1179[49,50] leading to a basal NO production and vasodilation.

Under conditions such as the absence of L-arginine or $BH_4$, or in the presence of the eNOS inhibitor $N^G$-monomethyl-L-arginine (L-NMMA), eNOS can undergo a process called eNOS uncoupling, whereby eNOS catalyzes an uncoupled NADPH oxidation leading to the formation of superoxide, instead of NO.[51–53] Superoxide can react very rapidly with NO, forming the powerful oxidant peroxynitrite[54] which can then readily modify protein residues and lipid moieties.[16,55]

### B. The Effect of Endothelial and Platelet NO on Platelet Reactivity

NO has both antithrombotic and vasodilatory effects. In the nondiseased blood vessel, the intact endothelium releases NO to inhibit platelet adhesion to the endothelium,[13,56] platelet activation, and platelet aggregation.[57–60] The vasodilatory effect of NO leads to smooth muscle relaxation.[10]

NO is not only important for inhibiting platelet aggregation and adhesion, but also for modulating platelet function itself. Platelet-derived NO is released both at rest[30] and during aggregation,[61–63] and has been linked to important autoregulatory functions in platelet reactivity. Shortly after activation, platelets release a large amount of NO (in the micromolar range) to prevent further platelet aggregation and adhesion to the growing thrombus.[63,64] By comparison, the NO released from platelets in the resting state is in the nanomolar range.[30]

NO inhibits platelet reactivity primarily by binding to the heme-containing enzyme soluble guanylyl cyclase (sGC), thus triggering a conformational change which increases the catalytic activity of sGC[65] and leads to a rapid increase in intracellular cyclic guanosine 5′-monophosphate (cGMP) levels formed from GTP.[66,67] This subsequently affects multiple signaling pathways[68,69] which include cGMP-dependent receptor proteins, such as cyclic nucleotide-gated cation channels, cGMP-regulated phosphodiesterases (PDE) that hydrolyze cyclic adenosine monophosphate (cAMP) and/or cGMP, and cGMP-dependent protein kinases.[70]

In the platelet, the increase in cGMP levels is accompanied by a decrease in intracellular $Ca^{2+}$ flux[71,72] caused by activation of cGMP-dependent protein kinase G,[15,73] which phosphorylates phospholamban and activates sarcoplasmic/ endoplasmic reticulum $Ca^{2+}$ ATPase (SERCA).[74,75] In addition, $Ca^{2+}$ levels are lowered by a decreased rate of $Ca^{2+}$ entry from the extracellular environment and inhibition of receptor-mediated $Ca^{2+}$ release from the dense tubular system.[76] The decrease in $Ca^{2+}$ levels inhibits the conformational change of GPIIb-IIIa into its active form[77] and thus decreases platelet association with fibrinogen.[78] The binding affinity of GPIIb-IIIa for fibrinogen is also decreased by cGMP-dependent inhibition of phosphoinositide 3-kinase activation[79] and by phosphorylation of the focal adhesion vasodilator-stimulated phosphoprotein (VASP) at Ser157 in a cGMP-dependent manner.[80] In addition, stimulation of cGMP-dependent protein kinases leads to the inhibition of phospholipase $A_2$- and C-mediated responses,[81] such as the inhibition of arachidonic acid release and inhibition of phospholipase C/G protein/receptor coupling.[73] Also, in a cGMP-dependent manner, the $TXA_2$ receptor becomes phosphorylated by cGMP-dependent protein kinase and is thus unable to mediate platelet activation.[82] Another cGMP-dependent mechanism preventing platelet adhesion to the endothelium is the downregulation of P-selectin expression by platelet NOS via an inhibitory effect on protein kinase C (PKC).[83]

Indirectly, cGMP can also increase intracellular cAMP levels by inhibiting the degradation of cAMP by phosphodiesterase III,[84,85] leading to protein kinase A (PKA) activation. It has been demonstrated that cGMP- and cAMP-elevating agents, such as prostacyclin (see Section III), have multiple synergistic interactions with respect to cyclic nucleotide generation/degradation, protein phosphorylation, and inhibition of platelet aggregation.[86,87]

NO-mediated platelet inhibition can also occur through cGMP-independent pathways such as modification of cellular or plasma proteins by S-nitrosylation of cysteine residues forming S-nitrosothiols.[69,88,89] The bioavailability of albumin thiols, or low-molecular-weight thiols, such as glutathione, was found to prolong the anti-platelet action of endothelial NO.[90,91] Transnitrosation reactions between protein thiol-bound NO and low-molecular-weight thiol-bound NO occur in vivo.[92,93] Extracellular NO in the nanomolar range is required for cGMP-independent inhibition of platelet activation, but plasma components may play an important role in the activation of cGMP-independent signaling by S-nitrosothiols or peroxynitrite generators.[94]

The outcome of NO-mediated effects are concentration dependent, such that NO synthesized in a nanomolar range is cytoprotective to maintain vascular hemostasis, whereas NO released in the micromolar range is considered to produce cytotoxic effects promoting vascular pathology.[15] A recent study showed that excessive NO production can inhibit $Ca^{2+}$ flux, prostacyclin production, and eNOS expression in endothelial cells.[95] Thus, the bioavailability of NO is regulated in at least two ways: reaction with various reactive oxygen species (ROS)[96,97] and desensitization of the NO/ cGMP system.[98]

As discussed above, under certain conditions eNOS can release the prothrombotic oxidant superoxide that can react rapidly with NO to form peroxynitrite.[54] NADPH oxidase has been identified as one primary source of superoxide production in platelets[97,99,100] and was found to play a role in platelet activation as a signaling mechanism.[100] Superoxide production by NADPH oxidase is decreased by NO-mediated PI3-kinase inhibition resulting in reversal of platelet aggregation.[101] This modulation of reactive oxygen and nitrogen species is a potential mechanism of regulating the bioavailability of NO at the site of platelet aggregation.[97] Upon activation, platelets generate superoxide via the glutathione cycle and lipoxygenase pathway[102] leading to a reduction in NO levels. Activation of the GPIIb-IIIa complex may contribute to lower NO levels as inhibition of GPIIb-IIIa increases platelet NO production,[103] decreases superoxide levels, and attenuates calcium flux.[101] Hydrogen peroxide and lipid hydroperoxides are additional forms of ROS, which affect the bioavailability of NO. The plasma enzyme gluthathione peroxidase protects both endothelial cells and platelets[104] from oxidative damage by reduction of ROS.[76]

## C. NO-Mediated Endothelial–Platelet Interactions and Thrombotic Disease

NO plays an important protective role in vascular hemostasis by suppressing thrombosis, atherosclerosis, and proliferation of vascular smooth muscle cells.[105] Diverse animal and human in vivo studies examining eNOS activity and NO production have reported the level of NO bioavailability and eNOS function during thrombotic disease states, as well as studied the effect of NO-donating or NO-enhancing substances in various pathological settings. Endothelial dysfunction or diminished NO bioavailability result in increased neutrophil adhesion to the endothelium and the initiation of atherosclerosis and thrombosis.[106–108] In atherosclerosis and hyperlipidemia, eNOS is dysfunctional and produces superoxide, which is implicated in endothelial dysfunction and impaired endothelium-dependent relaxations.[109]

Animal studies have helped elucidate the importance of NO in platelet-endothelial mediated homeostasis and vascular patency. Mice deficient in eNOS have decreased vascular reactivity and are hypertensive, but do not exhibit spontaneous thrombosis.[110] In eNOS-deficient mice, the lack of platelet-derived NO alters the *in vivo* hemostatic response by increasing platelet recruitment.[64] Conversely, eNOS deficiency is also associated with enhanced fibrinolysis due to lack of NO-dependent inhibition of the release of endothelial Weibel–Palade bodies,[111] which might partially explain the lack of enhanced thrombosis in eNOS-deficient mice.

Mice overexpressing the human eNOS gene were studied in a murine model of infarct-induced congestive heart failure.[112] In these mice both cardiac and pulmonary dysfunctions were attenuated and survival was drastically improved for the eNOS transgenic mice.[112] The genetic overexpression of eNOS in mice also attenuated myocardial infarction after ischemia/reperfusion injury but failed to significantly protect against postischemic myocardial contractile dysfunction.[113] Overexpression of eNOS was found to inhibit lesion formation in a mouse model of vascular remodeling[114] and when crossed with ApoE-deficient (hypercholesterolemic) mice atherosclerotic lesion formation was accelerated.[115]

In the human diseased blood vessel and during unstable coronary syndromes, release of both endothelial and platelet NO is impaired and thus contributes to thrombus formation. NO availability in the vascular system has been associated with various disease states and genetic variants. Risk factors for atherosclerosis, such as high cholesterol, male gender, family history, and age, have also been linked to this impaired endothelium-dependent vasodilation in coronary arteries.[116] NO deficiency has also been linked to an increasing number of cardiac and non-cardiac thrombotic disorders in humans. Activated platelets from patients with acute coronary syndromes produce significantly less NO as compared to patients with stable coronary artery disease.[117] Platelet activation increases in unstable coronary disease[118] and these platelets produce less NO compared to platelets from patients with stable coronary artery disease.[119] Furthermore, in human atherosclerotic coronary arteries, endothelial NO-dependent vessel dilation is impaired.[120] Endothelial dysfunction has also been associated with pulmonary hypertension leading to eventual pulmonary vascular hypertrophy and thrombosis.[121] Measurement of eNOS expression in the pulmonary vascular tissue of patients with pulmonary hypertension is decreased compared to healthy controls[122] and inhibition of eNOS leads to an increase in platelet deposition.[123] Impaired bioavailability of NO due to a deficiency in plasma glutathione peroxidase was found to be the cause of thrombotic childhood stroke.[124]

Pharmacological agents may have direct beneficial effects by improving the function of endothelial cells and platelets.[105,125] In a canine model of coronary artery stenosis, intravenous nitroglycerin infusion has been shown to inhibit platelet thrombus formation.[126] In a rat study, inhaled NO resulted in a lower incidence of collagen-induced platelet aggregation in small pulmonary vessels compared to controls.[127] The antithrombotic activity of NO was similar to the platelet GPIIb-IIIa antagonistic peptide G4120, suggesting that NO had a direct effect on inhibiting platelet activation.[127] Chronic treatment with L-arginine, a substrate for eNOS, inhibits formation of atherosclerotic lesions in animal models of atherosclerosis.[128,129]

Human *in vivo* studies have shown that modulating NO clinically affects platelet function. Oral supplementation of L-arginine in healthy adults inhibits ADP-dependent platelet aggregation when compared to placebo,[130] whereas inhibition of platelet eNOS enhances platelet aggregation and reduces bleeding times.[131] Furthermore, infusion of L-arginine reversed the vasoconstrictive effect of eNOS inhibitor L-NMMA in coronary arteries.[132]

The use of GPIIb-IIIa antagonists is known to reduce myocardial infarction and death in patients undergoing percutaneous coronary intervention and in patients with acute coronary syndromes (see Chapter 62).[133,134] Inhibition of GPIIb-IIIa in patients with known cardiovascular disease increases NO bioactivity.[135] NO donors have also been found to directly reverse GPIIb-IIIa activation of platelets,[136] although the therapeutic use of GPIIb-IIIa antagonists remains in flux (see Chapter 62).[137,138]

### D. Genetic Variants and Polymorphisms of Endothelial and Platelet eNOS

The study of genetic polymorphisms in the human eNOS gene[139] is important to a full understanding of the complex pathogenic process of atherosclerosis and arterial thrombosis. Numerous investigations have elucidated eNOS polymorphisms and linked them to various thrombotic diseases in association studies. The most common and thus extensively studied polymorphisms are the Glu298Asp polymorphism in exon 7, the T-786C polymorphism in the promoter region of the eNOS gene, and the intron 4b/a polymorphism. Table 13-2 provides an overview of these and other polymorphisms with regard to their association with thrombotic diseases.

Genetic variability due to heritable factors has a major influence on platelet aggregation[140] (see Chapter 14) and the

## Table 13-2: Polymorphisms of the eNOS Gene and Associated Disease States

| eNOS Polymorphism | Associated Disease/Population | Ref. |
|---|---|---|
| SNP T-786C in the promoter region | No association with ischemic heart disease (meta-analysis) | 148 |
| | No association with MI (France/Ireland) | 389 |
| | No association with MI (UK) | 390 |
| | No association with premature CAD (Australia) | 391 |
| | Association with CAD (Spain) | 392 |
| | Association with CAD (Italy) | 393 |
| | Association with carotid stenosis (Italy) | 394 |
| | Association with acute chest syndrome (African American) | 395 |
| | Association with hypertension (Canada) | 396 |
| Intron 4b/a | Association with ischemic heart disease (homozygous) (meta-analysis) | 148 |
| | No association with CAD (Germany) | 144 |
| | No association with premature CAD (Australia) | 391 |
| | No association with CAD/MI (Taiwan) | 397 |
| | Association with CAD/MI protection (Finland) | 145 |
| | Association with MI (Turkey) | 398 |
| | No association with hypertension (Australia) | 399 |
| SNP G894T in exon 7 (Glu298Asp) | Association with CAD + smoking (Australia) | 400 |
| | Association with ischemic heart disease (homozygous) (meta-analysis) | 148 |
| | No family-based association with ischemic heart disease (UK) | 401 |
| | No association with premature CAD (Australia) | 391 |
| | Association with premature CAD (China) | 402 |
| | Association with MI (Japan) | 403 |
| | Association with MI (homozygous) (UK) | 404 |
| | No association with MI (France/Ireland) | 389 |
| | No association with MI (UK) | 390 |
| | Association with carotid stenosis (Italy) | 405 |
| | Association with coronary spasm (Korea) | 406 |
| | No association with hypertension or systolic or diastolic blood pressure (Caucasian) | 407 |
| | Association with hypertension (Japan) | 408 |
| | No association with hypertension (Japan) | 409 |
| | No association with hypertension (France) | 410 |
| | No association with hypertension (Australia) | 399 |
| | Association with stroke (homozygous) (France) | 411 |
| Intron 13 CA Repeat | Association with CAD (Germany) | 412 |
| Polymorphism (VNTR) | No association with hypertension (France) | 413 |
| | Association with hypertension (Japan) | 414 |
| A-922G | No association with MI (UK) | 390 |
| G10-T intron in intron 23 | No association with hypertension (Japan) | 408 |
| | No association with hypertension (France) | 410 |

CAD, coronary artery disease; MI, myocardial infarction; SNP, single nucleotide polymorphism; VNTR, variable-number tandem repeat.

availability of plasma NO has been linked to several of these polymorphisms.[141,142]

Studies attempting to demonstrate an association of clinical coronary artery disease or acute myocardial infarction with eNOS polymorphisms are inconsistent and vary between ethnic groups (Table 13-2). Japanese patients homozygous for the intron 4 polymorphism were associated with an increased risk of myocardial infarction.[143] However, in a similar German population there was no increased risk of cardiovascular disease,[144] whereas a decreased risk of myocardial infarction was reported in Finnish men.[145] Part of this apparent confusion may lie in the various presentations of acute and chronic coronary artery disease, including coronary vasospasm, intra-arterial thrombosis, and chronic plaque formation or plaque rupture. There are also numerous difficulties associated with reporting genetic associations linked to complex outcomes (see Chapter 14);[146] for instance, many of the current trials involving eNOS polymorphisms are relatively small and likely underpowered, in addition to the possibly different distribution of eNOS polymorphisms among ethnic groups.[147]

In a recent meta-analysis, the effects of three of the most common eNOS genetic variations on the susceptibility to ischemic heart disease were examined:[148] Glu298Asp, T-786C, and intron 4. In this study,[148] only homozygosity for Asp298 and intron 4a alleles were associated with a 31% and 34%, respectively, increased risk of ischemic heart disease.

Besides conventional risk factors which may differ between ethnic groups,[149] other novel risk factors for atherosclerosis have to be taken into account, such as homocysteine, fibrinogen, impaired fibrinolysis, increased platelet reactivity, hypercoagulability, lipoprotein(a), small dense low-density lipoprotein cholesterol, and inflammatory-infectious markers.[149]

## III. Prostacyclin

### A. *Prostacyclin Biosynthesis and Characteristics of Endothelial Prostacyclin Synthase*

Prostaglandin $I_2$ ($PGI_2$) or prostacyclin is a derivative of the C-20 unsaturated fatty acid arachidonic acid (5,8,11,14-eicosatetraenoic acid)[150] which was discovered in 1976 as a substance inhibiting platelet secretion, platelet aggregation, and vasoconstriction.[151,152] With a half-life of about 3 min[153,154] prostacyclin is extremely labile under physiological conditions, but this is improved by binding to serum albumin to stabilize $PGI_2$ activity and enhance its receptor binding.[155] Prostacyclin's short lifetime has led to the development and experimental application of several prostacyclin mimetics.[156] The ones most commonly used are cicaprost,[157] iloprost,[158] and carbacyclin,[156,159,160] which are all analogs of $PGI_2$.[161] Cicaprost is often the analog of choice as iloprost is less selective and partially acts as an agonist for prostaglandin $E_1$ receptors.[156]

The biosynthesis of prostacyclin is part of the arachidonic acid metabolic pathway and takes place in a stepwise manner. Arachidonic acid is a member of the ω-6 series of essential fatty acids and is covalently bound at the *sn-2*-position in membrane phospholipids. It is released by phospholipase $A_2$ upon activation of the enzyme by an increase in intracellular $Ca^{2+}$ concentration. Arachidonic acid is further metabolized by at least two major enzyme complexes: (i) prostaglandin endoperoxide H (PGH) synthase, which is commonly known as cyclooxygenase (COX), and (ii) 5-lipoxygenase[162] (see Chapter 60). COX is a bifunctional enzyme with two related catalytic functions, a bis-oxygenase (cyclooxygenase) activity that catalyzes first the

formation of prostaglandin $G_2$ ($PGG_2$)[163] and a hydroperoxidase activity that subsequently catalyzes the two-electron reduction of $PGG_2$ to the unstable $PGH_2$.[164] $PGH_2$ is further converted by several specific synthases into prostanoids like $PGE_2$, $PGD_2$, $PGF_{2\alpha}$, prostacyclin, and $TXA_2$ (see Chapter 31) in a tissue-dependent manner.[162,165] Prostacyclin synthase (PGIS) in particular catalyzes the formation of prostacyclin from $PGH_2$. PGIS is a unique member of family 8 of the cytochrome P450 superfamily (CYP8A1).[166,167] This heme-binding enzyme exhibits features of the absorbance spectrum that are characteristic for cytochrome P450, but it has no mono-oxygenase activity. Instead, it catalyzes an isomerization reaction that does not require a reductase to initiate the catalytic activity.[168,169]

PGIS is widely distributed, predominantly in endothelial and smooth muscle cells.[170–175] The vascular endothelium constitutively expresses PGIS.[176] Platelets, however, lack the enzym PGIS, a membrane-bound enzyme of approximately 57 kDa, has one membrane anchor domain at the N-terminus[177] and a large cytoplasmic domain containing a substrate channel, heme-binding sites, and two Couet-motifs.[178,179] The enzyme is mainly localized in the membrane of the endoplasmic reticulum (ER)[180] whereas the heme-containing domain faces the cytoplasmic site of the ER.[178,181] The two isoforms of COX are also located in the ER membrane, though their active site is positioned at the ER lumenal side.[182] In addition, recent studies demonstrated the localization of PGIS in caveolae of human endothelial cell membranes, and its binding to caveolin-1–the primary coat protein of caveolae–does not reduce the enzymatic activity of PGIS.[183]

Comparison of the amino acid sequence of human PGIS with P450s shows that it has a significant sequence similarity to other P450s in the C-terminal region including the Cys-pocket near the helix L.[184] The invariant cysteine in this pocket at residue 441 (Cys441) is located in the heme-binding region and has been shown to constitute the fifth ligand of the heme iron.[185,186] Site-directed mutagenesis revealed that alanine and serine mutants of Cys441 resulted in a diminished enzyme activity (13%) without altering the expression level of PGIS.[184,187] A histidine mutant of Cys441 possessed no heme-binding ability or enzymatic activity.[188] Another study using molecular modeling-guided site-directed mutagenesis identified nine amino acids in the heme-binding region (Cys441, Leu112) and the substrate channel region (Ile67, Val76, Leu210, Pro355, Glu 360, Asp364, Leu384) that are important for catalytic activity but do not affect the expression level of PGIS.[187]

Previous studies suggested that PGIS is subject to multi-level regulation *in vivo*. One proposed mechanism is suicide inactivation by which PGIS undergoes a catalytic inactivation by its substrate $PGH_2$.[189] Another control mechanism has been described as selective inactivation of PGIS by tyrosine nitration at the active site due to exposure to micromolar peroxynitrite concentrations.[190] Tyrosine 430 (Tyr430) has been nitrated by peroxynitrite but substitution of Tyr430 by phenylalanine did not block the enzymatic activity, but rather sterically blocked access to the active site.[191] At the gene level, it has been previously established that variations in the PGIS gene sequence can affect the enzymatic activity (see Section III.E).

## B. The Platelet Prostacyclin Receptor — Biochemistry, Structure, and Function

Prostacyclin is a potent vasodilator, antithrombotic, and antiplatelet agent that mediates its effects through a specific membrane-bound receptor, the prostacyclin receptor (IP receptor). The receptor belongs to the prostanoid family of G protein-coupled membrane receptors (GPCR)[165] and is expressed on platelets[192,193] and smooth muscle cells.[194] IP receptor expression within the cardiovascular system is most abundant in the aorta.[195] It is also expressed in the atrium and ventricle of the heart indicating possible roles for prostacyclin in cardiac tissue,[196] whereas no IP expression is found in veins.[165]

The human prostacyclin receptor has a molecular weight of 41 to 83 kDa,[197,198] is located in the plasma membrane, and contains seven transmembrane domains with a short extracellular N-terminal region and an extended intracellular C-terminal region.[196] Prostaglandins have two structural features, a cyclopentane ring and side chains that are recognized by their receptor to stabilize ligand binding. In the IP receptor, specifically, transmembrane domains VI and VII play a role in distinct binding interactions with the PGI$_2$ side chains whereas transmembrane domains I and II confer broader binding functions, including recognition and interaction with the cyclopentane ring of prostacyclin.[150,199] The transmembrane domain I also contains a dimerization motif.[196]

The prostaglandins are not necessarily specific for an individual receptor. The binding pocket of the IP receptor can accommodate the cyclopentane rings of PGI$_2$, PGE$_1$, and PGE$_2$.[196] Prostacyclin analogs such as iloprost and cicaprost bind the receptor with the same affinity as prostacyclin. The rank order of agonist affinity for the cloned human prostacyclin receptor are: iloprost = cicaprost > PGE$_1$ > carbacyclin >> PGE$_2$ > PGD$_2$, PGF$_{2\alpha}$.[156,165,195,200] The cross-reactivity of IP ligands with several receptors for PGE$_2$ (EP) has been reported.[165,196] However, further elucidation of substrate specificity of the IP receptor and receptor subclassification as well as the evaluation of the role of endogenous prostacyclin has been hampered by the lack of a potent, selective IP receptor antagonist and highly selective agonists.[196,201,202]

The human prostacyclin receptor contains two potential N-glycosylation sites, one in the extracellular N-terminal domain at Asn7 and one in the first extracellular loop at Asn78,[196] and is expressed as a glycoprotein in human embryonic kidney (HEK 293) cells.[203] A functional study of the glycosylation sites in HEK and COS-7 cells revealed a greater degree of glycosylation at Asn78 than at Asn7, in addition to the finding that the extent of N-glycosylation might play a role in membrane localization, ligand binding, and signal transduction of the IP receptor.[204]

The C-terminus of the IP receptor contains a consensus sequence for isoprenylation, and is modified at Cys383 by a C-15 farnesyl isoprenoid chain.[205] This modification has no effect on the ligand binding characteristics, but is required for efficient coupling of the prostacyclin receptor to adenylyl cyclase as well as phospholipase C via the G proteins G$\alpha_s$ and G$\alpha_q$, respectively,[205] which are important effector proteins in the downstream signaling cascade of the prostacyclin receptor. In addition, it was established that the isoprenyl modification is necessary for efficient agonist-mediated receptor internalization[206] following receptor desensitization[207] and G protein uncoupling.[208] Additional studies revealed the palmitoylation of the intracellular C-terminal domain at residues Cys308 and Cys311[209] which, again, are not required for ligand binding, but play a crucial role in G$\alpha_s$/adenylyl cyclase coupling, whereby Cys308 is necessary for G$\alpha_q$ coupling and phospholipase C activation.[209] Several sites in the intracellular loops and the C-terminal domain of the IP receptor represent consensus sequences for phosphokinase A (PKA) and C (PKC) phosphorylation.[196,208] In particular, Ser328 was found to be a primary site for PKC phosphorylation, which is of critical importance for homologous regulation of the IP receptor.[208]

In addition to protein regulation via posttranslational modifications, the composition of the platelet membrane also plays a role in the expression of prostacyclin binding sites whereby an increase in cholesterol content lowers and an increase in phospholipids enhances the amount of bound prostacyclin.[210] Furthermore, the expression of the prostacyclin receptor was shown to be upregulated by proinflammatory and megakaryocytopoietic cytokines.[211]

A recent genetic analysis has revealed the first two naturally occurring polymorphisms within the coding sequence of the human prostacyclin receptor, Val25Met in the transmembrane domain I and Arg212His in the third intracellular loop.[212] Structure–function characterizations of these polymorphisms were performed at physiological pH (7.4) and an acidic pH (6.8) that would be encountered during stress such as renal, respiratory, or heart failure. In particular, the Arg212His polymorphism showed significant reduction in receptor binding affinity at low pH and in signal transduction activity at both pH values, whereas the Val-25Met polymorphism had no significant effects.[212] Lastly, a nuclear prostacyclin receptor has been identified which belongs to the peroxisome proliferator-activated receptor (PPAR) family (PPARδ)[213] and was suggested to be involved in angiogenesis.[179]

## C. Prostacyclin — Signaling Mechanisms and Platelet Inhibition

Prostacyclin is a very potent endogenous inhibitor of platelet aggregation, as well as a strong vasodilator that inhibits the growth of vascular smooth muscle cells.[151,156,170,172,214–218] It inhibits platelet activation induced by various stimulants, such as thrombin, collagen, ADP, TXA$_2$, or the calcium ionophore A23187[202] whereas it shows selectivity toward TXA$_2$-initiated platelet activation and its inhibitory potential decreases toward platelet aggregation in the following order: TXA$_2$ >> A23187 > thrombin > ADP.[219] Besides inhibiting platelet activation and limiting thrombus size,[170] prostacyclin and its mimetics can also prevent platelet[67] and leukocyte adhesion to endothelial cells.[220]

After release from the vessel wall, prostacyclin mediates its inhibitory effects via the high-affinity prostacyclin receptors on the platelet surface, thus causing an increase in cAMP levels and ultimately leading to inhibition of platelet activation and aggregation.[156,221,222] In particular, the binding of prostacyclin to its receptor induces a signaling cascade through coupling to the heterotrimeric G protein G$\alpha_s$[150,223] which then stimulates adenylate cyclase, assumed to be located in the dense tubular system,[224] and leads to an increase in cAMP levels.[221,222] Increased cAMP production activates phosphokinase A (PKA),[225] which causes the phosphorylation of several key proteins,[226,227] such as myosin light chain kinase (MLCK),[228] the platelet inositol 1,4,5-triphosphate receptor,[229] and vasodilator-stimulated phosphoprotein (VASP),[230] and then leads to the inhibition of several pathways: (i) Rho-kinase and MLCK activation and subsequent granule secretion, (ii) GPIIb-IIIa activation, and (iii) PKC activation and increase in intracellular calcium levels.[202,231]

Phosphorylated MLCK is inactive and has a reduced affinity for calmodulin which then reduces the amount of phosphorylated myosin.[232] The effect of this is a decreased platelet contractile activity, including secretion, and a decreased association of myosin with the platelet cytoskeleton, since only the phosphorylated form of myosin can bind to actin.[233] Through these mechanisms, prostacyclin is able to inhibit platelet activation and to cause disaggregation of existing platelet aggregates.[234]

Prostacyclin synthesis and its subsequent action is a dynamic process that is regulated on many levels. The binding of prostacyclin to its receptor and subsequent activation of adenylyl cyclase via G$\alpha_s$ is a concentration-dependent process. Also, it has become clear that, similar to other GPCR, the IP receptor may activate more than one G protein.[196] At higher ligand concentrations the IP receptor activates phospholipase C, most likely through G$\alpha_q$ coupling,[235] and causes calcium mobilization.[236] Furthermore, it has been proposed that the net effect on cAMP levels by prostacyclin, iloprost, and cicaprost depends on the ability of these agonists to stimulate as well as inactivate adenylyl cyclase.[237]

Prostacyclin receptor desensitization is another regulatory mechanism of PGI$_2$ responsiveness. It has been known for some time that elevated prostacyclin levels are accompanied by reduced binding to the IP receptor and by decreased responsiveness in human vascular disease, for example, myocardial infarction[238,239] and preeclampsia,[240] as well as during administration of prostacyclin and its analogs.[196,241] The desensitization process involves uncoupling of the receptor from the G protein, inhibition of adenylyl cyclase activity, and subsequent receptor internalization, which then results in a diminished response due to long-term exposure to agonists or high agonist concentrations.[207,208] Several studies have shown that desensitization and internalization of the human IP receptor takes place in platelets,[207,242,243] in cell cultures,[236,244] as well as *in vivo*.[245,246] PKC-mediated phosphorylation of the human IP receptor is a critical determinant of agonist-induced desensitization[208] whereas sequestration is independent of the PKC pathway and proceeds partially via an endocytotic pathway involving clathrin-coated vesicles.[236] In platelets, the response to desensitization is hyperresponsiveness and augmented platelet–endothelial cell adhesion in the presence of thrombin, which may result in an enhanced tendency toward thrombus formation.[238,243] A more recent study in platelets demonstrated that the short-term desensitization of the IP receptor is a reversible phenomenon whereby the receptors are not degraded but can be rapidly recycled in a functionally active form after agonist withdrawal.[247]

Another form of regulating IP receptor responsiveness is of heterologous nature and may involve receptor coexpression and colocalization, receptor cross-regulation through shared G proteins as well as proposed heterodimerization with other prostanoid receptors.[156,196,202,248,249] It has been shown that on platelets the IP receptor (a stimulatory receptor) is colocalized with the EP$_3$ receptor (an inhibitory receptor) to regulate cAMP production and to act as a buffering mechanism against a local increase in prostacyclin levels which otherwise would make the platelet refractory to activating signals.[250,251] This observation emphasizes the functional advantage of nonspecific agonists. It has been shown that low concentrations of prostacyclin stimulate the IP receptor and inhibit platelet activation, whereas high prostacyclin concentrations act on the EP$_3$ receptors to inhibit adenylate cyclase and restore the platelet's normal state.[202]

Heterodimerization of GPCR can substantially modify receptor function.[252,253] The formation of heterodimers between the prostacyclin receptor and the TXA$_2$ receptor $\alpha$ (TP$\alpha$) has been recently suggested.[249] Endogenously, both receptors are coexpressed, thus facilitating TP-mediated cAMP production that is independent of IP-mediated cAMP signaling. In the absence of the IP receptor, TP is largely inactive with regard to the activity of adenylyl cyclase.[249] These findings represent a new mechanism by which prostacyclin and its receptor may limit the prothrombotic effects of $TXA_2$.[249]

In general, eicosanoids are known to be important in preserving the dynamic equilibrium between thrombosis, hemostasis, and blood flow. In particular, maintenance of the balance between prostacyclin and $TXA_2$ appears to be a critical regulator for the maintenance of vessel integrity,[254] the interaction between platelets, and the endothelium[172,255] and cardiovascular processes.[202,256] Synthesized by platelets[257] and released after platelet activation,[258,259] $TXA_2$ is a platelet agonist and vasoconstrictor participating in a positive feedback loop that mediates additional platelet activation and aggregation (see Chapter 31). The actions of $PGI_2$ counteract the actions of $TXA_2$. Platelet activation and vasoconstriction induced by $TXA_2$ can be inhibited by prostacyclin.[156] Through increased platelet cAMP levels, prostacyclin blocks arachidonic acid formation[260] and thus prevents synthesis of prothrombotic prostaglandins and thromboxanes by the human platelet,[261,262] which further inhibits platelet activation.[263]

In addition, platelet-derived endoperoxides (like $TXA_2$) can act as substrate and thus upregulate prostacyclin synthesis in endothelial cells.[202,264,265] In general, there appear to be two mechanisms for prostacyclin synthesis by cultured endothelial cells: the first one involves prostacyclin synthesis from endogenous precursors, whereas the second one involves endoperoxides derived from stimulated platelets.[264] As discussed above, platelets release other antithrombotic substances, such as NO, that provide a negative feedback for thrombus formation. Indeed, low doses of NO and $PGI_2$ synergize in their antiaggregatory effect,[266,267] although not at the level of platelet adhesion to the endothelium.[267,268]

## D. The Relevance of Prostacyclin in Vivo

Prostacyclin as an antiplatelet and antithrombotic mediator plays an important role in the development of cardiovascular disease.[269] Dysfunctional prostacyclin activity has been implicated in the development of various cardiovascular diseases including thrombosis, myocardial infarction, stroke, myocardial ischemia, atherosclerosis, and hypertension.[165] Mice deficient in the PGIS gene are hypertensive and develop vascular disorders with vascular wall thickening and interstitial fibrosis, especially in the kidneys, demonstrating in vivo that prostacyclin is important in the hemostasis of blood vessels.[270] In vivo, inhibition or deficiency of PGIS causes a marked enhancement of white platelet thrombus formation[271] and plays a role in arterial thrombosis.[272,273] Several investigators have reported that abnormal prostacyclin synthesis or metabolism may be a risk factor for myo-

cardial and cerebral infarction.[274,275] Although prostacyclin reduced cerebral infarction in animal models,[276,277] many therapeutic trials of prostacyclin failed to show significant clinical improvement.[278–280] A decrease in prostacyclin production has been implicated in the pathogenesis of severe pulmonary hypertension[281] but not in human essential hypertension, although it has been reported to precede the clinical manifestation of pregnancy-induced hypertension.[282]

Because PGIS is upregulated by various cytokines, prostacyclin biosynthesis has been reported to be increased in the presence of atherosclerosis and platelet activation.[283] Therefore, it is possible that a subset of subjects with a less active promoter of the PGIS gene might be difficult to identify in the setting of established arteriosclerosis.[284] Prostacyclin-mediated protection of cardiomyocytes in vitro[285] and in vivo[201] has also been reported.

Delivery of PGIS in vivo can prevent proliferation and migration of smooth muscle cells, key features of restenosis and atherosclerosis.[286,287] The reduction of platelet prostacyclin binding observed in acute myocardial infarction might induce further intracoronary thrombotic events or a decrease in coronary blood flow due to vasospasm, particularly in combination with locally reduced prostacyclin synthesis in atheromatous plaques.[239]

Administration of prostacyclin mimetics, such as epoprostenol, in the clinical setting has been used to limit platelet aggregation, but a major drawback is the profound vascular dilatation they induce.[156] In pulmonary arterial hypertension, treatment with prostacyclin analogs, such as epoprostenol, treprostinil, iloprost, and beraprost, continues to play an important role despite the complications induced by their generally short half-lives and complicated drug delivery systems.[288]

Several investigators have generated mice deficient in the IP receptor and analyzed them in various disease settings.[201,289–293] Murata and coworkers found that IP receptor knockout mice display an increased thrombotic tendency and a reduction in inflammatory swelling and pain responses, but are not susceptible to hypertension.[289] Another group reported a protective effect of endogenously produced $PGI_2$ on cardiomyocytes during cardiac ischemia/reperfusion injuries which was lost after knockout of the IP receptor, an effect observed independently of its action on platelets and neutrophils.[201] Using mice that either overexpressed or lacked the IP receptor or the $TXA_2$ receptor,[291] eicosanoid-mediated interactions between platelets and endothelial cells were shown to play a role in thrombosis.[256] This study found that injury of the carotid artery leads to obstruction in the mice lacking the IP receptor and that this response was abolished in mice lacking the $TXA_2$.[291] In the IP null mice, the production of $TXA_2$ by platelets and components of the injured vessel wall had also increased.[291]

The tendency toward injury-induced thrombosis in IP$^{-/-}$ mice could be the result of either the observed fall in basal cAMP levels or a failure to respond to local accumulation of PGI$_2$.[292] The fact that thrombin is produced locally in response to vascular injury suggests that part of the response to injury is to generate an inhibitor of platelet activation (PGI$_2$) whose effects on platelet function must then be overcome to permit the formation of the platelet plug.[292] It was suggested that the increased tendency toward thrombosis in IP$^{-/-}$ mice after arterial injury is not caused by a generalized increase of agonist responsiveness in the circulating platelet population but rather by a loss of the regulatory effects of PGI$_2$, especially at the site of injury.[292] Furthermore, it has been reported that prostacyclin and its receptor do not work on a constitutive basis in regulating the systemic circulation, but rather work on demand in response to local stimuli.[165]

Clinical studies suggested that loss of the IP receptor may contribute to atherogenesis in patients with chronic spinal cord injury.[294] During spontaneous angina pectoris,[295] severe atherosclerosis,[296] and acute myocardial infarction,[297] the prostacyclin binding capacity and the number of IP receptors has been reported to decrease, but not in other patients with angiographically proven coronary artery disease and stable angina.[239,298] It has been shown that in myocardial infarction and unstable angina, biosynthesis of prostacyclin is considerably increased,[118,297] which could lead to agonist-induced receptor changes such as desensitization of the PGI$_2$ binding sites.[298]

The actions of prostacyclin and TXA$_2$ on platelets and the vessel wall are opposite to each other and their lack of balance is considered to be critical for thrombus formation[152] and in various occlusive vascular diseases including coronary heart disease.[249,299] During cardiac ischemia/reperfusion, the synthesis of PGI$_2$ and TXA$_2$ is significantly increased.[300,301] It has been reported that PGI$_2$ and its analogs attenuate cardiac ischemia/reperfusion injury when administered exogenously *in vivo*,[302–305] possibly due to their inhibitory effect on platelets and neutrophils.

Aspirin (acetylsalicylic acid) (see Chapter 60) has an antithrombotic effect[306] based on its preferential action on blocking COX in platelets and the endothelium by acetylating COX's active center.[307] Aspirin has also been reported to increase NO production in neutrophils[308,309] and in the arterial wall.[310] There are two different cyclooxygenases, COX-1 and COX-2.[162,163,311] COX-1 is involved in TXA$_2$ synthesis in platelets whereas COX-2 is involved in the synthesis of prostacyclin in endothelial cells.[162,312] Aspirin at low doses acetylates COX-1 in platelets and thus irreversibly blocks TXA$_2$ synthesis for their lifetime in the circulation. At the same low doses, aspirin has little effect on the synthesis of PGI$_2$. Thus, the overall effect of low-dose aspirin is a reduced risk of thrombosis.[162] Interestingly, the combined inhibition of both COX isoforms, but not the selective

COX-2 blockade, retarded atherogenesis in low-density lipoprotein receptor knockout mice.[313]

### E. Polymorphisms of the PGIS and Prostacyclin Receptor Genes

The PGIS gene is a candidate gene for cardiovascular disease and abnormalities in the gene may lead to altered vasodilation and platelet aggregation. The genomic organization of the PGIS gene has been investigated by a number of researchers.[314–316] Several polymorphisms in the PGIS gene have been detected and various association studies have been performed, primarily in the Japanese population.[284,316–326] Correlations between some of these polymorphisms, including single nucleotide polymorphisms (SNP), variable number of tandem repeat polymorphism (VNTR) and splicing mutations, and an increased risk for essential hypertension, myocardial infarction, and cerebral infarction have been found.[316,319,323–325] Two additional studies found polymorphisms in the French Caucasian population and two in African ethnic groups[327,328] (see Table 13-3).

Other mutations have been identified and include a nonsense mutation in exon 2 at codon 26 (CGA/TGA) of the PGIS gene. This polymorphism causes a large part of the PGIS mRNA to not be translated and is assumed to decrease the enzymatic activity of PGIS.[322] A VNTR polymorphism in the 5'-upstream promoter region has also been found.[324] This region contains binding sites for the transcription factors Sp1 and AP-2. The alleles vary in size from three to seven repeats of nine base pairs. Among Japanese subjects an association was found between this polymorphism and cerebral infarction.[324] This repeat polymorphism was found to influence the promoter activity of the PGIS gene in human endothelial cells.[284] In a study with almost 5000 Japanese subjects, the 3 and 4 repeats of the VTNR was found to be associated with increased systolic hypertension and higher pulse pressure.[284,329]

Another mutation includes a splicing variant in intron 9 that alters the translational reading frame of exon 10 and creates a premature stop codon.[318] The result is a truncated protein read with mismatch codons beginning from codon 403 and a protein lacking the heme-binding region. This mutation was found to be infrequent.[318,319] The SNP, 1117A $\rightarrow$ C, which is a silent mutation, was identified in exon 8 and found to be associated with risk for myocardial infarction in a Japanese population.[325]

In the human prostacyclin receptor gene a recent genetic analysis revealed the first two naturally occurring polymorphisms, which result in the amino acid mutations Val25Met and Arg212His.[212] The latter polymorph exhibited a significant decrease in signaling activation by the IP receptor;[212] however, no association study or other population study has been performed.

**Table 13-3: Polymorphisms of the Prostacyclin Synthase (PGIS) Gene and Associated Disease States**

| PGIS Polymorphism and Gene Product Modification | Associated Disease/Population | Ref. |
|---|---|---|
| Nonsense mutation in exon 2 at codon 26 (CGA/TGA); decrease in enzyme activity likely | Possibly linked to hypertension and cerebral infarction (Japanese population) | 322 |
| VNTR polymorphism in 5′-upstream promoter region; 3 to 7 repeats of 9 bp, increase in transcriptional activity with number of repeats; regulates prostacyclin production | Associated with risk of cerebral infarction (Japanese population) | 324 |
| | Not associated with essential hypertension (Japanese population) | 326 |
| | Not associated with development of chronic thromboembolic pulmonary hypertension (Japanese population) | 321 |
| | Repeats of 3 and 4 appear to be risk factor for higher pulse pressure and systolic hypertension (Japanese population) | 284 |
| VNTR polymorphism in 5′-upstream promoter region with 4 to 6 repeats of 9 bp; repeats of 6 show higher enzyme activity | Caucasian population | 327 |
| 9 VTNR polymorphisms in 5′-upstream promoter region with 3 to 9 repeats; single bp change in the 9 bp unit | No functional analysis performed, differences found in ethnic populations (French Caucasian, Tunisian, Gabonese) | 328 |
| Splicing site mutation in intron 9; truncated protein read with mismatch-codons from codon 403; PGIS lacks heme-binding region | Possibly causes hypertension complicated with renal dysfunction (Japanese population) | 318, 319 |
| CA/TG dinucleotide repeat polymorphism in intron 6; repeat size between 10 and 14 | Not associated with essential hypertension (Japanese population) | 316, 317 |
| SNP T-192G in the 5′-flanking region | Not associated with essential hypertension (Japanese population) | 323 |
| SNP 1117A→C in exon 8, no change in amino acid in codon 373, silent mutation | Associated with risk of myocardial infarction (Japanese population) | 325 |
| | Associated with 4 other mutations (see below); 70% frequency in Caucasian population; no phenotype/ association study | 327 |
| 1117A→C and 768G→A in exon 6; both silent mutations | Rare in Caucasian population | 327 |
| 1117A→C and 113C→T in exon 2; P38L substitution | Rare in Caucasian population | 327 |
| 1117A→C and 354T→A in exon 3; S118R substitution | Rare in Caucasian population | 327 |
| 1117A→C and 1135C→A in exon 8; R379S substitution, close to substrate binding region in PGIS | Rare in Caucasian population | 327 |
| Haplotype analysis of three polymorphisms: T-192G, VNTR, 1117A→C | Not associated with essential hypertension (Japanese population) | 320 |

bp, base pairs; PGIS, prostacyclin synthase; SNP, single nucleotide polymorphism; VNTR, variable-number tandem repeat.


## IV. CD39 (NTPDase-1)

### A. Endothelial CD39 — Biochemistry, Structure, and Function

Ecto-nucleoside triphosphate diphosphohydrolases (ENTPD-ases) are a family of membrane proteins that are ubiquitously expressed in eukaryotic cells and play a pivotal role in mediating platelet–endothelial interactions. These enzymes hydrolyze nucleoside 5′-di- and triphosphates in a $Ca^{2+}$- or $Mg^{2+}$-dependent manner.[330,331] Endothelial CD39

(NTPDase1), the major vascular ENTPDase, is a membrane-anchored glycoprotein with ecto-apyrase activity[332,333] and a cell type-dependent molecular weight of 70 to 95 kDa.[334] As an integral component of the endothelial cell surface, it is activated by its extracellular substrates in the bloodstream, adenosine triphosphate (ATP) and ADP, which are rapidly hydrolyzed into AMP.[330,335] An ecto-5′-nucleotidase (CD73) further converts AMP into adenosine.[336,337]

CD39 enzyme expression is found in a wide variety of vascular cells. The enzyme is expressed on human umbilical vein endothelial cells (HUVECs),[338,339] natural killer cells,[340] a subset of T cells,[338] activated B cells,[341] Epstein–Barr virus-transformed B cells,[332] megakaryocytes, and platelets.[334] HUVECs were shown to have a markedly higher enzymatic CD39 expression and activity than monocytes, natural killer cells, megakaryocytes, and platelets,[334] thus confirming that endothelial CD39 is the dominant contributor to hydrolysis of ADP and ATP in the vascular system.

Sequence analysis of soluble apyrase from potato tubers revealed 25% amino acid identity and 40% amino acid homology with human CD39.[342] Structural analysis of CD39 showed that the protein is membrane-anchored at its N- and C-terminus, whereas each terminus is composed of one transmembrane domain and a short cytoplasmic tail.[341] The middle of the protein forms a large extracellular loop containing a more central hydrophobic region.[330,343] This large extracellular domain has four apyrase conserved regions (ACR) which were suggested to contain the sites of catalytic activities.[342,344] The four ACR are highly conserved throughout the plant and animal kingdom suggesting their importance in the biological function of CD39.[342,345] A fifth ACR (termed ACR-5) has been described in the C-terminal region of the extracellular domain.[346,347] The ACR-4 in CD39 was suggested to contain the putative γ-phosphate binding motif, to be highly homologous with the actin-HSP70-hexokinase superfamily,[342] and to be involved in ATP hydrolysis.[344] ACR-1 has been proposed to be the β-phosphate binding domain by analogy with the same superfamily[342] and to play a role in ADP hydrolysis.[344] The enzyme's ADPase (but not ATPase) activity depends on the presence of divalent cations with $Ca^{2+}$ preferred over $Mg^{2+}$.[345] Heterologous interactions between both transmembrane domains of CD39 cause the tetramerization of the enzyme in the plasma membrane.[348] The tetrameric structure has a greater enzymatic activity than the monomeric form.[348]

Human CD39 has six potential N-linked glycosylation sites.[330,341] The extent of glycosylation is different in endothelial cells, platelets, and leukocytes.[334] The enzymatic activity of CD39 was reported to remain essentially unaltered by deglycosylation[343] after the protein is properly folded and targeted to the membrane surface.[349] Complete N-glycosylation was found to be essential for targeting the enzyme to its surface location, and only surface-expressed CD39 is enzymatically active.[350] Recently it was shown that glycosylation of CD39 in the endoplasmic reticulum, but not in the Golgi, is essential for the surface expression of an enzymatically functional CD39.[349] The glycosylation process in the endoplasmic reticulum is required for proper folding and intracellular trafficking of the enzyme.[349] CD39 also has several sites that may be modified by ectoprotein kinases,[351,352] a few potential phosphorylation sites for intracellular protein kinase C,[351] as well as one N-terminal palmitoylation site.[353] Oxidative modifications and proteolytic cleavages may modulate and regulate the enzymatic activity of CD39.[354,355]

Caveolae are specialized membrane invaginations that play a role in signal transduction events. CD39 is preferentially localized in caveolae of HUVECs and COS-7 cells,[356] a process which is mediated by S-palmitoylation at the residue Cys13 of the intracytoplasmic N-terminal region.[353] The activity of CD39 is cholesterol-dependent, whereby depletion or sequestering of membrane cholesterol results in inhibition of the enzymatic activity.[357] The absence of caveolin-1 and the subsequent loss of caveolae formation do not affect the enzymatic activity or the targeting of CD39 to the membrane.[357] Large aggregates of endogenous CD39 were found to colocalize with the membrane-anchored CD73 which associates with lipid rafts.[357]

Several studies performed mutation analysis of CD39 *in vitro* to elucidate the functional significance of single amino acids and enzyme tetramerization.[343–345] The removal of either of the C- or N-terminal transmembrane regions does not alter the biochemical activity of CD39.[343] However, intact ACR-1, ACR-4, and ACR-5 are necessary for maintaining the enzymatic activity of CD39. Native and mutant forms of CD39 that lack the transmembrane regions undergo multimerization by the formation of intermolecular disulfide bonds. In addition, limited tryptic cleavage of the intact CD39 protein results in two noncovalently membrane-associated fragments (56 and 27 kDa) that substantially augment the enzymatic activity.[343] In intact CD39 tetramers in the rat brain, the substitution of histidine at residue 59, located in ACR-1, with glycine (His59Gly) or serine (His59Ser) abolishes more than 90% of the ATPase activity but less than 50% of the ADPase activity, converting the enzyme into an ADPase with relative ADP:ATP hydrolysis rates of 6 : 1 or 8 : 1, respectively.[344] In contrast, the same substitutions in tetramers lacking either one transmembrane domain, in monomers lacking both transmembrane domains, or in detergent-solubilized full-length monomers, have no effect on ATPase activity and increase ADPase activity about twofold, which results in equal ATPase and ADPase activities. A substitution of asparagine at residue 61 with arginine (Asn61Arg) has a much smaller effect on the ADPase : ATPase ratio. These results suggest that CD39 uses different ATPase and ADPase mechanisms in different quaternary structures. His59 in ACR-1 plays a central role specifically in ATP hydrolysis in intact CD39 tetramers.[344]

Engineering of a recombinant soluble form of human CD39 (solCD39) allowed further functional studies of the membrane protein. Although solCD39 lacks both transmembrane regions and cytoplasmic domains, the enzymatic and biological properties of solCD39 and the full-length CD39 were found to be identical.[358] Using alanine scanning site-directed mutagenesis in the ACR of solCD39, loss-of-function mutations have been found. The study revealed that glutamic acid at residue 174 (Glu174) and serine at residue 218 (Ser218) are essential for the enzymatic activity of CD39.[345] Furthermore, aspartates 54 and 213 are involved in calcium utilization of solCD39.[359] Two splicing variants of CD39, type I and II, have been reported in the placenta.[360] Both differ from CD39 in the N-terminal intracellular region, whereas the type II variant also lacks ACR-5 and the C-terminal transmembrane domain.[360] Recently, the type II variant of CD39 has been identified in HUVECs and leukocytes.[9]

## B. The Effect of Endothelial CD39 on Platelet Reactivity

Nucleotides serve as intracellular energy sources as well as extracellular signaling molecules, and they can be released upon cellular activation or injury and induce biological responses through specific receptors.[331] Some of them can have pathological consequences; for example, the excessive ADP released from platelets and injured tissue have prothrombotic effects.[4]

Endothelial CD39 does not act on platelets *per se;* instead it reacts with the ADP released from activated platelets.[339] Upon activation, platelets release ADP from their dense granules, which leads to further activation and aggregation of platelets. By hydrolyzing ADP to AMP, CD39 prevents further prothrombotic platelet activation and mitigates against excessive platelet recruitment[335] with consequent restoration of platelets to the resting state.[361] Stimulation of platelets with collagen or low levels of thrombin is inhibited by the presence of CD39 in COS-7 cells that have been transfected with a human CD39 plasmid.[333] solCD39 strongly inhibits aggregation of human platelets induced by collagen, ADP, arachidonate, and thrombin receptor agonist peptide (TRAP).[362,363]

By performing site-directed mutagenesis within the four ACR in solCD39, it was shown that Glu174 and Ser218 are essential for inhibition of platelet reactivity.[345] The alanine mutation at Glu174 caused complete loss of the enzymatic activity and platelet aggregation. The alanine mutation at Ser218 resulted in 91% reduction of ADPase activity and 88% reduction of ATPase activity, whereas it was able to reverse platelet aggregation to a lesser extent than solCD39.[345]

The platelet-released ADP is transiently hydrolyzed to AMP and further broken down to adenosine by endothelial CD73 (ecto-5'-nucleotidase).[333,336] Adenosine, an antithrombotic and anti-inflammatory mediator, is itself a strong platelet inhibitor.[8] It blocks ADP-induced platelet activation and aggregation[364] by elevating intracellular cyclic AMP levels and blocking calcium mobilization,[365] a mechanism mediated by the adenosine $A_{2A}$ receptor on platelets,[366] which belongs to the class of purinergic P1 receptors.[367] A study of transgenic CD73$^{-/-}$ mice supports these findings by demonstrating a significant reduction in the occlusion time of the carotid artery, as well as a reduction in the tail bleed-

ing time in the CD73$^{-/-}$ mice compared to wild-type mice.[368] Platelet cAMP levels, but not cGMP levels, were also found to be significantly lower in the transgenic mice.[368] Thus, endothelial CD39 and CD73 act in concert to prevent platelet activation and aggregation, and neither NO nor prostacyclin is involved in their inhibitory activity.[335,364]

ADP and ATP are also substrates for the purinergic type 2 receptors (P2) (see Chapter 10). Several P2 receptors have been localized on megakaryocytes and platelets,[369–371] as well as on endothelial cells[372,373] (see Chapter 10). ATP is a competitive antagonist of ADP for P2Y receptors on human platelets.[374] The situation is further complicated by purinergic receptor desensitization where adenine nucleotides could also limit ADP-induced platelet activation[374] and cause platelet dysfunction.[375] The platelet purinergic receptor P2Y$_1$ plays an integral part in ADP function, whereby P2Y$_1$-deficient mice exhibited prolonged bleeding times and decreased platelet aggregation.[376,377] CD39 appears to modulate the functional expression of P2Y receptors. The coexpression of CD39 and P2 receptors in endothelial cells as well as platelets has been confirmed.[334] A more recent study observed that P2Y$_1$ and CD39 colocalize in the caveolae of endothelial cells.[378]

ADP and ATP released from activated platelets also affect the endothelium of the vessel wall; through various G protein-coupled P2 receptors both molecules stimulate the release of NO and prostacyclin from endothelial cells[373,379] via a phospholipase C-mediated pathway that leads to increased intracellular $Ca^{2+}$ and subsequent activation of eNOS and phospholipase $A_2$. This negative feedback mechanism can further limit the extent of platelet aggregation.[373]

The strong thromboregulatory potential of CD39 has been confirmed by experiments in which prostacyclin and NO production in the endothelium was blocked by treatment with aspirin and hemoglobin, respectively. However, the inhibition of platelet activation and aggregation was maintained due to the presence of CD39 in the endothelial cell membrane.[335,339,365] Although CD39 is therefore able to act independently of NO and prostacyclin, the presence of these thromboregulatory molecules enhanced the activity of CD39.[337]

Interestingly, activation of endothelial cells by proinflammatory cytokines like tumor necrosis factor α (TNFα) or exposure to oxidative stress leads to loss of the antithrombotic potential of CD39[333,354,355] due to loss of enzymatic activity, whereas the cell surface expression of CD39 is unaltered after TNFα treatment or oxidative stress.[354] The effect of cholesterol on the enzymatic activity of CD39 is noted in platelet aggregation studies in which cholesterol depletion of CD39-containing membranes caused a significant delay and a substantial decrease in platelet disaggregation.[357] Although the function of endothelial CD39 has been related to the inhibition of platelet activation and aggregation by ADP/ATP hydrolysis,[333,335] the function of CD39 on platelets and megakaryocytes has not yet been established.

## C. CD39 — In Vivo *Studies*

The clinical relevance of CD39 has been primarily established in animal studies. Recombinant solCD39 is able to inhibit platelet aggregation due to its enzymatic activity,[358] and intravenous injection into mice showed that solCD39 remained active over an extended period of time with an elimination phase half-life of almost two days.[358] solCD39 has been considered as a potential therapeutic agent in preventing platelet activation and thus blocking recruitment of additional platelets to a growing thrombus. It has been suggested that the extent of occlusion and vessel wall damage during and immediately after cardiovascular and cerebrovascular events, such as myocardial infarction and stroke, might be largely prevented.[358]

Several animal studies have investigated the therapeutic potential of human solCD39. In sympathetic nerve endings from guinea pig hearts, neuronal ATP enhances norepinephrine exocytosis. Treatment with solCD39 significantly attenuates norepinephrine release,[380] suggesting that CD39 could provide cardioprotection by reducing ATP-mediated norepinephrine release. In a murine stroke model driven by an excessive platelet recruitment, solCD39 reduced the extent of stroke without increasing intracerebral hemorrhage.[380]

Following reperfusion injury[381] or after exposure to oxidative stress during vascular inflammation,[354] the enzymatic and antithrombotic properties of endothelial CD39 are rapidly lost. Locally administered soluble apyrase (a soluble enzyme with identical function to CD39) was found to be beneficial in inhibiting platelet reactivity within the graft vasculature of transplanted organs.[382] Thus, targeted expression of a stable and active CD39 could be an effective therapeutic means to intervene in vascular inflammation[343] and transplantation-associated diseases.[383] Administration of solCD39 either before or 3 hours after arterial occlusion improves the neurological score and reduces the infarct size in a rat stroke model.[384] SolCD39 may therefore have potential for the treatment of focal ischemic stroke in humans.[384]

Studies with CD39 knockout mice have investigated the lack of this gene on cardiovascular properties and disease development, as studied by several groups in different settings and disease states.[337,375,385–388] Although CD39-deficient (CD39$^{-/-}$) mice and wild-type mice displayed no differences in their phenotype, the CD39$^{-/-}$ mice had substantially prolonged bleeding times and prolonged platelet plug formation times with minimally perturbed coagulation parameters.[375] However, no substantial differences were found in platelet activation. The platelet count was found to be 20% lower in CD39$^{-/-}$ mice compared to wild-type mice, whereas CD39$^{-/-}$ platelets had no morphological abnormalities. The interaction between platelet and injured vasculature *in vivo* was considerably reduced.[375] In agreement with the prolonged bleeding time and platelet plug formation, purified platelets from CD39$^{-/-}$ mice did not aggregate upon stimulation with ADP, collagen, and low doses of thrombin. Platelet dysfunction was reversible and correlated with purinergic receptor P2Y$_1$ desensitization, suggesting a dual role for CD39 in hemostasis and thrombotic reactions.[375] Platelets from wild-type mice did not express CD39,[375] whereas human platelets were reported to express low levels of CD39.[334] The CD39$^{-/-}$ mice were also very sensitive to ischemia/reperfusion injuries as they were incapable of producing adenosine at the local endothelial vessel wall layer. However, treatment with apyrase reversed the effect and protected the mice.[375] Using endothelial cell cultures from the CD39$^{-/-}$ mice, it was shown that both ADPase and ATPase activities were diminished.[375] This suggests that other nucleotidases associated with endothelial cells contribute only minimally to the hydrolysis of extracellular adenine nucleotides. Furthermore, wild-type endothelial cells completely inhibit the response of platelets to ADP, whereas endothelial cells from CD39$^{-/-}$ mice were not able to adequately inhibit platelet aggregation after ADP stimulation.[375]

The transgenic expression of human CD39 (hCD39) in mice has been investigated.[337] Under normal conditions, these mice have no spontaneous bleeding tendencies; however, platelet aggregation is impaired, bleeding times are prolonged, and the mice are resistant to systemic thromboembolism.[337] Paradoxically, the bleeding tendency of CD39$^{-/-}$ mice was similar to that of hCD39 transgenic mice, but this effect was secondary to purinergic receptor desensitization and could be corrected by prior administration of solCD39.[375] A different CD39-null mouse exhibited normal bleeding times but had increased cerebral infarct volumes and reduced postischemic perfusion, whereas solCD39 restored postischemic cerebral perfusion and rescued the mice from cerebral injury, suggesting a protective function of CD39 in stroke.[385]

Other animal models have been studied using CD39$^{-/-}$ mice, including intestinal ischemia and reperfusion injuries.[386] In CD39-null or -deficient mice, 80% died due to ischemia/ reperfusion injury as compared to wild-type mice. Apyrase supplementation protected all wild-type mice from intestinal ischemia-related death, but did not fully protect CD39$^{-/-}$ mice. Adenosine treatment failed to improve the survival rate. Platelet adherence to postcapillary venules in wild-type mice was significantly decreased and vascular integrity was well preserved following apyrase treatment. The potential of CD39 to maintain vascular integrity suggests potential pharmacological benefit in mesenteric ischemic injury.[386] Balloon injury in rabbit arteries decreases native CD39 activity which, however, can be augmented by adenovirus-mediated gene transfer of CD39.[387] Platelet deposition on the injured arterial surface is modest and not different between vessels treated either with CD39 or control vector. Studies of ApoE-deficient mice showed that deletion of CD39 accelerates the development of atherosclerotic lesions, a process reversed by solCD39 administration.[388] Furthermore, supplementation of solCD39 slows atherosclerosis in CD39 wild-type/apoE-deficient mice.[388] In conclusion, the deletion of CD39 renders the mice sensitive to vascular injury. Supplementation of CD39 by somatic gene transfer or administration of soluble CD39 has been shown to be advantageous in models of transplantation, inflammation, and atherosclerosis.

### D. Polymorphisms of Endothelial CD39

To date, polymorphisms in the human CD39 gene have not been reported.

## V. Summary

During platelet adhesion and activation, the endothelium responds to limit the size and growth of the thrombus. This response, known as endothelial thromboregulation, requires several forms of communication between endothelial cells and platelets. The three primary and functionally independent pathways involve NO, prostacyclin, and the ecto-nucleotidase CD39. Taken together, the endothelium utilizes these multiple pathways to regulate and prevent uncontrolled platelet activation, aggregation, and adhesion to the vessel wall. In addition, the platelet itself possesses several protective and counterregulatory mechanisms to balance its own prothrombotic actions and to stimulate endothelial antithrombotic factors and processes. As each of these pathways is defined, it is becoming more apparent that the underlying thromboregulatory mechanisms and enzymatic systems involved are interconnected and more complex than previously assumed. Furthermore, the future discovery of new mechanisms related to the established ones or the discovery of additional enzymes and pathways involved in thromboregulation appears plausible. It is also clear that disturbances of these balanced systems contribute to the pathophysiology of vascular diseases. Thus, gaining a better understanding of the complexity of early thromboregulation will likely lead to improved understanding and treatment of clinical thrombosis.

## References

1. Bloom, A. L. (1990). Physiology of blood coagulation. *Haemostasis, 20*(Suppl 1), 14–29.

2. Luscher, E. F., & Weber, S. (1993). The formation of the haemostatic plug — a special case of platelet aggregation. An experiment and a survey of the literature. *Thromb Haemost, 70,* 234–237.

3. Clemetson, K. J. (1999). Primary haemostasis: Sticky fingers cement the relationship. *Curr Biol, 9,* R110–R112.

4. Marcus, A. J., & Safier, L. B. (1993). Thromboregulation: Multicellular modulation of platelet reactivity in hemostasis and thrombosis. *FASEB J, 7,* 516–522.

5. Harker, L. A. (1994). Platelets and vascular thrombosis. *N Engl J Med, 330,* 1006–1007.

6. Marcus, A. J., Safier, L. B., Kaminski, et al. (1995). Principles of thromboregulation: Control of platelet reactivity in vascular disease. *Adv Prostaglandin Thromboxane Leukot Res. 23,* 413–418.

7. Ross R. (1993). The pathogenesis of atherosclerosis: A perspective for the 1990s. *Nature, 362,* 801–809.

8. Marcus, A. J., Broekman, M. J., Drosopoulos, et al. (2003). Heterologous cell–cell interactions: Thromboregulation, cerebroprotection and cardioprotection by CD39 (NTPDase-1). *J Thromb Haemost, 1,* 2497–2509.

9. Marcus, A. J., Broekman, M. J., Drosopoulos, J. H., et al. (2005). Role of CD39 (NTPDase-1) in thromboregulation, cerebroprotection, and cardioprotection. *Semin Thromb Hemost, 31,* 234–246.

10. Furchgott, R. F., & Zawadzki, J. V. (1980). The obligatory role of endothelial cells in the relaxation of arterial smooth muscle by acetylcholine. *Nature, 288,* 373–376.

11. Pryor, W. A., Lemercier, J. N., Zhang, H., et al. (1997). The catalytic role of carbon dioxide in the decomposition of peroxynitrite. *Free Radic Biol Med, 23,* 331–338.

12. Moncada, S., Palmer, R. M., & Higgs, E. A. (1991). Nitric oxide: Physiology, pathophysiology, and pharmacology. *Pharmacol Rev, 43,* 109–142.

13. Ignarro, L. J. (1989). Biological actions and properties of endothelium-derived nitric oxide formed and released from artery and vein. *Circ Res, 65,* 1–21.

14. Stamler, J. S., Singel, D. J., & Loscalzo, J. (1992). Biochemistry of nitric oxide and its redox-activated forms. *Science, 258,* 1898–1902.

15. Walford, G., & Loscalzo, J. (2003). Nitric oxide in vascular biology. *J Thromb Haemost, 1,* 2112–2118.

16. Radi, R., Beckman, J. S., Bush, K. M., et al. (1991). Peroxynitrite oxidation of sulfhydryls. The cytotoxic potential of superoxide and nitric oxide. *J Biol Chem, 266,* 4244–4250.

17. Rubanyi, G., & Vanhoutte, P. (1986). Superoxide anions and hyperoxia inactivate endothelium-derived relaxing factor. *Am J Physiol, 250,* H822–H827.

18. Rubanyi, G. M., & Vanhoutte, P. M. (1986). Oxygen-derived free radicals, endothelium, and responsiveness of vascular smooth muscle. *Am J Physiol, 250,* H815–H821.

19. Lacza, Z., Snipes, J. A., Zhang, J., et al. (2003). Mitochondrial nitric oxide synthase is not eNOS, nNOS or iNOS. *Free Radic Biol Med, 35,* 1217–1228.

20. Carreras, M. C., Franco, M. C., Peralta, J. G., et al. (2004). Nitric oxide, complex I, and the modulation of mitochondrial reactive species in biology and disease. *Mol Aspects Med, 25,* 125–139.

21. Haynes, V., Elfering, S., Traaseth, N., et al. (2004). Mitochondrial nitric-oxide synthase: Enzyme expression, characterization, and regulation. *J Bioenerg Biomembr, 36,* 341–346.

22. Ghafourifar, P., & Cadenas, E. (2005). Mitochondrial nitric oxide synthase. *Trends Pharmacol Sci, 26,* 190–195.

23. Pollock, J. S., Forstermann, U., Mitchell, J. A., et al. (1991). Purification and characterization of particulate endothelium-derived relaxing factor synthase from cultured and native bovine aortic endothelial cells. *Proc Natl Acad Sci USA, 88,* 10480–10484.

24. Nathan, C. (1992). Nitric oxide as a secretory product of mammalian cells. *FASEB J, 6,* 3051–3064.

25. Michel, T., & Feron, O. (1997). Nitric oxide synthases: Which, where, how, and why? *J Clin Invest, 100,* 2146–2152.

26. Shaul, P. W., North, A. J., Wu, L. C., et al. (1994). Endothelial nitric oxide synthase is expressed in cultured human bronchiolar epithelium. *J Clin Invest, 94,* 2231–2236.

27. German, Z., Chambliss, K. L., Pace, M. C., et al. (2000). Molecular basis of cell-specific endothelial nitric-oxide synthase expression in airway epithelium. *J Biol Chem, 275,* 8183–8189.

28. Mehta, J. L., Chen, L. Y., Kone, B. C., et al. (1995). Identification of constitutive and inducible forms of nitric oxide synthase in human platelets. *J Lab Clin Med, 125,* 370–377.

29. Sase, K., & Michel, T. (1995). Expression of constitutive endothelial nitric oxide synthase in human blood platelets. *Life Sciences, 57,* 2049–2055.

30. Zhou, Q., Hellermann, G., & Solomonson, L. (1995). Nitric oxide release from resting human platelets. *Thrombosis Res, 77,* 87–96.

31. Marletta, M. A., Yoon, P. S., Iyengar, R., et al. (1988). Macrophage oxidation of L-arginine to nitrite and nitrate: Nitric oxide is an intermediate. *Biochemistry, 27,* 8706–8711.

32. Bredt, D. S., & Snyder, S. H. (1994). Nitric oxide: A physiologic messenger molecule. *Annu Rev Biochem, 63,* 175–195.

33. Knowles, R. G., & Moncada, S. (1994). Nitric oxide synthases in mammals. *Biochem J, 298(Pt 2),* 249–258.

34. Marletta, M. A. (1994). Nitric oxide synthase: Aspects concerning structure and catalysis. *Cell, 78,* 927–930.

35. Nathan, C., & Xie, Q. W. (1994). Nitric oxide synthases: roles, tolls, and controls. *Cell, 78,* 915–918.

36. Alderton, W. K., Cooper, C. E., & Knowles, R. G. (2001). Nitric oxide synthases: Structure, function and inhibition. *Biochem J, 357,* 593–615.

37. Raman, C. S., Li, H., Martasek, P., et al. (1998). Crystal structure of constitutive endothelial nitric oxide synthase: A paradigm for pterin function involving a novel metal center. *Cell, 95,* 939–950.

38. Fischmann, T. O., Hruza, A., Niu, X. D., et al. (1999). Structural characterization of nitric oxide synthase isoforms reveals striking active-site conservation. *Nature Struct Biol, 6,* 233–242.

39. Crane, B. R., Rosenfeld, R. J., Arvai, A. S., et al. (1999). N-Terminal domain swapping and metal ion binding in nitric oxide synthase dimerization. *Embo J, 18,* 6271–6281.

40. Lamas, S., Marsden, P. A., Li, G. K., et al. (1992). Endothelial nitric oxide synthase: Molecular cloning and characterization of a distinct constitutive enzyme isoform. *Proc Natl Acad Sci USA, 89,* 6348–6352.

41. Robinson, L. J., & Michel, T. (1995). Mutagenesis of palmitoylation sites in endothelial nitric oxide synthase identifies a novel motif for dual acylation and subcellular targeting. *Proc Natl Acad Sci USA, 92,* 11776–11780.

42. Shaul, P. W. (2002). Regulation of endothelial nitric oxide synthase: Location, location, location. *Annu Rev Physiol, 64,* 749–774.

43. Feron, O., Belhassen, L., Kobzik, L., et al. (1996). Endothelial nitric oxide synthase targeting to caveolae. Specific interactions with caveolin isoforms in cardiac myocytes and endothelial cells. *J Biol Chem, 271,* 22810–22814.

44. Conrad, P. A., Smart, E. J., Ying, Y. S., et al. (1995). Caveolin cycles between plasma membrane caveolae and the Golgi complex by microtubule-dependent and microtubule-independent steps. *J Cell Biol, 131,* 1421–1433.

45. Shaul, P. W., Smart, E. J., Robinson, L. J., et al. (1996). Acylation targets endothelial nitric-oxide synthase to plasmalemmal caveolae. *J Biol Chem, 271,* 6518–6522.

46. Bucci, M., Gratton, J. P., Rudic, R. D., et al. (2000). In vivo delivery of the caveolin-1 scaffolding domain inhibits nitric oxide synthesis and reduces inflammation. *Nature Med, 6,* 1362–1367.

47. Fleming, I., & Busse, R. (2003). Molecular mechanisms involved in the regulation of the endothelial nitric oxide synthase. *Am J Physiol Regul Integr Comp Physiol, 284,* R1–R12.

48. Rubanyi, G. M., Romero, J. C., & Vanhoutte, P. M. (1986). Flow-induced release of endothelium-derived relaxing factor. *Am J Physiol, 250,* H1145–H1149.

49. Fulton, D., Gratton, J. P., McCabe, T. J., et al. (1999). Regulation of endothelium-derived nitric oxide production by the protein kinase Akt. *Nature, 399,* 597–601.

50. Dimmeler, S., Fleming, I., Fisslthaler, B., et al. (1999). Activation of nitric oxide synthase in endothelial cells by Akt-dependent phosphorylation. *Nature, 399,* 601–605.

51. Papapetropoulos, A., Rudic, R. D., & Sessa, W. C. (1999). Molecular control of nitric oxide synthases in the cardiovascular system. *Cardiovasc Res, 43,* 509–520.

52. Andrew, P. J., & Mayer, B. (1999). Enzymatic function of nitric oxide synthases. *Cardiovasc Res, 43,* 521–531.

53. Govers, R., & Rabelink, T. J. (2001). Cellular regulation of endothelial nitric oxide synthase. *Am J Physiol Renal Physiol, 280,* F193–F206.

54. Beckman, J., Beckman, T., Chen, J., et al. (1990). Apparent hydroxyl radical production by peroxynitrite: Implications for endothelial injury from nitric oxide and superoxide. *Proc Natl Acad Sci USA, 87,* 1620–1624.

55. Radi, R., Beckman, J. S., Bush, K. M., et al. (1991). Peroxynitrite-induced membrane lipid peroxidation: the cytotoxic potential of superoxide and nitric oxide. *Arch Biochem Biophys, 288,* 481–487.

56. de Graaf, J. C., Banga, J. D., Moncada, S., et al. (1992). Nitric oxide functions as an inhibitor of platelet adhesion under flow conditions. *Circulation, 85,* 2284–2290.

57. Azuma, H., Ishikawa, M., & Sekizaki, S. (1986). Endothelium-dependent inhibition of platelet aggregation. *Br J Pharmacol, 88,* 411–415.

58. Broekman, M. J., Eiroa, A. M., & Marcus, A. J. (1991). Inhibition of human platelet reactivity by endothelium-derived relaxing factor from human umbilical vein endothelial cells in suspension: Blockade of aggregation and secretion by an aspirin-insensitive mechanism. *Blood, 78,* 1033–1040.

59. Stamler, J., Mendelsohn, M. E., Amarante, P., et al. (1989). *N*-Acetylcysteine potentiates platelet inhibition by endothelium-derived relaxing factor. *Circulation Res, 65,* 789–795.

60. Cooke, J. P., Stamler, J., Andon, N., et al. (1990). Flow stimulates endothelial cells to release a nitrovasodilator that is potentiated by reduced thiol. *Am J Physiol, 259,* H804–H812.

61. Malinski, T., Radomski, M. W., Taha, Z., et al. (1993). Direct electrochemical measurement of nitric oxide released from human platelets. *Biochem Biophys Res Commun, 194,* 960–965.

62. Radomski, M. W., Palmer, R. M. J., & Moncada, S. (1990). An L-arginine/nitric oxide pathway present in human platelets regulates aggregation. *Proc Natl Acad Sci USA, 87,* 5193–5197.

63. Freedman, J. E., Loscalzo, J., Barnard, M. R., et al. (1997). Nitric oxide released from activated platelets inhibits platelet recruitment. *J Clin Invest, 100,* 350–356.

64. Freedman, J., Sauter, R., Battinelli, B., et al. (1999). Deficient platelet-derived nitric oxide and enhanced hemostasis in mice lacking the NOS3 gene. *Circ Res, 84,* 1416–1421.

65. Bellamy, T. C., & Garthwaite, J. (2002). The receptor-like properties of nitric oxide-activated soluble guanylyl cyclase in intact cells. *Mol Cell Biochem, 230,* 165–176.

66. Murad, F., Mittal, C. K., Arnold, W. P., et al. (1978). Guanylate cyclase: Activation by azide, nitro compounds, nitric oxide, and hydroxyl radical and inhibition by hemoglobin and myoglobin. *Adv Cyclic Nucleotide Res, 9,* 145–158.

67. Warner, T. D. (1996). Influence of endothelial mediators on the vascular smooth muscle and circulating platelets and blood cells. *Int Angiol, 15,* 93–99.

68. Beavo, J. A., & Brunton, L. L. (2002). Cyclic nucleotide research — still expanding after half a century. *Nature Rev Mol Cell Biol, 3,* 710–718.

69. Hanafy, K. A., Krumenacker, J. S., & Murad, F. (2001). NO, nitrotyrosine, and cyclic GMP in signal transduction. *Med Sci Monit, 7,* 801–819.

70. Feil, R., Lohmann, S. M., de Jonge, H., et al. (2003). Cyclic GMP-dependent protein kinases and the cardiovascular system: insights from genetically modified mice. *Circ Res, 93,* 907–916.

71. Moro, M., Russel, R., Cellek, S., et al. (1996). cGMP mediates the vascular and platelet action of nitric oxide: Confirmation using an inhibitor of the soluble guanylyl cylase. *Proc Natl Acad Sci USA, 93,* 1480–1485.

72. Rao, G. H., Krishnamurthi, S., Raij, L., et al. (1990). Influence of nitric oxide on agonist-mediated calcium mobilization in platelets. *Biochem Med Metab Biol, 43,* 271–275.

73. Walter, U., Geiger, J., Haffner, C., et al. (1995). Platelet–vessel wall interactions, focal adhesions, and the mechanism of action of endothelial factors. *Agents Actions Suppl, 45,* 255–268.

74. Nguyen, B. L., Saitoh, M., & Ware, J. A. (1991). Interaction of nitric oxide and cGMP with signal transduction in activated platelets. *Am J Physiol, 261,* H1043–H1052.

75. Trepakova, E. S., Cohen, R. A., & Bolotina, V. M. (1999). Nitric oxide inhibits capacitative cation influx in human platelets by promoting sarcoplasmic/endoplasmic reticulum $Ca^{2+}$-ATPase-dependent refilling of $Ca^{2+}$ stores. *Circ Res, 84,* 201–209.

76. Jin, R. C., Voetsch, B., & Loscalzo, J. (2005). Endogenous mechanisms of inhibition of platelet function. *Microcirculation, 12,* 247–258.

77. Michelson, A. D., Benoit, S. E., Furman, M. I., et al. (1996). Effects of nitric oxide/endothelium-derived relaxing factor on platelet surface glycoproteins. *Am J Physiol, 270,* H1640–H1648.

78. Mendelsohn, M. E., O'Neill, S., George, D., et al. (1990). Inhibition of fibrinogen binding to human platelets by *S*-nitroso-*N*-acetylcysteine. *J Biol Chem, 265,* 19028–19034.

79. Pigazzi, A., Heydrick, S., Folli, F., et al. (1999). Nitric oxide inhibits thrombin receptor-activating peptide-induced phosphoinositide 3-kinase activity in human platelets. *J Biol Chem, 274,* 14368–14375.

80. Horstrup, K., Jablonka, B., Honig-Liedl, P., et al. (1994). Phosphorylation of focal adhesion vasodilator-stimulated phosphoprotein at Ser157 in intact human platelets correlates with fibrinogen receptor inhibition. *Eur J Biochem, 225,* 21–27.

81. Radomski, M., & Moncada, S. (1993). Regulation of vascular homeostasis by nitric oxide. *Thromb Haemost, 70,* 36–41.

82. Wang, G. R., Zhu, Y., Halushka, P. V., et al. (1998). Mechanism of platelet inhibition by nitric oxide: In vivo phosphorylation of thromboxane receptor by cyclic GMP-dependent protein kinase. *Proc Natl Acad Sci USA, 95,* 4888–4893.

83. Murohara, T., Parkinson, S., Waldman, S., et al. (1995). Inhibition of nitric oxide biosynthesis promotes P-selectin expression in platelets. *Arterioscl Thromb Vasc Biol, 15,* 2068–2075.

84. Maurice, D. H., & Haslam, R. J. (1990). Molecular basis of the synergistic inhibition of platelet function by nitrovasodilators and activators of adenylate cyclase: Inhibition of cyclic AMP breakdown by cyclic GMP. *Mol Pharmacol, 37,* 671–681.

85. Bowen, R., & Haslam, R. J. (1991). Effects of nitrovasodilators on platelet cyclic nucleotide levels in rabbit blood; role for cyclic AMP in synergistic inhibition of platelet function by SIN-1 and prostaglandin E1. *J Cardiovasc Pharmacol, 17,* 424–433.

86. Stamler, J. S., Vaughan, D. E., & Loscalzo, J. (1989). Synergistic disaggregation of platelets by tissue-type plasminogen activator, prostaglandin E1, and nitroglycerin. *Circ Res, 65,* 796–804.

87. Schwarz, U. R., Walter, U., & Eigenthaler, M. (2001). Taming platelets with cyclic nucleotides. *Biochem Pharmacol, 62,* 1153–1161.

88. Stamler, J. S., Simon, D. I., Osborne, J. A., et al. (1992). S-Nitrosylation of proteins with nitric oxide: Synthesis and characterization of biologically active compounds. *Proc Natl Acad Sci USA, 89,* 444–448.

89. Stamler, J. S., Simon, D. I., Jaraki, O., et al. (1992). S-Nitrosylation of tissue-type plasminogen activator confers vasodilatory and antiplatelet properties on the enzyme. *Proc Natl Acad Sci USA, 89,* 8087–8091.

90. de Belder, A. J., MacAllister, R., Radomski, M. W., et al. (1994). Effects of S-nitroso-glutathione in the human forearm circulation: Evidence for selective inhibition of platelet activation. *Cardiovasc Res, 28,* 691–694.

91. Crane, M. S., Ollosson, R., Moore, K. P., et al. (2002). Novel role for low molecular weight plasma thiols in nitric oxide-mediated control of platelet function. *J Biol Chem, 277,* 46858–46863.

92. Scharfstein, J. S., Keaney, J. F., Jr., Slivka, A., et al. (1994). In vivo transfer of nitric oxide between a plasma protein-bound reservoir and low molecular weight thiols. *J Clin Invest, 94,* 1432–1439.

93. Liu, Z., Rudd, M. A., Freedman, J. E., et al. (1998). S-Transnitrosation reactions are involved in the metabolic fate and biological actions of nitric oxide. *J Pharmacol Exp Ther, 284,* 526–534.

94. Crane, M. S., Rossi, A. G., & Megson, I. L. (2005). A potential role for extracellular nitric oxide generation in cGMP-independent inhibition of human platelet aggregation: Biochemical and pharmacological considerations. *Br J Pharmacol, 144,* 849–859.

95. Takeuchi, K., Watanabe, H., Tran, Q. K., et al. (2004). Nitric oxide: Inhibitory effects on endothelial cell calcium signaling, prostaglandin I2 production and nitric oxide synthase expression. *Cardiovasc Res, 62,* 194–201.

96. Loscalzo, J. (2001). Nitric oxide insufficiency, platelet activation, and arterial thrombosis. *Circ Res, 88,* 756–762.

97. Krotz, F., Sohn, H. Y., & Pohl, U. (2004). Reactive oxygen species: Players in the platelet game. *Arterioscler Thromb Vasc Biol, 24,* 1988–1996.

98. Friebe, A., & Koesling, D. (2003). Regulation of nitric oxide-sensitive guanylyl cyclase. *Circ Res, 93,* 96–105.

99. Salvemini, D., Radziszewski, W., Mollace, V., et al. (1991). Diphenylene iodonium, an inhibitor of free radical formation, inhibits platelet aggregation. *Eur J Pharmacol, 199,* 15–18.

100. Chlopicki, S., Olszanecki, R., Janiszewski, M., et al. (2004). Functional role of NADPH oxidase in activation of platelets. *Antioxid Redox Signal, 6,* 691–698.

101. Clutton, P., Miermont, A., & Freedman, J. E. (2004). Regulation of endogenous reactive oxygen species in platelets can reverse aggregation. *Arterioscler Thromb Vasc Biol, 24,* 187–192.

102. Jahn, B., & Hansch, G. M. (1990). Oxygen radical generation in human platelets: Dependence on 12-lipoxygenase activity and on the glutathione cycle. *Int Arch Allergy Appl Immunol, 93,* 73–79.

103. Chakrabarti, S., Clutton, P., Varghese, S., et al. (2004). Glycoprotein IIb/IIIa inhibition enhances platelet nitric oxide release. *Thromb Res, 113,* 225–233.

104. Freedman, J. E., Frei, B., Welch, G. N., et al. (1995). Glutathione peroxidase potentiates the inhibition of platelet function by S-nitrosothiols. *J Clin Invest, 96,* 394–400.

105. Ignarro, L. J., & Napoli, C. (2004). Novel features of nitric oxide, endothelial nitric oxide synthase, and atherosclerosis. *Curr Atheroscler Rep, 6,* 281–287.

106. Cohen, R. A. (1995). The role of nitric oxide and other endothelium-derived vasoactive substances in vascular disease. *Prog Cardiovasc Dis, 38,* 105–128.

107. Harrison, D. G. (1994). Endothelial dysfunction in atherosclerosis. *Basic Res Cardiol, 89*(Suppl 1), 87–102.

108. Flavahan, N. A. (1992). Atherosclerosis or lipoprotein-induced endothelial dysfunction. Potential mechanisms underlying reduction in EDRF/nitric oxide activity. *Circulation, 85,* 1927–1938.

109. Kawashima, S., & Yokoyama, M. (2004). Dysfunction of endothelial nitric oxide synthase and atherosclerosis. *Arterioscler Thromb Vasc Biol, 24,* 998–1005.

110. Huang, P. L., Huang, Z., Mashimo, H., et al. (1995). Hypertension in mice lacking the gene for endothelial nitric oxide synthase. *Nature, 377,* 239–242.

111. Iafrati, M. D., Vitseva, O., Tanriverdi, K., et al. (2005). Compensatory mechanisms influence hemostasis in setting of eNOS deficiency. *Am J Physiol Heart Circ Physiol, 288,* H1627–H1632.

112. Jones, S. P., Greer, J. J., van Haperen, R., et al. (2003). Endothelial nitric oxide synthase overexpression attenuates congestive heart failure in mice. *Proc Natl Acad Sci USA, 100,* 4891–4896.

113. Jones, S. P., Greer, J. J., Kakkar, A. K., et al. (2004). Endothelial nitric oxide synthase overexpression attenuates myocardial reperfusion injury. *Am J Physiol Heart Circ Physiol, 286,* H276–H282.

114. Kawashima, S., Yamashita, T., Ozaki, M., et al. (2001). Endothelial NO synthase overexpression inhibits lesion formation in mouse model of vascular remodeling. *Arterioscler Thromb Vasc Biol, 21,* 201–207.

115. Ozaki, M., Kawashima, S., Yamashita, T., et al. (2002). Overexpression of endothelial nitric oxide synthase accelerates atherosclerotic lesion formation in apoE-deficient mice. *J Clin Invest, 110,* 331–340.

116. Vita, J. A., Treasure, C. B., Nabel, E. G., et al. (1990). Coronary vasomotor response to acetylcholine relates to risk factors for coronary artery disease. *Circulation, 81,* 491–497.

117. Freedman, J. E., Ting, B., Hankin, B., et al. (1998). Impaired platelet production of nitric oxide in patients with unstable angina. *Circulation, 98,* 1481–1486.

118. Fitzgerald, D. J., Roy, L., Catella, F., et al. (1986). Platelet activation in unstable coronary disease. *N Engl J Med, 315,* 983–989.

119. Freedman, J. E., Ting, B., Hankin, B., et al. (1998). Impaired platelet production of nitric oxide predicts presence of acute coronary syndromes. *Circulation, 98,* 1481–1486.

120. Bossaller, C., Habib, G. B., Yamamoto, H., et al. (1986). Impaired muscarinic endothelium-dependent relaxation and

cyclic guanosine 5′-monophosphate formation in atherosclerotic human coronary artery and rabbit aorta. *J Clin Invest, 79,* 170–174.

121. Cool, C. D., Stewart, J. S., Werahera, P., et al. (1999). Three-dimensional reconstruction of pulmonary arteries in plexiform pulmonary hypertension using cell-specific markers. Evidence for a dynamic and heterogeneous process of pulmonary endothelial cell growth. *Am J Pathol, 155,* 411–419.

122. Giaid, A., & Saleh, D. (1995). Reduced expression of endothelial nitric oxide synthase in the lungs of patients with pulmonary hypertension. *N Engl J Med, 333,* 214–221.

123. Stagliano, N., Zhao, W., Prado, R., et al. (1997). The effect of nitric oxide synthase inhibition on acute platelet accumulation and hemodynamic depression in a rat model of thromboembolic stroke. *J Cereb Blood Flow Metab, 17,* 1182–1190.

124. Freedman, J. E., Loscalzo, J., Benoit, S. E., et al. (1996). Decreased platelet inhibition by nitric oxide in two brothers with a history of arterial thrombosis. *J Clin Invest, 97,* 979–987.

125. Goumas, G., Tentolouris, C., Tousoulis, D., et al. (2001). Therapeutic modification of the L-arginine-eNOS pathway in cardiovascular diseases. *Atherosclerosis, 154,* 255–267.

126. Folts, J. D., Stamler, J., & Loscalzo, J. (1991). Intravenous nitroglycerin infusion inhibits cyclic blood flow responses caused by periodic platelet thrombus formation in stenosed canine coronary arteries. *Circulation, 83,* 2122–2127.

127. Nong, Z., Hoylaerts, M., Van Pelt, N., et al. (1997). Nitric oxide inhalation inhibits platelet aggregation and platelet-mediated pulmonary thrombosis in rats. *Circ Res, 81,* 865–869.

128. Candipan, R. C., Wang, B. Y., Buitrago, R., et al. (1996). Regression or progression. Dependency on vascular nitric oxide. *Arterioscler Thromb Vasc Biol, 16,* 44–50.

129. Aji, W., Ravalli, S., Szabolcs, M., et al. (1997). L-arginine prevents xanthoma development and inhibits atherosclerosis in LDL receptor knockout mice. *Circulation, 95,* 430–437.

130. Adams, M., Forsyth, C., Jessup, W., et al. (1995). Oral L-arginine inhibits platelet aggregation but does not enhance endothelium-dependent dilation in healthy young men. *J Am Coll Cardiol, 26,* 1054–1061.

131. Simon, D. I., Stamler, J. S., Loh, E., et al. (1995). Effect of nitric oxide synthase inhibition on bleeding time in humans. *J Cardiovasc Pharmacol, 26,* 339–342.

132. Quyyumi, A. A., Dakak, N., Andrews, N. P., et al. (1995). Nitric oxide activity in the human coronary circulation. Impact of risk factors for coronary atherosclerosis. *J Clin Invest, 95,* 1747–1755.

133. The CAPTURE Investigators (1997). Randomised placebo-controlled trial of abciximab before and during coronary intervention in refractory unstable angina: The CAPTURE Study. *Lancet, 349,* 1429–1435.

134. Kleiman, N. S., Lincoff, A. M., Ohman, E. M., & Harrington, R. A. (1998). Glycoprotein IIb/IIIa inhibitors in acute coronary syndromes: Pathophysiologic foundation and clinical findings. *Am Heart J, 136,* S32–S42.

135. Heitzer, T., Ollmann, I., Koke, K., et al. (2003). Platelet glycoprotein IIb/IIIa receptor blockade improves vascular nitric oxide bioavailability in patients with coronary artery disease. *Circulation, 108,* 536–541.

136. Keh, D., Gerlach, M., Kurer, I., et al. (1996). The effects of nitric oxide (NO) on platelet membrane receptor expression during activation with human alpha-thrombin. *Blood Coagul Fibrinolysis, 7,* 615–624.

137. Quinn, M. J., Byzova, T. V., Qin, J., et al. (2003). Integrin alphaIIbbeta3 and its antagonism. *Arterioscler Thromb Vasc Biol, 23,* 945–952.

138. Quinn, M. J., Plow, E. F., & Topol, E. J. (2002). Platelet glycoprotein IIb/IIIa inhibitors: recognition of a two-edged sword? *Circulation, 106,* 379–385.

139. Marsden, P. A., Heng, H. H., Scherer, S. W., et al. (1993). Structure and chromosomal localization of the human constitutive endothelial nitric oxide synthase gene. *J Biol Chem, 268,* 17478–17488.

140. O'Donnell, C. J., Larson, M. G., Feng, D., et al. (2001). Genetic and environmental contributions to platelet aggregation: the Framingham heart study. *Circulation, 103,* 3051–3056.

141. Tsukada, T., Yokoyama, K., Arai, T., et al. (1998). Evidence of association of the ecNOS gene polymorphism with plasma NO metabolite levels in humans. *Biochem Biophys Res Commun, 245,* 190–193.

142. Yoon, Y., Song, J., Hong, S. H., et al. (2000). Plasma nitric oxide concentrations and nitric oxide synthase gene polymorphisms in coronary artery disease. *Clin Chem, 46,* 1626–1630.

143. Ichihara, S., Yamada, Y., Fujimura, T., et al. (1998). Association of a polymorphism of the endothelial constitutive nitric oxide synthase gene with myocardial infarction in the Japanese population. *Am J Cardiol, 81,* 83–86.

144. Sigusch, H. H., Surber, R., Lehmann, M. H., et al. (2000). Lack of association between 27-bp repeat polymorphism in intron 4 of the endothelial nitric oxide synthase gene and the risk of coronary artery disease. *Scand J Clin Lab Invest, 60,* 229–235.

145. Kunnas, T. A., Ilveskoski, E., Niskakangas, T., et al. (2002). Association of the endothelial nitric oxide synthase gene polymorphism with risk of coronary artery disease and myocardial infarction in middle-aged men. *J Mol Med, 80,* 605–609.

146. Colhoun, H. M., McKeigue, P. M., & Davey Smith, G. (2003). Problems of reporting genetic associations with complex outcomes. *Lancet, 361,* 865–872.

147. Tanus-Santos, J. E., Desai, M., & Flockhart, D. A. (2001). Effects of ethnicity on the distribution of clinically relevant endothelial nitric oxide variants. *Pharmacogenetics, 11,* 719–725.

148. Casas, J. P., Bautista, L. E., Humphries, S. E., et al. (2004). Endothelial nitric oxide synthase genotype and ischemic heart disease: Meta-analysis of 26 studies involving 23028 subjects. *Circulation, 109,* 1359–1365.

149. Kullo, I. J., Gau, G. T., & Tajik, A. J. (2000). Novel risk factors for atherosclerosis. *Mayo Clin Proc, 75,* 369–380.

150. Kobayashi, T., Ushikubi, F., & Narumiya, S. (2000). Amino acid residues conferring ligand binding properties of prostaglandin I and prostaglandin D receptors. Identification by site-directed mutagenesis. *J Biol Chem, 275,* 24294–24303.

151. Moncada, S., Gryglewski, R., Bunting, S., et al. (1976). An enzyme isolated from arteries transforms prostaglandin endoperoxides to an unstable substance that inhibits platelet aggregation. *Nature, 263,* 663–665.

152. Gryglewski, R. J., Bunting, S., Moncada, S., et al. (1976). Arterial walls are protected against deposition of platelet thrombi by a substance (prostaglandin X) which they make from prostaglandin endoperoxides. *Prostaglandins, 12,* 685–713.

153. Cho, M. J., & Allen, M. A. (1978). Chemical stability of prostacyclin (PGI2) in aqueous solutions. *Prostaglandins, 15,* 943–954.

154. Dusting, G. J., Moncada, S., & Vane, J. R. (1978). Recirculation of prostacyclin (PGI2) in the dog. *Br J Pharmacol, 64,* 315–320.

155. Tsai, A. L., Hsu, M. J., Patsch, W., et al. (1991). Regulation of PGI2 activity by serum proteins: serum albumin but not high density lipoprotein is the PGI2 binding and stabilizing protein in human blood. *Biochim Biophys Acta, 1115,* 131–140.

156. Armstrong, R. A. (1996). Platelet prostanoid receptors. *Pharmacol Ther, 72,* 171–191.

157. Sturzebecher, C., Sr., Haberey, M., Muller, B., et al. (1985). Pharmacological profile of ZK 96480, a new chemically and metabolically stable prostacyclin analogue with oral availability and high PGI2 intrinsic activity. In Schror, K. (Ed.), *Prostaglandins and other eicosanoids in the cardiovascular system* (pp. 485–491). Basel: Karger.

158. Schror, K., Darius, H., Matzky, R., et al. (1981). The antiplatelet and cardiovascular actions of a new carbacyclin derivative (ZK 36 374) — equipotent to PGI2 in vitro. *Naunyn Schmiedebergs Arch Pharmacol, 316,* 252–255.

159. Whittle, B. J., Moncada, S., Whiting, F., et al. (1980). Carbacyclin — a potent stable prostacyclin analogue for the inhibition of platelet aggregation. *Prostaglandins, 19,* 605–627.

160. Whittle, B. J., & Moncada, S. (1985). Platelet actions of stable carbocyclic analogues of prostacyclin. *Circulation, 72,* 1219–1225.

161. Nickolson, R. C., Town, M. H., & Vorbruggen, H. (1985). Prostacyclin-analogs. *Med Res Rev, 5,* 1–53.

162. Parente, L., & Perretti, M. (2003). Advances in the pathophysiology of constitutive and inducible cyclooxygenases: Two enzymes in the spotlight. *Biochem Pharmacol, 65,* 153–159.

163. Marnett, L. J., Rowlinson, S. W., Goodwin, D. C., et al. (1999). Arachidonic acid oxygenation by COX-1 and COX-2. Mechanisms of catalysis and inhibition. *J Biol Chem, 274,* 22903–22906.

164. Smith, W. L., DeWitt, D. L., & Garavito, R. M. (2000). Cyclooxygenases: Structural, cellular, and molecular biology. *Annu Rev Biochem, 69,* 145–182.

165. Narumiya, S., Sugimoto, Y., & Ushikubi, F. (1999). Prostanoid receptors: Structures, properties, and functions. *Physiol Rev, 79,* 1193–1226.

166. Miyata, A., Hara, S., Yokoyama, C., et al. (1994). Molecular cloning and expression of human prostacyclin synthase. *Biochem Biophys Res Commun, 200,* 1728–1734.

167. Nelson, D. R., Koymans, L., Kamataki, T., et al. (1996). P450 superfamily: Update on new sequences, gene mapping, accession numbers and nomenclature. *Pharmacogenetics, 6,* 1–42.

168. Ullrich, V., Castle, L., & Weber, P. (1981). Spectral evidence for the cytochrome P450 nature of prostacyclin synthetase. *Biochem Pharmacol, 30,* 2033–2036.

169. Ullrich, V., & Hecker, M. (1990). A concept for the mechanism of prostacyclin and thromboxane A2 biosynthesis. *Adv Prostaglandin Thromboxane Leukot Res, 20,* 95–101.

170. Weksler, B. B., Marcus, A. J., & Jaffe, E. A. (1977). Synthesis of prostaglandin I2 (prostacyclin) by cultured human and bovine endothelial cells. *Proc Natl Acad Sci USA, 74,* 3922–3926.

171. Ingerman-Wojenski, C., Silver, M. J., Smith, J. B., et al. (1981). Bovine endothelial cells in culture produce thromboxane as well as prostacyclin. *J Clin Invest, 67,* 1292–1296.

172. Moncada, S. (1982). Eighth Gaddum Memorial Lecture. University of London Institute of Education, December 1980. Biological importance of prostacyclin. *Br J Pharmacol, 76,* 3–31.

173. Bunting, S., Moncada, S., & Vane, J. R. (1983). The prostacyclin–thromboxane A2 balance: pathophysiological and therapeutic implications. *Br Med Bull, 39,* 271–276.

174. Smith, W. L., DeWitt, D. L., & Allen, M. L. (1983). Bimodal distribution of the prostaglandin I2 synthase antigen in smooth muscle cells. *J Biol Chem, 258,* 5922–5926.

175. Oates, J. A., FitzGerald, G. A., Branch, R. A., et al. (1988). Clinical implications of prostaglandin and thromboxane A2 formation (1). *N Engl J Med, 319,* 689–698.

176. Spisni, E., Bartolini, G., Orlandi, M., et al. (1995). Prostacyclin (PGI2) synthase is a constitutively expressed enzyme in human endothelial cells. *Exp Cell Res, 219,* 507–513.

177. Lin, Y., Wu, K. K., & Ruan, K. H. (1998). Characterization of the secondary structure and membrane interaction of the putative membrane anchor domains of prostaglandin I2 synthase and cytochrome P450 2C1. *Arch Biochem Biophys, 352,* 78–84.

178. Ruan, K. H., So, S. P., Zheng, W., et al. (2002). Solution structure and topology of the N-terminal membrane anchor domain of a microsomal cytochrome P450: prostaglandin I2 synthase. *Biochem J, 368,* 721–728.

179. Massimino, M. L., Griffoni, C., Spisni, E., et al. (2002). Involvement of caveolae and caveolae-like domains in signalling, cell survival and angiogenesis. *Cell Signal, 14,* 93–98.

180. Smith, W. L. (1986). Prostaglandin biosynthesis and its compartmentation in vascular smooth muscle and endothelial cells. *Annu Rev Physiol, 48,* 251–262.

181. Deng, H., Huang, A., So, S. P., et al. (2002). Substrate access channel topology in membrane-bound prostacyclin synthase. *Biochem J, 362,* 545–551.

182. Ruan, K. H., Deng, H., Wu, J., et al. (2005). The N-terminal membrane anchor domain of the membrane-bound prostacyclin synthase involved in the substrate presentation of the coupling reaction with cyclooxygenase. *Arch Biochem Biophys, 435,* 372–381.

183. Spisni, E., Griffoni, C., Santi, S., et al. (2001). Colocalization prostacyclin (PGI2) synthase — caveolin-1 in endothelial cells and new roles for PGI2 in angiogenesis. *Exp Cell Res, 266,* 31–43.

184. Hatae, T., Hara, S., Yokoyama, C., et al. (1996). Site-directed mutagenesis of human prostacyclin synthase: Alteration of Cys441 of the Cys-pocket, and Glu347 and Arg350 of the EXXR motif. *FEBS Lett, 389,* 268–272.

185. Nelson, D. R., & Strobel, H. W. (1988). On the membrane topology of vertebrate cytochrome P-450 proteins. *J Biol Chem, 263,* 6038–6050.

186. Xia, Z., Shen, R. F., Baek, S. J., et al. (1993). Expression of two different forms of cDNA for thromboxane synthase in insect cells and site-directed mutagenesis of a critical cysteine residue. *Biochem J, 295(Pt 2),* 457–461.

187. Shyue, S. K., Ruan, K. H., Wang, L. H., et al. (1997). Prostacyclin synthase active sites. Identification by molecular modeling-guided site-directed mutagenesis. *J Biol Chem, 272,* 3657–3662.

188. Wada, M., Yokoyama, C., Hatae, T., et al. (2004). Purification and characterization of recombinant human prostacyclin synthase. *J Biochem (Tokyo), 135,* 455–463.

189. Wade, M. L., Voelkel, N. F., & Fitzpatrick, F. A. (1995). "Suicide" inactivation of prostaglandin I2 synthase: Characterization of mechanism-based inactivation with isolated enzyme and endothelial cells. *Arch Biochem Biophys, 321,* 453–458.

190. Zou, M., Martin, C., & Ullrich, V. (1997). Tyrosine nitration as a mechanism of selective inactivation of prostacyclin synthase by peroxynitrite. *Biol Chem, 378,* 707–713.

191. Schmidt, P., Youhnovski, N., Daiber, A., Balan, A., et al. (2003). Specific nitration at tyrosine 430 revealed by high resolution mass spectrometry as basis for redox regulation of bovine prostacyclin synthase. *J Biol Chem, 278,* 12813–12819.

192. Dutta-Roy, A. K., & Sinha, A. K. (1987). Purification and properties of prostaglandin E1/prostacyclin receptor of human blood platelets. *J Biol Chem, 262,* 12685–12691.

193. Tsai, A. L., Hsu, M. J., Vijjeswarapu, H., et al. (1989). Solubilization of prostacyclin membrane receptors from human platelets. *J Biol Chem, 264,* 61–67.

194. Jones, R. L., Qian, Y., Wong, H. N., et al. (1997). Prostanoid action on the human pulmonary vascular system. *Clin Exp Pharmacol Physiol, 24,* 969–972.

195. Nakagawa, O., Tanaka, I., Usui, T., et al. (1994). Molecular cloning of human prostacyclin receptor cDNA and its gene expression in the cardiovascular system. *Circulation, 90,* 1643–1647.

196. Smyth, E. M., & FitzGerald, G. A. (2002). Human prostacyclin receptor. *Vitam Horm, 65,* 149–165.

197. Ogawa, Y., Tanaka, I., Inoue, M., et al. (1995). Structural organization and chromosomal assignment of the human prostacyclin receptor gene. *Genomics, 27,* 142–148.

198. Leigh, P. J., Cramp, W. A., & MacDermot, J. (1984). Identification of the prostacyclin receptor by radiation inactivation. *J Biol Chem, 259,* 12431–12436.

199. Kobayashi, T., Kiriyama, M., Hirata, T., et al. (1997). Identification of domains conferring ligand binding specificity to the prostanoid receptor. Studies on chimeric prostacyclin/prostaglandin D receptors. *J Biol Chem, 272,* 15154–15160.

200. Boie, Y., Rushmore, T. H., Darmon-Goodwin, A., et al. (1994). Cloning and expression of a cDNA for the human prostanoid IP receptor. *J Biol Chem, 269,* 12173–12178.

201. Xiao, C. Y., Hara, A., Yuhki, K., et al. (2001). Roles of prostaglandin I(2) and thromboxane A(2) in cardiac ischemia-reperfusion injury: A study using mice lacking their respective receptors. *Circulation, 104,* 2210–2215.

202. Wise, H., Wong, Y. H., & Jones, R. L. (2002). Prostanoid signal integration and cross talk. *Neurosignals, 11,* 20–28.

203. Smyth, E. M., Nestor, P. V., & FitzGerald, G. A. (1996). Agonist-dependent phosphorylation of an epitope-tagged human prostacyclin receptor. *J Biol Chem, 271,* 33698–33704.

204. Zhang, Z., Austin, S. C., & Smyth, E. M. (2001). Glycosylation of the human prostacyclin receptor: Role in ligand binding and signal transduction. *Mol Pharmacol, 60,* 480–487.

205. Hayes, J. S., Lawler, O. A., Walsh, M. T., et al. (1999). The prostacyclin receptor is isoprenylated. Isoprenylation is required for efficient receptor-effector coupling. *J Biol Chem, 274,* 23707–23718.

206. Miggin, S. M., Lawler, O. A., & Kinsella, B. T. (2002). Investigation of a functional requirement for isoprenylation by the human prostacyclin receptor. *Eur J Biochem, 269,* 1714–1725.

207. Giovanazzi, S., Accomazzo, M. R., Letari, O., et al. (1997). Internalization and down-regulation of the prostacyclin receptor in human platelets. *Biochem J, 325(Pt 1),* 71–77.

208. Smyth, E. M., Li, W. H., & FitzGerald, G. A. (1998). Phosphorylation of the prostacyclin receptor during homologous desensitization. A critical role for protein kinase c. *J Biol Chem, 273,* 23258–23266.

209. Miggin, S. M., Lawler, O. A., & Kinsella, B. T. (2003). Palmitoylation of the human prostacyclin receptor. Functional implications of palmitoylation and isoprenylation. *J Biol Chem, 278,* 6947–6958.

210. Modesti, P. A., Abbate, R., Prisco, D., et al. (1990). Human prostacyclin platelet receptors and platelet lipid composition. *J Lipid Mediat, 2,* 309–315.

211. Sasaki, Y., Takahashi, T., Tanaka, I., et al. (1997). Expression of prostacyclin receptor in human megakaryocytes. *Blood, 90,* 1039–1046.

212. Stitham, J., Stojanovic, A., & Hwa, J. (2002). Impaired receptor binding and activation associated with a human prostacyclin receptor polymorphism. *J Biol Chem, 277,* 15439–15444.

213. Forman, B. M., Chen, J., & Evans, R. M. (1997). Hypolipidemic drugs, polyunsaturated fatty acids, and eicosanoids are ligands for peroxisome proliferator-activated receptors alpha and delta. *Proc Natl Acad Sci USA, 94,* 4312–4317.

214. Svensson, J., Hamberg, M., & Samuelsson, B. (1976). On the formation and effects of thromboxane A2 in human platelets. *Acta Physiol Scand, 98,* 285–294.

215. Moncada, S., & Vane, J. R. (1978). Pharmacology and endogenous roles of prostaglandin endoperoxides, thromboxane A2, and prostacyclin. *Pharmacol Rev, 30,* 293–331.

216. Moncada, S., & Vane, J. R. (1980). Prostacyclin in the cardiovascular system. *Adv Prostaglandin Thromboxane Res, 6,* 43–60.

217. Needleman, P., Turk, J., Jakschik, B. A., et al. (1986). Arachidonic acid metabolism. *Annu Rev Biochem, 55,* 69–102.

218. Pober, J. S., & Cotran, R. S. (1990). Cytokines and endothelial cell biology. *Physiol Rev, 70,* 427–451.

219. Manganello, J. M., Djellas, Y., Borg, C., et al. (1999). Cyclic AMP-dependent phosphorylation of thromboxane A(2) receptor-associated Galpha(13). *J Biol Chem, 274,* 28003–28010.

220. Panes, J., Perry, M., & Granger, D. N. (1999). Leukocyte–endothelial cell adhesion: avenues for therapeutic intervention. *Br J Pharmacol, 126,* 537–550.

221. Gorman, R. R., Bunting, S., & Miller, O. V. (1977). Modulation of human platelet adenylate cyclase by prostacyclin (PGX). *Prostaglandins, 13,* 377–388.

222. Tateson, J. E., Moncada, S., & Vane, J. R. (1977). Effects of prostacyclin (PGX) on cyclic AMP concentrations in human platelets. *Prostaglandins, 13,* 389–397.

223. Wise, H., & Jones, R. L. (2000). *Prostacyclin and its receptors.* New York: Kluwer Academic/Plenum Publishers.

224. Gonzalez-Utor, A. L., Sanchez-Aguayo, I., & Hidalgo, J. (1992). Cytochemical localization of K($^+$)-dependent $p$-nitrophenyl phosphatase and adenylate cyclase by using one-step method in human washed platelets. *Histochemistry, 97,* 503–507.

225. Seiss, W. (1989). Molecular mechanisms of platelet activation. *Physiol Rev, 69,* 58–178.

226. Fox, J. E., Say, A. K., & Haslam, R. J. (1979). Subcellular distribution of the different platelet proteins phosphorylated on exposure of intact platelets to ionophore A23187 or to prostaglandin E1. Possible role of a membrane phosphopolypeptide in the regulation of calcium-ion transport. *Biochem J, 184,* 651–661.

227. Kaser-Glanzmann, R., Gerber, E., & Luscher, E. F. (1979). Regulation of the intracellular calcium level in human blood platelets: Cyclic adenosine 3′,5′-monophosphate dependent phosphorylation of a 22,000 dalton component in isolated Ca$^{2+}$-accumulating vesicles. *Biochim Biophys Acta, 558,* 344–347.

228. Hathaway, D. R., Eaton, C. R., & Adelstein, R. S. (1981). Regulation of human platelet myosin light chain kinase by the catalytic subunit of cyclic AMP-dependent protein kinase. *Nature, 291,* 252–256.

229. Cavallini, L., Coassin, M., Borean, A., et al. (1996). Prostacyclin and sodium nitroprusside inhibit the activity of the platelet inositol 1,4,5-trisphosphate receptor and promote its phosphorylation. *J Biol Chem, 271,* 5545–5551.

230. Aszodi, A., Pfeifer, A., Ahmad, M., et al. (1999). The vasodilator-stimulated phosphoprotein (VASP) is involved in cGMP- and cAMP-mediated inhibition of agonist-induced platelet aggregation, but is dispensable for smooth muscle function. *EMBO J, 18,* 37–48.

231. Siess, W., & Lapetina, E. G. (1989). Prostacyclin inhibits platelet aggregation induced by phorbol ester or Ca$^{2+}$ ionophore at steps distal to activation of protein kinase C and Ca$^{2+}$-dependent protein kinases. *Biochem J, 258,* 57–65.

232. Conti, M. A., & Adelstein, R. S. (1981). The relationship between calmodulin binding and phosphorylation of smooth muscle myosin kinase by the catalytic subunit of 3′ : 5′ cAMP-dependent protein kinase. *J Biol Chem, 256,* 3178–3181.

233. Fox, J. E., & Phillips, D. R. (1982). Role of phosphorylation in mediating the association of myosin with the cytoskeletal structures of human platelets. *J Biol Chem, 257,* 4120–4126.

234. Body, S. C. (1996). Platelet activation and interactions with the microvasculature. *J Cardiovasc Pharmacol, 27*(Suppl 1), S13–S25.

235. Narumiya, S., & FitzGerald, G. A. (2001). Genetic and pharmacological analysis of prostanoid receptor function. *J Clin Invest, 108,* 25–30.

236. Smyth, E. M., Austin, S. C., Reilly, M. P., & FitzGerald, G. A. (2000). Internalization and sequestration of the human prostacyclin receptor. *J Biol Chem, 275,* 32037–32045.

237. Ashby, B. (1992). Comparison of iloprost, cicaprost and prostacyclin effects on cyclic AMP metabolism in intact platelets. *Prostaglandins, 43,* 255–261.

238. Darius, H., Veit, K., Binz, C., et al. (1995). Diminished inhibition of adhesion molecule expression in prostacyclin receptor desensitized human platelets. *Agents Actions Suppl, 45,* 77–83.

239. Jaschonek, K., Karsch, K. R., Weisenberger, H., et al. (1986). Platelet prostacyclin binding in coronary artery disease. *J Am Coll Cardiol, 8,* 259–266.

240. Klockenbusch, W., Hohlfeld, T., Wilhelm, M., et al. (1996). Platelet PGI2 receptor affinity is reduced in pre-eclampsia. *Br J Clin Pharmacol, 41,* 616–618.

241. Modesti, P. A., Fortini, A., Poggesi, L., et al. (1987). Acute reversible reduction of PGI2 platelet receptors after iloprost infusion in man. *Thromb Res, 48,* 663–669.

242. Alt, U., Leigh, P. J., Wilkins, A. J., et al. (1986). Desensitization of iloprost responsiveness in human platelets follows prolonged exposure to iloprost in vitro. *Br J Clin Pharmacol, 22,* 118–119.

243. Darius, H., Binz, C., Veit, K., et al. (1995). Platelet receptor desensitization induced by elevated prostacyclin levels causes platelet-endothelial cell adhesion. *J Am Coll Cardiol, 26,* 800–806.

244. Krane, A., MacDermot, J., & Keen, M. (1994). Desensitization of adenylate cyclase responses following exposure to IP prostanoid receptor agonists. Homologous and heterologous desensitization exhibit the same time course. *Biochem Pharmacol, 47,* 953–959.

245. Sinzinger, H., Silberbauer, K., Horsch, A. K., et al. (1981). Decreased sensitivity of human platelets to PGI2 during long-term intraarterial prostacyclin infusion in patients with

peripheral vascular disease — a rebound phenomenon? *Prostaglandins, 21,* 49–51.

246. MacDermot, J. (1986). Desensitization of prostacyclin responsiveness in platelets. Apparent differences in the mechanism in vitro or in vivo. *Biochem Pharmacol, 35,* 2645–2649.

247. Fisch, A., Tobusch, K., Veit, K., et al. (1997). Prostacyclin receptor desensitization is a reversible phenomenon in human platelets. *Circulation, 96,* 756–760.

248. Ashby, B. (1994). Interactions among prostaglandin receptors. *Receptor, 4,* 31–42.

249. Wilson, S. J., Roche, A. M., Kostetskaia, E., et al. (2004). Dimerization of the human receptors for prostacyclin and thromboxane facilitates thromboxane receptor-mediated cAMP generation. *J Biol Chem, 279,* 53036–53047.

250. Ortiz-Vega, S., & Ashby, B. (1997). Human prostacyclin receptor: cloning and co-expression with EP3 prostaglandin receptor. *Adv Exp Med Biol, 433,* 235–238.

251. Paul, B. Z., Ashby, B., & Sheth, S. B. (1998). Distribution of prostaglandin IP and EP receptor subtypes and isoforms in platelets and human umbilical artery smooth muscle cells. *Br J Haematol, 102,* 1204–1211.

252. Devi, L. A. (2001). Heterodimerization of G-protein-coupled receptors: Pharmacology, signaling and trafficking. *Trends Pharmacol Sci, 22,* 532–537.

253. Breitwieser, G. E. (2004). G protein-coupled receptor oligomerization: Implications for G protein activation and cell signaling. *Circ Res, 94,* 17–27.

254. Moncada, S., & Vane, J. R. (1979). Arachidonic acid metabolites and the interactions between platelets and blood-vessel walls. *N Engl J Med, 300,* 1142–1147.

255. Jorgensen, K. A. (1982). Studies on the biological balance between thromboxanes and prostacyclins in relation to the platelet–vessel wall interaction. *Dan Med Bull, 29,* 169–197.

256. Marcus, A. J., Broekman, M. J., & Pinsky, D. J. (2002). COX inhibitors and thromboregulation. *N Engl J Med, 347,* 1025–1026.

257. Coleman, R., Kennedy, I., Humphrey, P., et al. (1990). Prostanoids and their receptors. In J. Emmett (Ed.), *Comprehensive medicinal chemistry: Membranes and receptors* (Vol. 3, pp. 643–714). Oxford, UK: Pergamon Press.

258. Smith, W. L., DeWitt, D. L., Shimokawa, T., et al. (1990). Molecular basis for the inhibition of prostanoid biosynthesis by nonsteroidal anti-inflammatory agents. *Stroke, 21,* IV24–IV28.

259. Smith. W. L. (1992). Prostanoid biosynthesis and mechanisms of action. *Am J Physiol, 263,* F181–F191.

260. Minkes, M., Stanford, N., Chi, M. M., et al. (1977). Cyclic adenosine 3′,5′-monophosphate inhibits the availability of arachidonate to prostaglandin synthetase in human platelet suspensions. *J Clin Invest, 59,* 449–454.

261. Malmsten, C., Granstrom, E., & Samuelsson, B. (1976). Cyclic AMP inhibits synthesis of prostaglandin endoperoxide (PGG2) in human platelets. *Biochem Biophys Res Commun, 68,* 569–576.

262. Gerrard, J. M., Peller, J. D., Krick, T. P., et al. (1977). Cyclic AMP and platelet prostaglandin synthesis. *Prostaglandins, 14,* 39–50.

263. Nony, P., French, P., Girard, P., et al. (1996). Platelet-aggregation inhibition and hemodynamic effects of beraprost sodium, a new oral prostacyclin derivative: A study in healthy male subjects. *Can J Physiol Pharmacol, 74,* 887–893.

264. Marcus, A. J., Weksler, B. B., Jaffe, E. A., et al. (1980). Synthesis of prostacyclin from platelet-derived endoperoxides by cultured human endothelial cells. *J Clin Invest, 66,* 979–986.

265. Caughey, G. E., Cleland, L. G., Gamble, J. R., et al. (2001). Up-regulation of endothelial cyclooxygenase-2 and prostanoid synthesis by platelets. Role of thromboxane A2. *J Biol Chem, 276,* 37839–37845.

266. Radomski, M. W., Palmer, R. M., & Moncada, S. (1987). The anti-aggregating properties of vascular endothelium: Interactions between prostacyclin and nitric oxide. *Br J Pharmacol, 92,* 639–646.

267. Lidbury, P. S., Antunes, E., de Nucci, G., et al. (1989). Interactions of iloprost and sodium nitroprusside on vascular smooth muscle and platelet aggregation. *Br J Pharmacol, 98,* 1275–1280.

268. Radomski, M. W., Palmer, M. J., & Moncada, S. (1987). The role of nitric oxide and cGMP in platelet adhesion to vascular endothelium. *Biochem Biophys Res Commun, 148,* 1482–1489.

269. Vane, J. R., & Botting, R. M. (1995). Pharmacodynamic profile of prostacyclin. *Am J Cardiol, 75,* 3A–10A.

270. Yokoyama, C., Yabuki, T., Shimonishi, M., et al. (2002). Prostacyclin-deficient mice develop ischemic renal disorders, including nephrosclerosis and renal infarction. *Circulation, 106,* 2397–2403.

271. Bourgain, R. H. (1978). Inhibition of PGI2 (prostacyclin) synthesis in the arterial wall enhances the formation of white platelet thrombi in vivo. *Haemostasis, 7,* 252–255.

272. Pettigrew, L. C., & Wu, K. K. (1986). Platelet function and antiaggregant therapy in ischemic stroke. *Semin Neurol, 6,* 293–298.

273. Vane, J. (1985). Prostacyclin in the cardiovascular system in health and disease. In K. Schror (Ed.), *Prostaglandins and other eicosanoids in the cardiovascular system* (pp. 7–28). Basel: Karger.

274. Stein, R. W., Papp, A. C., Weiner, W. J., et al. (1985). Reduction of serum prostacyclin stability in ischemic stroke. *Stroke, 16,* 16–18.

275. Akopov, S., Darbinian, V., Grigorian, G., et al. (1993). Elevated velocity of prostacyclin degradation in blood as a possible risk factor in patients with cerebrovascular disorders. *Eur Neurol, 33,* 252–255.

276. Uchiyama-Tsuyuki, Y., Kawashima, K., Araki, H., et al. (1995). Prostacyclin analogue TTC-909 reduces memory impairment in rats with cerebral embolism. *Pharmacol Biochem Behav, 52,* 555–559.

277. Stier, C. T., Jr., Chander, P. N., Belmonte, A., et al. (1997). Beneficial action of beraprost sodium, a prostacyclin analog, in stroke-prone rats. *J Cardiovasc Pharmacol, 30,* 285–293.

278. Huczynski, J., Kostka-Trabka, E., Sotowska, W., et al. (1985). Double-blind controlled trial of the therapeutic effects of

prostacyclin in patients with completed ischaemic stroke. *Stroke, 16,* 810–814.

279. Martin, J. F., Hamdy, N., Nicholl, J., et al. (1985). Prostacyclin in cerebral infarction. *N Engl J Med, 312,* 1642.

280. Hsu, C. Y., Faught, R. E., Jr., Furlan, A. J., et al. (1987). Intravenous prostacyclin in acute nonhemorrhagic stroke: A placebo-controlled double-blind trial. *Stroke, 18,* 352–358.

281. Rubin, L. J. (1995). Pathology and pathophysiology of primary pulmonary hypertension. *Am J Cardiol, 75,* 51A–54A.

282. Fitzgerald, D. J., Entman, S. S., Mulloy, K., et al. (1987). Decreased prostacyclin biosynthesis preceding the clinical manifestation of pregnancy-induced hypertension. *Circulation, 75,* 956–963.

283. FitzGerald, G. A., Smith, B., Pedersen, A. K., et al. (1984). Increased prostacyclin biosynthesis in patients with severe atherosclerosis and platelet activation. *N Engl J Med, 310,* 1065–1068.

284. Iwai, N., Katsuya, T., Ishikawa, K., et al. (1999). Human prostacyclin synthase gene and hypertension: The Suita Study. *Circulation, 100,* 2231–2236.

285. Adderley, S. R., & Fitzgerald, D. J. (1999). Oxidative damage of cardiomyocytes is limited by extracellular regulated kinases 1/2-mediated induction of cyclooxygenase-2. *J Biol Chem, 274,* 5038–5046.

286. Numaguchi, Y., Naruse, K., Harada, M., et al. (1999). Prostacyclin synthase gene transfer accelerates reendothelialization and inhibits neointimal formation in rat carotid arteries after balloon injury. *Arterioscler Thromb Vasc Biol, 19,* 727–733.

287. Harada, M., Toki, Y., Numaguchi, Y., et al. (1999). Prostacyclin synthase gene transfer inhibits neointimal formation in rat balloon-injured arteries without bleeding complications. *Cardiovasc Res, 43,* 481–491.

288. Badesch, D. B., McLaughlin, V. V., Delcroix, M., et al. (2004). Prostanoid therapy for pulmonary arterial hypertension. *J Am Coll Cardiol, 43,* 56S–61S.

289. Murata, T., Ushikubi, F., Matsuoka, T., et al. (1997). Altered pain perception and inflammatory response in mice lacking prostacyclin receptor. *Nature, 388,* 678–682.

290. Ueno, A., Naraba, H., Ikeda, Y., et al. (2000). Intrinsic prostacyclin contributes to exudation induced by bradykinin or carrageenin: a study on the paw edema induced in IP-receptor-deficient mice. *Life Sci, 66,* PL155–160.

291. Cheng, Y., Austin, S. C., Rocca, B., et al. (2002). Role of prostacyclin in the cardiovascular response to thromboxane A2. *Science, 296,* 539–541.

292. Yang, J., Wu, J., Jiang, H., et al. (2002). Signaling through Gi family members in platelets. Redundancy and specificity in the regulation of adenylyl cyclase and other effectors. *J Biol Chem, 277,* 46035–46042.

293. Egan, K. M., Lawson, J. A., Fries, S., Koller, B., et al. (2004). COX-2-derived prostacyclin confers atheroprotection on female mice. *Science, 306,* 1954–1957.

294. Kahn, N. N., Bauman, W. A., & Sinha, A. K. (1996). Loss of high-affinity prostacyclin receptors in platelets and the lack of prostaglandin-induced inhibition of platelet-stimulated thrombin generation in subjects with

spinal cord injury. *Proc Natl Acad Sci USA, 93,* 245–249.

295. Neri Serneri, G. G., Modesti, P. A., Fortini, A., et al. (1984). Reduction in prostacyclin platelet receptors in active spontaneous angina. *Lancet, 2,* 838–841.

296. Betteridge, D. J., El Tahir, K. E., Reckless, J. P., et al. (1984). Platelets from diabetic subjects show diminished sensitivity to prostacyclin. *Eur J Clin Invest, 12,* 395–398.

297. Jaschonek, K., Weisenberger, H., Karsch, K. R., et al. (1984). Impaired platelet prostacyclin binding in acute myocardial infarction. *Lancet, 2,* 1341–1342.

298. Jaschonek, K., & Muller, C. P. (1988). Platelet and vessel associated prostacyclin and thromboxane A2/prostaglandin endoperoxide receptors. *Eur J Clin Invest, 18,* 18.

299. Majerus, P. W. (1983). Arachidonate metabolism in vascular disorders. *J Clin Invest, 72,* 1521–1525.

300. de Deckere, E. A., Nugteren, D. H., & Ten Hoor, F. (1977). Prostacyclin is the major prostaglandin released from the isolated perfused rabbit and rat heart. *Nature, 268,* 160–163.

301. Coker, S. J., Parratt, J. R., Ledingham, I. M., et al. (1981). Thromboxane and prostacyclin release from ischaemic myocardium in relation to arrhythmias. *Nature, 291,* 323–324.

302. Johnson III, G., Furlan, L. E., Aoki, N., et al. (1990). Endothelium and myocardial protecting actions of taprostene, a stable prostacyclin analogue, after acute myocardial ischemia and reperfusion in cats. *Circ Res, 66,* 1362–1370.

303. Simpson, P. J., Mickelson, J., Fantone, J. C., Gallagher, K. P., & Lucchesi, B. R. (1987). Iloprost inhibits neutrophil function in vitro and in vivo and limits experimental infarct size in canine heart. *Circ Res, 60,* 666–673.

304. Simpson, P. J., Mitsos, S. E., Ventura, A., et al. (1987). Prostacyclin protects ischemic reperfused myocardium in the dog by inhibition of neutrophil activation. *Am Heart J, 113,* 129–137.

305. Aherne, T., Price, D. C., Yee, E. S., et al. (1986). Prevention of ischemia-induced myocardial platelet deposition by exogenous prostacyclin. *J Thorac Cardiovasc Surg, 92,* 99–104.

306. Moncada, S., & Amezcua, J. L. (1979). Prostacyclin, thromboxane A2 interactions in haemostasis and thrombosis. *Haemostasis, 8,* 252–265.

307. Vane, J. R. (1971). Inhibition of prostaglandin synthesis as a mechanism of action for aspirin-like drugs. *Nature New Biol, 231,* 232–235.

308. Lopez-Farre, A., Caramelo, C., Esteban, A., et al. (1995). Effects of aspirin on platelet–neutrophil interactions. Role of nitric oxide and endothelin-1. *Circulation, 91,* 2080–2088.

309. De La Cruz, J. P., Blanco, E., & Sanchez de la Cuesta, F. (2000). Effect of dipyridamole and aspirin on the platelet-neutrophil interaction via the nitric oxide pathway. *Eur J Pharmacol, 397,* 35–41.

310. De La Cruz, J. P., Arrebola, M. M., Guerrero, A., et al. (2002). Influence of nitric oxide on the in vitro antiaggregant effect of ticlopidine. *Vasc Pharmacol, 38,* 183–186.

311. Smith, W. L., Garavito, R. M., & DeWitt, D. L. (1996). Prostaglandin endoperoxide H synthases (cyclooxygenases)-1 and -2. *J Biol Chem, 271,* 33157–33160.

312. McAdam, B. F., Catella-Lawson, F., Mardini, I. A., et al. (1999). Systemic biosynthesis of prostacyclin by cyclooxygenase (COX)-2: The human pharmacology of a selective inhibitor of COX-2. *Proc Natl Acad Sci USA, 96,* 272–277.

313. Pratico, D., Tillmann, C., Zhang, Z. B., et al. (2001). Acceleration of atherogenesis by COX-1-dependent prostanoid formation in low density lipoprotein receptor knockout mice. *Proc Natl Acad Sci USA, 98,* 3358–3363.

314. Yokoyama, C., Yabuki, T., Inoue, H., et al. (1996). Human gene encoding prostacyclin synthase (PTGIS): genomic organization, chromosomal localization, and promoter activity. *Genomics, 36,* 296–304.

315. Nakayama, T., Soma, M., Izumi, Y., et al. (1996). Organization of the human prostacyclin synthase gene. *Biochem Biophys Res Commun, 221,* 803–806.

316. Nakayama, T., Soma, M., & Kanmatsuse, K. (1997). Organization of the human prostacyclin synthase gene and association analysis of a novel CA repeat in essential hypertension. *Adv Exp Med Biol, 433,* 127–130.

317. Nakayama, T., Soma, M., Takahashi, Y., et al. (1997). Novel polymorphic CA/TG repeat identified in the human prostacyclin synthase gene. *Hum Hered, 47,* 176–177.

318. Nakayama, T., Soma, M., Watanabe, Y., et al. (2003). Splicing mutation of the prostacyclin synthase gene in a family associated with hypertension. *Adv Exp Med Biol, 525,* 165–168.

319. Nakayama, T., Soma, M., Watanabe, Y., et al. (2002). Splicing mutation of the prostacyclin synthase gene in a family associated with hypertension. *Biochem Biophys Res Commun, 297,* 1135–1139.

320. Nakayama, T., Soma, M., Haketa, A., et al. (2003). Haplotype analysis of the prostacyclin synthase gene and essential hypertension. *Hypertens Res, 26,* 553–557.

321. Amano, S., Tatsumi, K., Tanabe, N., et al. (2004). Polymorphism of the promoter region of prostacyclin synthase gene in chronic thromboembolic pulmonary hypertension. *Respirology, 9,* 184–189.

322. Nakayama, T., Soma, M., Rahmutula, D., et al. (1997). Nonsense mutation of prostacyclin synthase gene in a family. *Lancet, 349,* 1887–1888.

323. Nakayama, T., Soma, M., Rahmutula, D., et al. (2002). Association study between a novel single nucleotide polymorphism of the promoter region of the prostacyclin synthase gene and essential hypertension. *Hypertens Res, 25,* 65–68.

324. Nakayama, T., Soma, M., Rehemudula, D., et al. (2000). Association of 5′ upstream promoter region of prostacyclin synthase gene variant with cerebral infarction. *Am J Hypertens, 13,* 1263–1267.

325. Nakayama, T., Soma, M., Saito, S., et al. (2002). Association of a novel single nucleotide polymorphism of the prostacyclin synthase gene with myocardial infarction. *Am Heart J, 143,* 797–801.

326. Nakayama, T., Soma, M., Takahashi, Y., et al. (2001). Polymorphism of the promoter region of prostacyclin synthase gene is not related to essential hypertension. *Am J Hypertens, 14,* 409–411.

327. Chevalier, D., Cauffiez, C., Bernard, C., et al. (2001). Characterization of new mutations in the coding sequence and 5′-untranslated region of the human prostacyclin synthase gene (CYP8A1). *Hum Genet, 108,* 148–155.

328. Chevalier, D., Allorge, D., Lo-Guidice, J. M., et al. (2002). Sequence analysis, frequency and ethnic distribution of VNTR polymorphism in the 5′-untranslated region of the human prostacyclin synthase gene (CYP8A1). *Prostaglandins Other Lipid Mediat, 70,* 31–37.

329. Luft, F. C. (2000). Molecular genetics of human hypertension. *Curr Opin Nephrol Hypertens, 9,* 259–266.

330. Plesner, L. (1995). Ecto-ATPases: Identities and functions. *Int Rev Cytol, 158,* 141–214.

331. Zimmermann, H. (1999). Two novel families of ectonucleotidases: Molecular structures, catalytic properties and a search for function. *Trends Pharmacol Sci, 20,* 231–236.

332. Wang, T. F., & Guidotti, G. (1996). CD39 is an ecto-($Ca^{2+}$, $Mg^{2+}$)-apyrase. *J Biol Chem, 271,* 9898–9901.

333. Kaczmarek, E., Koziak, K., Sevigny, J., et al. (1996). Identification and characterization of CD39/vascular ATP diphosphohydrolase. *J Biol Chem, 271,* 33116–33122.

334. Koziak, K., Sevigny, J., Robson, S. C., Siegel, J. B., & Kaczmarek, E. (1999). Analysis of CD39/ATP diphosphohydrolase (ATPDase) expression in endothelial cells, platelets and leukocytes. *Thromb Haemost, 82,* 1538–1544.

335. Marcus, A. J., Safier, L. B., Hajjar, K. A., et al. (1991). Inhibition of platelet function by an aspirin-insensitive endothelial cell ADPase. Thromboregulation by endothelial cells. *J Clin Invest, 88,* 1690–1696.

336. Zimmermann, H. (1992). 5′-Nucleotidase: Molecular structure and functional aspects. *Biochem J, 285(Pt 2),* 345–365.

337. Dwyer, K. M., Robson, S. C., Nandurkar, H. H., et al. (2004). Thromboregulatory manifestations in human CD39 transgenic mice and the implications for thrombotic disease and transplantation. *J Clin Invest, 113,* 1440–1446.

338. Kansas, G. S., Wood, G. S., & Tedder, T. F. (1991). Expression, distribution, and biochemistry of human CD39. Role in activation-associated homotypic adhesion of lymphocytes. *J Immunol, 146,* 2235–2244.

339. Marcus, A. J., Broekman, M. J., Drosopoulos, J. H., et al. (1997). The endothelial cell ecto-ADPase responsible for inhibition of platelet function is CD39. *J Clin Invest, 99,* 1351–1360.

340. Dombrowski, K. E., Trevillyan, J. M., Cone, J. C., et al. (1993). Identification and partial characterization of an ectoATPase expressed by human natural killer cells. *Biochemistry, 32,* 6515–6522.

341. Maliszewski, C. R., Delespesse, G. J., Schoenborn, M. A., et al. (1994). The CD39 lymphoid cell activation antigen. Molecular cloning and structural characterization. *J Immunol, 153,* 3574–3583.

342. Handa, M., & Guidotti, G. (1996). Purification and cloning of a soluble ATP-diphosphohydrolase (apyrase) from potato tubers (*Solanum tuberosum*). *Biochem Biophys Res Commun, 218,* 916–923.

343. Schulte am Esch, J., 2nd, Sevigny, J., Kaczmarek, E., et al. (1999). Structural elements and limited proteolysis of CD39

influence ATP diphosphohydrolase activity. *Biochemistry, 38*, 2248–2258.

344. Grinthal, A., & Guidotti, G. (2000). Substitution of His59 converts CD39 apyrase into an ADPase in a quaternary structure dependent manner. *Biochemistry, 39*, 9–16.

345. Drosopoulos, J. H., Broekman, M. J., Islam, N., et al. (2000). Site-directed mutagenesis of human endothelial cell ecto-ADPase/soluble CD39: requirement of glutamate 174 and serine 218 for enzyme activity and inhibition of platelet recruitment. *Biochemistry, 39*, 6936–6943.

346. Vasconcelos, E. G., Nascimento, P. S., Meirelles, M. N., et al. (1993). Characterization and localization of an ATP-diphosphohydrolase on the external surface of the tegument of *Schistosoma mansoni. Mol Biochem Parasitol, 58*, 205–214.

347. Vasconcelos, E. G., Ferreira, S. T., Carvalho, T. M., et al. (1996). Partial purification and immunohistochemical localization of ATP diphosphohydrolase from *Schistosoma mansoni*. Immunological cross-reactivities with potato apyrase and *Toxoplasma gondii* nucleoside triphosphate hydrolase. *J Biol Chem, 271*, 22139–22145.

348. Wang, T. F., Ou, Y., & Guidotti, G. (1998). The transmembrane domains of ectoapyrase (CD39) affect its enzymatic activity and quaternary structure. *J Biol Chem, 273*, 24814–24821.

349. Zhong, X., Kriz, R., Kumar, R., et al. (2005). Distinctive roles of endoplasmic reticulum and golgi glycosylation in functional surface expression of mammalian E-NTPDase1, CD39. *Biochim Biophys Acta, 1723*, 143–150.

350. Zhong, X., Malhotra, R., Woodruff, R., et al. (2001). Mammalian plasma membrane ecto-nucleoside triphosphate diphosphohydrolase 1, CD39, is not active intracellularly. The N-glycosylation state of CD39 correlates with surface activity and localization. *J Biol Chem, 276*, 41518–41525.

351. Zimmermann, H., Braun, N., Kegel, B., et al. (1998). New insights into molecular structure and function of ectonucleotidases in the nervous system. *Neurochem Int, 32*, 421–425.

352. Kegel, B., Braun, N., Heine, P., et al. (1997). An ecto-ATPase and an ecto-ATP diphosphohydrolase are expressed in rat brain. *Neuropharmacology, 36*, 1189–1200.

353. Koziak, K., Kaczmarek, E., Kittel, A., et al. (2000). Palmitoylation targets CD39/endothelial ATP diphosphohydrolase to caveolae. *J Biol Chem, 275*, 2057–2062.

354. Robson, S. C., Kaczmarek, E., Siegel, J. B., et al. (1997). Loss of ATP diphosphohydrolase activity with endothelial cell activation. *J Exp Med, 185*, 153–163.

355. Krotz, F., Sohn, H. Y., Keller, M., et al. (2002). Depolarization of endothelial cells enhances platelet aggregation through oxidative inactivation of endothelial NTPDase. *Arterioscler Thromb Vasc Biol, 22*, 2003–2009.

356. Kittel, A., Kaczmarek, E., Sevigny, J., et al. (1999). CD39 as a caveolar-associated ectonucleotidase. *Biochem Biophys Res Commun, 262*, 596–599.

357. Papanikolaou, A., Papafotika, A., Murphy, C., et al. (2005). Cholesterol-dependent lipid assemblies regulate the activity of the ecto-nucleotidase CD39. *J Biol Chem, 280*, 26406–26414.

358. Gayle III, R. B., Maliszewski, C. R., Gimpel, S. D., et al. (1998). Inhibition of platelet function by recombinant soluble ecto-ADPase/CD39. *J Clin Invest, 101*, 1851–1859.

359. Drosopoulos, J. H. (2002). Roles of Asp54 and Asp213 in Ca$^{2+}$ utilization by soluble human CD39/ecto-nucleotidase. *Arch Biochem Biophys, 406*, 85–95.

360. Matsumoto, M., Sakurai, Y., Kokubo, T., et al. (1999). The cDNA cloning of human placental ecto-ATP diphosphohydrolases I and II. *FEBS Lett, 453*, 335–340.

361. Marcus, A. J., Safier, L. B., Broekman, M. J., et al. (1995). Thrombosis and inflammation as multicellular processes: Significance of cell–cell interactions. *Thromb Haemost, 74*, 213–217.

362. Marcus, A. J., Broekman, M. J., Drosopoulos, J. H., et al. (2001). Inhibition of platelet recruitment by endothelial cell CD39/ecto-ADPase: Significance for occlusive vascular diseases. *Ital Heart J, 2*, 824–830.

363. Marcus, A., Broekman, M., Drosopoulos, J., et al. (2001). Thromboregulation by endothelial cells: Significance for occlusive vascular diseases. *Arterioscler Thromb Vasc Biol, 21*, 178–182.

364. Kawashima, Y., Nagasawa, T., & Ninomiya, H. (2000). Contribution of ecto-5′-nucleotidase to the inhibition of platelet aggregation by human endothelial cells. *Blood, 96*, 2157–2162.

365. Cote, Y. P., Filep, J. G., Battistini, B., et al. (1992). Characterization of ATP-diphosphohydrolase activities in the intima and media of the bovine aorta: evidence for a regulatory role in platelet activation in vitro. *Biochim Biophys Acta, 1139*, 133–142.

366. Linden, J. (2001). Molecular approach to adenosine receptors: receptor-mediated mechanisms of tissue protection. *Annu Rev Pharmacol Toxicol, 41*, 775–787.

367. Burnstock, G. (2002). Purinergic signaling and vascular cell proliferation and death. *Arterioscler Thromb Vasc Biol, 22*, 364–373.

368. Koszalka, P., Ozuyaman, B., Huo, Y., et al. (2004). Targeted disruption of cd73/ecto-5′-nucleotidase alters thromboregulation and augments vascular inflammatory response. *Circ Res, 95*, 814–821.

369. Vial, C., Hechler, B., Leon, C., et al. (1997). Presence of P2X1 purinoceptors in human platelets and megakaryoblastic cell lines. *Thromb Haemost, 78*, 1500–1504.

370. Leon, C., Hechler, B., Vial, C., et al. (1997). The P2Y1 receptor is an ADP receptor antagonized by ATP and expressed in platelets and megakaryoblastic cells. *FEBS Lett, 403*, 26–30.

371. Daniel, J. L., Dangelmaier, C., Jin, J., et al. (1998). Molecular basis for ADP-induced platelet activation. I. Evidence for three distinct ADP receptors on human platelets. *J Biol Chem, 273*, 2024–2029.

372. Leon, C., Vial, C., Cazenave, J. P., et al. (1996). Cloning and sequencing of a human cDNA encoding endothelial P2Y1 purinoceptor. *Gene, 171*, 295–297.

373. Motte, S., Communi, D., Pirotton, S., et al. (1995). Involvement of multiple receptors in the actions of extracellular ATP: The example of vascular endothelial cells. *Int J Biochem Cell Biol, 27*, 1–7.

374. Fijnheer, R., Boomgaard, M. N., van den Eertwegh, A. J., et al. (1992). Stored platelets release nucleotides as inhibitors of platelet function. *Thromb Haemost, 68,* 595–599.

375. Enjyoji, K., Sevigny, J., Lin, Y., et al. (1999). Targeted disruption of cd39/ATP diphosphohydrolase results in disordered hemostasis and thromboregulation. *Nature Med, 5,* 1010–1017.

376. Fabre, J. E., Nguyen, M., Latour, A., et al. (1999). Decreased platelet aggregation, increased bleeding time and resistance to thromboembolism in P2Y1-deficient mice. *Nature Med, 5,* 1199–1202.

377. Leon, C., Hechler, B., Freund, M., et al. (1999). Defective platelet aggregation and increased resistance to thrombosis in purinergic P2Y(1) receptor-null mice. *J Clin Invest, 104,* 1731–1737.

378. Kittel, A., Csapo, Z. S., Csizmadia, E., Jackson, S. W., & Robson, S. C. (2004). Co-localization of P2Y1 receptor and NTPDase1/CD39 within caveolae in human placenta. *Eur J Histochem, 48,* 253–259.

379. Boeynaems, J. M., & Pearson, J. D. (1990). P2 purinoceptors on vascular endothelial cells: Physiological significance and transduction mechanisms. *Trends Pharmacol Sci, 11,* 34–37.

380. Marcus, A. J., Broekman, M. J., Drosopoulos, J. H., et al. (2003). Metabolic control of excessive extracellular nucleotide accumulation by CD39/ecto-nucleotidase-1: Implications for ischemic vascular diseases. *J Pharmacol Exp Ther, 305,* 9–16.

381. Candinas, D., Koyamada, N., Miyatake, T., et al. (1996). Loss of rat glomerular ATP diphosphohydrolase activity during reperfusion injury is associated with oxidative stress reactions. *Thromb Haemost, 76,* 807–812.

382. Koyamada, N., Miyatake, T., Candinas, D., et al. (1996). Apyrase administration prolongs discordant xenograft survival. *Transplantation, 62,* 1739–1743.

383. Robson, S. C., Wu, Y., Sun, X., et al. (2005). Ectonucleotidases of CD39 family modulate vascular inflammation and thrombosis in transplantation. *Semin Thromb Hemost, 31,* 217–233.

384. Belayev, L., Khoutorova, L., Deisher, T. A., et al. (2003). Neuroprotective effect of SolCD39, a novel platelet aggregation inhibitor, on transient middle cerebral artery occlusion in rats. *Stroke, 34,* 758–763.

385. Pinsky, D. J., Broekman, M. J., Peschon, J. J., et al. (2002). Elucidation of the thromboregulatory role of CD39/ectoapyrase in the ischemic brain. *J Clin Invest, 109,* 1031–1040.

386. Guckelberger, O., Sun, X. F., Sevigny, J., et al. (2004). Beneficial effects of CD39/ecto-nucleoside triphosphate diphosphohydrolase-1 in murine intestinal ischemia-reperfusion injury. *Thromb Haemost, 91,* 576–586.

387. Gangadharan, S. P., Imai, M., Rhynhart, K. K., et al. (2001). Targeting platelet aggregation: CD39 gene transfer augments nucleoside triphosphate diphosphohydrolase activity in injured rabbit arteries. *Surgery, 130,* 296–303.

388. Mazer, S., Fedarau, M., Liu, Y., et al. (2004). Deletion of endothelial ectoapyrase (CD39) promotes atherogenesis in hyperlipidemic mice. *Circulation, 110,* 79 (abstract).

389. Poirier, O., Mao, C., Mallet, C., et al. (1999). Polymorphisms of the endothelial nitric oxide synthase gene — no consistent association with myocardial infarction in the ECTIM study. *Eur J Clin Invest, 29,* 284–290.

390. Jeerooburkhan, N., Jones, L. C., Bujac, S., et al. (2001). Genetic and environmental determinants of plasma nitrogen oxides and risk of ischemic heart disease. *Hypertension, 38,* 1054–1061.

391. Granath, B., Taylor, R. R., van Bockxmeer, F. M., et al. (2001). Lack of evidence for association between endothelial nitric oxide synthase gene polymorphisms and coronary artery disease in the Australian Caucasian population. *J Cardiovasc Risk, 8,* 235–241.

392. Alvarez, R., Gonzalez, P., Batalla, A., et al. (2001). Association between the NOS3 (-786 T/C) and the ACE (I/D) DNA genotypes and early coronary artery disease. *Nitric Oxide, 5,* 343–348.

393. Rossi, G. P., Cesari, M., Zanchetta, M., et al. (2003). The T-786C endothelial nitric oxide synthase genotype is a novel risk factor for coronary artery disease in Caucasian patients of the GENICA study. *J Am Coll Cardiol, 41,* 930–937.

394. Ghilardi, G., Biondi, M. L., DeMonti, M., et al. (2002). Independent risk factor for moderate to severe internal carotid artery stenosis: T786C mutation of the endothelial nitric oxide synthase gene. *Clin Chem, 48,* 989–993.

395. Sharan, K., Surrey, S., Ballas, S., et al. (2004). Association of T-786C eNOS gene polymorphism with increased susceptibility to acute chest syndrome in females with sickle cell disease. *Br J Haematol, 124,* 240–243.

396. Hyndman, M. E., Parsons, H. G., Verma, S., et al. (2002). The T-786→C mutation in endothelial nitric oxide synthase is associated with hypertension. *Hypertension, 39,* 919–922.

397. Hwang, J. J., Tsai, C. T., Yeh, H. M., et al. (2002). The 27-bp tandem repeat polymorphism in intron 4 of the endothelial nitric oxide synthase gene is not associated with coronary artery disease in a hospital-based Taiwanese population. *Cardiology, 97,* 67–72.

398. Cine, N., Hatemi, A. C., & Erginel-Unaltuna, N. (2002). Association of a polymorphism of the ecNOS gene with myocardial infarction in a subgroup of Turkish MI patients. *Clin Genet, 61,* 66–70.

399. Benjafield, A. V., & Morris, B. J. (2000). Association analyses of endothelial nitric oxide synthase gene polymorphisms in essential hypertension. *Am J Hypertens, 13,* 994–998.

400. Wang, X. L., Sim, A. S., Badenhop, R. F., McCredie, R. M., & Wilcken, D. E. (1996). A smoking-dependent risk of coronary artery disease associated with a polymorphism of the endothelial nitric oxide synthase gene. *Nature Med, 2,* 41–45.

401. Spence, M. S., McGlinchey, P. G., Patterson, C. C., et al. (2004). Endothelial nitric oxide synthase gene polymorphism and ischemic heart disease. *Am Heart J, 148,* 847–851.

402. Jia, C. Q., Ning, Y., Liu, T. T., & Liu, Z. L. (2005). Association between G894T mutation in endothelial nitric oxide synthase gene and premature coronary heart disease. *Zhonghua Liu Xing Bing Xue Za Zhi, 26,* 51–53.

403. Shimasaki, Y., Yasue, H., Yoshimura, M., et al. (1998). Association of the missense Glu298Asp variant of the endothelial nitric oxide synthase gene with myocardial infarction. *J Am Coll Cardiol, 31,* 1506–1510.

404. Hingorani, A. D., Liang, C. F., Fatibene, J., et al. (1999). A common variant of the endothelial nitric oxide synthase (Glu298→Asp) is a major risk factor for coronary artery disease in the UK. *Circulation, 100,* 1515–1520.

405. Lembo, G., De Luca, N., Battagli, C., et al. (2001). A common variant of endothelial nitric oxide synthase (Glu298Asp) is an independent risk factor for carotid atherosclerosis. *Stroke, 32,* 735–740.

406. Chang, K., Baek, S. H., Seung, K. B., et al. (2003). The Glu298Asp polymorphism in the endothelial nitric oxide synthase gene is strongly associated with coronary spasm. *Coron Artery Dis, 14,* 293–299.

407. Wolff, B., Grabe, H. J., Schluter, C., et al. (2003). Endothelial nitric oxide synthase Glu298Asp gene polymorphism, blood pressure and hypertension in a general population sample. *J Hypertens, 23,* 1361–1366.

408. Miyamoto, Y., Saito, Y., Kajiyama, N., et al. (1998). Endothelial nitric oxide synthase gene is positively associated with essential hypertension. *Hypertension, 32,* 3–8.

409. Kishimoto, T., Misawa, Y., Kaetu, A., et al. (2004). eNOS Glu298Asp polymorphism and hypertension in a cohort study in Japanese. *Prev Med, 39,* 927–931.

410. Lacolley, P., Gautier, S., Poirier, O., et al. (1998). Nitric oxide synthase gene polymorphisms, blood pressure and aortic stiffness in normotensive and hypertensive subjects. *J Hypertens, 16,* 31–35.

411. Elbaz, A., Poirier, O., Moulin, T., et al. (2000). Association between the Glu298Asp polymorphism in the endothelial constitutive nitric oxide synthase gene and brain infarction. The GENIC Investigators. *Stroke, 31,* 1634–1639.

412. Stangl, K., Cascorbi, I., Laule, M., et al. (2000). High CA repeat numbers in intron 13 of the endothelial nitric oxide synthase gene and increased risk of coronary artery disease. *Pharmacogenetics, 10,* 133–140.

413. Bonnardeaux, A., Nadaud, S., Charru, A., et al. (1995). Lack of evidence for linkage of the endothelial cell nitric oxide synthase gene to essential hypertension. *Circulation, 91,* 96–102.

414. Nakayama, T., Soma, M., Takahashi, Y., et al. (1997). Association analysis of CA repeat polymorphism of the endothelial nitric oxide synthase gene with essential hypertension in Japanese. *Clin Genet, 51,* 26–30.

# CHAPTER 14

# Platelet Polymorphisms

**Vahid Afshar-Kharghan, K. Vinod Vijayan, and Paul F. Bray**

*Thrombosis Research Section, Department of Medicine, Baylor College of Medicine, Houston, Texas*

## I. Introduction

The first 30 years of investigations into platelet polymorphisms focused on the alloimmune thrombocytopenias. The nature of the alloantigen, the identification of the alloantibody, and the immune basis for antibody formation were discovered for those platelet membrane antigens most commonly responsible for alloimmune thrombocytopenia. Based on this knowledge, rational treatment strategies were developed for fetal and neonatal alloimmune thrombocytopenia (FNAIT) and posttransfusion purpura (PTP) (see Chapter 53). As molecular biologists uncovered the genetic bases for these polymorphisms, DNA testing became a useful addition to patient evaluation and management. The second phase of platelet polymorphism history began when investigators started to consider the genetic basis of common multifactorial diseases, such as vascular thrombosis. The impact of genetic variations as modifiers of disease processes led to the investigation of platelet polymorphisms as risk factors for arterial thrombosis and modifiers of hemorrhage. This new area of medicine presents unique challenges for the platelet researcher, such as the accurate detection of a small but real effect on the platelet and the clinical phenotype affected by the inherited sequence variation. Although in its infancy, this area of platelet biology promises to have a significant impact on future strategies for managing disorders such as myocardial infarction, stroke, and peripheral vascular disease. We have now entered a third phase comprising the intersection of information derived from the sequencing of the human genome with the study of inherited variations of platelet genes. This phase has identified large numbers of sequence variations in platelet genes via very different approaches than previously used by immunohematologists, using comprehensive genome searches for polymorphisms and resequencing specific genes. In the first half of this chapter we consider the genetic basis for platelet alloantigens and the evidence for alterations in protein antigenicity and function. The second half of the chapter covers the clinical consequences of these inherited variations.

The area of medicine and the field of investigation pertaining to platelet polymorphisms have their roots in disorders of alloimmune thrombocytopenias (see also Chapter 53). From the 1950s through the 1980s, the molecular understanding of alloimmune thrombocytopenias naturally focused on platelet antigens and the immune response to these antigens. The discovery of alloantibodies, presumably directed at novel platelet-specific antigens (alloantigens), were found to be the basis for immunopathologic conditions such as FNAIT, PTP, and some transfusion-refractory thrombocytopenias. Incompatibility for these antigens was shown to be necessary for developing alloimmune platelet destruction. Typically, the glycoprotein target of the alloantibody was discovered, and the genetic basis of these immunologic phenomena was inferred. For 40 years, serological methods utilizing well-characterized antisera and platelets of known phenotype were the only clinically useful means for typing a patient's platelets and detecting the presence of alloantibodies. These same reagents were used to characterize the pathophysiologic basis of the thrombocytopenia. By the early 1990s, most of the cDNAs for the major platelet membrane proteins were isolated and the genetic basis of the common antigenic variations were quickly uncovered. This enabled a shift in our thinking in this area of biology "upstream" from the antigen to the DNA, and throughout the 1990s numerous reports appeared which described different genotyping techniques permitting accurate definition of the DNA sequence variations responsible for the different alloantigens. These welcome advances in our understanding have come with at least one cost: confusion over nomenclature. However, a major advantage of readjusting the approach to this topic (by starting with DNA and not protein) is the improved ability to consider the rapidly evolving area of inherited platelet risk factors for arterial thrombosis, modifiers of hemorrhage, and platelet pharmacogenetics.

The nomenclature for the different platelet polymorphisms is confusing for a variety of reasons. The antigens were initially given names such as Pl$^A$ or Ko, derived from the names of thrombocytopenic patients with alloantibodies to the antigen. This led to the first source of confusion: multiple common names due to the independent naming of the same antigen by different investigators. An attempt to resolve this confusion resulted in a nomenclature called the Human Platelet Alloantigen (HPA) system.[1] This system considers each antigen independently and is based on the ability of specific antisera to recognize the relevant target. This nomenclature is useful when considering the allo-immune thrombocytopenias (see Chapter 53). However, as pointed out by Newman,[2] the HPA system may be misleading from a genetic point of view because it assigns different HPA names to the same haplotype. For example, there are two HPA names (HPA-1a and HPA-4a) for the single glycoprotein (GP) IIIa allele encoding Leu33 and Arg143. Viewed another way, many distinct haplotypes carry the Pl$^{A1}$ (Leu33) variant of GPIIIa, but all are called HPA-1a using the HPA nomenclature. Unfortunately, no existing nomenclature is able to describe haplotypes, which hinders the ability to think in genetic terms and to convey the idea that combinations of polymorphisms could provide more information and/or have different effects on protein function than each single polymorphism alone. Although this chapter will not resolve the confusion over nomenclature, this latter distinction will be emphasized. For purposes of clarity, we will arbitrarily italicize a common name when referring to DNA and genotypes. For example, GPIIIa will refer to the protein, whereas Pl$^{A2}$ will refer to the antigen on GPIIIa. We will use the standard nomenclature for genes (http://www.gene.ucl.ac.uk/nomenclature/), such that *ITGB3* will refer to the gene encoding GPIIIa and *Pl$^{A2}$* will be used to refer to the allele containing a C at nucleotide 1565 of *ITGB3*.

## II. The Molecular Basis for Platelet Polymorphisms

Polymorphisms are stable DNA sequence variations at a given locus that occur in greater than 1% of chromosomes in the population. Table 14-1 lists some additional basic genetic definitions relevant to the molecular approach to platelet polymorphisms and their functional and clinical consequences. The human genome displays a great deal of variability among individuals. It has been estimated that (1) human genetic sequence variations occur once every 1000 base pairs, (2) there are 200,000 amino acid-altering single nucleotide polymorphisms (SNPs) in the genome, (3) the typical gene has about four coding sequence polymorphisms, and (4) the typical person is heterozygous for 24,000 to 40,000 amino acid–altering substitutions.[3,4] A polymorphism in the coding sequence that alters an amino acid can

### Table 14-1: Basic Genetic Definitions

**Gene:** A unit of inheritance on a chromosome that has a function.

**Locus:** The position of a gene on a chromosome.

**Alleles:** The variant forms of a gene.

**Linkage disequilibrium:** Combinations of alleles occurring more frequently than would be predicted due to chance. Linkage disequilibrium (LD) of combinations of allele is likely due to infrequent crossover between these chromosomal locations during meiosis.

**Haplotype:** A unique combination of genetic variations from closely linked loci present on a single chromosome. (This combination can include several polymorphic sites on the same gene.)

**Genotype:** The genetic constitution of an individual at one or more loci.

**Polymorphisms:** Stable DNA sequence variations in which two or more alleles for a given locus occur in greater than 1% of chromosomes in the population.

**Mutations:** Heritable change in the genetic material, most often used to imply disease producing.

**Silent or synonomous sequence variations:** Genetic variations that do not change the amino acid sequence.

**Private alloantigen:** An alloantigen only present in a patient and his or her immediate family.

change the antigenicity or function of the protein it encodes. Genetic variations in noncoding sequences may alter the transcription or translation of a gene, thereby affecting the phenotype of the organism by changing the quantity, rather than the quality, of the gene product (quantitative trait locus). Altered antigenicity of the transmembrane glycoproteins expressed at moderate-to-high levels led to the ultimate identification of most known platelet polymorphisms. Thus, the most clinically relevant polymorphisms involve surface glycoprotein molecules that play a key role in platelet adhesion, activation, and aggregation. Table 14-2 lists the polymorphisms that are known to have the greatest clinical impact. The roles these glycoproteins play in platelet physiology are covered in Chapters 6–8. It is certain that there are polymorphisms in genes whose products are expressed in platelets (whether platelet-specific or not) that have yet to be identified, and many of these are likely to affect platelet biology.

The variant forms of a gene are called alleles. After many generations allele frequencies in large, randomly mating populations will tend to reach equilibrium. For a locus with two alleles, this equilibrium is represented by the Hardy-Weinberg equation, $p^2 + 2pq + q^2$, where $p$ is the allele frequency of one allele and $q$ is the allele frequency

**Table 14-2: Human Platelet Alloantigens**

| Gene/Protein | Common Alloantigen Name | Protein Change | Ref |
|---|---|---|---|
| *ITGB3*/GPIIIa | Pl^A, ZW (HPA-1) | Leu33Pro | 18 |
| | Pl^A1 (HPA-1a) | Leu33 | |
| | Pl^A2 (HPA-1b) | Pro33 | |
| | Pen, Yuk (HPA-4) | Arg143Gln | 45 |
| | Pen^a (HPA-4a) | Arg143 | |
| | Pen^b (HPA-4b) | Gln143 | |
| | Ca/Tu (HPA-6W) | Arg489Gln | 53 |
| | Ca/Tu^b | Arg489 | |
| | Ca/Tu^a (HPA-6bW) | Gln489 | |
| | Mo (HPA-7W) | Pro407Ala | 48 |
| | Mo^b | Pro407 | |
| | Mo^a (HPA-7bW) | Ala407 | |
| | Sr (HPA-8W) | Arg636Cys | 52 |
| | Sr^b | Arg636 | |
| | Sr^a (HPA-8bW) | Cys636 | |
| | La (HPA-10W) | Arg62Gln | 55 |
| | La^b | Arg62 | |
| | La^a (HPA-10bW) | Gln62 | |
| | Gro (HPA-11W) | Arg633His | 56 |
| | Gro^b | Arg633 | |
| | Gro^a (HPA-11bW) | His633 | |
| *ITGA2B*/GPIIb | Bak, Lek (HPA-3) | Ile843Ser | 67 |
| | Bak^a (HPA-3a) | Ile843 | |
| | Bak^b (HPA-3b) | Ser843 | |
| | Max (HPA-9W) | Val837Met | 72 |
| | Max^b | Val837 | |
| | Max^a (HPA-9bW) | Met837 | |
| GPIb-IX-V | Ko, Sib (HPA-2) | Thr145Met | 78 |
| *GP1BA* | Ko^b (HPA-2a) | Thr145 | |
| | Ko^a (HPA-2b) | Met145 | |
| *GP1BB* | Iy^a (HPA-12W) | Gly15Glu | 109 |
| | Iy^a- | Gly15 | |
| | Iy^a + (HPA-12bW) | Glu15 | |
| *ITGA2*/Integrin α₂ | Br, Zav (HPA-5) | Glu505Lys | 122 |
| | Br^b (HPA-5a) | Glu505 | |
| | Br^a (HPA-5b) | Lys505 | |
| | Sit (HPA-13W) | Thr799Met | 126 |
| | Sit^b | Thr799 | |
| | Sit^a (HPA-13bW) | Met799 | |

of the other. For example, when considering the alleles of the gene encoding the Leu33 (Pl^A1) and Pro33 (Pl^A2) isoforms of platelet GPIIIa, the possible genotypes are $Pl^{A1,A1}$, $Pl^{A1,A2}$, and $Pl^{A2,A2}$. If the allele frequencies of $Pl^{A1}$ and $Pl^{A2}$ in a given population are 0.86 and 0.14, respectively, the Hardy-Weinberg equation tells us that $Pl^{A1}$ (= p) and $Pl^{A2}$ (= q) alleles are in equilibrium if the frequencies of different genotypes are $(0.86)^2 + 2 (0.86)(0.14) + (0.14)^2$. In this case

one would expect $Pl^{A1,A1}$ to occur in 74% of the population, $Pl^{A1,A2}$ in 24% of the population, and $Pl^{A2,A2}$ in 2% of the population. The prevalence of different alleles varies among populations reflecting ethnic composition, geographic location, the length of time the genetic change has been in the population, etc. Relatively few studies have assessed allele frequencies for platelet polymorphisms among different ethnic groups. An exception is the study by Di Paola et al. who genotyped 1064 DNAs from 51 ethnic groups and found striking differences for some polymorphisms in the genes encoding GPIbα and integrin α2.[5] Future genetic epidemiology studies must consider this crucial issue of genetic admixture when evaluating thrombosis risk in patients.

### A. GPIIIa (Integrin β3); Gene: ITGB3

As the β subunit of both the GPIIb-IIIa receptor (integrin αIIbβ3) and integrin αvβ3, GPIIIa is the most abundantly expressed platelet membrane glycoprotein at ~80,000 copies (see Chapter 8).[6] Resting platelets can bind to immobilized fibrinogen and fibrin via GPIIb-IIIa. Platelet aggregation is mediated via binding of soluble fibrinogen or von Willebrand factor (VWF) to activated GPIIb-IIIa. Null or loss-of-function mutations of *ITGB3* result in the inherited bleeding disorder Glanzmann thrombasthenia (see Chapter 57).[7,8] *ITGB3* is located on chromosome 17, is 63 kb in length, and is organized into 15 exons.[9–12]

**1. The Pl^A (Leu33Pro, 1565T/C) Polymorphism.** Incompatibility for the Pl^A alloantigens is the most common cause of FNAIT and PTP in Caucasians (see Chapter 53).[13] The Pl^A antigen was first shown to reside on GPIIIa by Kunicki and Aster,[14] and further sublocalized to a 17-kD noncarbohydrate portion of the molecule.[15] With the cloning of the GPIIIa cDNA and gene[10,12,16,17] and the advent of the polymerase chain reaction technology (a godsend for molecular biologists studying a cell without a nucleus), investigators were able to apply molecular techniques in the field of platelet biology. Newman et al. identified a T→C nucleotide substitution at position 1565 in exon 2 of *ITGB3* that altered the codon for residue 33 from a Leu to a Pro.[18] They demonstrated that Leu33 was responsible for the Pl^A1 epitope; Pro33, for the Pl^A2 epitope. The prevalence of the *ITGB3* 1565T (Leu33, *Pl^A1*) and 1565C (Pro33, *Pl^A2*) alleles varies by ethnic background and geographical distribution. As shown in Table 14-3, the *Pl^A2* allele frequency is relatively common in white Northern European populations and virtually nonexistent in most Asian populations. This has obvious clinical consequences and is an important consideration in population studies. Between 1 and 3% of the *Pl^A2* alleles have a second nearby substitution in which the Leu at residue 40 is replaced by an Arg.[19,20] To date, the only known significance of this substitution is the additional recognition

### Table 14-3: Allele Frequency by Ethnic Group

| | | | Population/Gene Frequency | | |
|---|---|---|---|---|---|
| Gene | Polymorphic Alleles | | White | Black | East Asians |
| ITGB3 | Pl$^A$ | Pl$^{A1}$ | 0.84–0.89 (334–337) | 0.92 (334) | 0.99–1 (46,334,338,339) |
| | | Pl$^{A2}$ | 0.11–0.15 | 0.08 | 0–0.01 |
| ITGA2 | Br | Br$^a$ | 0.05–0.12 (334–337) | 0.21 (334) | 0.02–0.04 (47,334,338) |
| | | Br$^b$ | 0.87–0.95 | 0.79 | 0.95–0.97 |
| | 807C | | 0.54–0.65 (127,272,274) | | 0.61 (340) |
| | 807T | | 0.35–0.46 | | 0.38 |
| GP1BA | Ko | Ko$^a$ | 0.06–0.09 (334–337) | 0.11–0.18 (92,334) | 0.07–0.13 (46,47,334) |
| | | Ko$^b$ | 0.91–0.93 | 0.82–0.88 | 0.87–0.92 |
| | VNTR | A | 0.001 (89,92,93) | | 0.06–0.09 (258,341) |
| | | B | 0.07–0.1 | 0.19–0.2 (89,92) | 0.006–0.01 |
| | | C | 0.80–0.84 | 0.70–0.75 | 0.47–0.60 |
| | | D | 0.08–0.11 | 0.03–0.10 | 0.33–0.41 |

References indicated in parentheses.

site for the *MspI* restriction endonuclease, which has been used to distinguish the *Pl$^{A1}$* and *Pl$^{A2}$* alleles.

GPIIIa is a globular protein with many disulfide bonds (see Chapter 8), and the nature of the Pl$^A$ epitopes are likely to be complex and depend on secondary and tertiary structure, since synthetic peptides are unable to recreate the Pl$^{A1}$ antigen[21] and the epitope is sensitive to reducing agents.[14] The immunoreactivity of some, but not all, anti-Pl$^{A1}$ antibodies depends on disulfide bonds at the N-terminus of GPIIIa.[22]

There is conflicting evidence regarding the functional effects of the Pro33 substitution. Most anti-Pl$^{A1}$ antibodies block fibrinogen binding, and it has been proposed that more severe bleeding is seen in FNAIT with these antibodies than with antibodies to other alloantigens.[23] These antibodies will also completely inhibit the aggregation of *Pl$^{A1,A1}$* platelets, but only retard or partially inhibit aggregation of *Pl$^{A1,A2}$* platelets.[24] These studies with anti-Pl$^{A1}$ antisera suggest that position 33 may affect ligand binding despite its linear distance from the putative fibrinogen binding sites (amino acids 109–171 and 211–222) on GPIIIa.[25,26] Residue 33 of GPIIIa is located in a plexin-semaphorin-integrin (PSI) domain. These domains are conserved across species other than humans (including *C. elegans, G. gallus, X. laevus,* etc.), suggesting they serve a yet-to-be-identified function.[27] PSI domains are known to bind proteins and recent evidence indicates this region participates in integrin activation.[28]

Functional studies using platelets of known *Pl$^A$* genotype are difficult to compare because they have generally used small numbers of subjects and different assays. The largest study (the Framingham Offspring Study) included 1422 subjects and involved determination of the threshold concentration of epinephrine required to induce irreversible turbidometric platelet aggregation.[29] Significantly less epinephrine was needed to induce platelet aggregation in *Pl$^{A2}$*-expressing platelets ($p < 0.007$), and the effect showed an allele dosage effect. These differences between *Pl$^{A1}$* and *Pl$^{A2}$* platelets were observed with low-dose or subthreshold concentrations of agonists, and were lost with the high agonist concentrations typically used in clinical laboratories that are designed to detect platelet hypofunction and not hyperreactivity. *Pl$^{A2}$*-positive subjects have been shown to have significantly shorter bleeding times than *Pl$^{A1,A1}$* individuals.[30,31] A handful of other studies have also observed functional differences according to *Pl$^{A2}$* genotypes, but with inconsistent — and in several cases, possibly conflicting — results.[32–35] Additional work is needed to address the possibility that platelet functional differences by *Pl$^A$* genotype vary according to the agonist or the ligand for GPIIb-IIIa. To overcome the limitation caused by heterogeneity of human platelets, we generated cell lines overexpressing either the Pl$^{A1}$ or Pl$^{A2}$ form of GPIIb-IIIa, and observed increased adhesion, spreading, actin cytoskeletal reorganization, and migration in the Pl$^{A2}$ cells.[36,37] Increased outside-in signaling (phosphorylation) to extracellular signal-regulated kinase 2 and myosin light chain in the Pl$^{A2}$ cells, perhaps due to a reduced phosphatase activity, could account in part for the increased adhesive phenotype of the Pl$^{A2}$ cells.[38,39] Furthermore, shear stress augmented the adhesion of the Pl$^{A2}$ cells to fibrinogen.[40] Taken together, these studies indicate that the Pl$^{A2}$ polymorphism may not influence the initial events of platelet activation, but contributes to increased signaling and cytoskeletal changes, reinforcing platelet responses to post activation events, and stabilizing platelet–platelet or platelet–extracellular matrix interactions.

**2. The Pen (Arg143Gln, 526G/A) Polymorphism.** Pen (or Yuk) was the second alloantigen identified on GPIIIa that was associated with alloantibodies causing FNAIT and PTP.[41–44] Wang et al. demonstrated a G→A substitution in *Pen^{b,b}* individuals that resulted in a substitution of Arg143 by Gln.[45] When expressed in heterologous cells, the Arg143 form of GPIIIa bound only anti-Pen^a alloantibodies, and the Gln143 form of GPIIIa bound only anti-Pen^b alloantibodies. The prevalence of Pen^b is higher in Asians (1% among Japanese[46] and Koreans[47]) than in individuals of Northern European ancestry. (Note the contrast between Pen and Pl^A in this regard.) Anti-Pen^a alloantibodies are able to inhibit platelet aggregation, raising the possibility that residue 143 of GPIIIa may be important in this platelet function.[41] One study using cell lines expressing each Pen isoform revealed similar kinetics for soluble fibrinogen binding and similar binding to immobilized fibrinogen in static adhesion assays.[45] The outside-inside signaling after binding to fibrinogen was also similar between the Pen^a and Pen^b variants of GPIIIa. However, studies with platelets of known Pen genotypes have not been performed.

**3. Other *ITGB3* Variations.** A handful of additional low frequency genetic variations have been described in *ITGB3* that alter antigenicity without producing Glanzmann thrombasthenia. These less widely distributed genetic variations likely entered the human gene pool at a later time than *Pl^A* or *Pen*. All were identified because they caused FNAIT.[48–50]

Duv^a is a rare (<1% of population) alloantigen identified on GPIIIa and associated with alloantibodies causing severe FNAIT.[51] The Duv^{a+} antigen is due to a C-T substitution that is localized within the RGD binding domain of GPIIIa, resulting in a substitution of Thr140 by Ile. Heterologous expression in Cos-1 cells revealed that the Ile140 was responsible for the expression of Duv^{a+} epitope. Adhesion of Duv^{a+} expressing CHO cells to fibrinogen was inhibited by anti-Duv serum, suggesting that the residue 140 may be important for GPIIb-IIIa function. However, dithiothreitol-induced aggregation, adhesion to fibrinogen, and focal adhesion kinase phosphorylation were no different between the Duv^{a+} and Duv^{a−} expressing CHO cells.[51]

The Mo^{a,b} antigen is due to a C→G substitution resulting in the replacement of Pro407 by Ala.[48] Among 450 normal individuals in the Netherlands, only one had the *Mo^a* allele. This mutation was not associated with any significant bleeding disorder, but no functional studies on platelets of individuals with this mutation have been performed.

The Sr^{a,b} is a private alloantigen due to a C→T substitution at nucleotide 2004, resulting in the replacement of Arg636 by Cys.[52] This mutation is located in the cysteine-rich region of GPIIIa and creates an unpaired cysteine. This variant of GPIIIa has a higher molecular weight, probably secondary to different glycosylation of the amino

acid backbone, but platelets from one individual of each of the three possible genotypes showed no difference in the amount of GPIIIa on platelets or in standard platelet aggregometry.[52]

The Ca^{a,b} antigen is due to a G→A substitution at nucleotide 1564, resulting in the replacement of Arg489 by Gln.[53] Despite being rare, the Ca (Tu) variation has a wide distribution and is present in different ethnic groups.[53,54] Interestingly, three different codons (and therefore three different alleles) were found to encode the common Arg at 489.[53]

The La^a antigen is due to a G→A substitution resulting in the replacement of Arg62 by Gln,[55] and the Gro^{a,b} antigen is due to a 1996G→A substitution resulting in the replacement of Arg633 by His.[56] These variations are rare, but in some populations are more common than *Pen^b* and perhaps more common than some thrombasthenic alleles, demonstrating the arbitrariness of categorizing sequence variations as polymorphisms or mutations, or of categorizing their protein products as "major" platelet antigens (as in the case of Pen^b).

### B. GPIIb (Integrin αIIb); Gene: ITGA2B

As the α subunit of the GPIIb-IIIa receptor, GPIIb is also highly expressed on platelets at ~80,000 copies.[57] This protein has a molecular mass of 136 kDa and contains two subunits linked by a disulfide bond derived from a single precursor polypeptide (see Chapter 8).[58] The heavier GPIIbα subunit is about 125 kDa and the lighter GPIIbβ subunit is 23 kDa in its molecular mass. *ITGA2B* is located on chromosome 17,[9] is 17.2 kb, and contains 30 exons.[59]

**1. The Bak (Ile843Ser, 2622T/G) Polymorphism.** Antibodies against Bak^a and Bak^b cause alloimmune thrombocytopenias (see Chapter 53)[60–62] and van der Schoot and colleagues showed that the Bak^a epitope resides on GPIIb.[63] *Bak^a* is linked to *Pl^{A1},*[64] consistent with the physical linkage of *ITGA2B* and *ITGB3*.[65,66] Lyman et al. determined that the *Bak^{a,b}* polymorphism was associated with a T→G substitution at position 2622 of the GPIIb mRNA that caused an Ile→Ser change at residue 843,[67] and this change was shown to be responsible for the alloantibody reactivity.[68] There are several other intron sequence variations that are linked to the *Bak* polymorphisms and can be used as genetic markers for *Bak*.[69] Some posttranslational processing of Ser847, such as glycosylation and/or sialic acid addition, can be recognized by alloantibodies.[68,70,71] To date, there are no studies that have examined whether the Ile→Ser 843 change affects platelet or GPIIb-IIIa physiology.

**2. The Max (Val837Met, 2603G/A) Variation.** A low frequency variant of the *Bak^b* allele, termed *Max^a*, has been identified.[72] This allele contained a G→A substitution at position 2603 of the GPIIb mRNA that caused a Val→Met change at residue 837. Association studies suggest this allele is responsible for the Max[a] antigen, but alloantibody binding has not been demonstrated in transfection experiments of heterologous cells, and this amino acid change is not known to affect GPIIb function.

## C. GPIbα; Gene: GP1BA

The GPIb-IX-V complex is the second most abundant platelet receptor at ~25,000 copies per platelet (see Chapter 7).[73] The importance of this receptor is underscored by the inherited bleeding disorder Bernard–Soulier syndrome, wherein the receptor is absent or nonfunctional (see Chapter 57). In addition to its major function of tethering platelets to immobilized VWF, GPIb-IX-V also binds thrombin and mediates platelet–leukocyte interactions. The GPIb-IX-V complex is composed of four polypeptides: GPIbα, GPIbβ, GPIX, and GPV, each encoded by separate genes (see Chapter 7). These genes are similar in their organization, with the entire coding sequence located on a single exon (except the gene encoding GPIbβ GP1BB), which contains an intron 10 bases after the start of the coding sequence). Despite their structural similarities, the genes encoding the GPIb-IX-V complex are physically separated throughout the human genome. GP1BA is located on the short arm of chromosome 17,[74] GP1BB is on the long arm of chromosome 22,[75] and the genes for GPIX (GP9) and GPV (GP5) are located on the long arm of chromosome 3 (3q21 and 3q29).[76] GPIbα, the most important component of the GPIb-IX-V complex, contains the binding sites for VWF and thrombin and functions as a "tether" for platelet rolling on immobilized VWF.[77] GP1BA is about 6 kb and contains two exons, which are separated by a single intron, and the entire coding sequence is contained within the second exon.[74]

### 1. The Ko (Thr145Met, 524C/T) Polymorphism.
The Ko[78] [also called Sib[79] or HPA-2[78]] polymorphism was initially recognized because of its association with platelet transfusion refractoriness[80] and FNAIT,[81] and later was localized to GPIbα.[82,83] A G→A substitution at position 524 of the GPIbα mRNA was identified as responsible for a Thr (Ko[b])→Met (Ko[a]) change at residue 145, which in turn was responsible for alloantibody specificity.[79] The Ko[a] allele has a higher frequency in Africa than in other ethnic groups.[5] The Ko polymorphism is located in the N-terminus of GPIbα at the site of interaction with VWF and thrombin.[83] This polymorphism induces a conformational change in GPIbα that can be recognized by alloantibodies and monoclonal antibodies against GPIbα.[84] Furthermore, Ko alloantibodies can prevent ristocetin-induced platelet agglutination.[85] A study using platelets from a small number of subjects of known Ko genotypes failed to demonstrate dif-

ferences in ristocetin- or botrocetin-induced platelet agglutination.[86] Using recombinant peptides of the 45 Kd GPIbα globular head, the Thr145 form bound VWF with higher affinity than did the Met145 form under static conditions.[84] However, under flow conditions, the presence of methionine instead of threonine at residue 145 of GPIbα enhanced binding of GPIbα-expressing heterologous cells to immobilized VWF.[87]

### 2. The Variable Number of Tandem Repeat (VNTR) Alleles.
The existence of GPIbα molecules with different molecular weights was known for several years before the genetic basis of this variability was recognized. Moroi et al. described four different species of GPIbα designated as A, B, C, and D with molecular masses of 168, 162, 159, and 153 kDa, respectively.[88] The genetic basis for the variation in the GPIbα molecular weight was found to be a variable number of 39 nucleotide repeats encoding identical tandems of 13 amino acids in the mucin-like domain.[89] The A allele contains four repeats[90] and alleles B, C, and D have three, two, and one repeats, respectively.[89] Recently, an E variant of GP1BA with no repeat has been reported.[91] From an evolutionary point of view, the Ko[a] sequence variation may have originated on the B allele such that Ko[a] is present only on the B allele and the immediate evolutionary derivative of the B allele, the A allele. Due to the close physical proximity of these polymorphic sites (less than 1000 bp separate the Ko and VNTR sites), the frequency of genetic crossover during meiosis is expected to be low, and for most populations Ko[a] has been found only on the A and B VNTR alleles.[82,90]

The longest variant of GP1BA (the A allele) was initially believed to be restricted to Eastern Asians. In two reports, the A allele was reported in American Indians and Caucasians from Southern Spain, although with a lower frequency compared to the Japanese population.[92,93] Others have also reported the presence of Ko[a]/VNTR-C among African Americans[92] and Brazilians from different racial backgrounds.[91] Ko[b] was shown to be allelic with the A or B variant of the VNTR polymorphism in Southern Spain[93] and in Brazil,[91] and Ko[b] was shown to be allelic with the A or B variant of the VNTR polymorphism in 5.9% of the population in Southern Spain[93] and Brazil.[91] Several groups reported linkage disequilibrium between the Ko and VNTR polymorphisms of GP1BA.[92–94] There is some evidence to suggest the VNTR A allele induces a slower GPIbα-mediated rolling velocity.[87]

### 3. The Kozak (−5T/C) Polymorphism.
A third polymorphism in GP1BA is located in the noncoding region, five base pairs upstream of the initiation codon. This −5T→C polymorphism was first reported by Kaski et al.[95] and Corral et al.[93] The −5C variant is only found on the Ko[b] (Thr145) allele, whereas the T variant can be located on either the Ko[a] or Ko[b] alleles.[94,96] Di Paolo et al. found no

significant difference in the −5T/C allele frequencies among different ethnic groups,[5] such that population stratification ought to have relatively little effect on epidemiological studies of this SNP. The sequence surrounding the translation initiation site is important for efficient translation of eukaryotic mRNA,[97–99] raising the possibility that this −5T/C sequence polymorphism could affect GPIbα expression. Indeed, there is evidence that the −5C allele is associated with increased surface expression of GPIbα on platelets and heterologous cell lines, in a gene dosage dependent manner,[100] leading to its naming as the *Kozak* polymorphism (for Marilyn Kozak who first described the importance of this region of mRNA for translation).[98,99] However, several other studies have not shown a relationship between the *Kozak* polymorphism and platelet GPIb-IX-V levels.[101–103] It is equally controversial whether the *Kozak* polymorphism is associated with altered GPIbα-dependent platelet function. Cadroy et al. used a parallel plate flow chamber to show that −5C allele-positive platelets demonstrated greater deposition onto collagen compared to −5TT platelets under medium, but not high, flow shear rates.[104] However, Jilma-Stohlawetz et al. found faster closure times in the high shear platelet function analyzer (PFA)-100 with −5TT platelets than with −5TC platelets.[102]

**4. *GP1BA* Haplotypes.** A number of additional nucleotide sequence variations in *GP1BA* have been identified. It therefore becomes important to know the linkage disequilibrium among these different sequence variations, enabling construction of haplotypes that provide more genetic information. Di Paolo et al. genotyped 1064 DNAs from 51 ethnic groups for SNPs in *GP1BA* at positions −5 (Kozak), 1018 (Thr145Met, HPA-2), and 1285 (VNTR *A, B, C,* and *D*).[5] This information was integrated into genotype data publicly available for this region, and linkage disequilibrium and haplotype block information construed using standard algorithms. Compared to other geographic locales, the VNTR *B* allele was much more common in African and Israeli groups. In contrast to other reports,[92–94] and there was incomplete linkage disequilibrium for polymorphisms at 1018 (Thr145Met) and 1285 (VNTR),[5] no haplotype block could be identified using the three polymorphisms of *GP1BA*. Fig. 14-1 indicates these genetic relationships.

### D. *GPIbβ (Gene:* **GP1BB***) and the* Iyᵃ *Variation*

GPIbβ is covalently linked to GPIbα and is necessary for surface expression of the GPIb-IX-V complex (see Chapter 7).[105] *GP1BB* has two exons that are separated by a short intron,[106,107] and almost the entire coding sequence for *GP1BB* is located in the second exon. Kiefel et al. described a low frequency platelet GPIbβ alloantigen, called Iy, associated with FNAIT that was due to a substitution at amino

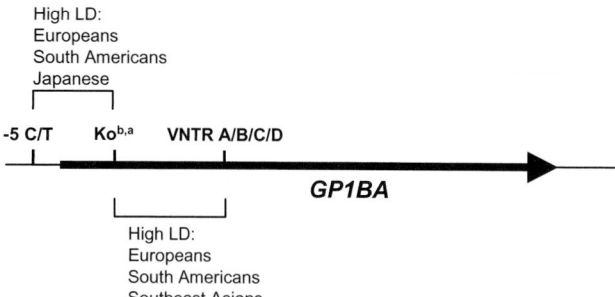

**Figure 14-1.** Schematic display of the best studied polymorphisms of *GP1BA*. The three polymorphic sites are shown in *GP1BA*. Displayed are regions of strong linkage disequilibrium (LD) that were observed for the indicated ethnic groups.[5] Haplotype summaries are not available, but would be quite cumbersome because of the VNTR 4 alleles.

acid residue 15.[108] Later, Sachs et al. showed that the Iyᵃ alloantigen results from a substitution at amino acid residue 15 (Gly15Glu).[109] This is a rare allele: Among 249 German subjects only one individual was Iyᵃ positive, and the Iyᵃ allele frequency in Japanese and Koreans was less than 0.0016.[110] The Iy polymorphism does not affect the expression of the GPIb-IX complex on platelets, and a study of ristocetin-induced aggregation with platelets from a single Iyᵃ positive subject was normal.[109]

### E. *GPIX and GPV; Genes:* **GP9** *and* **GP5**

GPIX, GPV, and GPIb associate through noncovalent interactions (see Chapter 7). Deficiency of GPIX decreases the expression of the GPIb-IX-V complex, as shown in several patients with Bernard–Soulier syndrome (see Chapter 57).[111] Studies with heterologous cell lines and a *GP5* knockout mouse model showed that although the presence of GPV was not required for the expression of GPIb-X-V,[112,113] GPV might play a role in regulating the binding of thrombin to the complex[114] or in modulating the stimulatory effect of thrombin on platelets.[115,116] There is a sequence variation in the 3′ untranslated region of *GP9*,[117] but it has no known functional consequence. Nine relatively common sequence variations have been identified in *GP5*, four of which change amino acids, with frequencies ranging from 1 to 4%.[117] No functional or antigenic alterations have been reported for these *GP5* variations.

### F. *Integrin α2 (Platelet GPIa); Gene:* **ITGA2**

Platelets contain several receptors for collagen, including integrin α2β1 (GPIa-IIa)[118] (see Chapter 6). Integrin α2 is 1181 amino acids long, has a molecular mass of about 130 kDa, and its mRNA has 5374 nucleotides.[119]

**1. The *Br* (Glu505Lys, 1648G/A) Polymorphism.** Incompatibility for the Br alloantigen is the second most common cause of FNAIT among Caucasians (see Chapter 53).[120] Initially, the Br alloantigens were shown to reside on the α2β1 complex,[121] and subsequently on the α2 subunit of the complex.[122] The *Br* polymorphism is caused by a G→A substitution at mRNA position 1648, and alters the Glu505 residue (Br[b]) to Lys505 (Br[a]).[122] Expressing each isoform in CHO cells confirmed that these amino acid changes were responsible for the Br alloantigens.[122] Studies on platelets from two *Br*[b,b] subjects and one *Br*[a,a] subject showed no difference in static adhesion to collagen type I, III, or V.[122] Expression of the *Br*[a] allele is associated with lower platelet surface levels of α2β1,[123,124] perhaps because it is in linkage disequilibrium with a −52T polymorphism in *ITGA2* that may cause a lower rate of α2 gene transcription.[125]

**2. The *Sit* (Thr799Met, 2531C/T) Sequence Variation.** Investigations of a case of FNAIT due to an alloantibody against integrin α2 identified a low frequency alloantigen called Sit.[126] This antigen was associated with a C→T point mutation at mRNA position 2531, resulting in a Thr799Met polymorphism, and anti-Sit[a] antisera reacted exclusively with Thr799 receptors. Studies on one subject homozygous for *Sit* Met799 allele showed reduced aggregation to low-dose collagen, but no difference in platelet adhesion to immobilized type I collagen.[126] Residue 799 is not within the collagen-binding I domain of the integrin α₂, and more work is needed to determine whether this potentially interesting, although rare, variant affects receptor function.

**3. The 807C/T Polymorphism.** Other *ITGA2* polymorphisms have been identified which do not alter amino acids. Initial studies focused on a limited number of polymorphic sites located in a contiguous region of exon 7 through exon 8 of *ITGA2*. Kritzik et al. proposed the existence of three alleles based on patterns of restriction digests.[123] Platelets from individuals with the 807T allele have higher levels of α2β1.[123,124,127,128] The 807C/T polymorphism is located in exon 7 and does not alter an amino acid and is not located in a usual region for regulating transcription, such that there is a relatively low likelihood it is responsible for the differences in receptor levels.

The 807C/T polymorphism has been associated with altered platelet adhesion to collagen using *in vitro* perfusion assays. Compared to platelets homozygous for 807C, platelets homozygous for 807T demonstrated a greater rate of attachment to surfaces coated with collagen type I.[123] Roest et al. confirmed the association of the 807C/T polymorphism with receptor density,[129] and demonstrated (1) that both plasma levels of VWF and α2β1 receptor density affected platelet adhesion to collagen under flow conditions, (2) that the type of collagen affected α2β1-dependent adhesion, and (3) a nonsignificant trend for both the 807C/T polymorphism and α2β1 expression to affect the collagen concentration required to induce turbidometric platelet aggregation. This was one of the very few studies of its kind to consider how differing demographics of the study subjects and differing levels of hemostatic factors (e.g., plasma and platelet VWF levels) affect platelet function based on genotype. A small study from Spain observed no association between the 807C/T genotype and differences in platelet P-selectin expression or platelet aggregation in response to thrombin and high-dose collagen.[127]

**4. The −52C/T and −92C/G Polymorphisms.** Two dimorphisms, −52C/T and −92C/G, within the *ITGA2* promoter in the 5′ regulatory region may influence the density of α2β1 receptors.[125,130] In a Caucasian population the −52T allele and −92G allele had a gene frequency of 0.35 and 0.15, respectively, and correlated with a reduced expression of platelet α2β1 receptors. This was due to decreased transcription of *ITGA2* resulting from a decreased binding of SP1 transcription factors. An additive negative influence in *ITGA2* transcription was observed when combined substitutions of −52T and −92G were used in a transfected megakaryocytic cell line.

**5. *ITGA2* Haplotypes.** Di Paola et al. identified four haplotype blocks for the *ITGA2* gene.[5] The −52 polymorphism was in linkage disequilibrium with most SNPs in haplotype blocks 3 and 4, and specifically with the SNP at 1648. The SNPs at 807 and 1648 were in strong linkage disequilibrium with one another and with most SNPs in blocks 3 and 4. Not surprisingly, there was a racial discrepancy for certain haplotypes. For example, the −52C/807C/1648A haplotype was rare in Caucasian and common in African populations. Fig. 14-2 displays the most common haplotypes of *ITGA2*.

### G. GPVI; Gene: GP6

*GP6* is located on chromosome 19q13.4[131] and codes for a immunoglobulin-like receptor family of molecular mass of 62 kDa. GPVI is a platelet collagen receptor and is physically associated with the Fc receptor γ chain, and serves a major signaling and platelet activation function (see Chapter 6). *GP6* has two primary alleles with frequencies of 0.85 and 0.13.[132] The T13254C polymorphism of *GP6* distinguishes these two alleles, and induces a Ser219Pro substitution.[133] The Pro219 isoform is associated with decreased GPVI receptors in normal subjects and a reduced thrombogenicity on collagen.[132,134]

### H. FcγRIIA; Gene: FCGR2A

Human blood platelets express a single Fc receptor, FcγRIIA (CD32), which has a low affinity for immunoglobulin (Ig)

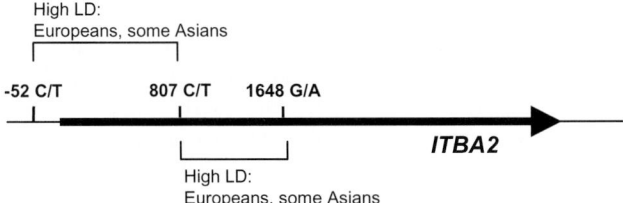

**Figure 14-2.** Schematic display of the best studied polymorphisms of *ITGA2*. The three polymorphic sites are shown in *ITGA2*. Displayed are regions of strong linkage disequilibrium (LD) that were observed for the indicated ethnic groups. The inserted table shows the most common haplotypes for this gene.[5] The remaining haplotype combinations are generally rare, although there are isolated exceptions in a few geographic regions.

molecules (see Chapter 6). Its size is ~40 kDa and estimates of copy number have ranged from 1300 to 5000 per platelet.[135,136] FcγRIIA is an integral membrane protein with two extracellular Ig-like domains and a cytoplasmic tail with an immunoreceptor tyrosine-based activation motif (ITAM).[137] A subclassification of FcγRIIA as high-responder (HR) or low-responder (LR) was based on the ability to induce T cell mitogenesis after binding mouse IgG₁.[138] The genetic basis for this functional polymorphism is a His131Arg polymorphism, corresponding to LR and HR, respectively, in the second Ig domain of FcγRIIA.[139] Compared to the His131 isoform, the Arg131 form binds little human IgG₂, but binds murine IgG₁ with a higher affinity.[140] The variation in interindividual copy number per platelet is not related to the His131Arg polymorphism.[141] We have shown that the His131Arg polymorphism of *FCGR2A* does not affect intrinsic platelet reactivity, but that some receptor- and ligand-induced binding site (RIBS and LIBS) antibodies preferentially bind to Arg131-positive platelets and are able to crosslink FcγRIIA and activate platelets.[142] For this reason, assays of platelet reactivity that use intact (Fc domain-containing) monoclonal antibodies against activation-dependent platelet epitopes may give artifactual results unless FcγRIIA is blocked or the His131Arg genotype is known and considered in the data analysis.[142]

### I. α₂A Adrenergic Receptor; Gene: ADRA2A

The α2A adrenergic receptor (A2AR) mediates epinephrine-induced platelet aggregation via a G protein-coupled response (see Chapter 6). This receptor is widely expressed in different organs, including the brain and kidney, in addition to platelets. In the blood-pressure-regulating region of the brain, the A2AR is the main inhibitory presynaptic receptor. In the kidney, it plays an important role in water and sodium excretion. *ADRA2A* is located in a single exon on chromosome 10.[143,144] The first polymorphism in *ADRA2A* identified two alleles by Southern-blot analysis of *Dra*I–digested genomic DNA: a less common allele had a 6.3-kb fragment; the more common allele, a 6.7-kb fragment. The presence of at least one 6.3-kb fragment in the genomic DNA was associated with increased blood pressure, especially in American blacks,[145] and increased platelet aggregation among normotensive individuals.[146] A second polymorphism has been identified in *ADRA2A* that results in the substitution of Asn251 by Lys.[147] Lys251 is a more common variation among African Americans compared to Caucasians. Although this polymorphism is not associated with hypertension, the Lys251 enhances signal transduction induced by ligand binding via an increased coupling of A2AR with the G$_i$ protein.

### J. Adenosine Diphosphate (ADP) Receptors

Platelets have two ADP receptors, P2Y₁ (gene: *P2RY1*) and P2Y₁₂ (gene: *P2RY12*), and an ATP receptor, P2X₁ (gene: *P2RX1*) (see Chapter 10). The P2Y₁ and P2Y₁₂ receptors mediate ADP-induced platelet aggregation via a G$_q$ and G$_i$ protein-coupled response, respectively. Four SNPs and a single nucleotide insertion (ins) polymorphism in *P2RY12* have been identified:[148] C139T, T744C, and the ins801A polymorphism are in the 5′ intronic sequence, while C34T and G52T polymorphisms are in exon 2 and do not alter amino acids. These polymorphisms are in complete linkage disequilibrium. A major haplotype H1, with an allelic frequency of 0.86, was defined as 139C/744T/no ins801A/52G. A minor haplotype H2, with an allelic frequency of 0.14, was represented as 139T/744C/ins801A/52T. The H2 haplotype was associated with greater ADP-induced platelet aggregation, perhaps via differences in the mechanisms regulating the cAMP inhibition.[148] Two synonymous SNPs have been identified for *P2RY1* that are in strong linkage disequilibrium.[149]

### K. Serotonin Receptor (5HT2A); Gene: HTR2A

5-Hydroxytryptamine (5-HT; serotonin) is transported by a 5-hydroxytryptamine transporter (5-HTT) and secreted from the dense granules of activated platelets (see Chapter 15). Serotonin binds to a G protein-coupled serotonin receptor (5HT₂A) on platelets and endothelial cells (see Chapter 6) to augment platelet aggregation and initiate vascular contraction, respectively. Two polymorphisms in *HTR2A*, namely, −1438A/G in the promoter region and T102C in exon 1 have been identified. Alleles 102T and 102C are in strong linkage disequilibrium with −1438A and −1438G, respectively. T allele frequency of T102C and A allele frequency of −1438A/G were 0.65 and 0.67. Subjects homozygous for 102T and −1438A demonstrated increased numbers of small sized (9–25 μm) platelet aggregates in response to serotonin.[150]

A 44-bp insertion (long)/deletion (short) polymorphism in the transcriptional initiation site has been described for the 5-HTT gene. The long allele has been associated with (a) higher promoter activity, (b) increased serotonin uptake in platelets,[151] and (c) increased platelet activation in elderly depressed subjects (see Chapter 44).[152]

### L. GPIV (CD36); Gene: CD36

The investigation of a Japanese patient who developed transfusion-refractory thrombocytopenia due to a platelet-specific antibody led to the discovery that 3 − 11% of the Japanese population do not express CD36 (GPIV).[153,154] The epitope on CD36 is called Nak, and individuals lacking CD36 are Nak[a−]. The frequency of the Nak[a−] phenotype among U.S. blood donors was reported to be about 0.3%,[155] but another study showed that this phenotype is more common among African Americans (2.5%) compared to whites (0%).[156] CD36 is a class B scavenger receptor expressed on monocyte/macrophages, platelets, endothelial cells, cardiac and skeletal muscle, and adipose tissue (see Chapter 6).[157] Two variants of CD36 deficiency are recognized: In type I both platelets and monocytes are CD36 deficient,[158] and in type II only platelets are affected.[159] The molecular basis of CD36 deficiency is due to a single base substitution (C478T) that replaces Pro90 with Ser and interrupts posttranslational modification of CD36 resulting in degradation of the CD36 precursor in the cytoplasm.[160] In patients with type II CD36 deficiency, monocytes carry both 478C and 487T alleles and platelets contain only the mRNA translated from the 487T variant of the gene. CD36 binds many ligands including native and oxidized lipoproteins, phospholipids, fatty acids including diacyglycerides,[161] collagen, and thrombospondin-1,[154] and mediates many physiologic and pathologic processes, including uptake of lipoprotein and fatty acids by cells, phagocytosis of apoptotic cells, and adhesion of malaria-infected red blood cells to the endothelium.

Nak[a−] platelets display abnormal aggregation to type V and fibrillar collagen, but aggregation is normal to other agonists.[162–165] Diaz-Ricart et al. showed that GPIV is important in the adhesive properties of platelets in flowing whole blood, particularly in their rapid adhesion to collagen.[166] Interestingly, Nak[a−] individuals show no evidence of a clinical hemostatic defect.[155] However, there are several reports on an association between CD36 deficiency and abnormalities in macrophage function, and the presence of metabolic disorders including insulin resistance and the metabolic syndrome.[157] Red blood cells infected with *Plasmodium falciparum* adhere to CD36 on endothelium; however, CD36 deficiency was reported to be associated with a more severe form of malaria with cerebral involvement.[167]

### M. P-Selectin; Gene: SELP

P-selectin is expressed on the surface of activated platelets and endothelial cells (see Chapter 12). It is encoded by a gene that extends over 50 Kb on chromosome 1 and comprises 17 exons. There are several polymorphisms reported in the coding region and the promoter of *SELP*.[168] Many of these polymorphisms are in linkage disequilibrium, and analysis of their frequencies among Caucasians determined the presence of two haplotype blocks in the structure of the P-selectin gene. Polymorphisms located in the promoter region of this gene form a haplotype block (5′ block) that is loosely associated with the second haplotype block (3′ block) containing polymorphisms in the coding region.[169] Among polymorphisms in the 3′ block, Thr715Pro has been shown to be associated with the level of soluble P-selectin in plasma. The Pro715 allele is associated with a lower plasma concentration of soluble P-selectin.[170–172] The frequency of the Pro715 allele is more common among whites (11.2%) compared to blacks (0.8%) and South Asians (3%).

### N. Other Platelet Polymorphisms

Gov[a,b] is an alloantigen system on human platelets responsible for refractory thrombocytopenia,[173,174] PTP, and FNAIT (see Chapter 53).[174,175] This alloantigen is expressed on CD109, a glycosylphosphatidylinositol-anchored protein (see Chapter 6).[176] Polymorphisms in several other platelet receptor genes have been identified that have not been associated with alloimmune responses. The noncoding sequence of the gene for the human protease-activated receptor 1 (PAR-1) (see Chapter 9) has several polymorphisms, including a C-T transition upstream of the transcriptional start site (−1426C/T), a 13-bp insertion at −506 (−506I/D), and an A-T transition 14 nucleotides upstream of exon 2 (IVS − 14A/T).[177] The IVS −14A/T, but not −506I/D or −1426C/T polymorphisms, was associated with PAR-1 expression levels in platelets. Carriers of the T allele at the IVS-14 position (allelic frequency 0.14) exhibited reduced PAR-1 expression, and reduced platelet aggregation and secretion in response to the PAR-1 activating peptide.[178]

Four nonsynonymous SNPs have been identified in platelet endothelial cell adhesion molecule-1 (PECAM-1), a member of the immunoglobulin family of adhesion receptors (see Chapter 11): Val80Met, Leu125Val, Ser563Asn, and Arg670Gly. The latter three are in strong linkage disequilibrium. A functional change in PECAM-1 was demonstrated for the Arg670Gly polymorphism, with the GG genotype exhibiting increased endothelial PECAM-1 phosphorylation and monocyte migration; no change in platelet aggregation was observed.[179]

c-mpl is a tyrosine kinase receptor for thrombopoietin on megakaryocytes and platelets that regulates platelet production (see Chapters 6 and 66). A SNP (G1238T) in exon 2 that induces an amino acid change (Lys39Asn) was reported in African American but not Caucasian patients with chronic myeloproliferative disorders.[180] The Lys39Asn polymorphism was associated with thrombocytosis and disrupted c-mpl expression in cell lines.[180] A C to A transversion at position 550 in the 5′ promoter region has also been identified, and the C allele was associated with higher platelet counts.[181] It is not known whether this SNP modulates c-mpl receptor levels.

In addition to polymorphisms in platelet plasma membrane receptor genes, polymorphisms in the genes of molecules involved in platelet activation (inside-out signaling), secretion, shape change, or postreceptor occupation events could affect platelet function. As an example, a single base substitution (C852T) in the $\beta_3$ subunit of heterotrimeric G proteins may modify the response of platelets to epinephrine.[182] The carriers of the 825T allele were observed to have enhanced platelet aggregation in response to epinephrine or combined epinephrine/ADP.[182] Compared to the Asn700 variant, the Ser700 variant was associated with increased platelet aggregation and thrombospondin expression on the platelet surface.[183] Megakaryocytes and platelets express estrogen receptor (ER) $\beta$, but not ER $\alpha$, and there are known gender differences in platelet function.[184] There are at least five polymorphisms in the ER $\beta$ gene,[185,186] two of which are associated with a functional consequence.[185,187] Finally, when considering how genetic variations will affect platelet biology, it will be important to consider functional polymorphisms in genes of nonplatelet origin whose products affect platelet function, such as endothelial cell nitric oxide synthase,[188–190] prion protein,[191] fibrinogen,[192] and VWF.[193,194]

## III. Clinical Consequences of Platelet Polymorphisms

### A. Alloimmune Thrombocytopenias

The alloimmune thrombocytopenias are the best established clinical disorders that result from platelet polymorphisms. These conditions — FNAIT and PTP — are discussed in detail in Chapter 53. The following sections will review other clinical aspects of platelet polymorphisms, the details of which are still evolving but which apply to much larger segments of the general population.

### B. Platelet Hyperreactivity

Intrinsic to the notion that platelet polymorphisms contribute to the development of clinical thrombosis, is that these variations increase platelet reactivity or adhesiveness. The heterogeneous results of platelet *in vitro* assays indicate that there is a range of platelet reactivity among different individuals. For example, some platelets require less epinephrine to induce aggregation than do other platelets and could be considered hyperreactive. Traditionally, clinical laboratories measure biochemical or functional parameters that are associated with clinical events. Elevations in creatine phosphokinase are used to diagnose acute myocardial infarction (MI), but similar biochemical (e.g., platelet factor 4, $\beta$-thromboglobulin, or thromboxane $A_2$) or platelet functional assays have never become routine in clinical coagulation laboratories, where platelet studies are designed almost solely to detect *hypo*function in hemorrhagic disorders. Nevertheless, based on our understanding of platelet physiology, it is possible to design assays of platelet *hyper*reactivity. However, such *in vitro* assays will only have clinical value when they are shown to correlate with clinical thrombosis. Attempts to correlate platelet reactivity with clinical thrombotic events have largely been cross-sectional, such that it has not been possible to assess whether hyperreactive platelets were (at least in part) responsible for the thrombosis, or whether the acute, stressful event caused expression of a marker of platelet reactivity. (Of course, a genetic risk factor would not be affected by such events.) Few prospective studies have attempted to correlate some parameter of platelet function with the development of an arterial thrombotic event. An increased platelet release reaction has been associated with a poor prognosis after MI.[195] This was a study of 22 MI patients and the association was seen only in heparinized, and not citrated, platelet-rich plasma. Trip et al. studied 149 survivors of MI, and found that those with spontaneous platelet aggregation (defined as >20% over 10 min) had a significantly higher risk of recurrent events.[196] However, aspirin use was not rigorously excluded and there were only 12 subjects who had events. In a large prospective study Martin et al. followed 1716 post-MI patients and found a significantly greater mean platelet volume (10.09 fL in cases vs. 9.72 fL in controls) was associated with an increased risk of recurrent ischemic events.[197] An even larger Canadian study found that among patients with documented coronary artery disease (CAD) who were taking aspirin for secondary prevention, urinary concentrations of 11-dehydro thromboxane $B_2$ predicted recurrent MI or thrombotic death (OR = 3.5; 95% CI = 1.7 – 7.4; $p < 0.001$).[198] However, all of these assays have limitations and there remains a need for better predictors of *in vitro* hyperreactivity that correlate with clinical thrombotic events. If genetic assays could predict platelet hyperreactivity they would offer certain advantages, since they could be performed long after a blood sample has been obtained and could be used in large population studies in which platelet assays cannot easily be performed.

## C. Platelet Polymorphisms as Risk Factors for Thrombosis

Platelets are generally believed to play a more important role in the development of arterial thrombi than venous thrombi. The higher arterial shear force pushes the low-density (compared to other blood cells) platelets to the lumen periphery where they contact the subendothelium after injury. In addition, platelet adhesion to subendothelial surfaces is shear-stress dependent, and high shear forces may induce platelet activation[199] and enhance the ligand-binding repertoire of GPIIb-IIIa to include VWF in addition to fibrinogen.[200] It is a reasonable assumption that polymorphisms altering function of platelets might change the propensity of an individual toward thrombotic or bleeding events. Most studies have examined patients with CAD, with fewer examining stroke, and fewer still peripheral artery disease. In this section, we will discuss platelet polymorphisms as genetic risk factors for arterial thrombosis. MI has received the most attention to date, and may serve as a model for thrombus formation in cerebral and peripheral arteries.

**1. Genetics and Myocardial Infarction.** Estimates of the effects of inheritance on MI have ranged from 20[201] to 80%.[202] Family studies provided the initial evidence that there is a genetic component to MI.[201–205] An especially powerful approach to assessing the influence of heredity on any phenotype is the use of twin studies, and three different large twin studies identified a genetic effect on MI risk that is lost at older ages.[206–208] Findings from a number of studies have suggested that the genetic risk may be greater in women than in men.[206,209,210] Until recently, the genetic component affecting the pathogenesis of MI has not extended far beyond considerations of lipid metabolism and the development of atherosclerosis. In the mid-1990s the possibility that gain-of-function genetic variants in platelet proteins might lead to a more subtle, prothrombotic phenotype gained momentum with the first association study between $Pl^{A2}$ and acute coronary syndrome (ACS). This notion was biologically plausible, although considering the multifactorial nature of ACS, it is generally believed that multiple genes affect the risk for ACS and the potential impact of a sequence variation in a platelet-related gene would be relatively small.[211] Nevertheless, considering the prevalence of CAD, the seriousness of ACS, and the prevalence of platelet polymorphisms, the overall potential prothrombotic effect of platelet polymorphisms may still be substantial.

**2. Pl^A2.** In 1996, $Pl^{A2}$-positivity was first reported to be twice as common in patients with ACS compared to in-

patients with similar traditional risk factors but who lacked CAD.[212] This genetic effect was most pronounced at a young age. Some subsequent studies supported the concept that $Pl^{A2}$ is a risk factor for arterial events,[210,212–223] findings consistent with hyperreactivity of $Pl^{A2}$-expressing GPIIb-IIIa receptors seen in platelet and cell line studies.[29,30,36] However, other studies did not find an association between $Pl^{A2}$ and ACS, including several that included large numbers of patients.[224–232] These various studies had fundamental differences in design, particularly in their choice of clinical phenotype and selection of control subjects, which in turn led to differences in patient demographics, such as age, gender, geographical origin, and the type of MI. This controversy is not surprising considering that the pathogenesis of ACS is complex and multifactorial. Furthermore, there are many errors made in interpreting association studies (thoughtfully reviewed by Cardon and Bell),[233] including publication bias against negative studies. The choice of control groups appears to be an especially important variable in determining whether an association was identified. With one exception,[214] if the $Pl^{A2}$ prevalence in the control group was less than 20%, an association between $Pl^{A2}$ and ACS was observed. If a true risk factor was overrepresented in the control group due to the improper inclusion of disease subjects, the effect of this risk factor may not be detected. This may be the case where the control groups that were more rigorously screened for the lack of CAD and/or better matched with the cases for other risk factors had a lower prevalence of the $Pl^{A2}$ allele.[212,215,216,218] Perhaps the most instructive studies have been two autopsy studies that found $Pl^{A2}$ was strongly associated with thrombus seen in the coronary arteries of subjects who died of acute MI.[222,223] This association was not seen in subjects >60 years of age. A prospective study of 592 patients with stable CAD found an increased risk for developing subsequent cardiac events in patients with the $Pl^{A2}$ polymorphism who smoked cigarettes.[234] A recent study in men with CAD noted an interaction between the $Pl^{A2}$ polymorphism and plasma fibrinogen levels in the determination of risk for cardiovascular events.[235]

GPIIIa is expressed in endothelial and smooth muscle cells, such that expression of the Pro33 isoform could modify the risk for atherosclerosis. Autopsy and angiography studies may be more useful for assessing this risk than are studies on ACS. Unlike the inconsistency in the MI studies, all seven of the published angiography studies found a higher prevalence of $Pl^{A2}$ in the patients with the most stenosis,[214,219,229,231,236–238] but this difference achieved statistical significance in only one of seven studies. Interestingly, postmortem angiography found significantly more stenosis in $Pl^{A1,A1}$ subjects.[222] The $Pl^{A2}$ polymorphism has also been linked with increased levels of cholesterol, triglyceride,[239] lipoprotein(a),[240] and plasma fibrinogen.[241] Although no known mechanism explains these associations, such variables should be considered in future epidemiologic studies. Much larger clinical studies and more basic research are needed to sort out the contributions of each of these markers to arterial thrombosis. Meta-analysis suggest that the *Pl*[A2] polymorphism presents a small but significant risk for cardiovascular disease, with higher risk in cohorts that are either younger or undergoing restenosis.[242,243]

The *Pl*[A2] polymorphism of GPIIIa has also been investigated for its association with stroke. The majority of studies have found no association with stroke.[224,228,244–246] Although some studies have found an association between *Pl*[A2] and stroke,[247–250] these tended to have a small number of patients and/or the association was only seen in post hoc subgroup analyses. Thus, the weight of the evidence does not support a link between *Pl*[A2] and stroke.

In summary, the Pl[A2] isoform of GPIIIa appears to cause a prothrombotic phenotype in platelets. Taken together, the epidemiologic data suggest *Pl*[A2] is not a risk for ACS in older, otherwise relatively healthy individuals,[224,232] but *Pl*[A2] positivity appears to confer a mild risk in younger individuals, and this risk may require interaction with environmental factors, especially cigarette smoking.[218] Of potential clinical significance are five studies suggesting a pharmacogenetic effect of the *Pl*[A2] genotype with aspirin,[30,31,35,251] pravastatin,[252] and clopidogrel.[253,254] Further work is needed to know whether this genotype should be considered when making therapeutic decisions.

**3. Bak.** There have only been a few studies testing for an association between the Bak polymorphism of *ITGA2B* and arterial thrombotic disorders. A German study found no association between *Bak* genotypes and the risk of thrombosis and restenosis after coronary stent placement.[255] A study from the UK found *Bak*[a] to be associated with mortality after stroke,[256] whereas a small study of young women from the U.S. found an association between *Bak*[b] and MI risk if other cardiovascular risk factors were present.[257] Given the heterogeneity of these study designs, additional investigations are necessary to address the risk of these *ITGA2B* alleles.

**4. GPIb Polymorphisms (Ko, VNTR, Kozak).** The presence of *Ko*[a] (Met145) has been associated with an increased risk of CAD in subjects younger than 60 years of age.[258] A study from Spain demonstrated that the *Ko* polymorphism of *GP1BA* is associated with an increased risk of CAD and cerebrovascular disease but not deep vein thrombosis.[259] Mikkelsson et al. performed an autopsy study on 700 middle-aged Finnish men who died suddenly or violently, and found that men with acute MI and coronary thrombosis were more likely to be carriers of the Met145 allele compared to those who died because of noncardiac causes (ORs of 2.0 and 2.6, respectively; $p < 0.005$ for both).[260] In men less than 55 years old, the overrepresentation of the Met145 allele among individuals with coronary thrombosis or victims of acute MI was even more pronounced (ORs of 9.2 and 5.6, respectively). In a third study, a higher frequency of the *Ko*[a] allele of *GP1BA* was observed among patients with cerebrovascular disease. The genotype effect was more significant in younger individuals (<60 years of age) and those without acquired cardiovascular risk factors.[255]

As mentioned above, the *Ko*[a] and VNTR B alleles are often linked. Several studies suggest that the *VNTR B/ Met145* allele may be a risk factor for CAD, MI, and stroke.[258–260] Due to a near complete linkage between *Ko* and *VNTR* polymorphisms, it is difficult to separate the effect of these polymorphisms as risk factors for arterial thrombotic events. However, two studies suggest that heterozygosity for the *GP1BA* VNTR polymorphism, rather than the absolute length of this protein, may be a risk factor.[261,262] We have shown that among African Americans the presence of the C/C VNTR genotype (vs. C/D or B/C genotypes) provided a protective effect against CAD.[261] In the second study, carriers of the C/D genotype for the *GP1BA* VNTR polymorphism had a twofold higher risk for MI compared to other genotypes.[255] It is unclear whether the Kozak (−5T/ C) polymorphism is associated with cardiovascular disease, with most studies finding no association.[94,101,103,263] In a few studies, carriage of the −5C allele was associated with acute coronary events,[264,265] but others have observed a protective effect.[96,96,266] However, there has been a consistent reporting of the −5C allele as a risk factor for ischemic cerebrovascular events.[267–269]

**5. The 807T/C Polymorphism of *ITGA2*.** The 807T allele has been associated with MI or stroke in some studies,[270–273] but not others.[127,274] A particularly large study of 2237 patients from Germany demonstrated that inheritance of the 807T allele of *ITGA2* represents a potent risk factor for nonfatal MI.[272] In this study and others,[273] the effect of the 807T allele was most pronounced at younger ages. A study from the Netherlands found the 807TT genotype to confer a potent risk in patients with other risk factors.[275] Only one study considered thrombotic complications after coronary stent placement, but no data were provided for a control group.[276] Thus, in younger individuals the 807T allele may represent a small risk factor for MI and stroke.

An investigation of the *Br* polymorphism included 2163 Caucasian males with angiographically proven CAD who were genotyped for G1648A polymorphism (*Br*[a,b]).[277] In the subgroup of individuals with few risk factors for CAD (patients with high apoAI/apoB and no history of active smoking), the presence of *Br*[a] (Glu505, 1648A) was associated with increased risk and extent of CAD.

In summary, the 807T allele of *ITGA2* is associated with a prothrombotic platelet phenotype,[123,124,127,128] and based on a small number of studies there is reason to believe it may be a risk for arterial thrombotic events in younger patients. As with several $Pl^{A2}$ studies, the 807T risk appears greatest in young smokers.

**6. P-Selectin.** The Pro715 allele was protective against MI in Europeans.[168,169] The presence of Asn290 and Asn562 in a haplotype background of Val599 and Thr715 increased the risk of MI in Europeans.[169] The frequency of 90 polymorphisms involving 56 genes was determined in 319 individuals who developed stroke and in 2092 control subjects in the Physician Health study followed for an average of 13.2 years.[278] The Val640Leu polymorphism in *SELP* (same as Val599Leu according to Herrman et al.)[168] was significantly associated with the risk of stroke ($p = 0.003$), whereas the Thr715Pro polymorphism was not.

**7. FcγRIIA.** The pathogenesis of heparin-induced thrombocytopenia (HIT) involves platelet factor 4, heparin, anti-heparin antibodies, and the platelet Fc receptor, FcγRIIA (see Chapter 48).[279,280] The requirement for each component was recently demonstrated in an animal model.[281] Because FcγRIIA can mediate platelet activation,[282–287] it was of interest to examine the association of the His131Arg polymorphism with clinical thrombosis in HIT. One study found no relationship.[288] A second, larger study[289] investigated 389 HIT patients, 351 patients with thrombocytopenia or thrombosis of non-HIT etiology, and 256 healthy controls. In a subgroup of 122 well-characterized HIT patients, the Arg131/Arg131 genotype was significantly overrepresented among patients with thrombotic events (37% vs. 17%; $p = 0.036$).

**8. Other Polymorphisms.** The 13254CC genotype (Ser219/Ser219) of *GP6* has been associated with an increased risk of MI,[133] despite being associated with less *in vitro* thrombogenicity on collagen.[132] Another study that recruited only men suffering from a fatal prehospital MI observed a correlation between the Ser219 allele and an increased coronary thrombosis and coronary stenosis.[290] The Thr249Ala polymorphism of *GP6* has also been associated with MI risk in Japanese,[291] whereas no association between stroke and the Gln317Leu polymorphism was found in a study from Australia.[292]

Few clinical studies have examined polymorphisms in the genes encoding PAR1 (gene is *F2R*), P-selectin, $P2Y_{12}$, $5HT_{2A}$, 5-HTT, GPVI, or PECAM-1. The −506I allele of *F2R* was not associated with MI risk, but was less frequent in males with venous thromboembolic disease than in control males.[293] The Pro715 allele of *SELP* was found to have a lower frequency among patients with MI than control subjects.[168] A single study has shown the *P2RY12* H2 haplotype was associated with peripheral arterial disease after controlling for confounding variables.[294] Polymorphisms in

the thrombospondin gene, notably Asn700Ser, have been linked to an increased risk of premature, familial heart attacks in some, but not all, studies.[295–297]

The TT genotype of the T102 polymorphism of *HTR2A* was reported to be associated with nonfatal MI in a Japanese population.[298] An association between the 5-HTT long/long genotype of the serotonin transporter gene and MI has been described in Japanese and European populations.[299,300] A Spanish study found no correlation between the risk of developing MI and the TT genotype of *HTR2A* or the 5-HTT long/long genotype.[301]

An emerging theme among the several studies that have examined the association of *PECAM-1* polymorphisms with CAD is that an increased risk for atherogenesis was associated with the haplotype 125Val/563Asn/670Gly,[179,302–305] whereas an increased risk for MI was associated with 125Leu/563Ser/670Arg.[179,306,307]

### D. Platelet Polymorphisms as Modifiers of Bleeding

An individual person's risk for thrombosis or bleeding is often presented as the sum total of factors that favor thrombosis (platelets, coagulation factors, etc.) and the factors that prevent thrombosis (anticoagulants, endothelium, etc.). For patients with inherited bleeding disorders like von Willebrand disease (VWD) or hemophilia, the hemostatic balance is shifted toward hemorrhage. It is not difficult to imagine that platelet genetic variants could modify the bleeding risk in these patients. To date, few studies have addressed this issue.

Kunicki et al. performed a detailed bleeding assessment of type 1 VWD patients from 14 families and tested for associations with nine SNPs in six platelet-related genes.[308] These investigators found that polymorphisms 807C, Ile843, and Pro219 in *ITGA2*, *ITGA2B*, and *GP6*, respectively, were associated with more bleeding. The finding with *ITGA2* is consistent with an earlier study showing 807C-positive patients more likely to be diagnosed with type I VWD,[309] although not all investigators have obtained similar data.[310]

Compared to individuals homozygous for $Pl^{A1}$, subjects carrying the $Pl^{A2}$ allele of *ITGB3* exhibit less bleeding after placement of ventricular assist devices,[311] after treatment with the GPIIb-IIIa antagonist orbofiban,[312] after subarachnoid hemorrhage,[313] and if they carry a diagnosis of Glanzmann thrombasthenia.[314] Thus, although the evidence is sparse, it suggests platelet polymorphisms modify the bleeding risk in a variety of congenital and acquired hemorrhagic conditions.

### E. Platelet Pharmacogenetics

**1. Aspirin.** The definition of aspirin resistance is controversial,[315] but is usually taken to mean a less than expected inhibition of platelet function by aspirin in an *in vitro* assay. Depending upon the assay, aspirin resistance has been reported to occur in 6 to 40% of individuals.[316,317] Variations in the COX-1 gene would seem obvious candidates to affect aspirin responsiveness, but although many SNPs have been identified, Hillarp et al. found these were not associated with aspirin sensitivity.[318] In general, GPIIIa $Pl^{A2}$-positive platelets are more sensitive to inhibition by aspirin,[30,35] but individuals expressing $Pl^{A2}$ show less activation of coagulation after aspirin than subjects lacking this variant.[319] Unfortunately, little attention has been given to the interaction between the $Pl^A$ polymorphism and aspirin in the many genetic studies of cardiovascular outcomes. Aspirin resistance has been reported to be significantly associated with genetic variation in the platelet surface ADP receptor gene P2Y$_1$.[149]

**2. Clopidogrel.** The antiplatelet activity of clopidogrel requires metabolism to an active form by the cytochrome P450, CYP3A4. Lau et al. have shown that some portion of the substantial variability in platelet inhibition by clopidogrel is due to individual variability in the activity of CYP3A4.[320] Although carriage of the H2 haplotype of *P2RY12* has been associated with enhanced ADP-induced platelet aggregation,[148] it does not seem to affect ADP aggregation after a loading dose of clopidogrel in patients undergoing PCI.[321] It has been reported that the 807T allele of *ITGA2* confers less sensitivity to clopidogrel inhibition of collagen-induced platelet aggregation.[322] However, this was a small study with many more statin users in the 807T-positive subjects. Since statin use has been reported to impair the antiplatelet effect of clopidogrel,[323] the interaction of the *ITGA2* 807T allele and clopidogrel needs confirmation.

**3. GPIIb-IIIa Antagonists.** There is substantial interindividual variability in the antiplatelet response to GPIIb-IIIa antagonists.[324,325] In general, GPIIb-IIIa antagonists have been shown to be less effective in $Pl^{A2}$-positive than $Pl^{A2}$-negative individuals,[30,326,327] although this effect may be both antagonist- and concentration-dependent.[326] Compared to $Pl^{A2}$-negative subjects, $Pl^{A2}$-positive patients who received the oral GPIIb-IIIa antagonist orbofiban had less bleeding.[312]

**4. GPIb-V-IX Antagonists.** Aurintricarboxylic acid blocks GPIbα-VWF binding and has been proposed for clinical use. This agent is more effective in blocking ristocetin-induced agglutination of VNTR *B*-positive platelets than VNTR *B*-negative platelets.[328]

**5. Statins.** Several studies have tested for an interaction between statins and the $Pl^A$ polymorphism, and found that the beneficial effects of statins for clinical CAD events were greater in the $Pl^{A2}$-positive patients.[254,329]

### F. Future Studies

The approaches for discovering the platelet genetic variations affecting hemostasis/thrombosis are the candidate gene method and genome scans.[330] Until now, investigators have only considered small numbers of candidate genes and usually only one SNP per gene, and it is virtually certain that there are many unidentified genes and genetic variations that participate in thrombosis. A more comprehensive candidate gene approach or the open-minded genome scan approach will likely be needed to move the field forward in a substantive manner. Candidate gene strategies with thousands of platelet genes and multiple SNPs in each gene are now a feasible approach for genetic association studies. Testing only SNPs that change amino acids is a strategy to reduce costs. Resequencing genes from well-defined phenotypes may be useful for identifying new variants that cause platelet hyper- or hyporeactivity.

The advantage of a genome scan is that it requires no presumptions about gene function and has a goal of defining loci/genes not previously known to contribute to the phenotype. Depending upon the ethnicity of the population, a comprehensive whole genome scan may require genotyping 220,000 to 570,000 haplotype-tag SNPs to characterize the whole genome.[331] Because the cost of genotyping continues to fall, these approaches are now feasible. With the rapid expansion in the list of identified SNPs and identification of haplotype blocks in the human genome, there is major interest in utilizing haplotypes, rather than single polymorphisms, in phenotype association studies[332,333] because of several advantages of haplotypes. First, the structure of the protein product of a gene is affected by interaction between all polymorphisms in that gene; thus it is more reasonable to study a combination of polymorphisms rather than several separate polymorphisms. Second, the structural units of genetics are haplotypes, not single polymorphisms, such that when linkage disequilibrium has been established, genotyping for haplotypes provides more information and is more cost-efficient. Third, the statistical power of analysis comparing haplotypes is greater than that of several single polymorphisms.[332]

## IV. Conclusions

The investigation of platelet polymorphisms has traveled a circuitous road in the past 50 years. The molecular bases for all of the common polymorphisms of surface antigens are known, but major challenges still exist. Better assays are needed for the detection of clinical platelet hyperfunction, and these assays must be prospectively correlated with clinical thrombosis. Larger *in vitro* studies with platelets are needed to address the potential functionality of these polymorphisms, and the effect of confounding variables must be considered. Caution is warranted in becoming overly optimistic about the significance of those polymorphisms for which single or few studies have yielded positive results. Perhaps polymorphisms of rate-limiting signaling molecules will produce more easily identifiable *ex vivo* and clinical effects. Additional studies from existing epidemiology databases may not be the best way to resolve the existing controversies. Rather, well-designed studies with an adequate sample size, proper control group selection, and very rigorous definitions of the clinical phenotype are needed. Larger genomic studies are needed — to define the extent of linkage disequilibrium between SNPs, determine haplotypes, tag SNPs, and utilize this information in broad candidate gene association studies and genomic scan approaches to discover new genes involved in hemostasis and thrombosis. Ultimately, the genomic profile of an individual's thrombotic risk factors, coupled with sound clinical judgment, should aid in decisions pertaining to invasive testing and therapy for arterial vascular disease.

# References

1. von dem Borne, A. E., & Decary, F. (1990). ICSH/ISBT Working Party on platelet serology. Nomenclature of platelet-specific antigens. *Vox Sang, 58,* 176.
2. Newman, P. J. (1994). Nomenclature of human platelet alloantigens: A problem with the HPA system? *Blood, 83,* 1447–1451.
3. Cargill, M., Altshuler, D., Ireland, J., et al. (1999). Characterization of single-nucleotide polymorphisms in coding regions of human genes. *Nature Genet, 22,* 231–238.
4. Halushka, M. K., Fan, J. B., Bentley, K., et al. (1999). Patterns of single-nucleotide polymorphisms in candidate genes for blood-pressure homeostasis. *Nat Genet, 22,* 239–247.
5. Di Paola, J., Jugessur, A., Goldman, T., et al. (2005). Platelet glycoprotein Ibalpha and integrin alphabeta polymorphisms: Gene frequencies and linkage disequilibrium in a population diversity panel. *J Thromb Haemost, 3,* 1511–1521.
6. Phillips, D. R., Charo, I. F., Parise, L. V., et al. (1988). The platelet membrane glycoprotein IIb-IIIa complex. *Blood, 71,* 831–843.
7. Bray, P. F., & Shuman, M. A. (1990). Identification of an abnormal gene for the GPIIIa subunit of the platelet fibrinogen receptor resulting in Glanzmann's thrombasthenia. *Blood, 75,* 881–888.
8. Nurden, A. T. (1999). Inherited abnormalities of platelets. *Thromb Haemost, 82,* 468–480.
9. Bray, P. F., Rosa, J. P., Johnston, G. I., et al. (1987). Platelet glycoprotein IIb. Chromosomal localization and tissue expression. *J Clin Invest, 80,* 1812–1817.
10. Zimrin, A. B., Gidwitz, S., Lord, S., et al. (1990). The genomic organization of platelet glycoprotein IIIa. *J Biol Chem, 265,* 8590–8595.
11. Wilhide, C. C., Jin, Y., Guo, Q., et al. (1997). The human integrin $b_3$ gene is 63 kb and contains a 5'-UTR sequence regulating expression. *Blood, 90,* 3951–3961.
12. Villa-Garcia, M., Li, L., Riely, G., et al. (1994). Isolation and characterization of a TATA-less promoter for the human $b_3$ integrin gene. *Blood, 83,* 668–676.
13. Newman, P. J., McFarland, J. G., & Aster, R. H. (1998). The alloimmune thrombocytopenias. In J. Loscalzo & A. I. Schafer (Eds.), *Thrombosis and hemorrhage* (pp. 599–615). Baltimore, MD: Williams & Wilkins.
14. Kunicki, T. J., & Aster, R. H. (1979). Isolation and immunologic characterization of the human platelet alloantigen, Pl^A1. *Mol Immunol, 16,* 353–360.
15. Newman, P. J., Martin, L. S., Knipp, M. A., et al. (1985). Studies on the nature of the human platelet alloantigen, PlA1: Localization to a 17,000-dalton polypeptide. *Mol Immunol, 22,* 719–729.
16. Fitzgerald, L. A., Steiner, B., Rall, S. C., et al. (1987). Protein sequence of endothelial glycoprotein IIIa derived from a cDNA clone. Identity with platelet glycoprotein IIIa and similarity to "integrin." *J Biol Chem, 262,* 3936–3939.
17. Rosa, J. P., Bray, P. F., Gayet, O., et al. (1988). Cloning of glycoprotein IIIa cDNA from human erythroleukemia cells and localization of the gene to chromosome 17. *Blood, 72,* 593–600.
18. Newman, P. J., Derbes, R. S., & Aster, R. H. (1989). The human platelet alloantigens, Pl^A1 and Pl^A2, are associated with a leucine33/proline33 amino acid polymorphism in membrane glycoprotein IIIa, and are distinguishable by DNA typing. *J Clin Invest, 83,* 1778–1781.
19. Unkelbach, K., Kalb, R., Breitfeld, C., et al. (1994). New polymorphism on platelet glycoprotein IIIa gene recognized by endonuclease Msp I: Implications for Pl^A typing by allele-specific restriction analysis. *Transfusion, 34,* 592–595.
20. Walchshofer, S., Ghali, D., Fink, M., et al. (1994). A rare leucine40/arginine40 polymorphism on platelet glycoprotein IIIa is linked to the human platelet antigen 1b. *Vox Sang, 67,* 231–234.
21. Flug, F., Espinola, R., Liu, L. X., et al. (1991). A 13-mer peptide straddling the leucine33/proline33 polymorphism in glycoprotein IIIa does not define the Pl^A1 epitope. *Blood, 77,* 1964–1969.
22. Valentin, N., Visentin, G. P., & Newman, P. J. (1995). Involvement of the cysteine-rich domain of glycoprotein IIIa in the expression of the human platelet alloantigen, Pl^A1: Evidence for heterogeneity in the humoral response. *Blood, 85,* 3028–3033.
23. van Leeuwen, E. F., Leeksma, O. C., Van Mourik, J. A., et al. (1984). Effect of the binding of anti-Zwa antibodies on platelet function. *Vox Sang, 47,* 280–289.
24. Kunicki, T. J. (1989). Biochemistry of platelet-associated isoantigens and alloantigens. In T. J. Kunicki & J. N. George (Eds.), *Platelet immunobiology: Molecular and clinical aspects* (pp. 99–120). Philadelphia, PA: Lippincott.
25. D'Souza, S. E., Ginsberg, M. H., Burke, T. A., et al. (1988). Localization of an Arg-Gly-Asp recognition site within an integrin adhesion receptor. *Science, 242,* 91–93.

26. Charo, I. F., Nannizzi, L., Phillips, D. R., et al. (1991). Inhibition of fibrinogen binding to GP IIb-IIIa by a GP IIIa peptide. *J Biol Chem, 266,* 1415–1421.

27. Bork, P., Doerks, T., Springer, T. A., et al. (1999). Domains in plexins: Links to integrins and transcription factors. *Trends Biochem Sci, 24,* 261–263.

28. Zang, Q., & Springer, T. A. (2001). Amino acid residues in the PSI domain and cysteine-rich repeats of the integrin $b_2$ subunit that restrain activation of the integrin $a_Xb_2$. *J Biol Chem, 276,* 6922–6929.

29. Feng, D., Lindpaintner, K., Larson, M. G., et al. (1999). Increased platelet aggregability associated with platelet GPIIIa PlA2 polymorphism: The Framingham Offspring Study. *Arterioscler Thromb Vasc Biol, 19,* 1142–1147.

30. Michelson, A. D., Furman, M. I., Goldschmidt-Clermont, P., et al. (2000). Platelet GP IIIa PlA polymorphisms display different sensitivities to agonists. *Circulation, 101,* 1013–1018.

31. Szczeklik, A., Undas, A., Sanak, M., et al. (2000). Short report: Relationship between bleeding time, aspirin and the PlA1/A2 polymorphism of platelet glycoprotein IIIa. *Br J Haematol, 110,* 965–967.

32. Lasne, D., Krenn, M., Pingault, V., et al. (1997). Interdonor variability of platelet response to thrombin receptor activation: Influence of PlA2 polymorphism. *Br J Haematol, 99,* 801–807.

33. Goodall, A. H., Curzen, N., Panesar, M., et al. (1999). Increased binding of fibrinogen to glycoprotein IIIa-proline33 (HPA-1b, PlA2, Zwb) positive platelets in patients with cardiovascular disease. *Eur Heart J, 20,* 742–747.

34. Meiklejohn, D. J., Urbaniak, S. J., & Greaves, M. (1999). Platelet glycoprotein IIIa polymorphism HPA 1b (PlA2): No association with platelet fibrinogen binding. *Br J Haematol, 105,* 664–666.

35. Andrioli, G., Minuz, P., Solero, P., et al. (2000). Defective platelet response to arachidonic acid and thromboxane A2 in subjects with PlA2 polymorphism of $b_3$ subunit (glycoprotein IIIa). *Br J Haematol, 110,* 911–918.

36. Vijayan, K. V., Goldschmidt-Clermont, P. J., Roos, C., et al. (2000). The PlA2 polymorphism of integrin $b_3$ enhances outside-in signaling and adhesive functions. *J Clin Invest, 105,* 793–802.

37. Sajid, M., Vijayan, K. V., Souza, S., et al. (2002). PlA polymorphism of integrin b3 differentially modulates cellular migration on extracellular matrix proteins. *Arterioscler Thromb Vasc Biol, 22,* 1984–1989.

38. Vijayan, K. V., Liu, Y., Dong, J. F., et al. (2003). Enhanced activation of mitogen-activated protein kinase and myosin light chain kinase by the Pro33 polymorphism of integrin b3. *J Biol Chem, 278,* 3860–3867.

39. Vijayan, K. V., Liu, Y., Sun, W., et al. (2005). The Pro33 isoform of integrin β3 enhances outside-in signaling in human platelets by regulating the activation of serine/threonine phosphatases. *J Biol Chem, 280,* 21756–21762.

40. Vijayan, K. V., Huang, T. C., Liu, Y., et al. (2003). Shear stress augments the enhanced adhesive phenotype of cells expressing the Pro33 isoform of integrin beta3. *FEBS Lett, 540,* 41–46.

41. Furihata, K., Nugent, D. J., Bissonette, A., et al. (1987). On the association of the platelet-specific alloantigen, Pena, with glycoprotein IIIa. Evidence for heterogeneity of glycoprotein IIIa. *J Clin Invest, 80,* 1624–1630.

42. Friedman, J. M., & Aster, R. H. (1985). Neonatal alloimmune thrombocytopenic purpura and congenital porencephaly in two siblings associated with a "new" maternal antiplatelet antibody. *Blood, 65,* 1412–1415.

43. Shibata, Y., Miyaji, T., Ichikawa, Y., et al. (1986). A new platelet antigen system, Yuka/Yukb. *Vox Sang, 51,* 334–336.

44. Simon, T. L., Collins, J., Kunicki, T. J., et al. (1988). Post-transfusion purpura associated with alloantibody specific for the platelet antigen, Pena. *Am J Hematol, 29,* 38–40.

45. Wang, R., Furihata, K., McFarland, J. G., et al. (1992). An amino acid polymorphism within the RGD binding domain of platelet membrane glycoprotein IIIa is responsible for the formation of the Pena/Penb alloantigen system. *J Clin Invest, 90,* 2038–2043.

46. Tanaka, S., Ohnoki, S., Shibata, H., et al. (1996). Gene frequencies of human platelet antigens on glycoprotein IIIa in Japanese. *Transfusion, 36,* 813–817.

47. Seo, D. H., Park, S. S., Kim, D. W., et al. (1998). Gene frequencies of eight human platelet-specific antigens in Koreans. *Transfus Med, 8,* 129–132.

48. Kuijpers, R. W., Simsek, S., Faber, N. M., et al. (1993). Single point mutation in human glycoprotein IIIa is associated with a new platelet-specific alloantigen (Mo) involved in neonatal alloimmune thrombocytopenia. *Blood, 81,* 70–76.

49. McFarland, J. G., Blanchette, V., Collins, J., et al. (1993). Neonatal alloimmune thrombocytopenia due to a new platelet-specific alloantibody. *Blood, 81,* 3318–3323.

50. Kroll, H., Kiefel, V., Santoso, S., et al. (1990). Sra, a private platelet antigen on glycoprotein IIIa associated with neonatal alloimmune thrombocytopenia. *Blood, 76,* 2296–2302.

51. Jallu, V., Meunier, M., Brement, M., et al. (2002). A new platelet polymorphism Duv(a+), localized within the RGD binding domain of glycoprotein IIIa, is associated with neonatal thrombocytopenia. *Blood, 99,* 4449–4456.

52. Santoso, S., Kalb, R., Kroll, H., et al. (1994). A point mutation leads to an unpaired cysteine residue and a molecular weight polymorphism of a functional platelet $b_3$ integrin subunit. The Sra alloantigen system of GPIIIa. *J Biol Chem, 269,* 8439–8444.

53. Wang, R., McFarland, J. G., Kekomaki, R., et al. (1993). Amino acid 489 is encoded by a mutational "hot spot" on the $b_3$ integrin chain: The CA/TU human platelet alloantigen system. *Blood, 82,* 3386–3391.

54. Tanaka, S., Taniue, A., Nagao, N., et al. (1996). Genotype frequencies of the human platelet antigen, Ca/Tu, in Japanese, determined by a PCR-RFLP method. *Vox Sang, 70,* 40–44.

55. Peyruchaud, O., Bourre, F., Morel-Kopp, M. C., et al. (1997). HPA-10w(b) (Laa): Genetic determination of a new platelet-specific alloantigen on glycoprotein IIIa and its expression in COS-7 cells. *Blood, 89,* 2422–2428.

56. Simsek, S., Folman, C., van der Schoot, C. E., et al. (1997). The Arg633His substitution responsible for the private

platelet antigen Gro[a] unravelled by SSCP analysis and direct sequencing. *Br J Haematol, 97,* 330–335.

57. Wagner, C. L., Mascelli, M. A., Neblock, D. S., et al. (1996). Analysis of GPIIb/IIIa receptor number by quantification of 7E3 binding to human platelets. *Blood, 88,* 907–914.

58. Bray, P. F., Rosa, J.-P., Lingappa, V. R., et al. (1986). Biogenesis of the platelet receptor for fibrinogen: Evidence for separate precursors for glycoproteins IIb and IIIa. *Proc Natl Acad Sci USA, 83,* 1480–1484.

59. Heidenreich, R., Eisman, R., Surrey, S., et al. (1990). Organization of the gene for platelet glycoprotein IIb. *Biochemistry, 29,* 1232–1244.

60. von dem Borne, A. E., von Riesz, E., Verheugt, F. W., et al. (1980). Bak[a], a new platelet-specific antigen involved in neonatal allo-immune thrombocytopenia. *Vox Sang, 39,* 113–120.

61. Keimowitz, R. M., Collins, J., Davis, K., et al. (1986). Posttransfusion purpura associated with alloimmunization against the platelet-specific antigen, Bak[a]. *Am J Hematol, 21,* 79–88.

62. Kickler, T. S., Herman, J. H., Furihata, K., et al. (1988). Identification of Bak[b], a new platelet-specific antigen associated with posttransfusion purpura. *Blood, 71,* 894–898.

63. van der Schoot, C. E., Wester, M., dem Borne, A. E., et al. (1986). Characterization of platelet-specific alloantigens by immunoblotting: Localization of Zw and Bak antigens. *Br J Haematol, 64,* 715–723.

64. Letellier, S. J., Hunter, J. B., & Aster, R. H. (1988). Probable genetic linkage between genes coding for platelet-specific antigens of the Pl[A] and Bak systems. *Am J Hematol, 29,* 139–143.

65. Bray, P. F., Barsh, G., Rosa, J. P., et al. (1988). Physical linkage of the genes for platelet membrane glycoproteins IIb and IIIa. *Proc Natl Acad Sci USA, 85,* 8683–8687.

66. Thornton, M. A., Poncz, M., Korostishevsky, M., et al. (1999). The human platelet alphaIIb gene is not closely linked to its integrin partner beta3. *Blood, 94,* 2039–2047.

67. Lyman, S., Aster, R. H., Visentin, G. P., et al. (1990). Polymorphism of human platelet membrane glycoprotein IIb associated with the Bak[a]/Bak[b] alloantigen system. *Blood, 75,* 2343–2348.

68. Goldberger, A., Kolodziej, M., Poncz, M., et al. (1991). Effect of single amino acid substitutions on the formation of the Pl[A] and Bak alloantigenic epitopes. *Blood, 78,* 681–687.

69. Ruan, J., Peyruchaud, O., Nurden, A., et al. (1998). Linkage of four polymorphisms on the a[IIb] gene. *Br J Haematol, 102,* 622–625.

70. Take, H., Tomiyama, Y., Shibata, Y., et al. (1990). Demonstration of the heterogeneity of epitopes of the platelet-specific alloantigen, Bak[a]. *Br J Haematol, 76,* 395–400

71. Djaffar, I., Vilette, D., Pidard, D., et al. (1993). Human platelet antigen 3 (HPA-3): Localization of the determinant of the alloantibody Lek[a] (HPA-3a) to the C-terminus of platelet glycoprotein IIb heavy chain and contribution of O-linked carbohydrates. *Thromb Haemost, 69,* 485–489.

72. Noris, P., Simsek, S., Bruijne-Admiraal, L. G., et al. (1995). Max[a], a new low-frequency platelet-specific antigen localized on glycoprotein IIb, is associated with neonatal alloimmune thrombocytopenia. *Blood, 86,* 1019–1026.

73. López, J. A., & Dong, J. F. (1997). Structure and function of the glycoprotein Ib-IX-V complex. *Curr Opin Hematol, 4,* 323–329.

74. Wenger, R. H., Wicki, A. N., Kieffer, N., et al. (1989). The 5′ flanking region and chromosomal localization of the gene encoding human platelet membrane glycoprotein Ibα. *Gene, 85,* 517–524.

75. Kelly, M. D., Essex, D. W., Shapiro, S. S., et al. (1994). Complementary DNA cloning of the alternatively expressed endothelial cell glycoprotein Ibβ (GPIbβ) and localization of the GPIbβ gene to chromosome 22. *J Clin Invest, 93,* 2417–2424.

76. Yagi, M., Edelhoff, S., Disteche, C. M., et al. (1995). Human platelet glycoproteins V and IX: Mapping of two leucine-rich glycoprotein genes to chromosome 3 and analysis of structures. *Biochemistry, 34,* 16132–16137.

77. Fredrickson, B. J., Dong, J. F., McIntire, L. V., et al. (1998). Shear-dependent rolling on von Willebrand factor of mammalian cells expressing the platelet glycoprotein Ib-IX-V complex. *Blood, 92,* 3684–3693.

78. Kuijpers, R. W. A. M., Faber, N. M., Cuypers, H. T. M., et al. (1992). NH2-terminal globular domain of human platelet glycoprotein Ibα has a methionine145/threonine145 amino acid polymorphism, which is associated with the HPA-2 (Ko) alloantigens. *J Clin Invest, 89,* 381–384.

79. Murata, M., Furihata, K., Ishida, F., et al. (1992). Genetic and structural characterization of an amino acid dimorphism in glycoprotein Ibα involved in platelet transfusion refractoriness. *Blood, 79,* 3086–3090.

80. Saji, H., Maruya, E., Fujii, H., et al. (1989). New platelet antigen, Sib[a], involved in platelet transfusion refractoriness in a Japanese man. *Vox Sang, 56,* 283–287.

81. Bizzaro, N., & Dianese, G. (1988). Neonatal alloimmune amegakaryocytosis. Case report. *Vox Sang, 54,* 112–114.

82. Ishida, F., Saji, H., Maruya, E., et al. (1991). Human platelet-specific antigen, Sib[a], is associated with the molecular weight polymorphism of glycoprotein Ibα. *Blood, 78,* 1722–1729.

83. Kuijpers, R. W., Ouwehand, W. H., Bleeker, P. M., et al. (1992). Localization of the platelet-specific HPA-2 (Ko) alloantigens on the N-terminal globular fragment of platelet glycoprotein Ibα. *Blood, 79,* 283–288.

84. Ulrichts, H., Vanhoorelbeke, K., Cauwenberghs, S., et al. (2003). von Willebrand factor but not alpha-thrombin binding to platelet glycoprotein Ibalpha is influenced by the HPA-2 polymorphism. *Arterioscler Thromb Vasc Biol, 23,* 1302–1307.

85. Li, C. Q., Garner, S. F., Davies, J., et al. (2000). Threonine-145/Methionine-145 variants of baculovirus produced recombinant ligand binding domain of GPIbα express HPA-2 epitopes and show equal binding of von Willebrand factor. *Blood, 95,* 205–211.

86. Mazzucato, M., Pradella, P., de A., et al. (1996). Frequency and functional relevance of genetic threonine145/methionine145 dimorphism in platelet glycoprotein Ibα in an Italian population. *Transfusion, 36,* 891–894.

87. Matsubara, Y., Murata, M., Hayashi, T., et al. (2005). Platelet glycoprotein Ibα polymorphisms affect the interaction with von Willebrand factor under flow conditions. *Br J Haematol, 128,* 533–539.

88. Moroi, M., Jung, S. M., & Yoshida, N. (1984). Genetic polymorphism of platelet glycoprotein Ib. *Blood, 64,* 622–629.

89. López, J. A., Ludwig, E. H., & McCarthy, B. J. (1992). Polymorphism of human glycoprotein Ibα results from a variable number of tandem repeats of a 13-amino acid sequence in the mucin-like macroglycopeptide region. Structure/function implications. *J Biol Chem, 267,* 10055–10061.

90. Ishida, F., Furihata, K., Ishida, K., et al. (1995). The largest variant of platelet glycoprotein Ibα has four tandem repeats of 13 amino acids in the macroglycopeptide region and a genetic linkage with methionine 145. *Blood, 86,* 1357–1360.

91. Ozelo, M. C., Costa, D. S., Siqueira, L. H., et al. (2004). Genetic variability of platelet glycoprotein Ibα gene. *Am J Hematol, 77,* 107–116.

92. Aramaki, K. M., & Reiner, A. P. (1999). A novel isoform of platelet glycoprotein Ibα is prevalent in African Americans. *Am J Hematol, 60,* 77–79.

93. Corral, J., Gonzalez-Conejero, R., Lozano, M. L., et al. (1998). New alleles of the platelet glycoprotein Ibα gene. *Br J Haematol, 103,* 997–1003.

94. Ishida, F., Ito, T., Takei, M., et al. (2000). Genetic linkage of Kozak sequence polymorphism of the platelet glycoprotein Ibα with human platelet antigen-2 and variable number of tandem repeat polymorphism, and its relationship with coronary artery disease. *Br J Haematol, 111,* 1247–1249.

95. Kaski, S., Kekomäki, R., & Partanen, J. (1996). Systematic screening for genetic polymorphism in human platelet glycoprotein Ibα. *Immunogenetics, 44,* 170–176.

96. Frank, M. B., Reiner, A. P., Schwartz, S. M., et al. (2001). The Kozak sequence polymorphism of platelet glycoprotein Ibα and risk of nonfatal myocardial infarction and nonfatal stroke in young women. *Blood, 97,* 875–879.

97. Kozak, M. (1981). Possible role of flanking nucleotides in recognition of the AUG initiator codon by eukaryotic ribosomes. *Nucleic Acids Res, 9,* 5233–5262.

98. Kozak, M. (1984). Point mutations close to the AUG initiator codon affect the efficiency of translation of rat preproinsulin *in vivo. Nature, 308,* 241–246.

99. Kozak, M. (1987). At least six nucleotides preceding the AUG initiator codon enhance translation in mammalian cells. *J Mol Biol, 196,* 947–950.

100. Afshar-Kharghan, V., Li, C. Q., Khoshnevis-Asl, M., et al. (1999). Kozak sequence polymorphism of the glycoprotein (GP) Ibα gene is a major determinant of the plasma membrane levels of the platelet GP Ib-IX-V complex. *Blood, 94,* 186–191.

101. Corral, J., Lozano, M. L., Gonzalez-Conejero, R., et al. (2000). A common polymorphism flanking the ATG initiator codon of GPIbα does not affect expression and is not a major risk factor for arterial thrombosis. *Thromb Haemost, 83,* 23–28.

102. Jilma-Stohlawetz, P., Homoncik, M., Jilma, B., et al. (2003). Glycoprotein Ib polymorphisms influence platelet plug formation under high shear rates. *Br J Haematol, 120,* 652–655.

103. Santoso, S., Zimmermann, P., Sachs, U. J., et al. (2002). The impact of the Kozak sequence polymorphism of the glycoprotein Ibα gene on the risk and extent of coronary heart disease. *Thromb Haemost, 87,* 345–346.

104. Cadroy, Y., Sakariassen, K. S., Charlet, J. P., et al. (2001). Role of four platelet membrane glycoprotein polymorphisms on experimental arterial thrombus formation in men. *Blood, 98,* 3159–3161.

105. López, J. A., Weisman, S., Sanan, D. A., et al. (1994). Glycoprotein (GP) Ibβ is the critical subunit linking GP Ibα and GP IX in the GP Ib-IX complex. Analysis of partial complexes. *J Biol Chem, 269,* 23716–23721.

106. Yagi, M., Edelhoff, S., Disteche, C. M., et al. (1994). Structural characterization and chromosomal location of the gene encoding human platelet glycoprotein Ibβ. *J Biol Chem, 269,* 17424–17427.

107. Lopez, J. A., Chung, D. W., Fujikawa, K., et al. (1988). The a and b chains of human platelet glycoprotein Ib are both transmembrane proteins containing a leucine-rich amino acid sequence. *Proc Natl Acad Sci USA, 85,* 2135–2139.

108. Kiefel, V., Vicariot, M., Giovangrandi, Y., et al. (1995). Alloimmunization against Iy, a low-frequency antigen on platelet glycoprotein Ib/IX as a cause of severe neonatal alloimmune thrombocytopenic purpura. *Vox Sang, 69,* 250–254.

109. Sachs, U. J., Kiefel, V., Bohringer, M., et al. (2000). Single amino acid substitution in human platelet glycoprotein Ibβ is responsible for the formation of the platelet-specific alloantigen Iy[a]. *Blood, 95,* 1849–1855.

110. Ishida, F., Ito, T., Santoso, S., et al. (1999). Low prevalence of a polymorphism of platelet membrane glycoprotein Ibβ associated with neonatal alloimmune thrombocytopenic purpura in Asian populations. *Int J Hematol, 69,* 54–56.

111. López, J. A., Andrews, R. K., Afshar-Kharghan, V., et al. (1998). Bernard–Soulier Syndrome. *Blood, 91,* 4397–4418.

112. Li, C. Q., Dong, J.-F., Lanza, F., et al. (1995). Expression of platelet glycoprotein (GP) V in heterologous cells and evidence for its association with GP Ibα in forming a GP Ib-IX-V complex on the cell surface. *J Biol Chem, 270,* 16302–16307.

113. Kahn, M. L., Diacovo, T. G., Bainton, D. F., et al. (1999). Glycoprotein V-deficient platelets have undiminished thrombin responsiveness and do not exhibit a Bernard–Soulier phenotype. *Blood, 94,* 4112–4121.

114. Dong, J.-F., Sae-Tung, G., & López, J. A. (1997). Role of glycoprotein V in the formation of the platelet high affinity thrombin-binding site. *Blood, 89,* 4355–4363.

115. Ramakrishnan, V., Reeves, P. S., DeGuzman, F., et al. (1999). Increased thrombin responsiveness in platelets from mice lacking glycoprotein V. *Proc Natl Acad Sci USA, 96,* 13336–13341.

116. Ramakrishnan, V., DeGuzman, F., Bao, M., et al. (2001). A thrombin receptor function for platelet glycoprotein Ib-IX unmasked by cleavage of glycoprotein V. *Proc Natl Acad Sci USA, 98,* 1823–1828.

117. Koskela, S., Kekomaki, R., & Partanen, J. (1998). Genetic polymorphism in human platelet glycoprotein GP Ib/IX/V

complex is enriched in GP V (CD42d). *Tissue Antigens, 52,* 236–241.

118. Watson, S. P., & Gibbins, J. (1998). Collagen receptor signalling in platelets: Extending the role of the ITAM. *Immunol Today, 19,* 260–264.

119. Takada, Y., & Hemler, M. E. (1989). The primary structure of the VLA-2/collagen receptor alpha 2 subunit (platelet GPIa): Homology to other integrins and the presence of a possible collagen-binding domain. *J Cell Biol, 109,* 397–407.

120. Mueller-Eckhardt, C., Kiefel, V., Grubert, A., et al. (1989). three hundred and forty-eight cases of suspected neonatal alloimmune thrombocytopenia. *Lancet, 1,* 363–366.

121. Santoso, S., Kiefel, V., & Mueller-Eckhardt, C. (1989). Immunochemical characterization of the new platelet alloantigen system $Br^a$/$Br^b$. *Br J Haematol 72,* 191–198.

122. Santoso, S., Kalb, R., Walka, M., et al. (1993). The human platelet alloantigens $Br^a$ and $Br^b$ are associated with a single amino acid polymorphism on glycoprotein Ia (integrin subunit $a_2$). *J Clin Invest, 92,* 2427–2432.

123. Kritzik, M., Savage, B., Nugent, D. J., et al. (1998). Nucleotide polymorphisms in the $a_2$ gene define multiple alleles that are associated with differences in platelet $a_2b_1$ density. *Blood, 92,* 2382–2388.

124. Kunicki, T. J., Kritzik, M., Annis, D. S., et al. (1997). Hereditary variation in platelet integrin $a_2b_1$ density is associated with two silent polymorphisms in the $a_2$ gene coding sequence. *Blood, 89,* 1939–1943.

125. Jacquelin, B., Tarantino, M. D., Kritzik, M., et al. (2001). Allele-dependent transcriptional regulation of the human integrin $a_2$ gene. *Blood, 97,* 1721–1726.

126. Santoso, S., Amrhein, J., Hofmann, H. A., et al. (1999). A point mutation $Thr_{799}Met$ on the $a_2$ integrin leads to the formation of new human platelet alloantigen $Sit^a$ and affects collagen-induced aggregation. *Blood, 94,* 4103–4111.

127. Corral, J., Gonzalez-Conejero, R., Rivera, J., et al. (1999). Role of the 807 C/T polymorphism of the $a_2$ gene in platelet GP Ia collagen receptor expression and function — effect in thromboembolic diseases. *Thromb Haemost, 81,* 951–956.

128. Corral, J., Rivera, J., Gonzalez-Conejero, R., et al. (1999). The number of platelet glycoprotein Ia molecules is associated with the genetically linked 807 C/T and HPA-5 polymorphisms. *Transfusion, 39,* 372–378.

129. Roest, M., Sixma, J. J., Wu, Y. P., et al. (2000). Platelet adhesion to collagen in healthy volunteers is influenced by variation of both $a_2b_1$ density and von Willebrand factor. *Blood, 96,* 1433–1437.

130. Jacquelin, B., Rozenshteyn, D., Kanaji, S., et al. (2001). Characterization of inherited differences in transcription of the human integrin $a_2$ gene. *J Biol Chem, 276,* 23518–23524.

131. Ezumi, Y., Uchiyama, T., & Takayama, H. (2000). Molecular cloning, genomic structure, chromosomal localization, and alternative splice forms of the platelet collagen receptor glycoprotein VI. *Biochem Biophys Res Commun, 277,* 27–36.

132. Joutsi-Korhonen, L., Smethurst, P. A., Rankin, A., et al. (2003). The low-frequency allele of the platelet collagen signaling receptor glycoprotein VI is associated with reduced functional responses and expression. *Blood, 101,* 4372–4379.

133. Croft, S. A., Samani, N. J., Teare, M. D., et al. (2001). Novel platelet membrane glycoprotein VI dimorphism is a risk factor for myocardial infarction. *Circulation, 104,* 1459–1463.

134. Best, D., Senis, Y. A., Jarvis, G. E., et al. (2003). GPVI levels in platelets: Relationship to platelet function at high shear. *Blood, 102,* 2811–2818.

135. McCrae, K. R., Shattil, S. J., & Cines, D. B. (1990). Platelet activation induces increased Fcg receptor expression. *J Immunol, 144,* 3920–3927.

136. King, M., McDermott, P., & Schreiber, A. D. (1990). Characterization of the Fcg receptor on human platelets. *Cell Immunol, 128,* 462–479.

137. Daeron, M. (1997). Fc receptor biology. *Annu Rev Immunol, 15,* 203–234.

138. Tax, W. J., Willems, H. W., Reekers, P. P., et al. (1983). Polymorphism in mitogenic effect of IgG1 monoclonal antibodies against T3 antigen on human T cells. *Nature, 304,* 445–447.

139. Warmerdam, P. A., van de Winkel, J. G., Gosselin, E. J., et al. (1990). Molecular basis for a polymorphism of human Fcg receptor II (CD32). *J Exp Med, 172,* 19–25.

140. Warmerdam, P. A., van de Winkel, J. G., Vlug, A., et al. (1991). A single amino acid in the second Ig-like domain of the human Fcg receptor II is critical for human IgG2 binding. *J Immunol, 147,* 1338–1343.

141. Looney, R. J., Anderson, C. L., Ryan, D. H., et al. (1988). Structural polymorphism of the human platelet Fc gamma receptor. *J Immunol, 141,* 2680–2683.

142. Chen, J., Dong, J. F., Sun, C., et al. (2003). Platelet FcgammaRIIA His131Arg polymorphism and platelet function: Antibodies to platelet-bound fibrinogen induce platelet activation. *J Thromb Haemost, 1,* 355–362.

143. Fraser, C. M., Arakawa, S., McCombie, W. R., et al. (1989). Cloning, sequence analysis, and permanent expression of a human $a_2$-adrenergic receptor in Chinese hamster ovary cells. Evidence for independent pathways of receptor coupling to adenylate cyclase attenuation and activation. *J Biol Chem, 264,* 11754–11761.

144. Kobilka, B. K., Matsui, H., Kobilka, T. S., et al. (1987). Cloning, sequencing, and expression of the gene coding for the human platelet $a_2$-adrenergic receptor. *Science, 238,* 650–656.

145. Lockette, W., Ghosh, S., Farrow, S., et al. (1995). Alpha 2-adrenergic receptor gene polymorphism and hypertension in blacks. *Am J Hypertens, 8,* 390–394.

146. Freeman, K., Farrow, S., Schmaier, A., et al. (1995). Genetic polymorphism of the $a_2$-adrenergic receptor is associated with increased platelet aggregation, baroreceptor sensitivity, and salt excretion in normotensive humans. *Am J Hypertens, 8,* 863–869.

147. Small, K. M., Forbes, S. L., Brown, K. M., et al. (2000). An Asn to Lys polymorphism in the third intracellular loop of the human $a_{2A}$-adrenergic receptor imparts enhanced agonist-promoted $G_i$ coupling. *J Biol Chem, 275,* 38518–38523.

148. Fontana, P., Dupont, A., Gandrille, S., et al. (2003). Adenosine diphosphate-induced platelet aggregation is associated with P2Y12 gene sequence variations in healthy subjects. *Circulation, 108,* 989–995.

149. Jefferson, B. K., Foster, J. H., McCarthy, J. J., et al. (2005). Aspirin resistance and a single gene. *Am J Cardiol, 95,* 805–808.

150. Shimizu, M., Kanazawa, K., Matsuda, Y., et al. (2003). Serotonin-2A receptor gene polymorphisms are associated with serotonin-induced platelet aggregation. *Thromb Res, 112,* 137–142.

151. Greenberg, B. D., Tolliver, T. J., Huang, S. J., et al. (1999). Genetic variation in the serotonin transporter promoter region affects serotonin uptake in human blood platelets. *Am J Med Genet, 88,* 83–87.

152. Whyte, E. M., Pollock, B. G., Wagner, W. R., et al. (2001). Influence of serotonin-transporter-linked promoter region polymorphism on platelet activation in geriatric depression. *Am J Psychiatry, 158,* 2074–2076.

153. Yanai, H., Chiba, H., Fujiwara, H., et al. (2000). Phenotype-genotype correlation in CD36 deficiency types I and II. *Thromb Haemost, 84,* 436–441.

154. Tomiyama, Y., Take, H., Ikeda, H., et al. (1990). Identification of the platelet-specific alloantigen, Nakᵃ, on platelet membrane glycoprotein IV. *Blood, 75,* 684–687.

155. Yamamoto, N., Ikeda, H., Tandon, N. N., et al. (1990). A platelet membrane glycoprotein (GP) deficiency in healthy blood donors. *Blood, 76,* 1698–1703.

156. Curtis, B. R., & Aster, R. H. (1996). Incidence of the Nak(a)-negative platelet phenotype in African Americans is similar to that of Asians. *Transfusion, 36,* 331–334.

157. Hirano, K., Kuwasako, T., Nakagawa-Toyama, Y., et al. (2003). Pathophysiology of human genetic CD36 deficiency. *Trends Cardiovasc Med, 13,* 136–141.

158. Kashiwagi, H., Tomiyama, Y., Kosugi, S., et al. (1994). Identification of molecular defects in a subject with type I CD36 deficiency. *Blood, 83,* 3545–3552.

159. Kashiwagi, H., Tomiyama, Y., Kosugi, S., et al. (1995). Family studies of type II CD36 deficient subjects: Linkage of a CD36 allele to a platelet-specific mRNA expression defect(s) causing type II CD36 deficiency. *Thromb Haemost, 74,* 758–763.

160. Kashiwagi, H., Tomiyama, Y., Honda, S., et al. (1995). Molecular basis of CD36 deficiency. Evidence that a 478C → T substitution (proline90 → serine) in CD36 cDNA accounts for CD36 deficiency. *J Clin Invest, 95,* 1040–1046.

161. Hoebe, K., Georgel, P., Rutschmann, S., et al. (2005). CD36 is a sensor of diacylglycerides. *Nature, 433,* 523–527.

162. McKeown, L., Vail, M., Williams, S., et al. (1994). Platelet adhesion to collagen in individuals lacking glycoprotein IV. *Blood, 83,* 2866–2871.

163. Yamamoto, N., Akamatsu, N., Yamazaki, H., et al. (1992). Normal aggregations of glycoprotein IV (CD36)-deficient platelets from seven healthy Japanese donors. *Br J Haematol, 81,* 86–92.

164. Kehrel, B., Kronenberg, A., Rauterberg, J., et al. (1993). Platelets deficient in glycoprotein IIIb aggregate normally to collagens type I and III but not to collagen type V. *Blood, 82,* 3364–3370.

165. Tandon, N. N., Ockenhouse, C. F., Greco, N. J., et al. (1991). Adhesive functions of platelets lacking glycoprotein IV (CD36). *Blood, 78,* 2809–2813.

166. Diaz-Ricart, M., Tandon, N. N., Carretero, M., et al. (1993). Platelets lacking functional CD36 (glycoprotein IV) show reduced adhesion to collagen in flowing whole blood. *Blood, 82,* 491–496.

167. Serghides, L., Smith, T. G., Patel, S. N., et al. (2003). CD36 and malaria: Friends or foes? *Trends Parasitol, 19,* 461–469.

168. Herrmann, S. M., Ricard, S., Nicaud, V., et al. (1998). The P-selectin gene is highly polymorphic: Reduced frequency of the Pro715 allele carriers in patients with myocardial infarction. *Hum Mol Genet, 7,* 1277–1284.

169. Tregouet, D. A., Barbaux, S., Escolano, S., et al. (2002). Specific haplotypes of the P-selectin gene are associated with myocardial infarction. *Hum Mol Genet, 11,* 2015–2023.

170. Barbaux, S. C., Blankenberg, S., Rupprecht, H. J., et al. (2001). Association between P-selectin gene polymorphisms and soluble P-selectin levels and their relation to coronary artery disease. *Arterioscler Thromb Vasc Biol, 21,* 1668–1673.

171. Carter, A. M., Anagnostopoulou, K., Mansfield, M. W., et al. (2003). Soluble P-selectin levels, P-selectin polymorphisms and cardiovascular disease. *J Thromb Haemost, 1,* 1718–1723.

172. Miller, M. A., Kerry, S. M., Dong, Y., et al. (2004). Association between the Thr715Pro P-selectin gene polymorphism and soluble P-selectin levels in a multiethnic population in South London. *Thromb Haemost, 92,* 1060–1065.

173. Kelton, J. G., Smith, J. W., Horsewood, P., et al. (1990). Govᵃ/ᵇ alloantigen system on human platelets. *Blood, 75,* 2172–2176.

174. Berry, J. E., Murphy, C. M., Smith, G. A., et al. (2000). Detection of Gov system antibodies by MAIPA reveals an immunogenicity similar to the HPA-5 alloantigens. *Br J Haematol, 110,* 735–742.

175. Bordin, J. O., Kelton, J. G., Warner, M. N., et al. (1997). Maternal immunization to Gov system alloantigens on human platelets. *Transfusion, 37,* 823–828.

176. Smith, J. W., Hayward, C. P., Horsewood, P., et al. (1995). Characterization and localization of the Govᵃ/ᵇ alloantigens to the glycosylphosphatidylinositol-anchored protein CDw109 on human platelets. *Blood, 86,* 2807–2814.

177. Arnaud, E., Nicaud, V., Poirier, O., et al. (2000). Protective effect of a thrombin receptor (protease-activated receptor 1) gene polymorphism toward venous thromboembolism. *Arterioscler Thromb Vasc Biol, 20,* 585–592.

178. Dupont, A., Fontana, P., Bachelot-Loza, C., et al. (2003). An intronic polymorphism in the PAR-1 gene is associated with platelet receptor density and the response to SFLLRN. *Blood, 101,* 1833–1840.

179. Elrayess, M. A., Webb, K. E., Bellingan, G. J., et al. (2004). R643G polymorphism in PECAM-1 influences transendothelial migration of monocytes and is associated with pro-

gression of CHD and CHD events. *Atherosclerosis, 177,* 127–135.

180. Moliterno, A. R., Williams, D. M., Gutierrez-Alamillo, L. I., et al. (2004). Mpl Baltimore: A thrombopoietin receptor polymorphism associated with thrombocytosis. *Proc Natl Acad Sci USA, 101,* 11444–11447.

181. Zeng, S. M., Murray, J. C., Widness, J. A., et al. (2004). Association of single nucleotide polymorphisms in the thrombopoietin-receptor gene, but not the thrombopoietin gene, with differences in platelet count. *Am J Hematol, 77,* 12–21.

182. Naber, C., Hermann, B. L., Vietzke, D., et al. (2000). Enhanced epinephrine-induced platelet aggregation in individuals carrying the G protein $b_3$ subunit 825T allele. *FEBS Lett, 484,* 199–201.

183. Narizhneva, N. V., Byers-Ward, V. J., Quinn, M. J., et al. (2004). Molecular and functional differences induced in thrombospondin-1 by the single nucleotide polymorphism associated with the risk of premature, familial myocardial infarction. *J Biol Chem, 279,* 21651–21657.

184. Faraday, N., Goldschmidt-Clermont, P. J., & Bray, P. F. (1997). Gender differences in platelet GPIIb-IIIa activation. *Thromb Haemost, 77,* 748–754.

185. Rosenkranz, K., Hinney, A., Ziegler, A., et al. (1998). Systematic mutation screening of the estrogen receptor b gene in probands of different weight extremes: Identification of several genetic variants. *J Clin. Endocrinol Metab, 83,* 4524–4527.

186. Tsukamoto, K., Inoue, S., Hosoi, T., et al. (1998). Isolation and radiation hybrid mapping of dinucleotide repeat polymorphism at the human estrogen receptor b locus. *J Hum Genet, 43,* 73–74.

187. Ogawa, S., Hosoi, T., Shiraki, M., et al. (2000). Association of estrogen receptor beta gene polymorphism with bone mineral density. *Biochem Biophys Res Commun, 269,* 537–541.

188. Wang, X. L., Sim, A. S., Badenhop, R. F., et al. (1996). A smoking-dependent risk of coronary artery disease associated with a polymorphism of the endothelial nitric oxide synthase gene. *Nat Med, 2,* 41–45.

189. Tsukada, T., Yokoyama, K., Arai, T., et al. (1998). Evidence of association of the ecNOS gene polymorphism with plasma NO metabolite levels in humans. *Biochem Biophys Res Commun, 245,* 190–193.

190. Hingorani, A. D., Liang, C. F., Fatibene, J., et al. (1999). A common variant of the endothelial nitric oxide synthase (Glu298 → Asp) is a major risk factor for coronary artery disease in the UK. *Circulation, 100,* 1515–1520.

191. Goldfarb, L. G., Petersen, R. B., Tabaton, M., et al. (1992). Fatal familial insomnia and familial Creutzfeldt-Jakob disease: Disease phenotype determined by a DNA polymorphism. *Science, 258,* 806–808.

192. Humphries, S. E., Cook, M., Dubowitz, M., et al. (1987). Role of genetic variation at the fibrinogen locus in determination of plasma fibrinogen concentrations. *Lancet, 1,* 1452–1455.

193. Harvey, P. J., Keightley, A. M., Lam, Y. M., et al. (2000). A single nucleotide polymorphism at nucleotide — 1793 in the von Willebrand factor (VWF) regulatory region is associated with plasma VWF:Ag levels. *Br J Haematol, 109,* 349–353.

194. Keightley, A. M., Lam, Y. M., Brady, J. N., et al. (1999). Variation at the von Willebrand factor (vWF) gene locus is associated with plasma vWF:Ag levels: Identification of three novel single nucleotide polymorphisms in the vWF gene promoter. *Blood, 93,* 4277–4283.

195. Heptinstall, S., Mulley, G. P., Taylor, P. M., et al. (1980). Platelet-release reaction in myocardial infarction. *Br Med J, 280,* 80–81.

196. Trip, M. D., Cats, V. M., van Capelle, F. J., et al. (1990). Platelet hyperreactivity and prognosis in survivors of myocardial infarction. *N Engl J Med, 322,* 1549–1554.

197. Martin, J. F., Bath, P. M., & Burr, M. L. (1991). Influence of platelet size on outcome after myocardial infarction. *Lancet, 338,* 1409–1411.

198. Eikelboom, J. W., Hirsh, J., Weitz, J. I., et al. (2002). Aspirin-resistant thromboxane biosynthesis and the risk of myocardial infarction, stroke, or cardiovascular death in patients at high risk for cardiovascular events. *Circulation, 105,* 1650–1655.

199. Kroll, M. H., Hellums, J. D., McIntire, L. V., et al. (1996). Platelets and shear stress. *Blood, 88,* 1525–1541.

200. Goto, S., Ikeda, Y., Saldivar, E., et al. (1998). Distinct mechanisms of platelet aggregation as a consequence of different shearing flow conditions. *J Clin Invest, 101,* 479–486.

201. Thordarson, O., & Fridriksson, S. (1979). Aggregation of deaths from ischaemic heart disease among first and second degree relatives of 108 males and 42 females with myocardial infarction. *Acta Med Scand, 205,* 493–500.

202. Rissanen, A. M., & Nikkila, E. A. (1977). Coronary artery disease and its risk factors in families of young men with angina pectoris and in controls. *Br Heart J, 39,* 875–883.

203. Slack, J., & Evans, K. A. (1966). The increased risk of death from ischaemic heart disease in first degree relatives of 121 men and 96 women with ischaemic heart disease. *J Med Genet, 3,* 329–357.

204. Rose, G. (1964). Familial patterns in ischaemic heart disease. *Br J Prev Soc Med, 18,* 75–80.

205. Rissanen, A. M. (1979). Familial occurrence of coronary heart disease: Effect of age at diagnosis. *Am J Cardiol, 44,* 60–66.

206. Marenberg, M. E., Risch, N., Berkman, L. F., et al. (1994). Genetic susceptibility to death from coronary heart disease in a study of twins. *N Engl J Med, 330,* 1041–1046.

207. Berg, K. (1982). The genetics of the hyperlipidemias and coronary artery disease. *Prog Clin Biol Res, 103* (Pt B), 111–125.

208. Sorensen, T. I., Nielsen, G. G., Andersen, P. K., et al. (1988). Genetic and environmental influences on premature death in adult adoptees. *N Engl J Med, 318,* 727–732.

209. Harvald, B., & Hauge, M. (1970). Coronary occlusion in twins. *Acta Genet Med Gemellol (Roma), 19,* 248–250.

210. Pastinen, T., Perola, M., Niini, P., et al. (1998). Array-based multiplex analysis of candidate genes reveals two independent and additive genetic risk factors for myocardial

infarction in the Finnish population. *Hum Mol Genet, 7,* 1453–1462.

211. Sing, C. F., Haviland, M. B., Templeton, A. R., et al. (1992). Biological complexity and strategies for finding DNA variations responsible for inter-individual variation in risk of a common chronic disease, coronary artery disease. *Ann Med, 24,* 539–547.

212. Weiss, E. J., Bray, P. F., Tayback, M., et al. (1996). A polymorphism of a platelet glycoprotein receptor as an inherited risk factor for coronary thrombosis. *N Engl J Med, 334,* 1090–1094.

213. Grove, E. L., Orntoft, T. F., Lassen, J. F., et al. (2004). The platelet polymorphism PlA2 is a genetic risk factor for myocardial infarction. *J Intern Med, 255,* 637–644.

214. Carter, A. M., Ossei-Gerning, N., Wilson, I. J., et al. (1997). Association of the platelet Pl^A polymorphism of glycoprotein IIb/IIIa and the fibrinogen Bb 448 polymorphism with myocardial infarction and extent of coronary artery disease. *Circulation, 96,* 1424–1431.

215. Zotz, R. B., Winkelmann, B. R., Nauck, M., et al. (1998). Polymorphism of platelet membrane glycoprotein IIIa: Human platelet antigen 1b (HPA-1b/Pl^A2) is an inherited risk factor for premature myocardial infarction in coronary artery disease. *Thromb Haemost, 79,* 731–735.

216. Garcia-Ribes, M., Gonzalez-Lamuno, D., Hernandez-Estefania, R., et al. (1998). Polymorphism of the platelet glycoprotein IIIa gene in patients with coronary stenosis. *Thromb Haemost, 79,* 1126–1129.

217. Araujo, F., Santos, A., Araujo, V., et al. (1999). Genetic risk factors in acute coronary disease. *Haemostasis, 29,* 212–218.

218. Ardissino, D., Mannucci, P. M., Merlini, P. A., et al. (1999). Prothrombotic genetic risk factors in young survivors of myocardial infarction. *Blood, 94,* 46–51.

219. Walter, D. H., Schachinger, V., Elsner, M., et al. (1997). Platelet glycoprotein IIIa polymorphisms and risk of coronary stent thrombosis. *Lancet, 350,* 1217–1219.

220. Kastrati, A., Schomig, A., Seyfarth, M., et al. (1999). Pl^A polymorphism of platelet glycoprotein IIIa and risk of restenosis after coronary stent placement. *Circulation, 99,* 1005–1010.

221. Zotz, R. B., Klein, M., Dauben, H. P., et al. (2000). Prospective analysis after coronary-artery bypass grafting: Platelet GP IIIa polymorphism (HPA-1b/Pl^A2) is a risk factor for bypass occlusion, myocardial infarction, and death. *Thromb Haemost, 83,* 404–407.

222. Mikkelsson, J., Perola, M., Laippala, P., et al. (1999). Glycoprotein IIIa Pl^A polymorphism associates with progression of coronary artery disease and with myocardial infarction in an autopsy series of middle-aged men who died suddenly. *Arterioscler Thromb Vasc Biol, 19,* 2573–2578.

223. Mikkelsson, J., Perola, M., Laippala, P., et al. (2000). Glycoprotein IIIa Pl^A1/A2 polymorphism and sudden cardiac death. *J Am Coll Cardiol, 36,* 1317–1323.

224. Ridker, P. M., Hennekens, C. H., Schmitz, C., et al. (1997). Pl^A1/A2 polymorphism of platelet glycoprotein IIIa and risks of myocardial infarction, stroke, and venous thrombosis. *Lancet, 349,* 385–388.

225. Herrmann, S. M., Poirier, O., Marques-Vidal, P., et al. (1997). The Leu33/Pro polymorphism (Pl^A1/Pl^A2) of the glycoprotein IIIa (GPIIIa) receptor is not related to myocardial infarction in the ECTIM Study. Étude Cas-Temoins de l'Infarctus du Myocarde. Thromb Haemost, 77, 1179–1181.

226. Mamotte, C. D., van Bockxmeer, F. M., & Taylor, R. R. (1998). Pl^A1/A2 polymorphism of glycoprotein IIIa and risk of coronary artery disease and restenosis following coronary angioplasty. *Am J Cardiol, 82,* 13–16.

227. Corral, J., Gonzalez-Conejero, R., Rivera, J., et al. (1997). HPA-1 genotype in arterial thrombosis — role of HPA-1b polymorphism in platelet function. *Blood Coagul Fibrinolysis, 8,* 284–290.

228. Kekomaki, S., Hamalainen, L., Kauppinen-Makelin, R., et al. (1999). Genetic polymorphism of platelet glycoprotein IIIa in patients with acute myocardial infarction and acute ischaemic stroke. *J Cardiovasc Risk, 6,* 1317.

229. Anderson, J. L., King, G. J., Bair, T. L., et al. (1999). Associations between a polymorphism in the gene encoding glycoprotein IIIa and myocardial infarction or coronary artery disease. *J Am Coll Cardiol, 33,* 727–733.

230. Hooper, W. C., Lally, C., Austin, H., et al. (1999). The relationship between polymorphisms in the endothelial cell nitric oxide synthase gene and the platelet GPIIIa gene with myocardial infarction and venous thromboembolism in African Americans. *Chest, 116,* 880–886.

231. Laule, M., Cascorbi, I., Stangl, V., et al. (1999). A1/A2 polymorphism of glycoprotein IIIa and association with excess procedural risk for coronary catheter interventions: A case-controlled study. *Lancet, 353,* 708–712.

232. Aleksic, N., Juneja, H., Folsom, A. R., et al. (2000). Platelet Pl^A2 allele and incidence of coronary heart disease: Results from the Atherosclerosis Risk In Communities (ARIC) Study. *Circulation, 102,* 1901–1905.

233. Cardon, L. R., & Bell, J. I. (2001). Association study designs for complex diseases. *Nature Rev Genet, 2,* 91–99.

234. Lopes, N. H., Pereira, A. C., Hueb, W., et al. (2004). Effect of glycoprotein IIIa PlA2 polymorphism on outcome of patients with stable coronary artery disease and effect of smoking. *Am J Cardiol, 93,* 1469–1472.

235. Boekholdt, S. M., Peters, R. J., de Maat, M. P., et al. (2004). Interaction between a genetic variant of the platelet fibrinogen receptor and fibrinogen levels in determining the risk of cardiovascular events. *Am Heart J, 147,* 181–186.

236. Durante-Mangoni, E., Davies, G. J., Ahmed, N., et al. (1998). Coronary thrombosis and the platelet glycoprotein IIIA gene Pl^A2 polymorphism. *Thromb Haemost, 80,* 218–219.

237. Gardemann, A., Humme, J., Stricker, J., et al. (1998). Association of the platelet glycoprotein IIIa Pl^A1/A2 gene polymorphism to coronary artery disease but not to nonfatal myocardial infarction in low risk patients. *Thromb Haemost, 80,* 214–217.

238. Bottiger, C., Kastrati, A., Koch, W., et al. (2000). HPA-1 and HPA-3 polymorphisms of the platelet fibrinogen receptor and coronary artery disease and myocardial infarction. *Thromb Haemost, 83,* 559–562.

239. Senti, M., Aubo, C., & Bosch, M. (1998). The relationship between smoking and triglyceride-rich lipoproteins is

modulated by genetic variation in the glycoprotein IIIa gene. *Metabolism, 47,* 1040–1041.

240. Joven, J., Simo, J. M., Vilella, E., et al. (1998). Lipoprotein(a) and the significance of the association between platelet glycoprotein IIIa polymorphisms and the risk of premature myocardial infarction. *Atherosclerosis, 140,* 155–159.

241. Senti, M., Aubo, C., Bosch, M., et al. (1998). Platelet glycoprotein IIb/IIIa genetic polymorphism is associated with plasma fibrinogen levels in myocardial infarction patients. The REGICOR Investigators. *Clin Biochem, 31,* 647–651.

242. Di Castelnuovo, A., de Gaetano, G., Donati, M. B., et al. (2001). Platelet glycoprotein receptor IIIa polymorphism Pl$^{A1/PIA2}$ and coronary risk: A meta-analysis. *Thromb Haemost, 85,* 626–633.

243. Wu, A. H., & Tsongalis, G. J. (2001). Correlation of polymorphisms to coagulation and biochemical risk factors for cardiovascular diseases. *Am J Cardiol, 87,* 1361–1366.

244. Meiklejohn, D. J., Vickers, M. A., Morrison, E. R., et al. (2001). In vivo platelet activation in atherothrombotic stroke is not determined by polymorphisms of human platelet glycoprotein IIIa or Ib. *Br J Haematol, 112,* 621–631.

245. Carlsson, L. E., Greinacher, A., Spitzer, C., et al. (1997). Polymorphisms of the human platelet antigens HPA-1, HPA-2, HPA-3, and HPA-5 on the platelet receptors for fibrinogen (GPIIb/IIIa), von Willebrand factor (GPIb/IX), and collagen (GPIa/IIa) are not correlated with an increased risk for stroke. *Stroke, 28,* 1392–1395.

246. van Goor, M. L., Gomez, G. E., Brouwers, G. J., et al. (2002). PLA1/A2 polymorphism of the platelet glycoprotein receptor IIb/IIIa in young patients with cryptogenic TIA or ischemic stroke. *Thromb Res, 108,* 63–65.

247. Wagner, K. R., Giles, W. H., Johnson, C. J., et al. (1998). Platelet glycoprotein receptor IIIa polymorphism P1A2 and ischemic stroke risk: The Stroke Prevention in Young Women Study. *Stroke, 29,* 581–585.

248. Carter, A. M., Catto, A. J., Bamford, J. M., et al. (1998). Platelet GP IIIa PlA and GP Ib variable number tandem repeat polymorphisms and markers of platelet activation in acute stroke. *Arterioscler Thromb Vasc Biol, 18,* 1124–1131.

249. Streifler, J. Y., Rosenberg, N., Chetrit, A., et al. (2001). Cerebrovascular events in patients with significant stenosis of the carotid artery are associated with hyperhomocysteinemia and platelet antigen-1 (Leu33Pro) polymorphism. *Stroke, 32,* 2753–2758.

250. Slowik, A., Dziedzic, T., Turaj, W., et al. (2004). A2 alelle of GpIIIa gene is a risk factor for stroke caused by large-vessel disease in males. *Stroke, 35,* 1589–1593.

251. Papp, E., Havasi, V., Bene, J., et al. (2005). Glycoprotein IIIA gene (PlA) Polymorphism and aspirin resistance: Is there any correlation? *Ann Pharmacother, 39,* 1013–1018.

252. Bray, P. F., Cannon, C., Goldschmidt-Clermont, P., et al. (1998). The Pl$^{A2}$ genetic alteration in the platelet GPIIb-IIIa receptor and the risk of recurrent events in post myocardial infarction patients taking pravastatin. *J Invest Med, 46,* 200A.

253. Angiolillo, D. J., Fernandez-Ortiz, A., Bernardo, E., et al. (2004). PlA polymorphism and platelet reactivity following clopidogrel loading dose in patients undergoing coronary stent implantation. *Blood Coagul Fibrinolysis, 15,* 89–93.

254. Walter, D. H., Schachinger, V., Elsner, M., et al. (2001). Statin therapy is associated with reduced restenosis rates after coronary stent implantation in carriers of the Pl$^{A2}$ allele of the platelet glycoprotein IIIa gene. *Eur Heart J, 22,* 587–595.

255. Bottiger, C., Kastrati, A., Koch, W., et al. (1999). Polymorphism of platelet glycoprotein IIb and risk of thrombosis and restenosis after coronary stent placement. *Am J Cardiol, 84,* 987–991.

256. Carter, A. M., Catto, A. J., Bamford, J. M., et al. (1999). Association of the platelet glycoprotein IIb HPA-3 polymorphism with survival after acute ischemic stroke. *Stroke, 30,* 2606–2611.

257. Reiner, A. P., Schwartz, S. M., Kumar, P. N., et al. (2001). Platelet glycoprotein IIb polymorphism, traditional risk factors and non-fatal myocardial infarction in young women. *Br J Haematol, 112,* 632–636.

258. Murata, M., Matsubara, Y., Kawano, K., et al. (1997). Coronary artery disease and polymorphisms in a receptor mediating shear stress-dependent platelet activation. *Circulation, 96,* 3281–3286.

259. Gonzalez-Conejero, R., Lozano, M. L., Rivera, J., et al. (1998). Polymorphisms of platelet membrane glycoprotein Ibα associated with arterial thrombotic disease. *Blood, 92,* 2771–2776.

260. Mikkelsson, J., Perola, M., Penttila, A., et al. (2001). Platelet glycoprotein Ibα HPA-2 Met/VNTR B haplotype as a genetic predictor of myocardial infarction and sudden cardiac death. *Circulation, 104,* 876–880.

261. Afshar-Kharghan, V., Matijevic-Aleksic, N., Ahn, C., et al. (2004). The variable number of tandem repeat polymorphism of platelet glycoprotein Ibα and risk of coronary heart disease. *Blood, 103,* 963–965.

262. Ozelo, M. C., Origa, A. F., Aranha, F. J., et al. (2004). Platelet glycoprotein Ibα polymorphisms modulate the risk for myocardial infarction. *Thromb Haemost, 92,* 384–386.

263. Croft, S. A., Hampton, K. K., Daly, M. E., et al. (2000). Kozak sequence polymorphism in the platelet GPIbalpha gene is not associated with risk of myocardial infarction. *Blood, 95,* 2183–2184.

264. Meisel, C., Afshar-Kharghan, V., Cascorbi, I., et al. (2001). Role of Kozak sequence polymorphism of platelet glycoprotein Ibalpha as a risk factor for coronary artery disease and catheter interventions. *J Am Coll Cardiol, 38,* 1023–1027.

265. Kenny, D., Muckian, C., Fitzgerald, D. J., et al. (2002). Platelet glycoprotein Ib alpha receptor polymorphisms and recurrent ischaemic events in acute coronary syndrome patients. *J Thromb Thrombolysis, 13,* 13–19.

266. Douglas, H., Michaelides, K., Gorog, D. A., et al. (2002). Platelet membrane glycoprotein Ibalpha gene -5T/C Kozak sequence polymorphism as an independent risk factor for the occurrence of coronary thrombosis. *Heart, 87,* 70–74.

267. Baker, R. I., Eikelboom, J., Lofthouse, E., et al. (2001). Platelet glycoprotein Ibα Kozak polymorphism is associated with an increased risk of ischemic stroke. *Blood, 98,* 36–40.

268. Sonoda, A., Murata, M., Ikeda, Y., et al. (2001). Stroke and platelet glycoprotein Ibalpha polymorphisms. *Thromb Haemost, 85,* 573–574.

269. Hsieh, K., Funk, M., Schillinger, M., et al. (2004). Vienna Stroke Registry. Impact of the platelet glycoprotein Iba Kozak polymorphism on the risk of ischemic cerebrovascular events: A case-control study. *Blood Coagul Fibrinolysis, 15,* 469–473.

270. Moshfegh, K., Wuillemin, W. A., Redondo, M., et al. (1999). Association of two silent polymorphisms of platelet glycoprotein Ia/IIa receptor with risk of myocardial infarction: A case-control study. *Lancet, 353,* 351–354.

271. Reiner, A. P., Kumar, P. N., Schwartz, S. M., et al. (2000). Genetic variants of platelet glycoprotein receptors and risk of stroke in young women. *Stroke, 31,* 1628–1633.

272. Santoso, S., Kunicki, T. J., Kroll, H., et al. (1999). Association of the platelet glycoprotein Ia C807T gene polymorphism with nonfatal myocardial infarction in younger patients. *Blood, 93,* 2449–2453.

273. Carlsson, L. E., Santoso, S., Spitzer, C., et al. (1999). The $a_2$ gene coding sequence T807/A873 of the platelet collagen receptor integrin $a_2b_1$ might be a genetic risk factor for the development of stroke in younger patients. *Blood, 93,* 3583–3586.

274. Croft, S. A., Hampton, K. K., Sorrell, J. A., et al. (1999). The GPIa C807T dimorphism associated with platelet collagen receptor density is not a risk factor for myocardial infarction. *Br J Haematol, 106,* 771–776.

275. Roest, M., Banga, J. D., Grobbee, D. E., et al. (2000). Homozygosity for 807 T polymorphism in $a_2$ subunit of platelet $a_2b_1$ is associated with increased risk of cardiovascular mortality in high-risk women. *Circulation, 102,* 1645–1650.

276. von Beckerath, N., Koch, W., Mehilli, J., et al. (2000). Glycoprotein Ia gene C807T polymorphism and risk for major adverse cardiac events within the first 30 days after coronary artery stenting. *Blood, 95,* 3297–3301.

277. Kroll, H., Gardemann, A., Fechter, A., et al. (2000). The impact of the glycoprotein Ia collagen receptor subunit A1648G gene polymorphism on coronary artery disease and acute myocardial infarction. *Thromb Haemost, 83,* 392–396.

278. Zee, R. Y., Cook, N. R., Cheng, S., et al. (2004). Polymorphism in the P-selectin and interleukin-4 genes as determinants of stroke: A population-based, prospective genetic analysis. *Hum Mol Genet, 13,* 389–396.

279. Warkentin, T. E. (1999). Heparin-induced thrombocytopenia: A ten-year retrospective. *Annu Rev Med, 50,* 129–147.

280. Warkentin, T. E. (1997). Heparin-induced thrombocytopenia. Pathogenesis, frequency, avoidance and management. *Drug Saf, 17,* 325–341.

281. Reilly, M. P., Taylor, S. M., Hartman, N. K., et al. (2000). Heparin-induced thrombocytopenia/thrombosis in a transgenic mouse model demonstrates the requirement for human platelet factor 4 and platelet activation through FcgRIIa. *Blood, 96,* 221a.

282. Bachelot, C., Saffroy, R., Gandrille, S., et al. (1995). Role of FcgRIIA gene polymorphism in human platelet activation by monoclonal antibodies. *Thromb Haemost, 74,* 1557–1563.

283. Rosenfeld, S. I., Looney, R. J., Leddy, J. P., et al. (1985). Human platelet Fc receptor for immunoglobulin G. Identification as a 40,000-molecular-weight membrane protein shared by monocytes. *J Clin Invest, 76,* 2317–2322.

284. Worthington, R. E., Carroll, R. C., & Boucheix, C. (1990). Platelet activation by CD9 monoclonal antibodies is mediated by the FcgII receptor. *Br J Haematol, 74,* 216–222.

285. Anderson, G. P., & Anderson, C. L. (1990). Signal transduction by the platelet Fc receptor. *Blood, 76,* 1165–1172.

286. Poole, A., Gibbins, J. M., Turner, M., et al. (1997). The Fc receptor g-chain and the tyrosine kinase Syk are essential for activation of mouse platelets by collagen. *EMBO J, 16,* 2333–2341.

287. Gross, B. S., Lee, J. R., Clements, J. L., et al. (1999). Tyrosine phosphorylation of SLP-76 is downstream of Syk following stimulation of the collagen receptor in platelets. *J Biol Chem, 274,* 5963–5971.

288. Arepally, G., McKenzie, S. E., Jiang, X. M., et al. (1997). FcgRIIA H/R 131 polymorphism, subclass-specific IgG anti-heparin/platelet factor 4 antibodies and clinical course in patients with heparin-induced thrombocytopenia and thrombosis. *Blood, 89,* 370–375.

289. Carlsson, L. E., Santoso, S., Baurichter, G., et al. (1998). Heparin-induced thrombocytopenia: New insights into the impact of the FcgRIIa-R-H131 polymorphism. *Blood, 92,* 1526–1531.

290. Ollikainen, E., Mikkelsson, J., Perola, M., et al. (2004). Platelet membrane collagen receptor glycoprotein VI polymorphism is associated with coronary thrombosis and fatal myocardial infarction in middle-aged men. *Atherosclerosis, 176,* 95–99.

291. Takagi, S., Iwai, N., Baba, S., et al. (2002). A GPVI polymorphism is a risk factor for myocardial infarction in Japanese. *Atherosclerosis, 165,* 397–398.

292. Cole, V. J., Staton, J. M., Eikelboom, J. W., et al. (2003). Collagen platelet receptor polymorphisms integrin alpha-2beta1 C807T and GPVI Q317L and risk of ischemic stroke. *J Thromb Haemost, 1,* 963–970.

293. Arnaud, E., Poirier, O., Aiach, M., et al. (2000). The −5061/D polymorphism of the thrombin receptor (PAR-1) gene is not related to myocardial infarction in the ECTIM study. The Étude Cas-Temoins de l'Infarctus du Myocarde. MONICA Members Group. *Thromb Haemost, 84,* 722–723.

294. Fontana, P., Gaussem, P., Aiach, M., et al. (2003). P2Y12 H2 haplotype is associated with peripheral arterial disease: A case-control study. *Circulation, 108,* 2971–2973.

295. Topol, E. J., McCarthy, J., Gabriel, S., et al. (2001). Single nucleotide polymorphisms in multiple novel thrombospondin genes may be associated with familial premature myocardial infarction. *Circulation, 104,* 2641–2644.

296. Yamada, Y., Izawa, H., Ichihara, S., et al. (2002). Prediction of the risk of myocardial infarction from polymorphisms in candidate genes. *N Engl J Med, 347,* 1916–1923.

297. Boekholdt, S. M., Trip, M. D., Peters, R. J., et al. (2002). Thrombospondin-2 polymorphism is associated with a reduced risk of premature myocardial infarction. *Arterioscler Thromb Vasc Biol, 22,* e24–e27.

298. Yamada, S., Akita, H., Kanazawa, K., et al. (2000). T102C polymorphism of the serotonin (5-HT) 2A receptor gene in patients with non-fatal acute myocardial infarction. *Atherosclerosis, 150,* 143–148.

299. Arinami, T., Ohtsuki, T., Yamakawa-Kobayashi, K., et al. (1999). A synergistic effect of serotonin transporter gene polymorphism and smoking in association with CHD. *Thromb Haemost, 81,* 853–856.

300. Fumeron, F., Betoulle, D., Nicaud, V., et al. (2002). Serotonin transporter gene polymorphism and myocardial infarction: Étude Cas-Temoins de l'Infarctus du Myocarde (ECTIM). *Circulation, 105,* 2943–2945.

301. Coto, E., Reguero, J. R., Alvarez, V., et al. (2003). 5-Hydroxytryptamine 5-HT2A receptor and 5-hydroxytryptamine transporter polymorphisms in acute myocardial infarction. *Clin Sci (Lond), 104,* 241–245.

302. Wenzel, K., Baumann, G., & Felix, S. B. (1999). The homozygous combination of Leu125Val and Ser563Asn polymorphisms in the PECAM1 (CD31) gene is associated with early severe coronary heart disease. *Hum Mutat, 14,* 545.

303. Gardemann, A., Knapp, A., Katz, N., et al. (2000). No evidence for the CD31 C/G gene polymorphism as an independent risk factor of coronary heart disease. *Thromb Haemost, 83,* 629.

304. Song, F. C., Chen, A. H., Tang, X. M., et al. (2003). Association of platelet endothelial cell adhesion molecule-1 gene polymorphism with coronary heart disease. *Di Yi Jun Yi Da Xue Xue Bao, 23,* 156–158.

305. Wei, H., Fang, L., Chowdhury, S. H., et al. (2004). Platelet-endothelial cell adhesion molecule-1 gene polymorphism and its soluble level are associated with severe coronary artery stenosis in Chinese Singaporean. *Clin Biochem, 37,* 1091–1097.

306. Sasaoka, T., Kimura, A., Hohta, S. A., et al. (2001). Polymorphisms in the platelet-endothelial cell adhesion molecule-1 (PECAM-1) gene, Asn563Ser and Gly670Arg, associated with myocardial infarction in the Japanese. *Ann NY Acad Sci, 947,* 259–269.

307. Listi, F., Candore, G., Lio, D., et al. (2004). Association between platelet endothelial cellular adhesion molecule 1 (PECAM-1/CD31) polymorphisms and acute myocardial infarction: A study in patients from Sicily. *Eur J Immunogenet, 31,* 175–178.

308. Kunicki, T. J., Federici, A. B., Salomon, D. R., et al. (2004). An association of candidate gene haplotypes and bleeding severity in von Willebrand disease (VWD) type 1 pedigrees. *Blood, 104,* 2359–2367.

309. Di Paola, J., Federici, A. B., Mannucci, P. M., et al. (1999). Low platelet a₂b₁ levels in type I von Willebrand disease correlate with impaired platelet function in a high shear stress system. *Blood, 93,* 3578–3582.

310. Pereira, J., Quiroga, T., Pereira, M. E., et al. (2003). Platelet membrane glycoprotein polymorphisms do not influence the clinical expressivity of von Willebrand disease type 1. *Thromb Haemost, 90,* 1135–1140.

311. Potapov, E. V., Ignatenko, S., Nasseri, B. A., et al. (2004). Clinical significance of PlA polymorphism of platelet GP IIb/IIIa receptors during long-term VAD support. *Ann Thorac Surg, 77,* 869–874.

312. O'Connor, F. F., Shields, D. C., Fitzgerald, A., et al. (2001). Genetic variation in glycoprotein IIb/IIIa (GPIIb/IIIa) as a determinant of the responses to an oral GPIIb/IIIa antagonist in patients with unstable coronary syndromes. *Blood, 98,* 3256–3260.

313. Iniesta, J. A., Gonzalez-Conejero, R., Piqueras, C., et al. (2004). Platelet GP IIIa polymorphism HPA-1 (PlA) protects against subarachnoid hemorrhage. *Stroke, 35,* 2282–2286.

314. Ghosh, K., Nair, S., Kulkarni, B., et al. (2003). Milder bleeding tendency in Glanzmann's thrombasthenia patients inheriting HPA-1b in the homozygous state. *J Thromb Haemost, 1,* 2255–2256.

315. Steinhubl, S. R., Varanasi, J. S., & Goldberg, L. (2003). Determination of the natural history of aspirin resistance among stable patients with cardiovascular disease. *J Am Coll Cardiol, 42,* 1336–1337.

316. Gum, P. A., Kottke-Marchant, K., Poggio, E. D., et al. (2001). Profile and prevalence of aspirin resistance in patients with cardiovascular disease. *Am J Cardiol, 88,* 230–235.

317. Buchanan, M. R., & Brister, S. J. (1995). Individual variation in the effects of ASA on platelet function: Implications for the use of ASA clinically. *Can J Cardiol, 11,* 221–227.

318. Hillarp, A., Palmqvist, B., Lethagen, S., et al. (2003). Mutations within the cyclooxygenase-1 gene in aspirin non-responders with recurrence of stroke. *Thromb Res, 112,* 275–283.

319. Undas, A., Brummel, K., Musial, J., et al. (2001). Pl(A2) polymorphism of beta(3) integrins is associated with enhanced thrombin generation and impaired antithrombotic action of aspirin at the site of microvascular injury. *Circulation, 104,* 2666–2672.

320. Lau, W. C., Gurbel, P. A., Watkins, P. B., et al. (2004). Contribution of hepatic cytochrome P450 3A4 metabolic activity to the phenomenon of clopidogrel resistance. *Circulation, 109,* 166–171.

321. von Beckerath, N., von Beckerath, O., Koch, W., et al. (2005). P2Y12 gene H2 haplotype is not associated with increased adenosine diphosphate-induced platelet aggregation after initiation of clopidogrel therapy with a high loading dose. *Blood Coagul Fibrinolysis, 16,* 199–204.

322. Angiolillo, D. J., Fernandez-Ortiz, A., Bernardo, E., et al. (2004). 807 C/T polymorphism of the glycoprotein Ia gene and pharmacogenetic modulation of platelet response to dual antiplatelet treatment. *Blood Coagul Fibrinolysis, 15,* 427–433.

323. Lau, W. C., Waskell, L. A., Watkins, P. B., et al. (2003). Atorvastatin reduces the ability of clopidogrel to inhibit platelet aggregation: A new drug–drug interaction. *Circulation, 107,* 32–37.

324. Bihour, C., Durrieu-Jais, C., Macchi, L., et al. (1999). Expression of markers of platelet activation and the interpatient variation in response to abciximab. *Arterioscler Thromb Vasc Biol, 19,* 212–219.

325. Holmes, M. B., Sobel, B. E., & Schneider, D. J. (1999). Variable responses to inhibition of fibrinogen binding induced by

tirofiban and eptifibatide in blood from healthy subjects. *Am J Cardiol, 84,* 203–207.

326. Rozalski, M., Boncler, M., Luzak, B., et al. (2005). Genetic factors underlying differential blood platelet sensitivity to inhibitors. *Pharmacol Rep, 57,* 1–13.

327. Wheeler, G. L., Braden, G. A., Bray, P. F., et al. (2002). Reduced inhibition by abciximab in platelets with the Pl$^{A2}$ polymorphism. *Am Heart J, 143,* 76–82.

328. Boncler, M. A., Golanski, J., Paczuski, R., et al. (2002). Polymorphisms of glycoprotein Ib affect the inhibition by aurintricarboxylic acid of the von Willebrand factor dependent platelet aggregation. *J Mol Med, 80,* 796–801.

329. Bray, P. F., Cannon, C. P., Goldschmidt-Clermont, P., et al. (2001). The platelet Pl$^{A2}$ and angiotensin-converting enzyme (ACE) D allele polymorphisms and the risk of recurrent events after acute myocardial infarction. *Am J Cardiol, 88,* 347–352.

330. Borecki, I. B., & Suarez, B. K. (2001). Linkage and association: Basic concepts. *Adv Genet, 42,* 45–66.

331. Hinds, D. A., Stuve, L. L., Nilsen, G. B., et al. (2005). Whole-genome patterns of common DNA variation in three human populations. *Science, 307,* 1072–1079.

332. Clark, A. G. (2004). The role of haplotypes in candidate gene studies. *Genet Epidemiol, 27,* 321–333.

333. Crawford, D. C., & Nickerson, D. A. (2005). Definition and clinical importance of haplotypes. *Annu Rev Med, 56,* 303–320.

334. Kim, H. O., Jin, Y., Kickler, T. S., et al. (1995). Gene frequencies of the five major human platelet antigens in African American, white, and Korean populations. *Transfusion, 35,* 863–867.

335. Kekomaki, S., Partanen, J., & Kekomaki, R. (1995). Platelet alloantigens HPA-1, -2, -3, -5 and -6b in Finns. *Transfus Med, 5,* 193–198.

336. Merieux, Y., Debost, M., Bernaud, J., et al. (1997). Human platelet antigen frequencies of platelet donors in the French population determined by polymerase chain reaction with sequence-specific primers. *Pathol Biol (Paris), 45,* 697–700.

337. Simsek, S., Faber, N. M., Bleeker, P. M., et al. (1993). Determination of human platelet antigen frequencies in the Dutch population by immunophenotyping and DNA (allele-specific restriction enzyme) analysis. *Blood, 81,* 835–840.

338. Santoso, S., Santoso, S., Kiefel, V., et al. (1993). Frequency of platelet-specific antigens among Indonesians. *Transfusion, 33,* 739–741.

339. Tanaka, S., Taniue, A., Nagao, N., et al. (1995). Simultaneous DNA typing of human platelet antigens 2, 3 and 4 by an allele-specific PCR method. *Vox Sang, 68,* 225–230.

340. Morita, H., Kurihara, H., Imai, Y., et al. (2001). Lack of association between the platelet glycoprotein Ia C807T gene polymorphism and myocardial infarction in Japanese. An approach entailing melting curve analysis with specific fluorescent hybridization probes. *Thromb Haemost, 85,* 226–230.

341. Ito, T., Ishida, F., Shimodaira, S., et al. (1999). Polymorphisms of platelet membrane glycoprotein Iba and plasma von Willebrand factor antigen in coronary artery disease. *Int J Hematol, 70,* 47–51.

# CHAPTER 15

# Platelet Secretion

## Guy L. Reed

*Cardiovascular Medicine and Research, Medical College of Georgia, Augusta, Georgia*

## I. Introduction

Platelets contain α-granules, dense granules, and lysosomes. These secretory organelles have unique molecular contents, ultrastructural patterns, kinetics of exocytosis, and genetic disorders. Platelet secretion or exocytosis releases molecules at sites of vascular injury to activate other cells or to facilitate cellular adhesion.[1–8] Platelets secrete molecules from intracellular granules that play central roles in hemostasis, thrombosis, and vascular remodeling.[9–12] Platelet secretion is increased in humans with cardiovascular diseases where it appears to accelerate the development of arteriosclerotic lesions through the release of secretory products.[12–20]

## II. Platelet Granules

The morphological structure of α-granules, dense granules, and lysosomes is discussed in detail in Chapter 3. The molecular content of platelet granules determines their morphology, density, size, and function. α-Granules and dense granules are found only in megakaryocytes and platelets, whereas lysosomes are ubiquitous. Each platelet contains an average of 80 α-granules, which are 200–500 nm in diameter.[21,22] α-Granules contain procoagulant molecules, fibrinolytic regulators, growth factors, chemokines, immunologic modulators, adhesion molecules, and other proteins (Table 15-1).[4,21,23–29] Some α-granule proteins are *"platelet-specific"* molecules that are synthesized only in megakaryocytes (e.g., platelet factor 4 and β-thromboglobulin, Table 15-1). Other molecules are *"platelet-selective"* (e.g., P-selectin, von Willebrand factor, and fibrinogen, Table 15-1) because they are synthesized (or endocytosed) by megakaryocytes and relatively few other cells.

Dense granules are nearly 10-fold less abundant than α-granules in human platelets.[30] They contain small molecules (e.g., ADP, serotonin, and Ca$^{2+}$) and comparatively few proteins (Table 15-2). Dense granules are slightly acidic (pH 6.1) and contain lysosomal membrane proteins such as CD63 (LAMP-3) and LAMP-2, but not LAMP-1.[31–34] P-selectin appears to be present in both α- and dense granules which may reflect dual sorting signals present in the molecule's cytoplasmic domain.[35,36]

The few primary and secondary lysosomes found in platelets are heterogeneous in appearance (Chapter 3).[5,37,38] Lysosomes in platelets, like those in other cells, contain acid hydrolases, cathepsins, and lysosomal membrane proteins (LAMP-1, LAMP-2, and CD63).[4,33,39,40]

## III. Mechanisms of Platelet Exocytosis

Platelet exocytosis occurs through mechanisms that are homologous to those used by other specialized secretory cells such as neurons (Fig. 15-1). Exocytosis involves reorganization of the actin structure, the movement of granules into close physical apposition with the plasma membrane, granule-plasma membrane fusion, and release of intracellular contents.[41–43] Platelet secretion occurs through a SNARE-dependent mechanism. The SNARE proteins are a group of structurally related molecules derived from three gene superfamilies (with over 30 members) that include the syntaxins, the VAMP, and SNAP-25-related genes.[44] Each SNARE protein contains a coiled-coil region or SNARE-motif with about 60 amino acids. Through this motif, SNARE proteins on different membranes form tight, stable oligomeric complexes *in trans* which have been shown to be sufficient in model membranes to catalyze membrane fusion.[45] *N*-ethylmaleimide sensitive factor (NSF), a Mg$^{2+}$-dependent ATPase, dissociates SNARE complexes and is required for exocytosis in neurons and neuroendocrinelike cells.[46–48] The interaction of the SNAREs is also regulated by a number of different proteins including the Sec1/Munc18 proteins. The Rab proteins may mediate vesicle docking and they may participate in secretory events.[49,50]

### Table 15-1: α-Granule Contents[4,21,23–29]

| | |
|---|---|
| Adhesion molecules | P-Selectin,[a] von Willebrand factor,[a] thrombospondin, fibrinogen,[a] integrin αIIb[b]β3, integrin αvβ3, fibronectin |
| Chemokines | Platelet basic protein[b,c] [platelet factor 4 and its variant[b] (CXCL4) and β-thromboglobulin[b]], CCL3 (MIP-1α), CCL5 (RANTES), CCL7 (MCP-3), CCL17, CXCL1 (growth-regulated oncogene-α), CXCL5 (ENA-78), CXCL8 (IL-8) |
| Coagulation pathway | Factor V,[a] multimerin,[a] factor VIII |
| Fibrinolytic pathway | α₂-Macroglobulin, plasminogen, plasminogen activator inhibitor 1 |
| Growth and angiogenesis | Basic fibroblast growth factor, epidermal growth factor, hepatocyte growth factor, insulin-like growth factor 1, transforming growth factor β, vascular endothelial growth factor-A, vascular endothelial growth factor-C, platelet-derived growth factor |
| Immunologic molecules | β1H Globulin, factor D, c1 inhibitor, IgG |
| Other proteins | Albumin, α₁-antitrypsin, Gas6, histidine-rich glycoprotein, high molecular weight kininogen, osteonectin protease nexin-II (amyloid beta-protein precursor) |

[a]These platelet *selective* proteins are synthesized or taken up by megakaryocytes-platelets and found in relatively few other cells.

[b]These platelet *specific* proteins are unique to platelets.

[c]Platelet basic protein undergoes proteolysis to yield platelet factor 4 and β-thromboglobulin-related proteins.

Note: Many other molecules have been identified in platelet releasate but their presence in granules has yet to be demonstrated.

### Table 15-2: Dense Granule Contents[4,31,138,139]

| | |
|---|---|
| Ions | Ca, Mg, P, pyrophosphate |
| Nucleotides | ATP, GTP, ADP, GDP |
| Membrane proteins | CD63 (granulophysin), LAMP 2 |
| Transmitters | Serotonin |

### A. Platelet SNARE Proteins Mediate Exocytosis

Platelets contain the SNARE proteins syntaxin 2, syntaxin 4, SNAP-23, SNAP-25, SNAP-29, VAMP 3, and VAMP 8.[51–60] The role of these SNARE proteins in platelet secretion has been established using selective membrane permeabilization techniques. NSF activity is required for platelet secretion, perhaps because it dissociates *cis*-SNARE complexes and prepares them for *trans*-complex formation.[53,56] Human platelet dense granule secretion appears to occur through a syntaxin-2-dependent mechanism and to require both SNAP-23 and VAMP 3.[56,58,60] However, mice lacking VAMP 3 are still capable of secretion; this may reflect species differences or functional complementation between different VAMP molecules.[61] In contrast, α-granule secretion requires syntaxin 4 and perhaps syntaxin 2, VAMP 3 and VAMP 8, and SNAP-23.[57–60]

### B. Regulatory Role of Sec1/Munc18 Proteins

The Sec1-Munc18 (SM) proteins are a conserved family of approximately 65-kDa molecules that play a central, but still enigmatic role in exocytosis in yeast, neurons, and special-ized secretory cells (reviewed in [62]). Three different genetic forms of the SM proteins exist in human platelets: Munc18a, Munc18b, and Munc18c.[52,63,64] Syntaxin 4-Munc18c complexes are readily detected in platelet membranes whereas syntaxin 2-Munc18c complexes are not.[60] Peptides that mimic sites in Munc18c are projected to interact with syntaxin 4 and promote dense granule exocytosis; monoclonal antibodies that block this interaction also promoted α-, dense, and lysosomal secretion.[63] Peptides that mimic a site on the Munc18 molecules not involved in interactions with the syntaxin molecules inhibited secretion.[64] When platelets are activated to secrete intracellular granules, there is dissociation of the Munc18c-syntaxin 4 complex.[63] Acute dissociation of the Munc18-syntaxin complex promotes Ca²⁺-induced exocytosis, indicating that Munc18-syntaxin complex formation *per se* has a regulatory effect on triggered secretion.[63]

### C. The Rab System and Platelet Secretion

The Rabs are an extensive family of low molecular weight GTPases.[50,65] The *gunmetal* mutation in mice provides evidence that the Rab pathway (Fig. 15-2) plays an important role in platelet granule development. *Gunmetal* results in a 75 to 80% reduction in the alpha subunit of the Rab geranylgeranyl complex (Step 2, Fig. 15-2) that attaches geranylgeranyl groups onto Rabs to insert them into membranes. As a result of the *gunmetal* mutation one or more Rabs in platelets are not delivered to their functional membrane sites. Mice with the *gunmetal* mutation, have reduced α- and dense granule contents and diminished platelet secretion (see below).[66–68]

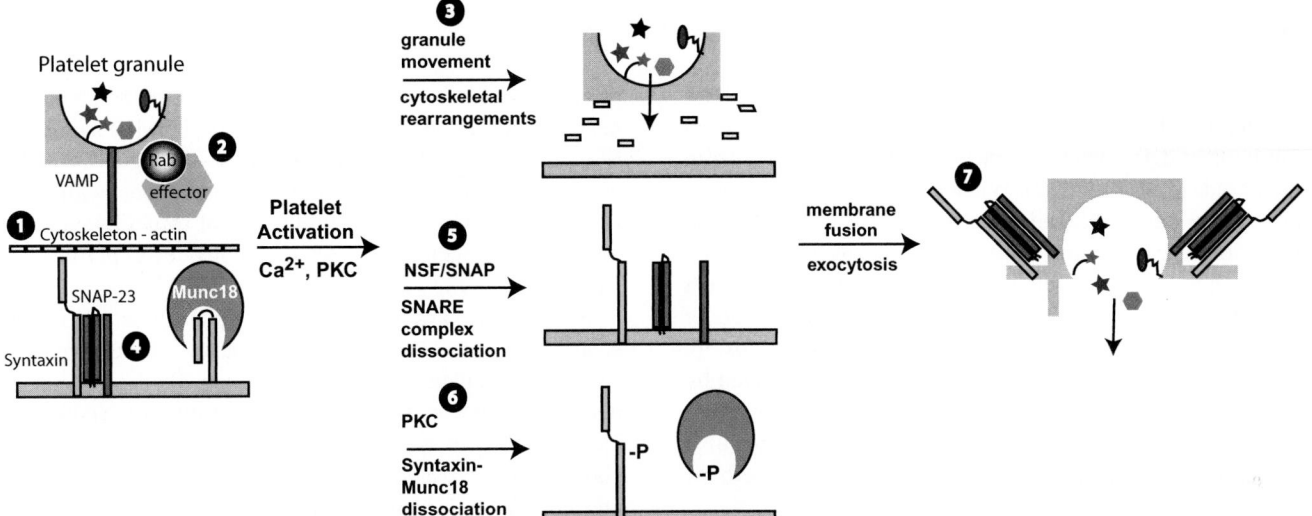

**Figure 15-1.** Platelet exocytosis model. Platelet secretion has important similarities to exocytosis in neurons and other cells, but involves a platelet-selective machinery that is uniquely coupled to signaling events triggered by cell activation. Platelets contain syntaxins 2 and 4, VAMPs 3 and 8, SNAP-23, and the Munc18 and Rab proteins.[51–60] In resting platelets, actin and other cytoskeletal elements may hinder exocytosis (Step 1) because when cells are activated, agents that affect actin assembly modulate secretion (Step 3).[85,100] Rabs 4 and 27a (Step 2) are implicated in platelet secretion, perhaps through interactions with effector proteins (Step 2) such as Munc13-4.[69,73] Resting platelets may also contain *cis*-SNARE complexes (Step 4) that block the nonspecific release of granules before platelet activation. This appears likely because NSF/SNAP activity, which dissociates SNARE complexes, is required for secretion (Step 5).[53] Munc18c binding to syntaxin (Step 4) prevents SNARE complex formation, but also may be necessary for priming the SNAREs for a *trans* interaction. Platelet secretion requires dissociation of Munc18c 6.[52,63,64] Munc18c is phosphorylated through a protein kinase C (PKC) mechanism when cells are activated to secrete by thrombin (Step 6).[52,63] Phosphorylation of Munc18c diminishes its binding interactions with syntaxin 4; the release of Munc18 may also prime syntaxin for SNARE interactions.[52,63] The formation of *trans*-SNARE complexes between the granule and plasma membrane (Step 7) are required for exocytosis.[56–60] Phosphorylation of syntaxin 4 and SNAP-23 that occurs during platelet activation may regulate interactions between the SNARE proteins and other molecules.[60,81] Increases in intracellular $Ca^{2+}$, acting through unidentified $Ca^{2+}$ sensor/transducer molecule(s), leads to fusion of the granule with the plasmalemma and exocytosis.

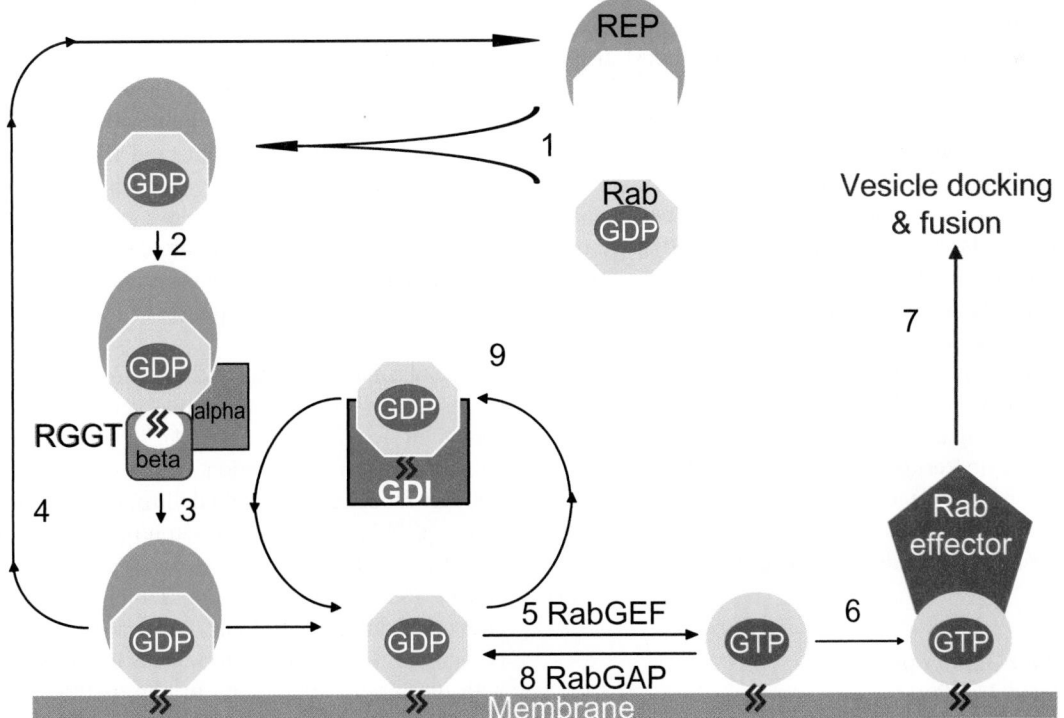

**Figure 15-2.** Rab cycle in membrane trafficking.[50,65] Newly synthesized GDP-bound Rabs interact with Rab escort protein (REP, Step 1). The Rab-GDP-REP complex then binds to the two-subunits (alpha, beta) of the Rab geranylgeranyltransferase (RGGT) which adds geranylgeranyl groups to the Rab protein (Step 2) that allow the Rab-GDP to be inserted by REP in the membrane (Step 3). Afterwards, REP dissociates to bind to another Rab protein (Step 4). In the membrane, a Rab guanine exchange factor (GEF), such as Vps39, "activates" the Rab by exchanging a GTP for a GDP (Step 5). The activated Rab then interacts with a specific effector molecule to mediate vesicle trafficking (Steps 6 and 7). The activated Rab GTP may be acted on by a RabGTPase activating protein (GAP, Step 8) to exchange GTP to GDP and "inactivate it." GDP dissociation inhibitor (GDI) can extract GDP-bound Rabs from the membrane to the cytosol (Step 9).

Additional evidence that Rabs play a critical role in platelet secretion has come from studies with GDP dissociation
inhibitor (GDI), which extracts GDP-bound Rabs from
membrane sites (Fig. 15-2). When a GDI isoform was introduced into permeabilized platelets it extracted membrane
Rabs to the cytosol and inhibited $Ca^{2+}$-induced α-granule
secretion.[69] Platelets contain a number of other Rabs such
as Rabs 1, 3b, 4, 6, 8, 11, 27, and 31, but their role in platelet
secretion or granule formation is still unknown.[70] Rabs 3b,
6, and 8 are phosphorylated when platelets are activated by
thrombin, suggesting that they may play a role in activation-
triggered secretion events.[71,72] Recent studies have also
implicated Rab27a in dense granule secretion — perhaps
through interactions with Munc13-4.[73]

# IV. Intracellular Signaling in Activation-Secretion Coupling

## A. $Ca^{2+}$ and Activated Protein Kinase C Are Critical Second Messengers in Secretion

Platelet secretion occurs after cells are activated by specific
ligands (e.g., thrombin, ADP, collagen, thromboxane $A_2$)
that interact with platelet membrane receptors, through $G_q$
protein-coupled or other mechanisms (see Chapter 16).[74]
Phosphatidylinositol 4,5 bisphosphate ($PIP_2$) is present in
platelet membranes and is cleaved to diacylglycerol and
inositol-1,4,5-trisphosphate ($IP_3$). Diacylglycerol activates
several forms of protein kinase C (PKC).[75] $IP_3$ increases
intracellular $[Ca^{2+}]$ from 40 to 100 nM to 2 to 10 mM, triggering granule secretion.[76–79] Increases in intracellular $[Ca^{2+}]$
are sufficient to induce platelet exocytosis, but PKC also
appears to interact synergistically with $Ca^{2+}$ to amplify
secretion.[80] Inhibitors of activated PKC block platelet secretion.[81–84] Unfortunately, the mechanisms through which elevations of intracellular $[Ca^{2+}]$ are coupled to secretion events
is poorly understood in platelets and is still not well-defined
in other secretory cells.

After platelet activation, several molecules are phosphorylated with kinetics that suggest that they are involved in
secretion. Myristoylated alanine-rich C kinase substrate
(MARCKS) may play a role at the interface between membranes and the cytoskeleton.[85] MARCKS binds tightly to
$PIP_2$ to protect it from degradation by phospholipase C.[3,86]
MARCKS is rapidly phosphorylated in platelets and a pseudo-
MARCKS substrate blocks dense granule secretion.[87]

## B. Intracellular Signaling and SNARE-Related Machinery

Interactions of platelet SNARE-related machinery may be
regulated by PKC-dependent phosphorylation events. When

platelets were stimulated by thrombin, Munc18c, syntaxin
4, and SNAP-23 became targets for phosphorylation.[52,60,64,81]
Munc18c phosphorylation proceeds through a PKC-
dependent process and it inhibits interactions with syntaxin.[52,64,63,64] Given that inhibition of Munc18-syntaxin
interactions enhanced platelet secretion,[63] PKC phosphorylation of Munc18c may be a trigger of exocytosis.

Syntaxin 4 plays a critical role in platelet secretion and
is also a target of intracellular kinases whose signaling
activity is coupled to cellular activation. Syntaxin 4 phosphorylation appears to be mediated through PKC and inhibits the binding of syntaxin 4 to SNAP-23.[81] Similarly, when
platelets are activated by thrombin to secrete, SNAP-23 is
phosphorylated largely on serine residues, through a PKC-
dependent mechanism, with kinetics that parallel the rate of
dense granule exocytosis and precede α-granule secretion.[60]
Mutants of SNAP-23 that mimic phosphorylation inhibited
syntaxin 4 interactions, which suggests that phosphorylation
may play a role in modulating SNARE-complex interactions
during membrane trafficking and fusion.[60]

These results provide evidence that extracellular activation can be coupled through intracellular PKC signaling to
modulate SNARE protein interactions involved in platelet
exocytosis. It remains to be elucidated how signaling to the
SNARE machinery is integrated with cell activation to regulate the platelet secretory process.

# V. Platelet Cytoskeleton and Secretion

As described in detail in Chapter 3, electron microscopy
studies of human platelets have shown that individual granules fuse with the plasma membrane that lines the surface-
connected canalicular system, and they have also shown
compound granule-granule-plasmalemma fusion, akin to
that seen in mast cells.[30,41,88] Dense granules have been identified in bovine platelets that are closely "anchored" to the
plasma membrane. These dense granules are most likely to
undergo exocytosis in response to increased intracellular
$Ca^{2+}$.[89]

The plasma membrane of the resting platelet is supported
by a spectrin-based skeleton that centralizes upon platelet
activation (see Chapters 3 and 4).[90] Platelet shape can occur
without triggering secretion, when platelets are activated
through receptors that signal through $G_i$-coupled mechanisms.[74,91,92] Still, the cytoskeleton may facilitate the development, targeting, and exocytosis of secretory granules.[93,94]
Agents that interfere with microtubules (e.g., colchicine,
antibodies) inhibit secretion.[95–97] The F-actin network, which
is thought to be a barrier to the fusion of exocytic vesicles
with the plasma membrane in chromaffin cells, may also
impede secretion in platelets.[43,94,98,99] Phalloidin and cytochalasin B (which prevent actin depolymerization) inhibit
dense granule secretion by ADP, whereas scinderin, a $Ca^{2+}$-

dependent F-actin severing protein, enhances dense granule secretion.[100]

## VI. Platelet Granule Development

Granule development is the first step in the platelet secretory process. Multivesicular bodies may be the precursor to α- and dense granules because they precede these organelles in young megakaryocytes.[101] Multivesicular bodies sort molecules during endocytosis[101] and have been found to contain proteins typical of α-granules (e.g., β-thromboglobulin, von Willebrand factor) as well as dense granules (CD63).[102] Type 1 multivesicular bodies have many internal vesicles (30–70 nm) that only contain the α-granule proteins β-thromboglobulin and von Willebrand factor. Type 2 multivesicular bodies contain internal vesicles (including the dense granule-lysosomal marker CD63) that are distributed in the periphery of the organelle; they also contain electron-dense material which includes β-thromboglobulin and von Willebrand factor.[101] As megakaryocytes enlarge, multivesicular bodies decline in number and α- and dense granules become numerous.[101,103] Subsequently, a demarcating membrane system develops and granules are transported into the developing proplatelet (Chapter 2).[101,103,104]

Although it was originally thought that α- and dense granules have completely distinct molecular cargo, recent studies suggest that there is trafficking through all membrane systems.[2] For example, dense granules and α-granules have detectable amounts of plasma membrane glycoproteins integrin αIIbβ3, P-selectin, and others.[35,105]

Proteins synthesized in the megakaryocyte (Table 15.1) may traffic to the developing α-granules in vesicles from the trans-Golgi apparatus.[106,107] Receptor-mediated endocytosis is another mechanism through which proteins such as fibrinogen (via integrin αIIbβ3) are trafficked to the α-granules.[108–112] Molecules taken up through endocytosis may be targeted to the α-granule by a clathrin-coated vesicle pathway that is nondegradative or by a clathrin-independent, degrading, endosomal-lysosomal route.[108,113–117]

There is less definitive information about dense granule development because there is not a specific protein marker for dense granules (CD63 is also found in lysosomes). Dense granule indicators (CD63 and serotonin) appear early in megakaryocytes at roughly the same time as α-granule formation but these nascent "dense" protovesicles have little identifiable cargo.[103,118] The specific transport and storage mechanisms responsible for loading the dense granule improve as the megakaryocyte and platelet develops, because it is only in mature cells that dense granules acquire their signature electron dense staining.[119,120]

Lysosomes are formed early in megakaryocyte maturation, even before α-granule development.[118] Endocytic trafficking to platelet lysosomes through endosomes appears to proceed through clathrin-independent pathways.[115]

## VII. Disorders of Platelet Secretion

Platelet storage pool disorders (SPD) are characterized by absent or empty granules and can be distinguished from conditions associated with impaired platelet secretion due to impaired cellular activation or signaling (see Chapter 57). Human α-SPD (Gray platelet syndrome) is characterized by markedly reduced α-granules, normal lysosomes, normal dense granules, and a variable tendency for bleeding.[9,121–123] Megakaryocytes from patients with α-SPD contain small abnormal vesicles that resemble a Golgi-associated, α-granule precursor.[9,124] Many of the megakaryocyte-synthesized, soluble α-granule proteins (such as β-thromboglobulin) are reduced in α-SPD platelets but are found in the plasma. This suggests that these molecules may be incorrectly targeted to a constitutive secretory pathway; alternatively there may be impaired α-granule storage mechanisms or premature granule release in these platelets.[124–127] In some α-SPD patients, increased amounts of the α-granule membrane protein P-selectin are found on the platelet membrane following cell activation,[121] which argues that the defect in some patients with α-SPD is in trafficking to the granule from the Golgi and not in exocytosis and membrane fusion. Interestingly, in some α-SPD patients, the Weibel–Palade body in endothelial cells is normal despite the close similarity of this organelle to the platelet α-granule.[128,129] In α-SPD, serotonin uptake is normal although some reduction in dense granule secretion has been reported.[126,130] In contrast to Hermansky–Pudlak syndrome (HPS, see below), only a small number of families with α-SPD have been described.[131,132] At the present time the arthrogryposis, renal dysfunction, and cholestasis (ARC) syndrome, which is due to mutations in *VPS33B*, is the only known genetic cause of α-SPD.[133] Few patients with combined α- and dense SPD have been reported and the molecular causes are unknown. One model of combined α- and dense SPD has been described in *gunmetal* mice (see above).[66–68]

Platelet dense SPDs are heterogeneous;[134] they may be an isolated defect or they may be associated with disorders such as Chediak–Higashi syndrome and HPS that affect pigmentation and other physiological parameters (see Chapter 57). HPS is an autosomal recessive condition characterized by lack of pigment in the eyes and skin (oculocutaneous albinism), deficient platelet dense granules, and variable accumulation of ceroid lipofuscin.[10] Although the molecular mechanisms underlying the pathophysiology of HPS are still not understood, there has been tremendous progress in identifying the genetic causes of HPS in humans and mice (Table 15-3).[135–137] A number of these gene products form complexes in the cell that are believed to mediate the

**Table 15-3: Genetic Mutations in Platelet Secretory Pool Disorders[135–137]**

| Human SPD | Mouse SPD | Affected Protein | Interacting Complex | Granules Affected |
|---|---|---|---|---|
| *HPS-1* | Pale ear | HPS1 | BLOC1 | D |
| *HPS-2* | Pearl | AP-3β3A | AP-3 complex | D |
| *HPS-3* | Cocoa | HPS3 | BLOC2 | D |
| *HPS-4* | Light ear | HPS4 | BLOC1 | D |
| *HPS-5* | Ruby eye 2 | HPS5 | BLOC2 | D |
| *HPS-6* | Ruby eye | HPS6 | BLOC2 | D |
| *HPS-7* | Sandy | HPS7 | BLOC1 | D |
| — | Buff | Vps33a | HOPS | D |
| — | Cappuccino | Cappucino | BLOC1 | D |
| — | Ashen | Rab27a | Myosin 5a | D |
| — | Gunmetal | RabGGTase α | RabGGTase-REP | A, D |
| — | Mocha | AP-3 δ | AP-3 complex | D |
| — | Muted | Muted | BLOC1 | D |
| — | Pallid | Palladin | BLOC1 | D |

Abbreviations: A, α-granule; D, dense granule; HPS, Hermansky-Pudlak syndrome; SPD, storage pool disease.

trafficking of molecules to vesicles or storage granules. These fascinating discoveries will provide important insights into the molecular pathways of granule development in megakaryocytes and contribute to a broader understanding of vesicle trafficking in other secretory cells.

## Acknowledgments

The author gratefully acknowledges the support of Aiilyan Houng, Sarah King, Janos Polgar, Michelle Bell, and the NIH (HL64057). Because of space limitations, it has not been possible to cite many key references in this field.

## References

1. Dale, G. L., Friese, P., Batar, P., et al. (2002). Stimulated platelets use serotonin to enhance their retention of pro-coagulant proteins on the cell surface. *Nature, 415,* 175–179.
2. Reed, G. L., Fitzgerald, M. L., & Polgar, J. (2000). Molecular mechanisms of platelet exocytosis: Insights into the "secrete" life of thrombocytes. *Blood, 96,* 3334–3342.
3. Flaumenhaft, R. (2003). Molecular basis of platelet granule secretion. *Arterioscler Thromb Vasc Biol, 23,* 1152–1160.
4. Fukami, M. H., Holmsen, H., Kowalska, M. A., et al. (2001). Platelet secretion. In R. W. Colman, J. Hirsch, V. J. Marder, A. W. Clowes, & J. N. George (Eds.), *Hemostasis and thrombosis: Basic principles and clinical practice* (4th.

ed., pp. 516–573). Philadelphia: Lippincott Williams & Wilkins.
5. Rendu, F., & Brohard-Bohn, B. (2001). The platelet release reaction: Granules' constituents, secretion and functions. *Platelets, 12,* 261–273.
6. Stenberg, P. E., McEver, R. P., Shuman, M. A., et al. (1985). A platelet alpha-granule membrane protein (GMP-140) is expressed on the plasma membrane after activation. *J Cell Biol, 101,* 880–886.
7. Berman, C. L., Yeo, E. L., Wencel-Drake, J. D., et al. (1986). A platelet alpha granule membrane protein that is associated with the plasma membrane after activation. Characterization and subcellular localization of platelet activation-dependent granule-external membrane protein. *J Clin Invest, 78,* 130–137.
8. Suzuki, H., Kaneko, T., Sakamoto, T., et al. (1994). Redistribution of alpha-granule membrane glycoprotein IIb/IIIa (integrin alpha IIb beta 3) to the surface membrane of human platelets during the release reaction. *J Electron Microsc (Tokyo), 43,* 282–289.
9. Weiss, H. J., Witte, L. D., Kaplan, K. L., et al. (1979). Heterogeneity in storage pool deficiency: Studies on granule-bound substances in 18 patients including variants deficient in alpha-granules, platelet factor 4, beta-thromboglobulin, and platelet-derived growth factor. *Blood, 54,* 1296–1319.
10. Hermansky, F., & Pudlak, P. (1959). Albinism associated with hemorrhagic diathesis and unusual pigmented reticular cells in the bone marrow. *Blood, 14,* 162.
11. Novak, E. K., Hui, S. W., & Swank, R. T. (1984). Platelet storage pool deficiency in mouse pigment mutations associated with seven distinct genetic loci. *Blood, 63,* 536–544.

12. Huo, Y., Schober, A., Forlow, S. B., et al. (2003). Circulating activated platelets exacerbate atherosclerosis in mice deficient in apolipoprotein E. *Nat Med, 9,* 61–67.

13. Packham, M. A., & Mustard, J. F. (1986). The role of platelets in the development and complications of atherosclerosis. *Sem Hematol, 23,* 8–26.

14. Ross, R., Glomset, J., & Harker, L. (1977). Response to injury and atherogenesis. *Am J Pathol, 86,* 675–684.

15. Blann, A. D., Lip, G. Y., Beevers, D. G., et al. (1997). Soluble P-selectin in atherosclerosis: A comparison with endothelial cell and platelet markers. *Thromb Haemost, 77,* 1077–1080.

16. Smith, F. B., Lowe, G. D., Fowkes, F. G., et al. (1993). Smoking, haemostatic factors and lipid peroxides in a population case control study of peripheral arterial disease. *Atherosclerosis, 102,* 155–162.

17. Davi, G., Romano, M., Mezzetti, A., et al. (1998). Increased levels of soluble P-selectin in hypercholesterolemic patients [see comments]. *Circulation, 97,* 953–957.

18. Rak, K., Beck, P., Udvardy, M., et al. (1983). Plasma levels of beta-thromboglobulin and factor VIII-related antigen in diabetic children and adults. *Thromb Res, 29,* 155–162.

19. Yamanishi, J., Sano, H., Saito, K., et al. (1985). Plasma concentrations of platelet-specific proteins in different stages of essential hypertension: Interactions between platelet aggregation, blood lipids and age. *Thromb Haemost, 54,* 539–543.

20. Ramsis, N., El-Hawary, A. A., & Ismail, E. (1998). Relation between carotid intima-media thickness, platelet surface activation and endothelial cell markers. *Haemostasis, 28,* 268–275.

21. Sixma, J. J., Slot, J. W., & Geuze, H. J. (1989). Immunocytochemical localization of platelet granule proteins. *Methods Enzymol, 169,* 301–311.

22. Harrison, P., Savidge, G. F., & Cramer, E. M. (1990). The origin and physiological relevance of alpha-granule adhesive proteins. *Br J Haematol, 74,* 125–130.

23. Beckstead, J. H., Stenberg, P. E., McEver, R. P., et al. (1986). Immunohistochemical localization of membrane and alpha-granule proteins in human megakaryocytes: Application to plastic-embedded bone marrow biopsy specimens. *Blood, 67,* 285–293.

24. Niewiarowski, S. (1977). Proteins secreted by the platelet. *Thromb Haemost, 38,* 924–938.

25. Schmaier, A. H. (1985). Platelet forms of plasma proteins: Plasma cofactors/substrates and inhibitors contained within platelets. *Sem Hematol, 22,* 187–202.

26. Handagama, P., Rappolee, D. A., Werb, Z., et al. (1990). Platelet alpha-granule fibrinogen, albumin, and immunoglobulin G are not synthesized by rat and mouse megakaryocytes. *J Clin Invest, 86,* 1364–1368.

27. Browder, T., Folkman, J., & Pirie-Shepherd, S. (2000). The hemostatic system as a regulator of angiogenesis. *J Biol Chem, 275,* 1521–1524.

28. Deuel, T. F. (1997). Protein granule factors. In F. von Bruchhausen & U. Walter (Eds.), *Platelets and their factors* (Vol. 126, pp. 247–296). New York: Springer.

29. Gear, A. R., & Camerini, D. (2003). Platelet chemokines and chemokine receptors: Linking hemostasis, inflammation, and host defense. *Microcirculation, 10,* 335–350.

30. White, J. G. (1969). The dense bodies of human platelets: Inherent electron opacity of the serotonin storage particles. *Blood, 33,* 598–606.

31. Holmsen, H., & Weiss, H. J. (1979). Secretable storage pools in platelets. *Annu Rev Med, 30,* 119–134.

32. Nishibori, M., Cham, B., McNicol, A., et al. (1993). The protein CD63 is in platelet dense granules, is deficient in a patient with Hermansky-Pudlak syndrome, and appears identical to granulophysin. *J Clin Invest, 91,* 1775–1782.

33. Israels, S. J., McMillan, E. M., Robertson, C., et al. (1996). The lysosomal granule membrane protein, LAMP-2, is also present in platelet dense granule membranes. *Thromb Haemost, 75,* 623–629.

34. Febbraio, M., & Silverstein, R. L. (1990). Identification and characterization of LAMP-1 as an activation-dependent platelet surface glycoprotein. *J Biol Chem, 265,* 18531–18537.

35. Israels, S. J., Gerrard, J. M., Jacques, Y. V., et al. (1992). Platelet dense granule membranes contain both granulophysin and P-selectin (GMP-140). *Blood, 80,* 143–152.

36. Norcott, J. P., Solari, R., & Cutler, D. F. (1996). Targeting of P-selectin to two regulated secretory organelles in PC12 cells. *J Cell Biol, 134,* 1229–1240.

37. Menard, M., Meyers, K. M., & Prieur, D. J. (1990). Demonstration of secondary lysosomes in bovine megakaryocytes and platelets using acid phosphatase cytochemistry with cerium as a trapping agent. *Thromb Haemost, 63,* 127–132.

38. Fukami, M. H., & Salganicoff, L. (1977). Human platelet storage organelles. A review. *Thromb Haemost, 38,* 963–970.

39. Ehrlich, H. P., & Gordon, J. L. (1976). *Proteinases in platelets* (Vol. 1). New York: Elsevier.

40. Dangelmaier, C. A., & Holmsen, H. (1980). Determination of acid hydrolases in human platelets. *Anal Biochem, 104,* 182–191.

41. White, J. G. (1974). Electron microscopic studies of platelet secretion. *Prog Hemost Thromb, 2,* 49–98.

42. Morgan, A. (1995). Exocytosis. *Essays Biochem, 30,* 77–95.

43. Flaumenhaft, R., Dilks, J. R., Rozenvayn, N., et al. (2005). The actin cytoskeleton differentially regulates platelet alpha-granule and dense-granule secretion. *Blood, 105,* 3879–3887.

44. Sollner, T., Whiteheart, S. W., Brunner, M., et al. (1993). SNAP receptors implicated in vesicle targeting and fusion [see comments]. *Nature, 362,* 318–324.

45. Weber, T., Zemelman, B. V., McNew, J. A., et al. (1998). SNAREpins: Minimal machinery for membrane fusion. *Cell, 92,* 759–772.

46. Hanson, P. I., Roth, R., Morisaki, H., et al. (1997). Structure and conformational changes in NSF and its membrane receptor complexes visualized by quick-freeze/deep-etch electron microscopy. *Cell, 90,* 523–535.

47. Haas, A. (1998). NSF — fusion and beyond. *Trends Cell Biol, 8,* 471–473.

48. Morgan, A., & Burgoyne, R. D. (1995). A role for soluble NSF attachment proteins (SNAPs) in regulated exocytosis in adrenal chromaffin cells. *EMBO J, 14,* 232–239.

49. Zerial, M., & McBride, H. (2001). Rab proteins as membrane organizers. *Nature Rev Mol Cell Biol, 2,* 107–117.

50. Seabra, M. C., Mules, E. H., & Hume, A. N. (2002). Rab GTPases, intracellular traffic and disease. *Trends Mol Med, 8,* 23–30.

51. Lemons, P. P., Chen, D., Bernstein, A. M., et al. (1997). Regulated secretion in platelets: Identification of elements of the platelet exocytosis machinery. *Blood, 90,* 1490–1500.

52. Reed, G. L., Houng, A. K., & Fitzgerald, M. L. (1999). Human platelets contain SNARE proteins and a Sec1p homologue that interacts with syntaxin 4 and is phosphorylated after thrombin activation: Implications for platelet secretion. *Blood, 93,* 2617–2626.

53. Polgar, J., & Reed, G. L. (1999). A critical role for *N*-ethylmaleimide-sensitive fusion protein (NSF) in platelet granule secretion. *Blood, 94,* 1313–1318.

54. Flaumenhaft, R., Croce, K., Chen, E., et al. (1999). Proteins of the exocytotic core complex mediate platelet alpha-granule secretion. Roles of vesicle-associated membrane protein, SNAP-23, and syntaxin 4. *J Biol Chem, 274,* 2492–2501.

55. Bernstein, A. M., & Whiteheart, S. W. (1999). Identification of a cellubrevin/vesicle associated membrane protein 3 homologue in human platelets. *Blood, 93,* 571–579.

56. Chen, D., Bernstein, A. M., Lemons, P. P., et al. (2000). Molecular mechanisms of platelet exocytosis: Role of SNAP-23 and syntaxin 2 in dense core granule release. *Blood, 95,* 921–929.

57. Lemons, P. P., Chen, D., & Whiteheart, S. W. (2000). Molecular mechanisms of platelet exocytosis: Requirements for alpha-granule release. *Biochem Biophys Res Commun, 267,* 875–880.

58. Polgar, J., Chung, S. H., & Reed, G. L. (2002). Vesicle-associated membrane protein three (VAMP-3) and VAMP-8 are present in human platelets and are required for granule secretion. *Blood, 100,* 1081–1083.

59. Feng, D., Crane, K., Rozenvayn, N., et al. (2002). Subcellular distribution of three functional platelet SNARE proteins: Human cellubrevin, SNAP-23, and syntaxin 2. *Blood, 99,* 4006–4014.

60. Polgar, J., Lane, W. S., Chung, S. H., et al. (2003). Phosphorylation of SNAP-23 in activated human platelets. *J Biol Chem, 278,* 44369–44376.

61. Schraw, T. D., Rutledge, T. W., Crawford, G. L., et al. (2003). Granule stores from cellubrevin/VAMP-3 null mouse platelets exhibit normal stimulus-induced release. *Blood, 102,* 1716–1722.

62. Rizo, J., & Sudhof, T. C. (2002). Snares and Munc18 in synaptic vesicle fusion. *Nature Rev Neurosci, 3,* 641–653.

63. Houng, A., Polgar, J., & Reed, G. L. (2003). Munc18-syntaxin complexes and exocytosis in human platelets. *J Biol Chem, 278,* 19627–19633.

64. Schraw, T. D., Lemons, P. P., Dean, W. L., et al. (2003). A role for Sec1/Munc18 proteins in platelet exocytosis. *Biochem J, 374,* 207–217.

65. Pfeffer, S. (2003). Membrane domains in the secretory and endocytic pathways. *Cell, 112,* 507–517.

66. Novak, E. K., Reddington, M., Zhen, L., et al. (1995). Inherited thrombocytopenia caused by reduced platelet production in mice with the gunmetal pigment gene mutation. *Blood, 85,* 1781–1789.

67. Stinchcombe, J. C., Barral, D. C., Mules, E. H., et al. (2001). Rab27a is required for regulated secretion in cytotoxic t lymphocytes. *J Cell Biol, 152,* 825–834.

68. Swank, R. T., Jiang, S. Y., Reddington, M., et al. (1993). Inherited abnormalities in platelet organelles and platelet formation and associated altered expression of low molecular weight guanosine triphosphate-binding proteins in the mouse pigment mutant gunmetal. *Blood, 81,* 2626–2635.

69. Shirakawa, R., Yoshioka, A., Horiuchi, H., et al. (2000). Small GTPase rab4 regulates Ca²⁺-induced alpha-granule secretion in platelets. *J Biol Chem, 275,* 33844–33849.

70. King, S., & Reed, G. (2002). Development of platelet secretory granules. *Sem Cell Dev Biol, 13,* 293.

71. Fitzgerald, M. L., & Reed, G. L. (1999). Rab6 is phosphorylated in thrombin-activated platelets by a protein kinase C-dependent mechanism: Effects on GTP/GDP binding and cellular distribution. *Biochem J, 342,* 353–360.

72. Karniguian, A., Zahraoui, A., & Tavitian, A. (1993). Identification of small GTP-binding rab proteins in human platelets: Thrombin-induced phosphorylation of rab3B, rab6, and rab8 proteins. *Proc Natl Acad Sci USA, 90,* 7647–7651.

73. Shirakawa, R., Higashi, T., Tabuchi, A., et al. (2004). Munc13-4 is a GTP-Rab27-binding protein regulating dense core granule secretion in platelets. *J Biol Chem, 279,* 10730–10737.

74. Brass, L. F. (2003). Thrombin and platelet activation. *Chest, 124,* 18S–25S.

75. Ware, J. A., & Chang, J. D. (1997). Protein kinase C and its interactions with other serine-threonine kinases. In F. von Bruchhausen & U. Walter (Eds.), *Platelets and their factors* (Vol. 126, pp. 247–296). New York: Springer.

76. Rink, T. J., Smith, S. W., & Tsien, R. Y. (1982). Cytoplasmic free Ca²⁺ in human platelets: Ca²⁺ thresholds and Ca-independent activation for shape-change and secretion. *FEBS Lett, 148,* 21–26.

77. Knight, D. E., Hallam, T. J., & Scrutton, M. C. (1982). Agonist selectivity and second messenger concentration in Ca²⁺-mediated secretion. *Nature, 296,* 256–257.

78. Authi, K. S., Evenden, B. J., & Crawford, N. (1986). Metabolic and functional consequences of introducing inositol 1,4,5-trisphosphate into saponin-permeabilized human platelets. *Biochem J, 233,* 707–718.

79. Watson, S. P., Ruggiero, M., Abrahams, S. L., et al. (1986). Inositol 1,4,5-trisphosphate induces aggregation and release of 5-hydroxytryptamine from saponin-permeabilized human platelets. *J Biol Chem, 261,* 5368–5372.

80. Walker, T. R., & Watson, S. P. (1993). Synergy between Ca²⁺ and protein kinase C is the major factor in determining the level of secretion from human platelets. *Biochem J, 289,* 277–282.

81. Chung, S. H., Polgar, J., & Reed, G. L. (2000). Protein kinase C phosphorylation of syntaxin 4 in thrombin-activated human platelets. *J Biol Chem, 2275,* 25286–25291.

82. Geanacopoulos, M., Turner, J., Bowling, K. E., et al. (1993). The role of protein kinase C in the initial events of platelet

activation by thrombin assessed with a selective inhibitor. *Thromb Res, 69,* 113–124.

83. Rozenvayn, N., & Flaumenhaft, R. (2003). Protein kinase C mediates translocation of type II phosphatidylinositol 5-phosphate 4-kinase required for platelet alpha-granule secretion. *J Biol Chem, 278,* 8126–8134.

84. Gerrard, J. M., McNicol, A., & Saxena, S. P. (1993). Protein kinase C, membrane fusion and platelet granule secretion. *Biochem Soc Trans, 21,* 289–293.

85. Arbuzova, A., Schmitz, A. A., & Vergeres, G. (2002). Cross-talk unfolded: MARCKS proteins. *Biochem J, 362,* 1–12.

86. Wang, J., Arbuzova, A., Hangyas-Mihalyne, G., et al. (2001). The effector domain of myristoylated alanine-rich C kinase substrate binds strongly to phosphatidylinositol 4,5-bisphosphate. *J Biol Chem, 276,* 5012–5019.

87. Elzagallaai, A., Rose, S. D., & Trifaro, J. M. (2000). Platelet secretion induced by phorbol esters stimulation is mediated through phosphorylation of MARCKS: A MARCKS-derived peptide blocks MARCKS phosphorylation and serotonin release without affecting pleckstrin phosphorylation. *Blood, 95,* 894–902.

88. Morgenstern, E., Neumann, K., & Patscheke, H. (1987). The exocytosis of human blood platelets. A fast freezing and freeze-substitution analysis. *Eur J Cell Biol, 43,* 273–282.

89. Morimoto, T., Ogihara, S., & Takisawa, H. (1990). Anchorage of secretion-competent dense granules on the plasma membrane of bovine platelets in the absence of secretory stimulation. *J Cell Biol, 111,* 79–86.

90. Barkalow, K. L., Italiano, J. E., Jr. Chou, D. E., et al. (2003). Alpha-adducin dissociates from F-actin and spectrin during platelet activation. *J Cell Biol, 161,* 557–570.

91. Otterdal, K., Pedersen, T. M., & Solum, N. O. (2001). Platelet shape change induced by the peptide YFLLRNP. *Thromb Res, 103,* 411–420.

92. Kometani, M., Sato, T., & Fujii, T. (1986). Platelet cytoskeletal components involved in shape change and secretion. *Thromb Res, 41,* 801–809.

93. Sasakawa, N., Ohara-Imaizumi, M., Okubo, S., et al. (2002). Roles of actin filaments and the actin-myosin interaction in the regulation of exocytosis in chromaffin cells. *Ann NY Acad Sci, 971,* 273–274.

94. Valentijn, K., Valentijn, J. A., & Jamieson, J. D. (1999). Role of actin in regulated exocytosis and compensatory membrane retrieval: Insights from an old acquaintance. *Biochem Biophys Res Commun, 266,* 652–661.

95. Menche, D., Israel, A., & Karpatkin, S. (1980). Platelets and microtubules. Effect of colchicine and D2O on platelet aggregation and release induced by calcium ionophore A23187. *J Clin Invest, 66,* 284–291.

96. Verhoeven, A. J., Mommersteeg, M. E., & Akkerman, J. W. (1986). Kinetics of energy consumption in human platelets with blocked ATP regeneration. *Int J Biochem, 18,* 985–990.

97. Berry, S., Dawicki, D. D., Agarwal, K. C., et al. (1989). The role of microtubules in platelet secretory release. *Biochim Biophys Acta, 1012,* 46–56.

98. Burgoyne, R. D., & Cheek, T. R. (1987). Reorganisation of peripheral actin filaments as a prelude to exocytosis. *Biosci Rep, 7,* 281–288.

99. Vitale, M. L., Seward, E. P., & Trifaro, J. M. (1995). Chromaffin cell cortical actin network dynamics control the size of the release-ready vesicle pool and the initial rate of exocytosis. *Neuron, 14,* 353–363.

100. Marcu, M. G., Zhang, L., Nau-Staudt, K., et al. (1996). Recombinant scinderin, an F-actin severing protein, increases calcium-induced release of serotonin from permeabilized platelets, an effect blocked by two scinderin-derived actin-binding peptides and phosphatidylinositol 4,5-bisphosphate. *Blood, 87,* 20–24.

101. Heijnen, H. F., Debili, N., Vainchencker, W., et al. (1998). Multivesicular bodies are an intermediate stage in the formation of platelet alpha-granules. *Blood, 91,* 2313–2325.

102. Youssefian, T., & Cramer, E. M. (2000). Megakaryocyte dense granule components are sorted in multivesicular bodies. *Blood, 95,* 4004–4007.

103. Cramer, E. M., Norol, F., Guichard, J., et al. (1997). Ultrastructure of platelet formation by human megakaryocytes cultured with the Mpl ligand. *Blood, 89,* 2336–2346.

104. Hartwig, J., & Italiano, J Jr. (2003). The birth of the platelet. *J Thromb Haemost, 1,* 1580–1586.

105. Youssefian, T., Masse, J. M., Rendu, F., et al. (1997). Platelet and megakaryocyte dense granules contain glycoproteins Ib and IIb-IIIa. *Blood, 89,* 4047–4057.

106. Cramer, E. M., Harrison, P., Savidge, G. F., et al. (1990). Uncoordinated expression of alpha-granule proteins in human megakaryocytes. *Prog Clin Biol Res, 356,* 131–142.

107. Hegyi, E., Heilbrun, L. K., & Nakeff, A. (1990). Immunogold probing of platelet factor 4 in different ploidy classes of rat megakaryocytes sorted by flow cytometry. *Exp Hematol, 18,* 789–793.

108. Handagama, P. J., George, J. N., Shuman, M. A., et al. (1987). Incorporation of a circulating protein into megakaryocyte and platelet granules. *Proc Natl Acad Sci USA, 84,* 861–865.

109. Handagama, P. J., Shuman, M. A., & Bainton, D. F. (1989). Incorporation of intravenously injected albumin, immunoglobulin G, and fibrinogen in guinea pig megakaryocyte granules. *J Clin Invest, 84,* 73–82.

110. Harrison, P., Wilbourn, B., Debili, N., et al. (1989). Uptake of plasma fibrinogen into the alpha granules of human megakaryocytes and platelets. *J Clin Invest, 84,* 1320–1324.

111. George, J. N. (1990). Platelet immunoglobulin G: Its significance for the evaluation of thrombocytopenia and for understanding the origin of alpha-granule proteins. *Blood, 76,* 859–870.

112. Handagama, P., Bainton, D. F., Jacques, Y., et al. (1993). Kistrin, an integrin antagonist, blocks endocytosis of fibrinogen into guinea pig megakaryocyte and platelet alpha-granules. *J Clin Invest, 91,* 193–200.

113. Morgenstern, E. (1982). Coated membranes in blood platelets. *Eur J Cell Biol, 26,* 315–318.

114. Behnke O. (1989). Coated pits and vesicles transfer plasma components to platelet granules. *Thromb Haemost, 62,* 718–722.

115. Behnke, O. (1992). Degrading and non-degrading pathways in fluid-phase (non-adsorptive) endocytosis in human blood platelets. *J Submicrosc Cytol Pathol, 24,* 169–178.

116. Klinger, M. H., & Kluter, H. (1995). Immunocytochemical colocalization of adhesive proteins with clathrin in human blood platelets: Further evidence for coated vesicle-mediated transport of von Willebrand factor, fibrinogen and fibronectin. *Cell Tissue Res, 279,* 453–457.

117. Stenberg, P. E., Pestina, T. I., Barrie, R. J., et al. (1997). The Src family kinases, Fgr, Fyn, Lck, and Lyn, colocalize with coated membranes in platelets. *Blood, 89,* 2384–2393.

118. Stenberg, P. E. (1986). Ultrastructural organization of maturing megakaryocytes. In R. F. Levine, N. Williams, J. Levin, & B. L. Evatt (Eds.), *Progress in clinical and biological research* (Vol. 215, pp. 373–386). New York: Alan R. Liss, Inc.

119. Wojenski, C. M., & Schick, P. K. (1993). Development of storage granules during megakaryocyte maturation: Accumulation of adenine nucleotides and the capacity for serotonin sequestration. *J Lab Clin Med, 121,* 479–485.

120. Cramer, E. M. (2001). Platelet and megakaryocytes: Anatomy and structural organization. In R. W. Colman, J. Hirsch, V. J. Marder, A. W. Clowes, & J. N. George (Eds.), *Hemostasis and thrombosis: Basic principles and clinical practice* (4th ed., pp. 411–428). Philadelphia: Lippincott Williams & Wilkins.

121. Rosa, J. P., George, J. N., Bainton, D. F., et al. (1987). Gray platelet syndrome. Demonstration of alpha granule membranes that can fuse with the cell surface. *J Clin Invest, 80,* 1138–1146.

122. Raccuglia, G. (1971). Gray platelet syndrome. A variety of qualitative platelet disorder. *Am J Med, 51,* 818–828.

123. Gerrard, J. M., Phillips, D. R., Rao, G. H., et al. (1980). Biochemical studies of two patients with the gray platelet syndrome. Selective deficiency of platelet alpha granules. *J Clin Invest, 66,* 102–109

124. Breton-Gorius, J., Vainchenker, W., Nurden, A., et al. (1981). Defective alpha-granule production in megakaryocytes from gray platelet syndrome: Ultrastructural studies of bone marrow cells and megakaryocytes growing in culture from blood precursors. *Am J Pathol, 102,* 10–19.

125. Cramer, E. M., Vainchenker, W., Vinci, G., et al. (1985). Gray platelet syndrome: Immunoelectron microscopic localization of fibrinogen and von Willebrand factor in platelets and megakaryocytes. *Blood, 66,* 1309–1316.

126. Levy-Toledano, S., Caen, J. P., Breton-Gorius, J., et al. (1981). Gray platelet syndrome: Alpha-granule deficiency. Its influence on platelet function. *J Lab Clin Med, 98,* 831–848.

127. Nurden, A. T., Kunicki, T. J., Dupuis, D., et al. (1982). Specific protein and glycoprotein deficiencies in platelets isolated from two patients with the gray platelet syndrome. *Blood, 59,* 709–718.

128. Gebrane-Younes, J., Cramer, E. M., Orcel, L., et al. (1993). Gray platelet syndrome. Dissociation between abnormal sorting in megakaryocyte alpha-granules and normal sorting in Weibel–Palade bodies of endothelial cells. *J Clin Invest, 92,* 3023–3028.

129. Hannah, M. J., Williams, R., Kaur, J., et al. (2002). Biogenesis of Weibel–Palade bodies. *Sem Cell Dev Biol, 13,* 313–324.

130. White, J. G. (1979). Ultrastructural studies of the gray platelet syndrome. *Am J Pathol, 95,* 445–462.

131. Falik-Zaccai, T. C., Anikster, Y., Rivera, C. E., et al. (2001). A new genetic isolate of gray platelet syndrome (GPS): Clinical, cellular, and hematologic characteristics. *Mol Genet Metab, 74,* 303–313.

132. Jantunen, E., Hanninen, A., Naukkarinen, A., et al. (1994). Gray platelet syndrome with splenomegaly and signs of extramedullary hematopoiesis: A case report with review of the literature. *Am J Hematol, 46,* 218–224.

133. Lo, B., Li, L., Gissen, P., et al. (2005). Requirement of VPS33B, a member of the Sec1/Munc18 protein family, in megakaryocyte and platelet α-granule biogenesis. *Blood, 106,* 4159–4166.

134. Weiss, H. J., Lages, B., Vicic, W., et al. (1993). Heterogeneous abnormalities of platelet dense granule ultrastructure in 20 patients with congenital storage pool deficiency. *Br J Haematol, 83,* 282–295.

135. Li, W., Rusiniak, M. E., Chintala, S., et al. (2004). Murine Hermansky–Pudlak syndrome genes: Regulators of lysosome-related organelles. *Bioessays, 26,* 616–628.

136. Gunay-Aygun, M., Huizing, M., & Gahl, W. A. (2004). Molecular defects that affect platelet dense granules. *Sem Thromb Hemost, 30,* 537–547.

137. Di Pietro, S. M., & Dell'angelica, E. C. (2005). The cell biology of Hermansky–Pudlak syndrome: Recent advances. *Traffic, 6,* 525–533.

138. Fukami, M. H. (1992). Isolation of dense granules from human platelets. *Methods Enzymol, 215,* 36–42.

139. Fukami, M. H. (1997). Dense granule factors. In F. von Bruchhausen & U. Walter (Eds.), *Platelets and their factors* (Vol. 126, pp. 419–432). New York: Springer.

# CHAPTER 16

# Signal Transduction During Platelet Plug Formation

**Lawrence F. Brass,[1] Timothy J. Stalker,[1] Li Zhu,[1] and Donna S. Woulfe[2]**

[1]*Departments of Medicine and Pharmacology, University of Pennsylvania, Philadelphia, Pennsylvania*
[2]*Department of Medicine, Thomas Jefferson College of Medicine, Philadelphia, Pennsylvania*

## I. Introduction

Platelet activation plays a central role in thrombus formation and in both benign and pathological responses to vascular injury. How platelet activation begins, and the degree to which it is essential, depends in part on where it occurs within the circulation and why. In the venous system, low flow rates and stasis permit the accumulation of activated coagulation factors and the local generation of thrombin largely without the benefit of platelets. Although venous thrombi contain platelets, the dominant cellular components are trapped red cells. In the arterial circulation, higher flow rates limit fibrin formation by washing out soluble clotting factors. Hemostasis in the arterial circulation requires platelets to accelerate thrombin formation, to form a physical barrier, and to provide a base upon which fibrin can accumulate. Hemostatic plugs and thrombi that form in the arterial circulation are therefore enriched in platelets as well as fibrin, giving them an appearance somewhat different from those formed in the venous circulation. The focus of this chapter is on the signaling events that underlie platelet activation and support stable thrombus formation. The data come from a variety of sources, including genetically altered mice. The emphasis will be on events as they are thought to occur in the arterial circulation.

The intracellular responses that support platelet activation involve a "toolkit" of signaling molecules that includes: the G protein-coupled receptors (GPCRs) that form the immediate response element for most soluble platelet agonists; the Ras superfamily members that regulate integrin activation and cytoskeletal reorganization; the phospholipases that are responsible for the hydrolysis of membrane phosphoinositides and the formation of prostanoids; the lipid kinases that replenish depleted pools of phosphoinositides; the protein tyrosine kinases which make possible the formation of signaling complexes; and the serine/threonine kinases that regulate the activity of other proteins by targeted phosphorylation events. Little or nothing on this list is unique to platelets. However, the signaling molecules within platelets collectively form a rapid response mechanism that makes it possible for moving platelets to be captured and transformed into noncirculating, tightly adherent cells. Such responsiveness is not without risk. When viewed from the perspective of biological evolution, human platelets are part of a beneficial system that reduces the risk of death following trauma or childbirth. The downside is the risk of unwanted platelet activation, particularly at sites of atherosclerotic disease in the coronary and cerebrovascular circulations.

## II. Stages in Platelet Plug Formation

Platelet activation can be divided into three overlapping stages: initiation, extension, and perpetuation (Fig. 16-1). *Initiation* can occur in more than one way. In the setting of trauma to the vessel wall, it may occur because circulating platelets are captured and then activated by exposed collagen and von Willebrand factor (VWF), forming a monolayer that supports thrombin generation and subsequent platelet aggregation. Key to these events is the presence of receptors on the platelet surface that can bind to collagen [integrin $\alpha 2\beta 1$ and glycoprotein (GP) VI] and VWF (GPIb$\alpha$ and $\alpha$IIb$\beta$3) and thereby initiate intracellular signaling. Platelet activation, particularly in thrombotic or inflammatory disorders, may also be initiated by thrombin, which activates platelets via GPCRs in the protease-activated receptor (PAR) family. Platelet activation may also occur because of the pathological activation of platelet FcR-IIA receptors, such as occurs in some patients receiving heparin (Chapter 48). *Extension* occurs when additional platelets are recruited and activated, sticking to each other and accumulating on top of the initial monolayer. Thrombin can play an important role at this point, as can secreted adenosine diphosphate (ADP) and released thromboxane $A_2$ (TXA$_2$). Each of these agonists is able to activate phospholipase C (PLC) in platelets, causing an increase in the cytosolic $Ca^{2+}$ concentration. The receptors that mediate this response are members of the superfamily of GPCRs. The signals that they engender support the activation of integrin $\alpha$IIb$\beta$3, making possible the cohesive

**A.** Initiation (capture, adhesion, activation)

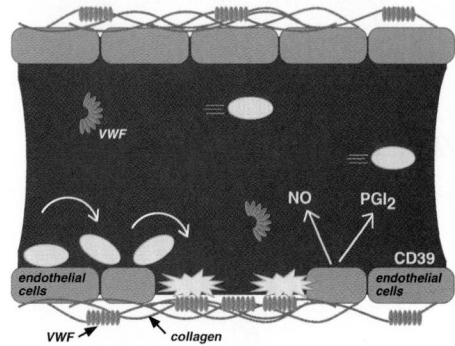

**B.** Extension (cohesion, secretion)

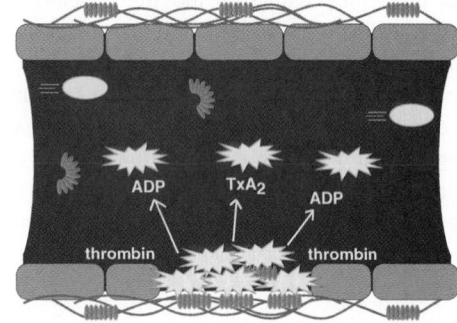

**C.** Perpetuation (stabilization)

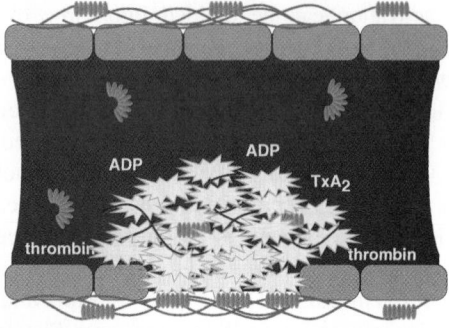

**Figure 16-1.** Steps in platelet plug formation. Prior to vascular injury, platelet activation is suppressed by endothelial cell-derived inhibitory factors. These include prostaglandin (PG) $I_2$ (prostacyclin), nitric oxide (NO), and CD39, an ADPase on the surface of endothelial cells that can hydrolyze trace amounts of ADP that might otherwise cause inappropriate platelet activation. (A) Initiation. The development of the platelet plug is initiated by thrombin and by the collagen-von Willebrand factor (VWF) complex, which captures and activates moving platelets. Platelets adhere and spread, forming a monolayer. (B) Extension. The platelet plug is extended as additional platelets are activated via the release or secretion of thromboxane $A_2$ (TXA$_2$), ADP, and other platelet agonists, most of which are ligands for G protein-coupled receptors on the platelet surface. Activated platelets stick to each other via bridges formed by the binding of fibrinogen, fibrin, or VWF to activated $\alpha$IIb$\beta$3. (C) Perpetuation. Finally, close contacts between platelets in the growing hemostatic plug, along with a fibrin meshwork (shown in red), help to perpetuate and stabilize the platelet plug.

interactions between platelets that are critical to the formation of the hemostatic plug. *Perpetuation* refers to the late events of platelet plug formation, when the intense, but often time-limited, signals arising from GPCRs may have faded. These late events help to stabilize the platelet plug and prevent premature disaggregation. Such events typically occur after aggregation has begun and are facilitated by close contacts between platelets. Examples include outside-in signaling through integrins and signaling through receptor tyrosine kinases that have membrane-bound ligands.

The platelet surface is crowded with receptors that are critical for hemostasis (Chapter 6). Those which are directly involved in binding to adhesive proteins such as collagen, VWF, and fibrinogen are expressed in the greatest numbers. There are approximately 80,000 copies of integrin $\alpha$IIb$\beta$3 (Chapter 8) and 25,000 copies of GPIb (Chapter 7) on the surface of human platelets. In contrast, receptors for thrombin, ADP, and TXA$_2$ range from a few hundred to a few thousand per platelet (Table 16-1). Receptor distribution is rarely uniform. Most platelet receptors and their associated effectors are mobile and tend to accumulate in cholesterol-enriched microdomains.[1–5] This may increase the efficiency with which platelets are activated. Lateral mobility may also allow adhesion and cohesion receptors to accumulate at sites of contact.

Platelets have been studied for more than a century. Initially, much of the work was descriptive, focusing on the macroscopic events of shape change, secretion, and aggregation. Subsequent studies on the molecular basis for these events have identified the receptors for most platelet agonists, shown that G proteins are the mediators for many signaling pathways within platelets, identified the integrin $\alpha$IIb$\beta$3 (also referred to as the GPIIb-IIIa complex) as the binding site for fibrinogen on the platelet surface, and begun the process of linking agonists and receptors to the machinery of cytoskeletal reorganization, granule exocytosis, and integrin activation. This information has proved to be useful for both a better understanding of platelet biology and the development of drugs that can prevent inappropriate platelet activation in the coronary and cerebrovascular circulations. Much is known, but much remains to be learned. The majority of studies on platelet signal transduction have been performed on human platelets, but methods that allow the genetic manipulation of mice combined with the establishment of increasingly accessible mouse models of thrombosis (Chapters 33 and 34) have made it possible to study platelet function *in vivo* — with at least some (although arguably not complete) relevance to platelet function in humans.

### A. Initiation of Platelet Activation

Although many molecules have been shown to cause platelet activation *in vitro*,[6] platelet activation *in vivo* is typically

**Table 16-1: G Protein-Coupled Receptors Expressed on Human Platelets**

| Agonist | Receptor | G Protein Families | Approximate Number of Copies per Platelet | References |
|---|---|---|---|---|
| Thrombin | PAR1 | $G_q$, $G_i$, $G_{12}$[a] | 2000 | 86, 90, 91, 210, 211 |
| | PAR4 | $G_q$, $G_{12}$[a] | ? | 91, 212[a] |
| ADP | $P2Y_1$ | $G_q$, $G_{12}$ | 150 | 57, 58, 213, 214 |
| | $P2Y_{12}$ | $G_i$ (particularly $G_{i2}$) | 600 | 62, 63 |
| TXA$_2$ | TPα/β | $G_q$, $G_{12}$ | 1000 | 117, 118, 120, 215–217 |
| Epinephrine | α$_{2A}$-adrenergic | $G_i$ (particularly $G_z$) | 300 | 107, 108, 218 |
| Vasopressin | V1 | $G_q$ | 100 | 219–222 |
| PAF | | $G_q$ | 200–2000 | 223, 224 |
| SDF-1 | CXCR4 | $G_i$ | ? | 225, 226 |
| PGI$_2$ | IP | $G_s$ | ? | 227 |

[a]The assignment of $G_q$ and $G_{12}$, but not $G_i$ family members to PAR4 is by inference. Selective PAR4 agonists cause platelet aggregation and shape change, but do not appear to inhibit adenylyl cyclase.[91,212] In contrast, PAR1 has been shown to couple to $G_i$ in platelet membrane preparation, but appears to do so less well in intact platelets.

Abbreviations: ADP, adenosine diphosphate; PAF, platelet activating factor; PAR, protease-activated receptor; PGI$_2$, prostaglandin I$_2$; SDF-1, stromal-derived factor 1; TXA$_2$, thromboxane A$_2$.

initiated by collagen and thrombin. This may take place at a site of acute vascular injury where platelet plug formation serves to stop bleeding, but the same platelet responses can be invoked in pathological states in which thrombin or other platelet-activating molecules are formed or exposed. Thrombin is generated at sites of vascular injury following the exposure of tissue factor sequestered within the vessel wall. In pathological states, tissue factor expression can be upregulated on the surface of endothelial cells or monocytes. It may also be present on circulating, monocyte-derived microvesicles that bind to activated platelets at sites of injury.[7–9] Signaling during platelet activation by thrombin will be considered in the next section. Here we will concentrate on the role of collagen.

Under static conditions, collagen is able to capture and activate platelets without the assistance of cofactors, but under the conditions of flow that exist in the arterial circulation VWF plays an essential role in supporting platelet adhesion and activation (see also Chapter 18). The collagen structural elements required for platelet activation include a basic quaternary structure.[10,11] Platelets can adhere to monomeric collagen, but require the more complex structure found in fibrillar collagen for optimal activation.[12] Four receptors for collagen have been identified on the surface of human and mouse platelets. Two bind directly to collagen (α2β1 and GPVI); the other two bind to collagen via VWF (αIIbβ3 and GPIb) (Fig. 16-2). Of these four receptors, GPVI is the most potent in terms of initiating signal generation. It is also the one that was identified most recently.[13] The structure of GPVI places it in the immunoglobulin

domain superfamily (see Chapter 6 for further structural details). The ability of GPVI to generate signals rests on a constitutive association with the Fc receptor γ-chain. Platelets from mice that lack either GPVI or the γ-chain have impaired responses to collagen, as do platelets in which GPVI has been depleted or blocked with an antibody.[14–17] Loss of the FcRγ chain affects collagen signaling in part because of loss of a necessary signaling element and in part because FcRγ is required for GPVI to reach the platelet surface. The α2β1 integrin also appears to be necessary for an optimal interaction with collagen (see also Chapter 18). Human platelets with reduced expression of α2β1 have impaired collagen responses,[18,19] as do mouse platelets that lack β1 integrins when the ability of these platelets to bind to collagen is tested at high shear.[17,20]

Signaling through GPVI can be studied in isolation with the snake venom protein, convulxin, or with synthetic "collagen-related" peptides (CRP), both of which bind to GPVI, but not to other collagen receptors. According to current models, collagen causes clustering of GPVI. This leads to the phosphorylation of FcRγ by tyrosine kinases in the Src family, creating a tandem phosphotyrosine motif recognized by the SH2 domains of Syk. Association of Syk with GPVI/γ-chain complex activates Syk and leads to the phosphorylation and activation of PLCγ$_2$. Loss of Syk impairs collagen responses.[16] PLCγ$_2$, like other forms of PLC in platelets, hydrolyzes phosphatidylinositol (PI)-4,5-P$_2$ to form 1,4,5-IP$_3$ and diacylglycerol. IP$_3$ opens Ca$^{2+}$ channels in the platelet dense tubular system, raising the cytosolic Ca$^{2+}$ concentration and triggering Ca$^{2+}$ influx across the platelet plasma mem-

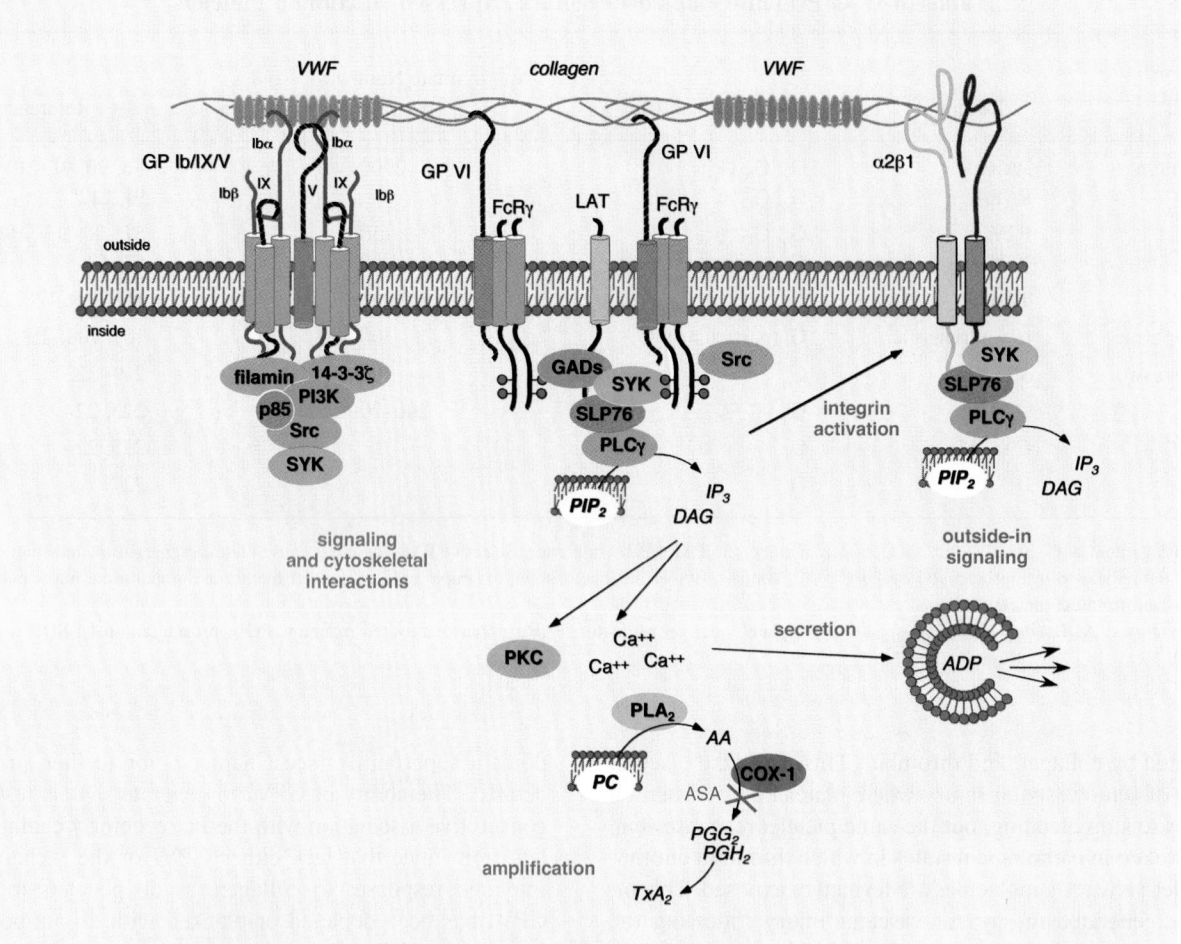

**Figure 16-2.** Platelet activation by collagen. Platelets use several different molecular complexes to support platelet activation by collagen. These include (1) VWF-mediated binding of collagen to the GPIb-IX-V complex and integrin αIIbβ3, (2) a direct interaction between collagen and both the integrin α2β1 and the GPVI/γ-chain complex. Clustering of GPVI results in the phosphorylation of tyrosine residues in the γ-chain, followed by the binding and activation of the tyrosine kinase, Syk. One consequence of Syk activation is the phosphorylation and activation of phospholipase Cγ, leading to phosphoinositide hydrolysis, secretion of ADP, and the production and release of TXA2. Abbreviations: ADP, adenosine diphosphate; AA, arachidonic acid; ASA, aspirin; COX-1, cyclooxygenase 1; DAG, diacylglycerol; GP, glycoprotein; PG, prostaglandin; PI3K, phosphatidylinositol 3-kinase; PIP2, phosphatidylinositol-4,5-bisphosphate; PKC, protein kinase C; PLA2, phospholipase A2; PLCγ, phospholipase Cγ; TXA2, thromboxane A2; VWF, von Willebrand factor; PC, phosphatidylcholine.

brane. The changes in the cytosolic $Ca^{2+}$ concentration that occur when platelets adhere to collagen under flow can be visualized in real time.[21,22] Diacylglycerol activates the more common protein kinase C (PKC) isoforms that are expressed in platelets, allowing the regulatory serine/threonine phosphorylation events that are needed for platelet activation.

Collectively, collagen receptors support the capture of fast-moving platelets at sites of injury, cause activation of the captured platelets, and stimulate the cytoskeletal reorganization that allows the previously discoid platelets to flatten out and adhere more closely to the exposed vessel wall. VWF supports this process in several ways. First, it increases the density of potential binding sites for collagen per platelet, because the number of copies of GPIbα and

αIIbβ3 greatly exceeds the number of copies of GPVI and α2β1. Since the form of VWF that supports platelet adhesion best is highly multimeric, VWF also increases the number of binding sites per collagen molecule. This increases the likelihood that platelets will encounter an available binding site and, once bound, will be able to increase the number of contacts with collagen by bringing additional receptors into play. It appears likely that only GPVI and GPIb are able to bind collagen and VWF, respectively, without prior platelet activation but once activation begins, α2β1 and αIIbβ3 are able to bind their respective ligands as well. Some of the integrin-activating signaling will occur downstream of GPVI, but there is growing evidence that the GPIb-IX-V complex can signal as well (see

Chapter 7), as can α2β1 and αIIbβ3 once they are engaged. Given all of this, it is perhaps not surprising that impairment of platelet receptors for collagen or loss of functional VWF produces defects in hemostasis that can range from modest to profound depending on the degree of impairment or the extent of VWF deficiency.

### B. Extension of the Platelet Plug

The formation of a platelet monolayer following the exposure of collagen and VWF is sufficient to initiate platelet plug formation, but is not sufficient to prevent bleeding. The next required step is the extension of the platelet plug in which additional circulating platelets are recruited and acquire the ability to stick to each other, a process that is technically termed cohesion but is commonly referred to as platelet aggregation. The recruitment of additional platelets beyond those in the collagen-bound monolayer is made possible by the local accumulation of agonists that are secreted or released from platelets, such as ADP and TXA₂, and by the local generation of thrombin, whose formation is accelerated on the surface of activated platelets (Fig. 16-1B). Contacts between platelets are maintained by a variety of molecular interactions, of which the most essential is the binding of fibrinogen (and VWF) to activated αIIbβ3 (Chapter 8). Circulating or locally secreted catecholamines can cause vasoconstriction, but also promote platelet activation by potentiating the effects of other platelet agonists. Inherited or acquired defects in platelet cohesion can produce significant risk of bleeding. Two obvious examples are Glanzmann thrombasthenia (Chapter 57), in which affected patients lack functional αIIbβ3 and therefore cannot bind fibrinogen, and the administration of αIIbβ3 (GPIIb-IIIa) antagonists (Chapter 62).

Most of the agonists that extend the platelet plug do so via GPCRs. The properties of these receptors make them particularly well suited for this task. First, most GPCRs bind their ligands with high affinity. Second, they are typically comprised of a single subunit, although receptor dimerization can affect the ligand preferences of the receptor. Third, they are constitutively associated with G proteins, reducing the time that would otherwise be required to recruit them into complexes.[23] Fourth, because they act as guanine nucleotide exchange factors, each occupied receptor can theoretically activate multiple G proteins and, in some cases, more than one class of G proteins. This allows amplification of a signal that might begin with a relatively small number of receptors. It also potentially allows each receptor to signal via more than one effector pathway. Finally, because several mechanisms exist that can limit the activation of GPCRs, platelet activation can be regulated, a property that may be useful when platelet activation is inappropriate and needs to be controlled.[24,25]

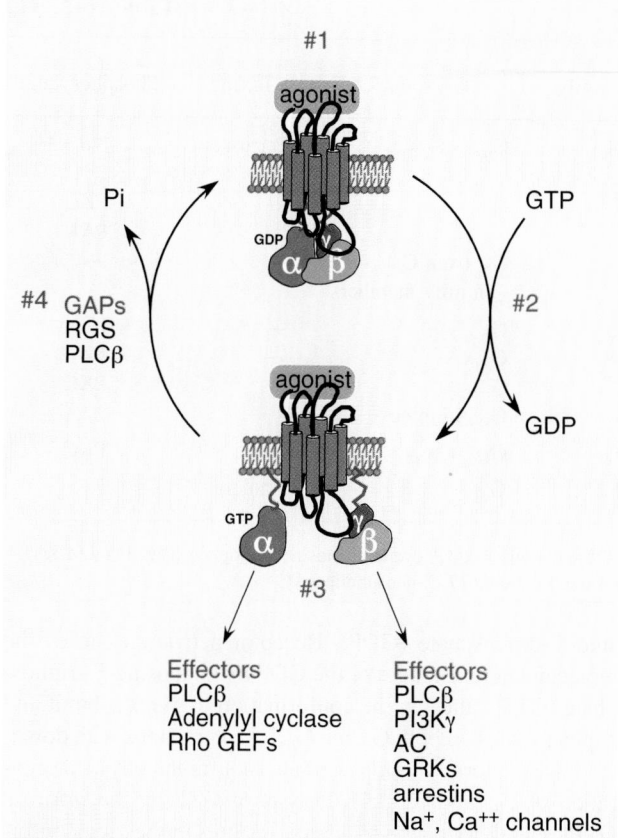

**Figure 16-3.** The G protein cycle. The binding of an agonist to a G protein-coupled receptor (#1), leads to the exchange of GTP for GDP on the G protein α subunit (#2) and the dissociation, but not necessarily the physical separation, of Gα from Gβγ. Both Gα and Gβγ can interact with effectors (#3). Proteins with GAP activity, including PLCβ as well as any of the RGS domain-containing proteins in platelets, promote the hydrolysis of GTP to GDP (#4), allowing reassociation of Gα and Gβγ (#1). Abbreviations: AC, adenylyl cyclase; GAP, GTPase activating protein; GDP, guanosine 5′-diphosphate; GPCR, G protein-coupled receptor; GRK, G protein-coupled receptor kinase; GTP, guanosine 5′-triphosphate; PI3K, phosphatidylinositol 3-kinase; PLCβ, phospholipase-β; RGS, regulator of G protein signaling; Rho GEFs refers collectively to proteins that are able to act as guanine nucleotide exchange factors for members of the Ras superfamily — an example is p115Rho-GEF, which can link G₁₂ to Rho.

GPCRs are comprised of a single polypeptide chain with seven transmembrane domains, an extracellular N-terminus, and an intracellular C-terminus.[26] Binding sites for agonists may involve the N-terminus, the extracellular loops, or a pocket formed by the transmembrane domains.[23] The G proteins that act as mediators for these receptors are αβγ heterotrimers (Fig. 16-3). The β subunit forms a propeller-like structure that is tightly, but not covalently, associated with the smaller γ subunit. The α subunit contains a guanine nucleotide-binding site that is normally occupied by guano-

**Table 16-2: $G_\alpha$ Proteins in Platelets**

| Family | G Protein | Toxin | Phosphorylated? | Effectors | Function | References |
|---|---|---|---|---|---|---|
| $G_i$ | $G_{i2\alpha} > G_{i3\alpha} \gg G_{i1\alpha}$ | Pertussis | No | Adenylyl cyclase, other effectors? | $\downarrow$ cAMP | 110 |
| | $G_{z\alpha}$ | — | PKC PAK? | Adenylyl cyclase, other effectors? | $\downarrow$ cAMP | 228–230 |
| | $G_{\beta\gamma}$ from $G_i$ family members | — | — | PLC PI 3K($\gamma$) | $\uparrow$ IP$_3$/DAG $\uparrow$ 3-PPIs | |
| $G_q$ | $G_{q\alpha}$ | — | No | PLC$\beta$ | $\uparrow$ IP$_3$/DAG | 231, 232 |
| | $G_{11\alpha}$ | — | No | PLC | $\uparrow$ IP$_3$/DAG | 231, 232 |
| | $G_{16\alpha}$ $G_{15\alpha}$ in mice | | PKC | PLC | $\uparrow$ IP$_3$/DAG | 233–236 |
| $G_{12}$ | $G_{12\alpha}$, $G_{13\alpha}$ | — | PKC | p115-RhoGEF? | Actin cytoskeleton | 35, 39, 119, 231, 237 |
| $G_s$ | $G_{s\alpha}$ | Cholera | No | Adenylyl cyclase | $\uparrow$ cAMP | 233 |

Abbreviations: cAMP, cyclic adenosine monophosphate; DAG, diacylglycerol; PAK, p21-activated kinase; PI 3 K, phosphatidylinositol 3-kinase; PKC, protein kinase C; PLC, phospholipase C.

sine 5′-diphosphate (GDP). Receptor activation causes the replacement (exchange) of the GDP by guanosine 5′-triphosphate (GTP), altering the conformation of the $\alpha$ subunit and exposing sites on both $G_\alpha$ and $G_{\beta\gamma}$ for interactions with downstream effectors.[27,28] This need not require the physical separation of $G_\alpha$ from $G_{\beta\gamma}$. Hydrolysis of GTP by the intrinsic GTPase activity of the $\alpha$ subunit restores the resting conformation of the heterotrimer, preparing it to undergo another round of activation and signaling. RGS (regulator of G protein signaling) proteins help to accelerate the hydrolysis of GTP by $\alpha$ subunits, shortening the time in which signaling occurs.[29] Cytosolic proteins with RGS domains can also serve as effectors. Prenylation of the $\gamma$-chain and acylation (with palmitate or myristate) of the $\alpha$ subunit help to anchor the complete heterotrimer to the plasma membrane.

Human platelets express at least 10 forms of $G_\alpha$ that fall into the $G_{s\alpha}$, $G_{i\alpha}$, $G_{q\alpha}$, and $G_{12\alpha}$ families (Table 16-2). This includes at least one $G_{s\alpha}$ family member, four $G_{i\alpha}$ family members ($G_{i1\alpha}$, $G_{i2\alpha}$, $G_{i3\alpha}$, and $G_{z\alpha}$), three $G_{q\alpha}$ family members ($G_{q\alpha}$, $G_{11\alpha}$, and $G_{16\alpha}$), and two $G_{12\alpha}$ family members ($G_{12\alpha}$ and $G_{13\alpha}$). Much has been learned about the critical role of G protein $\alpha$ subunits in platelets. Much less is known about the $G_{\beta\gamma}$ isoforms that are expressed in platelets and the selectivity of their individual contributions to platelet activation.

**1. G Protein-Mediated Signaling in Platelets: General Principles.** The signaling events that underlie platelet activation have not been fully identified. Large gaps remain in signaling maps that attempt to connect receptors on the platelet surface with the most characteristic platelet responses: shape change, aggregation, and secretion. In general terms, the events that have been described to date can be divided into three broad categories, all of which are required if robust aggregation and secretion are to

occur. The first category begins with the activation of PLC, which by hydrolyzing membrane phosphatidylinositol-4,5-bisphosphate (PIP$_2$) produces the second messengers needed to raise the cytosolic Ca$^{2+}$ concentration and activate protein serine/threonine kinases and lipid kinases. How PLC is activated varies with the agonist. Collagen activates PLC$\gamma$2 using a mechanism that depends on adaptor molecules and protein tyrosine kinases. Thrombin, ADP, and TXA$_2$ activate the PLC$\beta$ isoforms that are expressed in platelets using heterotrimeric G proteins in the $G_q$ and (less efficiently) the $G_i$ family as intermediaries. Regardless of which PLC isoform is activated, the subsequent rise in the Ca$^{2+}$ concentration triggers downstream events including integrin activation and the generation of TXA$_2$. Although it is possible to cause platelet activation and secretion *in vitro* without causing an increase in cytosolic Ca$^{2+}$, that is probably not how it happens *in vivo*. The pathways that comprise this first category of events are needed for shape change, aggregation, and secretion to occur.

The second broad category of necessary events involve monomeric G proteins in the Rho and Rac families, whose activation trigger the reorganization of the actin cytoskeleton that underlies filopodia and lamellopodia formation (Fig. 16-4). Along with changes in the platelet's circumferential microtubular ring, filopodia and lamellopodia formation are the essence of platelet shape change (see Chapter 4 for more details). When platelets in suspension are activated by soluble agonists, shape change typically precedes platelet aggregation. Most platelet agonists can trigger shape change, the notable exception being epinephrine. The soluble agonists (thrombin, ADP, and TXA$_2$) that trigger shape change typically act through receptors that are coupled to members of the $G_q$ and $G_{12}$ family. Epinephrine is unable to cause shape change because its receptors are coupled solely to the $G_i$ family member, $G_z$.

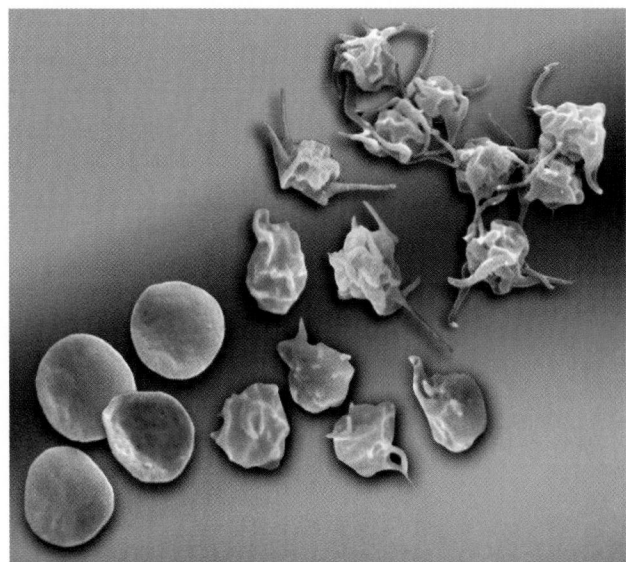

**Figure 16-4.** Platelet shape change and aggregation. Scanning electron micrographs of resting (lower left), partially activated (lower middle), and fully activated platelets (upper right), showing the accompanying shape changes, formation of filopodia and lamellipodia, and platelet aggregation. Image generously provided by John W. Weisel, Chandrasekaran Nagaswami, and Rustem I. Litvinov, Department of Cell and Developmental Biology, University of Pennsylvania School of Medicine.

The third category of required events for the platelet signaling map includes the suppression of cyclic adenosine monophosphate (cAMP) synthesis by adenylyl cyclase, particularly if the intracellular cAMP concentration has been raised above baseline by the action of endothelial cell-derived prostaglandin (PG) $I_2$ and nitric oxide (NO). Inhibition of cAMP formation relieves a block on platelet signaling that otherwise serves to limit inopportune platelet activation. The agonists that inhibit cAMP formation in platelets (ADP and epinephrine) do so by binding to receptors coupled to the $G_i$ family members, especially $G_{i2}$ (ADP) and $G_z$ (epinephrine). Suppression of adenylyl cyclase is clearly critical when cAMP levels are elevated. It is less clear that it is necessary under basal cAMP conditions.

*a. Platelet Activation In Vivo.* Although convenient for didactic purposes, slicing the events of platelet activation into individual pathways that are activated or suppressed by individual agonists is just that, a didactic convenience. Platelet activation at sites of vascular injury typically results from contact with more than one agonist. The initial injury may expose collagen, but it will also produce thrombin. Similarly, tissue and red cell damage will release ADP, as will activated platelets — which will also produce $TXA_2$ from membrane phospholipids. It is also important to remember that different agonists couple to different pathways with differing efficiencies and that it is the combined effect of multiple agonists

that produces the most robust platelet activation. Even thrombin, one of the most potent platelet agonists, relies on its ability to induce ADP and $TXA_2$ release to yield maximal platelet activation when used at lower concentrations *in vitro*. This is particularly important because endogenous antagonists of platelet activation, including $PGI_2$, antithrombin-III, NO, the *ecto*-ADPase CD39, and the diluting effects of continued blood flow, work against the accumulation of platelet agonists and limit the ability of platelets to respond.

As will be discussed in greater detail in the next section of this chapter, platelet activation may be invoked by a single agonist working through receptors coupled to multiple effectors, but more commonly it is produced by the combined effects of different agonists working through distinguishable receptors coupled to individual effectors. Platelets produce agonists as well as responding to agonists. This allows a degree of amplification that makes rapid platelet activation possible *in vivo*. Furthermore, once platelet aggregation begins, an additional wave of supportive signaling occurs as a result of outside-in signaling by integrins (see Chapter 17) and the contact-dependent activation of receptor tyrosine kinases on the platelet surface.

*b. $G_q$ and the Activation of Phospholipase C$\beta$.* Agonists whose receptors are coupled to $G_q$ can provide a strong stimulus for platelet activation by activating PLC$\beta$. PLC$\beta$ hydrolyzes membrane PI-4,5-$P_2$ to produce two second messengers: 1,4,5-$IP_3$ and diacylglycerol (Fig. 16-5). $IP_3$ formation triggers an increase in cytosolic $Ca^{2+}$. Diacylglycerol activates PKC. In resting platelets, the cytosolic $Ca^{2+}$ concentration is maintained at approximately 100 nM by limiting $Ca^{2+}$ influx and by pumping $Ca^{2+}$ out of the cytosol across the plasma membrane or into the dense tubular system. The latter is a closed membrane system within platelets (see Chapter 3) that has been thought to be derived from megakaryocyte smooth endoplasmic reticulum — although this conclusion needs to be revisited in view of recent information about the mechanisms underlying platelet formation by megakaryocytes (see Chapter 2).

The removal of $Ca^{2+}$ from the cytosol produces a steep $Ca^{2+}$ gradient across the plasma membrane that can be used to passively drive the movement of $Ca^{2+}$ back into the cytosol when platelets are activated. 1,4,5-$IP_3$ triggers the release of $Ca^{2+}$ from the dense tubular system by binding to receptors in the DTS membrane. In activated platelets, the cytosolic free concentration can exceed 1 µM (a 10-fold increase over baseline) with potent agonists like thrombin. Activation of PLC also releases membrane-associated proteins that bind to PI-4,5-$P_2$ via their pleckstrin homology (PH) domains. Different PLC$\beta$ isoforms can be activated by either $G_\alpha$ or $G_{\beta\gamma}$ (or both). PLC$\beta$-activating $\alpha$ subunits are typically derived from $G_q$ in platelets, but PLC can also be activated by $G_{\beta\gamma}$ derived from $G_i$ family members.

The rising $Ca^{2+}$ concentration in activated platelets is undoubtedly a trigger for numerous events, but one that has

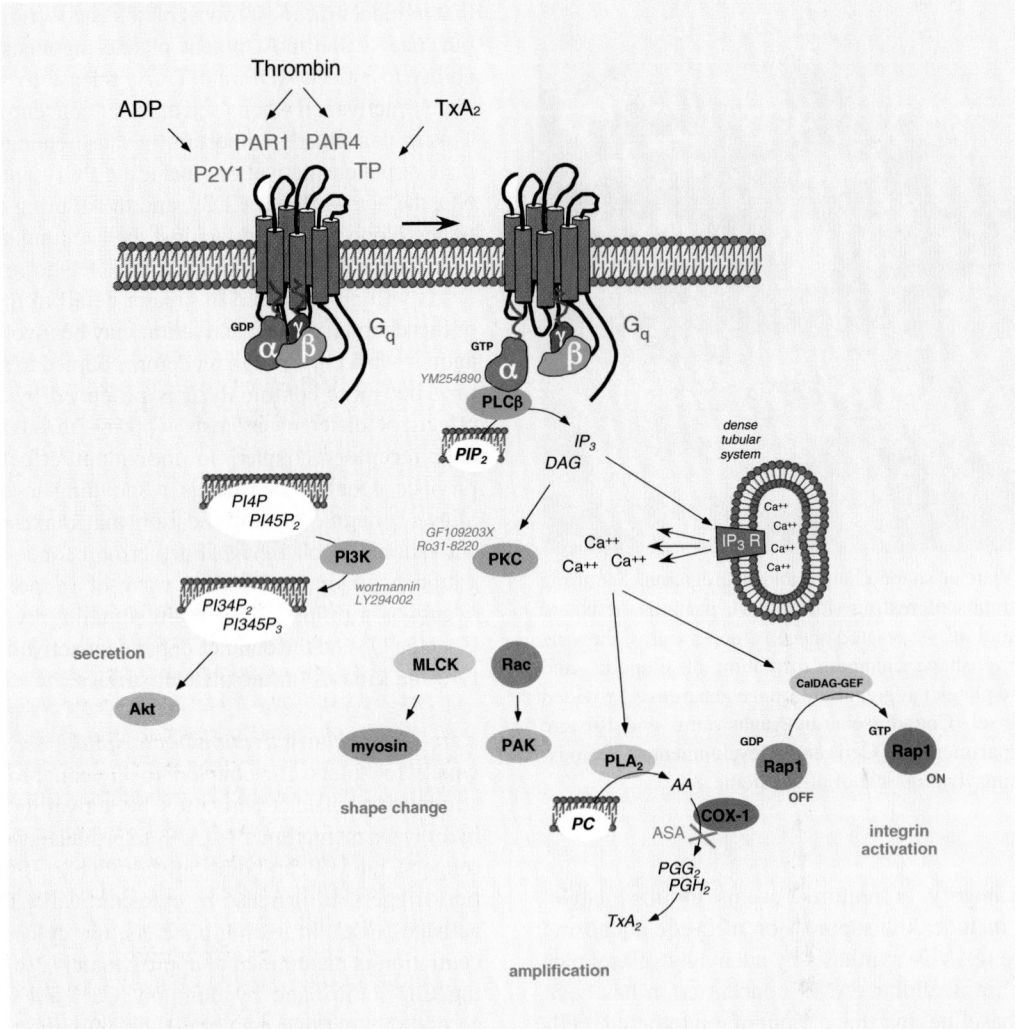

**Figure 16-5.** $G_q$ signaling in platelets. Agonists whose receptors are coupled to $G_q$ are able to activate PLCβ via $G_{q\alpha}$. The potency with which the activation occurs varies with the agonist, with thrombin and $TXA_2$ providing a stronger stimulus for PLCβ-mediated phosphoinositide hydrolysis than ADP. Thrombin activates two $G_q$-coupled receptors on human platelets, PAR1 and PAR4, which differ somewhat in the kinetics of PLC activation. YM254890 inhibits PLC activation by $G_q$.[205,206] GF109203X and Ro31-8220 are PKC inhibitors.[207] Wortmannin and LY294002 are PI3K inhibitors. Abbreviations: AA, arachidonic acid; ADP, adenosine diphosphate; ASA, aspirin; COX-1, cyclooxygenase 1; DAG, diacylglycerol; GDP, guanosine 5′-diphosphate; GTP, guanosine 5′-triphosphate; $IP_3$ R, receptor for 1,4,5-$IP_3$; MLCK, myosin light chain kinase; PI3K, phosphatidylinositol 3-kinase; PAK, p21-activated kinase; PAR, protease-activated receptor; PG, prostaglandin; PI3K, phosphatidylinositol 3-kinase; $PIP_2$, phosphatidylinositol-4,5-bisphosphate; PKC, protein kinase C; $PLA_2$, phospholipase $A_2$; PLCβ, phospholipase Cβ; $TXA_2$, thromboxane $A_2$; PC, phosphatidylcholine.

received recent attention is the $Ca^{2+}$-dependent activation of the Ras family member, Rap1B, via the guanine nucleotide exchange protein, Cal-DAG GEF.[30] Rap1B has been shown to be an important contributor to signaling pathways that converge on the activation of αIIbβ3 in platelets.[31–33] PKC isoforms are serine/threonine kinases. PKC has been identified as the cellular receptor for the lipid second messenger diacylglycerol, and it is therefore a key enzyme in the signaling events that follow activation of receptors coupled to PLC. PKC isoforms phosphorylate multiple cellular proteins

on serine and threonine residues. Activation of PKC by a phorbol ester such as phorbol myristate acetate is sufficient to cause integrin activation, granule secretion, and platelet aggregation in the absence of an increase in cytosolic $Ca^{2+}$.[34] How these responses are produced is not well understood.

*c. $G_q$, $G_{13}$ and the Reorganization of the Actin Cytoskeleton.* At least two effector pathways are involved in the reorganization of the actin cytoskeleton that accompanies platelet activation: $Ca^{2+}$-dependent activation of myosin

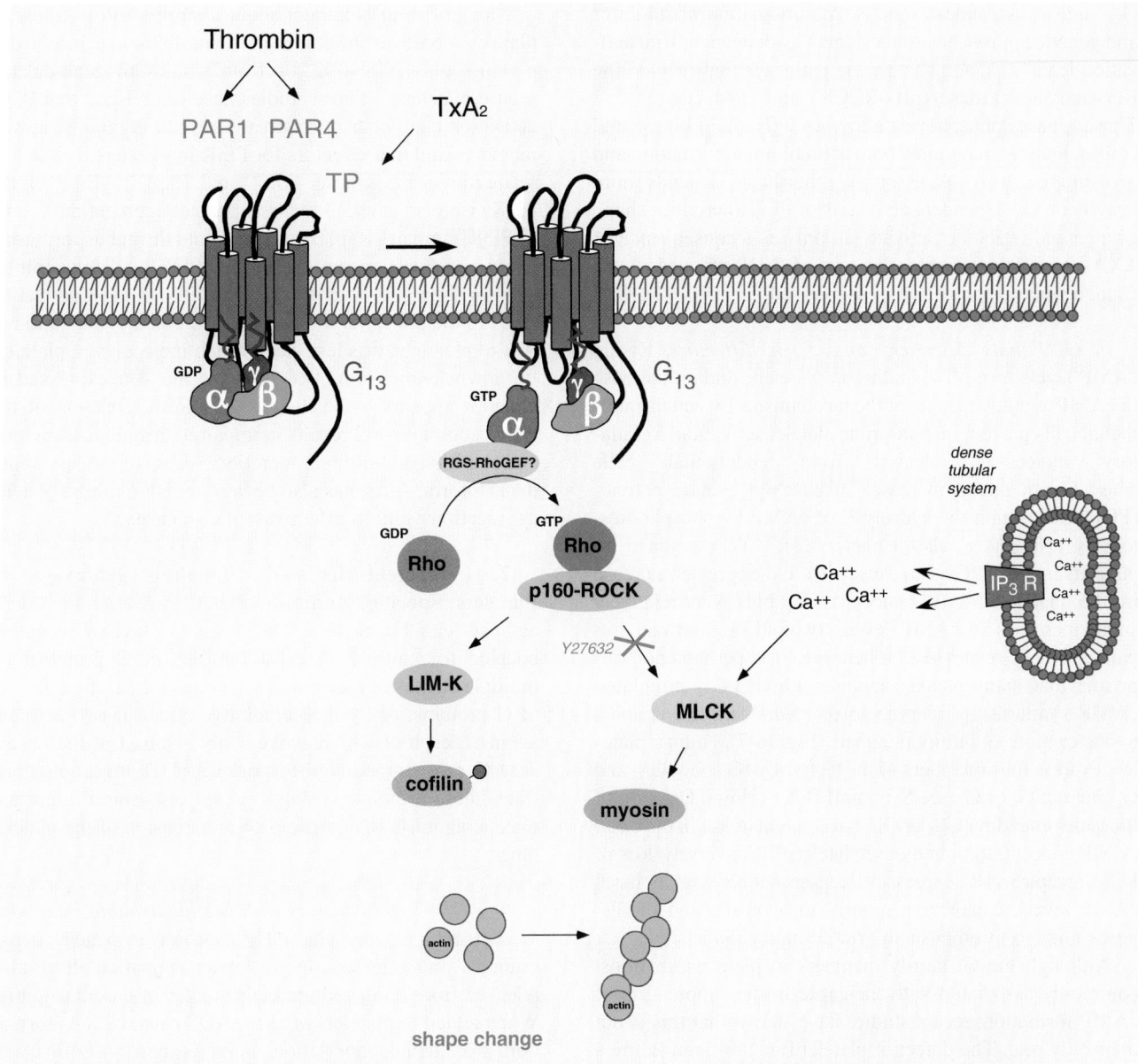

**Figure 16-6.** $G_{12/13}$ signaling in platelets. Agonists whose receptors are coupled to the $G_{12}$ family members expressed in platelets are able to trigger shape change, in part by Rho-dependent activation of kinases that include the Rho-activated kinase, p160 ROCK, and the downstream kinases, MLCK and LIM-K. Although $G_{12}$ and $G_{13}$ are both expressed, based on knockout studies, $G_{13}$ is the dominant $G_{12}$ family member in mouse platelets. Y27632 inhibits p160 ROCK.[208] Abbreviations: GDP, guanosine 5'-diphosphate; GTP, guanosine 5'-triphosphate; $IP_3$ R, receptor for 1,4,5-$IP_3$; MLCK, myosin light chain kinase; PAR, protease-activated receptor; TXA$_2$, thromboxane A$_2$.

light chain kinase downstream of $G_q$ family members and activation of low molecular weight GTP-binding proteins in the Rho family, which occurs downstream of $G_{12}$ family members (Figs. 16-5 and 16-6) (see also Chapter 4).[35,36] Several proteins having both $G_\alpha$-interacting domains and guanine nucleotide exchange factor (GEF) domains can link $G_{12}$ family members to Rho family members. One example is p115RhoGEF, but others exist as well.[37] With the exception of ADP, shape change persists in platelets from mice

that lack $G_{q\alpha}$[38] but is lost when $G_{13\alpha}$ expression is suppressed, alone or in combination with $G_{12\alpha}$.[39,40] Put somewhat differently, loss of $G_{q\alpha}$ expression prevents shape change when mouse platelets are activated by ADP.[38] Loss of $G_{13\alpha}$ expression (but not loss of $G_{12\alpha}$ expression alone) suppresses shape change in response to thrombin and TXA$_2$.[41]

These results indicate that for most platelet agonists, $G_{13}$ signaling is essential for shape change and that the events underlying shape change are invoked by a combination of

$G_{13}$- and $G_q$-dependent signals. A combination of inhibitor and genetic approaches suggest that $G_{13}$-dependent Rho activation leads to shape change via pathways that include the Rho-activated kinase (p160ROCK) and LIM-kinase.[35,42,43] These kinases phosphorylate myosin light chain kinase and cofilin, helping to regulate both actin filament formation and myosin (Fig. 16-6). ADP, on the other hand, depends more heavily on $G_q$-dependent activation of PLC to produce shape change and is able to activate $G_{13}$ only as a consequence of $TXA_2$ generation; hence the loss of ADP-induced shape change when $G_q$ signaling is suppressed.

*d. $G_i$ Family Members and Their Effectors.* Rising cAMP levels turn off signaling in platelets, and an increase in cAMP synthesis is one of the mechanisms by which endothelial cells prevent inappropriate platelet activation. Regulatory molecules released from endothelial cells cause $G_{s\alpha}$-mediated increases in adenylyl cyclase activity ($PGI_2$) and inhibit the hydrolysis of cAMP by phosphodiesterases (NO) (see also Chapter 13).[44] When added to platelets *in vitro,* $PGI_2$ can cause a 10-fold or greater increase in the platelet cAMP concentration, but even relatively small increases in cAMP levels (twofold or less) can impair thrombin responses.[45] Therefore, it is perhaps not surprising that many platelet agonists inhibit $PGI_2$-stimulated cAMP synthesis by binding to receptors that are coupled to one or more $G_i$ family members (Fig. 16-7). Human platelets express four members of the $G_i$ family ($G_{i1}$, $G_{i2}$, $G_{i3}$, and $G_z$) but not $G_o$ or $G_t$ (see Section II.B.2.c below). Deletion of the genes encoding $G_{i2\alpha}$ or $G_{z\alpha}$ causes an increase in the basal cAMP concentration in mouse platelets.[46] Conversely, loss of $PGI_2$ receptor (IP) expression causes a decrease in basal cAMP levels, enhances responses to agonists, and predisposes mice to thrombosis in arterial injury models.[46,47]

Although the $G_i$ family members in platelets are most commonly associated with their role in the suppression of cAMP formation, recent studies have shown that this is not their only role. The defect of platelet function seen in mice that are missing $G_{i\alpha}$ family members cannot be reversed by adding inhibitors of adenylyl cyclase,[46] and those same inhibitors also do not restore ADP responses in the presence of antagonists selective for the $G_i$-coupled ADP receptor.[48] In addition to providing $G_{\beta\gamma}$ heterodimers that can activate PLCβ, the downstream effectors for $G_i$ family members in platelets include PI 3-kinase, Src family members, and Rap1B (Figs. 16-3 and 16-7A).[33,49–51] The role of Rap1B in integrin activation has been discussed above. PI 3-kinases phosphorylate PI-4-P and PI-4,5-$P_2$ to produce PI-3,4-$P_2$ and PI-3,4,5-$P_3$. Human platelets express the α, β, γ, and δ isoforms of PI 3-kinase, each of which is comprised of a catalytic subunit and a regulatory subunit. The α, β, and δ isoforms are activated by binding to phosphorylated tyrosine residues. The PI3Kγ isoform is activated by $G_{\beta\gamma}$ derived from $G_i$ family members.

Much of what is known about the role of PI 3-kinase in platelets comes from studies with inhibitors such as wortmannin and LY294002, or from studies of gene-deleted mouse platelets.[52] Those studies have established that PI3K activation can occur downstream of both $G_q$ and $G_i$ family members and that effectors for PI3K in platelets include the serine/threonine kinase, Akt,[49] and Rap1B.[53] Loss of the PI3Kγ isoform causes impaired platelet aggregation.[54] Loss of PI3Kβ impairs Rap1B activation and thrombus formation *in vivo,* as does a recently described PI3Kβ-selective inhibitor.[53] Although it seems clear that PI 3-kinase activation leads to the phosphorylation and (presumably) activation of Akt in platelets, it is less clear which molecules in platelets lie downstream from Akt. Most of the Akt expressed in platelets appears to be the Akt2 isoform. Deletion of the gene encoding Akt2 results in impaired thrombus formation and stability, and inhibits secretion.[49] Loss of the less abundant isoform, Akt1, has also been reported to inhibit platelet aggregation[55] and to affect vascular integrity.[56]

**2. G Protein-Mediated Signaling Pathways in Platelets: Specific Examples.** Platelet activation by soluble agonists can be mediated by a single class of receptors coupled to multiple different families of G proteins, by multiple classes of receptors each coupled to a single family of G proteins, or by the cumulative effect of two or more agonists, each of which evokes only a subset of the necessary G protein-mediated responses.[57,58] This section illustrates these principles using as examples several important platelet agonists that participate in extension of the platelet plug.

*a. ADP: Two Receptors with Distinguishable Functions Mediated by $G_q$ and $G_{i2}$.* ADP is stored in platelet dense granules and released upon platelet activation. It is also released from damaged red cells at sites of vascular injury. When added to platelets *in vitro,* ADP causes $TXA_2$ formation, protein phosphorylation, an increase in cytosolic $Ca^{2+}$, shape change, aggregation, and secretion. It also inhibits cAMP formation. These responses are half-maximal at approximately 1 μM ADP. However, even at high concentrations, ADP is a comparatively weak activator of PLC. Instead, its utility as a platelet agonist rests more upon its ability to activate other pathways.[59,60] Human and mouse platelets express two distinct receptors for ADP, denoted $P2Y_1$ and $P2Y_{12}$ (which are discussed in detail in Chapter 10). Both receptors are members of the purinergic class of GPCRs (Table 16-1 and Fig. 16-8).[57,58,61–63] $P2Y_1$ receptors couple to $G_q$. $P2Y_{12}$ receptors couple to $G_i$ family members other than $G_z$ (Figs. 16-5 and 16-7A). Optimal activation of platelets by ADP requires activation of both receptors. The genes for $P2Y_1$ and $P2Y_{12}$ have been deleted in mice, producing effects that are consistent with those predicted by pharmacological studies on human platelets.[57,64] A third

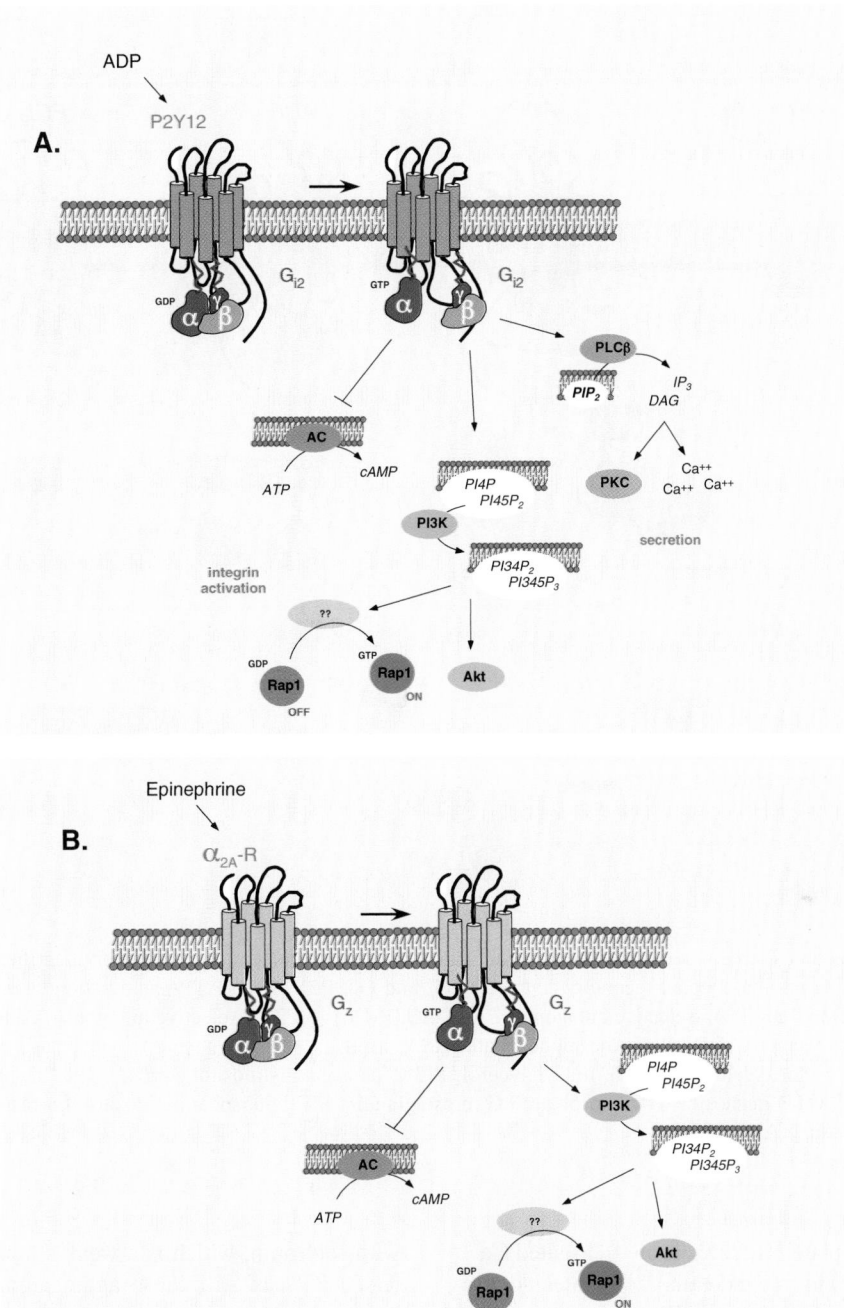

**Figure 16-7.** $G_i$ and $G_z$ signaling in platelets. (A) $G_{i2}$ is the predominant $G_i$ family member expressed in human platelets. In addition to inhibiting adenylyl cyclase (alleviating the repressive effects of cAMP), $G_{i2}$ couples $P2Y_{12}$ ADP receptors to PI 3-kinase, Akt phosphorylation, and Rap1B activation. Other effectors may exist as well. (B) The role of $G_z$ in platelets appears to be limited to interactions with $\alpha_{2A}$-adrenergic receptors and responses to epinephrine. Epinephrine appears to be primarily a potentiator of the effects of other platelet agonists, inhibiting cAMP formation and promoting the activation of Rap1B and PI 3-kinase. Abbreviations: AC, adenylyl cyclase; ADP, adenosine diphosphate; ATP, adenosine triphosphate; cAMP, cyclic adenosine monophosphate; DAG, diacylglycerol; PI3K, phosphatidylinositol 3-kinase; PLA$_2$, phospholipase A$_2$; PLC$\beta$, phospholipase C$\beta$; GDP, guanosine 5'-diphosphate; GTP, guanosine 5'-triphosphate; PIP$_2$, phosphatidylinositol-4,5-bisphosphate; PKC, protein kinase C.

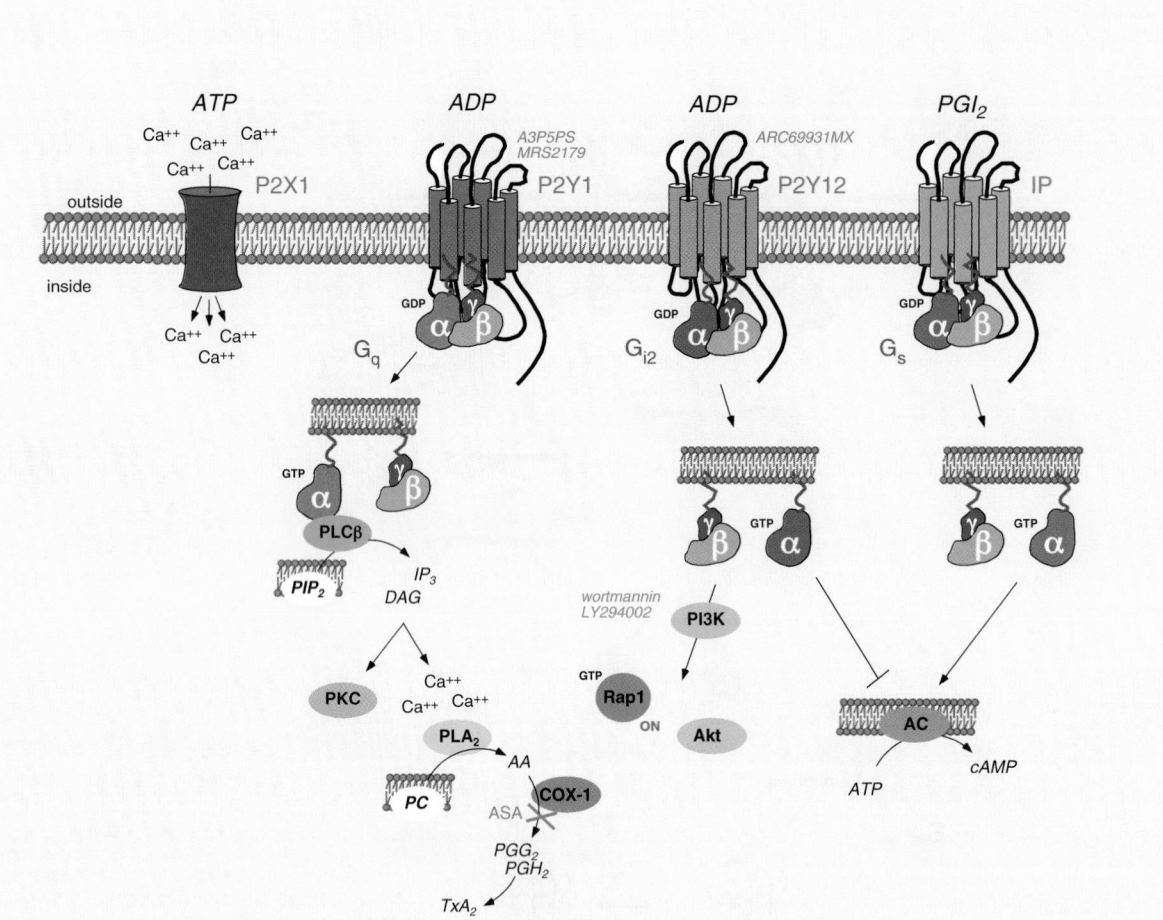

**Figure 16-8.** ADP receptors. Three receptors that can be activated by adenine nucleotides have been identified in platelets: P2Y$_1$ and P2Y$_{12}$ are ADP-activated G protein-coupled receptors coupled to G$_q$ and G$_{i2}$. P2X$_1$ is an ATP-gated cation channel that can allow Ca$^{2+}$ influx. A3P5PS and MRS2179 are P2Y$_1$-selective antagonists. ARC69931MX is a P2Y$_{12}$-selective antagonist. Abbreviations: AA, arachidonic acid; AC, adenylyl cyclase; ADP, adenosine diphosphate; ASA, aspirin; ATP, adenosine triphosphate; cAMP, cyclic adenosine monophosphate; COX-1, cyclooxygenase 1; DAG, diacylglycerol; PI3K, phosphatidylinositol 3-kinase; PLCβ, phospholipase Cβ; GDP, guanosine 5'-diphosphate; GTP, guanosine 5'-triphosphate; PG, prostaglandin; PIP$_2$, phosphatidylinositol-4,5-bisphosphate; PKC, protein kinase C; TXA$_2$, thromboxane A$_2$.

purinergic receptor on platelets, P2X$_1$, is an ATP-gated Ca$^{2+}$ channel (see Chapter 10 for more details).[65–68] Platelet dense granules contain ATP as well as ADP, and recent studies suggest that there are conditions in which P2X$_1$ activity is essential for platelet activation.[69–71] P2X$_1$ is also functional on megakaryocytes.[72]

When P2Y$_1$ is blocked or deleted, ADP is still able to inhibit cAMP formation, but its ability to cause an increase in cytosolic Ca$^{2+}$, shape change, and aggregation is greatly impaired — as is the case in platelets from mice that lack G$_{qα}$.[73] P2Y$_1^{-/-}$ mice have a minimal increase in bleeding time and show some resistance to thromboembolic mortality following injection of ADP, but no predisposition to spontaneous hemorrhage. Primary responses to platelet agonists other than ADP are unaffected and when combined

with serotonin, which is a weak stimulus for PLC in platelets, ADP can still cause aggregation of P2Y$_1^{-/-}$ platelets. Taken together, these results show that platelet P2Y$_1$ receptors are coupled to G$_{qα}$ and responsible for activation of PLC. P2Y$_1$ receptors can also activate Rac and the Rac effector, p21-activated kinase (PAK),[74] but do not appear to be coupled to G$_i$ family members.

P2Y$_{12}$ was independently identified by two groups of investigators.[62,63] As was predicted by inhibitor studies and by the phenotype of a patient lacking functional P2Y$_{12}$,[75] platelets from P2Y$_{12}^{-/-}$ mice do not aggregate normally in response to ADP.[76] P2Y$_{12}^{-/-}$ platelets retain P2Y$_1$-associated responses, including shape change and PLC activation, but lack the ability to inhibit cAMP formation in response to ADP. The G$_i$ family member associated with P2Y$_{12}$ appears

**Table 16-3: Platelet Phenotypes of $G_\alpha$ Knockout Mice**

| Family | $G_\alpha$ | Platelet Phenotype | References |
|---|---|---|---|
| $G_{i\alpha}$ | $G_{i2\alpha}$ | Reduced aggregation and inhibition of cAMP formation with ADP. Normal shape change. | 46, 77 |
| | $G_{i3\alpha}$ | No apparent effect. | 46 |
| | $G_{z\alpha}$ | Impaired aggregation and suppression of cAMP formation by epinephrine. Resistance to disseminated thrombosis initiated with collagen plus epinephrine, but not collagen plus ADP. Normal shape change. Increased bleeding time in one of two studies. | 78, 238 |
| $G_{q\alpha}$ | $G_{q\alpha}$ | Predisposed to bleed. Absent aggregation with all agonists tested, but normal shape change. Resistance to disseminated thrombosis initiated with collagen plus epinephrine. | 73 |
| | $G_{11\alpha}$ | No apparent effect on platelet function. | Our unpublished observations. |
| | $G_{15\alpha}$ | No apparent effect on platelet function. | Our unpublished observations; knockout reference is [239]. |
| $G_{12\alpha}$ | $G_{12\alpha}$ | No apparent effect on platelet function. | 36 |
| | $G_{13\alpha}$ | Embryonic lethal due to vascular defects, but decreased platelet shape change and aggregation when limited to hematopoietic cells. | 39, 240 |

to be primarily $G_{i2}$, because platelets from $G_{i2\alpha}^{-/-}$ mice have an impaired response to ADP,[46,77] whereas those lacking $G_{i3\alpha}$ or $G_{z\alpha}$ do not[46,78] (Table 16-3).

The identification of the receptors that mediate platelet responses to ADP, the development of antagonists that target each of the known receptors, and the successful knockouts of the genes encoding P2Y$_1$, P2Y$_{12}$, $G_{q\alpha}$, and $G_{i2\alpha}$ have brought an increased appreciation of the contribution of ADP to platelet plug formation *in vivo*. Absence of P2Y$_{12}$ produces a hemorrhagic phenotype in humans, albeit a relatively mild one.[62,75,79] Deletion of either P2Y$_1$ or P2Y$_{12}$ in mice (Table 16-4) prolongs the bleeding time and impairs platelet responses not only to ADP, but also to thrombin and TXA$_2$, particularly at low concentrations.[76,80,81] Since the receptors for thrombin and TXA$_2$ can cause robust activation of PLC, the contribution of ADP when thrombin or TXA$_2$ are present appears to be largely due to its ability to activate $G_{i2}$.[82] Two widely used antiplatelet agents, ticlopidine and clopidogrel, are antagonists of P2Y$_{12}$.[83] Both drugs block P2Y$_{12}$ by an irreversible mechanism involving the generation of a reactive metabolite in the liver (see Chapter 61 for details).

*b. Thrombin: Two Receptors with Overlapping Functions Mediated by $G_q$ and $G_{12/13}$.* Thrombin is able to activate platelets at concentrations as low as 0.1 nM (approximately 0.01 units per mL), making it the most potent activator of platelets. Although other platelet agonists can also cause phosphoinositide hydrolysis, none appear to be as efficiently coupled to PLC as thrombin. Within seconds of the addition of thrombin, the cytosolic Ca$^{2+}$ concentration increases 10-fold, triggering downstream Ca$^{2+}$-dependent

events, including the activation of phospholipase A$_2$. All of these responses, but not shape change, are abolished in platelets from mice lacking $G_{q\alpha}$.[73] Thrombin also activates Rho in platelets, leading to rearrangement of the actin cytoskeleton and shape change, responses that are greatly reduced or absent in mouse platelets that lack $G_{13\alpha}$.[41] Finally, thrombin is able to inhibit adenylyl cyclase activity in human platelets, either directly (*via* a $G_i$ family member) or indirectly (*via* released ADP).[84,85]

Platelet responses to thrombin are mediated by members of the protease-activated receptor (PAR) family of GPCRs. There are four members of this family (see Chapter 9 for more details). Three (PAR1, PAR3, and PAR4) can be activated by thrombin. PAR1 and PAR4 are expressed on human platelets; mouse platelets express PAR3 and PAR4 (Table 16-1). Receptor activation occurs when thrombin cleaves the extended N-terminus of each of these receptors, exposing a new N-terminus that serves as a tethered ligand.[86] Synthetic peptides based on the sequence of the tethered ligand domain of PAR1 and PAR4 are able to activate the receptors, mimicking at least some of the actions of thrombin. PAR3 was identified after gene ablation studies showed that platelets from mice lacking PAR1 were still responsive to thrombin (Table 16-4).[87] Approximately half of PAR1$^{-/-}$ mice die *in utero*, but this appears to be due to loss of receptor expression in the vasculature, rather than in platelets.[88] Whereas human PAR3 has been shown to signal in response to thrombin in transfected cells, PAR3 on mouse platelets appears to primarily serve to facilitate cleavage of PAR4 rather than to generate signals on its own.[89] PAR4 is expressed on human and mouse platelets and presumably accounts for the continued ability of platelets from PAR3

## Table 16-4: Platelet Phenotypes of GPCR Knockout Mice

| Agonist | Receptor | Phenotype | References |
|---|---|---|---|
| Thrombin | PAR1 | None (not expressed on mouse platelets). | 87, 241 |
| | PAR3 | Higher concentrations of thrombin required to elicit responses, presumably because of the loss of the normal contribution of PAR3 to PAR4 activation. | 91 |
| | PAR4 | Complete loss of thrombin responsiveness. | 96 |
| ADP | $P2Y_1$ | Impaired response to ADP. Absent increase in cytosolic $Ca^{2+}$ and shape change in response to ADP. Normal inhibition of cAMP formation by ADP. | 81, 242 |
| | $P2Y_{12}$ | Greatly diminished aggregation in response to ADP. Absent inhibition of adenylyl cyclase by ADP. Increased bleeding time. | 76 |
| $TXA_2$ | $TP\alpha/\beta$ | Prolonged bleeding time, absent aggregation in response to $TXA_2$ agonists, and delayed aggregation with collagen. | 122 |
| Epinephrine | $\alpha_{2A}$-adrenergic | The mice have been developed, but there are no reports on platelet function. | — |
| $PGI_2$ | IP | Increased aggregation in response to ADP. Increased rate of thrombosis following vascular injury. | 47 |

Abbreviations: ADP, adenosine diphosphate; cAMP, cyclic adenosine monophosphate; GPCR, G protein-coupled receptors; PAR, protease-activated receptor; $PGI_2$, prostaglandin $I_2$; $TXA_2$, thromboxane $A_2$.

knockout mice to respond to thrombin. Activation of PAR4 requires higher concentrations of thrombin than PAR1, apparently because it lacks the hirudin-like sequences that can interact with thrombin's anion-binding exosite and facilitate receptor cleavage.[89–92] Kinetic studies in human platelets suggest that thrombin signals first through PAR1 and subsequently through PAR4.[93,94]

There is little reason to doubt that PAR family members are sufficient to activate platelets. Peptide agonists for either PAR1 or PAR4 cause platelet aggregation and secretion.[86,91] Conversely, simultaneous inhibition of human PAR1 and PAR4 abolishes responses to thrombin,[95] as does deletion of the gene encoding PAR4 in mice.[96] Thus, PAR family members are necessary and the picture has emerged that thrombin responses in human platelets require PAR1 and PAR4 (Fig. 16-9), whereas those in mouse platelets are mediated by PAR3-facilitated cleavage of PAR4. Abundant evidence, including studies in knockout mice, shows that PAR1 and PAR4 are coupled to $G_q$ and $G_{13}$. There is conflicting evidence about whether they can also activate $G_i$ family members or whether $G_i$-dependent signaling in thrombin-activated platelets is entirely mediated by secreted ADP.

A requirement for PAR family members does not preclude the involvement of other participants in platelet responses to thrombin. One long-standing candidate is GPIb. GPIb is a heterodimer comprised of disulfide-linked $\alpha$ and $\beta$ subunits that exist in a complex with GPIX and GPV. Together they provide a binding site for VWF and an anchor for the platelet cytoskeleton (reviewed in reference [97] and Chapter 7). GPIb$\alpha$ has a high affinity thrombin binding site located within residues 268–287 GPIb$\alpha$.[98] Deletion or blockade of this site reduces platelet responses to thrombin, par-

ticularly at low thrombin concentrations[99–103] and has been shown to impair PAR1 cleavage on human platelets.[102] These observations suggest that GPIb may play a role in human platelets that is, at least in some respects, analogous to the role of PAR3 on mouse platelets: facilitating activation of a receptor that does most, if not all, of the actual signaling.

*c. Epinephrine: $G_z$-Mediated Responses That Potentiate the Effects of Other Agonists.* Compared to thrombin, epinephrine is a weak activator of human platelets when added on its own. Nonetheless, there are reports of human families in which a mild bleeding disorder is associated with impaired epinephrine-induced aggregation and reduced numbers of catecholamine receptors.[104,105] Platelet responses to epinephrine are mediated by $\alpha_{2A}$-adrenergic receptors (Table 16-1).[106–108] In both mice and humans, epinephrine is able to potentiate the effects of other agonists so that the combination is a stronger stimulus for platelet aggregation than either agonist alone. Potentiation is usually attributed to the ability of epinephrine to inhibit cAMP formation, but as is discussed below, there are clearly other effects as well. In contrast to other platelet agonists, epinephrine has no detectable direct effect on PLC and does not cause shape change, although it can trigger phosphoinositide hydrolysis indirectly by stimulating $TXA_2$ formation.[109]

Taken together, the reported effects of epinephrine on platelets suggest that platelet $\alpha_{2A}$-adrenergic receptors are coupled to $G_i$ family members, but not $G_q$ or $G_{12}$ family members (Fig. 16-7B). Gene deletion studies have helped to confirm and extend this conclusion. Human and mouse platelets express four members of the $G_i$ family. Immunoblots with specific antisera suggest that the relative level of

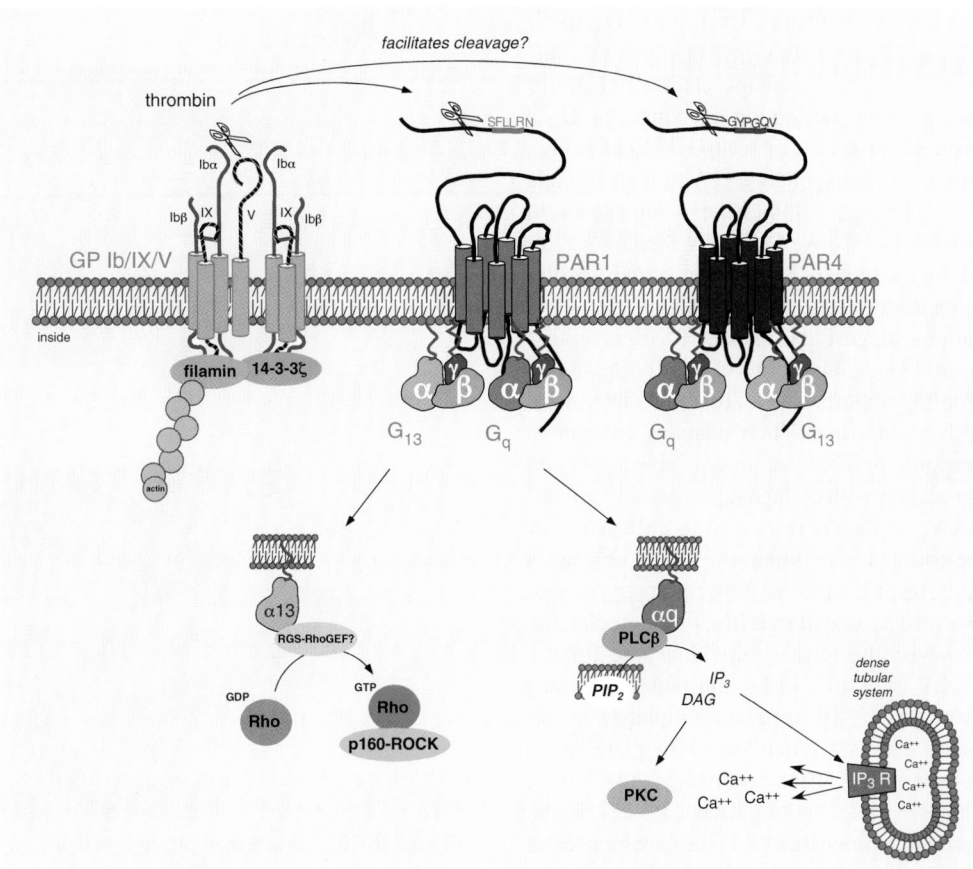

**Figure 16-9.** Thrombin receptors. Platelet responses to thrombin are mediated largely by members of the protease-activated receptor (PAR) family. Human platelets express PAR1 and PAR4, which collectively are coupled to $G_q$ and $G_{13}$-mediated effector pathways. Secretion of ADP acts as a further activator of $G_i$-mediated pathways via the receptor, $P2Y_{12}$. Cleavage of PAR1 by thrombin appears to be facilitated by the binding of thrombin to GPIbα in the GPIb-IX-V complex.

expression of the α subunits of the G proteins is $G_{i2\alpha} > G_{z\alpha} > G_{i3\alpha} \gg G_{i1\alpha}$ (Table 16-2).[110] Of the four, the sequence of $G_{z\alpha}$ is the most distinct. Where $G_{i1\alpha}$, $G_{i2\alpha}$, and $G_{i3\alpha}$ are >90% identical with each other and ubiquitously expressed, the amino acid sequence of $G_{z\alpha}$ is only 60% related to the others and $G_{z\alpha}$ has a limited range of expression. $G_{z\alpha}$ also has the slowest rate of intrinsic GTP hydrolysis and, lacking a critical cysteine residue near the N-terminus, is the only $G_i$ family member that is not a substrate for pertussis toxin. $G_{z\alpha}$ is also notable for being a substrate for PKC and PAK.[111–113] Knockout studies show that epinephrine responses in platelets are abolished when the gene encoding $G_{z\alpha}$ is removed by homologous recombination. Loss of $G_{i2\alpha}$ or $G_{i3\alpha}$ has no effect.[46,78] Therefore, it appears that in mouse platelets $\alpha_{2A}$-adrenergic receptors couple to $G_z$, but not $G_{i2}$ or $G_{i3}$. $G_z$ also appears to be responsible for the ability of epinephrine to activate Rap1B.[33,46]

*d. TXA₂ Receptor(s): Coupling to $G_q$ and $G_{12/13}$.* TXA₂ is produced from arachidonate in platelets by the aspirin-sensitive cyclooxygenase-1 (COX-1) pathway (see Chapter 31 for more details). When added to platelets *in vitro,* stable endoperoxide/thromboxane analogs such as U46619 cause shape change, aggregation, secretion, phosphoinositide hydrolysis, protein phosphorylation and an increase in cytosolic $Ca^{2+}$, while having little if any direct effect on cAMP formation. Similar responses are seen when platelets are incubated with exogenous arachidonate.[114] Once formed, TXA₂ can diffuse across the plasma membrane and activate other platelets (Fig. 16-1).[115] Like secreted ADP, release of TXA₂ amplifies the initial stimulus for platelet activation and helps to recruit additional platelets.[116] This process is effective locally, but is limited by the brief (~30 second) half-life of TXA₂ in solution, helping to confine the spread of platelet activation to the original area of injury.

Only one gene encodes TXA₂ receptors, but two splice variants (TPα and TPβ) are produced that differ in their cytoplasmic tails (Table 16-1). Human platelets express both.[117] Biochemical studies have shown that platelet TXA₂ receptors are physically associated with $G_{q\alpha}$ and $G_{13\alpha}$, and able to activate G$_{12}$ family members.[118–120] Loss of G$_{q\alpha}$ abolishes U46619-induced IP$_3$ formation and changes in cytosolic Ca$^{2+}$, but does not prevent shape change.[73] U46619 continues to cause guanine nucleotide exchange on G$_{12\alpha}$/G$_{13\alpha}$ in platelets from G$_{q\alpha}$$^{-/-}$ mice (Table 16-3).[35] Loss of G$_{13\alpha}$ abolishes TXA$_2$-induced shape change.[41] Although in cells other than platelets, TP$\alpha$ and TP$\beta$ have been shown to couple to pertussis toxin-sensitive G$_i$ family members,[121] in platelets the inhibitory effects of U46619 on cAMP formation appear to be mediated by secreted ADP.

These observations suggest that platelet TXA$_2$ receptors are coupled to G$_q$ and G$_{12/13}$, but not to G$_i$ family members. Loss of TXA$_2$ signaling impairs platelet function. TP$^{-/-}$ mice have a prolonged bleeding time. Their platelets are unable to aggregate in response to TXA$_2$ agonists (Table 16-4) and show delayed aggregation with collagen, presumably reflecting the role of TXA$_2$ in platelet responses to collagen.[122] A group of Japanese patients with impaired platelet responses to TXA$_2$ analogs have proved to be either homozygous or heterozygous for an R60L mutation in the first cytoplasmic loop of TP.[123] However, the most compelling case for the contribution of TXA$_2$ signaling in human platelets comes from the successful use of aspirin as an antiplatelet agent (see Chapter 60 for details). When added to platelets *in vitro*, aspirin abolishes TXA$_2$ generation (Figs. 16-2, 16-5, and 16-8) (see Chapter 31 for more details). It also blocks platelet activation by arachidonate and impairs responses to thrombin and ADP. The defect in thrombin responses appears as a shift in the dose/response curve, indicating that TXA$_2$ generation is supportive of platelet activation by thrombin, but not essential.

### C. Perpetuation: Contact-Dependent and Contact-Facilitated Events

Signaling from collagen receptors and GPCRs is responsible for the initiation and the extension of the platelet plug, but additional signaling events help to stabilize the platelet mass. These events are facilitated and, in some cases, made possible by the close contacts between platelets that can only occur once platelet aggregation begins. Electron micrographs show the close proximity of the plasma membranes of adjacent platelets within an aggregate, but do not show the adherens and tight junctions that are typical of contacts between endothelial cells (Fig. 16-10).[124–126] Estimates for the width of the gap between adjacent platelets range from as little as zero to as much as 50 nm,[127] making it possible for molecules on the surface of one platelet to bind to those on an adjacent platelet. This can be a direct interaction, as when one cell adhesion molecule binds to another in *trans*. It can also be an indirect interaction, such as occurs when multivalent adhesive proteins link activated αIIbβ3 on adjacent platelets. In either case, these interactions can theoreti-

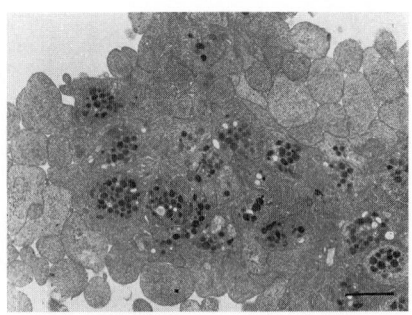

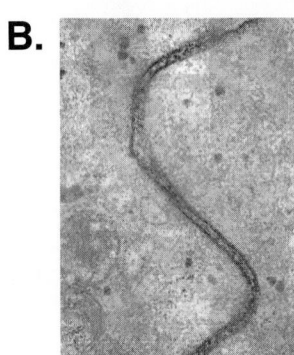

**Figure 16-10.** (A) Human platelets activated by ADP. (B) Close up of the junction between two activated platelets. Transmission electron micrographs by Alan T. Nurden and James G. White, with permission.[124,209]

cally provide both an additional adhesive force and a secondary source of intracellular signaling.

Molecules that are associated with junctions in other types of cells have been detected in platelets, but for the most part their roles in platelets are not entirely clear.[128] Although there is no evidence that platelets can form tight junctions or adherens junctions, activated platelets clearly come into close, stable contact with each other. Such contacts not only allow direct platelet–platelet interactions to occur, but also restrict the diffusion of plasma molecules into the gaps between platelets and prevent the escape of platelet activators from within the gaps. This might, for example, limit the access of plasmin to embedded fibrin, thereby helping to prevent premature dissolution of the hemostatic plug. It might also foster the accumulation of platelet activators, allowing higher concentrations to be reached locally and maintained.

Clot retraction is a platelet-dependent event in which a fibrin clot gradually pulls in upon itself, shrinking to a smaller volume after platelet activation and fibrin deposition are well under way. Clot retraction depends on the interaction between actin/myosin complexes and the cytoplasmic

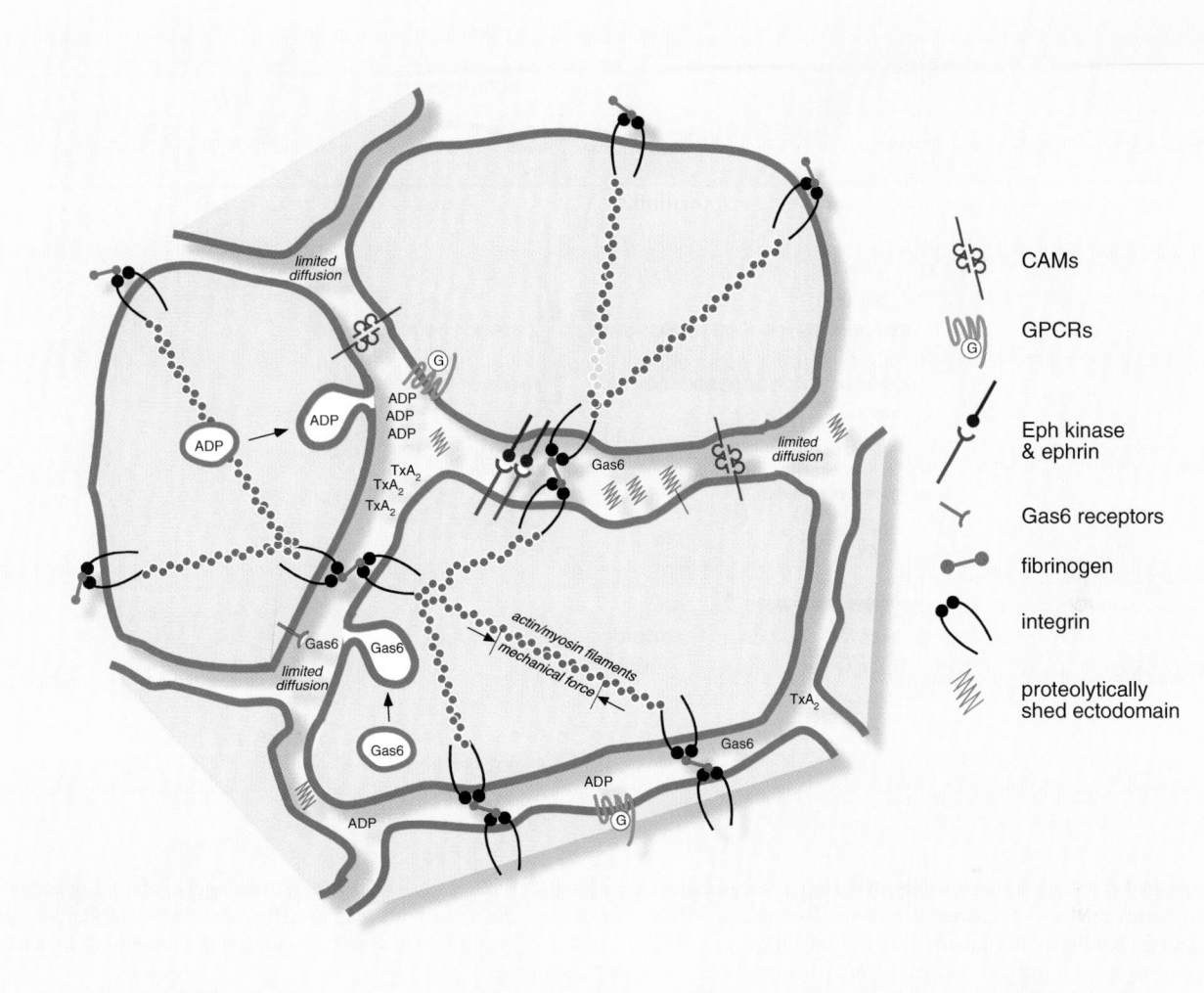

**Figure 16-11.** Contact-dependent and contact-facilitated events during thrombus formation. The onset of aggregation brings platelets into sufficiently close contact for integrins and other cell adhesion molecules to interact and for the activation of Eph receptor kinases by their cell surface ligands known as ephrins. The space between platelets also provides a protected environment in which soluble agonists for G protein-coupled receptors (ADP, thrombin, and TXA$_2$) and receptor tyrosine kinases (Gas-6), and the proteolytically shed bioactive exodomains of platelet surface proteins (CD40L) can accumulate. The mechanical forces generated by the contraction of actin/myosin filaments helps to compress the space between platelets, improving contacts and possibly increasing the concentration of soluble agonists. Abbreviations: ADP, adenosine diphosphate; CAMs, cell adhesion molecules; Gas6, growth-arrest specific gene 6; GPCRs, G protein-coupled receptors; TXA$_2$, thromboxane A$_2$.

domain of αIIbβ3, as well as the binding of fibrinogen or VWF to the extracellular domain of the integrin (see Chapter 4 for details). The absence of clot retraction is one of the hallmarks of platelets from patients with Glanzmann thrombasthenia, whose platelets lack αIIbβ3 (see Chapter 57). In the context of platelet–platelet interactions, clot retraction can be viewed as a mechanism for narrowing the gaps between platelets and increasing the local concentration of soluble ligands for platelet receptors (Fig. 16-11).[129]

Although there is currently evidence that a number of molecules in platelets participate in contact-dependent and

contact-facilitated interactions, this list is probably incomplete. Some of these molecules also play a role in the interaction of platelets with other types of cells, but the focus here will be on platelet–platelet interactions.

**1. Integrins, Adhesion, and Outside-In Signaling.** Outside-in signaling refers to the intracellular signaling events that occur downstream of activated integrins once ligand binding has occurred (see Chapter 17).[130] Integrin signaling depends in large part on the formation of protein complexes that link to the integrin cytoplasmic domain

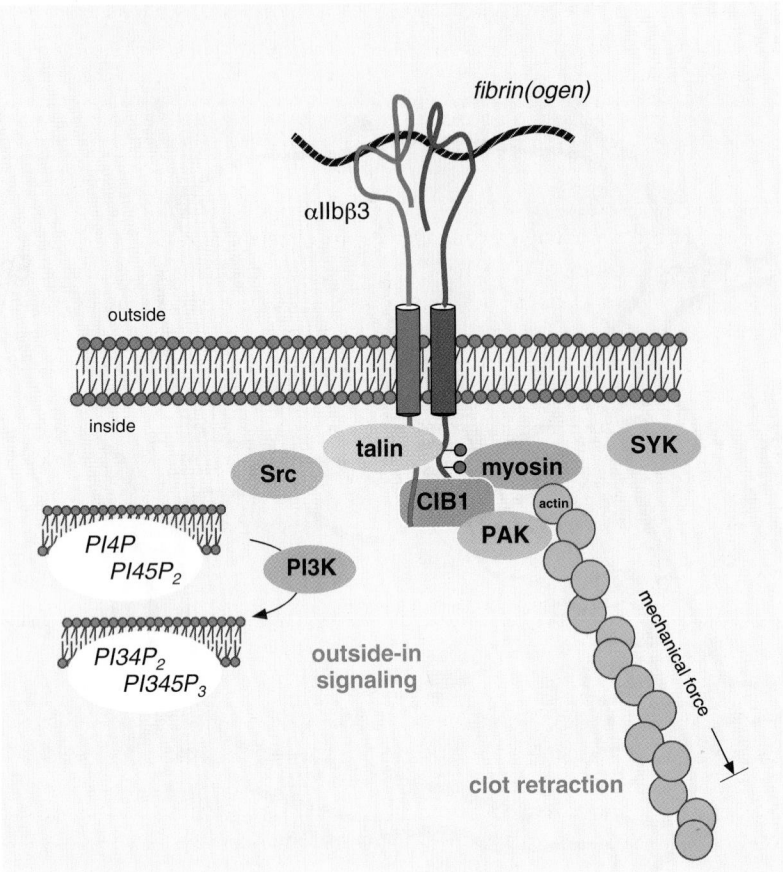

**Figure 16-12.** Outside-in signaling by αIIbβ3. Integrins in general and αIIbβ3 in particular are bidirectional signaling molecules as well as being receptors for adhesive proteins. Events downstream of platelet agonist receptors lead to αIIbβ3 activation (inside-out signaling) and the binding of the integrin to fibrinogen, fibrin, or VWF. Outside-in signaling occurs when the cytoplasmic domains of the activated integrin serves as a focus for formation of signaling and actin/myosin complexes.

(Fig. 16-12). Some of the protein–protein interactions that involve the cytoplasmic domains of αIIbβ3 help regulate integrin activation; others participate in outside-in signaling and clot retraction. Proteins that are capable of binding directly to the cytoplasmic domains of αIIbβ3 include β3-endonexin,[131] CIB1,[132] talin,[133] myosin,[134] Shc,[135] and the tyrosine kinases, Src[136] and Syk.[137,138] It is not always clear how these proteins bind to integrins and what the interaction contributes. Talin binding has been proposed to be one of the final events in the allosteric regulation of integrin activation.[133,139–141] CIB1 can directly interact with and regulate PAK, and therefore affect actin assembly.[142] Some interactions require the phosphorylation of tyrosine residues Y773 and Y785 (Y747 and Y759 in mice) in the β3 cytoplasmic domain. Shc, for example, requires Y773 phosphorylation;[135] myosin binding requires phosphorylation of both tyrosines.[134] Phosphorylation of the β3 cytoplasmic domain is thought to be mediated by Src family members and can require both activation of the integrin and its engagement with an adhesive protein.[143,144] Mutation of Y747 and Y759

to phenylalanine produces mice whose platelets tend to disaggregate and which show impaired clot retraction and a tendency to rebleed from tail bleeding time sites.[145] Fibrinogen binding to the extracellular domain of activated αIIbβ3 stimulates a rapid increase in the activity of Src family members and Syk. Studies of platelets from mice lacking these kinases suggest that these events are required for the initiation of outside-in signaling and for full platelet spreading, irreversible aggregation, and clot retraction.[143,144,146,147]

**2. Additional Cell Adhesion Molecules.** Platelet endothelial cell adhesion molecule-1 (PECAM-1; CD31) is perhaps best known for its high level of expression on endothelial cells where it accumulates at the junctions between cells. However, as its name suggests, PECAM-1 is also expressed on the surface of resting and activated platelets (see Chapter 11 for more details). PECAM-1 is a type-1 transmembrane protein with six extracellular Ig domains and an extended cytoplasmic domain of approximately 118 residues.[148] The most membrane-distal Ig domain is able to support homotypic interactions in *trans*. The C-terminus contains phosphorylatable tyrosine residues that represent tandem ITIM domains capable of binding the tyrosine phosphatases, SHP-2, and, possibly, SHP-1.[149] Loss of PECAM-1 expression in mice causes an increased responsiveness to collagen, consistent with a model in which platelet activation leads to tyrosine phosphorylation of PECAM-1, allowing SHP-1/2 to bind and bringing the phosphatase near its substrates, including the Fcγ chain partner of GPVI.[150,151] Loss of PECAM-1 expression has also now been shown to cause an increase in thrombus formation *in vivo*.[152] Taken together, these observations suggest that PECAM-1 may normally provide a braking effect on collagen-induced signaling and thereby help to prevent unwarranted platelet activation.

Junctional adhesion molecules (JAMs) are members of the CTX family of immunoglobulin (Ig) domain-containing, $Ca^{2+}$-independent cell adhesion molecules. Three have been identified: JAM-A (also known as JAM-1 and F11R), JAM-B (JAM-2, VE-JAM), and JAM-C (JAM-3).[153,154] Platelets express JAM-A and JAM-C. JAMs have an extracellular domain with two Ig domains, a single transmembrane region, and a short cytoplasmic tail that terminates in a binding site for cytosolic proteins with an appropriate PDZ domain. JAM-A localizes to tight junctions of endothelial and epithelial cells, and is also found on monocytes, neutrophils, and lymphocytes. JAM-C has been found on endothelial cells, lymphatic vessels, dendritic cells, and NK cells. JAM-A contributes to cell–cell adhesion by forming *trans* interactions involving the N-terminal Ig domain. However, JAMs also support heterotypic interactions, including binding to integrins via their membrane proximal Ig domain.[155,156] Platelet JAM-A was originally described as the antigen for a platelet-activating antibody.[157] Subsequent studies showed that the activating effects of the antibody are dependent on activation of platelet FcγRII receptors,[158] but several observations suggest that JAM-A could play a role in platelet–platelet signaling, including the ability of platelets to adhere and spread on immobilized recombinant JAM-A. A plausible model is that JAM-A localized to platelet–platelet and platelet–leukocyte contacts supports cohesive and signaling interactions between cells.[155,159] This model remains to be fully established.

In addition to JAM-A and JAM-C, two other CTX family members have been identified on platelets: ESAM and CD226. ESAM was first identified in endothelial cells, but it is also expressed by platelets, where it appears to be localized to α granules, emerging on the surface during platelet activation.[160,161] CD226 or DNAM-1 has also been identified on the surface of platelets.[162] CD226 consists of an extracellular domain with two Ig domains, a single transmembrane domain, and a cytoplasmic tail.[163] In addition to the integrin LFA-1 (αLβ2), a recent study identified the polio virus receptor (PVR, CD155) and the adherens junction protein,

nectin-2 as ligands for CD226.[164] The functional significance of these adhesion molecules in platelet biology has yet to be determined.

SLAM (signaling lymphocytic activation molecule; CD150) and CD84 are members of the CD2 family of homophilic adhesion molecules that have been studied extensively in lymphocytes, but have now been shown to be expressed in platelets as well.[165–167] The members of the family are type-1 membrane glycoproteins in the Ig superfamily. Notable differences among them are in the cytoplasmic domain, which supports binding interactions with a variety of adaptor/partner proteins. SLAM and CD84 are expressed on the surface of resting as well as activated platelets and become tyrosine phosphorylated during platelet activation, but only if aggregation is allowed to occur.[165] Mice that lack SLAM have a defect in platelet aggregation in response to collagen or a PAR4-activating peptide, but a normal response to ADP and a normal bleeding time. In a mesenteric vascular injury model, female (but not male) SLAM$^{-/-}$ mice showed a marked decrease in platelet accumulation.[165] The presence of SLAM, CD84, and two of their known adaptor proteins (SAP and EAT-2) in platelets provides a potential novel mechanism by which close contacts between platelets can help to support thrombus stability.

**3. Receptor Tyrosine Kinases: Contact-Dependent and Contact-Facilitated Signaling.** Direct contacts between platelets can promote signaling by more than one mechanism. In addition to signaling events that occur downstream from integrins and other cell adhesion molecules, there are receptors that interact *in trans* with cell surface ligands. One recently characterized example is the Eph kinases and their membrane-bound ligands, known as ephrins. Eph kinases are receptor tyrosine kinases with an intracellular kinase domain and a C-terminal binding site for cytosolic proteins with an appropriate PDZ domain. The ligands for Eph kinases are cell surface proteins known as ephrins that have either a glycophosphatidylinositol anchor (the ephrin A family) or a transmembrane domain (the ephrin B family). The cytoplasmic domains of the ephrin B family members are comprised of 90 to 100 residues with several phosphorylatable tyrosine residues[168] and, like the receptors, a PDZ target domain.[169–172] Contact between an ephrin-expressing cell and an Eph-expressing cell causes bidirectional signaling. Although Eph kinases and ephrins are best known for their role in neural development[173–175] and vasculogenesis,[176] human platelets express EphA4, EphB1, and ephrinB1.[177] Forced clustering of either EphA4 or ephrin B1 causes platelets to adhere to immobilized fibrinogen. Clustering of ephrinB1 also causes the activation of Rap1B and promotes platelet aggregation.[177,178] Blockade of Eph/ephrin interactions leads to reversible platelet aggregation at low agonist concentrations and limits the growth of platelet thrombi on collagen-coated surfaces under arterial flow conditions.[177,179] It also impairs β3 phosphorylation, which inhibits clot retraction.[179] EphA4 is constitutively associated with αIIbβ3 in both resting and activated platelets and colocalizes with the integrin at sites of contact between aggregated platelets. Collectively, these observations suggest a model in which the onset of aggregation brings platelets into close proximity and allows ephrinB1 to bind to EphA4 and EphB1. Signaling downstream of both the receptors and the kinases then promotes further integrin activation (by activating Rap1B) and integrin signaling (by promoting β3 phosphorylation).

A second example of a ligand/receptor tyrosine kinase interaction that is facilitated by platelet–platelet contacts is the binding of growth-arrest specific gene 6 (Gas-6) to its receptors. Gas-6 is a 75-kDa protein related to protein S and, like protein S, contains γ-carboxylated glutamic acid residues.[180] Gas-6 is expressed in a number of tissues and levels of Gas-6 expression are upregulated following vascular injury. In rodent platelets, Gas-6 is found in platelet α granules.[181,182] There is disagreement about whether human platelets contain Gas-6.[183] Secreted Gas-6 can serve as a ligand for the receptor tyrosine kinases, Tyro3, Axl, and Mer,[184,185] all of which are expressed on platelets.[182] Because Tyro3 family members have been shown to stimulate PI 3-kinase and PLCγ,[180] a reasonable hypothesis is that secreted Gas-6 can bind to its receptors on the platelet surface and cause signaling that promotes platelet plug formation and stability. Consistent with this hypothesis, platelets from Gas-6[−/−] mice were found to have an aberrant response to agonists in which aggregation terminates prematurely.[182] Furthermore, although the tail bleeding time of the Gas-6[−/−] mice was normal, the mice were resistant to thrombosis,[182] as are mice lacking any one of the three Gas-6 receptors. Platelets from the receptor-deleted mice also failed to aggregate normally in response to agonists.[186–188] Curiously, this appears to occur regardless of which of the receptors is suppressed. Biochemical studies showed that Gas-6 signaling promotes β3 phosphorylation and, therefore, clot retraction.[187] Secretion of Gas-6 into the spaces between platelets in a growing thrombus would be expected to allow it to achieve higher local concentrations and provide protection from being washed away.

**4. Shedding Surface Proteins into the Gaps between Platelets.** More than just an empty space, the gaps between platelets in a growing thrombus provide a safe harbor in which platelet-derived molecules can accumulate. In addition to the numerous proteins that are secreted from platelet α granules (Chapter 15),[189] activated platelets release ADP and TXA₂ and presumably continue to do so even after thrombus formation has begun. Platelets also shed a number of surface molecules, including GPIbα,[190] GPV,[191] GPVI,[192,193] and P-selectin.[194] Shedding of these proteins can be prevented by inhibitors of metalloproteases and in at least two cases (GPIbα and GPV) a role for a particular metalloprotease, ADAM17, has been established through studies on platelets from deficient mice.[190,191]

The advantages that the platelets derive from shedding surface proteins can sometimes only be surmised, but cleavage of at least one surface molecule, CD40 ligand (CD40L; CD154), gives rise to a soluble fragment that can stimulate platelets as well as other nearby cells. CD40L is a 33-kDa transmembrane protein that is present on the surface of activated platelets, but not resting platelets.[195–197] Its appearance on the platelet surface is followed by the gradual release of an 18-kDa exodomain fragment.[196] Both the surface-expressed and soluble form of CD40L (sCD40L) are trimers.[198] CD40L is a member of the tissue necrosis factor (TNF) family and platelet-derived sCD40L or activated, CD40L-expressing platelets can elicit responses from endothelial cells and monocytes that appear to be proatherogenic.[195,196,199–201] CD40L[−/−] platelets aggregate normally, but the growth of platelet plugs on collagen-coated surfaces under shear is impaired.[202,203] CD40L[−/−] mice show delayed occlusion following vascular injury and decreased thrombus stability.[202] The extracellular portion of CD40L includes a binding domain for the CD40L receptor, CD40, as well as a KGD (RGD in mice) integrin-recognition sequence. Platelets also express CD40.[196,202,204] Although the binding of sCD40L to activated platelets can be blocked by mutating the KGD sequence or blocking αIIbβ3, loss of CD40 has no apparent effect[202] — suggesting that the effects of CD40L on platelets are mediated by the integrin rather than by CD40.

## III. Conclusion

In summary, the intracellular signaling that underlies platelet activation is a dynamic process with different receptor and effector pathways dominant at different phases in the initiation, extension, and perpetuation of platelet plugs. The main objective throughout normal hemostasis is to activate integrin αIIbβ3 so that it can bind adhesive proteins, and then maintain it in an active (bound) state so that platelet plugs will remain stable long enough for wound healing to occur. In general, this requires the activation of PLC and PI3K-dependent pathways. It also involves the suppression of inhibitory mechanisms that are normally designed to prevent platelet activation, including the formation of cAMP by adenylyl cyclase. If the initial stimulus for platelet activation is the exposure of collagen and VWF, then αIIbβ3 activation is accomplished by a process that involves activation of PLCγ. If the initial stimulus is the generation of thrombin, as can be the case in pathological states, then a G protein-dependent mechanism results in a more rapid and robust activation of PLCβ — the same signaling mechanism that supports the recruitment of circulating platelets by secreted ADP and released $TXA_2$. Once $\alpha IIb\beta 3$ has been activated and platelet aggregation has occurred, then a third wave of signaling is facilitated by the close contacts between platelets within a hemostatic plug or thrombus.

# References

1. Simons, K., & Ikonen, E. (1997). Functional rafts in cell membranes. *Nature, 387,* 569–572.
2. Maguire, P. B., Foy, M., & Fitzgerald, D. J. (2005). Using proteomics to identify potential therapeutic targets in platelets. *Biochem Soc Trans, 33,* 409–412.
3. Lopez, J. A., del Conde, I., & Shrimpton, C. N. (2005). Receptors, rafts, and microvesicles in thrombosis and inflammation. *J Thromb Haexmost, 3,* 1737–1744.
4. Quinton, T. M., Kim, S., Jin, J., et al. (2005). Lipid rafts are required in Galpha(i) signaling downstream of the P2Y12 receptor during ADP-mediated platelet activation. *J Thromb Haemost, 3,* 1036–1041.
5. Locke, D., Chen, H., Liu, Y., et al. (2002). Lipid rafts orchestrate signaling by the platelet receptor glycoprotein VI. *J Biol Chem, 277,* 18801–18809.
6. Mustard, J. F., & Packham, M. A. (1970). Factors influencing platelet function: Adhesion, release, and aggregation. *Pharm Rev, 22,* 97–187.
7. Falati, S., Liu, Q., Gross, P., et al. (2003). Accumulation of tissue factor into developing thrombi in vivo is dependent upon microparticle P-selectin glycoprotein ligand 1 and platelet P-selectin. *J Exp Med, 197,* 1585–1598.
8. Gross, P. L., Furie, B. C., Merrill-Skoloff, G., et al. (2005). Leukocyte versus microparticle-mediated tissue factor transfer during arteriolar thrombus development. *J Leukoc Biol, 78,* 1318–1326.
9. Del Conde, I., Shrimpton, C. N., Thiagarajan, P., et al. (2005). Tissue-factor-bearing microvesicles arise from lipid rafts and fuse with activated platelets to initiate coagulation. *Blood, 106,* 1604–1611.
10. Santoro, S. A. (1986). Identification of a 160,000 dalton platelet membrane protein that mediates the initial divalent cation-dependent adhesion of platelets to collagen. *Cell, 46,* 913–920.
11. Brass, L. F., & Bensusan, H. B. (1974). The role of collagen quaternary structure in the platelet: Collagen interaction. *J Clin Invest, 54,* 1480–1487.
12. Brass, L. F., Faile, D., & Bensusan, H. B. (1976). Direct measurement of the platelet: Collagen interaction by affinity chromatography on collagen/Sepharose. *J Lab Clin Med, 87,* 525–534.
13. Clemetson, J. M., Polgar, J., Magnenat, E., et al. (1999). The platelet collagen receptor glycoprotein VI is a member of the immunoglobulin superfamily closely related to FcalphaR and the natural killer receptors. *J Biol Chem, 274,* 29019–29024.
14. Massberg, S., Gawaz, M., Gruner, S., et al. (2003). A crucial role of glycoprotein VI for platelet recruitment to the injured arterial wall in vivo. *J Exp Med, 197,* 41–49.
15. Kato, K., Kanaji, T., Russell, S., et al. (2003). The contribution of glycoprotein VI to stable platelet adhesion and thrombus formation illustrated by targeted gene deletion. *Blood, 102,* 1701–1707.
16. Poole, A., Gibbins, J. M., Turner, M., et al. (1997). The Fc receptor gamma-chain and the tyrosine kinase Syk are essential for activation of mouse platelets by collagen. *EMBO J, 16,* 2333–2341.
17. Nieswandt, B., Brakebusch, C., Bergmeier, W., et al. (2001). Glycoprotein VI but not alpha2beta1 integrin is essential for platelet interaction with collagen. *EMBO J, 20,* 2120–2130.
18. Nieuwenhuis, H. K., Akkerman, J. W. N., Houdijk, W. P. M., et al. (1985). Human blood platelets showing no response to collagen fail to express glycoprotein Ia. *Nature, 318,* 470–472.
19. Sixma, J. J., Van Zanten, G. H., Huizinga, E. G., et al. (1997). Platelet adhesion to collagen: An update. *Thromb Haemost, 78,* 434–438.
20. Kuijpers, M. J., Schulte, V., Bergmeier, W., et al. (2003). Complementary roles of glycoprotein VI and alpha2beta1 integrin in collagen-induced thrombus formation in flowing whole blood ex vivo. *FASEB J, 17,* 685–687.
21. Nesbitt, W. S., Giuliano, S., Kulkarni, S., et al. (2003). Intercellular calcium communication regulates platelet aggregation and thrombus growth. *J Cell Biol, 160,* 1151–1161.
22. Kulkarni, S., Nesbitt, W. S., Dopheide, S. M., et al. (2004). Techniques to examine platelet adhesive interactions under flow. *Methods Mol Biol, 272,* 165–186.
23. Hamm, H. E. (2001). How activated receptors couple to G proteins. *Proc Natl Acad Sci USA, 98,* 4819–4821.
24. Bünemann, M., Lee, K. B., Pals-Rylaarsdam, R., et al. (1999). Desensitization of G-protein-coupled receptors in the cardiovascular system. *Annu Rev Physiol, 61,* 169–192.
25. Penn, R. B., Pronin, A. N., & Benovic, J. L. (2000). Regulation of G protein-coupled receptor kinases. *Trends Cardiovasc Med, 10,* 81–89.
26. Palczewski, K., Kumasaka, T., Hori, T., et al. (2000). Crystal structure of rhodopsin: A G protein-coupled receptor. *Science, 289,* 739–745.
27. Lambright, D. G., Sondek, J., Bohm, A., et al. (1996). The 2.0 Å crystal structure of a heterotrimeric G protein. *Nature, 379,* 311–319.
28. Ford, C. E., Skiba, N. P., Bae, H. S., et al. (1998). Molecular basis for interactions of G protein betagamma subunits with effectors. *Science, 280,* 1271–1274.
29. Ross, E. M., & Wilkie, T. M. (2000). GTPase-activating proteins for heterotrimeric G proteins: Regulators of G protein signaling (RGS) and RGS-like proteins. *Annu Rev Biochem, 69,* 795–827.
30. Crittenden, J. R., Bergmeier, W., Zhang, Y., et al. (2004). CalDAG-GEFI integrates signaling for platelet aggregation and thrombus formation. *Nature Med, 10,* 982–986.
31. Bertoni, A., Tadokoro, S., Eto, K., et al. (2002). Relationships between Rap1b, affinity modulation of integrin $\alpha IIb\beta 3$, and the actin cytoskeleton. *J Biol Chem, 277,* 25715–25721.
32. Chrzanowska-Wodnicka, M., Smyth, S. S., Schoenwaelder, S. M., et al. (2005). Rap1b is required for normal platelet function and hemostasis in mice. *J Clin Invest, 115,* 680–687.

33. Woulfe, D., Jiang, H., Mortensen, R., et al. (2002). Activation of Rap1B by Gi family members in platelets. *J Biol Chem, 277,* 23382–23390.

34. Shattil, S. J., & Brass, L. F. (1987). Induction of the fibrinogen receptor on human platelets by intracellular mediators. *J Biol Chem, 262,* 992–1000.

35. Klages, B., Brandt, U., Simon, M. I., et al. (1999). Activation of G12/G13 results in shape change and Rho/Rho-kinase-mediated myosin light chain phosphorylation in mouse platelets. *J Cell Biol, 144,* 745–754.

36. Offermanns, S. (2001). In vivo functions of heterotrimeric G-proteins: Studies in Galpha-deficient mice. *Oncogene, 20,* 1635–1642.

37. Fukuhara, S., Chikumi, H., & Gutkind, J. S. (2001). RGS-containing RhoGEFs: The missing link between transforming G proteins and Rho? *Oncogene, 20,* 1661–1668.

38. Offermanns, S., Toombs, C. F., Hu, Y.-H., et al. (1997). Defective platelet activation in Galphaq-deficient mice. *Nature, 389,* 183–186.

39. Moers, A., Nieswandt, B., Massberg, S., et al. (2003). G(13) is an essential mediator of platelet activation in hemostasis and thrombosis. *Nature Med, 9,* 1418–1422.

40. Moers, A., Wettschureck, N., Gruner, S., et al. (2004). Unresponsiveness of platelets lacking both Galpha(q) and Galpha(13). Implications for collagen-induced platelet activation. *J Biol Chem, 279,* 45354–45359.

41. Moers, A., Nieswandt, B., Massberg, S., et al. (2003). G13 is an essential mediator of platelet activation in hemostasis and thrombosis. *Nature Med, 9,* 1418–1422.

42. Wilde, J. I., Retzer, M., Siess, W., et al. (2000). ADP-induced platelet shape change: An investigation of the signalling pathways involved and their dependence on the method of platelet preparation. *Platelets, 11,* 286–295.

43. Pandey, D., Goyal, P., Bamburg, J. R., et al. (2005). Regulation of LIM-kinase 1 and cofilin in thrombin-stimulated platelets. *Blood, 107,* 575–583.

44. Haslam, R. J., Dickinson, N. T., & Jang, E. K. (1999). Cyclic nucleotides and phosphodiesterases in platelets. *Thromb Haemost, 82,* 412–423.

45. Keularts, I. M. L. W., Van Gorp, R. M. A., Feijge, M. A. H., et al. (2000). alpha2A-adrenergic receptor stimulation potentiates calcium release in platelets by modulating cAMP levels. *J Biol Chem, 275,* 1763–1772.

46. Yang, J., Wu, J., Jiang, H., et al. (2002). Signaling through Gi family members in platelets — Redundancy and specificity in the regulation of adenylyl cyclase and other effectors. *J Biol Chem, 277,* 46035–46042.

47. Murata, T., Ushikubi, F., Matsuoka, T., et al. (1997). Altered pain perception and inflammatory response in mice lacking prostacyclin receptor. *Nature, 388,* 678–682.

48. Daniel, J. L., Dangelmaier, C., Jin, J. G., et al. (1999). Role of intracellular signaling events in ADP-induced platelet aggregation. *Thromb Haemost, 82,* 1322–1326.

49. Woulfe, D., Jiang, H., Morgans, A., et al. (2004). Defects in secretion, aggregation, and thrombus formation in platelets from mice lacking Akt2. *J Clin Invest, 113,* 441–450.

50. Lova, P., Paganini, S., Sinigaglia, F., et al. (2002). A Gi-dependent pathway is required for activation of the small GTPase Rap1B in human platelets. *J Biol Chem, 277,* 12009–12015.

51. Dorsam, R. T., Kim, S., Jin, J. G., et al. (2002). Coordinated signaling through both G12/13 and Gi pathways is sufficient to activate GPIIb/IIIa in human platelets. *J Biol Chem, 277,* 47588–47595.

52. Jackson, S. P., Yap, C. L., & Anderson, K. E. (2004). Phosphoinositide 3-kinases and the regulation of platelet function. *Biochem Soc Trans, 32,* 387–392.

53. Jackson, S. P., Schoenwaelder, S. M., Goncalves, I., et al. (2005). PI 3-kinase p110beta: A new target for antithrombotic therapy. Nature Med, *11,* 507–514.

54. Hirsch, E., Bosco, O., Tropel, P., et al. (2001). Resistance to thromboembolism in PI3Kgamma-deficient mice. *FASEB J, 15,* NIL307–NIL326.

55. Chen, J., De, S., Damron, D. S., et al. (2004). Impaired platelet responses to thrombin and collagen in AKT-1-deficient mice. *Blood, 104,* 1703–1710.

56. Chen, J., Somanath, P. R., Razorenova, O., et al. (2005). Akt1 regulates pathological angiogenesis, vascular maturation and permeability in vivo. *Nature Med, 11,* 1188–1196.

57. Daniel, J. L., Dangelmaier, C., Jin, J. G., et al. (1998). Molecular basis for ADP-induced platelet activation I. Evidence for three distinct ADP receptors on human platelets. *J Biol Chem, 273,* 2024–2029.

58. Jin, J. G., Daniel, J. L., & Kunapuli, S. P. (1998). Molecular basis for ADP-induced platelet activation II. The P2Y1 receptor mediates ADP-induced intracellular calcium mobilization and shape change in platelets. *J Biol Chem, 273,* 2030–2034.

59. Fisher, G. J., Bakshian, S., & Baldassare, J. J. (1985). Activation of human platelets by ADP causes a rapid rise in cytosolic free calcium without hydrolysis of phosphatidylinositol-4,5-bisphosphate. *Biochem Biophys Res Commun, 129,* 958–964.

60. Daniel, J. L., Dangelmaier, C. A., Selak, M., et al. (1986). ADP stimulates IP3 formation in human platelets. *FEBS Lett, 206,* 299–303.

61. Léon, C., Hechler, B., Vial, C., et al. (1997). The P2Y1 receptor is an ADP receptor antagonized by ATP and expressed in platelets and megakaryoblastic cells. *FEBS Lett, 403,* 26–30.

62. Hollopeter, G., Jantzen, H. M., Vincent, D., et al. (2001). Identification of the platelet ADP receptor targeted by antithrombotic drugs. *Nature, 409,* 202–207.

63. Zhang, F. L., Luo, L., Gustafson, E., et al. (2001). ADP is the cognate ligand for the orphan G protein-coupled receptor SP1999. *J Biol Chem, 276,* 8608–8615.

64. Jin, J. G., & Kunapuli, S. P. (1998). Coactivation of two different G protein-coupled receptors is essential for ADP-induced platelet aggregation. *Proc Natl Acad Sci USA, 95,* 8070–8074.

65. McKenzie, A. B., Mahout-Smith, M. P., & Sage, S. O. (1996). Activation of receptor-operated channels via P2X1 not P2T purinoreceptors in human platelets. *J Biol Chem, 271,* 2879–2881.

66. Vial, C., Hechler, B., Léon, C., et al. (1997). Presence of P2X1 purinoceptors in human platelets and megakaryoblastic cell lines. *Thromb Haemost, 78,* 1500–1504.

67. Sun, B., Li, J., Okahara, K., et al. (1998). P2X1 purinoceptor in human platelets — Molecular cloning and functional characterization after heterologous expression. *J Biol Chem, 273*, 11544–11547.

68. Mahaut-Smith, M. P., Ennion, S. J., Rolf, M. G., et al. (2000). ADP is not an agonist at P2X1 receptors: Evidence for separate receptors stimulated by ATP and ADP on human platelets. *Br J Pharmacol, 131*, 108–114.

69. Fung, C. Y., Brearley, C. A., Farndale, R. W., et al. (2005). A major role for P2X1 receptors in the early collagen-evoked intracellular $Ca^{2+}$ responses of human platelets. *Thromb Haemost, 94*, 37–40.

70. Mahaut-Smith, M. P., Tolhurst, G., & Evans, R. J. (2004). Emerging roles for P2X1 receptors in platelet activation. *Platelets, 15*, 131–144.

71. Hechler, B., Lenain, N., Marchese, P., et al. (2003). A role of the fast ATP-gated P2X1 cation channel in thrombosis of small arteries in vivo. *J Exp Med, 198*, 661–667.

72. Tolhurst, G., Vial, C., Leon, C., et al. (2005). Interplay between P2Y(1), P2Y(12), and P2X(1) receptors in the activation of megakaryocyte cation influx currents by ADP: Evidence that the primary megakaryocyte represents a fully functional model of platelet P2 receptor signaling. *Blood, 106*, 1644–1651.

73. Offermanns, S., Toombs, C. F., Hu, Y. H., et al. (1997). Defective platelet activation in Galphaq-deficient mice. *Nature, 389*, 183–186.

74. Soulet, C., Hechler, B., Gratacap, M. P., et al. (2005). A differential role of the platelet ADP receptors P2Y and P2Y in Rac activation. *J Thromb Haemost, 3*, 2296–2306.

75. Nurden, P. F., Savi, P., Heilmann, E., et al. (1995). An inherited bleeding disorder linked to a defective interaction between ADP and its receptor on platelets. Its influence on glycoprotein IIb-IIIa complex function. *J Clin Invest, 95*, 1612–1622.

76. Foster, C. J. (2001). Molecular identification and characterization of the platelet ADP receptor targeted by thienopyridine drugs using P2Yac-null mice. *J Clin Invest, 107*, 1591–1598.

77. Jantzen, H.-M., Milstone, D. S., Gousset, L., et al. (2001). Impaired activation of murine platelets lacking Galphai2. *J Clin Invest, 108*, 477–483.

78. Yang, J., Wu, J., Kowalska, M. A., et al. (2000). Loss of signaling through the G protein, Gz, results in abnormal platelet activation and altered responses to psychoactive drugs. *Proc Natl Acad Sci USA, 97*, 9984–9989.

79. Cattaneo, M., & Gachet, C. (1999). ADP receptors and clinical bleeding disorders. *Arterioscler Thromb Vasc Biol, 19*, 2281–2285.

80. Fabre, J. E., Nguyen, M. T., Latour, A., et al. (1999). Decreased platelet aggregation, increased bleeding time and resistance to thromboembolism in P2Y1-deficient mice. *Nature Med, 5*, 1199–1202.

81. Léon, C., Hechler, B., Freund, M., et al. (1999). Defective platelet aggregation and increased resistance to thrombosis in purinergic P2Y1 receptor-null mice. *J Clin Invest, 104*, 1731–1737.

82. Paul, B. Z. S., Jin, J. G., & Kunapuli, S. P. (1999). Molecular mechanism of thromboxane A2-induced platelet aggrega-

tion — Essential role for P2TAC and alpha2A receptors. *J Biol Chem, 274*, 29108–29114.

83. Bennett, J. S. (2001). Novel platelet inhibitors. *Annu Rev Med, 52*, 161–184.

84. Barr, A. J., Brass, L. F., & Manning, D. R. (1997). Reconstitution of receptors and GTP-binding regulatory proteins (G proteins) in Sf9 cells: A direct evaluation of selectivity in receptor-G protein coupling. *J Biol Chem, 272*, 2223–2229.

85. Kim, S., Foster, C., Lecchi, A., et al. (2002). Protease activated recepton 1 and 4 do not stimulate G(i) signaling pathways in the absence of secreted ADP and cause human platelet aggregation independantly of G(i) signaling. *Blood, 99*, 3629–3636.

86. Vu, T.-K. H., Hung, D. T., Wheaton, V. I., et al. (1991). Molecular cloning of a functional thrombin receptor reveals a novel proteolytic mechanism of receptor activation. *Cell, 64*, 1057–1068.

87. Connolly, A. J., Ishihara, H., Kahn, M. L., et al. (1996). Role of the thrombin receptor in development and evidence for a second receptor. *Nature, 381*, 516–519.

88. Griffin, C. T., Srinivasan, Y., Zheng, Y. W., et al. (2001). A role of thrombin receptor signaling in endothelial cells during embryonic development. *Science, 293*, 1666–1670.

89. Nakanishi-Matsui, M., Zheng, Y. W., Sulciner, D. J., et al. (2000). PAR3 is a cofactor for PAR4 activation by thrombin. *Nature, 404*, 609–610.

90. Xu, W.-F., Andersen, H., Whitmore, T. E., et al. (1998). Cloning and characterization of human protease-activated receptor 4. *Proc Natl Acad Sci USA, 95*, 6642–6646.

91. Kahn, M. L., Zheng, Y. W., Huang, W., et al. (1998). A dual thrombin receptor system for platelet activation. *Nature, 394*, 690–694.

92. Ishii, K., Gerszten, R., Zheng, Y. W., et al. (1995). Determinants of thrombin receptor cleavage. Receptor domains involved, specificity, and role of the P3 aspartate. *J Biol Chem, 270*, 16435–16440.

93. Covic, L., Gresser, A. L., & Kuliopulos, A. (2000). Biphasic kinetics of activation and signaling for PAR1 and PAR4 thrombin receptors in platelets. *Biochemistry, 39*, 5458–5467.

94. Shapiro, M. J., Weiss, E. J., Faruqi, T. R., et al. (2000). Protease-activated receptors 1 and 4 are shut off with distinct kinetics after activation by thrombin. *J Biol Chem, 275*, 25216–25221.

95. Kahn, M. L., Nakanishi-Matsui, M., Shapiro, M. J., et al. (1999). Protease-activated receptors 1 and 4 mediate activation of human platelets by thrombin. *J Clin Invest, 103*, 879–887.

96. Sambrano, G. R., Weiss, E. J., Zheng, Y.-W., et al. (2001). Role of thrombin signaling in platelets in hemostasis and thrombosis. *Nature, 413*, 74–78.

97. Lopez, J. A., Andrews, R. K., Afshar-Khargan, V., et al. (1998). Bernard–Soulier syndrome. *Blood, 91*, 4397–4418.

98. De Cristofaro, R., De Candia, E., Rutella, S., et al. (2000). The Asp272-Glu282 region of platelet glycoprotein Ibalpha interacts with the heparin-binding site of alpha-thrombin and protects the enzyme from the heparincatalyzed inhibition by antithrombin III. *J Biol Chem, 275*, 3887–3895.

99. De Marco, L., Mazzucato, M., Masotti, A., et al. (1991). Function of glycoprotein Ibalpha in platelet activation induced by alpha-thrombin. *J Biol Chem, 266*, 23776–23783.

100. Harmon, J. T., & Jamieson, G. A. (1988). Platelet activation by thrombin in the absence of the high affinity thrombin receptor. *Biochemistry, 27*, 2151–2157.

101. Mazzucato, M., De Marco, L., Masotti, A., et al. (1998). Characterization of the initial alpha-thrombin interaction with glycoprotein Ibalpha in relation to platelet activation. *J Biol Chem, 273*, 1880–1887.

102. De Candia, E., Hall, S. W., Rutella, S., et al. (2001). Binding of thrombin to glycoprotein Ib accelerates hydrolysis of PAR1 on intact platelets. *J Biol Chem, 276*, 4692–4698.

103. Dörmann, D., Clemetson, K. J., & Kehrel, B. E. (2000). The GPIb thrombin-binding site is essential for thrombin-induced platelet procoagulant activity. *Blood, 96*, 2469–2478.

104. Rao, A. K., Willis, J., Kowalska, M. A., et al. (1988). Differential requirements for platelet aggregation and inhibition of adenylate cyclase by epinephrine. Studies of a familial platelet alpha2-adrenergic receptor defect. *Blood, 71*, 494–501.

105. Tamponi, G., Pannocchia, A., Arduino, C., et al. (1987). Congenital deficiency of alpha2-adrenoreceptors on human platelets: Description of two cases. *Thromb Haemost, 58*, 1012–1016.

106. Newman, K. D., Williams, L. T., Bishopric, N. H., et al. (1978). Identification of alpha-adrenergic receptors in human platelets by 3H-dihydroergocryptine binding. *J Clin Invest, 61*, 395–402.

107. Kaywin, P., McDonough, M., Insel, P. A., et al. (1978). Platelet function in essential thrombocythemia: Decreased epinephrine responsivenesss associated with a deficiency of platelet alpha-adrenergic receptors. *N Engl J Med, 299*, 505–509.

108. Motulsky, H. J., & Insel, P. A. (1982). [3H]Dihydroergocryptine binding to alpha-adrenergic receptors of human platelets. A reassessment using the selective radioligands [3H] prazosin, [3H]yohimbine, and [3H]rauwolscine. *Biochem Pharmacol, 31*, 2591–2597.

109. Siess, W., Weber, P. C., & Lapetina, E. G. (1984). Activation of phospholipase C is dissociated from arachidonate metabolism during platelet shape change induced by thrombin or platelet-activating factor. Epinephrine does not induce phospholipase C activation or platelet shape change. *J Biol Chem, 259*, 8286–8292.

110. Williams, A., Woolkalis, M. J., Poncz, M., et al. (1990). Identification of the pertussis toxin-sensitive G proteins in platelets, megakaryocytes and HEL cells. *Blood, 76*, 721–730.

111. Casey, P. J., Fong, H. K. W., Simon, M. I., et al. (1990). Gz, a guanine nucleotide-binding protein with unique biochemical properties. *J Biol Chem, 265*, 2383–2390.

112. Lounsbury, K. M., Casey, P. J., Brass, L. F., et al. (1991). Phosphorylation of Gz in human platelets: Selectivity and site of modification. *J Biol Chem, 266*, 22051–22056.

113. Wang, J., Frost, J. A., & Ross, E. M. (1999). Reciprocal signaling between heterotrimeric G proteins and the p21-stimulated protein kinase. *J Biol Chem, 274*, 31641–31647.

114. Gerrard, J. M., & Carroll, R. C. (1981). Stimulation of protein phosphorylation by arachidonic acid and endoperoxide analog. *Prostaglandins, 22*, 81–94.

115. FitzGerald, G. A. (1991). Mechanisms of platelet activation: Thromboxane A2 as an amplifying signal for other agonists. *Am J Cardiol, 68*, 11B–15B.

116. Brass, L. F., Shaller, C. C., & Belmonte, E. J. (1987). Inositol 1,4,5-triphosphate-induced granule secretion in platelets. Evidence that the activation of phospholipase C mediated by platelet thromboxane receptors involves a guanine nucleotide binding protein-dependent mechanism distinct from that of thrombin. *J Clin Invest, 79*, 1269–1275.

117. Hirata, T., Ushikubi, F., Kakizuka, A., et al. (1996). Two thromboxane A2 receptor isoforms in human platelets — Opposite coupling to adenylyl cyclase with different sensitivity to Arg60 to Leu mutation. *J Clin Invest, 97*, 949–956.

118. Knezevic, I., Borg, C., & Le Breton, G. C. (1993). Identification of Gq as one of the G-proteins which copurify with human platelet thromboxane A2/prostaglandin H2 receptors. *J Biol Chem, 268*, 26011–26017.

119. Offermanns, S., Laugwitz, K.-L., Spicher, K., et al. (1994). G proteins of the G12 family are activated via thromboxane A2 and thrombin receptors in human platelets. *Proc Natl Acad Sci USA, 91*, 504–508.

120. Djellas, Y., Manganello, J. M., Antonakis, K., et al. (1999). Identification of Galpha13 as one of the G-proteins that couple to human platelet thromboxane A2 receptors. *J Biol Chem, 274*, 14325–14330.

121. Gao, Y., Tang, S., Zhou, S., et al. (2001). The thromboxane A2 receptor activates mitogen-activated protein kinase via protein kinase C-dependent Gi coupling and Src-dependent phosphorylation of the epidermal growth factor receptor. *J Pharmacol Exp Ther, 296*, 426–433.

122. Thomas, D. W., Mannon, R. B., Mannon, P. J., et al. (1998). Coagulation defects and altered hemodynamic responses in mice lacking receptors for thromboxane A2. *J Clin Invest, 102*, 1994–2001.

123. Higuchi, W., Fuse, I., Hattori, A., et al. (1999). Mutations of the platelet thromboxane A2 (TXA2) receptor in patients characterized by the absence of TXA2-induced platelet aggregation despite normal TXA2 binding activity. *Thromb Haemost, 82*, 1528–1531.

124. Humbert, M., Nurden, P., Bihour, C., et al. (1996). Ultrastructural studies of platelet aggregates from human subjects receiving clopidogrel and from a patient with an inherited defect of an ADP-dependent pathway of platelet activation. *Arterioscler Thromb Vasc Biol, 16*, 1532–1543.

125. White, J. G. (1972). Interaction of membrane systems in blood platelets. *Am J Pathol, 66*, 295–312.

126. White, J. G. (1988). Platelet membrane ultrastructure and its changes during platelet activation. *Prog Clin Biol Res, 283*, 1–32.

127. Skaer, R. J., Emmines, J. P., & Skaer, H. B. (1979). The fine structure of cell contacts in platelet aggregation. *J Ultrastruct Res, 69*, 28–42.

128. Elrod, J. W., Park, J. H., Oshima, T., et al. (2003). Expression of junctional proteins in human platelets. *Platelets, 14,* 247–251.

129. Tschumperlin, D. J., Dai, G., Maly, I. V., et al. (2004). Mechanotransduction through growth-factor shedding into the extracellular space. *Nature, 429,* 83–86.

130. Shattil, S. J., & Newman, P. J. (2004). Integrins: Dynamic scaffolds for adhesion and signaling in platelets. *Blood, 104,* 1606–1615.

131. Shattil, S. J., O'Toole, T., Eigenthaler, M., et al. (1995). beta3-Endonexin, a novel polypeptide that interacts specifically with the cytoplasmic tail of the integrin beta3 subunit. *J Cell Biol, 131,* 807–816.

132. Naik, U. P., Patel, P. M., & Parise, L. V. (1997). Identification of a novel calcium-binding protein that interacts with the integrin alphaIIb cytoplasmic domain. *J Biol Chem, 272,* 4651–4654.

133. Calderwood, D. A., Zent, R., Grant, R., et al. (1999). The talin head domain binds to integrin beta subunit cytoplasmic tails and regulates integrin activation. *J Biol Chem, 274,* 28071–28074.

134. Jenkins, A. L., Nannizzi-Alaimo, L., Silver, D., et al. (1998). Tyrosine phosphorylation of the beta3 cytoplasmic domain mediates integrin-cytoskeletal interactions. *J Biol Chem, 273,* 13878–13885.

135. Cowan, K. J., Law, D. A., & Phillips, D. R. (2000). Identification of Shc as the primary protein binding to the tyrosine-phosphorylated beta3 subunit of alphaIIbbeta3 during outside-in integrin platelet signaling. *J Biol Chem, 275,* 36423–36429.

136. Arias-Salgado, E. G., Lizano, S., Sarkar, S., et al. (2003). Src kinase activation by direct interaction with the integrin {beta} cytoplasmic domain. *Proc Natl Acad Sci USA, 100,* 13298–13302.

137. Gao, J., Zoller, K. E., Ginsberg, M. H., et al. (1997). Regulation of the pp72syk protein tyrosine kinase by platelet integrin alphaIIbbeta3. *EMBO J, 16,* 6414–6425.

138. Woodside, D. G., Obergfell, A., Leng, L., et al. (2001). Activation of Syk protein tyrosine kinase through interaction with integrin beta cytoplasmic domains. *Curr Biol, 11,* 1799–1804.

139. Calderwood, D. A., & Ginsberg, M. H. (2003). Talin forges the links between integrins and actin. *Nature Cell Biol, 5,* 694–697.

140. Ratnikov, B. I., Partridge, A. W., & Ginsberg, M. H. (2005). Integrin activation by talin. *J Thromb Haemost, 3,* 1783–1790.

141. Tadokoro, S., Shattil, S. J., Eto, K., et al. (2003). Talin binding to integrin beta tails: A final common step in integrin activation. *Science, 302,* 103–106.

142. Leisner, T. M., Liu, M., Jaffer, Z. M., et al. (2005). Essential role of CIB1 in regulating PAK1 activation and cell migration. *J Cell Biol, 170,* 465–476.

143. Philips, D. R., Prasad, K. S. S., Manganello, J., et al. (2001). Integrin tyrosine phosphorylation in platelet signaling. *Curr Opin Cell Biol, 13,* 546–554.

144. Phillips, D. R., Nannizzi-Alamio, L., & Prasad, K. S. S. (2001). beta3 tyrosine phosphorylation in alphaIIbbeta3 (platelet membrane GP IIb-IIIa) outside-in integrin signaling. *Thromb Haemost, 86,* 246–258.

145. Law, D. A., DeGuzman, F. R., Heiser, P., et al. (1999). Integrin cytoplasmic tyrosine motif is required for outside-in alphaIIbbeta3 signalling and platelet function. *Nature, 401,* 808–811.

146. Payrastre, B., Missy, K., Trumel, C., et al. (2000). The integrin alphaIIb/beta3 in human platelet signal transduction. *Biochem Pharmacol, 60,* 1069–1074.

147. Obergfell, A., Eto, K., Mocsai, A., et al. (2002). Coordinate interactions of Csk, Src and Syk kinases with alphaIIbbeta3 initiate integrin signaling to the cytoskeleton. *J Cell Biol, 157,* 265–275.

148. Newman, P. J., & Newman, D. K. (2003). Signal transduction pathways mediated by PECAM-1: New roles for an old molecule in platelet and vascular cell biology. *Arterioscler Thromb Vasc Biol, 23,* 953–964.

149. Newman, P. J. (1999). Switched at birth: A new family for PECAM-1. *J Clin Invest, 103,* 5–9.

150. Patil, S., Newman, D. K., & Newman, P. J. (2001). Platelet endothelial cell adhesion molecule-1 serves as an inhibitory receptor that modulates platelet responses to collagen. *Blood, 97,* 1727–1732.

151. Jones, K. L., Hughan, S. C., Dopheide, S. M., et al. (2001). Platelet endothelial cell adhesion molecule-1 is a negative regulator of platelet–collagen interactions. *Blood, 98,* 1456–1463.

152. Falati, S., Patil, S., Gross, P. L., et al. (2006). Platelet PECAM-1 inhibits thrombus formation in vivo. *Blood, 107,* 535–541.

153. Muller, W. A. (2003). Leukocyte-endothelial-cell interactions in leukocyte transmigration and the inflammatory response. *Trends Immunol, 24,* 327–334.

154. Bazzoni, G. (2003). The JAM family of junctional adhesion molecules. *Curr Opin Cell Biol, 15,* 525–530.

155. Ostermann, G., Weber, K. S., Zernecke, A., et al. (2002). JAM-1 is a ligand of the beta(2) integrin LFA-1 involved in transendothelial migration of leukocytes. *Nature Immunol, 3,* 151–158.

156. Santoso, S., Sachs, U. J., Kroll, H., et al. (2002). The junctional adhesion molecule 3 (JAM-3) on human platelets is a counterreceptor for the leukocyte integrin Mac-1. *J Exp Med, 196,* 679–691.

157. Kornecki, E., Walkowiak, B., Naik, U. P., et al. (1990). Activation of human platelets by a stimulatory monoclonal antibody. *J Biol Chem, 265,* 10042–10048.

158. Naik, U. P., Ehrlich, Y. H., & Kornecki, E. (1995). Mechanisms of platelet activation by a stimulatory antibody: Crosslinking of a novel platelet receptor for monoclonal antibody F11 with the Fc gamma RII receptor. *Biochem J, 310* (Pt 1), 155–162.

159. Babinska, A., Kedees, M. H., Athar, H., et al. (2002). F11-receptor (F11R/JAM) mediates platelet adhesion to endothelial cells: Role in inflammatory thrombosis. *Thromb Haemost, 88,* 843–850.

160. Hirata, K., Ishida, T., Penta, K., et al. (2001). Cloning of an immunoglobulin family adhesion molecule selectively expressed by endothelial cells. *J Biol Chem, 276,* 16223–16231.

161. Nasdala, I., Wolburg-Buchholz, K., Wolburg, H., et al. (2002). A transmembrane tight junction protein selectively expressed on endothelial cells and platelets. *J Biol Chem, 277,* 16294–16303.

162. Scott, J. L., Dunn, S. M., Jin, B., et al. (1989). Characterization of a novel membrane glycoprotein involved in platelet activation. *J Biol Chem, 264,* 13475–13482.

163. Kojima, H., Kanada, H., Shimizu, S., et al. (2003). CD226 mediates platelet and megakaryocytic cell adhesion to vascular endothelial cells. *J Biol Chem, 278,* 36748–36753.

164. Bottino, C., Castriconi, R., Pende, D., et al. (2003). Identification of PVR (CD155) and Nectin-2 (CD112) as cell surface ligands for the human DNAM-1 (CD226) activating molecule. *J Exp Med, 198,* 557–567.

165. Nanda, N., Andre, P., Bao, M., et al. (2005). Platelet aggregation induces platelet aggregate stability via SLAM family receptor signaling. *Blood, 106,* 3028–3034.

166. Krause, S. W., Rehli, M., Heinz, S., et al. (2000). Characterization of MAX.3 antigen, a glycoprotein expressed on mature macrophages, dendritic cells and blood platelets: Identity with CD84. *Biochem J, 346,* 729–736.

167. Martin, M., Romero, X., de la Fuente, M. A., et al. (2001). CD84 functions as a homophilic adhesion molecule and enhances IFN-gamma secretion: Adhesion is mediated by Ig-like domain 1. *J Immunol, 167,* 3668–3676.

168. Kalo, M. S., Yu, H. H., & Pasquale, E. B. (2001). In vivo tyrosine phosphorylation sites of activated ephrin-B1 and EphB2 from neural tissues. *J Biol Chem, 276,* 38940–38948.

169. Torres, R., Firestein, B. L., Dong, H. L., et al. (1998). PDZ proteins bind, cluster, and synaptically colocalize with Eph receptors and their ephrin ligands. *Neuron, 21,* 1453–1463.

170. Lin, D., Gish, G. D., Songyang, Z., et al. (1999). The carboxyl terminus of B class ephrins constitutes a PDZ binding motif. *J Biol Chem, 274,* 3726–3733.

171. Bruckner, K., Pablo Labrador, J., Scheiffele, P., et al. (1999). Ephrin B ligands recruit GRIP family PDZ adaptor proteins into raft membrane microdomains. *Neuron, 22,* 511–524.

172. Della Rocca, G. J., Van Biesen, T., Daaka, Y., et al. (1997). Ras-dependent mitogen-activated protein kinase activation by G protein-coupled receptors — Convergence of Gi- and Gq-mediated pathways on calcium/calmodulin, Pyk2, and Src kinase. *J Biol Chem, 272,* 19125–19132.

173. Kullander, K., & Klein, R. (2002). Mechanisms and function of Eph and ephrin signaling. *Nature Rev Mol Cell Biol, 3,* 475–486.

174. Holmberg, J., & Frisen, J. (2002). Ephrins are not only unattractive. *Trends Neurosci, 25,* 239–243.

175. Depaepe, V., Suarez-Gonzalez, N., Dufour, A., et al. (2005). Ephrin signalling controls brain size by regulating apoptosis of neural progenitors. *Nature, 435,* 1244–1250.

176. Adams, R. H., & Klein, R. (2000). Eph receptors and ephrin ligands: Essential mediators of vascular development. *Trends Cardiovasc Med, 10,* 183–188.

177. Prevost, N., Woulfe, D., Tanaka, T., et al. (2002). Interactions between Eph kinases and ephrins provide a mechanism to support platelet aggregation once cell-to-cell contact has occurred. *Proc Natl Acad Sci USA, 99,* 9219–9224.

178. Prevost, N., Woulfe, D. S., Tognolini, M., et al. (2004). Signaling by ephrinB1 and Eph kinases in platelets promotes Rap1 activation, platelet adhesion, and aggregation via effector pathways that do not require phosphorylation of ephrinB1. *Blood, 103,* 1348–1355.

179. Prevost, N., Woulfe, D. S., Jiang, H., et al. (2005). Eph kinases and ephrins support thrombus growth and stability by regulating integrin outside-in signaling in platelets. *Proc Natl Acad Sci USA, 102,* 9820–9825.

180. Melaragno, M. G., Fridell, Y.-W., & Berk, B. C. (1999). The Gas6/Axl system: A novel regulator of vascular cell function. *Trends Cardiovasc Med, 9,* 250–253.

181. Ishimoto, Y., & Nakano, T. (2000). Release of a product of growth arrest-specific gene 6 from rat platelets. *FEBS Lett, 466,* 197–199.

182. Angelillo-Scherrer, A., De Frutos, P. G., Aparicio, C., et al. (2001). Deficiency or inhibition of Gas6 causes platelet dysfunction and protects mice against thrombosis. *Nature Med, 7,* 215–221.

183. Balogh, I., Hafizi, S., Stenhoff, J., et al. (2005). Analysis of Gas6 in human platelets and plasma. *Arterioscler Thromb Vasc Biol, 25,* 1280–1286.

184. Stitt, T. N., Conn, G., Gore, M., et al. (1995). The anticoagulation factor protein and its relative, Gas6, are ligands for the Tyro3/Axl family of receptor tyrosine kinases. *Cell, 80,* 661–670.

185. Varnum, B. C., Young, C., Elliott, G., et al. (1995). Axl receptor tyrosine kinase stimulated by the vitamin K-dependent protein encoded by the growth-arrest-specific gene 6. *Nature, 373,* 623–626.

186. Chen, C., Li, Q., Darrow, A. L., et al. (2004). Mer receptor tyrosine kinase signaling participates in platelet function. *Arterioscler Thromb Vasc Biol, 24,* 1118–1123.

187. Angelillo-Scherrer, A., Burnier, L., Flores, N., et al. (2005). Role of Gas6 receptors in platelet signaling during thrombus stabilization and implications for antithrombotic therapy. *J Clin Invest, 115,* 237–246.

188. Gould, W. R., Baxi, S. M., Schroeder, R., et al. (2005). Gas6 receptors Axl, Sky and Mer enhance platelet activation and regulate thrombotic responses. *J Thromb Haemost, 3,* 733–741.

189. Coppinger, J. A., Cagney, G., Toomey, S., et al. (2004). Characterization of the proteins released from activated platelets leads to localization of novel platelet proteins in human atherosclerotic lesions. *Blood, 103,* 2096–2104.

190. Bergmeier, W., Piffath, C. L., Cheng, G., et al. (2004). Tumor necrosis factor-alpha-converting enzyme (ADAM17) mediates GPIbalpha shedding from platelets in vitro and in vivo. *Circ Res, 95,* 677–683.

191. Rabie, T., Strehl, A., Ludwig, A., et al. (2005). Evidence for a role of ADAM17 (TACE) in the regulation of platelet glycoprotein V. *J Biol Chem, 280,* 14462–14468.

192. Bergmeier, W., Rabie, T., Strehl, A., et al. (2004). GPVI down-regulation in murine platelets through metalloproteinase-dependent shedding. *Thromb Haemost, 91,* 951–958.

193. Stephens, G., Yan, Y., Jandrot-Perrus, M., et al. (2005). Platelet activation induces metalloproteinase-dependent GP

VI cleavage to down-regulate platelet reactivity to collagen. *Blood, 105,* 186–191.

194. Berger, G., Hartwell, D. W., & Wagner, D. D. (1998). P-Selectin and platelet clearance. *Blood, 92,* 4446–4452.

195. Henn, V., Slupsky, J. R., Gräfe, M., et al. (1998). CD40 ligand on activated platelets triggers an inflammatory reaction of endothelial cells. *Nature, 391,* 591–594.

196. Henn, V., Steinbach, S., Buchner, K., et al. (2001). The inflammatory action of CD40 ligand (CD154) expressed on activated human platelets is temporally limited by coexpressed CD40. *Blood, 98,* 1047–1054.

197. Hermann, A., Rauch, B. H., Braun, M., et al. (2001). Platelet CD40 ligand (CD40L) — subcellular localization, regulation of expression, and inhibition by clopidogrel. *Platelets, 12,* 74–82.

198. Locksley, R. M., Killeen, N., & Lenardo, M. J. (2001). The TNF and TNF receptor superfamilies: Integrating mammalian biology. *Cell, 104,* 487–501.

199. Mach, F., Schonbeck, U., Sukhova, G. K., et al. (1998). Reduction of atherosclerosis in mice by inhibition of CD40 signaling. *Nature, 394,* 200–203.

200. May, A. E., Kalsch, T., Massberg, S., et al. (2002). Engagement of glycprotein IIb/IIIa on platelets upregulates CD40L and triggers CD40L-dependent matrix degradation by endothelial cells. *Circulation, 106,* 2111–2117.

201. Danese, S., & Fiocchi, C. (2005). Platelet activation and the CD40/CD40 ligand pathway: Mechanisms and implications for human disease. *Crit Rev Immunol, 25,* 103–122.

202. Andre, P., Prasad, K. S., Denis, C. V., et al. (2002). CD40L stabilizes arterial thrombi by a β3 integrin-dependent mechanism. *Nature Med, 8,* 247–252.

203. Crow, A. R., Leytin, V., Starkey, A. F., et al. (2003). CD154 (CD40 ligand)-deficient mice exhibit prolonged bleeding time and decreased shear-induced platelet aggregates. *J Thromb Haemost, 1,* 850–852.

204. Inwald, D. P., McDowall, A., Peters, M. J., et al. (2003). CD40 is constitutively expressed on platelets and provides a novel mechanism for platelet activation. *Circ Res, 92,* 1041–1048.

205. Kawasaki, T., Taniguchi, M., Moritani, Y., et al. (2005). Pharmacological properties of YM-254890, a specific G(alpha)q/11 inhibitor, on thrombosis and neointima formation in mice. *Thromb Haemost, 94,* 184–192.

206. Takasaki, J., Saito, T., Taniguchi, M., et al. (2004). A novel Galphaq/11-selective inhibitor. *J Biol Chem, 279,* 47438–47445.

207. Goekjian, P. G., & Jirousek, M. R. (1999). Protein kinase C in the treatment of disease: Signal transduction pathways, inhibitors, and agents in development. *Curr Med Chem, 6,* 877–903.

208. Narumiya, S., Ishizaki, T., & Uehata, M. (2000). Use and properties of ROCK-specific inhibitor Y-27632. *Methods Enzymol, 325,* 273–284.

209. White, J. G., Krumwiede, M., & Escolar, G. (1999). EDTA induced changes in platelet structure and function: Influence on particle uptake. *Platelets, 10,* 327–337.

210. Brass, L. F., Vassallo, R. R., Jr., Belmonte, E., et al. (1992). Structure and function of the human platelet thrombin receptor: Studies using monoclonal antibodies against a defined epitope within the receptor N-terminus. *J Biol Chem, 267,* 13795–13798.

211. Ishihara, H., Connolly, A. J., Zeng, D., et al. (1997). Protease-activated receptor 3 is a second thrombin receptor in humans. *Nature, 386,* 502–508.

212. Faruqi, T. R., Weiss, E. J., Shapiro, M. J., et al. (2000). Structure-function analysis of protease-activated receptor 4 tethered ligand peptides — Determinants of specificity and utility in assays of receptor function. *J Biol Chem, 275,* 19728–19734.

213. Mills, D. C. B. (1996). ADP receptors on platelets. *Thromb Haemost, 76,* 835–856.

214. Gachet, C., Hechler, B., Léon, C., et al. (1997). Activation of ADP receptors and platelet function. *Thromb Haemost, 78,* 271–275.

215. Hirata, M., Hayashi, Y., Ushikubi, F., et al. (1991). Cloning and expression of cDNA for a human thromboxane A2 receptor. *Nature, 349,* 617–620.

216. Hanasaki, K., & Arita, H. (1988). Characterization of thromboxane A2/prostaglandin H2 (TXA2/PGH2) receptors of rat platelets and their interaction with TXA2/PGH2 receptor antagonists. *Biochem Pharmacol, 37,* 3923–3929.

217. Furci, L., Fitzgerald, D. J., & FitzGerald, G. A. (1991). Heterogeneity of prostaglandin H2/thromboxane A2 receptors: Distinct subtypes mediate vascular smooth muscle contraction and platelet aggregation. *J Pharmacol Exp Ther, 258,* 74–81.

218. Kobilka, B. K., Matsui, H., Kobilka, T. S., et al. (1987). Cloning, sequencing, and expression of the gene coding for the human platelet alpha2-adrenergic receptor. *Science, 238,* 650–656.

219. Inaba, K., Umeda, Y., Yamane, Y., et al. (1988). Characterization of human platelet vasopressin receptor and the relation between vasopressin-induced platelet aggregation and vasopressin binding to platelets. *Clin Endocrinol (Oxf), 29,* 377–386.

220. Siess, W., Stifel, M., Binder, H., et al. (1986). Activation of V1-receptors by vasopressin stimulates inositol phospholipid hydrolysis and arachidonate metabolism in human platelets. *Biochem J, 233,* 83–91.

221. Bichet, D. G., Arthus, M.-F., Barjon, J. N., et al. (1987). Human platelet fraction arginine-vasopressin: Potential physiological role. *J Clin Invest, 79,* 881–887.

222. Vittet, D., Cantau, B., Mathieu, M.-N., et al. (1988). Properties of vasopressin-activated human platelet high affinity GTPase. *Biochem Biophys Res Commun, 154,* 213–218.

223. Honda, Z., Nakamura, M., Miki, I., et al. (1991). Cloning by functional expression of platelet-activating factor receptor from guinea-pig lung. *Nature, 349,* 342–346.

224. Chao, W., & Olson, M. S. (1993). Platelet-activating factor: Receptors and signal transduction. *Biochem J, 292,* 617–629.

225. Kowalska, M. A., Ratajczak, J., Hoxie, J., et al. (1999). Megakaryocyte precursors, megakaryocytes and platelets express the HIV co-receptor CXCR4 on their surface: Determination of response to stromal-derived factor-1 by megakaryocytes and platelets. *Br J Haematol, 104,* 220–229.

226. Kowalska, M. A., Ratajczak, M. Z., Majka, M., et al. (2000). Stromal cell-derived factor-1 and macrophage-derived chemokine: two chemokines that activate platelets. *Blood, 96,* 50–57.

227. Vane, J. R., & Botting, R. M. (1995). Pharmacodynamic profile of prostacyclin. *Am J Cardiol, 75,* 3A–10A.

228. Gagnon, A. W., Manning, D. R., Catani, L., et al. (1991). Identification of Gzalpha as a pertussis toxin-insensitive G protein in human platelets and megakaryocytes. *Blood, 78,* 1247–1253.

229. Lounsbury, K. M., Casey, P. J., Brass, L. F. & Manning, D. R., Phosphorylation of GZ in human platelets: Selectivity and site of modification. *J Biol Chem, 266,* 22051–22056.

230. Carlson, K., Brass, L. F., & Manning, D. R. (1989). Thrombin and phorbol esters cause the selective phosphorylation of a G protein other than Gi in human platelets. *J Biol Chem, 264,* 13298–13305.

231. Brass, L. F., Hoxie, J. A., & Manning, D. R. (1993). Signaling through G proteins and G protein-coupled receptors during platelet activation. *Thromb Haemost, 70,* 217–223.

232. Shenker, A., Goldsmith, P., Unson, C. G., et al. (1991). The G protein coupled to the thomboxane A2 receptor in human platelets is a member of the novel Gq family. *J Biol Chem, 266,* 9309–9313.

233. Van Willigen, G., Donath, J., Lapetina, E. G., et al. (1995). Identification of alpha-subunits of trimeric GTP-binding proteins in human platelets by RT-PCR. *Biochem Biophys Res Commun, 214,* 254–262.

234. Giesberts, A. N., Van Ginneken, M., Gorter, G., et al. (1997). Subcellular localization of alpha-subunits of trimeric G-proteins in human platelets. *Biochem Biophys Res Commun, 234,* 439–444.

235. Tenailleau, S., Corre, J., & Hermouet, S. (1997). Specific expression of heterotrimeric G proteins G12 and G16 during human myeloid differentiation. *Exp Hematol, 25,* 927–934.

236. Aragay, A. M., & Quick, M. W. (1999). Functional regulation of Galpha16 by protein kinase C. *J Biol Chem, 274,* 4807–4815.

237. Offermanns, S., Hu, Y. H., & Simon, M. I. (1996). Galpha12 and Galpha13 are phosphorylated during platelet activation. *J Biol Chem, 271,* 26044–26048.

238. Kelleher, K. L., Matthaei, K. I., & Hendry, I. A. (2001). Targeted disruption of the mouse GZ-alpha gene: A role for GZ in platelet function? *Thromb Haemost, 85,* 529–532.

239. Davignon, I., Catalina, M. D., Smith, D., et al. (2000). Normal hematopoiesis and inflammatory responses despite discrete signaling defects in Galpha15 knockout mice. *Mol Cell Biol, 20,* 797–804.

240. Offermanns, S., Mancino, V., Revel, J.-P., et al. (1997). Vascular system defects and impaired cell chemokinesis as a result of Galpha13 deficiency. *Science, 275,* 533–536.

241. Darrow, A. L., Fung-Leung, W. P., Ye, R. D., et al. (1996). Biological consequences of thrombin receptor deficiency in mice. *Thromb Haemost, 76,* 860–866.

242. Fabre, J.-E., Nguyen, M., Latour, A., et al. (1999). Decreased platelet aggregation, increased bleeding time and resistance to thromboembolism in P2Y1-deficient mice. *Nature Med, 5,* 1199–1202.

# CHAPTER 17

# Outside-In Signaling by Integrin αIIbβ3

**Nicolas Prévost and Sanford J. Shattil**

*Department of Medicine, University of California San Diego, La Jolla, California*

## I. Introduction

The formation of a thrombus is dependent, in part, on molecular events that lead to the activation of platelets. Platelet activation is a prerequisite for most of the adhesive responses of this cell, and it is initiated by platelet contact with components of the extracellular matrix, such as collagen and von Willebrand factor (VWF), and by exposure to excitatory agonists, such as adenosine diphosphate (ADP) and thrombin. Integrin αIIbβ3 (Chapter 8), also known as the glycoprotein (GP) IIb-IIIa complex, is a major target of these initial activation events through a process known as "inside-out signaling." The conformational changes associated with the activation of integrin αIIbβ3 enable its high-affinity interaction with cognate ligands, particularly soluble fibrinogen and VWF, leading to the spreading and aggregation of platelets. In addition to being the primary mediator of platelet aggregation, αIIbβ3 functions as a signaling receptor capable of triggering a number of molecular and cellular events essential to or permissive of "post-ligand-binding" events that include cytoskeletal reorganization, formation and stabilization of large platelet aggregates, secretion, development of a procoagulant surface, and clot retraction. Signals emanating from ligand-occupied αIIbβ3 are known as "outside-in" signals. This chapter will focus on molecular events responsible for outside-in αIIbβ3 signaling, with one message being that this is a work-in-progress in the field of platelet biology.

## II. Structural Basis for Integrin αIIbβ3-Dependent Signaling

A platelet contains an average of 80,000 copies of αIIbβ3 on its surface, and there are additional pools of αIIbβ3 in α granule and open-canalicular membranes.[1,2] αIIbβ3 is a heterodimer composed of two noncovalently bound subunits: αIIb (1008 amino acid residues) and β3 (762 amino acid residues). Both subunits are type I transmembrane proteins with relatively short cytoplasmic tails (20 residues for

αIIb and 47 for β3). Activation of αIIbβ3, either by inside-out signals or ligand binding, is associated with conformational changes in the extracellular domain that lead to a switch between a "closed" low-affinity ligand-binding conformation and an "open" high-affinity conformation. These conformational changes are linked to a spatial separation of the transmembrane and cytoplasmic domains of αIIb from those of β3.[3,4] (See Chapter 8 for further details.)

### A. Platelet Responses Associated with Outside-In Signaling

Ligand binding to αIIbβ3 also triggers integrin oligomerization, or clustering.[5] This appears to be required for the assembly of a nascent signaling complex proximal to the cytoplasmic tails of αIIbβ3, and the growth of a larger actin-based signaling complex. Platelet aggregation, the central mechanism through which thrombi form, can be described in two phases: a primary reversible aggregation dependent on inside-out signals, and a secondary irreversible aggregation dependent on outside-in signals. Platelet adhesion to fibrinogen or VWF under static conditions offers an experimental setting in which to study outside-in αIIbβ3 signaling. Adhesion of platelets to their substrate, in the absence of added agonists, triggers morphological changes ranging from filopodial and lamellipodial extension to full cell spreading.[6-8] These changes reflect the ability of αIIbβ3 to anchor signaling complexes capable of regulating actin dynamics.[6,9-11] αIIbβ3-expressing CHO cells share similar adhesive and dynamic responses when layered on a fibrinogen matrix, making them a useful model system in which to study these processes.

### B. Changes in αIIbβ3 That May Promote Outside-In Signaling

**1. Separation of the α and β Cytoplasmic Tails.** There is increasing evidence that activation of integrin αIIbβ3 by

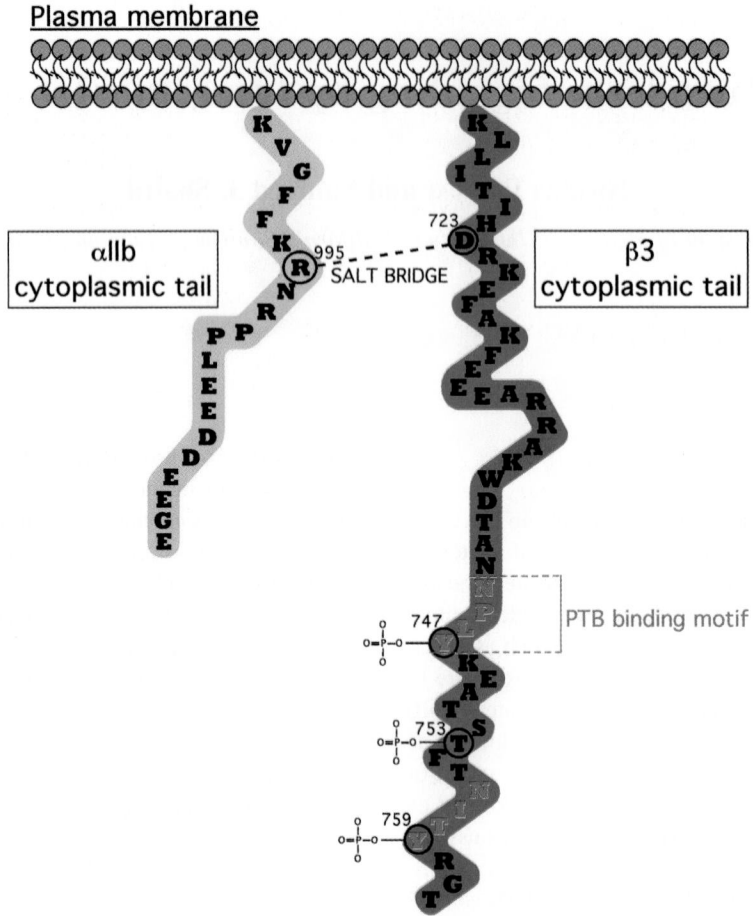

**Figure 17-1.** Regulatory motifs of the αIIb and β3 cytoplasmic tails. A salt bridge between residues Arg[995] on αIIb and Asp[723] on β3 may act as a structural constraint maintaining the integrin in its inactive state. The β3 tail bears an NPXY PTB binding motif that interacts with the PTB (FERM) domain of talin and possibly other proteins. The three known sites of phosphorylation of the β3 tail: Tyr[747], Tyr[759], and Thr[753] are indicated.

inside-out signals leads to dissociation of the αIIb and β3 transmembrane domains from each other.[12,13] In the low-affinity or unliganded state, the transmembrane domains of both subunits (αIIb Ile 966-Trp 988, β3 Ile 693-Trp 715) adopt the conformation of two interacting α-helices.[14,15] The association of the two subunits may be further reinforced by a salt bridge between the membrane-proximal residues αIIb Arg 995 and β3 Asp 723[16] (Fig. 17-1). Upon αIIbβ3 activation, a lateral separation of the transmembrane helices occurs, thus contributing to the regulation of the conformation and ligand affinity of the extracellular domain of the integrin.[17] The dissociation of the transmembrane and cytoplasmic domains of the α and β chains are promoted by inside-out signaling responses that culminate in talin recruitment to the β3 tail[18–21] and fibrinogen, VWF, or fibronectin binding.[13,17] Although conjectural, subunit separation might also facilitate tail interactions with specific intracellular proteins involved in outside-in signaling. For example, in CHO cells, disruption of the putative salt bridge between

αIIb Arg 995 and β3 Asp 723 promotes constitutive intracellular signaling downstream of αIIbβ3 as evidenced by tyrosine phosphorylation of focal adhesion kinase (FAK).[16]

**2. Integrin Clustering.** Another context in which dissociation of the α and β tails of integrin αIIbβ3 might be relevant to outside-in signaling is integrin clustering. Ligand binding to αIIbβ3 causes the integrin to cluster into oligomers in detergent solution,[22] in platelets,[23,24] and in CHO cells.[5,17] In vitro studies by Li et al. show that αIIb and β3 transmembrane domains, when combined in an artificial lipid environment, in the absence of any extracellular structural constraints, form homo-oligomers, suggesting that integrin clustering might entail a preliminary dissociation of the transmembrane α and β subunits.[25] While αIIbβ3 clustering may affect integrin affinity only minimally,[5,17,26] it appears necessary for many, if not all, outside-in signaling reactions downstream of αIIbβ3. For example, clustering of αIIbβ3 by the lectin concanavalin-A on the surface of platelets induces rapid tyrosine phosphorylation of Syk and phospholipase Cγ2 (PLCγ2) in an aggregation-independent manner.[27] These responses are dependent on αIIbβ3, since they are not observed with Glanzmann thrombasthenia platelets devoid of αIIbβ3. Furthermore, conditional clustering of αIIbβ3 in CHO cells using a chemical dimerizer strategy is sufficient to trigger activation of the protein tyrosine kinases, c-Src and Syk.[26,28]

## III. Effectors of Outside-In αIIbβ3 Signaling

### A. c-Src and Protein Phosphatases

c-Src is constitutively associated with β3 integrins in platelets,[10,29] an interaction that requires the SH3 domain of c-Src and the three carboxy-terminal residues $R^{760}GT^{762}$ of β3.[30] In unstimulated cells, the bulk of c-Src is believed to be maintained in an autoinhibited state by intramolecular interactions between the SH3 domain and the region linking the SH2 domain to the kinase domain, and between the SH2 domain and the carboxy-terminal phospho-$Tyr^{529}$.[31,32] Activation of c-Src is associated with unlatching of this assembled state, for example, by dephosphorylation of $Tyr^{529}$, and with phosphorylation of a tyrosine residue ($Tyr^{418}$) located within the activation loop of the kinase domain.[33,34] The monitoring of $Tyr^{529}$ dephosphorylation and $Tyr^{418}$ phosphorylation of c-Src has revealed that platelet adhesion to fibrinogen stimulates the activation of αIIbβ3-associated c-Src,[10] an observation that could be reproduced by conditional clustering of αIIbβ3 in CHO cells.[28]

In resting platelets, αIIbβ3 also associates with Csk, the protein kinase responsible for the phosphorylation of c-Src $Tyr^{529}$.[10,35] Fibrinogen binding to αIIbβ3 promotes the recruitment of protein tyrosine phosphatase 1B (PTP-1B) to the αIIbβ3/c-Src/Csk complex, leading to dissociation of Csk from the complex, dephosphorylation of c-Src $Tyr^{529}$, and c-Src activation[10,36] (Fig. 17-2). The role of PTP-1B in c-Src activation downstream of αIIbβ3 was further demonstrated by the use of platelets isolated from PTP-1B null mice. Fibrinogen binding to αIIbβ3 on the surface of PTP-1B[−/−] platelets elicited only minimal dissociation of Csk from the αIIbβ3/c-Src complex and little c-Src activation. In addition, PTP-1B[−/−] platelets exhibited impaired spreading on a fibrinogen matrix, and defective platelet thrombus formation in cremasteric arterioles exposed to laser injury.[36]

Protein phosphatases other than PTP-1B may contribute to integrin αIIbβ3 outside-in signaling. Protein phosphatase 1 (PP1) has been shown to interact with the membrane-proximal region of αIIb in resting platelets. αIIbβ3 engagement by fibrinogen and platelet aggregation triggered PP1 dissociation from αIIbβ3, PP1 activation, and phosphorylation of myosin light chain, a PP1 substrate[37] (Fig. 17-3).

### B. Syk

Syk has been identified as being one of the protein kinases activated downstream of αIIbβ3, both in platelets and CHO cells.[26,38–40] Studies performed in Syk[−/−] mice have shown that Syk is involved in αIIbβ3 outside-in but not inside-out signaling. The platelets from these mice undergo normal αIIbβ3 activation and primary aggregation[41] but are incapable of spreading on immobilized fibrinogen.[10] Fibrinogen binding to αIIbβ3 induces Syk association with the β3 tail (Fig. 17-3) and Syk tyrosine phosphorylation and activation. The latter two events are dependent on both c-Src activity and Syk autophosphorylation in response to integrin clustering.[10,39]

Syk participates in the phosphorylation of a number of substrates downstream of αIIbβ3 that are potentially relevant to cytoskeletal reorganization, including cortactin, Cbl, Vav1, and SLP-76[10,40,42] (Fig. 17-2) (see also Chapter 4). The adapter protein SLP-76 is of special interest to αIIbβ3-mediated functions, for the platelets of SLP-76 null mice spread poorly on fibrinogen and exhibit reduced phosphotyrosine responses downstream of αIIbβ3.[43] The molecular mechanisms through which Syk and SLP-76 contribute to integrin outside-in signaling may likely involve the SLP-76-associated proteins, Vav1 and Nck.[44] Vav1 is a guanine nucleotide exchange factor for Rac1 and Cdc42, and Nck is an adapter that recruits the Rac effector, Pak1, to plasma membrane adhesion sites.[45]

### C. Calcium and Integrin Binding Protein (CIB)

CIB was initially isolated through the yeast two-hybrid screening of a fetal liver cDNA library using the cytoplasmic tail of αIIb as bait. Further analysis revealed that CIB binds to αIIb, but not to αV, α2, α5, β1, or β3.[46] A minimal CIB binding motif has been characterized on αIIb, a 15-residue sequence that encompasses the membrane-proximal region of the cytoplasmic tail and part of the transmembrane domain.[47] This sequence includes $G^{991}FFKR^{995}$, a motif essential to the maintenance of αIIbβ3 in a low-affinity state.[16,48] The interaction of CIB with αIIb may[49,50] or may not[51] be influenced by binding of $Ca^{2+}$ to CIB. A preliminary report on the effects of CIB overexpression and knockdown in primary murine megakaryocytes suggests that one function of CIB is to negatively regulate agonist-induced activation of αIIbβ3.[52]

Platelet adhesion to fibrinogen induces CIB localization to filopodia, followed by redistribution to the periphery of spread platelets. Interestingly, activated Rac3, but not Rac1 and Rac2, interacts with CIB *in vitro* and in transfected CHO cells, and this interaction appears to promote cell spreading on fibrinogen[53] (Fig. 17-4). Furthermore, introduction of an αIIb tail peptide into platelets not only

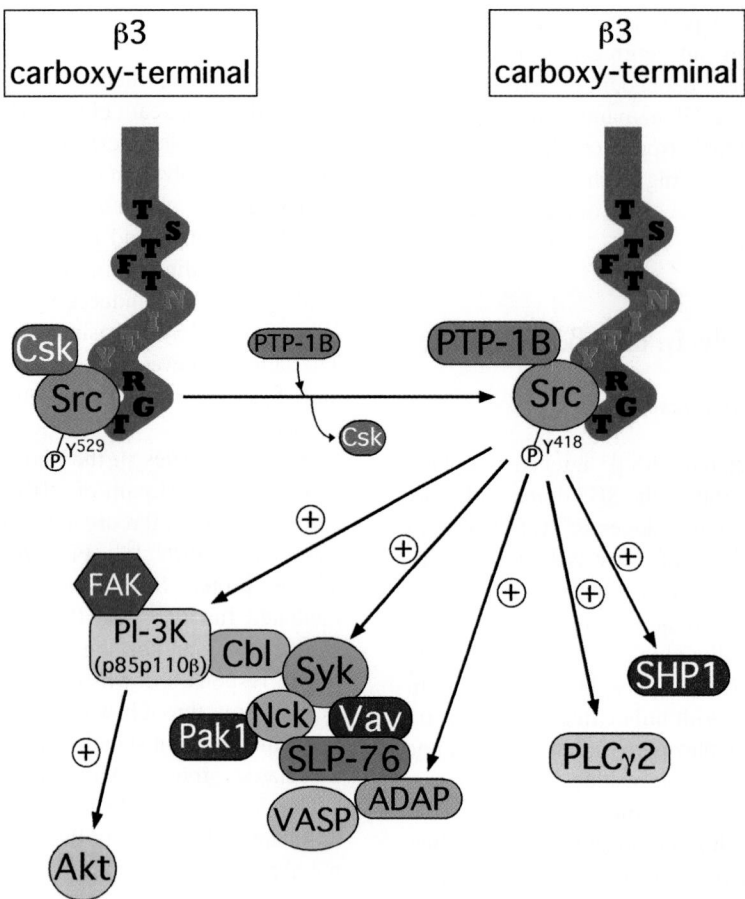

**Figure 17-2.** Early events associated with outside-in αIIbβ3 signaling. Activation of β3-bound c-Src requires the recruitment of PTP-1B and the displacement of Csk. c-Src activation initiates secondary signaling events through phosphorylation of multiple substrates, leading to actin reorganization.

prevents the CIB/αIIb interaction, but abolishes platelet spreading on fibrinogen.[54] CIB may therefore be involved in bidirectional αIIbβ3 signaling, although more work with platelets is needed to establish this with certainty.

### D. Tec Tyrosine Kinases

Human platelets express at least two Tec tyrosine kinases: Tec and Btk. Both kinases become activated and associate with the cytoskeleton during platelet aggregation. The tyrosine phosphorylation and activation of Tec and Btk require integrin αIIbβ3 engagement.[11,55–57] Activated Btk associates with the actin cytoskeleton upon platelet activation, a process regulated by αIIbβ3 and dependent on actin polymerization.[58] Nonetheless, human platelets deficient in Btk undergo normal shape change,[59] and Btk−/−/Tec−/− double knockout mouse platelets are reported to spread normally on fibrinogen.[60] However, spreading of Btk−/− mouse platelets may be reduced under certain conditions, suggesting some role in

coupling outside-in αIIbβ3 signals to the cytoskeleton (Soriani, A., Moran, B., de Virgilio, M., Kawakami, T., Altman, A., Lowell, C., Eto, K., & Shattil, S. J., unpublished observations).

### E. Protein Kinase C (PKC)

The PKC family of serine/threonine kinases acts as a regulator of integrin function in many cell types[61] and as a modulator of αIIbβ3 affinity in platelets.[62] Platelets express several PKC isoforms.[63–66] Whereas PKCα has been implicated in inside-out signaling,[67] PKCβ is necessary for outside-in signaling, as indicated by defective spreading of PKCβ−/− mouse platelets on fibrinogen.[66] Indeed, activated PKCβ inducibly associates with αIIbβ3 in response to fibrinogen binding to platelets, and this interaction appears to be mediated by the adapter, RACK1 (Fig. 17-3).[66] Further work is required to identify the relevant substrates of PKCβ during outside-in signaling and deter-

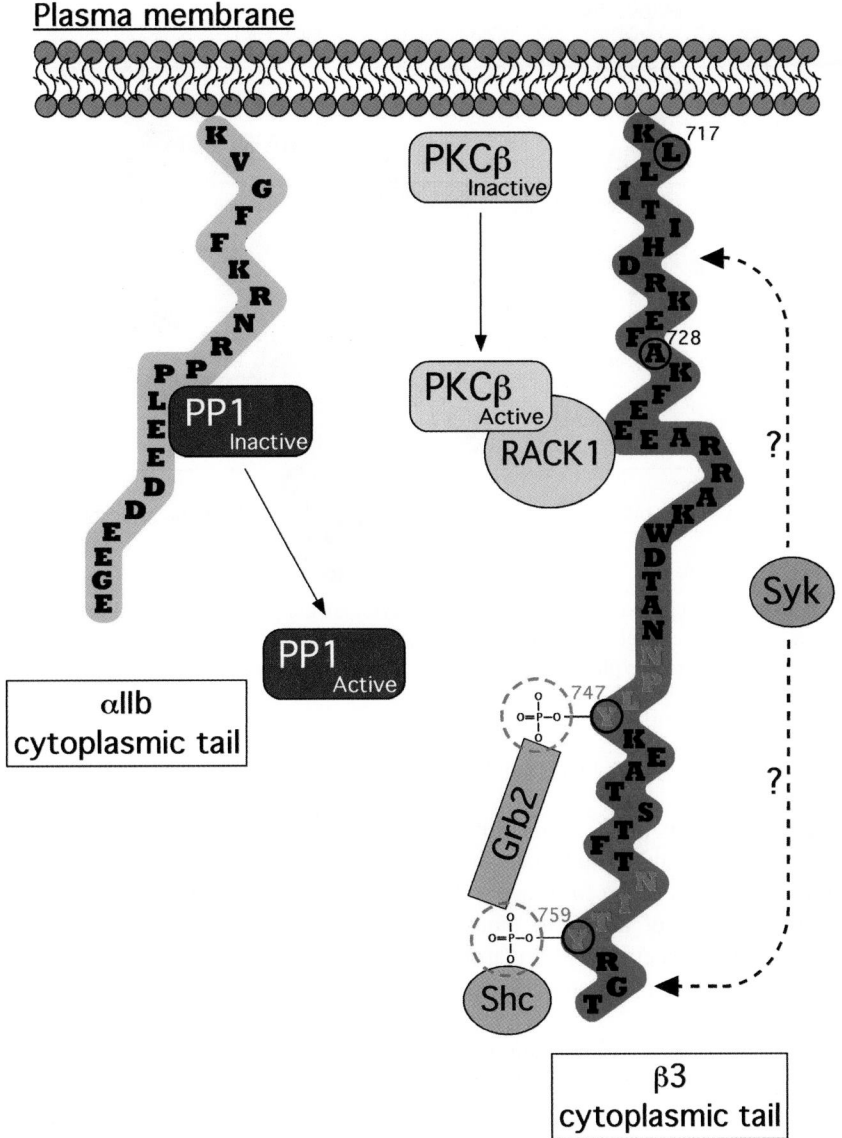

**Figure 17-3.** Examples of inducible or reversible protein interactions with the cytoplasmic tails of αIIbβ3. See text for discussion.

mine the potential roles of other PKC isoforms in this process.

### F. Integrin-Linked Kinase (ILK)

ILK is a serine-threonine protein kinase that directly interacts with the cytoplasmic tails of β1 and β3 integrins[68] and contributes to cell adhesion and spreading of adherent cells.[69–71] The effects of ILK may be mediated through its kinase function and adapter functions.[71] ILK coimmunoprecipitates with β3 and is activated upon agonist-induced platelet activation, but ascertainment of its precise role in platelet function requires further study.[72,73]

### G. Rho GTPases

Rho-GTPases regulate the formation of actin cytoskeletal structures such as filopodia, lamellipodia, stress fibers, focal complexes, and focal adhesions.[74] The pharmacological inactivation of RhoA in platelets has no detectable effect on the inside-out activation of αIIbβ3, platelet adhesion, or cell spreading, but it inhibits the formation of stress fibers and vinculin-rich patches.[7] In platelet aggregation studies, pretreatment with the αIIbβ3-blocking antibody, 7E3, resulted in inhibition of RhoA activation, demonstrating a relationship between integrin engagement and RhoA activation.[75] Activation of RhoA by αIIbβ3 is Src-dependent[76] and appears to be required for the stabilization of platelet

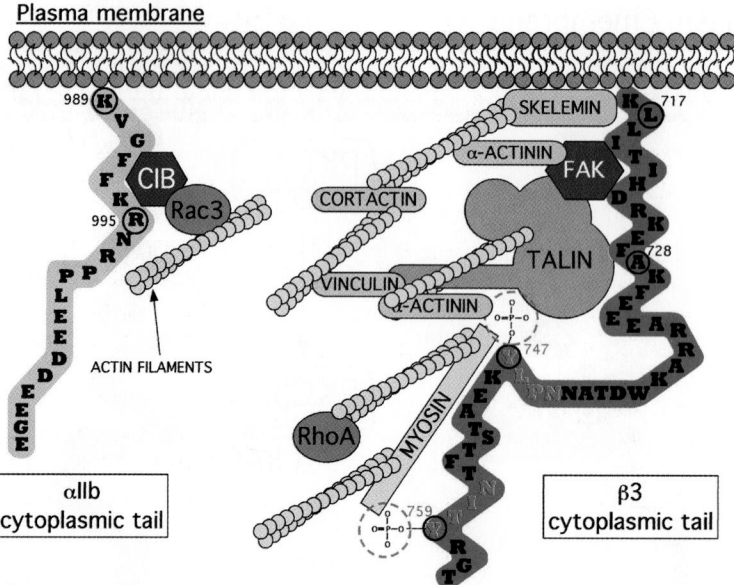

**Figure 17-4.** Assembly of actin-based cytoskeletal complexes to the cytoplasmic tails of αIIbβ3. Illustrated are proteins whose recruitment to actin-based complexes may be mediated by the membrane proximal region of β3 (e.g., skelemin) or modulated positively (myosin and α-actinin) or negatively (talin) by tyrosine phosphorylation of β3. CIB recruitment is mediated by residues Lys[989]-Arg[995] on αIIb.

adhesion and aggregation under high shear, but not static, conditions.[75]

Platelets express three Rac isoforms (1–3).[77] Lamellipodia assembly was shown to be regulated by Rac1 in a study addressing the effect of a dominant-negative form of Rac1 in permeabilized platelets.[78] Studies of knockout mouse platelets confirm a role for Rac1 in lamellipodia formation and in the formation of stable platelet aggregates under flow.[77]

Translocation of Cdc42 to the actin cytoskeleton of aggregated platelets requires αIIbβ3 activation and actin polymerization, suggesting that Cdc42 is regulated through αIIbβ3 outside-in signaling.[79] Based on work in CHO cells, one proposed trigger for the activation of Cdc42 is the release into the cytoplasm of the adapter protein 14-3-3ζ from its membrane-sequestering anchor GPIbα.[80] Whether this mechanism pertains to platelets is speculative.

### H. Actin-Binding Proteins

Early studies looking at the composition of detergent-insoluble, F-actin-rich fractions isolated from fibrinogen-bound platelets established that αIIbβ3 cosediments with a number of cytoskeletal proteins, including talin, spectrin, and vinculin.[81] There is evidence that three other actin-binding proteins: myosin, skelemin, and α-actinin — interact directly with the β3 tail (Figure 17-4). The interaction of myosin with β3 *in vitro* requires the phosphorylation of β3 tail Tyr[747] and Tyr[759],[82] and at least the phosphorylation of Tyr[747] in

platelets.[83] The C2 motifs 3-7 of skelemin were shown to interact directly with the membrane-proximal region K[716]LLITIHDRK[725] of β3 *in vitro*.[84]

The interaction of α-actinin with αIIbβ3 was initially demonstrated in a phospholipid vesicle environment.[85] Platelet adhesion to fibrinogen is associated with an increase in the phosphotyrosine content of α-actinin, a response that is dependent on αIIbβ3 engagement, cytosolic $Ca^{2+}$, and PKC activation.[86,87] α-actinin is a substrate for FAK, and phosphorylation of Tyr[12] by FAK is required for the optimal association of α-actinin with actin filaments.[88] The association of α-actinin with the platelet actin cytoskeleton also is dependent on inactivation of the protein tyrosine phosphatase SHP-1.[89]

Cortactin is a cortical actin-associated protein that becomes tyrosine-phosphorylated during agonist-induced platelet aggregation.[90] Tyrosine phosphorylation of cortactin by c-Src and/or Syk may modulate its ability, in concert with WASP/SCAR proteins, to nucleate branching actin filaments via the Arp2/3 complex[91] (see also Chapter 4).

## IV. Role of Phosphorylation of the αIIb and β3 Cytoplasmic Tails

Phosphorylation of the β3 integrin cytoplasmic tail occurs during agonist-induced platelet activation and aggregation. Whereas threonine/serine residues are believed to account for most of the phosphorylation sites in the β3 tail,[92,93] tyrosine phosphorylation has been shown to play a role in outside-in αIIbβ3 signaling. Of the three β3 phosphorylation sites that have been mapped, two are tyrosines — Tyr[747] and Tyr[759] [82] — and one is a threonine — Thr 753[94,95] (Fig. 17-1).

### A. Serine/Threonine Phosphorylation

The integrin β3 tail is serine- and threonine-phosphorylated in both resting and activated platelets. Treatment of platelets with thrombin or phorbol 12-myristate 13-acetate (PMA) causes an increase in serine/threonine phosphorylation,[96] a response that does not require fibrinogen binding.[93] Dephosphorylation of at least one threonine residue (Thr 753) might play a role in outside-in signaling. Thus, treatment of platelets with the protein seryl/threonyl phosphatase inhibitor calyculin A not only promotes phosphorylation of Thr 753, but also inhibits platelet adhesion and spreading on fibrinogen.[94] Any effects of Thr 753 phosphorylation on outside-in signaling might be explained by the presence of unfavorable negative charges within the tyrosine spacing motif Y[747]KEATSTFTNITY[759] of β3 (Fig. 17-1).

### B. Tyrosine Phosphorylation

Phosphorylation of residues Tyr[747] and Tyr[759] requires both fibrinogen binding to αIIbβ3 and platelet aggregation or spreading.[82,97] Another potential ligand for αIIbβ3 is CD40L, a transmembrane protein and member of the tumor necrosis factor (TNF) family expressed on the surface of activated platelets.[98] The soluble form of CD40L (sCD40L) is shed from the platelet surface upon platelet activation.[99] sCD40L binds specifically to αIIbβ3 and is capable of promoting platelet spreading when immobilized on a surface.[100] Treatment of platelets with sCD40L promotes the phosphorylation of Tyr[759] on β3.[101] It has been proposed that β3 Tyr[747] and Tyr[759] are substrates of the Fyn tyrosine kinase,[102,103] which like c-Src can bind directly to β3.[28]

"diYF" knock-in mice have been developed in which β3 tyrosine residues 747 and 759 have been mutated to phenylalanine. Their platelets are selectively impaired in outside-in αIIbβ3 signaling, resulting in decreased size and stability of ADP-induced platelet aggregates and decreased clot retraction[104] The latter may be explained by a loss of myosin anchorage to β3.[82,83] In addition, Tyr[745] is located within motifs recognized by the phosphotyrosine binding domains (PTB) of several signaling proteins.[105,106] For example, Shc binding to β3 tail residues requires the phosphorylation of Tyr[759] *in vitro* and *in vivo* (as demonstrated in mice expressing Y747, 759F β3), and this interaction promotes the adhesion-dependent tyrosine phosphorylation of Shc.[97,107] Tyrosine phosphorylation of the β3 Tyr[747] and Tyr[759] may also lead to recruitment of the adapter, Grb2, to the integrin in aggregated platelets.[97,103]

## V. Contribution of Phospholipids to Outside-In αIIbβ3 Signaling

Binding of fibrinogen to αIIbβ3 leads to the activation of two enzymes associated with phospholipid metabolism: phosphatidylinositol 3-kinase (PI 3-kinase) and PLCγ2. Platelets express several PI 3-kinase isoforms,[108–110] although p110β isoform appears to be the predominant one responsible for mediating αIIbβ3-induced PI 3-kinase responses in platelets.[111] Fibrinogen binding to αIIbβ3 activates PI 3-kinase, as demonstrated by production of PI(3,4)P2 and activation of the PI 3-kinase effector, Akt.[112] Pharmacological inhibition of PI 3-kinase is associated with reduced αIIbβ3-mediated lamellipodia formation, platelet spreading, and Ca[2+] mobilization.[113,114] Ligand binding to αIIbβ3 leads to the association of PI 3-kinase with several other αIIbβ3 effectors, such as FAK, Syk, and Cbl (Fig. 17-2), and the interaction with FAK is associated with increased FAK auto-phosphorylation.[115] PI 3-kinase might be involved in the recruitment of ILK to αIIbβ3 in platelets since PI 3-kinase inhibition abrogates the coimmunoprecipitation of these two proteins.[73]

PLCγ2 null mice have provided a model in which to study this isoform in αIIbβ3 signaling.[11] αIIbβ3 binding to fibrinogen promoted cytoplasmic Ca[2+] oscillations and lamellipodia formation independently of ADP and thromboxane A2 liberation in wild-type platelets, but these responses were abolished in PLCγ2[−/−] platelets. Tyrosine phosphorylation and activation of PLCγ2 downstream of αIIbβ3 requires Src activity.[11,116] Similar observations have been made in the context of platelet adhesion to fibronectin or shear-induced platelet adhesion to VWF.[116,117]

## References

1. Wagner, C. L., Mascelli, M. A., Neblock, D. S., et al. (1996). Analysis of GP IIb/IIIa receptor number by quantification of 7E3 binding to human platelets. *Blood, 88,* 907–914.
2. Woods, V. L., Wolff, L. E., & Keller, D. M. (1986). Resting platelets contain a substantial centrally located pool of glycoprotein IIb-IIIa complex which may be accessible to some but not other extracellular proteins. *J Biol Chem, 261,* 15242–15251.
3. Springer, T. A., & Wang, J.-H. (2004). The three-dimensional structure of integrins and their ligands, and conformational regulation of cell adhesion. *Adv Protein Chem, 68,* 29–63.
4. Vinogradova, O., Vaynberg, J., Kong, X., et al. (2004). Membrane-mediated structural transitions at the cytoplasmic face during integrin activation. *Proc Natl Acad Sci USA 101,* 4094–4099.
5. Buensuceso, C., De Virgilio, M., & Shattil, S. J. (2003). Detection of integrin αIIbβ3 clustering in living cells. *J Biol Chem, 278,* 15217–15224.

6. Hartwig, J. H., Kung, S., Kovacsovics, T., et al. (1996). D3 phosphoinositides and outside-in integrin signaling by glycoprotein IIb-IIIa mediate platelet actin assembly and filopodial extension induced by phorbol 12-myristate 13-acetate. *J Biol Chem, 271,* 32986–32993.

7. Leng, L., Kashiwagi, H., Ren, X.-D., et al. (1998). RhoA and the function of platelet integrin αIIbβ3. *Blood, 91,* 4206–4215.

8. Yuan, Y. P., Kulkarni, S., Ulsemer, P., et al. (1999). The von Willebrand factor-glycoprotein Ib/V/IX interaction induces actin polymerization and cytoskeletal reorganization in rolling platelets and glycoprotein Ib/V/IX-transfected cells. *J Biol Chem, 274,* 36241–36251.

9. Bearer, E. L., Prakash, J. M., & Li, Z. (2002). Actin dynamics in platelets. *Int Rev Cytol, 217,* 137–182.

10. Obergfell, A., Eto, K., Mocsai, A., et al. (2002). Coordinate interactions of Csk, Src, and Syk kinases with αIIbβ3 initiate integrin signaling to the cytoskeleton. *J Cell Biol, 157,* 265–275.

11. Wonerow, P., Pearce, A. C., Vaux, D. J., et al. (2003). A critical role for phospholipase Cγ2 in αIIbβ3-mediated platelet spreading. *J Biol Chem, 278,* 37520–37529.

12. Vinogradova, O., Velyvis, A., Velyviene, A., et al. (2002). A structural mechanism of integrin αIIbβ3 "inside-out" activation as regulated by its cytoplasmic face. *Cell, 110,* 587–597.

13. Luo, B. H., Springer, T. A., & Takagi, J. (2004). A specific interface between integrin transmembrane helices and affinity for ligand. *PLoS Biol, 2,* 776–786.

14. Takagi, J., Petre, B., Walz, T., et al. (2002). Global conformational rearrangements in integrin extracellular domains in outside-in and inside-out signaling. *Cell, 110,* 599–611.

15. Partridge, A. W., Liu, S., Kim, S., et al. (2005). Transmembrane domain helix packing stabilizes integrin αIIbβ3 in the low affinity state. *J Biol Chem, 280,* 7294–7300.

16. Hughes, P. E., Diaz-Gonzalez, F., Leong, L., et al. (1996). Breaking the integrin hinge — A defined structural constraint regulates integrin signaling. *J Biol Chem, 271,* 6571–6574.

17. Luo, B. H., Carman, C. V., Takagi, J., et al. (2005). Disrupting integrin transmembrane domain heterodimerization increases ligand binding affinity, not valency or clustering. *Proc Natl Acad Sci USA, 102,* 3679–3684.

18. Calderwood, D. A., Zent, R., Grant, R., et al. (1999). The talin head domain binds to integrin β subunit cytoplasmic tails and regulates integrin activation. *J Biol Chem, 274,* 28071–28074.

19. Patil, S., Jedsadayanmata, A., Wencel-Drake, J. D., et al. (1999). Identification of a talin-binding site in the integrin β3 subunit distinct from the NPLY regulatory motif of post-ligand binding functions. The talin n-terminal head domain interacts with the membrane-proximal region of the β3 cytoplasmic tail. *J Biol Chem, 274,* 28575–28583.

20. Calderwood, D. A., Yan, B., de Pereda, J. M., et al. (2002). The phosphotyrosine binding-like domain of talin activates integrins. *J Biol Chem, 277,* 21749–21758.

21. Tadokoro, S., Shattil, S. J., Eto, K., et al. (2003). Talin binding to integrin β cytoplasmic tails: A final common step in integrin activation. *Science, 302,* 103–106.

22. Hantgan, R. R., Lyles, D. S., Mallett, T. C., et al. (2003). Ligand binding promotes the entropy-driven oligomerization of integrin αIIbβ3. *J Biol Chem, 278,* 3417–3426.

23. Fox, J., Shattil, S. J., Kinlough-Rathbone, R., et al. (1996). The platelet cytoskeleton stabilizes the interaction between αIIbβ3 and its ligand and induces selective movements of ligand-occupied integrin. *J Biol Chem, 271,* 7004–7011.

24. Simmons, S. R., Sims, P. A., & Albrecht, R. M. (1997). αIIbβ3 redistribution triggered by receptor cross-linking. *Arterioscler Thromb Vasc Biol, 17,* 3311–3320.

25. Li, R., Babu, C. R., Lear, J. D., et al. (2001). Oligomerization of the integrin αIIbβ3: Roles of the transmembrane and cytoplasmic domains. *Proc Natl Acad Sci USA, 98,* 12462–12467.

26. Hato, T., Pampori, N., & Shattil, S. J. (1998). Complementary roles for receptor clustering and conformational change in the adhesive and signaling functions of integrin αIIbβ3. *J Cell Biol, 141,* 1685–1695.

27. Torti, M., Festetics, E. T., Bertoni, A., et al. (1999). Clustering of integrin αIIbβ3 differently regulates tyrosine phosphorylation of pp72[syk], PLCγ2 and pp125FAK in Concanavalin A-stimulated platelets. *Thromb Haemost, 81,* 124–130.

28. Arias-Salgado, E. G., Lizano, S., Sarker, S., et al. (2003). Src kinase activation by a novel and direct interaction with the integrin β cytoplasmic domain. *Proc Natl Acad Sci USA, 100,* 13298–13302.

29. Dorahy, D. J., Berndt, M. C., & Burns, G. F. (1995). Capture by chemical crosslinkers provides evidence that integrin αIIbβ3 forms a complex with protein tyrosine kinases in intact platelets. *Biochem J, 309,* 481–490.

30. Arias-Salgado, E. G., Lizano, S., Shattil, S. J., et al. (2005). Specification of the direction of adhesive signaling by the integrin β cytoplasmic domain. *J Biol Chem, 280,* 29699–29707.

31. Sicheri, F., Moarefi, I., & Kuriyan, J. (1997). Crystal structure of the Src family tyrosine kinase Hck. *Nature, 385,* 602–609.

32. Young, M. A., Gonfloni, S., Superti-Furga, G., et al. (2001). Dynamic coupling between the SH2 and SH3 domains of c-Src and Hck underlies their inactivation by C-terminal tyrosine phosphorylation. *Cell, 105,* 115–126.

33. Xu, W. Q., Harrison, S. C., & Eck, M. J. (1997). Three-dimensional structure of the tyrosine kinase c-Src. *Nature, 385,* 595–602.

34. Xu, W., Doshi, A., Lei, M., et al. (1999). Crystal structures of c-Src reveal features of its autoinhibitory mechanism. *Mol Cell, 3,* 629–638.

35. Okada, M., Nada, S., Yamanashi, Y., et al. (1991). CSK: A protein-tyrosine kinase involved in regulation of src family kinases. *J Biol Chem, 266,* 24249–24252.

36. Arias-Salgado, E. G., Haj, F., Dubois, C., et al. (2005). PTP-1B is an essential positive regulator of platelet integrin signaling. *J Cell Biol, 170,* 837–845.

37. Vijayan, K. V., Liu, Y., Li, T. T., et al. (2004). Protein phosphatase 1 associates with the integrin αIIb subunit and regulates signaling. *J Biol Chem, 279,* 33039–33042.

38. Clark, E. A., Shattil, S. J., Ginsberg, M. H., et al. (1994). Regulation of the protein tyrosine kinase, pp72[syk], by platelet

agonists and the integrin, αIIbβ3. *J Biol Chem, 46,* 28859–28864.

39. Gao, J., Zoller, K., Ginsberg, M. H., et al. (1997). Regulation of the pp72Syk protein tyrosine kinase by platelet integrin αIIbβ3. *EMBO J, 16,* 6414–6425.

40. Miranti, C., Leng, L., Maschberger, P., et al. (1998). Integrin-induced assembly of a Syk- and Vav1-regulated signaling pathway independent of actin polymerization. *Curr Biol, 8,* 1289–1299.

41. Law, D. A., Nannizzi-Alaimo, L., Ministri, K., et al. (1999). Genetic and pharmacological analyses of Syk function in αIIbβ3 signaling in platelets. *Blood, 93,* 2645–2652.

42. Saci, A., Rendu, F., & Bachelot-Loza, C. (2000). Platelet αIIbβ3 integrin engagement induces the tyrosine phosphorylation of Cbl and its association with phosphoinositide 3-kinase and Syk. *Biochem J, 351,* 669–676.

43. Judd, B. A., Myung, P. S., Leng, L., et al. (2000). Hematopoietic reconstitution of SLP-76 corrects hemostasis and platelet signaling through αIIbβ3 and collagen receptors. *Proc Natl Acad Sci USA, 97,* 12056–12061.

44. Fang, N., Motto, D. G., Ross, S. E., et al. (1996). Tyrosines 113, 128, and 145 of SLP-76 are required for optimal augmentation of NFAT promoter activity. *J Immunol, 157,* 3769–3773.

45. Bagrodia, S., & Cerione, R. A. (1999). Pak to the future. *Trends Cell Biol, 9,* 350–355.

46. Naik, U. P., Patel, P. M., & Parise, L. V. (1997). Identification of a novel calcium binding protein that interacts with the integrin αIIb cytoplasmic domain. *J Biol Chem, 272,* 4651–4654.

47. Barry, W. T., Boudignon-Proudhon, C., Shock, D. D., et al. (2002). Molecular basis of CIB binding to the integrin αIIb cytoplasmic domain. *J Biol Chem, 277,* 28877–28883.

48. O'Toole, T. E., Mandelman, D., Forsyth, J., et al. (1991). Modulation of the affinity of integrin αIIbβ3 (GPIIb-IIIa) by the cytoplasmic domain of αIIb. *Science, 254,* 845–847.

49. Shock, D. D., Naik, U. P., Brittain, J. E., et al. (1999). Calcium-dependent properties of CIB binding to the integrin αIIb cytoplasmic domain and translocation to the platelet cytoskeleton. *Biochem J, 342,* 729–735.

50. Tsuboi, S. (2002). Calcium integrin-binding protein activates platelet integrin αIIbβ3. *J Biol Chem, 277,* 1919–1923.

51. Vallar, L., Melchior, C., Plançon, S., et al. (1999). Divalent cations differentially regulate integrin αIIb cytoplasmic tail binding to β3 and to calcium- and integrin-binding protein. *J Biol Chem, 274,* 17257–17266.

52. Yuan, W., McFadden, A. W., Wang, Z., et al. (2003). CIB is an endogenous inhibitor of inside-out integrin αIIbβ3 activation in primary murine megakaryocytes. *Blood, 102,* 160a.

53. Haataja, L., Kaartinen, V., Groffen, J., et al. (2002). The small GTPase Rac3 interacts with the integrin-binding protein CIB and promotes integrin αIIbβ3-mediated adhesion and spreading. *J Biol Chem, 277,* 8321–8328.

54. Naik, U. P., & Naik, M. U. (2003). Association of CIB with GPIIb/IIIa during outside-in signaling is required for platelet spreading on fibrinogen. *Blood, 102,* 1355–1362.

55. Hamazaki, Y., Kojima, H., Mano, H., et al. (1998). Tec is involved in G protein-coupled receptor- and integrin-mediated signalings in human blood platelets. *Oncogene, 16,* 2773–2779.

56. Laffargue, M., Monnereau, L., Tuech, J., et al. (1997). Integrin-dependent tyrosine phoshorylation and cytoskeletal translocation of Tec in thrombin-activated platelets. *Biochem Biophys Res Commun, 238,* 247–251.

57. Laffargue, M., Ragab-Thomas, J. M., Ragab, A., et al. (1999). Phosphoinositide 3-kinase and integrin signalling are involved in activation of Bruton tyrosine kinase in thrombin-stimulated platelets. *FEBS Lett, 443,* 66–70.

58. Mukhopadhyay, S., Ramars, A. S. S., & Dash, D. (2001). Bruton's tyrosine kinase associates with the actin-based cytoskeleton in activated platelets. *J Cell Biochem, 81,* 659–665.

59. Bauer, M., Maschberger, P., Quek, L., et al. (2001). Genetic and pharmacological analyses of involvement of Src-family, Syk and Btk tyrosine kinases in platelet shape change. Src-kinases mediate integrin αIIbβ3 inside-out signalling during shape change. *Thromb Haemost, 85,* 331–340.

60. Atkinson, B. T., Ellmeier, W., & Watson, S. P. (2003). Tec regulates platelet activation by GPVI in the absence of Btk. *Blood, 102,* 3592–3599.

61. Ivaska, J., Kermorgant, S., Whelan, R., et al. (2003). Integrin–protein kinase C relationships. *Biochem Soc Trans, 31,* 90–93.

62. Shattil, S. J., & Newman, P. J. (2004). Integrins: Dynamic scaffolds for adhesion and signaling in platelets. *Blood, 104,* 1606–1615.

63. Grabarek, J., Raychowdhury, M., Ravid, K., et al. (1992). Identification and functional characterization of protein kinase C isozymes in platelets and HEL cells. *J Biol Chem, 267,* 10011–10017.

64. Baldassare, J. J., Henderson, P. A., Burns, D., et al. (1992). Translocation of protein kinase C isozymes in thrombin-stimulated human platelets. Correlation with 1,2-diacylglycerol levels. *J Biol Chem, 267,* 15585–15590.

65. Crosby, D., & Poole, A. W. (2003). Physical and functional interaction between protein kinase C delta and Fyn tyrosine kinase in human platelets. *J Biol Chem, 278,* 24533–24541.

66. Buensuceso, C. S., Obergfell, A., Soriani, A., et al. (2005). Regulation of outside-in signaling in platelets by integrin-associated protein kinase Cβ. *J Biol Chem, 280,* 644–653.

67. Tabuchi, A., Yoshioka, A., Higashi, T., et al. (2003). Direct demonstration of involvement of protein kinase Cα in the $Ca^{2+}$-induced platelet aggregation. *J Biol Chem, 278,* 26374–26379.

68. Hannigan, G. E., Leung-Hagesteijn, C., Fitz-Gibbon, L., et al. (1996). Regulation of cell adhesion and anchorage-dependent growth by a new β1-integrin-linked protein kinase. *Nature, 379,* 91–96.

69. Yamaji, S., Suzuki, A., Sugiyama, Y., et al. (2001). A novel integrin-linked kinase-binding protein, affixin, is involved in the early stage of cell-substrate interaction. *J Cell Biol, 153,* 1251–1264.

70. Nikolopoulos, S. N., & Turner, C. E. (2002). Molecular dissection of actopaxin-integrin-linked kinase-Paxillin interactions and their role in subcellular localization. *J Biol Chem, 277,* 1568–1575.

71. Wu, C. (2004). The PINCH-ILK-parvin complexes: Assembly, functions and regulation. *Biochim Biophys Acta, 1692,* 55–62.

72. Yamaji, S., Suzuki, A., Kanamori, H., et al. (2002). Possible role of ILK-affixin complex in integrin-cytoskeleton linkage during platelet aggregation. *Biochem Biophys Res Commun, 297,* 1324–1331.

73. Pasquet, J. M., Noury, M., & Nurden, A. T. (2002). Evidence that the platelet integrin αIIbβ3 is regulated by the integrin-linked kinase, ILK, in a PI3-kinase dependent pathway. *Thromb Haemost, 88,* 115–122.

74. Hall, A. (2005). Rho GTPases and the control of cell behaviour. *Biochem Soc Trans, 33,* 891–895.

75. Schoenwaelder, S. M., Hughan, S. C., Boniface, K., et al. (2002). RhoA sustains integrin αIIbβ3 adhesion contacts under high shear. *J Biol Chem, 277,* 14738–14746.

76. Salsmann, A., Schaffner-Reckinger, E., Kabile, F., et al. (2005). A new functional role of the fibrinogen RGD motif as the molecular switch that selectively triggers integrin αIIbβ3-dependent RhoA activation during cell spreading. *J Biol Chem, 280,* 33610–33619.

77. McCarty, O. J. T., Larson, M. K., Auger, J. M., et al. (2005). Rac1 is essential for platelet lamellipodia formation and aggregate stability under flow. *J Biol Chem, 280,* 39474–39484. doi:10.1074/jbc.M504672200.

78. Hartwig, J. H., Bokoch, G. M., Carpenter, C. L., et al. (1995). Thrombin receptor ligation and activated rac uncap actin filament barbed ends through phosphoinositide synthesis in permeabilized human platelets. *Cell, 82,* 643–653.

79. Dash, D., Aepfelbacher, M., & Siess, W. (1995). Integrin αIIbβ3-mediated translocation of CDC42Hs to the cytoskeleton in stimulated human platelets. *J Biol Chem, 270,* 17321–17326.

80. Bialkowska, K., Zaffran, Y., Meyer, S. C., et al. (2003). 14-3-3 zeta mediates integrin-induced activation of Cdc42 and Rac. Platelet glycoprotein Ib-IX regulates integrin-induced signaling by sequestering 14-3-3 zeta. *J Biol Chem, 278,* 33342–33350.

81. Fox, J. E. B. (1993). The platelet cytoskeleton. *Thromb Haemost, 70,* 884–893.

82. Jenkins, A. L., Nannizzi-Alaimo, L., Silver, D., et al. (1998). Tyrosine phosphorylation of the β3 cytoplasmic domain mediates integrin–cytoskeletal interactions. *J Biol Chem, 273,* 13878–13885.

83. Prevost, N., Woulfe, D. S., Jiang, H., et al. (2005). Eph kinases and ephrins support thrombus growth and stability by regulating integrin outside-in signaling in platelets. *Proc Natl Acad Sci USA, 102,* 9820–9825.

84. Reddy, K. B., Bialkowska, K., & Fox, J. E. B. (2001). Dynamic modulation of cytoskeletal proteins linking integrins to signaling complexes in spreading cells — Role of skelemin in initial integrin-induced spreading. *J Biol Chem, 276,* 28300–28308.

85. Otey, C. A., Pavalko, F. M., & Burridge, K. (1990). An interaction between α-actinin and the β1 integrin subunit in vitro. *J Cell Biol, 111,* 721–729.

86. Haimovich, B., Kaneshiki, M., & Ji, P. (1996). Protein kinase C regulates tyrosine phosphorylation of pp125FAK in platelets adherent to fibrinogen. *Blood, 87,* 152–161.

87. Izaguirre, G., Aguirre, L., Ji, P., et al. (1999). Tyrosine phosphorylation of α-actinin in activated platelets. *J Biol Chem, 274,* 37012–37020.

88. Izaguirre, G., Aguirre, L., Hu, Y. P., et al. (2001). The cytoskeletal/non-muscle isoform of α-actinin is phosphorylated on its actin-binding domain by the focal adhesion kinase. *J Biol Chem, 276,* 28676–28685.

89. Lin, S. Y., Raval, S., Zhang, Z., et al. (2004). The protein-tyrosine phosphatase SHP-1 regulates the phosphorylation of α-actinin. *J Biol Chem, 279,* 25755–25764.

90. Rosa, J. P., Artcanuthurry, V., Grelac, F., et al. (1997). Reassessment of protein tyrosine phosphorylation in thrombasthenic platelets: Evidence that phosphorylation of cortactin and a 64-kD protein is dependent on thrombin activation and integrin αIIbβ3. *Blood, 89,* 4385–4392.

91. Daly, R. J. (2004). Cortactin signalling and dynamic actin networks. *Biochem J, 382,* 13–25.

92. Van Willigen, G., Hers, I., Gorter, G., et al. (1996). Exposure of ligand-binding sites on platelet integrin αIIbβ3 by phosphorylation of the β3 subunit. *Biochem J, 314,* 769–779.

93. Hillery, C. A., Smyth, S. S., & Parise, L. V. (1990). Stoichiometry of glycoprotein IIIa phosphorylation in whole platelets and in vitro. *Blood, 76,* 459a.

94. Lerea, K. M., Cordero, K. P., Sakariassen, K. S., et al. (1999). Phosphorylation sites in the integrin β3 cytoplasmic domain in intact platelets. *J Biol Chem, 274,* 1914–1919.

95. Kirk, R. I., Sanderson, M. R., & Lerea, K. M. (2000). Threonine phosphorylation of the β3 integrin cytoplasmic tail, at a site recognized by PDK1 and Akt/PKB in vitro, regulates Shc binding. *J Biol Chem, 275,* 30901–30906.

96. Parise, L. V., Criss, A. B., Nannizzi, L., et al. (1990). Glycoprotein IIIa is phosphorylated in intact human platelets. *Blood, 75,* 2363–2368.

97. Law, D. A., Nannizzi-Alaimo, L., & Phillips, D. R. (1996). Outside-in signal transduction: αIIbβ3 (GP IIb-IIIa) tyrosine phosphorylation induced by platelet aggregation. *J Biol Chem, 271,* 10811–10815.

98. Henn, V., Slupsky, J. R., Gräfe, M., et al. (1998). CD40 ligand on activated platelets triggers an inflammatory reaction of endothelial cells. *Nature, 391,* 591–594.

99. Henn, V., Steinbach, S., Büchner, K., et al. (2001). The inflammatory action of CD40 ligand (CD154) expressed on activated human platelets is temporally limited by coexpressed CD40. *Blood, 98,* 1047–1054.

100. Andre, P., Prasad, K. S., Denis, C. V., et al. (2002). CD40L stabilizes arterial thrombi by a β3 integrin-dependent mechanism. *Nature Med, 8,* 247–252.

101. Prasad, K. S., Andre, P., He, M., et al. (2003). Soluble CD40 ligand induces β3 integrin tyrosine phosphorylation and triggers platelet activation by outside-in signaling. *Proc Natl Acad Sci USA, 100,* 12367–12371.

102. Phillips, D. R., Nannizzi-Alamio, L., & Prasad, K. S. S. (2001). β3 tyrosine phosphorylation in αIIbβ3 (platelet membrane GP IIb-IIIa) outside-in integrin signaling. *Thromb Haemost, 86,* 246–258.

103. Phillips, D. R., Prasad, K. S., Manganello, J., et al. (2001). Integrin tyrosine phosphorylation in platelet signaling. *Curr Opin Cell Biol, 13,* 546–554.

104. Law, D. A., DeGuzman, F. R., Heiser, P., et al. (1999). Integrin cytoplasmic tyrosine motif is required for outside-in αIIbβ3 signalling and platelet function. *Nature, 401,* 808–811.

105. Van der Geer, P., & Pawson, T. (1995). The PTB domain: A new protein module implicated in signal transduction. *Trends Biochem Sci, 20,* 277–280.

106. Calderwood, D. A., Fujioka, Y., de Pereda, J. M., et al. (2003). Integrin β cytoplasmic domain interactions with phosphotyrosine-binding domains: A structural prototype for diversity in integrin signaling. *Proc Natl Acad Sci USA, 100,* 2272–2277.

107. Cowan, K. J., Law, D. A., & Phillips, D. R. (2000). Identification of Shc as the primary protein binding to the tyrosine-phosphorylated β3 subunit of αIIbβ3 during outside-in integrin platelet signaling. *J Biol Chem, 275,* 36423–36429.

108. Selheim, F., Holmsen, H., & Vassbotn, F. S. (2000). PI 3-kinase signalling in platelets: The significance of synergistic, autocrine stimulation. *Platelets, 11,* 69–82.

109. Zhang, J., Banfic, H., Straforini, F., et al. (1998). A type II phosphoinositide 3-kinase is stimulated via activated integrin in platelets. A source of phosphatidylinositol 3-phosphate. *J Biol Chem, 273,* 14081–14084.

110. Rittenhouse, S. E. (1996). Phosphoinositide 3-kinase activation and platelet function. *Blood, 88,* 4401–4414.

111. Jackson, S. P., Schoenwaelder, S. M., Goncalves, I., et al. (2005). PI 3-kinase p110β: A new target for antithrombotic therapy. *Nature Med, 11,* 507–514.

112. Banfic, H., Tang, X. W., Batty, I. H., et al. (1998). A novel integrin-activated pathway forms PKB/Akt-stimulatory phosphatidylinositol 3,4-bisphosphate via phosphatidylinositol 3-phosphate in platelets. *J Biol Chem, 273,* 13–16.

113. Nesbitt, W. S., Kulkarni, S., Giuliano, S., et al. (2002). Distinct glycoprotein Ib/V/IX and integrin αIIbβ3-dependent calcium signals cooperatively regulate platelet adhesion under flow. *J Biol Chem, 277,* 2965–2972.

114. Goncalves, I., Nesbitt, W. S., Yuan, Y., et al. (2005). Importance of temporal flow gradients and integrin αIIββ3 mechanotransduction for shear activation of platelets. *J Biol Chem, 280,* 15430–15437.

115. Guinebault, C., Payrastre, B., Racaud-Sultan, C., et al. (1995). Integrin-dependent translocation of phosphoinositide 3-kinase to the cytoskeleton of thrombin-activated platelets involves specific interactions of p85-α with actin filaments and focal adhesion kinase. *J Cell Biol, 129,* 831–842.

116. Goncalves, I., Hughan, S. C., Schoenwaelder, S. M., et al. (2003). Integrin αIIbβ3-dependent calcium signals regulate platelet–fibrinogen interactions under flow. Involvement of phospholipase Cγ2. *J Biol Chem, 278,* 34812–34822.

117. McCarty, O. J., Zhao, Y., Andrew, N., et al. (2004). Evaluation of the role of platelet integrins in fibronectin-dependent spreading and adhesion. *J Thromb Haemost, 2,* 1823–1833.

# CHAPTER 18

# Platelet Thrombus Formation in Flowing Blood

**Brian Savage and Zaverio M. Ruggeri**

*Department of Molecular and Experimental Medicine, The Scripps Research Institute,*
*La Jolla, California*

## I. Introduction

The inner lining of normal blood vessels is composed of endothelial cells that form a surface resistant to the adhesion of circulating platelets. In areas where the endothelium is altered, however, or at sites of vascular damage where subendothelial or other extracellular matrices are exposed, firm platelet attachment rapidly occurs. This is a critical initial step in hemostasis and thrombosis, as well as in inflammatory and immunopathogenic responses,[1] requiring the concerted interaction of matrix proteins with platelet receptors ultimately leading to platelet activation and aggregation. The hemostatic response to vascular injury is contingent on the extent of damage, the specific matrix proteins exposed, and flow conditions. In this chapter, we focus first on the mechanisms responsible for the initial interaction of platelets with thrombogenic surfaces associated with vascular and tissue trauma. We then discuss the secondary events that, pursuant to the initial adhesion of platelets, lead to thrombus propagation and stabilization by mechanisms that mostly depend on the formation of interplatelet contacts.

Platelets are essential for normal hemostasis and their function is particularly relevant to arrest bleeding from the microarteriolar circulation, where shear stress is elevated.[2] In pathological conditions, platelets are a major contributor to arterial thrombosis, which typically occurs at sites of atherosclerosis with stenosis of the vessel lumen. Such vascular lesions result in shear stress values markedly higher than in the normal circulation.[3–5] For these reasons, the mechanisms that support platelet adhesion and aggregation under the constraints of elevated shear stress are of particular interest in the study of thrombus formation under flow.

## II. Initial Events in Thrombus Formation under Flow

The complex events that regulate platelet function are influenced by the flow of blood. The velocity of blood near the wall is lower than at the center of the vessel, and this difference creates a shearing effect between adjacent layers of fluid (Fig. 18-1).[6] Thus, shear is the consequence of the relative parallel motion between adjacent fluid laminae and is greatest at the luminal surface but decreases progressively toward the center of the vessel. The shear rate, a difference in flow velocity as a function of distance from the wall, is expressed in centimeters per second per centimeter or the equivalent inverse seconds ($s^{-1}$). Fluid shear stress is force per unit area and is expressed in Pascals (Pa), equivalent to one Newton per square meter ($N/m^2$), or in dynes per square centimeter ($1\,Pa = 10\,dynes/cm^2$). The shear rate is directly proportional to the shear stress and inversely proportional to the fluid viscosity. In blood, therefore, in which viscosity (force multiplied by time) is approximately $0.004\,Pa\cdot s$ (or $0.04\,dynes/cm^2 \cdot s$), a shear stress of $1\,dyne/cm^2$ corresponds to a shear rate of $25\,s^{-1}$. The highest wall shear rate in the normal circulation occurs in small arterioles of 10 to 50 μm in diameter, in which levels have been estimated to vary between 500 and $5000\,s^{-1}$.[2] Values up to 10 times higher have been calculated to occur at the tip of severe stenosis in atherosclerotic coronary arteries.[3–5]

### A. Platelet Adhesion to von Willebrand Factor (VWF)

The extracellular matrix components that react with platelets include different types of collagen, von Willebrand factor (VWF), fibronectin,[7] and other adhesive proteins such as laminin,[8] fibulin,[9] and thrombospondin.[10] Fibrinogen/fibrin[11] and vitronectin[12] are not synthesized by vascular wall cells, but must be considered as potentially relevant thrombogenic substrates because they become immobilized onto extracellular matrix at sites of injury. One can assume that all tissue components capable of interacting with platelets can contribute to the initiation of thrombus formation when exposed to blood, even though only a few may have essential roles. For example, exposure of whole blood to surfaces coated with vitronectin (or fibronectin) failed to

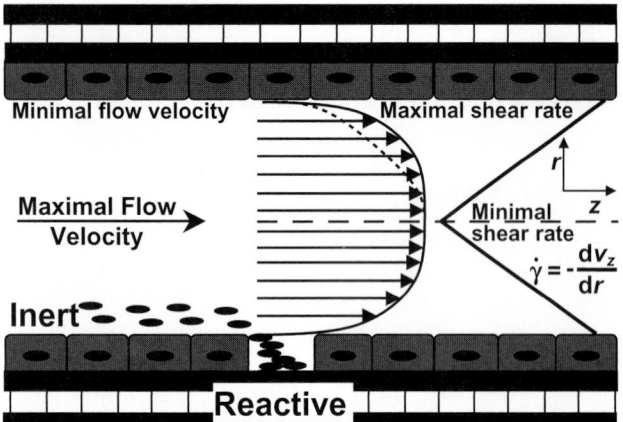

**Figure 18-1.** Schematic representation of blood flow in a vessel. Normal endothelial cells are nonreactive for platelets, but exposed subendothelial structures induce rapid platelet adhesion and aggregation. Blood flow in a cylindrical vessel can be visualized as a series of fluid layers (laminae) moving at different velocities. The laminae near the center of the vessel have greater velocities than those near the wall (depicted by arrows of different lengths). The corresponding velocity profile (solid line) is more blunted than the parabolic profile expected with a homogeneous suspension (dotted line) because of cell depletion in the boundary layer near the wall. The shear rate is the rate of change of velocity with respect to distance measured perpendicularly to the direction of flow. The negative sign indicates that the gradient is defined from the center (where velocity is maximal) to the wall (where velocity is minimal). (Adapted from Ruggeri and Saldivar,[6] with permission.)

result in adhesion under conditions that prompted rapid attachment to immobilized fibrinogen.[13] Experiments with purified proteins, although useful to establish specific mechanisms of action, may not reflect functions within the complex supramolecular assembly of extracellular matrices. The relative contribution of different adhesive interactions to the process of platelet adhesion to vascular surfaces, therefore, remains to be elucidated in detail.

Among potential substrates in the vessel wall, VWF through its A1 domain is uniquely required to mediate platelet adhesion under conditions of elevated shear stress,[14] a function effected by binding to glycoprotein (GP) Ibα in the platelet GPIb-IX-V receptor complex (see Chapter 7).[15] The constitutive subendothelial location of VWF[16] allows direct support of platelet adhesion when exposed to flowing blood.[17–19] Type VI collagen filaments have affinity for VWF,[20,21] and the two molecules are associated in the extracellular matrix.[22] Notwithstanding the presence of endogenous VWF, the binding of plasma VWF to exposed perivascular tissue is a key initial step in response to vascular injury.[17] The binding of soluble VWF to nonactivated platelets is tightly regulated to prevent aggregation in the circulation, but VWF immobilized onto a surface is highly reactive toward flowing platelets. This may be because cir-

culating VWF multimers are in a coiled conformation that mostly shields the A1 domain from interacting with platelets, whereas the molecule assumes an extended shape when bound to a substrate under the effect of shear stress.[23] Possible vascular wall binding sites for plasma VWF include collagen types I and III in deeper layers of the vessel wall, and microfibrillar collagen type VI in the subendothelial matrix.[22,24–26] The A3 domain of VWF is required for binding to human collagen types I and III,[27,28] whereas domain A1 may be a collagen type VI binding site.[25] Moreover, VWF multimers have a tendency to self-associate, such that immobilized VWF becomes the site for continuing recruitment of circulating multimers that enhance considerably the adhesive potential in areas of vascular lesion.[28] A relevant contribution in this regard may come from very large multimers locally released by stimulated endothelial cells, as these VWF molecules form high-strength bonds with GPIbα.[29]

Elucidation of the mechanisms involved in the initiation of thrombus formation has been facilitated by perfusion experiments that simulate relevant *in vivo* conditions. Real-time epifluorescence and confocal videomicroscopy have been used to examine how blood platelets interact with immobilized VWF and other adhesive substrates under flow conditions similar to those encountered in different blood vessels.[11,14] Thus, it has been shown that platelet adhesion to VWF is a two-stage dynamic process (Fig. 18-2), in which the initial tethering of platelets mediated by GPIbα is characterized by a succession of transient interactions resulting in constant motion, but with a substantially reduced velocity compared with that of free-flowing platelets adjacent to the surface.[11] VWF-mediated tethering is efficient even at shear rates in excess of $20,000\,s^{-1}$, which explains the essential role in support of the hemostatic function of platelets to prevent hemorrhage from wounded microarterioles, in which high shear rates are prevalent.[2] This property of VWF depends on the fact that bonds between the A1 domain and GPIbα form rapidly, even though their limited lifetime prevents stable adhesion. While kept in close proximity to the surface and in slow motion, platelets can form stabilizing bonds that would not occur directly in rapidly flowing blood, and become activated. These additional bonds that lead to stable adhesion may involve other domains of VWF, specifically the Arg–Gly–Asp sequence to which activated integrin αIIbβ3 (see Chapter 8) binds (Fig. 18-2), or other substrates associated with immobilized VWF (e.g., collagen), to which different specific platelet receptors bind.

## B. VWF-Initiated Signaling through GPIb-IX-V

VWF binding to GPIbα supports platelet tethering to a thrombogenic substrate, but the extent to which it contributes to subsequent activation is less well defined. In fact,

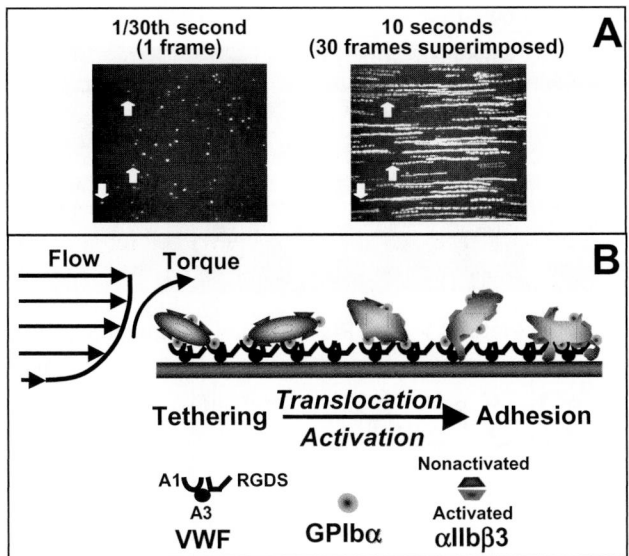

**Figure 18-2.** A. Time lapse analysis of platelet movement on immobilized VWF. Blood containing 40 μM D-phenylalanyl-L-prolyl-L-arginine chloromethyl ketone dihydrochloride to inhibit α-thrombin and prostaglandin $E_1$ to inhibit platelet activation was perfused through a parallel plate flow chamber at 37°C. The bottom surface of the chamber was coated with purified VWF, and the wall shear rate during perfusion was 1500 s$^{-1}$. Platelets were rendered fluorescent with mepacrine and visualized by epifluorescence videomicroscopy. Images were recorded in real time (30 frames per second). The left panel is a single frame showing a snapshot of individual platelets interacting with the VWF-coated surface. The right panel shows the superimposition of 30 consecutive frames at a sampling rate of three frames per second, corresponding to 10 s of observation. The occurrence of platelet translocation on VWF, mediated exclusively by GPIbα because platelet activation was blocked, is rendered by the streaking effect produced by moving platelets. (Adapted from Savage et al.,[11] with permission.) B. Schematic representation of the mechanism of platelet adhesion to immobilized VWF. The VWF A1–GPIbα bond forms rapidly and has high resistance to tensile stress, but a limited half-life. In the absence of other bonds, platelets detach at the tailing edge where tension is greatest and move forward with a rotational movement (rolling) resulting from the torque imposed by the flowing fluid. New VWF A1–GPIbα bonds form as different regions of the membrane of rolling platelets come close to the surface. Thus, platelets remain in contact with the substrate while translocating at low velocity. The initial tethering is followed by platelet activation and formation of stable bonds mediated by integrin αIIbβ3.

VWF in the vessel wall is associated with other potent platelet agonists, such as collagen (discussed later), and it is difficult to ascertain whether VWF-induced GPIbα signaling adds significantly to the overall activation response after initial adhesion. One problem is that, to date, it has not been possible to separate the adhesive and signaling functions of the receptor because the latter is dependent on the former,

which in turn is essential for the initial tethering that precedes any other response when shear rates are elevated. It seems reasonable to assume that if GPIbα signaling has a role in platelet thrombus formation, it may be limited to conditions of high shear stress[30] on surfaces with a high density of immobilized VWF.

Indirect evidence suggests that VWF-dependent signaling involves the cross-linking of GPIbα by multivalent VWF.[31] The observations that a GPIb-IX-V subpopulation is associated with two proteins — FcR γ-chain and FcγRI-IA[32] — that contain an immunoreceptor tyrosine-based activation motif (ITAM)[33,34] lends support to the cross-linking mechanism. FcγRIIa and GPIbα are within 10 nm of each other in the platelet membrane,[35] and both FcR γ-chain and FcγRIIA coimmunoprecipitate with the GPIb-IX-V complex.[32,35] On cross-linking, FcR γ-chain and FcγRIIA are phosphorylated within the cytoplasmic ITAM sequence by Src family tyrosine kinases,[36] thereby permitting the binding and autophosphorylation of the tyrosine kinase Syk. Activated Syk initiates signaling that leads to activation of phospholipase C-γ2 (PLC-γ2) and subsequent formation of inositol triphosphate (leading to $Ca^{2+}$ mobilization) and diacylglycerol (leading to activation of protein kinase C [PKC]).[32,34] There is also evidence that VWF A1 domain-induced signaling through GPIbα can activate αIIbβ3 independently of other receptors.[37] In particular, mouse platelets expressing human GPIbα but lacking the FcR γ-chain adhere to the dimeric VWF A1 domain and undergo $Ca^{2+}$ oscillations that are only slightly lower than normal, suggesting that the FcR γ-chain may contribute to maximal $Ca^{2+}$ elevations, but is not strictly required to initiate the response.

The cytoplasmic region of GPIbα is associated with actin binding protein and 14-3-3ζ,[38-40] which provide potential links to several intracellular signaling molecules including phosphatidylinositol (PI) 3-kinase, focal adhesion kinase, Src-related tyrosine kinases (Syk, Src, Fyn Lyn, and Yes), GTPase-activating protein, and tyrosine phosphatases (PTP-1B and SHPTP10).[15] 14-3-3ζ may regulate signaling through GPIb-IX-V, leading to activation of αIIbβ3.[41-43] Chimes hamster ovary (CHO) cells coexpressing GPIb-IX and αIIbβ3 were found to spread onto immobilized VWF and subsequently bind fibrinogen in an αIIbβ3 activation-dependent manner.[41] Both PI 3-kinase and PKC play important roles in this process, because their inhibition blocked cell spreading.[41] Moreover, deletion of the C-terminal 18 amino acids of GPIbα, where 14-3-3ζ binds, prevented cell spreading and fibrinogen binding to αIIbβ3. PI 3-kinase forms a constitutive complex with both GPIb-IX-V and 14-3-3ζ in resting platelets,[42] and may therefore play a key role in 14-3-3ζ signaling. Although dissociation of 14-3-3ζ from the GPIb-IX-V complex is seen after platelet activation,[42,43] it is unclear whether this is a direct effect of platelet activation or a process that precedes activation.

The engagement of GPIbα by surface-bound VWF elicits different kinds of transient cytoplasmic $Ca^{2+}$ elevations.[30,44] The first, designated a type α/β peak, is linked to the initial GPIbα-mediated platelet tethering to the VWF A1 domain and precedes stable adhesion (Fig. 18-3).[45] These spikes are independent of extracellular $Ca^{2+}$,[44] indicating that signals downstream of GPIbα may directly regulate $Ca^{2+}$ release from internal stores. Type α/β $Ca^{2+}$ peaks are partially inhibited by specific and direct blockage of the adenosine diphosphate (ADP) receptor $P2Y_1$, but not by apyrase, which degrades ADP enzymatically.[45] This indicates that GPIbα binding to the VWF A1 domain induces release of ADP that, in turn, acts rapidly (to the extent that it cannot be blocked by apyrase) through $P2Y_1$ to cause additional $Ca^{2+}$ release from intracellular stores, thereby reinforcing GPIbα-mediated α/β peaks.[45] Platelets that have established firm adhesion to VWF through αIIbβ3 exhibit sustained $Ca^{2+}$ oscillations, designated γ peaks, which precede platelet aggregation. Type γ peaks involve a transmembrane $Ca^{2+}$ flux operating downstream of αIIbβ3 signaling. Thus, the initial GPIbα interaction with the VWF A1 domain leads to a first level of αIIbβ3 activation sufficient for stable platelet adhesion to immobilized VWF, but not for binding soluble VWF or fibrinogen (required for aggregation). Progression to thrombus formation requires further αIIbβ3 activation that is contingent upon signal amplification associated with type γ $[Ca^{2+}]_i$ peaks (Fig. 18-3). PI 3-kinase inhibition or ADP removal by apyrase blocks γ peaks and prevents platelet aggregation,[30,46,47] and this ADP-dependent reinforcement of platelet activation is mediated by the $P2Y_{12}$ receptor.[45] Because only a small proportion of translocating platelets on immobilized VWF exhibit type γ $Ca^{2+}$ peaks,[30,45] it appears that the efficiency of this process is low, in agreement with the concept that immobilized VWF contributes to but is not sufficient for platelet activation after adhesion. Such a property may be important for a regulated initiation of thrombus formation to prevent the development of pathological thrombi.

## C. Platelet Adhesion to Collagen

Collagen of different types — but particularly I, III, and VI — is among the most thrombogenic component of the vessel wall responsible for the initiation of platelet adhesion. The influence of collagen structure on platelet adhesion and activation under flow is poorly understood. Images of surface-bound, insoluble type I collagen fibers and microfibrils obtained after pepsin solubilization reveal distinct features. Fibers display a characteristic banded pattern resulting from the regular staggering of collagen monomers with a 67-nm periodicity,[48] whereas pepsin-solubilized microfibrils lack this pattern and form spiraled structures (Fig. 18-4).[49] Spiral vascular collagen, formed by fibrils in

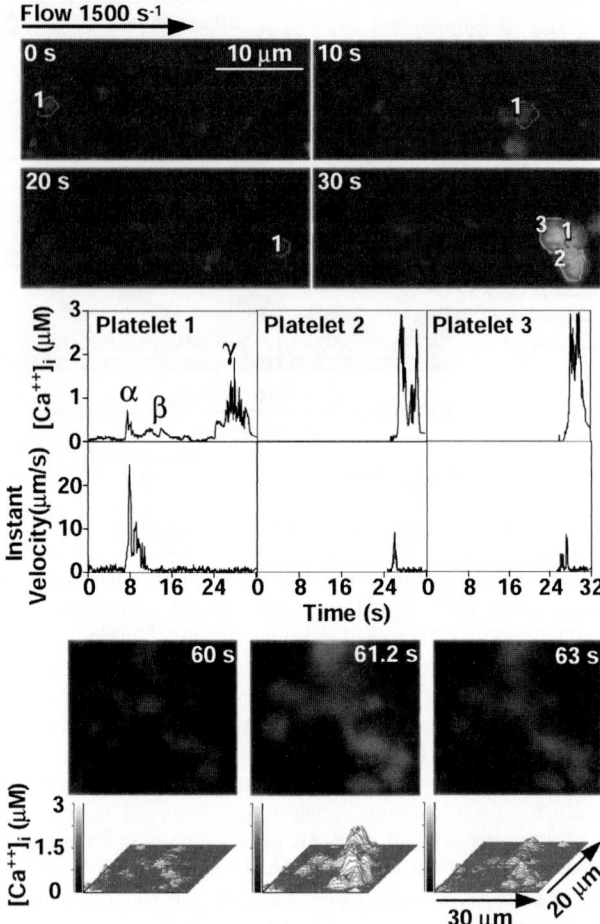

**Figure 18-3.** Real-time analysis of $[Ca^{++}]_i$ during platelet adhesion to immobilized VWF. Platelets loaded with $2 \times 10^7$/mL fluo-3 AM (1-[2-Amino-5-(2, 7-dichloro-6-hydroxy-3-oxo-9-xanthenyl) phenoxy]-2-(2-amino-5-methylphenoxy)ethane-N,$N$,$N'$,$N'$-tetraacetic acid, pentaacetoxymethyl ester) were suspended with washed erythrocytes in homologous plasma and perfused over immobilized VWF for 3 minutes at a shear rate of 1500 s⁻¹. The upper images show the sequence of events leading to aggregate formation. At 0 s platelet 1 appears in the view field, at 10 s it has moved in the direction of flow by approximately 20 μm, at 20 s it has moved by an additional few micrometers, and at 30 s it has remained stationary and two new platelets (2 and 3) adhere to it, forming a small aggregate. The diagrams in the middle of the figure show $[Ca^{++}]_i$ and the instant velocity of platelets 1, 2, and 3. Platelet 1 exhibits a period of rapid movement coincident with the appearance of transient $[Ca^{++}]_i$ peaks (α/β); a higher and longer lasting increase in $[Ca^{++}]_i$ (γ) develops while the platelet is stationary. Cytosolic $Ca^{++}$ oscillations appear also when platelets 2 and 3 arrest on the surface, without a clear sequence from α/β to γ. The lower images, captured between 60 s and 63 s after the appearance of platelet 1, show the long-lasting synchronous increase of $[Ca^{++}]_i$ in platelets forming a large aggregate. The three-dimensional diagrams below each image show the measurement of $[Ca^{++}]_i$ in all the platelets in the field. (From Mazzucato et al.,[30] with permission.)

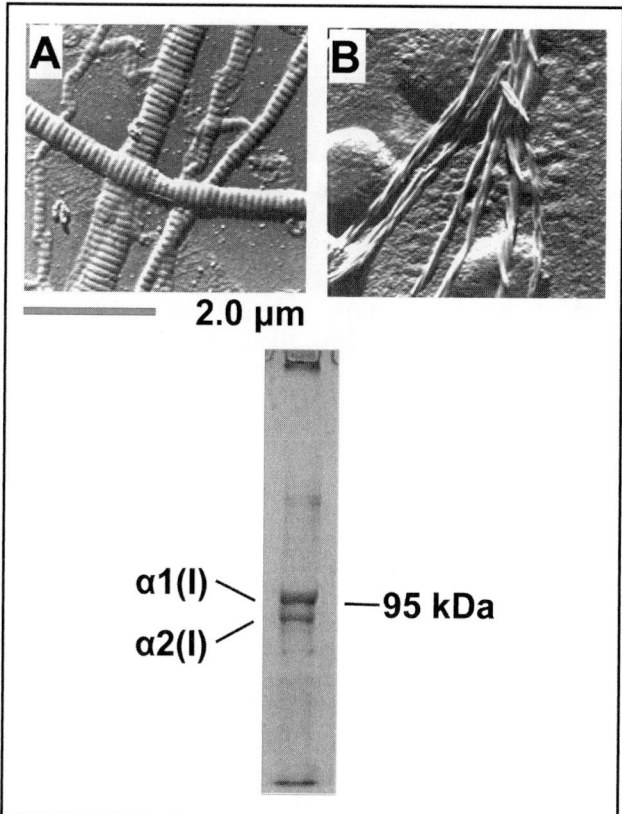

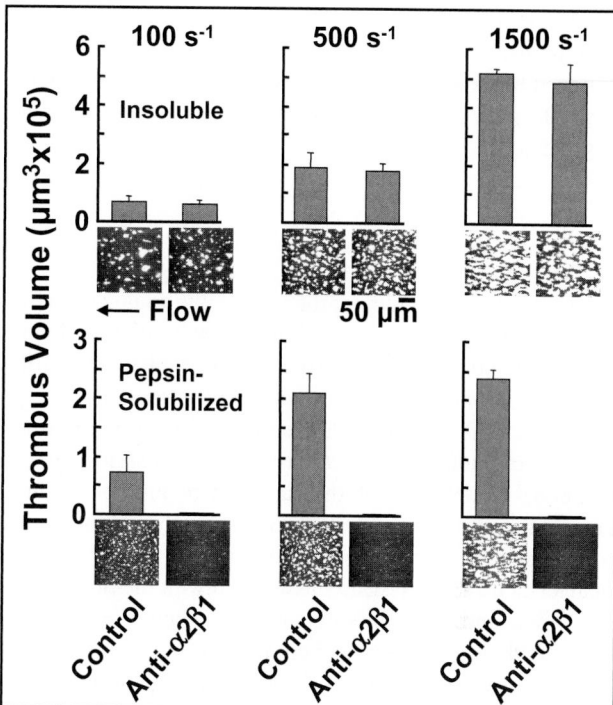

**Figure 18-4.** Structural features of native insoluble and pepsin-solubilized type I collagen. Surface replication analysis of insoluble type I collagen (A) displays a characteristic striated pattern that is absent from the smaller fibrils derived from pepsin-solubilized type I collagen (B), in which microfibril assemblies form a nonbanded spiral configuration. Polyacrylamide gel electrophoresis under denaturing and nonreducing conditions demonstrate the $\alpha_1(I)$ and $\alpha_2(I)$ type I collagen subunits as well as high-molecular weight multimers (lower panel). (From Savage et al.,[49] with permission.)

**Figure 18-5.** Role of integrin $\alpha2\beta1$ in platelet thrombus formation onto type I collagen. Blood containing D-phenylalanyl-L-prolyl-L-arginine chloromethyl ketone dihydrochloride to block $\alpha$-thrombin and mepacrine to render platelets fluorescent was perfused at different wall shear rates, as indicated, over distinct type I collagen preparations in a parallel plate flow chamber maintained at 37°C. Single-frame images were captured at 3 minutes after the beginning of blood flow at the indicated wall shear rates. Blood was either untreated (control) or treated with an anti-$\alpha2\beta1$ monoclonal antibody. Functional inhibition of $\alpha2\beta1$ prevented stable platelet adhesion and thrombus formation on pepsin-solubilized (nonbanded), but not on native insoluble, collagen (banded). The volume of platelet thrombi over an area of $102,236\,\mu m^2$ (mean ± SEM of four separate experiments with different blood donors) was measured 3 minutes after the beginning of blood flow by confocal optical sectioning in the z-axis (vertical). (From Savage et al.,[49] with permission.)

helical assembly, has been found in both normal and pathological tissues,[50] and is more abundant in veins than arteries. Immunohistochemical examination of matrix metalloproteinases (MMP-1, -2, and -3) revealed significant expression of MMP-1 in smooth muscle cells of the vascular wall where spiral collagen was abundant,[50] suggesting its origin from the degradation of collagen fibers. In experimental conditions, the banded pattern of collagen fibers can be reconstituted by dialyzing spiraled microfibrils at physiologic pH.[49] When immobilized, all these substrates are thrombogenic, but elicit platelet adhesion and aggregation through distinct mechanisms. This is best illustrated by the different consequences of integrin $\alpha2\beta1$ inhibition. Thrombus formation on spiraled microfibrils at all shear rates tested from 100 to $1500\,s^{-1}$ is completely dependent on this receptor, and its inhibition yields the transient surface translocation medi-

ated by the interaction of GPIb$\alpha$ with collagen-bound VWF. In contrast, thrombi of similar size form on banded collagen fibers of insoluble collagen regardless of $\alpha2\beta1$ inhibition (Fig. 18-5), even though initial platelet surface interactions may be delayed.[14] GPVI is the collagen receptor required for thrombus formation on banded collagen fibers, as shown by the defective function of GPVI$^{-/-}$ platelets from mice with a targeted gene deletion[51] (discussed later). Regardless of the collagen structure, inhibition of plasma VWF binding to collagen, or blockage of GPIb$\alpha$ on platelets, completely eliminates all initial platelet–surface interactions at $1500\,s^{-1}$ but not at or less than $500\,s^{-1}$.[14] Collagen and VWF, therefore, form a functional unit for thrombus formation in rapidly flowing blood, in which VWF contributes mainly to the capturing of platelets on the surface, and collagen to the establishment of stable bonds for firm adhesion as well as to the activation required for platelet aggregation.

### D. The Platelet Collagen Receptors

Several collagen-binding proteins on the platelet surface may mediate platelet adhesion under flow (see also Chapter 6). These include the integrin α2β1,[52–54] GPVI,[55,56] GPIV (CD36),[57–59] and the 65-kDa protein (p65) specific for type I collagen.[60] GPIV is unlikely to make a major contribution to collagen-induced platelet activation because it is absent in approximately 5% of the Japanese population without hemostatic impairment. In fact, GPIV may be essential only for platelet interaction with collagen type V.[61] In resting platelets, GPIV is physically associated with the Src kinases Yes, Fyn, and Lyn, although it has not been established whether they become activated after stimulation by collagen.[62] Likewise, it is not yet known whether binding of collagen to p65 can generate intracellular signals, although this receptor appears to be linked to the generation of nitric oxide.[63] Of the proposed platelet collagen receptors, therefore, only α2β1 and GPVI appear to have a defined role in thrombus formation.

**1. α2β1.** Integrin α2β1 corresponds to the platelet membrane GPIa-IIa, originally identified as the very late activation antigen-2 on activated T cells[64] and the class II extracellular matrix receptor on fibroblasts.[53] In contrast to other integrins, α2β1 expression varies by as much as an order of magnitude on normal platelets.[65] These differences, which are inherited, may modulate the initial phase of platelet adhesion to collagen, when rapid responses may be critical to counter inhibitory mechanisms during thrombus formation in vivo.[66] Indeed, blood perfusion experiments at the shear rate of 1500 s⁻¹ have shown delayed platelet adhesion and aggregation onto acid-soluble type I collagen fibrils when individuals with low α2β1 expression are compared with those with higher levels, but show equivalent thrombus formation within 3 to 5 minutes regardless of α2β1 density (Fig. 18-6).[66] The role of α2β1 in mediating platelet adhesion to collagen is well established, but the extent to which it generates intracellular signals that lead to platelet activation is less clear. It has been proposed that α2β1 binding to collagen facilitates the engagement of lower affinity receptors, such as GPVI, which are considered to play a crucial role in platelet activation. This could explain why antibodies and other reagents directed against α2β1 inhibit collagen-induced platelet activation, but the effect can be overcome with increasing concentrations of collagen.[67] Even so, evidence in support of α2β1 as a signaling receptor has been

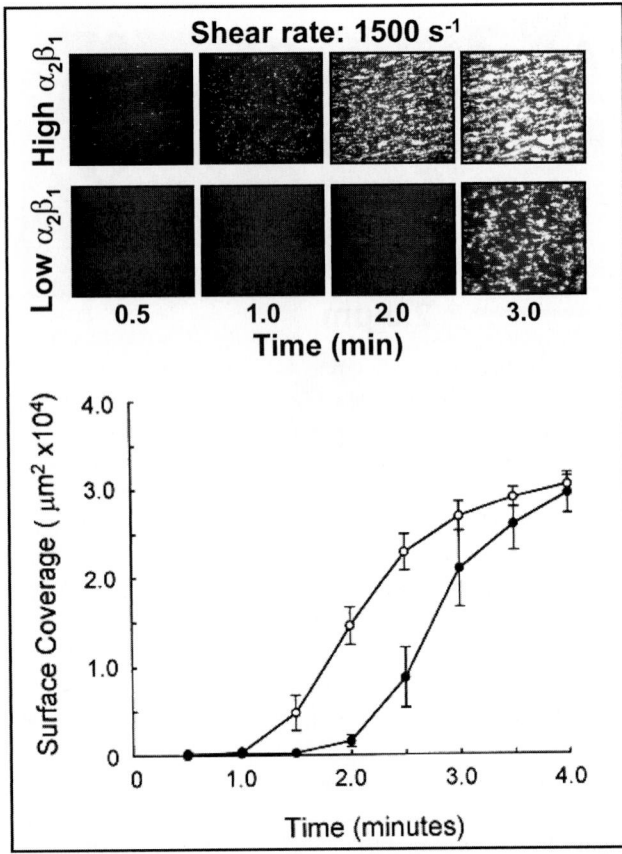

**Figure 18-6.** Effect of α2β1 density on initial platelet adhesion to type I collagen. Top. Images of platelet adhesion to solubilized human type I collagen at different times during blood perfusion with a wall shear rate of 1500 s⁻¹. The upper row shows the results with blood from an individual homozygous for α2 allele 1 and high α2β1 expression; the lower row shows blood from an individual homozygous for the α allele 2 and low α2β1 expression. Platelet α2β1 surface density was determined by flow cytometry and is expressed as normalized mean fluorescence intensity (nMFI). In the individuals shown here, the nMFI varied from 9.2 to 2.2. Bottom. Results of three paired comparisons (mean ± SEM) of time-dependent surface coverage by platelets. High- and low-density α2β1 individuals are represented by open and solid circles respectively. Differences in surface coverage at 1.5, 2.0, and 2.5 minutes are statistically significant ($p < 0.01$). Note that, after 4 minutes of flow, surface coverage by platelets is essentially equivalent regardless of α2β1 genotype. (Adapted from Kritzik et al.,[66] with permission.)

presented by many groups.[68–72] For example, an anti-α2 monoclonal antibody was found to inhibit collagen-induced platelet aggregation and tyrosine phosphorylation of Syk and PLC-γ2, although the effect could be overcome by increasing collagen concentrations.[73] Similarly, the snake venom jararhagin, which cleaves the α2 subunit inactivating the integrin, inhibits collagen-dependent phosphorylation of Syk and other proteins at low, but not higher, collagen concentrations, confirming that $\alpha2\beta1$ is not essential for collagen-induced activation.[67] In this context, a series of peptides based on the collagen structure and containing a motif that confers selectivity to $\alpha2\beta1$ failed to stimulate tyrosine phosphorylation.[74,75]

**2. GPVI.** This receptor plays a major role in collagen-induced platelet activation,[76] as illustrated by the lack of functional responses in human platelets deficient in GPVI[56,77] and confirmed by GPVI gene ablation in the mouse.[51] Collagen induces platelet activation through a tyrosine kinase-dependent pathway involving the tyrosine phosphorylation of PLC-$\gamma2$.[78,79] The demonstration that tyrosine phosphorylation of PLC-$\gamma2$ is lost in GPVI-deficient patients attests to the importance of GPVI in this signaling pathway.[80] The similarities in signaling between GPVI and the receptor for immune complexes, Fc$\gamma$RIIA, led to the demonstration that both stimulate Syk phosphorylation.[81] Mice lacking either the FcR $\gamma$-chain or Syk fail to aggregate or secrete dense granules in response to collagen,[82] indicating that collagen and immune receptors signal through common pathways involving both the FcR $\gamma$-chain and Syk. Indeed, the membrane expression of GPVI is abolished in the absence of the FcR $\gamma$-chain, consistent with the notion that coassembly of the two proteins in a noncovalent complex is an obligatory requirement for stable GPVI expression in the membrane.[83] Major tools for studying GPVI receptor signaling have been found in the snake venom toxin convulxin,[84,85] and in collagen-related peptides,[86] both potent platelet agonists selective for GPVI. Collagen-related peptides and convulxin mimic the pattern of tyrosine phosphorylation of the FcR $\gamma$-chain, Syk, and PLC-$\gamma2$.[87,88]

The potential role of GPVI in arterial thrombus formation has been evaluated by intravital fluorescence microscopy of injured mouse carotid arteries. Mouse platelets rendered deficient in GPVI by treatment with an antibody failed to aggregate on the damaged vessel wall, and a concurrent reduction in primary adhesion was also reported.[89] The latter finding could be the consequence of reduced platelet activation with inability to establish irreversible attachment after a normal initial tethering, which under high flow conditions is mediated by GPIb$\alpha$ binding to VWF immobilized onto exposed collagen.[14] Unexpected consequences of platelet treatment with an anti-GPVI antibody could account for the marked inhibition of adhesion observed in these studies, which is not in agreement with the results obtained in GPVI$^{-/-}$ mice generated by gene targeting methods.[51] In the latter, initial platelet adhesion to insoluble collagen type I fibrils is not decreased, but platelet spreading associated with irreversible adhesion is abolished (Fig. 18-7). Therefore, the defect observed in GPVI$^{-/-}$ mice with respect to thrombus formation on a fibrillar collagen surface indicates a severe abnormality of platelet activation but not of initial tethering.

## III. Secondary Events Leading to Thrombus Propagation under Flow

### A. Role of Soluble Agonists in Thrombus Propagation under Flow

Thrombogenic surfaces only interact with the initial layer of adherent platelets, but propagation of the resulting activating signals promotes the binding of soluble adhesive proteins and thrombus growth. These signals originate from the costimulation of platelet receptors by soluble agonists that become available at sites of vascular injury after the initial adhesion of platelets. ADP, $\alpha$-thrombin, and thromboxane $A_2$ (TXA$_2$) are of particular prominence in the process that turns $\alpha$IIb$\beta3$ into a high-affinity/avidity receptor capable of binding soluble adhesive proteins.[90–93] The activation of $\alpha$IIb$\beta3$ through diverse pathways is a prerequisite for platelet aggregation that can be induced by distinct platelet agonists. The platelet-activating effects of soluble agonists *in vivo* may be limited by their clearance from the luminal surface of a developing thrombus by flowing blood. Efficient platelet activation by soluble agonists is therefore only likely to occur in the immediate vicinity of platelet–platelet adhesion contacts where high local concentrations of soluble agonists can act synergistically to counter their rapid clearance (see also Chapter 16).

**1. ADP.** This agonist activates platelets through the G protein-coupled purinergic receptors, P2Y$_1$ and P2Y$_{12}$ (see also Chapters 10 and 16). P2Y$_1$ coupled to G$\alpha_q$ regulates PLC-$\beta$-mediated intracellular $Ca^{2+}$ elevation leading to platelet shape change and $\alpha$IIb$\beta3$ activation,[94–96] whereas P2Y$_{12}$ coupled to G$\alpha_i$ contributes to $\alpha$IIb$\beta3$ activation by suppressing cyclic adenosine monophosphate elevations.[97] P2Y$_1$ and P2Y$_{12}$ may have distinct and complementary roles in thrombus formation, with P2Y$_1$ initiating platelet aggregation and P2Y$_{12}$ providing a sustained activation response.[98–100] Studies using P2Y$_1$- and P2Y$_{12}$-deficient mice or selective antagonists of the two receptors have confirmed their crucial role in promoting thrombus formation under high shear stress conditions.[98,99,101–104] For example, arterial thrombosis was reduced in mice lacking P2Y$_1$ function, whereas venous thrombosis was only marginally inhibited.[102] Moreover, a synergistic inhibitory effect on thrombus formation was seen when both P2Y$_1$ and P2Y$_{12}$ were blocked concurrently.[102–104] These *in vivo* studies confirm previous experimental results on the role of ADP in shear-induced aggregation of human platelets.[105–107]

**2. $\alpha$-Thrombin.** Activated platelets provide a surface for $\alpha$-thrombin generation by promoting the assembly of procoagulant enzyme complexes.[108] In turn, $\alpha$-thrombin activates human platelets through distinct membrane receptors and plays a critical role in thrombus formation under

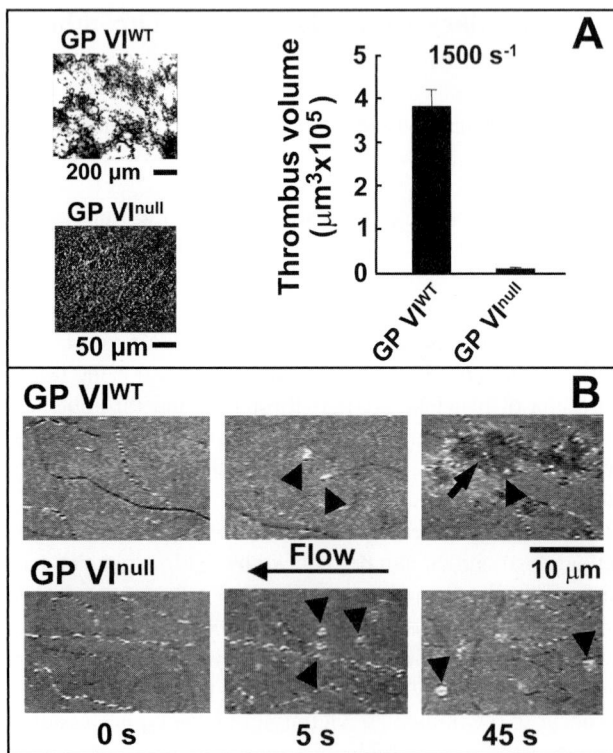

**Figure 18-7.** Role of GPVI in platelet thrombus formation onto insoluble type I collagen fibers. A. Blood was collected from anesthetized mice via a retro-orbital puncture using 40 U/mL heparin as an anticoagulant. Apyrase was added to a final concentration of 1.5 U/mL. Glass coverslips were coated with insoluble fibrillar type I collagen and placed in a parallel plate flow chamber maintained at 37°C. Mouse blood was treated with mepacrine for platelet visualization and was perfused through the chamber at $1500\,s^{-1}$ wall shear rate. The images on the left are single frames taken from a continuous recording and show the collagen-coated surface after 2.5 minutes of blood perfusion. In the top frame, thrombi formed by normal wild-type (WT) platelets are seen at a relatively low magnification. In the bottom frame, the surface exposed to GPVI[null] blood is seen at a higher magnification and exhibits complete coverage by single platelets but absence of thrombus formation. The graph on the right represents the volume of thrombi over an area of $102,236\,\mu m^2$ (mean ± SEM of three separate experiments) determined from confocal sections after perfusion of normal or GPVI[null] mouse blood. B. Analysis by reflection interference contrast microscopy of platelet–collagen interactions during flow. This microscopy technique allows direct visualization of the larger collagen fibrils at the bottom of the flow chamber (before blood perfusion; time, 0 s). After 5 s of perfusion with either normal WT or GPVI[null] blood, comparable numbers of single platelets (arrowheads) are seen interacting with the surface. Their shape is round, indicating that spreading after activation has not yet taken place. After 45 s, in the case of normal blood perfusion, the adherent platelets have become spread and occupy a larger portion of the surface. In contrast, in the case of GPVI[null] blood perfusion, platelets have the same morphology as after 5 s, indicating that they have not become activated and, thus, have not spread. The darker color of spread platelets compared with those that have not spread indicates a closer proximity to the collagen fibrils. The spread platelets represent the base of large thrombi attached to the collagen fibrils and protruding into the flow path. (Adapted from Kato et al.,[51] with permission.)

all shear stress conditions. The predominant α-thrombin-induced activation pathways are initiated by two protease-activated receptors (PARs), PAR-1[109,110] and PAR-4[109,111] (see also Chapters 9 and 16). Cleavage of PAR-1 leads to Gq-linked responses associated with the activation of the β-isoforms of PLC, PI 3-kinase, and RhoA/Rho kinases.[109,111,112] Cleavage of PAR-4 leads to platelet activation at a slower but more sustained rate than through PAR-1.[113–115]

GPIbα in the GPIb-IX-V complex is the major α-thrombin binding site on the platelet surface, accounting for as much as 90% of the total protease bound.[116] Yet, the functional significance of this interaction is still debated. Thrombin binding to GPIb-IX-V induces platelet adhesion and spreading, dense granule secretion, and aggregation.[117,118] Crystallographic studies have shown the potential for GPIbα to bind two separate thrombin molecules by interacting with both exosite I and II of the protease.[119,120] This may promote receptor cross-linking and aid in platelet activation.

GPV is a prominent α-thrombin substrate,[121] but the possible functional significance of its cleavage has only recently been surmised by evaluating the consequences of its absence from mouse platelets.[122] Despite the fact that platelets of GPV[−/−] mice are normal in size, express normal amounts of the modified GPIb-IX complex, and display normal function with regard to VWF binding, they have an enhanced response to thrombin.[122] The unexpected finding that GPV may negatively modulate the platelet response to thrombin led to the

description of a novel signaling mechanism.[123] These studies utilized proteolytically inactive thrombin, incapable of activating the PARs but capable of binding to GPIb-IX. Removal of GPV, by gene deletion or through cleavage by a low-dose thrombin pretreatment, permitted platelet activation by proteolytically inactive thrombin — a response that is completely lacking when GPV is present.[123] The conclusion was that thrombin can initiate platelet signaling upon interacting with GPIb-IX independently of proteolytic activity, but proteolysis is required to release the inhibitory effect of GPV in the receptor complex.

**3. TXA₂.** Stimulated platelets release arachidonic acid that cyclooxygenase converts to $TXA_2$, a platelet agonist and vasoconstrictor (see also Chapters 16 and 31). $TXA_2$ activates platelets by binding to a specific G-protein-coupled receptor linked to $G_q$ and $Gq_{12/13}$.[94,124] The best characterized

TXA$_2$ receptor signaling pathway through G$_q$ is linked to the activation of PLC-β[125] and the generation of diacylglycerol and inositol triphosphate,[126] which lead to intraplatelet Ca$^{2+}$ mobilization[126] and activation of PKC,[74] respectively. Signaling through G$_{12/13}$ regulates myosin light-chain phosphorylation through activation of the Rho/Rho kinase pathway.[124,127] These events precede alterations in actin formation, platelet cytoskeletal rearrangement, and inside-out signaling.[125] Although the role of TXA$_2$ in supporting the hemostatic function of platelets is clearly established, the antithrombotic effects of aspirin, which blocks TXA$_2$ production, can be overcome by increasing shear forces. For example, inhibition of TXA$_2$ generation has no effect on shear-induced platelet aggregation in a laminar flow field using a cone-and-plate viscometer.[106] Moreover, aspirin has no inhibitory effect on thrombus formation at pathological shear rates (>10,000 s$^{-1}$), but it inhibits thrombus growth at lower rates.[128–131]

## B. Role of Soluble Adhesive Proteins and Their Platelet Receptors in Thrombus Propagation under Flow

After the initial events that lead to adhesion and activation at sites of vascular injury, platelets bind soluble adhesive proteins and form a reactive surface for continuing platelet deposition. Subsequent thrombus growth is therefore strictly dependent on the formation of interplatelet bonds. Aggregation has typically been studied with agonist-stimulated platelets in suspension, without surface interactions, and under stirring conditions that create a turbulent flow with low shear rates.[132] These studies have led to the assumption that fibrinogen binding to αIIbβ3 represents the only interaction relevant for platelet aggregation.[133] This view has changed with the use of the cone-and-plate viscometer, in which platelets in suspension can be exposed to defined levels of shear stress in a laminar flow field. High shear stress by itself, without the addition of exogenous agonists, can lead to platelet aggregation that is mediated by VWF and its two membrane receptors.[105,134] Indeed, at elevated shear rates, typically in excess of 5000 s$^{-1}$, VWF binds specifically to platelets in a process that involves sequentially GPIbα and αIIbβ3.[105,135] Thus, fibrinogen and VWF, as well as GPIbα and αIIbβ3, have distinct but complementary roles in platelet aggregation depending on fluid dynamic conditions.[135] Aggregation of platelets in suspension, however, fails to reproduce the correct spatial and temporal sequence of events that occur during thrombus growth *in vivo*. Thus, to study the mechanisms that mediate homotypic platelet cohesion onto an immobilized reactive substrate, it is necessary to use experimental approaches that not only simulate hemodynamic conditions on a vascular surface, but also permit three-dimensional measurement in real time of the volume of growing thrombi.[136]

The role of different adhesion receptors in mediating interplatelet contacts has been analyzed with two-step perfusion experiments in which an aliquot of untreated blood was perfused over collagen and followed, without interruption, by blood containing specific inhibitors of GPIbα or αIIbβ3 (Fig. 18-8). These studies have shown that thrombus height is limited by GPIbα inhibition even when blood is initially perfused at shear rates permissive of adhesion to collagen without requiring the GPIbα–VWF interaction. This is because the shear rate increases at the tip of a growing thrombus where the flow channel narrows, such that a threshold may be reached at which only a rapidly forming bond can capture flowing platelets. Thus, the synergy between fibrinogen and VWF in supporting platelet aggregation depends on the recognized ability of each of these molecules to establish bonds with distinct adhesive properties. The initial function of VWF may depend primarily on its rapid rate of association with GPIbα,[11,14] which is particularly important when high flow rates increase the velocity differential between adjacent platelets. The subsequent interaction of multimeric VWF with both GPIbα and activated αIIbβ3[135,137] may temporarily stabilize interplatelet contacts and allow the ensuing permanent bridging mediated by fibrinogen binding across activated platelet membranes.[135] Therefore, adhesive interactions between platelets and a thrombogenic surface or between two platelets may depend essentially on the same mechanisms to oppose fluid dynamic forces generated by blood flow.

*Ex vivo* perfusion experiments indicate a synergistic role for fibrinogen and VWF in supporting platelet aggregation onto collagen fibrils. In the absence of fibrinogen, VWF-mediated thrombi grow rapidly at high shear rates, but are unstable; in the presence of VWF and fibrinogen, thrombi grow more slowly, but are stable.[136] In mice deficient in VWF but with normal fibrinogen levels, platelet adhesion at sites of experimental vascular lesion is delayed, but stable platelet aggregates eventually develop even though arterial occlusion is often impaired.[138] In mice selectively deficient in fibrinogen but with normal VWF levels, in contrast, platelet thrombi develop rapidly but detach from the subendothelial surface and embolize, causing vascular occlusion downstream of the lesion.[138] Thus, in agreement with the *ex vivo* perfusion studies outlined here, both VWF and fibrinogen are required to ensure stable aggregates. These results provide a plausible explanation for the altered hemostatic properties of platelets from patients with isolated congenital deficiency of either fibrinogen[139] or VWF.[140] Because neither protein by itself can sustain the full development of stable thrombi within the range of pathophysiologically relevant flow conditions, hemostasis cannot be normal unless both are present and functional. Paradoxically, platelets from mice deficient in both VWF and fibrinogen form thrombi at sites of vascular damage, even though often without progression to a stable occlusion.[138] However, fibrinogen/VWF-

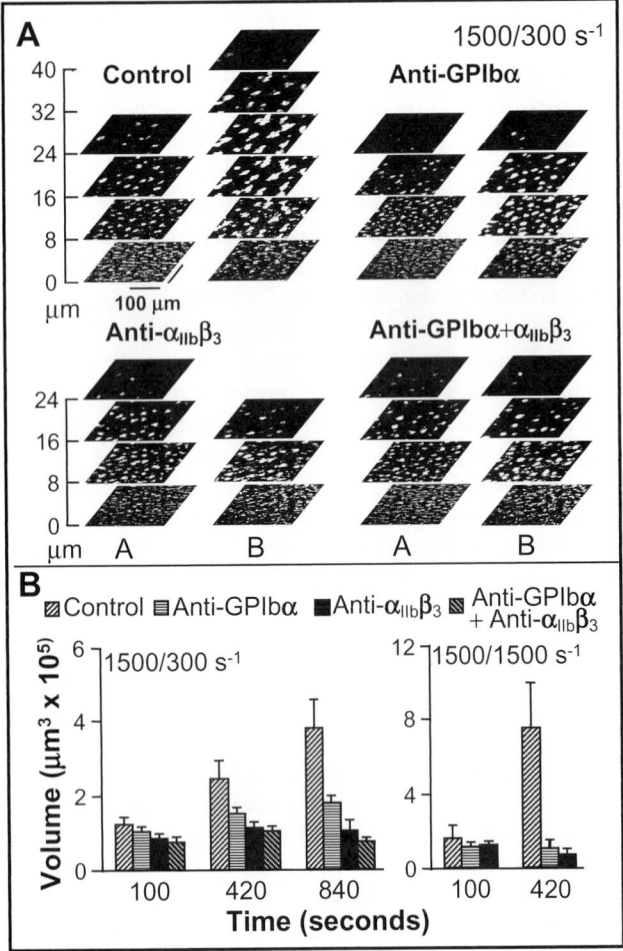

**Figure 18-8.** Role of GPIbα and αIIbβ3 in thrombus growth dependent on the formation of interplatelet bonds. In a two-stage experiment, untreated blood was perfused over collagen type I fibrils at 1500 s⁻¹ for 100 s, and the total volume of thrombi was measured from confocal sections. Blood flow was then continued without interruption at 300 s⁻¹ or 1500 s⁻¹ with blood containing either buffer (control) or monoclonal antibodies blocking GPIbα or αIIbβ3 as indicated. The total volume of thrombi was then determined at the indicated time points. A. Stacks of z sections labeled A show thrombi formed after 100 s of perfusion with blood containing no antibody at 1500 s⁻¹. B. Growth of thrombi after an additional 740 s of perfusion at 300 s⁻¹ with blood containing either buffer or the indicated antibodies (total perfusion time, 840 s). Total thrombus volume was measured at the indicated cumulative perfusion times in these two-stage experiments. Bars are identified according to the type of inhibitor used in the second stage, but perfusion for the first 100 s was always performed with untreated blood. The results are expressed as mean ± SEM for 4 to 10 experiments for the different conditions tested. Note that experiments at 1500 s⁻¹ were terminated after collecting z sections at 420 s, because thrombi had already reached a large volume in control blood. Experiments with the combination of anti-GPIbα and anti-αIIbβ3 monoclonal antibodies were not done at 1500 s⁻¹ because at this shear rate each individual antibody completely prevented thrombus growth compared with the volume attained after the initial perfusion for 100 s with untreated blood. (From Ruggeri et al.,¹³⁶ with permission.)

deficient mice that also lack the β3 integrin are unable to form thrombi under the same conditions, indicating that another β3 ligand may contribute to platelet cohesion. Experiments in mice with conditional depletion of plasma fibronectin indicate that this is the adhesive protein capable of contributing to thrombus growth and stability by acting as an interplatelet link across activated αIIbβ3 in synergy with VWF and fibrinogen.¹⁴¹

VWF in plasma or released by altered endothelial cells and/or activated platelets at sites of vascular injury has a potent prothrombotic effect by promoting both platelet adhesion and aggregation, as discussed earlier, particularly under high shear stress conditions. The largest multimers of VWF, which are known to have the tightest binding to GPIbα²⁹ and the most powerful prothrombotic function,¹⁴² are present inside cellular storage granules,¹⁴³,¹⁴⁴ but are not normally found in the circulation. The reason for this is the efficient processing of all secreted VWF by the metalloprotease ADAMTS-13,¹⁴⁵,¹⁴⁶ which cleaves one single peptide bond in the VWF subunit,¹⁴⁷ but in so doing reduces multimer size.¹⁴⁸ Absence of ADAMTS-13 results in a thrombotic microangiopathy (see Chapter 50),¹⁴² suggesting that the physiological function of the protease is to contain the activity of the largest VWF multimers to the sites where they are released from cells.¹⁴⁹ Recently, the results of *ex vivo* perfusion experiments have added to this concept by demonstrating that ADAMTS-13 can further cleave circulating VWF multimers while they mediate activation-independent interplatelet cohesion induced by elevated shear stress, resulting in a time-dependent dispersion of the aggregates.¹⁵⁰ In contrast, the protease appeared to have no effect, at least under the conditions studied, when thrombus formation was induced by blood exposure to a collagen surface. The likely explanation for this finding is that other adhesive molecules not processed by ADAMTS-13, such as fibrinogen and fibronectin (discussed earlier), contribute to maintaining the cohesion of aggregated platelets when they are fully activated. Thus, ADAMTS-13 may be a selective modulator of VWF-mediated platelet thrombus formation in pathological shear stress conditions. A process of this kind may be needed to avoid the propagation of platelet aggregates beyond the limits of a vascular lesion exposed to rapidly flowing blood.

## C. Mechanisms of Thrombus Stabilization under Flow

After platelet activation and aggregation have occurred in response to a vascular lesion, processes take place that consolidate the stability of the forming thrombus. These have been the focus of increasing attention in recent years. As mentioned previously in this chapter, one example of this function is the role that fibrinogen (likely after conversion to fibrin) plays in anchoring aggregated platelets to the site of vascular injury, thus preventing downstream embolization under the effects of flow.[138] Platelet aggregation in itself is not an irreversible process, which is an important characteristic allowing for the intervention of regulatory mechanisms capable of preventing excessive, and potentially dangerous, propagation of thrombosis. It is apparent that newly recruited platelets at the growing edge of a thrombus transmit feedback signals to aggregated platelets in deeper layers that are necessary to prevent disaggregation. The propagation of these signals throughout the thrombus is visible as recurrent cycles of intracytoplasmic $Ca^{2+}$ elevations.[151] Both ADP receptors, $P2Y_1$ and $P2Y_{12}$, are involved in the process, which can be demonstrated experimentally by two-stage perfusion experiments in which normal blood is first exposed to a thrombogenic surface, such as collagen type I fibers, followed by blood containing specific inhibitors. In such a manner, it has been shown that a thrombus can begin to disperse even after several minutes of growth if ADP activation pathways are blocked, intracytoplasmic $Ca^{2+}$ elevations are prevented, or inhibitors of $\alpha IIb\beta 3$ function are added.[152] These findings indicate that platelet activation must be sustained to keep adhesive ligands bound to $\alpha IIb\beta 3$ in a stable manner. It should be considered, however, that *ex vivo* experiments may overemphasize platelet-dependent mechanisms of this kind, because anticoagulants used to handle blood outside of vessels markedly suppress $\alpha$-thrombin generation and/or activity and reduce the deposition of fibrin that normally contributes to thrombus stability. Nevertheless, sustained platelet activation may be critical for maintaining the integrity of platelet aggregates, particularly in rapid arterial flow conditions and in early stages of thrombus development.

Several specific interactions have been identified that directly enhance the cohesion between platelets initially established through the well-known adhesive ligands (VWF, fibrinogen, fibronectin) and receptors (GPIb$\alpha$ and $\alpha IIb\beta 3$) mentioned earlier in this chapter. Two main mechanisms can be envisioned to operate in this regard, one resulting in the generation of signals that enhance or sustain platelet activation, and the other based on homophilic or heterophilic interactions that add adhesive strength to platelet aggregates. The hormone leptin is an example of the first type of action, in that it has been shown to reinforce the response of platelets to weak agonist stimulation and thereby contribute to the stability of platelet aggregation.[153] The effect of leptin on thrombus growth has been confirmed with *in vivo* experiments in obese mice, who lack leptin and exhibit delayed thrombotic occlusion with frequent embolization in injured arteries.[154] Moreover, inhibition of endogenous leptin has been found to protect mice from arterial and venous thrombosis.[155]

Cellular pathways involved in inflammatory and immune responses also appear to have prothrombotic effects. CD40 ligand (CD40L), a member of the tumor necrosis factor family of ligands, has been detected on the surface of activated platelets where it interacts with $\alpha IIb\beta 3$ through an integrin recognition sequence (KGD). In the absence of CD40L, platelet aggregates formed in experimental *in vivo* models of thrombosis are unstable, confirming the ability of this protein to support platelet–platelet cohesion under high shear stress conditions via an $\alpha IIb\beta 3$-dependent mechanism.[156]

Growth arrest-specific gene 6 product (Gas6; see also Chapter 16) is a vitamin K-dependent protein (homologous to the anticoagulant cofactor, protein S) whose absence or inactivation was reported to protect mice from fatal thromboembolism.[157] Gas6 is present in plasma and in the $\alpha$-granules of mouse platelets, from which it is secreted upon activation, but it may be absent in human platelets.[158] Because of the latter observation, the relevance of Gas6 in human platelet physiology is being reevaluated; it may depend on the function of plasma-derived Gas6 only. On the other hand, all three known Gas6 receptors (Axl, Sky, and Mer) are present on mouse and human platelets, and their inactivation or stimulation results in the expected inhibition or enhancement of agonist-induced platelet activation responses.[159] Thus, Gas6-dependent pathways may indeed represent an amplification mechanism for platelet aggregate stability.

Eph kinases and ephrins (see also Chapter 16) are families of membrane-bound molecules that interact with each other, exhibiting a major role in neuronal organization and as early markers for vascular commitment to arterial or venous development. Platelet-expressed Eph/ephrins include EphA4 and ephrinB1, which appear to contribute to "outside-in" signals originating from ligand-occupied $\alpha IIb\beta 3$ (see also Chapter 17). Inhibition of Eph/ephrin interactions results in a diminution of the volume of thrombi formed on a collagen type I surface, presumably as a consequence of decreased aggregate cohesion.[160]

A strategy has been proposed to identify molecules that are phosphorylated after the induction of platelet aggregation, because this could indicate a specific role in controlling thrombus stability.[161] Two studies have recently verified the validity of this hypothesis. In one, a novel membrane protein has been identified — platelet endothelial aggregation receptor 1 — that appears to signal secondary to $\alpha IIb\beta 3$-mediated platelet–platelet contacts.[161] Future studies will indicate the pathophysiological significance of this finding. In another study, evidence has been presented for the prothrombotic function of CD84 and SLAM (signaling lymphocyte activation molecule), both members of the same family of homophilic adhesion receptors. Immobilized CD84 can promote platelet microaggregation, and SLAM$^{-/-}$ mice exhibit decreased agonist-induced platelet aggregation, a defect confirmed by intravital studies of thrombus formation in injured vessels.[162] Because numerous cell-signaling and adhesive pathways potentially involved in controlling platelet thrombus integrity have already been identified, an obvious challenge for the future will be the integration of this analytical knowledge into a comprehensive representation of the mechanisms of platelet response to vascular injury.

## D. Analysis of Thrombus Formation by Intravital Microscopy

Studying the mechanisms of thrombus formation in the vasculature of a living animal offers the unique opportunity of obtaining a global view of the process, with the appropriate contribution of all pro- and antithrombotic factors[163] responding to lesions of endothelial and subendothelial structures (see Chapter 34).[164,165] Moreover, intravital experiments typically performed with videomicroscopy techniques avoid the problem of using anticoagulants, as required for the *ex vivo* perfusion of blood specimens through flow chambers. One can thus monitor concurrently the aggregation of platelets and deposition of fibrin,[166] and derive quantitative information that is temporally and spatially relevant to explain the process of thrombus formation (Fig. 18-9). Specific questions can be addressed in ever greater detail using this methodology, in particular with respect to the events that initiate and consolidate hemostatic and thrombotic responses. For example, intravital thrombosis models are providing new perspectives on how tissue factor is involved in the response to vascular injury,[167,168] and have revealed a still unexplained role of coagulation factor XII in the formation of platelet thrombi.[169] Factor XII (Hageman factor) is the initiator of the contact phase of coagulation, which is thought to be important for inducing clotting on negatively charged surfaces, but to have a limited, if not negligible, role *in vivo*. Factor XII deficiency in humans, for example, is not associated with a bleeding phenotype, and affected patients have no apparent defect of coagulation or platelet function. This is in contrast to the experimental results in mice with ablation of the factor XII gene, which exhibited a markedly decreased platelet response to a thrombogenic stimulus. The reasons why factor XII should perform a role in leading to α-thrombin and fibrin generation, and presumably platelet activation, that cannot be fulfilled by the tissue factor-initiated pathway of coagulation remain to be understood. Mechanisms relevant to the stability of platelet aggregates, which may be essential for the development of occluding arterial thrombi, can also be properly studied with *in vivo* models. In this manner, for example, it has been shown that plasminogen activator inhibitor 1 and its cofactor, vitronectin, are required for the integrity of thrombi that develop in response to vascular injury.[170]

Because of their relevance, the results of numerous studies based on intravital microscopy have been reported throughout this chapter. Nevertheless, no experimental method is without limitations, and conclusions from intravital animal experiments may not be directly applicable to human disease. This is particularly true with respect to

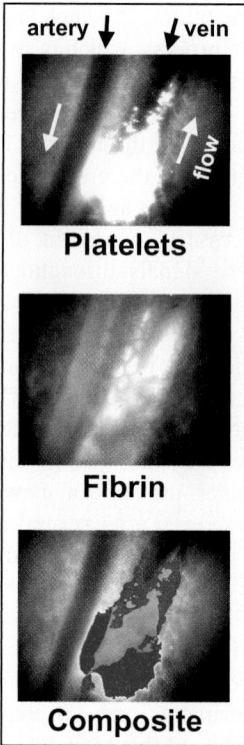

**Figure 18-9.** Images of an occluding thrombus formed in an injured mesenteric venule. A monoclonal antibody specifically interacting with mouse fibrin, and without cross-reactivity with fibrinogen, labeled with a red fluorochrome (Alexa 568) was mixed with homologous donor calcein-labeled platelets (4–5 × 10^6 platelets/g),[138] and the mixture was injected into a tail vein of a recipient mouse before injuring a mesenteric venule by application of a 5% ferric chloride solution over the vessel intima. Top. Image obtained in the green channel (platelets). Middle. Image obtained in the red channel (fibrin). The two original images shown here are on a gray scale (black and white) as viewed through a silicon intensified target (SIT) camera. Bottom. Superimposition of the top and middle images with pseudocolor rendition based on the original gray scale. Yellow indicates the areas where both platelets and fibrin are present. Note that the thrombus formed at the site of injury demonstrates incorporation of both calcein-labeled platelets and fibrin, and the distribution of these two components is not coincident. Both platelets and fibrin are present at the center of the thrombus mass, whereas platelets alone occupied the front edge (upstream of flow) and fibrin alone occupied the trailing edge (downstream of flow).

vascular disease, because experimental lesions in healthy vessels most likely represent a significantly different thrombogenic stimulus compared with pathologic atherosclerotic lesions. In fact, experimental evidence indicates that intravital models of thrombosis are greatly influenced by the methods used to cause injury in the target vessel, which may explain the variability of reported results (see Chapter 34). Understanding the mechanisms that modulate a thrombogenic response in relation to the nature of the initiating vascular lesion remains a challenge for both *ex vivo* and *in vivo* experimentation.

## IV. Conclusions

The last few years have witnessed major advances in our understanding of the mechanisms that support platelet thrombus formation in flowing blood. The results of *ex vivo* flow experiments and intravital microscopy studies have shed new light on the processes underlying hemostasis and thrombosis. Animal models with targeted gene deletions or mutations have already greatly contributed to these advances, and will provide even more insights in the future. Significant advances in genomics and proteomics (see Chapter 5) have already started to contribute additional information relevant to elucidating in ever finer detail the integrated processes that link platelet–substrate interactions and signaling pathways to thrombus growth and stability. Finally, the advent of improved drug development technologies is likely to permit the translation of this fundamental knowledge into more efficient therapeutic approaches to prevent excessive bleeding and thrombosis.

## References

1. Iannacone, M., Sitia, G., Isogawa, M., et al. (2005). Platelets mediate cytotoxic T lymphocyte-induced liver damage. *Nat Med, 11*, 1167–1169.
2. Tangelder, G. J., Slaaf, D. W., Arts, T., et al. (1988). Wall shear rate in arterioles in vivo: Least estimates from platelet velocity profiles. *Am J Physiol, 254*, H1059–H1064.
3. Mailhac, A., Badimon, J. J., Fallon, J. T., et al. (1994). Effect of an eccentric severe stenosis on fibrin(ogen) deposition on severely damaged vessel wall in arterial thrombosis. Relative contribution of fibrin(ogen) and platelets. *Circulation, 90*, 988–996.
4. Siegel, J. M., Markou, C. P., Ku, D. N., et al. (1994). A scaling law for wall shear rate through an arterial stenosis. *J Biomech Eng, 116*, 446–451.
5. Bluestein, D., Niu, L., Schoephoerster, R. T., et al. (1997). Fluid mechanics of arterial stenosis: Relationship to the development of mural thrombus. *Ann Biomed Eng, 25*, 344–356.
6. Ruggeri, Z. M., & Saldivar, E. (1998). Platelets, hemostasis and thrombosis. In Z. M. Ruggeri (Ed.), *Von Willebrand factor and the mechanisms of platelet function* (pp. 1–32). Berlin: Springer.
7. Beumer, S., Heijnen, H. F., IJsseldijk, M. J., et al. (1995). Platelet adhesion to fibronectin in flow: The importance of von Willebrand factor and glycoprotein Ib. *Blood, 86*, 3452–3460.
8. Hindriks, G., Ijsseldijk, M. J. W., Sonnenberg, A., et al. (1992). Platelet adhesion to laminin: Role of $Ca^{2+}$ and $Mg^{2+}$ ions, shear rate, and platelet membrane glycoproteins. *Blood, 79*, 928–935.
9. Godyna, S., Diaz–Ricart, M., & Argraves, W. S. (1996). Fibulin-1 mediates platelet adhesion via a bridge of fibrinogen. *Blood, 88*, 2569–2577.
10. Jurk, K., Clemetson, K. J., de Groot, P. G., et al. (2003). Thrombospondin-1 mediates platelet adhesion at high shear via glycoprotein Ib (GPIb): An alternative/backup mechanism to von Willebrand factor. *FASEB J, 17*, 1490–1492.
11. Savage, B., Saldivar, E., & Ruggeri, Z. M. (1996). Initiation of platelet adhesion by arrest onto fibrinogen or translocation on von Willebrand factor. *Cell, 84*, 289–297.
12. Asch, E., & Podack, E. (1990). Vitronectin binds to activated human platelets and plays a role in platelet aggregation. *J Clin Invest, 85*, 1372–1378.
13. Zaidi, T. N., McIntire, L. V., Farrell, D. H., et al. (1996). Adhesion of platelets to surface-bound fibrinogen under flow. *Blood, 88*, 2967–2972.
14. Savage, B., Almus–Jacobs, F., & Ruggeri, Z. M. (1998). Specific synergy of multiple substrate–receptor interactions in platelet thrombus formation under flow. *Cell, 94*, 657–666.
15. Andrews, R. K., Shen, Y., Gardiner, E. E., et al. (1999). The glycoprotein Ib-IX-V complex in platelet adhesion and signaling. *Thromb Haemost, 82*, 357–364.
16. Rand, J. H., Sussman, I. I., Gordon, R. E., et al. (1980). Localization of factor-VIII-related antigen in human vascular subendothelium. *Blood, 55*, 752–756.
17. Sakariassen, K. S., Bolhuis, P. A., & Sixma, J. J. (1979). Human blood platelet adhesion to artery subendothelium is mediated by factor VIII/von Willebrand factor bound to the subendothelium. *Nature, 279*, 636–638.
18. Stel, H. V., Sakariassen, K. S., de Groot, P. G., et al. (1985). Von Willebrand factor in the vessel wall mediates platelet adherence. *Blood, 65*, 85–90.
19. Turitto, V. T., Weiss, H. J., Zimmerman, T. S., et al. (1985). Factor VIII/von Willebrand factor in subendothelium mediates platelet adhesion. *Blood, 65*, 823–831.
20. Rand, J. H., Patel, N. D., Schwartz, E., et al. (1991). 150-kD von Willebrand Factor binding protein extracted from human vascular subendothelium is type VI collagen. *J Clin Invest, 88*, 253–259.
21. Denis, C., Baruch, D., Kielty, C. M., et al. (1993). Localization on von Willebrand factor binding domains to endothelial extracellular matrix and to type VI collagen. *Arterioscler Thromb, 13*, 398–406.
22. Rand, J. H., Glanville, R. W., Wu, X.-X., et al. (1997). The significance of subendothelial von Willebrand factor. *Thromb Haemost, 78*, 445–450.
23. Siediecki, C. A., Lestini, B. J., Kottke–Marchant, K., et al. (1996). Shear-dependent changes in the three-dimensional structure of human von Willebrand factor. *Blood, 88*, 2939–2950.
24. Sixma, J. J., van Zanten, G. H., Saelman, E. U., et al. (1995). Platelet adhesion to collagen. *Thromb Haemost, 74*, 454–459.
25. Mazzucato, M., Spessotto, P., Masotti, A., et al. (1999). Identification of domains responsible for von Willebrand factor type VI collagen interaction mediating platelet adhesion under high flow. *J Biol Chem, 274*, 3033–3041.

26. van der Plas, R. M., Gomes, L., Marquart, J. A., et al. (2000). Binding of von Willebrand factor to collagen type III: Role of specific amino acids in the collagen binding domain of vWF and effects of neighboring domains. *Thromb Haemost, 84,* 1005–1111.

27. Lankhof, H., van Hoeij, M., Schiphorst, M. E., et al. (1996). A3 domain is essential for interaction of von Willebrand factor with collagen type III. *Thromb Haemost, 75,* 950–958.

28. Savage, B., Sixma, J. J., & Ruggeri, Z. M. (2002). Functional self-association of von Willebrand factor during platelet adhesion under flow. *Proc Natl Acad Sci USA, 99,* 425–430.

29. Arya, M., Anvari, B., Romo, G. M., et al. (2002). Ultralarge multimers of von Willebrand factor form spontaneous high-strength bonds with the platelet glycoprotein Ib-IX complex: Studies using optical tweezers. *Blood, 99,* 3971–3977.

30. Mazzucato, M., Pradella, P., Cozzi, M. R., et al. (2002). Sequential cytoplasmic calcium signals in a two-stage platelet activation process induced by the glycoprotein Ibα mechanoreceptor. *Blood, 100,* 2793–2800.

31. Kasirer–Friede, A., Ware, J., Leng, L., et al. (2002). Lateral clustering of platelet GP Ib-IX complexes leads to up-regulation of the adhesive function of integrin $\alpha_{IIb}\beta_3$. *J Biol Chem, 277,* 11949–11956.

32. Falati, S., Edmead, C. E., & Poole, A. W. (1999). Glycoprotein Ib-V-IX, a receptor for von Willebrand factor, couples physically and functionally to the Fc receptor γ-chain, Fyn, and Lyn to activate human platelets. *Blood, 94,* 1648–1656.

33. Wu, Y., Suzuki–Inoue, K., Satoh, K., et al. (2001). Role of Fc receptor gamma-chain in platelet glycoprotein Ib-mediated signaling. *Blood, 97,* 3836–3845.

34. Torti, M., Bertoni, A., Canobbio, I., et al. (1999). Rap 1B and Rap2B translocation to the cytoskeleton by von Willebrand factor involves FcgammaII receptor-mediated protein tyrosine phosphorylation. *J Biol Chem, 274,* 13690–13697.

35. Sullam, P. M., Hyun, W. C., Szollosi, J., et al. (1998). Physical proximity and functional interplay of the glycoprotein Ib-IX-V complex and the Fc receptor FcgammaRIIA on the platelet plasma membrane. *J Biol Chem, 273,* 5331–5336.

36. Watson, S. P., Asazuma, N., Atkinson, B., et al. (2001). The role of ITAM- and ITIM-coupled receptors in platelet activation by collagen. *Thromb Haemost, 86,* 276–288.

37. Kasirer–Friede, A., Cozzi, M. R., Mazzucato, M., et al. (2004). Signaling through GP Ib-IX-V activates αIIbβ3 independently of other receptors. *Blood, 103,* 3403–3411.

38. Cunningham, J. G., Meyer, S. C., & Fox, J. E. B. (1996). The cytoplasmic domain of the α-subunit of glycoprotein (GP) Ib mediates attachment of the entire GP Ib-IX complex to the cytoskeleton and regulates von Willebrand factor-induced changes in cell morphology. *J Biol Chem, 271,* 11581–11587.

39. Andrews, R. K., & Fox, J. E. B. (1992). Identification of a region in the cytoplasmic domain of the platelet membrane glycoprotein Ib-IX complex that binds to purified actin-binding protein. *J Biol Chem, 267,* 18605–18611.

40. Du, X., Harris, S. J., Tetaz, T. J., et al. (1994). Association of a phospholipase A$_2$ (14-3-3 protein) with the platelet glycoprotein Ib-IX complex. *J Biol Chem, 269,* 18287–18290.

41. Gu, M., Xi, X., Englund, G. D., et al. (1999). Analysis of the roles of 14-3-3 in the platelet glycoprotein Ib-IX-mediated activation of integrin alpha (IIB)beta(3) using a reconstituted mammalian cell expression model. *J Cell Biol, 147,* 1085–1096.

42. Munday, A. D., Berndt, M. C., & Mitchell, C. A. (2000). Phosphoinositide 3-kinase forms a complex with platelet membrane glycoprotein Ib-IX-V complex and 14-3-3zeta. *Blood, 96,* 577–584.

43. Feng, S., Christodoulides, N., Resendiz, J. C., et al. (2000). Cytoplasmic domains of GpIbalpha and GpIbbeta regulate 14-3-3zeta binding to GpIb/IX/V. *Blood, 95,* 551–557.

44. Nesbitt, W. S., Kulkarni, S., Giuliano, S., et al. (2002). Distinct glycoprotein Ib/V/IX and integrin $\alpha_{IIb}\beta_3$-dependent calcium signals cooperatively regulate platelet adhesion under flow. *J Biol Chem, 277,* 2965–2972.

45. Mazzucato, M., Cozzi, M. R., Pradella, P., et al. (2004). Distinct roles of ADP receptors in von Willebrand factor-mediated platelet signaling and activation under high flow. *Blood, 104,* 3221–3227.

46. Yap, C. L., Anderson, K. E., Hughan, S. C., et al. (2002). Essential role for phosphoinositide 3-kinase in shear-dependent signaling between platelet glycoprotein Ib/V/IX and integrin $\alpha_{IIb}\beta_3$. *Blood, 99,* 151–158.

47. Goto, S., Tamura, N., Eto, K., et al. (2002). Functional significance of adenosine 5′-diphosphate receptor (P2Y$_{12}$) in platelet activation initiated by binding of von Willebrand factor to platelet Gp Ibα induced by conditions of high shear rate. *Circulation, 105,* 2531–2536.

48. Prockop, D. J., & Kivirikko, K. I. (1995). Collagens: Molecular biology, diseases and potentials for therapy. *Annu Rev Biochem, 64,* 403–434.

49. Savage, B., Ginsberg, M. H., & Ruggeri, Z. M. (1999). Influence of fibrillar collagen structure on the mechanisms of platelet thrombus formation under flow. *Blood, 94,* 2704–2715.

50. Ishii, T., & Asuwa, N. (1996). Spiraled collagen in the major blood vessels. *Mod Pathol, 9,* 843–848.

51. Kato, K., Kanaji, T., Russell, S., et al. (2003). The contribution of glycoprotein VI to stable platelet adhesion and thrombus formation illustrated by targeted gene deletion. *Blood, 102,* 1701–1707.

52. Santoro, S. A. (1986). Identification of a 160,000 dalton platelet membrane protein that mediates the initial divalent cation-dependent adhesion of platelets to collagen. *Cell, 46,* 913–920.

53. Kunicki, T. J., Nugent, D. J., Staats, S. J., et al. (1988). The human fibroblast class II extracellular matrix receptor mediates platelet adhesion to collagen and is identical to the platelet glycoprotein Ia-IIa complex. *J Biol Chem, 263,* 4516–4519.

54. Nieuwenhuis, H. K., Akkerman, J. W. N., Houdijk, W. P. M., et al. (1985). Human blood platelets showing no response to collagen fail to express surface glycoprotein Ia. *Nature, 318,* 470–472.

55. Moroi, M., Jung, S. M., Okuma, M., et al. (1989). A patient with platelets deficient in glycoprotein VI that lack both collagen-induced aggregation and adhesion. *J Clin Invest, 84,* 1440–1445.

56. Moroi, M., Jung, S. M., Shinmyozu, K., et al. (1996). Analysis of platelet adhesion to a collagen-coated surface under flow conditions: Involvement of glycoprotein VI in the platelet adhesion. *Blood, 88,* 2081–2092.

57. Asch, A. S., Liu, I., Briccetti, F. M., et al. (1993). Analysis of CD36 binding domains: Ligand specificity controlled by dephosphorylation of an ectodomain. *Science, 262,* 1436–1440.

58. Tandon, N. N., Kralisz, U., & Jamieson, G. A. (1989). Identification of glycoprotein IV (CD36) as a primary receptor for platelet-collagen adhesion. *J Biol Chem, 264,* 7576–7583.

59. Diaz–Ricart, M., Tandon, N. N., Carretero, M., et al. (1993). Platelets lacking functional CD36 (glycoprotein IV) show reduced adhesion to collagen in flowing whole blood. *Blood, 82,* 491–496.

60. Chiang, T. M., Rinaldy, A., & Kang, A. H. (1997). Cloning, characterization, and functional studies of a nonintegrin platelet receptor for type I collagen. *J Biol Chem, 100,* 514–529.

61. Kehrel, B., Kronenberg, A., Rauterberg, J., et al. (1993). Platelets deficient in glycoprotein IIIb aggregate normally to collagens type I and III but not to collagen type V. *Blood, 82,* 3364–3370.

62. Huang, M.- M., Bolen, J. B., Barnwell, J. W., et al. (1991). Membrane glycoprotein IC CD36 is physically associated with the Fyn, Lyn, and Yes protein-tyrosine kinases in human platelets. *Proc Natl Acad Sci USA, 88,* 7844–7848.

63. Chiang, T. M., Wang, Y. B., & Kang, E. S. (2000). Role of the recombinant protein of the platelet receptor for type I collagen in the release of nitric oxide during platelet aggregation. *Thromb Res, 100,* 427–432.

64. Pischel, K. D., Bluestein, H. G., & Woods, V. L. (1988). Platelet glycoprotein Ia, Ic, and IIa are physicochemically indistinguishable from the very late activation antigens adhesion-related proteins of lymphocytes and other cell types. *J Clin Invest, 81,* 505–513.

65. Kunicki, T. J., Orchekowski, R., Annis, D., et al. (1993). Variability of integrin $\alpha_2\beta_1$ activity on human platelets. *Blood, 82,* 2693–2703.

66. Kritzik, M., Savage, B., Nugent, D. J., et al. (1998). Nucleotide polymorphisms in the $\alpha_2$ gene define multiple alleles which are associated with differences in platelet $\alpha_2\beta_1$ density. *Blood, 92,* 2382–2388.

67. Watson, S. P. (1999). Collagen receptor signaling in platelets and megakaryocytes. *Thromb Haemost, 82,* 365–376.

68. Keely, P. J., & Parise, L. V. (1996). The $\alpha_2\beta_1$ integrin is a necessary co-receptor for collagen-induced activation of Syk and subsequent phosphorylation of phospholipase Cy2 in platelets. *J Biol Chem, 271,* 26668–26676.

69. Kamiguti, A. S., Hay, C. R., & Zuzel, M. (1996). Inhibition of collagen-induced platelet aggregation as the result of cleavage of $\alpha_2\beta_1$-integrin by the snake venom metalloproteinase jararhagin. *Biochem J, 320,* 635–641.

70. Kamiguti, A. S., Markland, F. S., Zhou, Q., et al. (1997). Proteolytic cleavage of the $\beta1$ subunit of platelet $\alpha_2\beta_1$ integrin by the metalloproteinase jararhagin compromises collagen-stimulated phosphorylation of pp72. *J Biol Chem, 272,* 32599–32605.

71. Huang, T.- F., Liu, C.- Z., & Yang, S.- H. (1995). Aggretin, a novel platelet-aggregation inducer from snake *Calloselasma rhodostoma* venom, activates phospholipase C by acting as a glycoprotein Ia/IIa agonist. *Biochem J, 309,* 1021–1027.

72. Teng, C. M., Ko, F. N., Tsai, I. H., et al. (1993). Trimucytin: A collagen-like aggregating inducer isolated from *Trimeresurus mucrosquamatus* snake venom. *Thromb Haemost, 69,* 286–292.

73. Coller, B. S., Beer, J. H., Scudder, L. E., et al. (1989). Collagen-platelet interactions: Evidence for a direct interaction of collagen with platelet GPIa/IIa and an indirect interaction with platelet GPIIb/IIIa mediated by adhesive proteins. *Blood, 74,* 182–192.

74. Rudd, C. E., Janssen, O., Cai, Y.- C., et al. (1996). Two-step TCRz/CD3-CD4 and CD28 signaling in T cells: SH2/SH3 domains, protein-tyrosine and lipid kinases. *Immunol Today, 15,* 225–234.

75. Knight, C. G., Morton, L. F., Onley, D. J., et al. (1998). Identification in collagen type I of an integrin $\alpha_2\beta_1$-binding site containing an essential GER sequence. *J Biol Chem, 273,* 33287–33294.

76. Nieswandt, B., & Watson, S. P. (2003). Platelet–collagen interaction: Is GPVI the central receptor? *Blood, 102,* 449–461.

77. Kehrel, B., Wierwille, S., Clemetson, K. J., et al. (1998). Glycoprotein VI is a major collagen receptor for platelet activation: It recognizes the platelet-activating quaternary structure of collagen, whereas CD36, glycoprotein IIb/IIIa, and von Willebrand factor do not. *Blood, 91,* 491–499.

78. Blake, R. A., Schieven, G. L., & Watson, S. P. (1994). Collagen stimulates tyrosine phosphorylation of phospholipase Cγ2 but not phospholipase Cγl in human platelets. *FEBS Lett, 353,* 212–216.

79. Daniel, J. L., Dangelmaier, C., & Smith, J. B. (1994). Evidence for a role for tyrosine phosphorylation of phospholipase Cγ2 in collagen-induced platelet cytosolic calcium mobilisation. *Biochem J, 302,* 617–622.

80. Ichinohe, T., Takayama, H., Ezumi, Y., et al. (1997). Collagen-stimulated activation of Syk but not c-Src is severely compromised in human platelets lacking membrane glycoprotein VI. *J Biol Chem, 272,* 63–68.

81. Yanaga, F., Poole, A., Asselin, J., et al. (1995). Syk interacts with tyrosine-phosphorylated proteins in human platelets activated by collagen and crosslinking of the FcγIIa receptor. *Biochem J, 311,* 471–478.

82. Poole, A., Gibbins, J. M., Turner, M., et al. (1997). The Fc receptor gamma-chain and the tyrosine kinase Syk are essential for activation of mouse platelets by collagen. *EMBO J, 16,* 2333–2341.

83. Tsuji, M., Ezumi, Y., Arai, M., et al. (1997). A novel association of Fc receptor γ-chain with glycoprotein VI and their

co-expression as a collagen receptor in human platelets. *J Biol Chem, 272,* 23528–23531.

84. Jandrot–Perrus, M., Lagrue, A. H., Okuma, M., et al. (1997). Adhesion and activation of human platelets induced by convulxin involves glycoprotein GI and integrin $\alpha_2\beta_1$. *J Biol Chem, 272,* 27035–27041.

85. Polgar, J., Clemetson, J. M., Kehrel, B. E., et al. (1997). Platelet activation and signal transduction by convulxin, a C-type lectin from *Crotalus durissus terrificus* (tropical rattlesnake) venom via the p62/GPVI collagen receptor. *J Biol Chem, 272,* 13576–13583.

86. Morton, L. F., Hargreaves, P. G., Farndale, R. W., et al. (1995). Integrin $\alpha_2\beta_1$-independent activation of platelets by simple collagen-like peptides: Collagen tertiary (triple-helical) and quaternary (polymeric) structures are sufficient alone for $\alpha_2\beta_1$-independent platelet reactivity. *Biochem J, 306,* 337–344.

87. Gibbins, J. M., Briddon, S., Shutes, A., et al. (1998). The p85 subunit of phosphatidylinositol 3-kinase associates with the Fc receptor γ-chain and linker for activator T cells (LAT) in platelets stimulated by collagen and convulxin. *J Biol Chem, 273,* 34437–34443.

88. Asselin, J., Gibbins, J. M., Achison, M., et al. (1997). A collagen-like peptide stimulates tyrosine phosphorylation of syk and phospholipase Cγ2 in platelets independent of the integrin $\alpha_2\beta_1$. *Blood, 89,* 1235–1242.

89. Massberg, S., Gawaz, M., Gruner, S., et al. (2003). A crucial role of glycoprotein VI for platelet recruitment to the injured arterial wall in vivo. *J Exp Med, 197,* 41–49.

90. Gralnick, H. R., Williams, S. B., & Coller, B. S. (1985). Asialo von Willebrand factor interactions with platelets. Interdependence of glycoproteins Ib and IIb/IIIa for binding and aggregation. *J Clin Invest, 75,* 19–25.

91. Weiss, H. J., Hawiger, J., Ruggeri, Z. M., et al. (1989). Fibrinogen-independent platelet adhesion and thrombus formation on subendothelium mediated by glycoprotein IIb-IIIa complex at high shear rate. *J Clin Invest, 83,* 288–297.

92. De Marco, L., Girolami, A., Russell, S., et al. (1985). Interaction of asialo von Willebrand factor with glycoprotein Ib induces fibrinogen binding to the glycoprotein IIb/IIIa complex and mediates platelet aggregation. *J Clin Invest, 75,* 1198–1203.

93. Ruggeri, Z. M., Fitzgerald, G. A., & Shattil, S. J. (1999). Platelet thrombus formation and antiplatelet therapy. K. R. Chien, J. L. Breslow, J. M. Leiden, R. D. Rosenberg, & C. E. Seidman (Eds.), In *Molecular basis of cardiovascular disease: A companion to Braunwald's heart disease* (pp. 566–589), W.B. Saunders Company, Philadelphia.

94. Offermanns, S. (2000). The role of heterotrimeric G proteins in platelet activation. *Biol Chem, 381,* 389–396.

95. Fabre, J.- E., Nguyen, M., Latour, A., et al. (1999). Decreased platelet aggregation, increased bleeding time and resistance to thromboembolism in P2Y$_1$-deficient mice. *Nat Med, 5,* 1199–1202.

96. Savage, B., Cattaneo, M., & Ruggeri, A. M. (2001). Mechanisms of platelet aggregation. *Curr Opin Hematol, 8,* 270–276.

97. Hollopeter, G., Jantzen, H.- M., Vincent, D., et al. (2001). Identification of the platelet ADP receptor targeted by antithrombotic drugs. *Nature, 409,* 202–206.

98. Nurden, A. T., & Nurden, P. (2003). Advantages of fast-acting ADP receptor blockade in ischemic heart disease. *Arterioscler Thromb Vasc Biol, 23,* 158–159.

99. Turner, N. A., Moake, J. L., & McIntire, L. V. (2001). Blockade of adenosine diphosphate receptors P2Y$_{12}$ and P2Y$_1$ is required to inhibit platelet aggregation in whole blood under flow. *Blood, 98,* 3340–3345.

100. Remijn, J. A., Wu, Y. P., Jeninga, E. H., et al. (2002). Role of ADP receptor P2Y(12) in platelet adhesion and thrombus formation in flowing blood. *Arterioscler Thromb Vasc Biol, 22,* 686–691.

101. Dorsam, R. T., & Kunapuli, S. P. (2004). Central role of the P2Y12 receptor in platelet activation. *J Clin Invest, 113,* 341–345.

102. Lenain, N., Freund, M., Leon, C., et al. (2003). Inhibition of localized thrombosis in P2Y1 deficient mice and rodents treated with MRS2179, a P2Y receptor antagonist. *J Thromb Haemost, 1,* 1144–1149.

103. Baurand, A., & Gachet, C. (2003). The P2Y1 receptor as a target for new antithrombotic drugs: A review of the P2Y1 antagonist MRS-2179. *Cardiovasc Drug Rev, 21,* 67–76.

104. Goto, S., Tamura, N., & Handa, S. (2002). Effects of adenosine 5-diphosphate (ADP) receptor blockade on platelet aggregation under flow. *Blood, 99,* 4644–4645.

105. Ikeda, Y., Handa, M., Kawano, K., et al. The role of von Willebrand Factor and fibrinogen in platelet aggregation under varying shear stress. *J Clin Invest, 87,* 1234–1240.

106. Moake, J. L., Turner, N. A., Stathopoulos, N. A., et al. (1988). Shear-induced platelet aggregation can be mediated by vWF released from platelets, as well as by exogenous large or unusually large vWF multimers, requires adenosine diphosphate, and is resistant to aspirin. *Blood, 71,* 1366–1374.

107. Chow, T. W., Hellums, J. D., Moake, J. L., et al. (1992). Shear stress-induced von Willebrand factor binding to platelet glycoprotein Ib initiates calcium influx associated with aggregation. *Blood, 80,* 113–120.

108. Monroe, D. M., Hoffman, M., & Roberts, H. R. (2002). Platelets and thrombin generation. *Arterioscler Thromb Vasc Biol, 22,* 1381–1389.

109. Kahn, M. L., Nakanishi–Matsui, M., Shapiro, M. J., et al. (1999). Protease-activated receptors 1 and 4 mediate activation of human platelets by thrombin. *J Clin Invest, 103,* 879–887.

110. Vu, T.-K. H., Hung, D. T., Wheaton, V. I., et al. (1991). Molecular cloning of a functional thrombin receptor reveals a novel proteolytic mechanism of receptor activation. *Cell, 64,* 1057–1068.

111. Xu, W.-F., Andersen, H., Whitmore, T. E., et al. (1998). Cloning and characterization of human protease-activated receptor 4. *Proc Natl Acad Sci U S A, 95,* 6642–6646.

112. Faruqi, T. R., Weiss, E. J., Shapiro, M. J., et al. (2000). Structure-function analysis of protease-activated receptor 4 tethered ligand peptides. Determinants of specificity and utility in assays of receptor function. *J Biol Chem, 275,* 19728–19734.

113. Coughlin, S. R. (2001). Protease-activated receptors in vascular biology. *Thromb Haemost, 86,* 298–307.

114. Shapiro, M. J., Weiss, E. J., Faruqi, T. R., et al. (2000). Protease-activated receptors 1 and 4 are shut off with distinct kinetics after activation by thrombin. *J Biol Chem, 275,* 25216–25221.

115. Covic, L., Gresser, A. L., & Kuliopulos, A. (2000). Biphasic kinetics of activation and signaling for PAR1 and PAR4 thrombin receptors in platelets. *Biochemistry, 39,* 5458–5467.

116. Mazzucato, M., De Marco, L., Masotti, A., et al. (1998). Characterization of the initial α-thrombin interaction with glycoprotein Ib α in relation to platelet activation. *J Biol Chem, 273,* 1880–1887.

117. Adam, F., Guillin, M. C., & Jandrot–Perrus, M. (2003). Glycoprotein Ib-mediated platelet activation. A signalling pathway triggered by thrombin. *Eur J Biochem, 270,* 2959–2970.

118. De Marco, L., Mazzucato, M., Masotti, A., et al. (1991). Function of glycoprotein Ibα in platelet activation induced by α-thrombin. *J Biol Chem, 266,* 23776–23783.

119. Celikel, R., McClintock, R. A., Roberts, J. R., et al. (2003). Modulation of α-thrombin function by distinct interactions with platelet glycoprotein Ibα. *Science, 301,* 218–221.

120. Dumas, J. J., Kumar, R., Seehra, J., et al.(2003). Crystal structure of the GPIbα-thrombin complex essential for platelet aggregation. *Science, 301,* 222–226.

121. Phillips, D. R., & Poh–Agin, P. (1977). Platelet plasma membrane glycoproteins. Identification of a proteolytic substrate for thrombin. *Biochem Biophys Res Commun, 75,* 940–947.

122. Ramakrishnan, V., Reeves, P. S., DeGuzman, F., et al. (1999). Increased thrombin responsiveness in platelets from mice lacking glycoprotein V. *Proc Natl Acad Sci U S A, 96,* 13336–13341.

123. Ramakrishnan, V., DeGuzman, F., Bao, M., et al. (2001). A thrombin receptor function for platelet glycoprotein Ib-IX unmasked by cleavage of glycoprotein V. *Proc Natl Acad Sci U S A, 98,* 1823–1828.

124. Klages, B., Brandt, U., Simon, M. I., et al. (1999). Activation of G12/G13 results in shape change and Rho/Rho-kinase-mediated myosin light chain phosphorylation in mouse platelets. *J Cell Biol, 144,* 745–754.

125. Huang, J. S., Ramamurthy, S. K., Lin, X., et al. (2004). Cell signalling through thromboxane A2 receptors. *Cell Signal, 16,* 521–533.

126. Berridge, M. J. (1993). Inositol trisphosphate and calcium signalling. *Nature, 361,* 315–325.

127. Bauer, M., Retzer, M., Wilde, J. I., et al. (1999). Dichotomous regulation of myosin phosphorylation and shape change by Rho-kinase and calcium in intact human platelets. *Blood, 94,* 1665–1672.

128. Alevriadou, B. R., Moake, J. L., Turner, N. A., et al. (1993). Real-time analysis of shear-dependent thrombus formation and its blockade by inhibitors of von Willebrand factor binding to platelets. *Blood, 81,* 1263–1276.

129. Barstad, R. M., Orvim, U., Hamers, M. J., et al. (1996). Reduced effect of aspirin on thrombus formation at high shear and disturbed laminar blood flow. *Thromb Haemost, 75,* 827–832.

130. Roald, H. E., Orvim, U., Bakkern, I., et al. (1994). Modulation of thrombotic responses in moderately stenosed arteries by cigarette smoking and aspirin ingestion. *Arterioscler Thromb, 14,* 617–621.

131. Barstad, R. M., Roald, H. E., Cui, Y., et al. (1994). A perfusion chamber developed to investigate thrombus formation and shear profiles in flowing native human blood at the apex of well-defined sciences. *Arterioscler Thromb, 14,* 1984–1991.

132. Born, G. V. R. (1962). Aggregation of blood platelets by adenosine diphosphate and its reversal. *Nature, 194,* 927–929.

133. Lefkovits, J., Plow, E. F., & Topol, E. J. (1995). Platelet glycoprotein IIb/IIIa receptors in cardiovascular medicine. *N Engl J Med, 332,* 1553–1559.

134. Peterson, D. M., Stathopoulos, N. A., Giorgio, T. D., et al. (1987). Shear-induced platelet aggregation requires von Willebrand factor and platelet membrane glycoproteins Ib and IIb-IIIa. *Blood, 69,* 625–628.

135. Goto, S., Ikeda, Y., Saldivar, E., et al. (1998). Distinct mechanisms of platelet aggregation as a consequence of different shearing flow conditions. *J Clin Invest, 101,* 479–486.

136. Ruggeri, Z. M., Dent, J. A., & Saldivar, E. (1999). Contribution of distinct adhesive interactions to platelet aggregation in flowing blood. *Blood, 94,* 172–178.

137. Goto, S., Salomon, D. R., Ikeda, Y., et al. (1995). Characterization of the unique mechanism mediating the shear-dependent binding of soluble von Willebrand factor to platelets. *J Biol Chem, 270,* 23352–23361.

138. Ni, H., Denis, C. V., Subbarao, S., et al. (2000). Persistence of platelet thrombus formation in arterioles of mice lacking both von Willebrand factor and fibrinogen. *J Clin Invest, 106,* 385–392.

139. Weiss, H. J., & Rogers, J. (1971). Fibrinogen and platelets in the primary arrest of bleeding: Studies in two patients with congenital afibrinogenemia. *N Engl J Med, 285,* 369–374.

140. Nichols, W. C., Cooney, K. A., Ginsburg, D., et al. (2003). Von Willebrand disease. In J. Loscalzo & A. I. Schafer (Eds.), *Thrombosis and hemorrhage* (Vol. 1., pp. 539–559). Philadelphia: Lippincott Williams and Wilkins.

141. Ni, H., Yuen, P. S., Papalia, J. M., et al. (2003). Plasma fibronectin promotes thrombus growth and stability in injured arterioles. *Proc Natl Acad Sci USA, 100,* 2415–2419.

142. Moake, J. L. (2002). Thrombotic thrombocytopenic purpura: the systemic clumping "plague." *Annu Rev Med, 53,* 75–88.

143. Wagner, D. D. (1993). The Weibel–Palade body: The storage granule for von Willebrand factor and P-selectin. *Thromb Haemost, 70,* 105–110.

144. Lopez–Fernandez, M. F., Ginsberg, M. H., Ruggeri, Z. M., et al. (1982). Multimeric structure of platelet factor VIII/von Willebrand factor. The presence of larger multimers and their reassociation with thrombin-stimulated platelets. *Blood, 60,* 1132–1138.

145. Zimmerman, T. S., Dent, J. A., Ruggeri, Z. M., et al. (1986). Subunit composition of plasma von Willebrand factor. Cleavage is present in normal individuals, increased in IIA and IIB von Willebrand disease, but minimal in variants with aberrant structure of individual oligomers (Types IIC, IID and IIE). *J Clin Invest, 77,* 947–951.

146. Levy, G. G., Nichols, W. C., Lian, E. C., et al. (2001). Mutations in a member of the ADAMTS gene family cause thrombotic thrombocytopenic purpura. *Nature, 413,* 488–494.

147. Dent, J. A., Berkowitz, S. D., Ware, J., et al. (1990). Identification of a cleavage site directing the immunochemical detection of molecular abnormalities in type IIA von Willebrand factor. *Proc Natl Acad Sci U S A, 87,* 6306–6310.

148. Dent, J. A., Galbusera, M., & Ruggeri, Z. M. (1991). Heterogeneity of plasma von Willebrand factor multimers resulting from proteolysis of the constituent subunit. *J Clin Invest, 88,* 774–782.

149. Dong, J.-F., Moake, J. L., Nolasco, L., et al. (2002). ADAMTS-13 rapidly cleaves newly secreted ultra-large von Willebrand factor multimers on the endothelial surface under flowing conditions. *Blood, 100,* 4033–4039.

150. Donadelli, R., Orje, J. N., Capoferri, C., et al. (2006). Size regulation of von Willebrand factor-mediated platelet thrombi by ADAMTS-13 in flowing blood. *Blood, 107,* 1943–1950.

151. Nesbitt, W. S., Giuliano, S., Kulkarni, S., et al. (2003). Intercellular calcium communication regulates platelet aggregation and thrombus growth. *J Cell Biol, 160,* 1151–1161.

152. Goto, S., Tamura, N., Ishida, H., et al. (2006). Dependence of platelet thrombus stability on sustained glycoprotein IIb/IIIa activation through ADP receptor stimulation and cyclic calcium signaling. *J Am Coll Cardiol, 47,* 155–162.

153. Giandomenico, G., Dellas, C., Czekay, R. P., et al. (2005). The leptin receptor system of human platelets. *J Thromb Haemost, 3,* 1042–1049.

154. Konstantinides, S., Schafer, K., Koschnick, S., et al. (2001). Leptin-dependent platelet aggregation and arterial thrombosis suggests a mechanism for atherothrombotic disease in obesity. *J Clin Invest, 108,* 1533–1540.

155. Konstantinides, S., Schafer, K., Neels, J. G., et al. (2004). Inhibition of endogenous leptin protects mice from arterial and venous thrombosis. *Arterioscler Thromb Vasc Biol, 24,* 2196–2201.

156. Andre, P., Prasad, K. S., Denis, C. V., et al. (2002). CD40L stabilizes arterial thrombi by a β₃ integrin-dependent mechanism. *Nat Med, 8,* 247–252.

157. Angelillo-Scherrer, A., de Frutos, P., Aparicio, C., et al. (2001). Deficiency or inhibition of Gas6 causes platelet dysfunction and protects mice against thrombosis. *Nat Med, 7,* 215–221.

158. Balogh, I., Hafizi, S., Stenhoff, J., et al. (2005). Analysis of Gas6 in human platelets and plasma. *Arterioscler Thromb Vasc Biol, 25,* 1280–1286.

159. Gould, W. R., Baxi, S. M., Schroeder, R., et al. (2005). Gas6 receptors Axl, Sky and Mer enhance platelet activation and regulate thrombotic responses. *J Thromb Haemost, 3,* 733–741.

160. Prevost, N., Woulfe, D. S., Jiang, H., et al. (2005). Eph kinases and ephrins support thrombus growth and stability by regulating integrin outside-in signaling in platelets. *Proc Natl Acad Sci USA, 102,* 9820–9825.

161. Nanda, N., Bao, B., Lin, H., et al. (2005). Platelet endothelial aggregation receptor 1 (PEAR1), a novel epidermal growth factor repeat-containing transmembrane receptor, participates in platelet contact-induced activation. *J Biol Chem, 280,* 24680–24689.

162. Nanda, N., Andre, P., Bao, M., et al. (2005). Platelet aggregation induces platelet aggregate stability via SLAM family receptor signaling. *Blood, 106,* 3028–3034.

163. Falati, S., Gross, P., Merrill-Skoloff, G., et al. (2002). Real-time in vivo imaging of platelets, tissue factor and fibrin during arterial thrombus formation in the mouse. *Nat Med, 8,* 1175–1180.

164. Mayadas, T. N., Johnson, R. C., Rayburn, H., et al. (1993). Leukocyte rolling and extravasation are severely compromised in P selectin-deficient mice. *Cell, 74,* 541–554.

165. Frenette, P. S., Johnson, R. C., Hynes, R. O., et al. (1995). Platelets roll on stimulated endothelium in vivo: An interaction mediated by endothelial P-selectin. *Proc Natl Acad Sci U S A, 92,* 7450–7454.

166. Sim, D., Flaumenhaft, R., Furie, B. C., et al. (2005). Interactions of platelets, blood-borne tissue factor, and fibrin during arteriolar thrombus formation in vivo. *Microcirculation, 12,* 301–311.

167. Hrachovinova, I., Cambien, B., Hafezi-Moghadam, A., et al. (2003). Interaction of P-selectin and PSGL-1 generates microparticles that correct hemostasis in a mouse model of hemophilia A. *Nat Med, 9,* 1020–1025.

168. Falati, S., Liu, Q., Gross, P., et al. (2003). Accumulation of tissue factor into developing thrombi in vivo is dependent upon microparticle P-selectin glycoprotein ligand 1 and platelet P-selectin. *J Exp Med, 197,* 1585–1598.

169. Renne, T., Pozgajova, M., Gruner, S., et al. (2005). Defective thrombus formation in mice lacking coagulation factor XII. *J Exp Med, 202,* 271–281.

170. Konstantinides, S., Schafer, K., Thinnes, T., et al. (2001). Plasminogen activator inhibitor-1 and its cofactor vitronectin stabilize arterial thrombi after vascular injury in mice. *Circulation, 103,* 576–583.

# CHAPTER 19

# Interactions Between Platelets and the Coagulation System

**Beth A. Bouchard, Saulius Butenas, Kenneth G. Mann, and Paula B. Tracy**

*Department of Biochemistry, University of Vermont College of Medicine, Burlington, Vermont*

## I. Overview of the Role of Platelets in Hemostasis

After disruption of the integrity of the endothelial cell lining of a blood vessel, circulating platelets function to localize, amplify, and sustain the coagulant response at the injury site. Platelets provide the first line of defense in plugging the leaky vessel by adhering to the site of injury. Platelet adhesion and activation are mediated primarily by their interactions with the subendothelial proteins von Willebrand factor (VWF) and collagen. Platelets bind to subendothelial VWF primarily via the platelet membrane glycoprotein (GP) Ib-IX-V complex, and to collagen via the integrin $\alpha2\beta1$ and GPVI.[1–3] For a full discussion of the platelet receptors regulating adhesion, see Chapters 6, 7, and 18.

Platelet adherence to collagen via GPVI, which is non-covalently associated with the Fc-receptor γ-chain, triggers the intracellular signaling events that result in platelet activation (Chapter 16),[4–6] an essential step in hemostasis. One platelet activation event, platelet aggregation, serves to prevent additional blood loss from a damaged vessel in the earliest stage of hemostasis. Platelet aggregation is a result of the cross-linking of activated integrin $\alpha$IIb$\beta3$ (GPIIb-IIIa; Chapter 8) molecules expressed on the surface of two different platelets by fibrinogen to form a temporary platelet plug that prevents additional blood loss from the leaky vessel.[7]

Platelet adherence and aggregation at a site of vascular injury serve not only to plug the damaged vessel temporarily, but also to localize subsequent procoagulant events to the injury site to prevent systemic activation of coagulation. Furthermore, activated platelets actively regulate the propagation of the coagulation reaction by (a) releasing a number of bioactive peptides and small molecules from their α- and dense granules that recruit additional platelets to the growing thrombus (Chapter 15), as well as participate directly in procoagulant events including a unique factor V molecule; (b) expressing specific, high-affinity membrane receptors for coagulation proteases, zymogens, and cofactors; and (c)

amplifying the initiating stimulus to lead to explosive thrombin generation. Platelets also function to sustain the procoagulant response by protecting the bound coagulant enzymes from inactivation/inhibition. This overall process provides for rapid fibrin polymerization at the wound site, transforming blood in the fluid phase to a solid phase plug.

## II. Enzyme Complex Formation Regulates the Coagulant Response

In this section, information is provided describing the macromolecular $Ca^{2+}$-dependent complexes that assemble on the damaged and/or activated cell membranes present at sites of vascular injury. The coagulation factors clearly associated with hemostatic risk (including tissue factor; factors V, VII, VIII, IX, and X; and prothrombin)[8] are discussed first.

Current literature supports the hypothesis that reactions of the blood coagulation cascade are propagated through the formation of discrete enzymatic complexes composed of a vitamin K-dependent serine protease and a nonenzymatic cofactor protein that are assembled on a membrane surface in a $Ca^{2+}$-dependent manner.[9,10] Under normal conditions, the blood coagulation response is initiated when subendothelial tissue factor, an integral membrane protein, is exposed/expressed to blood flow after damage to the intact endothelium.[11,12] Alternatively, tissue factor is also expressed under pathological conditions on the surface of activated monocytes or endothelial cells.[13–16] The serine protease, factor VIIa (activated factor VII), circulating in blood at subnanomolar concentrations,[17,18] binds to tissue factor to form extrinsic Xase, which activates the zymogens, factor IX and factor X, to their corresponding serine proteases, factor IXaβ and factor Xa[19–22] (Fig. 19-1). Subsequent to formation of a tissue factor/factor VIIa complex, the limited amounts of factor Xa produced assemble into prothrombinase through $Ca^{2+}$-dependent interactions with membrane-bound factor Va initially to generate picomolar amounts of thrombin.[23–25] Thrombin, once formed, amplifies, propagates, and sustains the coagulant response in several ways. Thrombin is a potent platelet agonist (Chapter 9), thereby recruiting more platelets to participate in and sustain coagulation. Thrombin activates additional factor V to factor Va, as well as factor VIII to factor VIIIa to ensure continued prothrombinase and intrinsic Xase formation and function respectively.[26–28] Thus, membrane-bound factor VIIIa binds factor IXaβ in a $Ca^{2+}$-dependent manner forming

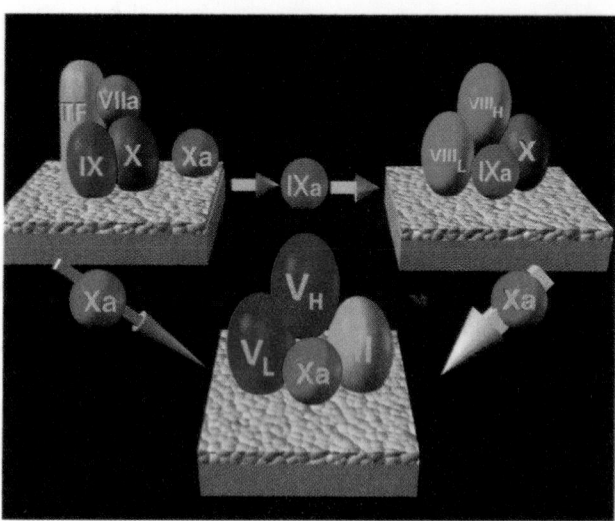

**Figure 19-1.** Physiologically relevant enzymatic complexes of hemostasis. Only those proteins known to be associated with hemorrhage or thrombosis are shown. Each procoagulant enzyme complex — extrinsic Xase, intrinsic Xase, and prothrombinase — is represented on a membrane surface. Each enzymatic complex is composed of a serine protease factor VIIa, factor IXaβ, or factor Xa bound to a cofactor protein, tissue factor (TF), factor VIIIa, or factor Va on an appropriate membrane surface. The vitamin K-dependent substrates for each complex — factor IX, factor X, and prothrombin — are also shown. $V_H$, factor Va heavy chain; $V_L$, factor Va light chain; $VIII_H$, factor VIIIa heavy chain; $VIII_L$, factor VIIIa light chain. (From Coagulation Explosion, Vermont Business Graphics, Kenneth G. Mann, 1997, with permission.) Oral presentation owned by Ken Mann.

intrinsic Xase,[29–32] which activates factor X 50 to 100 times faster than the factor VIIa–tissue factor complex (i.e., extrinsic Xase).[22,33,34] The thrombin, continuously produced via prothrombinase, further amplifies its own generation by activating factor XI[35] and continuing to activate additional platelets and factors V and VIII.[28,36] Thrombin also cleaves fibrinogen[37,38] and factor XIII[39,40] to form the insoluble, cross-linked fibrin clot.[41,42]

The significance of the individual components of each of these essential enzymatic complexes is best illustrated by the changes in catalytic efficiency that they display in proteolysis of their natural substrates. In the case of extrinsic Xase, the proteolytic activity of factor VIIa in the absence of the other components of the enzymatic complex is negligible.[20,21] Although the addition of $Ca^{2+}$ and phospholipids (a representative membrane surface) slightly increases the proteolytic activity of factor VIIa (Table 19-1),[20] only the presence of the cofactor, tissue factor, leads to the formation of an effective enzymatic complex. Similar observations have been made with the individual constituents of intrinsic Xase[34] and prothrombinase.[24,43] Deletion of either the membrane surface or the cofactor from each of the complexes essentially abolishes enzymatic activity. Kinetic studies of all the enzyme complexes involved in propagating or regulating the coagulant response indicate that when comparing the catalytic efficiency observed for each of the complete complexes versus those of the enzyme alone, a complete complex is $10^5$- to $10^9$-fold more efficient in catalyzing the proteolysis of its substrate (Table 19-2).[20,33]

A great deal of information has been obtained detailing the mechanisms that govern how the cofactors, proteases, and substrates interact, as well as the importance of an appropriately activated cell membrane surface in mediating these protein–protein interactions. In contrast, limited information has been obtained that details the structural nature of specific cell membrane receptors, which mediate the assembly of these various complexes on platelets. Anionic membrane phospholipids appear to make significant contributions to complex assembly and may, in some instances, constitute an important membrane binding site. However, several lines of evidence suggest that complex assembly

**Table 19-1: Activation of Factor X by Factor VIIa in the Presence of Various Cofactors of Extrinsic Xase**

| Cofactor | Concentration | $K_m$, μM | $k_{cat}$, min$^{-1}$ | $k_{cat}/K_m$, μM$^{-1}$·min$^{-1}$ |
|---|---|---|---|---|
| None | NA | >20 | $>1.5 \times 10^{-4}$ | ND |
| CaCl$_2$ | 2.5 mM | 2.10 | $1.0 \times 10^{-4}$ | $4.8 \times 10^{-5}$ |
| Phospholipid (PCPS) | 21 μM | 0.25 | 0.016 | 0.062 |
| Tissue factor | 9.4 pM | 0.23 | 186 | 885 |

NA, not applicable; ND, not determined; PCPS, phosphatidylcholine phosphatidylserine.
Adapted from Komiyama et al.,[20] with permission.

**Table 19-2: Kinetic Properties of the Vitamin K-Dependent Enzymes and Enzymatic Complexes**

| Enzyme | Substrate | $K_m$, µM | $k_{cat}$, min$^{-1}$ | $k_{cat}/K_m$, µM$^{-1}$ s$^{-1}$ | Efficiency Ratio |
|---|---|---|---|---|---|
| Factor VIIa | Factor IX | ND | ND | ND | |
| Factor VIIa/TF/PCPS/CaCl$_2$ | Factor IX | 0.016 | 91.9 | 5560 | |
| Factor VIIa | Factor X | 2.1 | $1.0 \times 10^{-4}$ | $4.8 \times 10^{-5}$ | |
| Factor VIIa/TF/PCPS/CaCl$_2$ | Factor X | 0.23 | 186 | 885 | $1.8 \times 10^7$ |
| Factor IXa | Factor X | 300 | 0.002 | $6.6 \times 10^{-6}$ | |
| Factor IXa/VIIIa/PCPS/CaCl$_2$ | Factor X | 0.063 | 500 | 7937 | $1.2 \times 10^9$ |
| Factor Xa | Factor II | 131 | 0.6 | $4.6 \times 10^{-3}$ | |
| Factor Xa/Va/PCPS/CaCl$_2$ | Factor II | 1.0 | 5016 | 5016 | $1.1 \times 10^6$ |
| Factor IIa | Protein C | 60 | 1.2 | 0.02 | |
| Factor IIa/Tm/PCPS/CaCl$_2$ | Protein C | 0.1 | 214 | 2140 | $1.1 \times 10^5$ |

ND, not determined; PCPS, phosphatidylcholine phosphatidylserine; TF, tissue factor; Tm, thrombomodulin.
Adapted from Komiyama et al.[20] and Mann et al.,[33] with permission.

requires membrane biochemical components other than lipids. These concepts are discussed in more detail in subsequent sections.

Proteins known to initiate the contact pathway (factor XII, prekallikrein, and high-molecular weight kininogen) are not shown in Figure 19-1, because patients with a hereditary deficiency of these proteins have no bleeding problems, despite their markedly prolonged partial thromboplastin time. The physiological role of factor XI is more difficult to assess, because the plasma level of factor XI activity alone is a poor predictor of a bleeding diathesis.[44–46] Some patients who are homozygous for factor XI deficiency are asymptomatic whereas others express a severe bleeding tendency. These disparate observations may be explained by the factor XI present in and released from platelets subsequent to their activation.[47] The known coagulant function of factor XIa is its proteolytic activation of factor IX to factor IXaβ.[48–51] Because thrombin effectively activates factor XI,[35,52,53] this positive feedback activation by thrombin may be important physiologically. The role of platelet-derived factor XI in the coagulant response is discussed in detail later.

Although the clear consequences of thrombin activation of factor XI have yet to be elucidated, thrombin also initiates many different cellular processes that may positively or negatively influence the normal hemostatic balance, fibrinolysis, and wound repair. Thrombin's interaction with the protein C system is essential for effective regulation of the coagulant response. Thrombin becomes a crucial anticoagulant by binding to its endothelial cell cofactor, thrombomodulin (Fig. 19-2). Protein Case assembled on the endothelial cell surface catalyzes the conversion of the zymogen protein C to the active serine protease, activated protein C.[54–56] Activated protein C effectively limits thrombin generation to the site of injury, primarily through the proteolytic inactivation of membrane-bound,

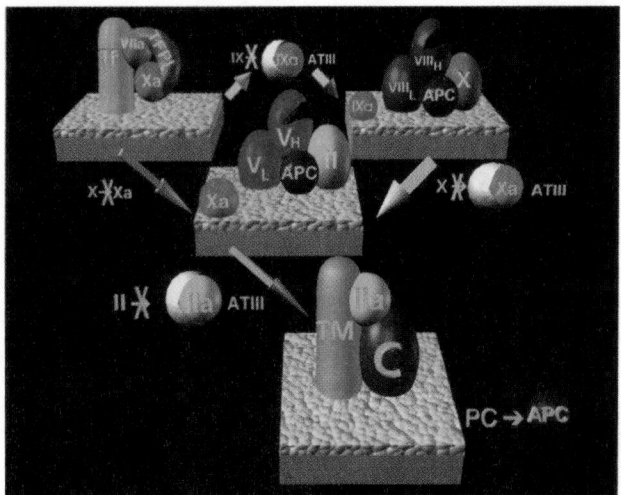

**Figure 19-2.** Downregulation of the procoagulant response. Thrombin (factor IIa) binds to thrombomodulin (TM) expressed by vascular cells and proteolytically activates protein C (PC) to activated protein C (APC). APC downregulates the procoagulant response by proteolytically inactivating membrane-bound, yet uncomplexed, factors VIIIa and Va. Antithrombin III (ATIII) stoichiometrically inhibits thrombin, factor Xa, and factor IXaβ, which are free in solution. Another stoichiometric inhibitor, tissue factor pathway inhibitor (TFPI), inhibits both factor Xa and the tissue factor–factor VIIa–factor Xa complex. (From "Coagulation Explosion," Vermont Business Graphics, Kenneth G. Mann, 1997, by permission.) Oral presentation owned by Ken Mann.

yet uncomplexed, factors V and Va.[57–60] Although factors VIII and VIIIa are also activated protein C substrates,[61,62] this mode of inactivation does not appear to be of any significance physiologically.[61] The inactivating cleavages of both factors Va and VIIIa are somewhat accelerated by protein S,[63] although the major function of this protein in the regulation of thrombin generation has not yet been clarified.[64–66]

The procoagulant processes are also attenuated by various stoichiometric inhibitors that inactivate serine proteases (Fig. 19-2). One of these inhibitors, antithrombin III, appears to be the most important quantitatively. Antithrombin III is able to neutralize effectively three of the serine proteases produced during the blood coagulation process (factors IXa and Xa, and thrombin). Effective antithrombin III inhibition of factors IXa and Xa requires that the proteases are free in solution, because complex formation with their respective cofactors protects them from inhibition.[67] Tissue factor pathway inhibitor, another stoichiometric inhibitor of blood coagulation (Fig. 19-2), neutralizes both factor Xa and the factor VIIa–tissue factor–factor Xa complex.[68–70] These proteolytic and stoichiometric inactivating or inhibitor processes are essential to terminating the coagulant response and to preventing systemic coagulation.

Adherent and activated platelets at the injury site provide the vast majority of membrane "receptors" that promote and regulate the assembly and function of intrinsic Xase and prothrombinase. They also store and, subsequent to their activation, release several proteins that are known or appear to play important roles in sustaining the coagulant response (i.e., factor Va and factor XI respectively). Furthermore, they sustain the generation of procoagulant activity by protecting the proteases and cofactors from inactivation/inhibition by various constitutive plasma inhibitors or formed inactivators.[57,58,60,71–74] Thus, when activated, platelets essentially function as procoagulant machines to sustain coagulant reactions at the site of vascular injury. The mechanisms regulating how platelets participate in amplifying and sustaining the coagulant response is the subject of this chapter.

## III. Model Systems Defining the Role of Platelets in Tissue Factor-Initiated Thrombin Generation

Several laboratories have been studying the dynamics of tissue factor-initiated blood coagulation processes *in vitro* under conditions resembling those that occur *in vivo*. In a synthetic plasma model, physiological concentrations of proteins and platelets generate thrombin subsequent to the addition of relipidated, recombinant tissue factor[26,28,75–82] or cells bearing tissue factor.[83–89] Tissue factor-initiated thrombin generation is studied also in minimally altered whole blood in which activation of the intrinsic pathway (contact pathway) is inhibited with corn trypsin inhibitor — a specific factor XIIa inhibitor.[27,90,91] These physiologically relevant models have all demonstrated that the tissue factor-initiated coagulant response is not uniform over time, but rather can be divided into two reasonably distinct inter-

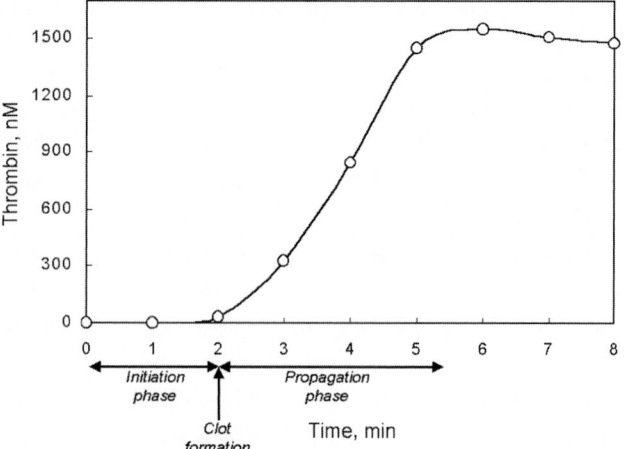

**Figure 19-3.** Thrombin generation over time in a synthetic plasma model system. Thrombin generation is initiated with the addition of 1.25 pM tissue factor to synthetic plasma, which contains plasma concentrations of factors VII/VIIa, IX, X, VIII, V, and prothrombin; as well as 200 μM phospholipids composed of 25% phosphatidylserine and 75% phosphatidylcholine. Thrombin generation is measured over time using chromogenic assays. Thrombin generation can be divided into the initiation phase and propagation phase, with clot formation occurring at the inception of the propagation phase. (Adapted from Butenas et al.,[28] with permission.)

vals: the initiation and propagation phases (Fig. 19-3).[28] During the initiation phase, nanomolar amounts of thrombin are produced as a result of the formation of small amounts of factor Xa catalyzed by the factor VIIa–tissue factor complex.[26–28] In addition, femto- to picomolar amounts of other coagulation enzymes are generated (i.e., factors IXa, Xa, and XIa), platelet activation begins, and clot formation occurs.[27,42,92] These events mark the inception of the propagation phase (Fig. 19-3). This phase is characterized by rapid and quantitative thrombin generation facilitated by the enhanced concentrations of factor Xa made available through the assembly and function of the factor VIIIa–factor IXa complex on the surface of activated platelets. By this point in the reaction, platelet activation is complete, and activation of factors Va and VIIIa is significant. As might be expected, the rate of thrombin generated in the propagation phase is limited by the rate at which factor Xa is generated.[26–28,42,92]

Evaluation of platelet effects in these model systems yields information consistent with clinical observations. Using a synthetic plasma system reconstituted with varying concentrations of platelets, duration of the initiation phase (clotting time) was normal, providing platelets were present at concentrations more than $0.25 \times 10^8$ cells/mL.[81] Higher platelet concentrations did not shorten the initiation phase. In contrast, the maximum rates of thrombin generation observed, as well as the maximum thrombin concentrations achieved, steadily increased as the platelet concentration increased from $0.25 \times 10^8$ cells/mL to normal physiological concentrations ($1.5–2 \times 10^8$ cells/mL).[81] The apparent divergence between the clotting time, and the rate and extent to which thrombin is generated is explained by the fact that the clotting time is dependent only upon the small amount of thrombin produced during the initiation phase. Most thrombin generation occurs subsequently during the propagation phase and is dependent upon the concentration of activated platelet membrane binding sites available for intrinsic Xase assembly and factor Xa generation — the events that control the amount of thrombin produced. This concept is supported by observations made in tissue factor-initiated clotting of blood from patients with hemophilia A, in which, again, factor Xa generation during the propagation phase should be, and is, severely compromised.[92]

Similarly, using samples from a chemotherapy patient who developed reduced platelet counts in a tissue factor-initiated whole blood assay, clotting times (initiation phase duration) were virtually unchanged for platelet concentrations exceeding $0.5 \times 10^8$/mL. However, when the platelet count decreased to $0.1 \times 10^8$/mL, significant prolongation of the clotting time was observed.[81] Clinical observations indicate these platelet concentrations mandate replacement therapy and are related to an increased risk of serious, spontaneous bleeding.[93]

Evaluation of the effects of platelet inhibitors in these model systems also yields important information that is applicable to clinical situations.[82] For example, the presence of the potent platelet inhibitor prostaglandin (PG) $E_1$ significantly prolongs not only the initiation phase of thrombin generation, but also suppresses the rate of thrombin generation during the propagation phase. $PGE_1$ appears to suppress formation of the platelet membrane binding sites required for complex enzyme assembly required for thrombin generation in both the initiation and propagation phases. Likewise, therapeutic concentrations of the $\alpha IIb\beta 3$ antagonists, abciximab (ReoPro) and eptifibatide (Integrilin), reduced the rate of thrombin generation during the propagation phase.[82] Thus, these inhibitors appear to influence significantly either the quantitative or qualitative expression of platelet-bound intrinsic Xase or prothrombinase enzyme complexes, consistent with observations made in other studies.[94,95] As suspected, these $\alpha IIb\beta 3$ antagonists appear to have a potent antithrombotic effect superimposed on their well-known antiaggregation characteristics (Chapter 62). Interestingly, the presence of the weaker platelet inhibitors, aspirin or dipyridamole (Persantine), at clinically relevant doses did not alter any aspect of tissue factor-initiated blood clotting.[82]

Platelets not only provide the membrane surface required for assembly and function of intrinsic Xase and prothrombinase, but also store factors V and XI that are released with platelet activation. The reconstituted models of blood coagulation allowed distinction of the relative importance of the plasma and platelet pools of these two coagulation factors.

Addition of plasma factors V or XI to a reconstituted model containing normal platelet concentrations was without effect on thrombin generation, suggesting that the platelet-derived pools of these coagulation factors were sufficient for effective clotting and consistent with several clinical observations, as discussed in detail in Section IV.[81]

## IV. Intrinsic Platelet Proteins Involved in the Coagulant Response

Four blood coagulation proteins, directly involved in the formation of a stable blood clot, are stored within platelets, primarily in platelet $\alpha$-granules or the cytosol. These proteins include factor V, factor XI, fibrinogen, and factor XIII. Of these, factor XIII and factor XI are synthesized by the platelet precursors — megakaryocytes — whereas the platelet-derived stores of factor V and fibrinogen are derived from the plasma pool after their endocytosis by megakaryocytes.

### A. *Factor V*

Radioimmunoassay and bioassay data indicate that platelets contain $\approx$ 4600–14,000 molecules of factor V per cell, which constitutes $\approx$ 18–25% of the total factor V pool in whole blood.[96] Platelet-derived factor V is stored in $\alpha$-granules in a complex with multimerin.[97] Multimerin consists of massive, disulfide-linked multimers that are millions of Daltons in size.[98–100] Multimerin specifically binds both factors V and Va through the light-chain portion of the molecules.[97] Multimerin immunodepletion of resting platelet lysates results in factor V removal and the loss of factor Va cofactor activity. Factor V is secreted from $\alpha$-granules upon platelet stimulation with a variety of agonists, including thrombin,[73,101] collagen,[102–104] epinephrine,[102] adenosine diphosphate (ADP),[102,103] arachidonic acid,[73] and the nonphysiological agonist calcium ionophore A23187.[43,73,104] Thrombin is the only known platelet agonist that affects both $\alpha$-granule release of factor V and full activation to its active cofactor factor Va. Factor V–multimerin complex dissociation is accomplished subsequent to platelet activation, and multimerin does not appear to regulate in any way the mechanisms by which platelet-derived factor Va functions in subsequent platelet procoagulant reactions.[97]

Clinical observations indicate that platelet-derived factor V plays a more important role in maintaining normal hemostasis than does its plasma counterpart,[105–107] and several studies demonstrate that platelet-derived factor V is physically and functionally distinct from its plasma counterpart. (The role of platelet-derived factor V in hemostasis is discussed in greater detail in Sections V and VI.) Platelet-derived factor V is released from $\alpha$-granules as a partially activated, proteolytically cleaved molecule (Mr = 74,000–

330,000) that exhibits substantial cofactor activity upon its release, which is increased two to three fold after activation by thrombin or factor Xa.[104] In contrast, plasma factor V circulates as a single-chain molecule (Mr = 330,000) with virtually no cofactor activity, which can be increased approximately 400-fold after activation with thrombin.[24] Furthermore, platelet-released factor V/Va appears to be a different substrate for various proteases, including factor Xa,[104,108] thrombin,[104,108] activated protein C,[109,110] and plasmin,[111] as well as platelet-derived kinases.[112,113]

Analyses of functional, platelet-derived factor V, purified from Triton X-100 platelet lysates using immunoaffinity chromatography with an antihuman factor V monoclonal antibody, confirmed several of these observations.[114] Purified, platelet-derived factor V was composed of a mixture of polypeptides ranging from approximately 40kDa to 330kDa, similar to that visualized by Western blotting of platelet lysates and releasates with antifactor V antibodies. Similar to platelet lysates, the purified, platelet-derived protein expressed significant levels of cofactor activity, such that its activation with thrombin resulted in only a two- to threefold increase in cofactor activity, yet led to expression of a specific activity nearly identical to that of purified plasma-derived factor Va. Furthermore, platelet-derived factor V/Va was two- to threefold more resistant to activated protein C-catalyzed inactivation than purified plasma-derived factor Va on the thrombin-activated platelet surface.

Similarly, purified platelet-derived factor V/Va is physically distinct from its plasma counterpart.[114] The heavy-chain subunit of platelet-derived factor Va contains only a fraction (approximately 10–15%) of the intrinsic phosphoserine present in the plasma-derived factor Va heavy chain and is resistant to phosphorylation at $Ser^{692}$ by either purified casein kinase 2 or thrombin-activated platelets. Resistance to phosphorylation is not the result of modification at $Ser^{692}$ as matrix-assisted laser desorption ionization (MALDI-TOF) analyses of tryptic digests confirm that it is unmodified at this site. In contrast, the platelet-derived factor V heavy chain is uniquely modified on $Thr^{402}$ with an N-acetylglucosamine or N-acetylgalactosamine. N-terminal sequencing and MALDI-TOF analyses of platelet-derived factor V/Va peptides identified the presence of a full-length heavy-chain subunit as well as a light-chain subunit formed by cleavage at $Tyr^{1543}$ rather than $Arg^{1545}$, accounting for the intrinsic levels of cofactor activity exhibited by the native platelet-derived cofactor. These collective data are the first to demonstrate physical differences between the two factor V cofactor pools. However, the effects, if any, of these differences on cofactor function remain to be determined.

Interestingly, even though platelet- and plasma-derived factor V are physically and functionally distinct, both cofactors are synthesized by the liver. Recently, it has been demonstrated unequivocally that platelet-derived factor V is acquired as a result of endocytosis by megakaryocytes from the plasma,[115–117] and not endogenous synthesis by megakaryocytes.[118,119] In these experiments, the released platelet-derived factor V pools from two patients heterozygous for factor $V^{Leiden}$, an inherited human polymorphism at the factor V gene locus, were analyzed after either a bone marrow or orthotopic liver transplant (OLT) from homozygous wild-type factor V donors. Western blotting analyses using an antibody that can distinguish wild-type factor V from factor $V^{Leiden}$ subsequent to activated protein C cleavage demonstrated that in both instances the platelet-derived factor V phenotype mirrored the plasma-derived factor V phenotype, and was independent of the megakaryocyte factor V gene.[115] These results were confirmed in another study in which a patient homozygous for factor $V^{Leiden}$ underwent OLT and received a wild-type liver.[117] Eighteen days posttransplant, Western blotting analysis indicated that the patient's platelets were acquiring wild-type factor V, consistent with the temporal differentiation of megakaryocytes and subsequent platelet production. Additional experiments performed with this patient 9 months after OLT indicated that the patient's total platelet-derived factor Va was completely cleaved by activated protein C at $Arg^{506}$, such that the factor Va cleavage pattern and rate of cleavage was identical to that observed with platelet-derived factor Va from wild-type controls. These data unequivocally demonstrate that the total platelet-derived factor V/Va pool originates from plasma via its endocytosis by megakaryocytes. In additional studies, Western blotting analyses of the patient's plasma- and platelet-derived factor V indicated that, although the plasma molecule retained its single-chain form, the platelet cofactor pool underwent activation and proteolysis subsequent to its endocytosis from plasma, appearing similar to that from a control individual. Furthermore, subsequent to the thrombin-catalyzed activation of the patient's platelets in the presence of $^{32}$P-adenosine triphosphate, analyses of endogenous platelet-derived factor Va, or added, purified platelet-derived factor Va, indicated that no labeled phosphate was incorporated into the heavy chain of the cofactor. In contrast, addition of purified plasma-derived factor V to the patient's platelets before their thrombin-catalyzed activation resulted in significant phosphate incorporation in the factor Va heavy chain presumably at $Ser^{692}$.

Because factor V uptake by platelets could not be demonstrated,[115] endocytosis of plasma-derived factor V by megakaryocytes was confirmed *in vitro* using megakaryocytes derived *ex vivo* from human CD34$^+$ stem cells and megakaryocytelike cell lines.[116] CD41$^+$ *ex vivo*-derived megakaryocytes showed little, if any, expression of endogenous factor V, suggesting that CD41$^+$ cells are incapable of factor V synthesis. In marked contrast, after incubation with plasma concentrations of factor V or factor IX, a significant number of CD41$^+$ cells were positive for factor V, but not factor IX, expression. Thus, a subpopulation of CD41$^+$ cells appears to be capable of specific factor V endocytosis. In addition, factor V uptake appears to be developmentally regulated during megakaryocyte growth and differentiation. Similar to observations made using *ex vivo*-derived megakaryocytes, subpopulations of the megakaryocytelike cell lines MEG-01 and CMK endocytosed plasma concentrations of factor V, as well as fibrinogen. Consistent with its ability to induce megakaryocyte differentiation, treatment with phorbol ester induced a substantial increase in the ability of these cells to endocytose factor V and fibrinogen.

Several observations suggest that factor V endocytosis by megakaryocytes proceeds through a specific and independent mechanism.[116] The amount of factor V and fibrinogen endocytosed did not merely reflect their disparate plasma concentrations. Factor IX, when present at its plasma concentrations, was not endocytosed under any of the conditions examined. Furthermore, another megakaryocytelike cell line, CHRF-288, endocytosed fibrinogen but not factor V. In addition, factor V endocytosis by CMK and MEG-01 cells is unaffected by the presence of plasma concentrations of other proteins such as fibrinogen, factor VIII, transferrin, and immunoglobulin (Ig) G. In support of this concept, subsequent to its endocytosis, factor V colocalized with endocytosed transferrin and fibrinogen, as well as with antibodies against α-adaptin and clathrin components of clathrin-coated vesicles, demonstrating that factor V endocytosis is clathrin dependent. Collectively, these observations suggest that megakaryocyte endocytosis of factor V is mediated via a specific, independent, and, most likely, receptor-mediated event.

Little is known about the mechanisms regulating the endocytosis, and the subsequent trafficking, of factor V by megakaryocytes. Subsequent to its endocytosis by megakaryocytelike cells and *ex vivo*-derived megakaryocytes, factor V undergoes specific proteolysis to yield fragments similar to those found in and isolated from platelet lysates,[120] suggesting that proteolysis occurs prior to platelet formation. Other studies by Hayward and colleagues[121] indicate that multimerin is not involved in endocytosis or intracellular trafficking or processing of factor V, because factor V can be stored and proteolytically processed normally in platelets deficient in multimerin.

### B. Fibrinogen

Immunofluorescence studies indicate that washed platelets contain both intracellular and membrane-associated fibrinogen.[122–126] Total platelet-derived fibrinogen accounts for 3 to 10%[126–128] of the total platelet protein (3–25 mg/10$^{11}$ platelets), with 25%[126] being found in α-granules.[129,130] Both the cytosolic and α-granular fibrinogen are released from platelets in response to various agonists, including thrombin, collagen, and ADP.[129–132]

In contrast to platelet-derived factors XIII and XI and VWF, platelet-derived fibrinogen is acquired exclusively by endocytosis from the plasma by platelets and megakaryocytes,[116,133–135] and not endogenous synthesis by megakaryocytes.[136–138] Although the mechanisms that regulate megakaryocyte and platelet endocytosis of fibrinogen are not entirely understood, several studies suggest that it occurs via clathrin-dependent, receptor-mediated endocytosis.

The receptor involved in fibrinogen endocytosis by megakaryocytes and platelets appears to be the integrin αIIbβ3, because platelets from individuals with type I Glanzmann thrombasthenia, which are deficient in αIIbβ3 (Chapter 57), have markedly reduced platelet fibrinogen levels.[139–142] Indeed, several studies suggest that endocytosis of fibrinogen is mediated by αIIbβ3 in both animal[143–145] and human models.[146,147] αIIbβ3-mediated fibrinogen endocytosis by platelets appears to be clathrin dependent, because both the receptor and the fibrinogen have been colocalized with clathrin in endocytic vesicles.[148,149] There is also evidence for the intracellular cycling of the αIIbβ3 pool in platelets.[150,151] Whether a similar mechanism occurs in megakaryocytes remains to be determined. Subsequent to its endocytosis by megakaryocytes, fibrinogen is trafficked to multivesicular bodies, which appear to be endocytic compartments that function as precursors to α-granules.[152] It is subsequently trafficked to mature α-granules.[133]

Other observations suggest that additional molecules are involved in fibrinogen endocytosis by megakaryocytes and platelets. For example, inhibition of fibrinogen binding to αIIbβ3 by its antagonist, abciximab, does not completely inhibit uptake of fibrinogen into the α-granules of unactivated platelets,[153] consistent with the observation that individuals lacking both αIIbβ3 and αVβ3 are slightly more deficient in fibrinogen.[142] Furthermore, an understanding of how αIIbβ3 regulates fibrinogen endocytosis is further complicated by the observation that soluble fibrinogen only binds the activated conformer of the integrin.[154] Consequently, it has been hypothesized that immobilized fibrinogen, such as that bound to another integrin (e.g., αVβ3) may bind to "unactivated" αIIbβ3, allowing for its subsequent uptake and trafficking to the α-granule.[143] Thus, even though fibrinogen endocytosis by megakaryocytes and platelets was documented almost 20 years ago, more studies are required to define this mechanism and its regulation in greater detail.

### C. Factor XIII

Factor XIII circulates in plasma as a 320-kDa heterotetramer consisting of two A-subunits and two B-subunits. The A-subunits contain the active site, calcium-binding site, and activation peptide that is removed as a result of factor XIIIa formation.[155,156] Platelets contain a large amount of the factor XIII zymogen in their cytoplasm. In contrast to the plasma molecule, human platelet-derived factor XIII is a homodimer consisting of two A-subunits.[157] The A-subunit of factor XIII is synthesized mainly by megakaryocytes.[158–160] However, it has also been shown to be synthesized by cells of the monocyte lineage[161–163] and, to a small extent, by hepatocytes.[164] It is unclear from which cell type the A-subunit is secreted into plasma. Interestingly, small amounts of both the A- and B-subunits of factor XIII have been detected in platelet α-granules, suggesting that either platelets or megakaryocytes are capable of taking up factor XIII from the plasma.[165] However, the mechanism by which this occurs is unknown. It is possible that factor XIII is simultaneously taken up from the plasma with fibrinogen,[165] which is known to be endocytosed by megakaryocytes and platelets, because much of the plasma-derived factor XIII has been demonstrated to circulate in complex with fibrinogen.[166–168]

The zymogen factor XIII is activated to the functional enzyme, factor XIIIa, by thrombin in a $Ca^{2+}$-dependent manner.[169] Platelet-derived factor XIII can also be activated to factor XIIIa by a slow, nonproteolytic mechanism by the elevated intracellular $Ca^{2+}$ concentrations that occur in platelets as a result of thrombin activation.[170] The platelet and plasma forms of factor XIIIa consist of the two A-subunits from which activation peptides have been removed.[157] It is a transglutaminase, and its primary function in plasma is to catalyze the formation of γ-glutamyl-ε-lysine bridges between the side chains of fibrin monomers to stabilize the clot.[171] The role of intraplatelet factor XIIIa in stabilization of the fibrin clot is less clear, because it does not appear to be secreted from platelets subsequent to their activation. Interestingly, in experiments in which platelets and monocytes are the only source of factor XIII, the clot contains fully cross-linked fibrin,[172] so there may be some effect of platelet-derived factor XIII in stabilizing the fibrin clot.[171] Other proposed functions of platelet-derived factor XIII include cross-linking of myosin filaments in the final stages of platelet aggregation,[173] or cross-linking of polymerizing fibrin to actin or fibronectin, which may be involved in clot retraction.[174]

## D. Factor XI

Individuals with plasma factor XI deficiency display heterogeneous bleeding tendencies with approximately 50% of these individuals displaying severe posttraumatic or postsurgical bleeding. The remaining individuals exhibit no bleeding diatheses. These observations may be explained by the presence of factor XI activity and antigen in platelets.[47]

Platelet-derived factor XI activity is enriched in the plasma membrane fraction.[175] Factor XI coagulant activity and antigen levels in washed platelet suspensions constitutes approximately 0.5% of the factor XI activity in normal plasma suggesting there are approximately 300 molecules per platelet.[175–179] As much as 50% of the total platelet factor XI is expressed on the surface of unactivated platelets.[180] Enhanced membrane expression of platelet factor XI can be achieved after platelet stimulation with a variety of platelet agonists. Agonist-induced expression of platelet factor XI activity parallels the expression of the activation-dependent conformer of αIIbβ3, rather than the release of α-granule constituents.[180] Platelets from individuals with severe plasma factor XI deficiency express normal amounts of both platelet factor XI pools — the pool constitutively expressed at the platelet membrane surface, and the pool released subsequent to activation.[180] In addition, well-washed platelets from both normal and severe plasma factor XI-deficient donors correct the clotting defect observed in factor XI-deficient plasma.[180] Thus, functionally active platelet factor XI is differentially expressed on platelet membranes, both constitutively and after activation, in the absence of detectable plasma factor XI.

Several reports indicate that factor XI is present in platelets in a form that is structurally different from plasma factor XI. Although plasma factor XI circulates as a disulfide-linked homodimer (Mr ≈ 143,000),[181–183] the platelet molecule has an apparent Mr value = 220,000 in the absence of disulfide bond reduction and an Mr value ≈ 55,000 after reduction.[175,179,184] Indeed, platelet factor XI appears to be an alternatively spliced product of the plasma factor XI gene, lacking exon V.[185] Molecular cloning of factor XI complementary DNA (cDNA) from a megakaryocyte library indicates the presence of a cDNA sequence identical to plasma factor XI with the exception that exon V is absent and exon IV is spliced to exon VI. The open reading frame is maintained without alteration of the amino acid sequence, except for the deletion of amino acids Ala91-Arg144 (encoded by exon V). *In situ* hybridization techniques demonstrate the presence of factor XI messenger RNA (mRNA; excluding exon V) in platelets, but not in other blood cells. Thus, platelet factor XI is expressed specifically within megakaryocytes and platelets in a form different from the plasma zymogen.[185] Alternatively, it may be the product of a second factor XI gene expressed specifically in megakaryocytes. Based on these combined observations, it has been suggested that platelet-derived factor XI represents a disulfide-linked tetramer of four 55,000-kDa subunits, each a truncated form of plasma factor XI lacking the region encoded by exon V. In contrast to these observations, two different laboratories detected factor XI mRNA in platelets and megakaryocytes that is identical to the liver mRNA.[186,187] However, the presence of identical factor XI gene products is not consistent with the reported discrepancy in the apparent Mr between the plasma and platelet form or the presence of platelet-derived factor XI in a subset of individuals with plasma factor XI deficiency. These discrepancies need to be resolved. The role of platelet-derived factor XIa in hemostasis is discussed in greater detail in Section V.

## V. Coagulation Reactions Supported by the Activated Platelet Surface

Through their constitutive expression of specific membrane binding sites on their surface, activated platelets can support the assembly and function of all the physiologically relevant coagulation complexes with the exception of the tissue factor–factor VIIa complex. Even though isolated, circulating human platelets from healthy individuals do not express tissue factor antigen or activity under either quiescent or activated conditions,[188] there is some recent evidence that suggests that platelets may recruit tissue factor-bearing microparticles into the growing thrombus under pathological conditions.[189,190] In these experiments, tissue factor-bearing microparticles generated by activated monocytes bound to and fused with activated platelets transferring both protein and lipid to the platelet surface. Transfer of tissue factor-positive microparticles was absolutely dependent upon the interaction of P-selectin expressed by the activated platelet with prostaglandin 1 expressed by the microparticle.[189,190] Similar observations have been made in mice.[191–193] In contrast to these observations, several investigators were unable to detect tissue factor activity in the circulating blood of healthy individuals when physiological concentrations of factor VIIa were used in the assay.[188,194–196] Furthermore, a recent study showed that a functional, membrane-bound tissue factor is essential for the initiation of thrombin generation but is not required for the propagation of the process, suggesting that there is no role for circulating tissue factor in blood coagulation.[197] Thus, a role for platelet-associated tissue factor in thrombus formation remains controversial.

### A. Factor XIa-Catalyzed Factor IXaβ Formation

Activated platelets provide a physiologically relevant surface for plasma factor XI activation by thrombin in the presence of prothrombin and calcium ions.[53,198] The requirement for this event appears to be prior complex formation of factor XI via its apple 1 domain to the kringle II domain of prothrombin.[199] Subsequent to platelet activation, this complex binds to the platelet surface through a $Ca^{2+}$-dependent prothrombin–platelet interaction. This binding event is postulated to cause a conformational change in factor XI, which exposes a binding site within the apple 3 domain[200] that mediates a direct, high-affinity interaction of factor XI with leucine-rich repeats of GPIbα[201] localized within membrane

rafts on the activated platelet surface (see also Section II.B.2 of Chapter 7).[202]

Studies utilizing a unique, cell-based tissue factor-initiated model system demonstrated that thrombin produced on the surface of activated platelets feeds back to activate factor XI efficiently on the activated platelet surface.[203] In that model, lipopolysaccharide-stimulated monocytes provided the tissue factor source, whereas the activated platelets provided the surface required for intrinsic Xase and prothrombinase assembly and function. These studies were recently confirmed by Baglia and Walsh,[198] who demonstrated that thrombin, once formed via prothrombinase, was the preferred physiological activator of plasma factor XI on the activated platelet surface. Activated platelets are absolutely required for thrombin–catalyzed factor XIa formation, suggesting that thrombin will not activate factor XI in the absence of an appropriately activated membrane surface.[198,203] Furthermore, the association of factor XI with GPIbα is absolutely dependent for its activation to factor XIa.[204] The ability of the formed factor XIa to enhance thrombin generation in the tissue factor-initiated model system is dependent on the presence of factor IX, suggesting that the factor XIa remains surface bound and recognizes factor IX as its normal macromolecular substrate.[203] Thus, these data suggest that factor XI can be activated by thrombin on the surface of activated platelets in amounts sufficient to enhance coagulation. In contrast, studies using a synthetic plasma model in which coagulation is initiated using 10 pM relipidated tissue factor suggest that factor XI has no effect on thrombin generation at physiological concentrations of coagulation proteins and platelets.[205]

Subsequent to its activation, factor XIa can bind to approximately 130 to 500 specific, high-affinity receptors (apparent $K_d \approx 0.8$ nM) expressed on the surface of activated platelets.[206] Because both factor XIa and factor IX[207] can bind to high-affinity, saturable receptors on activated platelets, this colocalization may serve to localize factor IX activation to the platelet surface, where factor VIIIa-dependent, factor IXa–catalyzed factor X activation occurs preferentially. Recently, Gailani and colleagues[208] proposed a model for factor IX activation by plasma-derived factor XIa on activated platelets. In this model, plasma-derived factor XIa, which is a disulfide-linked homodimer,[181–183] binds to activated platelets by one chain of the dimer, and to its substrate factor IX by the other chain to effect factor IXa formation.[208] More recent studies by Walsh and colleagues[209] suggest that factor XIa dimerization is not necessary for its optimal function in the presence of activated platelets.

In contrast to the plasma molecule, several fundamental questions regarding platelet factor XI remain unanswered. It is not known whether, or by what mechanism, platelet factor XI is activated. Furthermore, the substrate specificity of the putative active enzyme may, or may not, be factor IX. Platelet-derived factor XI does not appear to be a homodimer,[175,179,184] suggesting that factor IX may not be its physiological substrate, because the homodimeric structure of plasma-derived factor XI appears to be required for factor IX activation. Furthermore, exon V, lacking in platelet factor XI, encodes the portion of plasma factor XI, which contains the postulated binding site for its substrate factor IX.[210] Thus, if activated, platelet factor XIa might fail to activate factor IX because of disruption of the normal substrate binding site. Alternatively, it may activate factor IX through an entirely different binding mechanism.

Despite substantial effort by several laboratories, the physiological importance and role of platelet factor XI in maintaining normal hemostasis remains unknown. However, the observation that individuals with severe plasma factor XI deficiency, but no evidence of a bleeding diathesis, have normal amounts of platelet factor XI[175,184,211] suggests that platelet factor XI may represent the physiologically relevant pool. This concept is supported by studies performed in model systems in which plasma factor XI was without affect on blood coagulation when normal concentrations of platelets were present. In addition, clinical studies have detailed individuals who are severely compromised hemostatically and have normal plasma factor XI levels, but no detectable platelet factor XI.[212] Furthermore, individuals who have no detectable plasma or platelet factor XI also have serious bleeding problems.[176] Thus, to date, several fundamental questions regarding platelet-derived factor XI and its role in factor IX activation remain unanswered.

Factor IXaβ, produced either by the factor VIIa-tissue factor complex assembled on tissue factor expressing cells or at the activated platelet membrane surface via bound factor XIa, is the serine protease constituent of intrinsic Xase assembled on the activated platelet membrane. This enzymatic complex is essential for the formation of physiologically relevant concentrations of factor Xa on the activated platelet membrane.

## B. Intrinsic Xase Complex Assembly and Function

Although significant contributions have been made by other investigators,[213,214] the binding mechanisms by which intrinsic Xase assembles on the human activated platelet surface to activate factor X effectively has been elucidated principally by Walsh and colleagues.[207,215–230] Their working hypothesis[224] requires that, subsequent to platelet activation, three classes of receptors are expressed that are occupied by factors VIIIa, IXaβ, and X. Direct binding measurements of human factors IX and IXaβ to human, activated platelets indicate that factor IXaβ interacts with platelets significantly better than does the zymogen, factor IX, although both binding interactions are $Ca^{2+}$ dependent.[207,215] Factor IXaβ binds reversibly, specifically, and exclusively to approximately 350 high-affinity binding sites (apparent $K_d = 0.5$ nM)

and to an additional 300 binding sites with weaker affinity (apparent $K_d = 2.5$ nM), which can also be occupied by factor IX.[207] Factor IXaβ binds to its exclusive, high-affinity sites only in the presence of both factor VIIIa and factor X.[207,215] Thus, the cofactor and the substrate serve to define this factor IXaβ binding site and increase factor IXaβ's affinity for its specific platelet receptor fivefold.[215] Interestingly, factor IXaβ's affinity for its high-affinity sites is increased an additional fourfold in the presence of plasma concentrations of factor IX.[228] Occupation of these factor VIIIa-dependent, high-affinity receptor sites by factor IXaβ appears to be responsible for factor Xa generation for two reasons.[215,217,226] First, occupation of these sites correlates closely with the rate at which factor X is activated on the platelet surface.[215,217] Second, a peptide comprised of factor IX residues Q4 to Q11 completely inhibits factor IXaβ binding to the low-affinity site, yet is relatively ineffective in inhibiting factor VIIIa-dependent factor IXaβ–catalyzed factor X activation at the activated platelet surface.[226] As anticipated, the coordinate binding of factors VIIIa and IXaβ to activated platelets is required for physiologically relevant factor X activation at a site of injury.

Quantitation of factor VIIIa binding to platelets has been more difficult because of its inherent instability. Purified, stable factor VIIIa binds to activated platelets with a 10-fold higher affinity than does factor VIII.[214] Thus, the activation of factor VIII by thrombin appears to be required for its binding to platelets. Binding and kinetic analyses indicate that factor VIII binds with high affinity to approximately 484 sites on the activated platelet membrane surface with a $K_d$ value ≈ 3.7 nM.[229] Factor VIIIa interacts with approximately 300 to 500 additional, discrete sites on the platelet governed by a $K_d$ value ≈ 1.5 nM. The presence of factor IXaβ and factor X increases both the number and affinity of the binding sites for factors VIII and VIIIa.[229] Other kinetic analyses indicate that thrombin-activated or factor Xa-activated factor VIIIa incorporate into the complex bound to thrombin-activated platelets in an identical manner.[231] Nearly identical apparent $K_d$ (1 nM) and $k_{cat}$ values were obtained for both forms of the activated cofactor, as well as the protease, indicating that a functional complex consists of a 1 : 1 stoichiometric complex of factors VIIIa and IXaβ bound to the surface of thrombin-activated platelets.

Additional studies have reported the dual-labeled, coordinate binding of factor VIIIa with active site-blocked factor IXaβ on human, activated platelets.[230] These studies were performed in the presence of factors IX and X to define how factor VIIIa interacts with factor IXaβ bound to its exclusive, high-affinity binding sites. In contrast to the kinetic experiments, at saturation, 1400 molecules of factor VIIIa were bound simultaneously with 350 molecules of factor IXaβ. Both interactions were governed by $K_d$ values ≈ 0.5 nM — a binding affinity nearly identical to that observed in the kinetic studies.[215,229,231] These combined data suggest that human, activated platelets may express a subset of factor VIIIa binding sites that do not function in factor X activation, the significance of which needs to be established.

The substrate, factor X, has been shown to bind saturably, reversibly, and in a $Ca^{2+}$-dependent manner to approximately 16,000 sites ($K_d \approx 0.3\,\mu M$) on platelets activated with thrombin, but not ADP,[225] suggesting that strong platelet agonists are required to generate this binding site. Prothrombin and its $NH_2$ terminal gla-containing fragment, prothrombin fragment 1, were equipotent inhibitors of factor X binding ($K_i \approx 0.5\,\mu M$), whereas gla-domainless factor X or any other vitamin K-dependent protein did not compete for these factor X "zymogen binding sites." In direct binding studies, prothrombin was shown to bind to approximately 20,000 such sites, governed by a $K_d$ nearly identical to its $K_i$ ($0.5 \times 10^{-6}\,M$).[225] These "zymogen binding sites" are hypothesized to represent effective substrate binding sites that allow for rapid turnover of factor X by intrinsic Xase and prothrombin by prothrombinase respectively. Indeed, factor X bound to activated platelets via the zymogen binding site, and not factor X free in solution, appears to be the preferred substrate for platelet-bound factor IXaβ.[224] This concept was demonstrated by the ability of prothrombin fragment 1 (which displaces factor X from its zymogen binding site) to inhibit factor X activation by a platelet-bound, factor VIIIa/factor IXaβ complex. Thus, it has been hypothesized that the factor X zymogen binding site and the distinct, factor VIIIa-dependent, high-affinity factor IXaβ binding site appear to act in concert both to localize factor Xa formation to the site of platelet activation as well as to orient the substrate with the enzyme in a configuration that makes their interaction more efficient.[224]

The ability of prothrombin to bind to these same zymogen binding sites and displace factor X does not appear to alter factor X activation under physiological conditions, even though the prothrombin plasma concentration is 10 times greater than that of factor X.[224] The large excess of zymogen binding sites (approximately 16,000–20,000) requires that only a small fraction of the total number of binding sites needs to be occupied by factor X to saturate completely the factor VIIIa-dependent factor IXaβ binding sites ($\approx 300$ sites/platelet). Comparing the $K_d$ governing factor X binding with the zymogen binding sites ($\approx 0.32\,\mu M$) versus its plasma concentration ($0.15\,\mu M$), it is quite likely that subsequent to platelet activation, a sufficient number of binding sites would be occupied to effect factor X activation *in vivo*.

The binding of factor X and prothrombin to these "zymogen binding sites" is $Ca^{2+}$ dependent, requires the presence of their $NH_2$-terminal gla domain, and is completely inhibited by the lipid-binding protein annexin V.[225] The latter observation suggests that phospholipids present in the platelet membrane mediate at least part of these protein–membrane interactions. This observation, however, does not eliminate the possibility that specific protein receptors are also involved. Consistent with this premise is the observation that annexin V differentially inhibits factor IXaβ binding and function on activated platelets when compared with model systems of defined phospholipid vesicles.[222] For example, in the absence of factor VIIIa, the affinity of factor IXaβ for the activated platelet surface is 25-fold higher than its affinity for phospholipid vesicles. Additional studies demonstrated that factor IXaβ binding and functional interactions with negatively charged phospholipid vesicles are highly dependent upon electrostatic interactions (assessed by varying NaCl concentrations).[223] Similar factor IXaβ functional interactions with activated platelets were largely unaffected,[223] again suggesting that components in addition to phospholipids govern factor VIIIa-independent factor IXaβ binding. These observations, in addition to those discussed later concerning prothrombinase, indicate that other, as yet undefined, membrane components, in addition to anionic phospholipids, regulate procoagulant protein binding to cellular membrane surfaces.

The physiological significance of the binding of factors VIIIa and IXaβ to activated platelets to effect factor X activation is best exemplified by the kinetic parameters defining factor Xa formation.[217] Consider that in the absence of both thrombin-activated platelets and factor VIIIa, factor IXaβ–catalyzed factor X activation is defined by a $K_m \approx 80\,\mu M$, a factor X concentration 500 times greater than its plasma concentration ($0.15\,\mu M$). The binding of factor IXaβ to activated platelets decreases the $K_m$ value approximately 200-fold to $0.39\,\mu M$, leaving the $k_{cat}$ unchanged ($\approx 0.05$ per minute). The incorporation of factor VIIIa into the complex has dramatic consequences by decreasing the $K_m$ of the substrate to its plasma concentration and increasing the $k_{cat}$ 24,000-fold to 1240 per minute. Therefore, the functional consequences of the assembly of the entire complex on the activated platelet membrane surface is an almost unbelievable 13 million-fold increase in catalytic efficiency.

A substantial amount of information has been provided regarding the structural requirements for the effective and functional platelet binding of factor IXaβ, whereas virtually nothing is known for factor VIIIa. The active site of factor IXaβ is not involved in its factor VIIIa-dependent, high-affinity binding to platelets.[215] Active site-blocked factor IXaβ competitively inhibits both factor IXaβ binding and factor IXaβ–catalyzed factor X activation, with a $K_i$ identical to the $K_d$ governing factor IXa binding. Studies with factor IXa variants have also provided valuable insights. Studies with factor $IXa_{Alabama}$ ($Asp^{47} \rightarrow Gly$) indicate this residue is important for factor IXaβ interactions with platelet-bound factor VIIIa.[216] Factor $IXa_{Chapel\ Hill}$ ($Arg^{145} \rightarrow His$) binds only to the low-affinity factor IXaβ binding sites, indicating that the normal cleavage of factor IX by factor

XIa or by a factor VIIa–tissue factor complex is essential for its specific, high-affinity platelet interaction.[216] Studies have been done with chimeric proteins in which the first or second epidermal growth factor domain (EGF-1 or EGF-2 respectively) of factor IXaβ is replaced by that found in factor X. Although the EGF-1 domain does not appear to contain an important platelet binding element,[219] the EGF-2 domain is an essential structural feature of factor IXaβ required for its exclusive, high-affinity factor VIIIa-dependent binding.[221,227,232,233] Using recombinant factor IXa molecules with various point mutations within the EGF-2 domain, residues $Asn^{89}$, $Ile^{90}$, and $Val^{107}$ are essential for assembly of intrinsic tenase on activated platelets.[234] In studies with factor IXa molecules modified to examine the potential role of specific gla residues, factor VIIIa-dependent factor IXaβ binding to platelets was found to be mediated in part, but not exclusively, by the high-affinity $Ca^{2+}$ binding sites in the gla domain.[218] The factor IX residues G4-Q11 play no role in high-affinity, functional factor IXaβ binding, but are required for its binding to the lower affinity sites it shares with factor IX.[226]

The importance of the protein–protein binding interactions and protein membrane binding interactions regulating intrinsic Xase assembly and function, and hence the generation of physiologically relevant concentrations of factor Xa, is underscored by the significant bleeding diatheses accompanying factor VIII deficiency (hemophilia A) and factor IX deficiency (hemophilia B). The importance of factor Xa generated at the platelet surface is underscored by observations made in model systems, which demonstrate that the concentrations of factor Xa generated via extrinsic Xase on tissue factor expressing cells are insufficient to incorporate into prothrombinase assembled on activated platelets. Therefore, the effective channeling of factor Xa produced via platelet-bound intrinsic Xase to platelet-bound factor Va to form prothrombinase is required to facilitate explosive thrombin generation at the activated platelet surface.

## C. Prothrombinase Complex Assembly and Function

A great deal of what is known regarding events regulating the assembly of prothrombinase on the human platelet surface was first described in the bovine system.[235] The assembly of the human prothrombinase complex has been defined using several different experimental approaches. Equilibrium binding measurements of radiolabeled factors V, Va, and Xa to platelets have been reported.[43,72,73,236–238] Quantitation of the binding parameters governing functional factor Va and factor Xa platelet binding characteristics has been obtained by analyses of kinetics of prothrombin activation.[101,239,240] Inhibitory antibodies to factor V have demonstrated its crucial role in complex assembly.[72,239] Radiolabeled,

noninhibitory antifactor V monoclonal antibodies have been used to evaluate and quantitate the role of platelet-sequestered and released factor Va in factor Xa-dependent thrombin generation.[241] Fluorescently tagged noninhibitory monoclonal antibodies directed against human factor Va and factor Xa have allowed visualization of complex assembly at the activated platelet surface.[242–244] These combined studies indicate that the platelet actively participates in and regulates prothrombinase assembly, and functions through several distinct, yet related, mechanisms.

Studies detailing the assembly and stoichiometry of prothrombinase on human platelets is complicated by the fact that they contain a significant reserve of factor V,[96] as detailed in Section IV.A. Because the status of the internal factor V in platelet binding experiments after its release is difficult to determine quantitatively, analyses of kinetics of prothrombin activation have been used to calculate the stoichiometry and binding parameters governing functional interactions of factor Va and factor Xa with thrombin-activated platelets.[101,240] Those studies indicated clearly that thrombin activation of the platelets was required to effect the release and activation of the cofactor. The kinetic data also demonstrated that thrombin-activated platelets do not release factor Va in sufficient concentration to saturate all the platelet factor Va binding sites capable of participating in prothrombin activation. Thus, exogenous excess factor Va was added to ensure that factor Va was not a limiting component. Kinetic data, interpreted in terms of a 1 : 1 stoichiometry of the factor Va-to-factor Xa complex, indicate that factor Xa binds to approximately 3000 functional sites on the surface of thrombin-activated platelets with an apparent $K_d \approx 10^{-10}$ M and $k_{cat} = 1800$ mol thrombin per minute per molecule factor Xa bound.[101] For normal platelets, the storage and release of factor V from platelet α-granules prevent an analogous determination of the functional factor Va platelet binding sites. Therefore, factor Va titrations were performed using platelets from a factor V antigen-deficient individual.[101] Those studies confirmed that factor Va and factor Xa form a 1 : 1 stoichiometric complex on the surface of thrombin-activated platelets. A total of 4100 factor Va and 5100 factor Xa binding sites per platelet were calculated. Both isotherms were governed by the same apparent $K_d$ and expressed the same $k_{cat}$ per site. Thus, a compilation of all the data indicates that factor Va and factor Xa form a 1 : 1 stoichiometric complex at the activated platelet surface governed by a $K_d \approx 10^{-10}$ M and expressing a $k_{cat}$ per site of 30 per second.[101,240] Furthermore, factor Xa binding is absolutely dependent upon prebound factor Va, indicating that factor Va forms at least part of the receptor for factor Xa at the activated, human platelet membrane surface.

Platelet activation is required to mediate both factors Va and Xa binding.[241] A radioiodinated monoclonal antibody directed against human factor Va, which does not affect complex assembly and function, and binds with equal avidity to both platelet-released and plasma forms of factor Va, was used to monitor indirectly factor Va–platelet binding interactions, whereas factor Xa binding was measured directly using radiolabeled ligand. These studies demonstrated that factor Va and factor Xa bind to discrete binding sites on activated platelets. Perhaps more important, the expression of the binding sites is independently regulated as a function of the thrombin concentration used to effect platelet activation. Although thrombin concentrations as low as 1 nM effect maximal factor Va binding, 50 nM thrombin is required to effect maximal factor Xa binding. Thus, in response to low thrombin concentrations (0.05–1 nM), platelet factor Va release parallels factor Va binding (platelet- or plasma-derived), which is independent of and exceeds factor Xa binding on a molar basis ($\approx 2:1$). Even when maximal platelet factor Va binding is achieved ($\approx 6000$ sites/platelet), the bound factor Va is not competent to support factor Xa binding. Rather, factor Xa binding ($\approx 3000$ sites/platelet) is dependent upon both bound factor Va and the expression of an additional platelet binding site that is not expressed until thrombin concentrations exceeding 1 nM are used to effect platelet activation. As expected, factor Xa binding correlates with the proteolytic activation of prothrombin. Thus, it would appear that platelets modulate the assembly of prothrombinase at their surface through activation-dependent events regulating both factors Va and Xa binding. These binding events are independently regulated, arguing for the existence of activation-dependent "receptors" for factor Va and factor Xa at the human platelet surface.[241]

The concept that factor Xa interacts with a specific receptor on thrombin-activated platelets is supported by data indicating that neither factor X, factor IXaβ, nor prothrombin compete with factor Xa for platelet binding sites.[72] Indeed, monoclonal antibodies against effector cell protease receptor-1 (EPR-1), an integral membrane protein expressed by select leukocyte subpopulations[245] and endothelial cells,[246] which binds factor Xa in the absence of factor Va, detected an activation-dependent protein and inhibited prothrombinase–catalyzed thrombin generation on activated platelets.[241] Furthermore, a factor Xa interaction with activated platelets cannot be mimicked qualitatively by model systems using phospholipid vesicles of defined content. This concept is supported by studies with a factor X variant, Factor X$_{St. Louis}$, in which a glycine is substituted for a gla residue at position 7. Although the activated variant expresses near-normal activity (compared with wild-type) when assayed with factor Va and phospholipid vesicles,[247] the activated variant expresses less than 5% the activity of wild-type factor Xa when assayed with factor Va on activated platelets.[247] Similar, yet contrasting, observations were made with factor X variants in which aspartate was substituted for each of the glutamate residues normally subject to γ-carboxylation.[248] All the variants expressed much greater

activity when assayed with factor Va and activated platelets than with factor Va and vesicles containing the phospholipid components of the thrombin-activated platelet surface.

Several studies using synthetic vesicles of defined phospholipid content have demonstrated that anionic phospholipids are also important for both the assembly and function of prothrombinase.[249–252] The exposure of anionic phospholipids, primarily phosphatidylserine, subsequent to agonist-induced platelet activation has likewise been correlated to the expression of prothrombinase activity at the platelet surface.[253] Additional studies have shown that annexin V inhibits both prothrombinase and intrinsic Xase assembly on the activated platelet surface.[253–255] However, competition studies with factors Va and VIIIa indicate that the binding of these cofactors to activated platelets shows much more specificity than on synthetic phospholipid bilayers, suggesting that activated platelets may express specific factor Va and VIIIa receptors.[256] Thus, a compilation of all the studies discussed thus far indicate that the platelet receptor for factor Xa assembly into prothrombinase consists minimally of EPR-1, or an EPR-1-like protein (perhaps through a specific receptor), membrane-bound factor Va, and the surrounding phospholipid environment.

Despite the substantial effect that the substrate factor X has on the assembly of intrinsic Xase on the activated platelet surface, prothrombin is without effect on prothrombinase assembly.[101] However, as indicated previously, prothrombin binds to approximately 20,000 "zymogen binding sites" expressed on the surface of thrombin-activated platelets governed by an apparent $K_d$ equal to its plasma concentration ($\approx 0.15\,\mu M$).[225] This observation had been made previously, although the role these sites may play in presenting prothrombin to prothrombinase for substrate turnover had not been defined.[257] More recent studies have shown that prothrombin may bind to integrin αIIbβ₃ expressed on unactivated platelets even though previous studies had been unable to define such a binding interaction.[225,257] This binding interaction is dependent upon prothrombin's Arg-Gly-Asp (RGD) sequence and appears to accelerate prothrombin activation catalyzed by prothrombinase.[258] αIIbβ₃ expressed on activated platelets also binds prothrombin (see also Chapter 8); however, fibrinogen is an effective competitor of this interaction. These observations led to the suggestion that the circulating platelet may be prearmed with a small amount of prothrombin bound to αIIbβ₃ and that this prothrombin pool may play an important physiological role in the early events of thrombin generation and thrombus formation.[258]

Little is known regarding the structural features of factors Va and Xa that regulate complex assembly and function. Factor Va is a two-subunit protein consisting of an NH₂-terminal-derived heavy chain (Mr = 105,000) and a COOH-terminal derived light chain (Mr = 74,000).[259,260] The light chain, which binds to platelets independently of the heavy chain, appears to express the sites mediating both the platelet[238,240,261] and factor Xa binding interactions.[240,261] Although the heavy chain appears to associate with platelets only through a $Ca^{2+}$-dependent interaction with the light chain,[261] the heavy chain appears to play an important role in binding and positioning prothrombin for effective cleavage by factor Xa. However, active site-blocked factor Xa incorporates into prothrombinase in a manner identical to the active protease.[262,263]

Platelet-derived factor V appears to play a preeminent role in the maintenance of normal hemostasis. This contention is based on several clinical observations. Factor V Quebec describes a bleeding disorder that exhibits an autosomal dominant inheritance pattern and a severe, but delayed, bleeding diathesis after trauma, which has now been described in two separate families.[105,264] In one study, clinical laboratory evaluations of the affected individuals revealed a mild thrombocytopenia, mild reduction in plasma factor V levels, and an abnormal Stypven time when platelet-rich plasma, but not platelet-poor plasma, was used in the assay. The two individuals are severely deficient in platelet factor V levels (<5%), with near-normal plasma factor V levels ($\approx$70%). In the absence of added plasma factor Va, their thrombin-activated platelets cannot function in prothrombin activation, whereas their platelets function normally in the presence of added purified factor Va.[105] The platelet-specific factor V deficiency appears to be the result of a generalized defect that results in the proteolytic degradation of most, but not all, proteins localized to the $\alpha$-granules including multimerin, fibrinogen, VWF, and osteonectin.[265] Similar studies in the second family defined the same, if not identical, generalized proteolytic degradation of the $\alpha$-granule proteins.[264] Although the delayed bleeding diathesis that manifests in these two families may not be due solely to a platelet-associated factor V deficiency, the combined data suggest that platelet-derived factor Va is essential in maintaining stable and prolonged hemostasis after trauma.

The hemostatic requirement for platelet-derived factor Va is underscored by other studies. Platelet transfusions corrected the hemostatic defect of a congenital factor V-deficient individual for as long as 6 days, even though no detectable increase in plasma factor V activity was noted.[266] In other studies in individuals with varying degrees of congenital factor V deficiency, Miletich and colleagues[236] observed that each patient's washed platelets had decreased numbers of factor Xa binding sites, which reflected the clinical severity of his/her bleeding diathesis, whereas each patient's plasma factor V concentration was a poor predictor of their bleeding tendency. In addition, studies have been reported in which platelet transfusion is more effective than plasma transfusion in managing individuals with acquired factor V inhibitors.[267] Finally, Nesheim and coworkers[106] described an individual with immunodepleted plasma factor V resulting from an acquired IgG inhibitor, but normal

platelet factor V stores, who was not compromised hemostatically even after extensive surgical challenge.

## VI. Positive and Negative Effectors of Platelet Coagulant Activity

### A. Platelet Coagulation Protein "Receptors"

As discussed earlier, several lines of evidence suggest that the contribution of platelets to some coagulant reactions involve membrane biochemical components other than lipids. All the enzymatic complexes discussed earlier have been shown to assemble and function efficiently in model systems comprised of lipid vesicles of defined phospholipid. Several studies have tried to define the physicochemical properties of phospholipid mixtures required for enzymatic complex assembly (reviewed by Walsh[268]). The presence of phosphatidylserine is an absolute requirement that confirms the well-established positive effect of anionic phospholipids in coagulation complex assembly. However, the exposure of phosphatidylserine alone is not sufficient to accommodate intrinsic Xase and prothrombinase assembly and function because annexin V, which binds to phosphatidylserine, is unable to block complex assembly completely. Furthermore, although many of the vitamin K-dependent protein zymogens and proteases, as well as the plasma cofactors, share the same phospholipid binding sites, the binding sites on activated platelets for these same coagulation factors show substantially greater specificity. Progress in identifying specific platelet protein receptors mediating these binding interactions has been slow. However, two notable observations have been made recently.

EPR-1, or an EPR-1-like protein, an integral membrane protein that binds factor Xa specifically, is an important component of the platelet membrane required for factor Xa incorporation into prothrombinase.[241] In contrast to what has been observed in studies with phospholipid vesicles, saturation of the activated platelet with factor Va is not sufficient to support a factor Xa interaction. In addition, monoclonal antibodies directed against EPR-1 will coordinately inhibit both factor Va-dependent factor Xa binding to activated platelets and generation of thrombin. The observation that EPR-1 is also an integral component of prothrombinase assembled on pericytes is confirmation of its importance in supporting the coagulant response.[269]

Integrin $\alpha$IIb$\beta$3 expressed on unactivated platelets may bind prothrombin specifically.[258] Subsequent to platelet activation, however, the conformational change occurring in $\alpha$IIb$\beta$3 favors fibrinogen binding (see Chapter 8 for details). Thus, even though platelet activation is required absolutely to support prothrombinase assembly, the prothrombin originally bound to $\alpha$IIb$\beta$3 may play an important role in the very early stages of thrombin formation. Clearly more work

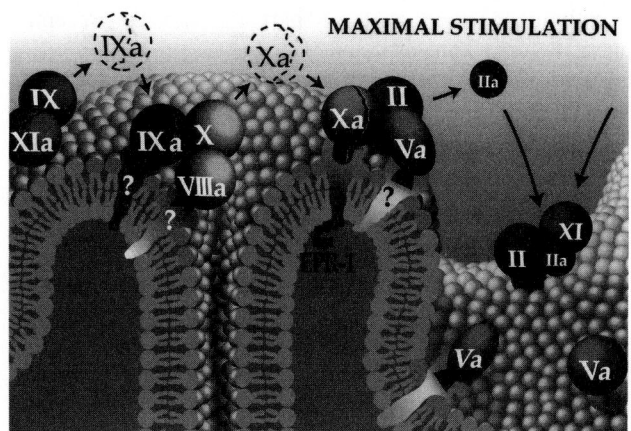

**MAXIMAL STIMULATION**

**Figure 19-4.** Hypothetical model for the propagation of a physiologically relevant coagulant response on the activated platelet surface. Subsequent to (a) tissue factor–factor VIIa-catalyzed formation of factor IXaβ, factor Xa, and thrombin; (b) platelet activation by thrombin (factor IIa); and (c) factor VIII activation to factor VIIIa by either thrombin or factor Xa; factors VIIIa and IXaβ interact with their hypothesized membrane receptors and anionic phospholipid to form functional intrinsic Xase, which channels factor Xa assembly into functional prothrombinase. Prothrombinase consists of platelet-released factor Va or thrombin-activated plasma factor Va binding to its putative receptor in combination with factor Xa binding to effector cell protease receptor-1 (EPR-1), or an EPR-1-like protein, as well as to anionic phospholipid. The thrombin produced via prothrombinase can activate factor XI bound to platelets to factor XIa, which subsequently activates factor IX to factor IXaβ to facilitate continued formation of intrinsic Xase. With the exception of EPR-1, all platelet membrane receptors shown as playing a part in procoagulant enzyme complex assembly are hypothetical (represented by ?) and remain to be established, as do their modes of expression. (Adapted from Tracy,[235] with permission.)

needs to be done in defining the platelet membrane components required for propagation of the coagulant response.

### B. Platelet Activation

Activation of platelets clearly qualifies as a key regulator of the coagulant response because the assembly of functional intrinsic Xase and prothrombinase is absolutely dependent upon platelet activation. In fact, with the exception of prothrombin binding to αIIbβ3 on human, unstimulated, platelets, every protein–platelet binding interaction shown in Figure 19-4 occurs only on an activated platelet membrane and is also required for physiologically relevant thrombin generation. Furthermore, the degree of activation required for complex assembly offers another point of regulation. Intrinsic Xase assembly and function on activated platelets occurs at low thrombin concentrations (0.1–2 nM),[207]

whereas 5 to 50 nM thrombin is required to support factor Xa incorporation into functional prothrombinase.[241]

Adherence of platelets to subendothelial components may also confer different degrees of regulation. Although no information is available regarding the effect of platelet adherence on intrinsic Xase, a few studies have been reported regarding prothrombinase. Two studies dealt with prothrombinase complex assembly[270] and/or function[271] on platelets adhered to immobilized VWF and collagen under static conditions. A third study investigated prothrombinase function on platelets adhered to fibrinogen under flow.[272] The combined data demonstrate that adherent and thrombin-stimulated adherent platelets support the equivalent assembly of a functional prothrombinase. Unlike platelets in suspension, additional platelet activation is not required for generation of procoagulant activity. However, in some instances, platelet activation of the adherent platelets results in the release of platelet microparticles (Chapter 20), although these membranous vesicles are incapable of participating in thrombin generation via prothrombinase.[271] All procoagulant activity remains associated with the adherent platelet population — an observation confirmed by other investigators.[273] Thus, it has been hypothesized that platelets adhered initially at a site of vascular damage may promote prothrombinase assembly before the generation of thrombin. This concept would be supported by the ability of the adherent platelets to release their stores of partially activated factor Va. The thrombin generated subsequently would induce activation of platelets in suspension, resulting in their recruitment into the growing platelet thrombus, as well as their activation to a "level," resulting in prothrombinase assembly and function at their membrane surface, which would sustain thrombin generation at the damaged site.[271]

### C. Procoagulant Platelet Subpopulations

Platelet activation alone may not be sufficient to define coagulation complex assembly and function. Several laboratories have identified discrete subpopulations of platelets that differentially regulate procoagulant enzyme complex and function.[242–244,274] When platelets are thrombin activated to a level consistent with expression of maximal prothrombinase activity, and were shown to express P-selectin at their membrane surface, factor Va and factor Xa colocalize to a subset (approximately 30–50%) of activated platelets, even though kinetic analyses of the same platelet population indicate that the system is saturated with factors Va and Xa.[242] As expected, no factor Xa binding is observed in the absence of factor Va binding. Furthermore, these platelets bind both platelet- and plasma-derived factor Va. Similar observations were made by London and colleagues[274] with respect to the intrinsic tenase complex. In these studies, maximal activation of platelets with thrombin or the protease-activated receptor 1 agonist peptide SFLLRN resulted in the generation of a small subpopulation of platelets (4–20%) capable of binding factor IXa. However, in contrast to observations made for prothrombinase, binding of factor IXa was not dependent upon bound factor VIIIa, because the size of the subpopulation was independent of the presence of this cofactor.

These observations extend those of Alberio and colleagues,[243] who demonstrated that a fraction of platelets express very high levels of surface-bound, platelet-derived factor Va after simultaneous activation with thrombin and the GPVI agonist convulxin, or activation with thrombin plus collagen. These platelets, referred to as COAT platelets, also preferentially bind factor Xa, as well as other α-granule proteins like thrombospondin and fibrinogen. Similar to the observations for platelet activated with thrombin alone,[274] COAT platelets can also bind coagulation proteins not present in α-granules.[243,244] Monroe and colleagues demonstrated in their cell-based model of blood coagulation that platelets activated by convulxin and thrombin can also bind factors VIIIa and IXa.[244] Furthermore, the same subpopulation of platelets that bound factors VIIIa and IXa also bound factors Va and Xa, consistent with previous observations that factor Xa formed on the platelet surface via intrinsic tenase channels into platelet prothrombinase. Subsequent analyses of COAT platelets demonstrated that serotonin released by dense granules covalently binds platelet-derived factor Va to the platelet surface via a transglutaminase-mediated reaction.[275] Furthermore, platelet-derived thrombospondin and fibrinogen may function as serotonin "binding sites" on the activated platelet.[276]

These combined observations demonstrate that subpopulations of activated platelets exist as defined by their ability to assemble active coagulation complexes, suggesting they are under distinct hematopoietic regulation.

### D. Complex Formation

Prothrombinase formation defines an important modulating event catalytically, because the complete complex is 300,000 times more efficient than free factor Xa acting on prothrombin in solution.[24,72] An implication of this significant rate increase is that any dissociation of components from the complex significantly and drastically decreases enzyme activity. The absence of the membrane surface from the complete catalyst results in a 1000-fold loss in catalytic efficiency, whereas deletion of factor Va from the complex results in a 10,000-fold decrease in the rate of thrombin generation.[24] The membrane surface is obligately required because, although contributing little to the intrinsic rate of catalysis, it provides an environment in which both the factor Va–factor Xa complex and substrate can coconcentrate.[277] Prothrombin appears to concentrate at the platelet

membrane surface in large part through the receptor-mediated mechanisms detailed earlier. Factor Va, by virtue of its membrane binding capacity, forms an obligate and high-affinity interaction with factor Xa, because factor Xa will not bind to activated platelets in the absence of factor Va. Concentration of reactants at the membrane surface results in high local concentrations of reagents that exceed the intrinsic $K_m$ of the reaction and significantly improve catalytic efficiency. In addition, factor Va enhances the turnover rate of the reaction by a mechanism that has not yet been rationalized.[278]

Complex formation also affords prothrombinase constituents protection from inhibitors and/or inactivators. In contrast to factor Xa free in solution, factor Xa incorporated into prothrombinase (i.e., complexed with membrane-bound factor Va) is protected from inhibition by antithrombin III and heparin.[72] Likewise, factor Va incorporated into prothrombinase is protected from proteolytic inactivation by activated protein C[109,110] — a process that is discussed in detail in the following section.

Intrinsic Xase is modulated by a variety of mechanisms similar to those observed with prothrombinase. Complex formation results in an approximate $2 \times 10^7$-fold increase in catalytic efficiency.[217] The binding of factors VIIIa and IXaβ to thrombin-activated platelets decreases the $K_m$ for factor X activation 2500-fold and enables factor VIIIa to increase the $k_{cat}$ approximately 7500-fold. Antithrombin III cannot inhibit factor IXaβ in the presence of factor VIIIa and platelets.[279] Similar to factor Va, complex formation of factor VIIIa with factor IXaβ modulates its activity. The dissociation of the A2 subunit from the heterotrimer factor VIIIa, which results in a complete loss of cofactor activity, is prevented by its association with factor IXaβ on a membrane surface,[280] as is its inactivation by APC.[62,281]

Because factor Va is required absolutely for physiologically relevant thrombin formation, significant variations in prothrombin activation can be accomplished via proteolytic, inactivating alterations in the cofactor catalyzed by activated protein C. Proteolysis and inactivation require that the factor Va is platelet[59,74] or membrane bound.[282] Early studies indicated that incubation of activated protein C with thrombin-activated platelets produced a parallel decrease in factor Xa binding and prothrombin activation, suggesting that after cleavage by activated protein C, factor Va does not bind factor Xa.[283,284] Hence, activated protein C-catalyzed inactivation of factor Va has a dramatic effect on thrombin generation.

Human platelets are able to promote and modulate the activated protein C-catalyzed inactivation of plasma- and platelet-derived factor Va, although these two cofactor pools are not equivalent substrates with respect to inactivation. Three cleavages in the heavy chain of the human factor Va molecule are required for efficient activated protein C-catalyzed inactivation: cleavage at Arg[506], followed by cleavages at Arg$^{306}$ and Arg$^{679}$.[282,285] Cleavage at Arg$^{506}$ is more rapid and precedes the membrane-dependent, inactivating cleavage at Arg$^{306}$.[282] The thrombin-activated, human platelet surface plays an important role in the activated protein C-catalyzed inactivation of platelet-derived factor Va and the factor Va variant, factor Va$^{Leiden}$.[109,110] Factor Va$^{Leiden}$ is derived by thrombin activation of factor V$^{Leiden}$, a variant in which Arg$^{506}$ has been replaced by glutamine, essentially removing the activated protein C cleavage site at position 506.[286] Thus, individuals homozygous (and in some instances, heterozygous) for this mutation suffer a poor anticoagulant response to activated protein C and are at increased risk for venous thrombosis.[287–289] The normal and mutant platelet cofactors are inactivated by activated protein C at near-identical rates; however, complete inactivation of both of the platelet-derived cofactors is never achieved,[109,110] because as much as 50% of the original activity remains. These results are in marked contrast to what has been observed in studies of the activated protein C-catalyzed inactivation of plasma-derived factor Va and factor Va$^{Leiden}$ on synthetic phospholipid vesicles.[290–292] In those studies, complete inactivation of the cofactors is always observed, although plasma factor Va$^{Leiden}$ is inactivated at a substantially slower rate than normal plasma factor Va. In platelet studies, greater residual activity of both platelet-derived factor Va and factor Va$^{Leiden}$ remain when using thrombin-activated platelets as the inactivating membrane surface, when contrasted with the near-complete inactivation when synthetic phospholipid vesicles are used as the inactivating surface.[110] Thus, activated platelets protect the platelet-derived cofactors from activated protein C-catalyzed inactivation.[109,110]

Activated platelets also protect the normal plasma-derived factor Va from inactivation by activated protein C by slowing the cleavage at Arg$^{506}$.[110] However, in contrast to the platelet-derived cofactor, plasma-derived factor Va can be completely inactivated by activated protein C, suggesting that the two different cofactor pools (plasma vs. platelet) are different substrates for APC.[110] These studies again confirm the importance of binding events in modulating prothrombinase activity. They also demonstrate that platelet and synthetic phospholipid membranes are not equivalent surfaces in regulating enzyme complex formation. Collectively, these studies indicate that platelets sustain procoagulant events by providing a membrane surface that delays cofactor inactivation and by releasing a cofactor molecule that displays an activated protein C-resistant phenotype.

It is interesting to consider that, at sites of arterial injury, the factor V$^{Leiden}$ mutation may not as readily predict arterial thrombosis, because the normal and variant platelet-derived cofactors are equally resistant to activated protein C at the activated platelet surface. In fact, the demonstrated role of the factor V$^{Leiden}$ mutation in predicting venous thrombosis[287–289] may reflect the documented resistance of the plasma-derived factor Va$^{Leiden}$ cofactor compared with normal, plasma-derived factor Va.[290–292] The plasma cofactor pool may play a more significant role in venous thrombosis because of the lack of a significant platelet contribution. In contrast, at the site of an arterial thrombus, the platelet-derived cofactor, as a result of its release and activation, may be present at a concentration exceeding 600 times that of the plasma cofactor.[106]

## VII. Summary

All the studies discussed in this chapter detailing the assembly, function, and regulation of intrinsic Xase and prothrombinase on the platelet surface underscore the significance of complex formation in controlling the coagulant response. Complex formation brings about a dramatic enhancement of catalytic efficiency that is significant to both the localization and modulation of procoagulant activity. When the complex is formed on adherent and aggregated platelets or leukocytes recruited to the site, the bound enzyme is localized with respect to the event for which coagulation is required, and the binding itself leads to significant amplification of the reaction rate. Complex formation also results in the resistance of such complexes to inhibition by the anticoagulants present in circulating blood, thereby limiting reactivity to the site where complex receptors are expressed. As a result, concentrated reactivity is generated and sustained at the site of vascular injury and platelet deposition. Furthermore, the formation and/or dissociation of the complex itself can be considered a regulatory event, because the only complex that is relevant on a biological time scale is the complete complex. Therefore, consider that any dissociation of factor Xa from membrane-bound factor Va will lead to an enormous decrease in reaction rate. As such, complex formation acts as a switch that can turn coagulation on or off.

## References

1. Hawiger, J. (2001). Adhesive interactions of blood cells and the vascular wall in hemostasis and thrombosis. In R. W. Colman, J. Hirsh, V. J. Marder, A. W. Clowes, & J. N. George (Eds.), *Hemostasis and thrombosis: Basic principles & clinical practice* (pp. 639–660). Philadelphia: Lippincott Williams & Wilkins.
2. Barnes, M. J., Knight, C. G., & Farndale, R. W. (1998). The collagen-platelet interaction. *Curr Opin Hematol, 5,* 314–320.
3. Nieswandt, B., Brakebusch, C., Bergmeier, W., et al. (2001). Glycoprotein VI but not alpha2beta1 integrin is essential for platelet interaction with collagen. *EMBO J, 20,* 2120–2130.
4. Gibbins, J. M., Okuma, M., Farndale, R., et al. (1997). Glycoprotein VI is the collagen receptor in platelets which

underlies tyrosine phosphorylation of the Fc receptor gamma-chain. *FEBS Lett, 413,* 255–259.

5. Nieswandt, B., Bergmeier, W., Schulte, V., et al. (2000). Expression and function of the mouse collagen receptor glycoprotein VI is strictly dependent on its association with the FcRgamma chain. *J Biol Chem, 275,* 23998–24002.

6. Tsuji, M., Ezumi, Y., Arai, M., et al. (1997). A novel association of Fc receptor gamma-chain with glycoprotein VI and their co-expression as a collagen receptor in human platelets. *J Biol Chem, 272,* 23528–23531.

7. Plow, E. F., & Shattil, S. J. (2001). Integrin a$_{IIb}$b$_3$ and platelet aggregation. In R. W. Colman, J. Hirsh, V. J. Marder, A. W. Clowes, & J. N. George (Eds.), *Hemostasis and thrombosis: Basic principles & clinical practice* (pp. 479–491). Philadelphia: Lippincott Williams & Wilkins.

8. Mann, K. G., Bovill, E. G., & Krishnaswamy, S. (1991). Surface-dependent reactions in the propagation phase of blood coagulation. *Ann N Y Acad Sci, 614,* 63–75.

9. Seegers, W. H. (1949). Activation of purified prothrombin. *Proc Soc Exptl Biol Med, 72,* 677–680.

10. Mann, K. G., Nesheim, M. E., Church, W. R., et al. (1990). Surface-dependent reactions of the vitamin K-dependent enzyme complexes. *Blood, 76,* 1–16.

11. Wilcox, J. N., Smith, K. M., Schwartz, S. M., et al. (1989). Localization of tissue factor in the normal vessel wall and in the atherosclerotic plaque. *Proc Natl Acad Sci U S A, 86,* 2839–2843.

12. Weiss, H. J., Turitto, V. T., Baumgartner, H. R., et al. (1989). Evidence for the presence of tissue factor activity on subendothelium. *Blood, 73,* 968–975.

13. Bevilacqua, M. P., Pober, J. S., Majeau, G. R., et al. (1984). Interleukin 1 (IL-1) induces biosynthesis and cell surface expression of procoagulant activity in human vascular endothelial cells. *J Exp Med, 160,* 618–623.

14. Bevilacqua, M. P., Pober, J. S., Majeau, G. R., et al. (1986). Recombinant tumor necrosis factor induces procoagulant activity in cultured human vascular endothelium: Characterization and comparison with the actions of interleukin 1. *Proc Natl Acad Sci U S A, 83,* 4533–4537.

15. Camera, M., Giesen, P. L., Fallon, J., et al. (1999). Cooperation between VEGF and TNF-alpha is necessary for exposure of active tissue factor on the surface of human endothelial cells. *Arterioscler Thromb Vasc Biol, 19,* 531–537.

16. Carlsen, E., Flatmark, A., & Prydz, H. (1988). Cytokine-induced procoagulant activity in monocytes and endothelial cells. Further enhancement by cyclosporine. *Transplantation, 46,* 575–580.

17. Morrissey, J. H., Macik, B. G., Neuenschwander, P. F., et al. (1993). Quantitation of activated factor VII levels in plasma using a tissue factor mutant selectively deficient in promoting factor VII activation. *Blood, 81,* 734–744.

18. Eichinger, S., Mannucci, P. M., Tradati, F., et al. (1995). Determinants of plasma factor VIIa levels in humans. *Blood, 86,* 3021–3025.

19. Silverberg, S. A., Nemerson, Y., & Zur, M. (1977). Kinetics of the activation of bovine coagulation factor X by components of the extrinsic pathway. Kinetic behavior of two-chain factor VII in the presence and absence of tissue factor. *J Biol Chem, 252,* 8481–8488.

20. Komiyama, Y., Pedersen, A. H., & Kisiel, W. (1990). Proteolytic activation of human factors IX and X by recombinant human factor VIIa: Effects of calcium, phospholipids, and tissue factor. *Biochemistry, 29,* 9418–9425.

21. Bom, V. J., & Bertina, R. M. (1990). The contributions of Ca2+, phospholipids and tissue-factor apoprotein to the activation of human blood-coagulation factor X by activated factor VII. *Biochem J, 265,* 327–336.

22. Lawson, J. H., & Mann, K. G. (1991). Cooperative activation of human factor IX by the human extrinsic pathway of blood coagulation. *J Biol Chem, 266,* 11317–11327.

23. Papahadjopoulos, D., & Hanahan, D. J. (1964). Observation of the interaction of phospholipids and certain clotting factors in prothrombin activator formation. *Biochim Biophys Acta, 90,* 436–439.

24. Nesheim, M. E., Taswell, J. B., & Mann, K. G. (1979). The contribution of bovine factor V and factor Va to the activity of prothrombinase. *J Biol Chem, 254,* 10952–10962.

25. Kalafatis, M., Swords, N. A., Rand, M. D., et al. (1994). Membrane-dependent reactions in blood coagulation: Role of the vitamin K-dependent enzyme complexes. *Biochim Biophys Acta, 1227,* 113–129.

26. Lawson, J. H., Kalafatis, M., Stram, S., et al. (1994). A model for the tissue factor pathway to thrombin. I. An empirical study. *J Biol Chem, 269,* 23357–23366.

27. Rand, M. D., Lock, J. B., van't Veer, C., et al. (1996). Blood clotting in minimally altered whole blood. *Blood, 88,* 3432–3445.

28. Butenas, S., van't Veer, C., & Mann, K. G. (1997). Evaluation of the initiation phase of blood coagulation using ultrasensitive assays for serine proteases. *J Biol Chem, 272,* 21527–21533.

29. Rosing, J., van Rijn, J. L., Bevers, E. M., et al. (1985). The role of activated human platelets in prothrombin and factor X activation. *Blood, 65,* 319–332.

30. Stern, D., Nawroth, P., Handley, D., et al. (1985). An endothelial cell-dependent pathway of coagulation. *Proc Natl Acad Sci U S A, 82,* 2523–2527.

31. Tans, G., Rosing, J., Thomassen, M. C., et al. (1991). Comparison of anticoagulant and procoagulant activities of stimulated platelets and platelet-derived microparticles. *Blood, 77,* 2641–2648.

32. Hugel, B., Socie, G., Vu, T., et al. (1999). Elevated levels of circulating procoagulant microparticles in patients with paroxysmal nocturnal hemoglobinuria and aplastic anemia. *Blood, 93,* 3451–3456.

33. Mann, K. G., Krishnaswamy, S., & Lawson, J. H. (1992). Surface-dependent hemostasis. *Semin Hematol, 29,* 213–226.

34. Ahmad, S. S., Rawala–Sheikh, R., & Walsh, P. N. (1992). Components and assembly of the factor X activating complex. *Semin Thromb Hemost, 18,* 311–323.

35. Gailani, D., & Broze, G. J., Jr. (1991). Factor XI activation in a revised model of blood coagulation. *Science, 253,* 909–912.

36. Pieters, J., Lindhout, T., & Hemker, H. C. (1989). In situ-generated thrombin is the only enzyme that effectively activates factor VIII and factor V in thromboplastin-activated plasma. *Blood, 74,* 1021–1024.

37. Bailey, K., Bettelheim, F. R., & Lorand, L. (1951). Action of thrombin in clotting of fibrinogen. *Nature, 167,* 233–234.

38. Mosesson, M. W. (1992). The assembly and structure of the fibrin clot. *Nouv Rev Fr Hematol, 34,* 11–16.

39. Lorand, L., & Konishi, K. (1964). Activation of the fibrin stabilizing factor of plasma by thrombin. *Arch Biochem Biophys, 105,* 58–67.

40. Naski, M. C., Lorand, L., & Shafer, J. A. (1991). Characterization of the kinetic pathway for fibrin promotion of alpha-thrombin-catalyzed activation of plasma factor XIII. *Biochemistry, 30,* 934–941.

41. Shen, L., & Lorand, L. (1983). Contribution of fibrin stabilization to clot strength. Supplementation of factor XIII-deficient plasma with the purified zymogen. *J Clin Invest, 71,* 1336–1341.

42. Brummel, K. E., Butenas, S., & Mann, K. G. (1999). An integrated study of fibrinogen during blood coagulation. *J Biol Chem, 274,* 22862–22870.

43. Miletich, J. P., Jackson, C. M., & Majerus, P. W. (1977). Interaction of coagulation factor Xa with human platelets. *Proc Natl Acad Sci U S A, 74,* 4033–4036.

44. Ragni, M. V., Sinha, D., Seaman, F., et al. (1985). Comparison of bleeding tendency, factor XI coagulant activity, and factor XI antigen in 25 factor XI-deficient kindreds. *Blood, 65,* 719–724.

45. Asakai, R., Chung, D. W., Davie, E. W., et al. (1991). Factor XI deficiency in Ashkenazi Jews in Israel. *N Engl J Med, 325,* 153–158.

46. Seligsohn, U. (1993). Factor XI deficiency. *Thromb Haemost, 70,* 68–71.

47. Walsh, P. N. (2001). Factor XI. In R. W. Colman, J. Hirsh, V. J. Marder, A. W. Clowes, & J. N. George (Eds.), *Hemostasis and thrombosis: Basic principles & clinical practice* (pp. 191–202). Philadelphia: Lippincott Williams & Wilkins.

48. Fujikawa, K., Legaz, M. E., Kato, H., et al. (1974). The mechanism of activation of bovine factor IX (Christmas factor) by bovine factor XIa (activated plasma thromboplastin antecedent). *Biochemistry, 13,* 4508–4516.

49. Di Scipio, R. G., Kurachi, K., & Davie, E. W. (1978). Activation of human factor IX (Christmas factor). *J Clin Invest, 61,* 1528–1538.

50. Osterud, B., Bouma, B. N., & Griffin, J. H. (1978). Human blood coagulation factor IX. Purification, properties, and mechanism of activation by activated factor XI. *J Biol Chem, 253,* 5946–5951.

51. Sinha, D., Seaman, F. S., & Walsh, P. N. (1987). Role of calcium ions and the heavy chain of factor XIa in the activation of human coagulation factor IX. *Biochemistry, 26,* 3768–3775.

52. Naito, K., & Fujikawa, K. (1991). Activation of human blood coagulation factor XI independent of factor XII. Factor XI is activated by thrombin and factor XIa in the presence of negatively charged surfaces. *J Biol Chem, 266,* 7353–7358.

53. Baglia, F. A., & Walsh, P. N. (1998). Prothrombin is a cofactor for the binding of factor XI to the platelet surface and for platelet-mediated factor XI activation by thrombin. *Biochemistry, 37,* 2271–2281.

54. Esmon, C. T., & Owen, W. G. (1981). Identification of an endothelial cell cofactor for thrombin-catalyzed activation of protein C. *Proc Natl Acad Sci U S A, 78,* 2249–2252.

55. Esmon, N. L., Owen, W. G., & Esmon, C. T. (1982). Isolation of a membrane-bound cofactor for thrombin-catalyzed activation of protein C. *J Biol Chem, 257,* 859–864.

56. Owen, W. G., & Esmon, C. T. (1981). Functional properties of an endothelial cell cofactor for thrombin-catalyzed activation of protein C. *J Biol Chem, 256,* 5532–5535.

57. Walker, F. J., Sexton, P. W., & Esmon, C. T. (1979). The inhibition of blood coagulation by activated protein C through the selective inactivation of activated factor V. *Biochim Biophys Acta, 571,* 333–342.

58. Nesheim, M. E., Canfield, W. M., Kisiel, W., et al. (1982). Studies of the capacity of factor Xa to protect factor Va from inactivation by activated protein C. *J Biol Chem, 257,* 1443–1447.

59. Suzuki, K., Stenflo, J., Dahlback, B., et al. (1983). Inactivation of human coagulation factor V by activated protein C. *J Biol Chem, 258,* 1914–1920.

60. Solymoss, S., Tucker, M. M., & Tracy, P. B. (1988). Kinetics of inactivation of membrane-bound factor Va by activated protein C. Protein S modulates factor Xa protection. *J Biol Chem, 263,* 14884–14890.

61. Fay, P. J., & Walker, F. J. (1989). Inactivation of human factor VIII by activated protein C: Evidence that the factor VIII light chain contains the activated protein C binding site. *Biochim Biophys Acta, 994,* 142–148.

62. Fay, P. J., Smudzin, T. M., & Walker, F. J. (1991). Activated protein C-catalyzed inactivation of human factor VIII and factor VIIIa. Identification of cleavage sites and correlation of proteolysis with cofactor activity. *J Biol Chem, 266,* 20139–20145.

63. Esmon, C. T. (1987). The regulation of natural anticoagulant pathways. *Science, 235,* 1348–1352.

64. Hackeng, T. M., van't Veer, C., Meijers, J. C., et al. (1994). Human protein S inhibits prothrombinase complex activity on endothelial cells and platelets via direct interactions with factors Va and Xa. *J Biol Chem, 269,* 21051–21058.

65. Koppelman, S. J., van't Veer, C., Sixma, J. J., et al. (1995). Synergistic inhibition of the intrinsic factor X activation by protein S and C4b-binding protein. *Blood, 86,* 2653–2660.

66. van't Veer, C., Butenas, S., Golden, N. J., et al. (1999). Regulation of prothrombinase activity by protein S. *Thromb Haemost, 82,* 80–87.

67. Olson, S. T., Bjork, I., & Shore, J. D. (1993). Kinetic characterization of heparin-catalyzed and uncatalyzed inhibition of blood coagulation proteinases by antithrombin. *Methods Enzymol, 222,* 525–559.

68. Broze, G. J., Jr., Warren, L. A., Novotny, W. F., et al. (1988). The lipoprotein-associated coagulation inhibitor that inhibits the factor VII-tissue factor complex also inhibits factor Xa: Insight into its possible mechanism of action. *Blood, 71,* 335–343.

69. Girard, T. J., Warren, L. A., Novotny, W. F., et al. (1989). Functional significance of the Kunitz-type inhibitory domains of lipoprotein-associated coagulation inhibitor. *Nature, 338,* 518–520.

70. Rapaport, S. I. (1991). The extrinsic pathway inhibitor: A regulator of tissue factor-dependent blood coagulation. *Thromb Haemost, 66,* 6–15.

71. Marciniak, E. (1973). Factor-Xa inactivation by antithrombin. 3. Evidence for biological stabilization of factor Xa by factor V-phospholipid complex. *Br J Haematol, 24,* 391–400.

72. Miletich, J. P., Jackson, C. M., & Majerus, P. W. (1978). Properties of the factor Xa binding site on human platelets. *J Biol Chem, 253,* 6908–6916.

73. Kane, W. H., Lindhout, M. J., Jackson, C. M., et al. (1980). Factor Va-dependent binding of factor Xa to human platelets. *J Biol Chem, 255,* 1170–1174.

74. Tracy, P. B., Nesheim, M. E., & Mann, K. G. (1983). Proteolytic alterations of factor Va bound to platelets. *J Biol Chem, 258,* 662–669.

75. van't Veer, C., & Mann, K. G. (1997). Regulation of tissue factor initiated thrombin generation by the stoichiometric inhibitors tissue factor pathway inhibitor, antithrombin-III, and heparin cofactor-II. *J Biol Chem, 272,* 4367–4377.

76. van't Veer, C., Golden, N. J., Kalafatis, M., et al. (1997). Inhibitory mechanism of the protein C pathway on tissue factor-induced thrombin generation. Synergistic effect in combination with tissue factor pathway inhibitor. *J Biol Chem, 272,* 7983–7994.

77. van't Veer, C., Kalafatis, M., Bertina, R. M., et al. (1997). Increased tissue factor-initiated prothrombin activation as a result of the Arg506→Gln mutation in factor V Leiden. *J Biol Chem, 272,* 20721–20729.

78. van't Veer, C., Golden, N. J., Kalafatis, M., et al. (1997). An in vitro analysis of the combination of hemophilia A and factor V (Leiden). *Blood, 90,* 3067–3072.

79. Butenas, S., van't Veer, C., & Mann, K. G. (1999). "Normal" thrombin generation. *Blood, 94,* 2169–2178.

80. van't Veer, C., Golden, N. J., & Mann, K. G. (2000). Inhibition of thrombin generation by the zymogen factor VII: Implications for the treatment of hemophilia A by factor VIIa. *Blood, 95,* 1330–1335.

81. Butenas, S., Branda, R. F., van't Veer, C., et al. (2001). Platelets and phospholipids in tissue factor-initiated thrombin generation. *Thromb Haemost, 86,* 660–667.

82. Butenas, S., Cawthern, K. M., van't Veer, C., et al. (2001). Antiplatelet agents in tissue factor-induced blood coagulation. *Blood, 97,* 2314–2322.

83. Monroe, D. M., Roberts, H. R., & Hoffman, M. (1994). Platelet procoagulant complex assembly in a tissue factor-initiated system. *Br J Haematol, 88,* 364–371.

84. Hoffman, M., Monroe, D. M., Oliver, J. A., et al. (1995). Factors IXa and Xa play distinct roles in tissue factor-dependent initiation of coagulation. *Blood, 86,* 1794–1801.

85. Monroe, D. M., Hoffman, M., & Roberts, H. R. (1996). Transmission of a procoagulant signal from tissue factor-bearing cell to platelets. *Blood Coagul Fibrinolysis, 7,* 459–464.

86. Hoffman, M., Monroe, D. M., & Roberts, H. R. (1996). Cellular interactions in hemostasis. *Haemostasis, 26,* 12–16.

87. Kjalke, M., Oliver, J. A., Monroe, D. M., et al. (1997). The effect of active site-inhibited factor VIIa on tissue factor-initiated coagulation using platelets before and after aspirin administration. *Thromb Haemost, 78,* 1202–1208.

88. Kjalke, M., Monroe, D. M., Hoffman, M., et al. (1998). Active site-inactivated factors VIIa, Xa, and IXa inhibit individual steps in a cell-based model of tissue factor-initiated coagulation. *Thromb Haemost, 80,* 578–584.

89. Hoffman, M., & Monroe, D. M., III. (2001). A cell-based model of hemostasis. *Thromb Haemost, 85,* 958–965.

90. Kirchhofer, D., Tschopp, T. B., & Baumgartner, H. R. (1995). Active site-blocked factors VIIa and IXa differentially inhibit fibrin formation in a human ex vivo thrombosis model. *Arterioscler Thromb Vasc Biol, 15,* 1098–1106.

91. Peyrou, V., Lormeau, J. C., Herault, J. P., et al. (1999). Contribution of erythrocytes to thrombin generation in whole blood. *Thromb Haemost, 81,* 400–406.

92. Cawthern, K. M., van't Veer, C., Lock, J. B., et al. (1998). Blood coagulation in hemophilia A and hemophilia C. *Blood, 91,* 4581–4592.

93. Beutler, E. (1993). Platelet tranfusions: The 20,000/mL trigger. *Blood, 81,* 1411–1413.

94. Ammar, T., Scudder, L. E., & Coller, B. S. (1997). In vitro effects of the platelet glycoprotein IIb/IIIa receptor antagonist c7E3 Fab on the activated clotting time. *Circulation, 95,* 614–617.

95. Reverter, J. C., Beguin, S., Kessels, H., et al. (1996). Inhibition of platelet-mediated, tissue factor-induced thrombin generation by the mouse/human chimeric 7E3 antibody. Potential implications for the effect of c7E3 Fab treatment on acute thrombosis and "clinical restenosis." *J Clin Invest, 98,* 863–874.

96. Tracy, P. B., Eide, L. L., Bowie, E. J., et al. (1982). Radioimmunoassay of factor V in human plasma and platelets. *Blood, 60,* 59–63.

97. Hayward, C. P., Furmaniak–Kazmierczak, E., Cieutat, A. M., et al. (1995). Factor V is complexed with multimerin in resting platelet lysates and colocalizes with multimerin in platelet alpha-granules. *J Biol Chem, 270,* 19217–19224.

98. Hayward, C. P., Warkentin, T. E., Horsewood, P., et al. (1991). Multimerin: A series of large disulfide-linked multimeric proteins within platelets. *Blood, 77,* 2556–2560.

99. Hayward, C. P., Bainton, D. F., Smith, J. W., et al. (1993). Multimerin is found in the alpha-granules of resting platelets and is synthesized by a megakaryocytic cell line. *J Clin Invest, 91,* 2630–2639.

100. Hayward, C. P., & Kelton, J. G. (1995). Multimerin. *Curr Opin Hematol, 2,* 339–344.

101. Tracy, P. B., Eide, L. L., & Mann, K. G. (1985). Human prothrombinase complex assembly and function on isolated peripheral blood cell populations. *J Biol Chem, 260,* 2119–2124.

102. Chesney, C. M., Pifer, D., & Colman, R. W. (1981). Subcellular localization and secretion of factor V from human platelets. *Proc Natl Acad Sci U S A, 78,* 5180–5184.

103. Vicic, W. J., Lages, B., & Weiss, H. J. (1980). Release of human platelet factor V activity is induced by both collagen and ADP and is inhibited by aspirin. *Blood, 56,* 448–455.

104. Monkovic, D. D., & Tracy, P. B. (1990). Functional characterization of human platelet-released factor V and its activation by factor Xa and thrombin. *J Biol Chem, 265,* 17132–17140.

105. Tracy, P. B., Giles, A. R., Mann, K. G., et al. (1984). Factor V (Quebec): A bleeding diathesis associated with a qualitative platelet factor V deficiency. *J Clin Invest, 74,* 1221–1228.

106. Nesheim, M. E., Nichols, W. L., Cole, T. L., et al. (1986). Isolation and study of an acquired inhibitor of human coagulation factor V. *J Clin Invest, 77,* 405–415.

107. Tracy, P. B., & Mann, K. G. (1987). Abnormal formation of the prothrombinase complex: Factor V deficiency and related disorders. *Hum Pathol, 18,* 162–169.

108. Monkovic, D. D., & Tracy, P. B. (1990). Activation of human factor V by factor Xa and thrombin. *Biochemistry, 29,* 1118–1128.

109. Camire, R. M., Kalafatis, M., Cushman, M., et al. (1995). The mechanism of inactivation of human platelet factor Va from normal and activated protein C-resistant individuals. *J Biol Chem, 270,* 20794–20800.

110. Camire, R. M., Kalafatis, M., Simioni, P., et al. (1998). Platelet-derived factor Va/Va Leiden cofactor activities are sustained on the surface of activated platelets despite the presence of activated protein C. *Blood, 91,* 2818–2829.

111. Conlon, S. J., Camire, R. M., Kalafatis, M., et al. (1997). Cleavage of platelet-derived factor Va by plasmin results in increased and sustained cofactor activity on the thrombin-activated platelet surface. *Thromb Haemost, 77,* 2507a.

112. Kalafatis, M., Rand, M. D., Jenny, R. J., et al. (1993). Phosphorylation of factor Va and factor VIIIa by activated platelets. *Blood, 81,* 704–719.

113. Rand, M. D., Kalafatis, M., & Mann, K. G. (1994). Platelet coagulation factor Va: The major secretory platelet phosphoprotein. *Blood, 83,* 2180–2190.

114. Gould, W. R., Silveira, J. R., & Tracy, P. B. (2004). Unique in vivo modifications of coagulation factor V produce a physically and functionally distinct platelet-derived cofactor: Characterization of purified platelet-derived factor V/Va. *J Biol Chem, 279,* 2383–2393.

115. Camire, R. M., Pollak, E. S., Kaushansky, K., et al. (1998). Secretable human platelet-derived factor V originates from the plasma pool. *Blood, 92,* 3035–3041.

116. Bouchard, B. A., Williams, J. L., Meisler, N. T., et al. (2005). Endocytosis of plasma-derived factor V by megakaryocytes occurs via a clathrin-dependent, specific membrane binding event. *J Thromb Haemost, 3,* 541–551.

117. Gould, W. R., Simioni, P., Silveira, J. R., et al. (2005). Megakaryocytes endocytose and subsequently modify human factor V in vivo to form the entire pool of a unique platelet-derived cofactor. *J Thromb Haemost, 3,* 450–456.

118. Christella, M., Thomassen, L. G., Castoldi, E., et al. (2003). Endogenous factor V synthesis in megakaryocytes contrib-

utes negligibly to the platelet factor V pool. *Haematologica, 88,* 1150–1156.

119. Veljkovic, D., Cramer, E. M., Alimardani, G., et al. (2003). Studies of alpha-granule proteins in cultured human megakaryocytes. *Thromb Haemost, 90,* 844–852.

120. Bouchard, B. A., Meisler, N. T., Bombard, J., et al. (2000). Endocytosis and proteolytic processing of plasma factor V by human CD34+-derived megakaryocytes and megakaryocyte-like cells. *Blood, 96,* 2688a.

121. Hayward, C. P., Weiss, H. J., Lages, B., et al. (2001). The storage defects in grey platelet syndrome and alphadelta-storage pool deficiency affect alpha-granule factor V and multimerin storage without altering their proteolytic processing. *Br J Haematol, 113,* 871–877.

122. Gokcen, M., & Yunis, E. (1963). Fibrinogen as part of platelet structure. *Nature (London), 200,* 590.

123. Nachman, R. L., Marcus, A. J., & Zucker–Franklin, D. (1964). Subcellular localization of platelet fibrinogen. *Blood, 24,* 853.

124. Nachman, R. L. (1965). Immunologic studies of platelet protein. *Blood, 25,* 703.

125. Castaldi, P. A., & Caen, J. (1965). Platelet fibrinogen. *J Clin Pathol, 18,* 579–585.

126. Nachman, R. L., Marcus, A. J., & Zucker–Franklin, D. (1967). Immunologic studies of proteins associated with subcellular fractions of normal human platelets. *J Lab Clin Med, 69,* 651–658.

127. Keenan, J. P. (1972). Platelet fibrinogen. I. Quantitation using fibrinogen sensitized tanned red cells. *Med Lab Technol, 29,* 71–79.

128. James, H. L., Bradford, H. R., & Ganguly, P. (1974). Quantitation of human platelet fibrinogen. *Circulation, 50,* II–284.

129. Lopaciuk, S., Lovette, K. M., McDonagh, J., et al. (1976). Subcellular distribution of fibrinogen and factor XIII in human blood platelets. *Thromb Res, 8,* 453–465.

130. Kaplan, K. L., Broekman, M. J., Chernoff, A., et al. (1979). Platelet alpha-granule proteins: Studies on release and subcellular localization. *Blood, 53,* 604–618.

131. Davey, M. G., & Luscher, E. F. (1968). Release reactions of human platelets induced by thrombin and other agents. *Biochim Biophys Acta, 165,* 490–506.

132. Koutts, J., Walsh, P. N., Plow, E. F., et al. (1978). Active release of human platelet factor VIII-related antigen by adenosine diphosphate, collagen, and thrombin. *J Clin Invest, 62,* 1255–1263.

133. Cramer, E. M., Debili, N., Martin, J. F., et al. (1989). Uncoordinated expression of fibrinogen compared with thrombospondin and von Willebrand factor in maturing human megakaryocytes. *Blood, 73,* 1123–1129.

134. Harrison, P., Wilbourn, B., Debili, N., et al. (1989). Uptake of plasma fibrinogen into the alpha granules of human megakaryocytes and platelets. *J Clin Invest, 84,* 1320–1324.

135. Handagama, P. J., Shuman, M. A., & Bainton, D. F. (1990). In vivo defibrination results in markedly decreased amounts of fibrinogen in rat megakaryocytes and platelets. *Am J Pathol, 137,* 1393–1399.

136. Handagama, P., Rappolee, D. A., Werb, Z., et al. (1990). Platelet alpha-granule fibrinogen, albumin, and immunoglobulin G are not synthesized by rat and mouse megakaryocytes. *J Clin Invest, 86,* 1364–1368.

137. Louache, F., Debili, N., Cramer, E., et al. (1991). Fibrinogen is not synthesized by human megakaryocytes. *Blood, 77,* 311–316.

138. Lange, W., Luig, A., Dolken, G., et al. (1991). Fibrinogen gamma-chain mRNA is not detected in human megakaryocytes. *Blood, 78,* 20–25.

139. Karpatkin, M., Howard, L., & Karpatkin, S. (1984). Studies of the origin of platelet-associated fibrinogen. *J Lab Clin Med, 104,* 223–237.

140. Disdier, M., Legrand, C., Bouillot, C., et al. (1989). Quantitation of platelet fibrinogen and thrombospondin in Glanzmann's thrombasthenia by electroimmunoassay. *Thromb Res, 53,* 521–533.

141. George, J. N., Caen, J. P., & Nurden, A. T. (1990). Glanzmann's thrombasthenia: The spectrum of clinical disease. *Blood, 75,* 1383–1395.

142. Coller, B. S., Seligsohn, U., West, S. M., et al. (1991). Platelet fibrinogen and vitronectin in Glanzmann thrombasthenia: Evidence consistent with specific roles for glycoprotein IIb/IIIA and alpha v beta 3 integrins in platelet protein trafficking. *Blood, 78,* 2603–2610.

143. Handagama, P., Scarborough, R. M., Shuman, M. A., et al. (1993). Endocytosis of fibrinogen into megakaryocyte and platelet alpha-granules is mediated by alpha IIb beta 3 (glycoprotein IIb-IIIa). *Blood, 82,* 135–138. [(1993). Erratum. *Blood, 82,* 2936.]

144. Handagama, P., Bainton, D. F., Jacques, Y., et al. (1993). Kistrin, an integrin antagonist, blocks endocytosis of fibrinogen into guinea pig megakaryocyte and platelet alpha-granules. *J Clin Invest, 91,* 193–200.

145. Suzuki, M., Kawakatsu, T., Nagata, H., et al. (1992). Effects of injected antibody against the platelet glycoprotein IIb/IIIa complex on monkey platelet fibrinogen. *Thromb Haemost, 67,* 578–581.

146. Harrison, P., Wilbourn, B., Cramer, E., et al. (1992). The influence of therapeutic blocking of Gp IIb/IIIa on platelet alpha-granular fibrinogen. *Br J Haematol, 82,* 721–728.

147. Nurden, P., Humbert, M., Piotrowicz, R. S., et al. (1996). Distribution of ligand-occupied alpha IIb beta 3 in resting and activated human platelets determined by expression of a novel class of ligand-induced binding site recognized by monoclonal antibody AP6. *Blood, 88,* 887–899.

148. Klinger, M. H., & Kluter, H. (1995). Immunocytochemical colocalization of adhesive proteins with clathrin in human blood platelets: Further evidence for coated vesicle-mediated transport of von Willebrand factor, fibrinogen and fibronectin. *Cell Tissue Res, 279,* 453–457.

149. Nurden, P., Poujol, C., Durrieu–Jais, C., et al. (1999). Labeling of the internal pool of GP IIb-IIIa in platelets by c7E3 Fab fragments (abciximab): Flow and endocytic mechanisms contribute to the transport. *Blood, 93,* 1622–1633.

150. Wencel–Drake, J. D. (1990). Plasma membrane GPIIb/IIIa. Evidence for a cycling receptor pool. *Am J Pathol, 136,* 61–70.

151. Wencel–Drake, J. D., Frelinger, A. L., Dieter, M. G., et al. (1993). Arg–Gly–Asp-dependent occupancy of GPIIb/IIIa by applaggin: Evidence for internalization and cycling of a platelet integrin. *Blood, 81,* 62–69.

152. Heijnen, H. F. G., Debili, N., Vainchencker, W., et al. (1998). Multivesicular bodies are an intermediate stage in the formation of platelet a-granules. *Blood, 91,* 2313–2325.

153. Schneider, D. J., Taatjes, D. J., Howard, D. B., et al. (1999). Increased reactivity of platelets induced by fibrinogen independent of its binding to the IIb-IIIa surface glycoprotein: A potential contributor to cardiovascular risk. *J Am Coll Cardiol, 33,* 261–266.

154. Smyth, S. S., Joneckis, C. C., & Parise, L. V. (1993). Regulation of vascular integrins. *Blood, 81,* 2827–2843.

155. Takahashi, N., Takahashi, Y., & Putnam, F. W. (1986). Primary structure of blood coagulation factor XIIIa (fibrinoligase, transglutaminase) from human placenta. *Proc Natl Acad Sci U S A, 83,* 8019–8023.

156. Ichinose, A., & Davie, E. W. (1988). Primary structure of human coagulation factor XIII. *Adv Exp Med Biol, 231,* 15–27.

157. Shuman, M. A., & Greenberg, C. S. (1986). Platelet regulation of thrombus generation. In D. R. Phillips & M. A. Shuman (Eds.), *Biochemistry of platelets* (pp. 319–346). Orlando, FL: Academic Press.

158. McDonagh, J., McDonagh, R. P., Jr., Delage, J. M., et al. (1969). Factor XIII in human plasma and platelets. *J Clin Invest, 48,* 940–946.

159. Kiesselbach, T. H., & Wagner, R. H. (1972). Demonstration of factor XIII in human megakaryocytes by a fluorescent antibody technique. *Ann N Y Acad Sci, 202,* 318–328.

160. Schwartz, M. L., Pizzo, S. V., Hill, R. L., et al. (1973). Human factor XIII from plasma and platelets: Molecular weights, subunit structures, proteolytic activation, and crosslinking of fibrinogen and fibrin. *J Biol Chem, 248,* 1395–1407.

161. Muszbek, L., Adany, R., Szegedi, G., et al. (1985). Factor XIII of blood coagulation in human monocytes. *Thromb Res, 37,* 401–410.

162. Henriksson, P., Becker, S., Lynch, G., et al. (1985). Identification of intracellular factor XIII in human monocytes and macrophages. *J Clin Invest, 76,* 528–534.

163. Kradin, R. L., Lynch, G. W., Kurnick, J. T., et al. (1987). Factor XIII A is synthesized and expressed on the surface of U937 cells and alveolar macrophages. *Blood, 69,* 778–785.

164. Kaczmarek, E., Liu, Y., Berse, B., et al. (1995). Biosynthesis of plasma factor XIII: Evidence for transcription and translation in hepatoma cells. *Biochim Biophys Acta, 1247,* 127–134.

165. Marx, G., Korner, G., Mou, X., et al. (1993). Packaging zinc, fibrinogen, and factor XIII in platelet alpha-granules. *J Cell Physiol, 156,* 437–442.

166. Loewy, A. G., Dahlberg, A., Dunathan, K., et al. (1961). Fibrinase II. Some physical properties. *J Biol Chem, 236,* 2634.

167. Loewy, A. G. (1972). Some thoughts on the state in nature, biosynthetic origin, and function of factor XIII. *Ann N Y Acad Sci, 202,* 41–58.

168. Bannerjee, D., & Mosesson, M. W. (1975). Characteristics of platelet protransglutaminase (factor XIII): Binding activity in human plasma. *Thromb Res, 7,* 323–327.

169. Curtis, C. G., Brown, K. L., Credo, R. B., et al. (1974). Calcium-dependent unmasking of active center cysteine during activation of fibrin stabilizing factor. *Biochemistry, 13,* 3774–3780.

170. Muszbek, L., Haramura, G., & Polgar, J. (1995). Transformation of cellular factor XIII into an active zymogen transglutaminase in thrombin-stimulated platelets. *Thromb Haemost, 73,* 702–705.

171. Loewy, A. G., McDonagh, J., Mikkola, H., et al. (2001). Structure and function of factor XIII. In R. W. Colman, J. Hirsh, V. J. Marder, A. W. Clowes, & J. N. George (Eds.), *Hemostasis and thrombosis: Basic principles & clinical practice* (pp. 233–247). Philadelphia: Lippincott Williams & Wilkins.

172. McDonagh, J., & McDonagh, R. P. (1972). Factor XIII from human platelets: Effect on fibrin cross-linking. *Thromb Res, 1,* 147.

173. Cohen, I., Young–Bandala, L., Blankenberg, T. A., et al. (1979). Fibrinoligase-catalyzed cross-linking of myosin from platelet and skeletal muscle. *Arch Biochem Biophys, 192,* 100–111.

174. Mosher, D. F. (1978). Cross-linking of plasma and cellular fibronectin by plasma transglutaminase. *Ann N Y Acad Sci, 312,* 38–42.

175. Lipscomb, M. S., & Walsh, P. N. (1979). Human platelets and factor XI. Localization in platelet membranes of factor XI-like activity and its functional distinction from plasma factor XI. *J Clin Invest, 63,* 1006–1014.

176. Walsh, P. N. (1972). The effects of collagen and kaolin on the intrinsic coagulant activity of platelets: Evidence for an alternative pathway in intrinsic coagulation not requiring factor XII. *Br J Haematol, 22,* 393–405.

177. Schiffman, S., Rapaport, S. I., & Chong, M. M. (1973). Platelets and initiation of intrinsic clotting. *Br J Haematol, 24,* 633–642.

178. Walsh, P. N., & Griffin, J. H. (1981). Contributions of human platelets to the proteolytic activation of blood coagulation factors XII and XI. *Blood, 57,* 106–118.

179. Schiffman, S., & Yeh, C. H. (1990). Purification and characterization of platelet factor XI. *Thromb Res, 60,* 87–97.

180. Hu, C. J., Baglia, F. A., Mills, D. C., et al. (1998). Tissue-specific expression of functional platelet factor XI is independent of plasma factor XI expression. *Blood, 91,* 3800–3807.

181. Heck, L. W., & Kaplan, A. P. (1974). Substrates of Hageman factor. I. Isolation and characterization of human factor XI (PTA) and inhibition of the activated enzyme by alpha 1-antitrypsin. *J Exp Med, 140,* 1615–1630.

182. Bouma, B. N., & Griffin, J. H. (1977). Human blood coagulation factor XI: Purification, properties, and mechanism of activation by activated factor XII. *J Biol Chem, 252,* 6432–6437.

183. Fujikawa, K., Chung, D. W., Hendrickson, L. E., et al. (1986). Amino acid sequence of human factor XI, a blood coagulation factor with four tandem repeats that are highly homolo-

gous with plasma prekallikrein. *Biochemistry, 25,* 2417–2424.

184. Tuszynski, G. P., Bevacqua, S. J., Schmaier, A. H., et al. (1982). Factor XI antigen and activity in human platelets. *Blood, 59,* 1148–1156.

185. Hsu, T. C., Shore, S. K., Seshsmma, T., et al. (1998). Molecular cloning of platelet factor XI, an alternative splicing product of the plasma factor XI gene. *J Biol Chem, 273,* 13787–13793.

186. Martincic, D., Kravtsov, V., & Gailani, D. (1999). Factor XI messenger RNA in human platelets. *Blood, 94,* 3397–3404.

187. Podmore, A., Smith, M., Savidge, G., et al. (2004). Real-time quantitative PCR analysis of factor XI mRNA variants in human platelets. *J Thromb Haemost, 2,* 1713–1719.

188. Butenas, S., Bouchard, B. A., Brummel–Ziedins, K. E., et al. (2005). Tissue factor activity in whole blood. *Blood, 105,* 2764–2770.

189. Rauch, U., Bonderman, D., Bohrmann, B., et al. (2000). Transfer of tissue factor from leukocytes to platelets is mediated by CD15 and tissue factor. *Blood, 96,* 170–175.

190. Del Conde, I., Shrimpton, C. N., Thiagarajan, P., et al. (2005). Tissue-factor-bearing microvesicles arise from lipid rafts and fuse with activated platelets to initiate coagulation. *Blood, 106,* 1604–1611.

191. Falati, S., Gross, P., Merrill–Skoloff, G., et al. (2002). Real-time in vivo imaging of platelets, tissue factor and fibrin during arterial thrombus formation in the mouse. *Nat Med, 8,* 1175–1181.

192. Falati, S., Liu, Q., Gross, P., et al. (2003). Accumulation of tissue factor into developing thrombi in vivo is dependent upon microparticle P-selectin glycoprotein ligand 1 and platelet P-selectin. *J Exp Med, 197,* 1585–1598.

193. Chou, J., Mackman, N., Merrill–Skoloff, G., et al. (2004). Hematopoietic cell-derived microparticle tissue factor contributes to fibrin formation during thrombus propagation. *Blood, 104,* 3190–3197.

194. Butenas, S., & Mann, K. G. (2004). Active tissue factor in blood? *Nat Med, 10,* 1155–1156. Reply. *Nat Med, 10,* 1156.

195. Berckmans, R. J., Neiuwland, R., Boing, A. N., et al. (2001). Cell-derived microparticles circulate in healthy humans and support low grade thrombin generation. *Thromb Haemost, 85,* 639–646.

196. Santucci, R. A., Erlich, J., Labriola, J., et al. (2000). Measurement of tissue factor activity in whole blood. *Thromb Haemost, 83,* 445–454.

197. Orfeo, T., Butenas, S., Brummel–Ziedins, K. E., et al. (2005). The tissue factor requirement in blood coagulation. *J Biol Chem, 280,* 42887–42896.

198. Baglia, F. A., & Walsh, P. N. (2000). Thrombin-mediated feedback activation of factor XI on the activated platelet surface is preferred over contact activation by factor XIIa or factor XIa. *J Biol Chem, 275,* 20514–20519.

199. Baglia, F. A., Badellino, K. O., Ho, D. H., et al. (2000). A binding site for the kringle II domain of prothrombin in the apple 1 domain of factor XI. *J Biol Chem, 275,* 31954–31962.

200. Baglia, F. A., Gailani, D., Lopez, J. A., et al. (2004). Identification of a binding site for glycoprotein Ibalpha in the apple 3 domain of factor XI. *J Biol Chem, 279,* 45470–45476.

201. Baglia, F. A., Shrimpton, C. N., Emsley, J., et al. (2004). Factor XI interacts with the leucine-rich repeats of glycoprotein Ibalpha on the activated platelet. *J Biol Chem, 279,* 49323–49329.

202. Baglia, F. A., Shrimpton, C. N., Lopez, J. A., et al. (2003). The glycoprotein Ib-IX-V complex mediates localization of factor XI to lipid rafts on the platelet membrane. *J Biol Chem, 278,* 21744–21750.

203. Oliver, J. A., Monroe, D. M., Roberts, H. R., et al. (1999). Thrombin activates factor XI on activated platelets in the absence of factor XII. *Arterioscler Thromb Vasc Biol, 19,* 170–177.

204. Baglia, F. A., Badellino, K. O., Li, C. Q., et al. (2002). Factor XI binding to the platelet glycoprotein Ib-IX-V complex promotes factor XI activation by thrombin. *J Biol Chem, 277,* 1662–1668.

205. Butenas, S., Dee, J. D., & Mann, K. G. (2003). The function of factor XI in tissue factor-initiated thrombin generation. *J Thromb Haemost, 1,* 2103–2111.

206. Sinha, D., Seaman, F. S., Koshy, A., et al. (1984). Blood coagulation factor XIa binds specifically to a site on activated human platelets distinct from that for factor XI. *J Clin Invest, 73,* 1550–1556.

207. Ahmad, S. S., Rawala–Sheikh, R., & Walsh, P. N. (1989). Comparative interactions of factor IX and factor IXa with human platelets. *J Biol Chem, 264,* 3244–3251.

208. Gailani, D., Ho, D., Sun, M. F., et al. (2001). Model for a factor IX activation complex on blood platelets: Dimeric conformation of factor XIa is essential. *Blood, 97,* 3117–3122.

209. Sinha, D., Marcinkiewicz, M., Lear, J. D., et al. (2005). Factor XIa dimer in the activation of factor IX. *Biochemistry, 44,* 10416–10422.

210. Baglia, F. A., Jameson, B. A., & Walsh, P. N. (1991). Identification and chemical synthesis of a substrate-binding site for factor IX on coagulation factor XIa. *J Biol Chem, 266,* 24190–24197.

211. Schiffman, S., Rimon, A., & Rapaport, S. I. (1977). Factor XI and platelets: Evidence that platelets contain only minimal factor XI activity and antigen. *Br J Haematol, 35,* 429–436.

212. Walsh, P. N., Mills, D. C., Pareti, F. I., et al. (1975). Hereditary giant platelet syndrome. Absence of collagen-induced coagulant activity and deficiency of factor-XI binding to platelets. *Br J Haematol, 29,* 639–655.

213. Neuenschwander, P., & Jesty, J. (1988). A comparison of phospholipid and platelets in the activation of human factor VIII by thrombin and factor Xa, and in the activation of factor X. *Blood, 72,* 1761–1770.

214. Saenko, E. L., Scandella, D., Yakhyaev, A. V., et al. (1998). Activation of factor VIII by thrombin increases its affinity for binding to synthetic phospholipid membranes and activated platelets. *J Biol Chem, 273,* 27918–27926.

215. Ahmad, S. S., Rawala–Sheikh, R., & Walsh, P. N. (1989). Platelet receptor occupancy with factor IXa promotes factor X activation. *J Biol Chem, 264,* 20012–20016.

216. Ahmad, S. S., Rawala–Sheikh, R., Monroe, D. M., et al. (1990). Comparative platelet binding and kinetic studies with normal and variant factor IXa molecules. *J Biol Chem, 265,* 20907–20911.

217. Rawala–Sheikh, R., Ahmad, S. S., Ashby, B., et al. (1990). Kinetics of coagulation factor X activation by platelet-bound factor IXa. *Biochemistry, 29,* 2606–2611.

218. Rawala–Sheikh, R., Ahmad, S. S., Monroe, D. M., et al. (1992). Role of gamma-carboxyglutamic acid residues in the binding of factor IXa to platelets and in factor-X activation. *Blood, 79,* 398–405.

219. Ahmad, S. S., Rawala–Sheikh, R., Cheung, W. F., et al. (1992). The role of the first growth factor domain of human factor IXa in binding to platelets and in factor X activation. *J Biol Chem, 267,* 8571–8576.

220. Ahmad, S. S., Rawala–Sheikh, R., Cheung, W. F., et al. (1994). High-affinity, specific factor IXa binding to platelets is mediated in part by residues 3–11. *Biochemistry, 33,* 12048–12055.

221. Ahmad, S. S., Rawala, R., Cheung, W. F., et al. (1995). The role of the second growth-factor domain of human factor IXa in binding to platelets and in factor-X activation. *Biochem J, 310,* 427–431. [(1995). Erratum. *Biochem J, 310,* 991.]

222. London, F., Ahmad, S. S., & Walsh, P. N. (1996). Annexin V inhibition of factor IXa-catalyzed factor X activation on human platelets and on negatively-charged phospholipid vesicles. *Biochemistry, 35,* 16886–16897.

223. London, F., & Walsh, P. N. (1996). The role of electrostatic interactions in the assembly of the factor X activating complex on both activated platelets and negatively-charged phospholipid vesicles. *Biochemistry, 35,* 12146–12154.

224. Scandura, J. M., & Walsh, P. N. (1996). Factor X bound to the surface of activated human platelets is preferentially activated by platelet-bound factor IXa. *Biochemistry, 35,* 8903–8913.

225. Scandura, J. M., Ahmad, S. S., & Walsh, P. N. (1996). A binding site expressed on the surface of activated human platelets is shared by factor X and prothrombin. *Biochemistry, 35,* 8890–8902.

226. Ahmad, S. S., Wong, M. Y., Rawala, R., et al. (1998). Coagulation factor IX residues G4-Q11 mediate its interaction with a shared factor IX/IXa binding site on activated platelets but not the assembly of the functional factor X activating complex. *Biochemistry, 37,* 1671–1679.

227. Wong, M. Y., Gurr, J. A., & Walsh, P. N. The second epidermal growth factor-like domain of human factor IXa mediates factor IXa binding to platelets and assembly of the factor X activating complex. *Biochemistry, 38,* 8948–8960.

228. London, F. S., & Walsh, P. N. (2000). Zymogen factor IX potentiates factor IXa-catalyzed factor X activation. *Biochemistry, 39,* 9850–9858.

229. Ahmad, S. S., Scandura, J. M., & Walsh, P. N. (2000). Structural and functional characterization of platelet receptor-mediated factor VIII binding. *J Biol Chem, 275,* 13071–13081.

230. Ahmad, S. S., London, F. S., & Walsh, P. N. (2003). Binding studies of the enzyme (factor IXa) with the cofactor (factor VIIIa) in the assembly of factor-X activating complex on the activated platelet surface. *J Thromb Haemost, 1,* 2348–2355.

231. McGee, M. P., Li, L. C., & Hensler, M. (1992). Functional assembly of intrinsic coagulation proteases on monocytes and platelets. Comparison between cofactor activities induced by thrombin and factor Xa. *J Exp Med, 176,* 27–35.

232. Wilkinson, F. H., London, F. S., & Walsh, P. N. (2002). Residues 88–109 of factor IXa are important for assembly of the factor X activating complex. *J Biol Chem, 277,* 5725–5733.

233. Wilkinson, F. H., Ahmad, S. S., & Walsh, P. N. (2002). The factor IXa second epidermal growth factor (EGF2) domain mediates platelet binding and assembly of the factor X activating complex. *J Biol Chem, 277,* 5734–5741.

234. Yang, X., Chang, Y. J., Lin, S. W., et al. (2004). Identification of residues Asn89, Ile90, and Val107 of the factor IXa second epidermal growth factor domain that are essential for the assembly of the factor X-activating complex on activated platelets. *J Biol Chem, 279,* 46400–46405.

235. Tracy, P. B. (2001). Role of platelets and leukocytes in coagulation. In R. W. Colman, J. Hirsh, V. J. Marder, A. J. Clowes, & J. N. George (Eds.), *Hemostasis and thrombosis: Basic principles of clinical practice* (4th ed., pp. 575–596). Philadelphia: Lippincott-Raven Publishers.

236. Miletich, J. P., Majerus, D. W., & Majerus, P. W. (1978). Patients with congenital factor V deficiency have decreased factor Xa binding sites on their platelets. *J Clin Invest, 62,* 824–831.

237. Miletich, J. P., Kane, W. H., Hofmann, S. L., et al. (1979). Deficiency of factor Xa-factor Va binding sites on the platelets of a patient with a bleeding disorder. *Blood, 54,* 1015–1022.

238. Kane, W. H., & Majerus, P. W. (1982). The interaction of human coagulation factor Va with platelets. *J Biol Chem, 257,* 3963–3969.

239. Tracy, P. B., Nesheim, M. E., & Mann, K. G. (1981). Coordinate binding of factor Va and factor Xa to the unstimulated platelet. *J Biol Chem, 256,* 743–751.

240. Tracy, P. B., Nesheim, M. E., & Mann, K. G. (1992). Platelet factor Xa receptor. *Methods Enzymol, 215,* 329–360.

241. Bouchard, B. A., Catcher, C. S., Thrash, B. R., et al. (1997). Effector cell protease receptor-1, a platelet activation-dependent membrane protein, regulates prothrombinase-catalyzed thrombin generation. *J Biol Chem, 272,* 9244–9251.

242. Feng, P., & Tracy, P. B. (1998). Not all platelets are equal procoagulants. *Blood, 92,* 1441a.

243. Alberio, L., Safa, O., Clemetson, K. J., et al. (2000). Surface expression and functional characterization of alpha-granule factor V in human platelets: Effects of ionophore A23187, thrombin, collagen, and convulxin. *Blood, 95,* 1694–1702.

244. Kempton, C. L., Hoffman, M., Roberts, H. R., et al. (2005). Platelet heterogeneity: Variation in coagulation complexes on platelet subpopulations. *Arterioscler Thromb Vasc Biol, 25,* 861–866.

245. Altieri, D. C., & Edgington, T. S. (1990). Identification of effector cell protease receptor-1. A leukocyte-distributed receptor for the serine protease factor Xa. *J Immunol, 145,* 246–253.

246. Nicholson, A. C., Nachman, R. L., Altieri, D. C., et al. (1996). Effector cell protease receptor-1 is a vascular receptor for coagulation factor Xa. *J Biol Chem, 271,* 28407–28413.

247. Rudolph, A. E., Mullane, M. P., Porche–Sorbet, R., et al. (1996). Factor X St. Louis II. Identification of a glycine substitution at residue 7 and characterization of the recombinant protein. *J Biol Chem, 271,* 28601–28606.

248. Larson, P. J., Camire, R. M., Wong, D., et al. (1998). Structure/function analyses of recombinant variants of human factor Xa: Factor Xa incorporation into prothrombinase on the thrombin-activated platelet surface is not mimicked by synthetic phospholipid vesicles. *Biochemistry, 37,* 5029–5038.

249. Higgins, D. L., Callahan, P. J., Prendergast, F. G., et al. (1985). Lipid mobility in the assembly and expression of the activity of the prothrombinase complex. *J Biol Chem, 260,* 3604–3612.

250. Bevers, E. M., Tilly, R. H., Senden, J. M., et al. (1989). Exposure of endogenous phosphatidylserine at the outer surface of stimulated platelets is reversed by restoration of aminophospholipid translocase activity. *Biochemistry, 28,* 2382–2387.

251. Bevers, E. M., Comfurius, P., & Zwaal, R. F. (1993). Mechanisms involved in platelet procoagulant response. *Adv Exp Med Biol, 344,* 195–207.

252. Kung, C., Hayes, E., & Mann, K. G. (1994). A membrane-mediated catalytic event in prothrombin activation. *J Biol Chem, 269,* 25838–25848.

253. Comfurius, P., Senden, J. M., Tilly, R. H., et al. (1990). Loss of membrane phospholipid asymmetry in platelets and red cells may be associated with calcium-induced shedding of plasma membrane and inhibition of aminophospholipid translocase. *Biochim Biophys Acta, 1026,* 153–160.

254. Thiagarajan, P., & Tait, J. F. (1990). Binding of annexin V/placental anticoagulant protein I to platelets. Evidence for phosphatidylserine exposure in the procoagulant response of activated platelets. *J Biol Chem, 265,* 17420–17423.

255. Thiagarajan, P., & Tait, J. F. (1991). Collagen-induced exposure of anionic phospholipid in platelets and platelet-derived microparticles. *J Biol Chem, 266,* 24302–24307.

256. Jenny, N. S., & Mann, K. G. (1995). Interactions of factor Va and factor VIIIa with adherent platelets and phospholipid bilayers. *Blood, 86,* 75a.

257. Slonosky, D., & Nesheim, M. E. (1988). Prothrombin binding to human activated platelets. *FASEB J, 2,* 5038a.

258. Byzova, T. V., & Plow, E. F. (1997). Networking in the hemostatic system. Integrin α β binds prothrombin and influences its activation. *J Biol Chem, 272,* 27183–27188.

259. Katzmann, J. A., Nesheim, M. E., Hibbard, L. S., et al. (1981). Isolation of functional human coagulation factor V by using a hybridoma antibody. *Proc Natl Acad Sci U S A, 78,* 162–166.

260. Kalafatis, M., Krishnaswamy, S., Rand, M. D., et al. (1993). Factor V. *Methods Enzymol, 222,* 224–236.

261. Tracy, P. B., & Mann, K. G. (1983). Prothrombinase complex assembly on the platelet surface is mediated through the 74,000-dalton component of factor Va. *Proc Natl Acad Sci U S A, 80,* 2380–2384.

262. Dahlback, B., & Stenflo, J. (1978). Binding of bovine coagulation factor Xa to platelets. *Biochemistry, 17,* 4938–4945.

263. Nesheim, M. E., Kettner, C., Shaw, E., et al. (1981). Cofactor dependence of factor Xa incorporation into the prothrombinase complex. *J Biol Chem, 256,* 6537–6540.

264. Hayward, C. P., Cramer, E. M., Kane, W. H., et al. (1997). Studies of a second family with the Quebec platelet disorder: Evidence that the degradation of the alpha-granule membrane and its soluble contents are not secondary to a defect in targeting proteins to alpha-granules. *Blood, 89,* 1243–1253.

265. Janeway, C. M., Rivard, G. E., Tracy, P. B., et al. (1996). Factor V Quebec revisited. *Blood, 87,* 3571–3578.

266. Borchgrevink, C. F., & Owren, P. A. (1961). The hemostatic effect of normal platelets in hemophilia and factor V deficiency. *Acta Med Scand, 170,* 743–746.

267. Chediak, J., Ashenhurst, J. B., Garlick, I., et al. (1980). Successful management of bleeding in a patient with factor V inhibitor by platelet transfusions. *Blood, 56,* 835–841.

268. Walsh, P. N. (1994). Platelet-coagulant protein interactions. In R. W. Colman, J. Hirsh, V. J. Marder, & E. W. Salzman (Eds.), *Hemostasis and thrombosis: Basic principles and clinical practice* (3rd ed., pp. 629–650). Philadelphia: Lippincott Company.

269. Bouchard, B. A., Shatos, M. A., & Tracy, P. B. (1997). Human brain pericytes differentially regulate expression of procoagulant enzyme complexes comprising the extrinsic pathway of blood coagulation. *Arterioscler Thromb Vasc Biol, 17,* 1–9.

270. Swords, N. A., & Mannm, K. G. (1993). The assembly of the prothrombinase complex on adherent platelets. *Arterioscler Thromb, 13,* 1602–1612.

271. Swords, N. A., Tracy, P. B., & Mann, K. G. (1993). Intact platelet membranes, not platelet-released microvesicles, support the procoagulant activity of adherent platelets. *Arterioscler Thromb, 13,* 1613–1622.

272. Billy, D., Briede, J., Heemskerk, J. W., et al. (1997). Prothrombin conversion under flow conditions by prothrombinase assembled on adherent platelets. *Blood Coagul Fibrinol, 8,* 168–174.

273. Hoffman, M., Monroe, D. M., & Roberts, H. R. (1992). Coagulation factor IXa binding to activated platelets and platelet-derived microparticles: A flow cytometric study. *Thromb Haemost, 68,* 74–78.

274. London, F. S., Marcinkiewicz, M., & Walsh, P. N. (2004). A subpopulation of platelets responds to thrombin- or SFLLRN-stimulation with binding sites for factor IXa. *J Biol Chem, 279,* 19854–19859.

275. Dale, G. L., Friese, P., Batar, P., et al. (2002). Stimulated platelets use serotonin to enhance their retention of procoagulant proteins on the cell surface. *Nature, 415,* 175–179.

276. Szasz, R., & Dale, G. L. (2002). Thrombospondin and fibrinogen bind serotonin-derivatized proteins on COAT-platelets. *Blood, 100,* 2827–2831.

277. Nesheim, M. E., Tracy, R. P., & Mann, K. G. (1984). "Clotspeed": A mathematical simulation of the functional properties of prothrombinase. *J Biol Chem, 259,* 1447–1453.

278. Rosing, J., Tans, G., Govers–Riemslag, J. W., et al. (1980). The role of phospholipids and factor Va in the prothrombinase complex. *J Biol Chem, 255,* 274–283.

279. Walsh, P. N., & Biggs, R. (1972). The role of platelets in intrinsic factor-Xa formation. *Br J Haematol, 22,* 743–760.

280. Lamphear, B. J., & Fay, P. J. (1992). Factor IXa enhances reconstitution of factor VIIIa from isolated A2 subunit and A1/A3-C1-C2 dimer. *J Biol Chem, 267,* 3725–3730.

281. Regan, L. M., Lamphear, B. J., Huggins, C. F., et al. (1994). Factor IXa protects factor VIIIa from activated protein C. Factor IXa inhibits activated protein C-catalyzed cleavage of factor VIIIa at Arg562. *J Biol Chem, 269,* 9445–9452.

282. Kalafatis, M., & Mann, K. G. (1993). Role of the membrane in the inactivation of factor Va by activated protein C. *J Biol Chem, 268,* 27246–27257.

283. Comp, P. C., & Esmon, C. T. (1979). Activated protein C inhibits platelet prothrombin-converting activity. *Blood, 54,* 1272–1281.

284. Dahlback, B., & Stenflo, J. (1980). Inhibitory effect of activated protein C on activation of prothrombin by platelet–bound factor Xa. *Eur J Biochem, 107,* 331–335.

285. Kalafatis, M., Rand, M. D., & Mann, K. G. (1994). The mechanism of inactivation of human factor V and human factor Va by activated protein C. *J Biol Chem, 269,* 31869–31880.

286. Bertina, R. M., Koeleman, B. P., Koster, T., et al. (1994). Mutation in blood coagulation factor V associated with resistance to activated protein C. *Nature, 369,* 64–67.

287. Dahlback, B., Carlsson, M., & Svensson, P. J. (1993). Familial thrombophilia due to a previously unrecognized mechanism characterized by poor anticoagulant response to activated protein C: Prediction of a cofactor to activated protein C. *Proc Natl Acad Sci U S A, 90,* 1004–1008.

288. Voorberg, J., Roelse, J., Koopman, R., et al. (1994). Association of idiopathic venous thromboembolism with single point-mutation at Arg506 of factor V. *Lancet, 343,* 1535–1536.

289. Svensson, P. J., & Dahlback, B. (1994). Resistance to activated protein C as a basis for venous thrombosis. *N Engl J Med, 330,* 517–522.

290. Kalafatis, M., Bertina, R. M., Rand, M. D., et al. (1995). Characterization of the molecular defect in factor VR506Q. *J Biol Chem, 270,* 4053–4057.

291. Kalafatis, M., Haley, P. E., Lu, D., et al. (1996). Proteolytic events that regulate factor V activity in whole plasma from normal and activated protein C (APC)-resistant individuals during clotting: An insight into the APC-resistance assay. *Blood, 87,* 4695–4707.

292. Kalafatis, M., & Mann, K. G. (1997). Factor V Leiden and thrombophilia. *Arterioscler Thromb Vasc Biol, 17,* 620–627.

# CHAPTER 20

# Platelet-Derived Microparticles

## Rienk Nieuwland and Augueste Sturk

*Department of Clinical Chemistry, Academic Medical Center, Amsterdam, The Netherlands*

## I. Introduction

Platelets play important roles in hemostasis. After vessel wall damage, platelets rapidly adhere, aggregate, and secrete compounds that activate other platelets. This results in the formation of a platelet plug (primary hemostasis). In addition, activated platelets support the formation of insoluble fibrin (coagulation, secondary hemostasis) that strengthens the platelet plug. In the 1940s it was known that platelets support clotting, because platelet-containing plasma clotted faster than platelet-poor plasma (PPP). High-speed centrifugation of PPP further prolonged the clotting time, suggesting that PPP contained another (subcellular) factor that facilitated clotting.[1,2] In 1967, Wolf[3] demonstrated the presence of "coagulant material in minute particulate form," which was present in "normal plasma, serum and fractions derived therefrom." This material originated from platelets and was initially called *platelet dust*.[3] The ability of platelets and platelet dust to facilitate coagulation was known in the early literature as platelet factor 3 activity. We now know that platelet dust consists of small vesicles that are mainly platelet derived (platelet-derived microparticles, or PMPs) and explain the coagulant activity of PPP. Especially during the last decade, there has been renewed interest in the potential role of PMP in health and disease.

## II. Microparticle Structure

### A. Size

Activated platelets release two types of membrane vesicles — PMP, which are budded from the surface membrane, and exosomes (Figs. 20-1 and 20-2). PMP are heterogeneous in size, ranging between 0.1 μm and 1.0 μm. Their upper size limit may therefore be almost that of a platelet. Exosomes are smaller on average than PMP and range in size between 40 nm and 100 nm. They are stored within α-granules and become released during the secretion response.[4]

### B. Phospholipids

The cell membrane that surrounds platelets is a (lipid) bilayer that contains different types of phospholipids, such as phosphatidylserine (PS), phosphatidylethanolamine (PE), phosphatidylcholine, and sphingomyelin. These different types of phospholipids are asymmetrically distributed in resting platelets. Uncharged phospholipids, such as phosphatidylcholine and sphingomyelin, are mainly present in the outer leaflet of the bilayer, whereas the inner leaflet contains the negatively charged aminophospholipids PS and PE.[5] Platelets and other cells contain specific transporters, such as aminophospholipid translocases, that actively maintain this asymmetrical phospholipid distribution within the membrane.[6] During activation, this asymmetrical distribution becomes disrupted, phospholipids are scrambled, and PS and PE become exposed on the cell surface.[7] PS is thought to originate mainly from the inner leaflet of the (cell) membrane and not from granule membranes.[8] Like activated platelets, PMP expose aminophospholipids. This exposure may facilitate their removal from the circulation, because phagocytic cells are equipped with PS-specific receptors that recognize and interact with PS-exposing cells, and, possibly, also with (P)MP.

### C. Glycoproteins

Platelets and PMP share glycoprotein (GP) receptors, such as GPIb (CD42b; see Chapter 7), platelet–endothelial cell adhesion molecule-1 (CD31; see Chapter 11), and the integrin αIIbβ3 (GPIIb-IIIa, CD41/CD61; see Chapter 8). In addition, subpopulations of PMPs can expose activation markers, including P-selectin (CD62P; see Chapter 12),[4] and may bind fibrinogen. The antigenic composition of PMP and their functions are dependent on the mechanisms underlying their release. For example, PMP released from platelets activated by collagen and thrombin expose integrin αIIbβ3 that binds fibrinogen, whereas PMP from complement C5b-9-activated platelets expose αIIbβ3, which does not bind fibrinogen.[9]

Recently, proteomics (see Chapter 5) was used to determine the proteome of plasma microparticles (MP).[10] Because plasma was used as a source of MP, one has to bear in mind that these plasma MP are not only of platelet origin. The proteome of plasma MP (1021–1055 protein spots) clearly differed from plasma (331–370 protein spots), and 30 proteins were identified in the plasma MP proteome that had never been reported before in such analyses of human plasma.[10] Additional studies will be necessary to establish the precise composition of MP from a single cell type, such as platelets.

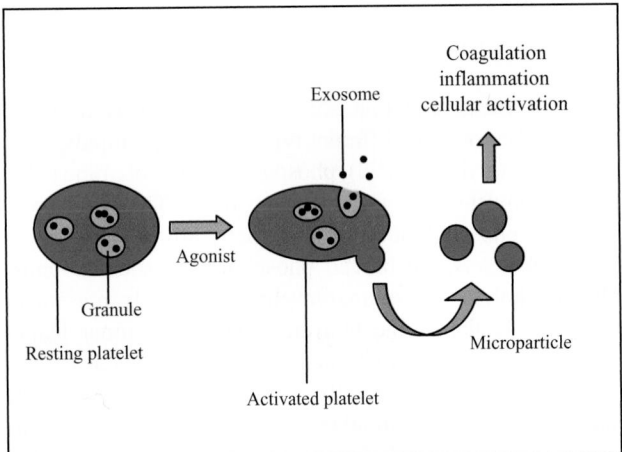

**Figure 20-1.** Platelet-derived microparticles (PMP) and exosomes. A *resting platelet* (left) contains α granules in which exosomes are stored. Physiological platelet agonists like thrombin, ADP or collagen, as well as non-physiological agonists such as calcium ionophores, activate the platelet (center). The *activated platelet* releases *microparticles* (right) which are 'blebbed' from the plasma membrane. When the activated platelet secretes its granule contents, granule membranes fuse with the plasma membrane and their contents, including the *exosomes,* are released.

# III.  Detection of Microparticles

## A.  Flow Cytometry

Currently, flow cytometry is the most widely used methodology to characterize cell-derived MP, including PMP. Using this technique, MP can be visualized directly (e.g., in whole blood; Fig. 20-3) or they can be isolated before labeling (Fig. 20-4). Particular antigens of interest exposed on (P)MP surfaces can be directly detected using fluorochrome-labeled antibodies (monoclonal antibodies [mAbs]) and/or annexin V. Because blood contains MP of different cellular origin, an mAb directed against a cell type-specific antigen is often included. In addition, other mAbs or annexin V can be included. Annexin V is a protein that binds with high affinity and specificity to exposed aminophospholipids in the presence of $Ca^{2+}$ ions.

As an example, PMP analysis in whole blood will be described. In the absence of an mAb against a platelet-specific antigen (Fig. 20-3A), neither platelets nor PMP can be reliably identified in whole blood when forward light scatter (FSC; which correlates with size) and side light scatter (SSC; which correlates with density) of all events detected by the flow cytometer are plotted. Upon addition of (fluorescein isothiocyanate [FITC]-labeled) anti-GPIb, platelets (region 2 [R2]) and PMP (R1) can be identified when FITC-positive events (FL-1 channel) are selected (Fig. 20-3B). For this experiment, blood was collected directly from the pericardial cavity of a patient undergoing cardiopulmonary bypass. This blood contains high numbers of PMP that are present in R1 of Fig. 20-3B. That R1 indeed contains PMP is illustrated in Fig. 20-3C and D. Fig. 20-3C shows an FSC-SSC dot plot of unstimulated platelet-rich plasma. Upon addition of calcium ionophore (A23187) in the presence of $Ca^{2+}$ ions, platelets (R2) release PMP (R1; Fig. 20-3D).

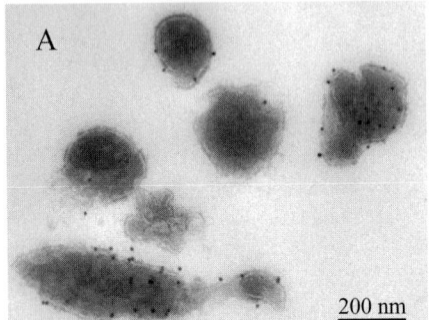
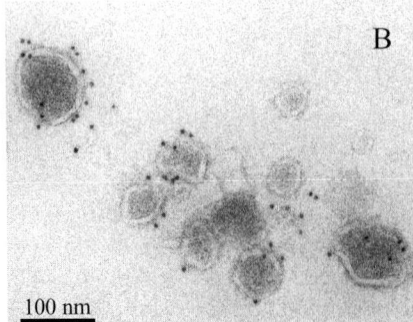
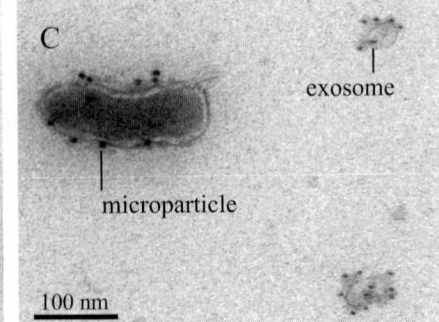

**Figure 20-2.** Identification of PMP and exosomes by electron microscopy. Platelets were stimulated by thrombin-receptor activating peptide. Microparticles and exosomes were isolated from platelet-free supernatants by centrifugation. Panels A and B show microparticles stained with anti-GPIb antibody (panel A; 10 nm gold particles) or annexin V (panel B; 10 nm gold). Panel C shows two exosomes stained with anti-CD63 (5 nm gold) and one microparticle stained with anti-GPIb (10 nm gold). Evidently, the exosomes are smaller than microparticles and differ in antigenic composition. These pictures were kindly provided by Dr. H. F. G. Heijnen (Department of Hematology, UMC Utrecht, The Netherlands).

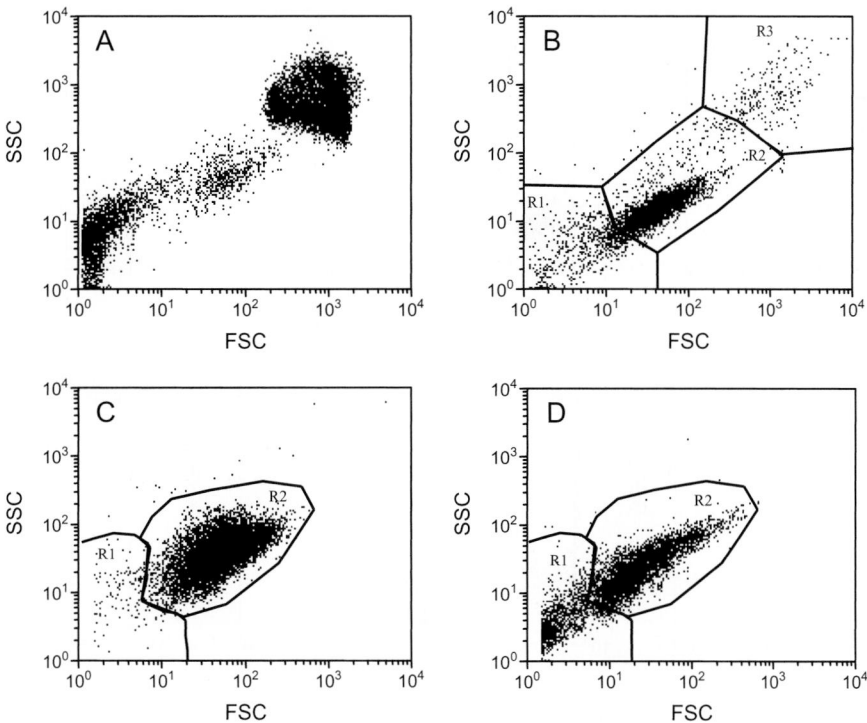

**Figure 20-3.** Detection of cell-derived MP in whole blood and in platelet-rich plasma. Panel A shows a representative flow cytometry dot plot in which all events in anticoagulated diluted whole blood are analyzed and plotted as forward scatter (FSC, which correlates with size) versus sideward scatter (SSC, which correlates with density). Most of the events are erythrocytes (in the upper right), but platelets (approximately in the center) and events smaller than platelets (lower left) cannot be reliably identified. Therefore, as shown in panel B, a fluorescein isothiocyanate (FITC)-labeled GPIb-specific antibody was added to blood obtained from the pericardial cavity of a patient undergoing cardiopulmonary bypass. It then becomes possible to select, on basis of fluorescence, platelets and platelet-derived events. From panel B it is apparent that in whole blood three GPIb-positive regions can be identified, arbitrarily defined as R (region) 1 to 3. Most of the GPIb-positive events are present in R2; these are the single platelets. The larger GPIb-positive events in R3 are platelet-platelet aggregates, platelet-erythrocyte aggregates, and platelet-leukocyte aggregates. Alternatively, these events could also represent PMP associated with erythrocytes or leukocytes. The R1 region contains the smallest size events, *i.e.* events smaller than platelets but positive for GPIb; these events are the PMP. That R1 contains PMP is confirmed in panels C and D. These panels show FSC versus SSC dot plots of platelet-rich plasma, before (C) and after (D) the addition of the calcium ionophore A23187. In the presence of A23187, PMP are released (panel D, R1).

Fig. 20-4 illustrates the detection of (P)MP from the blood of a healthy individual that were isolated by differential centrifugation from whole blood before labeling. Fig. 20-4A shows a typical FSC-SSC dot plot of such MP, almost all of which are within R1. (P)MP are stained with (phycoerythrin-labeled) annexin V in the presence and absence of $Ca^{2+}$ ions (Fig. 20-4C and B, respectively). Fig. 20-4D shows binding of a nonspecific control antibody (FITC labeled, isotype specific) to correct for binding to Fc receptors. From Fig. 20-4E (FITC labeled anti-GPIIIa [CD61]) and Fig. 20-4F (annexin V plus anti-GPIIIa), it is apparent that most MP bind anti-GPIIIa and annexin V. Thus, these MP are mainly of platelet origin and expose aminophospholipids.

Similarly, MP from other cells can be visualized by flow cytometry using mAbs directed against (e.g., glycophorin) A (erythrocytes) or vascular–endothelial cadherin (CD144, endothelial cells). By using a flow cytometric approach,

MPs can be identified and quantified in clinical samples.[11–14] Their presence and numbers differ between various clinical conditions and between body fluids. A striking example of the latter is the presence of high numbers of leukocytic MP and the almost complete absence of MP of platelet or erythrocyte origin in synovial samples of arthritic patients.[15]

Currently, there is no consensus between laboratories on blood collection, blood handling, and subsequent detection of MPs, either by flow cytometry or by other methods. Therefore, MP numbers, composition, and even function are likely to be affected by methodological differences between different laboratories.[16]

### B. Electron Microscopy

(P)MP and exosomes can also be visualized by electron microscopy. Fig. 20-2 shows PMP (Fig. 20-2A–C) and exo-

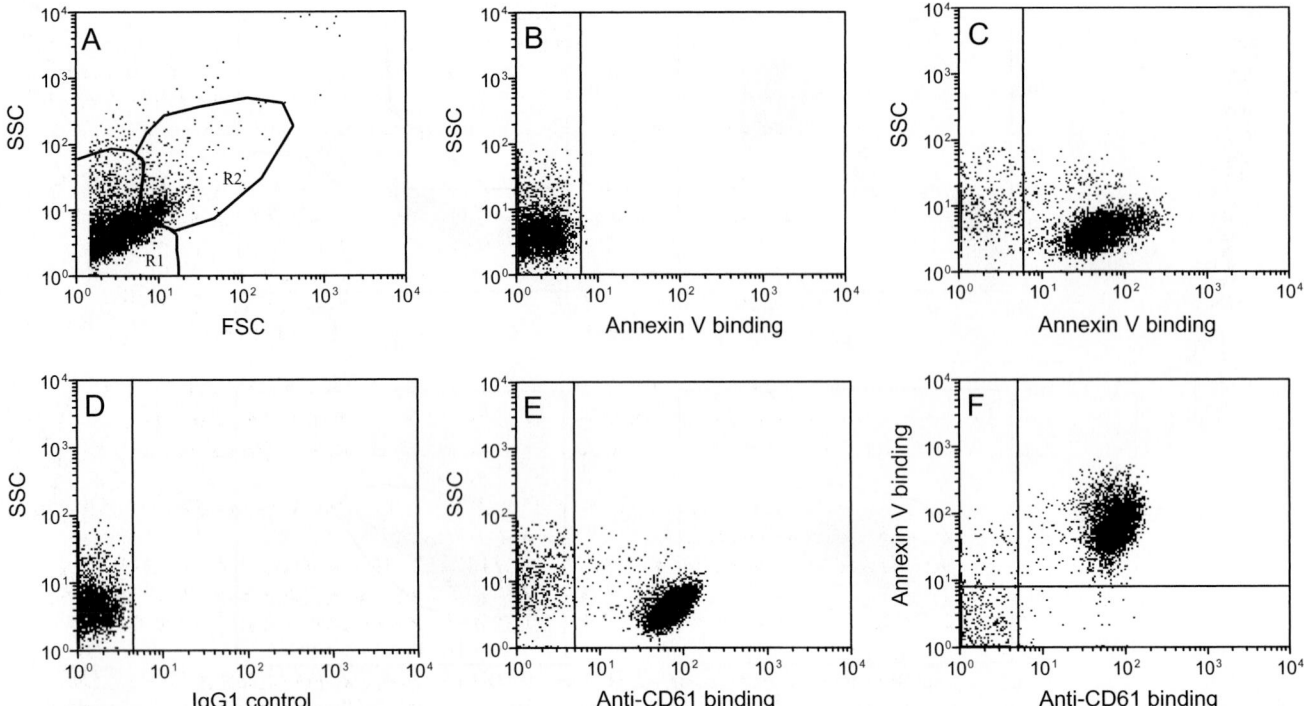

**Figure 20-4.** Detection of PMP by flow cytometry after isolation from plasma. MP were isolated from fresh cell-free plasma of a healthy individual by high-speed centrifugation, washed, and stained with FITC-labeled annexin V and a phycoerythrin-labeled GPIIIa (CD61)-specific antibody. Panel A shows the characteristic flow cytometric dot plot of MP. In panel B, annexin V was added in the absence of calcium ions, whereas in panel C the calcium ions were included. Because annexin V binds to negatively charged phospholipids such as PS only in the presence of calcium ions, in their absence only autofluorescence and/or background staining is observed (panel B). In panel D, a phycoerythrin-labeled IgG control antibody was added to correct for binding of antibodies via their Fc portion to Fc receptors on the microparticle surface. In panel E, an anti-CD61 antibody was added. Finally, in panel F both annexin V and anti-CD61 were added. Because different fluorescent probes were attached to annexin V (FITC, measured in the FL1-channel) and anti-CD61 (phycoerythrin, measured in the FL2-channel), a dot plot with the two fluorescent channels can be plotted (panel F) to show that most MP are of platelet origin and expose negatively charged phospholipids.

somes (Fig. 20-2C) isolated from supernatants of thrombin receptor-activating peptide-stimulated platelets. The PMP were labeled and analyzed by whole-mount immunoelectron microscopy as described elsewhere.[4] PMP were stained with anti-GPIb (Fig. 20-2A) or annexin V (Fig. 20-2B). Fig. 20-2C shows the simultaneous labeling of exosomes with anti-GP53 (CD63), and PMP with anti-GPIb. Fig. 20-2 shows that exosomes are smaller than PMP and differ in antigenic composition.

### C. Enzyme-Linked Immunosorbent Assay (ELISA)

MP can also be detected by ELISA. For example, by coating ELISA plates with annexin V or cell-specific mAbs, MP can be captured from plasma.[17] Other investigators capture PMP by coating ELISA plates with kistrin, a disintegrin for which integrin αIIbβ3 has a much higher affinity than for fibrinogen or Arginine-glycine-aspartic acid (RGD)-peptides.[18] Compared with flow cytometry, larger numbers of samples

can be analyzed more easily by ELISA. The main advantages of flow cytometry are its sensitivity, the ability to include combinations of mAbs and/or annexin V to study combinations of exposed antigens, and the identification of subpopulations.

## IV. Formation of Platelet Microparticles

Our current knowledge of the mechanisms underlying PMP formation is still fragmentary. PMP are formed upon platelet activation, during platelet aging and platelet destruction, and, possibly, directly from megakaryocytes.

### A. Platelet Activation

Activated platelets release PMP *in vitro* when agonists (first messengers), such as collagen and thrombin, bind to their surface receptors. These receptors then transduce signals across the cell membrane and generate changes in levels of intracellular second messengers such as $Ca^{2+}$, leading to the release of PMP. Alternatively, platelets release PMP when compounds are added that directly affect second messenger levels (e.g., calcium ionophores [A23187, ionomycin] and phorbol esters). PMP release also occurs when platelets are activated in response to high shear,[19,20] contact with surfaces of foreign bodies,[21] complement,[22] or low temperatures.[23]

Integrin αIIbβ3 is essential for the interaction between platelets (see Chapter 8). Several studies suggested that αIIbβ3 plays a prominent and possibly even essential role in PMP release,[24] but additional studies showed that (functional) αIIbβ3 is not essential. For example, platelets from patients with Glanzmann thrombasthenia (i.e., platelets lack functional αIIbβ3) release fewer PMP upon stimulation by thrombin or collagen than control platelets, but equal numbers of PMP during complement activation.[25]

One of the essential steps in platelet activation is the elevation of cytosolic calcium levels ($Ca^{2+}_i$). This elevation is a prerequisite for PMP release during activation, and results in the activation of several enzymes that are calcium-dependent for their activity, such as calpains and protein kinase C (PKC). Calpains are calcium-dependent proteases that facilitate PMP formation by degrading structural proteins including actin-binding protein, talin, and the heavy chain of myosin. The extent to which calpains contribute to vesiculation, however, is still not fully understood. Although complement C5b-9-activated platelets have increased levels of $Ca^{2+}_i$ and calpain is activated, the release of PMPs is not prevented by calpain inhibitors.[26–28] Increased levels of $Ca^{2+}_i$ are also a cofactor of various isoforms of PKC — serine–threonine kinases that phosphorylate a wide array of intracellular proteins, including signal transduction elements and structural proteins. Platelets release PMP upon direct activation of PKC by phorbol esters. Also, platelet agonists, such as thrombin and adenosine diphosphate, induce the release of PMP at least partially via activation of PKC.

Concurrent with the aminophospholipid exposure on platelets and PMP release, extensive tyrosine dephosphorylation of various signal transduction proteins occurs.[29] Because the extent of tyrosine phosphorylation directly affects the biological activity of many intracellular enzymes, tyrosine phosphorylation may be linked to the formation of PMP, but additional and more direct evidence is necessary to prove this.

### B. Platelet Aging

PMP are also formed during storage of platelets *in vitro*.[30] Although prolonged storage is often paralleled by platelet activation, and despite the fact that platelets are anucleate cells that cannot undergo full-blown programmed cell death, including DNA fragmentation (apoptosis), platelets do contain all proteins required for apoptosis, including cytochrome C, procaspase 9, procaspase 8, and procaspase 3. Addition of cytochrome C to platelet lysates results in formation of active caspase 3, illustrating that at least the so-called intrinsic pathway of apoptosis is functional. Thus far, however, studies of the possible contribution of apoptotic-like processes to PMP formation and other platelet responses, both during activation and aging (storage), have produced conflicting data.[27,31]

Recently, we demonstrated that MP from various cell types, including those from endothelial cells and platelets, contain caspase 3.[32] The presence of caspase 3 in MP from endothelial cells, however, was not related to the apoptotic status of the endothelial cells, which may imply that the presence of caspase 3 in MP is not necessarily linked to apoptosis. Surprisingly, we observed cleaved (130-kDa) Rho-associated kinase I (ROCK-I) in platelets during storage. This kinase, present as a 160-kDa protein in platelets and other cells, is a well-known sub-strate of caspase 3 that degrades ROCK-I to a 130-kDa form. This 130-kDa form directly contributes to the (constitutive) formation of "apoptotic blebs."[33] Whether ROCK-I is involved in the formation of PMP remains to be determined.

### C. Platelet Destruction

*In vitro*, complement C5b-9-treated platelets, particularly those from paroxysmal nocturnal hemoglobinuria (PNH) patients lacking the surface complement inhibitors CD55 and CD59, generate large numbers of PMP.[34] Although this PMP formation is accompanied by proteolytic degradation of structural proteins, inhibition of calpain only blocked this degradation, but not PMP formation. However, PMP formation still depends on calcium influx during these conditions, suggesting that other mechanisms contribute to PMP formation.

### D. Megakaryocytes

*In vivo*, some PMP may originate directly from megakaryocytes rather than from platelets. In clinical states such as immune thrombocytopenic purpura (ITP), in which platelet destruction is compensated for by increased platelet production (by megakaryocytes), hemostasis is often less impaired than predicted from the low platelet counts. It has been suggested that under such conditions PMP, directly released from megakaryocytes, support hemostasis. *In vitro*, cultured human bone marrow CD34+/CD38+ cells that matured to megakaryocyte progenitor cells not only produced platelets but also numerous PMP.[35] The contribution of megakaryocytes to the pool of PMP in the circulation is currently unknown.

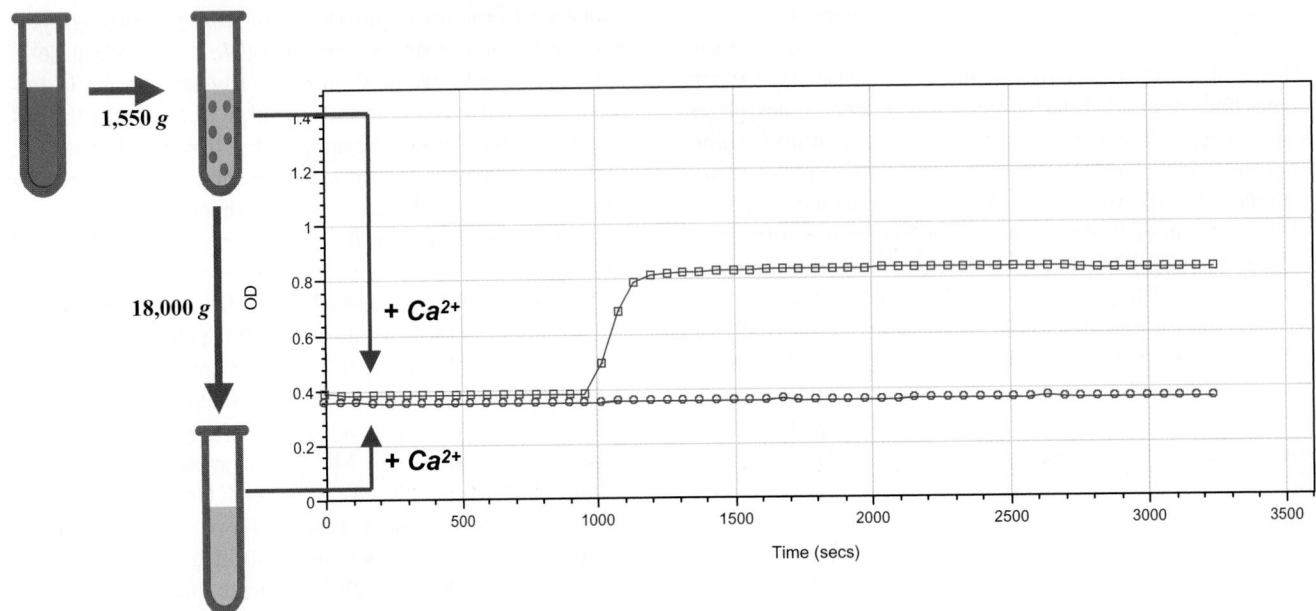

**Figure 20-5.** MP promote clot formation *in vivo*. Blood (citrate-anticoagulated) was collected from a healthy human individual. After centrifugation (20 minutes at 1550 × g), the MP-containing plasma was recalcified. In this experiment, the plasma started clotting after approximately 950 seconds. However, when the plasma was subjected to high-speed centrifugation (30 minutes at 18,000 × g to prepare MP-free plasma) before recalcification, no clotting occurred. This experiment illustrates that the presence of MP is essential for clotting of normal human plasma.

# V. Microparticle Function

## A. General

As discussed in Section I, the subcellular factor in PPP that elicited platelet factor 3 activity was identified as platelet dust or PMP. However, not all MP present in blood originate from platelets. In many studies, relatively low (compared with PMP) numbers of MP from erythrocytes, leukocytes, and endothelial cells have been identified, usually by flow cytometry. The numbers of MP from platelets and other cells in the circulation, and their composition and function are, at least in part, disease dependent. Although many functions have been attributed to MP directly isolated from blood or synovial fluid, the actual contribution of PMP to the observed (i.e., overall) responses are incompletely known. In the following sections, the (putative) functions of PMP are briefly discussed.

## B. Coagulation

Coagulation factors bind to exposed aminophospholipids via $Ca^{2+}$ ions. This binding facilitates the formation of tenase and prothrombinase complexes. PMP are enriched in binding sites for activated factor V (factor Va), factor VIIIa, and factor IXa, and provide the membrane surface for thrombin formation.[9,22]

MP in the plasma of healthy humans are mainly (>80–90%) of platelet origin.[13] Fig. 20-5 shows the clotting (i.e., fibrin formation) of recalcified (citrate-anticoagulated) healthy human PPP. When (P)MP are removed from the PPP by high-speed centrifugation, the (MP-depleted) plasma fails to clot after recalcification. This experiment illustrates that the presence of (P)MP is essential for clotting *in vitro*. MP-depleted plasma fails to clot even in the presence of artificially prepared PS-containing phospholipid vesicles. Thus, (P)MP are likely not only to support, but also to initiate, coagulation.

(P)MP isolated from human plasma samples may utilize both the extrinsic (tissue factor- and factor VIIa-dependent) and contact activation (intrinsic, factor XIIa-dependent) pathways.[12] The contribution of both pathways depended on the clinical conditions and could in some cases be attributed to the presence of particular populations of nonplatelet MP.[12]

Recently, tissue factor-exposing MP from a monocytic cell line were shown to fuse with activated platelets via a P-selectin–P-selectin glycoprotein ligand 1 (PSGL-1)-dependent mechanism *in vitro*,[36] suggesting that all membrane-associated coagulation events (including initiation) may occur on the platelet membrane surface. Previously, we reported that MP in plasma samples from patients with early-onset diabetes type II expose both antigens from platelets and leukocytes, suggesting that fusion of cells and MP or of MP themselves also occurs *in vivo*.[37]

## C. Inhibition of Coagulation

Activated protein C (APC) inhibits both factor Va and factor VIIIa, and (activated) platelets and PMP facilitate this inactivation.[38] APC resistance is associated with PMP,[39] and inverse correlations between levels of circulating (P)MP and prothrombin fragment $F_{1+2}$ or thrombin–antithrombin complexes have been reported.[13,40] Because such (P)MP fractions generate only minute amounts of thrombin, one may hypothesize that (P)MP are predominantly anticoagulant under these conditions.

## D. Adhesion

*In vitro*, PMP bind via integrin $\alpha IIb\beta 3$ to the subendothelial matrix. Once bound, PMP promote platelet and leukocyte adhesion under flow conditions.[41,42] In a rabbit model of arterial injury, PMP adhered mainly at the site of injury.[41] Evidently, PMP facilitate binding of blood cells to the (damaged) vessel wall.

## E. Carrier Function and Cell Activation

Secretory phospholipase $A_2$-treated PMP contain the "bioactive lipid" arachidonic acid (AA). This AA can be transferred to platelets and endothelial cells, which become activated.[43,44] PMP also contain other bioactive lipids, including platelet activating factor.[45] Thus, PMP can modulate the activation status of "target" cells. Interestingly, MP from preeclamptic patients or from myocardial infarction patients impaired nitric oxide-mediated vasodilatation,[46,47] suggesting that (P)MP may directly affect vascular tone *in vivo*. Whether this effect can be attributed solely to the delivery of AA, however, is unclear. Recently, in a rabbit artery model, PMP were also shown to contain cyclooxygenase that converted AA, produced by endothelial cells themselves, into the vasoconstrictor thromboxane (TX) $A_2$ and its stable metabolite $TXB_2$.[48] Thus, PMP may modulate the vascular tone via various pathways, including the delivery of AA and the production of $TXA_2$. PMP expose cell-specific adhesion receptors, such as P-selectin, and therefore may act as "long-range carriers" (third messengers) that deliver bioactive molecules to specific target cells. For example, P-selectin binds to PSGL-1 on monocytes, and this interaction initiates expression of tissue factor and cytokines.[49] Thus, PMP may not only support processes like coagulation directly, but also indirectly.

## F. Inflammation, Angiogenesis, and Cell Proliferation

Complement C5b-9-treated platelets release PMP *in vitro*.[22] Recently, we demonstrated the presence of complement activators (C-reactive protein and immunoglobulin [Ig] M) and activated complement components (C3bc and C4bc) on (P)MP from human plasma (É. Biró, A. Sturk, R. Nieuwland, unpublished observations). Whether this exposure contributes to complement activation *in vivo* or facilitates the removal of (P)MP from the circulation remains to be determined.

In a recent study, PMP were shown to support angiogenesis by promoting the survival, migration, and tube formation of endothelial cells.[50] PMP also stimulate the proliferation of vascular smooth muscle cells by a platelet-derived growth factor-independent mechanism.[51]

# VI. Clinical Disorders Associated with Microparticles

## A. Inherited Disorders

**1. Scott Syndrome and Castaman Syndrome.** In 1979, a case report was presented of a 34-year-old woman with a moderate to severe life-long bleeding disorder. No deficiencies of platelet adhesion, aggregation, secretion, or metabolism were found, but prothrombin consumption and platelet factor 3 activity were found to be impaired, suggesting an "isolated deficiency of platelet procoagulant activity."[52] This disease, now referred to as *Scott syndrome,* is an inherited disorder that is characterized by an impaired scrambling of phosphatidylserine,[53] an impaired release of PMPs,[54,55] a reduced number of binding sites for factor Va,[53] and a decreased ability to support tenase and prothrombinase complex formation.[9] This syndrome is not limited to humans; dogs from a single, inbred colony had a bleeding disorder with characteristic features of Scott syndrome.[56] More recently, another bleeding disorder (Castaman syndrome) was described in four patients from three unrelated families, in which prolonged bleeding times could be attributed to a deficiency of PMP formation.[57]

**2. Wiskott–Aldrich Syndrome (WAS).** WAS patients suffer from severe bleeding resulting from thrombocytopenia and aberrantly small platelets (see Chapter 54). These platelets are efficiently removed in the spleen, and splenectomy partially counteracts the hemorrhagic symptoms. WAS patients have a mutated WAS protein that is involved in signal-mediated cytoskeleton rearrangements. WAS platelets expose PS, which is believed to trigger macrophage-mediated phagocytosis in the spleen. Isolated WAS platelets release large numbers of PMP, and platelets have strongly elevated levels of $Ca^{2+}_i$ compared with control. The aberrant increase in $Ca^{2+}_i$ is thought to underlie both exposure of PS and release of PMP.[58]

## B. Acquired Disorders

### 1. Immune Mediated

*a. Immune Thrombocytopenic Purpura.* ITP is an autoimmune bleeding disorder in which platelets bind IgG autoantibodies and are removed by macrophages from the circulation (see Chapter 46). ITP patients have decreased platelet counts, but elevated numbers of PMP. Some patients with severe thrombocytopenia have hardly any hemorrhagic symptoms, whereas others with only a moderate thrombocytopenia may have frequent and serious hemorrhagic symptoms. It has been suggested that circulating PMP support hemostasis in the former patients. ITP patients without petechiae or mucosal bleeding have higher PMP numbers than patients with these symptoms, suggesting that PMP are involved in hemostatic protection. On the other hand, ITP patients with additional neurological complications have higher PMP levels than patients without these complications, suggesting that these patients are at risk for thromboembolic complication.[59,60] Because a large subpopulation of PMP from ITP patients, compared with control subjects, exposes P-selectin, platelet activation may underlie PMP formation in these patients.[61]

*b. Heparin-Induced Thrombocytopenia (HIT).* HIT (see Chapter 48) is a second well-known example of an acquired immunological disease associated with PMPs. Some patients who receive heparin as an anticoagulant produce antibodies directed against complexes of heparin and platelet factor 4. These antibodies, when bound to the heparin–platelet factor 4 complex, activate platelets via Fc receptors. The platelets aggregate, secrete their granule contents, and release PMP.[62] HIT patients have low platelet numbers, but elevated levels of procoagulant PMP compared with control subjects, and these elevated levels have been associated with thrombotic complications. When platelets from healthy individuals are incubated with HIT sera, they generate PMP in the presence of heparin. This release is paralleled by platelet activation, indicating that PMP formation in HIT is the result of platelet activation.[63,64]

*c. GPIIb-IIIa Antagonist-Induced Thrombocytopenia.* A third example of an autoantibody-induced thrombocytopenia with concurrent PMP formation was described recently. About 0.2% of patients treated with the GPIIb-IIIa (integrin $\alpha$IIb$\beta$3) antagonist eptifibatide develop acute thrombocytopenia (see Chapter 62). In a case report, a 55-year-old man was described who entered the hospital with symptoms of an acute coronary syndrome.[65] He developed acute thrombocytopenia in response to eptifibatide, and his PMP levels were elevated. When control blood was incubated with eptifibatide in the presence of either patient or control plasma, platelets bound IgG and released PMP only in the presence of patient plasma. Thus, the patient's IgG evidently binds to platelets in an eptifibatide-dependent fashion and thereby causes PMP formation. Despite the fact that GPIIb-IIIa antagonists are widely used as platelet antagonists, not only eptifibatide, but also some other GPIIb-IIIa antagonists promote fibrinogen binding and platelet aggregation in some patients, particularly in the lower range of pharmacological concentrations. These platelet antagonists are all thought to trigger conformational changes in the $\alpha$IIb$\beta$3 receptor that leads to IgG binding and platelet activation. Interestingly, despite the development of thrombocytopenia in these patients, the frequency of severe bleeding is low, which again suggests a role for PMP in hemostasis.[65]

### 2. Nonimmune Mediated

*a. Paroxysmal Nocturnal Hemoglobinuria (PNH).* PMP can be formed by platelet destruction. The best-known example is PNH, an acquired stem cell disorder, characterized clinically by hemolytic anemia and a hypercoagulable state that often leads to venous thrombosis in liver, abdominal organs, cerebrum, and skin. Blood cells of these patients partially or completely lack glycolipid-anchored membrane proteins. Two of these proteins, CD55 and CD59, are cell-surface complement inhibitors. Blood cells from these patients are highly sensitive to complement-induced lysis. *In vitro,* platelets from PNH patients generate approximately 10-fold more procoagulant PMP when incubated with the complement C5b-9 complex than control platelets. The elevated numbers of PMP in these patients have been associated with their hypercoagulable state.[34]

*b. Miscellaneous.* Evidence that (P)MP trigger coagulation directly *in vivo* is still scarce. Isolated MP from pericardial wound blood of patients undergoing cardiac surgery (i.e., MP mainly of platelet and erythrocyte origin) strongly initiate (tissue factor-mediated) thrombus formation in a rat venous stasis model.[66] Also PMP infused into rabbits shortened bleeding times.[67] In most studies, however, only associations between changes in levels of circulating (P)MP and the risk for thromboembolic events have been reported. Examples are meningococcal septic shock,[12] sepsis complicated with multiple-organ dysfunction,[40] endotoxemia, systemic vasculitis, several groups of acute cardiovascular diseases, patients with diabetes type I (but not type II),[14,37,60,68,69] Legionella pneumophila infection (Legionnaires' disease),[70] and a variety of other diseases including uremia,[71] sickle cell disease,[72] and thrombotic thrombocytopenic purpura.[73] The occurrence of PMP in such diseases has been extensively reviewed elsewhere.[74]

## VII. Future Developments

Bleeding abnormalities of patients with certain congenital and acquired platelet abnormalities can be corrected by the administration of plasma cryoprecipitate. Cryoprecipitate contains high numbers of PMP, which may account, at least in part, for its therapeutic effectiveness.[75] Furthermore, administration of PMP from outdated platelet concentrates shortened the bleeding time in thrombocytopenic rabbits, and these infusible PMP have been investigated as a substitute for platelet transfusion (see Chapter 70).[76] Thus, PMP administration may have promise as a hemostatic agent.

In many clinical conditions, elevated numbers of PMP are encountered and their presence is associated with an increased incidence of thromboembolic complications. In view of the ability of PMP to promote coagulation, to activate cells, and, most likely, to support inflammation, interference with excessive PMP release may become a therapeutic target.

Alternatively, PMP may contribute to the therapeutic effectiveness of medicines. For example, patients receiving factor VIIa (see Chapter 68) also have transiently elevated levels of PMP that may contribute to the therapeutic effectiveness of factor VIIa.[77] In the coming years, more collaboration between laboratories will be essential to standardize isolation procedures, to exclude artifacts, and to compare data obtained in different centers. In addition, more insight must be gained into the mechanisms underlying PMP release to enable the potential clinical modification of their release and/or functions. Comparative studies between platelets and PMP (e.g., by proteomics) will result in much more information on their formation and composition. Finally, more direct evidence will have to be obtained that PMP contribute directly to physiological and pathological processes *in vivo*.

Taken together, considering the rapidly increasing evidence for the pluripotent role of PMP *in vivo*, a more thorough investigation of their composition, occurrence, and function is essential for a better understanding of their contribution to coagulation, inflammation, and cellular activation.

# References

1. Chargaff, E., & West, R. (1946). The biological significance of the thromboplastic protein of blood. *J Biol Chem, 166*, 189–197.
2. O'Brian, J. R. (1995). The platelet-like-activity of serum. *Br J Haematol, 1*, 223–228.
3. Wolf, P. (1967). The nature and significance of platelet products in human plasma. *Br J Haematol, 13*, 269–288.
4. Heijnen, H. F., Schiel, A. E., Fijnheer, R., et al. (1999). Activated platelets release two types of membrane vesicles: Microvesicles by surface shedding and exosomes derived from exocytosis of multivesicular bodies and alpha-granules. *Blood, 94*, 3791–3799.
5. Schroit, A. J., & Zwaal, R. F. (1991). Transbilayer movement of phospholipids in red cell and platelet membranes. *Biochim Biophys Acta, 1071*, 313–329.
6. Bevers, E. M., Comfurius, P., Dekkers, D. W., et al. (1999). Lipid translocation across the plasma membrane of mammalian cells. *Biochim Biophys Acta, 1439*, 317–330.
7. Bevers, E. M., Comfurius, P., & Zwaal, R. F. (1983). Changes in membrane phospholipid distribution during platelet activation. *Biochim Biophys Acta, 736*, 57–66.
8. Chang, C. P., Zhao, J., Wiedmer, T., et al. (1993). Contribution of platelet microparticle formation and granule secretion to the transmembrane migration of phosphatidylserine. *J Biol Chem, 268*, 7171–7178.
9. Sims, P. J., Wiedmer, T., Esmon, C. T., et al. (1989). Assembly of the platelet prothrombinase complex is linked to vesiculation of the platelet plasma membrane. Studies in Scott syndrome: An isolated defect in platelet procoagulant activity. *J Biol Chem, 264*, 17049–17057.
10. Jin, M., Drwal, G., Bourgeois, T., et al. (2005). Distinct proteome features of plasma microparticles. *Proteomics, 5*, 1940–1952.
11. Nieuwland, R., Berckmans, R. J., Rotteveel–Eijkman, R. C., et al. (1997). Cell-derived microparticles generated in patients during cardiopulmonary bypass are highly procoagulant. *Circulation, 96*, 3534–3541.
12. Nieuwland, R., Berckmans, R. J., McGregor, S., et al. (2000). Cellular origin and procoagulant properties of microparticles in meningococcal sepsis. *Blood, 95*, 930–935.
13. Berckmans, R. J., Nieuwland, R., Böing, A. N., et al. (2001). Cell-derived microparticles circulate in healthy humans and support low grade thrombin generation. *Thromb Haemost, 85*, 639–646.
14. Combes, V., Simon, A. C., Grau, G. E., et al. (1999). In vitro generation of endothelial microparticles and possible prothrombotic activity in patients with lupus anticoagulant. *J Clin Invest, 104*, 93–102.
15. Berckmans, R. J., Nieuwland, R., Tak, P. P., et al. (2002). Cell-derived microparticles in synovial fluid from inflamed arthritic joints support coagulation exclusively via a factor VII-dependent mechanism. *Arthritis Rheum, 46*, 2857–2866.
16. Biró, É., Nieuwland, R., & Sturk, A. (2004). Measuring circulating cell-derived microparticles. *J Thromb Haemost, 2*, 1843–1844.
17. Aupeix, K., Hugel, B., Martin, T., et al. (1997). The significance of shed membrane particles during programmed cell death in vitro, and in vivo, in HIV-1 infection. *J Clin Invest, 99*, 1546–1554.
18. Miyamoto, S., Marcinkiewicz, C., Edmunds, L. H., et al. (1998). Measurement of platelet microparticles during cardiopulmonary bypass by means of captured ELISA for GPIIb/IIIa. *Thromb Haemost, 80*, 225–230.
19. Miyazaki, Y., Nomura, S., Miyake, T., et al. (1996). High shear stress can initiate both platelet aggregation and shedding of procoagulant containing microparticles. *Blood, 88*, 3456–3464.
20. Holme, P. A., Orvim, U., Hamers, M. J., et al. (1997). Shear-induced platelet activation and platelet microparticle formation at blood flow conditions as in arteries with a severe stenosis. *Arterioscler Thromb Vasc Biol, 17*, 646–653.
21. Gemmell, C. H., Ramirez, S. M., Yeo, E. L., et al. (1995). Platelet activation in whole blood by artificial surfaces: Iden-

tification of platelet-derived microparticles and activated platelet binding to leukocytes as material-induced activation events. *J Lab Clin Med, 125,* 276–287.

22. Sims, P. J., Faioni, E. M., Wiedmer, T., et al. (1988). Complement proteins C5b-9 cause release of membrane vesicles from the platelet surface that are enriched in the membrane receptor for coagulation factor Va and express prothrombinase activity. *J Biol Chem, 263,* 18205–18212.

23. Bode, A. P., & Knupp, C. L. (1994). Effect of cold storage on platelet glycoprotein Ib and vesiculation. *Transfusion, 34,* 690–696.

24. Gemmell, C. H., Sefton, M. V., & Yeo, E. L. (1993). Platelet-derived microparticle formation involves glycoprotein IIb-IIIa. Inhibition by RGDS and a Glanzmann's thrombasthenia defect. *J Biol Chem, 268,* 14586–14589.

25. Holme, P. A., Solum, N. O., Brosstad, F., et al. (1995). Stimulated Glanzmann's thrombasthenia platelets produced microvesicles. Microvesiculation correlates better to exposure of procoagulant surface than to activation of GPIIb-IIIa. *Thromb Haemost, 74,* 1533–1540.

26. Fox, J. E., Austin, C. D., Reynolds, C. C., et al. (1991). Evidence that agonist-induced activation of calpain causes the shedding of procoagulant-containing microvesicles from the membrane of aggregating platelets. *J Biol Chem, 266,* 13289–13295.

27. Shcherbina, A., & Remold–O'Donnell, E. (1999). Role of caspase in a subset of human platelet activation responses. *Blood, 93,* 4222–4231.

28. Wiedmer, T., Shattil, S. J., Cunningham, M., et al. (1990). Role of calcium and calpain in complement-induced vesiculation of the platelet plasma membrane and in the exposure of the platelet factor Va receptor. *Biochemistry, 29,* 623–632.

29. Pasquet, J. M., Dachary–Prigent, J., & Nurden, A. T. (1998). Microvesicle release is associated with extensive protein tyrosine dephosphorylation in platelets stimulated by A23187 or a mixture of thrombin and collagen. *Biochem J, 333(Pt 3),* 591–599.

30. Bode, A. P., Orton, S. M., Frye, M. J., et al. (1991). Vesiculation of platelets during in vitro aging. *Blood, 77,* 887–895.

31. Wolf, B. B., Goldstein, J. C., Stennicke, H. R., et al. (1999). Calpain functions in a caspase-independent manner to promote apoptosis-like events during platelet activation. *Blood, 94,* 1683–1692.

32. Abid Hussein, M. N., Nieuwland, R., Hau, C. M., et al. (2005). Cell-derived microparticles contain caspase 3 in vitro and in vivo. *J Thromb Haemost, 3,* 888–896.

33. Sebbagh, M., Renvoize, C., Hamelin, J., et al. (2001). Caspase-3-mediated cleavage of ROCK I induces MLC phosphorylation and apoptotic membrane blebbing. *Nat Cell Biol, 3,* 346–352.

34. Wiedmer, T., Hall, S. E., Ortel, T. L., et al. (1993). Complement-induced vesiculation and exposure of membrane prothrombinase sites in platelets of paroxysmal nocturnal hemoglobinuria. *Blood, 82,* 1192–1196.

35. Cramer, E. M., Norol, F., Guichard, J., et al. (1997). Ultrastructure of platelet formation by human megakaryocytes cultured with the Mpl ligand. *Blood, 89,* 2336–2346.

36. Del Conde, I., Shrimpton, C. N., Thiagarajan, P., et al. (2005). Tissue-factor-bearing microvesicles arise from lipid rafts and fuse with activated platelets to initiate coagulation. *Blood, 106,* 1604–1611.

37. Diamant, M., Tushuizen, M. E., Sturk, A., et al. (2004). Cellular microparticles: New players in the field of vascular disease? *Eur J Clin Invest, 34,* 392–401.

38. Tans, G., Rosing, J., Thomassen, M. C., et al. (1991). Comparison of anticoagulant and procoagulant activities of stimulated platelets and platelet-derived microparticles. *Blood, 77,* 2641–2648.

39. Taube, J., McWilliam, N., Luddington, R., et al. (1999). Activated protein C resistance: Effect of platelet activation, platelet-derived microparticles, and atherogenic lipoproteins. *Blood, 93,* 3792–3797.

40. Joop, K., Berckmans, R. J., Nieuwland, R., et al. (2001). Microparticles from patients with multiple organ dysfunction syndrome and sepsis support coagulation through multiple mechanisms. *Thromb Haemost, 85,* 810–820.

41. Merten, M., Pakala, R., Thiagarajan, P., et al. (1999). Platelet microparticles promote platelet interaction with subendothelial matrix in a glycoprotein IIb/IIIa-dependent mechanism. *Circulation, 99,* 2577–2582.

42. Forlow, S. B., McEver, R. P., & Nollert, M. U. (2000). Leukocyte–leukocyte interactions mediated by platelet microparticles under flow. *Blood, 95,* 1317–1323.

43. Barry, O. P., Pratico, D., Lawson, J. A., et al. (1997). Transcellular activation of platelets and endothelial cells by bioactive lipids in platelet microparticles. *J Clin Invest, 99,* 2118–2127.

44. Barry, O. P., Kazanietz, M. G., Pratico, D., et al. (1999). Arachidonic acid in platelet microparticles up-regulates cyclooxygenase-2-dependent prostaglandin formation via a protein kinase C/mitogen-activated protein kinase-dependent pathway. *J Biol Chem, 274,* 7545–7556.

45. Iwamoto, S., Kawasaki, T., Kambayashi, J., et al. (1996). Platelet microparticles: A carrier of platelet-activating factor? *Biochem Biophys Res Commun, 218,* 940–944.

46. Boulanger, C. M., Scoazec, A., Ebrahimian, T., et al. (2001). Circulating microparticles from patients with myocardial infarction cause endothelial dysfunction. *Circulation, 104,* 2649–2652.

47. VanWijk, M. J., Svedas, E., Boer, K., et al. (2002). Isolated microparticles, but not whole plasma, from women with pre-eclampsia impair endothelium-dependent relaxation in isolated myometrial arteries from healthy pregnant women. *Am J Obstet Gynecol, 187,* 1686–1693.

48. Pfister, S. L. (2004). Role of platelet microparticles in the production of thromboxane by rabbit pulmonary artery. *Hypertension, 43,* 428–433.

49. Celi, A., Pellegrini, G., Lorenzet, R., et al. (1994). P-selectin induces the expression of tissue factor on monocytes. *Proc Natl Acad Sci U S A, 91,* 8767–8771.

50. Kim, H. K., Song, K. S., Chung, J. H., et al. (2004). Platelet microparticles induce angiogenesis in vitro. *Br J Haematol, 124,* 376–384.

51. Weber, A., Koppen, H. O., & Schror, K. (2000). Platelet-derived microparticles stimulate coronary artery smooth

muscle cell mitogenesis by a PDGF-independent mechanism. *Thromb Res, 98,* 461–466.

52. Weiss, H. J., Vicic, W. J., Lages, B. A., et al. (1979). Isolated deficiency of platelet procoagulant activity. *Am J Med, 67,* 206–213.

53. Toti, F., Satta, N., Fressinaud, E., et al. (1996). Scott syndrome, characterized by impaired transmembrane migration of procoagulant phosphatidylserine and hemorrhagic complications, is an inherited disorder. *Blood, 87,* 1409–1415.

54. Dachary–Prigent, J., Pasquet, J. M., Fressinaud, E., et al. (1997). Aminophospholipid exposure, microvesiculation and abnormal protein tyrosine phosphorylation in the platelets of a patient with Scott syndrome: A study using physiologic agonists and local anaesthetics. *Br J Haematol, 99,* 959–967.

55. Bettache, N., Gaffet, P., Allegre, N., et al. (1998). Impaired redistribution of aminophospholipids with distinctive cell shape change during Ca2+-induced activation of platelets from a patient with Scott syndrome. *Br J Haematol, 101,* 50–58.

56. Brooks, M. B., Catalfamo, J. L., Brown, H. A., et al. (2002). A hereditary bleeding disorder of dogs caused by a lack of platelet procoagulant activity. *Blood, 99,* 2434–2441.

57. Castaman, G., Yu-Feng, L., Battistin, E., et al. (1997). Characterization of a novel bleeding disorder with isolated prolonged bleeding time and deficiency of platelet microvesicle generation. *Br J Haematol, 96,* 458–463.

58. Shcherbina, A., Rosen, F. S., & Remold–O'Donnell, E. (1999). Pathological events in platelets of Wiskott–Aldrich syndrome patients. *Br J Haematol, 106,* 875–883.

59. Jy, W., Horstman, L. L., Arce, M., et al. (1992). Clinical significance of platelet microparticles in autoimmune thrombocytopenias. *J Lab Clin Med, 119,* 334–345.

60. Lee, Y. J., Jy, W., Horstman, L. L., et al. (1993). Elevated platelet microparticles in transient ischemic attacks, lacunar infarcts, and multiinfarct dementias. *Thromb Res, 72,* 295–304.

61. Nomura, S., Komiyama, Y., Miyake, T., et al. (1994). Amyloid beta-protein precursor-rich platelet microparticles in thrombotic disease. *Thromb Haemost, 72,* 519–522.

62. Visentin, G. P., Ford, S. E., Scott, J. P., et al. (1994). Antibodies from patients with heparin-induced thrombocytopenia/thrombosis are specific for platelet factor 4 complexed with heparin or bound to endothelial cells. *J Clin Invest, 93,* 81–88.

63. Warkentin, T. E. (1996). Heparin-induced thrombocytopenia: IgG-mediated platelet activation, platelet microparticle generation, and altered procoagulant/anticoagulant balance in the pathogenesis of thrombosis and venous limb gangrene complicating heparin-induced thrombocytopenia. *Transfus Med Rev, 10,* 249–258.

64. Warkentin, T. E., Hayward, C. P., Boshkov, L. K., et al. (1994). Sera from patients with heparin-induced thrombocytopenia generate platelet-derived microparticles with procoagulant activity: An explanation for the thrombotic complications of heparin-induced thrombocytopenia. *Blood, 84,* 3691–3699.

65. Morel, O., Jesel, L., Chauvin, M., et al. (2003). Eptifibatide-induced thrombocytopenia and circulating procoagulant platelet-derived microparticles in a patient with acute coronary syndrome. *J Thromb Haemost, 1,* 2685–2687.

66. Biró, É., Sturk–Maquelin, K. N., Vogel, G. M., et al. (2003). Human cell-derived microparticles promote thrombus formation in vivo in a tissue factor-dependent manner. *J Thromb Haemost, 1,* 2561–2568.

67. McGill, M., Fugman, D. A., Vittorio, N., et al. (1987). Platelet membrane vesicles reduced microvascular bleeding times in thrombocytopenic rabbits. *J Lab Clin Med, 109,* 127–133.

68. Katopodis, J. N., Kolodny, L., Jy, W., et al. (1997). Platelet microparticles and calcium homeostasis in acute coronary ischemias. *Am J Hematol, 54,* 95–101.

69. Nomura, S., Suzuki, M., Katsura, K., et al. (1995). Platelet-derived microparticles may influence the development of atherosclerosis in diabetes mellitus. *Atherosclerosis, 116,* 235–240.

70. Larsson, A., Nilsson, B., & Eriksson, M. (1999). Thrombocytopenia and platelet microvesicle formation caused by *Legionella pneumophila* infection. *Thromb Res, 96,* 391–397.

71. Nomura, S., Shouzu, A., Nishikawa, M., et al. (1993). Significance of platelet-derived microparticles in uremia. *Nephron, 63,* 485.

72. Wun, T., Paglieroni, T., Rangaswami, A., et al. (1998). Platelet activation in patients with sickle cell disease. *Br J Haematol, 100,* 741–749.

73. Kelton, J. G., Moore, J. C., Warkentin, T. E., et al. (1996). Isolation and characterization of cysteine proteinase in thrombotic thrombocytopenic purpura. *Br J Haematol, 93,* 421–426.

74. Horstman, L. L., & Ahn, Y. S. (1999). Platelet microparticles: A wide-angle perspective. *Crit Rev Oncol Hematol, 30,* 111–142.

75. George, J. N., Pickett, E. B., & Heinz, R. (1986). Platelet membrane microparticles in blood bank fresh frozen plasma and cryoprecipitate. *Blood, 68,* 307–309.

76. Chao, F. C., Kim, B. K., Houranieh, A. M., et al. (1996). Infusible platelet membrane microvesicles: A potential transfusion substitute for platelets. *Transfusion, 36,* 536–542.

77. Proulle, V., Hugel, B., Guillet, B., et al. (2004). Injection of recombinant activated factor VII can induce transient increase in circulating procoagulant microparticles. *Thromb Haemost, 91,* 873–878.

# CHAPTER 21

# The Role of Platelets in Fibrinolysis

**Bradley A. Maron and Joseph Loscalzo**

*Department of Medicine, Brigham and Women's Hospital, Harvard Medical School, Boston, Massachusetts*

## I. Introduction

By convention, the process of fibrin formation and dissolution has long been conceptualized as three distinct pathways. The classification of separate intrinsic and extrinsic procoagulant pathways, together with the fibrinolytic pathway, provides a model for hemostasis and explains, from a biomolecular perspective, the complicated balance between clot formation and clot dissolution *in vivo*. More recently, however, there has been emphasis on the notion that "cross-talk" among these entities exists, underscoring the potential interrelatedness of all three pathways. Of the three pathways, the importance of fibrinolysis and its direct relationship to platelets has perhaps been underappreciated. New biomolecular components, linked to the regulation of fibrinolysis, have been identified and may offer potential clues to understanding and predicting the platelet-mediated predisposition to thrombus formation (or suppressed thrombolysis).

Thrombolysis and fibrinolysis have served as a major area of investigation for the development of therapeutic agents for acute and life-threatening thrombotic conditions. Since the time of its first clinical use in the early 1980s, tissue plasminogen activator (t-PA) has served as a key treatment for rapid thrombus dissolution.[1] Improving plasminogen-activating potency, enhancing fibrin specificity, and reducing its plasma clearance remain active goals in the development of new-generation thrombolytic and fibrinolytic agents.[2]

Thrombosis is the largest cause of death worldwide.[3] Traditional thinking and clinical subspecialization often narrow the focus of thrombosis to a specific disease entity or organ system. When considering the process of thrombus formation and breakdown in the context of vascular pathology, however, its universality is better appreciated. For example, in the United States, coronary artery disease is the leading cause of death, and cerebrovascular disease is the third leading cause of death.[4] Indeed, the hematologic and vascular aspects of thrombosis, rather than the target organ *per se*, together serve as the key determinants of thrombus formation and pathophysiology.

Thrombogenesis and fibrinolysis constitute important mechanistic components of other human physiological and pathological processes besides thromboembolic disease. Fibrinolysis has been observed in a variety of noncardiovascular processes; it is well understood that fertilization and even some aspects of embryogenesis are driven by fibrinolysis.[5] Specifically, tissue recession, as is the case in digit formation during embryogenesis, is predicated, at least in part, upon appropriate and site-directed fibrinolysis.[6] In turn, pathological processes such as tumor metastasis and sepsis (e.g., disseminated intravascular coagulation) provide helpful insight into mechanisms of aberrantly regulated (or suppressed) fibrinolysis.[7]

## II. Key Mediators

### A. Basic Principles

Fibrinolysis is a complex series of molecular events that work in concert to achieve thrombus dissolution. The biomolecular focal point of this cascade is plasminogen, which is converted to plasmin *in vivo* by t-PA. This single step, conversion of plasminogen to plasmin by t-PA, is critical to successful clot lysis and is also regulated by certain key proteins, including plasminogen activator inhibitor 1 (PAI-1). In addition, $\alpha$-2-antiplasmin inhibits activated plasmin to reduce its lytic activity.

In general, endothelial activation or injury results in exposure of blood elements to prothrombotic determinants contemporaneously for activation of fibrinolysis. Hemostasis and thrombosis place in opposition these two complementary functions. A review of the pathways and the localization of fibrinolytic mediators are central to its understanding.

### B. Localization

Primarily, the fibrin surface and the endothelial surface serve as the two major sites for active fibrinolysis. This fact

PLATELETS

been shown through experimental models identifying t-PA and urokinase-type plasminogen activator (u-PA) receptors at these two sites.[8,9] However, a broader distribution of receptor sites exists and includes platelets, neutrophils, monocytes, and membrane gangliosides and glycolipids.[5] Essentially, all major circulating blood cells, with the exception of the erythrocyte, bind plasminogen. Receptor density and affinity are not constant, and in some cases may change in response to prothrombotic states. For example, thrombin-stimulated platelets demonstrate greater receptor affinity and density (approximately 200,000 per cell) for plasminogen than the unstimulated platelet.[10] Receptor density and affinity are fluid concepts and therefore differ with physiological and pathological states. The endothelium, specifically, has certain receptor characteristics that favor fibrinolysis in the basal state, with more than 50% of plasminogen sites occupied on blood-exposed cells.[11] Conversion of plasminogen to plasmin is achieved with even greater efficacy when the binding sites are on a fibrin substrate. The presence of plasminogen receptors has also been identified in nonhematological cell types in vitro, including hepatocytes, fibroblasts,[5] and tumor cells in situ (e.g., acute promyelocytic leukemia).[12] The heterogeneous distribution of plasminogen binding sites, implicating the potential for fibrinolytic activity, may provide insight into other cellular phenomenon such as cell migration.[5]

## C. Plasminogen

Plasminogen circulates in abundant concentration in plasma and is synthesized primarily in the liver.[13] Plasminogen is a zymogen (preactive form) that is converted in vivo to plasmin by t-PA. It is a single-chain glycoprotein of 92 kDa, and contains approximately 2% carbohydrate.[14] The complete amino acid sequence shows the proenzyme comprises 790 amino acids, including a preactivation peptide that is 77 amino acids in length, and five homologous triple-loop structures called kringles. The kringle is an important determinant of fibrinolysis because it is the region of plasminogen that modulates binding of the zymogen to regulatory sites, α-2-antiplasmin, and cell-surface receptors. An activation site comprised of $Arg^{560}$-$Val^{561}$ is cleaved by a plasminogen activator (e.g., t-PA), which yields the active enzyme plasmin.[14,15]

The activated form of plasminogen, plasmin, is a two-chain molecule that consists of a heavy chain (A chain) and a light chain (B chain), the latter of which is enzymatically active.[14] Conversion of plasminogen to plasmin requires the appropriate conformation of t-PA and plasminogen. Much effort has been expended on understanding determinants of the molecular conformations of proteins that bind to plasminogen and facilitate its conversion to plasmin. Although not entirely understood, it appears that binding proteins in

the 40,000 to 50,000 molecular-mass range and the presence of a carboxyterminal lysine residue confer a binding advantage. Support for this claim includes the observation that activated annexin II contains an intact carboxyterminal lysine residue and is able to bind plasminogen, whereas annexin II absent this residue cannot. Similar to annexin, α-enolase and actin are other potential binding proteins.[5,16]

## D. t-PA

t-PA is derived primarily from vascular endothelial cells. It circulates at a relatively low concentration in plasma and has a half-life of 5 to 7 minutes. t-PA is a serine protease of 68 kDa synthesized as a single-chain molecule consisting of 527 amino acids.[14] The carboxyterminal region, comprising amino acids 276 to 527, constitutes the light or B chain, which possesses its catalytic site.[17] This catalytic site is composed of the catalytic triad $His^{322}$, $Asp^{371}$, and $Ser^{478}$. Hydrolysis of the $Arg^{275}$-$Ile^{276}$ bond by plasmin converts t-PA to a two-chain molecule in which the heavy and light chains are connected by a disulfide bond. Both two-chain and single-chain forms of the enzyme are catalytically active.[14]

## E. t-PA Receptors

t-PA receptors are widely distributed and overlap to some extent with plasminogen binding sites. t-PA receptors can be found on platelets, neutrophils, and monocytes.[5] These cell types have distinct properties, affording them a unique relationship with t-PA. For example, monocytes, when triggered by certain interleukins, are a source of t-PA production.[18] In addition, platelets have specific t-PA binding sites, underscoring their exclusive role in both thrombin formation and fibrinolysis. After platelet stimulation by thrombin, plasminogen binding sites increase in number and affinity.[10,19]

t-PA receptors are found on endothelial cells where their activity is enhanced by as much as 10-fold. Fibrin, even more than endothelial cells, favors t-PA binding and activity.[20,21] Hepatocytes are responsible for the clearance of t-PA, which occurs through discrimination of a mannose carbohydrate side chain found on the enzyme; hepatocytes additionally synthesize PAI-1, which binds and suppresses plasminogen activity.[5]

There are three classes of t-PA binding sites found on endothelial cells: high, intermediate, and low affinity. Receptor density is inversely proportional to receptor affinity. High-affinity receptors are PAI-1 dependent, whereas the other two categories are PAI-1 independent.[5] This is an interesting observation because it illustrates the presence of an intrinsic counterbalancing mechanism, on a cellular level, between prothrombotic and fibrinolytic tendencies. The intermediate affinity receptor sites function by noncompetitive mutual binding with plasminogen, and the mechanism by which t-PA binds to these receptors may be, at least in part, tubulin dependent.[22] Although these receptors bind t-PA noncompetitively, the concentration of plasminogen greatly supersedes the concentration of functional t-PA under normal physiological conditions.[5] It follows, therefore, that under conditions favorable for thrombolysis, either physiological or iatrogenic, t-PA concentrations favor plasminogen binding.

### F. u-PA

Single-chain u-PA is a 411 amino acid glycoprotein with a molecular mass of 54kDa.[23] It has three distinct structural components: an aminoterminal portion, a receptor binding region, and an epidermal growth factorlike domain. u-PA also contains a central kringle region and a carboxyterminal catalytic serine protease domain.[24] The kringle domain is similar to t-PA kringle 2, and a binding site for an anionic polysaccharide (such as heparin) has been localized to this region.[25]

### G. u-PA Receptors

As with t-PA receptor sites, the distribution of u-PA receptor sites is heterogeneous. Similarly, u-PA receptors are densely localized to the endothelial cell surface. Receptors additionally exist on neutrophils, monocytes, natural killer cells (NK), platelets, and certain tumor cells.[5] Receptors have a unique affinity for u-PA and other urokinase species, such as prourokinase, light-chain urokinase, and heavy-chain urokinase. These receptors are anchored by specific glycoproteins, whose absence or abnormal function has been correlated clinically to certain hematologic disorders, such as paroxysmal nocturnal hemoglobinuria (PNH).[26] In PNH, u-PA cannot bind to receptors and a prothrombotic state ensues. u-PA also binds cell-surface receptors complexed to its natural inhibitor PAI-1. In contrast to t-PA, which is degraded by hepatocytes, u-PA is internalized via the low density lipoprotein (LDL) receptor-related protein as a u-PA–PAI-1 complex.[5]

The binding, activation, and function of u-PA depends on multiple factors. For example, endothelial cell receptor binding of u-PA is dictated by both time and temperature, and is of high affinity.[5] Likewise, expression of u-PA receptors depends on various agonist and cytokine interactions. For example, thrombin has been shown to downregulate expression of u-PA receptors on human umbilical vein endothelial cells, whereas stimulation of endothelial cells with basic fibroblast growth factor leads to decreased receptor

affinity (but not density).[27,28] Furthermore, u-PA functions in an autocrine fashion, with conversion from prourokinase to u-PA serving to increase plasmin concentration on the endothelial cell surface. Receptor-bound u-PA, like receptor-bound t-PA, has been shown to contribute to cell migration, such as in tumor metastasis, angiogenesis, and even myeloid differentiation.[5]

### H. Thrombin Activatable Fibrinolysis Inhibitor (TAFI)

The coagulation cascade and fibrinolytic cascade oppose each other in function and purpose, but reflect one another with respect to design. Each pathway contains key elements that serve to accelerate enzymatic function and achieve either clot formation or lysis; however, each pathway contains inhibitory factors that serve as the framework for balancing the net physiological effect of these reactions. Key factors that suppress activity in the coagulation cascade include thrombomodulin and activated protein C; likewise, a symmetrical regulatory mechanism has recently been described in the fibrinolytic pathway. TAFI appears to serve as one major factor constituting antifibrinolytic regulation (Fig. 21-1).

Identification of TAFI began as an outgrowth of efforts to reconstitute the effects of activated protein C in a controlled laboratory system, which proved difficult.[29,30] Various

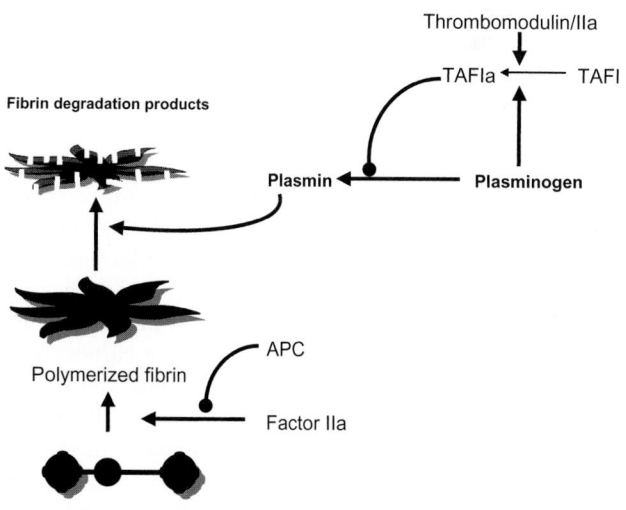

**Figure 21-1.** Activated thrombin activatable fibrinolysis inhibitor (TAFIa) is a key regulator of fibrinolysis. Activation is mainly triggered (straight arrow) by thrombomodulin in complex with activated factor II, but also by plasminogen. TAFIa blocks conversion (circle-headed line) of plasminogen to plasmin, which ultimately limits degradation of polymerized fibrin. In parallel fashion, the coagulation cascade utilizes activated protein C (APC), which downregulates factor IIa facilitation of fibrinogen conversion to fibrin.

experiments conducted by Bajzar[31] and colleagues indicated that thrombin alone did not intrinsically possess the magnitude of antifibrinolytic activity that may have been first believed. Thus, an intermediary (i.e., an additional biomolecular factor) to this process was suggested, and the protein TAFI was ultimately experimentally identified.[32]

TAFI is the preactive form of the protein that undergoes at least two enzymatic cleavages to attain its active state. TAFI activation is catalyzed by plasmin, trypsin, and thrombin, forming a 309 amino acid enzyme (TAFIa) that can be further cleaved to yield fragments of 24 kDa and 11 kDa.[33]

Paradoxically, thrombomodulin (in a complex with factor IIa) is a strong activator of TAFI; its formation occurs simultaneously with the formation of fibrin, as shown in experimental models using whole blood.[31] (Thrombomodulin is antithrombotic by virtue of its facilitation of protein C activation, but also promotes TAFI formation and thereby suppresses fibrinolysis.)

The mechanism by which TAFI functions has been elucidated in the following way. TAFIa concentrations on the order of 10 nM are sufficient to prolong clot lysis time. Because plasma TAFI concentrations are approximately 70 to 100 nM, it is parsimonious to conclude that TAFIa is efficacious in suppressing fibrinolytic activity.[34] Furthermore, t-PA activity is modulated, at least in part, by fibrin. The level of activity in the presence of fibrin is significantly greater than that in a fibrin-free environment. The basis for this effect of fibrin is that fibrin degradation exposes carboxyterminal lysine residues, which enhance the rate of plasmin formation when compared with that in the absence of fibrin. TAFIa works not by competitive site inhibition of either t-PA or plasmin, but rather as a carboxypeptidase eliminating exposed carboxyterminal lysine residues from catabolized fibrin.[32]

The identification and purification of TAFIa and the evaluation of its physiological role are relatively recent events. TAFI was purified by several groups in the late 1980s, but it was not until 1996 that Redlitz[35] and colleagues provided the first indication that TAFI was involved in the regulation of fibrinolysis in vivo. Inhibitors of carboxypeptidases, such as those found naturally in some foods (e.g., potato), may have inhibitory effects,[36] but a naturally occurring, specific inhibitor of TAFI has not yet been identified. TAFI is spontaneously degraded in a relatively rapid fashion, raising the possibility that its activity is substrate dependent, because the absence of a substrate results in its own degradation (i.e., without an endogenous, inhibitory regulator).[31]

The true framework for the physiological and pathophysiological functions of TAFI, for the most part, remains limited at this time. One potential clinical application of TAFI is pharmacological. Addition of TAFI antagonists to conventional thrombolytic agents might afford improved efficacy and specificity of clot lysis. Such an addition would complement current pharmacological strategies (e.g., synthetic engineered t-PA and its mutants) with an additional blocking component to the natural inhibitor of fibrinolysis.[31,32]

## I. α-2-Antiplasmin

Isolated from human plasma by two groups in 1976,[37,38] α-2-antiplasmin (also referred to as α-2 plasmin inhibitor) has long been studied for its inhibitory effects on plasmin. α-2 Antiplasmin is a single-chain glycoprotein with varying molecular mass between 60 kDa and 70 kDa. Although its nascent structure consists of 464 amino acids with aminoterminal methionine (Met-form), it loses its aminoterminal dodecapeptide in plasma as it is converted to the aminoterminal Asn-form.[39,40] Within the Asn-form, the active site consists of Arg376-Met377.[41] The function of α-2-antiplasmin is threefold: plasmin proteolysis, inhibition of plasminogen binding to fibrin, and cross-linking fibrin. α-2 Antiplasmin binds noncovalently to the lysine binding site of plasminogen, competitively inhibiting binding of plasminogen to fibrin.[42] It appears that α-2-antiplasmin synthesis is derived from a variety of human tissue types, although liver parenchymal cells are a predominate site.[43] α-2-Antiplasmin production from other sites, such as renal tissue, may be subject to hormonal influence (e.g., testosterone).[42] α-2-Antiplasmin binds to plasminogen competitively with other plasma proteins, such as fibrinogen.[44] As a consequence, α-2-antiplasmin functions most effectively in plasma (where binding is noncompetitive) as opposed to the fibrin surface.[45]

Fibrinolysis in vivo is associated with evidence of both profibrinolytic activity and a compensatory elevation in inhibitor regulators. Indeed, α-2-antiplasmin levels can be used to aid in determining the presence and quantification of profibrinolytic agents in certain pathological states. For example, in the Quebec platelet disorder (QPD), a rare bleeding disorder characterized by increased platelet α-granule proteolysis thought to be secondary to abnormal platelet stores of u-PA, elevated levels of plasmin–α-2-antiplasmin complexes are also noted.[46] In QPD platelets, the presence of high α-2-antiplasmin infers the presence of elevated plasminogen levels; however, an overall imbalance among zymogens, activators, and inactivators ultimately results in excessive and pathologic degradation of platelet α-granules, contributing to poor platelet function, an overall profibrinolytic state, and, consequently, abnormal bleeding in QPD.[46]

## J. Thrombomodulin

Thrombomodulin is a glycosylated type I transmembrane molecule of 557 amino acids. Its extracellular domain con-

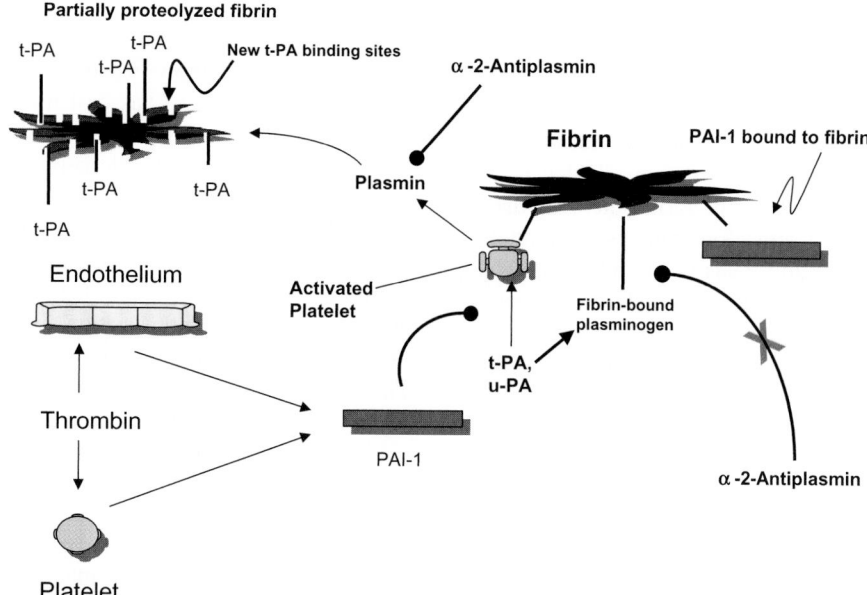

**Figure 21-2.** Plasminogen activator inhibitor-1 (PAI-1). Endothelial and platelet PAI-1 secretion is triggered by thrombin (straight arrow), which inhibits (circle-headed line) conversion of plasminogen to plasmin by tissue plasminogen activation (t-PA). The inhibitory effect of PAI-1 is decisively reduced when fibrin, plasmin, and t-PA exist as a ternary complex. In this complex, fibrin is also protected from alpha-2-antiplasmin (indicated by X), which can otherwise inhibit plasmin degradation. PAI-1 also exists bound to fibrin, maintaining its t-PA inhibiting properties.

tains an aminoterminal globular domain that participates in receptor endocytosis, regulation of tumor growth, and endothelial function in inflammation.[47] The extracellular portion also consists of six epidermal growth factor modules,[48] which serve as target sites for posttranslational modification.[49]

Thrombomodulin is integral to a wide array of pathophysiological processes, including thrombosis, oncogenesis, inflammation, and even embryogenesis.[47] Additionally, its role in fibrinolysis has recently been elucidated, as the thrombomodulin–thrombin complex appears to activate latent plasma carboxypeptidase B enzymes and TAFI (Fig. 21-1).[31,50] This proteolytic process may serve as a key intermediary step in TAFI activation and, therefore, thrombomodulin may indirectly affect clot lysis resistance and possibly regulation of complement activity.[51] Furthermore, experimental *in vitro* models have shown that u-PA may serve as one substrate for the thrombomodulin–thrombin complex,[52] raising the possibility that the full physiological range for thrombin activity has not been completely elucidated.

## K. PAI-1

PAI-1 is a glycoprotein composed of 379 amino acids with a molecular mass of 48 kDa. It binds in a 1 : 1 ratio with t-PA and u-PA, is cleared by the hepatocyte, and has a half-life

of approximately 30 minutes.[53] Thrombin triggers platelets and endothelial cells to secrete PAI-1, which serves as an antagonist to plasminogen, ultimately creating a prothrombotic state (Fig. 21-2).[54,55]

Given that fibrin cross-linking reduces plasminogen binding site availability, variability in fibrin concentration and fibrin cross-linking efficacy might promote thrombotic events. Indeed, one animal model of pulmonary embolism using ferrets has shown that fibrin–fibrin and α-2-antiplasmin cross-linking to fibrin mediated by factor XIII (a key protransglutaminase that facilitates cross-linking of fibrin) yields thrombi resistant to endogenous fibrinolysis.[56] Furthermore, although under normal circumstances fibrin-bound plasminogen converted to plasmin by t-PA yields fibrin degradation and, subsequently, new t-PA binding sites, this was not found to be the case in the ferret model; rather, inhibition of factor XIII in this model yielded increased thrombus lysing potential. It is likely that the mechanism underlying these observations involves α-2-antiplasmin, which is covalently bound to polymerizing fibrin by activated factor XIII.[56,57]

The previous example lends support to the view that pathological thrombosis principally represents a failure of fibrinolysis. The behavior of PAI-1 also supports this notion. Fibrinolysis occurs on the surface of the fibrin clot and, therefore, remains a local process. A ternary complex is formed among plasminogen, t-PA, and fibrin during this process, and it promotes the formation of plasmin and, subsequently, fibrin degradation.[58] When in this complex, plasmin is relatively protected from α-2-antiplasmin, and t-PA from PAI-1 (inhibition of t-PA by PAI-1 is decreased 80% in the presence of fibrin, likely secondary to reduced access of PAI-1 to the catalytic domain of fibrin-bound t-PA[59]). Despite the ternary conformation, PAI-1 does, however, have the capacity to bind fibrin and can partly suppress local plasminogen activation, perhaps by competitive binding.[53] Thus, although t-PA is protected from PAI-1 when in the ternary complex, overall there exists some potential for inhibition of t-PA by PAI-1 via PAI-1 binding to fibrin. Based on this observation, inhibition of PAI-1 should decrease thrombotic potential. In fact, administration of monoclonal antibodies against PAI-1 has been shown to increase fibrinolytic activity and decrease thrombus extension.[59]

Perhaps the most exciting clinical aspect of PAI-1 is its role as an independent risk factor for myocardial infarction (MI).[53] A variety of observational, prospective studies, including the Northwick Park Heart study and the European Concerted Action on Thrombosis and Disabilities study, have shown that elevated levels of PAI-1 are independently associated with increased risk for a myocardial event.[60,61] For example, high levels of PAI-1 are associated with high reinfarction rates among men younger than 45 who have sustained a prior MI. In addition, high PAI-1 levels predict future MI in patients with stable angina and have been associated with angiographic evidence of advanced coronary disease. A preponderance of clinical data show a relationship between elevated PAI-1 levels and reduced insulin sensitivity, and, likewise, patients with non-insulin dependent type II diabetes mellitus have higher levels of PAI-1 compared with those without diabetes.[62] This finding is corroborated by the observation that insulin treatment among diabetic patients decreases plasma PAI-1 levels.[63]

## III. Mechanisms of Fibrinolysis

### A. Basic Principles

Fibrinolysis is a complicated process, predicated on the conversion of inactive proenzymes to active forms, the ultimate goal of which is fibrin degradation. The central mediator of this process is plasmin. The primary substrate for physiological protein degradation by plasmin is the fibrin clot, keeping this process local.[14] This is in contrast to non-specific iatrogenic fibrinolysis triggered by administration of agents such as recombinant t-PA under specific clinical situations, including MI or active stroke. In these pathological situations, the nidus for thrombosis may evolve from both dysregulated thrombin formation and failure of fibrinolysis.

### B. Endothelium

Investigation of fibrinolysis and thrombosis in the setting of endothelial dysfunction has cast these processes in a new light, providing enhanced understanding of their biomolecular pathogenesis. Normal endothelium protects the organism from a prothrombotic state (see also Chapter 13).[64] Endothelial dysfunction leads to decreased nitric oxide availability, which normally serves to suppress platelet activation and promote vasodilation.[65] Thus, decreased nitric oxide production leads to vasoconstriction, enhanced platelet activation, and, subsequently, abnormal vessel wall morphology. The most striking clinical correlate of this process is coronary thrombus formation. In this model, atherosclerotic plaque rupture occurs secondary to endothelial dysfunction. Atherosclerotic plaques usually exist in quiescent, undisturbed vessels, and in such circumstances are clinically insignificant. Because tissue factor exists in the subendothelial matrix of atherosclerotic plaque, however, exposure of factor X to flowing blood occurs upon plaque rupture, triggering the coagulation cascade.[66] Furthermore, endothelial cells can upregulate PAI-1 levels and, because of the central role of plasmin in fibrinolysis, this mechanism may hold particular importance in the regulation of thrombosis.[5,64,65]

Recently, certain modifiable components of fibrinolysis *in vivo* have received attention. For example, bradykinin has been shown to stimulate t-PA release in human vasculature, and elevated angiotensin II levels (by infusion) appear to increase levels of PAI-1.[64,67] These findings are significant when considering the widespread use of angiotensin converting enzyme inhibitor medications. This pharmacotherapy has been shown to reduce angiotensin II levels *and* increase bradykinin levels. Additionally, there are some data that suggest that angiotensin converting enzyme inhibitors have been shown to reduce levels of PAI-1 antigen and activity.[68]

### C. Platelets

Platelets are paramount in hemostasis and the process of fibrin formation. In addition, although endothelial cells constitute the primary locus for thrombogenesis and fibrinolysis, these processes also occur on platelets and polymorphonuclear cells. Foremost, the platelet surface serves as a site for assembly of proteins of the plasminogen activator system[69] and, therefore, plays a critical role in fibrinolysis (Fig. 21-3).[70] Platelets secrete and bind plasmin. Of equal importance, however, is the privileged state afforded to platelet-bound plasmin, which cannot be inhibited by α-2-antiplasmin.[69] Plasmin-generating proteins assemble on the platelet surface, maximizing plasminogen efficacy; t-PA, plasmin, and plasminogen have the capacity to bind directly to the platelet surface. The relationship between platelets

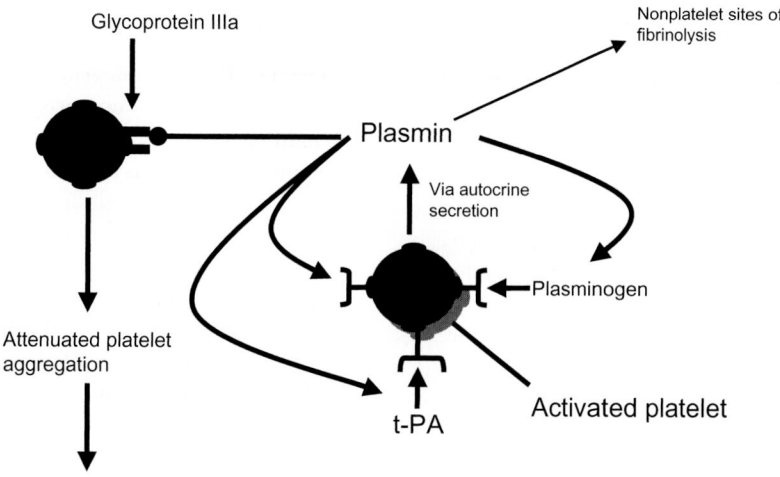

**Figure 21-3.** The role of the platelet in fibrinolysis. Upon activation, platelets secrete plasmin, which binds to the platelet surface in a concentration-dependent fashion (straight arrow). The presence of plasmin upregulates t-PA to platelet binding, and plasminogen to platelet binding (straight arrows). At the same time, the catalytic nature of fibrinolysis is accentuated by the cleavage of glycoprotein IIIa (circle-headed line), thereby inhibiting platelet function.

and plasmin is proportional to concentration (and is likely also time dependent), but varies among tissue types and experimental conditions. Nevertheless, it appears that at low levels of plasmin, platelets are activated, whereas high levels of plasmin induce platelet dysfunction.[69] These observations, of course, are relative, and it must be stated that platelets are conducive to plasminogen activation in both resting and active states.

Local platelet release of plasmin at high levels via autocrine secretion results in peak fibrinolysis and platelet activation. As a consequence, plasminogen and t-PA binding to the platelet is enhanced. Plasmin accentuates the catalytic nature of fibrinolysis by cleaving glycoprotein IIIa and rendering the platelet dysfunctional (Fig. 21-3).[70] Ultimately, the process of platelet aggregation, normally augmented in this process by the presence of fibrinogen, is attenuated and the hemostatic plug is disrupted. In contrast, PAI-1 also acting on the platelet surface negatively affects the platelet–plasminogen relationship, preventing early clot lysis.[71]

## IV. Platelet–Fibrin(ogen) Interactions

### A. Fibrin Clot

The relationship among fibrin, the platelet, and the endothelial surface is complex, but is fundamental in illustrating the mechanics of fibrinolysis. Vascular injury resulting in exposure of the subendothelial matrix to flowing blood leads to

platelet adhesion. This process is largely dependent on platelet recognition and binding to matrix components.[72] Platelet-specific glycoproteins, most of which are heterodimeric molecules composed of both $\alpha$ and $\beta$ subunits, enable this process. There are five platelet integrins involved in platelet binding, and each plays a key role in achieving hemostasis. Platelet adhesion first involves platelet surface glycoprotein (GP) Ib, existing in complex with GPIX and GPV, which binds von Willebrand factor (VWF) (discussed in detail in Chapter 7).[73,74] The GPIIb-IIIa complex (integrin $\alpha$IIb$\beta$3) supports both platelet adhesion and aggregation (discussed in detail in Chapter 8). Other integrins such as $\alpha$2$\beta$1 (and nonintegrin glycoproteins such as GPIV) are involved in platelet binding to collagen (see Chapters 6 and 18).[75,76] After adhesion, platelet activation is mediated by multiple factors, including components of exposed subendothelial matrix and the presence of thrombin.[72]

Fibrinogen, along with VWF, serves as the major ligand interconnecting platelets within the platelet plug, and is a large 340-kDa protein comprised of two sets of A$\alpha$, B$\beta$, and $\gamma$ chains.[77] The protein is arranged in symmetrical alignment with two lateral D domains and a central E domain. The D domain, specifically the $\gamma$ 408-411 residue (AGDV), plays a key role in the interaction between fibrin and platelets, which directly involves the GPIIb-IIIa platelet receptor (see Chapter 8).[77]

The relationship between fibrinogen and platelet binding is dynamic, and is influenced by the state of platelet activation. In the resting state, platelet–fibrinogen binding affinity is low, but upon activation increases considerably. Platelet activation generally involves one of two processes: *inside-*

*to-outside* or *outside-to-inside* signaling (see Chapters 16 and 17).[78,79] The former occurs after platelet binding and leads to platelet receptor site modification, which includes alteration in receptor conformation, ultimately increasing binding affinity. The latter, however, focuses primarily on postoccupancy events subsequent to VWF binding. The result of *inside-to-outside* signaling is modification in membrane structure, fluidity, intracellular ion levels, and possibly new epitope exposure.[79,80] These reactions are receptor specific and implicate potential (unwanted) secondary effects of therapeutic interventions that block platelet receptor activity.[79–81]

## B. Platelet-Rich Clots and Clot Retraction

Clot lysis is a highly variable process and is heavily influenced by the presence of platelets. A substantial body of literature has shown in both *in vivo* and *in vitro* models that platelet-rich clots are resistant to fibrinolysis.[81] Suppressed platelet-rich clot lysis is likely multifactorial, although two measurable indices have been elucidated: reduced t-PA-binding capacity and reduced lysis front velocity. Clot retraction, the primary step in thrombus removal *in vivo,* is likely important in suppressing clot lysis. This process involves conformational changes that decrease t-PA binding to the platelet. One possible explanation for the observed reduced rate of t-PA binding may be the high density of retracted fibrin fibers and the decreased structural availability of receptor sites. By varying clot structure *in vitro,* differential clot lysis rates have been observed.[81] Furthermore, reconstructed confocal micrograph images have demonstrated progressive deformation of the lysis front in a characteristic pattern, lending support to the hypothesis that fibrinolysis advances through clots via microchannels. Proliferation of (fingerlike) projections through the clot in a recanalizationlike pattern has been reported in several experimental venues.[82] For example, models of coronary artery thrombolysis in animal models and in whole blood clots demonstrate on a macroscopic level a similar recanalization pattern. Of note, addition of GPIIb-IIIa antagonists before clot formation has been shown in some studies to increase t-PA binding capacity, approaching that of platelet-poor clots.[81,82]

Platelet secretion of PAI-1 contributes to retarded clot lysis. Further underscoring the importance of clot retraction, however, *in vitro* models comparing PAI-1 resistant t-PA with wild-type t-PA have shown little difference in clot lysis time. An alternate factor that may affect clot lysis and account for this difference are α-2-antiplasmin levels.[83] Factors influencing clot retraction are not clearly defined, but likely involve interaction among the platelet surface integrin αIIbβ3 and fibrinogen, its primary ligand, in concert with platelet shape change.[81–83]

## C. GPIIb-IIIa Receptor

The GPIIb-IIIa (integrin αIIbβ3) receptor consists of a 136-kDa α subunit comprised of a heavy chain and a light chain (see Chapter 8 for details).[84] The light chain is transmembrane, whereas the heavy chain is entirely extracellular. The 92-kDa β subunit consists of a 762 amino acid polypeptide chain that is transmembrane.[85] GPIIb-IIIa is found exclusively on platelets and megakaryocytes; 70,000 to 90,000 receptors are expressed on each platelet in the resting state.[86] Although GPIIb-IIIa binding to fibrin(ogen) is the principle mechanism underlying platelet aggregation, other adhesive glycoproteins bind platelets via this site as well, including VWF and fibronectin.

Three fibrinogen binding sites exist that facilitate binding of GPIIb-IIIa during interaction with platelets: two Arg-Gly-Asp (RGD) sites and the carboxyterminal portion of the γ chain (see Chapter 8). Both RGD sites are located on the Aα chain at residues 95 to 97 and 572 to 574. The third site is fibrinogen specific and is composed of the 12 carboxyterminal residues of the γ chain (H12:HHLGGAKQAGDV).[83] Conflicting data exist from *in vitro* models regarding which receptor has the leading role in fibrin–platelet binding. Determining a single, major site would have obvious clinical implications in developing platelet–fibrin binding inhibitors. It appears, however, that no single site predominates; loss of one site can be compensated for by one or both of the other two sites to sustain fibrinogen–platelet binding. Furthermore, when all three sites are blocked, fibrinogen released from platelet α-granules may be sufficient for promoting platelet adhesion and subsequent clot retraction.[77] Nonetheless, in models that eliminate α-granule efficacy and the H12 and RGD receptors, clot retraction can still occur, suggesting that yet other sources facilitate fibrinogen–platelet binding. Candidates to serve this role include VWF, fibronectin, and thrombospondin, although there is as yet no consensus on the comparative efficacy of these ligands in this process.[74,83]

## D. GPIIb-IIIa Antagonists

The GPIIb-IIIa receptor plays a critical role in platelet aggregation and clot formation (see Chapter 8). This conclusion is based on experimental data and clinical observations. Indeed, patients with Glanzmann thrombasthenia, a genetic disease resulting in the absence or largely decreased functional state of the GPIIb-IIIa receptor, have persistent mucocutaneous bleeding (see Chapter 57).[87] GPIIb-IIIa antagonists (see Chapter 62 for details) include agents that have subtle mechanistic and physiological differences. Abciximab, a monoclonal antibody fragment, demonstrates high binding affinity for the GPIIb-IIIa receptor in either active or inactive states. Additional abciximab activity exists for binding integrin αvβ 3 and Mac-1.[88] In contrast, eptifibatide is a KGD (Lys-Gly-Asp) motif-specific agent, whereas tirofiban and lamifiban are RGD motif specific with relatively decreased binding affinity and a short half-life. The therapeutic goals for treatment are more than 80% receptor occupancy and more than 80% inhibition of platelet aggregation; however, a diverse range of factors prevents dose universalization among patients. For example, platelet quantity, platelet receptor availability, the basal state of platelet activation, spleen size, and comorbidity (e.g., diabetes mellitus) will affect efficacy and therapeutic dosing of these medications.[88]

Although the common goal of all GPIIb-IIIa antagonists is enhanced fibrinolysis, other potential outcomes exist. Ironically, prothrombotic states have been observed in patients undergoing fibrinolytic therapy. This observation may have a variety of explanations. First, via *outside-to-inside* signaling, activation of quiescent platelet receptors may occur, possibly superseding GPIIb-IIIa inhibition.[89] Second, GPIIb-IIIa antagonists do not affect platelet secretion, and compensatory platelet hypersecretion can disrupt the therapeutic balance between medication bioavailability and platelet receptor availability.[90]

Interestingly, trials that have examined oral GPIIb-IIIa antagonists do not support their use as primary fibrinolytic agents. Fluctuation in bioavailability, a confounding factor intrinsic to oral dosing, can (to some degree) account for this outcome.[89,90] One oral agent, however, has been shown to improve patient outcome in the setting of acute MI. Clopidogrel, an antagonist to the P2Y$_{12}$ receptor (see Chapter 61), when used with aspirin (cyclooxygenase inhibitor) can suppress both platelet activation and aggregation, and has been shown to be beneficial in patients with ST-segment elevation MI.[91]

## V. Surgery, Fibrinolysis, and the Platelet

Postoperative bleeding is a well-documented complication of cardiopulmonary bypass and coronary artery bypass grafting (CABG) surgery. The etiology of the hemorrhagic diathesis associated with bypass surgery has been the subject of empirical scrutiny for at least two decades.[92] Although its precise mechanism remains unresolved, it is likely, in part, related to the adverse effects of heparin on platelet function, resulting in abnormal fibrinolysis (see Chapter 59 for details).

CABG itself tends to alter hemodynamics, producing a state favoring bleeding.[9] Perhaps intuitively, extracorporeal bypass-induced thrombocytopenia was initially regarded as the cause for increased bleeding risk among CABG patients, although recent data suggest that such bleeding may not be clinically relevant.[93] Rather, the universal (and conventional) use of heparin to reduce thrombotic complications of cardiovascular surgery is a more likely cause of these hemorrhagic complications. The mechanism for heparin-induced postoperative bleeding is complex and multifactorial, and may include nonpharmacological factors such as hypothermia, which may contribute to increased bleeding risk by affecting platelet function via inhibition of thromboxane A$_2$ expression.[94]

Heparin binds to antithrombin III and catalyzes its interaction with various procoagulants, particularly thrombin, affecting hemostasis. Its impact on bleeding, however, is not exclusive to this effect. In preoperative patients, heparin increases bleeding time and reduces platelet thromboxane B$_2$ production. In addition, patients receiving heparin demonstrate abnormal fibrinolysis reflected by increased levels of plasmin and D-dimer.[92] One early theory explained these findings by implicating structural GPIIb-IIIa and GPIb platelet receptor (necessary for VWF and fibrinogen binding) differences between heparinized and nonheparinized patients undergoing CABG. This hypothesis, however, has been largely refuted by the use of flow cytometry, which demonstrated that surface expression of these platelet membrane receptors remains intact.[95]

Suppression of normal platelet function in CABG, therefore, must be accounted for by other mechanisms. One potential explanation is the inhibitory effect of heparin on thrombin, which is a potent platelet agonist.[96] In addition, inhibited VWF activity may also account for dysregulated platelet function. In an *in vitro* model examining the effect of heparin on VWF, Sobel and colleagues[97] showed that the administration of intravenous heparin reduced ristocetin-induced platelet agglutination (which depends upon VWF binding to platelet GPIb-IX-V) by more than one-half, supporting the notion that platelet dysfunction in CABG may, in part, be related to abnormal VWF binding to platelets.

Although these examples indirectly support fibrinolytic dysfunction as a consequence of heparinization, increased D-dimer and fibrinogen degradation products (FDPs) have also been observed in CABG patients, raising the possibility that heparin administration also directly affects fibrinolysis. Interestingly, the profibrinolytic properties of heparin have been recognized for more than 20 years.[96,98] Indeed, aprotinin administration during CABG prevents the formation of FDPs, reduces α-2-antiplasmin levels, and reduces bleeding time (see Chapter 59), all of which implicate a profibrinolytic tendency of heparin. Furthermore, preoperative CABG patients administered heparin have demonstrated an increase in plasmin and D-dimer levels, again supporting a link between heparin and fibrinolysis.[96] Of note, elevated t-PA levels after heparin administration *in vitro* do not necessarily indicate increased activity, because t-PA antigen detection does not correlate linearly with t-PA activity. There does appear, however, to be a direct relationship between plasmin activity and plasmin light-chain generation as detected by Western blot analysis, and there are data to suggest that heparin administration results in a 10- to 17-fold increase in plasminogen activity.[99] Upchurch and colleagues[99] showed that with increasing heparin doses administered to baboons, a significant increase in a specific FDP, fragment E, is observed in plasma. Because FDPs have been shown to interfere with platelet aggregation by competitively binding to platelet GPIIb–IIIa, the impact of increasing FDP levels after heparin administration may provide one possible link to its anticoagulant effect.[99] High plasmin, D-dimer, and FDP concentrations, all of which have been reported in CABG patients, have also been shown in other studies to interfere with platelet aggregation.[100]

The impact of heparin on platelet function nevertheless remains unresolved, because the bleeding diathesis accompanying cardiopulmonary bypass cannot be completely explained by the aforementioned mechanisms. For example, reversal of the heparin effect by protamine administration does not reduce bleeding time or decrease D-dimer levels, and, similarly, cessation of surgery does not affect the postoperative duration of elevated bleeding time.[96]

Of note, abnormal hemostasis is pervasive throughout the realm of vascular surgery and is not limited to CABG. For example, Haithcock and colleagues[101] reported reduced fibrinogen levels associated with invasive supraceliac aortic cross-clamp procedures. Aortic cross-clamping proximal to the celiac trunk is required for thoracoabdominal aortic aneurysm repair, and nearly one-quarter of such procedures results in a bleeding complication. Cohen and colleagues[102] were among the first groups to examine the effect of supraceliac cross-clamping on hemostasis. These investigators reported that 60 minutes of transaortic cross-clamping resulted in decreased platelet count and fibrinogen concentrations. Subsequent studies suggested that t-PA levels are elevated in such patients, although only t-PA antigen (and not activity) was measured.[103] More recently, Haithcock and colleagues[101] showed that cross-clamping resulted in elevated thrombin–antithrombin III complexes and t-PA activity compared with noninvasive controls. Although the mechanism is not entirely clear, it is possible that t-PA release may be induced by tissue ischemia secondary to cross-clamping. Ischemic tissue has been shown to release t-PA and upregulate gene expression for t-PA production in tissue distal to the primary site of ischemia.[104] Furthermore, procedures that compromise hepatic blood flow may adversely affect the metabolism of hemostatic factors, further creating an environment favorable for impaired hemostasis.[101]

# VI. Fibrinolytic Response to Inflammation and Infection

The link between inflammation and coagulation is well documented, although a basic understanding of this impor-

tant interaction has only recently evolved. Initially, hypotheses suggested that the bridge between the two pathways involved the contact system ("intrinsic pathway"), with microorganisms or endotoxins as the primary stimuli.[105] More recently, the concept of cross-talk between the inflammatory and coagulation pathways, mediated at a cellular level by cytokine release and signaling, has been advanced. Specifically, the tissue factor–factor VII interaction, in part stimulated by tumor necrosis factor-$\alpha$ (TNF-$\alpha$) and interleukin 1 (IL-1), constitutes a key regulatory event for mediation of normal response to injury, infection, or trauma.[105–108]

Disseminated intravascular coagulation (DIC) reflects this process in its unregulated form. DIC is characterized by widespread activation of the coagulation cascade, resulting in the intravascular formation of fibrin, and, ultimately, thrombotic occlusion of small and medium vessels.[7] In this scenario, overall hemostasis can be difficult to predict clinically, because procoagulant pathways are upregulated via cytokine signaling, and, alternatively, platelet and coagulation proteins can be depleted, resulting in hemorrhage. Diseases and iatrogenic triggers for DIC are heterogeneous and include trauma, malignancy, obstetrical complications, immunological disorders, and vascular malformations (e.g., giant hemangioma or Kasabach–Merritt syndrome).[7]

Independent of etiology, the pathogenesis of DIC involves three distinct mechanisms: increased generation of thrombin resulting in the systemic formation of fibrin, suppression of physiological anticoagulant mechanisms, and the delayed removal of fibrin as a consequence of impaired fibrinolysis.[7] Bacteremia and endotoxemia result in increased fibrinolytic activity, and although the precise mechanism underlying this relationship is unclear, it is likely related to plasminogen activators released from endothelial cells.[106] Consequently, the cytokine-mediated tissue factor and factor VII interaction shifts hemostasis toward hypercoagulability via activation of the "extrinsic" and contact pathways. Fibrinolytic activity, as a physiological response to a heightened prothrombotic state is overwhelmed by abnormal, sustained release of PAI-1.[107,108] In addition, downregulation of the endogenous anticoagulant protein C and S pathways[105] likely contributes to the net prothrombotic state. Sustained release of PAI-1 is cytokine mediated; experimental models have shown that high levels of TNF-$\alpha$ and IL-1 decrease free t-PA and increase PAI-1 production by endothelial cells.[109,110] In addition, TNF-$\alpha$ has been shown to stimulate PAI-1 production in the liver, kidney, lung, and adrenal glands of mice independent of DIC.[111] The role of platelets during DIC is generally confined to adverse consequences associated with thrombocytopenia. The mechanism for DIC-related thrombocytopenia is likely multifactorial, and, although it has been considered an immune-mediated process,[112] evidence to support alternative pathways exists that focus on platelet consumption in thrombi and hemophagocytosis in the bone marrow.[97] Furthermore, viral-mediated endothelial injury prompting platelet adherence to the vessel wall,[113] and direct interaction between platelets and certain viruses (e.g., Herpes simplex virus, cytomegalovirus, and others), appear to cause both thrombocytopenia and reduced platelet function.[114] Because platelets are directly involved in inflammation, however, their role as potential mediators in suppressing fibrinolysis has been investigated in greater detail. Indeed, in addition to thrombocytopenia, there are at least two distinct mechanisms by which platelets contribute to the dysregulation of hemostasis in DIC. First, endotoxins and cytokines stimulate platelet activation, which occurs via the sphingosine and platelet activating factor (PAF) pathways and at the level of the PAF receptor.[105] Activated platelets contribute to a thrombotic state by providing a suitable, phospholipid-based surface for the assembly of activated coagulation factors such as factors X and V, as well as prothrombin and calcium, which catalyze the formation of thrombin.[105] Second, platelet activation also leads to release of platelet α-granules, which contain significant quantities of PAI-1. Of note, mutations in the PAI-1 gene, specifically the 4G/5G polymorphism, has been shown to influence levels of PAI-1 and has also been linked to clinical outcome in certain diseases, such as meningococcemia. In that clinical scenario, patients with a PAI-1 mutation demonstrate elevated concentrations of PAI-1, and, therefore, enhanced suppression of fibrinolysis, conferring an increased mortality risk.[115,116]

The impact of coagulation and fibrinolytic abnormalities extends beyond the intravascular compartment, as well. For example, procoagulant activity has been observed in bronchoalveolar lavage samples from adult patients with acute respiratory distress syndrome (ARDS) and also less severe acute lung injury (ALI).[117] Although the trigger for ARDS differs from that of DIC by virtue of direct injury to the lung parenchyma (e.g., hyperoxia, inhaled agents), procoagulant and suppressed fibrinolytic activity have been observed similarly in both conditions. For example, bronchoalveolar lavage fluid in ARDS or ALI patients mimics serum samples from patients with DIC with evidence of elevated tissue factor–factor VII interaction and elevated PAI-1 levels.[118–120] Several weeks after the initial lung injury, however, procoagulant activity decreases whereas fibrinolytic activity remains suppressed, implicating one possible explanation for persistent fibrin deposition in alveoli[121] in the prolonged clinical course of ALI patients. Similar observations have been made in neonates with respiratory distress syndrome and early stages of bronchopulmonary dysplasia.[122]

## VII. Fibrinolysis, the Platelet, and Angiogenesis

Parallel to the biomolecular relationship between inflammation and coagulation, hemostasis and angiogenesis are inter-connected processes, linked by circulating and cell-associated mediators. Angiogenesis (see also Chapter 41) is the process of sprouting and developing new blood vessels from existing ones, and apart from the physiological processes of wound healing, and changes of the ovary and endometrium during menstruation, it generally connotes a pathological event.[123] The best understood pathologic variant of angiogenesis is tumor metastasis, in which the balance between pro- and antiangiogenic regulators is disrupted. This usually occurs as a consequence of blood vessel recruitment from an *in situ* tumor (stimulated by hypoxia), which triggers a cascade of events resulting in a tendency toward angiogenesis[124]; however, angiogenesis under physiological conditions is triggered by endothelial injury and subsequent initiation of the coagulation cascade. In this case, angiogenesis is promoted by increased vascular permeability, leading to extravasation of plasma glycoproteins, forming a meshwork for migrating endothelial cells,[125] in a process that is largely mediated by platelets.

Hemostatic upregulation of platelet function at the site of vessel wall injury triggers platelet α-granule secretion, resulting in the release of several positive and negative angiogenesis regulators (see also Chapters 15 and 41).[124,125] Among the best understood and most commonly found platelet-derived angiogenesis promoter in tumor cells is vascular endothelial growth factor A (VEGF-A).[126] VEGF-A, which has been shown to increase expression of thrombomodulin, PAI-1, and t-PA, is upregulated under hypoxic conditions in tumor cells. Additionally, VEGF-A may be the chief angiogenic protein in other pathological processes, such as adult and pediatric retinal neovascularization.[127]

Interestingly, hepatocyte growth factor serves as both a positive and negative angiogenic regulator — a phenotype dependent on alternative processing of the hepatocyte growth factor α chain generating antiangiogenic fragments comprised of either the first kringle domain (NK1) or the first two kringle domains (NK2).[123] This differential activity is important because development of a recombinant form of NK4 containing all four kringles serves as a potent inhibitor of tumor growth *in vivo,* a process partly mediated through its antiangiogenic properties.[128]

Other important platelet-derived antiangiogenic regulators include thrombospondin 1, transforming growth factor (TGF) β1, and platelet factor 4.[123] Thrombospondin 1 is the most abundant constituent of platelet α-granules and aids in platelet aggregation. Its role in angiogenesis is multifaceted, serving to interrupt growth factor and integrin signaling between endothelial cells and the fibrin clot, as well as suppressing motility and the chemokinetic response of endothelial cells.[129] Platelet factor 4 is unique to platelets and binds surface heparinlike glycosaminoglycans on endothelial cells, enhancing the antithrombotic activity of antithrombin III,[123] which, among several functions, inhibits the endothelial cell stimulatory activity of VEGF-A.[130] TGF-β1 appears to be antiangiogenic *in vitro* by inhibiting endothelial cell proliferation and migration, but *in vivo* TGF-β1 functions in a proangiogenic fashion, possibly via recruitment of macrophages, which have been shown to secrete endothelial growth factors.[123]

Platelets in the context of fibrinolysis constitute a potent regulatory mechanism of angiogenesis. The process of fibrinolysis provides abundant exposure to the vasculature of endogenous inhibitors of proteolysis, which regulate basement membrane degradation and matrix protein turnover.[123] For example, vascular endothelium and platelet-derived PAI-1 combines with vitronectin to inhibit vascular smooth muscle cell migration via cell surface integrin αvβ3 blockade.[131] Although elevated PAI-1 levels appear to be antiangiogenic *in vitro*, this does not necessarily translate into clinical benefit among cancer patients, in whom its increased expression is associated with an unfavorable outcome.[123] In addition to PAI-1, a role exists among other fibrinolytic determinants in the modification of angiogenesis *in vitro*. For example, α$_2$-macroglobulin binds VEGF, inhibiting interaction with its receptors,[132] and α$_2$-antiplasmin demonstrates endothelial cell growth suppression properties in the presence of angiopoietin 1.[133]

# References

1. Collen, D., & Lijnen, H. R. (2004). Tissue-type plasminogen activator: A historical perspective and personal account. *J Thromb Haemost, 2,* 541–546.

2. Lijnen, H. R., & Collen, D. (1991). Strategies for the improvement of thrombolytic agents. *Thromb Haemost, 66,* 88–110.

3. Deitcher, S. R., & Jaff, M. R. (2002). Pharmacologic and clinical characteristics of thrombolytic agents. *Rev Cardiovasc Med, 3*(S2), S25–S33.

4. Brott, T., & Bogousslavasky, J. (2000). Treatment of acute ischemic stroke. *N Engl J Med, 7,* 710–722.

5. Plow, E. F., Ugarova, T., & Miles, L. A. (1998). The interaction of the fibrinolytic system with the vessel wall. In J. Loscalzo & A. Schafer (Eds.), *Thrombosis and hemorrhage* (vol. 2, pp. 373–386). Baltimore: Williams and Wilkins.

6. Vassalli, J. D., Sappino, A. P., & Belin, D. (1991). The plasminogen activator/plasmin system [Review]. *J Clin Invest, 88,* 1067–1072.

7. Levi, M., & Cate, H. T. (1999). Disseminated intravascular coagulation. *N Engl J Med, 341*(8), 586–592.

8. Dudani, A. K., & Ganz, P. R. (1996). Endothelial cell surface actin serves as a binding site for plasminogen, tissue plasminogen activator and lipoprotein(a). *Br J Haemost, 95*(1), 168–178.

9. Nieuwenhuizen, W. (2001). Fibrin-mediated plasminogen activation. Fibrinogen: XVIth International Fibrinogen Workshop: Part III. Fibrinolysis and related subjects. *Ann N Y Acad Sci, 936,* 237–246.

10. Miles, L. A., & Plow, E. F. (1985). Binding and activation of plasminogen on the platelet surface. *J Biol Chem, 260,* 4303–4311.

11. Adelman, B., Rizk, A., & Hanners, E. (1988). Plasminogen interactions with platelets in plasma. *Blood, 72,* 1530–1535.

12. Kwaan, H. C., Wang, J., & Boggio, L. N. (2002). Abnormalities in hemostasis in acute promyelocytic leukemia. *Haematol Oncol, 20,* 33–41.

13. Raum, D., Marcus, D., Alper, C. A., et al. (1980). Synthesis of human plasminogen by the liver. *Science, 208,* 1036–1037.

14. Vaughan, D. E., & Declerck, P. J. Regulation of fibrinolysis. In J. Loscalzo & A. Schafer (Eds.), *Thrombosis and hemorrhage* (vol. 3, pp. 105–119). Baltimore: Williams and Wilkins.

15. Robbins, K. C., Summaria, L., Hsieh, B., et al. (167). The peptide chains of human plasmin. Mechanism of activation of human plasminogen to plasmin. *J Biol Chem, 242,* 2333–2342.

16. Cesarman, G. M., Guevara, C. A., & Hajjar, K. A. (1994). An endothelial cell receptor for plasminogen/tissue plasminogen activator (t-PA) II. Annexin II-mediated enhancement of t-PA-dependent plasminogen activation. *J Biol Chem, 269,* 21198–21203.

17. Davis, E. W., Ichinose, A., & Leytus, S. P. (1986). Structural features of the proteins participating in blood coagulation and fibrinolysis. *Cold Spring Harb Symp Quant Biol, 51,* 509–514.

18. Hart, P. H., Vitti, G. F., Burgess, D. R., et al. (1989). Human monocyte can produce tissue-type plasminogen activator. *J Exp Med, 169,* 1509–1514.

19. Peerschke, E. I., & López, J. A. (2003). Platelet membranes and receptors. In J. Loscalzo & A. Schafer (Eds.), *Thrombosis and hemorrhage* (vol. 3, pp. 161–186). Baltimore: Williams and Wilkins.

20. Hajjar, K. A., Hamel, N. M., Harpel, P. C., et al. (1987). Binding of tissue plasminogen activator cultured human endothelial cells. *J Clin Invest, 80,* 1712–1719.

21. Felez, J., Miles, L. A., Fabregas, P., et al. (1996). Characterization of cellular binding sites and interactive regions within reactants required for enhancement of plasminogen activation by t-PA on the surface of leukocytic cells. *Thromb Haemost, 76,* 577–584.

22. Beebe, D. P., Wood, L. L., & Moos, M. (1990). Characterization of tissue plasminogen activator binding proteins isolated from endothelial cells and other cell types. *Thromb Res, 59,* 339–350.

23. Holmes, W., Pennica, D., Blaber, M., et al. (1985). Cloning an expression of the gene for pro-urokinase in *Escherichia coli*. *Biotech, 3,* 923–929.

24. Riccio, A., Grimaldi, G., Verde, P., et al. (1985). The human urokinase–plasminogen activator gene and its promoter. *Nucl Acids Res, 13,* 2759–2771.

25. Li, X., Bokman, A., Llinas, M., et al. (1994). Solution structure of the kringle domain from urokinase-type plasminogen activator. *J Mol Biol, 235,* 1548–1559.

26. Ploug, M., Plesner, T., Ronne, E., et al. (1992). The receptor for urokinase-type plasminogen activator is deficient on peripheral blood leukocytes in individuals with paroxysmal nocturnal hemoglobinuria. *Blood, 79,* 1447–1455.

27. Miles, L. A., Levin, E. G., Plescia, J., et al. (1988). Plasminogen receptors, urokinase receptor, and their modulation on human endothelial cells. *Blood, 72,* 628–635.

28. Barthanan, E. S. (1992). Characterization and regulation of the urokinase receptor of human endothelial cells. *Fibrinolysis, 6,* 1–9.

29. Hendriks, D., Scharpe, S., van Sande, M., et al. (1989). Characterization of a carboxypeptidase in human serum distinct from carboxypeptidase N. *Clin Biochem, 27,* 277–285.

30. Campbell, W., & Okada, H. (1989). An arginine specific carboxypeptidase generated in blood during coagulation or inflammation which is unrelated to carboxypeptidase N or its subunits. *Biochem Biophys Res Commun, 162,* 933–939.

31. Bajzar, L. (2000). Thrombin activatable fibrinolysis inhibitor and an antifibrinolytic pathway. *Arterioscler Thromb Vasc Biol, 20,* 2511–2518.

32. Broze, G. J., & Higuchi, D. A. (1996). Coagulation-dependent inhibition of fibrinolysis, roles of carboxypeptidase-U and the premature lysis of clots from hemophilic plasma. *Blood, 88,* 3815–3823.

33. Eaton, D. L., Malloy, B. E., Tsai, S., et al. (1991). Isolation, molecular cloning, and partial characterization of a novel carboxypeptidase B from human plasma. *J Biol Chem, 269,* 21833–21838.

34. Bajzar, L., Manuel, R., & Nesheim, M. (1996). Purification and characterization of TAFI, a thrombin activatable fibrinolysis inhibitor. *J Biol Chem, 88,* 2093–2100.

35. Redlitz, A., Tan, A. K., Eaton, D. L., et al. (1995). Plasma carboxypeptidases as regulators of the plasminogen system. *J Clin Invest, 96,* 2534–2538.

36. Ryan, C. A. (1974). Purification and properties of a carboxypeptidase inhibitor from potatoes. *J Biol Chem, 249,* 5494–5499.

37. Moroi, M., & Aoki, N. A. (1976). Isolation and characterization of alpha 2-plasmin inhibitor from human plasma. A novel proteinase inhibitor which inhibits activator-induced clot lysis. *J Biol Chem, 251,* 5956–5965.

38. Wiman, B., & Collen, D. (1977). Purification and characterization of human antiplasmin, the fast-acting plasmin inhibitor in plasma. *Eur J Biochem, 78,* 19–26.

39. Bangert, K., Johnsen, A. H., Christensen, U., et al. (1993). Different Na-terminal forms of α-2-antiplasmin in human plasma. *Biochem J, 291,* 623–615.

40. Koyama, T., Koike, Y., Toyota, S., et al. (1994). Different NH2-terminal form with 12 additional residues of alpha 2-plasmin inhibitor from human and culture media of Hep G2 cells. *Biochem Biophys Res Commun, 200,* 417–422.

41. Holmes, W. E., Lijnen, H. R., Nelles, L., et al. (1987). α-2 Antiplasmin, a serine proteinase inhibitor (serpin). *J Biol Chem, 262,* 1659–1664.

42. Favier, R., Aoki, N., & de Moerloose, P. (2001). Congenital α-2 plasmin inhibitor deficiencies: A review. *Br J Haematol, 14,* 4–10.

43. Saito, H., Goodnough, L. T., Knowles, B. B., et al. (1982). Synthesis and secretion of α-2-antiplasmin inhibitor by established liver cell lines. *Proc Natl Acad Sci U S A, 79,* 5684–5687.

44. Wiman, B., Linjen, H., & Collen, D. (1979). On the specific interaction between the lysine-binding sites in plasmin and complementary sites in alpha2-anti-plasmin and in fibrinogen. *Bicohim Biophys Acta, 579,* 142–154.

45. Wiman, B., & Collen, D. (1978). On the kinetics of the reaction between human anti-plasmin and plasmin. *Eur J Biochem, 84,* 573–578.

46. Seth, P. M., Wlater, H. A., Kahr, M., et al. (2003). Intracellular activation of the fibrinolytic cascade in the Quebec platelet disorder. *Thromb Haemost, 90,* 293–298.

47. Weiler, H., & Isermann, B. H. (2003). Thrombomodulin. *J Thromb Haemost, 1,* 1515–1524.

48. Weisel, J. W., Naaswami, C., Young, T. A., et al. (1996). The shape of thrombomodulin and interactions with thrombin as determined by electron microscopy. *J Biol Chem, 271,* 31485–31490.

49. Bourin, M. C., Lundgren–Akerlund, E., & Indahl, U. (1990). Isolation and characterization of the glycosaminoglycan component of rabbit thrombomodulin proteoglycan. *J Biol Chem, 265,* 15424–15431.

50. Sakharaov, D. V., Plow, E. F., & Rijken, D. C. (1997). On the mechanism of the antifibrinolytic activity of plasma carboxypeptidase B. *J Biol Chem, 272,* 14477–144882.

51. Campbell, W., Okada, N., & Okada, H. (2001). Carboxypeptidase R is an inactivator of complement-derived inflammatory peptides and an inhibitor of fibrinolysis. *Immunol Rev, 180,* 162–167.

52. Schenk–Braat, E. T., Morser, J., & Riijken, D. C. (2001). Identification of the epidermal growth factor-like domains of thrombomodulin essential for the acceleration of thrombin-mediated inactivation of single chain urokinase-type plasminogen activator. *Eur J Biochem, 268,* 5562–5569.

53. Kohler, H. P., & Grant, P. J. (2000). Mechanisms of disease: Plasminogen-activator inhibitor type 1 and coronary artery disease. *N Engl J Med, 342*(24), 1792–1801.

54. Sprengers, E. D., & Kluft, C. (1987). Plasminogen activator inhibitors. *Blood, 69,* 381–387.

55. Kooistra, T., Sprengers, E. D., & van Hinsbergh, V. W. (1986). Rapid inactivation of the plasminogen-activator inhibitor upon secretion from cultured human endothelial cells. *Biochem J, 239,* 497–503.

56. Reed, G. L., & Houng, A. K. (1999). The contribution of activated factor XIII to fibrinolytic resistance in experimental pulmonary embolism. *Circulation, 99,* 299–304.

57. Shebuski, R. J., Sitko, G. R., Claremon, D. A., et al. (1990). Inhibition of factor XIIIa in a canine model of coronary thrombosis: Effect on reperfusion and acute reocclusion after recombinant tissue-type plasminogen activator. *Blood, 75,* 1455–1459.

58. Holaerts, M., Rijken, D. C., Lijnen, H. R., et al. (1982). Kinetics of the activation of plasminogen by human tissue plasminogen activator: Role of fibrin. *J Biol Chem, 257,* 2912–2919.

59. Keijer, J., Linders, M., van Zonneveld, A. J., et al. (1991). The interaction of plasminogen activator inhibitor 1 with plasminogen activators (tissue-type and urokinase-type) and fibrin: Localization of interaction sites and physiologic relevance. *Blood, 78,* 401–409.

60. Meade, T. W., Ruddock, V., Stirling, Y., et al. (1993). Fibrinolytic activity, clotting factors, and long-term incidence of ischaemic heart disease in the Northwick Park Heart Study. *Lancet, 342,* 1076–1079.

61. Juhan–Vague, I., Pyke, S. D., Alessi, M. C., et al. (1996). Fibrinolytic factors and the risk of myocardial infarction or sudden death in patients with angina pectoris. *Circulation, 94,* 2057–2063.

62. Juhan–Vague, I., Alessi, M., & Vague, P. (1991). Increased plasma plasminogen activator inhibitor I levels: A possible link between insulin resistance and atherothrombosis. *Diabetologia, 23,* 457–462.

63. Jain, S., Nagi, D., Slavin, B., et al. (1993). Insulin therapy in type 2 diabetic subjects suppresses plasminogen activator inhibitor (PAI-1) activity and proinsulin like molecules independently of glycaemic control. *Diabetes Med, 10,* 27–32.

64. Saigo, M., Hsue, P. Y., & Waters, D. D. (2004). Role of thrombotic and fibrinolytic factors in acute coronary syndromes. *Prog Cardiovasc Dis, 46,* 524–538.

65. Stamler, J., Loh, E., Roddy, M., et al. (1994). Nitric oxide regulates basal systemic and pulmonary vascular resistance in healthy humans. *Circulation, 89,* 2035–2040.

66. Ardissino, D., Merlini, P. A., Ariens, R., et al. (1997). Tissue factor antigen and activity in human coronary atherosclerotic plaque. *Lancet, 349,* 769–771.

67. Ridker, P. M., Gaboury, C. L., Conlin, P. R., et al. (1993). Stimulation of plasminogen activator inhibitor in vivo by infusion of angiotensin II: Evidence of a potential interaction between the rennin–angiotensin system and fibrinolytic function. *Circulation, 87,* 1969–1973.

68. Vaughan, D. E., Roueau, J. L., Ridker, P. M., et al. (1997). Effects of ramipril on plasma fibrinolytic balance in patients with acute anterior myocardial infarction. *Circulation, 96,* 442–447.

69. Loscalzo, J., Pasche, B., Ouimet, H., et al. (1995). Platelets and plasminogen activation. *Thromb Haemost, 74,* 291–293.

70. Loscalzo, J. (1989). Platelets and thrombolysis. *J Vasc Med Biol, 1,* 213–219.

71. Sprengers, E. D., Akkerman, J. W., & Jansenm, B. G. (1986). Blood platelet plasminogen activator inhibitor: Two different pools of endothelial cell type plasminogen activator inhibitor in human blood. *Thromb Haemost, 55,* 325–329.

72. Lefkovits, J., Plow, E. F., & Topol, E. J. (1995). Platelet glycoprotein IIb/IIIa receptors in cardiovascular medicine. *N Engl J Med, 332,* 1553–1559.

73. Kroll, M. H., Harris, T. S., Moake, J. L., et al. (1991). Von Willebrand factor binding to platelet GpIb initiates signals for platelet activation. *J Clin Invest, 88,* 1568–1573.

74. Fitzgerald, L. A., & Phillips, D. R. (1987). Platelet membrane glycoproteins. In R. Colman, J. Hirsh, V. Marder, & E. Salzman (Eds.), *Hemostasis and thrombosis: Basic principles of clinical practice* (vol. 2, pp. 572–593). Philadelphia: Lippincott.

75. Saelman, E. U., Nieuwenhuis, H. K., Hese, K. M., et al. (1994). Platelet adhesion to collagen types I through VIII under conditions of stasis and flow is mediated by GPIa/IIa. *Blood, 83,* 1244–1250.

76. Kunicki, T. J., Nugent, D. J., Staats, S. J., et al. (1988). The human fibroblast class II extracellular matrix receptor mediates platelet adhesion to collagen and is identical to the platelet glycoprotein Ia-IIa complex. *J Biol Chem, 263,* 4516–4519.

77. Rooney, M. M., Farrel, D. H., van Hemel, B. M., et al. (1988). The contribution of the three hypothesized integrin-binding sites in fibrinogen to platelet-mediated clot retraction. *Blood, 92,* 2374–2381.

78. Cierniewski, C. S., Byzova, T., Papierak, M., et al. (1999). Peptide ligands can bind to distinct sites in integrin alphaIIbbeta3 and elicit different functional responses. *J Biol Chem, 274,* 16923–16932.

79. Shattil, S. J., & Ginsberg, M. H. (1997). Integrin signaling in vascular biology. *J Clin Invest, 100,* S91–S95.

80. Rybak, M. E., & Renzulli, L. A. (1989). Ligand inhibition of the platelet glycoprotein IIb-IIIa complex function as a calcium channel in liposomes. *J Biol Chem, 264,* 14617–14620.

81. Collet, J. G., Montalescot, G., Lesty, C., et al. (2002). A structural and dynamic investigation of the facilitating effect of glycoprotein IIb/IIIa inhibitors in dissolving platelet-rich clots. *Circulation Res, 90,* 428–434.

82. Zidansek, A., Blinc, A., Lahanjar, G., et al. (1995). Finger-like lysing patterns of blood clots. *Biophys J, 69,* 803–809.

83. Braaten, J. V., Jerome, G. W., & Hantgan, R. R. (1994). Uncoupling fibrin from integrin receptors hastens fibrinolysis at the platelet–fibrin interface. *Blood, 83,* 982–993.

84. Poncz, M., Eisman, R., Heidenreich, R., et al. (1987). Structure of the platelet membrane glycoprotein IIb: Homology to the alpha subunits of vitronectin and fibronectin membrane receptors. *J Biol Chem, 262,* 8476–8482.

85. Phillips, D. R., Charo, I. F., Parise, L. V., et al. (1998). The platelet membrane glycoprotein IIb-IIIa complex. *Blood, 71,* 831–843.

86. Wagner, C. L., Mascelli, M. A., Neblok, D. S., et al. (1996). Analysis of GPIIb/IIIa receptor number by quantification of 7E3 binding to human platelets. *Blood, 88,* 907–914.

87. Endenburg, S. C., Hantgan, R. R., Sixma, J. J., et al. (1993). Platelet adhesion to fibrin(ogen). *Blood Coag Fibrinolys, 4,* 139.

88. Chew, D. P., & Moliterno, D. J. (2000). A critical appraisal of platelet glycoprotein IIb/IIIa inhibition. *J Am Coll Cardiol, 36,* 2028–2035.

89. Peter, K., Schwarz, M., Ylänne, J., et al. (1998). Induction of fibrinogen binding and platelet aggregation as a potential intrinsic property of various glycoprotein IIb/IIIa inhibitors. *Blood, 92,* 3240–3249.

90. McClure, M. W., Berkowitz, S. D., Sparapani, R., et al. (1999). Clinical significance of thrombocytopenia during a non-ST-elevation acute coronary syndrome. The platelet glycoprotein IIb/IIIa in unstable angina: Receptor suppression using integrilin therapy (PURSUIT) trial experience. *Circulation, 99,* 2892–2900.

91. Sabatine, M. S., Cannon, C. P., Gibson, C. M., et al. (2005). Addition of clopidogrel to aspirin and fibrinolytic therapy for myocardial infarction with ST-elevation. *N Engl J Med, 352,* 1179–1189.

92. Smith, B. R., Rinder, H. M., & Rinder, C. S. (2003). Interaction of blood and artificial surfaces. In J. Loscalzo & A. Shafer (Eds.), *Thrombosis and hemorrhage* (vol. 3, pp. 865–885). Baltimore: Williams and Wilkins.

93. Khuri, S. F., Wolfe, J. A., Josa, M., et al. (1992). Hematologic complications during and after cardiopulmonary bypass and their relationship to the bleeding time and nonsurgical blood loss. *J Thorac Cardiovasc Surg, 104*, 94–107.

94. Valeri, C. R., Habbaz, K., Khuri, S. F., et al. (1992). Effect of skin temperature on platelet function in patients undergoing extracorporeal bypass. *J Thorac Cardiovasc Surg, 104*, 108–116.

95. Kestin, A. S., Valeri, C. R., Khuri, S. F., et al. (1993). The platelet function defect of cardiopulmonary bypass. *Blood, 82*, 107–117.

96. Khuri, S. F., Valeri, R., Loscalzo, J., et al. (1995). Heparin causes platelet dysfunction and induces fibrinolysis before cardiopulmonary bypass. *Ann Thorac Surg, 60*, 1008–1014.

97. Sobel, M., Bird, K. E., Tyler–Cross, R., et al. (1996). Heparins designed to specifically inhibit platelet interactions with von Willebrand factor. *Circulation, 93*, 992–999.

98. Fareed, J., Walenga, J. M., Hoppensteadt, D. A., et al. (1985). Studies on the profibrinolytic actions of heparin and its fractions. *Semin Thromb Hemost, 11*, 199–207.

99. Upchurch, G. R., Valeri, R., Khuri, S. F., et al. (1996). Effect of heparin on fibrinolytic activity and platelet function in vivo. *Am J Physiol, 271*, H528–H534.

100. Ouimet, H., & Loscalzo, J. (1993). Reciprocating autocatalytic interactions between platelets and the plasminogen activation system. *Thromb Res, 70*, 355–364.

101. Haithcock, B. E., Shepard, A. D., Raman, S. B., et al. (2004). Activation of fibrinolytic pathways is associated with duration of supraceliac aortic cross-clamping. *J Vasc Surg, 40*(2), 325–333.

102. Cohen, J. R., Angus, L., Ahser, A., et al. (1987). Disseminated intravascular coagulation as a result of supraceliac clamping: Implications for thoracoabdominal aneurysm repair. *Ann Vasc Surg, 1*, 552–557.

103. Illig, K. A., Green, R. M., Ouriel, K., et al. (1997). Primary fibrinolysis during supraceliac aortic clamping. *J Vasc Surg, 25*, 244–254.

104. Schneiderman, J., Eguchi, Y., Adar, R., et al. (1994). Modulation of the fibrinolytic system by major peripheral ischemia. *J Vasc Surg, 10*, 516–524.

105. Levi, M., Keller, T. T., Gorp, E., et al. (2003). Infection and inflammation and the coagulation system. *Cardiovasc Res, 60*, 26–39.

106. Levi, M., ten Cate, H., Bauer, K. A., et al. (1994). Inhibition of endotoxin-induced activation of coagulation and fibrinolysis by pentoxifylline or by a monoclonal anti-tissue factor antibody in chimpanzees. *J Clin Invest, 93*, 114–120.

107. Biemond, B. J., Levi, M., ten Cate, H., et al. (1995). Plasminogen activator and plasminogen activator inhibitor I release during experimental endotoxemia in chimpanzees: Effect of interventions in the cytokine and coagulation cascade. *Clin Sci, 88*, 587–594.

108. Mesters, R. M., Florke, N., Ostermann, H., et al. (1996). Increased PAI levels predict outcome of leukocytopenic patients with sepsis. *Thromb Haemost, 75*, 902–907.

109. Schleef, R. R., Bevilacqua, M. P., Sawdey, M., et al. (1988). Cytokine activation of vascular endothelium. Effects on tissue-type plasminogen activator and type 1 plasminogen activator inhibitor. *J Biol Chem, 263*, 5797–5803.

110. Sawdey, M. S., & Loskutoff, D. J. (1991). Regulation of murine type 1 plasminogen activator inhibitor gene expression in vivo. Tissue specificity and induction by lipopolysaccharide, tumor necrosis factor alpha, and transforming growth factor-beta. *J Clin Invest, 88*, 1346–1353.

111. Yamamoto, K., & Loskutoff, D. J. (1996). Fibrin deposition in tissues from endotoxin-treated mice correlates with decreases in the expression of urokinase-type but not tissue-type plasminogen activator. *J Clin Invest, 97*, 2440–2451.

112. Levin, J. (1984). Bleeding with infectious disease. In O. Ratnof & C. Forbes (Eds.), *Disorders of hemostasis* (pp. 367–378). Grune and Stratton, Orlando.

113. Curwen, K. D., Gimbrone, Jr., M. A., & Handin, R. I. (1980). In vitro studies of thromboresistance: The role of prostacyclin (PGI2) in platelet adhesion to cultured normal and virally transformed human vascular endothelial cells. *Lab Invest, 42*, 366–374.

114. Halstead, S. B. (1989). Antibody, macrophages, dengue virus infection, shock, and hemorrhage: A pathogenetic cascade. *Rev Infect Dis, 11*(S4), S830–S839.

115. Hermans, P. W., Hibberd, M. L., Booy, R., et al. (1999). 4G/5G promoter polymorphism in the plasminogen-activator-inhibitor-1 gene and outcome of meningococcal disease. Meningococcal Research Group. *Lancet, 354*, 556–560.

116. Wesendorp, R. G., Hottenga, J. J., & Slagboom, P. E. (1999). Variation in plasminogen activator inhibitor-1 gene and risk of meningococcal septic shock. *Lancet, 354*, 561–563.

117. Idell, S. (2003). Coagulation, fibrinolysis, and fibrin deposition in acute lung injury. *Crit Care Med, 31*, S213–S220.

118. Idell, S., James, K., Levin, E., et al. (1989). Local abnormalities in coagulation and fibrinolytic pathways predispose to alveolar fibrin deposition in the adult respiratory distress syndrome. *J Clin Invest, 84*, 695–705.

119. Idell, S., James, K., Gillies, C., et al. (1989). Abnormalities of pathways of fibrin turnover in lung lavage of rats with oleic acid and bleomycin-induced lung injury support alveolar fibrin deposition. *Am J Pathol, 135*, 387–399.

120. Idell, S. (1995). Coagulation, fibrinolysis and fibrin deposition in lung injury and repair. In S. Phan & R. Thrall (Eds.), *Pulmonary fibrosis* (pp. 743–776). New York: Marcel Dekker.

121. Bertozzi, P., Astedt, B., Zenzius, L., et al. (1990). Depressed bronchoalveolar urokinase activity in patients with adult respiratory distress syndrome. *N Engl J Med, 322*, 890–897.

122. Viscardi, R. M., Broderick, K., Sun, C. C., et al. (1992). Disordered pathways of fibrin turnover in lung lavage of premature infants with respiratory distress syndrome. *Am Rev Respir Dis, 146*, 492–499.

123. Browder, T., Folkman, J., & Pirie–Shepherd, S. (2000). The hemostatic system as a regulator of angiogenesis. *J Biol Chem, 275*, 1521–1524.

124. Daly, M. E., Makris, A., Reed, M., et al. (2003). Hemostatic regulators of tumor angiogenesis: A source of antiangiogenic agents for cancer treatment? *J Nat Cancer Inst, 95*(22), 1660–1670.

125. Carmeliet, P. (2000). Mechanisms of angiogenesis and arteriogenesis. *Nat Med, 6*, 389–395.

126. Mohle, R., Green, D., Moore, M. A., et al. (1997). Constitutive production and thrombin-induced release of vascular endothelial growth factor by human megakaryocytes and platelets. *Proc Natl Acad Sci U S A, 94*, 663–668.

127. Folkman, J. (1995). Seminars in medicine of the Beth Israel Hospital, Boston: Clinical applications of research on angiogenesis. *N Engl J Med, 333*, 1757–1763.

128. Date, K., Matsumoto, K., Kubs, K., et al. (1998). Inhibition of tumor growth and invasion by a four-kringle antagonist (HGF/NK4) for hepatocyte growth factor. *Oncogene, 17*, 3045–3054.

129. Good, D. J., Polverini, P. J., Rastinejad, F., et al. (1990). A tumor suppressor-dependent inhibitor of angiogenesis is immunologically and functionally indistinguishable from a fragment of thrombospondin. *Proc Natl Acad Sci U S A, 87*, 6624–6628.

130. Gengrinovitch, S., Greenberg, S., Cohen, T., et al. (1995). Platelet factor 4 inhibits the mitogenic activity of VEGF-121 and VEGF 165 using several concurrent mechanisms. *J Biol Chem, 170*, 15059–15065.

131. Stefansson, S., & Lawrence, D. A. (1996). The serpin PAI-1 inhibits cell migration by blocking integrin $\alpha v \ \beta 3$ binding to vitronectin. *Nature, 383*, 441–443.

132. Soker, S., Svahn, C. M., & Neufeld, G. (1993). Vascular endothelial growth factor is inactivated by binding to alpha 2 macroglobulin and the binding is inhibited by heparin. *J Biol Chem, 268*, 7685–7691.

133. Kim, I., Kim, H. G., Moon, S. O., et al. (2000). Angiopoietin-1 induces endothelial cell sprouting through the activation of focal adhesion kinase and plasmin secretion. *Circ Res, 86*, 952–959.

# CHAPTER 22

# Platelet Function in the Newborn

## Sara J. Israels

*Department of Pediatrics and Child Health and the Manitoba Institute of Cell Biology, University of Manitoba,
Winnipeg, Manitoba, Canada*

## I. Introduction

The development of the hemostatic system is not complete at birth. Delayed maturation has been well documented for components of the coagulation and fibrinolytic systems. Thrombin generation and clot lysis are reduced significantly in neonates compared with older children and adults.[1–4] Similarly, there is a relative deficiency in neonatal platelet function.[5–8] Studies of neonatal platelets have been hampered in the past by the difficulty in obtaining adequate blood samples. As a result, most studies have relied on umbilical cord blood collected at the time of delivery, with very few studies on platelets collected directly from the neonate. New techniques have permitted the use of smaller sample volumes for the evaluation of specific aspects of neonatal platelet function — technology that can be adapted to the clinical setting.[9] The focus of this chapter is on the physiological differences between neonatal platelets and those of older children and adults. Inherited disorders of platelet function are discussed in Chapter 57. Neonatal thrombocytopenia is discussed in Chapter 52.

## II. Platelet Number and Size

Platelet counts in full-term newborns are not different from adult values. In premature infants, the mean count is lower, but still within the normal range for adults: 150 to $450 \times 10^9$/L.[10–12] In fetuses between 18 weeks and 30 weeks gestation, platelet numbers also are within this range, with no change in the mean of $250 \times 10^9$/L during this period.[13] The mean platelet volume averages 7 to 9 fL in both full-term and premature infants, similar to the adult normal range.[14,15] A single study reported a mean platelet volume that was greater in term than in preterm infants.[12]

## III. Platelet Production

The numbers of circulating megakaryocyte progenitors and mature megakaryocytes are increased in newborns and, more strikingly, in preterm infants.[16,17] In studies by Murray and colleagues,[17,18] the number of circulating megakaryocyte precursors correlated with both the number of bone marrow megakaryocyte precursors and the platelet count; thrombocytopenic newborns have fewer circulating megakaryocyte precursors.

Platelet production can be evaluated by measuring peripheral blood reticulated platelets. Newly synthesized platelets with increased RNA content are identified by staining with the fluorescent nucleic acid dye thiazole orange and are quantified by flow cytometry (see Chapter 30). A single study of reticulated platelets in fetal blood obtained by cordocentesis showed a 30-fold increase in the number of reticulated platelets in fetuses compared with their mothers.[19] The proportion of reticulated platelets in nonthrombocytopenic newborns is two to three times higher than the adult normal range.[20–24] A recent study has demonstrated an elevated reticulated platelet count in neonates of $2.7 \pm 1.6\%$ compared with adult counts of $1.1 \pm 0.5\%$.[22] Early studies in neonates reported conflicting results on the number of reticulated platelets in term versus preterm newborns, probably reflecting differences in methodology.[18,19] Saxonhouse and associates[22] demonstrated an inverse relationship between the reticulated platelet count and gestational age during the first week of life. In premature infants, the reticulated platelet count rose between days 0 and 5 before stabilizing at approximately 3% from days 7 to 28.

Thrombopoietin is detectable in the fetus as early as 24 weeks of gestation, and in the plasma of newborns of all gestations. In nonthrombocytopenic neonates, plasma thrombopoietin concentrations are comparable with or slightly higher than levels in older children and adults (approximately 150 pg/mL, with a very wide range).[18,25–28]

Most studies have reported similar plasma thrombopoietin concentrations in term and preterm infants, although two studies suggest a correlation with gestational age.[28,29] There is a significant difference in the *in vitro* sensitivity of neonatal and adult megakaryocyte progenitor cells to thrombopoietin: the proliferative response to thrombopoietin is several fold higher in cultured megakaryocyte progenitor cells from cord blood or neonatal marrow compared with megakaryocyte progenitor cells from adult marrow. Progenitor cells from preterm infants are more sensitive to exogenous thrombopoietin than those from term neonates.[18,30,31]

Reports of thrombopoietin levels in neonates with thrombocytopenia, resulting from a variety of causes including maternal gestational hypertension and sepsis, have been contradictory, probably because of differences in the study populations. There is, however, consensus that thrombocytopenic neonates have lower thrombopoietin concentrations than comparably thrombocytopenic adults.[32]

## IV. Platelet Structure

Platelet structure is described in detail in Chapter 3. The ultrastructure of neonatal platelets closely resembles that of adult platelets, although some immature forms may be found in the peripheral blood of neonates (Fig. 22-1).[33,34] In a single study, minor differences in neonatal platelet structure were described: fewer pseudopods, a less developed microtubular structure, and increased numbers of α-granules.[35] However, some of these reported differences may have been the result of issues with sample collection and processing.[36]

Receptors for adhesive proteins including the glycoprotein (GP) Ib-IX-V complex (receptor for von Willebrand factor [VWF]; see Chapter 7), integrins αIIbβ3 (the GPIIb-IIIa complex, receptor for fibrinogen, fibronectin, VWF; see Chapter 8), and α2β1 (the GPIa-IIa complex, receptor for collagen; see Chapters 6 and 18) are present on fetal and cord blood platelets.[37-39] Flow cytometric analyses indicate that the surface expression of αIIbβ3 is consistently lower on neonatal platelets than on those of older children and adults by 15 to 20%, whereas levels of GPIb-IX-V are similar in all age groups.[40-43]

Fewer α2-adrenergic receptors are found on neonatal platelets, accounting for the poor response of neonatal platelets to stimulation by epinephrine. Ligand-binding studies using the α-adrenergic antagonist dihydroergocryptine have demonstrated that the number of binding sites on neonatal platelets is approximately half the number on adult platelets. This difference was shown not to be the result of receptor occupancy by elevated levels of catecholamines in cord blood.[44] By 2 months of age, the number of α2-adrenergic receptors has reached adult levels.[45]

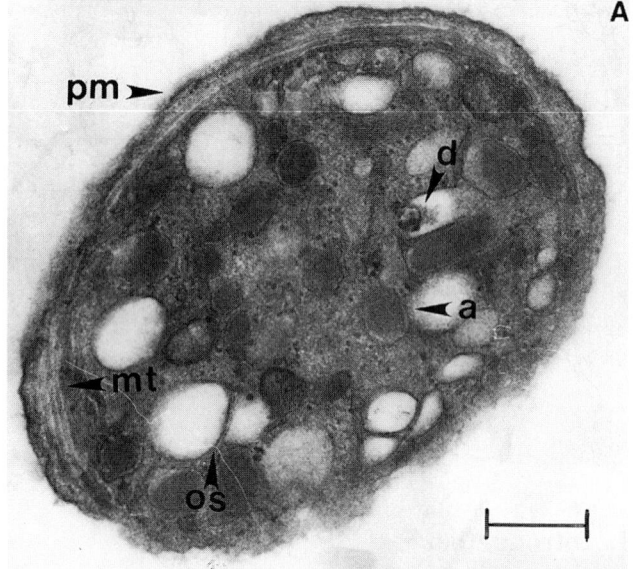

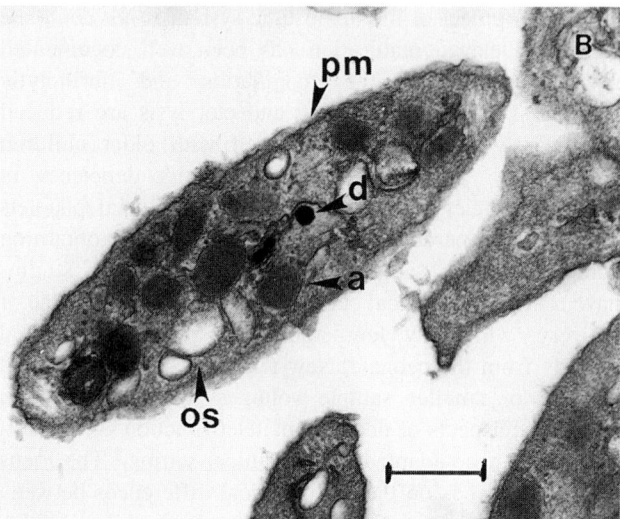

**Figure 22-1.** Cord blood platelet ultrastructure. A, B. Transmission electron micrographs of discoid platelets from cord blood. Platelets were washed in physiological buffer, fixed with glutaraldehyde, and embedded in Epon. Identified elements include plasma membrane (pm), surface-connected open canalicular system (os), microtubules (mt), α-granules (a), and dense granules (d). Bar = 0.5 μm.

The number of thromboxane receptors and their binding affinity are similar on platelets of adults, term, and premature infants.[46] The decreased response of neonatal platelets to stimulation with thromboxane analogs appears to be the result of differences in the signaling pathway downstream from the receptor (discussed later). Other agonist receptors have not been studied in neonatal platelets.

Early reports of decreased pools of nonmetabolic nucleotides in neonatal platelets consistent with a storage pool abnormality have not been confirmed.[47–49] More recent studies with neonatal and adult platelets have demonstrated comparable stores of nonmetabolic adenosine triphosphate and adenosine diphosphate (ADP),[50] and mepacrine staining.[51] The serotonin concentration in cord blood platelets is significantly lower than in adults, but increases rapidly within the first few weeks of life to levels similar to older children and adults.[52] Because serotonin uptake by the newborn platelet has been shown to be intact,[33] the decreased platelet serotonin at birth may be secondary to decreased plasma levels of serotonin *in utero,* which then increase with increased release of serotonin from the gut as feeding begins.[52,53] Release of granule contents from neonatal platelets, however, is reduced, as discussed later.

Evidence for a functional contractile cytoskeleton in neonatal platelets has been obtained indirectly by measuring the isometric contraction of platelet-rich plasma clots prepared from cord blood. Tension development and contraction of these clots are dependent upon fibrinogen binding to platelets and assembly of the actin–myosin cytoskeleton. The equivalent results obtained with adult and cord blood indicate that the components of the contractile cytoskeleton are present and functional in neonatal platelets.[54]

## V. Measurements of Platelet Function

The study of platelet function has been challenging as a result of the technical limitations of some of the standard methods when applied to neonates. The majority of studies have been performed on blood collected from the umbilical cord at delivery (rather than from the newborn infants), because this has allowed adequate specimen volumes for study. Difficulties with collection and processing have probably contributed to some of the inconsistency in reported findings. Newer techniques may help to address some of these problems. For example, thromboelastography, which offers a rapid global assessment of hemostasis in whole blood, providing assessment of clot formation, strength, and subsequent lysis, was developed originally in the 1960s but has gained renewed popularity, particularly for monitoring hemostasis in the surgical setting.[55] Two small thromboelastographic studies of newborns with gestational ages between 29 weeks and 40 weeks did not demonstrate defects in fibrin clot formation, the tensile strength of the clot, or in the rate of fibrinolysis when compared with adult control subjects.[56,57]

### A. The Bleeding Time

The bleeding time is an *in vivo* screening test for the interaction between platelets and the blood vessel wall (see Chapter 25). The devices used to measure bleeding times in adults and older children are not satisfactory for use in small infants. Andrew and colleagues,[58,59] using a device and technique modified to make a smaller skin incision (Surgicut Newborn; International Technidyne Co., Edison, NJ, USA), demonstrated that bleeding times in term newborns are shorter than bleeding times in adults. There is some evidence that the bleeding time is inversely proportional to gestational age.[60] The enhanced platelet–vessel wall interaction (i.e., the shorter bleeding time) in healthy neonates may be the result of multiple factors, including higher concentrations of circulating VWF with enhanced adhesive activity[1,2,61,62] and higher hematocrit levels.[63] In very low-birth weight infants, longer bleeding times associated with hematocrits less than 0.28 L/L improved after red cell transfusion.[64]

### B. Platelet Function Analyzer-100

The Platelet Function Analyzer-100 (PFA-100; Dade Behring, Deerfield, IL, USA) provides an *in vitro* assessment of primary hemostasis utilizing citrated whole blood (see Chapter 28).[65] This instrument measures the time (referred to as the *closure time*) required for a platelet plug to occlude an aperture (150 μm) in a membrane coated with 2 μg fibrillar type I collagen in response to either epinephrine 10 μg or ADP 50 μg. Platelets are activated by exposure to these agonists and by high shear stress as the blood is aspirated through the membrane aperture. Closure times in adults correlate with hematocrit, quantitative and functional levels of VWF, and the number and functional activity of platelets.[66–68] A series of studies now confirms that cord blood samples from term neonates have shorter closure times than samples from older children or adults (Fig. 22-2).[69–72] The shorter closure times correlate most consistently with the increased VWF concentration and activity in cord blood.[69–71]

### C. Cone and Platelet Analyzer

The cone and platelet analyzer is a device that measures platelet adhesion and aggregation under high shear conditions using a modified cone and plate viscometer (see Chapter 29).[73] Requiring only small volumes of citrated whole blood, the analyzer evaluates platelet surface coverage of extracellular matrix-coated plates and aggregate formation. Studies of umbilical cord and peripheral vein blood from healthy neonates demonstrated increased platelet adhesion and similar aggregate formation compared with adult control subjects, which correlated with plasma concentrations and functional activity of VWF, consistent with the studies of the bleeding time and the PFA-100 closure time.[39] Platelets from preterm infants demonstrated reduced adhesion and aggregation compared with full-term infants.[126]

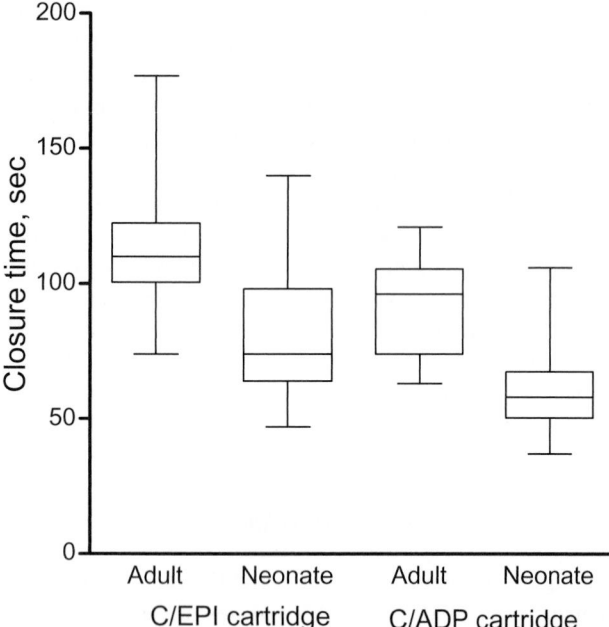

**Figure 22-2.** PFA-100 closure times. Box plots comparing PFA-100 closure times obtained with cartridges containing membranes coated with collagen/epinephrine (C/EPI) or collagen/ADP (C/ADP) in normal adults (n = 21) and cord blood samples from term neonates (n = 31). Mean closure times of cord blood are significantly shorter than mean closure times of adult samples (p < 0. 001). (Data from Israels and colleagues.[68])

Although not widely available, this instrument has the potential to evaluate neonatal platelet function in a variety of disease states and after exposure to antiplatelet drugs.

### D. Platelet Aggregation

Platelet aggregation (see Chapter 26) has been the most frequently used method to evaluate neonatal platelet function. The variability of results underlines the difficulties in obtaining cord blood samples that have not been activated during collection. Despite the technical challenges, platelet aggregation studies have revealed significant differences between neonates and adults (Fig. 22-3). In studies with a variety of agonists, including ADP, epinephrine, collagen, thrombin, and thromboxane analogs (e.g., U46619), the aggregation of neonatal platelets was decreased compared with adult platelets[33,37,75–78]; differences that were more marked in platelets of preterm infants.[46] The degree of aggregation is agonist specific. The poor response of neonatal platelets to epinephrine is the result of decreased numbers of $\alpha_2$-adrenergic receptors,[44] and the reduced response to collagen is the result of impairment of calcium mobilization. In contrast, aggregation in response to thrombin is more modestly decreased in neonatal platelets.

Ristocetin-induced aggregation of neonatal platelet-rich plasma is enhanced compared with that of adult control

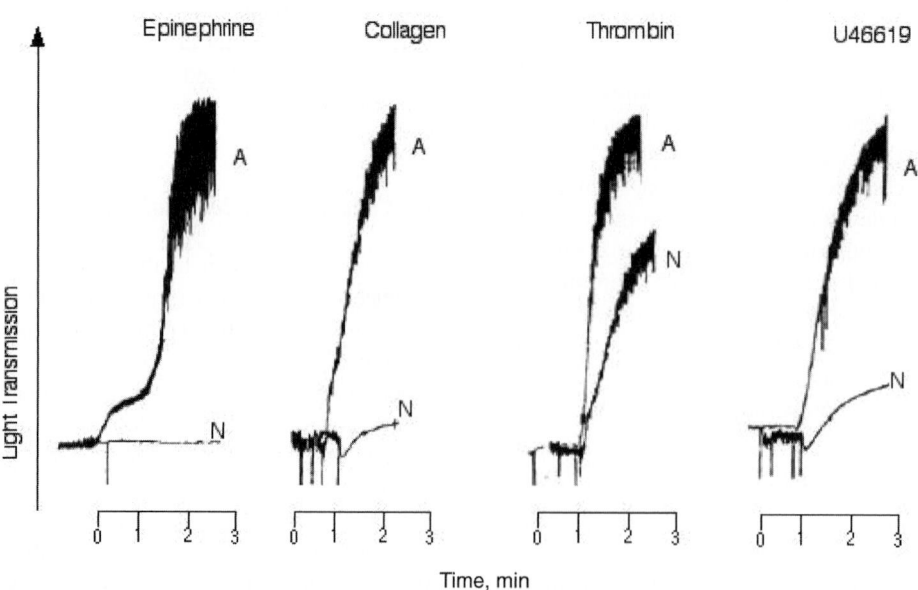

**Figure 22-3.** Aggregation responses of washed platelets. Platelets from umbilical cord blood (N) and adult control subjects (A) were washed and resuspended in physiological buffer, then stimulated with epinephrine 5 μM, collagen type 1 2 μg/mL, thrombin 0.5 U/mL, or thromboxane A₂ analogue U46619 1 μM.

subjects; neonatal platelet-rich plasma aggregates at lower concentrations of ristocetin (Fig. 22-4).[33,39,79] Moreover, ristocetin-induced aggregation of adult platelets is augmented by the addition of neonatal plasma (Fig. 22-4).[33,39,79] The agglutination of platelets by ristocetin depends on plasma VWF and its platelet receptor GPIb-IX-V (see Chapter 7). The increased response of neonatal platelet-rich plasma demonstrates the impact of the higher levels and enhanced activity of circulating neonatal VWF on platelet aggregation.[1,2,61,62,80]

### E. Flow Cytometry

Whole blood flow cytometry (see Chapter 30) has addressed many of the problems encountered in early studies of neonatal platelet activation. Only minute volumes of blood are required and there is minimal sample manipulation prior to analysis. With this technique, it has been possible to examine platelet responses in cord blood and in blood drawn from neonates at birth or on subsequent days, allowing changes in the postnatal platelet response to be monitored.[9,40,81] Using monoclonal antibodies directed at platelet activation markers, flow cytometric studies have confirmed that neonatal platelets from cord blood or from neonates on the first postnatal day are hyporeactive in comparison with adult platelets. Rajasekhar and associates[82,83] found reduced expression of activation markers on platelets from term and preterm (<30 weeks' gestation) infants after stimulation

with the agonists thrombin, ADP/epinephrine, and U46619. The surface expression of P-selectin and the exposure of the fibrinogen binding site on the integrin αIIbβ3 in response to these agents was lower in neonatal platelets compared with adult platelets, and the downregulation of GPIb-IX-V expression was blunted or absent.

The duration of the hyporeactivity identified in neonatal platelets varies significantly among the few available studies. Gatti and coworkers[81] demonstrated that by day 10 the *in vitro* neonatal platelet response resembles that of adult control subjects, suggesting that platelet hyporeactivity may be short-lived in healthy term neonates. Hezard and colleagues[42] described a gradual increase in platelet responsiveness during the first 15 years of childhood, although the most striking differences in response were between neonates and infants older than 1 month. In the setting of cardiopulmonary bypass, *in vivo* platelet activation (as measured by surface P-selectin expression) in infants up to 2 months of age was significantly less than that seen in older children.[84]

## VI. Platelet Activation during Delivery

It has been suggested that the hyporesponsiveness of cord blood platelets is a result of platelet activation and degranulation during labor and delivery. Evidence for this was provided by one study that found increased cord blood levels of thromboxane $B_2$ ($TXB_2$), β-thromboglobulin, and platelet

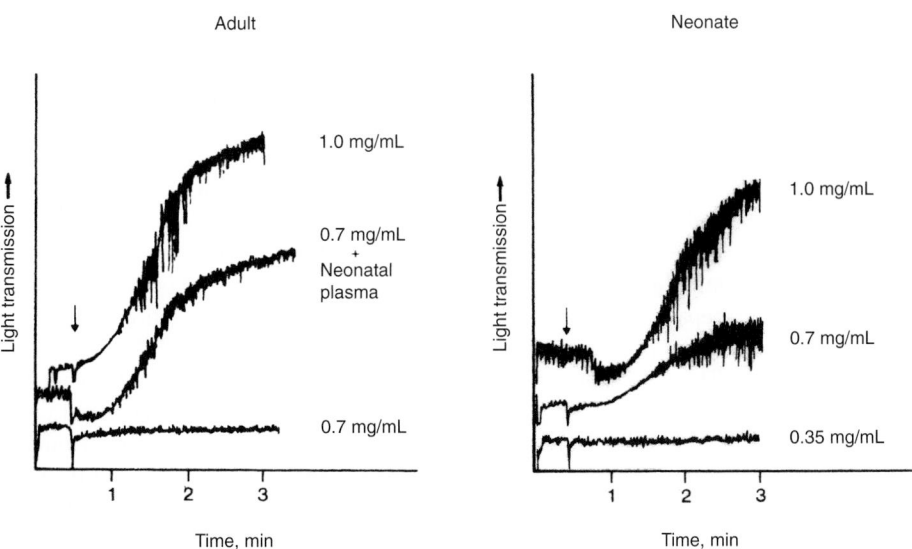

**Figure 22-4.** Ristocetin-induced platelet aggregation. Aggregation responses to varying concentrations of ristocetin (0.35–1.0 mg/mL) in platelet-rich plasma from adult controls and umbilical cord blood. A suspension of washed adult platelets in platelet-poor plasma from umbilical cord blood (Adult: neonatal plasma) demonstrated a response to ristocetin that was comparable to that seen in neonatal platelet-rich plasma (modified from Israels[79]).

factor 4, although no ultrastructural changes suggestive of degranulation were seen.[34] In more recent studies using whole blood flow cytometric measurements of activation markers (which are less prone to artifact), there was no increase in the number of circulating degranulated platelets in either the fetus or the neonate, and the mode of delivery did not influence these finding.[81-83,85-87] Thus, the hypo-responsiveness of neonatal platelets does not appear to be the result of partial activation or degranulation during labor or delivery.

## VII. Platelet Adhesion

*In vivo,* platelet adhesion is dependent on platelet adhesive receptors, the substrates for platelet attachment, and the presence of shear stress (see Chapter 18). The receptors for adhesive proteins — integrin $\alpha2\beta1$ (collagen), GPIb-IX-V (VWF), and integrin $\alpha IIb\beta3$ (fibrinogen, fibronectin, VWF) — are present on fetal and cord blood platelets.[37,38] Static adhesion of platelets to collagen substrates is similar for adult and neonatal platelets.[37] Studies of platelet deposition on extracellular matrix proteins under flow conditions demonstrated enhanced adhesion of cord blood platelets from term neonates compared with adult platelets.[39] This increased adhesion is mediated by the larger, more adhesive VWF multimers in neonatal plasma,[61,62] as can be demonstrated by the ability of umbilical cord plasma to promote aggregation of either adult or neonatal platelets at lower concentrations of ristocetin (0.6–0.7 mg/mL) than adult plasma (Fig. 22-4).[39,79]

The high-molecular weight VWF multimers in umbilical cord plasma are similar in size to multimers present in Weibel–Palade bodies of endothelial cells and $\alpha$-granules of platelets, and are responsible for increased ristocetin- and shear stress-induced binding to both adult and neonatal platelets.[88] Their presence in neonatal plasma is likely the result of the balance of several factors that contribute to both the size and concentration of VWF multimers in plasma, including endothelial cell production and secretion, and subsequent cleavage by the zinc metalloprotease ADAMTS-13 facilitated by shear stress-induced unfolding of the VWF molecule.[89,90] There have been discrepant reports on the concentration of ADAMTS-13 in neonatal plasma, ranging from 50 to 100% of adult values.[91-93] These differences may be related to assay methods. More significantly, neither the plasma concentration of VWF nor the percentage of ultra-large multimers correlates with the level of protease activity.[93] The increased adhesive activity of these larger multimers is a major contributor to the shorter bleeding times and PFA-100 closure times in healthy term neonates.[69-71,94]

Intrinsic platelet function also contributes to the adhesion of platelets in the setting of shear stress. Using a cone and plate(let) analyzer, Linder and colleagues[74] demonstrated significantly reduced platelet deposition on extracellular matrix-coated plates in blood from preterm infants compared with full-term infants, despite comparable levels and activity of VWF. Decreased platelet adhesion was similarly observed in sick neonates with respiratory distress syndrome or sepsis,[74,95] underlining the role of adhesive receptor-mediated signal transduction pathways in optimizing the adhesion process.

## VIII. Platelet Signal Transduction

Platelet signal transduction is discussed in detail in Chapter 16. In neonatal platelets, functional differences compared with adults have been demonstrated in the signaling pathways that involve $TXA_2$ synthesis, G protein-mediated responses, and intracellular calcium mobilization. The impaired platelet activation responses to a variety of agonists are the result of decreased receptor-mediated signal transduction.

$TXA_2$ is a potent endogenous platelet activator (see Chapter 31). It is produced from arachidonic acid in platelets through the aspirin-sensitive cyclooxygenase 1 pathway via endoperoxides, prostaglandin (PG) $G_2$ and $PGH_2$. Agonists such as collagen and low concentrations of thrombin require the participation of $TXA_2$ to produce maximum platelet activation. The amount of arachidonic acid released from membrane phospholipids is agonist dependent. High concentrations of thrombin stimulate greater arachidonic acid release from neonatal platelets than from adult platelets.[37,96,97] It has been suggested that this enhanced release of arachidonic acid from platelet phospholipids may be the result of the relative instability of the neonatal platelet membrane, possibly the result of vitamin E deficiency.[96,98] With agonists such as collagen and epinephrine, less arachidonic acid is released from neonatal platelets than from adult control platelets.[37,97] In addition, the proportion of arachidonic acid converted to $TXA_2$ in neonatal platelets is decreased. This difference can be overcome *in vitro.* Addition of more substrate arachidonic acid to neonatal platelets can compensate for the decreased rate of conversion. Under these conditions, the absolute $TXA_2$ production in neonatal platelets is similar to that of adult platelets.[46,97]

Platelet activation by the stable $TXA_2$ mimetic U46619 is decreased in both full-term and preterm neonatal platelets, although the number and affinity of the thromboxane receptors are equivalent to adult platelets.[46,82,83] The platelet thromboxane receptor is coupled through the heterotrimeric G protein Gq to phospholipase C$\beta$, which generates the intracellular second messengers inositol trisphosphate and diacylglycerol. These molecules stimulate intracellular calcium flux and protein kinase C activation, leading to platelet granule secretion and aggregation.[99] Many of the events downstream from phospholipase Cβ activation, including phosphoinositide metabolism,[37,100] intracellular calcium mobilization,[100,101] and protein kinase C activation,[37] are impaired in neonatal platelets. This appears to be the result of decreased GTPase activity of the α subunit of Gq in neonatal platelet membranes, leading to diminished signal transduction through phospholipase Cβ[100] and affecting the downstream events: calcium mobilization, granule secretion, and aggregation. With agonists that bypass surface receptor-mediated activation pathways, such as calcium ionophore or phorbol esters, there is comparable activation of neonatal and adult platelets.[37]

## IX. Platelet Secretion

The recruitment of platelets to the platelet plug and promotion of hemostasis is augmented by mediators such as ADP and serotonin released from platelet dense granules and adhesive proteins such as VWF and fibrinogen released from α-granules. Activation-induced secretion can be assessed by measuring the release of adenosine triphosphate,[102] [$^{14}$C]-serotonin,[46,48] or mepacrine[51] from dense granules, or by determining the surface expression of granule-specific markers such as P-selectin and CD63.[40,82,83] Agonist-induced secretion of platelet granule contents is reduced in both term and preterm neonates as a result of immature signal transduction pathways[51,82,83]; the granules and their contents are not deficient.[33,50]

## X. Platelet Procoagulant Activity

Phosphatidylserine and other negatively charged phospholipids are sequestered in the inner leaflet of the plasma membrane of resting platelets. Upon platelet activation, phosphatidylserine is relocated to the outer leaflet, where it provides a procoagulant surface that can bind coagulation factor zymogens (IX, X, and V, and prothrombin) and activated cofactors (Va and VIIIa), and promote assembly of the "tenase" complex and the "prothrombinase" complex (see Chapter 19). Phosphatidylserine is expressed on microparticles generated from platelet surface membranes after activation (see Chapter 20). Michelson and associates[103] examined the production and procoagulant activity of platelet-derived microparticles in whole blood from term and preterm neonates. In response to the calcium ionophore A23187, thrombin, or a combination of ADP and epinephrine, more microparticles were produced by both term and preterm platelets than by adult platelets under the same conditions. In response to each of these agonists, the procoagulant activity measured by factor V/Va binding was decreased in preterm platelets, but could be corrected by the

addition of factor V. These results suggest that neonatal platelets can generate a procoagulant surface after activation, but that low concentrations of circulating factor V limit the procoagulant activity of preterm platelets. In contrast, when platelets were stimulated by collagen, cord blood platelets showed significantly less procoagulant activity (as measured by thrombin generation) than adult platelets, likely related to impaired intracellular calcium mobilization and activation of the neonatal platelets by this agonist, as described earlier.[104]

## XI. Effects of Drugs on Fetal and Neonatal Platelet Function

### A. Cyclooxygenase Inhibitors

Inhibitors of platelet function have equivalent effects on neonatal and adult platelets, whether the platelets are exposed *in vivo* by maternal ingestion or *in vitro,* although some studies have shown an increased sensitivity of neonatal platelets (measured by aggregation) to aspirin and antihistamine.[76,105] Antenatal exposure to either aspirin or indomethacin is associated with increased clinical bleeding in newborns. Maternal aspirin ingestion within 5 days of delivery results in a significant risk of mucocutaneous bleeding in the newborn.[106,107] Indomethacin given to mothers for preterm labor resulted in an increased risk of intraventricular hemorrhage in their infants compared with gestation-matched control subjects whose mothers had not received indomethacin.[108] These cyclooxygenase 1 inhibitors cross the placenta and block thromboxane synthesis in the fetus and neonate, as well as in the mother. Use of low-dose aspirin (≤100 mg/day) for the management of pregnancy-induced hypertension is associated with less clinical bleeding than described earlier, although low-dose aspirin inhibits thromboxane generation in the fetus and neonate by more than 60%.[109–111] Thromboxane synthesis by neonatal platelets begins to recover 2 to 3 days after the last exposure to low-dose aspirin as unaffected platelets appear in the circulation.[111] Aspirin and other salicylates are excreted in breast milk and can be detected in the serum of breast-feeding infants whose mothers are taking aspirin regularly.[112]

### B. Ethanol

The *in vitro* effects of ethanol on neonatal platelets (inhibition of thromboxane formation) are comparable with those observed in adults.[113] Extrapolation from this study suggests that maternal ethanol intake will affect fetal and neonatal platelet function in a similar manner, but such data have not been reported.

## C. Nitric Oxide (NO)

Inhaled NO is used in the treatment of persistent pulmonary hypertension of the newborn. *In vitro,* NO inhibits platelet aggregation and adhesion by increasing levels of intracellular cyclic Guanosine monophosphate (GMP) with subsequent inhibition of activation pathways (see Chapter 13).[114] *Ex vivo* aggregation studies are complicated by the very short half-life of NO and, as a result, there are both positive and negative reports of its effects on neonatal platelet aggregation. The discrepancies are probably the result of differences in experimental methodology.[115–117] *In vitro* studies of platelet activation (measured by binding of the monoclonal antibody PAC-1 to integrin $\alpha IIb\beta 3$) in the presence of the NO donor SIN-1 demonstrated inhibition of agonist-induced activation by 70% in comparison with unexposed controls.[118] Bleeding times in infants receiving inhaled NO are prolonged, as they are in adults, but return to baseline within 24 hours of discontinuing the NO.[116] Although these studies have raised concerns about aggravating the risk of hemorrhage in sick neonates, an increased incidence of intracranial hemorrhage has not been reported in either term or preterm neonates who receive NO.[114,116,119]

## D. Selective Serotonin Reuptake Inhibitors (SSRIs)

The use of antidepressants that inhibit serotonin reuptake (SSRIs) in the settings of prenatal and postpartum depression has led to the investigation of SSRI exposure in the fetus and nursing newborn.[120] Similarities between the neuronal and platelet serotonin transporter has prompted studies of platelet serotonin as a surrogate marker for possible effects of these drugs on the developing nervous system. Cord blood platelet serotonin concentrations were significantly lower in newborn platelets exposed *in utero* to SSRIs than in unexposed control subjects.[52] These decreased levels in exposed newborns correlated with decreased levels in maternal platelets. In nursing infants of mothers taking SSRIs, the platelet serotonin concentrations were near normal, even though these drugs have been detected in the plasma of infants exposed through breast-feeding.[121]

## XII. Effects of Maternal Disease on Fetal and Neonatal Platelet Function

There is limited information about the effects of gestational diseases on fetal hemostasis, although there have been a few studies of platelet function in infants of diabetic mothers and the infants of mothers with pregnancy-induced hypertension. Prostaglandin synthesis (measured by formation of the metabolite malonyldialdehyde) and aggregation responses are increased in cord blood platelets from some diabetic

pregnancies.[122] Platelets from newborns of diabetic mothers also generate more thromboxane on day 1 than platelets from control newborns, but these differences resolve during the first week.[123] These findings are consistent with the increased thromboxane synthesis and hyperfunction in platelets of adult diabetics, and may contribute (with other factors such as polycythemia) to the increased incidence of thrombosis in infants of diabetic mothers.[124]

Platelets from infants of mothers with severe pregnancy-induced hypertension and/or placental insufficiency showed decreased platelet activation in response to thrombin compared with healthy neonatal control subjects.[40,125] However, it is difficult to determine whether some of these differences were related to the effects of maternal medications on neonatal platelet function rather than to the primary disease. In one study, increased *in vivo* platelet activation was detected in cord blood from pregnancies complicated by placental insufficiency.[125]

## XIII. Conclusions

The functional differences between neonatal cord blood platelets and those of older individuals are as follows:

- Decreased aggregation responses to agonists, including epinephrine, collagen, ADP, thrombin, and $TXA_2$ analogs
- Decreased agonist-induced exposure of the fibrinogen binding site on integrin $\alpha IIb\beta 3$
- Decreased agonist-induced granule secretion
- Decreased agonist-induced calcium mobilization
- Increased VWF-mediated platelet adhesion and agglutination
- Shorter bleeding times and PFA-100 closure times

*In vitro,* neonatal platelets demonstrate impaired activation in response to physiological agonists as a result of immature receptor-mediated signal transduction. These physiological differences can be influenced by antenatal exposure to platelet inhibitors or maternal disease states. Enhanced VWF-mediated platelet adhesion may compensate for other functional deficiencies in the first days of life.

Although it may appear that the immaturity of intrinsic platelet function has predisposed the newborn to an increased risk of hemorrhage, it is interesting to speculate on the potential benefit to the fetus of platelets that are relatively resistant to activation. No studies have addressed this question directly, but it is reasonable to consider that VWF ensures adequate hemostasis, whereas decreased platelet activation may protect the fetus and newborn from thrombotic risks associated with perinatal increases in circulating platelet activators.

# References

1. Andrew, M., Paes, B., Milner, R., et al. (1987). The development of the human coagulation system in the full term infant. *Blood, 70,* 165–172.
2. Andrew, M., Paes, B., Milner, R., et al. (1988). Development of the human coagulation system in the healthy premature infant. *Blood, 72,* 1651–1657.
3. Andrew, M., Schmidt, B., Mitchell, L., et al. (1990). Thrombin generation in newborn plasma is critically dependent on the concentration of prothrombin. *Thromb Haemost, 63,* 27–30.
4. Andrew, M., Brooker, L., Leaker, M., et al. (1990). Fibrin clot lysis by thrombolytic agents is impaired in newborns due to low plasminogen concentration. *Thromb Haemost, 68,* 325–330.
5. Blanchette, V. S., & Rand, M. L. (1997). Platelet disorders in newborn infants: Diagnosis and management. *Semin Perinatol, 21,* 53–62.
6. Kuhne, T., & Imbach, P. (1998). Neonatal platelet physiology and pathology. *Eur J Pediatr, 57,* 87–94.
7. Israels, S. J., Rand, M. L., & Michelson, A. D. (2003). Neonatal platelet function. *Semin Thromb Hemost, 29,* 363–371.
8. Saxonhouse, M. A., & Sola, M. C. (2004). Platelet function in term and preterm neonates. *Clin Perinatol, 31,* 15–28.
9. Michelson, A. D. (1998). Platelet function in the newborn. *Semin Thromb Hemost, 24,* 507–512.
10. Beverly, D. W., Inwood, M. J., Chance, G. W., et al. (1984). "Normal" haemostasis parameters: A study in a well-defined inborn population of preterm infants. *Early Hum Dev, 9,* 249–257.
11. Andrew, M., & Kelton, J. (1984). Neonatal thrombocytopenia. *Clin Perinatol, 11,* 359–391.
12. Patrick, C. H., Lazarchick, J., Stubbs, T., et al. (1987). Mean platelet volume and platelet distribution width in the neonate. *Am J Pediatr Hematol Oncol, 9,* 130–132.
13. Forestier, F., Daffos, F., Galacteros, F., et al. (1986). Hematological values of 163 normal fetuses between 18 and 30 weeks of gestation. *Pediatr Res, 20,* 342–346.
14. Kipper, S., & Sieger, L. (1982). Whole blood platelet volumes in newborn infants. *J Pediatr, 101,* 763–766.
15. Arad, I. D., Alpan, G., Sznajderman, S. D., et al. (1986). The mean platelet volume (MPV) in the neonatal period. *Am J Perinatol, 3,* 1–3.
16. Clapp, D. W., Baley, J. E., & Gerson, S. T. (1989). Gestational age-dependent changes in circulating hemopoietic stem cells in newborn infants. *J Lab Clin Med, 113,* 422–427.
17. Murray, N. A., & Roberts, I. A. G. (1995). Circulating megakaryocytes and their progenitors (BFU-MK and CFU-MK) in term and pre-term neonates. *Br J Haematol, 89,* 41–46.
18. Murray, N. A., Watts, T. L., & Roberts, I. A. G. (1998). Endogenous thrombopoietin levels and effect of recombinant human thrombopoietin on megakaryocyte precursors in term and preterm babies. *Pediatr Res, 43,* 148–151.
19. Jilma–Stohlawetz, P., Homoncik, M., Jilma, B., et al. (2001). High levels of reticulated platelets and thrombopoietin characterize fetal thrombopoiesis. *Br J Haematol, 112,* 466–468.
20. Peterec, S. M., Brennan, S. A., Rinder, H. M., et al. (1996). Reticulated platelet values in normal and thrombocytopenic neonates. *J Pediatr, 129,* 269–274.
21. Joseph, M. A., Adams, D., Maragos, J., et al.(1996). Flow cytometry of neonatal platelet RNA. *J Pediatr Hematol Oncol, 18,* 277–281.
22. Saxonhouse, M. A., Sola, M. C., Pastos, K. M., et al. (2004). Reticulated platelet percentages in term and preterm neonates. *J Pediatr Hematol Oncol, 26,* 797–802.
23. Kienast, J., & Schmitz, G. (1990). Flow cytometric analysis of thiazole orange uptake by platelets: A diagnostic aid in the evaluation of thrombocytopenic disorders. *Blood, 75,* 116–121.
24. Ault, K. A., Rinder, H., Mitchell, J., et al. (1992). The significance of platelets with increased RNA content (reticulated platelets): A measure of the rate of thrombopoiesis. *Am J Clin Pathol, 98,* 637–646.
25. Ishiguro, K., Nakahata, T., Matsubara, K., et al. (1999). Age-related changes in thrombopoietin in children: Reference interval for serum thrombopoietin level. *Br J Haematol, 106,* 884–888.
26. Watts, T. L., Murray, N. A., & Roberts I. A. G. (1999). Thrombopoietin has a primary role in the regulation of platelet production in preterm babies. *Pediatr Res, 46,* 28–32.
27. Sola, M. C., Calhoun, D. A., Hutson, A. D., et al. (1999). Plasma thrombopoietin concentrations in thrombocytopenic and non-thrombocytopenic patients in a neonatal intensive care unit. *Br J Haematol, 104,* 90–92.
28. Albert, T. S. E., Meng, Y. G., Simms, P., et al. (2000). Thrombopoietin in the thrombocytopenic term and preterm newborn. *Pediatrics, 105,* 1286–1291.
29. Paul, D. A., Leef, K. H., Taylor, S., et al. (2002). Thrombopoietin in preterm infants: Gestational age-dependent response. *J Pediatr Hematol Oncol, 24,* 304–309.
30. Traycoff, C. M., Abboud, M. R., Laver, J., et al. (1994). Human umbilical cord blood hematopoietic progenitor cells: Are they the same as their adult bone marrow counterparts? *Blood Cells, 20,* 382–391.
31. Sola, M. C., Du, Y., Hutson, A. D., et al. (2000). Dose–response relationship of megakaryocyte progenitors from the bone marrow of thrombocytopenic and non-thrombocytopenic neonates to recombinant thrombopoietin. *Br J Haematol, 110,* 449–453.
32. Sola, M. C., & Rimsza, L. M. (2002). Mechanisms underlying thrombocytopenia in the neonatal intensive care unit. *Acta Paediatr (Suppl), 438,* 66–73.
33. Ts'ao, C. H., Green, D., & Schultz, K. (1976). Function and ultrastructure of platelets of neonates: Enhanced ristocetin aggregation of neonatal platelets. *Br J Haematol, 32,* 225–233.
34. Suarez, C. R., Gonzalez, J., Menendez, C., et al. (1988). Neonatal and maternal platelets: Activation at time of birth. *Am J Hematol, 29,* 18–21.
35. Saving, K. L., Jennings, D. E., Aldag, J. C., et al. (1994). Platelet ultrastructure of high-risk premature infants. *Thromb Res, 73,* 371–384.

36. Saving, K. L., Aldag, J., Jennings, D., et al. (1991). Electron microscopic characterization of neonatal platelet ultrastructure: Effects of sampling techniques. *Thromb Res, 61,* 65–80.

37. Israels, S. J., Daniels, M., & McMillan, E. M. (1990). Deficient collagen-induced activation in the newborn platelet. *Pediatr Res, 27,* 337–343.

38. Gruel, Y., Boizard, B., Daffos, F., et al. (1986). Determination of platelet antigens and glycoproteins in the human fetus. *Blood, 68,* 488–492.

39. Shenkman, B., Linder, N., Savion, N., et al. (1999). Increased neonatal platelet deposition on subendothelium under flow conditions: The role of plasma von Willebrand factor. *Pediatr Res, 45,* 270–275.

40. Kuhne, T., Ryan, G., Blanchette, V., et al. (1996). Platelet-surface glycoproteins in healthy and preeclamptic mothers and their newborn infants. *Pediatr Res, 40,* 876–880.

41. Simak, J., Holada, K., Janota, J., et al. (1999). Surface expression of major membrane glycoproteins on resting and TRAP-activated neonatal platelets. *Pediatr Res, 46,* 445–449.

42. Hezard, N., Potron, G., Schlegel, N., et al. (2003). Unexpected persistence of platelet hyporeactivity beyond the neonatal period: A flow cytometric study in neonates, infants and older children. *Thromb Haemost, 90,* 116–123.

43. Schmugge, M., Rand, M. L., Bang, K. W. A., et al. (2003). The relationship of von Willebrand factor binding to activated platelet from healthy neonates and adults. *Pediatr Res, 54,* 474–479.

44. Corby, D. G., & O'Barr, T. P. (1981). Decreased alpha-adrenergic receptors in newborn platelets: Cause of abnormal response to epinephrine. *Dev Pharmacol Ther, 2,* 215–225.

45. Davidson Ward, S. L., Schuetz, S., Wachman, L., et al. (1991). Elevated plasma norepinephrine levels in infants of substance-abusing mothers. *Am J Dis Child, 145,* 44–48.

46. Israels, S. J., Odaibo, F. S., Robertson, C., et al. (1997). Deficient thromboxane synthesis and response in platelets from premature infants. *Pediatr Res, 41,* 218–223.

47. Corby, D. G., & Zuck, T. F. (1976). Newborn platelet dysfunction: A storage pool and release defect. *Thromb Haemost, 36,* 200–207.

48. Whaun, J. M. (1973). The platelet of the newborn infant: 5-Hydroxytryptamine uptake and release. *Thromb Diathes Haemorrh, 30,* 327–333.

49. Flachaire, E., Beney, C., Berthier, A., et al. (1990). Determination of reference values for serotonin concentration in platelet of healthy newborns, children, adults and elderly subjects by HPLC with electrochemical detection. *Clin Chem, 36,* 2117–2120.

50. Whaun, J. M. (1988). Platelet function in the neonate: Including qualitative platelet abnormalities associated with bleeding. In J. A. Stockman & C. Pochedly (Eds.), *Developmental and neonatal hematology* (pp. 131–144). New York: Raven Press.

51. Mankin, P., Maragos, J., Akhand, M., et al. (2000). Impaired platelet-dense granule release in neonates. *J Pediatr Hematol Oncol, 22,* 143–147.

52. Anderson, G. M., Czarkowski, K., Ravski, N., et al. (2004). Platelet serotonin in newborns and infants: Ontogeny, heritability, and effect of *in utero* exposure to selective serotonin reuptake inhibitors. *Pediatr Res, 56,* 418–422.

53. Anderson, G. M., Stevenson, J. M., & Cohen, D. J. (1987). Steady-state model for plasma free and platelet serotonin in man. *Life Sci, 41,* 1777–1785.

54. Israels, S. J., Gowen, B., & Gerrard, J. M. (1987). Contractile activity of neonatal platelets. *Pediatr Res, 21,* 293–295.

55. Traverso, C. I., Caprini, J. A., & Arcelus, J. I. (1995). The normal thromboelastogram and its interpretation. *Semin Thromb Hemost, 21*(suppl 4), 7–13.

56. Miller, B. E., Bailey, J. M., Mancuso, T. J., et al. (1997). Functional maturity of the coagulation system in children: An evaluation using thrombelastography. *Anesth Analg, 84,* 745–748.

57. Kettner, S. C., Pollak, A., Zimpfer, M., et al. (2004). Heparinase-modified thrombelastography in term and preterm neonates. *Anesth Analg, 98,* 1650–1652.

58. Andrew, M., Castle, V., Mitchell, L., et al. (1989). Modified bleeding time in the infant. *Am J Hematol, 30,* 190–191.

59. Andrew, M., Paes, B., Bowker, J., et al. (1990). Evaluation of an automated bleeding time device in the newborn. *Am J Hematol, 35,* 275–277.

60. Del Vecchio, A. (2002). Use of the bleeding time in the neonatal intensive care unit. *Acta Paediatr (Suppl), 438,* 82–86.

61. Katz, J. A., Moake, J. L., McPherson, P. D., et al. (1989). Relationship between human development and disappearance of unusually large von Willebrand factor multimers from plasma. *Blood, 73,* 1851–1858.

62. Weinstein, M. J., Blanchard, R., Moake, J. L., et al. (1989). Fetal and neonatal von Willebrand factor (vWF) is unusually large and similar to the vWF in patients with thrombotic thrombocytopenic purpura. *Br J Haematol, 72,* 68–72.

63. Gerrard, J. M., Docherty, J. C., Israels, S. J., et al. (1989). A reassessment of the bleeding time: Association of age, hematocrit, platelet function, von Willebrand factor, and bleeding time thromboxane $B_2$ with the length of the bleeding time. *Clin Invest Med, 12,* 165–171.

64. Sola, M. C., del Vecchio, A., Edwards, T. J., et al. (2001). The relationship between hematocrit and bleeding time in very low birth weight infants during the first week of life. *J Perinatol, 21,* 368–371.

65. Kundu, S. K., Heilmann, E. J., Sio, R., et al. (1996). Characterization of an *in vitro* platelet function analyzer, PFA-100®. *Clin Appl Thromb Hemost, 2,* 241–249.

66. Favalaro, E. J., Facey, D., & Henniker, A. (1999). Use of a novel platelet function analyzer (PFA-100™) with a high sensitivity to disturbances in von Willebrand factor to screen for von Willebrand's disease and other disorders. *Am J Hematol, 62,* 165–174.

67. Fressinaud, E., Veyradier, A., Truchaud, F., et al. (1998). Screening for von Willebrand disease with a new analyzer using high shear stress: A study of 60 cases. *Blood, 91,* 1325–1331.

68. Israels, S. J., Cheang, T., McMillan–Ward, E. M., et al. (1999). Comparison of the PFA-100® platelet function analyzer and the template bleeding time in evaluation of patients

with a clinical bleeding history [abstract]. *Blood, 94*(suppl 1), 452a.

69. Carcao, M. D., Blanchette, V. S., Dean, J. A., et al. (1998). The platelet function analyzer (PFA-100®): A novel *in vitro* system for evaluation of primary haemostasis in children. *Br J Haematol, 101,* 70–73.

70. Israels, S. J., Cheang, T., McMillan–Ward, E. M., et al. (2001). Evaluation of primary hemostasis in neonates with a new in vitro platelet function analyzer. *J Pediatr, 138,* 116–119.

71. Roschitz, B., Sudi, K., Kostenberger, M., et al. (2001). Shorter PFA-100 closure times in neonates than in adults: Role of red cells, white cells, platelets and von Willebrand factor. *Acta Paediatr, 90,* 664–670.

72. Boudewijns, M., Raes, M., Peeters, V., et al. (2003). Evaluation of platelet function on cord blood in 80 healthy term neonates using the platelet function analyser (PFA-100): Shorter in vitro bleeding times in neonates than adults. *Eur J Pediatr, 162,* 212–213.

73. Varon, D., Dardik, R., Shenkman, B., et al. (1997). A new method for quantitative analysis of whole blood platelet interaction with the extracellular matrix under flow conditions. *Thromb Res, 85,* 283–294.

74. Linder, N., Shenkman, B., Levin, E., et al. (2002). Deposition of whole blood platelets on extracellular matrix under flow conditions in preterm infants. *Arch Dis Child, 86,* F127–F130.

75. Mull, M. M., & Hathaway, W. E. (1970). Altered platelet function in newborns. *Pediatr Res, 4,* 229–237.

76. Corby, D. G., & Schulman, I. (1971). The effects of antenatal drug administration on aggregation of platelets of newborn infants. *J Pediatr, 79,* 307–313.

77. Corby, D. G., & O'Barr, T. P. (1981). Neonatal platelet function: A membrane-related phenomenon? *Haemostasis, 10,* 177–185.

78. Louden, K. A., Broughton Pipkin, F., Heptinstall, S., et al. (1994). Neonatal platelet reactivity and serum thromboxane $B_2$ production in whole blood: The effect of maternal low dose aspirin. *Br J Obstet Gynecol, 101,* 203–208.

79. Israels, S. J., & Andrew, M. (1994). In A. L. Bloom, C. D. Forbes, D. P. Thomas, & E. G. D. Tuddenham (Eds.), *Haemostasis and thrombosis* (3rd ed., pp. 1017–1055). Edinburgh: Churchill Livingstone.

80. Johnson, S. S., Montgomery, R. R., & Hathaway, W. E. (1981). Newborn factor VIII complex: Elevated activities in term infants and alterations in electrophoretic mobility related to illness and activated coagulation. *Br J Haemotol, 47,* 597–606.

81. Gatti, L., Guarneri, D., Caccamo, M. L., et al. (1996). Platelet activation in newborns detected by flow-cytometry. *Biol Neonate, 70,* 322–327.

82. Rajasekhar, D., Kestin, A. S., Bednarek, F. J., et al. (1994). Neonatal platelets are less reactive than adult platelets to physiological agonists in whole blood. *Thromb Haemost, 72,* 957–963.

83. Rajasekhar, D., Barnard, M. R., Bednarek, F. J., et al. (1997). Platelet hyporeactivity in very low birth weight neonates. *Thromb Haemost, 77,* 1002–1007.

84. Ichinose, F., Uezono, S., Muto, R., et al. (1999). Platelet hyporeactivity in young infants during cardiopulmonary bypass. *Anesth Analg, 88,* 258–262.

85. Nicolini, U., Guarneri, D., Gianotti, G. A., et al. (1994). Maternal and fetal platelet activation in normal pregnancy. *Obstet Gynecol, 83,* 65–69.

86. Grosshaupt, B., Muntean, W., & Sedlmayr, P. (1997). Hyporeactivity of neonatal platelets is not caused by preactivation during birth. *Eur J Pediatr, 156,* 944–948.

87. Irken, G., Uysal, K. M., Olgun, N., et al. (1998). Platelet activation during the early neonatal period. *Biol Neonate, 73,* 166–171.

88. Rehak, T., Cvirn, G., Gallistl, S., et al. (2004). Increased shear stress- and ristocetin-induced binding of von Willebrand factor to platelets in cord compared with adult plasma. *Thromb Haemost, 92,* 682–687.

89. Dong, J. F., Moake, J. L., Nolasco, L., et al. (2002). ADAMTS-13 rapidly cleaves newly secreted ultralarge von Willebrand factor multimers on the endothelial surface under flowing conditions. *Blood, 100,* 4033–4039.

90. Tsai, H. M. (2003). Shear stress and von Willebrand factor in health and disease. *Semin Thromb Hemost, 29,* 479–488.

91. Mannucci, P. M., Canciani, T., Forza, I., et al. (2001). Changes in health and disease of the metalloprotease that cleaves von Willebrand factor. *Blood, 98,* 2730–2735.

92. Tsai, H. M., Saroda, R., & Downes, K. A. (2003). Ultralarge von Willebrand factor multimers and normal ADAMTS 13 activity in the umbilical cord blood. *Thromb Res, 108,* 121–125.

93. Schmugge, M., Dunn, M. S., Amankwah, K. S., et al. (2004). The activity of the von Willebrand factor cleaving protease ADAMTS-13 in newborn infants. *J Thromb Haemost, 2,* 228–233.

94. Fischer, B. E., Thomas, K. B., & Dorner, F. (1998). Von Willebrand factor: Measuring its antigen or function? Correlation between the level of antigen, activity, and multimer size using various detection systems. *Thromb Res, 91,* 39–43.

95. Finkelstein, Y., Shenkman, B., Sirota, L., et al. (2002). Whole blood platelet deposition on extracellular matrix under flow conditions in preterm neonatal sepsis. *Eur J Pediatr, 161,* 270–274.

96. Stuart, M. J., & Allen, J. B. (1982). Arachidonic acid metabolism in the neonatal platelet. *Pediatrics, 69,* 714–718.

97. Stuart, M. J., Dusse, J., Clark, D. A., et al. (1984). Differences in thromboxane production between neonatal and adult platelets in response to arachidonic acid and epinephrine. Pediatr Res, 18, 823–826.

98. Stuart, M. J., & Oski, F. A. (1979). Vitamin E and platelet function. *Am J Pediatr Hematol Oncol, 1,* 77–82.

99. Brass, L. F., Manning, D. R., Cichowski, K., et al. (1997). Signaling through G proteins in platelets: To integrins and beyond. *Thromb Haemost, 78,* 581–589.

100. Israels, S. J., Cheang, T., Robertson, C., et al. (1999). Impaired signal transduction in neonatal platelets. *Pediatr Res, 45,* 687–691.

101. Gelman, B., Setty, B. N. Y., Chen, D., et al. (1996). Impaired mobilization of intracellular calcium in neonatal platelets. *Pediatr Res, 39,* 692–696.

102. Saving, K. L., Mankin, P., Maragos, J., et al. (2001). Association of whole blood aggregation response with immunogold-labeled glycoproteins in adult and neonatal platelets. *Thromb Res, 101,* 73–81.

103. Michelson, A. D., Rajasekhar, D., Bednarek, F. J., et al. (2000). Platelet and platelet-derived microparticle surface factor V/Va binding in whole blood: Differences between neonates and adults. *Thromb Haemost, 84,* 689–694.

104. Israels, S. J., & McMillan–Ward, E. M. (2003). Procoagulant activity of neonatal platelets. *J Thromb Haemost, 1* (Suppl 1), 1277.

105. Stuart, M. J., & Dusse, J. (1985). In vitro comparison of the efficacy of cyclooxygenase inhibitors on the adult versus neonatal platelet. *Biol Neonate, 47,* 265–269.

106. Bleyer, W. A., & Brekenridge, R. T. (1970). Studies on the detection of adverse drug reactions in the newborn II. The effects of prenatal aspirin on newborn hemostasis. *JAMA, 213,* 2049–2053.

107. Stuart, M. J., Gross, S. J., Elrad, H., et al. (1982). Effects of acetylsalicylic-acid ingestion on maternal and neonatal hemostasis. *N Engl J Med, 307,* 909–912.

108. Norton, M. E., Merrill, J., Cooper, B., et al. (1993). Neonatal complications after the administration of indomethacin for preterm labor. *N Engl J Med, 329,* 1602–1607.

109. Benigni, A., Gregorini, G., Frusca, T., et al. (1989). Effect of low-dose aspirin on fetal and maternal generation of thromboxane by platelets in women at risk for pregnancy-induced hypertension. *N Engl J Med, 321,* 357–362.

110. Duley, L., Henderson–Smart, D., Knight, M., et al. (2001). Antiplatelet drugs for prevention of pre-eclampsia and its consequences: Systematic review. *BMJ, 322,* 329–333.

111. Leonhardt, A., Bernert, S., Watzer, B., et al. (2003). Low-dose aspirin in pregnancy: Maternal and neonatal aspirin concentration and neonatal prostanoid formation. *Pediatrics, 111,* e77–e81.

112. Unsworth, J., d'Assis–Fonseca, A., & Beswick, D. T. (1987). Serum salicylate levels in a breast fed infant. *Ann Rheum, 46,* 638–639.

113. Ylikorkala, O., Halmesmaki, E., A., & Viinikka, L. (1987). Effect of ethanol on thromboxane and prostacyclin synthesis by fetal platelets and umbilical artery. *Life Sci, 41,* 371–376.

114. Cheung, P. Y., Salas, E., Schulz, R., et al. (1997). Nitric oxide and platelet function: Implications for neonatology. *Semin Perinatol, 21,* 409–417.

115. Varels, A. F., Runge, A., Ignano, L. J., et al. (1992). Nitric oxide and prostacyclin inhibit fetal platelet aggregation: A response similar to that observed in adults. *Am J Obstet Gynecol, 167,* 1599–1604.

116. George, T. N., Johnson, K. J., Bates, J. N., et al. (1998). The effect of inhaled nitric oxide therapy on bleeding time and platelet aggregation in neonates. *J Pediatr, 132,* 731–734.

117. Cheung, P. Y., Salas, E., Etches, P. C., et al. (1998). Inhaled nitric oxide and inhibition of platelet aggregation in critically ill neonates. *Lancet, 351,* 1181–1182.

118. Keh, D., Kurer, I., Dudenhausen, J. W., et al. (2001). Response of neonatal platelets to nitric oxide in vitro. *Intensive Care Med, 27,* 283–286.

119. Hoehn, T., Krause, M. F., & Buhrer, C. (2000). Inhaled nitric oxide in premature infants: A meta-analysis. *J Perinat Med, 28,* 7–13.

120. Koren, G., Matsui, D., Einarson, A., et al. (2005). Is maternal use of selective serotonin reuptake inhibitors in the third trimester of pregnancy harmful to neonates? *Can Med Assoc J, 172,* 1457–1459.

121. Epperson, C. N., Jatlow, P. I., Czarkowski, K., et al. (2003). Maternal fluoxetine treatment in the post-partum period: Effects on platelet serotonin and plasma drug levels in breast-feeding mother–infant pairs. *Pediatrics, 112,* e425–e429.

122. Stuart, M. J., Elrad, H., Graeber, J. E., et al. (1979). Increased synthesis of prostaglandin endoperoxides and platelet hyperfunction in infants of mothers with diabetes mellitus. *J Lab Clin Med, 94,* 12–26.

123. Kaapa, P., Knip, M., Viinikka, L., et al. (1986). Increased platelet thromboxane $B_2$ production in newborn infants of diabetic mothers. *Prostaglandins Leukotrienes Med, 21,* 299–304.

124. Oppenheimer, E. H., & Esterly, J. R. (1965). Thrombosis in the newborn: Comparison between infants of diabetic and nondiabetic mothers. *J Pediatr, 67,* 549–556.

125. Trudinger, B., Song, J. Z., Wu, Z. H., et al. (2003). Placental insufficiency is characterized by platelet activation in the fetus. *Obstet Gynecol, 101,* 975–981.

126. Levy-Shraga, Y., Maayan-Metzger, A., Lubetsky, A., et al. (2006). Platelet function of newborns as tested by cone and plate(let) analyzer correlates with gestationalage. *Acta Haematol, 115,* 152–156.

# PART TWO

# Tests of Platelet Function

# CHAPTER 23

# Clinical Tests of Platelet Function

## Paul Harrison and David Keeling

*Oxford Haemophilia Centre & Thrombosis Unit, Churchill Hospital, Oxford, United Kingdom*

## I. Introduction

Platelets play a pivotal role in both normal hemostasis and pathological bleeding and thrombosis.[1] Most platelet function tests have been traditionally utilized for the diagnosis and management of patients presenting with bleeding problems rather than thrombosis.[2] However, as platelets are now implicated in the development of atherothrombosis, which is the leading cause of mortality in the Western world,[3,4] new and existing platelet function tests are increasingly being used for monitoring the efficacy of antiplatelet drugs to treat these conditions and/or to try to identify patients at risk of arterial disease. Conversely, as increasing numbers of patients are being treated with antiplatelet drugs, there is an associated increased risk of bleeding, especially during trauma and surgical procedures. Platelet function tests are therefore also being increasingly proposed as presurgical/ perioperative tools to aid in the prediction of bleeding and for monitoring the efficacy of various types of prohemostatic therapies. This, coupled with the development of new, simpler tests and point-of-care (POC) instruments, has resulted in the increasing tendency of platelet function testing to be performed away from specialized hemostasis clinical or research laboratories, where the more traditional and complex tests are still performed.[5,6]

The following chapters in this book provide a detailed discussion of specific, well-characterized clinical tests of platelet function: the bleeding time (Chapter 25), platelet aggregation (Chapter 26), VerifyNow® (Chapter 27), the platelet function analyzer (PFA)-100® (Chapter 28), the Impact cone and plate(let) analyzer® (Chapter 29), flow cytometry (Chapter 30), and thromboxane generation (Chapter 31).

This chapter serves as an introduction to currently available clinical tests of platelet function, and discusses their relative advantages and limitations in clinical settings as well. Table 23-1 is a summary of platelet function tests and hemostatic tests that depend, to some extent, on platelets, and their clinical utility, advantages, and disadvantages.

## II. History of Platelet Function Testing and Overview of Currently Available Tests

The accurate enumeration of platelets in blood is important to eliminate thrombocytopenia as a potential cause of bleeding before any platelet function testing is performed. Although platelets were first identified as distinct corpuscles in the blood more than 120 years ago by microscopy,[7] the routine application of accurate platelet counting and blood smears to study platelet morphology was not widespread until the 1950s. Phase contrast microscopy was the first accurate method used to count platelets in lysed blood diluted within counting chambers.[8,9] Until very recently, this technique remained the gold standard reference method, even though cell counting was revolutionized by the invention of the Coulter or impedance principle in the 1950s.[10,11] A fully automated full blood count, including a platelet count, first became available in the 1970s, and it became possible, in addition, to measure other important parameters based upon distribution, including mean platelet volume (MPV), platelet distribution width, and platelet large cell ratio.[12-14] With the ultimate convergence of flow cytometric and aperture impedance principles, platelet counting can now also be performed by either optical (one-dimensional or two-dimensional scatter and fluorescence) or immunological methods (e.g., CD61 ImmunoPLT®), which may be more accurate in some samples, the latter providing the basis of a new, very precise flow cytometric reference method for platelet counting (i.e., the platelet-to-red blood cell ratio).[15-17] Details of all platelet counting instruments and reference methods are provided in Chapter 24 and elsewhere.[11]

**Table 23-1: An Alphabetical List of Many of the Currently Available Tests of Platelet Function, and Global Hemostatic Tests with a Significant Dependence on Platelet Number and Function**

| Name of Test | Principle | Advantages | Disadvantages | Frequency of Use | Clinical Applications |
|---|---|---|---|---|---|
| Adenine nucleotides | Measurement of total and released nucleotides by HPLC or luminescence | Sensitive | Sample preparation, assay calibration, extra equipment | Widely used in specialized labs, usually in conjunction with LTA | Diagnosis of storage and release defects |
| AspirinWorks® | Immunoassay of urinary 11-dehydrothromboxane $B_2$ | Measures stable thromboxane metabolite, dependent upon COX-1 activity | Indirect assay, not platelet specific, renal function dependent | Increasing use | Monitoring aspirin therapy and identifying poor responders at increased risk of thrombosis |
| Bleeding time | In vivo cessation of blood flow | In vivo test, physiological POC | Insensitive, invasive, scarring, high CV | Was widely used, now less popular | Screening test |
| Blood smear | Microscopic analysis of blood cell on glass slide | Diagnostic | Artifacts can occur | Widely used | Detection of abnormalities in platelet size, number and granules, and leukocyte inclusion bodies |
| Clot retraction | Measures platelet interaction with fibrin | Simple | Nonspecific | Not widely used | Detection of abnormalities in integrin $\alpha IIb\beta3$ and fibrinogen |
| Coaxial cylinder couette | Shear-induced platelet aggregation | Rapid, whole blood, no external activator | In development | In development | Monitoring antiplatelet therapy, detecting platelet hyperfunction |
| Combined aggregometry and luminescence | Combined WBA or LTA and nucleotide release | Monitors release reaction with secondary aggregation | Semiquantitative | Widely used in specialized labs, although less than LTA | Diagnosis of a wide variety of acquired and inherited platelet defects, diagnosis of storage and release defects |
| Electron microscopy | Ultrastructural analysis of platelets | Diagnostic | Expensive, specialized equipment | Only available in special units | Detection of granular and ultrastructural defects, whole-mount method can detect dense granular defects |
| Endogenous thrombin potential (ETP) assay | Global thrombin generation | Global test, measures total thrombin generating capacity within whole blood, PRP, and PPP | Requires fluorescent plate reader | Becoming popular in hemostasis laboratories, but mainly a research tool at present | Detection of clotting defects, monitoring prohemostatic therapy, measuring platelet procoagulant activity |

| Test | Principle | Advantages | Disadvantages | Experience | Clinical use |
|---|---|---|---|---|---|
| Flow cytometry | Measurement of platelet glycoproteins and activation markers by fluorescence | Whole blood test, small blood volumes, wide variety of tests | Specialized operator, expensive, samples prone to artifact unless carefully prepared | Widely used | Diagnosis of platelet glycoprotein defects, quantification of glycoproteins, detection of platelet activation *in vivo* or in response to agonists, monitoring antiplatelet therapy (e.g., VASP phosphorylation to monitor $P2Y_{12}$ inhibition) |
| Full blood count | Automated impedance and/or flow cytometry-based analysis of cells | Rapid, precise, provides platelet distribution and MPV | Less accuracy and precision at $<20 \times 10^9/L$ | Widely used | Abnormalities in platelet number, size, and distribution; immature platelet fraction now available |
| Gorog Thrombosis Test® | High shear-dependent platelet function and thrombolysis | Simple, global test, rapid, POC | Fresh nonanticoagulated blood required | Little widespread experience | Measurement of platelet function and thrombolysis |
| Hemostasis Analysis System® | Platelet contractile force, clot elastic modulus, and thrombin generation time | Rapid, simple, POC | Measures mainly clot properties | Used in surgery, cardiology, and hemostasis laboratories | Prediction of bleeding or thrombosis, monitoring rFVIIa therapy |
| Hemostatus® device | Platelet procoagulant activity | Simple, POC | Insensitive to aspirin and GPIb function | Used in surgery and cardiology | Prediction of bleeding |
| Ichor Plateletworks® | Platelet counting pre- and postactivation | Rapid, simple, POC, small blood volume | Indirect test measuring count after aggregation | Used in surgery and cardiology | Monitoring antiplatelet drugs, prediction of bleeding |
| Impact® cone and plate(let) analyzer | Quantification of high shear platelet adhesion/aggregation onto surface | Small blood volume required, high shear, rapid, simple, research (variable) and fixed versions available POC | Instrument not yet widely available | Little widespread experience, because only recently commercially available | Detection of inherited and acquired defects in primary hemostasis, detection of platelet hyperfunction, monitoring antiplatelet therapy |
| Kinetic aggregometer | Monitoring of fluorescent platelet accumulation within a thrombus forming in a collagen-coated capillary | Sensitive to amplification pathways | In development | In development | Monitoring antiplatelet therapy, detection of defects in primary hemostasis |
| Laser platelet aggregometer (PA-200) | Monitoring of aggregation using a laser | Detection of microaggregates, sensitive | Little widespread experience | Little widespread experience | Detection of platelet hyperfunction |
| Light transmission aggregometry | Low shear platelet-to-platelet aggregation in response to classic agonists | Gold standard | Time-consuming, sample preparation, expensive | Widely used in specialized labs | Diagnosis of a wide variety of acquired and inherited platelet defects |
| O'Brien filterometer | High-shear platelet function | Simple | Nonphysiological surface, requires blood counter | Not widely used anymore | Detection of defects in primary hemostasis |

**Table 23-1: An Alphabetical List of Many of the Currently Available Tests of Platelet Function, and Global Hemostatic Tests with a Significant Dependence on Platelet Number and Function—*Continued***

| Name of Test | Principle | Advantages | Disadvantages | Frequency of Use | Clinical Applications |
|---|---|---|---|---|---|
| PFA-100® | High-shear platelet adhesion and aggregation during formation of a platelet plug | Whole blood test, high shear, small blood volumes, simple, rapid, POC | Inflexible, VWF-dependent, Hct-dependent, insensitive to clopidogrel | Widely used | Detection of inherited and acquired defects in primary hemostasis, monitoring aspirin, monitoring DDAVP therapy |
| Platelet adhesion assay (PADA) | Adhesion and aggregation to polymer beads | Rapid | Requires counter and minishaker | Little widespread experience | Detection of platelet dysfunction |
| Platelet genome or transcriptome analysis | Total mRNA content by microarray technology | Detection of all mRNAs in platelets and MKs | Instability of mRNA, possible impurity of starting material | Research only at present | Rapid detection of SNPs and gene defects |
| Platelet reactivity index | Measurement of platelet aggregates in whole blood (modified Wu and Hoak method) | Simple, rapid, inexpensive | Requires blood counter, indirect test measuring count after aggregation | Little widespread experience | Monitoring antiplatelet therapy, detection of platelet hyperreactivity |
| Proteomic analysis | Measures total individual protein content | Sensitivity increasing | Loss of some membrane glycoproteins, possible impurity of starting material | Research only at present | Detection of protein deficiencies |
| Retention Test Homburg | Adhesion and aggregation resulting in retention within a filter during centrifugation | Simple, cheap | Requires blood counter and centrifuge | Aspirin insensitive, little widespread experience | Monitoring VWD therapy, detection of platelet hyperreactivity |
| Serum thromboxane B₂ | Immunoassay | Dependent upon COX-1 activity | Prone to artifact, not platelet specific | Widespread use | Monitoring aspirin therapy, detection of thromboxane production defects |
| Soluble platelet release markers (e.g., PF4, βTG, sCD40L, sCD62P, GPV) | Usually by ELISA | Relatively simple | Prone to artifact during blood collection and processing | Fairly widely used in research | Detection of *in vivo* platelet activation |

| Method | Principle | Type | Limitations | Usage | Application |
|---|---|---|---|---|---|
| Sonoclot® analyzer | Impedance detection of changes of viscoelastic properties of celite-activated blood sample | Global test, POC, platelet count and function dependent | Measures clot properties only | Widely used in surgery and anesthesiology | Prediction of surgical bleeding, aid to blood product usage |
| Thromboelastography (TEG® or ROTEM®) | Monitoring of rate and quality of clot formation | Global whole blood test, POC | Measures clot properties only, largely platelet independent unless platelet activators are used | Widely used in surgery and anesthesiology | Prediction of surgical bleeding, aid to blood product usage, monitoring rFVIIa therapy, new platelet mapping system can be used to monitor antiplatelet therapy (e.g., aspirin or clopidogrel) |
| Thrombotic status analyzer | High-shear-dependent platelet plug formation | Simple | Prototype | Little widespread experience | Detection of inherited and acquired defects in primary hemostasis |
| VerifyNow® | Fully automated platelet aggregometer to measure antiplatelet therapy | Simple, POC 3 test cartridges (aspirin, $P2Y_{12}$, and GPIIb-IIIa) | Inflexible, cartridges can only be used for single purpose | Increasing use | Monitoring antiplatelet therapy |
| WBA | Monitors changes in impedance in response to classic agonists | Whole blood test | Older instruments require electrodes to be cleaned and recycled | Widely used in specialized labs although less than LTA | Diagnosis of a wide variety of acquired and inherited platelet defects |

COX-1, cyclooxygenase 1; CV, coefficient of variation; DDAVP, 1-deamino-8-D-arginine vasopressin (desmopressin); ELISA, enzyme-linked immunoassay; GP, glycoprotein; Hct, hematocrit; HPLC, high-performance liquid chromatography; LTA, light transmission aggregometry; MKs, megakaryocytes; MPV, mean platelet volume; mRNA, messenger RNA; PFA-100, platelet function analyzer 100; PF4, platelet factor 4; POC, point of care; PPP, platelet-poor plasma; PRP, platelet-rich plasma; rFVIIa, recombinant activated factor VII; sCD40L, soluble CD40 ligand; sCD62P, soluble CD62P (P-selectin); SNPs, single nucleotide polymorphisms; βTG, β-thromboglobulin; VASP, vasodilator-stimulated phosphoprotein; VWD, von Willebrand disease; VWF, von Willebrand factor; WBA, whole blood aggregometry.

Platelet function testing began with the application of the *in vivo* bleeding time by Duke in 1910.[18] The bleeding time was further refined by the Ivy technique and the availability of commercial spring-loaded template disposable devices containing sterile blades (e.g., Simplate II®; Organon Technika Corporation, Durham, NC, USA) and was still regarded as the most useful screening test of platelet function until the early 1990s (Fig. 23-1).[2,19,20] During the last 10 to 15 years, the widespread use of the bleeding time has rapidly declined because its limitations have been recognized (discussed later) and other, less invasive, screening tests have become available.[21–23] The bleeding time is discussed in detail in Chapter 25.

Platelet aggregometry (light transmission aggregometry [LTA]) was invented in the 1960s and soon revolutionized the identification and diagnosis of primary hemostatic defects.[24,25] LTA is still regarded as the gold standard of platelet function testing and, by adding a panel of agonists at a range of concentrations to stirred platelets, it is possible to obtain a large amount of information about many different aspects of platelet function and biochemistry.[26] This test, often now coupled with the measurement of stored and releasable platelet nucleotide content, is still utilized in most laboratories for the identification and diagnosis of many platelet defects.[27] More recently, commercial aggregometers have become easier to use, with multichannel capability, simple automatic setting of 100% and 0% baselines, and computer operation and storage of results. For example, a new, fully computerized, eight-channel aggregometer has just become available (Fig. 23-2). Some instruments can simultaneously measure luminescence, to monitor the release reaction of dense granular nucleotides during secondary aggregation. For details on the principles and applications of platelet aggregometry, see Chapter 26. Although still considered the most useful diagnostic and research tool, LTA is relatively nonphysiological, because separated platelets are usually stirred under low shear conditions during the test and only form aggregates after addition of agonists — conditions that do not accurately mimic platelet adhesion, activation, and aggregation upon vessel wall damage. Also, conventional LTA using a full panel of agonists requires both large blood volumes and significant expertise both to perform the tests and to interpret the tracings. In response to the problems with the bleeding time and LTA, a number of alternative tests have been developed, including impedance whole blood aggregometry (WBA), a fully automated cartridge-based instrument (VerifyNow®) that measures platelet LTA in anticoagulated whole blood, and a variety of tests that attempt to simulate primary haemostasis *in vitro* (Table 23-1).

WBA provides a means to study platelet function within anticoagulated whole blood without any sample processing.[28] The test measures the change in resistance or impedance between two electrodes as platelets adhere and

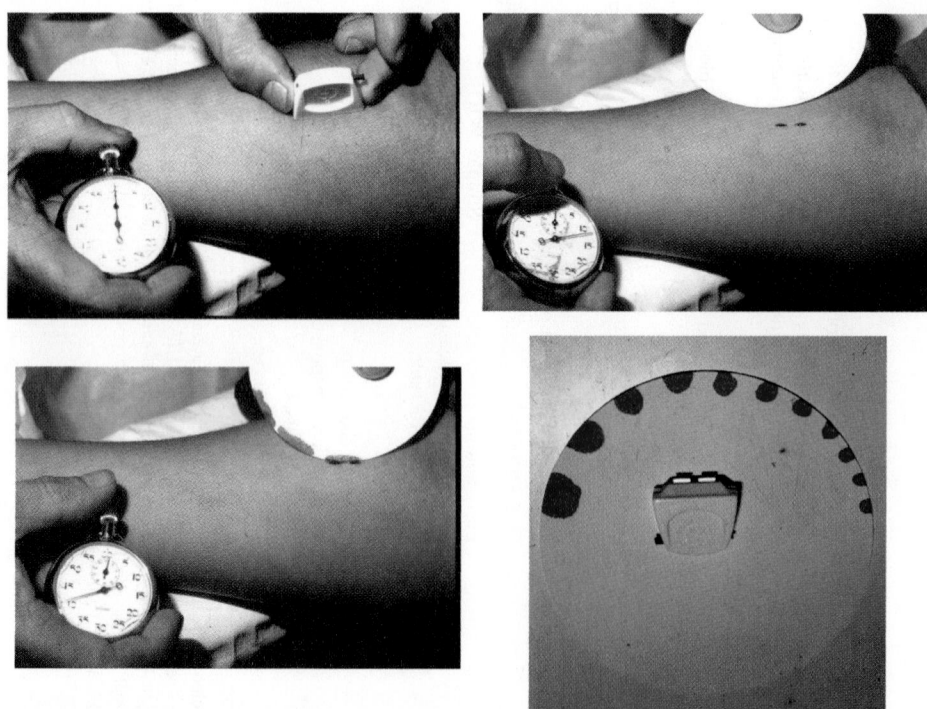

**Figure 23-1.** The *in vivo* bleeding time performed with a Simplate II® device. Horizontal cuts are made into a cleaned area of forearm skin, excess blood is blotted onto filter paper, and the time to cessation of bleeding is recorded. (Reproduced with kind permission of Prof. Sam Machin, Department of Haematology, University College Hospital, London, UK.)

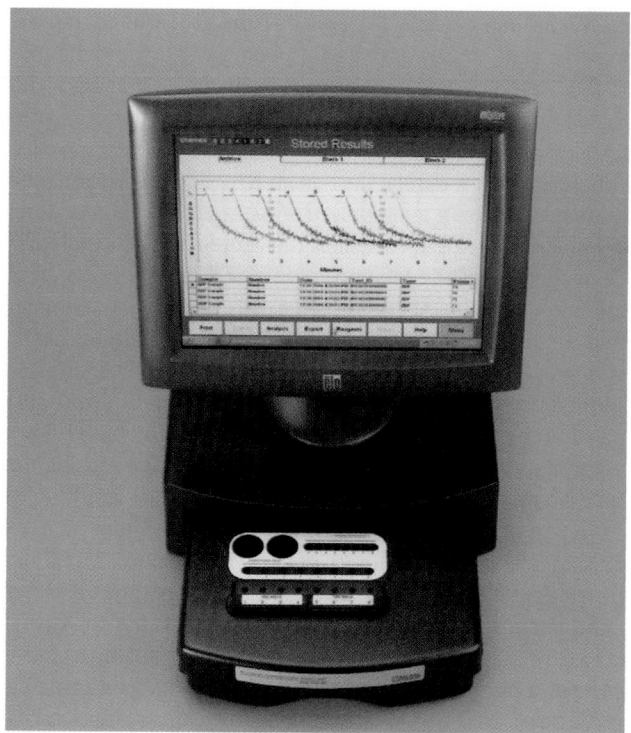

**Figure 23-2.** An example of a modern eight-channel platelet aggregometer. The model shown is the Biodata PAP-8E. (From Biodata and Biodis with permission.)

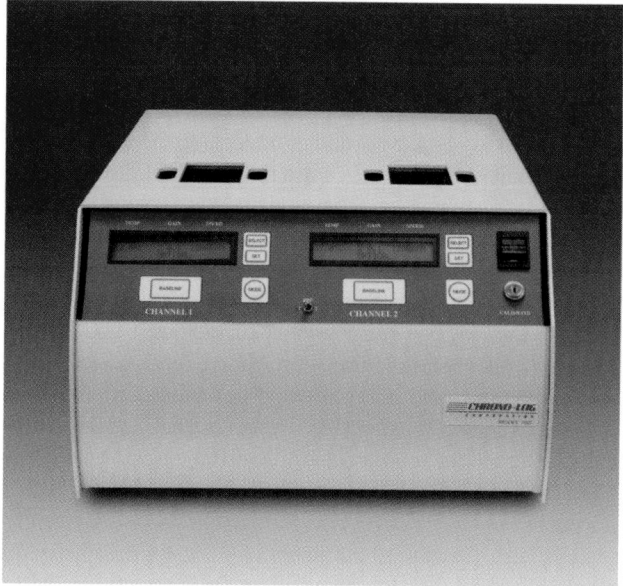

**Figure 23-3.** The Chrono-log Model 700 whole blood/optical two-channel lumiaggregometer. (From Chrono-log with permission.)

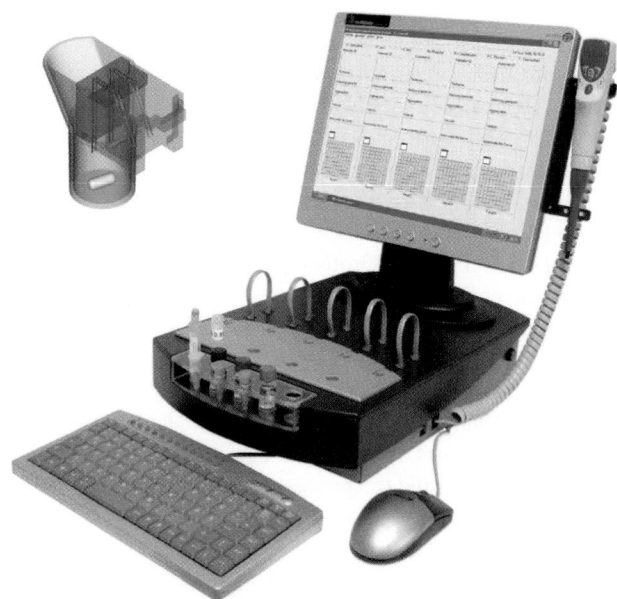

**Figure 23-4.** The Multiplate® multiple platelet function five-channel impedance analyzer. The inset illustrates the disposable cuvettes with electrodes and stir bar. (From Hart Biologicals with permission.)

aggregate in response to classic agonists. The original instrument was a two-channel device with luminescence capability. A new, fully computerized two- or four-channel instrument has now become available (Fig. 23-3). Although the latter instrument can also be used for LTA of platelet-rich plasma (PRP), WBA has many significant advantages, including the use of smaller sample volumes and the immediate analysis of samples without manipulation, loss of time, or potential loss of platelet subpopulations or platelet activation during centrifugation. The combined measurement of WBA with adenosine triphosphate (ATP) luminescence helps define the secondary aggregation response or release reaction and should theoretically be more sensitive than LTA alone for the rapid detection of storage/release disorders and defects in thromboxane production. The main disadvantage of the older WBA instruments was that the electrodes had to be carefully cleaned to remove platelet aggregates after the test. However, disposable electrodes are now available. Another new, five-channel, computerized WBA instrument (Multiple Platelet Function Analyzer or Multiplate® Dynobyte Medical, Munich, Germany) has disposable cuvettes/electrodes with a range of different agonists for different applications, including diagnosis and monitoring of antiplatelet therapy (Fig. 23-4).

The VerifyNow® (formerly known as the Ultegra Rapid Platelet Function Analyzer [or RPFA]) instrument (Fig.

27-1B in Chapter 27) is a fully automated POC test that was originally developed to monitor glycoprotein (GP) IIb-IIIa (integrin αIIbβ3) antagonists within a specialized self-contained cartridge (containing a platelet activator and fibrinogen-coated beads) that is inserted into the instrument at test initiation.[29–31] Blood sample tubes are then simply mixed prior to insertion onto the cartridge, which has been premounted onto the instrument. Aggregation in response to the agonist is monitored by light transmission through two duplicate reaction chambers in each cartridge, and the mean result is displayed and printed after a few minutes. Other specialized cartridges are now available for measuring platelet responses to either aspirin (VerifyNow® Aspirin) or clopidogrel and other $P2Y_{12}$ antagonists (VerifyNow® P2Y12). For full details on the VerifyNow® device, see Chapter 27. This instrument is a considerable advance, because the test is a fully automated POC test without the requirements of sample transport, time delays, or a specialized laboratory, and it can provide immediate information. It is also relatively expensive. However, the test is specifically designed for monitoring three different classes of antiplatelet drugs and cannot currently be used for any other purpose.

It is also possible to monitor platelet aggregometry in whole blood by a simple platelet counting technique. After addition of an agonist to anticoagulated, stirred whole blood, platelets aggregate and the platelet count decreases when compared with a control tube.[32–34] The Plateletworks® aggregation kits and Ichor full blood counter (Helena Biosciences) are simply based upon comparing platelet counts within a control ethylenediamine tetraacetic acid (EDTA) tube and after aggregation with either adenosine diphosphate (ADP) or collagen within citrated tubes.[35–38] The test correlates well with standard aggregometry[39] and can be used to monitor antiplatelet therapy,[40,41] potentially as a POC device. In the 1970s, Wu and Hoak[42] described a simple method for detecting circulating platelet aggregates. This method was refined by Grotemeyer into the platelet reactivity test and compares platelet counts within two blood tubes anticoagulated with EDTA and EDTA/fixative. The EDTA resolves the platelet aggregates that remain within the fixed sample.[43,44]

Because platelet LTA does not simulate physiological primary hemostasis, a number of tests have been developed that attempt to mimic the processes that occur during vessel wall damage. Many of these techniques have remained primarily research tools within expert laboratories because of their inherent complexity and technical difficulty. For example, the application of perfusion chambers to study platelet biology is discussed in Chapter 32. However, many simpler *in vitro* assays have been developed to measure platelet adhesion and retention of platelets on exposure to foreign surfaces. The original glass column platelet retention test was developed by Hellem.[45] Further modifications of this principle include the O'Brien filterometer,[46,47] the Homburg retention test,[48,49] and the platelet adhesion assay.[50] More recently, with significant advances in microscopy and digital imaging/processing it is now possible to perform real-time imaging of fluorescently labeled platelets and coagulation system components during thrombus formation

within animal models.[51–53] This has already resulted in some exciting new discoveries about platelets and the dynamics of their interaction with the vessel wall, each other, and with the coagulation system (see Chapter 34). A number of prototype instruments have been developed over the years, some of which remained as research tools (e.g., Thrombotic Status Analyser[54]) and some that were commercialized but are no longer available, such as the Clot Signature Analyser® developed from the Haemostatometer.[55–57] The inventor of the latter instrument has also recently developed a simpler four-channel instrument called the Gorog Thrombosis Test®, which is a global test of platelet function and fibrinolysis.[58] In this test, nonanticoagulated blood is added to disposable plastic tubes containing two ball bearings. The blood flows via gravity through the narrow gaps thus created inside the tubes. The high shear stress activates platelets, resulting in aggregation and thrombus formation. The flow rate of exiting blood is monitored with a light source, and the instrument displays both occlusion and lysis times.[58] Other commercially available instruments include the PFA-100® (Fig. 23-5; discussed in detail in Chapter 28) and the Impact® cone and plate(let) analyzer (Fig. 23-6; discussed in detail in Chapter 29). Both of these tests measure platelet adhesion and aggregation under conditions of high shear, in an attempt to simulate primary hemostatic mechanisms that are encountered *in vivo*.

The cone and plate(let) analyzer, originally developed by Varon and colleagues, monitors platelet adhesion and aggregation to a plate coated with collagen or extracellular

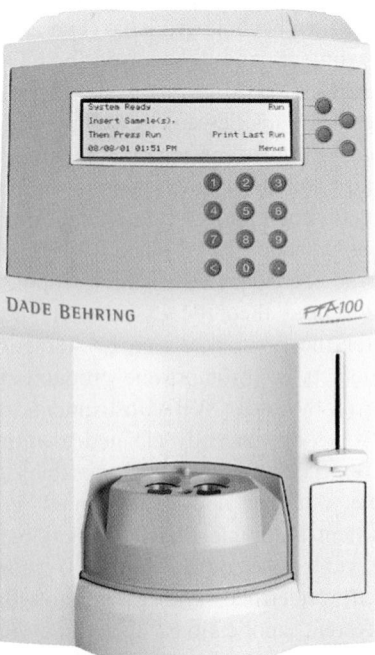

**Figure 23-5.** The PFA-100® instrument. (From Dade-Behring with permission.)

**Figure 23-6.** The DiaMed Impact® device. (From DiaMed with permission.)

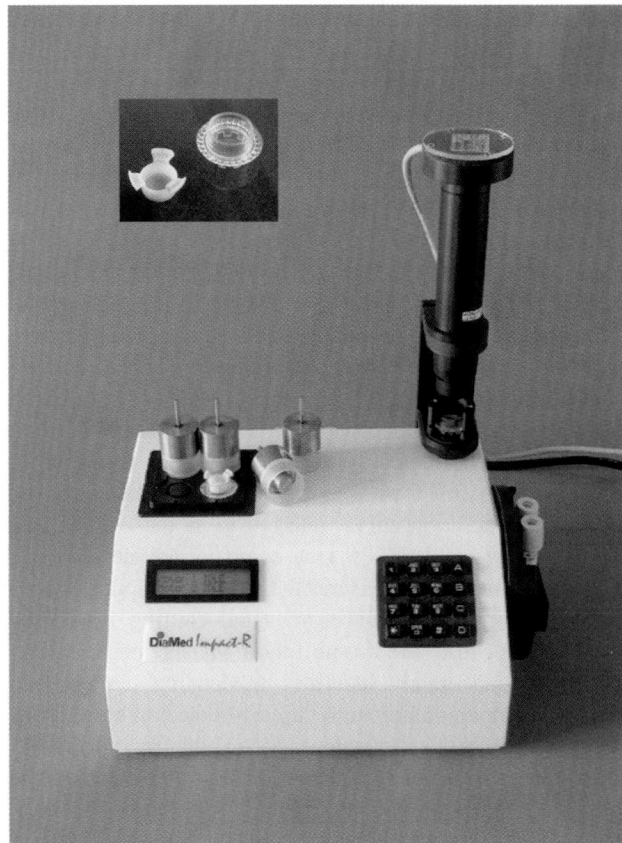

**Figure 23-7.** The DiaMed Impact-R® device. The cone and plate are shown in the inset. (From DiaMed with permission.)

matrix (ECM) under high shear conditions of 1800 s⁻¹.[59–61] In the commercial version of the device, the Impact® (DiaMed; Fig. 23-6), a plastic plate is utilized instead of a collagen or an ECM-coated surface. The test is now fully automated, simple to operate, uses a small quantity of blood (0.12 mL), and displays results in 6 minutes. The instrument contains a microscope and performs staining and image analysis of the platelets that have adhered and aggregated on the plate. The software permits storage of the images from each analysis, and records a number of parameters, including surface coverage, average size, and distribution histogram of adhered platelets. Preliminary data suggest the test can be used in the diagnosis of platelet defects and monitoring antiplatelet therapy. Because the test has only just become commercially available, widespread experience is limited. There is also a recently released research version of the instrument called the Impact-R® (Fig. 23-7), which requires some of the test steps to be manually performed, but also facilitates adjustment of the shear rate. Chapter 29 contains a full discussion of the Impact® cone and plate(let) analyzer.

The PFA-100® device (Fig. 23-5) has been available for a number of years and is now in widespread use within many laboratories, with more than 200 papers published on various clinical applications.[62,63] The test was originally developed as a prototype instrument called the Thrombostat 4000 by Kratzer and Born, and was further developed into the PFA-100® by Dade-Behring.[64,65] The PFA-100® measures the fall in flow rate as platelets within citrated whole blood are aspirated through a capillary and begin to seal a 150-μm aperture within a collagen-coated membrane. This reaction takes place contained within one of two types of disposable cartridge. The instrument records the time (closure time or CT) it takes to occlude the aperture, along with the total volume of blood used during the test. Maximal CTs that can be obtained are 300 seconds. Chapter 28 contains a full discussion of the PFA-100®.

Platelets contribute significantly to the generation of thrombin and the dynamics of blood clotting, including clot formation, clot retraction, and lysis (see Chapters 19 and 21). Clot retraction can be easily measured in whole blood or PRP within glass tubes after the addition of calcium. The

role of platelets in clot retraction was first described by
Hayem in the late 19th century, and Glanzmann famously
described patients with poor clot retraction or thrombasthenia in 1918, who were subsequently shown to be defective
in integrin αIIbβ3 (GPIIb-IIIa) (see Foreword and Chapter
57).[66] Modern tests are available that can study the role of
platelets in thrombin production, clot formation, and clot
retraction. For example, thrombin generation tests can be
used to measure thrombin generation in PRP and whole
blood.[67–69] However, early tests involved subsampling and
centrifugation steps to remove cells that would interfere
with the measurement. The recent development of fluorescent thrombin substrates has enabled the test to be utilized
in PRP and whole blood without the need for subsampling
and there is commercially available software (thrombino-
scope®) that can be used to calculate the area under the
thrombin generation curve, referred to as the *endogenous
thrombin potential* (ETP). One company has developed a
POC instrument that measures the influence of platelet activating factor on the kaolin activated clotting time. The
HemoStatus® test (Medtronic Blood Management, Parker,
CO, USA) accurately identifies a bleeding tendency in
thrombocytopenia and can be used to detect abnormalities
of, or the effects of antiplatelet drugs on, integrin αIIbβ3.[70–72]
There are also a number of instruments that measure the
physical properties of clot formation. Thromboelastogra-
phy® (TEG) was developed more than 50 years ago.[73–75]
Anticoagulated whole blood is incubated in a heated sample
cup in which a pin is suspended that is connected to a detector system. The cup oscillates 5° in each direction. In normal
anticoagulated blood, the pin is unaffected, but as the blood
clots, the motion of the cup is transmitted to the pin. Whole
blood or recalcified plasma can be used, with or without
activators of the tissue factor or contact factor pathways. The
instrument has been significantly upgraded to the TEG analyzer 5000 series (Fig. 23-8). The TEG is relatively rapid to
perform (<30 minutes), can be conducted in a POC fashion,
and provides various data relating to clot formation and lysis
(the lag time before the clot starts to form, the rate at which
clotting occurs, the maximal amplitude of the trace, and the
extent and rate of amplitude reduction). Rotational TEG
(ROTEG® or ROTEM®) is an adaptation of the TEG in
which the cup is stationary and the pin oscillates (Fig. 23-
9).[73,76] Unlike platelet function tests, TEG instruments have
been traditionally utilized within surgical and anesthesiology departments as POC tests for determining the risk of
bleeding and as a guide to transfusion requirements. More
recent developments include an expansion in the range of
activators to initiate aggregation rather than coagulation,
such as the platelet mapping system® using ADP and arachidonic acid, making the Haemoscope TEG theoretically
more sensitive to antiplatelet drugs than the conventional
TEG.[77,78] The Sonoclot analyzer (Sienco, Arvada, CO, USA)
also measures changes in the viscoelastic properties of a clot

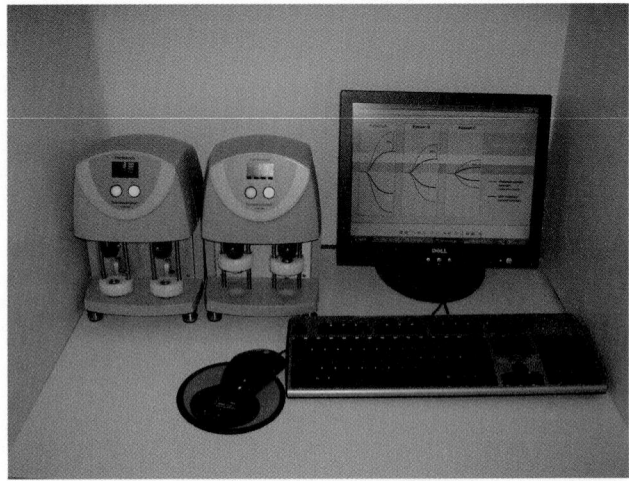

**Figure 23-8.** The Haemoscope TEG® instrument. (From Haemoscope with permission.)

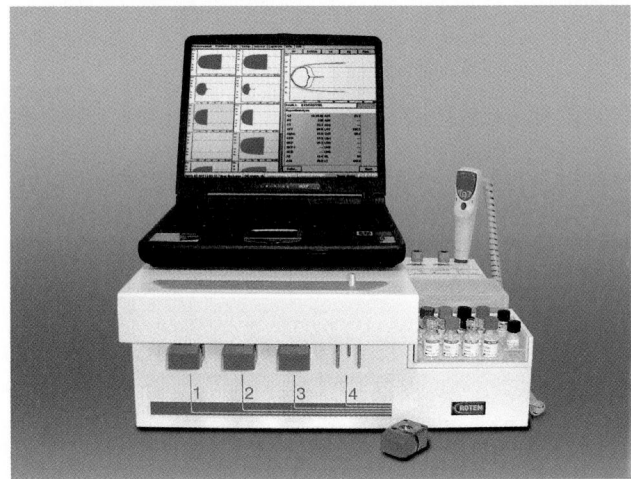

**Figure 23-9.** The ROTEM® gamma instrument. (From Pen-
tapharm with permission.)

in whole blood activated by celite.[79–81] There is some evidence to suggest that this test also provides an indication of
platelet number and function.

The Haemostasis Analysis System® (HAS) by Hemodyne
(Fig. 23-10) is based upon the original technique developed
by Carr and colleagues.[82–85] The HAS measures a number
of parameters in clotting blood, including platelet contractile force (PCF), clot elastic modulus, and thrombin generation time (TGT) in a small sample (700 µL) of whole
blood.

During the last 20 years, flow cytometric analysis of
platelets has also developed into a powerful and popular
means to study many aspects of platelet biology and function. Preferred modern methods now utilize diluted antico-

**Figure 23-10.** The Hemostasis Analysis System.® (From Hemodyne with permission.)

agulated whole blood incubated with a variety of reagents including antibodies and dyes that bind specifically to individual platelet proteins, granules, and lipid membranes.[86–88] Many of these reagents are now commercially available from many sources, enabling flow cytometric analysis of platelets to be widely performed. Flow cytometric analysis of platelet function is discussed in detail in Chapter 30.

# III. Clinical Utility of Platelet Function Testing

Most platelet function tests have been traditionally utilized either to screen for or diagnose platelet defects. Most traditional tests are not only difficult to perform, but are expensive, time-consuming, and require relatively large volumes of fresh blood. They are, therefore, usually performed within specialized hemostasis laboratories, often in close proximity to their associated clinics. Many of these tests are limited in their capacity to predict bleeding or thrombosis. These limitations have largely restricted their widespread clinical use within other disciplines (e.g., cardiology, stroke, and surgery). However, this is now beginning to change as simpler tests of platelet function become available that can

potentially be utilized as POC tests or at least within nonspecialized laboratories. There is much renewed interest in determining whether a measurement of platelet function and/or global hemostasis can reliably predict bleeding (e.g., peri- or postoperatively) and/or predict thrombosis within selected patients. With the increasing development of new classes of antiplatelet drugs (see Chapters 60–65) and the known heterogeneity in their biological effects between patients, it may also become useful to monitor an individual's response to antiplatelet therapy so that either the dosage and/or the type of drug or drugs administered can be titrated or optimized within individual patients to help control and minimize the risk of either thrombosis or bleeding.

This section discusses the currently known clinical utility of platelet function tests: screening tools, diagnostic tests, monitoring of antiplatelet therapy, predicting bleeding, predicting thrombosis, aids in transfusion medicine, and monitoring prohemostatic therapy.

## A. Screening Tools

When investigating a suspected bleeding disorder, clinicians must obtain a detailed clinical history and perform a physical examination (as described in detail in Chapter 45). If the patient is suspected to have a hemostatic defect, then the clinical history is usually very informative in the differentiation of a platelet defect from a coagulation defect (see Table 45-4 in Chapter 45). A typical panel of screening tests includes a full blood count; a blood smear (especially if there are abnormalities or flags from the blood counter); coagulation tests such as the activated partial thromboplastin time (APTT), prothrombin time (PT), and thrombin time (TT); the bleeding time or an *in vitro* equivalent (e.g., the PFA-100®, but see the important disclaimers mentioned later); and a von Willebrand factor (VWF) screen (ristocetin cofactor assay, VWF antigen, and factor VIII coagulant). The full blood count is essential in the workup of patients because modern blood counters can detect abnormalities in platelet number, platelet size distribution, or platelet volume (e.g., macrothrombocytopenia), and other problems with both red cells and white cells that may give rise to an acquired platelet defect (e.g., myelodysplasia). If abnormalities in either platelet count, size (MPV), or distribution are flagged by the instrument, then a blood smear should be examined to confirm any defects in platelet size and granule content, as well as any abnormalities in the red cells (e.g., schistocytes in thrombotic thrombocytopenic purpura/ hemolytic uremic syndrome) or leukocytes (e.g., neutrophil inclusions). Because von Willebrand disease (VWD) is the most common bleeding diathesis and presents with bleeding symptoms similar to (rarer) platelet defects as well, it can be eliminated or diagnosed by performing the three previously listed VWF tests. The APTT, PT, and TT screens will determine whether the patient has a coagulation defect. If so, the defect can be confirmed by specific assays for the appropriate coagulation factors.

Until the 1990s, the bleeding time was widely used as a screening tool of platelet function.[2] The bleeding time is simple, measures physiological hemostasis, including the role of the vessel wall components, and does not require expensive equipment or a specialized laboratory (Fig. 23-1). However, despite its relative simplicity, the bleeding time is poorly reproducible, invasive, insensitive to many mild platelet defects, time-consuming, and is not reproducible in repeat or consecutive testing. The bleeding time also does not correlate with the bleeding tendency, and an accurate bleeding history is a more valuable screening test.[21–23] However, in certain situations, the test may still be considered. The bleeding time is discussed in detail in Chapter 25.

The PFA-100® (Fig. 23-5) can be considered to be the equivalent of an "*in vitro* bleeding time" and can be utilized as a limited screening test for some primary hemostatic defects.[63,89,90] The test is simple, rapid, does not require substantial specialist training, and only requires 0.8 mL blood per cartridge. The PFA-100® is a global test of platelet (and VWF) function; it is not specific for any platelet defect, but is independent of the coagulation system. A key question is: Can the PFA-100® now be used as a screening test for platelet disorders and VWD? In other words, does a normal PFA-100® result exclude these disorders and obviate the need for further specialist tests? Although the test is highly sensitive for moderate and severe forms of VWD, there are conflicting data on the sensitivity and specificity to screen for platelet disorders.[90] An important general concept needs to be applied here; a normal test result needs to be interpreted in the light of clinical circumstances. If the pretest probability of a bleeding disorder is low, then a normal PFA-100® result will effectively exclude VWD or a platelet disorder. However, if the pretest probability of a bleeding disorder is high, then specific testing must be performed.[62] Clinicians are well used to applying this sort of Bayesian reasoning when interpreting a normal D-dimer test in suspected venous thromboembolism. Normal PFA-100® CTs do, however, exclude severe platelet defects (e.g., Glanzmann thrombasthenia or Bernard–Soulier syndrome) and moderate to severe VWD. Prolonged CTs can reflect VWD or a platelet defect, and abnormal results require further diagnostic evaluation. The PFA-100® is discussed in detail in Chapter 28.

The Impact® cone and plate(let) analyzer (Figs. 23-6 and 23-7) can detect a range of different platelet disorders and VWD, and may therefore have potential as a screening test. However, clinical experience is still limited with this device and there are few publications on the commercial form of this instrument. The Impact® is discussed in detail in Chapter 29.

## B. Diagnostic Tests

Most laboratories will utilize a panel of screening tests as discussed earlier to determine whether a bleeding patient has a clotting and/or a platelet defect including VWD. As indicated, if the screening tests are all normal but the clinical suspicion is strong for a platelet defect, it is imperative that a complete laboratory work up still be performed. The full diagnosis of any potential platelet disorder depends upon the use of a complex specialized panel of tests including platelet LTA or WBA, platelet nucleotides, and in some cases a measure of platelet release, flow cytometry or, rarely, electron microscopy.

Platelet LTA is still regarded as the gold standard for platelet function testing and remains the most useful technique for diagnosing a wide variety of platelet defects (see Chapter 26). Although most laboratories perform LTA with a panel of different concentrations of classic agonists (e.g., ADP, collagen, epinephrine, arachidonic acid, ristocetin), the exact range of agonists and their concentrations remain poorly standardized.[26,27] LTA is not 100% sensitive to storage pool and release defects, and so laboratories should ideally measure stored and released nucleotides simultaneously.[91,92] Either high-performance liquid chromatography or bioluminescence of lysed platelet preparations at standardized counts are the two most popular assays of platelet nucleotides, although flow cytometric analysis of mepacrine uptake and release from granules has also been used.[93,94] Some aggregometers are particularly useful because they can provide simultaneous measurement of ATP luminescence during the aggregation response and, as expected, demonstrate the release reaction during secondary aggregation. The advantage of this method is that any defects in storage or release can be simultaneously determined along with the LTA tracing.[95] Release reactions can also be measured by a variety of tests based upon the measurement of platelet granule constituents (e.g., uptake and release of radiolabeled serotonin, platelet factor 4, β-thromboglobulin, ADP/ATP) before and after degranulation with a strong platelet agonist. Many laboratories do not routinely assess release reactions and, if relying upon LTA alone, may therefore not detect all release defects, which according to some authors may be surprisingly common.[27,96]

Flow cytometry has now become a vital specialist tool not only for assessing platelet defects but for studying many aspects of platelet biology (see Chapter 30 for further details). Flow cytometry is most commonly used for the measurement of platelet surface glycoproteins and can definitively determine whether any individual glycoprotein is absent or reduced (e.g., integrin αIIbβ3 in Glanzmann thrombasthenia, GPIb-IX-V in Bernard–Soulier syndrome).[97] Modern assays can now provide accurate estimates of platelet glycoprotein densities and can therefore be useful for the diagnosis of heterozygous conditions. With the appropriate monoclonal antibodies, one can determine whether other key platelet surface receptors are deficient or low in number (e.g., the collagen receptors GPIa-IIa and GPVI, although defects in these receptors are very rare). However, some receptors have copy numbers near to or below the sensitivity of the instrument (i.e., 500 copies/platelet), so other techniques (e.g., radioligand binding assays in the case of the ADP receptor P2Y$_{12}$) may have to be used to specifically study an individual receptor.[98] Flow cytometry can also be used for studying platelet activation responses to agonists (e.g., ADP, thrombin, or thrombin receptor activating peptide) by measuring the exocytosis of granule markers (e.g., P-selectin, CD63) and changes in the conformation of the integrin αIIbβ3 (either directly via monoclonal antibody PAC-1 or indirectly via monoclonal antibodies to ligand-induced binding sites or receptor-induced binding sites).[86] It is therefore theoretically possible also to measure the platelet release reaction, but in both dense granular (e.g., storage pool disease) and α-granular disorders (gray platelet syndrome) some of the membrane markers may still be present and detectable after platelet activation.[99,100] Mepacrine uptake into resting and activated platelets can provide an alternative means to potentially diagnose these disorders.[93,94] Flow cytometry is also particularly useful for monitoring the exposure of negatively charged phospholipids and the generation of procoagulant microvesicles for the diagnosis of Scott syndrome and related disorders.[101,102] Full details of the flow cytometric analysis of platelets is provided in Chapter 30.

Various electron microscopic techniques also provide detailed analysis of the cytoplasmic organelles and granules of platelets. Most organelles are visualized in thin sections of fixed, plastic embedded platelets.[103] However, dense granules can also be easily identified as a single population in unfixed and unstained whole-mount preparations.[103] The latter technique is therefore particularly useful for studying platelets from patients with dense granular defects. Examples of platelet disorders diagnosed by electron microscopy are shown in Chapter 3 (Epstein syndrome in Fig. 3-28, Medich syndrome in Fig. 3-39, and gray platelet syndrome in Fig. 3-43) and Chapter 57 (May–Hegglin anomaly, Epstein syndrome, and gray platelet syndrome in Fig. 57-1).

### C. Monitoring Antiplatelet Therapy

The antiplatelet drug aspirin has traditionally been administered at a standard dose with no monitoring of effect, on the assumption that usual doses are two to three times that thought to be required to inhibit all cyclooxygenase 1 (COX-1) activity. However, the lack of a simple, convenient, reliable, and clinically relevant test of platelet function has meant that lack of effect in individual patients has gone

undetected. With the availability of other classes of antiplatelet drugs (e.g., thienopyridines, new P2Y$_{12}$ antagonists, and GPIIb-IIIa antagonists), there is now much interest in the potential utility of platelet function tests to monitor the efficacy of platelet inhibition. The development of GPIIb-IIIa antagonists in particular resulted in the development of a number of new assays to monitor a patient's response (e.g., VerifyNow® IIb/IIIa, flow cytometry of GPIIb-IIIa occupancy), mainly because of their narrow therapeutic window with associated increased risk of bleeding. This, coupled with the now well-studied but poorly defined phenomenon of "drug resistance" (i.e., the failure of a given antiplatelet drug or treatment to prevent an arterial thrombotic event), has led to an explosion of interest, research, and availability of a variety of tests that can potentially monitor an individual's response to antiplatelet therapy.[104] The question remains as to whether these tests are clinically useful, either to predict bleeding or thrombosis. Patient noncompliance with their therapy is also an important but relatively common confounding problem in many studies.[104]

It is well-known that there is considerable variation in the response of individuals (either patients or normal control subjects) to aspirin, clopidogrel, and GPIIb-IIIa antagonists as measured by various platelet function tests. Those individuals who respond poorly to a given drug are therefore termed *resistant*. However, this is a poorly defined phenomenon, and a precise definition of resistance should only relate to the action of a specific drug to inhibit its biochemical target.[105] Many platelet function tests are nonspecific (e.g., the PFA-100®) and they do not do this. Resistance may simply represent a natural biological variation in a given drug response or may be the result of specific or more complicated mechanisms.[106] Is resistance specific to an individual class of drug and related to its mechanism of action, or are there common inherited and/or acquired mechanisms that may influence an individual's response not to just one but potentially all antiplatelet drugs?[106] Whatever the mechanisms, the key question is whether any laboratory tests that detect either resistance or nonresponse predict clinical events. Until these links are firmly proved within large trials, then resistance in the laboratory cannot necessarily be ascribed as a cause of thrombosis. Therefore, except in research trials, it is still not yet clinically useful to test for resistance and change a patient's therapy on the basis of a laboratory test.[104,106,107] The following sections discuss the specific laboratory tests for the three main current choices of antiplatelet drug.

**1. Monitoring Aspirin.** Aspirin irreversibly inhibits COX-1 resulting in the inhibition of thromboxane A$_2$ (TXA$_2$) generation for the entire life span of the platelet.[108] (For more details on the pharmacology of aspirin, see Chapter 60.) Aspirin is an effective antiplatelet agent because it reduces the relative risk of major vascular events and vascular death by about 25% after ischemic stroke and acute coronary syndrome.[109] Regular low doses of aspirin (e.g., 81 mg/day) will result in more than 95% inhibition of thromboxane generation, as shown by arachidonic acid-induced platelet LTA. Therapeutic monitoring was therefore thought to be unnecessary. However, the antiplatelet properties of aspirin have been shown to vary between individuals, and recurrent events in some patients could be the result of "aspirin resistance" or aspirin nonresponsiveness.[105,106] The reported incidence of aspirin nonresponsiveness varies widely (between 5–60%), partly because there is no accepted standard definition based upon either clinical or laboratory criteria. There are also many possible mechanisms for aspirin resistance that have been discussed in detail elsewhere.[106,110] Recently it has been proposed that the term *aspirin resistance* should only be utilized as a description of the failure of aspirin to inhibit $TXA_2$ production, irrespective of a nonspecific test of platelet function.[105] This is because there are many other biochemical pathways that can potentially bypass COX-1 even if this enzyme is inhibited. Depending upon the test system used, "aspirin resistance" or (more correctly) an aspirin nonresponsiveness may therefore be detected even if COX-1 is fully blocked.[105] Recent studies also suggest that, in compliant patients, the incidence of aspirin resistance is rare using methods dependent on COX-1 activity.[77,111] Addition of *in vitro* aspirin to samples followed by retesting should also be an important consideration for testing compliance.[112]

Many tests have been used to assess the influence of aspirin on platelets and aspirin resistance, including arachidonic acid- and ADP-induced LTA (see Chapter 26), ADP- and collagen-induced impedance aggregation (see Chapter 26), VerifyNow® Aspirin (see Chapter 27), PFA-100® (see Chapter 28), Thromboelastography® (TEG–platelet mapping system®), flow cytometry using arachidonic acid stimulation (see Chapter 30), and serum and urinary thromboxane (see Chapter 31).[104] Tests should ideally be performed pre- and postdrug. Some of the tests have been claimed to be predictive of adverse clinical events.[104] However, the large majority of these studies are small and often statistically underpowered to answer completely whether each test can reliably predict the small number of adverse outcomes that were observed in each study.[106,110]

Although preliminary results from some studies could suggest that responses to aspirin should be monitored, there are additional problems in that LTA is time-consuming, difficult, and cannot realistically be performed on large numbers of patients in routine practice. However, the simpler tests of platelet function (e.g., PFA-100®, VerifyNow® Aspirin, TEG platelet mapping®, and urinary thromboxane) could offer the possibility of rapid and reliable identification of aspirin-nonresponsive patients, without the requirement of a specialized laboratory. The PFA-100® usually gives a prolongation in the collagen/epinephrine (CEPI) CT in

response to aspirin, with the collagen/ADP (CADP) CT usually remaining within the normal range.[113,114] A number of studies have observed that an appreciable number of both healthy subjects and patients are "aspirin resistant" or fail to respond in terms of prolongation of their CEPI CT in response to aspirin.[115–122] Because the PFA-100® is a global high-shear test of platelet function, many variables have been shown to influence the CT, including VWF levels, platelet count, and hematocrit levels.[63] In patients identified with "aspirin resistance" by the PFA-100®, a number of studies have shown that VWF levels are elevated in nonresponders compared with responders.[120,121,123] Because the CEPI CT is highly dependent upon VWF and other variables, pre- and postaspirin CTs should ideally be determined, because the true aspirin response may be masked by either short or prolonged CTs before the drug is given.[105] Also CADP CTs are lower in these patients, which may be caused by a combination of high VWF but also increased sensitivity to collagen and ADP as shown by LTA.[120,124,125] It is therefore possible that the apparent increased sensitivity of the PFA-100® to detecting aspirin nonresponsiveness is caused by a combination of these factors, resulting in the normalization of the CT despite adequate COX-1 blockade by aspirin. It is therefore not surprising that the incidence of aspirin nonresponders is reportedly much higher with the PFA-100® than other tests.[126,127] It is likely that the PFA-100® is detecting not only resistance (i.e., failure to inhibit COX-1) but also individuals who are able to give normal CEPI CTs despite adequate COX-1 blockade. The question remains whether either of these groups of patients are at increased risk of thrombosis. Preliminary data suggest that PFA-100® CEPI CTs were noninformative in patients with stable coronary artery disease, in contrast to LTA.[128–131] However, another study suggests that the PFA-100® could be informative,[132] and that shortened CTs with the CADP cartridge (which is not affected by aspirin) may also be predictive.[124,133–135] (For more details, see Section III.E.) Further large prospective studies on the PFA-100® are required.

The VerifyNow® Aspirin assay provides a true POC test for monitoring responses to aspirin (see Chapter 27). The test offers the possibility of rapid and reliable identification of aspirin resistance or nonresponsiveness without the requirement of a specialized laboratory or LTA. Indeed, the test has U.S. Food and Drug Administration approval for monitoring aspirin therapy and is being used by some cardiologists and general practitioners in the United States. The original VerifyNow® Aspirin cartridge contains fibrinogen-coated beads and a platelet activator (metallic cations and propyl gallate) to stimulate the COX-1 pathway and activate platelets.[136] Ideally, the test should produce similar results to those obtained by arachidonic acid-induced LTA. One study showed an 87% agreement with epinephrine-induced LTA.[137] Previous data comparing propyl gallate and other agonists by platelet aggregometry suggest that this agonist detects a lower number of responders in volunteers receiving either 100 mg or 400 mg aspirin.[136] A more recent study compared LTA with VerifyNow® Aspirin and PFA-100® in 100 stroke patients on low-dose aspirin therapy and demonstrated that aspirin nonresponsiveness was not only higher in both POC tests, but that agreement between the tests was poor and few patients were nonresponsive by all three tests.[126] Nevertheless, the VerifyNow® Aspirin test can potentially identify a correlation between aspirin nonresponders, adverse clinical outcomes, and aspirin dose.[138–141] Since the end of 2004, the VerifyNow® Aspirin cartridge has been modified and arachidonic acid has replaced propyl gallate as the principle agonist (see Chapter 27). Further studies are therefore warranted to relate adverse clinical outcomes to the new VerifyNow® Aspirin assay and to determine whether changing therapy based upon the result can also improve outcomes.

Because aspirin inhibits COX-1, measurement of $TXA_2$ and its metabolites either within serum or urine provides a potentially relatively simple way to monitor aspirin therapy. *In vivo*, $TXA_2$ is rapidly converted into the more stable and inert metabolite $TXB_2$, which is further metabolized to 11-dehydro $TXB_2$ — the major product found in urine. Measurement of $TXB_2$ by various immunoassays can facilitate an indirect assessment of the capacity of platelets to form $TXA_2$. Assays can be standardized so that $TXB_2$ is measured either within serum derived from whole blood clotted for 30 minutes at 37°C or in supernatants derived from PRP or purified platelets (with standardized platelet counts) activated by agonists to stimulate COX-1 activity. The metabolite 11-dehydro $TXB_2$ can also be measured within urine samples, and the assay is also commercially available as the AspirinWorks® test. This assay has the advantage that it is noninvasive, and one large study suggested that high levels of urinary 11-dehydro $TXB_2$ are associated with adverse clinical events in patients receiving low-dose aspirin.[142] Further details on $TXA_2$ measurements are provided in Chapter 31.

**2. Monitoring Clopidogrel.** Clopidogrel (Plavix) is a prodrug that is metabolized by cytochrome P450 in the liver to an active metabolite that specifically and irreversibly blocks the platelet ADP $P2Y_{12}$ receptor (as discussed in detail in Chapter 62).[143] Platelet inhibition by clopidogrel is both dose and time dependent, and patients are usually given a loading dose of 300 to 600 mg and then maintained on 75 mg/day. The CAPRIE trial showed that clopidogrel prevented more thrombotic vascular events than aspirin (RRR, 8.7%) in patients with known atherosclerosis.[144] The CURE trial showed aspirin plus clopidogrel was 20% more effective than aspirin alone in acute coronary syndromes,[145] but the MATCH study showed equivalence of aspirin plus clopidogrel with clopidogrel alone in patients with ischemic stroke or transient ischemic attack.[146] Combination therapy

is regarded as the gold standard during percutaneous coronary intervention (PCI).[147] However, interindividual variability in platelet response to clopidogrel has been observed,[148] and 5 to 10% of patients still experience acute or subacute thrombosis after coronary stent implantation.[105,149–151] The phenomenon of "clopidogrel resistance" has been estimated to be between 4% and 30%.[149] The definition of clopidogrel resistance is even more complex than aspirin resistance, because the physiological degree of inhibition detected by ADP-induced LTA can vary widely between individuals, especially because ADP can also activate platelets via a second receptor, $P2Y_1$, and there is interindividual variability of cytochrome P450 activity.[143] There is an inverse correlation between P450 3A4 activity and platelet aggregation, and other drugs can either promote or inhibit metabolism to a certain degree.[152] Preexisting variability in ADP responsiveness is also an important variable and may provide the explanation for response variability.[153] Many mechanisms of clopidogrel resistance have also been proposed, some of which are similar to aspirin.[104]

Laboratory responses to clopidogrel and other $P2Y_{12}$ inhibitors are largely based upon monitoring ADP-stimulated responses.[154] Platelets are stimulated with ADP, and responses are monitored using either LTA, the VerifyNow® P2Y12 assay, the TEG platelet mapping system® or flow cytometric analysis of activation-dependent markers (e.g., P-selectin, PAC-1), flow cytometric analysis of intracellular signaling by monitoring vasodilator-stimulated phosphoprotein (VASP), or Plateletworks®.[36,40,143,154] Ideally, responses are monitored pre- and postdrug, although this is not always possible in the clinical world. LTA using 5 μM or 20 μM ADP can be used to classify patients arbitrarily into three categories — nonresponders, intermediate responders, and responders — based upon measuring the change in (δ) aggregation at baseline and postdrug.[155] Nonresponders can be defined with a δ aggregation of less than 10%. Studies have shown that there is considerable variation in patient response to clopidogrel, and up to 30% of patients may be nonresponders. The largest analysis so far has found 4% of 544 patients to be hyporesponsive to clopidogrel.[148] More recent data suggest that a proportion of patients are probably underdosed and that a 600-mg loading dose significantly reduces the number of nonresponders when compared with 300 mg.[155–157] There is still the critical unresolved question regarding whether *in vitro* lack of responsiveness to clopidogrel correlates with the increased incidence of adverse events.

ADP-induced LTA is probably not very practical to test on large numbers of clinical samples outside a research setting. Also, because residual $P2Y_1$ function can potentially vary widely despite $P2Y_{12}$ inhibition, this could not only explain some of the heterogeneity observed with LTA, but it suggests that ADP alone may not be specific enough to measure the effect of clopidogrel and other $P2Y_{12}$

antagonists.[158] Despite these problems, Matetzsky and colleagues[159] found, in a small study, evidence that ADP-induced LTA predicted adverse events and that this assay also correlated with epinephrine-induced LTA and the cone and plate(let) analyzer. The VerifyNow® instrument was originally designed to overcome the major limitations of LTA and can be used as a POC test (see Chapter 27). The VerifyNow® P2Y12 cartridge has become available for monitoring clopidogrel and other $P2Y_{12}$ antagonists. The assay uses prostaglandin (PG) $E_1$ in addition to ADP to increase intracellular cyclic adenosine monophosphate, theoretically enhancing the sensitivity and specificity of the test for ADP-induced activation of platelets via $P2Y_{12}$.[160,161] $PGE_1$ should suppress the activation of platelets by $P2Y_1$. The VERITAS (Verify Thrombosis Risk Assessment) trial will determine whether the VerifyNow® $P2Y_{12}$ test is a reliable and sensitive measure for monitoring clopidogrel therapy, although the exact cutoff in this assay remains to be defined.

The combination of ADP and $PGE_1$ is also used in the flow cytometric-based VASP assay (BioCytex, Marseilles, France).[158,162] The principle of this assay is to measure the phosphorylation of VASP, which is theoretically proportional to the level of inhibition of the $P2Y_{12}$ receptor (see Chapter 30). Comparison of the VASP assay with LTA shows that the level of inhibition is higher in the flow cytometry assay, because nonspecific aggregation can occur via ADP stimulation of $P2Y_1$ during aggregation.[158] Recent data indeed show that the phosphorylation of VASP correlates with inhibition of LTA, but not platelet surface expression of P-selectin or the PFA-100® CT.[154] The latter test, with which variable results have been observed, is considered unsuitable for monitoring clopidogrel.[63,163–165] Theoretically the PFA-100® CADP cartridge may be more suitable for monitoring $P2Y_{12}$ antagonists than the CEPI cartridge, but both collagen activation and ADP acting through the $P2Y_1$ receptor, along with the high shear conditions, may be normally sufficient to largely overcome $P2Y_{12}$ blockade.[143] There may be also be a degree of time and dose dependence. It has also been observed that there is synergy with clopidogrel/aspirin combination therapy, demonstrated by the prolongation of both CADP and CEPI CTs.[166,167]

Assessment of platelet function by a variety of tests in correlation with clinical outcomes will also be necessary to define responsiveness to clopidogrel and other $P2Y_{12}$ antagonists. Preliminary data from the CREST (Clopidogrel Resistance and Stent Thrombosis) study by Gurbel and associates[168] show differences with VASP, LTA, and activated GPIIb-IIIa responsiveness to ADP between patients with and without subacute stent thrombosis (SAT).[168] Comparing data from patients with (n = 20) and without SAT (n = 100) suggest that clopidogrel response variability to ADP is significantly associated with an increased risk of SAT.[168] This, coupled with other studies on postdischarge and post-PCI events, suggests that high posttreatment

*ex vivo* reactivity to ADP may indeed be an important risk factor for adverse clinical events.[159,162,169]

Carefully controlled, large, randomized trials will be required to define an inadequate response to $P2Y_{12}$ inhibition for an individual test and to show that this correlates with adverse clinical events. Without such data, therapy should not be altered based upon the results of any of the tests that purport to determine responsiveness to a $P2Y_{12}$ antagonist. The new RESISTOR (Research Evaluation to Study Individuals Who Show Thromboxane or $P2Y_{12}$ Resistance) trial that is currently under way in 600 PCI patients may determine whether the level of $P2Y_{12}$ inhibition correlates with clinical outcome and whether changing therapy in resistant patients improves outcome.

The development and clinical application of thienopyridines such as clopidogrel has proved that the $P2Y_{12}$ inhibitor is an attractive target for the development of new drugs. Because thienopyridines are metabolized to their active derivatives by the liver, a number of direct antagonists have also been developed (e.g., cangrelor and AZD6140).[143] Some new thienopyridines (e.g., prasugrel) have also been developed that exhibit superior properties (e.g., higher efficacy, faster onset, and longer duration of action) over clopidogrel (see Chapter 61 for further details).[143] Because some of the observed interindividual heterogeneity of clopidogrel responsiveness may be caused by differences in liver metabolism, it will be interesting to determine whether the incidence of nonresponsiveness is lower or even eradicated with these new drugs, and whether high posttreatment reactivity to ADP remains a potentially significant problem.

**3. Monitoring GPIIb-IIIa Antagonists.** The identification of the importance of the GPIIb-IIIa complex (integrin $\alpha IIb\beta 3$) in mediating platelet aggregation (i.e., the final common pathway of platelet activation) suggested that this receptor would be another attractive target for antithrombotic therapy. The platelet GPIIb-IIIa antagonists (abciximab, tirofiban, and eptifibatide) have now become an important class of antiplatelet agents that are widely used for the prevention of thrombotic complications in patients undergoing PCI or presenting with acute coronary syndromes (see Chapter 62 for details). Early observations of the inhibition of thrombus formation within animal models not only established a strong correlation between the level of GPIIb-IIIa blockade and the prevention of thrombus formation, but demonstrated steep dose–response curves.[170,171] It became rapidly apparent that a certain level of GPIIb-IIIa inhibition was required for the optimal efficacy of GPIIb-IIIa antagonists. This strongly suggested that monitoring of platelet inhibition could be important in patients treated with these agents. Monitoring GPIIb-IIIa antagonists can be performed by a variety of tests, including LTA, WBA, flow cytometry, and radiolabeled antibody binding assays.[172] However, some of these tests are time-consuming, expensive, and are usually performed within specialized laboratories. Given the widespread clinical use of these GPIIb-IIIa antagonists in cardiology, there existed a demand for a simple, inexpensive, and rapid method that could be utilized as a POC test either at the bedside or in the clinic, so that the degree of GPIIb-IIIa blockade could also be potentially determined by nonspecialists. The VerifyNow® system (see Chapter 27 and the previous discussion in this chapter for a full description of the instrument) was originally developed to meet this demand. The assay principle was developed based upon experiments using fibrinogen-coated beads and thrombin receptor activating peptide, which facilitated the rapid visual analysis of the degree of GPIIb-IIIa blockade (see Fig. 27-1A in Chapter 27).[31] The basis of the VerifyNow® IIb/IIIa assay is that fibrinogen-coated beads will agglutinate in whole blood in direct proportion to the degree of platelet activation and GPIIb-IIIa exposure.[29] The presence of a GPIIb-IIIa antagonist will therefore decrease the amount of agglutination in proportion to the level of inhibition achieved (see Chapter 27 for further details).

Initial *in vitro* evaluations of the VerifyNow® IIb/IIIa assay demonstrated good correlations with either LTA in PRP or radiolabeled receptor binding assays.[29] Studies in patients receiving either abciximab or other GPIIb-IIIa antagonists also demonstrated good correlations with LTA.[173,174] A slightly modified Plateletworks® POC assay was recently reported to correlate more strongly than VerifyNow® IIb/IIIa with LTA in measuring platelet inhibition by GPIIb-IIIa antagonists.[37] GOLD (AU-Assessing Ultegra), a large prospective multicenter study, showed a significant association between the level of platelet inhibition by the VerifyNow® IIb/IIIa assay and clinical outcomes.[175] This suggests that the device has clinical utility, although no study has yet been performed to determine whether titration of GPIIb-IIIa therapy based upon the VerifyNow® IIb/IIIa test result decreases adverse events. The PFA-100 has also been utilized to monitor GPIIb-IIIa blockade and correlates well with LTA and receptor occupancy measurements.[176–178] Although many patients give nonclosure or less than a 300-second CT in the PFA-100 after GPIIb-IIIa antagonist treatment, one study suggests that failure to observe nonclosure may be associated with an increased risk of cardiac events.[178] This warrants further investigation.

### D. Predicting Bleeding

Approximately 3 to 5% of patients undergoing surgery have either an acquired or congenital platelet defect, or VWD.[179,180] Some types of surgery may result in a postoperative hypocoagulable state (e.g., cardiopulmonary bypass; see Chapter 59), even though most surgery is procoagulant. The situation may also be complicated by the use of pharmacological thromboprophylaxis. A key question is: Do preoperative and/or perioperative platelet function or global hemostatic tests predict either bleeding or thrombosis within individual patients? It has been estimated that more than 90% of all bleeding times performed in the United States are used preoperatively to try to predict bleeding risk. However, most data suggest that the bleeding time is uninformative in this regard in the absence of a clinical history of bleeding (see Chapter 25).

A separate use of perioperative POC tests might be to guide blood component transfusion in patients who are bleeding. The principle reasons for perioperative hemostatic failure are (a) a preexisting, undetected acquired or inherited bleeding disorder; (b) the procedure itself resulting in changes in hemostasis (e.g., cardiopulmonary bypass); (c) massive surgical blood loss; and (d) the coexistence of other pathologies. Management of potential bleeding or thrombotic episode therefore consists of (a) identifying patients at high risk of bleeding or thrombosis before surgery; (b) understanding the changes in hemostasis that may occur during the type surgery to be performed; (c) giving pharmacological treatment or transfusions as therapy either before, during, or after surgery; (d) reducing unnecessary transfusions and associated risks; and (e) monitoring hemostasis with one of a variety of POC tests, but also understanding their limitations.[181]

As with screening for inherited defects of primary and secondary hemostasis, a careful bleeding history, examination, and recent drug history are probably the key elements to the detection of a congenital or acquired platelet defect prior to a surgical procedure.[182] If the clinical suspicion of a hemostatic defect is strong, then the diagnosis can be confirmed using the traditional panel of platelet function and coagulation tests described earlier. The type and duration of surgery is relevant. Patients undergoing low-risk surgery may not require any screening, whereas those at moderate/high risk may require a preoperative check of platelet count, coagulation, and even platelet function.[181] Correct diagnosis would then facilitate the necessary preoperative normalization of hemostasis prior to surgery. The most common acquired defects encountered are those associated with antiplatelet drugs (e.g., aspirin and clopidogrel). There are conflicting data regarding whether continuing antiplatelet therapy will increase peri- and postoperative blood loss, whereas discontinuation of antiplatelet therapy may increase the risk of thrombosis.

There is currently increasing emphasis on using POC devices to monitor hemostasis during complex surgical procedures. An ideal test would be inexpensive, reliable, and rapid, and would provide a global assessment of both primary and secondary hemostasis and fibrinolysis. This is because traditional coagulation tests have been shown to correlate with either bleeding or thrombosis during surgery. Global tests such as TEG®, ROTEM®, and the Sonoclot® are therefore very popular as POC tests within surgical departments.[183-185] Use of these instruments in monitoring perioperative hemostasis has resulted in the development of algorithms for the appropriate transfusion of blood products and pharmacological intervention in high-risk patients. Some randomized studies have shown that such algorithms, if applied correctly, can not only decrease the frequency of transfusions, but can decrease postoperative bleeding.[186-190] Some of the new POC platelet function tests are also being increasingly used in surgical departments, including the PFA-100®, HAS®, VerifyNow®, HemoStatus® test, and Plateletworks®.[183,191] Conflicting reports on the utility of some of these instruments exist in the literature, so much further work is warranted to establish their potential promise in prediction of bleeding and monitoring of therapy.[71,179,183,191-197]

### E. Platelet Hyperfunction and Prediction of Thrombosis

The evidence for the role of platelets in arterial thrombosis is conclusive.[4] Given the availability of many new tests of platelet function, there is now much renewed interest in testing platelet hyperreactivity or enhanced platelet function and whether this may reliably predict increased risk of thrombosis. A simple, reliable test could be of great value in clinical practice if subjects at increased risk of thrombosis could be identified. Hyperreactivity may also modify bleeding risk in patients with either primary or secondary hemostatic defects. There have been few prospective studies on whether any existing platelet function test can predict thrombosis, especially stroke.[198] Two small studies have suggested that LTA could be predictive of cardiovascular events.[199,200] However, a much larger study in 740 patients followed for 16 years suggested that LTA with either ADP or epinephrine did not predict adverse events.[201] The Caerphilly Cohort Study of Heart Disease and Cognitive Decline studied three different tests of platelet function (LTA, WBA, and the O'Brien filterometer retention test) in a very large number (n = 2000) of older men.[198] In the first examination of the data, LTA using ADP was shown not to be predictive of ischemic heart disease.[202,203] After 10 years of extended follow-up there were approximately 200 patients with acute myocardial infarction (MI) and 100 with ischemic stroke, and neither LTA, WBA, nor the retention test were predictive of MI.[198] Paradoxically, one fifth of the men with the smallest response to ADP in LTA and WBA showed the highest incidence of ischemic stroke.[198] The Northwick Park study also demonstrated that both platelet count and LTA with either ADP or epinephrine showed no association with ischemic heart disease.[201] Although the bleeding time is now considered to be insensitive to the detection of mild platelet function defects, there is some evidence to suggest that the bleeding time can be significantly shorter in patients with

arterial thrombosis (e.g., MI and unstable angina) when compared with normal subjects.[204] Similar studies have also suggested a relationship between MPV, the bleeding time, and the risk of thrombosis.[205] However, the bleeding time was also used in a prospective study within a subset of 1319 men in the Caerphilly study, and the bleeding time did not predict MI or ischemic stroke.[206] Given these results and the inherent problems with the bleeding time (see Chapter 25), it is unlikely that this test has a role in detecting a hyperreactive phenotype.

LTA, although providing the gold standard for the detection and diagnosis of platelet hypofunction for the last 40 years, has not been used to study enhanced platelet responses except in the few trials mentioned here. LTA has been used to identify patients who exhibit spontaneous platelet aggregation (without addition of agonists), sometimes, rather unscientifically, referred to as *sticky platelet syndrome*.[207,208] Although there have been reports of clinical thrombosis in some of these hyperreactive individuals, there have been no prospective studies to assess whether LTA could be reliably used to predict thrombosis. The large Caerphilly and Northwick Park studies suggest that LTA cannot predict adverse events.[198] However a study by Yee and coworkers[209] has recently revisited this question. These investigators studied platelet hyperreactivity in 359 healthy individuals by LTA using a variety of agonists including ADP, epinephrine, collagen, collagen-related peptide, and ristocetin. The healthy subjects exhibited considerable interindividual variability, especially at submaximal dosages of agonists. Intra-individual variability was assessed within a subset of the group (n = 27) on four different occasions and showed that results were not only consistent but that there was agreement between different agonists. These assays may therefore be suitable for the detection of increased platelet reactivity,[209] although LTA cannot be widely performed on large numbers of clinical samples.

Do any of the newer platelet function tests reliably detect platelet hyperfunction and, if so, do they predict clinical outcomes? The PFA-100® CEPI cartridge has been shown to be potentially useful for monitoring responses to aspirin therapy as discussed earlier and in Chapter 28. The CADP cartridge, however, is largely insensitive to the detection of aspirin. It is apparent that patients with increased risk of thrombosis can sometimes present with shortened CADP CT.[124] The PFA-100® CT is dependent upon many variables, including VWF, platelet count, and hematocrit, and could be thought of as a global measure of platelet function. So, does a hyperfunctional phenotype have any potential clinical significance? Recent evidence suggests that CADP CTs are significantly shorter in patients with MI, particularly those with ST-segment elevation or when compared with unstable angina, stable coronary disease, and control subjects.[124] In the MI group, patients in the first quartile of the CTs on both CADP and CEPI exhibited a threefold increase in markers of myocardial damage (creatine kinase and troponin) over those patients in the highest quartile. This suggests that the CADP CT is related to the severity of MI. A follow-up study from the same group in 312 patients with acute coronary syndrome has also shown that both shorter CADP and CEPI CTs along with high ristocetin cofactor levels are associated with adverse events.[134] Patients with basal CADP CTs in the lowest quartile had a 26% event rate compared with to an 11% event rate in those patients with CADP CTs in the highest quartile (odds ratio, 2.4; $p = 0.023$).[134] A similar study in 300 stable cardiovascular disease patients also demonstrated that patients with a CADP CT ≤ 91 seconds had an event rate of 18.6% compared with 6.4% in patients with a CADP CT of more than 91 seconds (odds ratio, 3.34; 95% confidence interval, 1.59–7.03).[135] However, in the same group, LTA aggregometry using either arachidonic acid or four concentrations of ADP between 1 to 10 μM was not predictive. An earlier small study in 163 patients showed that there was a reduction in CADP CT (>15.5 seconds) in response to exercise in patients with coronary artery disease compared with an increase (of 12.5 seconds) in control subjects.[133] These studies suggest that the PFA-100® CADP CT may be useful for predicting adverse events in high-risk groups. Large prospective trials are therefore warranted, because a platelet function test that predicts MI would enable risk stratification and modulation of therapy.

A variety of platelet activation markers measured by flow cytometry and assay of plasma samples *ex vivo* also provides evidence for increased *in vivo* platelet activation in a variety of clinical conditions including acute coronary syndrome and stroke.[104,210] These markers could also potentially be used to determine optimal antiplatelet therapy. Whether these markers can be used to predict adverse events will need to be assessed in large prospective trials. For further details, see Chapter 30.

### F. Platelet Function Testing in Transfusion Medicine

Platelet function tests can be potentially used in transfusion medicine to (a) defer blood or platelet donors with defective platelet function, (b) study the platelet storage lesion and the quality of platelet concentrates, (c) determine whether recipients require a platelet transfusion, and (d) determine the efficacy of a platelet transfusion.[211] Again, advances in this area have been hampered by the past unavailability of reliable, simple tests that provide clinically relevant information to any of the potential applications mentioned earlier. Historically, platelet function tests have not been routinely utilized in any of the previous applications, although this is now becoming a very active area of research. In contrast, many institutions use platelet counting to monitor donors, concentrates, and recipients, and TEG® has also been extensively used as a guide to transfusion decision making. Given the cost of platelet concentrate production and the significant risks associated with transfusion (Chapter 69), relevant platelet function tests may not only be important for the quality control of platelet concentrates, but could also be a significant aid in the decision of whether a platelet transfusion should be given at all. Once a transfusion is given, then (ideally) the degree of improvement in platelet function could be monitored to aid in the future management of the recipient.

In current transfusion practice, donors of platelet concentrates are not routinely tested for platelet function. Normal platelet function is usually assessed by a lack of bleeding history in combination with routine questionnaires. Although this procedure will eliminate donors with severe platelet defects, it is also designed to question donors about recent drug intake that may interfere with normal platelet function (e.g., aspirin or other nonsteroidal antiinflammatory drugs). However, there is increasing evidence that a significant proportion of platelet donors fail to declare their intake of COX-1 inhibitors.[212,213] Although this effect will be masked or diluted within pooled buffy coat platelet concentrates, this will not be the case with single-donor plateletpheresis concentrates that are being now being prepared with increasing frequency. Although transfusion practice guidelines recommend that donors should be deferred from donating platelets if they have taken antiplatelet medication in the previous 3 to 10 days, current practice is failing to identify some of these donors. However, until there is evidence to show negative consequences of this either within the concentrate or the recipients, this will probably remain the case. Two recent studies have shown that a significant proportion of platelet donors exhibit an aspirinlike defect as detected by the PFA-100®.[212,213] The consequences of this are largely unexplored, but if the plateletpheresis concentrates remain defective for their entire shelf-life, then this may affect their clinical efficacy and ability to prevent bleeding in recipients. A cost-effective platelet function POC test to screen donors could theoretically provide a means to prevent defective concentrates from being transfused.

The platelet storage lesion has been identified and characterized using a variety of platelet function tests including LTA and flow cytometry.[211] Global tests that are dependent upon whole blood are not well suited to studying platelet concentrates, because the whole blood has to be reconstituted before testing can be performed (e.g., PFA-100® or WBA).[214] Platelet concentrates are currently not tested routinely, even though their function is known to diminish during storage. This is because platelet concentrates are thought to remain sufficiently clinically viable at room temperature for up to 5 days (see Chapter 69).

Platelet counting and, to a lesser extent, TEG® testing in the patient are often used to direct platelet transfusion decision making. In the stable untraumatized patient, the platelet transfusion threshold is usually now set at $10 \times 10^9$/L (see Chapter 69). This threshold may be higher (approximately $50 \times 10^9$/L) in patients undergoing surgery. One of the problems with many routine full-blood counters is that they become inaccurate after the platelet count is less than $20 \times 10^9$/L. A recent UK study suggests that many transfusion decisions are made based on inaccurate counts.[17,216] The data suggest that patients are being undertransfused, because platelet counts are often overestimated around typical transfusion thresholds by all available counting methods (except immunocounting). Improved quality control of platelet counts at this level is therefore warranted, along with improved calibration, in an attempt to rectify this problem.

Platelet function tests in the patient have not been used to direct transfusion decisions because it is often difficult to determine accurate platelet function in severe thrombocytopenia, especially using LTA. However, a recent small study suggests that the PFA-100® could also be used to direct transfusion decision making, even in patients with severe thrombocytopenia, and to monitor improved platelet function posttransfusion.[217] Larger trials are warranted with a variety of the newer tests to determine whether platelet function can be reliably monitored in recipients of platelet transfusions, and whether this improves their management and clinical outcome.

### G. Monitoring Prohemostatic Therapy

Although platelet transfusions are often used to treat clinical bleeding in both acquired and inherited platelet disorders, there are several other therapeutic agents that can also be utilized that are not associated with the risks of transfusion. These prohemostatic therapies that can be used to increase platelet number or function include platelet growth factors (see Chapter 66), desmopressin (see Chapter 67), and recombinant factor VIIa (rFVIIa; see Chapter 68). Platelet function tests could potentially be used to monitor these therapies. As with antiplatelet therapy, platelet function tests have not traditionally been used to monitor prohemostatic therapy, although this is beginning to change with the availability of simple-to-use and POC tests. The important question is whether any of these agents can be reliably monitored with tests that reliably predict clinical outcome (i.e., the prevention or cessation of bleeding).

Desmopressin (DDAVP) is a particularly useful agent because it mediates an immediate release of the high-molecular weight VWF stores from the vascular endothelium into the circulation. Desmopressin has been traditionally used in patients with mild hemophilia A and VWD for the prevention and treatment of bleeding. However, the clinical use of desmopressin has been extended to the treatment of congenital and acquired platelet defects and for the prevention of bleeding during surgery (see Chapters 59 and 67 for

details). Desmopressin has been shown to normalize and/or shorten the bleeding time in most cases of hemophilia A, mild VWD, and many types of congenital platelet defects (see Chapter 67). However, whether the effect of desmopressin on the bleeding time or any other platelet function test corresponds with in vivo prohemostatic efficacy is not well established. The high shear conditions of the PFA-100® should theoretically make it very sensitive for the assessment of the contribution of ultralarge VWF multimers released by desmopressin therapy.[63] The PFA-100® therefore offers a potential rapid test to discriminate between good and poor desmopressin responders.[218] Desmopressin significantly shortens both CADP and CEPI CT in healthy individuals. In type I VWD, prolonged CEPI and CADP CTs show correction after desmopressin-induced increases in plasma VWF, but not in patients with discordant or low levels of platelet VWF.[219,220] Direct comparison with the bleeding time suggests that the PFA-100® is more sensitive at detecting abnormalities in type I VWD that are correctable after desmopressin therapy.[220] In severe VWD, PFA-100® CT is very prolonged but often does not correct with VWF concentrate therapy, possibly because of abnormalities in concentrate multimer profile and/or lack of intraplatelet VWF.[221] Correction of prolonged CTs with increased plasma VWF levels after desmopressin therapy has also been reported in small studies of patients with storage pool disease and primary platelet secretion defects.[222] CT shortening also occurs in healthy individuals given desmopressin after antiplatelet agents.[223] Although the PFA-100® can thus detect the prohemostatic effects of desmopressin given to treat or prevent bleeding, there have been few studies designed to determine whether shortening of the CT predicts improved clinical outcomes with regard to hemorrhage.[90] However, in one prospective study, 254 of 5649 patients scheduled for surgery were identified with acquired or inherited defects of hemostasis.[179] Of these 254 patients, 90% responded to preoperative treatment with desmopressin as assessed by the PFA-100® and the nonresponders (10%) were given other hemostatic agents. The frequency of blood transfusion was significantly higher in those patients without preoperative correction of the PFA-100® when compared with control subjects.[180] The study suggests that the PFA-100® test could be used to monitor preoperative correction of hemostasis in abnormal patients. Further well-designed trials in this area are warranted, particularly comparing different tests of platelet function in relation to prediction of clinical bleeding and transfusion requirements during and after surgery.

rFVIIa (NovoSeven) has been successfully used to treat patients with inhibitors to factors VIII or IX. High-dose factor rFVIIa infusion is now becoming increasingly useful for the treatment of platelet disorders and thrombocytopenia, and as an alternative to platelet transfusion (see Chapter 68).[224] The mechanism of action of rFVIIa is interesting because its seems to bind to and enhance thrombin generation on the surface of platelets, resulting in the correction of a variety of hemostatic defects, especially when used at high doses.[225–227] Increased *in vitro* platelet adhesion and aggregation by rFVIIa-mediated thrombin formation has been recently observed and could potentially explain the therapeutic effects of rFVIIa in thrombocytopenic conditions and in patients with a normal platelet count.[228] rFVIIa has also been shown to result in increased *in vitro* activation of platelets detected by flow cytometry within whole blood.[229]

Because factor VII levels and shortening of the APTT and PT do not correlate with the *in vivo* response to rFVIIa, an appropriate laboratory test that correlates with clinical response is yet to be found. The question is whether any platelet function tests could be used to monitor the efficacy of rFVIIa therapy. Even though rFVIIa is now widely used to prevent bleeding, there are few reports on whether the bleeding time could be used to monitor therapy. An *in vivo* animal model reported that rFVIIa shortens the bleeding time, decreases *in vivo* bleeding, and does not promote thrombosis.[230] Given the problems with the bleeding time (see Chapter 25), are there any other platelet function tests that could be used to monitor rFVIIa? The PFA-100® could be insensitive to the effects of rFVIIa, because rFVIIa increases the amount of thrombin generation on the platelet surface, and this test is largely coagulation independent.[63] Indeed, rFVIIa does not appear to influence the PFA-100® CT in children with platelet disorders.[231] Perhaps the most promising tests for monitoring rFVIIa are global measures of hemostasis, such as platelet-dependent thrombin generation assays performed in PRP, TEG® and ROTEM®, and the HAS.[83] Increasing doses of rFVIIa have been shown to increase the ETP progressively *in vitro*, but with a variable plateau between patients at supratherapeutic concentrations.[232–234] The platelet-dependent ETP assay may therefore be useful for determining the optimal dose in each patient, because once a plateau is reached, higher doses of rFVIIa have no additional prohemostatic effect.[233] TEG® and ROTEM® have also shown promise, and changes in profiles correlate with clinical outcome.[235,236] Measurement of clot structure using the PCF and TGT parameters within the HAS instrument have also recently been shown to be potentially useful for monitoring rFVIIa.[237]

## IV. Summary and the Future of Platelet Function Testing

Up until the late 1980s, the only clinical platelet function tests that were available were the bleeding time, platelet LTA, and various biochemical assays.[2] These were mainly performed within specialized research and clinical laboratories. Around this time, many researchers began to utilize flow cytometry to study various aspects of platelet biology, and this soon led to the widespread commercial availability of reagents (antibodies and dyes). In a relatively short space of time, flow cytometry has thus become an important tool in both clinical and research laboratories (see Chapter 30). With the recent convergence of flow cytometric and impedance principles within many commercial full-blood counters, an increasing number of fully automated platelet parameters can now be determined within these analyzers without the need of a specialized operator. These parameters include immunological platelet counting (Chapter 24),[16,238] measurement of platelet activation, and determination of the immature platelet fraction.[239,240] Therefore, any newly identified, important clinical parameter that can be measured by flow cytometry could also be potentially measured within a fully automated, modern, full-blood counter that can also measure fluorescence and light scatter.

In the early 1990s, it was realized that the bleeding time was unreliable as a screening test despite its widespread use (see Chapter 25). Although LTA has become an indispensable gold standard test for the diagnosis of many platelet-related disorders, it is well recognized that it does not accurately simulate all aspects of platelet function, and its utility is significantly limited outside of specialized laboratories. Although many researchers were already utilizing flow chambers and microscopy to study platelet behavior under conditions that simulate *in vivo* conditions more accurately, these tests were restricted to specialized laboratories and therefore were not ideally suitable for routine clinical applications (see Chapters 18, 32, and 34). This, coupled with the limitations of both LTA and the bleeding time, paved the way for the development of a number of easy-to-use prototype platelet function analyzers. Not surprisingly, there has been a high casualty rate among the first generation of these instruments, because commercialization requires a large capital investment to overcome many hurdles in the development of a reliable clinical test. Nevertheless, a number of different prototypes have now been fully developed into commercial instruments that are widely available to both clinical and research laboratories. These include the PFA-100® (see Chapter 28), VerifyNow® (see Chapter 27), Impact® cone and plate(let) analyzer (see Chapter 29), HAS, Plateletworks®, and modifications of the TEG®. Some of these instruments have U.S. Food and Drug Administration approval for a variety of different applications. Although many of these tests have potential clinical utility, much further research is required to determine whether antiplatelet therapy should be routinely monitored and treatment adapted or titrated based upon the result of a platelet function test. Whether these tests can reliably predict thrombosis or bleeding and monitor prohemostatic therapy in various settings also needs to be determined. Large, randomized, controlled trials will be required to determine the true prognostic and therapeutic value of any existing or new platelet function test.

As stated by Coller in the Foreword to this book, it is ironic that intravital microscopy of platelets, originally described by Bizzozero in 1882, has now become a powerful research tool for studying the role of platelets in thrombosis.[7] With significant advances in microscopy and digital imaging/processing, it is now possible to perform real-time imaging of fluorescently labeled platelets and hemostatic system components during thrombus formation within animal models.[53] This has already resulted in many exciting discoveries about platelets and their dynamic interactions with the vessel wall and coagulation system (see Chapters 33 and 34). Future platelet function instruments could therefore be based upon studying the interaction of fluorescently tagged platelets with collagen-coated surfaces under flow conditions that closely mimic *in vivo* conditions. One such potential example is a kinetic platelet aggregometer that provides sensitive information about the rate of fluorescent platelet adhesion and thrombus stabilization in real time (Fig. 23-11).[241] This type of technology could potentially be developed into multichannel and miniaturized versions. Dynamic studies of platelet adhesion to protein-coated microparticles, and of platelet–platelet aggregation, can also be performed within homogeneous laminar shear flow over a range of 0 to $8000\,s^{-1}$ shear rates, within coaxial cylinder microcoquette devices. Such flow devices consist of two coaxial cylindrical chambers separated by a gap of typically 0.5 mm, with controlled rotation of the inner cylinder generating the desired shear rates. This technology may be particularly useful for studying the action of antiplatelet drugs under a range of shear conditions. Therefore both "closed" devices such as cone-plate and coquettes, as well as "flow-through" devices, are important candidates for future development.[242-246]

Advances in the definition of both the platelet genome and proteome have led to many recent interesting discoveries in platelet biology (see Chapter 5).[247] Utilization of these technologies within specifically designed arrays or chips could also potentially revolutionize not only the diagnosis of inherited defects in platelet function, but possibly also the rapid identification of individuals with increased risk of bleeding or thrombosis. The future of platelet function testing is therefore a particularly interesting area at present, because the development of new technologies and new instruments may have future utility in a variety of different clinical and laboratory settings.

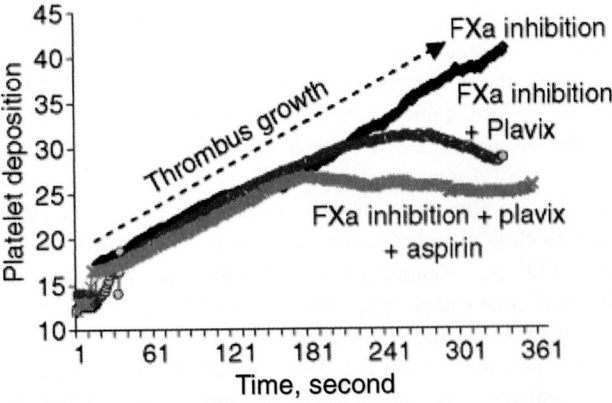

**Figure 23-11.** Synergism between aspirin and $P2Y_{12}$ inhibition in blocking thrombus growth. Blood from a control individual treated with clopidogrel, or an individual treated with aspirin and clopidogrel, was anticoagulated with a factor Xa (FXa) inhibitor, treated with rhodamine 6G to label platelets, and then perfused through a chamber coated with type III collagen at $1000\,s^{-1}$. The continuous accumulation of fluorescence was used to quantify platelet thrombus formation. (From Phillips et al.,[241] with permission.)

# References

1. George, J. N. (2000). Platelets. *Lancet, 355,* 1531–1539.
2. The British Society for Haematology (BCSH) Haemostasis and Thrombosis Task Force. (1988). Guidelines on platelet function testing. *J Clin Pathol, 41,* 1322–1330.
3. Gawaz, M., Langer, H., & May, A. E. (2005). Platelets in inflammation and atherogenesis. *J Clin Invest, 115,* 3378–3384.
4. Ruggeri, Z. M. (2002). Platelets in atherothrombosis. *Nat Med, 8,* 1227–1234.
5. Harrison, P. (2005). Platelet function analysis. *Blood Rev, 19,* 111–123.
6. Harrison, P. (2000). Progress in the assessment of platelet function. *Br J Haematol, 111,* 733–744.
7. de Gaetano, G. (2001). A new blood corpuscle: An impossible interview with Giulio Bizzozero. *Thromb Haemost, 86,* 973–979.
8. Brecher, G., & Cronkite, E. P. (1950). Morphology and enumeration of human blood platelets. *J Appl Physiol, 3,* 365–377.
9. Brecher, G., Schneiderman, M., & Cronkite, E. P. (1953). The reproducibility and constancy of the platelet count. *Am J Clin Pathol, 23,* 15–26.
10. Coulter, W. H. (1953). Means for counting particles suspended in a fluid. US patent no. 2,656,508.
11. Harrison, P., Briggs, C., & Machin, S. J. (2004). Platelet counting. *Methods Mol Biol, 272,* 29–46.
12. Mundschenk, D. D., Connelly, D. P., White, J. G., et al. (1976). An improved technique for the electronic measurement of platelet size and shape. *J Lab Clin Med, 88,* 301–315.
13. Bull, B. S., Schneiderman, M. A., & Brecher, G. (1965). Platelet counts with the Coulter counter. *Am J Clin Pathol, 44,* 678–688.
14. Bessman, J. D., Williams, L. J., & Gilmer, P. R., Jr. (1981). Mean platelet volume. The inverse relation of platelet size and count in normal subjects, and an artifact of other particles. *Am J Clin Pathol, 76,* 289–293.

15. Kunicka, J. E., Fischer, G., Murphy, J., et al. (2000). Improved platelet counting using two-dimensional laser light scatter. *Am J Clin Pathol, 114*, 283–289.

16. Harrison, P., Ault, K. A., Chapman, S., et al. (2001). An interlaboratory study of a candidate reference method for platelet counting. *Am J Clin Pathol, 115*, 448–459.

17. Harrison, P., Segal, H., Briggs, C., et al. (2005). Impact of immunological platelet counting (by the platelet/RBC ratio) on haematological practice. *Cytom B Clin Cytom, 67*, 1–5.

18. Duke, W. W. (1910). The relation of blood platelets to hemorrhagic disease. Description of a method for determining the bleeding time and the coagulation time and report of three cases of hemorrhagic disease relieved by blood transfusion. *JAMA, 55*, 1185–1192.

19. Harker, L. A., & Slichter, S. J. (1972). The bleeding time as a screening test for evaluation of platelet function. *N Engl J Med, 287*, 155–159.

20. Nilsson, I. M., Magnusson, S., & Borchgrevink, C. (1963). The Duke and Ivy methods for determination of the bleeding time. *Thromb Diath Haemorrh, 10*, 223–234.

21. Rodgers, R. P., & Levin, J. (1990). A critical reappraisal of the bleeding time. *Semin Thromb Hemost, 16*, 1–20.

22. Lind, S. E. (1991). The bleeding time does not predict surgical bleeding. *Blood, 77*, 2547–2552.

23. Peterson, P., Hayes, T. E., Arkin, C. F., et al. (1998). The preoperative bleeding time test lacks clinical benefit: College of American Pathologists' and American Society of Clinical Pathologists' position article. *Arch Surg, 133*, 134–139.

24. Born, G. V. (1962). Aggregation of blood platelets by adenosine diphosphate and its reversal. *Nature, 194*, 927–929.

25. O'Brien, J. M. (1962). Platelet aggregation. II. Some results from a new method of study. *J Clin Pathol, 15*, 452–481.

26. Zhou, L., & Schmaier, A. H. (2005). Platelet aggregation testing in platelet-rich plasma: Description of procedures with the aim to develop standards in the field. *Am J Clin Pathol, 123*, 172–183.

27. Moffat, K. A., Ledford–Kraemer, M. R., Nichols, W. L., et al. (2005). Variability in clinical laboratory practice in testing for disorders of platelet function: Results of two surveys of the North American Specialized Coagulation Laboratory Association. *Thromb Haemost, 93*, 549–553.

28. Cardinal, D. C., & Flower, R. J. (1980). The electronic aggregometer: A novel device for assessing platelet behavior in blood. *J Pharmacol Methods, 3*, 135–158.

29. Smith, J. W., Steinhubl, S. R., Lincoff, A. M., et al. (1999). Rapid platelet-function assay: An automated and quantitative cartridge-based method. *Circulation, 99*, 620–625.

30. Kereiakes, D. J., Mueller, M., Howard, W., et al. (1999). Efficacy of abciximab induced platelet blockade using a rapid point of care assay. *J Thromb Thrombolysis, 7*, 265–276.

31. Coller, B. S., Lang, D., & Scudder, L. E. (1997). Rapid and simple platelet function assay to assess glycoprotein IIb/IIIa receptor blockade. *Circulation, 95*, 860–867.

32. Fox, S. C., Burgess–Wilson, M., Heptinstall, S., et al. (1982). Platelet aggregation in whole blood determined using the Ultra-Flo 100 Platelet Counter. *Thromb Haemost, 48*, 327–329.

33. Heptinstall, S., Fox, S., Crawford, J., et al. (1986). Inhibition of platelet aggregation in whole blood by dipyridamole and aspirin. *Thromb Res, 42*, 215–223.

34. Glenn, J. R., White, A. E., Johnson, A., et al. (2005). Leukocyte count and leukocyte ecto-nucleotidase are major determinants of the effects of adenosine triphosphate and adenosine diphosphate on platelet aggregation in human blood. *Platelets, 16*, 159–170.

35. Carville, D. G., Schleckser, P. A., Guyer, K. E., et al. (1998). Whole blood platelet function assay on the ICHOR point-of-care hematology analyzer. *J Extra Corpor Technol, 30*, 171–177.

36. Craft, R. M., Chavez, J. J., Snider, C. C., et al. (2005). Comparison of modified thrombelastograph and Plateletworks whole blood assays to optical platelet aggregation for monitoring reversal of clopidogrel inhibition in elective surgery patients. *J Lab Clin Med, 145*, 309–315.

37. White, M. M., Krishnan, R., Kueter, T. J., et al. (2004). The use of the point of care Helena ICHOR/Plateletworks and the Accumetrics Ultegra RPFA for assessment of platelet function with GPIIB-IIIa antagonists. *J Thromb Thrombolysis, 18*, 163–169.

38. Lennon, M. J., Gibbs, N. M., Weightman, W. M., et al. (2004). A comparison of Plateletworks and platelet aggregometry for the assessment of aspirin-related platelet dysfunction in cardiac surgical patients. *J Cardiothorac Vasc Anesth, 18*, 136–140.

39. Nicholson, N. S., Panzer–Knodle, S. G., Haas, N. F., et al. (1998). Assessment of platelet function assays. *Am Heart J, 135*, S170–S178.

40. Mobley, J. E., Bresee, S. J., Wortham, D. C., et al. (2004). Frequency of nonresponse antiplatelet activity of clopidogrel during pretreatment for cardiac catheterization. *Am J Cardiol, 93*, 456–458.

41. Lakkis, N. M., George, S., Thomas, E., et al. (2001). Use of ICHOR/Plateletworks to assess platelet function in patients treated with GP IIb/IIIa inhibitors. *Catheter Cardiovasc Interv, 53*, 346–351.

42. Wu, K. K., & Hoak, J. C. (1974). A new method for the quantitative detection of platelet aggregates in patients with arterial insufficiency. *Lancet, ii*, 924–926.

43. Grotemeyer, K. H. (1991). The platelet reactivity test: A useful "by-product" of the blood-sampling procedure? *Thromb Res, 61*, 423–431.

44. Koscielny, J., Meyer, O., Kiesewetter, H., et al. (2004). [Platelet reactivity index by Grotemeyer.] *Hamostaseologie, 24*, 207–210.

45. Hellem, A. J. (1960). The adhesiveness of human blood platelets in vitro. *Scand J Clin Lab Invest, 12*(suppl), 1–117.

46. O'Brien, J. R., & Etherington, M. D. (1998). Antiglycoprotein Ib causes platelet aggregation: Different effects of blocking glycoprotein Ib and glycoprotein IIb/IIIa in the high shear filterometer. *Blood Coagul Fibrinolysis, 9*, 453–461.

47. O'Brien, J. R., & Salmon, G. P. (1987). Shear stress activation of platelet glycoprotein IIb/IIIa plus von Willebrand factor causes aggregation: Filter blockage and the long bleeding time in von Willebrand's disease. *Blood, 70*, 1354–1361.

48. Krischek, B., Morgenstern, E., Mestres, P., et al. (2005). Adhesion, spreading, and aggregation of platelets in flowing blood and the reliability of the retention test Homburg. *Semin Thromb Hemost, 31,* 449–457.

49. Krischek, B., Brannath, W., Klinkhardt, U., et al. (2005). Role of the retention test Homburg in evaluating platelet hyperactivity and in monitoring therapy with antiplatelet drugs. *Semin Thromb Hemost, 31,* 458–463.

50. Nowak, G., Wiesenburg, A., Schumann, A., et al. (2005). Platelet adhesion assay: A new quantitative whole blood test to measure platelet function. *Semin Thromb Hemost, 31,* 470–475.

51. Falati, S., Gross, P. L., & Merrill–Skoloff, G., et al. (2004). In vivo models of platelet function and thrombosis: Study of real-time thrombus formation. *Methods Mol Biol, 272,* 187–197.

52. Falati, S., Gross, P., Merrill–Skoloff, G., et al. (2002). Real-time in vivo imaging of platelets, tissue factor and fibrin during arterial thrombus formation in the mouse. *Nat Med, 8,* 1175–1181.

53. Furie, B., & Furie, B. C. (2005). Thrombus formation in vivo. *J Clin Invest, 115,* 3355–3362.

54. Gorog, D. A., & Kovacs, I. B. (1995). Thrombotic status analyser. Measurement of platelet-rich thrombus formation and lysis in native blood. *Thromb Haemost, 73,* 514–520.

55. Li, C. K., Hoffmann, T. J., Hsieh, P. Y., et al. (1998). The Xylum Clot Signature Analyzer: A dynamic flow system that simulates vascular injury. *Thromb Res, 92,* S67–S77.

56. Fricke, W., Kouides, P., Kessler, C., et al. (2004). A multicenter clinical evaluation of the Clot Signature Analyzer. *J Thromb Haemost, 2,* 763–768.

57. Gorog, P., & Ahmed, A. (1984). Haemostatometer: A new in vitro technique for assessing haemostatic activity of blood. *Thromb Res, 34,* 341–357.

58. Yamamoto, J., Yamashita, T., Ikarugi, H., et al. (2003). Gorog thrombosis test: A global in-vitro test of platelet function and thrombolysis. *Blood Coagul Fibrinolysis, 14,* 31–39.

59. Spectre, G., Brill, A., Gural, A., et al. (2005). A new point-of-care method for monitoring anti-platelet therapy: Application of the cone and plate(let) analyzer. *Platelets, 16,* 293–299.

60. Kenet, G., Lubetsky, A., Shenkman, B., et al. (1998). Cone and platelet analyser (CPA): A new test for the prediction of bleeding among thrombocytopenic patients. *Br J Haematol, 101,* 255–259.

61. Varon, D., Lashevski, I., Brenner, B., et al. (1998). Cone and plate(let) analyzer: Monitoring glycoprotein IIb/IIIa antagonists and von Willebrand disease replacement therapy by testing platelet deposition under flow conditions. *Am Heart J, 135,* S187–S193.

62. Harrison, P. (2005). The role of PFA-100 testing in the investigation and management of haemostatic defects in children and adults. *Br J Haematol, 130,* 3–10.

63. Jilma, B. (2001). Platelet function analyzer (PFA-100): A tool to quantify congenital or acquired platelet dysfunction. *J Lab Clin Med, 138,* 152–163.

64. Kundu, S. K., Heilmann, E. J., Sio, R., et al. (1996). Characterization of an in vitro platelet function analyzer: PFA-100. *Clin Appl Thromb Hemost, 2,* 241–249.

65. Kundu, S. K., Heilmann, E. J., Sio, R., et al. (1995). Description of an in vitro platelet function analyzer: PFA-100. *Semin Thromb Hemost, 21*(suppl 2), 106–112.

66. Caen, J. P., & Rosa, J. P. (1995). Platelet–vessel wall interaction: From the bedside to molecules. *Thromb Haemost, 74,* 18–24.

67. Gerotziafas, G. T., Depasse, F., Busson, J., et al. (2005). Towards a standardization of thrombin generation assessment: The influence of tissue factor, platelets and phospholipids concentration on the normal values of Thrombogram–Thrombinoscope assay. *Thromb J, 3,* 16.

68. Al, D. R., Peyvandi, F., Santagostino, E., et al. (2002). The thrombogram in rare inherited coagulation disorders: Its relation to clinical bleeding. *Thromb Haemost, 88,* 576–582.

69. Hemker, H. C., & Beguin, S. (1995). Thrombin generation in plasma: Its assessment via the endogenous thrombin potential. *Thromb Haemost, 74,* 134–138.

70. Coiffic, A., Cazes, E., Janvier, G., et al. (1999). Inhibition of platelet aggregation by abciximab but not by aspirin can be detected by a new point-of-care test, the hemostatus. *Thromb Res, 95,* 83–91.

71. Despotis, G. J., Levine, V., Saleem, R., et al. (1999). Use of point-of-care test in identification of patients who can benefit from desmopressin during cardiac surgery: A randomised controlled trial. *Lancet, 354,* 106–110.

72. Despotis, G. J., Levine, V., Filos, K. S., et al. (1996). Evaluation of a new point-of-care test that measures PAF-mediated acceleration of coagulation in cardiac surgical patients. *Anesthesiology, 85,* 1311–1323.

73. Luddington, R. J. (2005). Thromboelastography/thromboelastometry. *Clin Lab Haematol, 27,* 81–90.

74. Salooja, N., & Perry, D. J. (2001). Thromboelastography. *Blood Coagul Fibrinolysis, 12,* 327–337.

75. Hartert, H. (1951). [Thromboelastography, a method for physical analysis of blood coagulation.] *Z Gesamte Exp Med, 117,* 189–203.

76. Lang, T., Bauters, A., Braun, S. L., et al. (2005). Multicentre investigation on reference ranges for ROTEM thromboelastometry. Blood Coagul Fibrinolysis, 16, 301–310.

77. Tantry, U. S., Bliden, K. P., & Gurbel, P. A. (2005). Overestimation of platelet aspirin resistance detection by thrombelastograph platelet mapping and validation by conventional aggregometry using arachidonic acid stimulation. *J Am Coll Cardiol, 46,* 1705–1709.

78. Bowbrick, V. A., Mikhailidis, D. P., & Stansby, G. (2003). Value of thromboelastography in the assessment of platelet function. *Clin Appl Thromb Hemost, 9,* 137–142.

79. Hett, D. A., Walker, D., Pilkington, S. N., et al. (1995). Sonoclot analysis. *Br J Anaesth, 75,* 771–776.

80. LaForce, W. R., Brudno, D. S., Kanto, W. P., et al. (1992). Evaluation of the SonoClot Analyzer for the measurement of platelet function in whole blood. *Ann Clin Lab Sci, 22,* 30–33.

81. Chandler, W. L., & Schmer, G. (1986). Evaluation of a new dynamic viscometer for measuring the viscosity of whole blood and plasma. *Clin Chem, 32,* 505–507.

82. Carr, M. E., Jr., & Zekert, S. L. (1991). Measurement of platelet-mediated force development during plasma clot formation. *Am J Med Sci, 302,* 13–18.

83. Carr, M. E., Martin, E. J., Kuhn, J. G., et al. (2003). Onset of force development as a marker of thrombin generation in whole blood: The thrombin generation time (TGT). *J Thromb Haemost, 1,* 1977–1983.

84. Carr, M. E., Jr. (2003). Development of platelet contractile force as a research and clinical measure of platelet function. *Cell Biochem Biophys, 38,* 55–78.

85. Carr, M. E., Jr. (1995). Measurement of platelet force: The Hemodyne hemostasis analyzer. *Clin Lab Manage Rev, 9,* 312–318, 320.

86. Michelson, A. D., & Furman, M. I. (1999). Laboratory markers of platelet activation and their clinical significance. *Curr Opin Hematol, 6,* 342–348.

87. Schmitz, G., Rothe, G., Ruf, A., et al. (1998). European Working Group on clinical cell analysis: Consensus protocol for the flow cytometric characterisation of platelet function. *Thromb Haemost, 79,* 885–896.

88. Michelson, A. D. (1996). Flow cytometry: A clinical test of platelet function. *Blood, 87,* 4925–4936.

89. Francis, J. L., Francis, D., Larson, L., et al. (1999). Can the platelet function analyser (PFA-100) substitute for the bleeding time in routine clinical practice. *Platelets, 10,* 132–136.

90. Hayward, C. P., Harrison, P., Cattaneo, M., et al. (2006). Platelet function analyser (PFA-100) closure time in the evaluation of platelet disorders and platelet function. *J Thromb Haemost, 4,* 1–8.

91. Nieuwenhuis, H. K., Akkerman, J. W., & Sixma, J. J. (1987). Patients with a prolonged bleeding time and normal aggregation tests may have storage pool deficiency: Studies on 106 patients. *Blood, 70,* 620–623.

92. Israels, S. J., McNicol, A., Robertson, C., et al. (1990). Platelet storage pool deficiency: Diagnosis in patients with prolonged bleeding times and normal platelet aggregation. *Br J Haematol, 75,* 118–121.

93. Wall, J. E., Buijs–Wilts, M., Arnold, J. T.. et al. (1995). A flow cytometric assay using mepacrine for study of uptake and release of platelet dense granule contents. *Br J Haematol, 89,* 380–385.

94. Gordon, N., Thom, J., Cole, C., et al. (1995). Rapid detection of hereditary and acquired platelet storage pool deficiency by flow cytometry. *Br J Haematol, 89,* 117–123.

95. White, M. M., Foust, J. T., Mauer, A. M., et al. (1992). Assessment of lumiaggregometry for research and clinical laboratories. *Thromb Haemost, 67,* 572–577.

96. Cattaneo, M. (2003). Inherited platelet-based bleeding disorders. *J Thromb Haemost, 1,* 1628–1636.

97. Linden, M. D., Frelinger, A. L., III, Barnard, M. R., et al. (2004). Application of flow cytometry to platelet disorders. *Semin Thromb Hemost, 30,* 501–511.

98. Cattaneo, M., Zighetti, M. L., Lombardi, R., et al. (2003). Molecular bases of defective signal transduction in the plate-let P2Y12 receptor of a patient with congenital bleeding. *Proc Natl Acad Sci U S A, 100,* 1978–1983.

99. Mazurov, A. V., Vinogradov, D. V., Khaspekova, S. G., et al. (1996). Deficiency of P-selectin in a patient with grey platelet syndrome. *Eur J Haematol, 57,* 38–41.

100. McNicol, A., Israels, S. J., Robertson, C., et al. (1994). The empty sack syndrome: A platelet storage pool deficiency associated with empty dense granules. *Br J Haematol, 86,* 574–582.

101. Solum, N. O. (1999). Procoagulant expression in platelets and defects leading to clinical disorders. *Arterioscler Thromb Vasc Biol, 19,* 2841–2846.

102. Horstman, L. L., & Ahn, Y. S. (1999). Platelet microparticles: A wide-angle perspective. *Crit Rev Oncol Hematol, 30,* 111–142.

103. White, J. G. (2004). Electron microscopy methods for studying platelet structure and function. *Methods Mol Biol, 272,* 47–63.

104. Michelson, A. D. (2004). Platelet function testing in cardiovascular diseases. *Circulation, 110,* e489–e493.

105. Cattaneo, M. (2004). Aspirin and clopidogrel: Efficacy, safety, and the issue of drug resistance. *Arterioscler Thromb Vasc Biol, 24,* 1980–1987.

106. Michelson, A. D., Cattaneo, M., Eikelboom, J. W., et al. (2005). Aspirin resistance: Position paper of the Working Group on Aspirin Resistance. *J Thromb Haemost, 3,* 1309–1311.

107. Szczeklik, A., Musial, J., Undas, A., et al. (2005). Aspirin resistance. *J Thromb Haemost, 3,* 1655–1662.

108. Roth, G. J., & Majerus, P. W. (1975). The mechanism of the effect of aspirin on human platelets. I. Acetylation of a particulate fraction protein. *J Clin Invest, 56,* 624–632.

109. (2002). Collaborative meta-analysis of randomised trials of antiplatelet therapy for prevention of death, myocardial infarction, and stroke in high risk patients. *BMJ, 324,* 71–86.

110. Rocca, B., & Patrono, C. (2005). Determinants of the interindividual variability in response to antiplatelet drugs. *J Thromb Haemost, 3,* 1597–1602.

111. Schwartz, K. A., Schwartz, D. E., Ghosheh, K., et al. (2005). Compliance as a critical consideration in patients who appear to be resistant to aspirin after healing of myocardial infarction. *Am J Cardiol, 95,* 973–975.

112. Weber, A. A., Przytulski, B., Schanz, A., et al. (2002). Towards a definition of aspirin resistance: A typological approach. *Platelets, 13,* 37–40.

113. Homoncik, M., Jilma, B., Hergovich, N., et al. (2000). Monitoring of aspirin (ASA) pharmacodynamics with the platelet function analyzer PFA-100. *Thromb Haemost, 83,* 316–321.

114. Feuring, M., Schultz, A., Losel, R., et al. (2005). Monitoring acetylsalicylic acid effects with the platelet function analyzer PFA-100. *Semin Thromb Hemost, 31,* 411–415.

115. Andersen, K., Hurlen, M., Arnesen, H., et al. (2002). Aspirin non-responsiveness as measured by PFA-100 in patients with coronary artery disease. *Thromb Res, 108,* 37–42.

116. Sambola, A., Heras, M., Escolar, G., et al. (2004). The PFA-100 detects sub-optimal antiplatelet responses in patients on aspirin. *Platelets, 15,* 439–446.

117. Coakley, M., Self, R., Marchant, W., et al. (2005). Use of the platelet function analyser (PFA-100) to quantify the effect of low dose aspirin in patients with ischaemic heart disease. *Anaesthesia, 60,* 1173–1178.

118. Coma–Canella, I, Velasco, A., & Castano, S. (2005). Prevalence of aspirin resistance measured by PFA-100. *Int J Cardiol, 101,* 71–76.

119. Peters, A. J., Borries, M., Gradaus, F., et al. (2001). In vitro bleeding test with PFA-100-aspects of controlling individual acetylsalicylic acid induced platelet inhibition in patients with cardiovascular disease. *J Thromb Thrombolysis, 12,* 263–272.

120. Harrison, P., Mackie, I., Mathur, A., et al. (2005). Platelet hyper-function in acute coronary syndromes. *Blood Coagul Fibrinolysis, 16,* 557–562.

121. McCabe, D. J., Harrison, P., Mackie, I. J., et al. (2005). Assessment of the antiplatelet effects of low to medium dose aspirin in the early and late phases after ischaemic stroke and TIA. *Platelets, 16,* 269–280.

122. Alberts, M. J., Bergman, D. L., Molner, E., et al. (2004). Antiplatelet effect of aspirin in patients with cerebrovascular disease. *Stroke, 35,* 175–178.

123. Chakroun, T., Gerotziafas, G., Robert, F., et al. (2004). In vitro aspirin resistance detected by PFA-100 closure time: pivotal role of plasma von Willebrand factor. *Br J Haematol, 124,* 80–85.

124. Frossard, M., Fuchs, I., Leitner, J. M., et al. (2004). Platelet function predicts myocardial damage in patients with acute myocardial infarction. *Circulation, 110,* 1392–1397.

125. Macchi, L., Christiaens, L., Brabant, S., et al. (2002). Resistance to aspirin in vitro is associated with increased platelet sensitivity to adenosine diphosphate. *Thromb Res, 107,* 45–49.

126. Harrison, P., Segal, H., Blasbery, K., et al. (2005). Screening for aspirin responsiveness after transient ischemic attack and stroke: Comparison of 2 point-of-care platelet function tests with optical aggregometry. *Stroke, 36,* 1001–1005.

127. Gum, P. A., Kottke–Marchant, K., Poggio, E. D., et al. (2001). Profile and prevalence of aspirin resistance in patients with cardiovascular disease. *Am J Cardiol, 88,* 230–235.

128. Gum, P. A., Kottke–Marchant, K., Welsh, P. A., et al. (2003). A prospective, blinded determination of the natural history of aspirin resistance among stable patients with cardiovascular disease. *J Am Coll Cardiol, 41,* 961–965.

129. Steinhubl, S. R., Varanasi, J. S., & Goldberg, L. (2003). Determination of the natural history of aspirin resistance among stable patients with cardiovascular disease. *J Am Coll Cardiol, 42,* 1336–1337.

130. Jilma, B. (2004). Therapeutic failure or resistance to aspirin. *J Am Coll Cardiol, 43,* 1332–1333.

131. Eikelboom, J. W., & Hankey, G. J. (2003). Aspirin resistance: A new independent predictor of vascular events? *J Am Coll Cardiol, 41,* 966–968.

132. Grundmann, K., Jaschonek, K., Kleine, B., et al. (2003). Aspirin non-responder status in patients with recurrent cerebral ischemic attacks. *J Neurol, 250,* 63–66.

133. Lanza, G. A., Sestito, A., Iacovella, S., et al. (2003). Relation between platelet response to exercise and coronary angiographic findings in patients with effort angina. *Circulation, 107,* 1378–1382.

134. Fuchs, I., Frossard, M., Spiel, A., et al. (2005). Platelet hyperfunction and low response to aspirin predict re-events in patients with acute coronary syndromes during long term follow up. *J Thromb Haemost, 3*(suppl 1), abstract 2142.

135. Christie, D. J., Kottke–Marchant, K., & Gorman, R. (2005). High shear platelet function is associated with major adverse events in patients with stable cardiovascular disease (CVD) despite aspirin therapy. *J Thromb Haemost, 3*(suppl 1), abstract 2204.

136. Stejskal, D., Proskova, J., Petrzelova, A., et al. (2001). Application of cationic propyl gallate as inducer of thrombocyte aggregation for evaluation of effectiveness of antiaggregation therapy. *Biomed Pap Med Fac Univ Palacky Olomouc Czech Repub, 145,* 69–74.

137. Malinin, A., Spergling, M., Muhlestein, B., et al. (2004). Assessing aspirin responsiveness in subjects with multiple risk factors for vascular disease with a rapid platelet function analyzer. *Blood Coagul Fibrinolysis, 15,* 295–301.

138. Wang, J. C., Aucoin–Barry, D., Manuelian, D., et al. (2003). Incidence of aspirin nonresponsiveness using the Ultegra Rapid Platelet Function Assay-ASA. *Am J Cardiol, 92,* 1492–1494.

139. Chen, W. H., Lee, P. Y., Ng, W., et al. (2005). Relation of aspirin resistance to coronary flow reserve in patients undergoing elective percutaneous coronary intervention. *Am J Cardiol, 96,* 760–763.

140. Chen, W. H., Lee, P. Y., Ng, W., et al. (2004). Aspirin resistance is associated with a high incidence of myonecrosis after non-urgent percutaneous coronary intervention despite clopidogrel pretreatment. *J Am Coll Cardiol, 43,* 1122–1126.

141. Lee, P. Y., Chen, W. H., Ng, W., et al. (2005). Low-dose aspirin increases aspirin resistance in patients with coronary artery disease. *Am J Med, 118,* 723–727.

142. Eikelboom, J. W., Hirsh, J., Weitz, J. I., et al. (2002). Aspirin-resistant thromboxane biosynthesis and the risk of myocardial infarction, stroke, or cardiovascular death in patients at high risk for cardiovascular events. *Circulation, 105,* 1650–1655.

143. Gachet, C. (2006). Regulation of platelet functions by P2 receptors. *Annu Rev Pharmacol Toxicol. 46,* 277–300.

144. CAPRIE Steering Committee. (1996). A randomised, blinded, trial of clopidogrel versus aspirin in patients at risk of ischaemic events (CAPRIE). *Lancet, 348,* 1329–1339.

145. Mehta, S. R., Yusuf, S., Peters, R. J., et al. (2001). Effects of pretreatment with clopidogrel and aspirin followed by long-term therapy in patients undergoing percutaneous coronary intervention: The PCI-CURE study. *Lancet, 358,* 527–533.

146. Diener, H. C., Bogousslavsky, J., Brass, L. M., et al. (2004). Aspirin and clopidogrel compared with clopidogrel alone after recent ischaemic stroke or transient ischaemic attack in high-risk patients (MATCH): Randomised, double-blind, placebo-controlled trial. *Lancet, 364,* 331–337.

147. Savi, P., & Herbert, J. M. (2005). Clopidogrel and ticlopidine: P2Y12 adenosine diphosphate-receptor antagonists for the prevention of atherothrombosis. *Semin Thromb Hemost, 31,* 174–183.

148. Serebruany, V. L., Steinhubl, S. R., Berger, P. B., et al. (2005). Variability in platelet responsiveness to clopidogrel among 544 individuals. *J Am Coll Cardiol, 45,* 246–251.

149. Nguyen, T. A., Diodati, J. G., & Pharand, C. (2005). Resistance to clopidogrel: A review of the evidence. *J Am Coll Cardiol, 45,* 1157–1164.

150. Gurbel, P. A., Bliden, K. P., Hiatt, B. L., et al. (2003). Clopidogrel for coronary stenting: Response variability, drug resistance, and the effect of pretreatment platelet reactivity. *Circulation, 107,* 2908–2913.

151. Jaremo, P., Lindahl, T. L., Fransson, S. G., et al. (2002). Individual variations of platelet inhibition after loading doses of clopidogrel. *J Intern Med, 252,* 233–238.

152. Lau, W. C., Gurbel, P. A., Watkins, P. B., et al. (2004). Contribution of hepatic cytochrome P450 3A4 metabolic activity to the phenomenon of clopidogrel resistance. *Circulation, 109,* 166–171.

153. Michelson, A. D., Linden, M. D., Furman, M. I., et al. (2005). Pre-existent variability in platelet response to ADP accounts for clopidogrel resistance. *J Thromb Haemost, 3*(suppl 1), abstract P0955.

154. Geiger, J., Teichmann, L., Grossmann, R., et al. (2005). Monitoring of clopidogrel action: Comparison of methods. *Clin Chem, 51,* 957–965.

155. Gurbel, P. A., Bliden, K. P., Hayes, K. M., et al. (2005). The relation of dosing to clopidogrel responsiveness and the incidence of high post-treatment platelet aggregation in patients undergoing coronary stenting. *J Am Coll Cardiol, 45,* 1392–1396.

156. Muller, I., Seyfarth, M., Rudiger, S., et al. (2001). Effect of a high loading dose of clopidogrel on platelet function in patients undergoing coronary stent placement. *Heart, 85,* 92–93.

157. Gurbel, P. A., Bliden, K. P., Zaman, K. A., et al. (2005). Clopidogrel loading with eptifibatide to arrest the reactivity of platelets: Results of the Clopidogrel Loading with Eptifibatide to Arrest the Reactivity of Platelets (CLEAR PLATELETS) study. *Circulation, 111,* 1153–1159.

158. Aleil, B., Ravanat, C., Cazenave, J. P., et al. (2005). Flow cytometric analysis of intraplatelet VASP phosphorylation for the detection of clopidogrel resistance in patients with ischemic cardiovascular diseases. *J Thromb Haemost, 3,* 85–92.

159. Matetzky, S., Shenkman, B., Guetta, V., et al. (2004). Clopidogrel resistance is associated with increased risk of recurrent atherothrombotic events in patients with acute myocardial infarction. *Circulation, 109,* 3171–3175.

160. Fox, S. C., Behan, M. W., & Heptinstall S. (2004). Inhibition of ADP-induced intracellular Ca2+ responses and platelet aggregation by the P2Y12 receptor antagonists AR-C69931MX and clopidogrel is enhanced by prostaglandin E1. *Cell Calcium, 35,* 39–46.

161. Gachet, C., Cazenave, J. P., Ohlmann, P., et al. (1990). The thienopyridine ticlopidine selectively prevents the inhibitory effects of ADP but not of adrenaline on cAMP levels raised by stimulation of the adenylate cyclase of human platelets by PGE1. *Biochem Pharmacol, 40,* 2683–2687.

162. Barragan, P., Bouvier, J. L., Roquebert, P. O., et al. (2003). Resistance to thienopyridines: Clinical detection of coronary stent thrombosis by monitoring of vasodilator-stimulated phosphoprotein phosphorylation. *Catheter Cardiovasc Interv, 59,* 295–302.

163. Golanski, J., Pluta, J., Baraniak, J., et al. (2004). Limited usefulness of the PFA-100 for the monitoring of ADP receptor antagonists: In vitro experience. *Clin Chem Lab Med, 42,* 25–29.

164. Mueller, T., Haltmayer, M., Poelz, W., et al. (2003). Monitoring aspirin 100 mg and clopidogrel 75 mg therapy with the PFA-100 device in patients with peripheral arterial disease. *Vasc Endovasc Surg, 37,* 117–123.

165. Ziegler, S., Maca, T., Alt, E., et al. (2002). Monitoring of antiplatelet therapy with the PFA-100 in peripheral angioplasty patients. *Platelets, 13,* 493–497.

166. Raman, S., & Jilma, B. (2004). Time lag in platelet function inhibition by clopidogrel in stroke patients as measured by PFA-100. *J Thromb Haemost, 2,* 2278–2279.

167. Jilma, B. (2003). Synergistic antiplatelet effects of clopidogrel and aspirin detected with the PFA-100 in stroke patients. *Stroke, 34,* 849–854

168. Gurbel, P. A., Bliden, K. P., Samara, W., et al. (2005). Clopidogrel effect on platelet reactivity in patients with stent thrombosis: Results of the CREST Study. *J Am Coll Cardiol, 46,* 1827–1832.

169. Muller, I., Besta, F., Schulz, C., et al. (2003). Prevalence of clopidogrel non-responders among patients with stable angina pectoris scheduled for elective coronary stent placement. *Thromb Haemost, 89,* 783–787.

170. Coller, B. S., Folts, J. D., Smith, S. R., et al. (1989). Abolition of in vivo platelet thrombus formation in primates with monoclonal antibodies to the platelet GPIIb/IIIa receptor. Correlation with bleeding time, platelet aggregation, and blockade of GPIIb/IIIa receptors. *Circulation, 80,* 1766–1774.

171. Gold, H. K., Coller, B. S., Yasuda, T., et al. (1988). Rapid and sustained coronary artery recanalization with combined bolus injection of recombinant tissue-type plasminogen activator and monoclonal antiplatelet GPIIb/IIIa antibody in a canine preparation. *Circulation, 77,* 670–677.

172. Thompson, C. M., & Steinhubl, S. R. (1999). Monitoring of platelet function in the setting of GPIIb/IIIa inhibitor therapy. *Curr Interv Cardiol Rep, 1,* 270–277.

173. Wheeler, G. L., Braden, G. A., Steinhubl, S. R., et al. (2002). The Ultegra rapid platelet-function assay: Comparison to standard platelet function assays in patients undergoing percutaneous coronary intervention with abciximab therapy. *Am Heart J, 143,* 602–611.

174. Simon, D. I., Liu, C. B., Ganz, P., et al. (2001). A comparative study of light transmission aggregometry and automated bedside platelet function assays in patients undergoing percutaneous coronary intervention and receiving abciximab, eptifibatide, or tirofiban. *Catheter Cardiovasc Interv, 52,* 425–432.

175. Steinhubl, S. R., Talley, J. D., Braden, G. A., et al. (2001). Point-of-care measured platelet inhibition correlates with a reduced risk of an adverse cardiac event after percutaneous coronary intervention: Results of the GOLD (AU-Assessing Ultegra) multicenter study. *Circulation, 103,* 2572–2578.

176. Hezard, N., Metz, D., Nazeyrollas, P., et al. (2000). Use of the PFA-100 apparatus to assess platelet function in patients undergoing PTCA during and after infusion of c7E3 Fab in the presence of other antiplatelet agents. *Thromb Haemost, 83,* 540–544.

177. Madan, M., Berkowitz, S. D., Christie, D. J., et al. (2001). Rapid assessment of glycoprotein IIb/IIIa blockade with the platelet function analyzer (PFA-100) during percutaneous coronary intervention. *Am Heart J, 141,* 226–233.

178. Madan, M., Berkowitz, S. D., Christie, D. J., et al. (2002). Determination of platelet aggregation inhibition during percutaneous coronary intervention with the platelet function analyzer PFA-100. *Am Heart J, 144,* 151–158.

179. Koscielny, J., Ziemer, S., Radtke, H., et al. (2004). A practical concept for preoperative identification of patients with impaired primary hemostasis. *Clin Appl Thromb Hemost, 10,* 195–204.

180. Koscielny, J., von Tempelhoff, G. F., Ziemer, S., et al. (2004). A practical concept for preoperative management of patients with impaired primary hemostasis. *Clin Appl Thromb Hemost, 10,* 155–166.

181. Koh, M. B., & Hunt, B. J. (2003). The management of perioperative bleeding. *Blood Rev, 17,* 179–185.

182. de Moerloose, P. (1996). Laboratory evaluation of hemostasis before cardiac operations. *Ann Thorac Surg, 62,* 1921–1925.

183. Prisco, D., & Paniccia, R. Point-of-care testing of hemostasis in cardiac surgery. *Thromb J, 1,* 1.

184. Mallett, S. V., & Cox, D. J. (1992). Thrombelastography. *Br J Anaesth, 69,* 307–313.

185. Caprini, J. A., Traverso, C. I., & Arcelus, J. I. (1995). Perspectives on thromboelastography. *Semin Thromb Hemost, 21*(suppl 4), 91–93.

186. Shore–Lesserson L. (2005). Evidence based coagulation monitors: Heparin monitoring, thromboelastography, and platelet function. *Semin Cardiothorac Vasc Anesth, 9,* 41–52.

187. Shore–Lesserson, L., Manspeizer, H. E., DePerio, M., et al. (1999). Thromboelastography-guided transfusion algorithm reduces transfusions in complex cardiac surgery. *Anesth Analg, 88,* 312–319.

188. Spiess, B. D. (1995). Thromboelastography and cardiopulmonary bypass. *Semin Thromb Hemost, 21*(suppl 4), 27–33.

189. Nuttall, G. A., Oliver, W. C., Santrach, P. J., et al. (2001). Efficacy of a simple intraoperative transfusion algorithm for nonerythrocyte component utilization after cardiopulmonary bypass. *Anesthesiology, 94,* 773–781.

190. Royston, D., & von Kier, S. (2001). Reduced haemostatic factor transfusion using heparinase-modified thromboelastography during cardiopulmonary bypass. *Br J Anaesth, 86,* 575–578.

191. Despotis, G. J., Joist, J. H., & Goodnough, L. T. (1997). Monitoring of hemostasis in cardiac surgical patients: Impact of point-of-care testing on blood loss and transfusion outcomes. *Clin Chem, 43,* 1684–1696.

192. Avidan, M. S., Alcock, E. L., Da Fonseca, J., et al. (2004). Comparison of structured use of routine laboratory tests or near-patient assessment with clinical judgment in the management of bleeding after cardiac surgery. *Br J Anaesth, 92,* 178–186.

193. Poston, R., Gu, J., Manchio, J., et al. (2005). Platelet function tests predict bleeding and thrombotic events after off-pump coronary bypass grafting. *Eur J Cardiothorac Surg, 27,* 584–591.

194. Ostrowsky, J., Foes, J., Warchol, M., et al. (2004). Platelet-works platelet function test compared to the thromboelastograph for prediction of postoperative outcomes. *J Extra Corpor Technol, 36,* 149–152.

195. Hertfelder, H. J., Bos, M., Weber, D., et al. (2005). Perioperative monitoring of primary and secondary hemostasis in coronary artery bypass grafting. *Semin Thromb Hemost, 31,* 426–440.

196. Forestier, F., Coiffic, A., Mouton, C., et al. (2002). Platelet function point-of-care tests in post-bypass cardiac surgery: Are they relevant? *Br J Anaesth, 89,* 715–721.

197. Fattorutto, M., Pradier, O., Schmartz, D., et al. (2003). Does the platelet function analyser (PFA-100) predict blood loss after cardiopulmonary bypass? *Br J Anaesth, 90,* 692–693.

198. Elwood, P. C., Beswick, A., Pickering, J., et al. (2001). Platelet tests in the prediction of myocardial infarction and ischaemic stroke: Evidence from the Caerphilly Prospective Study. *Br J Haematol, 113,* 514–520.

199. Thaulow, E., Erikssen, J., Sandvik, L., et al. (1991). Blood platelet count and function are related to total and cardiovascular death in apparently healthy men. *Circulation, 84,* 613–617.

200. Trip, M. D., Cats, V. M., van Capelle, F. J., et al. (1990). Platelet hyperreactivity and prognosis in survivors of myocardial infarction. *N Engl J Med, 322,* 1549–1554.

201. Meade, T. W., Cooper, J. A., & Miller, G. J. (1997). Platelet counts and aggregation measures in the incidence of ischaemic heart disease (IHD). *Thromb Haemost, 78,* 926–929.

202. Elwood, P. C., Renaud, S., Beswick, A. D., et al. (1998). Platelet aggregation and incident ischaemic heart disease in the Caerphilly cohort. *Heart, 80,* 578–582.

203. Elwood, P. C., Renaud, S., Sharp, D. S., et al. (1991). Ischemic heart disease and platelet aggregation. The Caerphilly Collaborative Heart Disease Study. *Circulation, 83,* 38–44.

204. Kristensen, S. D., Bath, P. M., & Martin, J. F. (1990). Differences in bleeding time, aspirin sensitivity and adrenaline between acute myocardial infarction and unstable angina. *Cardiovasc Res, 24,* 19–23.

205. Martin, J. F., Bath, P. M., & Burr, M. L. (1991). Influence of platelet size on outcome after myocardial infarction. *Lancet, 338,* 1409–1411.

206. Elwood, P. C., Pickering, J., Yarnell, J., et al. (2003). Bleeding time, stroke and myocardial infarction: The Caerphilly prospective study. *Platelets, 14,* 139–141.

207. Mammen, E. F. (1999). Sticky platelet syndrome. *Semin Thromb Hemost, 25,* 361–365.

208. Mammen, E. F., Barnhart, M. I., Selik, N. R., et al. (1988). "Sticky platelet syndrome": A congenital platelet abnormality predisposing to thrombosis? *Folia Haematol Int Mag Klin Morphol Blutforsch, 115,* 361–365.

209. Yee, D. L., Sun, C. W., Bergeron, A. L., et al. (2005). Aggregometry detects platelet hyperreactivity in healthy individuals. *Blood, 106,* 2723–2729.

210. Ault, K. A., Cannon, C. P., Mitchell, J., et al. (1999). Platelet activation in patients after an acute coronary syndrome: Results from the TIMI-12 trial. Thrombolysis in myocardial infarction. *J Am Coll Cardiol, 33,* 634–639.

211. Cardigan, R., Turner, C., & Harrison, P. (2005). Current methods of assessing platelet function: Relevance to transfusion medicine. *Vox Sang, 88,* 153–163.

212. Harrison, P., Segal, H., Furtado, C., et al. (2004). High incidence of defective high-shear platelet function among platelet donors. *Transfusion, 44,* 764–770.

213. Jilma–Stohlawetz, P., Hergovich, N., Homoncik, M., et al. (2001). Impaired platelet function among platelet donors. *Thromb Haemost, 86,* 880–886.

214. Borzini, P., Lazzaro, A., & Mazzucco, L. (1999). Evaluation of the hemostatic function of stored platelet concentrates using the platelet function analyzer (PFA-100). *Haematologica, 84,* 1104–1109.

215. Hoffmeister, K. M., Josefsson, E. C., Isaac, N. A., et al. (2003). Glycosylation restores survival of chilled blood platelets. *Science, 301,* 1531–1534.

216. Segal, H. C., Briggs, C., Kunka, S., et al. (2005). Accuracy of platelet counting haematology analysers in severe thrombocytopenia and potential impact on platelet transfusion. *Br J Haematol, 128,* 520–525.

217. Salama, M. E., Raman, S., Drew, M. J., et al. (2004). Platelet function testing to assess effectiveness of platelet transfusion therapy. *Transfus Apheresis Sci, 30,* 93–100.

218. Favaloro, E. J., Kershaw, G., Bukuya, M., et al. (2001). Laboratory diagnosis of von Willebrand disorder (vWD) and monitoring of DDAVP therapy: Efficacy of the PFA-100 and vWF:CBA as combined diagnostic strategies. *Haemophilia, 7,* 180–189.

219. Fressinaud, E., Veyradier, A., Truchaud, F., et al. (1998). Screening for von Willebrand disease with a new analyzer using high shear stress: A study of 60 cases. *Blood, 91,* 1325–1331.

220. Fressinaud, E., Veyradier, A., Sigaud, M., et al. (1999). Therapeutic monitoring of von Willebrand disease: Interest and limits of a platelet function analyser at high shear rates. *Br J Haematol, 106,* 777–783.

221. Cattaneo, M., Federici, A. B., Lecchi, A., et al. (1999). Evaluation of the PFA-100 system in the diagnosis and therapeutic monitoring of patients with von Willebrand disease. *Thromb Haemost, 82,* 35–39.

222. Cattaneo, M., Lecchi, A., Agati, B., et al. (1999). Evaluation of platelet function with the PFA-100 system in patients with congenital defects of platelet secretion. *Thromb Res, 96,* 213–217.

223. Reiter, R. A., Mayr, F., Blazicek, H., et al. (2003). Desmopressin antagonizes the in vitro platelet dysfunction induced by GPIIb/IIIa inhibitors and aspirin. *Blood, 102,* 4594–4599.

224. Ghorashian, S., & Hunt, B. J. (2004). "Off-license" use of recombinant activated factor VII. *Blood Rev, 18,* 245–259.

225. Hoffman, M. (2003). A cell-based model of coagulation and the role of factor VIIa. *Blood Rev, 17*(suppl 1), S1–S5.

226. Hoffman, M., Monroe, D. M., & Roberts, H. R. (2002). Platelet-dependent action of high-dose factor VIIa. *Blood, 100,* 364–365.

227. Kjalke, M., Johannessen, M., & Hedner, U. (2001). Effect of recombinant factor VIIa (NovoSeven) on thrombocytopenia-like conditions in vitro. *Semin Hematol, 38,* 15–20.

228. Kjalke, M., Ezban, M., Monroe, D. M., et al. (2001). High-dose factor VIIa increases initial thrombin generation and mediates faster platelet activation in thrombocytopenia-like conditions in a cell-based model system. *Br J Haematol, 114,* 114–120.

229. Wilbourn, B., Harrison, P., Mackie, I. J., et al. (2003). Activation of platelets in whole blood by recombinant factor VIIa by a thrombin-dependent mechanism. *Br J Haematol, 122,* 651–661.

230. Fattorutto, M., Tourreau–Pham, S., Mazoyer, E., et al. (2004). Recombinant activated factor VII decreases bleeding without increasing arterial thrombosis in rabbits. *Can J Anaesth, 51,* 672–679.

231. Almeida, A. M., Khair, K., Hann, I., et al. (2003). The use of recombinant factor VIIa in children with inherited platelet function disorders. *Br J Haematol, 121,* 477–481.

232. Gerotziafas, G. T., Chakroun, T., Depasse, F., et al. (2004). The role of platelets and recombinant factor VIIa on thrombin generation, platelet activation and clot formation. *Thromb Haemost, 91,* 977–985.

233. Wegert, W., Harder, S., Bassus, S., et al. (2005). Platelet-dependent thrombin generation assay for monitoring the efficacy of recombinant factor VIIa. *Platelets, 16,* 45–50.

234. Tomokiyo, K., Nakatomi, Y., Araki, T., et al. (2003). A novel therapeutic approach combining human plasma-derived factors VIIa and X for haemophiliacs with inhibitors: Evidence of a higher thrombin generation rate in vitro and more sustained haemostatic activity in vivo than obtained with factor VIIa alone. *Vox Sang, 85,* 290–299.

235. Sorensen, B., & Ingerslev, J. (2004). Thromboelastography and recombinant factor VIIa: Hemophilia and beyond. *Semin Hematol, 41,* 140–144.

236. Yoshioka, A., Nishio, K., & Shima, M. (1996). Thromboelastogram as a hemostatic monitor during recombinant factor VIIa treatment in hemophilia A patients with inhibitor to factor VIII. *Haemostasis, 26*(suppl 1), 139–142.

237. Carr, M. E., Jr., Martin, E. J., Kuhn, J. G., et al. (2004). Monitoring of hemostatic status in four patients being treated with recombinant factor VIIa. *Clin Lab, 50,* 529–538.

238. Kunz, D., Kunz, W. S., Scott, C. S., et al. (2001). Automated CD61 immunoplatelet analysis of thrombocytopenic samples. *Br J Haematol, 112,* 584–592.

239. Abe, Y., Wada, H., Tomatsu, H., et al. (2005). A simple technique to determine thrombopoiesis level using immature platelet fraction (IPF). *Thromb Res.* doi: 10.1016/j.thromres.2005.09.007

240. Briggs, C., Kunka, S., Hart, D., et al. (2004). Assessment of an immature platelet fraction (IPF) in peripheral thrombocytopenia. *Br J Haematol, 126,* 93–99.

241. Phillips, D. R., Conley, P. B., Sinha, U., et al. (2005). Therapeutic approaches in arterial thrombosis. *J Thromb Haemost, 3,* 1577–1589.

242. Frojmovic, M. M. (1998). Platelet aggregation in flow: Differential roles for adhesive receptors and ligands. *Am Heart J Suppl, 135,* 5119–5131.

243. Frojmovic, M. M., Kasirer–Friede, A., Goldsmith, H., et al. (1997). Surface-secreted von Willebrand factor mediates aggregation of ADP-activated platelets at moderate shear stress: Facilitated by GPIb but controlled by GPIIb-IIIa. *Thromb Haemost, 76,* 568–576.

244. Bonnefoy, A., Hantgan, R., Legrand, C., et al. (2001). A model of platelet aggregation involving multiple interactions of thrombospondin-1, fibrinogen and GPIIbIIIa receptor. *J Biol Chem, 276,* 5605–5612.

245. Bonnefoy, A., Liu, Q., Gray, J. W., et al. (2001). Platelets in suspension require preactivation to adhere to immobilized fibrinogen. *Ann N Y Acad Sci, 936,* 459–463.

246. Chang, H. M., & Robertson, C. R. (1976). Platelet aggregation by laminar shear and Brownian motion. *Ann Biomed Eng, 4,* 151–183.

247. Macaulay, I. C., Carr, P., Gusnanto, A., et al. (2005). Platelet genomics and proteomics in human health and disease. *J Clin Invest, 115,* 3370–3377.

# CHAPTER 24

# Platelet Counting

**Carol Briggs,[1] Paul Harrison,[2] and Samuel J. Machin[1]**

[1]*Department of Haematology, University College London Hospital, London, United Kingdom*
[2]*Oxford Haemophilia Centre & Thrombosis Unit, Churchill Hospital, Oxford, United Kingdom*

## I. Introduction

Platelet counting is essential in both clinical hematology and platelet research laboratories. Accurate and precise enumeration of platelets is not only critical to assist in diagnosis and treatment of various clinical disorders but is a vital research tool particularly for standardizing counts within whole blood, platelet-rich plasma (PRP), or purified platelet preparations. The normal platelet count is widely quoted as $150-400 \times 10^9/L$ of whole blood. In this chapter, both the existing and newly proposed international reference methods for platelet counting will be discussed in detail, as they can be performed in any laboratory with suitable equipment. However, the full range of alternative automated methodologies will also be reviewed, so that readers will be able to understand their relative advantages and disadvantages and utilize the most appropriate platelet counting procedure within their laboratory. Quality control and assurance procedures will be briefly discussed.

The four main analytical procedures for platelet counting are: (1) manual counting using phase contrast microscopy, (2) impedance analysis, (3) optical light scatter/fluorescence analysis using various commercially available analyzers, and (4) immunoplatelet counting by flow cytometry. Although manual methods have been largely replaced by automated instrumentation, many research and nonspecialized laboratories interested in a reliable method for accurately counting platelets can still either perform manual counting or utilize small impedance analyzers if access to a larger automated blood counter is not possible.

Early methods to enumerate platelets in blood were usually inaccurate and irreproducible until the mid-twentieth century. In 1953, Brecher et al.[1] developed a manual phase contrast microscopy method, enabling platelets to be easily discriminated from lysed red cells within a counting chamber or hemocytometer. Although the development of the Coulter Principle in the 1950s (see Section III.A below) revolutionized blood counting, platelet counts were only added to the automated full blood count in the late 1970s. In early impedance analyzers, platelet counting could only be performed by analysis of PRP or purified platelet preparations and was thus prone to considerable error. Until the widespread availability of platelet counting as part of the full blood count, the majority of platelet counts were still performed manually *via* phase contrast microscopy. The manual phase count is still recognized as the gold standard or international reference method.[1,2] Thus, until very recently, the calibration of platelet counts on automated blood cell counters and quality control material was still routinely performed via the manual method by the majority of instrument manufacturers. However, the manual method is not only time-consuming, subjective, and tedious, but results in high levels of imprecision with typical interobserver coefficient of variations (CV) in the range of 10 to 25%.[3] At low platelet numbers, because fewer cells are counted, observed CVs increase proportionally. Although relatively imprecise, the manual method still offers a relatively inexpensive, simple, and viable means to enumerate platelets in the nonspecialized laboratory.

The introduction of automated full blood counters using impedance technology resulted in a dramatic improvement in precision, with typical CVs of less than 3%,[4] because much higher total numbers of platelets are counted.[5] However, impedance platelet counting methods still have significant limitations, despite their widespread use. One of the major problems is that cell size analysis cannot discriminate platelets from other similarly sized particles, such as small or fragmented red cells, immune complexes, etc.[6] These may be erroneously included in the platelet count, and in severely thrombocytopenic samples the number of interfering particles may even exceed the number of true platelets. Large or giant platelets may be excluded from the count on the basis of their size, because they cannot be resolved

475

from red cells. There may also be significant variation in the results obtained on different analyzers with the same sample due to differences in the method of analysis, linearity over the entire measuring range, and the number of events actually counted. More recently, multiple light scatter parameters and/or fluorescence, rather than impedance sizing alone, have been introduced for platelet counting in automated hematology analyzers. This has improved the ability of automated analyzers to discriminate platelets. Despite these newer methods, there are still occasional cases in which absolute accuracy of the platelet count remains a challenge. Thus, there has been renewed interest in the development of an improved reference procedure to enable optimization of automated platelet counting. This latter method utilizes specific monoclonal antibodies to platelet cell surface antigens (e.g., anti-CD41, -CD42, or -CD61) conjugated to a suitable fluorophore. By performing flow cytometric analysis of the ratio of fluorescent platelet events to nonfluorescent red cell events, a highly accurate and precise technique is now available for counting platelets in whole blood[3] (see Section III.E and III.F). This relatively new approach permits the possible implementation of a new international reference method to calibrate cell counters, assign values to calibrators, and to obtain a direct platelet count on a wide variety of pathological samples. This should hopefully lead to an improved accuracy of platelet counting in thrombocytopenia and facilitate further studies to establish whether current platelet transfusion thresholds (see Chapter 69) can be safely lowered without risk of bleeding. The method can also be adapted so that platelets can be quantified within purified preparations via the addition of precise numbers of fluorescent beads.[3]

## II. Manual Platelet Counting

Despite the widespread use of automated technology, manual counting of platelets is still widely performed in under-resourced laboratories and within research laboratories with no access to specialized instrumentation. It is also common practice to use manual counting methods in the clinical laboratory if the platelet count is low or there are atypical platelets present in the sample. The current international reference method for platelet counting is still performed by the standard manual method using phase contrast microscopy and was established by the International Committee for Standardisation in Haematology (ICSH).[2] Whole blood platelet counts are usually performed on ethylene diamine tetraacetic acid (EDTA)-anticoagulated blood obtained by standard clean venepuncture. To discriminate platelets from red cells, manual counting is usually performed by visual examination of diluted and lysed whole blood using a Neubauer counting chamber, which contains a precise volume of fluid.[2] Purified platelet preparations can also be

counted with this method. Specific methodological details of manual platelet counting are available.[2,7,8]

## III. Automated Platelet Counting

There are now several methods on commercial analyzers for counting platelets, including aperture impedance, optical scattering, and fluorescence. Table 24-1 lists some of the currently available large hematology analyzers which incorporate platelet counting. Normal platelets give a classical log-normal volume distribution curve, which is particularly useful as the basis for determining a valid platelet count (Fig. 24-1).

**Table 24-1: Examples of Currently Available Large Hematology Analyzers**

| Manufacturer | Instrument | Principle of Platelet Count |
|---|---|---|
| Abbott Diagnostics | CELL-DYN 4000 | Impedance, optical, and immunological |
| | CELL-DYN Sapphire | Impedance, optical, and immunological |
| ABX Diagnostics | Pentra-120 | Impedance |
| Bayer Corporation | ADVIA 120 | Optical |
| | ADVIA 2120 | Optical |
| Beckman Coulter | GEN-S | Impedance |
| | LH 750 | Impedance |
| Sysmex Corporation | SE-9500 | Impedance |
| | XE-2100 | Impedance and optical |

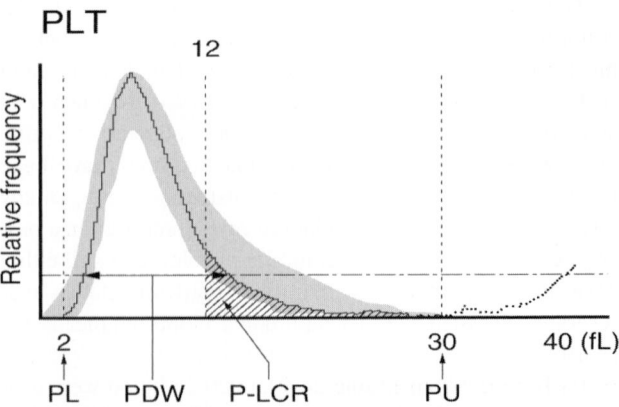

**Figure 24-1.** Typical platelet size distribution in an automated hematology analyzer. PL, lower discrimination for platelet size distribution. PDW, platelet distribution width. P-LCR, platelets-large cell ratio. PU, upper discrimination for platelet size distribution. (reprinted with permission from Sysmex corporation)

Other derived platelet parameters are highly dependent upon the individual technology and are influenced by the anticoagulant and delay time from sampling to analysis (e.g., EDTA-induced swelling). If mean platelet volume (MPV) is to be reliably measured, then the potential influence of anticoagulant on the MPV must be controlled for, either by using an alternative anticoagulant or standardizing the time delay between sampling and analysis. In impedance analyzers, MPV and platelet distribution width (PDW) are derived from the platelet distribution curve. Although these derived platelet parameters must be interpreted carefully, there is normally an established inverse relationship between MPV and the platelet count, which contributes to the maintenance of hemostatic function. There is also evidence that MPV is an important risk factor for acute myocardial infarction.[9]

Whichever automated method is used for platelet counting it must be demonstrated that it is precise, shows minimum fluctuation in repeated results on the same sample, and gives linear results over the entire analytical range. At high counts there is a growing probability of coincidence, two or more cells passing through the sensing zone at the same time, as well as possible sample carryover if a high count precedes a low count. With thrombocytopenic samples, it is important that added counts due to spurious signals caused by electronic noise are not included within the reported result. It is also desirable that there is minimal method variation between different analyzers; results obtained with different systems on the same sample should be comparable.

### A. Impedance Platelet Counting

Wallace Coulter first described the resistance detection method, usually referred to as the "Coulter Principle" or impedance method[10] (Fig. 24-2). In this method, biological cells are regarded as completely nonconductive resistivity particles. When a blood cell passes through an aperture (sensing zone) suspended in electrolyte solution, the change in electric impedance is detected. Each individual cell gives an impedance signal, which is proportional to the volume of the cell detected, and so this method can be used to size and count individual cells. This method was originally used for the counting of red cells and white cells; the first Coulter platelet counter required the use of PRP to avoid counting red cells as platelets. Many research laboratories still utilize small analyzers to count platelets in PRP or in purified platelet preparations (Table 24-2).

It was not until the 1970s that improvements in technology, including coincidence correction and hydrodynamic focusing, allowed the discrimination of platelets from red cells to enable an accurate platelet count to be obtained from a whole blood sample. Ideally, if cells pass through the

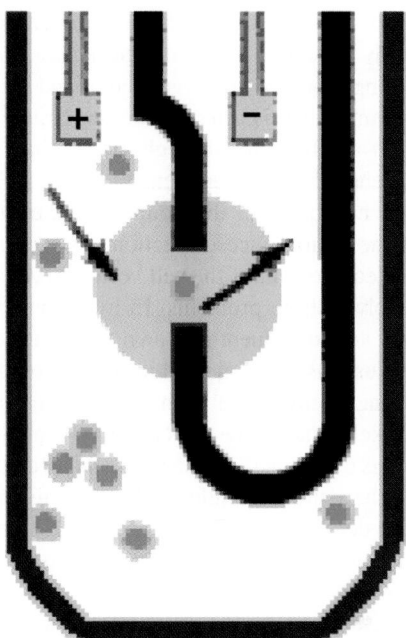

**Figure 24-2.** Electrical impedance method or Coulter Principle. See text for explanation.

**Table 24-2: Examples of Currently Available Compact Analyzers for Platelet Counting**

| Manufacturer | Instrument | Principle of Platelet Count |
|---|---|---|
| Abbott Diagnostics | CELL-DYN 1200 | Impedance |
| ABX Diagnostics | Pentra 60 | Impedance |
|  | Micros 60 | Impedance |
| Bayer Corporation | ADVIA 70 | Impedance |
| Beckman Coulter | A$^C$T | Impedance |
| Sysmex Corporation | KX21 | Impedance |
|  | pocH-100i | Impedance |

sensing zone one by one, the total number of detected cells is counted. However, simultaneous occupancy of the sensing zone by more than one particle occurs. This phenomenon is called "coincidence" and the resulting count error is known as the coincidence error. The magnitude of coincidence error increases with the concentration of cells suspended. For major hematology analyzers, by measuring the results from several samples of different concentrations, the coincidence correction formula can be established. The correction formula may be integrated into the analyzer's computer and the coincidence corrected result reported.

In order to minimize coincidence physically, the hydrodynamic focusing method has been developed for some analyzers. If two cells pass through the sensing zone together, the count may be corrected by the coincidence correction but a large single pulse will be generated and it is not possible to determine if this arises from one large cell or two small cells. If a cell passes through the sensing zone close to the wall, where high current density exists, an M-shaped pulse is generated; while the count result may be valid because of the coincidence correction, there is no way to correct the measurement of the cell volume. Hydrodynamic focusing resolves these problems. In hydrodynamic focusing, a steady flow of diluent is drawn through the aperture and the cell suspension is injected into this moving body of liquid in a fine stream close to the aperture entrance (Fig. 24-3). The likelihood of two cells passing through the aperture together is dramatically decreased and no cell goes near the wall or the entrance angle of the sensing zone where high current density exists. Hydrodynamic focusing produces a clear discrimination between red cells and platelets.

In the presently available Beckman Coulter analyzers (e.g., GEN S, LH 750), particles between 2 and 20 femtoliters (fL) are counted as platelets.[11] Pulses are obtained from three red cell/platelet orifices to obtain 64-channel size distribution histograms for each orifice. These histograms are

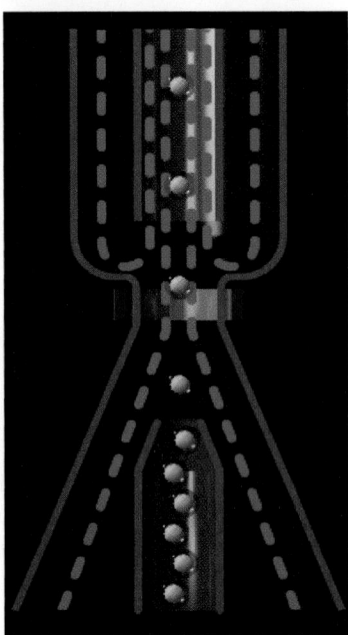

**Figure 24-3.** Schematic diagram of sheath flow and the principle of hydrodynamic focusing. Platelets and red cells are analyzed by a hydrodynamic focusing system, which eliminates potential errors of coincidence, recirculation, and stress changes associated with traditional methods of analysis. This results in more accurate platelet and red cell counts and sizing, even when cell counts are low or high. (Reprinted with permission from the Sysmex Corporation.)

smoothed and a high point and two low points are identified in the distribution. A log-normal curve is fitted to these points. The curves have a range of 0–70 fL and the platelet count and parameters are derived from this curve.

In the Sysmex counting systems (e.g., SE-9500, XE-2100), platelets are also counted by the orifice impedance method.[11] A platelet size distribution plot is produced using three thresholds. One is fixed at the 12 fL level and the other two are allowed to hunt the upper and lower ends of the platelet population between certain limits. The lower platelet size threshold may move between 2 and 6 fL, and the higher between 12 and 30 fL. The purpose of these thresholds is to endeavor to distinguish platelets from small red cells or red cell fragments at the upper end of the platelet population, and debris at the lower end. Analyzers using the standard impedance measurements are able (for most samples) to provide an accurate platelet count down to $20 \times 10^9$/L. Below this level, impedance analyzers become less accurate, due to decreasing statistical confidence, fewer events analyzed, and the increasing influence of background and plasma nonplatelet particulate matter. A major disadvantage of the electrical impedance method for counting platelets is the difficulty in distinguishing large platelets from extremely microcytic or fragmented red cells, even with the use of hydrodynamic focusing methods. False increases in the platelet count will occur when red cell or white cell fragments, microcytic red cells, immune complexes, bacteria, or cell debris are included in the reported platelet count.[6] False decreases in the count will occur in the presence of large platelets, platelet clumping (as seen with pseudo-thrombocytopenia by EDTA-dependent agglutinins), or platelet satellitism[6] (see Chapter 55).

### B. Optical Platelet Counting

More recently, optical light scatter methods have been introduced for platelet counting. In one-dimensional platelet analysis, platelets are counted and sized by a flow cytometry system in which the cells in a suitable diluent pass through a narrow beam of light (i.e., helium–neon laser). The illumination and light scatter by each cell is measured at a single angle (2–3°). This allows assessment of the number of electrical pulses generated in proportion to the number of cells and cell volume. In these automated systems, a series of algorithms, or a smoothing or fitting routine on the platelet volume histogram, is used to establish the validity of each platelet count.

To improve discrimination of platelets accurately from nonplatelet particles, two-dimensional laser light scatter was developed. The ADVIA 120 and 2120 analyzers (Bayer) use two-dimensional platelet analysis; volume and refractive index of effectively sphered individual platelets are simultaneously determined on a cell-by-cell basis by measuring two angles of laser light scatter at 2 to 3° and at 5 to 15°.[12] The two scatter measurements are converted to volume (platelet size) and refractive index (platelet density) values using the Mie theory of light scattering for homogenous spheres.[13] The platelet scatter cytogram map resolves volumes between 1 and 30 fL and refractive index values between 1.35 and 1.44. Large platelets, red cell fragments, red cell ghosts, microcytes, and cellular debris are distinguished. Platelets are identified within the map on the platelet scatter cytogram based on their volume and refractive index (1.35–1.40) (Fig. 24-4). Red cell fragments and microcytes with the same volume range have a greater refractive index than platelets and fall below and to the right of the grid, and red cell ghosts with a refractive index less than platelets fall above and to the left of the grid (Fig. 24-5). Large platelets with volumes between 30 and 60 fL are identified in the large platelet area of the red cell map. The reported two-dimensional platelet count is the sum of platelets and large platelets identified in the platelet and red cell scatter cytograms. Recent published data suggests that the two-dimensional platelet count improves the accuracy of the platelet count in thrombocytopenic samples.[3,12]

The CELL-DYN 4000 instrument also routinely reports an optical platelet count (as well as an impedance count) based on two light scatter parameters, intermediate light scatter (7°) and a wide angle scatter (90°). An algorithm is used to identify platelets using these two parameters to exclude, insofar as possible, nonplatelet particles while including all platelets. This is a two-dimensional analysis in which platelets must fall within a region that defines the correlation between the two light scatter parameters (a sloping window) and between a lower threshold and an upper discriminator (Fig. 24-5). The three-discriminator lines are set dynamically; the lower threshold is fixed. A simultaneous determination of the impedance platelet count is performed and discrepancies between the two counts generate an alert flag suggesting the presence of sample inter-ferences. The combination of two-dimensional optical analysis and flow impedance counting on the CELL-DYN 4000 has made a significant contribution to improving the accuracy and precision of platelet counting.[6]

## C. Optical Fluorescence Platelet Counting

An optical fluorescence platelet count has been introduced on the Sysmex XE-2100 analyzer, in addition to the tradi-

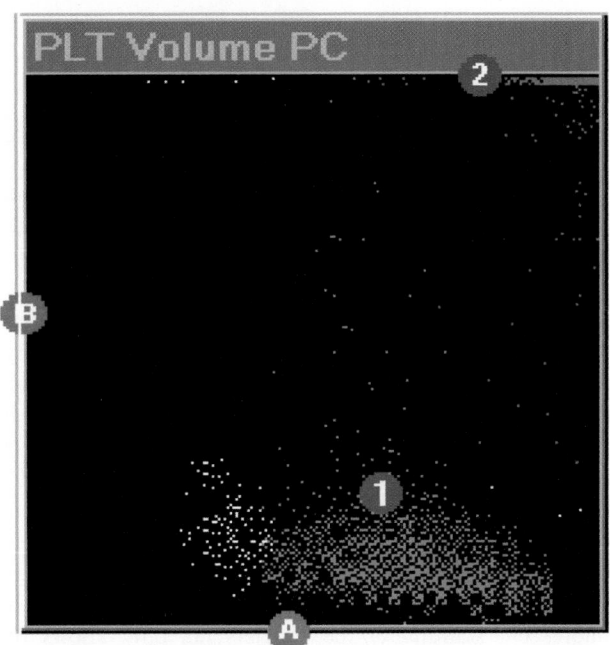

**Figure 24-4.** Identification of platelets by the Bayer ADVIA 120 hematology analyzer. The vertical axis (B) indicates low angle light scatter or cell volume. The horizontal axis (A) indicates high angle light scatter or refractive index. Particles in area 1 are platelets. Particles in area 2 are red cells.

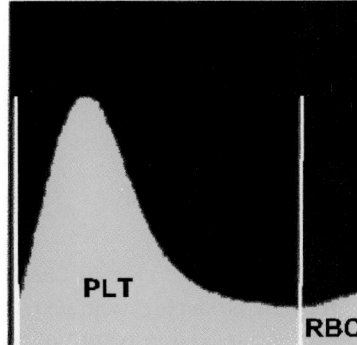

**Figure 24-5.** An example of a scattergram produced by the Abbott CELL DYN 4000, showing the optical platelet count and impedance platelet size distribution. See text for details.

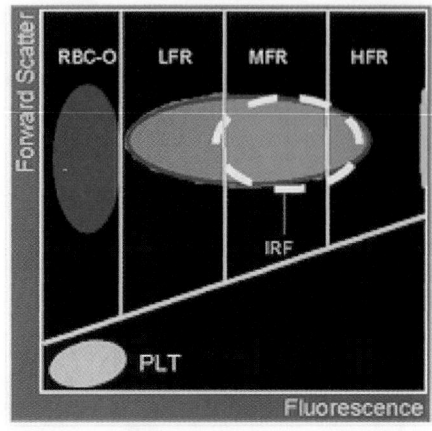 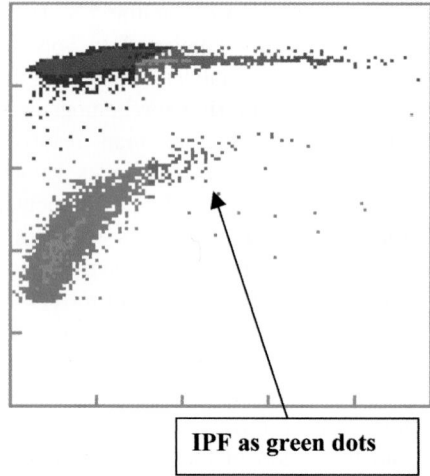

**Figure 24-6.** An example of a scattergram produced by the Sysmex XE-2100 hematology analyzer in both cartoon (left) and dot plot (right) formats. The vertical axis indicates forward scattered light or cell volume. The horizontal axis indicates fluorescence intensity. The scattergram is divided into a platelet (PLT) area, a mature red cell area (RBC-O), and the various immature reticulocyte fractions (IRF): LFR, MFR, and HFR. The immature platelet fraction (IPF) is represented as green dots in the dot plot format. (reprinted with permission from Sysmex Corporation)

tional impedance count.[14] The optical fluorescent platelet count is measured in the reticulocyte channel. A polymethine dye is used to stain the RNA/DNA of reticulated cells and platelet membranes and granules. This technology allows the simultaneous counting of the red cell reticulocytes, erythrocytes, and fluorescent platelets (Fig. 24-6). Within the flow cell, each single cell is passed through the light beam of a semiconductor diode laser. The fluorescence intensity of each cell is analyzed, which allows the separation of platelets from red cells and reticulocytes. The fluorescent staining of the platelets not only allows the exclusion of nonplatelet particles from the count, but also allows the inclusion of large or giant platelets.

The optical fluorescence count is more reliable at levels below $100 \times 10^9$/L and may allow more appropriate clinical decisions to be made, particularly with regard to platelet transfusions.[14] However, for samples from patients undergoing cytotoxic chemotherapy, the impedance count is sometimes more accurate. This is probably due to the erroneous staining of white cell fragments following apoptosis.[15] A switching algorithm has been designed on the XE-2100 to report the most accurate platelet count, either optical or impedance.[16]

### D. Quality Control for Automated Hematology Analyzers

Modern analyzers are very precise, but care still needs to be taken to ensure that they are producing accurate platelet counts. Most modern instruments tend to be precalibrated by the manufacturer but often perform adjustments according to the individual blood sample characteristics. Despite

improved calibration, each instrument requires regular maintenance and cleaning (according to the manufacturer's specifications) to ensure optimum performance. Each laboratory also should establish an in-house reference range for each measured cell type, including the platelet count. Quality control procedures should be performed regularly (e.g., daily) to check for accuracy. Each analyzer manufacturer produces quality control material that can be purchased to monitor performance of the instrument. The control can only usually be used on those manufacturers' instruments with their specific reagents. The control consists of treated stabilized human erythrocytes in an isotonic bacteriostatic medium, with the addition of a stabilized platelet-sized component and white blood cells or fixed erythrocytes to simulate blood cells. The controls are usually available with low, normal, or high levels of white cells, red cells, and platelets. Each control has assigned values and expected ranges. Expected ranges include variation between lots and between individual instruments, and represent 95% confidence limits for well-maintained instrument systems.

The United Kingdom National External Quality Assessment Service (UK NEQAS) is an external quality control service. Participating clinical laboratories are sent stabilized blood samples on a regular basis that they treat as a normal patient sample for testing. The samples are analyzed in each of the hematology laboratory's instruments and results are returned to NEQAS. NEQAS then provides a report that compares the participating laboratory's performance to that of all laboratories using the same test method or analyzer. Other countries have developed similar external quality controls. However, the lack of an internationally recognized standard for platelets necessitates the use of a consensus target value to establish limits of performance. As the many different blood cell counters available use a variety of technologies and diluents, they may respond in different ways to the stabilized blood used in the surveys. Performance is therefore individually assessed within instrument groups against the consensus target value.

### E. Immunological Platelet Counting

With the widespread availability of flow cytometers within hematology and research laboratories, a number of different groups began to investigate the applicability of this technology (Chapter 30) to accurately enumerate various cells within whole blood, including platelets. The principle of this methodology involves simply labeling EDTA-anticoagulated blood with a suitable antiplatelet monoclonal antibody, that has been fluorescently conjugated with, for example, fluorescein isothiocyanate (FITC). Because most flow cytometers cannot measure a fixed volume of sample, counting procedures involve indirect derivation of cell number using the ratio of fluorescent platelets to either added bead preparations or inherent number of red cells within the sample. Recently, a number of flow cytometric counting procedures were reviewed by the ICSH expert panel on cytometry.[3,17–19] The ICSH panel identified the variables and problems associated with this methodology which enabled the International Society of Laboratory Haematology (ISLH) task force panel to develop, evolve, and test a new candidate reference method with a multilaboratory study.[20,21] The preferred method simply derives the platelet count from the ratio of fluorescent platelets to red cells within the sample (Fig. 24-7). The main advantage of the red cell (RBC) ratio is that, providing the blood sample is well mixed and that coincident events (RBC/RBC and RBC/platelet) are eliminated by optimal dilution, the count obtained is not only accurate and precise but also independent of potential pipeting artifacts. The method is also superior to derived counts from bead ratios,[3] because these methods are dependent upon a stable bead preparation (with an accurate bead count) in combination with very accurate/precise pipeting. However, bead-derived platelet counts may be useful for simply counting platelets within purified preparations when the red cells have been removed.

### F. Automated Immunological Counting

With the more recent convergence of flow cytometry and analyzer technology, it became feasible not only to perform optical counting by light scatter and fluorescence but to simultaneously measure cells identified with fluorescent monoclonal antibodies. Currently, the only commercially available hematology analyzer that can measure antibody-labeled platelets is the Abbott CELL-DYN 4000 or CELL-

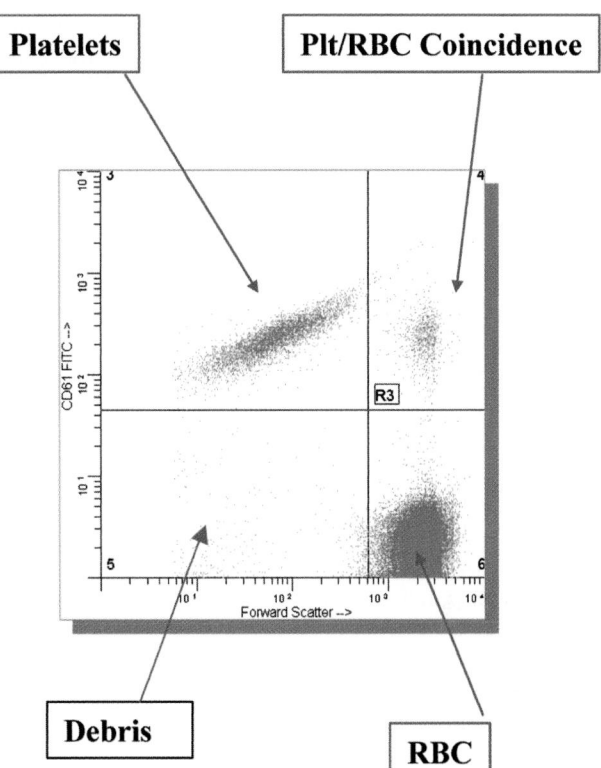

**Figure 24-7.** Immunological platelet counting. Flow cytometry scattergram of log fluorescence (CD62-FITC, FL1, vertical axis) versus log forward scatter (horizontal axis). The fluorescent platelets are clearly resolved from noise/debris, red cells (RBCs), and platelet (Plt)/RBC coincidence events.

DYN Sapphire. Unlike the flow cytometric method, the ImmunoPLT™ method is a fully automated procedure. It labels platelets within whole blood by using anti-CD61 antibodies contained within a lyophilized pellet inside special evacuated tubes (Becton Dickinson, San Jose, CA). During analysis, the CELL-DYN 4000 simply aspirates 56.3 μL of blood into the antibody-containing tube and performs a standard incubation. Final counting is performed within a fixed volume and includes PLT/RBC coincidence events but is hence not based upon a cell ratio. The method has been shown to provide an accurate platelet count, especially within thrombocytopenic samples.[22,23] As expected, it has also been recently shown to agree closely with immunocounting by flow cytometry.[23] The fully automated immunological technique has obvious advantages and will be very useful in clinical situations in which accurate platelet counts are required.

## IV. Conclusions

Many methods for counting platelets have been published, and the number of alternative methods is no doubt due to the difficulties in counting small cells which are activated easily, aggregate, and are also difficult to resolve from extraneous matter. For the research or nonspecialized laboratory setting, the manual count still offers the least expensive and easiest methodology if there is no access to a large hematology analyzer. Alternatively, some laboratories do invest in small impedance analyzers, which provide a rapid and precise way of counting platelets. The recent development of a new immunological platelet counting method allows laboratories with access to a flow cytometer to count platelets very accurately by ratioing to either red cell number (in whole blood) or to added bead preparations (in whole blood, PRP, or purified platelet preparations). Accurate and precise platelet counts in severely thrombocytopenic patients have become more important in recent years due to increased cytotoxic treatments resulting in prolonged thrombocytopenia, and the desire to reduce the frequency and threshold of platelet transfusions (Chapter 69). With the development of new automated platelet counting methods and two-dimensional analysis using light scatter or fluorescence, many of the limitations that exist with so called one-dimensional analyzers (e.g., impedance and single light scatter) are reduced. In two-dimensional analysis, platelets of a similar size to red cells should be included in the count and red cell fragments, cell debris, and other particulate excluded. Alternative platelet counting approaches using immunological markers to unequivocally identify platelets have still further improved the accuracy of the count. The flow cytometric method for counting platelets has been recommended as a potential reference method and has been the subject of review by the ICSH expert panel on cytometry.[3,17,19,20]

A fully automated immunological technique, as on the Abbott CELL-DYN 4000, has obvious advantages. Using the immunological platelet counting reference method, manufacturers of all hematology analyzers will now be able to calibrate the platelet count with more accuracy. External quality control programs (e.g., NEQAS and CAP) must now develop suitable stabilized and calibrated materials to assess the accuracy of counting in thrombocytopenia. These developments would lead to the reporting of reliable low platelet counts on which clinicians can base their treatment or transfusion decision making with confidence. Comparative studies with different analyzers and immunocounting will facilitate the re-evaluation of current platelet transfusion thresholds and may lead to the threshold being reduced from $10 \times 10^9$/L to $5 \times 10^9$/L, as has been proposed in the past.[24–28] A large multicenter study to compare the inaccuracy of platelet counts of current analyzers in severe thrombocytopenia (compared to a reference flow cytometric method) showed that most analyzers overestimated the count, which would result in undertransfusion of platelets at any set threshold.[29] This study highlights the inaccuracies of hematology analyzers in platelet counting and re-emphasizes the need for external quality control to improve analyzer calibration for samples with a low platelet count. It also suggests that the optimal thresholds for prophylactic platelet transfusions should be re-evaluated. Current methods of optical platelet counting may not be superior to impedance counts for all patient populations.[30] In the future, certain analyzers may introduce additional parameters from the platelet counting technology. A reliable method to quantify immature or reticulated platelets may be useful, providing correlation has taken place with well-characterized clinical situations.[31–33]

# References

1. Brecher, G., Schneiderman, M., & Cronkite, E. P. (1953). The reproducibility of the platelet count. *Am J Clin Path, 23,* 15–21.
2. England, J. M., Rowan, R. M., Bins, M., et al. (1998). Recommended methods for the visual determination of white cell and platelet counts. *WHO LAB, 88,* 1.
3. Harrison, P., Horton, A., Grant, D., et al. (2000). Immuno-platelet counting: A proposed new reference procedure. *Br J Haematol, 108,* 228–235.
4. Bentley, S. A., Johnson, A., & Bishop, C. A. (1993). A parallel evaluation of four automated hematology analysers. *Am J Clin Path, 100,* 626–632.
5. Bull, B. S., Schneiderman, M. A., & Brecher, G. (1965). Platelet counts with the Coulter counter. *Am J Clin Path, 44,* 678–688.
6. Ault, K. A. (1996). Platelet counting. Is there room for improvement? *Lab Haematol, 2,* 139–143.
7. Rowan, R. M. (1991). Platelet counting and assessment of platelet function. In J. A. Koepke (Ed.), *Practical laboratory haematology* (p. 157). New York: Churchill Livingstone.
8. Harrison, P., Briggs, C., & Machin, S. J. (2004). Platelet counting. *Methods Mol Biol, 272,* 29–46.
9. Martin, J. F., Bath, P. M., & Burr, M. L. (1991). Influence of platelet size on outcome after myocardial infarction. *Lancet, 338,* 1409–1411.
10. Coulter, W. H. (1953). Means for counting particles suspended in a fluid. U.S. Patent 2656508.
11. Patterson, K. (1997). Platelet parameters generated by automated blood counters, *CME Bull Haematol, 1,* 13–16.
12. Kunicka, J. E., Fischer, G., Murphy, J., et al. (2000). Improving platelet counting using two-dimensional laser light scatter. *Am J Clin Path, 114,* 283–289.
13. Tycko, D. H., Metz, M. H., & Epstein, E. A. (1985). Flow-cytometric light scattering measurement of red blood cell volume and haemoglobin concentration. *Appl Optics, 24,* 1355–1365.
14. Briggs, C., Harrison, P., Grant, D., et al. (2000). New quantitative parameters on a recently introduced automated blood cell counter — the XE 2100. *Clin Lab Haematol, 17,* 163–172.
15. van der Meer, W., Mackenzie, M. A., Dinnissen, J. W., et al. (2003). Pseudoplatelets: A retrospective study of their inci-

dence and interference with platelet counting. *J Clin Path, 56,* 772–774.

16. Briggs, C., Kunka, S., & Machin S. J. (2004). The most accurate platelet count on the XE-2100. Optical or impedance? *Clin Lab Haematol, 26,* 157–158.

17. Davis, B., & Bigelow, N. C. (1999). Indirect immunoplatelet counting by flow cytometry as a reference method for platelet count calibration. *Lab Haematol, 5,* 15–21.

18. Tanaka, C., Isii, T., & Fujimoto, K. (1996). Flow cytometric platelet enumeration utilising monoclonal antibody CD42a. *Clin Lab Haematol, 118,* 265–269.

19. Groner, W., Mayer, K., & Chapman, E. (1994). An indirect platelet count using platelet specific monoclonal antibody and flow cytometry can produce reliable platelet counts for assessing thrombocytopenia [abstract]. *Blood, 84,* (suppl 1) 687a.

20. International Council for Standardization in Haematology (ICSH) Expert Panel on Cytometry and International Society of Laboratory Haematology (ISLH). Task Force on Platelet Counting (2000). Platelet counting by the PLT/RBC ratio — a reference method, *Am J Clin Path, 115,* 460–464.

21. Harrison, P., Ault, K. A., Chapman, S. E., et al. (2000). An inter-laboratory study of a candidate reference method for platelet counting. *Am J Clin Path, 115,* 448–459.

22. Ault, K. A., Mitchell, J., Knowles, C., et al. (1997). Implementation of the immunological platelet count on a haematology analyser — the Abbott CELL-DYN 4000. *Lab Haematol, 3,* 125–128.

23. Kunz, D., Kunz, W. S., Scott, C. S., et al. (2001). Automated CD61 immunoplatelet analysis of thrombocytopenic samples. *Br J Haematol, 112,* 584–592.

24. Gmur, J., Burger, J., Schanz, U., et al. (1991). Safety of stringent prophylatic platelet transfusion policy for patients with acute leukaemia. *Lancet, 338,* 1223–1226.

25. Murphy, W. G. (1992). Prophylactic platelet transfusion in acute leukaemia. *Lancet, 339,* 120.

26. Springer, W., von Ruecker, A., & Dickerhoff, R. (1998). Difficulties in determining prophylactic transfusion thresholds of platelets in leukemia patients. *Blood, 92,* 2183–2184.

27. Ancliff, P. J., & Machin S. J. (1998). Trigger factors for prophylactic platelet transfusions. *Blood Rev, 12,* 234–238.

28. Norfolk, D. R., Ancliff, P. J., & Contreras, M. et al. (1998). Consensus Conference on Platelet Transfusion, Royal College of Physicians of Edinburgh, 27–28 November 1997. Synopsis of background papers, *Br J Haematol, 101,* 609–617.

29. Segal, H., Briggs, C., Kunka, S., et al. (2005). Accuracy of platelet counting haematology analysers in severe thrombocytopenia and potential impact on platelet transfusion. *Br J Haematol, 128,* 520–525.

30. Sandhaus, L. M., Osei, E. S., Agrawal, N. N., et al. (2002). Platelet counting by the Coulter LH 750, Sysmex XE 2100, and Advia 120: A comparative analysis using the RBC/platelet ratio reference method. *Am J Clin Path, 118,* 235–241.

31. Briggs, C., Kunka, S., Hart, Dan., et al. (2005). Assessment of an immature platelet fraction (IPF) in peripheral thrombocytopenia. *Br J Haematol, 126,* 93–99.

32. Kickler, T. S., Oguni, S., & Borowitz, M. J. (2006). A clinical evaluation of high fluorescent platelet fraction percentage in thrombocytopenia. *Am J Clin Pathol, 125,* 1–6.

33. Abe, Y., Wada, H., Tomatsu, H., et al. (2005). A simple technique to determine thrombopoiesis level using immature platelet fraction (IPF). *Thromb Res,* Epub ahead of print.

# CHAPTER 25

# The Bleeding Time

## Stuart E. Lind[1] and Carla D. Kurkjian[2]

[1]*Department of Medicine, University of Colorado Health Sciences Center, Denver, Colorado*
[2]*University of Oklahoma Health Sciences Center, Oklahoma City, Oklahoma*

## I. Introduction

Credit for the first description of the skin bleeding time is often given to a 1901 publication by the French physician, Milian.[1] One of the oldest tests still in clinical use, the bleeding time appears to be falling out of favor (at least for the time being) with clinicians.[2] Despite its shortcomings, the bleeding time is likely to continue to play an important role in the evolution of our knowledge of the blood and vasculature. It should therefore not be discarded, but be placed into the category of specialized tests that are utilized sparingly, but knowingly.

There are a number of lessons in the story of the bleeding time for both the clinician and the investigator. This chapter will review the history and shortcomings of the bleeding time as it has been used by clinicians. In addition, efforts will be made to point out some of the contributions of the bleeding time to our understanding of the complexities that underlie modern concepts of hemostasis and thrombosis. Those readers interested in more complete histories of thinking about platelet function in general, and the bleeding time in particular, should consult the several available reviews of the subject,[3,4] and the Foreword to this book.

## II. The Development of the Bleeding Time

Although credit[3] for the first description of the bleeding time is given to Milian's 1901 paper, Duke's[5] description of the earlobe bleeding time in 1910 had a greater influence upon medical thinking.[6] Not only did Duke show that the bleeding time (and the amount of blood shed) was increased in thrombocytopenia (thereby proving the primary function of blood platelets), he showed that a whole blood transfusion raised the platelet count, led to a cessation of clinical bleeding, and shortened the bleeding time to normal. (Only many decades later, as is discussed further below, would the beneficial effects of red cell transfusions upon hemostasis become commonly recognized.) Duke introduced the tech-

nique, still in use today, of touching a piece of filter paper to the edge of the bleeding time wound at fixed intervals (see Fig. 23-1 in Chapter 23). The analysis of the blood shed from a bleeding time wound has been useful for many investigators, who have quantified many parameters related to the amount and rate of bleeding, as well as the number and state of activation of platelets flowing from the bleeding time wound,[7–10] thereby generating important insights into the dynamics of hemostasis.

Ivy, a surgeon, popularized the next major development in the performance of the bleeding time.[11] Because physicians had recognized that patients with jaundice had bleeding problems, Ivy felt that a worthwhile test of hemostasis would detect preoperatively what he often found at the operating table. He noted that the Duke bleeding time was often normal in patients with jaundice, and he concluded that it "has been disappointing." Believing that the contraction of capillaries in a bleeding time wound could compensate for hemostatic impairment that would otherwise prolong the bleeding time, he sought to eliminate interpatient variability in capillary "tone." He therefore imposed a uniform "tone" upon the capillaries by applying a standard amount of pressure with a blood pressure cuff proximal to the site of the bleeding time wound, which he made on the arm. He noted that "*in a number of cases of jaundice, it was found that often, when the Duke's bleeding time was normal, the venous pressure bleeding time was definitely prolonged*" (emphasis his). Having demonstrated what he believed to be true, Ivy did not perform the statistical analysis that would be required of any new test introduced today. Ivy left it to his colleagues to provide more clinical detail about the bleeding time in different subgroups of jaundiced patients.[12]

Both the Duke and Ivy bleeding times were used in subsequent years. Eventually, the modified Ivy bleeding time won out over the earlobe method of Duke, though some hemostasis specialists continued to use the Duke bleeding time. Several factors were responsible for this shift. Probably of greatest importance was the (theoretical) appeal to clinicians of the modifications that were made over the years to "standardize" the performance of the test. Though Tocantins[13] had made modifications as early as 1936, it was the work of Mielke,[14] as well as the development of commercial instruments to perform the test, that most persuaded physicians that this modified Ivy method was preferred to the Duke earlobe puncture method. Of additional importance were anecdotal reports of earlobe wounds that would not stop bleeding in patients with von Willebrand disease, observations that may be of great physiological importance, as discussed further below.

The modified Ivy bleeding time became firmly established as the preferred method with the 1972 publication of a widely quoted paper which concluded that "the standardized bleeding time measures the overall hemostatic role of platelets *in vivo,* and is thus suitable for systematic screening."[15] This paper was conceptually appealing to many physicians because (1) it showed a quantitative relationship between the bleeding time and the platelet count; (2) it showed that patients with immune thrombocytopenia have bleeding times shorter than one would predict from their counts (which appeared to explain the clinical observation that they often had less bleeding than one would predict from their platelet count); and (3) that the bleeding time of uremic patients, who often suffered from severe mucosal bleeding (in the days before erythropoietin therapy), was longer than would be predicted from their platelet counts. With this information, and the apparent causal relationship between prolonged bleeding times and postcardiac bypass bleeding,[16] clinicians felt secure in using the bleeding time to screen a wide variety of patients, particularly in the preoperative setting.

Many years were to pass before the bleeding time was openly criticized in print. Privately, some discussed the apparent variability of the bleeding time in individual patients, although the variability may have been more biological than methodological in origin. (Abildgaard et al.,[17] for example, carefully showed that the bleeding time, factor VIII, von Willebrand factor antigen, and ristocetin cofactor levels may vary dramatically in individual patients studied on more than one occasion, indicating the importance of biological variability.) Thus, the definitive review article of the 1970s concerning the bleeding time provided a detailed appraisal of the various techniques of performing the bleeding time and countered criticisms of the test by stating that "... the bleeding time test will give reliable, reproducible results if the variables are carefully standardized and if the test is meticulously performed by an experienced operator." Although these authors seemed to champion the test, they were aware of its limitations, and concluded "... it seems to us that some of the literature on the bleeding time is not without a certain calorific gaseousness."[3]

The tide began to turn against the widespread use of the bleeding time with a 1984 review.[18] This paper noted some of the test's shortcomings and pointed out that there were few established guidelines to help the physician in evaluating the results of the test. It concluded that the database describing its performance as a diagnostic test was inadequate for the calculation of even the basic determinants of a test's performance characteristics (such as sensitivity, specificity, or predictive value) and that it should not be used as a routine preoperative test.

A more direct criticism of the widespread use of preoperative screening with the bleeding time began with a retrospective review of its use at a single institution. The authors[19] found that the bleeding time gave virtually no novel information beyond that which would have been obvious to a knowledgeable historian and observer. Picking up on this evolving theme, Burns and Lawrence stated in 1989 that "the clinician need not perform a BT (bleeding time) prior to routine general or cardiac surgery or minor surgical procedures."[20] That this view was not universally held is apparent from remarks made in an editorial that accompanied its publication.[21]

In the early 1990s, two reviews appeared that attacked the bleeding time head on. The first,[22] an encyclopedic review of 862 articles concerning the bleeding time, explored in great detail the various claims made on its behalf. Its compilation was so time-consuming that it only cited papers published through 1986. The authors moved beyond the descriptive analyses provided in prior reviews and attempted to quantify the diverse and disparate literature concerning its use with graphical analyses using receiver-operating characteristic (ROC) curves. Their analysis painted a damning picture of the test and showed that it performed poorly in many of the ways that clinicians had come to accept on faith.

The second[23] reviewed the literature between 1986 and 1990 and used published, clinical examples to explain to a wider audience of physicians and surgeons what had been apparent to knowledgeable subspecialists for many years. Subsequent analyses confirmed these conclusions,[10,24,25] and the message began to filter to the surgical community that the test was not the safety net that many hoped for.[26] The final blow to the bleeding time was delivered in the form of a position paper published in 1998 in the *Archives of Surgery* by leading clinical pathologists that concluded that the use of the bleeding time as a routine screening test was not warranted.[27]

Eventually, the message of these papers and those that followed them was heeded by the medical community, and the number of preoperative bleeding time tests began to fall. Documentation that medical care is not compromised by its retirement from the roster of active clinical tests is beginning to appear in the medical literature.[28] Discontinuation of the bleeding time test at a tertiary care center did not result in increased postprocedural bleeding complications, increased platelet or blood transfusions, or changes in clinicians' practices (i.e., the number of platelet aggregation studies ordered or the amount and dose of desmopressin [DDAVP] administered).[28]

## III. Clinical Uses of the Bleeding Time

The bleeding time has been used to (1) screen patients preoperatively, (2) find a cause of bleeding in an actively bleeding patient, (3) understand the cause of previous bleeding episodes, or (4) determine if an individual has a hereditary bleeding disorder.[23]

As discussed in virtually all of the recent articles cited above, the bleeding time is a poor preoperative screening test, and should not be used in a routine manner. While neurosurgeons, perhaps to a greater degree than other surgeons, live in fear of unexpected bleeding, there is no evidence in the literature that the bleeding time is more valuable in screening their patients, compared to other patients. One group went so far as to study the time required for bleeding to stop when an incision is made in the (rat) brain. They found that while aspirin prolongs the skin bleeding time, it does not prolong the brain bleeding time.[29]

There is little literature to help the clinician in deciding whether to perform a skin bleeding time when confronted with an actively bleeding patient. While some might argue for the performance of this test before undertaking a specific therapeutic maneuver, the empiric administration of desmopressin is so safe, and the performance of a bleeding time in rushed circumstances so questionable, that there is little to be gained by first performing the test. (Although there are reports[30] of thrombotic complications following desmopressin, they appear to be quite uncommon, except perhaps in the setting of thrombotic microangiopathy.[31,32])

The bleeding time may occasionally be useful in considering the cause of a prior bleeding event, particularly if the effect of desmopressin is ascertained on the same day. Clinicians must, however, always keep in mind the lessons learned from the study of the bleeding time, especially in the most common congenital disease affecting platelet function, von Willebrand disease: bleeding times vary from one day to the next in the same patient;[17] the bleeding time is normal in about 50% of patients with von Willebrand disease; the intra-and interobserver coefficient of variation of the bleeding time is about 15%;[10] and drugs, alcohol,[33] and (probably) vitamin and food supplements likely interact to alter the bleeding time, and may not be present at equivalent concentration on different days.

The clinician asked to confirm the presence of a congenital disorder of platelet function can expect, at best, confirmatory information from a bleeding time. If normal, one cannot exclude von Willebrand disease, and if abnormal, one cannot point to a specific disorder. Searching for other, pathognomonic findings of platelet dysfunction (such as

characteristic aggregation tracings or granule deficiencies on electron microscopy) is more likely to be of assistance. The use of the bleeding time to screen patients with significant personal and family bleeding histories has been disappointing. One major, recent series has demonstrated a sensitivity of only 33.5% in this patient population.[34] Finding that a prolonged bleeding time shortens with desmopressin may be of psychological benefit to all concerned, but many patients with von Willebrand disease can successfully undergo surgery even when their bleeding time is abnormal.

## IV. Contributions and Benefits of the Bleeding Time

As discussed above, the bleeding time led to an understanding of the function of platelets in the body. Those who might be inclined to abandon the bleeding time altogether, because it has not fulfilled the goal of finding a test suitable for mass screening of patients before surgery, risk overlooking the major contribution of the test: it has shed important light on the processes of hemostasis and thrombosis. Until such time as there are no remaining gaps in our knowledge, the bleeding time should remain in the armamentarium of those studying the biology of the blood and the vasculature. Areas where the bleeding time has been useful include the following.

### A. Transfusion Medicine

The bleeding time has been useful in studying the functionality of platelets collected for transfusion and, in the view of some, provides the only hard evidence that transfused platelets are functional. The bleeding time has been important in showing that hemostasis may be obtained without utilizing viable platelets, and may lead to the development of functional preparations of platelet membranes or even artificial platelets (Chapter 70).[35]

### B. Drug and Device Development

Virtually all antiplatelet drugs prolong the bleeding time and studies of the bleeding time are an integral component of new antiplatelet drug testing. Of interest is the observation that some drugs may counteract the effects of antiplatelet drugs on the bleeding time,[36] or prevent platelet dysfunction in circumstances where it would otherwise be expected, such as following cardiopulmonary bypass (reviewed in references 23 and 37). As the use of antiplatelet agents in combination has increased, the bleeding time has provided useful information regarding the potential for bleeding in this patient population.[38] For example, the demonstration that addition of the platelet inhibitor cilostazol did not induce further prolongation of the bleeding time in patients treated with aspirin and clopidogrel has clinical implications (Chapter 64).[39]

The bleeding time has also been important in defining drugs (such as desmopressin and estrogens) that improve hemostasis[40] and demonstrating the hemostatic benefits of hemodialysis for uremic patients.[41] In women with menorrhagia, prolonged bleeding times, and no defined coagulation factor deficiencies, nasal desmopressin has been shown to significantly reduce menstrual blood loss.[42] This finding has significant clinical utility, as it is estimated that 10% of women of reproductive age suffer from menorrhagia.

The bleeding time also provides insights into favorable interactions between drugs and clinical interventions. For example, a recent series demonstrated that although the bleeding time was shortened after 4 hours of dialysis, a further shortening was seen after administration of desmopressin.[43]

### C. Development of Concepts Regarding Important Physiological Interrelationships

**1. The Importance of Thrombin Generation and Platelet Function.** Although it is convenient to think of "thrombin generation" as being distinct from "platelet function," the connection between the two is evident from reports of prolonged bleeding times in some patients with the classic disease of thrombin generation, hemophilia (reviewed in Ref. 23). The importance of thrombin is also illustrated by the effects of heparin upon the bleeding time.[44]

**2. The Effect of Plasmin upon Platelet Function.** The ability of plasmin generation to affect the bleeding time is an underappreciated fact of clinical life that may be more important in explaining clinical bleeding problems in the group that Ivy was initially concerned with (those with cirrhosis) than are the prolonged clotting times that attract most clinicians' attention. That cirrhosis is a state of chronic hyperfibrinolysis has long been known, though the importance of fibrinolysis in determining the extent or timing[45] of clinical bleeding is still not fully appreciated. It is possible that the prolonged bleeding time observed in cirrhosis[46] is due, in part, to chronic activation of the fibrinolytic system. The study of the plasmin–platelet interaction was stimulated by the finding that acute activation of the fibrinolytic system by thrombolytic agents leads to changes in platelet function and the bleeding time,[47] and the report that aprotinin preserves platelet function[48] and prevents the normal prolongation of the bleeding time seen after cardiopulmonary bypass.

**3. The *In Vivo* Effects of Nitric Oxide upon Platelet Function.** The far-reaching importance of the body's nitric oxide generating systems (Chapter 13) extends, not surprisingly, to the bleeding time. Adults exposed to inhaled nitric oxide at 30 parts per million for 15 minutes demonstrated a prolongation of the bleeding time by 33%, with normalization of values within 1 hour of its discontinuation.[49] It now appears that at least part of the platelet dysfunction long associated with chronic renal failure is due to excessive nitric oxide production (Chapter 58).[50] Increased nitric oxide production may also explain part of the bleeding time prolongation seen in patients with cirrhosis.[51,52] Further, it appears that the beneficial effects of estrogen in shortening the bleeding time in renal failure result from effects upon nitric oxide production.[53,54]

**4. The Effect of Erythrocytes upon Primary Hemostasis.** The connection between the ability of platelets to close a wound and the hematocrit can be inferred in part from the 1910 paper of Duke,[5] in which he showed that a wound bled longer and more profusely when created in a patient with anemia than in a normal subject. This concept was re-explored in more detail in the 1980s and 1990s[55] after it was shown that the prolonged bleeding time of uremia could be shortened by red cell transfusions[56] or erythropoietin.[57] Further *in vitro* studies led to greater understanding of how platelets interact with cells other than endothelial cells, notably leukocytes[58] and erythrocytes[59] (especially sickled erythrocytes[60]). Recently, the importance of red cells in determining the bleeding time in normal subjects has been documented, a finding that may have implications for the management of surgical and/or thrombocytopenic patients.[61] The effect of increasing the hematocrit does appear to have a ceiling, since raising the hematocrit to more than 30% does not cause additional shortening of the bleeding time.[62]

**5. The Effect of Temperature upon Primary Hemostasis.** The importance of the temperature upon platelet function has been shown *in vivo* through the study of the bleeding time in human subjects.[63–65]

## V. Is There a Need for More Clinical Research Using the Bleeding Time?

The answer to this question is a qualified "yes," provided that those contemplating such research are aware of what has been written, and what new information is needed. Further retrospective studies are not needed to document that the performance of a bleeding time in the general preoperative setting is not helpful. It is possible that research into specialized populations, such as children undergoing tonsillectomy or women being evaluated for menorrhagia, might be useful, both for the patients and those seeking to study the bleeding time. Those investigators contemplating such an approach, however, are unlikely to generate useful information just by sending their patients off to a routine clinical laboratory, but must be willing to make the investment necessary to advance the field. Specific questions that might be addressed, in appropriate populations, are the following.

### A. How Should the Bleeding Time Be Performed?

Even if we believe that the skin bleeding time might have some use in predicting bleeding problems, it is not at all clear that the modified Ivy bleeding time, as presently performed, is the ideal way to measure it. Nilsson and colleagues showed that the greater "sensitivity" of the Ivy bleeding time did not correlate well with clinical bleeding.[66] These investigators[66] noted that patients with mild von Willebrand disease had prolonged Ivy bleeding times when the blood pressure cuff was applied, but normal ones without it, or when the Duke method was used. Since these patients usually do not have a serious bleeding disorder, the emphasis on a more "sensitive" test was apparently misplaced. (To previous generations of investigators, "sensitive" tests were those which were more likely to be abnormal, reflecting real or imagined problems. Modern investigators understand "sensitivity" to indicate how well a test detects a given disease state.) Thus, one area of investigation would be to

determine how different types of bleeding time tests perform in detecting von Willebrand disease or predicting surgical bleeding in specialized populations.

As noted above, investigators have studied the modified Ivy bleeding time (on the arm) with or without a blood pressure cuff. Reports of a thigh bleeding time[67] or a nonocclusive arm bleeding time[68] have been published. These modifications could be further studied to see if they are better predictors of bleeding.

### B. Can a Bleeding Time-Equivalent Be Performed In Vitro?

A number of methods have been devised in an attempt to parallel the bleeding time, using a blood sample and one of variety of types of machines (Chapter 23)[69] (e.g., the platelet function analyzer (PFA)-100) (Chapter 28) and the cone and plate(let) analyzer (Chapter 29). Efforts to evaluate the potential benefits of these tools should be informed by the lessons learned from the clinical evaluation of the bleeding time over the past 100 years. While testing to show that such evaluations parallel the findings of other tests is important, it does not follow that clinically useful information will be gained by applying the test to all individuals undergoing surgery or invasive procedures.[70] Although the PFA-100 appears to be able to detect von Willebrand disease (Table 25-1) or provide abnormal test results in selected patient

**Table 25-1: Detection of von Willebrand Disease**

| Reference No. | Number of Subjects | Sensitivity (%) | | |
| --- | --- | --- | --- | --- |
| | | PFA-100 Collagen–ADP Cartridge | PFA-100 Collagen–Epinephrine Cartridge | Bleeding Time |
| 77 | 60 | 100 | 97 | 66 |
| 78 | 52 | 87 | 88 | 65 |
| 79 | 32 | 73 | 84 | 48 |
| 80 | 41 | 83 | 79 | 27 |
| 81 | 30 | 61 | 65 | 21 |
| 82 | 43 | 56 | 78 | 75 |
| 83 | 34 | 61 | 65 | 21 |
| 84 | 12 | 80 | 100 | 17 |
| 85 | 53 | (94)[a] | (94)[a] | 58 |
| 34 | 26 | (62)[a] | (62)[a] | 42 |
| 86 | 54 | 78 | 96 | N/A |
| Mean | | 76 | 83 | 44 |

[a]Only overall sensitivity for PFA-100 reported.
PFA-100, platelet function analyzer 100; N/A, not available.

**Table 25-2: Frequency of Test Abnormalities in Selected Populations Studied with the PFA-100 and the Template Bleeding Time**

| Reference No. | Population | Number of Subjects | PFA-100 (% abnormal) | Bleeding Time (% abnormal) |
|---|---|---|---|---|
| 34 | Patients presenting with hereditary mucocutaneous hemorrhage | 148 | 29.7 | 35.8 |
| 78 | Congenital defects of platelet secretion | 17 | 41 | 47 |
| 87 | Patients on hemodialysis | 34 | 59/62[a] | 20 |
| 86 | Preoperative assessment in patients with positive bleeding history | 628 | 98/78[a] | 74 |

[a]Collagen–epinephrine cartridge and collagen–ADP cartridge, respectively.

groups (Table 25-2) with a frequency superior or at least comparable to that of the bleeding time, its specificity remains an important issue for future study (Chapter 28).

## VI. The Future of the Bleeding Time: Consider Heterogeneity

In virtually all studies that have looked, there is significant heterogeneity in the bleeding time[13] (or in its *in vitro* counterparts[71]) and its response to pharmacological challenge.[72,73] While this may reflect changes in the environment of the platelet (as occurs in von Willebrand disease[17]), it may also signify differences in the contents of the platelet α-granules[74] or the molecular makeup that distinguishes one individual from another. Recent studies, for example, have linked a polymorphism of integrin β3 (which accounts for the Pl$^{A1/A2}$ phenotype) with the bleeding time response to aspirin[75] and have shown that mice deficient in the adhesion molecule CD31 (PECAM-1) have prolonged bleeding times.[76] These studies illustrate that the variability of the bleeding time, for which many laboratory technicians have been unjustly maligned, may provide important insights into the genetics and cell biology of vascular function.

In summary, care must be taken to insure that the bleeding time is not used for purposes for which it is not well suited. However, it is important that specialists in platelet and vascular function are knowledgeable about the bleeding time, and continue to use it as a specialized tool for investigating platelet physiology *in vivo*.

## References

1. Milian, M. G. (1901). Technique pour l'etude clinique de la coagulation du sang. *Soc Med Hosp Paris, 18,* 777–779.

2. Lehman, C. M., & Rodgers, G. M. (2000). Abolition of the bleeding time test at an academic medical center. *Blood, 96,* 436a.

3. Bowie, E. J., & Owen, C. A., Jr. (1974). The bleeding time. *Prog Hemost Thromb, 2,* 249–271.

4. Spaet, T. H. (1980). Platelets: The blood dust. In M. M. Wintrobe (Ed.), *Blood, pure and eloquent,* pp. 548–571. New York: McGraw-Hill.

5. Duke, W. W. (1910). The relation of blood platelets to hemorrhagic disease. *JAMA, 60,* 1185–1192.

6. Brinkhous, K. M. (1983). W. W. Duke and his bleeding time test. A commentary on platelet function. *JAMA, 250,* 1210–1214.

7. Sutor, A. H., Bowie, E. J., & Owen, C. A., Jr. (1977). Quantitative bleeding time (hemorrhagometry). A review. *Mayo Clin Proc, 52,* 238–240.

8. Schwartz, B. S., Leis, L. A., & Johnson, G. J. (1979). In vivo platelet retention in human bleeding-time wounds. Ii. Effect of aspirin ingestion. *J Lab Clin Med, 94,* 574–584.

9. Abrams, C. S., Ellison, N., Budzynski, A. Z., et al. (1990). Direct detection of activated platelets and platelet-derived microparticles in humans. *Blood, 75,* 128–138.

10. De Caterina, R., Lanza, M., Manca, G., et al. (1994). Bleeding time and bleeding: An analysis of the relationship of the bleeding time test with parameters of surgical bleeding. *Blood, 84,* 3363–3370.

11. Ivy, A. C., Shapiro, P. F., & Melnick, P. (1935). The bleeding tendency in jaundice. *Surg Gynecol Obstet, 60,* 781–784.

12. McNealy, R. W., Shapiro, P. F., & Melnick, P. (1935). The effect of viosterol in jaundice. *Surg Gynecol Obstet, 60,* 785–801.

13. Tocantins, L. M. (1936). The bleeding time. *Am J Clin Path, 6,* 160–171.

14. Mielke, C. H., Jr., Kaneshiro, M. M., Maher, I. A., et al. (1969). The standardized normal Ivy bleeding time and its prolongation by aspirin. *Blood, 34,* 204–215.

15. Harker, L. A., & Slichter, S. J. (1972). The bleeding time as a screening test for evaluation of platelet function. *N Engl J Med, 287,* 155–159.

16. Harker, L. A., Malpass, T. W., Branson, H. E., et al. (1980). Mechanism of abnormal bleeding in patients undergoing cardiopulmonary bypass: Acquired transient platelet dysfunction associated with selective alpha-granule release. *Blood, 56,* 824–834.
17. Abildgaard, C. F., Suzucki, Z., Harrison, J., et al. (1980). Serial studies in von Willebrand's disease: Variability versus "variants." *Blood, 56,* 712–716.
18. Lind, S. E. (1984). Prolonged bleeding time. *Am J Med, 77,* 305–312.
19. Barber, A., Green, D., Galluzzo, T., er al. (1985). The bleeding time as a preoperative screening test. *Am J Med, 78,* 761–764.
20. Burns, E. R., & Lawrence, C. (1989). Bleeding time. A guide to its diagnostic and clinical utility [see comments]. *Arch Pathol Lab Med, 113,* 1219–1224.
21. Triplett, D. A. (1989). The bleeding time. Neither pariah or panacea [editorial; comment]. *Arch Pathol Lab Med, 113,* 1207–1208.
22. Rodgers, R. P., & Levin, J. (1990). A critical reappraisal of the bleeding time. *Sem Thromb Hemost, 16,* 1–20.
23. Lind, S. E. (1991). The bleeding time does not predict surgical bleeding. *Blood, 77,* 2547–2552.
24. Bernardi, M. M., Califf, R. M., Kleiman, N., et al. (1993). Lack of usefulness of prolonged bleeding times in predicting hemorrhagic events in patients receiving the 7E3 glycoprotein IIb/IIIa platelet antibody. The TAMI Study Group. *Am J Cardiol, 72,* 1121–1125.
25. Gewirtz, A. S., Miller, M. L., & Keys, T. F. (1996). The clinical usefulness of the preoperative bleeding time. *Arch Pathol Lab Med, 120,* 353–356.
26. De Rossi, S. S., & Glick, M. (1996). Bleeding time: An unreliable predictor of clinical hemostasis. *J Oral Maxillofac Surg, 54,* 1119–1120.
27. Peterson, P., Hayes, T. E., Arkin, C. F., et al. (1998). The preoperative bleeding time test lacks clinical benefit: College of American Pathologists' and American Society of Clinical Pathologists' position article. *Arch Surg, 133,* 134–139.
28. Lehman, C. M., Blaylock, R. C., Alexander, D. P., et al. (2001). Discontinuation of the bleeding time test without detectable adverse clinical impact. *Clin Chem, 47,* 1204–1211.
29. MacDonald, J. D., Remington, B. J., & Rodgers, G. M. (1994). The skin bleeding time test as a predictor of brain bleeding time in a rat model. *Thromb Res, 76,* 535–540.
30. Vermylen, J., & Peerlinck, K. (1995). Fatal complication of desmopressin. *Lancet, 346,* 447.
31. Stratton, J., Warwicker, P., Watkins, S., et al. (2001). Desmopressin may be hazardous in thrombotic microangiopathy. *Nephrol Dial Transplant, 16,* 161–162.
32. Overman, M., & Brass, E. (2004). Worsening of thrombotic thrombocytopenic purpura symptoms associated with desmopressin administration. *Thromb Haemost, 92,* 886–887.
33. Deykin, D., Janson, P., & McMahon, L. (1982). Ethanol potentiation of aspirin-induced prolongation of the bleeding time. *N Engl J Med, 306,* 852–854.
34. Quiroga, T., Goycoolea, M., Munoz, B., et al. (2004). Template bleeding time and PFA-100 have low sensitivity to screen patients with hereditary mucocutaneous hemorrhages: Comparative study in 148 patients. *J Thromb Haemost, 2,* 892–898.
35. Blajchman, M. A., & Lee, D. H. (1997). The thrombocytopenic rabbit bleeding time model to evaluate the in vivo hemostatic efficacy of platelets and platelet substitutes. *Transfus Med Rev, 11,* 95–105.
36. Herbert, J. M., Bernat, A., & Maffrand, J. P. (1993). Aprotinin reduces clopidogrel-induced prolongation of the bleeding time in the rat. *Thromb Res, 71,* 433–441.
37. Suzuki, Y., Hillyer, P., Miyamoto, S., et al. (1998). Integrilin prevents prolonged bleeding times after cardiopulmonary bypass. *Ann Thorac Surg, 66,* 373–381.
38. Payne, D. A., Hayes, P. D., Jones, C. I., et al. (2002). Combined therapy with clopidogrel and aspirin significantly increases the bleeding time through a synergistic antiplatelet action. *J Vasc Surg, 35,* 1204–1209.
39. Wilhite, D. B., Comerota, A. J., Schmieder, F. A., et al. (2003). Managing pad with multiple platelet inhibitors: The effect of combination therapy on bleeding time. *J Vasc Surg, 38,* 710–713.
40. Mannucci, P. M.(1998). Hemostatic drugs. *N Engl J Med, 339,* 245–253.
41. Butt, M. L., Shafi, T., Farooqi, I., et al. (1998). Effect of dialysis on bleeding time in chronic renal failure. *JPMA, J Pak Med Assoc, 48,* 242–244.
42. Edlund, M., Blomback, M., & Fried, G. (2002). Desmopressin in the treatment of menorrhagia in women with no common coagulation factor deficiency but with prolonged bleeding time. *Blood Coagul Fibrinolysis, 13,* 225–231.
43. Ulusoy, S., Ovali, E., Aydin, F., et al. (2004). Hemostatic and fibrinolytic response to nasal desmopressin in hemodialysis patients. *Med Princ Pract, 13,* 340–345.
44. Heiden, D., Mielke, C. H., Jr., & Rodvien, R. (1977). Impairment by heparin of primary haemostasis and platelet [14c]5-hydroxytryptamine release. *Br J Haematol, 36,* 427–436.
45. Piscaglia, F., Siringo, S., Hermida, R. C., et al. (2000). Diurnal changes of fibrinolysis in patients with liver cirrhosis and esophageal varices. *Hepatology, 31,* 349–357.
46. Violi, F., Leo, R., Vezza, E., et al. (1994). Bleeding time in patients with cirrhosis: Relation with degree of liver failure and clotting abnormalities. C.A.L.C. Group. Coagulation abnormalities in cirrhosis study group. *J Hepatol, 20,* 531–536.
47. Hirsch, D. R., Reis, S. E., Polak, J. F., et al. (1991). Prolonged bleeding time as a marker of venous clot lysis during streptokinase therapy. *Am Heart J, 122,* 965–971.
48. van Oeveren, W., Harder, M. P., Roozendaal, K. J., et al. (1990). Aprotinin protects platelets against the initial effect of cardiopulmonary bypass. *J Thorac Cardiovasc Surg, 99,* 788–796; discussion 796–787.
49. Hogman, M., Frostell, C., Arnberg, H., et al. (1993). Bleeding time prolongation and no inhalation. *Lancet, 341,* 1664–1665.
50. Noris, M., & Remuzzi, G. (1999). Uremic bleeding: Closing the circle after 30 years of controversies? *Blood, 94,* 2569–2574.
51. Hori, N., Okanoue, T., Mori, T., et al. (1996). Endogenous nitric oxide production is augmented as the severity advances

in patients with liver cirrhosis. *Clin Exp Pharmacol Physiol, 23,* 30–35.

52. Albornoz, L., Bandi, J. C., Otaso, J. C., et al. (1999). Prolonged bleeding time in experimental cirrhosis: Role of nitric oxide [see comments]. *J Hepatol, 30,* 456–460.

53. Zoja, C., Noris, M., Corna, D., et al. (1991). L-arginine, the precursor of nitric oxide, abolishes the effect of estrogens on bleeding time in experimental uremia. *Lab Invest, 65,* 479–483.

54. Noris, M., Todeschini, M., Zappella, S., et al. (2000). 17beta-estradiol corrects hemostasis in uremic rats by limiting vascular expression of nitric oxide synthases. *Am J Physiol Renal Physiol, 279,* F626–635.

55. Blajchman, M. A., Bordin, J. O., Bardossy, L., et al. (1994). The contribution of the haematocrit to thrombocytopenic bleeding in experimental animals. *Br J Haematol, 86,* 347–350.

56. Livio, M., Gotti, E., Marchesi, D., et al. (1982). Uraemic bleeding: Role of anaemia and beneficial effect of red cell transfusions. *Lancet, 2,* 1013–1015.

57. Moia, M., Mannucci, P. M., Vizzotto, L., et al. (1987). Improvement in the haemostatic defect of uraemia after treatment with recombinant human erythropoietin. *Lancet, 2,* 1227–1229.

58. Li N., Hu, H., Lindqvist, M., et al. (2000). Platelet-leukocyte cross talk in whole blood. *Arterioscler Vasc Biol, 20,* 2702–2708.

59. Valles, J., Santos, M. T., Aznar, J., et al. (1998). Erythrocyte promotion of platelet reactivity decreases the effectiveness of aspirin as an antithrombotic therapeutic modality: The effect of low-dose aspirin is less than optimal in patients with vascular disease due to prothrombotic effects of erythrocytes on platelet reactivity. *Circulation, 97,* 350–355.

60. Wun, T., Paglieroni, T., Field, C. L., et al. (1999). Platelet-erythrocyte adhesion in sickle cell disease. *J Investig Med, 47,* 121–127.

61. Valeri, C. R., Cassidy, G., Pivacek, L. E., et al. (2001). Anemia-induced increase in the bleeding time: Implications for treatment of nonsurgical blood loss. *Transfusion, 41,* 977–983.

62. Sola, M. C., del Vecchio, A., Edwards, T. J., et al. (2001). The relationship between hematocrit and bleeding time in very low birth weight infants during the first week of life. *J Perinatol, 21,* 368–371.

63. Valeri, C. R., Khabbaz, K., Khuri, S. F., et al. (1992). Effect of skin temperature on platelet function in patients undergoing extracorporeal bypass. *J Thorac Cardiovasc Surg, 104,* 108–116.

64. Kahn, H. A., Faust, G. R., Richard, R., et al. (1994). Hypothermia and bleeding during abdominal aortic aneurysm repair. *Ann Vasc Surg, 8,* 6–9.

65. Valeri, C. R., MacGregor, H., Cassidy, G., et al. (1995). Effects of temperature on bleeding time and clotting time in normal male and female volunteers. *Crit Care Med, 23,* 698–704.

66. Nilsson, I. M., Magnusson, S., & Borchgrevink, C. (1963). The Duke and Ivy methods for determination of the bleeding time. *Thrombosis Diath Haemorr, 10,* 223–234.

67. Liu, Y. K., Goldstein, D. M., Arora, K., et al. (1991). Thigh bleeding time as a valid indicator of hemostatic competency during surgical treatment of patients with advanced renal disease. *Surg Gynecol Obstet, 172,* 269–274.

68. Zeigler, Z. R. (1990). Non-occlusive bleeding times may improve the value of Ivy bleeding times. *Thromb Haemost, 63,* 371–374.

69. Harrison, P. (2000). Progress in the assessment of platelet function. *Br J Haematol, 111,* 733–744.

70. Lasne, D., Fiemeyer, A., Chatellier, G., et al. (2000). A study of platelet functions with a new analyzer using high shear stress (PFA-100) in patients undergoing coronary artery bypass graft. *Throm Haemostas, 84,* 794–799.

71. Marshall, P. W., Williams, A. J., Dixon, R. M., et al. (1997). A comparison of the effects of aspirin on bleeding time measured using the simplate method and closure time measured using the PFA-100, in healthy volunteers. *Br J Clin Pharmacol, 44,* 151–155.

72. Quick, A. J. (1966). Salicylates and bleeding: The aspirin tolerance test. *American J Med Sci, 252,* 265–269.

73. Fiore, L. D., Brophy, M. T., Lopez, A., et al. (1990). The bleeding time response to aspirin. Identifying the hyperresponder. *Am J Clin Pathol, 94,* 292–296.

74. Rodeghiero, F., Castaman, G., Ruggeri, M., et al. (1992). The bleeding time in normal subjects is mainly determined by platelet von Willebrand factor and is independent from blood group. *Thromb Res, 65,* 605–615.

75. Szczeklik, A., Undas, A., Sanak, M., et al. (2000). Relationship between bleeding time, aspirin and the PLA1/A2 polymorphism of platelet glycoprotein IIIa. *Br J Haematol, 110,* 965–967.

76. Mahooti, S., Graesser, D., Patil, S., et al. (2000). PECAM-1 (CD31) expression modulates bleeding time in vivo. *Am J Pathol, 157,* 75–81.

77. Fressinaud, E., Veyradier, A., Truchaud, F., et al. (1998). Screening for von Willebrand disease with a new analyzer using high shear stress: A study of 60 cases. *Blood, 91,* 1325–1331.

78. Cattaneo, M., Federici, A. B., Lecchi, A., et al. (1999). Evaluation of the PFA-100 system in the diagnosis and therapeutic monitoring of patients with von Willebrand disease. *Thromb Haemost, 82,* 35–39.

79. Kerenyi, A., Schlammadinger, A., Ajzner, E., et al. (1999). Comparison of PFA-100 closure time and template bleeding time of patients with inherited disorders causing defective platelet function. *Thromb Res, 96,* 487–492.

80. Dean, J. A., Blanchette, V. S., Carcao, M. D., et al. (2000). Von Willebrand disease in a pediatric-based population — comparison of type 1 diagnostic criteria and use of the PFA-100 and a von Willebrand factor/collagen-binding assay. *Thromb Haemost, 84,* 401–409.

81. Schlammadinger, A., Kerenyi, A., Muszbek, L., et al. (2000). Comparison of the O'Brien filter test and the PFA-100 platelet analyzer in the laboratory diagnosis of von Willebrand's disease. *Thromb Haemost, 84,* 88–92.

82. Wuillemin, W. A., Gasser, K. M., Zeerleder, S. S., et al. (2002). Evaluation of a platelet function analyser (PFA-100) in patients with a bleeding tendency. *Swiss Med Wkly, 132,* 443–448.

83. Posan, E., McBane, R. D., Grill, D. E., et al. (2003). Comparison of PFA-100 testing and bleeding time for detecting platelet

hypofunction and von Willebrand disease in clinical practice. *Thromb Haemost, 90,* 483–490.

84. Cariappa, R., Wilhite, T. R., Parvin, C. A., et al. (2003). Comparison of PFA-100 and bleeding time testing in pediatric patients with suspected hemorrhagic problems. *J Pediatr Hematol Oncol, 25,* 474–479.

85. Nitu-Whalley, I. C., Lee, C. A., Brown, S. A., et al. (2003). The role of the platelet function analyser (PFA-100) in the characterization of patients with von Willebrand's disease and its relationships with von Willebrand

factor and the ABO blood group. *Haemophilia, 9,* 298–302.

86. Koscielny, J., Ziemer, S., Radtke, H., et al. (2004). A practical concept for preoperative identification of patients with impaired primary hemostasis. *Clin Appl Thromb Hemost, 10,* 195–204.

87. Zupan, I. P., Sabovic, M., Salobir, B., et al. (2003). Utility of in vitro closure time test for evaluating platelet-related primary hemostasis in dialysis patients. *Am J Kidney Dis, 42,* 746–751.

# CHAPTER 26

# Platelet Aggregation

## Lisa K. Jennings and Melanie McCabe White

*Vascular Biology Center of Excellence and Department of Medicine, University of Tennessee
Health Science Center, Memphis, Tennessee*

## I. Introduction

Historically, studies of platelet function were performed in an effort to elucidate the basis of bleeding in patients with a history of epistaxes, bruising, gingival bleeding, etc. The original aggregometer described in 1962 by Born[1] consisted of an absorbptiometer and experiments evaluating platelet function were performed at room temperature. In the same year, O'Brien reported his results of aggregation studies using a photoelectric colorimeter run at three different temperatures.[2] This methodology has progressed to an instrument designed specifically for measuring platelet aggregation that is basically a spectrophotometer attached to a chart recorder or computer.

The current instrument is standardized for each subject using platelet-rich plasma (PRP) as the most opaque setting possible (0% aggregation) and autologous platelet-poor plasma (PPP) as the maximally transparent situation (100% aggregation). As platelets aggregate in response to the addition of an exogenous platelet agonist, the sample becomes more "clear" and an increase in light transmission through the test sample is recorded (Fig. 26-1a). Platelet aggregation response is calculated by dividing the distance from baseline to the maximal aggregation achieved by the distance from baseline to the theoretical 100% aggregation (Fig. 26-1b). The platelet aggregation pattern is classically thought of in terms of a primary response to the addition of an exogenous agonist, such as adenosine diphosphate (ADP), followed by a secondary response to the release of adenine nucleotides that are stored within the dense granules of platelets. These responses are often referred to as the first and second "wave" of aggregation (Fig. 26-2a). This biphasic response can be masked if high concentrations of agonists are added. With the agonist collagen, the aggregation pattern reflects the adhesion of platelets to the collagen fibrils and then the aggregation in response to the activation caused by that event (Fig. 26-2b). Acetylsalicylic acid

(aspirin) can totally inhibit the aggregation of platelets in response to low doses of collagen, but at higher doses aggregation will still occur.

Specialized aggregometers called lumiaggregometers assess the release of adenine nucleotides from platelet storage granules concomitant with the extent of aggregation response. This assay is adapted from a method originally described by Holmsen et al.[3] Lumiaggregation is a quantitative bioluminescent determination. It is based on the conversion of ADP released from the platelet dense granules to ATP which then reacts with firefly lantern extracts (luciferin and luciferase) generating adenyl-luciferon. Light is emitted when oxidation of adenyl-luciferin occurs. The light emitted is proportional to the nanomoles of ATP present in the aggregometer cuvette.[4,5] Studies of platelet secretion take the investigation of platelet function one step further than simple aggregation testing and allow the assessment not only of platelet aggregation but also release (Fig. 26-3). This is useful in the diagnosis of bleeding disorders such as storage pool disease and release defects especially in cases where the patients have clinical bleeding in the absence of the abnormal aggregation tracings usually associated with these disorders.[6,7] Secretion studies that provide a quantitative determination of second wave aggregation can also be useful in the investigation of platelet activation or inhibition. A more detailed methodology for assessing platelet aggregation and secretion can be found in published research and clinical laboratory procedures.[8]

In addition to the classic PRP system, platelet aggregation testing may be carried out in a whole blood system. The most common whole blood method is electrical impedance. This method can be used with either PRP or whole blood and measures an increase in impedance across electrodes placed in the anticoagulated blood as activated platelets accumulate on them.[9] Aggregometry recordings obtained by the electrical method do not discriminate two waves of platelet aggregation or correlate with platelet secretion as

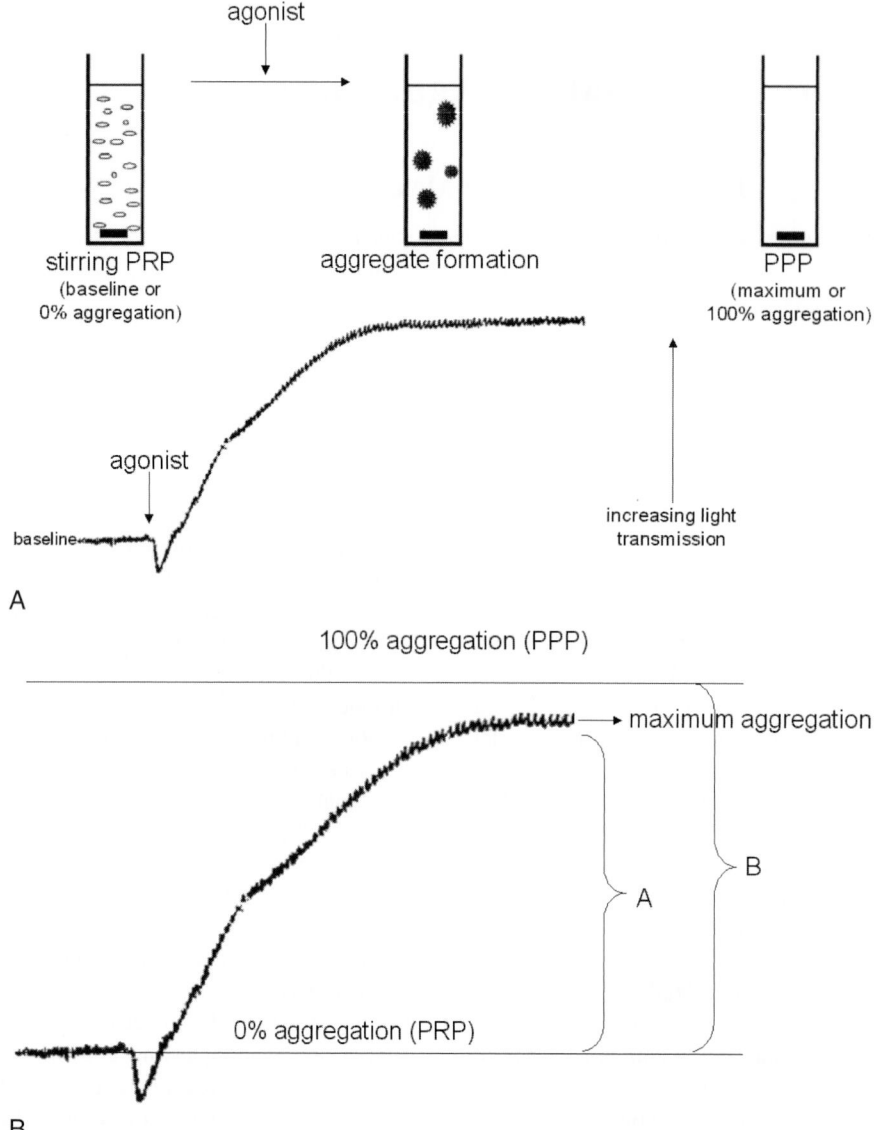

**Figure 26-1.** A. A representation of platelet response measured in an aggregometer cuvette. Stirring platelets in platelet-rich plasma (PRP) inhibits the transmission of light through the specimen. After an agonist is added, aggregation commences and the instrument measures an increase in light transmission. The maximal amount of light transmission possible for any sample is that seen with autologous platelet-poor plasma (PPP). Adapted from White and Jennings,[8] Fig 2-10, page 43. B. Calculation of percent platelet aggregation. Measure the distance between 0% aggregation and maximal aggregation (A). This value, divided by the distance between 0% aggregation and 100% aggregation (B) × 100 equals the % platelet aggregation. Abbreviations: PPP, platelet-poor plasma; PRP, platelet-rich plasma.

well as recordings obtained by the traditional optical method. In a study by Podczasy et al.,[10] when inhibitors of platelet function were added to PRP, both the rate and extent of aggregation were inhibited, but the main consequence on impedance was a decrease in its rate and not in its extent. Increases in impedance and secretion of ATP were also measured in whole blood after preincubation of an antibody specific for platelet surface glycoproteins. This study[10] showed that increases in impedance lagged several minutes behind the formation of platelet aggregates and the secretion

of platelet ATP. Therefore, this study and others suggest that there are advantages and disadvantages to both methods of measurement of platelet aggregation and that the parameters being measured must be clearly understood to properly interpret the results.

In recent years, several point-of-care technologies have been developed to assess platelet aggregation. These systems include the VerifyNow Rapid Platelet Function Analyzer (see Chapter 27) and the Plateletworks/ICHOR system (see Chapter 23).[11–13] VerifyNow is a unique platelet function

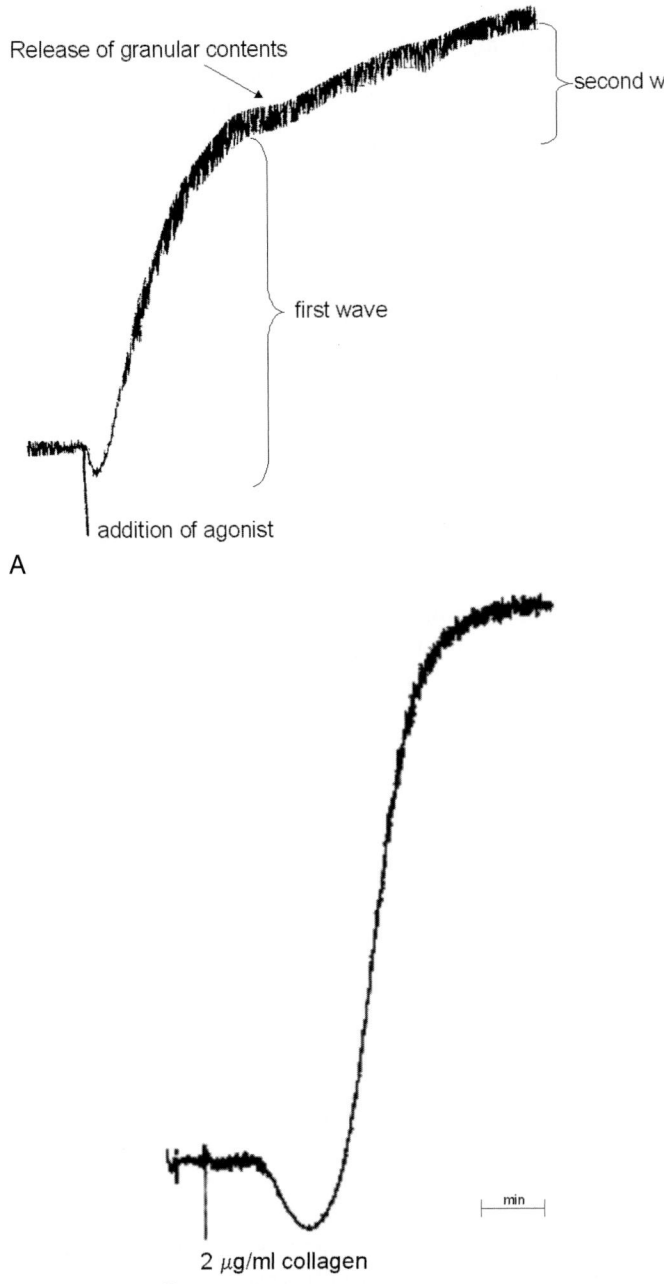

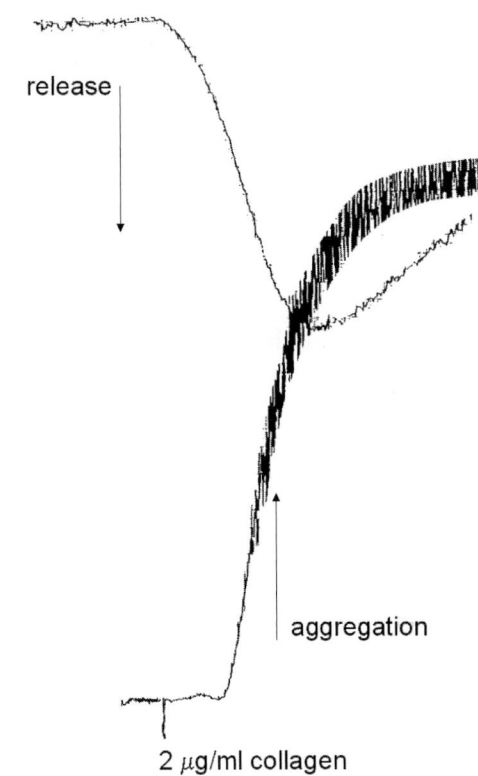

**Figure 26-3.** Platelet aggregation measured simultaneously with the release of ADP from platelet dense granules, as determined by lumiaggregometry.

**Figure 26-2.** A. A classic platelet aggregation pattern showing a first wave of aggregation in response to addition of an exogenous agonist and a second wave in response to release of adenine nucleotides from dense granules. B. Collagen-induced platelet aggregation showing the delayed aggregation in response to release of stored ADP.

assay system that relies upon the initial agglutination of platelets to fibrinogen-coated beads and platelet response to agonist exposure. This system measures the rate of platelet response but not the extent of aggregation response. Thus data obtained from this system are more similar to slope determinations in classical light transmission aggregometry. Recent data suggest that this system has certain limitations in assessing the level of platelet inhibition to the glycoprotein (GP) IIb-IIIa antagonists *ex vivo.*[13] New VerifyNow test cartridges for the assessment of aspirin and thienopyridine mediated inhibition of platelet function are currently under investigation. The Plateletworks/ICHOR system is another point-of-care platelet function system that measures the extent of platelet aggregation similar to that obtained with light transmission aggregometry. Initially the single platelet count in a whole blood sample is determined and, upon agonist exposure, the number of single platelets remaining in the sample is determined. These values are used to determine the percent of platelet aggregation. The Plateletworks/ICHOR, while still a point-of-care system, has similar results to that obtained with light transmission aggregometry and provides more flexibility in terms of agonist and anticoagulant choices.

The basic theory and assessment of platelet aggregation in PRP has changed very little from its origins. What has changed over time is the use of platelet aggregation to assess more than bleeding problems. The aggregometer has become a very useful tool in the assessment of inhibition of platelet aggregation by new anti-platelet therapies. Initially used to monitor inhibition by aspirin, light transmission aggregometry is now used to determine the pharmacodynamics of thienopyridines, antithrombins, and GPIIb-IIIa antagonists.[14,15]

## II. Variables of Platelet Aggregation Testing

In platelet aggregation testing, it is essential to develop normal ranges for the agonists being tested and to carefully control variables that might affect the test results.

### A. Venipucture

Blood from adult donors should be obtained using a 19 to 21-gauge needle and plastic syringe. In the case of pediatric patients, a smaller gauge needle such as a 23 to 25-gauge needle may be used. A single syringe, as opposed to the two syringe technique, may be used as long as the venipuncture is clean with no necessity for probing to find a vein.[16] Vacutainer™ (Becton Dickinson) collection of blood is not considered suitable for platelet aggregation measurements as increased responsiveness to low-dose ADP is typically observed from Vacutainer versus syringe-derived PRP (White and Jennings, unpublished observations). Until a Vacutainer is developed that does not increase the platelet activation or alter pharmacodynamic measurements, we recommend the use of a syringe.

### B. Anticoagulant

**1. Citrate.** Sodium citrate (0.102 M, 0.129 M citrate, buffered and nonbuffered) at a ratio of nine parts blood to one part anticoagulant is the anticoagulant typically chosen for platelet aggregation testing. Laboratories involved in aggregation testing do not use Vacutainers to obtain blood due to the concern that the platelets may become activated by the shear force of the vacuum. Instead, a plastic syringe and a butterfly needle are used to obtain the specimen. This practice is also helpful in assuring anticoagulant consistency because all of the above listed varieties of citrate anticoagulant come in blue stoppered Vacutainers. Some laboratories correct for a subject's hematocrit, especially if the hematocrit is very high or low, as the final plasma concentration of the anticoagulant can affect test results. Hardisty et al. showed that in individuals with a high hematocrit, more aggregating agent is necessary to produce an effect due to the decreased amount of free calcium available in the

plasma.[17] One can correct for hematocrit either by changing the amount of citrate added to a fixed amount of blood[16] or by changing the amount of whole blood added to a set amount of anticoagulant using the formula $(5/(1 - 0.\text{Hct}) = $ amount of whole blood to add to 1.0 mL anticoagulant). We have found the latter approach to be more convenient. Agonists such as ADP, collagen, and epinephrine, at moderate concentrations, show little or no difference in aggregation results between blood drawn into 0.102 M versus 0.129 M citrate. However, when testing in the lowest concentration ranges of agonist (e.g., ADP, where higher concentrations of citrate can result in lower aggregation responses), there may be a difference in the results obtained with the different citrate concentrations.

If the pharmacodynamics of the GPIIb-IIIa antagonists are to be evaluated, it is necessary to use a noncitrate-based anticoagulant. Studies have shown the binding of these antagonists to GPIIb-IIIa is calcium concentration dependent.[14,18] A citrate-based anticoagulant can enhance the amount of inhibition observed in *ex vivo* or *in vitro* testing. For general aggregation testing in which citrate anticoagulant is used, buffered citrates are preferable to the nonbuffered varieties because they help maintain the pH of the PRP, thus negating the possible effects of pH change. Figure 26-4 demonstrates the platelet aggregation response to ADP in three citrate anticoagulants: 0.102 M buffered citrate, 0.129 M buffered citrate, and ACD (acid-citrate-dextrose). The latter is only used for planned preparation of washed platelets and not for plasma-based aggregation assays.

**2. Heparin.** Heparin, which inhibits the generation and activity of thrombin via its complex with antithrombin III, can be used for platelet testing, but in many donors the PRP platelet count will be significantly lower when collected into heparin as compared to citrate. "Spontaneous" aggregation in the presence of heparin may be seen in a small percentage of the population. Heparin is therefore not an anticoagulant of choice for platelet aggregation testing.

**3. EDTA.** Because platelet aggregation is dependent on the presence of free calcium in the plasma, EDTA is not suitable for use in aggregation testing.

**4. PPACK.** D-phenylalanine-proline-arginine chloromethyl ketone (PPACK), an antithrombin, has become a familiar anticoagulant for platelet aggregation in systems that are being used to assess platelet inhibition by the GPIIb-IIIa antagonists. Because the Food and Drug Administration (FDA) approved antagonists (eptifibatide, abciximab, and tirofiban) have a calcium-dependent inhibition response, a nonchelating anticoagulant has to be used to avoid overestimation of inhibition during *ex vivo* testing.[14] The problems associated with heparin have made it a poor choice as an anticoagulant (see above). PPACK, which has the bene-

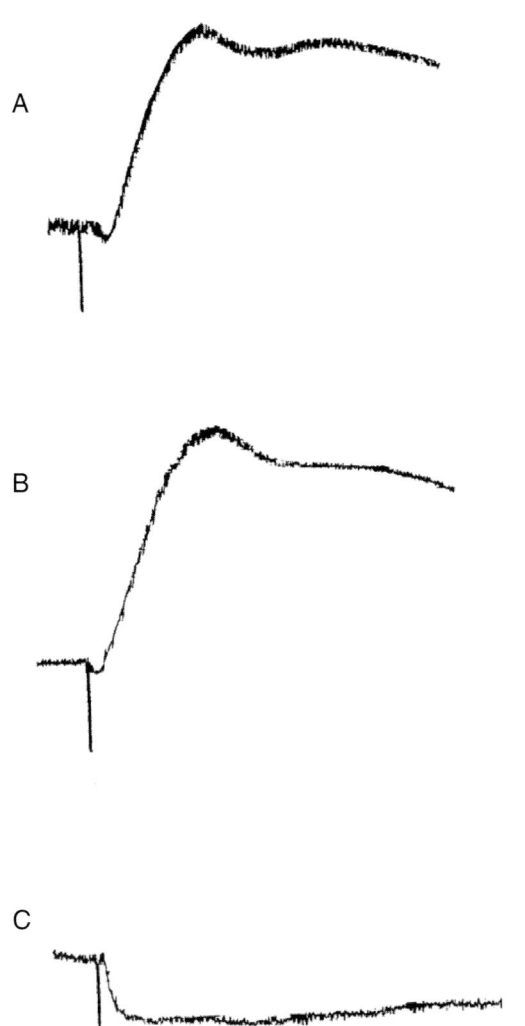

**Figure 26-4.** The response of platelets to 1 μM ADP in (A) 3.2% (0.102 M) buffered citrate, (B) 3.8% (0.129 M) buffered citrate, and (C) ACD anticoagulants. (Adapted from White and Jennings,[8] Fig 2-1, page 29.)

fits of not chelating calcium and therefore not exerting an effect on platelet function based on available plasma calcium, has filled the void. Unfortunately, PPACK is very expensive relative to the other anticoagulants listed and its anticoagulant effect can be short-lived depending on the concentration used. A final concentration of 1.6 mg per 10 mL whole blood (0.3 mM final concentration) prevents the specimen from clotting for several hours.

**5. ACD.** This anticoagulant brings the pH of the PRP to 6.5, and is, therefore, unsuitable for use in aggregation experiments (see Section II.F below). For washing or gel-filtering platelets, it is an excellent choice and avoids platelet aggregate formation in the centrifugation process.

**6. ACD-A.** This formulation of ACD keeps the pH of the PRP at 7.2 and may be acceptable for aggregation testing.[19]

If simultaneous measurements of aggregation and release are being made, it should be noted that anticoagulants such as PPACK and heparin cannot be used. The release reaction seen in citrated blood, which is used in the diagnosis of SPD and release defects, is not seen with the agonist ADP in blood anticoagulated with PPACK or heparin (Fig. 26-5). In our laboratory, the results of platelet aggregation testing performed in both PPACK and citrated blood were very similar at high concentrations of agonist. At lower concentrations of collagen, platelets from PPACK anticoagulated blood sometimes gave lower responses than platelets drawn into citrate. With very low-dose ADP (1 μM), lower responses are obtained in citrate than PPACK but at higher concentrations (5 μM) the response is greater in citrate. We find no response to 5 μM epinephrine in PPACK anticoagulant versus approximately 75% response when citrate anticoagulant is used.

### C. Glass versus Plastic Processing Tubes

The preparation of platelets for aggregation studies should always be carried out in either plastic or siliconized glass tubes. Uncoated glass can cause platelet activation and will, therefore, artifactually affect results. In our experience, polypropylene tubes are superior to either polystyrene or polycarbonate when used for platelet preparation and/or storage.

### D. Platelet Count Correction

There are a variety of opinions on whether or not it is necessary or beneficial to standardize the platelet count of the PRP used in platelet aggregation assays.[8,16,20] Because it has been reported that aggregation responses can vary in relation to platelet count, comparison of aggregation responses from donor to donor or in the case of multicenter studies necessitates standardization of the platelet count. This practice is recommended because established normal ranges of response to each agonist are typically carried out with adjusted PRP platelet counts.

### E. Red Blood Cell Contamination and Lipemia

Because platelet aggregation in PRP is based on optical transmission, the presence of any contaminating particles, such as red blood cells, or the presence of lipids can affect the ability of the aggregometer to measure platelet aggregates and can lead to a decrease in the percent aggregation.

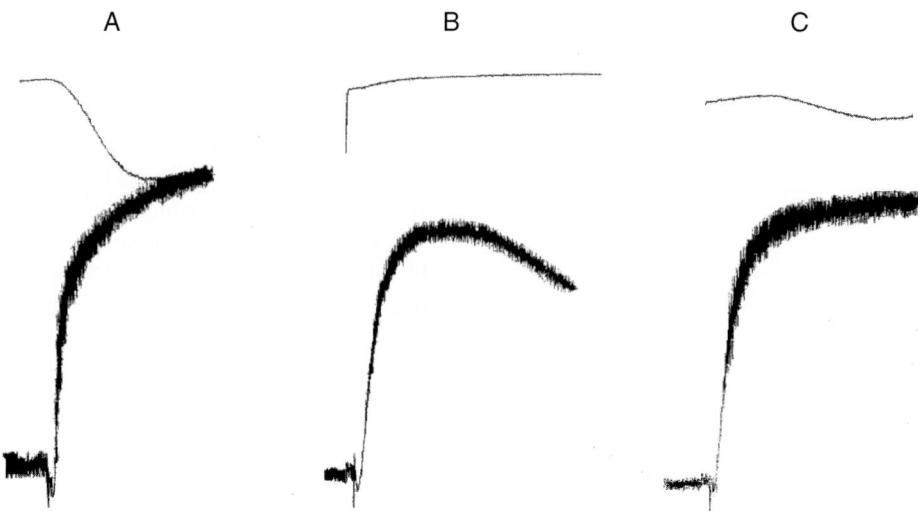

**Figure 26-5.** Platelet aggregation (lower tracing) and release (upper tracing) measured in (A) 0.102 M buffered citrate, (B) PPACK, and (C) heparin. Note the lack of release in platelets drawn into PPACK and the very diminished response in platelets drawn into heparin.

Red cell lysis can result in the release of ADP from the red cells, which in turn may cause the platelets to become refractory to the addition of exogenous ADP.

### F. pH

Platelet aggregation is pH sensitive and, therefore, when preparing a specimen for aggregation studies, pH must be maintained between 7.2 and 8.0. If the pH of the plasma drops below 6.4, no aggregation will occur and at a pH above 8.0, spontaneous aggregation can occur. If the pH approaches 10, inhibition of aggregation once again is evident. The change in pH of the plasma is mediated by the diffusion of $CO_2$ out of the plasma. As the $CO_2$ diffuses out of the plasma, the pH rises. To avoid this situation, PRP should be kept in a tube that minimizes the surface area exposed to the atmosphere (small diameter tubes), the tube should be kept capped, and the tube should be mixed as little as necessary.[16] Finally, although various buffers may not affect the pH, they may affect the platelet aggregation response. Usually, isotonic saline is the diluent of choice for agonists. Phosphate buffers in particular have the effect of lowering aggregation responses.

### G. Temperature

While platelet aggregation should be performed at 37°C to mimic the *in vivo* situation, the method used to store the platelets prior to and during aggregation studies is a matter of debate. Although it has been reported that platelets stored at room temperature are more sensitive than platelets stored

at 37°C, the aggregation response of platelets stored at 37°C is less stable. Studies have suggested that storage of platelet preparations at 4°C can prolong responsiveness, but exposure of platelets to cold temperatures can lead to a spontaneous aggregation response upon rewarming and stirring of the platelet suspension.[16] Furthermore, when the prechilled platelets have been incubated at warmer temperatures for longer periods, the agonist-induced aggregation response is higher compared to a control platelet sample. Thus, the choice of the "correct" method of storage for platelets to be used in aggregation testing is unclear; however, the most consistent results are those obtained with storage at room temperature in a capped tube.[16]

### H. Aggregometer Stir Speed

In order to aggregate, platelets must come in contact with each other. If an agonist is added to nonstirred platelets, they will become activated but will not aggregate. This situation produces a very typical aggregation tracing (Fig. 26-6). The optimal stir speed for any instrument is based on the height of the PRP column, the diameter of the aggregation cuvette, and the size of the stir bar used. Most aggregometer manufacturers recommend the optimal stir speed for their system. We have performed simultaneous studies on four different manufacturers' instruments (Helena Laboratories, Beaumont, TX; Chronolog Corporation, Havertown, PA; Bio/Data Corporation, Horsham, PA; and Payton Scientific Inc., Buffalo, NY) based on their recommended stir speed and found that the interinstrument reproducibility was excellent.

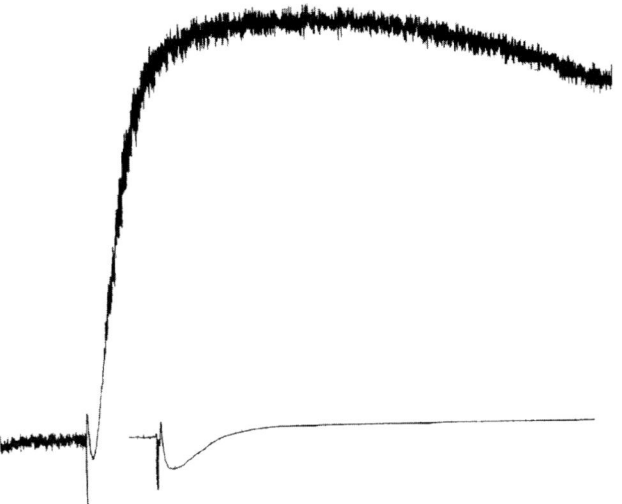

**Figure 26-6.** Platelets in PRP exposed to 10 μM ADP with (upper tracing) and without (lower tracing) a stir bar. Without platelet–platelet contact, shape change will occur but not aggregation.

## I. Time Frame of Platelet Aggregation

In assessing the effect of time after venipuncture on the responsiveness of platelets, aggregation studies were carried out with three concentrations of ADP (2, 5, and 10 μM) over a 5-hour period. When the platelets were tested immediately after preparation of PRP, the responses to all three concentrations were diminished. For the higher concentrations of ADP, the response reached a stable level when tested 30 minutes after preparation of PRP (White and Jennings, unpublished observations). It took 1 hour of "resting" for stable responses to be seen with all three concentrations of agonist. This stability remained for up to 3 hours after platelet preparation and then began to diminish at the lower concentration of ADP. While the responses to higher concentrations of agonist remained relatively stable for over 4 hours (White and Jennings, unpublished observations), it is our recommendation to complete the testing of platelet aggregation within 3 hours of the time the PRP is prepared.

## III. Platelet Agonists

### A. ADP

Concentrations of 1 to 10 μM ADP are typically used in the assessment of platelet aggregation. However, studies assessing the GPIIb-IIIa antagonists have primarily used 20 μM ADP. Lower ADP concentrations (1–3 μM) produce either a single (monophasic) aggregation response curve or a clearly biphasic curve. At lower concentrations of ADP, fibrinogen binding is usually reversible and platelets disaggregate. Higher ADP concentrations (typically 10 or 20 μM) can mask the biphasic response elicited by the release of endogenous ADP. This is still considered a biphasic response because ADP release has occurred, but it is not evident in the aggregation tracings. Aspirin will inhibit the ADP aggregation response observed with lower concentrations of agonist, due to the inhibition of the cyclooxygenase pathway and the release of granular constituents.

### B. Epinephrine

Epinephrine 5–10 μM is typically used in platelet aggregation testing. Epinephrine is the most erratic and unreliable of the agonists for platelet aggregation. Classically, a small first wave of response is seen, sometimes followed by a larger, full scale secondary response (Fig. 26-7). This second wave of aggregation, when present, is inhibited by aspirin (Fig. 26-8), nonsteroidal anti-inflammatory drugs (NSAIDs), antihistamines, some antibiotics, and many other prescription and over-the-counter compounds.

### C. Collagen

Collagen, either from bovine or equine tendon, is typically used at concentrations ranging from 1 to 5 μg/mL. Collagen is the strongest of the typical agonists used in the clinical laboratory. Collagen-induced platelet aggregation usually reflects a lag phase of approximately 1 minute, during which the platelets adhere to the collagen fibrils and undergo shape change and then release (Fig. 26-2b). The aggregation response measured is in fact the "second wave" of aggregation subsequent to platelet activation and release. At low concentrations of collagen, aspirin and other anti-platelet drugs may totally inhibit the aggregation response (Fig. 26-8).

### D. Arachidonic Acid

Arachidonic acid in a reaction with cyclooxygenase, is converted to thromboxane $A_2$, a potent platelet agonist. Aspirin inhibits the cyclooxygenase pathway and the aggregation response to arachidonic acid (see Chapter 60). Subjects who have taken aspirin or other anti-platelet drugs, or who have an intrinsic release defect or Glanzmann thrombasthenia, will have abnormal arachidonic acid-induced platelet aggregation. Patients with SPD should exhibit normal arachidonic acid-induced platelet aggregation.

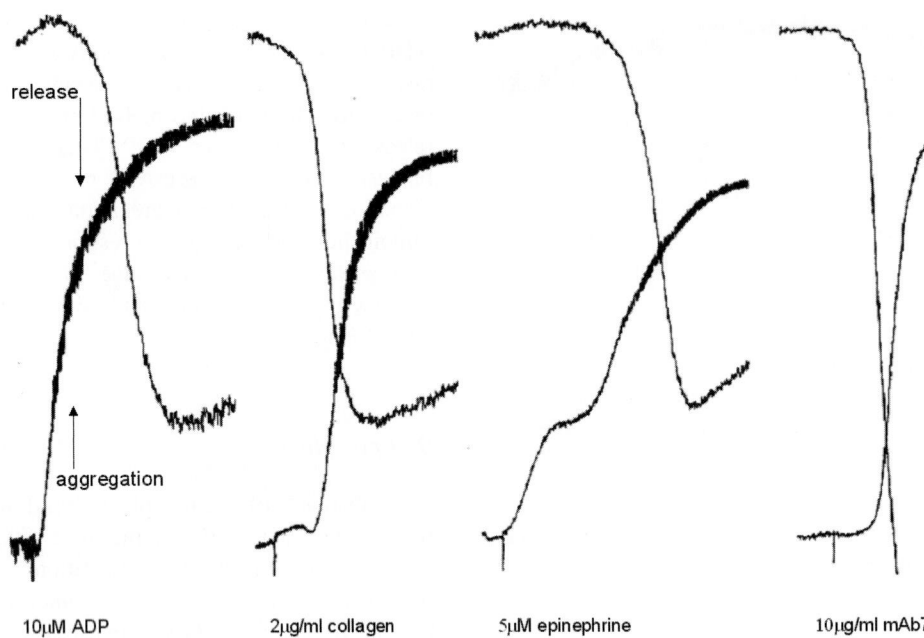

10μM ADP     2μg/ml collagen     5μM epinephrine     10μg/ml mAb7

**Figure 26-7.** Platelet aggregation and release tracings from a normal individual in response to ADP, collagen, epinephrine, and anti-CD9 mAb7 which activates via FCγRII/CD9 crosslinking. (Adapted from White and Jennings,[8] Fig 3-1, page 74.)

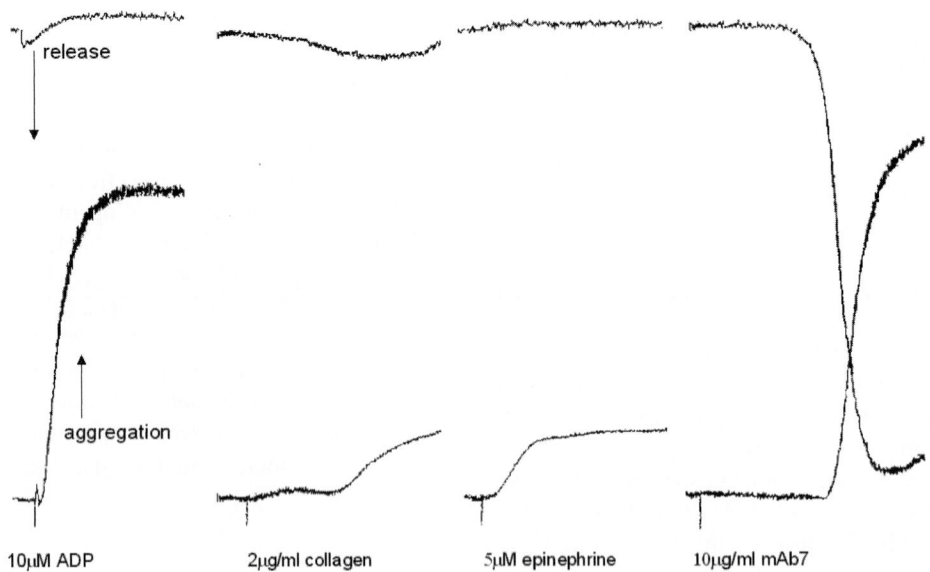

10μM ADP     2μg/ml collagen     5μM epinephrine     10μg/ml mAb7

**Figure 26-8.** Platelet aggregation and release tracings from an individual who had ingested aspirin the day prior to testing. Agonists used: ADP, collagen, epinephrine, and anti-CD9 mAb7 which activates via FCγRII/CD9 crosslinking. These tracings could also represent a patient with an intrinsic release defect. (Adapted from White and Jenning,[8] Fig 3-2, page 75.)

### E. Ristocetin

In the presence of normal platelets and a normal complement of von Willebrand factor (VWF) antigen, the antibiotic ristocetin, at a concentration of 1.5 mg/mL, causes a GPIb/VWF-dependent platelet agglutination. If abnormal agglutination in response to ristocetin is observed, von Willebrand disease or Bernard–Soulier syndrome (inherited lack of the GPIb-IX-V complex — the VWF receptor [see Chapters 7 and 57]) should be considered. Abnormal ristocetin-induced agglutination has also been reported in SPD.[21]

### F. Thrombin

Thrombin is a very potent platelet agonist (see Chapter 9). The fact that thrombin cleaves fibrinogen and leads to the formation of a clot makes it a very difficult agonist to use for platelet aggregation testing. The synthetic peptide Gly-Pro-Arg-Pro (GPRP) inhibits thrombin-induced fibrin polymerization (and therefore clot formation) while allowing thrombin-induced platelet aggregation.[21a] Alternatively, γ-thrombin can be used for aggregation in place of α-thrombin in a plasma system, but it is not readily available commercially. α-Thrombin at a concentration of 0.1 to 0.5 U/mL may be used to activate platelets in washed or gel-filtered platelet preparations.

### G. TRAP

Thrombin receptor activating peptide (TRAP) is a synthetic peptide that corresponds to the new N-terminal amino acid sequence of the "tethered ligand" generated after thrombin hydrolysis of the thrombin protease-activated receptor (PAR1)[22] (see Chapter 9). The addition of TRAP (10 μM) to platelets elicits the very strong activation response to thrombin without the complications of fibrinogen cleavage and clot formation. Most platelet defects reflect a normal aggregation response to TRAP except for Glanzmann thrombasthenia (see Chapter 57). TRAP is now being used in platelet aggregation testing to monitor the pharmacodynamic effects of new anti-platelet drugs which block fibrinogen binding to the platelet or target the platelet PAR receptors.

## IV. Trouble Shooting

Because there are no quality control (QC) kits currently available for aggregometers, it is important to be aware of problems in both sample preparation and instrument calibration that can cause inaccuracy in platelet aggregation results. One common problem encountered in aggregation testing is obtaining a result of 100% or greater. It is virtually impossible for platelets to aggregate to such an extent that 100% light transmission is achieved. If the sample clots, 100% light transmission may occur, but this is not a true measure of platelet aggregation. More often the problem is in the preparation of the PPP sample. If there are residual platelets in the PPP sample, then 100% or greater aggregation may be reported. If this situation occurs, a platelet count should be performed on the PPP. If the platelet count is less than $5 \times 10^9$/L in the PPP, then there may be a problem with the calibration of the instrument. To check this, set the PRP (baseline, 0%) and PPP (100%) limits and then put an extra PPP tube into the PRP well. The instrument should reflect 100% light transmission; if it does not there is a problem in the calibration

of the instrument. This test can also be useful in troubleshooting two channels that are set to run duplicates but do not give answers that agree. Occasionally, instruments that are set to read against a PPP standard will not read PPP as "100%." Unless the aggregometer manufacturer provides a calibration technique, this instrument will have to be serviced.

Often interlaboratory variability in platelet aggregation testing is a result of the manner in which the results are calculated. Whether the laboratory is manually calculating the results or the instrument performs the calculations automatically, error can be introduced by methodology. If an instrument automatically calculates the answers to an aggregation assay, it is important to be sure that it is reading the maximal aggregation at actual maximum rather than at a predetermined time. If time is used to determine maximum aggregation and disaggregation has occurred, the instrument will report out a falsely low aggregation value. Slopes that are automatically calculated should be reported as change per minute and the instrument must be checked to be certain that the line it has drawn for the calculation of slope is tangential to the actual aggregation and not to shape change. For a more complete description, see the detailed protocols by White and Jennings.[8]

## V. Medications That May Affect Platelet Aggregation

Many prescription and over-the-counter medications can affect platelet function (see Chapter 58). If platelet aggregation results are not clearly indicative of any classic defect but resemble partial defects, there is a significant likelihood that the patient has ingested aspirin, thienopyridines, or NSAIDs within the past week to 10 days and has simply forgotten or neglected to relate this fact. Because it is often difficult to get a precise drug history from patients coming to the laboratory, it is very important to reconfirm a defect before making the diagnosis of an intrinsic platelet function disorder.

### A. Antibiotics

Antibiotics that have a β-lactam ring structure, such as the penicillins and cephalosporins, may affect platelet function. The mechanism of action is postulated to be a membrane change that blocks receptor–agonist interactions or affects $Ca^{2+}$ influx.[23]

### B. Dipyridamole

Dipyridamole is a pyrimidopyrimidine that inhibits adenosine uptake into platelets, endothelial cells, and erythrocytes (see Chapter 63 for details). This inhibition causes an increase in the local concentration of adenosine that stimulates platelet adenylate cyclase and increases levels of cyclic 3′, 5′-adenosine monophosphate (cAMP). Elevation of cyclic AMP lowers the ability of platelets to be aggregated by platelet agonists such as platelet activating factor (PAF), collagen, and ADP.[24] The overall benefit of dipyridamole has been more evident on prosthetic surfaces. This drug as an extended release formulation is used in combination with low-dose aspirin as Aggrenox™ (Boehringer Ingelheim Pharmaceuticals, Inc., Ridgefield, CT).

## C. Fibrinolytics

Fibrinolysis and the formation of fibrin degradation products (FDPs) have been associated with a decrease in platelet aggregation. FDPs compete with fibrinogen for binding to the platelet membrane and interfere with platelet aggregation. One recent study[25] has shown that subjects treated with tenecteplase and alteplase, two newer fibrinolytic compounds, exhibited significant inhibition of platelet aggregation when tested in whole blood, traditional PRP aggregation systems, or point-of-care analyzers. In another study[26] comparing reteplase, alteplase, and streptokinase, inhibition of platelet aggregation was observed with all three treatments. The decrease in aggregation was most pronounced with streptokinase, followed by reteplase and then alteplase. Decreased levels of plasma fibrinogen and impaired binding of fibrinogen to GPIIb-IIIa correlated with the severity of the platelet aggregation defect with these agents.

## D. Dextran

Intravenous infusion of dextran can result in reduced platelet function. In peripheral arterial disease patients, it was recently shown that Dextran 40 reduced spontaneous and agonist-mediated platelet aggregation and the expression of activation markers such as P-selectin on the platelet surface.[27]

## E. Anesthetics

Anesthetics have been demonstrated to have an effect on the aggregation response of platelets and have been implicated in an increased risk of hemorrhagic complications.[29,30] Anesthetics such as lidocaine, dibucaine, cocaine, etc. have a direct effect on the platelet membrane. Cocaine added to platelets in vitro causes a reduction in fibrinogen binding to the activated GPIIb-IIIa receptor.[28] The concentration at which these anesthetics mediate a platelet inhibitory effect is typically at least an order of magnitude greater than that considered to have potentially lethal effects in vivo. There have been reports that patients undergoing general anesthesia not only had prolonged bleeding times but also had reduced aggregation response ex vivo to weak agonists. Although there are conflicting reports, data suggest that halothane, sevoflane, and propofol inhibit platelet function signaling pathways at clinically relevant concentrations.[29,30]

## F. Thrombin Inhibitors

Thrombin is a key regulator in the pathophysiology of acute coronary syndromes. It mediates the conversion of fibrinogen to fibrin, activates factor XIII that aids in clot stabilization, and is a potent agonist of platelets. Heparin is the most used clinical anti-thrombin, but it has many limitations with regard to unpredictable bioavailability and safety. The recent generation of direct thrombin inhibitors that act independently of antithrombin III can inhibit clot-bound thrombin as well as thrombin-induced platelet activation.[31]

## G. Thienopyridines

ADP is a key platelet agonist that functions by binding to G protein coupled receptors, $P2Y_1$ and $P2Y_{12}$ (see Chapter 10). The $P2Y_{12}$ receptor is the primary ADP receptor that mediates fibrinogen binding and sustained aggregation response.[32] The thienopyridines, ticlopidine and clopidogrel, irreversibly and covalently bind to this receptor and inhibit platelet aggregation by ADP on the order of 40 to 60% (see Chapter 61).[33] Platelet aggregation to other agonists is not inhibited but may be attenuated because released ADP contributes to the platelet aggregation response, particularly in conjunction with weak agonists.

## H. GPIIb-IIIa Antagonists

GPIIb-IIIa antagonists bind to the GPIIb-IIIa (integrin αIIbβ3) receptor and prevent the binding of fibrinogen or VWF to the activated platelet (see Chapter 62).[14,34] These agents, eptifibatide, abciximab, and tirofiban, are the most potent of the anti-platelet agents because when bound to GPIIb-IIIa, platelet aggregation to all agonists (e.g., ADP, collagen, TRAP) is inhibited significantly. Typically, the level of platelet inhibition by these agents is tested ex vivo using the PPACK anticoagulant and 20 μM ADP.[35]

Many other drugs have been identified that alter normal platelet function (see Chapter 58 for a comprehensive review).

# VI. Inherited Platelet Function Defects

## A. Storage Pool Disease

In contrast to normal platelet aggregation (Fig. 26-7), SPD and release defects are usually distinguished by an abnormal second wave of aggregation secondary to the absent or decreased release of adenine nucleotides from platelet dense granules (Figs. 26-8 and 26-9). However, SPD may have normal platelet aggregation.[6,7] The diagnosis of these patients is made using studies of adenine nucleotide release. SPD and release defects are discussed in detail in Chapters 15 and 57.

## B. Glanzmann Thrombasthenia

Patients with Glanzmann thrombasthenia[36] have a mutation in the GPIIb-IIIa receptor resulting in either the absence or deficiency of functional GPIIb-IIIa (see Chapter 57 for details). While the absence of GPIIb-IIIa yields a definitive diagnosis of Glanzmann thrombasthenia, variants have been described in which platelet GPIIb-IIIa is present, but a mutation renders it nonfunctional.[37] The diagnostic features are: absent platelet aggregation to ADP, collagen, epinephrine,

and thrombin (Fig. 26-10), and an abnormal clot retraction.

## C. von Willebrand Disease

Patients with von Willebrand disease exhibit bleeding similar to those with intrinsic platelet defects. The aggregation responses in these individuals are normal to all agonists with the exception of ristocetin. In the rare type IIb and platelet-type von Willebrand disease, there is increased ristocetin-induced platelet aggregation.[38]

## D. Bernard–Soulier Syndrome

Bernard–Soulier syndrome[39] patients have a deficiency of GPIb-IX-V which renders their platelets incapable of interacting with VWF (see Chapter 57 for details). This leads to a problem with platelet adhesion to injury-induced exposure of subendothelium. The platelet aggregation responses seen in Bernard–Soulier syndrome are normal to all agonists but ristocetin. If measuring only platelet aggregation in the diagnosis of a patient with a bleeding disorder, Bernard–Soulier could be mistaken for von Willebrand disease.

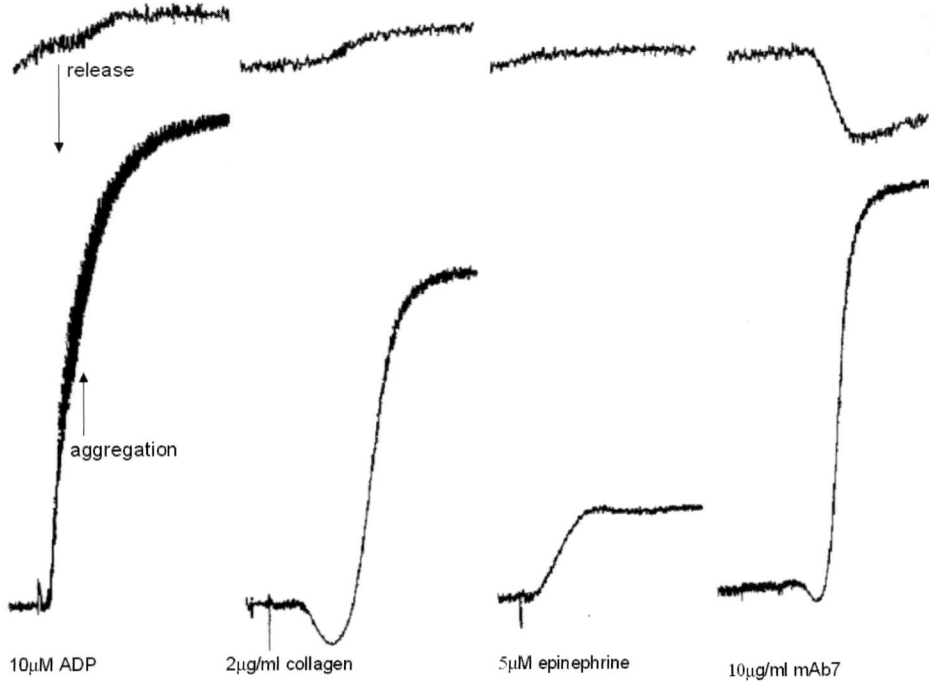

**Figure 26-9.** Platelet aggregation and release tracings from an individual with storage pool disease, in response to ADP, collagen, epinephrine, and anti-CD9 mAb7 which activates via FCγRII/CD9 crosslinking. Note that, compared to the patient with release defect (aspirin, Fig. 26-8), this patient exhibits a markedly decreased release in response to mAb7. (Adapted from White and Jennings,[8] Fig 3-3, page 76.)

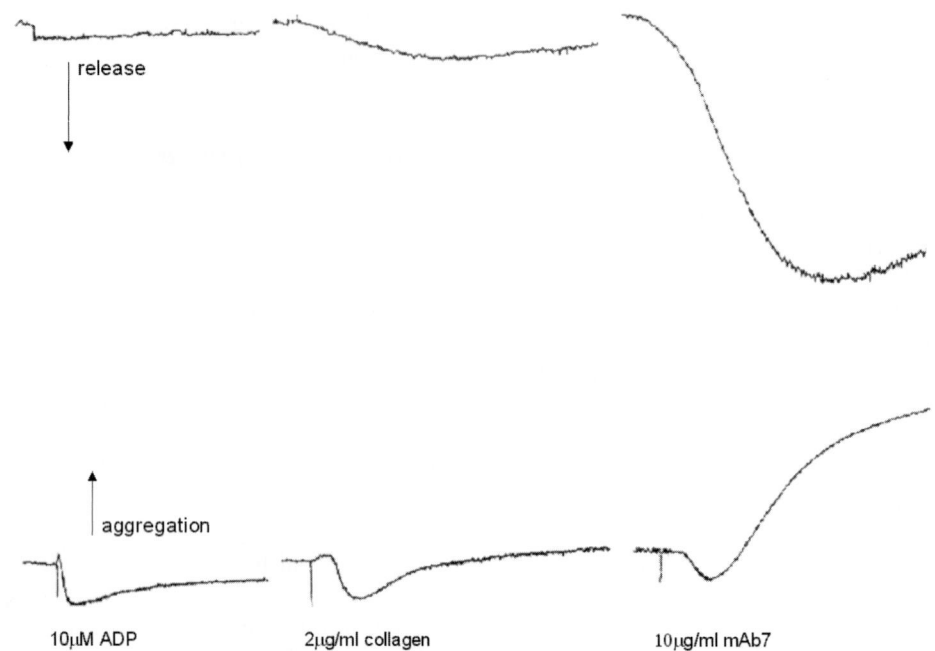

**Figure 26-10.** Platelet aggregation and release tracings from a patient with Glanzmann thrombasthenia. Agonists used: ADP, collagen, epinephrine, and anti-CD9 mAb7 which activates vial FCγRII/CD9 crosslinking. Due to the lack of GPIIb-IIIa expression, platelets are unable to bind fibrinogen and aggregate. Note that there is release in response to strong agonists, showing that, although the platelets are incapable of aggregating, activation can occur. (Adapted from White and Jennings,[8] Fig 3-4, page 77.)

However, flow cytometric analysis of GPIb-IX-V surface density is able to distinguish between these disorders (see Chapter 30).

# VII. Acquired Platelet Function Defects

Acquired platelet function defects occur in uremia, preleukemia and acute leukemias, myeloproliferative disorders, dysproteinemias, liver disease, and anti-platelet antibodies. These disorders are discussed in detail in Chapter 58.

# References

1. Born, G. V. R. (1962). Aggregation of blood platelets by adenosine diphosphate and its reversal. *Nature, 194,* 927–929.
2. O'Brien, J. R. (1962). Platelet aggregation: Part II. Some results of a new method. *J Clin Path, 15,* 452–455.
3. Holmsen, H., Holmsen, I., & Bernhardsen, A. (1966). Microdetermination of adenosine diphosphate and adenosine triphosphate in plasma with the firefly luciferase system. *Anal Biochem, 17,* 456–473.
4. Feinmann, R. D., Kubowsky, J., Charo, I., et al. (1977). The lumiaggregometer: A new instrument for simultaneous measurement of secretion and aggregation by platelets. *J Lab Clin Med, 90,* 125–129.
5. White, M. M., Foust, J. T., Mauer, A. M., et al. (1992). Assessment of lumiaggregometry for research and clinical laboratories. *Thromb Haemost, 67,* 572–577.
6. Nieuwenhuis, H. K., Akkerman, J.-W. N., & Sixma, J. J. (1987). Patients with prolonged bleeding time and normal aggregation tests may have storage pool deficiency: Studies on 106 patients. *Blood, 70,* 620–623.
7. Israels, S. J., McNicol, A., Robertson, C., et al. (1990). Platelet storage pool deficiency: Diagnosis in patients with prolonged bleeding times and normal platelet aggregation. *Br J Haematol, 75,* 118–121.
8. White, M. M., & Jennings, L. K. (1999). *Platelet protocols: Research and clinical laboratory procedures.* San Diego: Academic Press.
9. Ingeman-Wojenski, C., Smith J. B., & Silver, M. J. (1983). Evaluation of electrical aggregometry: Comparison with optical aggregometry, secretion of ATP, and accumulation of radiolabeled platelets. *J Lab Clin Med, 101,* 44–52.
10. Podczasy, J. J., Lee, J., & Vucenik, I. (1997). Evaluation of whole-blood lumiaggregation. *Clin Appl Thromb Hem, 3,* 190–195.
11. Steinhubl, S. R., Talley, J. D., Braden, G. A., et al. (2001). Point of care measured platelet inhibition correlates with a reduced risk of an adverse cardiac event after percutaneous coronary intervention. *Circulation, 103,* 2572–2578.
12. Kereiakes, D. J., Broderick, T. M., Roth, E. M., et al. (1999). Time course, magnitude and consistency of platelet inhibition by abciximab, tirofiban, or eptifibatide in patients with unsta-

ble angina pectoris undergoing percutaneous coronary intervention. *Am J Cardiol, 84,* 391–395.

13. White, M. M., Krishnan, R., Kueter, T. J., et al. (2005). The use of the point of care Helena ICHOR/Plateletworks and the Accumetrics Ultegra RPFA for assessment of platelet function with GPIIb-IIIa antagonists. *J Thromb Thrombolysis, 18,* 163–169.

14. Jennings, L. K., Jacoski, M. V., & White, M. M. (2002). The pharmacodynamics of parenteral glycoprotein IIb-IIIa Inhibitors. *J Intervent Cardiol, 15,* 45–60.

15. Tardiff, B. E., Jennings, L. K., Harrington, R. A., et al. (2001). Pharmacodynamics and pharmacokinetics of eptifibatide in patients with acute coronary syndromes: Prospective analysis from the Pursuit Trial. (Tardiff and Jennings are co-first authors.) *Circulation, 104,* 399–405.

16. Triplett, D. A., Harms, C. S., Newhouse, P., et al. (1978). *Platelet function: Laboratory evaluation and clinical application.* Chicago: American Society of Clinical Pathologists.

17. Hardisty, R. M., Hutton, R. A., Montgomery, D., et al. (1970). Secondary platelet aggregation: A quantitative study. *Br J Haematol, 19,* 307–319.

18. Phillips, D. R., Teng, W. S., Arfsten, A., et al. (1997). Effect of $Ca^{++}$ on Integrilin GPIIb-IIIa interactions: Enhanced GPIIb-IIIa binding and inhibition of platelet aggregation by reductions in the concentration of ionized calcium in plasma anticoagulated with citrate. *Circulation, 96,* 1488–1494.

19. Pignatelli, P., Pulcinelli, F. M., Ciatti, F., et al. (1995). Acid citrate dextrose (ACD) formula A as a new anticoagulant in the measurement of *in vitro* platelet aggregation. *J Lab Clin Anal, 9,* 138–140.

20. Luxembourg, M. H., Klaffling, C., Erbe, M., et al. (2005). Use of native or platelet count adjusted platelet rich plasma for platelet aggregation measurements. *J Clin Pathol, 58,* 747–750.

21. Pujol-Moix, N., Hernandez, A., Escolar, G., et al. (2000). Platelet ultrastructural morphometry for diagnosis of partial δ-storage pool disease in patients with mild platelet dysfunction and/or thrombocytopenia of unknown origin. A study of 24 cases. *Haematologica, 85,* 619–626.

21a. Michelson, A. D., Ellis, P., Barnard, M. R., et al. (1991). Downregulation of the platelet surface glycoprotein Ib-IX complex in whole blood stimulated by thrombin, ADP or an *in vivo* wound. *Blood, 77,* 770–779.

22.Coughlin, S. R. (2000). Thrombin signaling and protease activated receptors. *Nature, 407,* 258–264.

23. Burroughs, S. F., & Johnson, G. J. (1990). Beta-lactam antibiotic-induced platelet dysfunction: Evidence for irreversible inhibition of platelet activation in vitro and in vivo after prolonged exposure to penicillin. *Blood, 75,* 1473–1480.

24. Philp, R. B., & Lemieux, J. P. V. (1975). Interactions of dipyridamole and adenosine on platelet aggregation. *Nature, 221,* 1162–1164.

25. Serebruany, V. L., Malinin, A. I., Callahan, K. P., et al. (2003). Effect of tenecteplase versus alteplase on platelets during the first 3 hours of treatment for acute myocardial infarction: The Assessment of the Safety and Efficacy of a New Thrombolytic Agent (ASSENT-2) platelet substudy. *Am Heart J, 145,* 636–642.

26. Moser, M., Nordt, T., Peter, K., et al. (1999). Platelet function during and after thrombolytic therapy for acute MI with reteplase, alteplase or streptokinase. *Circulation, 100,* 1858–1864.

27. Robless, P., Okonko, D., Milhailidis, D. P., et al. (2004). Dextran 40 also reduces *in vitro* platelet aggregation in peripheral arterial disease. *Platelets, 15,* 215–222.

28. Jennings, L. K., White, M. M., Sauer, C. M., et al. (1993). Cocaine-induced platelet defects. *Stroke, 24,* 1352–1359.

29. Kozek-Langenecker, S. A. (2002). The effects of drugs used in anaesthesia on platelet membrane receptors and on platelet function. *Curr Drug Targets, 3,* 247–258.

30. Dordoni, P. L., Frassanito, L., Bruno, M. F., et al. (2004). *In vivo* and *in vitro* effects of different anaesthetics on platelet function. *Br J Haematol, 125,* 79–82.

31. Chen, M. S., & Lincoff, A. M. (2005). Direct thrombin inhibitors. *Curr Cardiol Rep, 7,* 355–359.

32. Conley, P. B., & Delaney, S. M. (2003). Scientific and therapeutic insights into the role of the platelet P2Y12 receptor in thrombosis. *Curr Opin Hematol, 10,* 333–338.

33. Gurbel, P. A., & Bliden, K. P. (2003). Durability of platelet inhibition by clopidogrel. *Am J Cardiol, 91,* 1123–1125.

34. Jennings, L. K. (2001). *Clinical trials of the parenteral glycoprotein IIb-IIIa inhibitors: The influence of dose selection on results.* A CME Booklet, Health Science Communications.

35. Saucedo, J. F., Wolford, D. C., Cook, S. L., et al. (2004). Comparative pharmacodynamic evaluation of eptifibatide and tirofiban HCl in patients undergoing coronary intervention—The TAM1 Study. *Am J Cardiol, 93,* 1270–1282.

36. Glanzmann, E. (1918). Heredititare hamorrhagische thrombasthenic: Ein Beitray zur pathologic der blutplatchen. *Jahrbuch fur Kinderheilkunde, 88,* 113.

37. Jackson, D. E., White, M. M., Jennings, L. K., et al. (1988). A Ser162-Leu mutation within glycoprotein (GP) IIIa (integrin β3) results in an unstable αIIbβ3 complex that retains partial function in a novel form of Type II Glanzmann's thrombasthenia. *Thromb Haemost, 80,* 42–48.

38. Schneppenheim, R. (2005). *Blood Coagul Fibrinolysis, Suppl 1,* S3–S10.

39. Bernard, J., & Soulier, J. P.(1948). Sur une nouvelle variete de dystrophie thrombocytaire hemorragipare congenitale. *Sem Hop Paris, 24,* 3217.

CHAPTER 27

# The VerifyNow System



**Steven R. Steinhubl**

*Linda and Jack Gill Heart Institute, University of Kentucky, Lexington, Kentucky*



## I. Introduction

Few pharmacotherapies have achieved such broad integration within virtually all aspects of cardiovascular clinical practice as have antiplatelet agents. Aspirin (Chapter 60) has been the mainstay of antiplatelet therapy for decades. More recently, the clinical benefit of additional platelet inhibitors such as the thienopyridines (Chapter 61) and the platelet glycoprotein (GP) IIb-IIIa antagonists (Chapter 62) have been well proven in a large number of clinical trials. Interestingly, despite their widespread use in millions of patients with diverse disease processes, antiplatelet agents are almost always administered as a standard dose, with no titration or monitoring. This practice is contrary to the more standard approach of titrating a pharmaceutical agent (e.g., an antihypertensive agent) to a specific, measurable effect (e.g., blood pressure).

One explanation for this "fixed dose" strategy for antiplatelet therapy has been the lack of a convenient, reproducible, clinically relevant measure of platelet function. Historically, the most widely utilized method for clinically evaluating platelet function has been the bleeding time (Chapter 25), which was first described over 100 years ago. However, bleeding time measurements are poorly standardized, labor intensive, subjective, and most importantly, have not been shown to correlate with clinical outcomes.[1,2] Light transmission turbidometric platelet aggregometry (Chapter 26) is considered the gold standard for determining the effectiveness of a wide variety of antiplatelet agents in a research setting, although just how translatable its results are into clinical outcomes has yet to be well proven. Its applicability in a patient care setting remains quite limited though by the requirement for significant technical expertise, prolonged sample preparation, and specialized equipment that restricts its use to hematology laboratories.

The VerifyNow® System, formerly known as the Ultegra Rapid Platelet Function Assay (RPFA) (Accumetrics, Inc., San Diego, CA), is a whole-blood, point-of-care assay developed to provide simple, rapid, and accurate determination of platelet function in any clinical setting. While initially designed to specifically monitor patients treated with a GPIIb-IIIa antagonist, further refinements have expanded its utility to include monitoring the effects of aspirin and thienopyridine treatment. This chapter will discuss how VerifyNow works, the correlation between measured platelet function using VerifyNow and other methods of monitoring platelet inhibition, and the association between the level of platelet inhibition as determined by VerifyNow and clinical outcomes among patients treated with antiplatelet agents.

## II. Mechanism

In early evaluations of the platelet GPIIb-IIIa receptor (integrin αIIbβ3) and monoclonal antibodies that might inhibit it, an assay incorporating fibrinogen-coated beads was utilized.[3,4] When these beads were manually mixed with whole blood and a thrombin receptor-activating peptide (iso-TRAP) the degree of platelet GPIIb-IIIa receptor blockade could be visually determined[5] (Fig. 27-1A). This simple but reproducible measure of platelet function was refined and automated into the current VerifyNow system (Fig. 27-1B).

The basis of the VerifyNow assay is that fibrinogen-coated polystyrene microparticles will agglutinate in whole blood in direct proportion to the degree of platelet activation and subsequent activated GPIIb-IIIa receptors (Fig. 27-2). Direct pharmacologic blockade of GPIIb-IIIa receptors with a GPIIb-IIIa antagonist or the prevention of their expression by the inhibition of arachidonic acid- or ADP-induced platelet activation diminishes agglutination in proportion to the degree of platelet inhibition achieved.

Blood samples are obtained in a tube containing either sodium citrate or another anticoagulant, e.g., PPACK (Phe-Pro-Arg chloromethyl ketone). The tube is inserted into a disposable plastic cartridge (Fig. 27-3) containing a lyophilized preparation of human fibrinogen-coated beads, a

509

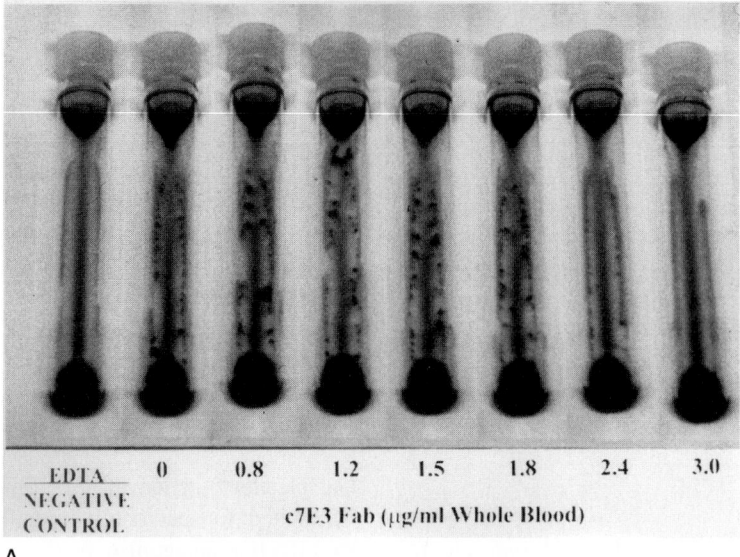

A

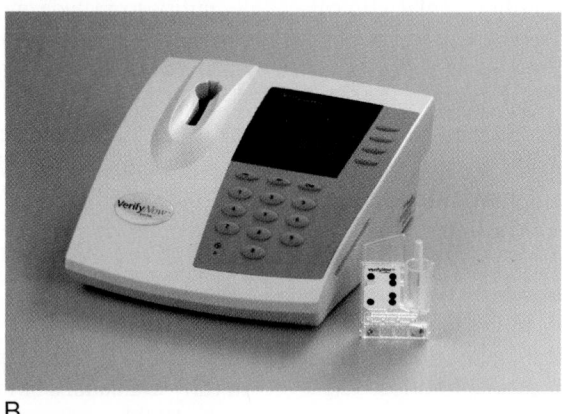

B

**Figure 27-1.** Evolution of the VerifyNow System. A. The original test-tube based assay showing decreasing agglutination with increasing concentrations of the GPIIb-IIIa antagonist monoclonal antibody c7E3 (photo courtesy of Dr. Barry S. Coller). B. The current VerifyNow system.

platelet activating agent, buffer, and preservative. The whole blood is automatically drawn into a sample channel containing the platelet agonist and fibrinogen-coated beads, and then mixed by the movement of a microprocessor-driven steel ball. The light absorbance of the sample is measured 16 times per second by an automated detector. As the platelets interact with the fibrinogen-coated beads resulting in agglutination, there is a progressive increase in light transmission. The rate of agglutination is quantified as the slope of the change of absorbance over a fixed time interval and measured in millivolts per 10 sec (mV/10 sec). The device then automatically displays this result in agonist-specific units.

Three types of cartridges are available: one for measuring the effects of aspirin, another for P2Y$_{12}$ inhibitors (clopidogrel and ticlopidine), and a third for GPIIb-IIIa antagonists (abciximab [ReoPro], eptifibatide [Integrilin],

and tirofiban [Aggrastat]). The cartridges differ primarily based on the platelet agonist contained in them. The same instrument can be used with all three cartridges.

## A. VerifyNow IIb/IIIa Assay

In determining the level of platelet inhibition in patients treated with a GPIIb-IIIa antagonist, the cartridge utilizes iso-TRAP ([iso-S]FLLRN) as the platelet agonist.[6] The final concentration of iso-TRAP is 4 μM. The rate of change in light transmittance as the TRAP-activated platelets bind to the fibrinogen-coated beads and fall out of solution is reported as platelet aggregation units (PAU). As the level of platelet inhibition in patients treated with GPIIb-IIIa antagonists is typically expressed in terms of a percentage inhibition compared with baseline, an individual patient's baseline

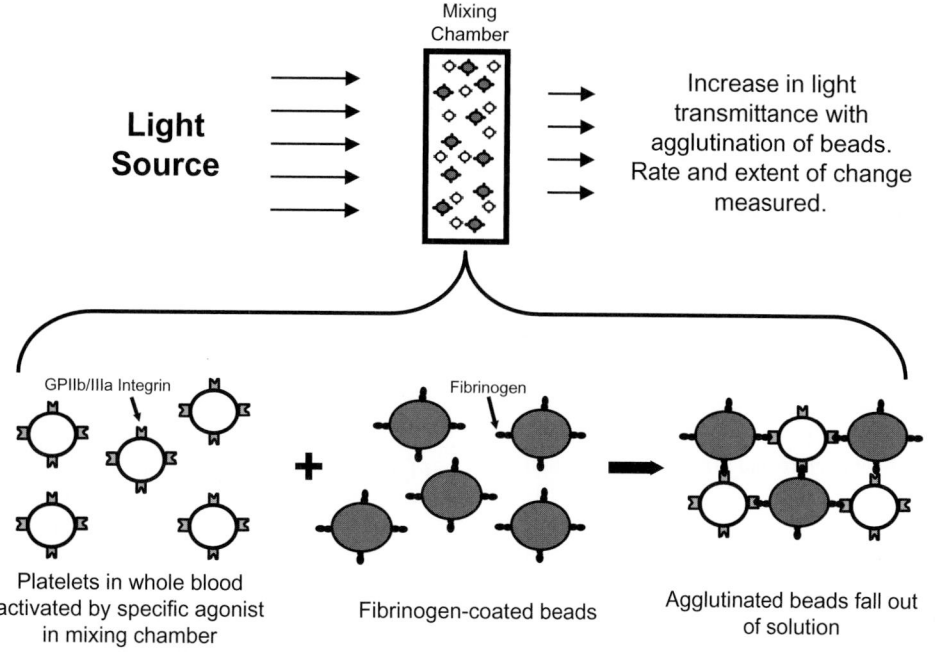

**Figure 27-2.** Diagram representing how platelet function is determined with the VerifyNow System. The mixing chamber contains a platelet agonist (thrombin receptor activating peptide [iso-TRAP], arachidonic acid, or ADP) and fibrinogen-coated beads. After anticoagulated whole blood is added to the mixing chamber, the platelets become activated. The activated GPIIb-IIIa receptors on the platelets bind via the fibrinogen on the beads and cause agglutination of the platelets and the beads. Light transmittance through the chamber is measured and increases as the agglutinated platelets and beads fall out of solution. Direct pharmacologic blockade of GPIIb-IIIa receptors with a GPIIb-IIIa antagonist, or the prevention of their expression by the inhibition of arachidonic acid- or ADP-induced platelet activation, diminishes agglutination in proportion to the degree of platelet inhibition achieved.

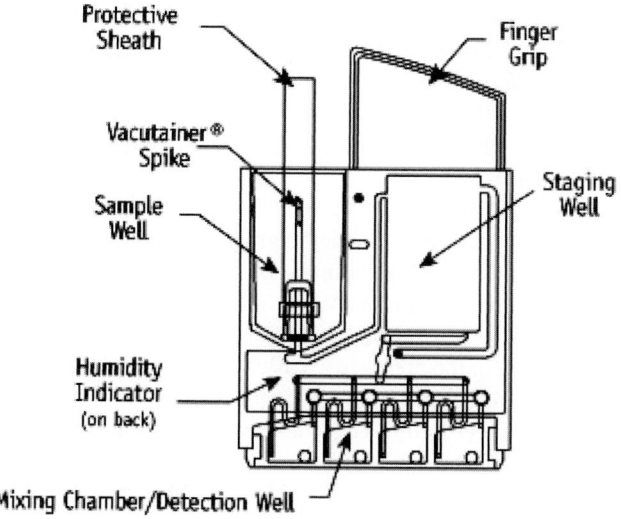

**Figure 27-3.** Schematic of the VerifyNow test cartridge.

(prior to treatment with GPIIb-IIIa antagonist) PAU is retained in memory and all additional specimens are reported as a raw PAU as well as a percentage of the baseline PAU. Therefore, during treatment with a GPIIb-IIIa antagonist, a patient's percent platelet function inhibition is

obtained by dividing the PAU determined after the initiation of therapy by the pretreatment baseline PAU (×100).

Because TRAP is a strong agonist for platelet activation and can overcome the antiplatelet effects of agents such as aspirin and clopidogrel, the results of the IIb/IIIa assay are not influenced by such concomitant antiplatelet therapies. Although it is recommended that samples be assayed within 1 hour of collection, the assay can be performed for up to 4 hours after sample collection.

### B. VerifyNow Aspirin Assay

The VerifyNow Aspirin Assay incorporates the agonist arachidonic acid (1 mM) to activate platelets. Arachidonic acid is converted into thromboxane $A_2$ depending on the level of platelet cyclooxygenase (COX) activity (Chapter 31), and this synthesized thromboxane $A_2$ leads to platelet activation and agglutination with the fibrinogen-coated beads. As this occurs, light transmittance increases and the rate of change in light transmittance is then reported as aspirin reaction units (ARU).

An individual's responsiveness to aspirin is not described as a percentage of baseline, but rather as just ARU and therefore a preaspirin sample is not required. A discrete

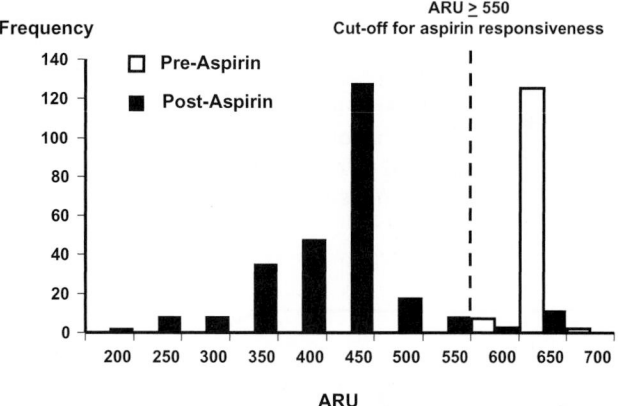

**Figure 27-4.** Frequency distribution of measured aspirin reaction units (ARU) in patients before and after aspirin ingestion.

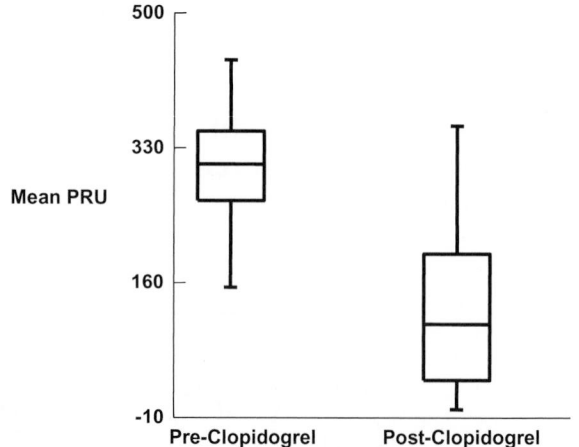

**Figure 27-5.** Histogram of mean P2Y12 reaction units (PRU) in 147 patients before and after clopidogrel treatment. Data are mean ± SEM. From the VerifyNow P2Y12 package insert.

cutoff for aspirin responsiveness was determined in clinical trials of individuals treated with and without aspirin and included over 400 samples. The reference range for preaspirin samples was found to be 620–672 ARU (2.5 to 97.5 percentile) (VerifyNow Aspirin Assay package insert). All patients not receiving aspirin had >550 ARU, and using this as a cutoff for the presence of aspirin provided a sensitivity of 91.4% and a specificity of 100% (Fig. 27-4). Based on this, an ARU ≥550 is defined as "aspirin resistance."

Unlike the IIb/IIIa assay, the VerifyNow aspirin assay results are influenced by a number of agents including other antiplatelet agents such as thienopyridines, phosphodiesterase inhibitors, and GPIIb-IIIa antagonists, as well as other inhibitors of COX such as nonsteroidal anti-inflammatory agents. Also, it is recommended that the blood sample incubate in its collection tube for at least 30 min (but not more than 4 hours) prior to the test's being run in order to allow the sample to adequately equilibrate with citrate.

### C. VerifyNow P2Y12 Assay

While the agonist utilized in the $P2Y_{12}$ assay is ADP at a concentration of 20 μM, a second agent, prostaglandin $E_1$ ($PGE_1$) is also added (22 nM) in order to suppress intracellular free calcium levels and thereby to reduce the platelet activation contribution from ADP-binding to its $P2Y_1$ receptor (Chapter 10).[7] Therefore, when a sample is stimulated with ADP, the rate of platelet agglutination with the fibrinogen-coated beads is influenced primarily by the degree of inhibition of the platelet $P2Y_{12}$ receptor, and this value is reported as P2Y12 Reaction Units (PRU). Studies in 147 patients before and after the $P2Y_{12}$ antagonist clopidogrel (either 24 hours after a 450-mg loading dose or after 7 days of 75 mg daily) showed a wide range of responses, but a

clear separation in the pre- and postclopidogrel values with a mean decrease of 185 PRU after clopidogrel (VerifyNow™ P2Y12 package insert) (Fig. 27-5).

Clopidogrel responsiveness is often expressed as a percentage of baseline, but because it is frequently impossible to obtain a predrug sample in a clinical setting, a unique mechanism for determining baseline platelet function in patients already being treated with a thienopyridine is incorporated in the $P2Y_{12}$ testing cartridge. In a separate channel, iso-TRAP is used as the agonist, and because it is unaffected by thienopyridine treatment a baseline value (Base) for platelet function is obtained simultaneously. Therefore, the VerifyNow P2Y12 assay reports patient results as PRU and percent inhibition calculated as PRU/ Base (×100).

The specificity of the VerifyNow P2Y12 assay was determined by evaluating platelet aggregation induced by ADP or ADP plus prostaglandin $E_1$ (ADP + $PGE_1$) in the presence of the specific $P2Y_{12}$ inhibitor 2-methylthio-AMP (2MeSAMP) in vitro using blood samples from 10 healthy volunteers. In the presence of 2MeSAMP, the average residual aggregation level of the 10 donors was 27% for ADP and 5% for ADP + $PGE_1$. There also was a strong agreement between ADP + $PGE_1$ aggregometry and the VerifyNow P2Y12 assay (93% vs. 95% average inhibition across all donors). The coefficient of variation for the test precision was less than 8%. The VerifyNow P2Y12 readings were not influenced by age, platelet count, hematocrit, fibrinogen, or cholesterol or triglyceride level.

The assay results can be influenced by treatment with GPIIb-IIIa antagonists and even phosphodiesterase inhibitors, but not aspirin. Before testing, blood should incubate in the collection tube for a minimum of 10 min (but not more than 4 hours) to equilibrate with citrate buffer.

## III. Correlation with Other Measures of Platelet Inhibition

### A. *VerifyNow IIb/IIIa Assay*

Initial evaluations of the VerifyNow IIb/IIIa assay found an excellent correlation between the assay and both turbidometric platelet aggregometry with platelet-rich plasma, and radiolabeled receptor binding assays.[6] Simultaneous measurement of platelet function determined by the Verify-Now IIb/IIIa assay and turbidometric platelet aggregometry induced by 20 μM ADP in blood samples treated with increasing doses of abciximab *in vitro* demonstrated a close correlation between the assays ($r^2 = 0.98$). Similarly, a close correlation was observed between the assay's results and the percentage of unblocked GPIIb-IIIa receptors assessed directly by radiolabeled binding ($r^2 = 0.96$) (Fig. 27-6).

To evaluate real world clinical utility, precision, and reliability of the VerifyNow IIb/IIIa assay, a study involving 192 patients undergoing percutaneous coronary intervention (PCI) at four centers with the use of the GPIIb-IIIa inhibitor abciximab was carried out.[8] All patients had platelet function monitored at three timepoints: (1) baseline, (2) within 1 hour of abciximab bolus administration, and (3) at 24 hours after the abciximab bolus. Abciximab was administered in the standard bolus and 12 hour intravenous infusion. Platelet function was determined at each time point using the VerifyNow IIb/IIIa assay, conventional turbidometric aggregometry and [$^{125}$I]abciximab receptor binding assay. Of the 125 evaluable patients in whom all three methods could be compared, the level of inhibition determined by the VerifyNow correlated well with both standard aggregometry ($r^2 = 0.87$) and with the receptor binding assay ($r^2 = 0.89$).

In a smaller study of 36 patients undergoing PCI and receiving an adjunctive GPIIb-IIIa antagonist, platelet function monitoring by light transmittance aggregometry and the VerifyNow IIb/IIIa assay was performed. Only when PPACK was utilized as the anticoagulant for platelet function testing in patients treated with eptifibatide and tirofiban did the results of light transmittance aggregometry and the VerifyNow correlate well ($r^2 = 0.73$).[9]

As part of the evaluation of the oral GPIIb-IIIa antagonist orbofiban, 176 patients underwent testing with both the VerifyNow IIb/IIIa assay and turbidometric aggregometry using ADP and TRAP as agonist.[10] Platelet inhibition was monitored 4 and 6 hours after orbofiban on days 1, 28, and 84 following randomized dose allocation. The results correlated well with both 20 μM ADP- ($r^2 = 0.87$) and 5 μM TRAP-induced ($r^2 = 0.81$) platelet aggregation, as well as with blood levels of orbofiban ($r^2 = 0.82$).

### B. *VerifyNow Aspirin Assay*

In the initial clinical study to directly compare the VerifyNow Aspirin Assay with standard platelet aggregometry using arachidonic acid as the agonist, a strong correlation was found ($r^2 = 0.88$) when platelet function was monitored prior to and after ingestion of 325 mg of aspirin in 31 individuals (VerifyNow Aspirin Assay package insert) (Fig. 27-7). A second study, involving 148 volunteers with multiple risk factors for coronary artery disease, compared an earlier version of the VerifyNow Aspirin Assay (which utilized cationic propyl gallate as the platelet agonist) with 5 μM epinephrine-induced aggregometry.[11] Platelet function was monitored prior to 325 mg of aspirin and then again between 2 and 30 hours after aspirin. The overall agreement

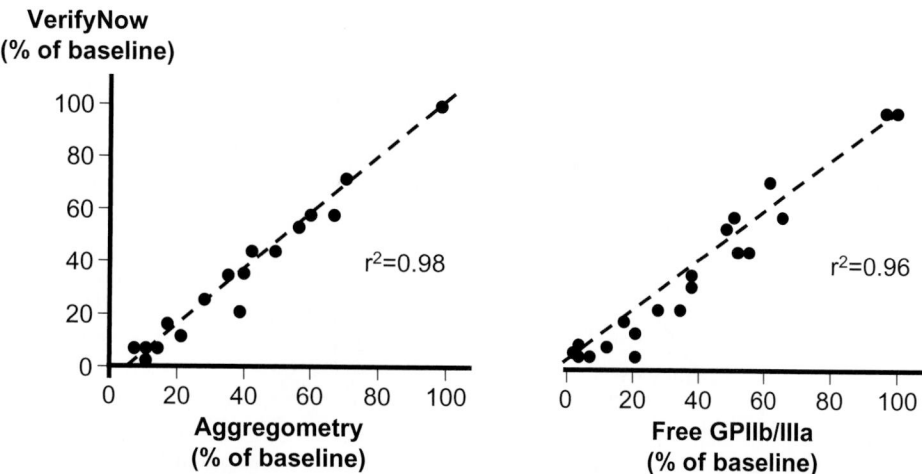

**Figure 27-6.** Correlation between the VerifyNow IIb/IIIa assay and turbidometric platelet aggregation in platelet-rich plasma with 20 μM ADP (left panel) and unblocked GPIIb-IIIa receptors as determined by radiolabeled receptor binding assay (right panel) in whole blood preincubated with increasing concentrations of unlabeled abciximab. Modified from Ref. 6 with permission.

between the two assays was 87% for determining aspirin response and the correlation between the two methods was 0.90.

This same older version of the VerifyNow Aspirin Assay was utilized in a study comparing it with light transmittance aggregometry using ADP and arachidonic acid as agonists, and another point-of-care device, the Platelet Function Analyzer (PFA)-100 (Chapter 28) in 100 patients treated with low-dose aspirin following a transient ischemic attack or an ischemic stroke.[12] Based on standard definitions, 17% of patients were identified as aspirin nonresponsive by the VerifyNow Aspirin Assay, 22% by the PFA-100, and only 5% by light transmittance aggregometry. Interestingly, these investigators found poor agreement between any one test with any other test of aspirin responsiveness.

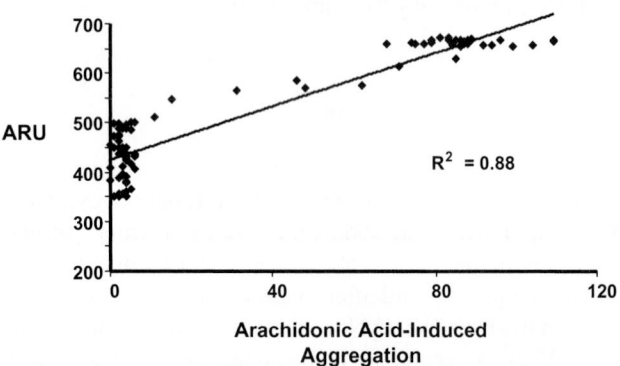

**Figure 27-7.** Correlation between VerifyNow Aspirin Assay, measured in aspirin reaction units (ARU) and arachidonic acid-induced platelet aggregation by light transmittance aggregometry in 71 patients before and after ingestion of 325 mg of aspirin.

## C. VerifyNow P2Y12 Assay

One clinical study has correlated the results of the VerifyNow P2Y12 assay with standard aggregometry. In the ISAR-CHOICE trial patients with known or suspected coronary artery disease were randomized in a double-blind fashion to a 300-, 600-, or 900-mg loading dose of clopidogrel with platelet function determined by optical aggregometry immediately before and 4 hours after the loading dose.[13] All patients also received aspirin. In a subset of 30 patients, platelet inhibition was also determined using the VerifyNow P2Y12 assay.[14] For both PRU and % P2Y12 inhibition, a strong correlation was found with 20 µM ADP and 5 µM ADP-induced platelet aggregation by optical aggregometry (Fig. 27-8). As seen in the graphs, low PRU values and higher values of percent $P2Y_{12}$ inhibition correlate poorly with conventional aggregometry. This is believed to be due to $P2Y_1$ activation by ADP in conventional aggregometry that is prevented in the VerifyNow P2Y12 assay due to the inclusion of $PGE_1$. Similar results have been found with other assays specific for the $P2Y_{12}$ receptor such as phosphorylation of the vasodilator-stimulated phosphoprotein (VASP).[15]

## IV. Association of Platelet Function Results and Clinical Outcomes

### A. VerifyNow IIb/IIIa Assay

The platelet GPIIb-IIIa antagonists are widely utilized for the prevention of thrombotic complications in patients

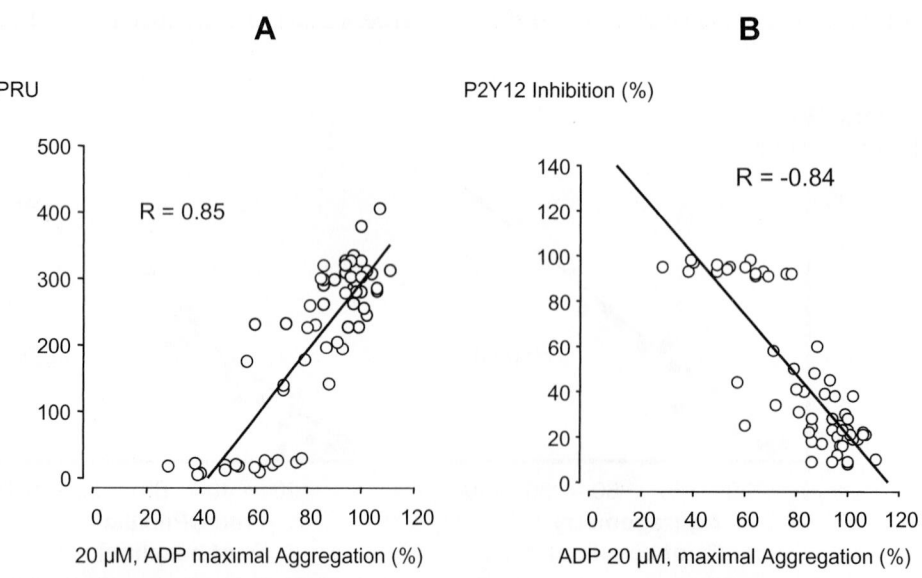

**Figure 27-8.** Correlation between the VerifyNow P2Y12 Assay results as P2Y12 reaction units (PRU) (A) or percent P2Y12 inhibition (B), and conventional ADP-induced platelet aggregometry in 30 patients prior to the initiation of clopidogrel therapy and 4 hours after a clopidogrel loading dose. Modified from Ref. 14 with permission.

undergoing PCI or who present with an acute coronary syndrome (Chapter 62). Because early studies performed in animals involving arterial injury and thrombus formation demonstrated a strong correlation between the level of GPIIb-IIIa receptor blockade and prevention of luminal thrombus formation, with a steep dose response, achievement of a specific level of GP IIb-IIIa inhibition was essential in their development.[16,17] Because of this, monitoring the level of platelet inhibition may be especially important in patients treated with these agents.

The first prospective study designed to correlate the level of platelet inhibition achieved with a GPIIb-IIIa antagonist and clinical outcomes was the GOLD (AU-Assessing Ultegra) study.[18] Five hundred patients at 13 centers undergoing a PCI with the use of a GPIIb-IIIa antagonist had serial platelet function evaluation using the VerifyNow IIb/IIIa Assay at 10 min, 1 hour, 8 hours, and 24 hours following the initiation of GPIIb-IIIa antagonist therapy. Major adverse cardiac events (MACE — composite of death, myocardial infarction, and urgent target vessel revascularization) were prospectively monitored and their incidence was correlated with the measured level of platelet inhibition at all time-points. Interestingly, one-quarter of all patients did not achieve ≥95% inhibition 10 min following the bolus and these patients experienced a significantly higher incidence of MACE (14.4% vs. 6.4%, $p = 0.006$) (Fig. 27-9A). Also, patients whose platelet function was <70% inhibited at 8 hours after the start of GPIIb-IIIa antagonist therapy had a MACE rate of 25% vs. 8.1% ($p = 0.009$) in those patients who were ≥70% inhibited (Fig. 27-9B). By multivariate analysis, platelet function inhibition ≥95% at 10 min following the start of GPIIb-IIIa antagonist therapy was associated with a significant decrease in the incidence of MACE (odds ratio 0.46, 95% confidence interval 0.22–0.96, $p = 0.04$).

This observed association between the level of platelet inhibition as determined by the VerifyNow IIb/IIIa Assay and clinical outcomes suggests the clinical utility of this device. Unfortunately no follow-up study has been conducted to date to confirm that titrating GP IIb/IIIa antagonist therapy based on measured levels of platelet inhibition will decrease adverse events, although one pilot study of 180 patients found similar outcomes in patients treated with a GPIIb-IIIa antagonist who received a second half-bolus of the agent if initial platelet inhibition was <90%.[19]

### B. VerifyNow Aspirin Assay

Although the assay has been available for a relatively short period of time, a number of clinical trials have evaluated the clinical utility of the VerifyNow Aspirin Assay. One of the earliest studies assessed the incidence and clinical predictors of aspirin responsiveness using the VerifyNow Aspirin Assay.[20] Among 422 patients self-reported to be taking

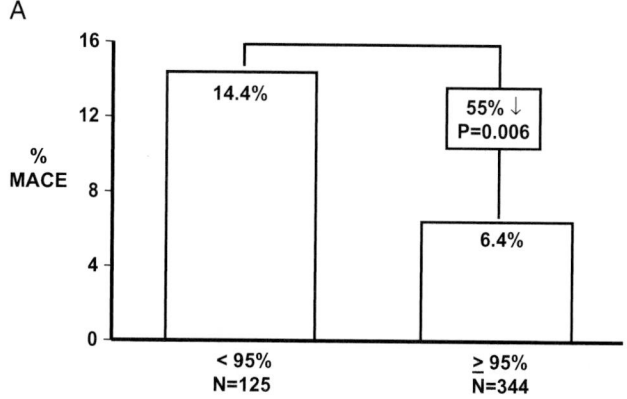

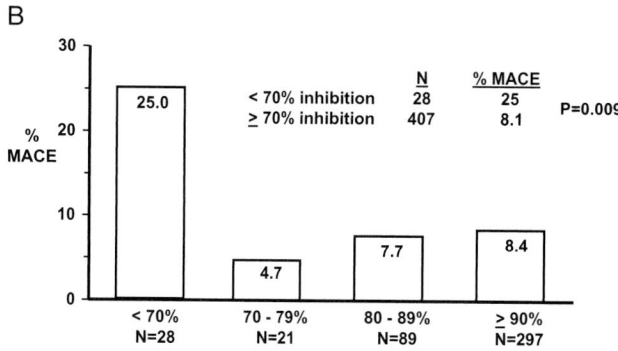

**Figure 27-9.** Incidence of major adverse cardiac events (MACE) in the GOLD (AU-Assessing Ultegra) study in relation to level of platelet function inhibition at 10 min following the GPIIb-IIIa inhibitor bolus (A), and at 8 hours following the bolus, during the infusion (B). Modified with permission from Ref. 18.

aspirin daily for at least 7 days, the investigators found 23% to be aspirin nonresponders using as a cutoff an ARU >550. Only a history of coronary artery disease was found to be an independent predictor of aspirin nonresponsiveness, whereas aspirin dose was not.

The clinical importance of being an aspirin nonresponder as determined by the VerifyNow Aspirin Assay was next studied in 151 patients scheduled for a nonurgent PCI.[21] All patients were being treated with daily aspirin between 80 and 325 mg for at least 1 week and received a 300-mg clopidogrel loading dose 12 to 24 hours prior to the planned PCI. Biomarkers of myonecrosis were measured at baseline and serially after the PCI. Aspirin responsiveness was also measured at baseline and prior to the clopidogrel loading dose. Twenty-nine (19.2%) patients were determined to be aspirin nonresponsive, and these individuals experienced a significantly higher incidence of myocardial necrosis (51.7%) measured by creatinine kinase-myocardial band (CK-MB) elevations than did aspirin responders (24.6%, $p = 0.006$ for comparison). All biomarkers of myonecrosis were signifi-

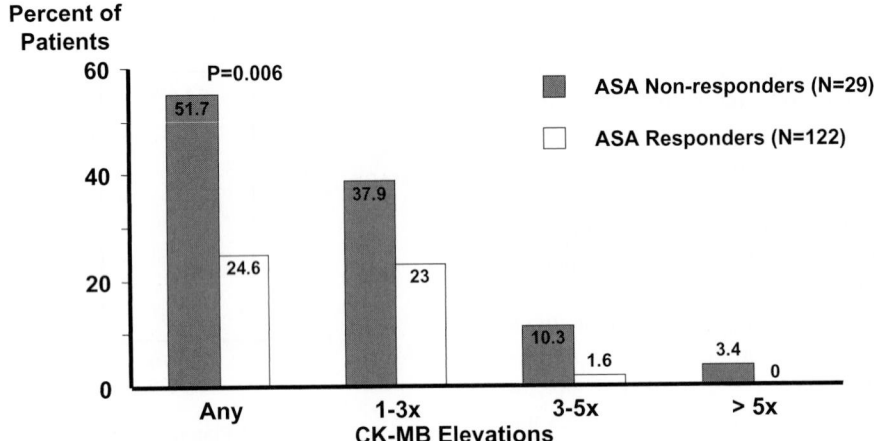

**Figure 27-10.** Incidence and magnitude of creatinine kinase-myocardial band (CK-MB) elevation in aspirin (ASA)-responsive and -nonresponsive patients, as determined by the VerifyNow Aspirin Assay, following a percutaneous coronary intervention. Modified with permission from Ref. 21.

cantly higher in the aspirin nonresponders, and the incidence of myocardial infarction, irrespective of the definition, was increased in nonresponders compared with responders (Fig. 27-10). Further study by the same investigators found that aspirin resistance was associated with impaired coronary flow reserve[22] and that, contrary to earlier studies with the VerifyNow Aspirin Assay, among 468 patients with coronary artery disease a low dose (≤100 mg) of aspirin was an independent predictor of aspirin nonresponsiveness.[23]

This same group has recently presented longer-term follow-up data in a stable population with coronary artery disease. They prospectively enrolled 422 patients who self-reported to be taking a daily aspirin (80 to 300 mg daily for at least 7 days) and followed these patients for approximately 1 year.[24] They identified approximately 27% of the patients to be aspirin nonresponsive and found these patients to have more than twice the risk of ischemic events than did aspirin-responsive patients.

While these early results support an association between aspirin responsiveness as determined by the VerifyNow Aspirin Assay and the risk of thrombotic events, prior to the determination of aspirin responsiveness becoming a routine part of the clinical care of patients treated with aspirin, further studies are needed to confirm that altering therapy — whether through changing the aspirin dose or adding an alternative agent — based on the results of the VerifyNow Aspirin Assay improves clinical outcomes.[25]

### C. VerifyNow P2Y12 Assay

While several studies measuring platelet function in individuals treated with the $P2Y_{12}$ inhibitor clopidogrel have demonstrated a relationship between the level of inhibition achieved and clinical outcomes,[26,27] no trial evaluating clinical outcomes has yet been reported utilizing the VerifyNow P2Y12 assay. However, one of the first trials designed to determine the benefit of alternative antiplatelet therapies based on measured platelet responsiveness to clopidogrel is currently in progress using the VerifyNow P2Y12 and Aspirin assays: the Research Evaluation to Study Individuals Who Show Thromboxane or $P2Y_{12}$ Receptor Resistance (RESISTOR) trial is a double-blind, multicenter trial involving 600 patients undergoing PCI.

## V. Conclusions

The VerifyNow system offers multiple advantages over conventional methods for monitoring platelet function inhibition which include: (1) use of whole blood, thus avoiding the need for sample preparation and eliminating variables in sample preparation, (2) semiautomated format, which avoids operator errors and subjective end point assessments, (3) rapid test completion, (4) digital readout, and (5) the ability to specifically determine the effects of the three major classes of clinically used antiplatelet agents (aspirin, thienopyridines, and GPIIb-IIIa antagonists). Importantly, the IIb/IIIa assay is still the only measure of platelet function that has been prospectively shown to correlate with clinical outcomes in patients treated with a GPIIb-IIIa antagonist,[18] and the aspirin assay has also been found to correlate with both short- and long-term clinical outcomes in aspirin-treated patients.[21] Other potential uses for the VerifyNow system currently being studied include preoperative assessment of platelet function in order to minimize bleeding risk and documentation of patient compliance with antiplatelet therapies.

While antiplatelet agents are used every day by more individuals than any other drug, they remain one of very few drug classes in which treatment is not titrated to the individual's response. Despite the long history of use of antiplatelet therapies, the measure of platelet inhibition that best correlates with clinical efficacy and safety remains unclear. Many of the limitations to the optimization of antiplatelet therapies stem from our inability to easily, accurately, and reproducibly measure their effects. The VerifyNow system offers the potential to overcome many of these limitations. Ongoing trials will be critical in defining its role in the clinical care of all patients treated with aspirin, $P2Y_{12}$ inhibitors, GPIIb-IIIa antagonists, or their combinations.

# References

1. Bernardi, M. M., et al. (1993). Lack of usefulness of prolonged bleeding time in predicting hemorrhagic events in patients receiving the 7E3 glycoprotein IIb/IIIa platelet antibody. *Am J Cardiol, 72,* 1121–1125.
2. Rogers, R., & Levin, J. (1990). A critical reappraisal of the bleeding time. *Sem Thromb Hemost, 16,* 1–20.
3. Coller, B. S. (1980). Interaction of normal, thrombasthenic, and Bernard-Soulier platelets with immobilized fibrinogen: Defective platelet-fibrinogen interaction in thrombasthenia. *Blood, 55,* 169–178.
4. Coller, B. S., et al. (1991). Monoclonal antibodies to platelet GPIIb/IIIa as antithrombotic agents. *Ann NY Acad Sci, 614,* 193–213.
5. Coller, B. S., Lang, D., & Scudder, L. E. (1997). Rapid and simple platelet function assay to assess glycoprotein IIb/IIIa receptor blockade. *Circulation, 95,* 860–867.
6. Smith, J. W., et al. (1999). Rapid platelet-function assay: An automated and quantitative cartridge-based method. *Circulation, 99,* 620–625.
7. Fox, S. C., Behan, M. W., & Heptinstall, S. (2004). Inhibition of ADP-induced intracellular $Ca^{2+}$ responses and platelet aggregation by the P2Y12 receptor antagonists AR-C69931MX and clopidogrel is enhanced by prostaglandin E1. *Cell Calcium, 35,* 39–46.
8. Wheeler, G. L., et al. (2002). The Ultegra rapid platelet-function assay: Comparison to standard platelet function assays in patients undergoing percutaneous coronary intervention with abciximab therapy. *Am Heart J, 143,* 602–611.
9. Simon, D. I., et al. (2001). A comparative study of light transmission aggregometry and automated bedside platelet function assays in patients undergoing percutaneous coronary intervention and receiving abciximab, eptifibatide, or tirofiban. *Cathet Cardiovasc Intervent, 52,* 425–432.
10. Theroux, P., et al. (1999). The Accumetrics Rapid Platelet Function Analyzer (RPFA) to monitor platelet aggregation during oral administration of a GPIIb/IIIa antagonist (abstract). *J Am Coll Cardiol, 33*(Suppl A), 330A.
11. Malinin, A., et al. (2004). Assessing aspirin responsiveness in subjects with multiple risk factors for vascular disease with a rapid platelet function analyzer. *Blood Coagul Fibrinolysis, 15,* 295–301.
12. Harrison, P., et al. (2005). Screening for aspirin responsiveness after transient ischemic attack and stroke: Comparison of 2 point-of-care platelet function tests with optical aggregometry. *Stroke, 36,* 1001–1005.
13. von Beckerath, N., et al. (2005). Absorption, metabolization, and antiplatelet effects of 300-, 600-, and 900-mg loading doses of clopidogrel. Results of the ISAR-CHOICE (Intracoronary Stenting and Antithrombotic Regimen: Choose Between 3 High Oral Doses for Immediate Clopidogrel Effect) Trial. *Circulation, 112,* 2946–2950.
14. von Beckerath, N., et al. (2005). Correlation between platelet response units measured with a point-of-care test and ADP-induced platelet aggregation assessed with conventional optical aggregometry. *Eur Heart J, 26*(Abstract Supplement), 479.
15. Aleil, B., et al. (2005). Flow cytometric analysis of intraplatelet VASP phosphorylation for the detection of clopidogrel resistance in patients with ischemic cardiovascular diseases. *J Thromb Haemost, 3,* 85–92.
16. Coller, B. S., et al. (1989). Abolition of in vivo platelet thrombus formation in primates with monoclonal antibodies to the platelet GPIIb/IIIa receptor. Correlation with bleeding time, platelet aggregation, and blockade of GPIIb/IIIa receptors. *Circulation, 80,* 1766–1774.
17. Gold, H. K., et al. (1988). Rapid and sustained coronary artery recanalization with combined bolus injection of recombinant tissue-type plasminogen activator and monoclonal antiplatelet GPIIb/IIIa antibody in a dog model. *Circulation, 77,* 670–677.
18. Steinhubl, S., et al. (2001). Point-of-care measured platelet inhibition correlates with a reduced risk of an adverse cardiac event following percutaneous coronary intervention. Results of the GOLD (AU-Assessing Ultegra) multicenter study. *Circulation, 103,* 1403–1409.
19. Kini, A. S., et al. (2002). Effectiveness of tirofiban, eptifibatide, and abciximab in minimizing myocardial necrosis during percutaneous coronary intervention (TEAM pilot study). *Am J Cardiol, 90,* 526–529.
20. Wang, J. C., et al. (2003). Incidence of aspirin nonresponsiveness using the Ultegra Rapid Platelet Function Assay-ASA. *Am J Cardiol, 92,* 1492–1494.
21. Chen, W. H., et al. (2004). Aspirin resistance is associated with a high incidence of myonecrosis after non-urgent percutaneous coronary intervention despite clopidogrel pretreatment. *J Am Coll Cardiol, 43,* 1122–1126.
22. Chen, W. H., et al. (2005). Relation of aspirin resistance to coronary flow reserve in patients undergoing elective percutaneous coronary intervention. *Am J Cardiol, 96,* 760–763.
23. Lee, P., et al. (2005). Low-dose aspirin increases aspirin resistance in patients with coronary artery disease. *Am J Med, 118,* 723–727.
24. Cheng, X., et al. (2005). Prevalence, profile, predictors, and natural history of aspirin resistance measured by the Ultegra

Rapid Platelet Function Assay-ASA in patients with coronary heart disease.

25. Michelson, A. D., et al. (2005). Aspirin resistance: Position paper of the Working Group on Aspirin Resistance. *J Thromb Haemost, 3*, 1309–1311.

26. Matetzky, S., et al. (2004). Clopidogrel resistance is associated with increased risk of recurrent atherothrombotic events in patients with acute myocardial infarction. *Circulation, 109*, 3171–3175.

27. Muller, I., et al. (2003). Prevalence of clopidogrel non-responders among patients with stable angina pectoris scheduled for elective coronary stent placement. *Thromb Haemost, 89*, 783–787.

# CHAPTER 28

# The Platelet Function Analyzer (PFA)-100

## John L. Francis

*Florida Hospital Center for Hemostasis and Thrombosis, Florida Hospital
Institute of Translational Research, Orlando, Florida*

## I. Platelet Function

The mechanisms by which platelets participate in hemostasis are complex and are comprehensively reviewed in Part One of this book. In brief, however, platelet activity can be divided into three phases:

1. Adhesion. Platelets readily adhere to foreign surfaces in a process mediated by binding of von Willebrand factor (VWF) to the glycoprotein (GP) Ib-IX-V complex on the platelet plasma membrane (Chapters 7 and 18). Most tests of platelet function do not assess this process.

2. Aggregation. This phase is initiated by many different agonists, including collagen, thrombin, adenosine diphosphate (ADP), and epinephrine that bind to the platelet surface membrane. *In vivo*, platelet aggregation is strongly dependent on fibrinogen binding to platelet integrin $\alpha IIb\beta 3$ (GPIIb-IIIa) (Chapter 8). In the clinical laboratory, this is usually assessed by platelet aggregometry (Chapter 26).

3. Secretion. During activation and aggregation, the contents of platelet granules are secreted into the exterior environment where they play an important role in the augmentation and propagation of the hemostatic plug (Chapter 15). Most hospital laboratories are not equipped to evaluate this aspect of platelet function. Importantly, platelets also act as a phospholipid-rich surface for subsequent coagulation factor assembly during the clotting process (Chapter 19). After fibrin formation, platelets also function to contract and strengthen the clot. Ideally, a screening test of primary hemostasis should be sensitive to all phases of platelet function.

## II. Disorders of Platelet Function

Platelets may be abnormal in number (thrombocytopenia or thrombocytosis) or function. If thrombocytosis is associated with malignancy (e.g., chronic myeloid leukemia or essential thrombocythemia), this process may also be accompanied by abnormal platelet function (Chapters 56 and 58). Reactive forms of thrombocytosis, in contrast, usually have normal platelet function. Functional defects may be inherited (Chapter 57) or, more commonly, are secondary to some other disease process or medication (Chapter 58). These conditions are comprehensively reviewed elsewhere in this book. However, for the purposes of this discussion, they can be classified according to the stage of platelet function affected by the abnormality.

### A. Adhesion Defects

Abnormalities of the GPIb-IX-V receptor that binds to VWF (Bernard–Soulier syndrome) are associated with reduced adhesion. von Willebrand disease (VWD) may also be classified as an adhesion defect, although the actual defect (reduced VWF activity) resides in the plasma.

### B. Aggregation Defects

Defects in integrin $\alpha IIb\beta 3$ (Glanzmann thrombasthenia) result in impaired aggregation.

### C. Secretion Defects

Defects that affect the secretion phase may be due to a deficiency of storage granules or to impairment of the release of granule contents. Examples include dense granule deficiency, gray platelet syndrome, and Hermansky–Pudlak syndrome (see Chapters 15 and 57).

The evaluation of a patient with a suspected platelet disorder will depend on the capability of the physician's local laboratory. Apart from the clinical and family history — a very important part of the evaluation (Chapter 45) — a platelet count and a bleeding time is often the only testing available. Larger institutions may offer platelet aggregation studies, while the specific evaluation of possible VWD is often left to the reference laboratory.

## III. The Bleeding Time

As discussed in detail in Chapter 25, the bleeding time is the time taken to stop bleeding after a wound is made to the skin. To standardize the variables associated with this test, an incision of standard length and depth is made using a commercially available spring-loaded device. Although simple in practice, the time to stop bleeding (typically 2–10 min) is a complex function of vascular properties and platelet number and function. In clinical practice, the bleeding time is used as a first-line screening test of platelet function, both for presurgical assessment and to detect disorders of primary hemostasis such as VWD, congenital or acquired platelet defects, and vascular disorders. It is also sometimes used to determine the response to treatment, for example, with DDAVP (1-deamino-8-D-arginine vasopressin) or platelet transfusions.[1]

Despite the apparent simplicity of the bleeding time, numerous technical variables and factors influence the test.[2] These include operator experience, temperature of the puncture site, skin depth, fat and capillary density, and the way that blood is "blotted" away from the incision. As a result, the bleeding time is poorly reproducible and dependent on the person performing the test. Even if these variables are well controlled, the technique has questionable sensitivity to factors known to affect platelet function, such as aspirin ingestion, congenital thrombocytopathies, and VWD. There is also strong evidence that the bleeding time does not correlate well with the bleeding tendency in individual patients or predict the likelihood of bleeding.[3] Indeed, an accurate bleeding history is considered to be a more valuable screening test.[4–6]

As extensively reviewed by Rodgers and Levin,[2] efforts to standardize the bleeding time have not improved the clinical utility of the test. The College of American Pathologists and the American Society of Clinical Pathologists identified three major shortcomings of the bleeding time[3]:

1. The bleeding time performs poorly as a presurgical screening test in that it does not identify patients who are likely to bleed excessively.

2. A normal bleeding time does not mean that a patient will not bleed.

3. The bleeding time does not reliably identify those subjects who have recently ingested antiplatelet drugs such as aspirin.

Because of the disadvantages of the bleeding time as a screening test for platelet dysfunction in clinical practice,

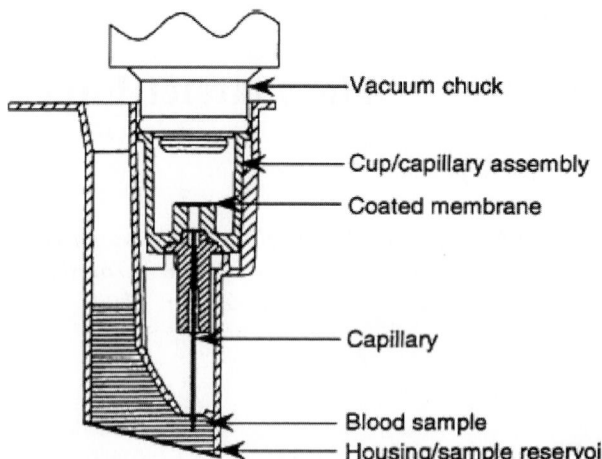

**Figure 28-1.** Schematic diagram of the Platelet Function Analyzer (PFA)-100. Citrated whole blood is placed into the sample reservoir. The blood is aspirated under vacuum through the membrane, which is coated with collagen and epinephrine (CEPI cartridge) or collagen and ADP (CADP cartridge). The time taken for platelets to occlude the aperture in the coated membrane and thereby stop flow through the system is recorded as the closure time (CT). The PFA-100 instrument is also shown in Fig. 23-5 of Chapter 23.

several alternative technologies have been developed. The most widely used at present is the Platelet Function Analyzer (PFA)-100® (Dade-Behring, Deerfield, IL).

## IV. PFA-100

### A. Principle

The PFA-100 test is performed by pipeting a small (0.8 mL) sample of citrated whole blood into a reservoir in a disposable cartridge. The blood sample is then drawn, under a constant vacuum (40 mbar), through a stainless steel capillary tube that activates the platelets via the shear force generated[7] (Fig. 28-1). The blood then passes through an aperture, 150 μm in diameter, in a membrane coated with collagen and epinephrine (CEPI) or collagen and ADP (CADP). Initial adhesion to this membrane is a function of the high shear rates (4000–5000 sec$^{-1}$) generated in the PFA-100 system and is mediated via a GPIb-VWF interaction. Subsequent release of platelet granule contents and the presence of epinephrine or ADP on the membrane result in platelet aggregation.[8]

By electron microscopy, adherent platelets can be seen on the membrane surface as early as 15 seconds. Typically, extensive adherence is evident by 45 seconds and aggregates intruding into the lumen are seen by 80 seconds. After approximately 110 seconds, the aperture is occluded, and blood flow through the membrane ceases.[9] This is referred to as the closure time (CT).[8] The current commercially available PFA-100 represents further development of a device known as the Thrombostat 4000.[10–12]

### B. Factors Affecting Closure Time

**1. Platelet Count.** CT in the PFA-100 is sensitive to both platelet number and function. There is an approximately linear relationship between CT and platelet count below $100 \times 10^9$/L.[9] However, in the presence of normal functionality, the CT is not abnormally prolonged until the platelet count falls below approximately $50 \times 10^9$/L, while nonclosure is virtually certain at platelet counts of $10 \times 10^9$/L.[13] The limitations posed by reduced platelet count may be partially overcome by reducing the diameter of the aperture in the PFA-100 instrument.[14] Thus, at very low platelet counts ($<30 \times 10^9$/L) an aperture of 100 μm gives better discrimination of abnormal platelet function than the standard 150 μm aperture. Unfortunately, these modifications are not commercially available for routine use.

**2. Hematocrit.** CT also increases with decreasing hematocrit.[9,10,15] This finding is compatible with the clinical observation that anemic patients are more likely to bleed and that a reduction of circulating red cell mass can impair platelet function *in vivo*.[16] The exact threshold of hematocrit at which CT is affected will vary between individuals because of differences in platelet functionality. As a rule of thumb, however, the CT will typically become abnormal when the hematocrit falls below 25% and will not be measurable when the hematocrit is below 10%.[13] Zupan et al.[17] reported a significant negative association between hematocrit values and CTs in patients on hemodialysis, estimating that a hematocrit above 35% was the level at which anemia no longer impacted hemostasis in these patients.

**3. Platelet Adhesive Receptors.** The importance of platelet GPIb and integrin αIIbβ3 is illustrated by the way in which increasing concentrations of blocking antibodies to these receptors progressively increase the CT.[9] Similar results are obtained with antibodies to the domains of VWF that bind collagen, GPIb and integrin αIIbβ3.[9] These experiments suggest that the PFA-100 is sensitive to disorders that affect these receptors. In contrast, antibodies to fibrinogen, the natural ligand for activated αIIbβ3, do not affect CT.[9] Deficiency of GPVI is associated with a bleeding phenotype, and allelic differences in this transmembrane glycoprotein may affect PFA-100 CT, as reported by Joutsi-Korhonen et al.[18] These authors speculated that individuals with the GPVI "bb" genotype may be more prone to bleeding when their hemostatic system is challenged.

The platelet integrin α2β1 (a collagen receptor) may also be a determinant of CT under certain circumstances. There are at least three alleles of the α2 gene, giving rise to low, intermediate, and high collagen binding phenotypes. In healthy individuals, these appear to have little influence on platelet function as measured by PFA-100.[19] However, in the presence of borderline levels of VWF activity, low platelet α2β1 density gives rise to less efficient platelet adhesion and a prolonged CT in the PFA-100.[20] It seems likely that the collagen binding capacity of platelets also assumes a more important role in determining CT in the presence of low platelet count, hematocrit, or VWF.

**4. von Willebrand Factor.** Plasma levels of VWF are an important determinant of CT in the PFA-100. The impact of low VWF activity is discussed in more detail below (Section IV.E.2) in the context of screening for VWD. Elevated VWF levels may shorten the CT,[21] an effect that may conspire to mask the tendency of mild platelet dysfunction, moderate thrombocytopenia, or anemia to prolong the CT. Haubelt et al.[22] recently concluded that plasma VWF levels modulate PFA-100 CT to a greater extent than platelet function.

**5. Age and Gender.** There are no significant differences in PFA-100 CT between males and females,[23,24] or between subjects aged more or less than 55 years. However, older males may have a tendency toward shorter values.[24] Neonates tend to have shorter CTs, which is more likely due to higher levels of multimeric VWF[25] than the higher hematocrit.[26] Children, however, tend to have CT values close to those of adults.[27–29] Specific reference ranges for PFA-100 tests on cord blood have been published.[30]

**6. Diurnal Variation.** Diurnal variation in platelet aggregation has been demonstrated using optical aggregometry in platelet-rich plasma, a finding that has been related to the fact that myocardial infarction (MI) is more common during the morning hours. Similarly, PFA-100 CT is shorter in the morning and gradually lengthens throughout the day, an effect that is more pronounced in the CEPI cartridge.[31] Thus, there appears to be a diurnal variation in platelet function studied under high shear, although whether this is enough to interfere with clinical interpretation of the PFA-100 test is not yet clear. However, it is probably wise to ensure that, for comparative purposes, blood for PFA-100 tests is collected at a similar time each day.

**7. Anticoagulant Concentration.** The concentration of sodium citrate used in blood collection tubes is an important determinant of CT. Test times are longer in 0.129 M (3.8%) citrate than in 0.109 M (3.2%), probably due to greater calcium chelation.[32] This effect is more pronounced in the CEPI cartridge, which provides less platelet stimulation compared to the ADP cartridge. The use of unbuffered sodium citrate may be associated with a higher frequency of flow obstructions — usually due to microaggregate formation — in the PFA-100 and should not be used for studies in this instrument.[32] As discussed in Section IV.F.1, the sodium citrate concentration may also affect the sensitivity of the PFA-100 to aspirin.

**8. Collection Tubes and Procedures.** As for all platelet function testing procedures, attention to obtaining a good quality venipuncture is critical. Collection of blood through a 21-gauge (or larger) needle is preferable, although using 23-gauge needles[27] or butterfly cannulae[33] does not appear to affect PFA-100 results. In the author's laboratory, the use of plastic collection tubes did not significantly affect the normal range for PFA-100 CT, although small tubes (for pediatric use) were associated with a greater degree of platelet activation (Francis, unpublished data). Transport of blood to a central laboratory via a pneumatic tube system may adversely affect the results of PFA-100 tests.[34] This may be due to platelet activation by ADP released from red cells, as well as to the formation of air bubbles that increase the frequency of flow obstructions in the PFA-100 instrument.[34]

**9. Time of Testing after Blood Collection.** CT in the PFA-100 appears to be stable up to approximately 5 hours after blood collection,[11,32] although the manufacturer currently recommends a 4-hour testing window. The author's (unpublished) experience is that normal blood is stable for up to 6 hours.

**10. Blood Group.** Patients of blood group O are well known to have lower VWF levels than non-O individuals.[35] As a result, PFA-100 CTs may be slightly longer in group O subjects.[22,36–38] Most, but not all,[22] reports have concluded that this difference is not significant, and at present there is little support for requiring blood group-specific normal ranges for the PFA-100.

## C. Establishing a Normal Range

Each laboratory must establish its own normal range for CT in both cartridge systems of the PFA-100. Normal volunteers must be carefully questioned about personal and family history of bleeding or easy bruising, and must be free from all drugs known to affect platelet function. Even among an apparently healthy blood donor population, a significant number will have prolonged closure in the CEPI cartridge, typically due to intake of aspirin-like agents.[39] In the author's experience, however, even close questioning may fail to reveal the recent ingestion of substances that can impair platelet function, and it is not unusual for a selected volunteer population to contain one or two cases of mild VWD that may skew the normal range. Ideally, "normality" should be established by parallel platelet aggregation studies and measurement of VWF activity. If this is not possible or

practical, the author suggests discarding those CT results from normal volunteers that fall outside the 95th percentile, as well as any that exhibit nonclosure in either test cartridge. Because of the effect of anticoagulant concentration, the normal range must be established using the same concentration of sodium citrate that is in routine use in the institution. Separate normal ranges for pediatric use are not feasible for most institutions and, in any event, do not differ significantly from the adult range.[29]

## D. Interpretation of Results

As described above, PFA-100 results are reported as CT (in seconds). Because the instrument only monitors flow through the system for 5 minutes, very abnormal platelet function may result in failure to achieve closure within this time. Such results are reported as CT >300 seconds. Normal results for both cartridges suggest that platelet function is normal. Typically, further investigations on such samples will reveal normal platelet aggregation responses and normal plasma levels of VWF. Nevertheless, a small number of patients may still have an underlying platelet function defect, and if the bleeding history is clear, such individuals will warrant further investigation.

Prolongation of the CEPI cartridge CT with normal results for CADP typically indicates a defect in thromboxane generation. This is most commonly seen following aspirin ingestion. However, as discussed in detail below, patients with storage pool or release defects may show a similar pattern. Because the use of aspirin-like medications is so common, it is not unusual to observe no closure (>300 sec) in the CEPI cartridge without clinical evidence of a bleeding disorder.

If the CT is significantly prolonged in both cartridges, it is likely that a significant defect of primary hemostasis is present, and further evaluation is certainly warranted. If both CEPI and CADP cartridges are routinely used in the evaluation, a small proportion of patients may be observed who have a normal CEPI result, but a prolonged CT with the CADP cartridge. The significance of an abnormality in the CADP cartridge alone is currently unclear.

Shortening of the PFA-100 CT is less well understood, but may often be associated with elevated VWF levels.[21] CT is shorter in patients with ST elevation MI,[40] not because of increased VWF, which is also high in patients with non-ST elevation MI,[40] but because of ADP-induced platelet activation in such patients.[41]

## E. Detection of Congenital Disorders of Primary Hemostasis

**1. Congenital Platelet Function Disorders.** In studies of normal individuals, and those known to have a platelet function defect, the sensitivity and specificity of the PFA-100 was similar to that of platelet aggregometry.[42] Indeed, for inherited defects of platelet function (mainly VWD), the PFA-100 may actually have greater sensitivity than platelet aggregation testing. In addition, the PFA-100 seems almost as sensitive to recent aspirin ingestion as platelet aggregometry.[42]

As predicted from the results of antibody inhibition experiments described above, the PFA-100 is sensitive to the presence of Bernard–Soulier syndrome (GPIb-IX-V deficiency) and Glanzmann thrombasthenia (integrin $\alpha IIb\beta 3$ deficiency).[13,42] Congenital absence of plasma fibrinogen does not prolong the CT.[13] The PFA-100 is usually abnormal in patients with $\delta$-storage pool deficiency and those with primary secretion defects.[13,43,44] Importantly, these abnormalities may affect closure only in the CEPI cartridge[44] (Table 28-1) and must therefore be distinguished from the more common "aspirin effect" (see Section IV.F.1). This is probably because, although patients with $\delta$-storage pool deficiency release less ADP, this is provided in the CADP cartridge, thereby normalizing the closure defect. Correction of the prolonged CT following DDAVP has been observed in these individuals[43] (Francis, unpublished observations). The performance of the PFA-100 in the detection of platelet secretion defects appears to be variable, and some patients give normal results.[45–47] Typically, though, only the milder defects are missed by the PFA-100.[47]

**2. VWD.** The diagnosis of VWD, especially mild type 1, is notoriously difficult. The bleeding time is normal in a significant proportion of these patients,[48] yet it is still often used as the primary screening test for this condition. Initial studies with the Thrombostat 4000 indicated that the sensitivity of this technology for VWD was very high[49] and more recent work with the PFA-100 has confirmed these findings. In one study, for example, the PFA-100 (CADP cartridge) detected all 58 known cases of VWD tested, while the CEPI cartridge was only slightly less sensitive (56/58).[48] The CT is inversely proportional to plasma VWF activity.[48] In another study, Favaloro et al.[50] reported 100% sensitivity for VWD in a panel of 47 patients with confirmed disease.

**Table 28-1: PFA-100 Results (Closure Time, sec) in Four Patients with Hermansky–Pudlak Syndrome (Francis, J.L., unpublished data)**

|           | CEPI Cartridge | CADP Cartridge |
|-----------|----------------|----------------|
| Patient 1 | 207            | 95             |
| Patient 2 | 248            | 87             |
| Patient 3 | 300            | 105            |
| Patient 4 | 289            | 168            |

Others, in contrast,[44,51,52] have reported a somewhat lower sensitivity (84–90% for CEPI and 65–88% for CADP), possibly due in part to differences in VWF activities in the type 2B patients in these studies. Despite these variations, the greater sensitivity to VWD of the PFA-100, compared to the bleeding time, has been consistent between reported studies.[53] The sensitivity of the bleeding time ranges from 15 to 66% depending on the severity of the VWD.[48,52] Variations in the integrin $\alpha 2\beta 1$ collagen receptor density may also partly account for differences in reported sensitivity, because this influences the CT in subjects with borderline VWF levels.[20]

The sensitivity of the PFA-100 to VWD has also been confirmed in pediatric patients by Dean et al.[54] In this study, all patients with type 2 or 3 VWD had prolonged CT with both cartridges, while 79% and 85%, respectively, of type 1 subjects gave abnormal results in the CEPI and CADP cartridges. In the pediatric population, the overall sensitivity of the PFA-100 was 90%, a diagnostic performance far superior to the bleeding time in the same patients.[54]

The PFA-100 clearly has utility in monitoring the response to DDAVP in patients with type 1 VWD in whom complete correction of the prolonged CT is typically obtained.[55–57] Although treatment with DDAVP normalizes CT in patients with type 1 "platelet-normal" or type 2M Vicenza VWD, it has no significant effects in patients with type 1 "platelet-low," type 1 "platelet-discordant" or type 2A VWD.[51,56] The utility of the PFA-100 as a marker of DDAVP responsiveness may be enhanced by combination with the VWF collagen binding activity assay.[56]

The PFA-100 may not be useful in assessing the response to some forms of VWF concentrate, because these may lack the high molecular weight multimers necessary for membrane closure under high shear conditions.[55] Thus, the CT in patients with DDAVP-unresponsive type 2 or type 3 disease remains prolonged.[51,55] This may also reflect the lack of platelet VWF in type 3 patients.[58] Despite its utility in screening for VWD, however, the PFA-100 appears to have little value in the characterization of severity and subtype of VWD.[38]

There has been considerable interest in using the PFA-100 as a screening test for defects of primary hemostasis in women with menorrhagia. This is the presenting symptom for the majority of women who undergo hysterectomy. About 20% of hysterectomies are performed for dysfunctional uterine bleeding that lacks a clearly defined anatomic cause. Unfortunately, most of these cases are not evaluated for underlying bleeding defects, despite ample evidence that platelet dysfunction and VWD are common in this patient group.[59–62] Indeed, the prevalence of VWD in white females with menorrhagia has been reported to be in the range 10 to 20%,[63–65] while in multiracial groups, platelet dysfunction may be even more common than VWD.[61] The PFA-100 appears to be a cost-effective and practical screening test for this patient population.[66]

## F. Detection of Acquired Platelet Function Disorders

**1. Drug-Induced Platelet Dysfunction.** Aspirin is by far the most commonly used drug that interferes with platelet function[67] (Chapter 60). Detection of the effect of full-dose aspirin on platelet function is readily achieved by the PFA-100 with a sensitivity of approximately 95%.[42] Generally, the CT will be prolonged with the CEPI cartridge, but normal (or near-normal) with the CADP cartridge.[68] This is in contrast to most "true" platelet defects, in which the CT is generally prolonged with both cartridges. Some individuals receiving aspirin, however, do not respond with a prolongation of the CT.[69] It is important to appreciate, however, that storage pool defects (see Section IV.E.1) may give rise to the "aspirin pattern" in the PFA-100.

The lack of effect of aspirin on the CT in some individuals[69] may not relate to a lack of sensitivity of the PFA-100, but the phenomenon of "aspirin resistance" (see Section IV.H). However, the ability of the PFA-100 to detect the aspirin effect appears to be partly dose-related. Among patients with coronary artery disease (CAD) taking 100 mg aspirin for at least 7 days, there was no statistically significant difference in CEPI or CADP values when compared to normal controls.[70] Only 31% of patients showed a prolongation of the CEPI time beyond the upper normal limit. Nevertheless, intake of 100 mg of aspirin elicits a more rapid onset of effect on the PFA-100 than 50 mg, which was only significant on days 3 and 4 of aspirin intake.[70] *In vitro*, the effect is rapid, nonclosure of the CEPI cartridge occurring within 3 minutes of exposure to aspirin.[71]

Healthy volunteers receiving a fractionated infusion of L-aspirin typically show an abnormal CEPI closure after 50 mg aspirin.[71] However, individual responses vary significantly, with some subjects showing nonclosure after just 30 mg, while others require more than 1000 mg to achieve a prolonged CT.[71] CADP closure remains relatively unaffected, despite escalating aspirin dose. Interestingly, the aspirin-induced prolongation of the CEPI cartridge correlates closely with the baseline values and is strongly dependent on VWF levels. Patients who do not achieve marked prolongation of the CEPI CT following aspirin do not have a significantly different degree of thromboxane $B_2$ synthesis compared to those with nonclosure in the PFA-100.[71]

Differences in published reports of the ability of the PFA-100 to detect aspirin, especially at relatively low doses, may depend on the concentration of sodium citrate anticoagulant used. In patients receiving low-dose (100 mg daily) aspirin, CTs return to normal more quickly (2 days) in blood collected into 0.106 M citrate than in samples drawn into 0.129 M (4 days).[72] In addition, there is a tendency for moderately prolonged CTs to normalize if the samples are not tested promptly (within 10 min) after blood collection.[72] Whether this is a real phenomenon, or is simply reflective of the transient platelet defect that occurs immediately after blood drawing, remains to be established.

Interestingly, although the aspirin effect is known to persist for the life of the platelet, the effect on the PFA-100 CT seems shorter lived. This seems to be true irrespective of the citrate concentration used. Thus, CTs normalize within 2 to 4 days following administration of low-dose[72] or full-dose[71] aspirin to normal volunteers, as well as in patients with CAD who stopped aspirin within 3 days of bypass surgery.[73] This is consistent with the lack of CT variation within the platelet count range of 100 to $400 \times 10^9$/L, suggesting that in most individuals, only about 25% of functional platelets (i.e., new platelets not affected by the last dose of aspirin) are needed for a normal CT.

Other nonsteroidal anti-inflammatory drugs have variable effects on the PFA-100. Indomethacin, for example, reduces platelet aggregation and prolongs the CT, at least in the CEPI cartridge. Inhibitors of cyclooxygenase (COX)-2, in contrast, have little or no effect.[74,75] CTs are slightly longer for patients taking aspirin plus ticlopidine compared to aspirin alone.[76] Diclofenac did not appear to prolong the CT, even though its effect was detectable by platelet aggregation with arachidonic acid.[77]

The effects on the PFA-100 CT of the antiplatelet drugs clopidogrel and GPIIb-IIIa antagonists are discussed in Sections IV.H.2 and IV.H.3.

**2. Pregnancy.** CTs decrease during pregnancy, becoming shortest during the second trimester and returning to control (nonpregnant) values at the onset of labor.[78] Interestingly, CTs also shorten in males injected with estrogen.[79]

**3. Renal and Liver Disease.** Patients with advanced renal and liver disease often develop a bleeding diathesis. The platelet component of the hemostatic abnormality is complex and may include abnormal adhesion, defective signal transduction, and acquired storage pool defects (Chapter 58). Many patients with end-stage renal disease or liver cirrhosis have abnormal CTs in the PFA-100, often in both cartridges.[80] In the author's laboratory (Francis, unpublished data), approximately 75% of patients with end-stage renal disease have a defect in the PFA-100. In 24 such patients, the mean ($\pm$ SD) CT for the CEPI and CADP cartridges were $200 \pm 73$ (normal <154) and $163 \pm 71$ (normal <109) seconds, respectively. However, much of this apparent defect seems to be due to reduced hematocrit, despite evidence of an intrinsic platelet defect using conventional aggregometry.[80–82]

**4. Coronary Artery Disease.** Although serum cardiac markers such as creatinine kinase-myocardial band (CK-MB) and troponin I reflect evidence of myocardial necrosis, and are specific markers for acute myocardial infarction, they provide no information about the pathophysiological state within the coronary vasculature. In a large 677 patient study, Linden et al.[83] demonstrated that indices of platelet activation, especially the PFA-100 CT, circulating monocyte-platelet aggregates, circulating neutrophil-platelet aggregates, and platelet surface activated integrin αIIbβ3, are increased in patients with angiographically demonstrated CAD. Furthermore, the degree of platelet activation by these markers reflected the severity of the acute thrombotic process, with the greatest degree of platelet activation occurring in patients with onset of MI <48 hours, followed by patients with MI onset >48 hours previously, then patients with no MI but unstable angina, followed by stable CAD and, finally, the least platelet activation in patients with no angiographically identifiable CAD.[83] Measurement of platelet activation, including PFA-100 CT and flow cytometric markers, may therefore constitute a potentially useful adjuvant for the early diagnosis of acute MI. Furthermore, Frossard et al.[40] found that PFA-100 CT was an independent predictor of myocardial damage as measured by CK-MB and troponin T in patients with ST-segment-elevation MI. Thus, measurement of platelet function with the PFA-100 may help in the risk stratification of patients presenting with MI.[40] Furthermore, in patients with stable angina, the PFA-100 CT may predict the presence or absence of coronary artery stenoses at angiography, thereby potentially avoiding further diagnostic investigations.[84] Another recent study[85] found that an enhanced platelet function after percutaneous coronary intervention (PCI) when measured under high shear rates by PFA-100 is an independent predictor of a worse clinical outcome, even during a short term follow-up, and may therefore help in patient risk stratification.

**5. Aortic and Mitral Valve Disease.** Francis[86] demonstrated that many individuals with significant aortic or mitral valve disease have a significantly abnormal CT, typically in both cartridges. In this study, 82% of patients with severe aortic or mitral valve disease had prolonged closure in both CEPI and CADP cartridges, compared to only 16% of patients with uncomplicated coronary artery disease (Fig. 28-2). Of potential clinical importance was the observation that three of six patients with nonclosure in the CADP cartridge bled excessively during and immediately following cardiopulmonary bypass (CPB) surgery. The etiology of this apparent platelet function defect in cardiac valve disease is not entirely clear, but may include pre existing high shear activation of platelets[87] and loss of high-molecular-weight VWF multimers, creating an acquired form of VWD.[88,89] Indeed, the degree of platelet dysfunction correlates closely with the degree of aortic stenosis and the reduction in high-molecular-weight VWF multimers.[89] CTs in the PFA-100 improved significantly, but often incompletely, after surgical implantation of prosthetic valves.[89]

**6. Cardiopulmonary Bypass.** Platelet dysfunction is often an important factor in blood loss during or after CPB

(Chapter 59). The PFA-100 may represent a rapid and simple test with which to identify platelet-related bleeding in this setting. Wahba et al.[90] found a significant correlation between blood loss and prolongation of the PFA-100 CT, although the practical value of this test to an individual patient was limited by the variability in individual results. Lasne and colleagues[73] reported that the CT is markedly prolonged 15 min after protamine neutralization, but shortens to less than preoperative values after 5 hours. Prolongation of the CT immediately after CPB is more pronounced with the CEPI cartridge. The increase in CT during CPB is partly due to hemodilution and thrombocytopenia and partly to other factors including loss of platelet membrane GPIb and GPIIb-IIIa[91] (see Chapter 59 for details). Despite the relationship of the CT to impaired platelet function, however, the PFA-100 results were not related to the bleeding risk in this study.[73]

In contrast, Slaughter et al.[92] showed that although CADP CTs increased significantly during CPB, they typically achieved near-baseline values 15 minutes after neutralization of heparin. Correction of hematocrit or supplementation with Humate P (factor VIII/VWF concentrate) had no effect, suggesting that neither hemodilution nor transient reductions in VWF activity could account for the abnormal PFA-100 results. In this study, the positive and negative predictive values for bleeding after CPB were 18% and 96%, respectively. These figures agree closely with the results obtained from more complex and time-consuming platelet aggregation studies.[93] Eighty percent of patients with an excessive postoperative chest tube output had prolonged CT in the PFA-100 when measured 15 minutes after protamine injection.

Raman and Silverman[94] reported that the PFA-100 may help identify patients that would benefit from platelet transfusion after open-heart surgery. In this small study,[94] 15/16 patients with bleeding who responded to platelet transfusion had abnormal PFA-100 CT 15 minutes after protamine. This was in contrast to nonbleeders and bleeders in whom platelet transfusion alone did not adequately stop the bleeding. Overall, this small study showed that the sensitivity of the PFA-100 for platelet-responders and nonresponders was 94% and 85%, respectively. Notably, the respective figures for the platelet count were only 56% and 55%.[94] A normal PFA CT in a bleeding patient following CPB suggests a surgical etiology and points to the need for surgical re-exploration.[95] Preoperative PFA-100 testing, however, does not appear to identify those patients at risk of intra- or postoperative bleeding[96] unless they give a history consistent with a bleeding disorder.[97]

Taken together, the available data on the use of the PFA-100 during CPB suggest that preoperative testing is unlikely to identify patients with an increased risk of bleeding, with the possible exception of a small number of previously unidentified patients with VWD or other mild platelet

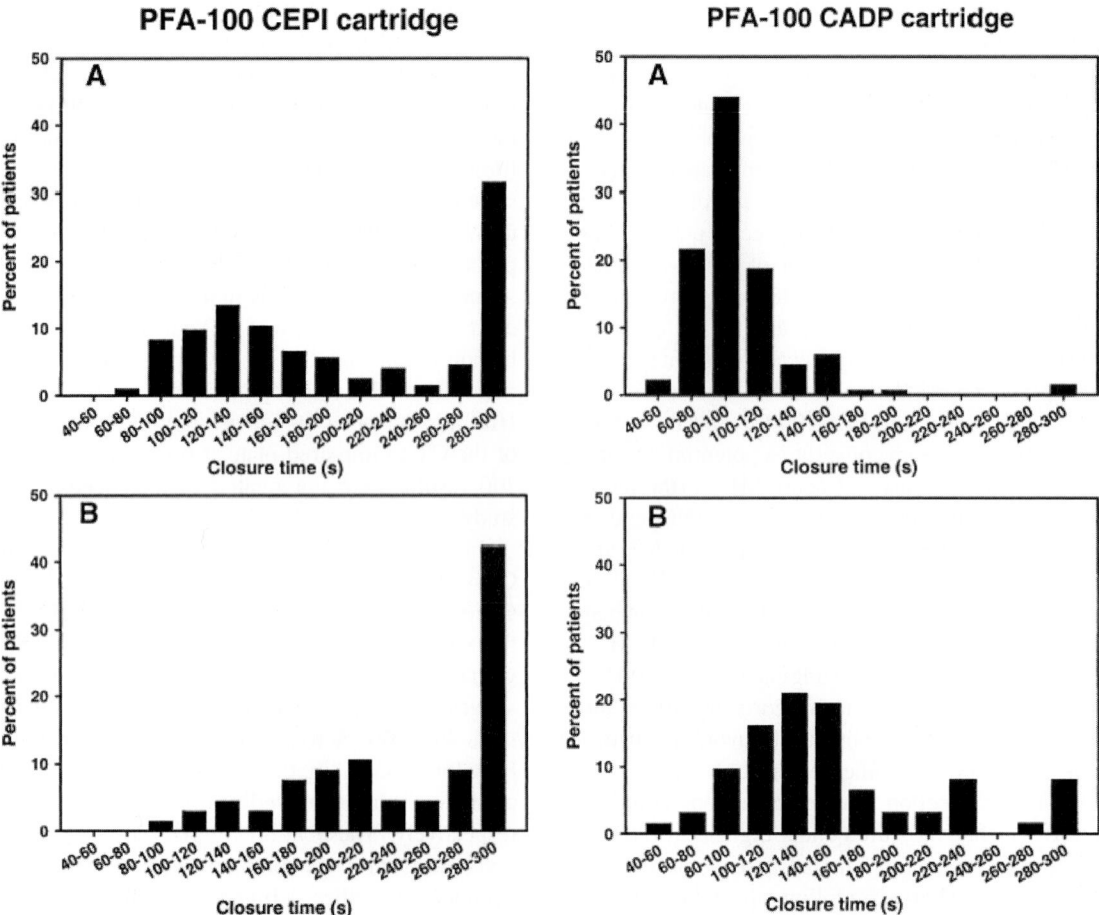

**Figure 28-2.** PFA-100 results in patients prior to undergoing uncomplicated coronary artery bypass compared with those having aortic and/or mitral valve replacement.[86] The left panels show the distribution of CTs (CEPI cartridge) for patients with coronary artery disease alone (A) and with severe aortic and/or mitral valve disease (B). The right panels show similar information for the CADP cartridge. The higher proportion of patients in the valvular disease group with abnormal platelet function, especially in the CADP cartridge, is evident.

function defects. However, among those who do develop bleeding complications, the PFA-100 may identify patients who are unlikely to benefit from platelet transfusions. Further studies in this area are warranted.

**7. Other Clinical Conditions.** Mimidis et al.[98] reported that the mean CT of the CADP cartridge (but not CEPI) was significantly shorter in patients with mild acute pancreatitis. Although these authors[98] point to the value of the PFA-100 as an indicator of increased platelet adhesiveness and aggregation potential in acute pancreatitis, it is possible that their results simply reflected an increase of VWF. Patients with essential thrombocythemia (Chapter 56) typically have a prolonged bleeding time following aspirin ingestion that has been attributed to platelet dysfunction. Such patients have a reduction in the level of high-molecular-weight VWF multimers, and are readily detected by the PFA-100, even without aspirin challenge.[99]

**8. Blood Donors.** Prospective platelet donors who have recently ingested aspirin are deferred. Because the use of antiplatelet medication is poorly identified by history alone, the PFA-100 has been proposed as a tool to screen potential platelet donors for abnormal platelet function. Several studies have demonstrated a significant incidence of platelet dysfunction in blood donor populations.[39,100] Among apheresis donors, about 20% had abnormal CEPI results, over half of which exhibited nonclosure.[100] Interestingly, the PFA-100 CT is prolonged following plateletpheresis, an abnormality that appears to be unrelated to the temporary decrease in platelet count that follows the procedure.[37] Plasmapheresis also leads to a mild platelet function defect as assessed by prolongation of PFA-100 CT.[101] Although aspirin ingestion is a common cause of platelet dysfunction in this population, it does not explain all the observed abnormalities, and recent intake of certain foods, including chocolate[102] and flavenol-rich cocoa[103] might be implicated.

### G. Comparison of the PFA-100 and the Bleeding Time as Tests of Platelet Dysfunction

As discussed above, the bleeding time is still widely used to evaluate platelet function, but has questionable sensitivity to thrombocytopenia and platelet dysfunction (Chapter 25). It is therefore not surprising that the PFA-100 has been compared to the bleeding time as a first line screening test for platelet dysfunction.[104] To address this question, the PFA-100 test was performed on 113 hospital inpatients in whom the bleeding time was ordered as part of their routine medical care. If the CEPI CT was abnormal, the PFA-100 was also performed using the CADP cartridge. If the bleeding time and PFA-100 agreed, no further action was taken. If the bleeding time and the PFA-100 CT gave discrepant results, whole blood platelet aggregation studies and a review of the patient's chart, including medication history, were performed. As shown in Table 28-2, the bleeding time and PFA-100 agreed in 84/113 (74%) of the patients studied.[104] Twenty-three patients had an abnormal PFA-100 despite a normal bleeding time. Platelet aggregation was abnormal in 20 (87%) of these. In the 29 discordant subjects, whole blood platelet aggregation supported the results of the PFA-100 test in 25 cases (86%). In summary, the PFA-100 detected significantly more abnormalities than the bleeding time in this unselected hospital population. The fact that these were indeed platelet function defects (albeit mainly drug-related) is indicated by the high degree of agreement (86%) between the PFA-100 test and platelet aggregation studies.

Most studies have reported that the PFA-100 is superior to the bleeding time in detecting VWD,[44,48,105] while others have suggested that the two tests have equivalent sensitivity.[106] The UK Haemophilia Center Doctors' Organization recommended against the use of the bleeding time in screening for VWD.[107] The relative sensitivities of the bleeding time and PFA-100 for VWD in 11 separate studies are shown in Table 25-1 of Chapter 25.

The relative performance of the PFA-100 and the bleeding time in the diagnosis of some congenital platelet function defects is less well established. In patients with Hermansky–Pudlak syndrome, for example, the bleeding time was slightly more sensitive than the PFA-100, although the number of patients studied was relatively small.[44] Quiroga et al.[108] reported that both tests performed relatively poorly in detecting hemostatic defects in patients with hereditary mucocutaneous hemorrhage (see also Table 25-2 in Chapter 25). In this study,[108] the PFA-100 performed better for detecting VWD — albeit with lower sensitivity than other reports — while the bleeding time detected more platelet secretion defects. Patients with a mild platelet function disorder according to platelet aggregation were more likely to have an abnormal bleeding time than PFA-100 CT.[106] In a pediatric population, the PFA-100 was significantly more sensitive to both VWD and qualitative platelet defects than the bleeding time.[109]

### H. Monitoring Antiplatelet Agents with the PFA-100

**1. Aspirin Resistance.** The ease of use of the PFA-100 and its sensitivity to platelet dysfunction suggest its application in monitoring antiplatelet therapy. In particular, its ability to detect the aspirin-induced platelet effect points to a role in assessing patient compliance[71] or aspirin resistance (see also Chapter 60). The latter syndrome has been the subject of recent intense research interest, although the term "aspirin resistance" remains poorly defined.[110] Thus, it is not clear whether this term should refer to the clinical failure of aspirin to protect individuals from arterial thrombotic events, or laboratory evidence that aspirin is not inhibiting platelet function.

Even among normal subjects taking aspirin, some individuals do not respond with a prolongation of the PFA-100 (CEPI) CT, even though their platelet aggregation in response to arachidonic acid is appropriately inhibited (Burr and Francis, unpublished data). As shown in Table 28-3, a significant proportion of patients taking aspirin for the prevention of MI and stroke appear "resistant" to the antiplatelet effect of aspirin when tested in the PFA-100. For most of these studies, outcome information has not been available, and thus the value of such testing is largely unconfirmed. There are, however, some suggestions that aspirin resistance according to the PFA-100 CT may have clinical significance. In one small (n = 53) study[111] of patients taking 100 mg aspirin daily, patients who remained free from recurrent cerebrovascular events all had prolonged CT, while those who suffered ischemic stroke or transient ischemic attack (TIA) had significantly shorter CT. Indeed, one third of the individual symptomatic patients had normal PFA-100 results, indicating aspirin resistance.

There is a discrepancy between the results of different tests of aspirin resistance, and between the PFA-100 and platelet aggregation studies in particular. The exact cause of aspirin resistance measured by the PFA-100 is unclear and in many patients it may be due to factors other than failure

### Table 28-2: Comparison of Bleeding Time (BT) and PFA-100 CT in 113 Hospital Inpatients[104]

|                  | BT Normal      | BT Abnormal    |
| ---------------- | -------------- | -------------- |
| PFA-100 normal   | 67<br>(59.3%)  | 6<br>(5.3%)    |
| PFA-100 abnormal | 23<br>(20.4%)  | 17<br>(15.0%)  |

**Table 28-3: Aspirin Resistance as Assessed by the PFA-100**

| Patient Population | n | Prevalence (%) | Reference |
|---|---|---|---|
| Stable CAD | 325 | 9.5 | 129 |
| Stable CAD | 50 | 20 | 130 |
| Stable CAD | 113 | 32 | 131 |
| Stable CAD | 100 | 27 | 132 |
| ACS | 75 | 40 | 132 |
| Stable angina | 98 | 30 | 133 |
| Unstable angina | 31 | 42 | 134 |
| Previous MI (aspirin alone) | 71 | 35 | 135 |
| Previous MI (aspirin + warfarin) | 58 | 40 | |
| NSTEMI | 38 | 26 | 41 |
| STEMI | 30 | 83 | |
| Peripheral arterial disease | 31 | 40 | 136 |
| Peripheral arterial disease | 116 | 20 | 120 |
| Peripheral arterial disease | 98 | 10 | 114 |
| Stroke | 31 | 16 | 123 |
| Stroke/TIA | 129 | 37 | 137 |
| CVD — early | 57 | 60 | 138 |
| CVD — convalescent | 46 | 43 | |
| Type II diabetes | 31 | 71 | 139 |
| Type II diabetes | 172 | 21 | 140 |
| Occluded saphenous vein post-CPB | 14 | 50 | 141 |

Abbreviations: ACS, acute coronary syndrome; CAD, coronary artery disease; CPB, cardiopulmonary bypass; CVD, cerebrovascular disease; MI, myocardial infarction; NSTEMI, non-ST elevation acute myocardial infarction; STEMI, ST elevation acute myocardial infarction; TIA, transient ischemic attack.

of aspirin to block COX activity. First, VWF levels may be higher in patients labeled as "poor responders."[112,113] As discussed in Sections IV.B.4 and IV.E.2, VWF may be the single most important determinant of CT in the PFA-100 and elevated plasma VWF concentrations might partially compensate for loss of the arachidonic acid pathway of platelet activation. This does not appear to be solely the result of an inflammatory response because elevated C-reactive protein levels in patients with peripheral arterial disease do not influence PFA-100 results.[114] Second, there is evidence that patients with ST elevation acute MI have markedly elevated plasma ADP levels that correlate with aspirin resistance as measured by the PFA-100.[41] Finally, genetic factors may play a role, as platelets homozygous for

the Pl[A1] allele appear to be less sensitive to the inhibitory action of low-dose aspirin.[115]

In summary, the etiology, clinical significance, and appropriate therapeutic approach to "aspirin resistance" as measured by the PFA-100 and other laboratory tests remain unclear. Therefore, in the absence of clinical trial data to guide patient management, it seems premature to recommend a change in antiplatelet therapy based on laboratory test results.[116]

**2. GPIIb-IIIa Antagonists.** The PFA-100 is extremely sensitive to the effects of GPIIb-IIIa antagonists. For example, abciximab (ReoPro) rapidly achieves nonclosure of both cartridge types.[8] The PFA-100 appears to be at least as useful as platelet aggregometry and flow cytometric assessment of GPIIb-IIIa receptor occupancy in patients undergoing PCI with GPIIb-IIIa antagonist therapy.[117,118] For most (>95%) patients, the PFA-100 CT remains above 300 seconds during PCI and platelet function has normalized in more than 70% of patients by 24 hours. Failure to observe nonclosure in the PFA-100 may be associated with an increased incidence of subsequent adverse cardiac events.[119]

**3. Clopidogrel.** Clopidogrel, an inhibitor of the platelet $P2Y_{12}$ ADP receptor, is a widely used antiplatelet therapy (Chapter 61). A test that could rapidly and accurately assess the effectiveness of such treatment could have clinical utility. Mueller et al.[120] studied 150 patients with peripheral arterial disease taking antiplatelet therapy. Patients taking 100 mg aspirin alone, or 100 mg aspirin plus 75 mg clopidogrel, had markedly prolonged CEPI CT compared to patients not on antiplatelet therapy. In contrast, those taking 75 mg clopidogrel alone had no prolongation of the CT. There were no significant differences in any of the treatment subgroups using the CADP cartridge. The PFA-100 failed to reliably demonstrate loss of platelet ADP responses in normal volunteers given clopidogrel (300-mg loading dose, followed by 75-mg maintenance dose) in contrast to ADP-induced platelet aggregation and the vasodilator-stimulated phosphoprotein (VASP) assay.[121] Thus, it appears that the PFA-100 is relatively insensitive to this commonly used dose of clopidogrel. In *in vitro* studies, the PFA-100 failed to detect the inhibitory effects of synthetic inhibitors of the $P2Y_1$ and $P2Y_{12}$ ADP receptors.[122] The apparently limited usefulness of the PFA-100 to monitor the effects of ADP receptor antagonists may be due to the relatively high concentration of ADP currently used in the PFA-100 system.

Although most studies have pointed to the relative insensitivity of the PFA-100 to ADP receptor antagonists, results from a small study in patients with ischemic stroke indicated that combined aspirin and clopidogrel may markedly prolong CT in the CADP cartridge.[123] This was despite the fact that monotherapy with either agent had no significant effect on the CADP CT. Why combination therapy exerted this synergistic effect is not clear. However, the fact that CT in the CADP cartridge was usually either normal or markedly prolonged (>300 sec) suggests that one single genetic or environmental factor might determine the response in a subset of patients.[124]

### I. Using the PFA-100 in Clinical Practice

In the clinical laboratory, as currently recommended by the manufacturer, it seems reasonable to use the CEPI cartridge as the initial screening test. If the CEPI time is normal, then it is probable that the patient does not have a significant platelet function defect. If the CEPI time is prolonged, then a CADP test should be performed. A normal CADP result is then strongly suggestive of recent aspirin ingestion. An abnormal CADP time should be followed by additional evaluation, especially conventional platelet aggregation tests, to elucidate the precise nature of the defect. If this approach is used, it is important to consider the patient's history in the interpretation of results. For example, an "aspirin pattern" in a patient with a history of bleeding, and who denies aspirin ingestion, might indicate a platelet storage pool defect. A small proportion of patients may have an abnormal CADP result despite a normal CT with the CEPI cartridge.[125] Such individuals would be missed with the recommended approach to testing, although the clinical significance of this finding is unknown. The PFA-100 can be used as part of an algorithmic approach for evaluating a bleeding disorder.[126]

In an analysis of over 2000 PFA-100 tests performed in the author's laboratory, approximately 55% had abnormal results, 25% had an "aspirin" pattern, and 30% showed a prolonged CT in both cartridges. This relatively high incidence of abnormal results probably reflects the high proportion of patients with cardiovascular disease in our practice, many of whom are taking aspirin and other medications. Prolongations of the CT in the CADP cartridge, however, are generally moderate, with only a minority (approximately 10%) of patients having a CT >200 seconds.

Ortel and colleagues[125] utilized the PFA-100 in a tertiary outpatient setting, further evaluating abnormal results with platelet aggregometry and measurement of VWF. Prolonged CT in the CADP and/or CEPI cartridges were found in 114/305 patients (37.3%), most of whom (n = 79) had isolated prolonged CEPI CTs, largely due to aspirin therapy. A few individuals had a prolonged CEPI time with normal aggregation studies—possibly due to δ-storage pool disease. Prolonged CADP CTs were most frequently due to qualitative platelet defects and/or decreased VWF levels. The latter accounted for 27% of patients with prolonged CADP. Prolonged CADP CTs were also seen in patients with sickle cell disease, in whom they were associated with a decreased

hematocrit.[125] Concurrent illnesses did not appear to reduce the ability of the PFA-100 to detect platelet defects or VWD, because the frequency of VWD in this study (2.3%) was consistent with the incidence of this condition reported elsewhere.

It has been estimated that over 90% of the bleeding times performed in the United States are done as part of a preoperative evaluation of hemostatic function, even though published data strongly suggest that such tests are not informative in the absence of a personal or family history of a bleeding disorder (see Chapter 25). Despite the greater sensitivity of the PFA-100 for platelet dysfunction and VWD, it is probably not cost effective to perform this test in a routine preoperative manner. Franchini et al.[127] screened 1342 patients scheduled for thyroid surgery, and found abnormal PFA-100 results in 37 (2.7%). Most of these defects were shown to be due to an acquired form of VWD, which is not uncommon in hypothyroidism. Only three individuals were subsequently proven to have congenital, type 1 VWD. In another large (n = 5649) prospective study,[97] preoperative patients were asked to complete a standardized questionnaire about their bleeding history. Of these patients, 628 (11.2%) gave a positive history, and abnormal platelet function was identified in almost 40% of these individuals by the PFA-100. In contrast, hemostatic testing of the 5021 patients with a negative bleeding history revealed only nine cases of lupus anticoagulant detected with the activated partial thromboplastin time (APTT). None of these patients had platelet dysfunction. This study[97] indicates that screening all surgical patients for hemostatic abnormalities is not cost effective; however, testing with PFA-100 in only those patients who provide a positive bleeding history will reveal a significant number with defective platelet function.

A recent review concluded that the PFA-100 should be considered an "optional" test in the routine evaluation of platelet dysfunction[128] on the grounds that it lacks sufficient sensitivity or specificity to be used as a screening tool for platelet disorders. While it is true that a prolonged CT does not distinguish between, for example, a platelet disorder and VWD, this author believes that the PFA-100 is nevertheless helpful in identifying those patients in whom further (and more complicated and expensive) laboratory tests might be indicated. This is especially useful for small hospital laboratories that lack the facilities for platelet aggregation testing and assays for VWF. In this regard, the PFA-100 can justifiably be regarded as the "sedimentation rate" of primary hemostasis.

## V. Conclusions

The PFA-100 has several advantages as a screening test of platelet function. Unlike many other platelet function tests, the PFA-100 is performed under conditions of shear that are known to have a physiologically important effect on platelet function (Chapter 18). The PFA-100 employs citrated whole blood and can therefore share samples collected for other coagulation tests. Blood samples are fairly stable at room temperature making transport from outlying sites feasible. The test provides more clinically useful information than the bleeding time and has greater sensitivity to congenital and acquired platelet function defects. The PFA-100 may also be cheaper than the bleeding time—especially if large numbers are performed. Finally, the PFA-100 is much quicker to perform than platelet aggregation studies, offering significant savings in technical time. Clinical data suggest that normal results in the PFA-100 should obviate the need for further platelet function testing unless the history indicates otherwise.

In conclusion, there is increasing interest in whole blood technology for the diagnosis of platelet function defects and several new instruments are becoming commercially available. Although many of these have significant research potential, application to the clinical laboratory requires that the test be rapidly performed, provide clinically relevant information, and be cost effective. The PFA-100 test meets these criteria.

## Acknowledgment

The author is grateful to Doug Christie, Ph.D., for critical review of the manuscript.

## References

1. Rodeghiero, F., Castaman, G., & Mannucci, P. M. (1991). Clinical indications for desmopressin (DDAVP) in congenital and acquired von Willebrand disease. *Blood Rev, 5,* 155–161.

2. Rodgers, R. P., & Levin, J. (1990). A critical reappraisal of the bleeding time. *Semin Thromb Hemost, 16,* 1–20.

3. Peterson, P., Hayes, T. E., Arkin, C. F., et al. (1998). The preoperative bleeding time test lacks clinical benefit: College of American Pathologists' and American Society of Clinical Pathologists' position article. *Arch Surg, 133,* 134–139.

4. Burns, E. R., Billett, H. H., Frater, R. W., et al. (1986). The preoperative bleeding time as a predictor of postoperative hemorrhage after cardiopulmonary bypass. *J Thorac Cardiovasc Surg, 92,* 310–312.

5. Burns, E. R., & Lawrence, C. (1989). Bleeding time. A guide to its diagnostic and clinical utility. *Arch Pathol Lab Med, 113,* 1219–1224.

6. Houry, S., Georgeac, C., Hay, J. M., et al. (1995). A prospective multicenter evaluation of preoperative hemostatic screening tests. The French Associations for Surgical Research. *Am J Surg, 170,* 19–23.

7. Dunkley, S., & Harrison, P. (2005). Platelet activation can occur by shear stress alone in the PFA-100 platelet analyser. *Platelets, 16,* 81–84.

8. Kundu, S. K., Heilmann, E. J., Sio, R., et al. (1995). Description of an in vitro platelet function analyzer—PFA-100. Semin *Thromb Hemost, 21*(Suppl 2), 106–112.

9. Kundu, S. K., Heilmann, E. J., Sio, R., et al. (1996). Characterization of an in vitro platelet function analyzer, PFA-100. *Clin Appl Thromb Hemost, 2,* 241–249.

10. Alshameeri, R. S., & Mammen, E. F. (1995). Performance characteristics and clinical evaluation of an in vitro bleeding time device—Thrombostat 4000. *Thromb Res, 79,* 275–287.

11. Alshameeri, R. S., & Mammen, E. F. (1995). Clinical experience with the Thrombostat 4000. *Semin Thromb Hemost, 21* (Suppl 2), 1–10.

12. Kratzer, M. A., Negrescu, E. V., Hirai, A., et al. (1995). The Thrombostat system. A useful method to test antiplatelet drugs and diets. *Semin Thromb Hemost, 21*(Suppl 2), 25–31.

13. Harrison, P., Robinson, M. S., Mackie, I. J., et al. (1999). Performance of the platelet function analyser PFA-100 in testing abnormalities of primary haemostasis. *Blood Coagul Fibrinolysis, 10,* 25–31.

14. Carcao, M. D., Blanchette, V. S., Stephens, D., et al. (2002). Assessment of thrombocytopenic disorders using the Platelet Function Analyzer (PFA-100). *Br J Haematol, 117,* 961–964.

15. Sohngen, D., Hattstein, E., Heyll, A., et al. (1995). Hematological parameters influencing the Thrombostat 4000. *Semin Thromb Hemost, 21*(Suppl 2), 20–24.

16. Anand, A., & Feffer, S. E. (1994). Hematocrit and bleeding time: An update. *South Med J, 87,* 299–301.

17. Zupan, I. P., Sabovic, M., Salobir, B., et al. (2003). Utility of in vitro closure time test for evaluating platelet-related primary hemostasis in dialysis patients. *Am J Kidney Dis, 42,* 746–751.

18. Joutsi-Korhonen, L., Smethurst, P. A., Rankin, A., et al. (2003). The low-frequency allele of the platelet collagen signaling receptor glycoprotein VI is associated with reduced functional responses and expression. *Blood, 101,* 4372–4379.

19. Luzak, B., Golanski, J., Rozalski, M., et al. (2003). Effect of the 807 c/t polymorphism in glycoprotein ia on blood platelet reactivity. *J Biomed Sci, 10,* 731–737.

20. Di Paola, J., Federici, A. B., Mannucci, P. M., et al. (1999). Low platelet alpha2beta1 levels in type I von Willebrand disease correlate with impaired platelet function in a high shear stress system. *Blood, 93,* 3578–3582.

21. Homoncik, M., Blann, A. D., Hollenstein, U., et al. (2000). Systemic inflammation increases shear stress-induced platelet plug formation measured by the PFA-100. *Br J Haematol, 111,* 1250–1252.

22. Haubelt, H., Anders, C., Vogt, A., et al. (2005). Variables influencing Platelet Function Analyzer-100 closure times in healthy individuals. *Br J Haematol, 130,* 759–767.

23. Bock, M., De Haan, J., Beck, K. H., et al. (1999). Standardization of the PFA-100(R) platelet function test in 105 mmol/l

buffered citrate: Effect of gender, smoking, and oral contraceptives. *AHA 6th scientific forum on quality care and outcomes in cardiovasc disease and stroke, Abstract,* p. 145. *Br J Haematol, 106,* 898–904.

24. Sestito, A., Sciahbasi, A., Landolfi, R., et al. (1999). A simple assay for platelet-mediated hemostasis in flowing whole blood (PFA-100): Reproducibility and effects of sex and age. *Cardiologia, 44,* 661–665.

25. Roschitz, B., Sudi, K., Kostenberger, M., et al. (2001). Shorter PFA-100 closure times in neonates than in adults: Role of red cells, white cells, platelets and von Willebrand factor. *Acta Paediatr, 90,* 664–670.

26. Israels, S. J., Cheang, T., McMillan-Ward, E. M., et al. (2001). Evaluation of primary hemostasis in neonates with a new in vitro platelet function analyzer. *J Pediatr, 138,* 116–119.

27. Carcao, M. D., Blanchette, V. S., Dean, J. A., et al. (1998). The Platelet Function Analyzer (PFA-100): A novel in-vitro system for evaluation of primary haemostasis in children. *Br J Haematol, 101,* 70–73.

28. Knofler, R., Weissbach, G., & Kuhlisch, E. (1998). Platelet function tests in childhood. Measuring aggregation and release reaction in whole blood. *Semin Thromb Hemost, 24,* 513–521.

29. Lippi, G., Manzato, F., Franchini, M., et al. (2001). Establishment of reference values for the PFA-100 platelet function analyzer in pediatrics. *Clin Exp Med, 1,* 69–70.

30. Boudewijns, M., Raes, M., Peeters, V., et al. (2003). Evaluation of platelet function on cord blood in 80 healthy term neonates using the Platelet Function Analyser (PFA-100); shorter in vitro bleeding times in neonates than adults. *Eur J Pediatr, 162,* 212–213.

31. Dalby, M. C., Davidson, S. J., Burman, J. F., et al. (2000). Diurnal variation in platelet aggregation with the PFA-100 platelet function analyser. *Platelets, 11,* 320–324.

32. Heilmann, E. J., Kundu, S. K., Sio, R., et al. (1997). Comparison of four commercial citrate blood collection systems for platelet function analysis by the PFA-100 system. *Thromb Res, 87,* 159–164.

33. Mani, H., Kirchmayr, K., Klaffling, C., et al. (2004). Influence of blood collection techniques on platelet function. *Platelets, 15,* 315–318.

34. Dyszkiewicz-Korpanty, A., Quinton, R., Yassine, J., et al. (2004). The effect of a pneumatic tube transport system on PFA-100 closure time and whole blood platelet aggregation. *J Thromb Haemost, 2,* 354–356.

35. Gill, J. C., Endres-Brooks, J., Bauer, P. J., et al. (1987). The effect of ABO blood group on the diagnosis of von Willebrand disease. *Blood, 69,* 1691–1695.

36. Moeller, A., Weippert-Kretschmer, M., Prinz, H., et al. (2001). Influence of ABO blood groups on primary hemostasis. *Transfusion, 41,* 56–60.

37. Boehlen, F., Michel, M., Reber, G., et al. (2001). Analysis of platelet donors function before and after thrombapheresis using the platelet function analyzer PFA-100. *Thromb Res, 102,* 49–52.

38. Nitu-Whalley, I. C., Lee, C. A., Brown, S. A., et al. (2003). The role of the platelet function analyser (PFA-100) in the characterization of patients with von Willebrand's disease and its relationships with von Willebrand factor and the ABO blood group. *Haemophilia, 9,* 298–302.

39. Harrison, P., Segal, H., Furtado, C., et al. (2004). High incidence of defective high-shear platelet function among platelet donors. *Transfusion, 44,* 764–770.

40. Frossard, M., Fuchs, I., Leitner, J. M., et al. (2004). Platelet function predicts myocardial damage in patients with acute myocardial infarction. *Circulation, 110,* 1392–1397.

41. Borna, C., Lazarowski, E., van Heusden, C., et al. (2005). Resistance to aspirin is increased by ST-elevation myocardial infarction and correlates with adenosine diphosphate levels. *Thromb J, 3,* 10.

42. Mammen, E. F., Comp, P. C., Gosselin, R., et al. (1998). PFA-100 system: A new method for assessment of platelet dysfunction. *Semin Thromb Hemost, 24,* 195–202.

43. Cattaneo, M., Lecchi, A., Agati, B., et al. (1999). Evaluation of platelet function with the PFA-100 system in patients with congenital defects of platelet secretion. *Thromb Res, 96,* 213–217.

44. Kerenyi, A., Schlammadinger, A., Ajzner, E., et al. (1999). Comparison of PFA-100 closure time and template bleeding time of patients with inherited disorders causing defective platelet function. *Thromb Res, 96,* 487–492.

45. Buyukasik, Y., Karakus, S., Goker, H., et al. (2002). Rational use of the PFA-100 device for screening of platelet function disorders and von Willebrand disease. *Blood Coagul Fibrinolysis, 13,* 349–353.

46. Harrison, C., Khair, K., Baxter, B., et al. (2002). Hermansky-Pudlak syndrome: Infrequent bleeding and first report of Turkish and Pakistani kindreds. *Arch Dis Child, 86,* 297–301.

47. Harrison, P., Robinson, M., Liesner, R., et al. (2002). The PFA-100®: A potential rapid screening tool for the assessment of platelet dysfunction. *Clin Lab Haematol, 24,* 225–232.

48. Fressinaud, E., Veyradier, A., Truchaud, F., et al. (1998). Screening for von Willebrand disease with a new analyzer using high shear stress: A study of 60 cases. *Blood, 91,* 1325–1331.

49. Weippert-Kretschmer, M., Witte, M., Budde, U., et al. (1995). The Thrombostat 4000. A sensitive screening test for von Willebrand's disease. *Semin Thromb Hemost, 21*(Suppl 2), 44–51.

50. Favaloro, E. J., Facey, D., & Henniker, A. (1999). Use of a novel platelet function analyzer (PFA-100) with high sensitivity to disturbances in von Willebrand factor to screen for von Willebrand's disease and other disorders. *Am J Hematol, 62,* 165–174.

51. Cattaneo, M., Federici, A. B., Lecchi, A., et al. (1999). Evaluation of the PFA-100 system in the diagnosis and therapeutic monitoring of patients with von Willebrand disease. *Thromb Haemost, 82,* 35–39.

52. Schlammadinger, A., Kerenyi, A., Muszbek, L., et al. (2000). Comparison of the O'Brien filter test and the PFA-100 platelet analyzer in the laboratory diagnosis of von Willebrand's disease. *Thromb Haemost, 84,* 88–92.

53. Favaloro, E. J. (2001). Appropriate laboratory assessment as a critical facet in the proper diagnosis and classification of

von Willebrand disorder. *Best Pract Res Clin Haematol, 14,* 299–319.

54. Dean, J. A., Blanchette, V. S., Carcao, M. D., et al. (2000). von Willebrand disease in a pediatric-based population—comparison of type 1 diagnostic criteria and use of the PFA-100 and a von Willebrand factor/collagen-binding assay. *Thromb Haemost, 84,* 401–409.

55. Fressinaud, E., Veyradier, A., Sigaud, M., et al. (1999). Therapeutic monitoring of von Willebrand disease: Interest and limits of a platelet function analyser at high shear rates. *Br J Haematol, 106,* 777–783.

56. Favaloro, E. J., Kershaw, G., Bukuya, M., et al. (2001). Laboratory diagnosis of von Willebrand disorder (vWD) and monitoring of DDAVP therapy: Efficacy of the PFA-100(R) and vWF : CBA as combined diagnostic strategies. *Haemophilia, 7,* 180–189.

57. Frank, R. D., Kunz, D., & Wirtz, D. C. (2002). Acquired von Willebrand disease—hemostatic management of major orthopedic surgery with high-dose immunoglobulin, desmopressin, and continuous factor concentrate infusion. *Am J Hematol, 70,* 64–71.

58. Meskal, A., Vertessen, F., Van der Planken, M., et al. (1999). The platelet function analyzer (PFA-100) may not be suitable for monitoring the therapeutic efficiency of von Willebrand concentrate in type III von Willebrand disease. *Ann Hematol, 78,* 426–430.

59. Kadir, R. A., Economides, D. L., Sabin, C. A., et al. (1998). Frequency of inherited bleeding disorders in women with menorrhagia. *Lancet, 351,* 485–489.

60. Dilley, A., Drews, C., Miller, C., et al. (2001). von Willebrand disease and other inherited bleeding disorders in women with diagnosed menorrhagia. *Obstet Gynecol, 97,* 630–636.

61. Philipp, C. S., Dilley, A., Miller, C. H., et al. (2003). Platelet functional defects in women with unexplained menorrhagia. *J Thromb Haemost, 1,* 477–484.

62. Saxena, R., Gupta, M., Gupta, P. K., et al. (2003). Inherited bleeding disorders in Indian women with menorrhagia. *Haemophilia, 9,* 193–196.

63. Edlund, M., Blombäck, M., Von Schoultz, B., et al. (1996). On the value of menorrhagia as a predictor for coagulation disorders. *Am J Hematol, 53,* 234–238.

64. Kadir, R. A., Economides, D. L., Sabin, C. A., et al. (1998). Frequency of inherited bleeding disorders in women with menorrhagia. *Lancet, 351,* 485–489.

65. Dilley, A., Drews, C., Lally, C., et al. (2002). A survey of gynecologists concerning menorrhagia: Perceptions of bleeding disorders as a possible cause. *J Womens Health Gend Based Med, 11,* 39–44.

66. James, A. H., Lukes, A. S., Brancazio, L. R., et al. (2004). Use of a new platelet function analyzer to detect von Willebrand disease in women with menorrhagia. *Am J Obstet Gynecol, 191,* 449–455.

67. Gewirtz, A. S., Miller, M. L., & Keys, T. F. (1996). The clinical usefulness of the preoperative bleeding time. *Arch Pathol Lab Med, 120,* 353–356.

68. Mammen, E. F., Alshameeri, R. S., & Comp, P. C. (1995). Preliminary data from a field trial of the PFA-100 system. *Semin Thromb Hemost, 21*(Suppl 2), 113–121.

69. Marshall, P. W., Williams, A. J., Dixon, R. M., et al. (1997). A comparison of the effects of aspirin on bleeding time measured using the Simplate method and closure time measured using the PFA-100, in healthy volunteers. *Br J Clin Pharmacol, 44,* 151–155.

70. Feuring, M., Haseroth, K., Janson, C. P., et al. (1999). Inhibition of platelet aggregation after intake of acetylsalicylic acid detected by a platelet function analyzer (PFA-100). *Int J Clin Pharmacol Ther, 37,* 584–588.

71. Homoncik, M., Jilma, B., Hergovich, N., et al. (2000). Monitoring of aspirin (ASA) pharmacodynamics with the platelet function analyzer PFA-100. *Thromb Haemost, 83,* 316–321.

72. von Pape, K. W., Aland, E., & Bohner, J. (2000). Platelet function analysis with PFA-100 in patients medicated with acetylsalicylic acid strongly depends on concentration of sodium citrate used for anticoagulation of blood sample. *Thromb Res, 98,* 295–299.

73. Lasne, D., Fiemeyer, A., Chatellier, G., et al. (2000). A study of platelet functions with a new analyzer using high shear stress (PFA 100) in patients undergoing coronary artery bypass graft. *Thromb Haemost, 84,* 794–799.

74. de Meijer, A., Vollaard, H., De Metz, M., et al. (1999). Meloxicam, 15 mg/day, spares platelet function in healthy volunteers. *Clin Pharmacol Ther, 66,* 425–430.

75. Homoncik, M., Malec, M., Marsik, C., et al. (2003). Rofecoxib exerts no effect on platelet plug formation in healthy volunteers. *Clin Exp Rheumatol, 21,* 229–231.

76. Kottke-Marchant, K., Powers, J. B., Brooks, L., et al. (1999). The effect of antiplatelet drugs, heparin, and preanalytical variables on platelet function detected by the platelet function analyzer (PFA-100). *Clin Appl Thromb Hemost, 5,* 122–130.

77. Munsterhjelm, E., Niemi, T. T., Syrjala, M. T., et al. (2003). Propacetamol augments inhibition of platelet function by diclofenac in volunteers. *Br J Anaesth, 91,* 357–362.

78. Suzuki, S., & Morishita, S. (1999). The relationship between the onset of labor mechanisms and the hemostatic system. *Immunopharmacology, 43,* 133–140.

79. Suzuki, S., Matsuno, K., & Kondoh, M. (1995). Primary hemostasis during women's life cycle measured by Thrombostat 4000. *Semin Thromb Hemost, 21*(Suppl 2), 103–105.

80. Escolar, G., Cases, A., Vinas, M., et al. (1999). Evaluation of acquired platelet dysfunctions in uremic and cirrhotic patients using the platelet function analyzer (PFA-100): Influence of hematocrit elevation. *Haematologica, 84,* 614–619.

81. Zupan, I. P., Sabovic, M., Salobir, B., et al. (2003). Utility of in vitro closure time test for evaluating platelet-related primary hemostasis in dialysis patients. *Am J Kidney Dis, 42,* 746–751.

82. Zupan, I. P., Sabovic, M., Salobir, B., et al. (2005). The study of anaemia-related haemostasis impairment in haemodialysis patients by in vitro closure time test. *Thromb Haemost, 93,* 375–379.

83. Linden, M. D., Furman, M. I., Michelson, A. D., et al. (2004). Indices of platelet activation as markers of recent myocardial infarction. *Circulation. 110,* III-450–III-451.

84. Lanza, G. A., Sestito, A., Iacovella, S., et al. (2003). Relation between platelet response to exercise and coronary angiographic findings in patients with effort angina. *Circulation, 107*, 1378–1382.

85. Gianetti, J., Parri, M. S., Sbrana, S., et al. (2006). Platelet activation predicts recurrent ischemic events after percutaneous coronary angioplasty: A 6 months prospective study. *Thromb Res.* [Epub ahead of print]

86. Francis, J. L. (2000). Platelet dysfunction detected at high shear in patients with heart valve disease. *Platelets, 11*, 133–136.

87. O'Brien, J. R., Etherington, M. D., Brant, J., et al. (1995). Decreased platelet function in aortic valve stenosis: High shear platelet activation then inactivation. *Br Heart J, 74*, 641–644.

88. Susen, S., Siaka, C., Caron, C., et al. (2000). Timing of reversal of acquired von Willebrand syndrome associated with aortic stenosis. *Blood, 96*, 89b(Abstract).

89. Vincentelli, A., Susen, S., Le Tourneau, T., et al. (2003). Acquired von Willebrand syndrome in aortic stenosis. *N Engl J Med, 349*, 343–349.

90. Wahba, A., Sander, S., & Birnbaum, D. E. (1998). Are in-vitro platelet function tests useful in predicting blood loss following open heart surgery? *Thorac Cardiovasc Surg, 46*, 228–231.

91. Wenger, R. K., Lukasiewicz, H., Mikuta, B. S., et al. (1989). Loss of platelet fibrinogen receptors during clinical cardiopulmonary bypass. *J Thorac Cardiovasc Surg, 97*, 235–239.

92. Slaughter, T. F., Sreeram, G., Sharma, A. D., et al. (2001). Reversible shear-mediated platelet dysfunction during cardiac surgery as assessed by the PFA-100 platelet function analyzer. *Blood Coagul Fibrinolysis, 12*, 85–93.

93. Ray, M. J., Hawson, G. A. T., Just, S. J. E., et al. (1994). Relationship of platelet aggregation to bleeding after cardiopulmonary bypass. *Ann Thorac Surg, 57*, 981–986.

94. Raman, S., & Silverman, N. A. (2001). Clinical utility of the platelet function analyzer (PFA-100) in cardiothoracic procedures involving extracorporeal circulation. *J Thorac Cardiovasc Surg, 122*, 190–191.

95. Cammerer, U., Dietrich, W., Rampf, T., et al. (2003). The predictive value of modified computerized thromboelastography and platelet function analysis for postoperative blood loss in routine cardiac surgery. *Anesth Analg, 96*, 51–57.

96. Fattorutto, M., Pradier, O., Schmartz, D., et al. (2003). Does the platelet function analyser (PFA-100) predict blood loss after cardiopulmonary bypass? *Br J Anaesth, 90*, 692–693.

97. Koscielny, J., Ziemer, S., Radtke, H., et al. (2004). A practical concept for preoperative identification of patients with impaired primary hemostasis. *Clin Appl Thromb Hemost, 10*, 195–204.

98. Mimidis, K., Papadopoulos, V., Kartasis, Z., et al. (2004). Assessment of platelet adhesiveness and aggregation in mild acute pancreatitis using the PFA-100 system. *JOP, 5*, 132–137.

99. Troost, M. M., & van Genderen, P. J. (2002). Excessive prolongation of the Ivy bleeding time after aspirin in essential thrombocythemia is also demonstrable in vitro in the high shear stress system PFA-100. *Ann Hematol, 81*, 353–354.

100. Jilma-Stohlawetz, P., Hergovich, N., Homoncik, M., et al. (2001). Impaired platelet function among platelet donors. *Thromb Haemost, 86*, 880–886.

101. Feuring, M., Gutfleisch, A., Ganschow, A., et al. (2001). Impact of plasmapheresis on platelet hemostatic capacity in healthy voluntary blood donors detected by the platelet function analyzer PFA-100. *Platelets, 12*, 236–240.

102. Paglieroni, T. G., Janatpour, K., Gosselin, R., et al. (2004). Platelet function abnormalities in qualified whole-blood donors: Effects of medication and recent food intake. *Vox Sang, 86*, 48–53.

103. Pearson, D., Paglieroni, T., Rein, D., et al. (2002). The effects of flavanol-rich cocoa and aspirin on ex vivo platelet function. *Thromb Res, 106*, 191.

104. Francis, J. L., Francis, D., Larson, L., et al. (1999). Can the Platelet Function Analyzer (PFA)-100 test substitute for the template bleeding time in routine clinical practice? *Platelets, 10*, 132–136.

105. Posan, E., McBane, R. D., Grill, D. E., et al. (2003). Comparison of PFA-100 testing and bleeding time for detecting platelet hypofunction and von Willebrand disease in clinical practice. *Thromb Haemost, 90*, 483–490.

106. Wuillemin, W. A., Gasser, K. M., Zeerleder, S. S., et al. (2002). Evaluation of a Platelet Function Analyser (PFA-100) in patients with a bleeding tendency. *Swiss Med Wkly, 132*, 443–448.

107. Laffan, M., Brown, S. A., Collins, P. W., et al. (2004). The diagnosis of von Willebrand disease: A guideline from the UK Haemophilia Centre Doctors' Organization. *Haemophilia, 10*, 199–217.

108. Quiroga, T., Goycoolea, M., Munoz, B., et al. (2004). Template bleeding time and PFA-100 have low sensitivity to screen patients with hereditary mucocutaneous hemorrhages: Comparative study in 148 patients. *J Thromb Haemost, 2*, 892–898.

109. Cariappa, R., Wilhite, T. R., Parvin, C. A., et al. (2003). Comparison of PFA-100 and bleeding time testing in pediatric patients with suspected hemorrhagic problems. *J Pediatr Hematol Oncol, 25*, 474–479.

110. Altman, R., Luciardi, H. L., Muntaner, J., et al. (2004). The antithrombotic profile of aspirin. Aspirin resistance, or simply failure? *Thromb J, 2*, 1.

111. Grundmann, K., Jaschonek, K., Kleine, B., et al. (2003). Aspirin non-responder status in patients with recurrent cerebral ischemic attacks. *J Neurol, 250*, 63–66.

112. Chakroun, T., Gerotziafas, G., Robert, F., et al. (2004). In vitro aspirin resistance detected by PFA-100 closure time: Pivotal role of plasma von Willebrand factor. *Br J Haematol, 124*, 80–85.

113. Harrison, P., Mackie, I., Mathur, A., et al. (2005). Platelet hyper-function in acute coronary syndromes. *Blood Coagul Fibrinolysis, 16*, 557–562.

114. Ziegler, S., Alt, E., Brunner, M., et al. (2005). Influence of systemic inflammation on the interpretation of response to antiplatelet therapy, Monitored by PFA-100. *Semin Thromb Hemost, 31*, 416–419.

115. Macchi, L., Christiaens, L., Brabant, S., et al. (2003). Resistance in vitro to low-dose aspirin is associated with platelet PlA1 (GP IIIa) polymorphism but not with C807T (GP

Ia/IIa) and C-5T Kozak (GP Ibalpha) polymorphisms. *J Am Coll Cardiol, 42,* 1115–1119.

116. Michelson, A. D., Cattaneo, M., Eikelboom, J. W., et al. (2005). Aspirin resistance: Position paper of the Working Group on Aspirin Resistance. *J Thromb Haemost, 3,* 1309–1311.

117. Hezard, N., Metz, D., Nazeyrollas, P., et al. (2000). Use of the PFA-100 apparatus to assess platelet function in patients undergoing PTCA during and after infusion of c7E3 Fab in the presence of other antiplatelet agents. *Thromb Haemost, 83,* 540–544.

118. Madan, M., Berkowitz, S. D., Christie, D. J., et al. (2001). Rapid assessment of glycoprotein IIb/IIIa block-ade with the platelet function analyzer (PFA-100) during percutaneous coronary intervention. *Am Heart J, 141,* 226–233.

119. Madan, M., Berkowitz, S. D., Christie, D. J., et al. (2002). Determination of platelet aggregation inhibition during percutaneous coronary intervention with the platelet function analyzer PFA-100. *Am Heart J, 144,* 151–158.

120. Mueller, T., Haltmayer, M., Poelz, W., et al. (2003). Monitoring aspirin 100 mg and clopidogrel 75 mg therapy with the PFA-100 device in patients with peripheral arterial disease. *Vasc Endovascular Surg, 37,* 117–123.

121. Geiger, J., Teichmann, L., Grossmann, R., et al. (2005). Monitoring of clopidogrel action: Comparison of methods. *Clin Chem, 51,* 957–965.

122. Golanski, J., Pluta, J., Baraniak, J., et al. (2004). Limited usefulness of the PFA-100 for the monitoring of ADP receptor antagonists—in vitro experience. *Clin Chem Lab Med, 42,* 25–29.

123. Grau, A. J., Reiners, S., Lichy, C., et al. (2003). Platelet function under aspirin, clopidogrel, and both after ischemic stroke: A case-crossover study. *Stroke, 34,* 849–854.

124. Jilma, B. (2003). Synergistic antiplatelet effects of clopidogrel and aspirin detected with the PFA-100 in stroke patients. *Stroke, 34,* 849–854.

125. Ortel, T. L., James, A. H., Thames, E. H., et al. (2000). Assessment of primary hemostasis by PFA-100 analysis in a tertiary care center. *Thromb Haemost, 84,* 93–97.

126. Kottke-Marchant, K., & Corcoran, G. (2002). The laboratory diagnosis of platelet disorders. *Arch Pathol Lab Med, 126,* 133–146.

127. Franchini, M., Zugni, C., Veneri, D., et al. (2004). High prevalence of acquired von Willebrand's syndrome in patients with thyroid diseases undergoing thyroid surgery. *Haematologica, 89,* 1341–1346.

128. Hayward, C. P. M., Harrison, P., Cattaneo, M., et al. (2006). Platelet function analyzer (PFA)-100 closure time in the evaluation of platelet disorders and platelet function. *J Thromb Haemost, 4,* 312–319.

129. Gum, P. A., Kottke-Marchant, K., Poggio, E. D., et al. (2001). Profile and prevalence of aspirin resistance in patients with cardiovascular disease. *Am J Cardiol, 88,* 230–235.

130. Christiaens, L., Macchi, L., Herpin, D., et al. (2002). Resistance to aspirin in vitro at rest and during exercise in patients with angiographically proven coronary artery disease. *Thromb Res, 108,* 115–119.

131. Coma-Canella, I., Velasco, A., & Castano, S. (2005). Prevalence of aspirin resistance measured by PFA-100. *Int J Cardiol, 101,* 71–76.

132. Hobikoglu, G. F., Norgaz, T., Aksu, H., et al. (2005). High frequency of aspirin resistance in patients with acute coronary syndrome. *Tohoku J Exp Med, 207,* 59–64.

133. Macchi, L., Christiaens, L., Brabant, S., et al. (2002). Resistance to aspirin in vitro is associated with increased platelet sensitivity to adenosine diphosphate. *Thromb Res, 107,* 45–49.

134. Crowe, B., Abbas, S., Meany, B., et al. (2005). Detection of aspirin resistance by PFA-100: Prevalence and aspirin compliance in patients with chronic stable angina. *Semin Thromb Hemost, 31,* 420–425.

135. Andersen, K., Hurlen, M., Arnesen, H., et al. (2002). Aspirin non-responsiveness as measured by PFA-100 in patients with coronary artery disease. *Thromb Res, 108,* 37–42.

136. Roller, R. E., Dorr, A., Ulrich, S., et al. (2002). Effect of aspirin treatment in patients with peripheral arterial disease monitored with the platelet function analyzer PFA-100. *Blood Coagul Fibrinolysis, 13,* 277–281.

137. Alberts, M. J., Bergman, D. L., Molner, E., et al. (2004). Antiplatelet effect of aspirin in patients with cerebrovascular disease. *Stroke, 35,* 175–178.

138. McCabe, D. J., Harrison, P., Mackie, I. J., et al. (2005). Assessment of the antiplatelet effects of low to medium dose aspirin in the early and late phases after ischaemic stroke and TIA. *Platelets, 16,* 269–280.

139. Watala, C., Golanski, J., Pluta, J., et al. (2004). Reduced sensitivity of platelets from type 2 diabetic patients to acetylsalicylic acid (aspirin)—its relation to metabolic control. *Thromb Res, 113,* 101–113.

140. Fateh-Moghadam, S., Plockinger, U., Cabeza, N., et al. (2005). Prevalence of aspirin resistance in patients with type 2 diabetes. *Acta Diabetol, 42,* 99–103.

141. Yilmaz, M. B., Balbay, Y., Caldir, V., et al. (2005). Late saphenous vein graft occlusion in patients with coronary bypass: Possible role of aspirin resistance. *Thromb Res, 115,* 25–29.

# CHAPTER 29

# Impact Cone and Plate(let) Analyzer

## David Varon[1] and Naphtali Savion[2]

[1]*Coagulation Unit, Hadassah Hebrew University Medical Center, Jerusalem, Israel*
[2]*Goldschleger Eye Research Institute, Sackler Faculty of Medicine, Tel-Aviv University, Tel-Aviv, Israel*

## I. Introduction

Blood platelets provide the primary defense mechanism in physiological hemostasis. Circulating platelets interact with and adhere to the exposed subendothelial extracellular matrix (ECM) under flow, thereby forming platelet aggregates leading to bleeding arrest. In order to better understand platelet physiology, it is essential to study their function under conditions including most of the physiological parameters playing a role in platelet adhesion and aggregation. The physiological milieu of platelets, including flow, red blood cells, and other blood components, regulates these interactions.

The role of flow conditions, operating in the vascular bed, in the interaction of platelet receptors with their ligands is well established (Chapter 18). Measurements of platelet function in whole blood samples under flow conditions may be achieved by various technologies, among them a parallel plate chamber (Chapter 32), a collagen-coated capillary (Chapter 28), and a cone and plate device (this chapter). The cone and plate apparatus induces laminar flow with uniform shear stress over the entire plate surface covered by a rotating cone. The shear force, at each point on the plate surface, is inversely related to the distance of the cone from the plate and is directly related to the angular velocity. At the specific cone angle used in the cone and plate apparatus, the decrease in shear force on the plate surface due to the increased radial distance is fully compensated by the increased angular velocity.[1] We have employed the cone and plate apparatus to develop a simple method for testing platelet adhesion and aggregation on ECM-coated plate or polystyrene surface, which is suitable for both basic studies and clinical applications.[2,3]

## II. Impact: The Cone and Plate(let) Analyzer (CPA) Technology

The Impact (DiaMed, Cressier, Switzerland) was designed in an attempt to establish a method that would test platelet function under close to physiological conditions. In addition, this system employs a very small blood volume, no blood processing, and it is a simple and easily performed testing procedure. Thus, the original CPA system tests whole blood platelet adhesion and aggregation on an ECM-coated plate, under physiological arterial flow conditions.[2] It has been further developed and the use of a polystyrene surface at which plasma proteins, in particular fibrinogen and von Willebrand factor (VWF), are immobilized and form a thrombogenic surface was introduced.[3] A 130 μL aliquot of whole blood in a sodium citrate solution (0.32%) is placed on the plastic and a defined shear rate (1800 sec$^{-1}$) applied using the cone and plate device (Fig. 29-1A). This is followed by staining and measuring the percentage of surface covered (SC) by the stained objects and the average size (AS) of the objects using an image analyzer. Under these conditions, only platelets, not other blood cells, adhere to the surface and form elongated aggregates aligned with the laminar flow lines (Fig. 29-1B). When normal blood is analyzed, platelet deposition is a shear- and time-dependent process, reaching maximal level within 2 min at high shear rate (1800 sec$^{-1}$).[3]

The effect of hematocrit on platelet adhesion and aggregation on ECM under high shear rate conditions was studied and the results are presented as the percent of control value of SC and AS obtained at 40% hematocrit (Fig. 29-2A). The extent of SC slightly decreased between 40 and 30% hematocrit followed by a sharp decrease below 30% hematocrit level, reaching about 10% of control SC in the absence of red blood cells (hematocrit = 0%). The mean AS decreased significantly as a function of the decrease in hematocrit level between 40 to 10%, reaching a size of about 24 μm$^2$ (about 30% of control) which represents a population of single adherent platelets. Further decrease in the hematocrit level to zero (platelet-rich plasma [PRP] fraction) did not further decrease the size of the ECM bound particles (Fig. 29-2A). These observations confirm that both platelet adhesion and aggregation under high shear rate conditions depend on the presence of red blood cells.

The effect of platelet count in whole blood on platelet deposition at high shear rate is shown in Fig. 29-2B.

PLATELETS

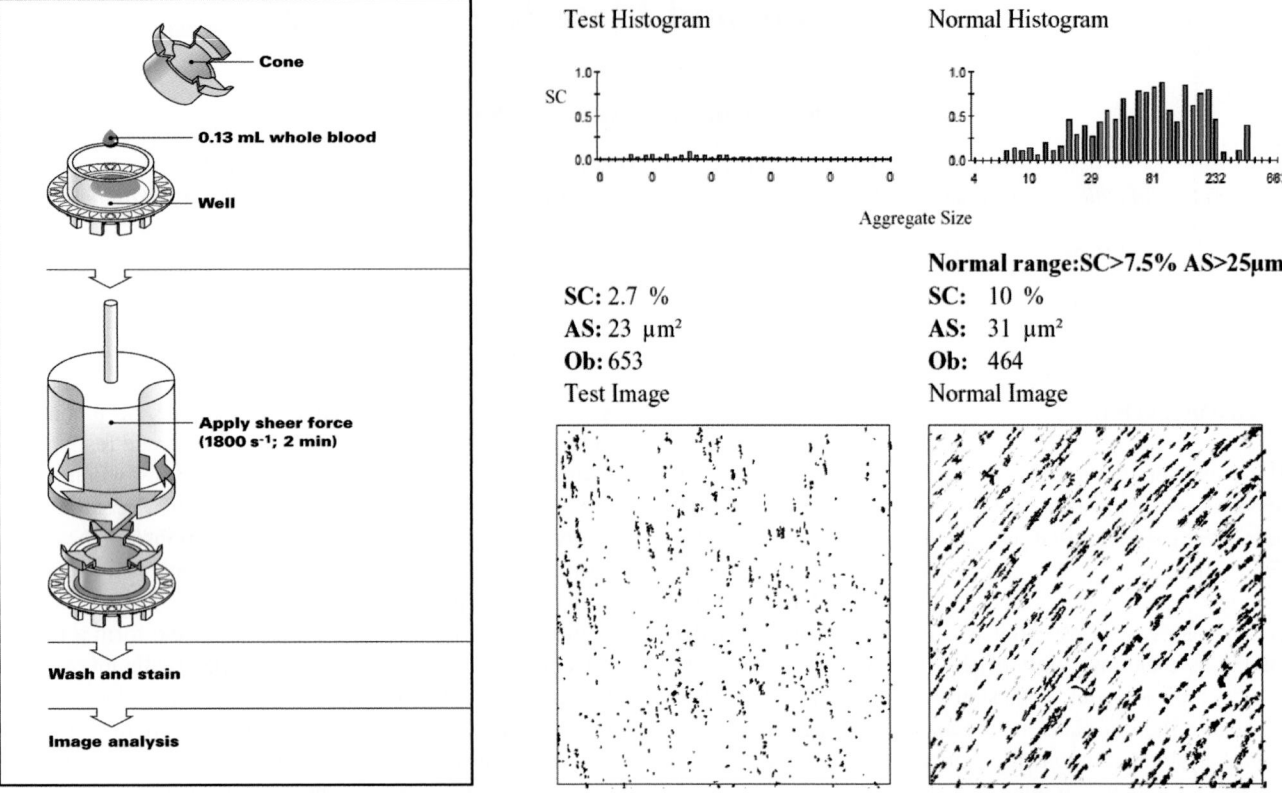

**Figure 29-1.**  The Impact cone and plate(let) analyzer (CPA) procedure. A. Citrated whole blood sample (0.13 mL) is applied onto a polystyrene well followed by application of the cone and then subjected to arterial shear flow (1800 $sec^{-1}$) for 2 min, and washed and stained by May Grünwald stain. Adherent platelets and platelet aggregates are evaluated by an image analyzer. B. The test results report include: (i) a particle size distribution histogram; (ii) the percentage of surface coverage (SC); (iii) the average size (AS) of the stained objects; (iv) the number of objects (Ob) per picture; and (v) a representative picture of the tested sample. For comparison, the results of a representative normal sample and the normal range are presented on the right side.

Reduction of platelet count resulted in a close to linear decrease in SC of platelet aggregates on the ECM. A marked decrease in aggregates AS was observed upon reduction of platelet count below $25 \times 10^9$/L, reaching the size of about 24 $\mu m^2$ which represents single adherent platelets.

Recently, the effects of above-normal hematocrit or platelet count on platelet function in the Impact were studied.[4] At normal hematocrit (30–45%), platelet adhesion on polystyrene surfaces (represented as the SC) increased linearly with platelet counts from $100 \times 10^9$/L to $1200 \times 10^9$/L. No significant change in the AS of platelet aggregates was noted over this range of platelet counts. The effect of high hematocrit (>45%) on platelet adhesion and aggregation was studied, and demonstrated a marked increase in platelet aggregate formation (AS) with decreased single platelet deposition. These morphological observations correlated with quantifiable increases of 30 to 50% in AS and slight increases (10–20%) in SC.[4] These results suggest that increase in the AS of platelet deposits (aggregate size) is probably a direct effect of increasing hematocrit but not of increasing platelet count.

## III.  Modifying the Impact for Testing Platelet Aggregation

Application of whole blood sample to the Impact system results in platelet adhesion to the well surface, followed by clustering of circulating platelets around adherent platelets associated with release reaction. The properties of circulating nonadherent platelets exposed to the released compounds (from adherent and activated platelets) were studied.[5] The results demonstrated that platelets in the suspension at the end of the Impact test form microaggregates, presumably in response to the released compounds, accompanied with a reduced single platelet count. This process was associated with transient adhesion refractoriness, as demonstrated by

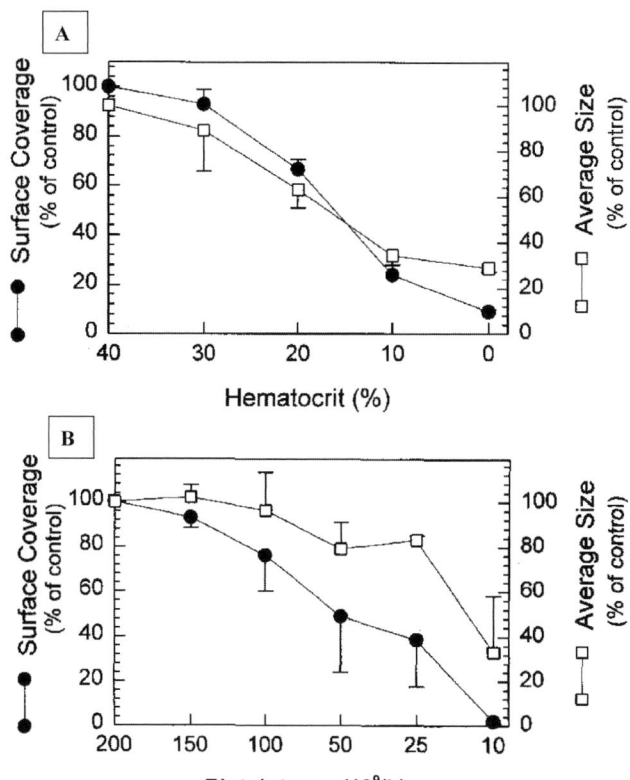

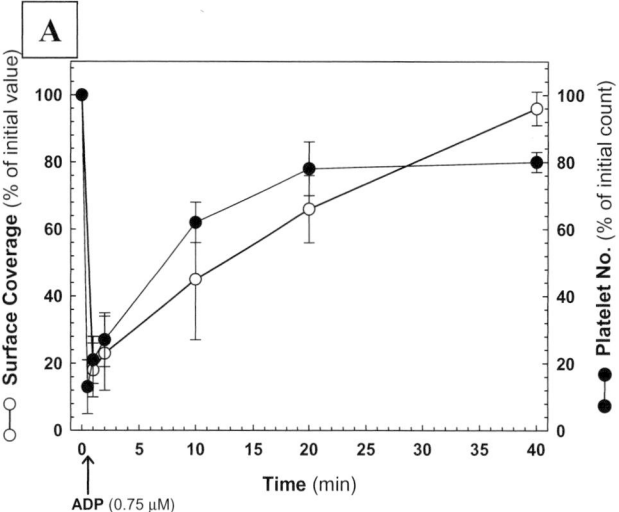

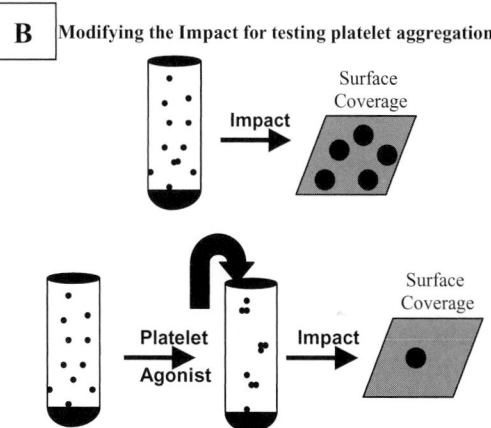

**Figure 29-2.** A. Effect of hematocrit in normal whole blood on SC and AS of ECM-bound objects. The results are the mean ± SD of three blood samples (each tested in duplicate) and expressed as percentage of control value obtained in a whole blood sample with 40% hematocrit and platelet count of $250 \times 10^9$/L. Control values were SC $20.1 \pm 1.4\%$ and AS $56 \pm 7$ $\mu m^2$. The experiment was performed at a high shear rate (1300 $sec^{-1}$). B. Effect of platelet count in normal whole blood on SC and AS of ECM-bound objects. The results are the mean ± SD of three blood samples (each tested in duplicate) and expressed as % of control value obtained in a whole blood sample containing $200 \times 10^9$/L platelets with 40% hematocrit. Control values were: SC $18.5 \pm 2.5\%$ and AS $51 \pm 9$ $\mu m^2$ performed at high shear rate (1300 $sec^{-1}$). Adapted from ref. 2 with permission.

the lack of platelet adhesion to the well surface following transfer of the blood sample at the end of the test to a new well for a second run. The adhesion refractoriness was dependent on platelet activation during the first run and was prevented by addition of apyrase (an adenosine diphosphate [ADP] scavenger) or ADP receptor inhibitor, suggesting a role for ADP in mediating this response. Furthermore, exposure of whole blood samples to suboptimal concentrations of ADP (0.75 μM) (Fig. 29-3A), or a thrombin receptor activating peptide (TRAP) (5 μM) for 1 min under gentle mixing conditions, resulted in a similar thrombocytopenia, platelet microaggregate formation, and adhesion refractoriness. Both microaggregate formation and adhesion

**Figure 29-3.** A. The modified Impact test: effect of ADP pretreatment on SC and platelet number in the suspension fraction. Whole blood samples were gently mixed (10 rpm) at room temperature in the presence of ADP (0.75 μM) for the indicated time intervals and then subjected to the Impact procedure. The SC was determined and presented as mean ± SD (*n* = 6). The initial SC determined for the blood samples before the addition of ADP (zero time) was considered to be 100%. In parallel, at the end of the mixing period with the agonist, at the time points indicated, the blood samples were fixed with 1% buffered formaldehyde and the platelet count determined. Each value was calculated as the percentage of the initial platelet count determined before the addition of the agonist and the mean ± SD (*n* = 6) is presented. Adapted from ref. 5. B. A schematic presentation of a modified Impact test for the measurement of platelet aggregation. (Upper) Testing a normal blood sample in the Impact results in aggregate formation on the well surface (normal SC). (Lower) However, incubating the blood sample under gentle mixing (10 rpm) for 1 min with any platelet agonist (e.g., ADP, arachidonic acid [AA], collagen, epinephrine) induces microaggregate formation in the suspension, resulting in reduced SC.

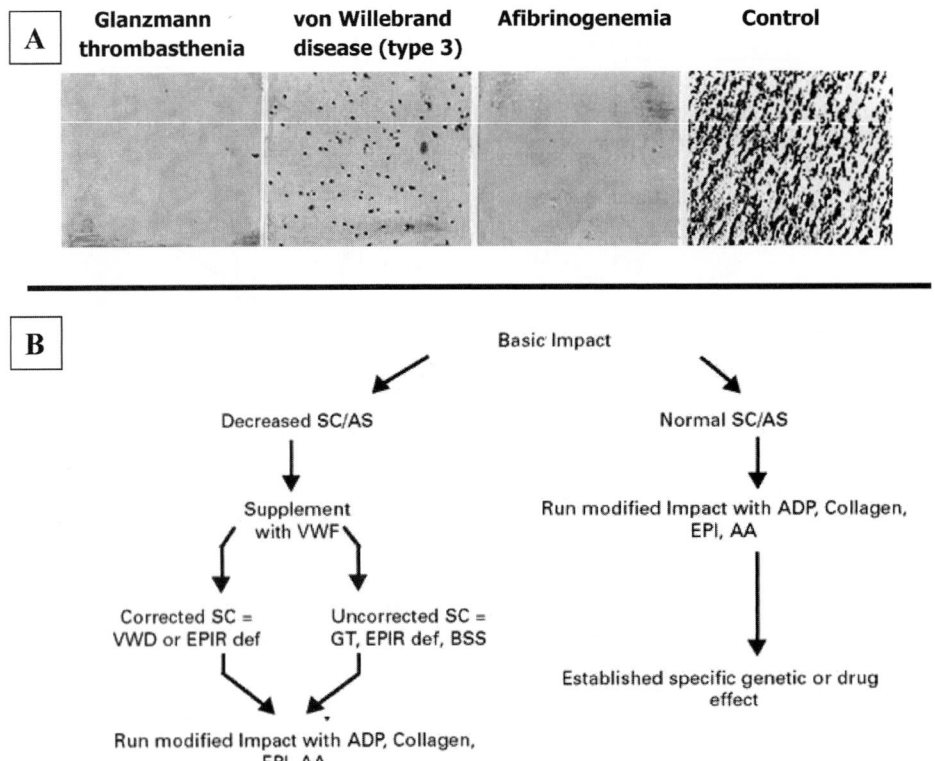

**Figure 29-4.** A. Impact results of blood samples from healthy donor, afibrinogenemia, von Willebrand disease (type 3), and Glanzmann thrombasthenia patients. Representative field of each well is presented. Adapted from ref. 3 with permission. B. Proposed flow chart scheme for evaluation of a patient with bleeding symptoms by the Impact method. Abbreviations: AA, arachidonic acid; ADP, adenosine diphosphate; AS, average size; BSS, Bernard-Soulier syndrome; EPI, epinephrine; EPIR def, epinephrine receptor deficiency; GT, Glanzmann thrombasthenia; SC, surface coverage; VWD, von Willebrand disease; VWF, von Willebrand factor.

refractoriness were found to be transient and after 20 to 40 min both recovered. Based on this phenomenon, it is now possible to perform a platelet aggregation test on a patient sample by a modified Impact test (Fig. 29-3B). In the modified assay, a patient blood sample is preincubated under gentle mixing (10 rpm), with a platelet agonist (e.g., ADP, thrombin, arachidonic acid [AA], epinephrine) at suboptimal concentration for 1 min and then subjected to the regular Impact test. A response to the agonist will result in platelet activation, microaggregate formation, and thrombocytopenia in the suspension and subsequently a reduced SC.

## IV. Testing Congenital Primary Hemostasis Abnormalities

A blood sample of a healthy donor analyzed by the Impact test demonstrates significant adhesion of platelets with elongated aggregates aligned with the laminar flow lines (Fig. 29-4A, Control). Samples from Glanzmann thrombasthenia (Chapter 57) and afibrinogenemia demonstrate complete

lack of adhesion to the polystyrene surface, indicating the crucial role of both the integrin $\alpha$IIb$\beta$3 (glycoprotein [GP] IIb-IIIa) receptor and fibrinogen in platelet adhesion under the test conditions. Moreover, a confirmation of the diagnosis of afibrinogenemia may be achieved by adding fibrinogen (0.25 mg/mL) to the blood sample prior to the Impact assay. Normalization of SC and AS following the fibrinogen addition indicated that the disorder was afibrinogenemia. Testing of von Willebrand disease (VWD) type III patient blood sample by the Impact assay results in reduced aggregate formation and adhesion of only single platelets (Fig. 29-4A). Conformation of VWD can be achieved by adding VWF (0.25 U/mL) to the blood sample prior to the Impact assay. Increased SC and AS following VWF addition suggests the diagnosis of VWD.

These studies have established the critical role of both plasma VWF and fibrinogen immobilization on the polystyrene surface in platelet adhesion to the polystyrene surface. This method was also employed in several studies in which the potential effect of a recombinant VWF fragment as an anti-thrombotic agent was assessed.[6-8] In these studies, the absolute dependency of the Impact test on the interaction of VWF with its receptor was demonstrated, confirming again that this interaction is required for the adhesion of platelets to the polystyrene plate under flow conditions.

The Impact technology may thus be used for a complete screening of bleeding patients. A proposed algorithm indicating the steps in this process is presented in Fig. 29-4B.

## V. Prediction of Bleeding in Cardiac Surgery Patients

The Impact method has been shown to be effective in the assessment of platelet function in thrombocytopenic patients and in predicting the risk of bleeding among these patients.[9] In this study, it was shown that SC and mean platelet volume were the best independent predictors of bleeding events among thrombocytopenic patients. Using this technology, we attempted to predict postoperative bleeding in cardiac surgery patients. Platelet dysfunction is a common finding during cardiac surgery, especially when a cardiopulmonary bypass pump is employed.[10,11] The circulation of blood through the foreign surfaces of extracorporeal circulation results in platelet activation and a decrease in platelet count and function. The result of platelet dysfunction is bleeding in most cases and thrombosis in others. We evaluated 50 patients undergoing cardiac surgery for preoperative platelet function as determined by the Impact and postoperative bleeding. Patients were divided into three groups according to the postoperative bleeding volume, with average bleeding volumes of 1277, 664, and 297 mL, respectively. A significant negative correlation between the preoperative SC and the postoperative bleeding volume was observed (Fig. 29-5). The univariate analysis model of this study also included other parameters, such as preoperative medications, patient age, urgency of operation, preoperative and postoperative platelet count, and blood transfusions. Except for the preoperative platelet function variables, AS and SC described above, the univariate analyses revealed no significant association between postoperative bleeding and any other preoperative factor incorporated in the study as a variable.[12,13] This finding may be important for the postoperative management of patients in the intensive care unit.

## VI. Monitoring GPIIb-IIIa Antagonist Therapy

The Impact is a good candidate for testing the response to GPIIb-IIIa antagonists, because it is absolutely dependent on the functional state of the GPIIb-IIIa and the GPIb-IX receptors and their ligands (fibrinogen and VWF). In an attempt

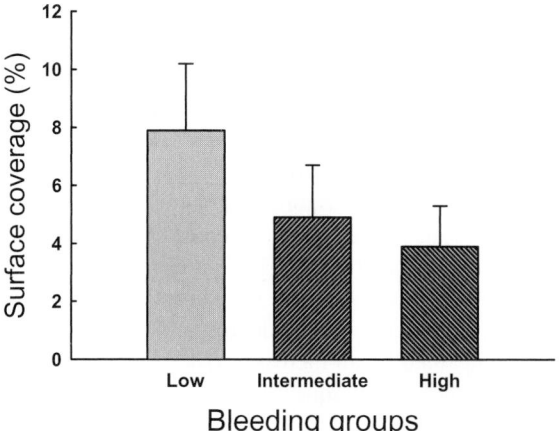

**Figure 29-5.** The postoperative bleeding volume of cardiac surgery patients ($n = 50$) was monitored and the patients were divided into the following three equal groups according to the severity of bleeding: low (297 ± 53 mL), intermediate (664 ± 217 mL), and high (1277 ± 270 mL). The preoperative SC was determined by the Impact and the mean ± SD for each bleeding group is presented. The differences between the groups were significant ($p < 0.001$).

to evaluate the potential application of the Impact method for this indication, we have studied the effect of several GPIIb-IIIa antagonists *in vitro* and found a dose response relationship between the SC and drug concentrations.[14] A good correlation with platelet aggregometry was also demonstrated. We then tested blood samples from patients undergoing percutaneous coronary intervention (PCI) in conjunction with the GPIIb-IIIa antagonist, abciximab. The results showed an almost 100% drop in the SC after bolus injection and a gradual recovery afterward. Additional experience with the Impact was gained in studying 16 patients receiving abciximab for carotid artery stenting.[15] In this study, comparable results were obtained with the Impact and the aggregometer. These two studies[14,15] also demonstrated a great variability in the degree of inhibition during the later phase after the GPIIb-IIIa antagonist therapy was discontinued. In another comparative study of patients undergoing GPIIb-IIIa antagonist therapy, the Impact was found to be superior to aggregation-based methods in monitoring the drug effect.[16] Furthermore, this method was successfully used for screening of a new GPIIb-IIIa antagonist.[17]

## VII. Testing Aspirin and Clopidogrel Effects

Adhesion of platelets under the Impact's testing conditions is extremely sensitive to the effect of GPIIb-IIIa antagonists (Section VI), but not to platelet activation blockers such as aspirin and clopidogrel. Both aspirin (Chapter 60) and clopidogrel (Chapter 61) are extensively used in the prevention and treatment of ischemic heart disease and stroke. Recently, it has been demonstrated that the pharmacological response to aspirin is not uniform in all treated individuals.[18] A correlation between the response to aspirin and clinical outcome has been suggested in several trials. Patients with "aspirin resistance" were found to be at an increased risk of developing cardiac event[19,20] and stroke.[21]

Similar to aspirin, variability in platelet response to clopidogrel has also recently been reported.[22] Moreover, we recently demonstrated that 25% of patients who had undergone primary PCI for acute myocardial infarction were laboratory "resistant" to clopidogrel, and this phenomenon was accompanied by an increased risk of recurrent cardiovascular events.[23] Consequently, evaluation of the pharmacological response to antiplatelet drugs is becoming an active field of research. Until recently, the lack of a reliable and yet practical method to detect patient response to platelet antagonists was a major obstacle in this field of research. This raised the need for a point-of-care method for wide-scale testing of platelet response to these drugs.

The modified Impact assay may be used for detecting the response of platelets to aspirin and clopidogrel. Pretreatment of blood sample with ADP (1.36 μM) or AA (0.275 mM) for 1 min is used to test the response to clopidogrel or aspirin, and a response to the drug results in SC higher than 3.2% and 2.5%, respectively.[24,25] When the SC is below the indicated values it may suggest that the platelets do not respond to the drug or the patient is not being treated.

The potential application of the modified Impact test to monitor the effect of aspirin in comparison to conventional aggregometry in 20 healthy volunteers treated with 100, 300, and 500 mg/day of aspirin was investigated.[24] As expected, platelet adhesion of the healthy volunteer samples measured *ex vivo* by standard Impact was unaffected by aspirin ingestion at any tested doses. In contrast, the decrease in SC in response to preincubation with AA was substantially inhibited during aspirin treatment (Fig. 29-6A). In contrast, performing the modified test with ADP as an agonist was insensitive to the effect of aspirin. Similarly, aspirin therapy significantly inhibited AA-induced platelet aggregation in a dose-independent manner reaching maximal effect at a dose of 100 mg of aspirin per day. Aspirin also inhibited ADP-induced platelet aggregation, but to a much lesser extent. A negative correlation was found between SC in the modified Impact and maximal aggregation in the aggregometer in monitoring the inhibitory effect of aspirin on platelets ($R^2 = 0.55$).

We also tested the potential application of this method in monitoring the response to aspirin among patients with acute coronary syndrome (ACS; mostly unstable angina). More than 100 patients were studied before and after PCI, which was accompanied by a loading dose of clopidogrel (300 mg).[25] The results demonstrated an unexpectedly low rate of response to aspirin as tested by both the aggregometer and the modified Impact on the day of admission for

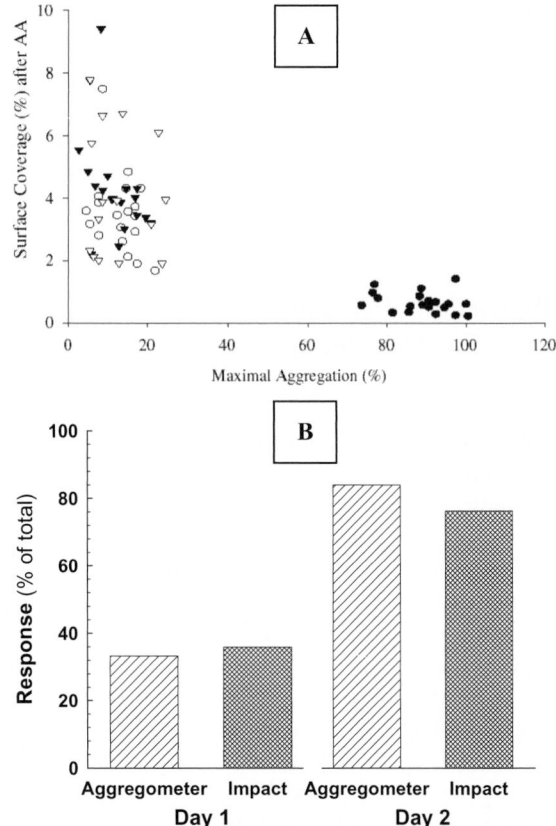

**Figure 29-6.** A. Correlation between the modified Impact (SC after AA) and platelet aggregometry (AA-induced maximal aggregation [%]) in detecting aspirin effect. Black circles, before aspirin treatment; open circles, aspirin 100 mg/day; black triangles, aspirin 300 mg/day; open triangles, aspirin 500 mg/day. Adapted from ref. 24 with permission. B. Response rate to aspirin among patients with acute coronary syndrome (ACS), before (day 1) and after (day 2) percutaneous coronary intervention (PCI) and a loading dose (300 mg) of clopidogrel. Maximal aggregation <25% in response to AA in the aggregometer and SC >2.5% in response to AA in the modified Impact were considered a response to aspirin. Note a significant increased rate of response on day 2 in comparison to the low response on day 1.

the procedure (Fig. 29-6B). In contrast, a dramatic shift in the response rate by both tests was observed one day after PCI and a loading dose of clopidogrel (Fig. 29-6B). A similar rate of response was also observed after 180 days when all patients were treated with aspirin only. This observation is supported by additional early reports in which cardiac and stroke patients were found to have a response to aspirin that was conversely correlated to disease severity.[26–28]

Testing the response to clopidogrel of ACS patients on day 4 of hospitalization demonstrated variability in the response by both assays, as reflected by the moderate change compared to the control group and the large standard deviation (Fig. 29-7A). In a preliminary study, a group of 19

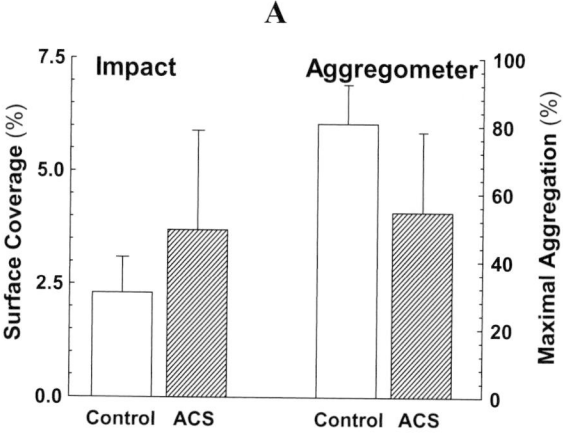

A

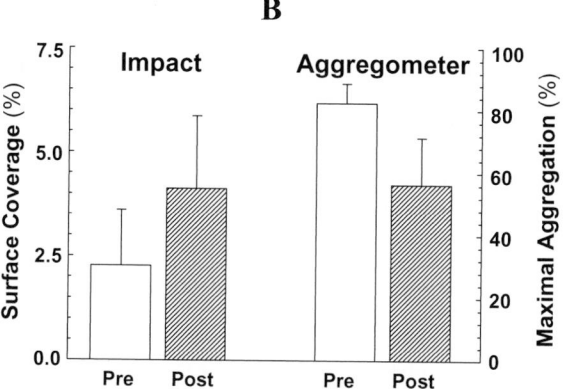

B

**Figure 29-7.** A. Blood samples of normal healthy volunteers (Control: $n = 46$) and ACS patients ($n = 259$) treated with aspirin and clopidogrel were subjected to the aggregometer test and the modified Impact test utilizing ADP as platelet agonist on day 4 of hospitalization. The results of the control group demonstrated maximal aggregation >55% and SC <3.2% representing the expected result of a donor not treated with clopidogrel. The results of the ACS group samples demonstrated maximal aggregation of about 55% and SC slightly higher than 3.2% both with very high standard deviation, representing a mixed population of responders and nonresponders to clopidogrel. B. Blood samples of ACS patients treated with aspirin and clopidogrel (loading dose 300 mg and daily dose 75 mg) were subjected to the aggregometer test with ADP as the agonist. Patients who were found to be nonresponders to clopidogrel (maximal aggregation >80%) ($n = 19$) were further treated by a loading dose of 600 mg and blood samples were collected 4 hours later. The results of the aggregometer and modified Impact with ADP before (pre) and following (post) the double dose are presented and demonstrate good response to the double dose of clopidogrel in both assays. Only three out of 19 patients did not respond to the double-dose treatment.

nonresponding patients (maximal aggregation >80%) selected from the group above were treated by a double loading dose of clopidogrel (600 mg). Following this treatment, the response to clopidogrel was tested by both assays and demonstrated a conversion of 13 of the 19 nonresponder patients to responders (Fig. 29-7B). These preliminary results suggest that both genetic factors and the thrombo-inflammatory state of the disease may have a significant effect on the pharmacological response to platelet anti-aggregant drugs. Future studies aiming to elucidate the mechanism of this phenomenon, as well as to develop strategies to improve response to the drugs and, most importantly, to improve clinical outcome, are warranted.

## VIII. Diagnosis of Thrombotic Microangiopathies

VWF stored in Weibel–Palade bodies of endothelial cells is rich in large and ultralarge multimers. Upon secretion, the VWF is rapidly, but partially cleaved by the metalloprotease ADAMTS-13 in the VWF A2 domain (Chapter 50).[29] Both the production of VWF and the ADAMTS-13 activity vary among individuals, thereby determining the interaction between VWF and the GPIb-IX-V complex on the platelets, especially under high shear stress (Chapter 7).[30,31] Severe deficiency of ADAMTS-13 leads to thrombotic thrombocytopenic purpura (TTP, Chapter 50). An inherited form of TTP (inTTP) is caused by a mutation in the ADAMTS-13 gene that results in deficiency of this VWF-cleaving protease.[32] The more common form of TTP, acquired TTP (acTTP), usually results from the development of autoantibodies against ADAMTS13.[33] TTP-mimicking syndromes occur in hemolytic uremic syndrome (HUS, Chapter 50).[34,35] ADAMTS-13 activity is usually measured under static condition in the presence of barium and urea.[29] This assay lacks the conditions that allow rapid cleavage of VWF multimers by the ADAMTS-13 *in vivo*.

Using the Impact, we recently showed that plasma from TTP patients induces excessive adhesion of normal whole blood platelets under high shear condition.[36] This test appeared to be specific for TTP, because mixing normal blood with plasma of patients with other thrombocytopenic conditions including HUS or with normal plasma, yielded decreased deposition of platelets which stemmed from dilution of the normal blood sample. Mixing plasma of TTP patients with normal blood yielded a 1.6- to 2-fold increase of platelet adhesion in a range of 1800 to 2500 sec$^{-1}$ shear rate compared to normal plasma. Maximal difference in platelet adhesion was observed under 2050 sec$^{-1}$. Under this shear rate, plasmas of five patients with inTTP and 11 patients with acTTP demonstrated indistinguishable effect on adhesion of normal whole blood platelets: increase by $77 \pm 19\%$ and $78 \pm 17\%$, respectively (Fig. 29-8A). The

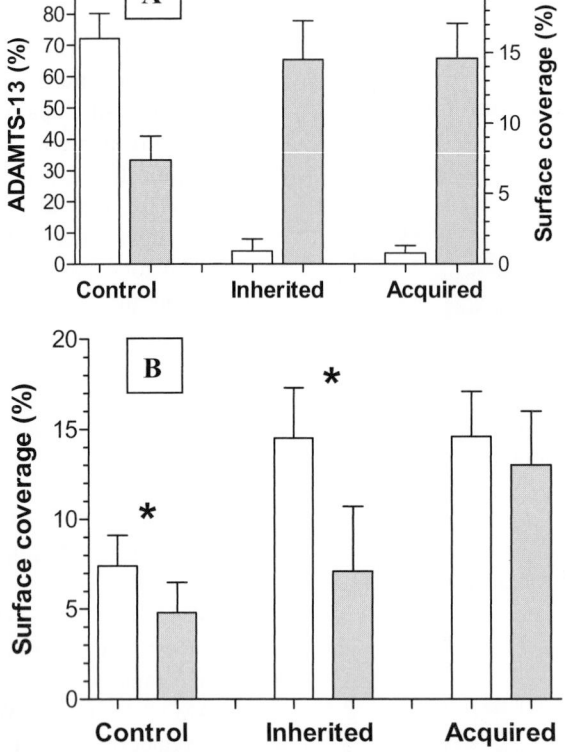

**Figure 29-8.** A. ADAMTS13 activity (white bars) in normal plasma ($n = 20$) and in plasma of inherited thrombotic thrombocytopenic purpura (inTTP) ($n = 5$) and acquired TTP (acTTP) ($n = 11$) evaluated by collagen-binding assay using fully multi-merized recombinant VWF as a substrate. Platelet adhesion (gray bars) under shear (2050 sec$^{-1}$) in mixed blood samples containing 150 μL normal blood and 50 μL plasma of the normal and patient groups indicated above is shown. The diluted blood samples were subjected to the Impact assay and surface coverage of adherent platelet particles was evaluated. B. Platelet adhesion under flow (2050 sec$^{-1}$) in mixed blood samples containing normal blood and either normal (Control; $n = 12$), inTTP (Inherited; $n = 5$) or acTTP (Acquired; $n = 11$) plasma. Before the Impact assay, the mixed blood samples were incubated for 10 min with phosphate-buffered saline (white bars) or BaCl$_2$ (10 mM; gray bars). The results are presented as mean ± SD. *$p < 0.05$.

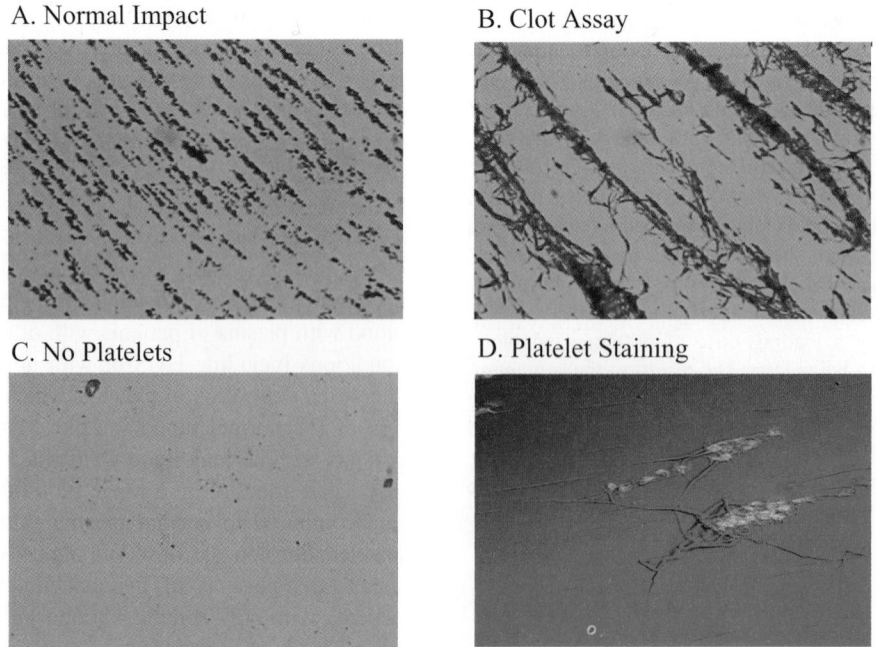

A. Normal Impact

B. Clot Assay

C. No Platelets

D. Platelet Staining

**Figure 29-9.** Testing platelet procoagulant effect under flow. A. A representative field of adherent platelets and aggregates in a typical Impact test. B. Fibrin clot formation upon addition of activated factor X. C. No fibrin clot is seen when the test is performed with a thrombocytopenic platelet-rich plasma sample, demonstrating the role of platelets in this process. D. A confocal microscope picture of a fibrin clot associated with adherent platelet aggregates. The fibrin fibers are seen only in association with adherent platelets.

values of ADAMTS13 activity measured by collagen-binding test were as follows: $72.2 \pm 8.0\%$ in controls, $4.2 \pm 3.8\%$ in patients with inTTP, and $3.5 \pm 2.4\%$ with acTTP (Fig. 29-8A). Addition of $BaCl_2$ to normal blood mixed with autologous plasma caused a $35 \pm 7\%$ reduction of SC compared to the mixture without $BaCl_2$ (Fig. 29-8B). Similarly, addition of $BaCl_2$ to the mixture of normal blood with inTTP plasma was followed by a $51 \pm 19\%$ reduction in SC. In contrast, addition of $BaCl_2$ to normal blood with acTTP plasma yielded only an $11 \pm 4\%$ decrease of SC, apparently due to the presence of ADAMTS13 inhibitor in acTTP plasma. Thus, the introduction of $BaCl_2$ in the Impact test may be useful for differentiation between inherited and acquired TTP.[37]

## IX. Testing Platelet Procoagulant Effect Under Flow

Activated platelets at the site of thrombus formation provide a negatively charged surface, upon which clotting factors assemble and accelerate the hemostatic process (Chapter 19). We therefore aimed to integrate the testing of clot formation into the framework of the Impact assay system. This new modification (Impact-C) is based on the ability of adherent platelets to very efficiently mediate clot formation. Clot initiators like thrombin, factor Xa, prothrombin time (PT), or partial thromboplastin time (PTT) reagents serve to promote fibrin clot formation on the surface of adherent and activated platelets in the Impact test.

Using the Impact-C, we established the feasibility of detecting clot formation after its initiation by PT/PTT reagents, factor Xa, or thrombin. Fig. 29-9 illustrates the principle of the test. In addition, this method was found effective in quantitative analysis of direct thrombin and factor Xa inhibitors using thrombin and factor Xa as clot initiators respectively.[38]

The Impact-C system is thus capable of quantitatively testing platelet-mediated clot formation under flow conditions in whole blood. This method can help in establishing a rapid diagnosis of a bleeding disorder resulting from a genetic or acquired deficiency of a clotting factor, may allow the detection of a prethrombotic (hypercoagulable) state related to increased clot formation, and may also provide a useful method for near-patient monitoring of the effect of various antithrombotic drugs, including new direct inhibitors of clotting factors.

## References

1. Einav, S., Dewey, C. F., & Hartenbaum, H. (1994). Cone and plate apparatus: A compact system for studying well characterized flow fields. *Exper Fluids, 16,* 196–202.

2. Varon, D., Dardik, R., Shenkman, S., et al. (1997). A new method for quantitative analysis of whole blood platelet interaction with extracellular matrix under flow conditions. *Thromb Res, 85,* 283–294.

3. Shenkman, B., Savion, N., Dardik, R., et al. (2000). Testing of platelet deposition on polystyrene surface under flow conditions by the cone and plate(let) analyzer: Role of platelet activation, fibrinogen and von Willebrand factor. *Thromb Res, 99,* 353–361.

4. Peerschke, E. I. B., Silver, R. T., Weksler, B., et al. (2004). *Ex vivo* evaluation of erythrocytosis-enhanced platelet thrombus formation using the cone and plate(let) analyzer: Effect of platelet antagonists. *Br J Haematol, 127,* 195–203.

5. Savion, N., Shenkman, B., Tamarin, I., et al. (2001). Transient adhesion refractoriness of circulating platelets under shear stress: The role of partial activation and microaggregate formation by suboptimal ADP concentration. *Br J Hematol, 112,* 1055–1061.

6. Gurevitz, O., Goldfarb, A., Hod, H., et al. (1998). Recombinant von Willebrand factor fragment AR545C inhibits platelet aggregation and enhances thrombolysis with rtPA in a rabbit thrombosis model. *Arterioscler Thromb Vasc Biol, 18,* 200–207.

7. Inbal, A., Gurevitz, O., Tamarin, I., et al. (1999). Unique antiplatelet effects of a novel S-nitroso derivative of a recombinant fragment of von Willebrand factor, AR545C: *In vitro* and *ex vivo* inhibition of platelet function. *Blood, 94,* 1693–1700.

8. Dardik, R., Varon, D., Eskaraev, R., et al. (2000). Recombinant fragment of von Willebrand factor AR545C inhibits platelets binding to thrombin and platelet adhesion of thrombin-treated endothelial cells. *Br J Haematol, 109,* 512–518.

9. Kenet, G., Lubetsky, A., Shenkman, B., et al. (1998). Cone and platelet analyser (CPA): A new test for the prediction of bleeding among thrombocytopenic patients. *Br J Haematol, 101,* 255–259.

10. Colman, R. W. (1990). Platelet and neutrophil activation in cardiopulmonary bypass. *Ann Thorac Surg, 49,* 32–34.

11. Weerasinghe, A., & Taylor, K. M. (1998). The platelets in cardiopulmonary bypass. *Ann Thorac Surg, 66,* 2145–2152.

12. Gerrah, R., Snir, E., Brill, A., et al. (2004). Platelet function changes as monitored by cone and plate(let) analyzer during beating heart surgery. *Heart Surg Forum, 7,* E191–E195.

13. Lubetsky, A., Jarrach, R., Brill, A., et al. (2002). Cone and plate(let) parameters as predictors of bleeding in patients undergoing cardiac surgery. *Blood, 100,* 2720.

14. Varon, D., Lashevski, I., Brenner, B., et al. (1998). Cone and plate(let) analyzer: Monitoring glycoprotein IIb/IIIa antagonists and von Willebrand disease replacement therapy by testing platelet deposition under flow conditions. *Am Heart J, 135,* 187–193.

15. Shenkman, B., Schneiderman, J., Tamarin, I., et al. (2001). Testing the effect of GPIIb-IIIa antagonist in patients undergoing carotid stenting: Correlation between standard aggregometry, flow cytometry and the cone and plate(let) analyzer (CPA) methods. *Thromb Res, 102,* 311–317.

16. Osende, J. I., Fuster, V., Lev, E. I., et al. (2001). Testing platelet activation with a shear-dependent platelet function test versus aggregation-based tests — Relevance for monitoring long-term glycoprotein IIb/IIIa inhibition. *Circulation, 103,* 1488–1491.

17. Wang, X. K., Dorsam, R. T., Lauver, A., et al. (2002). Comparative analysis of various platelet glycoprotein IIb/IIIa antagonists on shear-induced platelet activation and adhesion. *J Pharmacol Exp Ther, 303,* 1114–1120.

18. Helgason, C. M., Bolin, K. M., Hoff, J. A., et al. (1994). Development of aspirin resistance in persons with previous ischemic stroke. *Stroke, 25,* 2331–2336.

19. Gum, P. A., Kottke-Marchant, K., Welsh, P. A., et al. (2003). A prospective, blinded determination of the natural history of aspirin resistance among stable patients with cardiovascular disease. *J Am Coll Cardiol, 41,* 961–965.

20. Eikelboom, J. W., Hirsh, J., Weitz, J. I., et al. (2002). Aspirin-resistant thromboxane biosynthesis and the risk of myocardial infarction, stroke, or cardiovascular death in patients at high risk for cardiovascular events. *Circulation, 105,* 1650–1655.

21. Grotemeyer, K. H., Scharafinski, H. W., & Husstedt, I. W. (1993). Two-year follow-up of aspirin responder and aspirin non-responder. A pilot-study including 180 post-stroke patients. *Thromb Res, 71,* 397–403.

22. Wiviott, S. D., & Antman, E. M. (2004). Clopidogrel resistance: A new chapter in a fast-moving story. *Circulation, 109,* 3064–3067.

23. Matetzky, S., Shenkman, B., Guetta, V., et al. (2004). Clopidogrel resistance is associated with increased risk of recurrent atherothrombotic events in patients with acute myocardial infarction. *Circulation, 109,* 3171–3175.

24. Spectre, G., Brill, A., Gural, A., et al. (2005). A new point-of-care method for monitoring anti-platelet therapy: Application of the Cone and Plate(let) Analyzer. *Platelets 16,* 293–299.

25. Shenkman, B., Matetzky, S., Hod, H., et al. (2006). High rate of unresponsiveness to clopidogrel in patients with acute coronary syndrome (ACS) as determined by the Impact-R [cone and Plate(let) Analyzer]. *J Thromb Haemost, 3*(Suppl. 1), 2145 (abstr).

26. Tsabari, R., Schwammenthal, Y., Matetzky, S., et al. (2005). Low rate of responsiveness to aspirin in acute brain ischemia: Association with stroke severity and clinical outcome. *Stroke, 35,* 474 (abstr).

27. Frelinger, A. L., Linden, M. D., Furman, M. I., et al. (2004). Patients with coronary artery disease have an increased incidence of aspirin resistance: Association of PFA-100 closure time with clinical findings. *Blood, 104,* 514a (abstr).

28. Abdelrahman, N., Brill, A., Turetsky, N., et al. (2004). High rate of aspirin resistance among patients undergoing percutaneous interventions for unstable angina is associated with clinical outcome. *Blood, 104,* 515a (abstr).

29. Furlan, M. (1996). Willebrand factor: Molecular size and functional activity. *Ann Hematol, 72,* 341–348.

30. Manucci, P. M., Canciani, M. T., Forza, I., et al. (2001). Changes in health and disease of the metalloprotease that cleaves von Willebrand factor. *Blood, 98,* 2730–2735.

31. Dong, J. F., Whitelock, J., Bernardo, A., et al. (2004). Variations among normal individuals in the cleavage of endothelial-derived ultra-large von Willebrand factor under flow. *J Thromb Haemost, 2,* 1460–1466.

32. Levy, G. G., Nichols, W. C., Lian, E. C., et al. (2001). Mutations in a member of the ADAMTS gene family cause thrombotic thrombocytopenic purpura. *Nature, 413,* 488–494.

33. Tsai, H. M., & Lian, E. C. (1998). Antibodies to von Willebrand factor cleaving protease in acute thrombotic thrombocytopenic purpura. *N Engl J Med, 339,* 1585–1589.

34. Frishberg, Y., Obrig, T. G., & Kaplan, B. S. (1994). Hemolytic uremic syndrome. In M. A. Hollyday et al. (Eds.), *Pediatric nephrology* (3rd ed., pp. 871–889). Baltimore, MD: Williams and Wilkins.

35. Hunt, B. J., Lammle, B., Nevard, C. H. F., et al. (2001). Von Willebrand factor-cleaving protease in childhood diarrhea-associated haemolytic uraemic syndrome. *Thromb Haemost, 85,* 975–978.

36. Shenkman, B., Inbal, A., Tamarin, I., et al. (2003). Diagnosis of thrombotic thrombocytopenic purpura based on modulation by patient plasma of normal platelet adhesion under flow condition. *Br J Haematol, 120,* 597–604.

37. Shenkman, B., Budde, U., Angerhaus, D., et al. (2006). ADAMTS-13 effect on platelet adhesion under flow: A new method for differentiation between inherited and acquired thrombotic thrombocytopenic purpura. *Thromb Haemost,* in press.

38. Gitel, S., Brill, A., Savion, N., et al. (2006). Platelet mediated clot formation in whole blood under flow: A new screening method for coagulation disorders and for drug monitoring. *J Thromb Haemost, 3*(Suppl. 1), 2169 (abstr).

# CHAPTER 30

# Flow Cytometry

**Alan D. Michelson, Matthew D. Linden, Marc R. Barnard, Mark I. Furman, and A. L. Frelinger III**

*Center for Platelet Function Studies, Department of Pediatrics and Division of Cardiovascular Medicine, Department of Medicine, University of Massachusetts Medical School, Worcester, Massachusetts*

## I. Introduction

Flow cytometry, a remarkably versatile tool for the study of platelet function, encompasses multiple assays for multiple purposes (Table 30-1). Flow cytometry can be used to: (a) measure the activation state of circulating platelets and their reactivity (by activation-dependent changes in platelet surface antigens, leukocyte-platelet aggregation, and procoagulant platelet-derived microparticles); (b) diagnose specific disorders (Bernard–Soulier syndrome, Glanzmann thrombasthenia, storage pool disease, and heparin-induced thrombocytopenia [HIT]); (c) monitor antiplatelet agents; (d) monitor thrombopoiesis (by the number of young, "reticulated" platelets); (e) perform assays relevant to blood banking (quality control of platelet concentrates, identification of leukocyte contamination in platelet concentrates, immunophenotyping of platelet HPA-1a, detection of maternal and fetal anti-HPA-1a antibodies, and platelet cross-matching); (f) measure platelet-associated IgG; (g) measure the platelet count; and (h) perform other research assays (measurement of platelet survival and function *in vivo,* calcium flux, F-actin content, signal transduction, fluorescence resonance energy transfer, platelet recruitment, and bacteria–platelet interactions). All these applications of flow cytometry to the study of platelets are discussed in this chapter.

Flow cytometry rapidly measures the specific characteristics of a large number of individual cells. Before flow cytometric analysis, single cells in suspension are fluorescently labeled, typically with a fluorescently conjugated monoclonal antibody. In the flow cytometer, the suspended cells pass through a flow chamber and, at a rate of up to 2000 cells per second, through the focused beam of a laser. After the laser light activates the fluorophore at the excitation wavelength, detectors process the emitted fluorescence and light scattering properties of each cell. The intensity of the emitted light is directly proportional to the antigen density or the characteristics of the cell being measured.

Clinical studies that utilize flow cytometric assays of washed platelets or platelet-rich plasma are, like other assays of platelet function, potentially susceptible to artifactual *in vitro* platelets as a result of the obligatory separation procedures. The introduction of whole blood flow cytometry by Shattil et al.[1] was therefore a major step toward the application of flow cytometry to clinical settings. A typical schema of a sample preparation for whole blood flow cytometric analysis of platelets is shown in Fig. 30-1. The anticoagulant is usually buffered sodium citrate, although other anticoagulants can be used.[2] The purpose of the initial dilution is to minimize the formation of platelet aggregates.[2] A minimum of two monoclonal antibodies is used, each conjugated with a different fluorophore. A wide variety of fluorophores are available for antibody conjugation (e.g., phycoerythrin, fluorescein, peridinin chlorophyll protein [PerCP], phycoerythrin-Cy5, phycoerythrin-Texas Red [RED-670], allophycocyanin [APC]). The "test" monoclonal antibody (recognizing the antigen to be measured) is added at a saturating concentration. The "platelet identifier" monoclonal antibody (e.g., glycoprotein [GP] Ib-, GPIX-, integrin $\alpha$IIb-, or integrin $\beta$3-specific) is added at a near saturating concentration. Physiological agonists can be used in the assay, including thrombin, thrombin receptor-activating peptide (TRAP), adenosine diphosphate (ADP), collagen, the complement fraction C5b-9, and thromboxane $A_2$ analogs. Nonphysiologic agonists include phorbol myristate acetate and calcium ionophore A23187. Samples are stabilized by fixation, typically with a final concentration of 1% paraformaldehyde. The antibodies can be added after fixation, provided fixation does not interfere with antibody binding.[2,3] Samples are then analyzed in a flow cytometer. After identification of platelets both by their characteristic light scatter and by (for example) phycoerythrin positivity, binding of the (for example) fluorescein isothiocyanate (FITC)-conjugated test monoclonal antibody is determined by analyzing 5000 to 10,000 individual platelets. References 4 and 5 contain specific methodological protocols of whole blood flow cytometric assays of platelet function, together with a discussion of methodological issues.

**545**

### Table 30-1: Applications of Flow Cytometry to the Study of Platelets

**Measurement of Platelet Activation (Circulating Activated Platelets, Platelet Hyperreactivity, or Platelet Hyporeactivity)**
- Activation-dependent monoclonal antibodies
- Leukocyte–platelet aggregates
- Platelet-derived microparticles
- Platelet–platelet aggregates
- Shed blood

**Diagnosis of Specific Disorders**
- Bernard–Soulier syndrome
- Glanzmann thrombasthenia
- Storage pool disease
- Heparin-induced thrombocytopenia

**Monitoring of Antiplatelet Agents**
- Thienopyridines
- GPIIb-IIIa antagonists
- Aspirin

**Monitoring of Thrombopoiesis**
- Reticulated platelets

**Blood Bank Applications**
- Quality control of platelet concentrates
- Identification of leukocyte contamination in platelet concentrates
- Immunophenotyping of platelet HPA-1a
- Detection of maternal and fetal anti-HPA-1a antibodies
- Platelet cross-matching

**Platelet-Associated IgG**
- Immune thrombocytopenias
- Alloimmunization

**Platelet Counting**

**Other Research Applications**
- Platelet survival, tracking, and function *in vivo*
- Calcium flux
- F-actin
- Signal transduction
- Fluorescence resonance energy transfer
- Platelet recruitment
- Bacteria–platelet interactions

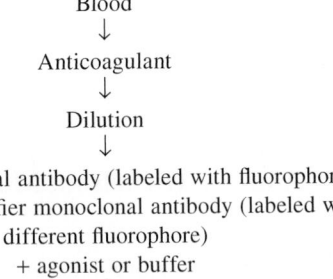

Blood
↓
Anticoagulant
↓
Dilution
↓
Test monoclonal antibody (labeled with fluorophore)
+ platelet identifier monoclonal antibody (labeled with different fluorophore)
+ agonist or buffer
↓
Fixation and dilution
↓
Analysis

**Figure 30-1.** A typical schema of sample preparation for analysis of platelets by whole blood flow cytometry.

nal antibodies directed against novel functional epitopes are developed, they can easily be incorporated into the assay. A subpopulation of as few as 1% partially activated platelets can be detected by whole blood flow cytometry.[10,11] Only minuscule volumes (approximately 5 μL) of blood are required[1,8] making whole blood flow cytometry particularly advantageous for neonatal studies.[12] The platelets of patients with profound thrombocytopenia can also be accurately analyzed. Finally, flow cytometric evaluation of platelets is easily adaptable to animal studies provided the antibody reagents are available.[13–15]

There are some disadvantages to flow cytometric analysis of platelet function. First, flow cytometers are expensive instruments to purchase and maintain. Second, for a clinical assay, sample preparation can be quite complicated, although the development of kits (e.g., BioCytex, Marseilles, France) has simplified some of the assays. Third, to avoid *ex vivo* platelet activation, blood samples should be processed within approximately 30 min of drawing for many assays.[1] For the evaluation of some platelet receptors, this time issue can be circumvented by immediate fixation.[2]

There are many advantages to flow cytometric analysis of platelet function. Platelets are directly analyzed in their physiological milieu of whole blood (including red cells and white cells, both of which affect platelet activation[6,7]). The minimal manipulation of the samples prevents artifactual *in vitro* activation and potential loss of platelet subpopulations.[1,8–10] Both the activation state of circulating platelets and the reactivity of circulating platelets can be determined. The flow cytometric method permits the detection of a spectrum of specific activation-dependent modifications in the platelet surface membrane. Furthermore, as new monoclo-

## II. Measurement of Platelet Activation

In the absence of an added exogenous platelet agonist, whole blood flow cytometry can determine the activation state of circulating platelets, as judged by the binding of an activation-dependent monoclonal antibody. In addition to this assessment of platelet function *in vivo*, inclusion of an exogenous agonist in the assay enables analysis of the reactivity of circulating platelets *in vitro*. In the latter application, whole blood flow cytometry is a physiological assay of platelet function in that an agonist results in a specific functional response by the platelets: a change in the surface expression

**Table 30-2: Activation-Dependent Monoclonal Antibodies (i.e., Antibodies That Bind to Activated but not Resting Platelets)**

| Activation-Dependent Platelet Surface Change | Prototypic Antibodies | References |
|---|---|---|
| *Conformational changes in integrins* | | |
| Activation-induced conformational change in integrin αIIbβ3 resulting in exposure of the fibrinogen binding site | PAC1 | 16 |
| Ligand-induced binding sites (LIBS) on integrin αIIbβ3 | PM 1.1, LIBS1, LIBS6 | 17,101,173 |
| Receptor-induced binding sites (RIBS) on bound fibrinogen | 2G5, 9F9, F26 | 18,32,174 |
| Activation-induced conformation change in integrin α2β1 resulting in exposure of the collagen binding site | IAC-1 | 175 |
| *Exposure of granule membrane proteins* | | |
| P-selectin (α-granules) | S12, AC1.2, 1E3 | 176–178 |
| GMP-33 (α-granules) | RUU-SP 1.77 | 179,180 |
| CD63 (lysosomes) | CLB-gran/12 | 181 |
| LAMP-1 (lysosomes) | H5G11 | 182 |
| LAMP-2 (lysosomes) | H4B4 | 183 |
| CD40 ligand | TRAP1 | 184 |
| Lectin–like oxidized LDL receptor-1 (LOX-1) | JTX68 | 185 |
| *Platelet surface binding of secreted platelet proteins* | | |
| Thrombospondin | P8, TSP-1 | 186,187 |
| Multimerin | JS-1 | 188,189 |
| *Development of a procoagulant surface[a]* | | |
| Factor V/Va binding | V237 | 27 |
| Factor X/Xa binding | 5224 | 190 |
| Factor VIII binding | 1B3 | 28 |

[a]Development of a procoagulant platelet surface can also be detected by the binding of annexin V to phosphatidylserine.[29]

of a physiological receptor (or other antigen or bound ligand), as determined by a change in the binding of a monoclonal antibody. In addition, as discussed below, whole blood flow cytometric enumeration of monocyte-platelet aggregates and procoagulant platelet-derived microparticles are also sensitive markers of *in vivo* platelet activation.

## A. Markers of Platelet Activation

**1. Activation-Dependent Monoclonal Antibodies.** Laboratory markers of platelet activation include activation-dependent conformational changes in integrin αIIbβ3 (the GPIIb-IIIa complex, CD41/CD61), exposure of granule membrane proteins, platelet surface binding of secreted platelet proteins, and development of a procoagulant surface (Table 30-2). The two most widely studied types of activation-dependent monoclonal antibodies are those directed against conformational changes in αIIbβ3 and those directed against granule membrane proteins.

Integrin αIIbβ3 is a receptor for fibrinogen and von Willebrand factor that is essential for platelet aggregation (see Chapter 8). Whereas most monoclonal antibodies directed against αIIbβ3 bind to resting platelets, monoclonal anti-

body PAC1 is directed against the fibrinogen binding site exposed by a conformational change in αIIbβ3 of activated platelets (Table 30-2).[16] Thus, PAC1 only binds to activated platelets, not to resting platelets. Other αIIbβ3-specific activation-dependent monoclonal antibodies are directed against either ligand-induced conformational changes in αIIbβ3 (ligand-induced binding sites, LIBS)[17] or receptor-induced conformational changes in the bound ligand (fibrinogen) (receptor-induced binding sites, RIBS)[18] (Table 30-2). Rather than αIIbβ3-specific monoclonal antibodies, fluorescein-conjugated fibrinogen can also be used in flow cytometric assays to detect the activated form of platelet surface αIIbβ3,[19,20] but the concentration of unlabeled plasma fibrinogen and unlabeled fibrinogen released from platelet α granules must also be considered in these assays.

The most widely studied type of activation-dependent monoclonal antibodies directed against granule membrane proteins are P-selectin (CD62P)-specific. P-selectin (see Chapter 12) is a component of the α granule membrane of resting platelets that is only expressed on the platelet surface membrane after α granule secretion. Therefore a P-selectin-specific monoclonal antibody only binds to degranulated platelets, not to resting platelets. The activation-dependent increase in platelet surface P-selectin is not reversible over

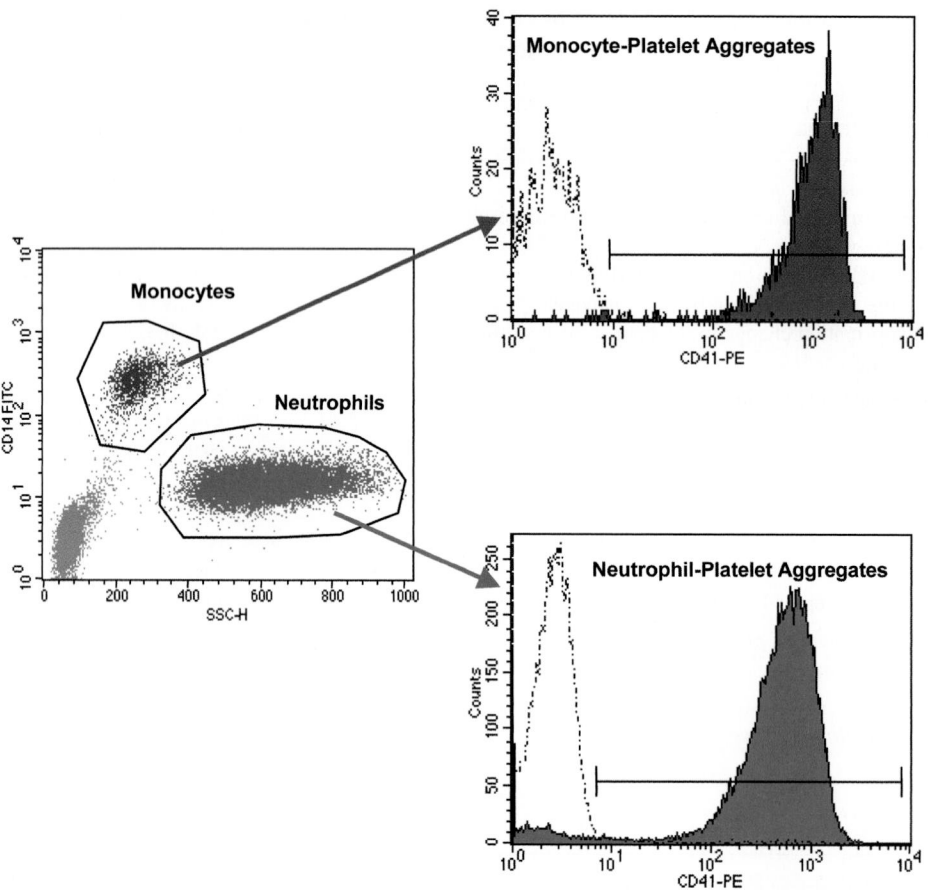

**Figure 30-2.** Whole blood flow cytometric analysis of monocyte–platelet aggregates and neutrophil–platelet aggregates in a normal donor after activation with thrombin receptor activating peptide (TRAP) 20 µM. Monocytes (blue) and neutrophils (green) were identified by their characteristic light scatter properties and the binding of FITC-conjugated CD14-specific monoclonal antibody TUK4 (left panel). Platelet-positive monocytes (i.e., monocyte–platelet aggregates) were identified by the binding of the phycoerythrin (PE)-conjugated αIIb (CD41)-specific monoclonal antibody 5B12 in the monocyte region (upper right panel). Platelet-positive neutrophils (i.e., neutrophil–platelet aggregates) were identified by the binding of the PE-conjugated αIIb (CD41)-specific monoclonal antibody 5B12 in the neutrophil region (lower right panel). SSC-H, side scatter height.

time *in vitro*.[21,22] However, *in vivo* circulating degranulated platelets rapidly lose their surface P-selectin, but continue to circulate and function.[14,23] Platelet surface P-selectin is therefore not an ideal marker for the detection of circulating degranulated platelets, unless (a) the blood sample is drawn immediately distal to the site of platelet activation, (b) the blood sample is drawn within 5 minutes of the activating stimulus, or (c) there is continuous activation of platelets. The length of time that other activation-dependent surface markers remain expressed on the platelet surface *in vivo* has not yet been definitively determined.

**2. Leukocyte–Platelet Aggregates.** P-selectin mediates the initial adhesion of activated platelets to monocytes and neutrophils via the P-selectin glycoprotein ligand 1 (PSGL-1) counter-receptor on the leukocyte surface (see Chapter 12). Monocyte–platelet and neutrophil–platelet

aggregates are readily identified by whole blood flow cytometry (Fig. 30-2).

Tracking of autologous infused biotinylated platelets in baboons by three-color whole blood flow cytometry enabled us[24] to directly demonstrate *in vivo* (Fig. 30-3) that (1) platelets degranulated by thrombin very rapidly (within 1 minute) form circulating aggregates with monocytes and neutrophils; (2) the percent of monocytes with adherent infused platelets is greater than the percent of neutrophils with adherent infused platelets; and (3) the *in vivo* half-life of detectable circulating monocyte–platelet aggregates (approximately 30 minutes) is longer than both the *in vivo* half-life of neutrophil–platelet aggregates (approximately 5 minutes) and the previously reported[14] rapid loss of surface P-selectin from nonaggregated infused platelets.

All these findings suggested that measurement of circulating monocyte–platelet aggregates may be a more

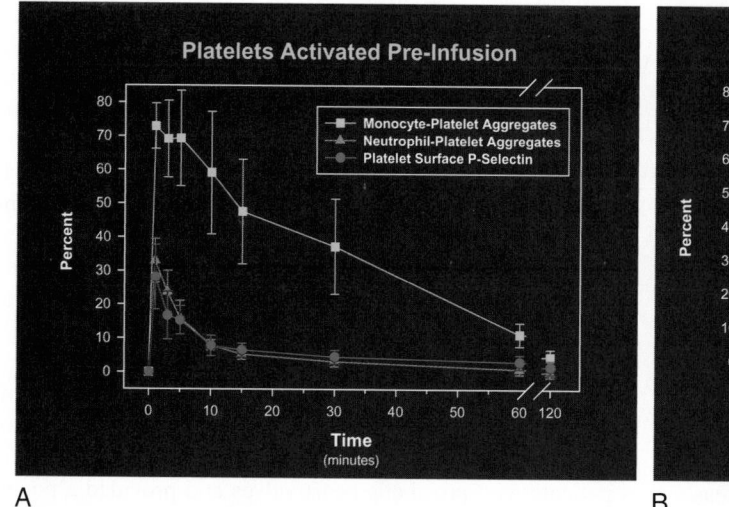
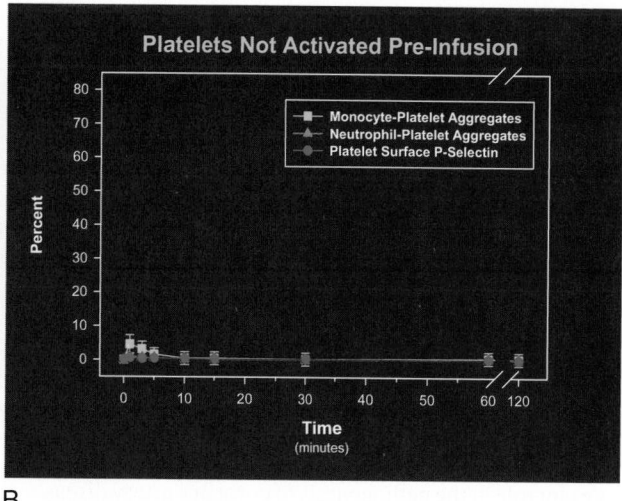

A                                                                                      B

**Figure 30-3.** Baboons were infused with autologous, biotinylated platelets that were (A) or were not (B) thrombin-activated preinfusion. Surface P-selectin on the infused platelets and participation of the infused platelets in circulating monocyte–platelet and neutrophil–platelet aggregates were determined by three-color whole blood flow cytometric analysis of peripheral blood samples drawn at the indicated time points. The "0" time point refers to blood samples taken immediately preinfusion. Platelet surface P-selectin is expressed as mean fluorescence intensity (MFI), as a percentage of the fluorescence of a preinfusion maximally activated (10 U/mL) thrombin control sample. Monocyte–platelet and neutrophil–platelet aggregates are expressed as the percent of all monocytes and neutrophils with adherent infused platelets. Data are mean ± SEM. (Reproduced with permission from ref. 24.)

sensitive indicator of *in vivo* platelet activation than either circulating neutrophil–platelet aggregates or circulating P-selectin-positive nonaggregated platelets. We therefore performed two clinical studies in patients with acute coronary syndromes.[24] First, after percutaneous coronary intervention (PCI), there was an increased number of circulating monocyte–platelet (and, to a lesser extent, neutrophil–platelet) aggregates, but not P-selectin-positive platelets, in peripheral blood. Second, of patients presenting to an Emergency Department with chest pain, patients with acute myocardial infarction had more circulating monocyte–platelet aggregates than patients without acute myocardial infarction and normal controls. However, circulating P-selectin-positive platelets were not increased in chest pain patients with or without acute myocardial infarction.[24]

In summary, we have demonstrated by five independent means—*in vivo* tracking of activated platelets in baboons (Fig. 30-3, upper panel),[24] human PCI,[24] human acute myocardial infarction,[24] stable coronary artery disease[25] (discussed in Section II.B1 below), and human chronic venous insufficiency[26] (discussed in Section II.B3. below)—that circulating monocyte–platelet aggregates are more sensitive markers of *in vivo* platelet activation than platelet surface P-selectin.

### 3. Platelet-Derived Microparticles.

As determined by flow cytometry, *in vitro* activation of platelets by some agonists (e.g., C5b-9, collagen/thrombin, and the calcium ionophore A23187) in the presence of extracellular calcium ions results in platelet-derived microparticles (defined by low forward angle light scatter and binding of a platelet-specific monoclonal antibody) that are procoagulant (determined by binding of monoclonal antibodies to activated factors V or VIII or by annexin V).[27–29] These findings suggest that procoagulant platelet-derived microparticles may have an important role in the assembly of the "tenase" and "prothrombinase" components of the coagulation system *in vivo*. A flow cytometric method for the direct detection of procoagulant platelet-derived microparticles in whole blood has been developed.[30] Platelet-derived microparticles are discussed in detail in Chapter 20.

### 4. Platelet–Platelet Aggregates.

Platelet–platelet aggregates can be measured by flow cytometry on the basis of light scattering properties. However, if the platelets are aggregated, the amount of antigen *per platelet* cannot be determined by flow cytometry.[14,31] This is because flow cytometry measures the amount of fluorescence per individual particle, irrespective of whether the particle is a single platelet or an aggregate of an unknown number of platelets. However, a rough estimate of aggregate size can be made by analyzing the increased platelet-specific fluorescence.

### 5. Shed Blood.

Because of the minuscule volumes of blood required, whole blood flow cytometry can be used to analyze the shed blood that emerges from a standardized bleeding time wound.[8,32–34] The time-dependent increase in the platelet surface expression of P-selectin in this shed blood reflects *in vivo* activation of platelets.[8,32–34] Immediate fixation (prior to antibody incubation) is obligatory in order to observe these time-dependent changes. The assay can be used to demonstrate deficient platelet reactivity in response to an *in vivo* wound, for example during cardiopulmonary bypass.[33] In addition, by tracking of infused platelets with biotin or PKH2 (see Section IX.A below), shed blood can be used to detect the functional participation of the infused platelets in *in vivo* platelet aggregates.[14]

## B. *Platelet Activation in Clinical Disorders*

**1. Acute Coronary Syndromes.** Platelets play an important role in the pathogenesis of coronary artery disease, including unstable angina and acute myocardial infarction (see Chapter 35). Whole blood flow cytometric studies have demonstrated circulating activated platelets, as determined by activation-dependent monoclonal antibodies, in patients with stable angina, unstable angina, and acute myocardial infarction.[25,35–37] In addition, as determined by activation-dependent monoclonal antibodies, PCI results in platelet activation in coronary sinus blood.[38,39]

Flow cytometric analysis of platelet activation-dependent markers can be used to determine optimal antiplatelet therapy in clinical settings (e.g., in acute coronary syndromes[40,41] and after coronary stenting).[42,43] Flow cytometric analysis of platelet activation markers before PCI can predict an increased risk of acute and subacute ischemic events after PCI.[44–47] Flow cytometrically detected exposure of LIBS is strongly associated with the development and progression of heart transplant vasculopathy.[48]

The Pl$^{A2}$ polymorphism of GPIIIa has been reported to be associated with ischemic coronary syndromes (Chapter 14). Flow cytometry has been used to demonstrate that (a) Pl$^{A2}$-positive platelets display a lower threshold for activation, and (b) platelets heterozygous for Pl$^A$ alleles show increased sensitivity to antiplatelet drugs.[49]

Circulating leukocyte–platelet aggregates are increased in stable coronary artery disease,[25,49] unstable angina,[35] acute myocardial infarction,[24,50–52] and cardiopulmonary bypass.[53] Circulating leukocyte-platelet aggregates also increase after PCI,[24] with a greater magnitude in patients experiencing late clinical events.[54] As discussed in Section II.A.2 above, circulating monocyte–platelet aggregates (but not neutrophil–platelet aggregates) are a more sensitive marker of *in vivo* platelet activation than platelet surface P-selectin in the clinical settings of stable coronary artery disease,[25] PCI,[24] and acute myocardial infarction.[24] Furthermore, circulating monocyte-platelet aggregates are an early marker of acute myocardial infarction.[52]

Platelet-derived microparticles are increased in acute coronary syndromes[55] and cardiopulmonary bypass.[32,56]

**2. Cerebrovascular Ischemia.** Platelets play an important role in the pathogenesis of ischemic cerebrovascular disease (Chapter 36). Increased circulating P-selectin-positive, CD63-positive, activated αIIbβ3-positive platelets, platelet-derived microparticles, and monocyte-platelet aggregates have been reported in acute cerebrovascular ischemia.[57–63] This platelet activation is evident 3 months after the acute event, suggesting the possibility of an underlying prothrombotic state.[59,61–64] Furthermore, increased expression of surface P-selectin on platelets is a risk factor for silent cerebral infarction in patients with atrial fibrillation.[65]

Platelet-derived microparticles are increased after transient ischemic attacks.[32,66] Increased platelet-derived microparticles and procoagulant activity occur in symptomatic patients with prosthetic heart valves and provided a potential pathophysiological explanation of cerebrovascular events in this patient group.[67]

Enhanced systemic platelet degranulation is associated with progression of intima-media thickness of the common carotid artery disease in patients with,[68] or without,[69] type 2 diabetes (as determined by platelet surface CD63 and CD40 ligand, and by platelet surface P-selectin, respectively).

**3. Peripheral Vascular Disease.** As discussed in detail in Chapter 37, circulating activated platelets and platelet hyperreactivity (as determined by P-selectin expression, platelet aggregates, and platelet-derived microparticle formation) are increased in patients with peripheral arterial disease compared with healthy volunteers.[70,71] Circulating monocyte– and neutrophil–platelet aggregates are significantly greater in the early postoperative period in patients with peripheral vascular disease who go on to develop later graft occlusion.[72]

With regard to peripheral venous disease, Peyton et al.[26] demonstrated an increased presence of monocyte–platelet aggregates in the lower extremity veins of patients with chronic venous stasis ulceration, as compared to control individuals without venous disease. Interestingly, these changes were present not only in blood drawn from the lower extremity veins of affected individuals, but also in blood drawn from arm veins, suggesting that the changes are systemic rather than localized to the lower extremities.[26] Powell et al.[73] further characterized these findings as being related to the presence of chronic venous disease rather than the presence of venous ulceration, because increased numbers of monocyte–platelet aggregates were noted in patients with all classes of venous disease, not just in patients with deep venous valvular insufficiency. Furthermore, increased levels of monocyte–platelet aggregates were noted even in patients with only superficial venous stasis disease manifested by the presence of varicose veins.

Even more intriguing is the fact that the number of monocyte–platelet aggregates remains elevated 6 weeks after

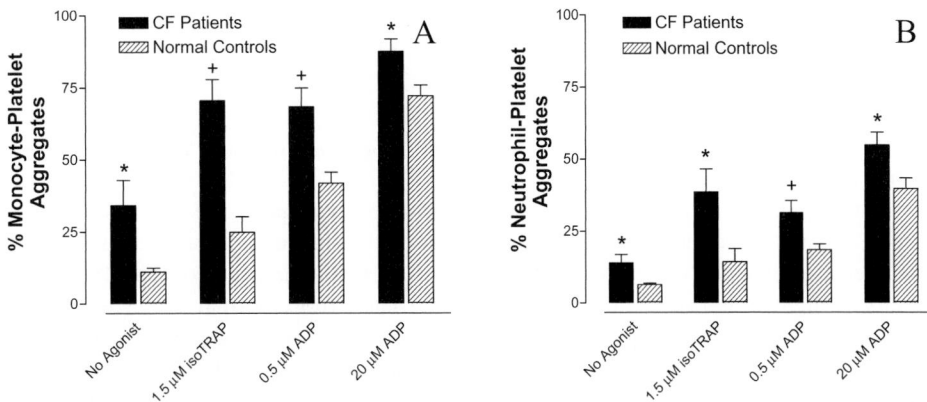

**Figure 30-4.** Cystic fibrosis (CF) patients have increased circulating monocyte– and neutrophil–platelet aggregates and increased platelet responsiveness to ADP and TRAP compared with healthy controls. Whole blood from CF patients and healthy controls was incubated with or without agonist and analyzed for (A) monocyte–platelet aggregates and (B) neutrophil–platelet aggregates. Data are mean ± SEM; n = 18. * $p < 0.05$ vs healthy controls. [+]$p < 0.01$ vs. healthy controls. (Reproduced with permission from ref. 78.)

total correction of the venous insufficiency by stripping of the abnormal veins, leaving normal venous physiology as documented by postoperative duplex scanning.[74] This finding suggests an underlying predisposition to the development of chronic venous disease in these patients, perhaps mediated by monocyte–platelet interactions.

**4. Other Clinical Disorders Associated with Platelet Hyperreactivity and/or Circulating Activated Platelets.** There are numerous other conditions in which whole blood flow cytometric measurement of platelet hyperreactivity, circulating activated platelets, and/or circulating leukocyte–platelet aggregates may prove to have a clinical role, including diabetes mellitus (discussed in Chapter 38),[68,75–77] cystic fibrosis (Fig. 30-4),[78] pre-eclampsia,[79,80] placental insufficiency,[81] migraine,[82] nephrotic syndrome,[83] hemodialysis,[84] sickle cell disease,[85] systemic inflammatory response syndrome,[86] septic multiple organ dysfunction syndrome,[87,88] antiphospholipid syndrome,[89] systemic lupus erythematosus,[89] rheumatoid arthritis,[89,90] inflammatory bowel disease,[91] myeloproliferative disorders,[92,93] and Alzheimer's disease.[94] High levels of circulating monocyte–platelet aggregates can predict rejection episodes after orthotopic liver transplantation.[95] Uremic patients with thrombotic events have higher numbers of circulating platelet-derived microparticles than those without thrombotic events.[96]

### C. Reduced Circulating Activated Platelets and Platelet Hyperreactivity

In addition to the detection of increased circulating activated platelets and platelet hyperreactivity discussed in Sections II.B above, whole blood flow cytometry may be useful in the clinical assessment of reduced circulating activated

platelets and platelet hyporeactivity — although there are few published studies in this area.

**1. Very Low Birth Weight Preterm Neonates.** Compared to adults, the platelets of very low birth weight preterm neonates are markedly hyporeactive to thrombin, ADP/epinephrine, and thromboxane $A_2$, as determined by flow cytometric detection of (a) the exposure of the fibrinogen binding site on αIIbβ3, (b) fibrinogen binding, (c) the increase in platelet surface P-selectin, and (d) the decrease in platelet surface GPIb.[12] Furthermore, procoagulant platelet-derived microparticles are decreased in preterm neonates compared with adults.[30] This platelet hyporeactivity of preterm neonates, as judged by both activation-dependent platelet surface changes and the generation of procoagulant platelet-derived microparticles, may contribute to the propensity of very low birth weight neonates to intraventricular hemorrhage.[97]

**2. Acute Myeloid Leukemia.** Prediction of hemorrhage in thrombocytopenic patients has been problematic because of the lack of laboratory markers or validated clinical assessment tools.[98] The clinical decision to prophylactically transfuse platelets is therefore usually based solely on the platelet count (see Chapter 69). However, low levels of expression of platelet surface P-selectin, as determined by whole blood flow cytometry, have recently been reported to be a prognostic marker for hemorrhage in acute myeloid leukemia.[98]

## III. Diagnosis of Specific Disorders

### A. Platelet Surface Glycoprotein Deficiencies

**1. Bernard–Soulier Syndrome.** Bernard–Soulier syndrome is an inherited deficiency of the GPIb-IX-V complex (see Chapters 54 and 57). Flow cytometric analysis with GPIb-, GPIX-, and GPV-specific monoclonal antibodies provides a rapid and simple means for the diagnosis of the homozygous and heterozygous states of Bernard–Soulier syndrome.[99] Whole blood flow cytometry allows analysis of platelets without attempting the technically difficult procedure of physically separating the giant Bernard–Soulier syndrome platelets from similarly-sized red and white blood cells. Because light scatter (especially forward light scatter) correlates with platelet size, light scatter gates may need to be adjusted in the flow cytometric analysis of giant platelet syndromes such as Bernard–Soulier syndrome. This adjustment may result in overlap of the light scatter of giant platelets with red and white blood cells. It is therefore essential to include in the assay a platelet-specific monoclonal antibody as a platelet identifier. For Bernard–Soulier syndrome platelets, this identifier antibody obviously cannot be GPIb-, GPIX-, or GPV-specific.

### 2. Glanzmann Thrombasthenia.

Glanzmann thrombasthenia is an inherited deficiency of integrin $\alpha IIb\beta 3$ (see Chapter 57). Flow cytometric analysis with $\alpha IIb$- and $\beta 3$-specific monoclonal antibodies provides a rapid and simple means for the diagnosis of the homozygous and heterozygous states of Glanzmann thrombasthenia.[99,100] In addition, a panel of activation-dependent monoclonal antibodies can be used to evaluate patients with defects in platelet aggregation,[101] secretion,[102] or procoagulant activity.[103]

### B. Storage Pool Disease

Inherited dense granule storage pool deficiency, a relatively common cause of a mild hemorrhagic diathesis, cannot be reliably diagnosed by standard platelet aggregometry.[104] The conventional method to establish the diagnosis of storage pool disease is to label the platelets with the fluorescent dye mepacrine and then measure platelet fluorescence by microscopy.[105] This assay is based on the selective binding of mepacrine to adenine nucleotides in dense granules. The assay is not ideal for the clinical laboratory because it is subjective and tedious, and examines only a small number of platelets. In contrast, dense granule storage pool deficiency can be accurately diagnosed by a simple, rapid, one-step flow cytometric assay in a clinical laboratory.[106,107] The method shows good correlation with fluorescent light microscopic methods, but improves the detection of the mepacrine-loaded platelets by quantitatively measuring fluorescence on a large number (5000) of platelets.[106,107] Acquired dense granule storage pool deficiency, which occurs in myeloproliferative disorders and end-stage renal failure, can also be diagnosed by flow cytometry.[106,108] An alternative approach to the flow cytometric diagnosis of storage pool disease is to measure intraplatelet serotonin (5-hydroxytryptamine) with a serotonin-specific monoclonal antibody.[109] Storage pool disease is discussed in detail in Chapters 15 and 57.

### C. Heparin-Induced Thrombocytopenia

Unlike normal sera and sera from patients with quinine- or quinidine-induced thrombocytopenia, sera from patients with HIT generate procoagulant platelet-derived microparticles from normal platelets.[110] This observation has been used to develop a rapid, specific, and sensitive flow cytometric assay for the diagnosis of HIT.[111] Alternative flow cytometric approaches to the diagnosis of HIT have been described.[112] HIT is discussed in detail in Chapter 48.

## IV. Monitoring of Antiplatelet Agents

The *in vivo* effect of the thienopyridines (clopidogrel and ticlopidine) on platelet function can be monitored by the vasodilator-stimulated phosphoprotein (VASP) assay (BioCytex, Marseilles, France).[113,114] The principle of this flow cytometric assay is shown in Fig. 30-5, and an example of

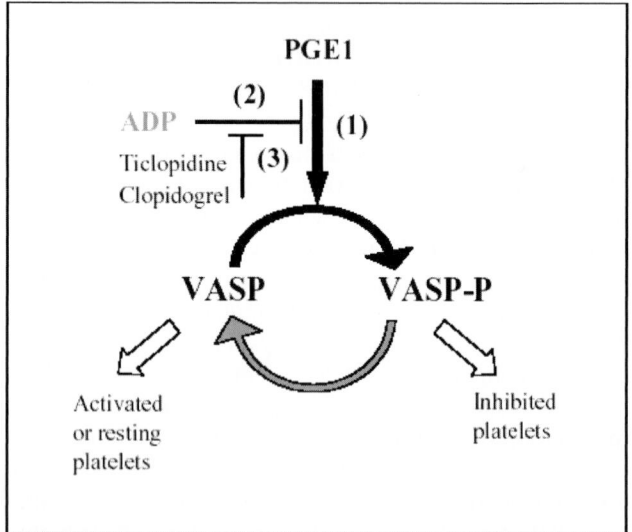

**Figure 30-5.** Monitoring thienopyridine therapy by vasodilator stimulated phosphoprotein (VASP) phosphorylation state, using a phosphorylation-specific monoclonal antibody. VASP is an intracellular platelet protein which is nonphosphorylated at basal state. VASP phosphorylation is regulated by the cyclic adenosine monophosphate (cAMP) cascade. Prostaglandin E1 (PGE1) activates this cascade (1), whereas it is inhibited by adenosine diphosphate (ADP) through the P2Y$_{12}$ receptor (2). Under the test conditions (i.e., the *in vitro* addition of ADP and PGE1), the degree of VASP phosphorylation correlates with the degree of P2Y$_{12}$ receptor inhibition by thienopyridines (clopidogrel or ticlopidine) (3). (From the BioCytex package insert.)

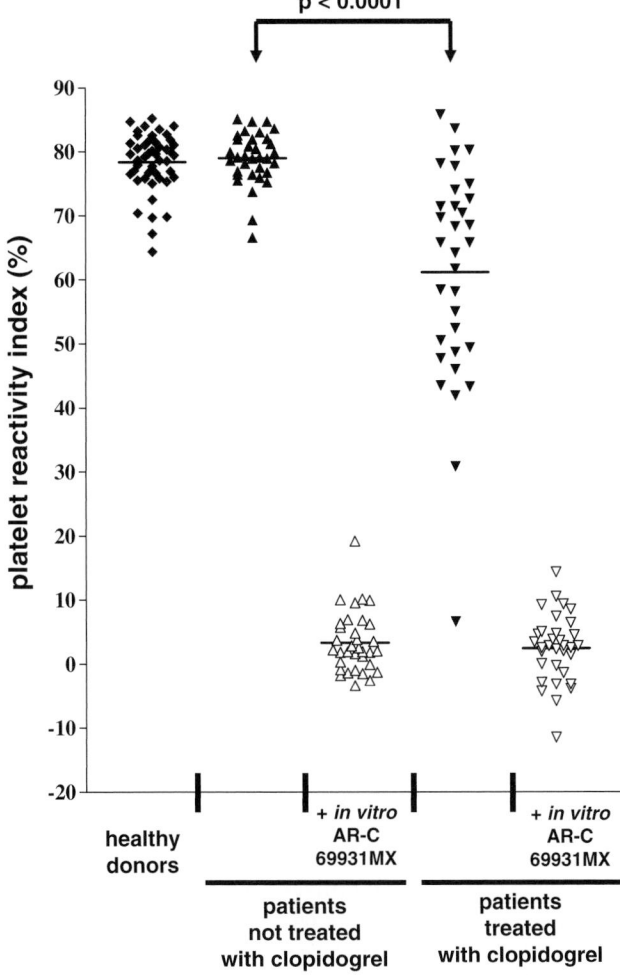

**Figure 30-6.** VASP phosphorylation (measured as platelet reactivity index) in healthy donors (n = 47) and in patients with ischemic cardiovascular diseases treated (n = 34) or not (n = 33) with clopidogrel. The empty symbols correspond to the platelet reactivity index in patients with ischemic cardiovascular diseases treated or not with clopidogrel after complementary *in vitro* inhibition of the P2Y$_{12}$ receptor by 10 μM AR-C69931MX. (Reproduced with permission from ref. 113.)

its use to measure clopidogrel response variability in patients with ischemic cardiovascular disease is shown in Fig. 30-6.

Flow cytometric methods can be used to monitor GPIIb-IIIa antagonists (abciximab, eptifibatide, and tirofiban) by measuring receptor occupancy by these drugs. These methods can be categorized as either direct or indirect. One type of direct method is a competitive binding assay with (a) biotinylated[115] or FITC-conjugated[116] GPIIb-IIIa antagonists; (b) blocking monoclonal antibodies[117,118]; (c) peptides with very high affinities for GPIIb-IIIa (disintegrins[119] or cyclic RGD peptides[120]); or, (d) fibrinogen binding after platelet activation (detected by a polyclonal antifibrinogen antibody).[121] Another type of direct method measures the binding of an antibody directed against the GPIIb-IIIa antagonist.[122,123] Indirect methods measure either (a) the GPIIb-IIIa antagonist-induced binding of an anti-LIBS antibody;[124] or, (b) platelet aggregation, as determined by light scatter.[115]

Through the use of arachidonic acid as the agonist, flow cytometry can also be used to monitor aspirin therapy.[125,126]

The monitoring of antiplatelet agents by platelet function tests is discussed in detail in Section III.B of Chapter 23.

## V. Monitoring of Thrombopoiesis

Whole blood flow cytometric methods have been developed for the identification of young platelets (i.e., those containing mRNA) by their staining with thiazole orange.[127–129] Because of the analogy to reticulocytes, these thiazole orange-positive platelets have been termed "reticulated platelets"[130] and have been used to monitor thrombopoiesis.[128] Thrombocytopenic patients whose bone marrow contains normal or increased numbers of megakaryocytes have significantly elevated proportions of circulating reticulated platelets.[127] In contrast, the proportion of reticulated platelets in thrombocytopenic patients with impaired platelet production (reduced bone marrow megakaryocytes) does not differ from normal controls and the absolute number of reticulated platelets is significantly lowered.[127] Measurement of reticulated platelets has been used as an aid in assessing bone marrow recovery after bone marrow transplantation.[131] In addition, measurement of reticulated platelets may be useful in evaluating both treatment response and thrombotic risk in patients with thrombocytosis.[132]

However, there are persistent methodological issues with the flow cytometric measurement of reticulated platelets. Because thiazole orange also binds to ADP and ATP (contained in dense granules), important controls in the flow cytometric measurement of reticulated platelets are the demonstration that thiazole orange staining is (a) abolished by pretreatment of the sample with RNAase, and (b) not abolished by pretreatment of the sample with thrombin. In fact, the higher thiazole orange signal of young platelets has been reported to be derived to a significant extent from their large volume and granule content,[133–135] leading to the suggestion that platelet degranulation with TRAP should be an initial part of the assay for reticulated platelets.[133] However, other investigators report that, under modified assay conditions, degranulation does not significantly change thiazole orange fluorescence and that RNAase demonstrates specificity of the thiazole orange staining.[129]

A new automated method to reliably quantify reticulated platelets, expressed as the immature platelet fraction (IPF), has been developed using the XE-2100 blood cell counter with upgraded software (Sysmex, Kobe, Japan).[136] The IPF is identified by flow cytometry techniques and the use of a nucleic acid-specific dye in the reticulocyte/optical platelet channel. The clinical utility of this parameter was established in the laboratory diagnosis of thrombocytopenia due to increased platelet destruction (e.g., in immune thrombocytopenic purpura).[136]

## VI. Blood Bank Applications

### A. *Quality Control of Platelet Concentrates*

Measurement of the platelet surface expression of P-selectin by flow cytometry is one of the most commonly applied measures of platelet activation in platelet concentrates stored in blood banks, and efforts have been made to standardize these measurements.[137] However, we[14] have demonstrated in a nonhuman primate model that infused, degranulated platelets rapidly lose surface P-selectin to the plasma pool, but continue to circulate and function *in vivo.* Thus, platelet surface P-selectin molecules, rather than degranulated platelets, are rapidly cleared. Our results[14] were subsequently independently confirmed by Berger et al.,[23] who found that the platelets of both wild-type and P-selectin knockout mice had identical life spans. When platelets were isolated, activated with thrombin, and reinjected into mice, the rate of platelet clearance was unchanged. The infused thrombin-activated platelets rapidly lost their surface P-selectin in circulation, and this loss was accompanied by the simultaneous appearance of a 100-kDa P-selectin fragment in the plasma.[23] Storage of platelets at 4°C caused a significant reduction in their life span *in vivo,* but again no significant differences were observed between the two genotypes. Thus, the results of Berger et al.[23] confirm that P-selectin does not mediate platelet clearance. Furthermore, in a thrombocytopenic rabbit kidney injury model, Krishnamurti et al.[138] reported that thrombin-activated human platelets (a) lose platelet surface P-selectin in the (reticuloendothelial system-inhibited) rabbit circulation; (b) survive in the circulation just as long as fresh human platelets; and, most importantly, (c) are just as effective as fresh human platelets at decreasing blood loss.

In summary, these studies[14,23,138] strongly suggest that the measurement of platelet surface P-selectin in platelet concentrates stored in the blood bank should not be used as a predictor of platelet survival or function *in vivo.* However, platelet surface P-selectin could still be a useful measure of quality control during processing, storage, and manipulation (filtration, washing).[137] This is because, in contrast to the situation *in vivo,*[14,23,138] the activation-dependent increase in platelet surface P-selectin is not reversible over time under standard blood banking conditions.[139] Quality control of platelet concentrates is discussed in detail in Chapter 69.

### B. *Other Blood Bank Applications*

Flow cytometry can also be used to identify leukocyte contamination of platelet concentrates (discussed in Chapter 69); immunophenotype platelet HPA-1a[140] and other polymorphisms (discussed in Chapters 14 and 53); detect maternal and fetal anti-HPA-1a antibodies[141] (discussed in Chapters 53); and crossmatch platelets, which may be useful for alloimmunized patients for whom HLA compatible platelets are not readily available[142] (discussed in Chapter 69).

## VII. Platelet-Associated IgG

Measurement of platelet-associated IgG by flow cytometry may be useful in immune thrombocytopenias[143,144] and alloimmunization.[145] A quantitative direct platelet immunofluorescence test for platelet-associated IgG has been proposed as a screening test for immune thrombocytopenias.[146]

## VIII. Platelet Count

### A. *Platelet Count in Humans*

The International Council for Standardization in Haematology and the International Society of Laboratory Hematology has recommended a flow cytometric reference method for platelet counting that utilizes erythrocyte counts, determined by an automated counter, as an internal reference standard.[147,148] Platelet counting is discussed in detail in Chapter 24.

### B. *Platelet Count in Mice*

With advances in manipulation of the mouse genome, murine models have become increasingly important in our understanding of platelet disorders. However, standard methods for murine platelet counting require relatively large volumes of blood and therefore serial platelet counts cannot be followed over time in the same mouse. To circumvent this problem, we recently described a rapid, reproducible, and accurate flow cytometric method to determine the number and activation state of circulating platelets from a single mouse over extended periods of time.[15] The method uses fluorescent staining of platelets in whole blood with a specific antibody and the addition of known numbers of fluorescent beads for standardization of the sample volume. Analysis of platelets obtained by tail bleeding indicated that this sampling procedure did not activate platelets, and only 5 μL of blood were required for platelet counting. Thus, the method can be used to follow the number and the activation state of circulating platelets from individual mice over extended periods of time and is applicable to a wide range of murine models of platelet disorders.[15]

## IX. Other Research Applications

### A. *Platelet Survival, Tracking, and Function* in Vivo

Multicolor whole blood flow cytometry can be used to track platelets *in vivo* and determine their survival and function (Fig. 30-3).[14,24,149–153] In three-color flow cytometry, the fluorescent labeling typically identifies (1) platelets in whole blood (e.g., by a phycoerythrin-conjugated GPIb-, GPIX-, αIIb-, or β3-specific monoclonal antibody), (2) the infused platelets (e.g., by prelabeling with PKH2, or biotin followed by the *ex vivo* addition of streptavidin-RED670), and (3) an activation-dependent monoclonal antibody (e.g., FITC-conjugated P-selectin-specific, or activated αIIb- or β3-specific, monoclonal antibody). By these means, the *in vivo* function of tracked, infused platelets can be determined at multiple time points by multiple independent assays including, for example, participation in platelet aggregation in response to an *in vivo* wound, exposure of the fibrinogen binding site on αIIbβ3, adherence to Dacron in an arteriovenous shunt, and generation of procoagulant platelet-derived microparticles.[14] In a recent study in humans, Hughes et al.[154] used two-color whole blood flow cytometry with a FITC-conjugated HLA-A2-specific monoclonal antibody to track and characterize transfused platelets by selectin donor/recipient pairs discrepant for HLA-A2. The measurement of platelet survival is discussed in detail in Chapter 33.

### B. *Monocyte and Neutrophil Procoagulant Activity with and without Surface-Adherent Platelets*

Monocytes and neutrophils form heterotypic aggregates with platelets via engagement of platelet surface P-selectin with leukocyte surface PSGL-1 (Chapter 12). The resultant intracellular signaling causes the leukocyte surface expression of tissue factor[155] and activation of leukocyte surface Mac-1 (integrin αMβ2, CD11b/CD18).[156,157] The activation-dependent conformational change in monocyte surface Mac-1[158,159] results in the binding of coagulation factor Xa and/or fibrinogen to Mac-1.[160–162] The platelet surface also binds coagulation factors via exposed negatively charged phospholipids such as phosphatidylserine (see Chapter 19).[29] Tissue factor from the vascular wall and platelets form a complex with activated coagulation factor VII, facilitating the activation of factor X. Tissue factor is a key component of monocyte surface procoagulant activity.[163]

We have developed whole blood flow cytometry assays to measure bound tissue factor, coagulation factor Xa,

fibrinogen, activated Mac-1 and CD11b on the surface of monocytes and neutrophils, allowing independent analysis of monocytes and neutrophils with and without surface-adherent platelets (Figs. 30-7 and 30-8).[164] These methods will be applicable to future *in vivo* and *in vitro* studies of pharmacological agents targeted to either coagulation and/or cellular activation processes. The monocyte and neutrophil surface binding of tissue factor (Fig. 30-8), coagulation factor Xa, and fibrinogen (Fig. 30-7) is mainly dependent on platelet adherence to the monocytes and neutrophils, whereas the monocyte and neutrophil surface expression of CD11b and activated Mac-1 is mainly independent of platelet adherence to the monocytes and neutrophils.[164]

### C. *Calcium Flux*

In washed platelet preparations, platelet-rich plasma, or whole blood, flow cytometry can be used to measure platelet calcium flux, an important platelet second messenger.[165]

### D. *F-Actin*

Platelet cytoskeletal rearrangement can be analyzed flow cytometrically by measuring F-actin content with NBD- or bodipy-phallicidin or FITC-phalloidin.[166,167]

### E. *Signal Transduction*

Flow cytometry can be used to investigate platelet signal transduction. For example, flow cytometry can be used to detect and quantify specific intracellular platelet protein phosphorylation using phosphorylation-specific monoclonal antibodies (Fig. 30-5).[168]

### F. *Fluorescence Resonance Energy Transfer*

Fluorescence resonance energy transfer (FRET) can be used to investigate the spatial separation or orientation of exoplasmic domains within receptor molecules,[169] and to detect and characterize antiplatelet antibodies directed against HLA class 1 molecules or platelet-specific glycoproteins.[170]

### G. *Platelet Recruitment*

A three-color flow cytometric method can be used for the simultaneous monitoring of two platelet populations that enables the study of the effects of stimuli (even very short-acting stimuli such as nitric oxide) on platelet recruitment.[171]

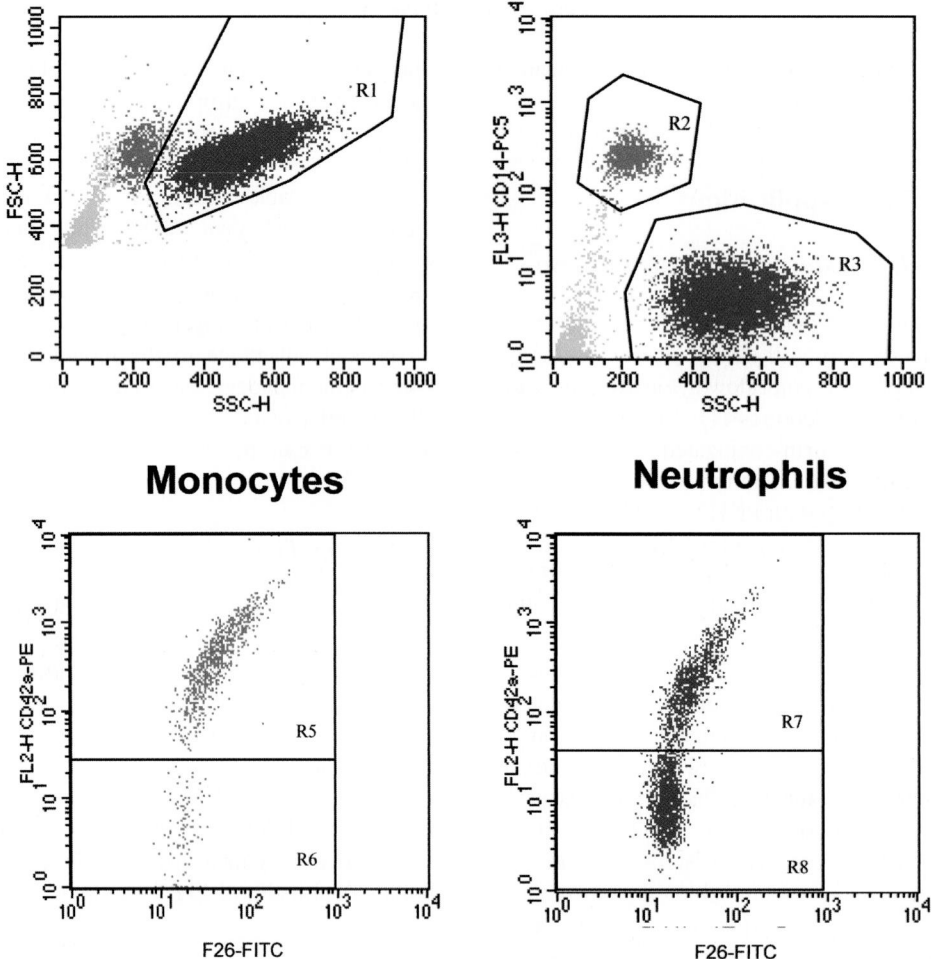

**Figure 30-7.** Assay of monocyte and neutrophil procoagulant activity with and without surface-adherent platelets. Upper panels: Monocytes and neutrophils were initially gated by a combination of light scattering and CD14 PECy5 fluorescence. Dim CD14, forward, and 90° light scatter define neutrophils (blue, R1 + R3). Bright CD14 and 90° light scatter define monocytes (red, R2). Forward and 90° light scatter increase significantly upon formation of heterotypic aggregates. Lower panels: In this collagen-stimulated example, monocytes (red) and neutrophils (blue) were further gated into CD42a-negative platelet-free (R6 and R8) and CD42a-positive platelet-bound (R5 and R7) subpopulations. Note monoclonal antibody F26 (directed against surface-bound fibrinogen) (x axis) increases with increased numbers of leukocyte-bound platelets (y axis). (Reproduced with permission from ref. 164.)

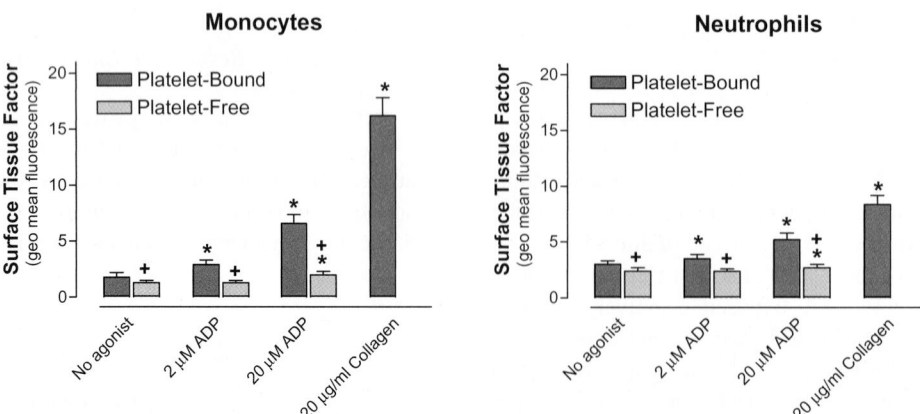

**Figure 30-8.** Surface expression of tissue factor on platelet-bound monocytes and neutrophils and platelet-free monocytes and neutrophils. These whole blood flow cytometric assays required preincubations at 37°C to allow physiological ligand binding prior to adding the test antibody.[164] Data are mean ± SEM, n = 10. *$p < 0.05$ by paired $t$ test for surface tissue factor with agonist compared with no agonist. +$p < 0.05$ by paired $t$ test for platelet-free monocytes compared with platelet-bound monocytes and for platelet-free neutrophils compared with platelet-bound neutrophils. There were insufficient platelet-free monocytes or neutrophils after collagen stimulation to generate these data points. (Reproduced with permission from ref. 164.)

## H. Bacteria–Platelet Interactions

The binding of platelets to other cells (e.g., bacteria, and the functional consequences of this binding on both cell types), can be studied by multicolor flow cytometry.[172]

## References

1. Shattil, S. J., Cunningham, M., & Hoxie, J. A. (1987). Detection of activated platelets in whole blood using activation-dependent monoclonal antibodies and flow cytometry. *Blood, 70,* 307–315.
2. Michelson, A. D., Barnard, M. R., Krueger, L. A., et al. (2000). Evaluation of platelet function by flow cytometry. *Methods, 21,* 259–270.
3. Michelson, A. D., Barnard, M. R., Benoit, S. E., et al. (1995). Characterization of platelet binding of blind panel mAb. In S. F. Schlossman, L. Boumsell, W. Gilks, et al. (Eds.), *Leucocyte typing.* (pp. 1207–1210). Oxford: Oxford University Press.
4. Krueger, L. A., Barnard, M. R., Frelinger, III A. L., et al. (2002). Immunophenotypic analysis of platelets. In J. P. Robinson, Z. Darzynkiewicz, P. N. Dean, et al. (Eds.), *Current protocols in cytometry.* (pp. 6.10.1–6.10.17). New York: John Wiley & Sons.
5. Barnard, M. R., Krueger, L. A., Frelinger, III A. L., et al. (2003). Whole blood analysis of leukocyte-platelet aggregates. In J. P. Robinson, Z. Darzynkiewicz, P. N. Dean, et al. (Eds.), *Current protocols in cytometry.* New York: John Wiley & Sons.
6. Santos, M. T., Valles, J., Marcus, A. J., et al. (1991). Enhancement of platelet reactivity and modulation of eicosanoid production by intact erythrocytes. A new approach to platelet activation and recruitment. *J Clin Invest, 87,* 571–580.
7. LaRosa, C. A., Rohrer, M. J., Rodino, L. J., et al. (1994). Human neutrophil cathepsin G is a potent platelet activator. *J Vasc Surg, 19,* 306–319.
8. Michelson, A. D., Ellis, P. A., Barnard, M. R., et al. (1991). Downregulation of the platelet surface glycoprotein Ib-IX complex in whole blood stimulated by thrombin, ADP or an in vivo wound. *Blood, 77,* 770–779.
9. Abrams, C., & Shattil, S. J. (1991). Immunological detection of activated platelets in clinical disorders. *Thromb Haemost, 65,* 467–473.
10. Michelson, A. D. (1994). Platelet activation by thrombin can be directly measured in whole blood through the use of the peptide GPRP and flow cytometry: Methods and clinical studies. *Blood Coagul Fibrinolysis, 5,* 121–131.
11. Kestin, A. S., Ellis, P. A., Barnard, M. R., et al. (1993). The effect of strenuous exercise on platelet activation state and reactivity. *Circulation, 88,* 1502–1511.
12. Rajasekhar, D., Barnard, M. R., Bednarek, F. J., et al. (1997). Platelet hyporeactivity in very low birth weight neonates. *Thromb Haemost, 77,* 1002–1007.
13. Michelson, A. D., Benoit, S. E., Barnard, M. R., et al. (1995). A panel of platelet mAb for the study of haemostasis and thrombosis in baboons. In S. F. Schlossman, L. Boumsell, W. Gilks, et al. (Eds.), *Leucocyte typing V.* (pp. 1230–1231). Oxford: Oxford University Press.
14. Michelson, A. D., Barnard, M. R., Hechtman, H. B., et al. (1996). In vivo tracking of platelets: Circulating degranulated platelets rapidly lose surface P-selectin but continue to circulate and function. *Proc Natl Acad Sci USA. 93,* 11877–11882.
15. Alugupalli, K. R., Michelson, A. D., Barnard, M. R., et al. (2001). Serial determinations of platelet counts in mice by flow cytometry. *Thromb Haemost, 86,* 668–671.
16. Shattil, S. J., Hoxie, J. A., Cunningham, M., et al. (1985). Changes in the platelet membrane glycoprotein IIb-IIIa complex during platelet activation. *J Biol Chem, 260,* 11107–11114.
17. Frelinger, A. L., Lam, S. C., Plow, E. F., et al. (1988). Occupancy of an adhesive glycoprotein receptor modulates expression of an antigenic site involved in cell adhesion. *J Biol Chem, 263,* 12397–12402.
18. Zamarron, C., Ginsberg, M. H., & Plow, E. F. (1990). Monoclonal antibodies specific for a conformationally altered state of fibrinogen. *Thromb Haemost, 64,* 41–46.
19. Faraday, N., Goldschmidt-Clermont, P., Dise, K., et al. (1994). Quantitation of soluble fibrinogen binding to platelets by fluorescence-activated flow cytometry. *J Lab Clin Med, 123,* 728–740.
20. Heilmann, E., Hynes, L. A., Burstein, S. A., et al. (1994). Fluorescein derivatization of fibrinogen for flow cytometric analysis of fibrinogen binding to platelets. *Cytometry, 17,* 287–293.
21. Michelson, A. D., Benoit, S. E., Kroll, M. H., et al. (1994). The activation-induced decrease in the platelet surface expression of the glycoprotein Ib-IX complex is reversible. *Blood, 83,* 3562–3573.
22. Ruf, A., & Patscheke, H. (1995). Flow cytometric detection of activated platelets: Comparison of determining shape change, fibrinogen binding, and P-selectin expression. *Semin Thromb Hemost, 21,* 146–151.
23. Berger, G., Hartwell, D. W., & Wagner, D. D. (1998). P-Selectin and platelet clearance. *Blood, 92,* 4446–4452.
24. Michelson, A. D., Barnard, M. R., Krueger, L. A., et al. (2001). Circulating monocyte-platelet aggregates are a more sensitive marker of in vivo platelet activation than platelet surface P-selectin: Studies in baboons, human coronary intervention, and human acute myocardial infarction. *Circulation, 104,* 1533–1537.
25. Furman, M. I., Benoit, S. E., Barnard, M. R., et al. (1998). Increased platelet reactivity and circulating monocyte-platelet aggregates in patients with stable coronary artery disease. *J Am Coll Cardiol, 31,* 352–358.
26. Peyton, B. D., Rohrer, M. J., Furman, M. I., et al. (1998). Patients with venous stasis ulceration have increased monocyte-platelet aggregation. *J Vasc Surg, 27,* 1109–1116.
27. Sims, P. J., Faioni, E. M., Wiedmer, T., et al. (1988). Complement proteins C5b-9 cause release of membrane vesicles from the platelet surface that are enriched in the membrane receptor for coagulation factor Va and express prothrombinase activity. *J Biol Chem, 263,* 18205–18212.

28. Gilbert, G. E., Sims, P. J., Wiedmer, T., et al. (1991). Platelet-derived microparticles express high affinity receptors for factor VIII. *J Biol Chem, 266,* 17261–17268.

29. Furman, M. I., Krueger, L. A., Frelinger, A. L., III et al. (2000). GPIIb-IIIa antagonist-induced reduction in platelet surface factor V/Va binding and phosphatidylserine expression in whole blood. *Thromb Haemost, 84,* 492–498.

30. Michelson, A. D., Rajasekhar, D., Bednarek, F. J., et al. (2000). Platelet and platelet-derived microparticle surface factor V/Va binding in whole blood: Differences between neonates and adults. *Thromb Haemost, 84,* 689–694.

31. Ault K. A., Rinder H. M., Mitchell J. G., et al. (1989). Correlated measurement of platelet release and aggregation in whole blood. *Cytometry, 10,* 448–455.

32. Abrams, C. S., Ellison, N., Budzynski, A. Z., et al. (1990). Direct detection of activated platelets and platelet-derived microparticles in humans. *Blood, 75,* 128–138.

33. Kestin, A. S., Valeri, C. R., Khuri, S. F., et al. (1993). The platelet function defect of cardiopulmonary bypass. *Blood, 82,* 107–117.

34. Michelson, A. D., MacGregor, H., Barnard, M. R., et al. (1994). Reversible inhibition of human platelet activation by hypothermia in vivo and in vitro. *Thromb Haemost, 71,* 633–640.

35. Ott, I., Neumann, F. J., Gawaz, M., et al. (1996). Increased neutrophil-platelet adhesion in patients with unstable angina. *Circulation, 94,* 1239–1246.

36. Coulter, S. A., Cannon, C. P., Ault, K. A., et al. (2000). High levels of platelet inhibition with abciximab despite heightened platelet activation and aggregation during thrombolysis for acute myocardial infarction: Results from TIMI (thrombolysis in myocardial infarction) 14. *Circulation, 101,* 2690–2695.

37. Schultheiss, H. P., Tschoepe, D., Esser, J., et al. (1994). Large platelets continue to circulate in an activated state after myocardial infarction. *Eur J Clin Invest, 24,* 243–247.

38. Scharf, R. E., Tomer, A., Marzec, U. M., et al. (1992). Activation of platelets in blood perfusing angioplasty-damaged coronary arteries. Flow cytometric detection. *Arterioscler Thromb, 12,* 1475–1487.

39. Langford, E. J., Brown, A. S., Wainwright, R. J., et al. (1994). Inhibition of platelet activity by S-nitrosoglutathione during coronary angioplasty. *Lancet, 344,* 1458–1460.

40. Langford, E. J., Wainwright, R. J., & Martin, J. F. (1996). Platelet activation in acute myocardial infarction and unstable angina is inhibited by nitric oxide donors. *Arterioscler Thromb Vasc Biol, 16,* 51–55.

41. Ault, K. A., Cannon, C. P., Mitchell, J., et al. (1993). Platelet activation in patients after an acute coronary syndrome: Results from the TIMI-12 trial. Thrombolysis in Myocardial Infarction. *J Am Coll Cardiol, 33,* 634–639.

42. Gawaz, M., Neumann, F. J., Ott, I., et al. (1996). Platelet activation and coronary stent implantation. Effect of antithrombotic therapy. *Circulation, 94,* 279–285.

43. Neumann, F. J., Gawaz, M., Dickfeld, T., et al. (1997). Antiplatelet effect of ticlopidine after coronary stenting. *J Am Coll Cardiol, 29,* 1515–1519.

44. Tschoepe, D., Schultheiss, H. P., Kolarov, P., et al. (1993). Platelet membrane activation markers are predictive for increased risk of acute ischemic events after PTCA. *Circulation, 88,* 37–42.

45. Gawaz, M., Neumann, F. J., Ott, I., et al. (1997). Role of activation-dependent platelet membrane glycoproteins in development of subacute occlusive coronary stent thrombosis. *Coron Artery Dis, 8,* 121–128.

46. Kabbani, S. S., Watkins, M. W., Ashikaga, T., et al. (2001). Platelet reactivity characterized prospectively: A determinant of outcome 90 days after percutaneous coronary intervention. *Circulation, 104,* 181–186.

47. Kabbani, S. S., Watkins, M. W., Ashikaga, T., et al. (2003). Usefulness of platelet reactivity before percutaneous coronary intervention in determining cardiac risk one year later. *Am J Cardiol, 91,* 876–878.

48. Fateh-Moghadam, S., Bocksch, W., Ruf, A., et al. (2000). Changes in surface expression of platelet membrane glycoproteins and progression of heart transplant vasculopathy. *Circulation, 102,* 890–897.

49. Michelson, A. D., Furman, M. I., Goldschmidt-Clermont, P., et al. (2000). Platelet GP IIIa Pl(A) polymorphisms display different sensitivities to agonists. *Circulation, 101,* 1013–1018.

50. Gawaz, M., Reininger, A., & Neumann, F. J. (1996). Platelet function and platelet-leukocyte adhesion in symptomatic coronary heart disease. Effects of intravenous magnesium. *Thromb Res, 83,* 341–349.

51. Neumann, F. J., Marx, N., Gawaz, M., et al. (1997). Induction of cytokine expression in leukocytes by binding of thrombin-stimulated platelets. *Circulation, 95,* 2387–2394.

52. Furman, M. I., Barnard, M. R., Krueger, L. A., et al. (2001). Circulating monocyte-platelet aggregates are an early marker of acute myocardial infarction. *J Am Coll Cardiol, 38,* 1002–1006.

53. Rinder, C. S., Bonan, J. L., Rinder, H. M., et al. (1992). Cardiopulmonary bypass induces leukocyte-platelet adhesion. *Blood, 79,* 1201–1205.

54. Mickelson, J. K., Lakkis, N. M., Villarreal-Levy, G., et al. (1996). Leukocyte activation with platelet adhesion after coronary angioplasty: A mechanism for recurrent disease? *J Am Coll Cardiol, 28,* 345–353.

55. Katopodis, J. N., Kolodny, L., Jy, W., et al. (1997). Platelet microparticles and calcium homeostasis in acute coronary ischemias. *Am J Hematol, 54,* 95–101.

56. George, J. N., Pickett, E. B., Saucerman, S., et al. (1986). Platelet surface glycoproteins. Studies on resting and activated platelets and platelet membrane microparticles in normal subjects, and observations in patients during adult respiratory distress syndrome and cardiac surgery. *J Clin Invest, 78,* 340–348.

57. Grau, A. J., Ruf, A., Vogt, A., et al. (1998). Increased fraction of circulating activated platelets in acute and previous cerebrovascular ischemia. *Thromb Haemost, 80,* 298–301.

58. Zeller, J. A., Tschoepe, D., & Kessler, C. (1999). Circulating platelets show increased activation in patients with acute cerebral ischemia. *Thromb Haemost, 81,* 373–377.

59. Meiklejohn, D. J., Vickers, M. A., Morrison, E. R., et al. (2001). In vivo platelet activation in atherothrombotic stroke is not determined by polymorphisms of human platelet glycoprotein IIIa or Ib. *Br J Haematol, 112,* 621–631.

60. Yamazaki, M., Uchiyama, S., & Iwata, M. (2001). Measurement of platelet fibrinogen binding and p-selectin expression by flow cytometry in patients with cerebral infarction. *Thromb Res, 104,* 197–205.

61. Cherian, P., Hankey, G. J., Eikelboom, J. W., et al. (2003). Endothelial and platelet activation in acute ischemic stroke and its etiological subtypes. *Stroke, 34,* 2132–2137.

62. Yip, H. K., Chen, S. S., Liu, J. S., et al. (2004). Serial changes in platelet activation in patients after ischemic stroke: Role of pharmacodynamic modulation. *Stroke, 35,* 1683–1687.

63. McCabe, D. J., Harrison, P., Mackie, I. J., et al. (2004). Platelet degranulation and monocyte-platelet complex formation are increased in the acute and convalescent phases after ischaemic stroke or transient ischaemic attack. *Br J Haematol, 125,* 777–787.

64. Marquardt, L., Ruf, A., Mansmann, U., et al. (2002). Course of platelet activation markers after ischemic stroke [see comment]. *Stroke, 33,* 2570–2574.

65. Minamino, T., Kitakaze, M., Sanada, S., et al. (1998). Increased expression of P-selectin on platelets is a risk factor for silent cerebral infarction in patients with atrial fibrillation: Role of nitric oxide. *Circulation, 98,* 1721–1727.

66. Lee, Y. J., Jy, W., Horstman, L. L., et al. (1994). Elevated platelet microparticles in transient ischemic attacks, lacunar infarcts, and multiinfarct dementias. *Thrombo Res, 72,* 295–304.

67. Geiser, T., Sturzenegger, M., Genewein, U., et al. (1998). Mechanisms of cerebrovascular events as assessed by procoagulant activity, cerebral microemboli, and platelet microparticles in patients with prosthetic heart valves. *Stroke, 29,* 1770–1777.

68. Fateh-Moghadam, S., Li, Z., Ersel, S., et al. (2005). Platelet degranulation is associated with progression of intima-media thickness of the common carotid artery in patients with diabetes mellitus type 2. *Arterioscler Thromb Vasc Biol, 25,* 1299–1303.

69. Koyama, H., Maeno, T., Fukumoto, S., et al. (2003). Platelet P-selectin expression is associated with atherosclerotic wall thickness in carotid artery in humans. *Circulation, 108,* 524–529.

70. Zeiger, F., Stephan, S., Hoheisel, G., et al. (2000). P-Selectin expression, platelet aggregates, and platelet-derived microparticle formation are increased in peripheral arterial disease. *Blood Coagul Fibrinolysis, 11,* 723–728.

71. Cassar, K., Bachoo, P., Ford, I., et al. (2003). Platelet activation is increased in peripheral arterial disease. *J Vasc Surg, 38,* 99–103.

72. Esposito, C. J., Popescu, W. M., Rinder, H. M., et al. (2003). Increased leukocyte-platelet adhesion in patients with graft occlusion after peripheral vascular surgery. *Thromb Haemost, 90,* 1128–1134.

73. Powell, C. C., Rohrer, M. J., Barnard, M. R., et al. (1999). Chronic venous insufficiency is associated with increased platelet and monocyte activation and aggregation. *J Vasc Surg, 30,* 844–851.

74. Rohrer, M. J., Claytor, R. B., Garnette, C. S. C., et al. (2002). Platelet-monocyte aggregates in patients with chronic venous insufficiency remain elevated following correction of reflux. *Cardiovasc Surg, 10,* 464–469.

75. Tschoepe, D., Roesen, P., Esser, J., et al. (1991). Large platelets circulate in an activated state in diabetes mellitus. *Semin Thromb Hemost, 17,* 433–438.

76. Angiolillo, D. J., Fernandez-Ortiz, A., Bernardo, E., et al. (2005). Platelet function profiles in patients with type 2 diabetes and coronary artery disease on combined aspirin and clopidogrel treatment. *Diabetes 54*(8), 2430–2435.

77. Hu, H., Li, N., Yngen, M., et al. (2004). Enhanced leukocyte-platelet cross-talk in Type 1 diabetes mellitus: Relationship to microangiopathy. *J Thromb Haemost, 2,* 58–64.

78. O'Sullivan, B. P., Linden, M. D., Frelinger, A. L., III, et al. (2005). Platelet activation in cystic fibrosis. *Blood, 105,* 4635–4641.

79. Janes, S. L., & Goodall, A. H. (1994). Flow cytometric detection of circulating activated platelets and platelet hyper-responsiveness in pre-eclampsia and pregnancy. *Clin Sci, 86,* 731–739.

80. Konijnenberg, A., van der Post, J. A., Mol, B. W., et al. (1997). Can flow cytometric detection of platelet activation early in pregnancy predict the occurrence of preeclampsia? A prospective study. *Am J Obstet Gynecol, 177,* 434–442.

81. Trudinger, B., Song, J. Z., Wu et al. (2003). Placental insufficiency is characterized by platelet activation in the fetus. *Obstet Gynecol, 101,* t-81.

82. Zeller, J. A., Frahm, K., Baron, R., et al. (2004). Platelet-leukocyte interaction and platelet activation in migraine: A link to ischemic stroke? *J Neurol Neurosurg Psychiatr, 75,* 984–987.

83. Sirolli, V., Ballone, E., Garofalo, D., et al. (2002). Platelet activation markers in patients with nephrotic syndrome. A comparative study of different platelet function tests. *Nephron, 91,* 424–430.

84. Gawaz, M. P., Mujais, S. K., Schmidt, B., et al. (1999). Platelet-leukocyte aggregates during hemodialysis: Effect of membrane type. *Artificial Organs, 23,* 29–36.

85. Wun, T., Cordoba, M., Rangaswami, A., et al. (2002). Activated monocytes and platelet-monocyte aggregates in patients with sickle cell disease. *Clin Lab Haematol, 24,* 81–88.

86. Ogura, H., Kawasaki, T., Tanaka, H., et al. (2001). Activated platelets enhance microparticle formation and platelet-leukocyte interaction in severe trauma and sepsis. *J Trauma-Injury Infection Crit Care, 50,* 801–809.

87. Gawaz, M., Dickfeld, T., Bogner, C., et al. (1997). Platelet function in septic multiple organ dysfunction syndrome. *Intensive Care Med, 23,* 379–385.

88. Russwurm, S., Vickers, J., Meier-Hellmann, A., et al. (2002). Platelet and leukocyte activation correlate with the severity of septic organ dysfunction. *Shock, 17,* 263–268.

89. Joseph, J. E., Harrison, P., Mackie, I. J., et al. (2001). Increased circulating platelet-leucocyte complexes and platelet activation in patients with antiphospholipid syndrome,

systemic lupus erythematosus and rheumatoid arthritis. *Br J Haemat, 115,* 451–459.

90. Bunescu, A., Seideman, P., Lenkei, R., et al. (2004). Enhanced Fcgamma receptor I, alphaMbeta2 integrin receptor expression by monocytes and neutrophils in rheumatoid arthritis: Interaction with platelets. *J Rheumatol, 31,* 2347–2355.

91. Danese, S., de la M. C., Sturm, A., et al. (2003). Platelets trigger a CD40-dependent inflammatory response in the microvasculature of inflammatory bowel disease patients. *Gastroenterology, 124,* 1249–1264.

92. Jensen, M. K., de Nully, B. P., Lund, B. V., et al. (2001). Increased circulating platelet-leukocyte aggregates in myeloproliferative disorders is correlated to previous thrombosis, platelet activation and platelet count. *Eur J Haematol, 66,* 143–151.

93. Villmow, T., Kemkes-Matthes, B., & Matzdorff, A. C. (2002). Markers of platelet activation and platelet-leukocyte interaction in patients with myeloproliferative syndromes. *Thromb Res, 108,* 139–145.

94. Sevush, S., Jy, W., Horstman, L. L., et al. (1998). Platelet activation in Alzheimer disease. *Arch Neurol, 55,* 530–536.

95. Vanacore, R., Guida, C., Urciuoli, P., et al. (2003). High levels of circulating monocyte-platelet aggregates can predict rejection episodes after orthotopic liver transplantation. *Transplant Proc, 35,* 1019.

96. Ando, M., Iwata, A., Ozeki, Y., et al. (2002). Circulating platelet-derived microparticles with procoagulant activity may be a potential cause of thrombosis in uremic patients. *Kidney Int, 62,* 1757–1763.

97. Setzer, E. S., Webb, I. B., Wassenaar, J. W., et al. (1982). Platelet dysfunction and coagulopathy in intraventricular hemorrhage in the premature infant. *J Pediatr, 100,* 599–605.

98. Leinoe, E. B., Hoffmann, M. H., Kjaersgaard, E., et al. (2005). Prediction of haemorrhage in the early stage of acute myeloid leukaemia by flow cytometric analysis of platelet function. *Br J Haematol, 128,* 526–532.

99. Michelson, A. D. (1987). Flow cytometric analysis of platelet surface glycoproteins: Phenotypically distinct subpopulations of platelets in children with chronic myeloid leukemia. *J Lab Clin Med, 110,* 346–354.

100. Jennings, L. K., Ashmun, R. A., Wang, W. C., et al. (1986). Analysis of human platelet glycoproteins IIb-IIIa and Glanzmann's thrombasthenia in whole blood by flow cytometry. *Blood, 68,* 173–179.

101. Ginsberg, M. H., Frelinger, A. L., Lam, S. C., et al. (1990). Analysis of platelet aggregation disorders based on flow cytometric analysis of membrane glycoprotein IIb-IIIa with conformation-specific monoclonal antibodies. *Blood, 76,* 2017–2023.

102. Lages, B., Shattil, S. J., Bainton, D. F., et al. (1991). Decreased content and surface expression of alpha-granule membrane protein GMP-140 in one of two types of platelet alpha delta storage pool deficiency. *J Clin Invest, 87,* 919–929.

103. Sims, P. J., Wiedmer, T., Esmon, C. T., et al. (1989). Assembly of the platelet prothrombinase complex is linked to vesiculation of the platelet plasma membrane. Studies in

Scott syndrome: An isolated defect in platelet procoagulant activity. *J Biol Chem, 264,* 17049–17057.

104. Nieuwenhuis, H. K., Akkerman, J. W., & Sixma, J. J. (1987). Patients with a prolonged bleeding time and normal aggregation tests may have storage pool deficiency: Studies on one hundred six patients. *Blood, 70,* 620–623.

105. Lorez, H. P., Richards, J. G., Da Prada, M., et al. (1979). Storage pool disease: Comparative fluorescence microscopical, cytochemical and biochemical studies on amine-storing organelles of human blood platelets. *Br J Haematol, 43,* 297–305.

106. Gordon, N., Thom, J., Cole, C., et al. (1995). Rapid detection of hereditary and acquired platelet storage pool deficiency by flow cytometry. *Br J Haematol, 89,* 117–123.

107. Wall, J. E., Buijs-Wilts, M., Arnold, J. T., et al. (1995). A flow cytometric assay using mepacrine for study of uptake and release of platelet dense granule contents. *Br J Haematol, 89,* 380–385.

108. Gawaz, M. P., Bogner, C., & Gurland, H. J. (1993). Flow-cytometric analysis of mepacrine-labelled platelets in patients with end-stage renal failure. *Haemostasis, 23,* 284–292.

109. Maurer-Spurej, E., Dyker, K., Gahl, W. A., et al. (2002). A novel immunocytochemical assay for the detection of serotonin in platelets. *Br J Haematol, 116,* 604–611.

110. Warkentin, T. E., Hayward, C. P., Boshkov, L. K., et al. (1994). Sera from patients with heparin-induced thrombocytopenia generate platelet-derived microparticles with procoagulant activity: An explanation for the thrombotic complications of heparin-induced thrombocytopenia. *Blood, 84,* 3691–3699.

111. Tomer, A. (1997). A sensitive and specific functional flow cytometric assay for the diagnosis of heparin-induced thrombocytopenia. *Br J Haematol, 98,* 648–656.

112. Gobbi, G., Mirandola, P., Tazzari, P. L., et al. (2004). New laboratory test in flow cytometry for the combined analysis of serologic and cellular parameters in the diagnosis of heparin-induced thrombocytopenia. *Cytometry B (Clin Cytometry), 58,* 32–38.

113. Aleil, B., Ravanat, C., Cazenave, J. P., et al. (2005). Flow cytometric analysis of intraplatelet VASP phosphorylation for the detection of clopidogrel resistance in patients with ischemic cardiovascular diseases. *J Thromb Haemost, 3,* 85–92.

114. Gurbel, P. A., Bliden, K. P., Samara, W., et al. (2005). Clopidogrel effect on platelet reactivity in patients with stent thrombosis. Results of the CREST study. *J Am Coll Cardiol, 46,* 1827–1832.

115. Konstantopoulos, K., Kamat, S. G., Schafer, A. I., et al. (1995). Shear-induced platelet aggregation is inhibited by in vivo infusion of an anti-glycoprotein IIb/IIIa antibody fragment, c7E3 Fab, in patients undergoing coronary angioplasty. *Circulation, 91,* 1427–1431.

116. Gawaz, M., Ruf, A., Neumann, F.-J., et al.(1998). Effect of glycoprotein IIb-IIIa receptor antagonism on platelet membrane glycoproteins after coronary stent placement. *Thromb Haemost, 80,* 994–1001.

117. Quinn, M., Deering, A., Stewart, M., et al. (1999). Quantifying GPIIb/IIIa receptor binding using 2 monoclonal antibod-

ies: Discriminating abciximab and small molecular weight antagonists. *Circulation, 99,* 2231–2238.

118. Hezard, N., Metz, D., Nazeyrollas, P., et al. (1999). Free and total platelet glycoprotein IIb/IIIa measurement in whole blood by quantitative flow cytometry during and after infusion of c7E3 Fab in patients undergoing PTCA. *Thromb Haemostas, 81,* 869–873.

119. Liu, C. Z., Hur, B. T., & Huang, T. F. (1996). Measurement of glycoprotein IIb/IIIa blockade by flow cytometry with fluorescein isothiocyanate-conjugated crotavirin, a member of disintegrins. *Thromb Haemost, 76,* 585–591.

120. Tsao, P. W., Bozarth, J. M., Jackson, S. A., et al. (1995). Platelet GPIIb/IIIa receptor occupancy studies using a novel fluoresceinated cyclic Arg-Gly-Asp peptide. *Thromb Res, 77,* 543–556.

121. Wittig, K., Rothe, G., & Schmitz, G. (1998). Inhibition of fibrinogen binding and surface recruitment of GpIIb/IIIa as dose-dependent effects of the RGD-mimetic MK-852. *Thromb Haemost, 79,* 625–630.

122. Mascelli, M. A., Lance, E. T., Damaraju, L., et al. (1998). Pharmacodynamic profile of short-term abciximab treatment demonstrates prolonged platelet inhibition with gradual recovery from GP IIb/IIIa receptor blockade. *Circulation, 97,* 1680–1688.

123. Peter, K., Kohler, B., Straub, A., et al. (2000). Flow cytometric monitoring of glycoprotein IIb/IIIa blockade and platelet function in patients with acute myocardial infarction receiving reteplase, abciximab, and ticlopidine: Continuous platelet inhibition by the combination of abciximab and ticlopidine. *Circulation, 102,* 1490–1496.

124. Jennings, L. K., & White, M. M. (1998). Expression of ligand-induced binding sites on glycoprotein IIb/IIIa complexes and the effect of various inhibitors. *Am Heart J, 135,* S179–S183.

125. Frelinger, A. L., Marchese, P. J., Barnard, M. R., et al. (2002). Flow cytometric assessment of aspirin and NSAID effects on platelet function and leukocyte-platelet aggregation [abstract]. *Blood, 100,* 704a.

126. Frelinger, A. L., Furman, M. I., Linden, M. D., et al. (2004). Residual arachidonic acid–induced platelet octivation via an adenosine diphosphate–dependent but cyclooxygenase and cyclooxygenase–2–independent pathway: A 700–patient study of aspirin resistance. *Circulation, 113,* 2888–2896.

127. Kienast, J., & Schmitz, G. (1990). Flow cytometric analysis of thiazole orange uptake by platelets: A diagnostic aid in the evaluation of thrombocytopenic disorders. *Blood, 75,* 116–121.

128. Bonan, J. L., Rinder, H. M., & Smith, B. R. (1993). Determination of the percentage of thiazole orange (TO)-positive, reticulated platelets using autologous erythrocyte TO fluorescence as an internal standard. *Cytometry, 14,* 690–694.

129. Matic, G. B., Chapman, E. S., Zaiss, M., et al. (1998). Whole blood analysis of reticulated platelets: Improvements of detection and assay stability. *Cytometry, 34,* 229–234.

130. Ault, K. A., Rinder, H. M., Mitchell, J., et al. (1992). The significance of platelets with increased RNA content (reticulated platelets). A measure of the rate of thrombopoiesis. *Am J Clin Pathol, 98,* 637–646.

131. Romp, K. G., Peters, W. P., & Hoffman, M. (1994). Reticulated platelet counts in patients undergoing autologous bone marrow transplantation: An aid in assessing marrow recovery. *Am J Hematol, 46,* 319–324.

132. Rinder, H. M., Schuster, J. E., Rinder, C. S., et al. (1998). Correlation of thrombosis with increased platelet turnover in thrombocytosis. *Blood, 91,* 1288–1294.

133. Robinson, M. S., Mackie, I. J., Khair, K., et al. (1998). Flow cytometric analysis of reticulated platelets: Evidence for a large proportion of non-specific labelling of dense granules by fluorescent dyes. *Br J Haematol, 100,* 351–357.

134. Balduini, C. L., Noris, P., Spedini, P., et al. (1999). Relationship between size and thiazole orange fluorescence of platelets in patients undergoing high-dose chemotherapy. *Br J Haematol, 106,* 202–207.

135. Robinson, M., Machin, S., Mackie, I., et al. (2000). In vivo biotinylation studies: Specificity of labelling of reticulated platelets by thiazole orange and mepacrine. *Br J Haematol, 108,* 859–864.

136. Briggs, C., Kunka, S., Hart, D., et al. (2004). Assessment of an immature platelet fraction (IPF) in peripheral thrombocytopenia. *Br J Haematol, 126,* 93–99.

137. Dumont, L. J., VandenBroeke, T., & Ault, K. A. (1999). Platelet surface P-selectin measurements in platelet preparations: An international collaborative study. Biomedical Excellence for Safer Transfusion (BEST) Working Party of the International Society of Blood Transfusion (ISBT). *Transfus Med Rev, 13*(1), 31–42.

138. Krishnamurti, C., Maglasang, P., & Rothwell, S. W. (1999). Reduction of blood loss by infusion of human platelets in a rabbit kidney injury model. *Transfusion, 39,* 967–974.

139. Holme, S., Sweeney, J. D., & Sawyer, S. (1997). The expression of P-selectin during collection, processing, and storage of platelet concentrates: Relationship to loss of in vivo viability. *Transfusion, 37,* 12–17.

140. Forsberg, B., Jacobsson, S., Stockelberg, D., et al. (1995). The platelet-specific alloantigen PlA1 (HPA-1a): A comparison of flow cytometric immunophenotyping and genotyping using polymerase chain reaction and restriction fragment length polymorphism in a Swedish blood donor population. *Transfusion, 35,* 241–246.

141. Marshall, L. R., Brogden, F. E., Roper, T. S., et al. (1994). Antenatal platelet antibody testing by flow cytometry — results of a pilot study. *Transfusion, 34,* 961–965.

142. Gates, K., & MacPherson, B. R. (1994). Retrospective evaluation of flow cytometry as a platelet crossmatching procedure. *Cytometry, 18,* 123–128.

143. Holme, S., Heaton, A., Kunchuba, A., et al. (1988). Increased levels of platelet associated IgG in patients with thrombocytopenia are not confined to any particular size class of platelets. *Br J Haematol, 68,* 431–436.

144. Tomer, A., Koziol, J., & McMillan, R. (2005). Autoimmune thrombocytopenia: Flow cytometric determination of platelet-associated autoantibodies against platelet-specific receptors. *J Thromb Haemost, 3,* 74–78.

145. Rosenfeld, C. S., & Bodensteiner, D. C. (1986). Detection of platelet alloantibodies by flow cytometry. Characterization and clinical significance. *Am J Clin Pathol, 85,* 207–212.

146. Hagenstrom, H., Schlenke, P., Hennig, H., et al. (2000). Quantification of platelet-associated IgG for differential diagnosis of patients with thrombocytopenia. *Thromb Haemost, 84,* 779–783.

147. Harrison, P., Ault, K. A., Chapman, S., et al. (2001). An interlaboratory study of a candidate reference method for platelet counting. *Am J Clin Pathol, 115,* 448–459.

148. International Council for Standardization in Haematology Expert Panel on Cytometry, International Society of Laboratory Hematology Task Force on Platelet Counting. (2001). Platelet counting by the RBC/platelet ratio method. A reference method. *Am J Clin Pathol, 115,* 460–464.

149. Peng, J., Friese, P., Heilmann, E., et al. (1994). Aged platelets have an impaired response to thrombin as quantitated by P-selectin expression. *Blood, 83,* 161–166.

150. Heilmann, E., Friese, P., Anderson, S., et al. (1993). Biotinylated platelets: A new approach to the measurement of platelet life span. *Br J Haematol, 85,* 729–735.

151. Heilmann, E., Hynes, L. A., Friese, P., et al. (1994). Dog platelets accumulate intracellular fibrinogen as they age. *J Cell Physiol, 161,* 23–30.

152. Dale, G. L., Friese, P., Hynes, L. A., et al. (1995). Demonstration that thiazole-orange-positive platelets in the dog are less than 24 hours old. *Blood, 85,* 1822–1825.

153. Dale, G. L., Wolf, R. F., Hynes, L. A., et al. (1996). Quantitation of platelet life span in splenectomized dogs. *Exp Hematol, 24,* 518–523.

154. Hughes, D. L., Evans, G., Metcalfe, P., et al. (2005). Tracking and characterisation of transfused platelets by two colour, whole blood flow cytometry. *Br J Haematol, 30*(5), 791–794.

155. Celi, A., Pellegrini, G., Lorenzet, R., et al. (1994). P-selectin induces the expression of tissue factor on monocytes. *Proc Natl Acad Sci USA, 91,* 8767–8771.

156. Evangelista, V., Manarini, S., Rotondo, S., et al. (1996). Platelet/polymorphonuclear leukocyte interaction in dynamic conditions: Evidence of adhesion cascade and cross talk between P-selectin and the beta 2 integrin CD11b/CD18. *Blood, 88,* 4183–4194.

157. Evangelista, V., Manarini, S., Sideri, R., et al. (1999). Platelet/polymorphonuclear leukocyte interaction: P-selectin triggers protein-tyrosine phosphorylation-dependent CD11b/CD18 adhesion: Role of PSGL-1 as a signaling molecule. *Blood, 93,* 876–885.

158. Altieri, D. C., & Edgington, T. S. (1988). A monoclonal antibody reacting with distinct adhesion molecules defines a transition in the functional state of the receptor CD11b/CD18 (Mac-1). *J Immunol, 141,* 2656–2660.

159. Hogg, N., Stewart, M. P., Scarth, S. L., et al. (1999). A novel leukocyte adhesion deficiency caused by expressed but nonfunctional beta2 integrins Mac-1 and LFA-1. *J Clin Invest, 103,* 97–106.

160. Altieri, D. C., Morrissey, J. H., & Edgington, T. S. (1988). Adhesive receptor Mac-1 coordinates the activation of factor X on stimulated cells of monocytic and myeloid differentiation: An alternative initiation of the coagulation protease cascade. *Proc Natl Acad Sci USA, 85,* 7462–7466.

161. Altieri, D. C., & Edgington, T. S. (1988). The saturable high affinity association of factor X to ADP-stimulated monocytes defines a novel function of the Mac-1 receptor. *J Biol Chem, 263,* 7007–7015.

162. Altieri, D. C., Bader, R., Mannucci, P. M., et al. (1988). Oligospecificity of the cellular adhesion receptor Mac-1 encompasses an inducible recognition specificity for fibrinogen. *J Cell Biol, 107,* 1893–1900.

163. Rauch, U., Bonderman, D., Bohrmann, B., et al. (2000). Transfer of tissue factor from leukocytes to platelets is mediated by CD15 and tissue factor. *Blood, 96,* 170–175.

164. Barnard, M. R., Linden, M. D., Frelinger, III, A. L., et al. (2005). Effects of platelet binding on whole blood flow cytometry assays of monocyte and neutrophil procoagulant activity. *J Thromb Haemost, 3,* 2563–2570.

165. do Ceu, M. M., Sansonetty, F., Goncalves, M. J., et al. (1999). Flow cytometric kinetic assay of calcium mobilization in whole blood platelets using Fluo-3 and CD41. *Cytometry, 35,* 302–310.

166. Cattaneo, M., Kinlough-Rathbone, R. L., Lecchi, A., et al. (1993). Fibrinogen-independent aggregation and deaggregation of human platelets: Studies of two afibrinogenemic patients. *Blood, 70,* 221–223.

167. LaRosa, C. A., Rohrer, M. J., Benoit, S. E., et al. (1994). Neutrophil cathepsin G modulates the platelet surface expression of the glycoprotein (GP) Ib-IX complex by proteolysis of the von Willebrand factor binding site on GPIbα and by a cytoskeletal-mediated redistribution of the remainder of the complex. *Blood, 84,* 158–168.

168. Schwarz, U. R., Geiger, J., Walter, U., et al. (1999). Flow cytometry analysis of intracellular VASP phosphorylation for the assessment of activating and inhibitory signal transduction pathways in human platelets—definition and detection of ticlopidine/clopidogrel effects. *Thromb Haemost, 82,* 1145–1152.

169. Sims, P. J., Ginsberg, M. H., Plow, E. F., et al. (1991). Effect of platelet activation on the conformation of the plasma membrane glycoprotein IIb-IIIa complex. *J Biol Chem, 266,* 7345–7352.

170. Koksch, M., Rothe, G., Kiefel, V., et al. (1995). Fluorescence resonance energy transfer as a new method for the epitope-specific characterization of anti-platelet antibodies. *J Immunol Methods, 187,* 53–67.

171. Freedman, J. E., Loscalzo, J., Barnard, M. R., et al. (1997). Nitric oxide released from activated platelets inhibits platelet recruitment. *J Clin Invest, 100,* 350–356.

172. Alugupalli, K. R., Michelson, A. D., Barnard, M. R., et al. (2001). Platelet activation by a relapsing fever spirochaete results in enhanced bacterium-platelet interaction via integrin alphaIIbbeta3 activation. *Mol Microbiol, 39,* 330–340.

173. Frelinger, A. L., Cohen, I., Plow, E. F., et al. (1990). Selective inhibition of integrin function by antibodies specific for ligand-occupied receptor conformers. *J Biol Chem, 265,* 6346–6352.

174. Gralnick, H. R., Williams, S. B., McKeown, L., et al. (1992). Endogenous platelet fibrinogen: Its modulation after surface expression is related to size-selective access to and confor-

mational changes in the bound fibrinogen. *Br J Haematol, 80,* 347–357.

175. Schoolmeester, A., Vanhoorelbeke, K., Katsutani, S., et al. (2004). Monoclonal antibody IAC-1 is specific for activated alpha2beta1 and binds to amino acids 199 to 201 of the integrin alpha2 I-domain. *Blood, 104,* 390–396.

176. Larsen, E., Celi, A., Gilbert, G. E., et al. (1989). PADGEM protein: A receptor that mediates the interaction of activated platelets with neutrophils and monocytes. *Cell, 59,* 305–312.

177. Stenberg, P. E., McEver, R. P., Shuman, M. A., et al. (1985). A platelet alpha-granule membrane protein (GMP-140) is expressed on the plasma membrane after activation. *J Cell Biol, 101,* 880–886.

178. Carmody, M. W., Ault, K. A., Mitchell, J. G., et al. (1990). Production of monoclonal antibodies specific for platelet activation antigens and their use in evaluating platelet function. *Hybridoma, 9,* 631–641.

179. Metzelaar, M. J., Heijnen, H. F., Sixma, J. J., et al. (1992). Identification of a 33-Kd protein associated with the alpha-granule membrane (GMP-33) that is expressed on the surface of activated platelets. *Blood, 79,* 372–379.

180. Damas, C., Vink, T., & Nieuwenhuis, H. K. (2001). The 33-kDa platelet alpha-granule membrane protein (GMP-33) is an N-terminal proteolytic fragment of thrombospondin. *Thromb Haemost, 86,* 887–893.

181. Nieuwenhuis, H. K., van Oosterhout, J. J., Rozemuller, E., et al. (1987). Studies with a monoclonal antibody against activated platelets: Evidence that a secreted 53,000-molecular weight lysosome-like granule protein is exposed on the surface of activated platelets in the circulation. *Blood, 70,* 838–845.

182. Febbraio, M., & Silverstein, R. L. (1990). Identification and characterization of LAMP-1 as an activation-dependent platelet surface glycoprotein. *J Biol Chem,* 18531–18537.

183. Silverstein, R. L., & Febbraio, M. (1992). Identification of lysosome-associated membrane protein-2 as an activation-dependent platelet surface glycoprotein. *Blood, 80,* 1470–1475.

184. Henn, V., Slupsky, J. R., Grafe, M., et al. (1998). CD40 ligand on activated platelets triggers an inflammatory reaction of endothelial cells. *Nature, 391,* 591–594.

185. Chen, M., Kakutani, M., Naruko, T., et al. (2001). Activation-dependent surface expression of LOX-1 in human platelets. *Biochem Biophys Res Commun,* 153–158.

186. Boukerche, H., & McGregor, J. L. (1988). Characterization of an anti-thrombospondin monoclonal antibody (P8) that inhibits human blood platelet functions. Normal binding of P8 to thrombin-activated Glanzmann thrombasthenic platelets. *Eur J Biochem, 171,* 383–392.

187. Aiken, M. L., Ginsberg, M. H., & Plow, E. F. (1987). Mechanisms for expression of thrombospondin on the platelet cell surface. *Semin Thromb Hemost, 13,* 307–316.

188. Hayward, C. P., Smith, J. W., Horsewood, P., et al. (1991). p-155, a multimeric platelet protein that is expressed on activated platelets. *J Biol Chem, 266,* 7114–7120.

189. Hayward, C. P., Bainton, D. F., Smith, J. W., et al. (1993). Multimerin is found in the alpha-granules of resting platelets and is synthesized by a megakaryocytic cell line. *J Clin Invest, 91,* 2630–2639.

190. Holme, P. A., Brosstad, F., & Solum, N. O. (1995). Platelet-derived microvesicles and activated platelets express factor Xa activity. *Blood Coagul Fibrinolysis, 6,* 302–310.

# CHAPTER 31

# Thromboxane Generation

**Tilo Grosser, Susanne Fries, and Garret A. FitzGerald**

*Institute for Translational Medicine and Therapeutics, Department of Pharmacology,
School of Medicine, University of Pennsylvania, Philadelphia, Pennsylvania*

## I. Introduction

A short-lived lipid mediator, thromboxane (Tx) A$_2$, is synthesized by activated platelets and released as a local signal, which amplifies activation and recruits additional platelets to the site of clot formation[1,2] TxA$_2$ is also a potent vasoconstrictor and stimulates mitogenesis, accelerating hemostasis and the proliferative response to vascular injury.[3-6] Derived from arachidonic acid (AA) by the sequential action of prostaglandin G/H synthase (PGHS)-1 and thromboxane synthase (TxAS) (Fig. 31-1), TxA$_2$ targets a specific G protein-coupled transmembrane thromboxane receptor (TP) that signals to facilitate dense granule secretion and integrin $\alpha_{IIb}\beta_3$ activation in platelets. Patients deficient in TxA$_2$ formation have a mild bleeding disorder,[7-16] as have those with a defective response to TxA$_2$,[17-19] such as an arginine[60] to leucine mutation of the TP which disrupts receptor-G protein interactions.[20]

Biosynthesis of TxA$_2$ is augmented in clinical syndromes of platelet activation, including unstable angina and myocardial infarction,[21] reperfusion following therapeutic thrombolysis,[22] severe peripheral vascular disease,[23] preeclampsia,[24] pulmonary hypertension,[25] sickle cell anemia,[26] systemic lupus erythematosus,[27] scleroderma,[28] and stroke.[29] Sustained suppression of platelet TxA$_2$ formation is sufficient to explain the reduction by aspirin of myocardial infarction and stroke by approximately 25%.[30]

The cellular capacity to generate TxA$_2$, as is the case with all eicosanoids, greatly exceeds actual biosynthetic rates in humans. This phenomenon constrains, in distinct ways, information which can be derived from measurements of the capacity to generate TxA$_2$, such as its hydrolysis product, TxB$_2$, in serum[31] and indices of TxA$_2$ biosynthesis, such as its major metabolites in plasma or urine. These are discussed in detail below.

## II. Thromboxane Biosynthesis

AA, the 20-carbon unsaturated fatty acid precursor of TxA$_2$, is cleaved from membrane phospholipids at the sn-2 ester binding site by the enzymatic activity of phospholipase A$_2$ (PLA$_2$). Group (G) IVA PLA$_2$, also referred to as 85-kD cytosolic PLA$_2$, is the main isoform expressed in platelets.[32] It possesses an $\alpha/\beta$ hydrolase and a regulatory C2 domain, which binds two Ca$^{2+}$ ions and is thought to mediate translocation of the enzyme from the cytosol to the phospholipid membrane in response to increments in intracellular calcium concentrations.[33] In addition to activation by calcium, a mechanism by which multiple platelet agonists initiate TxA$_2$ generation, GIVA PLA$_2$ can also be regulated by phosphorylation of serine-505,[34] mediated by p38 MAP kinase[35] and perhaps also by ERK1/2.[36] Von Willebrand factor-initiated adhesion of platelets through the glycoprotein (GP) Ib-V-IX complex is thought to stimulate AA release primarily through p38 MAP kinase.[37]

Free AA is acted on by membrane-anchored dimers of PGHS, a hemoprotein with two distinct catalytic activities. The cyclooxygenase (COX) activity, after which the enzyme is commonly named, incorporates two oxygen molecules into AA or alternate polyunsaturated fatty acid substrates, such as linoleic and eicosapentaenoic acid. This reaction forms a labile intermediate peroxide, PGG$_2$, from AA, which is reduced to the corresponding alcohol, PGH$_2$, by the enzyme's hydroperoxidase (HOX) activity (Fig. 31-1, ref. 38). Two evolutionarily conserved[39] PGHS isoforms exist, PGHS-1 and PGHS-2, of which only PGHS-1 is expressed in mature platelets.[40] Megakaryocytes and immature platelet forms released in clinical conditions of markedly accelerated platelet turnover also express PGHS-2,[41,42] but its role in platelet development and function has yet to be elucidated.[43] Inhibitors of the PGHS enzymes, nonsteroi-

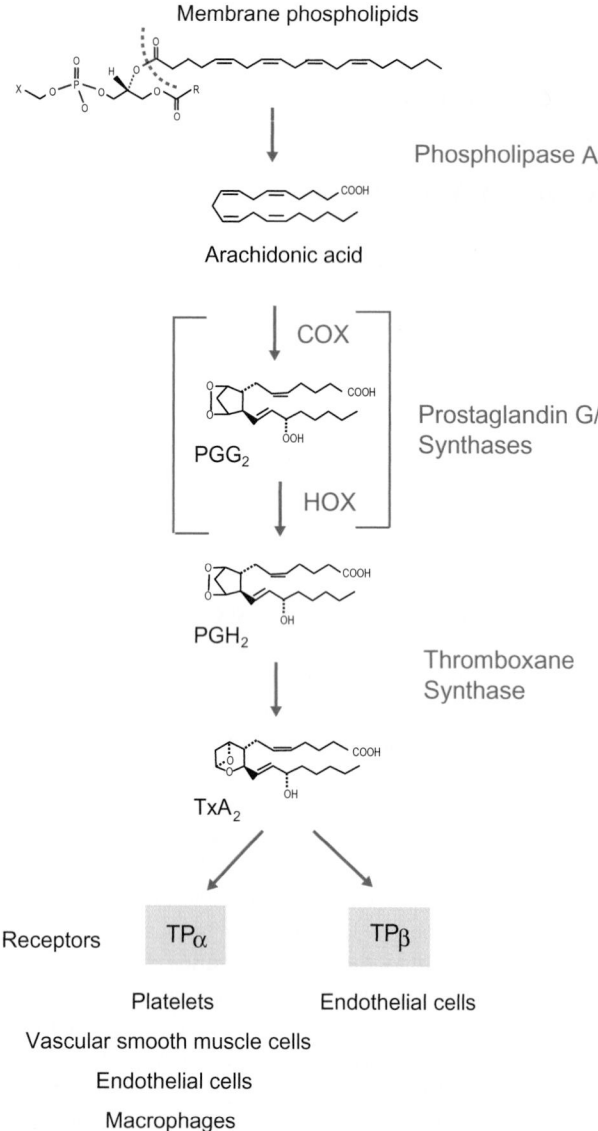

**Figure 31-1.** The biosynthetic pathway of thromboxane $A_2$. Abbreviations: COX, cyclooxygenase; HOX, hydroperoxidase; $PGG_2$, prostaglandin $G_2$; $PGH_2$, prostaglandin $H_2$; $TxA_2$, thromboxane $A_2$.

dal anti-inflammatory drugs (NSAIDs), and aspirin, block only the COX activity and are hence referred to as COX inhibitors. While the enzymatic characteristics of both isoforms are very similar, PGHS-2 is distinguished by a recess in the hydrophobic substrate binding channel. This has allowed the targeted design of molecules with cantilevered side chains that insert tightly into the active center of PGHS-2, but are too bulky to bind PGHS-1 with similar high affinity — NSAIDs selective for inhibition of COX-2 activity, such as the coxibs. As expected, these drugs depress platelet $TxA_2$ formation only marginally at high concentrations and have no effect on platelet function,[44,45] unlike many

traditional (t)NSAIDs, which inhibit both PGHS isoforms, depress $TxA_2$ formation during the dosing interval, and may inhibit platelet function transiently. Both tNSAIDs and NSAIDs selective for PGHS-2 inhibit the enzymes reversibly. Only aspirin acetylates serine-529 in PGHS-1 (or serine-516 in PGHS-2) covalently and inhibits enzymatic activity irreversibly. This unique feature, which sustains inhibition of platelet $TxA_2$ throughout the dosing interval, and the limited capacity of platelets for *de novo* protein synthesis, are thought to render aspirin the only COX inhibitor with proven cardioprotective activity (Chapter 60),[46] as new platelets have to be formed to restore function.

The unstable PGHS endoperoxide product $PGH_2$ is subject to further metabolism by isomerases and synthases which are expressed in a relatively tissue-specific manner. TxAS catalyzes the isomerization reaction of $PGH_2$ to $TxA_2$ in platelets (Fig. 31-1).[47,48] This membrane-bound hemoprotein has been assigned to the cytochrome (CYP) P450 5 superfamily as its sole member. TxAS also catalyzes the conversion of $PGH_2$ to 12-L-hydroxy-5,8,10-heptadecatrienoic acid (HHT) and malondialdehyde (MDA).[49] The 12-lipoxygenase (LOX) product 12-hydroxyeicosatetraenoic acid (HETE) is an even more abundant platelet AA product than $TxA_2$.[50,51] Although deletion of platelet 12-LOX accelerates thrombotic death caused by intravenous adenosine diphosphate (ADP) in mice,[52] its importance in the regulation of human platelet function is, like that of HHT and MDA, unknown.

## III. $TxA_2$ Signaling

The short half-life of $TxA_2$ of about 30 second in aqueous milieux limits its range of activity.[53] Thus, it acts as an autacoid or paracoid close to the site of its formation, rather than as a circulating hormone.[54] This is compatible with the role of $TxA_2$ as a diffusible mediator which is released in high local concentrations to activate the TP receptor on platelets and vascular cells close to the site of injury. Two human carboxyterminal TP splice variants exist, TPα and TPβ, resulting in sequence lengths of 343 and 407 amino residues that diverge distal to the seventh transmembrane domain. Orthologous murine splice variants have not been identified. While the exact biological roles of the two isoforms remain to be defined, their expression is apparently differentially regulated by two distinct promoters within the single human TP gene.[55] Both transcripts are present in human platelets, but only TPα protein has been detected in human platelets.[56] TPβ was first cloned from endothelial cells,[57] and the only functional difference to TPα detected so far is a restraining effect on mediators that stimulate angiogenesis,[58] while TPα apparently accelerates angiogenesis.[59] Differences in signal transduction pathways activated by the two splice variants have been noted,[60] as have differential rates of agonist-induced desensitization.[61] Pharmacological studies[62,63] have discriminated functionally distinct subpopulations among platelet TPs, which are not accounted for by the splice variants. Recent studies of potential relevance have revealed that platelet TPα may heterodimerize with other G protein-coupled receptors, such as the prostacyclin receptor (IP) to modify both agonist affinity and downstream signaling.[64]

TP receptors in various tissues have been shown to couple via $G_q$, $G_{11}$, $G_{12/13}$, and the tissue transglutaminase II, $G_h$, to activate phospholipase C (PLC)-dependent inositol phosphate generation, elevate intracellular calcium, or signal to the MAP kinase pathways.[65,66] TPα via $G_s$ and TPβ via $G_i$ may also increase or decrease, respectively, adenylyl cyclase activity in heterologous expression systems.[67] In platelets, the activated TPα couples predominantly to $G_q$ and $G_{12/13}$.[68,69] $G_q$ stimulates $PLC_β$ to increase intracellular calcium and activates protein kinase C-dependent pathways, which facilitate platelet aggregation, whereas $G_{12}/G_{13}$-mediated Rho/Rho-kinase-dependent regulation of myosin light chain phosphorylation participates in receptor-induced platelet shape change.[70] Both signals converge with responses to other platelet stimuli, which signal through the same G proteins, such as the thrombin receptors PAR1 and PAR4[71] or the ADP receptors $P2Y_1$ and $P2Y_{12}$,[72] consistent with the role of $TxA_2$ as an amplifying mediator.[2]

## IV. The Biological Activity of $TxA_2$

Rapid synthesis of $TxA_2$ during platelet stimulation, which is tightly controlled by GIVA $PLA_2$-dependent release of AA, is a mechanism to amplify the activation response and recruit additional platelets to the site of clotting. Thus, lack of $TxA_2$ signaling does not completely prevent clot formation *in vivo,* but it substantially restrains the aggregation response. Given the potential redundancy in the process of platelet activation and aggregation, low dose aspirin is remarkably effective in the secondary prevention of important vascular events (Chapter 60), but it is unsurprising that some patients suffer thrombotic events while being treated with aspirin. As previously mentioned, mutations in the human $TxA_2$ synthesis/response pathway may result in mild bleeding disorders, such as mucosal bleeding and easy bruising. These patients typically have prolonged bleeding times and defective platelet aggregation responses to AA/$TxA_2$ pathway agonists.[7-19] Mice deficient in $TxAS$[43] or $TP$[73] mimic these coagulation defects and show altered hemodynamic responses to hypertensive stimuli,[73,74] consistent with the role of $TxA_2$ in renal function and vasoconstriction. Studies in mice suggest that $TxA_2$ may contribute to the renovascular response to activation of the renin–angiotensin system.[74,75] $TxA_2$ derived from both platelet and nonplatelet sources may also accelerate remodeling of the blood vessel

wall in response to vascular injury[6] and lesion development during initiation of atherogenesis.[76-78] In this regard, $TxA_2$ is a major product of macrophage PGHS-2.[79,80]

$TxA_2$-induced platelet activation beyond the site of vascular injury is restrained by its limited half life,[81] while the activity of high local concentrations is moderated by rapid desensitization of its receptors.[82] Similarly, endogenous inhibitors of platelet function, including prostacyclin ($PGI_2$) and nitric oxide (NO), limit platelet activation *in vivo* (see Chapter 13).[6,83] Augmented $PGI_2$ biosynthesis in syndromes of platelet activation serves to constrain the effects of $TxA_2$ and other platelet agonists, vasoconstrictors, and stimuli to platelet activation. Indeed, the increase in the incidence of hypertension, myocardial infarction, and stroke in patients receiving the selective inhibitors of COX-2, rofecoxib, celecoxib, and valdecoxib, which is most parsimoniously explained by inhibition of PGHS-2 dependent $PGI_2$ formation, without the restraining effect of concomitant inhibition of platelet $TxA_2$ biosynthesis,[44,45] supports this concept.[84]

The relationship between the capacity of platelets to form $TxA_2$, however, and platelet function is nonlinear (Fig. 31-2).[85] A residual capacity to generate $TxA_2$ of 10% is sufficient to sustain fully $TxA_2$-dependent platelet aggregation.[86] Thus, antiplatelet drugs are thought to require permanent inhibition of more than 95 to 98% of the capacity of platelets to generate $TxA_2$ to afford cardioprotection.[87] So far, aspirin is the only compound to fall into this category throughout a typical internal dosing, as it inactivates platelet PGHS-1 irreversibly. Traditional NSAIDs, with the possible exception of naproxen in some individuals, which inhibit the enzyme reversibly, achieve a level of inhibition of $TxA_2$ synthesis that translates into inhibition of platelet function only transiently during the dosing interval.[88]

Exploration of the clinical efficacy as platelet inhibitors of compounds that inhibit selectively the activity of TxAS and/or block the TP was largely halted when the cardioprotective effect of low dose aspirin was recognized. Development of such drugs was initiated in the 1980s based on the hypothesis that they would cause less gastrointestinal adverse events than aspirin, which were attributed to its inhibition of platelet $TxA_2$ and mucosal $PGE_2$ and $PGI_2$ formation.[89] One cause for the failure to develop selective $TxA_2$ pathway inhibitors was the expected costs of therapy in comparison to generic aspirin. Other causes lay in the human pharmacology of these medications; selective inhibition of TxAS resulted in an accumulation of the $TxA_2$ precursor $PGH_2$, which, exerts via the TP receptor similar biological effects to $TxA_2$, including the activation of platelets. Indeed, TxAS inhibition failed consistently to inhibit platelet activity in human volunteers, despite marked depression of $TxA_2$ biosynthesis.[90] Similarly, selective antagonists of the TP receptor failed to achieve persistent and complete platelet inhibition, because of their potency at the doses used

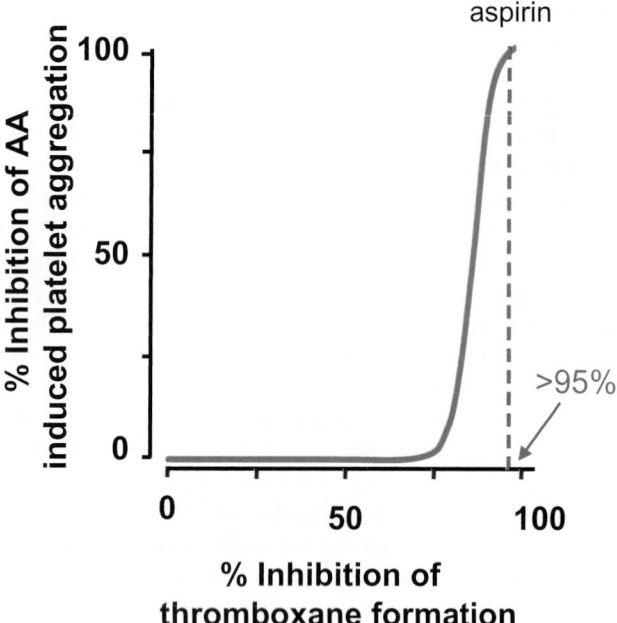

**Figure 31-2.** The nonlinear relationship between inhibition of platelet TxA$_2$ formation and inhibition of platelet function. Inhibition of more than 95 to 98% of the capacity of platelets to generate TxA$_2$ is necessary to afford inhibition of platelet function.[85] Thus, minor changes in the degree of inhibition of TxA$_2$ formation — as occur during the dosing interval of reversible inhibitors of PGHS — may translate into large changes in the degree of platelet inhibition. The cardioprotective effect of aspirin is attributed to its irreversible inactivation of platelet PGHS-1, which affords permanent inhibition of >95% TxA$_2$ formation (dashed line) throughout the dosing interval. (Reprinted with permission from ref. 84.)

and/or the kinetics of their interaction with the TP receptor.[3] It was also noted that inhibition of TxAS diverted a portion of the PGH$_2$ precursor to other prostanoid pathways evident as augmented biosynthesis of PGI$_2$ *in vivo*.[91,92] These observations triggered the development of combined inhibitors of TxAS and the TP receptor to exploit a potentially beneficial effect of elevating PGI$_2$ biosynthesis concurrent with platelet inhibition.[91,93] Randomized controlled trials designed to demonstrate advantages above aspirin therapy in settings of acute myocardial infarction and angioplasty restenosis were performed.[94–96] However, the initial compounds were relatively weak antagonists of the TP receptor and their platelet inhibitory effects were variable throughout the dosing interval and between individuals.[97–101] Development was stopped when these trials failed, and compounds with more favorable pharmacokinetic and dynamic properties have not been developed. However, new interest in TxA$_2$ pathway inhibitors as alternatives to low dose aspirin has been sparked by the perceived incidence of failure of aspirin to sufficiently inhibit platelet function[102] and the recognition that COX-

independent ligands may activate the TP receptor *in vivo* (see also Chapter 65).[103,104]

The incidence of this phenomenon, frequently referred to as aspirin "resistance" or "non-responsiveness," has been estimated to lie between 5 and 50%.[105–108] However, these estimates derive from small patient populations which were classified into responders and nonresponders based on the failure to meet arbitrarily selected response criteria in various platelet function tests. These criteria include alone or in combination. AA-induced platelet aggregation, cartridge-based bedside whole-blood clotting assays, measurement of platelet TxA$_2$ synthesis *ex vivo*, and quantitation of a stable urinary Tx metabolite. Currently, no combination of tests has been validated as reflecting reproducibly clinically important "resistance" that is specific to aspirin. One retrospective study has linked high levels of a urinary TxA$_2$ metabolite, 11 dehydro-TxB$_2$ (see below), in patients on low dose aspirin therapy to elevated cardiovascular risk.[109] Many factors may result in an insufficient pharmacological response to aspirin, and indeed to any medication (assuming the patient actually takes the drug as prescribed — nonadherence is probably the most frequent reason for treatment failure[110]). These include factors that affect bioavailability, interactions with other medications or nutritional compounds, and genetic variation in the biological target. Currently, no genetic variants of PGHS-1 have been linked to impaired inhibition of TxA$_2$ formation by aspirin. Traditional NSAIDs, which bind to both platelet PGHS-1 and PGHS-2, ibuprofen[88] naproxen,[111] and indomethacin,[112] may interfere with the capacity of aspirin to inhibit TxA$_2$ formation, most likely by impeding access of aspirin to the active site of PGHS-1.[88] Such a mechanism may not be restricted to these compounds and it would be expected to translate into clinical nonresponsiveness to aspirin — a lack of reduction of myocardial infarction and stroke.[113–115] Research is needed to identify criteria which reproducibly define resistance to aspirin, to assess its incidence in large populations, and to link these criteria to clinical outcome.

## V. Thromboxane Metabolism

The TxA$_2$ molecule is characterized by a tense bicyclic structure in which one carbon atom binds two oxygens. It is particularly susceptible to nucleophilic attack by H$_2$O, resulting in rapid hydrolysis (T$_{1/2}$ approximately 30 second) in biological fluids to TxB$_2$, a stable, biologically inactive compound. TxB$_2$ itself is rapidly cleared from the circulation with an estimated initial half-life of 7 minute and a plasma clearance of 52 mL × min$^{-1}$ × kg$^{-1}$ following infusion of TxB$_2$.[54] TxB$_2$ is extensively metabolized and only approximately 2.5% of TxB$_2$ is excreted unchanged in human urine

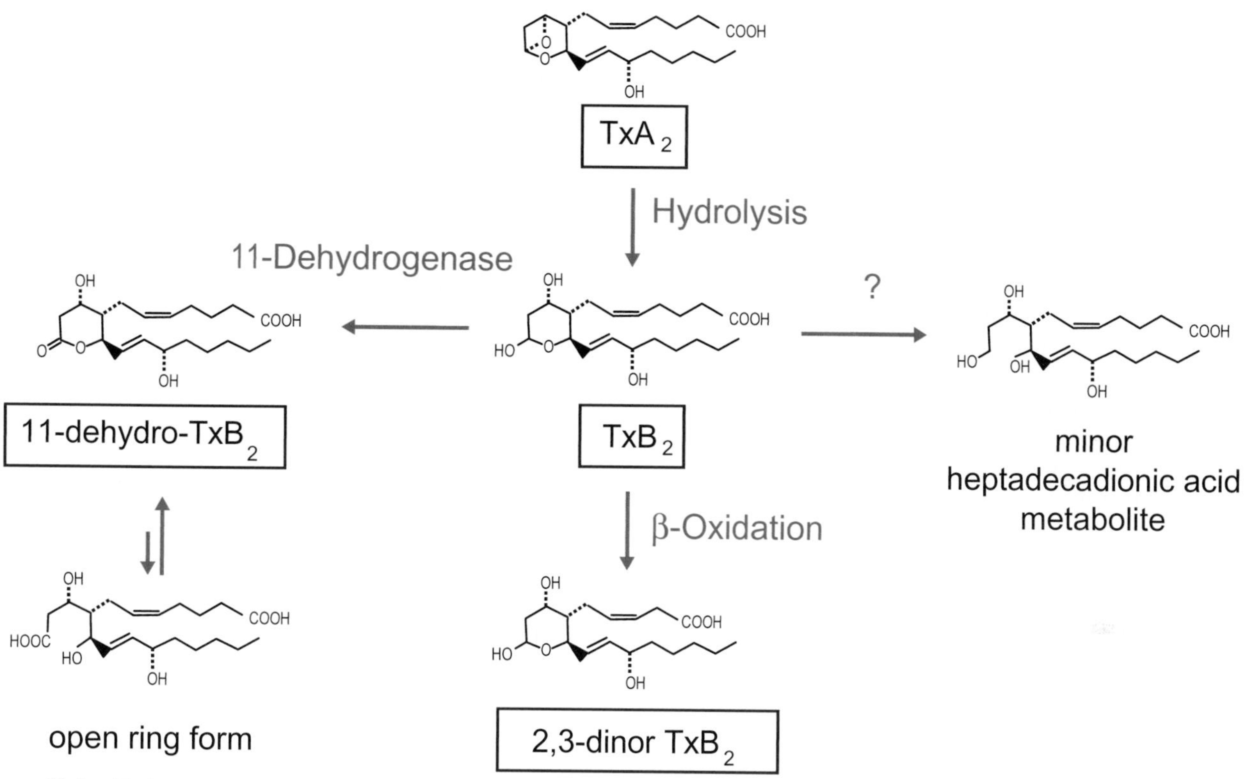

**Figure 31-3.** TxA$_2$ metabolism *in vivo*.

(Fig. 31-3).[81] Injection of radiolabeled TxB$_2$ into a human volunteer identified two major pathways of metabolism. One metabolic route results in the formation of the β-oxidation products, 2,3-dinor TxB$_2$ (23% of administered TxB$_2$), and 2,3,4,5 tetranor TxB$_2$ (5.3%) retaining the original TxB$_2$ hemiacetal ring.[81] The second includes dehydrogenation of the C-11 hemiacetal ring group and formation of a series of molecules with a δ-lactone ring structure, of which 11-dehydro TxB$_2$ is the major urinary product.[81] 11-dehydro TxB$_2$ itself undergoes extensive degradation by β and W oxidation, dehydrogenation of the C-15 alcohol group, and reduction of the $\Delta^5$ and $\Delta^{13}$ double bonds. Together, the δ-lactone products account for approximately one third of the urinary TxB$_2$ metabolites.[81] Subsequent development of quantitative methodology for the measurements of 2,3-dinor TxB$_2$ and 11-dehydro TxB$_2$ established the latter as the more abundant metabolite in human urine, while the former was more abundant in mice (see below). Interestingly, evidence has emerged that 11-dehydro-TxB$_2$ is not biologically inactive, but may act as an agonist of the PGD$_2$ receptor, DP2 (also named CRTH2). Activation of DP2 induces chemotaxis of eosinophils, basophils, and Th2 type T-cells.[116] While the DP2 receptor affinity for 11-dehydro TxB$_2$ is lower than for PGD$_2$, such activity may be relevant to the effects of TxA$_2$ in inflammation and immunity.[117] Finally, initial reduction of the hemiacetal ring represents a third minor route of TxB$_2$ metabolism (Fig. 31-3).[81]

## VI. Monitoring of TxA$_2$ Metabolites

Quantitation of TxB$_2$ by immunoassays[118] or mass spectrometry[119] permits assessment of the capacity of platelets to form TxA$_2$ *ex vivo*. Variations on this approach include measurement in serum derived from whole blood allowed to clot at 37°C for 30 minutes[120] or in platelet-rich plasma after standardized approaches to platelet activation.[121] These approaches provide essentially similar information and have been extremely useful in defining aspirin's dose-dependent inhibition of the capacity of platelets to form TxA$_2$[31,120] and validating the use of low doses of aspirin in cardiovascular prophylaxis.[89] They have also been used to discriminate the irreversible impact of aspirin on platelet TxA$_2$ formation compared to the transient impact of tNSAIDs, such as ibuprofen[88] or the more prolonged inhibition in some individuals receiving naproxen.[122] While TxB$_2$ proved to be a rational and tractable analyte in such capacity-related indices, its widespread analysis as an index of actual *in vivo* biosynthesis of TxA$_2$ proved fallible. Platelets have the capacity to form TxA$_2$ in concentrations of approximately 300 to 400 ng/mL blood,[31] while maximal endogenous concentrations have been estimated to be three orders of magnitude lower, in the range of 1 to 2 pg/mL.[54] Thus, because of the large discrepancy between capacity and actual biosynthesis, the variable contribution of *ex vivo* platelet activation during blood sampling to TxB$_2$ levels proved to be an insurmountable source of artifact.[123] Measurement of enzymatic metabolites, themselves not formed from TxB$_2$ in whole blood, bypassed this problem.

The two major TxB$_2$ metabolites, 2,3-dinor TxB$_2$ and 11-dehydro TxB$_2$, have been validated as such indices of TxA$_2$ biosynthesis *in vivo*.[54,123–125] Both quantitative immunoassays[125–127] and standardized stable isotope dilution mass spectrometry methods[128,129] have been developed. While initial detection by gas chromatography-mass spectrometry required a number of chemical derivatization and thin layer chromatography purification steps,[130] instrumentation has become sufficiently sensitive to assay the TxB$_2$ metabolites by liquid chromatography electrospray ionization tandem mass spectrometry (LC-ESI-MS/MS), reducing sample preparation to a single solid phase extraction.[131] Advantages of LC-ESI-MS/MS techniques include their excellent specificity and sensitivity. Advantages of commercial or noncommercial immunoassays include their high throughput and low costs per sample.

The half-life of 11-dehydro TxB$_2$ in the circulation is markedly longer (approximately 60 minute) than that of 2,3-dinor TxB$_2$ (15–17 min) and it is more abundant in urine.[129,132] Thus, 11-dehydro TxB$_2$ is thought to represent a more time-integrated — and perhaps less variable — index of TxA$_2$ formation *in vivo* than 2,3-dinor TxB$_2$.[132] Because neither metabolite is formed in the kidney, their detection in urine affords a noninvasive measure of systemic TxA$_2$ formation that is largely, but not exclusively,[133] reflective of platelet activation. Studies with low dose aspirin suggest that roughly 30% of TxA$_2$ urinary metabolites derive from extraplatelet sources,[133] although this contribution may change under pathological conditions. Clinical situations in which extraplatelet sources of TxA$_2$ may gain relevance would be expected to include particularly inflammatory conditions involving the accumulation of monocytes/macrophages.[79,80] Another such condition is the degranulation of mast cells in mastocytosis patients in whom urinary 11-dehydro TxB$_2$ is augmented in the absence of platelet activation.[134]

Quantitation of urinary Tx metabolites has permitted the assessment of platelet function *in vivo* in a number of diseases[21–29] and the clinical development of platelet inhibitors.[31,120] More rapid and sensitive detection afforded by novel mass spectrometry instrumentation, combined with accurate test protocols, would be expected to allow better assessment of drug effects in individuals within clinical trials and perhaps ultimately personalized therapy with platelet inhibitors.

# References

1. Hamberg, M., Svensson, J., & Samuelsson, B. (1975). Thromboxanes: A new group of biologically active compounds derived from prostaglandin endoperoxides. *Proc Natl Acad Sci USA, 72*, 2994–2998.

2. FitzGerald, G. A. (1991). Mechanisms of platelet activation: Thromboxane A$_2$ as an amplifying signal for other agonists. *Am J Cardiol, 68*, 11B–15B.

3. Saussy, D. L., Jr., Mais, D. E., Knapp, D. R., et al. (1985). Thromboxane A$_2$ and prostaglandin endoperoxide receptors in platelets and vascular smooth muscle. *Circulation, 72*, 1202–1207.

4. Ogletree, M. L., Smith, J. B., & Lefer, A. M. (1978). Actions of prostaglandins on isolated perfused cat coronary arteries. *Am J Physiol, 235*, H400–H406.

5. Grosser, T., Zucker, T. P., Weber, A. A., et al. (1997). Thromboxane A$_2$ induces cell signaling but requires platelet-derived growth factor to act as a mitogen. *Eur J Pharmacol, 319*, 327–332.

6. Cheng, Y., Austin, S. C., Rocca, B., et al. (2002). Role of prostacyclin in the cardiovascular response to thromboxane A$_2$. *Science, 296*, 539–541.

7. Malmsten, C., Hamberg, M., Svensson, J., et al. (1975). Physiological role of an endoperoxide in human platelets: Hemostatic defect due to platelet cyclo-oxygenase deficiency. *Proc Natl Acad Sci USA, 72*, 1446–1450.

8. Lagarde, M., Byron, P. A., Vargaftig, B. B., et al. (1978). Impairment of platelet thromboxane A$_2$ generation and of the platelet release reaction in two patients with congenital deficiency of platelet cyclo-oxygenase. *Br J Haematol, 38*, 251–266.

9. Nyman, D., Eriksson, A. W., Lehmann, W., et al. (1979). Inherited defective platelet aggregation with arachidonate as the main expression of a defective metabolism of arachidonic acid. *Thromb Res, 14*, 739–746.

10. Pareti, F. I., Mannucci, P. M., D'Angelo, A., et al. (1980). Congenital deficiency of thromboxane and prostacyclin. *Lancet, 1*, 898–901.

11. Wu, K. K., Le Breton, G. C., Tai, H. H., et al. (1981). Abnormal platelet response to thromboxane A$_2$. *J Clin Invest, 67*, 1801–1804.

12. Horellou, M. H., Lecompte, T., Lecrubier, C., et al. (1983). Familial and constitutional bleeding disorder due to platelet cyclo-oxygenase deficiency. *Am J Hematol, 14*, 1–9.

13. Ehara, H., Yoshimoto, T., Yamamoto, S., et al. (1988). Enzymological and immunological studies on a clinical case of platelet cyclooxygenase abnormality. *Biochim Biophys Acta, 960*, 35–42.

14. Nakanishi, K., Ikeda, K., Hato, T., et al. (1984). Platelet cyclo-oxygenase deficiency in a Japanese. *Scand J Haematol, 32*, 167–174.

15. Matijevic-Aleksic, N., McPhedran, P., & Wu, K. K. (1996). Bleeding disorder due to platelet prostaglandin H synthase-1 (PGHS-1) deficiency. *Br J Haematol, 92*, 212–217.

16. Dube, J. N., Drouin, J., Aminian, M., et al. (2001). Characterization of a partial prostaglandin endoperoxide H synthase-1 deficiency in a patient with a bleeding disorder. *Br J Haematol, 113*, 878–885.

17. Wu, K. K., Minkoff, I. M., Rossi, E. C., et al. (1981). Hereditary bleeding disorder due to a primary defect in platelet release reaction. *Br J Haematol, 47*, 241–249.

18. Defreyn, G., Machin, S. J., Carreras, L. O., et al. (1981). Familial bleeding tendency with partial platelet thromboxane synthetase deficiency: Reorientation of cyclic endoperoxide metabolism. *Br J Haematol, 49,* 29–41.

19. Fuse, I., Mito, M., Hattori, A., et al. (1993). Defective signal transduction induced by thromboxane $A_2$ in a patient with a mild bleeding disorder: Impaired phospholipase C activation despite normal phospholipase A2 activation. *Blood, 81,* 994–1000.

20. Hirata, T., Kakizuka, A., Ushikubi, F., et al. (1994). Arg60 to Leu mutation of the human thromboxane $A_2$ receptor in a dominantly inherited bleeding disorder. *J Clin Invest, 94,* 1662–1667.

21. Fitzgerald, D. J., Roy, L., Catella, F., et al. (1986). Platelet activation in unstable coronary disease. *N Engl J Med, 315,* 983–989.

22. Fitzgerald, D. J., Catella, F., Roy, L., et al. (1988). Marked platelet activation in vivo after intravenous streptokinase in patients with acute myocardial infarction. *Circulation, 77,* 142–150.

23. Reilly, I. A., Doran, J. B., Smith, B., et al. (1986). Increased thromboxane biosynthesis in a human preparation of platelet activation: Biochemical and functional consequences of selective inhibition of thromboxane synthase. *Circulation, 73,* 1300–1309.

24. Fitzgerald, D. J., Rocki, W., Murray, R., et al. (1990). Thromboxane $A_2$ synthesis in pregnancy-induced hypertension. *Lancet, 335,* 751–754.

25. Christman, B. W., McPherson, C. D., Newman, J. H., et al. (1992). An imbalance between the excretion of thromboxane and prostacyclin metabolites in pulmonary hypertension. *N Engl J Med, 327,* 70–75.

26. Foulon, I., Bachir, D., Galacteros, F., et al. (1993). Increased in vivo production of thromboxane in patients with sickle cell disease is accompanied by an impairment of platelet functions to the thromboxane $A_2$ agonist U46619. *Arterioscler Thromb, 13,* 421–426.

27. Pierucci, A., Simonetti, B. M., Pecci, G., et al. (1989). Improvement of renal function with selective thromboxane antagonism in lupus nephritis. *N Engl J Med, 320,* 421–425.

28. Reilly, I. A., Roy, L., & FitzGerald, G. A. (1986). Biosynthesis of thromboxane in patients with systemic sclerosis and Raynaud's phenomenon. *Br Med J (Clin Res Ed), 292,* 1037–1039.

29. Koudstaal, P. J., Ciabattoni, G., van Gijn, J., et al. (1993). Increased thromboxane biosynthesis in patients with acute cerebral ischemia. *Stroke, 24,* 219–223.

30. Antithrombolic Group (2002). Collaborative meta-analysis of randomised trials of antiplatelet therapy for prevention of death, myocardial infarction, and stroke in high risk patients. *Br Med J, 324,* 71–86.

31. Patrono, C., Ciabattoni, G., Pinca, E., et al. (1980). Low dose aspirin and inhibition of thromboxane $B_2$ production in healthy subjects. *Thromb Res, 17,* 317–327.

32. Kramer, R. M., Checani, G. C., Deykin, A., et al. (1986). Solubilization and properties of $Ca^{2+}$-dependent human platelet phospholipase A2. *Biochim Biophys Acta, 878,* 394–403.

33. Channon, J. Y., & Leslie, C. C. (1990). A calcium-dependent mechanism for associating a soluble arachidonoyl-hydrolyzing phospholipase A2 with membrane in the macrophage cell line RAW 264.7. *J Biol Chem, 265,* 5409–5413.

34. Lin, L. L., Wartmann, M., Lin, A. Y., et al. (1993). cPLA2 is phosphorylated and activated by MAP kinase. *Cell, 72,* 269–278.

35. Kramer, R. M., Roberts, E. F., Um, S. L., et al. (1996). p38 mitogen-activated protein kinase phosphorylates cytosolic phospholipase A2 (cPLA2) in thrombin-stimulated platelets. Evidence that proline-directed phosphorylation is not required for mobilization of arachidonic acid by cPLA2. *J Biol Chem, 271,* 27723–27729.

36. Garcia, A., Quinton, T. M., Dorsam, R. T., et al. (2005). Src family kinase and Erk-mediated thromboxane $A_2$ generation is essential for VWF/GPIb-induced fibrinogen receptor activation in human platelets. *Blood, 106,* 3410–3414.

37. Canobbio, I., Reineri, S., Sinigaglia, F., et al. (2004). A role for p38 MAP kinase in platelet activation by von Willebrand factor. *Thromb Haemost, 91,* 102–110.

38. Kulmacz, R. J., van der Donk, W. A., & Tsai, A. L. (2003). Comparison of the properties of prostaglandin H synthase-1 and -2. *Prog Lipid Res, 42,* 377–404.

39. Grosser, T., Yusuff, S., Cheskis, E., et al. (2002). Developmental expression of functional cyclooxygenases in zebrafish. *Proc Natl Acad Sci USA, 99,* 8418–8423.

40. Patrignani, P., Sciulli, M. G., Manarini, S., et al. (1999). COX-2 is not involved in thromboxane biosynthesis by activated human platelets. *J Physiol Pharmacol, 50,* 661–667.

41. Rocca, B., Secchiero, P., Ciabattoni, G., et al. (2002). Cyclooxygenase-2 expression is induced during human megakaryopoiesis and characterizes newly formed platelets. *Proc Natl Acad Sci USA, 99,* 7634–7639.

42. Zimmermann, N., Wenk, A., Kim, U., et al. (2003). Functional and biochemical evaluation of platelet aspirin resistance after coronary artery bypass surgery. *Circulation, 108,* 542–547.

43. Yu, I. S., Lin, S. R., Huang, C. C., et al. (2004). TXAS-deleted mice exhibit normal thrombopoiesis, defective hemostasis, and resistance to arachidonate-induced death. *Blood, 104,* 135–142.

44. McAdam, B. F., Catella-Lawson, F., Mardini, I. A., et al. (1999). Systemic biosynthesis of prostacyclin by cyclooxygenase (COX)-2: The human pharmacology of a selective inhibitor of COX-2. *Proc Natl Acad Sci USA, 96,* 272–277.

45. Catella-Lawson, F., McAdam, B., Morrison, B. W., et al. (1999). Effects of specific inhibition of cyclooxygenase-2 on sodium balance, hemodynamics, and vasoactive eicosanoids. *J Pharmacol Exp Ther, 289,* 735–741.

46. Patrono, C., Coller, B., Dalen, J. E., et al. (2001). Platelet-active drugs: The relationships among dose, effectiveness, and side effects. *Chest, 119*(1 Suppl), 39S–63S.

47. Needleman, P., Moncada, S., Bunting, S., et al. (1976). Identification of an enzyme in platelet microsomes which generates thromboxane $A_2$ from prostaglandin endoperoxides. *Nature, 261,* 558–560.

48. Hsu, P. Y., Tsai, A. L., Kulmacz, R. J., et al. (1999). Expression, purification, and spectroscopic characterization of human thromboxane synthase. *J Biol Chem, 274,* 762–769.

49. Diczfalusy, U., Falardeau, P., & Hammarstrom, S. (1977). Conversion of prostaglandin endoperoxides to C17-hydroxy acids catalyzed by human platelet thromboxane synthase. *FEBS Lett, 84,* 271–274.

50. Marcus, A. J., Safier, L. B., Ullman, H. L., et al. (1984). 12S,20-dihydroxyicosatetraenoic acid: A new icosanoid synthesized by neutrophils from 12S-hydroxyicosatetraenoic acid produced by thrombin- or collagen-stimulated platelets. *Proc Natl Acad Sci USA, 81,* 903–907.

51. Brash, A. R. (2001). Arachidonic acid as a bioactive molecule. *J Clin Invest, 107,* 1339–1345.

52. Johnson, E. N., Brass, L. F., & Funk, C. D. (1998). Increased platelet sensitivity to ADP in mice lacking platelet-type 12-lipoxygenase. *Proc Natl Acad Sci USA, 95,* 3100–3105.

53. Fitzpatrick, F. A., Bundy, G. L., Gorman, R. R., et al. (1978). 9,11-Epoxyiminoprosta-5,13-dienoic acid is a thromboxane A$_2$ antagonist in human platelets. *Nature, 275,* 764–766.

54. Patrono, C., Ciabattoni, G., Pugliese, F., et al. (1986). Estimated rate of thromboxane secretion into the circulation of normal humans. *J Clin Invest, 77,* 590–594.

55. Coyle, A. T., Miggin, S. M., & Kinsella, B. T. (2002). Characterization of the 5′ untranslated region of alpha and beta isoforms of the human thromboxane A$_2$ receptor (TP). Differential promoter utilization by the TP isoforms. *Eur J Biochem, 269,* 4058–4073.

56. Habib, A., FitzGerald, G. A., & Maclouf, J. (1999). Phosphorylation of the thromboxane receptor alpha, the predominant isoform expressed in human platelets. *J Biol Chem, 274,* 2645–2651.

57. Raychowdhury, M. K., Yukawa, M., Collins, L. J., et al. (1995). Alternative splicing produces a divergent cytoplasmic tail in the human endothelial thromboxane A$_2$ receptor. *J Biol Chem, 270,* 7011.

58. Ashton, A. W., Cheng, Y., Helisch, A., et al. (2004). Thromboxane A$_2$ receptor agonists antagonize the proangiogenic effects of fibroblast growth factor-2: Role of receptor internalization, thrombospondin-1, and alpha(v)beta3. *Circ Res, 94,* 735–742.

59. Daniel, T. O., Liu, H., Morrow, J. D., et al. (1999). Thromboxane A$_2$ is a mediator of cyclooxygenase-2-dependent endothelial migration and angiogenesis. *Cancer Res, 59,* 4574–4577.

60. Vezza, R., Habib, A., & FitzGerald, G. A. (1999). Differential signaling by the thromboxane receptor isoforms via the novel GTP-binding protein, Gh. *J Biol Chem, 274,* 12774–12779.

61. Parent, J. L., Labrecque, P., Driss Rochdi, M., et al. (2001). Role of the differentially spliced carboxyl terminus in thromboxane A$_2$ receptor trafficking: Identification of a distinct motif for tonic internalization. *J Biol Chem, 276,* 7079–7085.

62. Dorn, G. W., 2nd (1989). Distinct platelet thromboxane A$_2$/prostaglandin H$_2$ receptor subtypes. A radioligand binding study of human platelets. *J Clin Invest, 84,* 1883–1891.

63. Takahara, K., Murray, R., FitzGerald, G. A., et al. (1990). The response to thromboxane A$_2$ analogues in human platelets. Discrimination of two binding sites linked to distinct effector systems. *J Biol Chem, 265,* 6836–6844.

64. Wilson, S. J., Roche, A. M., Kostetskaia, E., et al. (2004). Dimerization of the human receptors for prostacyclin and thromboxane facilitates thromboxane receptor-mediated cAMP generation. *J Biol Chem, 279,* 53036–53047.

65. Narumiya, S., Sugimoto, Y., & Ushikubi, F. (1999). Prostanoid receptors: structures, properties, and functions. *Physiol Rev, 79,* 1193–1226.

66. Hata, A. N., & Breyer, R. M. (2004). Pharmacology and signaling of prostaglandin receptors: Multiple roles in inflammation and immune modulation. *Pharmacol Ther, 103,* 147–166.

67. Hirata, T., Ushikubi, F., Kakizuka, A., et al. (1996). Two thromboxane A$_2$ receptor isoforms in human platelets. Opposite coupling to adenylyl cyclase with different sensitivity to Arg60 to Leu mutation. *J Clin Invest, 97,* 949–956.

68. Djellas, Y., Manganello, J. M., Antonakis, K., et al. (1999). Identification of Galpha13 as one of the G-proteins that couple to human platelet thromboxane A$_2$ receptors. *J Biol Chem, 274,* 14325–14330.

69. Klages, B., Brandt, U., Simon, M. I., et al. (1999). Activation of G12/G13 results in shape change and Rho/Rho-kinase-mediated myosin light chain phosphorylation in mouse platelets. *J Cell Biol, 144,* 745–754.

70. Moers, A., Wettschureck, N., & Offermanns, S. (2004). G13-mediated signaling as a potential target for antiplatelet drugs. *Drug News Perspect, 17,* 493–498.

71. Coughlin, S. R. (2000). Thrombin signalling and protease-activated receptors. *Nature, 407,* 258–264.

72. Andre, P., Delaney, S. M., LaRocca, T., et al. (2003). P2Y12 regulates platelet adhesion/activation, thrombus growth, and thrombus stability in injured arteries. *J Clin Invest, 112,* 398–406.

73. Thomas, D. W., Mannon, R. B., Mannon, P. J., et al. (1998). Coagulation defects and altered hemodynamic responses in mice lacking receptors for thromboxane A$_2$. *J Clin Invest, 102,* 1994–2001.

74. Francois, H., Athirakul, K., Mao, L., et al. (2004). Role for thromboxane receptors in angiotensin-II-induced hypertension. *Hypertension, 43,* 364–369.

75. Athirakul, K., Kim, H. S., Audoly, L. P., et al. (2001). Deficiency of COX-1 causes natriuresis and enhanced sensitivity to ACE inhibition. *Kidney Int, 60,* 2324–2329.

76. Cyrus, T., Sung, S., Zhao, L., et al. (2002). Effect of low-dose aspirin on vascular inflammation, plaque stability, and atherogenesis in low-density lipoprotein receptor-deficient mice. *Circulation, 106,* 1282–1287.

77. Kobayashi, T., Tahara, Y., Matsumoto, M., et al. (2004). Roles of thromboxane A$_2$ and prostacyclin in the development of atherosclerosis in apoE-deficient mice. *J Clin Invest, 114,* 784–794.

78. Egan, K. M., Wang, M., Lucitt, M. B., et al. (2005). Cyclooxygenases, thromboxane, and atherosclerosis: Plaque destabilization by cyclooxygenase-2 inhibition combined

with thromboxane receptor antagonism. *Circulation, 111,* 334–342.

79. Brune, K., Glatt, M., Kalin, H., et al. (1978). Pharmacological control of prostaglandin and thromboxane release from macrophages. *Nature, 274,* 261–263.
80. Hla, T., & Neilson, K. (1992). Human cyclooxygenase-2 cDNA. *Proc Natl Acad Sci USA, 89,* 7384–7388.
81. Roberts, L. J., 2nd, Sweetman, B. J., & Oates, J. A. (1981). Metabolism of thromboxane B2 in man. Identification of 20 urinary metabolites. *J Biol Chem, 256,* 8384–8393.
82. Liel, N., Mais, D. E., & Halushka, P. V. (1988). Desensitization of platelet thromboxane A$_2$/prostaglandin H$_2$ receptors by the mimetic U46619. *J Pharmacol Exp Ther, 247,* 1133–1138.
83. FitzGerald, G. A., & Patrono, C. (2001). The coxibs, selective inhibitors of cyclooxygenase-2. *N Engl J Med, 345,* 433–442.
84. Grosser, T., Fries, S., & FitzGerald, G. A. (2006). Biological basis for the cardiovascular consequences of COX-2 inhibition: Therapeutic challenges and opportunities. *J Clin Invest, 116,* 4–15.
85. Reilly, I. A., & FitzGerald, G. A. (1983). Inhibition of thromboxane formation in vivo and ex vivo: Implications for therapy with platelet inhibitory drugs. *Blood, 69,* 180–186.
86. Di Minno, G., Silver, M. J., & Murphy, S. (1983). Monitoring the entry of new platelets into the circulation after ingestion of aspirin. *Blood, 61,* 1081–1085.
87. Patrono, C., Coller, B., FitzGerald, G. A., et al. (2004). Platelet-active drugs: The relationships among dose, effectiveness, and side effects: The Seventh ACCP Conference on Antithrombotic and Thrombolytic Therapy. *Chest, 126*(3 Suppl), 234S–264S.
88. Catella-Lawson, F., Reilly, M. P., Kapoor, S. C., et al. (2001). Cyclooxygenase inhibitors and the antiplatelet effects of aspirin. *N Engl J Med, 345,* 1809–1817.
89. Patrono, C. (1994). Aspirin as an antiplatelet drug. *N Engl J Med, 330,* 1287–1294.
90. Reilly, I. A., Doran, J. B., Smith, B., et al. (1986). Increased thromboxane biosynthesis in a human preparation of platelet activation: Biochemical and functional consequences of selective inhibition of thromboxane synthase. *Circulation, 73,* 1300–1309.
91. Fitzgerald, D. J., Fragetta, J., & FitzGerald, G. A. (1988). Prostaglandin endoperoxides modulate the response to thromboxane synthase inhibition during coronary thrombosis. *J Clin Invest, 82,* 1708–1713.
92. Morio, H., Hirai, A., Terano, T., et al. (1993). Effect of the infusion of OKY-046, a thromboxane A$_2$ synthase inhibitor, on urinary metabolites of prostacyclin and thromboxane A$_2$ in healthy human subjects. *Thromb Haemost, 69,* 276–281.
93. Gresele, P., Arnout, J., Deckmyn, H., et al. (1987). Role of proaggregatory and antiaggregatory prostaglandins in hemostasis. Studies with combined thromboxane synthase inhibition and thromboxane receptor antagonism. *J Clin Invest, 80,* 1435–1445.
94. RAPT Investigations (1994). Randomized trial of ridogrel, a combined thromboxane A$_2$ synthase inhibitor and thromboxane A$_2$/prostaglandin endoperoxide receptor antagonist, versus aspirin as adjunct to thrombolysis in patients with acute myocardial infarction. The Ridogrel Versus Aspirin Patency Trial (RAPT). *Circulation, 89,* 588–595.
95. van der Wieken, L. R., Simoons, M. L., Laarman, G. J., et al. (1995). Ridogrel as an adjunct to thrombolysis in acute myocardial infarction. *Int J Cardiol, 52,* 125–134.
96. Balsano, F., & Violi, F. (1993). Effect of picotamide on the clinical progression of peripheral vascular disease. A double-blind placebo-controlled study. The ADEP Group. *Circulation, 87,* 1563–1569.
97. Gresele, P., Deckmyn, H., Arnout, J., et al. (1989). Characterization of N,N′-bis(3-picolyl)-4-methoxy-isophtalamide (picotamide) as a dual thromboxane synthase inhibitor/thromboxane A$_2$ receptor antagonist in human platelets. *Thromb Haemost, 61,* 479–484.
98. Modesti, P. A., Cecioni, I., Colella, A., et al. (1994). Binding kinetics and antiplatelet activities of picotamide, a thromboxane A$_2$ receptor antagonist. *Br J Pharmacol, 112,* 81–86.
99. Hoet, B., Falcon, C., De Reys, S., et al. (1990). R68070, a combined thromboxane/endoperoxide receptor antagonist and thromboxane synthase inhibitor, inhibits human platelet activation in vitro and in vivo: A comparison with aspirin. *Blood, 75,* 646–653.
100. Hoet, B., Arnout, J., Van Geet, C., et al. (1990). Ridogrel, a combined thromboxane synthase inhibitor and receptor blocker, decreases elevated plasma beta-thromboglobulin levels in patients with documented peripheral arterial disease. *Thromb Haemost, 64,* 87–90.
101. Guth, B. D., Narjes, H., Schubert, H. D., et al. (2004). Pharmacokinetics and pharmacodynamics of terbogrel, a combined thromboxane A$_2$ recep-tor and synthase inhibitor, in healthy subjects. *Br J Clin Pharmacol, 58,* 40–51.
102. Neri Serneri, G. G., Coccheri, S., Marubini, E., et al. (2004). Picotamide, a combined inhibitor of thromboxane A$_2$ synthase and receptor, reduces 2-year mortality in diabetics with peripheral arterial disease: The DAVID study. *Eur Heart J, 25,* 1845–1852.
103. Audoly, L. P., Rocca, B., Fabre, J. E., et al. (2000). Cardiovascular responses to the isoprostanes iPF(2alpha)-III and iPE(2)-III are mediated via the thromboxane A$_2$ receptor in vivo. *Circulation, 101,* 2833–2840.
104. Tang, M., Cyrus, T., Yao, Y., et al. (2005). Involvement of thromboxane receptor in the proatherogenic effect of isoprostane F$_2$alpha-III: Evidence from apolipoprotein E- and LDL receptor-deficient mice. *Circulation, 112,* 2867–2874.
105. Helgason, C. M., Bolin, K. M., Hoff, J. A., et al. (1994). Development of aspirin resistance in persons with previous ischemic stroke. *Stroke, 25,* 2331–2336.
106. Buchanan, M. R., Schwartz, L., Bourassa, M., et al. (2000). Results of the BRAT study — a pilot study investigating the possible significance of ASA nonresponsiveness on the benefits and risks of ASA on thrombosis in patients undergoing coronary artery bypass surgery. *Can J Cardiol, 16,* 1385–1390.
107. Gum, P. A., Kottke-Marchant, K., Poggio, E. D., et al. (2001). Profile and prevalence of aspirin resistance in patients with cardiovascular disease. *Am J Cardiol, 88,* 230–235.

108. Gum, P. A., Kottke-Marchant, K., Welsh, P. A., et al. (2003). A prospective, blinded determination of the natural history of aspirin resistance among stable patients with cardiovascular disease. *J Am Coll Cardiol, 41,* 961–965.

109. Eikelboom, J. W., Hirsh, J., Weitz, J. I., et al. (2002). Aspirin-resistant thromboxane biosynthesis and the risk of myocardial infarction, stroke, or cardiovascular death in patients at high risk for cardiovascular events. *Circulation, 105,* 1650–1655.

110. Cotter, G., Shemesh, E., Zehavi, M., et al. (2004). Lack of aspirin effect: Aspirin resistance or resistance to taking aspirin? *Am Heart J, 147,* 293–300.

111. Capone, M. L., Sciulli, M. G., Tacconelli, S., et al. (2005). Pharmacodynamic interaction of naproxen with low-dose aspirin in healthy subjects. *J Am Coll Cardiol, 45,* 1295–1301.

112. Livio, M., Del Maschio, A., Cerletti, C., et al. (1982). Indomethacin prevents the long-lasting inhibitory effect of aspirin on human platelet cyclo-oxygenase activity. *Prostaglandins, 23,* 787–796.

113. MacDonald, T. M., & Wei, L. (2003). Effect of ibuprofen on cardioprotective effect of aspirin. *Lancet, 361,* 573–574.

114. Kurth, T., Glynn, R. J., Walker, A. M., et al. (2003). Inhibition of clinical benefits of aspirin on first myocardial infarction by nonsteroidal antiinflammatory drugs. *Circulation, 108,* 1191–1195.

115. Hudson, M., Baron, M., Rahme, E., et al. (2005). Ibuprofen may abrogate the benefits of aspirin when used for secondary prevention of myocardial infarction. *J Rheumatol, 32,* 1589–1593.

116. Bohm, E., Sturm, G. J., Weiglhofer, I., et al. (2004). 11-Dehydro-thromboxane $B_2$, a stable thromboxane metabolite, is a full agonist of chemo-attractant receptor-homologous molecule expressed on TH2 cells (CRTH2) in human eosinophils and basophils. *J Biol Chem, 279,* 7663–7670.

117. Thomas, D. W., Rocha, P. N., Nataraj, C., et al. (2003). Proinflammatory actions of thromboxane receptors to enhance cellular immune responses. *J Immunol, 171,* 6389–6395.

118. Fitzpatrick, F. A., Gorman, R. R., Mc Guire, J. C., et al. (1977). A radioimmunoassay for thromboxane $B_2$. *Anal Biochem, 82,* 1–7.

119. Fitzpatrick, F. A., Gorman, R. R., & Wynalda, M. A. (1977). Electron capture gas chromatographic detection of thromboxane $B_2$. *Prostaglandins, 13,* 201–208.

120. Patrignani, P., Filabozzi, P., & Patrono, C. (1982). Selective cumulative inhibition of platelet thromboxane production by low-dose aspirin in healthy subjects. *J Clin Invest, 69,* 1366–1372.

121. Siess, W., Roth, P., & Weber, P. C. (1981). Stimulated platelet aggregation, thromboxane $B_2$ formation and platelet sensitivity to prostacyclin — a critical evaluation. *Thromb Haemost, 45,* 204–207.

122. Capone, M. L., Tacconelli, S., Sciulli, M. G., et al. (2004). Clinical pharmacology of platelet, monocyte, and vascular cyclooxygenase inhibition by naproxen and low-dose aspirin in healthy subjects. *Circulation, 109,* 1468–1471.

123. Catella, F., Healy, D., Lawson, J. A., et al. (1986). 11-Dehydrothromboxane $B_2$: A quantitative index of thromboxane $A_2$ formation in the human circulation. *Proc Natl Acad Sci USA, 83,* 5861–5865.

124. Westlund, P., Granstrom, E., Kumlin, M., et al. (1986). Identification of 11-dehydro-$TXB_2$ as a suitable parameter for monitoring thromboxane production in the human. *Prostaglandins, 31,* 929–960.

125. Ciabattoni, G., Boss, A. H., Daffonchio, L., et al. (1987). Radioimmunoassay measurement of 2,3-dinor metabolites of prostacyclin and thromboxane in human urine. *Adv Prostaglandin Thromboxane Leukot Res, 17B,* 598–602.

126. Kumlin, M., & Granstrom, E. (1986). Radioimmunoassay for 11-dehydro-$TXB_2$: A method for monitoring thromboxane production in vivo. *Prostaglandins, 32,* 741–767.

127. Hayashi, Y., Shono, F., Yamamoto, S., et al. (1989). Radioimmunoassay of 11-dehydro-thromboxane $B_2$ using monoclonal antibody. *Adv Prostaglandin Thromboxane Leukot Res, 19,* 688–691.

128. Maas, R. L., Taber, D. F., & Roberts, L. J., 2nd. (1982). Quantitative assay of urinary 2,3-dinor thromboxane $B_2$ by GC-MS. *Methods Enzymol, 86,* 592–603.

129. Lawson, J. A., Patrono, C., Ciabattoni, G., et al. (1986). Long-lived enzymatic metabolites of thromboxane $B_2$ in the human circulation. *Anal Biochem, 155,* 198–205.

130. Murphy, R. C., & FitzGerald, G. A. (1994). Current approaches to estimation of eicosanoid formation in vivo. *Adv Prostaglandin Thromboxane Leukot Res, 22,* 341–348.

131. Fries, S., Grosser, T., Price, T. S., et al. (2005). Marked interindividual variability in the response to selective inhibitors of cyclooxygenase-2. *Gastroenterology,* Epub Sept 21, 2005.

132. Catella, F., Lawson, J. A., Fitzgerald, D. J., et al. (1987). Analysis of multiple thromboxane metabolites in plasma and urine. *Adv Prostaglandin Thromboxane Leukot Res, 17,* 611–614.

133. Catella, F., & FitzGerald, G. A. (1987). Paired analysis of urinary thromboxane $B_2$ metabolites in humans. *Thromb Res, 47,* 647–656.

134. Morrow, J. D., Oates, J. A., Roberts, L. J., 2nd, et al. (1999). Increased formation of thromboxane in vivo in humans with mastocytosis. *J Invest Dermatol, 113,* 93–97.

# CHAPTER 32

# Perfusion Chambers

## Philip G. de Groot and Jan J. Sixma

*Department of Clinical Chemistry and Haematology, University Medical Center, Utrecht, The Netherlands*

## I. Introduction

Over the past 35 years, shear forces generated by flowing blood have been recognized as having a significant impact on platelet adhesion and thrombus formation.[1] Detailed studies on the role of fluid dynamics on platelet deposition have shown that shear conditions are not only responsible for platelet transport to the vessel wall, but also determine which receptors on the platelets and which ligands in the vessel wall support platelet adhesion. From these observations it became clear that studies on platelet adhesion and thrombus formation were only relevant when performed under conditions of flow and therefore a large number of flow chambers have been devised.[2] These devices enable experiments under well-controlled conditions of exposure time, thrombogenic surface, and reproducible flow.

The first perfusion devices developed in the early 1970s were relatively large, requiring a substantial amount of anticoagulated blood to be circulated through the perfusion chambers. The large volume of blood limited the number of experiments that could be performed and made the model unattractive for studies with blood of patients and with recombinant proteins. The presence of an anticoagulant was another problem, especially when thrombus formation was studied. In addition, the recirculation of the blood through the device by a roller pump, which slightly activates platelets, could influence the results. In recent years, several strategies have been developed to overcome these disadvantages. Smaller perfusion chambers that only require 0.5 mL of blood for each experiment have been designed.[3,4] An *ex vivo* perfusion system has been developed in which non-anticoagulated blood is directly drawn from the antecubital vein through the perfusion chamber to better study the role of antithrombotic drugs.[5] Most of the systems have now replaced the roller pump by a syringe and an automated syringe pump. These modern flow systems have become essential tools for the investigation of the molecular mechanisms supporting platelet adhesion and thrombus formation. Furthermore, they are very valuable in studies on the effectiveness of various antithrombotic drugs and are suitable for studies of platelet function in patients with bleeding disorders.

## II. Influences of Shear Stress on Platelet Transport

To describe the influence of blood flow on platelet function, some basic aspects of blood rheology should be discussed. Flowing blood is considered laminar, which means that layers of blood pass each other at a different velocity, resulting in a parabolic velocity profile.[6] The velocity of the blood layers is highest in the middle of the vessel, gradually decreasing to almost zero at the vessel wall. The relevant fluid dynamic factors describing the influence of flow on platelet adhesion are shear rate ($\gamma$) and shear stress ($\tau$).[6] Shear rate ($sec^{-1}$) is defined as the velocity with which fluid layers pass each other (a flow gradient, $dV/dr$). Shear stress (dyn/$cm^2$) is defined as the tangential force per unit area exerted in the direction of flow. The highest flow gradient is near the vessel wall and the flow gradient near the axis of the vessel is almost absent. As a consequence, the shear rate and shear stress are almost zero in the middle of the vessel and maximal at the wall. Thus, shear stress and shear rate describe the flow characteristics near the vessel wall surface.

In the case of a liquid obeying Newton's Second Law, shear stress is linearly related to the shear rate, $\tau = \eta \cdot \gamma$, in which $\eta$ is the viscosity of the blood. Shear rates and shear stresses can be calculated for various parts of the vasculature from the known vascular diameter and volume flow rates (Table 32-1). The shear rates are low in large vessels but increase up to 5000 $second^{-1}$ in capillaries. The shear rates in stenosed arteries are much higher still. However, the calculations of shear rate and shear stress are based on a simplified model of steady flow and not on pulsatile flow. These calculations also assume laminar flow, while branching and abnormal vascular curvature will cause local turbulence. Furthermore, blood is not a Newtonian fluid: the

**Table 32-1: Shear Rates and Shear Stresses in the Vascular System**

| Vessel | Diameter (cm) | Wall Shear Rate (sec$^{-1}$) | Wall Shear Stress (dyne/cm$^2$) |
|---|---|---|---|
| Ascending aorta | 2.3–4.5 | 50–300 | 2–10 |
| Femoral artery | 0.5 | 350 | 10 |
| Small arteries | 0.03 | 1500 | 55 |
| Capillaries | 0.0006 | 2000–5000 | [a] |
| Large veins | 0.5–10 | 200 | 7 |
| Inferior vena cava | 2.0 | 50 | 2 |
| Stenosed arteries | 0.025 | 40,000 | 3000 |

The results presented are from refs. 1, 6, 93, and 94.
[a]Calculations are not meaningful because flow regime is totally different and bulk viscosity is not applicable.

viscosity of blood is not constant, but decreases with increasing shear rates, especially at shear rates below 50 sec$^{-1}$. The non-Newtonian flow properties of whole blood are due to the presence of red blood cells, which are by far the most numerous cells, occupying 40% of the volume. Nevertheless, the use of shear rates and shear stress allows a description of the effect of blood flow on platelet adhesion.[7,8]

One of the main factors determining platelet adhesion is the transport of platelets toward the vessel wall surface. This transport occurs by two processes: diffusion and convection. Diffusion is random and the result of Brownian motion of the platelets, representing the movement of material relative to the average fluid motion. Convection is the movement of platelets with the fluid and is the result of flow. Diffusion is a slow process in blood, whereas convection can be very rapid. In the absence of flow, platelet transport takes place by diffusion; in flowing blood convection forces predominate. With increasing shear rates, more platelets will be transported to the vessel wall. Platelet adhesion to subendothelial structures therefore increases with increasing shear rate, up to approximately 2000 sec$^{-1}$. At higher shear rates, platelet adhesion remains more or less unchanged until shear rates of 5000 sec$^{-1}$ and above.

Red cells have a profound influence on platelet distribution within a vessel. Platelet transport is driven by the shear rate of the blood in combination with the presence of red cells. The distribution of red cells in flowing blood is not homogeneous. At increasing shear rate, the red cells migrate toward the middle of the vessel. The smaller platelets are pushed aside toward the vessel surface due to collisions with the red cells. The rotation of the red cells further translocates the platelets to the wall. The diffusion of platelets in response to red cells perpendicular on the direction of flow can increase up to three orders of magnitude at the highest

shear rates. The consequence is that the local concentration of platelets near the vessel wall increases with increasing shear rate. Under physiological conditions, a thin red cell-free layer exists at the vessel wall. The width of this boundary layer with platelets depends on the shear rate. The concentration of platelets in the boundary layer is not only determined by the fluid dynamics but also by "consumption" of platelets via attachment to the vessel wall. The residence time of the platelets in the boundary layer is an important factor determining the initial steps of adhesion and thrombus formation. Thus, the increased platelet adhesion and thrombus formation observed with increasing shear rate is dependent on two mechanisms, convection in the direction of flow due to the flowing of blood and diffusion perpendicular to flow due to the presence of red cells.

Two rate-limiting mechanisms determine the number of platelets adhering at a certain shear rate. When the wall shear rate is low (<300 sec$^{-1}$), adhesion is controlled by mass transport of platelets. When the wall shear rate is high (>1300 sec$^{-1}$), the adhesion is controlled by the reaction kinetics of the interaction of the platelet membrane receptors and the adhesive proteins in the vessel wall. In the intermediate area, both processes contribute.

The transport of platelets by the red cells toward the boundary layer is not only dependent on the hematocrit; the rigidity and size of red cells also influence the transport. The rigidity of the red cells enhances the diffusion of platelets to the vessel wall.[9] Together with the plasma viscosity, the red cell rigidity determines the viscosity of the blood.[10] An increase in plasma viscosity will result in an increase in platelet deposition.[11] The increased blood viscosity found in patients with, for example, diabetes mellitus, may explain in part the increased thrombotic tendency in these patients (see Chapter 38).

## III. Perfusion Devices

Perfusion devices have been developed and improved during the past three decades. In this chapter we will focus on a number of selected models that have been used by various research groups. The cone and plate viscometer and the commercial device developed from this principle are discussed in detail in Chapter 29.

A perfusion system consists of a perfusion chamber, a roller pump or infusion pump, containers, and silastic tubing to connect chamber and pump. Perfusion chambers have been designed not only to mimic blood flow but also to provide a set of known rheological parameters in order to correlate platelet deposition with these parameters. As a consequence, flow chambers must fulfill various criteria. First, they must allow studies at shear rates throughout the whole pathophysiological range and allow the precise description of the rheological parameters. Second, the introduction of surfaces to which platelets should adhere must be easy. Third, a relatively simple assessment of the platelets interacting with these surfaces should be available. This also defines the limitations of flow chambers. Blood flow is pulsatile and the vessel walls can constrict or dilate. In flow chambers, blood flow is constant and the walls of the flow channel are rigid. *In vivo,* blood flows through branched vessels with expansions and obstructions; thus shear stresses on platelets continuously vary. The influence of sudden changes in shear stress on platelet function has been little studied.[12] Another difference between flow chambers and real blood vessels is that the vessel wall is covered with endothelial cells with powerful antiadhesive properties. Endothelial cells can secrete nitric oxide (NO) and prostacyclin, which keep platelets "dormant"[13] (see Chapter 13). Moreover, the ecto-ADPase CD39 on the endothelial cell surface removes the platelet agonist ADP from the circulation. In a perfusion set-up, platelets pass through tubing instead of through endothelial-cell covered vessels before encountering the thrombogenic surface. The chambers are thus not a real mimic of *in vivo* conditions, but models that allow experiments on platelet/vessel wall interactions under conditions of well-defined flow.

### A. Annular Flow Chambers

Baumgartner and colleagues developed the annular perfusion chamber in the early 1970s.[14] This model allowed studies of platelet adhesion and thrombus formation on deendothelialized rabbit or human arterial vessel wall segments. The vessel wall segments were everted on a central rod mounted in a cylinder and subsequently exposed to anticoagulated or nonanticoagulated blood. A large range of shear rates could be achieved by varying the blood flow or varying the distance between the inner wall of the outer cylinder and the surface of the vessel wall segment (= the width of the annulus). Morphometric techniques were developed for detailed quantification of the deposition of platelets and fibrin.[15] The annular perfusion chamber was the first model to study platelet adhesion under conditions of flow and with this model the importance of von Willebrand factor (VWF), glycoprotein (GP) Ib, and integrin αIIbβ3 (GPIIb-IIIa) for platelet deposition at higher shear rates was highlighted for the first time.[16–19] The disadvantage of this type of chamber is that the evaluation through cross sections of the vessel wall segment is very time-consuming and laborious. The model is, therefore, hardly used any more.

### B. Tubular Flow Chambers in Animals

The first cylindrical perfusion chamber was developed by Hanson et al.[20] The perfusion chamber was inserted into an extracorporeal arteriovenous shunt located between the carotid artery and the jugular vein in baboons. The wall of the chamber was coated with collagen type I and the blood flow was controlled with a pump or with a clamp distal to the flow device. The advantage of the use of these chambers is that vessel damage can be performed under controlled and reproducible conditions. Another advantage is that nonanticoagulated blood is used, which allows physiologic study of the efficacy of antithrombotic drugs. Moreover, the thrombotic surface is placed between real vessels covered with endothelial cells. When antithrombotic drugs are tested, the drugs are also subjected to the influence of the metabolic pathways in the baboon.[21–24] The disadvantage of this system is that it needs expensive experimental animals and therefore the number of experiments that can be performed is low. These types of chambers were also extensively used to study antithrombotic properties of graft materials.[25–27]

### C. Parallel Plate Chambers

The most commonly used flow chamber is the parallel plate perfusion chamber. Sakariassen et al. in 1983[28] and Hubbell and McIntyre in 1986[29] published the first examples of this type of perfusion chamber. A parallel plate perfusion chamber consists of a polycarbonate slab that can contain a glass coverslip (Fig. 32-1). The coverslip forms one side of the flow chamber. The major difference with the previously described chambers is that in this chamber the thrombogenic surface is introduced on the glass coverslip and subsequently exposed in a controlled way to flowing blood. The variation in shear rates is achieved in the same way as the annular perfusion chamber, by varying the flow rate or the dimensions of the chamber. The dimensions of a parallel plate flow chamber ($h$ = height, $d$ = width) in combination with the volumetric flow rate ($Q$) and the viscosity ($\mu$) determines the shear stress ($\tau$) in the chamber: $\tau = 6\,Q \cdot \mu \cdot h^{-2} \cdot d^{-2}$. The advantage of this system is that, after removal of the coverslip and staining, the platelet deposition on the coverslip can be analyzed mechanically with an imaging system.[30] The role of vessel wall components can be studied by coating or spraying purified proteins onto a coverslip. After culturing cells on the coverslip, extracellular matrices of these cells can easily be used as a thrombogenic surface.[31] For a personal view on the birth and coming of age of perfusion chambers, read the scientific memories of Sakariassen, Turitto, and Baumgartner.[32]

A parallel plate perfusion chamber set-up can easily be modified to answer specific questions. A cam-driven clamp flow oscillator that pinches the elastic tubing apical to the chamber can introduce pulsating flow.[33] Chambers containing a stenosis that narrows the lumen to 80% have been designed to study the effect of large changes in shear stresses on platelet deposition onto collagen surfaces.[34] The use of

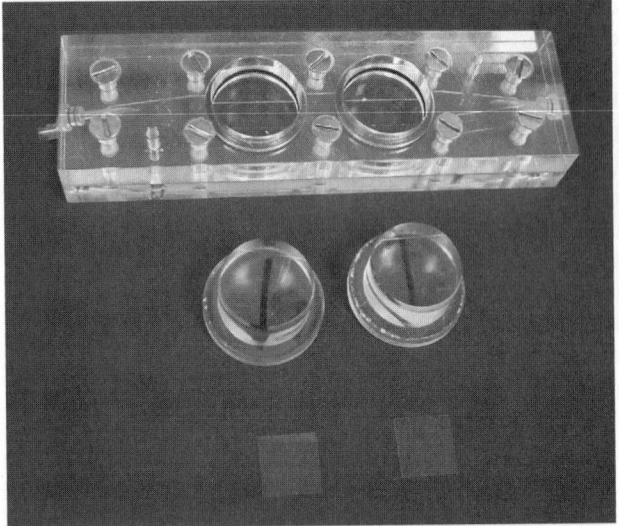

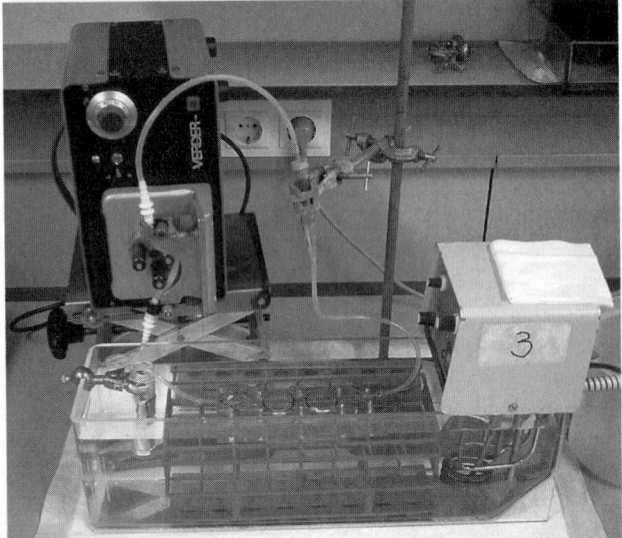

**Figure 32-1.** Parallel plate perfusion chamber. The blood is pumped from a container by a roller pump to a funnel and from the funnel through the perfusion chamber back into the container. The knobs that fit in the perfusion chamber contain a notch through which the blood flows. The glass coverslip contains a platelet-reactive surface and forms a roof on the notch.

an adhesive surface coated on a filter instead of coated on glass allows the local infusion of unstable components such as prostacyclin and NO, and thereby enables the study of the regulation of platelet deposition by endothelial cell products.[35]

Parallel plate perfusion chambers can also be combined with animal models. Perfusion chambers inserted in arteriovenous shunts between the carotid artery and the jugular vein improve the reproducibility of thrombus formation and the studies are performed with nonanticoagulated blood. This set-up is thus very suitable for studies on antithrombotic drugs and the thrombogenicity of artificial grafts. One study compared platelet thrombus formation and the effects of aspirin in parallel flow chambers mounted in an arteriovenous shunt in dogs with an *in vitro* set-up and found comparable results.[36]

Parallel plate perfusion chambers are still widely used. To decrease the required amount of blood, miniature perfusion chambers have now been developed that only require 0.3 mL of blood for a 5-minute perfusion at a shear rate of 300 second$^{-1}$ (Fig. 32-2).[3,4] The parallel perfusion chamber has allowed detailed studies on the role of various platelet membrane receptors, adhesive proteins in the vessel wall, plasma proteins, cations, hematocrit, thrombin formation, and shear rate on platelet adhesion and aggregate formation. In addition, the effects of a large number of drugs that inhibit platelet function have been analyzed. However, the major advantage of a parallel plate perfusion chamber nowadays is that, due to its flat design, it allows real-time analysis of platelet deposition[37,38] (see Chapter 18). The chamber can be easily modified to allow direct mounting on a microscope stage. With the help of a video camera and a CD recorder, direct visualization of platelet deposition and thrombus formation is possible. Not only is the end stage after a fixed time period measured, but information is now available on platelet rolling, platelet detachment, and the stabilization of thrombi under different flow conditions.[39–41]

### D. Cone and Plate Flow Devices

In most devices, a fluid is flowing over a static surface and the interaction of platelets in flowing blood with proteins on this static surface is studied. However, there are also devices in which the entire system is subjected to uniform flow. Rotational viscometers can produce constant and uniform shear stresses on all cells and proteins in a liquid suspension. This is different from the parallel plate chambers in which a shear stress gradient is created perpendicular to the vessel wall that mimics more the *in vivo* situation. In a parallel plate chamber the platelets in the flowing blood experience much lower shear stresses than the platelets in a cone and plate flow device. As higher shear stresses can activate platelets,[42] the difference in device might influence the results.

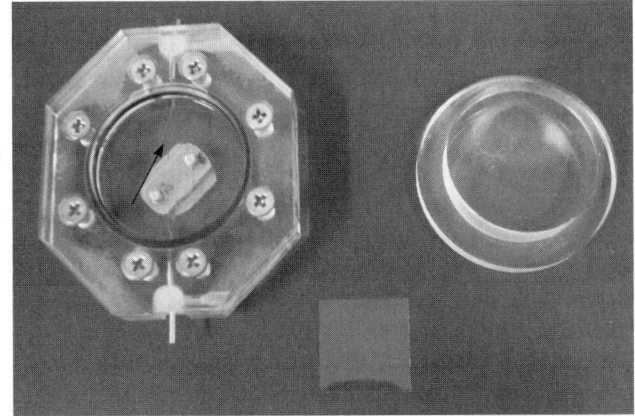

**Figure 32-2.** Parallel plate perfusion chamber for small amounts of blood. The differences from the "classic" perfusion chamber (Fig. 32-1) are the replacement of the roller pump by a syringe pump downstream of the perfusion chamber and the much smaller dimensions of the notch in the perfusion chamber. The arrow indicates the cavity through which the blood flows.

In cone and plate viscometers, the sample of interest is placed between a rotating cone and a stationary platen. The cone angle ranges from 0.3 to 1.0°. The shear rate in the device depends on the cone angle and the rotation rate of the cone.[43,44] The cone and plate viscometer is discussed in detail in Chapter 29.

### E. Stagnation Point Flow Chambers

In all the perfusion chambers previously described, the adhesive surface is placed parallel to the direction of flow. In a stagnation point flow chamber the adhesive surface is placed perpendicular to the flow, providing a maximal platelet/vessel wall contact.[45,46] The connective transport of platelets takes place to within the critical distance from the surface, eliminating a diffusion-controlled transport through the plasma boundary layer. This model can mimic flow profiles in branched and stenosed vessels. The model allows the study of platelet thrombus formation in areas with non-parallel streamlines.

### F. Capillary Perfusion Chambers

A glass capillary tube is coated with collagen or another purified matrix protein and perfused with blood. The advantage of this approach is that, due to the very small diameter of the tube, very high shear rates can be reached.[47] Capillary tubes can also be used in combination with animal models.[48]

## IV. Pumps

In the first chambers that were developed, the blood was recirculated through the perfusion chamber with the help of a roller pump. It was essential to use a roller pump that was not occlusive in order to prevent mechanical lysis of red cells.[49] The blood was pumped from a container through a perfusion chamber back into the container. Because a roller pump creates a pulsating flow, a funnel was introduced between the pump and the chamber to create a steady flow. The height between the funnel and the container determined the flow rate through the perfusion chamber. A better solution, feasible now that smaller chambers are available, is to use a syringe pump placed distally to the perfusion chamber.[50] The advantage is that blood is not circulated and it encounters the flow device after the chamber, reducing unwanted platelet activation to a minimum.

## V. Perfusate

The choice of the perfusate depends on the aim of the studies. Anticoagulated whole blood can be used for those studies in which the possible effects of formed thrombin can be neglected. The classic anticoagulant is 1/9 vol 108 mM trisodium citrate. Citrate blocks all coagulation activity. Platelet adhesion to collagen types I and III depends strongly on the presence of $Mg^{2+}$, making the use of citrated blood undesirable.[51] The anticoagulants of choice for collagen are low molecular weight heparin (LMWH, 20 U/mL) or direct thrombin inhibitors such as hirudin (20 U/mL) or PPACK (*o*-phenylalanyl-L-propyl-L-arginine chloromethyl ketone dihydrochloride, 40 μM). These two anticoagulants allow some thrombin formation and thus some fibrin formation on the surface. A disadvantage of heparin, and to a lesser extent LMWH, is that these anticoagulants influence platelet adhesion to purified VWF[52] and the interaction between the platelet receptors GPIb and P-selectin.[53] As an alternative, a synthetic pentasaccharide with factor Xa inhibitory activity (Arixtra, natriumfondaparinux, Sanofi Aventis) can be used (200 U/mL). Direct thrombin inhibitors are less effective coagulation inhibitors; they inhibit thrombin activity but they do not entirely prevent the formation of thrombin. Thrombin formed on the surface of a platelet can activate the platelet. In our laboratory, we now use a mixture of natriumfondaparinux and PPACK.

The major disadvantage of the use of anticoagulants is that the influence of thrombin formation on platelet deposition is excluded. If platelet adhesion is to be studied in the context of an active coagulation cascade, nonanticoagulated blood should be drawn directly from the antecubital vein of the donor through the perfusion chamber.[55] Although this method approaches the *in vivo* situation, a major disadvantage is that the analysis of platelet and fibrin deposition is very difficult, partly due to their massive deposition. Moreover, the experiments are not very reproducible, because of the activation of the coagulation cascade by the perfusion system itself (e.g., the tubing), the variable contamination of the collected blood with tissue factor, and the explosive character of thrombin formation. Also, skilled technicians are necessary. Another disadvantage of nonanticoagulated blood is that the manipulation of the blood is not simple; for example, to add inhibitors to the blood, a mixing device should be introduced between the donor and the chamber to allow proper mixing of the additives to the blood.[55-57] The mixing device is another risk of massive unwanted activation, and the device only functions when large amounts of blood (10 mL/min) are drawn through it.

An attractive method to study the effects of different plasma components is the use of reconstituted blood. Reconstituted blood consists of washed platelets (final concentration $2 \times 10^8$/mL) and washed red cells (40%) reconstituted in (patient) plasma or a buffer containing various purified proteins.

## VI. Surfaces

The ideal surface for platelet adhesion is a vessel wall segment, as used in the original annular perfusion chamber. However, the analysis of platelet deposition on vessel wall segments is difficult and laborious. Most studies are therefore performed with purified proteins, in particular collagen, or with the extracellular matrices of cultured cells. The extracellular matrix of cultured endothelial cells is a good alternative to vessel wall segments, in particular for the subendothelium.[57] The subendothelium is defined as the part of the vessel wall that lies between the internal elastic lamina and the endothelium. The characteristics of the extracellular matrix of endothelial cells mimic closely the composition of the subendothelium of a vessel wall. The extracellular matrix is a highly reactive surface for platelets

and its characteristics are well described. By stimulating the endothelial cells with, for example, phorbol myristate acetate, tissue factor expression in the matrix of the cells is induced and the reactivity of the extracellular matrix can be manipulated.[58]

Platelets adhere to a variety of purified proteins. Fibrillar collagen types I and III are the most potent inducers of platelet thrombi, because they not only support adhesion, but also induce aggregate formation. Most other adhesive proteins such as VWF, fibrinogen, thrombospondin, fibronectin, and laminin only support platelet adhesion. The different adhesive proteins show a large difference in shear rates for optimal platelet adhesion.[47] Coating a coverslip with a protein can take place via spraying of a protein solution or via adsorption. The technique of choice depends on the protein studied. For spraying of the protein on a coverslip with a commercial retouching airbrush, it is necessary to dissolve the proteins in a volatile buffer. Collagen is almost always sprayed because fibrillar collagen is an insoluble protein and should be coated from a suspension.

## VII. Methods to Quantify Platelet Deposition

Platelet deposition on a thrombogenic surface can be quantified as an end point measurement by morphometric methods, immunological methods,[59] or radioactive platelets.[60] Morphometric methods have important advantages, because platelet accumulation is directly observed. Irregular adhesion patterns are immediately noticed. Moreover, in most perfusion chambers only a part of the adhesive surface is exposed to the platelets. With chemical methods it may be difficult to discriminate between exposed and nonexposed surfaces. An additional advantage of morphometric methods is that it is easy to discriminate between platelet adhesion and thrombus formation. Using video cameras, real-time measurements are possible and very informative[37] (see Chapters 18 and 34). To visualize platelets with a video system, the platelets can be labeled by adding a fluorescent dye to the blood. The most commonly used dye is mepacrine (quinacrine dihydrochloride). Although this dye also labels leukocytes, these cells can easily be distinguished from platelets by their size. Moreover, leukocytes hardly adhere at shear rates used to study platelet adhesion. Mepacrine accumulates in the dense granules of the platelets and has no effect on platelet function in an aggregometer.[61] If the platelets are labeled with a fluorescent dye, the actual process of platelet adhesion can be recorded with the help of a confocal laser microscope coupled to a video imaging system. During the perfusion, the chamber is mounted on a fluorescence microscope equipped with a video camera and coupled to a video recorder. The evaluation of the videotapes showing real-time platelet adhesion has fundamentally changed our understanding of platelet adhesion. When platelets are labeled with other fluorescent probes such as Fluo-3 acetoxymethyl ester, real time analysis of $Ca^{2+}$-signaling when platelets adhere and aggregate is possible.[62]

## VIII. Influence of Shear on Platelet Reactivity

### A. Shear Rate Dependency of Platelet Adhesion

Shear influences platelet adhesion not only by mediating convective transport of platelets to the vessel wall. Platelet adhesion is dependent on adhesive proteins in the vessel wall that interact with specific receptors on the platelet membrane. Each adhesive protein appears to have a specific range of shear rates for which it supports platelet adhesion when tested as purified proteins[63] (Table 32-2). The most important adhesive proteins are collagen, VWF, and fibronectin. Laminin and thrombospondin may play a minor role. The shear rate dependency is determined by the affinity of platelet receptors for the specific adhesive proteins. When the shear stresses become too large, detachment of platelets occurs.[64] In a complex matrix such as subendothelium, adhesive proteins may act synergistically. However, as the different proteins in the subendothelium interact with each other, it is also possible that reactive domains in proteins are shielded off from interaction with platelets by interaction with other proteins.[65]

Platelet adhesion and aggregation are regulated by low molecular weight substances such as prostacyclin and NO that are synthesized and secreted by endothelial cells (see Chapter 13). Both the synthesis of prostacyclin and NO are

### Table 32-2: Characteristics of Platelet Adhesion to Purified Proteins

| Protein | Adhesion Receptor | Optimal Shear (sec$^{-1}$) | Cations |
|---|---|---|---|
| Fibrinogen | αIIbβ3 | 500–1000 | Not necessary |
| Fibronectin | α5β1 αIIbβ3 | 300 | Not necessary |
| Collagen type I | α2β1 GPVI | >4000 | $Mg^{2+}$ |
| Collagen type III | α2β1 GPVI | >4000 | $Mg^{2+}$ |
| Collagen type IV | α2β1 | 1600 | $Mg^{2+}$ |
| von Willebrand factor | GPIb αIIbβ3 | >4000 | Not necessary |
| Laminin | α6β1 | 500 | $Mg^{2+}$ |
| Vitronectin | αIIbβ3 | Very low | Not necessary |
| Thrombospondin | ? | 1500 | $Ca^{2+}$ |

upregulated by variations in shear stress exerted on the endothelial cell.[66–68]

### B. Simplified Model of Platelet Adhesion to Collagen under Flow (see also Chapter 18)

Platelet adhesion occurs in flowing blood. The presence of VWF is necessary for optimal platelet adhesion to deal with the shear forces exposed by the flowing blood on the platelets. At the site of vascular injury, when collagens become exposed to the circulating blood, VWF present in plasma binds via its A3 domain to collagens.[69] When VWF binds to collagen, it will change its conformation and the A1 domain, shielded off by the adjacent domains, becomes exposed.[70] The initial contact of platelets takes place via the interaction of the platelet receptor complex GPIb-IX-V with the A1 domain of immobilized VWD.[71] This unstable interaction facilitates tethering and rolling of platelets over the injured surface. By doing so, the platelets are slowed down.[37] Platelets roll over VWF in the direction of flow, driven by the shear forces.[72] A continuous loss of GPIb–VWF interactions at one side of the platelet and the formation of new interactions at the other side of the platelet support the rolling process.[73] The rolling of the platelets will finally end in firm attachment through the participation of other platelet membrane receptors, some of which become upregulated via inside-out signaling of the rolling platelets.[74] The adhesion of platelets to subendothelium involves, in addition to the GPIb-IX-V complex, the functioning of at least four other receptors: the collagen receptors α2β1[75] and GPVI,[76,77] the fibronectin receptor α5β1,[78] and the fibrinogen receptor (αIIbβ3).[79] Integrin αIIbβ3-mediated spreading of the platelets will follow the firm adhesion and is essential to enable the platelets to withstand the shear forces exerted by the flowing blood.[80] Before firm adhesion and spreading are established, some of the platelets will detach and return to the circulation.[81]

Spread platelets are themselves a new surface for additional platelets to adhere to and are the basis of a platelet aggregate.[82] Historically, platelet adhesion and platelet aggregate formation were considered as two separate stages of the formation of a thrombus. However, the similarities between adhesion and aggregation are now clearer.[83] For both adhesion and aggregation, circulating platelets must attach to an adhesive protein while shear forces are exposed on the platelet by the flowing blood. An adhered and subsequently spread platelet binds fibrinogen, and VWF from the circulation to integrin αIIbβ3 and GPIb, respectively, creating an ideal surface for the next platelet to adhere. Although platelet adhesion is dependent on additional receptors such as integrin α2β1, and aggregation is strongly supported by responses to thrombin and ADP, the fundamental mechanism for both processes, an interaction between a platelet receptor (αIIbβ3 and GPIb-IX-V) with its substrate (fibrinogen and VWF) in flowing blood, is similar.[84–86] This is highlighted by the observation that, for both adhesion and aggregation, the importance of VWF increases with increasing shear rate.

Recently, experiments with knockout mice have taught us that the formation of a stable thrombus is more complex than originally thought. Unexpected platelet and plasma proteins are essential to prevent embolization of a thrombus.[87–90] Perfusion chambers coupled to real-time monitoring are an ideal set-up to further study the mechanism of stable thrombus formation. The insertion of a Doppler device downstream of the flow chamber offers the possibility to quantify the fragmentation of newly formed thrombi.[91]

## IX. Future Directions

The use of flow models has provided important insights into the role of platelets in hemostasis and thrombus formation. Future studies on platelet function will also need flow models. The choice of the perfusion device will depend on the aims of the study. Perfusion studies may be improved in the near future by the systematic combination of confocal laser microscopy and real-time observations. Current systems are still relatively slow when assessing thrombus build-up, but faster systems are being developed and will allow an accurate description of thrombus growth and thrombus embolization in time. This may also be of importance for the study of the effects of local disturbances of flow.

Perfusion systems thus far have concentrated on laminar flow in rigid tubes. Physical techniques for dealing with turbulent flow are being developed, but it may take some time before they will become applicable to perfusion studies with blood. Theoretical description of these models will be difficult.[92] The problem of wall rigidity should also be solvable, by generating vessels that allow perfusion at physiological shear rates and in which the vessel wall is subject to similar deformation to that in the circulation.

There may be an increasing need for perfusion models in drug research. The current techniques of using receptor-based rational drug research will lead to drugs that have a tendency to work only in primates. The availability of experimental animals may decline and perfusion models may become attractive alternatives. Perfusion models can be adapted to bulk screening.

Whether perfusion studies will have a place in patient testing remains to be seen. Other than von Willebrand disease, there is no single clinical defect currently known that is only evident at higher blood shear rates, although there are theoretical reasons to believe that such conditions should exist. The increased sensitivity of the newer techniques of real-time perfusion in combination with confocal laser scanning microscopy may lead to the detection of subtle, but perhaps important, patient defects that have thus far escaped attention.

## References

1. Frederickson, B. J., & McIntyre, L. V. (2001). Rheology of thrombosis. In R. W. Colman, J. Hirsh, V. J. Marder, A. W. Clowes, & J. N. George (Eds.), *Hemostasis and thrombosis, basic principles and clinical practice* (4th ed.), (pp. 625–638). Philadelphia: Lippincott Williams & Wilkins.
2. de Groot, P. G., & Sixma, J. J. (1999). Platelet adhesion to the subendothelium under flow. In E. Dejana & M. Corada (Eds.), *Protocols for adhesion proteins: Methods in molecular biology* (Vol. 96, pp. 159–170). Totowa, NJ: Humana Press.
3. Kirchhofer, D., Tschopp, T. B., & Baumgartner, H. R. (1995). Active site-blocked factors VIIa and IXa differentially inhibit fibrin formation in a human ex vivo thrombosis model. *Arterioscler Thromb Vasc Biol, 15,* 1098–1106.
4. Sixma, J. J., de Groot, R. G., van Zanten, H., et al. (1998). A new perfusion chamber to detect platelet adhesion using a small volume of blood. *Thromb Res, 92,* S43–S46.
5. Sakariassen, K. S., Roald, H. E., & Salatti, J. A. (1992). Ex-vivo models for studying thrombosis: Special emphasis on shear rate dependent blood-collagen interactions. In N. H. Hwang (Ed.), *Advances in cardiovascular engineering* (pp. 151–174). New York: Plenum Press.
6. Goldsmith, H. L., & Turitto, V. T. (1986). Rheological aspects of thrombosis and haemostasis: Basic principles and applications. *Thromb Haemost, 5,* 415–435.
7. Turitto, V. T., & Baumgartner, H. R. (1975). Platelet deposition on subendothelium exposed to flowing blood. Mathematical analysis of physical parameters. *Trans Am Soc Artif Organs, 21,* 593–599.
8. Sakariassen, K. S., Hanson, S. R., & Cadroy, Y. (2001). Methods and models to evaluate shear-dependent and surface reactivity-dependent antithrombotic efficacy. *Thromb Res, 104,* 149–174.
9. Wang, N. H., & Keller, K. H. (1979). Solute transport induced by erythrocyte motion in shear flow. *Fed Proc, 30,* 1591–1596.
10. Van Breugel, H. H. F. I., de Groot, P. G., Heethaar, R. M., et al. (1992). Role of plasma viscosity in platelet adhesion. *Blood, 80,* 953–959.
11. Aarts, R. A. M. M., Heethaar, R. M., & Sixma, J. J. (1984). Red blood cell deformability influences platelet-vessel wall interaction in flowing blood. *Blood, 64,* 1228–1233.
12. Barstad, R. M., Roald, H. E., Cui, Y., et al. (1994). A perfusion chamber developed to investigate thrombus formation and shear profiles in flowing native human blood at the apex of well-defined stenoses. *Arterioscler Thromb, 14,* 1984–1991.
13. Pearson, J. D. (1999). Endothelial cell function and thrombosis. *Baillieres Best Pract Res Clin Haematol, 12,* 329–341.
14. Baumgartner, H. R., & Haudenschild, C. (1972). Adhesion of platelets to subendothelium. *Ann NY Acad Sci, 201,* 22–36.

15. Turitto, V. T., Weiss, H. J., & Baumgartner, H. R. (1983). Decreased platelet adhesion on vessel segments in von Willebrand's disease: A defect in initial platelet attachment. *J Lab Clin Med, 102,* 551–564.

16. Tschopp, T. B., Weiss, H. J., & Baumgartner, H. R. (1974). Decreased adhesion of platelets to subendothelium in von Willebrand's disease. *J Lab Clin Med, 83,* 296–300.

17. Weiss, H. J., Tschopp, T. B., Baumgartner, H. R., et al. (1974). Decreased-adhesion of giant (Bernard–Soulier) platelets to subendothelium. Further implications on the role of the von Willebrand factor in hemostasis. *Am J Med, 57,* 920–925.

18. Weiss, H. J., Tschopp, T. B., & Baumgartner, H. R. (1975). Impaired interaction (adhesion-aggregation) of platelets with the subendothelium in storage-pool disease and after aspirin ingestion. A comparison with von Willebrand's disease. *N Engl J Med, 293,* 619–623.

19. Sakariassen, K. S., Nievelstein, R. F., Coller, B. S., et al. (1986). The role of platelet membrane glycoproteins Ib and IIb-IIIa in platelet adherence to human artery subendothelium. *Br J Haematol, 63,* 681–691.

20. Hanson, S. R., Reidy, M. A., Hattori, A., et al. (1982). Pharmacologic modification of acute vascular graft thrombosis. *Scan Electron Microsc, 2,* 773–779.

21. Badimon, L., Badimon, J. J., Galvez, A., et al. (1986). Influence of arterial damage and wall shear rate on platelet deposition. Ex vivo study in a swine model. *Arteriosclerosis, 6,* 312–320.

22. Hanson, S. R., Griffin, J. H., Harker, L. A., et al. (1993). Antithrombotic effects of thrombin-induced activation of endogenous protein C in primates. *J Clin Invest, 92,* 2003–2012.

23. Badimon, L., Badimon, J. J., Galvez, A., et al. (1986). Influence of arterial damage and wall shear rate on platelet deposition. Ex vivo study in a swine model. *Arteriosclerosis, 6,* 312–320.

24. Cadroy, Y., Horbett, T. A., & Hanson, S. R. (1989). Discrimination between platelet-mediated and coagulation-mediated mechanisms in a model of complex thrombus formation in vivo. *J Lab Clin Med, 113,* 436–448.

25. Hanson, S. R., Kotze, H. F., Savage, B., et al. (1985). Platelet interactions with Dacron vascular grafts. A model of acute thrombosis in baboons. *Arteriosclerosis, 5,* 595–603.

26. Chen, C., Ofenloch, J. C., Yianni, Y. R., et al. (1998). Phosphorylcholine coating of PTFE reduces platelet deposition and neointimal hyperplasia in arteriovenous grafts. *J Surg Res, 77,* 119–125.

27. Roald, H. E., Barstad, R. M., Bakken, I. J., et al. (1994). Initial interactions of platelets and plasma proteins in flowing non-anticoagulated human blood with the artificial surfaces Dacron and PTFE. *Blood Coagul Fibrinol, 5,* 355–363.

28. Sakariassen, K. S., Aarts, R. A. M. M., de Groot, R. G., et al. (1983). A perfusion chamber developed to investigate platelet interaction in flowing blood with human vessel wall cells, their extracellular matrix, and purified components. *J Lab Clin Med, 102,* 522–536.

29. Hubbell, J. A., & McIntyre, L. V. (1986). Technique for visualisation and analysis of mural thrombogenesis. *Rev Sci Instrum, 57,* 892–896.

30. Sakariassen, K. S., Kuhn, H., Muggli, R., et al. (1988). Growth and stability of thrombi in flowing citrated blood: Assessment of platelet-surface interactions with computer assisted morphometry. *Thromb Haemost, 60,* 392–398.

31. De Groot, R G., & Sixma, J. J. (1993). Platelet adhesion to the vessel wall. Role of adhesive proteins and their receptors. *J Blood Rheol,* 777–782.

32. Sakariassen, K. S., Turitto, V. T., & Baumgartner, H. R. (2004). Recollections of the development of flow devices for studying mechanisms of hemostasis and thrombosis in flowing whole blood. *J Thromb Haemost, 2,* 1681–1690.

33. van Breugel, H. H. F. I., Sixma, J. J., & Heethaar, R. M. (1988). Effect of flow pulsatility on platelet adhesion to subendothelium. *Arteriosclerosis, 8,* 332–335.

34. Barstad, R. M., Kierulf, P., & Sakariassen, K. S. (1996). Collagen induced thrombus formation at the apex of eccentric stenoses — a time course study with non-anticoagulated human blood. *Thromb Haemost, 75,* 685–692.

35. de Graaf, J. C., Banga, J. D., Moncada, S., et al. (1992). Nitric oxide functions as an inhibitor of platelet adhesion under flow conditions. *Circulation, 85,* 2284–2290.

36. Roux, S. P., Sakariassen, K. S., Turitto, V. T., et al. (1991). Effect of aspirin and epinephrine on experimentally induced thrombogenesis in dogs. A parallelism between in vivo and ex vivo thrombosis models. *Arterioscler Thromb, 11,* 1182–1191.

37. Savage, B., Saldivar, E., & Ruggeri, Z. M. (1996). Initiation of platelet adhesion by arrest onto fibrinogen or translocation on von Willebrand factor. *Cell, 84,* 289–297.

38. Savage, B., Almus-Jacobs, F., & Ruggeri, Z. M. (1998). Specific synergy of multiple substrate-receptor interactions in platelet thrombus formation under flow. *Cell, 94,* 657–666.

39. Ruggeri, Z. M., Dent, J. A., & Saldivar, E. (1999). Contribution of distinct adhesive interactions to platelet aggregation in flowing blood. *Blood, 94,* 172–178.

40. Remijn, J. A., Wu, Y. P., Usseldijk, M. J. W., et al. (2001). Absence of fibrinogen in afibrinogemia patients results in large but loosely packed aggregates. *Thromb Haemost, 85,* 736–742.

41. Tsuji, S., Sugimoto, M., Miyata, S., et al. (1999). Real-time analysis of mural thrombus formation in various platelet aggregation disorders: Distinct shear-dependent roles of platelet receptors and adhesive proteins under flow. *Blood, 94,* 968–975.

42. Shankaran, H., Alexandridis, P., & Neelamegham, S. (2003). Aspects of hydrodynamic shear regulating shear-induced platelet activation and self-association of von Willebrand factor in suspension. *Blood, 101,* 2637–2645.

43. Brown, C. H., Leverett, L. B., Lewis, C. W., et al. (1975). Morphological, biochemical, and functional changes in human platelets subjected to shear stress. *J Lab Clin Med, 86,* 462–471.

44. Giorgio, T. D., & Hellums, J. D. (1988). A cone and plate viscometer for the continuous measurement of blood platelet activation. *Biorheology, 25,* 605–624.

45. Reininger, C. B., Reininger, A. J., Graf, J., et al. (1996). Real-time analysis of platelet adhesion under stagnation point flow conditions: The effect of red blood cells and glycoprotein IIb/IIIa receptor blockade. *J Vasc Invest, 2,* 1–11.

46. Reininger, A. J., Komdorfer, M. A., & Wurzinger, L. J. (1998). Adhesion of ADP-activated platelets to intact endothelium under stagnation point flow in vitro is mediated by the integrin alphaIIbbeta3. *Thromb Haemost, 79,* 998–1003.

47. Wu, Y. P., de Groot, P. G., & Sixma, J. J. (1997). Shear-stress-induced detachment of blood platelets from various surfaces. *Arterioscler Thromb Vasc Biol, 17,* 3202–3207.

48. Andre, P., Arbeille, B., Drouet, V., et al. (1996). Optimal antagonism of GPIIb/IIIa favors platelet adhesion by inhibiting thrombus growth. An ex vivo capillary perfusion chamber study in the guinea pig. *Arterioscler Thromb Vasc Biol, 16,* 56–63.

49. Kennedy, P. S., Ware, J., Horak, J. K., et al. (1981). Factors affecting the size of platelet aggregates in blood. *Thromb Haemost, 46,* 725–730.

50. Van Zanten, G. H., Saelman, E. U. M., Schut-Hese, K. M., et al. (1996). Platelet adhesion to collagen type IV under flow conditions. *Blood, 88,* 3862–3871.

51. Saelman, E. U. M., Nieuwenhuis, H. K., Hese, K. M., et al. (1993). Platelet adhesion to collagen types I through VIII under conditions of stasis and flow is mediated by Gpla-IIa (a2bl-integrin). *Blood, 83,* 1244–1250.

52. Wu, Y. P., van Breugel, H. E. L., Lankhof, H., et al. (1996). Platelet adhesion and von Willebrand Factor. *Arteriosc Thromb Vasc Biol, 16,* 611–620.

53. Romo, G. M., Dong, J. E., Schade, A. J., et al. (1999). The glycoprotein Ib-IX-V complex is a platelet counter receptor for P-selectin. *J Exp Med, 190,* 803–814.

54. Sakariassen, K. S., Roald, H. E., & Salatti, J. A. (1992). Ex-vivo models for studying thrombosis: Special emphasis on shear rate dependent blood-collagen interactions. In N. H. C. Huang, V. T. Turitto, & M. R. Yen (Eds.), *Advances in cardiovascular engineering* (pp. 151–174). New York: Plenum Press.

55. Kirchhofer, D., Tschopp, T. B., Hadvary, P., et al. (1994). Endothelial cells stimulated with tumor necrosis factor-alpha express varying amounts of tissue factor resulting in inhomogenous fibrin deposition in a native blood flow system. Effects of thrombin inhibitors. *J Clin Invest, 93,* 2073–2083.

56. Orvim, U., Barstad, R. M., Orning, L., et al. (1997). Antithrombotic efficacy of inactivated active site recombinant factor VIIa is shear dependent in human blood. *Arterioscler Thromb Vasc Biol, 17,* 3049–3056.

57. Vlodavski, I., Eldor, A., Hyam, R., et al. (1982). Platelet interaction with the extracellular matrix produced by cultured endothelial cells: A model to study the thrombogenicity of isolated subendothelial basal lamina. *Thromb Res, 28,* 60–65.

58. Zwaginga, J. J., Sixma, J. J., & de Groot, P. G. (1990). Activation of endothelial cells with various stimuli induces platelet thrombus formation on their matrix. Studies of a new ex vivo thrombosis model using low molecular weight heparin as anticoagulant. *Arteriosclerosis, 10,* 49–61.

59. van Zanten, G. H., de Graaf, S., Slootweg, P. J., et al. (1994). Increased platelet deposition on atherosclerotic coronary arteries. *J Clin Invest, 93,* 615–632.

60. Sakariassen, K. S., Bolhuis, R. A., & Sixma, J. (1980). Platelet adherence to subendothelium of human arteries in pulsatile and steady flow. *Thromb Res, 19,* 547–549.

61. Dise, C. A., Burch, J. W., & Goodman, D. B. P. (1982). Direct interaction of mepacrine with erythrocyte and platelet membrane phospholipids. *J Biol Chem, 257,* 4701–4704.

62. Siljander, P. R., Munnix, I. C., Smethurst, P. A., et al. (2004). Platelet receptor interplay regulates collagen-induced thrombus formation in flowing human blood. *Blood, 103,* 1333–1341.

63. Sixma, J. J., & de Groot, P. G. (1995). Regulation of platelet adhesion to the vessel wall. In H. J. Dengler, T. F. Lijscher, & E. Markward (Eds.), *Die gefasswand als pharmakologischer* (pp. 103–114). Stuttgard: Angriffspunkt Gustav Fisher.

64. Jen, C. J., Li, H. M., Wang, J. S., et al. (1996). Flow-induced detachment of adherent platelets from fibrinogen-coated surface. *Am J Physiol, 270,* H160–H166.

65. Agbanyo, F. R., Sixma, J. J., de Groot, P. G., et al. (1993). Thrombospondin-platelet interactions. Role of divalent cations, wall shear rate, and platelet membrane glycoproteins. *J Clin Invest, 92,* 288–296.

66. Resnick, N., Yahav, H., Khachigian, L. M., et al. (1997). Endothelial gene regulation by laminar shear stress. *Adv Exp Med Biol, 430,* 155–164.

67. Wu, K. K., & Thiagarajan, P. (1996). Role of endothelium in thrombosis and hemostasis. *Annu Rev Med, 47,* 315–331.

68. Nerem, R. M., Harrison, D. G., Taylor, W. R., et al. (1993). Hemodynamics and vascular endothelial biology. *J Cardiovasc Pharmacol, 11,* S6–SIO.

69. Sakariassen, K. S., Bolhuis, P. A., & Sixma, J. J. (1979). Human blood platelets adhesion to artery subendothelium is mediated by factor VIII-von Willebrand factor bound to the subendothelium. *Nature, 279,* 636–638.

70. Hulsten, J. J., deGroot, P. G., Silence, K., et al. (2005). A novel antibody that detects a gain-of-function phenotype of von Willebrand factor in ADAMTS-13 deficiency and von Willebrand disease type 2B. *Blood, 106,* 3035–3042.

71. Clemetson, K. J., & Clemetson, J. M. (1995). Platelet GPIb-V-IX complex. Structure, function, physiology and pathology. *Semin Thromb Hemost, 21,* 130–136.

72. Hafezi-Moghadam, A., Thomas, K. L., & Cornelssen, C. A. (2004). A novel mouse-driven ex vivo flow chamber for the study of leukocyte and platelet function. *Am J Physiol Cell Physiol, 286,* C876–C892.

73. Miura, S., Li, C. Q., Cao, Z., et al. (2000). Interaction of von Willebrand factor domain Al with platelet glycoprotein Ibalpha-(1–289). Slow intrinsic binding kinetics mediate rapid platelet adhesion. *J Biol Chem, 275,* 7539–7546.

74. Andrews, R. K., & Berndt, M. C. (2004). Platelet physiology and thrombosis. *Thromb Res, 114,* 447–453.

75. Nieuwenhuis, H. K., Akkerman, J. W. N., Houdijk, W. P. M., et al. (1985). Human blood platelets showing no response to collagen fail to express surface glycoprotein la. *Nature, 318,* 470–473.

76. Arai, M., Yamamoto, N., Moroi, M., et al. (1995). Platelets with 10% of the normal amount of GPVI have an impaired response to collagen that results in a mild bleeding tendency. *Br J Haemotol, 89,* 357–364.

77. Kehrel, B., Clementson, K. J., Anders, O., et al. (1998). Glycoprotein VI is a major collagen receptor for platelet activation: It recognizes the platelet-activating quartemary structure

of collagen, whereas CD36, glycoprotein IIb/IIIa and von Willebrand factor do not. *Blood, 91,* 491–496.

78. Beumer, S., Ijsseldijk, M. J. W., de Groot, R. G., et al. (1994). Platelet adhesion to fibronectin in flow: Dependence on surface concentration and shear rate, role of platelet membrane glycoproteins GPIIb/IIIa and VLA-5, and inhibition by heparin. *Blood, 84,* 3724–3733.

79. Weiss, H. J., Turitto, V. T., & Baumgartner, H. R. (1986). Platelet adhesion and thrombus formation on subendothelium in platelets deficient in glycoproteins IIb-IIIa, Ib, and storage granules. *Blood, 67,* 322–330.

80. Weiss, H. J., Turitto, V. T., & Baumgartner, H. R. (1991). Further evidence that glycoprotein IIb-IIIa mediates platelet spreading on subendothelium. *Thromb Haemost, 65,* 202–205.

81. Wu, Y. P., de Groot, P. G., & Sixma, J. J. (1997). Shear stress-induced detachment of blood platelets adhered to various adhesive surfaces under flow. *Arterioscler Thromb Vasc Biol, 17,* 3202–3207.

82. Kulkarni, S., Dopheide, S. M., Yap, C. L., et al. (2000). A revised model of platelet aggregation. *J Clin Invest, 105,* 783–791.

83. Ruggeri, Z. M. (2000). Old concepts and new developments in the study of platelet aggregation. *J Clin Invest, 105,* 699–701.

84. Ruggeri, Z. M., Dent, J. A., & Saldivar, E. (1999). Contribution of distinct adhesive interactions to platelet aggregation in flowing blood. *Blood, 94,* 172–178.

85. Wu, Y. P., Vink, T., Schiphorst, M., et al. (2000). Platelet thrombus formation on collagen at high shear rates is mediated by von Willebrand factor-GPIb interaction and inhibited by vWF-GPIIb:IIIa interaction. *Arterioscler Thromb Vasc Biol, 20,* 1661–1667.

86. Tsuji, S., Sugimoto, M., Miyata, S., et al. (1999). Real time analysis of mural thrombus formation in various platelet aggregation disorders: Distinct shear-dependent roles of platelet receptors and adhesive proteins under flow. *Blood, 94,* 968–975.

87. Wagner, D. D. (2005). New links between inflammation and thrombosis. *Arterioscler Thromb Vasc Biol, 25,* 1321–1326.

88. Kawai, T., Andrews, D., Colvin, R. B., et al. (2002). Thromboembolic complications after treatment with monoclonal antibody against CD40 ligand. *Nat Med, 6,* 114.

89. Andre, P., Prasad, K. S., Denis, C. V., et al. (2002). CD40L stabilizes arterial thrombi by a beta3 integrin–dependent mechanism. *Nat Med, 8,* 247–252.

90. Wu, Y. P., Bloemendal, H. J., Voest, E. E., et al. (2004). Fibrin-incorporated vitronectin is involved in platelet adhesion and thrombus formation through homotypic interactions with platelet-associated vitronectin. *Blood, 104,* 1034–1041.

91. Sukavaneshvar, S., Solen, K. A., & Mohammad, S. F. (2000). An in-vitro model to study device-induced thrombosis and embolism: Evaluation of the efficacy of tirofiban, aspirin, and dipyridamole. *Thromb Haemost, 83,* 322–326.

92. David, T., de Groot, P. G., & Walker, P. G. (2002). Boundary-layer type solutions for initial platelet activation and deposition. *J Thor Med, 4,* 95–108.

93. Strony, J., Beaudoin, A., Brands, D., et al. (1993). Analysis of shear stress and hemodynamic factors in a model of coronary artery stenosis and thrombosis. *Am J Physiol, 265,* H1787–H1796.

94. Slack, S. M., Cui, Yw., & Turitto, V. T. (1993). The effects of blood flow on coagulation and thrombosis. *Thromb Haemost, 76,* 129–134.

# CHAPTER 33

# Animal Models

## David H. Lee[1] and Morris A. Blajchman[2]

[1]Department of Medicine, Queen's University, Kingston, Ontario, Canada
[2]Department of Pathology and Molecular Medicine, McMaster University, Hamilton, Ontario, Canada

## I. Introduction

Over the past several decades, enormous advances have been made in understanding and manipulating platelet function *in vitro*. However, extending these observations and measuring the effects *in vivo* remains challenging. Animal models have the potential to fill the preclinical gap between *in vitro* and human studies in evaluating an intervention. In this setting, experimental animal models are only approximations to the human condition, but defining the limits of that approximation is difficult. They can also provide extremely useful insights into the biology of platelets in intact organisms, allowing a glimpse of how these anucleate cells, that are so often studied in isolation, fit into the orchestra of hemostasis, vascular biology, inflammation, and tissue modeling.

What makes a good experimental animal model? Above all, animal models should be reproducible and relevant to the research question being asked. Beyond this, the broad diversity of research questions makes it difficult to generate a stringent list of universal criteria of what makes a good animal model. As was recognized long ago, "models are different things to different people."[1] This chapter reviews the wide range of experimental animal models as simulations of platelet disorders or for evaluating *in vivo* platelet function and survival.

## II. Simulating Platelet Disorders in Animals

### A. *Disorders of Platelet Number*

#### 1. Methods of Inducing Experimental Thrombocytopenia

*a. Antiplatelet Antibodies.* For over a century, serum from animals immunized with platelets from another species has been infused into donor species to induce thrombocytopenia.[2–7] This method can reliably produce immediate severe thrombocytopenia, but the duration of action of a single dose of antiserum is usually short. Heterologous antiplatelet serum or commercially available polyclonal antiplatelet antibody has been used to produce immune thrombocytopenia in animals for the hemostatic evaluation of transfused lyophilized platelets,[8] fibrinogen-coated red cells,[9] and fibrinogen-coated albumin microcapsules.[10]

The use of antiplatelet antibodies to produce thrombocytopenia also provides a model for immune thrombocytopenic purpura (ITP). Aged (NZW × BXSB)F1 (W/B F1) hybrid mice spontaneously develop immune thrombocytopenia in the context of an antiphospholipid antibody-like syndrome.[11,12] Monoclonal antibody 6A6 derived from the splenocytes of (NZW × BXSB)F1 (W/B F1) mice induces isolated immune thrombocytopenia in normal mice,[13] producing a model of ITP that has been used in several studies investigating the pathobiology of ITP.[14–16] Heterologous antiplatelet sera[17] or commercially available monoclonal antibodies against mouse integrins αIIb or β3 can also be injected to simulate ITP in mice.[18–21] The propensity to bleed or develop acute systemic reactions upon injection with antiplatelet antibodies in mice appears to depend on the antigen specificity of such antibodies.[22]

*b. Ionizing Irradiation.* Another conventional method for inducing thrombocytopenia in animals is irradiation,[6,23–29] although the dosage must be carefully titrated to produce thrombocytopenia without excess morbidity and mortality. The platelet count nadir occurs 8 to 14 days after irradiation.[27,30,31] Circulating platelets at and prior to the nadir represent a cohort of "old" platelets, while platelets during the period of platelet count recovery represent "young" platelets that are more hemostatically effective.[32] Leukopenia also typically occurs several days after irradiation, often preceding the onset of thrombocytopenia. When bleeding is measured from a standardized incision, the site of incision should be shielded from potentially confounding irradiation-induced endothelial activation.

*c. Chemotherapy.* Like the use of irradiation, the challenge with the use of chemotherapeutic agents is to produce severe thrombocytopenia with minimal morbidity and mortality. Busulfan has been used by many investigators to experimentally induce thrombocytopenia in rabbits for the purpose of evaluating the hemostatic efficacy of platelet replacement therapy. Busulfan in doses of up to 70 mg/kg can be given to rabbits in one[4,33,34] or two injections.[5,10,35–37] Dual injection regimens have been used to produce thrombocytopenic rabbits for the hemostatic evaluation of thrombin-activated human platelets,[37] novel platelet products, and platelet substitutes.[10,36] Busulfan has also been used to produce thrombocytopenia in the rat to study the response of thrombopoietin to thrombocytopenia.[38]

Because thrombocytopenia is a dose-limiting toxicity of carboplatin, this agent has been used in the past decade to produce thrombocytopenia in primates,[39] dogs,[40] mice[41–44] and rats,[45,46] primarily for the evaluation of thrombopoietic agents.

*d. Combined Modalities.* A combination of irradiation followed by the administration of heterologous platelet antisera is used in a well-established rabbit model of severe thrombocytopenia for evaluating platelet replacement therapy.[6,31,32,47–51] Rabbits are exposed to 930 cGy from a $^{137}$Cs (cesium) source for 30 minute, with lead shielding of the ears for bleeding time studies. Heterologous platelet antiserum from immunized sheep is infused on day 8 after irradiation, producing severe thrombocytopenia (platelet count $<10 \times 10^9$/L) with 3 to 10% mortality.[31]

A combination of carboplatin and sublethal irradiation has been used to produce a murine model of life-threatening thrombocytopenic bleeding in which the control group has a 94% mortality and widespread hemorrhage.[52] This model was used to evaluate the efficacy of pegylated recombinant thrombopoietin in ameliorating lethal thrombocytopenia.

*e. Deficiency of Thrombopoietic Effectors.* Chronic thrombocytopenia of moderate severity in dogs has been produced by immunization with recombinant human thrombopoietin, due to the development of cross-reacting antibodies to canine thrombopoietin.[53] Knockout mice lacking c-Mpl, thrombopoietin, or conditional knockout mice lacking GATA-1, are also moderately thrombocytopenic.[54–57] In contrast, mice lacking the transcription factor NF-E2 are devoid of circulating platelets.[58,59]

*f. High Dose Estrogen.* A single injection of high dose estradiol has also been used to suppress thrombopoiesis without leukopenia in male dogs. The mean platelet count reached a nadir of $12 \times 10^9$/L on day 14, with recovery starting on day 17.[60]

*g. Cardiopulmonary Bypass Models.* Several animal models of cardiopulmonary bypass have also been developed to simulate the platelet dysfunction and thrombocytopenia that occurs clinically.[61–65] The efficacy of platelet products and substitutes has occasionally been tested in such models.[66]

*h. Thrombotic Thrombocytopenic Purpura.* Prior to the recognition of the pivotal role of ADAMTS13 in the pathogenesis of thrombotic thrombocytopenic purpura (TTP, Chapter 50), previous animal models of TTP were based on infusing animals with botrocetin, which initiates von Willebrand factor-mediated platelet agglutination, causing transient thrombocytopenia, microthrombi, and the immediate reduction of higher molecular weight multimers of von Willebrand factor.[67,68] The recent cloning, expression, and characterization of murine ADAMTS13[69] will likely serve as a first step toward the development of a genetically engineered mouse model of TTP.

*i. Heparin-Induced Thrombocytopenia.* Until the development of transgenic mice that express human components, heparin-induced thrombocytopenia (HIT, Chapter 48) has been difficult to simulate in animals. Platelet activation by the HIT antibody typically occurs through the platelet receptor for the Fc portion of IgG (FcγRIIA). However, many animal species, including mice, lack platelet FcγRIIA receptors. Furthermore, human HIT antibodies that recognize the complex of human platelet factor 4 (PF4)/heparin have very low cross-reactivity with murine PF4-heparin.[70,71] Notwithstanding these challenges, generation of an animal model by immunizing naive wild-type mice with anti-PF4/heparin antibodies from patient sera has been reported.[72]

Development of an animal model that contains the requisite components for HIT became possible with the use of transgenic mice. Different transgenic mouse lines that express human FcγRIIA on platelets[73] and human PF4[74] were crossbred to produce double transgenic mice expressing both human antigens. When these mice were injected with KKO, a platelet-activating murine monoclonal antibody that recognizes human PF4/heparin complexes,[71] they developed thrombocytopenia and thrombotic events. These effects were dependent on the presence of heparin, human FcγRIIA, human PF4, and KKO, thus recapitulating the clinicopathologic features of HIT.[75]

**2. Models of Thrombocytosis.** In contrast to thrombocytopenic animal models, there have been relatively few reports of animal models of thrombocytosis. While the examples listed below are all associated with increased platelet number, some are also associated with platelet dysfunction.

*a. Primary Thrombocytosis.* Thrombocytosis due to apparent myeloproliferative disorders can occur naturally in dogs[76,77] and horses, often without clinical sequelae.[78] Several methods of inducing constitutive thrombocytosis as part of a myeloproliferative disorder or conditions that resemble myeloproliferative disorders have been used. A myeloproliferative disorder associated with an elevated platelet count, splenomegaly, thrombosis, and hemorrhage can be produced by inoculating mice with a retrovirus carrying the middle T gene of the polyomavirus.[79] Mouse models producing the usual phenotypes of chronic myeloid leukemia have also been described.[80] However, preferential thrombocytosis and megakaryocytic hyperplasia with relatively modest granulocytosis has been reported in two models of BCR/ABL transgenic mice.[81,82] Thrombocytosis and increased megakaryocytopoiesis have also been observed in mice with mutations in hematopoietic regulatory elements p300[81] and c-Myb,[83] mice lacking the transcription factor Ikaros,[84] Lnk-null mice,[85] and transgenic mice that overexpress thrombopoietin.[86]

*b. Reactive Thrombocytosis.* Many of the biological response modifiers that drive reactive thrombocytosis can be administered pharmacologically to animals. For example, thrombocytosis in large and small animals occurs after the pharmacologic administration of thrombopoietin,[87–89] interleukin (IL)-6,[90–93] and IL-11.[94]

## B. Disorders of Platelet Function

### 1. Engineered Models of Platelet Disorders

*a. Knockout Mice.* The targeted inactivation of genes relevant to platelet development, function, and survival in the mouse has been a powerful tool for studying and modeling platelet biology. As such, they also have the potential to serve as disease models in which to evaluate therapies. Over the past decade, a large and ever-increasing number of different knockout mice have been reported, as summarized in Table 33-1. Many of these specific knockout mice are discussed in more detail in other chapters.

With the use of this technology, some challenges and surprises have emerged. First, homozygous deletion of a gene can be developmentally lethal. In order to study the null platelet phenotype in mature animals, the lethality problem can sometimes be bypassed by conditionally inactivating the gene of interest later in life using the Cre/loxP system.[100,171] Alternate strategies include transplanting hematopoietic stem cells from the fetal liver of null mutant mice prior to developmental collapse into lethally irradiated recipients,[132] or establishing transgenic mice that express a dominant-negative form of the protein of interest in megakaryocytes.[219]

Second, while mouse and human platelets are very similar, certain differences between the species have become more apparent from knockout studies. For example, protease activated receptors (PAR) 1 and PAR4 are the primary thrombin receptors on human platelets, but on mouse platelets PAR3 and PAR4 are the main receptors (Chapter 9),[111,220] Deletion of PAR1 in mice leads to a high rate of embryonic loss, but surviving mice have normal platelet responses to thrombin and do not have a bleeding phenotype.[113,114]

Third, surprising results have emerged on a few occasions. Paradoxically, instead of being predisposed to arterial thrombosis, mice deficient in endothelial nitric oxide synthase are protected from it, possibly due to the increased fibrinolysis associated with the genotype.[211] CD39/ATP-diphosphohydrolase knockout mice unexpectedly have impaired platelet function.[207] Finally, double knockout mice lacking both von Willebrand factor and fibrinogen unexpectedly are still able to form large occlusive thrombi in arterioles,[145] despite the belief that these ligands are crucial for platelet adhesion and thrombus formation. All of these interesting results serve as reminders of the complexity of hemostasis and provide researchers with more questions to be answered.

Finally, the different responses of phospholipase C (PLC) γ2 knockout mice to superficial and severe thrombotic stimuli serve as a reminder against drawing conclusions about the phenotype on the basis of a single model.[183]

### 2. Naturally Occurring Disorders of Platelet Function in Animals

*a. Disorders of Granule Formation and Trafficking.* The Hermansky–Pudlak syndrome (HPS) and Chediak–Higashi syndrome represent lysosome-related organelle disorders in which lysosomes, melanosomes, and platelet dense granules are affected (Chapters 15 and 57). Naturally occurring disorders resembling human Chediak–Higashi syndrome have been described in several animal species.[221–225] Of these, the beige mouse has been the most extensively characterized, and represents the most important animal model for Chediak–Higashi syndrome.[226–230]

There are at least 16 mouse mutants that serve as models for HPS,[231] many of which were originally identified among naturally occurring coat color mutations.[232] Many of these mouse mutant lines are now maintained on the same C57BL/6J genetic background, allowing comparisons between other mutants or the wild-type parent strain. Mutants that have similar coat colors have HPS mutations that cluster within common protein complexes (BLOC, or biogenesis of lysosome-related organelle complexes) involved in vesicle trafficking.[231] These mouse mutants have defective hemostasis associated with reduced or abnormal platelet dense granules.[226,233–235] The Fawn-hooded rat and Tester Moriyama rat are also models of HPS[236,237]; however, it appears that a common mutation in the Rab38 gene accounts for both.[238]

**Table 33-1: Mouse Knockout Models and Their Functional Effects on Platelets**

| Target of Gene Inactivation | | Functional Phenotype of Homozygous Knockout Mice in Comparison with Wild-Type Mice | References |
|---|---|---|---|
| Integrins | α2 | Delayed aggregation in response to fibrillar collagen; absent aggregation in response to digested soluble collagen; normal bleeding time | Holtkotter 2002[95] Chen 2002[96] |
| | αIIb | Absent aggregation in response to ADP and collagen; severe reduction in ability of platelets to bind fibrinogen; absent clot retraction; prolonged bleeding time; severe bleeding diathesis | Tronik-Le Roux 2000[97] |
| | αv | Embryonic lethal in 80%; live-born null mice are hemorrhagic, but platelet function not reported | Bader 1998[98] |
| | β1 | Embryonic lethal in complete knockout; conditional knockout permits survival with delayed aggregation in response to fibrillar collagen; absent aggregation in response to digested soluble collagen; normal bleeding time | Fassler 1995[99] Nieswandt 2001[100] |
| | β2 | Shortened platelet lifespan; enhanced platelet caspase activation; decreased localization of platelets to site of inflammation | Piquet 2001[101] |
| | β3 | Absent aggregation in response to ADP, thrombin, arachidonic acid, PMA; absent clot retraction; virtually absent fibrinogen binding; prolonged bleeding time; spontaneous hemorrhage; reduced thrombogenicity | Hodivala-Dilke 1999[102] Smyth 2001[103] |
| Leucine-rich repeat receptor family | GPIbα | Reduced platelet count; large platelets; prolonged bleeding time; "rescue" with transgenic expression of human GPIbα | Ware 2000[104] |
| | GPIbβ | Macrothrombocytopenia; large alpha granules; prolonged bleeding time; severe bleeding phenotype | Kato 2004[105] |
| | GPV | Normal platelet size; normal to increased responsiveness to thrombin; normal or shortened bleeding time; no bleeding diathesis; shorter time to occlusion and larger emboli in mesenteric thrombosis model | Ramakrishnan 1999[106] Kahn 1999[107] Ni 2001[108] |
| Thrombin receptors | PAR4 | Platelets unresponsive to thrombin; prolonged bleeding time; protection against experimental arterial thrombosis; *in vivo* effects are due to loss of PAR4 on hematopoietic cells rather than endothelium | Sambrano 2001[109] Hamilton 2004[110] |
| | PAR3 | Diminished and delayed platelet response to thrombin; bleeding time normal[111] or prolonged,[112] depending on method; protection against experimental arterial and venous thrombosis | Kahn 1998[111] Weiss 2002[112] |
| | PAR1 | Approximately 50% embryonic loss; null mice that survive have normal platelet response to thrombin, no bleeding tendency, and appear to be anatomically and histologically normal | Connolly 1996[113] Darrow 1996[114] |
| ADP receptors | P2Y1 | Absent to reduced aggregation in response to ADP; decreased sensitivity to other agonists; no spontaneous hemorrhage; prolonged bleeding time; relatively resistant to experimental thromboembolism | Leon 1999[115] Fabre 1999[116] |
| | P2Y12 | Absent to reduced aggregation in response to ADP; decreased sensitivity to other agonists; prolonged bleeding time; no spontaneous hemorrhage; prolonged time to occlusion in mesenteric artery injury model | Foster 2001[117] Andre 2003[118] |
| ATP receptor | P2X1 | Reduced aggregation in response to low dose collagen; normal bleeding time; no spontaneous hemorrhage | Hechler 2003[119] |
| Prostaglandin family receptors | TP (TXA2) receptor | Absent aggregation in response to U46619; delayed collagen-induced aggregation; normal ADP-induced aggregation; prolonged bleeding time | Thomas 1998[120] |

**Table 33-1: Mouse Knockout Models and Their Functional Effects on Platelets—*Continued***

| Target of Gene Inactivation | | Functional Phenotype of Homozygous Knockout Mice in Comparison with Wild-Type Mice | References |
|---|---|---|---|
| | IP (PGI$_2$, prostacyclin) receptor | Anti-aggregatory effect of prostacyclin analog abolished; failure of prostacyclin analogue to prolong normal bleeding time; predisposition to experimental arterial thrombosis | Murata 1997[121] |
| | EP3 (PGE$_2$) receptor | Loss of PGE$_2$-induced potentiation of aggregation; normal[122] or prolonged bleeding time[123]; resistance to experimental thrombosis | Fabre 2001[122] Ma 2001[123] |
| Immunoglobulin superfamily of receptors | GPVI | No aggregation in response to fibrillar collagen and convulxin; loss of *ex vivo* thrombus formation on surface-bound collagen; no major effect on bleeding time | Kato 2003[124] |
| | PECAM-1 (CD31) | Normal aggregation in response to ADP and thrombin; increased aggregation and thrombus size in response to collagen and VWF; variable effect on bleeding time; no effect on experimental thrombosis | Duncan 1999[125] Mahooti 2000[126] Jones 2001[127] Vollmar 2001[128] Rathore 2003[129] |
| | CD47 | Mild thrombocytopenia; increased sensitivity to experimental ITP; decreased aggregation in response to collagen; reduced adhesion of resting platelets to inflammatory vascular endothelium | Olsson 2005[16] Lagadec 2003[130] Chung 1999[131] |
| | FcR γ-chain | Resistant to ITP; platelet release and adhesion to collagen abolished; normal response to thrombin | Clynes 1995[14] Poole 1997[132] Kato 2003[124] |
| Tetraspanins | CD151 | Reduced aggregation in response to multiple agonists; normal adhesion; impaired clot retraction; prolonged bleeding time | Wright 2004[133] Lau 2004[134] |
| Chemokine receptors | CXCR4 | Fetal lethality; reduced embryonic hematopoiesis | Zou 1998[135] |
| | CCR4 | Normal resting platelet count; no effect on LPS-induced thrombocytopenia | Ma 1998[136] Chvatchko 2000[137] |
| Other receptors | P-selectin | Prolonged bleeding time; normal platelet survival; platelet rolling mediated by endothelial rather than platelet P-selectin | Frenette 1995[138] Subramaniam 1996[139] Berger 1998[140] |
| | A2a (adenosine) receptor | Enhanced ADP-induced platelet aggregation; insensitivity to anti-aggregating effects of adenosine analogues | Ledent 1997[141] |
| | CD36 (GPIV) | Platelet function not reported in CD36-deficient mice[142]; however, CD36-deficient rats lack VLDL-induced enhancement of collagen-induced aggregation[143] | Moore 2002[142] Englyst 2003[143] |
| Adhesive ligands | VWF | Prolonged bleeding time; spontaneous hemorrhage; reduced initial platelet deposition and occlusion rate in experimental thrombosis model | Denis 1998[144] Ni 2000[145] |
| | Fibrinogen α-chain | Absent aggregation; spontaneous bleeding; normal initial platelet deposition at site of arterial injury, but increased embolization in experimental thrombosis model | Suh 1995[146] Ni 2000[145] |
| | Vitronectin | Variable effects on platelet aggregation across studies; enhanced[147] and reduced[148,149] *in vivo* thrombogenicity reported | Fay 1999[147] Eitzman 2000[148] Reheman 2005[149] |
| | Fibronectin | Embryonic lethal in complete knockout; conditional knockout permits survival with normal collagen-induced aggregation, clot retraction, and bleeding time; delayed vessel occlusion in thrombosis model | George 1993[150] Sakai 2001[151] Ni 2003[152] |
| Matricellular and other secreted proteins | Thrombospondin 2 | Impaired aggregation in response to ADP; normal aggregation in response to thrombin and collagen; abnormal megakaryocyte structure; prolonged bleeding time; reduced thrombus formation following endothelial injury *in vivo* | Kyriakides 1998[153] Kyriakides 2003[154] |

DAVID H. LEE AND MORRIS A. BLAJCHMAN

**Table 33-1: Mouse Knockout Models and Their Functional Effects on Platelets—*Continued***

| Target of Gene Inactivation | | Functional Phenotype of Homozygous Knockout Mice in Comparison with Wild-Type Mice | References |
|---|---|---|---|
| | Thrombospondin 1 | Normal aggregation in response to thrombin; increased aggregation in response to collagen and ADP; smaller VWF multimer size in plasma, yet larger in releasate from activated platelets; normal bleeding time | Lawler 1998[155] Pimanda 2002[156] |
| | CD40 ligand (CD154) | Normal aggregation at low shear, reduced platelet aggregation at high shear; normal[157] or prolonged bleeding time[158]; delayed vessel occlusion after thrombogenic injury | Andre 2002[157] Crow 2003[158] |
| | Platelet factor 4 | Reduced aggregation in response to low thrombin concentration; normal bleeding time; reduced thrombosis in vascular injury model | Eslin 2004[159] |
| Thrombopoietic effectors | c-Mpl | Reduction in platelet count and bone marrow megakaryocyte count; no spontaneous hemorrhage | Gurney 1994[54] |
| | Thrombopoietin (Mpl-ligand) | Reduction in platelet count and bone marrow megakaryocyte count | de Sauvage 1996[55] |
| | IL-11Rα | Normal platelet count | Nandurker 1997[160] |
| | IL-6 | Normal platelet count | Bernad 1994[161] |
| | IL-3 | Normal platelet count | Gainsford 1998[162] |
| | Leukemia inhibitory factor | Normal platelet count | Escary 1993[163] |
| | Stat3 | Embryonic lethal | Takeda 1997[164] |
| | GATA-1 | Embryonic lethal in full knockout; targeted mutation produces thrombocytopenia and defective megakaryocyte maturation; spherocytic platelets; impaired aggregation in response to collagen; mild defect in P-selectin expression | Fujiwara 1996[165] Shivdasani 1997[56] Vyas 1999[57] Hughan 2005[166] |
| | c-Myb | Embryonic lethal | Mucenski 1991[167] |
| | Lnk | Thrombocytosis and increased megakaryocytopoiesis | Velazquez 2002[85] |
| | NF-E2 (p45 subunit) | Complete absence of circulating platelets; inability of megakaryocytes to form proplatelets | Shivdasani 1995[58] Lecine 1998[59] |
| | Ikaros | Thrombocytosis and increased splenic megakaryocytopoiesis | Lopez 2002[84] |
| G-proteins | $G_{\alpha i2}$ | Reduced aggregation and fibrinogen binding in response to ADP and low thrombin concentration; resistance to experimental thromboembolism | Jantzen 2001[168] |
| | $G_{\alpha z}$ | Normal aggregation in response to agonists when used alone; loss of potentiating effect of epinephrine on agonist-induced aggregation | Yang 2000[169] |
| | $G_{\alpha 12}$ | Viable with normal platelet aggregation and secretion; normal bleeding time; no apparent abnormality | Gu 2002[170] Moers 2003[171] |
| | $G_{\alpha 13}$ | Embryonic lethal in full knockout; conditional knockout characterized by decreased platelet aggregation to thrombin, U46619; normal aggregation to ADP; prolonged bleeding time; resistance to thrombosis induced by arterial injury | Offermanns 1997[172] Moers 2003[171] Moers 2004[173] |
| | $G_{\alpha q}$ | Absent aggregation in response to ADP, thrombin, collagen, U46619; prolonged bleeding time; resistance to experimental thromboembolism | Offermanns 1997[174] |
| | $G_{\alpha q}$ and $G_{\alpha 13}$ | Absent aggregation in response to ADP, thrombin, U46619; normal collagen Adhesion, but no aggregation in response to collagen | Moers 2004[173] |
| PI3-kinases | PI3K p110γ | Mild aggregation defect in response to ADP and U46619; loss of second wave of aggregation; normal response to thrombin; impaired platelet spreading; normal bleeding time; resistance to thromboembolism | Hirsch 2001[175] Li 2003[176] Lian 2005[177] |

**Table 33-1: Mouse Knockout Models and Their Functional Effects on Platelets—*Continued***

| Target of Gene Inactivation | | Functional Phenotype of Homozygous Knockout Mice in Comparison with Wild-Type Mice | References |
|---|---|---|---|
| | PI3K p110δ | Modest aggregation defect in response to CRP; normal adhesion under flow; no effect on thrombogenicity | Jackson 2005[178] Senis 2005[179] |
| | PI3K p85α | Reduced collagen-induced aggregation and spreading; normal responses to other agonists; normal bleeding time | Watanabe 2003[180] |
| PLC | PLCγ2 | Reduced collagen-induced aggregation and adhesion; normal aggregation to other agonists; prolonged bleeding time; partial resistance to thrombosis | Mangin 2003[181] Suzuki-Inoue 2003[182] Nonne 2005[183] |
| | PLCβ2/β3 | Mild aggregation defect in response to ADP, collagen, thrombin; impaired platelet spreading; resistance to thrombosis | Lian 2005[177] |
| Src family kinases | Lyn | Reduced thrombin-induced aggregation and secretion | Cho 2002[184] |
| | Fyn | Normal aggregation in response to thrombin | Cho 2002[184] |
| | Src | Normal aggregation in response to thrombin | Cho 2002[184] |
| Vav family GTP exchangers | Vav1 | Impaired aggregation in response to thrombin and CRP; no spontaneous hemorrhage | Pearce 2002[185] |
| | Vav2 | Normal aggregation in response to thrombin and CRP | Pearce 2002[185] |
| | Vav3 | Normal aggregation in response to collagen and CRP | Pearce 2004[186] |
| Tec family kinases | Btk | Reduced aggregation in response to collagen and CRP; normal aggregation in response to ADP; absent expression of P-selectin in response to CRP | Atkinson 2003[187] |
| | Tec | Reduced aggregation in response to collagen and CRP; normal aggregation in response to ADP; normal expression of P-selectin in response to CRP | Atkinson 2003[187] |
| Gas6 and its receptors | Gas6 | Impaired aggregation in response to ADP, collagen, U46619; resistant to arterial and venous thrombosis; normal blood loss after tail clipping; no spontaneous bleeding | Angelillo-Scherrer 2001[188] |
| | Tyro3 | Impaired aggregation to ADP, collagen, U46619; impaired clot retraction; impaired adhesion; normal bleeding time but increased blood loss; resistance to experimental thrombosis | Angelillo-Scherrer 2005[189] |
| | Axl | Impaired aggregation to ADP, collagen, U46619; impaired clot retraction; impaired adhesion; normal bleeding time but increased blood loss; resistance to experimental thrombosis | Angelillo-Scherrer 2005[189] |
| | Mer | Impaired aggregation to collagen, U46619; impaired[189] or normal[190] aggregation to ADP; impaired adhesion; normal bleeding time; increased blood loss; resistance to thrombosis | Chen 2004[190] Angelillo-Scherrer 2005[189] |
| Other signaling and adapter proteins | SLP-76 | Impaired aggregation and secretion in response to collagen; normal response to thrombin; normal bleeding time but spontaneous hemorrhages | Clements 1999[191] |
| | LAT (linker for activation of T-cells) | Reduced P-selectin expression and fibrinogen binding in platelets stimulated with CRP; normal P-selectin expression in thrombin-stimulated platelets | Pasquet 1999[192] |
| | cGMP-dependent protein kinase (PKG) | Enhanced or reduced platelet activity, depending on assay and conditions; prolonged bleeding time[193], yet enhancement of platelet adhesion in ischemia/reperfusion–induced injury[194] | Li 2003[193] Massberg 1999[194] Li 2004[195] |
| | Vasodilator-stimulated phosphoprotein (VASP) | Enhanced platelet activation in response to thrombin; increased platelet adhesion to vessel wall, unresponsive to inhibitory effect of NO-pretreatment of platelets | Hauser 1999[196] Massberg 2004[197] |
| | Akt-1 | Impaired[198] or normal[199] platelet aggregation and bleeding time | Chen 2004[198] Woulfe 2004[199] |
| | Akt-2 | Impaired aggregation in response to PAR4 agonist and U46619; normal bleeding time; resistant to thrombosis | Woulfe 2004[199] |

**Table 33-1: Mouse Knockout Models and Their Functional Effects on Platelets—*Continued***

| Target of Gene Inactivation | Functional Phenotype of Homozygous Knockout Mice in Comparison with Wild-Type Mice | References |
|---|---|---|
| | Syk | High embryonic/perinatal mortality; reduced aggregation and secretion in response to collagen; normal response to thrombin; normal bleeding time | Turner 1995[200] Poole 1997[132] |
| | Rap1b | 85% embryonic and perinatal lethality; reduced aggregation; prolonged bleeding time; resistance to thrombosis | Chrzanowska-Wodnicka 2005[201] |
| | SHIP1 | Enhanced platelet adhesion; increased resting P-selectin expression | Pasquet 2000[202] Maxwell 2004[203] |
| | Gads | Normal P-selectin expression in response to thrombin and convulxin | Judd 2002[204] |
| | Fps | Enhanced collagen-induced aggregation | Senis 2003[205] |
| Other proteins | COX-1 | Decreased aggregation in response to arachidonic acid | Langenbach 1995[206] |
| | CD39/ATP-diphosphohydrolase | Reduced aggregation in response to ADP, collagen, and thrombin; prolonged bleeding time; reduced thrombus formation in arterial injury model | Enjyoji 1999[207] |
| | CD79/ecto-5'-nucleotidase | Shortened bleeding time; shortened time to experimental carotid artery occlusion | Koszalka 2004[208] |
| | TXA(2) synthase | Prolonged bleeding time; aggregation markedly reduced in response to arachidonic acid, normal to ADP and thrombin | Yu 2004[209] |
| | eNOS | Shortened bleeding time; increased platelet recruitment; no effect on platelet P-selectin expression; paradoxical delay in occlusion in carotid injury model; increased fibrinolysis | Freedman 1999[210] Iafrati 2005[211] |
| | Gelsolin | Impaired shape change; prolonged bleeding time | Witke 1995[212] |
| | β1-tubulin | Defective proplatelet formation; thrombocytopenia; loss of discoid shape; normal aggregation and release; mild impairment of P-selectin expression; prolonged bleeding time | Schwer 2001[213] Italiano 2003[214] |
| | MYH9 | Embryonic lethal | Matsushita 2004[215] |
| | CDCrel-1 | Aggregation and granule release in response to subthreshold levels of collagen | Dent 2002[216] |
| | Tryptophan hydroxylase | Normal aggregation; prolonged bleeding time; reduced thrombosis | Walther 2003[217] |
| | Pituitary adenylate cyclase-activating polypeptide | Increased aggregation in response to collagen | Freson 2004[218] |

Abbreviations: cGMP, cyclic guanosine 3',5'-monophosphate; COX-1, cyclo-oxygenase 1; CRP, collagen-related peptide; eNOS, endothelial nitric oxide synthase; GP, glycoprotein; IL, interleukin; ITP, immune thrombocytopenic purpura; LPS, lipopolysaccharide; NO, nitric oxide; PAR, protease activated receptor; PECAM-1, platelet/endothelial adhesion molecule 1; PG, prostaglandin; PI3K, phosphoinositide 3-kinase; PLC, phospholipase C; PMA, phorbol myristate acetate; TX, thromboxane; VLDL, very low density lipoprotein; VWF, von Willebrand factor.

There are few descriptions of α-granule deficiency in animals. The Wistar Furth rat has a hereditary macrothrombocytopenia with platelets that are deficient in α-granule proteins, thus resembling human gray platelet syndrome[239,240] (Chapters 15 and 57). Platelet spreading is delayed and the bleeding time is prolonged.[241]

*b. Glanzmann Thrombasthenia.* Several naturally occurring cases of type I Glanzmann thrombasthenia have been described in dogs, and they are associated with a severe clinical bleeding phenotype.[242–244] A thrombasthenic variant has been described in a closed colony of Wistar rats,

characterized by a thrombasthenic aggregation profile and decreased integrin αIIbβ3 and fibrinogen binding.[245]

*c. Simmental Cattle Thrombocytopathy.* A severe hereditary bleeding diathesis associated with spontaneous hemorrhage has been described in Simmental cattle. Platelet aggregation in response to most agonists is markedly impaired to absent in affected animals, without evidence of a secretory defect or an ultrastructural abnormality.[246–250]

*d. Basset Hound Thrombocytopathy.* A thrombocytopathy has also been described in basset hounds in which platelets from affected animals have a thrombasthenic-like aggregation defect, but are able to bind fibrinogen normally in response to agonists. Affected animals have electrophoretically normal platelet membrane glycoproteins.[251–253]

## III. Measurement of Platelet Function and Survival in Animals

### A. Measurement of Bleeding Tendency

Measures of bleeding tendency are required when evaluating the efficacy of platelet replacement therapy for thrombocytopenia or the effects of interventions that are hypothesized to impair the hemostatic function of platelets. This section reviews the most commonly used approach of measuring bleeding from a standardized incision (usually the bleeding time or blood loss), but also discusses seldom-used methods to measure the spontaneous red cell losses associated with thrombocytopenia.

#### 1. Bleeding from Standardized Wounds

*a. Microvascular Ear Bleeding Time.* The most commonly used animal model for evaluating the hemostatic effectiveness of platelet products and substitutes is the microvascular ear bleeding time in thrombocytopenic rabbits. Several methodologic variations differ in the type of incision (template[10,254] or freehand[31]), or the method of removal of the shed blood (saline immersion,[31] saline irrigation,[29,255] or blotting with filter paper[10,36,254]).

The McMaster group has used the microvascular ear bleeding time extensively to evaluate platelet products, substitutes, and other agents in thrombocytopenic rabbits.[6,31,32,47,48,50,51,256,257] In this method, the ear is prewarmed in a 37°C saline bath containing a magnetic stirrer. A full thickness freehand incision is then made though the ear with a #11 scalpel blade at a site that is devoid of macroscopically visible vessels. The incised ear is immediately reimmersed into the stirred warm saline bath. The time taken for visible bleeding to cease is recorded. At least two bleeding time determinations are recorded for each rabbit and the mean is used for statistical analyses.[31] When evaluating human platelets (which normally have a very short survival in rabbits), ethyl palmitate is preinfused to block the reticuloendothelial system. This prolongs the survival of transfused human platelets to several hours, thereby permitting the assessment of their hemostatic function.[31,37]

The inverse relationship between the platelet count and microvascular ear bleeding time in thrombocytopenic rabbits is shown for endogenous young and old rabbit platelets in Fig. 33-1. A similar relationship between the platelet count and bleeding time has been reported by others.[5,167] When rabbits with profound thrombocytopenia are trans-

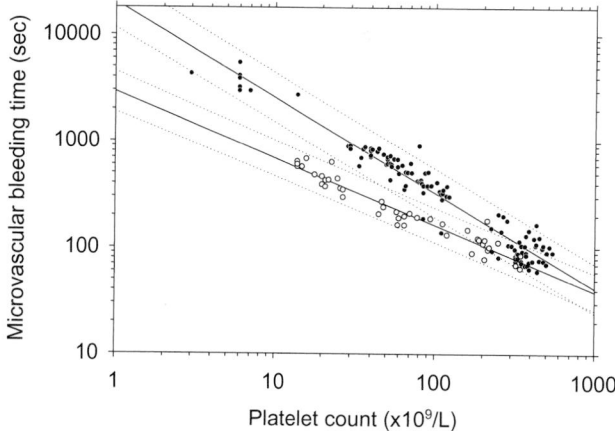

**Figure 33-1.** The relationship between the microvascular ear bleeding time and platelet count for old (closed circles) and young (open circles) rabbit platelets. The solid line and dotted lines represent the linear regression lines and 95% prediction limits for old and young platelets.

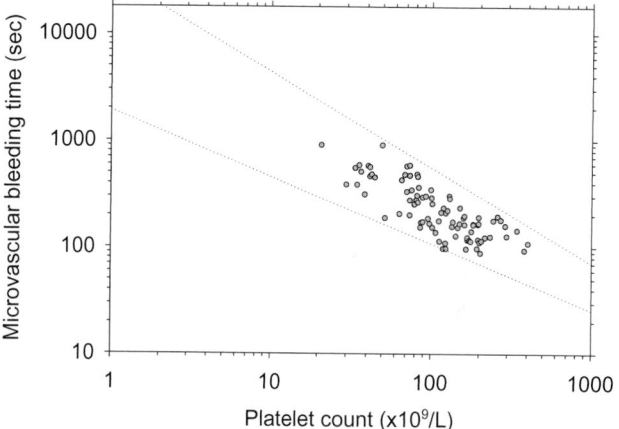

**Figure 33-2.** Fresh human platelets transfused into thrombocytopenic ethyl palmitate-treated rabbits correct the microvascular ear bleeding time as well as endogenous rabbit platelets. The platelet count prior to transfusion was $<10 \times 10^9$/L. The dotted lines represent the upper and lower 95% prediction limits for the linear regression lines derived from bleeding times for old and young endogenous rabbit platelets, respectively (see Fig. 33-1).

fused with human platelets, the bleeding times shorten in inverse proportion to the platelet count after transfusion (Fig. 33-2). Because the magnitude of the bleeding time prolongation increases with the severity of thrombocytopenia, the use of animals with severe thrombocytopenia may permit a hemostatic effect to be more easily detected. Furthermore, severe thrombocytopenia is the setting in which platelet replacement therapy is most commonly used clinically.

*b. Tail Bleeding Time.* The tail bleeding time is widely used in mice and rats, particularly when evaluating pharmacologic or genetic interventions (Table 33-1) that impair platelet hemostatic function. The most widely used method is a complete transection of the tail 1 to 5 mm from the tip. Alternate techniques include a template-assisted longitudinal incision on the dorsal part of the tail at a set distance from the tip,[258] a transverse incision over a lateral tail vein at a position where the diameter of the tail is 2.25 to 2.5 mm,[259] and a single puncture of the dorsal tail vein with a 23-gauge needle.[210] The rat tail bleeding time depends on a variety of variables that should all be controlled, including tail position (vertical or horizontal), temperature, method of incision (transection or longitudinal), air versus saline immersion, and the use of anesthesia or not.[258,260] In mice, the bleeding time also varies with the genetic strain.[259,261] While the bleeding time in humans and small rodents are both platelet dependent, the transection tail bleeding time in mice appears to be more sensitive to hemophilia and heparin than the skin bleeding time in humans.[262–266] In contrast, the longitudinal incision and lateral tail vein bleeding times appear to be less sensitive to impairment of coagulation than transection bleeding times.[259,265]

*c. Other Bleeding Time Models.* The rabbit jugular vein bleeding time is performed by puncturing the exposed jugular vein with a 23-gauge needle.[7,28] Because of the technical challenges in avoiding damage to the vessel and the disruption of nearby tributary vessels that result in bleeding near the puncture site, this method has largely been abandoned in favor of other measures of bleeding. The dog buccal mucosal bleeding time,[267–269] and a cuticle bleeding time methods in rats, mice, rabbits, and dogs have also been described.[245,270–272]

*d. Blood Loss Measurement from Standardized Incisions.* The volume of blood shed from a standardized wound into a water bath can be determined using $^{51}$Cr-labeled red cells[273] or colorimetrically.[51,109,274,275] Blood from standard injuries can also be determined by weighing the gauze into which the shed blood has been absorbed.[10]

A kidney injury model has been developed to assess the efficacy of transfused human platelets in rabbits.[37] In this model, anesthetized rabbits undergo laparotomy and the anterior pole of the kidney is removed with a cut that slices through the medulla. Blood loss is determined from the weight of gauze pads that have absorbed the shed blood. The increased hemorrhage observed in thrombocytopenic rabbits can be corrected with the transfusion of fresh human platelets, thrombin-activated fresh platelets, and platelets cryopreserved in dimethylsulfoxide (DMSO), but not by cold-stored platelets or platelets frozen without DMSO.[37,276]

*e. Limitations of Standardized Wound Approach.* Methods that employ standardized wounds have limitations that must be acknowledged. First, the measurement of the bleeding time and blood loss from standardized incisions when evaluating platelet replacement therapy in thrombocytopenic animals is based on the unproven assumption that these measurements actually predict thrombocytopenic bleeding. In animals, as in humans, the *in vivo* measurement of thrombocytopenic bleeding remains problematic.

Second, a prolonged bleeding time in one anatomic location using one technique does not necessarily predict a prolonged bleeding time in another location or in the same location using a different technique. For example, the rat ear bleeding time is prolonged by aspirin but the rat brain bleeding time is not affected.[277] Similarly, aspirin prolongs tail bleeding time when the tip is transected, but not when a longitudinal template incision is made.[265] Finally, mice lacking a fibrinogen motif had a prolonged cuticle bleeding time, but normal bleeding times from a full thickness skin incision.[270] Such discrepancies may reflect differences in the size of vessels that are disrupted by each type of standardized wound, or possibly local hemostatic differences.

Third, the wide range of bleeding times for any given platelet count limits the utility of the bleeding time as a predictor of the effective platelet count in individual animals. A similar conclusion has been made for the use of the bleeding time in humans (Chapter 25).[278] However, evaluation of platelet replacement therapy or interventions that inhibit platelet function involves the analysis of groups of animals rather than making predictions for individual animals. Differences in mean bleeding times between groups may still be discernable even though it may not be possible to attribute any single individual bleeding time value to any one group.

**2. Measurement of Spontaneous Red Cell Losses in Thrombocytopenic Animals.** It is believed that a minimum circulating platelet mass is required to maintain the integrity of the vascular endothelium, and that severe thrombocytopenia results in the generalized extravasation of red cells from the microvasculature into the extravascular compartment and lymph. Mucosal microhemorrhages may also result in loss of red cells into the gastrointestinal tract. Thus, red cell losses in thrombocytopenic animals may be measured without inflicting a standardized wound. In these models, the hemostatic efficacy of platelet replacement therapy is determined by the suppression of spontaneous red cell extravasation or hemorrhage that normally occurs during thrombocytopenia, rather than the effect on bleeding from an experimental incision.

*a. Lymph Red Cell Count.* The measurement of red cell content in lymph as a gauge of thrombocytopenic bleeding is based on the premise that vascular integrity is compromised by severe thrombocytopenia resulting in the generalized extravasation of red cells from the microvasculature into the extravascular compartment and into lymph. Studies performed half a century ago ascertained that the red cell count in thoracic duct lymph increased dramatically in animals with radiation-induced thrombocytopenia,[279] an effect that was immediately ameliorated by the transfusion of fresh platelets.[23,24,26] A similar but less robust relationship between thrombocytopenia and the red cell count in lymph obtained from the ear was observed in irradiated rabbits.[27] Unfortunately, such methods are technically difficult because the disruption of lymph flow and even small amounts of traumatic hemorrhage into lymph cannot be tolerated.[280] As a result, lymph red cell measurement is rarely used.

*b. Fecal $^{51}$Cr-Blood Loss.* Chromium-labeled fecal blood loss has been used to assess spontaneous thrombocytopenic bleeding from the gastrointestinal tract in patients[281] and rabbits.[282] Rabbits are infused with $^{51}$Cr-labeled red cells and the radioactivity in a 24-hour fecal sample is determined. This method has a reported sensitivity of $1 \times 10^{-4}$ mL of blood per gram of feces. The mean daily gastrointestinal blood loss in rabbits rendered severely thrombocytopenic by busulfan was 1.5 mL compared with 0.12 mL in control animals.[282]

## B. Measurement of Thrombotic Tendency

**1. Acute Platelet-Dependent Macrovascular Arterial Thrombosis.** The contribution of the platelet-rich thrombus to acute myocardial infarction, unstable angina, thrombotic stroke, and thrombosis following coronary interventions has led to the development of many different *in vivo* models of experimental thrombosis to evaluate the effectiveness of antithrombotic agents. Such tools have also been used to evaluate the *in vivo* effects of genetic and molecular interventions that target various receptors, signal transduction proteins, and other effector proteins relevant to platelets in knockout mice. There is an extensive body of literature in this area, and this section will focus on only some of the established techniques that have been used to evaluate platelet function. These models are primarily arterial or, in some cases, involve arteriovenous shunts. While venous stasis thrombosis models are affected by platelet reactivity, these models are more dependent on thrombin generation and fibrin deposition and will not be reviewed here.

*a. Folts Model.* The Folts model is the prototypical model for evaluating the *in vivo* antithrombotic properties of pharmacologic agents that inhibit *in vitro* platelet function. In this model, blood flow in the circumflex artery of dogs is continuously measured with a flow-meter probe and a plastic cylinder is placed around the vessel, producing approximately 70% stenosis and intimal damage is induced by compressing the artery.[283] Blood flow falls to zero as

platelets accumulate to produce an occlusive thrombus. Depending on the degree of stenosis and intimal damage, the thrombus will dislodge either spontaneously or by shaking the constrictor, temporarily restoring flow until platelets reaccumulate again to occlude the vessel. These cyclic flow reductions (CFRs) are repetitive and unchanging over hours which permits the testing of one or more doses of antithrombotic agents, using each animal as its internal control. Detailed technical considerations for this method using coronary or carotid arteries in dogs, pigs, monkeys, and rabbits have been summarized elsewhere.[284]

Histologically, the thrombi formed in the Folts model are platelet-rich, with some red cells, but are relatively fibrin-poor. CFRs can be reduced or abolished by aspirin[283] and a variety of other antiplatelet agents but not usually by heparin alone.[284–287] The similarities between the results from such animal studies and clinical studies cautiously suggest some generalizability to the human condition.[285] CFRs have been reported in humans[288,289] and the Folts model has been cited as a model for acute coronary syndromes in humans.[290] However, the thrombogenic lipid core of atherosclerotic plaques and the underlying risk factors for coronary artery disease in humans are not reproduced in these models.[285]

*b. Other Techniques for Producing Acute Macrovascular Arterial Thrombosis.* Acute platelet-rich thrombosis in the arteries of animals can be induced using electrical current,[51,291–296] topical ferric chloride,[145,147,297–299] and photochemical injury.[148,300–303]

Ferric chloride ($FeCl_3$) is frequently used in mice and other animals to induce platelet-rich arterial thrombi. Filter paper that is saturated with ferric chloride is applied topically to the exposed artery, inducing a transmural injury that denudes the endothelium and exposes the subendothelial matrix.[145] The rate of occlusive thrombus formation is dependent on the concentration of ferric chloride and size of the filter paper.[304,305]

Photochemical injury-induced thrombosis can be produced by the intravenous administration of Rose Bengal followed by exposure of the artery to green laser light, producing a reactive oxygen species that injures the endothelium.[306] Variables that should be kept constant to maximize reproducibility include the vein through which infusion occurs, the duration of the infusion, the distance of the light source, and the time interval between Rose Bengal administration and illumination.[305]

While ultrasonic Doppler or electromagnetic vascular probes are generally used to measure blood flow through the vessel of interest, arterial wall temperature distal to the thrombosis has also been used as a surrogate for blood flow.[294,296,297] Platelet deposition onto injured vessels or artificial materials can be measured objectively *in vivo* using $^{111}$In-labeled platelets.[307–309] Radiolabeled platelets are also a useful means to localize and quantitate experimentally induced embolic platelet thrombi.[302,310–312] Platelet-dependent thrombus deposition onto cotton or silk thread dwelling within vessels or prosthetic shunts is another end point that has been used to evaluate antithrombotic agents.[313–315]

The different models of arterial thrombosis vary in the degree of platelet and fibrin dependence as well as their susceptibility to thrombus prevention, lysis, and reocclusion.[286] These specific properties and nuances cannot be predicted and are primarily empirically derived. Some models, such as the Folts model, have been extensively characterized and have been widely accepted as a reproducible model of platelet-dependent thrombosis in the field of cardiology, while the detailed characteristics of some other animal models remain to be defined.

### 2. Acute Platelet-Dependent Microvascular Arterial Thrombosis

*a. Intravenous Administration of Platelet Agonists.* The transient decrease in circulating platelet count that occurs after intravenous infusion of platelet agonists was first recognized as a crude model of *in vivo* platelet aggregation decades ago.[316–318] The subsequent use of $^{51}$Cr- or $^{111}$In-labeled platelets permitted the simultaneous detection of accumulated platelet aggregates in the pulmonary circulation.[319] Such models have been used extensively to evaluate the *in vivo* efficacy of antithrombotic drugs.[280,320–323] More recently, intravenous infusion of ADP or collagen and epinephrine has been used as a measure of thrombosis in knockout mice (Table 33-1).[103,115,174,175,189] Because death commonly occurs from pulmonary embolism, mortality is used as an end point in many of these studies.

*b. Intravital Microscopy.* Intravital microscopy, the technique that Bizzozero used in guinea pigs over 120 years ago to first identify platelets anatomically and assign their roles in hemostasis and thrombosis[324] (see the Foreword to this book), has been reborn as a powerful method to study real-time platelet thrombus formation *in vivo* in animals (Chapter 34). This recent popularity has been made possible because of advances in microscopy, fluorescent probes, and image analysis software.

Arterioles (from 30 to 120 μm in diameter) in the cremaster muscle,[325,326] mesentery,[109,144,145,183,327] or, less commonly, the ear,[307,328] are examined by fluorescence videomicroscopy, although large vessel models have also been described.[329] Vessel injury is induced by ferric chloride,[109,144,145] laser applied through the microscope objective,[183,307,326] or, less commonly, by photochemical means.[307] Platelets are fluorescently labeled with calcein acetoxymethyl ester,[109,144,144,329] carboxyfluorescein diacetate succinimidyl ester,[217] or monoclonal antibody[326] to facilitate detection. Additional layers of data can be obtained by

the simultaneous imaging of fluorescently labeled tissue factor, fibrinogen,[326] and leukocytes[329] in the developing thrombus.

A variety of end points can be ascertained both visually and with the assistance of image analysis software. Thrombus formation, number and size of emboli, and vessel patency are all end points that can be quantified. Centerline erythrocyte velocity, flow, and shear can also be calculated. Composite images using confocal fluorescence and bright-field microscopy permit a comprehensive picture of the spatial–temporal dynamics of thrombus evolution in real-time and in real-life.[326] It is anticipated that as additional technical advances in optical sectioning such as two-photon microscopy[330,331] become incorporated into vascular biology research, even more exciting glimpses of how platelets function *in vivo* will appear.

Real time *in vivo* imaging of platelets during thrombus formation is discussed in more detail in Chapter 34.

## C. Platelet Survival

### 1. Isotopic Labeling.
Isotopic labeling of platelets with $^{111}$In or $^{51}$Cr represents the gold standard for performing studies of platelet recovery and survival. These methods have been well described for several species.[332–335] Dual isotopic labeling methods used in humans also permit the simultaneous measurement of survival for two platelet populations in animals.[336,337]

### 2. Nonisotopic Labeling.
In the past two decades, methods of fluorescent platelet labeling have been gaining popularity. Fluorescent labeling permits the investigator to conveniently identify circulating platelet populations using flow cytometry (Chapter 30) and simultaneously determine the expression of platelet surface antigens of interest in the identified platelet populations. Moreover, labeled platelets can be identified in thrombi using fluorescence microscopy (Chapter 34).

*a. Biotinylation.* A variety of nonisotopic methods for platelet labeling have been described, but biotinylation using *N*-hydroxysuccinimido biotin (NHS-biotin) is the most widely used. NHS-biotin covalently binds to protein lysine residues,[338] and biotinylated platelets can be readily identified by flow cytometry after the addition of a streptavidin-fluorochrome conjugate.

Whole blood[339] or washed platelets[340,341] can be biotinylated *ex vivo,* then infused into the recipient animal. Because of the large number of potential labeling sites, the density of biotinylation can be altered by using different concentrations of NHS-biotin during labeling. Thus, two levels of biotinylation can be used to track two platelet populations.[340] Platelet survival for dogs,[339] rabbits,[340] cows,[342] and baboons[343] are similar to those obtained in studies using radiolabeled platelets.

The technical ease of biotinylation also permits circulating platelets (and other cells exposed to the circulation) to be biotinylated *in vivo* by the intravenous infusion of NHS-biotin.[140,341,344–351] In dogs, a single bolus infusion of NHS-biotin results in the biotinylation of about 80% of platelets,[344] whereas the same total dose administered in two infusions separated by 30 minute increases the labeling efficiency to approximately 95%.[345,346] The *in vivo* biotinylation process is complete less than 30 minute after infusion.[347] Because megakaryocytes in the bone marrow are not labeled[344] and elution of NHS-biotin from labeled platelets does not appear to occur,[344] only the cohort of platelets present in the circulation at the time of intravenous biotinylation is labeled. Most[339,344,347,352] but not all[353] groups have found normal platelet responsiveness following biotinylation. Although this method permits *in vivo* platelet labeling, it does not permit calculations of recovery.

*b. Other Fluorescent Methods for Platelet Labeling.* Other fluorescent labeling options are available for platelet tracking and survival studies in animals. PKH2, PKH26, and PKH67 are dyes that possess a polar fluorescent head group and a lipophilic moiety that can be incorporated into the membrane of platelets and other cells.[354] PKH2 has been used as one component of a multicolor flow cytometric method to track baboon platelets and measure their function *in vivo* using cross-reacting monoclonal antibodies to human platelet antigens (P-selectin and the active conformation of integrin $\alpha IIb\beta 3$).[355] In this study, PKH2-labeled autologous baboon platelets were activated with thrombin *ex vivo* prior to their reinfusion. After reinfusion, PKH2-positive platelets shed their P-selectin but retained the ability to expose the fibrinogen-binding site of $\alpha IIb\beta 3$ in response to some agonists.

Rehydrated lyophilized platelets have been labeled with PKH26 to permit their identification by fluorescence microscopy after they have been incorporated into thrombi *in vivo*.[8] Rabbit platelets labeled with PKH26 and PKH67 have been reported to have lower recovery but similar survival to platelets labeled with biotin.[352]

In contrast to cell-surface labeling with biotin or PKH dyes, 5-chloromethyl fluorescein diacetate (CMFDA) has been used to internally label murine platelets, without altering aggregation and P-selectin expression in response to agonists. When infused into mice, CMFDA-labeled platelets have normal recovery and survival.[356] Calcein acetoxymethyl ester[144] or carboxyfluorescein diacetate succinimidyl ester[217] can be used to label platelets for intravital microscopy.

Finally, flow cytometry can be used to identify human platelets infused into ethyl palmitate-treated rabbits using monoclonal antibodies against human platelet antigens that do not cross react with rabbit antigens.[357]

## IV. Conclusions

Animal models for the evaluation of platelet products and substitutes, platelet survival, and interventions that modulate platelet reactivity have been an essential tool in the advancement of platelet biology and clinical medicine. Despite the challenges of platelet function measurement *in vivo,* the use of animal models has been ever increasing to meet these demands. Creative and novel approaches punctuate the literature, although many models remain incompletely understood. Generalizability of the results of most animal models to humans remains limited at this time. In the future, the utility of these models for evaluating platelet products, antithrombotic agents, and the study of human platelet biology will become clearer as parallels between the models and clinical reality either become consolidated or fail to materialize. For some experimental animal models, this process is already well under way.

## References

1. Frenkel, J. K. (1969). Choice of animal models for the study of disease processes in man. Introduction. *Fed Proc, 28,* 160–161.
2. Marino, F. (1905). Recherches sur les plaquettes du sang. *Compt Rend Soc Biol, 58,* 194–196.
3. Ledingham, J. O. G. (1914). The experimental production of purpura in animals by the introduction of anti-blood plate sera; a preliminary communication. *Lancet, 1,* 1673–1676.
4. Kitchens, C. S., & Weiss, L. (1975). Ultrastructural changes of endothelium associated with thrombocytopenia. *Blood, 46,* 567–578.
5. Bergqvist, D., & Arfors, K. E. (1973). Influence of platelet count on haemostatic plug formation and plug stability: An experimental study in rabbits with graded thrombocytopenia. *Thromb Diath Haemorrh, 30,* 586–596.
6. Blajchman, M. A., Senyi, A. F., Hirsh, J., et al. (1979). Shortening of the bleeding time in rabbits by hydrocortisone caused by inhibition of prostacyclin generation by the vessel wall. *J Clin Invest, 63,* 1026–1035.
7. Buchanan, M. R., Blajchman, M. A., Dejana, E., et al. (1979). Shortening of the bleeding time in thrombocytopenic rabbits after exposure of jugular vein to high aspirin concentration. *Prostaglandins Med, 3,* 333–342.
8. Read, M. S., Reddick, R. L., Bode, A. P., et al. (1995). Preservation of hemostatic and structural properties of rehydrated lyophilized platelets: Potential for long-term storage of dried platelets for transfusion. *Proc Natl Acad Sci USA, 92,* 397–401.
9. Agam, G., & Livne, A. A. (1992). Erythrocytes with covalently bound fibrinogen as a cellular replacement for the treatment of thrombocytopenia. *Eur J Clin Invest, 22,* 105–112.
10. Levi, M., Friederich, P. W., Middleton, S., et al. (1999). Fibrinogen-coated albumin microcapsules reduce bleeding in severely thrombocytopenic rabbits. *Nat Med, 5,* 107–111.

11. Oyaizu, N., Yasumizu, R., Miyama-Inaba, M., et al. (1988). (NZW × BXSB) F1 mouse: A new animal model of idiopathic thrombocytopenic purpura. *J Exp Med, 167,* 2017–2022.

12. Mizutani, H., Furubayashi, T., Kuriu, A., et al. (1990). Analyses of thrombocytopenia in idiopathic thrombocytopenic purpura-prone mice by platelet transfer experiments between (NZW × BXSB) F1 and normal mice. *Blood, 75,* 1809–1812.

13. Mizutani, H., Engelman, R. W., Kurata, Y., et al. (1993). Development and characterization of monoclonal anti-platelet autoantibodies from autoimmune thrombocytopenic purpura-prone (NZW × BXSB) F1 mice. *Blood, 82,* 837–844.

14. Clynes, R., & Ravetch, J. V. (1995). Cytotoxic antibodies trigger inflammation through Fc receptors. *Immunity, 3,* 21–26.

15. Samuelsson, A., Towers, T. L., & Ravetch, J. V. (2001). Anti-inflammatory activity of IVIG mediated through the inhibitory Fc receptor. *Science, 291,* 484–486.

16. Olsson, M., Bruhns, P., Frazier, W. A., et al. (2005). Platelet homeostasis is regulated by platelet expression of CD47 under normal conditions and in passive immune thrombocytopenia. *Blood, 105,* 3577–3582.

17. Alves-Rosa, F., Stanganelli, C., Cabrera, J., et al. (2000). Treatment with liposome-encapsulated clodronate as a new strategic approach in the management of immune thrombocytopenic purpura in a mouse model. *Blood, 96,* 2834–2840.

18. Crow, A. R., Song, S., Semple, J. W., et al. (2001). IVIg inhibits reticuloendothelial system function and ameliorates murine passive-immune thrombocytopenia independent of anti-idiotype reactivity. *Br J Haematol, 115,* 679–686.

19. Crow, A. R., Song, S., Freedman, J., et al. (2003). IVIg-mediated amelioration of murine ITP via FcgammaRIIB is independent of SHIP1, SHP-1, and Btk activity. *Blood, 102,* 558–560.

20. Song, S., Crow, A. R., Freedman, J., et al. (2003). Monoclonal IgG can ameliorate immune thrombocytopenia in a murine model of ITP: An alternative to IVIG. *Blood, 101,* 3708–3713.

21. Song, S., Crow, A. R., Siragam, V., et al. (2005). Monoclonal antibodies that mimic the action of anti-D in the amelioration of murine ITP act by a mechanism distinct from that of IVIg. *Blood, 105,* 1546–1548.

22. Nieswandt, B., Bergmeier, W., Rackebrandt, K., et al. (2000). Identification of critical antigen-specific mechanisms in the development of immune thrombocytopenic purpura in mice. *Blood, 96,* 2520–2527.

23. Woods, M. C., Gamble, F. N., Furth, J., et al. (1953). Control of the postirradiation hemorrhagic state by platelet transfusions. *Blood, 8,* 545–553.

24. Jackson, D. P., Sorensen, D. K., Cronkite, E. P., et al. (1959). Effectiveness of transfusions of fresh and lyophilized platelets in controlling bleeding due to thrombocytopenia. *J Clin Invest, 38,* 1689–1697.

25. Hjort, P. F., Perman, V., & Cronkite, E. P. (1959). Fresh, disintegrated platelets in radiation thrombocytopenia: Correction of prothrombin consumption without correction of bleeding. *Proc Soc Exp Biol Med, 102,* 31–35.

26. Roy, A. J., & Djerassi, I. (1972). Effects of platelet transfusions: Plug formation and maintenance of vascular integrity. *Proc Soc Exp Biol Med, 139,* 137–142.

27. Aursnes, I. (1973). Appearance of red cells in peripheral lymph during radiation-induced thrombocytopenia. *Acta Physiol Scand, 88,* 392–400.

28. Reimers, H. J., Kinlough-Rathbone, R. L., Cazenave, J. P., et al. (1976). In vitro and in vivo functions of thrombin-treated platelets. *Thromb Haemost, 35,* 151–166.

29. Bjornson, J., & Aursnes, I. (1977). The haemostatic effect of $^{51}$Cr-labelled blood platelets: An experimental study in the rabbit. *Scand J Haematol, 18,* 326–332.

30. Cronkite, E. P., Jacobs, G. J., Brecher, G., et al. (1952). The hemorrhagic phase of the acute radiation syndrome due to exposure of the whole body to penetrating ionizing radiation. *Am J Roentgenol Radium Ther Nucl Med, 67,* 796–804.

31. Blajchman, M. A., & Lee, D. H. (1997). The thrombocytopenic rabbit bleeding time model to evaluate the in vivo hemostatic efficacy of platelets and platelet substitutes. *Transfus Med Rev, 11,* 95–105.

32. Blajchman, M. A., Senyi, A. F., Hirsh, J., et al. (1981). Hemostatic function, survival, and membrane glycoprotein changes in young versus old rabbit platelets. *J Clin Invest, 68,* 1289–1294.

33. Evensen, S. A., Jeremic, M., & Hjort, P. F. (1968). Experimental thrombocytopenia induced by busulphan (Myleran) in rabbits: Extremely low platelet levels and intact plasma clotting system. *Thromb Diath Haemorrh, 19,* 570–577.

34. McGill, M., Fugman, D. A., Vittorio, N., et al. (1987). Platelet membrane vesicles reduced microvascular bleeding times in thrombocytopenic rabbits. *J Lab Clin Med, 109,* 127–133.

35. Kuter, D. J., & Rosenberg, R. D. (1995). The reciprocal relationship of thrombopoietin (c-Mpl ligand) to changes in the platelet mass during busulfan-induced thrombocytopenia in the rabbit. *Blood, 85,* 2720–2730.

36. Chao, F. C., Kim, B. K., Houranieh, A. M., et al. (1996). Infusible platelet membrane microvesicles: A potential transfusion substitute for platelets. *Transfusion, 36,* 536–542.

37. Krishnamurti, C., Maglasang, P., & Rothwell, S. W. (1999). Reduction of blood loss by infusion of human platelets in a rabbit kidney injury model. *Transfusion, 39,* 967–974.

38. Yang, C., Li, Y. C., & Kuter, D. J. (1999). The physiological response of thrombopoietin (c-Mpl ligand) to thrombocytopenia in the rat. *Br J Haematol, 105,* 478–485.

39. Schlerman, F. J., Bree, A. G., Kaviani, M. D., et al. (1996). Thrombopoietic activity of recombinant human interleukin 11 (rHuIL-11) in normal and myelosuppressed nonhuman primates. *Stem Cells, 14,* 517–532.

40. Case, B. C., Hauck, M. L., Yeager, R. L., et al. (2000). The pharmacokinetics and pharmacodynamics of GW395058, a peptide agonist of the thrombopoietin receptor, in the dog, a large-animal model of chemotherapy-induced thrombocytopenia. *Stem Cells, 18,* 360–365.

41. Leonard, J. P., Quinto, C. M., Kozitza, M. K., et al. (1994). Recombinant human interleukin-11 stimulates multilineage

hematopoietic recovery in mice after a myelosuppressive regimen of sublethal irradiation and carboplatin. *Blood, 83,* 1499–1506.

42. Ulich, T. R., del Castillo, J., Yin, S., et al. (1995). Megakaryocyte growth and development factor ameliorates carboplatin-induced thrombocytopenia in mice. *Blood, 86,* 971–976.

43. Abushullaih, B. A., Pestina, T. I., Srivastava, D. K., et al. (2001). A schedule of recombinant Mpl ligand highly effective at preventing lethal myelosuppression in mice given carboplatin and radiation. *Exp Hematol, 29,* 1425–1431.

44. Saitoh, M., Taguchi, K., Momose, K., et al. (2002). Recombinant human interleukin-11 improved carboplatin-induced thrombocytopenia without affecting antitumor activities in mice bearing Lewis lung carcinoma cells. *Cancer Chemother Pharmacol, 49,* 161–166.

45. Yonemura, Y., Kawakita, M., Miyake, H., et al. (1997). Effects of interleukin-11 on carboplatin-induced thrombocytopenia in rats and in combination with stem cell factor. *Int J Hematol, 65,* 397–404.

46. Ide, Y., Harada, K., Imai, A., et al. (1999). PEG-rHuMGDF ameliorates thrombocytopenia in carboplatin-treated rats without inducing myelofibrosis. *Int J Hematol, 70,* 91–96.

47. Wagner, S. J., Bardossy, L., Moroff, G., et al. (1993). Assessment of the hemostatic effectiveness of human platelets treated with aminomethyltrimethyl psoralen and UV A light using a rabbit ear bleeding time technique. *Blood, 82,* 3489–3492.

48. Ali, A. M., Warkentin, T. E., Bardossy, L., et al. (1994). Platelet concentrates stored for 5 days in a reduced volume of plasma maintain hemostatic function and viability. *Transfusion, 34,* 44–47.

49. Blajchman, M. A., Bordin, J. O., Bardossy, L., et al. (1994). The contribution of the haematocrit to thrombocytopenic bleeding in experimental animals. *Br J Haematol, 86,* 347–350.

50. Margolis-Nunno, H., Bardossy, L., Robinson, R., et al. (1997). Psoralen-mediated photodecontamination of platelet concentrates: Inactivation of cell-free and cell-associated forms of human immunodeficiency virus and assessment of platelet function in vivo. *Transfusion, 37,* 889–895.

51. Lee, D. H., Bardossy, L., Peterson, N., et al. (2000). *o*-Raffinose cross-linked hemoglobin improves the hemostatic defect associated with anemia and thrombocytopenia in rabbits. *Blood, 96,* 3630–3636.

52. Hokom, M. M., Lacey, D., Kinstler, O. B., et al. (1995). Pegylated megakaryocyte growth and development factor abrogates the lethal thrombocytopenia associated with carboplatin and irradiation in mice. *Blood, 86,* 4486–4492.

53. Dale, D. C., Nichol, J. L., Rich, D. A., et al. (1997). Chronic thrombocytopenia is induced in dogs by development of cross-reacting antibodies to the MpL ligand. *Blood, 90,* 3456–3461.

54. Gurney, A. L., Carver-Moore, K., de Sauvage, F. J., et al. (1994). Thrombocytopenia in c-Mpl-deficient mice. *Science, 265,* 1445–1447.

55. de Sauvage, F. J., Carver-Moore, K., Luoh, S. M., et al. (1996). Physiological regulation of early and late stages of megakaryocytopoiesis by thrombopoietin. *J Exp Med, 183,* 651–656.

56. Shivdasani, R. A., Fujiwara, Y., McDevitt, M. A., et al. (1997). A lineage-selective knockout establishes the critical role of transcription factor GATA-1 in megakaryocyte growth and platelet development. *EMBO J, 16,* 3965–3973.

57. Vyas, P., Ault, K., Jackson, C. W., et al. (1999). Consequences of GATA-1 deficiency in megakaryocytes and platelets. *Blood, 93,* 2867–2875.

58. Shivdasani, R. A., Rosenblatt, M. F., Zucker-Franklin, D., et al. (1995). Transcription factor NF-E2 is required for platelet formation independent of the actions of thrombopoietin/MGDF in megakaryocyte development. *Cell, 81,* 695–704.

59. Lecine, P., Villeval, J. L., Vyas, P., et al. (1998). Mice lacking transcription factor NF-E2 provide in vivo validation of the proplatelet model of thrombocytopoiesis and show a platelet production defect that is intrinsic to megakaryocytes. *Blood, 92,* 1608–1616.

60. Aranda, E., Pizarro, M., Pereira, J., et al. (1994). Accumulation of 5-hydroxytryptamine by aging platelets: Studies in a model of suppressed thrombopoiesis in dogs. *Thromb Haemost, 71,* 488–492.

61. Plachetka, J. R., Salomon, N. W., Larson, D. F., et al. (1980). Platelet loss during experimental cardiopulmonary bypass and its prevention with prostacyclin. *Ann Thorac Surg, 30,* 58–63.

62. Malpass, T. W., Hanson, S. R., Savage, B., et al. (1981). Prevention of acquired transient defect in platelet plug formation by infused prostacyclin. *Blood, 57,* 736–740.

63. Palatianos, G. M., Dewanjee, M. K., Robinson, R. P., et al. (1989). Quantitation of platelet loss with indium-111 labeled platelets in a hollow-fiber membrane oxygenator and arterial filter during extracorporeal circulation in a pig model. *ASAIO Trans, 35,* 667–670.

64. Hiramatsu, Y., Gikakis, N., Gorman, J. H., 3rd, et al. (1997). A baboon model for hematologic studies of cardiopulmonary bypass. *J Lab Clin Med, 130,* 412–420.

65. Nakamura, M., Toombs, C. F., Duarte, I. G., et al. (1998). Recombinant human megakaryocyte growth and development factor attenuates postbypass thrombocytopenia. *Ann Thorac Surg, 66,* 1216–1223.

66. Fischer, T. H., Merricks, E. P., Bode, A. P., et al. (2002). Thrombus formation with rehydrated, lyophilized platelets. *Hematology, 7,* 359–369.

67. Sanders, W. E., Read, M. S., Reddick, R. L., et al. (1988). Thrombotic thrombocytopenia with von Willebrand factor deficiency induced by botrocetin: An animal model. *Lab Invest, 59,* 443–452.

68. Sanders, W. E., Jr., Reddick, R. L., Nichols, T. C., et al. (1995). Thrombotic thrombocytopenia induced in dogs and pigs: The role of plasma and platelet vWF in animal models of thrombotic thrombocytopenic purpura. *Arterioscler Thromb Vasc Biol, 15,* 793–800.

69. Bruno, K., Volkel, D., Plaimauer, B., et al. (2005). Cloning, expression and functional characterization of the full-length murine ADAMTS13. *J Thromb Haemost, 3,* 1064–1073.

70. Ziporen, L., Li, Z. Q., Park, K. S., et al. (1998). Defining an antigenic epitope on platelet factor 4 associated with heparin-induced thrombocytopenia. *Blood, 92,* 3250–3259.

71. Arepally, G. M., Kamei, S., Park, K. S., et al. (2000). Characterization of a murine monoclonal antibody that mimics heparin-induced thrombocytopenia antibodies. *Blood, 95,* 1533–1540.

72. Blank, M., Cines, D. B., Arepally, G., et al. (1997). Pathogenicity of human anti-platelet factor 4 (PF4)/heparin in vivo: Generation of mouse anti-PF4/heparin and induction of thrombocytopenia by heparin. *Clin Exp Immunol, 108,* 333–339.

73. McKenzie, S. E., Taylor, S. M., Malladi, P., et al. (1999). The role of the human Fc receptor Fc gamma RIIA in the immune clearance of platelets: A transgenic mouse model. *J Immunol, 162,* 4311–4318.

74. Zhang, C., Thornton, M. A., Kowalska, M. A., et al. (2001). Localization of distal regulatory domains in the megakaryocyte-specific platelet basic protein/platelet factor 4 gene locus. *Blood, 98,* 610–617.

75. Reilly, M. P., Taylor, S. M., Hartman, N. K., et al. (2001). Heparin-induced thrombocytopenia/thrombosis in a transgenic mouse model requires human platelet factor 4 and platelet activation through FcgammaRIIA. *Blood, 98,* 2442–2447.

76. Hopper, P. E., Mandell, C. P., Turrel, J. M., et al. (1989). Probable essential thrombocythemia in a dog. *J Vet Intern Med, 3,* 79–85.

77. Bass, M. C., & Schultze, A. E. (1998). Essential thrombocythemia in a dog: Case report and literature review. *J Am Anim Hosp Assoc, 34,* 197–203.

78. Sellon, D. C., Levine, J. F., Palmer, K., et al. (1997). Thrombocytosis in 24 horses (1989–1994). *J Vet Intern Med, 11,* 24–29.

79. Fusco, A., Portella, G., Grieco, M., et al. (1988). A retrovirus carrying the polyomavirus middle T gene induces acute thrombocythemic myeloproliferative disease in mice. *J Virol, 62,* 361–365.

80. Van Etten, R. A. (2001). Models of chronic myeloid leukemia. *Curr Oncol Rep, 3,* 228–237.

81. Kasper, L. H., Boussouar, F., Ney, P. A., et al. (2002). A transcription-factor-binding surface of coactivator p300 is required for haematopoiesis. *Nature, 419,* 738–743.

82. Inokuchi, K., Dan, K., Takatori, M., et al. (2003). Myeloproliferative disease in transgenic mice expressing P230 Bcr/Abl: Longer disease latency, thrombocytosis, and mild leukocytosis. *Blood, 102,* 320–323.

83. Carpinelli, M. R., Hilton, D. J., Metcalf, D., et al. (2004). Suppressor screen in Mpl−/− mice: c-Myb mutation causes supraphysiological production of platelets in the absence of thrombopoietin signaling. *Proc Natl Acad Sci USA, 101,* 6553–6558.

84. Lopez, R. A., Schoetz, S., DeAngelis, K., et al. (2002). Multiple hematopoietic defects and delayed globin switching in Ikaros null mice. *Proc Natl Acad Sci USA, 99,* 602–607.

85. Velazquez, L., Cheng, A. M., Fleming, H. E., et al. (2002). Cytokine signaling and hematopoietic homeostasis are disrupted in Lnk-deficient mice. *J Exp Med, 195,* 1599–1611.

86. Zhou, W., Toombs, C. F., Zou, T., et al. (1997). Transgenic mice overexpressing human c-Mpl ligand exhibit chronic thrombocytosis and display enhanced recovery from 5-fluorouracil or antiplatelet serum treatment. *Blood, 89,* 1551–1559.

87. de Sauvage, F. J., Hass, P. E., Spencer, S. D., et al. (1994). Stimulation of megakaryocytopoiesis and thrombopoiesis by the c-Mpl ligand. *Nature, 369,* 533–538.

88. Lok, S., Kaushansky, K., Holly, R. D., et al. (1994). Cloning and expression of murine thrombopoietin cDNA and stimulation of platelet production in vivo. *Nature, 369,* 565–568.

89. Harker, L. A., Hunt, P., Marzec, U. M., et al. (1996). Regulation of platelet production and function by megakaryocyte growth and development factor in nonhuman primates. *Blood, 87,* 1833–1844.

90. Ishibashi, T., Kimura, H., Shikama, Y., et al. (1989). Interleukin-6 is a potent thrombopoietic factor in vivo in mice. *Blood, 74,* 1241–1244.

91. Hill, R. J., Warren, M. K., & Levin, J. (1990). Stimulation of thrombopoiesis in mice by human recombinant interleukin 6. *J Clin Invest, 85,* 1242–1247.

92. Asano, S., Okano, A., Ozawa, K., et al. (1990). In vivo effects of recombinant human interleukin-6 in primates: Stimulated production of platelets. *Blood, 75,* 1602–1605.

93. Stahl, C. P., Zucker-Franklin, D., Evatt, B. L., et al. (1991). Effects of human interleukin-6 on megakaryocyte development and thrombocytopoiesis in primates. *Blood, 78,* 1467–1475.

94. Neben, T. Y., Loebelenz, J., Hayes, L., et al. (1993). Recombinant human interleukin-11 stimulates megakaryocytopoiesis and increases peripheral platelets in normal and splenectomized mice. *Blood, 81,* 901–908.

95. Holtkotter, O., Nieswandt, B., Smyth, N., et al. (2002). Integrin alpha 2-deficient mice develop normally, are fertile, but display partially defective platelet interaction with collagen. *J Biol Chem, 277,* 10789–10794.

96. Chen, J., Diacovo, T. G., Grenache, D. G., et al. (2002). The alpha(2) integrin subunit-deficient mouse: A multifaceted phenotype including defects of branching morphogenesis and hemostasis. *Am J Pathol, 161,* 337–344.

97. Tronik-Le Roux, D., Roullot, V., Poujol, C., et al. (2000). Thrombasthenic mice generated by replacement of the integrin alpha(IIb) gene: Demonstration that transcriptional activation of this megakaryocytic locus precedes lineage commitment. *Blood, 96,* 1399–1408.

98. Bader, B. L., Rayburn, H., Crowley, D., et al. (1998). Extensive vasculogenesis, angiogenesis, and organogenesis precede lethality in mice lacking all alpha v integrins. *Cell, 95,* 507–519.

99. Fassler, R., & Meyer, M. (1995). Consequences of lack of beta 1 integrin gene expression in mice. *Genes Dev, 9,* 1896–1908.

100. Nieswandt, B., Brakebusch, C., Bergmeier, W., et al. (2001). Glycoprotein VI but not alpha2beta1 integrin is essential for platelet interaction with collagen. *EMBO J, 20,* 2120–2130.

101. Piguet, P. F., Vesin, C., & Rochat, A. (2001). Beta2 integrin modulates platelet caspase activation and life span in mice. *Eur J Cell Biol, 80,* 171–177.

102. Hodivala-Dilke, K. M., McHugh, K. P., Tsakiris, D. A., et al. (1999). Beta3-integrin-deficient mice are a model for Glanzmann thrombasthenia showing placental defects and reduced survival. *J Clin Invest, 103,* 229–238.

103. Smyth, S. S., Reis, E. D., Vaananen, H., et al. (2001). Variable protection of beta 3-integrin-deficient mice from thrombosis initiated by different mechanisms. *Blood, 98,* 1055–1062.

104. Ware, J., Russell, S., & Ruggeri, Z. M. (2000). Generation and rescue of a murine model of platelet dysfunction: The Bernard–Soulier syndrome. *Proc Natl Acad Sci USA, 97,* 2803–2808.

105. Kato, K., Martinez, C., Russell, S., et al. (2004). Genetic deletion of mouse platelet glycoprotein Ib beta produces a Bernard–Soulier phenotype with increased alpha-granule size. *Blood, 104,* 2339–2344.

106. Ramakrishnan, V., Reeves, P. S., DeGuzman, F., et al. (1999). Increased thrombin responsiveness in platelets from mice lacking glycoprotein V. *Proc Natl Acad Sci USA, 96,* 13336–13341.

107. Kahn, M. L., Diacovo, T. G., Bainton, D. F., et al. (1999). Glycoprotein V-deficient platelets have undiminished thrombin responsiveness and do not exhibit a Bernard–Soulier phenotype. *Blood, 94,* 4112–4121.

108. Ni, H., Ramakrishnan, V., Ruggeri, Z. M., et al. (2001). Increased thrombogenesis and embolus formation in mice lacking glycoprotein V. *Blood, 98,* 368–373.

109. Sambrano, G. R., Weiss, E. J., Zheng, Y. W., et al. (2001). Role of thrombin signalling in platelets in haemostasis and thrombosis. *Nature, 413,* 74–78.

110. Hamilton, J. R., Cornelissen, I., & Coughlin, S. R. (2004). Impaired hemostasis and protection against thrombosis in protease-activated receptor 4-deficient mice is due to lack of thrombin signaling in platelets. *J Thromb Haemost, 2,* 1429–1435.

111. Kahn, M. L., Zheng, Y. W., Huang, W., et al. (1998). A dual thrombin receptor system for platelet activation. *Nature, 394,* 690–694.

112. Weiss, E. J., Hamilton, J. R., Lease, K. E., et al. (2002). Protection against thrombosis in mice lacking PAR3. *Blood, 100,* 3240–3244.

113. Connolly, A. J., Ishihara, H., Kahn, M. L., et al. (1996). Role of the thrombin receptor in development and evidence for a second receptor. *Nature, 381,* 516–519.

114. Darrow, A. L., Fung-Leung, W. P., Ye, R. D., et al. (1996). Biological consequences of thrombin receptor deficiency in mice. *Thromb Haemost, 76,* 860–866.

115. Leon, C., Hechler, B., Freund, M., et al. (1999). Defective platelet aggregation and increased resistance to thrombosis in purinergic P2Y(1) receptor-null mice. *J Clin Invest, 104,* 1731–1737.

116. Fabre, J. E., Nguyen, M., Latour, A., et al. (1999). Decreased platelet aggregation, increased bleeding time and resistance to thromboembolism in P2Y1-deficient mice. *Nat Med, 5,* 1199–1202.

117. Foster, C. J., Prosser, D. M., Agans, J. M., et al. (2001). Molecular identification and characterization of the platelet ADP receptor targeted by thienopyridine antithrombotic drugs. *J Clin Invest, 107,* 1591–1598.

118. Andre, P., Delaney, S. M., LaRocca, T., et al. (2003). P2Y12 regulates platelet adhesion/activation, thrombus growth, and thrombus stability in injured arteries. *J Clin Invest, 112,* 398–406.

119. Hechler, B., Lenain, N., Marchese, P., et al. (2003). A role of the fast ATP-gated P2X1 cation channel in thrombosis of small arteries in vivo. *J Exp Med, 198,* 661–667.

120. Thomas, D. W., Mannon, R. B., Mannon, P. J., et al. (1998). Coagulation defects and altered hemodynamic responses in mice lacking receptors for thromboxane A2. *J Clin Invest, 102,* 1994–2001.

121. Murata, T., Ushikubi, F., Matsuoka, T., et al. (1997). Altered pain perception and inflammatory response in mice lacking prostacyclin receptor. *Nature, 388,* 678–682.

122. Fabre, J. E., Nguyen, M., Athirakul, K., et al. (2001). Activation of the murine EP3 receptor for PGE2 inhibits cAMP production and promotes platelet aggregation. *J Clin Invest, 107,* 603–610.

123. Ma, H., Hara, A., Xiao, C. Y., et al. (2001). Increased bleeding tendency and decreased susceptibility to thromboembolism in mice lacking the prostaglandin E receptor subtype EP(3). *Circulation, 104,* 1176–1180.

124. Kato, K., Kanaji, T., Russell, S., et al. (2003). The contribution of glycoprotein VI to stable platelet adhesion and thrombus formation illustrated by targeted gene deletion. *Blood, 102,* 1701–1707.

125. Duncan, G. S., Andrew, D. P., Takimoto, H., et al. (1999). Genetic evidence for functional redundancy of platelet/endothelial cell adhesion molecule-1 (PECAM-1): CD31-deficient mice reveal PECAM-1-dependent and PECAM-1-independent functions. *J Immunol, 162,* 3022–3030.

126. Mahooti, S., Graesser, D., Patil, S., et al. (2000). PECAM-1 (CD31) expression modulates bleeding time in vivo. *Am J Pathol, 157,* 75–81.

127. Jones, K. L., Hughan, S. C., Dopheide, S. M., et al. (2001). Platelet endothelial cell adhesion molecule-1 is a negative regulator of platelet-collagen interactions. *Blood, 98,* 1456–1463.

128. Vollmar, B., Schmits, R., Kunz, D., et al. (2001). Lack of in vivo function of CD31 in vascular thrombosis. *Thromb Haemost, 85,* 160–164.

129. Rathore, V., Stapleton, M. A., Hillery, C. A., et al. (2003). PECAM-1 negatively regulates GPIb/V/IX signaling in murine platelets. *Blood, 102,* 3658–3664.

130. Lagadec, P., Dejoux, O., Ticchioni, M., et al. (2003). Involvement of a CD47-dependent pathway in platelet adhesion on inflamed vascular endothelium under flow. *Blood, 101,* 4836–4843.

131. Chung, J., Wang, X. Q., Lindberg, F. P., et al. (1999). Thrombospondin-1 acts via IAP/CD47 to synergize with collagen in alpha2beta1-mediated platelet activation. *Blood, 94,* 642–648.

132. Poole, A., Gibbins, J. M., Turner, M., et al. (1997). The Fc receptor gamma-chain and the tyrosine kinase Syk are essential for activation of mouse platelets by collagen. *EMBO J, 16,* 2333–2341.

133. Wright, M. D., Geary, S. M., Fitter, S., et al. (2004). Characterization of mice lacking the tetraspanin superfamily member CD151. *Mol Cell Biol, 24,* 5978–5988.

134. Lau, L. M., Wee, J. L., Wright, M. D., et al. (2004). The tetraspanin superfamily member CD151 regulates outside-in integrin alphaIIbbeta3 signaling and platelet function. *Blood, 104,* 2368–2375.

135. Zou, Y. R., Kottmann, A. H., Kuroda, M., et al. (1998). Function of the chemokine receptor CXCR4 in haematopoiesis and in cerebellar development. *Nature, 393,* 595–599.

136. Ma, Q., Jones, D., Borghesani, P. R., et al. (1998). Impaired B-lymphopoiesis, myelopoiesis, and derailed cerebellar neuron migration in CXCR4- and SDF-1-deficient mice. *Proc Natl Acad Sci USA, 95,* 9448–9453.

137. Chvatchko, Y., Hoogewerf, A. J., Meyer, A., et al. (2000). A key role for CC chemokine receptor 4 in lipopolysaccharide-induced endotoxic shock. *J Exp Med, 191,* 1755–1764.

138. Frenette, P. S., Johnson, R. C., Hynes, R. O., et al. (1995). Platelets roll on stimulated endothelium in vivo: An interaction mediated by endothelial P-selectin. *Proc Natl Acad Sci USA, 92,* 7450–7454.

139. Subramaniam, M., Frenette, P. S., Saffaripour, S., et al. (1996). Defects in hemostasis in P-selectin-deficient mice. *Blood, 87,* 1238–1242.

140. Berger, G., Hartwell, D. W., & Wagner, D. D. (1998). P-selectin and platelet clearance. *Blood, 92,* 4446–4452.

141. Ledent, C., Vaugeois, J. M., Schiffmann, S. N., et al. (1997). Aggressiveness, hypoalgesia and high blood pressure in mice lacking the adenosine A2a receptor. *Nature, 388,* 674–678.

142. Moore, K. J., El Khoury, J., Medeiros, L. A., et al. (2002). A CD36-initiated signaling cascade mediates inflammatory effects of beta-amyloid. *J Biol Chem, 277,* 47373–47379.

143. Englyst, N. A., Taube, J. M., Aitman, T. J., et al. (2003). A novel role for CD36 in VLDL-enhanced platelet activation. *Diabetes, 52,* 1248–1255.

144. Denis, C., Methia, N., Frenette, P. S., et al. (1998). A mouse model of severe von Willebrand disease: Defects in hemostasis and thrombosis. *Proc Natl Acad Sci USA, 95,* 9524–9529.

145. Ni, H., Denis, C. V., Subbarao, S., et al. (2000). Persistence of platelet thrombus formation in arterioles of mice lacking both von Willebrand factor and fibrinogen. *J Clin Invest, 106,* 385–392.

146. Suh, T. T., Holmback, K., Jensen, N. J., et al. (1995). Resolution of spontaneous bleeding events but failure of pregnancy in fibrinogen-deficient mice. *Genes Dev, 9,* 2020–2033.

147. Fay, W. P., Parker, A. C., Ansari, M. N., et al. (1999). Vitronectin inhibits the thrombotic response to arterial injury in mice. *Blood, 93,* 1825–1830.

148. Eitzman, D. T., Westrick, R. J., Nabel, E. G., et al. (2000). Plasminogen activator inhibitor-1 and vitronectin promote vascular thrombosis in mice. *Blood, 95,* 577–580.

149. Reheman, A., Gross, P., Yang, H., et al. (2005). Vitronectin stabilizes thrombi and vessel occlusion but plays a dual role in platelet aggregation. *J Thromb Haemost, 3,* 875–883.

150. George, E. L., Georges-Labouesse, E. N., Patel-King, R. S., et al. (1993). Defects in mesoderm, neural tube and vascular development in mouse embryos lacking fibronectin. *Development, 119,* 1079–1091.

151. Sakai, T., Johnson, K. J., Murozono, M., et al. (2001). Plasma fibronectin supports neuronal survival and reduces brain injury following transient focal cerebral ischemia but is not essential for skin-wound healing and hemostasis. *Nat Med, 7,* 324–330.

152. Ni, H., Yuen, P. S., Papalia, J. M., et al. (2003). Plasma fibronectin promotes thrombus growth and stability in injured arterioles. *Proc Natl Acad Sci USA, 100,* 2415–2419.

153. Kyriakides, T. R., Zhu, Y. H., Smith, L. T., et al. (1998). Mice that lack thrombospondin 2 display connective tissue abnormalities that are associated with disordered collagen fibrillogenesis, an increased vascular density, and a bleeding diathesis. *J Cell Biol, 140,* 419–430.

154. Kyriakides, T. R., Rojnuckarin, P., Reidy, M. A., et al. (2003). Megakaryocytes require thrombospondin-2 for normal platelet formation and function. *Blood, 101,* 3915–3923.

155. Lawler, J., Sunday, M., Thibert, V., et al. (1998). Thrombospondin-1 is required for normal murine pulmonary homeostasis and its absence causes pneumonia. *J Clin Invest, 101,* 982–992.

156. Pimanda, J. E., Ganderton, T., Maekawa, A., et al. (2004). Role of thrombospondin-1 in control of von Willebrand factor multimer size in mice. *J Biol Chem, 279,* 21439–2148.

157. Andre, P., Prasad, K. S., Denis, C. V., et al. (2002). CD40L stabilizes arterial thrombi by a beta3 integrin-dependent mechanism. *Nat Med, 8,* 247–252.

158. Crow, A. R., Leytin, V., Starkey, A. F., et al. (2003). CD154 (CD40 ligand)-deficient mice exhibit prolonged bleeding time and decreased shear-induced platelet aggregates. *J Thromb Haemost, 1,* 850–852.

159. Eslin, D. E., Zhang, C., Samuels, K. J., et al. (2004). Transgenic mice studies demonstrate a role for platelet factor 4 in thrombosis: Dissociation between anticoagulant and antithrombotic effect of heparin. *Blood, 104,* 3173–3180.

160. Nandurkar, H. H., Robb, L., Tarlinton, D., et al. (1997). Adult mice with targeted mutation of the interleukin-11 receptor (IL11Ra) display normal hematopoiesis. *Blood, 90,* 2148–2159.

161. Bernad, A., Kopf, M., Kulbacki, R., et al. (1994). Interleukin-6 is required in vivo for the regulation of stem cells and committed progenitors of the hematopoietic system. *Immunity, 1,* 725–731.

162. Gainsford, T., Roberts, A. W., Kimura, S., et al. (1998). Cytokine production and function in c-Mpl-deficient mice: No physiologic role for interleukin-3 in residual megakaryocyte and platelet production. *Blood, 91,* 2745–2752.

163. Escary, J. L., Perreau, J., Dumenil, D., et al. (1993). Leukaemia inhibitory factor is necessary for maintenance of haematopoietic stem cells and thymocyte stimulation. *Nature, 363,* 361–364.

164. Takeda, K., Noguchi, K., Shi, W., et al. (1997). Targeted disruption of the mouse Stat3 gene leads to early embryonic lethality. *Proc Natl Acad Sci USA, 94,* 3801–3804.

165. Fujiwara, Y., Browne, C. P., Cunniff, K., et al. (1996). Arrested development of embryonic red cell precursors in mouse embryos lacking transcription factor GATA-1. *Proc Natl Acad Sci USA, 93,* 12355–12358.

166. Hughan, S. C., Senis, Y., Best, D., et al. (2005). Selective impairment of platelet activation to collagen in the absence of GATA1. *Blood, 105,* 4369–4376.

167. Mucenski, M. L., McLain, K., Kier, A. B., et al. (1991). A functional c-Myb gene is required for normal murine fetal hepatic hematopoiesis. *Cell, 65,* 677–689.

168. Jantzen, H. M., Milstone, D. S., Gousset, L., et al. (2001). Impaired activation of murine platelets lacking G alpha(i2). *J Clin Invest, 108,* 477–483.

169. Yang, J., Wu, J., Kowalska, M. A., et al. (2000). Loss of signaling through the G protein, Gz, results in abnormal platelet activation and altered responses to psychoactive drugs. *Proc Natl Acad Sci USA, 97,* 9984–9989.

170. Gu, J. L., Muller, S., Mancino, V., et al. (2002). Interaction of G alpha(12) with G alpha(13) and G alpha(q) signaling pathways. *Proc Natl Acad Sci USA, 99,* 9352–9357.

171. Moers, A., Nieswandt, B., Massberg, S., et al. (2003). G13 is an essential mediator of platelet activation in hemostasis and thrombosis. *Nat Med, 9,* 1418–1422.

172. Offermanns, S., Mancino, V., Revel, J. P., et al. (1997). Vascular system defects and impaired cell chemokinesis as a result of G alpha13 deficiency. *Science, 275,* 533–536.

173. Moers, A., Wettschureck, N., Gruner, S., et al. (2004). Unresponsiveness of platelets lacking both G alpha(q) and G alpha(13): Implications for collagen-induced platelet activation. *J Biol Chem, 279,* 45354–45359.

174. Offermanns, S., Toombs, C. F., Hu, Y. H., et al. (1997). Defective platelet activation in G alpha(q)-deficient mice. *Nature, 389,* 183–186.

175. Hirsch, E., Bosco, O., Tropel, P., et al. (2001). Resistance to thromboembolism in PI3Kgamma-deficient mice. *FASEB J, 15,* 2019–2021.

176. Li, Z., Zhang, G., Le Breton, G. C., et al. (2003). Two waves of platelet secretion induced by thromboxane A2 receptor and a critical role for phosphoinositide 3-kinases. *J Biol Chem, 278,* 30725–30731.

177. Lian, L., Wang, Y., Draznin, J., et al. (2005). The relative role of PLCbeta and PI3Kgamma in platelet activation. *Blood, 106,* 110–117.

178. Jackson, S. P., Schoenwaelder, S. M., Goncalves, I., et al. (2005). PI 3-kinase p110beta: A new target for antithrombotic therapy. *Nat Med, 11,* 507–514.

179. Senis, Y. A., Atkinson, B. T., Pearce, A. C., et al. (2005). Role of the p110delta PI 3-kinase in integrin and ITAM receptor signalling in platelets. *Platelets, 16,* 191–202.

180. Watanabe, N., Nakajima, H., Suzuki, H., et al. (2003). Functional phenotype of phosphoinositide 3-kinase p85alpha-null platelets characterized by an impaired response to GP VI stimulation. *Blood, 102,* 541–548.

181. Mangin, P., Nonne, C., Eckly, A., et al. (2003). A PLC gamma 2-independent platelet collagen aggregation requiring functional association of GPVI and integrin alpha2beta1. *FEBS Lett, 542,* 53–59.

182. Suzuki-Inoue, K., Inoue, O., Frampton, J., et al. (2003). Murine GPVI stimulates weak integrin activation in PLC-gamma2−/− platelets: Involvement of PLCgamma1 and PI3-kinase. *Blood, 102,* 1367–1373.

183. Nonne, C., Lenain, N., Hechler, B., et al. (2005). Importance of platelet phospholipase Cgamma2 signaling in arterial thrombosis as a function of lesion severity. *Arterioscler Thromb Vasc Biol, 25,* 1293–1298.

184. Cho, M. J., Pestina, T. I., Steward, S. A., et al. (2002). Role of the Src family kinase Lyn in TxA2 production, adenosine diphosphate secretion, Akt phosphorylation, and irreversible aggregation in platelets stimulated with gamma-thrombin. *Blood, 99,* 2442–2447.

185. Pearce, A. C., Wilde, J. I., Doody, G. M., et al. (2002). Vav1, but not Vav2, contributes to platelet aggregation by CRP and thrombin, but neither is required for regulation of phospholipase C. *Blood, 100,* 3561–3569.

186. Pearce, A. C., Senis, Y. A., Billadeau, D. D., et al. (2004). Vav1 and Vav3 have critical but redundant roles in mediating platelet activation by collagen. *J Biol Chem, 279,* 53955–53962.

187. Atkinson, B. T., Ellmeier, W., & Watson, S. P. (2003). Tec regulates platelet activation by GPVI in the absence of *Btk*. *Blood, 102,* 3592–3599.

188. Angelillo-Scherrer, A., de Frutos, P., Aparicio, C., et al. (2001). Deficiency or inhibition of Gas6 causes platelet dysfunction and protects mice against thrombosis. *Nat Med, 7,* 215–221.

189. Angelillo-Scherrer, A., Burnier, L., Flores, N., et al. (2005). Role of Gas6 receptors in platelet signaling during thrombus stabilization and implications for antithrombotic therapy. *J Clin Invest, 115,* 237–246.

190. Chen, C., Li, Q., Darrow, A. L., et al. (2004). Mer receptor tyrosine kinase signaling participates in platelet function. *Arterioscler Thromb Vasc Biol, 24,* 1118–1123.

191. Clements, J. L., Lee, J. R., Gross, B., et al. (1999). Fetal hemorrhage and platelet dysfunction in SLP-76-deficient mice. *J Clin Invest, 103,* 19–25.

192. Pasquet, J. M., Gross, B., Quek, L., et al. (1999). LAT is required for tyrosine phosphorylation of phospholipase cgamma2 and platelet activation by the collagen receptor GPVI. *Mol Cell Biol, 19,* 8326–8334.

193. Li, Z., Xi, X., Gu, M., et al. (2003). A stimulatory role for cGMP-dependent protein kinase in platelet activation. *Cell, 112,* 77–86.

194. Massberg, S., Sausbier, M., Klatt, P., et al. (1999). Increased adhesion and aggregation of platelets lacking cyclic guanosine 3′,5′-monophosphate kinase I. *J Exp Med, 189,* 1255–1264.

195. Li, Z., Zhang, G., Marjanovic, J. A., et al. (2004). A platelet secretion pathway mediated by cGMP-dependent protein kinase. *J Biol Chem, 279,* 42469–42475.

196. Hauser, W., Knobeloch, K. P., Eigenthaler, M., et al. (1999). Megakaryocyte hyperplasia and enhanced agonist-induced platelet activation in vasodilator-stimulated phosphoprotein knockout mice. *Proc Natl Acad Sci USA, 96,* 8120–8125.

197. Massberg, S., Gruner, S., Konrad, I., et al. (2004). Enhanced in vivo platelet adhesion in vasodilator-stimulated

phosphoprotein (VASP)-deficient mice. *Blood, 103,* 136–142.

198. Chen, J., De, S., Damron, D. S., et al. (2004). Impaired platelet responses to thrombin and collagen in AKT-1-deficient mice. *Blood, 104,* 1703–1710.

199. Woulfe, D., Jiang, H., Morgans, A., et al. (2004). Defects in secretion, aggregation, and thrombus formation in platelets from mice lacking Akt2. *J Clin Invest, 113,* 441–450.

200. Turner, M., Mee, P. J., Costello, P. S., et al. (1995). Perinatal lethality and blocked B-cell development in mice lacking the tyrosine kinase Syk. *Nature, 378,* 298–302.

201. Chrzanowska-Wodnicka, M., Smyth, S. S., Schoenwaelder, S. M., et al. (2005). Rap1b is required for normal platelet function and hemostasis in mice. *J Clin Invest, 115,* 680–687.

202. Pasquet, J. M., Quek, L., Stevens, C., et al. (2000). Phosphatidylinositol 3,4,5-trisphosphate regulates Ca(2+) entry via Btk in platelets and megakaryocytes without increasing phospholipase C activity. *EMBO J, 19,* 2793–2802.

203. Maxwell, M. J., Yuan, Y., Anderson, K. E., et al. (2004). SHIP1 and Lyn kinase negatively regulate integrin alpha IIb beta 3 signaling in platelets. *J Biol Chem, 279,* 32196–32204.

204. Judd, B. A., Myung, P. S., Obergfell, A., et al. (2002). Differential requirement for LAT and SLP-76 in GPVI versus T cell receptor signaling. *J Exp Med, 195,* 705–717.

205. Senis, Y. A., Sangrar, W., Zirngibl, R. A., et al. (2003). Fps/Fes and Fer non-receptor protein-tyrosine kinases regulate collagen- and ADP-induced platelet aggregation. *J Thromb Haemost, 1,* 1062–1070.

206. Langenbach, R., Morham, S. G., Tiano, H. F., et al. (1995). Prostaglandin synthase 1 gene disruption in mice reduces arachidonic acid-induced inflammation and indomethacin-induced gastric ulceration. *Cell, 83,* 483–492.

207. Enjyoji, K., Sevigny, J., Lin, Y., et al. (1999). Targeted disruption of CD39/ATP diphosphohydrolase results in disordered hemostasis and thromboregulation. *Nat Med, 5,* 1010–1017.

208. Koszalka, P., Ozuyaman, B., Huo, Y., et al. (2004). Targeted disruption of CD73/ecto-5′-nucleotidase alters thromboregulation and augments vascular inflammatory response. *Circ Res, 95,* 814–821.

209. Yu, I. S., Lin, S. R., Huang, C. C., et al. (2004). TXAS-deleted mice exhibit normal thrombopoiesis, defective hemostasis, and resistance to arachidonate-induced death. *Blood, 104,* 135–142.

210. Freedman, J. E., Sauter, R., Battinelli, E. M., et al. (1999). Deficient platelet-derived nitric oxide and enhanced hemostasis in mice lacking the NOSIII gene. *Circ Res, 84,* 1416–1421.

211. Iafrati, M. D., Vitseva, O., Tanriverdi, K., et al. (2005). Compensatory mechanisms influence hemostasis in setting of eNOS deficiency. *Am J Physiol Heart Circ Physiol, 288,* H1627–H1632.

212. Witke, W., Sharpe, A. H., Hartwig, J. H., et al. (1995). Hemostatic, inflammatory, and fibroblast responses are blunted in mice lacking gelsolin. *Cell, 81,* 41–51.

213. Schwer, H. D., Lecine, P., Tiwari, S., et al. (2001). A lineage-restricted and divergent beta-tubulin isoform is essential for the biogenesis, structure and function of blood platelets. *Curr Biol, 11,* 579–586.

214. Italiano, J. E., Jr., Bergmeier, W., Tiwari, S., et al. (2003). Mechanisms and implications of platelet discoid shape. *Blood, 101,* 4789–4796.

215. Matsushita, T., Hayashi, H., Kunishima, S., et al. (2004). Targeted disruption of mouse ortholog of the human MYH9 responsible for macrothrombocytopenia with different organ involvement: Hematological, nephrological, and otological studies of heterozygous KO mice. *Biochem Biophys Res Commun, 325,* 1163–1171.

216. Dent, J., Kato, K., Peng, X. R., et al. (2002). A prototypic platelet septin and its participation in secretion. *Proc Natl Acad Sci USA, 99,* 3064–3069.

217. Walther, D. J., Peter, J. U., Winter, S., et al. (2003). Serotonylation of small GTPases is a signal transduction pathway that triggers platelet alpha-granule release. *Cell, 115,* 851–862.

218. Freson, K., Hashimoto, H., Thys, C., et al. (2004). The pituitary adenylate cyclase-activating polypeptide is a physiological inhibitor of platelet activation. *J Clin Invest, 113,* 905–912.

219. Kirito, K., Osawa, M., Morita, H., et al. (2002). A functional role of Stat3 in in vivo megakaryopoiesis. *Blood, 99,* 3220–3227.

220. Kahn, M. L., Nakanishi-Matsui, M., Shapiro, M. J., et al. (1999). Protease-activated receptors 1 and 4 mediate activation of human platelets by thrombin. *J Clin Invest, 103,* 879–887.

221. Leader, R. W., Padgett, G. A., & Gorham, J. R. (1963). Studies of abnormal leukocyte bodies in the mink. *Blood, 22,* 477–484.

222. Padgett, G. A., Leader, R. W., Gorham, J. R., et al. (1964). The familial occurrence of the Chediak–Higashi syndrome in mink and cattle. *Genetics, 49,* 505–512.

223. Lutzner, M. A., Lowrie, C. T., & Jordan, H. W. (1967). Giant granules in leukocytes of the beige mouse. *J Hered, 58,* 299–300.

224. Kramer, J. W., Davis, W. C., Prieur, D. J., et al. (1975). An inherited disorder of Persian cats with intracytoplasmic inclusions in neutrophils. J *Am Vet Med Assoc, 166,* 1103–1104.

225. Cowles, B. E., Meyers, K. M., Wardrop, K. J., et al. (1992). Prolonged bleeding time of Chediak–Higashi cats corrected by platelet transfusion. *Thromb Haemost, 67,* 708–712.

226. Novak, E. K., Hui, S. W., & Swank, R. T. (1984). Platelet storage pool deficiency in mouse pigment mutations associated with seven distinct genetic loci. *Blood, 63,* 536–544.

227. Barbosa, M. D., Nguyen, Q. A., Tchernev, V. T., et al. (1996). Identification of the homologous beige and Chediak–Higashi syndrome genes. *Nature, 382,* 262–265.

228. Perou, C. M., Moore, K. J., Nagle, D. L., et al. (1996). Identification of the murine beige gene by YAC complementation and positional cloning. *Nat Genet, 13,* 303–308.

229. Ozaki, K., Fujimori, H., Nomura, S., et al. (1998). Morphologic and hematologic characteristics of storage pool defi-

ciency in beige rats (Chediak–Higashi syndrome of rats). *Lab Anim Sci, 48,* 502–506.

230. Introne, W., Boissy, R. E., & Gahl, W. A. (1999). Clinical, molecular, and cell biological aspects of Chediak–Higashi syndrome. *Mol Genet Metab, 68,* 283–303.

231. Li, W., Rusiniak, M. E., Chintala, S., et al. (2004). Murine Hermansky-Pudlak syndrome genes: Regulators of lysosome-related organelles. *Bioessays, 26,* 616–628.

232. Novak, E. K., & Swank, R. T. (1979). Lysosomal dysfunctions associated with mutations at mouse pigment genes. *Genetics, 92,* 189–204.

233. Novak, E. K., Sweet, H. O., Prochazka, M., et al. (1988). Cocoa: A new mouse model for platelet storage pool deficiency. *Br J Haematol, 69,* 371–378.

234. Swank, R. T., Reddington, M., Howlett, O., et al. (1991). Platelet storage pool deficiency associated with inherited abnormalities of the inner ear in the mouse pigment mutants muted and mocha. *Blood, 78,* 2036–2044.

235. Reddington, M., Novak, E. K., Hurley, E., et al. (1987). Immature dense granules in platelets from mice with platelet storage pool disease. *Blood, 69,* 1300–1306.

236. Tschopp, B., & Weiss, H. J. (1974). Decreased ATP, ADP and serotonin in young platelets of fawn-hooded rats with storage pool disease. *Thromb Diath Haemorrh, 32,* 670–677.

237. Hamada, S., Nishikawa, T., Yokoi, N., et al. (1997). TM rats: A model for platelet storage pool deficiency. *Exp Anim, 46,* 235–239.

238. Oiso, N., Riddle, S. R., Serikawa, T., et al. (2004). The rat ruby (R) locus is Rab38: Identical mutations in fawn-hooded and Tester–Moriyama rats derived from an ancestral long Evans rat sub-strain. *Mamm Genome, 15,* 307–314.

239. Jackson, C. W., Hutson, N. K., Steward, S. A., et al. (1988). The Wistar Furth rat: An animal model of hereditary macrothrombocytopenia. *Blood, 71,* 1676–1686.

240. Jackson, C. W., Hutson, N. K., Steward, S. A., et al. (1991). Platelets of the Wistar Furth rat have reduced levels of alpha-granule proteins: An animal model resembling gray platelet syndrome. *J Clin Invest, 87,* 1985–1991.

241. Stenberg, P. E., Barrie, R. J., Pestina, T. I., et al. (1998). Prolonged bleeding time with defective platelet filopodia formation in the Wistar Furth rat. *Blood, 91,* 1599–1608.

242. Boudreaux, M. K., Kvam, K., Dillon, A. R., et al. (1996). Type I Glanzmann's thrombasthenia in a Great Pyrenees dog. *Vet Pathol, 33,* 503–511.

243. Boudreaux, M. K., & Catalfamo, J. L. (2001). Molecular and genetic basis for thrombasthenic thrombopathia in otterhounds. *Am J Vet Res, 62,* 1797–1804.

244. Boudreaux, M. K., & Lipscomb, D. L. (2001). Clinical, biochemical, and molecular aspects of Glanzmann's thrombasthenia in humans and dogs. *Vet Pathol, 38,* 249–260.

245. Smith, S. V., Lumeng, L., Read, M. S., et al. (1996). Characterization of a new hereditary thrombopathy in a closed colony of Wistar rats. *J Lab Clin Med, 128,* 601–611.

246. Searcy, G. P., Sheridan, D., & Dobson, K. A. (1990). Preliminary studies of a platelet function disorder in Simmental cattle. *Can J Vet Res, 54,* 394–396.

247. Steficek, B. A., Thomas, J. S., Baker, J. C., et al. (1993). Hemorrhagic diathesis associated with a hereditary platelet disorder in Simmental cattle. *J Vet Diagn Invest, 5,* 202–207.

248. Steficek, B. A., Thomas, J. S., McConnell, M. F., et al. (1993). A primary platelet disorder of consanguineous Simmental cattle. *Thromb Res, 72,* 145–153.

249. Searcy, G. P., Frojmovic, M. M., McNicol, A., et al. (1994). Platelets from bleeding Simmental cattle mobilize calcium, phosphorylate myosin light chain and bind normal numbers of fibrinogen molecules but have abnormal cytoskeletal assembly and aggregation in response to ADP. *Thromb Haemost, 71,* 240–246.

250. Frojmovic, M. M., Wong, T., & Searcy, G. P. (1996). Platelets from bleeding Simmental cattle have a long delay in both ADP-activated expression of GpIIB-IIIA receptors and fibrinogen-dependent platelet aggregation. *Thromb Haemost, 76,* 1047–1052.

251. Patterson, W. R., Padgett, G. A., & Bell, T. G. (1985). Abnormal release of storage pool adenine nucleotides from platelets of dogs affected with basset hound hereditary thrombopathy. *Thromb Res, 37,* 61–71.

252. Patterson, W. R., Kunicki, T. J., & Bell, T. G. (1986). Two-dimensional electrophoretic studies of platelets from dogs affected with basset hound hereditary thrombopathy: A thrombasthenia-like aggregation defect. *Thromb Res, 42,* 195–203.

253. Patterson, W. R., Estry, D. W., Schwartz, K. A., et al. (1989). Absent platelet aggregation with normal fibrinogen binding in basset hound hereditary thrombopathy. *Thromb Haemost, 62,* 1011–1015.

254. Schmidt, K. G., Rasmussen, J. W., & Lorentzen, M. (1982). Function and morphology of 111-In-labelled platelets: In vitro, in vivo and ex vivo studies. *Haemostasis, 11,* 193–203.

255. Sokal, G. (1958). Étude de la transfusion de plaquettes experimental chez le lapin. *Proc VIIth Congr Int Soc Blood Transf,* 936–944.

256. Yen, R. C. K., Ho, T. W. C., & Blajchman, M. A. (1995). A new hemostatic agent: Thrombospheres shorten bleeding time in thrombocytopenic rabbits. *Thromb Haemost, 73,* 986.

257. Yen, R. K. C., Ho, T. W. C., & Blajchman, M. A. (1995). A novel approach to correcting the bleeding associated with thrombocytopenia. *Transfusion, 35,* 41S.

258. Dejana, E., Callioni, A., Quintana, A., et al. (1979). Bleeding time in laboratory animals. II — A comparison of different assay conditions in rats. *Thromb Res, 15,* 191–197.

259. Broze, G. J., Jr, Yin, Z. F., & Lasky, N. (2001). A tail vein bleeding time model and delayed bleeding in hemophiliac mice. *Thromb Haemost, 85,* 747–748.

260. Dejana, E., Villa, S., & de Gaetano, G. (1982). Bleeding time in rats: A comparison of different experimental conditions. *Thromb Haemost, 48,* 108–111.

261. Zumbach, A., Marbet, G. A., & Tsakiris, D. A. (2001). Influence of the genetic background on platelet function, microparticle and thrombin generation in the common laboratory mouse. *Platelets, 12,* 496–502.

262. Bi, L., Sarkar, R., Naas, T., et al. (1996). Further characterization of factor VIII-deficient mice created by gene targeting: RNA and protein studies. *Blood, 88,* 3446–3450.

263. Bi, L., Lawler, A. M., Antonarakis, S. E., et al. (1995). Targeted disruption of the mouse factor VIII gene produces a model of haemophilia A. *Nat Genet, 10,* 119–121.

264. Eyster, M. E., Gordon, R. A., & Ballard, J. O. (1981). The bleeding time is longer than normal in hemophilia. *Blood, 58,* 719–723.

265. Dejana, E., Quintana, A., Callioni, A., et al. (1979). Bleeding time in laboratory animals. III — Do tail bleeding times in rats only measure a platelet defect? (the aspirin puzzle). *Thromb Res, 15,* 199–207.

266. Tsakiris, D. A., Scudder, L., Hodivala-Dilke, K., et al. (1999). Hemostasis in the mouse (Mus musculus): A review. *Thromb Haemost, 81,* 177–188.

267. Jergens, A. E., Turrentine, M. A., Kraus, K. H., et al. (1987). Buccal mucosa bleeding times of healthy dogs and of dogs in various pathologic states, including thrombocytopenia, uremia, and von Willebrand's disease. *Am J Vet Res, 48,* 1337–1342.

268. Brooks, M., & Catalfamo, J. (1993). Buccal mucosa bleeding time is prolonged in canine models of primary hemostatic disorders. *Thromb Haemost, 70,* 777–780.

269. Sato, I., Anderson, G. A., & Parry, B. W. (2000). An interobserver and intraobserver study of buccal mucosal bleeding time in greyhounds. *Res Vet Sci, 68,* 41–45.

270. Holmback, K., Danton, M. J., Suh, T. T., et al. (1996). Impaired platelet aggregation and sustained bleeding in mice lacking the fibrinogen motif bound by integrin alpha IIb beta 3. *EMBO J, 15,* 5760–5771.

271. Tranholm, M., Rojkjaer, R., Pyke, C., et al. (2003). Recombinant factor VIIa reduces bleeding in severely thrombocytopenic rabbits. *Thromb Res, 109,* 217–223.

272. Giles, A. R., Tinlin, S., & Greenwood, R. (1982). A canine model of hemophilic (factor VIII:C deficiency) bleeding. *Blood, 60,* 727–730.

273. Carter, C. J., Kelton, J. G., & Hirsh, J. (1979). Comparison of the haemorrhagic effects of porcine and bovine heparin in rabbits. *Thromb Res, 15,* 581–586.

274. Fliedner, T. M., Sorensen, D. K., Bond, V. P., et al. (1958). Comparative effectiveness of fresh and lyophilized platelets in controlling irradiation hemorrhage in the rat. *Proc Soc Exp Biol Med, 99,* 731–733.

275. Rybak, M. E., & Renzulli, L. A. (1993). A liposome based platelet substitute, the plateletsome, with hemostatic efficacy. *Biomater Artif Cells Immobilization Biotechnol, 21,* 101–118.

276. Rothwell, S. W., Maglasang, P., Reid, T. J., et al. (2000). Correlation of in vivo and in vitro functions of fresh and stored human platelets. *Transfusion, 40,* 988–993.

277. MacDonald, J. D., Remington, B. J., & Rodgers, G. M. (1994). The skin bleeding time test as a predictor of brain bleeding time in a rat model. *Thromb Res, 76,* 535–540.

278. Rodgers, R. P., & Levin, J. (1990). A critical reappraisal of the bleeding time. *Semin Thromb Hemost, 16,* 1–20.

279. Ross, M. H., Furth, J., & Bigelow, R. R. (1952). Changes in cellular composition of the lymph caused by ionizing radiations. *Blood, 7,* 417–428.

280. May, G. R., Crook, P., Moore, P. K., et al. (1991). The role of nitric oxide as an endogenous regulator of platelet and neutrophil activation within the pulmonary circulation of the rabbit. *Br J Pharmacol, 102,* 759–763.

281. Slichter, S. J., & Harker, L. A. (1978). Thrombocytopenia: Mechanisms and management of defects in platelet production. *Clin Haematol, 7,* 523–539.

282. Giles, A. R., Greenwood. P., & Tinlin, S. (1984). A platelet release defect induced by aspirin or penicillin G does not increase gastrointestinal blood loss in thrombocytopenic rabbits. *Br J Haematol, 57,* 17–23.

283. Folts, J. D., Crowell, E. B., Jr, & Rowe, G. G. (1976). Platelet aggregation in partially obstructed vessels and its elimination with aspirin. *Circulation, 54,* 365–370.

284. Folts, J. (1991). An in vivo model of experimental arterial stenosis, intimal damage, and periodic thrombosis. *Circulation, 83,* IV3-IV14.

285. Folts, J. D., Schafer, A. I., Loscalzo, J., et al. (1999). A perspective on the potential problems with aspirin as an antithrombotic agent: A comparison of studies in an animal model with clinical trials. *J Am Coll Cardiol, 33,* 295–303.

286. Bush, L. R., & Shebuski, R. J. (1990). In vivo models of arterial thrombosis and thrombolysis. *FASEB J, 4,* 3087–3098.

287. Eidt, J. F., Allison, P., Noble, S., et al. (1989). Thrombin is an important mediator of platelet aggregation in stenosed canine coronary arteries with endothelial injury. *J Clin Invest, 84,* 18–27.

288. Eichhorn, E. J., Grayburn, P. A., Willard, J. E., et al. (1991). Spontaneous alterations in coronary blood flow velocity before and after coronary angioplasty in patients with severe angina. *J Am Coll Cardiol, 17,* 43–52.

289. Anderson, H. V., Kirkeeide, R. L., Krishnaswami, A., et al. (1994). Cyclic flow variations after coronary angioplasty in humans: Clinical and angiographic characteristics and elimination with 7E3 monoclonal antiplatelet antibody. *J Am Coll Cardiol, 23,* 1031–1037.

290. Ikeda, H., Koga, Y., Kuwano, K., et al. (1993). Cyclic flow variations in a conscious dog model of coronary artery stenosis and endothelial injury correlate with acute ischemic heart disease syndromes in humans. *J Am Coll Cardiol, 21,* 1008–1017.

291. Salazar, A. E. (1961). Experimental myocardial infarction: Induction of coronary thrombosis in the intact closed-chest dog. *Circ Res, 9,* 1351–1356.

292. Hladovec, J. (1971). Experimental arterial thrombosis in rats with continuous registration. *Thromb Diath Haemorrh, 26,* 407–410.

293. Romson, J. L., Haack, D. W., & Lucchesi, B. R. (1980). Electrical induction of coronary artery thrombosis in the ambulatory canine: A model for in vivo evaluation of antithrombotic agents. *Thromb Res, 17,* 841–853.

294. Bernat, A., Mares, A. M., Defreyn, G., et al. (1993). Effect of various antiplatelet agents on acute arterial thrombosis in the rat. *Thromb Haemost, 70,* 812–816.

295. Thiagarajan, P., & Benedict, C. R. (1997). Inhibition of arterial thrombosis by recombinant annexin V in a rabbit carotid artery injury model. *Circulation, 96,* 2339–2347.

296. Sugidachi, A., Asai, F., Tani, Y., et al. (1996). Occlusive thrombosis in the femoral artery of the rabbit: A pharmacological model for evaluating antiplatelet and anticoagulant agents. *Blood Coagul Fibrinolysis, 7,* 57–64.

297. Kurz, K. D., Main, B. W., & Sandusky, G. E. (1990). Rat model of arterial thrombosis induced by ferric chloride. *Thromb Res, 60,* 269–280.

298. Lockyer, S., & Kambayashi, J. (1999). Demonstration of flow and platelet dependency in a ferric chloride-induced model of thrombosis. *J Cardiovasc Pharmacol, 33,* 718–725.

299. Tanaka, T., Sato, R., & Kurimoto, T. (2000). Z-335, a new thromboxane A(2) receptor antagonist, prevents arterial thrombosis induced by ferric chloride in rats. *Eur J Pharmacol, 401,* 413–418.

300. Futrell. N., Millikan, C., Watson, B. D., et al. (1989). Embolic stroke from a carotid arterial source in the rat: Pathology and clinical implications. *Neurology, 39,* 1050–1056.

301. Futrell, N. (1991). An improved photochemical model of embolic cerebral infarction in rats. *Stroke, 22,* 225–232.

302. Dietrich, W. D., Dewanjee, S., Prado, R., et al. (1993). Transient platelet accumulation in the rat brain after common carotid artery thrombosis: An $^{111}$In-labeled platelet study. *Stroke, 24,* 1534–1540.

303. Yao, H., Ibayashi, S., Sugimori, H., et al. (1996). Simplified model of krypton laser-induced thrombotic distal middle cerebral artery occlusion in spontaneously hypertensive rats. *Stroke, 27,* 333–336.

304. Wang, X., & Xu, L. (2005). An optimized murine model of ferric chloride-induced arterial thrombosis for thrombosis research. *Thromb Res, 115,* 95–100.

305. Day, S. M., Reeve, J. L., Myers, D. D., et al. (2004). Murine thrombosis models. *Thromb Haemost, 92,* 486–494.

306. Saniabadi, A. R., Umemura, K., Matsumoto, N., et al. (1995). Vessel wall injury and arterial thrombosis induced by a photochemical reaction. *Thromb Haemost, 73,* 868–872.

307. Rosen, E. D., Raymond, S., Zollman, A., et al. (2001). Laser-induced noninvasive vascular injury models in mice generate platelet- and coagulation-dependent thrombi. *Am J Pathol, 158,* 1613–1622.

308. Lam, J. Y., Chesebro, J. H., Badimon, L., et al. (1993). Exogenous prostacyclin decreases vasoconstriction but not platelet thrombus deposition after arterial injury. *J Am Coll Cardiol, 21,* 488–492.

309. Olsen, S. B., Tang, D. B., Jackson, M. R., et al. (1996). Enhancement of platelet deposition by cross-linked hemoglobin in a rat carotid endarterectomy model. *Circulation, 93,* 327–332.

310. Vandenberg, B. F., Seabold, J. E., Conrad, G. R., et al. (1988). $^{111}$In-labeled platelet scintigraphy and two-dimensional echocardiography for detection of left atrial appendage thrombi studies in a new canine model. *Circulation, 78,* 1040–1046.

311. Hanson, S. R., Paxton, L. D., & Harker, L. A. (1986). Iliac artery mural thrombus formation: Effect of antiplatelet therapy on $^{111}$In-platelet deposition in baboons. *Arteriosclerosis, 6,* 511–518.

312. Dewanjee, M. K., Zhai, P., Hsu, L. C., et al. (1997). A new method for quantitation of platelet microthrombi and microemboli from cardiopulmonary bypass in organs using $^{111}$In labeled platelets. *ASAIO J, 43,* M701-M705.

313. Umetsu, T., & Sanai, K. (1978). Effect of 1-methyl-2-mercapto-5-(3-pyridyl)-imidazole (KC-6141), an antiaggregating compound, on experimental thrombosis in rats. *Thromb Haemost, 39,* 74–83.

314. Vogel, G. M., van Amsterdam, R. G., Zandberg, P., et al. (1997). Two new closely related rat models with relevance to arterial thrombosis — efficacies of different antithrombotic drugs. *Thromb Haemost, 77,* 183–189.

315. Herbert, J. M., Dol, F., Bernat, A., et al. (1998). The antiaggregating and antithrombotic activity of clopidogrel is potentiated by aspirin in several experimental models in the rabbit. *Thromb Haemost, 80,* 512–518.

316. Born, G. V., & Cross, M. J. (1963). Effect of adenosine diphosphate on the concentration of platelets in circulating blood. *Nature, 197,* 974–976.

317. Regoli, D., & Clark, V. (1963). Prevention by adenosine of the effect of adenosine diphosphate on the concentration of circulating platelets. *Nature, 200,* 546–548.

318. Nordöy, A., & Chandler, A. B. (1964). Platelet thrombosis induced by adenosine diphosphate in the rat. *Scand J Haematol, 17,* 16–25.

319. May, G. R., Herd, C. M., Butler, K. D., et al. (1990). Radioisotopic model for investigating thromboembolism in the rabbit. *J Pharmacol Methods, 24,* 19–35.

320. Page, C. P., Paul, W., & Morley, J. (1982). An in vivo model for studying platelet aggregation and disaggregation. *Thromb Haemost, 47,* 210–213.

321. Butler, K. D., Shand, R. A., & Wallis, R. B. (1986). The effects of modulation of prostanoid metabolism on the thoracic platelet accumulation induced by intravenous administration of collagen in the guinea-pig. *Thromb Haemost, 56,* 263–267.

322. Klee, A., & Seiffge, D. (1991). Evaluation of pulmonary accumulation of 51chromium-labelled rat platelets following intravenous application of ADP and collagen. *Thromb Haemost, 65,* 588–595.

323. Momi, S., Emerson, M., Paul, W., et al. (2000). Prevention of pulmonary thromboembolism by NCX 4016, a nitric oxide-releasing aspirin. *Eur J Pharmacol, 397,* 177–185.

324. Bizzozero, J. (1882). Ueber einen neuen formbestandtheil des blutes und dessen rolle bei der thrombose und der blutgerinnung. *Virchows Arch Pathol Anat Physiol, 90,* 261–332.

325. Baez, S. (1973). An open cremaster muscle preparation for the study of blood vessels by in vivo microscopy. *Microvasc Res, 5,* 384–394.

326. Falati, S., Gross, P., Merrill-Skoloff, G., et al. (2002). Real-time in vivo imaging of platelets, tissue factor and fibrin during arterial thrombus formation in the mouse. *Nat Med, 8,* 1175–1181.

327. Weichert, W., Pauliks, V., & Breddin, H. K. (1983). Laser-induced thrombi in rat mesenteric vessels and antithrombotic drugs. *Haemostasis, 13,* 61–71.

328. Roesken, F., Ruecker, M., Vollmar, B., et al. (1997). A new model for quantitative in vivo microscopic analysis of thrombus formation and vascular recanalisation: The ear of the hairless (hr/hr) mouse. *Thromb Haemost, 78,* 1408–1414.

329. Huo, Y., Schober, A., Forlow, S. B., et al. (2003). Circulating activated platelets exacerbate atherosclerosis in mice deficient in apolipoprotein E. *Nat Med, 9,* 61–67.

330. Denk, W., Strickler, J. H., & Webb, W. W. (1990). Two-photon laser scanning fluorescence microscopy. *Science, 248,* 73–76.

331. Piston, D. W. (2005). When two is better than one: Elements of intravital microscopy. *PLoS Biol, 3,* e207.

332. Hill-Zobel, R. L., Scheffel, U., McIntyre, P. A., et al. (1983). $^{111}$In oxine-labeled rabbit platelets: In vivo distribution and sites of destruction. *Blood, 61,* 149–153.

333. Kotze, H. F., Lotter, M. G., Badenhorst, P. N., et al. (1985). Kinetics of In-111-platelets in the baboon. II. In vitro distribution and sites of sequestration. *Thromb Haemost, 53,* 408–410.

334. Reimers, H. J., Buchanan, M. R., & Mustard, J. F. (1973). Survival of washed rabbit platelets in vivo. *Proc Soc Exp Biol Med, 142,* 1222–1225.

335. Winocour, P. D., Cattaneo, M., Somers, D., et al. (1982). Platelet survival and thrombosis. *Arteriosclerosis, 2,* 458–466.

336. Holme, S., Heaton, A., & Roodt, J. (1993). Concurrent label method with $^{111}$In and $^{51}$Cr allows accurate evaluation of platelet viability of stored platelet concentrates. *Br J Haematol, 84,* 717–723.

337. Wadenvik, H., & Kutti, J. (1991). The in vivo kinetics of $^{111}$In- and $^{51}$Cr-labelled platelets: A comparative study using both stored and fresh platelets. *Br J Haematol, 78,* 523–528.

338. Bayer, E. A., & Wilchek, M. (1990). Protein biotinylation. *Methods Enzymol, 184,* 138–160.

339. Heilmann, E., Friese, P., Anderson, S., et al. (1993). Biotinylated platelets: A new approach to the measurement of platelet life span. *Br J Haematol, 85,* 729–735.

340. Franco, R. S., Lee, K. N., Barker-Gear, R., et al. (1994). Use of bi-level biotinylation for concurrent measurement of in vivo recovery and survival in two rabbit platelet populations. *Transfusion, 34,* 784–789.

341. Rand, M. L., Wang, H., Bang, K. W., et al. (2004). Procoagulant surface exposure and apoptosis in rabbit platelets: Association with shortened survival and steady-state senescence. *J Thromb Haemost, 2,* 651–659.

342. Baker, L. C., Kameneva, M. V., Watach, M. J., et al. (1998). Assessment of bovine platelet life span with biotinylation and flow cytometry. *Artif Organs, 22,* 799–803.

343. Valeri, C. R., Macgregor, H., Giorgio, A., et al. (2005). Comparison of radioisotope methods and a non-radioisotope method to measure platelet survival in the baboon. *Transfus Apheresis Sci, 32,* 275–281.

344. Peng, J., Friese, P., Heilmann, E., et al. (1994). Aged platelets have an impaired response to thrombin as quantitated by P-selectin expression. *Blood, 83,* 161–166.

345. Heilmann, E., Hynes, L. A., Friese, P., et al. (1994). Dog platelets accumulate intracellular fibrinogen as they age. *J Cell Physiol, 161,* 23–30.

346. Dale, G. L., Friese, P., Hynes, L. A., et al. (1995). Demonstration that thiazole-orange-positive platelets in the dog are less than 24 hours old. *Blood, 85,* 1822–1825.

347. Ault, K. A., & Knowles, C. (1995). In vivo biotinylation demonstrates that reticulated platelets are the youngest platelets in circulation. *Exp Hematol, 23,* 996–1001.

348. Manning, K. L., Novinger, S., Sullivan, P. S., et al. (1996). Successful determination of platelet lifespan in C3H mice by in vivo biotinylation. *Lab Anim Sci, 46,* 545–548.

349. Gemmell, C. H., Yeo, E. L., & Sefton, M. V. (1997). Flow cytometric analysis of material-induced platelet activation in a canine model: Elevated microparticle levels and reduced platelet life span. *J Biomed Mater Res, 37,* 176–181.

350. Robinson, M., Machin, S., Mackie, I., et al. (2000). In vivo biotinylation studies: Specificity of labelling of reticulated platelets by thiazole orange and mepacrine. *Br J Haematol, 108,* 859–864.

351. Yamao, T., Noguchi, T., Takeuchi, O., et al. (2002). Negative regulation of platelet clearance and of the macrophage phagocytic response by the transmembrane glycoprotein SHPS-1. *J Biol Chem, 277,* 39833–39839.

352. Rand, M. L., Wang, H., Mody, M., et al. (2002). Concurrent measurement of the survival of two populations of rabbit platelets labeled with either two PKH lipophilic dyes or two concentrations of biotin. *Cytometry, 47,* 111–117.

353. Magnusson, S., Hou, M., Hallberg, E. C., et al. (1998). Biotinylated platelets have an impaired response to agonists as evidenced by in vitro platelet aggregation tests. *Thromb Res, 89,* 53–58.

354. Horan, P. K., Melnicoff, M. J., Jensen, B. D., et al. (1990). Fluorescent cell labeling for in vivo and in vitro cell tracking. *Methods Cell Biol, 33,* 469–490.

355. Michelson, A. D., Barnard, M. R., Hechtman, H. B., et al. (1996). In vivo tracking of platelets: Circulating degranulated platelets rapidly lose surface P-selectin but continue to circulate and function. *Proc Natl Acad Sci USA, 93,* 11877–11882.

356. Baker, G. R., Sullam, P. M., & Levin, J. (1997). A simple, fluorescent method to internally label platelets suitable for physiological measurements. *Am J Hematol, 56,* 17–25.

357. Rothwell, S. W., Maglasang, P., & Krishnamurti, C. (1998). Survival of fresh human platelets in a rabbit model as traced by flow cytometry. *Transfusion, 38,* 550–556.

# CHAPTER 34

# Real-Time *In Vivo* Imaging of Platelets During Thrombus Formation

**Christophe Dubois,[a] Ben Atkinson,[a] Barbara Furie, and Bruce Furie**

*Center for Hemostasis and Thrombosis Research, Beth Israel Deaconess Medical Center and Harvard Medical School, Boston, Massachusetts*

## I. Introduction

Real-time *in vivo* imaging makes it possible to initiate, observe, record, and quantitatively measure platelet thrombus formation within a living animal. The goal of using such a technique is to understand, in a physiological context, the relative involvement and contribution of platelets and other cells and proteins in thrombus formation.

The object of the majority of biological and biochemical research projects is to provide us with an understanding of how the components within a biological system interact to form a functioning unit. Prior experiments within the field of hemostasis and thrombosis, however, have generally been performed under highly simplified conditions in which the component under investigation is far removed from the whole system. As such there has always been a leap of faith as to whether any of the observations made in chemically defined, purified systems can be extrapolated to *in vivo* systems. So, as our questions get more specific and our technology and techniques more refined, the quest for conducting experiments in conditions closer to the physiological environment has been a high priority. The techniques involved in real time imaging of thrombus formation *in vivo* described here were developed to bring experiments in thrombosis and hemostasis into a living animal and thus significantly closer to intact physiology. In particular, these new developments have made it possible to test platelet function *in vivo*.

Intravital microscopy and the study of blood flow *in vivo* has been practiced for many decades.[1-7] In fact, Bizzozero[8] used intravital microscopy of guinea pig venules in 1882 to make his landmark discoveries about platelets (see Fig. 2 of the Foreword). However, until recently, with the development of improved cameras and computers, this field has always relied on the keen and unbiased eyesight of the investigator for accurate results.

The first step toward a method of studying thrombus formation and platelet function *in vivo* was the application of microscopy to conditions in which the blood is driven to flow within a coverslip bounded chamber, a "flow chamber" or a microcapillary tube.[9] The advantages of these systems are that conditions experienced by the platelets and other blood components can easily be controlled and altered, including: shear rate, fluid turbidity, cell count, and surface adhesion molecules. However, this is also the major disadvantage of such a system, because experiments cannot fully recreate the complex combination of conditions experienced by blood flowing *in vivo*.

The earliest forms of modern intravital microscopy were simple widefield systems using white light illumination and analog video camera capture technology. Thrombus formation could be recorded, presented, and to a certain extent, analyzed from the analog videos.[2,3] Unfortunately, analog video images cannot be subjected to detailed quantitative analysis other than simple two-dimensional measurements. However, these systems were instrumental in developing various tissue preparation techniques and different models of injury.

Recent advances in computer and digital camera technology have made possible the development of microscope systems that are both sensitive and fast enough to monitor thrombus formation *in vivo* and in real time. In addition, through the use of spinning Nipkow disk technology, confocal images can now also be acquired at rates comparable to video-frame rates.[10] Finally, the development of more powerful computer workstations that can handle rapid data collection and quickly store massive amounts of image information in digital format has enabled a recording of the acquired real-time images.

The more recent availability of techniques to genetically modify mice to generate transgenic, knockout, and knock-in mouse strains has allowed the investigation of the role of a single gene and its product in a specific physiologic context. This has in turn prompted the development of mouse models of human disease to study the role of individual proteins in

---

[a]These two authors made equivalent contributions.

different pathologic conditions, including thrombosis. For these reasons, much intravital microscopy has been focused on mouse models.

There are a number of different methods used to initiate experimental thrombosis. These include: electrical and mechanical destruction of the endothelium,[2,11–15] topical application of ferric chloride solution,[16–18] systemic infusion of Rose-Bengal followed by a low-power laser excitation,[19–25] and a higher power pulsed nitrogen laser that delivers an injury via the microscope optics to the vascular wall.[5,25–30] In order to monitor and quantitate subsequent platelet accumulation and thrombus formation, fluorescently labeled antibodies directed against cell-specific or soluble antigens are often used to study the role of specific cell populations and proteins in thrombus formation.

A full description of the components involved in the contemporary systems capable of real time *in vivo* imaging of thrombus formation and the methods that are used to generate and analyze the data are given in this chapter. However, because hardware, software, and user applications continue to evolve rapidly, the descriptions that follow are likely to change with continued improvements.

## II. Equipment and Methods

Our primary methods image the cremaster muscle microcirculation in a living mouse.[31–37] Although other vascular windows have been used in our laboratory, including the mesenteric microcirculation and the ear microcirculation, the cremaster muscle offers excellent blood vessel images because it is thin and transparent, contains many arterioles and venuoles, and grants a stable preparation for multiple hours of study. Wild-type mice and genetically altered mice are routinely employed, and thus provide important information about the role of particular gene products in thrombus formation. Mostly, we employ a laser-induced injury model.[25] Although many other injury models have been adapted to experiments in the mouse,[38] the laser injury is temporally and spatially defined and its magnitude adjustable. Confocal imaging employs laser excitation, a high-speed confocal scanner using Nipkow disk technology, and capture of intensified images on a high-speed digital charge-coupled device (CCD) camera. Widefield imaging illuminates the specimen with monochromatic light and captures intensified images on a high-speed digital CCD camera. The cameras are software-controlled, acquiring digital video images at a high frame rate. These images are analyzed in a computer workstation following data acquisition.

### A. The Microscope

Our system was developed around an Olympus AX70 microscope with a trinocular head (Olympus America, Inc.,

Melville, NY, USA). A binocular head is essential for easy switching between viewing through the eye pieces and the camera; in addition, a trinocular head is useful if more than one camera is regularly used or if a confocal head is present. Although the microscope has a full set of objectives, the intravital microscopy applications described here require mostly a 40× and a 60× water immersion lens (LUMPlanFl/IR), with numerical apertures of 0.80 and 0.90, respectively. Transmitted light under the specimen is delivered through a high-speed shutter (Uniblitz VS25, Vincent Associates, Rochester, NY, USA) with an electromagnetic actuator that can be operated manually (for example, during thrombus generation), or through computer control during image acquisition. In the latter case, the brightfield light can be shuttered on and off to allow brightfield image capture interlaced between fluorescent illumination. During widefield microscopy, fluorescent illumination is provided by a 175 watt xenon lamp that can supply light from 350 through 700 nm. In contrast to a mercury lamp, the xenon lamp provides similar intensity across the excitation spectrum. The excitation beam is controlled by a high-speed wavelength changer (DG-4, Sutter Instrument Company, Novato, CA, USA) that allows switching of the beam position via an electrogalvanometer-controlled mirror to pass through, in sequence, any of four positions, each with a specific excitation filter. Switching from one filter to the next is computer controlled and is effected in time intervals as short as 1.2 microsecond. For multicolor experiments such a wavelength changer can be equipped with excitation filters (360, 480, and 590 nm; Chroma, Brattleboro, VT, USA) matched to a triple band pass dichroic emission filter (DAPI/FITC/Cy5; Chroma) in the microscope. These components allow near-simultaneous collection of images in up to three fluorescence channels and a brightfield channel.

For confocal microscopy, we use a Yokogawa CSU-10 confocal scanner. Out-of-focus haze can be eliminated by confocal microscopy. However, the Nipkow disk and short exposure times required for real-time image acquisition greatly limit light, thus requiring light amplification with an intensifier. Intensifiers convert the light they receive into electrons which then pass down an electron multiplier tube and amplify the signal before being converted back to light on a phosphorous plate. The use of intensifiers results in a loss of resolution that degrades the image. This confocal scanner uses the same pinholes for entrance and exit light beams but is also equipped with a second rotating disk that contains approximately 20,000 pinholes, each with a microlens. The disks rotate together at 1800 rpm so that the light beams raster-scan the specimen. This technology allows the microscope to capture, in theory, up to 360 frames sec[-1]. In contrast, standard point-scanning confocal microscopes have a single pinhole and take about 0.5 to 1 second to acquire an image. An argon–krypton three-line laser (Melles Griot, Carlsbad, CA, USA) provides the fluorescent light for confocal microscopy, with excitation at 488, 568, and 647 nm. A filter wheel (Lambda L-10, Sutter Instrument Company, Novato, CA, USA) is mounted on the excitation source to enable delivery of monochromatic light.

To achieve three-dimensional (optical sectioning) image acquisition, the microscope objective is mounted on a piezoelectric driver (Physiks Instrumente International, Waldbronn, Germany), which is controlled by the computer and allows changes in the focal plane as rapidly as every 20 microsecond with movements as small as $0.1\ \mu m$ in the $z$ axis. A motorized focus drive also allows three-dimensional optical sectioning without the need for a piezoelectric collar. However, these drives have significantly longer movement times, thus reducing acquisition rates, an important consideration when recording a growing thrombus.

Our microscope is equipped with a motorized stage (ProScan 101, Prior Scientific, Cambridge, UK) specifically modified to accommodate our sample tray. The stage, which can be operated either via a manual joystick or through coordinates introduced into the computer, allows precise and reproducible movement of the animal preparation. Certain experiments require the observation of thrombi after an extended period of time (e.g., 15–30 min); with a motorized stage, a thrombus can be initiated, electronically tagged via its $xy$ coordinates, and returned to at a later time.

The central part of the trinocular stand is connected to a 12-bit CCD video camera through a Gen III intensifier (Videoscope, Sterling, VA, USA). The intensifier amplifies the signal that reaches the CCD chip, thus allowing shorter exposures for the same amount of available light. This increases the temporal resolution of the images, albeit at the expense of spatial resolution. The gain of the intensifier can be adjusted and is usually kept at the minimum value that allows sufficient time resolution. A higher intensifier gain significantly increases noise during image acquisition. We currently use two CCD cameras: a high-resolution camera (CoolSnap HQ, Roper Scientific, Tucson, AZ, USA) which captures images in $1390 \times 1024$ format (this camera has a high readout rate and pixel dimensions of $6.6 \times 6.6$ mm) and a high-speed camera (Sensicam, Cooke Corporation, Auburn Hill, MI, USA) with a resolution of $640 \times 480$ and pixel dimensions of 9.9 mm. While this latter camera has a lower spatial resolution, its smaller format allows a faster frame rate. Without binning, the Sensicam can operate at frame rates of 30 images $sec^{-1}$ and the CoolSnap HQ can operate at frame rates of 10 images $sec^{-1}$ during capture of a fluorescent image in a single channel. To increase the frame rate, the image acquisition software permits binning (i.e., the combination of signal from 4 ($2 \times 2$ binning) or 16 ($4 \times 4$ binning) adjacent pixels). Again, this combines light in adjacent pixels at the expense of resolution. Binning also dramatically reduces the file size for computer storage. We have found that $2 \times 2$ binning with the

high-speed camera, while keeping the light requirement low enough for most applications, still allows us to record images with an acceptable resolution and represents an optimal compromise for most of our experiments. A schematic representation of the microscope and its components is shown in Fig. 34-1.

The images captured by the CCD cameras are transferred to the random access memory of the computer. Imaging software (Slidebook, Intelligent Imaging Innovations, Denver, CO, USA) is used to control and coordinate the hardware components (such as the brightfield shutter, the high-speed wavelength changer, the piezoelectric driver, the CCD camera, image acquisition channel, etc.) during image acquisition and is also used for image analysis.

### B. Mouse Models

In theory, all veins and arteries could be used to study thrombus formation by intravital microscopy. However, some limits apply, including the difficulty in isolating the studied vasculature and the subsequent trauma to the animal and specific tissue, both of which could influence the blood flow/shear rate and the degree of inflammation. Furthermore, vessel caliber and organ bed may be important parameters for the experiment. These are some of the most important issues to consider in the choice of the mouse model used.

**1. General Procedure.** The animal is anesthetized with an intraperitoneal injection of ketamine (125 mg/kg), xylazine (12.5 mg/kg), and atropine sulfate (0.25 mg/kg), and the jugular vein is exposed under a dissecting microscope and cannulated. For some experiments, the femoral vein is used to introduce reagents into the circulation. When blood pressure is monitored continuously, a transducer is associated with the carotid artery. The trachea is always intubated to facilitate spontaneous respiration by the mouse. The mouse is then placed on a custom-designed Plexiglas tray where the appropriate vascular tissue is exteriorized. Anesthesia is maintained with pentobarbital (50 mg/kg) injected though the jugular vein as needed. The same access is also used to inject other agents, for example, antibodies, fluorochromes, and cells.

**2. Cremaster Muscle.** The cremaster muscle is one of the commonly used preparations to visualize the microcirculation in mice.[31,32,36] The original procedure for isolating this muscle was described by Baez in 1973.[1] Briefly, the cremaster muscle is exteriorized through an incision in the scrotum. After removal of the connective tissue, the muscle is pinned across a coverslip mounted within the tray. The muscle preparation is superfused throughout the experiment with preheated (37–38°C) bicarbonate-buffered saline

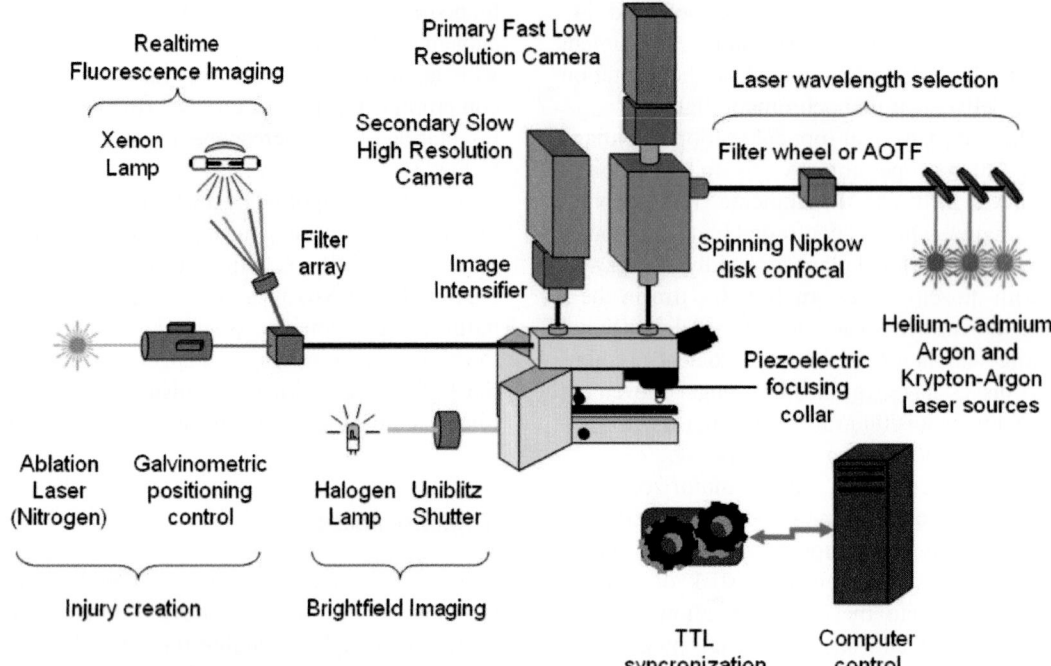

**Figure 34-1.**  Schematic diagram of the design of the confocal and widefield system for multichannel digital intravital microscopic imaging.

and aerated with 5% $CO_2$/95% $N_2$, and the mouse body temperature maintained on a heated water blanket. The animal is then moved to the microscope stage. Buffer superfusion is restarted immediately upon repositioning of the animal onto the microscope stage, and the buffer is used as the medium to immerse the lens of the microscope objective. The procedure described typically requires 5 to 7 minutes.[39,40]

The major advantages to studying the cremaster muscle are that the tissue is thin enough to record high-resolution brightfield images, the preparation contains many arterioles and venuoles, and exteriorization of the muscle does not traumatize the mouse. This muscle can be isolated in 2- to 25-week-old mice. However, arterioles and venules contained in the cremaster are too small to place a Doppler flow probe directly on them or to induce a mechanical injury by, for example, ligation of the artery.

**3.  Mesentery.**  Mice of 3 to 4 weeks are most often used, because larger mice tend to have a significant amount of fat that collects around the postcapillary venules and arteries. A cautery incision is made along the abdominal region of an anesthetized mouse. The mesenteric vascular bed is exteriorized by gently teasing it out of the abdominal cavity, then placing it on and pinning it across a coverslip on a Plexiglas viewing stage. The muscle preparation is superfused throughout the experiment as described previously and the animal is moved to the microscope stage.[6,12,16,27,41] Mesen-

teric arteries are large enough (about 100 μm) to perform some mechanical or ligation-induced injuries. However, only three to five different arteries can be studied in this model and the trauma induced by the exteriorization of the mesentery is greater than the isolation of the cremaster, such that experiments must be performed rapidly.[42,43] The blood shear rate in mesenteric arteries is lower than in cremaster arterioles (about 1300 $sec^{-1}$ versus 1500 $sec^{-1}$). Lastly, the thickness of the mesentery generates a greater degree of out-of-focus light, resulting in a subsequent drop in brightfield image resolution.

**4.  Ear.**  The hair on the ear of an anesthetized mouse is removed and the animal positioned in a plastic restraint on a microscope stage. The restraint is designed so one ear can be taped to a glass platform and blood flow through the vasculature of the ear can be visualized by transillumination. Although the architecture of the ear vasculature varies among mice, generally there are either one or two primary veins proximal to the head with a diameter of approximately 200 to 350 μm. This primary vein(s) is fed by secondary veins of approximately 150 to 250 μm, which in turn are fed by tertiary vessels of 100 to 200 μm. Although it is possible to define parameters for making reproducible injuries in the vein of the ear,[21,22,25] the elaboration of conditions to provoke reproducible injuries in the artery is problematic in this model. Indeed, the thicker wall of the artery and smaller diameter of the vessel disrupts the optical path of the laser beam used to induce the injury and prevents reliable focusing of the beam on the luminal surface of the artery.

**5. Carotid.** An incision is made with a scalpel directly over the right or left common carotid artery of an anesthetized mouse, and a segment of the artery is exposed using blunt dissection. The isolated artery is dissected free of surrounding tissue and pinned, and the animal is placed on the microscope stage. The advantage of this model is that the carotid artery is large enough to directly attach a Doppler flow probe connected to a flow meter for the measurement of the effect of vascular injury on the flow rate. The major inconvenience is that due to the tissue surrounding the carotid artery it is impossible to record brightfield images, although epifluorescent images can still be obtained.[14,17,18,23,44]

## C. Induction of Injury

Several methods have been proposed to induce thrombus formation in mice.[2,13–15]

**1. Mechanical Endothelial Denudation.** Endothelial denudation can be induced mechanically in large arteries (aorta, carotid artery, mesenteric artery). The damage can be produced indirectly on the vascular bed by compression or ligation of the vessel using a forceps or a surgical filament.[2,13–15] Alternatively, the luminal side of the vessel can be directly damaged by puncture of the vessel or, more subtly, using a guide-wire.[12,44] In this case, the artery is looped proximally and tied off distally. A transverse arteriotomy is made in the artery and a flexible angioplasty guide wire is introduced. Endothelial denudation injury of the artery is performed by wire withdrawal with rotating motion to ensure uniform and complete endothelial denudation. After removal of the wire, the artery is tied off and platelet–vessel wall interactions are studied. All these procedures result in exposure of the endothelial cell matrix.

**2. Ferric Chloride.** $FeCl_3$ is an oxidative chemical agent that penetrates the endothelium rapidly and induces removal of endothelial cells, leading to exposure of subendothelial matrix. $FeCl_3$-induced injury of the vascular bed results from superfusion of an $FeCl_3$ solution directed onto the isolated artery or by transient topical application of a filter paper saturated with a solution of $FeCl_3$ (Fig. 34-2A). The severity of the resultant injury is dependent on the concentration of $FeCl_3$ used and the duration of its application to the artery. Depending on the animal model studied, a wide range of conditions are used (e.g., a concentration of $FeCl_3$ of 5 to 50% applied to the artery for a few seconds to several minutes).[16–18,45]

**3. Rose Bengal/Photo Excitation.** Another chemical method to trigger thrombosis involves the intravenous injection of the photoreactive substance Rose Bengal (Fig. 34-2B) which rapidly accumulates in the membranes of endothelial and other cells. The subsequent exposure of an arterial segment to green light (540 nm) locally triggers a photochemical reaction of Rose Bengal, resulting in the formation of reactive oxygen species that damage the endothelium and induce the formation of occlusive thrombi.[23–25] Typically, mice are administered 40 μg/gm body weight of a 10 mg/mL solution of Rose Bengal dye dissolved in normal saline. Subsequent laser illumination induced mural clots in the target vessels near the site of illumination. The power (from 10 to 100 mW) and the duration (from a few seconds to several minutes) of irradiation by the laser of the Rose Bengal directly influences the severity of the induced injury. This technique has been used successfully to investigate thrombus formation in mouse ear vessels albeit only to produce qualitative data.[25]

**4. Laser-Induced Injury.** In our models, we use a nitrogen dye laser to injure the vessel wall.[5,25–30] The site of the vessel wall to be injured is established and several images of the field are captured. The laser beam is then aimed at the endothelial cell–blood interface through a crosshair in the ocular and bursts of laser light are delivered for a predefined time interval. The power and frequency of these bursts controlled by the software are empirically determined and depend on the thickness of the tissues and the desired size of the thrombus. It is usually possible to generate sufficiently accurate data by capturing a few images of the site to be injured prior to laser injury, to stop recording, and to injure the vessel wall by pulsing the laser manually, and then to record back at the site of injury (Fig. 34-2C).[25,32,36] In this method, proper barrier filters must be in place to avoid eye injury to the operator. Alternatively, laser-induced injuries can be performed "live" while recording with the imaging software. Care must be taken to assure that the emission filter being used absorbs light at the frequency of the laser, so as to avoid stray reflections that might otherwise damage the intensifier or camera. This method necessitates that the laser be targeted while looking at the computer screen rather than through the microscope oculars, and the software must support real-time video imaging directed to a monitor. Thrombi can be generated both in venules and arterioles. The number of thrombi that can be generated in a single experiment varies depending on the anatomy of the vessel tree in the specific mouse preparation. We routinely generate as many as 10 arteriolar thrombi over the course of a 60- to 90-minute experiment, moving upstream along the vessels.

To generate laser-induced thrombi, the system is fitted with a nitrogen ablation laser (MicroPoint, Photonic Instruments, St. Charles, IL, USA) which is introduced via the epi-illumination port and is focused on the specimen

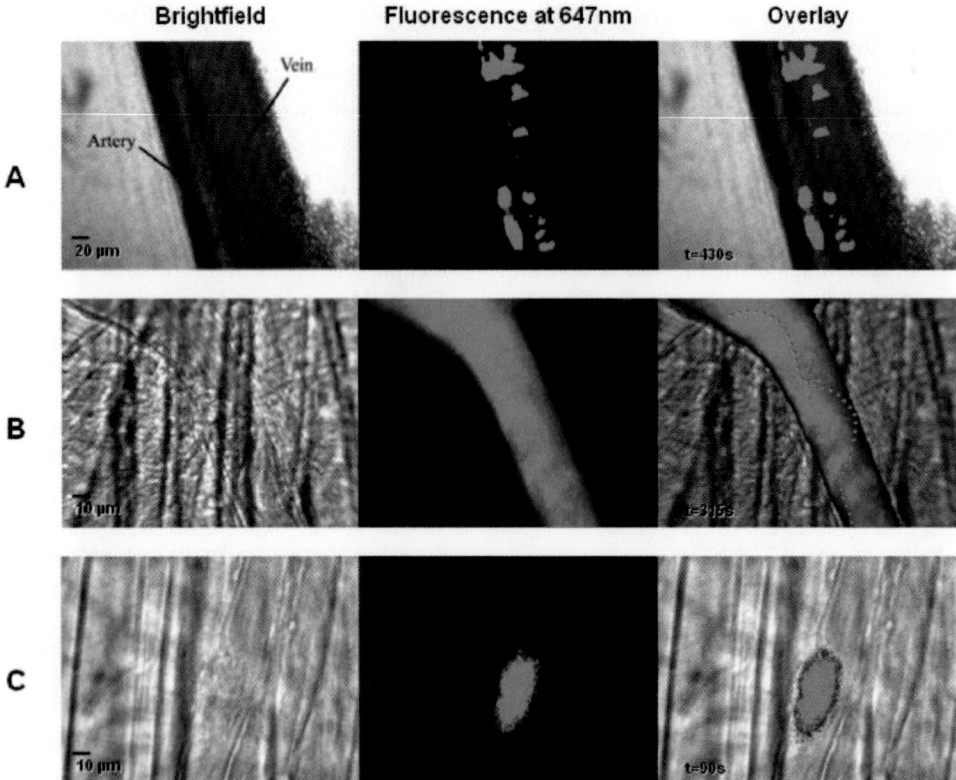

**Figure 34-2.** Intravital imaging of platelets in the developing thrombus of a living wild-type mouse following three different models of vascular injury. Brightfield, fluorescent, and composite (overlay) images of widefield microscopy are represented. The fluorescence image, recorded digitally, of platelets participating in growing thrombi is presented as the red pseudocolor. Images represent (A) growing thrombi formed 430 second after application of a 10% solution of ferric chloride for 5 minute to the adventitial surface of the mesentery, (B) 315 second after a Rose Bengal-induced injury in the cremaster microcirculation, and (C) 90 second after a laser-induced injury in the cremaster microcirculation.

through the microscope objectives. The output of the laser is at 337 nm but is subsequently tuned through a dye cell. We routinely use coumarin as dye, which emits at 440 nm. The laser delivers 4 nanosecond energy pulses, at typically three pulses sec[1], over a surface approximately 1 μm in diameter. The energy of the pulses can be controlled by the operator.

## III. Data Acquisition and Analysis

Digital images are captured with a Cooke Sensicam CCD camera in 640 × 480 format. The system is controlled using Slidebook (Intelligent Imaging Innovations, Denver, CO, USA). For each frame, the software provides a graphical representation (histogram) of the dynamic range of intensities within the image. Because images are captured at a 12-bit depth, $2^{12}$ (4096) levels of intensity are available. The histogram presents the number of pixels (y axis) at each level of intensity (x axis). To differentiate subtle changes in intensity, it is important to take advantage of the full dynamic range. This requires changing the amount of light that reaches the camera, which in turn is obtained by changing light

intensity, the intensifier gain, exposure time, and binning. If too much light reaches the CCD camera, all pixels with an intensity above 4096 will be recorded as equal; if not enough light is available, some pixels will not be distinguishable from background noise due to autofluorescence.

Many of our experiments involve the comparison of signals generated by specific antibodies with those generated by isotype-matched control antibodies or nonimmune IgG when polyclonal antibodies are used. More often than not, the optimized settings still are associated with some small signal observed in the control experiments. This has necessitated the development of a method to account for background signal.

Prior to the generation of the thrombus, but after the jugular injection of the fluorescent dye-labeled antibody, 30 image frames are recorded. For each thrombus a rectangular mask is defined to include a portion of the vessel upstream to the site of injury (Fig. 34-3). The maximum intensity (in pixels) contained on this mask is extracted for all frames (pre- and postinjury) for each thrombus. The mean value from the maximal intensity values in the mask for each frame is calculated and used as the threshold value to extract

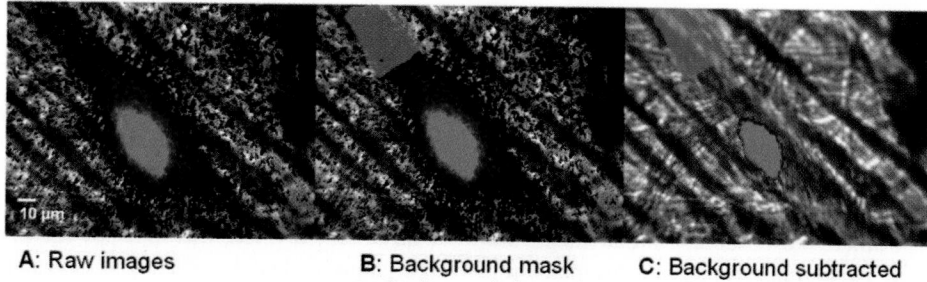

**A: Raw images**  **B: Background mask**  **C: Background subtracted**
  **marked on red channel**  **from red channel**

**Figure 34-3.** Data analysis of fluorescent signal. Images represent brightfield and fluorescent signal of platelets depicted in red after laser-induced injury on the cremaster artery. We have developed a method to account for background fluorescent signal (A). We draw a rectangular mask to include a portion of the vessel upstream to the site of injury (B), calculate the background signal for each frame (see text for details), and subtract it from the fluorescent signal (C).

the intensity values (in pixels) and the area corresponding to the specific signal. For each frame, the integrated fluorescence intensity is calculated by subtracting the background values contained in the mask from the specific intensity values by the following equation:

Integrated fluorescence intensity = [mean intensity value of the signal in the frame × area of the intensity value of the signal] − [mean intensity value of the background in the mask × area of the intensity value of the signal].

This calculation is done for all image frames captured in each thrombus. The final results (typically 30 thrombi, three mice) are expressed as the median of integrated fluorescence intensity (arbitrary units) over the time (sec).

## IV. Study of Thrombus Formation

Several approaches can be used to visualize the various components of a thrombus in the microcirculation. Each has advantages and disadvantages. The choice of the wavelength of the fluorochrome depends on the specific experimental design and must take into consideration background tissue autofluorescence, which is maximal at about 500 nm, and the presence of other fluorochromes so that crossover (i.e., "bleeding" of the signal generated by one channel into another), can be minimized or eliminated. When only two fluorescent labels are present, the use of Alexa 350 or 488 and Alexa 660 usually represents the best choice; when a third label is introduced, we use Alexa 350, Alexa 488, and either Alexa 660 or Cy5. Because the excitation spectrum of Alexa 640, Alexa 660, and Cy5 are broad, one has to consider that excitation of fluorochromes at shorter wavelengths may also excite these fluorochromes even though their excitation maxima are at longer wavelengths. For example, Alexa 567 can be excited at about 10% efficiency by illumination at 488 nm; therefore, Alexa 488 (or fluoroscein) is never used with Alexa 567 (or rhodamine).

### A. Kinetics of Platelet Accumulation

To follow platelet accumulation over time by intravital microscopy, we use a monoclonal antibody directed against the integrin αIIb (GPIIb, CD41) directly coupled to an Alexa fluorochrome (Molecular Probes). We prefer the Alexa series of fluorochromes, which offer the advantage of a low rate of photobleaching. Antibodies are labeled by incubating purified antibody with the label, followed by gel filtration to separate the labeled antibody from the excess label. A direct labeling of primary antibody reduces the number of proteins to be infused and therefore binding and nonspecific "trapping" of molecules within the thrombus. It also reduces the complexity of interpretation, specifically the concern about cross-species binding of secondary antibodies. Careful controls are therefore particularly relevant in this *in vivo* model in which nonspecific binding cannot be overcome by, for example, more stringent washing steps. Controls to consider are relevant nonimmune antibodies or isotype-matched controls and the relative fluorochrome labeling of each antibody, cell, or molecule used. The use of F(ab)₂ fragments to follow platelets is preferred to reduce the toxic effects on platelets of the antibody via the Fc receptor. We generally observe that upward of 70% of platelets are labeled *in vivo*, a value determined by flow cytometry of blood platelets removed from the mouse circulation.

To compare thrombus formation in wild-type control and treated or genetically deficient mice, we determine the integrated fluorescence intensity of the antibody directed against integrin αIIb after a laser-induced injury in several independent experiments. After a laser-induced injury on the cremaster muscle of wild-type mice we observed a rapid accumulation of platelets to the site of vascular injury. Platelet accumulation over time increases (net positive phase) and reaches a peak between approximately 70 and 100 seconds. After that time, the net platelet accumulation in the thrombus diminishes (net negative phase) and is followed by a stabilization of the thrombus (plateau phase) (Fig. 34-4A).

The kinetics of platelet accumulation in wild-type mice is different when the ferric chloride model is used on the mesenteric vasculature. In this model, platelets start to accumulate later in comparison with the laser model of injury. The thrombus grows and reaches a peak which corresponds to the occlusion of the artery. Finally, the thrombus is maintained at its maximal size for typically 35 minutes (Fig. 34-4B). In wild-type mice, we observe a broad range of platelet accumulation over time from one thrombus to another, even using the same experimental settings (e.g., concentration of the antibody/gram mouse, time of exposure of the fluorescent antibody, number of pulses and power of the dye laser, gain of the fluorescent intensifier). This vari-

ability is mainly due to the differences in the cremaster muscle from one mouse to another and to the focal plane in which the laser beam injures the vascular bed. Because of this variability, we typically induce 10 thrombi per mouse in three different mice and calculate the median signal accumulation over time. Two other parameters are also used to characterize the thrombus formation in wild-type and genetically defined mice: the time for the platelet thrombus to reach maximal peak size (Fig. 34-5A) and the maximal fluorescence intensity value at peak size for each thrombus (Fig. 34-5B). These three parameters are used to compare the kinetics of thrombus formation in different mouse genotypes or pharmacological situations or to determine whether

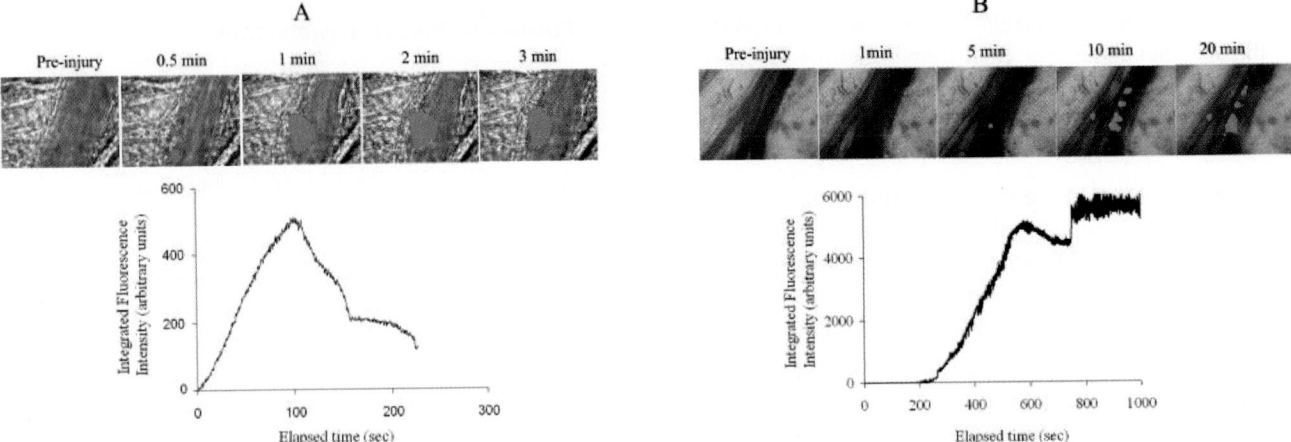

**Figure 34-4.** Kinetics of platelet accumulation after a laser- or a ferric chloride-induced injury. Incorporation of fluorescence signals into thrombi generated in wild-type mice was analyzed after infusion in the microcirculation of a CD41-specific antibody labeled with Alexa 647. Blood flow is from top to bottom. Upper panels, brightfield, fluorescence, and composite images after (A) laser-induced injury or (B) ferric chloride-induced injury. Lower panels represent integrated fluorescence intensity over time for representative thrombi after (A) laser- or (B) ferric chloride-induced injury. Note that the kinetics of thrombus formation in these two models of vascular injury are different.

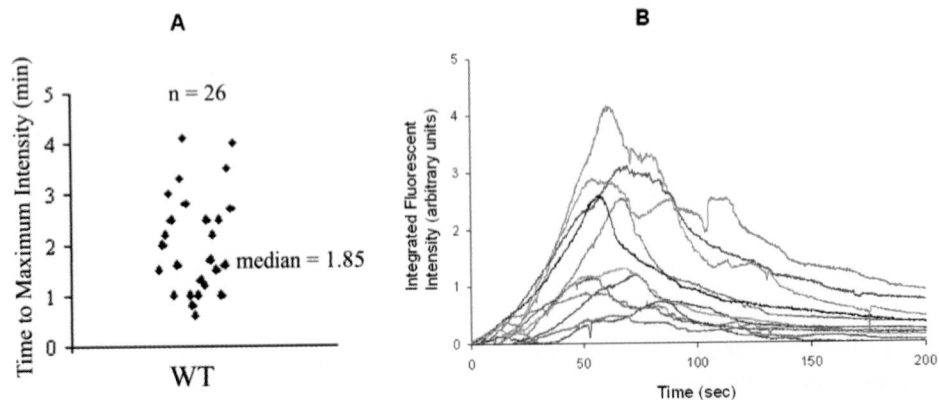

**Figure 34-5.** Thrombus formation in wild-type mice after laser-induced injury. Variation in thrombus size with each injury has led us to induce 7 to 10 thrombi per mouse in three different mice. We calculate the median integrated fluorescent intensity from these data. Panel A illustrates the variation in kinetics of platelet accumulation following 26 independent laser-induced thrombi in a wild-type (WT) mouse. Panel B represents the differences for 10 thrombi performed in three different wild-type mice to reach the maximal size after laser-induced injury.

an injection of a specific drug (e.g., directed against the integrin αIIbβ3-fibrinogen interaction) into a wild-type mouse can affect thrombus formation *in vivo*.

We have developed an alternative strategy to directly compare the participation in a growing thrombus of two different populations of platelets. For example, platelets treated with the calcium chelator BAPTA-AM and non-treated platelets can be simultaneously injected into a wild-type mouse to investigate the importance of calcium mobilization into platelets on their participation in the growing thrombus (Fig. 34-6A). In these experiments, blood is obtained by cardiac or hepatic portal vein puncture or carotid cannulation and collected in an acid–citrate–dextrose buffer. After centrifugation to obtain platelet-rich plasma, platelets are further purified by washing three times in Tyrode's buffer or by gel filtration. Platelets are then labeled *in vitro* with a fluorochrome. Platelet suspensions containing $3 \times 10^7$ platelets/mL are incubated with calcein acetoxymethyl ester (0.5 mg/mL) for 30 minutes. Once hydrolyzed inside the cell, this compound becomes fluorescent and charged, a characteristic that traps it inside the cell. Two different calcein dyes are used to label the two different platelet populations. We typically use calcein (monitored by excitation at 480 nm) and red–orange calcein (monitored by excitation at 570 nm). After an additional wash to eliminate the excess of calcein, the two platelet preparations at the same concentration are infused into the same mouse and laser-induced injuries are performed. Depending upon the particular experiment, the labeled platelets represent 0.5 to 10% of the total circulating platelets. Participation of calcein-labeled platelets and orange–red calcein-labeled platelets in the growing thrombus after a laser-induced injury is determined and the kinetics of labeled platelet accumulation are calculated as previously described (Fig. 34-6B).

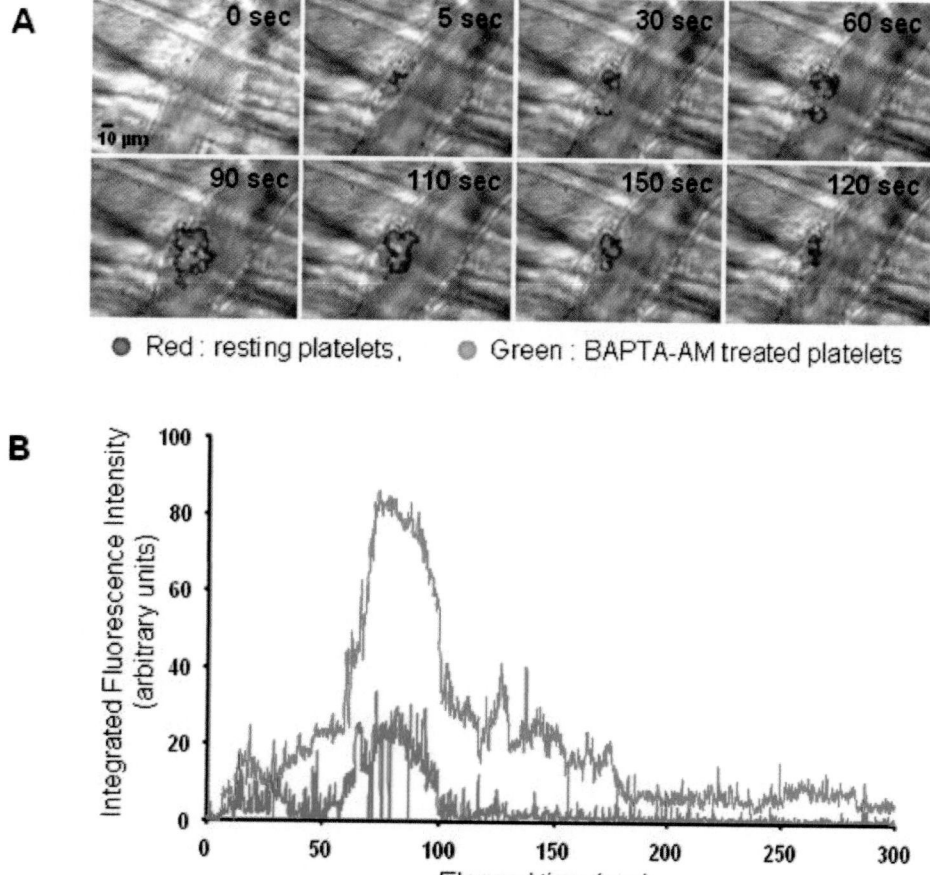

**Figure 34-6.** Accumulation of resting and BAPTA-AM-treated calcein-labeled platelets after laser-induced injury. Purified platelets were labeled with orange–red calcein or treated with the calcium chelator BAPTA-AM and calcein and infused into the circulation of a living wild-type mouse. A. Participation of calcein platelets, depicted in red, and orange–red BAPTA-AM treated platelets, depicted in green, in the growing thrombus following laser-induced injury was recorded by fluorescence microscopy. B. Medians of integrated fluorescence intensity of calcein platelets and orange–red calcein platelets from multiple experiments were calculated and compared to determine the importance of calcium mobilization into platelets on their participation in growing thrombi.

## B. Platelet Activation In Vivo

In some experiments the goal is to determine whether the defect observed in the kinetics of platelet accumulation to the site of vascular injury is due to a defect in platelet activation. We have developed two different strategies to determine *in vivo* the activation state of platelets participating in a growing thrombus: (1) monitoring the kinetics of calcium mobilization using fluorescent labeled platelets; (2) monitoring the exposure on the platelet surface of P-selectin.

### 1. Calcium Mobilization.

Calcium mobilization in platelets plays a critical role in inside-out and outside-in signaling, leading to the activation of the platelet integrins, platelet shape change, and granule secretion. The rise in cytosolic $Ca^{2+}$ levels can vary from basal levels of approximately 60 to 100 nM to micromolar levels in activated platelets, depending upon the potency of the agonist. To follow calcium mobilization in platelets *in vivo*, we loaded washed platelets with Fura 2-AM. Fura 2-AM is a calcium ion-binding fluorochrome used in living cells to measure cytosolic calcium mobilization. Fura 2-AM is characterized by distinct spectral properties in the presence of low and high concentrations of calcium ions. The binding constant, $K_d$, of Fura 2-AM for calcium is 0.14 μM, and this fluorochrome is therefore sensitive to changes in the intracellular calcium concentration where in the basal state calcium concentrations are significantly lower than 0.14 μM, while in the activated state calcium concentrations are significantly higher than 0.14 μM. Washed mouse platelets at a concentration of $1 \times 10^8$/mL are incubated in the dark at 37°C for 30 minutes. After washing the platelets to eliminate the excess of Fura-2, platelets are infused into a living mouse. Calcium binding to Fura-2 results in a shift in the absorbance maximum from 380 to 340 nm. Fura-2-loaded platelet accumulation in a growing thrombus is followed after excitation at 380 nm and calcium mobilization into these platelets is followed by excitation at 340 nm. Two different kinds of experiments are performed to characterize the kinetics of calcium mobilization *in vivo*. Firstly, if we infuse $1 \times 10^8$ of Fura-2 labeled platelets into a mouse only 1% of the total platelets will be labeled. This allows for the determination of the characteristics of binding and calcium mobilization of a single platelet (Fig. 34-7A). We have observed in the laser-induced injury model that a wildtype resting platelet first binds independently from calcium mobilization. Calcium flux into the platelet occurs 3 to 5 sec after the binding of the platelet to a growing thrombus. The duration of the calcium mobilization into a platelet directly correlates with the time that this platelet is bound to the thrombus. Secondly, if we wish to compare the kinetics of platelet binding and calcium mobilization between different genotypes of mice, we can infuse a greater amount of Fura-2 loaded platelets into the mouse and generate circulating platelets that are approximately 20% labeled. Using such conditions, multiple Fura-2 loaded platelets interact with the growing thrombus and the median Fura-2 loaded platelet accumulation and calcium mobilization from several thrombi performed in several mice can be calculated (Fig. 34-7B).

### 2. P-selectin Expression.

Platelet activation can also be monitored by P-selectin expression on the platelet surface. Expression of P-selection on the platelet surface is a consequence of α granule secretion (see Chapter 12). We study P-selectin expression on platelets incorporated into the developing thrombus by infusing into a living mouse a polyclonal rabbit antibody directed against P-selectin and a secondary anti-rabbit antibody coupled to an Alexa fluorochrome. In these experiments, we typically also infuse an antibody directed against integrin αIIb to monitor both platelet accumulation on the injured vascular bed and P-selectin expression on platelets over time. Using such conditions we detect P-selectin on the platelet surface 2 to 3 minutes after the laser-induced injury.[46] P-selectin first appears on platelets at the vessel wall–thrombus interface. The quantity of platelets expressing P-selectin increases over a time course of approximately 3 to 4 minutes and moves through the thrombus from the vessel wall to the luminal surface of the thrombus (Fig. 34-8). To compare the kinetics of platelet secretion over time as measured by P-selectin expression between two genetically different mice, we calculate as previously described both the median of platelet accumulation over time and the median of P-selectin expression to determine the relative quantity of P-selectin expressed per platelet.

## C. Thromboembolization

The thrombi that are generated in the mouse microcirculation undergo embolization, with small fragments of the thrombus breaking off and flowing downstream. To formally quantitate this phenomenon, we have established methods for detecting the image of the embolism as it floats downstream, and we can quantitate the fluorochrome in the emboli. Using a high image capture rate of over 100 frames per second, the integrated fluorescence intensity of each embolus is measured as a function of time. This high capture rate is achieved using a combination of $8 \times 8$ binning and limiting the image size to the specific region of interest. Multiple embolic events can be detected, and the amount of fluorochrome (proportional to the number of platelets) in each embolus is quantitated (Fig. 34-9). This allows formal analysis of a series of events observed previously using analogue videomicroscopy, but which could not be analyzed quantitatively.

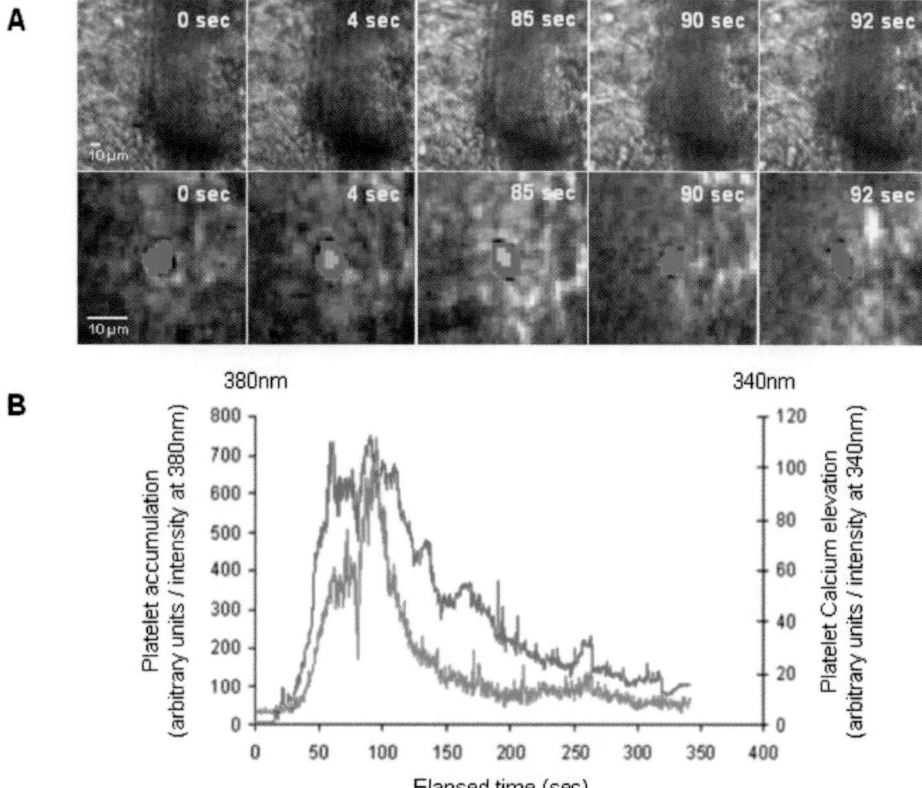

**Figure 34-7.** Intracellular calcium mobilization in platelets participating in thrombus formation *in vivo*. To characterize platelet activation *in vivo* following laser-induced injury, we have developed a method to follow intracellular calcium mobilization in platelets participating in growing thrombi. A. Representative kinetic of a single Fura-2 AM labeled platelet (depicted in green). Platelet mobilizes calcium (depicted in red, overlay in yellow) 2 to 5 seconds after binding to the growing thrombus. The platelet mobilizes calcium for approximately 80 second and then detaches from the thrombus. B. Median of Fura-2 labeled platelets participation (green) and calcium mobilization (red) in growing thrombi calculated from 18 thrombi performed in three different wild-type mice.

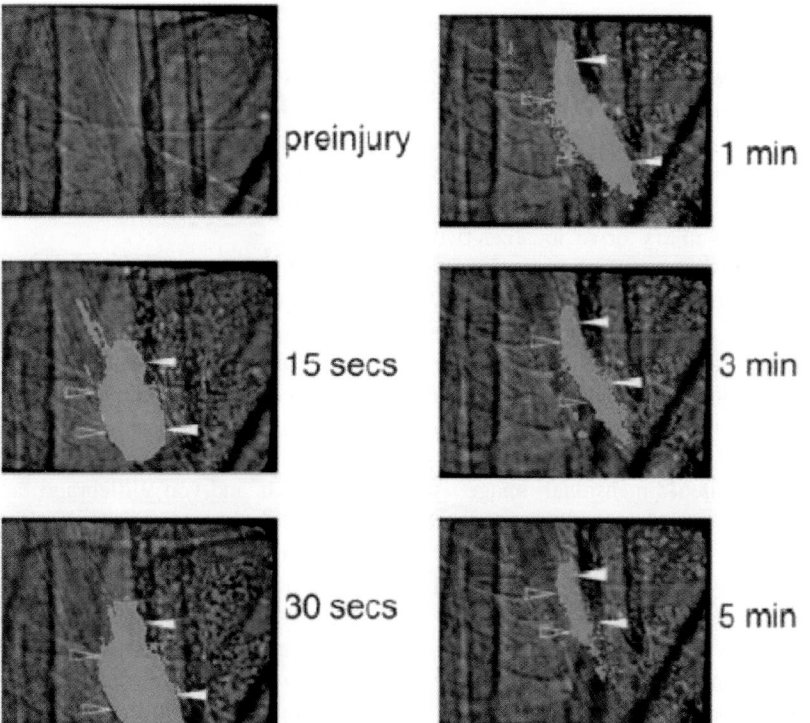

**Figure 34-8.** P-selectin expression in the developing arteriolar thrombus. High-speed confocal microscopy was used to generate an image through the center of the developing thrombus. A sequence of images of the confocal plane as a function of time, starting with initial injury, demonstrates a wave of P-selectin initiating at the vessel wall and ultimately extending to the thrombus–lumen interface. Platelets are labeled with Alexa 660-conjugated CD41 F$_{ab}$ fragments (red); P-selectin is labeled with Alexa 488-conjugated anti-P-selectin antibody (blue). Colocalization of P-selectin and platelets is displayed as magenta. Blood flow is from top to bottom. Thrombus–vessel wall interface (solid white arrowhead); thrombus–lumen interface (open white arrowhead).

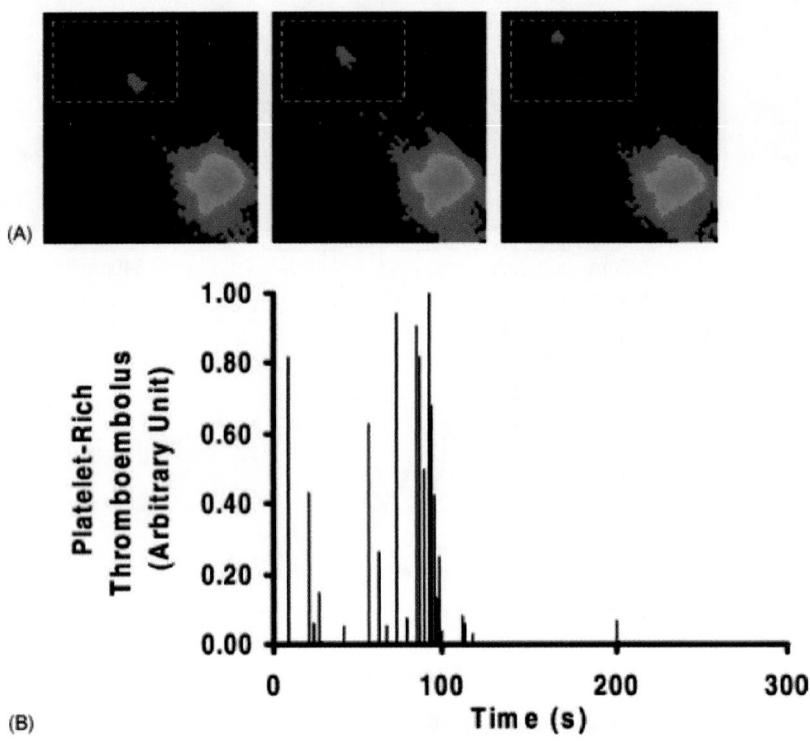

**Figure 34-9.** Quantitation of thrombus microembolization during thrombus formation. The total amount of thromboemboli originating from a thrombus has been quantitated by widefield fluorescence microscopy. A. In this series, a thromboembolus digitally recorded at 9-microsecond intervals is pictured in the dashed box. Thromboembolic events are captured by defining a subarray of pixels downstream of the thrombus with regard to blood flow. B. The integrated fluorescence of thromboemboli entering the defined subarray over a period of 300 seconds is recorded. Each bar represents a unique thromboembolic event.

### D. Confocal Microscopy: Thrombus Volume Determination and Colocalization of Proteins

High-speed confocal intravital microscopy has been used to perform detailed structural analysis of the growing thrombus generated by laser-induced injury or to localize two different components involved in thrombus formation *in vivo*. Using a piezoelectric driver on the microscope objective lens, image "slices" of 0.5 to 5 μm can be obtained at a high frame rate. These slices are stacked, thus generating an optical three-dimensional computer reconstruction of the thrombus (Fig. 34-10). The volume of thrombi formed *in vivo* can be determined by confocal microscopy and image analysis. In a thrombus, a three-dimensional image is obtained in which each component is labeled by a different fluorochrome. The total volume of the thrombus is obtained by image analysis. The boundary between the thrombus and the flowing blood is defined, and the thrombus surface structure rendered. Pixels included within this volume, including pixels defined by each of the fluorochromes reporting each of the thrombus components, are totaled to yield the number of "voxels." Because the dimension of a single pixel is known for a given magnification and thus the voxel volume

is known, totaling of the number of voxels within a thrombus yields the thrombus volume in cubic micrometers. To obtain the volume of a thrombus component, such as platelets, the number of voxels within the thrombus that are defined by the platelet-specific fluorochrome are totaled.

Confocal microscopy can also be used to determine the relative localization of two different components during *in vivo* thrombus formation.[47] Unlike widefield microscopy in which certain components above and below the z axis may appear colocalized although they are not adjacent, confocal microscopy images only components in a specific focal plane. For example, we have determined the colocalization of platelets and von Willebrand factor (VWF) *in vivo* after laser-induced injury. CD41-specific antibodies labeled with Alexa fluor 647 nm and directed against VWF labeled with Alexa fluor 488 nm were infused into a mouse. Confocal microscopy, using the piezoelectric driver on the microscope objective lens, was used to observe the accumulation of platelets and VWF at the site of a laser-induced injury, at different *z* planes over the growing thrombus. Confocal images of a growing thrombus were obtained by taking optical slices at 2-μm increments through approximately 70

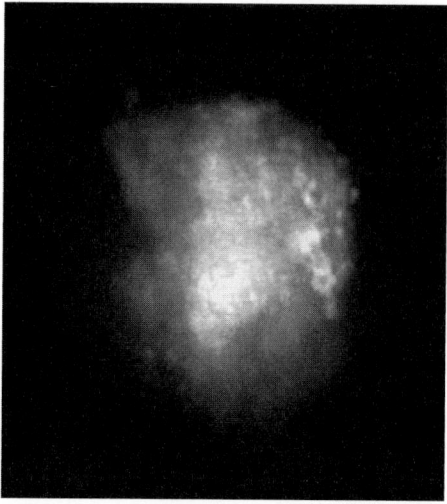

**Figure 34-10.** Three-dimensional image of a thrombus. An arteriole of the cremaster muscle was injured. Images of fluorescent platelets in the thrombus were obtained, as the working distances between the objective and the specimen were rapidly altered using the piezoelectric driver. At each confocal plane, a two-dimensional image was obtained. In this experiment, confocal *XY* images were stacked in sequential order with the imaging software to yield a three-dimensional image of the thrombus.

μm. Images were collected in both the 488- and 647-nm channels after an exposure time of 10 and 2 microseconds, respectively. Whereas widefield microscopy would suggest that VWF and platelets are colocalized in the growing thrombus (Fig. 34-11A), confocal microscopy reveals that most of the signal corresponding to VWF surrounds the growing thrombus and the platelets (Fig. 34-11B). Using the image analysis software, the signals corresponding to VWF and platelets at each *z* stack over the growing thrombus could be extracted and the percentage of platelets colocalized with VWF during thrombus formation was determined (Fig. 34-11C). The results indicate that over the time course of thrombus formation, few platelets were colocalized with VWF. A maximum of 25% of platelets colocalized with VWF 6 minutes after the laser-induced injury when the thrombus size was stable. Using the piezoelectric driver and the widefield microscope, we perform high-speed capture of fluorescent signals corresponding to platelets and fibrin through the growing thrombus. Using deconvolution techniques we eliminate the out-of-focus signal by reassigning the out-of-focus light back to its plane of origin using mathematical algorithms. Therefore, we can accurately determine the localization of platelets and fibrin over time in one *z* plane (Fig. 34-12).

## V. Conclusions

The development of real-time *in vivo* imaging has made possible a quantitative analysis of thrombus formation. It is now possible to observe, record, and subsequently measure both platelet and protein accumulation into the growing thrombus. Using fluorochrome-conjugated antibodies or direct labeling, proteins including tissue factor and fibrin have been imaged during thrombus formation. By quantitative analysis the kinetics of fluorochrome accumulation is determined. Using this analysis it is possible to show modification of platelet kinetics with various pharmacological agents or differing genotypes, thus revealing factors that could be important in the initiation and early stages of thrombus growth.

The addition of high-speed confocal capabilities to this technique has been crucial in facilitating the investigation of thrombus structure and homogeneity. Expanding uses of this technique have included the study of platelet activation states within the thrombus. Initially, the exposure of P-selectin, a known α granule release and platelet activation marker, was used to monitor platelet activation. However, expression of this marker only occurs at a late stage during thrombus formation. Therefore, $Ca^{2+}$ elevation within a platelet population or even single platelets has proven a useful early marker of platelet activation.

Hemostasis and thrombosis are highly complex processes that involve a large number of cellular and protein components. *In vivo* imaging of thrombus formation has enabled analysis of the role and relative contribution of each of these various components, including platelets, in the setting of an intact system in a living animal. These approaches have great potential for understanding thrombus formation and for investigating potential therapeutic strategies to inhibit pathogenic thrombus formation.

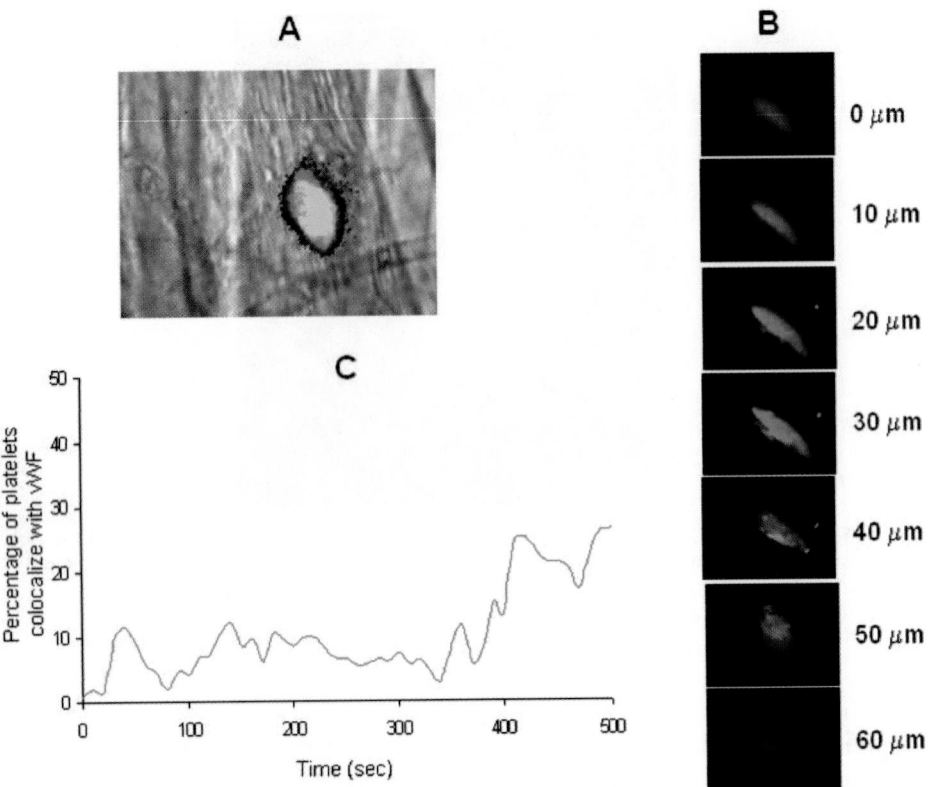

**Figure 34-11.** Colocalization of platelets and von Willebrand factor (vWF) in a growing thrombus. A. Widefield microscopy depicting platelets (red) and vWF (green) incorporation into a growing thrombus induced by a dye laser. Note that by using widefield microscopy one could conclude that platelets and vWF are colocalized. B. Confocal microscopy of platelet (red) and vWF (green) incorporation into a growing thrombus, 3 minutes after a laser-induced injury. Note that platelets are localized in the confocal $z$ planes 10 to 30 $\mu$m, whereas most of the signal corresponding to vWF is detected at the confocal $z$ planes 40 to 60 $\mu$m. C. Median curve representing percentage of platelets that colocalize with vWF over time.

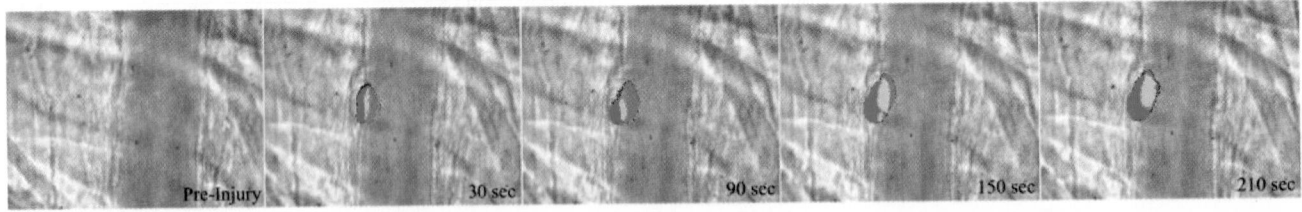

**Figure 34-12.** Reconstitution of colocalization of platelets (red) and fibrin (green) over time in one $z$ plane after deconvolution and projection techniques.

## References

1. Baez, S. (1973). An open cremaster muscle preparation for the study of blood vessels by in vivo microscopy. *Microvasc Res, 5,* 384–394.
2. Stockmans, F., Deckmyn, H., Gruwez, J., et al. (1991). Continuous quantitative monitoring of mural, platelet-dependent, thrombus kinetics in the crushed rat femoral vein. *Thromb Haemost, 65,* 425–431.
3. Seiffge, D., & Kremer, E. (1986). Influence of ADP, blood flow velocity, and vessel diameter on the laser-induced thrombus formation. *Thromb Res, 42,* 331–341.
4. Seiffge, D., & Kremer, E. (1986). Effects of propentofylline (HWA 285) on laser-induced thrombus formation in healthy and diseased rat mesenteric arterioles. *Curr Med Res Opin, 10,* 94–98.
5. Wiedeman, M. P. (1974). Vascular reactions to laser in vivo. *Microvasc Res, 8,* 132–138.

6. Araki, H., Muramoto, J., Nishi, K., et al. (1992). Heparin adheres to the damaged arterial wall and inhibits its thrombogenicity. *Circ Res, 71,* 577–584.

7. Rumbaut, R. E., Slaff, D. W., & Burns, A. R. (2005). Microvascular thrombosis models in venules and arterioles in vivo. *Microcirculation, 12,* 259–274.

8. Bizzozero, J. (1882). Uber einen neuen formbestandteil des blutes und dessen rolle bei der thrombose und blutgerinnung. *Virchows Arch, 90,* 261–332.

9. Cooke, B. M., Usami, S., Perry, I., et al. (1993). A simplified method for culture of endothelial cells and analysis of adhesion of blood cells under conditions of flow. *Microvasc Res, 45,* 33–45.

10. Inoue, S., & Inoue, T. (2002). Direct-view high-speed confocal scanner: The CSU-10. *Methods Cell Biol, 70,* 87–127.

11. Carmeliet, P., Moons, L., Stassen, J. M., et al. (1997). Vascular wound healing and neointima formation induced by perivascular electric injury in mice. *Am J Pathol, 150,* 761–776.

12. Broeders, M. A., Tangelder, G. J., Slaaf, D. W., et al. (1998). Endogenous nitric oxide protects against thromboembolism in venules but not in arterioles. *Arterioscler Thromb Vasc Biol, 18,* 139–145.

13. Stockmans, F., Stassen, J. M., Vermylen, J., et al. (1997). A technique to investigate mural thrombus formation in small arteries and veins: I. Comparative morphometric and histological analysis. *Ann Plast Surg, 38,* 56–62.

14. Busuttil, S. J., Drumm, C., Ploplis, V. A., et al. (2000). Endoluminal arterial injury in plasminogen-deficient mice. *J Surg Res, 91,* 159–164.

15. Roque, M., Fallon, J. T., Badimon, J. J., et al. (2000). Mouse model of femoral artery denudation injury associated with the rapid accumulation of adhesion molecules on the luminal surface and recruitment of neutrophils. *Arterioscler Thromb Vasc Biol, 20,* 335–342.

16. Denis, C., Methia, N., Frenette, P. S., et al. (1998). A mouse model of severe von Willebrand disease: Defects in hemostasis and thrombosis. *Proc Natl Acad Sci USA, 95,* 9524–9529.

17. Farrehi, P. M., Ozaki, C. K., Carmeliet, P., et al. (1998). Regulation of arterial thrombolysis by plasminogen activator inhibitor-1 in mice. *Circulation, 97,* 1002–1008.

18. Kurz, K. D., Main, B. W., & Sandusky, G. E. (1990). Rat model of arterial thrombosis induced by ferric chloride. *Thromb Res, 60,* 269–280.

19. Thorlacius, H., Vollmar, B., Seyfert, U. T., et al. (2000). The polysaccharide fucoidan inhibits microvascular thrombus formation independently from P- and L-selectin function in vivo. *Eur J Clin Invest, 30,* 804–810.

20. Mori, M., Sakata, I., Hirano, T., et al. (2000). Photodynamic therapy for experimental tumors using ATX-S10(Na), a hydrophilic chlorin photosensitizer, and diode laser. *Jpn J Cancer Res, 91,* 753–759.

21. Roesken, F., Vollmar, B., Rucker, M., et al. (1998). In vivo analysis of antithrombotic effectiveness of recombinant hirudin on microvascular thrombus formation and recanalization. *J Vasc Surg, 28,* 498–505.

22. Roesken, F., Ruecker, M., Vollmar, B., et al. (1997). A new model for quantitative in vivo microscopic analysis of throm-

bus formation and vascular recanalisation: The ear of the hairless (hr/hr) mouse. *Thromb Haemost, 78,* 1408–1414.

23. Kawasaki, T., Kaida, T., Arnout, J., et al. (1999). A new animal model of thrombophilia confirms that high plasma factor VIII levels are thrombogenic. *Thromb Haemost, 81,* 306–311.

24. Matsuno, H., Kozawa, O., Niwa, M., et al. (1999). Differential role of components of the fibrinolytic system in the formation and removal of thrombus induced by endothelial injury. *Thromb Haemost, 81,* 601–604.

25. Rosen, E. D., Raymond, S., Zollman, A., et al. (2001). Laser-induced noninvasive vascular injury models in mice generate platelet- and coagulation-dependent thrombi. *Am J Pathol, 158,* 1613–1622.

26. Kovacs, I. B., Sebes, A., Trombitas, K., et al. (1975). Proceedings: Improved technique to produce endothelial injury by laser beam without direct damage of blood cells. *Thromb Diath Haemorrh, 34,* 331.

27. Hovig, T., McKenzie, F. N., & Arfors, K. E. (1974). Measurement of the platelet response to laser induced microvascular injury. Ultrastructural studies. *Thromb Diath Haemorrh, 32,* 695–703.

28. Doutremepuich, F., Aguejouf, O., Imbault, P., et al. (1995). Effect of the low molecular weight heparin/nonsteroidal anti-inflammatory drugs association on an experimental thrombosis induced by laser. *Thromb Res, 77,* 311–319.

29. Rosenblum, W. I., Nelson, G. H., & Povlishock, J. T. (1987). Laser-induced endothelial damage inhibits endothelium-dependent relaxation in the cerebral microcirculation of the mouse. *Circ Res, 60,* 169–176.

30. Povlishock, J. T., & Rosenblum, W. I. (1987). Injury of brain microvessels with a helium-neon laser and Evans blue can elicit local platelet aggregation without endothelial denudation. *Arch Pathol Lab Med, 111,* 415–421.

31. Chou, J., Mackman, N., Merrill-Skoloff, G., et al. (2004). Hematopoietic cell-derived microparticle tissue factor contributes to fibrin formation during thrombus propagation. *Blood, 104,* 3190–3197.

32. Falati, S., Gross, P., Merrill-Skoloff, G., et al. (2002). Real-time in vivo imaging of platelets, tissue factor and fibrin during arterial thrombus formation in the mouse. *Nat Med, 8,* 1175–1181.

33. Falati, S., Gross, P. L., Merrill-Skoloff, G., et al. (2004). In vivo models of platelet function and thrombosis: Study of real-time thrombus formation. *Methods Mol Biol, 272,* 187–197.

34. Falati, S., Liu, Q., Gross, P., et al. (2003). Accumulation of tissue factor into developing thrombi in vivo is dependent upon microparticle P-selectin glycoprotein ligand 1 and platelet P-selectin. *J Exp Med, 197,* 1585–1598.

35. Falati, S., Patil, S., Gross, P. L., et al. (2005). Platelet PECAM-1 inhibits thrombus formation in vivo. *Blood.*

36. Celi, A., Merrill-Skoloff, G., Gross, P., et al. (2003). Thrombus formation: Direct real-time observation and digital analysis of thrombus assembly in a living mouse by confocal and widefield intravital microscopy. *J Thromb Haemost, 1,* 60–68.

37. Sim, D. S., Merrill-Skoloff, G., Furie, B. C., et al. (2004). Initial accumulation of platelets during arterial thrombus for-

mation in vivo is inhibited by elevation of basal cAMP levels. *Blood, 103,* 2127–2134.

38. Carmeliet, P., Moons, L., & Collen, D. (1998). Mouse models of angiogenesis, arterial stenosis, atherosclerosis and hemostasis. *Cardiovasc Res, 39,* 8–33.

39. Yang, J., Hirata, T., Croce, K., et al. (1999). Targeted gene disruption demonstrates that P-selectin glycoprotein ligand 1 (PSGL-1) is required for P-selectin-mediated but not E-selectin-mediated neutrophil rolling and migration. *J Exp Med, 190,* 1769–1782.

40. Ley, K. (1995). Gene-targeted mice in leukocyte adhesion research. *Microcirculation, 2,* 141–150.

41. van Gestel, M. A., Heemskerk, J. W., Slaaf, D. W., et al. (2002). Real-time detection of activation patterns in individual platelets during thromboembolism in vivo: Differences between thrombus growth and embolus formation. *J Vasc Res, 39,* 534–543.

42. Thorlacius, H., Raud, J., Xie, X., et al. (1995). Microvascular mechanisms of histamine-induced potentiation of leukocyte adhesion evoked by chemoattractants. *Br J Pharmacol, 116,* 3175–3180.

43. Gavins, F. N., & Chatterjee, B. E. (2004). Intravital microscopy for the study of mouse microcirculation in anti-inflammatory drug research: Focus on the mesentery and cremaster preparations. *J Pharmacol Toxicol Methods, 49,* 1–14.

44. Lindner, V., Fingerle, J., & Reidy, M. A. (1993). Mouse model of arterial injury. *Circ Res, 73,* 792–796.

45. Wang, X., & Xu, L. (2005). An optimized murine model of ferric chloride-induced arterial thrombosis for thrombosis research. *Thromb Res, 115,* 95–100.

46. Gross, P. L., Furie, B. C., Merrill-Skoloff, G., et al. (2005). Leukocyte-versus microparticle-mediated tissue factor transfer during arteriolar thrombus development. *J Leukoc Biol, 78,* 1318–1326.

47. Sim, D., Flaumenhaft, R., & Furie, B. (2005). Interactions of platelets, blood-borne tissue factor, and fibrin during arteriolar thrombus formation in vivo. *Microcirculation, 12,* 301–311.

PART THREE

# The Role of Platelets in Disease

# CHAPTER 35

# Atherothrombosis and Coronary Artery Disease

**Pascal J. Goldschmidt, Neuza Lopes, Lawrence E. Crawford, and Richard C. Becker**

*Department of Medicine, Cardiovascular Center for Genomic Science, Cardiovascular Thrombosis Center, Duke University Medical Center, Durham, North Carolina*

## I. Introduction

Atherosclerosis, the combined end-result of genetic modifiers, environmental factors, and spontaneous cellular–molecular events, selectively affects medium- and large-sized elastic and muscular arteries (Fig. 35-1). The coronary location of atherosclerosis results in its major clinical manifestation: coronary artery disease (CAD).[1–5] Atherosclerosis represents the premier human epidemic which emerged between the second and third millennia, an epidemic that affects all industrialized countries and is expanding worldwide.[6–10] Aging defines the advancement of atherosclerosis, both phenotypically and clinically. Men and women in their second and third decades of life contain early, preclinical lesions of atherosclerosis (fatty streaks), whereas most individuals in their seventh and eighth decades have advanced, complex atherosclerotic plaques (Fig. 35-2A and B).[11–17] Atherosclerosis can propagate uniformly across the arterial tree, usually starting within the distal (infrarenal) abdominal aorta, or can evolve in a mosaic fashion, with predilection for certain locations, such as coronary and extracranial vessels.

The extent of atherosclerotic burden can be predicted, to some degree, by the presence or absence of established risk factors: male gender, tobacco smoking, hypertension, dyslipidemia, diabetes mellitus, obesity, and sedentary lifestyle.[18–25] Unfortunately, because a majority of these factors develop or become evident at a time when atherosclerosis is already pathologically advanced, they fail to adequately identify individuals at risk for serious clinical events. Biological markers have evolved to assist with predicting an individual's risk for developing the thromboembolic complications of atherosclerosis. The levels of C-reactive protein (CRP), low-density lipoprotein (LDL), high-density lipoprotein (HDL), glycemia, homocysteinemia, and lipoprotein(a) [Lp(a)] have been associated either directly or inversely with the likelihood of a coronary thromboembolic event.[26–32] In contrast, most platelet-directed studies, whether currently available genetic tests (Chapter 14) or tests of platelet function (Chapter 23), lack strong predictive potential for reliably determining the development of coronary thrombosis. In selected conditions, some of these tests may be of clinical relevance.[33,34] A basis for the relative lack of predictiveness of traditional platelet function studies may be that platelet-dependent thrombosis represents just one of many biological processes participating in atherogenic activity and that, except in certain situations such as percutaneous coronary interventions (PCI) and coronary artery bypass grafting (CABG), the contribution of platelet thrombus is difficult to quantitate with currently available measurement tools.[35–38] The more direct, biology-based approach to platelet biomarkers may rest with an ability to characterize them in either a state of interaction with other cells (e.g., platelet–leukocyte aggregates [Chapter 30]) or transformation (e.g., microparticles [Chapter 20], apoptosis, and/or protein/mRNA expression [Chapter 5]).

In the setting of tissue injury, a *sine qua non* of atherogenesis, there is seamless continuity between the processes of thrombosis, fibrinolysis, and inflammation (Fig. 35-3).[39] The interface between thrombotic and inflammatory events might explain, at least in part, a widely recognized relationship between inflammatory biomarkers, such as CRP, and future thromboembolic events in several distinct vascular beds.[40] The subsequent phases in atherothrombosis include smooth muscle cell (SMC) proliferation (hyperplastic phase), growth of endothelial cells and fibroblasts, positive remodeling of the vessel wall, and, with continued provocative stimuli, negative remodeling, with eventual narrowing of the vessel's lumen (Fig. 35-3). Thus, atherosclerosis represents a series of highly specific cellular and molecular responses that can best be described, phenotypically in aggregate, as a chronic, arterial circulatory system-specific inflammatory disease.[1]

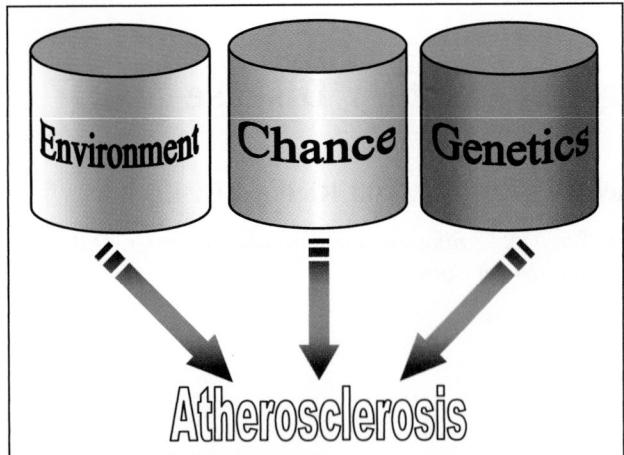

**Figure 35-1.** Factors contributing to atherosclerosis. The factors that contribute to atherosclerosis can be organized within three general classes: (i) factors related to the environment, such as cigarette smoking, a fat-rich diet, or *Chlamydia* infection; (ii) genetic factors that have been difficult to identify, given the multigenic complexity; and (iii) a stochastic component best defined as chance (e.g., a benign infection such as *Chlamydia pneumoniae,* if occurring at a time of defenselessness for the arterial tree, can accelerate atherogenesis substantially, whereas the same infection taking place while the arterial tree is stable could be harmless for coronary arteries).

## II. Atherogenesis: Endothelial Cell Contribution

Toxins and reactive oxygen species (ROS) generated by tobacco smoke, modified LDL particles, diabetes mellitus and resulting glycated products, homocysteine, and infectious organisms such as herpes viruses and *Chlamydia pneumoniae* can each cause endothelial injury and, in a susceptible host, trigger the progression of atherosclerosis.[41–47] Contrary to the appearance of static images of the endothelium, endothelial cells are constantly in motion, even when forming a confluent endothelium.[48] Such innate cellular motility is associated with the formation of small breaks between endothelial cells, and such intercellular gaps are rapidly sealed by adjacent cells. Nevertheless, these gaps represent an opportunity for particles, such as LDL-cholesterol, to migrate toward the subendothelial space.

Flow challenges *in vivo* using a pig model of a femoral arteriovenous (AV) shunt, have shown that endothelial gap permeability can be increased by changes of flow that are no greater than those expected following a fast walk or a jog.[49] Such rearrangement of endothelial cells might be the reason why small, dense LDL particles are more "vasotoxic" than larger lipoproteins. Indeed, their small size may provide greater access to the subendothelial space. The changes in endothelial call gap permeability caused by AV shunting

illustrate the importance of flow and shear to vessel wall homeostasis. Other examples of this phenomenon include the predilection for atherosclerosis within arterial regions that are directly downstream from a branching vessel (Fig. 35-4).[50] The molecular and cellular changes associated with the effect of shear stress appear to involve multiple pathways, including those responsible for the production of nitric oxide (NO), superoxide, and derived ROS.[51–53]

In the subendothelial space, LDL particles undergo transformation and may be oxidized, glycated (in the presence of hyperglycemia), or aggregated, and form complexes with proteoglycans; their modifications can attenuate immune tolerance and induce recognition by antigen-presenting cells residing in the vessel wall, such as dendritic cells and macrophages.[1] The accumulation of lipid particles in macrophages by means of scavenger receptors contributes to the formation of foam cells. Additionally, the presentation of neoantigens, derived from intracellular processing of engulfed lipid particles by the phagocytic cells, causes the activation of T and, subsequently, B lymphocytes. Immunoglobulins produced by B cells that target epitopes present on the surface of modified lipid particles and other antigens (e.g., those present on infectious agents that might have caused a past infection and whose antigens persist within the arterial wall [e.g., *Chlamydia pneumoniae* and herpes viruses]) may accumulate within the vessel wall. The presence of immunoglobulins, by means of activation of the Fc receptor of macrophages, induces the production of macrophage colony-stimulating factors (M-CSF) by the phagocytic cells, thereby enhancing the survival of new monocytes accessing the vessel wall (Fig. 35-5).[54–60]

## III. Inflammation in Atherothrombosis

Monocytes, the precursors of tissue macrophages, are produced by the bone marrow and engineered to circulate with a lifespan of 24 to 48 hours.[61] Once their time elapses, monocytes undergo apoptosis and are removed from circulation, unless they encounter a supportive environment that provides them with the survival factors required for longevity (Fig. 35-5).[62–64] Monocytes can be attracted to the arterial wall by dysfunctional endothelial cells via adhesive receptors,[65–67] such as selectins,[68,69] intercellular adhesion molecule (ICAM),[70,71] and vascular cell adhesion molecule-1 (VCAM)[72,73] that bind to integrins like $\alpha M\beta 2$ (CD11b/CD18, Mac-1) on the surface of monocytes, as well as many other molecules with well-defined roles.[74,75] Next, monocytes anchored to the arterial wall infiltrate the inner layers of the vessel wall by diapedesis through the action of monocyte chemoattractant protein 1 (MCP-1).[76] Activated platelets modulate chemotactic (MCP-1) and adhesive (ICAM-1) properties of endothelial cells via a nuclear factor (NF)-kappaβ-dependent mechanism, and these reactions repre-

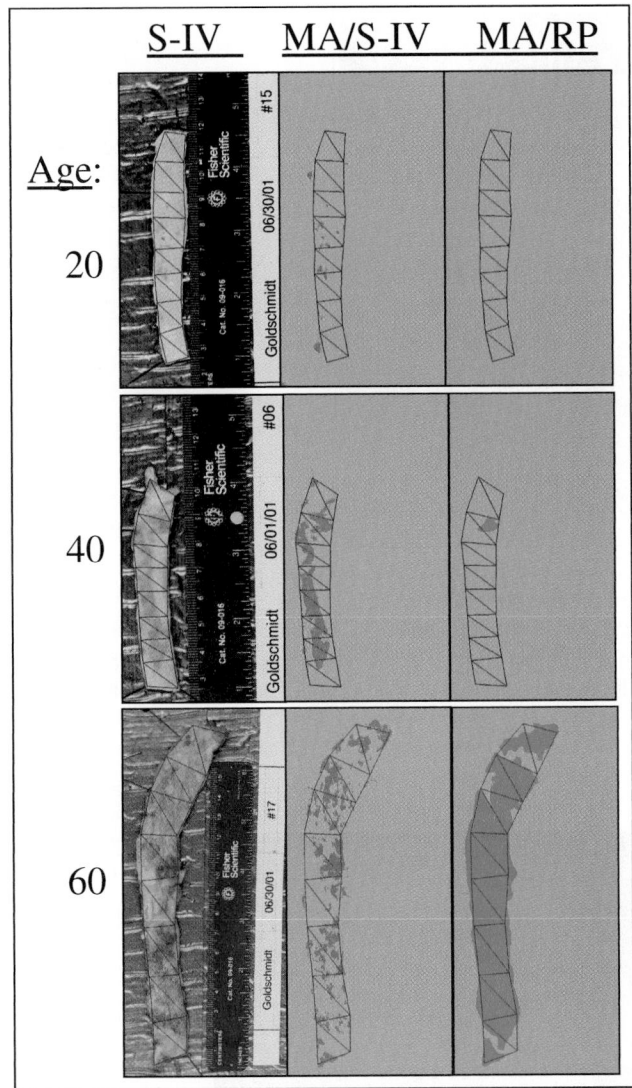

A

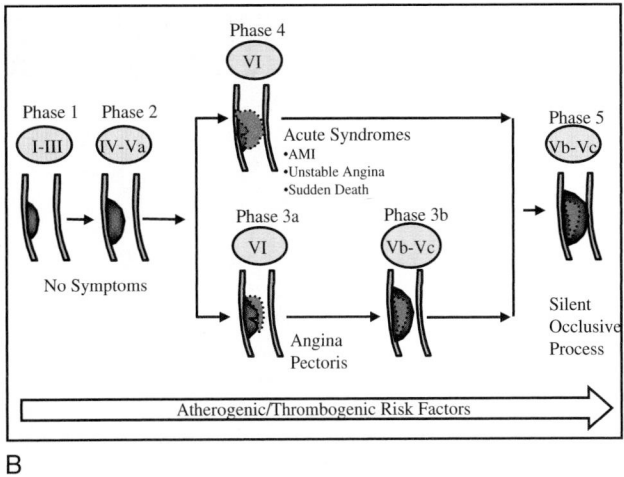

B

**Figure 35-2.** A. Progression of atherosclerosis with aging. Human arterial specimens corresponding to the proximal aorta harvested from heart donors show worsening lesions with aging. For each sample, the left panel displays Sudan-IV (S-IV) staining of the arterial tissue, which highlights early lesions of atherosclerosis called fatty streaks. The middle panel corresponds to a morphometric analysis (MA) of the S-IV stained tissue. The right panel is a morphometric analysis of more complex lesions: raised plaques (RP). Note the progression of these two types of lesions with aging. B. Progression of atherosclerosis, according to lesion characteristics and clinical presentation. The diagram displays the various phases of atherosclerosis, from uncomplicated atheroma (Phase 1) to occlusive lesions (Phase 5). Pathological findings corresponding to each phase have been classified according to the landmark Starry classification.[11] (Modified with permission from refs. 12 and 13.)

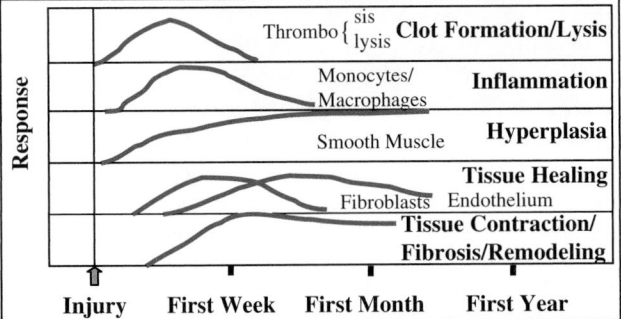

**Figure 35-3.** Sequential steps in the response of arteries to injurious agents and conditions. The pathobiologic link between thrombotic and inflammatory responses highlights the interaction between these two teleologically vital systems that can combine in potentially damaging ways. Narrowing of the vessel can develop during any one of these steps: a thrombus at the site of vessel injury, the proliferation of smooth muscle cells, or the fibrotic remodeling of the vessel wall.

sent important steps that contribute to early inflammatory events in atherogenesis[77–79]

Platelets express CD40 ligand (CD40L) upon activation both *in vitro* and *in vivo* (see Chapter 39).[80–82] CD40L on the surface of platelets can provoke endothelial cells to secrete chemokines (MCP-1) and express adhesion molecules. Interactions between platelets and endothelial cells generate the signals that recruit inflammatory cells to the site of vascular injury and promote the extravasation of leukocytes to the inner layers of the vessel wall.[80–82] Platelets accumulating at sites of arterial injury can also directly recruit monocytes to the vessel wall through the binding of monocyte surface P-selectin glycoprotein ligand-1 (PSGL-1) to platelet surface P-selectin (Chapter 12) and monocyte surface Mac-1 to the platelet surface glycoprotein (GP) Ib-IX-V complex (Chapters 7 and 12).[83,84] Monocyte–platelet aggregates have been identified in the peripheral blood of patients with CAD and may be markers of disease activ-

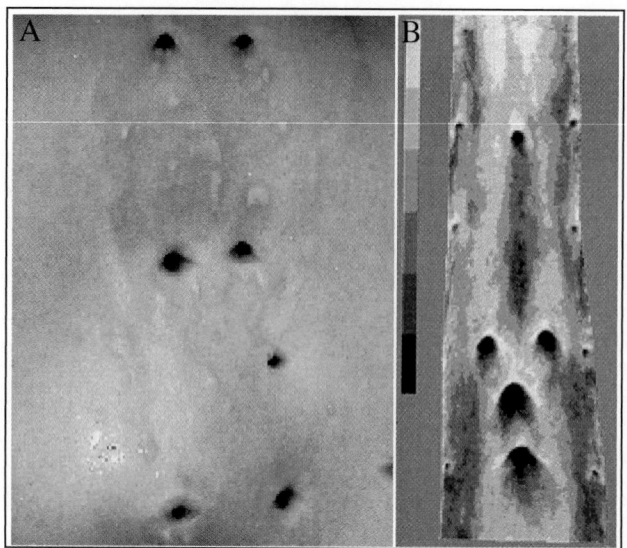

**Figure 35-4.** Predilection for development of atherosclerosis at certain sites in the arterial tree. A. A segment of thoracic human aorta is shown, with the take-off of intercostal arteries. Note the proximity of the lesions to the branching points. B. Summary slide from the Pathological Determinants of Atherosclerosis in Youth (PDAY) study,[50] where the distribution of atherosclerotic lesions is shown for thousands of specimens relative to landmark vessels within the human abdominal aorta. The prevalence of atheroma for each region is shown according to a scale from rare (light) to frequent (dark). (Reproduced with permission from Cornhill et al. [1990]. *Monogr Atheroscler, 15,* 13–19.)

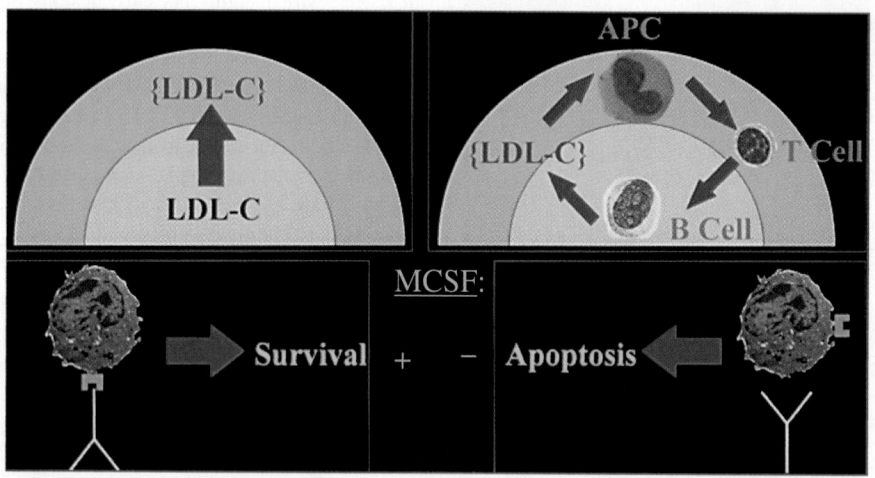

**Figure 35-5.** Basic inflammatory process driving atherosclerosis. Elevated low-density lipoprotein-cholesterol (LDL-C) level is an established risk factor for atherosclerosis. Modification of LDL-C {LDL-C} takes place within the arterial wall (e.g., by oxidation, aggregation). Alteration of LDL-C results in the generation of neoantigens, which can be processed by antigen presenting cells (APC) for activation of T lymphocytes, which in turn trigger the production of immunoglobulin by B lymphocytes. Immunoglobulins (Ig) accumulate within the vessel wall, where the targeted antigen is located. The $F_c$ portion of Ig stimulates the $F_c$-receptor on the surface of monocytes. Monocytes are produced by the bone marrow, and are engineered to survive 24 to 48 hours in the circulation, unless they encounter a supportive environment. Thus, apoptosis (programmed cell death) of monocytes prevents these cells from accumulating within the arterial wall. However, the presence of immunoglobulins targeted at {LDL-C}, or other antigens such as the remnants of infections like *Chlamydia pneumoniae,* via the $F_c$-receptor on the surface of monocytes, supports the survival of monocytes and their differentiation into macrophages. Such differentiation requires the autocrine production of macrophage-colony stimulating factor (M-CSF), in response to $F_c$-receptor engagement on the surface of monocytes infiltrating the arterial wall. Platelets, through their ability to accelerate the recruitment of monocytes to the arterial wall, contribute to this process. Hence, a vicious cycle is established, in which neoantigens like {LDL-C} trigger APC, which in turn trigger T cells, B cells, production of Ig, stabilization of macrophages, and consequent destruction by macrophages of the cells and structural proteins of the arterial wall.

ity.[85,86] However, the interaction of monocytes with molecular factors such as MCP-1 or Mac-1 is not sufficient to result in their survival. Additional stimuli must be present within the vessel wall to induce the production of survival factors such as M-CSF. M-CSF helps to support the survival of the recruited monocytes by differentiation into tissue macrophages.[87,88] Thus, a bioamplification loop can develop during which processed LDL particles trigger activation of the immune system, that in turn promotes differentiation, maturation, proliferation, and survival of macrophages within the vessel wall (Fig. 35-5). M-CSF-activated macrophages subsequently promote the inflammatory reaction by processing LDL particles and presenting neoantigens to helper T lymphocytes.[89–91] In support of this paradigm is the identification of immunoglobulins specific for oxidized LDL particles among patients with atherosclerosis. The activation and expansion of the macrophage population further accentuates the endothelial cell dysfunction.

ROS are pivotal mediators of numerous signaling pathways that underlie vascular inflammatory responses in atherosclerosis. In addition to the impact of oxidative stress on cellular aging and molecular damage, particularly DNA at the mitochondrial level, leukocyte-released $O_2^-$ may play a key role in platelet–leukocyte aggregation.[92] NAD(P)H oxidase in platelets undergoes protein kinase C activation[93–95] and contributes directly to tissue factor upregulation[96,97] and thrombosis.

Particulate matter air pollution through generation of ROS and accelerated NO destruction by superoxide contribute to endothelial injury, reduced vasodilatory reserve, and atherothrombosis.[98]

Intraplaque hemorrhage, a common event in the natural history of atherothrombosis, has been associated with accelerated atherosclerosis.[99] Although the mechanistic relationship between the events has not yet been fully elucidated, erythrocyte-derived free cholesterol and macrophage-stimulated phagocytosis of platelets and erythrocytes may provide requisite substrates for plaque growth.[100]

## IV. Platelets: An Essential Mediator in Atherothrombosis

Dysfunctional endothelial cells can change their activity substantially from their normal physiological state. For example, instead of forming a thromboresistant surface, dysfunctional endothelial cells develop prothrombotic activity with increased adhesiveness for platelets and leukocytes (Fig. 35-6) and secrete procoagulant compounds.[101–105] Endothelial cells also produce, when activated, vasoactive molecules, cytokines, and growth factors, which may induce vasoconstriction and migration and proliferation of smooth muscle cells.[106–108] Permeability of the endothelium increases

substantially upon becoming dysfunctional, further facilitating the migration of inflammatory cells from the blood stream to the inner layers of the vessel wall.[109,110] Chemoattracted macrophages release hydrolytic enzymes, cytokines, chemokines, and growth factors that promote, in synergy with factors produced by endothelial cells, further accumulation of macrophages, as well as migration and proliferation of smooth muscle cells.[111] The production of hydrolytic enzymes such as matrix metalloproteinase (MMP) promotes the degradation of the extracellular matrix proteins and remodeling of the vessel wall.[112]

The mechanism by which dysfunctional endothelial cells promote platelet thrombosis can be considered in two steps: (1) adhesion of platelets and (2) aggregation of platelets (Fig. 35-7).[113] Endothelial cells accumulate von Willebrand factor (VWF) within their Weibel–Palade bodies, which are secreted upon injury.[114–116] The process of Weibel–Palade body secretion upon disruption of endothelial cells involves the participation of the small G-protein Ral.[117] VWF released by endothelial cells in the extracellular space as oligomers, whose polymerization is a result of the oxidative cross-linking of VWF molecules within Weibel–Palade bodies of disrupted endothelial cells, represents a unique anchor for circulating platelets (Chapters 7 and 18).[118–121]

The adhesion of platelets to VWF is mediated by the GPIb-IX-V surface receptor on platelets (Chapters 7 and 18). Following the adhesion of platelets via the VWF/GPIb-IX-V interaction and via the interaction of collagen with GPVI and integrin $\alpha2\beta1$ (Chapter 18), the platelets are exposed to the microenvironment of dysfunctional endothelial cells. This environment provides platelets with a multitude of agonists, including collagen fibers, epinephrine, adenosine diphosphate (ADP), and thrombin.[122–125] Furthermore, dysfunctional endothelial cells fail to produce platelet antagonists such as NO[126] and prostacyclin (PGI$_2$),[127] which are normally constitutively produced by endothelial cells (see Chapter 13). The concurrent stimulation of platelets by various agonists results in substantial changes in platelet structure, secretion of granule contents including ADP, production of thromboxane A$_2$ via a cyclooxygenase 1-dependent mechanism (Chapter 31), and activation of the most abundant receptor on the surface of platelets, integrin $\alpha$IIb$\beta$3 (the GPIIb-IIIa complex). The various steps in the activation of $\alpha$IIb$\beta$3 are reviewed in Chapters 8 and 17. The aggregated platelets constitute an endovascular graft that promotes the formation of a thrombus and exacerbates the inflammatory reaction. Furthermore, activated platelets, by releasing their granules, provide the microenvironment with growth factors such as platelet-derived growth factor (PDGF) and proinflammatory molecules such as CD40L, which contribute to the migration and proliferation of smooth muscle cells and macrophages.[128,129]

As reviewed in earlier chapters, the primary role of platelets is to trigger hemostasis in a damaged vessel to maintain

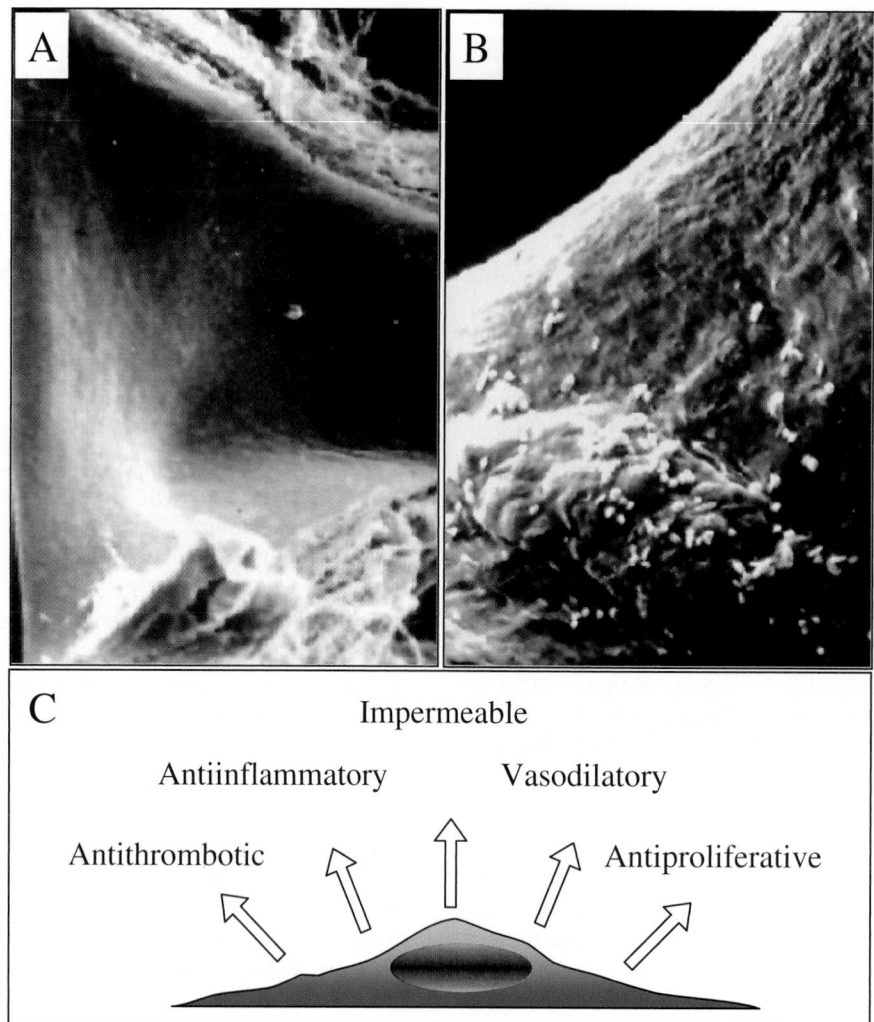

**Figure 35-6.** The endothelium, an adapting organ in the arterial wall. In their physiological state, endothelial cells are remarkably non-adhesive to platelets and other contributors to inflammation. However, in the presence of inflammatory cytokines and other noxious stimuli, the endothelial repertoire changes markedly. Thus, in physiological conditions, the endothelium is a remarkably smooth surface. A. Electron micrograph of a resected coronary vessel, showing a normal endothelium. B. Electron micrograph of an atherosclerotic coronary vessel; note the presence of adherent platelets and other inflammatory cells in the vicinity of an atherosclerotic plaque. (Both electron micrographs reproduced with permission from Ross (1986). *N Engl J Med, 314,* 488–500.) C. Endothelial cells normally exhibit properties of the best antithrombotic surface of any tissue, with antiinflammatory cells that provide an impermeable surface, with active vasodilatory and platelet inhibitory activity, and control of the proliferation of smooth muscle cells. Atherosclerotic endothelial cells, switching to a very different phenotype, promote the adhesion of platelets and the formation of a coronary thrombus, allow the passage of macromolecules and inflammatory cells, increase the arterial tone, and facilitate the proliferation of smooth muscle cells within the arterial wall.

vascular integrity. However, platelets have not been designed to differentiate between a disrupted vessel wall within, for example, a small digital vein and the atherosclerotic disruption of a coronary artery. As a consequence, the function of normal platelets is usually too efficient for the safety of patients with CAD, and potent antiplatelet drugs have been designed to reduce platelet function (Chapters 60–65).[130–135]

In addition to their role in thrombosis, platelets have recently been shown to have a direct role in the genesis of atherosclerosis, as reviewed in refs. 136 and 137, discussed in Chapter 39, and summarized in Fig. 39-9 in Chapter 39.

The relationship between thrombosis and atherosclerosis is complex. However, it has become increasingly clear that the recruitment of inflammatory cells to areas of endothelial

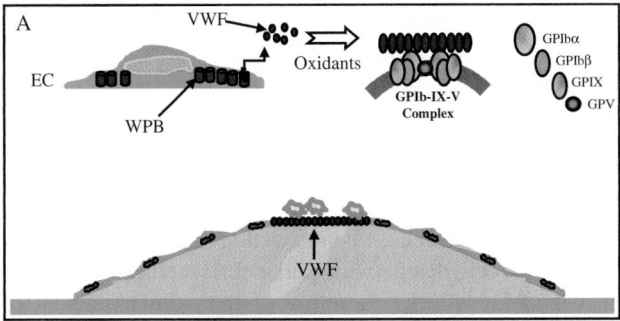

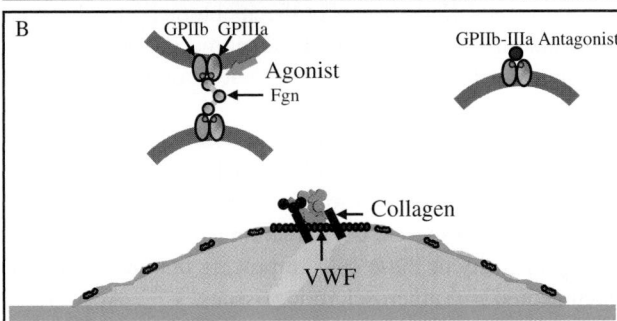

**Figure 35-7.** Key steps of platelet interaction with the arterial wall. A. Upon disruption of the endothelium on the surface of an atherosclerotic plaque, von Willebrand factor (VWF) is deposited. Surrounding endothelial cells (EC), activated by inflammatory and prothombotic cytokines, secrete VWF contained in their Weibel–Palade bodies (WPB). Within the WPB, VWF is oligomerized, probably as a consequence of oxidation, with formation of disulfide bridges between VWF molecules. Secreted VWF oligomers crosslink platelets on the surface of the atheroma. Binding of platelets to VWF is mediated by a unique receptor, the GPIb-IX-V complex (the stoichiometry of which is shown). This receptor allows for the anchoring of platelets to the arterial wall, in spite of the velocity of these cells in the arterial flow, and the shear exerted at the surface of coronary arteries. B. Adhered platelets need to undergo activation to allow for additional recruitment of platelets to the atherosclerotic plaque. In the presence of agonists like collagen, an abundant polymer of the vessel wall, platelets experience outside-in signaling, and consequently, the most abundant receptor on the surface of platelets, GPIIb-IIIa (integrin αIIbβ3), becomes activated and its affinity for fibrinogen (Fgn) and other extracellular ligands becomes substantially greater. Activated GPIIb-IIIa on the surface of platelets binds fibrinogen and, because fibrinogen has two binding sites for GPIIb-IIIa, fibrinogen can crosslink platelets together. The consequence is the formation of a platelet plug at the site of the damaged arterial wall. GPIIb-IIIa antagonists hinder the binding of fibrinogen and other ligands to GPIIb-IIIa and thereby inhibit the aggregation of platelets at the site of a disrupted plaque.

cell injury and existing atheromatous plaques is a constitutive phenomenon throughout the evolution of plaque growth. Accordingly, platelet–leukocyte interactions, as in thrombogenesis itself, represent an important pathobiologic theme in atherogenesis. Platelet–leukocyte adhesion and other interactions are discussed in detail in Chapter 12.

# V. The Atheromatous Plaque and Plaque Rupture

With expansion of the atherosclerotic inflammatory process and adaptation of vascular cells, which includes migration and proliferation of smooth muscle cells, a more complex lesion develops: the atherosclerotic plaque.[138,139] Plaque rupture is a seminal event in the pathophysiology of unstable coronary syndromes, including myocardial infarction (MI), unstable angina, and sudden cardiac death. Much attention, therefore, has been directed at understanding the mechanisms that render plaques vulnerable to rupture. The Q-wave MI (also called ST-elevation acute MI or STE-AMI) and other unstable ischemic coronary syndromes are known to be triggered either by the rupture or the ulceration of a coronary plaque.[140–148] Therefore, stabilization of coronary plaques has become a defined strategy in the management of patients with CAD. However, the lack of a clear understanding of the mechanism that renders plaques vulnerable to rupture has hindered progress in the field.

Large prospective, randomized, double-blind trials of 3-hydroxy-3-methylglutaryl-coenzyme A (HMG-CoA) reductase inhibitors, or statins, have demonstrated conclusively their ability to prevent MI, cardiac death, and stroke.[20,149–151] An important effect of this drug class is their well-established ability to reduce hepatocellular cholesterol synthesis, which augments the surface expression of LDL-receptor molecules, with consequent removal of LDL particles from the blood stream.[152] Instructively, even when total cholesterol, and specifically LDL cholesterol, has been entirely normalized, the probability of coronary thromboembolic events is, at best, reduced by 20 to 30%, a fact that underlines the importance of other contributors to atherosclerosis and associated MI and sudden cardiac death.[20,153–157]

The initial objective of secondary prevention trials among individuals with CAD was to achieve "plaque regression."[158] However, the degree of "reverse remodeling" remains insignificant for most patients in spite of significant reduction in the probability of coronary events, clearly indicating that plaque size, or the degree of lumen narrowing, can vary independently of event reduction.[159] In other words, statin drugs "stabilize" plaques, probably through combined anti-inflammatory, cellular recovery, and antiapoptotic effects.

Most theories of plaque instability implicate an imbalance between inflammatory cells and normal cellular constituents of the vessel wall, in particular between activated macrophages and smooth muscle cells.[160–171] Overproduction of metalloproteinases by activated macrophages promotes the destruction of extracellular matrix proteins (ECMP) such as collagen and elastin. Loss of ECMP production by damaged smooth muscle cells can also weaken the superstructure of the vessel wall.[172–177] In this setting, the disruption of an unstable plaque can be induced by mechanical forces such as shear stress and transmural vessel wall pressure.

A reliable pathologic marker for unstable plaques is the presence of a central or eccentric lipid core, which corresponds to the accumulation of cell debris and lipids, in particular cholesterol esters, which are mixed to form a liquid crystalline substance within the plaque. The constituents of the lipid core are highly thrombogenic, and therefore, contribute directly to clot formation upon plaque rupture.[178–180] The fibrous cap overlying the lipid core is stabilized by increased activity of PDGF, transforming growth factor β (TGF-β), interleukin-1, tumor necrosis factor α, osteopontin, and attenuated connective tissue degradation.[1]

Although the genesis of the atheromatous lipid core has remained enigmatic, statins have been shown to reliably reduce the size of the lipid core within lesions.[181,182] Statins can induce the restoration of a collagen matrix within the space (lipid core) previously emptied of such proteins and filled with cholesterol esters.[183] The restoration of a collagen matrix within the core of the plaque might represent an important effect of the statins that contributes to plaque stabilization. The observation of such a process of arterial tissue recovery, or "resuscitation," suggests that the matrix-free lipid core was formed at least in part because of the loss of the cells responsible for producing ECMP, a loss that can be successfully antagonized and even reversed through statin therapy.

Many disease processes involve in their pathogenesis the de novo reactivation of well-rehearsed cellular programs (e.g., programs that are required during development to ensure the optimal modeling of tissues by differentiated cells). In the genesis of the cardiovascular system, monocytes and macrophages can be viewed as artisans in growing tissues that precisely explore, drill, engulf debris, and destroy the remnants of blood vessels that have become idle. Their contribution is pivotal to the establishment of the functional human vascular tree as we know it. This well-crafted activity ceases once the system reaches a steady state—blood flow to downstream tissues is equal to the nutritive and oxygen requirements of such tissues. (No cells can be separated from a source of oxygen by a distance greater than 200 μm.)[184] Interestingly, these embryonic programs, active during cardiovascular genesis, can be reactivated upon tumor growth or following the thromboembolic occlusion of a feeding vessel.[176] The excessive activity of monocytes within the wall of atherosclerotic arteries is similar to that found during embryonic development. However, in the case of atherosclerosis, the destruction of blood vessels by macrophages is misguided and does not achieve its physiological mission. In fact, it is the arterial wall destruction that results in thrombosis and subsequent clinical events including heart failure or cardiovascular death. Furthermore, the elaboration of an angiogenic "factor" within the coronary plaque can contribute to plaque fragility.[185]

Smooth muscle cell death, whether as a result of apoptosis or necrosis, represents a highly regulated process within the vessel wall.[186–188] In the arterial wall, smooth muscle cells are surrounded by a cocoon that provides them with an optimal milieu for their survival. Physiological apoptosis of smooth muscle cells, when it occurs, is a highly organized process of cell death, with cell debris being specifically packaged for optimal recycling of the cellular content of apoptotic cells. Endocytosis of the apoptotic bodies by adjacent smooth muscle cells and resident macrophages ensures that the vessel wall swiftly recovers its protein and cellular biomass.

Activated macrophages and other vascular cells associated with atherogenesis produce metalloproteinases,[189–191] which disrupt the protein matrix supporting both endothelial and smooth muscle cells, thereby enhancing the apoptotic propensity of these cells, a process that has been well documented in atherosclerotic tissues.[192–196] Accelerated apoptosis of smooth muscle cells in the wall of damaged arteries recruits the Fas death pathway.[197,198] The Fas death pathway has also been implicated in the apoptosis of other cellular constituents of the vessel wall.[199–201]

Fas-mediated injury to medial smooth muscle cells could contribute substantially to triggering smooth muscle cell proliferation within the intima, where the balance between cell death and growth favors cell proliferation. Fas-ligand (Fas-L) can be provided by cytolytic T cells, but within atherosclerotic plaques the primary source of Fas-L is macrophages.[202] Interestingly, the induction of vascular smooth muscle cell killing by macrophages requires a tight connection via binding of Mac-1 to its ligand on SMCs (ICAM-1).[203] Furthermore, the presence of M-CSF is required for the death sentence to be carried out.[203–205] Other death pathways could also be involved in the natural history of atherosclerotic plaques that require a tight binding between macrophages and their SMC target.

Macrophages also undergo apoptosis. Accelerated apoptosis of multiple cells, including both SMCs and macrophages, results in the uptake of lipid-rich apoptotic bodies by adjacent cells, leading to arterial cells that are increasingly saturated with lipids.[206,207] An important feature of the apoptotic/necrotic death of cellular components of the vessel wall is the sometimes massive and synchronized occurrence of such deaths. In an experimental model of atherosclerosis, a multitude of cells belonging to the neointima can be found to sporadically undergo apoptosis/necrosis in a synchronized manner. The dying cells are rapidly removed by the blood flow when the process involves the surface of the neointima, creating an "ulcerated" scar, but when the process takes place within deeper layers of the atherosclerotic plaque, an acellular lipid core filled with cell debris and extracellular lipids can rapidly develop.[208] While the molecular mechanism for such a collective cellular suicide remains unknown, it might be relevant to plaque ulceration and/or rupture that is responsible for many thrombotic events. In other words, lipid cores may correspond to a form of "cemetery" containing cells that have undergone rapid and synchronized apoptosis/necrosis portending a heightened risk for plaque rupture and coronary thrombosis.[1,209]

Apoptosis, an evolutionary conserved and genetically regulated form of cell suicide, is an essential component for normal tissue development and homeostasis. Induced or provoked apoptosis is characterized by cell membrane shrinkage, active membrane blebbing, nuclear and cytoplasmic condensation, cellular fragmentation, and engulfment by neighboring cells.[210] When occurring in vascular endothelial cells, SMCs and/or macrophages contribute to plaque instability and atherothrombotic events.[211] Platelet apoptosis in response to chemical agonists and high shear stress triggers microparticle formation (Chapter 20) and activation of cytosolic caspase 3, contributing to the proatherogenic and prothrombotic environments of stenosed vessels.[212]

Constriction of coronary vessels (coronary vasospastic activity) can also trigger plaque rupture.[213] Vascular SMCs are mechanosensitive, constricting to elevations in transmural pressure.[214] Oxidant signaling regulates this myogenic response in arteries whose SMC expression of NADPH oxidase components is increased. Inhibition of NADPH oxidase prevents transmural pressure-induced generation of ROS and consequent vasoconstriction, which seems to be mediated by a hydrogen peroxide ($H_2O_2$)-dependent process. Angiotensin II, the principal product of the renin–angiotensin system, is a potent vasoconstrictor.[215] In addition to causing hypertension, it can contribute to plaque rupture by stimulating coronary vasospasm. Angiotensin II binds to specific receptors on SMCs. In addition to increasing intracellular calcium concentration, angiotensin II also stimulates the activity/expression of the NADPH complex.[216-218] Hence, the homeostasis of ROS, together with the production of NO and prostacyclin, appears central to the control of vascular tone. The lifesaving effect of angiotensin converting enzyme (ACE) inhibitors for patients with CAD[219-221] may therefore be due, at least in part, to the ability of these drugs to suppress coronary spasm and consequent plaque rupture. Other effects of ACE inhibitors are mediated by the ability of these drugs to inhibit SMC proliferation and consequent atheromatous plaque formation.

The most vulnerable plaque is filled with dead cellular debris, in particular SMCs that have died as a result of apoptosis, because of their interaction with monocytes/macrophages.[222] Elimination of the cells capable of producing extracellular matrix proteins within the plaque leaves a thin cap of vascular tissue exposed to mechanical constraints and likely to rupture, particularly in the presence of proteolytic enzymes such as MMPs. Plasminogen activator inhibitor 1 (PAI-1), whose expression is primarily controlled by TGF-β via Smad proteins, is the major physiological inhibitor of tissue/urokinase plasminogen activators (TPA/UPA).[223-225] TPA and UPA convert plasminogen into plasmin (Chapter 21). The latter is an important enzyme for the activation of MMPs that are responsible for the degradation of ECM. The ratio of PAI-1 to PAs also determines the net fibrinolytic activity in the plasma and tissue. Thus, elevated PAI-1 levels may enhance ECM accumulation and increase propensity for thrombus formation. Low PAI-1 concentration would favor the activation of MMPs, with weakening of the cap of atheromatous plaques and increased susceptibility for plaque rupture. Considering the multifaceted consequences of tilting the equilibrium between PAI-1 and PAs, it has been challenging to determine precisely the individual impact of these molecules on plaque rupture and thrombosis.[226-232]

## VI. Atherothrombosis and Endothelial Progenitor Cells

Vascular cell apoptosis may be crucial for the early development of atherosclerosis and thrombosis[233,234] As a consequence, certain antiapoptotic factors previously considered potentially harmful may actually promote vascular health[235] and resilience (i.e., the "youthful" capacity for rapid and complete self-repair). The resilience of injured endothelium has been related to the availability of endothelial progenitor cells (EPCs), as these cells can reendothelialize sites of vascular injury.[236] In turn, the number of circulating EPCs depends on the bone marrow's capacity both to produce these cells in response to factors such as vascular endothelial growth factor (VEGF), erythropoietin,[237] and insulin-like growth factor 1 (IGF-1),[235] and to release them in the blood stream. EPC mobilization requires endothelial NO synthase. Thus, NO (which is typically reduced in atherothrombotic states or in the presence of cardiovascular risk factors)[238] and progenitor-cell stimulating factors may conceivably contribute to avert atherothrombosis by promoting vascular cell resilience.

Although this hypothesis, focused on prompt and healthy vascular regeneration, requires rigorous testing, elements in its favor include the direct relation between preserved endothelial function and circulating EPCs[239]; the beneficial effects of VEGF against arterial restenosis[240]; the prediction of ischemic events by reduced circulating levels of IGF-1 and by renal impairment (a state of low IGF-1 and erythropoietin production;[241] and the lack of clinical benefit of somatostatin analogues (growth factor inhibitors) in patients with ischemic heart disease. Tissue injury and ischemia cause increased concentrations of VEGF,[242] thrombopoietin,[243] and IGF-1,[235] which may signal an underlying disorder, but may represent compensatory (bioattenuation) loops rather than active participation in the onset of disease. Perhaps the most convincing evidence can be derived from recent observations linking reduced numbers of circulating EPCs and the likelihood of future cardiovascular events (cardiovascular death, unstable angina, MI, revascularization, or ischemic stroke).[244]

The adherence of circulating EPCs to sites of vessel wall injury is a subject of intense current interest. Platelet microparticles adhere readily to vascular endothelial cells[245] and in addition to a role in vascular repair, may contribute to EPC survival, proliferation, and differentiation.[246]

## VII. Prevention of Coronary Arterial Plaque Rupture: Progress and Limitations

Although plaque rupture creates a risk for total or subtotal occlusion of the coronary vessel due to disrupted vessel wall fragments extending into the lumen of the vessel and to consequent thrombus formation, it is believed that most ruptures of atheromatous plaques go unrecognized clinically. Indeed, plaque ruptures can promote the removal of the cell debris and lipid material that constitute the semiliquid material within the core of the plaque. Plaque ruptures also allow the access of platelets to the inner layers of the lesion, resulting in the release of growth factors that stimulate the proliferation of residual SMCs that have the capacity to regenerate the extracellular matrix. Hence, once ruptured, the plaque might become more resistant to further mechanical disruption.

Such consequences of plaque rupture might explain, at least in part, the clinical success of PCI, during which the coronary plaque is ruptured under controlled conditions, usually with the placement of a stent to prevent the collapse and/or recoil of the vessel. Thrombus formation is limited by the concomitant use of potent thrombin and platelet inhibitors.[247–249] The intravenous platelet GPIIb-IIIa antagonists abciximab, tirofiban, and eptifibatide have been shown to be effective at reducing MI, urgent vessel revascularization with coronary artery bypass, and even death following PCI (Chapter 62).[250–252]

Immediately after the angioplasty procedure, with or without stent placement, a wave of smooth muscle cell apoptosis can be detected, followed by a prominent SMC proliferative reaction. The latter reaction can cause, in some cases, clinically significant restenosis.[253,254] This supports the concept that SMC proliferation, which is potentiated by the presence of platelets, follows plaque rupture and stabilizes the plaque, causing further unstable coronary events at that specific site of the atherosclerotic coronary vessel.

In addition to the administration of intravenous antithrombotic agents during PCI, such as the GPIIb-IIIa antag-

onists and intravenous drugs such as thrombolytics that can be administered to patients experiencing an STE-AMI, additional lifesaving drugs are available for oral administration to patients with CAD.[255–257] The antiplatelet agents aspirin (Chapter 60) and clopidogrel (Chapter 61), the lipid-lowering statins, the beta blockers, and the ACE inhibitors have all been shown to prevent MI and save lives of patients with CAD, and all of these drugs could interfere with the pathological process of plaque rupture and consequent thrombus formation.[258–260]

Despite our understanding of the biology of atherosclerosis and coronary thrombosis, the yearly rate of death for patients with established CAD remains greater than 5%.[20,149–158,261] Furthermore, patients can develop life-threatening CAD in the absence of any established risk factors.[262] For example, a famous young athlete, who died suddenly of the consequences of acute MI at age 28, had a risk profile that was clearly nonthreatening, according to criteria established by the Framingham study: absence of a smoking habit, hypertension, dyslipidemia (total cholesterol 189 mg/dL, triglycerides 116 mg/dL, HDL cholesterol 45 mg/dL, and LDL cholesterol 121 mg/dL), diabetes, and performance-enhancing substances.[263] The only notable risk factor for this patient, aside from his gender, was the presence of a positive family history for CAD, with a father who died of MI at age 52.[262]

Despite being able to train (with the goal of winning his third gold medal at the next Olympic Games) without experiencing angina pectoris, the coronary arteries of this young man were substantially affected by advanced atherosclerosis. His case, although exceptional (but not rare), illustrates a clear need for continued research of the genomics, proteomics, environmental contribution, and phenotypic expression of atherosclerosis. Further progress in preventing CAD and coronary plaque rupture will likely require additional characterization at the genetic and proteomic levels of the susceptibility to the atherosclerotic process, thereby supporting efficient targeting of preventive strategies.

## VIII. Genetics and Genomics of Atherothrombosis

The ensemble of genetic modifiers that enhances the impact of environmental factors on atherosclerosis represents the genetic susceptibility to this ailment. Individual modifiers often correspond to subtle variations in a gene coding sequence or regulatory regions outside coding sequences, usually in the form of single nucleotide polymorphisms (SNPs), that have a small to moderate impact on the function or expression level of the encoded protein.[264–267] However, the coincidental inheritance of unique combinations of susceptibility genes can have a dominant impact that fosters the pathogenesis of atherosclerosis.

Early-onset (premature) CAD, defined as age of onset less than 50 years, is known to have a particularly strong genetic component. A family history of CAD is one of the most robust risk factors for CAD, even after adjustment for environmental risks that may be shared within families.[268] Several studies have concluded that the relative risk (RR) of developing CAD in a first-degree relative (sibling) is between 3.8% and 12.1%, depending on the age of onset in the proband.[269] The risk of premature coronary heart disease in monozygotic and dizygotic twins has been measured.[270] For men, the relative risk of death from CAD before the age of 55 was 8.1% for monozygotic twins and 3.8% for dizygotic twins. For women, the relative risk of death from CAD before age 65 was 15.0% for monozygotic twins and 2.6% for dizygotic twins.[271]

In familial hypercholesterolemia, a condition in which premature development of atherosclerosis is precipitated by very high levels of LDL cholesterol, four types of gene defects have been identified[272]: (1) LDL-receptor gene defects (in excess of 600 different defects reported), (2) ApoB-100 mutations, (3) autosomal recessive hypercholesterolemia, and (4) sitosterolemia. Patients with Tangier disease have a defect in cellular cholesterol removal, which results in near-zero plasma levels of HDL cholesterol and in massive tissue deposition of cholesteryl esters.[273] The ABC1 transporter was identified as the flawed gene in Tangier disease through a combined strategy of gene expression microarray analysis, genetic mapping, and biochemical studies.[274–278] Homocysteinuria, due to a deficit in cystathionine β-synthase deficiency,[279] and Hurler syndrome,[280] due to a deficiency of L-iduronidase (a lysosomal hydrolase), are other examples of metabolic disorders with an accumulation of homocysteine or mucopolysaccharide, respectively, in which premature atherosclerosis represents a major cause of death. Thus, single gene defects can accelerate substantially the "sclerosis" of the arterial vessels. Interestingly, two individuals with familial hypercholesterolemia who share the same founder LDL-receptor defect and identical LDL cholesterol levels can experience substantially different fates.[281] This observation exemplifies the fact that atherosclerosis and its complications involve many genes, major and minor, that significantly impact upon the pathogenesis and in a way that takes cues from the environment. Perhaps due to such complexity, there is a general paucity of data related to genome-wide mapping of culprit genes for atherosclerosis.[282]

Although previous work has provided evidence for the contribution of single gene defects to cardiovascular disease, one anticipates that much of the variation in disease onset, progression, and severity within populations of patients is due to contributions of multiple loci (Fig. 35-8). Our understanding of ischemic heart disease has already become more sophisticated as recent discoveries have revealed the influence of gene variants on the development of atherosclerotic disease and arterial thrombosis. The role of platelets in unstable coronary syndromes and adverse events follow-

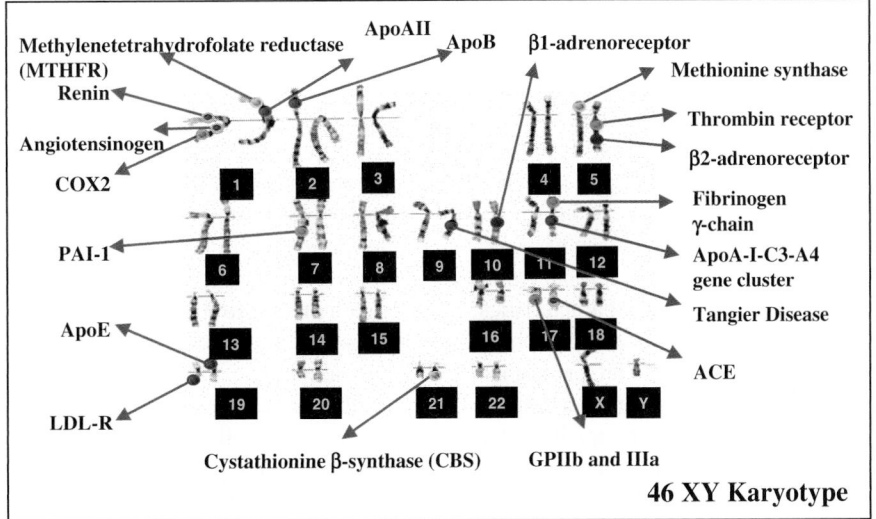

**Figure 35-8.** Coronary artery disease (CAD), a complex trait. Unlike monogenic disorders, for which a specific section of the human genome can be identified as the carrier for the culprit gene, in CAD, using a "linkage" approach, the culprit genes appear to be dispersed across the entire human genome. A normal 46 XY human karyotype is shown. Genes that have been linked to atherosclerosis, such as the LDL-receptor (LDL-R) cystathionine β-synthase (CBS), are marked on their respective chromosomes. Furthermore, genes whose product has been targeted successfully with drugs that improve survival of patients with CAD are also marked. Clearly, the genes that contribute to atherosclerosis are dispersed across the entire human genome and will require novel genomic techniques to be systematically identified.

ing coronary intervention is a typical example of the interplay between genetic and environmental factors. A number of gene polymorphisms within the subunits of platelet receptors and intracellular proteins (Chapter 14) may induce gain or loss of function, thereby predisposing some individuals to thrombotic events. Gain or loss of function may be secondary to polymorphisms that either encode a missense (the resulting codon provokes a change in amino acid) or modify the level of expression of the gene product. Additional alterations can involve SNPs at exon/intron junctions that modify splicing, or even SNPs falling within intronic regions that can affect the interaction between the primary nuclear RNA and RNA-binding proteins.[283–286]

There is ample controversy regarding individual platelet polymorphisms that contribute to an increased likelihood of ischemic events (reviewed in Chapter 14). For example, GPIIIa, together with GPIIb, constitutes the fibrinogen receptor (GPIIb-IIIa, or integrin $\alpha$IIb$\beta$3), whose engagement represents the final common pathway for platelet aggregation (Chapter 8).[287] Due to the substitution of a cytosine for a thymidine at position 1565 in exon 2 of the GPIIIa gene, the platelet antigen 2 (Pl$^{A2}$) variant displays a proline instead of a leucine at amino acid 33.[288] Pl$^{A2}$ has been implicated in arterial thrombosis and the development of unstable coronary syndromes. Specifically, a high prevalence of the Pl$^{A2}$ allele has been reported in patients with unstable angina or MI and in siblings of patients with a history of premature ischemic heart disease.[289,290] Mikkelsson et al. also reported a higher prevalence of Pl$^{A2}$ among victims of sudden cardiac death whose coronary arteries contained thrombi.[291] Still other studies have not confirmed the association between Pl$^{A2}$ and MI,[292–296] maintaining the controversy over the influence of the Pl$^{A2}$ polymorphism. Despite the variance in epidemiological data, most studies examining the molecular effect of the Pl$^{A2}$ polymorphism on GPIIb-IIIa function have been consistent, showing that the mutation results in increased platelet responsiveness.[297–300] Although it is appealing to attribute much of the disagreement between studies to design or patient populations, it is also important to recognize that the clinical diagnosis of acute coronary syndromes includes a broad range of phenotypic characteristics. Moreover, the relative impact of a platelet polymorphism such as Pl$^{A2}$ may vary according to multiple factors such as age, gender, smoking history, diabetes, and medications.

Although platelet polymorphisms are a promising addition to more established cardiovascular risk factors, identifying genetic variants as a single cause of cardiovascular disease would be an oversimplification; instead, the contribution of these polymorphisms should also be considered in the context of nongenetic factors. To date, no single trial has conclusively established the importance of platelet polymorphisms. Until a large database involving thousands of patients is able to satisfactorily reconcile divergent findings,

the controversy will remain. Considering inheritance patterns, the accuracy of genotyping in some cases, and the impact of environmental factors, the required sample size to have adequate power has been estimated to be several thousand patients.[301] To create such large databases, carefully supervised genetic sampling should therefore be encouraged as part of large clinical trials or even routine patient care.

Despite substantial research activity in the genetics of CAD, relatively little coordinated effort has been made to confirm the associated risks of specific gene polymorphisms. As a result, identifying risk associated with genetic discoveries and translating this knowledge into clinical practice will likely occur gradually rather than through one revolutionary discovery. To solve this problem, future research should include larger studies that adhere to the same standards as contemporary large-scale, randomized trials for drug development. Including mechanistic studies as part of clinical trials will also facilitate our understanding of the functional significance of the studied polymorphisms, help define the impact of these mutations on clinical outcomes, and validate case-control association studies. A spin-off effect of these studies might be a better definition of phenotypes for the group of syndromes that we have somewhat arbitrarily assembled under the generic name of acute coronary syndromes. Only through a larger collective effort will the role of platelet polymorphisms and other susceptibility genes for cardiovascular disease be defined.

Furthermore, large-scale efforts need to be developed to extend the panoply of genes already identified as important candidates for a role in atherosclerosis and coronary thrombosis. Thus far, most of the genes studied were selected on the basis of biological understanding of the disease process and clinical knowledge based on the efficacy of drugs that have been empirically defined as useful for patients with CAD. There is a need for nonbiased approaches to discovering new genes that participate in atherogenesis of coronary vessels and thrombosis. Genome-wide mapping of such genes based on linkage analyses, combined with expression studies on arterial tissue, performed on families of individuals with premature CAD presenting with various degrees of atherosclerosis and/or thrombosis, should help prioritize genes across the entire human genome in terms of their contribution to susceptibility or resistance to atherosclerosis and its coronary thromboembolic complications.[302]

Cross-species investigations could also lead to identification of important genes that contribute to atherosclerosis. Genetic models have been developed in the mouse, and the ApoE and LDL-receptor knockout models have already provided useful opportunities in our search for genetic modifiers that could interfere with the process of atherogenesis and thromboembolic complications.[303–307] Genes discovered to be important modifiers in mouse models can be tested in clinical trials for their relevance to human atherosclerosis. Advances will also be required in our ability to analyze the substantial masses of data that will be generated by large-scale genotyping or expression studies. Bioinformatic and statistical methods need to be upgraded to meet the demands of the future scale of investigation.

## IX. Clinical Aspects and Management of Coronary Atherothrombosis

As discussed in the preceding sections, there are three clinical stages that characterize the progression of CAD and its thromboembolic complications: preclinical, stable, and unstable. The following section discusses clinical aspects of these three stages of CAD and summarizes the recommendations for antiplatelet drug therapy and other treatments for CAD. The individual antiplatelet drugs and their clinical indications are discussed in detail in other chapters: aspirin (Chapter 60), clopidogrel and ticlopidine (Chapter 61), and GPIIb-IIIa antagonists (Chapter 62).

### A. Preclinical CAD

In industrialized countries, CAD is rampant but, in most of the population, not advanced enough to trigger clinical manifestations such as angina pectoris and other forms of chest pain related to coronary thrombosis. At this presymptomatic state, antiplatelet therapy is currently recommended only for men over the age of 50 or women over the age of 60 with at least one major risk factor for CAD: tobacco smoking, hypertension, diabetes mellitus, high cholesterol, or a family history of premature CAD.[308] For these patients, aspirin at a dose of 75 to 162 mg/day is recommended (Table 35-1). This recommendation assumes that there is no contraindication to aspirin (see Chapter 60 for a review of contraindications). Meticulous control of blood pressure is recommended for patients who are placed on aspirin, in order to reduce the risk of cerebral hemorrhage. It is also important to consider antithrombotic therapy in a context of global preventive measures for CAD events. Diet and 3-hydroxy-3-methylglutaryl-coenzyme A reductase inhibitors (statins) are recommended for these patients who have elevated cholesterol and, in particular, LDL cholesterol and are therefore at particular risk for plaque rupture and consequent coronary thrombosis.

Although the use of warfarin is more complex and costly than that of aspirin, and sometimes less safe, for those patients who are not candidates for receiving aspirin, warfarin is recommended (target international normalized ratio [INR] of ~1.5).[308,309] A combination of low-dose aspirin (75

–100 mg/day) and low-intensity warfarin (target INR of 1.3–1.5) may be considered for individuals who are at very high risk of cardiovascular events (those with multiple risk factors and extensive family history of premature CAD).[308,309] It is also possible that a combination of aspirin and clopidogrel could replace the aspirin–warfarin combination for patients at high risk or patients who have experienced an event while receiving aspirin. However, additional studies will be needed to support this approach in primary prevention. It is also important to note that patients with diabetes mellitus, but with no manifestation of CAD, carry a risk that is similar to that of patients with CAD, but no diabetes. Hence, preventive measures must be considered carefully in these patients.

### B. Stable CAD

The second stage of CAD consists of the development of stable coronary symptoms. Many patients have very predictable symptoms of angina pectoris upon reaching a certain level of exertion. All of these patients should receive oral aspirin at a dose of 75 to 162 mg daily (Table 35-1).[310] This recommendation includes both men and women, and in the case of a contraindication to aspirin, substitution of aspirin with clopidogrel (75 mg/day) should be recommended (for details on clopidogrel therapy, see Chapter 61).[311] Long-term aspirin therapy is preferred to the use of warfarin because of its simplicity, safety, and low cost. Patients with stable CAD should receive oral aspirin for life, together with comprehensive preventive efforts. Thus, patients are recommended to follow a diet and initiate a regular exercise routine to control their weight and improve their lipid profile. Most patients in this category are also candidates for statin therapy. At least two drugs in the statin group have been shown to save lives in patients with established CAD: simvastatin and pravastatin.[20,154] Furthermore, patients should also receive therapy with beta blockers, which have demonstrated capacity to improve $O_2$ mismatch, reduce symptoms of angina pectoris, and improve long-term survival for these patients.[312] The use of ACE inhibitors or angiotensin II-receptor blockers is recommended for control of blood pressure, aiming for a diastolic pressure of <85 mm Hg.[313] Some of the ACE inhibitors have been shown to improve survival in secondary prevention.[314,315] The use of nitrates and calcium-channel blockers should be individualized and based on specific needs of the patient. Importantly, neither of these two drugs has been shown to improve survival of patients with CAD.

This second stage of CAD usually corresponds to the presence of one or several lesions in the coronary vessels that reduce the lumen of the vessel by 75% or more (residual lumen of 25% or less of normal).[316] It is important to note

**Table 35-1: Guidelines for Antithrombotic Therapy in the Prevention and Treatment of Thrombosis among Patients with CAD**

|  | Aspirin[a] | Thienopyridine | GPIIb-IIIa Antagonist | Thrombolytics |
|---|---|---|---|---|
| Preclinical | II-A | — | — | — |
| Stable CAD | I-A[b] | I-A[c] | — | — |
| Non-STE-ACS | I-A | I-A[d] | I-A[e] | — |
| STE-MI | I-A | — | [f] | I-A |
| PCI | I-A | I-A | I-A | — |

CAD, coronary artery disease; NSTE-ACS, non-ST-elevation acute coronary syndrome; PCI, percutaneous coronary intervention; STE-MI, ST-elevation myocardial infarction.
[a]Vitamin K antagonist to an INR of 1.5 in very high-risk patients who can be monitored closely.
[b]I-A indicates a very strong recommendation based on the results of well-designed, randomized, clinical trials with consistent results.
[c]In stable CAD, a thienopyridine derivative (clopidogrel) is recommended (when surgical intervention [CABG] is unlikely) only for those patients who are unable to take aspirin.
[d]In unstable CAD/ACS, clopidogrel is recommended in addition to aspirin, unless the risk of bleeding is prohibitive. A loading dose of 300 mg, followed by 75 mg daily, is recommended.
[e]In NSTE-ACS, only the small-molecule GPIIb-IIIa antagonists (eptifibatide and tirofiban) are recommended. Abciximab failed to show benefit to patients with NSTE-ACS (in GUSTO-IV/ACS) in the absence of PCI.
[f]In STE-AMI, the only GPIIb-IIIa antagonist that has been tested in a phase III trial is abciximab which, when added to a thrombolytic, did not show improved survival. However, it is possible that the combination of abciximab and a thrombolytic could be beneficial for younger patients with anterior distribution of their MI. Abciximab may also reduce recurrent events in patients with STE-AMI, but such findings fall short of a I-A recommendation.
Adapted from Guyatt et al. (2004). *Chest, 126,* 179S–187S.

that, although we arbitrarily define a clear demarcation between clinical stages I, II and III, it is well-established that a number of "accidents" can take place within coronary vessels, triggered by plaque rupture and thrombus formation, that do not lead to sufficient narrowing of coronary vessels to result in angina symptoms or worse. Hence, the preventive maneuvers applied to patients with supposedly stable CAD should probably be as intense as those applied to patients who have experienced unstable syndromes.

### C. Unstable CAD

The third stage of CAD is characterized by instability (plaque and clinical expression). The clinical syndromes of this third stage can be divided into subgroups according to a hierarchy based on severity. The unstable clinical syndromes are divided into (1) unstable angina, (2) acute MI, and (3) sudden ischemic cardiac death. Unstable angina is characterized clinically by the development of angina symptoms at rest, in the absence of unusual exertion. However, such a strict definition can be misleading. Indeed, life-threatening destabilization of a coronary plaque with thrombus formation can manifest itself simply by acceleration in the rate of angina symptoms, with lowering of the exertion threshold for the development of chest pain (usually defined as crescendo angina). Particularly in women, excess shortness of breath, gastrointestinal symptoms, and fatigue can

represent more subtle manifestations of unstable angina. Once the patient reaches a hospital, confirmation of the diagnosis of unstable angina can be provided by dynamic changes in the ST-segment and T-wave of the electrocardiogram (EKG) (in particular, ST depression of more than 1 mm is a marker of severity). Elevated troponin is another indication of severity.[317]

MI corresponds to a clinical syndrome associated with myocardial necrosis. Hence, a landmark component of the diagnosis of MI is the elevation of creatine phosphokinase, in particular the MB isoform of this enzyme (CK-MB). The formation of new and significant Q-waves on EKG is another recognized proof of MI. The persistence of angina pain that is unusually severe is another landmark finding in patients with acute MI, and such pain is frequently associated with shortness of breath, diaphoresis, lightheadedness, nausea, and sometimes vomiting.[318]

There is an important dichotomy in the MI syndromes that is defined by a biomarker that was best established by the work of DeWood and colleagues,[319,320] who showed that the presence of consistent ST elevations on EKG for patients with ongoing MI is associated with a very specific status of the culprit coronary vessel. Thus, ST elevation of more than 1 mm (the biomarker in question) on the EKG of patients with ongoing MI identifies individuals with total occlusion of the culprit coronary vessel as the cause of the infarction process.[319] In contrast, MI that develops in the absence of ST elevation (no sustained ST changes or ST depression) indicates a condition in which a minority of the patients are found to have total occlusion of the culprit coronary vessel.[320,321] For the ST-elevation type of MI, with total occlusion of the culprit vessel, myocardial damage occurs, at least in part, as a result of acute ischemic injury at the epicenter of the infarction. Consequently, the myocardium in this region contains very few inflammatory cells and the area corresponds to a lesion that is called "mummified" by pathologists. Surrounding this lesion is an area where inflammatory cells are abundant and where necrosis and apoptosis may occur, at least in part as a consequence of the interaction of cardiomyocytes with inflammatory cells.

In contrast, in the non-ST-elevation type of MI, cardiomyocyte necrosis and apoptosis appear to be induced exclusively by the direct interaction of inflammatory cells with ischemic cardiomyocytes. This interaction occurs as a consequence of distal embolization of aggregates of platelets and phagocytes that originate from destabilized coronary plaques,[322] and possibly also of platelet–leukocyte aggregates that originate from the systemic circulation and have been detected in the blood of patients with unstable coronary lesions.[323]

Most importantly, patients with ST-elevation MI respond differently to therapy than those with non-ST-elevation MI. Patients with unstable angina and non-ST-elevation MI have been grouped into a single category of patients with "non-ST elevation acute coronary syndromes" (NSTE-ACS). The difference between unstable angina and MI in the absence of ST-elevation seems to reside in the intensity of the myocardial injury and, in general, a patient who experiences unstable angina symptoms for a period exceeding 15 to 20 minutes will eventually develop MI.

**1. Treatment of Non-ST-Elevation Acute Coronary Syndromes.** Patients with NSTE-ACS should receive aspirin at an initial dose of 160 to 325 mg/day followed by 75 to 162 mg/day indefinitely (Table 35-1).[324] The first tablet should be a nonenteric-coated aspirin tablet that the patient should chew in order to accelerate the impact of the drug on blood passing through the splanchnic circulation. Patients with acute coronary syndromes should also receive intravenous small molecule GPIIb-IIIa antagonists (either tirofiban or eptifibatide) (Table 35-1) (Chapter 62).[324] It is usually agreed that the more severe the condition of the patient, the stronger the indication for a GPIIb-IIIa antagonist. Markers of severity include dynamic ST depression on EKG of more than 1 mm and/or elevated levels of troponin (I or T). The infusion of tirofiban or eptifibatide should be continued for 48 to 72 hours or until PCI is performed (see following). Antithrombin therapy with unfractionated heparin, according to a weight-based dosing strategy to achieve an aPTT between 50 and 75 sec, is suggested. Low-molecular-weight heparin can also be used in this setting instead of unfractionated heparin. Heparin therapy should be main-

tained for at least 48 hours or until the unstable symptoms resolve.

Clopidogrel has also been shown to improve the outcome of patients with acute coronary syndromes.[325] Clopidogrel, an inhibitor of the P2Y$_{12}$ ADP receptor on platelets (Chapter 61), can reduce platelet aggregation induced by ADP. The active clopidogrel metabolite produced by the liver binds irreversibly to the P2Y$_{12}$ receptor.[326] In the Clopidogrel in Unstable Angina to Prevent Recurrent Events (CURE) trial, aspirin alone was compared to aspirin plus clopidogrel in patients with NSTE-ACS.[325] The addition of clopidogrel to aspirin significantly reduced the event rate (by 20%, $p < 0.001$) for these acute coronary syndrome patients. The benefits of the combination therapy were sustained over the entire duration of the study (mean of 9 months). As a consequence, clopidogrel is now indicated in the management of patients with acute coronary syndromes. An aspirin dose of 75 to 100 mg daily reduces the risk of hemorrhagic complications (compared to higher doses), while maintaining efficiency.[327] Interestingly, the establishment of clopidogrel for the management of patients with acute coronary syndromes might have an impact on other recommendations (presented earlier) for the management of patients with NSTE-ACS, including the appropriateness and timing of administration of GPIIb-IIIa antagonists. Furthermore, it is not clear at this time that recommendations for patients with NSTE-ACS regarding the combination of aspirin and clopidogrel can be extrapolated to patients with stable CAD or patients who are at high risk but do not have detectable manifestations of CAD. Nevertheless, the CURE findings indicate that blocking individual activation pathways of platelets has a synergistic effect rather than an overlapping effect in terms of patient benefit. Aspirin blocks the cyclo-oxygenase pathway (Chapter 60), whereas clopidogrel blocks the P2Y$_{12}$ receptor pathway (Chapters 61).

Additional important information is the fact that for patients with NSTE-ACS, the use of GPIIb-IIIa antagonists is particularly beneficial for patients who undergo early PCI (Chapter 62). For those patients who do not undergo early revascularization, the benefit of these drugs is present, but markedly less pronounced.

**2. Treatment of ST-Elevation Acute Myocardial Infarction.** A fundamental difference in the management of patients with NSTE-ACS versus patients with STE-MI relates to the response to intravenous fibrinolytic therapy. While the administration of TPA and other thrombolytic drugs is life-saving for patients with STE-MI, fibrinolytics may be detrimental to patients with NSTE-ACS.[328–332] This differential impact of thrombolytics on the two syndromes that lead to myocardial necrosis must be related to the pathophysiological differences that exist between these two coronary thromboembolic complications. In STE-MI, the complete occlusion of the culprit coronary vessel might explain the significant benefit provided by thrombolytic drugs that substantially accelerate the reperfusion of the culprit vessel. Because a large fraction of the myocardial necrosis in STE-AMI results from lack of blood flow ("mummified" region, see earlier), the accelerated restoration of TIMI-3 (Thrombolysis in Myocardial Infarction-3) flow in the culprit vessel is key to limiting myocardial damage.[333-335] In contrast, because complete occlusion of a coronary vessel is rarely the mechanism responsible for myocardial necrosis in NSTE-ACS, the potentially proaggregatory effect of thrombolytics toward platelets might explain, at least in part, the lack of benefit from thrombolytic drugs in patients with NSTE-ACS. Typically, the thrombolytic regimen for STE-MI consists of two boluses of 10 units of a drug such as reteplase administered 30 minutes apart.[336] Heparin administration is also recommended to achieve an aPTT between 50 and 75 seconds. Usually, a heparin bolus of 5000 U followed by 1000 U/hour infusion (for patients weighing 80 kg or more) or 800 U/hour infusion (for patients <80 kg) is state of the art.[329,330,336] Aspirin is given as in NSTE-ACS, with a nonenteric-coated tablet of 162 to 325 mg to chew followed by a daily dose of 75 to 162 mg (Table 35-1).

Primary angioplasty has been shown, when performed in a hospital by a cardiologist with substantial experience with this procedure, to be at least as good as thrombolytic therapy for patients with acute MI.[337] The combination of thrombolytics and a GPIIb-IIIa antagonist has been tested in the GUSTO V study.[338] In a double-blind, randomized, placebo-controlled study, full-dose reteplase was compared to abciximab bolus and infusion plus half-dose reteplase in patients with STE-AMI. Although there was no significant survival advantage of combination therapy versus thrombolytic alone, reinfarction was reduced significantly in the group receiving abciximab and reteplase. Such benefit was particularly pronounced in individuals with anterior MI and <75 years old. Combination therapy was associated with a significant increase in severe bleeding (0.5 versus 1.1%, $p < 0.0001$). Importantly, intracranial bleeds were not increased by the combination of abciximab and reteplase, compared to reteplase alone. The benefits in terms of recurrent events provided by combination therapy relative to thrombolytic therapy alone is instructive and will need additional studies to be extended to other GPIIb-IIIa antagonists besides abciximab.

Other benefits of abciximab that have not yet been reproduced with other GPIIb-IIIa antagonists include the long-term effect of abciximab on survival.[339] In three separate PCI trials, abciximab has been shown to improve long-term survival (Chapter 62). Such an effect was unexpected for a drug that is administered acutely IV for a short period. This observation contrasts with the lack of benefit of abciximab for patients experiencing NSTE-ACS and for whom early revascularization is not readily prescribed. For patients with NSTE-ACS, the small-molecule GPIIb-IIIa antagonists have shown a benefit and are indicated, whereas abciximab has not shown a benefit and is therefore not indicated for patients with NSTE-ACS (GUSTO IV-ACS).[340] The role of clopidogrel in the treatment of patients with STE-AMI has yet to be established, but emerging results when used in combination with fibrinolytic therapy are encouraging.[341]

### D. Percutaneous Coronary Intervention

PCI, particularly with stent placement, is a procedure that can induce substantial platelet activation. Antiplatelet agents have improved the prognosis of patients undergoing PCI. After a period of trial and error, selective antiplatelet agents have become the mainstay of antithrombotic therapy for patients undergoing PCI. Accordingly, aspirin (100–325 mg) and clopidogrel (loading dose 300 mg, maintenance dose 75 mg/day) are routinely used for patients undergoing PCI (Table 35-1).[342] Ticlopidine was previously the thienopyridine derivative of choice, but because clopidogrel was shown to be at least as effective as ticlopidine in preventing stent thrombosis and is a safer drug in terms of side effects (less neutropenia and thrombotic thrombocytopenic purpura [TTP]), clopidogrel has replaced ticlopidine for patients undergoing PCI (Chapter 61).

It is usually agreed that the GPIIb-IIIa antagonists are particularly indicated for patients at high risk (patients with NSTE-ACS with positive troponin and other factors that complicate the PCI procedure such as diabetes mellitus and advanced age).[342] For patients who undergo primary PCI for MI, abciximab is currently considered to be the GPIIb-IIIa antagonist of choice.

Patients undergoing PCI also need to receive heparin. Unfractionated heparin is administered to achieve an activated clotting time (ACT) of 250 to 300 seconds (HemoTec device) or 300 to 350 seconds (Hemochron device) with adjusted heparin boluses (60–100 IU/kg).[342] Early sheath removal reduces complications and should be performed when the ACT falls below 150 to 180 seconds. When abciximab is used, the heparin bolus should be reduced to 50 to 70 IU/kg to achieve a target ACT of >200 second (with either the HemoTec or Hemochron device).[342] The use of low-molecular-weight heparin or direct thrombin inhibitors has shown some promise in NSTE-ACS and STE-MI,[343] but does not thus far provide patients with obvious advantages compared to unfractionated heparin during PCI.

Finally, restenosis is a frequent complication following PCI and there is little evidence that antiplatelet therapy affects the rate of restenosis. Nevertheless, because patients undergoing PCI are at risk for recurrent thromboembolic events, the use of aspirin, and perhaps a combination of aspirin and clopidogrel, is recommended indefinitely.

Patients receiving Drug Eluting Stents (DES) should receive a combination of aspirin (75–100 mg daily) and clopidogrel (75 mg daily) for a minimum of 3 months with sirolimus DES and 6 months with paclitaxel DES. The risk of late stent thrombosis after discontinuation is a concern.

# References

1. Ross, R. (1999). Atherosclerosis—an inflammatory disease. *N Engl J Med, 340,* 115–132.
2. Ross, R., & Fuster, V. (1996). The pathogenesis of atherosclerosis. In V. Fuster, R. Ross, & E. J. Topol (Eds.), *Atherosclerosis and coronary artery disease* (Vol. I., pp. 441–460). Philadelphia: Lippincott-Raven.
3. Ross, R. (1993). The pathogenesis of atherosclerosis: A perspective for the 1990s. *Nature 362,* 801–809.
4. Badimon, J. J., Fuster, V., Chesebro, J., et al. (1993). Coronary atherosclerosis: A multifactorial disease. *Circulation, 87*(suppl. II), 3–16.
5. Libby, P. (2000). Coronary artery injury and the biology of atherosclerosis: Inflammation, thrombosis, and stabilization. *Am J Cardiol, 86,* 3J–8J.
6. Faergeman, O. (2001). The atherosclerosis epidemic: Methodology, nosology, and clinical practice. *Am J Cardiol, 88*(suppl), 4E–7E.
7. Tunstall-Pedoe, H., Kuulasmaa, K., Mahonen, M., et al. (1999). Contribution of trends in survival and coronary-event rates to changes in coronary heart disease mortality: 10-year results from 37 WHO MONICA project populations. Monitoring trends and determinants in cardiovascular disease. *Lancet, 353*(9164), 1547–1557.
8. Murray, C., & Lopez, A. D. (1997). Alternative projections of mortality and disability by cause 1990–2020. Global Burden of Disease Study. *Lancet, 349,* 1498–1504.
9. Breslow, J. L. (1997). Cardiovascular disease burden increases, NIH funding decreases. *Nat Med, 3,* 600–601.
10. Braunwald, E. (1997). Shattuck Lecture — cardiovascular medicine at the turn of the millennium: Triumphs, concerns, and opportunities. *N Engl J Med, 337,* 1360–1369.
11. Stary, H. C. (2000). Natural history and histological classification of atherosclerotic lesions: An update. *Arterioscler Thromb Vasc Biol, 20,* 1177–1178.
12. Fuster, V., Badimon, L., Badimon, J. J., et al. (1992). The pathogenesis of coronary artery disease and the acute coronary syndromes (1). *N Engl J Med, 332*(4), 242–250.
13. Fuster, V., Badimon, L., Badimon, J. J., et al. (1992). The pathogenesis of coronary artery disease and the acute coronary syndromes (2). *N Engl J Med, 332*(5), 310–318.
14. Stary, H. C., Chandler, A. B., Glagov, S., et al. (1994). A definition of initial, fatty streak, and intermediate lesions of atherosclerosis: A report from the Committee on Vascular Lesions of the Council on Arteriosclerosis, American Heart Association. *Circulation, 89,* 2462–2478.
15. Stary, H. C., Blankenhorn, D. H., Chandeler, A. B., et al. (1995). A definition of advanced arteriosclerotic lesions and a classification of arteriosclerosis. *Circulation, 92,* 1355–1374.
16. Zaman, A. G., Helft, G., Worthley, S. G., et al. (2000). The role of plaque rupture and thrombosis in coronary artery disease. *Atherosclerosis, 149,* 251–326.
17. Napoli, C., D'Armiento, F. P., Mancini, F. P., et al. (1997). Fatty streak formation occurs in human fetal aortas and is greatly enhanced by maternal hypercholesterolemia: Intimal accumulation of low density lipoprotein and its oxidation precede monocyte recruitment into early atherosclerotic lesions. *J Clin Invest, 100,* 3280–3290.
18. Worthley, S. G., Osende, J. I., Helft, G., et al. (2001). Coronary artery disease: Pathogenesis and acute coronary syndromes. *Mt Sinai J Med, 68,* 167–181.
19. National Cholesterol Education Program. (1993). Second report of the Expert Panel on Detection, Evaluation, and Treatment of High Blood Cholesterol in Adults (Adult Treatment Panel II). Bethesda, MD.: National Heart, Lung, and Blood Institute. (NIH publication no. 93–3095),
20. Scandinavian Simvastatin Survival Study Group. (1994). Randomised trial of cholesterol lowering in 4444 patients with coronary heart disease: The Scandinavian Simvastatin Survival Study (4S). *Lancet, 344,* 1383–1389.
21. Goto, Jr., A. M. (1999). Lipid management in patients at moderate risk for coronary heart disease: Insights from the Air Force/Texas Coronary Atherosclerosis Prevention Study (AFCAPS/TexCAPS). *Am J Med, 107,* 36S–39S.
22. Aronson, D., Rayfield, E. J., & Chesebro, J. H. (1997). Mechanisms determining course and outcome of diabetic patients who have had acute myocardial infarction. *Ann Intern Med, 132,* 296–306.
23. Sytkowski, P. A., Kannel, W. B., & D'Agostino, R. B. (1990). Changes in risk factors and the decline in mortality from cardiovascular disease. The Framingham Heart Study. *N Engl J Med, 322,* 1635–1641.
24. Anonymous. (1990). Mortality after $10^{1}/_2$ years for hypertensive participants in the Multiple Risk Factor Intervention Trial. *Circulation, 82,* 1616–1628.
25. Castelli, W. P. (1996). Lipids, risk factors and ischaemic heart disease. *Atherosclerosis, 124,* S1–S9.
26. McCully, K. S. (1969). Vascular pathology of homocysteinemia: Implications for the pathogenesis of arteriosclerosis. *Am J Pathol, 56,* 111–128.
27. Nehler, M. R., Taylor, Jr., L. M., & Porter, J. M. (1997). Homocysteinemia as a risk factor for atherosclerosis: A review. *Cardiovasc Surg, 6,* 559–567.
28. Nygard, O., Nordrehaug, J. E., Refsum, H., et al. (1997). Plasma homocysteine levels and mortality in patients with coronary artery disease. *N Engl J Med, 337,* 230–236.
29. Malinow, M. R. (1995). Plasma homocyst(e)ine and arterial occlusive diseases: A mini-review. *Clin Chem, 41,* 173–176.
30. Levenson, J., Giral, P., Razavian, M., et al. (1995). Fibrinogen and silent atherosclerosis in subjects with cardiovascular risk factors. *Arterioscler Thromb Vasc Biol, 15,* 1323–1328.
31. Hopkins, P. N., Hunt, S. C., Schreiner, P. J., et al. (1998). Lipoprotein(a) interactions with lipid and non-lipid risk factors in patients with early onset coronary artery disease: Results from the NHLBI Family Heart Study. *Atherosclerosis, 141,* 333–345.
32. Djurovic, S., & Berg, K. (1997). Epidemiology of Lp(a) lipoprotein: Its role in atherosclerotic/thrombotic disease. *Clin Genet, 52,* 281–292.

33. Steinhubl, S. R., Talley, J. D., Braden, G. A., et al. (2001). Point-of-care measured platelet inhibition correlates with a reduced risk of an adverse cardiac event after percutaneous coronary intervention: Results of the GOLD (AU-Assessing Ultegra) multicenter study. *Circulation, 103,* 2572–2578.

34. Ridker, P. M. (1999). Evaluating novel cardiovascular risk factors: Can we better predict heart attacks? *Ann Intern Med, 130,* 933–937.

35. Libby, P. (1995). Molecular basis in the acute coronary syndromes. *Circulation, 91,* 2844–2850.

36. Ross, R. (1981). Atherosclerosis—a problem of the biology of arterial wall cells and their interactions with blood components. *Arteriosclerosis, 1,* 293–311.

37. Lincoff, A. M., Kereiakes, D. J. Mascelli, M. A., et al. (2001). Abciximab suppresses the rise in levels of circulating inflammatory markers after percutaneous coronary revascularization. *Circulation, 104,* 163–167.

38. Fernandez-Ortiz, A., Badimon, J. J., Falk, E., et al. (1994). Characterization of the relative thrombogenicity of atherosclerotic plaque components: Implications for consequences of plaque rupture. *J Am Coll Cardiol, 23,* 1562–1569.

39. Libby, P. (2001). Current concepts of the pathogenesis of the acute coronary syndromes. *Circulation, 104,* 365–372.

40. Ridker, P. M., Cushman, M., Stampfer, M. J., et al. (1997). Inflammation, aspirin, and the risk of cardiovascular disease in apparently healthy men. *N Engl J Med, 336,* 973–979.

41. Morel, D. W., Hessler, J. R., & Chisholm, G. M. (1983). Low density lipoprotein cytotoxicity induced by free radical peroxidation of lipid. *J Lipid Res, 24,* 1070–1076.

42. Eberhardt, R. T., Forgione, M. A., Cap, A., et al. (2000). Endothelial dysfunction in a murine model of mild hyperhomocyst(e)inemia. *J Clin Invest, 106*(4), 483–491.

43. Libby, P., Egan, D., & Skarlatos, S. (1997). Roles of infectious agents in atherosclerosis and restenosis: An assessment of the evidence and need for future research. *Circulation, 96,* 4095–4103.

44. Gupta, S., Leatham, E. W., Carrington, D., et al. (1997). Elevated Chlamydia pneumoniae antibodies, cardiovascular events, and azithromycin in male survivors of myocardial infarction. *Circulation, 96,* 404–407.

45. Hajjar, D. P., Fabricant, C. G., Minick, C. R., et al. (1986). Virus-induced atherosclerosis: Herpesvirus infection alters aortic cholesterol metabolism and accumulation. *Am J Pathol, 122,* 62–70.

46. Nicholson, A. C., & Hajjar, D. P. (1998). Herpesviruses in atherosclerosis and thrombosis: Etiologic agents or ubiquitous bystanders? *Arterioscler Thromb Vasc Biol, 18,* 339–348.

47. Danesh, J., Collins, R., & Peto, R. (1997). Chronic infections and coronary heart disease: Is there a link? *Lancet, 350,* 430–436.

48. Gimbrone, M. A. (1999). Vascular endothelium, hemodynamic forces, and atherogenesis. *Am J Pathol, 155,* 1–5.

49. Henderson, J. M., Aukerman, J. A., Clingan, P. A., et al. (1999). Effect of alterations in femoral artery flow on abdominal vessel hemodynamics in swine. *Biorheology, 36,* 257–326.

50. Strong, J. P., Herderick, G.T, Cornhill, J. F., et al. (1999). Prevalence and extent of atherosclerosis in adolescents and young adults: Implications of prevention from the pathobiological determinants of atherosclerosis in youth study. *J Am Med Assoc, 281,* 727–735.

51. Griendling, K. K., & Alexander, R. W. (1997). Oxidative stress and cardiovascular disease. *Circulation, 96,* 3324–3325.

52. Resnick, N., Yahav, H., Schubert, S., et al. (2000). Signalling pathways in vascular endothelium activated by shear stress: Relevance to atherosclerosis. *Curr Opin Lipidol, 11,* 176–177.

53. Topper, J. N., Cai, J., Falb, D., et al. (1996). Identification of vascular endothelial genes differentially responsive to fluid mechanical stimuli: Cyclooxygenase-2, manganese superoxide dismutase and endothelial cell nitric oxide synthase are selectively up-regulated by steady laminar shear stress. *Proc Natl Acad Sci USA, 93,* 10417–10422.

54. Lusis, A. J. (2000). Atherosclerosis. *Nature, 407,* 233–241.

55. Libby, P., & Hansson, G. K. (1991). Involvement of the immune system in human atherogenesis: Current knowledge and unanswered questions. *Lab Invest, 64,* 5–15.

56. Khoo, J. C., Miller, E., McLoughlin, P., et al. (1988). Enhanced macrophage uptake of low density lipoprotein after self-aggregation. *Arteriosclerosis, 8,* 348–358.

57. Khoo, J. C., Miller, E., Pio, F., et al. (1992). Monoclonal antibodies against LDL further enhance macrophage uptake of LDL aggregates. *Arterioscler Thromb, 12,* 1258–1326.

58. Munn, D. H., & Cheung, N. K. (1989). Antibody-dependent antitumor cytotoxicity by human monocytes cultured with recombinant macrophage colony-stimulating factor. Induction of efficient antibody-mediated antitumor cytotoxicity not detected by isotope release assays. *J Exp Med, 170,* 511–532.

59. Marsh, C. B., Winnard, A. V., Mazzaferri, Jr., E. L., et al. (1997). Immune complexes induce monocyte maturation through MCSF production. *Am J Resp Crit Care Med, 155,* A682.

60. Tushinski, R. J., Oliver, I. T., Guilbert, L. J., et al. (1982). Survival of mononuclear phagocytes depends on a lineage specific growth factor that the differentiated cells selectively destroy. *Cell, 28,* 71–81.

61. Marsh, C. B., Pomerantz, R. P., Parker, J. M., et al. (1999). Regulation of monocyte survival in vitro by deposited IgG: Role of macrophage colony-stimulating factor. *J Immunol, 162,* 6217–6225.

62. Faruqi, R. M., & DiCorleto, P. E. (1993). Mechanisms of monocyte recruitment and accumulation. *Br Heart J, 69,* S19–S29.

63. Adams, D. O., & Hamilton, T. A. (1984). The cell biology of macrophage activation. *Annu Rev Immunol, 2,* 283–318.

64. Muller, W. A., Weigl, S. A., Deng, X., et al. (1993). PECAM-1 is required for transendothelial migration of leukocytes. *J Exp Med, 178,* 449–460.

65. Libby, P. (2000). Changing concepts of atherogenesis. *J Int Med, 247,* 349–358.

66. Springer, T. A., & Cybulsky, M. I. (1996). Traffic signals on endothelium for leukocytes in health, inflammation, and

atherosclerosis. In V. Fuster, R. Ross, & E. J. Topol (Eds.), *Atherosclerosis and coronary artery disease* (Vol. 1, pp. 511–538). Philadelphia: Lippincott-Raven.

67. O'Brien, K., McDonald, T. O., Chait, A., et al. (1996). Neovascular expression of E-selectin, intracellular adhesion molecule-1, and vascular cell adhesion molecule-1 in human atherosclerosis and their relation to intimal leukocyte content. *Circulation. 93,* 672–682.

68. Allen, S., Khan, S., Al-Mohanna, F., et al. (1998). Native low density lipoproteins-induced calcium transients trigger VCAM-1 and E-selectin expression in cultured human vascular endothelial cells. *J Clin Invest, 101,* 1064–1075.

69. Printseva, O.-Y., Peclo, M. M., & Gown, A. M. (1992). Various cell types in human atherosclerotic lesions express ICAM-1. *Am J Pathol, 140,* 889–896.

70. Davies, M. J., Gordon, J. L., Gearing, A. J. H., et al. (1993). The expression of the adhesion molecules ICAM-1, VCAM1, PECAM, and E-selectin in human atherosclerosis. *J Pathol, 171,* 223–229.

71. Bobryshev, Y. V., Lord, R. S. A., Rainer, S. P., et al. (1996). VCAM-1 expression and network of VCAM-1 positive vascular dendritic cells in advanced atherosclerotic lesions of carotid arteries and aortas. *Acta Histochem, 98,* 185–194.

72. Braun, M., Pietsch, P., Felix, S. B., et al. (1995). Modulation of intracellular adhesion molecule-1 and vascular cell adhesion molecule-1 on human coronary smooth muscle cells by cytokines. *J Mol Cell Cardiol, 27,* 2571–2579.

73. Corbi, A. L., Kishimoto, T. K., Miller, L. J., et al. (1988). The human leukocyte adhesion glycoprotein Mac-1 (complement receptor type 3, CD11b) alpha subunit. Cloning, primary structure, and relation to the integrins, von Willebrand factor and factor B. *J Biol Chem, 323,* 12403–12411.

74. Bombeli, T., Schwartz, B. R., & Harlan, J. M. (1998). Adhesion of activated platelets to endothelial cells: Evidence for a GPIIb-IIIa-dependent bridging mechanism and novel roles for endothelial intercellular adhesion molecule 1 (ICAM-1), $(\alpha)v(\beta)3$ integrin, and GPIb$(\alpha)$. *J Exp Med, 187,* 329–339.

75. Shyy, Y. J., Wickham, L. L., Hagan, J. P., et al. (1993). Human monocyte colony-stimulating factor stimulates the gene expression of monocyte chemotactic protein-1 and increases the adhesion of monocytes to endothelial monolayers. *J Clin Invest, 92,* 1745–1751.

76. Dickfeld, T., Lengyel, E., May, A. E., et al. (2001). Transient interaction of activated platelets with endothelial cells induces expression of monocytes-chemoattractant protein-1 via a p38 mitogen-activated protein kinase mediated pathway. Implication for atherogenesis. *Cardiovasc Res, 49*(1), 189–199.

77. Weyrich, A. S., McIntyre, T. M., McEver, R. P., et al. (1995). Monocyte tethering by P-selectin regulates monocyte chemotactic protein-1 and tumor necrosis factor-alpha secretion. Signal integration and NF-kappa β translocation. *J Clin Invest, 95,* 2297–2303.

78. Glass, C. K., & Witztum, J. L. (2001). Atherosclerosis: The road ahead. *Cell, 104,* 503–516.

79. Hollenbaugh, D., Mischel-Petty, N., Edwards, C. P., et al. (1995). Expression of functional CD40 by vascular endothelial cells. *J Exp Med, 182,* 33–40.

80. Henn, V., Slupsky, J. R., Grafe, M., et al. (1998). CD40 ligand on activated platelets triggers an inflammatory reaction of endothelial cells. *Nature, 391*(6667), 591–594.

81. Gawaz, M., Neumann, F. J., Dickfeld, T., et al. (1998). Activated platelets induce monocyte chemotactic protein-1 secretion and surface expression of intercellular adhesion molecule-1 on endothelial cells. *Circulation, 98*(12), 1164–1171.

82. Becker, B. F., Heindl, B., Kupatt, C., et al. (2000). Endothelial function and hemostasis. *Z Kardiol, 89,* 160–167.

83. Simon, D. I., Chen, Z., Xu, H., et al. (2000). Platelet glycoprotein Ibalpha is a counterreceptor for the leukocyte integrin Mac-1 (CD11b/CD18). *J Exp Med, 17,* 193–204.

84. Furman, M. I., Benoit, S. E., Barnard, M. R., et al. (1998). Increased platelet reactivity and circulating monocyte-platelet aggregates in patients with stable coronary artery disease. *J Am Coll Cardiol, 31,* 352–358.

85. Michelson, A. D., Barnard, M. R., Krueger, L. A., et al. (2001). Circulating monocyte-platelet aggregates are a more sensitive marker of in vivo platelet activation than platelet surface P-selectin: Studies in baboons, human coronary intervention, and human acute myocardial infarction. *Circulation, 104,* 1533–1537.

86. Clinton, S. K., Underwood, R., Hayes, L., et al. (1992). Macrophage colony-stimulating factor gene expression in vascular cells and in experimental and human atherosclerosis. *Am J Pathol, 140,* 301–316.

87. Rajavashisth, T. B., Andalibi, A., Territo, M. C., et al. (1990). Induction of endothelial cell expression of granulocyte and macrophage colony-stimulating factors by modified low-density lipoproteins. *Nature, 344,* 254–257.

88. Witztum, J. L. (1997). Immunological response to oxidized LDL. *Atherosclerosis, 131*(suppl.), S9–S11.

89. van der Wal, A. C., Das, P. K., Bentz van de Berg, D., et al. (1989). Atherosclerotic lesions in humans: In situ immunophenotypic analysis suggesting an immune mediated response. *Lab Invest, 61,* 166–170.

90. Hansson, G. K., & Libby, P. (1996). The role of the lymphocyte. In V. Fuster, R. Ross, & E. J. Topol, (Eds.), *Atherosclerosis and coronary artery disease* (Vol. 1, pp. 557–568). Philadelphia: Lippincott-Raven.

91. Bazzoni, G., Dejana, E., & Del Maschio, A. (1991). Platelet-neutrophil interactions. Possible relevance in the pathogenesis of thrombosis and inflammation. *Haematologica, 76,* 491–499.

92. Leo, R., Pratico, D., Iuliano, L., et al. (1997). Platelet activation by superoxide anion and hydroxyl radicals intrinsically generated by platelets that had undergone anoxia and then reoxygenated. *Circulation, 95,* 885–891.

93. Tajima, M., & Sakagami, H. (2000). Tetrahydrobiopterin impairs the action of endothelial nitric oxide via superoxide derived from platelets. *Br J Pharmacol, 131,* 958–64.

94. Seno, T., Inoue, N., Gao, D., et al. (2001). Involvement of NADH/NADPH oxidase in human platelet ROS production. *Thromb Res, 103,* 399–409.

95. Gorlach, A., Brandes, R. P., Bassus, S., et al. (2000). Oxidative stress and expression of p22phox are involved in the up-regulation of tissue factor in vascular smooth muscle cells in response to activated platelets. *FASEB J, 14,* 1518–1528.

96. Herkert, O., Diebold, I., Brandes, R. P., et al. (2002). NADPH oxidase mediates tissue factor-dependent surface procoagulant activity by thrombin in human vascular smooth muscle cells. *Circulation, 105,* 2030–2036.

97. O'Neill, M. S., Veves, A., Zanobetti, A., et al. (2005). Diabetes enhances vulnerability to participate air pollution-associated impairment in vasoreactivity and endothelial function. *Circulation, 111,* 2913–2920.

98. Takaya, N., Yuan, C., Chu, B., et al. (2005). Presence of intraplaque hemorrhage stimulated progression of carotid atherosclerosis plaques. *Circulation, 111,* 2768–2775.

99. Kockx, M. M., Cromheeke, K. M., Knaapen, M. W., et al. (2003). Phagocytosis and macrophage activation associated with hemorrhagic microvessels in human atherosclerosis. *Arterioscler Thromb Vasc Biol, 23,* 440–446.

100. Gimbrone, Jr., M. A. (1999). Vascular endothelium, hemodynamic forces, and atherogenesis. *Am J Pathol, 155,* 1–5.

101. Lin, M. C., Almus-Jacobs, F., & Chen, H. H., et al. (1997). Shear stress induction of the tissue factor gene. *J Clin Invest, 99,* 737–744.

102. Nagel, T., Resnick, N., Atkinson, W. J., et al. (1994). Shear stress selectively upregulates intercellular adhesion molecule-1 expression in cultured human vascular endothelial cells. *J Clin Invest, 94,* 885–891.

103. Traub, O., & Berk, B. C. (1998). Laminar shear stress: Mechanisms by which endothelial cells transduce an atheroprotective force. *Arterioscler Thromb Vasc Biol, 18,* 677–685.

104. Navab, M., Berliner, J. A., Watson, A. D., et al. (1996). The Yin and Yang of oxidation in the development of the fatty streak: A review based on the 1994 George Lyman Duff Memorial Lecture. *Arterioscler Thromb Vasc Biol, 16,* 831–842.

105. Berk, B. C., Alexander, R. W., Brock, T. A., et al. (1986). Vasoconstriction: A new activity for platelet-derived growth factor. *Science, 232,* 87–90.

106. Ross, R., & Glomset, J. A. (1963). Atherosclerosis and the arterial smooth muscle cell: Proliferation of smooth muscle is key event in the genesis of the lesions of atherosclerosis. *Science, 180,* 1332–1339.

107. Filonzi, E. L., Zoellner, H., Stanton, H., et al. (1993). Cytokine regulation of granulocyte-macrophage colony stimulating factor and macrophage-colony stimulating factor production in human arterial smooth muscle cells. *Atherosclerosis, 99,* 241–252.

108. Forgione, M. A., Leopold, J. A., & Loscalzo, J. (2000). Roles of endothelial dysfunction in coronary artery disease. *Curr Opin Cardiol, 15,* 409–415.

109. Moreno, P. R., Falk, E., & Palacios, I. F., et al. (1994). Macrophage infiltration in acute coronary syndromes. Implications for plaque rupture. *Circulation, 90,* 775–778.

110. Rosenfeld, M. E., & Ross, R. (1990). Macrophage and smooth muscle cell proliferation in atherosclerotic lesions of WHHL and comparably hypercholesterolemic fat-fed rabbits. *Arteriosclerosis, 10,* 680–687.

111. Shah, P. K., Falk, E., Badimon, J. J., et al. (1995). Human monocyte-derived macrophages induce collagen breakdown in fibrous caps of atherosclerotic plaques: Potential role of matrix-degrading metalloproteinases and implications for plaque rupture. *Circulation, 92,* 1565–1569.

112. Ginsberg, M. H., Loftus, J., & Plow, E. F. (1988). Platelets and the adhesion receptor superfamily. *Prog Clin Biol Res, 283,* 171–195.

113. Andrews, R. K., Shen, Y., Gardiner, E. E., et al. (2000). Platelets adhere to and translocate on von Willebrand factor presented by endothelium in stimulated veins. *Blood 96,* 3322–3328.

114. Berndt, M. C., Shen, Y., Dopheide, S. M., et al. (2001). The vascular biology of the glycoprotein Ib-IX-V complex. *Thromb Haemost, 86,* 178–188.

115. Sadler, J. E. (1998). Biochemistry and genetics of von Willebrand factor. *Annu Rev Biochem, 67,* 395–424.

116. de Leeuw, H. P., Fernandez-Borja, M., Reits, E. A., et al. (2001). Small GTP-binding protein Ral modulates regulated exocytosis of von Willebrand factor by endothelial cells. *Arterioscler Thromb Vasc Biol, 21,* 899–904.

117. Shen, Y., Romo, G. M., Dong, J. F., et al. (2000). Requirement of leucine-rich repeats of glycoprotein (GP) Ibalpha for shear-dependent and static binding of von Willebrand factor to the platelet membrane GP Ib-IX-V complex. *Blood, 95*(3), 903–910.

118. Arribas, M., & Cutler, D. F. (2000). Weibel–Palade body membrane proteins exhibit differential trafficking after exocytosis in endothelial cells. *Traffic, 1,* 783–793.

119. Romo, G. M., Dong, J. F., Schade, A. J., et al. (1999). The glycoprotein Ib-IX-V complex is a platelet counterreceptor for P-selectin. *J Exp Med, 190,* 803–814.

120. Sun, R. J., Muller, S., Wang, X., et al. (2000). Regulation of von Willebrand factor of human endothelial cells exposed to laminar flow: An *in vitro* study. *Clin Hemorheol Microcirc, 23*(1), 1–11.

121. Gibbins, J. M., Okuma, M., Farndale, R., et al. (1997). Glycoprotein VI is the collagen receptor in platelets which underlies tyrosine phosphorylation of the Fc receptor γ chain. *FEBS Lett, 413,* 255–259.

122. Alexander, R. W., Cooper, B., & Handin, R. I. (1978). Characterization of the human platelet α-adrenergic receptor. Correlation of $^3$H dihydroergocryptine binding with aggregation and adenylate cyclase inhibition. *J Clin Invest, 61,* 1136–1144.

123. Hollopeter, G., Jantzen, H. M., Vincent, D., et al. (2001). Identification of the platelet ADP receptor targeted by antithrombotic drugs. *Nature, 409,* 202–207.

124. Vu, T. K., Hung, D. T., Wheaton, V. I., et al. (1991). Molecular cloning of a functional thrombin receptor reveals a novel proteolytic mechanism of receptor activation. *Cell, 64,* 1057–1068.

125. Freedman, J. E., Loscalzo, J., Benoit, S. E., et al. (1996). Decreased platelet inhibition by nitric oxide in two brothers with a history of arterial thrombosis. *J Clin Invest, 97,* 979.

126. Ware, J. A., & Heistad, D. D. (1993). Platelet-endothelium interactions. N *Engl J Med, 328,* 628.

127. Henn, V., Steinbach, S., Buchner, K., et al. (2001). The inflammatory action of CD40 ligand (CD154) expressed on activated human platelets is temporally limited by coexpressed CD40. *Blood, 98,* 1047–1054.

128. Ross, R., Masuda, J., Raine, E. W., et al. (1990). Localization of PDGF-B protein in macrophages in all phases of atherogenesis. *Science, 248,* 1009–1012.

129. Badimon, J. J., Meyer, B., Feigen, L. P., et al. (1997). Thrombosis triggered by severe arterial lesions is inhibited by oral administration of a glycoprotein IIb/IIIa antagonist. *Eur J Clin Invest, 27,* 568–574.

130. Lincoff, A. M., Califf, R. M., & Topol, E. J. (2000). Platelet glycoprotein IIb/IIIa receptor blockade in coronary artery disease. *J Am Coll Cardiol, 35,* 1103–1115.

131. Topol, E. J., Moliterno, D. J., & Hermann, H. C., & TARGET investigators. (2001). Comparison of two platelet glycoprotein IIb/IIIa inhibitors, Tirofiban and abciximab, for the prevention of ischaemic events with percutaneous coronary revascularization. *N Eng J Med, 44*(25), 1937–1939.

132. Ridker, P. M., Cushman, M., Stampfer, M. J., et al. (1997). Inflammation, aspirin, and the risk of cardiovascular disease in apparently healthy men. *N Engl J Med, 336,* 973–979.

133. The ESPRIT Investigators. (2000). Novel dosing regimen of eptifibatide in planned coronary stent implantation (ESPRIT): A randomized, placebo-controlled trial. Enhanced Suppression of the Platelet IIb/IIIa Receptor with Integrilin Therapy. *Lancet, 356,* 2037–2044.

134. Boersma, E., Harrington, R. A., Moliterno, D. J., et al. (2002). Platelet glycoprotein IIb/IIIa inhibitors in acute coronary syndromes: A meta-analysis of all major randomised clinical trials. *Lancet, 359,* 189–198.

135. Gawaz, M., Langer, H., & May, A. E. (2005). Platelets in inflammation and atherogenesis. *J Clin Invest, 115,* 3378–3384.

136. Weber, C. (2005). Platelets and chemokines in atherosclerosis. Partners in crime. *Circ Res, 96,* 612–616.

137. Lee, R. T., & Libby, P. (1997). The unstable atheroma. *Arterioscler Thromb Vasc Biol, 17,* 1859–1867.

138. Farb, A., Burke, A. P., Tang, A. L., et al. (1996). Coronary plaque erosion without rupture into a lipid core. A frequent cause of coronary thrombosis in sudden coronary death. *Circulation, 93,* 1354–1363.

139. Schwartz, R. S. (1998). Pathophysiology of restenosis: Interaction of thrombosis, hyperplasia, and/or remodeling. *Am J Cardiol, 81*(7A), 14E–17E.

140. Levine, G. N., Chodos, A. P., & Loscalzo, J. (1995). Restenosis following coronary angioplasty: Clinical presentations and therapeutic options. *Clin Cardiol, 18*(12), 693–703.

141. Davies, M. J., et al. (1984). Thrombosis and acute coronary-artery lesions in sudden cardiac ischemic death. *N Engl J Med, 310,* 1137–1140.

142. Gutstein, D. E., & Fuster, V. (1999). Pathophysiology and clinical significance of atherosclerotic plaque rupture. *Cardiovasc Res, 41,* 323–333.

143. Falk, E., Shah, P. K., & Fuster, V. (1995). Coronary plaque disruption. *Circulation, 92,* 657–671.

144. Zhou, J., Chew, M., Ravn, H. B., et al. (1999). Plaque pathology and coronary thrombosis in the pathogenesis of acute coronary syndromes. *Scand J Clin Lab Invest, 230*(Suppl.), 3–11.

145. Falk, E., Shah, P. K., & Fuster, V. (1996). Pathogenesis of plaque disruption. In V. Fuster, R. Ross, & E. J. Topol (Eds.), *Atherosclerosis and coronary artery disease* (Vol. 2, pp. 492–510). Philadelphia: Lippincott-Raven.

146. Bouch, D. C., & Montgomery, G. L. (1970). Cardiac lesions in fatal cases of recent myocardial ischaemia from a coronary care unit. *Br Heart J, 32,* 795–803.

147. Falk, E., Shah, P. K., & Fuster, V. (1995). Coronary plaque disruption. *Circulation, 92,* 657–671.

148. Shepherd, J., Cobbe, S. M., Ford, I., et al. (1995). Prevention of coronary heart disease with pravastatin in men with hypercholesterolemia. West of Scotland Coronary Prevention Study Group. *N Engl J Med, 333,* 1301–1307.

149. Plehn, J. F., Davis, B. R., Sacks, F. M., et al. (1999). Reduction of stroke after myocardial infarction with Pravastatin in the Cholesterol and Recurrent Events (CARE) study. *Circulation, 99,* 216–223.

150. Sacks, F. M., Pfeffer, M. A., Moye, L. A., et al. (1996). The effect of Pravastatin on coronary events after myocardial infarction in patients with average cholesterol levels. *N Engl J Med, 335*(14), 1001–1009.

151. Endo, A. L. (1992). The discovery and development of HMG-CoA reductase inhibitors. *J Lipid Res, 33,* 1569.

152. Downs, J. R., Clearfield, M., Weis, S., et al. (1998). Primary prevention of acute coronary events with lovastatin in men and women with average cholesterol levels: Results of AFCAPS/TexCAPS. Air Force/Texas Coronary Atherosclerosis Prevention Study. *JAMA, 279,* 1615–1622.

153. LIPID Study Group. (1998). Prevention of cardiovascular events and death with pravastatin in patients with coronary heart disease and a broad range of initial cholesterol levels. The Long-Term Intervention with Pravastatin in Ischaemic Disease. *N Engl J Med, 339,* 1349–1357.

154. Maron, D. J., Fazio, S., & Linton, M. F. (2000). Current perspectives on statins. *Circulation, 101,* 207.

155. Redersen, T. R., Kjekshus, J., Berg, K., et al. (1996). Cholesterol-lowering and the use of health care resources: Results of the Scandinavian Sinvastatin Survival Study (4S). *Circulation, 93,* 1796–1802.

156. Canner, P. L., Berge, K. G., Wenger, N. K., et al. (1986). Fifteen-year mortality in Coronary Drug Project Patients: Long term benefit with niacin. *J Am Coll Cardiol, 8,* 1245–1255.

157. Brown, B. G., Zhao, X. Q., Sacco, D. E., et al. (1993). Lipid lowering and plaque regression. New insights into prevention of plaque disruption and clinical events in coronary disease. *Circulation, 87,* 1781–1791.

158. Sacks, F. M., Moye, L. A., Davis, B. R., et al. (1998). Relationship between plasma LDL concentrations during treatment with pravastatin and recurrent coronary events in the Cholesterol and Recurrent Events trial. *Circulation, 97,* 1446–1452.

159. Moreno, P. R., Falk, E., Palacios, I. F., et al. (1994). Macrophage infiltration in acute coronary syndromes: Implications for plaque rupture. *Circulation, 90,* 775–778.

160. Van der Wal, A. C., Becker, A. E., van der Loos, C. M., et al. (1994). Site of intimal rupture or erosion of thrombosed coronary atherosclerotic plaques is characterized by an inflammatory process irrespective of the dominant plaque morphology. *Circulation, 84,* 36–44.

161. Van der Wal, A. C., Becker, A. E., Tigges, A. J., et al. (1994). Fibrous and lipid-rich atherosclerotic plaques are part of interchangeable morphologies related to inflammation. *Coron Artery Dis, 5,* 463–469.

162. Davies, M. J. (1990). A macro and micro view of coronary vascular insult in ischemic heart disease. *Circulation, 82*(Suppl. II), II-38–II-46.

163. Kovanen, P. T., Kaartinen, M., Paavonen, T., et al. (1995). Infiltrates of activated mast cells at the site of coronary atheromatous erosion or rupture in myocardial infarction. *Circulation, 92,* 1084–1088.

164. Ross, R. (1993). Atherosclerosis: A defense mechanism gone awry. *Am J Pathol, 143,* 987–1002.

165. Frid, M. G., Aldashev, A. A., Dempsey, E. C., et al. (1997). Smooth muscle cells isolated from discrete compartments of the mature vascular media exhibit unique phenotypes and distinct growth capabilities. *Circ Res, 81,* 940–952.

166. Davies, M. J. (2001). Going from immutable to mutable atherosclerotic plaques. *Am J Cardiol, 88*(suppl), 2F–9F.

167. Babaev, V. R., Bobryshev, Y. V., Stenina, O. V., et al. (1990). Heterogeneity of smooth muscle cells in atheromatous plaque of human aorta. *Am J Pathol, 136,* 1031–1042.

168. Smith, R. E., Hogaboam, C. M., Strieter, R. M., et al. (1997). Cell-to-cell and cell-to-matrix interactions mediate chemokine expression: An important component of the inflammatory lesion. *J Leukoc Biol, 62,* 612–619.

169. Bauriedel, G., Hutter, R., Welsch, U., et al. (1999). Role of smooth muscle cell death in advanced coronary primary lesions: Implications for plaque instability. *Cardiovasc Res, 41,* 480–488.

170. Shan, P. K., Falk, E., Badimon, J. I., et al. (1995). Human monocyte-derived macrophages induce collagen breakdown in fibrous caps of atherosclerotic plaques. Potential role of matrix-degrading metalloproteinases and implications for plaque rupture. *Circulation, 130,* 386–396.

171. Davies, M. J., Richardson, P., Woolf, N., et al. (1993). Risk of thrombosis in human atherosclerotic plaque: Role of extracellular lipid, macrophages, and smooth muscle cell content. *Br Heart J, 69,* 377–381.

172. Wesley, R. B. II., Meng, X., Godin, D., et al. (1998). Extracellular matrix modulates macrophage functions characteristic to atheroma: Collagen type I enhances acquisition of resident macrophage traits by human peripheral blood monocytes in vitro. *Arterioscler Thromb Vasc Biol, 18,* 432–440.

173. Moss, M. L., Jin, S.-L. C., Milla, M. E., et al. (1997). Cloning of a disintegrin metalloproteinase that processes precursor tumour necrosis factor-α. *Nature, 385,* 733–736.

174. Dollery, C. M., McEwan, J. R., & Henney, A. M. (1995). Matrix metalloproteinases and cardiovascular disease. *Circ Res, 77,* 863–868.

175. Herren, B., Raines, E. W., & Ross, R. (1997). Expression of a disintegrin-like protein in cultured human vascular cells and in vivo. *FASEB J, 11,* 173–180.

176. Black, R. A., Rauch, C. T., Kozlosky, C. J., et al. (1997). A metalloproteinase disintegrin that releases tumor-necrosis factor-α from cells. *Nature, 385,* 729–733.

177. Lendon, C. L., Davies, M. J., Born, G. V., et al. (1991). Atherosclerotic plaque caps are locally weakened when macrophages density is increased. *Atherosclerosis, 87,* 87–90.

178. Toschi, V., Gallo, R., Lettino, M., et al. (1997). Tissue factor modulates the thrombogenicity of human atherosclerotic plaques. *Circulation, 95,* 594–599.

179. Penn, M. S., & Topol, E. J. (2001). Tissue factor, the emerging link between inflammation, thrombosis, and vascular remodeling. *Circ Res, 89,* 1–2.

180. Dupuis, J. (2001). Mechanisms of acute coronary syndromes and the potential role of statins. *Atheroscler Suppl, 2*(1), 9–14.

181. Jukema, J. W., Bruschke, A. V. G., van Boren, A. J., et al. (1995). Effects of lipid-lowering by pravastatin on progression and regression of coronary artery disease in symptomatic men with normal or moderately elevated cholesterol levels: The Regression Growth Evaluation Statin Study (REGRESS). *Circulation, 91,* 2528–2540.

182. La Rosa, J. C. (2001). Pleiotropic effects of statins and their clinical significance. *Am J Cardiol, 88,* 291–293.

183. Folkman, J. (1998). Is tissue mass regulated by vascular endothelial cells? Prostate as the first evidence. *Endocrinology, 139,* 441–442.

184. Moulton, K. S. (2001). Plaque angiogenesis and atherosclerosis. *Curr Atheroscler Rep, 3,* 225–233.

185. Geng, Y. J., Wu, Q., Muszynski, M., et al. (1996). Apoptosis of vascular smooth muscle cells induced by in vitro stimulation with interferon-γ, tumor necrosis factor-α, and interleukin-1 β. *Arterioscler Thromb Vasc Biol, 16,* 19–27.

186. Kockx, M. M., & Herman, A. G. (1998). Apoptosis in atherogenesis: Implications for plaque destabilization. *Eur Heart J, 19*(Suppl. G), G23–G28.

187. Isner, J., Kearney, M., Bortman, S., et al. (1995). Apoptosis in human atherosclerosis and restenosis. *Circulation, 91,* 2703–2711.

188. Galis, Z., Sukhova, G., Kranzhofer, M., et al. (1995). Macrophage foam cells from experimental atheroma constitutively produce matrix-degrading proteinases. *Proc Natl Acad Sci USA, 92,* 402–406.

189. Galis, Z. S., Sukhova, G. K., Lark, M. W., et al. (1994). Increased expression of matrix metalloproteinases and matrix degrading activity in vulnerable regions of human atherosclerotic plaques. *J Clin Invest, 94,* 2493–2503.

190. Newby, A. C., Southgate, K. M., & Davies, M. (1994). Extracellular matrix degrading metalloproteinases in the pathogenesis of arteriosclerosis. *Basic Res Cardiol, 89*(Suppl. 1), 59–70.

191. Koyama, H., Raines, E. W., Bornfeldt, K. E., et al. (1996). Fibrillar collagen inhibits arterial smooth muscle proliferation through regulation of Cdk2 inhibitors. *Cell, 87,* 1069–1078.

192. Dollery, C. M., McEwan, J. R., & Henney A. M. (1995). Matrix metalloproteinases and cardiovascular disease. *Circ Res, 77,* 863–868.

193. Assoian, R. K., & Marcantonio, E. E. (1996). The extracellular matrix as a cell cycle control element in atherosclerosis and restenosis. *J Clin Invest, 98,* 2436–2439.

194. Mercurius, K. O., & Morla, A. O. (1998). Inhibition of vascular smooth muscle cell growth by inhibition of fibronectin matrix assembly. *Circ Res, 82,* 548–556.

195. Frisch, S. M., Vuori, K., Ruoslahti, E., et al. (1996). Control of adhesion-dependent cell survival by focal adhesion kinase. *J Cell Biol, 134,* 793–799.

196. Geng, Y. J., Henderson, L. E., Levesque, E. B., et al. (1997). Fas is expressed in human atherosclerotic intima and promotes apoptosis of cytokine-primed human vascular smooth muscle cells. *Arterioscler Thromb Vasc Biol, 17,* 2200–2208.

197. Moldovan, N. I., Qian, Z., Dong, C., et al. (1998). Fas-mediated apoptosis in accelerated graft arteriosclerosis. *Angiogenesis, 2,* 245–254.

198. Dong, C., Wilson, J. E., Winters, G. L., et al. (1996). Human transplant coronary artery disease: Pathological evidence for Fas-mediated apoptotic cytotoxicity in allograft arteriopathy. *Lab Invest, 74,* 921–931.

199. Perlman, H., Maillard, L., Krasinski, K., et al. (1997). Evidence for the rapid onset of apoptosis in medial smooth muscle cells after balloon injury. *Circulation, 95,* 981–987.

200. Sata, M., & Walsh, K. (1998). Oxidized LDL activates Fas-mediated endothelial cell apoptosis. *J Clin Invest, 102,* 1682–1689.

201. Bremner, T. A., Chatterjee, D., Han, Z., et al. (1999). THP-1 monocytic leukemia cells express Fas ligand constitutively and kill Fas-positive Jurkat cells. *Leuk Res, 23,* 865–870.

202. Seshiah, P. N., Kereiakes, D. J., Vasudevan, S. S., et al. (2002). Activated monocytes induce smooth muscle cell death: Role of macrophage colony-stimulating factor and cell contact. *Circulation, 105,* 174–180.

203. Qiao, J.-H., Tripathi, J., Mishra, N. K., et al. (1997). Role of macrophage colony-stimulating factor in atherosclerosis: Studies of osteopetrotic mice. *Am J Pathol, 150,* 1687–1699.

204. Simon, D. I., Xu, H., Ortlepp, S., et al. (1997). 7E3 monoclonal antibody directed against the platelet glycoprotein IIb/IIIa cross-react with the leukocyte integrin Mac-1 and blocks adhesion to fibrinogen and ICAM-1. *Arterioscler Thromb Vasc Biol, 17,* 528–535.

205. Nakashima, Y., Plump, A. S., Raines, E. W., et al. (1994). ApoE-deficient mice develop lesions of all phases of atherosclerosis throughout the arterial tree. *Arterioscler Thromb, 14,* 133–140.

206. Lee, R. T., & Libby, P. (1997). The unstable atheroma. *Arterioscler Thromb Vasc Biol, 17,* 1859–1867.

207. Belanger, A. J., Scaria, A., Lu, H., et al. (2001). Fas ligand/Fas-mediated apoptosis in human coronary artery smooth muscle cells: Therapeutic implications of fratricidal mode of action. *Cardiovasc Res, 51,* 749–761.

208. Geng, Y. J. (2001). Biologic effect and molecular regulation of vascular apoptosis in atherosclerosis. *Curr Atheroscler Rep, 3,* 234–242.

209. Kerr, J. R. F., Wyllie, A. H., & Curie, A. R. (1972). Apoptosis: A basic biological phenomenon with wide-ranging implications in tissue kinetics. *Br J Cancer, 26,* 239–257.

210. Tabas, I. (2004). Apoptosis and plaque destabilization in atherosclerosis: The role of macrophage apoptosis induced by cholesterol. *Cell Death Differ, 11,* S12-S16.

211. Leytin, V., Allen, D., Mykhaylov, S., et al. (2005). Pathologic high shear stress induces apoptosis events in human platelets. *Biochem Biophys Res Comm, 320,* 303–310.

212. Bogaty, P., Haeckett, D., Davies, G., et al. (1994). Vasoreactivity of the culprit lesion in unstable angina. *Circulation, 90,* 5–11.

213. Nowicki, P. T., Flavahan, S., Hassanain, H., et al. (2001). Redox signaling of the arteriolar myogenic response. *Circ Res, 89,* 114–116.

214. Chobanian, A. V., & Dzau, V. J. (1996). Renin angiotensin system and atherosclerotic vascular disease. In V. Fuster, R. Ross, & E. J. Topol (Eds.), *Atherosclerosis and coronary artery disease* (Vol. 1, pp. 237–242). Philadelphia: Lippincott-Raven.

215. Rajagopalan, S., Kurz, S., Munzel, T., et al. (1996). Angiotensin II-mediated hypertension in the rat increases vascular superoxide production via membrane NADH/NADPH oxidase activation. Contribution to alterations of vasomotor tone. *J Clin Invest, 97,* 1916–1923.

216. Schmidt-Ott, K. M., Kagiyama, S., & Phillips, M. I. (2000). The multiple actions of angiotensin II in atherosclerosis. *Regul Pept, 93,* 65–77.

217. Griendling, K. K., Minieri, C. A., Ollerenshaw, J. D., et al. (1994). Angiotensin II stimulates NADH and NADPH oxidase activity in cultured vascular smooth muscle cells. *Circ Res, 74,* 1141–1148.

218. Weiss, D., Sorescu, D., & Taylor, W. R. (2001). Angiotensin II and atherosclerosis. *Am J Cardiol, 87,* 25C–32C.

219. Achard, J., Fournier, A., Mazouz, H., et al. (2001). Protection against ischemia: A physiological function of the renin-angiotensin system. *Biochem Pharmacol, 62,* 321–271.

220. Shlipak, M. G., Browner, W. S., Noguchi, H., et al. (2001). Comparison of the effects of angiotensin converting-enzyme inhibitors and beta blockers on survival in elderly patients with reduced left ventricular function after myocardial infarction. *Am J Med, 110,* 425–433.

221. Mallat, Z., & Tedgui, A. (2001). Current perspective on the role of apoptosis in atherothrombotic disease. *Circ Res, 88,* 998–1003.

222. Brown, N. J., & Vaughan, D. E. (2000). Prothrombotic effects of angiotensin. *Adv Intern Med, 45,* 419–429.

223. Dong, C., Zhu, S., Alvarez, R. J., et al. (2001). Angiotensin II induces PAI-1 expression through MAP kinase-dependent, but TGF beta and PI3 kinase-independent pathway. *J Heart Lung Transplant, 20,* 232–227.

224. Dong, C., Zhu, S., Yoon, W., et al. (2001). Upregulation of PAI-1 is mediated through TGF beta/SMAD pathway in transplant arteriopathy. *J Heart Lung Transplant, 20,* 219.

225. Fukao, H., Ueshima, S., Okada, K., et al. (2000). Binding of mutant tissue-type plasminogen activators to human endo-

thelial cells and their extracellular matrix. *Life Sci, 66,* 2473–2487.

226. Newby, D. E., McLeod, A. L., Uren, N. G., et al. (2001). Impaired coronary tissue plasminogen activator release is associated with coronary atherosclerosis and cigarette smoking: Direct link between endothelial dysfunction and atherothrombosis. *Circulation, 103,* 1936–1941.

227. Hevey, D., McGee, H. M., Fitzgerald, D., et al. (2000). Acute psychological stress decreases plasma tissue plasminogen activator (tPA) and tissue plasminogen activator/plasminogen activator inhibitor-1 (tPA/PAI-1) complexes in cardiac patients. *Eur J Appl Physiol, 83,* 344–348.

228. Wiman, B., Andersson, T., Hallqvist, J., et al. (2000). Plasma levels of tissue plasminogen activator/plasminogen activator inhibitor-1 complex and von Willebrand factor are significant risk markers for recurrent myocardial infarction in the Stockholm Heart Epidemiology Program (SHEEP) study. *Arterioscler Thromb Vasc Biol, 20,* 2019–2023.

229. Yamada, S., Yamada, R., Ishii, A., et al. (1996). Evaluation of tissue plasminogen activator and plasminogen activator inhibitor-1 levels in acute myocardial infarction. *J Cardiol, 27,* 171–178.

230. Falcone, D. J., McCaffrey, T. A., & Vergilio, J. A. (1991). Stimulation of macrophage urokinase expression by polyanions is protein kinase C–dependent and requires protein and RNA synthesis. *J Biol Chem, 266,* 22726–22732.

231. Chavakis, T., Kanse, S. M., Yutzy, B., et al. (1998). Vitronectin concentrates proteolytic activity on the cell surface and extracellular matrix by trapping soluble urokinase receptor–urokinase complexes. *Blood, 91,* 2305–2312.

232. Okura, Y., Brink, M., Itabe, H., et al. (2000). Oxidized low-density lipoprotein is associated with apoptosis of vascular smooth muscle cells in human atherosclerotic plaques. *Circulation, 28,* 2680–2686.

233. Hutter, R., Sauter, B. V., Reis, E. D., et al. (2003). Decreased reendothelialization and increased neointima formation with endostatin overexpression in a mouse model of arterial injury. *Circulation, 107,* 1658–1663.

234. Conti, E., Carrozza, C., Capoluongo, E., et al. (2004). IGF-1 as a vascular protective factor. *Circulation, 110,* 2260–2265.

235. Szmitko, P. E., Fedak, P. W. M., Weisel, R. D., et al. (2003). Endothelial progenitor cells. New hope for a broken heart. *Circulation, 107,* 3093–3100.

236. Heeschen, C., Aicher, A., Lehmann, R., et al. (2003). Erythropoietin is a potent physiologic stimulus for endothelial progenitor cells. *Blood, 102,* 1340–1346.

237. Aicher, A., Heeschen, C., Mildner-Rihm, C., et al. (2003). Essential role of endothelial nitric oxide synthase for mobilization of stem and progenitor cells. *Nat Med, 9,* 1370–1376.

238. Hill, J. M., Zalos, G., Halcox, J. P., et al. (2003). Circulating endothelial progenitor cells, vascular function, and cardiovascular risk. *N Engl J Med, 348,* 593–600.

239. Walter, D. H., Cejna, M., Diaz-Sandoval, L., et al. (2004). Local gene transfer of phVEGF-2 plasmid by gene-eluting stents. An alternative strategy for inhibition of restenosis. *Circulation, 110,* 36–45.

240. Andreotti, F., Crea, F., & Conti, E. (2004). Heart-kidney interactions in ischemic syndromes. *Circulation, 109,* e31–32.

241. Shintani, S., Murohara, T., Ikeda, H., et al. (2001). Mobilization of endothelial progenitor cells in patients with acute myocardial infarction. *Circulation, 103,* 2776–2779.

242. van der Meer, P., Voors, A. A., Lipsic, E., et al. (2004). Erythropoietin in cardiovascular diseases. *Eur Heart J, 25,* 285–291.

243. Schmidt-Lucke, C., Rssig, L., Fichtlscherer, S., et al. (2005). Reduced number of circulating endothelial progenitor cells predicts future cardiovascular events. *Circulation, 111,* 2981–2987.

244. Liu, B., Liao, C., Chen, J., et al. (2003). Significance of increasing adhesion of cord blood hematopoietic cells and a new method: Platelet microparticles. *Am J Hematol, 74,* 216–217.

245. Kim, H. K., Song, K. S., Chung, J. H., et al. (2004). Platelet microparticles induce angiogenesis in vitro. *Br J Haematol, 124,* 376–384.

246. The EPISTENT Investigators. (1998). Randomised placebo-controlled and balloon-angioplasty-controlled trial to assess safety of coronary stenting with use of platelet glycoprotein-IIb/IIIa blockade. Evaluation of Platelet IIb/IIIa Inhibitor for Stenting. *Lancet, 352,* 87–92.

247. Neumann, F. J., Schomig, A. (1998). Glycoprotein IIb/IIIa receptor blocked with coronary stent placement. *Semin Inv Cardiol, 3,* 81–90.

248. Anderson, K. M., Califf, R. M., Stone, G. W., et al. (2001). Long-term mortality benefits with Abciximab in patients undergoing percutaneous coronary intervention. *J Am Coll Cardiol, 37*(8), 2059–2065.

249. Topol, E. (2001). Recent advances in anticoagulant therapy for acute coronary syndromes. *Am Heart J, 142*(2 Pt 2), S22–S29.

250. Lincoff, A. M., Tcheng, J. E., Califf, R. M., et al., for the EPILOG Investigators. (1999). Sustained suppression of ischemic complications of coronary intervention by platelet GPIIb/IIIa blockade with abciximab: One year outcome in the EPILOG Trial. *Circulation, 99,* 1951–1958.

251. Topol, E. J., Mark, D. B., Lincoff, A. M., et al. (1999). Outcomes at 1 year and economic implications of platelet glycoprotein IIb/IIIa blockade in patients undergoing coronary stenting: Results from a multicentre randomised trial. EPISTENT Investigators. Evaluation of Platelet IIb/IIIa Inhibitor for Stenting. *Lancet, 354,* 2019–2024.

252. O'Shea, J. C., Buller, C. E., Cantor, W. J., et al. (2002). Long-term efficacy of platelet glycoprotein IIb/IIIa integrin blockade with eptifibatide in coronary stent intervention. *JAMA, 287,* 618–621.

253. Schwartz, R. S. (1998). Pathophysiology of restenosis: Interaction of thrombosis, hyperplasia, and/or remodeling. *Am J Cardiol, 81,* 14E–17E.

254. Levine, G. N., Chodos, A. P., & Loscalzo, J. (1995). Restenosis following coronary angioplasty: Clinical presentations and therapeutic options. *Clin Cardiol, 18,* 693–703.

255. Topol, E. J., & The GUSTO V Investigators. (2001). Reperfusion therapy for acute myocardial infarction with

fibrinolytic therapy or combination reduced fibrinolytic therapy and platelet glycoprotein IIb/IIIa inhibition: The GUSTO V randomised trial. *Lancet, 357,* 1905–1914.

256. Aronow, H. D., Topol, E. J., Roe, M. T., et al. (2001). Effect of lipid-lowering therapy on early mortality after acute coronary syndromes: An observational study. *Lancet, 357,* 1063–1068.

257. Fuster, V. V. (1999). Current issues of lipid-lowering and antithrombotic therapy. *Curr Cardiol Rep, 1,* 173–174.

258. Peterson, J. G., Topol, E. J., Sapp, S. K., et al. (2000). Evaluation of the effects of aspirin combined with angiotensin-converting enzyme inhibitors in patients with coronary artery disease. *Am J Med, 109,* 371–377.

259. Helft, G., Osende, J. I., Worthley, S. G., et al. (2000). Acute antithrombotic effect of a front-loaded regimen of clopidogrel in patients with atherosclerosis on aspirin. *Arterioscler Thromb Vasc Biol, 20,* 2316–2321.

260. Andrews, T. C., Raby, K., Pyrola, K., et al. (1997). Effect of cholesterol reduction on myocardial ischemia in patients with coronary disease. *Circulation, 95,* 324–328.

261. MacIntrye, K., Stewart, S., Capewell, S., et al. (2001). Gender and survival: A population-based study of 201,114 men and women following a first acute myocardial infarction. *J Am Coll Cardiol, 38,* 729–735.

262. Goldschmidt-Clermont, P. J, Shear, W. S., Schwartzberg, J., et al. (1996). Clues to the death of an Olympic champion. *Lancet, 347,* 1833.

263. Wilson, P. W., Castelli, W. P., & Kannel, W. B. (1987). Coronary risk predictor in adults (The Framingham Heart Study). *Am J Cardiol, 59,* 91G–94G.

264. Murata, M., Kawano, K., Matsubara, Y., et al. (1998). Genetic polymorphisms and risk of coronary artery disease. *Sem Thromb Hemost, 24,* 245–250.

265. Boerwinkle, E., Elsworth, D. L., Hallman, D. M., et al. (1996). Genetic analysis of atherosclerosis: A research paradigm for the common diseases. *Hum Mol Genet, 5,* 1405–1410.

266. Halushka, M. K., Fan, J. B., Benley, K., et al. (1999). Patterns of single nucleotide polymorphisms in candidate genes for BP homeostasis. *Nature Genet, 22,* 239–247.

267. Lockhart, D. J., & Winzeler, E. A. (2000). Genomics gene expression and DNA arrays. *Nature, 404,* 827–836.

268. Hopkins, P. N., Williams, R. R., Kuida, H., et al. (1998). Family history as an independent risk factor for incident coronary artery disease in a high-risk cohort in Utah. *Am J Cardiol, 62,* 703–707.

269. Nora, J. J., Lortscher, R. H., Spangler, R. D., et al. (1980). Genetic-epidemiologic study of early-onset ischemic heart disease. *Circulation, 61,* 503–508.

270. Beng, K. (1984). Twin studies on coronary heart disease and its risk factors. *Ac Genet Med Gemellol, 33,* 349–361.

271. Marenberg, M. E., Rish, N., Berkman, L. F., et al. (1994). Genetic susceptibility to death from coronary heart disease in a study of twins. *N Engl J Med, 330,* 1141–1146.

272. Goldstein, J. L., & Brown, M. S. (2001). Molecular medicine. The cholesterol quartet. *Science, 292,* 1310–1312.

273. Dong, C., Nevins, J. R., & Goldschmidt-Clermont, P. J. (2001). ABCA1 single nucleotide polymorphisms. Snipping at the pathogenesis of atherosclerosis. *Circ Res, 88,* 855–857.

274. Bertolini, S., Pisciotta, L., Seri, M., et al. (2001). A point mutation in ABC1 gene in a patient with severe premature coronary heart disease and mild clinical phenotype of Tangier disease. *Atherosclerosis, 154,* 599–605.

275. Lutucuta, S., Ballantyne, C. M., Elghannam, H., et al. (2001). Novel polymorphisms in promoter region of ATP-binding cassette transporter gene and plasma lipids, severity, progression, and regression of coronary atherosclerosis and response to therapy. *Circ Res, 88,* 969–973.

276. Oram, J. F., & Lawn, R. M. (2001). ABC1: The gatekeeper for eliminating excess tissue cholesterol. *J Lipid Res, 42,* 1173–1179.

277. Lapicka-Bodzioch, K., Bodzioch, M., Krull, M., et al. (2001). Homogeneous assay based on 52 primer sets to scan for mutations of the ABC1 gene and its application in genetic analysis of a new patient with familial high-density lipoprotein deficiency syndrome. *Biochim Biophys Acta, 27,* 42–48.

278. Orso, E., et al. (2000). Transport of lipids from Golgi to plasma membrane is defective in Tangier disease patients and ABC-1-deficient mice. *Nature Genet, 24,* 192–196.

279. Mudd, S. H., Skovby, F., Levy, H. L., et al. (1985). The natural history of homocystinuria due to cystathionine (β)-synthase deficiency. *Am J Hum Genet, 37,* 1–31.

280. Hinek, A., & Wilson, S. E. (2000). Impaired elastogenesis in Hurler disease: Dermatan sulfate accumulation linked to deficiency in elastin-binding protein and elastic fiber assembly. *Am J Pathol, 156*(3), 925–938.

281. Hobbs, H. H., Russell, D. W., Brown, M. S., et al. (1990). The LDL receptor locus in familial hypercholesterolemia: Mutational analysis of a membrane protein. *Annu Rev Genet, 24,* 133–170.

282. Faber, B., Cleutjens, K., Niessen, R., et al. (2001). Identification of genes potentially involved in rupture of human atherosclerotic plaques. *Circ Res, 89,* 547–554.

283. Carrill, M., et al. (1999). Characterization of single-nucleotide polymorphism in coding regions of human genes. *Nature Genet, 22,* 231–238.

284. Doevendans, P. A., Jukema, W., Spiering, W., et al. (2001). Molecular genetics and gene expression in atherosclerosis. *Int J Cardiol, 80,* 161–172.

285. Loots, G. G., Locksley, R. M., Blankespoor, C. M., et al. (2000). Identification of a coordinate regulator of interleukins 4, 13 and 5 by cross-species sequence comparisons. *Science, 288,* 136–140.

286. Chien, K. R. (2000). Genomic circuits and the integrative biology of cardiac disease. *Nature, 407,* 227–232.

287. Coller, B. S. (1997). Platelet GPIIb/IIIa antagonists: The first anti-integrin receptor therapeutics. *J Clin Invest, 100,* S57–S60.

288. Newman, P. J., Derbes, R., & Aster, R. H. (1989). The human platelet alloantigens, PLA1 and PLA2, are associated with a leucine33/proline33 amino acid polymorphism in membrane glycoprotein IIIa, and are distinguishable by DNA typing. *J Clin Invest, 83,* 1778–1781.

289. Weiss, E. J., Bray, P. F., Tayback, M., et al. (1996). A polymorphism of a platelet glycoprotein receptor as an inherited risk factor for coronary thrombosis. *N Engl J Med, 334,* 1090–1094.

290. Goldschmidt-Clermont, P. J., Coleman, L. D., Pham, Y. M., et al. (1999). Higher prevalence of GPIIIa Pl$^{a2}$ polymorphism in siblings of patients with premature coronary heart disease. *Arch Pathol Lab Med, 123,* 1223–1229.

291. Mikkelsson, J., Perola, M., Laippala, P., et al. (1999). Glycoprotein IIIa/Pl$^A$ polymorphism associates with progression of coronary artery disease and with myocardial infarction in an autopsy series of middle-aged men who died suddenly. *Arterioscler Thromb Vasc Biol, 19,* 2573–2578.

292. Bray, P. F. (1999). Integrin polymorphisms as risk factors for thrombosis. *Thromb Haemost, 82,* 337–344.

293. Hermann, S. M., Poirier, O., Marques-Vidal, P., et al. (1997). The leu33/Pro polymorphism (Pl$^{A1}$/Pl$^{A2}$) of the glycoprotein IIIa (GPIIIa) receptor is not related to myocardial infarction in the ECTIM Study: Estude Cas-Teoins de Infarctus du Myocarde. *Thromb Haemost, 77,* 1179–1181.

294. Laule, M., Cascorbi, I., Stangl, V., et al. (1999). A1/A2 polymorphism of glycoprotein IIIa and association with excess procedural risk for coronary catheter interventions: A case-controlled study. *Lancet, 353,* 708–712.

295. Ridker, P. M., Hennekens, C. H., Schmitz, C., et al. (1997). Pl$^{A1/A2}$ polymorphism of platelet glycoprotein IIIa and risks of myocardial infarction, stroke and venous thrombosis. *Lancet, 349,* 385–388.

296. Aleksic, N., Juneja, H., Folson, A. R., et al. (2000). Platelet Pl$^{A2}$ allele and incidence of coronary heart disease: Results from the Atherosclerosis Risk in Communities (ARIC) Study. *Circulation, 103,* 1901–1905.

297. Michelson, A. D., Furman, M. I., Goldschmidt-Clermont, P., et al. (2000). Platelet GP IIIa Pl(A) polymorphisms display different sensitivities to agonists. *Circulation, 101,* 1013–1018.

298. Feng, D., Lindpaintner, K., Larson, M. G., et al. (1999). Increased platelet aggregability associated with platelet GP IIIa Pl$^{A2}$ polymorphism: The Framingham Offspring Study. *Arterioscler Thromb Vasc Biol, 19,* 1142–1147.

299. Vijayan, K. V., Goldschmidt-Clermont, P. J., Roos, C., et al. (2000). The Pl(A2) polymorphism of integrin β3 enhances outside-in signaling and adhesive functions. *J Clin Invest, 105,* 793–802.

300. Kandzari, D. E., & Goldschmidt-Clermont, P. J. (2001). Platelet polymorphisms and ischemic heart disease: Moving beyond traditional risk factors. *J Am Coll Cardiol, 38,* 1028–1032.

301. Winkelmann, B. R., Hager, J., Kraus, W. E., et al. (2000). Genetics of coronary heart disease: Current knowledge and research principles. *Am Heart J, 140*(Suppl.), S11–S32.

302. Risch, N. J. (2000). Searching for genetic determinants in the new millennium. *Nature, 405,* 847–856.

303. Knoweles, J. W., & Maeda, N. (2000). Genetic modifiers of atherosclerosis in mice. *Arterioscler Thromb Vasc Biol, 20*(11), 2336–2345.

304. Chien, K. R. (1996). Genes and physiology: Molecular physiology in genetically engineered animals. *J Clin Invest, 97,* 901–909.

305. Plump, A. S., Smith, J. D., Hayek, T., et al. (1992). Severe hypercholesterolemia and atherosclerosis in apolipoprotein E-deficient mice created by homologous recombination in ES cells. *Cell, 71,* 343–353.

306. Nakashima, Y., Plump, A. S., Raines, E. W., et al. (1994). ApoE-deficient mice develop lesions of all phases of atherosclerosis throughout the arterial tree. *Arterioscler Thromb, 14,* 133–140.

307. de Villiers, W. J. S., Smith, J. D., Miyata, M., et al. (1998). Macrophage phenotype in mice deficient in both macrophage colony-stimulating factor (op) and apolipoprotein E. *Arterioscler Thromb Vasc Biol, 18,* 631–640.

308. Patrono, C., Coller, B., FitzGerald, G. A., et al. (2004). Platelet-active drugs. The relationships among dose, effectiveness, and side effect. *Chest, 126,* 234S–264S.

309. Ansell, J., Hirsh, J., et al. (2004). The pharmacology and management of Vitamin K antagonists. *Chest, 126,* 2045–2335.

310. Gibbons, R. J., Abrams, J., Chatterjee, K., et al. (2002). ACC/AHA Guideline update for the management of patients with chronic stable angina. www.americanheart.org.

311. CAPRIE Steering Committee. (1996). A randomized, blinded trial of clopidogrel versus aspirin in patients at risk of ischemic events (CAPRIE). *Lancet, 348*(9038), 1329–1339.

312. Olsson, G., Wikstrand, J., Warnold, I., et al. (1992). Metoprolol-induced reduction in postinfarction mortality: Pooled results from five double-blind randomized trials. *Eur Heart J, 13,* 28–32.

313. Yusuf, S., Sleight, P., Pogue, J., et al. (2000). Effects of an angiotensin-converting-enzyme inhibitor, ramipril, on cardiovascular events in high-risk patients. The Heart Outcomes Prevention Evaluation Study Investigators. *N Engl J Med, 342,* 145–153.

314. Pfeffer, M. A., Braunwald, E., Moye, L. A., et al. (1992). Effect of captopril on mortality and morbidity in patients with left ventricular dysfunction after myocardial infarction. Results of the survival and ventricular enlargement trial. The SAVE Investigators. *N Engl J Med, 327,* 669.

315. Yusulf, S., Pepine, C. J., Garces, C., et al. (1992). Effect of enalapril on myocardial infarction and unstable angina in patients with low ejection fractions. *Lancet, 340,* 1173–1178.

316. Walter, B. F. (1989). The eccentric coronary atherosclerotic plaque: Morphologic observations and clinical relevance. *Clin Cardiol, 12,* 14–20.

317. Hamm, C. W., & Braunwald, E. (2000). A classification of unstable angina revisited. *Circulation, 102,* 118–122.

318. Thygesen, K. A., & Alpert, J. S. (2001). The definitions of acute coronary syndrome, myocardial infarction, and unstable angina. *Curr Cardiol Rep, 3*(4), 328–272.

319. DeWood, M. A., Spores, J., Notske, R., et al. (1980). Prevalence of total coronary occlusion during the early hours of transmural myocardial infarction. *N Engl J Med, 303,* 897–902.

320. DeWood, M. A., Stifter, W. F., Simpson, C. S., et al. (1986). Coronary arteriographic findings soon after non-Q-wave myocardial infarction. *N Engl J Med, 315,* 417–423.

321. Boldenheimer, M. M., Bankas, V. S., Trout, R. D., et al. (1979). Relationship between myocardial fibrosis and epicar-

dial and surface electrocardiographic Q-waves in man. *J Eletrocardiol, 12,* 205.

322. Mak, K. H., Challapalli, R., Eisenberg, M. J., et al. (1997). Effect of platelet glycoprotein IIb/IIIa receptor inhibition on distal embolization during percutaneous revascularization of aortocoronary saphenous vein grafts. EPIC Investigators. Evaluation of IIb/IIIa platelet receptor antagonist 7E3 in Preventing Ischemic Complications. *Am J Cardiol, 80,* 985–988.

323. Furman, M. I., Kereiakes, D. J., Krueger, L. A., et al. (2001). Leukocyte-platelet aggregation, platelet surface P-selectin, and platelet surface glycoprotein IIIa after percutaneous coronary intervention: Effects of dalteparin or unfractionated heparin in combination with abciximab. *Am Heart, 142,* 790–798.

324. Braunwald, E., Altman, E. M., Beasle, J. W., et al. (2000). ACC/AHA guidelines for the management of patients with unstable angina and non-ST-segment elevation myocardial infarction: Executive summary and recommendations: A report of the American College of Cardiology/American Heart Association Task Force on Practice Guidelines (Committee on Management of Patients with Unstable Angina). *Circulation, 102,* 1193–1209.

325. Yusuf, S., Zhao, F., Mehta, S. R., et al. (2001). Effects of clopidogrel in addition to aspirin in patients with acute coronary syndromes without ST-segment elevation. *N Engl J Med, 345,* 494–502.

326. Quinn, M. J., & Fitzgerald, D. J. (1999). Ticlopidine and clopidogrel. *Circulation, 100,* 1667–1672.

327. Peters, R., Mehta, S., Fox, K., et al. (2003). For the Clopidogrel in Unstable Angina to Prevent Recurrent Events (CURE) Trial Investigators. *Circulation, 108,* 1682–1687.

328. TIMI Study Group. (1989). Comparison of invasive and conservative strategies after treatment with intravenous tissue plasminogen activator in acute myocardial infarction. Results of thrombolysis in myocardial infarction (TIMI) phase II trial. *N Engl J Med, 320,* 618.

329. ISSIS-3 Collaborative Group. (1992). A randomized comparison of streptokinase vs tissue plasminogen activator vs anistreplase and of aspirin plus heparin vs aspirin alone among 41,299 cases of suspected acute myocardial infarction. *Lancet, 339,* 753.

330. Gusto Investigators. (1993). An international randomized trial comparing four thrombolytic strategies for acute myocardial infarction. *N Engl J Med, 329,* 673.

331. Langer, A., Goodman, S. G., Topol, E. J., et al. (1996). Late Assessment of Thrombolytic Efficacy (LATE) Study: Prognostic in patients with non-Q wave myocardial infarction. *J Am Coll Cardiol, 27,* 1333.

332. Roberts, R. (1996). La difference: Long-term benefit of one thrombolytic over another. *Circulation, 94,* 1203.

333. GUSTO III Investigators. (1997). A comparison of reteplase with alteplase for acute myocardial infarction. *N Engl J Med, 337,* 1118.

334. Reiner, J. S., Lundergan, C. F., Fung, A., et al. (1996). Evolution of early TIMI 2 flow after thrombolysis for acute myocardial infarction. GUSTO-1 Angiographic Investigators. *Circulation, 94,* 2441–2446.

335. van de Werf, F., Cannon, C. P., Luyten, A., et al. (1999). Safety assessment of single-bolus administration of TNK tissue-plasminogen activator in acute myocardial infarction: The ASSENT-1 trial. The ASSENT Investigators. *Am Heart J, 137,* 786.

336. Hampton, J. R., Schroder, R., Wilcox, R. G., et al. (1995). Randomised, double-blind comparison of reteplase double-bolus administration with streptokinase in acute myocardial infarction (INJECT): Trial to investigate equivalence. *Lancet, 346,* 329.

337. Grines, C. L., Browne, K. F., Marco, J., et al. (1993). A comparison of immediate angioplasty with thrombolytic therapy for acute myocardial infarction. The Primary Angioplasty in Myocardial Infarction Study Group. *N Engl J Med, 328,* 673–679.

338. Topol, E. J., & The GUSTO Investigators. (2001). Reperfusion therapy for acute myocardial infarction with fibrinolytic therapy or combination reduced fibrinolytic therapy and platelet glycoprotein IIb/IIIa inhibition: The GUSTO V randomized trial. *Lancet, 357*(9272), 1905–1914.

339. Chan, A. W., & Moliterno, D. J. (2001). Defining the role of abciximab for acute coronary syndromes: Lessons from CADILLAC, ADMIRAL, GUSTO IV, GUSTO V, and TARGET. *Curr Opin Cardiol, 16,* 375–383.

340. Simoons, M. L., & GUSTO IV-ACS Investigators. (2001). Effect of glycoprotein IIb/IIIa receptor blocker abciximab on outcome in patients with acute coronary syndromes without early coronary revascularisation: The GUSTO IV-ACS randomized trial. *Lancet, 357,* 1915–1924.

341. Sabatine, M., Cannon, C., Gibson, M., et al., for the CLARITY-TIMI 20 Investigators. (2005). *N Engl J Med, 352.*

342. Smith, S. C., Felamen, T. E., Hirshfeld, J. W., et al. (2005). ACC/AHA/SCAI 2005 guideline update for percutaneous coronary intervention. www.americanheart.org.

343. Anonymous. (2001). Efficacy and safety of tenecteplase in combination with enoxaparin, abciximab, or unfractionated heparin: The ASSENT-3 randomised trial in acute myocardial infarction. *Lancet, 358,* 605–613.

# CHAPTER 36

# Central Nervous System Ischemia

## Gregory J. del Zoppo

*Department of Molecular and Experimental Medicine, The Scripps Research Institute, La Jolla, California*

## I. Introduction

Platelets contribute to the evolution of ischemic lesions in the central nervous system (CNS), and their normal function is required for protection from the hemorrhagic consequences of ischemic injury. A role for activated platelets in the development of focal ischemia was suspected by Denny–Brown,[1] Russell,[2] and Hollenhorst[3] in the 1960s. Migrating refractile bodies observed in cerebral and retinal arteries in patients during and before focal cerebral ischemia,[2,3] and platelet-containing thrombi seen in cerebral arteries during craniotomy,[4,5] supported the participation of platelets in ischemic events of the CNS. Occlusions of brain-supplying and cerebral arteries (whose composition may include platelets) have been visualized by angiography in patients with recent cerebral ischemia.[6–8] In support of those observations is evidence from short-term studies demonstrating prothrombin activation, plasminogen activation, and alteration of fibrinolysis during ischemic cerebrovascular events. Hence, considerable circumstantial evidence now links platelet–fibrin interactions with cerebral ischemic events. Recent interest in stroke risk reduction by statins highlights alterations (downregulation) in platelet activation.[9] For example, platelet activation is decreased in atorvastatin-treated wild-type mice, but not endothelial nitric oxide synthase (eNOS) knockout mice following middle cerebral artery occlusion (MCA:O).[10] Several clinical trials have now demonstrated a significant reduction of CNS ischemic events (stroke) in patients receiving statins.[11–15]

Both experimental and clinical observations implicate activation of circulating platelets in the development of the ischemic lesion at at least four levels (Table 36-1). An interrelationship between platelet activation at sites of atherosclerosis in large brain-supplying arteries and the downstream microvasculature exists.[16,17] As discussed in detail in this chapter, abrogation of platelet activation is a useful intervention in altering the frequency, course, and outcome of CNS ischemic events.

## II. Ischemic Cerebrovascular Disease

Stroke refers to a syndrome of fixed symptomatic or transient focal neurological deficits. These result from atherothrombotic events (40–57%), thromboembolism (16–23%), subarachnoid hemorrhage (10–19%), intracerebral hemorrhage (4–18%), lacunae (14%), or other vascular causes.[18–20] Atherothrombotic stroke refers to focal neurological events resulting from the stenosis or occlusion of intracranial arteries or of the principal extracranial brain-supplying arteries due to *in situ* arterial thrombosis (e.g., at sites of atherosclerosis) or artery-to-artery emboli.

Cerebrovascular ischemic events are still classified according to the temporal aspects of neurologic symptoms as transient ischemic attacks (TIAs) (which often last only minutes, but not more than 24 hours), reversible ischemic neurologic deficits (RINDs) or minor stroke (leaving persistent dysfunction), and completed stroke. Acute interventions refer to approaches applied within 6 hours of symptom onset.

Focal cerebral ischemia in the carotid artery territory results primarily from *in situ* thrombosis or emboli originating from the heart, aortic arch, or the carotid artery, while vertebrobasilar arterial ischemia frequently results from *in situ* thrombus on atheromata in the basilar artery or emboli from the vertebral arteries.

Unusually, emboli can reach the CNS from the peripheral venous circulation via a patent foramen ovale (PFO). Although it has been suggested that PFO plays little role in stroke etiology,[21] there is reasonable evidence that the presence of a PFO is associated with a significantly higher rate of recurrent cerebral ischemic events (12 vs. 5%/patient/year compared with patients without PFO, $p < 0.03$).[22] Among 630 patients presenting with ischemic stroke, the prevalence of PFO was approximately 34%, but was somewhat higher in the subgroup with cryptogenic stroke (39%).[23]

Selective angiographic studies during the early hours of ischemic stroke have documented a high frequency of

**Table 36-1: Platelet Activation during Focal
Cerebral Ischemia**

1. Platelet–thrombus formation on atheromata of brain-supplying arteries.

2. Platelet accumulation within microvessels of the ischemic regions.

3. Normal platelet function required to prevent hemorrhage within the ischemic region.

4. Potential vascular effects of platelet release products.

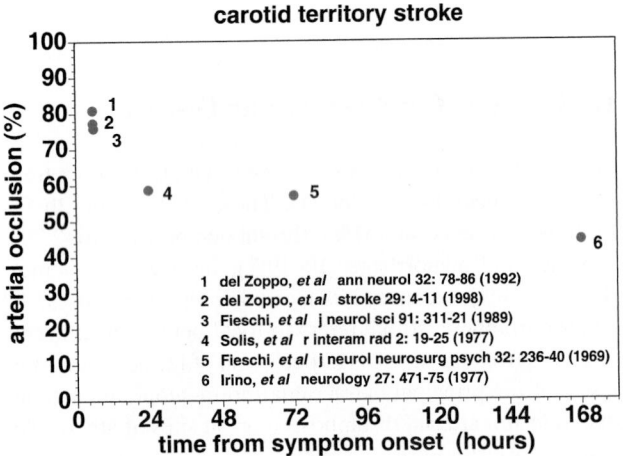

**Figure 36-1.** Frequency of occlusion within the carotid artery system in patients with ischemic stroke, relative to the time from symptom onset (see text).

atherothrombotic lesions or arterial occlusions (Fig. 36-1).[7,8,24–26] In the carotid artery territory, completed strokes occur in 40 to 75% of individuals with one or more premonitory TIAs,[27–29] with a prevalence of approximately 30% per year.[30] Among individuals with a premonitory TIA, about 50% may have a stroke within the first year.[30] Approximately 64% of individuals with TIAs show evidence of infarction on their initial CT scan.[31] Patients suffering their first stroke are at risk for recurrence, often within the same vascular territory. The 5-year cumulative incidence of secondary stroke was 42% among males in one prospective follow-up study.[19] A separate study described a 7-year cumulative incidence of 32% for recurrent stroke.[32] Again, the highest recurrence rates for stroke are within the first year following the initial event, suggesting an ongoing active vascular pathology and continuous platelet activation.[30,33,34] These data indicate that, for the most part, the processes underlying TIAs are continuous. This is supported by the frequent appearance of high-intensity transcranial signals (HITS) in the hemispheric circulation in the setting of TIAs.[35–40]

Both neurologic presentation and outcome depend on the stroke subtype, which has been well-demonstrated by Wityk et al.[41] Improvements in neurologic status among individuals surviving stroke without a therapeutic intervention are commonly observed. Early stroke-related deterioration or mortality is in part a function of edema formation, injury volume, and secondary illness (e.g., pneumonia, cardiovascular ischemia). The clinical measures of injury and its course do not readily allow one to distinguish the hemostatic (or thrombotic) basis for the stroke symptoms, however.

Nonetheless, considerable indirect evidence exists for the participation of activated platelets, coagulation factors, and the fibrinolytic system during thrombotic strokes and episodes of transient cerebral ischemia.[42–46] Spontaneous platelet aggregation and circulating platelet aggregates have been reported in individuals with atherothrombotic and thromboembolic cerebral ischemia. Markers of platelet activation can increase significantly in the setting of ischemic stroke, as demonstrated by β-thromboglobulin (β-TG) and platelet factor 4 (PF4) (Table 36-2), platelet aggregation (Table 36-3), and flow cytometry (Chapter 30).

## III. Platelet Activation in the Development of Focal Cerebral Ischemia

Circumstantial support for the participation of platelets in cerebral ischemia derives from evidence of platelet activation in the peripheral circulation, platelet accumulation on carotid artery atheromata and their embolization, and the accumulation of platelets within microvessels of the ischemic core in experimental models. Injury within the cerebrovascular tree is central to these events.

### A. Systemic Evidence of Platelet Activation

Measures of platelet activation can be abnormal in stroke patients (Tables 36-2 and 36-3).[33,42,45–50] Several reviews have addressed platelet dysfunction and the interactions of platelets, leukocytes, and endothelial cells during cerebral ischemia.[51,52] Independent studies have indicated that platelet survival times (and $t_{1/2}$) are significantly shortened in patients with evidence of transient cerebral ischemic events or sustained ischemic lesions.[53–57] Plasma levels of the platelet α-granule components PF4, β-TG, and thrombospondin (TSP), released into plasma upon platelet activation, are increased in patients suffering stroke (Table 36-2).[50,58–60] Alterations in platelet integrin αIIbβ3 (glycoprotein [GP] IIb-IIIa) and GPIV have been seen in some patients presenting with ischemic stroke[60] (Chapter 30). Iwamoto et al.[58] have shown that the ratios of β-TG concentration in the internal jugular vein to the antecubital vein are elevated in

## Table 36-2: Platelet Activation Studies in Stroke

| Source | Patients (*n*) | Groups | Age (years) | Platelet Count | β-TG | PF4 |
|---|---|---|---|---|---|---|
| Hoogendijk[292] | 186 | Stroke + TIA | 64 | Normal | ↑ | — |
| | 176 | Control | 35 | Normal | 0 | — |
| Cella[45] | 24 | Stroke (old) | — | — | 0 | — |
| | 103 | Control | 35.1 (M), 39.6 (F) | | | |
| Fisher[293] | 24 | Stroke | 59.4 ± 18.5 | — | ↑ | ↑ |
| | 20 | TIA | 60.2 ± 12.7 | — | ↑ | ↑ |
| | 40 | Control (younger) | 29.8 – 5.8 | — | | |
| | 15 | Control (age-matched) | 58.3 ± 7.0 | — | | |
| De Boer[63] | 36 | Stroke + heparin | — | — | 0 | — |
| | 31 | Stroke + placebo | — | — | 0 | — |
| | 80 | Control | — | — | | |
| Taomoto[294] | 70 | Stroke, acute | — | — | ↑ | — |
| | 80 | Stroke, chronic | — | — | ↑ | — |
| | 117 | TIA + RIND | — | — | ↑ | — |
| | 136 | Cerebral atherosclerosis | — | — | ↑ | — |
| | 39 | Control | 28.7 | — | — | — |
| Lane[72] | 66 | Stroke | 69 | — | ↑ | — |
| | 16 | Control | 69 | — | | — |
| Shah[62] | 13 | Stroke (thromboembolic) | 62.5 | — | ↑ | ↑ |
| | 10 | Stroke (cardioembolic) | 68.7 | — | ↑ | 0 |
| | 11 | TIA | 60.9 | — | 0 | 0 |
| | 10 | Lacunes | 59.2 | — | 0 | 0 |
| | 14 | Uncertain | 66.4 | — | ↑ | 0 |
| | 20 | Control, young | 35.7 | — | | |
| | 15 | Control, elderly | 65.2 | — | | |
| Landi[295] | 70 | Stroke | 67.7 | Normal | ↑ | — |
| | 45 | Control | 66.2 | Normal | | — |
| Fisher[296] | 85 | Stroke + TIA, acute | 51.7 | — | ↑ | ↑ |
| | 57 | Stroke + TIA, follow-up | 52.0 | — | ↑ | ↑ |
| | 18 | Control, nonvascular disease | 53.7 | — | | |
| | 44 | Control, normal | 53.0 | — | | |
| Feinberg[297] | 39 | Stroke (2 weeks) | — | — | ↑ | ↑ |
| | 37 | Control | — | — | | |
| Iwamoto[58] | 56 | Atherothrombotic | 77.5 ± 10.3 | — | ↑ | — |
| | 31 | Cardioembolic | 72.8 ± 9.2 | — | ↑ | — |
| | 62 | Lacunar | 75.6 ± 9.1 | — | ↑ | — |
| | 7 | TIA | 76.7 ± 3.1 | — | 0 | — |
| | 30 | Binswanger's disease | 74.7 ± 9.0 | — | ↑ | — |
| | 25 | Control, nonstroke | 80.2 ± 5.9 | — | — | — |
| | 25 | Control, healthy | 75.4 ± 9.1 | — | — | — |

β-TG, β-thromboglobulin; F, female; M, male; PF4, platelet factor 4; RIND, reversible ischemic neurologic deficit; TIA, transient ischemic attack; 0, not different from control; ↑, significantly greater than control.

the chronic phase following lacunar stroke and in athero-thrombotic stroke, suggesting that platelet activation can occur in the ischemic brain independent of vascular pathology. In addition, *in vitro* studies by Konstantopoulos et al.[61] demonstrated that platelets obtained from patients with atherosclerotic stroke exhibit increased shear-induced aggregation compared with those from healthy persons. Among reports which described evidence of platelet α-granule β-TG and PF4 release, considerable variability in outcome was observed between ischemic stroke patients and their control groups. The observations of Shah et al. implied that strokes of different etiologies (e.g., lacunar vs. thromboembolic

## Table 36-3: Platelet Aggregation Studies in Stroke

| Source | Patients ($n$) | Groups | Age (years) | Platelet Aggregation | ADP | Epinephrine | Collagen |
|---|---|---|---|---|---|---|---|
| Couch[298] | 18 | Stroke + TIA | <61 (38–61) | ↑ | — | — | — |
| | 18 | Control | Age-matched | | — | — | — |
| | 21 | Stroke + TIA | ≥61 (62–89) | 0 | — | — | — |
| | 21 | Control | Age-matched | | — | — | — |
| Dougherty[42] | | Acute (<10 days) | | | | | |
| | 53 | Stroke | | ↑ | | | |
| | 29 | TIA | | ↑ | | | |
| | | Chronic (10 days–6 weeks) | | | | | |
| | 34 | Stroke or TIA | | 0 | | | |
| | 30 | Control | | | | | |
| | | Acute (<10 days) | | | | | |
| | 44 | Stroke or TIA | | | ↑ | ↑ | 0 |
| | | Chronic (10 days–6 weeks) | | | | | |
| | 33 | Stroke or TIA | | | 0 | 0 | 0 |
| | 20 | Control | | | | | |
| Konstantopoulos[61] | 15 | Atherosclerotic stroke | 62 ± 11 | ↑[a] | — | — | — |
| | 8 | Lacunar | 57 ± 11 | 0 | — | — | — |
| | 11 | Control | 62 ± 11 | | — | — | — |

[a]Shear-induced platelet aggregation.

Abbreviations: ADP, adenosine diphosphate; TIA, transient ischemic attack; 0, not different from control; ↑, significantly greater than control.

stroke) involved platelet activation of different degrees,[62] but the details were at odds with the findings of Iwamoto et al.[58] β-TG levels were higher in patients with thromboembolic and cardioembolic stroke than control.[62] Furthermore, disparities among reports examining β-TG or PF4 release in patients suffering TIAs or stroke can be found.[45,58,59,63] β-TG levels were not elevated in patients with "old CVA disease."[45] These studies carry concerns regarding the technical reliability of platelet granule release measurements (e.g., platelet activation during venipuncture) and the nature of their respective control groups.

Flow cytometric analysis of activation-dependent changes in platelet surface receptors (Chapter 30) circumvents the methodological problems of the measurement of plasma β-TG and PF4. Increased circulating P-selectin (CD62P)-positive, CD63-positive, and activated αIIbβ3-positive platelets have been reported in acute cerebrovascular ischemia.[50,64,65] This platelet activation is evident 3 months after the acute event. But there is no evidence that it antedates the ischemic event in these studies. Nonetheless, increased expression of surface P-selectin on platelets is a risk factor for silent cerebral infarction in patients with atrial fibrillation.[66] Platelet-derived microparticles (Chapter 20) have been reported to be increased in TIAs.[67] Increased platelet-derived microparticles and procoagulant activity occur in symptomatic patients with prosthetic heart valves and provide a potential pathophysiological explanation of cerebrovascular events in this patient group.[68]

Evidence of thromboxane (TX) A$_2$ generation, demonstrated through stable urinary excretion of metabolites of TXB$_2$ as measures of platelet activation (see Chapter 31), has been documented in all ischemic stroke subtypes.[33,34,49,69] Patrono and colleagues have demonstrated that platelet activation occurs repeatedly during the 48 hours following ischemic stroke and can also accompany TIAs and intracerebral hemorrhage.[33,34] More recently, measurement of platelet surface P-selectin has further confirmed significant platelet activation early following ischemic stroke.[70] Those studies, in aggregate, emphasize the continuous activation of platelets (especially) during all forms of nonlacunar stroke.

In addition, a number of studies have described increased levels of fibrinopeptide A (FPA), fibrinogen B-β peptides, or D-dimer in patients with "early stroke," or in later phases, indicating the generation of thrombin during cerebral ischemia.[71] Several studies of symptomatic ischemic cerebrovascular disease have reported accelerated activated partial thromboplastin times (aPTT), decreased levels of tissue plasminogen activator (t-PA), increased levels of plasma fibrinogen and serum fragment E levels (FgE, E monomer), and increased levels of soluble fibrin monomer.[55,63,72] Increased levels of FPA and D-dimer levels also have been documented for up to 4 weeks after stroke, suggesting the presence of ongoing fibrin formation and fibrinolysis.[46] Those findings most probably represented consequences, rather than the cause, of cerebral infarction; however, no prospective population-based studies have been performed to test this notion. Despite some inconsistencies, these studies overall imply the coincidence of platelet activation and thrombin generation. Reports of cerebral thromboembolic events in young individuals with hyperhomocystinemia, protein C deficiency, and familial plasminogen activator deficiency also implicate altered hemostatic (and platelet) function as predisposing factors in thrombus formation and dissolution in stroke and support a primary role for arterial thrombosis in the pathogenesis of ischemic stroke.[73]

Scintigraphic imaging of platelets labeled with [111]In indicate active platelet accumulation on carotid artery atheromata.[74] Of importance to patient management, such imaging studies confirm that four-vessel cerebral angiography can produce vascular injury with sites of platelet accumulation.[75] This has been confirmed by the presence of platelet activation products following cerebral angiography.[76]

The persistence of stroke risk following a single event and the appearance of continuous platelet activation, thrombin generation, and plasmin activity suggests targets for antiplatelet/antithrombotic intervention. This notion is underscored by the meta-analyses of the Antiplatelet Trialists' Collaboration[77] and the Cochrane Collaboration,[78] which support the hypothesis that antiplatelet agents can reduce the incidence of cerebral ischemic events in the setting of cerebrovascular activation.

### B. Platelet Activation Associated with Brain-Supplying Arteries

The primary vascular prothrombotic condition associated with cerebral ischemia is atherosclerosis of the carotid and vertebral arteries and the aortic arch. Platelet adhesion, activation and accumulation contributes to local flow disturbance, and the downstream embolization of the platelet-fibrin thrombi.[79,80] The evolving complexity of atheromata in the carotid artery bulb may be due in part to platelet activation products and local flow disturbances.[81–85] One proposed hypothesis is that the configuration of the carotid bulb and the flow divider separating the internal and external artery limbs causes turbulence and high-shear stresses which promote platelet activation, and the secretion of factors responsible for myointimal growth (e.g., platelet-derived growth factor).[16,81,82] The participation of platelets and inflammatory cells in the growing atheroma has been summarized by Ross.[16,17]

Platelet-fibrin thrombi on atheromata within the extracranial portion of the carotid artery or in the aortic arch can embolize downstream, predominantly into the middle (MCA) and anterior (ACA) cerebral arterial territories. Emboli have been observed in the retinal circulation as

Hollenhorst plaques,[3] or directly at operation.[4,5] A similar process can send emboli from atheromata at the subclavian artery origins of the vertebral arteries' junctions into the basilar artery. Cardiac source emboli originate from left ventricular mural thrombi formed during myocardial ischemia, atrial thrombi formed in association with (nonvalvular) atrial fibrillation, valvular injury during rheumatic disease, or from prosthetic valves. There is, however, little information regarding the actual composition of such thrombi, although they contain platelets and fibrin. It is presumed that the processes of platelet accumulation *in situ* on atheromata in the basilar artery are identical to those of atherosclerotic coronary arteries which lead to their occlusion and myocardial ischemia (Chapter 35).[86] Thrombosis of small, penetrating cerebral arteries contributes to lipohyalinosis and the formation of lacunae, which is less common than atherothrombotic stroke.[87]

Platelet accumulation into complex thrombi formed on atheromata at the proximal internal carotid and vertebral artery origins are responsible for both transient (i.e., TIAs) and persistent symptoms of focal ischemia. Internal carotid artery atheromata have been shown by duplex ultrasonographic studies to generate embolic material to the downstream cerebral microcirculation.[36–38] HITS in part result from platelet–fibrin thromboemboli, which may contain lipid materials derived from the carotid lesion or microbubbles caused by cavitation of blood from the lesion or more proximal sources (e.g., injected gases).[88,89] The relative importance of activated platelets in carotid artery thromboemboli has been addressed in part by the Clopidogrel and Aspirin for Reduction of Emboli in Symptomatic Carotid Stenosis (CARESS) trial.[90] Asymptomatic HITS were significantly reduced by exposure to the antiplatelet combination aspirin + clopidogrel or aspirin alone.[90] Rare observations at postmortem have confirmed the presence of cholesterol crystals in the downstream microvasculature of the ischemic territory in selected patients with proximal carotid artery atheromata in the symptomatic vasculature.[91,92] Limited experiments in rabbits have indicated that pulverized atheromatous materials from human carotid endarterectomy specimens can occlude cerebral microvessels and initiate local thrombosis when injected into the proximal common carotid artery.[93] It is presumed that similar obstructions are produced by platelet–fibrin emboli from atherosclerotic lesions in humans.

### C. Experimental Modeling of Platelet–Fibrin Embolization

Several experimental settings have been constructed to explore the possible roles of platelet-derived thrombi in focal cerebral ischemia. Early attempts to generate platelet emboli to mimic TIAs in small animals involved local intracarotid artery infusion of platelet proaggregatory substances, including adenosine diphosphate (ADP),[94] arachidonate,[95,96] human atheroma,[93] and thrombin. In only a few studies did poor behavioral outcome follow sustained injury of the microvascular bed in a dose-concentration-dependent fashion.[96] Evidence of platelet-containing occlusions of the dependent microvessels was observed in the platelet activation studies,[94,95] while arteriolar lesions typical of atheroembolic microvascular occlusion followed infusion of pulverized atheromata.[93] Limitations of this experimental platelet activation approach lay in the difficulties in adjusting ADP and arachidonate dose-effects, their nonpathophysiologic nature, and the uncertainty of the injury.[93,97]

A second approach (in rabbits, hamsters, and rats) modeled carotid artery injury and stenosis, by constructing an extrinsic subocclusive constriction of the artery with or without accompanying endothelial cell injury. Cyclic platelet activation, accumulation, and embolization were generated under conditions quite similar to those developed by Folts and used for the coronary artery (discussed in Section III.B.1.a of Chapter 33).[98–101] The cyclic platelet thrombus formation and embolization could be accelerated by injection of thrombin upstream of the stenosis.[102] Near complete abrogation of cyclic platelet embolization followed exposure to the antiplatelet agent aspirin,[98–101] while complete elimination of cyclic flow variations was produced in one study by pharmacologic inhibition of $TXA_2$ synthase and the $TXA_2$ receptor.[99] The effects of cyclic embolization on cerebral function or regional cerebral blood flow were not studied in those models; however, a model of similar intent, in which the platelet embolus generator consisted of an externalized silastic loop in the carotid arterial system, has been described.[103] All models of this type require anesthesia which can alter the local cerebral blood flow distributions. They point to the difficulties of modeling focal cerebral ischemia and generating brain lesions by activated platelets. However, separate studies imply that the platelet–fibrin interaction is central to carotid artery thrombosis during flow restriction.

Two other models generate activated platelets in the cerebral circulation. Dietrich and colleagues have used Rose Bengal-enhanced laser injury to the carotid artery to cause platelet emboli in the rat.[104,105] Platelet-containing occlusions accumulated in the downstream microvasculature, suggesting embolization, and there was evidence of blood–brain barrier disruption.[104] In a separate study, [111]In-labeled platelets impacted in both the ipsilateral and the contralateral hemispheres.[50] Platelet activation occurs together with leukocyte activation in a number of settings. Akopov and colleagues demonstrated that 4-β-phorbol-12 β-myristate-13 α-acetate (PMA) provoked intravascular leukocyte and platelet aggregation and regional ischemic injury with blood–brain barrier disruption.[106] Products of leukocyte activation, rather than platelet release (e.g., serotonin), were

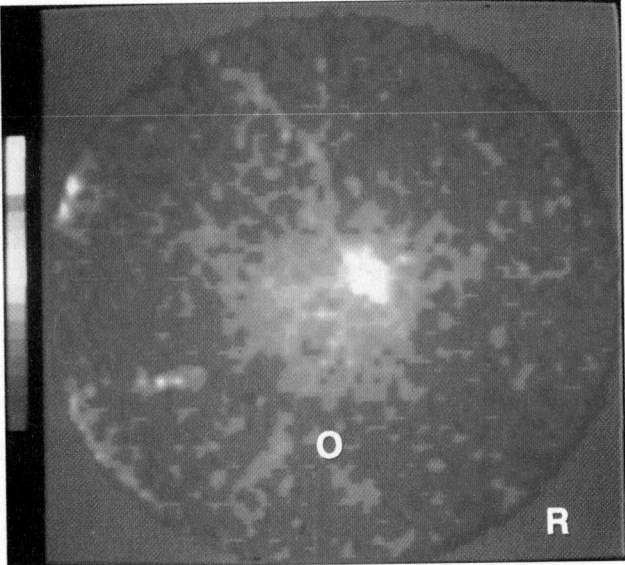

**Figure 36-2.** Deposition of [111]In-labeled autologous platelets in the ischemic basal ganglia by 1 hour reperfusion, following 3 hours middle cerebral artery occlusion in a nonhuman primate. Reproduced with permission from ref. 109.

apparently involved in PMA-stimulated cerebral arterial constriction.[107] Both model systems have limited utility in exploring the precise roles of platelet activation in the ischemic territory, but suggest that platelet-containing aggregates might lead to local ischemia.

### D. Platelet Activation within the Cerebral Microvasculature

Other experimental approaches which generate focal cerebral ischemia by MCA:O have shown that platelets are involved in growing ischemic lesions.[108–110] MCA:O results in accumulation of activated platelets within microvessels in the focal ischemic core (Fig. 36-2). del Zoppo et al. first described the deposition of autologous [111]In-labeled platelets infused prior to MCA:O within the ischemic region early following reperfusion in the nonhuman primate.[109] Platelet accumulation and occlusions within the microvessels were significantly reduced by treatment with the antiplatelet/anticoagulant combination ticlopidine/heparin. Okada et al. subsequently demonstrated time-dependent accumulation of activated platelets within the microvasculature up to 24 hours after MCA:O by the anti-human monoclonal antibody (MoAb) LJ-P4 directed against the platelet integrin αIIbβ3 receptor.[108] Platelet accumulation coincided with the appearance of nonvascular P-selectin antigen within ischemic microvessels.[108] Electron microscopic studies confirmed the presence of aggregates of degranulated platelets interspersed with fibrin in association with leukocytes within the microvasculature of the ischemic region, providing direct evidence that platelet activation occurs in the ischemic region

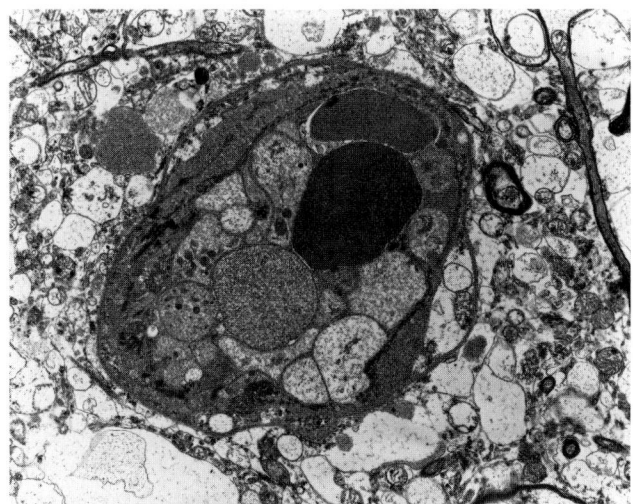

**Figure 36-3.** Impact of activated platelets and fibrin, together with neutrophils in a noncapillary microvessel in the ischemic basal ganglia, under similar conditions to Fig. 36-2.[111]

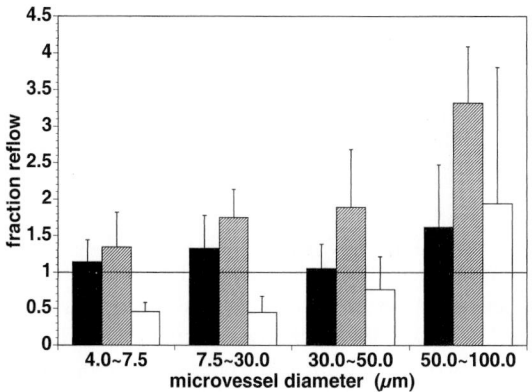

**Figure 36-4.** Microvessel patency in the ischemic basal ganglia of nonhuman primates in response to inhibition of the platelet integrin $\alpha IIb\beta3$-fibrin interaction by the RGD-containing nonapeptide TP9201. Clinically significant hemorrhage was associated with inhibition at the $IC_{80}$. Open columns, placebo; gray columns, TP9201 at the $IC_{30}$; black columns, TP9201 at the $IC_{80}$. (Modified from ref. 110.)

(Fig. 36-3).[111] Those observations are in accord with the known contribution of neutrophils to microvascular "no-reflow" following MCA:O, and the interrelationship of leukocyte and platelet activation during thrombosis.[108,111] Three reports have demonstrated that interruption of the platelet–fibrinogen interaction can abrogate "no-reflow" instigated by reperfusion during cerebral ischemia.[110,112–114] Garcia and colleagues observed the accumulation of platelets with microvessels in the peripheral border of the expanding ischemic lesion in the Wistar rat following MCA:O.[115] Entry of unactivated platelets into the ischemic field from patent channels was interpreted as a major source for the platelets which accumulated in response to ischemia. Whether platelet activation occurred in response to regional changes in flow, thrombin generation associated with endothelial barrier permeability, leukocyte activation, or other stimuli is not yet known. Also, it is unknown whether subtle alterations in endothelial cell antithrombotic properties can occur. A central issue is whether the accumulation of activated platelets within the ischemic microvascular bed is a secondary event or contributes directly to the evolving ischemic lesion and, if so, in what manner.

### E. Secondary Effects of Platelet Activation within the CNS

Several features of platelet activation have been explored for their potential to augment the injury initiated by ischemia. Dense granule secretion products have been implicated in local vasomotor alterations following platelet aggregation.[76,116] One hypothesis examined in detail is that serotonin modulates vasoreactivity in normal and atherosclerotic arteries and contributes to cerebral ischemia.[117–120] In stroke-prone hypertensive rats (SHRSP), abrogation of vasodilation by platelet products could contribute to cerebral injury.[118–120] In the setting of atherosclerotic lesions upstream, serotonin can decrease blood flow, an effect which resolves with regression of the atherosclerotic lesions.[121] Welch and colleagues have supported the hypothesis that serotonin (as released during platelet activation) is neurotoxic.[122–124] Glutamate is a neurotransmitter which in excessive quantity can be toxic to neurons. During ischemic stroke plasma glutamate levels increase, while platelet glutamate release and uptake are decreased.[125] These changes are compatible with platelet activation. However, whether they contribute to neuron toxicity during ischemic stroke is unclear.

### F. Contributions of Microvessel Obstruction to Neurological Outcome

Platelet accumulation contributes to the focal "no-reflow" phenomenon, to microvascular obstruction, and to extension of the ischemic lesion.[108,110,126] Blockade of platelet activation and of the platelet–fibrin(ogen) interaction have been shown to alter focal "no-reflow."[109,110,112] Specifically, inhibition of the integrin $\alpha IIb\beta3$–fibrin interaction by the RGD-containing nonapeptide TP9201 significantly reduced microvessel occlusion within capillaries and larger microvessels (7.5–30.0 $\mu m$ diameter).[110] Dose-dependent induction of intraparenchymal hemorrhage was significant at the $IC_{80}$ for ADP-induced platelet aggregation, but not at the $IC_{30}$, although microvessel patency was preserved at both concentrations (Fig. 36-4).[110] The effect of preserving microvascular patency on neuron viability was not studied. Inhibition of the integrin αIIbβ3–fibrin interaction in a murine MCA: O model was associated with a differential reduction in the injury zone, but also with dose-dependent significant hemorrhage.[112] Intravenous weight-adjusted doses of the organic integrin αIIbβ3 inhibitor SDZ GPI 562 produced a significant reduction in ischemic injury volume at 24 hours. The salutary effect on injury volume correlated with inhibition of [111]In-labeled platelet accumulation at 24 hours and a reduction in tissue fibrin deposition. The latter findings are consistent with the effects of anti-tissue factor strategies during MCA:O in the nonhuman primate.[127,128] However, it could not be excluded that GPI 562 at the high doses also affected the course of tissue injury by independent effects. Hemorrhage confounded the surgical procedures and increased dose-dependently at 24 hours with GPI 562. No experimental test in this setting of abciximab (the humanized Fab fragment of the monoclonal antibody 7E3) has been reported so far.

These results imply that normal platelet function is required to prevent or reduce detectable parenchymal hemorrhage during evolving cerebral infarction. Furthermore, platelet activation and deposition within the microvasculature during focal cerebral ischemia (a) contributes significantly to microvascular occlusion and focal "no-reflow" and (b) may participate in the extension of injury, at least in rodent models. Separate experiments have demonstrated that inhibitions of the platelet integrin αIIbβ3 can prevent carotid artery thrombosis caused by electrical or mechanical injury.[129–132] The principal clinical hypothesis, formally untested, is that platelet–fibrin(ogen) interactions, if blocked in the acute moments of ischemic stroke, may contribute to clinical (behavioral) improvement. However, a critical awareness of the differential risk of significant hemorrhage and its impact on clinical efficacy is required with these antiplatelet agents.

### G. Genetic Manipulation of Platelet Adherence and CNS Injury

P-selectin, expressed by platelets and activated endothelial cells, interacts with its counterreceptor P-selectin glycoprotein ligand 1 (PSGL-1) on leukocytes,[133,134] allowing the interaction of these three cell types under activating conditions (Chapter 12). Platelet–leukocyte interactions modulate thrombus formation, and can promote leukocyte adhesion.[134] Homozygous deletion of P-selectin in mice has been associated with a significant reduction in infarct volume following MCA:O.[135] However, this study[135] did not provide an indication of the relative contributions of platelet and leukocyte activation to injury. The reduction in injury could also be explained by a reduction in leukocyte adhesion, as suggested by separate studies.[136–138]

Polymorphisms of platelet membrane proteins have been linked to stroke risk in small studies, but the evidence is not strong [65,69,139–143] (see Chapter 14 for details). It has been suggested that the PL[A1/2] polymorphism of integrin β3 is associated with atherothrombotic stroke in patients under 50 years.[139] Several recent studies suggest that the Kozak T/C polymorphism of the GPIb gene is associated with first ever stroke.[140,141,144]

### H. Summary

Experimental studies implicate platelets in the development of microvascular "no-reflow" tissue injury in the ischemic territory and in protection from hemorrhagic transformation following focal ischemia. These roles for activated platelets are in addition to their observed contributions to vascular lesions and their embolization from atherosclerotic brain-supplying arteries. Three relevant clinical settings in which platelet function contributes to cerebrovascular ischemia can be considered: (a) hemorrhagic transformation and its augmentation by antithrombotic agents, (b) the development of parenchymal injury in the acute setting, and (c) secondary prevention of additional cerebrovascular events.

## IV. Hemorrhagic Transformation

Intracerebral hemorrhage can result from overt vascular rupture, or occur in the setting of secondary injury following focal cerebral ischemia. Hemorrhage accounts for approximately 10% of strokes, and hemorrhage in the neuroaxis is a risk of the use of all antithrombotic agents.[18,145,146] In otherwise normal individuals, neuropathologic changes probably underlie the very low, but significant, frequency of spontaneous intracranial hemorrhage (e.g., 0.02%) in individuals with acute myocardial ischemia.[147]

Normal platelet function appears to be necessary to maintain the integrity of the cerebrovascular beds and to prevent clinically detectable hemorrhage. Spontaneous intracerebral hemorrhage has been observed in some disorders of altered platelet function. In severe thrombocytopenia accompanying refractory immune thrombocytopenic purpura (ITP), symptomatic intracerebral hemorrhage not infrequently causes demise.[148,149] In the absence of known CNS disease, unexpected cerebral hemorrhage and post-traumatic hemorrhage have been reported in von Willebrand disease, platelet storage pool disease, and myelodysplastic syndrome. But no prospective comparisons of the frequency of cerebral hemorrhage in patients with these disorders who suffer ischemic stroke have appeared. The incidence of intracerebral hemorrhage in patients with severe uremia appears to be low. However, as in the above hematologic conditions, the risk would be expected to be increased in the setting of a pathological intracerebral event. Spontaneous intracerebral hemorrhage can be observed in primary disorders of coagulation (e.g., inherited or acquired deficiencies of factors VIII and IX), most often accompanying trauma, but sometimes the mode of presentation of rare deficiencies of factors VII[150,151] and XIII.[152]

Hemorrhagic infarction (HI) and parenchymal hemorrhage (PH) are common accompaniments of untreated ischemic stroke, occurring in 50 to 70% of patients.[153–159] Symptomatic intracerebral hemorrhage consists of either HI, PH, or both — although HI is more commonly asymptomatic.[6] Postmortem studies have described HI as a spectrum from scattered petechiae to more confluent hemorrhage within the regions of infarction.[153–155] In CT studies, HI has been observed in 10 to 43% of nonanticoagulated patients with acute cerebral infarction, an incidence somewhat lower than that found at autopsy.[159,160] HI usually appears within 2 to 4 days, but it also occurs during the first few hours after the onset of stroke-related symptoms.[156,161,162] Most commonly, HI occurs in association with cardioembolic stroke, rather than *in situ* atherothrombotic vascular occlusions.[158] In contrast, PH is a homogeneous, circumscribed mass of blood, often associated with a shift of midline structures. Many reports of PH in patients with cerebral embolism are associated with anticoagulant treatment.[163–166]

The mechanisms underlying the appearance of HI and PH in cerebrovascular ischemia, as well as their relative risks, are not known. It has been hypothesized that hemorrhage results from fragmentation and distal migration of thromboemboli, thereby exposing ischemic arteries to systemic pressure. These arteries may rupture, producing hemorrhage into the infarct.[153] This concept has been broadened by angiographic studies of thrombolytic agents in individuals with acute thrombotic stroke,[25,167–169] and the potential contribution of degradation of the (micro)vascular basal lamina to petechial hemorrhage.[170] However, hemorrhagic transformation despite persistent occlusion of the primary artery suggests that hemorrhage can occur from other vascular sources as well (e.g., collateral channels).[153,154]

Antithrombotic agents increase the frequency of symptomatic hemorrhage associated with ischemic stroke, most probably by directly interfering with vascular thrombus generation within the affected vessel(s) or by rapidly degrading fibrin(ogen) (e.g., plasminogen activator). At therapeutic doses, plasminogen activators, anticoagulants, and antiplatelet agents increase intracerebral hemorrhage over their respective untreated cohorts in descending order of frequency.

Antiplatelet agents increase the incidence of symptomatic hemorrhage during ischemic stroke. The International Stroke Trial (IST), a 3 × 2 factorial comparison of aspirin, heparin delivered subcutaneously at two doses, and placebo, prospectively demonstrated a significant 0.1% increase in the incidence of symptomatic intracerebral hemorrhage in the aspirin arms compared to the placebo arm of the study.[171] This differential increased risk associated with aspirin was confirmed by the Chinese Acute Stroke Trial (CAST).[172,173] Both trials entered patients into the study within 48 hours of symptom onset, but there was no differentiation on the basis of stroke subtype.

A prospective open phase II study of the αIIbβ3 antagonist abciximab in patients presenting with ischemic stroke within 24 hours of symptom onset has been reported.[174] Despite the appearance of hemorrhage at the $IC_{80}$ for two different integrin αIIbβ3 antagonists,[110,112] dosing at the $IC_{80}$ for abciximab produced no apparent change in hemorrhagic risk from control,[174] although (a) the cohort sizes were small, and (b) binding of this pan-β3 integrin inhibitor to other vascular integrin receptors (e.g., integrin αVβ3) which appeared in the ischemic core could not be excluded.[175] A dose-dependent increase in the incidence of symptomatic intracerebral hemorrhage was observed with aspirin in the UK-TIA Study.[176] The European Stroke Prevention Study-2 (ESPS-2) trial of aspirin, dipyridamole, aspirin/dipyridamole in sustained release, and placebo also demonstrated an increase in intracerebral hemorrhage with aspirin.[177] Hence, antiplatelet agents, and particularly aspirin for which the most information exists, can increase the frequency of symptomatic hemorrhage associated with ischemic stroke. These clinical conditions support the importance of normal platelet function to limit the risk of intracerebral hemorrhage.

## V. Antiplatelet Interventions in Cerebral Ischemia

Agents that inhibit any event or step essential for platelet activation and adhesion, platelet aggregation, or platelet–vascular reactivity can function as antiplatelet agents.[178–180] Antiplatelet agents have been used to limit the consequences of recent transient or fixed neurologic deficits. They are also used as prophylactic strategies to reduce the risk of second events following an initial CNS ischemic event (i.e., secondary prevention). Antiplatelet agents tested in various settings of cerebral ischemia have included aspirin, dipyridamole, sulfinpyrazone, suloctidil, ticlopidine, clopidogrel, prostacyclin, and platelet integrin αIIbβ3 antagonists. In comparison, anticoagulants and plasminogen activators have been used successfully to treat arterial thrombosis leading to acute cerebral ischemia or for prevention of thromboembolic events originating because of atrial fibrillation, acute myocardial infarction (MI), or cardiac valve injury.

Updated recommendations for the use of antiplatelet agents (and other antithrombotics) in ischemic stroke are made at regular intervals in conjunction with the American

College of Chest Physicians (ACCP) Consensus Conference on Antithrombotic and Thrombolytic Therapy.[181] Methodologic issues related to stroke intervention trials have been discussed elsewhere.[182] Attempts to evaluate large comparative studies of antiplatelet agents in stroke have been made by a meta-analysis. The utility of these antiplatelet agents and the potential roles of more novel agents have been reviewed.[183,184]

The most recent Antiplatelet Trialists' Collaboration has reported the effect of sustained antiplatelet treatment in 21 prospective trials of patients with a history of stroke or TIA, and seven trials of patients with completed stroke.[77] In the most recent formulation, antiplatelet therapy was associated with $25 \pm 5$ (per 1000) fewer nonfatal strokes in patients with prior stroke or TIA ($2p < 0.00001$) and $4 \pm 2$ (per 1000) fewer nonfatal strokes among the patients treated for completed stroke ($2p = 0.003$). Those results were "driven" primarily by the outcome of a few prospective trials.[176,185,186] In this type of analysis, individual contributions from each antiplatelet agent are obscured, as are anatomic considerations, stroke subtypes, embolus or thrombus source, and the location, volume, and other characteristics of the cerebrovascular ischemia for the overall outcome and persistence of symptoms. Nonetheless, these studies demonstrate that specific antiplatelet agents can reduce the risk of ischemic stroke following TIAs or a signal stroke.

Recently, interest has arisen about the pleiotropic effects of $\beta$-hydroxy-$\beta$-methylglutaryl coenzyme A (HMG-CoA) reductase inhibitors (statins), which have significant effects on the severity and incidence of ischemic stroke in high-risk patients. Among patients treated with a statin following their MI in the CARE trial there was a 31% significant reduction in subsequent stroke, while a post hoc analysis of the 4S trial indicated a similar benefit.[11,12] Subsequently, meta-analyses and smaller prospective studies have underscored the benefit to stroke frequency[187,188] and severity[13,14] of statin use. These findings have led to the promulgation of guidelines supporting the use of statins in patients suffering stroke[15] and their increased usage.[189] Direct effects of statins on the behavior of cerebral tissue in animal models have been summarized.[190] A direct effect of cerivastatin on platelets to reduce thrombin generation has been reported,[191] suggesting the possibility of an antiplatelet effect of this class of agent. Whether statin use will have an impact on cerebrovascular events is the thrust of the ongoing Stroke Prevention by Aggressive Reduction of Cholesterol Levels (SPARCL) trial.

## A. Transient Ischemic Attacks (TIAs)

**1. Aspirin.** Aspirin interferes with platelet function by irreversible acetylation and inactivation of platelet cyclooxygenase (COX-1)[192–194] (Chapter 60). In doing so, aspirin inhibits production of $TXA_2$. Aspirin inhibits collagen-stimulated adhesion at low shear stress, but has little effect on adhesion or aggregation at high shear stress.[195,196] Despite these observations, aspirin has been employed in both primary and secondary prevention of thrombotic processes occurring at arterial flow rates. In most trials involving cerebral ischemia, aspirin doses (300–1300 mg/day) have exceeded those at which the antiplatelet effect occurs, thereby also contributing to a potential vascular effect.

A role for antiplatelet agents in individuals with transient ischemia was first suggested by the reduction in transient monocular blindness (amaurosis fugax) by aspirin in selected patients.[197,198] Aspirin had a similar effect on the frequency of TIAs in 26 individuals with complex alterations of plaques at the carotid artery bifurcation.[199] That aspirin can affect the sequelae of transient ischemia, including completed stroke, has been shown by level I, level II, and level III trials (Table 36-4). The Aspirin in Transient Ischemic Attacks (AITIA) study randomized individuals with a 3-month history of TIAs who were not considered candidates for carotid endarterectomy to aspirin (1300 mg/day) or placebo.[200] There was a significant decrease in the combined outcome events of recurrent TIAs, cerebral/retinal infarction, and death among those receiving aspirin ($n = 88$) compared to placebo ($n = 90$) at 6 months, but there was no reduction by life-table analysis survival at the 24-month follow-up. No difference in the incidence of stroke between the two groups was observed, and post hoc subgroup analyses did not reveal any meaningful relationships. In the subsequent Danish Cooperative Study, there was a less substantial decrease in disabling stroke and death among TIA patients who received aspirin (1000 mg/day, $n = 101$) compared to those who received placebo ($n = 102$) after a mean follow-up of 2.1 years.[201] Both studies, however, suffered from limited power to distinguish principal outcome events of stroke frequency.

The Canadian Cooperative Study Group (CCSG) trial was a randomized, double-blind (level I), four-arm study comparing aspirin (1300 mg/day), sulfinpyrazone (800 mg/day), and the combination of aspirin and sulfinpyrazone with placebo among patients with a history of TIAs.[202] It demonstrated a significant decrease in the incidence of stroke and death in the patients who received aspirin, after a mean follow-up of 2.2 years. Sulfinpyrazone was not associated with benefit, and the risk reduction for stroke and death was most significant for males who received aspirin. In the CCSG trial, 66% of the ischemic episodes were in the carotid artery territory alone.

In the three-arm, randomized, double-blind (level I) AICLA trial, a significantly reduced incidence of stroke, MI, and death was attributed to aspirin.[186] Patients presenting with a history of TIAs were randomized to aspirin (990 mg/day, $n = 198$), a combination of aspirin (990 mg/

## Table 36-4: Antiplatelet Agents for Transient Ischemic Attacks (±Stroke)

| Study | Year | Agent | Dose (per day in mg) | Patients (n) | Follow-up (years) | Stroke (n) | Mortality (n) Vascular[a] | Nonvascular | Total |
|---|---|---|---|---|---|---|---|---|---|
| AITIA Study[200] | 1977 | ASA | 1300 | 88 | 0.5 | 10 | 3 | — | 3 |
| | | Placebo | | 90 | | 12 | 6 | 1 | 7 |
| Canadian Cooperative Study Group[202] | 1978 | ASA/placebo 1 | 1300/— | 144 | 2.2 | 22 | 4 | 0 | 4 |
| | | Sulfinpyrazone/placebo 2 | 800/— | 156 | | 29 | 6 | 3 | 9 |
| | | ASA/sulfinpyrazone | 1300/800 | 146 | | 14 | 4 | 2 | 6 |
| | | Placebo 1/placebo 2 | —/— | 139 | | 20 | 8 | 2 | 10 |
| Danish Cooperative Study[201] | 1983 | ASA | 1000 | 101 | 2.1 | 11 | 6 | 1 | 7 |
| | | Placebo | | 102 | | 18 | 6 | 1 | 7 |
| AICLA[186] | 1983 | ASA/dipyridamole | 990/225 | 202 | 3.0 | 18 | 5 | 3 | 8 |
| | | ASA | 990 | 198 | | 17 | 4 | 6 | 10 |
| | | Placebo | | 204 | | 31 | 4 | 3 | 7 |
| UK-TLA Study Group[176] | 1988 | ASA | 1200 | 815 | 4.0 | 66 | 84 | 27 | 111 |
| | | ASA | 300 | 806 | | 68 | 84 | 22 | 106 |
| | | Placebo | | 814 | | 88 | 86 | 36 | 122 |
| Dutch TIA Trial Study Group[203] | 1991 | ASA | 283 | 1576 | 2.6 | 109 | 107 | 44 | 151 |
| | | ASA | 30 | 1555 | | 90 | 105 | 55 | 160 |
| SALT[204] | 1991 | ASA | 75 | 676 | 2.7 | 93 | 48 | 13 | 61 |
| | | Placebo | | 684 | | 112 | 50 | 19 | 39 |
| Acheson et al.[209] | 1969 | Dipyridamole | 400–800 | 85/69[a] | 2.1 | 5/7[b] | — | — | 13 |
| | | Placebo | | 84/70[a] | | 7/4[b] | — | — | 9 |
| European Stroke Prevention Study Group[185] | 1987 | ASA/dipyridamole | 975/225 | 1250 | 2.0 | 114[c] | 69 | 39 | 108 |
| | | Placebo | | 1250 | | 184[c] | 100 | 56 | 156 |
| American–Canadian Cooperative Study Group[210] | 1985 | ASA/dipyridamole | 1300/300 | 448 | 1.5 | 53 | 29 | 17 | 46 |
| | | ASA | 1300 | 442 | | 60 | 29 | 9 | 38 |
| Matius-Guiu et al.[211] | 1987 | ASA/dipyridamole | 50/300 | 115 | 1.75 | 3 | 1 | 1 | 2 |
| | | Dipyridamole | 400 | 71 | | 3 | 2 | 1 | 3 |
| Roden et al.[215] | 1981 | Sulfinpyrazone | 800 | 39[c] | 0.33 | 1 | 1 | — | 1 |
| | | Placebo | | 39[c] | | 3 | 1 | — | 1 |
| Candelise et al.[299] | 1982 | Sulfinpyrazone | 800 | 61 | 2.0 | 2 | 1 | — | — |
| | | ASA | 1000 | 63 | | 2 | 0 | — | — |
| Hass et al.[218] | 1989 | Ticlopidine | 500 | 1529 | 2.0–6.0 | 172 | 120 | 55 | 175 |
| | | ASA | 1300 | 1540 | | 212 | 116 | 80 | 196 |

ASA, aspirin.

[a]Stroke, myocardial infection, and peripheral vascular events.

[b]First evaluation/second evaluation.

[c]Intention-to-treat analysis.

[d]Cross-over format.

For data on ESPS-2, see Table 36-5 and text.

day) and dipyridamole (225 mg/day, n = 202), or placebo (n = 204).

In the subsequent UK-TIA trial, 2435 patients with TIAs or minor ischemic stroke were randomized to "high-dose" aspirin (1200 mg/day, n = 815), "low-dose" aspirin (300 mg/day, n = 806), or placebo (n = 814).[176] The incidence of non-fatal MI, nonfatal major stroke, and vascular or nonvascular death was significantly reduced by 18% at a mean follow-up of 4 years in patients who received aspirin (300–1200 mg/day, n = 1621). No aspirin dose-response for the combined outcome events was seen. Gastrointestinal side effects were more common with the high-dose aspirin regimen, however.

Nine deaths were attributed to intracranial hemorrhage in the aspirin-treated groups, whereas only one occurred in the placebo group.

Another study (the Dutch TIA Trial Study Group) of aspirin dose response on nonfatal stroke, nonfatal MI, and vascular death found no difference in outcome between individuals with TIAs or minor strokes treated with aspirin at 30 mg/day ($n = 1555$) or 283 mg/day ($n = 1576$) at a mean follow-up of 2.6 years.[203] Six fatal intracerebral hemorrhages occurred in the low-dose group and nine in the high-dose group.

Confirming the benefit of low-dose aspirin, the Swedish Aspirin Low-Dose Trial (SALT) Collaborative Group reported an 18% reduction in the risk of stroke and death among 676 individuals with TIA who began aspirin (75 mg/day) within 1 to 4 months of their initial symptoms as compared to 684 individuals who received placebo.[204] A 16 to 20% reduction in risk of TIAs, frequent stroke, and MI was also observed, thereby confirming the positive impact of low-dose aspirin on the reduction of the risk of stroke and death following a TIA.

In summary, four level I clinical trials have indicated a benefit from aspirin on early stroke and vascular mortality in patients with a history of recent TIAs or minor ischemic events. Low-dose aspirin (50 mg/day) is sufficient, although 325 mg/day is commonly employed. The pathophysiologic basis for the relative benefit seen among aspirin-treated individuals, however, has not been explored.

**2. Aspirin and Dipyridamole.** Aspirin has been used in combination with dipyridamole (Chapter 63) in the treatment of TIAs, based upon a positive interaction between the agents in both preclinical and platelet survival studies of clinical vascular disease.[56,179,205,206] In combination with dipyridamole, aspirin can normalize the decreased platelet survival associated with various arterial thrombotic disorders.[207] That dipyridamole can act together with aspirin to normalize platelet survival *in vivo* has led to the use of combination aspirin/dipyridamole therapy for carotid artery TIAs. The efficacy of this combination, however, has been challenged.[205,208] Dipyridamole (400–800 mg/day) alone did not significantly change the incidence of stroke or related mortality compared with placebo in a single double-blind, randomized trial.[209] Because of the limited power of that study, however, a clinical benefit from dipyridamole might have been missed. A second comparison of aspirin/dipyridamole ($n = 202$) to placebo ($n = 204$), derives from the three-arm AICLA study.[186] Significant benefit for patients with TIAs occurred in those who received aspirin/dipyridamole for the combined outcome events of stroke, MI, and mortality. For that comparison, however, a type II error could not be excluded. No difference in stroke or mortality was reported by the American-Canadian Cooperative Study Group at 1.5 years follow-up among patients who received

aspirin (1300 mg/day) and dipyridamole (300 mg/day) in comparison to those who received aspirin (1300 mg/day) alone.[210] The results of a small, open, nonrandomized trial of 115 patients treated with the combination low-dose aspirin (50 mg/day) with dipyridamole (300 mg/day) suggested no differential benefit over dipyridamole (400 mg/day) alone for the combined outcome events of TIA, stroke, and mortality at a mean follow-up of 1.75 years in 71 patients who had at least one atherothrombotic cerebral event.[211]

In contrast, two randomized, double-blind controlled studies demonstrated that the combination of aspirin/dipyridamole could provide a benefit over placebo. The European Stroke Prevention Study (ESPS) Group compared the combination of aspirin (975 mg/day) with dipyridamole (225 mg/day) to placebo in equal patient groups ($n = 1250$ each).[185] A 33% reduction in the risk of stroke and death was associated with the combination treatment. The more recent ESPS-2 study examined the relative efficacy of time-release dipyridamole (400 mg/day), aspirin (50 mg/day), the combination of aspirin/time-release dipyridamole, and placebo to limit stroke, death, or both in individuals presenting with TIAs or stroke in a $2 \times 2$ factorial design.[212] Aspirin alone and dipyridamole alone (relative risk reductions [RRR] for stroke, 15.8 and 17.7%, respectively) and the combination (RRR, 36.7%) were superior to placebo at 2 year follow-up. For patients with a history of TIAs or recent stroke, the combination was superior to no treatment, but was not substantially different from aspirin alone for the outcomes of stroke, MI, and mortality. This low-dose aspirin/extended-release dipyridamole combination is used for secondary prevention.

The potential benefit of combination aspirin/extended-release dipyridamole relative to clopidogrel in reducing the risk of ischemic stroke has not been formally tested. The ongoing PROFESS study is designed to compare the efficacy (and safety) of aspirin/dipyridamole to aspirin/clopidogrel and the angiotensin receptor blocker (ARB) telmisartan to placebo in a $2 \times 2$ factorial design in the prevention of recurrent stroke.

**3. Sulfinpyrazone.** Sulfinpyrazone, a phenylbutazone derivative, has uricosuric properties *in vivo* and inhibits platelet aggregation *in vitro* by inhibiting COX and by other as yet poorly defined mechanisms.[213] When used alone, sulfinpyrazone can normalize decreased platelet survival times associated with TIAs, prosthetic heart valves, and coronary artery disease.[178] Following a report that sulfinpyrazone could significantly reduce episodes of amaurosis fugax and TIAs,[214] the Canadian Cooperative Study Group compared aspirin, sulfinpyrazone, and placebo for the outcomes of stroke and mortality. However, no significant difference in the outcome was observed at 2.2 years follow-up between sulfinpyrazone and placebo.[202] In a separate double-blind, crossover trial, sulfinpyrazone (800 mg/day) produced a nonsignificant reduction in transient cerebral ischemic events and the incidence of stroke and mortality at 4-month follow-up.[215] In yet another study, individuals with TIAs treated with sulfinpyrazone (800 mg/day) had a higher incidence of stroke, MI, and vascular death than those treated with aspirin (1000 mg/day) at 11-month follow-up.[215a] Considering the small number of patients entered into each study, a type II error could not be excluded. Based on those studies, however, sulfinpyrazone is not usually used in the treatment of TIAs.

**4. Ticlopidine.** Ticlopidine,[216,217] a potent thienopyridine antiplatelet agent, is a P2Y$_{12}$ ADP receptor antagonist that is discussed in detail in Chapter 61. Ticlopidine was tested in patients presenting with TIAs or minor stroke, and in secondary stroke prevention.[218,219] In one study, patients with transient ischemic symptoms were randomized to either ticlopidine (500 mg/day) or aspirin (1300 mg/day).[218] In the "intention-to-treat" analysis, ticlopidine was associated with a significant 12% reduction in risk of stroke and death from any cause ($p = 0.048$), and with a similar 21% reduction in the risk of secondary outcomes of stroke and stroke-related death ($p = 0.024$). Leukopenia, which was reversible, occurred in 0.8% of individuals taking ticlopidine, while gastrointestinal symptoms were more common in patients receiving aspirin. Intracerebral hemorrhage occurred equally in each group. A growing worry was engendered by reports of thrombotic thrombocytopenic purpura (TTP, Chapter 50) with the use of ticlopidine.[220,221] However, the incidence of TTP and/or leukopenia may be lower with the use of clopidogrel, a related P2Y$_{12}$ antagonist (Chapter 61).

**5. Summary.** In summary, aspirin reduces the risk of stroke, MI, and mortality in individuals with a history of TIAs or minor stroke. This risk reduction is not obviously dose dependent.[176] The combination of low-dose aspirin and dipyridamole in extended release also produces a substantial risk reduction.[212,222] Ticlopidine was somewhat more effective than aspirin in preventing stroke and mortality after TIAs in one study, but systemic side effects have limited its use in the CNS (see following).[218]

### B. Completed Stroke

Focal cerebral ischemic events, MI, death, or vascular-related death following ("secondary" to) the signal stroke occur at a rate of 10 to 12% per year.[219,223,224] Antithrombotic interventions with antiplatelet agents have proven successful in preventing and reducing these secondary ischemic events following completed stroke. A group of level I studies have suggested a benefit of aspirin and ticlopidine for secondary prevention (Table 36-5).

**1. Aspirin.** Much as in the setting of TIAs, aspirin can provide protective benefit from additional cerebral ischemic events. The Swedish Cooperative Study found no difference in the outcomes of recurrent stroke or death between individuals treated early (≤3 weeks) after completed stroke who were assigned to receive either aspirin (1500 mg/day) or placebo.[223] To further examine the effects of simple antithrombotic regimens on secondary stroke in patients with completed stroke, the International Stroke Trial (IST) and the Chinese Acute Stroke Trial (CAST) were performed. IST randomized 19,436 individuals with presumed ischemic stroke within 48 hours of the onset of symptoms in a $3 \times 2$ factorial design to receive placebo, aspirin alone (300 mg/day), subcutaneous heparin alone (low dose, 10,000 IU/day; or medium dose, 25,000 IU/day), or both aspirin and heparin for 14 days.[171] Aspirin was associated with an overall significant reduction in total recurrent ischemic strokes within 14 days compared to control ($p < 0.001$), with no significant excess of intracranial hemorrhages. At 6 months, a modest decrease in the risk of death or dependency was seen, with a significant 0.1% increase in intracranial hemorrhage. Heparin was associated with a much greater increase in symptomatic intracranial hemorrhage, which negated any benefit in efficacy. CAST also entered patients within 48 hours of symptom onset, and demonstrated a modest but significant reduction in recurrent ischemic stroke and a reduction in the prescribed combined outcomes.[172,173] Both trials together indicate that aspirin is associated with a small but significant reduction in the incidence of recurrent stroke and mortality in the first weeks poststroke when applied to an unselected group of ischemic stroke patients (in whom stroke subgroups were not defined). The mean follow-up in this study was 2 years (Table 36-5).

**2. Prostacyclin.** Prostacyclin (prostaglandin [PG] I$_2$), a product of endothelial cell arachidonic acid metabolism, inhibits platelet aggregation both *in vitro* and *in vivo* by elevation of platelet cytoplasmic cyclic adenosine monophosphate (cAMP) levels (see Chapter 13 for details). Small studies of PGI$_2$, iloprost, or other stable prostacyclin analogues in patients with completed stroke have failed to demonstrate therapeutic efficacy.[225–228] However, these studies were generally underpowered for their prescribed outcome. In addition, one level I, prospective controlled trial demonstrated significant worsening by 2 weeks among patients who received PGI$_2$ within 24 hours of stroke-related symptoms compared to placebo.[229] So far, there has been no evidence of benefit with PGI$_2$ in completed stroke, in part because of inadequate sample sizes and side effects.[230] However, experimental data imply the possibility of benefit.[231,232]

**3. Suloctidil.** Suloctidil inhibits platelet aggregation induced by collagen and thrombin, serotonin leakage, and

**Table 36-5: Antiplatelet Agents for Completed Stroke**

| Study | Year | Agent | Dose (per day in mg) | Patients (n) | Follow-up (years) | Stroke (n) | Mortality (n) Vascular[a] | Nonvascular | Total |
|---|---|---|---|---|---|---|---|---|---|
| Swedish Cooperative Study[223] | 1987 | ASA | 1500 | 253 | 2.0 | 32 | 27 | 7 | 34 |
| | | Placebo | | 252 | | 32 | 25 | 12 | 37 |
| Cent et al.[224] | 1985 | Suloctidil | 600 | 218 | 1.7 | 29[b] | 4[b] | 9[b] | 13[b] |
| | | Placebo | | 220 | | 28[b] | 14[b] | 11[b] | 25[b] |
| Blakeley[234] | 1979 | Sulfinpyrazone | 800 | 145 | — | — | — | — | 25 |
| | | Placebo | | 145 | | — | — | — | 28 |
| Canadian–American Ticlopidine Study[219] | 1989 | Ticlopidine | 500 | 525 | 2.0 | 54 | 17[b] | 13[b] | 30[b] |
| | | Placebo | | 528 | | 89 | 29[b] | 8[b] | 38[b] |
| CAPRIE[237] | 1996 | Clopidogrel | 75 | 3233 | 1.9 | 315 | 102 | — | — |
| | | ASA | 325 | 3198 | | 338 | 102 | — | — |
| European Stroke Prevention Study-2[177] | 1997 | Dipyridamole/ASA[d] | 400/50 | 1650 | 2.0 | 157 | 105 | 180 | 285 |
| | | Dipyridamole | 400 | 1654 | | 211 | 118 | 170 | 288 |
| | | ASA | 50 | 1649 | | 206 | 109 | 173 | 282 |
| | | Placebo | — | 1649 | | 250 | 117 | 85 | 202 |
| IST[171] | 1997 | ASA[e] | 300 | 9720 | 0.5 | 362[f] | 855 | 17 | 872 |
| | | No ASA[e] | — | 9715 | | 452[f] | 896 | 13 | 909 |
| CAST[173] | 1997 | ASA | 160 | 10,554 | 0.08[g] | 335 | 243 | 100[h] | 343 |
| | | No ASA | — | 10,552 | | 351 | 283 | 115[h] | 398 |

ASA, aspirin.
[a]Stroke, myocardial infarction, and peripheral vascular events.
[b]Eligible events only, excluding events >28 days after study drug was permanently discontinued.
[c]Stroke subgroup.
[d]Sustained release.
[e]Factorial design (include heparin ± ASA).
[f]14-day outcomes.
[g]4-week outcomes.
[h]Unknown cause.

other mechanisms leading to reduced thrombus extension.[233] Decreased fibrinogen level and whole blood viscosity, a hypocholesterolemic effect, and peripheral dilatation also occur. In a single study, no significant difference was seen between suloctidil (600 mg/day, n = 218) and placebo (n = 220) for the outcomes of subsequent first stroke, MI, or vascular death at the 1.7 year follow-up.[224] Side effects, including reversible hepatitis, resulted in individual withdrawal.

**4. Sulfinpyrazone.** In one limited study, sulfinpyrazone (800 mg/day) did not significantly reduce the mortality in individuals treated within 6 months of a clinically presumed atherothrombotic or thromboembolic stroke, although a trend favoring sulfinpyrazone was suggested.[234]

**5. Ticlopidine and Clopidogrel.** Studies with the thienopyridines in completed stroke have been complex. Clopid-

ogrel and ticlopidine are thienopyridines, which irreversibly inhibit binding of ADP to its platelet $P2Y_{12}$ receptor [235,236] (see Chapter 61 for details). Studies of dose equivalence as measured on the basis of *ex vivo* platelet aggregation confirm the efficacy of single daily dosing with clopidogrel. Gent et al. tested the effect of ticlopidine (500 mg/day) on the incidence of second stroke following initial thromboembolic stroke in the Canadian American Ticlopidine Study (CATS).[219] CATS demonstrated a significant decrease in the combined outcomes of stroke, MI, and vascular mortality after a mean follow-up of 2 years in patients entered within 1 to 16 weeks of a first completed stroke who received ticlopidine (n = 525) over placebo (n = 528). Side effects related to ticlopidine, which included reversible leukopenia, diarrhea, or rash, occurred in 8% of patients.

Given the more favorable safety profile of clopidogrel (see Chapter 61), a broader study of that agent in ischemic disorders was undertaken. The Clopidogrel versus Aspirin in Individuals at Risk of Ischaemic Events (CAPRIE) study was a prospective, randomized, blinded comparison of clopidogrel (75 mg/day, $n = 9577$) versus aspirin (325 mg/day, $n = 9566$) in patients with a signal MI less than 35 days old, thrombotic peripheral artery disease (PAD), or ischemic stroke (including lacunes).[237] An 8.7% relative reduction in the risk of the combined outcomes of ischemic stroke, MI, and vascular-related death at 1 year was observed in favor of clopidogrel ($p = 0.043$). No differences in the incidence of clinically significant neutropenia or in adverse experiences were noted. A nearly equivalent frequency of intracranial hemorrhage was noted between the two treatment arms which was greater than placebo (clopidogrel, 0.33%; aspirin, 0.47%). The benefits of clopidogrel in that study were primarily "driven" by the favorable outcomes observed with PAD. No specific advantage of clopidogrel for stroke was observed. However, in current clinical practice clopidogrel (or aspirin/dipyridamole) has replaced ticlopidine. The Management of Atherothrombosis with Clopidogrel in High-Risk Patients (MATCH) trial compared the efficacy of clopidogrel/aspirin with aspirin alone for ischemic events including stroke. The addition of clopidogrel to aspirin increased the absolute risk reduction by an insignificant 1% over aspirin alone, but a worrisome increase in major hemorrhage with the combination was seen.[238,239] The Clopidogrel in Unstable Angina to Prevent Recurrent Events (CURE) trial indicated that both aspirin and the combination clopidogrel/aspirin generated an aspirin dose-dependent increase in major hemorrhage.[240,241]

**6. GPIIb-IIIa Antagonists.** The acute use of GPIIb-IIIa (integrin $\alpha IIb\beta 3$) antagonists in acute ischemic stroke or other cerebrovascular disorders is so far experimental. A phase II study of the humanized monoclonal antibody abciximab given early in 74 ischemic stroke patients, and several animal model studies of polypeptide or organic molecule GPIIb-IIIa antagonists in acute MCA:O have been reported.[110,112,174] Concerns regarding hemorrhagic risk are relevant.[242] In this phase II study, the pan-integrin $\beta 3$ inhibitor abciximab at the purported $IC_{50}$ did not apparently produce significant intracerebral hemorrhage or apparent improvement among patients with symptoms of ischemic stroke treated within 24 hours of the event.[174] Recently, the prospective blinded phase III controlled study of abciximab in early ischemic stroke (Abciximab Emergent Stroke Treatment Trial-II, AbeSTT-II), at a dose applied for cardiac indications, was terminated because of significant safety (hemorrhage) concerns.[242a] This study followed a second phase II trial that had suggested benefit in the cohort receiving abciximab.[242b] Hence, there is no current indication for the use of this agent in stroke.

Other reports of the use of abciximab have been confined to a limited number of situations involving angioplasty or stent placement within large brain-supplying arteries (e.g., vertebrobasilar territory or carotid artery) of patients with cerebral ischemic symptoms.[242c,242d] The use of this agent in this setting has been evaluated prospectively.[242e] At the time of this writing, a single prospective randomized trial of acute intervention recombinant tissue plasminogen activator (rt-PA) with or without the GPIIb-IIIa antagonist eptifibatide is in progress (the Combined Approach to Lysis Utilizing Eptifibatide and rt-PA in Acute Ischemic Stroke [CLEAR Stroke Trial]).

**7. Other Agents with Antiplatelet Effect.** In addition to agents with direct antiplatelet effect, statins and other pharmacologic classes employed in patients with risk of stroke can alter platelet function. Select statins may have direct antiplatelet activity.[191,243,244] In addition, a recent report suggests that the angiotensin II receptor inhibitor (valsartan) can significantly inhibit platelet aggregation *in vitro* under controlled conditions.[245] Whether this is a class effect is unknown.

**8. Summary.** In summary, on the basis of its salutary effects on stroke outcome in individuals with TIA, aspirin is recommended for secondary prevention. Ticlopidine produces a significant further reduction in subsequent stroke, MI, or vascular-related death over aspirin in individuals with an initial completed stroke. On the basis of its similar mode of action to ticlopidine, clopidogrel has been substituted in clinical practice. In general, for patients with a noncardioembolic ischemic stroke, secondary prevention (when there is no contraindication) should be provided with aspirin (50–325 mg/day), aspirin/extended-release dipyridamole, or clopidogrel (75 mg/day). The latter recommendation is based upon data with ticlopidine. For patients allergic to aspirin, clopidogrel has been recommended over ticlopidine.[181] For secondary prevention, aspirin/extended release dipyridamole is superior to aspirin in direct comparison.[212]

## VI. Carotid Artery Atherothrombotic Disease

Atheromas within the aorta and brain-supplying carotid, vertebral, and basilar arteries generate a local prothrombotic state. The carotid artery bifurcation is a predilection site for atheromatous change, which generates platelet–fibrin thromboemboli primarily to the MCA and ACA territories. Endarterectomy of the extracranial atherosclerotic portion of the carotid artery in selected patients, with or without patch angioplasty, can resolve the local abnormality of vascular flow, but cerebral ischemic symptoms can recur.

The results of both the North American Symptomatic Carotid Endarterectomy Trial (NASCET) and the MRC

European Carotid Surgery Trial (ECST) demonstrated significant survival benefit and symptomatic relief from endarterectomy over accepted medical (i.e., antiplatelet agent) therapy for symptomatic carotid artery stenosis of 70 to 99%, in which surgical risk is low.[246,247] Both trials were undertaken to resolve controversies arising from the liberal use of this surgical procedure in individuals with TIAs.

Trials of adjunctive antiplatelet treatment have sought to alter the incidence of stroke and death or to decrease the incidence of carotid artery restenosis after carotid endarterectomy. Restenosis can occur early due to thrombotic occlusion at the sites of surgical anastomoses, or late due to myointimal hyperplasia.[248] One prospective study failed to show that the combination of aspirin with dipyridamole could decrease postendarterectomy restenosis at the surgical site.[248] In another prospective, double-blind trial, patients who underwent carotid endarterectomy were randomized to aspirin (1300 mg/day, $n = 65$) or to placebo ($n = 60$) within 5 days of the procedure.[249] The incidence of stroke or death at 6 months was greater in those receiving placebo, but the difference did not reach statistical significance. In the Danish Very Low-Dose Aspirin trial, no significant difference in the outcomes of stroke, MI, or vascular death was seen at 2.1 years follow-up between individuals receiving aspirin (50–100 mg/day, $n = 150$) or placebo ($n = 151$) within 1 to 12 weeks after carotid endarterectomy.[250] In a retrospective examination, Kretschmer et al. found significantly prolonged survival in patients undergoing carotid endarterectomy who had received aspirin (1500 mg/day) compared to no adjunctive treatment.[251] This study[251] formed the basis for a prospective comparison of presurgical aspirin (1000 mg/day, $n = 32$) versus no treatment ($n = 34$).[251a] Survival was prolonged in the aspirin-treated group, although cerebral events were equally frequent in both. The small number of individuals in all three trials, however, may have obscured any potential real benefit to cerebrovascular outcome in the active arms. At this time, there are no results of level I trials to support a recommendation for a specific adjunctive antiplatelet therapy after endarterectomy. However, in endarterectomy trials, and in clinical practice, patients undergoing endarterectomy are continued on aspirin after the operation.

# VII. Cerebral Embolism from a Cardiac Source

Cardiac sources of embolic occlusion of CNS arteries derive from mural thrombi associated with left ventricular dyskinesis in myocardial ischemia, prosthetic valves and rheumatic valvular vegetations, and thromboembolism in atrial fibrillation (AF). For most cardioembolic events, there is a role for anticoagulation in both primary and secondary prevention. The use of antiplatelet agents is more limited in this setting.

## A. Cardioembolic Stroke in Acute Myocardial Infarction

The incidence of systemic thromboembolism, including stroke, following MI varies from 1 to 3% per year, but it may be as high as 3.7% during the first month after MI.[252-257] The number of cerebrovascular events after recovery from MI can be reduced with the use of long-term anticoagulation, as indicated by two level I trials.[252-254,256-258] In practice, routine use of anticoagulants does not occur because of the higher incidence of serious hemorrhagic events, including intracerebral hemorrhages. Evidence that antiplatelet agents might be effective in decreasing the incidence of post-MI cerebral events is limited. The CAPRIE study suggested that clopidogrel could reduce the combined events of second MI, stroke, or PAD following an initial MI, although the benefit for stroke was not significant.[237]

### B. Cardioembolic Stroke in Atrial Fibrillation

AF can be classified either as nonvalvular (i.e., nonrheumatic) or as AF associated with valvular dysfunction. AF is a significant risk factor for thromboembolic stroke and stroke recurrence, and it may be first observed during stroke in up to 30% of patients.[28,259-261] The two-year incidence of stroke in individuals with chronic nonvalvular AF is 6.2 to 7.6%.[260,262]

In general, aspirin is not employed for primary prevention in nonvalvular AF. However, the evidence is interesting. The Stroke Prevention in Atrial Fibrillation (SPAF) study randomized 1244 individuals to warfarin, aspirin, or placebo (group 1) if they were eligible for warfarin, or to aspirin or placebo (group 2, double-blind) if they were not.[263] SPAF-1 was terminated at a mean follow-up of 1.13 years, when the warfarin and aspirin arms of group 1 were shown to have a significant combined 81% risk reduction for the outcomes of ischemic stroke and systemic embolism. In group 2 there was relative benefit of aspirin over control (3–6 events/year vs. 6.3 events/year, $p = 0.02$). The relative benefit of warfarin over aspirin was not reported, but 10.9% of individuals randomly assigned to warfarin were withdrawn because of drug intolerance. Nonsignificant differences between the aspirin treatment and control arms of the Atrial Fibrillation Aspirin and Anticoagulation (AFASAK)-1 study, the European Atrial Fibrillation Trial (EAFT), and ESPS-2 were observed in patients with nonvalvular AF.[177,264,265] Aspirin was associated with a 21% RRR in annual stroke events (aspirin treated = 6.3, control = 8.1%, $p = 0.05$) in an individual patient combined analysis of AFASAK-1, EAFT, and SPAF-1.[266] That conclusion was supported by a broader meta-analysis.[267] Heterogeneity in the results could not be excluded.[268]

Concern regarding the risk of intracranial hemorrhage associated with warfarin fueled other approaches using aspirin or adjusted low-dose oral anticoagulation.[269,270] The SPAF-II trial tested the relative efficacies of warfarin and aspirin in patients with nonvalvular AF and suggested a modest, but not significant, decrease in ischemic stroke events with warfarin over aspirin in the ≤75-year group (33%) and >5-year group (27%). There was a significantly greater frequency of major hemorrhagic events with warfarin in the older cohort (>75 years) than in the younger cohort (*p* = 0.008). However, both AFASAK-1 and the EAFT demonstrated that warfarin could significantly reduce the annual stroke rate compared to aspirin.[264]

In the SPAF-III trial, patients received either adjusted dose warfarin (international normalized ratio [INR] 2.0–3.0; *n* = 523) or low-intensity warfarin with aspirin (325 mg/day) to maintain an INR of between 1.2 and 1.5 (*n* = 521).[270] After a mean follow-up of 1.1 years SPAF-III was terminated when the annual disabling stroke rate in the low-intensity warfarin/aspirin combination group exceeded that of adjusted-dose warfarin (5.6 vs. 1.7%, respectively; *p* = 0.0007). This led to the early termination of the AFASAK-2 study of moderate risk patients.[271,272] Recently, a prospective randomized comparison of the antiplatelet agent clopidogrel with warfarin for nonvalvular AF was terminated when the clopidogrel treatment group suffered a 75% increased incidence of stroke compared to the control group.[272a] The basis for this trial was not clearly stated.

Nonetheless, for high-risk AF patients (with a prior TIA/ stroke or systemic embolus, history of hypertension, poor left ventricular function, age >75 years, rheumatic mitral valve disease, or a prosthetic valve) the recommended therapy is adjusted dose warfarin anticoagulation at a target INR of 2.5 (range 2.0–3.0) rather than aspirin.[273]

### C. Cardiac Valvular Disease

Cerebral embolism can be associated with rheumatic valvular disease, both mechanical and xenograft prosthetic cardiac valves and calcified mitral annuli. The embolic risk is greatly increased by coexistent AF.[274–276] Coulshed et al. recorded a 3.7 and 1.9% yearly incidence of systemic embolism from untreated rheumatic mitral stenosis and mitral insufficiency, respectively.[277] The incidence of systemic embolism from prosthetic heart valves is greater for mechanical devices than for xenograft prostheses.[278–280]

**1. Prosthetic Mechanical Cardiac Valves.** Antiplatelet agents and anticoagulants have been used as prophylaxis against the embolic complications of mechanical valves.[276] The combination of warfarin and aspirin in one level I study was associated with a decreased incidence of cerebral thromboembolism in patients with mechanical prosthetic cardiac valves with fewer hemorrhagic events.[276,281] Generally, all individuals with mechanical prosthetic valves should be treated with long-term anticoagulation at a target INR of 3.5 (range, 3.0–4.5).[282,283] Antiplatelet agents can increase the protection against thromboemboli afforded by anticoagulants alone, but the risk of hemorrhagic complications may increase. Two level I studies demonstrated reduced thromboembolic risk in individuals receiving combination anticoagulation with aspirin over anticoagulation alone.[284,285] The doses of aspirin used were in the range of 500 to 1000 mg/day. In a separate randomized double-blind trial, the addition of dipyridamole (400 mg/day) to warfarin produced a significant reduction in thromboembolism or death over warfarin and placebo, a finding subsequently supported by an open trial.[286,287] A meta-analysis of trials combining dipyridamole with oral anticoagulation indicated a decrease in the frequency of fatal and nonfatal thromboemboli.[276,289] The addition of dipyridamole to aspirin reduced the frequency of thromboembolic events, and major hemorrhagic complications were significantly fewer (*p* = 0.001). To date there is no reported experience with combination low-dose aspirin/long-acting dipyridamole for prosthetic cardiac valves.

**2. (Bio)Prosthetic Xenograft Cardiac Valves.** The lower incidence of thromboembolic events with xenograft valve prostheses as compared to that with mechanical prostheses implies that antiplatelet prophylaxis may have some benefit with lower hemorrhagic risk. Two prospective, nonrandomized trials of warfarin versus aspirin (1000 or 500 mg/day) have examined this notion.[290,291] In the first study, patients with AF had fewer embolic events with aspirin (*n* = 135) compared to warfarin (*n* = 151).[290] In the second level III study, individuals who received treatment every 2 days with 500 mg aspirin following placement of mitral valve bioprostheses were less likely to have embolic events than those who received 1000 mg/day.[291] Thus, with a mitral valve bioprosthesis, aspirin may provide some protection against embolic events; however, this conclusion has not been rigorously tested. Generally accepted clinical practice requires anticoagulation of individuals with a bioprosthesis for the initial three postoperative months, when the thromboembolic risk is greatest. The addition of aspirin to adequate dose oral anticoagulation may increase the incidence or severity of hemorrhagic complications.

## VIII. Conclusions

Platelets participate in the development of ischemic stroke and recurrent stroke at four levels: (a) Platelet-containing thrombi originating from atherosclerotic lesions of the brain-supplying arteries or the aortic arch can occlude the artery or embolize downstream to alter regional cerebral blood flow. (b) Within the early moments following the onset of focal cerebral ischemia, activated platelets accumulate within the microvasculature of the ischemic regions. Microvessel occlusions can be reduced by blockade of platelet–fibrin(ogen) interactions in experimental systems with evidence of potential benefit. (c) Intact platelet function is required to prevent significant hemorrhage within the ischemic region. (d) The involvement of platelets with large artery lesions leading to stroke suggests a role for antiplatelet agents in prophylaxis or secondary prevention.

Antiplatelet agents reduce the risk for the combined outcomes of subsequent stroke, MI, and mortality in individuals with a history of recent TIAs. Aspirin is the most widely used antiplatelet agent in this setting. When TIAs are associated with carotid artery stenosis of 70 to 99%, carotid endarterectomy and aspirin are superior to aspirin alone in preventing subsequent stroke and stroke-related mortality. While the combination of aspirin and dipyridamole was no better than aspirin alone for TIAs in one study, aspirin/extended-release dipyridamole significantly reduced the risk of stroke, death, or both compared to either agent alone or to placebo in patients with a history of TIAs or stroke. For secondary prevention of recurrent stroke in individuals presenting with a stable, recent focal cerebral deficit, ticlopidine and clopidogrel are beneficial, and aspirin/dipyridamole also may be active. But these strategies have not been compared directly. It is intriguing to consider that the stroke-risk reduction of certain statins could be due to their individual antiplatelet effects. Although as yet untested in the acute setting, the possibility that safe inhibition of platelet aggregation can reduce subsequent injury in ischemic stroke has an experimental basis.

# References

1. Denny–Brown, D. (1960). Recurrent cerebrovascular episodes. *Arch Neurol, 2,* 194–210.
2. Russell, R. W. R. (1961). Observations on the retinal blood vessels in monocular blindness. *Lancet, 2,* 1422–1428.
3. Hollenhorst, R. W. (1966). Vascular status of patients who have cholesterol emboli in the retina. *Am J Ophthalmol, 77,* 1159–1165.
4. Barnett, H. J. M. (1979). The pathophysiology of transient cerebral ischemic attacks: Therapy with platelet antiaggregants. *Med Clin North Am, 63,* 649–679.
5. Barnett, H. J. M. (1978). Delayed cerebral ischemic episodes distal to occlusion of major cerebral arteries. *Neurology, 28,* 769.
6. del Zoppo G. J., Poecm, K., Pessin, M. S., et al. (1992). Recombinant tissue plasminogen activator in acute thrombotic and embolic stroke. *Ann Neurol, 32,* 78–86.
7. Fieschi, C., Argentino, C., Lenzi, G. L., et al. (1989). Clinical and instrumental evaluation of patients with ischemic stroke within the first six hours. *J Neurol Sci, 91,* 311–321.
8. del Zoppo, G. J., Higashida, R. T., Furlan, A. J., et al. (1998). PROACT: A phase II randomized trial of recombinant pro-urokinase by direct arterial delivery in acute middle cerebral artery stroke. *Stroke, 29,* 4–11.
9. Delanty, N., & Vaughan, C. J. (1997). Vascular effects of statins in stroke. *Stroke, 28,* 2315–2320.
10. Laufs, U., Gertz, K., Huang, P., et al. (2000). Atorvastatin upregulates type III nitric oxide synthase in thrombocytes, decreases platelet activation, and protects from cerebral ischemia in normocholesterolemic mice. *Stroke, 31,* 2442–2449.
11. Sacks, F. M., Pfeffer, M. A., Moye, L. A., et al. (1996). The effect of pravastatin on coronary events after myocardial infarction in patients with average cholesterol levels. *N Eng J Med, 335,* 1001–1009.
12. Scandinavian Simvastatin Survival Study Group. (1994). Randomised trial of cholesterol lowering in 4444 patients with coronary heart disease: The Scandinavian Simvastatin Survival Study 4S. *Lancet, 344,* 1383–1389.
13. Greisenegger, S., Mullner, M., Tentschert, S., et al. (2004). Effect of pretreatment with statins on the severity of acute ischemic cerebrovascular events. *J Neurol Sci, 221,* 5–10.
14. Marti-Fabregas, J., Gomis, M., Arboix, A., et al. (2004). Favorable outcome of ischemic stroke in patients pretreated with statins. *Stroke, 35,* 1117–1121.
15. Stroke Council AHAASA. (2004). Statins after ischemic stroke and transient ischemic attack: An advisory statement from the Stroke Council, American Heart Association and American Stroke Association. *Stroke, 35,* 1023.
16. Ross, R. (1993). The pathogenesis of atherosclerosis: A perspective for the 1990s. *Nature, 362,* 801–809.
17. Ross, R. (1999). Atherosclerosis—An inflammatory disease. *N Engl J Med, 340,* 115–126.
18. Mohr, J. P., Caplan, L. R., Melski, J. W., et al. (1978). The Harvard Cooperative Stroke Registry: A prospective registry of patients hospitalized with stroke. *Neurology, 28,* 754–762.
19. Sacco, R. L., Wolf, P. A., Kannel, W. B., et al. (1982). Survival and recurrence following stroke: The Framingham Study. *Stroke, 13,* 290–295.
20. Mohr, J. P., & Barnett, H. J. M. (1986). Classification of ischemic strokes. In H. J. M. Barnett, B. M. Stein, J. P. Mohr, et al. (Eds.), *Stroke: Pathophysiology, diagnosis and management* (Vol. 1, pp. 281–291). New York: Churchill Livingstone.
21. Jones, E. F., Calafiore, P., Donnan, G. A., et al. (1994). Evidence that patent foramen ovale is not a risk factor for cerebral ischemia in the elderly. *Am J Cardiol, 74,* 596–599.
22. Cujec, B., Mainra, R., & Johnson, D. H. (1999). Prevention of recurrent cerebral ischemic events in patients with patent foramen ovale and cryptogenic strokes or transient ischemic attacks. *Can J Cardiol, 15,* 57–64.
23. Rodriguez, C. J., Homma, S., Sacco, R. L., et al. (2003). Race-ethnic differences in patent foramen ovale, atrial septal aneurysm, and right atrial anatomy among ischemic stroke patients. *Stroke, 34,* 2097–2102.

24. Solis, O. J., Roberson, G. R., Taveras, J. M., et al. (1977). Cerebral angiography in acute cerebral infarction. *Revist Interam Radiol, 2,* 19–25.
25. rt-PA Acute Stroke Study Group. (1991). An open safety/efficacy trial of rt-PA in acute thromboembolic stroke: Final report. *Stroke, 22,* 153.
26. Irino, T., Taneda, M., & Minami, T. (1977). Angiographic manifestations in post-recanalized cerebral infarction. *Neurology, 27,* 471–475.
27. Marshall, J. (1964). The natural history of transient ischemic cerebrovascular attacks. *Q J Med, 33,* 309–324.
28. Wolf, P. A., Kannel, W. B., McGee, D. L., et al. (1983). Duration of atrial fibrillation and imminence of stroke: The Framingham Study. *Stroke, 14,* 664–667.
29. Wolf, P. A., Kannel, W. B., & McGee, D. L. (1986). Prevention of ischemic stroke: Risk factors. In H. J. M. Barnett, J. P. Mohr, & B. M. Stein (Eds.), *Stroke: Pathophysiology, diagnosis, and management* (p. 983). New York: Churchill-Livingstone.
30. Whisnant, J. P., Matsumoto, N., & Elveback, L. R. (1973). Transient cerebral ischemic attacks in a community, Rochester, Minnesota, 1955 through 1969. *Mayo Clin Proc, 48,* 194–198.
31. Caplan, L. R. (1983). Are terms such as completed stroke or RIND of continued usefulness? *Stroke, 14,* 431–433.
32. Schmidt, E. V., Smirnov, V. E., & Ryabova, V. S. (1988). Results of the seven-year prospective study of stroke patients. *Stroke, 19,* 942–949.
33. van Kooten, F., Ciabattoni, G., Patrono, C., et al. (1994). Evidence for episodic platelet activation in acute ischemic stroke. *Stroke, 25,* 278–281.
34. van Kooten, F., Ciabattoni, G., Koudstaal, P. J., et al. (1999). Increased platelet activation in the chronic phase after cerebral ischemia and intracerebral hemorrhage. *Stroke, 30,* 546–549.
35. Spencer, M. P., Thomas, G. I., Nicholls, S. C., et al. (1990). Detection of middle cerebral artery emboli during carotid endarterectomy using transcranial Doppler sonography. *Stroke, 21,* 415–423.
36. Markus, H. S., Thomson, N., & Brown, M. M. (1995). Asymptomatic cerebral embolic signals in symptomatic and asymptomatic carotid artery disease. *Brain, 118,* 1005–1011.
37. Tong, D., & Albers, G. (1995). Transcranial Doppler-detected microemboli in patients with acute stroke. *Stroke, 26,* 1588–1592.
38. Sliwka, U., Job, F. P., Wissuwa, D., et al. (1996). Occurrence of transcranial Doppler high intensity signals in patients with potential cardiac source of embolism. *Stroke, 26,* 2067–2070.
39. Sliwka, U., Lingnau, A., Stohlmann, W.-D., et al. (1997). Prevalence and time course of microembolic signals in patients with acute stroke. *Stroke, 27,* 1844–1849.
40. Batista, P., Oliveira, V., & Ferro, J. M. (1999). The detection of microembolic signals in patients at risk of recurrent cardioembolic stroke: Possible therapeutic relevance. *Cerebrovasc Dis, 9,* 314–319.
41. Wityk, R. J., Pessin, M. S., Kaplan, R. F., et al. (1994). Serial assessment of acute stroke using the NIH stroke scale. *Stroke, 25,* 362–365.
42. Dougherty, J. H., Levy, D. E., & Weksler, B. B. (1977). Platelet activation in acute cerebral ischemia. *Lancet, 3,* 821–824.
43. Mettinger, K. L., Nyman, D., Kjellin, K.-G., et al. (1979). Factor VIII related antigen, anti-thrombin III, spontaneous platelet aggregation and plasminogen activator in ischemic cerebrovascular disease. *J Neurol Sci, 41,* 31–38.
44. deBoer, A. C. (1982). Plasma beta-thromboglobulin and serum fragment E in acute partial stroke. *Br J Haematol, 50,* 327–344.
45. Cella, G., Zahavi, J., de Haas, H. A., et al. (1979). β-thromboglobulin, platelet production time, and platelet function in vascular disease. *Br J Haematol, 43,* 127–136.
46. Feinberg, W. M., Bruck, D. C., Ring, M. E., et al. (1989). Hemostatic markers in acute stroke. *Stroke, 20,* 592–597.
47. Feinberg, W. M., & Bruck, D. C. (1991). Time course of platelet activation following acute ischemic stroke. *J Stroke Cerebrovasc Dis, 1,* 124–128.
48. Feinberg, W. M., Pearce, L. A., Hart, R. G., et al. (1999). Markers of thrombin and platelet activity in patients with atrial fibrillation: Correlation with stroke among 1531 participants in the stroke prevention in atrial fibrillation III study. *Stroke, 30,* 2547–2553.
49. Fisher, M., & Zipser, R. (1985). Increased excretion of immunoreactive thromboxane B2 in cerebral ischemia. *Stroke, 16,* 10–14.
50. Zeller, J. A., Tschoepe, D., & Kessler, C. (1999). Circulating platelets show increased activation in patients with acute cerebral ischemia. *Thromb Haemost, 81,* 373–377.
51. Akopov, S., Sercombe, R., & Seylaz, J. (1996). Cerebrovascular reactivity: Role of endothelium/platelet/leukocyte interactions. *Cerebrovasc Brain Metab Rev, 8,* 11–94.
52. Weksler, B. B. (1995). Hematologic disorders and ischemic stroke. *Curr Opin Neurol, 8,* 38–44.
53. Uchiyama, S., Takeuchi, M., Osawa, M., et al. (1983). Platelet function in thrombotic cerebrovascular disorders. *Stroke, 14,* 511–517.
54. Pramsohler, B., Lupattelli, G., Scholz, H., et al. (1990). Platelet scintigraphy and survival in juvenile stroke patients. *Prog Clin Biol Res, 355,* 71–80.
55. Uchiyama, S., Yamazaki, M., Hara, Y., et al. (1997). Alterations of platelet, coagulation, and fibrinolysis markers in patients with acute ischemic stroke. *Semin Thromb Hemost, 23,* 535–541.
56. Harker, L. A., & Slichter, S. J. (1974). Arterial and venous thromboembolism. *Diath Haemorrh, 31,* 188–203.
57. Sinzinger, H., Silberbauer, K., Fitscha, P., et al. (1982). Wertigkeit des Nachweises atherosklerotischer Lasionen mit markierten autologen Thrombozyten. *Acta Med Aust, 9,* 181–184.
58. Iwamoto, T., Kubo, H., & Takasaki, M. (1995). Platelet activation in the cerebral circulation in different subtypes of ischemic stroke and Binswanger's disease. *Stroke, 26,* 52–56.

59. Hamann, G. F., Okada, Y., & del Zoppo, G. J. (1996). Hemorrhagic transformation and microvascular integrity during focal cerebral ischemia/reperfusion. *J Cereb Blood Flow Metab, 16*, 1373–1378.

60. Legrand, C., Woimant, F., Haguenau, M., et al. (1991). Platelet surface glycoprotein changes in patients with cerebral ischemia. *Nouvelle Revue Francaise d'hematology, 33*, 497–499.

61. Konstantopoulos, K., Grotta, J. C., Sills, C., et al. (1995). Shear-induced platelet aggregation in normal subjects and stroke patients. *Thromb Haemost, 74*, 1329–1334.

62. Shah, A. B., Beamer, N., & Coull, B. M. (1985). Enhanced *in vivo* platelet activation in subtypes of ischemic stroke. *Stroke, 16*, 643–647.

63. de Boer, A. C., Turpie, A. G., Butt, R. W., et al. (1982). Plasma beta thromboglobulin and serum fragment E in acute partial stroke. *Br J Haematol, 50*, 327–334.

64. Grau, A. J., Boddy, A. W., Dukovic, D. A., et al. (2004). CAPRIE Investigators. Leukocyte count as an independent predictor of recurrent ischemic events. *Stroke, 35*, 1147–1152.

65. Meiklejohn, D. J., Vickers, M. A., Morrison, E. R., et al. (2001). *In vivo* platelet activation in atherothrombotic stroke is not determined by polymorphisms of human platelet glycoprotein IIIa or Ib. *Br J Haematol, 112*, 621–631.

66. Minamino, T., Kitakaze, M., Sanada, S., et al. (1998). Increased expression of P-selectin on platelets is a risk factor for silent cerebral infarction in patients with atrial fibrillation: Role of nitric oxide. *Br J Haematol, 98*, 1721–1727.

67. Lee, Y. J., Jy, W., Horstman, L. L., et al. (1994). Elevated platelet microparticles in transient ischemic attacks, lacunar infarcts, and multiinfarct dementias. *Thromb Res, 72*, 295–304.

68. Geiser, T., Sturzenegger, M., Genewein, U., et al. (1998). Mechanisms of cerebrovascular events as assessed by procoagulant activity, cerebral microemboli, and platelet microparticles in patients with prosthetic heart valves. *Stroke, 29*, 1770–1777.

69. van Kooten, F., Ciabattoni, G., Patrono, C., et al. (1997). Platelet activation and lipid peroxidation in patients with acute ischemic stroke. *Stroke, 28*, 1557–1563.

70. Yip, H. K., Chen, S. S., Liu, J. S., et al. (2005). Serial changes in platelet activation in patients after ischemic stroke: Role of pharmacodynamic modulation. *Stroke, 35*, 1683–1687.

71. Barnett, H. J. M. (1987). Medical treatment of transient ischemia. In J. H. Wood (Ed.), *Cerebral blood flow physiologic and clinical aspects* (p. 595). New York: McGraw-Hill.

72. Lane, D. A., Wolff, S., Ireland, H., et al. (1983). Activation of coagulation and fibrinolytic systems following stroke. *Br J Haematol, 53*, 655–658.

73. Reiner, A. P., Siscovick, D. S., & Rosendaal, F. R. (2001). Hemostatic risk factors and arterial thrombotic disease. *Thromb Haemost, 85*, 584–595.

74. Manca, G., Parenti, G., Bellina, R., et al. (2001). 111In-platelet scintigraphy for the noninvasive detection of carotid plaque thrombosis. *Stroke, 32*, 719–727.

75. Fisher, M., Sandler, R., & Weiner, J. M. (1985). Delayed cerebral ischemia following arteriography. *Stroke, 16*, 431–434.

76. Schror, K., & Braun, M. (1990). Platelets as a source of vasoactive mediators. *Stroke, 21*(12 Suppl.), IV32–IV35.

77. Antiplatelet Trialists' Collaboration. (2002). Collaborative meta-analysis of randomised trials of antiplatelet therapy for prevention of death, myocardial infarction, and stroke in high-risk patients. *Br Med J, 324*, 71–86.

78. Gubitz, G., Sandercock, P., & Counsell, C. (2000). Antiplatelet therapy for acute ischaemic stroke. Cochrane Database of Systematic Reviews computer file. [2.], CD000029.

79. Nag, S., & Robertson, D. M. (1987). Gross pathologic anatomic. In J. H. Wood (Ed.), *Cerebral blood flow* (pp. 59–74). New York: McGraw-Hill.

80. Frijns, C. J., Kappelle, L. J., van Gijn, J., et al. (1997). Soluble adhesion molecules reflect endothelial cell activation in ischemic stroke and in carotid atherosclerosis. *Stroke, 28*, 2214–2218.

81. Mondy, J. S., Lindner, V., Miyashiro, J. K., et al. (1997). Platelet-derived growth factor ligand and receptor expression in response to altered blood flow in vivo. *Circ Res, 81*, 320–327.

82. Myllarniemi, M., Calderon, L., Lemstrom, K., et al. (1997). Inhibition of platelet-derived growth factor receptor tyrosine kinase inhibits vascular smooth muscle cell migration and proliferation. *FASEB J, 11*, 1119–1126.

83. Spence, J. D., Perkins, D. G., Kline, R. L., et al. (1984). Hemodynamic modification of aortic atherosclerosis. Effects of propranolol versus hydralazine in hypertensive hyperlipidemic rabbits. *Atherosclerosis, 50*, 325–333.

84. Steinman, D. A., Poepping, T. L., Tambasco, M., et al. (2000). Flow patterns at the stenosed carotid bifurcation: Effect of concentric versus eccentric stenosis. *Ann Biomed Engineering, 28*, 415–423.

85. Kawasaki, T., Dewerchin, M., Lijnen, H. R., et al. (2001). Mouse carotid artery ligation induces platelet-leukocyte-dependent luminal fibrin, required for neointima development. *Circ Res, 88*, 159–166.

86. Caplan, L. R., & Tettenborn, B. (1992). Embolism in the posterior circulation. In R. Bergeur & L. R. Caplan (eds.), *Vertebrobasilar arterial disease* (pp. 52–65). St. Louis: Quality Medical Publishing.

87. Fisher, C. M. (1969). The arterial lesions underlying lacunes. *Acta Neuropathol Berl, 12*, 1–15.

88. Potthast, K., Erdonmez, G., Schnelke, C., et al. (2000). Origin and appearance of HITS induced by prosthetic heart valves: An *in vitro* study. *Int J Artif Organs, 23*, 441–445.

89. Rambod, E., Beizaie, M., Shusser, M., et al. (1999). A physical model describing the mechanism for formation of gas microbubbles in patients with mitral mechanical heart valves. *Ann Biomed Engineering, 27*, 774–792.

90. Markus, H. S., Droste, D. W., Kaps, M., et al. (2005). Dual antiplatelet therapy with clopidogrel and aspirin in symptomatic carotid stenosis evaluated using Doppler embolic signal detection: The Clopidogrel and Aspirin for Reduction of Emboli in Symptomatic Carotid Stenosis CARESS. Trial. *Circulation, 111*, 2233–2240.

91. Kealy, W. F. (1978). Atheroembolism. *J Clin Pathol, 31,* 984–989.

92. Lammie, G. A., Sandercock, P. A., & Dennis, M. S. (1999). Recently occluded intracranial and extracranial carotid arteries. Relevance of the unstable atherosclerotic plaque. *Stroke, 30,* 1319–1325.

93. Jeynes, B. J. (1986). An assessment of the extent and distribution of experimentally induced cerebral atheroembolic vascular occlusions and infarcts. *Artery, 14,* 35–42.

94. Fieschi, C., Battistini, N., Volanto, F., et al. (1975). Animal model of TIA: An experimental study with intracarotid ADP infusion in rabbits. *Stroke, 6,* 617–621.

95. Furlow, T. W., Jr., & Bass, N. H. (1975). Stroke in rats produced by carotid injection of sodium arachidonate. *Science, 197,* 658–660.

96. Passero, S., Battistini, N., & Fieschi, C. (1981). Platelet embolism in rabbit brain. *Stroke, 12,* 781–786.

97. del Zoppo, G. J. (1990). Relevance of focal cerebral ischemia models. Experience with fibrinolytic agents. *Stroke, 21,* IV-155–IV-160.

98. Roux, S., Carteaux, J. P., Hess, P., et al. (1994). Experimental carotid thrombosis in the guinea pig. *Thromb Haemost, 71,* 252–256.

99. Golino, P., Ambrosio, G., Pascucci, I., et al. (1992). Experimental carotid stenosis and endothelial injury in the rabbit: An *in vivo* model to study intravascular platelet aggregation. *Thromb Haemost, 67,* 302–305.

100. Ito, T., Matsuno, H., Kozawa, O., et al. (1999). Comparison of the antithrombotic effects and bleeding risk of fractionated aurin tricarboxylic acid and the GPIIb/IIIa antagonist GR144053 in a hamster model of stenosis. *Thromb Res, 95,* 49–61.

101. Matsuno, H., Kozawa, O., Niwa, M., et al. (1998). Effect of GR144053, a fibrinogen-receptor antagonist, on thrombus formation and vascular patency after thrombolysis by t-PA in the injured carotid artery of the hamster. *J Cardiovasc Pharmacol, 32,* 191–197.

102. Quaknine-Orlando, B., Samama, C. M., Riou, B., et al. (1999). Role of the hematocrit in a rabbit model of arterial thrombosis and bleeding. *Anesthesiology, 90,* 1454–1461.

103. Kessler, C., Kelly, A. B., Suggs, W. D., et al. (1992). Induction of transient neurological dysfunction in baboons by platelet microemboli. *Stroke, 23,* 697–702.

104. Dietrich, W. D., Prado, R., Halley, M., et al. (1993). Microvascular and neuronal consequences of common carotid artery thrombosis and platelet embolization in rats. *J Neuropathol Exp Neurol, 52,* 351–360.

105. Dietrich, W. D., Dewanjee, S., Prado, R., et al. (1993). Transient platelet accumulation in the rat brain after common carotid artery thrombosis. An 111-In-labeled platelet study. *Stroke, 24,* 1534–1540.

106. Akopov, S., Ghazarian, A., & Gabrielian, E. (1992). Effects of nimodipine and nicergoline on cerebrovascular injuries induced by activation of platelets and leukocytes *in vivo*. *Arch Int Pharmacodyn Ther, 318,* 66–75.

107. Akopov, S. E., Sercombe, R., & Seylaz, J. (1994). Endothelial dysfunction in cerebral vessels following carotid artery infusion of phorbol ester in rabbits: The role of polymorphonuclear leukocytes. *J Cereb Blood Flow Metab, 14,* 1078–1087.

108. Okada, Y., Copeland, B. R., Mori, E., et al. (1994). P-selectin and intercellular adhesion molecule-1 expression after focal brain ischemia and reperfusion. *Stroke, 25,* 202–211.

109. del Zoppo, G. J., Copeland, B. R., Harker, L. A., et al. (1986). Experimental acute thrombotic stroke in baboons. *Stroke, 17,* 1254–1265.

110. Abumiya, T., Fitridge, R., Mazur, C., et al. (2000). Integrin $\alpha_{IIb}\beta_3$ inhibitor preserves microvascular patency in experimental acute focal cerebral ischemia. *Stroke, 31,* 1402–1410.

111. del Zoppo, G. J., Schmid-Schönbein, G. W., Mori, E., et al. (1991). Polymorphonuclear leukocytes occlude capillaries following middle cerebral artery occlusion and reperfusion in baboons. *Stroke, 22,* 1276–1284.

112. Choudhri, T. F., Hoh, B. L., Zerwes, H. G., et al. (1998). Reduced microvascular thrombosis and improved outcome in acute murine stroke by inhibiting GP IIb/IIIa receptor-mediated platelet aggregation. *J Clin Invest, 102,* 1301–1310.

113. Moriguchi, A., Maeda, M., Mihara, K., et al. (2005). FK419, a novel nonpeptide GPIIb/IIIa antagonist, restores microvascular patency and improves outcome in the guinea-pig middle cerebral artery thrombotic occlusion model: Comparison with triofiban. *J Cereb Blood Flow Metab, 25,* 75–86.

114. Moriguchi, A., Mihara, K., Aoki, T., et al. (2004). Restoration of middle cerebral artery thrombosis by novel glycoprotein IIb/IIIa antagonist FK419 in guinea pig. *Eur J Pharmacol, 498,* 179–188.

115. Garcia, J. H., Liu, K. F., Yoshida, Y., et al. (1994). Influx of leukocytes and platelets in an evolving brain infarct Wistar rat. *Am J Pathol, 144,* 188–199.

116. Rosenblum, W. I. (1990). Control of brain microcirculation by endothelium. *Keio J Med, 39,* 137–141.

117. Mayhan, W. G., & Faraci, F. M. (1990). Cerebral vasoconstrictor responses to serotonin during chronic hypertension. *Hypertension, 15,* 872–876.

118. Mayhan, W. G., Faraci, F. M., & Heistad, D. D. (1988). Responses of cerebral arterioles to adenosine 5′-diphosphate, serotonin, and the thromboxane analogue U-46619 during chronic hypertension. *Hypertension, 12,* 556–561.

119. Heistad, D. D., Breese, K., & Armstrong, M. L. (1987). Cerebral vasoconstrictor responses to serotonin after dietary treatment of atherosclerosis: Implications for transient ischemic attacks. *Stroke, 18,* 1068–1073.

120. Mayhan, W. G., Faraci, F. M., & Heistad, D. D. (1987). Impairment of endothelium-dependent responses of cerebral arterioles in chronic hypertension. *Am J Physiol, 253,* H1435–H1440.

121. Sobey, C. G., Faraci, F. M., Piegors, D. J., et al. (1996). Effect of short-term regression of atherosclerosis on reactivity of carotid and retinal arteries. *Stroke, 27,* 927–933.

122. Joseph, R., Tsering, C., Grunfeld, S., et al. (1991). Platelet secretory products may contribute to neuronal injury. *Stroke, 22,* 1448–1451.

123. Joseph, R., Tsering, C., & Welch, K. M. (1992). Study of platelet-mediated neurotoxicity in rat brain. *Stroke, 23,* 394–398.

124. Joseph, R., Tsering, C., Grunfeld, S., et al. (1992). Further studies on platelet-mediated neurotoxicity. *Brain Res, 577,* 268–275.

125. Aliprandi, A., Longoni, M., Stanzani, L., et al. (2005). Increased plasma glutamate in stroke patients might be linked to altered platelet release and uptake. *J Cereb Blood Flow Metab, 25,* 513–519.

126. Okada, Y., Copeland, B. R., Fitridge, R., et al. (1994). Fibrin contributes to microvascular obstructions and parenchymal changes during early focal cerebral ischemia and reperfusion. *Stroke, 25,* 1847–1854.

127. del Zoppo, G. J., Yu, J.-Q., Copeland, B. R., et al. (1992). Tissue factor localization in non-human primate cerebral tissue. *Thromb Haemost, 68,* 642–647.

128. Thomas, W. S., Mori, E., Copeland, B. R., et al. (1993). Tissue factor contributes to microvascular defects following cerebral ischemia. *Stroke, 24,* 847–853.

129. Mousa, S. A., DeGrado, W. F., Mu, D. X., et al. (1996). Oral antiplatelet, anthithrombotic efficacy of DMP 728, a novel platelet GPIIb/IIIa antagonist. *Circulation, 93,* 537–543.

130. Rebello, S. S., Driscoll, E. M., & Lucchesi, B. R. (1997). TP-9201, a glycoprotein IIb/IIIa platelet receptor antagonist, prevents rethrombosis after successful arterial thrombolysis in the dog. *Stroke, 289,* 1789–1796.

131. Mousa, S. A., Mu, D. X., & Lucchesi, B. R. (1997). Prevention of carotid artery thrombosis by oral platelet GPIIb/IIIa antagonist in dogs. *Stroke, 28,* 830–835.

132. Rote, W. E., Nedelman, M. A., Mu, D. X., et al. (1994). Chimeric 7E3 prevents carotid artery thrombosis in cynomolgus monkeys. *Stroke, 25,* 1223–1232.

133. Frenette, P. S., Denis, C. V., Weiss, L., et al. (2000). P-selectin glycoprotein ligand 1 (PSGL-1). is expressed on platelets and can mediate platelet-endothelial interactions in vivo. *J Exp Med, 191,* 1413–1422.

134. Forlow, S. B., McEver, R. P., & Nollert, M. U. (2000). Leukocyte-leukocyte interactions mediated by platelet microparticles under flow. *Blood, 95,* 1317–1323.

135. Connolly, E. S., Jr., Winfree, C. J., Prestigiacomo, C. J., et al. (1997). Exacerbation of cerebral injury in mice that express the P-selectin gene: Identification of P-selectin blockade as a new target for the treatment of stroke. *Circ Res, 81,* 304–310.

136. Vasthare, U. S., Heinel, L. A., Rosenwasser, R. H., et al. (1990). Leukocyte involvement in cerebral ischemia and reperfusion injury. *Surg Neurol, 33,* 261–265.

137. Chopp, M., Zhang, R. L., Chen, H., et al. (1994). Postischemic administration of an anti-Mac-1 antibody reduces ischemic cell damage after transient middle cerebral artery occlusion in rats. *Stroke, 25,* 869–875.

138. Walder, C. E., Green, S. P., Darbonne, W. C., et al. (1997). Ischemic stroke injury is reduced in mice lacking a functional NADPH oxidase. *Stroke, 28,* 2252–2258.

139. Carter, A. M., Catto, A. J., Bamford, J. M., et al. (1998). Platelet GP IIIa PlA and GP Ib variable number tandem repeat polymorphisms and markers of platelet activation in acute stroke. *Arterioscler Thromb Vasc Biol, 18,* 1124–1131.

140. Baker, R. I., Eikelboom, J., Lofthouse, E., et al. (2001). Platelet glycoprotein Ib alpha Kozak polymorphism is associated with an increased risk of ischemic stroke. *Blood, 98,* 36–40.

141. Frank, M. B., Reiner, A. P., Schwartz, S. M., et al. (2001). The Kozak sequence polymorphism of platelet glycoprotein Iba alpha and risk of nonfatal myocardial infarction and nonfatal stroke in young women. *Blood, 97,* 875–879.

142. Reiner, A. P., Kumar, P. N., Schwartz, S. M., et al. (2000). Genetic variants of platelet glycoprotein receptors and risk of stroke in young women. *Stroke, 31,* 1628–1633.

143. Carter, A. M., Catto, A. J., Bamford, J. M., et al. (1999). Association of the platelet glycoprotein IIb HPA-3 polymorphism with survival after acute ischemic stroke. *Stroke, 30,* 2606–2611.

144. Hsieh, K., Funk, M., Schillinger, M., et al. (2004). Vienna Stroke Registry. Impact of the platelet glycoprotein Ib alpha Kozak polymorphism on the risk of ischemic cerebrovascular events: A case-control study. *Blood Coagul Fibrinolysis, 15,* 469–473.

145. Bogousslavsky, J., Van Melle, G., & Regli, F. (1988). The Lausanne Stroke Registry: Analysis of 1000 consecutive patients with first stroke. *Stroke, 19,* 1083–1092.

146. Foulkes, M. A., Wolf, P. A., Price, T. R., et al. (1988). The stroke data bank: Design, methods, and baseline characteristics. *Stroke, 19,* 547–554.

147. del Zoppo, G. J., & Mori, E. (1992). Hematologic causes of intracerebral hemorrhage and their treatment. In H. H. Batjer (Ed.), *Spontaneous intracerebral hemorrhage* (Neurosurg Clin North Am, Vol. 33, pp. 637–658). Saunders, Philadelphia.

148. Figueroa, M., Gehlsen, J., Hammond, D., et al. (1993). Combination chemotherapy in refractory immune thrombocytopenic purpura. *N Engl J Med, 328,* 1226–1229.

149. McMillan, R. (2001). Long-term outcomes after treatment for refractory immune thrombocytopenic purpura. *N Eng J Med, 344,* 1402–1403.

150. Matthay, K. K., Koerper, M. A., & Ablin, A. R. (1979). Intracranial hemorrhage in congenital factor VII deficiency. *J Pediatr, 94,* 413–415.

151. Ragni, M. V., Lewis, J. H., Spero, J. A., et al. (1981). Factor VII deficiency. *Am J Hematol, 10,* 79–88.

152. Petri, M., Ellman, L., & Carey, R. (1983). Acquired factor XIII deficiency with chronic myelomonocytic leukemia. *Ann Intern Med, 99,* 638–639.

153. Fisher, M., & Adams, R. D. (1951). Observations on brain embolism with special reference to the mechanism of hemorrhagic infarction. *Neuropathol Exp Neurol, 10,* 92–94.

154. Fisher, C. M., & Adams, R. D. (1987). Observations on brain embolism with special reference to hemorrhage infarction. In A. J. Furlan (Ed.), *The heart and stroke. Exploring mutual cerebrovascular and cardiovascular issues* (pp. 17–36). New York: Springer-Verlag.

155. Jörgensen, L., & Torvik, A. (1969). Ischaemic cerebrovascular diseases in an autopsy series. Part 2. Prevalence, location,

pathogenesis, and clinical course of cerebral infarcts. *J Neurol Sci, 9,* 285–320.

156. Lodder, J., Krijne-Kubat, B., & Broekman, J. (1986). Cerebral hemorrhagic infarction at autopsy: Cardiac embolic cause and the relationship to the cause of death. *Stroke, 17,* 626–629.

157. Lodder, J. (1984). CT-detected hemorrhagic infarction: Relation with the size of the infarct, and the presence of midline shift. *Acta Neurol Scand, 70,* 329–335.

158. Yamaguchi, T., Minematsu, K., Choki, J., et al. (1984). Clinical and neuroradiological analysis of thrombotic and embolic cerebral infarction. *Jpn Circ J, 48,* 50–58.

159. Hornig, C. R., Dorndorf, W., & Agnoli, A. L. (1986). Hemorrhagic cerebral infarction: A prospective study. *Stroke, 17,* 179–185.

160. Okada, Y., Yamaguchi, T., Minematsu, K., et al. (1989). Hemorrhagic transformation in cerebral embolism. *Stroke, 20,* 598–603.

161. Eisenberger, H. M., & Suddith, R. L. (1979). Cerebral vessels have the capacity to transport sodium and potassium. *Science, 206,* 1083–1085.

162. Hart, R. G. (1986). Cerebral Embolism Study Group: Timing of hemorrhagic transformation of cardioembolic stroke. In T. Stober, K. Schimrigk, D. Ganten, et al. (Eds.), *Central nervous system control of the heart* (pp. 229–232). Boston: Martinus Nijhoff Publishing.

163. Cerebral Embolism Study Group. (1984). Immediate anticoagulation of embolic stroke: Brain hemorrhage and management options. *Stroke, 15,* 779–789.

164. Drake, M. E., & Shin, C. (1983). Conversion of ischemic to hemorrhagic infarction by anticoagulant administration. Report of two cases with evidence from serial computed tomographic brain scans. *Arch Neurol, 40,* 44–46.

165. Meyer, J. S., Gilroy, J., Barnhart, M. I., et al. (1963). Therapeutic thrombolysis in cerebral thromboembolism. *Neurology, 13,* 927–937.

166. Babikian, V. L., Kase, C. S., Pessin, M. S., et al. (1989). Intracerebral hemorrhage in stroke patients anticoagulated with heparin. *Stroke, 29,* 1500–1503.

167. Mori, E., Tabuchi, M., Yoshida, T., et al. (1988). Intracarotid urokinase with thromboembolic occlusion of the middle cerebral artery. *Stroke, 19,* 802–812.

168. del Zoppo, G. J., Ferbert, A., Otis, S., et al. (1988). Local intra-arterial fibrinolytic therapy in acute carotid territory stroke: A pilot study. *Stroke, 19,* 307–313.

169. Mori, E., Yoneda, Y., Tabuchi, M., et al. (1992). Intravenous recombinant tissue plasminogen activator in acute carotid artery territory stroke. *Neurology, 42,* 976–982.

170. Hamann, G. F., Okada, Y., Fitridge, R., et al. (1995). Microvascular basal lamina antigens disappear during cerebral ischemia and reperfusion. *Stroke, 26,* 2120–2126.

171. International Stroke Trial Collaborative Group. (1997). The International Stroke Trial (IST).: A randomised trial of aspirin, subcutaneous heparin, both, or neither among 19,435 patients with acute ischaemic stroke. *Lancet, 349,* 1569–1681.

172. Chen, Z. M., Sandercock, P., Pan, H. C., et al. (2000). Indications for early aspirin use in acute ischemic stroke: A combined analysis of 40,000 randomized patients from the Chinese acute stroke trial and the international stroke trial. On behalf of the CAST and IST collaborative groups. *Stroke, 31,* 1240–1249.

173. CAST (Chinese Acute Stroke Trial) Collaborative Group. (1997). CAST: A randomised placebo-controlled trial of early aspirin use in 20,000 patients with acute ischaemic stroke. *Lancet, 349,* 1641–1649.

174. The Abciximab in Ischemic Stroke Investigators. (2000). Abciximab in acute ischemic stroke: A randomized, double-blind, placebo-controlled, dose-escalation study. *Stroke, 31,* 601–609.

175. Abumiya, T., Lucero, J., Heo, J. H., et al. (1999). Activated microvessels express vascular endothelial growth factor and integrin $\alpha_v\beta_3$ during focal cerebral ischemia. *J Cereb Blood Flow Metab, 19,* 1038–1050.

176. UK-TIA Study Group. (1988). United Kingdom transient ischemic attack UK-TIA. aspirin trial: Interim results. *Br Med J, 296,* 316–320.

177. Diener, H., Cunha, L., Forbes, C., et al. (1996). European Stroke Prevention Study 2. Dipyridamole and acetylsalicylic acid in the secondary prevention of stroke. *J Neurol Sci, 143,* 1–13.

178. Kelton, J. G. (1983). Antiplatelet agents: Rationale and results. *Clin Haematol, 12,* 311–354.

179. Harker, L. A. (1986). Antiplatelet drugs in the management of patients with thrombotic disorders. *Semin Thromb Hemost, 12,* 134–155.

180. Fuster, V., Badimon, L., Badimon, J., et al. (1987). Drugs interfering with platelet functions: Mechanisms and clinical relevance. In M. Verstraete, J. Vermylen, R. Lijnen, et al. (eds.), *Thrombosis and haemostesis* (pp. 349–418). Leuven, Belgium: Leuven University Press.

181. Albers, G. W., Amarenco, P., Easton, J. D., et al. (2004). Antithrombotic and thrombolytic therapy for ischemic stroke. *Chest, 126,* 483S–512S.

182. del Zoppo, G. J. (1998). Antithrombotic interventions in cerebrovascular disease. In J. Loscalzo, & A. I. Schafer (Eds.), *Thrombosis and hemorrhage* (pp. 1279–1307). Baltimore: Williams and Wilkins.

183. Easton, J. D. (2001). Future perspectives for optimizing oral antiplatelet therapy. *Cerebrovasc Dis, 2,* 23–28.

184. Weksler, B. B. (2000). Antiplatelet agents in stroke prevention, combination therapy: Present and future. *Cerebrovasc Dis, 5,* 41–48.

185. European Stroke Prevention Study Group. (1987). The European Stroke Prevention Study ESPS. Principal end-points. *Lancet, 2,* 1351–1354.

186. Bousser, M. G., Eschwege, E., Haguenau, M., et al. (1983). "AICLA" controlled trial of aspirin and dipyridamole in the secondary prevention of athero-thrombotic cerebral ischemia. *Stroke, 14,* 5–14.

187. Crouse, J. R., Byington, R. P., Hoen, H. M., et al. (1997). Reductase inhibitor monotherapy and stroke prevention. *Arch Intern Med, 157,* 1305–1310.

188. Blauw, G. J., Lagaay, A. M., Smelt, A. H. M., et al. (1997). Stroke, statins, and cholesterol: A meta-analysis of randomized, placebo-controlled, double-blind trials

with HMG-CoA reductase inhibitors. *Stroke, 28,* 946–950.

189. Yoon, S. S., Dambrosia, J., Chalela, J., et al. (2004). Rising statin use and effect on ischemic stroke outcome. *BMC Med, 2,* 4.

190. Vaughan, C. J., & Delanty, N. (1999). Neuroprotective properties of statins in cerebral ischemia and stroke. *Stroke, 30,* 1969–1973.

191. Puccetti, L., Bruni, F., Di Renzo, M., et al. (1999). Hypercoagulable state in hypercholesterolemic subjects assessed by platelet-dependent thrombin generation: *In vitro* effect of cerivastatin. *Eur Rev Med Pharm Sci, 3,* 197–204.

192. Roth, G. J., Stanford, N., & Majerus, P. W. (1975). Acetylation of prostaglandin synthetase by aspirin. *Proc Natl Acad Sci USA, 72,* 3073–3076.

193. Preston, F. E., Whipps, S., Jackson, C. A., et al. (1981). Inhibition of prostaglandin and platelet thromboxane $A_2$ after low dose aspirin. *N Engl J Med, 304,* 76–79.

194. Clarke, R. J., Mayo, G., Price, P., et al. (1991). Suppression of thromboxane $A_2$ but not of systemic prostacyclin by controlled-release aspirin. *N Engl J Med, 325,* 1137–1141.

195. Baumgartner, H. R., Tschopp, T. B., & Weiss, H. J. (1977). Platelet interaction with collagen fibrils in flowing blood II. Impaired adhesion-aggregation in bleeding disorders. A comparison with subendothelium. *Thromb Haemost, 37,* 17–28.

196. Moake, J. L., Turner, N. A., Stathopoulos, N. A., et al. (1988). Shear-induced platelet aggregation can be mediated by vWF released from platelets, as well as by exogenous large or unusually large vWF multimers, requires adenosine diphosphate and is resistant to aspirin. *Blood, 71,* 1366–1374.

197. Mundall, J., Quintero, P., von Kaulla, K. N., et al. (1972). Transient monocular blindness and increased platelet aggregability treated with aspirin. *Neurology, 22,* 280–285.

198. Harrison, M. J. G., Marshall, J., Meadows, J. C., et al. (1971). Effect of aspirin in amaurosis fugax. *Lancet, 2,* 743–744.

199. Dyken, M. L., Kolar, O. J., & Jones, F. H. (1973). Differences in the occurrence of carotid transient ischemic attacks associated with antiplatelet aggregation therapy. *Stroke, 4,* 732–736.

200. Fields, W. S., Lemak, N. A., Frankowski, R. F., et al. (1977). Controlled trial of aspirin in cerebral ischemia. *Stroke, 8,* 301–314.

201. Sorenson, P. S., Pedersen, H., Marquardsen, J., et al. (1983). Acetylsalicylic acid in the prevention of stroke in patients with reversible cerebral ischemic attacks. A Danish Cooperative Study. *Stroke, 14,* 15–22.

202. Canadian Cooperative Study Group. (1978). A randomized trial of aspirin and sulfinpyrazone in threatened stroke. *N Engl J Med, 299,* 53–59.

203. Dutch TIA Trial Study Group. (1991). A comparison of two doses of aspirin 30 mg versus 283 mg a day in patients after a transient ischemic attack or minor ischemic stroke. *N Engl J Med, 325,* 1261–1266.

204. SALT Collaborative Group. (1991). Swedish Aspirin Low-Dose Trial (SALT). of 75 mg aspirin as secondary prophylaxis after cerebrovascular ischemic events. *Lancet, 338,* 1345–1349.

205. Fitzgerald, G. A. (1987). Dipyridamole. *N Engl J Med, 316,* 1247–1257.

206. Hirsch, L. (1986). Anticoagulant and platelet antiaggregant agents. In H. J. M. Barnett, J. P. Mohr, B. M. Stein, et al. (Eds.), *Stroke: Pathophysiology, diagnosis and management* (Vol. 2, pp. 925–966). New York: Churchill Livingston.

207. Harker, L. A., & Slichter, S. J. (1970). Studies of platelet and fibrinogen kinetics in patients wtih prosthetic heart valves. *N Engl J Med, 283,* 1302–1305.

208. Ranhosky, A. (1987). Dipyridamole letter. *N Engl J Med, 317,* 1734.

209. Acheson, J., Danta, G., & Hutchinson, E. C. (1969). Controlled trial of dipyridamole in cerebral vascular disease. *Br Med J, 1,* 614–615.

210. American–Canadian Cooperative Study Group. (1985). Persantine aspirin trial in cerebral ischemia. Part II. Endpoint results. *Stroke, 16,* 406–415.

211. Matias-Guiu, J., Davalos, A., Pico, M., et al. (1987). Low-dose acetylsalicylic acid (ASA). plus dipyridamole versus dipyridamole alone in the prevention of stroke in patients with reversible ischemic attacks. *Acta Neurol Scand, 76,* 413–421.

212. ESPS-2 Working Group. (1992). Second European Stroke Prevention Study. *J Neurol, 239,* 299–301.

213. Wiley, J. S., Chesterman, C. N., Morgan, F. J., et al. (1979). The effect of sulphinpyrazone on the aggregation and release reactions of human platelets. *Thromb Res, 14,* 23–33.

214. Evans, G. (1972). Effect of drugs that suppress platelet surface interaction on incidence of amaurosis fugax and transient cerebral ischemia. *Surg Forum, 23,* 239–241.

215. Roden, S., Low-Beer, T., Carmalt, M., et al. (1981). Transient cerebral ischemic attacks — Management and prognosis. *Postgrad Med J, 57,* 275–278.

215a. Candelise, L., Landi, G., Perrone, P., et al. (1982). A randomized trial of aspirin and sulfinpyrazone in patients with TIA. *Stroke, 13,* 175–179.

216. McTairoh, D., Faulds, D., & Goa, K. L. (1990). Ticlopidine. An updated review of its pharmacology and therapeutic use in platelet-dependent disorders. *Drugs, 40,* 238–259.

217. Panak, E., Maffrand, J. P., Picard-Faire, C., et al. (1983). Ticlopidine: A promise for the prevention and treatment of thrombosis and its complications. *Haemostasis, 12*(Suppl I), 1–54.

218. Hass, W. K., Easton, J. D., Adams, H. P., Jr., et al. (1989). A randomized trial comparing ticlopidine hydrochloride with aspirin for prevention of stroke in high-risk patients. *N Engl J Med, 321,* 501–507.

219. Gent, M., Blakely, J. A., Easton, J. D., et al. (1989). The Canadian-American Ticlopidine Study (CATS) in thromboembolic stroke. *Lancet, 1,* 1215–1220.

220. Tsai, H. M., Rice, L., Sarode, R., et al. (2000). Antibody inhibitors to von Willebrand factor metallogproteinase and increased binding of von Willebrand factor to platelets in ticlopidine-associated thrombotic thrombocytopenic purpura. *Ann Intern Med, 132,* 794–799.

221. Bennett, C. L., Davidson, C. J., Raisch, D. W., et al. (1999). Thrombotic thrombocytopenic purpura associated with

ticlopidine in the setting of coronary artery stents and stroke prevention. *Arch Intern Med, 159,* 2524–2538.

222. The European Stroke Prevention Study ESPS-2. Working Group. (1996). Secondary stroke prevention: Aspirin/dypyridamole combination is superior to either agent alone and to placebo [abstract]. *Stroke, 27,* 195.

223. Britton, M., Helmers, C., & Samuelsson, K. (1987). High-dose acetylsalicylic acid after cerebral infarction. A Swedish Cooperative Study. *Stroke, 18,* 325–334.

224. Gent, M., Blakeley, J. A., Hachinski, V., et al. (1985). A secondary prevention, randomized trial of suloctidil in patients with a recent history of thromboembolic stroke. *Stroke, 16,* 416–424.

225. Gryglewski, R. J., Nowak, S., Kostka-Trabka, E., et al. (1983). Treatment of ischaemic stroke with prostacyclin. *Stroke, 14,* 197–202.

226. Huczynski, J., Kostka-Trabka, E., Sotowska, W., et al. (1985). Double-blind controlled trial of the therapeutic effects of prostacyclin in patients with completed ischaemic stroke. *Stroke, 16,* 810–814.

227. Martin, J. F., Hamdy, N., Nicholl, J., et al. (1985). Double-blind controlled trial of prostacyclin in cerebral infarction. *Stroke, 16,* 386–390.

228. Bath, P. M., & Bath, F. J. (2000). Prostacyclin and analogues for acute ischaemic stroke. *Cochrane Database Syst Rev, 2,* CD000177.

229. Hsu, C. Y., Faught, R. E., Jr., Furlan, A. J., et al. (1987). Intravenous prostacyclin in acute nonhemorrhagic stroke: A placebo-controlled double-blind trial. *Stroke, 18,* 352–358.

230. Bath, P. M. (2004). Prostacyclin and analogues for acute ischaemic stroke. *Cochrane Database Syst Rev* computer file, CD0001777.

231. Tanaka, K., Gotoh, F., Fukuuchi, Y., et al. (1988). Stable prostacyclin analogue preventing microcirculatory derangement in experimental cerebral ischemia in cats. *Stroke, 19,* 1267–1274.

232. Steir, C. T., Jr., Chander, P. N., Belmonte, A., et al. (1997). Beneficial action of beraprost sodium, a prostacyclin analog, in stroke-prone rats. *J Cardiovasc Pharmacol, 30,* 285–293.

233. Gurewich, O., & Lipinski, B. (1976). Evaluation of antithrombotic properties of suloctidil in comparison with aspirin and dipyridamole. *Thromb Res, 9,* 101–108.

234. Blakeley, J. A. (1979). A prospective trial of sulfinpyrazone and survival after thrombotic stroke [abstract]. *Thromb Haemost, 42,* 382.

235. Herbert, J. M., Frechel, D., Vallee, E., et al. (1993). Clopidogrel, a novel antiplatelet and antithrombotic agent. *Cardiovasc Drug Rev, 11,* 180–198.

236. Savi, P., Laplace, M. C., Maffrand, J. P., et al. (1994). Binding of [3H]-2-methylthio ADP to rat platelets: Effect of clopidogrel and ticlopidine. *J Pharmacol Exp Ther, 269,* 772–777.

237. CAPRIE Steering Committee. (1996). A randomised, blinded, trial of clopidogrel versus aspirin in patients at risk of ischaemic events CAPRIE. *Lancet, 348,* 1329–1339.

238. Diener, H.-C., Bogousslavsky, J., Brass, L. M., et al. (2004). Aspirin and clopidogrel compared with clopidogrel alone after recent ischaemic stroke or transient ischaemic attack in high-risk patients (MATCH): Randomised, doubled-blind, placebo-controlled trial. *Lancet, 364,* 331–337.

239. Diener, H.-C., Bogousslavsky, J., Brass, L. M., et al. (2004). Management of atherothrombosis with clopidogrel in high-risk patients with recent transient ischaemic attack or ischaemic stroke (MATCH): Study design and baseline data. *Cerebrovasc Dis, 17,* 253–261.

240. Peters, R. J., Mehta, S. R., Fox, K. A. A., et al. (2003). Effects of aspirin dose when used alone or in combination with clopidogrel in patients with acute coronary syndromes: Observations from the Clopidogrel in Unstable angina to prevent Recurrent Events (CURE). study. *Circulation, 108,* 1682–1687.

241. Mehta, S. R., & Yusuf, S. (2000). The Clopidogrel in Unstable angina to prevent Recurrent Events (CURE). trial programme; Rationale, design and baseline characteristics including a meta-analysis of the effects of thienopyridines in vascular disease. *Eur Heart J, 21,* 2033–2041.

242. Schror, K. (1995). Antiplatelet drugs. A comparative review. *Drugs, 50,* 7–28.

242a. Patient enrollment permanently discontinued in investigational clinical trial of ReoPro for acute ischemic stroke. October 28, 2005 < http://www.centocor.com/news/news_103105.jsp > .

242b. Abciximab Emergent Stroke Treatment Trial AbESTT. Investigators. (2005). Emergency administration of abciximab for treatment of patients with acute ischemic stroke: Results of a randomized phase 2 trial. *Stroke, 36,* 880–890.

242c. Qureshi, A. I., Suri, M. F., Khan, J., et al. (2000). Abciximab as an adjunct to high-risk carotid or vertebrobasilar angioplasty: Preliminary experience. *Neurosurgery, 46,* 1316–1324.

242d. Tong, F. C., Cloft, H. J., Joseph, G. J., et al. (2000). Abciximab rescue in acute carotid stent thrombosis. *Am J Neuroradiol, 21,* 1750–1752.

242e. Hofman, R., Kerschner, K., Steinwender, C., et al. (2002). Abciximab bolus injection does not reduce cerebral iscemic complications of elective carotid artery stenting: A randomized study. *Stroke, 33,* 725–727.

243. Puccetti, L., Pasqui, A. L., Pastorelli, M., et al. (2002). Time-dependent effect of statins on platelet function in hypercholesterolemia. *Eur J Clin Invest, 32,* 901–908.

244. Schror, K., Lobel, P., & Steinhagen-Thiessen, E. (1989). Simvastatin reduces platelet thromboxane formation and restores normal platelet sensitivity against prostacyclin in type IIa hypercholesterolemia. *Eicosanoids, 2,* 39–45.

245. Serebruany, V. L., Malinin, A. I., Lowry, D. R., et al. (2004). Effects of valsartan and valeryl 4-hydroxy valsartan on human platelets: A possible additional mechanism for clinical benefits. *J Cardiovasc Pharmacol, 43,* 677–684.

246. Northern American Symptomatic Carotid Endarterectomy Trial Collaborators. (1991). Beneficial effect of carotid endarterectomy in symptomatic patients with high-grade carotid stenosis. *N Engl J Med, 325,* 445–453.

247. European Carotid Surgery Trialists' Collaborative Group. (1991). MRC European Carotid Surgery Trial: Interim results

for symptomatic patients with severe (70–99%) or with mild (0–29%) carotid stenosis. *Lancet, 337,* 1235–1243.

248. Harker, L. A., Bernstein, E. F., Dilley, R. B., et al. (1992). Failure of aspirin plus dipyridamole to prevent restenosis after carotid endarterectomy. *Ann Intern Med, 116,* 731–736.

249. Fields, W. S., Lemak, N. A., Frankowski, R. F., et al. (1978). Controlled trial of aspirin in cerebral ischemia. Part II. Surgical group. *Stroke, 9,* 309–319.

250. Boysen, G., Sorensen, P. S., Juhler, M., et al. (1988). Danish very low dose aspirin after carotid endarterectomy trial. *Stroke, 19,* 1211–1215.

251. Kretschmer, G., Pratschner, T., Prager, M., et al. (1990). Antiplatelet treatment prolongs survival after carotid bifurcation endarterectomy. *Ann Surg, 211,* 317–322.

251a. Pratschner, T., Kretschmer, G., Prager, M., et al. (1990). Antiplatelet therapy following carotid bifurcation endarterectomy. Evaluation of a controlled clinical trial. Prognostic significance of histologic plaque examination on behalf of survival, *Eur J Vasc Surg, 4,* 285–289.

252. Harvald, B., Hilden, T., & Lund, E. (1962). Long-term anticoagulant therapy after myocardial infarction. *Lancet, 2,* 626–630.

253. Cooperative Study. (1965). Long-term anticoagulant therapy after myocardial infarction. *JAMA, 193,* 929–934.

254. Loeliger, E. A., Hensen, A., Kroes, F., et al. (1967). A double blind trial of long-term anticoagulant treatment after myocardial infarction. *Acta Med Scand, 182,* 549–567.

255. Breddin, K., Loew, D., Lechner, K., et al. (1980). The German-Austrian Aspirin Trial: A comparison of acetylsalicyclic acid, placebo, and phenprocoumon in secondary prevention of myocardial infarction. *Circulation, 62*(Supp 1), V63–V72.

256. Sixty-Plus Reinfarction Study Research Group. (1980). A double-blind trial to assess long-term anticoagulant therapy in elderly patients after myocardial infarction. *Lancet, 2,* 989–994.

257. Smith, P., Arnesen, H., & Holme, I. (1990). The effect of warfarin on mortality and reinfarction after myocardial infarction. *N Engl J Med, 323,* 147–152.

258. Breddin, K., Loew, D., Lechner, K., et al. (1980). Secondary prevention of myocardial infarction: A comparison of acetylsalicylic acid, placebo, and phenprocoumen. *Haemostasis, 9,* 325–344.

259. Hinton, R. C., Kistler, J. P., Fallon, J. T., et al. (1977). Influence of etiology of atrial fibrillation on incidence of systemic embolism. *Am J Cardiol, 40,* 509–513.

260. Wolf, P. A., Abbott, R. D., & Kannel, W. B. (1987). Atrial fibrillation: A major contributor to stroke in the elderly. The Framingham Study. *Arch Intern Med, 147,* 1561–1564.

261. Chesebro, J. H., Fuster, V., & Halperin, J. L. (1990). Atrial fibrillation—Risk marker for stroke. *N Engl J Med, 323,* 1556–1558.

262. Sherman, D. A., Goldman, L., Whiting, R. B., et al. (1984). Thromboembolism in patients with atrial fibrillation. *Arch Neurol, 41,* 708–710.

263. Stroke Prevention in Atrial Fibrillation Study Group Investigators. (1990). Preliminary report of the Stroke Pre-

vention in Atrial Fibrillation Study. *N Engl J Med, 322,* 863–868.

264. Petersen, P., Boysen, G., Godtfredsen, J., et al. (1989). Placebo-controlled randomized trial of warfarin and aspirin to prevention of thromboembolic complications in chronic atrial fibrillation. The Copenhagen AFASAK Study. *Lancet, 1,* 175–179.

265. EAFT European Atrial Fibrillation Trial Study Group. (1993). Secondary prevention in non-rheumatic atrial fibrillation after transient ischaemic attack or minor stroke. *Lancet, 342,* 1255–1262.

266. Atrial Fibrillation Investigators. (1997). The efficacy of aspirin in patients with atrial fibrillation: Analysis of pooled data from three randomized trials. *Arch Intern Med, 157,* 1237–1240.

267. Hart, R. G., Benavente, O., McBride, R., et al. (1999). Antithrombotic therapy to prevent stroke in patients with atrial fibrillation: A meta-analysis. *Ann Intern Med, 131,* 492–501.

268. Segal, J. B., McNamara, R. L., Miller, M. R., et al. (2000). Prevention of thromboembolism in atrial fibrillation: A meta-analysis of trials of anticoagulants and antiplatelet drugs. *J Gen Intern Med, 15,* 56–67.

269. Stroke Prevention in Atrial Fibrillation Investigators. (1994). Warfarin versus aspirin for prevention of thromboembolism in atrial fibrillation: Stroke Prevention in Atrial Fibrillation II study. *Lancet, 343,* 687–691.

270. Stroke Prevention in Atrial Fibrillation Investigators. (1996). Adjusted-dose warfarin versus low-intensity, fixed-dose warfarin plus aspirin for high-risk patients with atrial fibrillation: Stroke prevention in atrial fibrillation III randomised clinical trial. *Lancet, 348,* 633–638.

271. Gullov, A. L., Koefoed, B. G., Petersen, P., et al. (1998). Fixed mini-dose warfarin and aspirin alone and in combination versus adjusted-dose warfarin for stroke prevention in atrial fibrillation: Second Copenhagen Atrial Fibrillation, Aspirin, and Anticoagulation Study: The AFASAK-2 study. *Arch Intern Med, 158,* 1513–1521.

272. Gullov, A. L., Koefoed, B. G., & Petersen, P. (1999). Bleeding during warfarin and asirin therapy in patients with atrial fibrillation: The AFASAK-2 Study. *Arch Intern Med, 159,* 1322–1328.

272a. News Release: Plavix inferior to standard drug in stroke study. [Reuters] November 15, 2005.

273. Singer, D. E., Albers, G. W., Dalen, J. E., et al. (2004). Antithrombotic therapy in atrial fibrillation. *Chest, 126,* 429S–456S.

274. Szekely, P. (1964). Systemic embolism and anticoagulant prophylaxis in rheumatic heart disease. *Br Med J, 1,* 1209–1212.

275. Levine, H. J., Pauker, S. G., & Salzman, E. W. (1989). Antithrombotic therapy in valvular heart disease. *Chest, 95,* 98S–106S.

276. Salem, D. N., Stein, P. D., Al-Ahmad, A., et al. (2004). Antithrombotic therapy in valvular heart disease — native and prosthetic: The Seventh ACCP Conference on Antithrombotic and Thrombolytic Therapy. *Chest, 126,* 457S–482S.

277. Coulshed, N., Epstein, E. J., McKendrick, C. S., et al. (1970). Systemic embolism in mitral valve disease. *Br Med J, 32,* 26–34.
278. Kopf, G. S., Hammond, G. L., Geha, A. S., et al. (1987). Long-term performance of the St. Jude medical valve: Low incidence of thromboembolism and hemorrhagic complications with modest doses of warfarin. *Circulation, 76*(Suppl), 132–136.
279. Barnhorst, D. A., Oxman, H. A., Connolly, D. C., et al. (1975). Long-term follow-up of isolated replacement of the aortic or mitral valve with the Starr-Edwards prosthesis. *Am J Cardiol, 35,* 228–233.
280. Turpie, A. G. G., Gunstensen, J., Hirsh, J., et al. (1988). Randomized comparison of two intensities of oral anticoagulant therapy after tissue heart valve placement. *Lancet, 1,* 1242–1245.
281. Turpie, A. G. G., Gent, M., Laupacis, A., et al. (1993). Comparison of aspirin with placebo in patients treated with warfarin after heart-valve replacement. *N Engl J Med, 329,* 524–529.
282. Loeliger, E. A., Poller, L., Samama, M., et al. (1985). Questions and answers on prothrombin time standardization in oral anticoagulant control. *Thromb Haemost, 54,* 515–517.
283. Loeliger, E. A. (1992). Therapeutic target values in oral anticoagulation—Justification of Dutch policy and a warning against the so-called moderate-intensity regimens. *Ann Hematol, 64,* 60–65.
284. Dale, J., Myhre, E., Storstein, O., et al. (1977). Prevention of arterial thromboembolism with acetylsalicylic acid. A controlled clinical study in patients with aortic ball valves. *Am Heart J, 94,* 101–111.
285. Altman, R., Boullon, F., Rouvier, J., et al. (1976). Aspirin and prophylaxis of thromboembolic complications in patients with substitute heart valves. *J Thorac Cardiovasc Surg, 72,* 127–129.
286. Sullivan, J. M., Harken, D. E., & Gorlin, R. (1969). Effect of dipyridamole on the incidence of arterial emboli after cardiac valve replacement. *Circulation, 39/40,* I-149–I-153.
287. Sullivan, J. M., Harken, D. E., & Gorlin, R. (1971). Pharmacologic control of thromboembolic complications of cardiac valve replacement. *N Engl J Med, 284,* 1391–1394.
288. Deleted in proof.
289. Pouleur, H., & Buyse, M. (1995). Effects of dipyridamole in comobination with anticoagulant therapy on survival and thromboembolic events in patients with prosthetic heart valves: A meta-analysis of the randomized trials. *J Thorac Cardiovasc Surg, 110,* 463–472.
290. Nuñez, L., Aguado, M. G., Celemin, D., et al. (1982). Aspirin or coumadin as the drug of choice for valve replacement with porcine bioprosthesis. *Ann Thorac Surg, 33,* 355–358.
291. Nuñez, L., Aguado, M. G., Larrea, J. L., et al. (1984). Prevention of thromboembolism using aspirin after mitral valve replacement with porcine bioprosthesis. *Ann Thorac Surg, 37,* 84–87.
292. Hoogendijk, E. M., Jenkins, C. S., van Wijk, E. M., et al. (1979). Spontaneous platelet aggregation in cerebrovascular disease. II. Further characterisation of the platelet defect. *Thromb Haemost, 41,* 512–522.
293. Fisher, M., Levine, P. H., Fullerton, A. L., et al. (1982). Marker proteins of platelet activation in patients with cerebrovascular disease. *Arch Neurol, 39,* 692–695.
294. Taomoto, K., Asada, M., Kanazawa, Y., et al. (1983). Usefulness of the measurement of plasma beta-thromboglobulin beta-TG in cerebrovascular disease. *Stroke, 14,* 518–524.
295. Landi, G., D'Angelo, A., Boccardi, E., et al. (1987). Hypercoagulability in acute stroke: Prognostic significance. *Neurology, 37,* 1667–1671.
296. Fisher, M., & Francis, R. (1990). Altered coagulation in cerebral ischemia, platelet, thrombin, and plasmin activity. *Arch Neurol, 47,* 1075–1079.
297. Feinberg, W. M. (1995). Coagulation. In L. R. Caplan (Ed.), *Brain ischemia. Basic concepts and clinical relevance* (pp. 85–96). Berlin, Heidelberg, New York: Springer-Verlag.
298. Couch, J. R., & Hassanein, R. S. (1976). Platelet aggregation, stroke, and transient ischemic attack in middle-aged and elderly patients. *Neurology, 26,* 888–895.
299. Candelise, L., Landi, G., Perrone, P., et al. (1982). A randomized trial of aspirin and sulfinpyrazone in patients with TIA. *Stroke, 13,* 175–179.

# CHAPTER 37

# Peripheral Vascular Disease

## Michael Sobel

*Department of Surgery, University of Washington School of Medicine, Seattle, Washington*

## I. Overview of Peripheral Arterial Disease

Approximately 5 million to 10 million adults in the United States have peripheral arterial disease (PAD), which is defined as atherosclerotic occlusive disease of the extremities. The atherosclerotic manifestations of PAD commonly affect the aortoiliac regions and the arteries of the lower extremities, although the arteries of the upper extremities may also be affected. The anatomic turns and bends of extremity arteries, and the resulting rheology and hemodynamics of blood flow, can explain, in part, the stereotypical sites at which critical atherosclerotic stenoses develop: at the aortic and iliac bifurcations, at the adductor canal of the superficial femoral artery, and at the trifurcation of the popliteal artery. Against this background of hemodynamic factors, the major risk factors for PAD play dominant roles in the development of atherosclerosis in these vascular beds — namely, cigarette smoking, diabetes, hyperlipidemia, hypertension, advancing age, and hyperhomocysteinemia.[1–3]

Above the age of 55, the prevalence of PAD may range from 12 to 30%, depending on the exact age group and criteria used.[4–6] Men and women are equally affected. The symptom complex of intermittent claudication is the hallmark of PAD: muscular pain and cramping or burning induced by exercise and relieved by rest. An ankle/arm index (AAI) of <0.9 is considered diagnostic of PAD, yet only one third to one half of such patients with a reduced AAI will have classical symptoms of claudication. In their review of the topic, Burns and colleagues[2] very succinctly described the significance of PAD: "Peripheral arterial disease is a marker for systemic atherosclerosis; the risk to the limb in claudication is low, but the risk to life is high" (p. 584). Considering that intermittent claudication is the leg-equivalent of angina pectoris, the natural history of PAD carries a relatively better prognosis for the affected vascular bed, compared to primary coronary artery disease. Over a 5-year period, approximately 15% of patients with intermittent claudication will progress to limb loss or require interventions for limb-threatening ischemia.[7–9] This may be why the diagnosis and aggressive treatment of PAD in primary care have lagged behind those of coronary or cerebrovascular disease.[4,10] In spite of this seemingly moderate prognosis for the legs, PAD patients have the same high risk of stroke, myocardial infarction, or vascular death as do those with established coronary or cerebrovascular disease. Even asymptomatic PAD patients without other overt manifestations of atherosclerosis have a 30% 5-year risk of suffering major cardiovascular events or death.[2,8,11,12] Thus, patients with PAD can be expected to share all of the derangements of platelet function associated with coronary or cerebrovascular disease (Chapters 35 and 36).

Most of the acute thromboembolic events seen in coronary and cerebrovascular syndromes (Chapters 35 and 36) have parallels in the peripheral arterial vasculature, although the pathophysiology of these events in PAD has mostly been inferred from the other vascular trees. Acute thrombosis at the site of a critically stenotic vessel may convert intermittent claudication to ischemic limb pain at rest (comparable to acute coronary syndromes). Arterio-arterial embolism of platelet thrombi and plaque debris from an iliac or femoral stenosis ("blue toe syndrome") is analogous to the transient ischemic attack or stroke from carotid stenosis. The repertoire of invasive treatments in PAD is also generally the same as that in other atherosclerotic beds: thrombolysis, percutaneous balloon angioplasty, stenting, open surgical endarterectomy, or vascular bypass with autogenous conduits. Lower-extremity arterial disease stands out as one of the few areas where prosthetic vascular grafts are also currently feasible. Perhaps because of the lower clinical profile of PAD, the roles of platelets and antithrombotic adjuncts have not been subjected to the same intense scrutiny as they have been in coronary artery disease. As a consequence, many of the evidence-based guidelines for antiplatelet therapy in PAD have been inferred from studies of coronary or cerebrovascular disease, without the benefit of clinical trials specific to PAD.

## II. Platelet-Vessel Wall Interactions in PAD

### A. *Platelet Interactions with the Atherosclerotic Vessel Wall*

Originally highlighted by Ross's "injury-repair" hypotheses,[13] platelets have been implicated in both the pathogenesis and progression of atherosclerotic lesions. Circulating platelets can be thought of as the forward scouts of the arterial vasculature, not only stopping leaks, but also inciting an inflammatory response. Platelets deliver platelet-derived growth factor (PDGF) and other mitogenic factors to sites of vascular injury (Chapters 15 and 41). Platelets, *via* their receptor P-selectin, bind to neutrophils and monocytes, *via* their counter-receptor P-selectin glycoprotein 1 ligand (PSGL-1), thereby recruiting and activating these inflammatory cells at the injury site (Chapter 12). P-selectin appears to be a crucial factor in the development of early atherosclerotic lesions, at least in animal models[14] (Chapter 39). Binding between platelet glycoprotein (GP) Ib and the integrin $\alpha M\beta 2$ (Mac-1 CD11b/CD18) expressed on leukocytes may also support the formation of heterotypic platelet-leukocyte aggregates (Chapters 7 and 12).[15,16] Platelet release of soluble CD40 ligand (sCD40L) is also implicated in the recruitment and extravasation of inflammatory cells (Chapter 39). Both endothelial and immune cells express the CD40 receptor, whose ligation induces inflammatory responses.[17,18] Patient cohorts with PAD or atherosclerosis have been found to have more activation of these platelet inflammatory pathways, including elevated expression of P-selectin, increased platelet-monocyte aggregates, and sCD40L levels (Table 37-1).

The loss of natural inhibitory factors by the diseased vessel wall also contributes to the heightened level of platelet activation in PAD. As a mature atherosclerotic plaque develops, endothelial dysfunction leads to reduced production of nitric oxide (NO) and prostacyclin $I_2$ (PGI$_2$, prostacyclin), which are physiologic platelet inhibitors (Chapter 13). Shear-induced platelet aggregation can occur at turbulent atherosclerotic stenoses or anastomoses, mediated by platelet GPIb and von Willebrand factor.[19–21] Tissue factor (and hence increased thrombin generation) can also be observed in the native plaque.

### B. *Platelet Responses to Vascular Grafts and Angioplasty*

**1. Vascular Grafts.** Prosthetic grafts are frequently used to bypass occluded or aneurysmal arteries. For the replacement of large-caliber, high-flow arteries (i.e., aorto-iliac-femoral), prosthetic grafts have excellent long-term patency and durability. For the femoral-popliteal position, autogenous venous grafts have a superior longevity over prosthetic grafts, although prosthetic grafting of these medium-sized arteries is still common when an autogenous conduit is lacking or other surgical risk factors are considered. The patency of prosthetic grafts in small-caliber tibial vessels is generally poor. Only two basic prosthetic materials are currently in use: expanded polytetrafluoroethylene (ePTFE) and Dacron polyester. Endovascular grafts are essentially specialized stents lined with a thin Dacron or ePTFE fabric.

All prosthetic grafts are platelet reactive. Radiolabeled platelet studies in humans and animals have shown that grafts actively accumulate platelets immediately upon contact with the blood. Platelet uptake is highest in the perioperative period, and human platelet half-life is shortened.[22,23] Thrombocytopenia from graft consumption, however, is extremely unusual and is generally a sign of complicating infection or disseminated intravascular coagulation. By 6 months, platelet half-life normalizes, but the platelet reactivity of the graft remains 50 to 100% higher than a reference artery.[24–26] Over a period of years, the graft's platelet reactivity continues to diminish, but even 5- to 9-year-old Dacron grafts still show an active accumulation of platelets, particularly at the anastomoses.[25] Neither modifications of the weave nor coatings seem to modify platelet accumulation on Dacron.[27,28] Acutely, ePTFE grafts are less platelet reactive than Dacron,[29,30] but at 6 months differences between ePTFE and Dacron graft reactivity are not discernible. Carbon-coated ePTFE may be less platelet-reactive,[31] and heparin-coated Dacron grafts may have improved patency.[32]

Compared to prosthetic grafts, vein grafts have minimal platelet reactivity. Yet even vein grafts may not retain the thromboresistance of a healthy blood vessel. The trauma of harvesting autogenous vein grafts can cause transient sloughing of endothelium and loss of endothelial-dependent relaxation.[33] Likewise, chronically arterialized vein grafts undergo pathologic changes of intimal thickening and medial smooth muscle hyperplasia, which can lead to a more platelet-reactive endothelium.[34]

Despite all the research on blood-prosthetic interfaces, the clinical differences between available graft materials are relatively small. In current practice, host biology, flow dynamics, and technical factors are the dominant influences on long-term graft patency, rather than the intrinsic platelet reactivity of the artificial surface. The holy grail in peripheral vascular surgery is a small-diameter prosthetic graft onto which a natural endothelial lining can grow and thus confer all the antithrombotic functions of a healthy artery. Attempts have been made to "seed" endothelium onto prostheses or modify their porosity to promote capillary ingrowth. Unfortunately, the results have been mixed[35–38] and complicated by a significant increase in platelet reactivity (and thrombogenicity) immediately after seeding.[39]

**Table 37-1: Evidence of Pathologic Platelet Reactivity in PAD**

| Measure of Platelet Function | Author | N | Population | Results | Comparison Population |
|---|---|---|---|---|---|
| Spontaneous platelet aggregation | Robless et al.[113] | 20 | PAD, no DM | Increased | AMHC |
| Thrombin responses | Riba et al.[70] | 15 | PAD, normal homocysteine | No difference (Fg binding) | AMHC |
| | Riba et al.[70] | 24 | PAD, high homocysteine | Increased (Fg binding) | AMHC |
| | Ejim et al.[114] | 13 | PAD | Increased (Fn binding) | AMHC |
| | Klinkhardt et al.[88] | 24 | PAD | Increased (P-sel) | HC |
| | Zeiger et al.[115] | 50 | PAD | Increased (P-sel) | ASMC |
| ADP responses | Robless et al.[113] | 20 | PAD, no DM | No difference | AMHC |
| | Keating et al.[116] | 26 | PAD | Increased (P-sel) | CAD or CVD |
| | Keating et al.[116] | 26 | PAD | No difference (Fg binding) | CAD or CVD |
| | Riba et al.[70] | 15 | PAD, normal homocysteine | No difference (Fg binding, P-sel) | AMHC |
| | Riba et al.[70] | 24 | PAD, high homocysteine | Increased (Fg binding, P-sel) | AMHC |
| | Zahavi and Zahavi[60] | 35 | PAD, no DM/hyperlipidemia | Increased (aggregation) | ASMHC |
| | Klinkhardt et al.[88] | 24 | PAD | No difference (P-sel) | HC |
| | Zeiger et al.[115] | 50 | PAD | Increased (P-sel) | ASMC |
| Collagen responses | Robless et al.[113] | 20 | PAD, no DM | Increased | AMHC |
| | Zahavi and Zahavi[60] | 35 | PAD, no DM/hyperlipidemia | No difference | ASMHC |
| | Walters et al.[117] | 31 | PAD | Increased | ASMHC |
| Basal P-selectin expression | Robless et al.[113] | 20 | PAD, no DM | No difference | AMHC |
| | Keating et al.[116] | 26 | PAD | No difference | CAD or CVD |
| | Riba et al.[70] | 39 | PAD | No difference | AMHC |
| | Klinkhardt et al.[88] | 7 | PAD | Increased | HC |
| | Zeiger et al.[115] | 50 | PAD | Increased | ASMC |
| Platelet-leukocyte aggregates | Klinkhardt et al.[88] | 24 | PAD | Increased | HC |
| | Esposito et al.[41] | 46 | PAD, perioperative | Increased | Patent vs. thrombosed |
| Basal GPIIb-IIIa activation | Robless et al.[113] | 20 | PAD, no DM | Increased (PAC-1binding) | AMHC |
| | Riba et al.[70] | 39 | PAD | No difference (Fg binding) | AMHC |
| Markers of chronic platelet activation | Trifiletti et al.[59] | 13 | PAD | Increased $\beta$TG | AMHC |
| | Zahavi and Zahavi[60] | 35 | PAD, no DM/hyperlipidemia | Increased $\beta$TG, $TXB_2$; shorter survival | ASMHC |
| | Sinzinger et al.[119] | 12 | PAD | Shorter survival | ASMC |
| | Daví et al.[62] | 64 | PAD | Increased urinary $TXB_2$ | ASMHC |
| Platelet responses to natural inhibitors | Riba et al.[70] | 24 | PAD, high homocysteine | Reduced (NO) | AMHC |
| | Riba et al.[70] | 15 | PAD, normal homocysteine | No difference (NO) | AMHC |
| | Zahavi and Zahavi[60] | 35 | PAD, no DM/hyperlipidemia | No difference ($PGI_2$) | ASMHC |
| Shear-induced platelet aggregation | Walters et al.[117] | 31 | PAD | Increased | ASMHC |
| Platelet receptor density | Gosk-Bierska et al.[120] | 64 | PAD | Increased GPIIb-IIIa; Increased GPIb | ASMHC |

Abbreviations: AMHC, age-matched healthy controls; ASMC, age- and sex-matched controls; ASMHC, age- and sex-matched healthy controls; $\beta$TG, $\beta$-thromboglobulin; CAD, coronary artery disease; CVD, cerebrovascular disease; DM, diabetes mellitus; Fg, fibrinogen; Fn, fibronectin; GP, glycoprotein; HC, healthy controls; NO, nitric oxide; PAD, peripheral arterial disease; $PGI_2$, prostaglandin $I_2$; P-sel, P-selectin; $TXB_2$, thromboxane $B_2$.

The responsiveness of the host's platelets can significantly influence long-term graft patency, regardless of the graft type. The relative amount of platelet uptake on the graft,[22] preoperative platelet aggregability,[40] and the level of circulating platelet-leukocyte aggregates[41] are strong predictors of graft failure. Many studies have shown that antiplatelet therapy will significantly reduce platelet accumulation on a fresh prosthetic graft.[22,39,42,43] Based on these studies, as well as an even larger body of clinical trials, pre- and postoperative aspirin therapy is now considered a standard adjunct in bypass grafting.[44]

**2. Percutaneous Angioplasty.** Although the technical fundamentals are similar for all vascular beds, percutaneous transluminal balloon angioplasty (PTA) of peripheral arteries involves a broader range of arterial sizes, flow, and patency rates than does coronary angioplasty, for example. Dilatation of high flow iliac arteries have low acute failure rates and excellent long-term patency, while the tibial arteries have lower velocities of flow than comparably sized coronaries and have poorer patency rates. Studies of peripheral artery PTA have documented both local and systemic platelet activation.[45–48] Radionuclide imaging studies reveal the acute accumulation of platelets at the site of injury,[49,50] and the platelets passing through the site show signs of activation.[51] Unlike vascular grafting, the platelet reactivity of these sites appears to subside more quickly, and some patients will have an unreactive angioplasty site within days.[49,50]

## III. Characteristics of Platelet Function in PAD

Ever since PAD was recognized as a distinct clinical entity, researchers have suspected that the platelets of patients with PAD were hyperaggregable, or hyperreactive. The earliest studies focused on the theory that patients with systemic atherosclerosis had circulating platelet aggregates or other manifestations of chronic, *in vivo* activation of their platelets. Wu and Hoak,[52] Salzman, and others[53,54] used simple but elegant *ex vivo* methods to measure the presence of circulating platelet aggregates and their stickiness, from carefully obtained samples of peripheral venous blood. They indeed found elevated levels of platelet aggregates associated with a variety of arterial pathologies, including stroke and myocardial infarction. As platelet activation progresses, platelet-specific proteins are released from α-granules (e.g., β-thromboglobulin, platelet factor 4 [Chapter 15]), and eicosanoid synthesis yields free thromboxane $A_2$ (Chapter 31). Thus, the plasma levels of these proteins and the stable metabolite thromboxane $B_2$ have also been measured in venous blood as markers for *in vivo* platelet activation. Elevated levels have been associated with unstable angina,

pathologic heart valves, and PAD.[55–60] However, concerns remain whether these methods for measuring the *in vivo* activation of platelets really reflect the physiologic state of circulating platelets or instead reflect a lower threshold to activation by the venipuncture and processing of the blood.[61] Still, there is also supporting evidence of *in vivo* platelet activation, based on the observations of elevated levels of urinary secretion of thromboxane metabolites and shortened platelet survival times[60,62] in PAD.

Another approach to estimating the level of *in vivo* platelet activation has been to measure the expression of activation-dependent receptors or markers on the platelet surface, usually by flow cytometry (Chapter 30). The risks of artifactual platelet activation are minimized by the use of whole blood methods, and flow cytometry also offers the advantage of specificity by focusing on individual receptors or pathways. For example, during platelet activation, P-selectin and CD40 ligand (CD40L) are more abundantly expressed on the platelet surface, and the platelet integrin αIIbβ3 (GPIIb-IIIa) undergoes conformational changes that can be detected by monoclonal antibodies such as PAC-1. Activated platelets shed microparticles (Chapter 20), which can be measured by flow cytometry, and platelets release soluble forms of P-selectin and CD40L, which are proinflammatory, procoagulant proteins (Chapter 39) that can be assayed in plasma.

Finally, Michelson and others have refined flow cytometric methods for measuring the presence of heterotypic platelet-monocyte aggregates, which may be one of the most sensitive markers for *in vivo* platelet activation.[63] The dominant receptor/ligand pair involved in platelet-leukocyte recruitment is platelet P-selectin and leukocyte PSGL-1 (Chapter 12). Platelet-leukocyte cross-talk also involves the interplay between the leukocyte integrin αMβ2 (Mac-1, CD11b/CD18) and platelet GPIb.[15,64] The CD40L/CD40 ligand/receptor pair is another important avenue for activated platelets to recruit inflammatory cells.[18,47,65]

Table 37-1 summarizes a range of platelet function studies that have been performed specifically on patients with PAD. Although the studies are flawed by small cohorts and heterogeneous methods, they support the contention that the platelets of PAD patients manifest a higher level of chronic activation *in vivo* and that they have lower threshold sensitivities to soluble agonists. Studies that employed low concentrations of agonists more frequently found significant differences between PAD patients and controls, while higher concentrations of agonists did not always distinguish healthy and diseased groups. Basal platelet expression of P-selectin has not been found to be consistently elevated in PAD. Some authors have observed increased platelet receptor density and increased expression of activation-dependent epitopes (Table 37-1). *In vitro* sensitivity to ADP and thrombin are generally increased in PAD. These observations in PAD patients are in keeping with the wealth of data supporting increased platelet activation in patients with symptomatic coronary or cerebrovascular disease (see Chapters 35 and 36).

Understanding the causes for platelet activation in PAD is a real "chicken or egg" conundrum. Which came first, the platelet dysfunction or the atherosclerosis? Epidemiological data suggest that the presence of hyperactive platelets in younger, healthy volunteers is an independent predictor of later vascular complications.[66,67] Alternatively, could the burden of dysfunctional endothelium and stenotic atherosclerotic vessels be the prime cause for activating platelets and lowering their threshold to stimulation? Diabetes, smoking, and hyperlipidemia are all independently associated with increased platelet reactivity (see Chapters 35, 36, 38, and refs.[67-69]).

Apart from these factors, is there an independent relationship between platelet hyperreactivity and PAD? Riba et al.[70] examined cohorts with PAD and found increased platelet activation only in those with elevated levels of homocysteine. Davi and colleagues[62] studied urinary excretion of thromboxane metabolites and found that diabetes, hypercholesterolemia, and hypertension — but not PAD per se — accounted for their significant increases in thromboxane excretion. These investigators[62] also noted, as have others, the close correlation between the degree of platelet activation and the incidence of subsequent vascular ischemic events.

From a clinician's point of view, the "chicken or egg" question about the causes of platelet hyperresponsiveness in PAD may be moot. The individual presenting with PAD can be assumed, on average, to have more reactive platelets than age-matched healthy controls, and this person's platelets' prothrombotic, proinflammatory, and proatherogenic effects will contribute to increased vascular morbidity and mortality if untreated. The evidence is clear that these patients should be chronically treated with platelet-inhibiting drugs for primary prevention of vascular ischemic events.

## IV. Platelet-Mediated Therapy in PAD

### A. Control of Metabolic Factors

Controlling modifiable risk factors should be the bulwark of treatment for patients with PAD. Diabetes, hyperlipidemia, smoking, hyperhomocysteinemia, and hypertension are all implicated in the prothrombotic and proinflammatory platelet phenotype seen in PAD. Each of these factors has been independently associated with exaggerated platelet reactivity. These are considered modifiable risk factors, as correction of these physiologic and metabolic derangements may help normalize platelet function.

Beyond its conventional lipid-lowering effects, statin therapy appears to reduce the prothrombotic and proinflam-

matory diathesis mediated in part by platelets.[71,72] The mechanisms of statin effects may include salutary increases in NO production.[73] A retrospective study suggested that statin therapy may be associated with improved patency of vein bypass grafts.[74] Cipollone et al. found that sCD40L and platelet and prothrombotic markers were elevated in hypercholesterolemia and were reduced by statin treatment.[75] Improved glycemic control in diabetics also helps normalize platelet function[76] (Chapter 38). It is well established that smoking causes an up-regulation of platelet reactivity and platelet-mediated proinflammatory markers.[68,77,78] Thus, control of these risk factors can be considered an indirect (yet effective) form of both antithrombotic and antiinflammatory therapy.

### B. Platelet Inhibitors for PAD

All patients with PAD should be considered for chronic, long-term treatment with a platelet-inhibiting drug. Because of the relative paucity of clinical trials devoted exclusively to PAD, many of the benefits of antiplatelet therapy in PAD have been inferred from large trials encompassing a broader range of atherosclerotic diseases. Summaries of the clinical data and evidence-based guidelines have been published in the *Cochrane Review*,[79] by the American College of Chest Physicians,[44] and others.[80-82]

**1. Antiplatelet Therapy to Retard the Progression of Peripheral Atherosclerotic Occlusive Disease.** It is difficult to know whether antiplatelet therapy can slow the progression of atherosclerosis per se. The episodic, symptomatic worsening of PAD is often caused by subclinical thromboses at critical stenoses, rather than pure progression of atherosclerosis. Accordingly, the antithrombotic effects of aspirin may retard the progression of symptoms in PAD, as has been observed in a number of placebo-controlled trials.[83-86] But the growing evidence linking platelets, inflammation, and atherosclerosis (see Chapters 35 and 39) suggests that aspirin alone is insufficient to slow atherosclerosis. For example, aspirin does not effectively block the formation of platelet-leukocyte aggregates,[87,88] while clopidogrel[88,89] and combinations of platelet inhibitors do.[90] Cilostazol, a selective phosphodiesterase inhibitor that is both a vasodilator and a platelet inhibitor, has been approved by the Food and Drug Administration (FDA) for use in the United States for the improvement of walking distance in patients with intermittent claudication (see Chapter 64).

**2. Primary or Secondary Prevention of Stroke, Myocardial Infarction, or Vascular Death.** In patients with PAD, a much more compelling case can be made for lifelong antiplatelet therapy to prevent future ischemic events and to reduce overall vascular mortality. The choices for chronic platelet inhibitor therapy currently boil down to aspirin, clopidogrel, or combined therapy. Chapters 60 and 61 review the pharmacology of aspirin and clopidogrel, respectively, in detail. Consensus guidelines recognize the optimal dosage of aspirin to be 80 to 325 mg daily. In its meta-analysis of studies of PAD, the Antithrombotic Trialists' Collaboration showed a 23% reduction in vascular morbidity and mortality by antiplatelet therapy.[82] A recent *Cochrane* review[79] of high-risk vascular patients found that either aspirin or thienopyridines (clopidogrel or ticlopidine) significantly reduced the incidence of vascular ischemic events. Thienopyridine therapy conferred a marginally (approximately 1%), albeit statistically significant, lower absolute incidence of vascular end points, compared to aspirin. Dipyridamole, used as an adjunct to aspirin, does not appear to offer any additive benefit, at least in PAD. The CAPRIE trial also found a marginal benefit (0.5% absolute risk reduction) of clopidogrel over aspirin.[91] Subgroup analysis of the data yielded some perplexing results. Aspirin and clopidogrel were equally effective for the subgroups with symptomatic coronary or cerebrovascular disease, but clopidogrel was clearly superior in patients enrolled for the diagnosis of PAD.

Why should PAD patients benefit more from clopidogrel? One possibility is that the members of the PAD cohort were generally undertreated for their modifiable risk factors, a neglect that is common in clinical practice.[4] With less aggressive therapy of hyperlipidemia and hypertension, for example, the small pharmacologic advantage of ADP $P2Y_{12}$ receptor blockade by clopidogrel may be more clinically significant. A genetic explanation is also possible — Fontana et al. found an association between PAD and a gain-of-function polymorphism in the $P2Y_{12}$ receptor.[65] Further studies are necessary to determine if PAD patients have unique responses to aspirin/clopidogrel or if the differences are related to the undertreatment of modifiable risk factors. Combined therapy with aspirin and clopidogrel has not been studied in patients with PAD.

**3. Improving the Patency of Peripheral Bypass Grafts.** All PAD patients should be on long-term aspirin (or clopidogrel, depending on how one views the previously discussed current controversy) to prevent myocardial infarction and stroke. Hence, a discussion of the optimal antiplatelet therapy after bypass grafting should start with baseline antiplatelet therapy (usually aspirin). Randomized, placebo-controlled trials have clearly shown the benefit of aspirin or thienopyridines in improving the patency of small-diameter (i.e., infrainguinal) bypasses, especially prosthetic bypasses.[44,81,92] The Antiplatelet Trialists' Collaboration found a highly significant difference in the absolute rates of occlusion of peripheral arterial interventions with antiplatelet therapy (16% versus 25% in controls),[81] which was also supported by a *Cochrane* review.[93]

The value of antiplatelet therapy for vein grafts is still inconclusive. After the native endothelium of a vein graft has recovered from the initial perioperative trauma, vein grafts generally have minimal intrinsic thrombogenicity. Late graft failures still occur, but these are mainly due to neointimal hyperplasia or progression of atherosclerosis in the inflow and outflow tracts. This may explain why some trials have failed to show a benefit from aspirin in improving the long-term patency of vein grafts.[94]

For prosthetic grafts, two major trials are in conflict regarding the optimal antithrombotic therapy. In both the Veterans Affairs multicenter cooperative trial[95] and the Dutch Bypass Oral Anticoagulants or Aspirin Study group,[96] aspirin alone yielded similar cumulative patency rates in prosthetic grafts (68–70% at 2 years). Yet the Veterans Affairs trial found that aspirin *plus* warfarin improved the patency rate over aspirin alone, while the Dutch group found that warfarin alone (*without aspirin*) was inferior to aspirin alone. One interpretation of these studies is that prosthetic grafts, by virtue of their continuing platelet reactivity and thrombogenicity, may benefit more from antiplatelet than from anticoagulant therapy. This controversy is still unsettled, and the value of ADP receptor antagonists has not been studied. As a minimum, for prosthetic grafts, long-term aspirin therapy is probably beneficial, but patients may benefit further from supplemental antithrombotic agents.

**4. Improving the Patency of Stents and Angioplasties.** Based on the wealth of evidence supporting antiplatelet therapy in coronary angioplasty (reviewed in Chapter 35 and reference[97]), one would expect that antiplatelet therapy for angioplasty (PTA) in the peripheral arteries would also offer a clear advantage. But the clinical evidence supporting antiplatelet therapy in peripheral angioplasty is equivocal.[98] Only a few placebo-controlled trials have been conducted, with conflicting results.[44,98] The main confounding factor in peripheral PTA studies is that the procedures are quite heterogeneous: interventions at both high and low risk for thrombosis; iliac versus tibial; angioplasty of focal stenoses versus recanalization of occluded arteries. As in peripheral arterial bypass, the host's platelet reactivity may predict later occlusion. Mueller and colleagues found that claudicators who were aspirin resistant were more prone to occlusion of their angioplasty site,[99] and others have noted an association between excessive platelet uptake at the site of injury and premature closure.[100,101] Cassar et al. measured markers of platelet activation by flow cytometry before and after lower extremity PTA.[102] Not surprisingly, more significant inhibition of platelet function was encountered in patients treated with two different inhibitors, compared to aspirin alone. Disappointingly, this study did not measure the impact of platelet inhibition on clinical outcomes.[102] From other experimental data and by inference from coronary trials, it is logical that periprocedural antiplatelet therapy should reduce the risks of thrombosis at sites of peripheral angioplasty.

## V. Platelets in Venous Disease

The initiation and progression of venous thromboembolism (VTE) appear to be more dependent on the activation of the plasma coagulation cascade than they do on the activation of platelets. Aspirin may reduce deep venous thrombosis and pulmonary embolism in high risk surgical or medical patients by approximately one third. The Antiplatelet Trialists' Collaboration meta-analysis[103] showed an absolute reduction from 35 to 25% in the incidence of VTE, and another large trial showed similar reductions.[104] However, more recent trials and evidence-based reviews advise against aspirin for primary thromboprophylaxis against VTE,[105,106] mainly due to the superiority of other methods. Prophylaxis with low-molecular-weight heparins or modern external pneumatic compression devices are generally more effective than aspirin in high-risk patients in head-to-head trials[105] and can reduce the risks of VTE in untreated patients by one half to two thirds.

Even though antiplatelet therapy is not the optimal strategy for venous thromboprophylaxis, that does not acquit platelets from their important role in the inflammatory component of venous thrombosis. Venous thrombi initiate a biochemical and cellular inflammatory response that dictates, in part, the speed of resolution of a thrombus, as well the later vein wall fibrosis and valve destruction.[107] Using genetic manipulations of P-selectin in animal models, Wakefield and colleagues convincingly showed that P-selectin is responsible for both the early recruitment of leukocytes to the thrombus and later fibrosis.[108,109] Blockade of P-selectin with an oral inhibitor reduced thrombus weight in mice, but paradoxically it increased the acute inflammatory cell infiltrate in the vein wall.[110] Pharmacological strategies to modulate the platelet's recruitment of leukocytes to venous thrombi may hold promise for reducing the need for anticoagulants, speeding the resolution of thrombi, and even preventing fibrosis.

Chronic venous insufficiency and the postphlebitic syndrome are both manifestations of chronic venous hypertension. This venous hypertension leads to poor local capillary perfusion, extravasation of plasma exudate, and the transmigration of inflammatory cells. Studies of patients with and without venous ulceration by Michelson's group[111,112] suggest that chronic venous insufficiency is also marked by an increase in systemic platelet activation and a lower agonist threshold to form platelet-leukocyte aggregates. Thus, venous insufficiency joins peripheral arterial disease as a condition associated with chronic activation of platelets and inflammatory pathways.

## VI. Conclusions

Peripheral arterial disease, although relatively benign in its outlook for the legs, is a serious prognostic marker for future stroke, myocardial infarction, and vascular death. Patients with PAD can be expected to have chronically activated and hyperreactive platelets. Lifelong antiplatelet therapy (aspirin or clopidogrel) is recommended. The early diagnosis of PAD and aggressive medical management of modifiable risk factors (hyperlipidemia, diabetes, hypertension, smoking) may also modulate the prothrombotic and proinflammatory diathesis observed in these patients. Antiplatelet therapy is also indicated for peripheral bypass grafting, percutaneous angioplasty, and stenting, although the optimal antiplatelet and antithrombotic regimens have not been clearly defined.

For the prevention of venous thromboembolism, antiplatelet agents are somewhat effective, but they are not the optimal therapy for prevention as better prophylactic strategies exist. Chronic venous insufficiency is also associated with circulating platelet-leukocyte aggregates and lower agonist thresholds to platelet activation.

## References

1. Hiatt, W. R. (2001). Medical treatment of peripheral arterial disease and claudication. *N Engl J Med, 344,* 1608–1621.
2. Burns, P., Gough, S., & Bradbury, A. W. (2003). Management of peripheral arterial disease in primary care. *BMJ, 326,* 584–588.
3. Meijer, W. T., Grobbee, D. E., Hunink, M. G., et al. (2000). Determinants of peripheral arterial disease in the elderly: The Rotterdam study. *Arch Int Med, 160,* 2934–2938.
4. Hirsch, A. T., Criqui, M. H., Treat-Jacobson, D., et al. (2001). Peripheral arterial disease detection, awareness, and treatment in primary care. *JAMA, 286,* 1317–1324.
5. Criqui, M. H., Fronek, A., Barrett-Connor, E., et al. (1985). The prevalence of peripheral arterial disease in a defined population. *Circulation, 71,* 510–515.
6. Meijer, W. T., Hoes, A. W., Rutgers, D., et al. (1998). Peripheral arterial disease in the elderly: The Rotterdam study. *Arterioscler Thromb Vasc Biol, 18,* 185–192.
7. Mohler, E. R. I. (2003). Peripheral arterial disease: Identification and implications. *Arch Int Med, 163,* 2306–2314.
8. Alving, B. M., Francis, C. W., Hiatt, W. R., et al. (2003). Consultations on patients with venous or arterial diseases. *Hematology, 2003,* 540–558.
9. Stewart, K. J., Hiatt, W. R., Regensteiner, J. G., & Hirsch, A. T. (2002). Exercise training for claudication. *N Engl J Med, 347,* 1941–1951.
10. Donnelly, R., & Yeung, J. M. (2002). Management of intermittent claudication: The importance of secondary prevention. *Eur J Vasc Endovasc Surg, 23,* 100–107.
11. Hiatt, W. R. (2002). Pharmacologic therapy for peripheral arterial disease and claudication. *J Vasc Surg, 36,* 1283–1291.

12. McDermott, M. (2002). Peripheral arterial disease: Epidemiology and drug therapy. *Am J Geriatr Cardiol, 11,* 258–266.

13. Ross, R. (1999). Atherosclerosis — an inflammatory disease. *N Eng J Med, 340,* 115–126.

14. Smyth, S. S., Reis, E. D., Zhang, W., et al. (2001). Beta(3)-integrin-deficient mice but not P-selectin-deficient mice develop intimal hyperplasia after vascular injury — correlation with leukocyte recruitment to adherent platelets 1 hour after injury. *Circulation, 103,* 2501–2507.

15. Wang, Y., Sakuma, M., Chen, Z., et al. (2005). Leukocyte engagement of platelet glycoprotein Ibalpha via the integrin Mac-1 is critical for the biological response to vascular injury. *Circulation, 112,* 2993–3000.

16. Ehlers, R., Ustinov, V., Chen, Z., et al. (2003). Targeting platelet-leukocyte interactions: Identification of the integrin Mac-1 binding site for the platelet counter receptor glycoprotein Ibalpha. *J Exp Med, 198,* 1077–1088.

17. Aukrust, P., Damay, J. K., & Solum, N. O. (2004). Soluble CD40 ligand and platelets: Self-perpetuating pathogenic loop in thrombosis and inflammation? *J Am Coll Cardiol, 43,* 2326–2328.

18. Henn, V., Slupsky, J. R., Grafe, M., et al. (1998). CD40 ligand on activated platelets triggers an inflammatory reaction of endothelial cells. *Nature, 391,* 591–594.

19. Andrews, R. K., & Berndt, M. C. (2004). Platelet physiology and thrombosis. *Thromb Res, 114,* 447–453.

20. Savage, B., Sixma, J. J., & Ruggeri, Z. M. (2002). Functional self-association of von Willebrand factor during platelet adhesion under flow. *Proc Natl Acad Sci USA, 99,* 425–430.

21. Ruggeri, Z. M. (1997). Mechanisms initiating platelet thrombus formation. *Thromb Haemost, 78,* 611–616.

22. Goldman, M., Hall, C., Dykes, J., et al. (1983). Does 111 indium-platelet deposition predict patency in prosthetic arterial grafts? *Br J Surg, 70,* 635–638.

23. Wadenvik, H., Örtenwall, P., Kutti, J., et al. (1988). Splenic platelet kinetics and platelet production after major reconstructive vascular surgery. *Acta Med Scand, 223,* 147–152.

24. McCollum, C. N., Kester, R. C., Rajah, S. M., et al. (1981). Arterial graft maturation: The duration of thrombotic activity in Dacron aortobifemoral grafts measured by platelet and fibrinogen kinetics. *Br J Surg, 68,* 61–64.

25. Goldman, M., Norcott, H. C., Hawker, R. J., et al. (1982). Platelet accumulation on mature Dacron grafts in man. *Br J Surg, 69,* Suppl: S38–S40.

26. Goldman, M., McCollum, C. N., Hawker, R. J., et al. (1982). Dacron arterial grafts: The influence of porosity, velour, and maturity on thrombogenicity. *Surgery, 92,* 947–952.

27. Robicsek, F., Duncan, G. D., Anderson, C. E., et al. (1987). Indium 111-labeled platelet deposition in woven and knitted Dacron bifurcated aortic grafts with the same patient as a clinical model. *J Vasc Surg, 5,* 833–837.

28. Bearn, P. E., McCollum, C. N., & Greenhalgh, R. M. (1993). The influence of collagen and albumen presealants on knitted Dacron grafts. *Eur J Vasc Surg, 7,* 271–276.

29. Callow, A. D., Connolly, R., O'Donnell, T. F., Jr., et al. (1982). Platelet–arterial synthetic graft interaction and its modification. *Arch Surg, 117,* 1447–1455.

30. Wakefield, T. W., Shulkin, B. L., Fellows, E. P., et al. (1989). Platelet reactivity in human aortic grafts: A prospective, randomized midterm study of platelet adherence and release products in Dacron and polytetrafluoroethylene conduits. *J Vasc Surg, 9,* 234–243.

31. Tsuchida, H., Cameron, B. L., Marcus, C. S., et al. (1992). Modified polytetrafluoroethylene: Indium 111-labeled platelet deposition on carbon-lined and high-porosity polytetrafluoroethylene grafts. *J Vasc Surg, 16,* 643–649.

32. Devine, C., & McCollum, C. (2004). Heparin-bonded Dacron or polytetrafluoroethylene for femoropopliteal bypass: Five-year results of a prospective randomized multicenter clinical trial. *J Vasc Surg, 40,* 924–931.

33. Zerkowski, H. R., Knocks, M., Konerding, M. A., et al. (1993). Endothelial damage of the venous graft in CABG: Influence of solutions used for storage and rinsing on endothelial function. *Eur J Cardiothorac Surg, 7,* 376–382.

34. Davies, M. G., & Hagen, P. O. (1995). Pathophysiology of vein graft failure: A review. *Eur J Vasc Endovasc Surg, 9,* 7–18.

35. Zilla, P., Fasol, R., Deutsch, M., et al. (1987). Endothelial cell seeding of polytetrafluoroethylene vascular grafts in humans: A preliminary report. *J Vasc Surg, 6,* 535–541.

36. Ortenwall, P., Wadenvik, H., & Risberg, B. (1989). Reduced platelet deposition on seeded versus unseeded segments of expanded polytetrafluoroethylene grafts: Clinical observations after a 6-month follow-up. *J Vasc Surg, 10,* 374–380.

37. Jensen, N., Lindblad, B., & Bergqvist, D. (1994). Endothelial cell seeded dacron aortobifurcated grafts: Platelet deposition and long-term follow-up. *J Cardiovasc Surg* (Torino), *35,* 425–429.

38. Kohler, T. R., Stratton, J. R., Kirkman, T. R., et al. (1992). Conventional versus high-porosity polytetrafluoroethylene grafts: Clinical evaluation. *Surgery, 112,* 901–907.

39. Allen, B. T., Long, J. A., Clark, R. E., et al. (1984). Influence of endothelial cell seeding on platelet deposition and patency in small-diameter Dacron arterial grafts. *J Vasc Surg, 1,* 224–233.

40. Saad, E. M., Kaplan, S., el-Massry, S., et al. (1993). Platelet aggregometry can accurately predict failure of externally supported knitted Dacron femoropopliteal bypass grafts. *J Vasc Surg, 18,* 587–594.

41. Esposito, C. J., Popescu, W. M., Rinder, H. M., et al. (2003). Increased leukocyte-platelet adhesion in patients with graft occlusion after peripheral vascular surgery. *Thromb Haemost, 90,* 1128–1134.

42. Pumphrey, C. W., Chesebro, J. H., Dewanjee, M. K., et al. (1983). In vivo quantitation of platelet deposition on human peripheral arterial bypass grafts using indium-111-labeled platelets: Effect of dipyridamole and aspirin. *Am J Cardiol, 51,* 796–801.

43. Stratton, J. R., & Ritchie, J. L. (1986). Reduction of indium-111 platelet deposition on Dacron vascular grafts in humans by aspirin plus dipyridamole. *Circulation, 73,* 325–330.

44. Clagett, G. P., Sobel, M., Jackson, M. R., et al. (2004). Antithrombotic therapy in peripheral arterial occlusive disease:

The Seventh ACCP Conference on Antithrombotic and Thrombolytic Therapy. *Chest, 126* (2 suppl), 609s–626s.

45. Ludemann, J., Schulte, K. L., Hader, O., et al. (2001). Leukocyte/endothelium activation and interactions during femoral percutaneous transluminal angioplasty. *Vasc Surg, 35*, 293–301.

46. Barani, J., Gottsater, A., Mattiasson, I., & Linblad, B. (2002). Platelet and leukocyte activation during aortoiliac angiography and angioplasty. *Eur J Vasc Endovasc Surg, 23*, 220–225.

47. Blann, A. D., Tan, K. T., Tayebjee, M. H., et al. (2005). Soluble CD40L in peripheral artery disease — relationship with disease severity, platelet markers and the effects of angioplasty. *Thromb Haemost, 93*, 578–583.

48. Buchholz, A. M., Bruch, L., & Schulte, K. L. (2003). Activation of circulating platelets in patients with peripheral arterial disease during digital subtraction angiography and percutaneous transluminal angioplasty. *Thromb Res, 109*, 13–22.

49. Miller, D. D., Rivera, F. J., Garcia, O. J., et al. (1992). Imaging of vascular injury with 99mTc-labeled monoclonal antiplatelet antibody S12. Preliminary experience in human percutaneous transluminal angioplasty. *Circulation, 85*, 1354–1363.

50. Minar, E., Ehringer, H., Ahmadi, R., et al. (1987). Platelet deposition at angioplasty sites and platelet survival time after PTA in iliac and femoral arteries: Investigations with indium-111-oxine labelled platelets in patients with ASA (1.0 g/day)-therapy. *Thromb Haemost, 58*, 718–723.

51. Scharf, R. E., Tomer, A., Marzec, U. M., et al. (1992). Activation of platelets in blood perfusing angioplasty-damaged coronary arteries: Flow cytometric detection. *Arterioscler Thromb, 12*, 1475–1487.

52. Wu, K. K., & Hoak, J. C. (1975). Increased platelet aggregates in patients with transient ischemic attacks. *Stroke, 6*, 521–524.

53. Salzman, E. W. (1963). Measurement of platelet adhesiveness. *J Lab Clin Med, 62*, 724–735.

54. Zucker, M. B., Brownlea, S., & McPherson, J. (1987). Insights into the mechanism of platelet retention in glass bead columns. *Ann N Y Acad Sci, 516*, 398–406.

55. Sobel, M., Salzman, E. W., Davies, G. C., et al. (1981). Circulating platelet products in unstable angina pectoris. *Circulation, 63*, 300–306.

56. Smitherman, T. C., Milam, M., Woo, J., et al. (1981). Elevated beta thromboglobulin in peripheral venous blood of patients with acute myocardial ischemia: Direct evidence for enhanced platelet reactivity in vivo. *Am J Cardiol, 48*, 395–402.

57. Sobel, M., Gervin, C. A., Qureshi, G. D., et al. (1987). Coagulation responses to heparin in the ischemic limb: Assessment of thrombin and platelet activation during vascular surgery. *Circulation, 76* (suppl III), 8–13.

58. Donaldson, M. C., Matthews, E. T., Hadjimichael, J., et al. (1987). Markers of thrombotic activity in arterial disease. *Arch Surg, 122*, 897–900.

59. Trifiletti, A., Barbera, N., Pizzoleo, M. A., et al. (1995). Hemostatic disorders associated with arterial hypertension and peripheral arterial disease. *J Cardiovasc Surg* (Torino), *36*, 483–485.

60. Zahavi, J., & Zahavi, M. (1985). Enhanced platelet release reaction, shortened platelet survival time and increased platelet aggregation and plasma thromboxane B2 in chronic obstructive arterial disease. *Thromb Haemost, 53*, 105–109.

61. Kohanna, F. H., Smith, M. H., & Salzman, E. W. (1984). Do patients with thromboembolic disease have circulating platelet aggregates? *Blood, 64*, 205–209.

62. Davi, G., Gresele, P., Violi, F., et al. (1997). Diabetes mellitus, hypercholesterolemia, and hypertension but not vascular disease per se are associated with persistent platelet activation in vivo: Evidence derived from the study of peripheral arterial disease. *Circulation, 96*, 69–75.

63. Michelson, A. D., Barnard, M. R., Krueger, L. A., et al. (2001). Circulating monocyte-platelet aggregates are a more sensitive marker of *in vivo* platelet activation than platelet surface P-selectin: Studies in baboons, human coronary intervention, and human acute myocardial infarction. *Circulation, 104*, 1533–1537.

64. Simon, D. I., Xu, H., Ballantyne, C. M., et al. (1999). Platelet glycoprotein Ibalpha is an adhesive ligand for the leukocyte integrin mac-1 (CD11b/CD18). *Circulation, 100*, I-1746.

65. Fontana, P., Gaussem, P., Aiach, M., et al. (2003). P2Y(12)H2 haplotype is associated with peripheral arterial disease: A case-control study. *Circulation, 108*, 2971–2973.

66. O'Donnell, C. J., Larson, M. G., Feng, D. L., et al. (2001). Genetic and environmental contributions to platelet aggregation: The Framingham Heart Study. *Circulation, 103*, 3051–3056.

67. Breddin, H. K., Lippold, R., Bittner, M., et al. (1999). Spontaneous platelet aggregation as a predictive risk factor for vascular occlusions in healthy volunteers? Results of the HAPARG Study. *Atherosclerosis, 144*, 211–219.

68. Lassila, R., & Laustiola, K. E. (1992). Cigarette smoking and platelet-vessel wall interactions. *Prostaglandins Leukot Essent Fatty Acids, 46*, 81–86.

69. Semb, A. G., van, W. S., Ueland, T., et al. (2003). Raised serum levels of soluble CD40 ligand in patients with familial hypercholesterolemia: Downregulatory effect of statin therapy. *J Am Coll Cardiol, 41*, 275–279.

70. Riba, R., Nicolaou, A. A., Troxler, M., et al. (2004). Altered platelet reactivity in peripheral vascular disease complicated with elevated plasma homocysteine levels. *Atherosclerosis, 175*, 69–75.

71. Gaddam, V., Li, D. Y., & Mehta, J. L. (2002). Antithrombotic effects of atorvastatin — an effect unrelated to lipid lowering. *J Cardiovasc Pharmacol Ther, 7*, 247–253.

72. Calabro, P., & Yeh, E. T. (2005). The pleiotropic effects of statins. *Curr Op Cardiol, 20*, 541–546.

73. Hognestad, A., Aukrust, P., Wergeland, R., et al. (2004). Effects of conventional and aggressive statin treatment on markers of endothelial function and inflammation. *Clin Cardiol, 27*, 199–203.

74. Abbruzzese, T. A., Havens, J., Belkin, M., et al. (2004). Statin therapy is associated with improved patency of

autogenous infrainguinal bypass grafts. *J Vasc Surg, 39,* 1178–1185.

75. Cipollone, F., Mezzetti, A., Porreca, E., et al. (2002). Association between enhanced soluble CD40L and prothrombotic state in hypercholesterolemia: Effects of statin therapy. *Circulation, 106,* 399–402.

76. Ferroni, P., Basili, S., Falco, A., & Davi, G. (2004). Platelet activation in type 2 diabetes mellitus. *J Thromb Haemost, 2,* 1282–1291.

77. Lu, J. T., & Creager, M. A. (2004). The relationship of cigarette smoking to peripheral arterial disease. *Rev Cardiovasc Med, 5,* 189–193.

78. Harding, S. A., Sarma, J., Josephs, D. H., et al. (2006). Upregulation of the CD40/CD40 ligand dyad and platelet-monocyte aggregation in cigarette smokers. *Circulation, 109,* 1926–1929.

79. Hankey, G. J., & Dunbabin, D. W. (2000). Thienopyridine derivatives (ticlopidine, clopidogrel) versus aspirin for preventing stroke and other serious vascular events in high vascular risk patients. *Cochrane Database of Systemic Review,* 2000CD001246.

80. Antiplatelet Trialists' Collaboration. (1994). Collaborative overview of randomised trials of antiplatelet therapy — I: Prevention of death, myocardial infarction, and stroke by prolonged antiplatelet therapy in various categories of patients. Antiplatelet Trialists' Collaboration. *BMJ, 308,* 81–106.

81. Antiplatelet Trialists' Collaboration. (1994). Collaborative overview of randomised trials of antiplatelet therapy — II: Maintenance of vascular graft or arterial patency by antiplatelet therapy. *BMJ, 308,* 159–168.

82. Antithrombotic Trialists' Collaboration. (2002). Collaborative meta-analysis of randomised trials of antiplatelet therapy for prevention of death, myocardial infarction, and stroke in high risk patients. *BMJ, 324,* 71–86.

83. Hess, H., Mietaschik, A., & Deichsel, G. (1985). Drug-induced inhibition of platelet function delays progression of peripheral occlusive arterial disease: A prospective double-blind arteriographically controlled trial. *Lancet, 1,* 416–419.

84. Libretti, A., & Catalano, M. (1986). Treatment of claudication with dipyridamole and aspirin. *Int J Clin Pharmacol Res, 6,* 59–60.

85. Giansante, C., Calabrese, S., Fisicaro, M., et al. (1990). Treatment of intermittent claudication with antiplatelet agents. *J Int Med, 18,* 400–407.

86. Goldhaber, S. Z., Manson, J. E., Stampfer, M. J., et al. (1992). Low-dose aspirin and subsequent peripheral arterial surgery in the Physicians' Health Study. *Lancet, 340,* 143–145.

87. Li, N., Hu, H., & Hjemdahl, P. (2003). Aspirin treatment does not attenuate platelet or leukocyte activation as monitored by whole blood flow cytometry. *Thromb Res, 111,* 165–170.

88. Klinkhardt, U., Bauersachs, R., Adams, J., et al. (2003). Clopidogrel but not aspirin reduces P-selectin expression and formation of platelet-leukocyte aggregates in patients with atherosclerotic vascular disease. *Clin Pharmacol Ther, 73,* 232–241.

89. Evangelista, V., Manarini, S., Dell'Elba, G., et al. (2005). Clopidogrel inhibits platelet-leukocyte adhesion and platelet-dependent leukocyte activation. *Thromb Haemost, 94,* 568–577.

90. Zhao, L., Bath, P., & Heptinstall, S. (2001). Effects of combining three different antiplatelet agents on platelets and leukocytes in whole blood in vitro. *Br J Pharmacol, 134,* 353–358.

91. CAPRIE Steering Committee. (1996). A randomised, blinded, trial of clopidogrel versus aspirin in patients at risk of ischaemic events (CAPRIE). *Lancet, 348,* 1329–1339.

92. Becquemin, J. P. (1997). Effect of ticlopidine on the long-term patency of saphenous vein bypass grafts in the legs. *N Engl J Med, 337,* 1726–1731.

93. Dorffler-Melly, J., Buller, H. R., Koopman, M. M., et al. (2003). Antiplatelet agents for preventing thrombosis after peripheral arterial bypass surgery (Review). *Cochrane Database Syst Rev (3),* CD000535.

94. McCollum, C., Alexander, C., Kenchington, G., et al. (1991). Antiplatelet drugs in femoropopliteal vein bypasses: A multicenter trial. *J Vasc Surg, 13,* 150–162.

95. Johnson, W. C., Williford, W. O., & Members of the Department of Veterans Affairs Cooperative Study #362. (2002). Benefits, morbidity, and mortality associated with long-term administration of oral anticoagulant therapy to patients with peripheral arterial bypass procedures: A prospective randomized study. *J Vasc Surg, 35,* 413–421.

96. Efficacy of oral anticoagulants compared with aspirin after infrainguinal bypass surgery (The Dutch Bypass Oral Anticoagulants or Aspirin Study): A randomised trial. (2000). *Lancet, 355,* 346–351.

97. Popma, J. J., Berger, P., Ohman, E. M., et al. (2004). Antithrombotic therapy during percutaneous coronary intervention: The Seventh ACCP Conference on Antithrombotic and Thrombolytic Therapy. *Chest, 126,* 576S–599.

98. Watson, H. R., & Bergqvist, D. (2000). Antithrombotic agents after peripheral transluminal angioplasty: A review of the studies, methods and evidence for their use. *Eur J Vasc Endovasc Surg, 19,* 445–450.

99. Mueller, M. R., Salat, A., Stangl, P., et al. (1997). Variable platelet response to low-dose ASA and the risk of limb deterioration in patients submitted to peripheral arterial angioplasty. *Thromb Haemost, 78,* 1003–1007.

100. Nyamekye, I., Costa, D., Raphael, M., et al. (1997). Thrombosis and restenosis after peripheral angioplasty: Does acute 111indium-platelet accumulation predict angioplasty outcome? *Eur J Vasc Endovasc Surg, 13,* 388–393.

101. Poskitt, K. R., Harwood, A., Scott D. J. A., et al. (1991). Failure of peripheral arterial balloon angioplasty: Does platelet deposition play a role? *Eur J Vasc Surg, 5,* 541–547.

102. Cassar, K., Ford, I., Greaves, M., et al. (2004). Randomised clinical trial of the antiplatelet effects of aspirin-clopidogrel combination versus aspirin alone after lower limb angioplasty. *Br J Surg, 92,* 159–165.

103. Collaborative overview of randomised trials of antiplatelet therapy — III: Reduction in venous thrombosis and pulmo-

nary embolism by antiplatelet prophylaxis among surgical and medical patients. *BMJ, 308,* 235–246.

104. Prevention of pulmonary embolism and deep vein thrombosis with low dose aspirin: Pulmonary Embolism Prevention (PEP) trial. (2000). *Lancet, 355,* 1295–1302.

105. Gent, M., Hirsh, J., Ginsberg, J. S., et al. (1996). Low-molecular-weight heparinoid Orgaran is more effective than aspirin in the prevention of venous thromboembolism after surgery for hip fracture. *Circulation, 93,* 80–84.

106. Geerts, W. H., Pineo, G. F., Heit, J. A., et al. (2004). Prevention of venous thromboembolism: The Seventh ACCP Conference on Antithrombotic and Thrombolytic Therapy. *Chest, 126,* 338S–3400.

107. Wakefield, T. W., & Henke, P. K. (2005). The role of inflammation in early and late venous thrombosis: Are there clinical implications? *Sem Vasc Surg, 18,* 118–129.

108. Thanaporn, P., Myers, D. D., Wrobleski, S. K., et al. (2003). P-selectin inhibition decreases post-thrombotic vein wall fibrosis in a rat model. *Surgery, 134,* 365–371.

109. Myers, D. D., Hawley, A. E., Farris, D. M., et al. (2003). P-selectin and leukocyte microparticles are associated with venous thrombogenesis. *J Vasc Surg, 38,* 1075–1089.

110. Myers, D. D., Jr., Rectenwald, J. E., Bedard, P. W., et al. (2005). Decreased venous thrombosis with an oral inhibitor of P-selectin. *J Vasc Surg, 42,* 329–336.

111. Peyton, B. D., Rohrer, M. J., Furman, M. I., et al. (1998). Patients with venous stasis ulceration have increased monocyte-platelet aggregation. *J Vasc Surg, 27,* 1109–1115.

112. Powell, C. C., Rohrer, M. J., Barnard, M. R., et al. (1999). Chronic venous insufficiency is associated with increased platelet and monocyte activation and aggregation. *J Vasc Surg, 30,* 844–851.

113. Robless, P. A., Okonko, D., Lintott, P., et al. (2003). Increased platelet aggregation and activation in peripheral arterial disease. *Eur J Vasc Endovasc Surg, 25,* 16–22.

114. Ejim, O. S., Powlin, M. J., Dandona, P., et al. (1990). A flow cytometric analysis of fibronectin binding to platelets from patients with peripheral vascular disease. *Thromb Res, 58,* 519–524.

115. Zeiger, F., Stephan, S., Hoheisel, G., et al. (2000). P-Selectin expression, platelet aggregates, and platelet-derived microparticle formation are increased in peripheral arterial disease. *Blood Coagul Fibrinolys, 11,* 723–728.

116. Keating, F. K., Whitaker, D. A., Kabbani, S. S., et al. (2004). Relation of augmented platelet reactivity to the magnitude of distribution of atherosclerosis. *Am J Cardiol, 94,* 725–728.

117. Walters, T. K., Mitchell, D. C., & Wood, R. F. (1993). Low-dose aspirin fails to inhibit increased platelet reactivity in patients with peripheral vascular disease. *Br J Surg, 80,* 1266–1268.

118. Shankar, V. K., Chaudhury, S. R., Uthappa, M. C., et al. (2004). Changes in blood coagulability as it traverses the ischemic limb. *J Vasc Surg, 39,* 1033–1042.

119. Sinzinger, H., Virgolini, I., & Fitscha, P. (1990). Platelet kinetics in patients with atherosclerosis. *Thromb Res, 57,* 507–516.

120. Gosk-Bierska, I., Adamiec, R., & Szuba, A. (2003). Platelets' glycoproteins and their ligands in patients with intermittent claudication. *Angiology, 22,* 164–171.

# CHAPTER 38

# Diabetes Mellitus

**Bernd Stratmann,[1] Barbara Menart,[2] and Diethelm Tschoepe[1]**

[1]*Heart and Diabetes Centre NRW, Ruhr-University Bochum, Bod Oeynhausen, Germany*
[2]*German Diabetes Centre, Heinrich Heine University, Düsseldorf, Germany*

## I. Introduction

The majority of patients with diabetes mellitus dies from vascular complications, usually acute thrombotic events superimposed on atherosclerotic lesions in the arterial circulation. Therefore diabetes is accepted as an independent risk factor for cardiovascular events and is in fact a vascular disease. The risk for coronary artery disease (CAD), stroke, and peripheral arterial disease (PAD) is increased from twofold to fourfold by diabetes.[1,2]

This chapter reviews the evidence that in patients with diabetes mellitus

- platelet emboli cause microvascular occlusion
- platelets amplify atherogenesis
- platelets control the ischemic cascade of persons with atheroma ("thrombogenesis")
- platelet functions are hyperreactive
- platelet hyperreactivity is conditioned from megakaryocytes and is enhanced by metabolic derangements
- platelet hyperreactivity is only partly normalized by metabolic control but can be effectively treated by antiplatelet agents

## II. Clinical Background

With improved glucose control, metabolic complications in diabetes sharply decrease. The excess mortality of patients with diabetes is predominantly determined by vascular events.[3–5] More than 75% of patients with diabetes die from vascular complications, which in turn account for the majority of direct and indirect medical and social costs of the disease.[6–11] The results of the Multiple Risk Factor Intervention Trial (MRFIT) show that classical cardiovascular risk factors are also operative in diabetic patients, but diabetes acts as an independent risk factor for cardiovascular end points. Therefore, all subjects with type 2 diabetes qualify for a secondary prevention treatment.[1,2] The increased cardiovascular risk starts with impaired glucose tolerance. As a seminal component of the so-called metabolic syndrome, insulin resistance plays a significant role in the development of vascular dysfunction.[12] An increase in fasting insulin level predicts the development of hypertension in nonobese nondiabetic persons. Progressive diabetes further negatively affects the outcome of patients admitted for acute coronary syndromes both in the natural course and following revascularization procedures.[7,13,14] The pathophysiology of diabetic micro- and macroangiopathy may be partly dependent on the classification (type 1 or 2) of the disease, but both types of diabetes are associated with a hypercoagulable state driven by platelet hyperreactivity.[15–21] This suggests that metabolism and classical cardiovascular risk factors translate into vascular damage by the disturbed interaction of platelets with small and large arteries, consecutively causing damage of the dependent organs (retina, kidney, heart, brain, etc.). The same mechanism can be assumed for the relatively poor outcome for diabetic patients from revascularization maneurves such as percutaneous coronary intervention (PCI). Patients with diabetes experience more post-PCI complications and have decreased infarct-free survival.[22–32] Consistent with this hypothesis, in addition to metabolic control, antiplatelet therapy has proven to be particularly effective for these patients in primary, secondary, and tertiary prevention of vascular thrombotic disease.[33–35]

## III. Pathophysiological Contributions of Platelets to Diabetic Vascular Disease

Ischemia and tissue hypoxemia represent a common pathogenic principle that accounts for the sequence of events in the pathogenesis of diabetic vascular complications. Patients with diabetes frequently have hypercoagulable blood, evidenced by increased content of plasmatic coagulators, depressed fibrinolysis, reduced thromboresistance, and platelet hyperreactivity.[16,17,20] These alterations lower the coagulation threshold in the arterial circulation where occlusive thrombi induce hypoxic organ damage (e.g., acute coronary syndromes or overt myocardial infarction).[26,36-38] Occlusion of the capillaries by platelet or mixed platelet-leukocyte emboli may cause sustained occlusion of the functional microcirculation even without any acute clinical symptoms. Basal membrane thickening and microaneurysms, together with capillary occlusions, contribute to the pathological basis for diabetic small vessel disease.[30,39,40] Clinically, microangiopathic alterations precede acute infarction, and this adds to the severe impairment of cardiac performance following myocardial infarction.[25,41,42] The pathophysiological development of atherothrombotic end points in the diabetic environment starts with endothelial dysfunction and is mediated by accelerated atherosclerosis. A key feature of this atheromatous phenotype is the reduced resistance of the plaque connective tissue cap toward vessel wall stressors such as pressure or erosion of the covering endothelial layer. There is activation of luminal adhesion molecules, resulting in the adherence of leukocytes and/or platelets.[43,44] These changes favor repeated occlusive thrombosis, which ultimately triggers the gross infarct.[23,26,27,36]

Platelets are the universal hemostatic interface between the vessel wall, liquid phase coagulation, and fibrinolysis, leading to a physiologically balanced hemostasis at sites of vascular injury. In arterial thrombosis, their role includes embolization of the microvasculature ("embolic showering"), amplification of the atherosclerotic lesion, and thrombogenesis. Thus, it appears clear that hyperactive platelets are one of the main determinants of the prethrombotic state in diabetes mellitus.[45,46]

The immediate aggregation response of platelets after exposure to agonists such as vessel wall components (e.g., glycated collagen) or fluid phase procoagulants (e.g., thrombin) occurs when circulating platelets encounter a ruptured atherosclerotic plaque *in vivo*. Initially, circulating resting platelets recognize vessel wall matrix ligands such as von Willebrand factor or collagen by specific constitutive membrane glycoprotein receptors at low affinity sufficient for loose adhesion (Chapter 18). This binding reaction triggers a signal flux ("outside-in" signaling, Chapter 17), which results in ion currents, protein kinase activation, polymerization of the cytoskeleton, and arachidonic acid metabolism (Chapter 16). These intraplatelet events ultimately lead to integrin activation ("inside-out" signaling): the constitutive integrin αIIbβ3 (glycoprotein [GP] IIb-IIIa) changes its steric conformation, exposing a high-affinity binding site for RGD (arginine-glycine-asparagine)-containing ligands such as fibrinogen (see Chapter 8 for details). This reaction results in cross-bridging of activated platelets by binding of the polyvalent ligand fibrinogen.[47]

During the process of platelet activation, platelets react by changing their morphology, aggregating, secreting various proteins, and metabolizing arachidonic acid that is rapidly converted to prostaglandins and lipoxygenase products. Constituents from platelet α-granule stores such as platelet factor 4[48] and β-thromboglobulin[49-51] are released, and platelet cell surface activation markers like P-selectin,[52,53] CD63, and CD40 ligand (CD40L)[54-57] are up-regulated (Chapter 30). Activated platelets show an increased adhesiveness and aggregation in response to collagen, thrombin, and platelet-activating factor (PAF). This may lead to increased microembolism in the capillaries and local progression of preexisting vascular lesions.[15,58] Abnormalities in platelet function may exacerbate the progression of atherosclerosis and the consequences of plaque rupture. The abnormal metabolic state that accompanies diabetes causes arterial dysfunction. Relevant abnormalities include chronic hyperglycemia, dyslipidemia, and insulin resistance, which engender adverse metabolic events within the endothelial cell. Activation of these systems impairs endothelial function, augments vasoconstriction, increases inflammation, and promotes thrombosis.

In diabetes, the complex regulation of platelet activity is altered toward

- increased reactivity and adhesion[18,59,60]
- amplified agonist-receptor coupling[61,62]
- increased capacity for prostanoid generation[63-65]
- decreased capacity for nitric oxide (NO) generation[66-68]
- enhanced generation of reactive oxygen species[69,70]
- resistance to NO and prostacyclin[71,72]
- increased cytosolic calcium mobilization[73-75]
- increased α-granule content with concomitantly increased release[40]
- increased platelet volume[76]
- increased numbers of GPIb and integrin αIIbβ3 receptors[77]
- increased membrane protein glycation[78]
- altered membrane fluidity[79,80]
- increased binding of adhesive RGD-protein ligands (e.g., fibrinogen)[81,83]
- increased content and release of plasminogen activator inhibitor-1[83,84]

Some, if not all, of these findings are linked to functional or ultrastructural changes in diabetes mellitus that may constitutively contribute to an increased platelet functional response. However, increased platelet volume and glycoprotein receptor enhancement together provide an important aspect to the understanding of platelet hyperactivity in diabetes mellitus. It is generally accepted that in diabetes the peripheral platelet volume distribution is broadened and shifted toward larger platelets.[76,85] These changes mirror the findings with nondiabetics with acute myocardial infarction, in whom the mean peripheral platelet volume is increased and the distribution width is broadened.[86,87] Martin et al. reported that the increased megakaryocyte DNA content observed in patients with coronary atherosclerosis is a signature of an underlying prethrombotic state, with predictive value for an increased risk of acute coronary events.[88,89] In diabetics with myocardial infarction who have complicated follow-up and excess mortality, Hendra et al. reported a further increase in the platelet volume distribution width.[90] Similar to acute coronary syndromes, in diabetes the clinical significance of platelet hyperactivity may be derived from the Platelet Aggregation as a Risk factor in Diabetics (PARD) study, which showed that the incidence of macrovascular end points in male diabetic patients was strongly related to the degree of spontaneous platelet aggregation.[91–93]

The assumption of a constitutive platelet hyperreactivity in diabetes does not necessarily lessen the importance of metabolic derangements such as hyperglycemia as a platelet activating stimulus.[94] Hyperglycemia-dependent metabolic changes are clearly linked to a secondary increased production of proaggregatory thromboxane and thrombin.[32,65,95] Proaggregatory protein kinase C activation in human platelets is closely linked to short-term hyperglycemia.[96] Ceriello et al. showed that the postprandial state induces coagulation activation, presumably by means of hyperglycemia-dependent generation of reactive oxygen species, which is a known platelet stimulus.[97–102] A similar mechanism has been assumed for postprandial lipemia, which has been shown to induce *in vivo* platelet activation.[103–106]

Prolonged exposure to hyperglycemia is therefore recognized as one primary causal factor in the pathogenesis of diabetic complications.[107–109] Hyperglycemia induces a large number of alterations in vascular tissue that potentially promote accelerated atherosclerosis. Currently, three major mechanisms have emerged that encompass most of the pathological alterations observed in the vasculature of diabetic animals and humans:

1. nonenzymatic glycation of proteins and lipids (formation of advanced glycation end products [AGEs])

2. oxidative stress

3. protein kinase C activation

Importantly, these mechanisms are not independent. For example, hyperglycemia-induced oxidative stress promotes the formation of AGEs and protein kinase C activation.[110] The effects of hyperglycemia are often irreversible and lead to progressive cellular dysfunction.[111] One of the important mechanisms responsible for the accelerated atherosclerosis

in diabetes is the nonenzymatic reaction between glucose and proteins or lipoproteins in arterial walls, collectively known as Maillard, or browning reaction.[112] Glucose forms chemically reversible early glycosylation products with reactive amino groups of circulating or vessel wall proteins (Schiff bases), which subsequently rearrange to form the more stable Amadori-type early glycosylation products. Equilibrium levels of Schiff base and Amadori products (the best known of which is hemoglobin $A_{1C}$) are reached in hours and weeks, respectively.[113] Some of the early glycosylation products on long-lived proteins (e.g., vessel wall collagen) continue to undergo a complex series of chemical rearrangements to form AGEs.[113] Once formed, AGE-protein adducts are stable and virtually irreversible. Although AGEs constitute a large number of chemical structures, carboxymethyllysine-protein adducts are the predominant AGEs present *in vivo*.[114,115] AGEs accumulate continuously on long-lived vessel wall proteins with aging and at an accelerated rate in diabetes.[113] The degree of nonenzymatic glycation is determined mainly by the glucose concentration and the time of exposure.[114] However, another critical factor to the formation of AGEs is the tissue microenvironment redox potential. Thus, in situations in which the local redox potential has been shifted to favor oxidant stress, the formation of AGEs increases substantially.[110,116–120] AGEs from either endogenous formation or exogenous food sources or tobacco smoke induce platelet activation via RAGE (receptor of AGEs)-mediated downstream signaling, or by direct effects on vascular and matrix proteins.[108,114,121,122] Food AGEs have impressive effects on flow-mediated vasodilatation in diabetic patients. About 10% of exogenously delivered AGEs are absorbed. Whereas healthy subjects excrete 30% of these AGEs, diabetic patients have an AGE clearance of below 5% — which therefore results in more AGE-induced vascular damage. Interestingly, these alterations have been linked to thickening of the intima-media complex as a hallmark of early atherosclerosis. By these mechanisms, acute hyperglycemia could be a thrombogenic plaque stressor that fosters the evolution of acute ischemic syndromes.[24,123]

## IV. The Diabetic Thrombocytopathy and the Stem Cell Hypothesis

As platelets are not supplied with all elements of the transcriptional apparatus, most constitutive proteins have to be synthesized in megakaryocytes and packed into platelet precursors, which are afterward shed into the peripheral circulation.[87,124–126] Increase in the surface expression of both GPIb and integrin αIIbβ3 on diabetic platelets may be regarded as a molecular basis for the increased functional behavior of these platelets, since both glycoproteins act as specific receptors for cytoadhesive proteins in the adhesion and aggregation process (Chapters 7 and 8).[77] DiMinno et al. reported increased fibrinogen binding to platelets from diabetic patients, and αIIbβ3 has been shown to be involved in fibrinogen uptake.[82,127] The reported increase in both glycoproteins in diabetes exceeds the expected increase with larger platelet volume, but rather represents a change on the level of expression or synthesis.[77] These surface glycoproteins are megakaryocytic lineage markers that are not synthesized by platelets.[124,128] These results provide indirect evidence for a primary change at the level of megakaryocytic thrombopoiesis.[76,77] Presumably, this could result in an increased peripheral thrombotic mass.

It is well known that megakaryocytes react to environmental changes such as dyslipoproteinemia.[129,130] Thus, an endocrine imbalance, with consecutive disturbances of the intermediate metabolism as present in the diabetic state, seems to be important. Insulin influences the size and maturation status of megakaryocytes in culture, which indicates a possible influence of insulin on the endomitotic cycle of megakarocytes.[131] Investigation of the megakaryocyte-platelet system both at the time of diabetes manifestation and under the conditions of long-term insulin substitution was studied in adult, genetically related BB rats, a model for immunogenic type 1-like diabetes mellitus. Significant qualitative and quantitative changes were detectable both in platelets and megakaryocytes when simultaneous DNA and lineage specific integrin αIIbβ3 staining patterns were evaluated.[132] The enhanced recruitment of progenitor cells along with consumption of mature higher ploid megakaryocytes was suggested as an explanation for the increased platelet count and size, as well as megakaryocytic expression of αIIbβ3, at the onset of diabetes and after insulin therapy. Enhanced megakaryocyte maturation, together with the release of a high number of larger platelets, thus reflects a superimposed insulin effect that could be mediated by various interleukins.[133–136]

The finding that peripheral platelets already circulate in an activated state in newly diagnosed and even in prediabetic (immunologically affected) patients suggests a chronic condition of ongoing platelet consumption.[137] The bone marrow may react with increased release considering that the peripheral platelet does not appear severely affected in human studies, suggesting increased turnover.[138] The interleukins associated with the inflammatory islet cell destruction of diabetes mellitus are well known to affect the megakaryocyte compartment and could account for the observed differences in early diabetic animals. The systemic increase in tumor necrosis factor (TNF)-α in recent-onset diabetic animals was considered as a hallmark of this assumption.[85,139] Furthermore, it was shown that elevations of interleukin (IL)-6 coincide with increased megakaryocyte ploidy in diabetic patients with atherosclerotic disease.[126] The initial recruitment of progenitor cells occurring together with megakaryocyte consumption and the subsequent acceleration of megakaryocyte maturation in terms of ploidy,

size, and αIIbβ3 expression occurring in insulin-treated diabetic animals could also reflect thrombopoietin-dependent reactions modified by the presence of the discussed immune mediators and insulin.[131–136,140–142] Currently, no data are available with regard to thrombopoietin levels in diabetes, but an increase could also account for a state of sustained hyperreactivity of the circulating platelets.[143]

In summary, the observations in spontaneously diabetic BB rats and available human data suggest that there is a primary alteration at the level of bone marrow (megakaryocytic) stem cells appearing in a thrombopoietin-dependent manner. However, the interplay of the discussed differential biological mediators on the process of stem cell differentiation, clonal expansion, megakaryocyte maturation, and process of platelet formation in the diabetic state remains incompletely defined (Fig. 38-1).

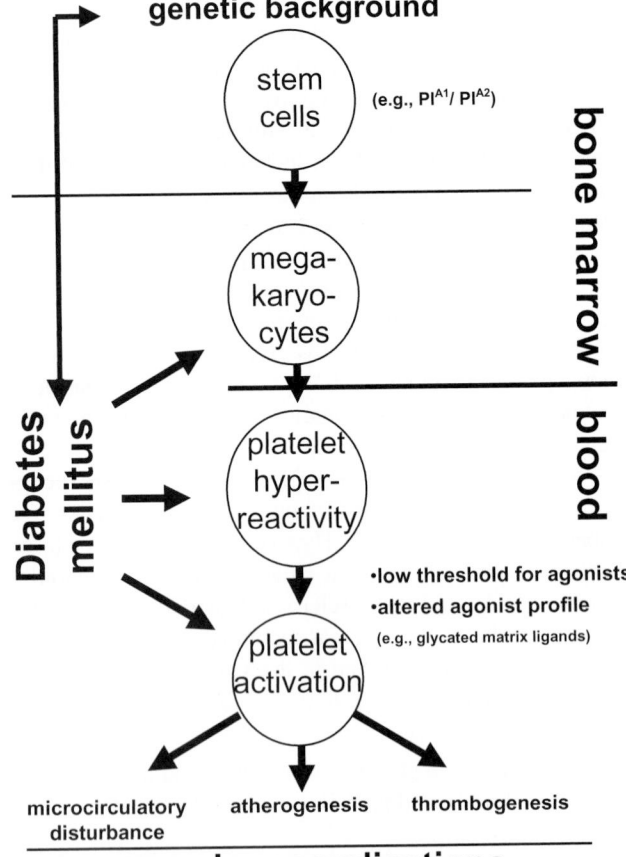

**Figure 38-1.** Networking mechanisms between stem cells and the evolution of clinical end points in diabetic vascular complications. Diabetes is a systemic disease. The stem cell hypothesis proposes that genetic preconditioning occurs at the megakaryocyte level and with thrombopoietic dependency, eventually producing hyperreactive platelets with increased levels of glycoprotein receptors involved in adhesion to the vascular endothelium and cross-linking of fibrinogen in the aggregation process.

Another line of argument addresses changes within the megakaryocytic genetic code that results in functional changes in the platelet proteome. The single nucleotide polymorphism PI[A1/A2] of the fibrinogen receptor integrin $\alpha IIb\beta 3$, which is thrombosis-associated (see Chapter 14), has been investigated as a surrogate of such an assumption. The relative frequency of the PI[A2] allele is increased in patients with diabetes but, in contrast to the nondiabetic population, this increase appears to be independent from clinical atherosclerosis.[144] This result confirms the hypothesis of a primary risk condition that operates by megakaryocyte lineage stem cells and fits with the so-called common soil hypothesis of type 2 diabetes.[145] Also consistent with the concept of genetic preconditioning, a polymorphism of the platelet collagen receptor integrin $\alpha 2\beta 1$ (GPIa-IIa) was reported to be associated with the development of diabetic retinopathy.[146]

Bone marrow transplantation can result in the transfer of insulin-dependent diabetes mellitus between HLA-identical relatives, thereby demonstrating that stem cells can transfer the "diabetes message."[147] The association of other autoimmune diseases with defects in hematopoietic stem cells has been described.[148] Given the possible genetic background for diabetes, this suggests that the genetic code could operate by stem cell-derived cellular mechanisms. Whether this occurs independently or involves the megakaryocytic differentiation series from the beginning remains an open issue.

## V. Platelet Activation

The previously described constitutive alterations lower the threshold for intravascular platelet activation in the diabetic state. There is a remarkable increase in plasma concentrations of the platelet-specific release proteins like platelet factor 4 and $\beta$-thromboglobulin in type 1 diabetes patients, independent from the state of metabolic control despite intensive therapy.[17,40] The interpretation of plasma concentrations of platelet factor 4 and $\beta$-thromboglobulin is, however, hampered by poorly predictable elimination kinetics through renal clearing or binding at the endothelial interface, both of which must be assumed to be changed under diabetic conditions. Flow cytometry-based assays characterize the functional status of freshly fixed platelets derived from the structural changes of the outer platelet membrane following the "inside-out" signaling events (Chapter 30).[149] Adhesion molecules within the internal membranes of platelet granules, such as the lysosomal GP53 (CD63) or the $\alpha$-granule P-selectin (CD62P, Chapter 12), become exocytosed and stored molecules such as thrombospondin become rebound to platelet plasma membrane receptors (CD36, GPIV). This intermediate platelet condition is referred to as thrombogenic transformation and indicates activation of the platelet that still circulates prior to functional consumption

in the aggregation cascade.[150] Flow cytometric analysis of these activation-dependent changes in platelet surface glycoproteins permits the classification of each platelet under investigation (Chapter 30). Subjects at various stages of diabetes proved to have increased numbers of P-selectin- and CD63-positive platelets compared to healthy controls.[17] Whereas such a finding was anticipated in line with the so-called response-to-injury theory in subjects with clinical detectable angiopathy, newly diagnosed insulin-dependent diabetes patients also show clearly increased levels of circulating activated platelets with exposed adhesion molecules.[151,152] Moreover, this activation was not related to the improvement of glycemic control with intensified insulin therapy. Most strikingly, P-selectin-positive platelets can also be found in metabolically healthy first-degree relatives of clinically manifest type 1 diabetic patients.[153] Recently Véricel et al. were also able to show that basal thromboxane $B_2$ is significantly increased in resting platelets from both type 1 and type 2 diabetic patients and that platelet hyperactivation was detectable in well-controlled diabetic patients without complications.[154]

P-selectin is a specific receptor for the attachment of leukocytes to activated platelets and endothelial cells (Chapter 12), which can occur under inflammatory conditions (including diabetes) as well as with platelet activation in response to vascular lesion sites.[152,153,155–157] Thus, activation-dependent molecules form a nidus for the adherence and coactivation of inflammatory leukocytes expressing surface counterligands such as the sialic acid containing oligosaccharide P-selectin glycoprotein ligand 1 (PSGL-1) or the CD11/CD18 complex (Chapter 12).[44] On the other hand, such activated platelets generate thrombin and initiate inflammatory endothelial responses with resultant leukocyte migration and further induction of a procoagulant state.[4,157–159]

By flow cytometry (Chapter 30), the confirmation of platelet activation being related to the subpopulation of large, potentially hyperactive platelets is possible. There was a clear increase in binding sites for activation markers with platelet size, but this increase was more enhanced for platelets from diabetic patients similar to that predicted from the results with myocardial infarction patients.[17,38] Large platelets circulate in an activated state in diabetes mellitus, which suggests that the increased aggregatory potential of such platelets lowers their threshold for activation.[85] The latter probably contributes to the increased incidence of acute cardiovascular events in diabetes mellitus, but activated platelets could also be considered to be potentially pathogenic in the prediabetic phase without apparent metabolic alterations. Interestingly, the diabetic state in turn was also detectable above random probability from flow cytometric list mode platelet data by an expert system (CLASSIF1) independently from metabolic indices.[160]

Features of the insulin resistance syndrome, including abdominal obesity, are associated with increased C-reactive protein levels.[161–164] Davi and others have reported that bio-active products of lipid peroxidation, such as the F2-isoprostane 8-iso prostaglandin (PG) F2[165,166] are enhanced and may contribute to persistent platelet activation in the setting of hypercholesterolemia,[166] diabetes mellitus,[167] and severe hyperhomocysteinemia.[168] It is suggested that the expanded abdominal fat depot may be responsible for a low-grade inflammatory state[169] by providing a source of increased production of IL-6, which is a potent stimulus of C-reactive protein synthesis by the liver.[170] Further evidence for a cause-and-effect relationship between the obese state and persistent platelet activation was obtained through a short-term, diet-induced weight-loss program. This demonstrated that a 10% reduction in body weight obtained through a successful program over a 12-week period was associated with more than a 50% reduction in thromboxane biosynthesis. This led to a normalization of this noninvasive index of platelet activation.

The findings of Davi et al. suggest that a low-grade inflammatory state associated with abdominal adiposity may be the primary trigger of thromboxane-dependent platelet activation mediated, at least partly, through enhanced lipid peroxidation.[168] Several mechanisms are likely to amplify and sustain the relationship between systemic inflammation and platelet activation, such as a direct proinflammatory effect of C-reactive protein, the effects of F2-isoprostanes on inflammatory gene expression, and the synthesis and release of inflammatory mediators from activated platelets like transforming growth factor-β, thromboxane $A_2$, $PGE_2$, and CD40L (Chapter 39).[171–175] Thus, there seems to be a vicious cycle of self-accelerating platelet activation in diabetes.[171]

## VI. Leukocyte-Platelet Cross-Talk in Diabetes

Pathophysiological processes like inflammation or thrombosis show multicellular activation involving endothelial cells, leukocytes, and platelets (see Chapters 12 and 39). The complex interactions between these cells' "cross-talk" is influenced by several mediators. Platelets influence the process of leukocyte activation, chemotaxis, and phagocytosis.[175] Platelet-released adenine nucleotides and platelet-derived growth factor (PDGF) may induce leukocyte degranulation. Adherent platelets,[176] platelet-derived microparticles (Chapter 20),[177] and platelet-released substances, such as PDGF, platelet factor 4, and thromboxane $A_2$, enhance leukocyte rolling and adhesion. Furthermore, PDGF is chemotactic and enhances phagocytosis by neutrophils and monocytes. Superoxide formation by neutrophils may be promoted by platelets bound to neutrophils or platelet-released ADP, whereas intact platelets may inhibit neutrophil superoxide production.[178] Leukocyte chemotaxis, adhesion, and superoxide generation is inhibited by

P-selectin and NO released from platelets.[179,180] Thus, platelets and platelet-derived products influence leukocyte function in many ways.

On the other hand, leukocytes influence platelet adhesion directly or through leukocyte-released superoxide. Platelet aggregation and secretion are induced by superoxide, PAF, metalloproteases like cathepsin G, and neutrophil elastase. Platelets and leukocytes may form platelet-leukocyte aggregates, initially via platelet surface-expressed P-selectin and its leukocyte counter-receptor PSGL-1 and subsequently via fibrinogen bridging between platelet integrin αIIbβ3 and leukocyte CD11b/CD18, as well as other mechanisms.

Li and coworkers were able to demonstrate platelet-leukocyte cross-talk in a whole blood flow cytometry assay.[181] Leukocytes activated with N-formyl-methionyl-leucyl-phenylalanine (fMLP) induced platelet activation with increased expression of platelet P-selectin. The effects may be related to the generation of PAF, 5-lipoxygenase production, and superoxide. The collagen-induced activation of platelets led to a significant leukocyte activation, which was monitored by the expression of CD11b on leukocytes. During platelet-leukocyte cross-talk, P-selectin and integrin αIIbβ3 take part in cellular signaling.[181]

The complex pathological process of diabetic angiopathy involves atherosclerosis, inflammation, and thrombosis with platelet and leukocyte dysfunction playing a central role. In diabetic microangiopathy, increased platelet and leukocyte activation and heterotypic aggregation of both are present.[39,182–185] Leukocytes from diabetic subjects exhibit enhanced adhesion molecule expression,[146] increased aggregability, impaired cell deformability,[187] decreased chemotaxis,[186] and altered superoxide anion production.[188] Hu et al. evaluated platelet-leukocyte cross-talk and its contribution to platelet and leukocyte dysfunction and microangiopathy in diabetes mellitus type 1 patients.[189] Whereas basal single platelet P-selectin and leukocyte CD11b expression levels were similar in diabetic patients and healthy subjects, circulating platelet-leukocyte aggregates and plasma elastase were elevated in diabetic patients. Upon treatment with the thromboxane $A_2$ analog U46619, a marked increase in platelet surface P-selectin expression and platelet-leukocyte aggregation in patients was detectable. The leukocyte-specific agonist fMLP induced more marked CD11b expression in diabetic patients with microangiopathy. Platelet-leukocyte cross-talk induced by U46619 showed no difference between patients and healthy subjects. fMLP evoked marked leukocyte activation, which subsequently caused mild platelet P-selectin expression. This leukocyte-platelet cross-talk was more pronounced in patients than in healthy subjects. Enhanced leukocyte-platelet cross-talk was correlated to platelet hyperreactivity among diabetic patients with microangiopathy only. Hu et al.[189] found that patients with diabetes mellitus had elevated plasma levels of elastase, which can in turn induce platelet activation. Other soluble mediators of platelet-leukocyte cross-talk (e.g., PAF and superoxide anions) are enhanced in type 1 diabetes and could therefore act in concert.[189] Platelet-monocyte aggregation was observed in diabetics with microangiopathy, which may facilitate tissue factor transfer between monocytes and platelets[190,191] resulting in increased thrombin generation. The fact that basal P-selectin levels did not differ significantly, but platelet-leukocyte aggregates were distinguishable in healthy subjects and patients, led to the concept that platelet-leukocyte aggregates may be a more sensitive marker of platelet activation *in vivo* — as previously reported by Michelson et al.[192] Hu et al.[189] supposed that leukocyte-platelet cross-talk may also prime platelets to be more sensitive to activation, which was demonstrated by *in vitro* activation with U46619 in the patients with microangiopathy.

Activated human platelets express and secrete CD40L and may interact with CD40-expressing cells like macrophages and vascular structural cells.[193–196] Expression of cyclooxygenase-2, prostaglandins, adhesion molecules, and cytokines like IL-6 and tissue factor occurs upon activation with CD40 and leads to further platelet activation.[197,198] Elevated CD40L levels in blood are associated with acute coronary syndromes and stroke and predict an increased cardiovascular risk in healthy populations.[199,200]

In summary, type 1 diabetes is associated with platelet and leukocyte hyperactivity, enhanced platelet-leukocyte aggregation, and enhanced leukocyte-platelet cross-talk, especially in patients with microangiopathy. The enhanced leukocyte-platelet cross-talk may contribute to diabetic platelet hyperreactivity and to the microvascular complications in type 1 diabetes mellitus.

## VII. Therapeutic Consequences

The normalization of metabolism is considered to prevent late complications of diabetes mellitus and also in patients with the metabolic syndrome. However, even advanced strategies for achieving near-normoglycemic control in the Diabetes Control and Complications Trial (DCCT) failed to significantly reduce the vascular mortality of the patients, despite clear evidence for reduced platelet thromboxane production and thrombin generation with metabolic improvement.[95,201] There is a lack of association between better acute metabolic control and reduced platelet adhesion molecule expression, and cellular perturbations may persist despite the return of normoglycemia (the so-called memory effect).[151,202] Thus, persistent rather than transient acute metabolic changes are of pivotal importance in the pathogenesis of diabetic complications. It remains questionable whether hypercoagulability in diabetes, including the hyperactivity of the megakaryocyte-platelet system, can be normalized by strict and effective near-normoglycemia

alone.[94,203] On the other hand, insulin was found to inhibit platelet aggregation by augmentation of the NO-dependent generation of cyclic guanosine-3′,5′-monophosphate (cGMP) and cyclic adenosine-3′,5′-monophosphate (cAMP) second messengers.[204,205] The clinical relevance of such findings, at least for type 2 patients, is still under debate.[206,207] Generally, it appears reasonable to use all antiplatelet effects of a broader risk factor intervention strategy including antihypertensive, lipid-lowering, and glucose-lowering drugs plus nutrition modification.[208–210]

Antiplatelet therapy should be expected to be of particular benefit, since platelets appear to be strongly involved in the evolution of the final ischemic end points.[17] Epidemiological data confirm the rational indication for additional antiplatelet therapy in diabetic patients, at least when a vascular high-risk condition can be indentified.[211] The majority of studies shows a remarkable positive effect of antiplatelet therapy on the predefined end points that constituted the macrocirculation as well as retinopathy and nephropathy.[35,212,213] The publication of the nonocular end points in the early treatment retinopathy study (ETDRS) confirmed a positive primary preventive influence on myocardial infarction similar to the results of the American Physician Health Study in nondiabetic patients.[211,214,215] Antiplatelet therapy in various protocols (e.g., aspirin, ticlopidine, clopidogrel, and picotamid) has been extensively evaluated longitudinally and the results, together with the meta-analytical data from the large-scale, population-based studies, confirm the indication for this kind of additional therapy, particularly in patients with diabetes who should be considered to be of vascular risk from the beginning and are particularly responsive to antiplatelet interventions.[33,35,216,217] The American Diabetes Association (ADA) publishes position statements every year regarding aspirin therapy in patients with diabetes mellitus.[218] Enteric-coated aspirin in dosages of 81 to 325 mg/day should be used as a secondary prevention strategy in men and women with evidence of a large vessel disease (myocardial infarction, vascular bypass, stroke, transient ischemic attack, peripheral vascular disease, claudication, or angina pectoris). Aspirin should be used as a primary prevention strategy in high-risk patients with diabetes mellitus. In cases of aspirin allergy, bleeding tendency, anticoagulant therapy, recent gastrointestinal bleeding, or clinical hepatic disease, clopidogrel should be used instead of aspirin. Patients treated with aspirin should be older than 21 years.

Aspirin (see Chapter 60 for further details) is underutilized as a primary prevention strategy in patients with diabetes. It is apparent that this inexpensive and well-tolerated preventive medication should be added to cardioprotective strategies to lower the risk for cardiac events in diabetic patients who are at high risk. The results of several clinical studies strongly support this treatment regimen — even as a combination therapy with other antiplatelet agents.

A very powerful proof of principle comes from studies with coronary revascularization using PCI and stents in which iatrogenic plaque rupture is used to obtain vessel patency. The outcome from this type of intervention is strongly linked to platelet activation, and parenteral administration of GPIIb-IIIa antagonists (Chapter 62) has been shown to improve the outcome among the general population.[47] In this setting of tertiary prevention, patients with diabetes have a significantly worse prognosis, but the administration of GPIIb-IIIa antagonists has neutralized the excess event rates and mortality.[28,29,219–221] These results are consistent with the specific platelet pathophysiology in diabetes, with increased numbers of integrin $\alpha$IIb$\beta$3 (GPIIb-IIIa) receptors contributing to platelet hyperreactivity.

The diagnostic identification of patients with an activated platelet system (e.g., by flow cytometry, see Chapter 30) may help to better discriminate those patients at high risk.[222] In addition, flow cytometry and other types of platelet function testing (Chapter 23) may allow better control of therapeutic efficacy, particularly in populations with pathophysiological platelet alterations such as diabetes mellitus.[223]

However, it is clear that antiplatelet therapy must be superimposed on effective metabolic adjustment in order to avoid the glucose toxicity pathway to late complications. Furthermore, it remains a fascinating challenge to reflect on a potentially protective effect of antiplatelet agents against ischemic $\beta$-cell destruction.

# References

1. Stamler, J., Vaccaro, O., Neaton, J. D., et al. (1993). Diabetes, other risk factors, and 12-yr cardiovascular mortality for men screened in the Multiple Risk Factor Intervention Trial. *Diabetes Care, 16,* 434–444.
2. Haffner, S. M., Lehto, S., Ronnemaa, T., et al. (1998). Mortality from coronary heart disease in subjects with type 2 diabetes and in nondiabetic subjects with and without prior myocardial infarction. *N Engl J Med, 339,* 229–234.
3. American Diabetes Association. (1998). Role of cardiovascular risk factors in prevention and treatment of macrovascular disease in diabetes. *Diabetes Care, 12,* 573–579.
4. Diabetes Epidemiology Research International Mortality Study Group. (1998). International evaluation of cause-specific mortality and IDDM. *Diabetes Care, 14,* 55–60.
5. Eastman, R. C., & Keen, H. (1997). The impact of cardiovascular disease on people with diabetes: The potential for prevention. *Lancet, 350,* SI29–32.
6. U.K. Prospective Diabetes Study Group. U.K. (1995). Prospective diabetes study 16. Overview of 6 years' therapy of type II diabetes: A progressive disease. *Diabetes, 44,* 1249–1258.
7. The DECODE study group. (1999). Glucose tolerance and mortality: Comparison of WHO and American Diabetes Association diagnostic criteria. The DECODE study group. European Diabetes Epidemiology Group. Diabetes Epidemiology: Collaborative analysis of diagnostic criteria in Europe. *Lancet, 354,* 617–621.
8. Brand, F. N., Abbott, R. D., & Kannel, W. B. (1998). Diabetes, intermittent claudication, and risk of cardiovascular events. The Framingham Study. *Diabetes, 38,* 504–509.
9. Calles-Escandon, J., Garcia-Rubi, E., Mirza, S., et al. (1999). Type 2 diabetes: One disease, multiple cardiovascular risk factors. *Cor Art Dis, 10,* 23–30.
10. Bransome, E. D., Jr. (1992). Financing the care of diabetes mellitus in the U.S.: Background, problems, and challenges. *Diab Care, 15,* S1–5.
11. Jacobs, J., Sena, M., & Fox, N. (1991). The cost of hospitalisation for the late complications of diabetes in the United States. *Diabetic Medicine, 8,* S23–29.
12. Osei, K. (1999). Insulin resistance and systemic hypertension. *Am J Cardiol, 84,* 33J–36J.
13. Malmberg, K., Yusuf, S., Gerstein, H. C., et al. (2000). Impact of diabetes on long-term prognosis in patients with unstable angina and non-Q-wave myocardial infarction: Results of the OASIS (Organisation to Assess Strategies for Ischemic Syndromes) Registry. *Circulation, 102,* 1014–1019.
14. McGuire, D. K., Emanuelsson, H., Granger, C. B., et al. (2000). Influence of diabetes mellitus on clinical outcomes across the spectrum of acute coronary syndromes: Findings from the GUSTO-IIb study. GUSTO IIb Investigators. *Eur Heart J, 21,* 1750–1758.
15. Tschoepe, D., Roesen, P., Schwippert, B., et al. (1993). Platelets in diabetes: The role in the hemostatic regulation in atherosclerosis. *Semin Thromb Hemost, 19,* 122–128.
16. Colwell, J. A. (1993). Vascular thrombosis in type II diabetes mellitus. *Diabetes, 42,* 8–11.
17. Ostermann, H., & van de Loo, J. (1986). Factors of the hemostatic system in diabetic patients: A survey of controlled studies. *Haemostasis, 16,* 386–416.
18. Sobol, A. B., & Watala, C. (2000). The role of platelets in diabetes-related vascular complications. *Diabetes Res Clin Pract, 50,* 1–16.
19. Winocour, P. D. (1992). Platelet abnormalities in diabetes mellitus. *Diabetes, 41,* 26–31.
20. Banga, J. D., & Sixma, J. J. (1986). Diabetes mellitus, vascular disease and thrombosis. *Clin Haematol, 15,* 465–492.
21. Kannel, W. B., D'Agostino, R. B., Wilson, P. W., et al. (1990). Diabetes, fibrinogen, and risk of cardiovascular disease: The Framingham experience. *Am Heart J, 120,* 672–676.
22. Fitzgerald, D. J., Roy, L., Catella, F., et al. (1986). Platelet activation in unstable coronary disease. *N Engl J Med, 315,* 983–989.
23. Fuster, V., Fayad, Z. A., & Badimon, J. J. (1999). Acute coronary syndromes: Biology. *Lancet, 353,* SII5–9.
24. Folsom, A. R., Eckfeldt, J. H., Weitzman, S., et al. (1994). Relation of carotid artery wall thickness to diabetes mellitus, fasting glucose and insulin, body size, and physical activity. Atherosclerosis Risk in Communities (ARIC) Study Investigators. *Stroke, 25,* 66–73.

25. Jacoby, R. M., & Nesto, R. W. (1992). Acute myocardial infarction in the diabetic patient: Pathophysiology, clinical course and prognosis. *J Am Coll Cardiol, 20,* 736–744.

26. Silva, J. A., Escobar, A., Collins, T. J., et al. (1995). Unstable angina: A comparison of angioscopic findings between diabetic and nondiabetic patients. *Circulation, 92,* 1731–1736.

27. Yeghiazarians, Y., Braunstein, J. B., Askari, A., & Stone, P. H. (2000). Unstable angina pectoris. *N Engl J Med, 342,* 101–114.

28. Van Belle, E., Bauters, C., Hubert, E., et al. (1997). Restenosis rates in diabetic patients: A comparison of coronary stenting and balloon angioplasty in native coronary vessels. *Circulation, 96,* 1454–1460.

29. Detre, K. M., Guo, P., Holubkov, R., et al. (1999). Coronary revascularisation in diabetic patients: A comparison of the randomised and observational components of the Aypass Angioplasty Revascularisation Investigation (BARI). *Circulation, 99,* 633–640.

30. Brooks, A. M., Hussein, S., Chesterman, C. N., et al. (1983). Platelets, coagulation and fibrinolysis in patients with diabetic retinopathy. *Thromb Haemost, 49,* 123–127.

31. van der Loo, B., & Martin, J. F. (1999). A role for changes in platelet production in the cause of acute coronary syndromes. *Arterioscler Thromb Vasc Biol, 19,* 672–679.

32. Davi, G., Ciabattoni, G., Consoli, A., et al. (1999). *In vivo* formation of 8-iso-prostaglandin f2alpha and platelet activation in diabetes mellitus: Effects of improved metabolic control and vitamin E supplementation. *Circulation, 99,* 224–229.

33. Antiplatelet Trialists' Collaboration Collaborative overview of randomised trials of antiplatelet therapy-I: Prevention of death, myocardial infarction, and stroke by prolonged antiplatelet therapy in various categories of patients. (1994). *Br Med J, 308,* 81–106.

34. Colwell, J. A. (1992). Antiplatelet drugs and prevention of macrovascular disease in diabetes mellitus. *Metabolism, 41,* 7–10.

35. Colwell, J. A. (1991). Clinical trials of antiplatelet agents in diabetes mellitus: Rationale and results. *Semin Thromb Hemost, 17,* 439–444.

36. Mizuno, K., Satomura, K., Miyamoto, A., et al. (1992). Angioscopic evaluation of coronary-artery thrombi in acute coronary syndromes. *N Engl J Med, 326,* 287–291.

37. Tschoepe, D., & Roesen, P. (1998). Heart disease in diabetes mellitus: A challenge for early diagnosis and intervention. *Exp Clin Endocrinol Diabetes, 106,* 16–24.

38. Schultheiss, H. P., Tschoepe, D., Esser, J., et al. (1994). Large platelets continue to circulate in an activated state after myocardial infarction. *Eur J Clin Invest, 24,* 243–247.

39. Rauch, U., Ziegler, D., Piolot, R., et al. (1999). Platelet activation in diabetic cardiovascular autonomic neuropathy. *Diabetic Med, 16,* 848–852.

40. Tschoepe, D., Ostermann, H., Huebinger, A., et al. (1990). Elevated platelet activation in type I diabetics with chronic complications under long-term near-normoglycemic control. *Haemostasis, 20,* 93–98.

41. Sobel, B. E., Woodcock-Mitchell, J., Schneider, D. J., et al. (1998). Increased plasminogen activator inhibitor type 1 in coronary artery atherectomy specimens from type 2 diabetic compared with nondiabetic patients: A potential factor predisposing to thrombosis and its persistence. *Circulation, 97,* 2213–2221.

42. Jelesoff, N. E., Feinglos, M., Granger, C. B., et al. (1997). Outcomes of diabetic patients following acute myocardial infarction: A review of the major thrombolytic trials. *J Cardiovasc Risk, 4,* 100–111.

43. Tschoepe, D. (1996). Adhesion molecules influencing atherosclerosis. *Diabetes Res Clin Prac, 30S,* 19–24.

44. Tschoepe, D., Rauch, U., & Schwippert, B. (1997). Platelet-leukocyte cross-talk in diabetes mellitus. *Horm Metab Res, 29,* 631–635.

45. Bauer, K. A., & Rosenberg, R. D. (1987). The pathophysiology of the prethrombotic state in humans: Insights gained from studies using markers of hemostatic system activation. *Blood, 70,* 343–350.

46. Schafer, A. I. (1985). The hypercoagulable states. *Ann Inter Med, 102,* 814–828.

47. Lefkovits, J., Plow, E. F., & Topol, E. J. (1985). Platelet glycoprotein IIb/IIIa receptors in cardiovascular medicine. *N Engl J Med, 332,* 1553–1559.

48. Jafri, S. M., Ozawa, T., Mammen, E., et al. (1993). Platelet function, thrombin and fibrinolytic activity in patients with heart failure. *Eur Heart J, 14,* 205–212.

49. Hughes, A., Daunt, S., Vass, G., & Wickes, J. (1982). *In vivo* platelet activation following myocardial infarction and acute coronary ischaemia. *Thromb Haemost, 48,* 133–135.

50. Kjeldsen, S. E., Lande, K., Gjesdal, K., et al. (1987). Increased platelet release reaction in 50-year-old men with essential hypertension: Correlation with atherogenic cholesterol fractions. *Am Heart J, 113,* 151–155.

51. Rossi, E., Casali, B., Regolisti, G., et al. (1998). Increased plasma levels of platelet-derived growth factor (PDGF-BB + PDGF-AB) in patients with never-treated mild essential hypertension. *Am J Hyperten, 11,* 1239–1243.

52. Ritchie, J. L., Alexander, H. D., & Rea, I. M. (2000). Flow cytometry analysis of platelet p-selectin expression in whole blood — methodological considerations. *Clin Lab Haematol, 22,* 359–363.

53. Holmes, M. B., Kabbani, S. S., Terrien, C. M., et al. (2001). Quantification by flow cytometry of the efficacy of and interindividual variation of platelet inhibition induced by treatment with tirofiban and abciximab. *Coron Artery Dis, 12,* 245–253.

54. Chakhtoura, E. Y., Shamoon, F. E., Haft, J. I., et al. (2000). Comparison of platelet activation in unstable and stable angina pectoris and correlation with coronary angiographic findings. *Am J Cardiol, 86,* 835–859.

55. Hagberg, I. A., & Lyberg, T. (2000). Blood platelet activation evaluated by flow cytometry: Optimised methods for clinical studies. *Platelets, 11,* 137–150.

56. Lee, Y., Lee, W. H., Lee, S. C., et al. (1999). CD40L activation in circulating platelets in patients with acute coronary syndrome. *Cardiology, 92,* 11–16.

57. Hermann, A., Rauch, B. H., Braun, M., et al. (2001). Platelet CD40 ligand (CD40L)-subcellular localisation, regulation of

expression, and inhibition by clopidogrel. *Platelets, 12,* 74–82.

58. Packham, M. A., & Mustard, J. F. (1986). The role of platelets in the development and complications of atherosclerosis. *Semin Hematol, 23,* 8–26.

59. Dittmar, S., Polanowska-Grabowska, R., & Gear, A. R. (1994). Platelet adhesion to collagen under flow conditions in diabetes mellitus. *Thromb Res, 74,* 273–283.

60. Knobler, H., Savion, N., Shenkman, B., et al. (1998). Shear-induced platelet adhesion and aggregation on subendothelium are increased in diabetic patients. *Thromb Res, 90,* 181–190.

61. Winocour, P. D., Kinlough-Rathbone, R. L., & Mustard, J. F. (1986). Pathways responsible for platelet hypersensitivity in rats with diabetes. I. Streptozocin-induced diabetes. *J Lab Clin Med, 107,* 148–153.

62. Fukuda, K., Ozaki, Y., Satoh, K., et al. (1997). Phosphorylation of myosin light chain in resting platelets from NIDDM patients is enhanced: Correlation with spontaneous aggregation. *Diabetes, 46,* 488–493.

63. Halushka, P. V., Lurie, D., & Colwell, J. A. (1997). Increased synthesis of prostaglandin-E-like material by platelets from patients with diabetes mellitus. *N Engl J Med, 297,* 1306–1310.

64. Halushka, P. V., Rogers, R. C., Loadholt, C. B., et al. (1981). Increased platelet thromboxane synthesis in diabetes mellitus. *J Lab Clin Med, 97,* 87–96.

65. Davi, G., Catalano, I., Averna, M., et al. (1990). Thromboxane biosynthesis and platelet function in type II diabetes mellitus. *N Engl J Med, 322,* 1769–1774.

66. Oskarsson, H. J., & Hofmeyer, T. G. (1996). Platelets from patients with diabetes mellitus have impaired ability to mediate vasodilation. *J Am Coll Cardiol, 27,* 1464–1470.

67. Rabini, R. A., Staffolani, R., Fumelli, P., et al. (1998). Decreased nitric oxide synthase activity in platelets from IDDM and NIDDM patients. *Diabetologia, 41,* 101–104.

68. Martina, V., Bruno, G. A., Trucco, F., et al. (1998). Platelet cNOS activity is reduced in patients with IDDM and NIDDM. *Thromb Haemost, 79,* 520–522.

69. Tannous, M., Rabini, R. A., Vignini, A., et al. (1999). Evidence for iNOS-dependent peroxynitrite production in diabetic platelets. *Diabetologia, 42,* 539–544.

70. Schaeffer, G., Wascher, T. C., Kostner, G. M., et al. (1999). Alterations in platelet Ca2+ signalling in diabetic patients is due to increased formation of superoxide anions and reduced nitric oxide production. *Diabetologia, 42,* 167–176.

71. Roesen, P., Schwippert, B., Kaufmann, L., et al. (1994). Expression of adhesion molecules on the surface of activated platelets is not diminished by PGI$_2$-analogues and an NO (EDRF)-donor: A comparison between platelets of healthy and diabetic subjects. *Platelets, 5,* 45–52.

72. Anfossi, G., Mularoni, E. M., Burzacca, S., et al. (1998). Platelet resistance to nitrates in obesity and obese NIDDM, and normal platelet sensitivity to both insulin and nitrates in lean NIDDM. *Diabetes Care, 21,* 121–126.

73. Vicari, A. M., Taglietti, M. V., Pellegatta, F., et al. (1996). Deranged platelet calcium hemostasis in diabetic patients with end-stage renal failure: A possible link to increased cardiovascular mortality? *Diabetes Care, 19,* 1062–1066.

74. Schaeffer, G., Wascher, T. C., Kostner, G. M., et al. (1999). Alterations in platelet Ca2+ signalling in diabetic patients is due to increased formation of superoxide anions and reduced nitric oxide production. *Diabetologia, 42,* 167–176.

75. Gill, J. K., Fonseca, V., Dandona, P., et al. (1999). Differential alterations of spontaneous and stimulated 45Ca(2+) uptake by platelets from patients with type I and type II diabetes mellitus. *J Diabetes Complications, 13,* 271–276.

76. Tschoepe, D., Langer, E., Schauseil, P., et al. (1989). Increased platelet volume–sign of impaired thrombopoiesis in diabetes mellitus. *Klin Wochenschr, 67,* 253–259.

77. Tschoepe, D., Roesen, P., Kaufmann, L., et al. (1990). Evidence for abnormal platelet glycoprotein expression in diabetes mellitus. *Eur J Clin Invest, 20,* 166–170.

78. Cohen, I., Burk, D. L., Fullerton, R., et al. (1991). Nonenzymatic glycation of platelet proteins in diabetic patients. *Semin Thromb Haemost, 17,* 426–432.

79. Mazzanti, L., Rabini, R. A., Fumelli, P., et al. (1997). Altered platelet membrane dynamic properties in type 1 diabetes. *Diabetes, 46,* 2069–2074.

80. Watala, C., Boncer, M., Golanski, J., et al. (1998). Platelet membrane lipid fluidity and intraplatelet calcium mobilisation in type 2 diabetes mellitus. *Eur J Haematol, 61,* 319–326.

81. DiMinno, G., Silver, M. J., Cerbone, A. M., et al. (1985). Increased binding of fibrinogen to platelets in diabetes: The role of prostaglandins and thromboxane. *Blood, 65,* 156–162.

82. DiMinno, G., Silver, M. J., Cerbone, A. M., et al. (1986). Platelet fibrinogen binding in diabetes mellitus: Differences between binding to platelets from nonretinopathic and retinopathic diabetic patients. *Diabetes, 35,* 182–185.

83. Jokl, R., Laimins, M., Klein, R. L., et al. (1994). Platelet plasminogen activator inhibitor 1 in patients with type II diabetes. *Diabetes Care, 17,* 818–823.

84. Jokl, R., Klein, R. L., Lopes-Virella, M. F., et al. (1995). Release of platelet plasminogen activator inhibitor 1 in whole blood is increased in patients with type II diabetes. *Diabetes Care, 18,* 1150–1155.

85. Tschoepe, D., Roesen, P., Esser, J., et al. (1991). Large platelets circulate in an activated state in diabetes mellitus. *Semin Thromb Hemost, 17,* 433–438.

86. Kristensen, S. D. (1992). The platelet-vessel wall interaction in experimental atherosclerosis and ischemic heart disease with special reference to thrombopoiesis. *Dan Med Bull, 39,* 110–127.

87. Martin, J. F., Trowbridge, E. A., Salmon, G., et al. (1983). The biological significance of platelet volume: Its relationship to bleeding time, platelet thromboxane B2 production and megakaryocyte nuclear DNA concentration. *Thromb Res, 32,* 443–460.

88. Bath, P. M., Gladwin, A. M., Carden, N., et al. (1994). Megakaryocyte DNA content is increased in patients with coronary artery atherosclerosis. *Cardiovascular Res, 28,* 1348–1352.

89. Martin, J. F., Bath, P. M., & Burr, M. L. (1991). Influence of platelet size on outcome after myocardial infarction. *Lancet, 338,* 1409–1411.

90. Hendra, T. J., Oswald, G. A., & Yudkin, J. S. (1988). Increased mean platelet volume after acute myocardial infarction relates to diabetes and to cardiac failure. *Diabetes Res Clin Pract, 5,* 63–69.

91. Breddin, H. K., Krzywanek, H. J., Althoff, P., et al. (1986). Spontaneous platelet aggregation and coagulation parameters as risk factors for arterial occlusions in diabetics: Results of the PARD study. *Int Angiol, 5,* 181–195.

92. Iwase, E., Tawata, M., Aida, K., et al. (1998). A cross-sectional evaluation of spontaneous platelet aggregation in relation to complications in patients with type II diabetes mellitus. *Metabolism, 47,* 699–705.

93. Gray, R. P., Hendra, T. J., Patterson, D. L., et al. (1993). Spontaneous; platelet aggregation in whole blood in diabetic and nondiabetic survivors of acute myocardial infarction. *Thromb Haemost, 70,* 932–936.

94. Roshan, B., Tofler, G. H., Weinrauch, L. A., et al. (2000). Improved glycemic control and platelet function abnormalities in diabetic patients with microvascular disease. *Metabolism, 49,* 88–91.

95. Aoki, I., Shimoyama, K., Aoki, N., et al. (1996). Platelet-dependent thrombin generation in patients with diabetes mellitus: Effects of glycemic control on coagulability in diabetes. *J Am Coll Cardiol, 27,* 560–566.

96. Assert, R., Scherk, G., Bumbure, A., et al. (2001). Regulation of protein kinase C by short term hyperglycemia in human platelets *in vivo* and *in vitro. Diabetologia, 44,* 188–195.

97. Ceriello, A., Taboga, C., Tonutti, L., et al. (1996). Post-meal coagulation activation in diabetes mellitus: The effect of acarbose. *Diabetologia, 39,* 469–73.

98. Catella-Lawson, F., & Fitzgerald, G. A. (1996). Oxidative stress and platelet activation in diabetes mellitus. *Diabetes Res Clin Pract, 30,* S13–18.

99. Rabini, R. A., Staffolani, R., Martarelli, D., et al. (1999). Influence of low density lipoprotein from insulin-dependent diabetic patients on platelet functions. *J Clin Endocrinol Metab, 84,* 3770–3774.

100. Ceriello, A., Giacomello, R., Stel, G., et al. (1995). Hyperglycemia-induced thrombin formation in diabetes: The possible role of oxidative stress. *Diabetes, 44,* 924–928.

101. Ceriello, A. (1993). Coagulation activation in diabetes mellitus: The role of hyperglycemia and therapeutic prospects. *Diabetologia, 36,* 1119–1125.

102. Mezzetti, A., Cipollone, F., & Cuccurullo, F. (2000). Oxidative stress and cardiovascular complications in diabetes: Isoprostanes as new markers on an old paradigm. *Cardiovasc Res, 47,* 475–488.

103. Antiplatelet Trialists' Collaboration. (1994). Collaborative overview of randomised trials of antiplatelet therapy–II: Maintenance of vascular graft or arterial patency by antiplatelet therapy. *Br Med J, 308,* 159–68.

104. Miller, G. J. (1988). Postprandial lipemia and haemostatic factors. *Atherosclerosis, 141,* S47–51.

105. Broijersen, A., Karpe, F., Hamsten, A., et al. (1998). Alimentary lipemia enhances the membrane expression of platelet P-selectin without affecting other markers of platelet activation. *Atherosclerosis, 137,* 107–113.

106. Keaney, J. F., Jr., & Loscalzo, J. (1999). Diabetes, oxidative stress, and platelet activation. *Circulation, 99,* 189–191.

107. Ulrich, P., & Cerami, A. (2001). Protein glycation, diabetes, and aging. *Recent Prog Horm Research, 56,* 1–21.

108. Cerami, A., & Ulrich, P. (2001). Pharmaceutical intervention of advanced glycation endproducts. *Novartis Foundat Symp, 235,* 202–212.

109. Brownlee, M. (1996). Advanced glycation end products in diabetic complications. *Curr Opin Endocrinol Diabetes, 3,* 291–297.

110. Brownlee, M. (2001). Biochemistry and molecular cell biology of diabetic complications. *Nature, 414,* 813–820.

111. Schmidt, A. M., Hori, O., Brett, J., et al. (1994). Cellular receptors for advanced glycation end products: Implications for induction of oxidant stress and cellular dysfunction in the pathogenesis of vascular lesions. *Arterioscler Thromb, 14,* 1521–1528.

112. Wautier, J. L., & Guillausseau, P. J. (2001). Advanced glycation end products, their receptors and diabetic angiopathy. *Diabetes Metab, 27,* 535–542.

113. Brownlee, M. (2000). Negative consequences of glycation. *Metabolism, 49,* 9–13.

114. Shinohara, M., Thornalley, P. J., Giardino, I., et al. (1998). Overexpression of glyoxalase-I in bovine endothelial cells inhibits intracellular advanced glycation endproduct formation and prevents hyperglycemia-induced increases in macromolecular endocytosis. *J Clin Invest, 101,* 1142–1147.

115. Thornalley, P. J. (2003). Glyoxalase I—structure, function and a critical role in the enzymatic defence against glycation. *Biochem Soc Trans, 31,* 1343–1348.

116. Brownlee, M., Cerami, A., & Vlassara, H. (1998). Advanced glycosylation end products in tissue and the biochemical basis of diabetic complications. *N Engl J Med, 318,* 1315–1321.

117. Szmitko, P. E., Wang, C. H., Weisel, R. D., et al. (2003). New markers of inflammation and endothelial cell activation: Part I. *Circulation, 108,* 1917–1923.

118. Vlassara, H., Fuh, H., Donnelly, T., et al. (1995). Advanced glycation endproducts promote adhesion molecule (VCAM-1, ICAM-1) expression and atheroma formation in normal rabbits. *Mol Med, 1,* 447–456.

119. Nawroth, P. P., Bank, I., Handley, D., et al. (1986). Tumor necrosis factor/cachectin interacts with endothelial cell receptors to induce release of interleukin 1. *J Exp Med, 163,* 1363–1375.

120. Montgomery, K. F., Osborn, L., Hession, C., et al. (1991). Activation of endothelial-leukocyte adhesion molecule 1 (ELAM-1) gene transcription. *Proc Nat Acad Sci USA, 88,* 6523–6527.

121. Hangaishi, M., Taguchi, J., Miyata, T., et al. (1998). Increased aggregation of human platelets produced by advanced glycation end products *in vitro. Biochem Biophys Res Commun, 248,* 285–292.

122. Koschinsky, T., Schwippert, B., Ruetter, R., et al. (2000). Food- and serum-advanced glycation endproducts induce activation of human platelets. *Diabetologia, 43,* A272.

123. Temelkova-Kurktschiev, T. S., Koehler, C., Henkel, E., et al. (2000). Postchallenge plasma glucose and glycemic spikes are more strongly associated with atherosclerosis than fasting glucose or HbA1c level. *Diabetes Care, 23,* 1830–1834.

124. Bray, P. F., Rosa, J. P., Johnston, G. I., et al. (1987). Platelet glycoprotein IIb. Chromosomal localisation and tissue expression. *J Clin Invest, 80,* 1812–1817.

125. Brown, A. S., & Martin, J. F. (1994). The megakaryocyte platelet system and vascular disease. *Eur J Clin Invest, 24,* S9–15.

126. Schroer, K., Roesen, P., Tschoepe, D. (Eds.). (1994). Megakaryocytes and platelets in cardiovascular diseases. Proceedings of a workshop held at the 27th meeting of the European Society of Clinical Investigation, Heidelberg, Germany, 15 April 1993. *Eur J Clin Invest, 24,* S1–52.

127. Handagama, P., Scarborough, R. M., Shuman, M. A., et al. (1993). Endocytosis of fibrinogen into megakaryocyte and platelet alpha-granules is mediated by alpha IIb beta 3 (glycoprotein IIb-IIIa). *Blood, 82,* 135–138.

128. Giles, H., Smith, R. E., & Martin, J. F. (1994). Platelet glycoprotein IIb-IIIa and size are increased in acute myocardial infarction. *Eur J Clin Invest, 24,* 69–72.

129. Dupont, H., Dupont, M. A., Larrue, J., et al. (1987). Megakaryopoiesis disturbances in atherosclerotic rabbits. *Atherosclerosis, 63,* 15–26.

130. Schick, P. K., Williams-Gartner, K., & He, X. L. (1990). Lipid composition and metabolism in megakaryocytes at different stages of maturation. *J Lipid Res, 31,* 27–35.

131. Watanabe, Y., Kawada, M., & Kobayashi, B. (1987). Effect of insulin on murine megakaryocytopoiesis in a liquid culture system. *Cell Struct Funct, 12,* 311–316.

132. Tschoepe, D., Schwippert, B., Schettler, B., et al. (1992). Increased GPIIB/IIIA expression and altered DNA-ploidy pattern in megakaryocytes of diabetic BB-rats. *Eur J Clin Invest, 22,* 591–598.

133. Burstein, S. A. (1994). Effects of interleukin 6 on megakaryocytes and on canine platelet function. *Stem Cells, 12,* 386–393.

134. Caux, C., Saeland, S., Favre, C., et al. (1990). Tumor necrosis factor-alpha strongly potentiates interleukin-3 and granulocyte-macrophage colony-stimulating factor-induced proliferation of human CD34+ hematopoietic progenitor cells. *Blood, 75,* 2292–2298.

135. Hussain, M. J., Peakman, M., Gallati, H., et al. (1996). Elevated serum levels of macrophage-derived cytokines precede and accompany the onset of IDDM. *Diabetologia, 39,* 60–69.

136. Ryffel, B., Car, B. D., Woerly, G., et al. (1994). Long-term interleukin-6 administration stimulates sustained thrombopoiesis and acute-phase protein synthesis in a small primate — the marmoset. *Blood, 83,* 2093–2102.

137. Winocour, P. D. (1994). Platelet turnover in advanced diabetes. *Eur J Clin Invest, 24,* S34–37.

138. Rinder, H. M., Schuster, J. E., Rinder, C. S., et al. (1998). Correlation of thrombosis with increased platelet turnover in thrombocytosis. *Blood, 91,* 1288–1294.

139. Rothe, H., Fehsel, K., & Kolb, H. (1990). Tumour necrosis factor alpha production is upregulated in diabetes prone BB rats. *Diabetologia, 33,* 573–575.

140. Broudy, V. C., Lin, N. L., & Kaushansky, K. (1995). Thrombopoietin (c-mpl ligand) acts synergistically with erythropoietin, stem cell factor, and interleukin-11 to enhance murine megakaryocyte colony growth and increases megakaryocyte ploidy *in vitro. Blood, 85,* 1719–1726.

141. Bruno, E., Miller, M. E., & Hoffman, R. (1989). Interacting cytokines regulate in vitro human megakaryocytopoiesis. *Blood, 73,* 671–677.

142. Lok, S., & Foster, D. C. (1994). The structure, biology and potential therapeutic applications of recombinant thrombopoietin. *Stem Cells, 12,* 586–598.

143. Kojima, H., Hamazaki, Y., Nagata, Y., et al. (1995). Modulation of platelet activation in vitro by thrombopoietin. *Thromb Haemost, 74,* 1541–1545.

144. Strapatsakis, S., Ferber, P., Faerber, K., et al. (1999). Reduced binding of anti-fibrinogen-receptor beta 3 subunit (CD61) monoclonal antibody SZ21 indicates altered receptor phenotype in PlAX7A2 genotype patients with diabetes mellitus. *Diabetes, 48,* A135.

145. Stern, M. P. (1995). Diabetes and cardiovascular disease: The "common soil" hypothesis. *Diabetes, 44,* 369–374.

146. Matsubara, Y., Murata, M., Maruyama, T., et al. (2000). Association between diabetic retinopathy and genetic variations in alpha2beta1 integrin, a platelet receptor for collagen. *Blood, 95,* 1560–1564.

147. Lampeter, E. F., Homberg, M., Quabeck, K., et al. (1993). Transfer of insulin-dependent diabetes between HLA-identical siblings by bone marrow transplantation. *Lancet, 341,* 1243–1244.

148. Ikehara, S., Kawamura, M., Takao, F., et al. (1990). Organ-specific and systemic autoimmune diseases originate from defects in hematopoietic stem cells. *Proc Nat Acad Sci USA, 87,* 8341–8344.

149. Tschoepe, D., Spangenberg, P., Esser, J., et al. (1990). Flow-cytometric detection of surface membrane alterations and concomitant changes in the cytoskeletal actin status of activated platelets. *Cytometry, 11,* 652–656.

150. Tschoepe, D., & Schwippert, B. (1997). Platelet flow cytometry-adhesive proteins in Handbook of Experimental Pharmacology: Platelets and Their Factors, F. von Bruchhausen & U. Walter, Eds. (pp. 619–643). Heidelberg: Springer.

151. Tschoepe, D., Driesch, E., Schwippert, B., et al. (1985). Exposure of adhesion molecules on activated platelets in patients with newly diagnosed IDDM is not normalised by near-normoglycemia. *Diabetes, 44,* 890–894.

152. Jilma, B., Fasching, P., Ruthner, C., et al. (1996). Elevated circulating P-selectin in insulin dependent diabetes mellitus. *Thromb Haemost, 76,* 328–332.

153. Tschoepe, D., Driesch, E., Schwippert, B., et al. (1997). Activated platelets in subjects at increased risk of IDDM. DENIS Study Group. Deutsche Nikotinamid Interventionsstudie. *Diabetologia, 40,* 573–577.

154. Vericel, E., Januel, C., Carreras, M., et al. (2004). Diabetic patients without vascular complications display enhanced

basal platelet activation and decreased antioxidant status. *Diabetes, 53,* 1046–1051.

155. Lorenzi, M., & Cagliero, E. (1991). Pathobiology of endothelial and other vascular cells in diabetes mellitus. *Diabetes, 40,* 653–659.

156. Frenette, P. S., & Wagner, D. D. (1996). Adhesion molecules–Part II: Blood vessels and blood cells. *N Engl J Med, 335,* 43–45.

157. Nomura, S., Shouzu, A., Omoto, S., et al. (2000). Significance of chemokines and activated platelets in patients with diabetes. *Clin Exp Immunol, 121,* 437–443.

158. Gawaz, M., Neumann, F. J., Dickfeld, T., et al. (1998). Activated platelets induce monocyte chemotactic protein-1 secretion and surface expression of intercellular adhesion molecule-1 on endothelial cells. *Circulation, 98,* 1164–1171.

159. Gawaz, M., Brand, K., Dickfeld, T., et al. (2000). Platelets induce alterations of chemotactic and adhesive properties of endothelial cells mediated through an interleukin-1-dependent mechanism. Implications for atherogenesis. *Atherosclerosis, 148,* 75–85.

160. Valet, G., Valet, M., Tschoepe, D., et al. (1993). White cell and thrombocyte disorders: Standardised, self-learning flow cytometric list mode data classification with the CLASSIF1 program system. *Ann N Y Acad Sci, 677,* 233–251.

161. Brook, R. D., Bard, R. L., Rubenfire, M., et al. (2001). Usefulness of visceral obesity (waist/hip ratio) in predicting vascular endothelial function in healthy overweight adults. *Am J Cardiol, 88,* 1264–1269.

162. Yudkin, J. S., Stehouwer, C. D., Emeis, J. J., et al. (1999). C-reactive protein in healthy subjects: Associations with obesity, insulin resistance, and endothelial dysfunction. *Arterioscler Thromb Vasc Biol, 19,* 972–978.

163. Hak, A. E., Stehouwer, C. D., Bots, M. L., et al. (1999). Associations of C-reactive protein with measures of obesity, insulin resistance, and subclinical arteriosclerosis in healthy, middle-aged women. *Arterioscler Thromb Vasc Biol, 19,* 1986–1991.

164. Lemieux, I., Pascot, A., Prud'homme, D., et al. (2001). Elevated C-reactive protein. *Arterioscler Thromb Vasc Biol, 21,* 961–967.

165. Morrow, J. D., Hill, K. E., Burk, R. F., et al. (1990). A series of prostaglandin F2-like compounds are produced *in vivo* in humans by a non-cyclooxygenase, free radical-catalyzed mechanism. *Proc Nat Acad Sci USA, 87,* 9383–9387.

166. Davì, G., Alessandrini, P., Mezzetti, A., et al. (1997). *In vivo* formation of 8-epi-prostaglandin F2 is increased in hypercholesterolemia. *Arterioscler Thromb Vasc Biol, 17,* 3230–3235.

167. Davì, G., Ciabattoni, G., Consoli, A., et al. (1999). *In vivo* formation of 8-iso-prostaglandin F2 and platelet activation in diabetes mellitus. *Circulation, 99,* 224–229.

168. Davì, G., Guagnano, M. T., Ciabattoni, G., et al. (2002). Platelet activation in obese women: Role of inflammation and oxidant stress. *JAMA, 288,* 2008–2014.

169. Davì, G., Di Minno, G., Coppola, A., et al. (2001). Oxidative stress and platelet activation in homozygous homocystinuria. *Circulation, 104,* 1124–1128.

170. Yudkin, J. S., Kumari, M., Humphries, S. E., et al. (2000). Inflammation, obesity, stress and coronary heart disease. *Atherosclerosis, 148,* 209–214.

171. Heinrich, P. C., Castell, J. V., & Andus, T. (1990). Interleukin-6 and the acute phase response. *Biochem J, 265,* 621–626.

172. Pasceri, V., Willerson, J. T., & Yeh, E. T. (2000). Direct proinflammatory effect of C-reactive protein on human endothelial cells. *Circulation, 102,* 2165–2168.

173. Meade, E. A., McIntyre, T. M., Zimmerman, G. A., et al. (1999). Peroxisome proliferators enhance cyclooxygenase-2 expression in epithelial cells. *J Biol Chem, 274,* 8328–8334.

174. Lindemann, S., Tolley, N. D., Dixon, D. A., et al. (2001). Activated platelets mediate inflammatory signaling by regulated interleukin 1 synthesis. *J Cell Biol, 154,* 485–490.

175. Bazzoni, G., Dejana, E., & Del Maschio, A. (1991). Platelet-neutrophil interactions: Possible relevance in the pathogenesis and inflammation. *Haematologica, 76,* 491–499.

176. Kuijper, P. H., Gallardo, T. H., Lammers, J. W., et al. (1997). Platelet and fibrin deposition at the damaged vessel wall: Cooperative substrates for neutrophil adhesion under flow conditions. *Blood, 89,* 166–175.

177. Barry, O. P., Pratico, D., Savani, R. C., et al. (1998). Modulation of monocyte-endothelial cell interactions by platelet microparticles. *J Clin Invest, 102,* 136–144.

178. Nagata, K., Tsuji, T., Todoroki, N., et al. (1993). Activated platelets induce superoxide anion release by monocytes and neutrophils through p-selectin (CD62). *J Immunol, 151,* 3267–3273.

179. Bath, P. M. W., Hassall, D. G., Gladwin, A.-M., et al. (1991). Nitric oxide and prostacyclin: Divergence of inhibitor effects on monocyte chemotaxis and adhesion to endothelium in vitro. *Arterioscler Thromb, 11,* 254–260.

180. Gamble, J. R., Skinner, M., Berndt, M. C., et al. (1990). Prevention of activated neutrophil adhesion to endothelium by soluble adhesion protein GMP-140. *Science, 249,* 414–417.

181. Li, N., Hu, H., Lindqvist, M., et al. (2000). Platelet-leukocyte cross talk in whole blood. *Arterioscler Thromb Vasc Biol, 20,* 2702–2708.

182. Tschoepe, D., Rauch, U., & Schwippert, B. (1997). Platelet-leukocyte cross-talk in diabetes mellitus. *Horm Metab Res, 29,* 631–635.

183. Alessandrini, P., McRae, J., Feman, S., et al. (1988). Thromboxane biosynthesis and platelet function in type I diabetes mellitus. *N Engl J Med, 319,* 208–212.

184. Joussen, A. M., Poulaki, V., Qin, W., et al. (2002). Retinal vascular endothelial growth factor induces intercellular adhesion molecule-1 and endothelial nitric oxide synthase expression and initiates early diabetic retinal leukocyte adhesion *in vivo. Am J Pathol, 160,* 501–509.

185. Kaplar, M., Kappelmayer, J., Veszpremi, A., et al. (2001). The possible association of *in vivo* leukocyte-platelet heterophilic aggregate formation and the development of diabetic angiopathy. *Platelets, 12,* 419–422.

186. Delamaire, M., Maugendre, D., Moreno, M., et al. (1997). Impaired leucocyte functions in diabetic patients. *Diabet Med, 14,* 29–34.

187. Linderkamp, O., Ruef, P., Zilow, E. P., et al. (1999). Impaired deformability of erythrocytes and neutrophils in children with newly diagnosed insulin-dependent diabetes mellitus. *Diabetologia, 42,* 865–869.

188. Wierusz-Wysocka, B., Wykretowicz, A., Byks, H., et al. (1993). Polymorphonuclear neutrophils adherence, superoxide anion (O$_2^-$) production and HBA1 level in diabetic patients. *Diabetes Res Clin Pract, 21,* 109–114.

189. Hu, H., Li, N., Yngen, M., et al. (2004). Enhanced leukocyte-platelet cross-talk in type 1 diabetes mellitus: Relationship to microangiopathy. *J Thromb Haemost, 2,* 58–64.

190. Rauch, U., Bonderman, D., Bohrmann, B., et al. (2000). Transfer of tissue factor from leukocytes to platelets is mediated by CD15 and tissue factor. *Blood, 96,* 170–175.

191. Mueller, I., Klocke, A., Alex, M., et al. (2003). Intravascular tissue factor initiates coagulation via circulating microvesicles and platelets. *FASEB J, 17,* 476–478.

192. Michelson, A. D., Barnard, M. R., Krueger, L. A., et al. (2001). Circulating monocyte-platelet aggregates are a more sensitive marker of *in vivo* platelet activation than platelet surface P-selectin: Studies in baboons, human coronary intervention, and human acute myocardial infarction. *Circulation, 104,* 1533–1537.

193. Phipps, R. P. (2000). Atherosclerosis: The emerging role of inflammation and the CD40-CD40 ligand system. *Proc Nat Acad Sci USA, 97,* 6930–6932.

194. Phipps, R. P., Kaufmann, J., & Blumberg, N. (2001). Platelet derived CD154 (CD40 ligand) and febrile responses to transfusion. *Lancet, 357,* 2023–2024.

195. Danese, S., de la Motte, C., Sturm, A., et al. (2003). Platelets trigger a CD40-dependent inflammatory response in the microvasculature of inflammatory bowel disease patients. *Gastroenterology, 124,* 1249–1264.

196. Henn, V., Slupsky, J. R., Grafe, M., et al. (1998). CD40 ligand on activated platelets triggers an inflammatory reaction of endothelial cells. *Nature, 391,* 591–594.

197. Mach, F., Schonbeck, U., & Libby, P. (1998). CD40 signaling in vascular cells: A key role in atherosclerosis? *Atherosclerosis, 137,* S89–S95.

198. Linton, M. F., & Fazio, S. (2002). Cyclooxygenase-2 and atherosclerosis. *Cur Opin Lipidol, 13,* 497–504.

199. Heeschen, C., Dimmeler, S., Hamm, C. W., et al. (2003). CAPTURE Study Investigators. Soluble CD40L in acute coronary syndromes. *N Engl J Med, 348,* 1104–1111.

200. Schonbeck, U., Varo, N., Libby, P., et al. (2001). Soluble CD40L and cardiovascular risk in women. *Circulation, 104,* 2266–2268.

201. Tschoepe, D. (1995). The activated megakaryocyte-platelet-system in vascular disease: Focus on diabetes. *Semin Thromb Hemost, 21,* 152–160.

202. Knobl, P., Schernthaner, G., Schnack, C., et al. (1994). Haemostatic abnormalities persist despite glycaemic improvement by insulin therapy in lean type 2 diabetic patients. *Thromb Haemost, 71,* 692–697.

203. Abraira, C., Colwell, J. A., Nuttall, F., et al. (1998). A critical issue. Intensive insulin treatment and macrovascular disease. *Diabetes Care, 21,* 669–671.

204. Trovati, M., Massucco, P., Mattiello, L., et al. (1994). Insulin increases guanosine-3′,5′-cyclic monophosphate in human platelets: A mechanism involved in the insulin anti-aggregating effect. *Diabetes, 43,* 1015–1019.

205. Trovati, M., Anfossi, G., Massucco, P., et al. (1997). Insulin stimulates nitric oxide synthesis in human platelets and, through nitric oxide, increases platelet concentrations of both guanosine-3′,5′-cyclic monophosphate and adenosine-3′,5′-cyclic monophosphate. *Diabetes, 46,* 742–749.

206. Wagner, B., Fasching, P., Schneider, B., et al. (1996). Platelet and endothelial cell function in patients with insulin-dependent diabetes mellitus with excellent, good, or poor metabolic control. *Microvasc Res, 52,* 183–187.

207. Coppola, L., Verrazzo, G., La Marca, C., et al. (1997). Effect of insulin on blood rheology in non-diabetic subjects and in patients with type 2 diabetes mellitus. *Diabet Med, 14,* 959–963.

208. Wood, D., De Backer, G., Faergeman, O., et al. (1998). Prevention of coronary heart disease in clinical practice: Recommendations of the Second Joint Task Force of European and other Societies on Coronary Prevention. *Atherosclerosis, 140,* 199–270.

209. O'Keefe, J. H., Jr., Miles, J. M., Harris, W. H., et al. (1999). Improving the adverse cardiovascular prognosis of type 2 diabetes. *Mayo Clin Proc, 74,* 171–180.

210. Gaede, P., Vedel, P., Parving, H. H., et al. (1999). Intensified multifactorial intervention in patients with type 2 diabetes mellitus and microalbuminuria: The Steno type 2 randomised study. *Lancet, 353,* 617–622.

211. Antiplatelet Trialists' Collaboration Secondary prevention of vascular disease by prolonged antiplatelet treatment. (1988). *Br Med J* (Clinical Research Edition), *296,* 320–331.

212. The DAMAD Study Group. (1989). Effect of aspirin alone and aspirin plus dipyridamole in early diabetic retinopathy: A multicenter randomised controlled clinical trial. *Diabetes, 38,* 491–498.

213. The TIMAD Study Group. (1990). Ticlopidine treatment reduces the progression of nonproliferative diabetic retinopathy. *Arch Ophthalmol, 108,* 1577–1583.

214. Early Treatment Diabetic Retinopathy Study Research Group. (1991). Effects of aspirin treatment on diabetic retinopathy. ETDRS report number 8. *Ophthalmology, 98,* 757–765.

215. Physicians' Health Study Research Group. (1989). Final report on the aspirin component of the ongoing Physicians' Health Study. Steering Committee of the Physicians' Health Study Research Group. *N Engl J Med, 321,* 129–135.

216. Patrono, C., & Davi, G. (1993). Antiplatelet agents in the prevention of diabetic vascular complications. *Diabetes Metabol Rev, 9,* 177–188.

217. Rolka, D. B., Fagot-Campagna, A., & Narayan, K. M. (2001). Aspirin use among adults with diabetes: Estimates from the Third National Health and Nutrition Examination Survey. *Diabetes Care, 24,* 197–201.

218. American Diabetes Association Position Statement: Aspirin Therapy in Diabetes. (2004). *Diabetes Care, 27*(S1), 687-S71.

219. Marso, S. P., Lincoff, A. M., Ellis, S. G., et al. (1999). Optimizing the percutaneous interventional outcomes for patients with diabetes mellitus: Results of the EPISTENT (evaluation of platelet IIb/IIIa inhibitor for stenting trial) diabetic substudy. *Circulation, 100,* 2477–2484.

220. Bhatt, D. L., Marso, S. P., Lincoff, A. M., et al. (2000). Abciximab reduces mortality in diabetics following percutaneous coronary intervention. *J Am Coll Cardiol, 35,* 922–928.

221. Kleiman, N. S., Lincoff, A. M., Kereiakes, D. J., et al. (1998). Diabetes mellitus, glycoprotein IIb/IIIa blockade, and heparin: Evidence for a complex interaction in a multi-center trial. EPILOG Investigators. *Circulation, 97,* 1912–1920.

222. Michelson, A. D., & Furman, M. I. (1999). Laboratory markers of platelet activation and their clinical significance. *Curr Opin Hematol, 6,* 342–348.

223. Steinhubl, S. R., Kottke-Marchant, K., Moliterno, D. J., et al. (1999). Attainment and maintenance of platelet inhibition through standard dosing of abciximab in diabetic and non-diabetic patients undergoing percutaneous coronary intervention. *Circulation, 100,* 1977–1982.

# CHAPTER 39

# Inflammation

**Wolfgang Bergmeier and Denisa D. Wagner**

*CBR Institute for Biomedical Research and Department of Pathology,
Harvard Medical School, Boston, Massachusetts*

## I. Introduction

Inflammation is characterized by a multitude of interactions between leukocytes, endothelial cells, and platelets. Irrespective of its etiology, inflammation causes endothelial activation. Activated endothelial cells express cell adhesion molecules such as P- and E-selectin, which mediate leukocyte rolling, the first step in the adhesion cascade leading to leukocyte extravasation.[1,2] Rolling leukocytes become activated by chemokines such as monocyte chemoattractant protein 1 (MCP-1) and RANTES (*r*egulated upon *a*ctivation, *n*ormal *T* cell *e*xpressed and presumably *s*ecreted), also presented by the endothelium. Subsequently, leukocytes, through their integrins, bind firmly to other endothelial adhesion molecules of the immunoglobulin superfamily, such as intercellular adhesion molecule-1 (ICAM-1) and vascular adhesion molecule-1 (VCAM-1). Following firm adhesion, leukocytes migrate across the endothelial barrier. The process of diapedesis (transmigration) involves various adhesion receptors expressed on both leukocytes and endothelial cells, such as platelet-endothelial cell adhesion molecule 1 (PECAM-1, Chapter 11), CD99, vascular endothelial (VE)-cadherin, as well as junctional adhesion molecule (JAM)-A, -B, and -C (reviewed in [3]).

Platelets are rapidly recruited to sites of injury and infection. Although best known for their key role in hemostasis, increasing evidence shows that platelets actively promote the inflammatory process. One piece of direct evidence for this concept originates from studies in platelet-depleted animals challenged by inflammatory stimuli. Platelet depletion leads to impaired (a) leukocyte infiltration in mouse models of chronic allergic lung inflammation,[4,5] (b) tumor necrosis factor (TNF)-α-induced leukocyte recruitment in the brain vasculature,[6] and (c) atherosclerosis in apoE-deficient mice.[7] Circulating platelets may affect inflammation by providing a bridge between endothelium and leukocytes, by transcellular metabolism with leukocytes, or via their secretory products (discussed later). In turn, inflammation leads to an imbalance between the procoagulant and anticoagulant properties of the endothelium. TNF-α, the first pro-inflammatory cytokine released at the site of infection, promotes a procoagulant state by inhibiting the synthesis of the anticoagulant protein C[8] and by stimulating tissue factor production by endothelial cells and monocytes.[9] At concentrations found in lethal or sub-lethal sepsis, however, TNF-α strongly inhibits thrombus formation in a model of ferric chloride-induced thrombosis (Fig. 39-1) and it prolongs the bleeding time in mice.[10] The antithrombotic effect of TNF-α is not mediated by a platelet TNF receptor but through the release of nitric oxide (NO), a strong inhibitor of platelet activation (Chapter 13),[11] via stimulation of inducible NO synthase by TNF receptors in the vessel wall. Thus, in the early phases of infection, when leukocyte recruitment is of primary importance to prevent bacterial spread, the formation of thrombi obstructing blood flow and leukocyte transmigration is not desirable. *In vitro* studies also show that platelets support neutrophil transmigration, but migration is impeded when the thickness of the platelet surface exceeds three layers of platelets.[12]

## II. Platelet Interactions with the Endothelium

In normal physiologic conditions, circulating platelets do not interact with nonactivated endothelium (Chapter 13). Endothelial denudation, however, induces immediate platelet adhesion and aggregation at the site of injury.[13] In addition, resting platelets are capable of adhering to stimulated endothelial cells.

### A. Platelet Rolling

As for leukocytes, the first step in platelet adhesion to stimulated endothelial cells is platelet rolling (Figs. 39-2 and 39-3). In small mesenteric venules at shear rates of approximately 500 s$^{-1}$, wild-type (WT) platelets show reduced rolling on stimulated/inflamed endothelium of P-selectin-

713

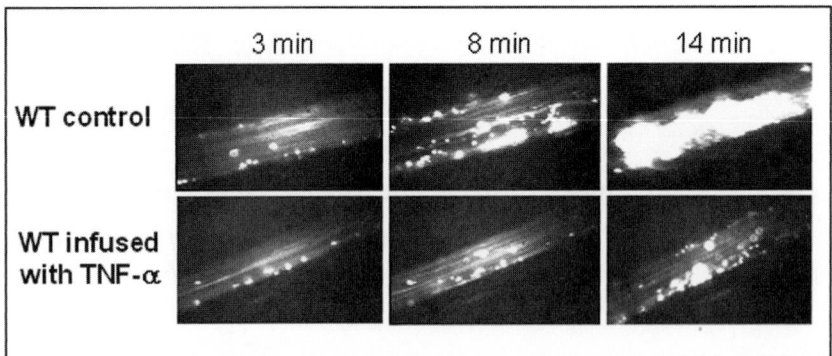

**Figure 39-1.**   Effect of TNF-α on thrombus formation *in vivo*. Thrombus formation in response to ferric chloride-induced vascular injury was visualized in arterioles of control (upper) or TNF-α-treated mice (lower). The times after ferric chloride application are indicated. No significant difference in initial platelet adhesion (white dots) to the injured vessel was observed between the two groups (3 minutes). The appearance of thrombi was delayed by TNF-α infusion (8 minutes). In addition, thrombi in treated mice were unstable and often did not grow to occlusive size, leaving, after 14 minutes, a still patent vessel, whereas the control arterioles occluded. Reproduced with permission.[10]

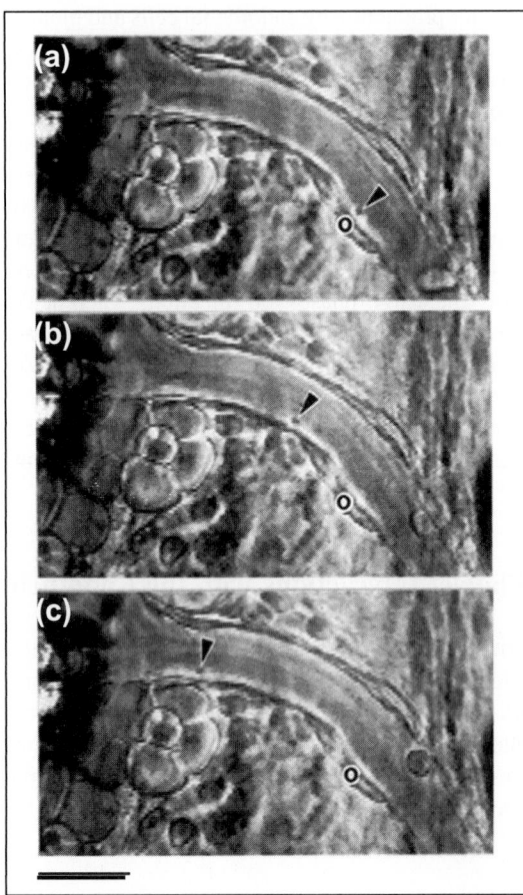

**Figure 39-2.**   Phase-contrast intravital microscopy showing a platelet rolling on a mesenteric venule after stimulation of Weibel–Palade body release by the calcium ionophore A23187. "0" indicates the location of the platelet at time 0. Arrowheads point toward the rolling platelet at 0 sec (a), 1.58 sec (b), and 2.96 sec (c). Arrow indicates a rolling leukocyte for size comparison. (Bar = 30 μm.) Reproduced with permission.[14]

or E-selectin-deficient mice, indicating that either selectin can serve as a ligand for circulating platelets (see also Chapter 12).[14,15] Endothelial P-selectin has also been identified as the major adhesion molecule for platelet adhesion to stimulated endothelium in a model of ischemia-reperfusion injury.[16] Platelets from mice lacking both fucosyl transferases (FucTs), FucT IV and FucT VII, show normal rolling on endothelium, indicating that, in contrast to leukocytes, α1,3-fucosylation of the platelet ligand is not required.[15] Two platelet receptors for endothelial P-selectin have been identified so far: P-selectin glycoprotein ligand 1 (PSGL-1) and glycoprotein (GP) Ibα.[17,18] These two receptors show striking structural similarities (see Chapters 7 and 12), but whereas PSGL-1 requires carbohydrate modification and the presence of calcium for its interaction with P-selectin, GPIbα requires neither. Blockage of the N-terminal domains of PSGL-1 by monoclonal antibodies has been shown to reduce platelet rolling on endothelial P-selectin both *in vitro* and *in vivo*. The same has been demonstrated for GPIbα *in vitro*. For a more detailed discussion of the molecular characteristics of P-selectin interactions with its ligands, see Chapter 12.

In endothelial cells, P-selectin is stored in Weibel–Palade bodies.[19,20] Upon stimulation of endothelial cells, P-selectin is rapidly translocated from the membranes of these storage granules to the plasma membrane. von Willebrand factor (VWF), an important ligand for platelet receptors GPIbα (Chapter 7) and integrin αIIbβ3 (Chapter 8), is also stored in Weibel–Palade bodies.[21] Resting platelets have been shown to translocate in a stop-and-go motion on VWF-coated coverslips under flow conditions *in vitro*[22] and on calcium ionophore- or histamine-treated mouse mesenteric venules at low shear conditions (80–100 s$^{-1}$) *in vivo*.[23] This transient adhesion of platelets to VWF is mediated by the

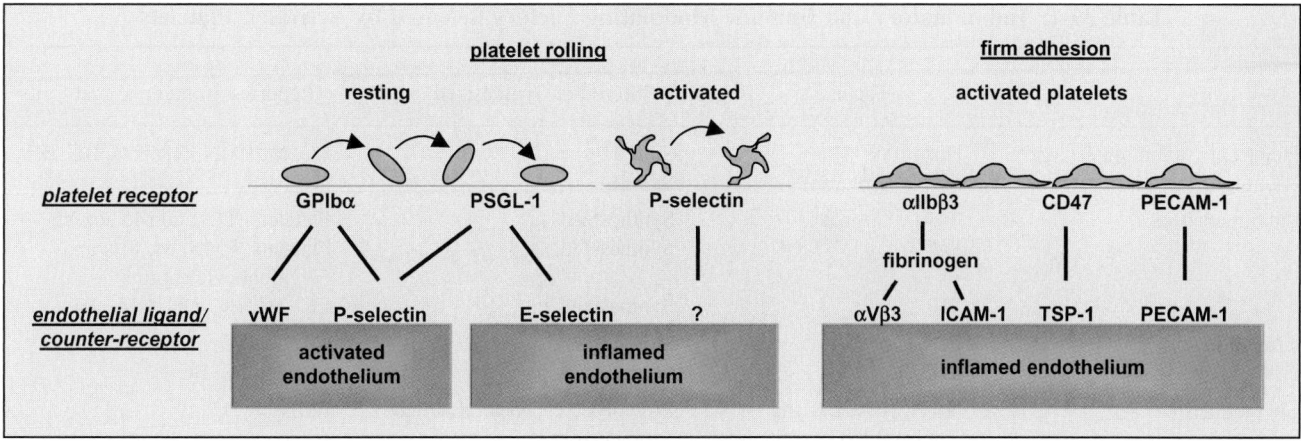

**Figure 39-3.** Schematic representation showing adhesion molecules involved in platelet rolling along activated or inflamed endothelium (left) and firm adhesion to inflamed endothelium (right). The endothelial ligand mediating rolling of activated platelets is unknown. Abbreviations: ICAM-1, intercellular adhesion molecule-1; PECAM-1, platelet-endothelial cell adhesion molecule 1; PSGL-1, P-selectin glycoprotein ligand 1; TSP-1, thrombospondin 1; vWF, von Willebrand factor.

GPIbα subunit of the VWF receptor complex (GPIb-IX-V) and is independent of integrin αIIbβ3. Interestingly, an interaction between VWF and P-selectin has been proposed, making it possible that VWF is anchored to the endothelium by binding to P-selectin originating from the same granule.[24] The transient adhesion process is terminated by the cleavage of VWF multimers by ADAMTS13,[25,26] the enzyme deficient in patients with thrombotic thrombocytopenic purpura (Chapter 50).[27]

### B. Platelet Adhesion to Endothelial Cells

The specific function of platelet rolling on stimulated endothelial cells is not known, but it certainly leads to an increased recruitment of platelets to sites of vascular injury/inflammation. To modulate the antithrombotic, anticoagulant, and antiinflammatory properties of endothelial cells, platelets may have to be fully activated and firmly adherent to the vasculature. The receptors implicated in firm platelet-endothelial cell adhesion are not yet well established. *In vitro* studies demonstrated that the adhesion of activated platelets to activated endothelial cells involves platelet integrin αIIbβ3 as well as endothelial ICAM-1 and integrin αVβ3.[28,29] Other studies suggested the participation of VWF secreted from endothelial cells[30] and platelet CD47 binding to thrombospondin-1.[31] *In vivo,* platelet adhesion to ischemic, procoagulant endothelial cells is facilitated by fibrinogen bridging ICAM-1 on endothelium and integrin αIIbβ3 on platelets.[32] In a different study, endothelial PECAM-1 was shown to contribute to platelet adhesion at a site of injured but not denuded endothelium (Fig. 39-3).[33]

### C. Activated Platelets Induce a Proinflammatory Phenotype in Endothelial Cells

The ability of activated platelets to modulate the properties of endothelial cells has been demonstrated repeatedly *in vitro*. CD40 ligand (CD40L), a member of the TNF-α family, is rapidly expressed on the platelet surface upon activation.[34–36] CD40L induces endothelial cell activation through binding to CD40. Subsequently, endothelial cells up-regulate the surface expression of various adhesion molecules and secrete chemokines, such as MCP-1 (CCL-2) and interleukin-8 (IL-8, CXCL8).[34] It has further been shown that CD40 engagement by platelet CD40L induces tissue factor expression on endothelial cells,[37,38] which could lead to a more procoagulant state of the vasculature. The role of membrane-expressed and soluble CD40L as modulators of inflammation and hemostasis is discussed in more detail in Section V.B.

In addition to signaling through direct cell-cell contacts, platelets secrete a variety of soluble inflammatory mediators with strong effects on the endothelium (Table 39-1). For example, IL-1β released from activated platelets induces endothelial MCP-1,[39,40] thus increasing neutrophil adhesion to endothelium. RANTES, a powerful chemoattractant for inflammatory cells, is also deposited on inflamed endothelium by activated platelets[41,42] and platelet microparticles.[43]

## III. Platelet Interactions with Leukocytes

Activated platelets in the circulation are prone to bind leukocytes, such as monocytes, polymorphonuclear cells (PMN), eosinophils, basophils, and T cells, to form platelet-

**Table 39-1:  Inflammatory and Immune Modulating Factors Released by Activated Platelets**

|  | Factor | Stored or Synthesized | Reported Immune Target Cells |
|---|---|---|---|
| Pleiotropic inflammatory and immune modulators | Histamine | Stored | EC, M, PMN, NK, TC, BC, E |
|  | 5-HT (serotonin) | Stored | M, Mφ |
| Lipid mediators | TxA$_2$ | Synthesized | Platelets, TC, and Mφ subsets |
|  | PAF | Synthesized | Platelets, PMN, M, Mφ, and lymphocyte subsets |
|  | S1P | Synthesized | EC, TC, BC, DC, NK, Mφ |
| Growth factors | PDGF | Stored | M, Mφ, TC |
|  | TGF-β | Stored | M, Mφ, TC, BC |
| Chemokines | NAP2 (CXCL7) and related β-TG variants | Proteolytic cleavage of stored precursors | PMN |
|  | PF4 (CXCL4) | Stored | PMN |
|  | GRO-α (CXCL1) | Stored | PMN |
|  | ENA-78 (CXCL5) | Stored | PMN |
|  | RANTES (CCL5) | Stored | M, E, B, NK, TC, and DC subsets |
|  | MIP-1α (CCL3) | Stored | M, E, B, NK, and DC subsets |
|  | MCP-3 (CCL7) | Stored | M, B, NK cell, and DC subsets |
| Cytokines | IL-1β | Synthesized | M, EC, DC, and Mφ subsets |
|  | HMGB1 | Stored | Mφ, PMN, EC |

Abbreviations: Factors: β-TG, β-thromboglobulin; ENA-78, epithelial neutrophil activating protein-78; GRO-α, growth-regulating oncogene-α; 5-HT, 5 hydroxytryptamine (serotonin); HMGB1, high mobility group box 1; IL-1β, interleukin-1β; MCP-3, monocyte chemotactic protein-3; MIP-1α, macrophage inflammatory protein-1α; NAP2, neutrophil activating peptide 2; PAF, platelet-activating factor; PDGF, platelet-derived growth factor; PF4, platelet factor 4; S1P, sphingosine 1-phosphate; TGF-β, transforming growth factor β; TxA$_2$, thromboxane A$_2$. Target cells: B, basophil; BC, B lymphocyte; DC, dendritic cell; E, eosinophil; EC, endothelial cell; M, monocyte; Mφ, macrophage; NK, natural killer cell; PMN, polymorphonuclear leukocyte (neutrophil); TC, T lymphocyte.
Adapted from ref. 72.

leukocyte aggregates.[44,45] The interaction between activated platelets and leukocytes can occur in two ways: (a) leukocyte adhesion to activated platelets that are themselves adherent to endothelial cells or components of the extracellular matrix, and (b) intravascular formation of heterotypic platelet-leukocyte aggregates. In both situations, the initial adhesion of leukocytes depends on platelet P-selectin and leukocyte PSGL-1[42,44,46,47] (see Chapter 12). P-selectin binding to PSGL-1 may also participate in the firm adhesion of leukocytes to adherent platelets, as it provides signals for the activation of integrins αLβ2 (LFA-1) and αMβ2 (Mac-1) on the leukocyte surface.[48–50] Several platelet counter-receptors for Mac-1 have been identified *in vitro:* junctional adhesion molecule-3 (JAM-3),[51] GPIbα,[52,53] and fibrinogen bound to integrin αIIbβ3[54] or ICAM-2.[55] Platelet ICAM-2 is also a counterreceptor for integrin αLβ2 (Fig. 39-4).[56] Furthermore, binding of platelet P-selectin to PSGL-1 can induce the transcription of chemokines, translation of mRNA for urokinase plasminogen activator receptor, and phosphatidyl serine exposure in leukocytes.[57–61]

Platelet-leukocyte aggregates are seen rolling on inflamed endothelium.[14] Physiological flow may support their interaction, as it leads to a high density of these cell types around the vessel periphery.[62,63] Burger and Wagner[64] demonstrated the importance of platelet P-selectin in the formation of atherosclerotic lesions in mice, as lesion size and maturation were significantly decreased in chimeric apoE-deficient mice expressing endothelial but not platelet P-selectin when compared with mice expressing P-selectin on both cell types (Fig. 39-5). Supporting the key role of platelet P-selectin in inflammation, infusion of activated WT but not P-selectin-deficient platelets into mice promotes monocyte adhesion to atherosclerotic lesions in apoE-deficient mice,[42] as well as leukocyte infiltration into lungs in a model of allergic asthma.[5,65] Activated platelets also enhance the accumulation of virus-specific cytotoxic T cells at sites of inflammation in the liver.[66] As the atherosclerosis model shows, the interaction of activated platelets with leukocytes is essential for the delivery of chemokines such as RANTES and platelet factor 4 to the monocyte surface.[42] These chemokines in turn induce integrin activation in monocytes and thus change the adhesiveness of the cells toward the inflamed endothelium.

Results from our group show that infusion of activated platelets into healthy mice induces a transient formation of

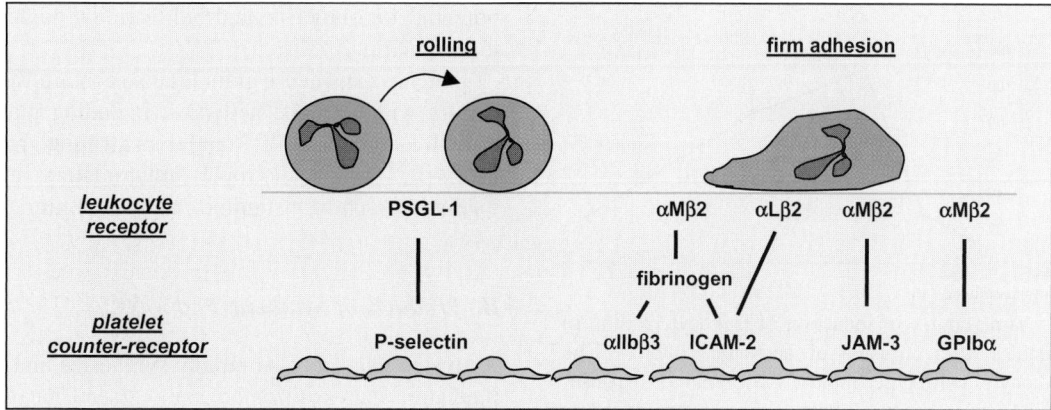

**Figure 39-4.** Adhesion molecules involved in leukocyte adhesion to activated platelets.

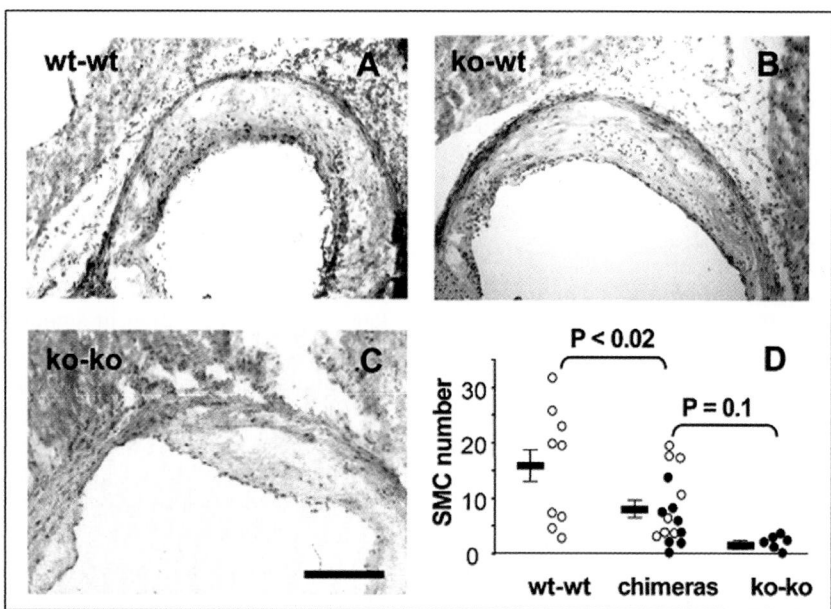

**Figure 39-5.** Immunohistochemical staining for α-actin shows positive SMCs in atherosclerotic lesions of apoE$^{-/-}$ mice chimeric for the expression of P-selectin, generated by bone marrow transplantation. The P-selectin genotype in platelets and endothelium of chimeric mice is indicated in the order of platelets-endothelium: (A) wt-wt, (B) ko-wt, (C) ko-ko animals; bar = 0.2 mm. (D) α-actin-positive SMCs within lesions were counted on 5 aortic sinus sections 80 μm apart. Black dots (•) correspond to animals that lack endothelial P-selectin (i.e., wt-ko, ko-ko). Bars show the mean ± SEM for each group. The atherosclerotic lesions of the chimeric animals appear to have an intermediate content of SMCs, indicating that both endothelial and platelet P-selectin contribute to lesion maturation. Abbreviations: ko, knockout; SMCs, smooth muscle cells; wt, wild type. Reproduced with permission.[64]

platelet-leukocyte aggregates leading to endothelial activation and leukocyte rolling.[67] These studies demonstrate that activated platelets induce Weibel–Palade body release, thereby systemically changing the vasculature to a proinflammatory state. Subsequent leukocyte rolling depends on leukocyte PSGL-1 and endothelial P-selectin. Platelet P-selectin also plays a key role, probably due to its importance in the formation of platelet-leukocyte aggregates (Fig. 39-6).

## IV. Platelet-Derived Inflammatory and Immune-Modulating Factors

Platelets are a rich source of inflammatory and immune-modulating factors, which can be rapidly released upon cellular activation. These factors, the so-called platelet secretome, have been extensively reviewed.[68–73] A summary of the best-characterized inflammatory and immune-

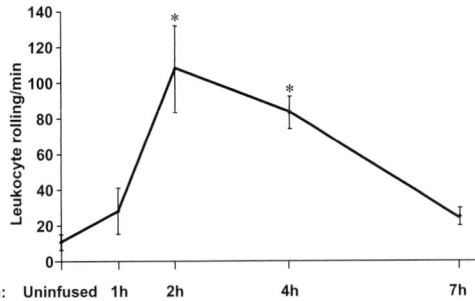

**Figure 39-6.** Time course of induction of leukocyte rolling in mesenteric venules after infusion of activated platelets. The number of leukocytes rolling per minute was observed by phase-contrast intravital microscopy. Infusion of activated platelets induced a significant increase in leukocyte rolling compared with uninjected mice after 2 hours and 4 hours. Leukocyte rolling decreased by 7 hours after infusion. * $p < 0.01$. Average ± SEM. Reproduced with permission.[67]

modulating factors released from activated platelets and their reported immune target cells is given in Table 39-1. Platelets also release specific antimicrobial factors, which are reviewed in detail in Chapter 40.

### A. Inflammatory and Immune-Modulating Factors Stored in Platelet Granules

Activated platelets release chemokines of the CXC and CC classes that are stored in processed or precursor forms in α granules.[70,71] In fact, platelets have been suggested as a unique delivery cell for chemokines, as unexpectedly high levels of these mediators have been detected at sites of platelet activation.[71,74] CXC chemokines like platelet factor 4 (PF4) or β-thromboglobulin (β-TG) play a crucial role in the early inflammatory response, as they are known modulators of PMN function.[70] CC chemokines like RANTES, macrophage inflammatory protein-1α (MIP-1α), and monocyte chemotactic protein-3 (MCP-3) have multiple immune and inflammatory activities, including the activation of innate and adaptive immune effector cells and the induction of chemokine and cytokine synthesis.[41,58,59,75,76] As shown *in vivo*, platelets deposit chemokines like RANTES on the surface of endothelial cells, thus enabling circulating monocytes to be activated and to arrest.[41,42,77] Platelets themselves express various chemokine receptors such as CXCR4, CCR1, CCR3, and CCR4 on their surface (see Chapter 6), and these receptors provide costimulatory signaling during platelet activation with low doses of well-established platelet agonists.[74,78–80] Three chemokines in particular activate platelets *in vitro*: CXCL12 (SDF-1), secreted from endothelial cells and present in atherosclerotic plaques as well as CCL17 (TARC) and CCL22 (MDC), which are produced by monocytes and macrophages. Thus, platelets have the

potential for autocrine and bidirectional paracrine chemokine signaling.

Platelet α and dense granules also contain various growth factors with immune activities, including platelet-derived growth factor (PDGF)[81,82] and transforming growth factor-β (TGF-β),[83] and pleiotropic inflammatory and immune mediators such as histamine[82] and serotonin.[84]

### B. Products of Synthetic Pathways

Activated platelets also rapidly synthesize and secrete lipid mediators such as thromboxane $A_2$ (see Chapter 31), sphingosine 1-phosphate (S1P), and platelet-activating factor (PAF), which have the potential to modulate multiple cellular functions in innate and adaptive immune effector cells.[85–90] Furthermore, activated platelets translate mRNA, which was passed on to them by their parent megakaryocytes.[72] For example, interleukin-1β (IL-1β) is synthesized from mRNA by activated platelets, explaining the previously described IL-1 activity in these cells.[40,91–97] IL-1β is a well-known proinflammatory cytokine with the ability to stimulate the expression of various genes associated with inflammation.[98] Platelet IL-1β has been shown to induce the up-regulation of adhesion receptors and the secretion of cytokines/chemokines in endothelial cells,[94,96,97] to stimulate cytokine production in smooth muscle cells,[95] and to promote IL-1-dependent growth in a T cell line.[93] To express its proinflammatory activity, IL-1β has to be associated with the plasma membrane of activated platelets. IL-1β activity is also released in association with platelet microparticles, which have been shown to modify cellular responses in leukocytes and endothelial cells.[40,99]

## V. Inflammatory Receptors Modulating Hemostasis and Thrombosis

### A. Procoagulant Activity of P-Selectin

The first demonstration that P-selectin can influence fibrin deposition was made by Palabrica and colleagues, who demonstrated that inhibition of P-selectin blocks leukocyte adhesion to platelets adherent on a thrombogenic graft and that, surprisingly at the time, this was accompanied by reduced deposition of fibrin on the graft.[100] Further evidence came from studies in mice deficient in P-selectin, which showed prolonged tail bleeding times as well as increased hemorrhage in experimental hemorrhagic lesions when compared to WT animals.[101] Unexpectedly, a procoagulant state was observed in mice with elevated levels of soluble P-selectin (P-selectin$^{\Delta CT}$),[102] indicating that P-selectin is not only involved in directing fibrin deposition at sites of injury but also systemically increases the coagulation potential.

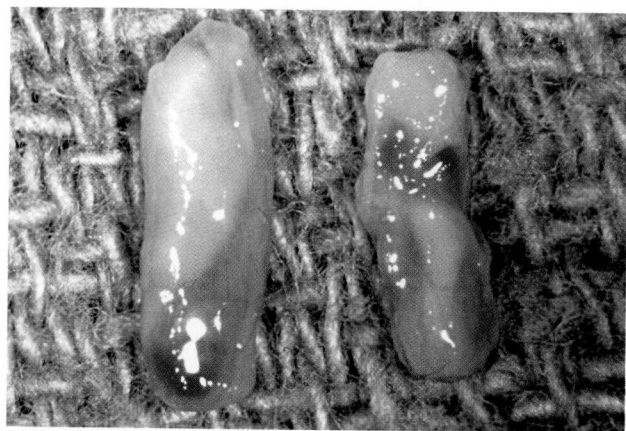

**Figure 39-7.** Venous thrombogenesis in wild-type mice and mice expressing high levels of soluble P-selectin (ΔCT mice). Thrombosis was induced by ligation of the inferior vena cava. Animals were sacrificed six days after the ligation, and thrombi were isolated. Photograph of representative thrombi is shown. ΔCT thrombus with a significantly higher mass is on the left; wild-type thrombus on the right. Photograph was provided by Thomas Wakefield (University of Michigan). Reproduced with permission.[104]

P-selectin-dependent fibrin deposition was further detected in platelet-rich thrombi *ex vivo*[102] and *in vivo* (Fig. 39-7).[103,104] The mechanism of how P-selectin potentiates coagulation appears to be dependent on a large population of tissue factor-containing leukocyte-derived microparticles, which were detected in plasma from P-selectin[ΔCT] but not WT mice. These microparticles represent what is now called the "blood-borne tissue factor," described by Nemerson and colleagues in 1999.[105] The formation of such microparticles appears to be a direct consequence of the P-selectin/PSGL-1 interaction, as the infusion of P-selectin-Ig induced microparticle formation in WT but not PSGL-1-deficient mice.[106] P-selectin, either purified or on activated platelets, was also shown to induce phosphatidyl serine expression on monocytes, a cellular change important for the assembly of coagulation complexes.[61] Interestingly, P-selectin, through binding to PSGL-1, also mediates capture of microparticles on activated platelets, thus concentrating the procoagulant activity at the thrombus site.[107,108] The importance of blood-borne tissue factor for normal thrombus formation may depend on the quality of the injury. In situations where high concentrations of tissue-derived tissue factor is exposed to blood, the blood-borne tissue factor may be less important than in small puncture wounds or in situations where thrombi form in the absence of vessel injury (for example, in deep vein thrombosis).[109,110]

*In vivo,* activated platelets and inflamed endothelium rapidly shed their P-selectin into the circulation.[64,111,112] The amount of P-selectin shed from activated platelets or endothelial cells reflects the extent of the injury and proportion-

ally increases the animal's procoagulant activity. This injury-related microparticle response occurs with a delay of a few hours and is likely to be more important for prolonged, fibrin-dependent thrombus stabilization than for initial plug formation.

Elevated plasma concentrations of soluble P-selectin may have pathological consequences. Elevated levels of plasma P-selectin are found in thrombotic consumptive platelet disorders, such as disseminated intravascular coagulation, thrombotic thrombocytopenic purpura, and heparin-induced thrombocytopenia.[113–116] They were further described for the recurrent arterial thrombotic processes of angina,[117] restenosis,[118] myocardial infarction,[119] and infective endocarditis[120] and have been suggested to be a predictive marker for future cardiovascular events.[121,122]

Although many aspects of microparticle formation are unclear, it has become evident that they play an important role in inflammation, coagulation, and vascular function (see also Chapter 20).[123,124] In coagulation, platelet-derived microparticles expressing P-selectin probably induce tissue factor on monocytes, leukocyte-derived microparticles induce tissue factor expression on endothelial cells,[125,126] and endothelial microparticles, which might also express P-selectin, were shown to induce tissue factor expression on monocytes.[127] *In vitro,* tissue factor-bearing microparticles can fuse with activated platelets, thereby transferring procoagulant activity to the platelet membrane. Successful fusion again requires surface-expressed phosphatidyl serine and PSGL-1.[128] Thus, the role of P-selectin in coagulation and thrombus stabilization may be just as important as its well-recognized function in inflammation.

### B. Platelet Surface CD40L and Soluble CD40L

CD40L expressed on activated platelets can generate signals for the recruitment and extravasation of leukocytes at sites of inflammation, because it induces, through engagement of CD40, the secretion of chemokines and the expression of adhesion receptors in endothelial cells.[34] CD40L also provides a powerful link between platelets and the immune system. Elzey et al.[36,129] have shown that CD40L expressed on activated platelets induces dendritic cell maturation and B cell isotype switching. In addition, platelet CD40L augments CD8[+] T cell responses *in vitro* and *in vivo*.[36,129] In turn, the interaction between CD40L and CD40 induces shedding of CD40L and thus leads to markedly increased levels of the soluble receptor (sCD40L) in plasma.[35] While the proinflammatory activity of sCD40L is controversial,[35,130] sCD40L modulates platelet function *in vitro* and *in vivo*.[131,132] Mice deficient in CD40L, but not mice deficient in CD40, developed highly unstable thrombi in an *in vivo* model of arterial thrombosis, suggesting that CD40L binds to a receptor other than CD40 on platelets (Fig. 39-8). Further studies demon-

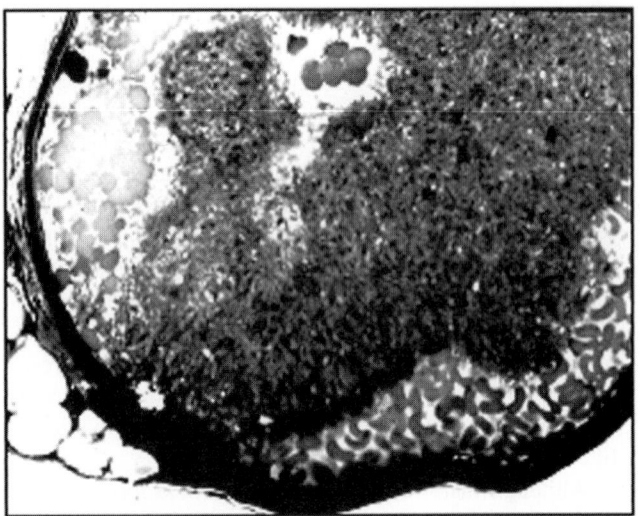 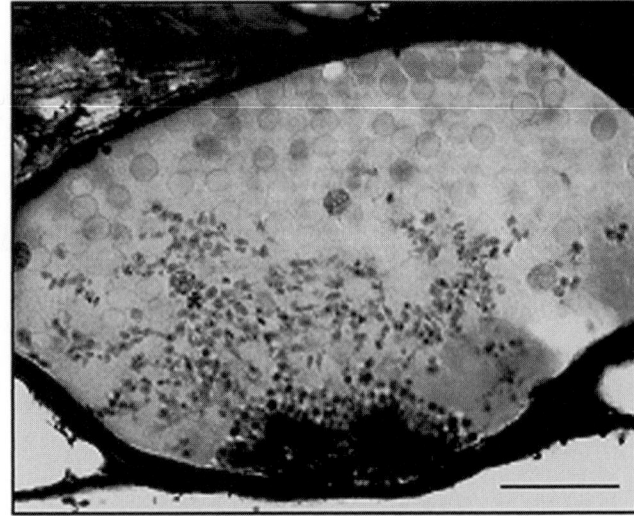

**WT**                    **CD40L⁻/⁻**

**Figure 39-8.**   Effect of CD40L-deficiency on the property of thrombi formed 30 minutes after FeCl₃-induced injury. Wild-type (left) arterioles revealed a dense, packed platelet-rich thrombus, whereas a loosely packed luminal thrombus on top of a dense mural thrombus developed in the CD40L⁻/⁻ (right) arteriole. Asterisk indicates luminal thrombus. Scale bar, 20 μm. Reproduced with permission.[131]

strated that sCD40L through a KGD sequence binds and activates platelet integrin αIIbβ3, leading to β3 phosphorylation. This interaction allows sCD40L to restore thrombus stability when infused into CD40L-deficient mice.[131,133]

## VI. Platelets in Inflammatory Disease

Platelet activation, via the multitude of mechanisms just described, results in a proinflammatory state. There is therefore a platelet-mediated proinflammatory component to the numerous common diseases in which there is an increase in circulating activated platelets or an increase in activatable platelets: acute coronary syndromes (see Chapter 35),[134–136] ischemic cerebrovascular disease (see Chapter 36),[137–139] peripheral arterial occlusive disease (see Chapter 37),[140,141] peripheral venous disease (see Chapter 37),[142,143] diabetes mellitus (see Chapter 38),[144] myeloproliferative disorders (see Chapter 56),[145,146] Alzheimer's disease (see Chapter 43),[147] preeclampsia,[148,149] placental insufficiency,[150] nephrotic syndrome,[151] hemodialysis,[152] sickle cell disease,[153] systemic inflammatory response syndrome,[154] septic multiple organ dysfunction syndrome,[155,156] antiphospholipid syndrome,[157] systemic lupus erythematosus,[157] rheumatoid arthritis,[157] inflammatory bowel disease,[158] Kawasaki disease,[159] and asthma.[160]

Cystic fibrosis (CF) is a recently characterized example of the role of platelet-mediated inflammation in disease and its potential therapeutic implications.[161] CF is caused by a mutation of the gene encoding the cystic fibrosis transmem-

brane conductance regulator (CFTR). Platelet function was examined in CF patients because lung inflammation is an important part of this disease. CF patients have an increase in circulating activated platelets and platelet reactivity, as determined by monocyte-platelet aggregation, neutrophil-platelet aggregation, and platelet surface P-selectin.[161] This increased platelet activation in CF is the result of both a plasma factor(s) and an intrinsic platelet mechanism via (a) cAMP/adenylate cyclase or (b) increased Mead acid, but not via platelet CFTR. These findings may account, at least in part, for the beneficial effects of ibuprofen in CF.[161]

## VII. Conclusions

Thrombosis and inflammation are intricately linked (Fig. 39-9). Although highly specialized for their role in hemostasis, platelets are also innate inflammatory cells, carrying inflammatory and antimicrobial activities. In addition, platelets or platelet-derived microparticles modulate leukocyte adhesion/extravasation at sites of inflammation by (a) inducing a proinflammatory, proadhesive state in endothelial cells and leukocytes and (b) providing a bridge between the endothelium and leukocytes. This enables leukocytes to firmly attach to the vessel wall and finally to transmigrate into the subendothelial tissue. In turn, inflammatory mediators such as TNF-α lead to an imbalance between the procoagulant and anticoagulant properties of the endothelium, thereby affecting both platelet function and coagulation and thus thrombus formation. The identification of the proco-

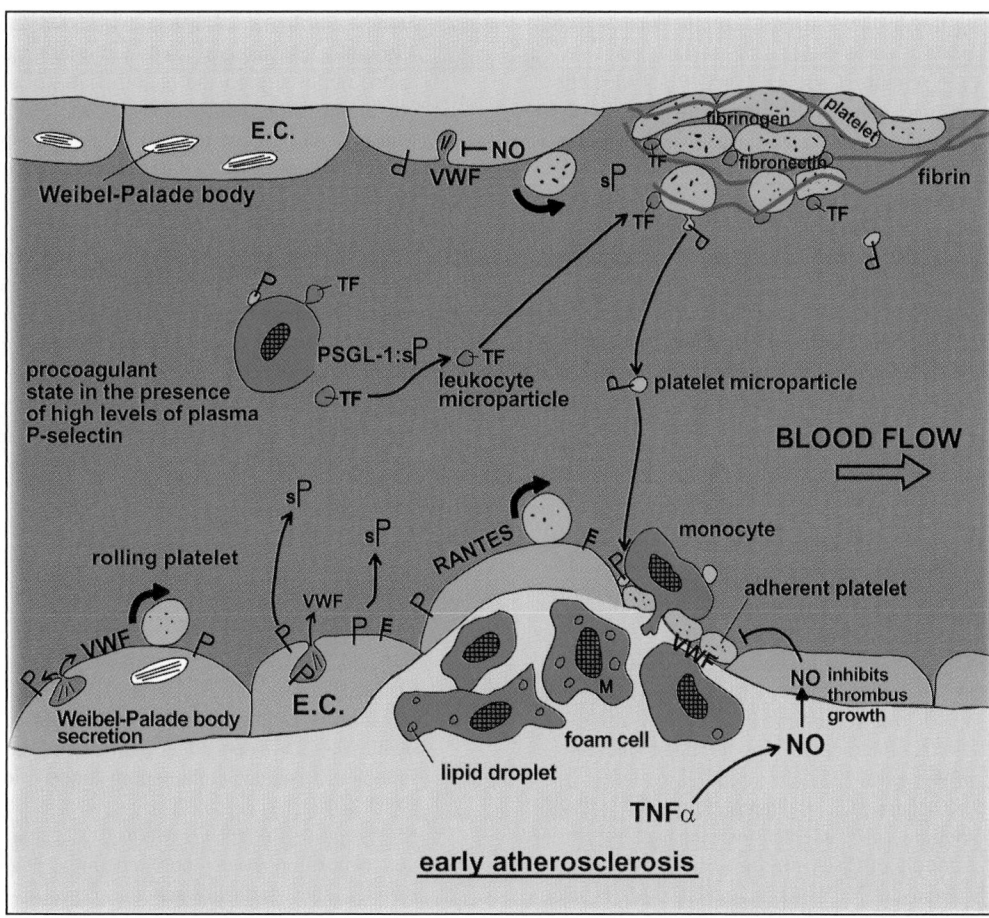

**Figure 39-9.** Schematic representation of the role of platelets in inflammation/atherosclerosis and thrombosis. Abbreviations: E, E-selectin; EC, endothelial cell; M, macrophage; NO, nitric oxide; P, transmembrane form of P-selectin; RANTES, regulated upon activation, normal T cell expressed and presumably secreted; sP, soluble P-selectin; TF, tissue factor; TNF-α, tumor necrosis factor α; VWF, von Willebrand factor. Reproduced with permission.[162]

agulant properties of soluble P-selectin, a key receptor in the adhesive interaction of endothelial cells and platelets with leukocytes, provides another example of the intricate relationship between thrombosis and inflammation.

## Acknowledgment

We thank Alan Michelson for writing Section VI of this chapter.

## References

1. Butcher, E. C. (1991). Leukocyte-endothelial cell recognition: Three (or more) steps to specificity and diversity. *Cell, 67,* 1033–1036.

2. Springer, T. A. (1994). Traffic signals for lymphocyte recirculation and leukocyte emigration: The multistep paradigm. *Cell, 76,* 301–314.

3. Muller, W. A. (2003). Leukocyte-endothelial-cell interactions in leukocyte transmigration and the inflammatory response. *Trends Immunol, 24,* 326–334.

4. Pitchford, S. C., Yano, H., Lever, R., et al. (2003). Platelets are essential for leukocyte recruitment in allergic inflammation. *Jour Allergy Clin Immunol, 112,* 109–118.

5. Pitchford, S. C., Momi, S., Giannini, S., et al. (2005). Platelet P-selectin is required for pulmonary eosinophil and lymphocyte recruitment in a murine model of allergic inflammation. *Blood, 105,* 2074–2081.

6. Carvalho-Tavares, J., Hickey, M. J., Hutchison, J., et al. (2000). A role for platelets and endothelial selectins in tumor necrosis factor-alpha-induced leukocyte recruitment in the brain microvasculature. *Circ Res, 87,* 1141–1148.

7. Massberg, S., Brand, K., Gruner, S., et al. (2002). A critical role of platelet adhesion in the initiation of atherosclerotic lesion formation. *J Exp Med, 196,* 887–896.

8. Yamamoto, K., Shimokawa, T., Kojima, T., et al. (1999). Regulation of murine protein C gene expression in vivo: Effects of tumor necrosis factor-alpha, interleukin-1, and

transforming growth factor-beta. *Thromb Haemost, 82,* 1297–1301.

9. Kirchhofer, D., Tschopp, T. B., Hadvary, P., et al. (1994). Endothelial cells stimulated with tumor necrosis factor-alpha express varying amounts of tissue factor resulting in inhomogenous fibrin deposition in a native blood flow system: Effects of thrombin inhibitors. *J Clin Invest, 93,* 2073–2083.

10. Cambien, B., Bergmeier, W., Saffaripour, S., et al. (2003). Antithrombotic activity of TNF-alpha. *J Clin Invest, 112,* 1589–1596.

11. Loscalzo, J. (2001). Nitric oxide insufficiency, platelet activation, and arterial thrombosis. *Circ Res, 88,* 756–762.

12. Diacovo, T. G., Roth, S. J., Buccola, J. M., et al. (1996). Neutrophil rolling, arrest, and transmigration across activated, surface-adherent platelets via sequential action of P-selectin and the beta 2-integrin CD11b/CD18. *Blood, 88,* 146–157.

13. Clemetson, K. J. (1999). Primary haemostasis: Sticky fingers cement the relationship. *Curr Biol, 9,* R110–112.

14. Frenette, P. S., Johnson, R. C., Hynes, R. O., et al. (1995). Platelets roll on stimulated endothelium *in vivo:* An interaction mediated by endothelial P-selectin. *Proc Nat Acad Sci, 92,* 7450–7454.

15. Frenette, P. S., Moyna, C., Hartwell, D. W., et al. (1998). Platelet-endothelial interactions in inflamed mesenteric venules. *Blood, 91,* 1318–1324.

16. Massberg, S., Enders, G., Leiderer, R., et al. (1998). Platelet-endothelial cell interactions during ischemia/reperfusion: The role of P-selectin. *Blood, 92,* 507–515.

17. Frenette, P. S., Denis, C. V., Weiss, L., et al. (2000). P-Selectin glycoprotein ligand 1 (PSGL-1) is expressed on platelets and can mediate platelet-endothelial interactions *in vivo. J Exp Med, 191,* 1413–1422.

18. Romo, G. M., Dong, J. F., Schade, A. J., et al. (1999). The glycoprotein Ib-IX-V complex is a platelet counterreceptor for P-selectin. *J Exp Med, 190,* 803–814.

19. Bonfanti, R., Furie, B. C., Furie, B., et al. (1989). PADGEM (GMP140) is a component of Weibel-Palade bodies of human endothelial cells. *Blood, 73,* 1109–1112.

20. McEver, R. P., Beckstead, J. H., Moore, K. L., et al. (1989). GMP-140, a platelet alpha-granule membrane protein, is also synthesized by vascular endothelial cells and is localized in Weibel-Palade bodies. *J Clin Invest, 84,* 92–99.

21. Wagner, D. D., Olmsted, J. B., & Marder, V. J. (1982). Immunolocalization of von Willebrand protein in Weibel-Palade bodies of human endothelial cells. *J Cell Biol, 95,* 355–360.

22. Savage, B., Saldivar, E., & Ruggeri, Z. M. (1996). Initiation of platelet adhesion by arrest onto fibrinogen or translocation on von Willebrand factor. *Cell, 84,* 289–297.

23. Andre, P., Denis, C. V., Ware, J., et al. (2000). Platelets adhere to and translocate on von Willebrand factor presented by endothelium in stimulated veins. *Blood, 96,* 3322–3328.

24. Padilla, A., Moake, J. L., Bernardo, A., et al. (2004). P-selectin anchors newly released ultralarge von Willebrand factor multimers to the endothelial cell surface. *Blood, 103,* 2150–2156.

25. Dong, J. F., Moake, J. L., Nolasco, L., et al. (2002). ADAMTS-13 rapidly cleaves newly secreted ultralarge von Willebrand factor multimers on the endothelial surface under flowing conditions. *Blood, 100,* 4033–4039.

26. Motto, D. G., Chauhan, A. K., Zhu, G., et al. (2005). Shigatoxin triggers thrombotic thrombocytopenic purpura in genetically susceptible ADAMTS13-deficient mice. *J Clin Invest, 115,* 2752–2761.

27. Levy, G. G., Nichols, W. C., Lian, E. C., et al. (2001). Mutations in a member of the ADAMTS gene family cause thrombotic thrombocytopenic purpura. *Nature, 413,* 488–494.

28. Li, J. M., Podolsky, R. S., Rohrer, M. J., et al. (1996). Adhesion of activated platelets to venous endothelial cells is mediated via GPIIb/IIIa. *J Surg Res, 61,* 543–548.

29. Bombeli, T., Schwartz, B. R., & Harlan, J. M. (1998). Adhesion of activated platelets to endothelial cells: Evidence for a GPIIbIIIa-dependent bridging mechanism and novel roles for endothelial intercellular adhesion molecule 1 (ICAM-1), alphavbeta3 integrin, and GPIbalpha. *J Exp Med, 187,* 329–339.

30. Etingin, O. R., Silverstein, R. L., & Hajjar, D. P. (1993). von Willebrand factor mediates platelet adhesion to virally infected endothelial cells. *Proc Nat Acad Sci USA, 90,* 5153–5156.

31. Lagadec, P., Dejoux, O., Ticchioni, M., et al. (2003). Involvement of a CD47-dependent pathway in platelet adhesion on inflamed vascular endothelium under flow. *Blood, 101,* 4836–4843.

32. Massberg, S., Enders, G., Matos, F. C., et al. (1999). Fibrinogen deposition at the postischemic vessel wall promotes platelet adhesion during ischemia-reperfusion in vivo. *Blood, 94,* 3829–3838.

33. Rosenblum, W. I., Nelson, G. H., Wormley, B., et al. (1996). Role of platelet-endothelial cell adhesion molecule (PECAM) in platelet adhesion/aggregation over injured but not denuded endothelium in vivo and *ex vivo. Stroke, 27,* 709–711.

34. Henn, V., Slupsky, J. R., Grafe, M., et al. (1998). CD40 ligand on activated platelets triggers an inflammatory reaction of endothelial cells. *Nature, 391,* 591–594.

35. Henn, V., Steinbach, S., Buchner, K., et al. (2001). The inflammatory action of CD40 ligand (CD154) expressed on activated human platelets is temporally limited by coexpressed CD40. *Blood, 98,* 1047–1054.

36. Elzey, B. D., Tian, J., Jensen, R. J., et al. (2003). Platelet-mediated modulation of adaptive immunity: A communication link between innate and adaptive immune compartments. *Immunity, 19,* 9–19.

37. Miller, D. L., Yaron, R., & Yellin, M. J. (1998). CD40L-CD40 interactions regulate endothelial cell surface tissue factor and thrombomodulin expression. *J Leukoc Biol, 63,* 373–379.

38. Slupsky, J. R., Kalbas, M., Willuweit, A., et al. (1998). Activated platelets induce tissue factor expression on human umbilical vein endothelial cells by ligation of CD40. *Thromb Haemost, 80,* 1008–1014.

39. Gawaz, M., Neumann, F. J., Dickfeld, T., et al. (1998). Activated platelets induce monocyte chemotactic protein-1 secretion and surface expression of intercellular adhesion

molecule-1 on endothelial cells. *Circulation, 98,* 1164–1171.

40. Lindemann, S., Tolley, N. D., Dixon, D. A., et al. (2001). Activated platelets mediate inflammatory signaling by regulated interleukin 1beta synthesis. *J Cell Biol, 154,* 485–490.

41. Schober, A., Manka, D., von Hundelshausen, P., et al. (2002). Deposition of platelet RANTES triggering monocyte recruitment requires P-selectin and is involved in neointima formation after arterial injury. *Circulation, 106,* 1523–1529.

42. Huo, Y., Schober, A., Forlow, S. B., et al. (2003). Circulating activated platelets exacerbate atherosclerosis in mice deficient in apolipoprotein E. *Nat Med, 9,* 61–67.

43. Mause, S. F., von Hundelshausen, P., Zernecke, A., et al. (2005). Platelet microparticles: A transcellular delivery system for RANTES promoting monocyte recruitment on endothelium. *Arterioscler Thromb Vasc Biol, 25,* 1512–1518.

44. Hamburger, S. A., & McEver, R. P. (1999). GMP-140 mediates adhesion of stimulated platelets to neutrophils. *Blood, 75,* 550–554.

45. Rinder, C. S., Bonan, J. L., Rinder, H. M., et al. (1992). Cardiopulmonary bypass induces leukocyte-platelet adhesion. *Blood, 79,* 1201–1205.

46. Larsen, E., Celi, A., Gilbert, G. E., et al. (1989). PADGEM protein: A receptor that mediates the interaction of activated platelets with neutrophils and monocytes. *Cell, 59,* 305–312.

47. Buttrum, S. M., Hatton, R., & Nash, G. B. (1993). Selectin-mediated rolling of neutrophils on immobilized platelets. *Blood, 82,* 1165–1174.

48. Blanks, J. E., Moll, T., Eytner, R., et al. (1998). Stimulation of P-selectin glycoprotein ligand-1 on mouse neutrophils activates beta 2-integrin mediated cell attachment to ICAM-1. *Eur J Immunol, 28,* 433–443.

49. Evangelista, V., Manarini, S., Sideri, R., et al. (1999). Platelet/polymorphonuclear leukocyte interaction: P-selectin triggers protein-tyrosine phosphorylation-dependent CD11b/CD18 adhesion: Role of PSGL-1 as a signaling molecule. *Blood, 93,* 876–885.

50. Yago, T., Tsukuda, M., & Minami, M. (1999). P-selectin binding promotes the adhesion of monocytes to VCAM-1 under flow conditions. *J Immunol, 163,* 367–373.

51. Santoso, S., Sachs, U. J., Kroll, H., et al. (2002). The junctional adhesion molecule 3 (JAM-3) on human platelets is a counterreceptor for the leukocyte integrin Mac-1. *J Exp Med, 196,* 679–691.

52. Ehlers, R., Ustinov, V., Chen, Z., et al. (2003). Targeting platelet-leukocyte interactions: Identification of the integrin Mac-1 binding site for the platelet counter receptor glycoprotein Ibalpha. *J Exp Med, 198,* 1077–1088.

53. Wang, Y., Sakuma, M., Chen, Z., et al. (2005). Leukocyte engagement of platelet glycoprotein Ibalpha via the integrin Mac-1 is critical for the biological response to vascular injury. *Circulation, 112,* 2993–3000.

54. Weber, C., & Springer, T. A. (1997). Neutrophil accumulation on activated, surface-adherent platelets in flow is mediated by interaction of Mac-1 with fibrinogen bound to alphaIIbbeta3 and stimulated by platelet-activating factor. *J Clin Invest, 100,* 2085–2093.

55. Kuijper, P. H., Gallardo Tores, H. I., Lammers, J. W., et al. (1998). Platelet associated fibrinogen and ICAM-2 induce firm adhesion of neutrophils under flow conditions. *Thromb Haemost, 80,* 443–448.

56. Weber, K. S., Alon, R., & Klickstein, L. B. (2004). Sialylation of ICAM-2 on platelets impairs adhesion of leukocytes via LFA-1 and DC-SIGN. *Inflammation, 28,* 177–188.

57. McIntyre, T. M., Prescott, S. M., Weyrich, A. S., et al. (2003). Cell-cell interactions: Leukocyte-endothelial interactions. *Curr Opin Hematol, 10,* 150–158.

58. Weyrich, A. S., McIntyre, T. M., McEver, R. P., et al. (1995). Monocyte tethering by P-selectin regulates monocyte chemotactic protein-1 and tumor necrosis factor-alpha secretion: Signal integration and NF-kappa B translocation. *J Clin Invest, 95,* 2297–2303.

59. Weyrich, A. S., Elstad, M. R., McEver, R. P., et al. (1996). Activated platelets signal chemokine synthesis by human monocytes. *J Clin Invest, 97,* 1525–1534.

60. Mahoney, T. S., Weyrich, A. S., Dixon, D. A., et al. (2001). Cell adhesion regulates gene expression at translational checkpoints in human myeloid leukocytes. *Proc Nat Acad Sci USA, 98,* 10284–10289.

61. del Conde, I., Nabi, F., Tonda, R., et al. (2005). Effect of P-selectin on phosphatidylserine exposure and surface-dependent thrombin generation on monocytes. *Arterioscler Thromb Vasc Biol, 25,* 1065–1070.

62. Goldsmith, H. L., Lichtarge, O., Tessier-Lavigne, M., et al. (1981). Some model experiments in hemodynamics: VI. Two-body collisions between blood cells. *Biorheology, 18,* 531–555.

63. Tangelder, G. J., Slaaf, D. W., Tierlinck, H. C., et al. (1982). Localization within a thin optical section of fluorescent blood platelets flowing in a microvessel. *Microvasc Res, 23,* 214–230.

64. Burger, P. C., & Wagner, D. D. (2003). Platelet P-selectin facilitates atherosclerotic lesion development. *Blood, 101,* 2661–2666.

65. Ulfman, L. H., Joosten, D. P., van Aalst, C. W., et al. (2003). Platelets promote eosinophil adhesion of patients with asthma to endothelium under flow conditions. *Am J Respir Cell Mol Biol, 28,* 512–519.

66. Iannacone, M., Sitia, G., Isogawa, M., et al. (2005). Platelets mediate cytotoxic T lymphocyte-induced liver damage. *Nat Med, 11,* 1167–1169.

67. Dole, V. S., Bergmeier, W., Mitchell, H. A., et al. (2005). Activated platelets induce Weibel-Palade-body secretion and leukocyte rolling *in vivo:* Role of P-selectin. *Blood, 106,* 2334–2339.

68. Weyrich, A. S., Lindemann, S., & Zimmerman, G. A. (2003). The evolving role of platelets in inflammation. *J Thromb Haemost, 1,* 1897–1905.

69. Elstad, M. R., McIntyre, T. M., Prescott, S. M., et al. (1995). The interaction of leukocytes with platelets in blood coagulation. *Curr Opin Hematol, 2,* 47–54.

70. Brandt, E., Ludwig, A., Petersen, F., et al. (2000). Platelet-derived CXC chemokines: Old players in new games. *Immunol Rev, 177,* 204–216.

71. Boehlen, F., & Clemetson, K. J. (2001). Platelet chemokines and their receptors: What is their relevance to platelet storage and transfusion practice? *Transfus Med, 11,* 403–417.

72. Weyrich, A. S., & Zimmerman, G. A. (2004). Platelets: Signaling cells in the immune continuum. *Trends Immunol, 25,* 489–495.

73. Weber, C. (2005). Platelets and chemokines in atherosclerosis: Partners in crime. *Circ Res, 96,* 612–616.

74. Clemetson, K. J., Clemetson, J. M., Proudfoot, A. E., et al. (2000). Functional expression of CCR1, CCR3, CCR4, and CXCR4 chemokine receptors on human platelets. *Blood, 96,* 4046–4054.

75. Kameyoshi, Y., Schroder, J. M., Christophers, E., et al. (1994). Identification of the cytokine RANTES released from platelets as an eosinophil chemotactic factor. *Int Arch Allergy Immunol, 104,* Suppl 1:49–51.

76. Klinger, M. H., Wilhelm, D., Bubel, S., et al. (1995). Immunocyto-chemical localization of the chemokines RANTES and MIP-1 alpha within human platelets and their release during storage. *Int Arch Allergy Immunol, 107,* 541–546.

77. von Hundelshausen, P., Weber, K. S., Huo, Y., et al. (2001). RANTES deposition by platelets triggers monocyte arrest on inflamed and atherosclerotic endothelium. *Circulation, 103,* 1772–1777.

78. Abi-Younes, S., Sauty, A., Mach, F., et al. (2000). The stromal cell-derived factor-1 chemokine is a potent platelet agonist highly expressed in atherosclerotic plaques. *Circ Res, 86,* 131–138.

79. Abi-Younes, S., Si-Tahar, M., & Luster, A. D. (2001). The CC chemokines MDC and TARC induce platelet activation via CCR4. *Thromb Res, 101,* 279–289.

80. Gear, A. R., Suttitanamongkol, S., Viisoreanu, D., et al. (2001). Adenosine diphosphate strongly potentiates the ability of the chemokines MDC, TARC, and SDF-1 to stimulate platelet function. *Blood, 97,* 937–945.

81. Antoniades, H. N., & Scher, C. D. (1997). Radioimmunoassay of a human serum growth factor for Balb/c-3T3 cells: Derivation from platelets. *Proc Nat Acad Sci USA, 74,* 1973–1977.

82. Mannaioni, P. F., Di Bello, M. G., & Masini, E. (1997). Platelets and inflammation: Role of platelet-derived growth factor, adhesion molecules and histamine. *Inflamm Res, 46,* 4–18.

83. Assoian, R. K., & Sporn, M. B. (1986). Type beta transforming growth factor in human platelets: Release during platelet degranulation and action on vascular smooth muscle cells. *J Cell Biol, 102,* 1217–1223.

84. Salganicoff, L., Hebda, P. A., Yandrasitz, J., et al. (1975). Subcellular fractionation of pig platelets. *Biochim Biophys Acta, 385,* 394–411.

85. Prescott, S. M., Zimmerman, G. A., Stafforini, D. M., et al. (2000). Platelet-activating factor and related lipid mediators. *Annu Rev Biochem, 69,* 419–445.

86. Rocca, B., & FitzGerald, G. A. (2002). Cyclooxygenases and prostaglandins: Shaping up the immune response. *Int Immunopharmacol, 2,* 603–630.

87. Thomas, D. W., Rocha, P. N., Nataraj, C., et al. (2003). Proinflammatory actions of thromboxane receptors to enhance cellular immune responses. *J Immunol, 171,* 6389–6395.

88. Lee, H., Lin, C. I., Liao, J. J., et al. (2004). Lysophospholipids increase ICAM-1 expression in HUVEC through a Gi- and NF-kappaB-dependent mechanism. *Am J Physiol Cell Physiol, 287,* C1657–1666.

89. Ozaki, H., Hla, T., & Lee, M. J. (2003). Sphingosine-1-phosphate signaling in endothelial activation. *J Atheroscler Thromb, 10,* 125–131.

90. Goetzl, E. J., Wang, W., McGiffert, C., et al. (2004). Sphingosine 1-phosphate and its G protein-coupled receptors constitute a multifunctional immunoregulatory system. *J Cell Biochem, 92,* 1104–1114.

91. O'Neill, L. A., & Greene, C. (1998). Signal transduction pathways activated by the IL-1 receptor family: Ancient signaling machinery in mammals, insects, and plants. *J Leukoc Biol, 63,* 650–657.

92. Lindemann, S., Tolley, N. D., Eyre, J. R., et al. (2001). Integrins regulate the intracellular distribution of eukaryotic initiation factor 4E in platelets: A checkpoint for translational control. *J Biol Chem, 276,* 33947–33951.

93. Hawrylowicz, C. M., Santoro, S. A., Platt, F. M., et al. (1989). Activated platelets express IL-1 activity. *J Immunol, 143,* 4015–4018.

94. Kaplanski, G., Porat, R., Aiura, K., et al. (1993). Activated platelets induce endothelial secretion of interleukin-8 *in vitro* via an interleukin-1-mediated event. *Blood, 81,* 2492–2495.

95. Loppnow, H., Bil, R., Hirt, S., et al. (1998). Platelet-derived interleukin-1 induces cytokine production, but not proliferation of human vascular smooth muscle cells. *Blood, 91,* 134–141.

96. Gawaz, M., Brand, K., Dickfeld, T., et al. (2000). Platelets induce alterations of chemotactic and adhesive properties of endothelial cells mediated through an interleukin-1-dependent mechanism. Implications for atherogenesis. *Atherosclerosis, 148,* 75–85.

97. Hawrylowicz, C. M., Howells, G. L., & Feldmann, M. (1991). Platelet-derived interleukin 1 induces human endothelial adhesion molecule expression and cytokine production. *J Exp Med, 174,* 785–790.

98. Dinarello, C. A. (2002). The IL-1 family and inflammatory diseases. *Clin Exp Rheumatol, 20,* S1–S13.

99. Barry, O. P., Pratico, D., Savani, R. C., et al. (1998). Modulation of monocyte-endothelial cell interactions by platelet microparticles. *J Clin Invest, 102,* 136–144.

100. Palabrica, T., Lobb, R., Furie, B. C., et al. (1992). Leukocyte accumulation promoting fibrin deposition is mediated *in vivo* by P-selectin on adherent platelets. *Nature, 359,* 848–851.

101. Subramaniam, M., Frenette, P. S., Saffaripour, S., et al. (1996). Defects in hemostasis in P-selectin-deficient mice. *Blood, 87,* 1238–1242.

102. Andre, P., Hartwell, D., Hrachovinova, I., et al. (2000). Pro-coagulant state resulting from high levels of soluble P-selectin in blood. *Proc Nat Acad Sci USA, 19,* 13835–13840.

103. Sullivan, V. V., Hawley, A. E., Farris, D. M., et al. (2003). Decrease in fibrin content of venous thrombi in selectin-deficient mice. *J Surg Res, 109,* 1–7.

104. Myers, D. D., Hawley, A. E., Farris, D. M., et al. (2003). P-selectin and leukocyte microparticles are associated with venous thrombogenesis. *J Vasc Surg, 38,* 1075–1089.

105. Giesen, P. L., Rauch, U., Bohrmann, B., et al. (1999). Blood-borne tissue factor: Another view of thrombosis. *Proc Nat Acad Sci USA, 96,* 2311–2315.

106. Hrachovinova, I., Cambien, B., Hafezi-Moghadam, A., et al. (2003). Interaction of P-selectin and PSGL-1 generates microparticles that correct hemostasis in a mouse model of hemophilia A. *Nat Med, 9,* 1020–1025.

107. Rauch, U., Bonderman, D., Bohrmann, B., et al. (2000). Transfer of tissue factor from leukocytes to platelets is mediated by CD15 and tissue factor. *Blood, 96,* 170–175.

108. Falati, S., Liu, Q., Gross, P., et al. (2003). Accumulation of tissue factor into developing thrombi *in vivo* is dependent upon microparticle P-selectin glycoprotein ligand 1 and platelet P-selectin. *J Exp Med, 197,* 1585–1598.

109. Chou, J., Mackman, N., Merrill-Skoloff, G., et al. (2004). Hematopoietic cell-derived microparticle tissue factor contributes to fibrin formation during thrombus propagation. *Blood, 104,* 3190–3197.

110. Day, S. M., Reeve, J. L., Pedersen, B., et al. (2005). Macrovascular thrombosis is driven by tissue factor derived primarily from the blood vessel wall. *Blood, 105,* 192–198.

111. Michelson, A. D., Barnard, M. R., Hechtman, H. B., et al. (1996). *In vivo* tracking of platelets: Circulating degranulated platelets rapidly lose surface P-selectin but continue to circulate and function. *Proc Nat Acad Sci USA, 93,* 11877–11882.

112. Berger, G., Hartwell, D. W., & Wagner, D. D. (1998). P-Selectin and platelet clearance. *Blood, 92,* 4446–4452.

113. Blann, A. D., Dobrotova, M., Kubisz, P., et al. (1995). von Willebrand factor, soluble P-selectin, tissue plasminogen activator and plasminogen activator inhibitor in atherosclerosis. *Thromb Haemost, 74,* 626–630.

114. Smith, A., Quarmby, J. W., Collins, M., et al. (1999). Changes in the levels of soluble adhesion molecules and coagulation factors in patients with deep vein thrombosis. *Thromb Haemost, 82,* 1593–1599.

115. Chong, B. H., Murray, B., Berndt, M. C., et al. (1994). Plasma P-selectin is increased in thrombotic consumptive platelet disorders. *Blood, 83,* 1535–1541.

116. Wu, G., Li, F., Li, P., et al. (1993). Detection of plasma alpha-granule membrane protein GMP-140 using radiolabeled monoclonal antibodies in thrombotic diseases. *Haemostasis, 23,* 121–128.

117. Ikeda, H., Takajo, Y., Ichiki, K., et al. (1995). Increased soluble form of P-selectin in patients with unstable angina. *Circulation, 92,* 1693–1696.

118. Tsakiris, D. A., Tschopl, M., Jager, K., et al. (1999). Circulating cell adhesion molecules and endothelial markers before and after transluminal angioplasty in peripheral arterial occlusive disease. *Atherosclerosis, 142,* 193–200.

119. Ikeda, H., Nakayama, H., Oda, T., et al. (1994). Soluble form of P-selectin in patients with acute myocardial infarction. *Coron Artery Dis, 5,* 515–518.

120. Korkmaz, S., Ileri, M., Hisar, I., et al. (2001). Increased levels of soluble adhesion molecules, E-selectin and P-selectin, in patients with infective endocarditis and embolic events. *Eur Heart J, 22,* 874–878.

121. Ridker, P. M., Buring, J. E., & Rifai, N. (2001). Soluble P-selectin and the risk of future cardiovascular events. *Circulation, 103,* 491–495.

122. Hillis, G. S., Terregino, C., Taggart, P., et al. (2002). Elevated soluble P-selectin levels are associated with an increased risk of early adverse events in patients with presumed myocardial ischemia. *Am Heart J, 143,* 235–241.

123. Satta, N., Toti, F., Feugeas, O., et al. (1994). Monocyte vesiculation is a possible mechanism for dissemination of membrane-associated procoagulant activities and adhesion molecules after stimulation by lipopolysaccharide. *J Immunol, 153,* 3245–3255.

124. VanWijk, M. J., VanBavel, E., Sturk, A., et al. (2003). Microparticles in cardiovascular diseases. *Cardiovasc Res, 59,* 277–287.

125. Mesri, M., & Altieri, D. C. (1998). Endothelial cell activation by leukocyte microparticles. *J Immunol, 161,* 4382–4387.

126. Mesri, M., & Altieri, D. C. (1999). Leukocyte microparticles stimulate endothelial cell cytokine release and tissue factor induction in a JNK1 signaling pathway. *J Biol Chem, 274,* 23111–23118.

127. Sabatier, F., Roux, V., Anfosso, F., et al. (2002). Interaction of endothelial microparticles with monocytic cells *in vitro* induces tissue factor-dependent procoagulant activity. *Blood, 99,* 3962–3970.

128. Del Conde, I., Shrimpton, C. N., Thiagarajan, P., et al. (2005). Tissue-factor-bearing microvesicles arise from lipid rafts and fuse with activated platelets to initiate coagulation. *Blood, 106,* 1604–1611.

129. Elzey, B. D., Grant, J. F., Sinn, H. W., et al. (2005). Cooperation between platelet-derived CD154 and CD4+ T cells for enhanced germinal center formation. *J Leukoc Biol, 78,* 80–84.

130. Damas, J. K., Otterdal, K., Yndestad, A., et al. (2004). Soluble CD40 ligand in pulmonary arterial hypertension: Possible pathogenic role of the interaction between platelets and endothelial cells. *Circulation, 110,* 999–1005.

131. Andre, P., Prasad, K. S., Denis, C. V., et al. (2002). CD40L stabilizes arterial thrombi by a beta3 integrin — dependent mechanism. *Nat Med, 8,* 247–252.

132. Vishnevetsky, D., Kiyanista, V. A., & Gandhi, P. J. (2004). CD40 ligand: A novel target in the fight against cardiovascular disease. *Ann Pharmacother, 38,* 1500–1508.

133. Prasad, K. S., Andre, P., He, M., et al. (2003). Soluble CD40 ligand induces beta3 integrin tyrosine phosphorylation and triggers platelet activation by outside-in signaling. *Proc Nat Acad Sci USA, 100,* 12367–12371.

134. Furman, M. I., Benoit, S. E., Barnard, M. R., et al. (1998). Increased platelet reactivity and circulating monocyte-platelet aggregates in patients with stable coronary artery disease. *J Am Coll Cardiol, 31,* 352–358.

135. Furman, M. I., Barnard, M. R., Krueger, L. A., et al. (2001). Circulating monocyte-platelet aggregates are an early marker of acute myocardial infarction. *J Am Coll Cardiol, 38,* 1002–1006.

136. Michelson, A. D., Barnard, M. R., Krueger, L. A., et al. (2001). Circulating monocyte-platelet aggregates are a more sensitive marker of *in vivo* platelet activation than platelet surface P-selectin: Studies in baboons, human coronary intervention, and human acute myocardial infarction. *Circulation, 104,* 1533–1537.

137. Grau, A. J., Ruf, A., Vogt, A., et al. (1998). Increased fraction of circulating activated platelets in acute and previous cerebrovascular ischemia. *Thromb Haemost, 80,* 298–301.

138. Zeller, J. A., Tschoepe, D., & Kessler, C. (1999). Circulating platelets show increased activation in patients with acute cerebral ischemia. *Thromb Haemost, 81,* 373–377.

139. Meiklejohn, D. J., Vickers, M. A., Morrison, E. R., et al. (2001). *In vivo* platelet activation in atherothrombotic stroke is not determined by polymorphisms of human platelet glycoprotein IIIa or Ib. *Br J Haematol, 112,* 621–631.

140. Zeiger, F., Stephan, S., Hoheisel, G., et al. (2000). P-Selectin expression, platelet aggregates, and platelet-derived microparticle formation are increased in peripheral arterial disease. *Blood Coagul Fibrinolysis, 11,* 723–728.

141. Cassar, K., Bachoo, P., Ford, I., et al. (2003). Platelet activation is increased in peripheral arterial disease. *J Vasc Surg, 38,* 99–103.

142. Peyton, B. D., Rohrer, M. J., Furman, M. I., et al. (1998). Patients with venous stasis ulceration have increased monocyte-platelet aggregation. *J Vasc Surg, 27,* 1109–1115; discussion 1115–1106.

143. Powell, C. C., Rohrer, M. J., Barnard, M. R., et al. (1999). Chronic venous insufficiency is associated with increased platelet and monocyte activation and aggregation. *J Vasc Surg, 30,* 844–851.

144. Tschoepe, D., Roesen, P., Esser, J., et al. (1991). Large platelets circulate in an activated state in diabetes mellitus. *Semin Thromb Hemost, 17,* 433–438.

145. Jensen, M. K., de Nully Brown, P., Lund, B. V., et al. (2001). Increased circulating platelet-leukocyte aggregates in myeloproliferative disorders is correlated to previous thrombosis, platelet activation and platelet count. *Eur J Haematol, 66,* 143–151.

146. Villmow, T., Kemkes-Matthes, B., & Matzdorff, A. C. (2000). Markers of platelet activation and platelet-leukocyte interaction in patients with myeloproliferative syndromes. *Thromb Res, 108,* 139–145.

147. Sevush, S., Jy, W., Horstman, L. L., et al. (1998). Platelet activation in Alzheimer disease. *Arch Neurol, 55,* 530–536.

148. Janes, S. L., & Goodall, A. H. (1994). Flow cytometric detection of circulating activated platelets and platelet hyper-responsiveness in pre-eclampsia and pregnancy. *Clin Sci (Lond), 86,* 731–739.

149. Konijnenberg, A., van der Post, J. A., Mol, B. W., et al. (1997). Can flow cytometric detection of platelet activation early in pregnancy predict the occurrence of preeclampsia? A prospective study. *Am J Obstet Gynecol, 177,* 434–442.

150. Trudinger, B., Song, J. Z., Wu, Z. H., et al. (2003). Placental insufficiency is characterized by platelet activation in the fetus. *Obstet Gynecol, 101,* 975–981.

151. Sirolli, V., Ballone, E., Garofalo, D., et al. (2002). Platelet activation markers in patients with nephrotic syndrome: A comparative study of different platelet function tests. *Nephron, 91,* 424–430.

152. Gawaz, M. P., Mujais, S. K., Schmidt, B., et al. (1999). Platelet-leukocyte aggregates during hemodialysis: Effect of membrane type. *Artif Organs, 23,* 29–36.

153. Wun, T., Cordoba, M., Rangaswami, A., et al. (2002). Activated monocytes and platelet-monocyte aggregates in patients with sickle cell disease. *Clin Lab Haematol, 24,* 81–88.

154. Ogura, H., Kawasaki, T., Tanaka, H., et al. (2001). Activated platelets enhance microparticle formation and platelet-leukocyte interaction in severe trauma and sepsis. *J Trauma, 50,* 801–809.

155. Gawaz, M., Dickfeld, T., Bogner, C., et al. (1997). Platelet function in septic multiple organ dysfunction syndrome. *Intensive Care Med, 23,* 379–385.

156. Russwurm, S., Vickers, J., Meier-Hellmann, A., et al. (2002). Platelet and leukocyte activation correlate with the severity of septic organ dysfunction. *Shock, 17,* 263–268.

157. Joseph, J. E., Harrison, P., Mackie, I. J., et al. (2001). Increased circulating platelet-leucocyte complexes and platelet activation in patients with antiphospholipid syndrome, systemic lupus erythematosus and rheumatoid arthritis. *Br J Haematol, 115,* 451–459.

158. Collins, C. E., Cahill, M. R., Newland, A. C., et al. (1994). Platelets circulate in an activated state in inflammatory bowel disease. *Gastroenterology, 106,* 840–845.

159. Burns, J. C., Glode, M. P., Clarke, S. H., et al. (1984). Coagulopathy and platelet activation in Kawasaki syndrome: Identification of patients at high risk for development of coronary artery aneurysms. *J Pediatr, 105,* 206–211.

160. Sullivan, P. J., Jafar, Z. H., Harbinson, P. L., et al. (2000). Platelet dynamics following allergen challenge in allergic asthmatics. *Respiration, 67,* 514–517.

161. O'Sullivan, B. P., Linden, M. D., Frelinger, A. L., 3rd, et al. (2005). Platelet activation in cystic fibrosis. *Blood, 105,* 4635–4641.

162. Wagner, D. D., & Burger, P. C. (2003). Platelets in inflammation and thrombosis. *Arterioscler Thromb Vasc Biol, 23,* 2131–2137.

# CHAPTER 40

# Antimicrobial Host Defense

**Michael R. Yeaman and Arnold S. Bayer**

*Department of Medicine, David Geffen School of Medicine at University of California, Los Angeles (UCLA), Los Angeles Biomedical Research Institute, LAC-Harbor UCLA Medical Center, Torrance, California*

## I. Introduction

Platelets have long been recognized for their role in maintaining hemostasis and contributing to wound healing. Yet platelets have historically been underappreciated for their multiple and integral functions in antimicrobial host defense. Structural and functional features continue to reinforce the concept that platelets have explicit functions in antimicrobial host defense. Platelets express a diverse array of constitutive and inducible membrane receptors. Thus, platelets have key attributes of sentinel cells that are highly sensitive and responsive to a broad spectrum of agonists associated with microbial contamination, infection, or tissue inflammation. Upon activation, platelets transform from quiescent discoid forms to amoeboid cells that recognize and adhere to microbial pathogens or ligands displayed by tissues injured by trauma or infection.

Platelets are now well recognized for their role(s) in antimicrobial immunity. In this regard, platelets share many structural and functional archetypes with leukocytes known to participate in immunoprotection against infection. For example, platelet activation leads to the display of membrane receptors that respond to soluble and particulate agonists generated in the setting of tissue injury or infection. Once activated, platelets respond in specific ways that emphasize their likely multiple roles in antimicrobial host defense, including a dramatic increase in their metabolic status, and change from discoid to amoeboid structure; vectored motility toward sites of infection; display of receptors mediating increased adhesivity to injured or infected tissues; projection of pseudopodia that may promote interiorization or clearance of microbial pathogens from the blood stream; production of antimicrobial oxygen species such as superoxide, peroxide, and hydroxyl radicals; cytoskeletal rearrangements facilitating granule mobilization and organization; and degranulation and release of preformed and processed antimicrobial molecules, including peptides. In

these ways, platelets exhibit hallmark structural and functional characteristics reflective of cells that function in host defense against infection. Furthermore, a body of evidence suggests that platelet ligands and degranulation products recruit professional phagocytes to sites of trauma or infection and potentiate the antimicrobial mechanisms of these cells. Studies have characterized the structural and mechanistic features of platelet microbicidal proteins and kinocidins (microbicidal chemokines). These findings indicate that these effector molecules are multifunctional, bridging molecular and cellular innate host defense against infection. Therefore, platelets likely play significant roles in antimicrobial host defense through both direct and indirect mechanisms.

## II. Mammalian Platelets are Multipurpose Inflammatory Cells

Invertebrate hemocytes serve as multipurpose inflammatory cells (Chapter 1). These cells prevent loss of hemolymph following tissue injury, interact with and clear microorganisms introduced by trauma, and initiate the process of wound repair and tissue remodeling.[1,2] In higher organisms, distinct cell types, thrombocytes and leukocytes, have classically been viewed as mediating hemostasis and inflammation, respectively.[3,4] Mammalian platelets represent further specialization. Although one principal function of platelets is undoubtedly the maintenance of hemostasis, mammalian platelets have retained features of inflammatory cells likely corresponding to multiple functions in antimicrobial host defense.

As described in detail in Chapters 1–3, mammalian platelets are small (2–4 μm), short-lived discoid cells derived from the megakaryocyte lineage.[5–9] Although platelets are devoid of nuclei, young platelets perform limited translation using stable megakaryocyte mRNA templates.[10–12] Platelets contain three distinct cytoplasmic granule types (Chapters 3 and 15). Dense δ (δ) granules predominantly store mediators of vascular tone, including serotonin, adenosine diphosphate (ADP), calcium, and phosphate.[3,5,13–15] In contrast, alpha (α) granules contain proteins that are important in hemostatic functions, including adhesion (e.g., fibrinogen, thrombospondin, vitronectin, laminin, and von Willebrand factor), modulation of coagulation (e.g., plasminogen and $\alpha_2$-plasmin inhibitor), and endothelial cell repair (e.g., platelet-derived growth factor [PDGF], permeability factor, and transforming growth factors α and β [TGF-α and TGF-β]).[3,13–15] As will be discussed in detail here, platelet α-granules also contain an arsenal of microbicidal proteins. Lysosomal λ (λ) granules contain enzymes that mediate thrombus dissolution.[3,5,14] These distinct platelet granules are subject to differential release, dependent upon agonist specificity and potency. For example, low levels of thrombin or ADP induce δ and α degranulation, whereas λ granules are not secreted until these agonists are present at much higher concentrations.[14–17] Based on their collective structural features, platelets may be viewed as vehicles that recognize and respond to agonists and ligands expressed at sites of endovascular damage or infection and release a diverse array of bioactive molecules in these settings.[3,15–18]

### A. Platelets Recognize Endothelium Damaged by Trauma or Microbial Colonization

Changes in vascular endothelial tone resulting from trauma or infection may prompt platelet responses. Platelets exhibit specific receptors that sense agonists and bind to ligands associated with activated endothelial cells or exposed subendothelial stroma resulting from injury or infection. Ligands recognized by platelet membrane glycoprotein (GP) receptors include collagen (GPVI and GPIa-IIa [VLA-2]), fibronectin (GPIc-IIa [VLA-5]), von Willebrand factor (GPIb-IX-V), laminin (GPIc-IIa [VLA-6]), vitronectin (integrin αVβ3), and thrombin (Chapters 6, 7, 9, and 18).[14,19,20,21] Contact with blood prompts the subendothelial stroma to release tissue factor.[3,22–24] Tissue factor facilitates an ensuing cascade of proteolytic reactions catalyzing thrombin generation.[3,22–24] Thrombin is among the most potent of all platelet agonists (Chapter 9).[14,19] As a result of platelet activation, P-selectin (CD62P) is exposed on the platelet plasma membrane (Chapter 12), and the platelet fibrinogen receptor (integrin αIIbβ3 [GPIIb-IIIa]) changes conformation to its activated form (Chapter 8).[3,14,19] Activated platelets undergo a rapid shape change from discoid to amoeboid form and exhibit extensive microtubule assembly and granule reorganization[25,26] (Chapters 3 and 4). Degranulation may ensue, liberating, among other agonists,

ADP from dense granules and generating thromboxane $A_2$ and platelet activating factor (PAF) through activation of membrane phospholipase $A_2$.[3,14–16] These potent agonists activate subsequent waves of platelets at sites of tissue injury or microbial colonization.[5,27] Thus, platelets recognize and respond rapidly to sites of endothelial activation resulting from trauma or microbial colonization.

### B. Platelets Accumulate at Sites of Infected Endovascular Lesions

Platelets are the earliest and most abundant cells that accumulate at sites of vascular infection.[28–38] Such endovascular infections include infective endocarditis, suppurative thrombophlebitis, mycotic aneurysm, septic endarteritis, catheter and dialysis access site infections, and infections of vascular prostheses and stents.[34,38] The rapid and numerically significant presence of platelets at these sites has been well established. In 1886, Osler reported perhaps the earliest observations of platelets rapidly coating filaments placed in animal veins.[39] More recently, Cheung and Fischetti[40] and others[33,38] have shown that platelets are the first cells to adhere to indwelling vascular catheters. Furthermore, platelets rapidly deposit on cardiac valve prostheses and endovascular stents.[38] In addition, platelets are the earliest and most abundant cells in endocarditis vegetations in rabbits and humans.[28–31,41] Thus, platelets rapidly target foreign surfaces, as well as sites of vascular endothelium that have been damaged or colonized by microorganisms.

## III. Early Studies Were Interpreted to Suggest Platelets Promote Infection

The natural history of endovascular infection, particularly when involving highly pathogenic organisms such as *Staphylococcus aureus*, has conventionally been viewed as follows: (a) adherence of hematogenous pathogens to normal or predisposed vascular endothelium (e.g., rheumatic heart disease); (b) endothelial cell expression of agonists or ligands that promote platelet activation, deposition, and platelet–fibrin matrix formation[3,5,42,43]; and (c) further deposition of circulating platelets in response to successive waves of agonists (e.g., thromboxane $A_2$, ADP, PAF) or ligands (e.g., integrin αIIbβ3) associated with secondary platelet activation.[14,20,21] Less virulent pathogens, including viridans streptococci, are believed to require altered or damaged endothelial surfaces to initiate endovascular infection through pathways that involve: (a) fibrinogen-mediated adherence to intact or abnormal vascular endothelium[5,23,24,32]; (b) vascular endothelial cell expression of platelet ligands and agonists (noted earlier); and (c) further platelet–fibrin deposition and evolution of the lesion.[20,21,32–38] In either case, platelets represent the most rapid and abundant inflammatory cells that respond to endothelial cell damage or microbial colonization. Thus, platelets may be viewed as the earliest of opportunities for inflammatory responses to intercede in microbial pathogenesis and effect antimicrobial host defense against induction and evolution of endovascular infection.

It is clear that platelets are present at sites of infection involving the vascular endothelium. The historic interpretation of their role in these settings has contended that platelets promote the establishment and evolution of endovascular infection.[28–31,35,44–46] For example, studies by several investigators have suggested that platelets facilitate microbial adhesion to fibrin matrices or endothelial cells *in vitro*.[28,30,46] Additionally, more recent data have demonstrated that both bacterial and fungal pathogens can aggregate human and rabbit platelets *in vitro*.[47–50] Herzberg et al. have suggested that streptococcal binding to and aggregation of platelets may be associated with increased virulence of these strains in experimental animal models of endocarditis.[51] Furthermore, platelets and microorganisms interact through specific ligands[48,49] (discussed later), suggesting pathogens may exploit platelets as adhesive sites via molecular mimicry. Platelet aggregation has also been suggested to be detrimental to the host, because massive endovascular vegetations are often associated with clinical complications such as emboli and infarcts.[35,52] Some investigators have suggested that platelet aggregation and internalization of microorganisms may protect pathogens from exposure to antibiotics or clearance by neutrophils or other leukocytes.[3,53]

This evidence demonstrates that platelets interact with microorganisms in response to endovascular infection. However, there are no studies that substantiate the concept that platelets inherently facilitate microbial pathogenesis (pathogen survival, endothelial cell penetration, or dissemination into deeper tissue parenchyma) or inhibit the immune response. On the contrary, cells that promote infection would likely be highly disadvantageous from an evolutionary perspective.

## IV. Platelets Likely Play Multiple and Key Roles in Antimicrobial Host Defense

A compelling body of evidence now exists indicating that platelets defend against infection. Thus, platelet interactions with microorganisms likely serve an important role in antimicrobial host defense. The following discussion focuses on observations that point to key functions for platelets in limiting the establishment and progression of infection *in vivo*.

### A. Platelets Exhibit Structural and Functional Archetypes of Antimicrobial Effector Cells

The concept that platelets may protect against infection is supported by a number of specific observations related to platelet structure and function (Fig. 40-1). For example, platelets share common surface antigens with professional phagocytes such as neutrophils and monocytes. These include the 40 kDa FcγRII receptor,[54] the Fcε receptor for IgE,[55] the C-reactive protein receptor,[56] and the thrombospondin receptor CD36 (platelet GPIV).[57] Platelets are also known to express the complement CR3 receptor,[58] receptors for $C_{3a}$ and $C_{5a}$ generated from either classical or alternative pathway complement fixation, and to respond to cytokines such as tumor necrosis factor-α (TNF-α), interleukin-1 (IL-1), and IL-6 in a manner similar to leukocytes.[59,60] Clemetson et al. have shown that human platelets express functional cystine–cystine chemokine receptors (CCR) CCR1, CCR3, CCR4, and cystine-X-cystine chemokine receptors (CXCR), such as CXCR4.[61] These findings further implicate platelets in coordinated navigation to sites of tissue injury related to infection. Also, like professional phagocytes, the platelet cytoplasm is rich in granules that contain bioactive molecules.[3,5,20,21] These relationships demonstrate that platelets possess structural features consistent with a role in antimicrobial host defense.

Platelets also perform functions that reflect their similarities to cell-mediated immune effector cells. *In vitro* and in experimental animal models, platelets have been shown to accumulate at sites enriched in microbial stimuli such as N-formyl-methionyl-leucyl-phenylalanine (N-*f*-met-leu-phe)[62] or complement proteins $C_{3a}$ and $C_{5a}$.[3,58,63] In this regard, platelets exhibit positive chemotactic responses to stimuli present in the setting of infection. Platelets are capable of interacting directly and indirectly with a broad variety of microbial pathogens both *in vitro* and *in vivo* (discussed later). Platelets can internalize microorganisms into phagosome-like vacuoles, likely enhancing pathogen clearance from the blood stream.[53,64,65] Platelets stimulated with microbial pathogens or inflammatory stimuli generate oxidative molecules, such as superoxide anion, hydrogen peroxide, hydroxyl radical, and lipoperoxides.[66,67] Thus, platelets respond to these stimuli by producing antimicrobial oxygen metabolites. Similar to neutrophils and macrophages, granules within activated platelets are mobilized by microtubule assembly and are subsequently secreted (Chapters 3 and 4). However, in contrast to leukocyte degranulation, directed toward intracellular phagolysosomes, platelet degranulation liberates the majority of granule contents to the extracellular milieu. Numerous investigators have reported that platelets inactivate, inhibit the growth of, or kill viral, bacterial, and fungal pathogens (discussed later). In the presence of IgE or C-reactive protein, platelets exhibit

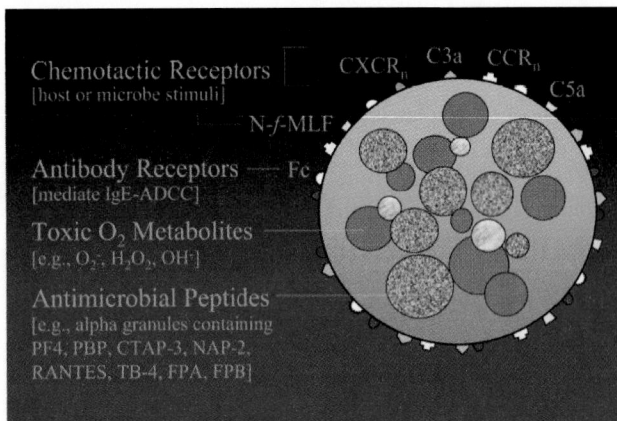

**Figure 40-1.** Platelet antimicrobial archetypes. Platelets exhibit structural and functional properties indicative of their multiple roles in antimicrobial host defense, including the following: *chemotactic receptors* — host (e.g., CXC or CC chemokines, complement fixation products C3a or C5a) or microbe (e.g., N-*f*MetLeuPhe sequences) components serve as positive chemoattractants that guide platelet navigation toward sites of microbial infection; *antibody receptors* — receptors (e.g., FcγRII) for Fc domains of antibody that has recognized antigens) may promote platelet interaction with microorganisms or toxins thereof, yielding opsonization, neutralization, or antibody-dependent cell cytotoxicity (ADCC); *toxic oxygen metabolites* — derivatives of oxygen that have potent antimicrobial activity (including superoxide, peroxide, and hydroxyl radicals) are generated from activated platelets in response to microbial components, as well as host cell-derived inflammatory stimuli, antibody recognition, and soluble agonists (e.g., platelet activating factor [PAF]); *antimicrobial peptides* — platelets are now known to release an array of proteins and peptides that exert direct antimicrobial activity, many of which are microbicidal chemokines (e.g., platelet factor 4 [PF4], platelet basic protein [PBP], connective tissue activating peptide-3 [CTAP-3], neutrophil activating peptide-2 [NAP-2], "released upon activation, normal T cell expressed and secreted" [RANTES], thymosin-beta-4 [TB-4], and fibrinopeptides A and B [FPA and FPB]). Collectively, these structural and functional features of platelets reflect their antimicrobial archetypes similar to those in leukocytes known to act in host defense against infection.

cytotoxic capabilities against microfilariae and *Schistosoma*, respectively[55,58,68,69] (as will be outlined later). Taken together, these studies strongly support the concept that platelets possess explicit antimicrobial functions.

The structures and functions of platelets as they relate to antimicrobial host defense have been the focus of reviews.[20,21] It is intriguing to consider the likelihood that mammalian platelets have retained these antimicrobial features over the course of their phylogenetic history. For example, many invertebrates and insects possess hemocytes that maintain hemostasis and simultaneously serve as the professional phagocytes for these organisms (see Chapter 1 for a full discussion). From this perspective, it is highly likely that

platelets represent efficient, multipurpose antimicrobial host defense cells. The following discussion will focus on the key structural and functional attributes that mediate the multiple antimicrobial functions of platelets: (a) Platelets respond rapidly to sites of tissue injury, which also have a high likelihood for contamination by microbial pathogens. (b) Platelets directly interact with many microbial pathogens, and platelets are activated by soluble stimuli and microbial components present at these sites or in the blood stream. (c) Activated platelets express inducible receptors, and thus become more adhesive for ligands expressed by injured endothelial cells and subendothelial stroma. (d) Platelet degranulation products integrate into the intrinsic and extrinsic coagulation cascades that prevent exsanguination due to the initial wound or injury. (e) Degranulation of activated platelets also releases a diverse array of potent antimicrobial peptides that exert direct microbicidal activities in this local setting. (f) Many of these antimicrobial peptides are also chemokines that simultaneously recruit leukocytes to respond to these sites in antimicrobial host defense. (g) Activated platelets express ligands (e.g., P-selectin) that assist in leukocyte recognition and adhesion to injured or infected tissues. (h) The microbicidal chemokines released from platelets likely interact with microorganisms, resulting in activation and potentiation of the ensuing antimicrobial mechanisms of recruited leukocytes. Thus, platelets play key and multifunctional roles that significantly contribute to antimicrobial host defense.

### B. Platelets Recognize and Interact Directly and Indirectly with Microbial Pathogens

**1. Platelets Interact with Viruses.** There is a compelling body of evidence substantiating the fact that viruses interact with and are internalized by megakaryocytes and platelets.[70–73] For example, Friend leukemia,[74,75] influenza,[76] measles,[3,73] Newcastle disease,[77] vaccinia,[73] and herpes[3,73] viruses enter megakaryocytes and platelets. Additionally, megakaryocytic cell lines have been infected with human immunodeficiency virus (HIV) *in vitro*.[70,71,78,79] Vaccinia virus appears to adhere to human platelets via electrostatic interactions and prompts platelets to undergo δ degranulation.[3,73] Some investigators have reported that viral replication occurs within the megakaryocyte cytoplasm, likely due to the vigorous metabolic status of these cells.[73,78,79] Although platelets lack significant macromolecular biosynthesis, several studies have documented viral budding from platelet internal membranes.[72,73] Subsequent to internalization, viral antigens appear on the surface of the platelet. For example, influenza virus hemagglutinin is expressed as a membrane-bound complex on the platelet surface.[3,73,76]

The potential consequences of platelet interactions with viral pathogens is an intriguing, yet poorly understood, phenomenon. For example, thrombocytopenia is frequently associated with viral infection[72,73] and appears to occur via one of two primary mechanisms: (a) increased lysis or destruction of megakaryocytes or platelets due to viral-induced injury or (b) destruction of virus-infected platelets by immune mechanisms. Brown and Axelrad[80] have shown that megakaryocytes are significantly depleted within days subsequent to infection of mice with Friend leukemia virus. Likewise, Oski and Naiman[81] reported that the number of human megakaryocytes reaches a nadir 3 days following vaccination with live measles vaccine. In each of these cases, primary thrombocytopenia was observed within 5 days after the initial viral challenge. Thus, primary thrombocytopenia associated with acute viral infection is believed to occur due to viral injury of megakaryocytes or platelets rather than from secondary immune clearance of infected cells. Along these lines, myxoviral neuraminidase has been demonstrated to reduce platelet life span by cleaving platelet membrane sialic acid.[76,77] Newcastle virus likewise appears to disrupt the platelet membrane, leading to lysis.[77,82] Ultrastructural studies have also shown that megakaryocytes and platelets of patients with acquired immunodeficiency syndrome (AIDS) exhibit HIV-induced nuclear and intracytoplasmic damage, respectively[70,71,79] (Chapter 47). Alternatively, viruses that do not proliferate within the megakaryocyte, such as human parvovirus-19, may nonetheless interfere with platelet production, yielding thrombocytopenia.[83] Additionally, viruses adsorbed onto, or internalized within, platelets may prompt aggregation and degranulation, facilitating a subsequent coagulation cascade and further consumption of platelets.

Immune thrombocytopenia (ITP) may also result from viral infection (Chapter 46). In contrast to primary thrombocytopenia, which commonly occurs within the first few days of infection, viral-induced ITP typically emerges 7 to 10 days after the primary symptoms of the infection have been manifested. Therefore, in many cases, ITP is manifested beginning approximately 14 to 20 days subsequent to the initial viral infection. The molecular or physiological mechanisms responsible for ITP have not been well defined. Anti-viral IgG appears on the surface of platelets of virus-infected patients.[73,84–87] The mechanisms for this adsorption of IgG onto platelets involve either binding of antibody to viral antigens expressed on the platelet membrane, antibody-virus complexes binding to platelets, or both. Supranormal levels of IgG are associated with platelets from patients with ITP as compared with healthy individuals.[72,73] Regardless of the precise mechanisms, platelet-associated IgG and platelet quantity exhibit an inverse relationship.[72,73] This effect is likely due to both viral-infected megakaryocytes and platelet accumulation in the spleen and bone marrow.[75] Viral antigens, viral-platelet complexes, and excessive levels of platelet surface-associated IgG are believed to mediate platelet destruction and clearance via the reticuloendothelial

system. In addition, platelets that exhibit sufficiently high levels of surface-bound IgG or viral antigens may be destroyed via activation of the classical or alternative pathway of complement fixation, respectively.[87] Dramatic examples of ITP resulting from immune-mediated platelet damage and thrombocytopenia from direct platelet and megakaryocyte injury may be observed in hemorrhagic viral infections such as dengue fever or hantavirus infection.[88]

**2. Platelets Interact with Bacteria.** Mammalian platelets interact directly and indirectly with bacterial pathogens *in vitro* and *in vivo*. Perhaps the most systematic studies performed in this regard are those in a series by Clawson and White.[53,89–91] In these studies, a broad spectrum of bacterial pathogens were shown to be capable of binding to, aggregating, and inducing secretory degranulation of platelets. In general, the sequence of events associated with platelet-bacterial interaction proceeds through successive and distinct phases: (a) contact, (b) shape change, (c) early aggregation, and (d) irreversible aggregation.[89] Over the evolution of this process, platelets undergo a shape change from smooth discoid cells to amoeboid forms with pseudopodia.[20,21,90] This structural change typically occurs prior to the onset of platelet aggregation. Stimulation by bacterial cells or antigens may also lead to changes within the platelet cytoplasm. For example, coincident with shape transition, bacterial-stimulated platelets exhibit microtubule reorganization, such that granules are mobilized and relocated from the perimeter to the center of the cytoplasm.[53,90] Granule centralization immediately precedes ensuing secretion of platelet granules.[20,21,90] Additionally, studies by Saluk–Juszczak et al. demonstrate that endotoxin from *Proteus mirabilis* strains stimulates platelets to generate superoxide radicals.[92]

Bacterial pathogens display considerable variation in their ability to adhere to, aggregate, and degranulate platelets. Clawson and White demonstrated that *Staphylococcus aureus, Streptococcus pyogenes, Enterococcus faecalis,* and *Escherichia coli* rapidly bind to and aggregate human platelets *in vitro*.[53,89–91] However, in the same studies, *Staphylococcus epidermidis* and *Streptococcus pneumoniae* (type 24) failed to stimulate irreversible platelet aggregation.[89] Other investigators have shown that an even more diverse group of bacterial pathogens may interact with and aggregate platelets *in vitro*, including *Fusobacterium*,[93] *Listeria*,[94] *Mycobacterium*,[95] *Pseudomonas*,[96] *Salmonella*,[97,98] and *Yersinia*.[99] Key differences in bacterial interaction with platelets can be observed in the overall magnitude of aggregation and in the delay period between contact and irreversible aggregation phases. For example, *S. aureus* and *St. pyogenes* are equivalent in their rapid and complete platelet aggregation, while *E. coli* and *Ent. faecalis* achieve considerably less rapid and vigorous platelet aggregation.[53,89]

Several phenotypic parameters appear to influence the ability of bacteria to stimulate platelet responsiveness. Herzberg et al. found that *viridans streptococci* bind directly to the platelet surface in a manner that was reduced by exposing the organisms to proteases.[100,101] For example, it is possible that *Streptococcus sanguis* adheres to platelets specifically through its expression of a 150-kDa two-domain adhesin.[102] Sullam et al. have used flow cytometry to study the interaction of *Streptococcus* species with rabbit and human platelets *in vitro*.[103] Bayer et al. and Yeaman et al. have also examined the molecular basis for the interaction between *S. aureus* and platelets using aggregometry complemented with flow cytometry.[104,105] These techniques have shown that streptococci and staphylococci may bind directly to washed human or rabbit platelets *in vitro*. Of note, the bacterium-to-platelet ratio has a direct relationship to the velocity and extent with which platelet aggregation occurs.[103–105] In addition, these organisms rapidly adhere to platelets via a mechanism that is saturable and reversible,[103,105] suggestive of a receptor–ligand interaction. However, it is also possible that bacteria and platelets interact though multiple ligands, or by nonspecific interactions such as charge or hydrophobicity. Alugupalli and colleagues have reported that *Borrelia* appears to interact with human platelets through the integrin αIIbβ3 receptors exhibited on resting platelets.[106] Moreover, prostacyclin (prostaglandin I₂), an inhibitor of platelet activation, reduced the ability of the organism to adhere to platelets through this receptor. In a related mechanism, Siboo et al. demonstrated that clumping factor A mediates binding of *S. aureus* to human platelets *in vitro*.[107] Additionally, Bensing et al. identified two genetic loci in *Streptococcus mitis* that mediate binding to human platelets *in vitro*.[108] These findings, enabled through the use of contemporary molecular techniques, suggest that the interactions between pathogen and platelet occur through specific receptor–ligand interactions.

Recently, Youssefian et al. found that platelets can actively phagocytose microorganisms.[245] *Staphylococcus aureus* and HIV internalization by platelets corresponded to morphological evidence of platelet activation. In turn, platelet activation enhanced the capacity for bacterial internalization. Interestingly, immunolabeling revealed that the engulfing vacuoles and the open canalicular system were composed of distinct antigens. Of particular importance, phagocytic vacuoles containing internalized *S. aureus* were shown to fuse with α-granules, which are known to contain platelet microbicidal proteins. Endocytic vacuoles were identified in close proximity to the plasma membrane and contained HIV particles. Examination of platelets from a patient with AIDS and high viremia suggested that HIV endocytosis may also occur *in vivo*.

Along with their specific interactions, it is becoming clear that different bacterial pathogens possess multiple and distinct mechanisms through which they interact with platelets. For example, in addition to their direct adhesion to platelets, staphylococci and streptococci likely utilize plasma cofactors to facilitate their interplay with platelets. Staphylococci utilize fibrinogen bridging to adhere to endothelial cells and platelets.[109,110] In contrast, fibronectin does not appear to be involved in staphylococcal interaction with platelets,[110] but may facilitate *S. aureus* adhesion to fibrin clots.[111] Bayer et al. found that *S. aureus* may induce rabbit platelet aggregation through a fibrinogen-dependent mechanism; however, this mechanism appeared to be independent of integrin αIIbβ3.[104] Hermann et al. suggested that *S. aureus* exploits thrombospondin to adhere to activated platelets and extracellular matrices.[112] Hawiger et al. and other investigators have demonstrated that the interaction among staphylococcal protein-A, IgG, and the platelet Fc receptor yields *S. aureus*-mediated damage of platelets.[113,114] Moreover, the fact that staphylococcal peptidoglycan may activate platelets[115] has been suggested as a mechanism leading to thrombocytopenia, complement activation, and disseminated intravascular coagulation. Furthermore, lipopolysaccharide rich in lipid-A amplifies the interaction among platelets, IgG, and microorganisms.[116] This phenomenon may contribute to the observation that thrombocytopenia often accompanies Gram-negative infection and endotoxemia.[3,116]

Studies conducted by Sullam et al. have shown that viridans-group streptococcal aggregation of human platelets may require direct bacterium-to-platelet binding and plasma cofactors.[48,103,117] For example, the aggregation of human platelets by some streptococci requires IgG specific to the organism, which presumably interacts with the 40-kDa FcγRII receptor on the platelet surface.[117] More recently, Siboo et al. found that a serine-rich protein, *SraP*, significantly mediates adhesion of *S. aureus* to platelets.[246] In addition, this means of binding appears to be relevant *in vivo*, as a mutant deficient in *SraP* had reduced virulence in a rabbit model of infective endocarditis. Staphylococcal *SraP* is estimated to be 227 kDa and, like its streptococcal homolog *GspB*, is predicted to contain an atypical N-terminal signal sequence, two serine-rich repeat regions (srr1 and srr2) separated by a nonrepeat region, and a C-terminal cell wall anchoring motif (LPDTG). Similarly, Zimmerman et al. showed that *St. pneumoniae* aggregates platelets only in the presence of specific antipneumococcal antibody *in vitro*.[118] Ford and coworkers,[119] found that platelet aggregation by *St. sanguis* is facilitated by fibrinogen interaction through the platelet integrin αIIbβ3 receptor. Likewise, platelet aggregation and degranulation by *St. pyogenes* appears to be mediated by fibrinogen,[120] and Johnson and Bowie reported that group C streptococci use von Willebrand factor to interact indirectly with platelets.[121] Other investigators have shown that monoclonal antibodies recognizing 170–230 kDa platelet aggregation-associated antigens (PAAPs) impede the binding of streptococci via 87 or 150 kDa ligands.[122] Niemann et al. and Herrmann et al. have documented key roles for soluble fibrin in *S. aureus* adhesion to platelets *in vitro*[247,248] and surface bound fibronectin in platelet-enhanced host defenses.[249] Loughman and coworkers have also suggested that specific immunoglobulin against *S. aureus* clumping factor (*ClfA*) may exploit platelet interactions.[250] Collectively, these studies underscore multiple ways platelets may interact with bacteria directly and indirectly through a variety of ligands and bridging molecules.

**3. Platelets Interact with Fungi.** The interaction of platelets with pathogenic fungi remains an area that requires much more attention. This view is underscored by the emergence of fungi such as *Candida albicans* as predominant blood stream isolates recovered in the nosocomial setting.[123–126] For example, the fact that many pathogenic fungi are angiotrophic and have a propensity to enter and exploit the vascular compartment suggests that interaction with platelets may influence their potential for pathogenesis. Several fungal organisms are capable of adhering to and aggregating mammalian platelets. As an example, Maisch and Calderone have shown that *C. albicans* and *C. stellatoidea* adhere to and aggregate platelets and platelet–fibrin matrices *in vitro*.[45,127] Importantly, *C. albicans* also binds to these biomatrices in animal models of infective endocarditis.[30] Robert et al. demonstrated that *C. albicans* may exploit multiple adhesins unique to its germ tube surface to interact with platelet integrin αIIbβ3 receptors.[128] Likewise, Klotz et al. suggested that binding to platelets enhances the adhesion of *C. albicans* to cultured vascular endothelial cells.[46] It is possible that pathogenic fungi, with their tendency to cause endovascular infections, may circumvent the antimicrobial functions of platelets and exploit these cells to adhere to vascular endothelium.

These findings illustrate the fact that although fungi interact with platelets, the role of this interaction in fungal pathogenesis remains to be determined. Christin et al. have shown that *Aspergillus fumigatus* interacts with, and evokes antifungal responses from, human platelets.[129] The consequences of this interaction cause platelets to attach to *Aspergillus* hyphae, yielding platelet activation and degranulation and ensuing damage of the organism.[129] In addition, platelets stimulated by this mold potentiate killing of *Aspergillus* by neutrophils *in vitro*. Thus, platelets appear to exert direct antifungal responses and magnify the antifungal mechanisms of leukocytes (see also the following discussion).

**4. Platelets Interact with Protozoa.** The interaction of platelets with protozoal pathogens is perhaps the best studied

of platelet interactions with microorganisms. The role of platelets as cytotoxic effector cells in antimicrobial host defense has been thoroughly demonstrated by several investigators through a body of studies. For example, human platelets interact with a diverse variety of protozoa, including *Schistosoma mansoni*,[55] microfilariae such as *Dipetalonema viteae* and *Brugia malayi*,[130,131] *Toxoplasma gondii*,[132,133] *Trypanosoma cruzi*, and *T. musculi*,[134,135] as well as *Plasmodium falciparum* and *P. vivax*.[136] Platelet interactions with these pathogens are believed to require specific IgE and corresponding platelet surface IgE receptors. Parallel studies utilizing radiolabeled IgE demonstrated that platelets possess approximately 1000 high-affinity IgE receptors and interact with protozoa specifically through this receptor.[137] Immunoglobulin E-rich serum from patients infected with *S. mansoni* activates platelets to mediate cytotoxic activities against this pathogen.[55]

Complementary studies have also demonstrated that immune sera or IgE monoclonal antibody may prompt cytotoxic anti-*S. mansoni* activities of platelets from uninfected animals.[138] The fact that the immune sera alone does not result in protozoal killing substantiated the role of platelets as effector cells integral to antibody-dependent antiprotozoal activity.[3,138] Thus, platelets likely contribute to antimicrobial host defense against protozoal infection via mechanisms that include antibody-dependent cell cytotoxicity (ADCC; also discussed later). Related studies have also shown that TNF-α, TNF-β, and interferon-γ (IFN-γ) promote platelet-mediated antiprotozoal effects.[59,139] Platelet activity suppressive lymphokine (PASL) is released from mitogen-stimulated CD8+ T lymphocytes, reducing the IgE-dependent cytotoxic activities of platelets.[140] Collectively, these data support the concept that platelets function in antiprotozoal host defense as modulated by counteracting cytokines that may be present in the setting of infection.

# V. Platelets Contain Antimicrobial Effector Molecules That Contribute to Host Defense

The relationship between platelet structure and function supports their integral roles in antimicrobial host defense. Yet, until recently, the molecular effectors contributing to these activities had not been established. Since the mid-1990s, contemporary techniques in protein chemistry and molecular biology have been applied to identify and resolve characteristics of these antimicrobial constituents. Thus, it is clear that platelets contain and elaborate a battery of antimicrobial peptides that significantly contribute to antimicrobial host defense.

Direct and indirect interactions with microbial pathogens provide a theme consistent with a role for platelets in antimicrobial host defense. A considerable number of previous investigations have focused on determining the antimicrobial capacity of mammalian platelets. In addition to the oxidative and IgE-mediated effects outlined earlier, platelet antimicrobial effects are likely associated with their release of antimicrobial proteins and peptides. In 1887, Fodor reported heat-stable bactericidal capacity of serum, termed beta-lysin, distinguishing it from heat-labile or alpha-lysin complement proteins.[141] Gengou showed that β-lysin bactericidal activity within serum was derived from cells involved in the clotting of blood and was independent of complement.[142] This was perhaps the first study to implicate platelets directly in antimicrobial host defense. In 1938, Tocantins authored a comprehensive review of the studies to that point examining the immune functions of platelets.[4] Later, Hirsch showed that platelets, not other leukocytes, reconstituted the bactericidal activity of rabbit serum.[143] Subsequently, Jago and Jacox,[144] Weksler,[145] Kahn et al.[146] Czuprynski and Balish,[94] and Miragliotta et al.[147] have shown that platelets exert *in vitro* bacteriostatic or bactericidal effects against a variety of pathogens, including bacillus, staphylococcus, listeria, and salmonella. In addition to these direct antibacterial effects, platelets likely amplify antimicrobial activities of professional phagocytes such as neutrophils, monocytes, and macrophages (discussed later).

Many investigators have worked to isolate and characterize the platelet-specific molecule(s) associated with the antimicrobial action of serum.[148–155] In the earliest of these studies, Myrvik reported that two platelet-derived agents in serum were associated with killing of *Bacillus subtilis*.[151] Jago and Jacox[144] and Myrvik and Leake[152] corroborated this finding by demonstrating that two heat-stable proteins from mammalian platelets correspond with serum bactericidal action. Based on these data, Johnson et al., Donaldson et al., and others[153,154] isolated distinct β-lysins from rabbit serum that exerted specific killing of *S. aureus* or *B. subtilis*, respectively. In similar studies, Weksler and Nachman resolved two cationic proteins from platelets, with molecular masses of 10 and 40 kDa, that were bactericidal *in vitro* versus *B. subtilis* and *S. aureus* strains.[155] Each of these studies reported that the antimicrobial effector molecules from platelets were small and basic, with masses ranging from 6 to 40 kDa. Tew et al. later found that one or more β-lysin proteins were released from rabbit platelets stimulated with thrombin.[156] Dankert et al.[157,158] also suggested that platelets release platelet-associated bactericidal substances (PABS) following thrombin stimulation. Carroll and Martinez[148–150] described the isolation, microcompositional analysis, microbicidal spectrum, and mechanisms of action of an antibacterial peptide (PC-III) from rabbit serum. Darveau et al. described peptides related to human platelet factor-4 (PF4) that exerted antimicrobial activity in combination with conventional antibiotics against Gram-negative bacteria.[159]

## A. Platelet Microbicidal Proteins

The molecular characterization of antimicrobial effector polypeptides from platelets has been achieved. Yeaman et al. isolated and characterized proteins that likely significantly contribute to the antimicrobial effects of platelets. For example, platelet microbicidal proteins, or PMPs, have been isolated from supernatants of rabbit platelets following thrombin stimulation.[160,161] The fact that thrombin is generated at sites of endovascular infection and is a potent stimulant of platelet degranulation may be relevant to PMP release in this setting.[22–24] Reversed-phase high performance liquid chromatography (RP-HPLC) has revealed that rabbit platelets contain a family of distinct PMPs; at least two predominant thrombin-inducible PMPs (tPMPs 1 and 2) are liberated from platelets upon thrombin-stimulation[160,161] (discussed later). As in the case of human platelet antimicrobial peptides (discussed later), it is possible that tPMPs represent either native or processed forms of PMPs isolated from unstimulated rabbit platelets. Mass spectroscopy has shown that PMPs and tPMPs range in size from 6 to 9 kDa.[20,21,161] Amino acid compositional analyses show that these polypeptides are cationic and are rich in basic residues lysine, arginine, and histidine (total content approximately 25%). The antimicrobial activities of PMPs and tPMPs are heat stable and inhibitable by anionic adsorption.[160,161] Molecular mass, a conspicuous presence of lysine, and the lack of a disulfide array integrating six cystine residues differentiate PMPs or tPMPs from the well-characterized neutrophil defensins.[20,21,161,163] Moreover, PMPs and tPMPs are distinguishable from platelet lysozyme in mass, amino acid composition, and antimicrobial activity.[161] PMPs and tPMPs exert strong antimicrobial activities *in vitro* against bacteria and fung (see below).[20,21,161] Azizi et al.[164] and Bayer et al.[165] also found that pathogens or purified staphylococcal α-toxin prompt PMP release from rabbit platelets *in vitro*.

## B. Multiple Genetic Lineages of PMPs

Whereas PMPs and tPMPs were isolated from rabbit platelets, analogous peptides were isolated soon thereafter from human platelets. Tang et al. isolated and identified antimicrobial peptides from human platelets that appear to be functional analogues of rabbit PMPs and tPMPs.[166] Currently, a broad collection of antimicrobial peptides is known to be released from human platelets following thrombin stimulation. These human peptides are structural and functional analogues of rabbit PMPs, and include PF4, platelet basic protein (PBP) and its derivatives, connective tissue activating peptide-3 (CTAP-3 and a derivative, neutrophil activating peptide-2, NAP-2), RANTES ("released upon activation, normal T cell expressed and secreted"), thymosin-β-4 (Tβ-4), and fibrinopeptides A and B (FP-A and

FP-B). Based on the preceding discussion, phylogenetically there are five distinct genetic lineages of mammalian PMPs: (a) PF4 and variants, (b) PBP and its proteolytic derivatives CTAP-3 and NAP-2, (c) RANTES, (d) Tβ-4, and (e) fibrinopeptides (Fig. 40-1). While PF4, PBP, CTAP-3, and NAP-2 are α- or CXC-chemokines, RANTES is a β- or CC-chemokine. Findings demonstrate the direct antimicrobial functions of these human platelet proteins, even though their structures had been characterized previously. Krijgsveld et al.[168] found that carboxy-terminal diamino acid truncated versions of NAP-2 or CTAP-3 exerted antimicrobial activity *in vitro*. These peptides have been termed thrombocidins 1 and 2, respectively. It is interesting to note that FP-A and FP-B were not detectable in total protein extracts of platelets. This finding suggests these PMPs are generated as a result of thrombin stimulation of platelets. Conceivably, such peptides may be generated via thrombin cleavable (e.g., Arg-Gly sites) in platelet fibrinogen.[250] Thus, thrombin (a serine protease), platelet-derived proteases, proteases generated by tissue injury, phagocytes (e.g., cathepsin G), or inflammation (e.g., plasmin) may process precursor proteins and generate multiple forms of antimicrobial peptides from platelets. Alternatively, microbial proteases may also modulate the bioactivity of PMPs and kinocidins. For example, Sieprawska–Lupa et al.[252] have shown that *Staphylococcus aureus* exoenzymes aureolysin and V8 protease cleave the antimicrobial peptide LL-37. Thus, it is possible that PMPs and kinocidins respond to microbial virulence factors such as proteases by deploying domains with optimized antimicrobial efficacy versus cognate pathogens in contexts of infection[253] (also discussed later). Collectively, these relationships may be crucial to the multifunctional roles of platelets in antimicrobial host defense.[253,255]

By convention, the term *platelet microbicidal proteins* (PMPs) is used to encompass PMPs, thrombin-induced PMPs (tPMPs), thrombocidins, or other antimicrobial polypeptides from platelets. Comparisons of the structures and functions of PMPs with antimicrobial peptides from other sources are reviewed elsewhere.[20,21,161,166–171,253–256]

Detailed molecular characterization of structure-activity relationships has revealed interesting new insights into the repertoire of PMPs and kinocidins.[257] For example, amino acid sequencing and mass spectrometry showed that distinct N-terminal polymorphism variants of rabbit PMP-1 from unactivated or thrombin-stimulated platelets, respectively, arise from a common PMP-1 propeptide. Cloning from bone marrow and characterization of its cDNA identified that PMP-1 is translated as a 106-amino acid precursor and processed to 73 (8053 Da) and 72 residue (7951 Da) variants. Analysis of sequence motifs revealed PMP-1 to be a member of the PF4 protein family. Moreover, phylogeny and three-dimensional structures confirmed that PMP-1 has greatest homology with PF4. Therefore, structural homology and antimicrobial properties established the identity of PMP-1 as a rabbit analogue of human PF4 (Fig. 40-2).

## C. Multifunctional Antimicrobial Peptides from Platelets: Kinocidins

Studies described earlier have demonstrated that several human and other mammalian PMPs are also chemokines. Important reports by Cole et al.[258] and Yang et al.[259] have subsequently reinforced this concept. Reflecting their dual complementary functions, chemokines that also exert direct microbicidal activities have been termed *kinocidins*.[252,259] Studies have identified and characterized antimicrobial efficacies of kinocidins from nonplatelet sources as well.[258,259] Therefore, kinocidins also have known chemotactic functions that recruit and activate antimicrobial mechanisms of leukocytes and lymphocytes. Yeaman and Yount have proposed how PMPs and other antimicrobial peptides are optimized to defend against infection of specific physiologic and anatomic contexts or contexts that change over time in the face of infection. This paradigm of immunocoordination has been termed the AEGIS model of antimicrobial peptides: archetype effectors governing immune syntax.[253] A principal aspect of this model is based on the concept of antimicrobial peptide immunorelativity (Fig. 40-3). This concept is a functional concatenation of molecular regulation and physicochemical properties of host defense molecules, in particular anatomic or physiologic context, optimized to defend against cognate pathogens in context. Hence, PMPs and kinocidins likely act in consort with leukocytes and their antimicrobial armamentarium to protect the host against invasive infection. The AEGIS model of immunocoordination is supported by multiple lines of investigation as illustrated in the following examples.

**1. Kinocidins Have Complementary and Compounded Host Defense Functions.** Studies suggest[253,257,260] how kinocidins contain distinguishable functional motifs that facilitate their complementary antimicrobial functions. Direct antimicrobial actions include rapid microbicidal effects on microbial targets. Indirect effects are less immediate but no less important. Leukocyte chemotaxis to sites of infection exemplifies an indirect antimicrobial effect of platelet kinocidins. Still other functions may further amplify the compounded roles of these molecules in host defense. For example, proteomic and functional data indicate that protease inhibitor, opsonic, and other complementary functions can be integrated along with microbicidal activity within specific disulfide-stabilized antimicrobial peptide configurations.[260] Elegant studies from the laboratory of Richard Gallo[261] suggested that certain human cathelicidins also encode microbicidal and protease inhibitor activities. Alternatively, prior investigations of classical defensins have

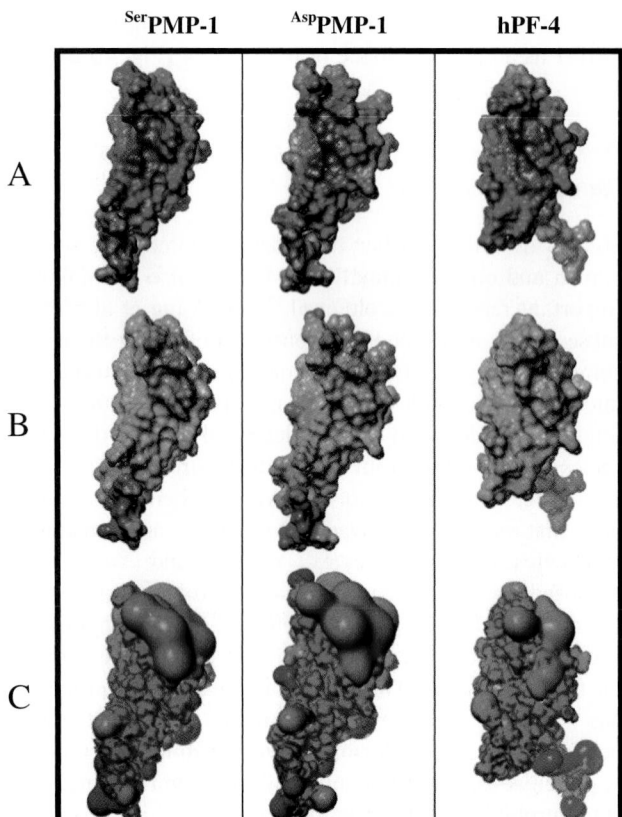

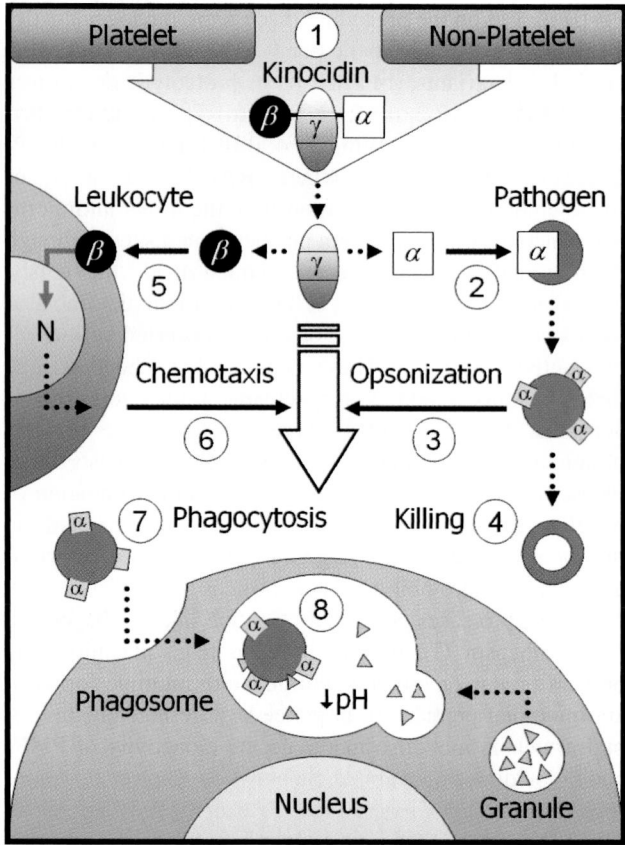

**Figure 40-2.** Comparative molecular models of rabbit PMP-1 variants and human PF4. PMP-1 structures were predicted from homology- and energy-based space filling and molecular surface models. The conserved C-terminal helices are shown at the front and top of each of the structures, which are aligned in identical orientations. The flexible N-terminal tails (bottom) are shown in the most common conformations. A. Hydrophobicity projected onto peptide solvent accessible surface area (most hydrophilic, blue; most hydrophobic, brown; intermediate values are green). Note the hydrophilic propensity of the N-terminal region and hydrophobic interior. B. Electrostatic surface (most positive surface, red; most negative surface, violet; intermediate values follow spectral colors). Note the segregation of surface change in each molecule. C. Electrostatic (Coulombic) fields. Each field is contoured at 30 kcal/mol for negative (blue) or positive (red) electrostatic energy. All molecules have a net positive electrical potential and behave much like a macro-cation. The cationic domain is clearly evident by red protrusions in the C-terminal regions (top) of each peptide. Note how the presence or absence of the N-terminal serine residue influences the electrostatic field in the C-terminal globular domain of the PMP-1 variants. The structural organization and conservation in these kinocidins is consistent with the presence of biochemically distinct functional domains separated in space (e.g., the anionic N-terminal chemokine domains versus the cationic C-terminal microbicidal domain). These models support the hypothesis that kinocidins such as PMP-1 and hPF4 are multidomain and multifunctional effectors of that bridge molecular (e.g., direct microbicidal function) and cellular (e.g., potentiation of neutrophils) innate immunity. (Modified with permission.[257])

**Figure 40-3.** AEGIS model of antimicrobial host defense. Kinocidins are microbicidal chemokines having characteristic multidomain structures: α, alpha-helical domain associated with direct microbicidal activity; β, beta-sheet or extended domain containing the chemokine motif (e.g., CXC or CC); and γ, gamma core motif of antimicrobial polypeptides.[252,256,259] The multiple and complementary functions encoded within kinocidins are deployed as coordinated in space and time (immune syntax): ① hallmark signals of infection evoke host elaboration of kinocidins in cognate physiologic, anatomic, and microbiologic contexts; ② affinity for microbial cells concentrates native kinocidins or by deployment of autonomous functional domain(s) due to proteolytic cleavage of the native kinocidin under specific conditions in context (e.g., associated with microbial target); ③ complementary functions of modules may include opsonization or ④ direct microbicidal activity upon the pathogen target; ⑤ if needed, diffusion or processing of kinocidins or their active domains ⑥ govern chemonavigation of leukocytes to contexts of infection and potentiate their antimicrobial mechanisms; ⑦ in turn, pathogens predecorated by opsonic kinocidin domains are more efficiently phagocytosed by recruited leukocytes; ⑧ once within the restricted context of a leukocyte phagolysosome, the microbicidal functions of kinocidin microbicidal domains may intensify as pH descends, augmenting the intrinsic oxidative (e.g., superoxide, chloramines) or nonoxidative microbicidal (defensins or other granule constituents) mechanisms of the leukocyte. Note that the AEGIS model suggests that ensuing cellular mechanisms may only be needed if preceding molecular mechanisms fail to control pathogen(s). (Modified with permission.[253])

revealed some of these molecules to be capable of leukocyte chemoattraction, consistent with this hypothesis.

## 2. Multiple Functions Encoded in Kinocidins Are Likely Deployed by Activation in Relevant Contexts of Infection.

The structures of kinocidins suggest multiple integrated functions in host defense. Functional decompression of multiple antimicrobial properties integrated in such molecules has been hypothesized to proceed through a co-ordinated process:[253,260] (a) signals of infection elicit host elaboration of a kinocidin within its cognate physiological/anatomic context; (b) where intensified, the kinocidin or its antimicrobial determinant(s) exerts direct microbicidal activity; (c) proteolytic cleavage may yield a protease inhibitory function of molecular domains, preserving antimicrobial integrity of others; (d) diffusion and proteolytic processing establishes gradients of kinocidins or their chemotactic determinants that govern leukocyte chemonavigation; (e) decoration by specific opsonic motifs promotes pathogen phagocytosis by arriving leukocytes; and (f) domains with enhanced microbicidal activity under acidic conditions are potentiated within the leukocyte phagolysosomes. Evidence substantiates each step in this orchestrated pathway. This model further suggests that latter steps deploy in inverse proportion to success of former events: phagocytic responses are only necessary if molecular effectors fail to control pathogen(s) directly. In this way, kinocidins likely contribute to host defense through the coordinated deployment of direct and indirect antimicrobial functions optimized to context.

## D. Elaboration and Maturation of PMPs and Kinocidins

The contributions of PMPs and kinocidins to platelet-mediated host defense against infection relate to mechanisms that prompt the mobilization or elaboration of these antimicrobial peptides. Studies have shown that PMPs and kinocidins are released from platelets exposed to thrombin, staphylococcal α-toxin, or microbial pathogens themselves, including viridans group streptococci, *S. aureus,* and *C. albicans.* As discussed, thrombin is a potent stimulus for the release of PMPs and kinocidins from rabbit and human platelets.

Contemporary studies have also provided molecular mechanisms by which platelets detect and respond to tPMP-1-susceptible (ISP479C; tPMP-1$^S$) or -resistant (ISP479R; tPMP-1$^R$) strains of *S. aureus.*[172] At platelet-to-*S. aureus* ratios ≥1000:1, anti-*S. aureus* effects corresponded to the tPMP-1 susceptibility phenotype of the challenge organism, with greater killing of the tPMP-1$^S$ versus tPMP-1$^R$ strain. Analytic RP-HPLC of staphylocidal supernatants resulting from these platelet–*S. aureus* interactions confirmed tPMP-

1 was released from *S. aureus*-stimulated platelets. Below ratios of 1000:1, equivalent survival of ISP479C or ISP479R strains was seen and did not differ from respective controls. A panel of specific platelet inhibitors was then used to probe pathways associated with platelet antistaphylococcal responses. Apyrase (inhibitor of extracellular ADP), ticlopidine (specific inhibitor of platelet P2Y$_{12}$ ADP receptors), suramin (inhibitor of platelet P2X and P2Y ADP receptors), and pyridoxyl 5′-phosphate derivative (PPND; inhibitor of platelet P2X ADP receptors) each interfered with platelet antistaphylococcal responses. However, specific inhibition of platelet β-adrenergic (yohimbine), phospholipase C (e.g., propranolol), cyclo-oxygenase-1 (COX-1) (indomethacin), or thromboxane A$_2$ (SQ29548) pathways each failed to impede platelet anti-*S.aureus* responses. Collectively, this pattern of results demonstrated that platelet release of PMPs in response to *S. aureus* occurs via an active, rapid, and direct mechanism, amplified through autocrine pathways where platelet ADP release triggers successive waves of platelet degranulation via the platelet P2X and P2Y$_{12}$ adenosine nucleotide receptor array. Likewise, Sharma et al. found that outer membrane constituents of *Porphyromonas gingivalis* potently activator murine platelet aggregation and degranulation.[173] These studies indicate that certain microbial pathogens may use molecular mimicry to exploit the platelet as an adhesive surface. In such cases, it would be anticipated that pathogens would possess mechanisms by which they circumvent the antimicrobial responses of platelets.

Numerous studies have shown that PF4 (a known platelet kinocidin, discussed earlier) levels in plasma increase dramatically (fourfold to sixfold) in the setting of microbial challenge *in vivo*. For example, PF4 plasma levels increase during acute cytomegaloviremia,[174] bacterial septicemia,[175] streptococcal nephritis,[176] candidiasis,[177] and malaria.[178] Similarly, Wilson et al. have shown that endotoxin prompts marked increases in circulating levels of soluble P-selectin, another indicator of platelet degranulation.[179] Interestingly, unlike most other markers of inflammation studied, plasma concentrations of soluble P-selectin progressively increased for up to 8 hours following endotoxin injection in human volunteers. Taken together, these findings substantiate the concept that the antimicrobial polypeptides are released from platelets in response to relevant signals of inflammation and infection, as wells as microorganisms themselves.

Like rabbit PMPs, human peptides exert potent and synergistic *in vitro* microbicidal activities against *B. subtilis, S. aureus, S. epidermidis, E. coli, C. albicans,* and *Cryptococcus neoformans*[20,21,164,165] (see the discussion that follows). The microbicidal effects of PMPs and platelet kinocidins are achieved in nanomolar to low micromolar concentrations (1 to 5 μg/mL).[160–167] In addition, these peptides are active against microbial pathogens in physiological ranges of pH (5.5 to 8.0) and appear to act synergistically against microbial pathogens *in vitro*.[166,167] Thus, the antimicrobial activities of PMPs and kinocidins observed *in vitro* are relevant to conditions that exist *in vivo*. Furthermore, structural similarities between rabbit and human PMPs likely relate to their functional congruence and provide opportunities to study relevant rabbit models (e.g., infective endocarditis) in the investigation of the role of human platelets and PMPs in antimicrobial host defense. These findings also imply that the existence of antimicrobial peptides is conserved among mammalian platelets. It is becoming clear that there are two predominant phylogenetic lineages of antimicrobial proteins and peptides from mammalian platelets: (a) those represented by PF4 and (b) those represented by PBP and its derivatives CTAP-3 and NAP-2.[20,21,160,166–168] It should be emphasized that, along with potent and direct antimicrobial effects, many of these proteins are chemokines. For example, PF4, PBP, CTAP-3, and NAP-2 are α- or CXC-chemokines, whereas RANTES represents a β- or CC-chemokine. Such chemokines recruit and potentiate the antimicrobial mechanisms of neutrophils, monocytes, and macrophages, contributing to direct and indirect mechanisms of antimicrobial host defense (discussed later).

# VI. Platelet Antimicrobial Peptides Likely Function via Multiple Mechanisms of Action

Antimicrobial effects of PMPs and kinocidins on whole microbial cells, protoplasts, and lipid bilayers *in vitro* have been investigated using multiple approaches, including genetic methods, biophysical techniques, and ultrastructural studies.[180–186] Generally, the results of these investigations reinforce a fundamental theme: PMPs and kinocidins initially target and perturb their microbial target cell membranes, but additional steps are involved in eventual lethal effects. In *S. aureus*, cytoplasmic membrane permeabilization occurs immediately, but membrane depolarization does not necessarily follow. Thereafter, cytoplasmic membrane condensation occurs, corresponding to cell wall hypertrophy after exposure to PMPs for 60 to 90 minutes.[180] Typically, perturbations in cell ultrastructure precede significant bactericidal and bacteriolytic effects of these peptides. Fungal pathogens are affected in a similar manner by tPMP-1 *in vitro*.[162] It is notable that *S. aureus* protoplasts exhibit tPMP-1 susceptibility or resistance characteristics corresponding to the strain phenotype from which they were prepared, suggesting that antimicrobial effects do not necessarily depend on the presence of the cell wall.[181,184] Rather, tPMP-1 appears to achieve antistaphylococcal effects through mechanisms that initially cause voltage-dependent membrane permeabilization.[181,184–186]

Results from these studies indicate that PMPs and kinocidins initially target and perturb microbial cell membranes.

Rapidly following exposure to peptides such as tPMP-1, perturbations of the *S. aureus* cell membrane are evident by ultrastructure analysis. Thus, like other endogenous antimicrobial peptides, tPMP-1 initially targets the cell membrane in its lethal pathway. Respiratory electron chain *S. aureus* mutants (e.g., menadione auxotrophs or "small colony variants") exhibit defects in normal transmembrane electric potential (Δψ). These strains demonstrate reduced susceptibility to killing by tPMP-1 (and other antimicrobial peptides) as compared to their parental counterparts with normal Δψ. Menadione supplementation of these small colony variants restores the Δψ to parental levels and restores tPMP-1 susceptibility.[180] Likewise, tPMP-1 mediated membrane permeability is also reduced in Δψ mutants and is reconstituted by menadione. Related studies[185] confirm that the overall proton motive force (Δψ and ΔpH) is important in tPMP-1-induced microbicidal action. Exposure of even very low concentrations of tPMP-1 (e.g., 1 μg/mL) to model planar lipid bilayers causes membrane permeabilization and bilayer dysfunction in a voltage-dependent manner, with maximal effects induced at a transnegative voltage orientation relative to the site of addition of the peptide. Well-defined, voltage-gated pores formed by defensins in similar model membranes are not formed by tPMP-1. Staphylocidal activity of tPMP-1 declines *in vitro* under conditions mitigating microbial energetics (e.g., stationary phase or low temperature).

An interesting paradigm is emerging from contemporary studies of membrane biochemistry in bacteria exposed to PMPs. Cytoplasmic membrane fluidity of tPMP-1-resistant strains of *S. aureus* is distinct from that of genetically related tPMP-1-susceptible counterparts. In addition, some findings[262–264] offer new evidence for interrelationships of the staphylococcal cell wall and cell membrane in antimicrobial peptide susceptibility or resistance. For example, the *mprF* gene in *Staphylococcus aureus* leads to increased phospholipid lysinylation (cationic charge) and reduced susceptibility to antimicrobial peptides *in vitro* ostensibly through electrostatic repulsion.[265] In contrast, defects in the *tagO* gene governing cell wall teichoic acid synthesis do not appear to influence tPMP-1 susceptibility but do mitigate endothelial cell binding and attenuate virulence in the rabbit model of endocarditis.[264] Likewise, membrane permeabilization is involved, but it is not exclusive as a mechanism of tPMP-1 staphylocidal activity.[262] Interesting results also reveal that the net cationic charge of the extracellular facet of the *S. aureus* cell membrane also affects susceptibility to PMPs and kinocidins. For example, Mukhopadhyay et al.[263] found a correlation between asymmetry of the outer membrane leaflet and orientation of lysylphosphatidylglycerol (LPG) constituents to its exterior to lower *S. aureus* susceptibility to tPMP-1 or other antimicrobial peptides *in vitro*. As LPG is the predominant basic lipid species, asymmetric localization to the outer membrane leaflet likely enhances positive surface charge and may repel cationic peptides.

Although PMPs and kinocidins cause rapid dysfunction of target microbial cell membranes, the 1 to 2-hour delay between initial exposure and microbicidal effect implicate other, likely intracellular, targets of action for these peptides. Xiong et al.[183] found that preexposure of tPMP-1$^S$ *S. aureus* to tetracycline, a 30S ribosomal subunit inhibitor, significantly decreased the ensuing staphylocidal effect of tPMP-1 over a concentration range of 0.16 to 1.25 μg/mL. In these studies, preexposure to novobiocin (DNA gyrase subunit B inhibitor) or azithromycin, quinupristin, or dalfopristin (inhibit 50S ribosomal subunits) mitigated the staphylocidal effect of tPMP-1 over a concentration range of 0.31 to 1.25 μg/mL. These data suggested that tPMP-1 exerts anti-*S. aureus* activities, in part, through mechanisms involving inhibition of macromolecular synthesis. Studies by Xiong et al.[266] provide further support for the hypothesis that PMPs and kinocidins access and disrupt intracellular pathways. In tPMP-1$^S$ *S. aureus* strains, purified tPMP-1 caused significant reduction in DNA and RNA synthesis temporally corresponding to the extent of staphylocidal activity. In contrast, tPMP-1 exerted strikingly reduced inhibition of macromolecular synthesis in the isogenic tPMP-1$^R$ counterpart, mirroring reduced staphylocidal effects. However, tPMP-1 caused equivalent degrees of protein synthesis inhibition in these strains. For example, preexposure of tPMP-1-susceptible strains with antibiotics that block either 50S ribosome-dependent protein synthesis or DNA gyrase subunit B actions completely inhibit tPMP-1-induced microbicidal effects. Moreover, *S. aureus* strains deficient in their autolytic pathway are also less susceptible to tPMP-1, suggesting an important role for this system in the overall lethal mechanism of this peptide.[266] Defective autolysin functions have also been implicated as a resistance mechanism in *S. aureus* against cationic antimicrobial peptides, including PMPs and kinocidins.[267–269] Collectively, a body of evidence indicates that PMP or kinocidin antimicrobial efficacy involves intracellular targets and autolytic pathways subsequent to initial membrane perturbation.

Beyond direct microbicidal effects, PMPs and kinocidins also cause prolonged postexposure growth-inhibitory effects against staphylococci, similar to those of cell wall-active antibiotics such as oxacillin and vancomycin[190] (discussed later). Thus, exposure of microorganisms to PMPs or kinocidins and antibiotics or antifungal agents exhibit favorable antimicrobial interactions. Moreover, such growth-inhibitory activities are observed against tPMP-1 susceptible and resistant strains. Likewise, preexposure of *S. aureus* or *C. albicans* to tPMP-1 reduces the capacity of these pathogens to adhere to platelets *in vitro,* and this effect can be amplified by other anti-infective agents.[193,194]

Antimicrobial mechanisms of PMPs and kinocidins appear to be distinctive, if not unique, as compared with classical antimicrobial peptides. For example, the staphylocidal mechanisms of PMP-2 and tPMP-1 action against tPMP-susceptible (tPMP$^S$) and tPMP-resistant (tPMP$^R$) *S. aureus* differ from those of human defensin NP-1 (hNP-1) in the following parameters: involvement of membrane potential (Δψ), permeabilization, and bactericidal activity. For example, PMP-2 rapidly permeabilizes and depolarizes the tPMP$^S$ strain; for this peptide, the extent of permeabilization was observed to be inversely related to pH.[180] On the contrary, tPMP-1 does not appear to depolarize the tPMP$^S$ strain, but it does permeabilize this strain in a manner directly correlated with pH. Depolarization, permeabilization, and killing of the tPMP$^R$ strain by PMP-2 and tPMP-1 are significantly less than with the tPMP$^S$ strain.[180] However, culture in the presence of menadione reconstitutes the tPMP$^R$ Δψ to a level equivalent to the tPMP$^S$ strain and is associated with increased depolarization due to PMP-2, but not tPMP-1. Accordingly, reconstitution of Δψ also enhances permeabilization and killing of the tPMP$^R$ strain due to tPMP-1 or PMP-2. Thus, the mechanisms of PMP-2 and tPMP-1 action appear to differ, involving rapid and pH-dependent membrane permeabilization, with or without membrane depolarization, respectively. These effects differ from hNP-1 or the cationic antibacterial agents protamine or gentamicin.[180]

The antimicrobial functions of PMPs and kinocidins are likely important in preventing or limiting infection of relevant biomatrices. Mercier et al.[188] studied the ability of platelets to limit the colonization and proliferation of *S. aureus* in an *in vitro* model of infective endocarditis. In time-kill studies, platelet activation by thrombin 30 minutes prior to bacterial inoculation correlated with a significant bactericidal effect against tPMP-1$^S$, but not tPMP-1$^R$ *S. aureus*. In the *in vitro* infective endocarditis model, thrombin activation significantly inhibited proliferation of the tPMP-1$^S$ strain within simulated infective endocarditis vegetations as compared to the isogenic tPMP-1$^R$ strain ($p < 0.05$). These outcomes were observed despite there being no detectable difference between the *S. aureus* strains in their adhesion to or initial colonization of simulated vegetations. Collectively, these data indicate that platelets limit intravegetation proliferation of tPMP-1$^S$ but not tPMP-1$^R$ *S. aureus* strains. Reinforcing this point, synthetic peptidomimetics of antimicrobial determinants in platelet kinocidins exert significantly greater antimicrobial efficacy in human blood and plasma *ex vivo* than in artificial medium, even after 2 hours incubation in the biomatrices prior to inoculation of the target organisms.[270] These findings underscore the view that platelets play crucial antimicrobial host defense roles in preventing or limiting endovascular infections due to tPMP-1$^S$ pathogens.

## VII. Platelet Antimicrobial Peptides Potentiate the Actions of Conventional Antibiotics

Platelet products are known to interact with conventional antibiotics. Asensi and Fierer demonstrated that β-lysin and ampicillin synergistically interact *in vitro* to inhibit *Listeria monocytogenes*.[189] Likewise, PMPs and kinocidins potentiate conventional antibiotics against other organisms, including *S. aureus*. For example, against inocula representing *S. aureus* densities present within early, developing, and established endocarditis vegetations *in vitro*, tPMP-1 exerts potent synergistic bactericidal effects in combination with antistaphylococcal antibiotics such as oxacillin and vancomycin.[190] A combined antibiotic-peptide amplification of antistaphylococcal activity was observed regardless of whether strains had tPMP^S or tPMP^R phenotypes.[190] Additionally, staphylocidal effects of tPMP-1 are augmented by pretreatment of *S. aureus* with either penicillin or vancomycin.[183] Moreover, sublethal exposure of tPMP-1 exerted a prolonged growth inhibitory effect against *S. aureus*, consistent with classic "postantibiotic effect" phenomena of vancomycin and oxacillin.[190]

Dhawan et al.[191,192] compared vancomycin prophylaxis and treatment outcomes in experimental infective endocarditis caused by isogenic ISP479C (tPMP-1^S) and ISP479R (tPMP-1^R) *S. aureus* strain pairs differing in tPMP-1 susceptibility *in vitro*. Vancomycin therapy (selected for its intrinsically slow bactericidal activity) reduced ISP479C, but not ISP479R, densities in vegetations as compared with controls ($p < 0.01$). However, vancomycin prophylaxis yielded no difference in ensuing infectivity or severity proliferation of the two challenge strains. By comparison, treatment with oxacillin (a more rapid staphylocidal agent than vancomycin) enhanced the clearance of both tPMP-1^S and tPMP-1^R cells from vegetations; the extent of clearance was greater for tPMP-1^S cells. Collectively, these data suggest the *S. aureus* tPMP-1^R phenotype *in vitro* corresponds with a negative effect on antimicrobial therapy, but not prophylaxis, in experimental infective endocarditis. These results may be explained, in part, by the requirement for microbicidal effects in the treatment of established infective endocarditis, whereas prophylactic efficacy apparently relies more on growth inhibitory and antiadhesion effects.

Studies by Mercier et al.[271] also indicate that PMPs or kinocidins and antibiotics interact synergistically in limiting the evolution of simulated human infective endocarditis vegetations. *S. aureus* strains differing in intrinsic susceptibility to PMP antibiotics were studied in the presence and absence of vancomycin or nafcillin in an *in vitro* model of infective endocarditis: ISP479C (thrombin-induced PMP-1 [tPMP-1] susceptible; nafcillin and vancomycin susceptible), ISP479R (tPMP-1 resistant; nafcillin and van-

comycin susceptible), and GISA-NJ (tPMP-1 intermediate-susceptible; vancomycin intermediate-susceptible). Platelets were introduced and activated with thrombin 30 minutes prior to inoculation with *S. aureus*. Activated platelets alone, or in combination with antibiotics, inhibited the proliferation of ISP479C in model fluid or simulated vegetations over the initial 4-hour period ($p < 0.05$ versus controls). Nafcillin regimens exerted inhibitory effects beyond 4 hours against ISP479C in both model phases. By comparison, activated platelets inhibited GISA-NJ proliferation in vegetations, but not in model fluid. The combination of platelets plus nafcillin or vancomycin significantly inhibited proliferation of the GISA-NJ strain in vegetations as compared to platelets or antibiotics alone. In contrast, platelets did not significantly alter the antistaphylococcal efficacies of nafcillin or vancomycin against the PMP-resistant strain ISP479R. These data support the hypothesis that beneficial antimicrobial effects result from the interaction among platelets, PMPs or kinocidins, and antiinfective agents against antibiotic-susceptible or -resistant staphylococci.

## VIII. PMPs Likely Modulate Pathogen Interactions with Platelets and Endothelial Cells

The fact that PMPs and kinocidins interact with microbial surfaces is consistent with the concept that these peptides may alter microorganism interactions with host cells. For platelet adhesion and aggregation, and susceptibility to tPMP-1, heterogeneity among *S. aureus* clinical isolates exists.[105,193] Whereas a significant, positive correlation is observed between platelet adherence and aggregation among such strains, there was no correlation of either platelet adherence or aggregation with *in vitro* susceptibility to tPMP-1.[105,193] These data indicate that susceptibility to tPMP-1 may be independent from other platelet-microorganism interactions.

As described previously, *S. aureus* adhesion to platelets *in vitro* is rapid, saturable, and reversible, corresponding to a receptor–ligand interaction. Modified Scatchard analyses suggest that the number of binding sites per platelet vary somewhat for distinct *S. aureus* strains.[105] Furthermore, the binding of individual *S. aureus* cells to platelets appears to be influenced more by the number of binding sites on platelets than on platelet-binding affinities of bacterial cells.[105] Taken together, these observations suggest that platelet binding by these organisms is specific. Proteolytic treatment failed to alter *S. aureus* binding to rabbit platelets *in vitro*. However, exposure to periodate or to tPMP-1 significantly reduced staphylococcal adherence to rabbit platelets.[105,194] These findings suggest that *S. aureus* adhesion to platelets may be multimodal, involving both tPMP-1-sensitive and carbohydrate surface ligands. Parallel investigations have shown that *in vitro* exposure to tPMP-1, alone or in combination with conventional antibiotics, significantly reduces adhesion of tPMP$^S$ or tPMP$^R$ *S. aureus* to platelets *in vitro*.[194] This effect was observed despite antibiotics alone (ampicillin, sulbactam, combination) increasing bacterial adhesion to platelets *in vitro*.[194]

Several lines of evidence indicate that platelets protect against endovascular infections such as infective endocarditis. It is likely that a principal mechanism of this platelet role is the release of PMPs in response to agonists generated at sites of endovascular infection. Superimposed on their direct microbicidal effects, PMPs and kinocidins likely interfere with *S. aureus*-induced platelet aggregation *in vitro*.[104] For example, exposure of tPMP$^S$ or tPMP$^R$ *S. aureus* strains to sublethal concentrations of tPMP-1 reduced both the velocity and magnitude of platelet aggregation by *S. aureus*.[104,193,194] Whether tPMP-1 achieves this effect by altering bacterium-to-platelet binding or other mechanisms has not been determined. Filler et al. have shown that platelets protect human umbilical vein endothelial cells (HUVECs) from *in vitro* injury due to *C. albicans*.[195] In these studies,[51]chromium-release from HUVECs due to a tPMP-1$^S$ *C. albicans* strain was reduced by 45% in the presence of a platelet-to-fungus ratio of 20:1. Moreover, HUVEC protection by platelets was associated with a 37% reduction in germ tube length in *C. albicans* after a 2-hour exposure. In contrast, damage of HUVECs due to an isogenic tPMP-1$^R$ strain was not inhibited by platelets.

## IX. Platelets Modulate Complement Activation

Platelets interact with complement proteins to amplify or suppress complement fixation through both the classical and alternative pathways. As described previously, platelets express surface receptors for $C_{3b}$ and $C_{5b}$ proteins bound to antigen surfaces. Additionally, human platelets have been shown to mediate the generation of terminal attack ($C_5$:$C_6$-$C_9$) complexes.[196–198] In turn, platelet surface complement receptors subsequently bind to microorganisms exhibiting these complexes. Platelet proteases may also cleave $C_5$ to $C_{5a}$, thereby yielding a positive chemotactic gradient for recruitment of cell-mediated immune effort cells.[199] Activation of complement cascades also stimulates platelet activation, phospholipase activity, and degranulation.[199,200] Likewise, prostaglandin and other eicosanoid metabolite synthesis and release is triggered in platelets by complement proteins $C_5$-$C_9$.[196] The generation of oxidative radicals by platelets has also been linked to stimulation by C-reactive protein,[201] $C_{3b}$ and $C_5$-$C_9$,[202] INF-γ,[203] or TNF-β.[204] Each of these soluble stimuli could conceivably lead to increased

direct platelet antimicrobial functions or enhanced platelet interactions with leukocytes (discussed later). Similarly, complement alternative pathway factor D exists within platelet α-granules and is released from thrombin-stimulated platelets.[205] Thus, platelet activation may be integrally linked to activation of the alternative pathway of complement fixation. Alternatively, disorders in complement systems may negatively affect the antimicrobial host defense functions of platelets. For example, Zucker et al. have shown that deficiencies of complement proteins $C_3$, $C_5$, $C_6$, or $C_7$ are associated with diminished platelet responses in patients.[206] Moreover, platelets may also downmodulate complement activation. For example, platelet decay accelerating factor appears to suppress $C_3$ convertase activity of $C_{4b2a}$ in classical complement fixation.[207] Similarly, platelet factor H may mitigate alternative pathway complement fixation.[206]

It is also interesting to note that some microbial pathogens may exploit the interaction between platelets and complement proteins in disease pathogenesis. For example, Nguyen et al. reported that *S. aureus* protein A recognizes the platelet gC1qR/p33 receptor.[208] Moreover, there appear to be specific structural requirements in protein A necessary to mediate gC1qR and IgG Fc binding. Investigations using a truncated gC1q receptor mutant (lacking amino acids 74 to 95) implicated that the protein A binding domain localizes beyond the amino-terminal α-helix of the gC1q receptor, which contains binding sites for the globular heads of C1q.[208] These data suggest that the platelet gC1q receptor may be exploitable as a binding site for staphylococcal protein A, providing an additional mechanism for *S. aureus* adhesion to sites of vascular injury and thrombosis. As described earlier, it would be expected that organisms using such a virulence strategy would also be capable of circumventing the known antimicrobial mechanisms of platelets (e.g., antimicrobial peptides, oxygen radicals).

## X. Platelets Potentiate the Antimicrobial Functions of Leukocytes

Interactions with leukocytes provide an additional mechanism by which platelets contribute to antimicrobial host defense. For example, an array of bioactive molecules released from activated platelets are chemoattractants for monocytes and neutrophils (Chapters 12, 15, and 39). These stimuli include PF4, CTAP-3, NAP-2, RANTES, PAF, PDGF, and eicosanoids such as 12-HETE.[209–216] Subcutaneous injection of PF4 or PDGF prompts a rapid neutrophil infiltration in experimental animal models.[217] Furthermore, intravenous injection of PAF into guinea pigs yields eosinophil infiltration within peribronchiolar tissues.[218] Thrombin-activated platelets bind avidly to human monocytes and neutrophils, whereas unactivated platelets do not.[219] One mechanism for this interaction is attributable to receptor-mediated affinity to thrombospondin released from, and rebound to, the platelet surface.[218,219] However, the primary mechanism of the adherence of activated platelets to neutrophil and monocytes is via platelet surface P-selectin binding to its constitutively expressed neutrophil and monocyte counterreceptor PSGL-1 (Chapter 12). Additionally, the GPIb-IX-V and integrin αIIbβ3 receptors facilitate monocyte and neutrophil adhesion to tissues[219,220] (Chapter 12).

Molecules liberated from activated monocytes or neutrophils may, in turn, counteractivate platelets (Chapters 12 and 39). For example, oxygen metabolites, myeloperoxidase, and halides generated by leukocytes may prompt rapid platelet degranulation.[3,68,221] Likewise, neutrophil or monocyte-derived PAF has a strong activating effect on platelets, triggering shape change, inducible receptor expression (e.g., activated integrin αIIbβ3), and granule organization corresponding to rapid platelet activation.[3,14] Neutrophil leukotrienes C4, D4, or E4 may also amplify platelet aggregation and degranulation, alone or in combination with epinephrine or thrombin.[222] It is important to emphasize that monocytes exposed to bacterial components generate tissue factor, which elicits thrombin production and subsequent platelet activation.[22-24,223] In turn, thrombin stimulation prompts platelets to release kinocidins that have chemokine functions.[20,21,160,166-168] Thus, cross-talk between agonists and stimuli generated by leukocytes and platelets enhances the interplay of these cells in antimicrobial host defense.

As detailed earlier, PF4, CTAP-3, NAP-2, and RANTES represent what can now be termed microbicidal chemokines, or kinocidins. This nomenclature reflects the fact that these proteins and peptides exert direct microbicidal activities and mediate leukocyte navigation to sites of infection. Moreover, it is highly likely that the antimicrobial mechanisms of leukocytes are augmented against organisms exposed to these molecules.[20,21] Thus, in concert with recruiting monocytes and neutrophils to sites of infection, platelet activation and degranulation is associated with the potentiation of leukocyte antimicrobial functions. For example, Mandell and Hook[98] have shown that activated platelets facilitate phagocytosis of *Salmonella* by mouse peritoneal macrophages. Other investigators have shown that PF4 amplifies neutrophil fungicidal activities *in vitro*.[211,213] Serotonin released from platelet dense granules increases the adherence of neutrophils to damaged vascular endothelial cells *in vitro*.[224] Platelet thromboxane A₂ has also been shown to enhance neutrophil adhesive and phagocytic properties.[225] Monocyte-derived IL-6 has been shown to induce cytotoxicity of human platelets and leukocytes to *Schistosoma mansoni* larvae *in vitro*.[59] Christin et al.[129] demonstrated that platelets and neutrophils act synergistically *in vitro* to damage and kill *Aspergillus*. Taken together, these findings underscore the likelihood that plate-

let interactions with leukocytes are important for optimal host defense against infection.

PMPs and kinocidins bridge molecular and cellular innate immunity. Individually, PMPs and kinocidins likely have distinct functions in antimicrobial host defense. Moreover, coordination of these distinct functions appears to afford compound roles, bridging molecular and cellular aspects of innate immunity (Fig. 40-3). For example, platelet accumulation at sites of infection, intensified PMP and kinocidin concentration in these settings, and their affinity for pathogens likely converge to facilitate direct antimicrobial host defense.[20,166,253-261] Kinocidins almost certainly play key immunoenhancing roles, analogous to nonplatelet kinocidins (e.g., IL-8) which have direct antimicrobial efficacy (Yount & Yeaman, unpublished). For example, like PF4, IL-8 is rapidly elaborated regionally or even systemically in response to microbial stimuli or infection but exerts local effects that target sites of infection. The immune coordinating roles of platelets and their products offers a rich field of investigation yet to be fully explored. For example, Cocchi et al.[272] demonstrated that the platelet kinocidin RANTES suppresses HIV proliferation or pathogenesis via direct antiviral effects or modulation of T-cell function. Finally, platelet kinocidins are also members of the intercrine family of chemokines termed alarmins.[273] Thus, PMPs and kinocidins appear to bridge molecular and cellular innate immunity through at least two key roles in antimicrobial host defense: (a) direct inhibition or killing of pathogens at sites of infection and (b) recruitment and amplification of leukocyte antimicrobial mechanisms. In this sense, platelets, PMPs, and kinocidins are part of a greater system of host surveillance and homeostasis that integrates defense against infection.

## XI. Platelets Are Integral to Antimicrobial Host Defense *In Vivo*

As outlined earlier, platelets are activated rapidly and in significant quantity in response to soluble mediators and surface ligands expressed by infected vascular and cardiac endothelium. Moreover, platelets respond to, interact directly with, and are activated by microbial pathogens themselves. Activated platelets are known to release a group of antimicrobial polypeptides (e.g., PMPs, kinocidins, and thrombocidins) that contribute to the antimicrobial functions of platelets. As described, platelets also generate antimicrobial oxygen radicals, facilitate complement fixation on microorganisms, internalize and assist in clearing pathogens from the blood stream, effect antibody-dependent cell cytotoxic actions against microorganisms, and potentiate the antimicrobial mechanisms of neutrophils.[20,21,255]

Because of their immediate response to damaged endothelium and the microbial pathogens responsible for its cause, platelets accumulate in the local proximity of infection. Thus, aggregated platelets represent a significant component of infected foci. Early findings that platelets are present in infected cardiac vegetations in infective endocarditis were interpreted to suggest that platelets promote the initiation or evolution of infective endocarditis.[28–31] Yet no data have been reported to substantiate a promotional role for platelets in this or any infection. On the contrary, platelet responses in these settings, resulting in the elaboration of multiple antimicrobial functions, is perhaps the strongest evidence that platelets function to prevent the establishment and progression of infective endocarditis. In this regard, platelets as inflammatory host defense cells would be expected to target and accumulate at these sites of infection.

However, strategic virulence properties of some microbial pathogens may circumvent the antimicrobial functions of platelets. These pathogens likely gain an advantage in pathogenesis of endovascular infections.[226–229] Thus, pathogens capable of resisting the antimicrobial functions of platelets may exploit exhausted platelets in the evolution of infective endocarditis or other vascular infections. For example, Dhawan et al. and Fowler et al. (see below) have implicated this likelihood in pathogenesis and therapy of experimental and human staphylococcal infective endocarditis.[228,229] Similarly, Fields et al.[230] and Groisman et al.[231] were the first to show that resistance to antimicrobial defensins from neutrophils was associated with enhanced virulence in experimental animal models (discussed later).

Related series of studies have strongly implicated the efficacy of PMPs and kinocidins as being critical to host defense against invasive infection in humans. An initial investigation, focused on a cohort of 60 bacteremic isolates from a single medical center, revealed that *S. aureus* strains from patients with infective endocarditis or vascular catheter infections were less susceptible *in vitro* to low levels of tPMP-1 than isolates associated with soft tissue abscesses.[229] These observations suggested that tPMP-1 is an important host defense effector and that resistance to this peptide may provide invading microbes with a survival advantage at sites of endovascular damage. Subsequent studies by the same group[274] found that methicillin-resistant *S. aureus* (MRSA) with dysfunctions in their accessory gene regulator (*agr*) and reduced tPMP-1 susceptibility *in vitro* have a greater propensity to cause persistent bacteremia than MRSA strains lacking these phenotypes. More recently, Sakoulas et al.[268] examined potential relationships among *agr* function, vancomycin exposure, and abnormal autolysis in development of the glycopeptide intermediate susceptible *S. aureus* (GISA) phenotype and low-level *in vitro* resistance to tPMP-1 in isogenic laboratory-derived and clinical *S. aureus* isolates. For both laboratory and clinical strains examined *in vitro*, vancomycin exposure correlated with modest increases in vancomycin MICs and slight reductions in killing by

tPMP, but it did not correlate to a change in autolysin profiles in *agr*-intact prototype strains. On the contrary, vancomycin exposure of *agr*-deficient strains yielded a hetero-GISA phenotype and was linked to defective lysis and reduced *in vitro* killing by tPMP. Collectively, these findings underscore the view that PMPs and kinocidins are integral components of host defense in relevant contexts of invasive infection. Moreover, they point to potential adaptive responses to antibiotics that may render organisms less susceptible to these host defense molecules.

Another strategy of microbial exploitation of platelets may be exemplified in viridans group streptococci. Through molecular mimicry of structural domains of collagen important in hemostasis, *Streptococcus sanguis* triggers platelets to aggregate *in vitro*.[232] Rapidly following inoculation into experimental animals, an aggregation-positive strain of *St. sanguis* caused increased blood pressure, intermittent electrocardiographic abnormalities, and changes in blood catecholamine concentration. These effects were associated with acute thrombocytopenia and accumulation of [111]indium-labeled platelets in the lungs. Moreover, platelet aggregation rendered thrombi that these investigators suggested were responsible for the observed hemodynamic changes, acute pulmonary hypertension, and cardiac abnormalities. In contrast, a *St. sanguis* strain incapable of inducing platelet aggregation failed to yield such effects. Thus, it would be predicted that the aggregation-positive *St. sanguis* strain may resist the antimicrobial properties of platelets, exploiting these cells as adhesive surfaces in the pathogenesis of infective endocarditis.

Despite pathogen countermeasures to overcome the antimicrobial mechanisms of platelets, evidence from a variety of investigators broadly supports the burgeoning view that platelets are integral to host defense against infection *in vivo*. This evidence integrates both *in vitro* and *in vivo* studies, providing a complementary perspective of platelets in this role. For example, Wu and colleagues studied the relationship between *in vitro* staphylococcal susceptibility to the platelet-derived antimicrobial peptide tPMP-1 and the clinical source of bacteremic isolates *in vivo*.[226] For isolates of *S. aureus* and *S. epidermidis*, they observed a significant correlation between the infective endocarditis source and diminished susceptibility to tPMP-1 *in vitro*. These data suggest that tPMP-susceptible (tPMP^S) organisms have a reduced propensity to cause endovascular infection in humans as compared with their tPMP-resistant (tPMP^R) counterparts. Similar observations have correlated *Salmonella* resistance to defensins, antimicrobial peptides present in neutrophils, with increased virulence.[230,231] In related studies, Fowler et al. examined the relationship between *S. aureus* infective endocarditis and *in vitro* resistance to tPMP-1.[229] These investigators evaluated the *in vitro* tPMP-1 susceptibility phenotype of *S. aureus* isolates from 58 prospectively-identified patients with definite infective endocarditis. On multivariate analyses, *S. aureus* infective endocarditis complicating an infected intravascular device was significantly more likely to be caused by a tPMP-1ᴿ strain (*p* = 0.02). Among the *S. aureus* strains studied, no correlations were detected between *in vitro* tPMP-1ᴿ phenotype and the severity of infective endocarditis. This work[229] supports the concept that *in vitro* resistance to tPMP-1 in clinical strains of *S. aureus* is associated with important clinical characteristics of endovascular infections due to this pathogen *in vivo*.

The role of platelets in antimicrobial host defense has also been demonstrated *in vivo* using two complementary approaches. For example, Sullam et al.[233] used an experimental animal model to examine the role of platelets in defense against infective endocarditis *in vivo*. In this study, a tPMPˢ strain of viridans streptococci was used to induce endocarditis in animals either with normal platelet counts or those with selective immune thrombocytopenia. Of note, there were no differences in these groups of animals regarding other leukocyte quantity or quality or complement activity. Animals with thrombocytopenia exhibited significantly higher streptococcal densities in vegetations as compared with their counterparts with normal platelet counts.[233] Reports of Dankert similarly support the concept that platelets are active in host defense against infective endocarditis.[157] These data substantiate the concept that platelets and PMPs are integral to host defense mechanisms that limit the establishment or evolution of endovascular infections.

In humans, thrombocytopenia has been shown to be a significant, independent indicator of worsened morbidity and mortality in patients undergoing cytotoxic cancer chemotherapy as well as non-neoplastic conditions, including infection.[234–238] In the absence of neutropenia, thrombocytopenia has also been positively correlated with increased incidence and severity of lobar pneumonia in elderly individuals.[235] In addition, antiplatelet agents in the experimental endocarditis model have been shown to significantly increase levels of bacteremia and mortality in some experimental animal models.[238] Moreover, neutropenia in the setting of a normal platelet count does not diminish host defense against endovascular infection *in vivo*.[239,240] Collectively, the majority of data suggest that the platelet-microbe interaction is beneficial in attenuating infection *in vivo*.

Studies of several human diseases have also implicated platelets as being crucial to antimicrobial host defense against infection. For example, Chang et al.[237] conducted an important study of liver transplant recipients in which the impact of thrombocytopenia on morbidity and mortality due to infection was evaluated. In several measures of outcome, thrombocytopenia was found to be a significant and independent predictor associated with increased infection and related morbidity and mortality. For example, nadir

platelet counts were found to be significantly lower in non-survivors as compared with survivors (16 versus $36 \times 10^9$/L; *p* = 0.0001). Nearly half (43%) of the patients with nadir platelet counts of $\leq 30 \times 10^9$/L encountered a major infection within 30 days of transplantation, as compared with only 17% of patients with nadir platelet counts exceeding this threshold (*p* = 0.04). Similarly, fungal infections occurred in 14% of those patients exhibiting nadir platelet counts below $30 \times 10^9$/L, versus 0% in those with nadir platelet counts above this limit (*p* = 0.06). Of note, all patients with fungal infections had nadir platelet counts below this threshold prior to the presentation of fungal infection. Such low nadir platelet counts preceded the first major infection by a median of 7 days in these studies. The results of this study strongly suggest that persistent thrombocytopenia portends a worsened outcome in liver transplant recipients due to infection. Moreover, thrombocytopenia appears to precede infection and may be associated with subgroups of transplant patients susceptible to early major infections. Similarly, Kirkpatrick et al.[236] showed thrombocytopenia to be a significant, independent predictor of mortality in pneumococcal meningitis.

Studies also underscore how antiplatelet therapies may enhance host defense against endovascular infection by reducing the adhesive functions of platelets without interfering with their antimicrobial responses. For example, Nicolau and coworkers found that aspirin administered prophylactically reduces the extent of *S. aureus* infective endocarditis in the rabbit model.[241] This study was interpreted as showing that platelets suppressed the induction phase of endovascular infection. However, given that aspirin is a global inhibitor of cyclo-oxygenase function (e.g., in endothelial cells as well as in platelets), these findings remain difficult to interpret. The studies of Kupferwasser et al.[242] shed light on this situation. These investigators demonstrated that acetylsalicylic acid (ASA) exerted direct antimicrobial activities against *S. aureus* in an experimental animal model.[242] In these studies, ASA at 8 mg/kg/day (but not at 4 or 12 mg/kg/day) was associated with significant decreases in vegetation weight (*p* < 0.05), echocardiographic-verified vegetation size (*p* < 0.001), vegetation (*p* < 0.05) and renal *S. aureus* densities, and renal embolic lesions (*p* < 0.05) as compared with untreated controls. Diminished aggregation also resulted when platelets were preexposed to ASA or when *S. aureus* was preexposed to salicylate (*p* < 0.05). Moreover, *S. aureus* adhesion to sterile vegetations (*p* < 0.05), platelets in suspension (*p* < 0.05), fibrin matrices (*p* < 0.05), or fibrin–platelet matrices (*p* < 0.05) were also significantly reduced following bacteria exposure to salicylate. Thus, a likely scenario is that aspirin or salicylates suppress the ability of *S. aureus* to adhere to platelets, and perhaps the counteradhesive properties of platelets, but do not impede the antimicrobial responses of platelets, such as their ability to release antimicrobial peptides such as PMPs and kinocidins. In follow-up studies, this group of investigators detailed how platelet activation, PMPs, and kinocidins likely play integral roles in host defense against cardiovascular infection *in vivo.*[275,276]

Abundant evidence also exists in support of the contention that PMPs and kinocidins participate in the antimicrobial mechanisms of platelets *in vivo.* For example, Yeaman et al.[227] have demonstrated that susceptibility to tPMP-1 negatively influences the establishment and evolution of *Candida albicans* infection.[227] For example, in the experimental rabbit model, tPMP-1$^S$ *C. albicans* exhibits significantly less proliferation in cardiac vegetations as compared with a related tPMP-1$^R$ strain. Moreover, the tPMP-1$^S$ *C. albicans* strain exhibited dramatically reduced incidence of splenic dissemination as compared with its tPMP-1$^R$ counterpart. Furthermore, studies indicate that PMPs and kinocidins likely amplify the antifungal activities of fluconazole *in vivo.*[277] Similarly, Dhawan et al.[228] have found that *in vitro* phenotypic resistance to tPMP-1 is correlated with enhanced virulence in experimental endocarditis due to *S. aureus.* Following simultaneous challenge of animals with both strains, significantly higher densities of the tPMP-1$^R$ strain were present in vegetations ($p < 0.0001$), kidneys ($p < 0.0001$), and spleens ($p < 0.0001$), as compared with those for the tPMP-1$^S$ strain. Importantly, there were no differences in the ability of these strains to adhere to platelet–fibrin matrices or in their clearance from the blood stream. Thus, these results suggest that tPMP-1 disproportionately limits the intravegetation survival, proliferation, and hematogenous dissemination of a tPMP-1$^S$ *S. aureus* strain in experimental endocarditis. Alternatively, the tPMP-1$^R$ phenotype confers a selective advantage associated with the enhanced progression of this infection. These studies are consistent with the concept that antimicrobial functions of platelets are due, at least in part, to elaboration of PMPs and kinocidins at sites of microbial colonization of vascular endothelium. Altered platelet binding or elaboration of toxins or other virulence factors may also correlate with changes in virulence of *S. aureus* in infective endocarditis.[107,108,165,226–229,243] Thus, it is possible that certain organisms are capable of exploiting the platelet as an adhesive surface, if they are also able to circumvent the antimicrobial properties of platelets.

These findings are consistent with platelet responses seen in the setting of human infection. For example, studies by Mavrommatis et al.[244] demonstrate that there are two distinguishable levels of platelet and coagulation cascade response to Gram-negative infection in humans. In the initial state, uncomplicated sepsis is associated with increases in blood levels of fibrinopeptide A and PF4, two known antimicrobial peptides released from activated platelets. Increases in the release of these polypeptides were temporally associated

with a reduction in platelet count, suggesting that activated platelets are cleared following degranulation. However, in severe sepsis, and particularly in septic shock, both platelet number and responsivity are reduced (as measured by markers of platelet activation), and coagulation factors are depleted. These consequences indicate that in such cases of profound sepsis, platelet responses to microbial challenge are likely overwhelmed.

## XII. Future Studies

Evidence continues to accumulate reinforcing the concept that platelets play a key and multifaceted role in host defense against infection. Much of this evidence points to an integral role of antimicrobial proteins and peptides (e.g., PMPs and kinocidins) released from platelets for this purpose. These observations suggest that the antimicrobial activities of platelets likely extend beyond the vascular compartment. These observations also emphasize the importance of studying human diseases and animal models to elucidate the influence of platelet quantity or quality on the prevention or limitation of infection.

Studies utilizing increasingly specific platelet inhibitors may provide important insights into platelet function in human infections. In addition, results from these investigations could lead to new therapeutic interventions in human infections. For example, therapeutic modalities may be identified that could potentially be used to stimulate platelet-mediated host defense in leukopenic patients. Additionally, the observation that antimicrobial peptides from platelets appear to amplify the action of conventional antimicrobial agents could also be exploited to improve the efficacies of conventional antibiotics. Alternatively, PMPs and kinocidins are already providing templates for the development of novel antimicrobial agents that act against pathogens exhibiting multiple antibiotic resistance. Thus, the fact that platelets are now known to release chemokines that exert direct antimicrobial effects (i.e., kinocidins), as well as recruit and potentiate the antimicrobial mechanisms of leukocytes, could provide new strategies and approaches to prevention or therapy of infectious diseases.

## XIII. Summary

Mammalian platelets are unique and multipurpose inflammatory cells that exhibit archetypal features indicative of their roles in antimicrobial host defense. When stimulated in the context of tissue injury or infection, platelets elaborate a diverse array of bioactive molecules that contribute to a wide variety of key host responses. In addition to their roles in modulating hemostasis, vascular tone, tissue adhesiveness, vascular permeability, wound healing, and tissue regeneration, platelets deliver potent antimicrobial molecules to sites of tissue injury or infection. These molecules are now known to include PMPs and kinocidins that exert direct antimicrobial effects and potentiate the antimicrobial mechanisms of leukocytes. Evidence that platelets, PMPs, and kinocidins play key roles in host defense against infection includes the following conclusions: (a) Platelets rapidly respond to soluble stimuli generated at sites of endovascular trauma and chemonavigate to chemotactic stimuli associated with microbial pathogens. (b) Platelets are the earliest and most predominant cells at sites of microbial colonization of vascular endothelium. (c) Platelets possess surface receptors and cytoplasmic granules consistent in structure and function with their antimicrobial roles. (d) Platelets interact directly and indirectly with and internalize microbial pathogens, thus promoting pathogen clearance from the blood stream and limiting their hematogenous dissemination. (e) Viral, bacterial, fungal, and protozoal pathogens are damaged or killed by activated platelets *in vitro*. (f) Platelets are capable of promoting complement fixation in the presence of microbial pathogens. (g) Platelets release microbicidal proteins (e.g., PMPs and kinocidins) when stimulated with microorganisms or host platelet agonists associated with infection *in vitro*. (h) Platelets generate toxic oxygen metabolites which likely contribute to their antimicrobial activity. (i) Platelets and leukocytes interact synergistically to exert enhanced antimicrobial functions *in vitro*. (j) Thrombocytopenia increases susceptibility to and severity of certain infections, and PMP[S] pathogens produce less severe infections *in vivo* as compared with their PMP[R] counterparts. Thus, platelets, PMPs, and kinocidins appear to play key and multifunctional roles in antimicrobial host defense.

## Acknowledgment

Many individuals have contributed to this chapter through their insights and efforts: Nannette Yount, Eric Brass, Paul Sullam, Steve Projan, Yi-Quan Tang, Bill Welch, Rich Proctor, Michael Selsted, Tomas Ganz, Robert Lehrer, Alan Waring, Dick Diamond, Elizabeth Simons, Renee Mercier, Chip Chambers, Gordon Archer, Ambrose Cheung, Scott Filler, Loren Miller, Ashraf Ibrahim, Matthew Fu, Don Sheppard, Iri Kupferwasser, Deborah Kupferwasser, Yan-Qiong Xiong, Kimberly Gank, Robert Dietz, Julie Gerberding, Kay Mitzel, and especially Jack Edwards. Their contributions, and those of others too numerous to list here, are appreciated.

M.R.Y and A.S.B. were supported, in part, through grants AI39001, AI39108, AI48031, RR13004, and RR14587 from the National Institutes of Health.

## References

1. Nachum, R., Watson, S. W., Sullivan, J. D., Jr., et al. (1980). Antimicrobial defense mechanisms in the horseshoe crab, *Limulus polyphemus:* Preliminary observations with heat-derived extracts of *Limulus* amoebocyte lysate. *J Invert Pathol, 32,* 51–58.
2. Maluf, N. S. R. (1939). The blood of arthropods. *Quart Rev Biol, 14,* 149–191.
3. Weksler, B. B. (1992). Platelets. In J. I. Gallin, I. M. Goldstein, & R. Snyderman (Eds.), *Inflammation: Basic principles and clinical correlates,* 2nd ed. (pp. 543–557). New York: Raven Press.
4. Tocantins, L. M. (1938). The mammalian blood platelet in health and disease. *Medicine, 17,* 155–257.
5. White, J. G. (1972). Platelet morphology and function. In W. J. Williams, E. Beutler, A. J. Erslev, & R. W. Rundles (Eds.), *Hematology* (pp. 1023–1039). New York: McGraw-Hill.
6. Marcus, A. J. (1969). Platelet function. *New Eng J Med, 280,* 1213, 1278, and 1330.
7. Heyssel, R. M. (1961). Determination of human platelet survival utilizing $^{14}$C-labelled serotonin. *J Clin Invest, 40,* 2134–2138.
8. Murphy, E. A., Robinson, G. A., & Rowsell, A. (1967). The pattern of platelet disappearance. *Blood, 30,* 26–31.
9. Harker, L. A., & Finch, C. A. (1969). Thrombokinetics in man. *J Clin Invest, 48,* 963–969.
10. Booyse, F., & Rafelson, M. E. (1967). Stable messenger RNA in the synthesis of contractile protein in human platelets. *Biochem Biophys Acta, 145,* 188–192.
11. Booyse, F., & Rafelson, M. E. (1967). *In vitro* incorporation of amino acids into the contractile protein of human blood platelets. *Nature, 215,* 283–285.
12. Warshaw, A. L., Laster, L., & Shulman, N. R. (1967). Protein synthesis by human platelets. *J Biol Chem, 242,* 2094–2100.
13. Day, H. J., Ang, G. A. T., & Holmsen, H. (1972). Platelet release reaction during clotting of native human platelet rich plasma. *Proc Soc Exp Med Biol, 139,* 717–721.
14. Colman, R. W. (1991). Receptors that activate platelets. *Proc Soc Exp Biol Med, 197,* 242–248.
15. MacFarlane, D. E., & Mills, D. C. B. (1975). The effects of ATP on platelets: Evidence against the central role of released ADP in primary aggregation. *Blood, 46,* 309–314.
16. MacFarlane, D. E., Walsh, P. N., Mills, D. C. B., et al. (1975). The role of thrombin in ADP-induced platelet aggregation and release: A critical evaluation. *Br J Haematol, 30,* 457–464.
17. Davies, T. A., Fine, R. E., Johnson, R. J., et al. (1993). Non-age related differences in thrombin responses by platelets from male patients with advanced Alzheimer's disease. *Biochem Biophys Res Comm, 194,* 537–543.
18. Day, H. J., & Rao, A. K. (1986). Evaluation of platelet function. *Semin Hematol, 23,* 89–101.
19. Parmentier, S., Kaplan, C., Catimel, B., et al. (1990). New families of adhesion molecules play a vital role in platelet functions. *Immunol Today, 11,* 225–227.
20. Yeaman, M. R. (1997). The role of platelets in antimicrobial host defense. *Clin Infect Dis, 25,* 951–968.

21. Yeaman, M. R., & Bayer, A. S. (1999). Antimicrobial peptides from platelets. *Drug Resistance Updates, 2,* 116–126.

22. Bancsi, M. J. L. F., Thompson, J., & Bertina, R. M. (1994). Stimulation of monocyte tissue factor expression in an *in vitro* model of bacterial endocarditis. *Infect Immun, 62,* 5669–5672.

23. Drake, T. A., & Pang, M. (1988). *Staphylococcus aureus* induces tissue factor expression in cultured human cardiac valve endothelium. *J Infect Dis, 157,* 749–756.

24. Drake, T. A., Rodgers, G. M., & Sande, M. A. (1984). Tissue factor is a major stimulus for vegetation formation in enterococcal endocarditis in rabbits. *J Clin Invest, 73,* 1750–1753.

25. White, J. G. (1987). Views of the platelet at rest and at work. *Ann NY Acad Sci, 509,* 156–176.

26. White, J. G., & Sauk, J. J. (1984). Microtubule coils in spread blood platelets. *Blood, 64,* 470–478.

27. Zucker, M. B. (1980). The functioning of blood platelets. *Sci Am, 242,* 70–89.

28. Scheld, W. M., Valone, J. A., & Sande, M. A. (1978). Bacterial adherence in the pathogenesis of infective endocarditis: Interaction of dextran, platelets, & fibrin. *J Clin Invest, 61,* 1394–1404.

29. Durack, D. T. (1975). Experimental bacterial endocarditis. IV. Structure and evolution of very early lesions. *J Clin Pathol, 45,* 81–89.

30. Calderone, R. A., Rotondo, M. F., & Sande, M. A. (1978). *Candida albicans* endocarditis: Ultrastructural studies of vegetation formation. *Infect Immun, 20,* 279–289.

31. Durack, D. T., Beeson, P. B., & Petersdorf, R. G. (1973). Experimental bacterial endocarditis. III. Production of progress of the disease in rabbits. *Br J Exp Pathol, 54,* 142–151.

32. Sullam, P. M., Drake, T. A., & Sande, M. A. (1985). Pathogenesis of endocarditis. *Am J Med, 78* (suppl 6B), 110–115.

33. Ferguson, D. J. P., McColm, A. A., Savage, T. J., et al. (1986). A morphological study of experimental rabbit staphylococcal endocarditis and aortitis. I. Formation and effect of infected and uninfected vegetations on the aorta. *Br J Exp Pathol, 67,* 667–678.

34. Scheld, W. M., & Sande, M. A. (1995). Endocarditis and intravascular infections. In G. L. Mandell, J. E. Bennet, & R. Dolin (Eds.), *Principles and practice of infectious diseases,* 5th ed. (pp. 740–782). New York: Churchill Livingstone.

35. Clawson, C. C. (1977). Role of platelets in the pathogenesis of endocarditis. In Infectious Endocarditis. *Am Heart Assoc Monograph, 52,* 24–27.

36. Roberts, W. C., & Buchbinder, N. A. (1972). Right-sided valvular infective endocarditis: A clinicopathologic study of 12 necropsy patients. *Am J Med, 53,* 7–19.

37. Buchbinder, N. A., & Roberts, W. C. (1972). Left-sided valvular infective endocarditis: A study of 45 necropsy patients. *Am J Med, 53,* 20–35.

38. Vinter, D. W., Burkel, W. E., Wakefield, T. W., et al. (1984). Radioisotope-labeled platelet studies and infection of vascular grafts. *J Vasc Surg, 6,* 921–923.

39. Osler, W. (1886). On certain problems in the physiology of the blood corpuscles. *Med News, 48,* 365–370, 393–399, and 421–425.

40. Cheung, A. L., & Fischetti, V. A. (1990). The role of fibrinogen in staphylococcal adherence to catheters *in vitro*. *J Infect Dis, 161,* 1177–1186.

41. Roberts, W. C., & Buchbinder, N. A. (1972). Right-sided valvular infective endocarditis: A clinicopathologic study of 12 necropsy patients. *Am J Med, 53,* 7–19.

42. Piguet, P. F., Vesin, C., Ryser, J. E., et al. (1993). An effector role for platelets in systemic and local lipopolysaccharide-induced toxicity in mice, mediated by a CD11a- and CD54a-dependent interaction with endothelium. *Infect Immum, 61,* 82–87.

43. Smith, C. W. (1993). Leukocyte-endothelial cell interaction. *Sem Hematol, 30,* 45–55.

44. Durack, D. T., Beeson, P. B., & Petersdorf, R. G. (1973). Experimental bacterial endocarditis. III. Production of progress of the disease in rabbits. *Br J Exp Pathol, 54,* 142–151.

45. Maisch, P. A., & Calderone, R. A. (1980). Adherence of *Candida albicans* to a fibrin-platelet matrix formed *in vitro*. *Infect Immun, 27,* 650–656.

46. Klotz, S. A., Harrison, J. L., & Misra, R. P. (1989). Aggregated platelets enhance adherence of *Candida* yeasts to endothelium. *J Infect Dis, 160,* 669–677.

47. Yeaman, M. R., Norman, D. C., & Bayer, A. S. (1992). *Staphylococcus aureus* susceptibility to thrombin-induced platelet microbicidal protein is independent of platelet adherence or aggregation *in vitro*. *Infect Immun, 60,* 2368–2374.

48. Sullam, P. M., Valone, F. H., & Mills, J. (1987). Mechanisms of platelet aggregation by viridans group streptococci. *Infect Immun, 55,* 1743–1750.

49. Robert, R., Senet, J. M., Mahaza, C., et al. (1992). Molecular basis of the interaction between *Candida albicans*, fibrinogen, and platelets. *J Mycol Med, 2,* 19–25.

50. Herzberg, M. C., Brintzenhofe, K. L., & Clawson, C. C. (1983). Aggregation of human platelets and adhesion of *Streptococcus sanguis*. *Infect Immun, 39,* 1457–1469.

51. Herzberg, M. C., Gong, K., MacFarlane, G. D., et al. (1990). Phenotypic characterization of *Streptococcus sanguis* virulence factors associated with bacterial endocarditis. *Infect Immun, 58,* 515–522.

52. Nicolau, D. P., Freeman, C. D., Nightingale, C. H., et al. (1993). Reduction of bacterial titers by low-dose aspirin in experimental aortic valve endocarditis. *Infect Immun, 61,* 1593–1595.

53. Clawson, C. C., & White, J. G. (1971). Platelet interaction with bacteria. II. Fate of bacteria. *Am J Pathol, 65,* 381–398.

54. Rosenfeld, S. I., Looney, R. J., & Leddy, J. P. (1985). Human platelet Fc receptor of immunoglobulin G. *J Clin Invest, 76,* 2317–2322.

55. Joseph, M., & Aurialt, C. (1983). A new function for platelets: IgE-dependent killing of schistosomes. *Nature, 303,* 810–812.

56. Bout, D., Joseph, M., & Ponet, M. (1986). Rat resistance to schistosomiasis: Platelet-mediated cytotoxicity induced by C-reactive protein. *Science, 231,* 153–156.

57. Ochenhouse, C. F., Magowan, C., & Chulay, J. D. (1989). Activation of monocytes and platelets by monoclonal anti-

bodies or malaria-infected erythrocytes binding to CD36 surface receptor *in vitro*. *J Clin Invest, 84,* 468–475.

58. Cosgrove, L. J., de Apice, A. J., Haddad, A., et al. (1987). CR3 receptors on platelets and it role in the prostaglandin metabolic pathway. *Immunol Cell Biol, 65,* 453–460.

59. Pancré, V., Monte, D., Delanoye, A., et al. (1990). Interleukin-6, the main mediator of interaction between monocytes and platelets in killing of *Schistosoma mansoni*. *Eur Cytokine Net, 1,* 15–19.

60. Peng, J., Friese, P., George, J. N., et al. (1994). Alteration of platelet function in dogs mediated by interleukin-6. *Blood, 83,* 398–403.

61. Clemetson K. J., Clemetson, J. M., Proudfoot, A. E., et al. (2000). Functional expression of CCR1, CCR3, CCR4, and CXCR4 chemokine receptors on human platelets. *Blood, 96,* 4046–4054.

62. Bureau, M. F., DeClerck, F., Lefort, J., et al. (1992). Thromboxane A2 accounts for bronchoconstriction, but not for platelet sequestration and microvascular albumin exchanges induced by f-met-leu-phe in the guinea pig lung. *J Pharmacol Exp Ther, 260,* 832–8840.

63. Coyle, A. J., & Vargaftig, B. B. (1995). Animal models for investigating the allergic and inflammatory properties of platelets. In M. Joseph, (Ed.), *Immunopharmacology of platelets*. (pp. 21–30). London: Academic Press.

64. Jaff, M. S., McKenna, D., & McCann, S. R. (1985). Platelet phagocytosis: A probable mechanism of thrombocytopenia in *Plasmodium falciparum* infection. *J Clin Pathol, 38,* 1318–1319.

65. Movat, H. Z., Weiser, W. J., Glynn, M. F., et al. (1965). Platelet phagocytosis and aggregation. *J Cell Biol, 27,* 531–543.

66. Joseph, M. (1995). Generation of free radicals by platelets. In M. Joseph, (Ed.), *Immunopharmacology of platelets*. (pp. 209–225). London: Academic Press.

67. Kitagawa, S., Fujisawa, H., Kametani, F., et al. (1992). Generation of active oxygen species in blood platelets: Spin trapping studies. *Free Radical Res Commun, 15,* 319–324.

68. Bout, D., Joseph, M., & Ponet, M. (1986). Rat resistance to schistosomiasis: Platelet-mediated cytotoxicity induced by C-reactive protein. *Science, 231,* 153–156.

69. Haque, A., Cuna, W., Bonnel, B., et al. (1985). Platelet-mediated killing of larvae from different filarial species in the presence of *Dipetalonema viteae*-stimlated IgE antibodies. *Parsitol Immunol, 7,* 517–526.

70. Zucker-Franklin, D., Termin, C. S., & Cooper, M. C. (1989). Structural changes in the megakaryocytes of patients infected with the human immune deficiency virus (HIV-I). *Am J Pathol, 134,* 1295–1303.

71. Zucker-Franklin, D., Seremetis, S., & Zheng, Z. Y. (1990). Internalization of human immunodeficiency virus type I and other retroviruses by megarkaryocytes and platelets. *Blood, 75,* 1920–1923.

72. Zucker-Franklin, D, (1995). Platelets in viral infections. In M. Joseph (Ed.), *Immunopharmacology of platelets*. (pp. 137–149). London: Academic Press.

73. Bik, T., Sarow, I., & Livne, A. (1982). Interaction between virus and human blood platelets. *Blood,* 482–495.

74. De Harven, E., & Friend, C. (1960). Further electron microscopic studies of a mouse leukemia induced by cell-free filtrates. *J Biophys Biochem Cytol, 7,* 747–754.

75. Dalton, A. J., Law, L. W., Moloney, J. B., et al. (1961). An electron microscopic study of a series of murine lymphoid neoplasms. *J Natl Cancer Inst (USA), 27,* 747–791.

76. Terada, H., Baldini, M., Ebbe, S., et al. (1966). Interaction of influenza virus with blood platelets. *Blood, 28,* 213–228.

77. Turpie, A. G., Chernesky, M. A., Larke, R. P., et al. (1973). Effect of Newcastle disease virus on human or rabbit platelets: Aggregation and loss of constituents. *Lab Invest, 28,* 575–580.

78. Sakaguchi, M., Sato, T., & Groopman, J. E. (1991). Human immunodeficiency virus infection of megakaryocytic cells. *Blood, 78,* 481–485.

79. Monté, D., Groux, H., Raharinivo, B., et al. (1992). Productive human immunodeficiency virus-1 infection of megakaryocytic cells is enhanced by tumor necrosis factor-α. *Blood, 79,* 2670–2679.

80. Brown, W. M., & Axelrad, A. A. (1976). Effect of Friend leukemia virus on megakaryocytes and platelets in mice. *Int J Cancer, 18,* 764–773.

81. Oski, F. A., & Naiman, J. L. (1966). Effect of live measles vaccine on the platelet count. *N Engl J Med, 275,* 352–356.

82. Danon, D., Jerushalmy, Z., & de Vries, A. (1959). Incorporation of influenza virus in human blood platelets *in vitro*: Electron microscopic observation. *Virology, 9,* 719–722.

83. Srivastava, A., Bruno, E., Bridell, R., et al. (1990). Parvovirus B19-induced perturbation of human megakaryocytopoiesis *in vitro*. *Blood, 76,* 1997–2004.

84. Myllya, G., Vaheri, A., Vesikari, T., et al. (1969). Interaction between human blood platelets, viruses, and antibodies. IV. Post-rubella thrombocytopenic purpura and platelet aggregation by rubella antigen-antibody interaction. *Clin Exp Immunol, 4,* 323–332.

85. Dixon, R., Rosse, W., & Ebbert, L. (1973). Quantitative determination of antibody in idiopathic thrombocytopenic purpura: Correlation of serum and platelet-bound antibody with clinical response. *N Engl J Med, 292,* 230–236.

86. Lurhuma, A. Z., Riccomi, H., & Masson, P. L. (1977). The occurrence of circulating immune complexes and viral antigens in idiopathic thrombocytopenic purpura. *Clin Exp Immunol, 28,* 49–55.

87. Sissons, J. G., Oldstone, M. B., & Schreiber, R. D. (1980). Antibody-dependent activation of the alternative complement pathway by measles virus-infected cells. *Proc Natl Acad Sci USA, 77,* 559–562.

88. Boonpucknavig, N., & Nimmanitya, S. (1979). Demonstration of dengue antibody complexes on the surface of platelets from patients with dengue hemorrhagic fever. *Am J Trop Med Hyg, 28,* 881–884.

89. Clawson, C. C., & White, J. G. (1971). Platelet interaction with bacteria. I. Reaction phases and effects of inhibitors. *Am J Pathol, 65,* 367–380.

90. Clawson, C. C. (1973). Platelet interaction with bacteria. III. Ultrastructure. *Am J Pathol, 70,* 449–472.

91. Clawson, C. C., Rao, G. H. R., & White, J. G. (1975). Platelet interaction with bacteria. IV. Stimulation of the release reaction. *Am J Pathol, 81,* 411–420.

92. Saluk-Juszczak, J., Wachowicz, B., & Kaca, W. (2000). Endotoxins stimulate generation of superoxide radicals and lipid peroxidation in blood platelets. *Microbios, 103,* 17–25.

93. Forrester, L. J., Campbell, B. J., Berg, J. N., et al. (1985). Aggregation of platelets by *Fusobacterium necrophorum. J Clin Microbiol, 22,* 245–249.

94. Czuprynski, C. J., & Balish, E. (1981). Interaction of rat platelets with *Listeria monocytogenes. Infect Immun, 33,* 103–108.

95. Copley, A. L., Maupin, B., & Balea, T. (1959). The agglutinant and adhesive behaviour of isolated human and rabbit platelets in contact with various strains of mycobacteria. *Acta Tubercul Scand, 37,* 151–161.

96. Kessler, C. M., Nussbaum, E., & Tuazon, C. U. (1987). *In vitro* correlation of platelet aggregation with occurrence of disseminated intravascular coagulation and subacute bacterial endocarditis. *J Lab Clin Med, 109,* 647–652.

97. Timmons, S., Huzoor-Akbar, A., Grabarek, J., et al. (1986). Mechanism of human platelet activation by endotoxic glycolipid-bearing mutant Re595 of *Salmonella minnesota. Blood, 68,* 1015–1023.

98. Mandell, G. L., & Hook, E. W. (1969). The interaction of platelets, *Salmonella,* and mouse peritoneal macrophages. *Proc Soc Exp Biol Med, 132,* 757–759.

99. Simmonet, M., Triadou, P., Frehel, C., et al. (1992). Human platelet aggregation by *Yersinia pseudotuberculosis* is mediated by invasin. *Infect Immun, 60,* 366–373.

100. Herzberg, M. C., Brintzenhofe, K. L., & Clawson, C. C. (1983). Aggregation of human platelets and adhesion of *Streptococcus sanguis. Infect Immun, 39,* 1457–1469.

101. Herzberg, M. C., Brintzenhofe, K. L., & Clawson, C. C. (1983). Cell-free released components of *Streptococcus sanguis* inhibit human platelet aggregation. *Infect Immun,* 394–401.

102. Gong, K., & Herzberg, M. C. (1997). *Streptococcus sanguis* expresses a 150-kilodalton two-domain adhesin: Characterization of several independent adhesin epitopes. *Infect Immun, 65,* 3815–3821.

103. Sullam, P. M., Payan, D. G., Dazin, P. F., et al. (1990). Binding of viridans group streptococci to human platelets: A quantitative analysis. *Infect Immun, 58,* 3802–3806.

104. Bayer, A. S., Sullam, P. M., Ramos, M., et al. (1995). *Staphylococcus aureus* induces platelet aggregation via a fibrinogen-dependent mechanism which is independent of principal platelet glycoprotein IIb/IIIa fibrinogen binding domains. *Infect Immun, 63,* 3634–3641.

105. Yeaman, M. R., Sullam, P. M., Dazin, P. F., et al. (1992). Characterization of *Staphylococcus aureus*-platelet binding by quantitative flow cytometric analysis. *J Infect Dis, 166,* 65–73.

106. Alugupalli, K. R., Michelson, A. D., Barnard, M. R., et al. (2001). Platelet activation by a relapsing fever spirochaete results in enhanced bacterium-platelet interaction via integrin alpha II/beta III activation. *Mol Microbiol, 39,* 330–340.

107. Siboo, I. R., Cheung, A. L., Bayer, A. S., et al. (2001). Clumping factor A mediates binding of *Staphylococcus aureus* to human platelets. *Infect Immun, 69,* 3120–3127.

108. Bensing, B. A., Rubens, C. E., & Sullam, P. M. (2001). Genetic loci of *Streptoccus mitis* that mediate binding to human platelets. *Infect Immun, 69,* 1373–1380.

109. Cheung, A. L., Krishnan, M., Jaffe, E. A., et al. (1991). Fibrinogen acts as a bridging molecule in the adherence of *Staphylococcus aureus* to cultured human endothelial cells. *J Clin Invest, 87,* 2236–2245.

110. Clawson, C. C., White, J. G., & Herzberg, M. C. (1980). Platelet interaction with bacteria. VI. Contrasting the role of fibrinogen and fibronectin. *Am J Hematol, 9,* 43–53.

111. Toy, P. T. C. Y., Lai, L-W., Drake, T. A., et al. (1985). Effect of fibronectin on adherence of *Staphylococcus aureus* to fibrin thrombi *in vitro. Infect Immun, 48,* 83–86.

112. Herrmann, M., Suchard, S. J., Boxer, L. A., et al. (1991). Thrombospondin binds to *Staphylococcus aureus* and promotes staphylococcal adherence to surfaces. *Infect Immun, 59,* 271–288.

113. Hawiger, J., Streackley, S., Hammond, D., et al. (1979). Staphylococcal-induced human platelet injury mediated by protein A and immunoglobulin G Fc receptor. *J Clin Invest, 64,* 931–937.

114. Pfueller, S. L., & Cosgrove, L. J. (1980). Staphylococci-induced human platelet injury. *Thromb Res, 19,* 733–735.

115. Spika, J. S., Peterson, P. K., Wilkinson, B. J., et al. (1982). Role of peptidoglycan from *Staphylococcus aureus* in leukopenia, thrombocytopenia, and complement activation associated with bacteremia. *J Infect Dis, 146,* 227–234.

116. Ginsberg, M. H., & Henson, P. M. (1978). Enhancement of platelet response to immune complexes and IgG aggregates by lipid A-rich bacterial lipopolysaccharides. *J Exp Med, 147,* 207–218.

117. Sullam, P. M., Jarvis, G. A., & Valone, F. H. (1988). Role of immunoglobulin G in platelet aggregation by viridans group streptococci. *Infect Immun, 56,* 2907–2911.

118. Zimmerman, T. S., & Spregelberg, H. L. (1975). Pneumococcus-induced serotonin release from human platelets. *J Clin Invest, 56,* 828–834.

119. Ford, I., Douglas, C. W. I., Preston, S. E., et al. (1993). Mechanisms of platelet aggregation by *Streptococcus sanguis,* a causative organism in infective endocarditis. *Brit J Hematol, 84,* 85–100.

120. Kurpiewski, G. E., Forrester, L. J., Campbell, B. J., et al. (1983). Platelet aggregation by *Streptococcus pyogenes. Infect Immun, 39,* 704–708.

121. Johnson, C. M., & Bowie, E. J. W. (1992). Pigs with von Willebrand disease may be resistant to experimental infective endocarditis. *J Lab Clin Med, 120,* 553–558.

122. Gong, K., Wen, D. Y., Ouyang, T., et al. (1995). Platelet receptors for the *Streptococcus sanguis* adhesin- and aggregation-associated antigens are distinguished by anti-idiotypical monoclonal antibodies. *Infect Immun, 63,* 3628–3633.

123. Sedwitz, M. M., Hye, R. J., & Stabile, B. E. (1988). The changing epidemiology of pseudoaneurysm: Therapeutic implications. *Arch Surg, 123,* 473–476.

124. Kaye, D. (1985). The changing pattern of infective endocarditis. *Am J Med, 78* (suppl 6B), 157–162.

125. Edwards, J. E., Jr. (1991). Invasive *Candida* infections: Evolution of a fungal pathogen. *N Engl J Med, 324,* 1060–1062.

126. Horn, R., Wong, B., Kiehn, T. E., et al. (1985). Fungemia in a cancer hospital: Changing frequency, earlier onset, and results of therapy. *Rev Infect Dis, 7,* 646–655.

127. Maisch, P. A., & Calderone, R. A. (1981). Role of surface mannans in the adherence of *Candida albicans* to fibrin-platelet clots formed *in vitro. Infect Immun, 32,* 92–97.

128. Robert, R., Senet, J. M., Mahaza, C., et al. (1992). Molecular basis of the interaction between *Candida albicans*, fibrinogen, and platelets. *J Mycol Med, 2,* 19–25.

129. Christin, L., Wysong, D. R., Meshulam, T., et al. (1997). Mechanisms and target sites of damage in killing of Candida albicans hyphae by human polymorphonuclear neutrophils. *J Infect Dis, 176,* 1567–1578.

130. Haque, A., Cuna, W., Bonnel, B., et al. (1985). Platelet-mediated killing of larvae from different filarial species in the presence of *Dipetalonema viteae*-stimulated IgE antibodies. *Parasite Immunol, 7,* 517–526.

131. Pancré, V., Cesbron, J. Y., Auriault, C., et al. (1988). IgE dependent killing of *Brugia malayi* microfilariae by human platelets and its modulation by T cell products. *Int Arch Allergy Appl Immunol, 85,* 483–486.

132. Ridel, P. R., Auriault, C., Darcy, F., et al. (1988). Protective role of IgE in immunocompromised rat toxoplasmosis. *J Immunol, 141,* 978–983.

133. Yong, E. C., Chi, E. Y., Fritsche, T. R., et al. (1991). Human platelet-mediated cytotoxicity against *Toxoplasma gondii*: Role of thromboxane. *J Exp Med, 173,* 65–78.

134. Viens, P., Dubois, R., & Kingshavn, P. A. L. (1983). Platelet activity in immune lysis of *Trypanosoma musculi J Parasitol, 13,* 527–530.

135. Umekita, L. F., & Mota, I. (1989). *In vitro* lysis of sensitized *Trypanosoma cruzi* by platelets: Role of C$_{3b}$ receptors. *Parasite Immunol, 11,* 561–563.

136. Peyron, F., Polack, B., Lamotte, D., et al. (1989). *Plasmodium falciparum* growth inhibition by human platelets *in vitro. Parasitol, 99,* 317–322.

137. Capron, M., Jouault, T., Prin, L., et al. (1986). Functional study of a monoclonal antibody to IgE Fc receptor (FcE R$_2$) of eosinophils, platelets, and macrophages. *J Exp Med, 164,* 72–89.

138. Verwaerde, C., Joseph, M., Capron, M., et al. (1987). Fuctional properties of a rat monoclonal IgE antibody specific for *Schistosoma mansoni. J Immunol, 138,* 4441–4446.

139. Monté, D., Wietzerbin, J., Pancré, V., et al. (1991). Identification and characterization of a functional receptor for interferon-γ on a megakaryocytic cell line. *Blood, 78,* 2062–2069.

140. Pancré, V., Auriault, C., Joseph, M., et al. (1986). A suppressive lymphokine of platelet cytotoxic function. *J Immunol, 137,* 585–591.

141. Fodor, J. (1887). Die fahigkeit des blutes bakterien zu vernichten. *Dtsch Med Wochenschr, 13,* 745–747.

142. Gengou, O. (1901). De l'origine de l'axenine de serums normaux. *Ann Inst Pasteur* (Paris), *15,* 232–245.

143. Hirsch, J. G. (1960). Comparative bactericidal activities of blood serum and plasma serum. *J Exp Med, 112,* 15–22.

144. Jago, R., & Jacox, R. F. (1961). Cellular source and character of a heat-stable bactericidal property associated with rabbit and rat platelets. *J Exp Med, 113,* 701–709.

145. Weksler, B. B. (1971). Induction of bactericidal activity in human platelets. *Clin Res, 19,* 434–435.

146. Kahn, R. A., & Flinton, L. J. (1974). The relationship between platelets and bacteria. *Blood, 44,* 715–721.

147. Miragliotta, G., Lafata, M., & Jirilla, E. (1988). Antibacterial activity mediated by human platelets. *Agents Actions, 25,* 401–406.

148. Carroll, S. F., & Martinez, R. J. (1981). Antibacterial peptide from normal rabbit serum. I. Isolation from whole serum, activity, and microbiologic spectrum. *Biochem, 20,* 5973–5981.

149. Carroll, S. F., & Martinez, R. J. (1981). Antibacterial peptide from normal rabbit serum. II. Compositional microanalysis. *Biochem, 20,* 5981–5987.

150. Carroll, S. F., & Martinez, R. J. (1981). Antibacterial peptide from normal rabbit serum. III. Inhibition of microbial electron transport. *Biochem, 20,* 5988–5994.

151. Myrvik, Q. N. (1956). Serum bactericidins active against Gram-positive bacteria. *Ann NY Acad Sci, 66,* 391–400.

152. Myrvik, Q. N., & Leake, E. S. (1960). Studies on antibacterial factors in mammalian tissues and fluids. IV. Demonstration of two non-dialyzable components in the serum bactericidin system for *Bacillus subtilis. J Immunol, 84,* 247–250.

153. Johnson, F. B., & Donaldson, D. M. (1968). Purification of staphylocidal β-lysin from rabbit serum. *J Bacteriol, 96,* 589–595.

154. Donaldson, D. M., & Tew, J. G. (1977). β-lysin of platelet origin. *Bacteriol Rev, 41,* 501–512.

155. Weksler, B. B., & Nachman, R. L. (1971). Rabbit platelet bactericidal protein. *J Exp Med, 134,* 1114–1130.

156. Tew, J. G., Roberts, R. R., & Donaldson, D. M. (1974). Release of β-lysin from platelets by thrombin and by a factor produced in heparinized blood. *Infect Immun, 9,* 179–186.

157. Dankert, J. (1988). *Role of platelets in early pathogenesis of viridans group streptococcal endocarditis.* Ph.D. Dissertation, University of Groningen, The Netherlands.

158. Dankert, J., van der Werff, J., Zaat, S. A. J., et al. (1995). Involvement of bactericidal factors from thrombin-stimulated platelets in clearance of adherent viridans streptococci in experimental infective endocarditis. *Infect Immun, 63,* 663–671.

159. Darveau, R. P., Blake, J., Seachord, C. L., et al. (1992). Peptide related to the carboxy-terminus of human platelet factor IV with antibacterial activity. *J Clin Invest, 90,* 447–455.

160. Yeaman, M. R., Puentes, S. M., Norman, D. C., et al. (1992). Partial purification and staphylocidal activity of thrombin-

induced platelet microbicidal protein. *Infect Immun, 60,* 1202–1209.

161. Yeaman, M. R., Tang, Y-Q., Shen, A. J., et al. (1997). Purification and *in vitro* activities of rabbit platelet microbicidal proteins. *Infect Immun, 65,* 1023–1031.

162. Yeaman, M. R., Ibrahim, A., Filler, S. G., et al. (1993). Thrombin-induced rabbit platelet microbicidal protein is fungicidal *in vitro. Antimicrob Agents Chemother, 37,* 546–553.

163. Ganz, T., Selsted, M. E., & Lehrer, R. I. (1990). Defensins. *Eur J Haematol, 44,* 1–8.

164. Azizi, N., Li, C., Shen, A. J., et al. (1996). *Staphylococcus aureus* elicits release of platelet microbicidal proteins *in vitro.* Abstract no. 866, 36th Interscience Conference on Antimicrobial Agents and Chemotherapy, New Orleans, LA.

165. Bayer, A. S., Ramos, M. D., Menzies, B. E., et al. (1997). Hyperproduction of alpha-toxin by *Staphylococcus aureus* results in paradoxically reduced virulence in experimental endocarditis: A host defense role for platelet microbicidal proteins. *Infect Immun, 65,* 4652–4660.

166. Tang, Y. Q., Yeaman, M. R., & Selsted, M. E. (2002). Antimicrobial proteins from human platelets. *Infect Immun, 70,* 6524–6533.

167. Tang, Y. Q., Yeaman, M. R., & Selsted, M. E. (1995). Microbicidal and synergistic activities of human platelet factor-4 (hPF-4) and connective tissue activating peptide-3 (CTAP-3). *Blood, 86,* 556a.

168. Krijgsveld, J., Zaat, S. A., Meeldijk, J., et al. (2000). Thrombocidins, microbicidal proteins from human blood platelets, are C-terminal deletion products of CXC chemokines. *J Biol Chem, 275,* 20374–20381.

169. Zasloff, M. (1992). Antibiotic peptides as mediators of innate immunity. *Curr Opin Immunol, 4,* 3–8.

170. Bowman, H. G. (1991). Antibacterial peptides: Key components in immunity. *Cell, 65,* 205–211.

171. Taylor, R. (1993). Drugs R Us: Finding new antibiotics at the endogenous pharmacy. *J NIH Res, 5,* 59–63.

172. Trier, D. A., Bayer, A. S., & Yeaman, M. R. (2000). *Staphylococcus aureus* elicits antimicrobial responses from platelets via an ADP-dependent pathway. Abstract 1010, 40th ICAAC, American Society for Microbiology. Toronto, Canada.

173. Sharma, A., Novak, E. K., Sojar, H. T., et al. (2000). *Porphyromonas gingivalis* platelet aggregation activity: Outer membrane vesicles are potent activators of murine platelets. *Oral Microbiol Immunol, 15,* 393–396.

174. Srichaikul, T., Nimmanitya, S., Sripaisarn, T., et al. (1989). Platelet function during acute phase of dengue hemorrhagic fever. *Southeast Asian J Trop Med Pub Health, 20,* 19–25.

175. Lorenz, R., & Brauer, M. (1988). Platelet factor 4 (PF4) in septicaemia. *Infection, 16,* 273–276.

176. Mezzano, S., Burgos, M. E., Ardiles, L., et al. (1992). Glomerular localization of platelet factor 4 in streptococcal nephritis. *Nephron, 61,* 58–63.

177. Yamamoto, Y., Klein, T. W., & Friedman, H. (1997). Involvement of mannose receptor in cytokine interleukin-1 beta (IL-1 beta), IL-6, and granulocyte macrophage colony stim-

ulating factor responses, but not in chemokine macrophage inflammatory protein-1 beta (MIP-1 beta), MIP-2, and KC responses, caused by attachment of *Candida albicans* to macrophages. *Infect Immun, 65,* 1077–1082.

178. Essien, E. M., & Ebhota, M. I. (1983). Platelet secretory activities in acute malaria (*Plasmodium falciparum*) infection. *Acta Haematol, 70,* 183–188.

179. Wilson, M., Blum, R., Dandona, P., et al. (2001). Effects in humans of intravenously administered endotoxin on soluble cell-adhesion molecule and inflammatory markers: A model of human diseases. *Clin Exp Pharmacol Physiol, 28,* 376–380.

180. Yeaman, M. R., Bayer, A. S., Koo, S. P., et al. (1998). Platelet microbicidal proteins and neutrophil defensin disrupt the *Staphylococcus aureus* cytoplasmic membrane by distinct mechanisms of action. *J Clin Invest, 101,* 178–187.

181. Koo, S. P., Bayer, A. S., Sahl, H. G., et al. (1996). Staphylocidal action of thrombin-induced platelet microbicidal protein is not solely dependent on transmembrane potential. *Infect Immun, 64,* 1070–1074.

182. Wu, T. M., Yeaman, M. R., Nast, C., et al. (1996). Ultrastructural evidence that platelet microbicidal protein (PMP) targets the bacterial cell membrane. Abstract no. A72, 96th General Meeting of the American Society for Microbiology, New Orleans, LA.

183. Xiong, Y. Q., Yeaman, M. R., & Bayer, A. S. (1999). *In vitro* antibacterial activities of platelet microbicidal protein and neutrophil defensin against *Staphylococcus aureus* are influenced by antibiotics differing in mechanism of action. *Antimicrob Agents Chemother, 43,* 1111–1117.

184. Koo, S. P., Yeaman, M. R., & Bayer, A. S. (1996). Staphylocidal action of thrombin-induced platelet microbicidal protein is influenced by microenvironment and target cell growth phase. *Infect Immun, 64,* 3758–3764.

185. Koo, S. P., Bayer, A. S., Kagan, B. L., et al. (1999). Membrane permeabilization by thrombin-induced PMP-1 is modulated by transmembrane voltage polarity and magnitude. *Infect Immun, 67,* 2475–2481.

186. Koo, S. P., Yeaman, M. R., & Bayer, A. S. (1996). Cell membrane is a principal target for the staphylocidal action of platelet microbicidal protein. Abstract no. 171, 34th Annual Meeting of the Infectious Diseases Society of America, New Orleans, LA.

187. Xiong, Y. Q., Yeaman, M. R., & Bayer, A. S. (2000). Thrombin-induced platelet microbicidal proteins (tPMPs) inhibit macromolecular synthesis in *Staphylococcus aureus.* Abstract A111, 100th General Meeting of the American Society for Microbiology, Los Angeles, CA.

188. Mercier, R. C., Rybak, M. J., Bayer, A. S., et al. (2000). Influence of platelets and platelet microbicidal protein susceptibility on the fate of *Staphylococcus aureus* in an *in vitro* model of infective endocarditis. *Infect Immun, 68,* 4699–4705.

189. Asensi, V., & Fierer, J. (1991). Synergistic effect of human lysozyme plus ampicillin or β-lysin on the killing of *Listeria monocytogenes. J Infect Dis, 163,* 574–578.

190. Yeaman, M. R., Norman, D. C., & Bayer, A. S. (1992). Platelet microbicidal protein enhances antibiotic-induced

killing of and post-antibiotic effect in *Staphylococcus aureus*. *Antimicrob Agents Chemother, 36,* 1665–1670.

191. Dhawan, V. K., Yeaman, M. R., & Bayer, A. S. (1999). Influence of *in vitro* susceptibility phenotype against thrombin-induced platelet microbicidal protein on treatment and prophylaxis outcomes of experimental *Staphylococcus aureus* endocarditis. *J Infect Dis, 180,* 1561–1568.

192. Dhawan, V. K., Bayer, A. S., & Yeaman, M. R. (2000). Thrombin-induced platelet microbicidal protein susceptibility phenotype influences the outcome of oxacillin prophylaxis and therapy of experimental *Staphylococcus aureus* endocarditis. *Antimicrob Agents Chemother, 44,* 3206–3209.

193. Yeaman, M. R., Sullam, P. M., Dazin, P. F., et al. (1994). Platelet microbicidal protein alone and in combination with antibiotics reduces *Staphylococcus aureus* adherence to platelets *in vitro. Infect Immun, 62,* 3416–3423.

194. Yeaman, M. R., Sullam, P. M., Dazin, P. F., et al. (1994). Fluconazole and platelet microbicidal protein inhibit *Candida* adherence to platelets *in vitro. Antimicrob Agents Chemother, 38,* 1460–1465.

195. Filler, S. G., Joshi, M., Phan, Q. T., et al. (1999). Platelets protect vascular endothelial cells from injury due to *Candida albicans* Abstract 2163, 39th ICAAC Meeting, American Society for Microbiology, San Francisco, CA.

196. Polley, M. J., Nachman, R. L., & Weksler, B. B. (1981). Human complement in the arachidonic acid transformation pathway in platelets. *J Exp Med, 153,* 257–268.

197. Polley, M. J., & Nachman, R. L. (1983). Human platelet activation by C3a and C3a des-arg. *J Exp Med, 158,* 603–615.

198. Zimmerman, T. S., & Kolb, W. P. (1976). Human platelet-initiated formation and uptake of the $C_5$-$C_9$ complex of human complement. *J Clin Invest, 57,* 203–211.

199. Weksler, B. B., & Coupal, C. E. (1973). Platelet dependent generation of chemotactic activity in serum. *J Exp Med, 137,* 1419–1429.

200. Hansch, G. M., Gemsa, D., & Resch, K. (1985). Induction of prostanoid synthesis in human platelets by the late complement components $C_{5b}$-$C_9$ and channel forming antibiotic nystatin: Inhibition of the reacylation of liberated arachidonic acid. *J Immunol, 135,* 1320–1324.

201. Simpson, R. M., Prancan, A., Izzi, J. M., et al. (1982). Generation of thromboxane $A_2$ and aorta-contracting activity from platelets stimulated with modified C-reactive protein. *Immunol, 47,* 193–202.

202. Henson, P. M., & Ginsberg, M. H. (1981). Immunological reactions of platelets. In J. L. Gordon (Ed.), *Platelets in biology and pathology, 2* (pp. 265–308). Amsterdam: Elsevier.

203. Pancré, V., Joseph, M., Cesbron, J. Y., et al. (1987). Induction of platelet cytotoxic functions by lymphokines: Role of interferon gamma. *J Immunol, 138,* 4490–4495.

204. Damonville, M., Wietzerbin, J., Pancré, V., et al. (1988). Recombinant tumor necrosis factors mediate platelet cytotoxicity to *Schistosoma mansoni* larvae. *J Immunol, 140,* 3962–3965.

205. Kenney, D. M., & Davis, A. E., III. (1981). Association of alternative complement pathway components with human

blood platelets: Secretion and localization of factor D and beta 1H globin. *Clin Immunol Immunopathol, 21,* 351–363.

206. Zucker, M. B., Grant, R. A., Alper, C. A., et al. (1974). Requirement for complement components and fibrinogen in the zymosan-induced release reaction of human blood platelets. *J Immunol, 113,* 1744–1755.

207. Devine, D. V., Siegel, R. S., & Rosse, W. F. (1987). Interactions of the platelets in paryoxysmal nocturmal hemoglobinuria with complement. *J Clin Invest, 79,* 131–137.

208. Nguyen, T., Ghebrehiwet, B., & Peerschke, E. I. (2000). *Staphylococcus aureus* protein A recognizes platelet gC1qR/p33: A novel mechanism for staphylococcal interactions with platelets. *Infect Immun, 68,* 2061–2068.

209. Clark-Lewis, I., Dewald, B., Geiser, T., et al. (1993). Platelet factor-4 binds to interleukin-8 receptors and activates neutrophils when its N-terminus is modified with Glu-Leu-Arg. *Proc Natl Acad Sci USA, 90,* 3574–3577.

210. Drake, W. T., Lopes, N. N., Fenton, J. W., II, et al. (1992). Thrombin enhancement of interleukin-1 and tumor necrosis factor-$\alpha$ induced polymorphonuclear leukocyte migration. *Lab Invest, 67,* 617–627.

211. Schroder, J. M., Sticherling, M., Persoon, N. L., et al. (1990). Identification of a novel platelet-derived neutrophil-chemotactic polypeptide with structural homology to platelet factor-4. *Biochem Biophys Res Comm, 172,* 898–904.

212. Schall, T. J., Bacon, K., Toy, K. J., et al. (1990). Selective attraction of monocytes and T-lymphocytes of the memory phenotype by cytokine RANTES. *Nature, 347,* 669–671.

213. Deuel, T. F., Senior, R. M., Chang, D., et al. (1981). Platelet factor-4 is chemotactic for neutrophils and monocytes. *Proc Natl Acad Sci USA, 78,* 4548–4587.

214. Tzeng, D. Y., Deuel, T. F., Huang, J. S., et al. (1985). Platelet-derived growth factor promotes polymorphonuclear leukocyte activation. *Blood, 4,* 1123–1128.

215. Nachman, R. L., & Weksler, B. B. (1972). The platelet as an inflammatory cell. *Ann NY Acad Sci, 201,* 131–137.

216. Lellouch-Tubiana, A., Lefort, J., Simon, M-T., et al. (1988). Eosinophil recruitment into guinea pig lungs after PAF-acether and allergen administration: Modulation by prostacyclin, platelet depletion, and selective antagonists. *Am Rev Respir Dis, 137,* 948–954.

217. Jungi, T. W., Spycher, M. O., Nydegger, V. E., et al. (1986). Platelet-leukocyte interactions: Selective binding of thrombin-stimulated platelets to human monocytes, polymorphonuclear leukocytes and related cell lines. *Blood, 67,* 629–636.

218. Silverstein, R., & Nachman, R. L. (1987). Thrombospondin mediates the interaction of stimulated platelets with monocytes. *J Clin Invest, 79,* 867–874.

219. Burns, G. F., Cosgrove, L., Triglia, T., et al. (1986). The IIb-IIIa glycoprotein complex that mediates platelet aggregation is directly implicated in leukocyte adhesion. *Cell, 45,* 269–280.

220. Simon, D. I., Chen, Z., Xu, H., et al. (2000). Platelet glycoprotein ibalpha is a counterreceptor for the leukocyte integrin Mac-1 (CD11b/CD18). *J Exp Med, 192,* 193–204.

221. Clark, R. A., & Klebanoff, S. J. (1979). Myeloperoxidase-mediated platelet release reaction. *J Clin Invest, 63,* 177–183.

222. Mehta, P., Mehta, J., & Lawson, D. (1986). Leukotrienes potentiate the effects of epinephrine and thrombin on human platelet-aggregation. *Thromb Res, 41,* 731–738.

223. Schwartz, B. F., & Monroe, M. C. (1986). Human platelet aggregation is initiated by peripheral blood mononuclear cells exposed to bacterial lipopolysaccharide *in vitro. J Clin Invest, 78,* 1136–1141.

224. Boogaerts, M. A., Yamada, O., Jacob, H. S., et al. (1982). Enhancement of granulocyte-endothelial cell adherence and granulocyte-induced cytotoxicity by platelet release products. *Proc Natl Acad Sci USA, 79,* 7019–7023.

225. Spagnuolo, P. J., Ellner, J. J., & Hassid, A. (1980). Thromboxane A$_2$ mediates augmented polymorphonuclear leukocyte adhesiveness.

226. Wu, T., Yeaman, M. R., & Bayer, A. S. (1994). *In vitro* resistance to platelet microbicidal protein correlates with endocarditis source among staphylococcal isolates. *Antimicrob Agents Chemother, 38,* 729–732.

227. Yeaman, M. R., Soldan, S. S., Ghannoum, M., et al. (1996). Resistance to platelet microbicidal protein results in the increased severity of experimental *Candida albicans* endocarditis. *Infect Immun, 64,* 1379–1384.

228. Dhawan, V. K., Bayer, A. S., & Yeaman, M. R. (1998). *In vitro* resistance to thrombin-induced platelet microbicidal protein is associated with enhanced progression and hematogenous dissemination in experimental *Staphylococcus aureus* infective endocarditis. *Infect Immun, 66,* 3476–3479.

229. Fowler, V. G., McIntyre, L. M., Yeaman, M. R., et al. (2000). *In vitro* resistance to thrombin-induced platelet microbicidal protein in isolates of *Staphylococcus aureus* from endocarditis patients correlates with an intravascular device source. *J Infect Dis, 182,* 1251–1254.

230. Fields, P. L., Groisman, E. A., & Heffron, F. (1989). A *Salmonella* locus that controls resistance to micro-bicidal proteins from phagocytic cells. *Science, 243,* 1059–1062.

231. Groisman, E. A., Parra-Lopez, C., Salcedo, M., et al. (1992). Resistance to host antimicrobial peptides in necessary for *Salmonella* virulence. *Proc Natl Acad Sci USA, 89,* 11939–11943.

232. Meyer, M. W., Gong, K., & Herzberg, M. C. (1998). *Streptococcus sanguis*-induced platelet clotting in rabbits and hemodynamic and cardiopulmonary consequences. *Infect Immun, 66,* 5906–5914.

233. Sullam, P. M., Frank, U., Yeaman, M. R., et al. (1993). Effect of thrombocytopenia on the early course of streptococcal endocarditis. *J Infect Dis, 168,* 910–914.

234. Viscoli, C., Bruzzi, P., Castagnola, E., et al. (1994). Factors associated with bacteraemia in febrile, granulocytopenic patients: The International Antimicrobial Therapy Cooperative Group (IATCG) of the European Organization for Research and Treatment of Cancer (EORTC). *Eur J Cancer, 30,* 430–437.

235. Feldman, C., Kallenbach, J. M., Levy, H., et al. (1991). Comparison of bacteraemic community-acquired lobar pneumonia due to *Streptococcus pneumoniae* and *Klebsiella pneumoniae* in an intensive care unit. *Respiration, 58,* 265–270.

236. Kirkpatrick, B., Reeves, D. S., & MacGowan, A. P. (1994). A review of the clinical presentation, laboratory features, antimicrobial therapy and outcome of 77 episodes of pneumococcal meningitis occurring in children and adults. *J Infect, 29,* 171–182.

237. Chang, F. Y., Singh, N., Gayowski, T., et al. (2000). Thrombocytopenia in liver transplant recipients: Predictors, impact on fungal infections, and role of endogenous thrombopoietin. *Transplantation, 69,* 70–75.

238. Korzweniowski, O. M., Scheld, W. M., Bithell, T. C., et al. (1979). The effect of aspirin on the production of experimental *Staphylococcus aureus* endocarditis [abstract]. In Program Abstracts of the 19th Interscience Conference on Antimicrobial Agents and Chemotherapy (Boston). Washington, DC: American Society for Microbiology.

239. Berney, P., & Francioli, P. (1990). Successful prophylaxis of experimental streptococcal endocarditis with single-dose amoxicillin administered after bacterial challenge. *J Infect Dis, 161,* 281–285.

240. Yersin, B. R., Glauser, M. P., & Freedman, L. R. (1982). Effect of nitrogen mustard on natural history of right-sided streptococcal endocarditis in rabbits: Role for cellular host defenses. *Infect Immun, 35,* 320–325.

241. Nicolau, D. P., Freeman, C. D., Nightingale, C. H., et al. (1993). Reduction of bacterial titers by low-dose aspirin in experimental aortic valve endocarditis. *Infect Immun, 61,* 1593–1595.

242. Kupferwasser, L. I., Yeaman, M. R., Shapiro, S. M., et al. (1999). Acetylsalicylic acid reduces vegetation bacterial density, hematogenous bacterial dissemination, and frequency of embolic events in experimental *Staphylococcus aureus* endocarditis through antiplatelet and antibacterial effects. *Circulation, 99,* 2791–2797.

243. Sullam, P. M., Bayer, A. S., Foss, W. M., et al. (1996). Diminished platelet binding *in vitro* by *Staphylococcus aureus* is associated with reduced virulence in a rabbit model of infective endocarditis. *Infect Immun, 64,* 4915–4921.

244. Mavrommatis, A. C., Theodoridis, T., Orfanidou, A., et al. (2000). Coagulation system and platelets are fully activated in uncomplicated sepsis. *Crit Care Med, 28,* 451–457.

245. Youssefian, T., Drouin, A., Masse, J. M., et al. (2002). Host defense role of platelets: Engulfment of HIV and *Staphylococcus aureus* occurs in a specific subcellular compartment and is enhanced by platelet activation. *Blood, 99,* 4021–4029.

246. Siboo, I. R., Chambers, H. F., & Sullam, P. M. (2005). Role of SraP, a serine-rich surface protein of *Staphylococcus aureus* in binding to human platelets. *Infect Immun, 73,* 2273–2280.

247. Niemann, S., Spehr, N., Van Aken, H., et al. (2004). Soluble fibrin is the main mediator of *Staphylococcus aureus* adhesion to platelets. *Circulation, 110,* 193–200.

248. Herrmann, M., Lai, Q. J., Albrecht, R. M., et al. (1993). Adhesion of *Staphylococcus aureus* to surface-bound

platelets: Role of fibrinogen-fibrin and platelet ligands. *J Infect Dis, 167,* 312–322.

249. Hermann, M., Jaconi, M. E., Dahlgren, C., et al. (1990). Neutrophil bactericidal activity against *Staphylococcus aureus* adherent on biological surfaces: Surface-bound extracellular matrix proteins activate intracellular killing by oxygen-dependent and -independent mechanisms. *J Clin Invest, 86,* 942–951.

250. Loughman, A., Fitzgerald, J. R., Brennan, M. P., et al. (2005). Roles for fibrinogen, immunoglobulin and complement in platelet activation promoted by *Staphylococcus aureus* clumping factor A. *Mol Microbiol, 57,* 804–818.

251. Turner, R. B., Liu, L., Sazonova, I. Y., et al. (2002). Structural elements that govern substrate specificity of the clot-dissolving enzyme plasmin. *J Biol Chem, 277,* 33068–33074.

252. Sieprawska-Lupa, M., Mydel, P., Krawczyk, K., et al. (2004). Degradation of human antimicrobial peptide LL-37 by *Staphylococcus aureus*-derived proteinases. *Antimicrob Agents Chemother, 48,* 4673–4679.

253. Yeaman, M. R., & Yount, N. Y. (2005). Code among chaos: Immunorelativity and the AEGIS model of antimicrobial peptides. *ASM News, 71,* 21–27.

254. Klinger, M. H., & Jelkmann, W. (2002). Role of blood platelets in infection and inflammation. *J Interferon Cytokine Res, 22,* 913–922.

255. Yeaman, M. R., & Bayer, A. S. (2000). *Staphylococcus aureus*, platelets, and the heart. *Current Infect Dis Reports, 2,* 281–298.

256. Yeaman, M. R. (2004). Antimicrobial peptides from platelets in defense against cardiovascular infections. In D. A. Devine & R. E. W. Hancock (Eds.), Mammalian host defense peptides. (pp. 279–322). Cambridge University Press: *Adv Mol Cell Micro.*

257. Yount, N. Y., Gank, K. D., Xiong, Y. Q., et al. (2004). Platelet microbicidal protein-1: Structural themes of a multifunctional antimicrobial peptide. *Antimicrob Agents Chemother, 48,* 4395–4404.

258. Cole, A. M., Ganz, T., Liese, A. M., et al. (2001). Cutting edge: IFN-inducible ELR- CXC chemokines display defensin-like antimicrobial activity. *J Immunol, 167,* 623–627.

259. Yang, D., Chen, Q., Hoover, D. M., et al. (2003). Many chemokines including CCL20/MIP-3α display antimicrobial activity. *J Leukoc Biol, 74,* 448–455.

260. Yount, N. Y., & Yeaman, M. R. (2004). Multidimensional signatures in antimicrobial peptides. *Proc Natl Acad Sci USA, 101,* 7363–7368.

261. Zaiou, M., Nizet, V., & Gallo, R. L. (2003). Antimicrobial and protease inhibitory functions of the human cathelicidin (hCAP18/LL-37) prosequence. *J Invest Dermatol, 120,* 810–816.

262. Xiong, Y. Q., Mukhopadhyay, K., Yeaman, M. R., et al. (2005). Functional interrelationships between cell membrane and cell wall in antimicrobial peptide-mediated killing of *Staphylococcus aureus*. *Antimicrob Agents Chemother, 49,* 3114–3121.

263. Mukhopadhyay, K., Whitmire, W., Xiong, Y. Q., et al. (2005). *In vitro* resistance of *Staphylococcus aureus* to thrombin-induced platelet microbicidal protein is associated with alterations in cytoplasmic membrane fluidity, fatty acid and phospholipids composition, and phospholipids asymmetry. Abstract A-002, 105th American Society for Microbiology, Atlanta, GA.

264. Weidenmaier, C., Peschel, A., Xiong, Y. Q., et al. (2005). Lack of wall teichoic acids in *Staphylococcus aureus* leads to reduced interactions with endothelial cells and to attenuated virulence in a rabbit model of endocarditis. *J Infect Dis, 191,* 1771–1777.

265. Kristian, S. A., Durr, M., Van Strijp, J. A., et al. (2003). MprF-mediated lysinylation of phospholipids in *Staphylococcus aureus* leads to protection against oxygen-independent neutrophil killing. *Infect Immun, 71,* 546–549.

266. Xiong, Y. Q., Bayer, A. S., & Yeamanm, M. R. (2002). Inhibition of *Staphylococcus aureus* intracellular macromolecular synthesis by thrombin-induced platelet microbicidal proteins. *J Infect Dis, 185,* 348–356.

267. Xiong, Y. Q., Bayer, A. S., Yeaman, M. R., et al. (2004). Impact of *sarA* and *agr* in *Staphylococcus aureus* upon fibronectin-binding protein A gene expression and fibronectin adherence capacity *in vitro* and in experimental infective endocarditis. *Infect Immun, 72,* 1832–1836.

268. Sakoulas, G., Eliopoulos, G. M., Fowler, V. G., et al. (2005). *Staphylococcus aureus* accessory gene regulator (*agr*) dysfunction and vancomycin exposure are associated with autolysin defect, glycopeptide intermediate susceptibility (GISA), and reduced antimicrobial peptide susceptibility phenotypes *in vitro*. *Antimicrob Agents Chemother, 49,* 2687–2692.

269. Ginsburg, I. (1988). The biochemistry of bacteriolysis: Facts, paradoxes, and myths. *Microbiol Sci, 5,* 137–142.

270. Yeaman, M. R., Gank, K. D., Bayer, A. S., et al. (2002). Synthetic peptides that exert antimicrobial activities in whole blood and blood-derived matrices. *Antimicro Agents Chemother, 46,* 3883–3891.

271. Mercier, R. C., Dietz, R. M., Mazzola, J. L., et al. (2004). Beneficial influence of platelets on antibiotic efficacy in an *in vitro* model of *Staphylococcus aureus* endocarditis. *Antimicrob Agents Chemother, 48,* 2551–2557.

272. Cocchi, F., DeVico, A. L., G-Demo, A., et al. (1995). Identification of RANTES, MIP-1α, and MIP-1β as the major HIV-suppressive factors produced by CD8+ T cells. *Science, 270,* 1811–1815.

273. Oppenheim, J. J., & Yang, D. (2005). Alarmins: Chemotactic activators of immune responses. *Curr Opin Immunol, 17,* 359–365.

274. Fowler, V. G., Jr., Sakoulas, G., McIntyre, L. M., et al. (2004). Persistent bacteremia due to methicillin-resistant Staphylococcus aureus infection is associated with agr dysfunction and low-level *in vitro* resistance to thrombin-induced platelet microbicidal protein. *J Infect Dis, 190,* 1140–1149.

275. Kupferwasser, L. I., Yeaman, M. R., Shapiro, S. M., et al. (2002). *In vitro* susceptibility to thrombin-induced platelet microbicidal protein is associated with reduced disease progression and complication rates in experimental *Staphylococcus aureus* endocarditis: Microbiological, histopatho-

logic, and echocardiographic analyses. *Circulation, 105,* 746–752.

276. Kupferwasser, L. I., Yeaman, M. R., Nast, C. C., et al. (2003). Salicylic acid attenuates virulence in endovascular infections by targeting global regulatory pathways in *Staphylococcus aureus. J Clin Invest, 112,* 222–233.

277. Yeaman, M. R., Cheng, D., Desai, B., et al. (2004). Susceptibility to thrombin-induced platelet microbicidal protein is associated with increased fluconazole efficacy in experimental *Candida albicans* endocarditis. *Antimicrob Agents Chemother, 48,* 3051–3056.

# CHAPTER 41

# Angiogenesis

## Alexander Brill and David Varon

*Coagulation Unit, Hadassah Medical Center, Jerusalem, Israel*

## I. Introduction

Platelets are key players in the initial stages of the hemostatic process leading to the arrest of bleeding. However, data indicate that, in addition to hemostasis, platelets are involved in other pathophysiological processes, such as atherosclerosis (Chapter 35) and tumor metastasis (Chapter 42).[1,2] The fact that platelets contain numerous biologically active molecules, which can be delivered to the endothelium and tissues upon platelet adhesion to the injured site of the vasculature, allows platelets to affect fundamental processes of vascular biology, including angiogenesis. The scope of this chapter is to summarize the existing information and hypotheses regarding the regulatory role of platelets in blood vessel development and to discuss their potential use as a tool to manipulate this process by diminishing deleterious angiogenesis and promoting beneficial angiogenesis.

## II. Platelets and Endothelial Cells: Historical Background

The first scientific evidence suggesting that platelets affect vascular endothelium in a way that constitutes a basis for new vessel development was reported in the late 1960s. It was demonstrated that perfusion of organs with platelet-depleted plasma caused instability of the endothelial layer, parenchymal degeneration, and hemorrhages. The addition of platelets markedly reduced this injurious effect.[3] In subsequent animal experiments, thrombocytopenia was associated with higher vascular permeability for blood cells and plasma constituents that appeared to result from large gaps between endothelial cells (EC).[4,5] (It is now known that an increase in permeability of the vessel wall establishes the first step of the angiogenic process and constitutes the essential element in the mechanism of action of different pro-angiogenic growth factors, such as vascular endothelial growth factor [VEGF].) At the same time, other studies have suggested that the presence of platelets in the medium promoted growth of EC.[6]

Observations aimed at the elucidation of the mechanisms underlying platelet effects on EC were initially, to a large extent, descriptive. Thus, Maca et al.[7] described a shape change of EC after contact with platelets from the regular polygonal form to a more elongated and bipolar one. Furthermore, intact EC seeded in a low amount form clusters with significant gaps between them. However, platelet-treated EC tend to distribute more consistently. This could suggest stimulation of a nondirectional migration. This hypothesis was further supported by the observation that EC filled a scratch through a confluent cell layer more rapidly in the presence of platelets than without them. Furthermore, investigation of the mitogenic action of platelets on EC showed that two intraplatelet agents mimic the platelet effect: ADP and serotonin.[6–8]

As a result of this research, vascular biologists in the 1960s and 1970s considered platelet interactions with the vascular wall more as trophic or nutritious than as related to new vessel development. Nevertheless this research created the foundation for the development of the current concepts of platelet involvement in the angiogenic response.

## III. Angiogenesis-Related Intraplatelet Compounds

Platelets contain three types of granules: α, dense, and lysosomes (Chapters 3 and 15). The following important angiogenesis-regulating agents are present in circulating platelets (mainly in the α-granules) and are released upon platelet activation and secretion.

### A. Angiogenesis Activators

**1. Vascular Endothelial Growth Factor (VEGF).** VEGF, one of the most potent angiogenesis stimulators, is present in platelets at a concentration of about 0.56 pg per

$10^6$ cells.[9] Two isoforms of VEGF are presented in platelets: VEGF-A and VEGF-C. The major isoform of VEGF-A found in platelets is $VEGF_{165}$, but intraplatelet mRNA of $VEGF_{189}$ and $VEGF_{121}$ has also been identified.[10] VEGF promotes permeability of the vessel wall and serves as a chemoattractant for EC that operate predominantly at the initial stage of the angiogenic response.

**2. Platelet-Derived Growth Factor (PDGF).** PDGF is a 28- to 31-kDa dimer that consists of two chains (A and B) in all possible combinations (AA, AB, and BB).[11] The PDGF isoforms bind two types of receptors, designated α and β. The α-receptor binds both A and B cytokine subunits, but the β-receptor binds only the B chain. This allows each isoform to engage in a specific complex of receptors.[12] Activation of specific PDGF receptors results in proliferation and movement of target cells, predominantly fibroblasts and smooth muscle cells, recruiting them to the angiogenic site to contribute to the maturation of blood vessels. Therefore, PDGF operates mainly at the advanced stages of the angiogenic process.

**3. Fibroblast Growth Factors (FGFs).** FGFs are a family of cytokines that possess a strong mitogenic and chemoattractant activity for EC and participate in their recruitment and proliferation. Platelets contain FGF-2[13] and bFGF,[14] two of the most potent activators of angiogenesis.

**4. Epidermal Growth Factor (EGF).** This angiogenesis agonist binds its specific receptors (EGFR, designated also as Erb1 through 4) that are expressed on various cells, especially mesenchimal and epithelial lineages.[15] Downstream signaling of ErbB receptors includes activation of Ras and MAP kinases that transfer the signal to the nucleus, thereby triggering cell proliferation. Stimulation of endothelial cells with EGF results in increased tubule formation, cell division, and movement.[16,17] *In vivo* neutralization of EGFR in a tumor model resulted in down-regulation of VEGF and bFGF expression and a decrease in microvessel density.[18] In addition, EGF can also augment the proangiogenic effect of other cytokines (e.g., insulin-like growth factor-II).[19]

The association of plasma EGF with platelets was demonstrated in the mid-1980s.[20] Platelets contain EGF in their α-granules in a concentration of approximately 105 pg per $10^8$ cells.[21] The cytokine is released from platelets during platelet aggregation. Platelets obtain EGF from megakaryocytes, where it is synthesized, rather than take it up from the environment.[22] Platelets are also reported to transport ERG precursor on their surface.[23]

**5. Hepatocyte Growth Factor (HGF).** HGF is a potent angiogenesis inducer and its receptor, c-met, is abundantly expressed on both endothelial cells and vascular smooth muscle cells.[24] As determined by an *in vivo* animal model, injection of either intramural or intravenous recombinant HGF facilitates angiogenesis in the ischemic heart, resulting in improved cardiac performance.[25,26] At the cellular level, HGF is a potent mitogen for different cell types including endothelial cells.[27] Mechanisms of its effects include stimulating of secretion of MMP-1, VEGF, HGF itself and its receptor, c-met, in endothelial cells.[28] While the whole HGF molecule possesses strong proangiogenic activity, alternative processing of the HGF α-chain mRNA produces anti-angiogenic fragments.[29]

HGF has been isolated from platelets.[30,31] The role of platelet-derived HGF in angiogenesis becomes even more complex, taking into account findings that HGF is a potent platelet antagonist that markedly reduces thrombin-induced platelet aggregation and transition of integrin αIIbβ3 to its activated state via Met receptors on the platelet surface.[32]

**6. Insulin-Like Growth Factor (IGF).** IGF is a polypeptide of 7.5 kDa that has structural homology with insulin. Receptors for this important angiogenesis agonist are present on endothelial cells in different tissues.[33–35] Both insulin and IGF stimulate VEGF mRNA synthesis in endothelial cells.[36] IGF facilitates endothelial cell motility and tubule formation[37] and produces angiogenic response in the rat aortic ring *ex vivo* model.[38] Moreover, it has been demonstrated that plasmid IGF application can stimulate angiogenesis in regenerating muscle and increase blood flow and angiogenesis in the diabetic limb.[39] Another structural homolog of IGF-1, IGF-2, possesses significant proangiogenic activity.[40]

Both IGF-I and II are stored in platelet α-granules and released upon thrombin stimulation.[41,42] Platelets obtain IGF from megakaryocytes, which both produce the cytokine and take it up from the environment.[43] Platelets carry also IGF receptors on their surface.[44]

**7. Angiopoietin.** This family of cytokines includes four members that bind the Tie2 tyrosine kinase receptor. Angiopoietin-1 (Ang-1) is a major ligand for Tie2, and while it does not induce endothelial cell proliferation, it can induce their migration, tubule formation, sprouting, and survival.[45] This suggests, in conjunction with experiments on Ang-1-null mice in which severe defects in vessel development and maturation were found, that Ang-1 is a potent proangiogenic agent.

Angiopoietin is constitutively expressed in megakaryocytes and platelets.[46] The platelet content of Ang-1 is increased in patients with breast cancer, indicating that this platelet-derived cytokine may play a role in cancer angiogenesis (see also Chapter 42).[47]

**8. Platelet Phospholipids.** This group of agents includes lysophosphatidic acid, sphingosine 1-phosphate (SPP), and phosphatidic acid. The first evidence of SPP influence on EC was reported in 1994.[48] Later it was reported that these substances, operating via the endothelium differentiation gene (EDG) receptor family, induce a strong *in vitro* and *in vivo* angiogenic response and stimulate chemotaxis, proliferation, and capillary tube formation by EC.[49,50]

**9. CD40 Ligand (CD40L, CD154).** This is a cytoplasmic intraplatelet molecule expressed on the membrane surface upon platelet activation (see Chapter 39). CD40L binding to CD40 on EC promotes various biological responses that include the development of new blood vessels,[51] partially due to platelet activating factor (PAF) synthesis.[52]

**10. Matrix Metalloproteinases (MMPs).** MMPs are a family of zinc-dependent endopeptidases capable of cleaving different components of extracellular matrix (ECM) and basement membrane.[53,54] Platelets contain MMP-1, MMP-2, and MMP-9,[55,56] which may be released upon activation and support new vessel development by assisting EC to migrate through the surrounding tissues.

**11. Heparanase.** Heparanase is an endoglycosidase cleaving heparan sulfate, both in the ECM and on the cell membrane. Heparanase facilitates angiogenesis by releasing angiogenic mediators normally associated with heparan sulfate that increases their bioavailability and participation in blood vessel development in wound healing, tumor growth, and metastasis.[57,58]

### B. Angiogenesis Inhibitors

**1. Angiostatin.** This potent inhibitor is produced by cleavage of plasminogen by several enzymes, such as metalloproteinases, urokinase, and tissue-type plasminogen activator. Intact platelets contain large amounts of angiostatin, which is released upon platelet activation.[59] Moreover, the intraplatelet plasminogen level is decreased during platelet aggregation, which indicates that platelets may not only secrete angiostatin but also synthesize it *de novo*.

**2. Thrombospondin-1 (TSP1).** TSP1, an endogenous lectin,[60] is a potent inhibitor of EC proliferation and capillary tube formation. It binds CD36 on the endothelial surface and activates the signaling cascade leading to stimulation of caspase-3 and increased EC apoptosis.[61] Furthermore, TSP1 prevents angiogenesis by displacement of VEGF from its complex with heparan sulfate and by direct binding of VEGF.[62] Suppression of TSP1 expression mediates a proangiogenic effect of factor XIII.[63] In very high concentrations, TSP1 can also promote an angiogenic response.[64]

**3. Platelet Factor 4 (PF4, CXCL4).** PF4 exerts its angiostatic effect via inhibition of binding of different growth factors to glycosaminoglycans on the EC surface.[65] Several other pathways also exist. In particular, PF4 interferes with signaling of the FGF receptor, suppresses angiogenin-induced actin polymerization, abrogates EC entry to the cell cycle,[66] and directly hampers FGF function.[67] A nonallelic variant of PF4 (CXCL4), designated CXCL4L1, a strong angiogenesis inhibitor, has recently been described in platelets.[68]

**4. Endostatin.** Endostatin is a short fragment of collagen XIII, a potent inhibitor of the angiogenic properties of EC *in vitro* and tumor vascularization *in vivo*.[69] Interestingly, the mechanisms of endostatin and VEGF release from platelets form a reciprocal relationship, since the activation of protease-activated receptor (PAR) 1 receptors on platelets results in stimulation of VEGF release and suppression of endostatin liberation, whereas PAR4 exerts the inverse effect.

**5. Transforming Growth Factor β (TGF-β).** The effects of TGF-β on angiogenesis are complex and apparently contradictory. *In vitro*, TGF-β reduces both the proliferation and migration of endothelial cells, but promotes tubule formation in three-dimensional collagen gels.[70] In contrast, *in vivo*, TGF-β promotes angiogenesis at the sites of injection and also exerts a proangiogenic effect in the rabbit cornea assay or on the chick chorioallantoic membrane.[70] These results are in accordance with the described stimulation of endothelial cell proliferation in response to TGF-β[71] and TGF-β-mediated endothelial cell survival during *in vitro* angiogenesis.[72] However, it was also demonstrated that TGF-β strongly inhibits angiogenesis in the arterial tree of the developing quail chorioallantoic membrane.[73]

Therefore, the effect of TGF-β on the angiogenic process potentially depends on various factors, such as cell type, experimental approach (*in vitro* or *in vivo*), and experimental model and design. It may, however, be supposed that the direct effect of TGF-β on endothelium is predominantly antiangiogenic, whereas *in vivo* it may promote angiogenesis indirectly via recruitment of macrophages releasing proangiogenic growth factors.

Platelets contain TGF-β[74] as a latent high molecular weight complex.[75] The cytokine is released from platelets upon their degranulation[76] and is activated by a furin-like convertase, which is also released upon platelet activation.[77]

**6. Tissue Inhibitors of Metalloproteinases (TIMPs).** TIMPs hinder the angiogenic process via neutralization of the activity of different MMPs. Platelets contain TIMP-4 (12 to 16 ng per $10^8$ platelets), a very potent MMP inhibitor, and to a lesser extent, TIMP-1 (<1 ng per $10^8$ platelets).[78]

ALEXANDER BRILL AND DAVID VARON

The intracellular localization of TIMP-4 coincides with that of MMP-2 and the inhibitor is released upon platelet activation by thrombin or collagen.

## IV. Angiogenesis-Related Properties of Platelets: Experimental Data

Development of novel capillaries involves recruitment of EC at earlier stages and smooth muscle cells with pericytes at later stages of vessel maturation. A successful angiogenic process requires EC to pass through several stages: activation, migration through tissues to the site of the provisional vessel branch, proliferation, and formation of tube-like structures that constitute the core of the future capillary. The latter process is believed to reflect EC differentiation. Each stage can be evaluated separately *in vitro*. Nevertheless, angiogenesis is a complex process that requires successful accomplishment of all stages. A delay in one stage may result in useless activation of the others.

Pipili–Synetos et al.[79] demonstrated a platelet effect on endothelial angiogenic properties *in vitro*. Platelets promote tubule formation by human umbilical vein endothelial cells (HUVEC) in a number- and time-dependent manner and adhere to EC along the formed capillaries. Notably, platelet releasate does not mimic the effect of whole platelets, indicating that it is not intraplatelet cytokines that mediate the effect, but rather the interaction between platelet surface glycoproteins and endothelial receptors. This hypothesis is further supported by the abolishment of the platelet effect by neuraminidase, which strips sialic acid from membrane glycoproteins to prevent their interaction with their ligands. Interestingly, this observation differs from a study reported in 1977 in which a direct platelet effect on EC behavior could not be reversed by neuraminidase.[7] Platelet involvement in the process of angiogenesis and tumor metastasis via the combined action of integrin $\alpha V\beta 3$ (glycoprotein [GP] IIb-IIIa) on platelets and integrin $\alpha V\beta 3$ on endothelium has also been demonstrated under *in vivo* conditions (see also Chapter 42).[80]

Platelets promote EC proliferation and support their survival under conditions of endothelial cell growth factor (ECGF) deprivation.[13] Interestingly, this effect does not require contact between platelets and EC, because separation of platelets and endothelium does not significantly reduce it. However, platelet activation appears to be necessary for endothelial proliferation and survival, because fixed or aspirin-pretreated platelets fail to exert the effects. The concerted action of two potent proangiogenic cytokines, VEGF and FGF-2, mediates this effect, which is associated with the cytokine-dependent amplification of extracellular regulated kinase (ERK) phosphorylation. Another important proangiogenic feature rendered to EC by platelet releasate, but independent of ERK, is stimulation of the

production and secretion of MMP-2, whose involvement in angiogenesis is well established.[81] Furthermore, adherence of whole platelets to HUVEC promotes the expression of membrane type-1 matrix metalloproteinase (MT1-MMP) and release of MMP-2 and MMP-9.[82]

Platelet surface-expressed CD40L is an interesting pathway in which platelets render a procoagulant and proangiogenic phenotype to EC. Binding of platelet CD40L to its receptor, CD40 on EC, results in activation of tissue factor expression and down-regulation of thrombomodulin.[83] Tissue factor, in turn, is involved in stimulation of blood vessel growth[84] and may be one more link between platelets and angiogenesis.

The studies cited earlier investigated the separate properties of endothelium associated with the angiogenic process and their modification by platelets. These differ from our work in which the question of how platelets affect angiogenesis *in toto* was addressed.[85] Indeed, given that platelets contain a cocktail of angiogenic agonists and antagonists (Section III), the question becomes, which effect will predominate? As Fig. 41-1 shows, platelets induce strong number-dependent sprouting of blood vessels in the rat aortic ring model. This effect was reduced by inhibiting either VEGF or bFGF, whereas combined inhibition of VEGF, bFGF, and PDGF completely abolished the angiogenic response. Similar results have been obtained in an *in vivo* model of subcutaneous Matrigel injection into mice (Fig. 41-2).[85] This *in vivo* approach suggested the proangiogenic importance of platelet heparanase. Consequently, the overall effect of platelets is angiogenesis stimulation via the concerted action of intraplatelet cytokines. Antiangiogenic agents are also operative in platelets, since neutralization of PF4 by a specific monoclonal antibody resulted in additional amplification of sprouting.

Rhee et al.[86] report similar findings. In particular, these investigators demonstrated that platelets induce proliferation and sprouting of microvascular EC in fibrin gel due to the synergistic action of VEGF and bFGF. Therefore, these two cytokines are major players in platelet-mediated angiogenesis, although numerous other molecules are involved. Further studies are required to dissect out the exact role of the molecules involved.

For quite a long time, it was thought that platelets obtain biologically active molecules either from their ancestors — megakaryocytes — or by an active recruitment from plasma. However, recent studies indicate another option: platelets may synthesize cytokines from cytoplasmic mRNA, as shown for Bcl-3 and interleukin (IL)-1$\beta$.[87–89] Moreover, it is now clear that platelets synthesize a wide variety of still unrecognized regulatory proteins in a signal-dependent manner.[55] The possibility of activation-regulated translation, and production of novel regulatory cytokines that depend on environmental requirements, makes platelets a more potent tool in vascular physiology and, in particular, angiogenesis.

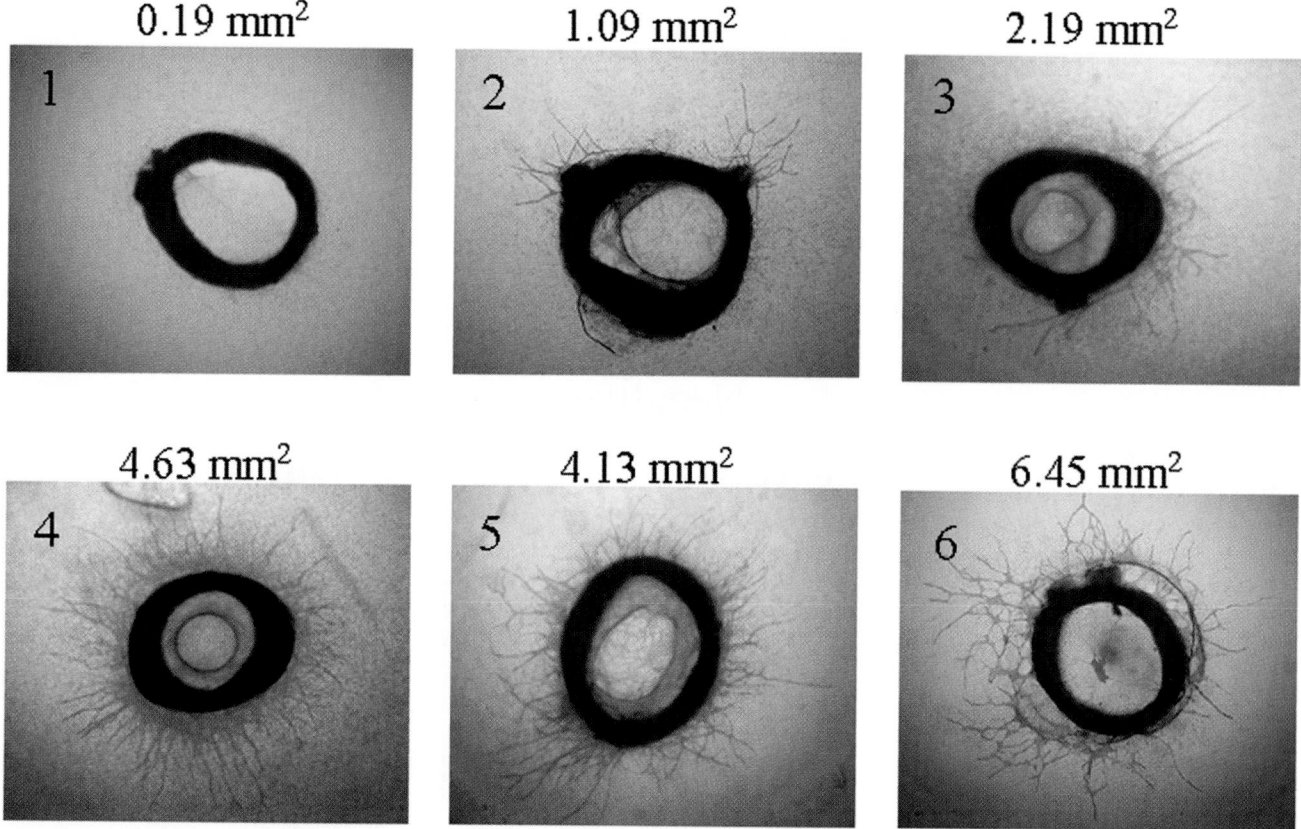

**Figure 41-1.** Platelets induce angiogenesis *ex vivo* in the rat aortic ring model. (1) Negative control. (2) through (5) $0.2 \times 10^5/\mu L$ to $1 \times 10^6/\mu L$ platelets. (6) VEGF + FGF (50 ng/mL of each). Area covered with the sprouts of blood vessels is indicated above each picture. Abbreviations: VEGF, vascular endothelial growth factor; FGF, fibroblast growth factors. (Adapted with permission.[85])

## V. Platelet-Derived Microparticles: Possible Mediators of the Angiogenic Response

Microparticles (MP) are small plasma membrane fragments (0.05–1 μm) shed from cells upon their activation or apoptosis.[90] The basic mechanism of MP formation is disruption of the mechanisms supporting asymmetry of phospholipids between two membrane layers. Platelet-derived microparticles (PMP) constitute the major pool of MP circulating in the blood (see Chapter 20 for a full discussion).[91] Platelets release two types of microvesicles: plasma surface membrane derived PMP and exosomes derived from α-granules.[92] Most studies have investigated plasma surface membrane-derived PMPs or do not address this difference.

PMP can express and transfer functional receptors, such as integrin αIIbβ3, P-selectin, CXCR4, PAR1, tissue necrosis factor (TNF)-RI, TNF-RII, CD95, and CD40L[93–95] and enhance engraftment of hematopoietic stem/progenitor cells.[96] PMP also promote monocyte adhesion to endothelium and stimulate cyclooxygenase (COX)-2 expression in monocytes and EC.[91] It was recently established that PMP not only passively carry various proteins and receptors but also chemoattract hematopoietic cells and stimulate their adhesion, survival, and proliferation.[95] Moreover, PMP activate intracellular signaling pathways, such as ERK, PI3-kinase/Akt, and STAT proteins. These effects could only partially be overturned by heat and trypsin, which suggests that in addition to intraparticle agents of protein nature, lipid components of PMP could also be involved.

Evidence that PMP might affect blood vessel development has recently been shown. PMP promote proliferation of EC.[97] This effect is mediated by a concerted action of VEGF, FGF-2, and a lipid component of PMP, most likely sphingosine 1-phosphate. This lipid constituent seems to mediate the antiapoptotic and chemotactic effects of PMP on EC, as well as PMP-induced tubule formation.

Circulating tumor cells are able to activate platelets and induce platelet aggregation.[98] Consequently, it may be expected that the concentration of PMP at sites of tumor cell accumulation may reach a high level and affect surrounding endothelium. However, considering that the effects of PMP on separate properties of EC are, to a large extent, activatory, could they trigger neovascularization? We recently demonstrated that PMP induce sprouting both *in vitro* and *in vivo* to a degree comparable with that of whole platelets.[99]

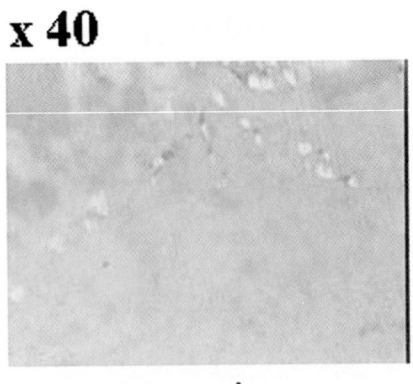

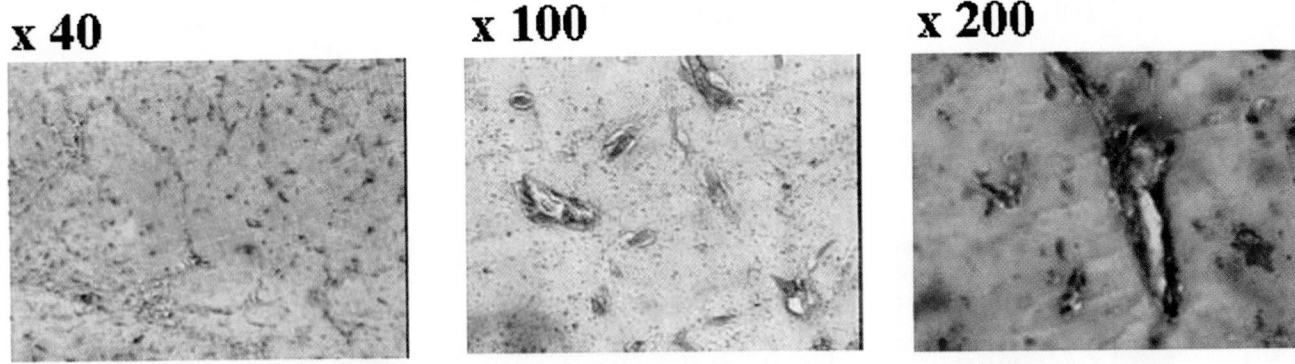

**Figure 41-2.** Platelets promote an angiogenic response *in vivo*. Matrigel was mixed with platelets and injected subcutaneously into Sabra mice. After 6 days, mice were sacrificed; matrigel was removed and used for preparation of paraffin-embedded sections and staining for von Willebrand factor (dark brown color) as a marker for endothelial cells. Upper panel: negative control (no added platelets). Lower three panels: test (added platelets). The respective magnifications are indicated above each picture. (Adapted with permission.[85])

Moreover, application of PMP may have a therapeutic potential. As shown in Fig. 41-3, intramyocardial injection of PMP markedly elevated the number of novel capillaries in ischemic heart muscle. Thus, PMP may further support the platelet-induced angiogenic response, operating relatively independently of platelets. Alternatively, PMP may be a tool or a mediator through which platelets exert their proangiogenic effect, given that various released molecules, traditionally considered as soluble, are in fact bound to MP.[100]

## VI. Platelets and Angiogenesis in Diseases

### A. *Platelets in Regulation of Tumor Angiogenesis and Metastasis (see also Chapter 42)*

It is known that neither primary tumors nor tumor metastases can grow larger than 3 mm in the absence of an additional blood supply.[101] Development of new capillaries in the tumor is essential, not only for oxygen and metabolite

supply but also for the access of paracrine regulatory factors necessary for the facilitation of tumor growth.[102] The hemostatic system as a whole, and platelets in particular, significantly affect the balance between inducers and inhibitors of angiogenesis maintained in the blood. Disturbance in this balance could therefore lead to further tumor expansion.

In a pioneering study that related hemostatic defects to cancer development, Trousseau[103] described approximately 200 patients in whom thrombophlebitis was associated with developing cancer. The first evidence that suggested platelet involvement in tumor development was reported by Gasic et al.[104] These investigators demonstrated a strong dependency of lung metastases on platelet count. Complete platelet depletion almost eliminated the development of metastasis. The association of platelet aggregability with cancer metastatic spread was confirmed in several subsequent studies.[105,106] Pinedo et al.[107] hypothesized that platelets may play a significant role in tumor-induced angiogenesis. This hypothesis has been supported by experimental and clinical data. A significant number of cancer patients have

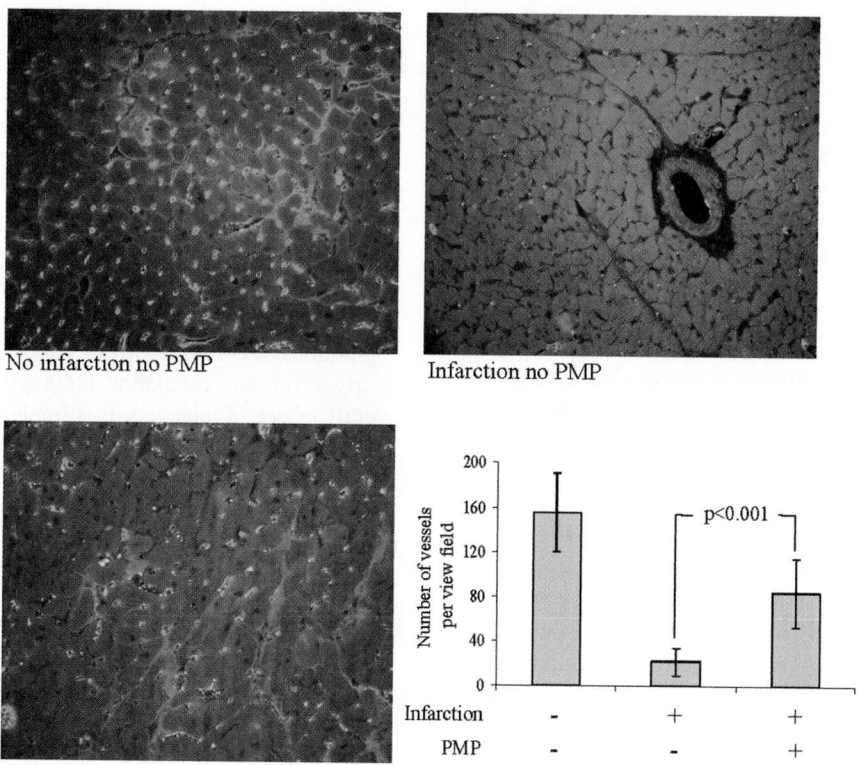

**Figure 41-3.** PMPs induce postischemic revascularization of myocardium. Myocardial infarction was modeled by ligation of left coronary artery of rats. PMPs (250 μg/mL protein) were injected into the border region of the infarction. Three weeks thereafter, bovine serum albumin (BSA) conjugated with fluorescein isothiocyanate (FITC) was injected into the left atrium. One minute later, the animals were sacrificed and their hearts were prepared into paraffin-embedded sections. Samples were photographed with objective ×20. Light green color: blood vessels stained with BSA-FITC. (Adapted with permission.[99])

high platelet counts, which is associated with a poor prognosis.[108,109] Other studies have shown that platelet cross-talk with tumor cells is mutual. Thus, breast cancer cells stimulate platelets, resulting in liberation of lysophosphatidic acid, which in turn promotes tumor progression.[110] Moreover, in an experimental animal model, platelet inhibition by the GPIIb-IIIa antagonist eptifibatide suppressed breast cancer metastasis.[110]

Tumor vasculature is very complex with numerous branches that result in a disorganized flow.[111] Turbulence, blood flow disturbances, and extremely high shear stress, as well as stasis, could lead to platelet activation and subject EC and surrounding tissues to platelet granular content. Moreover, this interaction could accelerate due to positive feedback, since platelets exhibit an increased tissue factor-mediated adhesion to VEGF-activated EC.[112]

The impact of platelets on the serum level of angiogenic mediators in cancer patients has been studied.[113,114] Platelets constitute a major source of serum VEGF and contain elevated levels of angiopoietin-1 in cancer patients.[47,115] Moreover, platelets may mediate a proangiogenic shift in the balance between pro- and antiangiogenic compounds

(VEGF versus TSP1) in these patients.[116] However, the amount of VEGF in platelets may reflect the VEGF level in the tumor.[114] Notably, platelets accumulate tumor-derived angiogenic factors (e.g., VEGF), preventing elevation of their concentration in the serum. Therefore, the concentration and ratio of angiogenic factors found in platelets could be considered as a diagnostic tool to distinguish a very small VEGF-producing tumor that is unidentifiable by other methods.[117]

PMP may also play a role in cancer growth and angiogenesis, although existing data are rather scanty. It has been reported that PMP may serve as chemoattractants to several lung cancer cell lines, activating phosphorylation of ERK and expression of MT1-MMP.[118] PMP also stimulate proliferation and adhesion of cancer cells to fibrinogen and EC. Furthermore, we recently observed induction of invasive properties associated with increased MMP-2 production by a prostate cancer cell line after incubation with PMP (Brill et al., unpublished data). These findings suggest that PMP may indeed play an auxiliary role in tumor growth and metastasis by promotion of tumor-associated angiogenesis.

## B. Platelets and Angiogenesis in Wound and Ulcer Healing

The diverse repertoire of growth factors present in platelet granules resulted in studies related to their beneficial effect in wound healing. Knighton et al.[119] reported positive results of such treatment in nonhealing cutaneous ulcers among 49 patients. Later, the same group, as well as others, reported an accelerated healing effect of the local application of platelets in different types of injuries.[120–122] Platelets, topically applied to a wound, stimulated granulation tissue production in experimental animals.[123] Impressive results were also observed when highly concentrated platelet components were applied to treat macular holes: anatomic closure of the holes was achieved in 18 out of 19 patients.[124] Successful use of platelet supernatant in several types of pathologies, such as treatment of diabetic ulcers, was also reported in several clinical trials.[125]

These findings suggest that platelets may be a useful therapeutic adjunct for tissue repair. However, the role of circulating platelets in this process is still poorly understood and data on this issue are rather controversial. It has been reported that experimental induction of gastric ulcers in rats is associated with an increase in serum VEGF accompanied by a diminished endostatin level.[126] This phenomenon was interpreted as a physiologic curative response, apparently linked to platelets, since ticlopidine (a platelet $P2Y_{12}$ ADP receptor inhibitor) impaired ulcer-related angiogenesis and healing and reversed the changes in VEGF and endostatin levels. The effect of ticlopidine could be reproduced by immunoreduction of platelets. In another study, platelets accumulated in the retinal vasculature in the course of diabetic retinopathy and exerted a beneficial effect by preventing a blood-retinal barrier breakdown.[127] These findings suggest that under certain conditions, platelet accumulation may be a healing factor in the local microcirculatory bed and not a handicap for blood flow. Significantly, it has been established that VEGF concentration in the blood near an *in vivo* thrombus is elevated by about three-fold during the first minutes after plug formation.[128] Moreover, VEGF secreted from platelets upon their activation accumulates inside the thrombus.[129] Unlike elevation of the serum level of angiogenesis-related compounds, this phenomenon may underlie the development of a targeted angiogenic response.

Nevertheless, the opposite view is also supported by data. It has been reported that the release of TGF-β from degranulating platelets is not essential at different stages of wound consolidation.[130] In another study, mice deficient in PAR1 (thrombin receptor expressed on platelets and other cell types) had intact parameters of wound repair, such as time to closure, tensile strength, wound histology, and hydroxyproline/DNA content of wound implants.[131] Finally, Szpaderska et al.[132] demonstrated that platelets are not directly involved in such essential processes in wound repair

as angiogenesis and collagen synthesis. At the same time, platelets clearly played a role in wound inflammation since the number of macrophages and T cells in wounds increased in mice deficient of platelets.

## VII. Conclusion

Both blood platelets and PMP contain a wide variety of biological regulators that include pro- and antiangiogenic compounds. It is now clear that platelets are highly involved in triggering and regulating the angiogenic response. The overall effect of platelets is stimulatory on blood vessel development. Two important corollaries are that (1) the global effect of antiplatelet therapy may expand beyond hemostasis; thus antiplatelet drugs could be effective in conditions such as cancer and inflammation by inhibiting angiogenesis; (2) local application of platelets or PMPs could be useful in developing novel therapeutic strategies that target angiogenesis-related conditions.

## References

1. Huo, Y., & Ley, K. F. (2004). Role of platelets in the development of atherosclerosis. *Trend Cardiovas Med, 14*, 18–22.
2. Camerer, E., Qazi, A. A., Duong, D. N., et al. (2004). Platelets, protease-activated receptors, and fibrinogen in hematogenous metastasis. *Blood, 104*, 397–401.
3. Gimbrone, M. A., Jr., Aster, R. H., Cotran, R. S., et al. (1969). Preservation of vascular integrity in organs perfused *in vitro* with a platelet-rich medium. *Nature, 222*, 33–36.
4. Gore, I., Takada, M., & Austin, J. (1970). Ultrastructural basis of experimental thrombocytopenic purpura. *Arch Pathol, 90*, 197–205.
5. Kitchens, C. S., & Weiss, L. (1975). Ultrastructural changes of endothelium associated with thrombocytopenia. *Blood, 46*, 567–578.
6. Saba, S. R., & Mason, R. G. (1975). Effects of platelets and certain platelet components on growth of cultured human endothelial cells. *Thromb Res, 7*, 807–812.
7. Maca, R. D., Fry, G. L., Hoak, J. C., et al. (1977). The effects of intact platelets on cultured human endothelial cells. *Thromb Res, 11*, 715–727.
8. Fratkin, J. D., Cancilla, P. A., & DeBault, L. E. (1980). Platelet factor and cerebral vascular endothelium: Platelet-induced mitogenesis. *Thromb Res, 19*, 473–483.
9. Banks, R. E., Forbes, M. A., Kinsey, S. E., et al. (1998). Release of the angiogenic cytokine vascular endothelial growth factor (VEGF) from platelets: Significance for VEGF measurements and cancer biology. *Br J Cancer, 77*, 956–964.
10. Mohle, R., Green, D., Moore, M. A., et al. (1997). Constitutive production and thrombin-induced release of vascular endothelial growth factor by human megakaryocytes and platelets. *Proc Nat Acad Sci USA, 94*, 663–668.

11. Mannaioni, P. F., Di Bello, M. G., & Masini, E. (1997). Platelets and inflammation: Role of platelet-derived growth factor, adhesion molecules and histamine. *Inflamm Res, 46,* 4–18.

12. Heldin, C. H. (1997). Simultaneous induction of stimulatory and inhibitory signals by PDGF. *FEBS Lett, 410,* 17–21.

13. Pintucci, G., Froum, S., Pinnell, J., et al. (2002). Trophic effects of platelets on cultured endothelial cells are mediated by platelet-associated fibroblast growth factor-2 (FGF-2) and vascular endothelial growth factor (VEGF). *Thromb Haemost, 88,* 834–842.

14. Brunner, G., Nguyen, H., Gabrilove, J., et al. (1993). Basic fibroblast growth factor expression in human bone marrow and peripheral blood cells. *Blood, 81,* 631–638.

15. Wells, A. (1999). EGF receptor. *Int J Biochem Cell Biol, 31,* 637–643.

16. Okamura, K., Morimoto, A., Hamanaka, R., et al. (1992). A model system for tumor angiogenesis: Involvement of transforming growth factor-alpha in tube formation of human microvascular endothelial cells induced by esophageal cancer cells. *Biochem Biophys Res Commun, 186,* 1471–1479.

17. Kim, H. S., Shin, H. S., Kwak, H. J., et al. (2003). Betacellulin induces angiogenesis through activation of mitogen-activated protein kinase and phosphatidylinositol 3'-kinase in endothelial cell. *FASEB J, 17,* 318–320.

18. Kedar, D., Baker, C. H., Killion, J. J., et al. (2002). Blockade of the epidermal growth factor receptor signaling inhibits angiogenesis leading to regression of human renal cell carcinoma growing orthotopically in nude mice. *Clin Cancer Res, 8,* 3592–3600.

19. Lee, Y. M., Bae, M. H., Lee, O. H., et al. (2004). Synergistic induction of *in vivo* angiogenesis by the combination of insulin-like growth factor-II and epidermal growth factor. *Oncology Report, 12,* 843–848.

20. Oka, Y., & Orth, D. N. (1983). Human plasma epidermal growth factor/beta-urogastrone is associated with blood platelets. *J Clin Invest, 72,* 249–259.

21. Nakamura, T., Kasai, K., Banba, N., et al. (1989). Release of human epidermal growth factor from platelets in accordance with aggregation *in vitro*. *Endocrinology Jpn, 36,* 23–28.

22. Ben-Ezra, J., Sheibani, K., Hwang, D. L., et al. (1990). Megakaryocyte synthesis is the source of epidermal growth factor in human platelets. *Am J Pathol, 137,* 755–759.

23. Valcarce, C., Bjork, I., & Stenflo, J. (1999). The epidermal growth factor precursor: A calcium-binding, beta-hydroxyasparagine containing modular protein present on the surface of platelets. *Eur J Biochem, 260,* 200–207.

24. Nakamura, Y., Morishita, R., Higaki, J., et al. (1995). Expression of local hepatocyte growth factor system in vascular tissues. *Biochem Biophys Res Commun, 215,* 483–488.

25. Wang, Y., Ahmad, N., Wani, M. A., et al. (2004). Hepatocyte growth factor prevents ventricular remodeling and dysfunction in mice via Akt pathway and angiogenesis. *J Mol Cell Cardiol, 37,* 1041–1052.

26. Yamaguchi, T., Sawa, Y., Miyamoto, Y., et al. (2005). Therapeutic angiogenesis induced by injecting hepatocyte growth factor in ischemic canine hearts. *Surg Today, 35,* 855–860.

27. Shima, N., Itagaki, Y., Nagao, M., et al. (1991). A fibroblast-derived tumor cytotoxic factor/F-TCF (hepatocyte growth factor/HGF) has multiple functions *in vitro*. *Cell Biol Int Rep, 15,* 397–408.

28. Tomita, N., Morishita, R., Taniyama, Y., et al. (2003). Angiogenic property of hepatocyte growth factor is dependent on upregulation of essential transcription factor for angiogenesis, ets-1. *Circulation, 107,* 1411–1417.

29. Browder, T., Folkman, J., & Pirie-Shepherd, S. (2000). The hemostatic system as a regulator of angiogenesis. *J Biol Chem, 275,* 1521–1524.

30. Russell, W. E., McGowan, J. A., & Bucher, N. L. (1984). Partial characterization of a hepatocyte growth factor from rat platelets. *J Cell Physiol, 119,* 183–192.

31. Nakamura, T., Teramoto, H., & Ichihara, A. (1986). Purification and characterization of a growth factor from rat platelets for mature parenchymal hepatocytes in primary cultures. *Proc Nat Acad Sci USA, 83,* 6489–6493.

32. Pietrapiana, D., Sala, M., Prat, M., et al. (2005). Met identification on human platelets: Role of hepatocyte growth factor in the modulation of platelet activation. *FEBS Let, 579,* 4550–4554.

33. Fiorelli, G., Formigli, L., Zecchi Orlandini, S., et al. (1996). Characterization and function of the receptor for IGF-I in human preosteoclastic cells. *Bone, 18,* 269–276.

34. Spoerri, P. E., Ellis, E. A., Tarnuzzer, R. W., et al. (1998). Insulin-like growth factor: Receptor and binding proteins in human retinal endothelial cell cultures of diabetic and non-diabetic origin. *Growth Horm IGF Res, 8,* 125–132.

35. Kobayashi, T., & Kamata, K. (2002). Short-term insulin treatment and aortic expressions of IGF-1 receptor and VEGF mRNA in diabetic rats. *Am J Physiol Heart Circ Physiol, 283,* H1761–1768.

36. Miele, C., Rochford, J. J., Filippa, N., et al. (2000). Insulin and insulin-like growth factor-I induce vascular endothelial growth factor mRNA expression via different signaling pathways. *J Biol Chem, 275,* 21695–21702.

37. Shigematsu, S., Yamauchi, K., Nakajima, K., et al. (1999). IGF-1 regulates migration and angiogenesis of human endothelial cells. *Endocrinol J, 46,* Suppl:S59–62.

38. Nicosia, R. F., Nicosia, S. V., & Smith, M. (1994). Vascular endothelial growth factor, platelet-derived growth factor, and insulin-like growth factor-1 promote rat aortic angiogenesis in vitro. *Am J Pathol, 145,* 1023–1029.

39. Rabinovsky, E. D., & Draghia-Akli, R. (2004). Insulin-like growth factor I plasmid therapy promotes *in vivo* angiogenesis. *Mol Therapy, 9,* 46–55.

40. Lee, O. H., Bae, S. K., Bae, M. H., et al. (2000). Identification of angiogenic properties of insulin-like growth factor II in *in vitro* angiogenesis models. *Br J Cancer, 82,* 385–391.

41. Karey, K. P., Marquardt, H., & Sirbasku, D. A. (1989). Human platelet-derived mitogens. I. Identification of insulin-like growth factors I and II by purification and N alpha amino acid sequence analysis. *Blood, 74,* 1084–1092.

42. Karey, K. P., & Sirbasku, D. A. (1989). Human platelet-derived mitogens. II. Subcellular localization of insulinlike growth factor I to the alpha-granule and release in response to thrombin. *Blood, 74,* 1093–1100.

43. Chan, K., & Spencer, E. M. (1998). Megakaryocytes endocytose insulin-like growth factor (IGF) I and IGF-binding protein-3: A novel mechanism directing them into alpha granules of platelets. *Endocrinology, 139,* 559–565.

44. Hartmann, K., Baier, T. G., Loibl, R., et al. (1989). Demonstration of type I insulin-like growth factor receptors on human platelets. *J Recept Res, 9,* 181–198.

45. Metheny-Barlow, L. J., & Li, L. Y. (2003). The enigmatic role of angiopoietin-1 in tumor angiogenesis. *Cell Res, 13,* 309–317.

46. Li, J. J., Huang, Y. Q., Basch, R., et al. (2001). Thrombin induces the release of angiopoietin-1 from platelets. *Thromb Haemost, 85,* 204–206.

47. Caine, G. J., Lip, G. Y., & Blann, A. D. (2004). Platelet-derived VEGF, Flt-1, angiopoietin-1 and P-selectin in breast and prostate cancer: Further evidence for a role of platelets in tumour angiogenesis. *Ann Med, 36,* 273–277.

48. Natarajan, V., Jayaram, H. N., Scribner, W. M., et al. (1994). Activation of endothelial cell phospholipase D by sphingosine and sphingosine-1-phosphate. *Am J Respir Cell Mol Biol, 11,* 221–229.

49. English, D., Welch, Z., Kovala, A. T., et al. (2000). Sphingosine 1-phosphate released from platelets during clotting accounts for the potent endothelial cell chemotactic activity of blood serum and provides a novel link between hemostasis and angiogenesis. *FASEB J, 14,* 2255–2265.

50. English, D., Garcia, J. G., & Brindley, D. N. (2001). Platelet-released phospholipids link haemostasis and angiogenesis. *Cardiovasc Res, 49,* 588–599.

51. Mach, F., Schonbeck, U., Fabunmi, R. P., et al. (1999). T lymphocytes induce endothelial cell matrix metalloproteinase expression by a CD40L-dependent mechanism: Implications for tubule formation. *Am J Pathol, 154,* 229–238.

52. Russo, S., Bussolati, B., Deambrosis, I., et al. (2003). Platelet-activating factor mediates CD40-dependent angiogenesis and endothelial-smooth muscle cell interaction. *J Immunol, 171,* 5489–5497.

53. Nguyen, M., Arkell, J., & Jackson, C. J. (2001). Human endothelial gelatinases and angiogenesis. *Int J Biochem Cell Biol, 33,* 960–970.

54. Mott, J. D., & Werb, Z. (2004). Regulation of matrix biology by matrix metalloproteinases. *Curr Opin Cell Biol, 16,* 558–564.

55. Weyrich, A. S., Lindemann, S., & Zimmerman, G. A. (2003). The evolving role of platelets in inflammation. *J Thromb Haemost, 1,* 1897–1905.

56. Sheu, J. R., Fong, T. H., Liu, C. M., et al. (2004). Expression of matrix metalloproteinase-9 in human platelets: Regulation of platelet activation in *in vitro* and *in vivo* studies. *Br J Pharmacol, 143,* 193–201.

57. Elkin, M., Ilan, N., Ishai-Michaeli, R., et al. (2001). Heparanase as mediator of angiogenesis: Mode of action. *FASEB J, 15,* 1661–1663.

58. Goldshmidt, O., Zcharia, E., Abramovitch, R., et al. (2002). Cell surface expression and secretion of heparanase markedly promote tumor angiogenesis and metastasis. *Proc Nat Acad Sci USA, 99,* 10031–10036.

59. Jurasz, P., Alonso, D., Castro-Blanco, S., et al. (2003). Generation and role of angiostatin in human platelets. *Blood, 102,* 3217–3223.

60. Jaffe, E. A., Leung, L. L., Nachman, R. L., et al. (1982). Thrombospondin is the endogenous lectin of human platelets. *Nature, 295,* 246–248.

61. Jimenez, B., Volpert, O. V., Crawford, S. E., et al. (2000). Signals leading to apoptosis-dependent inhibition of neovascularization by thrombospondin-1. *Nat Med, 6,* 41–48.

62. Gupta, K., Gupta, P., Wild, R., et al. (1999). Binding and displacement of vascular endothelial growth factor (VEGF) by thrombospondin: Effect on human microvascular endothelial cell proliferation and angiogenesis. *Angiogenesis, 3,* 147–158.

63. Dardik, R., Solomon, A., Loscalzo, J., et al. (2003). Novel proangiogenic effect of factor XIII associated with suppression of thrombospondin 1 expression. *Arterioscler Thromb Vasc Biol, 23,* 1472–1477.

64. Dawson, D. W., & Bouck, N. (1999). *Thrombospondin as an inhibitor of angiogenesis* (pp. 185–203). In B. A. Teicher (Ed.), Totowa, NJ: Humana Press.

65. Bikfalvi, A. (2004). Platelet factor 4: An inhibitor of angiogenesis. *Semin Thromb Hemost, 30,* 379–385.

66. Daly, M. E., Makris, A., Reed, M., et al. (2003). Hemostatic regulators of tumor angiogenesis: A source of antiangiogenic agents for cancer treatment? *J Nat Cancer Inst, 95,* 1660–1673.

67. Perollet, C., Han, Z. C., Savona, C., et al. (1998). Platelet factor 4 modulates fibroblast growth factor 2 (FGF-2) activity and inhibits FGF-2 dimerization. *Blood, 91,* 3289–3299.

68. Struyf, S., Burdick, M. D., Proost, P., et al. (2004). Platelets release CXCL4L1, a nonallelic variant of the chemokine platelet factor-4/CXCL4 and potent inhibitor of angiogenesis. *Circ Res, 95,* 855–857.

69. Marneros, A. G., & Olsen, B. R. (2001). The role of collagen-derived proteolytic fragments in angiogenesis. *Matrix Biol, 20,* 337–345.

70. Roberts, A. B., & Sporn, M. B. (1989). Regulation of endothelial cell growth, architecture, and matrix synthesis by TGF-beta. *Am Review Respir Dis, 140,* 1126–1128.

71. Iruela-Arispe, M. L., & Sage, E. H. (1993). Endothelial cells exhibiting angiogenesis *in vitro* proliferate in response to TGF-beta 1. *J Cell Biochem, 52,* 414–430.

72. Vinals, F., & Pouyssegur, J. (2001). Transforming growth factor beta1 (TGF-beta1) promotes endothelial cell survival during in vitro angiogenesis via an autocrine mechanism implicating TGF-alpha signaling. *Mol Cell Biol, 21,* 7218–7230.

73. Parsons-Wingerter, P., Elliott, K. E., Farr, A. G., et al. (2000). Generational analysis reveals that TGF-beta1 inhibits the rate of angiogenesis *in vivo* by selective decrease in the number of new vessels. *Microvas Res, 59,* 221–232.

74. Assoian, R. K., Komoriya, A., Meyers, C. A., et al. (1983). Transforming growth factor-beta in human platelets: Identification of a major storage site, purification, and characterization. *J Biol Chem, 258,* 7155–7160.

75. Pircher, R., Jullien, P., & Lawrence, D. A. (1986). Beta-transforming growth factor is stored in human blood plate-

lets as a latent high molecular weight complex. *Biochem Biophys Res Commun, 136,* 30–37.

76. Assoian, R. K., & Sporn, M. B. (1986). Type beta transforming growth factor in human platelets: Release during platelet degranulation and action on vascular smooth muscle cells. *J Cell Biol, 102,* 1217–1223.

77. Blakytny, R., Ludlow, A., Martin, G. E., et al. (2004). Latent TGF-beta1 activation by platelets. *J Cell Physiol, 199,* 67–76.

78. Radomski, A., Jurasz, P., Sanders, E. J., et al. (2002). Identification, regulation and role of tissue inhibitor of metalloproteinases-4 (TIMP-4) in human platelets. *Br J Pharmacol, 137,* 1330–1338.

79. Pipili-Synetos, E., Papadimitriou, E., & Maragoudakis, M. E. (1998). Evidence that platelets promote tube formation by endothelial cells on matrigel. *Br J Pharmacol, 125,* 1252–1257.

80. Trikha, M., Zhou, Z., Timar, J., et al. (2002). Multiple roles for platelet GPIIb/IIIa and alphavbeta3 integrins in tumor growth, angiogenesis, and metastasis. *Cancer Res, 62,* 2824–2833.

81. Jackson, C. (2002). Matrix metalloproteinases and angiogenesis. *Curr Opin Nephrol Hypertens, 11,* 295–299.

82. May, A. E., Kalsch, T., Massberg, S., et al. (2002). Engagement of glycoprotein IIb/IIIa (alpha(IIb)beta3) on platelets upregulates CD40L and triggers CD40L-dependent matrix degradation by endothelial cells. *Circulation, 106,* 2111–2117.

83. Slupsky, J. R., Kalbas, M., Willuweit, A., et al. (1998). Activated platelets induce tissue factor expression on human umbilical vein endothelial cells by ligation of CD40. *Thromb Haemost, 80,* 1008–1014.

84. Belting, M., Ahamed, J., & Ruf, W. (2005). Signaling of the tissue factor coagulation pathway in angiogenesis and cancer. *Arterioscler Thromb Vasc Biol,* ••.

85. Brill, A., Elinav, H., & Varon, D. (2004). Differential role of platelet granular mediators in angiogenesis. *Cardiovascular Res, 63,* 226–235.

86. Rhee, J. S., Black, M., Schubert, U., et al. (2004). The functional role of blood platelet components in angiogenesis. *Thromb Haemost, 92,* 394–402.

87. Weyrich, A. S., Dixon, D. A., Pabla, R., et al. (1998). Signal-dependent translation of a regulatory protein, Bcl-3, in activated human platelets. *Proceedings of the Nat Acad Sci USA, 95,* 5556–5561.

88. Pabla, R., Weyrich, A. S., Dixon, D. A., et al. (1999). Integrin-dependent control of translation: Engagement of integrin alphaIIbbeta3 regulates synthesis of proteins in activated human platelets. *J Cell Biol, 144,* 175–184.

89. Lindemann, S., Tolley, N. D., Dixon, D. A., et al. (2001). Activated platelets mediate inflammatory signaling by regulated interleukin 1beta synthesis. *J Cell Biol, 154,* 485–490.

90. Freyssinet, J. M. (2003). Cellular microparticles: What are they bad or good for? *J Thromb Haemost, 1,* 1655–1662.

91. Diamant, M., Tushuizen, M. E., Sturk, A., et al. (2004). Cellular microparticles: New players in the field of vascular disease? *Eur J Clin Invest, 34,* 392–401.

92. Heijnen, H. F., Schiel, A. E., Fijnheer, R., et al. (1999). Activated platelets release two types of membrane vesicles: Microvesicles by surface shedding and exosomes derived from exocytosis of multivesicular bodies and alpha-granules. *Blood, 94,* 3791–3799.

93. Barry, O. P., Pratico, D., Savani, R. C., et al. (1998). Modulation of monocyte-endothelial cell interactions by platelet microparticles. *J Clin Invest, 102,* 136–144.

94. Barry, O. P., & FitzGerald, G. A. (1999). Mechanisms of cellular activation by platelet microparticles. *Thromb Haemost, 82,* 794–800.

95. Baj-Krzyworzeka, M., Majka, M., Pratico, D., et al. (2002). Platelet-derived microparticles stimulate proliferation, survival, adhesion, and chemotaxis of hematopoietic cells. *Exp Hematol, 30,* 450–459.

96. Janowska-Wieczorek, A., Majka, M., Kijowski, J., et al. (2001). Platelet-derived microparticles bind to hematopoietic stem/progenitor cells and enhance their engraftment. *Blood, 98,* 3143–3149.

97. Kim, H. K., Song, K. S., Chung, J. H., et al. (2004). Platelet microparticles induce angiogenesis *in vitro. Br J Haematol, 124,* 376–384.

98. Jurasz, P., Alonso-Escolano, D., & Radomski, M. W. (2004). Platelet-cancer interactions: Mechanisms and pharmacology of tumour cell-induced platelet aggregation. *Br J Pharmacol, 143,* 819–826.

99. Brill, A., Dashevsky, O., Rivo, J., et al. (2005). Platelet-derived microparticles induce angiogenesis and stimulate post-ischemic revascularization. *Cardiovas Res, 67,* 30–38.

100. Horstman, L. L., Jy, W., Jimenez, J. J., et al. (2004). New horizons in the analysis of circulating cell-derived microparticles. *Keio J Med, 53,* 210–230.

101. Folkman, J. (1971). Tumor angiogenesis: Therapeutic implications. *N Engl J Med, 285,* 1182–1186.

102. Wojtukiewicz, M. Z., Sierko, E., Klement, P., et al. (2001). The hemostatic system and angiogenesis in malignancy. *Neoplasia, 3,* 371–384.

103. Trousseau, A. (1865). *Phlegmasia alba dolens. Clinique Medicale de L'Hotel-Dieu Paris* (pp. 94–96). London: New Sydenham Society.

104. Gasic, G. J., Gasic, T. B., & Stewart, C. C. (1968). Antimetastatic effects associated with platelet reduction. *Proc Nat Acad Sci USA, 61,* 46–52.

105. Karpatkin, S., Pearlstein, E., Salk, P. L., et al. (1981). Role of platelets in tumor cell metastases. *Annals of N Y Acad Sci, 370,* 101–118.

106. Radomski, M. W., Jenkins, D. C., Holmes, L., et al. (1991). Human colorectal adenocarcinoma cells: Differential nitric oxide synthesis determines their ability to aggregate platelets. *Cancer Res, 51,* 6073–6078.

107. Pinedo, H. M., Verheul, H. M., D'Amato, R. J., et al. (1998). Involvement of platelets in tumour angiogenesis? *Lancet, 352,* 1775–1777.

108. Pedersen, L. M., & Milman, N. (1996). Prognostic significance of thrombocytosis in patients with primary lung cancer. *Eur Respir J, 9,* 1826–1830.

109. Ikeda, M., Furukawa, H., Imamura, H., et al. (2002). Poor prognosis associated with thrombocytosis in patients with gastric cancer. *Ann Surg Oncol, 9,* 287–291.

110. Boucharaba, A., Serre, C. M., Gres, S., et al. (2004). Platelet-derived lysophosphatidic acid supports the progression of osteolytic bone metastases in breast cancer. *J Clin Invest, 114*, 1714–1725.

111. Jain, R. K. (1997). Delivery of molecular and cellular medicine to solid tumors. *Adv Drug Deliv Rev, 26*, 71–90.

112. Verheul, H. M., Jorna, A. S., Hoekman, K., et al. (2000). Vascular endothelial growth factor-stimulated endothelial cells promote adhesion and activation of platelets. *Blood, 96*, 4216–4221.

113. Verheul, H. M., & Pinedo, H. M. (1998). Tumor growth: A putative role for platelets? *Oncologist, 3*, II.

114. Poon, R. T., Lau, C. P., Cheung, S. T., et al. (2003). Quantitative correlation of serum levels and tumor expression of vascular endothelial growth factor in patients with hepatocellular carcinoma. *Cancer Res, 63*, 3121–3126.

115. Gunsilius, E., Petzer, A., Stockhammer, G., et al. (2000). Thrombocytes are the major source for soluble vascular endothelial growth factor in peripheral blood. *Oncology, 58*, 169–174.

116. Gonzalez, F. J., Rueda, A., Sevilla, I., et al. (2004). Shift in the balance between circulating thrombospondin-1 and vascular endothelial growth factor in cancer patients: Relationship to platelet alpha-granule content and primary activation. *Int J Biol Markers, 19*, 221–228.

117. Klement, G., Kikuchi, L., Kieran, M., et al. (2004). Early tumor detection using platelet uptake of angiogenesis regulators. *Blood, 104*, 239a.

118. Janowska-Wieczorek, A., Wysoczynski, M., Kijowski, J., et al. (2005). Microvesicles derived from activated platelets induce metastasis and angiogenesis in lung cancer. *Int J Cancer, 113*, 752–760.

119. Knighton, D. R., Ciresi, K. F., Fiegel, V. D., et al. (1986). Classification and treatment of chronic nonhealing wounds. Successful treatment with autologous platelet-derived wound healing factors (PDWHF). *Ann Surg, 204*, 322–330.

120. Knighton, D. R., Ciresi, K., Fiegel, V. D., et al. (1990). Stimulation of repair in chronic, nonhealing, cutaneous ulcers using platelet-derived wound healing formula. *Surg Gynecol Obstet, 170*, 56–60.

121. Hiraizumi, Y., Transfeldt, E. E., Kawahara, N., et al. (1993). *In vivo* angiogenesis by platelet-derived wound-healing formula in injured spinal cord. *Brain Res Bull, 30*, 353–357.

122. Mazzucco, L., Medici, D., Serra, M., et al. (2004). The use of autologous platelet gel to treat difficult-to-heal wounds: A pilot study. *Transfusion, 44*, 1013–1018.

123. Ksander, G. A., Sawamura, S. J., Ogawa, Y., et al. (1990). The effect of platelet releasate on wound healing in animal models. *J Am Acad Dermatol, 22*, 781–791.

124. Gehring, S., Hoerauf, H., Laqua, H., et al. (1999). Preparation of autologous platelets for the ophthalmologic treatment of macular holes. *Transfusion, 39*, 144–148.

125. Steed, D. L., Goslen, J. B., Holloway, G. A., et al. (1992). Randomized prospective double-blind trial in healing chronic diabetic foot ulcers: CT-102 activated platelet supernatant, topical versus placebo. *Diabetes Care, 15*, 1598–1604.

126. Ma, L., Elliott, S. N., Cirino, G., et al. (2001). Platelets modulate gastric ulcer healing: Role of endostatin and vascular endothelial growth factor release. *Proc Nat Acad Sci USA, 98*, 6470–6475.

127. Yamashiro, K., Tsujikawa, A., Ishida, S., et al. (2003). Platelets accumulate in the diabetic retinal vasculature following endothelial death and suppress blood-retinal barrier breakdown. *Am J Pathol, 163*, 253–259.

128. Weltermann, A., Wolzt, M., Petersmann, K., et al. (1999). Large amounts of vascular endothelial growth factor at the site of hemostatic plug formation *in vivo. Arterioscler Thromb Vasc Biol, 19*, 1757–1760.

129. Arisato, T., Hashiguchi, T., Sarker, K. P., et al. (2003). Highly accumulated platelet vascular endothelial growth factor in coagulant thrombotic region. *J Thromb Haemost, 1*, 2589–2593.

130. Koch, R. M., Roche, N. S., Parks, W. T., et al. (2000). Incisional wound healing in transforming growth factor-beta1 null mice. *Wound Repair Regen, 85*, 179–191.

131. Connolly, A. J., Suh, D. Y., Hunt, T. K., et al. (1997). Mice lacking the thrombin receptor, PAR1, have normal skin wound healing. *Am J Pathol, 151*, 1199–1204.

132. Szpaderska, A. M., Egozi, E. I., Gamelli, R. L., et al. (2003). The effect of thrombocytopenia on dermal wound healing. *J Invest Dermatol, 120*, 1130–1137.

# CHAPTER 42

# Tumor Growth and Metastasis

## Mary Lynn Nierodzik and Simon Karpatkin

*Division of Hematology, New York University School of Medicine, New York, New York*

## I. Introduction

The requirement of platelets for experimental pulmonary metastasis and the ability of tumor cells to aggregate platelets was first recognized by Gasic and coworkers in 1968, who demonstrated impaired experimental pulmonary metastasis in mice following platelet depletion.[1] These observations prompted numerous studies that have demonstrated a role for platelets in the hematogenous dissemination of animal tumors: many tumor cells require platelets for the development of metastasis;[1–5] ultrastructural studies have demonstrated arrested tumor emboli surrounded by platelets;[6–9] several tumor cells induce thrombocytopenia *in vivo*[1–3] and aggregate platelets *in vitro;*[2,4,10–18] and a correlation exists between the ability of some tumor cells to aggregate platelets *in vitro* and their requirement for metastasis *in vivo.*[2,4]

The rationale for the platelet requirement in tumor metastasis can be explained by the following observations:

1. Platelets interact with certain tumor emboli which prolong tumor survival in the circulation.[6–9] Indeed it has been shown that >98% of labeled tumor cells injected intravenously into mice disappear within 24 hours.[19]

2. Platelet-tumor emboli induce downstream ischemic endothelial damage, which exposes an adhesive subendothelial matrix (fibronectin, vitronectin, von Willebrand factor [VWF], laminin) for tumor cell binding and arrest.[20]

3. Sequestration of tumor cells by platelets protects tumor cells from immunologic host surveillance by impeding natural killer cell-mediated elimination of tumor cells.[21,22]

4. Platelets secrete tumor cell growth factors, such as platelet-derived growth factor (PDGF).[15,23–25]

5. Platelets secrete angiogenesis growth factors, such as vascular endothelial growth factor (VEGF) and angio-

poietin-1,[26–28] which are required for tumor angiogenesis, growth, and metastasis.

6. Platelets secrete permeability factors such as VEGF,[26] which could theoretically facilitate tumor cell invasion by penetration of the vessel wall.

7. Platelet-derived lysophosphatidic acid (LPA) promotes tumor cell proliferation by binding its $LPA_1$ receptor on breast (MDA-BO2) and ovarian (CHO) cancer cells. LPA also contributes to bone metastasis by stimulating the release of the bone osteoclast resorption stimulators, interleukin (IL)-8, and IL-6.[29]

8. Activated platelets generate thrombin on their surface, which can stimulate platelet-tumor adhesion as well as tumor growth and metastasis (discussed later).

## II. Platelet-Tumor Aggregation

In the early 1980s, considerable attention was addressed to the mode of tumor-induced platelet aggregation because it was hypothesized that interruption of this mechanism could lead to inhibition of tumor metastasis. Tumor cells aggregate platelets in heparinized platelet-rich plasma following a prolonged lag period of 1 to 3 minutes in a platelet aggregometer (Fig. 42-1). Citrated plasma is ineffective with most tumors examined, suggesting a role of divalent cation in this interaction. Three mechanisms of tumor-induced platelet aggregation were identified:[30] (a) a requirement for serum complement activation, a stable plasma cofactor other than fibrinogen, divalent cation, and the sialo-lipo-protein vesicular component of the tumor membrane (seen in murine SV40 transformed 3T3 fibroblasts and possibly a rat renal sarcoma, PW20); (b) a requirement for the generation of thrombin and a phospholipid component of the plasma membrane (seen in human colon carcinomas, LoVo, and HCT8 and an anaplastic murine cell line HUT-20); (c) a requirement for a trypsin-sensitive surface protein for activity (seen in a hamster-human melanoma, HM29, a

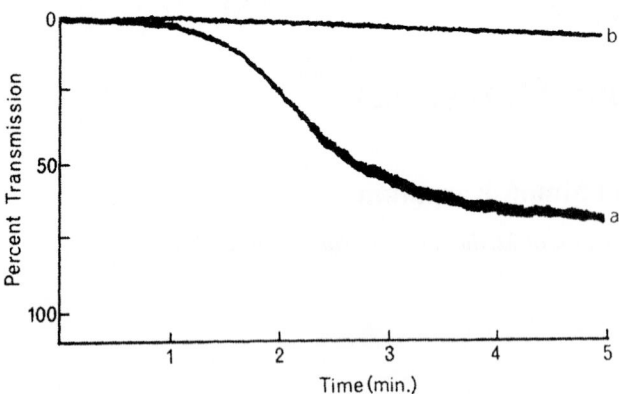

**Figure 42-1.** Induction of rabbit platelet aggregation in heparinized platelet-rich plasma (PRP) by $10^6$ intact SV3T3 cells before (a) and following (b) treatment of monolayer cells with 0.25 mL of 0.25% trypsin for 5 min at 37°C. Trypsin was neutralized with 0.5 mL of 0.25% soybean trypsin inhibitor prior to the preparation of the cell suspension for the aggregation assay. Then 0.05 mL of cells was added to 0.4 mL of PRP. Data taken from 1 of 5 experiments with similar results.

murine melanoma, B16F10, and colon carcinoma, CT26) (Fig. 42-1).

Inhibition of experimental tumor metastasis with anti-platelet aggregating agents has been controversial, because even the specific agents aspirin and ticlopidine as well as prostaglandin $I_2$ were unsuccessful.[6,16,25,31–35] These disappointing observations prompted a reevaluation of Gasic's original observation, which was confirmed employing three different tumor cell lines (CT26, Lewis lung, and B16 amelanotic melanoma). The new experiments demonstrated that the platelet requirement was early (within the first 6 hours) and that human platelets could reconstitute metastasis in mice protected by the induction of thrombocytopenia.[36]

## III. Platelet-Tumor Adhesion

Agents previously employed to inhibit tumor-induced platelet aggregation had little effect on platelet adhesion. Studies were therefore designed to measure the adhesion of tumor cells to a "lawn" of platelets on a plastic microtiter plate, demonstrating the requirement of platelet glycoprotein (GP) IIb-IIIa (integrin $\alpha IIb\beta 3$), fibronectin, VWF, and the RGDS domain of the adhesive proteins for adhesion (Table 42-1). Indeed, experimental murine tumor metastasis with CT26, B16a, and T241 Lewis bladder CA could be blocked 45% to 65% *in vivo* with an anti-VWF antibody (without inducing thrombocytopenia), and a monoclonal antibody against platelet GPIIb-IIIa could block the platelet reconstitution effect in mice protected from pulmonary metastasis with thrombocytopenia[35] (Table 42-2).

**Table 42-1:** **Effect of Anti-Platelet Monoclonal Antibodies 10E5 (Anti-GPIIb-IIIa), 6D1 (Anti-GPIb), 3B2 (Anti-IIb), Polyclonal Anti-Fibronectin Antibody, Polyclonal Anti-von Willebrand Factor Antibody, and Peptide RGDS on the Adhesion of CT26 and HCT8 Tumor Cells to Platelets *in vitro*\***

| n | Treatment | Tumor Cells × $10^3$ Bound to Platelets | Δ% | p |
|---|---|---|---|---|
| | | *CT26 Cells* | | |
| 10 | Buffer | 10.9 ± 3.8 | | |
| 10 | RGDS | 4.0 ± 0.7 | −64 | <0.001 |
| 8 | 10E5 | 4.0 ± 1.4 | −63 | =0.04 |
| 6 | 6D1 | 9.8 ± 3.9 | −10 | >0.1 |
| 6 | 3B2 | 10.5 ± 4.2 | −4 | >0.1 |
| 8 | Anti-fibronectin antibody | 2.0 ± 0.3 | −82 | <0.001 |
| 5 | Non-immune IgG | 7.7 ± 1.8 | | |
| 5 | Anti-mouse VWF antibody | 1.9 ± 1.0 | −75 | =0.01 |
| | | *HCT8 Cells* | | |
| 6 | Buffer | 7.4 ± 1.4 | | |
| 6 | RGDS | 2.3 ± 0.6 | −69 | =0.01 |
| 6 | 10E5 | 2.5 ± 0.9 | −66 | =0.03 |
| 6 | Non-immune IgG | 7.6 ± 1.6 | | |
| 6 | Anti-fibronectin antibody | 3.0 ± 0.8 | −60 | =0.01 |
| 6 | Anti-human VWF antibody | 1.4 ± 0.4 | −81 | =0.005 |

*1 × $10^3$ stractan-separated platelets were incubated with flat-bottomed plastic microtiter plates for 24 hrs at 4°, followed by addition of 0.01 M phosphate-buffered saline, pH 7.4 (PBS) + 1% bovine serum albumin (BSA) for 1 hr at 37° to block "free adherent" sites on the plastic. Buffer + BSA or F(ab')₂ fragments of monoclonal antibody 10E5, 6D1 or 3B2 (25 μg/mL) or the peptide RGDS (25 μg/mL) or rabbit anti-human fibronectin IgG or rabbit anti-human von Willebrand factor IgG (6 μg/mL) or rabbit anti-mouse von Willebrand factor IgG(37 μg/mL) or non-immune rabbit IgG (6 or 37 μg/mL), respectively, were then added for 1 hr at 37°, washed, followed by the addition of 1 × $10^5$ mouse CT26 or human HCT8 cells for 1 hr at 37° in PBS-0.9 mM CaCl₂. Non-adherent tumor cells were washed away, and the adherent tumor cells released into the supernatant with trypsin-EDTA. Tumor cells were then counted directly in a counting chamber under phase microscopy. Background adhesion of tumor cells to microtiter wells (0–10% of control values) was subtracted from tumor cells adherent to platelets. Platelet-adherent tumor cells ± SEM is given. n refers to number of experiments. Platelet preparations were human, except for the CT26 cell line experiment with anti-mouse von Willebrand factor, where mouse platelets were employed.

**Table 42-2: Effect of Monoclonal Antibody 10E5 on Reconstitution of Murine Metastases with Human Platelets After Inhibition of Metastases by Induction of Thrombocytopenia***

| | | | Mean Total Nodule Volume/mm³ | | | |
|---|---|---|---|---|---|---|
| Expt. | n | Control | Anti-Platelet Antibody | Anti-Platelet Ab + Platelets | Anti-Platelet Ab + Platelets + 10E5 | % Inhibition of Reconstitution |
| 1 | 45 | 61.9 ± 8.7 | 37.1 ± 8.7 | 61.1 ± 13.8 | 36.8 ± 6.4 | 100 |
| 2 | 50 | 10.4 ± 1.7 | 3.0 ± 0.6 | 10.6 ± 2.8 | 3.6 ± 0.5 | 92 |
| 3 | 20 | 31.0 ± 10.5 | 7.4 ± 3.0 | 16.3 ± 3.2 | 10.4 ± 2.9 | 66 |
| 4 | 30 | 16.6 ± 4.5 | 4.9 ± 1.2 | 14.1 ± 4.7 | 11.2 ± 5.4 | 32 |
| 5 | 52 | 18.0 ± 3.2 | 6.3 ± 1.3 | 9.8 ± 1.6 | 7.2 ± 1.4 | 74 |
| Mean | | 27.3 ± 5.0 | 12.4 ± 3.0 | 23.0 ± 5.3 | 14.0 ± 3.1 | 77 |

*Control mice (Column 3) received 10 μL of non-immune mouse serum 6 hr prior to the intravenous injection of 50,000 CT26 tumor cells. Thrombocytopenic mice (Column 4) received 100 μL of rabbit anti-mouse platelet serum 6 hr prior to injection of CT26 cells. The reconstituted group of mice (Column 5) received $5 \times 10^8$ human platelets 5½ hr after injection of anti-platelet antibody and 30 min prior to the injection of tumor cells. The mice reconstituted with 10E5-treated platelets (Column 6) were identical to the third group except for preincubation of human platelets with 2.5 μg of F(ab')₂ fragments of 10E5 for 30 min at room temperature prior to their injection. Mean refers to the "weighted" mean for the number of experiments. n refers to the total number of animals in each experiment. SEM is given. % Inhibition of Reconstitution refers to the difference between column 6 and column 5 divided by the difference between column 5 and column 4.

Soluble fibrin monomer also augments platelet-tumor cell adherence *in vitro* and *in vivo* and also requires platelet GPIIb-IIIa, as well as tumor ICAM-1.[37]

A study employing human colon carcinomas LS174HT and COLO205 has extended these observations for adhesion of tumor cells to platelets under dynamic flow conditions. This group has demonstrated a two-phase sequential process for tumor adhesion to platelets under flow: platelet P-selectin mediates tumor cell tethering and rolling followed by stable adhesion initiated by GPIIb-IIIa and VWF, via an RGD-dependent mechanism. HCT8 cells failed to react with P-selectin tethered minimally to platelets under flow conditions, despite extensive adhesion under static conditions.[38] Similar observations have been made by others who have demonstrated the presence of the P-selectin receptors, PSGL-1 (P-selectin glycoprotein ligand-1) and CD24 in KS breast CA cells,[39,40] the presence of distinct ligands for selectins on colon carcinoma mucin-type glycoproteins, and the attenuation of tumor growth and metastasis in P-selectin deficient mice in association with absence, *in vivo*, of platelet tumor aggregates.[41]

## IV. Effect of Thrombin on Platelet-Tumor Adhesion and Pulmonary Metastasis

Thrombin both activates and enhances exposure of GPIIb-IIIa on the platelet membrane surface (see Chapter 8). Thrombin also induces the release of platelet fibronectin and VWF onto the platelet surface. We therefore hypothesized

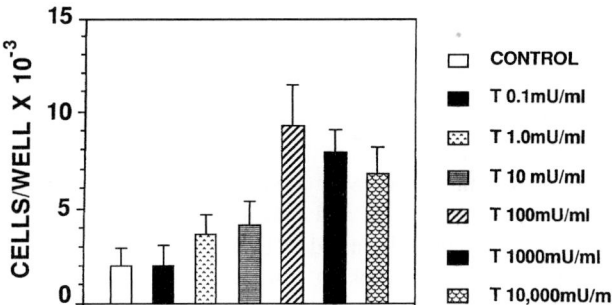

**Figure 42-2.** Effect of thrombin treatment of platelets on adhesion of HM54 melanoma cells to platelets. Stractan-separated platelets ($3 \times 10^7$) were applied to microtiter plates at 22°C for 1 hour, followed by the addition of thrombin (T) at the designated concentrations (in milliunits per milliliter). After 18 hours of overnight incubation at 4°C, the platelets were washed and blocked with PBS-1% BSA, followed by the addition of $1 \times 10^5$ HM54 melanoma cells in PBS-0.9 mM MgCl₂ + 0.9 mM CaCl₂ for 1 hour at 37°C. Nonadherent cells were removed by washing. Adherent cells were removed with trypsin-EDTA and enumerated by phase microscopy (n = 5).

that thrombin might activate tumor-platelet adhesion *in vitro* and metastasis *in vivo*. Figure 42-2 demonstrates the exquisite sensitivity of platelet-tumor adhesion to thrombin. Platelets treated with as little as 0.1 milliunits of thrombin 15 hours prior to the addition of tumor cells enhanced tumor adhesion, with an optimal twofold to fourfold enhancement obtained at 10 to 100 milliunits of thrombin. Similar results were obtained with four additional tumor cell lines: HCT8

**Table 42-3: Effect of Thrombin on Pulmonary Metastases of CT26 Colon Carcinoma and
B16a Amelanotic Melanoma Tumor Cells**

| Expt. | n | Group | Mean No. of Nodules per Lung | Mean Nodule Volume per Lung | Tumor Mass | Fold Increase | p |
|---|---|---|---|---|---|---|---|
| | | | CT26 Colon Carcinoma | | | | |
| 1 | 19 | Control | 0.68 ± 0.17 | 9.1 ± 3.8 | 6.2 | 4.3 | =0.007 |
| | 20 | Thrombin | 1.95 ± 0.40 | 13.6 ± 4.4 | 26.5 | | |
| | | | B16a Amelanotic Carcinoma | | | | |
| 2 | 7 | Control | 2.1 ± 0.6 | 3.3 ± 1.3 | 6.9 | | =0.006 |
| | 6 | Thrombin | 11.3 ± 2.9 | 41.7 ± 6.0 | 471.2 | 68 | =0.016 |

| Expt. | n | Group | Mean No. of Nodules per Section (μm²) | Mean Surface Area per Histologic Section | Total Tumor Area per Histologic Section | Fold Increase | p |
|---|---|---|---|---|---|---|---|
| 3 | 19 | Control | 0.37 ± 0.17 | 11.8 ± 6.0 | 3.1 | | <0.001 |
| | 17 | Thrombin | 3.71 ± 0.60 | 330 ± 71 | 1237.5 | 413 | <0.001 |

50,000 viable CT26 colon carcinoma cells plus thrombin 500 milliunits per mL and 25,000 B16a amelanotic cells plus thrombin 250 milliunits per mL were injected intravenously into BALB/c and C57B1/6J mice, respectively. The number and volume of pulmonary metastases were enumerated macroscopically (expts. 1 and 2) and microscopically (expt. 3) on day 28. SEM is given. The first p value refers to the difference between control and thrombin mean number of nodules per lung; the second value refers to the difference between mean nodule volume (or mean surface area per histologic section) per lung.

human colon carcinoma and murine, B16a, KLN205 squamous cell, and hamster.[42]

This thrombin effect could be inhibited by high-affinity PPACK thrombin (phenylanyl-L-propyl-L-arginine chloromethyl ketone), hirudin, and DAPA (dansylarginine N-)3-ethyl-1,5 pentanediyl) amide, but not by low-affinity TLCK-thrombin (N-P-tosyl-L-lysine chloromethyl ketone). Further experiments demonstrated that the thrombin-induced effect on enhanced adhesion, as with naïve platelets, also required GPIIb-IIIa, fibronectin, VWF, and the RGDS domain of fibronectin and VWF, because agents inhibiting the receptor or ligand also inhibited tumor cell adhesion.[42]

*In vivo* experiments with thrombin substantiated these *in vitro* observations. When 0.2 to 0.5 units of thrombin were injected together with tumor cells intravenously (a dose titered not to reduce the platelet count), pulmonary metastases increased 4- to 413-fold with two different syngeneic tumor cell lines (CT26 and B16a)[42] (Table 42-3).

## V. Effect of Thrombin-Treated Tumor Cells on Adhesion to Naïve Platelets and Endothelial Cells *in Vitro* and Pulmonary Metastasis *in Vivo*

Because of the dramatic increase *in vivo* of tumor metastasis when thrombin was administered intravenously together with tumor cells, its direct effect on tumor cells was tested.

Figure 42-3 demonstrates a threefold increased adhesion of thrombin-treated tumor cells to adherent platelets. When thrombin-treated platelets and thrombin-treated tumor cells were added together, no additive effect was noted.[43] This would suggest that thrombin-treated tumor cells and thrombin-treated platelets were operating through the same mechanism. Five of seven tumor cell lines from three different species can be activated by thrombin (B16a, HCT-8, Lewis Lung, HM29, and SK-Mel-28 human melanoma) to adhere to platelets. Others have reported similar results.[44,45] Further studies revealed that the thrombin effect was maximum after 1 hour of incubation and not inhibited by the protein synthesis inhibitor, cyclohexamide. In addition, the thrombin effect on tumor cells could not be inhibited by DAPA, indicating that thrombin induced a secondary event.[43]

Thrombin-activated tumor cells (SKMel-28 and HM29) also enhance their adhesion to bovine aortic and capillary endothelial cells 2.1- to 2.3-fold.[46] Similar observations have been made by others employing flow conditions in which thrombin increased the adhesion of human melanoma 397 cells to endothelial cells 2.2-fold and adhesion was blocked by antibodies to GPIIb-IIIa as well as P-selectin;[47] another study demonstrated that ⁵¹Cr-labeled tumor cells (HELA or HT29) have greatly increased adhesion to endothelial cells in culture in the presence of both platelets and thrombin compared to platelets or thrombin alone.[48]

Of particular interest was the observation that thrombin-treated melanoma tumor cells studied also required a

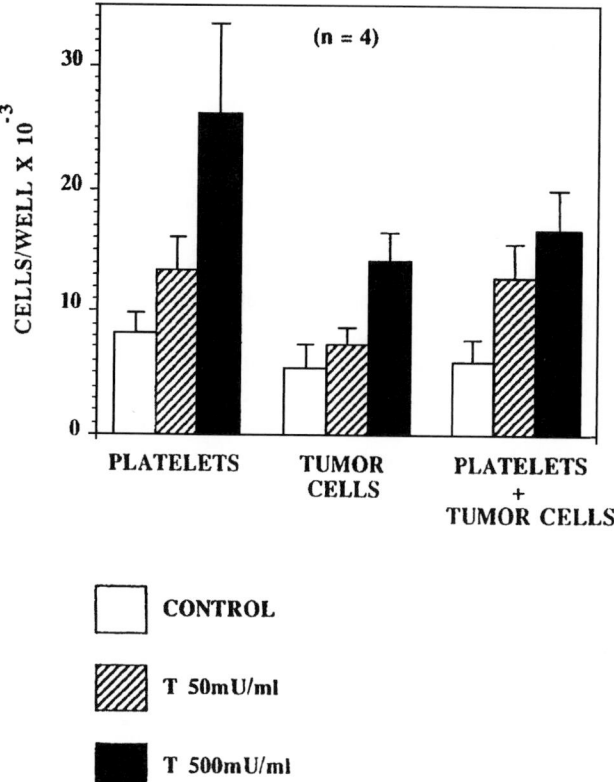

**Figure 42-3.** Effect of thrombin-treated platelets versus thrombin-treated tumor cells on tumor cell-platelet adhesion. Platelets ($3 \times 10^7$) were applied to microtiter plates. (*Left*) Platelets were treated with thrombin or buffer overnight and then incubated with $1 \times 10^5$ naïve HM29 cells in PBS-BSA-MgCa. (*Middle*) Naïve platelets were incubated with $1 \times 10^5$ HM29 cells previously treated with buffer or thrombin for 1 hour at 37°C, followed by washing. (*Right*) Thrombin-treated platelets were incubated with thrombin-treated HM29 cells as in left and middle. In all experiments, cells were washed with buffer, eluted with trypsin-EDTA, and enumerated under phase microscopy. Columns, mean of four experiments performed in triplicate; bars, SEM.

"GPIIb-IIIa-like" receptor on their surface (inhibited by anti-GPIIb-IIIa and RGDS).[49] The presence of a GPIIb-IIIa-like receptor on tumor cells has been supported by the observations of others.[49-54] M3Dau melanoma cells react with platelets, bind a GPIIb-IIIa-specific monoclonal antibody, synthesize GPIIb-IIIa-like glycoproteins, and do not react with Glanzmann thrombasthenia platelets (congenital absence of GPIIb-IIIa).[49] When preincubated with anti-GPIIb-IIIa antibody, M3Dau tumor cell growth was dramatically inhibited following subcutaneous implantation into nude mice.[49] Sixteen of 21 human malignant melanoma frozen biopsies reacted with an anti-GPIIb-IIIa antibody, whereas negative results were obtained with 15 benign human melanoma specimens and 73 of 75 other human carcinomas.[52] Monoclonal and polyclonal antibodies against GPIIb-IIIa reacted with two cell lines derived from human

tumors (MS751 cervical CA and a human colon CA) and were localized to the plasma membrane. When pretreated with the antibodies, both tumor cells were impaired in their ability to aggregate platelets, as well as adhere to fibronectin.[51] GPIIb-IIIa has been found on human melanoma cells WM983B, 983A, and 35 in its inactive conformation (lack of reactivity with monoclonal antibody PAC1). Stimulation with a protein kinase C activator enhances binding to PAC1 and immobilized fibronectin (blocked by RGD peptide) and stimulated invasion through a fibronectin monolayer (blocked by PAC1).[54] GPIIb-IIIa has also been reported in 17 tumor cell lines of different histological origin: skin, blood, lung, liver, kidney, cervix, colon, bladder, breast, and prostate.[53] However, GPIIb-IIIa was not found in another study employing a human melanoma SK-Mel-28 cell line.[46]

*In vivo* experiments with thrombin-treated tumor cells (CT26, B16F1, and B16F10), followed by washing, enhanced experimental pulmonary metastases 10- to 156-fold.[43] It is likely that thrombin may be inducing effects on tumor cells other than increased adhesion, such as induction of growth proto-oncogenes/oncogenes as well as metastasis. For example, thrombin has been shown to enhance the synthesis and secretion of the angiogenesis growth factors VEGF and angiopoietin-2[55,56] (added to its effect on secretion of both factors from platelets) in human primary FS4 fibroblasts as well as prostate DU145 and megakaryocyte CHRF cells.

The thrombin protease activated receptor 1 (PAR1) was present on nine of nine tumor cells examined by RT-PCR. Seven of 11 lines respond to the PAR1 thrombin receptor activation peptide (TRAP), SFLLRN, by a twofold to threefold enhanced adhesion to platelets.[57] Two TRAP-treated murine cell lines, B16F10 and CT26, enhance their experimental pulmonary metastasis 17- to 320-fold, despite having no effect on enhanced adhesion to platelets. Others have observed similar results.[44,45] The enhanced effect on pulmonary metastasis is proportional to PAR1 receptor density[58] on tumor cells. Activation of PAR1 has been shown to stimulate chemokinesis of melanoma and prostate tumor cells.[59] Thus, thrombin stimulates other tumor growth/metastatic promoting properties, as well as adhesion.

Genetic confirmation of the role of circulating platelets, platelet activation, and fibrinogen has been studied in a B16 melanoma model in NF-E2$^{-/-}$ knockout mice with virtually no circulating platelets, PAR4$^{-/-}$ knockout mice (the major thrombin receptor in mice) with platelets that fail to respond to thrombin, and fibrinogen Fib$^{-/-}$ knockout mice.[5,22] Marked reduction in tail vein pulmonary metastasis was seen in all three groups (6%, 14%, and 24% of wild-type, respectively). Of considerable interest are the observations obtained with Gαq$^{-/-}$ platelet knockout mice, a critical G protein required for platelet activation. Experimental and spontaneous pulmonary metastasis obtained with syngeneic Lewis Lung as well as B16-BL6 tumor cells were decreased by approximately 100-fold and approximately 60-fold B16-BL6, respectively.[22] However, Gαq[-/-] platelets had no effect on subcutaneous tumor growth *in vivo*. In addition, a possible link between NK cell function and platelets was suggested from the observation that enhanced experimental pulmonary metastasis in the absence of NK cells was no different with Gαq platelets versus control platelets.[22]

Several genes have been reported to be induced in cells following thrombin treatment in specific cell lines: urokinase-type plasminogen activators in PC-3 prostate cancer cells,[60] PDGF and E-selectin in HUVEC,[61] monocyte chemotactic protein 1 in monocytes,[62,63] atrial natriuretic factor in myocardium,[64] and tissue factor in endothelial cells.[65] A precedent for thrombin-induced enhanced oncogenesis is supported by the following observations. Thrombin can act as a mitogenic agent for mesenchymal tissue: fibroblasts, endothelial cells, and smooth muscle cells.[66–71] Thrombin can induce the early response oncogene, c-*fos* in endothelial cells,[72] and CHRF-288 megakaryocytes.[73] Indeed, thrombin generates some of the same signals induced by oncogenes.

## VI. Discussion

These data clearly show a role for platelets, platelet GPIIb-GPIIIa (integrin αIIbβ3), platelet P-selectin, fibronectin, fibrin, VWF, and thrombin in experimental pulmonary murine metastasis. It is likely that other adhesive ligands such as laminin,[74–79] vitronectin,[50,80] type IV collagen,[74,75,81] and thrombospondin,[82,83] as well as other integrin receptors[74,78] (α3β1,[46] α5β1,[46] αvβ3[46,80]) are also involved in tumor adhesion,[50,78,80,83–85] platelet-tumor interaction,[83,85] and metastases.[75,76,82,85]

The role of a required GPIIb-IIIa-like integrin on the platelet, tumor cell, or other structure is supported by additional *in vivo* data. In experimental pulmonary metastases, the pentapeptide GRGDS blocked B16F10 melanoma metastases by 97% without impairing cellular tumorigenicity. This was shown to be due to inhibition of retention of tumor cells in the vasculature. Five times more cells were retained in control mice at 7 hours. This difference increased with time to parallel the effect of pulmonary colonization at 2 weeks.[86] Similar results were noted in a second report in which platelets appeared not to be responsible, that is, thrombocytopenia did not inhibit the RGDS effect[87] — suggesting that tumor cells may bind via an RGDS mechanism to structures other than platelets. The disintegrin, Albolabrin, an RGD-containing peptide inhibited attachment of B16F10 mouse melanoma to fibronectin or laminin when immobilized on plastic *in vitro* and inhibited experimental pulmonary metastasis by 90% *in vivo*.[78]

Extrapolating from these data, it is possible that low-grade thrombin generation may be harmful to some patients with malignancies, because it may predispose to metastatic progression of the lesion. Indeed, there is abundant evidence that many tumor cells activate the coagulation system with generation of thrombin.[30,88–94] Low-grade intravascular coagulation — as diagnosed by increased fibrinogen turnover,[95] increased plasma levels of fibrinogen/fibrin-related antigen,[95] and increased plasma levels of fibrinopeptide A — has been observed in most patients with solid tumors.[96–98] One study noted elevated fibrinopeptide A levels in 60% of patients at the time of presentation, with increasing levels noted with disease progression. Persistent elevation was associated with a poor prognosis.[98] In this respect, it is of interest that r-hirudin, a potent antithrombin agent, inhibited experimental pulmonary metastasis of B16F10 melanoma cells by 95% to 98%, with the optimum effect occurring when given 2 to 15 minutes before intravenous tumor inoculation.[99] Similar results have been noted with human xenografts, where r-hirudin appears to inhibit tumor implantation, seeding, and spontaneous metastasis rather than growth.[100] Thus, thrombin activates an early component of the metastatic event. However, this does not rule out the possibility that thrombin may also be activating those tumor cells that have survived within the circulation after the initial rapid disappearance of the vast majority of tumor cells.[19]

Thus, it is suggested that antithrombin and antiplatelet adhesive agents may be helpful in the prevention of tumor metastasis and that thrombin may awaken dormant tumor cells in the host.[101]

## VII. Summary

Platelets and thrombin play significant roles in tumor cell adhesion *in vitro* and metastasis *in vivo*. Adhesion of tumor cells to platelets is inhibited by agents inhibiting platelet GPIIb-IIIa (integrin αIIbβ3) receptor occupancy as well as GPIIb-IIIa ligands. Initial tethering of tumor cells to platelets is via P-selectin. *In vivo* murine experimental pulmonary metastasis (tail vein injection) is inhibited by antibody-induced induction of thrombocytopenia and is reconstituted by simultaneous injection of human platelets. Preincubation of human platelets with a GPIIb-IIIa-specific monoclonal antibody inhibits reconstitution of metastasis. Thrombin activates tumor cell adhesion to platelets by activating platelets as well as tumor cells. Activated platelets release angiogenesis growth factors VEGF and angiopoietin-1 as well as other growth factors (e.g., PDGF). Thrombin-activated tumor cells also enhance their adhesion to endothelial cells as well as adhesive ligands fibronectin and VWF by a GPIIb-IIIa-like receptor. Experimental pulmonary metastasis is enhanced 4- to 400-fold by preinfusion of thrombin into mice or 10- to 160-fold by prior treatment of tumor cells with thrombin. The *in vitro* and *in vivo* effects of thrombin are mimicked by the thrombin receptor activation peptides SFLLRNPNDKYEPF and SFLLRN and are

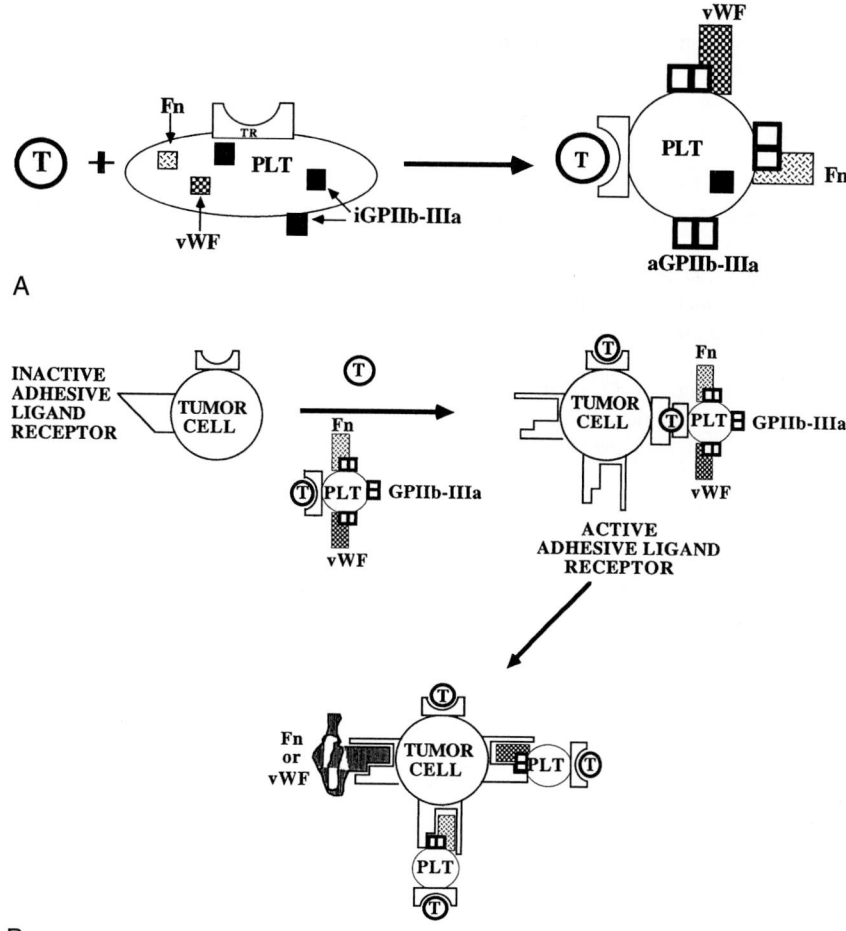

**Figure 42-4.** Proposed schematic diagram of platelet-tumor adhesion. (A) Thrombin binds to a receptor(s) on platelets, which converts GPIIb-IIIa to its conformationally active form, induces the expression of additional GPIIb-IIIa on the platelet surface, and stimulates the release of platelet fibronectin and von Willebrand factor (vWF), which then binds to conformationally active GPIIb-IIIa. (B) Thrombin or thrombin bound to a platelet receptor(s) binds to PAR1 on tumor cells, initiating the activation of a tumor "GPIIb-IIIa-like" receptor, which forms a bridge with fibronectin, vWF and, possibly, fibrin on the activated platelet surface. Tumor cells, like platelets, bind to sub-endothelial matrix fibronectin and vWF. Abbreviations: aGPIIb-IIIa, activated GPIIb-IIIa; Fn, fibronectin; iGPIIb-IIIa, inactivated GPIIb-IIIa; PLT, platelet; T, thrombin; vWF, von Willebrand factor.

proportional to tumor PAR1 receptor density. Nine of nine tumor cell lines have the seven transmembrane-spanning thrombin receptors detected by the polymerase chain reaction. Thus, both platelets and thrombin contribute to experimental tumor metastasis by fostering and enhancing platelet GPIIb-IIIa interaction with tumor cells (Fig. 42-4). Because many tumor cells generate thrombin, it is proposed that tumor cells may autoactivate a metastatic phenotype.

# References

1. Gasic, G., Gasic, T., & Stewart, C. (1968). Antimetastatic effects associated with platelet reduction. *Proc Nat Acad Sci USA, 61,* 46–52.

2. Gasic, G. J., Gasic, T. B., Galanti, N., et al. (1973). Platelet-tumor-cell interactions in mice: The role of platelets in the spread of malignant disease. *Int J Cancer, 11,* 704–718.

3. Hilgard, P. (1973). The role of blood platelets in experimental metastases. *Br J Cancer, 28,* 429–435.

4. Pearlstein, E., Salk, P. L., Yogeeswaran, G., et al. (1980). Correlation between spontaneous metastatic potential, platelet-aggregating activity of cell surface extracts, and cell surface sialylation in 10 metastatic-variant derivatives of a rat renal sarcoma cell line. *Proc Nat Acad Sci USA, 77,* 4336–4339.

5. Camerer, E., Qazi, A., Duong, D., et al. (2004). Platelets, protease-activated receptors, and fibrinogen in hematogenous metastasis. *Blood, 104,* 397–401.

6. Hilgard, P., & Gordon-Smith, E. (1974). Microangiopathic haemolytic anemia and experimental tumor-cell emboli. *Br J Haematol, 26,* 651–659.

7. Jones, J., Wallace, A., & Fraser, E. (1991). Sequence of events in experimental and electron microscopic observations. *J Nat Cancer Inst, 46,* 493–504.

8. Sindelar, W., Tralka, T., & Ketcham, A. (1975). Electron microscope observations on formation of pulmonary metastases. *J Surg Res, 18,* 137–161.

9. Warren, B., & Vales, O. (1972). The adhesion of thromboplastic tumor emboli to vessel walls *in vivo. Br J Haematol, 53,* 301–313.

10. Fitzpatrick, F. A., & Stringfellow, D. A. (1979). Prostaglandin D2 formation by malignant melanoma cells correlates inversely with cellular metastatic potential. *Proc Nat Acad Sci USA, 76,* 1765–1769.

11. Gasic, G. J., Boettiger, D., Catalfamo, J. L., et al. (1978). Aggregation of platelets and cell membrane vesiculation by rat cells transformed *in vitro* by Rous sarcoma virus. *Cancer Res, 38,* 2950–2955.

12. Gasic, G. J., Gasic, T. B., & Jimenez, S. A. (1977). Effects of trypsin on the platelet-aggregating activity of mouse tumor cells. *Thromb Res, 10,* 33–45.

13. Gasic, G. J., Gasic, T. B., & Jimenez, S. A. (1977). Platelet aggregating material in mouse tumor cells: Removal and regeneration. *Lab Invest, 36,* 413–419.

14. Gasic, G. J., Koch, P. A., Hsu, B., et al. (1976). Thrombogenic activity of mouse and human tumors: Effects on platelets, coagulation, and fibrinolysis, and possible significance for metastases. *Z Krebs forsch Klin Onkol Cancer Res Clin Oncol, 86,* 263–277.

15. Hara, Y., Steiner, M., & Baldini, M. (1980). Platelets as a source of growth-promoting factor(s) for tumor cells. *Cancer Res, 40,* 1212–1216.

16. Karpatkin, S., Smerling, A., & Pearlstein, E. (1980). Plasma requirement for the aggregation of rabbit platelets by an aggregating material derived from SV40-transformed 3T3 fibroblasts. *J Lab Clin Med, 96,* 994–1001.

17. Marcum, J. M., McGill, M., Bastida, E., et al. (1980). The interaction of platelets, tumor cells, and vascular subendothelium. *J Lab Clin Med, 96,* 1046–1053.

18. Pearlstein, E., Cooper, L. B., & Karpatkin, S. (1979). Extraction and characterization of a platelet-aggregating material from SV40-transformed mouse 3T3 fibroblasts. *J Lab Clin Med, 93,* 332–344.

19. Fidler, I. (1970). Metastases: Quantitative analysis of distribution and fate of tumor emboli labeled with [125]I-5-iodo-2'-deoxyuridine. *J Nat Cancer Inst, 45,* 773–782.

20. Warren, B. (1984). The microinjury hypothesis and metastasis. In *Hemostatic mechanisms and metastasis* (p. 56). Boston: M. Nijhoff.

21. Nieswandt, B., Hafner, M., Echtenacher, B., & Mannel, D. (1999). Lysis of tumor cells by natural killer cells in mice is impeded by platelets. *Cancer Res, 59,* 1295–1300.

22. Palumbo, J. S., Talmage, K. E., Massari, J. V., et al. (2005). Platelets and fibrin(ogen) increase metastatic potential by impeding natural killer cell-mediated elimination of tumor cells. *Blood, 105,* 178–185.

23. Cowan, D., & Graham, J. (1982). *Effect of platelet growth factor(s) on growth of human tumor colonies* (pp. 249–268). New York: R. Alan, Liss.

24. Eastman, B., & Sirbasku, D. (1981). Platelet derived growth factor(s) for a hormone responsive rat mammary tumor cell line. *J Cell Physiol, 97,* 17–28.

25. Kepner, N., & Lipton, A. (1981). A mitogenic factor for transformed fibroblasts from human platelets. *Cancer Res, 41,* 430–432.

26. Mohle, R., Green, D., Moore, M., et al. (1997). Constitutive production and thrombin-induced release of VEGF by human megakaryocytes and platelets. *Proc Nat Acad Sci USA, 94,* 663–668.

27. Li, J.-J., Huang, Y.-Q., Basch, R., et al. (1998). Thrombin induces the release of angiopoietin-1 from platelets. *Thromb Haemost, 85,* 204–206.

28. Maloney, J., Sillimon, C., Ambruso, D., et al. (1998). *In vitro* release of VEGF during platelet aggregation. *Am J Physicians, 275,* H1054.

29. Boucharaba, A., Serre, C. M., Gres, S., et al. (2004). Platelet-derived lysophosphatidic acid supports the progression of osteolytic bone metastases in breast cancer. *J Clin Invest, 114,* 1714–1725.

30. Lerner, W., Pearlstein, E., Ambrogio, C., et al. (1983). A new mechanism for tumor-induced platelet aggregation: Comparison with mechanism shared by other tumors with possible pharmacologic strategy toward prevention of metastases. *Int J Cancer, 31,* 463–469.

31. Gasic, G., & Gasic, T. (1962). Removal of sialic acid from the cell coat in tumor cells and vascular endothelium, and its effects on metastasis. *Proc Nat Acad Sci USA, 48,* 1172–1177.

32. Miller, O. V., Aiken, J. W., Shebuski, R. J., et al. (1980). 6-keto-prostaglandin E1 is not equipotent to prostacyclin (PGI2) as an antiaggregatory agent. *Prostaglandins, 20,* 391–400.

33. Tamao, Y., Hara, H., Kikumoto, R., et al. (1981). The effect of the synthetic thrombin inhibitor, no. 805, on an experimental DIC in rabbits. *Thromb Haemos, 46,* ••.

34. Karpatkin, S., Ambrogio, C., & Pearlstein, E. (1984). Lack of effect of *in vivo* prostacyclin on the development of pulmonary metastases in mice following intravenous injection of CT26 colon carcinoma, Lewis lung carcinoma, or B16 amelanotic melanoma cells. *Cancer Res, 44,* 3880–3883.

35. Karpatkin, S., Pearlstein, E., Ambrogio, C., et al. (1988). Role of adhesive proteins in platelet tumor interaction *in vitro* and metastasis formation *in vivo. J Clin Invest, 81,* 1012–1019.

36. Pearlstein, E., Ambrogio, C., & Karpatkin, S. (1984). Effect of anti-platelet antibody on the development of pulmonary metastases following injection of CT26 colon adenocarcinoma, Lewis lung carcinoma and B16 amelanotic melanoma tumor cells in mice. *Cancer Res, 44,* 3884–3887.

37. Biggerstaff, J., Seth, N., Amirkhosravi, A., et al. (1999). Soluble fibrin augments platelet/tumor cell adherence *in vitro* and *in vivo,* and enhances experimental metastasis. *Clin Exp Metastasis, 17,* 723–730.

38. McCarty, O., Mousa, S., Bray, P., et al. (2000). Immobilized platelets support human colon carcinoma cell tethering, rolling, and firm adhesion under dynamic flow conditions, *96,* ••.

39. Aigner, S., Ramos, C., Hafezi-Moghadam, A., et al. (1998). CD24 mediates rolling of breast carcinoma cells on P-selectin. *FASEB J, 12,* 1241–1251.

40. Aigner, S., Sthoeger, M., Fogel, M., et al. (1997). CD24, a mucin-type glycoprotein, is a ligand for P-selectin on human tumor cells, *89,* 3385–3395.

41. Kim, Y., Borsig, L., Varki, N., et al. (1998). P-selectin deficiency attenuates tumor growth and metastasis. *Proc Nat Acad Sci USA, 95,* 9325–9330.

42. Nierodzik, M., Plotkin, A., Kajumo, F., et al. (1991). Thrombin stimulates tumor-platelet adhesion *in vitro* and metastasis *in vivo. J Clin Invest, 87,* 229–236.

43. Nierodzik, M., Kajumo, F., & Karpatkin, S. (1992). Effect of thrombin treatment of tumor cells on adhesion of tumor cells to platelets *in vitro* and metastasis *in vivo. Cancer Res, 52,* 3267–3272.

44. Wojtukiewicz, M., Tang, D., Ben-Josef, E., et al. (1995). Solid tumor cells express functional tethered ligand thrombin receptor. *Cancer Res, 55,* 698–700.

45. Wojtukiewicz, M., Tang, D., Ciarelli, J., et al. (1993). Thrombin increases the metastatic potential of tumor cells. *Int J Cancer, 54,* 793–806.

46. Klepfish, A., Greco, M. A., & Karpatkin, S. (1993). Thrombin stimulates melanoma tumor-cell binding to endothelial cells and subendothelial matrix. *Int J Cancer, 53,* 978–982.

47. Dardik, R., Savion, N., Kaufmann, Y., et al. (1998). Thrombin promotes platelet-mediated melanoma cell adhesion to endothelial cells under flow conditions: Role of platelet glycoproteins P-selectin and GPIIb-IIIa. *Br J Cancer, 77,* 2069–2075.

48. Helland, I., Klemensten, B., & Jorgensen, L. (1997). Addition of both platelets and thrombin in combination accelerates tumor cells to adhere to endothelial cells *in vitro. In Vitro Cell Dev Biol Anim, 33,* 182–186.

49. Boukerche, H., Berthier-Vergnes, O., Tabone, E., et al. (1989). Platelet-melanoma cell interaction is mediated by the glycoprotein IIb-IIIa complex. *Blood, 74,* 658–663.

50. Cheresh, D., & Spiro, R. (1989). Biosynthetic and functional properties of an arg-gly-asp-directed receptor involved in human melanoma cell attachment to vitronectin, fibrinogen and von Willebrand factor. *J Biol Chem, 262,* 17703–17711.

51. Grossi, I., Hatfield, J., FitzGerald, L., et al. (1989). Presence of cytoadhesions (IIb-IIIa-like glycoproteins) on human metastatic melanomas but not on benign melanocytes. *Am J Clin Pathol, 92,* 495–499.

52. McGregor, B. C., McGregor, J. L., Weiss, L. M., et al. (1989). Presence of cytoadhesins (IIb-IIIa-like glycoproteins) on human metastatic melanomas but not on benign melanocytes. *Am J Clin Pathol, 92,* 495–499.

53. Chen, Y., Trikha, M., Gao, X., et al. (1997). Ectopic expression of platelet integrin alphaIIb beta3 in tumor cells from various species and histological origin. *Int J Cancer, 72,* 642–648.

54. Trikha, M., Timar, J., Lundy, S., et al. (1997). The high affinity alphaIIb beta3 integrin is involved in invasion of human melanoma cells. *Cancer Res, 57,* 2522–2528.

55. Huang, Y. Q., Li, J. J., Hu, L., et al. (2001). Thrombin induces increased expression and secretion of VEGF from human FS4 fibroblasts, DU145 prostate cells and CHRF megakaryocytes. *Thromb Haemost, 86,* 1094–1098.

56. Huang, Y. Q., Li, J. J., Hu, L., et al. (2002). Thrombin induces increased expression and secretion of angiopoietin-2 from human umbilical vein endothelial cells. *Blood, 99,* 1646–1650.

57. Nierodzik, M. L., Bain, R. M., Liu, L.-X., et al. (1996). Presence of the seven transmembrane thrombin receptor on human tumour cells: Effect of activation on tumour adhesion to platelets and tumour tyrosine phosphorylation. *Br J Haematol, 92,* 452–457.

58. Nierodzik, M., Chen, K., Takeshita, K., et al. (1998). Protease-activated receptor 1 (PAR-1) is required and rate-limiting for thrombin-enhanced experimental pulmonary metastasis. *Blood, 92,* 3694–3700.

59. Shi, X., Gangadharan, B., Brass, L. F., et al. (2004). Protease-activated receptors (PAR1 and PAR2) contribute to tumor cell motility and metastasis. *Molecules Cancer Res, 2,* 395–402.

60. Yoshida, E., Verrusio, E. N., Mihara, H., et al. (1994). Enhancement of the expression of urokinase-type plasminogen activator from PC-3 human prostate cancer cells by thrombin. *Cancer Res, 54,* 3300–3304.

61. Shankar, R., de la Motte, C. A., Poptic, E. J., et al. (1994). Thrombin receptor-activating peptides differentially stimulate platelet-derived growth factor production, monocytic cell adhesion, and E-selectin expression in human umbilical vein endothelial cells. *J Biol Chem, 269,* 13936–13941.

62. Colotta, F., Sciacca, F. L., Sironi, M., et al. (1994). Expression of monocyte chemotactic protein-1 by monocytes and endothelial cells exposed to thrombin. *Am J Pathol, 144,* 975–985.

63. Grandaliano, G., Valente, A. J., & Abboud, H. E. (1994). A novel biologic activity of thrombin: Stimulation of monocyte chemotactic protein production. *J Exp Med, 179,* 1737–1741.

64. Glembotski, C. C., Irons, C. E., Krown, K. A., et al. (1993). Myocardial alpha-thrombin receptor activation induces hypertrophy and increases atrial natriuretic factor gene expression. *J Biol Chem, 268,* 20646–20652.

65. Bartha, K., Brisson, C., Archipoff, G., et al. (1993). Thrombin regulates tissue factor and thrombomodulin mRNA levels and activities in human saphenous vein endothelial cells by distinct mechanisms. *J Biol Chem, 268,* 421–429.

66. Carney, D., Glenn, K., & Cunningham, D. (1978). Conditions which affect initiation of animal cell division by trypsin and thrombin. *J Cell Physiol, 95,* 13–22.

67. Carney, D., Stiernberg, J., & Fenton, J. (1984). Initiation of proliferative events by human α-thrombin requires both receptor binding and enzymatic activity. *J Cell Biochem, 26,* 181–195.

68. Chen, L., & Buchanan, J. (1975). Mitogenic activity of blood components. I. Thrombin and prothrombin. *Proc Nat Acad Sci USA, 72,* 131–135.

69. Glenn, K., Carney, D., Fenton, J., et al. (1980). Thrombin active site regions required for fibroblast receptor binding

and initiation of cell division. *J Biol Chem, 255,* 6609–6616.

70. Gospodarowicz, D., Brown, K., Birdwell, C., et al. (1978). Control of proliferation of human vascular endothelial cells: Characterization of the response of human umbilical vein endothelial cells to fibroblast growth factor, epidermal growth factor, and thrombin. *J Cell Biol, 77,* 774–788.

71. Morris, D., Ward, J., & Carney, D. (1992). Thrombin promotes cell transformation in Balb 3T3/A 31-1-13 cells. *Carcinogenesis, 13,* 67–73.

72. Lampugnani, M. G., Colotta, F., Polentarutti, N., et al. (1990). Thrombin induces c-fos expression in cultured human endothelial cells by a Ca2(+)-dependent mechanism. *Blood, 76,* 1173–1180.

73. Dorn, G. W., 2nd, & Davis, M. G. (1992). Thrombin, but not thromboxane, stimulates megakaryocytic differentiation in human megakaryoblastic leukemia cells. *J Pharmacol Exp Ther, 262,* 1242–1247.

74. Kramer, R., McDonald, K., Crowley, E., et al. (1989). Melanoma cell adhesion to basement membrane mediated by integrin-related complexes. *Cancer Res, 49,* 393–402.

75. Chan, B., Matsuura, N., & Takada, Y. (1991). *In vitro* and *in vivo* consequences of VLA-2 expression on rhabdomyosarcoma cells. *Science, 251,* 1600–1602.

76. Iwamoto, Y., Robey, F., & Graf, J. (1987). YIGSR, a synthetic laminin pentapeptide, inhibits experimental metastasis formation. *Science, 238,* 1132–1134.

77. Kramer, R., McDonald, K., & Vu, M. (1989). Human melanoma cells express a novel integrin receptor for laminin. *J Biol Chem, 264,* 15642–15649.

78. Soszka, T., Knudsen, K., Beviglia, L., et al. (1991). Inhibition of murine melanoma cell-matrix adhesion and experimental metastasis by albolabrin, an RGD-containing peptide isolated from the venom of *Trimeresarus* albolabris. *Exp Cell Res, 196,* 6–12.

79. Terranova, V., Williams, J., & Liotta, L. (1984). Modulation of metastatic activity of melanoma cells by laminin and fibronectin. *Science, 226,* 982–984.

80. Cheresh, D., Smith, W., Cooper, H., et al. (1989). A novel vitronectin receptor integrin (avbx) is responsible for distinct adhesive properties of carcinoma cells. *Cell, 57,* 59–69.

81. Kramer, R., & Marks, N. (1989). Identification of integrin collagen receptors on human melanoma cells. *J Biol Chem, 264,* 4684–4688.

82. Castler, V., Varan, J., & Fligiel, S. (1991). Anti-sense reduction in thrombospondin reverses the malignant phenotype of a human squamous carcinoma. *J Clin Invest, 87,* 1883–1888.

83. Roberts, D., Sherwood, J., & Ginsburg, V. (1987). Platelet thrombospondin mediates attachment and spreading of human melanoma cells. *J Cell Biol, 104,* 131–139.

84. Kramer, A., Keitel, T., Winkler, K., et al. (1997). Molecular basis for the binding promiscuity of an anti-p24 (HIV-1) monoclonal antibody. *Cell, 91,* 799–809.

85. Tuszynski, G. P., Gasic, T. B., Rothman, V. L., et al. (1987). Thrombospondin, a potentiator of tumor cell metastasis. *Cancer Res, 47,* 4130–4133.

86. Humphries, M., Olden, K., & Yamada, K. (1986). A synthetic peptide from fibronectin inhibits experimental metastases of murine melanoma cells. *Science, 467,* 467–470.

87. Humphries, M., Yamada, K., & Olden, K. (1988). Investigation of the biological effect of anti-cell adhesive synthetic peptides that inhibit experimental metastasis of B16F10 murine melanoma cells. *J Clin Invest, 81,* 782–790.

88. Bastida, E., Ordinas, A., Escolar, G., et al. (1984). Tissue factor in microvesicles shed from U87 MG human glioblastoma cells induces coagulation, platelet aggregation and thrombogenesis. *Blood, 64,* 177–184.

89. Curatolo, L., Colucci, M., Cambini, A., et al. (1979). Evidence that cells from experimental tumors can activate coagulation factor X. *Br J Haematol, 40,* 228–233.

90. Dvorak, H., Van deWater, L., Bitzer, A., et al. (1983). Procoagulant activity associated with plasma membrane vesicles shed by cultured tumor cells. *Cancer Res, 43,* 4334–4342.

91. Gordon, S., & Cross, B. (1981). A factor X-activating cysteine protease from malignant tissue. *J Clin Invest, 67,* 1665–1671.

92. Honn, K., Sloane, B., & Cavanaugh, P. (1988). Role of the coagulation system in tumor-cell-induced platelet aggregation and metastasis. *Haemostasis, 18,* 37–46.

93. Kadish, J., Wenc, K., & Dvorak, H. (1983). Tissue factor activity of normal and neoplastic cells: Quantitation and species specificity. *J Nat Cancer Inst, 70,* 551–557.

94. Pearlstein, E., Ambrogio, C., Gasic, G., et al. (1981). Inhibition of platelet aggregating activity of two human adenocarcinomas of the colon and an anaplastic murine tumor with a specific thrombin inhibitor: Dansylarginine N-(3-ethyl-1,5-pentanediyl)amide. *Cancer Res, 41,* 4535–4539.

95. Yoda, Y., & Abe, T. (1981). Fibrinopeptide A (FPA) level and fibrinogen kinetics in patients with malignant disease. *Thromb Haemost, 46,* 706–709.

96. Merskey, C., Johnson, A., Harris, J., et al. (1980). Isolation of fibrinogen-fibrin related antigen from human plasma by immune-affinity chromatography: Its characterization in normal subjects and in defibrinating patients with abruptio placentae and disseminated cancer. *Br J Haematol, 44,* 655–670.

97. Peuscher, F., Cleton, F., Armstrong, L., et al. (1980). Significance of plasma fibrinopeptide A (FPA) in patients with malignancy. *J Lab Clin Med, 96,* 5–14.

98. Rickles, F., Edwards, R., Barb, C., et al. (1983). Abnormalities of blood coagulation in patients with cancer: Fibrinopeptide, a generation and tumor growth. *Cancer, 51,* 301–307.

99. Esumi, N., Fan, D., & Fidler, I. (1981). Inhibition of murine melanoma experimental metastasis by recombinant-desulfatohirudin, a highly specific thrombin inhibitor. *Cancer Res, 51,* 4549–4556.

100. Lee, M., Campbell, W., Huang, Y.-Q., et al. (2000). Hirudin inhibits human tumor implantation and metastasis in nude mice. *Blood, 96,* 818A.

101. Karpatkin, S. (2005). Hypercoagulability preceding cancer. Does hypercoagulability awaken dormant tumor cells in the host? *J Thromb Haemost, 3,* 577–580.

# CHAPTER 43

# Alzheimer's Disease

## Qiao-Xin Li and Colin L. Masters

*Department of Pathology, University of Melbourne, and Mental Health Research Institute of Victoria, Parkville, Victoria, Australia*

## I. Introduction

Alzheimer's disease (AD) is the most common cause of progressive cognitive decline in the aging human population. The disease is definitively diagnosed by the presence of extracellular amyloid deposits in the form of plaques and congophilic angiopathy, as well as intracellular neurofibrillary tangles, in postmortem brains. The amyloid deposits consist mostly of self-aggregating 40 to 43 amino acid residue peptides, Aβ, which are proteolytically derived from a family of 695 to 770 amino acid transmembrane glycoproteins, the amyloid precursor proteins (APP). The molecular pathogenesis of the disease remains unclear, but progress toward identifying therapeutic targets is gaining pace. Studies carried out on human platelets have shed light on APP and Aβ metabolism and function. Platelets contain large amounts of APP protein at a level comparable to that of brain and Aβ peptides to a lesser extent. Because platelets can generate Aβ and soluble forms of APP by mechanisms similar to those observed in neuronal cells, platelets may be useful in monitoring responses to therapeutic interventions targeted directly or indirectly at APP/Aβ metabolism. The platelet APP isoform ratio has been shown to correlate with the progression of clinical symptoms and cognitive decline in AD, and this ratio could potentially be useful as an adjunctive clinical value in the diagnosis of AD, monitoring disease progression, and therapeutic response. Other pathophysiological changes, including in membrane fluidity, calcium homeostasis, and glutamate uptake, have also been observed in platelets. Circulating Aβ has deleterious effects on cerebral vessels and peripheral endothelial cells. These systemic abnormalities in the periphery may reflect pathological changes in the brain and potentially serve as a diagnostic marker for AD. Therefore, the study of platelets may help to understand the molecular pathogenesis of AD and reveal biomarkers for the disease.

## II. The Amyloid Precursor Proteins and Alzheimer's Disease

### A. APP Structure and Function

AD is characterized pathologically by the presence of extracellular amyloid plaques, cerebrovascular amyloid, and intracellular neurofibrillary tangles in postmortem brains.[1,2] The main protein component in the amyloid is the self-aggregating 40 to 43 amino acid residue peptides, Aβ, which is proteolytically derived from a 695 to 770 amino acid protein, APP (Fig. 43-1). APP is part of a family of type I transmembrane glycoproteins[3] consisting of at least ten isoforms, which are generated by alternative mRNA splicing of exons 7 (KPI), 8 (Ox-2), and 15 of the APP gene on chromosome 21. The two major APP isoforms are the Kunitz-type protease inhibitor domain-containing isoform (APP-KPI⁺) and the Kunitz-type protease inhibitor domain-lacking isoform (APP-KPI⁻). Under normal conditions, the APP-KPI⁻ (APP695) isoform is the predominant form produced by neurons. The APP-KPI⁺ isoforms (including APP751 and APP770)[4] are the predominant forms expressed in non-neuronal cells,[5,6] including platelets.[7] An increased ratio of APP-KPI⁺ relative to APP-KPI⁻ is observed in the AD brain.[8] The APP-KPI⁺ isoforms may also be more amyloidogenic, as overexpression of APP751 in cultured cells results in increased Aβ peptide secretion.[9] The exclusion of exon 15 creates L-APP isoforms, which are expressed in the lymphocyte/monocyte lineage and non-neuronal cells within the central nervous system (CNS), but not in neurons.[10,11] The L-APP isoforms contain an attachment site for chondroitin sulfate proteoglycans, and the binding of a glycosaminoglycan chain close to the Aβ sequence of APP may affect the processing of APP and the production of Aβ.[12] Other members of the APP family are the amyloid precursor-like proteins APLP1 and

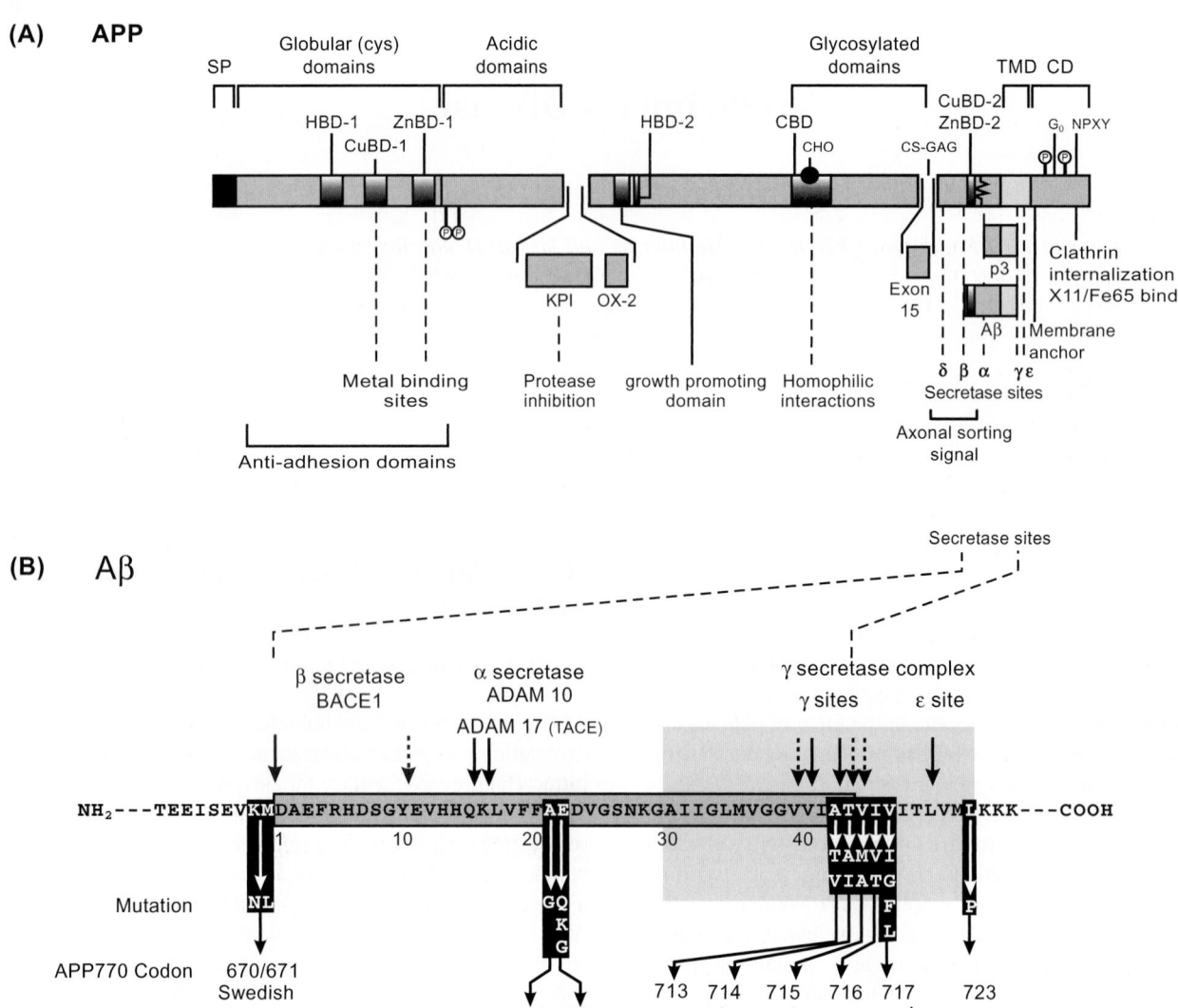

**Figure 43-1.** Schematic illustration of the structure and proteolytic processing of amyloid protein precursor (APP770). A. APP domains and their putative functions. HBD: heparin binding domain; CuBD: copper binding domain; ZnBD: Zn binding domain; CBD: collagen binding domain; CHO: carbohydrate attachment site; P: phosphorylation site; TMD: transmembrane domain; CD: carboxyl domain. Alternatively spliced exons are the KPI (Kunitz protease inhibitor domain), OX-2 (exon with homology to OX-2 domain), and exon 15 (splicing out exon 15 creates a chondroitin sulfate attachment site [CS-GAG]). B. Aβ sequence. The arrows below the sequence indicate APP pathogenic mutations, most of which lead to familial early-onset Alzheimer's disease. The arrows above show the secretase cleavage sites.

APLP2; however, these do not contain the amyloidogenic Aβ sequence.[13,14]

APP is synthesized in the endoplasmic reticulum, posttranslationally modified in the Golgi (glycosylation, sulfation, and phosphorylation), and trafficked to the cell surface. A subset of APP is then endocytosed from the cell surface and processed by the endosomal-lysosomal pathway.[15] As illustrated in Figure 43-1, APP contains a globular domain (including heparin-, zinc-, and copper-binding sites), an acidic domain, a KPI domain, and a glycosylation domain, which may be involved in dimerization.[16] The cytoplasmic C-terminal domain of APP contains transduction and internalization signals, which interact with a range of adaptor proteins including the $G_0$ protein,[17] the brain proteins Fe65and X11,[18,19] JIP (c-Jun N-terminal kinase [JNK]-interacting protein) families,[20] and Dab1 (disabled-1).[21] Interaction with Fe65 and X11 may be involved in modulating the trafficking and processing of APP, as X11 stabilizes APP and decreases Aβ peptide secretion, while Fe65 promotes cell type-dependent Aβ secretion.[22] Upon phosphorylation, the APP C-terminus interacts with other adaptor proteins including ShcA and Grb2 via the PTB (phosphotyrosine binding) or SH2 (Src homology 2) domain, playing a role in cell signaling.[23] The APP N-terminal extracellular domain also interacts with apoA-I.[24] Cell surface APP can act as a receptor involved in cell-cell or cell-matrix interactions and is shown to bind to sulfated proteoglycans, laminin, collagen, or integrin-like receptors.[16,25-27] In addition, APP has growth-promoting and cell-adhesive properties and can protect neurons against excitotoxic or ischemic insults by stabilizing intracellular calcium.[28-32] The Cu/Zn binding sites in the APP N-terminal domain and in the Aβ domain have been shown to involve physiological metal regulation.[33,34] The physiological function of APP has been further addressed by gene knockout experiments. Single knockout mice of APP$^{-/-}$, APLP1$^{-/-}$, APLP2$^{-/-}$ are viable and fertile; however, double knockouts of APP$^{-/-}$/APLP2$^{-/-}$ and APLP1$^{-/-}$/APLP2$^{-/-}$ are perinatally lethal. Because APLP1$^{-/-}$/APP$^{-/-}$ double knockout mice are viable, APP and APLP2 could serve overlapping and nonredundant functions *in vivo*.[35-37]

## B. APP Processing and Aβ Production

APP is cleaved within the Aβ sequence by α-secretase, which belongs to the metalloprotease family of Adamlysins. Candidates include ADAM 10[38] and ADAM 17[39] (Fig. 43-1B). Cleavage by α-secretase generates soluble N-terminal fragments of 100 to 130 kDa (sAPPα) and an approximately 10 kDa membrane-associated C-terminal fragment (C83, p3CT) that contains the C-terminal part of Aβ and the C-terminus of APP.[40,41] An alternative pathway involves cleavage of APP at the amino terminus of Aβ by β-secretase, identified as the membrane-anchored aspartyl protease BACE1,[42] to yield sAPPβ and the A4CT (or C99) fragment, which extends from the Aβ N-terminus to the C-terminus of APP. Subsequent cleavage of A4CT within its transmembrane region by γ-secretase generates Aβ(1-40), Aβ(1-42/43) (at the γ-cleavage site), and APP intracellular domain (AICD) (at the ε-cleavage site), which has potential transcriptional activity that resembles the Notch intracellular domain (NICD).[43-46] Accumulating evidence suggests that the presenilins, which have been genetically linked to early-onset familial forms of AD, contain γ-secretase activity within a complex composed of presenilin 1 or 2, nicastrin,

PEN-2, and APH-1.[47,48] Both sAPP and Aβ are secreted by cells and can be detected in plasma and cerebrospinal fluid.[40,41,49-52] APP is also a substrate of caspases that release the last 31 residues from the C-terminus.[53] Utilization of various APP processing pathways is regulated by signal transduction (reviewed in ref.[54]) and determined by cell type. In non-neuronal cells such as platelets, the "α-secretase" cleavage is the dominant APP processing pathway,[55] whereas in cultured hippocampal neurons and neuronal cell lines, a higher portion of APP is processed by the "β/γ-secretases" pathway generating relatively greater amounts of Aβ.[49,56] Characterization of the γ-secretase activity is of great importance, as this enzyme controls the production of the more amyloidogenic Aβ(42/43), which appears to increase the rate of amyloid deposition.[57-59] Presenilins are expressed in platelets,[60,61] but the expression of other partners of the putative γ-secretase complex such as nicastrin, APH-1, and PEN-2 in platelets is still to be demonstrated. Both ADAM 10 (α-secretase) and BACE1 (β-secretase) have been found in platelets,[62] indicating a resemblance between the processing of APP in platelets and in brain.

Although Aβ appears to be ubiquitously produced, the clinically important Aβ deposition occurs only in the brain. This suggests that other factors specific to the CNS may be involved in promoting the deposition of Aβ or preventing its clearance. Under normal conditions, approximately 90% of secreted Aβ consists of the Aβ(1-40) peptide, and approximately 10% consists of longer Aβ(1-42/43) peptides. Although the Aβ(1-42/43) peptides are minor Aβ products, these longer variants are more amyloidogenic *in vitro*.[57] Aβ(1-42/43) initiates Aβ deposition and plaque formation *in vivo*, as immunocytochemistry studies show that Aβ(1-42/43) is selectively deposited in all types of amyloid plaques.[58,59] Therefore, the amount of Aβ(1-42/43) in the brain appears to play an important role in the initiation of Aβ deposition. The relative amount of Aβ(1-40) versus Aβ(1-42/43) in peripheral tissues is not clear, because the Aβ(1-42/43) levels are too low to be quantified accurately.[50]

## C. The Genetics of Alzheimer's Disease

Mutations in the APP gene near or within the Aβ sequence and in the multitransmembrane proteins presenilin 1 and 2 have been identified that lead to early-onset of familial AD (FAD) (Fig. 43-1B) (see review[63] and an updated list of mutations at www.alzforum.org/res/com/mut/app/). The phenotypes of all these mutations result in aberrant Aβ production, leading to increased secretion of all Aβ isoforms and/or to increased ratios of Aβ(1-42/43)/Aβ(1-40).[63-65] This finding supports the hypothesis that the development of AD neuropathology is mostly due to the toxic properties of aggregated Aβ fibrils or soluble oligomers,[66] although the molecular mechanisms involved in neuronal degeneration and the progression of dementia in AD are still unclear. Aberrant Aβ metabolism is consistently detected in the plasma of carriers of AD mutations and of the media conditioned by the fibroblasts of human carriers,[67] as well as in the brains of transgenic mice carrying some of these AD mutations.[68–73] These studies have consolidated the theory that Aβ deposition is involved in the pathogenesis of AD, as the only common factor known to link these different mutations in different proteins is the change in Aβ metabolism. The relationships between these proteins have not been fully characterized, although it was shown that APP and presenilins interact closely[74,75] and that the presenilins are essential components of the γ-secretase complex activity.[76,77]

Despite advances in the genetics of early-onset forms of AD, the vast majority of AD cases are sporadic. Several factors (such as estrogen deficiency, cerebrovascular disease, cholesterol, oxidants, and metal ions) have been implicated in the pathogenesis of sporadic AD; however, the processes by which these factors influence disease development in sporadic cases are still poorly understood. Studies on the pathogenic mutations that lead to early-onset FAD are helping to characterize the molecular mechanisms underlying the pathogenesis of sporadic AD. One of the major risk factors that leads to an increased likelihood of developing AD is the inheritance of one or two apoE4 (apolipoprotein E4) alleles. However, it should be emphasized that possession of apoE4 alleles is only a risk factor, because not all apoE4 carriers will develop AD.[65]

## III. The Proposed Functions of Platelet APP and Aβ

### A. Platelet APP and Aβ

As platelets are a readily accessible source of human tissue, many groups have utilized platelets as an *ex vivo* system to study APP processing and function. Platelets have the highest APP levels of all peripheral tissues, these levels being comparable to the total APP levels in the brain.[78] Dietary factors, such as zinc and copper levels, can affect platelet APP levels.[79] Our studies indicate that platelets possess the α-, β-, and γ-secretase activities and produce similar APP fragments to neurons: sAPPα, the amyloidogenic fragment A4CT (C99), the soluble sAPPβ isoform, and the Aβ peptide.[50,55] α-Secretase cleavage appears to be the dominant pathway in platelets, as the sAPPα levels are much higher than Aβ levels,[55] in contrast to cells of neuronal origin in which the β-secretase pathway is dominant. The levels of ADAM 10 (an α-secretase candidate) is reduced, paralleling reduced levels of sAPPα, and BACE1 (β-secretase) is altered in AD platelets.[62,80] Unlike neurons

that produce significant amounts of Aβ(1-42) peptide, platelets produce mainly the Aβ(1-40) peptide. Full-length APP is also proteolytically cleaved by a calcium-dependent cysteine protease during platelet activation.[81,82]

Although platelets contain mRNA for APP695/751/770, APP in platelets consists mostly of the sAPPα-KPI$^+$ isoforms (APP770 and APP751) and do not contain detectable amounts of L-APP isoforms or APLP proteins.[83] In platelets, 90% of APP corresponds to the soluble isoform, and 10% of APP consists of full-length APP (APP$_{FL}$-KPI$^+$) and C-terminally truncated membrane-associated APP (APP$_{Mem}$-KPI$^+$),[7] whereas in the brain only approximately 50% of APP is in its soluble form.[84] Activation of platelets by thrombin increases the surface expression of APP by up to three-fold, suggesting that APP-KPI$^+$ may regulate hemostatic protease inhibitory activity on the platelet surface.[7]

Platelets are the primary source of APP in the circulation, producing greater than 90% of the circulating APP[85] and up to 90% of the Aβ in circulation.[50,86] The sAPP (sAPPα, sAPPβ) and Aβ can be released by agents that induce platelet degranulation, including the physiological agonists thrombin and collagen, or nonphysiological agonists such as the calcium ionophore A23187 or ionomycin.[50,78,86,87] This is consistent with the finding that α granules are one of the cellular locations for APP[88] and Aβ. The release of APP from activated platelets is cyclooxygenase (COX) independent but protein kinase C dependent, whereas the release of Aβ is independent of both COX and PKC activities.[89] The released sAPP isoforms potently inhibit the intrinsic coagulation factor XIa and IXa,[90,91] suggesting a role in hemostasis. The inhibition of factor XIa by APP has a $K_i$ of $450 \pm 50$ pM, and this inhibition is enhanced by heparin and zinc.[92] Co-crystal structural analysis of the factor XIa catalytic domain and the APP KPI domain has identified critical amino acid residues within the KPI domain in factor XIa inhibition.[93] Our work shows that recombinant sAPP inhibits platelet aggregation and secretion induced by ADP or adrenaline via the arachidonic acid pathway, indicating that platelet degranulation may result in negative feedback regulation during platelet activation.[94] These data, together with the findings that APP possesses growth factor activity,[30–32] suggest a physiological function for platelet-derived APP in wound repair and in the "microenvironmental" regulation of the coagulation cascade. The effects of Aβ on platelet function appear to be in direct contrast to the sAPP inhibition of platelet aggregation: Aβ has been shown to augment ADP-dependent platelet aggregation and support platelet adhesion.[95,96] Therefore, a balance of the levels of sAPP and Aβ may be important in hemostasis. A similar balance between APP and Aβ may also be important in the brain, as Aβ is thought to be toxic to neurons by disrupting $Ca^{2+}$ homeostasis, whereas sAPP has been shown to protect neurons against excitotoxic insults by stabilizing the intracellular $Ca^{2+}$ concentration.[29]

## B. Aβ in Endothelial Cell Function

Aβ peptides are capable of constricting the microvasculature via mechanisms that involve the release of endothelin and reactive oxygen species. These effects can be reversed using appropriate inhibitors in rat skin,[97] isolated blood vessels, or cerebral blood vessels.[98–101] The impeded vascular effects of Aβ have also been observed in the cerebral cortex vessels[102,103] and peripheral blood vessels[104] of transgenic mice overexpressing Aβ, where it causes loss of vasodilation in response to acetylcholine and increased contractility in response to vasoconstrictors. This vasoconstriction effect is likely due to the toxic Aβ oligomers (present at 5 μM of Aβ40 peptide), which act on the endothelium to reduce vasodilator output, as suggested in a study that utilized aortic rings from Sprague–Dawley rats.[105] Soluble Aβ induces endothelial nitric oxide (NO) dysfunction by inhibiting endothelial nitric oxide synthase (eNOS) enzymatic activity via alteration of intracellular $Ca^{2+}$ homeostasis and the PKC signaling pathway.[106] The vascular damage by circulating Aβ may thus be an early event in the development of the AD pathology. In an extension to these studies, we have further shown that in human skin, peripheral endothelial alterations can be detected early in the course of the disease and can potentially be applied as a diagnostic marker in patients with mild cognitive symptoms or those with early clinical evidence of AD (Khalil, LoGiudice, Khodr, et al., manuscript in preparation). These studies indicate that Aβ is able to induce endothelial dysfunction, and this effect may represent the link between vascular and neuronal pathophysiological factors involved in AD.

## C. APP and Aβ in Plasma

Although the origin of plasma APP is uncertain, our studies suggest that platelets are the major source due to their high concentration of APP (30 nM) compared to other cells in the circulation. Plasma contains low concentrations (approximately 10 pM) of sAPP-KPI+ when blood is carefully collected with minimal platelet activation,[107] as compared to 60 pM using normal blood collection techniques.[85] The theory that most plasma APP originates from platelets is supported by the study of APP levels in platelets and the plasma of a patient with gray platelet syndrome, in which platelets have characteristically low α granule contents (see Chapter 57). Reduction of APP in the platelets of gray platelet syndrome is associated with a similar reduction in plasma APP levels.[7] Our and other studies also suggest that platelets contribute to plasma Aβ levels. Serum contains approximately twofold more APP and Aβ than plasma; this is consistent with the release of APP and Aβ by platelets during activation.[50] Elevation of plasma Aβ levels has also been observed in AD.[108,109] The APP levels in plasma are rela-

tively low when compared to platelet APP levels (approximately 10-fold lower), suggesting that APP has a short half-life once secreted into the plasma. This is supported by a turnover rate of approximately 7 hours for APP observed in the gene-targeted mouse expressing the human Aβ sequence.[110]

The origin of Aβ deposited in the amyloid plaques in AD brain is unclear, although it is most likely to be produced locally by neurons. Brain endothelial cells can also process APP via γ-secretase-like activity. Furthermore, human smooth muscle cells can internalize and accumulate Aβ.[111] It is also possible that Aβ in the circulation contributes to the deposits in the brain. The mechanism involved in the delivery of Aβ across the blood–brain barrier is not clear; however, it has been suggested that Aβ can be taken into the brain as a complex with $HDL_3$ and VHDL in association with apoJ[112,113] or apoE.[114] Another pathway of transporting circulating Aβ across the blood–brain barrier into the CNS is by the interaction of Aβ with the receptor for advanced glycation end products (RAGE)-bearing cells in the vessel wall, a process that is accompanied by Aβ-induced vasoconstriction.[115] Increased penetration of soluble APP has also been observed in cultured human blood–brain barrier endothelial cells from AD subjects compared to endothelial cells from age-matched controls.[116] Therefore, circulating Aβ may contribute to cerebrovascular amyloid (congophilic angiopathy, one of the pathological features of AD) and neuro/peripheral vascular dysfunction.[117]

# IV. Platelets and Alzheimer's Disease

Abnormalities in the brain may be reflected in the function and morphology of platelets, as demonstrated for Parkinson's disease[118] and Huntington's disease.[119] As abnormalities of platelet metabolism have also been observed in AD, these can be used as potential peripheral diagnostic tools for AD.

## A. Platelet APP/Aβ and AD

Altered APP metabolism has been observed in AD platelets, which may reflect abnormal APP metabolism in the brain. Platelets possess all the appropriate biochemical machinery to generate Aβ.[50,55,62,86] The ratio of the 120 to 130 kDa APP isoforms to the 110 kDa is lower in AD patients compared to age-matched cognitively normal controls, non-AD Parkinson's disease, and hemorrhagic stroke subjects.[120,121] This AD-specific alteration of APP isoforms in platelets correlates with the progression of clinical symptoms and cognitive decline, suggesting that this assay can be used as a peripheral biochemical marker for AD. The presence of different proportions of APP isoforms may be related to altered APP processing in AD patients, as shown in studies of platelets derived from patients with severe AD.[122] Alternatively, the declining ratio of APP isoforms in platelets may be a result of increased release of the 120 to 130 kDa species into plasma during platelet activation. This is consistent with our finding of an elevation of the 130 kDa sAPP species in the plasma of AD patients with moderate to severe dementia when compared to age-matched controls.[123] Platelets of advanced AD patients also exhibit abnormal signal transduction as they appear to be hyperacidified upon thrombin activation.[124] High cholesterol affects platelet APP isoform ratio, as hypercholesterolemic AD patients have a lower APP isoform ratio than normocholesterolemic AD patients.[125] Abnormal platelet Aβ secretion is also observed under hypercholesterolemic conditions.[126] Interestingly, AD patients receiving the anticholesterol drug statin have an increased APP isoform ratio, correlating with reduced cholesterol levels.[127] The platelet APP isoform ratio is also increased by donepezil (an acetylcholinesterase inhibitor-based AD drug) and influenced by apoE genotype in mild to moderate AD patients.[128,129] Cholinesterase inhibitor treatment in AD also rescues impaired APP metabolism by increasing α-secretase and decreasing β-secretase activities in platelets.[130]

### B. Platelet Membrane Structure and AD

Because the generation of Aβ involves a cleavage of APP within the membrane, the structure of the membrane is important to Aβ production and therefore to the pathogenesis of AD. Membrane phospholipid is altered in affected brain regions in AD.[131] Several studies that compared AD platelets with normal control platelets have found differences in platelet membrane structures. Abnormalities in platelet membrane fluidity (with a 50% increase shown in AD patients) have been demonstrated in most of the studies, and this change is not reported for platelets derived from other dementias, for example, multi-infarct dementia.[132–134] The membrane fluidity may be linked to the genetic background in the AD subgroup.

### C. Calcium Homeostasis, Oxidative Stress, Glutamate Uptake, and Protein Kinase C

Altered calcium homeostasis is implicated in AD pathogenesis. In platelets of patients with early stages of AD, the basal level of cytosolic calcium $[Ca^{2+}]$ in the absence of extracellular $Ca^{2+}$ is lower but is markedly increased in the presence of extracellular $Ca^{2+}$ compared to age-matched and vascular dementia controls.[135,136] Oxidative stress has also

been implicated in the progression of AD.[137] Using cells (cybrids) transformed by mitochondria isolated from platelets of AD or control subjects, an increase of reactive oxygen species (ROS) production and basal calcium concentrations has been demonstrated in the AD cybrids.[138] In addition, cytochrome c oxidase activity is decreased in mitochondria isolated from AD platelets.[139] This indicates that mitochondrial dysfunction contributes to the pathogenesis of sporadic AD by ROS overproduction and ATP underproduction, and this dysfunction is also reflected in mitochondria from platelets. This is consistent with the observation of systemic disruption in modulation of oxidative stress in AD, as AD platelets have increased levels of thiobarbituric acid-reactive substances (TBARS) and increased activities of superoxide dismutase (SOD), nitric oxide synthase (NOD), and Na,K-ATPase.[140] Glutamate uptake is decreased by 40% in AD platelets compared to platelets obtained from control or multi-infarct dementia subjects[141] due to a decrease of glutamate transporter EAAT1 expression,[142] indicating glutamatergic involvement in AD. Phospholipase A(2) controls phospholipid metabolism and modulates APP metabolism in brain. In AD platelets, phospholipase A(2) activity is reduced and correlates with impaired cognitive performance.[143] The cytosolic level of type II protein kinase C is also altered in platelets derived from severe AD subjects,[144] indicating that abnormalities in signal transduction occur in the peripheral tissues of AD subjects.

## V. Conclusions

Studies indicate that AD-related abnormalities in the CNS might be reflected in the circulation. Abnormal APP metabolism can increase Aβ concentration leading to Aβ deposition. APP and Aβ are produced by peripheral tissues as well as by the CNS; however, the mechanism by which amyloid deposition occurs only in the brain remains unclear. This may be due to APP metabolism by different preferred pathways leading to altered concentrations of Aβ, the presence of factor(s) that may accelerate or prevent the deposition of Aβ, or the more effective removal of Aβ from peripheral tissues. The information drawn from platelet/peripheral studies may help to develop a therapeutic strategy for AD and reveal potential peripheral markers for this disease. Studies so far have shown that APP isoform ratios might be useful in the diagnosis of AD and in monitoring the response to cholinesterase inhibitor treatment. The peripheral endothelial vascular responses in AD patients may also prove to be a potential diagnostic marker. As human platelets produce all forms of APP and secrete amyloidogenic Aβ peptide, platelets may be useful in monitoring responses to therapeutic interventions directed at APP metabolism, such as BACE and γ-secretase inhibitors, or the clearance of amyloid by Aβ immunization.

## Acknowledgments

This work was supported in part by a program grant from the National Health and Medical Research Council of Australia (#208978). We thank Drs. Genevieve Evin and Su San Mok for their critical reading of the manuscript.

## References

1. Glenner, G. G., & Wong, C. W. (1984). Alzheimer's disease: Initial report of the purification and characterization of a novel cerebrovascular amyloid protein. *Biochem Biophy Res Commun, 120,* 885–890.

2. Masters, C. L., Simms, G., Weinman, N. A., et al. (1985). Amyloid plaque core protein in Alzheimer's disease and Down syndrome. *Proc Nat Acad Sci USA, 82,* 4245–4249.

3. Kang, J., Lemaire, H., Unterbeck, A., et al. (1987). The precursor of Alzheimer's disease amyloid A4 protein resembles a cell-surface receptor. *Nature, 325,* 733–736.

4. Tanzi, R. E., McClatchey, A. I., Lamperti, E. D., et al. (1983). Protease inhibitor domain encoded by an amyloid protein precursor mRNA associated with Alzheimer's disease. *Nature, 331,* 528–530.

5. Golde, T. E., Estus, S., Usiak, M., et al. (1990). Expression of β amyloid protein precursor mRNAs: Recognition of a novel alternatively spliced form and quantitation in Alzheimer's disease using PCR. *Neuron, 4,* 253–267.

6. Mönning, U., König, G., Banati, R. B., et al. (1992). Alzheimer βA4-amyloid protein precursor in immunocompetent cells. *J Biol Chem, 267,* 23950–23956.

7. Li, Q. X., Berndt, M. C., Bush, A. I., et al. (1994). Membrane-associated forms of the βA4 amyloid protein precursor of Alzheimer's disease in human platelet and brain: Surface expression on the activated human platelet. *Blood, 84,* 133–142.

8. Moir, R. D., Lynch, T., Bush, A. I., et al. (1998). Relative increase in Alzheimer's disease of soluble forms of cerebral Aβ amyloid protein precursor containing the Kunitz inhibitory domain. *J Biol Chem, 273,* 5013–5019.

9. Ho, L., Fukuchi, K., & Younkin, S. G. (1996). The alternatively spliced Kunitz protease inhibitor domain alters amyloid beta protein precursor processing and amyloid beta protein production in cultured cells. *J Biol Chem, 271,* 30929–30934.

10. König, G., Mönning, U., Czech, C., et al. (1992). Identification and differential expression of a novel alternative splice isoform of the βA4 amyloid precursor protein (APP) mRNA in leukocytes and brain microglial cells. *J Biol Chem, 267,* 10804–10809.

11. Sandbrink, R., Masters, C. L., & Beyreuther, K. (1994). βA4-amyloid protein precursor mRNA isoforms without exon 15 are ubiquitously expressed in rat tissues including brain, but not in neurons. *J Biol Chem, 269,* 1510–1517.

12. Pangalos, M. N., Efthimiopoulos, S., Shioi, J., & Robakis, N. K. (1995). The chondroitin sulfate attachment site of appican is formed by splicing out exon 15 of the amyloid precursor gene. *J Biol Chem, 270,* 10388–10391.

13. Sprecher, C. A., Grant, F. J., Grimm, G., et al. (1993). Molecular cloning of the cDNA for a human amyloid precursor protein homolog: Evidence for a multigene family. *Biochemistry, 32,* 4481–4486.

14. Wasco, W., Brook, J. D., & Tanzi, R. E. (1993). The amyloid precursor-like protein (APLP) gene maps to the long arm of human chromosome 19. *Genomics, 15,* 237–239.

15. Selkoe, D. J. (1998). The cell biology of beta-amyloid precursor protein and presenilin in Alzheimer's disease. *Trends Cell Biol, 8,* 447–453.

16. Beher, D., Hesse, L., Masters, C. L., et al. (1996). Regulation of amyloid protein precursor (APP) binding to collagen and mapping of the binding sites on APP and collagen type I. *J Biol Chem, 271,* 1613–1620.

17. Brouillet, E., Trembleau, A., Galanaud, D., et al. (1999). The amyloid precursor protein interacts with Go heterotrimeric protein within a cell compartment specialized in signal transduction. *J Neurosci, 19,* 1717–1727.

18. Tomita, S., Ozaki, T., Taru, H., et al. (1999). Interaction of a neuron-specific protein containing PDZ domains with Alzheimer's amyloid precursor protein. *J Biol Chem, 274,* 2243–2254.

19. Zambrano, N., Buxbaum, J. D., Minopoli, G., et al. (1997). Interaction of the phosphotyrosine interaction/phosphotyrosine binding-related domains of Fe65 with wild-type and mutant Alzheimer's beta-amyloid precursor proteins. *J Biol Chem, 272,* 6399–6405.

20. Inomata, H., Nakamura, Y., Hayakawa, A., et al. (2003). A scaffold protein JIP-1b enhances amyloid precursor protein phosphorylation by JNK and its association with kinesin light chain 1. *J Biol Chem, 278,* 22946–22955.

21. Homayouni, R., Rice, D. S., Sheldon, M., & Curran, T. (1999). Disabled-1 binds to the cytoplasmic domain of amyloid precursor-like protein 1. *J Neurosci, 19,* 7507–7515.

22. King, G. D., & Scott Turner, R. (2004). Adaptor protein interactions: Modulators of amyloid precursor protein metabolism and Alzheimer's disease risk? *Exp Neurol, 185,* 208–219.

23. Russo, C., Venezia, V., Repetto, E., et al. (2005). The amyloid precursor protein and its network of interacting proteins: Physiological and pathological implications. *Brain Res Brain Res Rev, 48,* 257–264.

24. Koldamova, R. P., Lefterov, I. M., Lefterova, M. I., et al. (2001). Apolipoprotein A-I directly interacts with amyloid precursor protein and inhibits A beta aggregation and toxicity. *Biochemistry, 40,* 3553–3560.

25. Ghiso, J., Rostagno, A., Gardella, J. E., et al. (1992). A 109-amino-acid C-terminal fragment of Alzheimer's-disease amyloid precursor protein contains a sequence, -RHDS-, that promotes cell adhesion. *Biochem J, 288,* 1053–1059.

26. Small, D. H., Nurcombe, V., Moir, R., et al. (1992). Association and release of the amyloid protein precursor of Alzheimer's disease from chick brain extracellular matrix. *J Neurosci, 12,* 4143–4150.

27. Williamson, T. G., Mok, S. S., Henry, A., et al. (1999). Secreted glypican binds to the amyloid precursor protein of Alzheimer's disease. *J Biol Chem, 271*, 31215–31221.

28. Clarris, H. J., Cappai, R., Heffernan, D., et al. (1997). Identification of heparin-binding domains in the amyloid precursor protein of Alzheimer's disease by deletion mutagenesis and peptide mapping. *J Neurochem, 68*, 1164–1172.

29. Mattson, M. P., Barger, S. W., Cheng, B., et al. (1993). β-Amyloid precursor protein metabolites and loss of neuronal $Ca^{2+}$ homeostasis in Alzheimer's disease. *Trends Neurosci, 16*, 409–414.

30. Milward, E., Papadopoulos, R., Fuller, S. J., et al. (1992). The amyloid protein precursor of Alzheimer's disease is a mediator of the effects of nerve growth factor on neurite outgrowth. *Neuron, 9*, 129–137.

31. Saitoh, T., Sundsmo, M., Roch, J. M., et al. (1989). Secreted form of amyloid-β protein precursor is involved in the growth regulation of fibroblasts. *Cell, 58*, 615–622.

32. Small, D. H., Nurcombe, V., Reed, G., et al. (1994). A heparin-binding domain in the amyloid protein precursor of Alzheimer's disease is involved in the regulation of neurite outgrowth. *J Neurosci, 14*, 2117–2127.

33. Maynard, C. J., Cappai, R., Volitakis, I., et al. (2002). Overexpression of Alzheimer's disease amyloid-beta opposes the age-dependent elevations of brain copper and iron. *J Biol Chem, 277*, 44670–44676.

34. White, A. R., Reyes, R., Mercer, J. F., et al. (1999). Copper levels are increased in the cerebral cortex and liver of APP and APLP2 knockout mice. *Brain Res, 842*, 439–444.

35. Heber, S., Herms, J., Gajic, V., et al. (2000). Mice with combined gene knock-outs reveal essential and partially redundant functions of amyloid precursor protein family members. *J Neurosci, 20*, 7951–7963.

36. von Koch, C. S., Zheng, H., Chen, H., et al. (1997). Generation of APLP2 KO mice and early postnatal lethality in APLP2/APP double KO mice. *Neurobiol Aging, 18*, 661–669.

37. Zheng, H., Jiang, M. H., Trumbauer, M. E., et al. (1995). β-amyloid precursor protein-deficient mice show reactive gliosis and decreased locomotor activity. *Cell, 81*, 525–531.

38. Lammich, S., Kojro, E., Postina, R., et al. (1999). Constitutive and regulated alpha-secretase cleavage of Alzheimer's amyloid precursor protein by a disintegrin metalloprotease. *Proc Nat Acad Sci USA, 96*, 3922–3927.

39. Buxbaum, J. D., Liu, K. N., Luo, Y., et al. (1998). Evidence that tumor necrosis factor alpha converting enzyme is involved in regulated alpha-secretase cleavage of the Alzheimer amyloid protein precursor. *J Biol Chem, 273*, 27765–27767.

40. Esch, F. S., Keim, P. S., Beattie, E. C., et al. (1990). Cleavage of amyloid β peptide during constitutive processing of its precursor. *Science, 248*, 1122–1124.

41. Weidemann, A., König, G., Bunke, D., et al. (1989). Identification, biogenesis and localization of precursors of Alzheimer's disease A4 amyloid protein. *Cell, 57*, 115–126.

42. Vassar, R., & Citron, M. (2000). Abeta-generating enzymes: Recent advances in beta- and gamma-secretase research. *Neuron, 27*, 419–422.

43. Gu, Y., Misonou, H., Sato, T., et al. (2001). Distinct intramembrane cleavage of the beta-amyloid precursor protein family resembling gamma-secretase-like cleavage of Notch. *J Biol Chem, 276*, 35235–35238.

44. Sastre, M., Steiner, H., Fuchs, K., et al. (2001). Presenilin-dependent gamma-secretase processing of beta-amyloid precursor protein at a site corresponding to the S3 cleavage of Notch. *EMBO Rep, 2*, 835–841.

45. Weidemann, A., Eggert, S., Reinhard, F. B., et al. (2002). A novel epsilon-cleavage within the transmembrane domain of the Alzheimer amyloid precursor protein demonstrates homology with Notch processing. *Biochemistry, 41*, 2825–2835.

46. Yu, C., Kim, S. H., Ikeuchi, T., et al. (2001). Characterization of a presenilin-mediated amyloid precursor protein carboxyl-terminal fragment gamma: Evidence for distinct mechanisms involved in gamma-secretase processing of the APP and Notch1 transmembrane domains. *J Biol Chem, 276*, 43756–43760.

47. Kopan, R., & Ilagan, M. X. (2004). Gamma-secretase: Proteasome of the membrane? *Nat Rev Molec Cell Biol, 5*, 499–504.

48. Iwatsubo, T. (2004). Alzheimer's disease: Focus on beta-amyloid and gamma-secretase. *Tanpakushitsu Kakusan Koso, 49*, 1086–1090.

49. Fuller, S. J., Storey, E., Li, Q. X., et al. (1995). Intracellular production of βA4 amyloid of Alzheimer's disease: Modulation by phosphoramidon and lack of coupling to the secretion of the amyloid precursor. *Biochemistry, 34*, 8091–8098.

50. Li, Q.-X., Whyte, S., Tanner, J. E., et al. (1998). Secretion of Alzheimer's disease Aβ amyloid peptide by activated human platelets. *Lab Invest, 78*, 461–469.

51. Seubert, P., Vigo-Pelfrey, C., Esch, F., et al. (1992). Isolation and quantification of soluble Alzheimer's β-peptide from biological fluids. *Nature, 359*, 325–327.

52. Suzuki, N., Cheung, T. T., Cai, X. D., et al. (1994). An increased percentage of long amyloid β protein secreted by familial amyloid β protein precursor (βAPP717) mutants. *Science, 264*, 1336–1340.

53. Ayala-Grosso, C., Ng, G., Roy, S., et al. (2002). Caspase-cleaved amyloid precursor protein in Alzheimer's disease. *Brain Pathol, 12*, 430–441.

54. Gandy, S., & Petanceska, S. (2000). Regulation of Alzheimer beta-amyloid precursor trafficking and metabolism. *Biochim Biophys Acta, 1502*, 44–52.

55. Li, Q.-X., Cappai, R., Evin, G., et al. (1998). Products of the Alzheimer's disease amyloid precursor protein generated by β-secretase are present in human platelets, and secreted upon degranulation. *Am J Alzheimer Dis, 13*, 236–244.

56. Simons, M., De Strooper, B., Multhaup, G., et al. (1996). Amyloidogenic processing of the human amyloid precursor protein in primary cultures of rat hippocampal neurons. *J Neurosci, 16*, 899–908.

57. Jarrett, J. T., Berger, E. P., & Lansbury, P. T., Jr. (1993). The carboxy terminus of the β amyloid protein is critical for the seeding of amyloid formation: Implications for the pathogenesis of Alzheimer's disease. *Biochemistry, 32*, 4693–4697.

58. Mann, D. M. A., Iwatsubo, T., Ihara. Y., et al. (1996). Predominant deposition of amyloid-$\beta_{42(43)}$ in plaques in cases of Alzheimer's disease and hereditary cerebral hemorrhage associated with mutations in the amyloid precursor protein gene. *Am J Pathol, 148,* 1257–1266.

59. Gravina, S. A., Ho, L., Eckman, C. B., et al. (1995). Amyloid β protein (A β) in Alzheimer's disease brain: Biochemical and immunocytochemical analysis with antibodies specific for forms ending at Aβ40 or Aβ42(43). *J Biol Chem, 270,* 7013–7016.

60. Evin, G., Zhu, A., Holsinger, R. M., et al. (2003). Proteolytic processing of the Alzheimer's disease amyloid precursor protein in brain and platelets. *J Neurosci Res, 74,* 386–392.

61. Vidal, R., Ghiso, J., Wisniewski, T., et al. (1996). Alzheimer's presenilin 1 gene expression in platelets and megakaryocytes: Identification of a novel splice variant. *FEBS Letters, 393,* 19–23.

62. Colciaghi, F., Marcello, E., Borroni, B., et al. (2004). Platelet APP, ADAM 10 and BACE alterations in the early stages of Alzheimer disease. *Neurology, 62,* 498–501.

63. Hardy, J. (1997). Amyloid, the presenilins and Alzheimer's disease. *Trends Neurosci, 20,* 154–159.

64. Kwok, J. B. J., Li, Q.-X., Hallupp, M., et al. (1998). Novel familial early-onset Alzheimer's disease mutation (Leu-723Pro) in amyloid precursor protein (APP) gene increases production of 42(43) amino-acid isoform of amyloid-beta peptide. *Neurobiol Aging, 19* (Supp 2), S91:S378 (abs).

65. Selkoe, D. J. (2001). Alzheimer's disease: Genes, proteins, and therapy. *Physiol Rev, 81,* 741–766.

66. Walsh, D. M., Klyubin, I., Fadeeva, J. V., et al. (2002). Naturally secreted oligomers of amyloid β protein potently inhibit hippocampal long-term potentiation *in vivo. Nature, 416,* 535–539.

67. Scheuner, D., Eckman, C., Jensen, M., et al. (1996). Secreted amyloid b-protein similar to that in the senile plaques of Alzheimer's disease is increased *in vivo* by the presenilin 1 and 2 and APP mutations linked to familial Alzheimer's disease. *Nat Med, 2,* 864–870.

68. Citron, M., Westaway, D., Xia, W. M., et al. (1997). Mutant presenilins of Alzheimer's disease increase production of 42-residue amyloid β-protein in both transfected cells and transgenic mice. *Nat Med, 3,* 67–72.

69. Duff, K., Eckman, C., Zehr, C., et al. (1996). Increased amyloid-β42(43) in brains of mice expressing mutant presenilin 1. *Nature, 383,* 710–713.

70. Hsiao, K., Chapman, P., Nilsen, S., et al. (1996). Correlative memory deficits, Aβ elevation, and amyloid plaques in transgenic mice. *Science, 274,* 99–102.

71. Li, Q.-X., Maynard, C., Cappai, R., et al. (1999). Intracellular accumulation of detergent-soluble amyloidogenic Aβ fragment of Alzheimer's disease precursor protein in the hippocampus of aged transgenic mice. *J Neurochem, 72,* 2479–2487.

72. Chishti, M. A., Yang, D. S., Janus, C., et al. (2004). Early-onset amyloid deposition and cognitive deficits in transgenic mice expressing a double mutant form of amyloid precursor protein 695. *J Biol Chem, 276,* 21562–21570.

73. Jankowsky, J. L., Fadale, D. J., Anderson, J., et al. (2004). Mutant presenilins specifically elevate the levels of the 42 residue beta-amyloid peptide in vivo: Evidence for augmentation of a 42-specific gamma secretase. *Human Mol Genet, 13,* 159–170.

74. Weidemann, A., Paliga, K., Dürrwang, U., et al. (1997). Formation of stable complexes between two Alzheimer's disease gene products — presenilin-2 and β-amyloid precursor protein. *Nat Med, 3,* 328–332.

75. Xia, W., Zhang, J., Perez, R., et al. (1997). Interaction between amyloid precursor protein and presenilins in mammalian cells: Implications for the pathogenesis of Alzheimer disease. *Pro Nat Acad Sci USA, 94,* 8208–8213.

76. De Strooper, B., Saftig, P., Craessaerts, K., et al. (1998). Deficiency of presenilin-1 inhibits the normal cleavage of amyloid precursor protein. *Nature, 391,* 387–390.

77. Wolfe, M. S., Xia, W., Ostaszewski, B. L., et al. (1999). Two transmembrane aspartates in presenilin-1 required for presenilin endoproteolysis and gamma-secretase activity. *Nature, 398,* 513–517.

78. Bush, A. I., Martins, R. N., Rumble, B., et al. (1990). The amyloid precursor protein of Alzheimer's disease is released by human platelets. *J Biol Chem, 265,* 15977–15983.

79. Davis, C. D., Milne, D. B., & Nielsen, F. H. (2000). Changes in dietary zinc and copper affect zinc-status indicators of postmenopausal women, notably, extracellular superoxide dismutase and amyloid precursor proteins. *Am J Clin Nutr, 71,* 781–788.

80. Colciaghi, F., Borroni, B., Pastorino, L., et al. (2002). [alpha]-Secretase ADAM10 as well as [alpha]APPs is reduced in platelets and CSF of Alzheimer disease patients. *Mol Med, 8,* 67–74.

81. Li, Q.-X., Evin, G., Small, D. H., et al. (1995). Proteolytic processing of Alzheimer's disease amyloid protein precursor in human platelets. *J Biol Chem, 270,* 14140–14147.

82. Chen, M., Durr, J., & Fernandez, H. L. (2000). Possible role of calpain in normal processing of beta-amyloid precursor protein in human platelets. *Biochem Biol Res Commun, 273,* 170–175.

83. Li, Q.-X., Whyte, S., Birchall, I., et al. (1995). The amyloid protein precursor of Alzheimer's disease in human platelets and kidney. In P. Zatta & M. Nicolini (Eds.). *Non-neuronal cells in Alzheimer's disease* (pp. 62–70). Singapore: World Scientific.

84. Moir, R. D., Martins, R. N., Small, D. H., et al. (1992). Human brain βA4 amyloid protein precursor (APP) of Alzheimer's disease: Purification and partial characterization. *J Neurochem, 59,* 1490–1498.

85. Van Nostrand, W. E., Schmaier, A. H., Farrow, J. S., et al. (1991). Protease nexin-2/amyloid β-protein precursor in blood is a platelet-specific protein. *Biochem Biological Res Commun, 175,* 15–21.

86. Chen, M., Inestrosa, N. C., Ross, G. S., & Fernandez, H. L. (1995). Platelets are the primary source of amyloid β-peptide in human blood. *Biochem Biological Res Commun, 213,* 96–103.

87. Smith, R. P., & Broze, G. J., Jr. (1992). Characterization of platelet-releasable forms of β-amyloid precursor proteins: The effect of thrombin. *Blood, 80,* 2252–2260.

88. Van Nostrand, W. E., Schmaier, A. H., Farrow, J. S., et al. (1990). Protease nexin-II (amyloid β-protein precursor): A platelet α-granule protein. *Science, 248,* 745–748.

89. Skovronsky, D. M., Lee, V. M., & Pratico, D. (2001). Amyloid precursor protein and amyloid beta peptide in human platelets: Role of cyclooxygenase and protein kinase C. *J Biol Chem, 276,* 17036–17043.

90. Schmaier, A. H., Dahl, L. D., Hasan, A. A., et al. (1995). Factor IXa inhibition by protease nexin-2/amyloid β-protein precursor on phospholipid vesicles and cell membranes. *Biochemistry, 34,* 1171–1178.

91. Smith, R. P., Higuchi, D. A., & Broze, G. J., Jr. (1990). Platelet coagulation factor XIa-inhibitor, a form of Alzheimer amyloid precursor protein. *Science, 248,* 1126–1128.

92. Komiyama, Y., Murakami, T., Egawa, H., et al. (1992). Purification of factor XIa inhibitor from human platelets. *Thromb Res, 66,* 397–408.

93. Navaneetham, D., Jin, L., Pandey, P., et al. (2005). Structural and mutational analyses of the molecular interactions between the catalytic domain of factor XIa and the Kunitz protease inhibitor domain of protease nexin 2. *J Biol Chem, 280,* 36165–36175.

94. Henry, A., Li, Q.-X., Galatis, D., et al. (1998). Inhibition of platelet activation by the Alzheimer's amyloid precursor protein. *Br J Haematol, 103,* 402–415.

95. Kowalska, M. A., & Badellino, K. (1994). Beta-amyloid protein induces platelet aggregation and supports platelet adhesion. *Biochem Biophys Res Commun, 205,* 1829–1835.

96. Wolozin, B. L., Maheshwari, S., Jones, C., et al. (1998). β-Amyloid augments platelet aggregation: Reduced activity of familial angiopathy-associated mutants. *Mol Psychiatry, 3,* 500–507.

97. Khalil, Z., Chen, H., & Helme, R. D. (1996). Mechanisms underlying the vascular activity of beta-amyloid protein fragment (beta A(4)25–35) at the level of skin microvasculature. *Brain Res, 736,* 206–216.

98. Thomas, T., McLendon, C., Sutton, E. T., et al. (1997). Beta-amyloid-induced cerebrovascular endothelial dysfunction. *Ann N Y Acad Sci, 826,* 447–451.

99. Thomas, T., Sutton, E. T., Hellermann, A., et al. (1997). Beta-amyloid-induced coronary artery vasoactivity and endothelial damage. *J Cardiovasc Pharmacol, 30,* 517–522.

100. Thomas, T., McLendon, C., Sutton, E. T., et al. (1997). Cerebrovascular endothelial dysfunction mediated by beta-amyloid. *Neuroreport, 8,* 1387–1391.

101. Crawford, F., Suo, Z., Fang, C., et al. (1997). Beta-amyloid peptides and enhancement of vasoconstriction by endothelin-1. *Ann N Y Acad Sci, 826,* 461–462.

102. Iadecola, C., Zhang, F., Niwa, K., et al. (1999). SOD1 rescues cerebral endothelial dysfunction in mice overexpressing amyloid precursor protein. *Nat Neurosci, 2,* 157–161.

103. Suo, Z., Humphrey, J., Kundtz, A., et al. (1998). Soluble Alzheimer's β-amyloid constricts the cerebral vasculature in vivo. *Neurosci Lett, 257,* 77–80.

104. Khalil, Z., Poliviou, H., Maynard, C. J., et al. (2002). Mechanisms of peripheral microvascular dysfunction in transgenic mice overexpressing the Alzheimer's disease amyloid Abeta protein. *J Alzheimer Dis, 4,* 467–478.

105. Smith, C. C., Stanyer, L., & Betteridge, D. J. (2004). Soluble beta-amyloid (A beta) 40 causes attenuation or potentiation of noradrenaline-induced vasoconstriction in rats depending upon the concentration employed. *Neurosci Lett, 367,* 129–132.

106. Gentile, M. T., Vecchione, C., Maffei, A., et al. (2004). Mechanisms of soluble beta-amyloid impairment of endothelial function. *J Biol Chem, 279,* 48135–48142.

107. Whyte, S., Wilson, N., Currie, J., et al. (1997). Collection and normal levels of the amyloid precursor protein in plasma. *Ann Neurol, 41,* 121–124.

108. Mayeux, R., Tang, M. X., Jacobs, D. M., et al. (1999). Plasma amyloid β-peptide 1-42 and incipient Alzheimer's disease. *Ann Neurol, ••,* 412–416.

109. Matsubara, E., Ghiso, J., Frangione, B., et al. (1999). Lipoprotein-free amyloidogenic peptides in plasma are elevated in patients with sporadic Alzheimer's disease and Down's syndrome. *Ann Neurol, 45,* 537–541.

110. Savage, M. J., Trusko, S. P., Howland, D. S., et al. (1998). Turnover of amyloid beta-protein in mouse brain and acute reduction of its level by phorbol ester. *J Neurosci, 18,* 1743–1752.

111. Urmoneit, B., Prikulis, I., Wihl, G., et al. (1997). Cerebrovascular smooth muscle cells internalize Alzheimer amyloid beta protein via a lipoprotein pathway: Implications for cerebral amyloid angiopathy. *Lab Invest, 77,* 157–166.

112. Koudinov, A., Matsubara, E., Frangione, B., et al. (1994). The soluble form of Alzheimer's amyloid beta protein is complexed to high density lipoprotein 3 and very high density lipoprotein in normal human plasma. *Biochem Biophys Res Commun, 205,* 1164–1171.

113. Zlokovic, B. V., Martel, C. L., Matsubara, E., et al. (1996). Glycoprotein 330/megalin: Probable role in receptor-mediated transport of apolipoprotein J alone and in a complex with Alzheimer disease amyloid beta at the blood-brain and blood-cerebrospinal fluid barriers. *Pro Nat Acad Sci USA, 93,* 4229–4234.

114. Martel, C. L., Mackic, J. B., Matsubara, E., et al. (1997). Isoform-specific effects of apolipoproteins E2, E3, and E4 on cerebral capillary sequestration and blood-brain barrier transport of circulating Alzheimer's amyloid beta. *J Neurochem, 69,* 1995–2004.

115. Deane, R., Du Yan, S., Submamaryan, R. K., et al. (2003). RAGE mediates amyloid-beta peptide transport across the blood-brain barrier and accumulation in brain. *Nat Med, 9,* 907–913.

116. Davies, T. A., Billingslea, A. M., Long, H. J., et al. (1998). Brain endothelial cell enzymes cleave platelet-retained amyloid precursor protein. *J Lab Clin Med, 132,* 341–350.

117. Zlokovic, B. V., Deane, R., Sallstrom, J., et al. (2005). Neurovascular pathways and Alzheimer amyloid beta-peptide. *Brain Pathol, 15,* 78–83.

118. Ferrarese, C., Zoia, C., Pecora, N., et al. (1999). Reduced platelet glutamate uptake in Parkinson's disease. *J Neural Transm, 106,* 685–692.

119. Reilmann, R., Rolf, L. H., & Lange, H. W. (1997). Huntington's disease: N-methyl-D-aspartate receptor coagonist glycine is increased in platelets. *Exp Neurol, 144,* 416–419.

120. Di Luca, M., Pastorino, L., Bianchetti, A., et al. (1998). Differential level of platelet amyloid beta precursor protein isoforms: An early marker for Alzheimer disease. *Arch Neurol, 55,* 1195–1200.

121. Baskin, F., Rosenberg, R. N., Iyer, L., et al. (2000). Platelet APP isoform ratios correlate with declining cognition in AD. *Neurology, 54,* 1907–1909.

122. Davies, T. A., Long, H. J., Tibbles, H. E., et al. (1997). Moderate and advanced Alzheimer's patients exhibit platelet activation differences. *Neurobiol Aging, 18,* 155–162.

123. Bush, A. I., Whyte, S., Thomas, L. D., et al. (1992). An abnormality of plasma amyloid protein precursor in Alzheimer's disease. *Ann Neurol, 32,* 57–65.

124. Davies, T. A., Fine, R. E., Johnson, R. J., et al. (1993). Non-age related differences in thrombin responses by platelets from male patients with advanced Alzheimer's disease. *Biochem Biophys Res Commun, 194,* 537–543.

125. Borroni, B., Colciaghi, F., Lenzi, G. L., et al. (2003). High cholesterol affects platelet APP processing in controls and in AD patients. *Neurobiol Aging, 24,* 631–636.

126. Smith, C. C., Hyatt, P. J., Stanyer, L., et al. (2001). Platelet secretion of beta-amyloid is increased in hypercholesterolaemia. *Brain Res, 896,* 161–164.

127. Baskin, F., Rosenberg, R. N., Fang, X., et al. (2003). Correlation of statin-increased platelet APP ratios and reduced blood lipids in AD patients. *Neurology, 60,* 2006–2007.

128. Borroni, B., Colciaghi, F., Pastorino, L., et al. (2001). Amyloid precursor protein in platelets of patients with Alzheimer disease: Effect of acetylcholinesterase inhibitor treatment. *Arch Neurol, 58,* 442–446.

129. Borroni, B., Colciaghi, F., Pastorino, L., et al. (2002). ApoE genotype influences the biological effect of donepezil on APP metabolism in Alzheimer disease: Evidence from a peripheral model. *Eur Neuropsychopharmacol, 12,* 195–200.

130. Zimmermann, M., Borroni, B., Cattabeni, F., et al. (2005). Cholinesterase inhibitors influence APP metabolism in Alzheimer disease patients. *Neurobiol Dis, 19,* 237–242.

131. Farooqui, A. A., Rapoport, S. I., & Horrocks, L. A. (1997). Membrane phospholipid alterations in Alzheimer's disease: Deficiency of ethanolamine plasmalogens. *Neurochem Res, 22,* 523–527.

132. Swiderek, M., Kozubski, W., & Watala, C. (1997). Abnormalities in platelet membrane structure and function in Alzheimer's disease and ischaemic stroke. *Platelets, 8,* 125–133.

133. Zubenko, G. S., Kopp, U., Seto, T., et al. (1999). Platelet membrane fluidity individuals at risk for Alzheimer's disease: A comparison of results from fluorescence spectroscopy and electron spin resonance spectroscopy. *Psychopharmacology (Berl), 145,* 175–180.

134. Kozubski, W., Swiderek, M., Kloszewska, I., et al. (2002). Blood platelet membrane fluidity and the exposition of membrane protein receptors in Alzheimer disease (AD) patients — preliminary study. *Alzheimer Dis Assoc Disord, 16,* 52–54.

135. Ripovi, D., Platilova, V., Strunecka, A., et al. (2000). Cytosolic calcium alterations in platelets of patients with early stages of Alzheimer's disease. *Neurobiol, Aging, 21,* 729–734.

136. Ripova, D., Platilova, V., Strunecka, A., et al. (2004). Alterations in calcium homeostasis as biological marker for mild Alzheimer's disease? *Physiol Res, 53,* 449–452.

137. Barnham, K. J., Masters, C. L., & Bush, A. I. (2004). Neurodegenerative diseases and oxidative stress. *Nat Rev Drug Discov, 3,* 205–214.

138. Sheehan, J. P., Swerdlow, R. H., Miller, S. W., et al. (1997). Calcium homeostasis and reactive oxygen species production in cells transformed by mitochondria from individuals with sporadic Alzheimer's disease. *J Neurosci, 17,* 4612–4622.

139. Cardoso, S. M., Proenca, M. T., Santos, S., et al. (2004). Cytochrome c oxidase is decreased in Alzheimer's disease platelets. *Neurobiol Aging, 25,* 105–110.

140. Kawamoto, E. M., Munhoz, C. D., Glezer, I., et al. (2005). Oxidative state in platelets and erythrocytes in aging and Alzheimer's disease. *Neurobiol Aging, 26,* 857–864.

141. Ferrarese, C., Begni, B., Canevari, C., et al. (2000). Glutamate uptake is decreased in platelets from Alzheimer's disease patients. *Ann Neurol, 47,* 641–643.

142. Zoia, C., Cogliati, T., Tagliabue, E., et al. (2004). Glutamate transporters in platelets: EAAT1 decrease in aging and in Alzheimer's disease. *Neurobiol Aging, 25,* 149–157.

143. Gattaz, W. F., Forlenza, O. V., Talib, L. L., et al. (2004). Platelet phospholipase A(2) activity in Alzheimer's disease and mild cognitive impairment. *J Neural Transm, 111,* 591–601.

144. Matsushima, H., Shimohama, S., Tanaka, S., et al. (1994). Platelet protein kinase C levels in Alzheimer's disease. *Neurobiol Aging, 15,* 671–674.

# CHAPTER 44

# Psychiatric Disorders

## George N. M. Gurguis

*Department of Psychiatry, University of Texas Southwestern Medical Center, Dallas, Texas*

## I. Introduction

Since the 1960s, an extensive body of investigation has employed platelets as a peripheral model of central nervous system (CNS) neuronal function to investigate the neurobiology of psychiatric disorders. The major forces driving these studies are the inherent difficulty of directly accessing brain function *in vivo* and the substantial similarities between platelets and central neurons.[1–5] However, as the rationale for extrapolating platelet findings to CNS function has been challenged (i.e., changes found in platelets might not necessarily mirror changes in the CNS), the rationale for platelet studies has shifted to using abnormalities in platelet receptor-signal transduction systems as peripheral markers in psychiatric disorders.

Virtually no single symptom is pathognomonic of a specific psychiatric syndrome. Psychiatric nosology depends primarily on identifying syndromes wherein a group of symptoms and signs cluster to make a clinical syndrome. Therefore, a peripheral marker that proves to be associated with a specific syndrome would have important clinical implications diagnostically or as a measure of disease process. Furthermore, it is clear that even a single diagnostic syndrome is neurobiologically heterogeneous. Combined with a syndrome-based diagnostic system, a biological marker in platelets can, in theory, be used to further subcategorize psychiatric disorders into more neurobiologically homogeneous subgroups.

A related aim has been to identify platelet abnormalities as markers that may predict, or correlate with, treatment outcome. It has long been observed that patients with the same diagnosis do not respond similarly to the same psychotropic medicine. Hypothetically, this differential treatment response may be a function of the neurobiological heterogeneity of psychiatric disorders (i.e., differences in underlying neurochemical processes). Thus, the use of psychotropic drugs — for example, antidepressants — as dissecting pharmacological tools to identify platelet functional abnormalities that are associated with response or nonresponse to a particular treatment may also have substantial clinical utility.

Platelets possess several receptors and second messenger systems (see Chapters 6 and 16) that make them an attractive peripheral model of monoaminergic neurons. Briefly, $\alpha_2$-adrenergic receptors ($\alpha_2$AR) have long been identified in platelet membranes.[6] They are of the $\alpha_{2A}$ (or C10) subtype,[7,8] similar to brain $\alpha_{2A}$AR. $\alpha_2$AR are coupled to the adenylyl cyclase (AC) second messenger system through $G_i$ protein, which mediates its inhibition.[9] $\beta$-adrenergic receptors ($\beta$AR) have also been identified in platelet membranes,[10–12] although they have not been used in psychiatric research as extensively as $\alpha_{2A}$AR. Platelet $\beta$AR are of the $\beta_2$ subtype, while both $\beta_1$ and $\beta_2$ subtypes exist in the CNS. Both subtypes mediate stimulation of AC through $G_s$ protein.

Platelet membranes also possess serotonin$_2$ receptors (5-$HT_{2A}$) that are similar to brain 5-$HT_{2A}$ receptors and are coupled to phospholipase C (PLC) through $G_{q/11}$ proteins.[13] 5-$HT_{2A}$ receptor stimulation mediates the conversion of phosphatidylinositol 4,5-bisphosphate ($PIP_2$) to inositol 1,4,5-triphosphate ($IP_3$) and diacylglycerol (DAG). DAG mediates intracellular $Ca^{++}$ release and the activation of protein kinase C (PKC).[14,15] By agonist displacement of $^3$H-ketanserine binding experiments, we demonstrated that platelet 5-$HT_{2A}$ receptors, like brain 5-$HT_{2A}$ receptors, display two affinity states upon agonist stimulation as a function of their coupling to $G_{q/11}$.[16] This method provides for a more detailed investigation of platelet 5-$HT_{2A}$ receptor coupling to $G_{q/11}$ protein, beyond mere examination of total receptor density.

In addition, platelets have a serotonin transporter (5-HTTP) and vesicular monoamine transporters ($VMAT_2$) that are identical to those in the brain,[17–19] and peripheral serotonin platelets mirror cerebral neuronal serotonin changes.[20] Platelet serotonin content reflects 5-HTTP function.[21,22] The presence of a similar norepinephrine or dopamine transporter in platelets that are analogues of the CNS transporter remains controversial. Platelets have a monoamine oxidase enzyme (MAO) of the B subtype, while both

MAO A and B exist in the brain. Platelets have protein kinases that have become lately the focus of research into the neurobiology of psychiatric syndromes. Both protein kinase A (PKA) and PKC modulate adrenergic and serotonergic receptor-G protein interactions, respectively. In addition, imidazoline$_1$ (I$_1$) receptors,[23,24] benzodiazepine receptors,[25] glutamate receptors,[26] and dopamine receptors[27] have all been identified in platelets and, in the majority of cases, mediate functions similar to their counterparts in the brain. Finally, platelets possess amyloid precursor protein isoforms similar to those in the brain and have been used to investigate pathophysiologic mechanisms in Alzheimer's disease, resulting in a sizable body of literature.[28] Alzheimer's disease is discussed in Chapter 43.

These characteristics of blood platelets and the advancement of the monoamine hypothesis of mood disorders, the hypothesis of noradrenergic dysregulation in anxiety disorders, the dopamine hypothesis in schizophrenia, and neuronal mechanisms of action of psychotropic medications have spurred a rich research using platelets as a tool. This chapter reviews major studies that have used platelets as a research tool in anxiety, mood and related disorders (e.g., alcoholism, eating disorders, and premenstrual dysphoric disorder), schizophrenia, and childhood psychiatric disorders.

## II. Anxiety Disorders

### A. α$_2$AR Studies

Anxiety disorders constitute a group of related disorders in which anxiety and avoidance or inhibited behaviors represent the hallmark symptoms. Panic disorder (PD) is characterized by recurrent discrete episodes of intense anxiety, associated with autonomic hyperarousal, anticipatory fear of recurrence of panic attacks, and, in some patients, avoidance of specific situations or phobias. Increased noradrenergic function has been associated with intense fear, as demonstrated in primates.[29] As autoreceptors, α$_{2A}$AR regulate norepinephrine release. Abnormal α$_{2A}$AR function was hypothesized to occur in anxiety disorders. Specifically, decreased platelet α$_2$AR density would be consistent with enhanced norepinephrine release and a hyperadrenergic state. This was first reported in PD, in generalized anxiety disorder (GAD),[30–34] and in borderline personality disorder,[35] a disorder characterized by a substantial component of anxiety and vulnerability to depression. Downregulation of α$_{2A}$AR in PD or GAD, however, was not replicated by other investigators who reported normal platelet α$_2$AR density in PD.[36–38] In fact, increased α$_{2A}$AR density was reported in PD[39] in a large cohort study. Methodological differences (e.g., drug washout period, ligand used, pH of

platelet-rich plasma) or population differences (e.g., age, gender) may have contributed to the discrepancies between these studies.

These studies measured the maximum binding capacity of α$_{2A}$AR but did not investigate functional aspects of the receptor, such as α$_{2A}$AR-mediated signal transduction or α$_{2A}$AR interaction with G$_i$ protein. α$_{2A}$AR exhibit high- and low-affinity states upon agonist stimulation, as a function of their coupling with G$_i$ protein.[9,40] The high-affinity state of the receptor constitutes the transitory agonist-receptor-G$_i$ tertiary complex, which precedes the activation of G$_{i\alpha}$. In a comprehensive study from our laboratory,[41] we found increased α$_{2A}$AR density in both affinity states in PD. The increase was more pronounced in the high-affinity state as shown by the significantly higher percentage of receptors in the high-affinity state (%R$_H$), an index of coupling to G$_i$ protein. This is also consistent with low epinephrine EC$_{50}$ for inhibition of platelet AC[36] and enhanced norepinephrine-mediated platelet aggregation[39] in PD. Increased G$_i$ coupling was observed in PD patients with high levels of depressive symptoms. We also demonstrated positive correlations between coupling measures and severity of anxiety and depressive symptoms.[41]

We further investigated the relationship between antidepressant treatment response and α$_{2A}$AR function in PD.[41] Imipramine, a tricyclic antidepressant, did not induce changes in any of α$_{2A}$AR density consistent with other studies.[39,42–54] However, higher pretreatment α$_{2A}$AR density and coupling to G$_i$ predicted a better treatment outcome.[41] These results underscore the adaptive nature of increased α$_{2A}$AR density in PD. They also suggest that normal or increased α$_{2A}$AR coupling may be a prerequisite for antidepressant response or a positive treatment outcome, as we shall see in our discussion of α$_{2A}$AR in major depressive disorder (MDD) (discussed later). In addition, we have found in the same PD patients that plasma norepinephrine and epinephrine levels were not different from those in normal controls (Gurguis et al., unpublished data). Thus, upregulation of α$_{2A}$AR occurs in the presence of normal agonist levels, which suggests abnormal α$_{2A}$AR agonist-mediated regulation of gene expression of α$_{2A}$AR in PD. Increased α$_{2A}$AR may also represent adaptation to increased βAR density[55,56] and higher PKA catalytic subunit,[57] which has been reported in PD, thus balancing the increased stimulatory input into the adenylyl cyclase system. This notion of adaptive upregulation in α$_2$AR density in PD is also supported by a recent study reporting increased α$_2$AR density response to psychological stress, and significant positive correlation between α$_2$AR density and severity of anxiety.[58]

Posttraumatic stress disorder (PTSD) is an anxiety disorder that develops following exposure to a severe, psychologically traumatic event. It is characterized by reliving the traumatic event in the form of flashbacks or nightmares, hypervigilance, autonomic hyperarousal, and exaggerated startle responses. PTSD may be related to PD as both syndromes share a common symptomatology. In our hands, approximately 40% of PTSD patients also met diagnostic criteria for PD.[59] Similar to PD, we found upregulation of $\alpha_{2A}AR$ in PTSD, which was also observed in receptor density in the high-affinity state, although differences in coupling measures in PTSD did not reach statistical significance, perhaps due to the small number of subjects. A decrease in the platelet $^3$H-rauwolscine low-affinity binding site, presumably related to $\alpha_{2A}AR$, has been reported.[60] However, the significance of this finding is unclear because rauwolscine, being an antagonist, does not induce a high-affinity tertiary complex with $G_{i\alpha}$ protein. Our findings of increased $\alpha_{2A}AR$ density in PTSD are consistent with enhanced epinephrine-mediated inhibition of forskolin-stimulated AC,[61] although this latter finding could not be replicated.[62] Upregulation of platelet $\alpha_{2A}AR$ was also reported in obsessive compulsive disorder (OCD),[63] and in OCD patients with motor tic.[64] Thus, $\alpha_{2A}AR$ upregulation appears to be a biological marker common to anxiety disorders. Downstream serotonergic input into the inhibitory AC pathway has been investigated in PD. Serotonin inhibition of AC was decreased in PD patients prior to treatment and was reversed after paroxetine treatment, concomitant with clinical improvement.[65]

## B. MAO Studies

Platelet MAO activity has been investigated in anxiety disorders, with increased MAO activity reported in PD.[66–69] Such an increase would suggest decreased monoamine levels and would be consistent with the established therapeutic efficacy of MAO inhibitors in anxiety and depressive disorders. Two other studies,[70,71] however, found no differences in MAO activity in PD compared to normal controls but did find significant correlations between MAO activity and severity of anxiety. A third study found decreased MAO activity in PD.[72] Low MAO-B activity was also reported in PTSD,[73,74] particularly in soldiers with lower ego strength and high neuroticism scores.[73] However, this result was not replicated in a study that reported normal 5-HT content and MAO enzyme activity in PTSD.[75] Both measures also did not correlate with measures of severity of anxiety. Methodological differences — such as gender, phase of menstrual cycle, diagnostic specificity, sampling time, and particularly substrate type — may account for the discrepant results. Serotonin is a primary substrate for MAO A but not MAO B. This, together with the inconsistent findings on MAO activity in anxiety disorders, calls for more controlled and well-conducted studies on the role of MAO in anxiety disorders.

## C. Serotonin Transporter and Serotonin$_2$ Receptor Studies

Platelet serotonin transporter (5-HTTP) has been investigated in anxiety disorders. Similar to findings in MDD, reduced $[^3H]$-imipramine and $[^3H]$-paroxetine binding has been reported in PD.[76–81] However, normal $^3$H-imipramine binding in PD has also been reported.[38,82–84] Ligand specificity (paroxetine may be a more specific ligand for 5-HTTP), among other methodological differences, seems to account for the inconclusive results in PD. Indeed, the actual values of $B_{max}$ or $K_d$ are hard to compare across these studies. A relationship between the serotonin transporter and $^{14}$C-serotonin uptake has been demonstrated.[21,22] Reduced density of platelet 5-HTTP may be consistent with decreased platelet serotonin uptake in PD.[85] Again, this finding is inconsistent with both normal serotonin content[86] and increased platelet serotonin uptake[86–88] reported in PD. In one study,[88] 5-HT uptake correlated positively with inhibition of aggression in PD. No differences either in the density of 5-HTTP sites,[74,89] in serotonin uptake,[90] or serotonin content[75,91] were found in PTSD. Only one study reported decreased 5-HTTP sites in PTSD,[92] which was associated with a trend for positive treatment outcome.[93] Normal transporter density was reported in social phobia.[83]

In summary, abnormal platelet 5-HTTP density or serotonin uptake has not been unequivocally established in PD, social phobia, GAD, or PTSD. The discrepancy between studies on 5-HTTP in anxiety disorders is intriguing given that 5-HTTP polymorphism is reported to account for 7% to 9% of inherited variance in anxiety-related personality traits in individuals and their siblings.[94] Finally, few studies reported decreased serotonin-mediated platelet aggregation in PD.[39,95,96]

5-HT$_{2A}$ receptor density was upregulated in PD when measured using $[^3H]$-ketanserin[39,95] but normal when measured using $[^3H]$-lysergic acid diethylamide ($[^3H]$-LSD).[97] Normal $[^3H]$-LSD binding decreased after citalopram treatment. Normal 5-HT$_{2A}$ and 5-HT$_{1A}$ receptor density was also observed in PD,[98] despite increased phosphorylation of receptors, suggesting decreased serotonergic receptor function. This decrease was also reflected in lower 5-HT-mediated ERK$_{1/2}$ phosphorylation in PD. Thus, there is converging evidence[65,98] of decreased 5-HT$_{2A}$ receptor function, but not density, in PD.

## D. Serotonergic Studies in OCD

Obsessive compulsive disorder (OCD) is characterized by repetitive thoughts associated with intense anxiety that is relieved by acting on ritualistic behavioral repertoires designed to mitigate the anxiety. Although one investigation[99] has challenged the association between abnormal serotonin function OCD spectrum disorders, the serotonergic system remains perhaps the most extensively studied system in OCD and may bear more relevance to this psychiatric disorder compared to other anxiety disorders. Several studies have targeted peripheral indices of serotonergic function. Downregulation of platelet 5-HTTP was reported in OCD and in OCD-spectrum disorders in six studies.[100–106] Five other studies found normal 5-HTTP density in OCD.[107–111] Although treatment had no effect on 5-HTTP density in MDD patients with comorbid OCD, treatment nonresponders had lower transporter density than responders before treatment.[109] Similarly, the literature is inconsistent on serotonin uptake kinetics in OCD. Normal or low serotonin uptake,[100,104,112] normal platelet serotonin content,[113,114] and normal MAO activity[113] were found in OCD. Both serotonin content and MAO activity were decreased following clomipramine and treatment.[113] Finally, platelet 5-HT$_{2A}$ receptor density was normal in OCD[115] and decreased after treatment with serotonin reuptake inhibitors.[106] Studies showed that PKC may play a role in the pathophysiology of OCD. PKC inhibits 5-HT uptake. Decreased 5-HT uptake suggests increased PKC activity.[112,116] Treatment with serotonin reuptake inhibitors normalized increased IP$_3$ platelet content[106] and increased 5-HT reuptake.[116] Opposite to findings in PD, decreased PKA catalytic subunit, PKA activity,[117] and phosphorylated Rap 1[118] were reported in OCD, suggesting altered cyclic adenosine monophosphate (cAMP) signaling.

## III. Mood Disorders

Mood disorders are a group of disorders in which the basic psychopathology is one of abnormal mood (i.e., severe elation, depression, or swings between either poles). Changes in cognitive, behavioral, motivational, and vegetative functions occur in the context of a primary disturbance in mood. Unipolar recurrent MDD and bipolar disorder (BPD) are the most commonly investigated mood disorders. Recently, abnormal signal transduction mechanisms and the effects of antidepressants and mood stabilizers on these mechanisms have been the focus of research in mood disorders.

### A. Signal Transduction Studies in Mood Disorders

G proteins play a crucial role in signal transduction between membrane receptors and second messenger systems (see Chapter 16). Increased platelet G$_s$ protein levels were reported in BPD[119,120] and in bipolar II disorder.[120] This increase was found in both manic and euthymic BPD patients, suggesting that increased G$_s$ protein levels may be a trait marker in BPD.[120] Increased platelet G$_s$ is consistent

with similar findings in postmortem brains from BPD subjects.[121,122] Whereas G$_{i1/2}$ levels were normal in one study,[119] they were increased in another[120] from the same research group. Increased platelet cAMP-dependent protein kinase level in euthymic BPD patients[123] suggests abnormal cAMP signaling in BPD. This finding was later replicated by other studies. cAMP-dependent PKA, Rap 1 levels, and cAMP-dependent phosphorylation of Rap 1 were increased in BPD.[124,125] Low PKA regulatory subunit type I and high C subunit was reported in BPD psychotically depressed patients.[126] This is in contrast with decreased PKA regulatory subunit type II and Rap 1 levels in unipolar depression.[127] However, increased PKA levels in BPD are difficult to reconcile with decreased cAMP-dependent protein kinase in postmortem brains from BPD patients.[128]

Platelet 5-HT$_{2A}$ receptors are coupled via G$_{q/11}$ to PLC, which in turn activates PKC. G$_{q/11}$ levels were normal in BPD patients.[119] Platelet 5-HT$_{2A}$ receptors are increased in BPD, particularly among suicidal patients.[129] High free intracellular Ca$^{++}$ concentration was reported in both manic and depressed bipolar patients,[130–133] and 5-HT-mediated platelet aggregation was increased in mania.[134] Serotonin-mediated intracellular Ca$^{++}$ mobilization was also elevated in BPD in comparison to other psychiatric disorders.[135] These findings, collectively, suggest enhanced platelet 5-HT$_{2A}$ receptor sensitivity in BPD independent of receptor density. Increased serotonin-mediated Ca$^{++}$ response predicted favorable response to mood stabilizing agents in bipolar patients.[136] An interesting body of literature on the role of PI, PLC, and PKC in BPD has emerged. Increased PKC translocation was found in the manic state.[137,138] This was observed in response to serotonin, thrombin, or phorbol myristate acetate (a PKC activator). Platelet membrane PIP$_2$ was increased in bipolar depressed patients.[139] Sautosporine (a PKC inhibitor) failed to inhibit Ca$^{++}$ mobilization in BPD, suggesting increased PKC activity.[140] Lithium decreased PKC activation through post-5-HT$_{2A}$ receptor mechanisms.[137] Chronic lithium treatment decreased platelet membrane PIP$_2$,[141] and decreased cytosolic PKC α isozyme and membrane PIP$_2$ in BPD,[142] and lithium levels were positively correlated with stimulated Ca$^{++}$ levels.[143] Only one study[144] reported low PI-PLC, PKC isozymes but high myristoylated alanine-rich C kinase substrate (MARKS) in BPD but not in unipolar depression. Future investigations are needed to elicit which of these mechanisms is phase dependent (manic versus depressed) and which mechanisms are trait related. These findings also may have significant implications for developing and testing new therapeutic agents that specifically target this system.

Quantitative studies of G protein revealed increased platelet G$_i$ protein levels in MDD patients.[145,146] However, α$_{2A}$AR agonist-dependent activation of G$_α$ was decreased.[147] G$_i$ levels were decreased after antidepressant treatment, although the two available studies did not clarify whether the decrease was observed in all patients regardless of treatment outcome[145] or only in treatment responders.[146] The decrease in $G_i$ levels may be consistent with desensitization of $\alpha_{2A}AR$ after antidepressant treatment, which has been shown in basic neuroscience studies. Functionally, agonist $\alpha_2AR$-mediated activation of $G_{i\alpha}$ GTPase activity, which terminates inhibition of AC by the $G_{i\alpha}$ subunit, was normal in MDD.[148]

## B. Adenylyl Cyclase Studies

The role of AC in MDD has been rather problematic. Decreased prostaglandin (PG) $E_1$-stimulated and decreased norepinephrine-mediated inhibition of $PGE_1$-stimulated AC responses were found in MDD,[149,150] demonstrating a dissociation between upregulation of $\alpha_{2A}AR$ density and decreased $\alpha_{2A}AR$-mediated inhibition of AC. However, four other investigations found no differences in $PGE_1$-, sodium fluoride-, 5-guanylylimidodiphosphate-(Gpp(NH)p), or forskolin-stimulated AC or $\alpha_{2A}AR$-mediated inhibition of AC.[151-154] A well-conducted study[155] found decreased basal and forskolin-stimulated AC activity in MDD patients. In that study, treatment had no effect on AC activity, consistent with basic studies showing lack of effects of antidepressants on AC activity, although changes in relationship to treatment response were not examined. Decreased platelet AC activity in MDD would be consistent with similar findings in postmortem brains from suicide victims.[156]

## C. $\alpha_2AR$ Studies

Platelet $\alpha_2AR$ have been studied in MDD, because decreased noradrenergic function was hypothesized in MDD. Increased $\alpha_{2A}AR$ density in MDD was reported,[157] which is consistent with decreased synaptic availability of norepinephrine or decreased noradrenergic transmission. Upregulation of platelet $\alpha_2AR$ was later inconsistently replicated. Some studies found upregulation in $\alpha_2AR$,[46,49,157-167] while others reported normal $\alpha_{2A}AR$ density.[42,43,47,48,50,52,53,168-172,173] Three studies reported decreased $\alpha_2AR$ density in MDD.[174-176] Either $^3H$-clonidine (a partial $\alpha_2AR$ agonist) or $^3H$-yohimbine (an $\alpha_2AR$ antagonist) was used as a ligand in most studies. Inconsistencies among studies were explained by differences in the ligands used. It was speculated that $\alpha_2AR$ upregulation might exist in the high-affinity state that is labeled by clonidine, which could not be demonstrated using yohimbine.

The pitfalls of this assumption were addressed in a study from our laboratory,[178] which suggested that clonidine, as a partial agonist, did not induce full formation of the high-affinity state of $\alpha_2AR$ or did not saturate that site. In an exhaustive study from our laboratory,[179] we demonstrated

increased platelet $\alpha_2AR$ density in MDD, particularly that in the high-conformational state. Despite $\alpha_2AR$ upregulation, $\alpha_2AR$-G protein coupling was normal. In addition, we found that treatment with the tricyclic antidepressant imipramine was associated with desensitization of $\alpha_2AR$ through shifts in $\alpha_2AR$ density from the high- to the low-affinity state in the absence of changes in total $\alpha_2AR$ density. This lack of treatment effect on total $\alpha_2AR$ density in MDD is consistent with the majority of studies.[42-44,48,50-54,172] We also demonstrated that normal $\alpha_2AR$ coupling to $G_i$ was associated with a positive treatment outcome, whereas decreased coupling predicted a negative treatment outcome.[179] This is consistent with findings that desipramine inhibition of $\beta AR$-mediated activation of AC was lacking after inactivation of $Gi\alpha$, $Go\alpha$ using pertussis toxin and indicates the need for intact inhibitory input into the adenylyl cyclase.[180] Similarly, another report showed that desipramine and roboxetine treatment of $G_{z\alpha}$ (which belongs to the $G_i$ and $G_o$ family) knockout mice did not shorten the immobilization time in the forced swim test as observed in the wild type.[181] Finally, the severity of somatic anxiety symptoms in depressed patients was correlated with $\alpha_2AR$ density in the high-affinity state.

The similarities in $\alpha_2AR$ binding characteristics between PD and MDD lead us to propose that upregulation of platelet $\alpha_2AR$ may be a common biological marker between these antidepressant-responsive disorders. Another aspect of $\alpha_{2A}AR$ upregulation in MDD is reflected in their mediation of phosphoinositide signaling. $\alpha_{2A}AR$ mediate phosphoinositide hydrolysis in normal subjects.[182] Several studies found enhanced epinephrine-mediated phosphoinositide signaling in MDD,[183,184] with antidepressants and lithium inhibiting phosphoinositol hydrolysis.[185] The GTPase activity of $G_{i\alpha}$, which terminates inhibition of AC, was normal in MDD patients.[148]

These results implicate abnormal platelet PKA and PKC in MDD.[179] Another study reported low PKA regulatory subunit type I and high catalytic subunit in psychotic unipolar depression.[126,127] Low regulatory subunit II and Rap1 levels were low in MDD.[127] Unlike BPD, there were no differences in PI-PLC, PKC activity, or in levels of various isozymes between MDD patients and controls.[144]

## D. Imidazoline$_I$ Receptor Studies

Clonidine, para-aminoclonidine, UK 14303, and idazoxan were used as ligands in several studies investigating platelet $\alpha_{2A}AR$ function in mood or anxiety disorders. These ligands were later found to also bind to platelet $I_1$ receptors.[23,24,186] $I_1$ receptors inhibit catecholamine release in the brainstem and sympathetic nerve terminals. Upregulation of $I_1$ receptors in MDD may further contribute to decreased noradrenergic transmission mediated through upregulation of $\alpha_{2A}AR$.

Platelet $I_1$ receptors are increased in MDD, but not in GAD.[187] These findings were confirmed using solubilized receptors[188] and are consistent with increased $I_1$ receptors in the frontal cortex from postmortem brains of suicide victims.[188] Similar increases were not observed in euthymic BP patients.[189] Positive correlations between immunoreactivities of $I_1$ receptors and $G_{q/11}$, $G_{i2}$, and $G\beta$ suggest that the $I_1$ receptor couples the phosphoinositide pathway in the platelet.[190] Upregulation of $I_1$ receptors in MDD has been replicated[191] and patients with highest $I_1$ pretreatment density responded worst to bupropion treatment. Downregulation of the $I_1$ site was correlated with plasma bupropion levels and was observed at a daily dose of 450 mg, but it was not related to treatment response. Finally, $I_1$ receptor density was increased in postmenopausal compared to premenopausal women, who were normalized after 60 days of estrogen replacement therapy,[192] perhaps suggesting an association between sex-steroid hormone changes and mood changes in menopausal women.

### E. Serotonin Transporter Studies

The platelet 5-HTTP has been the subject of extensive investigation in MDD and to a lesser extent in BPD. Most studies have reported decreased 5-HTTP in MDD,[81,193–212] as well as in depressed or euthymic BPD patients.[213–215] Fewer studies did not find differences in 5-HTTP density.[45,110,216–230] Decreased transporter density was further confirmed in a meta-analytic study.[231–233] Other studies investigated the kinetics of 5-HT uptake by measuring the rate constant ($K_m$) and maximum velocity ($V_{max}$) of $^{14}C$ 5-HT uptake. These studies showed decreased $V_{max}$,[234–240] while others showed increased $K_m$ in bipolar or depressed, suicidal patients.[241,242] Fewer studies did not find differences.[230,243–246] The $V_{max}/K_m$ ratio of serotonin uptake correlates with platelet serotonin content in normal controls, but not in depressed patients,[247] and high HAM-D scores were observed in patients with low net reuptake. Low platelet 5-HT concentration was reported in MDD.[248] Interestingly in that study, platelet 5-HT concentration was correlated with plasma and prolactin levels in healthy controls, but this relationship was not found in depressed patients. Cortisol has been reported to increase serotonin uptake in cortical neuronal cells *in vitro,* due to promotion of serotonin transporter.[249] Hypercortisolemia exists in a subgroup of patients with major depression. Low platelet serotonin concentration or transporter density may thus suggest dysregulation in the relationship between the hypothalamic-pituitary-adrenocortical axis and the serotonergic system. Decreased platelet serotonin uptake has been associated with somatic symptoms in MDD,[250,251] and abnormal 5-HTTP processing or alternative splicing of the gene encoding the transporter was reported in somatoform disorder.[252] Decreased $V_{max}$ is consistent with decreased 5-HTTP density. Pretreatment platelet serotonin concentration

predicted treatment response to serotonin reuptake inhibitors,[253,254] and high pretreatment $K_m$ predicted good response to nortriptyline and was further increased after treatment.[255] Interestingly, it was demonstrated that deletion/insertion polymorphism regulates the 5-HTTP gene transcription. Long and short variants of 5-HTTP promoter have differing transcriptional efficiencies. Significant decrease in the $V_{max}$ of 5-HTTP was associated with the l/l genotype.[238] Low $V_{max}$ of 5-HT reuptake reflects low expression of the 5-HTTP in platelets and perhaps in the brain.[256] Finally, another study showed enhanced platelet vesicular monoamine transporter in MDD suggesting enhancement of vesicular storage and low monoamine turnover in depression.[19]

Decreased platelet 5-HTTP density or serotonin reuptake in MDD is intriguing, because antidepressants block the transporter site. Treatment with various antidepressants, electroconvulsive therapy, and lithium was associated with increased platelet 5-HTPP density or function in most[45,199,208,209,213,220,228,257–261] but not all[198,200,205,210,211,222,226,259] studies. Moreover, this increase was associated with either positive treatment outcome or remission of the depressive episode, suggesting that decreased 5-HTTP may be a state marker in MDD. Blockade of serotonin uptake by antidepressants is only one of many mechanisms of action of antidepressants. It is unclear which mechanisms are specifically related to their antidepressant effects.

### F. Serotonin$_{2A}$ Receptor Studies

Modulation of serotonin neurotransmission by antidepressants prompted the investigation of serotonin receptor function in MDD. Increased platelet 5-HT$_{2A}$ receptor density in MDD has been found using [$^3$H]-ketanserin, [$^3$H]- or [$^{125}$I]-LSD, or [$^{125}$I]-1-(2,5-dimethoxy-4-iodophenyl)-2-aminopropane (DOI) as ligands,[262–271] consistent with increased 5-HT$_{2A}$ receptor density in the frontal cortex from depressed subjects or suicide victims,[272–275] although a few studies found normal 5-HT$_{2A}$ receptors in MDD.[276,277] Upregulation of 5-HT$_{2A}$ receptors may pertain either to depressed female patients[266,270] or to suicidal patients.[262,268,278,279] A study investigating the association between 5-HT$_{2A}$ gene C102T polymorphism found that the TT genotype was associated with higher platelet 5-HT$_{2A}$ density in a healthy population, but this effect was not associated with the pathophysiology of mood disorders.[279] 5-HT$_{2A}$ supersensitivity may exist regardless of changes in receptor density. 5-HT$_{2A}$ receptor sensitivity was found to be correlated with behavioral changes.[280] Increased 5-HT$_{2A}$ receptor sensitivity as reflected by agonist-mediated phosphoinositide hydrolysis, aggregation, or Ca$^{++}$ responses in MDD has been reported[185,281–285] (see the section on platelet activation studies). Increased serotonin-mediated intraplatelet Ca$^{++}$ mobilization was specifically found in depressed cancer patients,[286] although another study found 5-HT$_{2A}$ receptor-mediated Ca$^{++}$ response was normal and not related to severity of depression or suicidality in cancer patients.[287] Taken together, the majority of studies seems to indicate either increased 5-HT$_{2A}$-receptor density or sensitivity in MDD.

The reported effects of antidepressants, electroconvulsive therapy (ECT), or lithium on 5-HT$_{2A}$ receptors have been inconsistent. Studies reported decreased platelet 5-HT$_{2A}$ density or sensitivity.[263,265,288–290] Lithium reduced serotonin-mediated inhibition of AC, which correlated with clinical improvement,[291] and imipramine reduced serotonin-mediated platelet aggregation upon remission of depressive symptoms.[292] Other studies, however, did not find changes[262,267] or found increased receptor density.[276,293,294] In a study in normal controls, fluvoxamine treatment decreased 5-HT$_{2A}$ receptors in the first week; however, density was not different from pretreatment values by the end of the fourth week,[271] suggesting perhaps that the duration of treatment or timing of repeating the assay could have contributed to this inconsistency. Mechanisms underlying increased 5-HT$_{2A}$ receptor density or sensitivity in MDD, or treatment-induced changes, remain unclear and warrant further investigation.

## G. MAO Studies

Platelet MAO enzyme activity in MDD and BPD was the subject of extensive investigations in the 1980s. Increased MAO enzyme activity would be consistent with decreased monoamine levels in depression, and hence with the monoamine hypothesis of depression. Increased platelet MAO activity was reported in MDD, particularly in patients with endogenous depression or those with lack of suppression of cortisol secretion after dexamethasone administration (dexamethasone escape or nonsuppression).[295–301] Low platelet MAO activity was observed in patients who attempted suicide and was associated with low urinary MHPG and high plasma cortisol levels.[302] However, other studies reported increased MAO in BPD,[303,304] or normal MAO in MDD.[70,305] Despite these inconsistencies, what remains clinically significant is that (a) inhibition of MAO is affected not only by MAO inhibitors, but also by tricyclic antidepressants[306,307] and serotonin reuptake inhibitors,[308] and (b) at least 60 to 80% inhibition of MAO enzyme activity is needed for a positive clinical outcome with MAO inhibitors.[306,309,310] Interest in MAO activity in mood disorders has long since waned.

## H. Platelet Activation Studies and Risk for Thromboembolic Events

A large number of studies have investigated platelet activation in MDD. For a long time, a relationship between stress, hostility, depression, and risk for cardiovascular diseases

has been observed. Depression is a risk factor for cardiovascular diseases.[311,312] Stress increases plasma levels of β-thromboglobulin, which is released from platelet alpha granules upon platelet activation (see Chapter 15). Hostility was associated with a greater increase in plasma β-thromboglobulin.[313] Platelet procoagulant activity, both basal and in response to orthostatic physiologic challenge, was high in MDD[314,315] and decreased after paroxetine treatment.[315] Platelet glycoprotein (GP) Ib receptors were significantly increased in depressed patients[316,317] similar to poststroke subjects. Higher plasma levels of platelet factor 4 (also released from platelet α granules upon platelet activation) and β-thromboglobulin were observed in depressed patients with ischemic heart disease compared to non-depressed patients with ischemic heart disease and healthy controls.[318] MDD patients have augmented serotonin-mediated calcium uptake[284] and increased thrombin-stimulated, but not NaFl-stimulated, IP, which was not changed after desipramine treatment,[319] and increased thrombin-mediated intracellular calcium mobilization, which remained increased after electroconvulsive therapy.[320] Depressed patients had increased activation-dependent surface glycoprotein receptors, P-selectin and GP-53, in response to strenuous physical stress.[321] Enhanced Ca$^{++}$-mediated platelet aggregation and secretion cascades were reported in MDD patients, particularly those with high anxiety.[322] Glutamate receptor supersensitivity further contributes to enhanced intercellular Ca$^{++}$ responses.[323] As summarized in one review, depression is associated with increased susceptibility to serotonin-mediated platelet activation, upregulation or supersensitivity of 5-HT$_{2A}$, and downregulation of 5-HTTP, and these changes may contribute to an increased risk for thromboembolic events and cardiovascular disease.[324] Increased α$_2$AR-mediated aggregation and delayed desensitization of aggregation responses were both observed in MDD and anxiety disorders.[39,151,159] G$_z$ knockout mice have significantly decreased platelet aggregation responses to epinephrine, ADP, serotonin, and the combination of epinephrine and collagen.[181] α$_2$AR couples G$_z$ at high epinephrine concentrations. It is likely that increased G$_z$ function or levels may underlie increased platelet activation in depression. This is an exciting possibility because, as discussed earlier, G$_z$ proteins appear to be a prerequisite for the mechanisms of action of antidepressants. Treatment with antidepressants such as paroxetine and sertraline and electroconvulsive therapy was associated with normalization of platelet secretion and activation,[289,315,325,326] and mood stabilizing agents normalized serotonin-mediated intraplatelet calcium mobilization.[327] Collectively, these studies have strong clinical significance in view of their direct relevance to clinical management and to the need for early identification and treatment of depression, particularly among the medically ill patients.

However, a few intriguing findings of apparently the opposite nature were also reported in MDD. Serotonin-mediated platelet aggregation was normal[328] or decreased in MDD.[329–332] This decrease was not found after recovery from depression, suggesting a state-dependent change.[330] Also, plasma from depressed patients was found to reduce platelet aggregation in response to ADP, ADP/serotonin, epinephrine, sodium fluoride, calcium ionophore A23187, and phorbol-12-myristate-13-acetate in controls.[332] Because multiple mechanisms are involved in platelet aggregation, it is unclear if these latter studies are necessarily inconsistent with the prior investigations. In any event, the high risk for thromboembolic events and mortality rates in depressed patients remains strongly supported.

## IV. Alcoholism

Alcohol dependence or alcoholism is a disorder of excessive alcohol intake despite the patient's awareness of the negative physical, mental, and social consequences. Although strong familial patterns have been observed, genetic associations have not been conclusively identified. Clinically, alcoholism is a complicated, multifaceted disorder that entails the interaction of a psychoactive substance with genetic and environmental factors. Moreover, comorbidity with other psychiatric disorders, such as anxiety disorders, mood disorders, and other substance abuse disorders, further complicates the interpretation of research findings in this area.

### A. G Protein Studies

Alcohol dependence has been associated with increased $G_{\alpha i}$ levels and $G_{i2\alpha}$ mRNA in both actively drinking and abstinent alcoholics.[333,334] $G_{s\alpha}$ levels were not different, but $G_{s\alpha}$ mRNA levels were increased.[334] Nonalcoholic men with a family history of alcoholism have increased levels of $G_{s\alpha}$, suggesting that this may be a marker for high risk of alcoholism.[335]

### B. Adenylyl Cyclase Studies

Several studies focused on the investigation of AC in alcoholism. Acute ethanol exposure increases AC activity. However, chronic alcoholism was associated with decreased cesium fluoride-, Gpp(NH)p-, and PGE$_1$-stimulated AC.[336–343] Ethanol withdrawal was associated with increased stimulatory input and decreased inhibitory input to AC.[336] Increased platelet AC activity was reported in recently abstinent alcoholics and in alcoholics with positive family history of alcoholism,[344,345] but it decreased after 4 days of abstinence. Inhibition of MAO by ethanol and decreased cesium fluoride-stimulated AC correctly identified 75% of alcoholics.[342] Decreased Gpp(NH)p- and ethanol-stimulated AC

was observed during withdrawal and normalized shortly thereafter, only to decrease with long-term abstinence.[343] Both Gpp(NH)p$^-$ and Gpp(NH)p$^+$ ethanol-stimulated AC were decreased in male alcoholics with a family history of alcoholism.[341] Similar findings were also seen in alcoholic women.[339] Both reduced Gpp(NH)p$^-$ and forskolin-stimulated AC were found in alcoholics with poor capacity for abstinence, regardless of comorbidity with major depression.[338] Low AC response to sodium fluoride stimulation was also found in children of alcoholics and children with multiple alcoholic family members.[346] Finally, cesium fluoride-stimulated AC activity was proposed to reflect a single gene effect.[337] This latter finding remains to be replicated. In summary, these findings suggest decreased AC activity or AC-G protein interaction in alcoholism. In addition, these studies identified molecular abnormalities in individuals with high risk for alcoholism.

### C. Serotonergic Studies

Ethanol decreases MAO enzyme activity. Platelet MAO activity is low in alcoholics.[347–352] Low MAO activity was also seen in abstinent alcoholics, was more pronounced in type II alcoholics (those with younger age of onset, high heritability, and higher incidence of social/legal problems)[347,350,353] and alcoholics with a positive family history of alcoholism,[351] and was proposed as a state marker for alcoholism.[352] Another study reported an association between low platelet MAO activity and homozygosity of the long allele CAAA of the transcription factor AP-2β in both genders.[354] Low MAO activity was associated with personality characteristics such as impulsivity and novelty seeking behavior.[355–357] Low platelet MAO is consistent with brain postmortem studies in alcoholics.[358,359] Although others found normal MAO activity in abstinent alcoholics,[360] ethanol-mediated inhibition of MAO was more pronounced.[342] MAO in abstinent alcoholic women was increased.[339] Some of the inconsistency between studies may be in part due to differences in the time elapsed since the last drink[361] or to nicotine dependence.[362] A neuroendocrine study showed decreased D$_2$ receptor function was associated with low platelet MAO-B activity.[363] The relationship between low MAO-B and the Taq1 A1 allele of the D$_2$ receptor gene in socially stable middle-aged Caucasian men with high alcohol consumption was significant only when subjects who met diagnostic criteria for alcohol dependence were included.[364] The relationship between platelet MAO, personality, and alcoholism has been reviewed.[365] However, the specificity of low MAO to the pathophysiology of alcoholism remains to be firmly established, as low MAO has also been reported in other psychiatric disorders.

Finally, although increased platelet 5-HT uptake was reported in early onset alcoholics,[366] the majority of studies reported low platelet 5-HT uptake. Low blood serotonin and low serotonin uptake in platelets from alcoholics was reported.[367,368] One study reported low platelet 5-HT concentration in alcoholics of both genders, with female alcoholics having lower levels than male alcoholics,[369] and comorbidity with PTSD was associated with higher 5-HT levels than comorbidity with other anxiety or depressive disorders. There was no correlation between 5-HT uptake and density of the serotonin transporter in platelets from alcoholics, but low 5-HT uptake was associated with the LL allele of the 5-HTTLPR region of the serotonin transporter gene,[370] but not the SS allele.[371] Decreased 5-HT$_{2A}$-mediated phosphoinositide hydrolysis was only observed shortly after withdrawal, not after long-term abstinence,[368] suggesting the short-term effects of ethanol.

## V. Eating Disorders

Eating disorders are characterized by abnormal impulse control over food intake. While anorectics are too restrictive, bulimics display binging episodes associated with loss of control over the amount of food ingested. Loss of control seems to exist in other aspects of bulimics' lives. Both disorders are associated with a higher incidence of MDD. Dieting, which resulted in a 4 to 5% weight loss in volunteers, was associated with increased platelet $\alpha_{2A}$AR and 5-HT$_{2A}$ receptor density but had no effect on 5-HTTP.[372] Serotonin-induced increase in intracellular calcium levels was enhanced by fasting.[373] Platelet MAO was low in most studies of bulimia and anorexia but was not different between the two groups.[374–377] Similar to alcoholics, platelet MAO activity was inversely correlated with measures of impulsivity in bulimics.[375,376] Serotonin uptake was normal in anorexia,[378,379] but 5-HTTP density was decreased in both anorexia and bulimia nervosa[378,380–385] and was inversely correlated with measures of impulsivity.[381] Curiously, low $^3$H-paroxetine binding bulimia was not related to clinical variables such as binging, depression,[382] or history of childhood abuse.[383] Decreased 5-HTTP density was associated with evidence for decreased central serotonin receptor function as reflected in blunted prolactin response to m-CPP. This emphasizes the need for a multisystem approach for investigating the pathophysiology of eating disorders.[384,385] 5-HT$_{2A}$ receptors were increased[386] in eating disorders. Platelet aggregation was enhanced,[387] and $\alpha_{2A}$AR density was increased in anorexia.[388] Increased $\alpha_{2A}$AR was associated with increased PGE$_1$-stimulated cAMP and norepinephrine-mediated inhibition of PGE$_1$-stimulated cAMP in anorexia. In the same study, $\alpha_{2A}$AR density normalized after a 10% weight gain.[388] βAR-mediated $^{45}$Ca$^{++}$ uptake was decreased,[389] but $\alpha_{2A}$AR-mediated $^{45}$Ca$^{++}$ uptake into platelets of anorexia patients was increased, consistent with upregulation of $\alpha_{2A}$AR density. It is unclear if several of these abnormalities are a result of abnormal metabolism (i.e., secondary to weight gain or loss). Alternately, upregulation of $\alpha_{2A}$AR may represent a vulnerability marker for depression in patients with eating disorders.

## VI. Premenstrual Dysphoric Disorder

Premenstrual changes in mood and anxiety levels have long been recognized in some women. The symptomatology of the severe form of premenstrual dysphoric disorder (PMDD) shares characteristics with mood and anxiety disorders.[390,391] Changes in platelet MAO activity in relationship to phase of menstrual cycle were inconsistently reported in normal women.[392–395] Similarly, inconsistent results were found in PMDD patients.[395,396] Only one[397] of six investigations[397–402] reported changes in $\alpha_{2A}$AR density in relationship to the phase of the menstrual cycle in healthy women. Increase in $\alpha_{2A}$AR density occurs during pregnancy. Failure of platelet $\alpha_{2A}$AR to downregulate in the postpartum period was found to be associated with "maternity blues."[403] One study reported increased $\alpha_{2A}$AR density in PMDD, with a small trend for differences between phases of the menstrual cycle.[404] In a study from our laboratory, we did not find any differences in $\alpha_{2A}$AR density, agonist affinity, or coupling to G$_i$ protein in PMDD.[405] Furthermore, there were no changes in any of the $\alpha_{2A}$AR binding parameters between phases of the menstrual cycle. However, we found multiple, significant correlations between $\alpha_{2A}$AR coupling and severity of anxiety and depression in the follicular phase in healthy controls, but not in PMDD patients. $\alpha_{2A}$AR density in the follicular phase predicted severity of symptoms in the luteal phase in PMDD patients, perhaps suggesting that changes in receptor function may lag behind hormonal changes during menstrual cycle and pointing out the need to investigate PMDD over the course of consecutive cycles. The lack of changes in $\alpha_{2A}$AR density in relation to phase of menstrual cycle was later replicated.[406] If increased $\alpha_{2A}$AR density represents a vulnerability marker for depression, as seen in other psychiatric disorders, it is peculiar that no similar changes were observed in PMDD. Future studies in PMDD should focus on the effects of sex steroid hormones on the regulation of monoaminergic receptor regulation, G protein expression and function, and signal transduction mechanisms.

## VII. Schizophrenia

Schizophrenia is a severe psychiatric disorder that tends to run a chronic debilitating course with significant deterioration in function overtime. Despite the existence of multiple subtypes, schizophrenia is characterized by perceptual abnormalities (such as auditory or visual hallucinations), strongly held erroneous beliefs or delusions (such as delusions of persecution or reference), abnormal thought processes, or severely disorganized behavior. Schizophrenia is associated with severe impairment of psychosocial functioning and downward socioeconomic drift.

## A. cAMP Signaling Studies

Few studies have examined platelet $\alpha_{2A}AR$ in schizophrenia. Two studies reported increased $\alpha_{2A}AR$ density in schizophrenic patients.[407,408] Decreased norepinephrine-mediated inhibition of cAMP and norepinephrine-mediated inhibition of $PGE_1$-stimulated cAMP were also found in schizophrenic patients despite increased $\alpha_{2A}AR$ density, suggesting a dissociation between receptor density and receptor-mediated function. Another study found decreased $\alpha_{2A}AR$-mediated function despite normal $\alpha_{2A}AR$ density.[409] Finally, two studies from the same laboratory reported decreased $\alpha_{2A}AR$ density in schizophrenia.[410,411] A further decrease in $\alpha_{2A}AR$ density was seen in treatment responders after chlorpromazine, while no change was seen in patients lacking clinical response.[410] Except for one study,[412] platelet $PGE_1$-stimulated cAMP responses were also decreased in schizophrenics (either during exacerbation or in remission),[413–415] suggesting decreased PGE receptor sensitivity. Such decreased sensitivity would imply enhanced catecholaminergic transmission, particularly dopaminergic transmission. Also, lower cAMP responses mediated either through $\alpha_{2A}AR$ or PGE receptors led to the investigation of platelet cAMP-dependent protein kinase. Decreased platelet AC regulatory subunit type I and type II was found in schizophrenia, but density of the catalytic subunit was normal.[416] In addition, forskolin-stimulated AC was higher in schizophrenics and was further enhanced by PKC-dependent phosphorylation.[417] [$^3$H]cAMP binding to PKA and $G_s$ and $G_i$ levels were increased in postmortem brains in schizophrenia.[418] There were no differences, however, in Rap1 or phosphorylated Rap1 levels in schizophrenia[118] or in epinephrine-stimulated GTPase activity, which terminates the $G_\alpha$ subunit-mediated signal.[148] Collectively, these results suggest abnormal cAMP signaling as part of the pathophysiology of schizophrenia.

## B. Serotonergic Studies

The indolamine hypothesis of schizophrenia derives support from the fact that hallucinogens exert their effects partly through modulating central serotonergic transmission. Serotonin studies in schizophrenia investigated platelet serotonin content, 5-HTTP, serotonin reuptake, and $5\text{-HT}_{2A}$ receptors. High platelet and plasma serotonin content has been inconsistently reported in schizophrenia. Platelet 5-HT content was increased in schizophrenics in comparison to controls and depressed patients and was not correlated with plasma cortisol or prolactin levels as observed in controls.[248] Both decreased[419–422] and normal[423] platelet serotonin reuptake were also reported in schizophrenia. Platelet serotonin content was low in treated schizophrenic patients and was not different between positive and negative symptom schizophrenics.[424] 5-HTTP density was normal in schizophrenia[425,426] and was not related to aggression,[427] although another study reported increased 5-HTTP density as a state maker in aggressive schizophrenics.[428] Increased 5-HTTP density in drug-free schizophrenics was decreased after antipsychotic therapy.[429] Dissociation between increased platelet serotonin content and normal serotonin uptake has been observed in other medical conditions (e.g., cirrhosis) and may be part of the pathophysiology of schizophrenia.[430] Finally, one study reported increased dopamine uptake into platelet dense granules in schizophrenic patients.[431] However, the methodology is controversial[432] and the exact mechanism underlying platelet dopamine uptake remains unclear.

By the use of various ligands, increased platelet $5\text{-HT}_{2A}$ receptor density has been reported in schizophrenic patients.[129,433–437] This upregulation is consistent with a similar finding in postmortem brains of schizophrenic patients.[438] Upregulation of $5\text{-HT}_{2A}$ receptors may also be consistent with studies showing increased basal and agonist-mediated hydrolysis of $PIP_2$ and DAG, and phosphatidic acid accumulation in schizophrenia.[439,440] However, some studies did not find differences in pre- or posttreatment $IP_3$,[436] while another found decreased serotonin-mediated platelet aggregation response, which increased after treatment with antipsychotics[441] as did $5\text{-HT}_{2A}$ receptor density.[129,436] These findings suggest that upregulation of $5\text{-HT}_{2A}$ receptor density does not translate into enhanced IP activation. Treatment of schizophrenics with antipsychotics was found to decrease platelet $5\text{-HT}_{2A}$ receptor density or receptor-mediated changes in $Ca^{++}$[434,437,442,443] or increase $5\text{-HT}_{2A}$ receptor density.[435,436,444–447] Upregulation of platelet $5\text{-HT}_{2A}$ receptors in schizophrenia may not be consistent with studies showing decreased[448,449] or normal[450] brain $5\text{-HT}_{2A}$. It is unclear if differences in the ligands used could have contributed to such inconsistency or if it is due to dissociation between $5\text{-HT}_{2A}$ receptor density and function. Alternately, receptor density in the high-conformational state, which couples $G_{q/11}$ protein, may correspond more accurately with receptor-mediated function. To date, most studies have only measured total receptor density.

## C. MAO Studies

Several factors have contributed to investigation of MAO activity in schizophrenia. Subjects with low platelet MAO appeared to have a higher incidence of psychopathology, drug abuse, and family psychiatric history compared to subjects with high platelet MAO activity.[357,451–453] Genetic control accounted for 70 to 80% of the variation in MAO activity.[298,454] First-degree relatives of schizophrenics with low MAO activity also have lower MAO activity than controls,[455] and parent-offspring and sibling-sibling correlations[456] strengthened the argument for genetic control of MAO. However, environmental factors could not be ruled out,[457–460] and genetic mutations have not been detected.[461] Finally, MAO activity in normal volunteers with low MAO activity and in chronic schizophrenics was positively correlated with plasma prolactin levels, suggesting that platelet MAO activity reflects central monoaminergic neuronal activity in the tuberoinfundibular dopaminergic system.[462] Low MAO activity would thus predict increased central monoamine levels including dopamine, which has been implicated in the pathophysiology of schizophrenia. In this context, platelet vesicular monoamine transporter ($VMAT_2$) density was increased in schizophrenia, further suggesting hyperactivity of monoaminergic transmission[463] may be part of the pathophysiology of schizophrenia or an adaptive response to treatment.

Low MAO activity in schizophrenia has been reported in numerous investigations.[297,304,464–483] Low MAO among schizophrenic patients correlated with prognosis[471,481] and exists in patients with chronic schizophrenia,[469,478,479,482] in patients with Schiederian hallucinations,[472] in patients with active delusions and hallucinations,[478] in patients with high score positive and low score negative symptoms,[476] in males,[477] and in treated schizophrenics.[424] No relationship was found to severity of illness.[467] These studies contrast with many other studies that failed to find decreased MAO activity in schizophrenic patients.[303,484–496] The sources contributing to this striking inconsistency have been discussed in detail[497,498] but include differences in the substrate employed, the use of saturating concentrations, the method used for separating the platelets, gender, self-reported mood, and, probably most important, the heterogeneity of the disorder itself. It is likely that low MAO activity exists in a subgroup of schizophrenic patients; however, the inconsistency in existing studies leaves unclear the exact role of MAO in the pathophysiology of schizophrenia.

### D. Platelet Activation, Metabolic, and Other Miscellaneous Investigations

Antipsychotic drugs have been shown to depress platelet aggregation responses *in vitro*.[499] Curiously, while drug-free schizophrenic patients had normal aggregation responses and thrombin-stimulated GTPase activity of $G_q$ in platelets,[148] treatment with neuroleptics increased aggregation responses to serotonin, an effect that faded after four weeks of treatment. These observations suggest a drug-disease

interaction. The half-life of platelets in schizophrenics was similar to that observed in aging and significantly shorter than in controls.[500] A significant increase in collagen-induced platelet aggregation[501] in both drug-free and medicated schizophrenics may reflect decreased membrane fluidity. $PLA_2$ activity was increased in schizophrenia and decreased after neuroleptic treatment,[502,503] suggesting an accelerated breakdown of membrane phospholipids in schizophrenia.[502] Such an abnormal membrane environment is likely to decrease dopamine-mediated signal transmission but also similarly to decrease other membrane receptor functions. Decreased arachidonic acid incorporation into platelet membrane lipids in schizophrenics was reported.[504] This decrease was more prominent in schizophreniform disorder and to a lesser extent in chronic schizophrenics,[505,506] thus providing another line of evidence for decreased membrane fluidity and enhanced phosphatidylinositol turnover. However, it has not been established whether enhanced phosphatidylinositol turnover in schizophrenia is secondary to enhanced acetylcholine, serotonin, or dopamine receptor function.

Phosphatidylcholine and phosphatidylethanolamine levels were decreased and lysophosphatidylcholine was increased in schizophrenia, suggesting an increased breakdown of phospholipids in schizophrenia.[507] Antipsychotic treatment was associated initially with a decrease in lysophosphatidylcholine, only to increase again with chronic treatment. Serum brain-derived neurotrophic factor (BDNF), one of several neurotrophic factors that regulate neuronal development and synaptic plasticity, was decreased in schizophrenic patients.[508] Decreased cerebral blood flow and altered energy metabolism has been reported in various brain regions in schizophrenia. Platelet mitochondrial complex I provides a good model for investigating energy metabolism in schizophrenia. Increased complex I activity was associated with positive psychotic symptoms and was positively correlated with their severity, whereas its decrease was associated with residual or negative symptoms.[509] Platelet nitrite content in schizophrenics was not altered, but polymorphonuclear leukocyte nitric oxide (NO) synthesis was decreased, which might contribute to oxidative stress.[510] Chronic overrelease of 2-arachindonoyl-glycerol has been hypothesized to be causally related to high arousal and cognitive deficits in schizophrenia.[511] These studies collectively provide several lines of evidence implicating other cellular processes in the pathophysiology of schizophrenia, besides the role of the classic monoamine receptors which have been typically pursued in previous investigations.

Finally, hypoglutamatergic function has been hypothesized in schizophrenia and may also pertain to other psychotic states. Platelet NMDA receptor supersensitivity, as reflected in enhanced $Ca^{++}$ response to glutamate, was reported in schizophrenia and psychotically depressed patients.[512] Excitatory NMDA receptors and the subsequent intracellular $Ca^{++}$ release regulate several transcription factors and further investigations may provide evidence for abnormal gene expression or function. Finally, platelet peripheral benzodiazepine receptors were approximately 30% lower and were correlated with overtly aggressive behavior, hostility, and anxiety in schizophrenic patients, but were not correlated with subtypes of illness.[279] In summary, it is clear that the pathophysiology of schizophrenia is complex and involves multiple systems and processes.

## VIII. Childhood Developmental and Psychiatric Disorders

### A. Studies in Autism

Autistic disorder is a pervasive developmental disorder characterized with qualitative impairment in social interaction, communication, and restricted behavioral repertoires. Hyperserotonemia (increased blood or platelet serotonin levels) has been reported in autism in all studies[513–521] except one.[522] Platelet-poor plasma 5-HT levels were low in autistic patients and were inversely correlated with severity of overt aggression.[523] Hyperserotonemia was identified in 40 to 70% of subjects[516,520] and was found more frequently in probands with affected relatives than in those with unaffected relatives.[518] A familial transmission or genetic effect was also suggested by the correlation of blood serotonin levels in affected probands and their parents.[513] Furthermore, increased platelet serotonin uptake was found in infantile autism and appeared to be genetically determined.[524] Platelet 5-HT levels were significantly increased in patients with autism and pervasive developmental disorder, but not in mental retardation, and showed a bimodal distribution with a hyperserotonemic subgroup.[525] This biphasic pattern was also observed in response to a challenge with a carbohydrate-rich meal (which influences peripheral serotonin levels), after which autistic patients had a hyperserotonemic response, whereas normal controls had a linear increase in peripheral serotonin levels.[526] A study that investigated polymorphism both in the promoter region LPR and in the intron 2 region VNTR of the serotonin transporter gene found no relationship between gene polymorphism and 5-HT blood levels or susceptibility to autism.[527]

Increased 5-HT platelet uptake may be consistent with increased $^3H$-paroxetine binding in autism[528,529] and low platelet-poor plasma serotonin. However, [$^3H$]-imipramine binding was normal in two other studies.[425,530] MAO activity was decreased in pervasive developmental disorder.[531] Thus, hyperserotonemia may be secondary to increased serotonin uptake or decreased metabolism of serotonin by MAO, but it is less likely to be due to increased density of the serotonin transporter. One study investigated serotonin uptake in relationship to the serotonin transporter promoter gene polymorphism and reported high uptake rate $V_{max}$ was associated with the ll genotype and found a correlation between $V_{max}$ and $B_{max}$ of the transporter density, but there was no association between the ll genotype and hyperserotonemia in autism.[532] Platelet $5\text{-HT}_{2A}$ receptors were decreased in hyperserotonemic children.[515] Another study did not find differences in affected children, their siblings, or their parents in comparison to controls, but found an inverse correlation between plasma norepinephrine levels and $5\text{-HT}_{2A}$ receptor density, suggesting heterologous regulation of $5\text{-HT}_{2A}$ receptors by plasma norepinephrine levels.[533] Thus, there is evidence to support abnormal serotonin uptake or metabolism in autism.

### B. Studies in Attention Deficit Hyperactivity Disorder

Attention deficit hyperactivity disorder (ADHD) is characterized by a persistent pattern of decreased attention and hyperactivity or impulsivity that exceeds acknowledged norms in children. The locus ceruleus is involved in attention. However, only a small trend for decreased $\alpha_{2A}AR$ density was found in ADHD.[534] Increased platelet serotonin was found particularly in impulsive ADHD children.[535] MAO activity was decreased in hyperactive children with high impulsivity scores and was not changed after d-amphetamine treatment or related to clinical response.[534] 5-HTTP density was also normal and was not changed after methylphenidate treatment.[536] Moreover, platelet $5\text{-HT}_{2A}$ receptor density was normal in ADHD.[537]

### C. Studies in Conduct and Impulse Control Disorders

Subjects with conduct disorder (CD) display a pattern of behavior that reflects disregard for the rights of others and social norms. Subjects with conduct disorder often commit impulsive acts. Increased blood and platelet serotonin levels were reported in impulsive and conduct disorder adolescents.[535,538] A high incidence of impulse control disorder was found in patients with carcinoid syndrome whose plasma tryptophan levels were low and who were negatively correlated with urinary 5-HIAA.[539] It was suggested that serotonin production by the tumor depleted the tryptophan pool, which is essential for brain serotonin. Similar to ADHD, low MAO was found in CD,[355] which was inversely correlated with impulsivity[540] and with criminal behavior in delinquent juveniles.[541] One study reported a correlation between criminal recidivism, high T3, and low MAO activity, which was stable over time.[542] Two profiles or subtypes were found in subjects with criminal behavior: one was characterized with low MAO enzyme activity, cluster B traits, and drug abuse, whereas the other did not present these characteristics,[543] perhaps identifying some differences between conduct disorder and the later developments of antisocial personality disorder and criminal activity. However, two studies did not replicate these findings.[544,545] 5-HTTP was decreased in CD and ADHD and showed a significant inverse correlation with measures of impulsivity, sensation seeking, and aggression.[546,547] Decreased serotonin uptake was correlated with aggression in schizophrenics and CD adolescents.[237] Low 5-HT uptake rate ($V_{max}$) was significantly correlated with parent-rated aggression scores in conduct disorder boys.[548] 5-HT$_{2A}$ receptor density was decreased in boys raised with harsh parental physical punishment who were at risk for delinquency.[549] Collectively, these observations suggest an overlap between the phenomenology and neurobiology, as reflected in serotonergic function, between ADHD and CD.

### D. Studies in Childhood Anxiety and Depressive Disorders

Platelet markers in childhood depressive disorders have been investigated. Similar to depression in adults, decreased 5-HTTP density was reported in depressed and enuretic children/adolescents in the majority of studies,[238,550–553] with some exceptions.[554,555] Significant seasonal variations in 5-HTTP, reported in some studies in normal adults and also observed in a study in normal adolescents, were absent in suicidal adolescents.[556] Increased platelet $\alpha_{2A}AR$ was reported in juvenile depressed patients, resembling the findings in depressed adults.[554]

Fewer platelet studies have been performed on anxiety disorders in children. Platelet MAO was positively correlated with self-ratings of anxiety in children,[557] reminiscent of similar findings in adults. Peripheral benzodiazepine binding in platelets from children with anxiety and PTSD was decreased.[558] 5-HTTP density was decreased in adolescents with OCD.[103,104] Thus, there are some similarities between children and adults with respect to platelet studies in anxiety and depressive disorders.

## IX. Conclusions

This chapter has summarized three decades of psychiatric studies that have used platelets as a research tool. Despite some inconsistencies, certain platelet abnormalities have been identified, including abnormal G protein levels and function in mood disorders; upregulation of $\alpha_{2A}AR$ in anxiety, depressive disorders, and related disorders with high risk for depression; enhanced platelet procoagulant activity in depression; decreased density of the serotonin transporter and increased sensitivity of 5-HT$_{2A}$ receptors in

MDD; and decreased AC function and related G protein abnormalities in subjects at high risk for alcoholism. Further research is required in these areas to clarify the underlying molecular mechanisms associated with these abnormalities so that a well-integrated picture of the pathophysiology of each psychiatric disorder emerges.

## Acknowledgments

The author thanks Ms. Shirley Campbell and Jennea Augsbury and the library staff at the Dallas VA Medical Center, without whose assistance this review would not have been possible.

## References

1. Abrams, W. B., & Solomon, H. M. (1969). The human platelet as a pharmacological model for the adrenergic neuron: The uptake and release of norepinephrine. *Clin Pharmacol Ther, 10,* 702–709.
2. Pletscher, A., Affolter, H., Cesura, M., et al. (1984). In H. G. Schlossberger, W. Kochen, & B. Linzen (Eds.), *Progress in tryptophan and serotonin release* (pp. 231–239). Berlin: W. Steinhart de Gruyter and Company.
3. Sneddon, J. M. (1973). Blood platelets as a model for monoamine-containing neurones. *Prog Neurobiol, 1,* 1598.
4. Stahl, S. M. (1977). The human platelet: A diagnostic and research tool for the study of biogenic amines in psychiatric and neurologic disorders. *Arch Gen Psychiatry, 34,* 509–516.
5. Wirz-Justice, A. (1988). Platelet research in psychiatry. *Experientia, 44,* 145–152.
6. García-Sevilla, J. A., Hollingsworth, P. J., & Smith, C. B. (1981). Alpha-2 adrenoreceptors on human platelets: Selective labeling by [³H] clonidine and [³H] yohimbine and competitive inhibition by antidepressant drugs. *Eur J Pharmacol, 74,* 329–341.
7. Bylund, D. B., Blaxall, H. S., Iversen, L. J., et al. (1992). Pharmacological characteristics of $\alpha_2$-adrenergic receptors: Comparison of pharmacologically defined subtypes with subtypes identified by molecular cloning. *Mol Pharmacol, 42,* 1–5.
8. Ordway, G. A., Jaconetta, S. M., & Halaris, A. E. (1993). Characterization of subtypes of alpha-2 adrenoceptors in the human brain. *J Pharmacol, 264,* 967–976.
9. Simonds, W. F., Goldsmith, P. K., Codina, J., et al. (1989). $G_{i2}$ mediates $\alpha_2$-adrenergic inhibition of adenylyl cyclase in platelet membranes: *In situ* identification with $G_\alpha$ C-terminal antibodies. *Proc Natl Acad Sci U S A, 86,* 7809–7813.
10. Cook, N., Nahorski, S. R., & Barnett, D. B. (1985). (−)-[¹²⁵I]Pindolol binding to the human platelet β-adrenoceptor: Characterisation and agonist interactions. *Eur J Pharmacol, 113,* 247–254.
11. Wang, X. L., & Brodde, O.-E. (1985). Identification of a homogeneous class of β₂-adrenoceptors in human platelets

by $(-)$-$^{125}$I-iodopindolol binding. *J Cyclic Nucleotide Protein Phosphor Res, 10,* 439–450.

12. Winther, K., Klysner, R., Geisler, A., et al. (1985). Characterization of human platelet beta-adrenoceptors. *Thromb Res, 40,* 757–767.

13. Smrcka, A. V., Hepler, J. R., Brown, K. O., et al. (1991). Regulation of polyphosphoinositide-specific phospholipase C activity by purified $G_q$. *Science, 251,* 804–807.

14. Conn, P. J., & Sanders-Bush, E. (1986). Regulation of serotonin-stimulated phosphoinositide hydrolysis: Relation to the serotonin 5-HT$_2$ binding site. *J Neurosci, 6,* 3669–3675.

15. Ivins, K. J., & Molinoff, P. B. (1990). Serotonin-2 receptors coupled to phosphoinositide hydrolysis in a clonal cell line. *Mol Pharmacol, 37,* 622–630.

16. Gurguis, G. N. M., Phan, S. P., & Blakeley, J. E. (1998). Characteristics of agonist displacement of $^3$H-ketanserin binding to platelet 5-HT$_{2A}$ receptors: Implications for psychiatric research. *Psychiatry Res, 80,* 227–238.

17. Hoffman, B. J., Mezey, E., & Brownstein, M. J. (1991). Cloning of a serotonin transporter affected by antidepressants. *Science, 254,* 579–580.

18. Lesch, K. P., Wolozin, B. L., Murphy, D. L., et al. (1993). Primary structure of the human platelet serotonin uptake site: Identity with the brain serotonin transporter. *J Neurochem, 60,* 2319–2322.

19. Zucker, M., Aviv, A., Shelef, A., et al. (2002). Elevated platelet vesicular monoamine transporter density in untreated patients diagnosed with major depression. *Psychiatry Res, 3,* 251–256.

20. Bianchi, M., Moser, C., Lazzarini, C., et al. (2002). Forced swimming test and fluoxetine treatment: In vivo evidence that peripheral 5-HT in tat platelet-rich plasma mirrors cerebral extracellular 5-HT levels, whilst 5-HT in isolated platelets mirrors neuronal 5-HT changes. *Exp Brain Res, 2,* 191–197.

21. Maguire, K., Tuckwell, V., Pereira, A., et al. (1993). Significant correlation between $^{14}$C-5-HT uptake by and $^3$H-paroxetine binding to platelets from healthy volunteers. *Biol Psychiatry, 34,* 356–360.

22. Cheetham, S. C., Viggers, J. A., Slater, N. A., et al. (1993). [$^3$H]Paroxetine binding in rat frontal cortex strongly correlates with [$^3$H]5-HT uptake: Effect of administration of various antidepressant treatments. *Neuropharmacology, 32,* 737–743.

23. Piletz, J. E., & Sletten, K. (1993). Nonadrenergic imidazoline binding sites on human platelets. *J Pharmacol Exp Ther, 267,* 1493–1502.

24. Piletz, J. E., Andorn, A. C., Unnerstall, J. R., et al. (1991). Binding of [$^3$H]-p-aminoclonidine to $\alpha_2$-adrenoceptor states plus a non-adrenergic site on human platelet plasma membranes. *Biochem Pharmacol, 42,* 569–584.

25. Gavish, M., Weizman, A., Karp, L., et al. (1986). Decreased peripheral benzodiazepine binding sites in platelets of neuroleptic-treated schizophrenics. *Eur J Pharmacol, 121,* 275–279.

26. Berk, M., Plein, H., & Csizmadia, T. (1999). Supersensitive platelet glutamate receptors as a possible peripheral marker in schizophrenia. *Int Clin Psychopharmacol, 14,* 119–122.

27. Sethi, B. B., Kumar, P., Agarwal, A. K., et al. (1986). $^3$H-Spiperone binding in platelet membranes: A possible biological marker for schizophrenia. *Acta Psychiatr Scand, 73,* 186–190.

28. Cattabeni, F., Colciaghi, F., & Di Luca, M. (2004). Platelets provide human tissue to unravel pathogenic mechanisms in Alzheimer disease. *Prog Neuropsychopharmacol Biol Psychiatry, 5,* 763–770.

29. Redmond, D. E., Jr. (1987). In H. Y. Meltzer (Ed.), *Psychopharmacology: The third generation of progress* (pp. 967–975). New York: Raven Press.

30. Cameron, O. G., Smith, C. B., Hollingsworth, P. J., et al. (1984). Platelet $\alpha_2$-adrenergic receptor binding and plasma catecholamines. *Arch Gen Psychiatry, 41,* 1144–1148.

31. Cameron, O. G., Smith, C. B., Lee, M. A., et al. (1990). Adrenergic status in anxiety disorders: Platelet alpha$_2$-adrenergic receptor binding, blood pressure, pulse, and plasma catecholamines in panic and generalized anxiety disorder patients and in normal subjects. *Biol Psychiatry, 28,* 3–20.

32. Cameron, O. G., Smith, C. B., Nesse, R. M., et al. (1996). Platelet $\alpha_2$-adrenoreceptors, catecholamines, hemodynamic variables, and anxiety in panic patients and their asymptomatic relatives. *Psychosom Med, 58,* 289–301.

33. Albus, M., Bondy, B., & Ackenheil, M. (1986). Adrenergic receptors on blood cells: Relation to the pathophysiology of anxiety. *Clin Neuropharmacol, 9,* 359–361.

34. Sevy, S., Papadimitriou, G. N., Surmont, D. W., et al. (1989). Noradrenergic function in generalized anxiety disorder, major depressive disorder, and healthy subjects. *Biol Psychiatry, 25,* 141–152.

35. Southwick, S. M., Yehuda, R., Giller, E. L., et al. (1990). Altered platelet $\alpha_2$-adrenergic receptor binding sites in borderline personality disorder. *Am J Psychiatry, 147,* 1014–1017.

36. Charney, D. S., Innis, R. B., Duman, R. S., et al. (1989). Platelet alpha-2-receptor binding and adenylate cyclase activity in panic disorder. *Psychopharmacology, 98,* 102–107.

37. Norman, T. R., Kimber, N. M., Judd, F. K., et al. (1987). Platelet $^3$H-rauwolscine binding in patients with panic attacks. *Psychiatry Res, 22,* 43–48.

38. Nutt, D. J., & Fraser, S. (1987). Platelet binding studies in panic disorder. *J Affect Disord, 12,* 7–11.

39. Butler, J., O'Halloran, A., & Leonard, B. E. (1992). The Galway study of panic disorder II: Changes in some peripheral markers of noradrenergic and serotonergic function in DSM III-R panic disorder. *J Affect Disord, 26,* 89–100.

40. Kim, M. H., & Neubig, R. R. (1987). Membrane reconstitution of high-affinity $\alpha_2$ adrenergic agonist binding with guanine nucleotide regulatory proteins. *Biochemistry, 26,* 3664–3672.

41. Gurguis, G. N. M., Antai-Otong, D., Vo, S. P., et al. (1999). Adrenergic receptor function in panic disorder: I. Platelet $\alpha_2$ receptors: $G_i$ protein coupling, effects of imipramine, and relationship to treatment outcome. *Neuropsychopharmacology, 20,* 162–176.

42. Bhatia, S. C., Hsieh, H. H., Theesen, K. A., et al. (1991). Platelet alpha-2 adrenoreceptor activity pre-treatment and post-treatment in major depressive disorder with melancholia. *Res Commun Chem Pathol Pharmacol, 74,* 47–57.

43. Campbell, I. C., McKernan, R. M., Checkley, S. A., et al. (1985). Characterization of platelet alpha$_2$ adrenoceptors and measurement in control and depressed subjects. *Psychiatry Res, 14,* 17–31.

44. Cooper, S. J., Kelly, J. G., & King, D. J. (1985). Adrenergic receptors in depression: Effects of electroconvulsive therapy. *Br J Psychiatry, 147,* 23–29.

45. Healy, D. T., Paykel, E. S., Whitehouse, A. M., et al. (1991). Platelet [$^3$H]imipramine and $\alpha_2$-adrenoceptor binding in normal subjects during desipramine administration and withdrawal. *Neuropsychopharmacology, 4,* 117–124.

46. Kaneko, M., Kanno, T., Honda, K., et al. (1992). Platelet alpha-2 adrenergic receptor binding and plasma free 3-methoxy-4-hydroxyphenyl-ethylene glycol in depressed patients before and after treatment with mianserin. *Neuropsychobiology, 25,* 14–19.

47. Katona, C. L. E., Theodorou, A. E., Davies, S. L., et al. (1989). [$^3$H]Yohimbine binding to platelet $\alpha_2$-adrenoceptors in depression. *J Affect Disord, 17,* 219–228.

48. Lenox, R. H., Ellis, J. E., Van Riper, D. A., et al. (1983). In E. Usdin, M. Goldstein, A. J. Friedhoff, & A. Georgotas, (Eds.), Frontiers in neuropsychiatric research (pp. 331–356). London: Macmillan Press.

49. Pandey, G. N., Janicak, P. G., Javaid, J. I., et al. (1989). Increased $^3$H-clonidine binding in the platelets of patients with depressive and schizophrenic disorders. *Psychiatry Res, 28,* 73–88.

50. Pimoule, C., Briley, M. S., Gay, C., et al. (1983). $^3$H-Rauwolscine binding in platelets from depressed patients and healthy volunteers. *Psychopharmacology, 79,* 308–312.

51. Siever, L. J., Kafka, M. S., Insel, T. R., et al. (1983). Effect of long-term clorgyline administration on human platelet alpha-adrenergic receptor binding and platelet cyclic AMP responses. *Psychiatry Res, 9,* 37–44.

52. Stahl, S. M., Lemoine, P. M., Ciaranello, R. D., et al. (1983). Platelet alpha$_2$-adrenergic receptor sensitivity in major depressive disorder. *Psychiatry Res, 10,* 157–164.

53. Werstiuk, E. S., Auffarth, S. E., Coote, M., et al. (1992). Platelet $\alpha_2$-adrenergic receptors in depressed patients and healthy volunteers: The effects of desipramine. *Pharmacopsychiatry, 25,* 199–206.

54. Wolfe, N., Gelenberg, A. J., & Lydiard, R. B. (1989). Alpha$_2$-adrenergic receptor sensitivity in depressed patients: Relation between $^3$H-yohimbine binding to platelet membranes and clonidine-induced hypotension. *Biol Psychiatry, 25,* 382–392.

55. Gurguis, G. N. M., Turkka, J., George, D., et al. (1997). β-Adrenoreceptor coupling to G$_s$ protein in alcohol dependence, panic disorder, and patients with both conditions. *Neuropsychopharmacol, 16,* 69–76.

56. Gurguis, G. N. M., Blakeley, J. E., Antai-Otong, D., et al. (1999). Adrenergic receptor function in panic disorder. II. Neutrophil β$_2$ receptors: G$_s$ protein coupling, effects of imip-

57. Tardito, D., Zanardi, R., Racagni, G., et al. (2002). The protein kinase A in platelets from patients with panic disorder. *Eur Neuropsychopharmacol, 5,* 483–487.

58. Maes, M., Van Gastel, A., Delmeire, L., et al. (2002). Platelet alpha2-adrenoceptor density in humans: Relationships to stress induced anxiety, psychosthenic constitution, gender and stress-induced changes in the inflammatory response system. *Psychol Med, 5,* 919–928.

59. Gurguis, G. N. M., Andrews, R., Antai-Otong, D., et al. (1999). Platelet $\alpha_2$-adrenergic receptor coupling efficiency to G$_i$ protein in subjects with post-traumatic stress disorder and normal controls. *Psychopharmacology, 141,* 258–266.

60. Perry, B. D., Giller, E. L., Jr., & Southwick, M. S. (1987). Altered platelet alpha-2-adrenergic binding sites in post-traumatic stress disorder. *Am J Psychiatry, 144,* 1511–1512.

61. Weizmann, R., Gur, E., Laor, N., et al. (1994). Platelet adenylate cyclase activity in Israeli victims of Iraqi scud missile attacks with post-traumatic stress disorder. *Psychopharmacology, 114,* 509–512.

62. Kohn, Y., Newman, M. E., Lerer, B., et al. (1995). Absence of reduced platelet adenylate cyclase activity in Vietnam veterans with PTSD. *Biol Psychiatry, 37,* 205–208.

63. Lee, M. A., Cameron, O. G., Gurguis, G. N. M., et al. (1990). Alpha$_2$-adrenoreceptor status in obsessive-compulsive disorder. *Biol Psychiatry, 27,* 1083–1093.

64. Marazziti, D., Baroni, S., Masala, I., et al. (2004). Platelet alpha2-adrenoreceptors in obsessive-compulsive disorder. *Neuropsychobiology, 2,* 81–83.

65. Dell'Osso, L., Carmassi, C., Palego, L., et al. (2004). Serotonin-mediated cyclic AMP inhibitory pathway in platelets of patients affected by panic disorder. *Neuropsychobiology, 1,* 28–36.

66. Flaskos, J., Theophilopoulos, N., & George, A. J. (1989). Platelet monoamine oxidase activity and 5-hydroxytryptamine uptake in agoraphobic patients. *Br J Psychiatry, 155,* 680–685.

67. Gorman, J., Liebowitz, M. R., Fyer, A. J., et al. (1985). Platelet monoamine oxidase activity in patients with panic disorder. *Biol Psychiatry, 20,* 852–857.

68. Mathew, R. J., Ho, B. T., Kralik, P., et al. (1981). Catecholamines and monoamine oxidase activity in anxiety. *Acta Psychiatr Scand, 63,* 245–252.

69. Yu, P. H., Bowen, R. C., Davis, B. A., et al. (1983). A study of the catabolism of trace amines in mentally disordered individuals with particular reference to agoraphobic patients with panic attacks. *Prog in Neuropsychopharmacol Biol Psychiatry, 7,* 611–615.

70. Kahn, A., Lee, E., Dager, S., et al. (1986). Platelet MAO-B activity in anxiety and depression. *Biol Psychiatry, 21,* 847–849.

71. Norman, T. R., Acevedo, A., McIntyre, I. M., et al. (1988). A kinetic analysis of platelet monoamine oxidase activity in patients with panic attacks. *J Affect Disord, 15,* 127–130.

72. Balon, R., Rainey, J. M., Pohl, R., et al. (1987). Platelet monoamine oxidase activity in panic disorder. *Psychiatry Res, 22,* 37–41.

73. Kozaric-Kovacic, D., Ljubin, T., Rutic-Puz, L., et al. (2000). Platelet monoamine oxidase, ego strength, and neuroticism is soldiers with combat-related current posttraumatic stress disorder. *Croat Med J, 1,* 76–80.

74. Cicin-Sain, L., Mimica, N., Hranilovic, D., et al. (2000). Posttraumatic stress disorder an platelet serotonin measures. *J Psychiatr Res, 2,* 155–161.

75. Pivac, N., Muck-Seler, D., Sagud, M., et al. (2002). Platelet serotonergic markers in posttraumatic stress disorder. *Prog Neuropsychopharmacol Biol Psychiatry, 6,* 1193–1198.

76. Faludi, G., Tekes, K., & TΛthfalusi, L. (1994). Comparative study of platelet $^3$H-paroxetine and $^3$H-imipramine binding in panic disorder patients and healthy controls. *J Psychiatry Neurosci, 19,* 109–113.

77. Lewis, D. A., Noyes, R. Jr., Coryell, W., et al. (1985). Tritiated imipramine binding to platelets is decreased in patients with agoraphobia. *Psychiatry Res, 16,* 1–9.

78. Marazziti, D., Rotondo, A., Placidi, G. F., et al. (1988). Imipramine binding in platelets of patients with panic disorder. *Pharmacopsychiatry, 2,* 47–49.

79. Marazziti, D. (1989). Imipramine binding in panic disorder. *Pharmacopsychiatry, 22,* 128–129.

80. Marazziti, D., Rossi, A., Dell'Osso, L., et al. (1999). Decreased platelet $^3$H-paroxetine binding in untreated panic disorder patients. *Life Sci, 65,* 2735–2741.

81. Pecknold, J. C., Chang, H., Fleury, D., et al. (1987). Platelet imipramine binding in patients with panic disorder and major familial depression. *J Psychiatr Res, 21,* 319–326.

82. Innis, R. B., Charney, D. S., & Heninger, G. R. (1987). Differential $^3$H-imipramine platelet binding in patients with panic disorder and depression. *Psychiatry Res, 21,* 33–41.

83. Stein, M. B., Delaney, S. M., Chartier, M. J., et al. (1995). [$^3$H]Paroxetine binding to platelets of patients with social phobia: Comparison to patients with panic disorder and healthy volunteers. *Biol Psychiatry, 37,* 224–228.

84. Uhde, T. W., Berrettini, W. H., Roy-Byrne, P. P., et al. (1987). Platelet [$^3$H]imipramine binding in patients with panic disorder. *Biol Psychiatry, 22,* 52–58.

85. Pecknold, J. C., Suranyi-Cadotte, B., Chang, H., et al. (1988). Serotonin uptake in panic disorder and agoraphobia. *Neuropsychopharmacology, 1,* 173–176.

86. McIntyre, I. M., Judd, F. K., Burrows, G. D., et al. (1989). Serotonin in panic disorder: Platelet uptake and concentration. *Int Clin Psychopharmacol, 4,* 1–6.

87. Norman, T. R., Judd, F. K., Gregory, M., et al. (1986). Platelet serotonin uptake in panic disorder. *J Affect Disord, 11,* 69–72.

88. Neuger, J., Wistedt, B., Aberg-Wistedt, A., et al. (2002). Effect of citalopram treatment on relationship between platelet serotonin functions and the Karlinska scales of personality in panic patients. *J Clin Psychopharmacol, 4,* 400–405.

89. Maguire, K., Norman, T., Burrows, G., et al. (1998). Platelet paroxetine binding in post-traumatic stress disorder. *Psychiatry Res, 77,* 1–7.

90. Mellman, T. A., & Kumar, A. M. (1994). Platelet serotonin measures in posttraumatic stress disorder. *Psychiatry Res, 53,* 99–101.

91. Muck-Seler, D., Pivac, N., Jakoljevic, M., et al. (2003). Platelet 5-HT concentration and comorbid depression in war veterans with and without posttraumatic stress disorder. *J Affect Disord, 2,* 171–179.

92. Arora, R. C., Fichtner, C. G., O'Connor, F., et al. (1993). Paroxetine binding in the blood platelets of post-traumatic stress disorder patients. *Life Sci, 53,* 919–928.

93. Fichtner, C. G., Arora, R. C., O'Connor, F. L., et al. (1994). Platelet paroxetine binding and fluoxetine pharmacotherapy in posttraumatic stress disorder: Preliminary observations on a possible predictor of clinical treatment response. *Life Sci, 54,* 39–44.

94. Lesch, K. P., Bengel, D., Heils, A., et al. (1996). Association of anxiety-related traits with a polymorphism in the serotonin transporter gene regulatory region. *Science, 274,* 1527–1531.

95. Butler, J., Tannion, M., O'Rourke, D., et al. (1988). In R. H. Belmaker, M. Sandler, & A. Dahlstrom (Eds.), *Progress in catecholamine research part C: Clinical aspects* (pp. 399–407). New York: Alan R. Liss.

96. Nugent, D. F., Mannion, L., & Leonard, B. E. (1996). The Galway study of panic disorder IV: Changes in platelet serotonin uptake in a 5–6 year follow up study of DSM III-R panic disorder. *J Serotonin Res, 4,* 259–265.

97. Neuger, J., Wistedt, B., Sinner, B., et al. (2000). The effect of citalopram treatment on platelet serotonin function in panic disorders. *Int Clin Psychopharmacol, 15,* 83–91.

98. Martini, C., Trincavelli, M. L., Tuscano, D., et al. (2004). Serotonin-mediated phosphorylation of extracellular regulated kinases in platelets of patients with panic disorder versus controls. *Neurochem Int, 4,* 475–483.

99. Cath, D. C., Spinhoven, P., Landman, A. D., et al. (2001). Psychopathology and personality characteristics in relation to blood serotonin in Tourette's syndrome and obsessive-compulsive disorder. *J Psychopharmacol, 2,* 111–119.

100. Bastani, B., Arora, R. C., & Meltzer, H. Y. (1991). Serotonin uptake and imipramine binding in the blood platelets of obsessive-compulsive disorder patients. *Biol Psychiatry, 30,* 131–139.

101. Marazziti, D., Hollander, E., Lensi, P., et al. (1992). Peripheral markers of serotonin and dopamine function in obsessive-compulsive disorder. *Psychiatry Res, 42,* 41–51.

102. Marazziti, D., Dell'Osso, L., Presta, S., et al. (1999). Platelet [$^3$H]paroxetine binding in patients with OCD-related disorders. *Psychiatry Res, 89,* 223–228.

103. Sallee, F. R., Richman, H., Beach, K., et al. (1996). Platelet serotonin transporter in children and adolescents with obsessive-compulsive disorder of Tourette's syndrome. *J Am Acad Child Adolesc Psychiatry, 35,* 1647–1656.

104. Weizman, A., Carmi, M., Hermesh, H., et al. (1986). High-affinity imipramine binding and serotonin uptake in platelets of eight adolescents and ten adult obsessive-compulsive patients. *Am J Psychiatry, 143,* 335–339.

105. Weizman, A., Mandel, A., Barber, Y., et al. (1992). Decreased platelet imipramine binding in Tourette syndrome children with obsessive-compulsive disorder. *Biol Psychiatry, 31,* 705–711.

106. Delorme, R., Chabane, N., Callebert, J., et al. (2004). Platelet serotonergic predictors of clinical improvement in obsessive-compulsive disorder. *J Clin Psychopharmacol, 1,* 18–23.

107. Black, D. W., Kelly, M., Myers, C., et al. (1990). Tritiated imipramine binding in obsessive-compulsive volunteers and psychiatrically normal controls. *Biol Psychiatry, 27,* 319–327.

108. Kim, S. W., Dysken, M. W., Pandey, G. N., et al. (1991). Platelet $^3$H-imipramine binding sites in obsessive-compulsive behavior. *Biol Psychiatry, 30,* 467–474.

109. Nelson, E. C., Sheline, Y. I., Bardgett, M. E., et al. (1995). Platelet paroxetine binding in major depressive disorder with and without comorbid obsessive-compulsive disorder. *Psychiatry Res, 58,* 117–125.

110. Pfohl, B., Black, D., Noyes, R. Jr., et al. (1990). A test of the tridimensional personality theory: Association with diagnosis and platelet imipramine binding in obsessive-compulsive disorder. *Biol Psychiatry, 28,* 41–46.

111. Vitiello, B., Shimon, H., Behar, D., et al. (1991). Platelet imipramine binding and serotonin uptake in obsessive-compulsive patients. *Acta Psychiatr Scand, 84,* 29–32.

112. Marazziti, D., Masala, I., Rossi, A., et al. (2000). Increased inhibitory activity of protein kinase C on the serotonin transporter in OCD. *Neuropsychobiology, 4,* 171–177.

113. Flament, M. F., Rapoport, J. L., Murphy, D. L., et al. (1987). Biochemical changes during clomipramine treatment of childhood obsessive-compulsive disorder. *Arch Gen Psychiatry, 44,* 219–225.

114. Hanna, G. L., Yuwiler, A., & Cantwell, D. P. (1991). Whole blood serotonin in juvenile obsessive-compulsive disorder. *Biol Psychiatry, 29,* 738–744.

115. Pandey, S. C., Kim, S. W., Davis, J. M., et al. (1993). Platelet serotonin-2 receptors in obsessive-compulsive disorder. *Biol Psychiatry, 33,* 367–372.

116. Marazziti, D., Dell'Osso, L., Masala, I., et al. (2002). Decreased inhibitory activity of PKC in OCD patients after six months of treatment. *Psychoneuroendocrinology, 7,* 769–776.

117. Perez, J., Tardito, D., Ravizza, L., et al. (2000). Altered cAMP-dependent protein kinase A in platelets of patients with obsessive-compulsive disorder. *Am J Psychiatry, 157,* 284–286.

118. Tardito, D., Maina, G., Tura, G. B., et al. (2001). The cAMP-dependent protein kinase substrate Rap1 in platelets from patients with obsessive-compulsive disorder or schizophrenia. *Eur Neuropsychopharmacol, 3,* 221–225.

119. Manji, H. K., Chen, G., Shimon, H., et al. (1995). Guanine nucleotide-binding proteins in bipolar affective disorder. *Arch Gen Psychiatry, 52,* 135–144.

120. Mitchell, P. B., Manji, H. K., Chen, G., et al. (1997). High levels of $G_{s\alpha}$ in platelets of euthymic patients with bipolar affective disorder. *Am J Psychiatry, 154,* 218–223.

121. Young, L. T., Li, P. P., Kish, S. J., et al. (1991). Postmortem cerebral cortex $G_s$ alpha-subunit levels are elevated in bipolar affective disorder. *Brain Res, 553,* 323–326.

122. Young, L. T., Li, P. P., Kish, S. J., et al. (1993). Cerebral cortex $G_s$ alpha protein levels and forskolin-stimulated cyclic AMP formation are increased in bipolar affective disorder. *J Neurochem, 61,* 890–898.

123. Perez, J., Tardito, D., Mori, S., et al. (1999). Abnormalities of cyclic adenosine monophosphate signaling in platelets from untreated patients with bipolar disorder. *Arch Gen Psychiatry, 56,* 248–253.

124. Tardito, D., Mori, S., Racagni, G., et al. (2003). Protein kinase A activity in platelets from patients with bipolar disorder. *J Affect Disord, 1–3,* 249–253.

125. Perez, J., Tardito, D., Mori, S., et al. (2000). Altered Rap1 endogenous phosphorylation and levels in platelets from patients with bipolar disorder. *J Psychiatr Res, 2,* 99–104.

126. Perez, J., Tardito, D., Racagni, G., et al. (2002). cAMP signaling pathway in depressed patients with psychotic features. *Mol Psychiatry, 2,* 208–212.

127. Perez, J., Tardito, D., Racagni, G., et al. (2001). Protein kinase A and Rap1 levels in platelets of untreated patients with major depression. *Mol Psychiatry, 1,* 44–49.

128. Rahman, S., Li, P. P., Young, T., et al. (1997). Reduced [$^3$H] cyclic amp binding in postmortem brain from subjects with bipolar affective disorder. *J Neurochem, 68,* 297–304.

129. Pandey, G. N., Pandey, S. C., Ren, X., et al. (2003). Serotonin receptors in platelets of bipolar and schizoaffective patients: Effect of lithium treatment. *Psychopharmacology, 2,* 115–123.

130. Dubovsky, S. L., Christiano, J., Daniell, L. C., et al. (1989). Increased platelet intracellular calcium concentration in patients with bipolar affective disorder. *Arch Gen Psychiatry, 46,* 632–638.

131. Dubovsky, S. L., Murphy, J., Thomas, M., et al. (1992). Abnormal intracellular calcium ion concentration in platelets and lymphocytes of bipolar patients. *Am J Psychiatry, 149,* 118–120.

132. Dubovsky, S. L., Thomas, M., Hijazi, A., et al. (1994). Intracellular calcium signalling in peripheral cells of patients with bipolar affective disorder. *Eur Arch Psychiatry Clin Neurosci, 243,* 229–234.

133. Kusumi, I., Koyama, T., & Yamashita, I. (1992). Thrombin-induced platelet calcium mobilization is enhanced in bipolar disorders. *Biol Psychiatry, 32,* 731–734.

134. Meagher, J., & Leonard, B. E. (1994). Increased 5HT-induced platelet aggregation in mania: Lack of change following neuroleptic medication. *Hum Psychopharmacol, 9,* 337–342.

135. Suzuki, K., Kusumi, I., Sasaki, Y., et al. (2001). Serotonin-induced platelet intracellular calcium mobilization in various psychiatric disorders: Is it specific to bipolar disorder? *J Affect Disord, 2–3,* 291–296.

136. Kusumi, I., Suzuki, K., Sasaki, Y., et al. (2000). Treatment response in depressed patients with enhanced Ca mobilization stimulated by serotonin. *Neuropsychopharmacology, 6,* 690–696.

137. Friedman, E., Wang, H. -Y., Levinson, D., et al. (1993). Altered platelet protein kinase C activity in bipolar

affective disorder, manic episode. *Biol Psychiatry, 33,* 520–525.

138. Wang, H.-Y., Markowitz, P., Levinson, D., et al. (1999). Increased membrane-associated protein kinase C activity and translocation in blood platelets from bipolar affective disorder patients. *J Psychiatr Res, 33,* 171–179.

139. Soares, J. C., Dippold, C. S., Wells, K. F., et al. (2001). Increased platelet membrane phosphatidylinositol-4,5-bisphosphate in drug-free depressed bipolar patients. *Neurosci Lett, 1–2,* 150–152.

140. Suzuki, K., Kusumi, I., Akimoto, T., et al. (2003). Altered 5-HT-induced calcium response in the presence of staurosporin in blood platelets from bipolar disorder patients. *Neuropsychopharmacology, 6,* 1210–1214.

141. Soares J. C., Mallinger, A. G., Dippold, C. S., et al. (2000). Effects of lithium on platelet membrane phosphoinositides in bipolar disorder patients: A pilot study. *Psychopharmacology, 1,* 12–16.

142. Soares, J. C., Chen, G., Dippold, C. S., et al. (2000). Concurrent measures of protein kinase C and phosphoinositides in lithium-treated bipolar patients and healthy individuals: A preliminary study. *Psychiatry Res, 2,* 109–118.

143. El Khoury, A., Petterson, U., Kallner, G., et al. (2002). Calcium homeostasis in long-term lithium-treated women with bipolar affective disorder. *Prog Neuropsychopharmacol Biol Psychiatry, 6,* 1063–1069.

144. Pandey, G. N., Dwivedi, Y., SridharaRao, J., et al. (2002). Protein kinase C and phospholipase C activity and expression of their specific isozymes is decreased and expression of MARCKS is increased in platelets of bipolar but not in unipolar patients. *Neuropsychopharmacology, 2,* 216–228.

145. García-Sevilla, J. A., Walzer, C., Busquets, X., et al. (1997). Density of guanine nucleotide-binding proteins in platelets of patients with major depression: Increased abundance of the $G_{\alpha i2}$ subunit and down-regulation by antidepressant drug treatment. *Biol Psychiatry, 42,* 704–712.

146. Karege, F., Bouvier, P., Stepanian, R., et al. (1998). The effect of clinical outcome on platelet G proteins of major depressed patients. *Eur Neuropsychopharmacol, 8,* 89–94.

147. Karege, F., Bouvier, P., Stepanian, R., et al. (1996). Change in platelet GTP-binding in drug-free depressed patients. *Hum Psychopharmacol, 11,* 115–121.

148. Odagaki, Y., & Koyama, T. (2002). Epinephrine- and thrombin-stimulated high-affinity GTPase activity in platelet membranes from patients with psychiatric disorders. *Psychiatry Res, 2,* 111–119.

149. Kafka, M. S., Nurnberger, J. I., Siever, L., et al. (1986). Alpha2-adrenergic receptor function in patients with unipolar and bipolar affective disorders. *J Affect Disord, 10,* 163–169.

150. Siever, L. J., Kafka, M. S., Targum, S., et al. (1984). Platelet alpha-adrenergic binding and biochemical responsiveness in depressed patients and controls. *Psychiatry Res, 11,* 287–302.

151. García-Sevilla, J. A., Padró, D., Giralt, T., et al. (1990). $\alpha_2$-Adrenoceptor-mediated inhibition of platelet adenylate cyclase and induction of aggregation in major depression. *Arch Gen Psychiatry, 47,* 125–132.

152. Murphy, D. L., Donnelly, C., & Moskowitz, J. (1974). Catecholamine receptor function in depressed patients. *Am J Psychiatry, 131,* 1389–1391.

153. Newman, M. E., Lerer, B., Lichtenberg, P., et al. (1992). Platelet adenylate cyclase activity in depression and after clomipramine and lithium treatment: Relation to serotonergic function. *Psychopharmacology, 109,* 231–234.

154. Wang, Y.-C., Pandey, G. N., Mendels, J., et al. (1974). Platelet adenylate cyclase responses in depression: Implications for a receptor defect. *Psychopharmacologia, 36,* 291–300.

155. Menninger, J. A., & Tabakoff, B. (1997). Forskolin-stimulated platelet adenylyl cyclase activity is lower in persons with major depression. *Biol Psychiatry, 42,* 30–38.

156. Cowburn, R. F., Marcusson, J. O., Eriksson, A., et al. (1994). Adenylyl cyclase activity and G-protein subunit levels in postmortem frontal cortex of suicide victims. *Brain Res, 633,* 297–304.

157. García-Sevilla, J. A., Zis, A. P., Hollingsworth, P. J., et al. (1981). Platelet $\alpha_2$-adrenergic receptors in major depressive disorder: Binding of tritiated clonidine before and after tricyclic antidepressant drug treatment. *Arch Gen Psychiatry, 38,* 1327–1333.

158. Doyle, M. C., George, A. J., Ravindran, A. U., et al. (1985). Platelet $\alpha_2$-adrenoreceptor binding in elderly depressed patients. *Am J Psychiatry, 142,* 1489–1490.

159. García-Sevilla, J. A., Guimón, J., García-Vallejo, P., et al. (1986). Biochemical and functional evidence of supersensitive platelet $\alpha_2$-adrenoceptors in major affective disorder: Effect of long-term lithium carbonate treatment. *Arch Gen Psychiatry, 43,* 51–57.

160. García-Sevilla, J. A., Udina, C., Fuster, M. J., et al. (1987). Enhanced binding of [³H](-) adrenaline to platelets of depressed patients with melancholia: Effect of long-term clomipramine treatment. *Acta Psychiatr Scand, 75,* 150–157.

161. Piletz, J. E., & Halaris, A. (1988). Super high affinity ³H-para-aminoclonidine binding to platelet adrenoceptors in depression. *Prog Neuropsychopharmacol Biol Psychiatry, 12,* 541–553.

162. Piletz, J. E., Halaris, A., Saran, A., et al. (1990). Elevated ³H-para-aminoclonidine binding to platelet purified plasma membranes from depressed patients. *Neuropsychopharmacology, 3,* 201–210.

163. Piletz, J. E., Halaris, A., Saran, A., et al. (1991). Desipramine lowers tritiated para-aminoclonidine binding in platelets of depressed patients. *Arch Gen Psychiatry, 48,* 813–820.

164. Smith, C. B., & Zelnik, T. C. (1983). Alpha-2 adrenergic receptors and the mechanism of action of antidepressants. *Behav Brain Sci, 4,* 559.

165. Smith, C. B., Hollingsworth, P. J., García-Sevilla, J. A., et al. (1983). Platelet alpha2 adrenoreceptors are decreased in number after antidepressant therapy. *Prog Neuropsychopharmacol Biol Psychiatry, 7,* 241–247.

166. Takeda, T., Harada, T., & Otsuki, S. (1989). Platelet $^3$H-clonidine and $^3$H-imipramine binding and plasma cortisol level in depression. *Biol Psychiatry, 26,* 52–60.

167. Werstiuk, E. S., Coote, M., Griffith, L., et al. (1996). Effects of electroconvulsive therapy on peripheral adrenoceptors, plasma, noradrenaline, MHPG and cortisol in depressed patients. *Br J Psychiatry, 169,* 758–765.

168. Braddock, L., Cowen, P. J., Elliott, J. M., et al. (1986). Binding of yohimbine and imipramine to platelets in depressive illness. *Psychol Med, 16,* 765–773.

169. Daiguji, M., Meltzer, H. Y., Tong, C., et al. (1981). $\alpha_2$-Adrenergic receptors in platelet membranes of depressed patients: No change in number or $^3$H-yohimbine affinity. *Life Sci, 29,* 2059–2064.

170. Georgotas, A., Schweitzer, J., McCue, R. E., et al. (1987). Clinical and treatment effects on $^3$H-clonidine and $^3$H-imipramine binding in elderly depressed patients. *Life Sci, 40,* 2137–2143.

171. Theodorou, A. E., Hale, A. S., Davies, S. L., et al. (1986). Platelet high affinity adrenoceptor binding sites labelled with the agonist [$^3$H]UK-14,304, in depressed patients and matched controls. *Eur J Pharmacol, 126,* 329–332.

172. Theodorou, A. E., Lawrence, K. M., Healy, D., et al. (1991). Platelet $\alpha_2$-adrenoceptors, defined with agonist and antagonist ligands, in depressed patients, prior to and following treatment. *J Affect Disord, 23,* 99–106.

173. Marazziti, D., Baroni, S., Masala, I., et al. (2001). Correlation between platelet alpha(2)-adrenoreceptors and symptom severity in major depression. *Neuropsychobiology, 3,* 122–125.

174. Wolfe, N., Cohen, B. C., & Gelenberg, A. J. (1987). Alpha$_2$-adrenergic receptors in platelet membranes of depressed patients: Increased affinity for $^3$H-yohimbine. *Psychiatry Res, 20,* 107–116.

175. Carstens, M. E., Engelbrecht, A. H., Russell, V. A., et al. (1986). Alpha$_2$-adrenoceptor levels on platelets of patients with major depressive disorders. *Psychiatry Res, 18,* 321–331.

176. Maes, M., Van Gastel, A., Delmeire, L., et al. (1999). Decreased platelet alpha-2 adrenoceptor density in major depression: Effects of tricyclic antidepressants and fluoxetine. *Biol Psychiatry, 45,* 278–284.

177. Mendlewicz, J., Hirsch, D., Sevy, S., et al. (1989). Alpha-2-adrenoreceptor binding as a possible vulnerability marker for affective disorders. *Neuropsychobiology, 22,* 61–67.

178. Gurguis, G. N. M., Vo, S. P., Blakeley, J. E., et al. (1999). Characteristics of norepinephrine and clonidine displacement of $^3$H-yohimbine binding to platelet $\alpha_2$-adrenergic receptors in healthy volunteers. *Psychiatry Res, 85,* 305–314.

179. Gurguis, G. N. M., Vo, S. P., Griffith, J. M., et al. (1999). Platelet $\alpha_{2A}$-adrenoceptor function in major depression: G$_i$ protein coupling, effects of imipramine and relationship to treatment outcome. *Psychiatry Res, 89,* 73–95.

180. Okada, F., Tokumitsu, Y., & Ui, M. (1988). Possible involvement of pertussis toxin substrates (G$_i$, G$_o$) in desipramine-induced refractoriness of adenylate cyclase in cerebral cortices of rats. *J Neurochem, 51,* 194–199.

181. Yang, J., Wu, J., Kowalska M. A., et al. (2000). Loss of signaling through the G protein, G$_z$, results in abnormal platelet activation and altered responses to psychoactive drugs. *Proc Natl Acad Sci USA, 97,* 9984–9989.

182. Rigatti, B. W., Paleos, G. A., et al. (1991). Simultaneous effects of the platelet 5-HT$_2$ and alpha$_2$-adrenergic receptor populations on phosphoinositide hydrolysis. *Life Sci, 50,* 169–180.

183. Karege, F., Bouvier, P., Rudolph, W., et al. (1996). Platelet phosphoinositide signaling system: An overstimulated pathway to depression. *Biol Psychiatry, 39,* 697–702.

184. Mori, H., Koyama, T., & Yamashita, I. (1991). Platelet alpha-2 adrenergic receptor-mediated phosphoinositide responses in endogenous depression. *Life Sci, 48,* 741–748.

185. Rehavi, M., Jerushalemi, Z., Aviv, A., et al. (1993). Interaction between antidepressants and phosphoinositide signal transduction system in human platelets. *Biol Psychiatry, 33,* 40–44.

186. Göthert, M., & Molderings, G. J. (1991). Involvement of presynaptic imidazoline receptors in the $\alpha_2$-adrenoceptor-independent inhibition of noradrenaline release by imidazoline derivatives. *Naunyn-Schmiedebergs Arch Pharmacol, 343,* 271–282.

187. Piletz, J. E., Halaris, A., Nelson, J., et al. (1996). Platelet I$_1$-imidazoline binding sites are elevated in depression but not generalized anxiety disorder. *J Psychiatr Res, 30,* 147–168.

188. García-Sevilla, J. A., Escribá, P. V., Sastre, M., et al. (1996). Immunodetection and quantitation of imidazoline receptor proteins in platelets of patients with major depression and in brains of suicide victims. *Arch Gen Psychiatry, 53,* 803–810.

189. García-Sevilla, J. A., Escribá, P. V., Ozaita, A., et al. (1998). Density of imidazoline receptors in platelets of euthymic patients with bipolar affective disorder and in brains of lithium-treated rats. *Biol Psychiatry, 43,* 616–618.

190. García-Sevilla, J. A., Escribá, P. V., Busquets, X., et al. (1996). Platelet imidazoline receptors and regulatory G proteins in patients with major depression. *Neuroreport, 8,* 169–172.

191. Halaris, A., Zhu, H., Ali, J., et al. (2002). Down-regulation of platelet imidazoline-1 binding sites after bupropion treatment. *Int J Neuropsychopharmacol, 1,* 37–46.

192. Piletz, J. E., & Halbreich, U. (2000). Imidazoline and alpha(2a)-adrenoceptor binding sites in postmenopausal women before and after estrogen replacement therapy. *Biol Psychiatry, 9,* 932–939.

193. Baron, M., Barkai, A., Gruen, R., et al. (1983). $^3$H-imipramine platelet binding sites in unipolar depression. *Biol Psychiatry, 18,* 1403–1409.

194. Briley, M. S., Langer, S. Z., Raisman, R., et al. (1980). Tritiated imipramine binding sites are decreased in platelets of untreated depressed patients. *Science, 209,* 303–305.

195. DeMet, E. M., Reist, C., Bell, K. M., et al. (1991). Decreased seasonal mesor of platelet $^3$H-imipramine binding in depression. *Biol Psychiatry, 29,* 427–440.

196. Egrise, D., Desmedt, D., Schoutens, A., et al. (1983). Circannual variations in the density of tritiated imipramine binding sites on blood platelets in man. *Neuropsychobiology, 10,* 101–102.

197. Iny, L. J., Pecknold, J., Suranyi-Cadotte, B. E., et al. (1994). Studies of a neurochemical link between depression, anxiety, and stress from [³H]imipramine and [³H]paroxetine binding on human platelets. *Biol Psychiatry, 36,* 281–291.

198. Langer, S. Z., & Raisman, R. (1983). Binding of [³H]imipramine and [³H]desipramine as biochemical tools for studies in depression. *Neuropharmacology, 22,* 407–413.

199. Langer, S. Z., Sechter, D., Loo, H., et al. (1986). Electroconvulsive shock therapy and maximum binding of platelet tritiated imipramine binding in depression. *Arch Gen Psychiatry, 43,* 949–952.

200. Mårtensson, B., Wågner, A., Beck, O., et al. (1991). Effects of clomipramine treatment on cerebrospinal fluid monoamine metabolites and platelet ³H-imipramine binding and serotonin uptake and concentration in major depressive disorder. *Acta Psychiatr Scand, 83,* 125–133.

201. Nemeroff, C. B., Knight, D. L., Krishnan, R. R., et al. (1988). Marked reduction in the number of platelet-tritiated imipramine binding sites in geriatric depression. *Arch Gen Psychiatry, 45,* 919–923.

202. Nemeroff, C. B., Knight, D. L., Franks, J., et al. (1994). Further studies on platelet serotonin transporter binding in depression. *Am J Psychiatry, 151,* 1623–1625.

203. Paul, S. M., Rehavi, M., Skolnick, P., et al. (1981). Depressed patients have decreased binding of tritiated imipramine to platelet serotonin "transporter." *Arch Gen Psychiatry, 38,* 1315–1317.

204. Poirier, M.-F., Benkelfat, C., Loo, H., et al. (1986). Reduced $B_{max}$ of [³H]-imipramine binding to platelets of depressed patients free of previous medication with 5HT uptake inhibitors. *Psychopharmacology, 89,* 456–461.

205. Raisman, R., Sechter, D., Briley, M. S., et al. (1981). High-affinity ³H-imipramine binding in platelets from untreated and treated depressed patients compared to healthy volunteers. *Psychopharmacology, 75,* 368–371.

206. Rosel, P., Arranz, B., Vallejo, J., et al. (1999). Altered [³H]imipramine and 5-HT2 but not [³H]paroxetine binding sites in platelets from depressed patients. *J Affect Disord, 52,* 225–233.

207. Suranyi-Cadotte, B. E., Wood, P. L., Nair, N. P. V., et al. (1982). Normalization of platelet [³H]imipramine binding in depressed patients during remission. *Eur J Pharmacol, 85,* 357–358.

208. Suranyi-Cadotte, B. E., Wood, P. L., Schwartz, G., et al. (1983). Altered platelet ³H-imipramine binding in schizoaffective and depressive disorders. *Biol Psychiatry, 18,* 923–927.

209. Suranyi-Cadotte, B. E., Quirion, R., Nair, N. P. V., et al. (1985). Imipramine treatment differentially affects platelet ³H-imipramine binding and serotonin uptake in depressed patients. *Life Sci, 36,* 795–799.

210. Tanimoto, K., Maeda, K., & Terada, T. (1985). Alteration of platelet [³H]imipramine binding in mildly depressed patients

211. Theodorou, A. E., Katona, C. L. E., Davies, S. L., et al. (1989). ³H-imipramine binding to freshly prepared platelet membranes in depression. *Psychiatry Res, 29,* 87–103.

212. Wågner, A., Åberg-Wistedt, A., Åsberg, M., et al. (1985). Lower ³H-imipramine binding in platelets from untreated depressed patients compared to healthy controls. *Psychiatry Res, 16,* 131–139.

213. Baron, M., Barkai, A., Gruen, R., et al. (1986). Platelet [³H]imipramine binding in affective disorders: Trait versus state characteristics. *Am J Psychiatry, 143,* 711–717.

214. Baron, M., Barkai, A., Gruen, R., et al. (1987). Platelet ³H-imipramine binding and familial transmission of affective disorders. *Neuropsychobiology, 17,* 182–186.

215. Nankai, M., Yoshimoto, S., Narita, K., et al. (1986). Platelet [³H]imipramine binding in depressed patients and its circadian variations in healthy controls. *J Affect Disord, 11,* 207–212.

216. Berrettini, W. H., Nurnberger, J. I. Jr., Post, R. M., et al. (1982). Platelet ³H-imipramine binding in euthymic bipolar patients. *Psychiatry Res, 7,* 215–219.

217. Carstens, M. E., Engelbrecht, A. H., Russell, V. A., et al. (1986). Imipramine binding sites on platelets of patients with major depressive disorder. *Psychiatry Res, 18,* 333–342.

218. D'Hondt, P. D., Maes, M., Leysen, J. E., et al. (1994). Binding of [³H]paroxetine to platelets of depressed patients: Seasonal differences and effects of diagnostic classification. *J Affect Disord, 32,* 27–35.

219. Gentsch, C., Lichtsteiner, M., Gastpar, M., et al. (1985). ³H-imipramine binding sites in platelets of hospitalized psychiatric patients. *Psychiatry Res, 14,* 177–187.

220. Healy, D., Theodorou, A. E., Whitehouse, A. M., et al. (1990). ³H-imipramine binding to previously frozen platelet membranes from depressed patients, before and after treatment. *Br J Psychiatry, 157,* 208–215.

221. Lawrence, K. M., Falkowski, J., Jacobson, R. R., et al. (1993). Platelet 5-HT uptake sites in depression: Three current measures using [³H]imipramine and [³H]paroxetine. *Psychopharmacology, 110,* 235–239.

222. Lawrence, K. M., Katona, C. L. E., Abou-Saleh, M. T., et al. (1994). Platelet 5-HT uptake sites, labelled with [³H]paroxetine, in controls and depressed patients before and after treatment with fluoxetine or lofepramine. *Psychopharmacology, 115,* 261–264.

223. Mellerup, E. T., Plenge, P., & Rosenberg, R. (1982). ³H-imipramine binding sites in platelets from psychiatric patients. *Psychiatry Res, 7,* 221–227.

224. Mellerup, E. T., & Plenge, P. (1988). Imipramine binding in depression and other psychiatric conditions. *Acta Psychiatr Scand, 78,* Suppl. 345, 61–68.

225. Nankai, M., Yamada, S., Yoshimoto, S., et al. (1994). Platelet ³H-paroxetine binding in control subjects and depressed patients: Relationship to serotonin uptake and age. *Psychiatry Res, 51,* 147–155.

226. Ozaki, N., Rosenthal, N. E., Mazzola, P., et al. (1994). Platelet [³H]paroxetine binding, 5-HT-stimulated Ca²⁺ response, correlates with disease severity. *Biol Psychiatry, 20,* 329–352.

and 5-HT content in winter seasonal affective disorder. *Biol Psychiatry, 36,* 458–466.

227. Ravindran, A. V., Chudzik, J., Bialik, R. J., et al. (1994). Platelet serotonin measures in primary dysthymia. *Am J Psychiatry, 151,* 1369–1371.

228. Tollefson, G. D., Heiligenstein, J. H., Tollefson, S. L., et al. (1996). Is there a relationship between baseline and treatment-associated changes in [³H]-IMI platelet binding and clinical response in major depression? *Neuropsychopharmacology, 14,* 47–53.

229. Whitaker, P. M., Warsh, J. J., Stancer, H. C., et al. (1984). Seasonal variation in platelet ³H-imipramine binding: Comparable values in control and depressed populations. *Psychiatry Res, 11,* 127–131.

230. El Khoury, A., Johnson, L., Aberg-Wistedt, A., et al. (2001). Effects of long-term lithium treatment on monoaminergic functions in major depression. *Psychiatry Res, 1–2,* 33–44.

231. Ellis, P. M., & Salmond, C. (1994). Is platelet imipramine binding reduced in depression? A meta-analysis. *Biol Psychiatry, 36,* 292–299.

232. Owens, M. J., & Nemeroff, C. B. (1994). Role of serotonin in the pathophysiology of depression: Focus on the serotonin transporter. *Clin Chem, 40,* 288–295.

233. Schneider, L. S., & Lyness, S. A. (1995). Is platelet imipramine binding reduced in depression? A meta-analysis by Ellis and Salmond. *Biol Psychiatry, 38,* 775–776.

234. Healy, D., O'Halloran, A., Carney, P. A., et al. (1986). Platelet 5-HT uptake in delusional and nondelusional depressions. *J Affect Disord, 10,* 233–239.

235. Meltzer, H. Y., Arora, R. C., Baber, R., et al. (1981). Serotonin uptake in blood platelets of psychiatric patients. *Arch Gen Psychiatry, 38,* 1322–1326.

236. Modai, I., Zemishlany, Z., & Jerushalmy, Z. (1984). 5-Hydroxytryptamine uptake by blood platelets of unipolar and bipolar depressed patients. *Neuropsychobiology, 12,* 93–95.

237. Modai, I., Apter, A., Meltzer, M., et al. (1989). Serotonin uptake by platelets of suicidal and aggressive adolescent psychiatric inpatients. *Neuropsychobiology, 21,* 9–13.

238. Nobile, M., Begni, B., Giorda, R., et al. (1999). Effects of serotonin transporter promoter genotype on platelet serotonin transporter functionality in depressed children and adolescents. *J Am Acad Child Adolesc Psychiatry, 38,* 1396–1402.

239. Rausch, J. L., Janowsky, D. S., Risch, S. C., et al. (1986). A kinetic analysis and replication of decreased platelet serotonin uptake in depressed patients. *Psychiatry Res, 19,* 105–112.

240. Stahl, S. M., Woo, D. J., Mefford, I. N., et al. (1983). Hyperserotonemia and platelet serotonin uptake and release in schizophrenia and affective disorders. *Am J Psychiatry, 140,* 26–30.

241. Meagher, J. B., O'Halloran, A., Carney, P. A., et al. (1990). Changes in platelet 5-hydroxytryptamine uptake in mania. *J Affect Disord, 19,* 191–196.

242. Roy, A. (1999). Suicidal behavior in depression: Relationship to platelet serotonin transporter. *Neuropsychobiology, 39,* 71–75.

243. Fišar, Z., Paclt, I., Anders, M., et al. (1998). Platelet serotonin uptake in depressed patients during treatment. *Homeostasis, 38,* 172–173.

244. Franke, L., Schewe, H. -J., Müller, B., et al. (2000). Serotonergic platelet variables in unmedicated patients suffering from major depression and healthy subjects: Relationship between 5HT content and 5HT uptake. *Life Sci, 67,* 301–315.

245. Healy, D., O'Halloran, A., Carney, P. A., et al. (1986). Variations in platelet 5-hydroxytryptamine in control and depressed populations. *J Psychiatr Res, 20,* 345–353.

246. Mellerup, E. T., Bolwig, T. G., Dam, H., et al. (1995). The serotonin transporter in depressed patients: A qualitative study based on dissociation kinetics. *Biol Psychiatry, 38,* 699–700.

247. Franke, L., Schewe, H. J., Muller, B., et al. (2000). Serotonergic platelet variables in unmedicated patients suffering from major depression and healthy subjects: Relationship between 5HT content and 5HT uptake. *Life Sci, 3,* 301–305.

248. Muck-Seler, D., Pivac, N., Mustapic, M., et al. (2004). Platelet serotonin and plasma prolactin and cortisol in healthy depressed and schizophrenic women. *Psychiatry Res, 3,* 217–226.

249. Tafet, G. E., Toister-Achituv, M., & Shinitzky, M. (2001). Enhancement of serotonin uptake by cortisol: A possible link between stress and depression. *Cogn Affect Behav Neurosci, 4,* 388–393.

250. Franke, L., Schewe, H. J., Uebelhack, R., et al. (2003). Platelet-5HT uptake and gastrointestinal symptoms in patients suffering from major depression. *Life Sci, 4,* 521–531.

251. Pivac, N., Muck-Seler, D., Barisic, I., et al. (2001). Platelet serotonin concentration in dialysis patients with somatic symptoms of depression. *Life Sci, 21,* 2423–2433.

252. Belous, A. R., Ramamoorthy, S., Blakely, R. D., et al. (2001). The state of the serotonin transporter protein in the platelets of patients with somatoform disorders. *Neurosci Behav Physiol, 2,* 185–189.

253. Goodnick, P. J., & Kumar, A. M. (2000). Pretreatment platelet 5-HT concentration predicts the short-term response to paroxetine in major depression. *Biol Psychiatry, 9,* 846–847.

254. Castrogiovanni, P., Blardi, P., De Lalla, A., et al. (2003). Can serotonin and fluoxetine levels in plasma and platelets predict clinical response in depression? *Psychopharmacol, 2,* 102–108.

255. Rausch, J. L., Moeller, F. G., & Johnson, M. E. (2003). Initial platelet serotonin (5-HT) transport kinetics predict nortriptyline treatment outcome. *J Clin Psychopharmacol, 2,* 138–144.

256. Koert, C. E., Spencer, G. E., van Minnen, J., et al. (2001) Functional implications of neurotransmitter expression during axonal regeneration: Serotonin, but not peptides, auto-regulate axon growth of an indentified central neuron. *J Neurosci, 21,* 5597–5606.

257. Arora, R. C., & Meltzer, H. Y. (1983). Effects of amoxapine on serotonin uptake in human blood platelets of depressed patients and normal controls. *Psychiatry Res, 9,* 29–36.

258. Braddock, L. E., Cowen, P. J., Elliott, J. M., et al. (1984). Changes in the binding to platelets of [³H]imipramine and [³H]yohimbine in normal subjects taking amitriptyline. *Neuropharmacology, 23,* 285–286.

259. Suranyi-Cadotte, B. E., Lafaille, F., Schwartz, G., et al. (1985). Unchanged platelet ³H-imipramine binding in normal subjects after imipramine administration. *Biol Psychiatry, 20,* 1237–1240.

260. Wågner, A., Åberg-Wistedt, A., Åberg, M., et al. (1987). Effects of antidepressant treatments on platelet tritiated imipramine binding in major depressive disorder. *Arch Gen Psychiatry, 44,* 870–877.

261. Waldmeier, P. C., Graf, T., Germer, M., et al. (1993). Serotonin uptake inhibition by the monoamine oxidase inhibitor brofaromine. *Biol Psychiatry, 33,* 373–379.

262. Bakish, D., Cavazzoni, P., Chudzik, J., et al. (1997). Effects of selective serotonin reuptake inhibitors on platelet serotonin parameters in major depressive disorder. *Biol Psychiatry, 41,* 184–190.

263. Biegon, A., Weizman, A., Karp, L., et al. (1987). Serotonin 5-HT2 receptor binding on blood platelets — a peripheral marker for depression? *Life Sci, 41,* 2485–2492.

264. Biegon, A., Grinspoon, A., Blumenfeld, B., et al. (1990). Increased serotonin 5-HT$_2$ receptor binding on blood platelets of suicidal men. *Psychopharmacology, 100,* 165–167.

265. Butler, J., & Leonard, B. E. (1988). The platelet serotonergic system in depression and following sertraline treatment. *Int Clin Psychopharmacol, 3,* 343–347.

266. Hrdina, P. D., Bakish, D., Chudzik, J., et al. (1995). Serotonergic markers in platelets of patients with major depression: Upregulation of 5-HT$_2$ receptors. *J Psychiatry Neurosci, 20,* 11–19.

267. Hrdina, P. D., Bakish, D., Ravindran, A., et al. (1997). Platelet serotonergic indices in major depression: Up-regulation of 5-HT$_{2A}$ receptors unchanged by antidepressant treatment. *Psychiatry Res, 66,* 73–85.

268. Pandey, G. N., Pandey, S. C., Janicak, P. G., et al. (1990). Platelet serotonin-2 receptor binding sites in depression and suicide. *Biol Psychiatry, 28,* 215–222.

269. Rao, M. L., Hawellek, B., Papassotiropoulos, A., et al. (1998). Upregulation of the platelet serotonin$_{2A}$ receptor and low blood serotonin in suicidal psychiatric patients. *Neuropsychobiology, 38,* 84–89.

270. Serres, F., Azorin, J.-M., Valli, M., et al. (1999). Evidence for an increase in functional platelet 5-HT$_{2A}$ receptors in depressed patients using the new ligand [¹²⁵I]-DOI. *Eur Psychiatry, 14,* 451–457.

271. Spigset. O., & Mjörndal, T. (1997). The effect of fluvoxamine on serum prolactin and serum sodium concentrations: Relation to platelet 5-HT$_{2A}$ receptor status. *J Clin Psychopharmacol, 17,* 292–297.

272. Arango, V., Underwood, M. D., & Mann, J. J. (1992). Alterations in monoamine receptors in the brain of suicide victims. *J Clin Psychopharmacol, 12,* 8S–12S.

273. Arora, R. C., & Meltzer, H. Y. (1989). Serotonergic measures in the brains of suicide victims: 5-HT$_2$ binding sites in the frontal cortex of suicide victims and control subjects. *Am J Psychiatry, 146,* 730–736.

274. Mann, J. J., Stanley, M., McBride, A., et al. (1986). Increased serotonin$_2$ and β-adrenergic receptor binding in the frontal cortices of suicide victims. *Arch Gen Psychiatry, 43,* 954–959.

275. Yates, M., Leake, A., Candy, J. M., et al. (1990). 5HT$_2$ receptor changes in major depression. *Biol Psychiatry, 27,* 489–496.

276. McBride, P. A., Brown, R. P., DeMeo, M., et al. (1994). The relationship of platelet 5-HT$_2$ receptor indices to major depressive disorder, personality traits, and suicidal behavior. *Biol Psychiatry, 35,* 295–308.

277. Neuger, J., El Khoury, A., Kjellman, B. F., et al. (1999). Platelet serotonin functions in untreated major depression. *Psychiatry Res, 85,* 189–198.

278. Alda, M., & Hrdina, P. D., (2000). Distribution of platelet 5-HT (2A) receptor densities in suicidal and non suicidal depressives and control subjects. *Pyschiatry Res, 3,* 273–277.

279. Khait, V. D., Huang, Y. Y., Zalsman, G., et al. (2005). Association of serotonin 5-HT2A receptor binding and the T102C polymorphism in depressed and healthy Caucasian subjects. *Neuropsychopharmacology, 1,* 166–172.

280. Smith, R. L., Barrett, R. J., & Sanders-Bush, E. (1990). Adaptation of brain 5HT$_2$ receptors after mianserin treatment: Receptor sensitivity, not receptor binding, more accurately correlates with behavior. *J Pharmacol Exp Ther, 254,* 484–488.

281. Eckert, A., Gann, H., Riemann, D., et al. (1993). Elevated intracellular calcium levels after 5-HT$_2$ receptor stimulation in platelets of depressed patients. *Biol Psychiatry, 34,* 565–568.

282. Konopka, L. M., Cooper, R., & Crayton, J. W. (1996). Serotonin-induced increases in platelet cytosolic calcium concentration in depressed, schizophrenic, and substance abuse patients. *Biol Psychiatry, 39,* 708–713.

283. Mikuni, M., Kusumi, I., Kagaya, A., et al. (1991). Increased 5-HT$_2$ receptor function as measured by serotonin-stimulated phosphoinositide hydrolysis in platelets of depressed patients. *Prog Neuropsychopharmacol Biol Psychiatry, 15,* 49–61.

284. Plein, H., Berk, M., Eppel, S., et al. (2000). Augmented platelet calcium uptake in response to serotonin stimulation in patients with major depression measured using MN$^{2+}$ influx and $^{45}$CA$^{2+}$ uptake. *Life Sci, 66,* 425–431.

285. Struneck•, A., Ripov•, D., and Ha•kovec, L. (1987). Incorporation of [³²P]-orthophosphate into phosphoinositides in platelets of depressive patients before and after 10-day lithium administration. *Med Sci Res, 15,* 197–198.

286. Uchitomi, Y., Kugaya, A., Akechi, T., et al. (2001). Three sets of diagnostic criteria for major depression and correlations with serotonin-induced platelet calcium mobilization in cancer patients. *Psychopharmacology, 2,* 244–248.

287. Uchitomi, Y., Kuyaga, A., Akechi, T., et al. (2002). Lack of association between suicidal ideation and enhanced platelet 5-HT2A receptor-mediated calcium mobilization in cancer patients with depression. *Biol Psychiatry, 12,* 1159–1165.

288. Biegon, A., Essar, N., Israeli, M., et al. (1990). Serotonin 5-HT2 receptor binding on blood platelets as a state dependent

marker in major affective disorder. *Psychopharmacology, 102,* 73–75.

289. Plein, H., & Berk, M. (2000). Changes in the platelet intracellular calcium response to serotonin in patients with major depression treated with electroconvulsive therapy: State or trait marker. *Int Clin Psychopharmacol, 15,* 93–98.

290. Stahl, S. M., Hauger, R. L., Rausch, J. L., et al. (1993). Downregulation of serotonin receptor subtypes by nortriptyline and adinazolam in major depressive disorder: Neuroendocrine and platelet markers. *Clin Neuropharmacol, 16,* (Suppl. 3), S19–S31.

291. Januel, D., Massot, O., Poirer, M. F., et al. (2002). Interaction of lithium with 5-HT(1B) receptors in depressed unipolar patients treated with clomipramine and lithium versus clomipramine. *Psychiatry Res, 2–3,* 117–124.

292. Gomez-Gil, E., Gasto, C., Carretero, M., et al. (2004). Decrease of the platelet 5-HT2A receptor function by long-term imipramine treatment in endogenous depression. *Hum Psychopharmacol, 4,* 251–258.

293. Massou, J. M., Trichard, C., Attar-Levy, D., et al. (1997). Frontal 5HT$_{2A}$ receptors studied in depressive patients during chronic treatment by selective serotonin reuptake inhibitors. *Psychopharmacology, 133,* 99–101.

294. Stain-Malmgren, R., Tham, A., & Åberg-Wistedt, A. (1998). Increased platelet 5-HT$_2$ receptor binding after electroconvulsive therapy in depression. *J ECT, 14,* 15–24.

295. Alexopoulos, G. S., Young, R. C., Lieberman, K. W., et al. (1987). Platelet MAO activity in geriatric patients with depression and dementia. *Am J Psychiatry, 144,* 1480–1483.

296. Hartong, E. G. T. M., Goekoop, J. G., Pennings, E. J. M., et al. (1985). DST results and platelet MAO activity. *Br J Psychiatry, 147,* 730–731.

297. Orsulak, P. J., Schildkraut, J. J., Schatzberg, A. F., et al. (1978). Differences in platelet monoamine oxidase activity in subgroups of schizophrenic and depressive disorders. *Biol Psychiatry, 13,* 637–647.

298. Reveley, M. A., Reveley, A. M., Clifford, C. A., et al. (1983). Genetics of platelet MAO activity in discordant schizophrenic and normal twins. *Brit J Psychiatry, 142,* 560–565.

299. Samson, J. A., Gudeman, J. E., Schatzberg, A. F., et al. (1985). Toward a biochemical classification of depressive disorders — VIII. Platelet monoamine oxidase activity in subtypes of depressions. *J Psychiatr Res, 19,* 547–555.

300. Schatzberg, A. F., Rothschild, A. J., Gerson, B., et al. (1985). Toward a biochemical classification of depressive disorders. IX.: DST results and platelet MAO activity. *Br J Psychiatry, 146,* 633–637.

301. White, K., Shih, J., Fong, T.-L., et al. (1980). Elevated platelet monoamine oxidase activity in patients with nonendogenous depression. *Am J Psychiatry, 137,* 1258–1259.

302. Tripodianakis, J., Markianos, M., Sarantidis, D., et al. (2000). Neurochemical variables in subjects with adjustment disorder after suicide attempts. *Eur Psychiatry, 3,* 190–195.

303. Belmaker, R. H., Ebbesen, K., Ebstein, R., et al. (1976). Platelet monoamine oxidase in schizophrenia and manic-depressive illness. *Brit J Psychiatry, 129,* 227–232.

304. Edwards, D. J., Spiker, D. G., Kupfer, D. J., et al. (1978). Platelet monoamine oxidase in affective disorders. *Arch Gen Psychiatry, 35,* 1443–1446.

305. Rogeness, G. A., Mitchell, E. L., Custer, G. J., et al. (1985). Comparison of whole blood serotonin and platelet MAO in children with schizophrenia and major depressive disorder. *Biol Psychiatry, 20,* 270–275.

306. Raft, D., Davidson, J., Wasik, J., et al. (1981). Relationship between response to phenelzine and MAO inhibition in a clinical trial of phenelzine, amitriptyline and placebo. *Neuropsychobiology, 7,* 122–126.

307. Sullivan, J. L., Zung, W. W. K., Stanfield, C. N., et al. (1978). Clinical correlates of tricyclic antidepressant-mediated inhibition of platelet monoamine oxidase. *Biol Psychiatry, 13,* 399–407.

308. Silver, H., Odnopozov, N., Jahjah, N., et al. (1995). Chronic fluvoxamine treatment reduces platelet MAO activity in medicated chronic schizophrenics. *Hum Psychopharmacol, 10,* 321–326.

309. Davidson, J., McLeod, M. N., & White, H. L. (1978). Inhibition of platelet monoamine oxidase in depressed subjects treated with phenelzine. *Am J Psychiatry, 135,* 470–472.

310. Georgotas, A., Mann, J., & Friedman, E. (1981). Platelet monoamine oxidase inhibition as a potential indicator of favorable response to MAOI's in geriatric depressions. *Biol Psychiatry, 16,* 997–1001.

311. Aromaa, A., Raitasalo, R., Reunanen, A., et al. (1994). Depression and cardiovascular diseases. *Acta Psychiatr Scand, 337,* 77–82.

312. Nemeroff, C. B., & Musselman, D. L. (2000). Are platelets the link between depression and ischemic heart disease? *Am Heart J, 140,* S57–S62.

313. Markovitz, J. H., Matthews, K. A., Kiss, J., et al. (1996). Effects of hostility on platelet reactivity to psychological stress in coronary heart disease patients and healthy controls. *Psychosom Med, 58,* 143–149.

314. Musselman, D. L., Tomer, A., Manatunga, A. K., et al. (1996). Exaggerated platelet reactivity in major depression. *Am J Psychiatry, 153,* 1313–1317.

315. Musselman, D. L., Marzec, U. M., Manatunga, A., et al. (2000). Platelet reactivity in depressed patients treated with paroxetine. *Arch Gen Psychiatry, 57,* 875–882.

316. Cassidy, E. M., Walsh, M. T., O'Connor, R., et al. (2003). Platelet surface glycoprotein expression in post-stroke depression: A preliminary study. *Psychiatry Res, 2,* 175–181.

317. Walsh, M. T., Dinan, T. G., Condren, R. M., et al. (2002). Depression is associated with an increase in the expression of the platelet adhesion receptor glycoprotein Ib. *Life Sci, 26,* 3155–3165.

318. Laghrissi-Thode, F., Wagner, W. R., Pollock, B. G., et al. (1997). Elevated platelet factor 4 and β-thromboglobulin plasma levels in depressed patients with ischemic heart disease. *Biol Psychiatry, 42,* 290–295.

319. Pandey, G. N., Ren, X., Pandey, S. C., et al. (2001). Hyperactive phosphoinositide signaling pathway in platelets of

depressed patients: Effect of desipramine treatment. *Psychiatry Res, 1–2,* 23–32.

320. Berk, M., & Plein, H. (2000). Platelet supersensitivity to thrombin stimulation in depression: A possible mechanism for the association with cardiovascular mortality. *Clin Neuropharmacology, 23,* 182–185.

321. Lederbogen, F., Baranyai, R., Gilles, M., et al. (2004). Effect of mental and physical stress on platelet activation markers in depressed patients and healthy subjects: A pilot study. *Psychiatry Res, 1–2,* 55–64.

322. Delisi, S. M., Konapka, L. M., O'Connor, F. L., et al. (1998). Platelet cytosolic calcium responses to serotonin in depressed patients and controls: Relationship to symptomatology and medication. *Biol Psychiatry, 43,* 327–334.

323. Berk, M., Plein, H., & Ferreira, D. (2001). Platelet glutamate receptor supersensitivity in major depressive disorder. *Clin Neuropharmacol, 3,* 129–132.

324. Schins, A., Honig, A., Crijins, H., et al. (2003). Increased coronary events in depressed cardiovascular patients: 5-HT2A receptor as missing link? *Psychosom Med, 5,* 729–737.

325. Markovitz, J. H., Shuster, J. L., Chitwood, W. S., et al. (2000). Platelet activation in depression and effects of sertraline treatment: An open-label study. *Am J Psychiatry, 157,* 1006–1008.

326. Serebruany, V. L., O'Connor, C. M., & Gurbel, P. A. (2001). Effect of selective serotonin reuptake inhibitors or platelets in patients with coronary artery disease. *Am J Cardiol, 12,* 1398–1400.

327. Kusumi, I., Suzuki, K., Sasaki, Y., et al. (2000). Treatment response in depressed patients with enhanced Ca mobilization stimulated by serotonin. *Neuropsychopharmacology, 6,* 690–696.

328. Gomez-Gil, E., Gasto, C., Diaz-Ricart, M., et al. (2002). Platelet 5-HT2A-receptor-mediated induction of aggregation is not altered in major depression. *Hum Psychopharmacol, 8,* 419–424.

329. Keely, H., Ring, M., Dunne, A. M. M., et al. (1996). Platelet serotonin aggregation in depression. *Ir J Psychol Med, 13,* 97–99.

330. McAdams, C., & Leonard, B. E. (1992). Changes in platelet aggregatory responses to collagen and 5-hydroxytryptamine in depressed, schizophrenic and manic patients. *Int Clin Psychopharmacol, 7,* 81–85.

331. Nugent, D. F., Dinan, T. G., & Leonard, B. E. (1994). Alteration by a plasma factor(s) of platelet aggregation in unmedicated unipolar depressed patients. *J Affect Disord, 31,* 61–66.

332. Nugent, D. F., Dinan, T. G., & Leonard, B. E. (1995). Further characterization of the inhibition of platelet aggregation by a plasma factor(s) in unmedicated unipolar depressed patients. *J Affect Disord, 33,* 227–231.

333. Lichtenberg-Kraag, B., May, T., Schmidt, L. G., et al. (1995). Changes of G-protein levels in platelet membranes from alcoholics during short-term and long-term abstinence. *Alcohol Alcohol, 30,* 455–464.

334. Waltman, C., Levine, M. A., McCaul, M. E., et al. (1993). Enhanced expression of the inhibitory protein $G_{i2\alpha}$ and decreased activity of adenylyl cyclase in lymphocytes of abstinent alcoholics. *Alcohol Clin Exp Res, 17,* 315–320.

335. Wand, G. S., Waltman, C., Martin, C. S., et al. (1994). Differential expression of guanosine triphosphate binding proteins in men at high and low risk for the future development of alcoholism. *J Clin Invest, 94,* 1004–1011.

336. Anokina, I. P., Kogan, B. M., Nickel, B., et al. (1989). Basal activities of adenylate and guanylate cyclase in lymphocytes and platelets of alcoholics. *BioMed Biochim Acta, 48,* 593–596.

337. Devor, E. J., Cloninger, C. R., Hoffman, P. L., et al. (1991). A genetic study of platelet adenylate cyclase activity: Evidence for a single major locus effect in fluoride-stimulated activity. *Am J Hum Genet, 49,* 372–377.

338. Ikeda, H., Menninger, J. A., & Tabakoff, B. (1998). An initial study of the relationship between platelet adenylyl cyclase activity and alcohol use disorder criteria. *Alcohol Clin Exp Res, 22,* 1057–1064.

339. Lex, B. W., Ellingboe, J., LaRosa, K., et al. (1993). Platelet adenylate cyclase and monoamine oxidase in women with alcoholism or a family history of alcoholism. *Harvard Rev Psychiatry, 1,* 229–237.

340. Parsian, A., Todd, R. D., Cloninger, C. R., et al. (1996). Platelet adenylyl cyclase activity in alcoholics and subtypes of alcoholics. WHO/ISBRA Study Clinical Centers. *Alcohol Clin Exp Res, 20,* 745–751.

341. Saito, T., Katamura, Y., Ozawa, H., et al. (1994). Platelet GTP-binding protein in long-term abstinent alcoholics with an alcoholic first-degree relative. *Biol Psychiatry, 36,* 495–497.

342. Tabakoff, B., Hoffman, P. L., Lee, J. M., et al. (1988). Differences in platelet enzyme activity between alcoholics and nonalcoholics. *N Engl J Med, 318,* 134–139.

343. Tsuchiya, F., Sirasaka, T., Ikeda, H., et al. (1987). Platelet adenylate cyclase activity in alcoholics. *Jpn J Alcohol Stud Drug Depend, 22,* 366–372.

344. Hoffman, P. L., Glanz, J., & Tabakoff, B., WHO/ISBRA Study on State and Trait Markers of Alcohol Use and Dependence Investigators. (2002). Platelet adenylyl cyclase activity as a state or trait marker in alcohol dependence: Results of the WHO/ISBRA Study on State and Trait Markers of Alcohol Use and Dependence. *Alcohol Clin Exp Res, 7,* 1078–1087.

345. Menninger, J. A., Baron, A. E., Conigrave, K. M., et al. (2000). Platelet adenylyl cyclase activity as a trait marker of alcohol dependence. WHO/ISBRA Collaborative Study Investigators. International Society for Biomedical Research on Alcoholism. *Alcohol Clin Exp Res, 6,* 810–821.

346. Ratsma, J. E., Gunning, W. B., Leurs, R., et al. (1999). Platelet adenylyl cyclase activity as a biochemical trait marker for predisposition to alcoholism. *Alcohol Clin Exp Res, 23,* 600–604.

347. Pandey, G. N., Fawcett, J., Gibbons, R., et al. (1988). Platelet monoamine oxidase in alcoholism. *Biol Psychiatry, 24,* 15–24.

348. Sullivan, J. L., Baenziger, J. C., Wagner, D. L., et al. (1990). Platelet MAO in subtypes of alcoholism. *Biol Psychiatry, 27,* 911–922.

349. Tabakoff, B., & Hoffman, P. L. (1988). Genetics and biological markers of risk for alcoholism. *Public Health Rep, 103,* 690–698.

350. Demir, B., Ucar, G., Ulug, B., et al. (2002). Platelet monoamine oxidase activity in alcoholism subtypes: Relationship to personality traits and executive functions. *Alcohol Alcohol, 6,* 597–602.

351. Berggren, U., Eriksson, M., Fahlke, C., et al. (2002). Platelet monoamine oxidase B in family history positive and family history negative type 1 alcohol-dependent subjects. *Alcohol Alcohol, 6,* 577–580.

352. Coccini, T., Castoldi, A. F., Gandini, C., et al. (2002). Platelet monoamine oxidase B activity as a state marker for alcoholism trend over time during withdrawal and influence of smoking and gender. *Alcohol Alcohol, 6,* 566–572.

353. Crabb, D. W. (1990). Biological markers for increased risk of alcoholism and for quantitation of alcohol consumption. *J Clin Invest, 85,* 311–315.

354. Damberg, M., Garpenstrand, H., Berggard, C., et al. (2000). The genotype of human transcription factor AP-2beta is associated with platelet monoamine oxidase B activity. *Neurosci Lett, 3,* 204–206.

355. Alm, P. O., af Klinteberg, B., Humble, K., et al. (1996). Psychopathy, platelet MAO activity and criminality among former juvenile delinquents. *Acta Psychiatr Scand, 94,* 105–111.

356. von Knorring, L., Oreland, L., & von Knorring, A.-L. (1987). Personality traits and platelet MAO activity in alcohol and drug abusing teenage boys. *Acta Psychiatr Scand, 75,* 307–314.

357. Yehuda, R., Edell, W. S., & Meyer, J. S. (1987). Platelet MAO activity and psychosis proneness in college students. *Psychiatry Res, 20,* 129–142.

358. Gottfries, C. G., Oreland, L., Wiberg, A., et al. (1985). Lowered monoamine oxidase activity in brains from alcoholic suicides. *J Neurochem, 25,* 667–673.

359. Oreland, L., Wiberg, A., Winblad, B., et al. (1983). The activity of monoamine oxidase A & B in brains from chronic alcoholics. *J. Neurol Transmission, 56,* 73–83.

360. Berggren, U., Eriksson, M., Fahlke, C., et al. (2002). Platelet monoamine oxidase-B activity in type 1 alcohol-dependent subjects in sustained full remission. *Alcohol Alcohol, 4,* 340–343.

361. Berggren, U., Fahlke, C., & Balldin, J. (2000). Transient increase in platelet monoamine oxidase b activity during early abstinence in alcoholics: Implications for research *Alcohol Alcohol, 4,* 377–380.

362. Whitfield, J. B., Pang, D., Bucholz, K. K., et al. (2000). Monoamine oxidase: Associations with alcohol dependence, smoking and other measures of psychopathology. *Psychol Med, 2,* 443–454.

363. Berggren, U., Fahlke, C., & Balldin, J. (2000). Alcohol-dependent patients with neuroendocrine evidence for reduced dopamine d(2) receptor function have decreased platelet monoamine oxidase-B activity. *Alcohol Alcohol, 2,* 210–211.

364. Eriksson, M., Berggren, U., Blennow, K., et al. (2000) Alcoholics with the dopamine receptor DRD2 A1 allele have lower platelet monoamine oxidase-B activity than those with the A2 allele: A preliminary study. *Alcohol Alcohol, 5,* 493–498.

365. Oreland, L. (2004). Platelet monoamine oxidase, personality and alcoholism: The rise, fall and resurrection. *Neurotoxicology, 1–2,* 79–89.

366. Javors, M., Tiouririne, M., & Prihoda, T. (2000). Platelet serotonin uptake is higher in early-onset than in late-onset alcoholics. *Alcohol Alcohol, 4,* 390–393.

367. Kent, T. A., Campbell, J. L., Pazdernik, T. L., et al. (1985). Blood platelet uptake of serotonin in men alcoholics. *J Stud Alcohol, 46,* 357–359.

368. Simonsson, P., Berglund, M., Oreland, L., et al. (1992). Serotonin-stimulated phosphoinositide hydrolysis in platelets from post-withdrawal alcoholics. *Alcohol Alcohol, 27,* 607–612.

369. Pivac, N., Muck-Seler, D., Mustapic, M., et al. (2004). Platelet serotonin concentration in alcoholic subjects. *Life Sci, 5,* 521–531.

370. Javors, M. A., Seneviratne, C., Roache, J. D., et al. (2005). Platelet serotonin uptake and paroxetine binding among allelic genotypes of the serotonin transporter in alcoholics. *Prog Neuropsychopharmacol Biol Psychiatry, 1,* 7–13.

371. Preuss, U. W., Soyka, M., Bahlmann, M., et al. (2000). Serotonin transporter gene regulatory region polymorphism (HTTLPR), [3H]paroxetine binding in healthy control subjects and alcohol-dependent patients and their relationships to impulsivity. *Psychiatry Res, 1,* 51–61.

372. Goodwin, G. M., Fraser, S., Stump, K., et al. (1987). Dieting and weight loss in volunteers increases the number of $\alpha_2$-adrenoceptors and 5-HT receptors on blood platelets without effect on [³H]imipramine binding. *J Affect Disord, 12,* 267–274.

373. Sudo, N., Sogawa, H., Komaki, G., & Kubo, C. (1997). The serotonin-induced elevation of intracellular $Ca^{2+}$ in human platelets is enhanced by total fasting. *Biol Psychiatry, 41,* 618–620.

374. Biederman, J., Rivinus, T. M., Herzog, D. B., et al. (1984). Platelet MAO activity in anorexia nervosa patients with and without a major depressive disorder. *Am J Psychiatry, 141,* 1244–1247.

375. Carrasco, J. L., Diaz-Marsá, M., Hollander, E., et al. (2000). Decreased platelet monoamine oxidase activity in female bulimia nervosa. *Eur Neuropsychopharmacol, 10,* 113–117.

376. Diaz-Marsa, M., Carrasco, J. L., Hollander, E., et al. (2000). Decreased platelet monoamine oxidase activity in female anorexia nervosa. *Acta Psychiatr Scand, 101,* 226–230.

377. Hallman, J., Sakurai, E., & Oreland, L. (1990). Blood platelet monoamine oxidase activity, serotonin uptake and release rates in anorexia and bulimia patients and in healthy controls. *Acta Psychiatr Scand, 81,* 73–77.

378. Weizman, R., Carmi, M., Tyano, S., et al. (1986). High affinity [3H]imipramine binding and serotonin uptake to platelets of adolescent females suffering from anorexia nervosa. *Life Sci, 38,* 1235–1242.

379. Zemishlany, Z., Modai, I., Apter, A., et al. (1987). Serotonin (5-HT) uptake by blood platelets in anorexia nervosa. *Acta Psychiatr Scand, 75,* 127–130.

380. Steiger, H., Leonard, S., Kin, N. Y., et al. (2000). Childhood abuse and platelet tritiated-paroxetine binding in bulimia nervosa: Implications of borderline personality disorder. *J Clin Psychiatry, 61,* 428–435.

381. Steiger, H., Young, S. N., Kin, N. M., et al. (2001). Implications of impulsive and affective symptoms for serotonin function in bulimia nervosa. *Psychol Med, 1,* 85–95.

382. Ramacciotti, C. E., Coli, E., Paoli, R., et al. (2003). Serotonergic activity measured by platelet [3H]paroxetine binding in patients with eating disorders. *Psychiatry Res, 1,* 33–38.

383. Steiger, H., Gauvin, L., Israel, M., et al. (2001). Association of serotonin and cortisol indices with childhood abuse in bulimia nervosa. *Arch Gen Psychiatry, 9,* 837–843.

384. Steiger, H., Gauvin, L., Israel, M., et al. (2004). Serotonin function, personalitytrait variations, and childhood, abuse in women with bulimia-spectrum eating disorders. *J Clin Psychiatry, 6,* 830–837.

385. Steiger, H., Israel, M., Gauvin, L., et al. (2003). Implications of compulsive and impulsive traits for serotonin status in women with bulimia nervosa. *Pyschiatry Res, 3,* 219–229.

386. Spigset, O., Andersen, T., Hagg, S., et al. (1999). Enhanced platelet serotonin $5\text{-}HT_{2A}$ receptor binding in anorexia nervosa and bulimia nervosa. *Eur Neuropsychopharmacol, 9,* 469–473.

387. Luck, P., Mikhailidis, D. P., Dashwood, M. R., et al. (1983). Platelet hyperaggregability and increased alpha-adrenoceptor density in anorexia nervosa. *J Clin Endocrinol Metab, 57,* 911–914.

388. Heufelder, A., Warnhoff, M., & Pirke, K. M. (1985). Platelet $\alpha_2$-adrenoceptor and adenylate cyclase in patients with anorexia nervosa and bulimia. *J Clin Endocrinol Metab, 61,* 1053–1060.

389. Gill, J., Desouza, V., Wakeling, A., et al. (1992). Differential changes in $\alpha$- and $\beta$-adrenoceptor linked [$^{45}Ca^{2+}$] uptake in platelets from patients with anorexia nervosa. *J Clin Endocrinol Metab, 74,* 441–446.

390. Yonkers, K. A. (1997). The association between premenstrual dysphoric disorder and other mood disorders. *J Clin Psychiatry, 58,* 1–7.

391. Yonkers, K. A. (1997). Anxiety symptoms and anxiety disorders: How are they related to premenstrual disorders? *J Clin Psychiatry, 58,* 62–67.

392. Belmaker, R. H., Murphy, D. L., Wyatt, R. J., et al. (1974). Human platelet monoamine oxidase changes during the menstrual cycle. *Arch Gen Psychiatry, 31,* 553–556.

393. Gilmore, N. J., Robinson, D. S., Nies, A., et al. (1971). Blood monoamine oxidase levels in pregnancy and during the menstrual cycle. *J Psychosom Res, 15,* 215–220.

394. Poirier, M. -F., Loo, H., Dennis, T., et al. (1985). Platelet monoamine oxidase activity and plasma 3,4-dihydroxyphenylethylene glycol levels during the menstrual cycle. *Neuropsychobiology, 14,* 165–169.

395. Ashby, C. R. Jr., Carr, L. A., Cook, C. L., et al. (1988). Alteration of platelet serotonergic mechanisms and monoamine oxidase activity in premenstrual syndrome. *Biol Psychiatry, 24,* 225–233.

396. Rapkin, A. J., Buckman, T. D., Sutphin, M. S., et al. (1988). Platelet monoamine oxidase B activity in women with premenstrual syndrome. *Am J Obstet Gynecol, 159,* 1536–1540.

397. Jones, S. B., Bylund, D. B., Rieser, C. A., et al. (1983). $\alpha_2$-adrenergic receptor binding in human platelets: Alterations during the menstrual cycle. *Clin Pharmacol Ther, 34,* 90–96.

398. Peters, J. R., Elliot, J. M., & Grahame-Smith, D. G. (1979). Effect of oral contraceptives on platelet noradrenaline and 5-hydroxytryptamine receptors and aggregation. *Lancet, 2,* 933–936.

399. Rosen, S. G., Berk, M. A., Popp, D. A., et al. (1984). $\beta_2$- and $\alpha_2$-adrenergic receptors and receptor coupling to adenylate cyclase in human mononuclear leukocytes and platelets in relation to physiological variations of sex steroids. *J Clin Endocrinol Metab, 58,* 1068–1076.

400. Stowell, L. I., McIntosh, C. J., Cooke, R., et al. (1988). Adrenoceptor and imipramine receptor binding during the menstrual cycle. *Acta Psychiatr Scand, 78,* 366–368.

401. Sundaresan, P. R., Madan, M. K., Kelvie, S. L., et al. (1985). Platelet alpha-2 adrenoceptors and the menstrual cycle. *Clin Pharmacol Ther, 37,* 337–342.

402. Theodorou, A. E., Mistry, H., Davies, S. L., et al. (1987). Platelet $\alpha_2$-adrenoceptor binding and function during the menstrual cycle. *J Psychiatr Res, 21,* 163–169.

403. Metz, A., Stump, K., Cowen, P. J., et al. (1983) Changes in platelet $\alpha_2$-adrenoceptor binding post partum: Possible relation to maternity blues. *Lancet, 1,* 495–498.

404. Halbreich, U., Piletz, J. E., Carson, S., et al. (1993). Increased imidazoline and $\alpha_2$-adrenergic binding in platelets of women with dysphoric premenstrual syndromes. *Biol Psychiatry, 34,* 676–686.

405. Gurguis, G. N. M., Yonkers, K. A., Phan, S. P., et al. (1998). Adrenergic receptors in PMDD: I. Platelet $\alpha_2$ receptors: $G_i$ protein coupling, phase of menstrual cycle and prediction of luteal phase symptom severity. *Biol Psychiatry, 44,* 600–609.

406. Piletz, J. E., Andrew, M., Zhu, H., et al. (1998). $\alpha_2$-adrenoceptors and $I_1$-imidazoline binding sites: Relationship with catecholamines in women of reproductive age. *J Psychiatr Res, 32,* 55–64.

407. Kafka, M. S., van Kammen, D. P., Kleinman, J. E., et al. (1980). Alpha-adrenergic receptor function in schizophrenia, affective disorders and some neurological diseases. *Commun Psychopharmacol, 4,* 477–486.

408. Pandey, G. N., Pandey, S. C., & Davis, J. M. (1990). Peripheral adrenergic receptors in affective illness and schizophrenia. *Pharmacol Toxicol, 66,* 13–36.

409. Kafka, M. S., van Kammen, D. P., & Bunney, W. E. (1979). Reduced cyclic AMP production in the blood platelets from schizophrenic patients. *Am J Psychiatry, 136,* 685–687.

410. Rice, H. E., Smith, C. B., Silk, K. R., et al. (1984). Platelet alpha$_2$-adrenergic receptors in schizophrenic patients before and after phenothiazine treatment. *Psychiatry Res, 12,* 69–77.

411. Rosen, J., Silk, K. R., Rice, H. E., et al. (1985). Platelet alpha-2-adrenergic dysfunction in negative symptom schizophrenia: A preliminary study. *Biol Psychiatry, 20,* 539–545.

412. Pandey, G. N., Garver, D. L., Tamminga, C., et al. (1977). Postsynaptic supersensitivity in schizophrenia. *Am J Psychiatry, 134,* 518–522.

413. Kanof, P. D., Johns, C., Davidson, M., et al. (1986). Prostaglandin receptor sensitivity in psychiatric disorders. *Arch Gen Psychiatry, 43,* 987–993.

414. Rotrosen, J., Miller, A. D., Mandio, D., et al. (1978). Reduced PGE₁ stimulated ³H-cAMP accumulation in platelets from schizophrenics. *Life Sci, 23,* 1989–1996.

415. Rotrosen, J., Miller, A. D., Mandio, D., et al. (1980). Prostaglandins, platelets, and schizophrenia. *Arch Gen Psychiatry, 37,* 1047–1054.

416. Tardito, D., Tura, G. B., Bocchio, L., et al. (2000). Abnormal levels of cAMP-dependent protein kinase regulatory subunits in platelets from schizophrenic patients. *Neuropsychopharmacology, 23,* 216–219.

417. Natsukari, N., Kuluga, H., Baker, I., et al. (1997). Increased cyclic AMP response to forskolin in Epstein-Barr virus-transformed human B-lymphocytes derived from schizophrenics. *Psychopharmacology, 130,* 235–241.

418. Nashino, N., Kitamura, N., Hashimoto, T., et al. (1993). Increase in [³H]cAMP binding sites and decrease in Gᵢₐ and Gₒₐ immunoreactivities in left temporal cortices from patients with schizophrenia. *Brain Res, 615,* 41–49.

419. Lingjærde, O. (1983). Serotonin uptake and efflux in blood platelets from untreated and neuroleptic-treated schizophrenics. *Biol Psychiatry, 18,* 1345–1356.

420. Modai, I., Rotman, A., Munitz, H., et al. (1979). Serotonin uptake by blood platelets of acute schizophrenic patients. *Psychopharmacology, 64,* 193–195.

421. Rotman, A., Zemishlany, Z., Munitz, H., et al. (1982). The active uptake of serotonin by platelets of schizophrenic patients and their families: Possibility of a genetic marker. *Psychopharmacology, 77,* 171–174.

422. Sankar, D. V. S. (1977). Uptake of 5-hydroxytryptamine by isolated platelets in childhood schizophrenia and autism. *Neuropsychobiology, 3,* 234–239.

423. Arora, R. C., & Meltzer, H. Y. (1982). Serotonin uptake by blood platelets of schizophrenic patients. *Psychiatry Res, 6,* 327–333.

424. Kaneda, Y., Fujii, A., & Nagamine, I. (2001). Platelet serotonin concentrations in medicated schizophrenic patients. *Prog Neuropsychopharmacol Biol Psychiatry, 5,* 983–992.

425. Weizman, A., Gonen, N., Tyano, S., et al. (1987). Platelet [³H]imipramine binding in autism and schizophrenia. *Psychopharmacology, 91,* 101–103.

426. Wood, P. L., Suranyi-Cadotte, B. E., Nair, N. P. V., et al. (1983). Lack of association between [³H]imipramine binding sites and uptake of serotonin in control, depressed and schizophrenic patients. *Neuropharmacology, 22,* 1211–1214.

427. Maguire, K., Cheung, P., Crowley, K., et al. (1997). Aggressive behaviour and platelet ³H-paroxetine binding in schizophrenia. *Schizophr Res, 23,* 61–67.

428. Modai, I., Gibel, A., Rauchverger, B., et al. (2000). Paroxetine binding in aggressive schizophrenic patients. *Psychiatry Res, 1,* 77–81.

429. Govitrapong, P., Mukda, S., Turakitwanakan, W., et al. (2002). *Neurochem Int, 4,* 209–216.

430. Ahtee, L., Briley, M., Raisman, R., et al. (1981). Reduced uptake of serotonin but unchanged ³H-imipramine binding in the platelets of cirrhotic patients. *Life Sci, 29,* 2323–2329.

431. Shah, N. S., Burch, E. A. Jr., Pressley, L. C., et al. (1982). Platelet uptake of dopamine in patients with schizophrenia. *Res Commun Psychol Psychiatr Behav, 7,* 489–492.

432. Malmgren, R. (1984). Platelets and biogenic amines 1. Platelets are poor investigative models for dopamine re-uptake. *Psychopharmacology, 84,* 480–485.

433. Arora, R. C., & Meltzer, H. Y. (1993). Serotonin₂ receptor binding in blood platelets of schizophrenic patients. *Psychiatry Res, 47,* 111–119.

434. Govitrapong, P., Chagkutip, J., Turakitwanakan, W., et al. (2000). Platelet 5-HT₂ₐ receptors in schizophrenic patients with and without neuroleptic treatment. *Psychiatry Res, 96,* 41–50.

435. Pandey, S. C., Sharma, R. P., Janicak, P. G., et al. (1993). Platelet serotonin-2 receptors in schizophrenia: Effects of illness and neuroleptic treatment. *Psychiatr Res, 48,* 57–68.

436. Arranz, B., Rosel, P., Sarro, S., et al. (2003). Altered platelet serotonin 5-HT2A receptor density but not second messenger inositol triphosphate levels in drug-free schizophrenic patients. *Psychiatry Res, 2,* 165–174.

437. Govitrapong, P., Chagkutip, J., Turakitwanakan, W., et al. (2000). Platelet 5-HT(2A) receptors in schizophrenic patients with and without neuroleptic treatment. *Psychiatry, 1,* 41–50.

438. Joyce, J. N., Shane, A., Lexow, N., et al. (1993). Serotonin uptake sites and serotonin receptors are altered in the limbic system of schizophrenics. *Neuropsychopharmacology, 8,* 315–336.

439. Das, I., Essali, M. A., de Belleroche, J., et al. (1994). Elevated platelet phosphatidylinositol bisphosphate in medicated schizophrenics. *Schizophrenia Res, 12,* 265–268.

440. Kaiya, H., Nishida, A., Imai, A., et al. (1989). Accumulation of diacylglycerol in platelet phosphoinositide turnover in schizophrenia: A biological marker of good prognosis? *Biol Psychiatry, 26,* 669–676.

441. Yao, J. K., van Kammen, D. P., Moss, H. B., et al. (1996). Decreased serotonergic responsivity in platelets of drug-free patients with schizophrenia. *Psychiatry Res, 63,* 123–132.

442. Singh, A. N., Barlas, C., Saeedi, H., et al. (2003). Effect of loxapine on peripheral dopamine-like and serotonin receptor in patients with schizophrenia. *J Psychiatry Neurosci, 1,* 39–47.

443. Ereshefsky, L., Riesenman, C., True, J. E., et al. (1996). Serotonin-mediated increase in cytosolic [Ca⁺⁺] in platelets of risperidone-treated schizophrenia patients. *Psychopharmacol Bull, 32,* 101–106.

444. Grahame-Smith, D. G., Geaney, D. P., Schachter, M., et al. (1988). Human platelet 5-hydroxytryptamine receptors: Binding of [³H]-lysergic acid diethylamide (LSD). Effects of chronic neuroleptic and antidepressant drug administration. *Experientia, 44,* 142–145.

445. Pandey, G. N., Pandey, S. C., Isaac, L., et al. (1992). Effect of electroconvulsive shock therapy on 5-HT₂ and α₁-adrenoceptors and phosphoinositide signalling system in rat brain. *Eur J Pharmacol, 226,* 303–310.

446. Paul, I. A., Duncan, G. E., Mueller, R. A., et al. (1991). Neural adaptation in response to chronic imipramine and electroconvulsive shock therapy: Evidence for separate mechanisms. *Eur J Pharmacol, 205,* 135–143.

447. Schachter, M., Geaney, D. P., Grahame-Smith, D. G., et al. (1985). Increased platelet membrane [³H]-LSD binding in patients on chronic neuroleptic treatment. *Br J Clin Pharmacol, 19,* 453–457.

448. Arora, R. C., & Meltzer, H. Y. (1991). Serotonin₂ (5-HT₂) receptor binding in the frontal cortex of schizophrenic patients. *J Neural Transm Gen Sect, 85,* 19–29.

449. Bennett, J. P., Enna, S. J., Bylund, D. B., et al. (1979). Neurotransmitter receptors in frontal cortex of schizophrenics. *Arch Gen Psychiatry, 36,* 927–934.

450. Whitaker, P. M., Crow, T. J., & Ferrier, I. N. (1981). Tritiated LSD binding in frontal cortex in schizophrenia. *Arch Gen Psychiatry, 38,* 278–280.

451. Buchsbaum, M. S., Murphy, D. L., Coursey, R. D., et al. (1978). Platelet monoamine oxidase, plasma dopamine-beta-hydroxylase and attention in a "biochemical high risk" sample. *J Psychiatr Res, 14,* 215–224.

452. Coursey, R. D., Buchsbaum, M. S., & Murphy, D. L. (1979). Platelet MAO activity and evoked potentials in the identification of subjects biologically at risk for psychiatric disorders. *Brit J Psychiatry, 134,* 372–381.

453. Coursey, R. D., Buchsbaum, M. S., & Murphy, D. L. (1980). Psychological characteristics of subjects identified by platelet MAO activity and evoked potentials as biologically at risk for psychopathology. *J Abnorm Psychol, 89,* 151–164.

454. Oxenstierna, G., Edman, G., Iselius, L., et al. (1986). Concentrations of monoamine metabolites in the cerebrospinal fluid of twins and unrelated individuals — a genetic study. *J Psychiatr Res, 20,* 19–29.

455. Berrettini, W. H., Benfield, T. C., Schmidt, A. O., et al. (1980). Platelet monoamine oxidase in families of chronic schizophrenics. *Schizophr Bull, 6,* 235–237.

456. Pandey, G. N., Dorus, E., Shaughnessy, R., et al. (1979). Genetic control of platelet monoamine oxidase activity: Studies on normal families. *Life Sci, 25,* 1173–1178.

457. Baron, M., Levitt, M., Gruen, R., et al. (1984). Platelet monoamine oxidase activity and genetic vulnerability to schizophrenia. *Am J Psychiatry, 141,* 836–842.

458. Baron, M., Risch, N., Levitt, M., et al. (1985). Genetic analysis of platelet monoamine oxidase activity in families of schizophrenic patients. *J Psychiatr Res, 19,* 9–21.

459. Propping, P., & Friedl, W. (1979). Platelet monoamine oxidase activity in first-degree relatives of schizophrenic patients. *Psychopharmacology, 65,* 265–272.

460. Wyatt, R. J., Saavedra, J. M., Belmaker, R., et al. (1973). The dimethyltryptamine-forming enzyme in blood platelets: A study in monozygotic twins discordant for schizophrenia. *Am J Psychiatry, 130,* 1359–1361.

461. Belmaker, R. H., Ebstein, R., Rimon, R., et al. (1976). Electrophoresis of platelet monoamine oxidase in schizophrenia and manic-depressive illness. *Acta Psychiatr Scand, 54,* 67–72.

462. Kleinman, J. E., Potkin, S., Rogol, A., et al. (1979). A correlation between platelet monoamine oxidase activity and plasma prolactin concentrations in man. *Science, 206,* 479–481.

463. Zucker, M., Valevski, A., Weizman, A., et al. (2002). Increased platelet vesicular monoamine transporter density in adult schizophrenia patients. *Eur Neuropsychopharmacol, 4,* 343–347.

464. Becker, R. E., & Shaskan, E. G. (1977). Platelet monoamine oxidase activity in schizophrenic patients. *Am J Psychiatry, 134,* 512–517.

465. Berger, P. A., Ginsburg, R. A., Barchas, J. D., et al. (1978). Platelet monoamine oxidase in chronic schizophrenic patients. *Am J Psychiatry, 135,* 95–99.

466. Berrettini, W. H., & Vogel, W. H. (1978). Evidence for an endogenous inhibitor of platelet MAO in chronic schizophrenia. *Am J Psychiatry, 135,* 605–607.

467. Berrettini, W. H., Vogel, W. H., & Clouse, R. (1977). Platelet monoamine oxidase in chronic schizophrenia. *Am J Psychiatry, 134,* 805–806.

468. Domino, E. F., & Khanna, S. S. (1976). Decreased blood platelet MAO activity in unmedicated chronic schizophrenic patients. *Am J Psychiatry, 133,* 323–326.

469. Friedhoff, A. J., Miller, J. C., & Weisenfreund, J. (1978). Human platelet MAO in drug-free and medicated schizophrenic patients. *Am J Psychiatry, 135,* 952–955.

470. Giller, E. L. Jr., Bierer, L., Rubinow, D., et al. (1980). Platelet MAO Vmax and Km in chronic schizophrenic subjects. *Am J Psychiatry, 137,* 97–98.

471. Gruen, R., Baron, M., Levitt, M., et al. (1982). Platelet MAO activity and schizophrenic prognosis. *Am J Psychiatry, 139,* 240–241.

472. Meltzer, H. Y., Arora, R. C., Jackman, H., et al. (1980). Platelet monoamine oxidase and plasma amine oxidase in psychiatric patients. *Schizophr Bull, 6,* 213–219.

473. Murphy, D. L., & Wyatt, R. J. (1972). Reduced monoamine oxidase activity in blood platelets from schizophrenic patients. *Nature* (London), *238,* 225–226.

474. Murphy, D. L., Donnelly, C. H., Miller, L., et al. (1976). Platelet monoamine oxidase in chronic schizophrenia. *Arch Gen Psychiatry, 33,* 1377–1381.

475. Murphy, D. L., Belmaker, R., Carpenter, W. T., et al. (1977). Monoamine oxidase in chronic schizophrenia: Studies of hormonal and other factors affecting enzyme activity. *Br J Psychiatry, 130,* 151–158.

476. Paik, I.-H., Lee, C., & Kim, C.-E. (1988). Platelet MAO in schizophrenics: Relationship to symptomatology and neuroleptics. *Biol Psychiatry, 23,* 89–93.

477. Rose, R. M., Castellani, S., Boeringa, J. A., et al. (1986). Platelet MAO concentration and molecular activity: II. Com-

parison of normal and schizophrenic populations. *Psychiatry Res, 17,* 141–151.

478. Schildkraut, J. J., Herzog, J. M., Orsulak, P. J., et al. (1976). Reduced platelet monoamine oxidase activity in a subgroup of schizophrenic patients. *Am J Psychiatry, 133,* 438–440.

479. Sullivan, J. L., Cavenar, J. O. Jr., Stanfield, C. N., et al. (1978). Reduced MAO activity in platelets and lymphocytes of chronic schizophrenics. *Am J Psychiatry, 135,* 597–598.

480. Tachiki, K. H., Buckman, T. D., Eiduson, S., et al. (1986). Phosphatidylserine inhibition of monoamine oxidase in platelets of schizophrenics. *Biol Psychiatry, 21,* 59–68.

481. Van Kammen, D. P., Marder, S. R., Murphy, D. L., et al. (1978). MAO activity, CSF amine metabolites, and drug-free improvement in schizophrenia. *Am J Psychiatry, 135,* 567–569.

482. Wyatt, R. J., Potkin, S. G., & Murphy, D. L. (1979). Platelet monoamine oxidase activity in schizophrenia: A review of the data. *Am J Psychiatry, 136,* 377–385.

483. Zureick, J. L., & Meltzer, H. Y. (1988). Platelet MAO activity in hallucinating and paranoid schizophrenics: A review and meta-analysis. *Biol Psychiatry, 24,* 63–78.

484. Becker, R. E., & Giambalvo, C. T. (1982). Endogenous modulation of monoamine oxidase in schizophrenic and normal humans. *Am J Psychiatry, 139,* 1567–1570.

485. Bond, P. A., Cundell, R. L., & Falloon, I. R. H. (1979). Monoamine oxidase (MAO) of platelets, plasma, lymphocytes and granulocytes in schizophrenia. *Brit J Psychiatry, 134,* 360–365.

486. Campbell, M., Friedman, E., Green, W. H., et al. (1976). Blood platelet monoamine oxidase activity in schizophrenic children and their families: A preliminary study. *Neuropsychobiology, 2,* 239–246.

487. Carpenter, W. T. Jr., Murphy, D. L., & Wyatt, R. J. (1975). Platelet monoamine oxidase activity in acute schizophrenia. *Am J Psychiatry, 132,* 438–441.

488. Fleissner, A., Seifert, R., Schneider, K., et al. (1987). Platelet monoamine oxidase activity and schizophrenia — a myth that refuses to die? *Eur Arch Psychiatry Neurol Sci, 237,* 8–15.

489. Gattaz, W. F., & Beckman, H. (1981). Platelet MAO activity and personality characteristics: A study in schizophrenic patients and normal individuals. *Acta Psychiatr Scand, 63,* 479–485.

490. Groshong, R., Baldessarini, R. J., Gibson, A., et al. (1978). Activities of types A and B MAO and catechol-o-methyltransferase in blood cells and skin fibroblasts of normal and chronic schizophrenic subjects. *Arch Gen Psychiatry, 35,* 1198–1205.

491. Jackman, H. L., & Meltzer, H. Y. (1983). Kinetic constants of platelet monoamine oxidase in schizophrenia. *Am J Psychiatry, 140,* 1044–1047.

492. Joseph, M. H., Owen, F., Baker, H. F., et al. (1977). Platelet serotonin concentration and monoamine oxidase activity in unmedicated chronic schizophrenic and in schizoaffective patients. *Psychol Med, 7,* 159–162.

493. Karson, C. N., Kleinman, J. E., Berman, K. F., et al. (1983). An inverse correlation between spontaneous eye-blink rate and platelet monoamine oxidase activity. *Br J Psychiatry, 142,* 43–46.

494. Mann, J., & Thomas, K. M. (1979). Platelet monoamine oxidase activity in schizophrenia: Relationship to disease, treatment, institutionalization and outcome. *Br J Psychiatry, 134,* 366–371.

495. Owen, F., Bourne, R., Crow, T. J., et al. (1976). Platelet monoamine oxidase in schizophrenia. *Arch Gen Psychiatry, 33,* 1370–1373.

496. White, H. L., McLeod, M. N., & Davidson, J. R. T. (1976). Platelet monoamine oxidase activity in schizophrenia. *Am J Psychiatry, 133,* 1191–1193.

497. Jackman, H. L., & Meltzer, H. Y. (1980). Factors affecting determination of platelet monoamine oxidase activity. *Schizophr Bull, 6,* 259–266.

498. Wise, C. D., Potkin, S. G., Bridge, T. P., et al. (1980). Sources of error in the determination of platelet monoamine oxidase: A review of methods. *Schizophr Bull, 6,* 245–253.

499. Dinan, T. G. (1987). Neuroleptic effects on platelet aggregation: A study in normal volunteers and schizophrenics. *Psychol Med, 17,* 875–881.

500. Kessler, A., Shinitzky, M., & Kessler, B. (1995). Number of platelet dense granules varies with age, schizophrenia and dementia. *Dementia, 6,* 330–333.

501. Yao, J. K., van Kammen, D. P., Gurklis, J., et al. (1994). Platelet aggregation and dense granule secretion in schizophrenia. *Psychiatry Res, 54,* 13–24.

502. Gattaz, W. F., Schmitt, A., & Maras, A. (1995). Increased platelet phospholipase $A_2$ activity in schizophrenia. *Schizophr Res, 16,* 1–6.

503. Noponen, M., Sanfilipo, M., Samanich, K., et al. (1993). Elevated $PLA_2$ activity in schizophrenics and other psychiatric patients. *Biol Psychiatry, 34,* 641–649.

504. Yao, J. K., van Kammen, D. P., & Gurklis, J. A. (1996). Abnormal incorporation of arachidonic acid into platelets of drug-free patients with schizophrenia. *Psychiatry Res, 60,* 11–21.

505. Demisch, L., Gerbaldo, H., Gebhart, P., et al. (1987). Incorporation of $^{14}$C-arachidonic acid into platelet phospholipids of untreated patients with schizophreniform or schizophrenic disorders. *Psychiatry Res, 22,* 275–282.

506. Demisch, L., Heinz, K., Gerbaldo, H., et al. (1992). Increased concentrations of phosphatidylinositol (PI) and decreased esterification of arachidonic acid into phospholipids in platelets from patients with schizoaffective disorders or atypic phasic psychoses. *Prostaglandins Leukot Essent Fatty Acids, 46,* 47–52.

507. Schmitt, A., Maras, A., Petroianu, G., et al. (2001). Effects of antipsychotic treatment on membrane phospholipid metabolism in schizophrenia. *J Neural Transm, 8–9,* 1081–1090.

508. Toyooka, K., Asama, K., Watanabe, Y., et al. (2002). Decreased levels of brain-derived neurotrophic factor in serum of chronic schizophrenic patients. *Psychiatry Res, 3,* 249–257.

509. Dror, N., Klein, E., Karry, R., et al. (2002). State-dependent alterations in mitochondrial complex I activity in platelets:

A potential peripheral marker for schizophrenia. *Mol Psychiatry, 9,* 995–1001.

510. Srivastava, N., Barthwal, M. K., Dalal, P. K., et al. (2001). Nitrite content and antioxidant enzyme levels in the blood of schizophrenia patients. *Psychopharmacology, 2,* 140–145.

511. Pryor, S. R. (2000). Is platelet release of 2-arachidonoylglycerol a mediator of cognitive deficits? An endocannabinoid theory of schizophrenia and arousal. *Med Hypotheses, 6,* 494–501.

512. Berk, M., Plein, H., & Belsham, B. (2002). The specificity of platelet glutamate receptor supersensitivity in psychotic disorders. *Life Sci, 25,* 2427–2432.

513. Abramson, R. K., Wright, H. H., Carpenter, R., et al. (1989). Elevated blood serotonin in autistic probands and their first-degree relatives. *J Autism Dev Disord, 19,* 397–407.

514. Anderson, G. M., Freedman, D. X., Cohen, D. J., et al. (1987). Whole blood serotonin in autistic and normal subjects. *J Child Psychol Psychiatry, 28,* 885–900.

515. Cook, E. H., Arora, R. C., Anderson, G. M., et al. (1993). Platelet serotonin studies in hyperserotonemic relatives of children with autistic disorder. *Life Sci, 52,* 2005–2015.

516. Levy, P. Q., & Bicho, M. P. (1997). Platelet serotonin as a biological marker of autism. *Acta Med Port, 10,* 927–931.

517. McBride, P. A., Anderson, G. M., Hertzig, M. E., et al. (1998). Effects of diagnosis, race, and puberty on platelet serotonin levels in autism and mental retardation. *J Am Acad Child Adolesc Psychiatry, 37,* 767–776.

518. Piven, J., Tsai, G. C., Nehme, E., et al. (1991). Platelet serotonin, a possible marker for familial autism. *J Autism Dev Disord, 21,* 51–59.

519. Rolf, L. H., Haarman, F. Y., Grotemeyer, K. H., et al. (1993). Serotonin and amino acid content in platelets of autistic children. *Acta Psychiatr Scand, 87,* 312–316.

520. Schain, R. J., & Freedman, D. X. (1961). Studies on 5-hydroxyindole metabolism in autistic and other mentally retarded children. *J Pediatr, 58,* 315–320.

521. Schmahmann, J. J. (1994). In M. L. Bauman & T. L. Kemper (Eds.), *The Neurobiology of Autism* (pp. 195–226). Baltimore, MD: Johns Hopkins University Press.

522. Geller, E., Yuwiler, A., Freeman, B. J., et al. (1988). Platelet size, number, and serotonin content in blood of autistic, childhood schizophrenic, and normal children. *J Autism Dev Disord, 18,* 119–126.

523. Spivak, B., Golubchik, P., Mozes, T., et al. (2004). Low platelet-poor plasma levels of serotonin in adult autistic patients. *Neuropsychobiology, 2,* 157–160.

524. Katsui, T., Okuda, M., Usuda, S., et al. (1986). Kinetics of $^3$H-serotonin uptake by platelets in infantile autism and developmental language disorder (including five pairs of twins). *J Autism Dev Disord, 16,* 69–76.

525. Mulder, E. J., Anderson, G. M., Kema, I. P., et al. (2004). Platelet serotonin levels in persuasive developmental disorders and mental retardation: Diagnostic group differences, within-group distribution, and behavioral correlates. *J Am Acad Child Adolesc Psychiatry, 4,* 491–499.

526. Vered, Y., Golubchik, P., Mozes, T., et al. (2003). The platelet-poor plasma 5-HT response to carbohydrate rich meal administration in adult autistic patients compared with normal control. *Hum Psychopharmacol, 5,* 395–399.

527. Betancur, C., Corbex, M., Spielewoy, C., et al. (2002). *Mol Psychiatry, 1,* 67–71.

528. Marazziti, D., Muratori, F., Cesari, A., et al. (2000). Increased density of the platelet serotonin transporter in autism. *Pharmacopsychiatry, 5,* 165–168.

529. Marazziti, D., Muratori, F., Cesari, A., et al. (2000). Increased density of the platelet serotonin transporter in autism. *Pharmacopsychiatry, 33,* 165–168.

530. Anderson, G. M., Minderaa, R. B., van Benthem, P. P., et al. (1984). Platelet imipramine binding in autistic subjects. *Psychiatry Res, 11,* 133–141.

531. Filinger, E. J., Garcia-Cotto, M. A., Vila, S., et al. (1987). Possible relationship between pervasive developmental disorders and platelet monoamine oxidase activity. *Braz J Med Biol Res, 20,* 161–164.

532. Anderson, G. M., Gutknecht, L., Cohen, D. J., et al. (2002). Serotonin transporter promoter variants in autism: Functional effects and relationship to platelet hyperserotonemia. *Mol Psychiatry, 8,* 831–836.

533. Perry, B. D., Cook, E. H. Jr., Leventhal, B. L., et al. (1991). Platelet 5-HT2 serotonin receptor binding sites in autistic children and their first-degree relatives. *Biol Psychiatry, 30,* 121–130.

534. Shekim, W. O., Bylund, D. B., Alexson, J., et al. (1986). Platelet MAO and measures of attention and impulsivity in boys with attention deficit disorder and hyperactivity. *Psychiatry Res, 18,* 179–188.

535. Askenazy, F., Caci, H., Myquel, M., et al. (2000). Relationship between impulsivity and platelet serotonin content in adolescents. *Psychiatry Res, 94,* 19–28.

536. Weizman, A., Bernhout, E., Weitz, R., et al. (1988). Imipramine binding to platelets of children with attention deficit disorder with hyperactivity. *Biol Psychiatry, 23,* 491–496.

537. Pornnoppadol, C., Friesen, D. S., Haussler, T. S., et al. (1999). No difference between platelet serotonin–5-HT(2A) receptors from children with and without ADHD. *J Child Adolesc Psychopharmacol, 9,* 27–33.

538. Unis, A. S., Cook, E. H., Vincent, J. G., et al. (1997). Platelet serotonin measures in adolescents with conduct disorder. *Biol Psychiatry, 42,* 553–559.

539. Russo, S., Boon, J. C., Kema, I. P., et al. Patients with carcinoid syndrome exhibit symptoms of aggressive. *Psychosom Med, 3,* 422–425.

540. Verkes, R. J., Van der Mast, R. C., Kerkhof, A. J., et al. (1998). Platelet serotonin, monoamine oxidase activity, and [$^3$H]paroxetine binding related to impulsive suicide attempts and borderline personality disorder. *Biol Psychiatry, 43,* 740–746.

541. Alm, P. O., Alm, M., Humble, K., et al. (1994). Criminality and platelet monoamine oxidase activity in former juvenile delinquents as adults. *Acta Psychiatr Scand, 89,* 41–45.

542. Stalenheim, E. G. (2004). Long-term validity of biological markers of psychopathy and criminal recidivism: Follow-up 6–8 years after forensic psychiatric investigation. *Psychiatry Res, 3,* 281–291.

543. Longato-Stadler, E., af Klinteberg, B., Garpenstrand, H., et al. (2002). Personality traits and platelet monoamine oxidase activity in a Swedish male criminal population. *Neuropsychobiology, 4,* 202–208.

544. Stoff, D. M., Pollock, L., Vitiello, B., et al. (1987). Reduction of (3H)-imipramine binding sites on platelets of conduct-disordered children. *Neuropsychopharmacology, 1,* 55–62.

545. Merenakk, L., Harro, M., Kiive, E., et al. (2003). Association between substance use, personality traits, and platelet MAO activity in preadolescents and adolescents. *Addict Behav, 8,* 1507–1514.

546. Stoff, D. M., Friedman, E., Pollock, L., et al. (1989). Elevated platelet MAO is related to impulsivity in disruptive behavior disorders. *J Am Acad Child Adolesc Psychiatry, 28,* 754–760.

547. Patkar, A. A., Gottheil, E., Berrettini, W. H., et al. Relationship between platelet serotonin uptake sites and measures of impulsivity, aggression, and craving among African-American cocaine abusers. *Am J Addict, 5,* 432–447.

548. Stadler, C., Schmeck, K., Nowraty, I., et al. (2004). Platelet 5-HT uptake in boys with conduct disorder. *Neuropsychobiology, 3,* 244–251.

549. Pine, D. S., Wasserman, G. A., Coplan, J., et al. (1996). Platelet serotonin 2A (5HT2A) receptor characteristics and parenting factors for boys at risk for delinquency: A preliminary report. *Am J Psychiatry, 153,* 538–544.

550. Ambrosini, P. J., Metz, C., Arora, R. C., et al. (1992). Platelet imipramine binding in depressed children and adolescents. *J Am Acad Child Adolesc Psychiatry, 31,* 298–305.

551. Sallee, F. R., Hilal, R., Dougherty, D., et al. (1998). Platelet serotonin transporter in depressed children and adolescents: [3]H-paroxetine platelet binding before and after sertraline. *J Am Acad Child Adolesc Psychiatry, 37,* 777–784.

552. Weizman, A., Carel, C., Tyano, S., et al. (1985). Decreased high affinity 3H-imipramine binding in platelets of enuretic children and adolescents. *Psychiatry Res, 14,* 39–46.

553. Weizman, R., Carmi, M., Tyano, S., et al. (1986). Reduced 3H-imipramine binding but unaltered 3H-serotonin uptake in platelets of adolescent enuretics. *Psychiatry Res, 19,* 37–42.

554. Carstens, M. E., Engelbrecht, A. H., Russell, V. A., et al. (1988). Biological markers in juvenile depression. *Psychiatry Res, 23,* 77–88.

555. Rehavi, M., Weizman, R., Carel, C., et al. (1984). High-affinity 3H-imipramine binding in platelets of children and adolescents with major affective disorders. *Psychiatry Res, 13,* 31–39.

556. Pine, D. S., Trautman, P. D., Shaffer, D., et al. (1995). Seasonal rhythm of platelet [3H]imipramine binding in adolescents who attempt suicide. *Am J Psychiatry, 152,* 923–925.

557. Pliszka, S. R., Rogeness, G. A., & Medrano, M. A. (1988). DBH, MHPG, and MAO in children with depressive, anxiety, and conduct disorders: Relationship to diagnosis and symptom ratings. *Psychiatry Res, 24,* 35–44.

558. Soreni, N., Apter, A., Weizman, A., et al. (1999). Decreased platelet peripheral-type benzodiazepine receptors in adolescent inpatients with repeated suicide attempts. *Biol Psychiatry, 46,* 484–488.

# PART FOUR

# Disorders of Platelet Number and Function

# CHAPTER 45

# The Clinical Approach to Disorders of Platelet Number and Function

## Alan D. Michelson

*Center for Platelet Function Studies, Departments of Pediatrics, Medicine, and Pathology,*
*University of Massachusetts Medical School, Worcester, Massachusetts*

## I. Introduction

This chapter discusses the clinical approach to patients with disorders of platelet number and function (Tables 45-1 through 45-3). As in all of clinical medicine, the history and physical examination are the keys to unlocking the diagnosis. Table 45-4 lists the characteristic clinical features that differentiate primary hemostatic disorders (thrombocytopenia, platelet function defects, and von Willebrand disease [VWD]) from coagulation disorders (e.g., hemophilia). Three of these clinical features are particularly helpful. First, petechiae are a strong pointer toward a primary hemostatic disorder and away from a coagulation disorder. Second, hemarthroses are a strong pointer toward a coagulation disorder and away from a primary hemostatic disorder. Third, in the setting of an injury, immediate excessive bleeding suggests a primary hemostatic disorder (because the initial platelet plug does not form correctly), whereas delayed bleeding suggests a coagulation disorder (because the lack of a well-formed fibrin clot results in gradual breakdown of the initial platelet plug).

The clinical approaches to thrombocytopenia during pregnancy and in the neonatal period are discussed separately in Chapters 51 and 52, respectively. The clinical approach to thrombocytosis is discussed in Chapter 56.

## II. Clinical History

### A. Bleeding

Spontaneous bleeding, or excessive bleeding during or after an injury, suggests an underlying hemostatic disorder. The clinician must obtain a history of the patient's responses to hemostatic challenges: surgery (including circumcision), dental procedures (especially tooth extractions), trauma, injections, menses, labor and delivery, and tooth brushing. Immediate bleeding suggests a primary hemostatic disorder,

whereas delayed bleeding suggests a coagulation disorder (Table 45-4).

Easy bruising is common in hemostatic disorders but does not help distinguish disorders of platelet number and function from other disorders of hemostasis. However, epistaxes, gingival bleeding, and menorrhagia are all more common in primary hemostatic disorders than in coagulation disorders.

The age of onset of the excessive bleeding may help to distinguish inherited bleeding disorders from acquired bleeding disorders.

### B. Gender and Family History

Some disorders of platelet number and function are X-linked recessive (e.g., Wiskott–Aldrich syndrome [Chapters 54 and 57]) and therefore present almost exclusively in males. Other disorders of platelet number and function are autosomal recessive (e.g., Bernard–Soulier syndrome and Glanzmann thrombasthenia) or autosomal dominant (e.g., May–Hegglin anomaly) (Chapters 54 and 57). Hemophilia A (factor VIII deficiency) and hemophilia B (factor IX deficiency) are common X-linked recessive coagulation disorders that occur almost exclusively in males.

### C. Medications

The history of the patient's intake of medications is important. Thrombocytopenia can result from drugs, including heparin (Chapter 48), quinidine, quinine, sulfonamides, rifampin, penicillin, vancomycin, gold salts, procainamide, valproic acid, carbamazepine, chlorothiazide, abciximab, tirofiban, and eptifibatide (Chapter 49). Furthermore, a platelet function defect can be caused by many drugs, including aspirin, nonsteroidal anti-inflammatory agents, ADP receptor antagonists (clopidogrel, ticlopidine), glycoprotein (GP) IIb-IIIa antagonists (abciximab, tirofiban,

## Table 45-1: Causes of Thrombocytopenia in the Absence of Leukopenia or Anemia

**Increased Platelet Destruction**

Immune thrombocytopenic purpura
Disseminated intravascular coagulation
Heparin-induced thrombocytopenia
Other drug-induced thrombocytopenias
Systemic lupus erythematosus
HIV-1-related thrombocytopenia
Thrombotic thrombocytopenic purpura/hemolytic-uremic
  syndrome
Common variable immunodeficiency
Post-transfusion purpura
Type 2B von Willebrand disease
Platelet-type von Willebrand disease
Wiskott–Aldrich syndrome/X-linked thrombocytopenia

**Decreased Platelet Production**

**Normal-sized platelets**
Amegakaryocytic thrombocytopenia
Amegakaryocytic thrombocytopenia with radio-ulnar
  synostosis
Thrombocytopenia with absent radii (TAR) syndrome
Autosomal dominant thrombocytopenia with linkage to
  human chromosome 10
Familial platelet disorder with predisposition to myeloid
  malignancy
GATA-1 mutation of X-linked thrombocytopenia and
  thalassemia
Paris–Trousseau thrombocytopenia/Jacobsen syndrome

**Large platelets**
Bernard–Soulier syndrome/DiGeorge syndrome (velo-
  cardio-facial syndrome)
Benign Mediterranean macrothrombocytopenia
MYH9-related disease (May–Hegglin anomaly, Sebastian
  syndrome, Fechtner syndrome, Epstein syndrome)
Gray platelet syndrome
Montreal platelet syndrome
Macrothrombocytopenia with platelet expression of
  glycophorin A
Macrothrombocytopenia with platelet $\beta 1$ tubulin Q43P
  polymorphism

**Sequestration**

Hypersplenism
Kasabach–Merritt syndrome

**Increased Platelet Destruction and Hemodilution**

Extracorporeal perfusion

## Table 45-2: Causes of Platelet Function Defects

**Acquired**

Uremia
Myeloproliferative disorders
  Essential thrombocythemia
  Polycythemia vera
  Chronic myeloid leukemia
  Agnogenic myeloid metaplasia
Acute leukemias and myelodysplastic syndromes
Dysproteinemias
Extracorporeal perfusion
Acquired von Willebrand disease
Acquired storage pool deficiency
Antiplatelet antibodies
Liver disease
Drugs and other agents

**Inherited**

**Platelet adhesion defects**
Bernard–Soulier syndrome
von Willebrand disease
**Agonist receptor defects**
Integrin $\alpha 2\beta 1$ (collagen receptor) deficiency
GPVI (collagen receptor) deficiency
P2Y$_{12}$ (ADP receptor) deficiency
Thromboxane A$_2$ receptor deficiency
**Signaling pathway defects**
G$_{\alpha q}$ deficiency
Phospholipase C-$\beta_2$ deficiency
Cyclooxygenase deficiency
Thromboxane synthetase deficiency
Lipoxygenase deficiency
Defects in calcium mobilization
**Secretion defects**
Storage pool disease
Hermansky–Pudlak syndrome
Chediak–Higashi syndrome
Gray platelet syndrome
Quebec syndrome
Wiskott–Aldrich syndrome
**Aggregation defects**
Glanzmann thrombasthenia
Congenital afibrinogenemia
**Platelet-coagulant protein interaction defects**
Scott syndrome

**Table 45-3: Causes of Thrombocytosis**

| Primary Thrombocytosis |
| --- |

Essential thrombocythemia
Chronic myeloproliferative disorders (including polycythemia vera, myelofibrosis with myeloid metaplasia)
Chronic myeloid leukemia
Myelodysplastic syndrome
Congenital thrombocytosis

| Reactive Thrombocytosis |
| --- |

Infection
Rebound thrombocytosis (e.g., after chemotherapy or immune thrombocytopenic purpura)
Tissue damage (e.g., surgery)
Chronic inflammation
Malignancy
Renal disorders
Hemolytic anemia
Iron deficiency
Asplenia (post-splenectomy, post-infarction, or congenital)

**Table 45-4: Characteristic Clinical Features That Differentiate Primary Hemostatic Disorders from Coagulation Disorders**

|  | Primary Hemostatic Disorder | Coagulation Disorder |
| --- | --- | --- |
| Prototypic Disorders | Thrombocytopenia Platelet function defect von Willebrand disease[a] | Hemophilia |
| Bleeding | Immediate | Delayed |
| Petechiae | Yes | No |
| Hemarthroses | No | Yes |
| Intramuscular hematomas | Uncommon | Common |
| Epistaxes | Common | Uncommon |
| Menorrhagia | Common | Uncommon |

[a]In the uncommon type 3 von Willebrand disease, the factor VIII coagulant level is low enough for the clinical features to be those of a combined primary hemostatic and coagulation disorder.

eptifibatide), dipyridamole, cilostazol, antimicrobial agents (β-lactam antibiotics, nitrofurantoin, hydroxychloroquine, miconazole), cardiovascular drugs (propranolol, nitroprusside, nitroglycerin, furosemide, calcium channel blockers), ethanol, and food items and supplements (omega-3 fatty acids, vitamin E, onions, garlic, ginger, cumin, turmeric, clove, black tree fungus, Ginko) (Chapter 58).

### D. Medical History

A history of renal disease, liver disease, myeloproliferative disorder, acute leukemia, myelodysplastic syndrome, or dysproteinemia suggests an acquired platelet function defect (Table 45-2 and Chapter 58).

### E. Transfusion History

The sudden development of profound thrombocytopenia 5 to 10 days after the transfusion of blood products (not necessarily platelet concentrates) strongly suggests the diagnosis of posttransfusion purpura (Chapter 53).

### F. Social-Sexual History

Homosexual persons, intravenous narcotic addicts, and their sexual partners are at risk for HIV-1 infection and the subsequent development of chronic immune thrombocytopenic purpura. The thrombocytopenia can present in the presence or absence of AIDS (Chapter 47).

## III. Physical Examination

### A. Hemorrhage

Petechiae strongly suggest a primary hemostatic disorder rather than a coagulation disorder (Table 45-4). Petechiae can also signal vasculitis or, in a sick patient, bacteremia (including meningococcemia) and disseminated intravascular coagulation (DIC). In contrast, hemarthroses strongly suggest a coagulation disorder rather than a primary hemostatic disorder (Table 45-4). Intramuscular hematomas are more typical of a coagulation disorder than of a primary hemostatic disorder.

### B. Other Physical Signs

Splenomegaly may indicate hypersplenism (e.g., Gaucher disease) or malignancy. Fluctuating global or focal neurological signs suggest thrombotic thrombocytopenic purpura (Chapter 50). Bilateral absent radii are observed in the thrombocytopenia with absent radii (TAR) syndrome (see Figure 54-2 in Chapter 54). Eczema in a boy suggests the possibility of Wiskott–Aldrich syndrome (Chapters 54 and 57). Giant hemangiomas suggest Kasabach–Merritt syndrome (Chapter 52). Oculocutaneous albinism is observed in Hermansky–Pudlak syndrome (Chapters 15 and 57) and Chediak–Higashi syndrome (Chapters 15 and 57). Deafness and cataracts suggest Fechtner syndrome (Chapters 54 and

57). Deafness and hypertension occur in Epstein syndrome (Chapters 54 and 57). Cardiac defects, delayed development, otorhinolaryngological manifestations, psychiatric disturbances, and mental retardation suggest the velo-cardio-facial syndrome that is associated with Bernard–Soulier syndrome (Chapters 54 and 57).

## IV. Clinical Tests

### A. Platelet Count

The causes of a low platelet count in the absence of leukopenia or anemia are shown in Table 45-1. Aplastic anemia and infiltrative disorders of the bone marrow, including leukemia and solid tumor metastases, do not usually cause an isolated thrombocytopenia of major degree in the absence of any leukopenia or anemia.

Pseudothrombocytopenia (Chapter 55), the result of platelet clumping *ex vivo,* can cause erroneously low platelet counts reported by electronic counters. The platelet clumping can be observed in the peripheral blood smear by light microscopy (see Figures 55-1 through 55-7 in Chapter 55). In most cases, drawing the blood into citrate anticoagulant rather than the standard EDTA will circumvent, and therefore diagnose, the problem of pseudothrombocytopenia.

The causes of a high platelet count are shown in Table 45-3.

### B. Blood Smear

**1. Platelets.** Large platelets may be observed in immune thrombocytopenic purpura (ITP, Chapter 46). Giant platelets are observed in essential thrombocythemia and other causes of primary thrombocytosis (Table 45-3; also see Figure 56-1 in Chapter 56) and in inherited giant platelet syndromes (including Bernard–Soulier syndrome, May–Hegglin anomaly, Fechtner syndrome, Sebastian syndrome, Epstein syndrome, and Montreal platelet syndrome) (Table 45-1 and see Figure 54-3 in Chapter 54). Agranular, gray platelets are observed in the gray platelet syndrome (Chapters 54 and 57). Although the platelets are small in Wiskott–Aldrich syndrome (Chapters 54 and 57), this may not be appreciated on conventional blood smears.

**2. Leukocytes.** Myeloblasts or lymphoblasts suggest the diagnosis of leukemia or myeloproliferative disorder. Döhle bodies are observed in the neutrophil cytoplasm in May–Hegglin anomaly (see Figure 54-3 in Chapter 54). Neutrophil inclusions are also observed in Sebastian syndrome (Chapters 54 and 57). Giant cytoplasmic granules are observed in the neutrophils, monocytes, and lymphocytes in Chediak–Higashi syndrome (Chapters 15 and 57), but these granules may be more easily recognized in leukocyte precursors in the bone marrow rather than in leukocytes in the peripheral blood.

**3. Erythrocytes.** Red cell fragmentation (schistocytes) suggests thrombotic thrombocytopenic purpura, hemolytic-uremic syndrome, or DIC (see Figure 50-1 in Chapter 50). Spherocytes may be observed in Evans syndrome (Chapter 46).

### C. Platelet Function Tests

Chapter 23 discusses clinical tests of platelet function. In addition, the following chapters in this book provide a detailed discussion of specific, well-characterized clinical tests of platelet function: the bleeding time (Chapter 25), platelet aggregation (Chapter 26), VerifyNow (Chapter 27), the platelet function analyzer (PFA)-100 (Chapter 28), the Impact cone and plate(let) analyzer (Chapter 29), flow cytometry (Chapter 30), and thromboxane generation (Chapter 31).

### D. Other Tests

Bone marrow aspiration with or without biopsy may be useful in the diagnosis of leukemia and other malignancies, aplastic anemia, ITP (especially in adults), amegakaryo-cytic thrombocytopenia, and Gaucher disease. Furthermore, bone marrow examination may be necessary in the evaluation of thrombocytosis, if reactive thrombocytosis (Table 45-3) has not been confirmed by history, physical examination, or noninvasive laboratory tests (as discussed in Chapter 56).

Other tests used in the specific diagnosis of disorders of platelet number and function are discussed in the context of the differential diagnosis (see Section V) and in the other chapters of this book (Chapters 46 through 59) that discuss these disorders in detail.

## V. Differential Diagnosis

The presence of a disorder of platelet number or function can usually be determined from the clinical history, physical examination (Table 45-4), and complete blood count (including examination of the blood smear). The following subsections provide an overview of the clinical approach to establishing the specific cause of the thrombocytopenia or platelet function defect.

## A. Thrombocytopenia

Although the list of causes of thrombocytopenia may appear dauntingly long (Table 45-1), the diagnosis of the specific cause of thrombocytopenia in a patient is often straightforward. ITP (Chapter 46) is a diagnosis of exclusion in a patient with acquired thrombocytopenia: a normal hemoglobin level, white blood cell count and differential, and blood smear (except for the thrombocytopenia and possibly enlarged platelets), in the absence of hepatosplenomegaly, lymphadenopathy, abnormalities of the radii, or other underlying disease. Anemia in the presence of ITP may signify either significant bleeding or autoimmune hemolytic anemia (Evans syndrome, Chapter 46). DIC can essentially be excluded if a thrombocytopenic patient is generally well without a clear precipitating cause for DIC (e.g., sepsis, hypoxia, or shock). If suspected, DIC can be diagnosed by D-dimer and fibrin split products. Heparin-induced thrombocytopenia (Chapter 48) or other drug-induced thrombocytopenia (Chapter 49) can be excluded or suspected based on the clinical history. Systemic lupus erythematosus (SLE)-related thrombocytopenia (Chapter 46) may be initially difficult to diagnose if the thrombocytopenia precedes other manifestations of SLE. HIV-1-related thrombocytopenia (Chapter 47) may be suspected from the clinical history, and HIV can be tested for if this diagnosis is not already known. Patients with thrombotic thrombocytopenic purpura or hemolytic-uremic syndrome, although listed in Table 45-1, usually have anemia and fragmented red cells on blood smear (see Figure 50-1 in Chapter 50). In addition, blood urea nitrogen and serum creatinine are elevated in hemolytic-uremic syndrome. Common variable immunodeficiency is diagnosed by measuring quantitative serum immunoglobulins. Post-transfusion purpura (Chapter 53) is strongly suggested by the sudden development of profound thrombocytopenia 5 to 10 days after the transfusion of blood products. Type 2B VWD and platelet-type VWD, both rare causes of thrombocytopenia, can be suspected by increased ristocetin-induced platelet agglutination at low ristocetin concentrations.[1] The absent radii in the TAR syndrome (see Figure 54-2 in Chapter 54) are all too evident by examination of the patient. Amegakaryocytic thrombocytopenia (Chapter 54) is diagnosed by examination of the bone marrow. Giant platelet syndromes (Chapters 54 and 57) are evident on peripheral blood smear (see Figure 54-3 in Chapter 54). Wiskott–Aldrich syndrome (Chapters 54 and 57) can be suspected in a male patient with thrombocytopenia and eczema. In hypersplenism, which can occur in any patient with an enlarged spleen from any cause, the thrombocytopenia is usually not severe and anemia and leukopenia are often present. The giant hemangiomas of Kasabach–Merritt syndrome (Chapter 52) are often evident on physical examination.

## B. Platelet Function Defects

The list of specific causes of platelet function defects is also long (Table 45-2). However, acquired defects are the most common and the underlying medical condition is usually evident. With the exception of VWD[1] and, to a lesser extent, storage pool disease (Chapters 15 and 57), inherited platelet function defects are all rare. These rare disorders (Chapter 57) usually require specialized research laboratories to make a specific diagnosis, with the exception of Bernard–Soulier syndrome and Glanzmann thrombasthenia, which are easily diagnosed by platelet aggregometry (Chapter 26) and flow cytometry (Chapter 30).

Platelet aggregation (Chapter 26) is useful for the diagnosis of platelet function defects. In Glanzmann thrombasthenia (Chapter 57), the platelets do not aggregate in response to adenosine diphosphate (ADP), epinephrine, arachidonic acid, or collagen, but do agglutinate in response to ristocetin. In Bernard–Soulier syndrome (Chapters 54 and 57), the platelets do not agglutinate in response to ristocetin but do aggregate in response to all other agonists. Patients with storage pool disease or defects in platelet secretion (Chapters 15 and 57) show decreased platelet aggregation or absence of the second wave of aggregation in response to ADP, epinephrine, arachidonic acid, or collagen. Some patients with storage pool disease cannot be reliably diagnosed by standard platelet aggregometry; the diagnosis is made by measuring the granule adenine nucleotides by biochemical assays or by labeling the platelets with the fluorescent dye mepacrine and then measuring platelet fluorescence by microscopy or flow cytometry (Chapter 30).

## C. von Willebrand Disease

A primary hemostatic disorder in the absence of thrombocytopenia or morphological evidence of a platelet function disorder may require testing for VWD. VWD is not primarily a disorder of platelet function but a disorder of von Willebrand factor (VWF).[1] Because of the key role of VWF in platelet adhesion via its binding to the GPIb-IX-V complex (Chapters 7 and 18), VWD results, however, in a platelet adhesion defect. Therefore, VWD usually presents clinically as a primary hemostatic disorder (Table 45-4), which, without laboratory testing, may be clinically indistinguishable from disorders of platelet number and function. VWF has a second major role: it is the carrier molecule for factor VIII. In the uncommon setting of type 3 VWD, the factor VIII coagulant activity is low enough for the clinical features to be those of a combined primary hemostatic disorder and coagulation disorder (Table 45-4). VWD is usually diagnosed by a decrease in plasma levels of ristocetin cofactor activity, VWF antigen, and factor VIII coagulant activity. However, not all patients with VWD have a decrease in the plasma levels of all three of these parameters.[1]

## VI. Specific Disorders

The specific disorders of platelet number and function (Tables 45-1 through 45-3), including their pathophysiology, clinical features, differential diagnosis, definitive diagnostic tests, and treatment, are discussed in detail in the chapters that follow (Chapters 46 through 59).

## Reference

1. Sadler, J. E. (2005). New concepts in von Willebrand disease. *Annual Rev Med, 56,* 173–191.

# CHAPTER 46

# Immune Thrombocytopenic Purpura

**James B. Bussel**

*Department of Pediatrics, Weill Medical College of Cornell University, New York Presbyterian Hospital*
*New York, New York*

## I. Introduction

The definition of immune thrombocytopenic purpura (ITP)[1-5] used in this chapter is modified from that proposed by the Practice Guidelines of the American Society of Hematology for ITP.[3] ITP is defined as thrombocytopenia: (a) in the presence of a normal hemoglobin level, white blood cell count and differential, and blood smear and (b) the absence of other causes of thrombocytopenia. However, the fulfillment of part b, "the absence of other causes of thrombocytopenia," is often not straightforward and may require considerable experience. Whereas marked hepatosplenomegaly or lymphadenopathy would direct the clinician to leukemia and abnormalities of the radial ray to the thrombocytopenia with absent radii (TAR) syndrome or Fanconi anemia, other, more subtle immune or nonimmune thrombocytopenias such as the MYH9-related diseases (e.g., May–Hegglin anomaly; see Chapters 54 and 57) or possible drug-induced thrombocytopenias may be difficult to diagnose. Non-immune inherited thrombocytopenias have recently been reviewed.[6,7]

The pathogenesis of ITP is well known to be accelerated platelet destruction as a result of antiplatelet autoantibodies. However, it has become clear that there is an important component of reduced platelet production, mediated by these antibodies. Initially, studies of platelet production showed in one case a normal rate of platelet production[8] and in the other only a threefold increase.[9] Subsequently, platelet survival studies using autologous chromium-labeled platelets suggested a platelet half-life as short of minutes. When Indium[111] became available, studies with autologous platelets demonstrated considerably longer platelet survivals in ITP (i.e., 2 to 3 days). In 1994, several different laboratories cloned thrombopoietin (TPO) (see Chapter 66), and shortly thereafter a number of studies demonstrated that the serum levels of TPO in patients with ITP were normal.[10,11] More recently, two independent investigations showed that antibodies in the sera of patients with ITP were capable of inhibiting the maturation and differentiation of immature megakaryocyte precursors to megakaryocytes.[12,13] Finally,

at least one study suggested that even though there were increased numbers of megakaryocytes in the marrow in patients with ITP, many of these were apoptotic and thus not making platelets.[14] Thus, a considerable body of work demonstrates that many (especially refractory) patients with ITP might not be making platelets very well.

Disease severity can be extrapolated from the degree of clinical hemorrhage as well as the platelet response to treatment(s). Fortunately, ITP has a relatively infrequent rate of severe bleeding including, but not limited to, intracranial hemorrhage.[3,15,16] This makes treatment and prevention of bleeding difficult to assess. Attempts to quantify bleeding are under development. The World Health Organization (WHO) bleeding scale,[17] used in trials involving chemotherapy-induced thrombocytopenia, has limited applicability to ITP because it depends too heavily on interventions such as transfusions that are rarely used in ITP.

There are a number of complicated/complicating issues in ITP:

1. There is no sensitive or specific diagnostic test. Platelet antibody testing, as currently available, cannot be used to include or exclude the diagnosis of ITP in the ambiguous cases for which there is the greatest need for such a test.[3,18]

2. The natural history and approach to management is different in children and adults, although certain similarities in both have become apparent.[3-5]

3. Whereas the great majority of cases considered to be ITP are in fact autoantibody mediated, some cases represent subtle forms of bone marrow failure (i.e., myelodysplastic syndrome). These cases, if the thrombocytopenia is severe, may be "selected" to appear among the cases of ITP that are the most refractory to treatment, because treatments that may be effective in cases that are autoimmune in origin often have little or no effect in patients with nonimmune forms of thrombocytopenia. There may also be cytotoxic T cell-mediated thrombocytopenia, the specific clinical features of which remain unknown.

4. There are a lack of comparative treatment studies in ITP, other than those performed at the time of diagnosis in children with acute ITP.

These issues and others make it difficult to scientifically determine the optimal approach to the diagnosis and treatment of individual patients with ITP, although this has recently been reviewed.[19]

This chapter on ITP is divided into the following sections: incidence, etiology, diagnosis, and treatment.

## II. Incidence

The incidence of ITP remains uncertain. The incidence has been estimated by comparing the number of cases of ITP to the incidence of better defined diseases such as leukemia and hemophilia.[24] Another approach has been to review discharge diagnoses from hospital admissions.[25] These estimates have approximated an incidence of 1 case of ITP per 66–100,000 population.[3] These and other studies have suggested that there is a 2 to 3 : 1 female to male predominance, consistent with the generally recognized increased incidence of autoimmune disease in women and a peak age of onset of ITP in women of child-bearing age. However, studies from Denmark[26] and Newcastle[27] accrued "all" cases of ITP in a selected part of the country and found no female predominance and an increasing incidence of ITP in the elderly population over the age of 60 years. Whether a fraction of these cases in the over-60 population were impending myelodysplasia is unknown. In prepubertal children, the peak age incidence is 2 to 4 years, and girls and boys are equally affected.[3]

There are a number of ancillary issues to consider in evaluating this information. First, the higher the "allowable" platelet count in making the diagnosis of ITP, the greater will be the incidence of ITP. Second, the diagnosis of ITP is currently a clinical one made by exclusion. Therefore, how often the diagnosis of ITP is made may vary according to the practices of a region, a hospital, or individual physicians, including whether the majority of the population obtains a yearly platelet count on routine visits to physicians. One study discovered that 50% of 185 adults with ITP being followed at tertiary care centers had had their initial thrombocytopenia discovered when they were asymptomatic and on a routine visit to their primary care physician.[28] Also, there may be a bias toward recognizing ITP in women because platelet counts may be routinely obtained during visits to obstetrician-gynecologists, especially during pregnancy.

An issue that impacts the prevalence of refractory ITP is the relatively good longevity of affected patients. If there are a certain number of adults who fail splenectomy each year, a relatively small percentage (probably 3% at most) die

within 1 year.[29] Some of these deaths are not directly ITP-related (i.e., they are not hemorrhagic). Therefore, if there are 1000 to 2000 adults with ITP who fail splenectomy each year in the United States, the number of adults who have failed splenectomy but are still alive and still require treatment is steadily increasing. This group is augmented by those patients with initial response to splenectomy who then relapse after years in remission (discussed later).[30,31] Conversely, there may be delayed "responders" whose ITP improves years after onset.

## III. Etiology

### A. Infection

Infections have a clear role in the initiation of many cases of ITP, at least in childhood. Virtually any virus has been identified immediately prior to the onset of ITP.[3,4] However, the infection is thought to serve merely as an initiator (by a mechanism still incompletely understood) of the disease, and viral persistence is not thought to be required for the development or persistence of ITP. A cross-reactive antibody generated to the virus also may react with an antigen on the platelet membrane, as has been suggested for several cases of varicella.[32] It is less clear if this pathogenesis is specifically related to cases of acute but not chronic ITP. Another proposed mechanism is that the viral infection disrupts the immunoregulatory network in a transient but strategic way that allows an antiplatelet autoantibody to be produced. This may be in association with a predominance of T Helper 2 (TH2) cytokines that represent an individual predisposition to autoimmunity.[33] Finally, in some cases the platelet count may, unknown to the patient or physician, already be reduced prior to the viral infection. In these cases, the viral infection may create a relatively slight further reduction in the platelet count that results in hemorrhagic symptomatology and diagnosis. Cooper and Bussel provide further details.[34]

In certain situations, the role of infection appears to be different: the chronicity and severity of ITP may be strongly linked to the persistence of the underlying infection. The most obvious example is HIV-related thrombocytopenia (see Chapter 47), in which there is persistent infection and the presence of antibody does not clear the infection. A percentage of HIV-infected patients develop thrombocytopenia, which is thought to result from a combination of autoimmune and platelet production-inhibiting pathophysiologies (see Chapter 47); this has become far less of a problem as a result of highly active antiretroviral therapy (HAART). Similarly, hepatitis C results in persistent viral infection, and a percentage of these patients develop thrombocytopenia.[35] The thrombocytopenia in hepatitis C infection has not been studied as extensively as HIV-related thrombocytopenia. However, recent work has suggested that these patients may bleed at higher platelet counts than classical ITP patients as a result of vascular damage mediated by cryoglobulins.[36] Furthermore, hypersplenism, low TPO levels if the liver is severely damaged, disseminated intravascular coagulation (DIC), and the effects of interferon treatment may all contribute to the hepatitis C-induced thrombocytopenia. Neither cytomegalovirus[37] nor HTLV1[38] have been demonstrated to be associated with ITP-like thrombocytopenia, although some anecdotal experience has suggested that a small number of highly refractory cases of ITP are exacerbated by CMV.

Whether *Helicobacter pylori* has an important role in the etiology or persistence of ITP remains unclear despite more than 50 publications in the past 10 years on this association.[39–43] Studies from Italy and Japan suggest that H. pylori may contribute to both the etiology and persistence of ITP in a number of cases;[41,42] however, investigations in other countries have not seen equivalent effects. Persistence has been demonstrated by eradicating the H. pylori and demonstrating an increase in the platelet count. One metanalysis[40] suggested that patients whose ITP is both milder and less pretreated are more likely to respond to eradication of H. pylori with a sustained platelet increase. The breath test and stool antigen analysis are both superior to antibody testing in the diagnosis of H. pylori infection.

### B. Genetic Predisposition

It has been hypothesized that individuals may have an inherited predisposition to ITP. An infection or medication may result in the development of ITP in these individuals but not other individuals. Against this hypothesis, however, is the fact that only quite infrequently is there more than one case of ITP in a family. Also unresolved is the percentage of patients with ITP who would have one or more specific predispositions to develop ITP if all known predispositions were tested (discussed later).

A component of the etiology of ITP may include immune response genes that predispose patients to develop ITP. One approach to understanding this component is to consider the associated diseases linked with ITP. A common theme of immune dysregulation, not defined in its specifics, is associated with many of the diseases linked to ITP. The most obvious example is an inability to make antibodies normally. Chronic lymphocytic leukemia (CLL), common variable immunodeficiency (CVID, hypogammaglobulinemia), IgA deficiency, and IgG$_2$ deficiency have all been linked to ITP with an incidence far higher than would be expected in the normal population.[44–47] The mechanism of development of ITP may vary, but one study suggested that autoantibodies may be restricted to only one Vh gene family.[48] The hypothesis is that an inability to make antibodies normally

(i.e., to use the full Vh gene repertoire) may result in overutilization of this autoantibody-generating Vh gene family. Of note, Bruton's X-linked agammaglobulinemia is not associated with ITP, suggesting that abnormal T or B cells are integral to the process of ITP. Although antibody deficiency with normal immunoglobulin levels may also be related to ITP, this has not been studied, at least in part because the widespread use of intravenous immunoglobulin (IVIG) makes it difficult to assess serum IgG antibody levels. These etiologies of ITP could be the result of abnormalities of B cells, but they could equally be explained by abnormal T cell interactions with B cells.

Alternatively, the T cell may be more directly implicated in ITP. The DiGeorge syndrome is a "pure" abnormality of T cells and is also associated with ITP.[49] Another pathophysiology could be failure of apoptosis of autoreactive T lymphocytes — that is, fas (CD95) or fas pathway deficiency genes such as those seen in the autoimmune lymphoproliferative Canale–Smith syndrome.[50,51] ITP and autoimmune hemolytic anemia are common in these patients. However, few if any cases of autoimmune lymphoproliferative syndrome (ALPS) have been identified in which the primary clinical manifestation is the ITP, without the associated splenomegaly and lymphadenopathy, which distinguish ALPS from ITP. Another area requiring further exploration is HLA typing and T cell Vβ repertoire selection. Early studies suggested that the "autoimmune" haplotype B8DR3 was increased in this patient population.[52,53] A more recent study in an inbred Asian population identified a link to failure to respond to splenectomy.[54] However, detailed studies of sufficient numbers of specific patient populations (e.g., splenectomy failures) have not yet been conducted. At least one ITP patient has been reported who overexpressed a certain T-cell receptor idiotype.[55] One study of T-cell spectratyping suggested that patients failing to respond to splenectomy have more abnormal T-cell clones than those patients who responded to splenectomy (11 versus 3, $p < 0.05$).[56] Rare cases of monoclonal T cell, and especially natural killer cell, populations have also been thought to create a thrombocytopenia analogous to "T killer cell" or large granular lymphocytosis with neutropenia.[57]

With regard to cytokine levels, one hypothesis is that patients with chronic ITP tend to produce a "TH2" response.[33] This can be observed either in the unstimulated state or, for example, following the administration of IV anti-D or IVIG. The implication is that the patient with chronic ITP has T cells that are primed for a certain response, unlike those of either normals or of other ITP patients with acute disease who are not predisposed to develop chronic disease.

Defects in the classical pathway of complement may operate via failure to rapidly clear immune complexes, resulting in longer persistence of an "enhanced" immunogen (part of an immune complex). Low C4 has been associated with ITP and low levels of all of the components of the classical pathway have been linked to systemic lupus erythematosus (SLE).[58] Studies of Fc receptors in patients with ITP have not been conclusive, but it appears that polymorphisms of these receptors may contribute to response to treatments.[59] Another approach has been to study heavy chain gene usage in autoantibodies.[48] Antiplatelet antibodies to glycoprotein (GP) Ib-IX were shown to share a single idiotype such that they could be identified by an anti-idiotypic antibody,[60] but there has been no follow-up of this finding.

In summary, a number of possible etiologies, postinfectious and immune predispositions, may underlie the apparent considerable heterogeneity among patients with ITP. Future studies need to target why certain people develop ITP, whereas the vast majority do not. These differences would also be important in characterizing this heterogeneity into several subtypes. These subtypes could potentially better define prognosis and therapeutic strategies. The apparent wide heterogeneity of affected patients could also result from the interaction of modifying genes rather than from multiple etiologies.

## IV. Diagnosis

As indicated in Section I, there is no diagnostic blood test (e.g., antiplatelet antibodies) that reliably and consistently includes or excludes ITP. The diagnosis of ITP therefore continues to be made on clinical grounds. An algorithm for the diagnosis of ITP could proceed in three steps:

1. Distinguish ITP from systemic disorders causing thrombocytopenia as part of their pathophysiology. The most often considered alternative diagnoses in this category are leukemia, bone marrow failure states, and infections.
2. Distinguish immune thrombocytopenia from nonimmune thrombocytopenia. Considering the latter involves identifying a positive family history, abnormalities of platelet size or morphology on blood smear (e.g., Bernard–Soulier syndrome [see Chapters 54 and 57]), or the presence of findings associated with specific thrombocytopenic syndromes such as renal disease, hearing loss, or absent radii (see Chapter 54).
3. Distinguish primary ITP from secondary ITP. Although in practice this overlaps with step 1, it is conceptually separate. It involves distinguishing primary, "idiopathic" ITP from ITP secondary to other diseases such as SLE, HIV, CLL, and CVID. In addition, heparin-induced thrombocytopenia (Chapter 48) and other drug-induced thrombocytopenias (Chapter 49) must also be considered; the latter are often particularly difficult to include or exclude.

A number of studies have clarified the approach to diagnosing ITP in children. One approach has been to review cases in which the bone marrow was examined to make the diagnosis of ITP (i.e., normal or increased megakaryocytes and no other diagnostic abnormalities) and compare the clinical features of the patients to the results of the bone marrow examination.[61] A second approach has been to review the presenting features of patients enrolled in childhood leukemia cooperative group studies to explore if any patients resembling ITP were eventually discovered to have leukemia.[62] The conclusions from these studies are as follows. If a child with usually severe thrombocytopenia has

1. the onset of bleeding symptoms and signs,
2. an otherwise normal complete blood count (CBC) including hemoglobin, red cell indices, white cell count, and blood smear, or
3. a normal physical examination except for signs of hemorrhage, including not only the absence of hepatosplenomegaly and lymphadenopathy but also normal radial rays (forearms and thumbs),

then the likelihood is very high that the diagnosis is ITP. In particular, such a child has a vanishingly small to nil chance of having leukemia. The primary diagnostic issue to consider in such cases is the differential diagnosis of isolated thrombocytopenia. Although primary ITP is by far the most common diagnosis, secondary ITP, familial/inherited thrombocytopenia (Chapter 54), and the rare cases of hypomegakaryocytic or amegakaryocytic thrombocytopenia are all possible. If the bleeding symptoms and signs develop suddenly and there is no family history of bleeding or thrombocytopenia, then the statistical likelihood that such a case will be primary ITP, already very high, increases to almost 100%. These diagnostic considerations point to the lack of the utility of a bone marrow examination alone to make the diagnosis of ITP in the typical case.

The diagnostic approach of the ASH guidelines requiring no laboratory testing other than a complete CBC with review of the smear is appropriate in a well child who is on no medications and whose only medical history of any note may be a viral upper respiratory infection 1 to 4 weeks prior to presentation. However, even in children, many features may complicate the diagnosis in certain cases. These include (a) the presence of hepatosplenomegaly in a post-Epstein–Barr virus case of ITP; (b) the existence of anemia of unclear etiology, possibly as a result of, for example, thrombocytopenia-induced epistaxis or menorrhagia (if the MCV is low suggesting iron deficiency anemia, this supports the diagnosis of ITP); or (c) the presence, especially in a female adolescent, of a number of features of "lupoid" ITP that are insufficient to make the formal diagnosis of SLE. In addition, the author has seen one case in which a patient had two normal bone marrow examinations, had steroids used to manage the presumed ITP, and ultimately was discovered to have leukemia-lymphoma after she no longer responded to steroids and a third marrow was performed. In these atypical cases of ITP, it is very important to proceed with additional testing, including bone marrow examination. If bone marrow hypoplasia is a consideration in the differential diagnosis, then it is mandatory to perform a bone marrow biopsy in addition to an aspirate. In the future, when the test becomes more readily available, an alternative to a bone marrow examination, when it is performed to determine the number of megakaryocytes in the marrow, may be plasma TPO, as plasma TPO levels appear to inversely correlate very well with the bone marrow megakaryocyte content rather than the platelet count.[63] Specifically, the TPO level is low (normal) in ITP, even in the setting of severe thrombocytopenia, whereas it is 10- to 20-fold increased in amegakaryocytic states. Testing of this latter setting has been retrospective and included patients with aplastic anemia and amegakaryocytic thrombocytopenia, but not large numbers of patients with leukemia.

In contrast to the diagnostic considerations in outpatient children described earlier, the diagnosis of ITP may be considerably more difficult in adult patients. For example, an elderly patient on 5 to 10 medications who is hospitalized, has a number of simultaneous medical problems including possibly infection, and then develops a progressive thrombocytopenia is a far more difficult diagnostic challenge. Because no single test confirms the diagnosis of ITP, testing is usually performed, as much as possible, to exclude other diagnoses. Examples of tests, in addition to a bone marrow examination, include but are not limited to the following:

1. Reticulocyte count, to help exclude both thrombotic thrombocytopenic purpura (TTP, see Chapter 50) and Evans syndrome (autoimmune anemia in association with ITP). (Other testing for Evans syndrome would include a direct Coombs test and a lactic dehydrogenase. Other testing for TTP would include serum ADAMTS13 levels, although TTP would be unlikely in the absence of fragmented red cells on the blood smear.)

2. Antinuclear antibody (ANA) and antidouble stranded DNA antibody to investigate the possibility of SLE.

3. Prothrombin time, partial thromboplastin time, fibrinogen, and D-dimer to consider DIC or other nonimmune consumption.

4. Radiologic studies (e.g., CT scan) to search for a lymphoma or other malignancy.

Measurement of antiplatelet antibodies is not listed, because no study has conclusively demonstrated that any method of platelet antibody testing sensitively and specifically identi-fies ITP in the ambiguous cases in which this test would be helpful. In addition to TPO levels, other testing that may be of use in the future includes reticulated (i.e., young) platelets.[64] These have been studied in a number of settings, because the percentage of reticulated platelets can be determined by flow cytometry (see Chapter 30) and is now available as a routine part of the CBC on the Sysmex XE2100 autoanalyzer. However, confirmation that the level of reticulated platelets provides clinically relevant information in ITP remains unclear. A similar parameter is that of "large platelets" available on a number of autoanalyzers including the Bayer Advia 120. The latter has been shown to predict response to IVIG and to IV anti-D suggesting that it is associated with the degree of platelet turnover.[65]

Despite all of the possibilities discussed here, it is not possible to be certain of the diagnosis of ITP in some cases. However, the response to treatments may be useful diagnostically. Patients whose thrombocytopenia is at least temporarily improved by IVIG or IV anti-D are those with ITP, whereas those whose platelets do not improve may or may not have ITP depending on the degree of their lack of response.[68,69] The diagnostic utility of anti-D infusion is greater at higher doses (i.e., 75 µg/kg).[70] Substantial (undefined by any clinical study but probably $>30-50 \times 10^9$/L) platelet responses to either treatment are more useful to include the diagnosis of ITP in responders than their absence is to exclude the diagnosis in nonresponders. However, if the platelet count does not change at all or actually goes down with IVIG or IV anti-D, ITP can probably be excluded. Response to steroids and to splenectomy are less specific to ITP. Importantly, response to treatment, as well as the testing described earlier, does not necessarily distinguish patients with secondary ITP. Therefore, very rarely a platelet response may be seen to IVIG in the setting, for example, of undiagnosed lymphoma. If a patient has a sustained rise in platelet count following a platelet transfusion, then ITP is not the diagnosis.

Testing for secondary ITP may facilitate the identification of the ITP itself if the workup provides evidence of other forms of autoimmunity. For example, if a patient with SLE, or merely with a positive ANA or lupus anticoagulant, develops thrombocytopenia, this implies that the thrombocytopenia is (secondary) ITP. A positive direct Coombs test is useful in the same way, as is the presence of thyroid disease.

In summary, the diagnosis of ITP requires, at a minimum, a complete history including that of the family, a physical examination, and a CBC with a review of the peripheral smear. Any suspicion generated by these data should lead to appropriate additional testing (e.g., CT scan, bone marrow examination, or family platelet counts). Ongoing medications should be changed to alternatives. Finally, response to all treatments should be carefully monitored as they may confirm or reject the diagnosis.

# V.  Treatment

The platelet count is the traditional surrogate marker for the risk of serious bleeding and therefore the need for treatment. However, variability in the bleeding tendency in patients with ITP has not been well studied and may significantly fluctuate over time in individual patients. This variability may depend on the fluctuation of the platelet count or may depend on changes in other factors. Because virtually all platelet function tests (except flow cytometry) require a normal or near-normal platelet count, these tests (e.g., platelet aggregation or platelet function analyzer [PFA]-100) will not be informative for the patient with a count of 10–$30 \times 10^9$/L. The only well-described platelet function abnormality in patients with ITP is the very rare case of acquired Glanzmann thrombasthenia, as a result of autoantibodies interfering with fibrinogen binding to platelet integrin αIIbβ3 (GPIIb-IIIa).[71] Whether other cases of ITP may have a more moderate inhibition of platelet function for other pathophysiological reasons, such as acquired von Willebrand disease, remains to be delineated. The bleeding time has not been linked to bleeding in either ITP or in other disorders and therefore its use is no longer encouraged as a monitor of treatment[73] (see Chapter 25).

Several bleeding scores have been developed,[17,74] and one or more of these may be useful in the future for monitoring the response to therapy in ITP. In addition, studies have been performed of the organic psychosocial disabilities associated with ITP, in addition to the studies of bleeding or the risk of bleeding. Disabilities such as depression and serial semiquantitative assessments of bleeding will become increasingly important in the future in the management of ITP and the assessment of therapeutic effects.

## A.  Methods to Rapidly Increase the Platelet Count

Most clinicians, especially in pediatrics, consider that excessive mucosal bleeding (i.e., wet purpura) suggests a greater likelihood of developing serious internal bleeding (e.g., in the central nervous system). Preliminary unpublished data also suggest that, in children, the presence by history or on examination of bleeding beyond petechiae and ecchymoses may be associated with a greater risk of intracranial hemorrhage. Studies of intracranial hemorrhage suggest that these fortunately rare events occur almost exclusively when the platelet count is less than 10 to $20 \times 10^9$/L.[73] Most treatment regimens in children and adults currently use 20 to $30 \times 10^9$/L as the cutoff below which treatment is initiated.[3–5] This particular issue is more extensively discussed, but not better resolved, in pediatric patients with ITP. In these patients, the dissension is whether patients may not need to be treated even at very low counts.[24] Conversely, in adult patients with

ITP, there is universal agreement to treat the "typical" patient with a platelet count of $\leq 30 \times 10^9$/L.

If a decision is made to treat the newly diagnosed patient with ITP, there are a number of methods to rapidly increase the platelet count (Table 46-1). The gold standard has been IVIG, which increases the platelet count significantly faster at a dose of 1 gm/kg than at lower daily doses and is as fast (IV anti-D 75 µg/kg, discussed later) or faster than any other treatment.[68,75] No ITP study has, however, compared IVIG 2 gm/kg/infusion in 1 day (the Kawasaki disease regimen) with 1 gm/kg, nor has IVIG alone been compared with IVIG at the same dose combined with high-dose IV steroids (discussed later).

Primarily in terms of childhood ITP, a number of studies have looked at the efficacy of high-dose steroids, both IV and PO, in increasing the platelet count. Doses have ranged from 5 to 30 mg/kg/day of IV methylprednisolone up to 1 gm total for 1 to 3 days or oral dexamethasone 30 to 50 mg/day or 24 to 28 mg/m².[76] In addition, uncontrolled studies have implied that higher doses of prednisone, 4 mg/kg/day rather than 1 to 2 mg/kg/day, increase the platelet count more rapidly. One drawback of many of these studies is that they were performed at the time of diagnosis of children with acute ITP. Many of these patients will have a spontaneous increase in platelet count without treatment, making it difficult to judge the true effect of treatment and to compare the effects of different treatments. In adult patients, the use of high-dose dexamethasone, 40 mg/day × 4 days, with variable repeats, has been reported to result in an approximately 50% incidence of lasting response, so that no further treatment is required.[77] However, this initial study requires confirmation. Multiple mechanisms for the beneficial effects of steroids have been reported, but it is unclear which one(s) predominates. The most important effect of steroids may be to increase the rate of platelet production, possibly via interference with antibody-mediated binding to megakaryocytes.[21] However, steroids also inhibit Fc-mediated phagocytosis and may accelerate clearance of autoantibodies via their effects on FcRn.[23,78]

Another option to increase the platelet count acutely in Rh-positive, direct Coombs-negative ITP patients (approximately 80 to 85% of ITP patients in the United States) is to treat them with IV anti-D at a dose of 75 µg/kg. In 1991, then standard doses of anti-D (40 to 50 µg/kg) were shown to require 72 hours on average to demonstrate a significant platelet increase from baseline.[69] This was confirmed by a 1994 study in children with acute ITP in which anti-D 25 µg/kg administered on 2 consecutive days was slower to increase the platelet count than were prednisone 1 to 2 mg/kg or IVIG 0.8 to 2.0 gm/kg.[79] In 2001, a randomized study in adults (the report also included a pilot study in children) demonstrated that using a higher dose of anti-D, 75 µg/kg instead of 50 µg/kg, increased the platelet count overnight, unlike the lower dose.[80] These findings demonstrated for the

## Table 46-1: Methods to Rapidly Increase the Platelet Count in ITP

**Standard Options**

"Gold standards"
  (a) IVIG: 1 gm/kg
  (b) IV anti-D 75 μg/kg (in Rh-positive, direct Coombs test-negative patients who have not been splenectomized)
High-dose steroids
  (a) IV methylprednisolone 5–30 mg/kg/day (limit 1 gm) for 1 to 3 days
  (b) PO dexamethasone 40 mg/day × 4 days
  (c) PO prednisone 4 mg/kg/day

**Other Options**

Vincristine 0.03 mg/kg/IV push.
Plasmapheresis (usually as part of combination therapy)

**Emergency Options**

Combination treatments
  (a) IVIG + IV methylprednisolone 30 mg/kg
  (b) (a) + IV anti-D 75 μg/kg
  (c) (a) + vincristine 0.03 mg/kg
  (d) all four treatments together
Platelet transfusion
  (a) Bolus
  (b) Continuous
  (c) (Best) Bolus + continuous

first time that IV anti-D can be used instead of IVIG in Rh-positive, nonsplenectomized ITP patients in order to rapidly increase the platelet count. A study in children at diagnosis of ITP directly compared IVIG at 800 mg/kg with IV anti-D at 75 μg/kg, as well as at 50 μg/kg, and showed that when IV anti-D was infused at a dose of 75 μg/kg, it was equivalent to IVIG in that both treatments increased the platelet count rapidly to a high level in a very high proportion of patients with ITP.[81] We believe that these patients should also be required to be direct antiglobulin test-negative to avoid the risk of hemolysis.[82] Hemolysis is the primary toxicity of anti-D and can result in transient anemia or fever-chill reactions to infusion.[69] The anemia is typically an average decrease of 1 to 2 gm/dL with return to baseline in 3 weeks,[83] but very rare cases of more severe, intravascular, hemolysis have been reported,[70] and these may result in DIC or renal failure.[82,84]

Other single agents that potentially could be effective include vincristine or plasmapheresis (Table 46-1), but these are less well documented as single treatments. Furthermore, plasmapheresis is cumbersome and probably requires plasma or IVIG replacement to be effective in this setting, and vin-cristine has substantial neurotoxicity with weekly use. Both, however, especially vincristine because of its ease of administration, can be valuable adjuncts in difficult cases.

Platelet transfusions can be useful if an acute increase in platelet count is required for several hours while awaiting the effect of a more definitive or lasting treatment. High platelet doses may need to be tried, and anecdotal experience suggests that in the setting of life-threatening bleeding in a patient refractory to treatment, a bolus followed by a continuous platelet infusion may be the optimal approach. Conversely, platelets should rarely be given and are almost never appropriate in the setting of a low platelet count with diffuse petechiae, ecchymoses, and oral blisters, but an otherwise well patient.

Fortunately, in most cases of ITP, a "rapid" (overnight) increase in platelet count is not mandatory, and standard treatment (e.g., oral prednisone) is sufficient. A common approach is to administer another treatment, such as IV anti-D or IVIG, and simultaneously initiate prednisone in order to shorten the duration of hospitalization and yet maintain the platelet count after it increases. However, whether prednisone or another treatment is used first, there are cases in which the initial treatment is ineffective. In addition, some patients, typically those with a very low platelet count ($<10 \times 10^9$/L), have diffuse clinical bleeding beyond petechiae and ecchymoses. These patients appear to be at higher risk of life-threatening bleeding and may be less likely than typical patients to respond to single treatments.[85] An urgent need for surgery is another setting in which a substantial overnight platelet increase is required, but if there is a past history of rapid response to a particular treatment, these cases may not be difficult to manage.

Whereas single agents may be effective in as many as 90% of typical cases, in cases refractory to single agent(s), combining IVIG, IV methylprednisolone, vincristine and/or IV anti-D has met with very good success, even in patients unresponsive to IVIG and steroids alone.[86] The basis for successful combination treatment appears to be the different mechanisms of effect of each component. Vincristine's effect is not clear beyond its binding to tubulin. This could impact both platelets (by decreasing antiplatelet antibody binding) and macrophages (by inhibiting phagocytosis). Clinical studies have suggested that the mechanism of the so-called Fc receptor blockade is different following IVIG and IV anti-D.[69] Data in mice suggest that the immediate effect of IVIG is mediated by an upregulation of FcRIIb, the inhibitory FcR.[87] A separate, slower effect may exist *via* saturation of FcRn, with a resulting decrease in the level of autoantibody. In contrast, IV anti-D seems to act by blocking FcRIIa and FcRIIIa.[59] Therefore, IVIG and IV anti-D may exert their acute clinical effects differently, and the combination (with steroids and possibly vincristine) would have additive, if not synergistic, effects on the platelet count, as indeed is often observed in patients.

A critical assumption underlying the use of these treatments is that the patient is producing and destroying platelets. If accelerated platelet destruction is an important component of the pathophysiology, then interfering with it will rapidly and substantially increase the platelet count. The directly proportional relationship of numbers of circulating large platelets to the 24-hour response to IVIG and to IV anti-D seems to confirm this hypothesis. However, some patients with ITP apparently have more markedly decreased platelet production than usual, which makes them relatively impervious to this treatment approach. The decreased platelet production in these patients appears to be mediated by antimegakaryocytic antibodies.[12,13] The normal or increased numbers of megakaryocytes seen in the marrow may be either damaged or undergoing apoptosis (or both).[14] Additional information is required to identify the clinical settings in which markedly decreased platelet production occurs (both reticulated platelets and the number of large platelets need to be measured), which in turn may clarify the optimal clinical approach to these difficult patients.

Thrombopoietin and, especially, megakaryocyte growth and development factor (MGDF) were developed for chemotherapy-induced thrombocytopenia and were planned to be later used for the stimulation of platelet production in ITP (see Chapter 66 for details). However, these two agents were discontinued from clinical trials when patients developed antibodies to MGDF that cross-reacted with native TPO resulting in lasting thrombocytopenia.[88] Subsequently, newer TPO-like agents were developed that have no biochemical resemblance to TPO per se and yet effectively stimulated megakaryocytopoiesis. These agents have been first used in trials in ITP (because of the limited efficacy in chemotherapy-induced thrombocytopenia of the earlier agents) and are currently demonstrating remarkable efficacy with apparently minimal toxicity.[89–91] Preliminary results suggest that these agents, if administered in a high enough dose, will result in substantial platelet increases in the great majority of even refractory patients.[92,93] In the trials of AMG531, the agent most frequently used to date, mild headaches have been the only common side effect. However, there has been at least one case of reticulin fibrosis, raising the question as to how frequently this may occur and its clinical significance. In animals repeatedly administered high doses of thrombopoietic agents, reticulin fibrosis has been seen as well; however, in both the animal studies and in the first described case in a patient, this fibrosis was reversible upon discontinuation of treatment.[94]

## B. Maintenance

Once a patient with ITP has been treated and the platelet count increases, follow-up therapy is not well defined. Approximately half of all children with ITP do not require further therapy.[4,5] Other children will spontaneously improve in the months that follow. Maintenance treatments include repeated IV anti-D, IVIG, and steroids. Irrespective of the treatment, only 5 to 20% of children with ITP will have the important chronic disease more than 6 to 12 months from diagnosis. Two recent trials have demonstrated the lasting effects of rituximab (anti-CD20) in 30 to 50% of difficult chronic ITP patients.[95,96]

Adults are known not to have as high a rate of spontaneous improvement as children, although views on this are slowly changing. Three studies have suggested that if treatment is given to thrombocytopenic adults to support the platelet count while waiting up to 1 to 2 years from diagnosis, then at least 50% of these patients will no longer require any treatment, allowing them to indefinitely avoid splenectomy. These studies used three different approaches in four arms suggesting, but not proving, that it is the natural history of the patient rather than a specific effect of a treatment that is responsible for the improvement. These studies include (a) the dexamethasone study cited previously,[77] (b) our study of repeated infusions of IV anti-D in which more than 50% of patients were able to discontinue all treatments,[59] and (c) a controlled trial comparing prednisone to IV anti-D in which more than 50% of patients in both arms were able to discontinue all treatment usually in less than 1 year.[97] In addition, many studies have shown that 10 to 30% of adults with ITP improve with an initial course of prednisone, and platelet counts will slowly normalize without further therapy.[69] Therefore, while the previous standard of care had been to proceed with splenectomy after an initial course of prednisone for 1 to 2 months, this no longer seems appropriate. The rationale for proceeding with splenectomy had been that further improvement was unlikely and that improvement (i.e., to a stable platelet count of $30–50 \times 10^9/L$) if it occurred with medical therapy, was likely to be temporary and insufficient. There are, however, few data describing the rate of relapse or occurrence of hemorrhagic emergencies in these "stable patients."

Increasingly clear, however, is that prolonged steroid treatment is associated with marked toxicity. In both children and adults, the toxicity of prednisone therapy acutely is primarily the mood swings and sleeplessness, which make dealing with a new disease more difficult. Studies of quality of life support the concept that these are real issues and not "minor complaints" that should be brushed aside.[98] Other steroid-induced complications such as hypertension, diabetes, gastritis, peptic ulceration, and infections are problematic although they can often be managed with additional drug therapy. Osteoporosis is the most common, serious long-term complication in women taking "only" 10 to 20 mg of prednisone per day for years. However, use of agents such as fosamax probably reduces the rate of this complication as well. Fortunately, steroid-induced cataract formation seems to be quite infrequent.

Therefore, it is our firm recommendation that if a patient is given an initial course of prednisone of several months, this not be continued unless the dose is ≤10 mg/day. Even this dose may result in complications in some patients. In addition to repeated infusions of anti-D administered on an as-needed basis, whenever the platelet count falls to <20 to $30 \times 10^9$/L, other treatments that can be used (there are no controlled comparative trials) include rituximab, danazol, azathioprin, and the thrombopoietic agents (although these are still in clinical trials). The advantage of rituximab is that it appears that approximately one third of adults will have a response, usually to a normal platelet count, that will last longer than a year if not indefinitely. Danazol is effective in some patients and generally not toxic, but the percentage of "cures" with prolonged use is not defined. Azathioprin also appears to have minor toxicity; both it and danazol may take up to 3 to 4 months to have an effect. We find that both together are effective in a high percentage of patients. The thrombopoietic agents appear likely to receive frequent usage in this patient group in the future pending further confirmatory studies.[91]

An issue that has been increasingly clarified is that a stable platelet count greater than $30 \times 10^9$/L requires no treatment.[3] Data that intracranial hemorrhage almost never occurs at a count >$20 \times 10^9$/L and clinical experience with patients at counts of 30 to $50 \times 10^9$/L who are off treatment support $30 \times 10^9$/L in preference to $50 \times 10^9$/L as the threshold in this regard. The choice of the threshold for instituting treatment needs, however, to be individualized based on many factors including bleeding tendency, risks undergone in the activities of daily living, and the chronic disabling effects of ITP. The choice of this threshold will clearly impact the ability to discontinue treatment and the ability to avoid splenectomy. Further, even if a threshold of $30 \times 10^9$/L is used as an indicator for treatment, virtually all forms of physical activity are to be encouraged including skiing, bicycle riding, baseball/softball, and working out at a gym. Probably only tackle football and lacrosse definitely require higher cutoffs for the institution of treatment.

### C. Splenectomy

The following are not well characterized with regard to splenectomy in ITP (a) which patients will respond to it; (b) the long-term outcome with regard to the ITP; and (c) the adverse consequences in the future, if any, of having a splenectomy, other than a small risk of postsplenectomy sepsis. Kojouri et al. summarized the current state of knowledge in these areas.[30]

Prediction of the outcome of splenectomy in ITP is uncertain. One general theme has been that response to splenectomy is more likely in patients who respond to other treatments, such as steroids or IVIG. However, data exploring the predictive value of these treatments individually are controversial, with some reports in favor of a relationship of responses[100] and some reports demonstrating no relationship.[101] It is likely that patients who do not increase their platelet count at all in response to steroids, IVIG, and IV anti-D are less likely to respond to splenectomy. Conversely, patients who respond to all these medical treatments, but not lastingly, may be more likely to respond to splenectomy. The difficulty in predicting splenectomy success or failure in the individual may exist because the spleen not only serves as the most sensitive filter of IgG antibody-coated platelets but also may be an important site of synthesis of the antiplatelet antibody.

There is a clear consensus that having Evans syndrome (i.e., autoimmune hemolytic anemia and ITP) is a poor prognostic factor for response to splenectomy,[86,102] although this is better documented in children than in adults. Whether or not a positive direct antiglobulin test in a patient with ITP without hemolysis, but which elution shows to be a panagglutinin, is predictive of a poor response to splenectomy remains unknown. Some patients with active "vasculitic" SLE do not respond well to splenectomy.[103] Finally, hypogammaglobulinemic patients with ITP will have an increased risk of postsplenectomy sepsis, will have an increased risk of infectious complications at the time of surgery, and are the most likely to benefit from the repeated administration of IVIG, with resolution of the ITP in a number of cases. In summary, among patients contemplating splenectomy, relatively few patients (those with hemolytic not just serologic Evans syndrome, active SLE, and hypogammaglobulinemia) can be considered poor candidates for splenectomy.

The other factor that has been studied with regard to a patient's response to splenectomy is the platelet life span and especially the site of platelet destruction. It has been proposed that patients with low platelet turnover are unlikely to respond to splenectomy.[104] It has been more strongly suggested that patients whose platelets are primarily destroyed in the liver will not respond well to splenectomy.[105]

The long-term beneficial outcome of splenectomy in ITP has been reported in many series of patients to be in the range of 65%, including the relapse rate in the years after an initially successful splenectomy. Our recent study suggests that with median 7.5 year follow-up and a minimum follow-up of 5 years, the long-term response rate (defined as platelet count always >$50 \times 10^9$/L with no postsplenectomy treatment) is 62%.[106] Results from another study are similar,[107] and meta-analysis also suggests a similar percentage.[30] In these studies, no assessed clinical variable predicted late relapse.

Laparoscopic splenectomy has advantages, especially more rapid recovery from surgery and reduced scarring. The ability of the patient to return to work is enhanced, and there does not seem to be an increased rate of missed accessory spleens. Certain patients may be ineligible, including those with markedly enlarged spleens.[108,109]

The long-term consequences of splenectomy remain uncertain, especially in adults. Children have a very low but finite rate of postsplenectomy sepsis, which, as in adults, can be reduced by pneumococcal and other vaccinations that must be repeated every 5 to 7 years and by immediate IV antibiotics in the setting of fever. The value of prophylactic antibiotics is probably overestimated, as studies have shown little to no benefit of penicillin prophylaxis in patients with sickle cell disease (who have autosplenectomy secondary to infarction) over the age of 5 years.[110] Completely unresolved are other possible consequences of splenectomy, such as accelerated atherosclerosis and pulmonary hypertension. These were not observed in our long-term follow-up study of 56 patients, but 80% of the patients in this study were women and only 2 males over the age of 50 underwent splenectomy. Finally, one highly controversial study has suggested that splenectomy may lead to dementia.[111]

### D. Splenectomy Failures

There are no clear answers from controlled trials regarding the choice of treatment in splenectomy failures in patients with ITP.[112–114] Table 46-2 lists the options for managing such refractory patients. No single therapy has an efficacy rate greater than 30% in ITP patients who have failed splenectomy.[115] Therefore, it seems appropriate to use multiple treatments in a combination approach.[86,116] Although combination approaches may have more toxicity, the response rate appears to be better and faster than single-agent therapy, and the toxicity can be minimized by selecting agents whose toxicities are not additive. The following is the approach taken by

**Table 46-2: Approaches to Treatment of Refractory ITP**

Conventional Single Agents

  **Pro:** "The known"
  **Con:** Low efficacy rate, slow onset

Combination Chemotherapy Protocols

  **Pro:** Better, faster response
  **Con:** More toxicity

Experimental Approaches May Have the Following Factors

  Better response rates (TPO-like agents)
  Lasting effects (Rituximab)
  Less toxicity (especially if monoclonal antibody)?

our center. It is supported by our single-arm pilot data,[86,102] but it has not been validated in comparative studies.

For combination treatments designed to increase the platelet count acutely, we use IVIG 1 gm/kg and IV methylprednisolone 30 mg/kg up to 1 gram. Additional treatments include vincristine 1 to 1.5 mg IV push and IV anti-D 75 μg/kg. Two, three, or all four agents can be infused together. These combinations can be infused in less than 8 hours in an experienced outpatient department. The intent is to use them whenever the platelet count is $<20 \times 10^9/L$, while awaiting the onset of the effect of the steady-state treatment (discussed later). In responders, at least 70% of these refractory splenectomized patients, the combination is usually required only two to four times at most.[86]

Combination treatments designed to maintain a stable increase in the platelet count usually involve two oral medications whose time until onset of effect (i.e., an increase in the platelet count) may be as long as 2 to 4 months. The medications most commonly used are danazol (400 to 800 mg/day) and azathioprine (2 mg/kg/day).[117,118] Before these drugs are used, liver function tests must be documented to be normal. This choice of drugs reflects their relatively low toxicity and the desire to reserve cyclophosphamide for true emergency situations. Danazol may cause mild facial hair growth and acne but relatively little edema; it will suppress menses completely, which may be an advantage. Azathioprine may cause mild diarrhea and leukopenia, but the latter rarely occurs at the stated dose and can be used as the dose-limiting toxicity. Clinically relevant immunosuppression (infections, induction of malignancy) has not been reported with this agent in patients with ITP.

Other medications can be used (Table 46-3) including cyclosporine, mycophenolate mofetil, dapsone, colchicine, and others for which there are less data in ITP. Long-term prednisone treatment is often deleterious, especially in women because of their greater susceptibility to osteoporosis. Ongoing studies involve the thrombopoietic agents, which appear to be potent and safe in preliminary trials (noted earlier). Another promising agent is anti-CD20 (rituximab), which is widely used because of the hematology community's experience with it in the management of lymphoma and because of its apparent potential to provide a curative effect in approximately one third of patients. Autologous bone marrow transplantation has also been tried with some success in very refractory patients; the NIH criteria requires a need for frequent transfusions.[119] There are many unanswered questions regarding the latter, including patient selection and the optimal conditioning regimen.

### E. Management of ITP during Pregnancy

The management of ITP during pregnancy is discussed in Chapter 51. The management of neonatal autoimmune

**Table 46-3: Refractory ITP: Experimental Treatment Choices**

| More Promising | Less Promising |
| --- | --- |
| Thrombopoietic agents | Interleukin-11 (toxic and minimally effective) |
| Rituximab (Anti-CD20) | Interferon (good long-term for hepatitis C-ITP) |
| Vincristine-IVIG-methylprednisolone-danazol-azathioprine | Plasmapheresis — also staph A |
| Cyclosporine/mycophenolate | Other chemotherapeutic agents |
| CHOP type chemo (cyclophosphamide-based) | Accessory splenectomy — only appropriate if remission to initial splenectomy lasted >2 years |
| Autologous bone marrow transplantation | Anti-D |
| | Alemtuzumab |

thrombocytopenia as a result of maternal ITP is discussed in Chapter 52.

## VI. Summary

In summary, ITP is usually an autoantibody-mediated disease that results in often severe thrombocytopenia. ITP has the virtue of having an easily obtainable and highly specific surrogate marker for hemorrhage: the platelet count. Fortunately, even at low platelet counts, serious hemorrhage (e.g., intracranial) is rare. Difficulties in diagnosis include the lack of a specific platelet antibody test. Other tests (e.g., plasma TPO levels, reticulated platelet counts, and the number of large platelets) are being developed to provide complementary pathophysiological information. It is still difficult to individualize treatment or prognosticate in many patients with ITP unless they demonstrate a specific feature, such as having Evans syndrome or hypogammaglobulinemia. Current management is therefore largely based on an individual patient's degree of bleeding and response to previous treatments.

There are a number of standard treatments. Methods to acutely increase the platelet count are numerous and generally effective, including high-dose steroids, IVIG, and higher dose, 75 μg/kg, IV anti-D. In addition, there appears to be a clear benefit of combining agents in difficult patients. Maintenance of the platelet count prior to splenectomy, or when splenectomy (the only curative treatment in more than 50% of patients) has failed, is the subject of ongoing studies but the upfront treatments and especially the novel thrombopoi-

etic agents seem useful. Studies with anti-CD20 suggest that this is another "curative" treatment, but longer term follow up is required to define this more precisely.

Areas ripe for further research in the field of ITP include specific antiplatelet antibody testing, better understanding of the autoimmune response, better ability to predict bleeding, easier distinction of ITP and non-ITP thrombocytopenia, specific treatments of autoimmunity, and distinction of high- and low-risk ITP patients. In addition, comparative trials of different agents, especially in adults, would be useful.

## Acknowledgment

Supported in part by the ITP Society of the Children's Blood Foundation. The assistance of Chris Yurchuck is gratefully acknowledged.

## References

1. Harrington, W. J., Minnich, V., Hollingsworth, J. W., et al. (1951). Demonstration of a thrombocytopenic factor in the blood of patients with thrombocytopenic purpura. *J Lab Clin Med, 38*, 1.
2. McMillan, R. (1981). Chronic idiopathic thrombocytopenic purpura. *N Engl J Med, 304*, 1135–1147.
3. George, J. N., Woolf, S. H., Raskob, E., et al. (1996). Idiopathic thrombocytopenic purpura: The recommendations of the American Society of Hematology. A practice guideline developed by explicit methods for the American Society of Hematology. *Blood, 88*, 3–40.
4. Bussel, J. B., & Cines, D. (in press). Immune thrombocytopenic purpura, neonatal alloimmune thrombocytopenia and post-transfusion purpura. *Hematology: Basic principles and practice* (4th ed). In R. Hoffman, E. J. Benz, S. Shattil, B. Furie, H. J. Cohen, & L. E. Silberstein (Eds.), Churchill-Livingston.
5. Cines, D., & Blanchette, V. (2002). Immune thrombocytopenic purpura. *N Engl J Med, 346*, 995–1008.
6. Drachman, J. G. (2004). Inherited thrombocytopenia: When a low platelet count does not mean ITP. *Blood, 103*, 390–398.
7. Cines, D. B., Bussel, J. B., McMillan, R. B., et al. (2004). Congenital and acquired thrombocytopenia. *Hematology (Am Soc Hematol Educ Program)*, 390–406.
8. Branehog, I., Kutti, J., & Weinfeld, A. (1974). Platelet survival and platelet production in idiopathic thrombocytopenic purpura (ITP). *Br J Haematol, 27*, 127–143.
9. Harker, L. A. (1970). Thrombokinetics in idiopathic thrombocytopenic purpura. *Br J Haematol, 19*, 95–104.
10. Emmons, R. V., Reid, D. M., Cohen, R. L., et al. (1996). Human thrombopoietin levels are high when thrombocytopenia is due to megakaryocyte deficiency and low when due to increased platelet destruction. *Blood, 87*, 4068–4071.

11. Kappers-Klunne, M. C., de Haan, M., Struijk, P. C., et al. (2001). Serum thrombopoietin levels in relation to disease status in patients with immune thrombocytopenic purpura. *Br J Haematol, 115,* 1004–1006.

12. Chang, M., Nakagawa, P. A., Williams, S. A., et al. (2003). Immune thrombocytopenic purpura (ITP) plasma and purified ITP monoclonal autoantibodies inhibit megakaryocytopoiesis *in vitro*. *Blood, 102,* 887–895.

13. McMillan, R., Wang, L., Tomer, A., et al. (2004). Suppression of *in vitro* megakaryocyte production by antiplatelet autoantibodies from adult patients with chronic ITP. *Blood, 103,* 1364–1369.

14. Houwerzijl, E. J., Blom, N. R., van der Want, J. J., et al. (2004). Ultrastructural study shows morphologic features of apoptosis and para-apoptosis in megakaryocytes from patients with idiopathic thrombocytopenic purpura. *Blood, 103,* 500–506.

15. Cortelazzo, S., Finazzi, G., Buelli, M., et al. (1991). High risk of severe bleeding in aged patients with chronic idiopathic thrombocytopenic purpura. *Blood, 77,* 31–33.

16. Schattner, E., & Bussel, J. B. (1994). Mortality in immune thrombocytopenic purpura: Report of seven cases and consideration of prognostic indicators. *Am J Hematol, 46,* 120–126.

17. Miller, A. B., Hoogstraten, B., Staquet, M., et al. (1981). Reporting results of cancer treatment. *Cancer, 47,* 207–214.

18. Arnott, J., Horsewood, P., & Kelton, J. G. (1987). Measurement of platelet-associated IgG in animal models of immune and non thrombocytopenia. *Blood, 69,* 1294–1299.

19. Cines, D., & Bussel, J. (2005). How I treat idiopathic thrombocytopenic purpura (ITP). *Blood, 106,* 2244–2251.

20. Ballem, P. J., Segal, G. M., Stratton, J. R., et al. (1987). Mechanisms of thrombocytopenia in chronic autoimmune thrombocytopenia of both impaired platelet production and increased platelet clearance. *J Clin Invest, 80,* 33–40.

21. Gernsheimer, T., Stratton, J., Ballem, P., et al. (1989). Mechanisms of response to treatment in autoimmune thrombocytopenic purpura. *N Eng J Med, 320,* 974–980.

22. Hersh, J. (2000). Mathematical analysis of the relative contributions of decreased production—a peripheral destruction in idiopathic thrombocytopenic purpura and implication. *J Theor Biol, 203,* 153–162.

23. Bussel, J. B. (2000). Fc receptor blockade and immune thrombocytopenic purpura. *Semin Hematol, 37,* 261–266.

24. Buchanan, G. R. (1987). The nontreatment of childhood idiopathic thrombocytopenic purpura. *Eur J Pediatr, 146,* 107–112.

25. Simpson, K. N., Coughlin, C. M., Eron, J., et al. (1998). Idiopathic thrombocytopenia purpura: Treatment patterns and an analysis of cost associated with intravenous immunoglobulin and anti-D therapy. *Sem Hematol, 35* (Suppl 1), 58–64.

26. Erederiksen, H., & Schmidt, K. (1999). The incidence of idiopathic thrombocytopenic purpura in adults increases with age. *Blood, 94,* 909–913.

27. Neylon, A. J., Saunders, P. W. G., Howard, M. R., et al. (2003). Clinically significant newly presenting autoimmune thrombocytopenic purpura in adults: A prospective study of a population-based cohort of 245 patients. *Br J Haematol, 122,* 966–974.

28. Aledort, L. M., Hayward, C., Chen, M. G., et al. (2004). Prospective screening of 205 patients with ITP including diagnosis, serological markers, and the relationship of platelet counts, endogenous thrombopoietin, and circulating antithrombopoietin antibodies. *Am J Hematol, 76,* 205–213.

29. Cohen, Y. C., Djulbegovic, B., Shamai-Lubovitz, O., et al. (2000). The bleeding risk and natural history of idiopathic thrombocytopenic purpura in patients with persistent low platelet counts. *Arch Intern Med, 160,* 1630–1638.

30. Kojouri, K., Vesely, S. K., Terrell, D. R., et al. (2004). Splenectomy for adult patients with idiopathic thrombocytopenic purpura: A systematic review to assess long-term platelet count responses, prediction of response, and surgical complications. *Blood, 104,* 2623–2634.

31. Schwartz, J., Eldor, A., Gillis, A., et al. (2003). Long-term follow-up after splenectomy performed for immune thrombocytopenic purpura (ITP). *Am J Hematol, 72,* 94–98.

32. Mayer, J. L., & Beardsley, D. S. (1996). Varicella-associated thrombocytopenia: Autoantibodies against platelet surface glycoprotein V. *Pediatr Res, 40,* 615–619.

33. Semple, J., Allen, D., Hoggarth, M., et al. (1998). *In vivo* actions of anti-D in children with chronic autoimmune thrombocytopenic purpura (AITP). *Blood,* [abstract] *92*(10), (Suppl 1) 720.

34. Cooper, N., & Bussel, J. B. (in press). The pathogenesis of ITP. *Br J Haematol.*

35. Rajan, S. K., & Liebman, H. A. (2000). Treatment of hepatitis C-related thrombocytopenia with interferon α. *Blood,* [abstract] *96*(11), 1090.

36. Rajan, S. K., Espina, B. M., & Liebman, H. A. (2005). Hepatitis C virus-related thrombocytopenia: Clinical and laboratory characteristics compared with chronic immune thrombocytopenic purpura. *Br J Haematol, 129,* 818–824.

37. Levy, A. S., & Bussel, J. (2004). Immune thrombocytopenic purpura: Investigation of the role of cytomegalovirus infection. *Br J Haematol, 126,* 622–623.

38. Calleja, E., Klein, R., & Bussel, J. (1999). Is it necessary to test patients with immune thrombocytopenic purpura (ITP) for seropositivity to HTLV-1? *Am J Hematol, 61,* 94–97.

39. Franchini, M., & Veneri, D. (2004). Helicobacter pylori infection and immune thrombocytopenic purpura: An update. *Helicobacter, 9,* 342–346.

40. Bussel, J. B., Ostrow, A., & Segal, J. B. (2005). Response to treatment of helicobacter pylori in patients with immune thrombocytopenic purpura: A systematic review and meta-analysis. *Blood, 106,* 2151.

41. Gasbarrini, A., Francechi, F., Tartaglione, R., et al. (1998). Regression of autoimmune thrombocytopenia after eradication of Helicobacter pylori. *Lancet, 352,* 878.

42. Emilia, G., Longo, G., Luppi, M., et al. (2001). Helicobacter pylori edication can induce platelet recovery in idiopathic thrombocytopenic purpura. *Blood, 97,* 812–814.

43. Lingwood, C. A. (1999). Glycolipid receptors for verotoxin and helicobacter pylori: Role in pathology. *Biochim Biophys Acta, 1455,* 375–386.

44. Kiehl, L. F., & Ketchum, L. H. (1998). Autoimmune disease and chronic lymphocytic leukemia: Autoimmune hemolytic anemia, pure red aplasia, and autoimmune thrombocytopenia. *Sem Oncol, 25,* 80–97.

45. Cunningham-Rundles, C., & Bodian, C. (1999). Common variable immunodeficiency: Clinical and immunological features of 248 patients. *Clin Immun, 92,* 34–48.

46. Khalifa, A. S., Take, H., Cejka, J., et al. (1974). Immunoglobulins in acute leukemia in children. *J Pediatr, 85,* 788–791.

47. Bussel, J., Morell, A., & Skvaril, F. (1986). IgG2 deficiency in autoimmune cytopenias. *Monographs in Allergy, 20,* 116–118.

48. Roark, J. H., Bussel, J. B., Cines, D. B., et al. (2002). Genetic analysis of autoantibodies in ITP reveals evidence of clonal expansion and somatic mutation. *Blood, 100,* 1388–1398.

49. Pinchas-Hamiel, O., Mandel, M., Engelberg, S., et al. (1994). Immune hemolytic anemia, thrombocytopenia and liver disease in a patient with DiGeorge syndrome. *Isr J Med Sci, 30,* 530–532.

50. Drappa, J., Vaishnaw, A. K., Sullivan, K. E., et al. (1996). Fas gene mutation in the Canale-Smith syndrome, and inherited lymphoprofilerative disorder associated with autoimmunity. *N Engl J Med, 335,* 1643–1649.

51. Straus, S. E., Lenardo, M., Puck, J. M., et al. (1997). The Canale-Smith syndrome. *N Engl J Med, 336,* 1457–1458.

52. el-Khateeb, M. S., Awidi, A. S., Tarawneh, M. S., et al. (1986). HLA antigens, blood groups and immunoglobulin levels in idiopathic thrombocytopenic purpura. *Acta Haematol, 76,* 110–114.

53. Helmerhorst, F. M., Nijenhuis, L. E., de Lange, G. G., et al. (1982). HLA antigens in idiopathic thrombocytopenic purpura. *Tissue Antigens, 20,* 372–379.

54. Kuwana, M., Kaburaki, J., & Ikeda, Y. (1998). Autoreactive T cells to platelet GPIIb-IIIa in immune thrombocytopenic purpura. *J Clin Invest, 102,* 1393–1402.

55. Posnett, D. N., Gottlieb, A., Bussel, J. B., et al. (1988). T cell antigen receptors in autoimmunity. *J Immunol, 141,* 1963–1969.

56. Fogarty, P. F., Rick, M. E., Zeng, W., et al. (2003). T cell receptor VB repertoire diversity in patients with immune thrombocytopenia following splenectomy. *Clin Exper Immunol, 133,* 461–466.

57. Englehard, D., Warner, J. L., Kapoor, N., et al. (1986). Effect of intravenous immune globulin on natural killer cell activity: Possible association with autoimmune neutropenia and idiopathic thrombocytopenia. *J Pediatr, 108,* 77–81.

58. Navratil, I. S., Korb, L. C., & Ahearn, J. M. (1999). Systemic lupus erythematosus and complement deficiency: Clues to a novel role for the classical complement pathway in the maintenance of immune tolerance. *Immunopharmacology, 42,* 47–52.

59. Cooper, N., Heddle, N., de Haas, M., et al. (2004). IV anti-D and IVIG achieve acute platelet increases by different mechanisms: Modulation of cytokine and platelet responses to IV anti-D by FcgRIIa and FcgRIIIa polymorphisms. *Br J Haematol, 124,* 511–518.

60. Brashem-Stein, C., Nugent, D. J., & Bernstein, I. D. (1988). Characterization of an antigen expressed on activated human T cells and platelets. *J Immunol, 40,* 2330–2333.

61. Calpin, C., Dick, P., Poon, A., et al. (1998). Is bone marrow aspiration needed in acute childhood idiopathic thrombocytopenic purpura to rule out leukemia? *Arch Ped Adol Med, 152,* 345.

62. Yenicesu, I., Sanli, C., & Gurgey, A. (2000). Idiopathic thrombocytopenic purpura in acute lymphocytic leukemia. *Pediatr Hematol Oncol, 17,* 719–720.

63. Emmons, R. V., Reid, D. M., Cohen, R. L., et al. (1996). Human thrombopoietin levels are high when thrombocytopenia is due to megakaryocyte deficiency and low when due to increased platelet destruction. *Blood, 87,* 4068–4071.

64. Rinder, H. M., Munz, U. J., Ault, K. A., et al. (1993). Reticulated platelets in the evaluation of thrombopoietic disorders. *Arch Pathol Lab Med, 117,* 606–610.

65. Michel, M., Kreidel, F., Chapman, E. S., et al. (2005). Prognostic relevance of large-platelet counts in patients with immune thrombocytopenic purpura. *Haematologica, 90,* 1715–1716.

66. Beer, J. H., Buchi, L., & Steiner, B. (1994). Glycocalicin: A new assay — the normal plasma levels and its potential usefulness in selected diseases. *Blood, 83,* 691–702.

67. Cooper, N., Woloski, B. M. R., Fodero, E. M., et al. (2002). Does treatment with intermittent infusions of IV anti-D allow a proportion of adults with recently diagnosed immune thrombocytopenic purpura (ITP) to avoid splenectomy? *Blood, 99,* 1922–1927.

68. Bussel, J. B., Goldman, A., Imbach, P., et al. (1985). Treatment of acute idiopathic thrombocytopenia of childhood with intravenous infusions of gammaglobulin. *J Pediatr, 106,* 886.

69. Bussel, J. B., Graziano, J. N., Kimberly, R. P., et al. (1991). Intravenous anti-D treatment of immune thrombocytopenic purpura: Analysis of efficacy, toxicity, and mechanism of effect. *Blood, 77,* 1884–1893.

70. Newman, G. C., Novoa, M. V., Fodero, E. M., et al. (2001). A dose of 75 ug/kg/d of i.v. anti-D increases the platelet count more rapidly and for a longer period of time than 50 µg/kg/d in adults with immune thrombocytopenic purpura. *Br J Haematol, 112,* 1076–1078.

71. Niessner, H., Clemetson, K. J., Panzer, S., et al. (1986). Acquired thrombasthenia due to GPIIb-IIIa-specific platelet autoantibodies. *Blood, 68,* 571–586.

72. Rodgers, R. P., & Levin J. (1990). A critical reappraisal of the bleeding time. *Sem Thromb Hemost, 16,* 1–20.

73. Woerner, S. J., Abildgaard, C. F., & French, B. N. (1981). Intracranial hemorrhage in children with idiopathic thrombocytopenic purpura. *Pediatrics, 67,* 453–460.

74. Buchanan, G. R., & Adix, L. (2002). Grading of hemorrhage in children with idiopathic thrombocytopenic purpura. *J Pediatr, 141,* 683–688.

75. Bussel, J. B., & Szatrowski, T. P. (1995). Uses of intravenous gammaglobulin in immune hematologic disease. *Immunol Invest, 24,* 451–456.

76. Andersen, J. C. (1994). Response of resistant idiopathic thrombocytopenic purpura to pulsed high-dose dexamethasone therapy. *N Engl J Med, 330*, 1560–1564.

77. Cheng, Y., Wong, R. S., Soo, Y. O., et al. (2003). Initial treatment of immune thrombocytopenic purpura with high-dose dexamethasone. *N Engl J Med, 349*, 831–836.

78. Bussel, J. B. (2002). Another interaction of the FcR system with IVIG. *Thromb Haemost, 88*, 890–891.

79. Blanchette, V., Imbach, P., Andrew, M., et al. (1994). Randomised trial of intravenous immunoglobulin G, intravenous anti-D, and oral prednisone in childhood acute immune thrombocytopenic purpura. *Lancet, 334*, 703–707.

80. Newman, G. C., Novoa, M. V., Fodero, E. M., et al. (2001). A dose of 75 ug/kg/d of IV anti-D increases the platelet count more rapidly and for a longer period of time than 50 ug/kg/d in adults with immune thrombocytopenic purpura. *Brit J Haematol, 112*, 1076–1078.

81. Tarantino, M. D., Young, G., Bertolone, S. J., et al. (in press). A single dose of anti-D immune globulin at 75 μg/kg is as effective as IVIg at rapidly raising the platelet count in newly diagnosed immune thrombocytopenic purpura (ITP) in children. *J Pediatr.*

82. Gaines, A. R. (2000). Acute onset hemoglobinemia and/or hemoglobinuria and sequelae following $RH_0(D)$ immune globulin intravenous administration in immune thrombocytopenic purpura patients. *Blood, 95*, 2523–2529.

83. Scaradavou, A., Woo, B., Woloski, B. M. R., et al. (1997). Intravenous anti-D treatment of immune thrombocytopenic purpura: Experience in 272 patients. *Blood, 89*, 2689–2700.

84. Gaines, A. R. (2005). Disseminated intravascular coagulation associated with acute hemoglobinemia or hemoglbinuria following $RH_0(D)$ immune globulin intravenous administration for immune thrombocytopenic purpura. *Blood, 106*, 1532–1537.

85. Medeiros, D., & Buchanan, G. R. (1998). Major hemorrhage in children with idiopathic thrombocytopenic purpura: Immediate response to therapy and long-term outcome. *J Pediatr, 133*, 334–339.

86. Gururangan, S., & Bussel, J. B. (1998). Combination immunotherapy for patients with refractory ITP or Evan's syndrome (ES) [abstract]. *Blood, 92* (suppl 1), 3345.

87. Samuelsson, A., Towers, T. L., & Ravetch, J. V. (2001). Anti-inflammatory activity of IVIG mediated through the inhibitory Fc receptor. *Science, 291*, 484.

88. Li, J., Yang, C., Xia, Y., et al. (2001). Thrombocytopenia caused by the development of antibodies to thrombopoietin. *Blood, 98*, 3241–3248.

89. Bussel, J. B., George, J. N., Kuter, D. J., et al. (2003). An open-label, dose-finding study evaluating the safety and platelet response of a novel thrombopoietic protein (AMG531) in thrombocytoepnic adult patients with immune thrombocytopenic purpura (ITP). *Blood, 102*, 293.

90. Kuter, D., Bussel, J. B., Aledort, L. M., et al. (2004). A phase 2 placebo controlled study evaluating the platelet response and safety of weekly dosing with a novel thrombopoietic protein (AMG531) in thrombocytopenic adult patients with immune thrombocytopenic purpura (ITP). *Blood, 104*, 511.

91. Bussel, J. B., Kuter, D. J., George, J. N., et al. (2005). Long-term dosing of AMG531 is effective and well-tolerated in thrombocytopenic patients with immune thrombocytopenic purpura. *Blood, 106*, 220.

92. Jenkins, J., Nicholl, R., Williams, D., et al. (2004). An oral, non-peptide, small molecule thrombopoietin receptor agonist increases platelet counts in healthy subjects. *Blood, 104*, 2916.

93. McMillan, R., Bussel, J. B., George, J. N., et al. (in press). Self-reported health-related quality of life in adults with chronic immune thrombocytopenic purpura. *N Engl J Med.*

94. Kakumitsu, H., Kamezaki, K., Shimoda, K., et al. (2005). Transgenic mice overexpressing murine thrombopoietin develop myelofibrosis and osteosclerosis. *Leuk Res, 29*, 761–769.

95. Wang, J., Wiley, J., Luddy, R., et al. (2005). Chronic immune thrombocytopenia purpura (ITP) in children: Assessment of rituximab treatment. *J Pediatr, 146*, 217–221.

96. Bennett, C. M., Rogers, Z. R., Kinnamon, D. D., et al. (2005). Prospective phase I/II study of rituximab in childhood and adolescent chronic immune thrombocytopenic purpura. *Blood*, (doi 10.1182/Blood-2005-08-3518.)

97. George, J. N., Raskob, G. E., Vesely, S. K., et al. (2003). Initial management of immune thrombocytopenic purpura in adults: A randomized controlled trial comparing intermittent anti-D with routine care. *Am J Hematol, 74*, 161–169.

98. McMillan, R., Bussel, J. B., George, J. N., et al. (2006). Self-reported health-related quality of life in adults with chronic immune thrombocytopenic purpura. Manuscript submitted.

99. Simpson, K. N., Coughlin, C. M., Eron, J., et al. (1998). Idiopathic thrombocytopenia purpura: Treatment patterns and an analysis of cost associated with intravenous immunoglobulin and anti-D intravenous. *Sem Hematol, 35*, 58–64.

100. Schneider, P., Wehmeier, A., Schneider, W., et al. (1997). High-dose intravenous immune globulin and the response to splenectomy in patients with idiopathic thrombocytopenic purpura. *N Engl J Med, 337*, 1087–1089.

101. Bussel, J. B., Kaufmann, C. P., Ware, R. E., et al. (2001). Do the acute platelet responses of patients with immune thrombocytopenic purpura to anti-D and to IV gammaglobulin predict response to subsequent splenectomy. *Am J Hematol, 67*, 27–33.

102. Scaradavou, A., & Bussel, J. (1995). Evans syndrome: Result of a pilot study utilizing a multiagent treatment protocol. *J Pediatr Hematol Oncol, 17*, 290–295.

103. Hall, S., McCormick, J. L., Jr., Greipp, P. R., et al. (1985). Splenectomy does not cure the thrombocytopenia of systemic lupus erythematosus. *Ann Intern Med, 102*, 325–328.

104. Siegel, R. S., Rae, J. L., Barth, S., et al. (1989). Platelet survival and turnover: Important factors in predicting response to splenectomy in immune thrombocytopenic purpura. *Am J Hematol, 30*, 206–212.

105. Najean, Y., Rain, D., & duFour, V. (1991). The sequestration of 111-In-labelled autologous platelets and the efficiency of splenectomy. *Nouv Rev Franc d'Hematol, 33*, 449–450.

106. Schwartz, J., Eldor, A., Gillis, S., et al. (2003). Long term follow-up of splenectomy for immune thrombocytopenic purpura (ITP). *Am J Hematol, 72*, 94–98.

107. Portielje, J. E. A., Westendorp, R. G. J., Kluin-Nelemans, H. C., et al. (2001). Morbidity and mortality in adults with idiopathic thrombocytopenic purpura. *Blood, 97,* 2549–2554.

108. Harold, K. L., Schlinkert, R. T., Mann, D. K., et al. (1999). Long-term results of laparoscopic splenectomy for immune thrombocytopenic purpura. *Mayo Clin Proc, 74,* 37–39.

109. Marcaccio, M. J. (2000). Laparoscopic vs. open splenectomy. *Sem Hematol, 37,* 267–274.

110. Pai, V. B., & Nahata, M. C. (2000). Duration of penicillin prophylaxis in sickle cell anemia: Issues and controversy. *Pharmacotherapy, 20,* 110–117.

111. Ahn, Y. S., Horstman, L. L., Jy, W., et al. (2002). Vascular dementia in patients with immune thrombocytopenic purpura. *Thromb Res, 107,* 337–344.

112. Berchtold, P., & McMillan, R. (1989). Therapy of chronic idiopathic thrombocytopenic purpura in adults. *Blood, 74,* 2309–2317.

113. McMillan, R. (1997). Therapy for adults with refractory chronic immune thrombocytopenic purpura. *Ann Intern Med, 126,* 307–314.

114. Bussel, J. B. (2000). Overview of idiopathic thrombocytopenic purpura: New approach to refractory patients. *Sem Oncol, 27*(suppl 12), 91–98.

115. Reiner, A., Gernsheimer, T., & Slichter, S. J. (1995). Pulse cyclophosphamide therapy for refractory autoimmune thrombocytopenic purpura. *Blood, 85,* 351–358.

116. Figueroa, M., Gehlsen, J., Hammond, D., et al. (1993). Combination chemotherapy in refractory immune thrombocytopenic purpura. *N Engl J Med, 328,* 1226–1229.

117. Rocha, R., Horstman, L., Ahn, Y. S., et al. (1991). Danazol therapy for cyclic thrombocytopenia. *Am J Hematol, 36,* 140–143.

118. Moake, J. L., Rudy, C. K., Troll, J. H., et al. (1985). Therapy of chronic relapsing thrombotic thrombocytopenic purpura with prednisone and azathioprine. *Am J Hematol, 20,* 73–79.

119. Huhn, R. D., Fogarty, P. F., Nakamura, R., et al. (2003). High-dose cyclophosphamide with autologous lymphocyte-depleted peripheral blood stem cell (PBSC) support for treatment of refractory chronic autoimmune thrombocytopenia. *Blood, 101,* 71–77.

# CHAPTER 47

# HIV-1-Related Thrombocytopenia

**Michael A. Nardi, Zongdong Li, and Simon Karpatkin**

*Division of Hematology, New York University Medical School, New York, New York*

## I. Introduction

Human immonodeficiency virus (HIV)-1 seropositive individuals (homosexuals,[1] intravenous (IV) narcotic addicts,[2] hemophiliacs,[3] and their heterosexual partners[4]) develop chronic immune thrombocytopenia purpura (ITP). The thrombocytopenia can present in association with AIDS or in its absence. The disease is clinically indistinguishable from classic autoimmune thrombocytopenic purpura (ATP) seen predominantly in females, with respect to increased megakaryocytes in the bone marrow, peripheral destruction of antibody-coated platelets or impaired platelet production, negative antinuclear antibody, and response to prednisone, IV gamma globulin, or splenectomy. The disease is different from classic ATP with respect to the markedly increased platelet-associated IgG, $C_3C_4$, the presence of circulating immune complexes, and the male predominance. Studies reveal the presence of a unique anti-GPIIIa49–66 antibody capable of fragmenting platelets in the absence of complement by the induction of reactive oxygen species.

## II. Historical Descriptions

### A. ITP in Homosexual Individuals

Shortly after the first recognition of Kaposi's sarcoma in sexually active homosexuals at the Bellevue Hospital Oncology Clinic of New York University Medical Center,[5] an "epidemic" of thrombocytopenia was recognized in the same cohort of patients at the same institution in November 1980.[1] The syndrome was clinically indistinguishable from classic ATP, which is seen predominantly in females. Eleven severe cases were first described with a mean platelet count of $16 \pm 3 \times 10^9$/L, as well as two mild cases. All patients had increased megakaryocytes in the bone marrow, no splenomegaly, negative antinuclear antibody titers, and no clinical disorders known to cause thrombocytopenia. Other findings included decreased helper/suppressor T-cell ratios, elevated platelet-bound IgG, elevated circulating immune complexes, relative lymphopenia, and elevated γ globulin levels. Both the patient group as well as a control homosexual group (normal complete blood count) had similar increased exposure to herpes simplex virus, cytomegalovirus (CMV), Epstein-Barr virus (EBV), hepatitis A and B viruses, and similar histories for the use of recreational drugs: marijuana, amyl nitrate, and cocaine.[1] HLA-typing for A-, B-, C-, and D-loci revealed no significant associations.

### B. ITP in Narcotic Addicts

In November 1982 (2 years later), a similar epidemic of chronic ITP was noted in 70 IV narcotic addicts treated at Bellevue Hospital (female to male ratio: 1 to 3), with mean platelet counts of $53 \pm 4$ (range 13 to 140) $\times 10^9$/L.[2] These patients were or had been chronic IV abusers of heroin, cocaine, or both for a mean duration of 10 years. Thirty-three patients had stopped taking IV drugs for an average of 21 months, indicating that the thrombocytopenia was not caused by acute exposure to narcotics. Other possible causes of thrombocytopenia such as classic ATP, hypersplenism, chronic active hepatitis, or AIDS had been ruled out. Elevated platelet-bound IgG, $C_3C_4$, and serum circulating immune complexes were also noted.

### C. ITP in Patients with Hemophilia

In 1983, Ratnoff et al.[3] reported the development of ITP in five patients with hemophilia (factor VIII deficiency) with a mean platelet count of $34 \pm 29$ (range 8 to 82) $\times 10^9$/L). These patients had been multiply-transfused with lyophilized factor VIII concentrates for 5 to 9 years. This syndrome was clinically indistinguishable from classic ATP. All five patients had elevated platelet-bound IgG levels. These patients also presented with hypergammaglobulinemia, elevated circulating immune complexes, lymphopenia, decreased helper/suppressor T-cell ratios, and positive serology for HIV-1.

### D. Heterosexually Transmitted HIV-1-Related ITP

In 1987, six cases of HIV-1-related ITP were recognized in "nonrisk" individuals who had acquired the disorder through heterosexual contact.[4] The disease was transmitted three times by male narcotic addicts to nonaddicted female partners. One of the other three patients was a 64-year-old white woman whose contact was her husband who had received an HIV-1-contaminated blood transfusion for a coronary bypass. This patient developed thrombocytopenia approximately 2 years after resumption of sexual intercourse. All six patients were HIV-1 positive by ELISA and immunoblot and had platelet-immunologic profiles easily distinguishable from patients with classic ATP but indistinguishable from that noted for HIV-1-related ITP patients. Thus, these observations were the first to note that HIV-1-related ITP can be transmitted to the heterosexual community. It can be disguised as classic ATP, which responds to prednisone, splenectomy, or IV gamma globulin. HIV-1-related ITP is now part of the differential diagnosis of unexplained thrombocytopenia. A careful social-sexual history is mandatory for diagnosing patients.

### E. Thrombotic Thrombocytopenia Purpura (TTP)

Eight cases of TTP associated with HIV-1 infection were reported in 1988.[6–8] All eight cases had been recognized since 1985, relatively late in the development of the HIV-1 epidemic. Five cases reported from New York University Medical Center[7] from 1985 to 1987 represented one third of all cases of TTP diagnosed at the institution. When this rate of incidence, 5 of 15 patients, was compared with the incidence of patients hospitalized with AIDS during the same period (3560 of 171,210), the difference was significant by chi square analysis, indicating that the apparent association is not caused by ascertainment bias. The mechanism of HIV-1-related TTP remains to be established.

## III. Incidence and Prognosis

In the early description of the epidemic, sexually active homosexual patients presented with ITP in the presence as well as absence of AIDS. Although many patients developed ITP early after HIV-1 infection, at least 11 other ITP patients had not developed AIDS 4 to 7.5 years (mean 5.2 years) after HIV-1 seropositivity.[9] Nine patients studied reverted to normal platelet counts 5 to 27 months (median 10 months) after the diagnosis of thrombocytopenia in the absence of treatment (i.e., splenectomy or steroids). One of these patients presented with a platelet count of $11 \times 10^9$/L.

HIV-1-related thrombocytopenia can occur without clinical symptoms and increases in incidence with duration of illness and the development of AIDS. The incidence of thrombocytopenia in an asymptomatic group of 26 seropositive British homosexual men was 0% in 1987.[10] This incidence rose to 8% in 59 patients with non-AIDS HIV-1-related symptoms and 30% in 20 patients with AIDS; the incidence in 461 homosexuals in Amsterdam in 1992 was 16%.[11] The incidence of thrombocytopenia ($<100 \times 10^9$/L) in 380 HIV-infected Italian (Milan) narcotic addicts was 20.8%, 5.3% with platelet counts $<30 \times 10^9$/L.[12] The incidence in 299 drug abusers in Amsterdam in 1992 was 37%.[11] The incidence of thrombocytopenia in 750 HIV-1-infected patients with hemophilia A, B, or von Willebrand disease in 1984 was 4.5%.[13] The incidence of thrombocytopenia ($<100 \times 10^9$/L) from a multicenter hemophilia cohort study of 961 HIV-1-infected patients in 1997 was 19%. The incidence rose in the first 5 years, with the risk increasing to 50% in older patients 1 year after the development of AIDS.[14] Surveillance data for thrombocytopenia from a longitudinal study of 30,214 HIV-infected patients from January 1990 to August 1996 was 1.7% in patients without clinical symptoms, 3.1% in patients with a CD4 count $<200$ cells/mm$^3$ without AIDS, and 8.7% with AIDS.[15]

An analysis of 44 HIV-1-ITP patients studied for a period of 844 person-months revealed no greater risk for the development of AIDS than a similar nonthrombocytopenic seropositive cohort over a period of 36 months.[16]

## IV. Treatment

The rationale for treatment has been to use the lowest possible corticosteroid dose capable of raising the platelet count to safe levels, generally $>25 \times 10^9$/L with absence of significant purpura. Prednisone is administered at a dose of 30 to 40 mg/day for 1 to 2 weeks and rapidly tapered to a maintenance dose of 10 to 15 mg/day. The patient is then observed for 10 months, if possible, which is the median time for spontaneous reversion to a normal platelet count in 18% of our patients. If the patient has not reverted to a safe platelet count by this time, or requires more than 10 to 15 mg of prednisone per day to maintain such a platelet count, splenectomy is performed. Of 41 homosexual patients, 35 had a moderate ($>50 \times 10^9$/L) to excellent ($>100 \times 10^9$/L) response while on corticosteroids (30 relapsed following cessation of corticosteroids). All of the 10 patients in our group had an excellent response to splenectomy that has persisted;[9] 10 of 15 responded in the group of Abrams et al.;[17] in a third group, all 7 responded with a mean follow-up time of 14 months.[18] Similar results have been observed with narcotic addicts.[19] Of our homosexual patients, 26 did not require treatment for their thrombocytopenia. There is no rigorous evidence that corticosteroid treatment or splenectomy contributes to the development of AIDS. Prednisone therapy has been complicated, however, by the development of oral candidiasis as well as activation of latent herpes simplex virus.

The efficacy of azidodeoxythymidine (AZT) in the treatment of ITP in homosexuals is often dramatic.[20] There is an exponential rise in platelet count within the first 1 to 2 weeks of AZT treatment (200 mg every 6 hours), with the maintenance of normal or higher counts for 7 weeks (duration of study). Hematocrits decreased in an exponential manner during the same period of time. These results were confirmed by the Swiss Group,[21] who treated 10 patients. After discontinuation of AZT, the platelet count remained elevated for more than 4 weeks in three of five patients. These observations on the relatively rapid rise in platelet count following AZT treatment may therefore suggest that HIV-1 infection of bone marrow precursor cells, megakaryocytes, or stromal cells may be inhibiting platelet production (discussed later). Similar results have been reported with the new antiretroviral drugs: indinavir, saquinavir, and ritonavir. Twenty-three patients treated with the drugs increased their platelet counts from 63 to 125 × 10⁹/L and decreased their viral load 74-fold during 6 months of observation with this treatment. No correlation was noted with CD4.[22]

IV gamma globulin (1 to 2 gm/kg for 2 to 5 days) is usually effective in the treatment of HIV-related thrombocytopenia; however, its duration is transient, 7 to 10 days.[23–26] Six of eight homosexual patients had a significant rise in their platelet count, as did all of three narcotic addicts and all of eight hemophiliacs.[23] Another group reported a good to excellent initial response in 12 of 17 patients (71%).[24] Patients refractory to prednisone will occasionally respond to gamma globulin given with steroids.

## V. Mechanisms of Thrombocytopenia

The mechanisms of thrombocytopenia in chronic ITP associated with homosexuals (HSITP), narcotic addiction (NITP), and hemophilia (HITP) have been studied extensively by our group[1,2,27–34] in patients with early onset HIV-1-infection in the absence of AIDS in whom the mechanism appears to be increased platelet destruction[35] and compared with those of classic ATP.

### A. Increased Platelet Destruction

Autologous platelet survival studies performed in 13 HSITP patients with a mean platelet count of <33 × 10⁹/L demonstrated a platelet survival of less than 1 day (normal survival time, 8 to 10 days). Platelet sequestration was exclusively to predominantly splenic in 85% of the patients studied. Calculated blood flow and mean transit time were normal, ruling out hypersplenism.[36] Autologous platelet survival studies performed in 19 NITP patients with a mean platelet

count of <31 × 10⁹/L were <1 day with no evidence of splenic sequestration.[36] These data support an increased peripheral platelet destruction mechanism for the pathophysiology. Indeed, the rapid improvement following treatment with steroids or splenectomy in most early-onset HIV-1-infected patients is strongly suggestive of this pattern. However, others have reported impaired platelet production, as well as decreased platelet survival (discussed next).

### B. Decreased Platelet Production

Kinetic data on the rapid rise in platelet count (as early as 1 week) following AZT administration suggested that impaired megakaryocyte production, platelet release, or both may also be involved.[20,21] The rapid rise in platelet count following the use of AZT precludes a prior change in antiplatelet antibody levels and suggests that this drug either blocks reticuloendothelial function or prevents HIV-1-induced damage of megakaryocytes, which otherwise results in ineffective thrombopoiesis or stimulates platelet production. A precedent for impaired thrombopoiesis in patients with classic ATP has been reported by Ballem et al.,[37] who noted normal to decreased platelet production in 73% of patients with mild to moderately severe classic ATP despite decreased platelet survival. Because of increased megakaryocytes in their marrow, the authors concluded that thrombopoiesis was ineffective. Three kinetic reports suggest that impaired thrombopoiesis may also be operative in patients with HSITP. In one study, platelet production was normal or decreased in 7 of 14 patients despite a decreased platelet survival.[38] Another group studied platelet survival and turnover in 19 HSITP patients treated with and without AZT.[39] Both groups had a decreased platelet survival of about 50% of normal. Platelet turnover (production) was significantly decreased (approximately half the normal rate) in the untreated group (13 patients), compared with normal or increased turnover in the treated patients (6 patients). Two patients were studied before and after AZT administration. Their rise in platelet count was associated with a threefold to sixfold rise in platelet turnover, with no change in platelet survival. Thus, AZT appears to exert its effect by enhancing platelet production, suggesting that the thrombocytopenia may also be related to impairment of the production or function of megakaryocytes. A third study of six patients revealed a combination of shortened autologous platelet survival by two thirds normal, doubling of splenic platelet sequestration and ineffective delivery of platelets to the peripheral blood despite a sixfold increase in serum thrombopoietin concentration and a threefold increase in megakaryocyte mass.[40] These observations are supported by a study of hematopoietic progenitor cells in 15 patients with AIDS or HIV-1 infection, three of whom were thrombocytopenic.[41] These investigators reported a significant impairment of growth of bone marrow megakaryocyte colony-forming units, as well as granulocyte, erythrocyte, and macrophage precursors in comparison with normal controls, which was corrected by depletion of T cells and reversed by readdition of autologous T cells. Because most patients have normal to increased megakaryocytes in their marrow, it is likely that the impairment is caused by ineffective thrombopoiesis (platelet production and release). The mechanism of inhibition of thrombopoiesis is currently unresolved.

An unlikely possibility is the direct infection of megakaryocytes with HIV-1.[42,43] Megakaryocytes have CD4[44,45] as well as CXCR4,[46,47] known receptors for HIV-1gp120, and various megakaryocyte lines are infectable with HIV-1.[45,48] However, reports of HIV-1 infection of wild-type human megakaryocytes reveal little to no infectivity (1–5% of cells) compared to T cells or macrophages, with therefore questionable clinical significance.[42] There is now abundant evidence from several laboratories that CD34[+] stem/precursor hematopoietic cells are incapable of being infected *in vivo* or *in vitro* with HIV-1.[49–58] It is therefore logical to suggest that HIV-1-infected stromal cells may be responsible for the well-documented impaired megakaryocytopoiesis of HIV-1-infected patients.[41,51,58,59] This could be via induction of inhibitory cytokines. This is supported by impaired colony-forming units-megakaryocytes (CFU-MK) in 15 HIV-1-seropositive patients compared to control bone marrows,[40,41] the morphological abnormalities detected in the megakaryocytes of HIV-1-ITP patients,[60] decreased megakaryocyte precursors,[41,51,58,59] and impaired megakaryocyte survival due to increased apoptosis.[59] Several cytokines have been shown to inhibit megakaryocyte growth and differentiation: transforming growth factor (TGF) β-1,[61–65] platelet factor 4,[66] tissue necrosis factor (TNF) α,[67,68] α and γ interferon,[69] and interleukin 4.[70] Of interest is the observation that HIV-1 *tat* protein, released by infected cells, stimulates macrophage production of TGFβ-1,[65] one of the most potent negative regulators of hematopoiesis, resulting in a dose-dependent inhibition of CFU-MK. An alternative or additional mechanism could be autoimmune destruction of megakaryocytes or their precursor by antiplatelet/megakaryocyte antibody. These apparently contradictory conclusions regarding decreased production versus increased destruction of platelets can be reconciled, perhaps by the observations of Najean and Rain who performed platelet kinetic studies on HIV-1-ITP patients with and without AIDS. They concluded that patients with AIDS are more likely to have decreased platelet production, whereas patients with early onset HIV-1 infection are more likely to have increased peripheral destruction of platelets.[35]

Immunologic measurements of platelet-associated IgG, $C_3C_4$, and serum circulating immune complexes (CIC), determined by the PEG method in HSITP, NITP, and HITP reveal strikingly elevated values compared to the values

**Table 47-1: Immune Thrombocytopenic Purpura in Homosexuals (HSITP) Compared with Classic Autoimmune Thrombocytopenic Purpura (ATP)**[a]

| | Platelet IgG (ng/10$^6$) | Platelet Complement (ng/10$^6$) | Serum Immune Complexes (mg/mL) |
|---|---|---|---|
| **HSITP** | | | |
| Mean ± SEM | 31.5 ± 7.6 | 7.7 ± 1.9 | 0.96 ± 0.10 |
| No. high/total[‡] | 29/33 | 16/20 | 21/24 |
| **Homosexual controls (HIV-sero-pos and -neg)** | | | |
| Mean ± SEM | 5.0 ± 0.6 | 1.9 ± 0.08 | 0.66 ± 0.15 |
| No. high/total | 14/28 | 6/28 | 7/8 |
| **Homosexual controls (HIV-sero-neg)** | | | |
| Mean ± SEM | 2.9 ± 0.2 | 1.1 ± 0.38 | 0.23 ± 0.08 |
| No. high/total | 0/4 | 0/4 | 0/4 |
| **Classic ATP** | | | |
| Mean ± SEM | 8.4 ± 1.6 | 1.9 ± 0.09 | 0.32 ± 0.02 |
| No. high/total | 19/22 | 6/23 | 0/5 |
| Controls ±2 SD | 1.5 ± 1.3 | 1.5 ± 0.7 | 0.34 ± 0.06 |

[a]Mean (±SEM) platelet counts for HSITP and patients with ATP were $62 \pm 6$ and $73 \pm 7 \times 10^9$/L, respectively, for measurements of platelet IgG and complement levels.
[‡]Values represent the number of subjects in whom elevated levels were found/the number of subjects studied.

found in classic ATP patients — that is, 2.6- to 3.8-fold greater IgG, 2.2- to 3.9-fold greater $C_3C_4$, 2.4- to 7.3-fold greater PEG-CIC. Intermediate elevations of these measurements were noted in nonthrombocytopenic HIV-1-seropositive individuals of the same cohorts, as presented in Table 47-1.[2,27,33] These data suggested that circulating immune complexes may be responsible for the elevated platelet levels and that no relationship exists between the platelet count and the platelet-associated IgG or $C_3C_4$. This proved to be the case. Isolated CIC were shown to bind to normal platelets in a concentration-dependent manner; indeed the level of serum CIC correlated with platelet bound IgG, r = 0.5, $p < 0.05$, n = 17. However, unlike the situation with classic ATP, there was no inverse relationship between platelet count and platelet associated IgG, r = −0.2, $p > 0.1$, n = 16 (similar results were reported by others); nor was the platelet count inversely related to CICs, r = −0.2, $p > 0.1$, n = 20. Therefore, although the platelet immunologic/serum CIC profile is of value diagnostically in differentiating the thrombocytopenia from classic ATP, it is of no value in predicting platelet count or response to therapy.

Serum antiplatelet binding has been reported, employing an indirect semiquantitative platelet suppression immunofluorescence test in homosexual patients, with impaired binding to Glanzmann thrombasthenia platelets (devoid of platelet glycoprotein [GP] IIb-IIIa[71]). However, these authors did not prove F(ab')$_2$ binding. A second group, although unable to elute antiplatelet antibody from NITP patients, was able to detect plasma antiplatelet GPIb-IX and GPIIb-IIIa antibody utilizing the monoclonal antibody-specific immobilization of platelet antigens (MAIPA) technique on normal platelets preincubated with patient plasma in 43% of 45 patients.[72] Our group has been unable to detect F(ab')$_2$ binding in HSITP patients. However, we have detected F(ab')$_2$ binding in NITP and HITP patients.[2,27] The reasons for these differences remains to be determined; however, a likelihood is the presence of anti-idiotype blocking antibody (discussed next).

## VI. Composition of Immune Complexes

### A. Anti-F(ab')$_2$ Antibodies

Studies designed to analyze platelets and immune complexes for viral antigens suggested the presence of anti-antibodies. Fixed washed platelets or eluates of HSITP patients revealed absence of EBV, CMV, herpes simplex virus (HSV), or adenovirus (ADE) viral antigens on their platelets, despite the presence of viral antibodies for these antigens in their sera. Similar findings were noted in fixed immune complexes of these patients. Viral antibodies against EBV, CMV, HSV, and rubeola virus were noted in these complexes and correlated with their presence in sera. No such correlation was noted with preparations from healthy control subjects whose immune complexes were negative in the presence of positive antiviral serum titers.[34]

Following this lead, anti-F(ab')$_2$ antibodies were sought and found in 9 of 12 homosexual patients and 6 of 6 narcotic addicts, which correlated with immune complex levels, $r = 0.83$, $p < 0.01$, $n = 16$. In contrast, in 6 patients with classic ATP, anti-F(ab')$_2$ antibodies were not detectable against F(ab')$_2$ fragments of subjects from autologous, homologous ATP, or normal control groups.[34]

### B. Anti-HIV-1gp120 and Anti-idiotype Antibodies

Analysis of serum PEG-CIC as well as platelet acid eluates of HIV-1-ITP, HSITP, and NITP patients has revealed the presence of anti-HIV-1gp120 antibody and its anti-idiotype. HIV-1 antigen or proviral DNA was not detectable. Approximately 50% of eluted platelet IgG contained anti-HIV-1gp120 antibody.[28] The anti-idiotype antibody was both

nonblocking as well as internal image blocking. Internal image blocking antibody was demonstrated by first looking for anti-CD4 antibody (receptor for gp120), which was found in 30% of HIV-1-ITP patients. Affinity-purified anti-CD4 (Ab2) bound to F(ab')$_2$ fragments of affinity-purified anti-HIV-1gp120 (Ab1) on solid phase radioimmunoassay. Binding could be blocked by rCD4 as well as rgp120 Ag, and anti-HIV-1gp120 (Ab1) inhibited the binding of anti-CD4 (Ab2) to rCD4. Ab3 (anti-HIV-1gp120) could be produced *in vivo* by immunizing mice with Ab2, and binding of Ab3 to rgp120 could be blocked with rCD4.[73]

### C. Anti-GPIIIa Antibody *in Vitro and* in Vivo

PEG-ICs of HIV-1-ITP patients were purified on protein A columns, dissociated in acid, separated by gel filtration on an acidified G200 column, and affinity-purified on anti-IgG and anti-IgM columns. Both IgG and IgM reactivity was found against HIV-1gp120, HIV-1gp24 CD4 receptor, and human platelets. Only the IgM reacted with purified Fc fragments, indicating rheumatoid factor (RF) reactivity. The IgG antiplatelet reactivity was shown to be specific for platelets by demonstrating reactivity with F(ab')$_2$ fragments.[29] Antiplatelet IgG reactivity was affinity purified on fixed platelets and the eluate incubated with platelet lysates containing $^{125}$I-labeled surface platelet membrane. The immune complex precipitate was then analyzed on SDS-PAGE in the presence and absence of reducing agent. An approximately 100-kDa predominant band was noted, which paradoxically increased its molecular weight upon reduction, demonstrating the physical characteristics of platelet GPIIIa. This was proven with specific monoclonal antibodies against platelet GPIIIa, simultaneously incubated with platelet lysate.[29] Further studies revealed specificity for amino acids 49–66 on GPIIIa, with an inverse curvilinear relationship between patient serum (as well as affinity-purified anti-GPIIIa49–66) and platelet count (Fig. 47-1). The *in vivo* pathophysiologic relevance of these *in vitro* observations was substantiated by demonstrating that approximately 25 μg of affinity-purified antiplatelet IgG$_1$ of HIV-1-ITP PEG-ICs induced dramatic thrombocytopenia in recipient mice (which have 83% homology with human GPIIIa and Fc receptors for human IgG$_1$). *In vivo* specificity was substantiated by demonstrating prevention or reversal of the anti-GPIIIa-induced thrombocytopenia with an albumin conjugate of GPIIIa49–66 (Fig. 47-2).[32]

The presence of this immunodominant autoimmune GPIIIa49–66 epitope appears to be unique, because classic ATP appears to have multiple epitopes within platelet membrane receptors GPIIb-IIIa and GPIb. The observations in HIV-1-ITP could be related to cross-reactivity with HIV-1 antigens. This is suggested from the data of Bettaieb et al.,

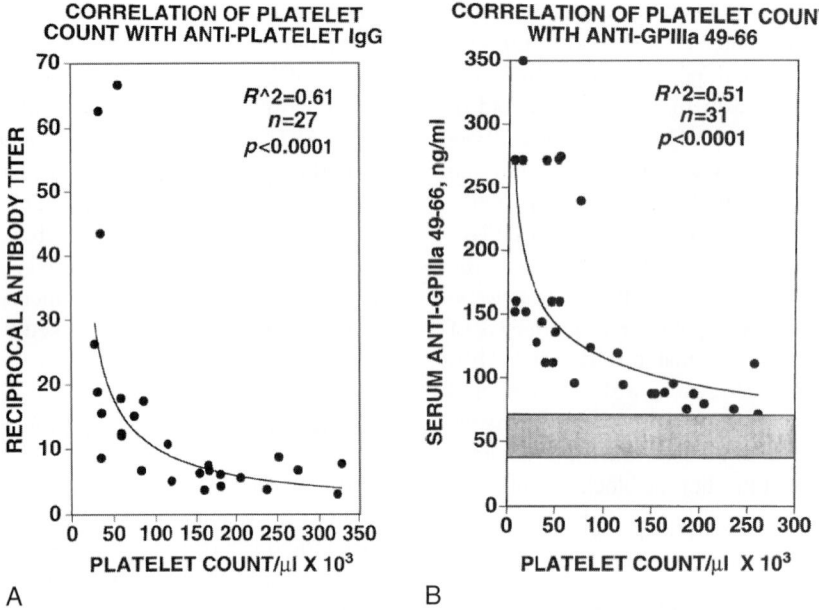

**Figure 47-1.** Correlation of platelet count with antiplatelet IgG and anti-GPIIIa-(49–66). (A.) Antiplatelet IgG was extracted from PEG-ICs, affinity-purified with fixed platelets, and tested for its reactivity with washed platelets adsorbed to a microtiter plate via a dilution determination of antibody titer, as described. A best-fit curvilinear power regression curve was selected by computer analysis (Macintosh Delta Graph Pro 3.5) with $R^2 = 0.61$, $F(1,25) = 40.2$, and $p < 0.0001$. (B.) Anti-GPIIIa-(49–66) concentration was measured in serum by using a standard curve of affinity-purified anti-GPIIIa-(49–66) versus purified peptide. Ordinate represents nanograms of antibody per ml of sera, diluted 1 : 4, reacting with GPIIIa-(49–66) adsorbed to microtiter plates. Gray horizontal band represents mean ±2 SD of 10 control subjects tested. A similar best-fit curvilinear power regression curve was selected with $R^2 = 0.51$, $F(1,29) = 30.0$, and $p < 0.0001$. Reproduced with permission from ref. [32].

who noted cross-reactivity with HIV-1gp120 and platelet GPIIIa,[74] as well as our report of anti-HIV-1gp120 on platelets of HIV-1-ITP patients,[28] and the work of Gonzalez-Conejero et al. who noted cross-reactivity with HIV-1gp120 of serum-platelet reactive eluates with GPIIb-IIIa and GPIb-IX in 20% of 45 drug abusers.[72] Also, Chia et al.[75] described cross-reactive HIV-1p24 and HIV-1gp120 antibodies in alkaline and acid eluates of serum antibodies bound to an HIV sepharose 4B affinity column. Indeed 48% of alkaline-eluted antibody bound to platelets as well as p24.

### D. Rheumatoid Factor Antibodies

Elevated CD5+ B cells have been reported in patients and mice with autoimmune diseases as well as in HIV-1-ITP.[76] CD5+ B cells generally produce low-affinity antibodies, which are predominantly RF of the IgM class against Fc of IgG. Because PEG-ICs of these patients contain high concentrations of IgM, RF-IgG complexes were sought within their serum PEG-ICs. Purified PEG-IC IgM was shown to react with purified PEG-IC IgG of these patients, relocating approximately 50% of the IgG preincubated with IgM to the void volume region of a G200 gel filtration column.[29] Thus, PEG-ICs of HIV-1-ITP patients contain RF-IgG complexes as well as anti-HIV-1gp120-anti-idiotype complexes.

### E. Anti-GPIIIa Antibody and Anti-Idiotype Antibody

Serum from HIV-1-ITP patients have little anti-GPIIIa49–66 antibody compared to purified serum IgG, suggesting the presence of blocking antibody. To investigate this theory, crude serum, purified serum IgG, and PEG-IC IgG were isolated and studied for anti-GPIIIa49–66 reactivity. This revealed approximately 150-fold greater antibody reactivity in purified serum IgG and approximately 4000-fold greater reactivity in PEG-IC IgG. This result was explained by the presence of anti-idiotype Ab2 (both IgG and IgM) sequestered within the PEG-IC. The IgM anti-idiotype is predominantly blocking antibody as demonstrated by specificity for F(ab')₂ fragments of anti-GPIIIa49–66 and inhibition with peptide GPIIIa49–66, not with a control peptide. The IgM anti-idiotype antibody is not polyreactive. In a group of HIV-1-infected patients (nonthrombocytopenic as well as thrombocytopenic), a positive correlation was noted between platelet count and anti-idiotype IgM (r = 0.7, p = 0.0001, n = 22). The in vivo relevance of the IgM anti-idiotype antibody was tested by inducing thrombocytopenia (approximately 30% baseline) in mice with anti-GPIIIa49–66. Preincubation of the anti-GPIIIa49–66 antibody with the IgM anti-idiotype (molar ratio 1 : 7) improved the platelet count to 80% of normal, and thrombocytopenia could be reversed 4 hours after induction of thrombocytopenia.[31] Thus, CICs of

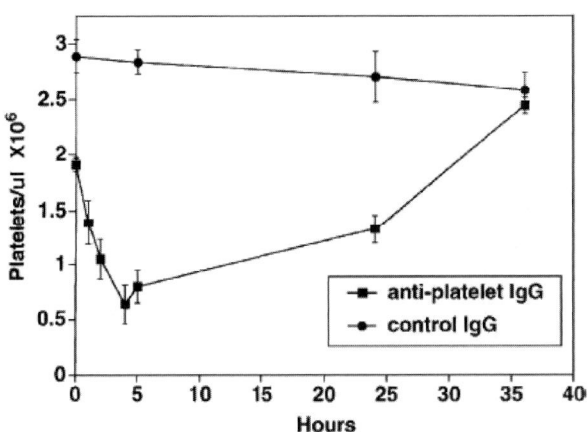

**EFFECT OF HUMAN ANTI-PLATELET GPIIIa ON MURINE PLATELETS**

A

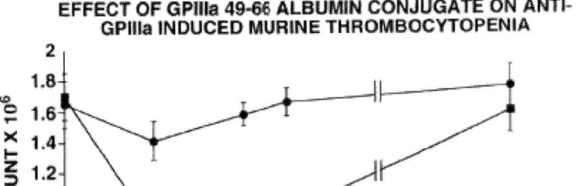

**EFFECT OF GPIIIa 49-66 ALBUMIN CONJUGATE ON ANTI-GPIIIa INDUCED MURINE THROMBOCYTOPENIA**

B

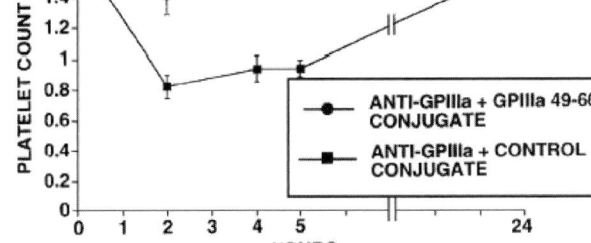

**REVERSAL OF ANTI-GPIIIa INDUCED THROMBOCYTOPENIA WITH GPIIIa 49-66 ALBUMIN CONJUGATE**

C

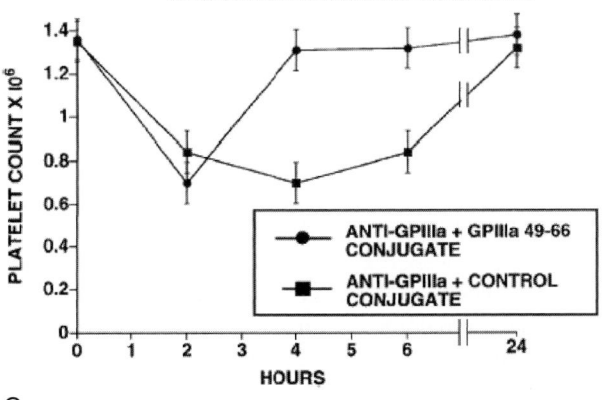

**Figure 47-2.** *In vivo* induction of thrombocytopenia in mice injected with affinity-purified anti-platelet IgG of HIV-1-ITP patients and its reversal with GPIIIa-(49–66). (A.) Thirty micrograms of antiplatelet IgG of patients 6 and 11 or purified control human IgG was injected intraperitoneally into eight experimental and six control mice, respectively. (B.) Twelve mice were injected intraperitoneally with 25 µg of antiplatelet IgG of patient 11, as noted previously. Seven of these mice were simultaneously injected in the opposite flank with 226 nmol of GPIIIa-(49–66)-albumin conjugate, whereas five were simultaneously injected with the irrelevant scrambled GPIIIa-albumin conjugate (CGGGARVLE-DRP) at the same concentration. (C.) Ten mice were injected intraperitoneally with 50 µg of antiplatelet IgG of patients 10 and 11, as prevously. At 2 hours, six of these mice were injected with GPIIIa-(49–66)-albumin conjugate, whereas four were injected with the irrelevant-scrambled GPIIIa-albumin conjugate. Reproduced with permission from ref. [32].

of HIV-1-ITP patients. They also offer an explanation for the presence of IgG on platelets of HIV-1-infected patients with normal platelet counts, namely the presence of sufficient anti-idiotype antibody to neutralize the thrombocytopenia.

### F. Complement-Independent, Peroxide-Induced Anti-GPIIIa49–66 Platelet Lysis

Recognition of the presence of platelet fragments within the serum PEG-ICs stimulated a search for antibody-dependent platelet lysis. These studies revealed the presence of a unique anti-GPIIIa49–66 Ab in HIV-1-ITP patients capable of fragmenting platelets *in vitro* (Figs. 47-3 and 47-4) and *in vivo* and inducing thrombocytopenia in the absence of complement.[77] Fragmentation is induced through oxidation of platelets by activation of a platelet NADPH oxidase pathway in concert with platelet 12-lipoxygenase.[78] Neither NADPH oxidase-deficient platelets [(p47phox-/-) or (gp91phox-/-)][77,78] nor 12-lipoxygenase-deficient platelets undergo antibody-induced oxidase fragmentation and thrombocytopenia. Antibody-activated platelet 12-lipoxygenase acts upstream of NADPH oxidase by producing a relatively stable oxidative product, 12(S)-HETE. Purified 12(S)-HETE mimics the effect of anti-GPIIIa49–66.[78]

### G. Molecular Mimicry to HIV-1 Peptides Determines Anti-GPIIIa49–66 Antibody

Molecular mimicry describes antigen sharing between host and parasite. Because of the immunodominant specificity of this antibody, we chose to examine whether molecular mimicry might be the mechanism for HIV-1-ITP.[79] Using a filamentous phage surface display 7-mer library, we screened

HIV-1-infected patients contain blocking IgM anti-idiotype antibody against anti-GPIIIa, which regulates their serum reactivity *in vitro* and level of thrombocytopenia *in vivo*.

These observations along with the presence of platelet cross-reactive idiotype-anti-idiotype complexes (i.e., HIV-1gp120) as well as the presence of RF offer an explanation for the markedly elevated immune complex/Ig on platelets

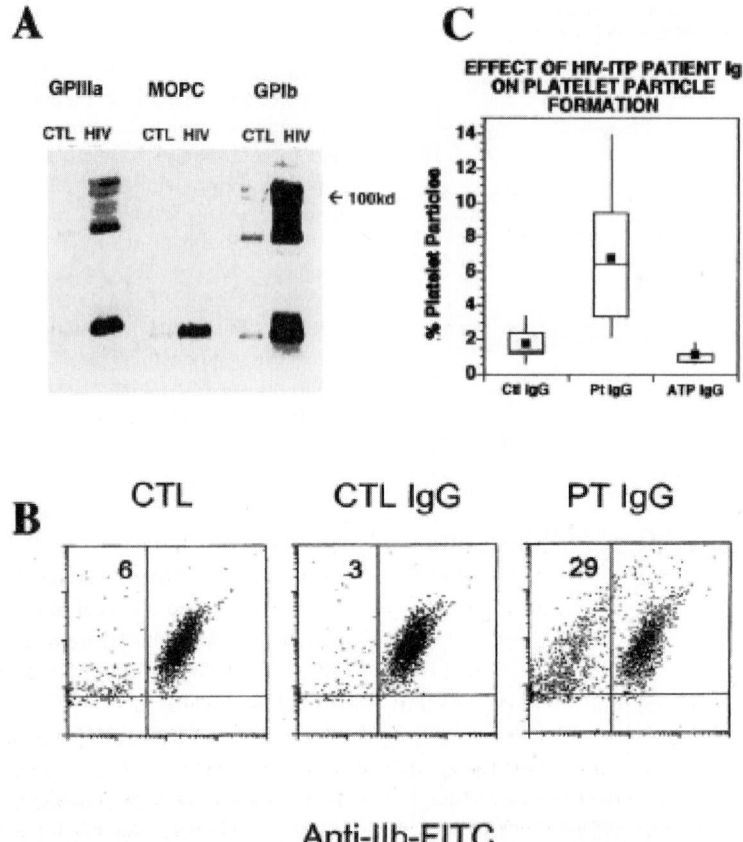

**Anti-IIb-FITC**

**Figure 47-3.** *In vitro* platelet fragmentation induced by anti-GPIIIa49–66 antibody. (A.) Immunoblot of PEG-IC of control and HIV-1-ITP patients employing anti-GPIIIa, anti-GPIbα, and control mouse monoclonal antibody, MoPC. CTL refers to control. (B.) Flow cytometry of platelet *in vitro* particle formation. Gel-filtered platelets were labeled with anti-GPIIb-FITC monoclonal antibody, washed, and then treated with various agents: buffer, control IgG, and patient anti-GPIIIa49–66 for 4 hours at 37°C. Three panels represent: CTL, buffer alone; CTL IgG, IgG isolated from control PEG-IC; and PT IgG, IgG isolated from HIV-1-ITP patient PEG-IC. *X*-axis refers to fluorescence; *Y*-axis to forward scatter. Numbers in left upper quadrant refer to the percentage of antibody-induced particles with decreased fluorescence attributed to platelet fragmentation. A highly reactive anti-GPIIIa49–66 IgG is shown. (C.) Distribution of percentage of platelet particle formation in control *versus* HIV-1-ITP versus ATP patients at 1 hour incubation. Results shown are for IgG isolated from PEG-IC's of 12 control, 16 HIV-1-ITP, and 5 ATP patients. Box plot. Mean is shown by the solid black box; median by the horizontal line in the large open box; 25th and 75th percentiles by the lower and upper border of the large open box from which spread of the data from the position of the median can be assessed. Whiskers include 99% of a Gaussian distribution. Reproduced with permission from ref. [77].

for antibody-reactive peptides capable of inhibiting platelet lysis and oxidation *in vitro*. Several phage-peptide clones were identified. Five shared close sequence similarity with GPIIIa49–66, as expected. However, 10 were molecular mimics with close sequence similarity to HIV-1-proteins nef, gag, env, and pol. Seven were synthesized as 10-mers from their known HIV-1 sequence and were found to inhibit antibody-induced platelet oxidation/fragmentation *in vitro*. Rabbit antibodies raised against the peptides mimicked the reaction of HIV-1-infected serum on HIV-1 antigens by immunoblot, demonstrating reactivity with intact proteins. Rabbit antibody induced platelet oxidative fragmentation *in vitro* and thrombocytopenia *in vivo* when passively transferred into mice (Figs. 47-5 and 47-6). Two of the peptides

shared a known epitope region[80] with HIV-1 protein nef (amino acid position 60–73) and were derived from a variant region of the protein known to be linear and shown to be antigenic and T-cell responsive by others.[81] This region maps in the predicted similar sequence region between nef and HLA class II.[80]

These data support the concept that other such variant mutants or nonconserved regions of HIV-1 may also be responsible in other patients. Indeed, it has been suggested that HIV is continuously undergoing mutations in the nonconserved region to mimic the host protein and escape immune surveillance. It is likely that such mutations may contribute to the susceptibility of thrombocytopenia in patients infected with HIV-1.

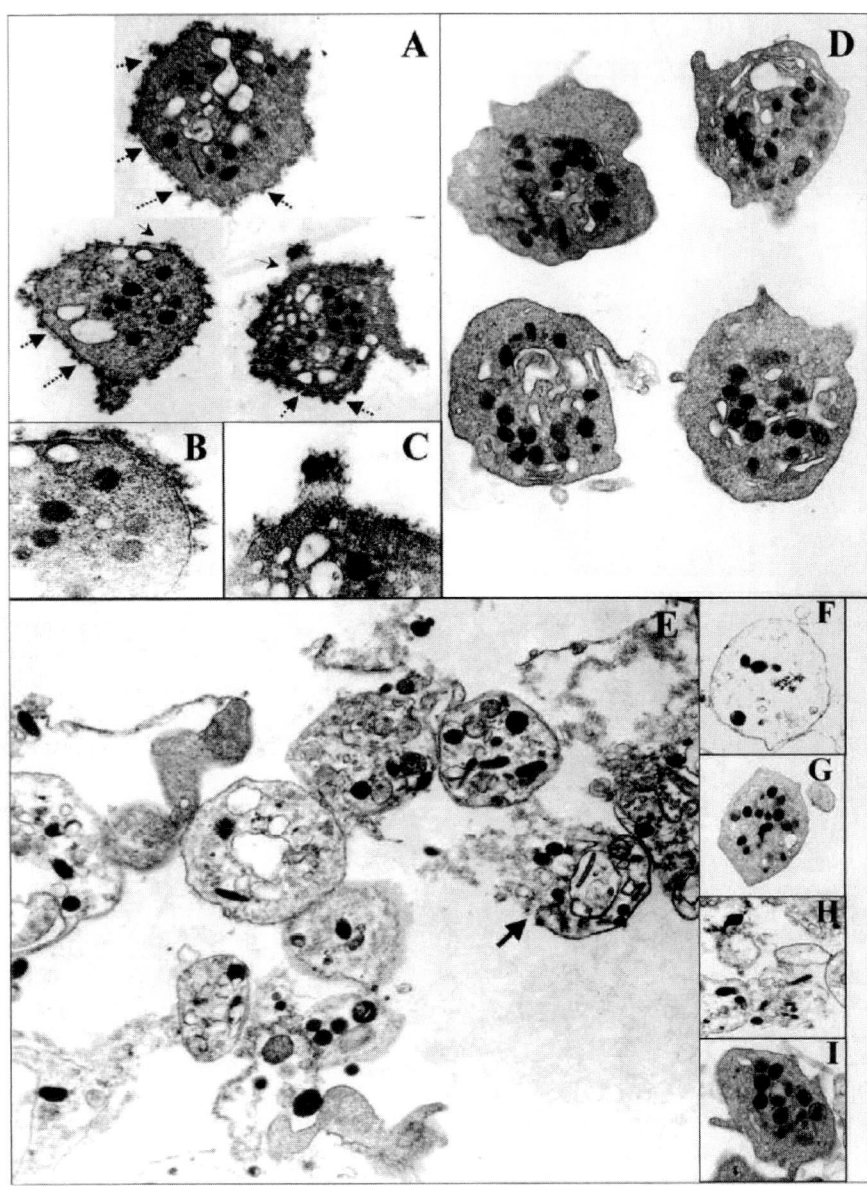

**Figure 47-4.** Electron microscopy of damaged platelets treated with anti-GPIIIa49–66 antibody. (A.) Patient sample at 1 hour showing platelets with a fuzzy material attached to the outer surface of the cell membranes (dotted arrows). Gaps are noted in the cell membranes with leakage of cytoplasmic content (arrows). These areas are demonstrated at higher magnification in (B) and (C). (E.) Patient sample at 4 hours showing degenerating platelets and disintegration of the cell membrane (arrow). Swollen platelet (F), platelet fragments (H), and occasional normal platelets (G and I) are also shown. None of these changes were present in IgG controls at 4 hours (D). Original magnifications: (A), (D–I): ×4000. (B) ×50,000. (C) ×40,000. Reproduced with permission from ref. [77].

## EFFECT OF RABBIT ANTI-HIV PEPTIDE
## Ab ON PLATELET FRAGMENTATION

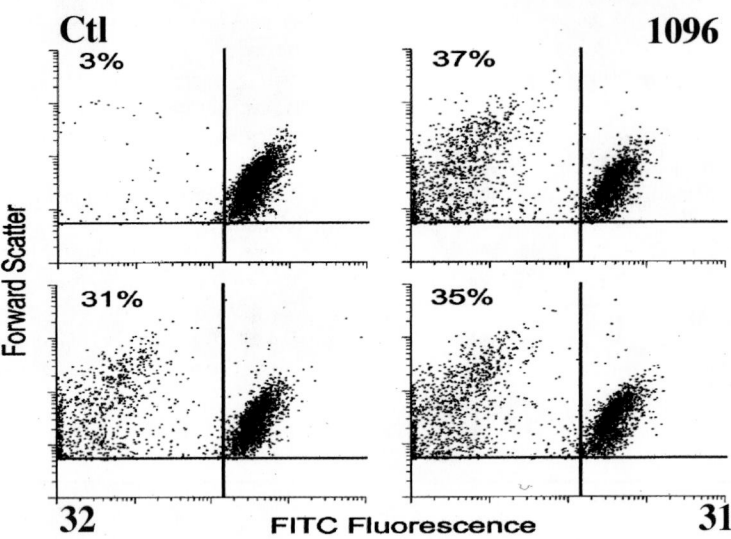

**Figure 47-5.** Effect of rabbit antibodies raised against HIV-1 10 mer peptides or 7 mer phagepeptides on platelet fragmentation. Gel-filtered platelets were incubated with control rabbit IgG (upper left quadrant) or rabbit antibody against peptide batches 1096 (upper right quadrant), 1098 (lower left), and 1099 (lower right) at 37°C for 4 hours. Abbreviation: FITC, fluorescein isothiocyanate. Reproduced with permission from ref. [79].

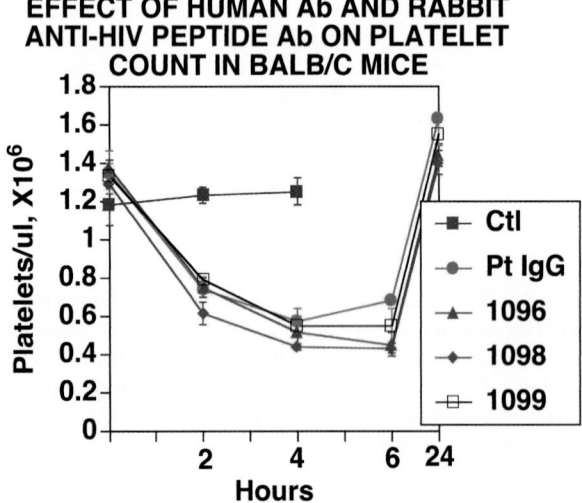

**Figure 47-6.** Effect of rabbit antibody against HIV-1 peptides on induction of thrombocytopenia in mice. Purified control IgG (CtI), patient IgG (Pt IgG), and rabbit Ab (25 μg) against mimicry HIV-1 peptide batches 1096, 1098, and 1099 were injected intraperitoneally into Balb/c mice, and platelet counts followed for 24 hours. Mean ± SEM, n = 4. Reproduced with permission from ref. 79.

## References

1. Morris, L., Distenfeld, A., Amorosi, E., et al. (1982). Autoimmune thrombocytopenic purpura in homosexual men. *Ann Int Med, 96,* 714–717.
2. Savona, S., Nardi, M. A., & Karpatkin, S. (1985). Thrombocytopenic purpura in narcotics addicts. *Ann Int Med, 102,* 737–741.
3. Ratnoff, O. D., Menitove, J. E., Aster, R. H., et al. (1983). Coincident classic hemophilia and "idiopathic" thrombocytopenic purpura in patients under treatment with concentrates of anti-hemophilic factor (factor VIII). *N Engl J Med, 308,* 439–442.
4. Karpatkin, S., Nardi, M., & Hymes, K. B. (1988). Immunologic thrombocytopenic purpura after heterosexual transmission of human immunodeficiency virus (HIV). *Ann Intern Med, 109,* 190–193.
5. Hymes, K. B., Cheung, T., Greene, J. B., et al. (1981). Kaposi's sarcoma in homosexual men: A report of eight cases. *Lancet, 2,* 598–600.
6. Jokela, J., Flynn, T., & Henry, K. (1987). Thrombotic thrombocytopenic purpura in a human immunodeficiency virus (HIV)-seropositive homosexual man. *Am J Hematol, 25,* 341–••.
7. Leaf, A. N., Raphael, B., Hochster, H., et al. (1988). Thrombotic thrombocytopenic purpura associated with HIV-1 infection. *Ann Intern Med, 109,* 194–197.

8. Nair, J. M. G., Bellevue, R., Bertoni, M., et al. (1988). TTP in patients with the AIDS-related complex. *Ann Intern Med, 109*, 209–••.

9. Walsh, C., Kriegel, R., Lennette, E., et al. (1985). Thrombocytopenia in homosexual patients: Prognosis, response to therapy, and prevalence of antibody to the retrovirus associated with the acquired immunodeficiency syndrome. *Ann Intern Med, 103*, 542–545.

10. Murphy, M. F., Metcalfe, P., & Waters, A. H. (1987). Incidence and mechanism of neutropenia and thrombocytopenia in patients with human immunodeficiency virus infection. *Brit J Haematol, 66*, 337–340.

11. Mientjes, C. H. C., Van Ameijden, E. J. C., Mulder, J. W., et al. (1992). Prevalence of thrombocytopenia in HIV-infected and non-HIV-infected drug abusers and homosexuals. *Brit J Haematol, 82*, 615–619.

12. Landonio, G., Gall, M., Nosari, A., et al. (1998). HIV-related severe thrombocytopenia in intravenous drug abusers: Prevalence, response to therapy in a medium term follow-up and pathogenetic evaluation. *AIDS, 4*, 29–34.

13. Auch, D., Budde, U., Hammerstein, U., et al. (1987). FcR-mediated clearance in thrombopenic and non-thrombopenic patients with hemophilia A and possible relation of thrombopenia to HIV seropositivity. *Eur J Haematol, 39*, 440–••.

14. Ehmann, W. C., Rabkin, C. S., Eyster, M. E., et al. (1997). Thrombocytopenia in HIV-infected and uninfected hemophiliacs. *Am J Hematol, 54*, 296–300.

15. Sullivan, P. S., Hanson, D. L., Chu, S. Y., et al. (1997). Surveillance for thrombocytopenia in persons infected with HIV: Results from the multistate Adult and Adolescent Spectrum of Disease Project. *JAIDS, 14*, 374–379.

16. Holzman, R. S., Walsh, C. M., & Karpatkin, S. (1987). Risk for the acquired immunodeficiency syndrome among thrombocytopenic and non-thrombocytopenic homosexuals nonseropositive for human immunodeficiency virus. *Ann Intern Med, 106*, 383–386.

17. Abrams, D. I., Kiprov, D. D., Goedert, J. J., et al. (1986). Antibodies to human T lymphotropic virus type III and development of the acquired immunodeficiency syndrome in homosexual men presenting with immune thrombocytopenia. *Ann Intern Med, 105*, 47–••.

18. Goldsweig, H. G., Grossman, R., & William, D. (1986). Thrombocytopenia in homosexual men. *Am J Hematol, 21*, 243–••.

19. Hobollah, J. J., Kim, E. H., & Dumont, A. E. (1989). Thrombocytopenic purpura in parental drug abusers: Response to splenectomy. *Surg Gynecol Obstet, 168*, 497–500.

20. Hymes, K. B., Greene, J. B., & Karpatkin, S. (1988). The effect of azidothymidine on HIV-related thrombocytopenia. *N Engl J Med, 318*, 516–517.

21. Swiss Group for Clinical Studies on AIDS. (1988). Zidovudine for the treatment of thrombocytopenia associated with HIV. A prospective study. *Ann Intern Med, 109*, 718.

22. Arranz Caso, J. A., Sanchez Mingo, C., & Garcia Tena, J. (1999). Effect of highly active antiretroviral therapy on thrombocytopenia in patients with HIV infection [letter]. *New Engl J Med, 341*, 1239–1240.

23. Bussell, J., & Haimi, J. (1988). Isolated thrombocytopenia in patients infected with HIV: Treament with intravenous gammaglobulin. *Am J Hematol, 28*, 79–••.

24. Oskendler, E., Bierling, P., Farcet, J.-P., et al. (1987). Response to therapy in 37 patients with HIV-related thrombocytopenic purpura. *Brit J Haematol, 66*, 491–••.

25. Pollak, A. N., Janinis, J., & Green, D. (1988). Successful intravenous immunoglobulin therapy for human immunodeficiency virus-associated thrombocytopenia. *Arch Intern Med, 148*, 695–••.

26. Rarick, M. U., Montgomery, T., Groshen, S., et al. (1991). Intravenous immunoglobulin in treatment of human immunodeficiency virus-related thrombocytopenia. *Am J Hematol, 38*, 261–266.

27. Karpatkin, S., & Nardi, M. A. (1988). On the mechanism of thrombocytopenia in hemophiliacs multiply transfused with AHF concentrates. *J Lab Clin Med, 111*, 441–448.

28. Karpatkin, S., & Nardi, M. A. (1992). Autoimmune anti-HIV-1gp120 antibody with anti-idiotype-like activity in sera and immune complexes of HIV-1 related immunologic thrombocytopenia (HIV-1-ITP). *J Clin Invest, 89*, 356–364.

29. Karpatkin, S., Nardi, M. A., & Hymes, K. B. (1995). Sequestration of anti-platelet GPIIIa antibody in rheumatoid factor-immune complexes of human immunodeficiency virus 1 thrombocytopenic patients. *Proc Natl Acad Sci USA, 92*, 2263–2267.

30. Karpatkin, S., Nardi, M. A., Lennette, E. T., et al. (1988). Anti-human immunodeficiency virus type 1 antibody complexes on platelets of seropositive thrombocytopenic homosexuals and narcotic addicts. *Proc Natl Acad Sci USA, 85*, 9763–9767.

31. Nardi, M., & Karpatkin, S. (2000). Antiidiotype antibody against platelet anti-GPIIIa contributes to the regulation of thrombocytopenia in HIV-1-ITP patients. *J Exper Med, 191*, 2093–2100.

32. Nardi, M. A., Liu, L.-X., & Karpatkin, S. (1997). GPIIIa (49–66) is a major pathophysiologically-relevant antigenic determinant for anti-platelet GPIIIa of HIV-1-related immunologic thrombocytopenia (HIV-1-ITP). *Proc Natl Acad Sci USA, 94*, 7589–7594.

33. Walsh, C. M., Nardi, M. A., & Karpatkin, S. (1984). On the mechanism of thrombocytopenic purpura in sexually-active homosexual men. *New Engl J Med, 311*, 635–639.

34. Yu, J.-R., Lennette, E. T., & Karpatkin, S. (1986). Anti-F(ab')2 antibodies in thrombocytopenic patients at risk for Acquired Immunodeficiency Syndrome. *J Clin Invest, 77*, 1756–1761.

35. Najean, Y., & Rain, J.-D. (1994). The mechanism of thrombocytopenia in patients with HIV. *J Lab Clin Med, 123*, 415.

36. Bel-Ali, Z., Dufour, V., & Najean, Y. (1987). Platelet kinetics in human immunodeficiency virus induced thrombocytopenia. *Am J Hematol, 26*, 229–••.

37. Ballem, P. J., Segal, G. M., Stratton, J. R., et al. (1987). Mechanisms of thrombocytopenia in chronic autoimmune thrombocytopenic purpura: Evidence of both impaired platelet production and increased platelet clearance. *J Clin Invest, 80*, 33–40.

38. Siegel, R. S., Rae, J. L., & Kessler, C. M. (1986). Immune thrombocytopenic purpura in HTLV-III men. *Blood, 68*(Suppl 1), 134A.

39. Ballem, P., Belzberg, A., Devine, D., et al. (1992). Kinetics studies of the mechanism of thrombocytopenia in patients with human immunodeficiency virus infection. *New Engl J Med, 327,* 1779–1784.

40. Cole, J. L., Marzec, U. M., Gunthel, C. J., et al. (1998). Ineffective platelet production in thrombocytopenic human immunodeficiency virus-infected patients. *Blood, 91,* 3239–3246.

41. Stella, C. C., Ganser, A., & Hoezler, D. (1987). Defective *in vitro* growth of the hemopoietic progenitor cells in the acquired immunodeficiency syndrome. *J Clin Invest, 80,* 286.

42. Chelucci, C., Federico, M., Guerriero, R. G. M., et al. (1998). Productive human immunodeficiency virus-1 infection of purified megakaryocytic progenitors/precursors and maturing megakaryocytes. *Blood, 91,* 1225–1234.

43. Zucker-Franklin, D., & Cao, Y. (1989). Megakaryocytes of human immunodeficiency virus-infected individuals express viral RNA. *Proc Natl Acad Sci USA, 86,* 5595.

44. Basch, R., Kouri, Y. H., & Karpatkin, S. (1990). Expression of CD4 by human megakaryocytes. *Proc Natl Acad Sci USA, 87,* 8085–8089.

45. Kouri, Y. H., Borkowsky, W., Nardi, M., et al. (1993). Human megakaryocytes have a CD4 molecule capable of binding human immunodeficiency virus-1. *Blood, 81,* 2664–2670.

46. Kowalska, M. A., Ratajczak, J., Hoxie, J., et al. (1999). Megakaryocyte precursors, megakaryocytes and platelets express the HIV co-receptor CXCR4 on their surface: Determination of response to stromal-derived factor-1 by megakaryocytes and platelets. *Brit J Haematol, 104,* 220–229.

47. Riviere, C., Subra, F., Cohen-Solal, K., et al. (1999). Phenotypic and functional evidence for the expression of CXCR4 receptor during megakaryocytopoiesis. *Blood, 93,* 1511–1523.

48. Sakaguchi, M., Sato, T., & Groopman, J. (1991). Human immunodeficiency virus infection of megakaryocytic cells. *Blood, 77,* 481.

49. Bahner, I., Kearns, K., Couthino, S., et al. (1997). Infection of human marrow stroma by HIV-1 is both required and sufficient for HIV-1 induced hematopoietic suppression *in vitro:* Demonstration by gene modification of primary human stroma. *Blood, 90,* 1787–1798.

50. Davis, B. R., Marx, J. C., Johnson, C. E., et al. (1991). Absent or rare HIV-1 infection of bone marrow stem/progenitor cells *in vivo. J Virol, 65,* 1985.

51. Louache, F., Henri, A., Bettaieb, A., et al. (1992). Role of human immunodeficiency virus replication in defective *in vitro* growth of hematopoietic progenitors. *Blood, 80,* 2991.

52. Moses, A., Nelson, J., & Bagby, G. C., Jr. (1998). The influence of human immunodeficiency virus 1 on hematopoiesis. *Blood, 91,* 1479–1491.

53. Moses, A. V., Williams, S., Heneveld, M. L., et al. (1996). Human immunodeficiency virus infection of bone marrow endothelium reduces induction of stromal hematopoietic growth factors. *Blood, 87,* 919–••.

54. Scadden, D. T., Zeira, M., Woon, A., et al. (1990). Human HIV-1 infection of human bone marrow stromal fibroblasts. *Blood, 76,* 317.

55. Steinberg, H. N., Anderson, J., Crumpacker, C. S., et al. (1993). HIV infection of the BS-1 human stromal cell line: Effect on murine hematopoiesis. *Virology, 193,* 524–••.

56. von Laer, D., Hufert, F. T., Fenner, T. E., et al. (1990). CD34⁺ hematopoietic progenitor cells are not a major reservoir of the HIV-1 virus. *Blood, 76,* 1281.

57. Von Laer, D., Hufert, F. T., Fenner, T. E., et al. (1990). Progenitor cells are not a major reservoir of the human immunodeficiency virus. *Blood, 76,* 1281–••.

58. Zauli, G., Carla-Re, M., Davis, B., et al. (1992). Impaired *in vitro* growth of purified (CD34⁺) hematopoietic progenitors in human immunodeficiency virus-1 seropositive thrombocytopenic individuals. *Blood, 79,* 2680.

59. Zauli, G., Catani, L., Gibellini, D., et al. (1996). Impaired survival of bone marrow GPIIb/GPIIIa+ megakaryocytic cells as an additional pathogenesis mechanism of HIV-1-related thrombocytopenia. *Brit J Haematol, 92,* 711–••.

60. Zucker-Franklin, D., Termin, C. S., & Cooper, M. C. (1989). Structural change in the megakaryocytes of patients infected with human immune deficiency virus (HIV-1). *Am J Pathol, 134,* 1295.

61. Carlo-Stella, C., Ganser, A., & Hoelzer, D. (1987). Defective in vitro growth of the hematopoietic progenitor cells in the acquired immunodeficiency syndrome. *J Clin Invest, 80,* 286–••.

62. Ishibashi, T., Miller, S. L., & Burstein, S. A. (1987). Type β transforming growth factor is a potent direct inhibitor of murine megakaryocytopoiesis *in vitro. Blood, 69,* 1737.

63. Kuter, D. J., Gminski, D. M., & Rosenberg, R. D. (1992). Transforming growth factor b inhibits megakaryocyte growth and endomitosis. *Blood, 79,* 619.

64. Mitjavila, M. T., Vinci, G., Villeval, J. L., et al. (1988). Human platelet alpha granules contain a non-specific inhibitor of megakaryocyte colony formation: Its relationship to type b transforming growth factor (TGF-b). *J Cell Physiol, 134,* 93.

65. Zauli, G., Davis, B., Re, M. C., et al. (1992). Tat protein stimulates production of TGFb-1 by marrow macrophages: A potential mechanism for human HIV-1-induced hematopoietic suppression. *Blood, 80,* 3036.

66. Gewirtz, A. M., Calabretta, B., Rucinski, B., et al. (1989). Inhibition of human megakaryocytopoiesis *in vitro* by platelet factor 4 (PF4) and a synthetic COOH-terminal PF4 peptide. *J Clin Invest, 83,* 1477.

67. Guarini, A., Sanavio, F., Novarino, A., et al. (1991). Thrombocytopenia in acute leukemia patients treated with IL-2: Cytolytic effect of LAK on megakaryocytic progenitors. *Brit J Haematol, 79,* 451.

68. Volkers, B., Ganser, A., Greher, J., et al. (1989). Effect of tumor necrosis factor on human hematopoietic precursor cells. *Onkolgie, 12,* 109.

69. Ganser, A., Carlo-Stella, C., Greher, J., et al. (1987). Effect of recombinant interferons alpha and gamma on human bone marrow-derived megakaryocytic progenitor cells. *Blood, 70,* 1171.

70. Sonoda, Y., Kuzuyama, Y., Tanaka, S., et al. (1993). Human IL-4 inhibits proliferation of megakaryocyte progenitor cells in culture. *Blood, 81,* 624.

71. Van Der Lelie, J., Lange, J. M. A., ••, J. J. E., et al. (1987). Autoimmunity against blood cells in human immunodeficiency virus (HIV) infection. *Brit J Haematol, 67,* 109–••.

72. Gonzalez-Conejero, R., Rivera, J., Rosillo, M. C., et al. (1996). Association of autoantibodies against platelet glycoproteins Ib/IX and IIb/IIIa and platelet-reactive anti-HIV antibodies in thrombocytopenic narcotic addicts. *Br J Haematol, 93,* 464–471.

73. Karpatkin, S., Nardi, M. A., & Kouri, Y. (1992). Internal image anti-idiotype HIV-1gp120 antibody in human immunodeficiency virus 1 (HIV-1)-seropositive individuals with thrombocytopenia. *Proc Natl Acad Sci USA, 89,* 1487–1491.

74. Bettaieb, A., Oksenhendler, E., Duedari, N., et al. (1996). Cross-reactive antibodies between HIV-gp120 and platelet gpIIIa (CD61) in HIV-related immune thrombocytopenic purpura. *Clin Exp Immunol, 103,* 19–23.

75. Chia, W. K., Blanchette, V., Mody, M., et al. (1998). Characterization of HIV-1-specific antibodies and HIV-1-crossreactive antibodies to platelets in HIV-1-infected haemophiliac patients. *Br J Haematol, 103,* 1014–1022.

76. Kouri, Y., Basch, R. S., & Karpatkin, S. (1992). B-cell subsets and platelet counts in HIV-1 seropositive subjects. *Lancet, 339,* 1445–1446.

77. Nardi, M., Tomlinson, S., Greco, M., et al. (2001). Complement-independent, peroxide induced antibody lysis of platelets in HIV-1-related immune thrombocytopenia. *Cell, 106,* 551–561.

78. Nardi, M., Feinmark, S., Liang, H., et al. (2004). Complement-independent Ab-induced peroxide lysis of platelets requires 12-lipoxygenase and a platelet NADPH oxidase pathway. *J Clin Invest, 113,* 973–980.

79. Li, Z., Nardi, M. A., & Karpatkin, S. (2005). Role of molecular mimicry to HIV-1 peptides in HIV-1-related immunologic thrombocytopenia. *Blood, 106,* 572–576.

80. Schneider, T., Harthus, H. P., Hildebrandt, P., et al. (1991). Epitopes of the HIV-1-negative factor (nef) reactive with murine monoclonal antibodies and human HIV-1-positive sera. *AIDS Res Hum Retroviruses, 7,* 37–44.

81. Kiepiela, P., Leslie, A. J., Honeyborne, I., et al. (2004). Dominant influence of HLA-B in mediating the potential co-evolution of HIV and HLA. *Nature, 432,* 769–775.

# CHAPTER 48

# Heparin-Induced Thrombocytopenia

## Beng H. Chong

*Department of Medicine, St. George Clinical School, University of New South Wales, Kogarah, Australia*

## I. Introduction

Heparin is a mixture of highly sulfated glycosaminoglycans of different chain lengths. The molecular weight of the heparin chains varies from 1800 to 30,000 Daltons. Unfractionated heparin and low molecular weight heparin (LMWH) are both effective anticoagulants and are widely used in the prevention and treatment of venous and arterial thromboembolic disorders. In this chapter, the term *heparin* will be used to denote unfractionated heparin. Heparin is also used as an anticoagulant in extra-corporeal circuits such as in cardiopulmonary bypass and renal dialysis. Pentasaccharide (comprising the five sugar moieties in heparin that mediate its anticoagulant activity) may soon replace heparin and LMWH in some of these indications.

Heparin and LMWH have potential side effects including bleeding and thrombocytopenia. Heparin-induced thrombocytopenia (HIT) has aroused considerable interest in the medical community. HIT is not only common but is potentially serious and life-threatening.[1] It occurs in 1 to 5% of patients receiving heparin and is associated with a high risk of thromboembolic complications, which may lead to limb gangrene, limb amputation, and even death.[2] Studies have provided new insights into the immune mechanism of this condition. However, most physicians find the diagnosis of HIT difficult, particularly in the presence of coexisting thrombocytopenic conditions. Effective drugs for its treatment have been available in North America and Europe for several years,[3,4] yet many clinicians are still unfamiliar with the use of these drugs. It is not surprising that the number of litigation cases involving patients with HIT has increased. This chapter discusses the incidence, pathophysiology, clinical features, diagnosis, and treatment of HIT and attempts to clarify some of the controversial issues.

## II. Historical Aspects of HIT

In the 1940s, several investigators reported on the ability of heparin to cause thrombocytopenia.[5-7] They demonstrated that a single injection of heparin administered to mice or dogs resulted in an immediate but transient drop in the platelet count. In 1962, Gollub and Ulin reported the same phenomenon in man,[8] but Davey and Lander later suggested that it was an *in vitro* artifact.[9] These early reports did not view the thrombocytopenia caused by heparin to be a clinically significant problem.

Thrombocytopenia associated with the clinical use of heparin was first reported by Natelson and coworkers in 1969.[10] They described a 78-year-old man who developed severe thrombocytopenia after treatment with heparin for 10 days. Although this man had disseminated intravascular coagulation (DIC), the thrombosis in this patient was attributed to his prostate carcinoma and not to heparin administration. The link between thrombosis and thrombocytopenia induced by heparin was not made at that time.

In 1958, Weismann and Tobin observed thrombotic complications during heparin therapy,[11] but they did not mention the presence of thrombocytopenia in their patients. They described a series of 10 patients who developed occlusion of lower limb arteries 7 to 15 days after commencement of heparin administration. Weismann and Tobin believed that heparin anticoagulation disrupted preexisting aortic thrombi leading to emboli being lodged in the limb arteries distally. They did not consider the possibility that heparin might have been the direct cause of the arterial thrombosis. A few years later, Roberts and Rosato reported a similar series of patients with heparin-associated thrombosis.[12]

The link between thrombocytopenia and thrombosis induced by heparin was first noticed by Rhodes and his colleagues in 1973.[13] They reported two patients who developed thrombocytopenia during heparin therapy. The patients also had myocardial infarction, petechiae, and heparin resistance. The thrombocytopenia resolved after cessation of heparin but promptly recurred upon rechallenge with heparin. These investigators also provided *in vitro* evidence for the immune basis of HIT. They demonstrated a heparin-dependent platelet aggregating factor in the patients' sera and suggested that the aggregating factor was in the IgG fraction on fractionation of one patient's serum.

During the latter half of the 1970s, there were many reports of patients with heparin-associated thrombocytopenia and thrombosis.[14–16] A report by Towne and coworkers stated that the pale thrombi consisting of fibrin and aggregated platelets were characteristic of this syndrome. They coined the term, "white clot syndrome."[17]

In these reports, the platelet counts in some patients returned to normal levels only after heparin was withdrawn and a heparin-dependent platelet aggregating factor was often detectable in the patients' sera. However, in other patients, the thrombocytopenia resolved even with continuation of heparin.[15,18,19] It was the latter patients who led some investigators to hold the view that HIT might not be mediated by an immune mechanism and that there might be another explanation (e.g., DIC) for the patients' thrombocytopenia. In 1974, Klein and Bell reported two patients with heparin-associated thrombocytopenia in whom they also found evidence of DIC.[18] Bell and his colleagues subsequently conducted the first prospective study to investigate the incidence of thrombocytopenia in patients receiving heparin.[19] They found a surprisingly high incidence of 31%. Sixteen of 52 patients developed thrombocytopenia, and some of these patients had hypofibrinogenemia and fibrin degradation products but no heparin-related platelet antibodies in their plasma. The investigators believed that a contaminating thromboplastin-like material in heparin was the cause of the DIC. There was much confusion then regarding the etiology of HIT. Some investigators believed that it was caused by an immune mechanism, whereas others favored a nonimmune mechanism.

## III. Nomenclature

Different terms have been used in the past for this heparin adverse effect. Investigators who believed that the condition was caused directly by heparin itself used the term *heparin-induced thrombocytopenia (HIT)*, while others who did not share this view frequently used the term *heparin-associated thrombocytopenia (HAT)*. Later, when the condition was seen not only with different heparin preparations but also with different heparin fragments (e.g., LMWHs produced by different fractionation processes), it became clear that it was heparin itself that was the cause of the clinical syndrome and not a contaminating substance from the manufacturing process. The term *heparin-associated thrombocytopenia* became less frequently used, and it is now largely replaced by *heparin-induced thrombocytopenia or HIT*.

From the confusion in the 1970s, the concept of two distinct clinical types of HIT emerged. In 1982, Chong et al. described two groups of patients with HIT.[20] In the first group, the patients were asymptomatic and had mild thrombocytopenia of early onset with no detectable heparin-dependent IgG platelet aggregating factor in their plasma.

In the second group, the patients had severe thrombocytopenia of delayed onset and frequently suffered thromboembolic complications. A plasma platelet aggregating factor was detectable in patients of the second group. The clinical course of the patients in the first group and the absence of a heparin-dependent antibody were consistent with a nonimmune mechanism. In contrast, the delayed onset of the thrombocytopenia and the presence of a heparin-dependent platelet antibody in the patients of the second group strongly suggested an immune mechanism. Hence, the term *nonimmune HIT* was used for the patients in the first group and *immune HIT* for patients in the second group. In a subsequent review, Chong and Berndt proposed the term "Type I" HIT for the patients with the early onset nonimmune HIT and "Type II" HIT for the patients with the delayed onset immune HIT.[21] These terms later became popular.

## IV. Pathogenesis

### A. *Type I HIT (Nonimmune HIT)*

The mechanism of type I HIT is distinct from that of type II HIT.[1,20] The mechanism of type I HIT is probably related to the platelet proaggregating effect of heparin. Heparin binds to platelets and causes mild cell activation and formation of tiny platelet aggregates.[22,23] Heparin also enhances the proplatelet aggregating effect of ADP and collagen.[24] This effect is more pronounced with hyperactive platelets of patients with infection, peripheral vascular disease, and anorexia nervosa.[25,26] Addition of heparin *in vitro* to platelets of patients with these conditions may lead to spontaneous platelet aggregation. Heparin-induced platelet aggregation is mediated by fibrinogen and its platelet receptor, integrin $\alpha IIb\beta 3$ (the glycoprotein [GP] IIb-IIIa complex).[23] The binding of heparin to platelets can be inhibited by heparin-binding proteins such as antithrombin and fibronectin[23,24] but not by platelet factor 4 (PF4).

It is quite conceivable that when heparin is administered to patients, particularly those with hyperactive platelets, it may cause mild platelet aggregation *in vivo*. The aggregated platelets are then cleared by the reticuloendothelial system. This may explain the modest drop in the platelet counts that occurs during the first 4 days of heparin administration. In the patients with hyperactive platelets or the infected patients with circulating immune complexes and bacterial products, heparin may cause more intense platelet aggregation and more severe thrombocytopenia. There are published data in support of this mechanism for type I HIT. In a randomized prospective study, Blahut et al. observed that patients with septicemia had a significant fall in their platelet counts when they were given intravenous (IV) heparin with or without antithrombin. In contrast, the control patients who received antithrombin alone showed no decrease in their platelet counts.[27] Furthermore, Balduini et al. found a more frequent and pronounced fall in platelet counts in patients receiving heparin after streptokinase for treatment of acute myocardial infarction, compared with the controls who were given only streptokinase.[28]

However, in some individuals (e.g., postsurgery patients), other concomitant clinical factors such as hemodilution may also account for the fall in platelet count. In these patients it is unclear whether heparin, the concomitant factor, or both are the cause of the thrombocytopenia.

### B. Type II HIT (Immune HIT)

Type II HIT is mediated by an antibody that induces platelet activation only in the presence of heparin. Consequently it is termed a *heparin-dependent antibody.* The target antigen of the antibody was identified in 1992 by Amiral et al. to be a complex formed by heparin and PF4.[29] PF4 is a positively charged tetrameric protein (molecular weight 35 kDa) found in the α granules of platelets and megakaryocytes. It belongs to the CXC chemokine family. The plasma PF4 concentration is very low in the resting state, but it rises when PF4 is secreted into the plasma during platelet activation or heparin infusion.

Heparin binds PF4 with high affinity. PF4-heparin complexes bind to platelets via the heparin binding sites.[30] Binding of heparin to PF4 induces a conformational change in PF4 exposing certain cryptic epitopes, to which the HIT antibody binds.[1] Other polysaccharides and negatively charged compounds (e.g., polyvinylsulfonate) can also bind PF4 and induce the same conformational change.[31] The binding of heparin/heparin fragments to PF4 is dependent on the composition, chain length (>12 to 14 oligosaccharide units), and the degree of sulfation of the heparin.[32,33] LMWHs are much shorter in length than heparin and have a lower affinity for PF4. They are therefore less antigenic and less likely to cause immune HIT. The pentasaccharide is even smaller and shorter; it does not bind PF4 and may possibly not cause HIT.

The PF4-heparin complex is formed at stoichiometric concentrations of heparin and PF4. Excess heparin causes dispersion of the PF4-heparin complex. The therapeutic plasma heparin concentration is 0.2 to 0.4 I.U./mL or 100 to 200 nM,[34] but the plasma drug level is lower if it is given for prophylaxis. In most clinical settings, heparin is present in excess in plasma. The plasma concentration of PF4 rises after an IV injection of heparin because heparin displaces PF4 from endothelial cells. The plasma concentration of PF4 reaches a maximum level of 8 nM, which is much below the stoichiometric concentration of PF4 that is optimal for PF4-heparin complex formation.[35] Therefore, it requires, in addition, platelet activation (with resultant release of α granule PF4) for the plasma PF4 concentration to rise to levels

compatible for the formation of PF4-heparin complex. Such levels are reached in patients undergoing cardiac surgery or orthopedic surgery. Not surprisingly, anti-PF4-heparin antibodies are frequently produced in these patients.[36,37]

Unlike the quinine-dependent antibody, which binds to an antigenic restricted site on the target protein,[38] the heparin-dependent antibody is polyclonal and binds multiple neoepitopes on PF4-heparin complex.[39] Li et al. and Zeporen et al. identified two antibody binding sites on the "P" surface of the PF4 tetramer.[40,41] We have identified several other antibody sites on the same surface of the PF4 tetramer (unpublished data).[39] These data, together with the finding by Rauova et al. that PF4 and heparin form ultralarge complexes of a molecular weight of 670 kDa or greater,[42] suggest that the heparin-dependent antibody can potentially bind bivalently to epitopes on the P surfaces of two adjacent PF4 tetramers in a large complex. It is not surprising that the antibody binds the antigenic complex with high affinity ($K_d$ 13 to 31 nM).[43] The PF4-heparin-antibody complex then binds to the platelets via their FcγR IIa receptors.[44–46] Cross-linking of the Fc receptors leads to platelet activation with the consequent release of platelet granular constituents (including PF4) and platelet microparticles, generation of thromboxane $A_2$, and ultimately platelet aggregation (Fig. 48-1).[20,47,48] Platelet activation releases PF4 from platelet α granules, and more PF4-heparin complexes are formed and become bound to the platelet surface, thereby allowing more heparin-dependent antibodies to bind (Fig. 48-2).[46] A chain reaction is thus set in place that results in intense platelet activation and formation of platelet aggregates.[48] Generation of platelet microparticles and other procoagulant materials leads to activation of blood coagulation pathways, thrombin generation, and thrombus formation.[47,49] These processes may account for the hypercoagulable state and the frequent occurrence of thrombotic complications in HIT. The thrombocytopenia may be attributed to the clearance of activated

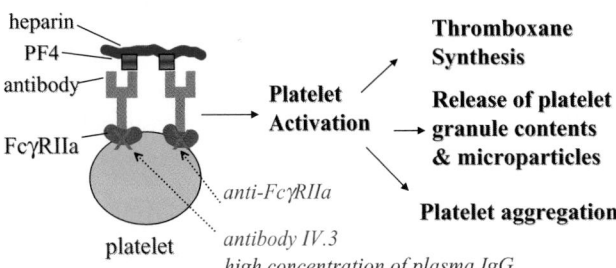

**Figure 48-1.** Platelet activation by heparin-antibody-PF4 complex. Heparin-IgG-PF4 complex binds to FcγRIIa receptors on the platelet, causing platelet activation and consequent release of platelet granule contents and platelet microparticles, thromboxane $A_2$ synthesis, and ultimately platelet aggregation. Binding of the immune complex to platelet Fc receptors can be blocked by anti-FcγRIIa antibody, IV.3, or a high concentration of plasma IgG.

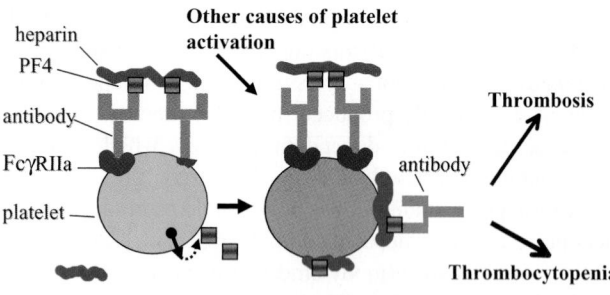

**Figure 48-2.** Antibody-PF4-heparin-platelet interaction. Binding of heparin-IgG-PF4 complex to platelet FcγRIIa receptors leads to platelet activation and an increase in PF4-heparin complexes on the activated platelet surface. HIT antibodies then bind to PF4-heparin complexes on the platelet surface via their Fab domain. This process ultimately results in thrombocytopenia and thrombosis.

or PF4-heparin-IgG–coated platelets by the reticuloendothelial system. Platelets are also consumed during thrombus formation.

**1. HIT Antibody Subclass.** In most patients, the antibody belongs to the IgG subclass. Some IgG subclasses are more common than others (IgG$_1$ > IgG$_3$ > IgG$_2$ > IgG$_4$). IgG antibodies may occur together with IgA or IgM antibodies.[50] In only a very small percentage of patients, only IgA, IgM, or both are present. As platelets do not possess Fc receptors for IgA or IgM, heparin-dependent antibodies of these two subclasses bind PF4/heparin on the platelet surface via their Fab domain. Some experts believe that IgA and IgM may occasionally cause thrombocytopenia or thrombosis, but others are not convinced that they do.[51,52]

**2. FcγRIIA Polymorphism.** The only Fc receptor that platelets express is the FcγRIIA receptor. There is an Arg/His polymorphism at amino acid 131 of human FcγRIIA receptors (see Chapter 14). Platelets bearing FcγRIIA$_{His131}$ receptors are activated more strongly by human IgG$_2$ antibodies. Three studies[53–55] found an overrepresentation of FcγRIIA$_{His131}$ among patients with HIT, suggesting that IgG$_2$ might be a pathogenetically important IgG subclass of HIT antibodies. However, two subsequent studies[56,57] showed no correlation with either FcγRIIA$_{His131}$ or FcγRIIA$_{Arg131}$, but a large study by Carlsson et al. found the reverse correlation.[58] This controversy remains unresolved and it is currently uncertain whether the FcγRIIA$_{Arg131/His131}$ polymorphism plays a role in the pathogenesis of HIT.

**3. Other Autoantigens Besides PF4.** In very occasional patients with type II HIT, the antigen is interleukin 8 (IL-8) or neutrophil-activating peptide 2 (NAP-2)[59] rather than PF4. These chemokines also belong to the CXC family

and they share considerabley structural homology with PF4. In contrast to anti-PF4-heparin antibodies, the binding of anti-IL-8 and anti-NAP-2 antibodies to their respective antigens is heparin independent. These antibodies tend to occur in HIT patients with comorbid conditions including autoimmune diseases, cancer, infection, and trauma.

**4. Animal Model.** The findings of a study using a transgenic mouse animal model confirm the critical role of PF4 and FcγR IIa receptors in the pathogenesis of HIT and HIT-associated thrombosis *in vivo*.[60] In this study, administration of heparin (20 U/day) and a murine monoclonal antihuman PF4 (hPF4)-heparin IgG antibody for 5 days to mice transgenic for both hPF4 and human FcγR IIa (hFcgRIIa) receptors resulted in severe thrombocytopenia. Whereas heparin administered at 50 U/mL together with the anti-hPF4-heparin antibody in the hFcγRIIa/hPF4 mice resulted in shock and formation of thrombi in multiple organs, administration of heparin and isotype control antibody to the hPF4/hFcγRIIa transgenic animals had no effect. Administration of heparin plus the antibody to mice transgenic for hPF4 alone or hFcγRIIa alone did not cause thrombocytopenia or thrombosis.

**5. Role of Endothelial Cells and Monocytes in the Pathogenesis of HIT.** The HIT antibody can bind to PF4 attached to endogenous heparan sulfate on endothelial cells, causing immunoinjury to these cells.[61,62] Although this has been considered to be another mechanism for the thrombosis in HIT, the transgenic mouse study[60] calls this hypothesis into question. It is known that binding of the HIT antibody to the endothelial cells is independent of FcγR IIa and exogenous heparin.[61,62] The antibody should bind to PF4-heparan or PF4-heparin on endothelial cells in the hPF4 transgenic mice when anti-hPF4-heparin antibody is administered. This should cause thrombocytopenia or thrombosis, but neither occurs. These observations suggest that binding of the heparin-dependent antibody to hPF4 on the endothelial cells may not be involved in the pathogenesis of thrombocytopenia or thrombosis in HIT.

The HIT antibodies can also bind heparin independently to PF4/proteoglycan (chrondroitin sulfate) complexes on monocytes. Binding of HIT antibodies causes activation of monocytes inducing the cells to express tissue factor and generate procoagulant activity.[63] Currently, it remains unclear whether monocyte activation is a significant factor in the pathogenesis of thrombosis in HIT.

## V. Frequency of HIT

### A. Type I HIT

Most prospective studies that investigated the frequency of HIT, particularly those in the 1970s and 1980s, did not differentiate between type I and type II HIT, and they probably included HIT of both types.[15,19,64-69] Lee and Warkentin attempted to differentiate type I and type II patients by the timing of the reported fall in platelet count and by the presence or absence of a positive *in vitro* HIT antibody test.[70] The resultant frequencies of type I and type II HIT are shown in Tables 48-1 and 48-2, respectively. These studies

### Table 48-1: Frequency of Type I HIT in Medical Patients

| Investigators | No. of Patients | No. (%) with Thrombocytopenia[a] |
|---|---|---|
| Bell et al. (1976)[19] | 52 | 13 (25%) |
| Nelson et al. (1978)[15] | 37 | 9 (24.3%) |
| Gallus et al. (1980)[64] | 166 | 2 (1.2%) |
| Holm et al. (1980)[66] | 90 | 1 (1.1%) |
| Eika et al. (1980)[67] | 120 | 2 (1.7%) |
| Cipolle et al. (1983)[69] | 211 | 4 (1.9%) |
| Powers et al. (1984)[68] | 131 | 3 (2.3%) |
| Kakkasseril et al. (1985)[65] | 142 | 5 (3.5%) |

[a]The number of patients in each study considered to have type I HIT on the basis of the timing of the fall in platelet count and the absence of a positive *in vitro* HIT test. In these studies, heparin was given IV for the treatment of thrombosis. Adapted from ref. 70.

### Table 48-2: Frequency of Type II HIT in Medical Patients

| Investigators | No. of Patients | No. (%) with thrombocytopenia[a] |
|---|---|---|
| Malcom et al. (1979)[77] | 66 | 1 (1.5%) |
| Powers et al. (1979)[76] | 120 | 2 (1.7%) |
| Holm et al. (1980)[66] | 90 | 0 (0%) |
| Gallus et al. (1980)[64] | 166 | 3 (1.8%) |
| Cipolle et al. (1983)[69] | 211 | 7 (3.3%) |
| Power et al. (1984)[68] | 131 | 2 (1.5%) |
| Green et al. (1984)[74] | 89 | 2 (2.2%) |
| Kakkassail et al. (1986)[65] | 142 | 4 (2.8%) |
| Bailey et al. (1986)[73] | 43 | 1 (2.3%) |
| Monreal et al. (1989)[85] | 89 | 2 (2.2%) |
| Rao et al. (1989)[86] | 94 | 0 (0%) |

[a]Included in each study were only patients considered to have type II HIT on the basis of delayed onset of thrombocytopenia, a positive *in vitro* HIT test, or other clinical criteria. In these studies, heparin was given IV for the treatment of thrombosis.
Adapted from ref. 70.

recruited mainly medical patients receiving IV heparin for treatment of thromboembolism. The initially reported that frequencies of HIT were about 25%, suggesting that type I HIT was very common at that time.[15,19] Subsequent studies reported much lower frequencies (about 1–4%) of type I HIT in medical patients.[64-69]

In contrast, the frequency of type I HIT is higher in surgical patients. Warkentin et al.[36] found that patients who underwent total hip replacement surgery and were given subcutaneous heparin at prophylactic doses had an incidence of type I HIT of 28% (<4 days of heparin administration). In the same study, a frequency of type I HIT of 29% was observed in the patients given LMWH.[36] The authors of this study suggested that in these postoperative patients, other unrelated factors such as perioperative hemodilution might have also contributed to the early onset thrombocytopenia.

### B. Type II HIT

The incidence of type II HIT (Table 48-2) is dependent on factors such as the type of heparin used and the clinical setting in which the drug is used.

**1. Types of Heparin.** Five clinical studies have compared the frequency of HIT in patients receiving bovine heparin with that in patients given porcine heparin.[68,69,72-74] These studies found a higher frequency of HIT in the patients receiving bovine heparin (Table 48-3). Schmitt and Adelman[75] examined the pooled data from the more reliable studies using a stricter criteria for diagnosis of HIT. They estimated an incidence of thrombocytopenia of 2.9% for bovine heparin and 1.1% for porcine heparin.[75] The increased

### Table 48-3: Frequency of Type II HIT in Medical Patients Receiving Bovine or Porcine Heparin

| Investigators | Frequency of Type II HIT (%)[a] | |
|---|---|---|
| | Bovine Heparin | Porcine Heparin |
| Ansell et al. (1980)[72] | 4/21 (19.0%) | 0/22 (0%) |
| Cipolle et al. (1983)[69] | 6/100 (6.0%) | 1/111 (0.9%) |
| Powers et al. (1984)[68] | 2/65 (3.1%) | 0/66 (0%) |
| Green et al. (1984)[74] | 2/45 (4.4%) | 0/44 (0%) |
| Bailey et al. (1986)[73] | 1/21 (4.8%) | 0/22 (0%) |

[a]Included in each study were only patients considered to have type II HIT on the basis of delayed onset of thrombocytopenia, a positive *in vitro* HIT test, or other clinical criteria. The study by Bell and Royall[71] was excluded because it contained patients with type I HIT.
Adapted from ref. 70.

tendency for bovine heparin to induce HIT has been attributed to its higher degree of sulfation, which renders it better able to form an antigenic complex with PF4.

## 2. Types of Patients

*a. Medical Patients.* Prospective studies in the 1980s showed that the frequency of immune HIT in medical patients was low when they were given therapeutic doses of porcine heparin for treatment of venous or arterial thrombosis.[64–66, 68,69,73,74,76,77,85,86] The studies reported frequencies of HIT that ranged from 0 to 3.3% (Table 48-2). These results were confirmed by two other studies.[78,79] Medical patients who receive prophylactic doses of subcutaneous porcine heparin may also have a low incidence of HIT, because 10 studies in the 1980s with a total of 527 patients found only four such cases of HIT (0.7%).[64,77,81–88] A study by Girolami et al. found a somewhat higher incidence of 1.4% (5 out of 360 patients).[78]

*b. Surgical Patients.* In contrast, surgical patients who received prophylactic doses of porcine heparin have much higher frequency of immune HIT. Two groups of surgical patients have been prospectively studied: postoperative orthopedic[36,89,90] and cardiac/vascular surgical patients.[91–94] Warkentin et al.[94] found that in patients who underwent hip replacement surgery the frequency of immune HIT was 2.7% if the platelet count cutoff for the diagnosis of HIT was $<150 \times 10^9$/L or 5% if a more liberal diagnostic criterion was used (i.e., a decrease in platelets of >50% of the baseline count). There were many more patients (7.8%) who developed anti-PF4/heparin antibodies (detected by the $^{14}$C-serotonin release assay [SRA]) than patients who developed thrombocytopenia.[36] Other investigators[89,90] obtained similar results.

Four studies have shown cardiac surgery patients to have an even greater tendency than orthopedic patients to form anti-PF4/heparin antibodies. Prior to surgery, about 20% of the patients had antibodies detectable by enzyme immunoassay (EIA) and 5% by SRA, probably as a result of prior heparin exposure during cardiac catheterization.[95,96] Postoperatively, even more patients developed the antibodies. Fifty percent of patients had antibodies postoperatively detectable by EIA and 13 to 20% by SRA. If heparin reexposure was avoided or restricted to only 1 to 3 days postoperatively, no patients developed immune HIT. However, in one study the patients were allowed reexposure to heparin for a significant period of time postoperatively (10 or more days), and a significant number of these patients developed immune HIT (3.8%) and thrombosis (1.3%).[93] Nevertheless, the incidence of HIT is still slightly lower than that in orthopedic patients. It is uncertain why this is so, given the greater tendency in the cardiac surgery patients to form anti-PF4/heparin antibodies.

In both cardiac surgery and orthopedic surgery patients, the frequency of immune HIT is very low if they are given LMWH postoperatively. In orthopedic surgery patients, 2.2% of the patients given LMWH developed the heparin-dependent antibody (detectable by SRA) and less than 1% developed immune HIT. These frequencies are much lower than those receiving heparin (noted earlies).[36] In patients treated with fondaparinux (Arixtra), a synthetic saccharide or ultra-low heparin fragment (pentasaccharide), the frequency of HIT antibody development is also very low.[70] Fondaparinux does not cross-react with HIT antibody.[97] To date, approximately 5000 patients have been given this drug for thromboprophylaxis and treatment in clinical trials, and yet no episodes of HIT have occurred.[97–99] However, we should reserve judgment of this issue until further confirmation when fondaparinux is widely used in clinical practice.

In cardiac surgery patients who received heparin during the procedure but LMWH postoperatively, the frequency of antibody formation (about 25%) is similar to that for those receiving heparin postoperatively, but the incidence of type II HIT was much lower: 0% with LMWH versus 3.8% with heparin.[93]

*c. Obstetric Patients.* The frequency of HIT in obstetric patients is very low, even though these patients tend to receive heparin for prolonged periods of time. Fausett et al. found no patients with HIT among 244 pregnant women who received heparin.[100] Lepercq and coworkers reported no cases of HIT in 624 pregnancies treated with LMWH.[101] This is surprising considering that women are reported to be more prone to develop HIT.[102]

**3. Heparin Flushes and Heparin-Coated Devices.** Heparin given in minute doses as flushing of intravascular catheters can induce the formation of HIT antibodies and has been implicated in causing HIT in anecdotal case reports.[103,104] The use of heparin-coated pulmonary catheters has also been reported to cause HIT in occasional patients.[105]

**4. Duration of Heparin Therapy.** Because type II HIT occurs 5 or more days after commencement of heparin treatment, a lower incidence of immune HIT would be expected if heparin therapy is given for fewer than 5 days. Traditionally, warfarin was started 5 to 7 days after commencement of heparin in the treatment of venous thromboembolism (VTE). As it takes a few days for warfarin to become therapeutic, heparin was frequently given for 10 days or more. However, with the change to early commencement of warfarin for the past 10 to 15 years, the duration of heparin treatment for VTE has been substantially reduced, and the incidence of immune HIT has correspondingly decreased (A. Gallus, personal communication).

# VI. Clinical Features of HIT

## A. Type I HIT

The thrombocytopenia in type I HIT is mild and usually occurs in the first 4 days of heparin administration. The platelet counts often drop to a level between $100 - 150 \times 10^9/L$ and rarely to below $80 \times 10^9/L$.[20] The platelet counts may return to normal despite continuation of heparin. When rechallenged with heparin after resolution of the thrombocytopenia, the platelet counts usually do not fall. In postoperative patients, type I HIT may be indistinguishable from the fall in platelet count due to hemodilution.[106] Patients with HIT type I usually remain asymptomatic; it is not associated with thrombosis or bleeding.

## B. Type II HIT

### 1. Thrombocytopenia

*a. Onset of Thrombocytopenia.* In type II HIT, the platelet count usually falls gradually starting 5 to 10 days after the initiation of heparin therapy, and it may not reach thrombocytopenic levels until several days later (classical onset HIT). In some cases, the onset of thrombocytopenia occurs abruptly before day 5 of heparin therapy, and this is now referred to as *rapid-onset HIT*.[51] In these cases, the patients have prior exposure to heparin within the past 100 days or so. Because they already have the HIT antibody when they are again given heparin, thrombocytopenia promptly occurs. In very occasional patients, thrombocytopenia occurs several days after cessation of heparin. The term *delayed-onset HIT* has been coined for these cases.[107,108] These patients often have high-titer HIT antibody that may be heparin independent.

*b. Severity of Thrombocytopenia.* The thrombocytopenia in type II HIT is usually moderately severe with a median nadir of $60 \times 10^9/L$. Occasionally it may drop to below $10 \times 10^9/L$.[20] In some patients, the platelet count may fall more than 50%, but the platelet nadir may still be above $150 \times 10^9/L$ if their initial baseline counts were high.[106] The clinical picture of type II HIT is unlike that of quinine/quinidine-induced thrombocytopenia in which the thrombocytopenia is almost invariably severe ($<10 \times 10^9/L$) and its onset is sudden.[38]

*c. Resolution and Recurrence of Thrombocytopenia.* Unlike type I HIT, the thrombocytopenia in type II HIT persists until heparin withdrawal. After cessation of heparin, it usually takes 5 to 7 days for the platelet count to rise to normal or frequently above normal levels, but occasionally it may take longer (up to 30 days). In this situation, there

may be conditions/factors (e.g., severe sepsis) that suppress the bone marrow, thereby preventing immediate platelet recovery. Upon rechallenge with heparin when the heparin-dependent antibody is still present (usually within 60 to 100 days of stopping heparin), the thrombocytopenia promptly recurs. Heparin has been administered to patients with type II HIT months or years after the acute event when the heparin-dependent antibody was no longer present and the thrombocytopenia in most cases did not recur. This was particularly true when heparin was given as a "one-off" treatment, as a bolus, or as an infusion for a short duration such as during a surgical procedure (e.g., cardiopulmonary bypass surgery).[109]

### 2. Thrombotic Complications.
Despite the presence of thrombocytopenia, thrombotic complications are common in patients with type II HIT, occurring in up to 50 to 60% of patients.[70,78] Even in patients with severe thrombocytopenia ($<20 \times 10^9/L$), bleeding rarely occurs. The thrombotic complications can be venous, arterial, or microcirculation thrombosis. They are characteristically severe and extensive, and they frequently occur at unusual sites. The frequency and type of thrombosis vary with the patient population.[110,111] HIT-associated thrombosis is common in patients already at high risk of developing thrombosis, such as patients who have undergone hip replacement surgery. In this patient population, 20 to 30% of patients are normally expected to develop venous thrombosis postoperatively even with heparin prophylaxis, but the thrombosis rate increases sharply to about 76% when the patients develop HIT.[36]

Postoperative patients, who are prone to venous thromboembolism, are more likely to suffer venous rather than arterial thrombosis when they develop HIT.[36, 111] In contrast, arterial thrombotic complications are more common in HIT patients who have cardiovascular disease or have undergone vascular surgery.

*a. Venous Thrombosis.* Lower limb deep vein thrombosis (DVT) is a common HIT-associated thrombotic complication. Bilateral DVT and pulmonary embolus are comparatively more common in patients with HIT than in patients without HIT.[106] The venous thrombosis in HIT can be extensive and result in limb gangrene or *phlegmasia cerulea dolens*, a condition extremely uncommon in patients without HIT. Development of this condition in HIT is frequently associated with coumarin treatment, particularly in the presence of supratherapeutic INRs (INR > 4).[112,113] It is believed that the coumarin-induced protein C and protein S deficiency is the underlying cause of the venous limb gangrene. This usually occurs in patients who do not receive treatment with an anticoagulant (danaparoid or hirudin) that inhibits thrombin generation.

Venous thrombosis in HIT can occur at other sites, including upper limb veins (often associated with the insertion of central venous catheters),[114] cerebral dural sinuses,[115,116] and adrenal veins.[117,118] Cerebral dural sinus thrombosis often occurs when there is another hypercoagulable state (e.g., pregnancy or myeloproliferative disorder) besides HIT. It is a potentially serious condition that can lead to stroke and even death.[119] Patients with HIT appear to have an unusual predilection for unilateral or bilateral adrenal vein thrombosis leading to adrenal hemorrhage.[117] This condition should be suspected when patients with HIT develop abdominal pain and hypotension.[118]

*b. Arterial Thrombosis.* Arterial thrombosis in HIT usually involves the distal aorta and lower limb arteries causing limb ischemia, which may lead limb gangrene and leg amputation.[11–13] Less commonly, arterial thrombosis in HIT results in thrombotic stroke or acute myocardial infarction. Occasionally, occlusions of brachial, mesenteric, and renal arteries have been described in patients with HIT, resulting in limb gangrene, bowel infarction, and renal infarction, respectively.

*c. Micro-Circulation Thrombosis.* DIC occurs in 5 to 10% of patients with type II HIT.[18, 106] Increased fibrin degradation products are usually detectable, but hypofibrinogenemia may not be present because in patients with HIT the baseline plasma fibrinogen levels are often elevated as an acute phase reactant. The decline in plasma fibrinogen concentration due to DIC may only become obvious upon serial determinations. DIC may lead to acute renal failure and other organ failure. Painful digital infarctions may occur in HIT with or without DIC.

*d. Other HIT-Associated Thrombotic Complications.* Other HIT-associated thrombotic complications include thrombosis of intravascular prostheses and vascular fistulae, and clot formation in renal hemodialysis circuits. The unexpected occurrence of clot formation in these devices should alert clinicians to the possibility of HIT.

Skin necrosis may occur at heparin or LMWH injection sites.[120,121] Its occurrence is usually associated with the presence of a heparin-dependent antibody but may occur in the absence of thrombocytopenia. Skin lesions may sometimes be just erythematous skin induration without necrosis. Substituting heparin with LMWH can lead to more lesions, but a change to danaparoid will avoid further lesions or recurrence.

**3. Heparin Resistance.** Heparin resistance in HIT was first described by Rhodes and colleagues.[13] In the typical case of heparin resistance, the activated partial thromboplastin time (APTT) used to monitor heparin therapy does not become adequately prolonged and will not reach therapeutic levels despite progressive increases in the heparin dose. Heparin resistance is probably due to high circulating

levels of PF4 and other heparin-binding proteins and can also occur with acute massive thrombosis.

**4. Acute Systemic Reactions.** Acute systemic reactions, as indicated by clinical features such as fever, chills, tachycardia, flushing, headache, chest pain, and dyspnea, may rapidly follow an IV bolus of heparin administered to a patient with the heparin-dependent antibody.[122,123] Acute amnesia, pulmonary arrest, or cardiac arrest may also occur.[124] These reactions are often accompanied by a transient drop in the platelet count, which may be missed if a blood count is not performed soon after the event.

## VII. Diagnosis

### A. *Clinical Diagnosis*

A high index of suspicion is necessary to recognize HIT. HIT should be suspected whenever thrombocytopenia occurs in a patient receiving heparin. However, in such a patient there are also other possible causes for the thrombocytopenia. Therefore, before the diagnosis of HIT is made, the clinical setting in which the thrombocytopenia occurs and serology test results should be carefully analyzed. A chart showing serial platelet counts and the dates of administration of heparin and other concomitant drugs can be very helpful in the diagnostic process.

The following clinical diagnostic criteria have been proposed for the clinical diagnosis of HIT:[1,2]

1. The thrombocytopenia occurs during heparin administration.

2. Other possible causes of thrombocytopenia are excluded.

3. The thrombocytopenia resolves after cessation of heparin.

Criteria 1 and 2 are applicable for both type I and type II HIT. If the thrombocytopenia resolves despite the continuation of heparin, it is type I HIT, but if the platelet counts return to normal only after heparin withdrawal, it is more likely to be type II HIT. The time of onset of the thrombocytopenia is crucially important in distinguishing type I from type II HIT. In the former, the onset of thrombocytopenia occurs during the first 4 days of heparin administration. In contrast, the onset of thrombocytopenia occurs between day 5 and day 10 of heparin administration in type II HIT, unless there is prior exposure to the drug in the previous 100 days as in "rapid-onset HIT".[106] After 10 days of heparin therapy, type II HIT is less likely but may still occur. It is important that the overall clinical picture of the patient is taken into consideration. For example, the presence of an unusual thrombotic complication (e.g., warfarin-induced venous gangrene) would further favor the diagnosis of type II HIT. Whenever possible, the diagnosis of immune HIT should be confirmed by a positive serological test. However, serological tests should not be performed to screen for HIT as the heparin-dependent antibody can be present in patients without HIT.[79,98,99]

### B. Laboratory Diagnosis of HIT

After a clinical diagnosis of HIT has been made, every attempt should be made to confirm the diagnosis. Two types of serology tests are available: functional tests and immunoassays.

**1. Functional Tests.** Reaction of the HIT antibody with the PF4-heparin complex and platelets leads to platelet activation, which results in a range of platelet changes including release of $\alpha$ granules and dense bodies, generation of platelet microparticles, alterations in the platelet membrane, and finally platelet aggregation (Fig. 48-1). Any of these platelet changes can be measured as an end point in the functional tests. The main differences among the functional tests are the following:

1. The platelet end point used
2. The use of washed platelets or platelet-rich plasma (PRP)

These differences (discussed later) can affect the sensitivity and specificity of the test. The commonly used functional tests are the [14]C-serotonin release assay (SRA), the heparin-induced platelet-activation assay (HIPA), and the platelet aggregation test (PAT).[125,126]

*a. [14]C-serotonin Release Assay (SRA).* This functional assay was first described by Sheridan et al.[127] ACD-anticoagulated PRP from two normal donors are combined and incubated at 37°C for 30 minutes with [14]C-serotonin. After the isotope has been taken up by the platelets, they are washed once in calcium-free and magnesium-free Tyrode's solution containing apyrase. Apyrase degrades the ADP released during platelet washing and hence prevents its accumulation, which can render the platelets refractory to subsequent stimulation by ADP. Finally the platelets are spun down and then resuspended in a calcium- and magnesium-containing Tyrode's buffer at a platelet concentration of $300 \times 10^9$/L. Washed platelet suspension (75 μL) is added to microtiter wells containing patient serum/plasma (20 μL) and heparin/buffer (5 μL). A range of heparin concentrations is customarily used: 0 U/mL (i.e., buffer only), 0.1 U/mL, 0.3 U/mL, and 100 U/mL. Several controls are also included: a normal serum (negative control) and a

weak HIT serum (positive control) are substituted for the test serum, and an Fc receptor inhibitor control is used, in which 0.3 U/mL heparin and the monoclonal antibody IV.3 are added instead of heparin. The microtiter plate containing the reaction mixtures is placed in a platelet shaker for 1 hour and thereafter the reaction is stopped by the addition of EDTA/phosphate-buffered saline (PBS). After centrifugation to pellet the platelets, 50 μL of the supernatant is transferred to tubes containing scintillation fluid for measurement of [14]C-serotonin released during the platelet-HIT antibody interaction. [14]C-serotonin release is deemed to have occurred if more than 20% serotonin release is detected. A positive test is one in which the test serum induces [14]C-serotonin release at therapeutic concentrations of heparin (0.1 to 0.3 U/mL) but not at high heparin concentration (100 U/mL) and not in the presence of IV.3 monoclonal antibody. Serotonin can also be measured by a nonradioactive method.[128]

*b. Heparin-Induced Platelet Activation (HIPA) Test.* The HIPA test is similar to the SRA. Both use washed platelets. In HIPA,[129] however, washed platelets from four normal donors are used, and platelets from each donor are prepared and tested separately. Also in the HIPA test, hirudin (1 U/mL) is added to the washing buffer instead of apyrase. The platelets are incubated with patient serum/plasma and heparin/buffer in microtiter wells. The platelets are stirred using two stainless steel spheres and a magnetic stirrer at approximately 500 rpm for 45 minutes at room temperature. Like SRA, a range of heparin concentrations is used: 0 U/mL, 0.2 U/mL, and 100 U/mL. In one recent modification, LMWH (reviparin) 0.2 U/mL is used instead of heparin 0.2 U/ml. The reaction mixture in each microtiter well is inspected by eye every 5 minutes for evidence of platelet aggregation, which is the end point in this test. A test serum is considered positive if it induces aggregation of platelets of at least 2 out of 4 donors in the presence of the low heparin or LMWH concentration (0.2 U/mL) and not the high heparin concentration (100 U/mL). Aggregation occurring in the presence of both the low and high heparin concentrations is considered an indeterminate result.

*c. Platelet Aggregation Test (PAT).* In this test, citrated PRP from a normal donor is used. PRP (300 μL) is mixed with patient serum/plasma (150 μL) and heparin/saline (50 μL) in a cuvette and stirred by a small magnetic bar at 37°C for 30 minutes.[130] Some laboratories prefer to use an equal volume of normal PRP and test serum or plasma. A range of heparin concentrations is also used in this test: 0 U/mL (saline), 0.5 U/mL or 1.0 U/mL, and 100 U/mL. In addition, a negative control (normal serum instead of test serum) and a positive control (a known weak HIT serum instead of test serum) are also included. Platelet aggregation is the end point, but unlike the HIPA test, it is measured quantitatively using a platelet aggregometer. Platelet aggregation of more than 20% is considered a positive end point. A positive result is one in which platelet aggregation occurs in the presence of low heparin concentrations (0.5 or 1.0 U/mL) but it is partially or completely inhibited by the high concentration of heparin (100 U/mL).

### d. Other Less Widely Used Functional Tests

i. ATP Release Test. ATP release from platelet-dense granules during platelet activation induced by the HIT antibody can be measured using a lumiaggregometer. ATP gives a flash of light in the presence of a luciferin-luciferase reagent, and this can be detected.[131]

ii. Platelet-Derived Microparticle Generation. Reaction of the HIT antibody with washed platelets and heparin generates microparticles. The fluorescein-labeled anti-GPIb monoclonal antibody binds to the platelet-derived microparticles, which can be detected using a flow cytometer.[132]

iii. Detection of Platelet Membrane Changes Using Flow Cytometry. The HIT antibody causes platelet activation and induces changes in the platelet membrane. Tomer described a PRP-based assay, which uses flow cytometry to detect fluorescein-labeled recombinant annexin V that binds to anionic phospholipids expressed on the surface of activated platelets.[133] Similarly P-selectin is expressed on the surface of activated platelets and can be measured by means of a fluorescein-labeled anti-P-selectin antibody using flow cytometric analysis.[134]

### e. Factors That Can Affect Functional Assays

i. Heat-Inactivation of Patient Serum or Plasma. Trace amounts of thrombin may be present in the test serum or plasma samples and it may cause platelet activation, thus interfering with the test results. Heating the samples at 56°C for 30 minutes will inactivate the contaminating thrombin. However, overheating the samples sometimes occurs, and this can generate aggregated IgG, which may also cause platelet activation.

ii. Heparin Concentrations. The concentration of heparin in the assay is critically important. Low heparin concentrations of 0.1 to 0.3 U/mL are optimal for the formation of the antigenic complex (heparin-PF4 complex) in washed platelets-based assays.[127] For PRP-based assays, the optimal heparin concentrations are 0.5 to 1.0 U/mL.[106] To maximize the sensitivity of the assay, these concentrations should be used. On the other hand, high concentrations of heparin such as 10 to 100 U/mL disrupt the heparin-PF4 complex, suppress platelet activation induced by the HIT antibody,[127,130] and result in a negative reaction.

iii. Variable Reactivity of Platelets from Different Donors to HIT Antibody. Platelets from different donors vary considerably in their reactivity to the HIT antibody.[130] Platelets with poor reactivity may give false negative results with weak HIT antibodies. Different laboratories have adopted different approaches to overcome this problem, (e.g., using platelets from four donors to minimize the variability of platelets).[129] In general, platelet donors are not easily available, and it may not be possible to obtain platelets from four donors for the test on a regular basis. The author's laboratory takes the approach of identifying a number of "good responders" among the laboratory staff, and they are called on in rotation to be platelet donors for the test. With this approach, the platelets of only one donor each time need to be used. It is important that the donor is asked to abstain from food or drugs (e.g., aspirin), which may impair platelet function. In addition, we always use one or more weak HIT positive sera as a positive control to ensure good platelet reactivity.

f. Interpretation of Test Results. In most functional assays, a positive result is one in which a positive reaction occurs with the low heparin concentration(s) and a negative reaction with the high heparin concentration. Adoption of this two-point system significantly enhances the specificity of the assay. However, in PRP-based assays, heparin at 100 U/mL may not always completely suppress the HIT antibody-induced platelet activation. As there are heparin binding proteins in PRP, a higher concentration of heparin (e.g., 200 to 500 U/mL) may be required or a partial inhibition at 100 U/mL of heparin may be acceptable.

A positive reaction in the absence of heparin occurs with a small proportion of HIT sera. This could be due to contaminating heparin in the test samples. If so, the reaction should still be positive with low concentrations of heparin (0.1 to 1 U/mL) and suppressed by high heparin concentration, (i.e., 10 to 100 U/mL). When this occurs, it is also considered a positive result. At other times, the reaction is positive with all heparin concentrations (i.e., buffer, low, and high heparin concentration). This is an indeterminate result. This could be due to the presence in the serum or plasma of an HLA alloantibody, immune complexes, aggregated IgG, or an unidentified platelet-activating factor. As aggregated IgG can be generated inadvertently by overheating the sample, the assay should be repeated with another properly heat-inactivated sample, and a clear-cut result may be obtained.

Some laboratories include IV.3, an anti-FcγRIIa monoclonal antibody, as a control because IV.3 is known to inhibit HIT antibody-mediated platelet activation (Fig. 48-1). However, inhibition by IV.3 is not entirely specific for the HIT antibody as it also inhibits platelet activation provoked by other agonists in serum such as aggregated IgG, immune complexes, and HLA alloantibodies (B. Chong, unpublished data).

*g. Washed Platelet versus PRP Assays.* Platelet washing is a time-consuming procedure that requires experience and care to avoid excessive platelet activation. The mild background platelet activation generated during the procedure can be advantageous in HIT antibody assays. However, in inexperienced hands excessive platelet activation can occur, and this will lead to erroneous results. Washed platelet-based assays (SRA and HIPA) are best performed in specialized or referral laboratories. On the other hand, PRP-based assays (e.g., PAT) are technically less demanding and can be performed in a nonspecialist clinical laboratory.

*h. Sensitivity.* In general, washed platelet-based assays are more sensitive than PRP-based assays for detection of the HIT antibody. The reasons for this are as follows:

1. The background platelet activation enhances the HIT antibody-induced platelet changes such as serotonin release (SRA) and platelet aggregation (HIPA). PF4 released during the procedure-induced platelet activation increases the availability of PF4 for antigenic complex formation.

2. The normal IgG concentration in washed platelet suspensions is much lower than that in PRP and hence there is less IgG competing with the PF4-heparin-HIT antibody complex for binding to platelet Fc receptors (FcγRIIa). On the other hand, IgG in high concentration in PRP decreases the HIT antibody-induced platelet activation.[130]

3. Apyrase, used in SRA, stops ADP accumulation and thus prevents the platelets from becoming refractory to subsequent ADP stimulation, which occurs during the second phase of HIT-antibody induced activation.

In the author's experience, the most important factor affecting the sensitivity of the HIT antibody functional assays is the selection of platelets from responsive donors for the assays. Using highly responsive platelets, the sensitivity of a PRP-based assay (PAT) can approach that of the washed platelet-based assays (SRA and HIPA).[130]

*i. Specificity.* The most important factor that can influence the specificity of the functional assays is the use of the two-point system to interpret test results. Agonists in the patient sera or plasma, which can cause false positive results, usually induce platelet activation in the presence of both low and high heparin concentrations. Only HIT antibody provokes platelet activation at low but not high heparin concentration. The author believes that if the two-point system is used, the specificity of washed platelet-based assays and PRP-based tests will be essentially the same. However, some experts believe that the former assays have a superior

specificity because they can be performed in a larger scale with duplicates or triplicates and with many more controls.

Some experts place emphasis on the inhibition of the reaction by IV.3, the anti-FcγRIIa monoclonal antibody. They require inhibition by IV.3 before they would consider a test result as positive. This author is less enthusiastic about the inhibition by IV.3. Although IV.3 inhibits HIT antibody-induced platelet activation, it also blocks platelet activation induced by agonists that frequently cause false positive results such as immune complexes, aggregated IgG, and HLA-related alloantibodies.

*j. Conclusion.* Both types of functional assays have advantages and disadvantages. In experienced hands, the washed platelet-based assays are more sensitive and possibly more specific. However, they are technically demanding and labor intensive and hence may be more suitable for specialist or referral laboratories. In the nonspecialist clinical laboratories, the technicians may not have the experience or training to perform platelet washing properly and a PRP-based assay may be more appropriate. Every effort must be made to use highly responsive platelets and one or more weak HIT antibodies as positive controls. Without these measures, the sensitivity of PRP-based assays can be low, possibly less than 40%. The two-point system or a modified two-point system (discussed earlier) should be used for maximum specificity.

**2. Immunoassays.** Immunoassays for HIT antibodies are based on detection of the antibody binding to the antigenic complex. Four types of immunoassays have been described: (a) solid-phase enzyme immunoassay (EIA), (b) PF4-polyvinylsulfonate immunoassay, (c) fluid-phase EIA, and (d) particle gel assay. A brief description of each method follows, but for description of the methods in detail, the reader is referred to previous reports of these assays.[29,31,135–137]

*a. Solid-Phase Enzyme Immunoassay.[113]* PF4 and heparin are coated to microtiter wells, which are "blocked" with blocking solution (e.g., buffer containing 20% fetal calf serum). Test or control serum (1 in 50 dilution) is added in duplicate and incubated for 1 hour at room temperature. Alkaline phosphatase-conjugated with goat antihuman immunoglobulin is added, followed by the addition of the substrate, p-nitrophenyl phosphate in 1 M diethanolamine buffer. The microtiter wells are washed with PBS-Tween 20 buffer in between each of the preceding steps. After incubation in the dark, the reaction is stopped with 1 N sodium hydroxide. Absorbance is read at 405 nm. The cutoff level is set at three standard deviations above the mean of absorbance readings of a large number of normal sera (e.g., 50).

*b. PF4-Polyvinylsulfonate Antigen Assay.* Polyvinylsulfonate (PVS), a negatively charged compound, can induce exposure of the antigenic epitopes when it binds to PF4. PVS-PF4 complex can be used to coat the microtiter wells in the solid phase EIA instead of heparin-PF4.[31] Indeed, a commercial PVS-PF4 assay kit is now available for detection of the HIT antibody. This assay has the advantage that the PVS-PF4 complex is stable for a long period of time.

*c. Fluid-Phase EIA.* PF4 (5% biotinylated) is mixed with an optimal concentration of heparin, and the antigen mixture is incubated for 1 hour with diluted serum/plasma (1/50 or 1/10). Subsequently the antigen-antibody mixture is incubated in duplicate in microfuge tubes containing Protein G Sepharose, which has been blocked with 1% bovine serum albumin (BSA). The Sepharose beads containing biotin-PF4-heparin-antibody complex are separated from unbound antigen by centrifugation and washing. The amount of antibody-antigen immobilized to the beads is determined by the addition of streptavidin conjugated to horse radish peroxidase and TMB peroxidase substrate. After the addition of 0.6 M $H_2SO_4$ to stop color development, the supernatant is transferred to microtiter wells and the absorbance at 450 nm is measured by a microtiter plate reader.[136]

The fluid phase assay has several advantages over the solid phase assays. Because the antigen (PF4) is in fluid phase, the denaturing of the protein inherent in solid-phase assays is avoided. The fluid phase assay is more sensitive because the very low background allows higher concentrations of test serum to be employed. The low background is due to the fact that serum is not added directly to the microtiter wells, thus avoiding the nonspecific binding of normal IgG to the plastic. Unlike solid phase assays, the antigen is not in contact with a plastic surface, cryptic epitopes of PF4 are not exposed in fluid phase, and nonspecific reaction with the antibody does not occur. This and the low background result in a lower rate of false positive reaction. The lack of antibody binding in the absence of heparin is particularly advantageous in cross-reaction studies with different heparin and heparinoid drugs. The main disadvantages of this assay are that it is time consuming and it does not detect IgM or IgA antibodies.

*d. Particle Gel Immunoassay.[137]* This is also known as H/PF4-PaGIA assay (DiaMed, Switzerland). It utilizes the ID microcolumn system widely used for red cell serology testing. In this assay, 10 µL of patient serum and 20 µL of PF4-heparin coated polystyrene microbeads are added onto the top of the gel microcolumn in a test card. After incubation for 5 minutes, the card is centrifuged. In the presence of a strong HIT antibody, the microbeads are agglutinated and remain on top of the gel microcolumn. In the absence of the antibody, the microbeads remain free and sediment to the bottom. Weak antibodies form small microbead aggregates that remain in the middle of the gel microcolumn. The results are read visually. The test is technically easy and can be performed quickly. The sensitivity is between that of solid phase EIA and the functional tests[137] (T. Brighton, personal communication). Risch and colleagues found that among 42 postoperative cardiac surgery patients who did not have HIT, more patient sera tested positive than in the solid phase EIA (69% versus 26%), suggesting that the particle gel assay detected more clinically insignificant antibodies.[138] This surprising finding requires confirmation by further studies.

**3. Functional Tests versus Immunoassays.** Because the antigen used in the immunoassays is PF4-heparin or PF4-PVS complex, immunoassays cannot detect the rare HIT antibodies with specificity for the minor antigens (IL-8 and NAP-2). Commercial EIA kits detect HIT antibodies of all major subclasses: IgG, IgM, and IgA. Identifying IgM and IgA antibodies, considered by some experts to be clinically insignificant, may be a disadvantage. On the other hand, functional assays are dependent on platelet activation that occurs when IgG antibody-antigen complexes bind to platelet Fcγ receptors. As platelets do not have Fc receptors for IgA and IgM, functional assays cannot detect HIT antibodies of these two subclasses. Hence, the immunoassay and the functional assay complement each other in the laboratory diagnosis of HIT.

However, the two types of assays do differ in sensitivity and specificity. Immunoassays are more sensitive than the functional assays (Fig. 48-3). The immunoassays can detect HIT antibody in the sera that have been diluted 200-fold, but the functional assays generally are unable to detect the antibody in such diluted sera. Immunoassays are more likely to detect the weak and clinically insignificant HIT antibodies, such as those in postcardiac surgery patients without thrombocytopenia.[98,99] On the other hand, the functional assays are less sensitive but they tend to detect HIT antibodies that are more clinically significant.[36,93] In the patients with a high probability of HIT, a sensitive assay such as an immunoassay is more useful, but in patients with an intermediate or low probability of HIT, a more specific test is more clinically useful.

## VIII. Management

Managing a patient with HIT depends on the clinical type of HIT and the clinical circumstances of the patient, such as whether there is a thrombosis that requires prompt anticoagulant treatment or whether the patient needs to undergo cardiac surgery immediately.

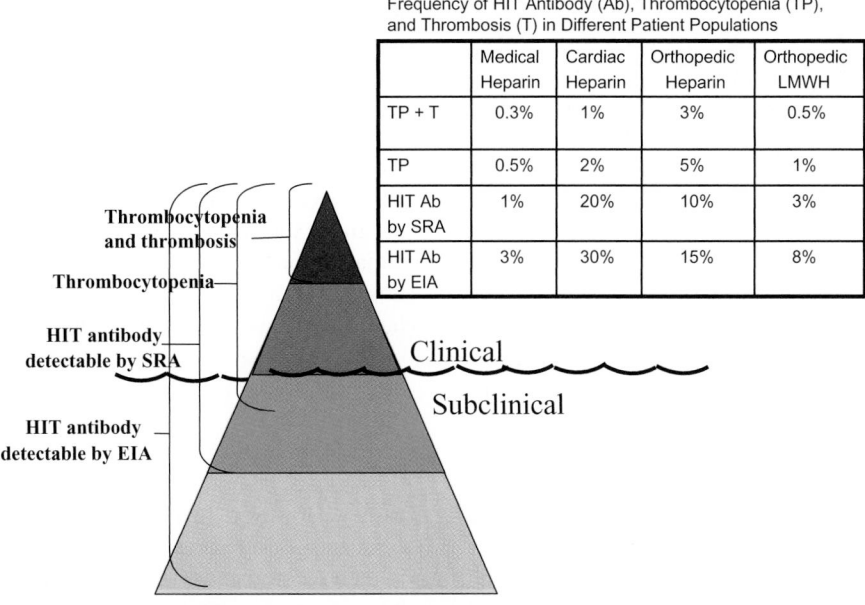

Figure 48-3. "Iceberg" model of HIT. Among patients treated with heparin or LMWH, a proportion of patients develop HIT antibody detectable by enzyme immunoassay (EIA). Only a proportion of EIA positive patients have HIT antibodies detectable by [14]C-serotonin release assay (SRA), a less sensitive assay than EIA. Among SRA-positive individuals, only a proportion has thrombocytopenia. About 50% of the thrombocytopenic patients have thrombosis. The actual proportions of these subclinical and clinical manifestations of HIT vary with the different patient populations, as shown in the table insert.

### A. Type I HIT (Nonimmune HIT)

In type I HIT, the thrombocytopenia is modest and the patient remains asymptomatic. No specific treatment is usually required. However, differentiation from type II HIT can be difficult in some cases, as the thrombocytopenia in type II patients with previous exposure to heparin can occur early. A decision to continue or withdraw heparin should be made after careful evaluation of the patient's clinical picture and laboratory test results. If heparin is continued, the patient's condition and platelet count should be closely monitored.

### B. Type II HIT (Immune HIT)

Once a clinical diagnosis of type II HIT is made, heparin should be stopped and an alternative anticoagulant started immediately. As HIT is a hypercoagulable state, serious thrombosis can occur or an existing thrombosis can progress very rapidly and cause devastating sequelae such as limb gangrene and death. One should not wait for the HIT antibody test result before commencing treatment. If the diagnosis is initially uncertain, it will often become clearer with subsequent clinical events or availability of the HIT antibody test result. The patient's management needs to be regularly reviewed and the appropriate changes made if the

patient's condition is unstable and labile. The specific therapeutic approach will vary with the patient's clinical situation.

**1. Treatment of Venous/Arterial Thrombosis.** After cessation of heparin, patients with a venous or arterial thrombosis should be treated with a rapidly acting anticoagulant that inhibits thrombin or its generation. Three drugs — danaparoid,[139–141] lepirudin,[142,143] and argatroban[144] — have been found to be effective in the treatment of thrombosis associated with HIT. Treatment with one of these drugs should continue for at least 5 days (see the upcoming discussion). For long-term treatment of the thrombosis, an oral anticoagulant such as warfarin is needed.[1,3] However, warfarin causes a decrease in plasma protein C and protein S levels. These changes, together with thrombin generation when the acute thrombosis is not yet controlled, can lead to further progression of thrombosis and result in venous limb gangrene as reported by Warkentin et al.[112] These investigators cautioned against starting warfarin too early. Commencement of warfarin should be delayed, particularly in patients with severe or extensive thrombosis, until after the acute thrombosis/HIT is under control, as judged clinically or as indicated by the return of platelet count to normal or baseline level. In addition, there must be at least a few days of overlap between stopping the parental anticoagulant (danaparoid, lepirudin, or argatroban) and the

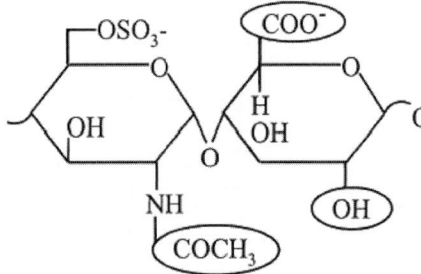

**Figure 48-4.** Disaccharide structure of danaparoid.

commencement of warfarin. After the acute treatment, warfarin should be continued for at least 6 months.

Thrombolytic therapy (streptokinase or recombinant-tissue plasminogen activator) is required in patients who have a massive pulmonary embolus with hemodynamic instability or in patients who have severe DVT with impending gangrene.[145–147] Timely administration of thrombolytic agent followed by danaparoid, lepirudin, or argatroban could be life- or limb-saving. Despite the presence of severe thrombocytopenia, thrombolytic treatment does not usually cause serious bleeding.

Occasionally, surgical intervention may be necessary. Embolectomy may be indicated in a patient with a lower limb arterial thrombosis and limb ischemia.[148] Insertion of an inferior vena filter may be required in a patient with uncontrolled recurrent pulmonary emboli or in a patient with venous thromboembolism who is unable to receive effective anticoagulation.[148]

*a. Danaparoid (Organan).* Danaparoid is a mixture of glycosaminoglycans isolated from porcine intestinal mucosa. It consists of mainly heparan sulfate (84%) and dermatan sulfate (12%) (Fig. 48-4) and has a mean molecular weight of about 6000 Daltons.[139,149] Danaparoid does not contain any heparin fragments. It exerts its anticoagulant effects primarily by inhibition of factor Xa, and it has only minimal antithrombin activity. Danaparoid's antifactor Xa activity is mediated by antithrombin and its antithrombin activity by antithrombin and heparin cofactor II.[149] Danaparoid has been shown to specifically inhibit platelet activation induced by the HIT antibody, a unique feature not seen with any other anticoagulant.[139] In animal models, danaparoid has been shown to have a high benefit (antithrombotic) to risk (bleeding) ratio.[149] Danaparoid is well absorbed after subcutaneous administration and has a bioavailability of almost 100%. It has a long plasma half life of about 25 hours.[150] Danaparoid is mainly excreted via the kidneys. It will accumulate in plasma in patients with impaired renal function and its dose has to be adjusted accordingly. Danaparoid's anticoagulant effect is only minimally neutralized by protamine.[151]

In a compassionate-use program, more than 460 patients with HIT-associated thrombosis have been treated with dan-

aparoid over a 10-year period. The success rate was greater than 90% in this series of patients.[4,140] In a prospective randomized comparative study, danaparoid has been shown to be more effective than dextran 70 in the treatment of HIT-related venous and arterial thrombosis.[152] In the prospective study, treatment with either drug was not associated with any bleeding complications. In the compassionate-use program, the patients received an IV bolus of 2500 U of danaparoid, followed by an infusion of 400 U/hour for 4 hours, then 300 U/hour for 4 hours, and then 150 to 200 U/hour for at least 5 days. The drug was also given subcutaneously at a dose of 2250 U twice daily. A slightly lower dose of danaparoid was used in the prospective study (Table 48-4).

Usually in the treatment of an uncomplicated case of thrombosis, laboratory monitoring of danaparoid is not necessary. However, in patients with HIT, the thrombosis is frequently severe and extensive and laboratory monitoring is then recommended. A plasma antifactor Xa level should be performed at least once, preferably 12 to 24 hours after commencement of danaparoid treatment to ensure that the level is within the therapeutic range of 0.5 to 0.8 antifactor Xa U/mL. Danaparoid does not significantly prolong the APTT or the INR. The APTT and INR should therefore not be used to monitor danaparoid therapy.[151] Other clinical situations that dictate laboratory monitoring of danaparoid treatment are (a) when the patient is clinically unstable, (b) in the presence of renal impairment, and (c) when the patient is overweight (>100 kg) or underweight (<50 kg).

i. Cross-Reactivity of HIT Antibody with Danaparoid. In 5 to 10% of patients with HIT, the antibody cross-reacts with danaparoid *in vitro* when testing is performed with SRA or PAT.[1,153] The cross-reactivity rate is higher (about 50%) if a more sensitive test (e.g., fluid-phase EIA) is used.[36] However, *in vitro* cross-reactivity with danaparoid appears to have no clinical significance. We have evaluated 21 patients with HIT and found that patients with and without *in vitro* cross-reactivity responded equally well to danaparoid treatment, except one patient who was refractory not only to danaparoid but also to several antithrombotic drugs including the anti-GPIIb-IIIa antagonist, abciximab.[136] Warkentin also studied 29 patients with HIT who were treated with danaparoid and found no difference in the clinical outcomes among those with and without *in vitro* cross-reactivity with the drug as detected by the SRA.[154] However, there have been occasional anecdotal reports of patients with HIT who showed *in vitro* cross-reactivity with danaparoid, and treatment with danaparoid led to unfavorable clinical outcomes.[155] In summary, *in vitro* cross-reactivity of the HIT antibody with danaparoid is not associated with unfavorable clinical outcomes in most HIT patients. True *in vivo* cross-reactivity is a rare event in the patients with demonstrated *in vitro* cross-reactivity. Pretreatment testing for

## Table 48-4: Danaparoid Dosing Schedules

| Clinical Indications | General Dosing Schedules[a] |
|---|---|
| Venous thromboembolism prophylaxis | 750 U SC bid |
| Arterial thromboembolism prophylaxis<br><br>then | (i)  1250 U SC bid, or<br>(ii)  infusion schedule 1 : 2500 U IV bolus followed by 400 U/h × 4 h, 300 U/h × 4 h,<br><br>     150–200 U/h.<br>Target plasma anti-Xa level: 0.5 to 0.8 U/mL |
| Venous thromboembolism treatment | (i)  1250 U SC bid (subacute event), or<br>(ii)  infusion schedule 1 (acute event) |
| Arterial thromboembolism treatment | (i)  1250 U SC bid (subacute event), or<br>(ii)  infusion schedule 1 (acute event) |
| Catheter patency | 750 U in 500 mL saline, then 50 mL per port or 50 to 100 mL flush |
| Hemodialysis/Hemofiltration | See Table 48-5 |
| PCI, IABP, cardiac catheterization | 2500 U IV bolus |
| Peripheral arterial surgery | 5000 U IV bolus |
| Cardiopulmonary bypass | (i)  125 U/kg IV bolus after thoracotomy<br>(ii)  2 U/mL priming fluid into bypass machine<br>(iii)  7 U/kg/h IV infusion after bypass hookup; discontinue 45 min before stopping bypass |

Abbreviations: bid, twice daily; IABP, intraaortic balloon pump; IV, intravenously; PCI, percutaneous coronary intervention; SC, subcutaneously; tid,

## Table 48-5: Anticoagulation of Patients with HIT during Hemodialysis and Hemofiltration

| Drug | Procedure | Drug dosing schedule |
|---|---|---|
| Danaparoid | Intermittent hemodialysis (HD) | 3750 U IV bolus before first and second dialyses<br>3000 U before third dialysis, then 2500 U before subsequent dialyses with a<br>   target plasma anti-Xa level of <0.3 U/mL |
| | Hemofiltration/continuous HD | 2500 U IV bolus followed by 600 U/h × 4 h, 400 U/h × 4 h, then 200 to 400 U/h<br>   with target plasma anti-Xa level of 0.5 to 1.0 U/mL |
| Lepirudin | Intermittent HD | 0.08 to 0.15 mg/kg bolus<br>Target APPT ratio: 2 to 3 |
| | Hemofiltration/continous HD | 0.005 mg/kg/h infusion<br>Target APPT ratio: 1.5 to 2.5 |
| Argatroban | Intermittent HD | 0.1 mg/kg bolus<br>Infusion schedule: 0.1 to 0.2 mg/kg/h<br>Target APPT ratio: 1.5 to 3.0 |

Adapted from ref. 206.

cross-reactivity is probably unnecessary as it will delay commencement of treatment and will consequently expose the patient to the risk of thrombosis extension or progression. However, it may be prudent to keep a pretreatment serum sample for testing later in the rare event that there is *in vivo* cross-reactivity, as indicated by the persistence of thrombocytopenia and worsening of thrombosis on treatment with danaparoid.

*b. r-Hirudin (Lepirudin).* Hirudin is an anticoagulant produced by the salivary glands of the medicinal leech, *Hirudo medicinalis.* It is a 65-amino-acid polypeptide with molecular weight of about 7000 Daltons.[156] Hirudin is a direct antithrombin that inhibits both fluid phase and clot-bound thrombin. It does not bind plasma proteins other than thrombin.[157,158] Hirudin has a short plasma half-life of 1 to 2 hours and is excreted predominantly by the kidneys.[158]

Recombinant hirudin (r-hirudin) is now commercially available as lepirudin (Refludan, HBW 023) and desirudin (Revasc, CGP 39393). Lepirudin has been shown to be effective in the treatment of thrombotic disorders, especially in patients with thrombosis associated with HIT.[142,143] There were three prospective nonrandomized clinical trials (HAT-1, HAT-2, HAT-3),[142,143,159] which showed that HIT patients treated with lepirudin had more favorable clinical outcomes than historical control patients treated with a variety of drugs including danaparoid, phenprocouman, aspirin, LMWH, or no treatment. In each of these studies, both HIT patients with thrombosis and without thrombosis were included. The event rates of death, new thromboembolic complications (TECs), and limb amputations at Day 35 in HAT-1 were 11.4%, 18.4%, and 5.7%, respectively in the lepirudin-treated patients and 27.8%, 32.1%, and 8.2%, respectively, in the historical controls. The event rate for each of these end points was lower in the lepirudin arm than in the control arm, but the difference did not reach statistically significance. However, the frequency of the combined end points (deaths, new TECs, and amputations) in the lepirudin-treated patients was 25.4%, which was significantly different from the incidence of 52.1% in the historical controls. Similar results were obtained in the two subsequent studies (HAT-2 and HAT-3).[143,159] A meta-analysis of HAT-1 and HAT-2 showed that in HIT patients with thrombosis, the combined end point of new thrombosis, limb amputation, and death in lepirudin-treated patients (n = 113) was lower than in the historical controls (n = 91), 21.3% versus 47.8%; $p = 0.004$.[160] The more favorable result in the lepirudin-treated patients was attributed to the substantial reduction of new thromboses in these patients. A meta-analysis of HAT-1, HAT-2, and HAT-3 showed that in HIT patients without thrombosis at diagnosis, even lower incidences of new thromboses, limb amputations, and deaths in patients treated with lepirudin were reported (2.7%, 2.7%, and 4.5%, respectively).[160] In the three HAT studies, there were more bleeding events in the lepirudin-treated patients. A meta-analysis of HAT-1, HAT-2, and HAT-3 studies revealed a cumulative bleeding incidence of 42.0% in patients treated with lepirudin compared with 23.6% in the control group ($p = 0.001$).[160]

i. Lepirudin Dosing in HAT-1, HAT-2, and HAT-3 Studies. In the HAT-1, HAT-2, and HAT-3 studies,[142,143,159] the patients received lepirudin for 2 to 10 days according to the following dosing regimens:

1. Patients with known TECs: 0.4 mg/kg IV bolus followed by infusion of 0.15 mg/kg/h
2. Patients with known TECs who also received thrombolytic therapy: 0.2 mg/kg IV bolus followed by infusion of 0.1 mg/kg/h

3. Patients without thrombosis who were receiving lepirudin for prophylaxis: infusion of 0.1 mg/kg/h

In view of the high bleeding incidence observed in the HAT studies, it is now recommended that the IV bolus be omitted except in patients with serious or extensive thrombosis.

ii. Laboratory Monitoring. Laboratory monitoring of lepirudin with APTT is required, particularly in patients with renal failure, because the drug tends to accumulate in these patients. In lepirudin-treated patients, both treatment efficacy and bleeding risk are influenced by the APTT. For a reasonable balance between maximizing efficacy and minimizing bleeding risk, the infusion rate of lepirudin should be adjusted to keep the APTT ratio within 1.5 to 2.5 times baseline. A meta-analysis of the HAT studies showed that efficacy is optimum when the APTT ratio was 1.5 to 2.5, but bleeding risk is greater than the control.[160] With an APTT ratio of more than 2.5, there was an even greater risk of bleeding without a further increase in efficacy. In patients with a high risk of bleeding, it is reasonable to keep the APTT ratio within 1.5 to 2.0.

iii. Other Safety Outcomes. In about 40% of patients, antihirudin antibodies are generated, usually from day 5 of treatment, peaking at days 9 to 10. The antibodies further prolong the APTT, implying that they enhance the anticoagulant effect of the drug. The dose of the drug should be adjusted accordingly.[161]

Allergic reactions such as skin rash, pruritus, urticaria, hot flushes, fever, and chills have been reported in lepirudin-treated patients. A serious but rare complication is anaphylaxis, which can be fatal. The incidence of anaphylaxis is estimated to be 0.015% on first exposure and 0.16% on reexposure. All anaphylactic reactions occurred within minutes of IV bolus of lepirudin administration, with fatal anaphylaxis occurring only in patients who had previous exposure to the drug.[160]

c. Argatroban. Argatroban is a synthetic direct thrombin inhibitor. It is a small molecule with a molecular weight of about 526 Daltons.[162] Its chemical structure is shown in Fig. 48-5. Like hirudin, argatroban can inhibit both fluid-phase and thrombus-bound thrombin.[163] Unlike the hirudin-thrombin interaction, the argatroban-thrombin interaction is reversible. Unlike lepirudin, argatroban is metabolized in the liver and excreted primarily in the feces, probably by biliary secretion.[164] In the plasma, argatroban is 54% protein bound. It has a short plasma half-life of 39 to 51 minutes.[165] Like danaparoid and hirudin, there is no specific antidote for argatroban. In addition to HIT, argatroban has been used to treat other thrombotic disorders.[144]

In HIT, three prospective multicenter studies have been carried out: ARG-911, ARG-915, and ARG-915X.[144] ARG-

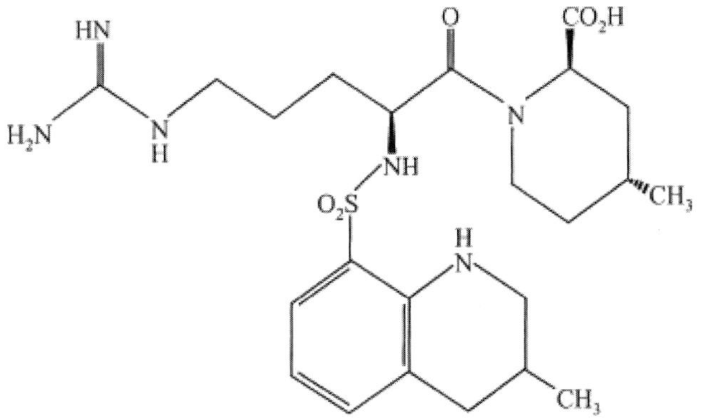

**Figure 48-5.** Chemical structure of argatroban.

911 is a historical controlled study. ARG-915 is a follow-up study using the same historical control group, and ARG-915X is a phase III extension study that allowed physicians to have continued access to the drug while it was under regulatory review. Altogether 754 patients in the three studies received argatroban for treatment of HIT and HIT-associated thrombosis syndrome (HITTS). In the first study (ARG-911), the incidence of the composite end point (death, amputation, and new thrombosis) was significantly reduced in argatroban-treated patients with HIT (isolated thrombocytopenia, n = 147) when compared to the historical control patients (25.6% versus 38.8%, p = 0.014).[144] However, the frequency of the composite end point in argatroban-treated patients with HITTS (n = 46) was not significantly different from that of the controls (43.8% versus 56.5%, p = 0.13). In each arm of the study, argatroban therapy significantly reduced new thrombosis and death due to thrombosis compared with the controls. It did not decrease all-cause mortality or limb amputation. However, bleeding events were similar between the argatroban-treated patients and the controls (major bleeding incidence: 6.9% versus 6.7%). The other adverse effects of the drug are diarrhea (11%) and pain (9%). Similar results were obtained in the other two studies.

i. Dosing and Monitoring. The patients in the three studies received argatroban IV initially at 2 μg/kg/min.[144] In patients with moderate liver impairment, an initial dose of 0.5 μg/kg/min is recommended. The dose should be adjusted after 2 hours and then daily to keep the APTT between 1.5 and 3.0 times baseline. If the APTT is subtherapeutic, the dose is increased by 0.5 μg/kg/min in most patients or 0.2 μg/kg/min in patients with liver impairment. Argatroban treatment is usually continued for up to 14 days or until recovery of the thrombocytopenia and adequate anticoagulation has been provided by warfarin or another agent. As argatroban prolongs the prothrombin time (PT) and increases the INR, during argatroban/warfarin overlapping therapy the INR reflects the combined effects of both

drugs. Serious thrombotic events were observed in patients with HIT when argatroban was prematurely discontinued when the INR was 2 to 3 during overlapping therapy. After stopping argatroban, the INR became subtherapeutic (<2). It may be more appropriate to stop argatroban when the INR is 4 and to repeat the INR a few hours later to ensure that the INR remains in the therapeutic range of 2 to 3.[144]

*d. Danaparoid versus Lepirudin versus Argatroban.* Danaparoid, lepirudin, and argatroban are rapid-acting anticoagulants capable of inhibiting thrombin generation or directly inhibiting thrombin. These are desirable properties for the treatment of HIT. There are now a large number of patients with HIT who have been treated with each of these drugs. In fact, in prospective comparative clinical trials and noncomparative studies, all three drugs have been shown to be highly effective in controlling thrombosis associated with HIT. Only danaparoid has been evaluated in a prospective randomized study and was found to be more efficacious than the control drug, dextran 70.[152] Argatroban and lepirudin have only been shown to be effective drugs for treatment of HIT in historical controlled studies.[142–144] No head-to-head clinical trial has yet been carried out between any two of these three drugs. In terms of efficacy, there is no convincing data to show that one drug is better than the other. Each drug has advantages and disadvantages. Danaparoid and argatroban do not cause an increase in bleeding,[144,152] but lepirudin was found in the three clinical trials (HAT-1, HAT-2, and HAT-3) to give rise to more bleeding events when compared with the historical controls.[142,143] There is no effective antidote for any of the three drugs, but in the bleeding patients the plasma drug levels will fall more quickly in the case of argatroban and lepirudin as they have much shorter half-lives. Danaparoid and lepirudin are excreted via the kidneys,[151,158] and the drugs will accumulate in patients with renal failure. On the other hand, argatroban is metabolized in the liver and will accumulate in patients with liver failure.[164] In a small proportion (about 7%) of HIT patients, the antibody cross-reacts *in vitro* with danaparoid, but this does not result in adverse clinical outcomes[136,154] except in the rare patients. In about 50% of HIT patients, treatment with lepirudin leads to formation of antihirudin antibodies, which prolongs the APTT but does not cause bleeding, provided the APTT ratio is closely monitored and not allowed to exceed three.[161]

The choice of drug may be determined by its availability, approval status by the drug regulatory authority of the particular country, and the clinical circumstances. Argatroban is approved in the United States for both thromboprophylaxis and treatment of HIT. Lepirudin is approved for treatment of HIT in the United States and the European Union. Danaparoid is approved for thromboprophylaxis and treatment of HIT in most European Union countries but not in the United States.

*e. Other Anticoagulants.* Besides these three drugs, there are other effective anticoagulants (such as bivalirudin and fondaparinux) that have been used in only a limited number of HIT patients. Each drug has some attractive features and may in the future be employed more widely in the management of HIT.

i. Bivalirudin. Bivalirudin (Angiomax, formerly known as hirulog), a hirudin analogue, is a small synthetic 20-amino-acid peptide (molecular mass, 2180 Daltons) and is a specific, direct inhibitor of thrombin.[166] Unlike lepirudin, it binds reversibly to thrombin. Bivalirudin has a short plasma half-life of 25 to 36 min. It is cleared by both renal mechanisms (20%) and proteolytic cleavage in plasma (80%) independent of organ functions. Bivalirudin's clearance rate is decreased in patients with moderate and severe renal impairment (by about 45% and 70%, respectively) and its plasma half-life is prolonged (3.5 hours), requiring dose adjustment in these patients.[167] It prolongs APTT, activated clotting time (ACT), and PT linearly in a dose-dependent manner. In general, the ACT is used to monitor bivalirudin in patients undergoing percutaneous coronary intervention (PCI), ecarin clotting time (ECT) during on-pump cardiac surgery, and APTT in HIT patients treated for non-PCI indications.

Bivalirudin has been used for treatment of HIT in only a limited numbers of patients, initially in a few individual patients and later in three patient series. In the first series of 39 patients, bivalirudin was used for DVT/PE treatment and several other indications.[168] In the second, Francis et al. treated 40 HIT patients with bivalirudin (32 with isolated HIT and 16 with HIT and thrombosis).[169] The antithrombotic efficacy was acceptable and bleeding occurred only in a few patients. Francis et al. employed a mean bivalirudin infusion rate of 0.165 mg/kg/h aiming for a target APTT of 1.5 to 2.5 times baseline.[169] Bivalirudin was given for a mean of 8.7 days, overlapping with warfarin in 35 patients. The third series, the ATBAT trial, was a prospective, open-label study designed to examine the efficacy and safety of bivalirudin in patients with acute or previous HIT undergoing PCI.[166] In addition, there is a large body of clinical experience of bivalirudin use in non-HIT patients with unstable angina undergoing PCI or off-pump coronary artery bypass surgery (OPCAB; see the upcoming discussion). The drug is not yet approved for the treatment of venous thromboembolism in HIT patients in the United States or Europe. Its short half-life, unique clearance mechanism, and low immunogenicity may prove attractive for the management of HIT in the future,[166] particularly for the indications of PCI and OPCAB.

ii. Fondaparinux. Fondaparinux (Arixtra) is a new synthetic pure anti-Xa compound that is identical to the pentasaccharide sequence in heparin and LMWH.[170] It has a high affinity for antithrombin III, through which it mediates its anticoagulant effect. Clinical trials have shown that fondaparinux is highly effective in thromboprophylaxis in patients undergoing orthopedic surgery[96] and in the initial treatment of venous thromboembolism.[97,171] For the latter indication, it is given by subcutaneous administration, once or twice daily without the need of laboratory monitoring.[171] Its proven antithrombotic efficacy, its good risk/benefit ratio, and its lack of cross-activity with the anti-PF4/heparin antibodies in HIT[95] may make it a very attractive drug for the treatment of HIT in the future, particularly in patients with DVT or PE. However, the clinical experience with fondaparinux in treatment of HIT is very limited.[172–174]

*f. Adjunctive Therapies.* Adjunctive therapies in HIT include antiplatelet agents (aspirin, dipyridamole, GPIIb-IIIa antagonists, high-dose IV IgG, and plasmapheresis[1,4,51]). Thrombolytic therapy, embolectomy, and insertion of inferior vena cava filters were discussed in Section VIII.B.1.

i. Antiplatelet Agents. Aspirin and dipyridamole have been used in the treatment of HIT with variable success.[175] The experience with GPIIb-IIIa inhibitors and thienopyridines in the treatment of HIT is very limited. Although these drugs should not be used alone as a first-line treatment of HIT, their use in conjunction with an effective anticoagulant may be beneficial in patients at high risk of arterial thrombosis.

ii. High-Dose IV IgG. IV IgG has been described in anecdotal reports to result in a rapid rise in platelet counts in patients with HIT.[176,177] Although IgG in high concentration *in vitro* can inhibit HIT antibody-induced platelet activation,[130] it has no anticoagulant effect. IV IgG should only be used an adjunctive therapy together with an effective anticoagulant in selected cases of HIT, such as patients with very severe thrombocytopenia or serious thrombosis refractory to a first-line drug.

iii. Plasmapheresis. Plasmapheresis has been reported to be successful in the treatment of some patients with severe HIT. This procedure not only removes the circulating HIT antibody, PF4, and activated clotting factors, but it also replenishes the depleted plasma anticoagulant proteins, such as antithrombin, protein C, and protein S (if fresh plasma is used as the replacement fluid).[178–180] Plasmapheresis can be a useful adjunctive therapy in patients with severe life- or limb-threatening thrombosis.

iv. LMWH. In a high percentage of patients with HIT (78%–88%), the antibody cross-reacts with LMWH.[153] Some reports described successful outcomes with LMWH treatment of HIT, but in other reports, LMWH therapy resulted in poor outcomes including persistent thrombocytopenia, development of new thrombosis, limb amputation, and even death.[181–183] With the availability of safer and more effective anticoagulants (danaparoid, lepirudin, and argatroban), LMWH can no longer be recommended for treatment of HIT. Its use should be strongly discouraged.

v. Platelet Transfusion. Even in patients with severe thrombocytopenia, bleeding rarely occurs. Prophylactic platelet transfusions are not recommended for the treatment of HIT. Platelet transfusion can precipitate an acute thrombotic event in patients with circulating HIT antibody.

## 2. Type II HIT with No Clinically Obvious Thrombosis (Isolated Thrombocytopenia).
Type II HIT is a hypercoagulable state. Even in HIT patients without clinically overt thrombosis, there is clear evidence of *in vivo* platelet activation and activation of blood coagulation. Previous studies have reported the presence of elevated plasma levels of β-thromboglobulin, P-selectin, and platelet microparticles, as well as increased thrombin-antithrombin complexes.[48,49,112,184] In addition, the plasma levels of natural anticoagulant proteins (e.g., antithrombin and protein C) in HIT patients are reduced. In one retrospective study, 52.8% of HIT patients with isolated thrombocytopenia at the time of diagnosis developed thrombosis during the subsequent 30-day follow-up period.[185] The rate of developing thrombosis soon after stopping heparin is very high, estimated to be 10% and 18% at 1 and 2 days after heparin withdrawal. These high rates of thrombosis (which may include serious and fatal thrombotic events) have also been observed by others.[142,143] It is not adequate to just stop heparin in the HIT patient with isolated thrombocytopenia;[186] these patients require prompt treatment with rapid-acting anticoagulants such as danaparoid, lepirudin, or argatroban.[137,143,144]

## 3. Treatment of Type II HIT in Special Situations

*a. On-Pump Cardiac Surgery.* A patient with acute or recent type II HIT still has circulating heparin-dependent antibodies. If the patient requires cardiac surgery, heparin should not be used during cardiopulmonary bypass (CPB). An alternative anticoagulant such as danaparoid,[139] lepirudin,[160] or bivalirudin[166] should be used. During surgery, the anticoagulant effect should be closely monitored by antifactor Xa for danaparoid or ECT for lepirudin. (Higher doses of lepirudin are used during CPB than in the treatment of thrombosis. ECT but not APTT still has a linear relationship with the higher plasma concentrations of lepirudin. Therefore, the ECT should be used for monitoring plasma lepirudin levels during CPB.) There is a high risk of postoperative bleeding, particularly with danaparoid. This may be due to its long plasma half-life and the lack of an antidote to neutralize its anticoagulant activity at the conclusion of surgery. With bivalirudin, which has a very short-half life, there was

no increase in perioperative bleeding when it was used for CPB in several HIT patients.[166] Argatroban has been used in a few patients during CPB with mixed outcomes. A safe and effective dose of argatroban for CPB surgery has not yet been established.[144]

Alternatively, if there is no urgency for the cardiac surgery, it should be delayed until the heparin-dependent antibody is no longer detectable by a sensitive assay (e.g., SRA). In this setting, heparin can be used exclusively during CPB with or without antiplatelet drugs (GPIIb-IIIa antagonists or epoprostenol),[189–194] but it should be deliberately avoided before and after surgery. Likewise, patients with a past history of HIT can use heparin exclusively for CPB, provided the heparin-dependent antibody is no longer detectable.

*b. Off-Pump Coronary Artery Bypass (OPCAB) Surgery.* A lesser degree of anticoagulation is needed in off-pump coronary artery bypass (OPCAB) surgery. This should lead to less perioperative bleeding. Indeed danaparoid, bivalirudin, and argatroban[195,196] have been used successfully in patients with acute HIT during OPCAB surgery, with acceptable perioperative blood loss. The most encouraging result was from a prospective trial involving 100 non-HIT patients who were randomly assigned to receive bivalirudin or heparin with protamine reversal. The trial showed that patients who received bivalirudin had better graft flow with no increase in perioperative blood loss.[197] A trial in HIT patients is under way. A prospective randomized trial in 71 non-HIT patients comparing danaparoid and heparin with protamine reversal showed slightly increased but still acceptable bleeding in danaparoid patients.[198]

*c. Percutaneous Coronary Intervention (PCI).* Argatroban has been evaluated in three multicenter prospective studies (ARG-216, ARG-310, and ARG-311) in patients with HIT undergoing PCI[144] and bivalirudin in one study (the ATBAT trial).[199] In the argatroban studies, the 71 patients were given an IV bolus of 350 μg/kg and then 25 μg/kg/min infusion titrated to an ACT of 300 to 450 seconds during PCI. Satisfactory outcome occurred in 94.5% of patients and adequate anticoagulation in 97.8%. Only one patient (1%) experienced a major nonfatal periprocedural bleed. Similar results were observed in the bivalirudin study. In the ATBAT trial, bivalirudin was given to 52 patients as 1.0 mg/kg IV bolus and 2.5 mg/kg/h infusion or 0.75 mg/kg bolus and 1.75 mg/kg/h infusion for 4 hours. Procedural success (TIMI grade 3 flow and <50% stenosis) was achieved in 98% of patients, and clinical success (absence of death, emergency bypass surgery, or Q-wave infarction) was achieved in 96%. One high-dose patient had major bleeding (1.9%).[199] Danaparoid[137] and lepirudin[200] have also been used for PCI in a few anecdotal HIT patients. In addition, recombinant hirudin was used in clinical studies for PCI in non-HIT patients with good results.[201] In the United States, only argatroban has been approved for use as an anticoagulant in patients with HIT undergoing PCI.

*d. Hemodialysis.* In patients with type II HIT who require hemodialysis, several alternative anticoagulants are now available and they include danaparoid,[137,140,141] lepirudin,[202,203] and argatroban.[204,205] The dosages of the drugs in hemodialysis are given in Table 48-5. Danaparoid and lepirudin are excreted mainly in the kidneys and the drugs therefore tend to accumulate in the patients undergoing dialysis. Careful laboratory monitoring is necessary, and the dose of the drug may need to be reduced with time. Because argatroban is not excreted renally, it has an advantage in patients undergoing dialysis. In some patients with HIT who are undergoing hemodialysis with argatroban, the platelets become activated, resulting in clotting in the extracorporeal circuit.[205] In these cases, aspirin may be added to inhibit the platelet activation and clotting.

Other approaches include hemodialysis without an anticoagulant or with other agents such as vitamin K antagonists, dermatan sulfate, prostacyclin, or aspirin. For more detailed information, the reader is referred to specific reviews of this topic.[137,141,206]

*e. Pregnant Women and Children.* HIT occasionally occurs in pregnant women. At least 13 women have been treated with danaparoid for type II HIT during pregnancy, and they did not suffer serious adverse effects.[137] In contrast, only one pregnant patient with HIT has been reported to be treated with lepirudin.[207] In preclinical studies, lepirudin in high doses induced increased mortality in pregnant animals. Argatroban has not been reported to have been used in pregnant women with HIT.[144] Danaparoid, which does not cross the placenta to affect the fetus, is at present the drug of choice for the treatment of type II HIT in pregnancy. The dose of danaparoid used in pregnancy is the same as that used in nonpregnant subjects.

Children usually require higher weight-adjusted doses of danaparoid than adults. Six children have been treated with danaparoid for various indications including renal dialysis and cardiac surgery.[137] Five children have been treated with lepirudin — two in HAT-1 and HAT-2 studies and three in anecdotal reports.[142,143,208] No standardized lepirudin dosing protocol is currently available for the treatment of children. A few neonates have been treated with argatroban.[208]

# References

1. Chong, B. H. (1995). Heparin induced thrombocytopenia. *Br J Haematol, 89,* 431–439.
2. Chong, B. H. (1988). Heparin induced thrombocytopenia. *Blood Reviews, 2,* 108–114.
3. Warketin, T. E., Chong, B. H., & Greinacher, A. (1998). Heparin induced thrombocytopenia: Towards consensus. *Thromb Haemost, 79,* 1–7.
4. Ortel, T. L., & Chong, B. H. (1998). New treatment options for heparin-induced thrombocytopenia. *Sem Hematol, 35,* 4:5:3–8.
5. Copley, A. L., & Robb, T. P. (1942). Studies on platelets III: The effect of heparin *in vivo* on the platelet count in mice and dogs. *Am J Clin Pathol, 12,* 563–570.
6. Quick, A. J., Shanberge, J. N., & Stefanini, M. (1948). The effect of heparin on platelets *in vivo. J Lab Clin Med, 33,* 1424–1430.
7. Fidlar, E., & Jaques, L. B. (1948). The effect of commercial heparin on the platelet count. *J Lab Clin Med, 33,* 1410–1413.
8. Gollub, S., & Ulin, A. W. (1962). Heparin-induced thrombocytopenia in man. *J Lab Clin Med, 59,* 430–435.
9. Davey, M. G., & Lander, H. (1968). Effect of injected heparin on platelet levels in man. *J Lab Clin Pathol, 21,* 55–99.
10. Natelson, E. A., Lynch, E. C., Alfrey, Jr., et al. (1969). Heparin-induced thrombocytopenia: An unexpected response to treatment of consumption coagulopathy. *Ann Intern Med, 71,* 1121–1125.
11. Weismann, R. E., & Tobin, R. W. (1958). Arterial embolism occurring during systemic heparin therapy. *AMA Archives of Surgery, 76,* 219–227.
12. Roberts, B., & Rosato, E. F. (1964). Heparin-A cause of arterial emboli? *Surgery, 55,* 803–808.
13. Rhodes, G. R., Dixon, R. H., & Silver, D. (1973). Heparin-induced thrombocytopenia with thrombotic and hemorrhagic manifestations. *Surg Gynecol Obstet, 136,* 409–416.
14. Fratantoni, J. C., Pollet, R., & Gralnick, H. R. (1975). Heparin-induced thrombocytopenia: Confirmation of diagnosis with *in vitro* methods. *Blood, 45,* 395–401.
15. Nelson, J. C., Lerner, R. G., Goldstein, R., et al. (1978). Heparin-induced thromboctyopenia. *Arch Intern Med, 138,* 548–552.
16. Cimo, P. L., Moake, J. L., Weinger, R. S., et al. (1979). Heparin-induced thrombocytopenia: Association with a platelet aggregating factor and arterial thromboses. *Am J Hematol, 6,* 125–133.
17. Towne, J. B., Bernhard, V. M., Hussey, C., et al. (1979). White clot syndrome, peripheral vascular complications of heparin therapy. *Arch Surg, 114,* 372–377.
18. Klein, H. G., & Bell, W. R. (1974). Disseminated intravascular coagulation during heparin therapy. *Intern Med, 80,* 477–478.
19. Bell, W. R., Tomasulo, P. A., Alving, B. M., et al. (1976). Thrombocytopenia occurring during the administration of heparin: A prospective study in 52 patients. *Ann Intern Med, 85,* 155–160.
20. Chong, B. H., Pitney, W. R., & Castaldi, P. A. (1982). Heparin-induced thrombocytopenia: Association of thrombotic complication with a heparin-dependent IgG antibody which induced platelet aggregation, release and thromboxane synthesis. *Lancet, ii,* 1246–1248.
21. Chong, B. H., & Berndt, M. C. (1989). Heparin-induced thrombocytopenia. *Blut, 58,* 53–57.

22. Salzman, E. W., Rosenberg, R. D., Smith, M. H., et al. (1980). Effect of heparin and heparin fractions on platelet aggregation. *J Clin Invest, 65,* 64–73.

23. Chong, B. H., & Ismail, F. (1989). The mechanism of heparin-induced platelet aggregation. *Eur J Haematol, 43,* 245–251.

24. Holmer, E., Lindahl, U., Bäckström, G., et al. (1980). Anticoagulant activities and effects on platelets of a heparin fragment with high affinity for antithrombin. *Thromb Res, 18,* 861–869.

25. Mikhaildis, D. P., Barradas, M. A., Jeremy, J. Y., et al. (1985). Heparin-induced platelet aggregation in anorexia nervosa and in severe peripheral vascular disease. *Eur J Clin Invest, 15,* 313–319.

26. Burgess, J. K., & Chong, B. H. (1997). The platelet proaggregating and potentiating effects of unfractionated heparin, low molecular weight heparin and heparinoid in intensive care patients and healthy controls. *Eur J Haematol, 58,* 270–285.

27. Blahut, B., Kramar, H., Vinazzer, H., et al. (1985). Substitution of antithrombin III in shock and DIC: A randomized study. *Thromb Res, 39,* 81–89.

28. Balduini, C. L., Noris, P., Bertolino, G., et al. (1993). *Thromb Haemost, 69,* 522–523.

29. Amiral, J., Bridey, F., Dreyfus, M., et al. (1992). Platelet factor 4 complexed to heparin is the target for antibodies generated in heparin induced thrombocytopenia [letter]. *Thromb Haemost, 68,* 95–96.

30. Horne, M. K. III., & Hutchinson, K. J. (1998). Simultaneous binding of heparin and platelet factor-4 to platelets: Further insights into the mechanism of heparin-induced thrombocytopenia. *Am J Hematol, 58,* 24–30.

31. Visentin, G. P., Moghaddam, M., Collins, J. L., et al. (1997). Antibodies associated with heparin-induced thrombocytopenia (HIT) report conformational changes in platelet factor 4 (PF4) induced by linear polyanionic compounds [abstr.]. *Blood, 90,* (suppl 1), 460a.

32. Horne, M. K. III., & Chao, E. S. (1990). The effect of molecular weight on heparin binding to platelets. *Br J Haematol, 74,* 306–312.

33. Greinacher, A., Alban, S., Dummel, V., et al. (1995). Characterization of the structural requirements for a carbohydrate based anticoagulant with a reduced risk of inducing the immunological type of heparin-associated thrombocytopenia. *Thromb Haemost, 74,* 886–892.

34. Horne, M. K., III. (2000). In T. E. Warkentin & A. Greinacher (Eds.), *Heparin-induced thrombocytopenia* (pp. 113–126). New York: Basel, Marcel Dekker.

35. Dawes, J., Pumphrey, C. W., McLaren, K. M., et al. (1982). The *in vivo* release of human platelet factor 4 by heparin. *Thromb Res, 27,* 65–76.

36. Warkentin, T. E., Levine, M. N., Hirsh, J., et al. (1995). Heparin-induced thrombocytopenia in patients treated with low-molecular-weight heparin or unfractionated heparin. *N Engl J Med, 332,* 1330–1335.

37. Visentin, G. P., Malik, M., Cyganiak, K. A., et al. (1996). Patients treated with unfractionated heparin during open heart surgery are at high risk to form antibodies reactive with heparin: Platelet factor 4 complexes. *J Lab Clin Med, 128,* 376–383.

38. Burgess, J. K., Lopez, J. A., Berndt, M. C., et al. (1998). Quinine-dependent antibodies bind a restricted set of epitopes on the glyooprotein 1b-1x complex: Charactarization of the epitopes. *Blood, 92,* 2366–2373.

39. Chee, M., Vun, & Chong, B. H. (unpublished data).

40. Li, Z. Q., Park, K. S., Sachais, B. S., et al. (2002). Defining a second epitope for heparin-induced thrombocytopenia/thrombosis antibodies using KKO, a murine HIT — like monoclonal antibody. *Blood, 99,* 1230–1236.

41. Zeporen, L., Li, Z. Q., Park, K. S., et al. (1998). Defining an antigenic epitope on platelet factor 4 associated with heparin-induced thrombocytopenia. *Blood, 92,* 3250–3259.

42. Rauova, L., Poncz, M., McKenzie, S. E., et al. (2005). Ultra large complexes of PF4 and heparin are central to the pathogenesis of heparin-induced thrombocytopenia. *Blood, 105,* 131–138.

43. Newman, P., & Chong, B. H. (1999). Heparin-induced thrombocytopenia: The major antigenic epitope is on the modified platelet factor 4 and not on heparin. *Br J Haematol, 107,* 303–309.

44. Kelton, J. G., Sheridan, D., Santos, A., et al. (1998). Heparin-induced thrombocytopenia: Laboratory studies. *Blood, 72,* 925–930.

45. Chong, B. H., Ismail, F., Chesterman, C. N., et al. (1989). Heparin-induced thrombocytopenia: Mechanism of interaction of the heparin-dependent antibody with platelets. *Br J Haematol, 73,* 235–240.

46. Newman, P., & Chong, B. H. (2000). Heparin-induced thrombocytopenia: New evidence for the dynamic binding of purtied anti-pf4-heparin antibodies to platelets and the resulting activation. *Blood, 96,* 182–187.

47. Warkentin, T. E., Hayward, C. P. M., Boshkov, L. K., et al. (1994). Sera from patients with heparin-induced thrombocytopenia generated platelet-derived microparticles with procoagulant acitivity: An explanation for the thrombotic complications of heparin-induced thrombocytopenia. *Blood, 84,* 3691–3699.

48. Chong, B. H., Grace, C. S., & Rosenberg, M. C. (1981). Heparin-induced thrombocytopenia: Effects of heparin platelet antibody on platelets. *Br J Haematol, 49,* 531–540.

49. Warkentin, T. E. (1996). Heparin-induced thrombocytopenia: IgG-mediated platelet activation, platelet microparticle generation, and altered procoagulant/anticoagulant balance in the pathogenesis of thrombosis and venous limb gangrene complicating heparin-induced thrombocytopenia. *Transfus Med Rev, 10,* 249–258.

50. Amiral, J., Wolf, M., Fischer, A. M., et al. (1996). Pathogenicity of IgA and/or IgM antibodies to heparin-PF4 complexes in patients with heparin-induced thrombocytopenia. *Br J Haematol, 92,* 954–959.

51. Warkentin, T. E. (2003). Heparin-induced thrombocytopenia: Pathogenesis and management. *Br J Hematol, 121,* 1365–2141.

52. Amiral, J., Peynaud-Debayle, E., Wolf, M., et al. (1996). Generation of antibodies to heparin-PF4 complexes without thrombocytopenia in patients treated with unfractionated or

low molecular weight heparin. *Am J Hematol, 52,* 90–95.

53. Brandt, J. T., Isenhart, C. E., Osborne, J. M., et al. (1995). On the role of platelet FcγRIIa phenotype in heparin-induced thrombocytopenia. *Thromb Haemost, 74,* 1564–1572.

54. Burgess, J. K., Lindeman, R., Chesterman, C. N., et al. (1995). Single amino acid mutation of Fcγ receptor is associated with the development of heparin-induced thrombocytopenia. *Br J Haematol, 91,* 761–766.

55. Denomme, G. A., Warkentin, T. E., Horsewood, P., et al. (1997). Activation of platelets by sera containing IgG1 heparin-dependent antibodies: An explanation for the predominance of the FcγRIIa "low responder" (His$_{131}$) gene in patients with hepain-induced thrombocytopenia. *J Lab Clin Med, 130,* 278–284.

56. Arepally, G., McKenzie, S. E., Jiang, X-M., et al. (1997). FcγRIIA H/R$^{131}$ polymorphism, subclass-specific IgG anti-heparin/platelet factor 4 antibodies and clinical course in patients with heparin-induced thrombocytopenia and thrombosis. *Blood, 89,* 370–375.

57. Bachelot-Loza, C., Saffroy, R., Lasne, D., et al. (1998). Importance of the FcγRIIa-Arg/His-131 polymorphism in heparin-induced thrombocytopenia diagnosis. *Thromb Haemost, 79,* 523–528.

58. Carlsson, L. E., Santoso, S., Baurichter, G., et al. (1998). Heparin-induced thrombocytopenia: New insights into the impact of the FcγRIIa-R-H$_{131}$ polymorphism. *Blood, 92,* 1526–1531.

59. Amiral, J., Marfaing-Koka, A., Wolf, M., et al. (1996). Presence of auto-antibodies to interleukin-8 or neutrophil-activating peptide-2 in patients with heparin-associated thrombocytopenia. *Blood, 88,* 410–416.

60. Reilly, M. P., Taylor, S. M., Hartman, N. K., et al. (2001). Heparin-induced thrombocytopenia/thrombosis in a transgenic mouse model demonstrates the requirement for human platelet factor 4 and platelet activation through Fcγ R II A. *Blood, 98,* 2442–2447.

61. Cines, D. B., Tomaski, A., & Tannenbaum, S. (1987). Immune endothelial-cell injury in heparin-associated thrombocytopenia. *N Engl J Med, 316,* 581–589.

62. Visentin, G. P., Ford, S. E., Scott, J. P., et al. (1994). Antibodies from patients with heparin-induced thrombocytopenia/thrombosis are specific for platelet factor 4 complexed with heparin or bound to endothelial cells. *J Clin Invest, 93,* 81–88.

63. Arepally, G. M., & Mayer, I. M. (2001). Antibodies from patients with heparin-induced thrombocytopenia stimulate monocytic cells to express tissue factor and secrete interleukin-8. *Blood, 98,* 1252–1254.

64. Gallus, A. S., Goodall, K. T., Beswick, W., et al. (1980). Heparin-associated thrombocytopenia: A case report and prospective study. *Aust N Z J Med, 10,* 25–31.

65. Kakkasseril, J. S. Cranley, J. J., Panke, T., et al. (1985). Heparin-induced thrombocytopenia: A prospective study of 142 patients. *J Vasc Surg, 2,* 382–384.

66. Holm, H. A., Eika, C., & Laake, K. (1980). Thrombocytes and treatment with heparin from porcine mucosa. *Scand J Haematol, 36* (suppl 1), 81–84.

67. Eika, C., Godal, H. C., Laake, K., et al. (1980). Low incidence of thrombocytopenia during treatment with hog

mucosa and beef lung heparin. *Scand J Haemost, 25,* 19–24.

68. Powers, P. J., Kelton, J. G., & Carter, C. J. (1984). Studies of the frequency of heparin-associated thrombocytopenia. *Thromb Res, 33,* 439–443.

69. Cipolle, R. J., Rodvold, K. A., Seifert, R., et al. (1983). Heparin-associated thrombocytopenia: A prospective evaluation of 211 patients. *Ther Drug Monit, 5,* 205–211.

70. Lee, D. H., & Warkentin, T. E. (2004). In T. E. Warkentin & A. Greinacher (Eds.), *Heparin-induced thrombocytopenia* (3rd ed., pp. 107–148). New York: Basel, Marcel Dekker.

71. Bell, W. R., & Royall, R. M. (1980). Heparin-associated thrombocytopenia: A comparison of three heparin preparations. *N Engl J M, 303,* 902–907.

72. Ansell, J., Stepchuk, N., Kumar, R., et al. (1980). Heparin-induced thrombocytopenia: A prospective study. *Thromb Haemost, 43,* 61–65.

73. Bailey, R. T., Jr., Ursick, J. A., Heirn, K. L., et al. (1986). Heparin-associated thrombocytopenia: A prospective comparison of bovine lung heparin intestinal heparin. *Drug Intell Clin Pharm, 20,* 374–378.

74. Green, D., Martin, G. J., Shoichet, S. H., et al. (1984). Thrombocytopenia in a prospective, randomized, double-blind trial of bovine and porcine heparin. *Am J Med Sci, 288,* 60–64.

75. Schmitt, B. P., & Adelman, B. (1993). Heparin-associated thrombocytopenia: A critical review and a pooled analysis. *Am J Med Sci, 305,* 208–215.

76. Powers, P. J., Cuthbert, D., & Hirsh, J. (1979). Thrombocytopenia found uncommonly during heparin therapy. *JAMA, 241,* 2396–2397.

77. Malcolm, I. D., Wigmore, T. A., & Steinbrecher, U. P. (1979). Heparin-associated thrombocytopenia: Low frequency in 104 patients treated with heparin of intestinal mucosal origin. *Can Med Assoc J, 120,* 1086–1088.

78. Girolami, B., Prandoni, P., Stefani, P. M., et al. (2003). The incidence of heparin-induced thrombocytopenia in hospitalised medical patients treated with subcutaneous unfractionated heparin: A prospective cohort study. *Blood, 101,* 2955–2959.

79. Amiral, J., Peynaud-Debayle, E., Wolf, M., et al. (1998). Generation of antibodies to heparin-PF4 complexes without thrombocytopenia in patients treated with unfractionated or low-molecular weight heparin. *Am J Hematol, 52,* 90–95.

80. Ansell, J. E., Price, J. M., Shah, S., et al. (1985). Heparin-induced thrombocytopenia: What is its real frequency? *Chest, 88,* 878–882.

81. Saffle, J. R., Russon, J. Jr., Dukes, G. E., et al. (1980). The effect of low-dose heparin therapy on serum platelet and transaminase levels. *J Surg Res, 28,* 297–305.

82. Olin, J., & Graor, R. (1981). Heparin-associated thrombocytopenia. *N Eng J Med, 304,* 609.

83. Romeril, K. R., Anzimlt, C. M. H., Hamer J. W., et al. (1982). Heparin-induced thrombocytopenia: Case reports and a prospective study. *N Z Med J, 95,* 267–269.

84. Weitberg, A. B., Spremulli, E., & Cummings, F. J. (1982). Effect of low-dose heparin on the platelet count. *South Med J, 75,* 190–192.

85. Monreal, M., Lafoz, E., Salvador, R., et al. (1989). Adverse effects of three different forms of heparin therapy: Thrombocytopenia, increased transaminases, and hyperkalemia. *Eur J Clin Pharmacol, 37,* 415–418.

86. Rao, A. K., White, G. C., Sherman, L., et al. (1989). Low incidence of thrombocytopenia with porcine muccossal heparin. A prospective multicentre study. *Arch Intern Med, 149,* 1285–1288.

87. Johnson, R. A., Lazarus, K. H., & Henry, D. H. (1984). Heparin-induced thrombocytopenia: A prospective study. *Am J Hematol, 17,* 349–353.

88. Ayars, G. H., & Tikoff, G. (1980). Incidence of thrombocytopenia in medical patients on "mini-dose" heparin prophylaxis. *Am Heart J, 99,* 816 (letter).

89. Leyvraz, P. F., Bachmann, F., Hoek, J., et al. (1991). Prevention of deep vein thrombosis after hip replacement: Randomised comparison between unfractionated heparin and low molecular weight heparin. *Br Med J, 303,* 543–548.

90. Ganzer, D., Gutezeit, A., Mayer, G., et al. (1997). Thromboembolieprophylaxe als Auslöser Thrombembolischer Komplicationen. Eine Untersuchung zur Inzidenz de Heparininduzierten Thrombozytopenie (HIT) Type II. *Z Orthop, 135,* 543–549.

91. Louridas, G. (1991). Heparin-induced thrombocytopenia. *S Afr J Surg, 29,* 50–52.

92. Trossaert, M., Gaillard, A., Commin, P. L., et al. (1998). High incidence of anti-heparin/platelet factor 4 antibodies after cardiopulmonary bypass. *Br J Haematol, 101,* 653–655.

93. Pouplard, C., May, M. A., Iochmann, S., et al. (1999). Antibodies to platelet factor 4-heparin after cardiopulmonary bypass in patients anticoagulated with unfractionated heparin or a low-molecular-weight heparin: Clinical implications for heparin-induced thrombocytopenia. *Circulation, 99,* 2530–3536.

94. Warkentin, T. E., Sheppard, J. A., Horsewood, P., et al. (2000). Impact of the patient population on the risk for heparin-induced thrombocytopenia. *Blood, 96,* 1703–1708.

95. Visentin, G. P., Mali, M., Cyganiak, K. A., et al. (1996). Patients treated with unfractionated heparin during open heart surgery are at high risk to form antibodies reactive with heparin platelet factor 4 complexes. *J Lab Clin Med, 128,* 376–383.

96. Bauer, T. L., Arepally, G., Konkle, B. A., et al. (1997). Prevalence of heparin-associated antibodies without thrombosis in patients undergoing cardiopulmonary bypass surgery. *Circulation, 95,* 1242–1246.

97. Savi, P., Chong, B. H., Greinacher, A., et al. (2005). Effect of fondaparinux on platelet activation in the presence of heparin-dependent antibodies: A blinded comparative multicentre study with unfractinated heparin. *Blood, 105,* 139–144.

98. Turpie, A. G. G., Gallus, A. S., & Hoek, J. A. (2001). For the pentasaccharide investigators: A synthetic pentasaccharide for the prevention of deep-vein thrombosis after total hip replacement. *N Engl J Med, 344,* 619–625.

99. The Rembrandt Investigators. (2000). Treatment of proximal deep-vein thrombosis with a novel synthetic compound (SR90107A/ORG31540) with pure anti-factor Xa activity: A phase 2 evaluation. *Circulation, 102,* 2726–2731.

100. Fausett, M. B., Vogtlander, M., Lee, R. M., et al. (2001). Heparin-induced thrombocytopenia is rare in pregnancy. *Am J Obstet Gynecol, 185,* 148–152.

101. Lepercq, J., Conard, J., Borel-Derlon, A., et al. (2001). Venous thromboembolism during pregnancy: A retrospective study of enoxaparin safety in 624 pregnancies. *Br J Obstet Gynaecol, 108,* 1134–1140.

102. Warkentin, T. E., & Sigouin, C. S. (2002). Gender and risk of immune heparin-induced thrombocytopenia. *Blood, 100,* 17a (abstract).

103. Heeger, P. S., & Backstrom, J. T. (1986). Heparin flushes and thrombocytopenia [letter]. *Ann Intern Med, 105,* 143.

104. Rama, B. N., Haake, R. E., Bander, S. J., et al. (1991). Heparin-flush associated thrombocytopenia-induced hemorrhage: A case report. *Nebr Med J, 76,* 392–394.

105. Laster, J., & Silver, D. (1988). Heparin-coated catheters and heparin-induced thrombocytopenia. *J Vasc Surg, 7,* 667–672.

106. Warkentin, T. E. (2004). In T. E. Warkentin & A. Greinacher (Eds.), *Heparin-induced thrombocytopenia* (3rd ed., pp. 105–106). New York: Basel, Marcel Dekker.

107. Rice, L., Attisha, W. K., Drexler, A., et al. (2002). Delayed-onset heparin-induced thormbocytopenia. *Ann Int Med, 136,* 210–215.

108. Warkentin, T. E., & Kelton, J. G. (2001). Delayed-onset heparin-induced thrombocytopenia and thrombosis. *Ann Int Med, 135,* 502–506.

109. Warkentin, T. E., & Kelton, J. G. (1998). Timing of heparin-induced thrombocytopenia (HIT) in relation to previous heparin use: Absence of an anamnestic immune response, and implications for repeat heparin use in patients with a history of HIT [abstr.]. *Blood, 92* (suppl 1), 182a.

110. Warkentin, T. E., Sheppard, J. I., Horsewood, P., et al. (2000). Impact of the patient population on the risk for heparin-induced thrombocytopenia. *Blood, 96,* 1703–1708.

111. Boshkov, L. K., Warkentin, T. E., Hayward, C. P. M., et al. (1993). Heparin-induced thrombocytopenia and thrombosis: Clinical and laboratory studies. *Br J Haematol, 84,* 322–328.

112. Warkentin, T. E., Elavathil, L. J., Hayward, C. P. M., et al. (1997). The pathogenesis of venous limb gangrene associated with heparin-induced thrombocytopenia. *Ann Intern Med, 127,* 804–812.

113. Srinivasan, A. F., Rice, L., Bartholomew, J. R., et al. (2004). Warfarin-induced skin necrosis and venous limb gangrene in the setting of heparin-induced thrombocytopenia. *Arch Intern Med, 164,* 66–70.

114. Warkentin, T. E., & Hong, A. P. (1998). Upper limb deep venous thrombosis and central venous catheter complicating immune heparin-induced thrombocytopenia (HIT). *Blood, 101,* 3049–3051.

115. Van der Weyden, M. B., Hunt, H., McGrath, K., et al. (1983). Delayed-onset heparin-induced thrombocytopenia: A potentially malignant syndrome. *Med J Aust, 2,* 132–135.

116. Kyritsis, A. P., Williams, E. C., & Schutta, H. S. (1990). Cerebral venous thrombosis due to heparin-induced thrombocytopenia. *Stroke, 21,* 1503–1505.

117. Arthur, C. K., Grant, S. J. B., Murray, W. K., et al. (1985). Heparin-associated acute adrenal insufficiency. *Aust NZ J Med, 15,* 454–455.

118. Bleasel, J. F., Rasko, J. E. J., Rickard, K. A., et al. (1992). Acute adrenal insufficiency secondary to heparin-induced thrombocytopenia-thrombosis syndrome. *Med J Aust, 157,* 192–193.

119. Meyer-Lindenberg, A., Quenzel, E-M., Bierhoff, E., et al. (1997). Fatal cerebral venous sinus thrombosis in heparin-induced thrombocytopenia. *Eur Neurol, 37,* 191–192.

120. Celoria, G. M., Steingart, R. H., Banson, B., et al. Coumarin skin necrosis in a patient with heparin-induced thrombocytopenia — a Case Report. *Angiology, 39,* 915–920.

121. Cohen, G. R., Hall, J. C., Yeast, J. D., et al. (1988). Heparin-induced cutaneous necrosis in a postpartum patient. *Obste Gynecol, 72,* 498–499.

122. Popov, D., Zarrabi, M. H., Foda, H., et al. (1997). Pseudopulmonary embolism: Acute respiratory distress in the syndrome of heparin-induced thrombocytopenia. *Am J Kidney Dis, 29,* 449–452.

123. Warkentin, T. E., Soutar, R. L., Panju, A., et al. (1992). Acute systemic reactions to intravenous bolus heparin therapy: Characterization and relationship to heparin-induced thrombocytopenia [abstr]. *Blood, 80* (suppl 1), 160a.

124. Warkentin, T. E., Hirte, H. W., Anderson, D. R., et al. (1994). Transient global amnesia associated with acute heparin-induced thrombocytopenia. *Am J Med, 97,* 489–491.

125. Chong, B. H. (1995). Diagnosis, treatment and pathophysiology of immune-mediated thrombocytopenia. *Critical Rev Oncol Hematol, 20,* 271–296.

126. Chong, B. H., & Eisbacher, M. (1998). Pathophysiology and laboratory testing of heparin-induced thrombocytopenia. *Sem in Haematol, 35,* 3–8.

127. Sheridan, D., Carter, C., & Kelton, J. G. (1986). A diagnostic test for heparin-induced thrombocytopenia. *Blood, 67,* 27–30.

128. Schnell, M. K., Giordano, K. J., Henry, M., et al. (1998). Diagnosis of heparin-induced thrombocytopenia (HIT): Comparison of methods. *Transfusion, 38* (suppl.), 98S.

129. Greinacher, A., Michels. J., Kiefel, V., et al. (1991). A rapid and sensitive test for diagnosing heparin-associated thrombocytopenia. *Thromb Haemost, 66,* 734–736.

130. Chong, B. H., Burgess, J., & Ismail, F. (1993). The clinical usefulness of the platelet aggregation test for the diagnosis of heparin-induced thrombocytopenia. *Thromb Haemost, 69,* 344–350.

131. Stewart, M. W., Etches, W. S., Boshkov, L. K., et al. (1995). Heparin-induced thrombocytopenia: An improved method of detection based on lumi-aggregometry. *Br J Haematol, 91,* 173–177.

132. Lee, D. P., Warkentin, T. E., Denomme, G. A., et al. (1996). A diagnostic test for heparin-induced thrombocytopenia: Detection of platelet microparticles using flow cytometry. *Br J Haematol, 95,* 724–731.

133. Tomer, A. (1997). A sensitive and specific functional flow cytometric assay for the diagnosis of heparin-induced thrombocytopenia. *Br J Haematol, 98,* 648–656.

134. Jy, W., Mao, W. W., Horstman, L. L., et al. (1999). A flow cytometric assay of platelet activation markers P-selectin (CD62P) distinguishes heparin-induced thrombocytopenia (HIT) from HIT with thrombosis (HITT). *Thromb Haemost, 82,* 1255–1259.

135. Amiral, J., Bridey, F., Wolf, M., et al. (1995). Antibodies to macromolecular platelet factor 4-heparin complexes in heparin-induced thrombocytopenia: A study of 44 cases. *Thromb Haemost, 73,* 21–28.

136. Newman, P. M., Swanson, R. L., & Chong, B. H. (1998). Heparin-induced thrombocytopenia: IgG binding to PF4-heparin complexes in the fluid phase and cross-reactivity with low molecular weight heparin and heparinoid. *Thromb Haemost, 80,* 292–297.

137. Eichler, P., Raschke, R., Lubenow, N., et al. (2002). The new ID-heparin/PF4 antibody test for rapid detection of heparin-induced antibodies in comparison with functional and antigenic assays. *Br J Haematol, 116,* 887–891.

138. Risch, L., Bertschmann, W., Heijnen, I. A. F. M., et al. (2003). A differentiated approach to assess the diagnostic usefulness of a rapid particle gel immunoassay for the detection of antibodies against heparin-platelet factor 4 in cardiac surgery patients. *Blood Coagul Fibinolysis, 14,* 99–106.

139. Chong, B. H., & Magnani, H. (2004). In T. E. Warkentin & A. Greinacher (Eds.), *Heparin-induced thrombocytopenia* (3rd ed., pp. 371–396). New York: Basel, Marcel Dekker.

140. Chong, B. H., & Magnani, H. N. (1992). Orgaran in heparin-induced thrombocytopenia. *Haemostasis, 22,* 85–91.

141. Magnani, H. N. (1993). Heparin-induced thrombocytopenia. (HIT): An overview of 230 patients treated with orgaran (Org. 10172). *Thromb Haemost, 70,* 554–561.

142. Greinacher, A., Völpel, H., Janssens, U., et al. (1999). Recombinant hirudin (Lepirudin) provides safe and effective anticoagulation in patients with the immunologic type of heparin-induced thrombocytopenia: A prospective study. *Circulation, 99,* 73–80.

143. Greinacher, A., Janssens, U., Berg, G., et al. (1999). Lepirudin (recombinant hirudin) for parenteral anticoagulation in patients with heparin-induced thrombocytopenia. *Circulation, 100,* 587–593.

144. Lewis, B. E., & Hursting, M. J. (2004). In T. E. Warkentin & A. Greinacher (Eds.), *Heparin-induced thrombocytopenia* (3rd ed., pp. 437–474). New York: Basel, Marcel Dekker.

145. Cohen, J. I., Cooper, M. R., & Greenberg, G. S. (1985). Streptokinase therapy of pulmonary emboli with heparin-associated thrombocytopenia. *Arch Intern Med, 145,* 1725–1726.

146. Fiessinger, J. N., Aiach, M., Rocanto, M., et al. (1984). Critical ischemia during heparin-induced thrombocytopenia: Treatment by intra-arterial streptokinase. *Thromb Res, 33,* 235–238.

147. Cummings, J. M., Mason, T. J., Chomka, E. V., et al. (1986). Fibrinolytic therapy of acute myocardial infarction in the heparin thrombosis syndrome. *Am Heart J, 112,* 407–409.

148. Sobel, M., Adelman, B., Szentpeterey, S., et al. (1988). Surgical management of heparin-associated thrombocytopenia: Strategies in the treatment of venous and arterial thromboembolism. *J Vasc Surg, 8,* 395–401.

149. Meuleman, D. G. (1992). Orgaran (Org. 10172): Its pharmacological profile in experimental models. *Haemostasis, 22,* 58–65.

150. Stiekema, J. C., Eijnand, H. P., van Dinther, T. G., et al. (1989). Safety and pharmocokinetics of the low molecular weight heparinoid org. 10172 administered to healthy elderly volunteers. *Br J Clin Pharmacol, 27,* 39–48.

151. Danhof, M., de Boer, A., Magnani, H. N., et al. (1992). Pharmacokinetic considerations on orgaran (org. 10172). *Hemostasis, 22,* 73–84.

152. Chong, B. H., Gallus, A. S., Cade, J. F., et al. (2001). Prospective randomised open-label comparison of danaparoid with Dextran 70 in the treatment of heparin-induced thormbocytopneia with thrombosis: A clinical outcome study. *Thromb Haemos, 86,* 1170–1175.

153. Vun, C. H., Evans, S., & Chong, B. H. (1996). Cross-reactivity study of low molecular weight heparinoid in heparin-induced thrombocytopenia. *Thromb Res, 81,* 525–523.

154. Warkentin, T. E. (1996). Danaparoid (organan) for the treatment of heparin-induced thrombocytopenia (HIT) and thrombosis: Effects on *in vivo* thrombin and cross-linked fibrin generation and evaluation of the clinical significance of vitro cross-reactivity of danaparoid for HIT-IgG [abstr.]. *Blood, 88,* 626a.

155. Tardy/Poncet, B., Mahul, P., Beraud, A. M., et al. (1995). Failure of organan therapy in a patient with a previous heparin-induced thrombocytopenia. *Br J Haematol, 90,* 69–70.

156. Clore, G. M., Sukumaran, D. K., Niges, M., et al. (1987). The conformation of hirudin in solution: A study using nuclear magnetic resonance, distance geometry and restrained molecular dynamics. *EMBO J, 6,* 529–537.

157. Weitz, J. I., Hudoba, M., Massel, D., et al. (1990). Clot bound thrombin is protected from inhibition by heparin-antithrombin III but is susceptible to inactivation by antithrombin III-independent inhibitors. *J Clin Invest, 86,* 385–391.

158. Glusa, E. (1998). Pharmacology and therapeutic applications of hirudin, a new anticoagulant. *Kidney Int, 53,* 54–56.

159. Eichler, P., Lubenow, N., & Greinacher, A. (2002). Results of the third prospective study of treatment with lepirudin in patients with heparin-induced thrombocytopenia (HIT) [abstr]. *Blood, 100* (suppl 1), 704a.

160. Greinacher, A. (2004). In T. E. Warkentin & A. Greinacher (Eds.), *Heparin-induced thrombocytopenia* (3rd ed., pp. 397–437). New York: Basel, Marcel Dekker.

161. Eichler, P., Friesen, H, J., Lubenow, N., et al. (2000). Anti-hirudin antibodies in patients with heparin-induced thrombocytopenia treated with lepirudin: Incidence, effects of aPTT, and clinical relevance. *Blood, 96,* 2372–2378.

162. Okamoto, S., & Okunomiya-Hijikata, A. (1993). Synthetic selective inhibitors of thrombin. *Methods, 222,* 328–340.

163. Berry, C. N., Girardot, C., Lecoffre, C., et al. (1994). Effects of the synthetic thrombin inhibitor argatroban on fibrin- or clot-incorporated thrombin: Comparison with heparin and recombinan hirudin. *Thromb Haemost, 72,* 381–386.

164. Izawa, O., Katsuki, M., Komatsu, T., et al. (1986). Pharmacokinetic studies of argatroban (MD-805) in human: Concentrations of argatroban and its metabolites in plasma, urine and feces during and after drip intravenous infusion. *Jap Pharmacol Ther, 14,* 251–263.

165. Swan, S. K., St. Peter, J. V., Lambrecht, L. J., et al. (2000). Comparison of anticoagulant effects and safety of argatroban and heparin in healthy subjects. *Pharmacotherapy, 20,* 756–770.

166. Bartholomew, J. (2004). In T. E. Warkentin & A. Greinacher (Eds.), *Heparin-induced thrombocytopenia* (3rd ed., pp. 475–507). New York: Basel, Marcel Dekker.

167. Robson, R., White, H., Aylward, P., et al. (2002). Bivalirudin pharmacokinetics and pharmacodynamics: Effect of renal function, dose and gender. *Clin Pharmacol Ther, 71,* 433–439.

168. Chamberlin, J. R., Lewis B., Leya, F., et al. (1994). Successful treatment of heparin-associated thrombocytopenia and thrombosis using hrulog. *Can J Cardiol, 11,* 511–514.

169. Francis, J., Drexler, A., Gwyn, G., et al. (2003). Bivalirubin, a direct thrombin inhibitor, in the treatment of heparin-induced thrombocytopenia (abstr). *Thromb Haemost, 1* (suppl 1), 1909.

170. Boneu, B., Necciari, J., Cariou, R., et al. (1995). Pharmaco-kinetics and tolerance of the natural pentasacchardie (ORG31540/SR90107A) with high affinity to antithrombin III in man. *Thromb Haemost, 74,* 1468–1473.

171. Buller, H. R., Davidson, B. L., Decousus, H., et al. (2003). Subcutaneous fondaparinux versus intravenous unfractionated heparin in the initial treatment of pulmonary embolism. *N Engl J Med, 349,* 1695–1702.

172. Kovacs, M. J. (2005). Successful treatment of heparin induced thrombocytopenia. *Thromb Haemost, 93,* 999–1000.

173. D'Amico, E. A., Villaca, P. R., Gualandro, S. F., et al. (2003). Successful use of arixtra in a patient with paroxysmal nocturnal hemoglobinuria, Budd–Chiari syndrome and heparin-induced thrombocytopenia [letter]. *J Thromb Haemost, 1,* 2452–2453.

174. Parody, R., Oliver, A., Souto, J. C., et al. (2003). Fondaparinux (ARIXTRA) as an alternative anti-thrombotic prophylaxis when there is hypersensitivity to low molecular weight and unfractionated heparins. *Haematologica, 88,* 32.

175. Gruel, Y., Lermusiaux, P., Lang, M., et al. (1991). Usefulness of antiplatelet drugs in the management of heparin-associated thrombocytopenia and thrombosis. *Ann Vasc Surg, 5,* 552–555.

176. Frame, J. N., Mulvey, K. P., Phares, J. C., et al. (1989). Correction of severe heparin-associated thrombocytopenia with intravenous immunoglobulin. *Ann Intern Med, 111,* 946–947.

177. Grau, E., Linares, M., Alaso, M. A., et al. (1992). Heparin-induced thrombocytopenia-response to intravenous immunoglobulin *in vivo* and *in vitro*. *Am J Hematol, 39,* 312–312.

178. Bouvier, J. L., Lefevre, P., Villain, P., et al. (1988). Treatment of serious heparin-induced thrombocytopenia by plasma exchange: Report on four cases. *Thromb Res, 51,* 335–336.

179. Nand, S., & Robinson, J. A. (1988). Plasmapheresis in the management of heparin-associated thrombocytopenia with thrombosis. *Am J Hematol, 28,* 204–206.

886 BENG H. CHONG

180. Thorp, D., Canty, A., Whiting, J., et al. (1990). Plasma exchange and heparin-induced thrombocytopenia. *Prog Clin Biol Res, 337,* 521–522.

181. Leroy, J., Leclerc, M. H., Delahousse, B., et al. (1985). Treatment of heparin-associated thrombocytopenia and thrombosis with low molecular weight heparin (C.Y. 216). *Semin Thromb Hemost, 11,* 326–329.

182. Vitoux, J. F., Mathieu, J. F., Roncato, M., et al. (1986). Heparin-associated thrombocytopenia: Treatment of low molecular weight heparin. *Thromb Haemost, 55,* 37–39.

183. Gouault-Heilmann, M., Huet, Y., Adnot, S., et al. (1987). Low molecular weight heparin fractions as an alternative therapy in heparin-induced thrombocytopenia. *Haemostasis, 17,* 134–140.

184. Chong, B. H., Murray, B., Berndt, M. C., et al. (1994). Plasma P-selectin is increased in thrombotic consumptive platelet disorders. *Blood, 83,* 1535–1541.

185. Warkentin, T. E., & Kelton, J. G. (1996). A 14-year study of heparin-induced thrombocytopenia. *Am J Med, 101,* 502–507.

186. Wallis, D. E., Workman, D. L., Lewis, B. E., et al. (1999). Failure of early heparin cessation as treatment for heparin-induced thrombocytopenia. *Am J Med, 106,* 629–635.

187. Reiss, F. C., Poetzch, B., & Mueller-Berghaus, G. (1997). In R. Pifarré (Ed.), *New anticoagulants for the Cardiovascular Patient* (pp. 197–222). Philadelphia: Hanley & Belfus.

188. Vasquez, J. C., Vichiendilokkul, A., Mahmood, S., et al. (2002). Anticoagulation with bivalirudin during cardio pulmonary bypass in cardiac surgery. *Ann Thorac Surg, 74,* 2177–2179.

189. Makhoul, R. G., McCann, R. L., Austin, E. H., et al. (1987). Management of patients with heparin-associated thrombocytopenia and thrombosis requiring cardiac surgery. *Ann Thorac Surg, 43,* 617–621.

190. Laster, J., Elfrin, R., & Silver, D. (1989). Re-exposure to heparin of patients with heparin-associated antibodies. *Vas Surg, 9,* 677–681.

191. Kappa, J. R., Fisher, C. A., Todd, B., et al. (1990). Intraoperative management of patients with heparin-induced thrombocytopenia. *Ann Thoracic Surg, 49,* 714–722.

192. Koster, A., Hansen, R., & Kuppe, H., et al. (2000). Recombinant hirudin as an alternative for anticoagulation during cardiopulmonary bypass in patients with heparin-induced thrombocytopenia type II: A 1-year experience in 57 patients. *J Cardiothorac Vasc Anesth, 14,* 243–248.

193. Koster, A., Meyer, O., & Fischer, T., et al. (2001). One-year experience with the platelet glycoprotein IIb/IIIa antagonist tirofiban and heparin during cardiopulmonary bypass in patients with heparin-induced thrombocytopenia type II. *J Thorac Cardiovasc Surg, 122,* 1254–1255.

194. Mertzlufft, F., Kuppe, H., & Koster, A. (2000). Management of urgent high-risk cardiopulmonary bypass in patients with heparin-induced thrombocytopenia type II and coexisting disorders of renal function: Use of heparin and epoprostenol combined with on-line monitoring of platelet function. *J Cardothoac Vasc Anesth, 14,* 304–308.

195. Warkentin, T. E., Dunn, G. L., & Cybulsky, I. J. (2001). Off-pump coronary artery bypass grafting for acute heparin-induced thrombocytopenia. *Ann Thorac Surg, 72,* 1730–1732.

196. Kieta, D. R., McCammon, A. T., & Homan, W. T., et al. (2003). Hemostatic analysis of a patient undergoing off-pump coronary artery bypass surgery with argatroban anticoagulation. *Anesth Analg, 96,* 956–958.

197. Merry, A. F., Raudkivi, P. J., & Middleton, N. G., et al. (2004). Bivalirudin versus heparin and procamine in off pump coronary artery bypass surgery. *Ann Thorac Surg, 77,* 925–931.

198. Carrier, M., Robitaille, D., & Perrault, L. P., et al. (2003). Heparin vs. danaparoid in off-pump coronary bypass grafting: Results of a prospective randomized clinical test. *J Thorac Cardiovasc Surg, 125,* 325–329.

199. Mahaffey, K. W., Lewis, B. E., Wildermann, N. M., et al. (2003). The anticoagulant therapy with bivalirudin to assist in the performance of percutaneous coronary intervention in patients with heparin-induced thrombocytopenia (ATBAT) study: Main results. *J Invasive Cardiol, 15,* 611–616.

200. Manfredi, J. A., Wall, R. P., & Sane, D. C., et al. (2001). Lepirudin as a safe alternative for effective anticoagulation in patients with know heparin-induced thrombocytopenia undergoing a percutaneous coronary intervention: Case reports. *Cathet Cardiovasc Interv, 52,* 468–472.

201. Rupprecht, H. J., Terres, W., & Ozbek, C., et al. (1995). Recombinant hirudin (HBW 023) prevents troponin T release after coronary angioplasty in patients with unstable angina. *J Am Coll Cardiol, 26,* 1637–1642.

202. Van Wyk, V., Badenhorst, P. N., Luus, H. G., et al. (1995). A comparison between the use of recombinant hirudin and heparin during hemodialysis. *Kidney Int, 48,* 1338–1343.

203. Nowak, G., Bucha, E., Brauns, I., et al. (1997). Anticoagulation with r-hirudin in regular application of r-hirudin in a haemodialysis patient. *Wien Klin Wochenschr, 109,* 354–358.

204. Matsuo, T., Kario, K., Kodama, K., et al. (1992). Clinical application of the synthetic thrombin inhibitor, argatroban (MD-805). *Semin Thromb Hemost, 18,* 155–160.

205. Koide, M., Yamamoto, S., Matsuo, M., et al. (1995). Anticoagulation for heparin-induced thrombocytopenia with spontaneous platelet aggregation in a patient requiring haemodialysis. *Nephrol Dial Transplant, 10,* 2137–2140.

206. Fischer, K-G. (2004). In T. E. Warkentin & A. Greinacher (Eds.), *Heparin-induced thrombocytopenia* (3rd ed., pp. 509–530). New York: Basel, Marcel Dekker.

207. Greinacher, A., & Warkentin, T. E. (2004). In T. E. Warkentin & A. Greinacher (Eds.), *Heparin-induced thrombocytopenia* (3rd ed., pp. 335–370). New York: Basel, Marcel Dekker.

208. Klenner, A. F., & Greinacher, A. (2004). In T. E. Warkentin & A. Greinacher (Eds.), *Heparin-induced thrombocytopenia* (3rd ed., pp. 553–571). New York: Basel, Marcel Dekker.

# CHAPTER 49

# Drug-Induced Thrombocytopenia

**Richard H. Aster**

*Blood Research Institute, BloodCenter of Wisconsin, Medical College of Wisconsin, Milwaukee, Wisconsin*

## I. Introduction

Many medications are capable of causing clinically significant thrombocytopenia by one of two mechanisms: inhibition of megakaryocyte proliferation and platelet production or destruction of platelets in the peripheral blood. In the former situation, all hematopoietic elements are usually affected, leading to pancytopenia. However, a few agents appear to be relatively specific for megakaryocytes. In the second group of conditions, antibodies are usually the cause of platelet destruction, but a few drugs appear to act directly on platelets without involvement of the immune system. Drugs can cause antibody-mediated platelet destruction by at least six different mechanisms. These mechanisms and the drugs implicated in each are reviewed in detail in this chapter. Heparin-induced thrombocytopenia (HIT) is a common and sometimes serious side effect of this widely used anticoagulant. Because the clinical presentation and pathogenesis of HIT is complex and unique, it is discussed separately in Chapter 48. Additional information about drug-induced thrombocytopenia can be found in other reviews.[1–7]

## II. Drug-Induced Suppression of Platelet Production

### A. Chemotherapeutic Agents

Most chemotherapeutic and immunosuppressive agents cause thrombocytopenia when used aggressively.[8] Cytosine arabinoside, 6-mercaptopurine, methotrexate, busulfan, cyclophosphamide, *cis*-platinum, and many others routinely cause myelosuppression, leading to thrombocytopenia when given at sufficient doses, whereas vinca alkaloids are much less likely to do so. Although all myeloid elements are usually affected, thrombocytopenia often limits the dose of a chemotherapeutic agent that can be administered. A risk model based on various clinical and laboratory findings that

may be helpful in identifying patients at high risk to develop profound thrombocytopenia has been described.[9] Recombinant interleukin (IL)-11 has been approved for the treatment and amelioration of chemotherapy-induced thrombocytopenia[10] (Chapter 66), but its cost-effectiveness has been questioned.[11] If eventually approved for this purpose, thrombopoietin may be useful[12,13] (Chapter 66). Granulocyte-macrophage colony-stimulating factor (GM-CSF),[14] granulocyte colony-stimulating factor (G-CSF),[15] and IL-3[16] appear to be relatively ineffective. Chemoprotective agents that may be capable of moderating the effects of chemotherapy on platelet production are in various stages of development.[17,18] Myelotoxicity leading to severe thrombocytopenia also occurs in patients treated with antiviral agents such as zidovudine used for the treatment of infection with human immunodeficiency virus (HIV).[19]

Anecdotal reports describe patients who became profoundly thrombocytopenic while taking chemotherapeutic agents such as actinomycin D, oxaliplatin, suramin, cyclosporine A, and aminoglutethimide, despite the absence of general myelosuppression and the presence of adequate numbers of megakaryocytes in the bone marrow. Findings made in such cases are consistent with the possibility that chemotherapeutic and immunosuppressive drugs occasionally cause immune thrombocytopenia of the "quinine type" (Section IV.B.2).

### B. Idiosyncratic Drug-Induced Marrow Aplasia

Hundreds of medications have been implicated as possible causes of aplastic anemia.[20] This complication can develop after weeks or months of regular or intermittent treatment and appears not to be dose related. The pathogenesis of this type of drug hypersensitivity is poorly understood.[21] Drugs most often implicated include anticonvulsants, sulfonamides, gold salts used in treatment of rheumatoid arthritis, and nonsteroidal anti-inflammatory drugs (NSAIDs).[22]

## C. *Drugs That Affect Megakaryocytopoiesis*

**1. Anagrelide.** Anagrelide is a quinazolin compound capable of causing isolated thrombocytopenia without a significant effect on other myeloid elements.[23] Its mechanism of action is not fully understood, but *in vivo* and *in vitro* studies indicate that the drug interferes with megakaryocyte maturation by several possible mechanisms.[24] This property of anagrelide has made it useful for treatment of thrombocythemia in patients with myeloproliferative syndromes[23,25] (Chapter 56). The therapeutic : toxic ratio of anagrelide is relatively low, and instances of profound thrombocytopenia in patients taking the drug have been described.[26]

**2. Alcohol.** Patients who ingest large quantities of alcohol for months or years sometimes present with severe thrombocytopenia.[27] Vitamin deficiencies or splenomegaly secondary to alcoholic cirrhosis are causative in some instances. However, *in vitro* and *in vivo* studies indicate that ethanol can exert a relatively specific effect on megakaryocytopoiesis in some individuals. Thrombocytopenia associated with chronic alcoholism is usually mild, but patients with extremely low platelet counts and hemorrhagic symptoms have been described.[28] Concomitant inhibition of platelet function by ethanol may contribute to bleeding symptoms in some patients[29] (Chapter 58). Upon withdrawal of alcohol, platelet levels usually return to normal within a few weeks but mild thrombocytopenia sometimes persists for a longer period of time.

**3. Estrogens.** Large doses of estrogenic hormones cause megakaryocytic aplasia and profound thrombocytopenia in several species of animals.[30] Only sporadic reports of thrombocytopenia with megakaryocytic hypoplasia in human patients treated with estrogen have been published.[31] Thus, it appears that the human species is relatively resistant to estrogen-induced suppression of megakaryocyte maturation. However, it has been suggested that estrogen therapy can worsen the severity of disease in women with autoimmune thrombocytopenia.[32]

**4. Interferons.** Mild to moderate thrombocytopenia is a recognized side effect of interferon therapy.[33–36] The drop in platelet levels tends to be more severe in patients with preexisting mild thrombocytopenia.[36] Severe life-threatening thrombocytopenia after interferon-α and pegylated interferon-2b therapy for chronic hepatitis C infection has been described.[37,38] Interferons α and γ inhibit megakaryocyte proliferation *in vitro*[36] and inhibit stem cell proliferation and differentiation in the bone marrow.[39] This effect may explain thrombocytopenia seen in patients given interferons, but its molecular basis has not yet been defined.

**5. Valproic Acid.** Mild thrombocytopenia occurs in up to one third of adults treated with valproate for epilepsy.[40]

This complication appears to be dose related[41] and sometimes resolves despite continuation of treatment.[42] Elderly patients experience thrombocytopenia more often than younger ones.[40] Clinical and *in vitro* studies suggest that valproate acts to inhibit megakaryocyte development and platelet production.[43–45] However, instances of acute, profound thrombocytopenia with severe bleeding have been described,[46,47] suggesting that drug-induced antibodies may cause platelet destruction in rare individuals.

**6. Imatinib Mesylate.** Thrombocytopenia, sometimes severe, occurs in a significant fraction of patients with chronic myelogenous leukemia, especially those in an accelerated phase of the disease during treatment with imatinib mesylate (Gleevec).[48,49] Responses to IL-11[49] and G-CSF[50] have been described.

**7. Linezolid.** Linezolid is an oxazolidinone antibiotic used in the treatment of certain drug-resistant bacterial infections. Up to one third of patients given this drug for more than 10 days develop thrombocytopenia,[51,52] usually mild but occasionally severe enough to require platelet transfusion.[51] Evidence suggests that reduced platelet levels are a consequence of dose-dependent marrow suppression of unknown mechanism.[53]

## III. Drug-Induced Platelet Destruction by Nonimmune Mechanisms

### A. *Desmopressin*

Desmopressin (DDAVP) promotes release of von Willebrand factor (VWF) from endothelial cells and is useful for the treatment of some patients with von Willebrand disease (VWD) and certain disorders of platelet function[54] (Chapter 67). In patients with the type 2B variant of VWD, characterized by a VWF mutant with high affinity for the platelet VWF receptor, however, injection of DDAVP can cause mild and occasionally severe thrombocytopenia.[55] DDAVP and cryoprecipitate can also induce thrombocytopenia in patients with "platelet-type" VWD, characterized by a mutant VWF receptor on platelets[56] (Chapter 57). Several reports suggest DDAVP can be deleterious in patients with thrombotic thrombocytopenic purpura (TTP).[57,58]

### B. *Hematopoietic Growth Factors and Leukokines*

GM-CSF triggers an acute drop in platelet levels in some patients.[59,60] Animal studies have provided evidence that GM-CSF causes destruction of platelets in the spleen and liver, possibly the result of activation of the macrophage-monocyte system in these organs.[61,62] A similar effect has been attributed to macrophage colony-stimulating factor (M-CSF).[63]

### C. Tumor Necrosis Factor α/Interferon γ

Treatment of Ewing's sarcoma with tumor necrosis factor α (TNF-α) combined with interferon γ was associated with a rapid drop in platelet levels by more than 90% in a small number of patients.[64] An effect of this cytokine combination on endothelium, leading to platelet-endothelial interaction and increased clearance of platelets, was postulated as the responsible mechanism.

### D. IL-2

A decrease in platelet levels, usually mild but sometimes profound, is a common side effect of high-dose IL-2 given for treatment of cancer.[65,66] Megakaryocytes appear to be unaffected and platelet destruction is thought to result from the direct action of IL-2 on platelets by an unknown mechanism.[65]

### E. Porcine Factor VIII

Mild but occasionally severe thrombocytopenia occurs in most patients with hemophilia treated with porcine factor VIII.[67] These patients sometimes experience concomitant symptoms of hypersensitivity such as fever, urticaria, and chills, and markers indicative of activation can be found on circulating platelets.[68] Whether platelet-activating immune complexes produced by the reaction of antibodies with the xenoprotein are responsible has not been established.

### F. Protamine Sulfate

Thrombocytopenia, often mild, occurs in up to one third of patients given protamine for the purpose of neutralizing heparin after surgery and has been reproduced by infusion of protamine into dogs.[69] A direct effect of heparin-protamine complexes on circulating platelets has been postulated as the cause.[70]

### G. Amrinone

Amrinone is a type III phosphodiesterase inhibitor used for short-term treatment of heart failure. About one third of patients infused with amrinone develop mild thrombocytopenia within a few hours of starting the drug.[71,72] The N-acetyl metabolite of amrinone may be the agent that causes

platelet destruction.[72] The responsible mechanism has not been identified.

## IV. Drug-Induced Platelet Destruction by Immune Mechanisms

### A. Background

Many drugs are capable of inducing immune thrombocytopenia. This section considers pathogenetic mechanisms, drug and drug metabolites implicated as triggers, clinical aspects, laboratory diagnosis, and treatment. The responsible mechanisms are summarized in Table 49-1.

### B. Mechanisms by Which Drugs Cause Immune Thrombocytopenia

**1. Induction of Hapten-Dependent Antibodies.** Under normal circumstances, only macromolecules induce antibodies, but it has long been known that low molecular weight compounds (haptens) covalently linked to proteins can trigger the formation of immunoglobulins specific for the low molecular weight substance itself. When drug-induced immune thrombocytopenia (DITP) was first characterized clinically, it was assumed that it was caused by hapten-dependent antibodies raised against the drug covalently linked to one or more protein constituents of the platelet membrane. However, the antibodies could not, in general, be inhibited by high concentrations of soluble drugs as would be expected of a hapten-dependent antibody, and it was rarely possible to use platelets preincubated with drugs and then washed as targets for antibody detection. Exceptions are penicillin and penicillin derivatives, which can link covalently and spontaneously to cell membrane proteins[73] and which appear to trigger hapten-dependent antibodies capable of causing thrombocytopenia on rare occasions. As with penicillin-induced immune hemolytic anemia,[73] patients treated with high doses of penicillin for several weeks may be more likely to develop thrombocytopenia.[74,75] Patients with established penicillin sensitivity may be more prone than others to experience DITP after treatment with penicillin derivatives,[76] but this relationship is not fully established.[77] In some patients who develop thrombocytopenia after treatment with penicillin or penicillin derivatives, the responsible antibodies may not be hapten-dependent but similar to those implicated in patients with "quinine-type" DITP (see Section IV.B.2).

**2. Quinine-Type Drug-Induced Immune Thrombocytopenia.** Profound, often life-threatening thrombocytopenia following treatment with quinine has been recognized as a clinical entity since the early 20th century. "Quinine-

**Table 49-1: Mechanisms of Drug-Induced Immune Thrombocytopenia**

| Type | Mechanism | Examples | References |
|---|---|---|---|
| Hapten-induced antibody | Drug binds covalently to membrane glycoprotein and functions as hapten to induce antibody response. | Penicillin | 74 |
| "Quinine-type" thrombocytopenia | Drug binds noncovalently to platelet membrane glycoprotein to produce a "compound" epitope or induce a conformational change for which antibody is specific. | Quinidine, quinine, sulfonamide antibiotics | 84, 86, 89 |
| Autoantibody induction | Drug induces true autoantibodies that bind to platelet membrane glycoproteins without need for added drug. | Gold salts, procaine amide | 99, 103 |
| Immune complex mediated | Drug binds to a normal protein to form immunogenic complexes; antibody reacts with these complexes to form immune complexes that activate platelets *via* Fc receptors. | Heparin | See Chapter 48 |
| Fibrinogen-receptor antagonist (FRA) | Drug binds to the platelet fibrinogen receptor (GPIIb-IIIa) to induce conformational changes (LIBS) that are recognized by antibody. | Tirofiban, eptifibatide | 108, 109, 114 |
| Drug specific | Antibody recognizes a platelet-specific monoclonal antibody bound to its target. | Abciximab | 121, 124, 128 |

type" DITP is associated with a remarkable class of antibodies that react with platelets only when the drug is present in soluble form.[78,79] Although many drugs appear to be capable of inducing such antibodies, quinine, quinidine, and sulfonamide derivatives are more prone to do so than others. The unique features of quinine-type antibodies are (a) they bind tightly to platelet membrane targets at the highest concentration of drug that can be achieved *in vitro*; (b) they generally do not bind to platelets pretreated with the sensitizing drug and then washed; (c) they react with glycoproteins present on human or primate platelets but not with those of other species;[78] and (d) the reaction is reversible — that is, both the antibodies and the drug dissociate from the target if sensitized platelets are suspended in buffer in the absence of the drug.[80] It is apparent that the behavior of these antibodies is quite different from that expected of hapten-dependent antibodies.

How the sensitizing drug promotes binding of a drug-dependent antibody (DDAb) to a specific site on the platelet membrane without covalently linking to the target or the antibody is not yet known. For some years, it was thought that the drug interacts with the antibody to form immune complexes, which somehow associate with platelets and cause their destruction. This hypothetical mechanism was referred to as the "immune complex" or "innocent bystander" concept and is still cited in some publications as the likely explanation for quinine-type immune thrombocytopenia. However, more recent studies have failed to support this concept. Evidence against the immune complex hypothesis includes failure to demonstrate the postulated immune complexes *in vitro,* failure to show that antibody binding is optimal at a particular drug/antibody ratio as would be

expected of an immune complex, and the finding that drug-mediated binding of antibody to its target takes place by way of the Fab immunoglobulin domain rather than Fc.[81,82] Thus, binding of immune complexes to platelets as an explanation for quinine-type DITP is now untenable.

Several alternative mechanisms for binding of DDAb to their targets have been proposed (Fig. 49-1). One holds that the sensitizing drug interacts with one or more platelet membrane glycoproteins to create an epitope, possibly comprised of the drug plus peptide, that is directly recognized by antibody. Weak interaction between the drug and the protein would not be surprising because most drugs that cause quinine-type DITP contain hydrophobic structural elements that could bind to complementary domains on protein. Close interaction between the drug and peptides on the target protein could create a "compound epitope" for which antibody is specific (Fig. 49-1B). This possibility is consistent with the observation that binding of a quinine-dependent antibody to platelets causes trapping of quinine in the antibody-glycoprotein complex and thus resistance of the drug to removal from the sensitized platelets by repeated washing.[80] Another possibility is that drug binding induces conformational changes elsewhere in the target protein that are recognized by antibody (Fig. 49-1A). Alternatively, the drug could react with the Fab region of the antibody and reconfigure it to create specificity for one or more sites on target glycoproteins, but no experimental support has been advanced to support this possibility.

Antibodies induced by quinine or its diastereoisomer, quinidine,[83–85] and many other drugs[86–90] are specific for epitopes on the platelet glycoprotein (GP) Ib-IX complex or the GPIIb-IIIa complex (integrin $\alpha$IIb$\beta$3). Some DDAb react

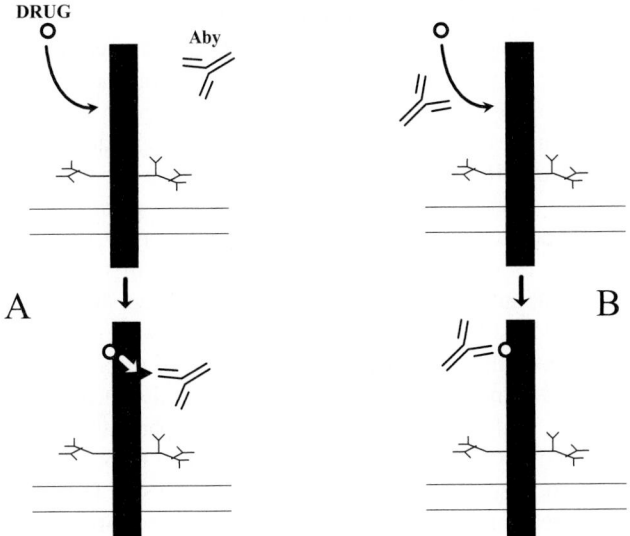

**Figure 49-1.** Two mechanisms by which drugs may promote binding of drug-induced antibodies to platelet membranes. A. Drug binds to glycoprotein (vertical dark bar), possibly through hydrophobic interaction. This induces a conformational change that creates an epitope elsewhere in the glycoprotein for which antibody (Aby) is specific. B. Drug binds to protein, creating a combinatorial epitope (part drug, part protein) for which antibody is specific. Reprinted with permission.[5]

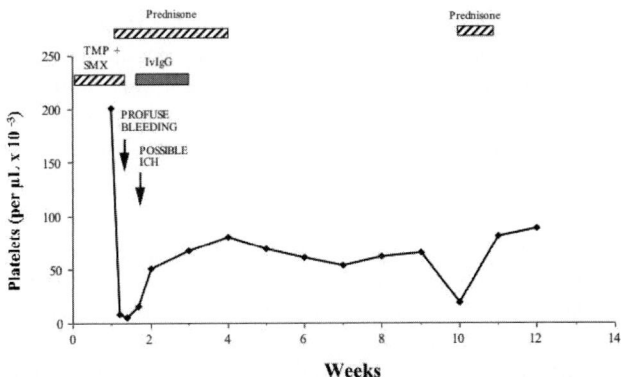

**Figure 49-2.** Chronic autoimmune thrombocytopenia in a patient who presented initially with sulfamethoxazole (SMX)-dependent, platelet-reactive antibodies. SMX-dependent antibodies were identified in acute phase serum together with GPIIb-IIIa-specific non-drug-dependent autoantibodies. Persistent non-drug-dependent antibodies reactive with autologous platelets were identified during weeks 1, 5, and 9. Abbreviations: ICH, intracranial hemorrhage; IVIgG, intravenous gamma globulin; TMP, trimethoprim. Reprinted with permission.[102]

with GPIIb or GPIIIa alone, whereas others require the intact heterodimer for binding.[84] The same appears to be true of antibodies specific for the GPIb-IX complex.[83,85] Most sulfonamide-induced DDAbs recognize only intact GPIIb-IIIa.[86] GPIIb-IIIa is also the preferred target for DDAbs induced by the macrolide antibiotic teicoplanin.[91] Platelet/endothelial cell adhesion molecule-1 (PECAM-1, CD31) has been implicated as a target for drug-dependent antibodies in a group of patients sensitive to the antithyroid drug carbimazole.[92]

An important step toward characterizing the mechanism by which drugs promote binding of DDAb to their targets would be characterization of the precise epitopes for which they are specific. Several studies suggest that a restricted region of GPIX is a favored target.[83,85,87,89,93] Recently, short peptide sequences on GPIbα[94] and GPIIIa[95] were identified as binding sites for quinine-dependent antibodies. Further studies along these lines may enable molecular models of drug-dependent antibody binding to be developed.

**3. Induction of Autoantibodies by Drugs.** Autoimmune hemolytic anemia is a well-characterized and relatively common side effect of the antihypertensive medication alpha methyldopa[96] and occurs occasionally in patients treated with levodopa, procainamide, and other drugs.[97] Whether drugs can cause autoimmune thrombocytopenia (AITP) is less certain. The most convincing evidence for

this possibility comes from studies of patients given gold salts for treatment of rheumatoid arthritis,[98,99] in whom AITP, usually responsive to corticosteroids, splenectomy, and intravenous gamma globulin, occurs with a frequency of about 1%. In such patients, thrombocytopenia usually occurs weeks or months after gold injection[99] but can develop acutely.[100] The associated autoantibodies were originally thought to be specific for GPIIb-IIIa,[99] but a more recent report provides evidence that most of these autoantibodies recognize GPV.[101]

Anecdotal reports of AITP following treatment with levodopa, procainamide, and other drugs have also appeared.[102–104] Patients with DITP and documented drug-dependent antibodies sometimes also make autoantibodies, usually short-lived, that bind to platelets without any requirement for the drug to be present.[105] The clinical course of a patient with sulfamethoxazole-induced thrombocytopenia who developed both transient drug-dependent and sustained autoantibodies after treatment with sulfamethoxazole[102] is illustrated in Fig. 49-2.

How gold salts and other drugs might induce platelet-reactive autoantibodies is unknown. Heavy metals can trigger autoimmunity in certain animal models,[106] and the suggestion has been made that gold might perturb the macrophage processing of platelet membrane glycoproteins in such a way that "cryptic peptides" not ordinarily seen by the immune system are generated and trigger an autoimmune response in some individuals.[102]

**4. Drug-Induced Platelet Destruction by Immune Complexes.** As discussed in Section IV.B.2, it was formerly thought that quinine-type immune thrombocytopenia was caused by immune complexes created by the interaction of antibodies and drugs for which they are specific, but evidence gathered more recently makes this very unlikely. Accordingly, the terms "immune complex" and "innocent bystander" used to describe mechanisms that cause DITP should be discarded. The one clear exception is heparin-induced immune thrombocytopenia (HITP). Although molecular pathogenesis of HITP is still incompletely understood, it is generally agreed that immune complexes created by the interaction of heparin, platelet factor 4 (a basic platelet α-granule protein), and antibody are of key importance in the pathogenesis. HITP is discussed in detail in Chapter 48.

### 5. Thrombocytopenia Induced by Ligand-Mimetic Fibrinogen Receptor Antagonists.

Ligand-mimetic fibrinogen receptor antagonists (FRA) are a group of antithrombotic agents that bind specifically to the RGD recognition site on GPIIb-IIIa and thereby block the binding of fibrinogen to this receptor[107] (Chapter 62). Two FRAs, tirofiban and eptifibatide, are currently approved for clinical use by intravenous infusion. Eptifibatide is a cyclic heptapeptide, and tirofiban is a synthetic ligand-mimetic compound that mimics the tripeptide sequence arginine-glycine-aspartic acid (RGD). Acute, severe thrombocytopenia occurs in 0.1 to 0.5% of patients given these drugs for the first time.[107,108] Systemic symptoms such as chills, fever, and hypotension sometimes accompany the onset of thrombocytopenia.[109–111] Most patients recover uneventfully in a few days, but fatal bleeding has been described.[109,112]

Accumulating evidence indicates that acute platelet destruction in patients given a drug of this class is caused by antibodies that recognize GPIIb-IIIa when complexed with a ligand-mimetic drug.[109,113,114] Remarkably, these antibodies can be naturally occurring, accounting for the observation that platelet destruction can occur within a few hours of the first exposure to one of these medications.[109,114] The molecular basis for the binding of these antibodies to GPIIb-IIIa occupied by a drug of this class is not yet understood. Upon reacting with the integrin, RGD and RGD-mimetic drugs induce conformational changes leading to expression of epitopes (ligand-induced binding sites, or LIBS) recognized by certain murine monoclonal antibodies.[115,116] It has been speculated, but not yet confirmed, that antibodies associated with FRA-induced thrombocytopenia may be human analogs of these LIBS-specific monoclonals.[108,117,118] Why the antibodies are found naturally in a subset of normal individuals and whether they serve any physiologic purpose are interesting unresolved questions. It appears that the incidence of thrombocytopenia in patients given ligand-mimetic agents can be reduced by screening for the presence of antibody before treatment,[114] but whether this would be cost-effective is uncertain.

### 6. Thrombocytopenia Induced by Abciximab.

Abciximab is a chimeric (human/mouse) Fab fragment that is specific for an epitope on GPIIIa close to a fibrinogen recognition site and blocks binding of fibrinogen to the activated integrin[119,120] (Chapter 62). Like the ligand-mimetic GPIIb-IIIa inhibitors, abciximab reduces the incidence of complications following coronary angioplasty. Acute thrombocytopenia occurs within a few hours of starting treatment in about 1% of patients given abciximab for the first time[121,122] and in about 5% of those given the drug a second time.[123] About 15% of patients given the drug twice within 30 days experience this complication.[123] The onset of thrombocytopenia is sometimes accompanied by fever, dyspnea, hypotension, and even anaphylaxis.[108,124] Most patients recover within a few days without incident, although thrombocytopenia occasionally persists for several weeks.[125] However, life-threatening bleeding symptoms, including intracranial hemorrhage, have been described.[124,126]

Studies in patients developing acute, severe thrombocytopenia after a second exposure to abciximab demonstrated the presence of strong IgG or IgM antibodies that react with abciximab-coated platelets.[124] Whether these antibodies actually cause thrombocytopenia was called into question by the finding that similar immunoglobulins are present in normal individuals.[124] However, "dangerous" antibodies capable of causing thrombocytopenia appear to recognize murine sequences incorporated into the chimeric abciximab molecule that confer specificity for GPIIIa, whereas apparently benign antibodies found in normal subjects are specific for the papain cleavage site at the C-terminus of the abciximab Fab fragment.[124] A testing algorithm that enables "dangerous" antibodies to be distinguished from naturally occurring and apparently "benign" antibodies has been described.[124] Why one type of antibody causes thrombocytopenia whereas the other does not is presently unresolved. Acute thrombocytopenia occurring after *first exposure* to abciximab is less well characterized but, like platelet destruction in patients given a ligand-mimetic GPIIb-IIIa antagonist (Section IV.B.5), may be caused by naturally occurring antibodies similar to those found in patients previously treated with the drug.[7,108]

Although abciximab-associated thrombocytopenia usually develops within a few hours of starting treatment, the drop in platelet levels may occur 5 to 10 days later (Fig. 49-3). After being infused, abciximab remains detectable on circulating platelets for up to two weeks, apparently because the drug can transfer from one platelet to another.[127] "Delayed thrombocytopenia" after abciximab appears to be caused by abciximab-specific antibodies newly produced in response to the infused drug, which react with abciximab-coated platelets still present in the circulation and cause their destruction.[128] Because this complication can be life threatening and usually occurs after hospital discharge in patients taking other platelet inhibitors such as aspirin and clopido-

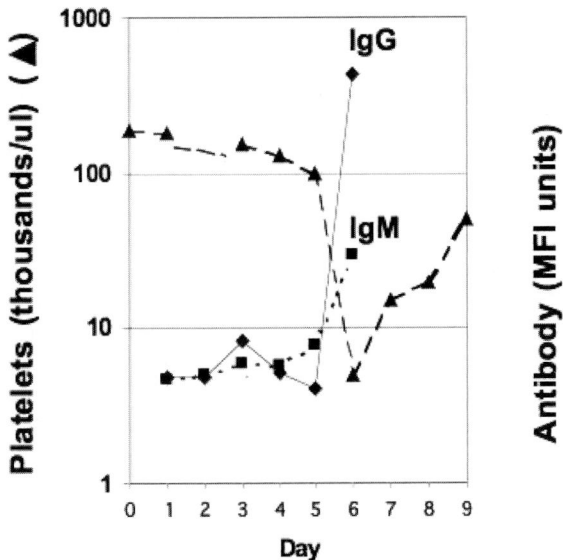

**Figure 49-3.** "Delayed thrombocytopenia" in a patient after treatment with abciximab. Abciximab was administered on day 0. Platelet levels remained above 100,000/µL until day 6, when profound thrombocytopenia and bleeding symptoms developed. At this time, IgG (diamonds) and IgM (squares) antibodies reactive with abciximab-coated platelets became detectable for the first time. Antibody activity (right ordinate) is expressed as mean platelet fluorescence intensity (MFI) in arbitrary units. Reprinted with permission.[128]

## Table 49-2: Criteria Used to Evaluate Causative Relationships in Drug-Induced Thrombocytopenic Purpura

| Criterion | Description |
|---|---|
| 1 | (a) Therapy with the candidate drug preceded thrombocytopenia, and |
| | (b) Recovery from thrombocytopenia was complete and sustained after therapy with the drug was discontinued. |
| 2 | (a) The candidate drug was the only drug used before the onset of thrombocytopenia or |
| | (b) Other drugs were continued or reintroduced after discontinuation of therapy with the candidate drug with a sustained normal platelet count. |
| 3 | Other causes for thrombocytopenia were excluded. |
| 4 | Reexposure to the candidate drug resulted in recurrent thrombocytopenia. |

**Level of Evidence**

| | |
|---|---|
| I | Definite: Criteria 1, 2, 3, and 4 are met. |
| II | Probable: Criteria 1, 2, and 3 are met. |
| III | Possible: Criterion 1 is met. |
| IV | Unlikely: Criterion 1 is not met. |

Reproduced with permission.[4]

grel, it may be prudent to advise patients given abciximab to watch for petechiae and other signs of bleeding after leaving the hospital.[128]

True thrombocytopenia following treatment with abciximab should be distinguished from pseudothrombocytopenia — an *in vitro* artifact caused by the clumping of platelets in blood from an abciximab-treated platelet anticoagulated with EDTA.[129,130] Pseudothrombocytopenia can usually be diagnosed by showing that the platelet count is normal in blood anticoagulated with citrate. It is also useful to demonstrate that platelet aggregates are present in a film prepared from EDTA blood and absent in one prepared from a finger stick. Pseudothrombocytopenia is discussed in detail in Chapter 55.

## C. Drugs Implicated as Causes of Immune Thrombocytopenia

Hundreds of drugs are thought to cause immune thrombocytopenia. In many cases, the association is supported only by one or two case reports and the connection may have been coincidental. George and coworkers[131] undertook a retrospective analysis of all English-language reports of DITP and identified 48 drugs they considered to be "definitely established" as causative on the basis of meeting four criteria (Table 49-2).[4] Fifteen other drugs that met three of these criteria were identified as "probable" causes of thrombocytopenia. Whether antibodies were detected that reacted with platelets in the presence of an implicated drug was not considered as a diagnostic criterion for two reasons: it was felt that available assays were lacking in sensitivity and it was argued that drug-dependent antiplatelet antibodies are sometimes detected in patients who do not develop thrombocytopenia. Since the 1990s or so, methods for detecting platelet-reactive allo-, auto-, and drug-dependent antibodies have improved significantly. Moreover, failure of drug-dependent antibodies to correlate with thrombocytopenia has been described only in patients receiving heparin (Chapter 48) or abciximab (Section IV.B.6). In the opinion of the writer, demonstration of an immunoglobulin that binds to normal platelets in the presence of a suspect drug (other than heparin) provides very strong evidence that thrombocytopenia was drug induced, provided the assay is done by an experienced laboratory. Unfortunately, the converse does not hold — drug-dependent antibodies are sometimes not detected in patients with a history strongly suggestive of DITP. Possible reasons for this are discussed in Section IV.D.

Table 49-3 lists "Level I" and "Level II" drugs identified as "definite" or "probable" causes of thrombocytopenia using the criteria of George and coworkers[131] (Table 49-2)

**Table 49-3: Drugs Implicated as Causes of Immune Thrombocytopenia**

| Medication | Documentation Level I/II[a] | DDAb Detected? | Frequency of Occurrence[b] |
|---|---|---|---|
| Abciximab | I | Yes | +++ |
| Acetaminophen | I | Yes | + |
| Acetylsalicylic acid | — | No | + |
| Alprenolol | I | No | + |
| Aminoglutethimide | I | No | + |
| Aminosalicylic acid | I | No | + |
| Amiodarone | I | No | + |
| Amoxicillin | — | Yes | + |
| Amphotericin B | I | No | + |
| Ampicillin | II | No | + |
| Amrinone | I | No | + |
| Captopril | II | No | + |
| Carbamazepine | II | Yes | ++ |
| Cefazolin | — | Yes | + |
| Cefotetan | — | Yes | + |
| Ceftriaxone | — | Yes | + |
| Cephalothin | I | No | + |
| Chlorothiazide | I | No | ++ |
| Chlorpromazine | I | No | + |
| Chlorpropamide | II | No | + |
| Cimetidine | I | No | + |
| Clopidogrel | — | No | + |
| Danazol | I | No | + |
| Deferoxamine | I | No | + |
| Diatrizoate meglumine (Hypaque) | I | No | + |
| Diazepam | I | Yes | + |
| Diazoxide | I | No | + |
| Diclofenac | I | Yes | + |
| Difluoromethylornithine | I | No | + |
| Digoxin | I | No | + |
| Diltiazem | — | No | + |
| Eptifibatide | — | Yes | ++ |
| Ethambutol | I | No | + |
| Fluconazole | II | No | + |
| Glibenclamide | II | No | + |
| Gold | II | No | ++ |
| Gold salts | — | No | +++ |
| Haloperidol | I | No | + |
| Heparin | — | Yes | +++ |
| Hydrochlorothiazide | II | No | + |
| Ibuprofen | II | Yes | + |
| Interferon-alfa | I | No | + |

**Table 49-3: Drugs Implicated as Causes of Immune Thrombocytopenia—*Continued***

| Medication | Documentation Level I/II[a] | DDAb Detected? | Frequency of Occurrence[b] |
|---|:---:|:---:|:---:|
| Iopanoic acid (Telepaque) | I | No | + |
| Isoniazid | I | No | + |
| Levamisole | I | No | + |
| Levofloxacin | — | Yes | + |
| Lithium | I | No | + |
| Meclofenamate | I | No | + |
| Methicillin | I | No | + |
| Methyldopa | I | No | + |
| Minoxidil | I | No | + |
| Nalidixic acid | I | No | + |
| Naphazoline | I | No | + |
| Naproxen | — | Yes | + |
| Nitroglycerin | I | No | + |
| Novobiocin | I | No | + |
| Oxaprozin | — | Yes | + |
| Oxprenolol | I | No | + |
| Oxyphenbutazone | II | No | + |
| Oxytetracycline | II | No | + |
| Penicillin | — | No | + |
| Phenytoin | II | Yes | + |
| Piperacillin | I | No | + |
| Procainamide | II | No | ++ |
| Quinidine | I | Yes | +++ |
| Quinine | I | Yes | +++ |
| Ranitidine | II | Yes | + |
| Rifampin | I | Yes | ++ |
| Simvastatin | — | No | + |
| Sulfamethoxazole | I | Yes | ++ |
| Sulfasalazine | I | No | + |
| Sulfisoxazole | I | Yes | ++ |
| Sulindac | II | No | + |
| Suramin | — | Yes | + |
| Tamoxifen | I | No | + |
| Thiothixene | I | No | + |
| Tirofiban | — | Yes | ++ |
| Tolmetin | I | No | + |
| Trimethoprim | I | Yes | + |
| Valproic acid | — | No | ++ |
| Vancomycin | I | Yes | ++ |

[a]Criteria of George et al. (1998).[4]
[b]+++, common (>25 cases reported); ++ rare (6–24 cases reported); + very rare (1–5 cases reported).
DDAb, drug-dependent antibody.

and other drugs considered by the author to be established as causes of DITP on the basis of laboratory testing and other criteria. Reports describing DITP induced by many of these drugs are cited in review articles.[4,131–135] A website containing useful reference material can be accessed at http://moon.ouhsc.edu/jgeorge.

Although thrombocytopenia associated with the use of chemotherapeutic and immunosuppressive drugs is usually the result of hematosuppression (Section II.A), evidence suggests that this class of drugs can sometimes cause acute platelet destruction by immune mechanisms.[136–141] Drug-induced immune thrombocytopenia should be considered in patients on chemotherapy who have a sudden drop in platelet levels, especially if the bone marrow contains normal numbers of megakaryocytes.

## D. Sensitivity to Drug Metabolites as a Cause of Immune Thrombocytopenia

Although detection of an antibody reactive with platelets in the presence of a suspect drug provides strong evidence for DITP, it is not uncommon for negative results to be obtained in patients with a strongly suggestive history. One reason for this is that a metabolite of the implicated drug, formed *in vivo*, can be the sensitizing agent, leading to the production of antibodies that can be detected only using that metabolite in the reaction mixture, even when its structure is very similar to the native compound. This has been documented convincingly in a number of patients with drug-induced immune hemolytic anemia.[142–145] In such patients, it is often possible to show that a metabolite is the sensitizing agent by using urine or serum from an individual taking the suspect drug as the source of "metabolite" for *in vitro* testing.[142,146] Fewer studies to characterize drug metabolites capable of inducing DITP have been conducted to date. In published reports, acetaminophen sulfate,[147,148] a glycine conjugate of para-amino salicylic acid (PAS),[149] the N-4 acetyl congener of sulfamethoxazole,[150] a glucuronide conjugate of naproxen,[148] and unidentified metabolites of ibuprofen[151] and diclofenac[152] have been shown to promote binding to platelets of antibodies from patients with drug-induced thrombocytopenia. It is possible that further studies to define responsible drug metabolites could greatly improve laboratory diagnosis in patients with drug-induced immune cytopenias.

## E. Clinical Presentation

Most drugs listed in Table 49-3 are rare causes of thrombocytopenia. Because quinine and sulfamethoxazole are among the most common triggers for quinine-type thrombocytopenia, any patient presenting with unexplained acute

thrombocytopenia should be asked about exposure to these drugs. It is important to recognize that acutely ill patients may not give an accurate account of drug exposure.[153,154] With respect to quinine, inquiries should be made about ingestion of soft drinks, mixers,[155] and aperitifs[156] that contain the drug. Recurrent, severe thrombocytopenia caused by surreptitious intake of quinidine[157] and quinine[158] has been described. Quinine used to cut heroin can cause sensitization and thrombocytopenia in addicts.[159] Even topical medications have been implicated in a few instances.[160,161]

In almost all cases of acute DITP, the patient will have taken the sensitizing drug within 1 day of the onset of symptoms. Gold salts, which are injected in depot form and can remain in the body for long periods, are an exception to this rule. Upon taking the provocative drug, sensitized patients often experience a feeling of warmth, flushing, and, subsequently, chills. Syncope is not uncommon. Acute thrombocytopenia occurs within a few hours of drug ingestion, but bleeding, typically characterized by petechial hemorrhages, purpura, and blood in the urinary and gastrointestinal tracts, may become apparent only after a delay of 12 to 24 hours. Hemorrhagic bullae in the buccal mucosa are indicative of a very low platelet count, usually less than $5 \times 10^9$/L. It is not unusual for a patient with DITP to report to a dentist because of bleeding gums after brushing teeth. If further exposure to the responsible drug is avoided, bleeding symptoms usually subside within 1 or 2 days and platelet counts return to normal within 3 to 4 days. For unknown reasons, thrombocytopenia occasionally persists for 1 or 2 weeks. Patients with drug-induced thrombocytopenia sometimes simultaneously produce drug-dependent antibodies specific for erythrocytes or neutrophils[162,163] and may present with hemolytic anemia or neutropenia in addition to thrombocytopenia. Patients sensitive to quinine[164,165] and perhaps other drugs[166–168] sometimes present with the hemolytic-uremic syndrome (HUS, microangiopathic hemolytic anemia, renal failure, and thrombocytopenia; see Chapter 50). It has been hypothesized that cell-mediated endothelial cell injury may cause the renal dysfunction in such cases.[163] This type of presentation should not be confused with HUS seen after treatment with certain chemotherapeutic and immunosuppressive agents, which appears to have a quite different etiology.[169]

## F. Laboratory Diagnosis

Various assays are capable of identifying antibodies that react with normal platelets in the presence of serum from a patient with drug-induced immune thrombocytopenia,[86,170,171] but these are available only in specialized reference laboratories. Antibodies induced by quinine, quinidine, and sulfamethoxazole are usually easy to detect, but negative results do not rule out a drug-induced etiology. Further studies are needed to determine the extent to which use of drug metabolites in testing (Section IV.D) will increase the diagnostic yield. In general, serologic reactions are highly drug specific and patients with documented DITP can safely be given another drug having the same pharmacologic effect. When a drug not confirmed in the laboratory to be the cause of thrombocytopenia is critical for a patient's welfare, a diagnostic clinical challenge can be undertaken. For safety reasons, however, it is important that very small amounts of the drug be initially given and that the platelet count be carefully monitored.[172] To be meaningful, an *in vivo* challenge should be undertaken as soon as possible after recovery from the thrombocytopenic episode, because drug-dependent antibodies may decline and disappear over time.

### G. Prognosis and Treatment

When a patient suspected of having acute drug-induced immune thrombocytopenia is encountered, drugs currently being taken should be discontinued until the sensitizing drug can be identified. As noted earlier, it is generally safe to substitute drugs with similar pharmacologic actions. Intracranial hemorrhage is a significant risk in patients who present with severe thrombocytopenia and widespread bleeding.[173,174] Fatal intrapulmonary hemorrhage has also been described.[175] Glucocorticoids have not been shown to be helpful but are often given because of the suspicion that the patient may have AITP. It should be apparent to the clinician that a good "response" to corticosteroids may simply be a consequence of stopping the sensitizing drug and does not imply that the patient has AITP. Failure to recognize this can lead to repeated hospital admissions and even unnecessary surgery in patients who actually have DITP.[154]

It is reasonable to give platelet transfusions to patients with severe bleeding symptoms, especially if the suspect drug is a fibrinogen receptor antagonist capable of inhibiting the function of platelets remaining in the circulation. Plasma exchange or intravenous gamma globulin can be considered in acutely ill patients.[176–179] Once established, sensitivity to drugs causing immune thrombocytopenia is usually permanent. Accordingly, patients should avoid reexposure.

# References

1. Garratty, G. (1993). Immune cytopenia associated with antibiotics. *Transf Med Rev, 7,* 255–267.
2. Pedersen-Bjergaard, U., Andersen, M., & Hansen, P. B. (1997). Drug-induced thrombocytopenia: Clinical data on 309 cases and the effect of corticosteroid therapy. *Eur J Clin Pharmacol, 52,* 183–189.
3. Pedersen-Bjergaard, U., Andersen, M., & Hansen, P. B. (1998). Drug-specific characteristics of thrombocytopenia caused by non-cytotoxic drugs. *Eur J Clin Pharmacol, 54,* 701–706.
4. George, J. N., Raskob, G. E., Shah, S. R., et al. (1998). Drug-induced thrombocytopenia: A systematic review of published case reports. *Ann Intern Med, 129,* 886–890.
5. Aster, R. (1997). Response of thrombocytes to toxic injury. In I. G. Sipes, A. J. Gandolfi, & C. A. McQueen (Eds.), *Comprehensive toxicology. Vol. IV* (pp. 263–268). Elsevier Science.
6. Wazny, L. D., & Ariano, R. E. (2000). Evaluation and management of drug-induced thrombocytopenia in the acutely ill patient. *Pharmacotherapy, 20,* 292–307.
7. Aster, R. H. (2005). Drug-induced immune cytopenias. *Toxicology, 209,* 149–153.
8. Carey, P. J. (2003). Drug-induced myelosuppression: Diagnosis and management. *Drug Saf, 26,* 691–706.
9. Blay, J. Y., Le Cesne, A., Mermet, C., et al. (1998). A risk model for thrombocytopenia requiring platelet transfusion after cytotoxic chemotherapy. *Blood, 92,* 405–410.
10. Recombinant interleukin-11 for chemotherapy-induced thrombocytopenia. (1998). *Med Lett Drugs Ther, 40,* 77–78.
11. Cantor, S. B., Elting, L. S., Hudson, D. V., Jr., et al. (2003). Pharmacoeconomic analysis of oprelvekin (recombinant human interleukin-11) for secondary prophylaxis of thrombocytopenia in solid tumor patients receiving chemotherapy. *Cancer, 97,* 3099–3106.
12. Vadhan-Raj, S. (2000). Clinical experience with recombinant human thrombopoietin in chemotherapy-induced thrombocytopenia. *Semin Hematol, 37,* 28–34.
13. Vadhan-Raj, S., Patel, S., Bueso-Ramos, C., et al. (2003). Importance of predosing of recombinant human thrombopoietin to reduce chemotherapy-induced early thrombocytopenia. *J Clin Oncol, 21,* 3158–3167.
14. Stone, R. M., Berg, D. T., George, S. L., et al. (1995). Granulocyte-macrophage colony-stimulating factor after initial chemotherapy for elderly patients with primary acute myelogenous leukemia. Cancer and Leukemia Group B. *N Engl J Med, 332,* 1671–1677.
15. Michel, G., Landman-Parker, J., Auclerc, M. F., et al. (2000). Use of recombinant human granulocyte colony-stimulating factor to increase chemotherapy dose-intensity: A randomized trial in very high-risk childhood acute lymphoblastic leukemia. *J Clin Oncol, 18,* 1517–1524.
16. Newland, A. C. (1996). Is interleukin 3 active in anticancer drug-induced thrombocytopenia? *Cancer Chemother Pharmacol, 38,* Suppl:S83–S88.
17. Phillips, K. A., & Tannock, I. F. (1998). Design and interpretation of clinical trials that evaluate agents that may offer protection from the toxic effects of cancer chemotherapy. *J Clin Oncol, 16,* 3179–3190.
18. Budd, G. T., Ganapathi, R., Wood, L., et al. (1999). Approaches to managing carboplatin-induced thrombocytopenia: Focus on the role of amifostine. *Semin Oncol, 26,* 41–50.
19. Koch, M. A., Volberding, P. A., Lagakos, S. W., et al. (1992). Toxic effects of zidovudine in asymptomatic human immu-

nodeficiency virus-infected individuals with CD4+ cell counts of 0.50×10(9)/L or less: Detailed and updated results from protocol 019 of the AIDS Clinical Trials Group. *Arch Intern Med, 152,* 2286–2292.

20. Kaufman, D. W., Kelly, J. P., Jurgelon, J. M., et al. (1996). Drugs in the aetiology of agranulocytosis and aplastic anaemia. *Eur J Haematol Suppl, 60,* 23–30.

21. Young, N. S. (1996). Immune pathophysiology of acquired aplastic anaemia. *Eur J Haematol Suppl, 60,* 55–59.

22. Wiholm, B. E., & Emanuelsson, S. (1996). Drug-related blood dyscrasias in a Swedish reporting system, 1985–1994. *Eur J Haematol Suppl, 60,* 42–46.

23. Dingli, D., & Tefferi, A. (2005). A critical review of anagrelide therapy in essential thrombocythemia and related disorders. *Leuk Lymphoma, 46,* 641–650.

24. Wang, G., Franklin, R., Hong, Y., et al. (2005). Comparison of the biological activities of anagrelide and its major metabolites in haematopoietic cell cultures. *Br J Pharmacol, 146,* 324–332.

25. Silverstein, M. N., & Tefferi, A. (1999). Treatment of essential thrombocythemia with anagrelide. *Semin Hematol, 36,* 23–25.

26. McCune, J. S., Liles, D., & Lindley, C. (1997). Precipitous fall in platelet count with anagrelide: Case report and critique of dosing recommendations. *Pharmacotherapy, 17,* 822–826.

27. Girard, D. E., Kumar, K. L., & McAfee, J. H. (1987). Hematologic effects of acute and chronic alcohol abuse. *Hematol Oncol Clin North Am, 1,* 321–334.

28. Gewirtz, A. M., & Hoffman, R. (1986). Transitory hypomegakaryocytic thrombocytopenia: Aetiological association with ethanol abuse and implications regarding regulation of human megakaryocytopoiesis. *Br J Haematol, 62,* 333–344.

29. Mikhailidis, D. P., Barradas, M. A., & Jeremy, J. Y. (1990). The effect of ethanol on platelet function and vascular prostanoids. *Alcohol, 7,* 171–180.

30. Aranda, E., Pizarro, M., Pereira, J., et al. (1994). Accumulation of 5-hydroxytryptamine by aging platelets: Studies in a model of suppressed thrombopoiesis in dogs. *Thromb Haemost, 71,* 488–492.

31. Cooper, B. A., & Bigelow, F. S. (1960). Thrombocytopenia associated with the administration of diethylstilbestrol in man. *Ann Intern Med, 52,* 907–909.

32. Onel, K., & Bussel, J. B. (2004). Adverse effects of estrogen therapy in a subset of women with ITP. *J Thromb Haemost, 2,* 670–671.

33. Rakela, J., Wood, J. R., Czaja, A. J., et al. (1990). Long-term versus short-term treatment with recombinant interferon alfa-2a in patients with chronic hepatitis B: A prospective, randomized treatment trial. *Mayo Clin Proc, 65,* 1330–1335.

34. Martin, T. G., & Shuman, M. A. (1998). Interferon-induced thrombocytopenia: Is it time for thrombopoietin? *Hepatology, 28,* 1430–1432.

35. Toccaceli, F., Rosati, S., Scuderi, M., et al. (1998). Leukocyte and platelet lowering by some interferon types during viral hepatitis treatment. *Hepatogastroenterology, 45,* 1748–1752.

36. Ganser, A., Carlo-Stella, C., Greher, J., et al. (1987). Effect of recombinant interferons alpha and gamma on human bone marrow-derived megakaryocytic progenitor cells. *Blood, 70,* 1173–1179.

37. Sevastianos, V. A., Deutsch, M., Dourakis, S. P., et al. (2003). Pegylated interferon-2b-associated autoimmune thrombocytopenia in a patient with chronic hepatitis C. *Am J Gastroenterol, 98,* 706–707.

38. Fujii, H., Kitada, T., Yamada, T., et al. (2003). Life-threatening severe immune thrombocytopenia during alpha-interferon therapy for chronic hepatitis C. *Hepatogastroenterology, 50,* 841–842.

39. Sata, M., Yano, Y., Yoshiyama, Y., et al. (1997). Mechanisms of thrombocytopenia induced by interferon therapy for chronic hepatitis B. *J Gastroenterol, 32,* 206–210.

40. Trannel, T. J., Ahmed, I., & Goebert, D. (2001). Occurrence of thrombocytopenia in psychiatric patients taking valproate. *Am J Psychiatry, 158,* 128–130.

41. Kaufman, K. R., & Gerner. R. (1998). Dose-dependent valproic acid thrombocytopenia in bipolar disorder. *Ann Clin Psychiatry, 10,* 35–37.

42. Eastham, R. D., & Jancar, J. (1980). Sodium valproate and platelet counts. *Br Med J, 280,* 186.

43. May, R. B., & Sunder, T. R. (1993). Hematologic manifestations of long-term valproate therapy. *Epilepsia, 34,* 1098–1101.

44. Brichard, B., Vermylen, C., Scheiff, J. M., et al. (1994). Haematological disturbances during long-term valproate therapy. *Eur J Pediatr, 153,* 378–380.

45. Gesundheit, B., Kirby, M., Lau, W., et al. (2002). Thrombocytopenia and megakaryocyte dysplasia: An adverse effect of valproic acid treatment. *J Pediatr Hematol Oncol, 24,* 589–590.

46. Proulle, V., Masnou, P., Cartron, J., et al. (2000). GPIaIIa as a candidate target for anti-platelet autoantibody occurring during valproate therapy and associated with preoperative bleeding. *Thromb Haemost, 83,* 175–176.

47. Sleiman, C., Raffy, O., Roue, C., et al. (2000). Fatal pulmonary hemorrhage during high-dose valproate monotherapy. *Chest, 117,* 613.

48. Miyazawa, K., Nishimaki, J., Katagiri, T., et al. (2003). Thrombocytopenia induced by imatinib mesylate (Glivec) in patients with chronic myelogenous leukemia: Is 400 mg daily of imatinib mesylate an optimal starting dose for Japanese patients? *Int J Hematol, 77,* 93–95.

49. Ault, P., Kantarjian, H., Welch, M. A., et al. (2004). Interleukin 11 may improve thrombocytopenia associated with imatinib mesylate therapy in chronic myelogenous leukemia. *Leuk Res, 28,* 613–618.

50. Marin, D., Marktel, S., Foot, N., et al. (2003). Granulocyte colony-stimulating factor reverses cytopenia and may permit cytogenetic responses in patients with chronic myeloid leukemia treated with imatinib mesylate. *Haematologica, 88,* 227–229.

51. Attassi, K., Hershberger, E., Alam, R., et al. (2002). Thrombocytopenia associated with linezolid therapy. *Clin Infect Dis, 34,* 695–698.

52. Orrick, J. J., Johns, T., Janelle, J., et al. (2002). Thrombocytopenia secondary to linezolid administration: What is the risk? *Clin Infect Dis, 35,* 348–349.

53. Gerson, S. L., Kaplan, S. L., Bruss, J. B., et al. (2005). Hematologic effects of linezolid: Summary of clinical experience. *Antimicrob Agents Chemother, 46,* 2723–2726.

54. Mannucci, P. M. (1997). Desmopressin (DDAVP) in the treatment of bleeding disorders: The first 20 years. *Blood, 90,* 2515–2521.

55. Casonato, A., Pontara, E., Dannhaeuser, D., et al. (1994). Re-evaluation of the therapeutic efficacy of DDAVP in type IIB von Willebrand's disease. *Blood Coagul Fibrinolysis, 5,* 959–964.

56. Takahashi, H., Okada, K., Abe, S., et al. (1987). Platelet aggregation induced by cryoprecipitate infusion in platelet-type von Willebrand's disease. *Thromb Res, 46,* 255–262.

57. Stratton, J., Warwicker, P., Watkins, S., et al. (2001). Desmopressin may be hazardous in thrombotic microangiopathy. *Nephrol Dial Transplant, 16,* 161–162.

58. Overman, M., & Brass, E. (2004). Worsening of thrombotic thromboctyopenic purpura symptoms associated with desmopressin administration. *Thromb Haemost, 92,* 886–887.

59. Bukowski, R. M., Murthy, S., McLain, D., et al. (1993). Phase I trial of recombinant granulocyte-macrophage colony-stimulating factor in patients with lung cancer: Clinical and immunologic effects. *J Immunother, 13,* 267–274.

60. Tortajada, C., Garcia, F., Miro, J. M., et al. (2000). Severe thrombocytopenia related to granulocyte-macrophage colony-stimulating factor (rHU-GM-CSF). *An Med Interna, 17,* 671.

61. Nash, R. A., Burstein, S. A., Storb, R., et al. (1995). Thrombocytopenia in dogs induced by granulocyte-macrophage colony-stimulating factor: Increased destruction of circulating platelets. *Blood, 86,* 1765–1775.

62. Abramsm, K., Yunusov, M. Y., Slichter, S., et al. (2003). Recombinant human macrophage colony-stimulating factor-induced thrombocytopenia in dogs. *Br J Haematol, 121,* 614–622.

63. Baker, G. R., & Levin, J. (1998). Transient thrombocytopenia produced by administration of macrophage colony-stimulating factor: Investigations of the mechanism. *Blood, 91,* 89–99.

64. Michelmann, I., Bockmann, D., Nurnberger, W., et al. (1997). Thrombocytopenia and complement activation under recombinant TNF alpha/IFN gamma therapy in man. *Ann Hematol, 74,* 179–184.

65. Paciucci, P. A., Mandeli, J., Oleksowicz, L., et al. (1990). Thrombocytopenia during immunotherapy with interleukin-2 by constant infusion. *Am J Med, 89,* 308–312.

66. Fleischmann, J. D., Shingleton, W. B., Gallagher, C., et al. (1991). Fibrinolysis, thrombocytopenia, and coagulation abnormalities complicating high-dose interleukin-2 immunotherapy. *J Lab Clin Med, 117,* 76–82.

67. Gringeri, A., Santagostino, E., Tradati, F., et al. (1991). Adverse effects of treatment with porcine factor VIII. *Thromb Haemost, 65,* 245–247.

68. Chang, H., Mody, M., Lazarus, A. H., et al. (1998). Platelet activation induced by porcine factor VIII (Hyate:C). *Am J Hematol, 57,* 200–205.

69. Wakefield, T. W., Bouffard, J. A., Spaulding, S. A., et al. (1987). Sequestration of platelets in the pulmonary circulation as a consequence of protamine reversal of the anticoagulant effects of heparin. *J Vasc Surg, 5,* 187–193.

70. Al-Mondhiry, H., Pierce, W. S., & Basarab, R. M. (1985). Protamine-induced thrombocytopenia and leukopenia. *Thromb Haemost, 53,* 60–64.

71. Kinney, E. L., Ballard, J. O., Carlin, B., et al. (1983). Amrinone-mediated thrombocytopenia. *Scand J Haematol, 31,* 376–380.

72. Ross, M. P., Allen-Webb, E. M., Pappas, J. B., et al. (1993). Amrinone-associated thrombocytopenia: Pharmacokinetic analysis. *Clin Pharmacol Ther, 53,* 661–667.

73. Garratty, G., & Petz, L. D. (1975). Drug-induced immune hemolytic anemia. *Am J Med, 58,* 398–407.

74. Murphy, M. F., Riordan, T., Minchinton, R. M., et al. (1983). Demonstration of an immune-mediated mechanism of penicillin-induced neutropenia and thrombocytopenia. *Br J Haematol, 55,* 155–160.

75. Lang, R., Lishner, M., & Ravid, M. (1991). Adverse reactions to prolonged treatment with high doses of carbenicillin and ureidopenicillins. *Rev Infect Dis, 13,* 68–72.

76. Parker, J. C., & Barrett, D. A., 2nd. (1971). Microangiopathic hemolysis and thrombocytopenia related to penicillin drugs. *Arch Intern Med, 127,* 474–477.

77. Christie, D. J., Lennon, S. S., Drew, R. L., et al. (1998). Cefotetan-induced immunologic thrombocytopenia. *Br J Haematol, 70,* 423–426.

78. Shulman, N. R., & Reid, D. M. (1993). Mechanisms of drug-induced immunologically mediated cytopenias. *Transfus Med Rev, 7,* 215–229.

79. Aster, R. H. (1999). Drug-induced immune thrombocytopenia: An overview of pathogenesis. *Semin Hematol, 36,* 2–6.

80. Christie, D. J., & Aster, R. H. (1982). Drug-antibody-platelet interaction in quinine- and quinidine-induced thrombocytopenia. *J Clin Invest, 70,* 989–998.

81. Christie, D. J., Mullen, P. C., & Aster, R. H. (1985). Fab-mediated binding of drug-dependent antibodies to platelets in quinidine- and quinine-induced thrombocytopenia. *J Clin Invest, 75,* 310–314.

82. Smith, M. E., Reid, D. M., Jones, C. E., et al. (1987). Binding of quinine- and quinidine-dependent drug antibodies to platelets is mediated by the Fab domain of the immunoglobulin G and is not Fc dependent. *J Clin Invest, 79,* 912–917.

83. Chong, B. H., Du, X. P., Berndt, M. C., et al. (1991). Characterization of the binding domains on platelet glycoproteins Ib-IX and IIb/IIIa complexes for the quinine/quinidine-dependent antibodies. *Blood, 77,* 2190–2199.

84. Visentin, G. P., Newman, P. J., & Aster, R. H. (1991). Characteristics of quinine- and quinidine-induced antibodies specific for platelet glycoproteins IIb and IIIa. *Blood, 77,* 2668–2676.

85. Lopez, J. A., Li, C. Q., Weisman, S., et al. (1995). The glycoprotein Ib-IX complex-specific monoclonal antibody SZ1

binds to a conformation-sensitive epitope on glycoprotein IX: Implications for the target antigen of quinine/quinidine-dependent autoantibodies. *Blood, 85,* 1254–1258.

86. Curtis, B. R., McFarland, J. G., Wu, G. G., et al. (1994). Antibodies in sulfonamide-induced immune thrombocytopenia recognize calcium-dependent epitopes on the glycoprotein IIb/IIIa complex. *Blood, 84,* 176–183.

87. Gentilini, G., Curtis, B. R., & Aster, R. H. (1998). An antibody from a patient with ranitidine-induced thrombocytopenia recognizes a site on glycoprotein IX that is a favored target for drug-induced antibodies. *Blood, 92,* 2359–2365.

88. Pereira, J., Hidalgo, P., Ocqueteau, M., et al. (2000). Glycoprotein Ib/IX complex is the target in rifampicin-induced immune thrombocytopenia. *Br J Haematol, 110,* 907–910.

89. Asvadi, P., Ahmadi, Z., & Chong, B. H. (2003). Drug-induced thrombocytopenia: Localization of the binding site of GPIX-specific quinine-dependent antibodies. *Blood, 102,* 1670–1677.

90. Grossjohann, B., Eichler, P., Greinacher, A., et al. (2004). Ceftriaxone causes drug-induced immune thrombocytopenia and hemolytic anemia: Characterization of targets on platelets and red blood cells. *Transfusion, 44,* 1033–1040.

91. Garner, S. F., Campbell, K., Smith, G., et al. (2005). Teicoplanin-dependent antibodies: Detection and characterization. *Br J Haematol, 129,* 279–281.

92. Kroll, H., Sun, Q. H., & Santoso, S. (2000). Platelet endothelial cell adhesion molecule-1 (PECAM-1) is a target glycoprotein in drug-induced thrombocytopenia. *Blood, 96,* 1409–1414.

93. Burgess, J. K., Lopez, J. A., Gaudry, L. E., et al.. (2000). Rifampicin-dependent antibodies bind a similar or identical epitope to glycoprotein IX-specific quinine-dependent antibodies. *Blood, 95,* 1988–1992.

94. Burgess, J. K., Lopez, J. A., Berndt, M. C., et al. (1998). Quinine-dependent antibodies bind a restricted set of epitopes on the glycoprotein Ib-IX complex: Characterization of the epitopes. *Blood, 92,* 2366–2373.

95. Peterson, J. A., Nyree, C. E., Newman, P. J., et al. (2003). A site involving the "hybrid" and PSI homology domains of GPIIIa (beta 3-integrin subunit) is a common target for antibodies associated with quinine-induced immune thrombocytopenia. *Blood, 101,* 937–942.

96. Worlledge, S. M. (1973). Immune drug-induced hemolytic anemias. *Semin Hematol, 10,* 327–344.

97. Petz, L. D. (1993). Drug-induced autoimmune hemolytic anemia. *Transfus Med Rev, 7,* 242–254.

98. Adachi, J. D., Bensen, W. G., Kassam, Y., et al. (1987). Gold induced thrombocytopenia: twelve cases and a review of the literature. *Semin Arthritis Rheum, 16,* 287–293.

99. von dem Borne, A. E., Pegels, J. G., van der Stadt, R. J., et al. (1986). Thrombocytopenia associated with gold therapy: A drug-induced autoimmune disease? *Br J Haematol, 63,* 509–516.

100. Levin, M. D., van T Veer, M. B., de Veld, J. C., et al. (2003). Two patients with acute thrombocytopenia following gold administration and five-year follow-up. *Neth J Med, 61,* 223–225.

101. Garner, S. F., Campbell, K., Metcalfe, P., et al. (2002). Glycoprotein V: The predominant target antigen in gold-induced autoimmune thrombocytopenia. *Blood, 100,* 344–346.

102. Aster, R. H. (2000). Can drugs cause autoimmune thrombocytopenic purpura? *Semin Hematol, 37,* 229–238.

103. Landrum, E. M., Siegert, E. A., Hanlon, J. T., et al. (1994). Prolonged thrombocytopenia associated with procainamide in an elderly patient. *Ann Pharmacother, 28,* 1172–1176.

104. Giner, V., Rueda, D., Salvador, A., et al. (2003). Thrombocytopenia associated with levodopa treatment. *Arch Intern Med, 163,* 735–736.

105. Lerner, W., Caruso, R., Faig, D., et al. (1985). Drug-dependent and non-drug-dependent antiplatelet anti-body in drug-induced immunologic thrombocytopenic purpura. *Blood, 66,* 306–311.

106. Griem, P., & Gleichmann, E. (1995). Metal ion induced autoimmunity. *Curr Opin Immunol, 7,* 831–838.

107. Topol, E. J., Byzova, T. V., & Plow, E. F. (1999). Platelet GPIIb-IIIa blockers. *Lancet, 353,* 227–231.

108. Aster, R. H. (2005). Immune thrombocytopenia caused by glycoprotein IIb/IIIa inhibitors. *Chest, 127,* 53S–59S.

109. Bougie, D. W., Wilker, P. R., Wuitschick, E. D., et al. (2002). Acute thrombocytopenia after treatment with tirofiban or eptifibatide is associated with antibodies specific for ligand-occupied GPIIb/IIIa. *Blood, 100,* 2071–2076.

110. Rezkalla, S. H., Hayes, J. J., Curtis, B. R., et al. (2003). Eptifibatide-induced acute profound thrombocytopenia presenting as refractory hypotension. *Catheter Cardiovasc Interv, 58,* 76–79.

111. Morel, O., Jesel, L., Chauvin, M., et al. (2003). Eptifibatide-induced thrombocytopenia and circulating procoagulant platelet-derived microparticles in a patient with acute coronary syndrome. *J Thromb Haemost, 1,* 2685–2687.

112. Coons, J. C., Barcelona, R. A., Freedy, T., et al. (2005). Eptifibatide-associated acute, profound thrombocytopenia. *Ann Pharmacother, 39,* 368–372.

113. Brassard, J. A., Curtis, B. R., Cooper, R. A., et al. (2002). Acute thrombocytopenia in patients treated with the oral glycoprotein IIb/IIIa inhibitors xemilofiban and orbofiban: Evidence for an immune etiology. *Thromb Haemost, 88,* 892–897.

114. Seiffert, D., Stern, A. M., Ebling, W., et al. (2003). Prospective testing for drug-dependent antibodies reduces the incidence of thrombocytopenia observed with the small molecule glycoprotein IIb/IIIa antagonist roxifiban: Implications for the etiology of thrombocytopenia. *Blood, 101,* 58–63.

115. Frelinger, A. L., 3rd, Du, X. P., Plow, E. F., et al. (1991). Monoclonal antibodies to ligand-occupied conformers of integrin alpha IIb beta 3 (glycoprotein IIb-IIIa) alter receptor affinity, specificity, and function. *J Biol Chem, 266,* 17106–17111.

116. Jennings, L. K., & White, M. M. (1998). Expression of ligand-induced binding sites on glycoprotein IIb/IIIa complexes and the effect of various inhibitors. *Am Heart J, 135,* S179–S183.

117. Madan, M., & Berkowitz, S. D. (1999). Understanding thrombocytopenia and antigenicity with glycoprotein IIb-IIIa inhibitors. *Am Heart J, 138,* 317–326.

118. Cines, D. B. (1998). Glycoprotein IIb/IIIa antagonists: Potential induction and detection of drug-dependent antiplatelet antibodies. *Am Heart J, 135,* S152–S159.

119. Coller, B. S. (1997). GPIIb/IIIa antagonists: Pathophysiologic and therapeutic insights from studies of c7E3 Fab. *Thromb Haemost, 78,* 730–735.

120. Artoni, A., Li, J., Mitchell, B., et al. (2004). Integrin beta3 regions controlling binding of murine mAb 7E3: Implications for the mechanism of integrin alphaIIbbeta3 activation. *Proc Natl Acad Sci USA, 101,* 13114–13120.

121. Berkowitz, S. D., Sane, D. C., Sigmon, K. N., et al. (1998). Occurrence and clinical significance of thrombocytopenia in a population undergoing high-risk percutaneous coronary revascularization: Evaluation of c7E3 for the Prevention of Ischemic Complications (EPIC) Study Group. *J Am Coll Cardiol, 32,* 311–319.

122. Jubelirer, S. J., Koenig, B. A., & Bates, M. C. (1999). Acute profound thrombocytopenia following C7E3 Fab (Abciximab) therapy: Case reports, review of the literature and implications for therapy. *Am J Hematol, 61,* 205–208.

123. Dery, J. P., Braden, G. A., Lincoff, A. M., et al. (2004). Final results of the ReoPro readministration registry. *Am J Cardiol, 93,* 979–984.

124. Curtis, B. R., Swyers, J., Divgi, A., et al. (2002). Thrombocytopenia after second exposure to abciximab is caused by antibodies that recognize abciximab-coated platelets. *Blood, 99,* 2054–2059.

125. Lown, J. A., Hughes, A. S., & Cannell, P. (2004). Prolonged profound abciximab associated immune thrombocytopenia complicated by transient multispecific platelet antibodies. *Heart, 90,* e55.

126. Moshiri, S., Di Mario, C., Liistro, F., et al. (2001). Severe intracranial hemorrhage after emergency carotid stenting and abciximab administration for postoperative thrombosis. *Catheter Cardiovasc Interv, 53,* 225–228.

127. Mascelli, M. A., Lance, E. T., Damaraju, L., et al. (1998). Pharmacodynamic profile of short-term abciximab treatment demonstrates prolonged platelet inhibition with gradual recovery from GP IIb/IIIa receptor blockade. *Circulation, 97,* 1680–1688.

128. Curtis, B. R., Divgi, A., Garritty, M., et al. (2004). Delayed thrombocytopenia after treatment with abciximab: A distinct clinical entity associated with the immune response to the drug. *J Thromb Haemost, 2,* 985–992.

129. Christopoulosm, C. G., & Machin, S. J. (1994). A new type of pseudothrombocytopenia: EDTA-mediated agglutination of platelets bearing Fab fragments of a chimaeric antibody. *Br J Haematol, 87,* 650–652.

130. Sane, D. C., Damaraju, L. V., Topol, E. J., et al. (2000). Occurrence and clinical significance of pseudothrombocytopenia during abciximab therapy. *J Am Coll Cardiol, 36,* 75–83.

131. George, J., & Rizva, M. (2005). Thrombocytopenia. In E. Beutler, M. Lichtman, B. Coller, T. Kipps, & U. Seligsohn (Eds.), *Williams hematology* (6th ed., pp. 1479–1494). New York: McGraw Hill.

132. Hibbard, A. B., Medina, P. J., & Vesely, S. K. (2003). Reports of drug-induced thrombocytopenia. *Ann Intern Med, 138,* 239.

133. Rizvi, M. A., Kojouri, K., & George, J. N. (2001). Drug-induced thrombocytopenia: An updated systematic review. *Ann Intern Med, 134,* 346.

134. Majhail, N. S., & Lichtin, A. E. (2002). What is the best way to determine if thrombocytopenia in a patient on multiple medications is drug-induced? *Cleve Clin J Med, 69,* 259–262.

135. Li, X., Hunt, L., & Vesely, S. K. (2005). Drug-induced thrombocytopenia: An updated systematic review. *Ann Intern Med, 142,* 474–475.

136. Tisdale, J. F., Figg, W. D., Reed, E., et al. (1996). Severe thrombocytopenia in patients treated with suramin: Evidence for an immune mechanism in one. *Am J Hematol, 51,* 152–157.

137. Bozec, L., Bierling, P., Fromont, P., et al. (1998). Irinotecan-induced immune thrombocytopenia. *Ann Oncol, 9,* 453–455.

138. Fernandez, M. J., Llopis, I., Pastor, E., et al. (2003). Immune thrombocytopenia induced by fludarabine successfully treated with rituximab. *Haematologica, 88,* ELT02.

139. Khatua, S., Nair, C. N., & Ghosh, K. (2004). Immune-mediated thrombocytopenia following dactinomycin therapy in a child with alveolar rhabdomyosarcoma: The unresolved issues. *J Pediatr Hematol Oncol, 26,* 777–779.

140. Pamuk, G. E., Donmez, S., Turgut, B., et al. (2005). Rituximab-induced acute thrombocytopenia in a patient with prolymphocytic leukemia. *Am J Hematol, 78,* 81.

141. Curtis, B. R. (in press). Immune-mediated thrombocytopenia resulting from sensitivity to oxaliplatin. *Am J Hematol.*

142. Salama, A., Mueller-Eckhardt, C., Kissel, K., et al. (1984). *Ex vivo* antigen preparation for the serological detection of drug-dependent antibodies in immune haemolytic anaemias. *Br J Haematol, 58,* 525–531.

143. Salama, A., Gottsche, B., & Mueller-Eckhardt, C. (1991). Autoantibodies and drug- or metabolite-dependent antibodies in patients with diclofenac-induced immune haemolysis. *Br J Haematol, 77,* 546–549.

144. Bougie, D., Johnson, S. T., Weitekamp, L. A., et al. (1997). Sensitivity to a metabolite of diclofenac as a cause of acute immune hemolytic anemia. *Blood, 90,* 407–413.

145. Cunha, P. D., Lord, R. S., Johnson, S. T., et al. (2000). Immune hemolytic anemia caused by sensitivity to a metabolite of etodolac, a nonsteroidal anti-inflammatory drug. *Transfusion, 40,* 663–668.

146. Mueller-Eckhardt, C., & Salama, A. (1990). Drug-induced immune cytopenias: A unifying pathogenetic concept with special emphasis on the role of drug metabolites. *Transfus Med Rev, 4,* 69–77.

147. Eisner, E. V., & Shahidi, N. T. (1992). Immune thrombocytopenia due to a drug metabolite. *N Engl J Med, 287,* 376–381.

148. Bougie, D., & Aster, R. (2001). Immune thrombocytopenia resulting from sensitivity to metabolites of naproxen and acetaminophen. *Blood, 97,* 3846–3850.

149. Eisner, E. V., & Kasper, K. (1972). Immune thrombocytopenia due to a metabolite of para-aminosalicylic acid. *Am J Med, 53,* 790–796.

150. Kiefel, V., Santoso, S., Schmidt, S., et al. (1987). Metabolite-specific (IgG) and drug-specific antibodies (IgG, IgM) in two cases of trimethoprim-sulfamethoxazole-induced immune thrombocytopenia. *Transfusion, 27,* 262–265.

151. Meyer, T., Herrmann, C., Wiegand, V., et al. (1993). Immune thrombocytopenia associated with hemorrhagic diathesis due to ibuprofen administration. *Clin Invest, 71,* 413–415.

152. Meyer, O., Hoffmann, T., Aslan, T., et al. (2003). Diclofenac-induced antibodies against RBCs and platelets: Two case reports and a concise review. *Transfusion, 43,* 345–349.

153. Kojouri, K., Perdue, J. J., Medina, P. J., et al. (2000). Occult quinine-induced thrombocytopenia. *J Okla State Med Assoc, 93,* 519–521.

154. Reddy, J. C., Shuman, M. A., & Aster, R. H. (2004). Quinine/quinidine-induced thrombocytopenia: A great imitator. *Arch Intern Med, 164,* 218–220.

155. Belkin, G. A. (1967). Cocktail purpura: An unusual case of quinine sensitivity. *Ann Intern Med, 66,* 583–586.

156. Siroty, R. R. (1976). Purpura on the rocks — with a twist. *JAMA, 235,* 2521–2522.

157. Reid, D. M., & Shulman, N. R. (1988). Drug purpura due to surreptitious quinidine intake. *Ann Intern Med, 108,* 206–208.

158. Abraham, R., & Whitehead, S. (1998). Factitious quinine-induced thrombocytopenia. *Med J Aust, 168,* 19–20.

159. Christie, D. J., Walker, R. H., Kolins, M. D., et al. (1983). Quinine-induced thrombocytopenia following intravenous use of heroin. *Arch Intern Med, 143,* 1174–1175.

160. Khaleeli, A. A. (1976). Quinaband-induced thrombocytopenic purpura in a patient with myxoedema coma. *Br Med J, 2,* 562–563.

161. Maloley, P. A., Nelson, E., Montgomery, H. A., et al. (1990). Severe reversible thrombocytopenia resulting from butoconazole cream. *Dicp, 24,* 143–144.

162. Zeigler, Z., Shadduck, R. K., Winkelstein, A., et al. (1979). Immune hemolytic anemia and thrombocytopenia secondary to quinidine: *In vitro* studies of the quinidine-dependent red cell and platelet antibodies. *Blood, 53,* 396–402.

163. Stroncek, D. F., Vercellotti, G. M., Hammerschmidt, D. E., et al.. (1992). Characterization of multiple quinine-dependent antibodies in a patient with episodic hemolytic uremic syndrome and immune agranulocytosis. *Blood, 80,* 241–248.

164. Gottschall, J. L., Elliot, W., Lianos, E., et al. (1991). Quinine-induced immune thrombocytopenia associated with hemolytic uremic syndrome: A new clinical entity. *Blood, 77,* 306–310.

165. Kojouri, K., Vesely, S. K., & George, J. N. (2001). Quinine-associated thrombotic thrombocytopenic purpurahemolytic uremic syndrome: Frequency, clinical features, and long-term outcomes. *Ann Intern Med, 135,* 1047–1051.

166. Wolf, B., Conradty, M., Grohmann, R., et al. (1989). A case of immune complex hemolytic anemia, thrombocytopenia, and acute renal failure associated with doxepin use. *J Clin Psychiatry, 50,* 99–100.

167. Juang, Y. C., Tsao, T. C., Chiang, Y. C., et al. (1992). Acute renal failure and severe thrombocytopenia induced by rifampicin: Report of a case. *J Formos Med Assoc, 91,* 475–476.

168. Coates, A. S., Childs, A., Cox, K., et al. (1992). Severe vascular adverse effects with thrombocytopenia and renal failure following emetogenic chemotherapy and ondansetron. *Ann Oncol, 3,* 719–722.

169. Dlott, J. S., Danielson, C. F., Blue-Hnidy, D. E., et al. (2004). Drug-induced thrombotic thrombocytopenic purpura/hemolytic uremic syndrome: A concise review. *Ther Apher Dial, 8,* 102–111.

170. Visentin, G. P., Wolfmeyer, K., Newman, P. J., et al. (1990). Detection of drug-dependent, platelet-reactive antibodies by antigen-capture ELISA and flow cytometry. *Transfusion, 30,* 694–700.

171. McFarland, J. G. (1993). Laboratory investigation of drug-induced immune thrombocytopenias. *Transfus Med Rev, 7,* 275–287.

172. Shulman, N. R. (1958). Immunoreactions involving platelets. IV. Studies on the pathogenesis of thrombocytopenia in drug purpura using test doses of quinidine in sensitized individuals; their implications in idiopathic thrombocytopenic purpura. *J Exp Med, 107,* 711–729.

173. Glass, J. T., Williams, J. P., Mankad, V. N., et al. (1989). Intracranial hemorrhage associated with quinidine induced thrombocytopenia. *Ala Med, 59,* 21, 24–25.

174. Freiman, J. P. (1990). Fatal quinine-induced thrombocytopenia. *Ann Intern Med, 112,* 308–309.

175. Fireman, Z., Yust, I., & Abramov, A. L. (1981). Lethal occult pulmonary hemorrhage in drug-induced thrombocytopenia. *Chest, 79,* 358–359.

176. Howrie, D. L., Schwinghammer, T. L., & Wollman, M. (1989). Use of i.v. immune globulin for treatment of phenytoin-induced thrombocytopenia. *Clin Pharm, 8,* 734–737.

177. Ray, J. B., Brereton, W. F., & Nullet, F. R. (1990). Intravenous immune globulin for the treatment of presumed quinidine-induced thrombocytopenia. *Dicp, 24,* 693–695.

178. Pourrat, O. (1994). Treatment of drug-related diseases by plasma exchanges. *Ann Med Interne* (Paris), *145,* 357–360.

179. Herrington, A., Mahmood, A., & Berger, R. (1994). Treatment options in sulfamethoxazole-trimethoprim-induced thrombocytopenic purpura. *South Med J, 87,* 948–950.

# CHAPTER 50

# Thrombotic Thrombocytopenic Purpura and the Hemolytic-Uremic Syndrome

**Joel L. Moake**

*Department of Medicine, Baylor College of Medicine and Biomedical Engineering Laboratory,*
*Rice University, Houston, Texas*

## I. Introduction

The combination of thrombocytopenia, hemolysis, and erythrocyte fragmentation (schistocytosis) is characteristic of thrombotic thrombocytopenic purpura (TTP) and the hemolytic-uremic syndrome (HUS). Systemic (in TTP) or predominantly renal (in HUS) microvascular occlusions are the essential pathophysiologic events in these disorders; consequently, the entities are often called "thrombotic microangiopathies."

## II. Clinical Presentations of the Thrombotic Microangiopathies

*TTP* is the most severe thrombotic microangiopathy. It is characterized by systemic platelet aggregation, organ ischemia, profound thrombocytopenia (with increased bone marrow megakaryocytes), and fragmentation of erythrocytes (Fig. 50-1).[1] The red blood cell fragmentation occurs, presumably, as blood flows through turbulent areas of the microcirculation partially occluded by platelet aggregates. Schistocytes, or "split" red cells, appear in the peripheral blood smear (>1% of total red cells)[2] as an indication of the microangiopathic hemolytic anemia. Serum levels of lactate dehydrogenase (LDH) are extremely elevated, mostly as a consequence of hemolysis and the leakage of LDH from ischemic or necrotic tissue cells.[3]

The systemic platelet clumping in TTP is often associated with blood platelet counts below $20 \times 10^9$/L. Occlusive ischemia of the brain or the gastrointestinal tract is common, and renal dysfunction may occur. In the past, a pentad of signs and symptoms was often associated with TTP: thrombocytopenia, microangiopathic hemolytic anemia, neurologic abnormalities, renal failure, and fever.[4] In current clinical practice, thrombocytopenia, schistocytosis, and an impressively elevated serum LDH value are sufficient to suggest the diagnosis of TTP.[1] Coagulation studies are usually normal.

*HUS* is a thrombotic microangiopathy that predominantly involves the glomerular microvasculature and causes acute renal failure along with thrombocytopenia, schistocytosis, and elevated serum LDH.

A type of thrombotic microangiopathy with clinical similarities to either HUS or TTP occurs in some patients weeks to months after their exposure to the following: *mitomycin C;* inhibitors of the $Ca^{2+}$-activated phosphatase, calcineurin *(cyclosporine or tacrolimus [FK 506]); quinine;* combinations of *chemotherapeutic agents; total-body irradiation;* or allogeneic bone marrow, kidney, liver, heart, or lung *transplantation.*[5–8] The microvascular thrombi in these entities may be either predominantly renal or systemic.

## III. Thrombotic Thrombocytopenic Purpura

In 1924, Dr. Eli Moschcowitz of New York City described the abrupt onset and rapid progression of petechiae, pallor, paralysis, coma, and death in a 16-year-old girl.[9,10] Terminal arterioles and capillaries were occluded in this unfortunate teenager by "hyaline" thrombi. In later years, these were determined to consist mostly of platelets and to be unassociated with perivascular inflammation or overt endothelial cell damage or desquamation. Moschcowitz suspected a "powerful poison which had both agglutinative and hemolytic properties"[10] as the cause of this disastrous illness, which he believed to be a new disease.

### A. *Types of TTP*

TTP is often associated with defective function of the plasma von Willebrand factor (VWF)-cleaving metalloprotease (ADAMTS-13). These types of the disorder include familial, acquired idiopathic, thienopyridine-related, and pregnancy-associated TTP (Table 50-1).

*Familial TTP* is rare. Study of this disease has been, nevertheless, of extraordinary importance. It usually (but not always) appears initially in infancy or childhood and

PLATELETS

## Table 50-1: ADAMTS-13 Deficiency and TTP

| Cause of Absent ADAMTS-13 Plasma Activity[a] | Clinical Presentation |
|---|---|
| *ADAMTS-13 mutations* | *Familial TTP; chronic relapsing TTP[b]* |
| *Autoantibodies against ADAMTS-13* | *Acquired idiopathic TTP* |
| Transient | Single episode TTP |
| Recurrent | Recurrent (intermittent) TTP |
| Thienopyridine-associated (ticlopidine/clopidogrel) | Thienopyridine-associated TTP |
| *ADAMTS-13 transient production or survival (?) defect* | Acquired idiopathic TTP (some patients?) |
| | Pregnancy-associated TTP[c] |

[a]<5% of normal. [b]Usually begins in infancy/childhood, but may be delayed. [c]Autoantibodies may also be present.

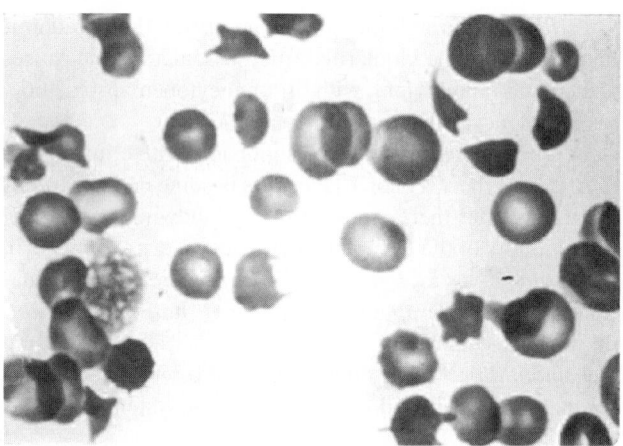

**Figure 50-1.** "Split" red cells (schistocytes) in the peripheral blood smear of a patient with TTP. Schistocytes are characteristic of TTP, HUS, and other thrombotic microangiopathies.

then recurs in some individuals as "chronic relapsing TTP" episodes at regular intervals of about 3 weeks.[1,11–13] In contrast, *acquired idiopathic ("out-of-the-blue") TTP* has become a commonly recognized disorder that occurs in adults and older children.[1,14–17] Following successful treatment of acquired idiopathic TTP, recurrent episodes at irregular intervals occur in 11 to 36% of patients. A small fraction of patients treated for arterial thrombosis with the platelet adenosine diphosphate (ADP) P2Y$_{12}$ receptor-inhibiting *thienopyridine drugs, ticlopidine* (Ticlid) or *clopidogrel* (Plavix), develop TTP within a few weeks after the initiation or therapy (see also Chapter 61).[1,18–21] TTP occurs occasionally in pregnancy (especially the last trimester) or immediately postpartum.[22–25]

### B. Etiology and Pathophysiology of the ADAMTS-13-Deficient Types of TTP

In 1982, *"unusually large" (UL) VWF multimers* were found in plasma samples taken repeatedly from four patients with chronic relapsing TTP. The ULVWF multimers were proposed to be the "agglutinative" substances in this rare disorder.[12] This report concluded that patients with chronic relapsing TTP have a defect in the "processing" of ULVWF multimers that makes them susceptible to periodic relapses.[12] Convincing evidence for ULVWF multimers as the cause of platelet clumping in TTP has accumulated in the subsequent years.[11,16,26–29]

Brief comments on VWF biochemistry and physiology are included here for orientation. Monomers of VWF (280,000 Daltons) are linked by disulfide bonds into multimers with varying molecular masses that range into the millions of Daltons.[30] Multimers of VWF are constructed within megakaryocytes and endothelial cells and are stored within platelet α granules and endothelial cell Weibel–Palade bodies. Most plasma VWF multimers are derived from endothelial cells. Both endothelial cells and platelets produce VWF multimers larger than the multimers in normal plasma.[30] These ULVWF multimers bind more efficiently than the largest plasma VWF multimers to the glycoprotein (GP) Ibα components of the platelet GPIb-IX-V receptors.[31,32] The initial attachment of ULVWF multimers to GPIbα receptors,[32] and subsequently to activated platelet integrin αIIbβ3 (GPIIb-IIIa complexes), induces platelet adhesion and aggregation *in vitro* in the presence of elevated levels of fluid shear stress.[27,31] After retrograde secretion by endothelial cells, ULVWF multimers become entangled in subendothelial collagen, thereby maximizing the VWF-mediated adhesion of blood platelets to any subendothelium exposed by vascular damage and endothelial cell desquamation. An efficient "processing activity"[12,26] in normal plasma prevents the highly adhesive ULVWF multimers, which are also secreted antegrade into the vessel lumen, from persisting in the bloodstream.

The VWF "processing activity" is now known to be a specific VWF-cleaving metalloprotease in normal plasma that prevents the persistence in the circulation of ULVWF multimers.[13,33,34] The enzyme degrades ULVWF multimers by cleaving 842Tyr-843Met peptide bonds in susceptible A2

## ADAMTS-13

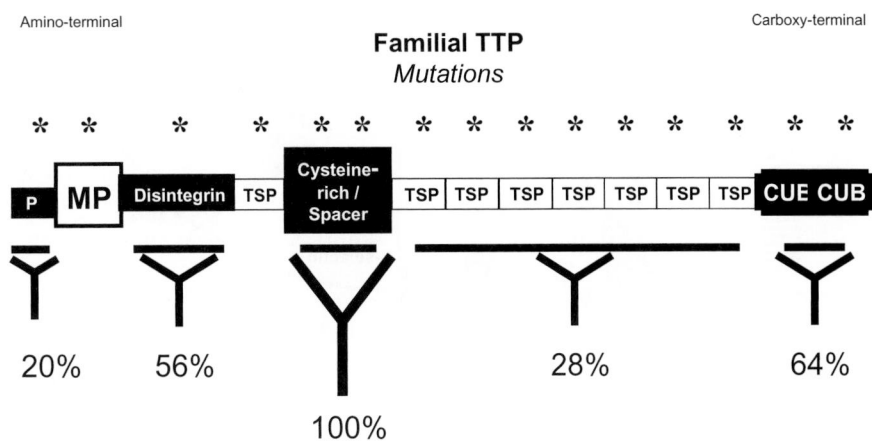

**Figure 50-2.** Domain structure of ADAMTS-13, the plasma VWF-cleaving metalloprotease. P, propeptide; MP, metalloprotease (proteolytic) domain; TSP, thrombospondin-1-like domain (eight are present); CUB, two nonidentical domains containing peptide segments similar to *c*omplement components, C1r/C1s, a sea *u*rchin protein, and a *b*one morphogenic protein. * Indicates the location of mutations in familial TTP patients that affect secretion or function of the enzyme. The percentages of polyclonal autoantibodies directed against specific domains of the enzyme in 25 patients with acquired idiopathic TTP are indicated.[55]

domains of VWF monomeric subunits.[35–37] The VWF-cleaving metalloprotease is number 13 in a family of 18 distinct ADAMTS-type enzymes identified to date that share structural similarities.[38,39] "ADAMTS-13" is *a d*isintegrin *a*nd *m*etalloprotease with eight *t*hrombo*s*pondin-1-like domains. More precisely, plasma ADAMTS-13 (Fig. 50-2) is composed of an amino-terminal reprolysin-type metalloprotease domain followed by a disintegrin domain, a thrombospondin-1-like domain, a cysteine-rich domain containing an arginine-glycine-aspartate (RGD) sequence, a spacer domain, seven additional thrombospondin-1-like domains, and two nonidentical CUB-type domains at the carboxyl-terminal end of the molecule. (CUB domains contain peptide sequences similar to complement subcomponents C1r/C1s; embryonic sea urchin protein egf; and bone morphogenic protein-1)[40]. ADAMTS-13 is a $Zn^{2+}$- and $Ca^{2+}$-requiring 190,000 Dalton glycosylated protein that is encoded on chromosome 9q34 and produced in hepatic stellate and vascular endothelial cells.[204] ADAMTS-13 is inhibited *in vitro* by EDTA; therefore, functional assays of the enzyme are usually performed using citrate-plasma.[13,33–39,41,42] Plasma anticoagulated with heparin, chloromethylketones (e.g., phenylalanine-proline-aspartate-chloromethylketone [PPACK]), hirudin, or other direct thrombin inhibitors would also probably be satisfactory for testing.

ULVWF multimers are cleaved by ADAMTS-13 as they are secreted as long "strings" from stimulated endothelial cells (Fig. 50-3A).[43,44] The ULVWF multimeric strings may be anchored in the endothelial cell membrane to P-selectin molecules that are secreted concurrently with the ULVWF multimers from Weibel–Palade bodies.[45] Included among the agents that stimulate endothelial cells to secrete ULVWF multimers are the proinflammatory cytokines, tumor necrosis factor (TNF) α, interleukin (IL)-8 and IL-6 (in complex with the IL-6 receptor),[46] and the Shiga toxins (discussed in Section IV on HUS). One of the repeated CUB domains at the carboxyl-terminal end of each ADAMTS-13 enzyme, as well as one or more of the thrombospondin-1-like domains along the length of the molecule, may modulate the binding of ADAMTS-13 to ULVWF multimers as they are secreted by endothelial cells.[47–49] Specifically, ADAMTS-13 enzymes may attach under flowing conditions to accessible A3 domains in the monomeric subunits of ULVWF multimers[43] and then cleave Tyr 842-843 Met peptide bonds in adjacent A2 domains (Fig. 50-3B). Partial unfolding of emerging ULVWF multimers by fluid shear stress may increase the efficiency of ADAMTS-13 attachment to ULVWF multimers, followed by ULVWF cleavage.[35,44] The VWF A1 domain may exert a negative influence on VWF cleavage by ADAMTS-13, because platelet GPIbα binding to the VWF A1 domain renders the adjacent VWF A2 domain more susceptible to proteolysis by ADAMTS-13.[50]

Failure to degrade ULVWF multimers has long been suspected to cause the familial and acquired idiopathic types of TTP or to predispose an individual to these disor-

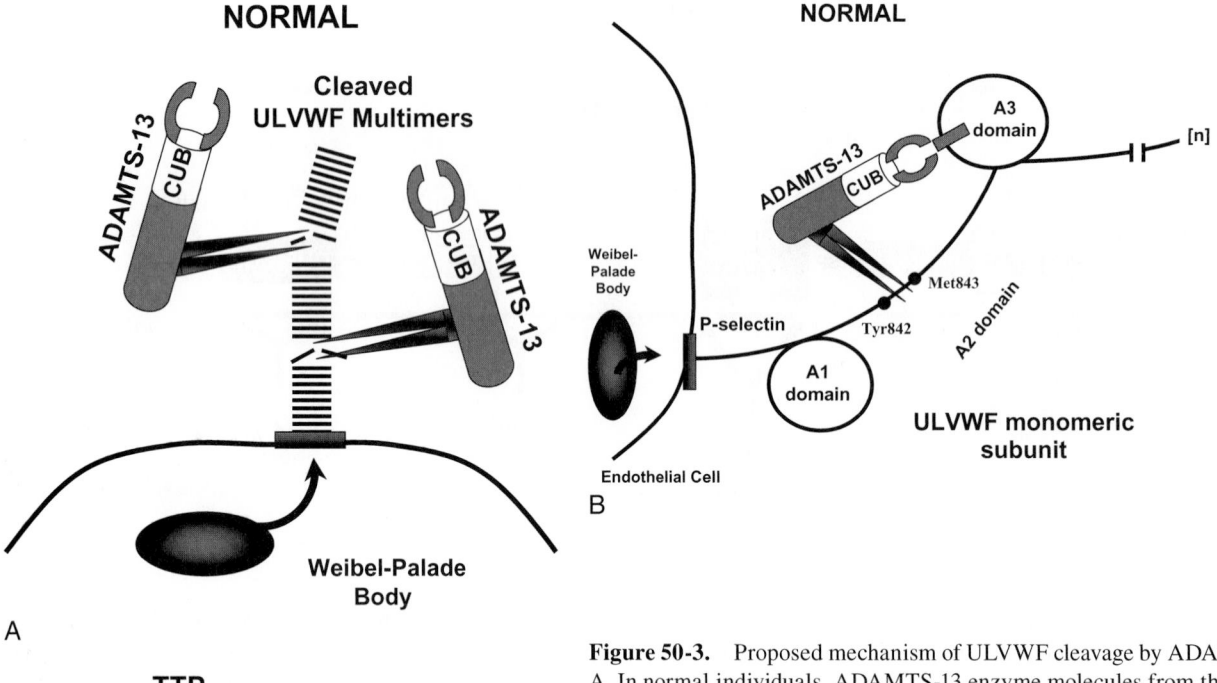

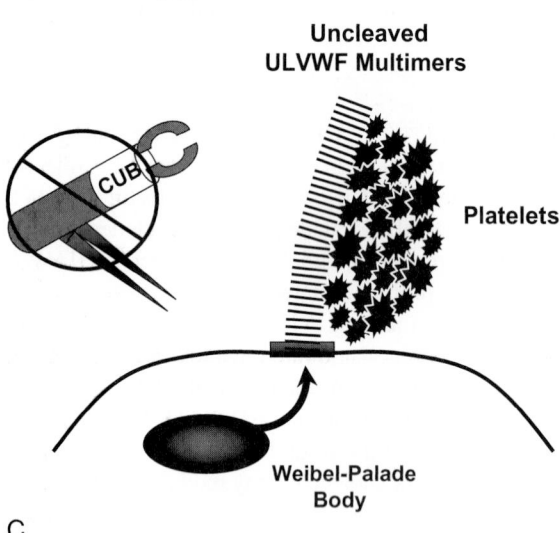

**Figure 50-3.** Proposed mechanism of ULVWF cleavage by ADAMTS-13. A. In normal individuals, ADAMTS-13 enzyme molecules from the plasma attach to, and then cleave, ULVWF multimers that are secreted in long "strings" from stimulated endothelial cells. B. In normal individuals, the ULVWF multimeric strings may be anchored in the endothelial cell membrane to P-selectin molecules that are secreted concurrently with the ULVWF multimers from Weibel–Palade bodies. Each ADAMTS-13 molecule may dock, possibly by utilizing its C-terminal CUB domain(s), to exposed A3 domains in ULVWF monomeric subunits. The attached ADAMTS-13 molecules then cleave Tyr 842-843 Met peptide bonds in adjacent A2 domains of ULVWF monomeric subunits. The smaller VWF forms that circulate after cleavage do not induce the adhesion and aggregation of platelets during normal blood flow. C. Absent or severely reduced activity of ADAMTS-13 in patients with TTP prevents the timely cleavage of ULVWF multimeric strings as they are secreted from endothelial cells. Uncleaved ULVWF multimers induce the adhesion and subsequent aggregation of platelets in flowing blood. Initial platelet adherence is via platelet GPIbα onto the exposed A1 domains of uncleaved ULVWF strings (not shown). TTP is caused by familial deficiencies of ADAMTS-13 activity caused by ADAMTS-13 gene mutations or acquired autoantibody-induced defects of ADAMTS-13 activity (or survival).

ders (Fig. 50-3C).[12,51] Critical experiments verifying this concept were reported in 1997 and 1998. In 1997, four patients were described with chronic relapsing TTP who had a deficiency of VWF-cleaving protease activity (ADAMTS-13) in plasma.[13] Because no inhibitor of the enzyme was detected, the deficiency was ascribed to an abnormality in the production, survival, or function of the protease. The following year, the pathogenesis of the more common acquired idiopathic type of TTP was elucidated.[33,34,52] Acquired idiopathic patients have little, if any, plasma VWF-cleaving protease activity during acute episodes; however, the activity often increases upon recovery. Although

the plasma assays in the 1997/1998 studies were "nonphysiologic" (as discussed in Section III.C), they were, nonetheless, innovative and informative. IgG autoantibodies against components of the enzyme probably accounted for the lack of protease activity in most of the acquired idiopathic TTP patients reported in 1998.[33,34,52] The reasons for the transient immune dysregulation, as well as for the selective antigenic targeting of the VWF-cleaving protease, is not yet known.

Patients with familial, chronic relapsing TTP almost always have ULVWF multimers in their plasma.[12,51] ULVWF multimers are also detectable using a sensitive gel electrophoresis method in some patient plasma samples during acute episodes of acquired idiopathic TTP, but not after recovery.[51] These findings were explained in 1997 and 1998[13,33,34,52] by investigators who demonstrated that there was a chronic absence from plasma of the VWF-cleaving protease (ADAMTS-13) in familial, chronic relapsing TTP, and transient inhibition of the enzyme during acute episodes of acquired idiopathic TTP.

Most patients with familial TTP have less than about 5% of normal ADAMTS-13 activity in their plasma, regardless of whether the plasma is obtained during or after acute episodes — provided that they have not recently received plasma infusions. Most patients with acquired idiopathic types of TTP have less than about 5% of normal activity of ADAMTS-13 in their plasma only during acute TTP episodes.[1,13,33,34,52,53] Severe deficiency of ADAMTS-13 activity in TTP patient plasma correlates with a failure to cleave ULVWF multimers as they emerge from the surface of endothelial cells (Fig. 50-3C).[44] As a consequence, ULVWF multimers secreted by endothelial cells remain anchored to the cells in long strings.[44] The anchoring may be via P-selectin molecules, which have transmembrane domains and are secreted along with ULVWF multimers from the Weibel–Palade bodies of endothelial cells.[45] (P-selectin molecules are predominantly retained in the cell membrane as the Weibel–Palade contents are secreted.) Passing platelets adhere via their GPIbα receptors to these long uncleaved ULVWF multimeric strings.[44] (Platelets do not adhere to the smaller VWF forms that circulate after cleavage of ULVWF multimers.[30]) Many additional platelets subsequently aggregate under flowing conditions, probably via their activated integrin αIIbβ3, onto the ULVWF multimeric strings to form large, potentially occlusive platelet thrombi[44,54] (Fig. 50-3C).

ULVWF multimeric strings are capable of detaching from endothelial cells in the absence of ADAMTS-13 activity, the presence of fluid shear stress, and the increasing torque generated as platelets adhere and aggregate onto the ULVWF strings.[44] The detached ULVWF-platelet strings may "embolize" to microvessels downstream and contribute to organ ischemia. The formation of ULVWF-platelet thrombi and emboli may account for the following observations: (a) ULVWF multimers are chronically detected in the plasma of familial TTP patients, as well as in the plasma of some patients with acquired idiopathic TTP during acute episodes;[1,12,26,51] (b) increased VWF antigen is found, by flow cytometry, on platelets during episodes of familial or acquired TTP;[55] and (c) abundant VWF antigen (but not fibrinogen) is observed by immuno-histochemistry on platelet occlusive lesions in TTP.[29]

Plasma ADAMTS-13 activity is almost always absent or severely reduced in familial TTP patients,[13,56,57] as a consequence of homozygous (or doubly heterozygous) mutations in each of the two ADAMTS-13 9q34 genes.[1,22,53,58,59] Mutations in familial TTP have been detected all along the gene,

in regions encoding different domains (Fig. 50-2).[1,22,53,58,59] In severe familial deficiency of ADAMTS-13 activity, episodes of TTP usually commence in infancy or childhood (Table 50-1). In some of these patients, however, overt TTP episodes do not develop for years (e.g., during a first pregnancy),[22] if ever. This latter observation suggests that *in vivo* ADAMTS-13 activity on ULVWF multimers emerging from stimulated endothelial cells may exceed the plasma enzyme activity measured by *in vitro* nonphysiologic assays. Additionally, or alternatively, accentuated secretion of ULVWF multimers by endothelial cells induced by estrogen or proinflammatory cytokines[46] may be required to provoke TTP episodes in some patients with severe plasma ADAMTS-13 deficiency.

In some infants with less than 5% ADAMTS-13 and neonatal onset of familial chronic relapsing TTP, transient or progressive renal failure is a prominent component of the disorder.[60] These patients clinically resemble two patients described in 1960 by Schulman et al.[12,61] and in 1978 by Upshaw,[12,62] and, consequently, this pediatric subgroup is sometimes said to have "Upshaw-Schulman syndrome."

Many patients with acquired idiopathic TTP have absent or severely reduced plasma ADAMTS-13 activity during an initial episode, as well as during any later recurrence.[33,34,52] ADAMTS-13 activity increase in these patients following recovery from either a single or a recurrent episode (Table 50-1). IgG antibodies (presumably autoantibodies) that inhibit plasma ADAMTS-13 activity can be detected in 44 to 94% of patients using the nonphysiologic techniques currently available.[22,25,33,34,52,63] These results suggest the presence of a transient, or intermittently recurrent, defect of immune regulation in many patients who have acquired idiopathic TTP associated with transient, or recurrent, ADAMTS-13 deficiency. Antibodies that inhibit plasma ADAMTS-13 have also been demonstrated in a few patients with ticlopidine or clopidogrel-associated TTP.[19,21] It is not yet known if there is a transient, severe defect of ADAMTS-13 production or survival in patients with acquired TTP who do not have detectable autoantibodies against the enzyme. Alternatively, failure to detect autoantibodies in some patients may reflect the limited sensitivity of the test systems in current use.

In one study[64] of polyclonal autoantibodies against ADAMTS-13 in 25 acquired TTP patients, the epitope targets always included the cysteine-rich/spacer domain sequence and were exclusively directed against the cysteine-rich/spacer domain sequence in 3 of the 25 patients (Fig. 50-2). The other 22 autoantibodies reacted with the cysteine-rich/spacer domain sequence plus either the CUB domains (64%), the metalloprotease/disintegrin-like/1st thrombospondin-1-like domain sequence (56%), or the second through eighth thrombospondin-1-like domain sequence (28%). The propeptide region was also identified by 20% of autoantibodies,[64] indicating that removal of the propeptide is not required for secretion of active enzyme.[65] Autoantibodies either inhibit the activity of ADAMTS-13 or decrease its survival.

The propensity to produce ADAMTS-13 autoantibodies is almost certainly genetically determined. This is emphasized by the demonstration of acquired idiopathic TTP caused by IgG autoantibodies against ADAMTS-13 in twin sisters.[66]

Relapses occur in 23 to 44% of patients with acquired idiopathic TTP,[22,25,67,68] often in the first year after the initial episode.[22] These relapsing patients usually have acquired idiopathic TTP with severe plasma ADAMTS-13 deficiency that is often due to the presence of autoantibodies against ADAMTS-13.[22] In a few instances, pregnancy-related TTP episodes have been caused by autoantibodies against ADAMTS-13.[22] The risk of recurrent TTP during any subsequent pregnancies is controversial, with estimates of possible recurrence (per woman) ranging from 26 to 73%.[22]

Plasma ADAMTS-13 activity in healthy adults ranges from approximately 50 to 178% using currently available static, nonphysiologic assays. Activity is often reduced below normal in liver disease, disseminated malignancies,[69] chronic metabolic and inflammatory conditions, pregnancy, and in newborns.[70] With the exception of those peripartum women who develop overt TTP,[22,25] the ADAMTS-13 activity in these conditions is not reduced to the extremely low values (<5% of normal) found in most patients with familial or acquired idiopathic TTP.

### C. Assays of Plasma ADAMTS-13

Most clinical series reported to date have depended on non-physiologic laboratory estimates of plasma ADAMTS-13 activity.[22,63,71] The divalent cation-dependent disappearance of large plasma-type VWF multimers in the presence of test citrate-plasma is evaluated directly (using porous sodium dodecyl sulfate [SDS]-agarose gel electrophoresis)[13,33,34,52] or indirectly (using residual collagen-binding or ristocetin cofactor activity).[63,70–72] The conditions are static (i.e., not under flowing conditions). The assay is often performed using low-ionic strength buffer, alkaline pH, a denaturing agent (urea or guanidine), $Ba^{2+}$, and a prolonged incubation time (from a few hours to 24 hours). In normal samples, large plasma-type VWF multimers are cleaved by ADAMTS-13 at peptide bond Tyr842-843Met in the A2 domain of susceptible VWF monomeric subunits.[35–37] VWF fragments of 170,000 and 140,000 Daltons (or dimers of these fragments), indicating specific cleavage of VWF monomers at 842–843, can also be detected using less-porous SDS-acrylamide gel electrophoresis. Cleavage by ADAMTS-13 of plasma VWF, VWF purified from plasma, or ULVWF multimers occurs under these harsh in vitro conditions in the presence of dilutions of normal plasma, but not plasma obtained from patients with familial TTP or from patients during an episode of acquired idiopathic TTP.[13,33,34,52]

A different type of test system is capable of observing directly in vitro under flowing conditions the capacity of plasma ADAMTS-13 to cleave ULVWF multimeric "strings" as they are secreted from stimulated human endothelial cells (Fig. 50-3).[44] The endothelial cell-based, rapid ULVWF multimeric "string" cleavage assay is performed under flowing conditions that are more nearly physiologic, although not entirely analogous to in vivo events (e.g., the endothelial cells are cultured human umbilical vein endothelial cells [HUVECs]). The assay requires only minutes to complete. ULVWF multimers secreted from endothelial cells are the substrates for ADAMTS-13, and no extraneous chemicals other than citrate are present (specifically, no $Ba^{2+}$, urea, or guanidine). An important limitation on the widespread use of this technique is that the equipment required is expensive and complicated.

To cleave plasma-type VWF in vitro, the ADAMTS-13 molecule must contain all domains that extend from the amino-terminal metalloprotease through the spacer region. These include the metalloprotease, disintegrin, first thrombospondin-1-like, cysteine-rich, and spacer domains (Fig. 50-2).[73,74] Although the other domains are not absolutely required for VWF proteolysis in vitro, cleavage of purified VWF and of ULVWF multimeric strings is modulated by a carboxyl-terminal CUB domain.[48,49]

More recently described assays use diluted test plasma samples to measure the cleavage by ADAMTS-13 activity of the Tyr842-843Met peptide bond in recombinant A2 domain proteins.[75–77] This type of assay requires only a few hours to complete and may be a useful screening test for clinical laboratories. In vivo, however, the interaction between ADAMTS-13 molecules and ULVWF multimeric strings is likely to be a complex structural interplay of the various domains that comprise the ADAMTS-13 enzyme and ULVWF multimeric substrate.

Any of these assays can be modified to screen for the presence of inhibitors to ADAMTS-13 autoantibodies by incubating normal plasma with patient plasma and then determining if the ADAMTS-13 activity in normal plasma has been diminished.

### D. TTP Unassociated with Severe ADAMTS-13 Deficiency

Some patients develop the characteristic clinical manifestations of TTP without either overt associated conditions or plasma ADAMTS-13 deficiency (at least as measured by techniques currently available).[25,67,68,78,79] The etiology of TTP in this subgroup of patients, who have a higher mortality rate than patients with severe ADAMTS-13 deficiency, is unknown. In some of these patients, any possible relationship between ADAMTS-13 deficiency and TTP is clouded by the transfusion of normal blood products containing ADAMTS-13 before the testing of patient plasma for enzyme activity.[22]

Bone marrow transplantation-associated thrombotic microangiopathy is not usually associated with an absence or severely reduced level of plasma ADAMTS-13 activity.[33,67,80,81] The explanation for VWF abnormalities in the plasma of the few chemotherapy/transplantation-associated thrombotic microangiopathy patients studied[82] is not known. For example, it is not known if endothelial cell injury contributes to the pathophysiology of this entity. Additional discussion of thrombotic microangiopathies associated with these and other conditions is presented in Section V.

## E. Therapy of TTP

In 1977, before there was any inkling of the underlying etiology of TTP, Byrnes and Khurana[83] reported that relapses in one young adult patient could be prevented or reversed by the *infusion* of only a few units of fresh-frozen plasma or its cryoprecipitate-poor fraction (cryosupernatant), without concurrent plasmapheresis. It was shown in 1985 that the processing of ULVWF multimers was restored in patients with *familial, chronic relapsing TTP* by transfusing fresh-frozen plasma, cryosupernatant (plasma depleted of VWF-rich cryoprecipitate),[26,28] or solvent/detergent-treated plasma.[11] Tsai and Lian[34] and Furlan et al.[33] demonstrated in 1998 that these plasma products contain functionally active ADAMTS-13 (Table 50-2).

Infants or young children with familial TTP produce inadequate quantities, or functionally defective forms, of ADAMTS-13.[1,13,56] The reason why the infusion of normal ADAMTS-13 only about every 3 weeks (Table 50-2) prevents TTP episodes in these patients is not known. The plasma $t_{1/2}$ of the infused ADAMTS-13 activity is relatively long (about 2 days).[56] The functional $t_{1/2}$ of the enzyme may be even longer, as a result of ADAMTS-13 docking and cleaving one ULVWF multimeric string after another as each string is secreted from endothelial cells.[43,44,48,84]

The sequence of the 190,000 Dalton ADAMTS-13 has been determined, and the enzyme has been partially purified from normal human plasma fractions.[37-39] Recombinant, active ADAMTS-13 has also been prepared.[85] As a consequence, purified or recombinant ADAMTS-13 may soon be developed for therapeutic use in TTP. A plasma level of ADAMTS-13 of only about 5% of normal is sufficient to prevent or truncate TTP episodes in most patients.[1,41,53,57,67] Gene therapy, consequently, may eventually be a practical approach to providing more lasting remissions in children with familial, chronic relapsing TTP.

Adults and older children with acquired idiopathic TTP episodes associated with ADAMTS-13 deficiency require

**Table 50-2: TTP Therapy**

| ADAMTS-13 Present In |
| --- |
| Fresh-frozen plasma |
| Cryoprecipitate-poor plasma (cryosupernatant) |
| [Solvent-detergent plasma] |

| Plasma Infusion |
| --- |
| Familial (chronic relapsing) TTP |
| Every 3 weeks |

| Plasma Infusion with Plasmapheresis (Plasma Exchange) |
| --- |
| Acquired TTP |
| Daily |

| Acquired Idiopathic TTP — Other Therapy |
| --- |
| Glucocorticoids |
| Transfusion: |
|   Red cells |
|   Platelets only for emergency bleeding |
| No aspirin |
| Rituximab (anti-CD20 on B-lymphocytes) |
| Splenectomy |
| [Vincristine?] |
| [Cytoxan, cyclosporine?] |

daily plasma exchange (Table 50-2). Plasma exchange combines plasmapheresis (which may remove circulating ULVWF-platelet strings, agents that stimulate endothelial cells to secrete ULVWF multimers, and autoantibodies against ADAMTS-13) and the infusion of fresh-frozen plasma or cryosupernatant (both containing uninhibited ADAMTS-13). Both solvent-detergent-treated plasma[34,86] and methylene blue/light-treated plasma[86] (for inactivation of lipid envelope viruses) also contain active ADAMTS-13; however, protein S activity is below normal in solvent-detergent plasma.[86]

Plasma exchange allows about 80 to 90% of acquired, "out-of-the-blue" TTP patients to survive an episode.[14,17,22] Patients with acquired idiopathic TTP who have severe ADAMTS-13 deficiency often respond more effectively and rapidly to this procedure than patients diagnosed with TTP who have normal levels of ADAMTS-13. This is illustrated by a study of 38 consecutive patients diagnosed with TTP at one hospital.[87] Of the 10 patients who did not respond to daily plasma exchange, 8 had normal plasma ADAMTS-13 (activity was absent in the other 2). In contrast, of the 28 patients who responded to plasma exchange, 25 had absent or severely reduced plasma ADAMTS-13 (activity was normal in the other 3). Of the 25 severe ADAMTS-13-

deficient TTP patients who responded to plasma exchange, 15 required only seven exchanges to attain platelet counts above $150 \times 10^9$/L and normal serum LDH values.

Lower titers of ADAMTS-13 autoantibodies may be associated with better responses to plasma exchange procedures.[68,88,89] In association with plasma exchange, production of ADAMTS-13 autoantibodies may be suppressed by high-dose *glucocorticoids*,[17] 4 to 8 weekly doses of *rituximab* (monoclonal antibody against CD20 on B-lymphocytes),[90–93] rituximab combined with cyclophosphamide,[22] or (most radically) *splenectomy* (Table 50-2).[94–96]

Recovery from TTP is not usually associated with persistent, overt organ damage;[14,17] however, some compromise of cognitive function may be detectable subsequently by careful testing.[22] Although almost all TTP recurrences can be quickly recognized by blood counts and LDH measurements, several disturbing exceptions have been reported. For example, three women who had recovered from previous episodes of TTP subsequently had symptoms of stroke *without thrombocytopenia,* but with absent ADAMTS-13 and symptomatic response to plasma exchanges.[93,97]

## IV. The Hemolytic-Uremic Syndrome

HUS, initially described by Gasser and colleagues in 1955,[98] is the constellation of acute renal failure, thrombocytopenia, intravascular hemolytic anemia with schistocytosis, and elevated serum LDH. With the exception of familial types of the disorder, HUS almost always occurs as a single episode. The thrombocytopenia, hemolytic anemia, schistocytosis, and LDH elevations are often less extreme in HUS than during TTP episodes. However, renal dysfunction is severe in HUS, in contrast to TTP. Proteinuria and hematuria are frequently present, and dialysis is often required. The hemolytic anemia may occasionally be exacerbated by erythropoietin deficiency.[99]

Chronic renal failure, oliguria, and hypertension can be long-term complications of HUS, but are uncommon in patients who recover from episodes of TTP. In HUS, microvascular platelet adhesion/aggregation and fibrin polymer formation is predominantly renal (and intraglomerular), although peritubular capillaries and even other organs may sometimes be involved.[98,100,101] The occasional extrarenal manifestations in HUS can obscure the distinction between HUS and TTP. Coagulation abnormalities are not consistently observed in HUS, indicating that any activation of coagulation may be a secondary, limited phenomenon.

### A. Types of HUS (Table 50-3)

*"Classical" HUS* is usually acquired by the inadvertent ingestion of enterohemorrhagic E. coli (e.g., serotype O157:

**Table 50-3: Types of HUS**

Diarrhea-Associated (D+)

Shiga toxins 1 and 2 (enterohemorrhagic *Escherichiae coli;* e.g., *E. coli O157:H7*)
Shiga toxin (*Shigella dysenteriae*)

Unassociated with diarrhea (D−)

**Familial complement-regulation defects**
  Plasma factor H deficiency
  Membrane cofactor protein (MCP) deficiency
  Plasma factor I deficiency
**Familial intracellular cobalamin reduction defects**

H7) and preceded by bloody diarrhea.[1,102] *Familial types of HUS* are caused by inadequate functional activity of one of the regulatory proteins for the alternative complement pathway, *plasma factor H, membrane-cofactor protein,*[103] or *factor I;*[104] or defective intracellular *cobalamin reduction/cofactor function.*[105]

### B. Diarrhea-Associated HUS (D+ HUS)

**1. Etiology and Pathogenesis.** HUS frequently occurs following an episode of hemorrhagic colitis/diarrhea. Among the offending microbes are *Shigella dysenteriae* or, more frequently, certain enterohemorrhagic serotypes of *Escherichiae coli* (O157:H7).[101,102,106–108] These bacteria produce the exotoxins, Shiga toxin (Stx; made by *S. dysenteriae*) or Stx-1 and Stx-2 (made by enterohemorrhagic *E. coli*).[107–111]

About a week after an episode of bloody diarrhea caused by enterohemorrhagic E. coli 0157:H7, HUS occurs in 9 to 30% of infected children.[101,102,106] About 10% of these HUS patients die, and another 25% have some residual renal dysfunction.[112,113] Infections with other enterohemorrhagic *Escherichia coli* serotypes (026:H11, 0113:H21),[107,108] *Shigella dysenteriae,* and occasionally other bacteria (*Streptococcus pneumoniae, Clostridium difficile*)[114,115] also cause HUS in children and adults. In some regions (e.g., Buenos Aires and Calgary), enterohemorrhagic *E. coli* infections are endemic, and HUS is a common cause of acute renal failure in children.[101]

The E. coli serotypes that cause HUS share the capability of producing Stx-1 and/or Stx-2.[107–110] Those E. coli serotypes that produce Stx-2 may be more likely to cause HUS in humans;[107–110] however, the injection of either Stx-1 or Stx-2 leads to HUS in primates.[116–118]

Stx is a 70,000 Dalton exotoxin protein encoded in S. dysenteriae DNA.[111] Stx and Stx-1 are structurally similar,

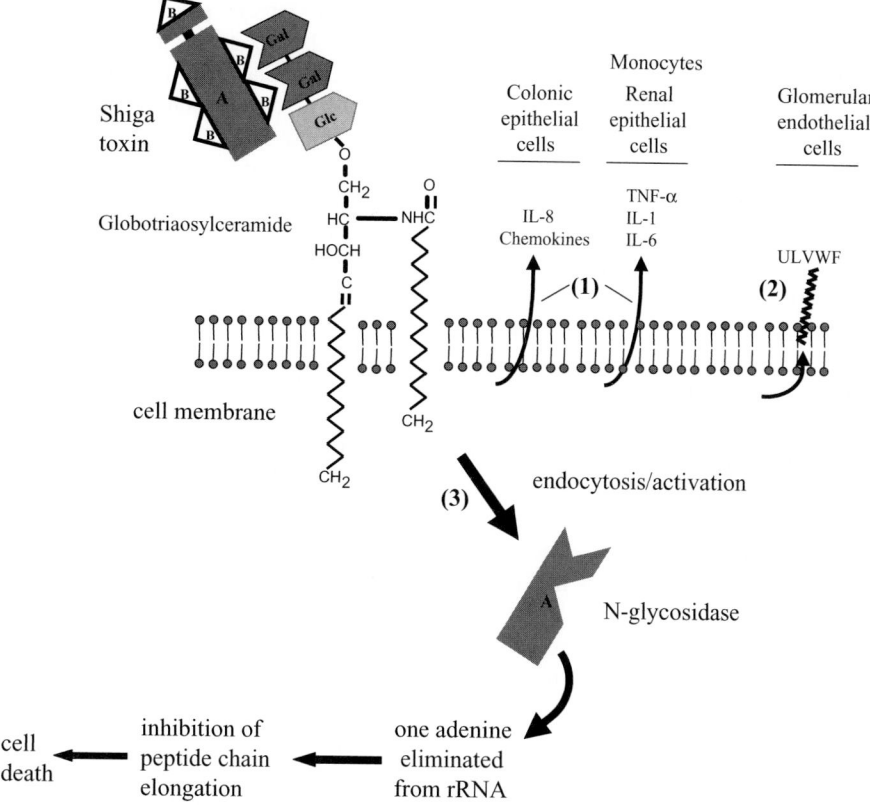

**Figure 50-4.** Shiga toxin and HUS. The B subunits of Shiga toxin molecules (or Shiga toxin-1 or Shiga toxin-2) attach to specific disaccharides of globotriaosylceramide (Gb₃) receptors in the membranes of some human cell types. These include colonic epithelial cells, monocytes, platelets, glomerular and tubular epithelial cells, renal mesangial cells, and glomerular and cerebrovascular endothelial cells. Initial Shiga toxin (or Shiga toxin-1 or Shiga toxin-2) binding to Gb₃ on these various cells stimulates epithelial cells and monocytes to secrete cytokines (and chemokines) (**1**), activates platelets, and stimulates endothelial cells to secrete ULVWF multimeric strings (**2**). The exposure of tissue factor on endothelial cells and epithelial cells may also be increased by Shiga toxin (or Shiga toxin-1 or Shiga toxin-2) binding to Gb₃. After the A subunit of Shiga toxin (or Shiga toxin-1 or Shiga toxin-2) is internalized, it is converted to an active glycosidase that cleaves one adenine from 28S ribosomal RNA subunits, arrests peptide chain elongation, and results in cell death (apoptosis) (**3**).

whereas Stx-1 and Stx-2 are only 53 to 56% homologous.[1,119] Stx-1 and Stx-2 are closely related exotoxins that are encoded by bacteriophage DNA. This bacteriophage DNA is incorporated into the genome of a restricted number of *E. coli* serotypes.[111] Stx is composed of one A subunit (33,000 Daltons) and five B ("binding") subunits (7,700 Daltons each) (Fig. 50-4).[111] Structural genes for the A and B subunits are adjacent to one another in the DNA of *S. dysenteriae* and in the incorporated bacteriophage DNA of enterohemorrhagic *E. coli* serotypes. The B subunits of Stx are capable of binding with high affinity to galactose α1,4β, galactose disaccharides in membrane globotriaosylceramide (Gb₃) receptors (Fig. 50-4).[111,120] Gb₃ (or Gb3-like) receptors are present in the membranes of human renal glomerular, mesangial, and tubular cells (where it may be most effectively expressed); colonic and cerebral microvascular endothelial cells; and peripheral blood leukocytes and platelets.[1,111,121–126,] The Gb₃ receptors associate with cholesterol and ganglioside G_{M1} (the cholera toxin receptor)

to form mobile lipid "rafts" floating in outer plasma membranes.[127]

*S. dysenteriae* capable of producing Stx, or *E. coli* serotypes that produce pathogenetic Stx-1 and -2, may contaminate water, meat, milk, cheese, or other insufficiently cooked or pasteurized food.[101,128] Cattle are a major reservoir of *E. coli* 0157:H7 and a few other Stx 1- and Stx-2-producing serotypes. In contrast to humans, the animals may remain well because their vascular endothelial cells lack the Gb₃ receptors necessary for the binding of *E. coli* Stx.[129] Stx is extremely stable and can remain active in the dust and structures of contaminated buildings for months.[130]

In humans, enterohemorrhagic bacteria adhere to mucosal epithelial cells of the terminal ileum and colon and then efface, invade, replicate, and destroy the cells. Stx-1 and -2 (possibly together with the bacterial lipopolysaccharide, endotoxin)[101,111] cause damage to the underlying tissue and vasculature and produce bloody diarrhea. This injurious process is potentiated by invading neutrophils, which are recruited into the damaged colon and activated by IL-8, granulocyte-stimulating factor (G-CSF),[131] and the C-X-C chemokines. These latter include GRO (growth related oncogene) proteins-α, -β and -γ and ENA (epithelial cell-derived neutrophil activating protein)-78.[132,133] IL-8, the GRO/ENA chemokines, and G-CSF are secreted by colonic epithelial cells in response to colonic cell contact with the H7 flagellin structures on enterohemorrhagic E. coli O157: H7.[134] The level of circulating G-CSF correlates with the likelihood of developing HUS in association with enteric infection.[131] Shiga toxin from S. dysenteriae, or Stx-1 and -2 from enterohemorrhagic E. coli, then enter the intestinal circulation and travel through the bloodstream on the surface of platelets, neutrophils, or monocytes.[124,135] The toxins attach to $Gb_3$ molecules on renal glomerular capillary endothelial cells, mesangial cells, and tubular epithelial cells.[136–139] Stx (especially in association with endotoxin lipopolysaccharide) stimulates the release of tumor necrosis factor (TNF)-α, IL-1, and IL-6 from monocytes and renal glomerular and tubular epithelial cells (Fig. 50-4).[118,135,139–141] These latter three cytokines upregulate the membrane expression of $Gb_3$ in renal, cerebral, and other microvascular endothelial cells by promoting the synthesis of enzymes responsible for the production of $Gb_3$ molecules.[111,142–144] Additional Stx molecules then attach to the increased number of $Gb_3$ receptors in glomerular, and sometimes other, endothelial cell membranes. It is likely that the relatively specific renal toxicity of Stx-1 and Stx-2 are the result of their transport via cells in the bloodstream to $Gb_3$ receptors concentrated in the membranes of glomerular endothelial (and other renal) cells.

Platelet adhesion, and subsequent aggregation, atop Stx-stimulated renal glomerular microvascular endothelial cells may initiate the thrombotic microangiopathy in HUS.[1,142,145,146] There is evidence in vitro, and in vivo in primates, that VWF may be important in this platelet endothelial cell adhesion.[116–118] VWF was not, however, found in the renal thrombi of 3 out of 4 patients with Stx-associated HUS thrombi examined in one small study.[147] Plasma ADAMTS-13 levels are usually normal in Stx-induced HUS, in comparison to the absent or severely reduced levels in most patients with TTP.[1,33,67,147–149]

Nanomolar concentrations of Stx-1 or Stx-2 stimulate the secretion of ULVWF multimeric strings from both HUVECs and glomerular microvascular endothelial cells (GMVECs) ex vivo.[146] The Stx stimulation of glomerular endothelial cells may account for the elevated levels of VWF found in the plasma of most patients with diarrhea-associated HUS.[51,150–153]

ULVWF strings remain attached to the endothelial cells via P-selectin molecules secreted concurrently with ULVWF from Weibel–Palade bodies.[45] Normal platelets adhere to the ULVWF strings via their GPIbα-IX-V receptors as the platelets flow through the "thicket" of secreted ULVWF

strings. Extensive platelet attachment to ULVWF strings secreted by Stx-stimulated glomerular endothelial cells in vivo may explain the thrombocytopenia and initiation of thrombotic obstruction in the renal microvasculature of HUS patients.[134] Thrombocytopenia and intrarenal thrombosis are even more likely under the higher flow/shear conditions of the glomerular microcirculation. At higher levels of shear stress (20 to 40 dynes/$cm^2$), activation of platelets attached to the ULVWF strings occurs and results in the formation of platelet aggregates.[43,44,54]

Cleavage and elimination of the ULVWF-platelet strings by plasma ADAMTS-13 is delayed by several minutes in the presence of nanomolar concentrations of either Stx-1 or Stx-2.[146] Stx 1- or Stx-2-induced ULVWF string secretion, ULVWF-platelet string formation, and the Stx-1 or Stx-2-related delay in ULVWF-platelet string cleavage may contribute to the initiation of platelet thrombus formation atop Stx-stimulated glomerular endothelial cells. These events may exceed the counteractive ULVWF-platelet string cleaving capacity of circulating ADAMTS-13, even though the plasma levels of ADAMTS-13 are usually normal. In addition to these effects on endothelial cells, Stx binds to, and activates, platelets via $Gb_3$ or $Gb_3$-like receptors.[124] All of these processes in combination may provoke platelet adhesion and aggregation onto the ULVWF multimeric strings protruding from Stx-stimulated glomerular endothelial cell surfaces (Fig. 50-5A) and, therefore, promote glomerular occlusion by platelet-fibrin thrombi, thrombotic microangiopathy, acute renal failure, and HUS.

The effects of Stx-1 and Stx-2 are likely to be potentiated by the "cytokine storm" that accompanies enterohemorrhagic E. coli infection. TNF-α, IL-8, and IL-6 (in complex with the soluble IL-6 receptor) all stimulate endothelial cell secretion of ULVWF multimeric strings, accompanied by the rapid adhesion of flowing platelets.[46] Of these three cytokines, IL-6 has effects similar to Stx-1 and Stx-2 in that it both stimulates ULVWF multimeric string secretion (and platelet adherence) and impairs the rate of ULVWF-platelet string cleavage by ADAMTS-13.[46,146]

In vivo observations in a primate model of Stx-induced HUS reinforce these pathophysiologic conclusions. The intravenous injection into baboons of Stx-1 caused increased VWF expression in glomerular and peritubular capillaries and increased circulating levels of TNF-α, IL-8, and IL-6.[117] Furthermore, the intravenous injection of Stx-2 into the animals was associated with increased IL-6 levels.[116] The outcome in the baboons injected with either Stx-1 or Stx-2 was glomerular thrombotic microangiopathy and HUS.[116–118]

Exposure to nanomolar concentrations of these toxins for minutes causes rapid and profuse secretion of ULVWF multimeric strings from glomerular endothelial cells. In contrast, exposure of endothelial cells to Stx-1 or Stx-2 for more than 1 hour results in progressive cell injury and death, as

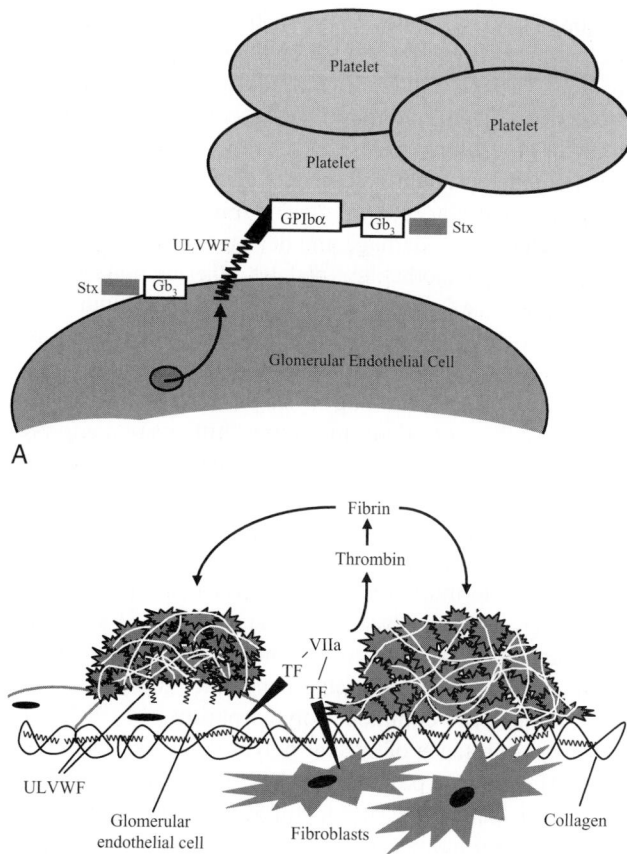

A

B

**Figure 50-5.** Proposed mechanisms of platelet adhesion/aggregation and fibrin formation in HUS. A. Platelets activated by Shiga toxin, Shiga toxin-1, or Shiga toxin-2 (abbreviated as Stx in the figure) may adhere via GPIbα in their GPIbα-IX-V complexes and aggregate on ULVWF multimers secreted by Stx-stimulated glomerular endothelial cells. This may be especially likely if ADAMTS-13 activity is impaired by Stx-1, Stx-2, or IL-6. Gb₃, globotriaosylceramide receptors for Stx. B. Stx-mediated endothelial cell death and desquamation may expose subendothelial ULVWF multimers intertwined with collagen. Platelets from flowing blood in the renal microcirculation may adhere, become activated, and aggregate on the exposed ULVWF/collagen. Local tissue factor (TF) exposure and factor VII binding and activation may occur on fibroblasts, invading phagocytic cells and a top simulated or injured endothelial and epithelial cells. This initiation of coagulation results in thrombin generation and fibrin polymer production, and it contributes to the formation of occlusive platelet-fibrin thrombi.

a result of the following events. The A subunit of Stx undergoes endocytosis, partial proteolysis, and disulfide bond reduction (Fig. 50-4). These processes generate an active intracellular enzyme (27,000 Daltons) that cleaves an adenine from 28 S-ribosomal RNA,[111] inhibits peptide chain elongation and protein synthesis, and causes cell death. This renal damage is potentiated by the monocytes and neutro-

phils that invade glomeruli in response to IL-8 and G-CSF, and to the renal production of monocyte chemoattractant protein-1 (MCP-1).[131,143,154] IL-8-activated neutrophils release oxygen-derived free radicals, $H_2O_2$, elastase, and other proteases that magnify renal damage.[155–157] Neutrophilia makes irreversible renal injury more likely in HUS.

Endothelial cell death and desquamation may also promote platelet adhesion via GPIbα to exposed subendothelial collagen-bound ULVWF multimers.[158] Subsequent fibrinogen binding to activated platelet integrin αIIbβ3 is known to induce the aggregation of platelets onto those initially adherent under high flow/high shear conditions,[158] as in the glomerular microcirculation.

Increased exposure of tissue factor and activated factor VII binding on perturbed cell surfaces, thrombin generation, and fibrin polymer formation occur in HUS.[159] The signals transduced rapidly by Stx-1 or Stx-2 attachment to Gb₃ include nuclear factor κB (NFκB), activator protein (AP)-1,[160] and several protein kinases.[161] Generation of NFκB and AP-1 activates the tissue factor promoter region of chromosome 1 and increases cell exposure of tissue factor.[160] Binding and activation of factor VII then leads to thrombin generation and fibrin polymer formation that contribute to the intraglomerular platelet-fibrin thrombi characteristic of *E. coli*-associated HUS (Fig. 50-5B).[1,159,162]

The local intraglomerular thrombin generation (and fibrin formation) may also be potentiated by Stx-disruption of the interaction between thrombin and thrombomodulin. The result is interference with the protein C/S coagulation control pathway and impaired local renal control of coagulation. Stx, in the presence of TNF-α, suppresses the expression of thrombomodulin on the surface of glomerular microvascular endothelial cells.[163] Thrombin must bind to cell surface thrombomodulin in order to activate protein C for binding to protein S and cleavage of activated factors V and VIII.[163,164]

**2. Therapy.** In mildly affected children with anuria for less than 24 hours, proper management of fluid and electrolytes is usually sufficient. Otherwise, the duration of oliguria correlates inversely with the likelihood of recovery. Renal failure is often more severe in adults. Dialysis or, ultimately, renal transplantation may be required. Plasma infusion or exchange (presumably to remove Stx and cytokines) has been tried with equivocal results.[165,166]

Antimotility agents increase the risk of developing HUS during *E. coli* 0157:H7 infection. Antimicrobial agents promote the release of Stx-1 and Stx-2 from *E. coli* 0157: H7,[167] and antibiotic therapy may, paradoxically, also increase the risk of HUS in infected children. Normal individuals generate neutralizing antibodies against Stx in response to infection with Stx-producing microbes,[101,168] but the extent of the protection provided is not known.

A large trial of a Stx-binding agent administered for 1 week orally to children with HUS failed to influence the course of the disease.[169] It is not yet known if anticoagulant therapy is useful or safe in human HUS.[159] Heparin at therapeutic levels was, however, ineffective in a baboon model of Stx-induced HUS.[170] It has been claimed that angiotensin-converting enzyme (ACE) inhibitors are "renoprotective" in HUS-related hypertension.[171]

### C. HUS Unassociated with Diarrhea (D⁻ HUS)

**1. Familial Complement Regulation Defects.** Familial HUS accounts for 5 to 10% of all cases of HUS. The mortality rate (about 50%) is much higher than that in typical childhood HUS (about 10%). About half of survivors have relapses, and almost 40% require chronic dialysis.[172] Among familial HUS patients who receive kidney allografts, about 15% lose function in the engrafted kidney within 1 month.[173]

*a. Inadequate Plasma Factor H* (Fig. 50-6). Between 13% and 30% of familial HUS patients have a deficiency or defect of complement factor H.[103,172,174–179] The factor H gene is encoded within a complement-related gene cluster on chromosome 1q that includes membrane cofactor protein (discussed later).[103,176,180] Factor H is a 155,000 Dalton plasma protein containing 20 tandem homologous short consensus repeats (SCRs) or "sushi" domains of 60 amino acids each.[181] Factor H normally protects host cells from accidental damage by the alternative complement pathway. C3bBb, the C3 convertase of the alternative pathway, amplifies the gen-

eration of C3b molecules on the surface of susceptible cells. Factor H regulates this process by displacing Bb from C3b, thereby exposing C3b to cleavage and inactivation by factor I (the C3b-cleaving protease). A deficiency or dysfunction of factor H results in excessive activation of C3 to C3b that may potentiate autoantibody-mediated or immune complex-induced glomerular injury.[172,182] The result may be glomerular endothelial cell damage and desquamation; exposure of glomerular subendothelium; platelet adhesion and aggregation; local tissue factor exposure with factor VII binding and activation; thrombin generation and fibrin polymer formation; and acute, relapsing, or progressive renal failure.

Point mutations resulting in single amino acid substitutions, as well as deletions and frame shifts resulting in premature stop-codons, have been identified in both patients and family members. Most mutations occur in the factor H gene encoding for the short consensus repeat, or "sushi" domain, number 20 of the factor H carboxy-terminus.[182,183] It is via this carboxy-terminal region that factor H binds to C3b, as well as glycosaminoglycans (e.g., heparan).[181–184] Mutations are usually heterozygous (occasionally doubly heterozygous or homozygous), with variable penetrance that may reflect association with common polymorphisms.[175,181–183,185] The concurrent presence of several polymorphic variants of the factor H gene may promote the development of HUS.[175]

In some patients, serum factor H antigenic levels are 10 to 50% of normal because of aberrant protein folding and decreased secretion, serum C3 is low, and HUS develops at a young age.[183,186] In other patients, the factor H antigenic value is normal (or even elevated), but half (or occasionally more) of the circulating protein is functionally abnormal. The clinical onset of factor H-deficient HUS may be delayed until later in childhood (or even adulthood) in these patients, and may be precipitated by infection or pregnancy.[183,185,186]

Measurement of the serum ratio of the antigen level of an inactive metabolite of C3b, C3d, to total C3 antigen (i.e., C3d/C3) may be a more accurate method of screening for overactivation of C3 than measurement of total serum C3 antigen alone. An elevated C3d/C3 antigen ratio, along with a normal C4 antigen level (C4 is in the classical, but not the alternative, complement pathway), indicates overactivation of the alternative complement pathway.[103,187] A functional assay of factor H activity is under development.[188]

One report[189] suggests that occasional patients may develop factor H deficiency as a result of the production of autoantibodies against the protein.

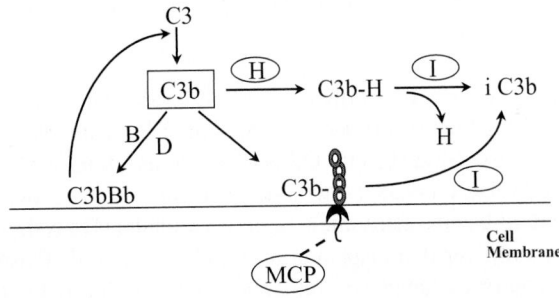

**Figure 50-6.** Control of the alternative complement pathway by plasma factor H membrane cofactor protein and plasma factor I. C3b, activated complement factor C3. B, complement factor B, which is activated to Bb by complement factor D. C3bBb, active C3 convertase. H, plasma factor H, which displaces Bb from C3b and allows its inactivation to iC3b by plasma complement factor I (C3b-cleaving protease). MCP, membrane cofactor protein, which functions on cell membranes as factor H functions (predominantly) in plasma. Heterozygous mutations in factor H, less commonly MCP, or least commonly factor I, cause increased susceptibility to a familial, recurrent type of HUS.

i. Therapy of Factor H Deficiency. The infusion of normal FFP containing factor H in familial factor H-deficient HUS patients has not been completely successful in preventing relapses or progressive renal disease.[176–178,190] Plasma exchange plus intravenous (IV) IgG may have been successful in restoring renal function in one case.[180] It was suggested that the IV IgG may adsorb C3b, and the C3b-bound IgG may then be removed by the plasmapheresis component of the plasma exchanges. Purified or recombinant factor H may eventually be developed for use in deficient patients. Transplantation of normal liver tissue may restore plasma factor H levels to normal, but it is a dangerous procedure.[191] Recurrences of HUS frequently destroy transplanted kidneys in factor H deficiency.[192]

*b. Inadequate Membrane Cofactor Protein (MCP) (Fig. 50-6).* MCP is a protein with four identical external C3b/Bb-interactive domains, a transmembrane domain, and an intracellular cytoplasmic domain. MCP is expressed on many cell types, especially renal endothelial cells.[103,193,194] It is encoded, along with factor H, within the complement-related gene cluster on chromosome 1q.[194] MCP impairs Bb attachment to C3b on the surfaces of renal endothelial (and other) cells. The formation of the alternative pathway C3b/Bb convertase that activates C3 to C3b is, thereby, suppressed. MCP, the membrane analogue of circulating plasma factor H, functions as a cofactor for factor I by displacing Bb from the C3b/Bb complex and exposing C3b to factor I-mediated proteolysis.[103,184,187,194] MCP function can be measured on the surface of blood mononuclear cells.[194]

Both heterozygous and homozygous forms of MCP-deficiency have been described.[194] HUS caused by MCP-deficiency may appear early in life and progress rapidly to end-stage renal failure or may be clinically mild. This is not entirely determined by homozygosity versus heterozygosity.[194] The differences may be influenced by plasma factor H level in individual patients, which is under separate genetic control (higher level of factor H = milder disease?).[187] Patients with MCP defects may have low serum C3 antigenic levels or an elevated C3d/C3 antigen ratio, along with normal serum C4 antigen, as in factor H deficiency.

i. Therapy of MCP Deficiency. There is no known specific treatment for MCP-deficient HUS. Dialysis and renal transplantation can be used to treat end-stage renal disease. MCP is expressed robustly on the surface of normal renal endothelial cells. Renal transplantation may, consequently, have a more favorable outcome and a decreased likelihood of recurrence in HUS caused by MCP-deficiency compared to the HUS due to factor H deficiency.[187]

*c. Inadequate Factor I (C3b-Cleaving Protease) Activity (Fig. 50-6).* One report[104] described a familial type of D- HUS in one infant and two young adults caused by heterozygous defects in the serine protease active site of factor I. In contrast to factor H and MCP genes, which are on chromosome 1, factor I is encoded on chromosome 4.

**2. Familial Intracellular Cobalamin Reduction Defects.** A rare type of HUS is the result of defective intracellular reduction of cobalamin $Co^{3+}$ to $Co^{2+}$, and then $Co^+$, the active ionic form.[105] As a consequence, there is reduced formation of adenosyl-cobalamin ($Co^+$) and methyl-

cobalamin ($Co^+$), or occasionally only the latter alone. Adenosyl-cobalamin ($Co^+$) is the cofactor for methylmalonyl CoA mutase, and methyl-cobalamin ($Co^+$) is the cofactor for methonine synthetase. Failure of these two enzyme to function causes methylmalonyl aciduria and hyperhomocysteinemia, respectively.[105,195,196] The precise gene/protein abnormalities responsible for the several different clinical and chemical phenotypes are unknown. A chromosome 8;19 translocation has been reported in two siblings and their asymptomatic father.

Cobalamin-deficient HUS usually occurs during the first year of life,[195,197] although in a few patients it has not appeared until later in childhood.[105,196] There are diffuse glomerular endothelial cell swelling, intravascular thrombi, and renal tubular atrophy.[105]

Although serum and red cell folate, along with serum $B_{12}$ and transcobalamin levels, are normal,[105,195–197] megaloblastic changes are present in peripheral blood and bone marrow. Associated problems may include failure to thrive or developmental delay; hypotonia or spasticity; micro- or hydrocephalus; psychosis, dementia, or seizures; absence of gastric parietal and chief cells; hepatic insufficiency; cardiomyopathy; and interstitial lung disease.

*a. Therapy of Intracellular Cobalamin Defects.* The course of cobalamin-deficient HUS is severe, chronic, and fatal from multiorgan failure unless frequent (every 1 to 3 days) high-dose (1 mg) subcutaneous hydroxycobalamin injections commence soon after the disease is recognized. Cyanocobalamin is ineffective.[105] These supraphysiological doses of parenteral vitamin $B_{12}$ somehow induce the generation of enough intracellular cobalamin ($Co^+$) to activate cobalamin ($Co^+$)-dependent enzymes. The concurrent administration of oral betaine may lower serum homocysteine levels via activation of betaine-homocysteine methyltransferase. Supplemental folic acid may be useful to counteract the effect of methyl-tetrahydrofolate "trapping."[105]

**3. Severe ADAMTS-13 Deficiency.** As mentioned in Section III.B, in some infants with less than 5% ADAMTS-13 and neonatal onset of chronic relapsing TTP, transient or progressive renal failure is a more prominent component of the disorder (Upshaw–Schulman syndrome)[60] than it is in most familial TTP patients.

## V. Other Types of Thrombotic Microangiopathies

Exposure to mitomycin C, cyclosporine, FK 506 (tacrolimus), quinine, gemcitabine, combinations of chemotherapeutic agents, or total-body irradiation has been associated with thrombotic microangiopathy weeks to months after exposure.[5,198,199] Thrombi may be predominantly renal

("HUS"), or systemic ("TTP"), and have been reported after allogeneic bone marrow,[5] kidney, liver, heart, or lung transplantion.[6] Bone marrow or stem cell transplantation-associated thrombotic microangiopathy is not usually associated with an absence or severely reduced level of plasma ADAMTS-13 activity.[25,80] It is not known if free radicals or other drug metabolites cause initial damage to endothelial cells. It is also not known if the delayed development of thrombotic microangiopathy is related to therapy-associated alterations in antigenic expression on renal and other types of endothelial cells, with subsequent antiendothelial cell antibody production.

In quinine-induced immune thrombocytopenia, patients produce antibodies against epitopes of platelet GPIb-IX-V or integrin $\alpha$IIb$\beta$3 that have been antigenically altered by the attachment of quinine. A subgroup of these patients also develop thrombotic microangiopathy,[7,199] possibly because the antibodies cross-react with quinine-altered $\beta$3 molecules on endothelial cell membranes.

### A. Therapy of These Other Types of Thrombotic Microangiopathies

Discontinuation of any putative disease-inducing drug should occur immediately. Patients in this "pathogenesis unknown" category of thrombotic microangiopathies often respond poorly to plasma exchange.[1,5] It has been demonstrated, however, that some of these patients improve in association with exchange.[25] It is, therefore, appropriate to commence daily plasma exchange procedures and continue for a sufficient period to determine effectiveness. Several patients with HUS-like thrombotic microangiopathy after solid organ transplantation have been reported to respond to plasma exchange plus IV IgG.[200,201] Plasma adsorption over Staphylococcal protein A columns has been reported to be useful in thrombotic microangiopathy as a result of mitomycin C exposure.[202]

## VI. Overview of Therapy for TTP Versus HUS

Microvascular platelet aggregation occurs in both TTP and HUS. If platelet aggregation is systemic, and especially if the central nervous system is involved, the disorder is usually called TTP. If platelet aggregation is predominantly confined to the renal circulation, the diagnosis is often HUS. Severe renal involvement in a "TTP" patient (detectable renal abnormalities occur in 50–75% of episodes),[14,17] or extrarenal manifestations in a patient with "HUS," may obscure clinical boundaries between the two entities.[14,17,203]

In patients with ADAMTS-13-deficient types of TTP, ADAMTS-13 activity measured on a pretreatment citrate-plasma sample is usually in the 0 to 5% range.[1,13,19,21,33,34,52,56,88] In contrast, plasma ADAMTS-13 is not severely reduced (or is normal) in some patients diagnosed with "TTP" and in HUS or other thrombotic microangiopathies.[33,67,70,147] Quantification of ADAMTS-13 activity may eventually allow patients with the ADAMTS-13 types of TTP to be identified precisely using plasma ADAMTS-13 assays. Assays of ADAMTS-13 are not generally or rapidly available currently and so are not yet capable of influencing emergency clinical decisions. This situation is likely to change with the implementation of newer and more rapid assay systems. It must be emphasized, however, that the correlation between pretreatment values obtained using the various new ADAMTS-13 testing methods and clinical responsiveness to plasma exchange awaits additional observation and experience. Currently, therefore, an adult patient who has an acquired syndrome that is either "TTP" or "HUS" should be presumed to have TTP, and plasma exchange should begin as soon as possible and continue until the clinical situation is clarified.[1,16,22,203]

## Acknowledgement

Support for the work of JLM provided by NIH grant P50–HL65967 and the Mary R. Gibson Foundation.

## References

1. Moake, J. L. (2002). Thrombotic microangiopathies. *N Engl J Med, 347,* 589–600.
2. Burns, E. R., Lou, Y., & Pathak, A. (2004). Morphologic diagnosis of thrombotic thrombocytopenic purpura. *Am J Hematol, 75,* 18–21.
3. Cohen, J. A., Brecher, M. E., & Bandarenko, N. (1998). Cellular source of serum lactate dehydrogenase elevation in patients with thrombotic thrombocytopenic purpura. *J Clin Apheresis, 13,* 16–19.
4. Amorosi, E. L., & Ultmann, J. E. (1966). Thrombotic thrombocytopenic purpura: Report of 16 cases and review of the literature. *Medicine, 45,* 139–159.
5. Moake, J. L., & Byrnes, J. J. (1996). Thrombotic microangiopathies associated with drugs and bone marrow transplantation. *Hematol/Oncol Clinics North America, 10,* 485–497.
6. Singh, N., Gayowski, T., & Marino, I. R. (1996). Hemolytic uremic syndrome in solid-organ transplant recipients. *Transplant Internat, 9,* 68–75.
7. Gottschall, J. L., Elliot, W., Lianos, E., et al. (1991). Quinine-induced immune thrombocytopenia associated with hemolytic-uremic syndrome: A new clinical entity. *Blood, 77,* 306–310.
8. Kojouri, K., Vesely, S., & George, J. N. (2001). Quinine-associated thrombotic thrombocytopenic purpura-hemolytic uremic syndrome: Frequency, clinical features, and long-term outcomes. *Ann Int Med, 135,* 1047–1051.

9. Moschcowitz, E. (1924). Hyaline thrombosis of the terminal arterioles and capillaries: A hitherto undescribed disease. *Proc NY Pathol Soc, 24,* 21–24.
10. Moschcowitz, E. (1925). An acute febrile pleiochromic anamia with hyaline thrombosis of the terminal arterioles and capillaries. *Arch Intern Med, 36,* 89–93.
11. Moake, J., Chintagumpala, M., Turner, N., et al. (1994). Solvent/detergent-treated plasma suppresses shear-induced platelet aggregation and prevents episodes of thrombotic thrombocytopenic purpura. *Blood, 84,* 490–497.
12. Moake, J. L., Rudy, C. K., Troll, J. H., et al. (1982). Unusually large plasma factor VIII: von Willebrand factor multimers in chronic relapsing thrombotic throbocytopenic purpura. *N Engl J Med, 307,* 1432–1435.
13. Furlan, M., Robles, R., Solenthaler, M., et al. (1997). Deficient activity of von Willegrand factor-cleaving protease in chronic relapsing thrombotic thrombocytopenic purpura. *Blood, 89,* 3097–3103.
14. Rock, G., Sumak, K., Buskard, N., et al. (1991). Comparison of plasma exchange with plasma infusion in the treatment of thrombotic thrombocytopenic purpura. *N Eng J Med, 325,* 393–397.
15. Shumak, K. H., Rock, G. A., & Nair, R. C. (1995). Late relapses in individuals successfully treated for thrombotic thrombocytopenic purpura. Canadian Apheresis Group. *Ann Intern Med, 122,* 569–572.
16. Byrnes, J. J., & Moake, J. L. (1986). Thrombotic thrombocytopenic purpura and the hemolytic-uremic syndrome: Evolving concepts of pathogenesis and therapy. *Clinics Haematol, 15,* 413–442.
17. Bell, W. R., Braine, H. G., Ness, P. M., et al. (1991). Improved survival in thrombotic thrombocytopenic purpura-hemolytic-uremic syndrome clinical experience in 108 patients. *New Eng J Med, 325,* 398–403.
18. Bennett, C. L., Weinberg, P. D., Rozenberg, B.-D. K., et al. (1998). Thrombotic thrombocytopenic purpura associated with ticlopidine: A review of 60 cases. *Ann Int Med, 128,* 541–544.
19. Bennett, C. L., Connors, J. M., Carwile, J. M., et al. (2000). Thrombotic thrombocytopenic purpura associated with clopidogrel. *New Engl J Med, 342,* 1773–1777.
20. Zakarija, A., Bandarenko, N., Pandey, D. K., et al. (2004). Clopidogrel-associated TTP: An update of pharmacovigilanc efforts conducted by independent researchers, and the Food and Drug Administration. *Stroke, 35,* 533–537.
21. Tsai, H.-M., Rice, L., Sarode, R., et al. (2000). Antibody inhibitors to von Willebrand factor metalloproteinase and increased von Willebrand factor-platelet binding in ticlopidine-associated thrombotic thrombocytopenic purpura. *Ann Int Med, 132,* 794–799.
22. Sadler, J. E., Moake, J. L., Miyata, T., et al. (2004). Recent advances in thrombotic thrombocytopenic purpura. *Hematology (Am Soc Hematol Educ Program),* 407–423.
23. McMinn, J. R., & George, J. N. (2001). Evaluation of women with clinically suspected thrombotic thrombocytopenic purpura-hemolytic uremic syndrome during pregnancy. *J Clin Apheresis, 16,* 202–209.
24. Neame, P. D. (1980). Immunologic and other factors in thrombotic thrombocytopenic purpura (TTP). *Semin Thromb Hemost, 6,* 416–429.
25. Vesely, S. K., George, J. N., Lammle, B., et al. (2003). ADAMTS13 activity in thrombotic thrombocytopenic purpura-hemolytic uremic syndrome: Relation to presenting features and clinical outcomes in a prospective cohort of 142 patients. *Blood, 102,* 60–68.
26. Moake, J. L., Byrnes, J. J., Troll, J. H., et al. (1985). Effects of fresh-frozen plasma and its cryosupernatant fraction on von Willebrand factor multimeric forms in chronic relapsing thrombotic thrombocytopenic purpura. *Blood, 65,* 1232–1236.
27. Moake, J. L., Turner, N. A., Stathopoulos, N. A., et al. (1988). Shear-induced platelet aggregation can be mediated by vWF released from platelets, as well as by exogenous large or unusually large vWF multimers, requires adenosine diphosphate, and is resistant to aspirin. *Blood, 71,* 1366–1374.
28. Frangos, J. A., Moake, J. L., Nolasco, L., et al. (1989). Cryosupernatant regulates accumulation of unusually large vWF multimers from endothelial cells. *Am J Physiol, 256,* H1635–1644.
29. Asada, Y., Sumiyoshi, A., Hayashi, T., et al. (1985). Immunochemistry of vascular lesions in thrombotic thromocytopenic purpura, with special reference to factor VIII related antigen. *Throm Res, 38,* 469–479.
30. Ruggeri, Z. M. (2000). Developing basic and clinical research on von Willebrand factor and von Willebrand disease. *Thromb Haemost, 84,* 147–149.
31. Moake, J. L., Turner, N. A., Stathopoulos, N. A., et al. (1986). Involvement of large plasma von Willebrand factor (vWF) multimers and unusually large vWF forms derived from endothelial cells in shear stress-induced platelet aggregation. *J Clin Invest, 78,* 1456–1461.
32. Arya, M., Anvari, B., Romo, G. M., et al. (2002). Ultra-large multimers of von Willebrand factor form spontaneous high-strength bonds with the platelet GP Ib-IX complex: Studies using optical tweezers. *Blood, 99,* 3971–3977.
33. Furlan, M., Robles, R., Galbusera, M., et al. (1998). von Willebrand factor-cleaving protease in thrombotic thrombocytopenic purpura and hemolytic-uremic syndrome. *N Eng J Med, 339,* 1578–1584.
34. Tsai, H. M., & Lian, E. C.-Y. (1998). Antibodies of von Willebrand factor cleaving protease in acute thrombotic thrombocytopenic purpura. *N Eng J Med, 339,* 1585–1594.
35. Tsai, H. M., Sussman, I. I., & Nagel, R. L. (1994). Shear stress enhances the proteolysis of von Willebrand factor in normal plasma. *Blood, 83,* 2171–2179.
36. Tsai, H. M. (1996). Physiologic cleavage of von Willebrand factor by a plasma protease is dependent on its confirmation and requires calcium ion. *Blood, 87,* 4235–4244.
37. Furlan, M., Robles, R., & Lammle, B. (1996). Partial purification and characterization of a protease from human plasma cleaving von Willebrand factor to fragments produced by *in vivo* proteolysis. *Blood, 87,* 4223–4234.
38. Fujikawa, K., Suzuki, H., McMullen, B., et al. (2001). Purification of von Willebrand factor-cleaving protease and its

identification as a new member of the metalloproteinase family. *Blood, 98,* 1662–1666.

39. Zheng, X., Chung, C., Takayama, T. K., et al. (2001). Structure of von Willebrand factor cleaving protease (ADAMTS13), a metaloprotease involved in thrombotic thrombocytopenic purpura. *J Biol Chem, 276,* 41059–41063.

40. Bork, P., & Beckmann, G. (1993). The CUB domain — a widespread module in developmentally regulated proteins. *J Mol Biol, 231,* 530–545.

41. Barbot, J., Costa, E., Guerra, M., et al. (2001). Ten years of prophylactic treatment with fresh-frozen plasma in a child with chronic relapsing thrombotic thrombocytopenic purpura as a result of a congenital deficiency of von Willebrand factor-cleaving protease. *Br J Haematol, 113,* 649–651.

42. Chung, D. W., & Fujikawa, K. (2002). Processing of von Willebrand factor by ADAMTS-13. *Biochem, 41,* 11065–11070.

43. Dong, J.-F., Moake, J. L., Bernardo, A., et al. (2003). ADAMTS-13 metalloprotease interacts with the endothelial cell-derived ultra-large von Willebrand factor. *J Biol Chem, 278,* 29633–29639.

44. Dong, J.-F., Moake, J. L., Nolasco, L., et al. (2002). ADAMTS-13 rapidly cleaves newly secreted ultralarge von Willebrand factor multimers on the endothelial surface under flowing conditions. *Blood, 100,* 4033–4039.

45. Padilla, A., Moake, J. L., Bernardo, A., et al. (2004). P-selectin anchors newly released ultralarge von Willebrand factor multimers to the endothelial cell surface. *Blood, 103,* 2150–2156.

46. Bernardo, A., Ball, C., Nolasco, L., et al. (2004). Effects of inflammatory cytokines on the release and cleavage of the endothelial cell-derived ultra-large von Willebrand factor multimers under flow. *Blood, 104,* 100–106.

47. Bernardo, A., Nolasco, L., Ball, C., et al. (2003). Peptides from the C-terminal regions of ADAMTS-13 specifically block cleavage of ultra-large von Willebrand factor multimers on the endothelial surface under flow. *J Thrombosis Haemostasis, 1,* July, Abstract #OC405.

48. Tao, Z., Peng, Y., Nolasco, L., et al. (2005). Recombinant CUB–1 domain poly peptide inhibits the cleavage of ULVWF strings by ADAMTS–13 under flowing conditions. *Blood, 106,* 4139–4145.

49. Majerus, E. M., Anderson, P. J., & Sadler, J. E. (2005). Binding of ADAMTS13 to von Willebrand factor. *J Biol Chem, 280,* 21773–21778.

50. Nishio, K., Anderson, P. J., Zheng, X. L., et al. (2004). Binding of platelet glycoprotein Ibalpha to von Willebrand factor domain A1 stimulates the cleavage of the adjacent domain A2 by ADAMTS13. *Proc Natl Acad Sci USA, 101,* 10578–10583.

51. Moake, J. L., & McPherson, P. D. (1989). Abnormalities of von Willebrand factor multimers in thrombotic thrombocytopenic purpura and hemolytic-uremic syndrome. *Am J Med, 87*(3N), 9N–15N.

52. Furlan, M., Robles, R., Solenthaler, M., et al. (1998). Acquired deficiency of von Willebrand factor-cleaving protease in a patient with thrombotic thrombocytopenic purpura. *Blood, 91,* 2839–2846.

53. Bianchi, V., Robles, R., Alberio, L., et al. (2002). Von Willebrand factor-cleaving protease (ADAMTS13) in thrombotic thrombocytopenic disorders: A severely deficient activity is specific for thrombotic thrombocytopenic purpura. *Blood, 100,* 710–713.

54. Bernardo, A., Ball, C., Nolasco, L., et al. (2005). Platelets adhered to endothelial cell-bound ultra-large von Willebrand factor strings support leukocyte tethering and rolling under high shear stress. *J Thromb Haemost, 3,* 562–570.

55. Chow, T. W., Turner, N. A., Chintagumpala, M., et al. (1998). Increased von Willebrand factor binding to platelets in single episode and recurrent types of thrombotic thrombocytopenic purpura. *Am J Hematol, 57,* 293–302.

56. Furlan, M., Robles, R., Morselli, B., et al. (1999). Recovery and half-life of von Willebrand factor-cleaving protease after plasma therapy in patients with thrombotic thrombocytopenic purpura. *Thromb Haemost, 81,* 8–13.

57. Allford, S. L., Harrison, P., Lawrie, A. S., et al. (2000). Von Willebrand factor-cleaving protease in congenital thrombotic thrombocytopenic purpura. *Br J Haematol, 111,* 1215–1222.

58. Levy, G. A., Nichols, W. C., Lian, E. C., et al. (2001). Mutations in a member of the ADAMTS gene family cause thrombotic thrombocytopenic purpura. *Nature, 413,* 488–494.

59. Pimanda, J. E., Maekawa, A., Wind, T., et al. (2004). Congenital thrombotic thrombocytopenic purpura in association with a mutation in the second CUB domain of ADAMTS13. *Blood, 103,* 627–629.

60. Veyradier, A., Obert, B., Haddad, E., et al. (2003). Severe deficiency of the specific von Willebrand factor-cleaving protease (ADAMTS 13) activity in a subgroup of children with atypical hemolytic uremic syndrome. *J Pediatr, 142,* 310–317.

61. Schulman, I., Pierce, M., Lukens, A., et al. (1960). Studies on thrombopoiesis. I. A factor in normal human plasma required for platelet production; chronic thrombocytopenia due to its deficiency. *Blood, 16,* 943–957.

62. Upshaw, J. D., Jr. (1978). Congenital deficiency of a factor in normal plasma that reverses microangiopathic hemolysis and thrombocytopenia. *N Engl J Med, 298,* 1350–1352.

63. Veyradier, A., & Girma, J. P. (2004). Assays of ADAMTS-13 activity. *Semin Hematol, 41,* 41–47.

64. Klaus, C., Plaimauer, B., Studt, J. D., et al. (2004). Epitope mapping of ADAMTS13 autoantibodies in acquired thrombotic thrombocytopenic purpura. *Blood, 103,* 4514–4519.

65. Majerus, E. M., Zheng, X., Tuley, E. A., et al. (2003). Cleavage of the ADAMTS13 propeptide is not required for protease activity. *J Biol Chem, 278,* 46643–46648.

66. Studt, J. D., Kremer Hovinga, J. A., Radonic, R., et al. (2004). Familial acquired thrombotic thrombocytopenic purpura: ADAMTS-13 inhibitory autoantibodies in identical twins. *Blood, 103,* 4195–4197.

67. Veyradier, A., Obert, B., Houllier, A., et al. (2001). Specific von Willebrand factor — cleaving protease in thrombotic microangiopathies: A study of 111 cases. *Blood, 98,* 1765–1762.

68. Zheng, X. L., Kaufman, R. M., Goodnough, L. T., et al. (2004). Effect of plasma exchange on plasma ADAMTS13 metalloprotease activity, inhibitor level, and clinical outcome in patients with idiopathic and non-idiopathic thrombotic thrombocytopenic purpura. *Blood, 103,* 4043–4049.
69. Oleksowicz, L., Bhagwati, N., & DeLoen-Fernandez, M. (1999). Deficient activity of von Willebrand's factor-cleaving protease in patients with disseminated malignancies. *Cancer Res, 59,* 2244–2250.
70. Mannucci, P. M., Canciani, M. T., Forza, I., et al. (2001). Changes in health and disease of the metalloprotease that cleaves von Willebrand factor. *Blood, 98,* 2730–2735.
71. Tripodi, A., Chantarangkul, V., Bohm, M., et al. (2004). Measurement of von Willebrand factor cleaving protease (ADAMTS-13): Results of an international collaborative study involving 11 methods testing the same set of coded plasmas. *J Thromb Haemost, 2,* 1601–1609.
72. Rick, M. E., Moll, S., Taylor, M. A., et al. (2002). Clinical use of a rapid collagen-binding assay for von Willebrand factor cleaving protease in patients with thrombotic thrombocytopenic purpura. *Thromb Haemost, 88,* 598–604.
73. Tao, Z., Wang, Y., Choi, H., et al. (2005). Cleavage of ultralarge multimers of von Willebrand factor by C-terminal-truncated mutants of ADAMTS-13 under flow. *Blood, 106,* 141–143.
74. Zheng, X., Nishio, K., Majerus, E. M., et al. (2003). Cleavage of von Willebrand requires the spacer domain of the metalloprotease ADAMTS13. *J Biol Chem, 278,* 30136–30141.
75. Kokame, K., Matsumoto, M., Fujimura, Y., et al. (2004). VWF73, a region from D1596 to R1668 of von Willebrand factor, provides a minimal substrate for ADAMTS-13. *Blood, 103,* 607–612.
76. Whitelock, J. L., Nolasco, L., Bernardo, A., et al. (2004). ADAMTS-13 activity in plasma is rapidly measured by a new ELISA method that uses recombinant VWF-A2 domain as substrate. *J Thromb Haemost, 2,* 485–491.
77. Zhou, W., & Tsai, H. M. (2004). An enzyme immunoassay of ADAMTS13 distinguishes patients with thrombotic thrombocytopenic purpura from normal individuals and carriers of ADAMTS13 mutations. *Thromb Haemost, 91,* 806–811.
78. Raife, T., Atkinson, B., Montgomery, R., et al. (2004). Severe deficiency of VWF-cleaving protease (ADAMTS13) activity defines a distinct population of thrombotic microangiopathy patients. *Transfusion, 44,* 146–150.
79. Mori, Y., Wada, H., Gabazza, E. C., et al. (2002). Predicting response to plasma exchange in patients with thrombotic thrombocytopenic purpura with measurement of vWF-cleaving protease activity. *Transfusion, 42,* 572–580.
80. van der Plas, R. M., Schiphorst, M. E., Huizinga, E. G., et al. (1999). von Willebrand factor proteolysis is deficient in classic, but not in bone marrow transplantation-associated thrombotic thrombocytopenic purpura. *Blood, 93,* 3798–3802.
81. Elliott, M. A., Nichols, W. L., Plumhoff, E. A., et al. (2003). Posttransplantation thrombotic thrombocytopenic purpura:

A single-center experience and a contemporary review. *Mayo Clin Proc, 78,* 421–430.
82. Charba, D., Moake, J. L., Harris, M. A., et al. (1993). Abnormalities of von Willebrand factor multimers in drug-associated thrombotic microangiopathies. *Am J Hematol, 42,* 268–277.
83. Byrnes, J. J., & Khurana, M. (1977). Treatment of thrombotic thrombocytopenic purpura with plasma. *New Engl J Med, 297,* 1386–1389.
84. Reiter, R. A., Knobl, P., Varadi, K., et al. (2003). Changes in von Willebrand factor-cleaving protease (ADAMTS13) activity after infusion of desmopressin. *Blood, 101,* 946–948.
85. Plaimauer, B., Zimmermann, K., Volkel, D., et al. (2002). Cloning, expression, and functional characterization of the von Willebrand factor-cleaving protease (ADAMTS13). *Blood, 100,* 3626–3632.
86. Yarranton, H., Lawrie, A. S., Purdy, G., et al. (2004). Comparison of von Willebrand factor antigen, von Willebrand factor-cleaving protease and protein S in blood components used for treatment of thrombotic thrombocytopenic purpura. *Transfus Med, 14,* 39–44.
87. Abassi, E., Yawn, D., Leveque, C., et al. (2004). Correlation of ADAMTS-13 activity with response to plasma exchange in patients diagnosed with thrombotic thrombocytopenic purpura. *Blood, 104,* 242a.
88. Tsai, H. M. (2000). High titers of inhibitors of von Willebrand factor-cleaving metalloproteinase in a fatal case of acute thrombotic thrombocytopenic purpura. *Am J Hematol, 65,* 251–255.
89. Tsai, H. M., Li, A., & Rock, G. (2001). Inhibitors of von Willebrand factor-cleaving protease in thrombotic thrombocytopenic purpura. *Clin Lab, 47,* 387–392.
90. Gutterman, L. A., Kloster, B., & Tsai, H. M. (2002). Rituximab therapy for refractory thrombotic thrombocytopenic purpura. *Blood Cells Mol Dis, 28,* 385.
91. Chemnitz, J., Draube, A., Scheid, C., et al. (2002). Successful treatment of severe thrombotic thrombocytopenic purpura with the monoclonal antibody rituximab. *Am J Hematol, 71,* 105.
92. Reff, M., Carner, K., Chambers, K., et al. (1994). Depletion of B cells in vivo by a chimeric mouse human monoclonal antibody to CD20. *Blood, 83,* 435–445.
93. Tsai, H. M., & Shulman, K. (2003). Rituximab induces remission of cerebral ischemia caused by thrombotic thrombocytopenic purpura. *Eur J Haematol, 70,* 183–185.
94. Thompson, C. E., Damon, L. E., Ries, C. A., et al. (1992). Thrombotic microangiopathies in the 1980s: Clinical features, response to treatment, and the impact of the human immunodeficiency virus epidemic. *Blood, 80,* 1890–1895.
95. Crowther, M. A., Heddle, N., Hayward, C. P. M., et al. (1996). Splenectomy done during hematologic remission to prevent relapse in patients with thrombotic thrombocytopenic purpura. *Ann Int Med, 125,* 294–296.
96. Kremer Hovinga, J. A., Studt, J. D., Demarmels Biasiutti, F., et al. (2004). Splenectomy in relapsing and plasma-

refractory acquired thrombotic thrombocytopenic purpura. *Haematologica, 89,* 320–324.

97. Downes, K. A., Yomtovian, R., Tsai, H. M., et al. (2004). Relapsed thrombotic thrombocytopenic purpura presenting as an acute cerebrovascular accident. *J Clin Apheresis, 19,* 86–89.

98. Gasser, C., Gautier, E., Steck, A., et al. (1955). [Hemolytic-uremic syndrome: Bilateral necrosis of the renal cortex in acute acquired hemolytic anemia.]. *Schweiz Med Wochenschr, 85,* 905–909.

99. Exeni, R., Donato, H., Rendo, P., et al. (1998). Low levels of serum erythropoietin in children with endemic hemolytic uremic syndrome. *Pediatr Nephrol, 12,* 226–230.

100. Kaplan, B. S., & Proesmans, W. (1987). The hemolytic uremic syndrome of childhood and its variants. *Semin Hematol, 24,* 148–160.

101. Cleary, T. G. (1988). Cytotoxin producing Escherichia coli and the hemolytic uremic syndrome. *Pediatr Clin North Am, 35,* 485–501.

102. Karmali, M. A., Petric, M., Lim, C., et al. (1985). The association between idiopathic hemolytic uremic syndrome and infection by verotoxin-producing Escherichia coli. *J Infect Dis, 151,* 775–782.

103. Bonnardeaux, A., & Pichette, V. (2003). Complement dysregulation in haemolytic uraemic syndrome. *Lancet, 362,* 1514–1515.

104. Fremeaux-Bacchi, V., Dragon-Durey, M. A., Blouin, J., et al. (2004). Complement factor I: A susceptibility gene for atypical haemolytic uraemic syndrome. *J Med Genet, 41,* e84.

105. Van Hove, J. L., Van Damme-Lombaerts, R., Grunewald, S., et al. (2002). Cobalamin disorder Cbl-C presenting with late-onset thrombotic microangiopathy. *Am J Med Genet, 111,* 195–201.

106. Banatvala, N., Griffin, P. M., Greene, K. D., et al. (2001). The United States National Progressive Hemolytic Uremic Syndrome Study: Microbiologic, serologic, clinical, and epidemiologic findings. *J Inf Dis, 183,* 1063–1070.

107. Brett, K. N., Hornitzky, M. A., Bettelheim, K. A., et al. (2003). Bovine non-O157 Shiga toxin 2-containing Escherichia coli isolates commonly possess stx2-EDL933 and/or stx2vhb subtypes. *J Clin Microbiol, 41,* 2716–2722.

108. Misselwitz, J., Karch, H., Bielazewska, M., et al. (2003). Cluster of hemolytic-uremic syndrome caused by Shiga toxin-producing Escherichia coli O26:H11. *Pediatr Infect Dis J, 22,* 349–354.

109. Matussek, A., Lauber, J., Bergau, A., et al. (2003). Molecular and functional analysis of Shiga toxin-induced response patterns in human vascular endothelial cells. *Blood, 102,* 1323–1332.

110. Eklund, M., Leino, K., & Siitonen, A. (2002). Clinical Escherichia coli strains carrying stx genes: Stx variants and stx-positive virulence profiles. *J Clin Microbiol, 40,* 4585–4593.

111. Obrig, T. G. (1992). Pathogenesis of Shiga toxin (verotoxin)-induced endothelial cell injury. In B. S. Kaplan, R. S. Trompeter, & J. L. Moake (Eds.), *Hemolytic-uremic syndrome and thrombotic thrombocytopenic purpura* (pp. 405–419). New York: Marcel Dekker.

112. Thorpe, C. M. (2004). Shiga toxin-producing Escherichia coli infection. *Clin Infect Dis, 38,* 1298–1303.

113. Garg, A. X., Suri, R. S., Barrowman, N., et al. (2003). Long-term renal prognosis of diarrhea-associated hemolytic uremic syndrome: A systematic review, meta-analysis, and meta-regression. *JAMA, 290,* 1360–1370.

114. Mbonu, C. C., Davison, D. L., El-Jazzar, K. M., et al. (2003). Clostridium difficile colitis associated with hemolytic-uremic syndrome. *Am J Kidney Dis, 41,* E14.

115. Cochran, J. B., Panzarino, V. M., Maes, L. Y., et al. (2004). Pneumococcus-induced T-antigen activation in hemolytic uremic syndrome and anemia. *Pediatr Nephrol, 19,* 317–321.

116. Siegler, R. L., Obrig, T. G., Pysher, T. J., et al. (2003). Response to Shiga toxin 1 and 2 in a baboon model of hemolytic uremic syndrome. *Pediatr Nephrol, 18,* 92–96.

117. Pysher, T. J., Siegler, R. L., Tesh, V. L., et al. (2002). von Willebrand factor expression in a Shiga toxin-mediated primate model of hemolytic uremic syndrome. *Pediatr Dev Pathol, 5,* 472–479.

118. Siegler, R. L., Pysher, T. J., Lou, R., et al. (2001). Response to Shiga toxin-1, with and without lipopolsaccharide, in a primate model of hemolytic uremic syndrome. *Am J Nephrol, 21,* 420–425.

119. Fraser, M. E., Fujinaga, M., Cherney, M. M., et al. (2004). Structure of shiga toxin type 2 (Stx2) from Escherichia coli O157:H7. *J Biol Chem, 279,* 27511–27517.

120. Lindberg, A. A., Brown, J. E., Stromberg, N., et al. (1987). Identification of the carbohydrate receptor for Shiga toxin produced by Shigella dysenteriae type I. *J Biol Chem, 262,* 1779–1785.

121. Simon, M., Cleary, T. G., Hernandez, J. D., et al. (1998). Shiga toxin 1 elicits diverse biologic responses in mesangial cells. *Kidney Int, 54,* 1117–1127.

122. Kiyokawa, N., Taguchi, T., Mori, T., et al. (1998). Induction of apoptosis in normal human renal tubular epithelial cells by Escherichia coli Shiga toxins 1 and 2. *J Infect Dis, 178,* 178–184.

123. Uchida, H., Kiyokawa, N., Horie, H., et al. (1999). The detection of shiga toxin in the kidney of a patient with hemolytic uremic syndrome. *Pediatr Res, 45,* 133–137.

124. Karpman, D., Papadopoulou, D., Nilsson, K., et al. (2001). Platelet activation by Shiga toxin and circulatory factors as a pathogenetic mechanism in the hemolytic uremic syndrome. *Blood, 97,* 3100–3108.

125. Karmali, M. A. (2004). Infection by Shiga toxin-producing Escherichia coli: An overview. *Mol Biotechnol, 26,* 117–122.

126. Cooling, L. L., Walker, K. E., Gille, T., et al. (1998). Shiga toxin binds human platelets via globotriaosylceramide (Pk antigen) and a novel platelet glycosphingolipid. *Infect Immun, 66,* 4355–4366.

127. Kovbasnjuk, O., Edidin, M., & Donowitz, M. (2001). Role of lipid rafts in Shiga toxin 1 interaction with the apical surface of Caco-2 cells. *J Cell Sci, 114,* 4025–4031.

128. Deschenes, G., Casenave, C., Grimont, F., et al. (1996). Cluster of cases of haemolytic uraemic syndrome due to unpasteurized cheese. *Pediatr Nephrol, 10,* 203–205.

129. Pruimboom-Brees, I. M., Morgan, T. W., Ackermann, M. R., et al. (2000). Cattle lack vascular receptors for Escherichia coli O157:H7 Shiga toxins. *Proc Nat Acad Sci USA, 97,* 10325–10329.

130. Varma, J. K., Greene, K. D., Reller, M. E., et al. (2003). An outbreak of Escherichia coli O157 infection following exposure to a contaminated building. *JAMA, 290,* 2709–2712.

131. Proulx, F., Toledano, B., Phan, V., et al. (2002). Circulating granulocyte colony-stimulating factor, C-X-C, and C-C chemokines in children with Escherichia coli O157:H7 associated hemolytic uremic syndrome. *Pediatr Res, 52,* 928–934.

132. Thorpe, C. M., Hurley, B. P., Lincicome, L. L., et al. (1999). Shiga toxins stimulate secretion of interleukin-8 from intestinal epithelial cells. *Infect Immunol, 67,* 5985–5993.

133. Thorpe, C. M., Smith, W. E., Hurley, B. P., et al. (2001). Shiga toxins induce, superinduce, and stabilize a variety of C-X-C chemokine mRNAs in intestinal epithelial cells, resulting in increased chemokine expression. *Infect Immunol, 69,* 6140–6147.

134. Moxley, R. A. (2004). Escherichia coli O157:H7: An update on intestinal colonization and virulence mechanisms. *Anim Health Res Rev, 5,* 15–33.

135. Foster, G. H., & Tesh, V. L. (2002). Shiga toxin 1-induced activation of c-Jun NH(2)-terminal kinase and in the human monocytic cell line THP-1: Possible involvement in the production of TNF-alpha. *Leukoc Biol, 71,* 107–114.

136. Williams, J. M., Boyd, B., Nutikka, A., et al. (1999). A comparison of the effects of verocytotoxin-1 on primary human renal cell cultures. *Toxicol Lett, 105,* 47–57.

137. Hughes, A. K., Stricklett, P. K., & Kohan, D. E. (1998). Cytotoxic effect of Shiga toxin-1 on human proximal tubule cells. *Kidney Int, 54,* 426–437.

138. Adler, S., & Bollu, R. (1998). Glomerular endothelial cell injury mediated by Shiga-like toxin-1. *Kidney Blood Press Res, 21,* 12–21.

139. Hughes, A. K., Stricklett, P. K., & Kohan, D. E. (2001). Shiga toxin-1 regulation of cytokine production by human glomerular epithelial cells. *Nephron, 88,* 14–23.

140. Eisenhauer, P. B., Chaturvedi, P., Fine, R. E., et al. (2001). Tumor necrosis factor alpha incrases human cerebral endothelial cell Gb3 and sensitivity to Shiga toxin. *Infect Immunol, 69,* 1889–1894.

141. Siegler, R. L., Pysher, T. J., Tesh, V. L., et al. (2001). Response to single and divided doses of Shiga toxin-1 in a primate model of hemolytic uremic syndrome. *J Am Soc Nephrol, 12,* 1458–1467.

142. Morigi, M., Galbusera, M., Binda, E., et al. (2001). Verotoxin-1 induced up-regulation of adhesive molecules renders microvascular endothelial cells thrombogenic at high shear stress. *Blood, 98,* 1828–1835.

143. Ray, P. E., & Liu, X. H. (2001). Pathogenesis of Shiga toxin-induced hemolytic uremic syndrome. *Pediatr Nephrol, 16,* 823–839.

144. Stricklett, P. K., Hughes, A. K., Ergonul, Z., et al. (2002). Molecular basis for up-regulation by inflammatory cytokines of Shiga toxin 1 cytotoxicity and globotriaosylceramide expression. *J Infect Dis, 186,* 976–982.

145. Moake, J. L. (1994). Haemolytic-uraemic syndrome: Basic science. *Lancet, 343,* 393–397.

146. Nolasco, L., Turner, N., Bernardo, A., et al. (2005). Hemolytic-uremic syndrome — associated Shiga toxins promote endothelial cell secretion and impair ADAMTS-13 cleavage of unusually large von Willebrand factor multimers. *Blood, 106,* 4199–4209.

147. Tsai, H.-M., Chandler, W. L., Sarode, R., et al. (2001). von Willebrand factor and von Willebrand factor-cleaving metalloprotease activity in Escherichia coli O157:H7-associated hemolytic uremic syndrome. *Pediatr Res, 49,* 653–659.

148. Hunt, B. J., Lammle, B., Nevard, C. H., et al. (2001). von Willebrand factor-cleaving protease in childhood diarrhoea-associated haemolytic uraemic syndrome. *Thromb Haemost, 85,* 975–978.

149. Gerritsen, H. E., Turecek, P. L., Schwarz, H. P., et al. (1999). Assay of von Willebrand factor (vWF)-cleaving protease based on decreased collagen binding affinity of degraded vWF: A tool for the diagnosis of thrombotic thrombocytopenic purpura (TTP). *Thromb Haemost, 82,* 1386–1389.

150. Sutor, A. H., Thomas, K. B., Prufer, F. H., et al. (2001). Function of von Willebrand factor in children with diarrhea-associated hemolytic-uremic syndrome (D + HUS). *Semin Thromb Hemost, 27,* 287–292.

151. Moake, J. L., Byrnes, J. J., Troll, J. H., et al. (1984). Abnormal VIII: von Willebrand factor patterns in the plasma of patients with the hemolytic-uremic syndrome. *Blood, 64,* 592–598.

152. Galbusera, M., Benigni, A., Paris, S., et al. (1999). Unrecognized pattern of von Willebrand factor abnormalities in hemolytic uremic syndrome and thrombotic thrombocytopenic purpura. *J Am Soc Nephrol, 10,* 1234–1241.

153. Gordjani, N., & Sutor, A. H. (1998). Coagulation changes associated with the hemolytic uremic syndrome. *Semin Thromb Hemost, 24,* 577–582.

154. van Setten, P. A., van Hinsbergh, V. W. M., van den Heuvel, L. P. W. J., et al. (1998). Monocyte chemoattractant protein-1 and interleuken-8 levels in urine and serum of patients with hemolytic uremic syndrome. *Pediatr Res, 43,* 759–767.

155. Milford, D. V., Taylor, C. M., Rafaat, F., et al. (1989). Neutrophil elastases and haemolytic uraemic syndrome. *Lancet, 2,* 1153.

156. Fitzpatrick, M. M., Shah, V., Trompeter, R. S., et al. (1992). Interleukin-8 and polymorphoneutrophil leukocyte activation in hemolytic-uremic syndrome of childhood. *Kidney Int, 42,* 951–956.

157. Walters, M. D., Matthei, I. U., Kay, R., et al. (1989). The polymorphunuclear leukocyte count in childhood haemolytic uraemic syndrome. *Pediatr Nephrol, 3,* 130–134.

158. Alevriadou, B. R., Moake, J. L., Turner, N. A., et al. (1983). Real-time analysis of shear-dependent thrombus formation and its blockade by inhibitors of von Willebrand factor binding to platelets. *Blood, 81,* 1263–1276.

159. Chandler, W. L., Jelacic, S., Boster, D. R., et al. (2002). Prothrombotic coagulation abnormalities preceding the hemolytic-uremic syndrome. *N Engl J Med, 346,* 23–32.

160. Ishii, H., Takada, K., Higuchi, T., et al. (2000). Verotoxin-1 induces tissue factor expression in human umbilical vein endothelial cells through activation of NF-kappaB/Rel and AP-1. *Thromb Haemost, 84,* 712–721.

161. Takenouchi, H., Kiyokawa, N., Taguchi, T., et al. (2004). Shiga toxin binding to globotriaosyl ceramide induces intracellular signals that mediate cytoskeleton remodeling in human renal carcinoma-derived cells. *J Cell Sci, 117,* 3911–3922.

162. Tarr, P. I., Gordon, C. A., & Chandler, W. L. (2005). Shiga-toxin-producing Escherichia coli and haemolytic uraemic syndrome. *Lancet, 365,* 1073–1086.

163. Fernandez, G. C., Te Loo, M. W., van der Velden, T. J., et al. (2003). Decrease of thrombomodulin contributes to the procoagulant state of endothelium in hemolytic uremic syndrome. *Pediatr Nephrol, 18,* 1066–1068.

164. Williams, J. M., & Taylor, C. M. (2005). Decrease of thrombomodulin contributes to the procoagulant state of endothelium in haemolytic uraemic syndrome. *Pediatr Nephrol, 20,* 243; author reply 244.

165. Misiani, R., Appiani, A. C., Edefonti, A., et al. (1982). Haemolytic uraemic syndrome: Therapeutic effect of plasma infusion. *Br Med J, 285,* 1304–1306.

166. Sheth, K. J., Gill, J. C., Hanna, J., et al. (1988/1989). Failure of fresh frozen plasma infusions to alter the course of hemolytic-uremic syndrome. *Child Nephrol Urol, 9,* 38–41.

167. Grif, K., Dierich, M. P., Karch, H., et al. (1998). Shiga toxin released from enterohemorrhagic Escherichia coli 0157 following exposure to subinhibitory concentrations of antimicrobial agents. *Eur J Clin Microbiol Infect Dis, 17,* 761–766.

168. Karmali, M. A., Mascarenhas, M., Petric, M., et al. (2003). Age-specific frequencies of antibodies to Escherichia coli verocytotoxins (Shiga toxins) 1 and 2 among urban and rural populations in southern Ontario. *J Infect Dis, 188,* 1724–1729.

169. Trachtman, H., Cnaan, A., Christen, E., et al. (2003). Effect of an oral Shiga toxin-binding agent on diarrhea-associated hemolytic uremic syndrome in children: A randomized controlled trial. *JAMA, 290,* 1337–1344.

170. Siegler, R. L., Pysher, T. J., Tesh, V. L., et al. (2002). Prophylactic heparinization is ineffective in a primate model of hemolytic uremic syndrome. *Pediatr Nephrol, 17,* 1053–1058.

171. Van Dyck, M., & Proesmans, W. (2004). Renoprotection by ACE inhibitors after severe hemolytic uremic syndrome. *Pediatr Nephrol, 19,* 688–690.

172. Noris, M., Ruggenenti, P., Perna, A., et al. (1999). Hypocomplementemia discloses genetic predisposition to hemolytic uremic syndrome and thrombotic thrombocytopenic purpura: Role of factor H abnormalities. *J Am Soc Nephrol, 10,* 281–293.

173. Muller, T., Sikora, P., Offnerr, G., et al. (1998). Recurrence of renal disease after kidney transplantation in children: Twenty-four years of experience in a single center. *Clin Nephrol, 49,* 82–90.

174. Pichette, V., Querin, S., Schurch, W., et al. (1994). Familial hemolytic-uremic syndrome and homozygous factor H deficiency. *Am J Kidney Dis, 24,* 936–941.

175. Caprioli, J., Castelletti, F., Bucchioni, S., et al. (2003). Complement factor H mutations and gene polymorphisms in haemolytic uraemic syndrome: The C-257T, the G2881T polymorphisms are strongly associated with the disease. *Hum Mol Genet, 12,* 3385–3395.

176. Warwicker, P., Goodship, T. H. J., Donne, R. L., et al. (1998). Genetic studies into inherited and sporadic hemolytic uremic syndrome. *Kidney Internat, 53,* 836–844.

177. Ohali, M., Shalev, H., Schlesinger, M., et al. (1998). Hypocomplementemic autosomal recessive hemolytic uremic syndrome with decreased factor H. *Pedriatr Nephrol, 12,* 619–624.

178. Rougier, N., Kazatchkine, M. D., Rougier, J.-P., et al. (1998). Human complement factor H deficiency associated with hemolytic uremic syndrome. *J Am Soc Nephrol, 9,* 2318–2326.

179. Landau, D., Shalev, H., Levy-Finer, G., et al. (2001). Familial hemolytic uremic syndrome associated with complement factor H deficiency. *J Pediatr, 138,* 412–417.

180. Stratton, J. D., & Warwicker, P. (2002). Successful treatment of factor H-related haemolytic uraemic syndrome. *Nephrol Dial Transplant, 17,* 684–685.

181. Perez-Caballero, D., Gonzalez-Rubio, C., Gallardo, M. E., et al. (2001). Clustering of missence mutations in the C-terminal region of factor H in atypical hemolytic uremic syndrome. *Am J Hum Genet, 68,* 478–484.

182. Zipfel, P. F. (2001). Hemolytic uremic syndrome: How do factor H mutants mediate endothelial damage. *Trends Immunol, 22,* 345.

183. Caprioli, J., Bettinaglio, P., Zipfel, P. F., et al. (2001). The molecular basis of familial hemolytic uremic syndrome: Mutation analysis of factor H gene reveals a hot spot in short consensus repeat 20. *J Am Soc Nephrol, 12,* 297–307.

184. Manuelian, T., Hellwage, J., Meri, S., et al. (2003). Mutations in factor H reduce binding affinity to C3b and heparin and surface attachment to endothelial cells in hemolytic uremic syndrome. *J Clin Invest, 111,* 1181–1190.

185. Richards, A., Buddles, M. R., Donne, R. L., et al. (2001). Factor H mutations in hemolytic uremic syndrome cluster in exons 18-20, a domain important for host cell recognition. *Am J Hum Genet, 68,* 485–490.

186. Taylor, C. M. (2001). Hemolytic-uremic syndrome and complement factor H deficiency: Clinical aspects. *Sem Thromb Hemost, 27,* 185.

187. Noris, M., Brioschi, S., Caprioli, J., et al. (2003). Familial haemolytic uraemic syndrome and an MCP mutation. *Lancet, 362,* 1542–1547.

188. Sanchez-Corral, P., Gonzalez-Rubio, C., Rodriguez De Cordoba, S., et al. (2004). Functional analysis in serum from atypical hemolytic uremic syndrome patients reveals impaired protection of host cells associated with mutations in factor H. *Mol Immunol, 41,* 81–84.

189. Dragon-Durey, M. A., Loirat, C., Cloarec, S., et al. (2005). Anti-factor H autoantibodies associated with atypical hemolytic uremic syndrome. *J Am Soc Nephrol, 16,* 555–563.

190. Warwicker, P., Donne, R. L., Goodship, J. A., et al. (1999). Familial relapsing haemolytic uraemic syndrome and complement factor H deficiency. *Nephrol Dial Transplant, 14,* 1229–1233.

191. Cheong, H. I., Lee, B. S., Kang, H. G., et al. (2004). Attempted treatment of factor H deficiency by liver transplantation. *Pediatr Nephrol, 19,* 454–458.
192. Loirat, C., & Niaudet, P. (2003). The risk of recurrence of hemolytic uremic syndrome after renal transplantation in children. *Pediatr Nephrol, 18,* 1095–1101.
193. Liszewski, M. K., Leung, M., Cui, W., et al. (2000). Dissecting sites important for complement regulatory activity in membrane cofactor protein (MCP; CD46). *J Biol Chem, 275,* 37692–37701.
194. Richards, A., Kemp, E. J., Liszewski, M. K., et al. (2003). Mutations in human complement regulator, membrane cofactor protein (CD46), predispose to development of familial hemolytic uremic syndrome. *Proc Natl Acad Sci USA, 100,* 12966–12971.
195. Baumgartner, E. R., Wick, H., Maurer, R., et al. (1979). Congenital defect in intracellular cobalamin metabolism resulting in homocysteinuria and methylmalonic aciduria. I. Case report and histopathology. *Helv Paediatr Acta, 34,* 465–482.
196. Brunelli, S. M., Meyers, K. E., Guttenberg, M., et al. (2002). Cobalamin C deficiency complicated by an atypical glomerulopathy. *Pediatr Nephrol, 17,* 800–803.
197. Russo, P., Doyon, J., Sonsino, E., et al. (1992). A congenital anomaly of vitamin B12 metabolism: A study of three cases. *Hum Pathol, 23,* 504–512.
198. Venat-Bouvet, L., Ly, K., Szelag, J. C., et al. (2003). Thrombotic microangiopathy and digital necrosis: Two unrecognized toxicities of gemcitabine. *Anticancer Drugs, 14,* 829–832.
199. Kojouri, K., Vesely, S. K., & George, J. N. (2001). Quinine-associated thrombotic thrombocytopenic-hemolytic uremic syndrome: Frequency, clinical features, and long-term outcomes. *Ann Intern Med, 135,* 1047–1051.
200. Banerjee, D., Kupin, W., & Roth, D. (2003). Hemolytic uremic syndrome after multivisceral transplantation treated with intravenous immunoglobulin. *J Nephrol, 16,* 733–735.
201. Gatti, S., Arru, M., Reggiani, P., et al. (2003). Successful treatment of hemolytic uremic syndrome after liver-kidney transplantation. *J Nephrol, 16,* 586–590.
202. Korec, S., Schein, P. S., Smith, F. P., et al. (1986). Treatment of cancer-associated hemolytic-uremic syndrome with staphylococcal protein A immunoperfusion. *J Clin Oncol, 4,* 210–215.
203. George, J. N. (2000). How I treat patients with thrombotic thrombocytopenic purpura-hemolytic uremic syndrome. *Blood, 96,* 1223–1229.
204. Turner, N., Nolasco, L., Dong, J.-F., et al. (2006). Human endothelial cells synthesize and release ADAMTS–13. *J Thromb Haemost, 4,* 1396–1404.

# CHAPTER 51

# Thrombocytopenia in Pregnancy

## Keith R. McCrae

*Departments of Medicine and Pathology, Case School of Medicine/University Hospitals of Cleveland, Cleveland, Ohio*

## I. Introduction

Thrombocytopenia in pregnant patients may result from many causes, including "incidental" thrombocytopenia; immune thrombocytopenic purpura (ITP); hypertensive disorders such as preeclampsia; the syndrome of hemolysis, elevated liver enzymes, and low platelets (HELLP); uncommon disorders such as thrombotic thrombocytopenic purpura (TTP) and the hemolytic uremic syndrome (HUS); and acute fatty liver of pregnancy, among others (Table 51-1). Because each of these disorders may have a distinct pathogenesis, the mechanisms by which thrombocytopenia develops may differ. The most common causes of thrombocytopenia in pregnancy are incidental thrombocytopenia and thrombocytopenia associated with hypertensive disorders. Though in each of these the degree of thrombocytopenia is generally mild, this does not diminish the importance of accurate diagnosis in facilitating optimal patient management.

The clinical manifestations of some causes of pregnancy-associated thrombocytopenia may overlap so extensively that it may be difficult, if not impossible, to distinguish them from each other. However, because the management of these disorders is critically dependent on defining the cause as accurately as possible, familiarity with the more common causes of thrombocytopenia in this population is essential for physicians who care for pregnant patients. In this chapter, the goal will be to provide an overview of these disorders and highlight clinical findings that may be useful in distinguishing them. For additional discussions of pregnancy-associated thrombocytopenia, the reader is referred to related reviews.[1–6]

## II. Specific Causes of Pregnancy-Associated Thrombocytopenia

### A. Incidental Thrombocytopenia of Pregnancy

**1. Definition and Clinical Characteristics.** "Incidental" thrombocytopenia, referred to by some authors as "gestational" thrombocytopenia, is the most common cause of thrombocytopenia in pregnancy, affecting 6 to 7% of pregnant women and accounting for 75 to 80% of all cases of pregnancy-associated thrombocytopenia.[4,7–11] Women with this disorder usually develop mild thrombocytopenia in the late second or third trimester, with platelet counts generally remaining above approximately $100 \times 10^9$/L. Although there is no absolute minimal platelet count below which incidental thrombocytopenia may be excluded, the likelihood of incidental thrombocytopenia diminishes as the platelet count falls below this value, and most experts consider other diagnoses (e.g., ITP) equally if not more likely when platelet counts fall below approximately $70 \times 10^9$/L.[7]

Patients with gestational thrombocytopenia are otherwise healthy, with no history of ITP or other autoimmune disorders. Serologic studies such as antinuclear antibody tests or antiphospholipid antibodies are generally negative,[7] and physical examination does not reveal hypertension or findings associated with other causes of pregnancy-induced thrombocytopenia.

**2. Platelet Counts in Normal Pregnancy: Implications for Incidental Thrombocytopenia.** The pathogenesis of incidental thrombocytopenia is uncertain, but it may involve hemodilution or accelerated platelet clearance in the placental circulation.[7,9] In at least some patients, the diagnosis of incidental thrombocytopenia likely reflects a physiological decrease in the platelet count associated with pregnancy. Several large, population-based studies have demonstrated that in uncomplicated pregnancies, the platelet count decreases by approximately 10% by term; in one study,[12] the mean platelet count in 6770 pregnant patients near term was $213 \times 10^9$/L, compared to $248 \times 10^9$/L in non-pregnant controls. The 2.5th percentile in the pregnant group, used to define the lower limit of a "normal" platelet count, was $116 \times 10^9$/L.[12] Similar results were obtained in another study involving 4382 pregnant women, in which the 5th percentile for the platelet count was $123 \times 10^9$/L.[13] These changes in mean platelet counts did not reflect excessively low platelet counts in a small group of outliers, but instead

**Table 51-1:  Causes of Thrombocytopenia in Pregnancy**

| Pregnancy Specific | Not Pregnancy Specific |
|---|---|
| Incidental thrombocytopenia of pregnancy | Immune thrombocytopenic purpura |
| Preeclampsia/eclampsia | Thrombotic microangiopathies |
| HELLP syndrome |   Thrombotic thrombocytopenic purpura |
| Acute fatty liver of pregnancy |   Hemolytic uremic syndrome |
| | Systemic lupus erythematosus |
| | HIV infection |
| | Other viral infections (e.g., CMV, EBV) |
| | Antiphospholipid antibodies |
| | Disseminated intravascular coagulation (DIC) |
| | Bone marrow dysfunction |
| | Nutritional deficiencies |
| | Drug-induced thrombocytopenia |
| | Type IIB von Willebrand disease |
| | Congenital |
| | Hypersplenism |

Abbreviations: CMV, cytomegalovirus; EBV, Epstein–Barr virus; HELLP, hemolysis, elevated liver enzymes, and low platelets; HIV, human immunodeficiency virus.

a shift to the left in the histogram of the platelet count distribution (Fig. 51-1).[12] In these studies, 7.3%[13] and 11.6%[12] of pregnant women had platelet counts below the lower limit for a normal platelet count of $150 \times 10^9$/L, as defined for nonpregnant populations, and were thus considered thrombocytopenic. In another study of 2263 pregnant women, the mean platelet count of the 1357 women considered normal was $225 \times 10^9$/L, and 8.3% of these individuals had platelet counts below $150 \times 10^9$/L.[8] This relatively high incidence of "thrombocytopenia" in pregnant patients likely contributes significantly to the fact that incidental thrombocytopenia is the second most common hematologic complication of pregnancy, following anemia.[10] However, it is likely that there are additional mechanisms that contribute to incidental thrombocytopenia in pregnant patients, most of which are not well understood.

**3. Management.** Patients with incidental thrombocytopenia are not at increased risk for poor pregnancy outcomes or delivery of thrombocytopenic offspring,[4,7–10,14] and for this reason evaluation of an otherwise healthy pregnant woman with mild thrombocytopenia occurring beyond the mid-second trimester may be limited to careful assessment for the presence of hypertension or proteinuria.[7,15] However, offspring of individuals with suspected incidental thrombocytopenia should have an umbilical cord platelet count determined in case the maternal thrombocytopenia actually resulted from mild ITP with resultant neonatal thrombocytopenia (discussed later).

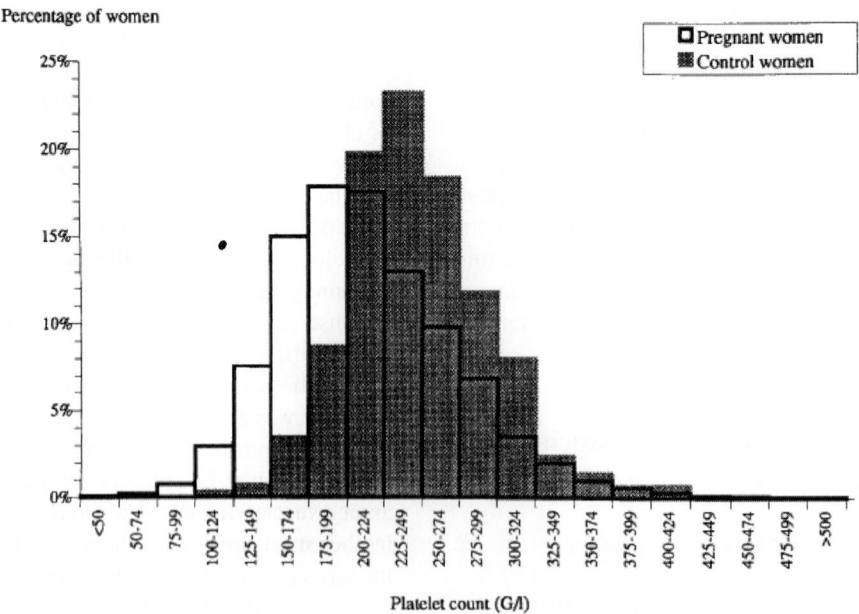

**Figure 51-1.**  Histogram of platelet counts of pregnant and nonpregnant women. Reproduced with permission.[12]

By definition, incidental thrombocytopenia remits after the pregnancy ends, often within several days, but almost always within 2 months.[7,13,16] Though little evidence exists, one report suggests that incidental thrombocytopenia may recur in some individuals during subsequent pregnancies.[16]

### B. Immune Thrombocytopenia Purpura

**1. Incidence in Pregnancy.** ITP is characterized by the presence of circulating antiplatelet glycoprotein antibodies that cause accelerated clearance of platelets by the reticulo-endothelial system, primarily the spleen (see Chapter 46 for further details).[17–20] Reports suggest that, in some patients, these antibodies may also affect megakaryocytes, inhibiting platelet production.[21,22] ITP affects approximately 0.1 to 1.0 of every 1000 pregnancies and accounts for 3 to 5% of cases of pregnancy-associated thrombocytopenia; thus, it is approximately 50-fold to 100-fold less common than incidental thrombocytopenia of pregnancy.[18] However, ITP is the most common cause of isolated thrombocytopenia occurring early in pregnancy, particularly in the first trimester,[1,4,9,23] and should be strongly suspected in a patient in this setting.

**2. Clinical and Laboratory Diagnosis.** Patients with ITP, whether pregnant or not, may present in a variety of ways. Those with more severe thrombocytopenia may be symptomatic and experience petechiae, purpura, or bleeding manifestations such as epistaxis or gingival bleeding. Those with mild thrombocytopenia are most often asymptomatic, with thrombocytopenia detected on routine monitoring of the complete blood count. There are no "gold-standard" laboratory studies for ITP, and the diagnosis of this disorder, like incidental thrombocytopenia of pregnancy, remains one of exclusion (Chapter 46).[24] Therefore, given the numerous causes of thrombocytopenia in pregnancy, the diagnosis of ITP requires careful consideration in pregnant patients. A history of thrombocytopenia preceding the pregnancy is particularly helpful in diagnosing ITP,[25] and a history of autoimmune disease or the presence of more severe thrombocytopenia (platelet count $<50 \times 10^9$/L) also makes the diagnosis more likely.[26] Increased levels of platelet surface-associated IgG are relatively nonspecific and may be increased in incidental thrombocytopenia as well as in other pregnancy-associated thrombocytopenic disorders.[27] However, contemporary antiplatelet glycoprotein antibody assays, though less sensitive, may be diagnostically useful if positive, although they do not exclude ITP if the tests are negative.[7,13,17,20]

ITP associated with mild thrombocytopenia may be indistinguishable from incidental thrombocytopenia. Although ITP may occur at any time during pregnancy, from a practical perspective, a platelet count below $100 \times$ $10^9$/L in the first trimester, with continued decline as pregnancy progresses, is most consistent with this disorder,[9,28] whereas mild isolated thrombocytopenia first noted in the second or third trimester that is not associated with hypertension or proteinuria most often results from incidental thrombocytopenia of pregnancy.

**3. Management of the Pregnant Patient with ITP.** Although antiplatelet glycoprotein antibodies in patients with ITP may cross the placenta and in some cases induce neonatal thrombocytopenia (see Chapter 52), therapy for pregnant patients with ITP should be dictated by the maternal platelet count and clinical manifestations such as the degree of bleeding or bruising. Throughout the first and second trimesters, patients with a platelet count above approximately $30 \times 10^9$/L, who are not bleeding, generally do not require treatment.[15,20,25,29] As term approaches, more aggressive therapy may be indicated to raise the platelet count to a level that allows for safe administration of epidural anesthesia and minimizes the risk of hemorrhage during delivery. Although there is no level 1 evidence defining the exact platelet count at which such procedures are safe, most experts consider a platelet count of $>50 \times 10^9$/L adequate for either vaginal delivery or cesarean section,[20,25] assuming that the coagulation system is otherwise normal. However, although epidural anesthesia has been administered successfully in anecdotal patients with very low platelet counts,[30] most experts recommend that a platelet count $>80 \times 10^9$/L should be achieved if epidural anesthesia is to be employed.[6,15] Hence, in many cases treatment can be deferred through the first and second trimesters of pregnancy, but initiated during the late third trimester 1 to 3 weeks prior to the expected date of delivery.

Although many physicians employ corticosteroids as first-line therapy for ITP in pregnant patients, these agents are associated not only with their usual toxicities but also with unique pregnancy-associated toxicities such as gestational diabetes, accelerated osteoporosis, pregnancy-induced hypertension and perhaps premature rupture of the fetal membranes, abruption, and prematurity.[20,25,28] These considerations have led some to advocate the use of high-dose (2 gm/kg) intravenous immunoglobulin (IV Ig) as initial therapy for pregnant women with ITP.[23] However, although approximately 70% of both pregnant and nonpregnant patients with ITP respond to IV Ig, these responses are often of short duration (3 weeks), and multiple courses of therapy may be required to maintain an adequate platelet count throughout pregnancy. Thus, although the optimal first-line therapy for ITP in pregnancy remains controversial, it is agreed that corticosteroids should be used judiciously and IV Ig considered if patients require prolonged therapy with an unacceptably high maintenance dose of corticosteroid (>7.5 mg/day of prednisone).[6] One approach that has been advocated suggests that if treatment is not needed until the late third trimester, and thus the time for development of steroid toxicity is expected to be limited, corticosteroids should be considered; however, if therapy must be initiated earlier in pregnancy and the expectation for prolonged treatment exists, then therapy should be initiated with IV Ig.[25] Intravenous anti-D has been used successfully and safely in 8 Rh(D)-positive pregnant patients with intact spleens and a median gestational age of 34 weeks.[31] Six of these eight patients responded, with one complete and five partial responses (coincident therapy was administered in four patients). Seven babies born to these patients were Rh(D)-positive, and three had positive direct antiglobulin tests (DATs) at the time of delivery (although one resulted from an ABO incompatibility). One of the neonates had jaundice, but a negative DAT, and none of the neonates had evidence of hemolytic anemia. Thus, despite the positive DAT results in a fraction of these neonates, it is likely that the majority of the anti-D immunoglobulin is adsorbed to circulating maternal red cells, thus minimizing its transplacental passage and its potential to cause hydrops in the fetus. Despite these promising initial observations, however, a larger experience is required to unequivocally establish the safety of this agent in pregnancy.[1] The significantly greater convenience of treatment with anti-D immunoglobulin might offer an advantage over IV Ig in pregnant patients who require maintenance therapy over an extended period.

Pregnant patients who are refractory to corticosteroids or IV Ig may either be treated medically or undergo splenectomy. Splenectomy is associated with an initial response rate of 75 to 85% in both pregnant and nonpregnant patients,[6] although there are currently no parameters that can be used to predict the response to splenectomy.[32] The second trimester has classically been considered as the optimal time for splenectomy, as procedures earlier in pregnancy may be associated with a higher incidence of premature labor, and at later points in pregnancy the surgical field may be obscured by the gravid uterus.[6] However, it has been argued that the gestational age at which splenectomy is performed may not be as important as once believed,[33] and hence the issue remains unsettled. Laparoscopic splenectomy may be performed safely during pregnancy and may be associated with a somewhat lower complication rate than open procedures.[32]

Additional medical therapy may be used either prior to splenectomy or in patients in whom splenectomy is unsuccessful. Some individuals with refractory ITP may respond to high-dose "pulse" corticosteroids (methylprednisolone, 1 gm IV) with or without IV Ig (2 gm/kg). For those who do not respond to this approach, additional immunosuppressive agents may be considered.[6] Most experts suggest the avoidance of potentially teratogenic agents such as Danazol, cyclophosphamide, or vinca alkaloids, although a limited amount of data suggests that azathioprine may be reasonably well tolerated in pregnant patients.[6] Rituximab induces complete or partial remissions in approximately 50% of nonpregnant patients with ITP,[34] but there is only anecdotal

experience with its use in pregnancy,[35] and thus its safety has not been established in this setting.

**4. Management of Thrombocytopenia in the Newborn of a Mother with ITP.** Due to the transplacental passage of antiplatelet antibodies, maternal ITP may cause neonatal thrombocytopenia (see also Chapter 52). Approximately 10% of the offspring of mothers with ITP will be born with a platelet count $<50 \times 10^9/L$, and 1 to 5% will be severely thrombocytopenic with platelet counts $<20 \times 10^9/L$.[36] However, in contrast to neonatal alloimmune thrombocytopenia (see Chapter 53), platelet counts rarely fall below $10 \times 10^9/L$, and although 25 to 50% of severely thrombocytopenic neonates experience bleeding during delivery, intracranial hemorrhage is rare (<1.0%).[14,20,37] Unfortunately, there is no noninvasive means to determine which neonates will be born thrombocytopenic, because neither the maternal platelet count, results of platelet antibody tests, response of maternal ITP to therapy, or a number of other parameters predict neonatal thrombocytopenia.[9] In fact, the best predictor of neonatal thrombocytopenia appears to be a previous history of neonatal thrombocytopenia in a sibling.[28,38]

The inability to determine the fetal platelet count noninvasively has led to debate concerning the use of percutaneous umbilical cord blood sampling (PUBS) to determine the fetal platelet count, with the aim of delivering neonates with platelet counts $<50 \times 10^9/L$ by cesarean section and avoiding potential cranial trauma during traversal of the birth canal. However, PUBS is associated with a neonatal morbidity (bleeding or fetal bradycardia requiring emergent cesarean section) of approximately 1%, which is equal to or greater than that of fetal intracranial hemorrhage during vaginal delivery.[36] Moreover, although an irreducibly low risk of fetal intracranial hemorrhage exists with any delivery method, there is no evidence that this risk is reduced by caesarean section.[39] Based on these considerations, most perinatologists now advocate the use of cesarean section for maternal indications only.[40]

The risk of neonatal intracranial hemorrhage in the offspring of patients with ITP may actually be greatest during the first few days after delivery, as platelet counts may fall further over this period. Thus, all offspring of mothers with ITP should have a cord platelet count obtained, as well as close monitoring of the platelet count during the first postpartum week. A cranial sonogram should also be considered for severely thrombocytopenic infants.[6] Treatment of severe neonatal thrombocytopenia with IV Ig is generally effective, and life-threatening bleeding should be treated with platelet transfusion and IVIG (see Chapter 52).[25,37]

### C. Preeclampsia and the HELLP Syndrome

**1. Preeclampsia: Incidence and Definition.** Preeclampsia affects approximately 6% of all pregnancies, most often those of primigravidas less than 20 or greater than 30 years of age,[9,41] and accounts for 17.6% of maternal deaths in the United States.[42] National data indicate that a woman's risk for developing preeclampsia increases by 30% for each additional year of age beyond 34.[43]

The classification of hypertensive disorders of pregnancy has undergone several revisions and may be particularly confusing to nonobstetricians. Diagnostic criteria for these disorders have been defined by the NIH Working Group on High Blood Pressure in Pregnancy (2000).[44] This family of disorders consists of (a) chronic hypertension, which includes patients with hypertension preceding, but exacerbated by pregnancy; (b) preeclampsia-eclampsia (discussed further later); (c) preeclampsia superimposed on chronic hypertension, the category that usually includes patients with the most severe illness; (d) gestational hypertension, defined as hypertension developing after 20 weeks of pregnancy but not associated with proteinuria (this classification may include transient cases, in which the hypertension resolves by 12 weeks postpartum); and (e) chronic hypertension, in which hypertension persists for more than 12 weeks postpartum.[44,45]

The specific diagnostic criteria for preeclampsia include hypertension (blood pressure of ≥140/90 mm Hg) and proteinuria (≥300 mg/24 hours, or at least 1+ on a dipstick urine analysis) developing after 20 weeks of gestation and noted on at least two separate occasions at least 6 hours apart. Patients with severe preeclampsia have a higher blood pressure (≥160/110 mm Hg) and more proteinuria (≥5 gm/24 hours).[46] These patients have evidence of end organ damage and may display a number of additional manifestations including headache, visual disturbances, pulmonary edema, right upper quadrant or epigastric pain, oligohydramnios, and fetal intrauterine growth retardation.[45] Eclampsia is defined by the occurrence of grand mal seizures occurring antenatally or within 7 days of delivery in patients with preeclampsia. Eclampsia complicates approximately 1 in every 2000 to 3000 pregnancies in the United States.[47]

**2. Preeclampsia: Genetic Associations.** A genetic component contributes to the development of preeclampsia, although the specific genetic associations have remained elusive. Evidence has shown that women with a family history of preeclampsia possess a threefold greater risk of developing the disorder than those that lack such a history. Moreover, women with severe preeclampsia are more likely to have a mother as opposed to a mother-in-law who has had preeclampsia.[43] Interestingly, paternal genetic factors are also involved, suggesting a fetal genotypic influence; both men and women who were the products of a pregnancy complicated by preeclampsia are significantly more likely than control men and women to have a child who is also the product of a pregnancy associated with preeclampsia.[48] Moreover, changed paternity is a risk factor for the development of preeclampsia in multiparous women,[49,50] although

one study suggests that this association may reflect a longer duration between pregnancies.[51] Finally, a history of prolonged cohabitation between parents appears to be associated with a decreased risk of preeclampsia.[52,53]

Despite these associations, a genetic basis for preeclampsia has not been defined at the molecular level.[53,54] One study genotyped 657 women affected by preeclampsia and their families for single nucleotide polymorphisms (SNPs) in the genes encoding angiotensinogen, angiotensin receptors, factor V Leiden, methylene tetrahydrofolate reductase (MTHFR), nitric oxide synthase, and tumor necrosis alpha (TNFα), but found that none of these genetic variants conferred an increased risk of disease.[55] A systematic review of the immunogenetics of preeclampsia concluded that the HLA-DR locus (particularly DR4) was a correlate of preeclampsia, though it was unclear whether any specific HLA allele, haplotype, or susceptibility gene in linkage disequilibrium with the HLA region was responsible.[56] Given the complex issues associated with maternal immune tolerance of the fetus, it is evident that additional studies designed to assess the effect of maternal-fetal HLA sharing on risk of preeclampsia are sorely needed.[56]

**3. Preeclampsia: Pathogenesis.** Preeclampsia is a disease that involves damage to vascular endothelium, and it is therefore not surprising that prominent risk factors for this disorder may reflect underlying vascular disease. These include preexisting hypertension, diabetes and insulin resistance, as well as increased homocysteine concentrations.[53] Testosterone levels and African-American ethnicity are additional risk factors.[53]

Several reports suggest that thrombophilic conditions may represent risk factors for preeclampsia. Of the acquired thrombophilic conditions, available evidence most strongly supports a role for antiphospholipid antibodies, which have been reported to be associated with a syndrome of early onset, severe preeclampsia.[57–59] An association of preeclampsia with maternally inherited thrombophilia has been suggested by some studies,[60–63] but not others.[64–67] Fetal inherited thrombophilias have also been suggested as contributing to the pathogenesis of preeclampsia, but no clear relationship between the most common fetal inherited thrombophilias, such as factor V Leiden, the prothrombin G20210A gene mutation, or MTHFR heterozygosity and maternal preeclampsia, has been demonstrated.[66,68] Likewise, elevated maternal factor VIII levels do not appear to influence the onset, progression, or resolution of preeclampsia.[69]

Though the clinical manifestations of preeclampsia usually do not appear until the third trimester, the pathophysiology of this disorder involves deficient remodeling of the maternal uterine vasculature early in pregnancy.[70] In normal pregnancy, placental cytotrophoblasts invade through the superficial decidual tissue and remodel the maternal uterine spiral arteries to the depth of the myometrium.[71] In contrast, cytotrophoblast invasion is shallow in preeclampsia, often limited to the superficial decidua.[72–74] Cytotrophoblast cells appear to modulate their phenotype during uterine invasion, with the more invasive cells switching their pattern of adhesion molecule expression to one that more closely reflects a vascular phenotype (i.e., integrin αVβ3, VE-cadherin), whereas less invasive trophoblast cells that remain in the placental villi express more E-cadherin and αVβ6.[74,75] It has been suggested that in preeclampsia, invasive cytotrophoblasts retain the "pro-adhesive" pattern of adhesion molecule expression characteristic of less invasive trophoblast cells, leading to diminished remodeling of the maternal uterine spiral arteries and progressive fetoplacental hypoxia as pregnancy progresses.[74] Isolated trophoblasts from preeclamptic patients may also be deficient in their expression of other invasion-related factors such as matrix metalloproteases.[76]

Studies have suggested that preeclampsia may also result from deficient activity of vascular endothelial cell growth factor (VEGF), which plays an important role in placental development. This may occur as a consequence of either increased production of soluble VEGF receptor 1, which binds and neutralizes VEGF-A and placental growth factor, or decreased expression of the membrane bound form of this receptor (VEGFR-1) in the placental bed.[77] Similar findings have been reported with regard to placental growth factor and VEGFR-1.[78] Regardless, the gene expression profiles of placentae from preeclamptic patients have been shown to resemble those of placentae from high-altitude pregnancies or from first-trimester placental explants cultured under hypoxic conditions, providing strong, if indirect evidence for hypoxia in preeclamptic placentae.[79] Fetoplacental hypoxia is thought to lead to the release or activation of factors, such as endothelin 1 and reactive oxygen species,[80–83] that may mediate the systemic manifestations of preeclampsia and induce endothelial damage or dysfunction.

**4. Preeclampsia: Thrombocytopenia and Microangiopathic Hemolytic Anemia.** Approximately 15 to 50% of patients with preeclampsia develop thrombocytopenia, the extent of which is generally proportional to the severity of disease.[84] The pathogenesis of preeclampsia-associated thrombocytopenia is uncertain and likely multifactorial. The elevated levels of thromboxane $A_2$ metabolites in the urine of preeclamptic patients, as well as the increased plasma levels of the platelet α-granule proteins β-thromboglobulin and platelet factor 4, support the argument that platelet activation is a major cause of accelerated platelet clearance in this disorder,[9,45] even if a definitive mechanism by which this occurs has not yet been defined. Although only the most severe cases of preeclampsia are associated with a coagulation profile suggestive of disseminated intravascular coagulation (DIC) (prolonged prothrombin time, elevated fibrin degradation products, decreased fibrinogen level), the plasma of most patients with preeclampsia contains increased levels of thrombin-antithrombin complexes, and approximately 40% of these plasmas contain increased levels of fibrin D-dimers, suggesting that subclinical activation of the coagulation system occurs in a substantial portion of patients with preeclampsia.[2,9,84,85] Thus, increased generation of thrombin may be one mechanism that promotes platelet activation, with accelerated clearance of activated platelets.[2] Platelets may also be stimulated through contact with activated endothelium or exposed subendothelium underlying the injured placental vasculature. Platelet adhesion and activation may be promoted by the elevated levels of von Willebrand factor (VWF),[86,87] as well as other adhesive proteins such as cellular fibronectin[88] that circulate in the plasma of many patients with preeclampsia.

Patients with preeclampsia also display microangiopathic hemolytic anemia (MAHA) due to red cell fragmentation, presumably due to shearing of red cells on fibrin strands in the microvasculature or placental circulation. However, the number of schistocytes on the peripheral blood film, a marker of MAHA, is relatively low compared to disorders in which MAHA is more prominent, such as HELLP, TTP, or HUS (discussed later).

Preeclampsia-associated thrombocytopenia may occasionally precede other clinical manifestations of the syndrome.[89] Thus, preeclampsia must be included in the differential diagnosis of a falling platelet count first noted in the third trimester of pregnancy, and the patient should be monitored for the development of this disorder if no other cause of thrombocytopenia can be identified.[2,9,84]

**5. HELLP: Definition and Association with Preeclampsia.** The HELLP (hemolysis, elevated liver function tests, low platelets) syndrome is considered by some investigators to be a variant of preeclampsia, because it may develop in 20% of patients with severe preeclampsia.[90] The criteria for diagnosis of HELLP are not entirely consistent among various reports, with authors from different institutions adopting similar, though somewhat different guidelines.[91,92] In general, HELLP is defined by (a) hemolysis, (b) elevated liver enzymes, and (c) low platelets. Hemolysis takes the form of MAHA, which is generally more severe than that occurring in preeclampsia. The extent of hepatic enzyme elevation required for a diagnosis of HELLP is another area of disagreement, but most experts accept an aspartate aminotransferase (AST) level ≥70 U/L (or approximately two times the upper limit of the normal laboratory value).[93] Likewise, disagreement exists concerning the degree of thrombocytopenia necessary for a diagnosis of HELLP, although the majority of reports accept a platelet count <100 × 10⁹/L. Finally, whereas elevated values of lactic dehydrogenase (LDH) have not been documented in some series of patients with HELLP, most authors consider an LDH value of ≥600 U/L as a necessary part of the syndrome. One group has classified HELLP into three subpopulations (referred to by some as the "Mississippi" classification) based on the severity of thrombocytopenia. Patients with class 1 HELLP have a platelet nadir below $50 \times 10^9$/L, patients with class 2 HELLP have a nadir between $51 \times 10^9$/L and $100 \times 10^9$/L, and patients with class 3 HELLP have a nadir between $101 \times 10^9$/L and $150 \times 10^9$/L.[94] This classification scheme is useful in predicting the rapidity of postpartum recovery, particularly of platelets, and the need for plasma exchange. Some authors consider patients who have some, but not all, manifestations of HELLP to have partial HELLP syndrome. For example, those with all the manifestations except MAHA would be referred to as having "ELLP."[92] However, hemolysis may be transient and episodic, and schistocytes may not appear increased on the peripheral blood film in all cases. The serum haptoglobin has been suggested to be a sensitive marker of hemolysis in patients with HELLP,[95] as this has been reported to be reduced in 85 to 97% of cases of the syndrome.

### 6. HELLP: Diagnosis and Clinical Characteristics.

HELLP has been reported to affect 0.17 to 0.85% of all live births and occurs most commonly in a slightly older population than preeclampsia, with a mean age of 25 years.[95] The percentage of nulliparous patients ranges from 52 to 81%, a significantly lower proportion than those with preeclampsia.[95] HELLP syndrome, particularly when it develops early in pregnancy,[96,97] may be associated with antiphospholipid antibodies in some patients, although only small series have been reported.[98,99] Patients with HELLP characteristically present with nonspecific complaints including nausea, fatigue, and malaise. Right-upper-quadrant and epigastric pain is the most common symptom, occurring in 86 to 92% of patients; in occasional patients, right-upper-quadrant pain may precede the onset of liver enzyme abnormalities.[95] Thus, it is not surprising that many patients with HELLP may be mistakenly diagnosed with a primary gastrointestinal disorder, particularly as only 50 to 70% of these patients have hypertension at the time of presentation, and the differential diagnosis in a patient with such manifestations is wide (Table 51-2). Patients may have accompanying edema, and 5 to 15% have no or minimal proteinuria; 15% may have neither hypertension nor proteinuria.[91,95] Two thirds of patients develop HELLP antepartum, most commonly between 27 and 37 weeks,[100] although in one series, 15% of patients had evidence of HELLP between the 17th and 26th weeks of gestation.[93] One third of patients present with HELLP postpartum, usually within several days after delivery,[100] with approximately 6% of these individuals having no signs of preeclampsia prior to delivery.[95]

The classic liver lesion in HELLP syndrome consists of periportal or focal parenchymal necrosis, with periportal hemorrhage and fibrin deposition in the liver sinusoids. Dissection of necrosis into the liver capsule may lead to

**Table 51-2: Differential Diagnosis of the HELLP Syndrome**

Gastrointestinal
  Acute fatty liver of pregnancy
  Cholelithiasis
  Nephrolithiasis
  Pancreatitis
  Peptic ulcer disease
  Appendicitis
  Hepatitis
  Gastroenteritis
Hematologic
  Idiopathic thrombocytopenic purpura (ITP)
  Thrombotic thrombocytopenic purpura (TTP)
  Hemolytic-uremic syndrome (HUS)
Other
  Hyperemesis gravidarum
  Pyelonephritis
  Diabetes insipidus

the development of subcapsular hemorrhage and eventual hepatic rupture, the most feared complication of HELLP, which may be a terminal event.[100]

### 7. HELLP: Maternal Morbidity.

Like preeclampsia, HELLP is associated with significant maternal and neonatal morbidity.[101] It has been suggested by some studies that the maternal outcomes of patients with HELLP are worse than those with severe preeclampsia, but this has not been a universal finding.[92] In many cases, patients with partial or full-blown HELLP are analyzed together, making the assignment of outcomes problematic, as patients with the latter category appear to have significantly worse outcomes.[92] Patients with class I HELLP have the greatest morbidity, which, in addition to hepatic necrosis, may include acute renal failure, pulmonary edema, abruptio placentae, intracerebral bleeding, and retinal detachment.[92] Visual defects that develop in the course of HELLP may in some cases be permanent.[102]

### 8. HELLP: Thrombocytopenia.

The thrombocytopenia that accompanies HELLP is generally more severe than that of preeclampsia. It has been reported that the rate of fall of the platelet count is a predictor of the eventual severity of HELLP, with patients whose platelet counts decrease by $>50 \times 10^9$/L per day having a significantly higher probability of developing class 1 or 2 HELLP.[103] There appears to be a correlation between the extent of thrombocytopenia and the degree of liver dysfunction in patients with HELLP, as suggested by the HELLP classification scheme described earlier. The platelet nadir is usually reached approximately 24 hours postpartum,[95] with normalization occurring within 6 to 11 days.[104] As with preeclampsia, the thrombocytopenia of the HELLP syndrome likely reflects a multifactorial pathogenesis. Activation of platelets in contact with damaged or disrupted liver endothelium is a likely scenario, as is consumption of platelets secondary to subclinical thrombin generation. Clinically apparent DIC complicates approximately 6% of cases of HELLP, as opposed to 1 to 2% of patients with preeclampsia,[105] although significant decreases in the fibrinogen level and elevated levels of fibrin degradation products are not generally seen until the platelet count is $<50 \times 10^9/L$.[100]

**9. HELLP: Neonatal Implications.** HELLP is associated with a 10 to 20% fetal mortality, as well as substantial morbidity, that has been attributed primarily to placental ischemia leading to profound fetal hypoxia.[106,107] Some neonates may have mild thrombocytopenia, although this may not be apparent until after delivery. The pathogenesis of neonatal thrombocytopenia, although often attributable to routine causes such as sepsis and prematurity, may also involve deficient platelet production.[84] This may result, in part, from relatively low erythropoietin levels and impaired megakaryocyte responses to erythropoietin in premature neonates.[84]

**10. Management of Preeclampsia and the HELLP Syndrome.** The first step in the management of patients with preeclampsia or HELLP is stabilization of the patient, including correction of fluid and electrolyte abnormalities and any apparent coagulopathy, with the goal of preparing the patient for the definitive treatment — delivery of the fetus. Intravenous magnesium sulfate is administered as seizure prophylaxis, and acute hypertension is most commonly treated with hydralazine.[108] Although conservative management of patients with evidence of mild preeclampsia may be attempted until fetal lung maturity is achieved or until 38 weeks of gestation, emergent therapy is usually employed for patients with severe preeclampsia, with or without superimposed HELLP.[45,54] After maternal stabilization, the fetal status should be assessed and patients with severe preeclampsia or HELLP who are beyond 34 weeks gestation should undergo urgent delivery.[100] If evidence of fetal lung maturity is not present, betamethasone should be administered to promote this, and delivery should be attempted after an additional 48 hours.[42,100] Although some have advocated more conservative management, there is no evidence that fetal or maternal outcomes have been improved by this approach.[91,109] Most patients managed conservatively continue to deteriorate and thus may be exposed to additional complications such as placental abruption, acute renal failure, pulmonary edema, eclampsia, or maternal or fetal death.[91,92] The presence of HELLP or severe preeclampsia is not an absolute indication for cesarean section, and vaginal delivery may be employed in the absence of obstetric contraindications.

Though platelet survival is reduced in patients with preeclampsia or HELLP, platelet transfusion may be used to raise the platelet count in order to allow epidural anesthesia or cesarean section. Although there is no definitive study defining what is an adequate platelet count for either of these procedures, most experts consider a platelet count of $>50 \times 10^9/L$ to be adequate for cesarean section, whereas a platelet count between $70 \times 10^9/L$ to $100 \times 10^9/L$ has been recommended for epidural anesthesia.[6,15,110] DIC should be considered in patients with bleeding, a prolonged prothrombin time, and elevated levels of fibrin(ogen) degradation products or decreased fibrinogen. DIC is managed through the use of fresh frozen plasma to normalize the prothrombin time, whereas severe hypofibrinogenemia, though uncommon, may be treated with cryoprecipitate.

Although a recent prospective, randomized placebo-controlled trial demonstrated that postpartum administration of corticosteroids did not benefit patients with severe preeclampsia who did not have HELLP,[111] there is significant support for the use of corticosteroids in patients with HELLP. Though no placebo-controlled studies have been performed, several small randomized trials have demonstrated that antepartum administration of corticosteroids, primarily dexamethasone, leads to a more rapid recovery of biochemical laboratory values and improvement of the platelet count in patients with the HELLP syndrome.[112] Retrospective analyses have also suggested that antepartum corticosteroid administration increases the probability of successful labor induction and candidacy for regional anesthesia.[113,114] Postpartum administration of corticosteroids has also been shown to lead to a more rapid resolution of HELLP syndrome.[115] Yet whether corticosteroids have a significant influence on maternal or fetal morbidity or mortality remains uncertain, as the number of patients included in these studies has been small, and mortality is an uncommon end point.

**11. Postpartum HELLP.** Although the manifestations of preeclampsia or the HELLP syndrome, including the associated thrombocytopenia, generally remit within several days after delivery, occasional patients may develop these disorders postpartum or experience progressive disease after delivery. Progressive disease may be associated with multiorgan dysfunction and significant morbidity. The management of such individuals is not well defined. In a retrospective study, Martin et al.[115] reported that steroids hastened the resolution of postpartum HELLP in 43 women who received them in comparison to 237 women who did not. Plasma exchange with fresh frozen plasma has also been employed in individuals with postpartum HELLP. In an initial study of 7 patients with postpartum HELLP persisting for at least 72 hours after delivery, Martin et al.[116] reported that plasma exchange led to sustained increases in the platelet count, and a decrease in LDH. In a subsequent study, 18 patients with postpartum HELLP were divided into two groups: group 1 included 9 women with persistent HELLP syndrome more than 72 hours after delivery; group 2 included 9 women with worsening HELLP syndrome and evidence of organ dysfunction at any time after delivery.[117] Although all group 1 patients responded to plasma exchange, responses in group 2 were inconsistent, and 2 deaths occurred in this group. Another research group compared the outcomes of 29 consecutive patients with HELLP treated with plasma exchange and found a decreased length of stay in the intensive care unit, a more rapid improvement in the platelet count and biochemical abnormalities, and significantly decreased mortality in the patients treated with plasma exchange compared to a group of historical controls.[118]

**12. Recurrent HELLP in Subsequent Pregnancies: Incidence and Prevention.** The question of risk of recurrence of preeclampsia or HELLP in subsequent pregnancies is of major concern to women who have experienced these disorders. In a study of primigravid women whose pregnancies were complicated by severe preeclampsia or eclampsia, second pregnancies were complicated by mild preeclampsia in 19.5%, severe preeclampsia in 29.5%, and eclampsia in 1.4%.[106] The risk declined in subsequent pregnancies yet remained significantly higher than control patients who had never experienced preeclampsia. The risk of recurrent HELLP derived from several reports has been collated by Barton et al.[92] (Table 51-3). In patients who have experienced a pregnancy complicated by HELLP, the risk of recurrent HELLP in subsequent pregnancies ranges from 3 to 19%, with a risk of preeclampsia of approximately 20%.

Given the high incidence and morbidity associated with preeclampsia, extensive effort has been devoted to the identification of strategies to prevent recurrent disease. Based on the premise that platelets play a primary role in the pathogenesis of preeclampsia, the ability of aspirin to inhibit the development of preeclampsia has been studied, but little if any beneficial effects have been documented.[119] More

recently, intensive interest has been focused on the use of antioxidants for prevention of preeclampsia based on the possible pathogenic role of placentally derived reactive oxygen species in the pathogenesis of this disorder.[120] A randomized, placebo-controlled trial comparing vitamin E and C supplementation of patients at high risk for preeclampsia demonstrated a marked reduction in the relative risk of developing this disorder in treated patients.[121] However, despite several smaller studies with similar observations, one meta-analysis concluded that currently "data are too few to say if vitamin E supplementation either alone or in combination with other supplements is beneficial during pregnancy."[122] Additional information concerning the utility of antioxidants in this setting should be forthcoming from the results of ongoing trials.

## D. Thrombotic Thrombocytopenic Purpura and the Hemolytic-Uremic Syndrome (see also Chapter 50)

**1. TTP: Incidence and Clinical Diagnosis.** TTP is a rare disorder with an incidence of 6.5 cases per million annually[123] and a peak incidence in the fourth decade. The disorder is more common in females (female/male ratio 3 : 2).[124,125] Other risk factors include African ancestry and obesity.[126,127]

TTP is defined by a classic pentad of symptoms that includes MAHA, thrombocytopenia, neurologic symptoms, fever, and renal dysfunction.[125] However, whereas only 40% of patients display the entire pentad of symptoms, 75% present with the triad of MAHA, neurologic symptoms, and thrombocytopenia.[125,128] Neurological symptoms generally predominate in TTP and may range from headache and confusion to seizures or coma.[125,128–130] Thrombocytopenia is often severe, with platelet counts below $20 \times 10^9$/L.[125,128,129] Renal dysfunction is generally mild in patients with TTP, usually limited to hematuria and proteinuria.[130] MAHA and compensatory reticulocytosis are central features of TTP and result from shearing of red cells in the

### Table 51-3: Pregnancy Outcome after HELLP in Normotensive Women

| Reference | Number of Women | Number of Subsequent Pregnancies | Preeclampsia (percentage of subsequent pregnancies) | HELLP (percentage of subsequent pregnancies) |
|---|---|---|---|---|
| Sibai et al.[186] | 139 | 192 | 19 | 3 |
| Sullivan et al.[187] | 122 | 161 | 23 | 19 |
| Van Pampus et al.[188] | 77 | 92 | 16 | 2 |
| Chames et al.[189a] | 40 | 42 | 52 | 6 |

[a]HELLP ≤28 weeks in previous pregnancy.
Adapted from ref.[92].

microvasculature. Markedly increased levels of LDH result from hemolysis and tissue ischemia.[131]

## 2. HUS: Clinical Manifestations and Comparison with TTP.

Patients with HUS can display the same pentad of symptoms as patients with TTP, although in contrast to TTP, renal failure generally predominates and CNS symptoms are less pronounced.[132] There are several variants of HUS, including the childhood form associated with gastroenteritis caused by E. coli and related organisms and induced by bacterially derived verotoxins,[133] as well as inherited forms of HUS associated with deficiencies of complement regulatory proteins, specifically factor H (see Chapter 50 for details).[134,135] However, the most common type of HUS that occurs in pregnancy is "sporadic" HUS. This form of HUS is not associated with a prodrome of bloody diarrhea and thus is sometimes referred to as D(-) HUS.[135] Sporadic HUS is also sometimes referred to as "malignant nephrosclerosis" or "irreversible postpartum renal failure."[136,137] Thrombocytopenia in patients with sporadic HUS is generally less severe than in classic TTP, although up to 60% of patients require dialysis. Despite these differences, the symptoms of TTP and HUS may overlap to such an extent that in some cases these disorders may be impossible to distinguish from one another, and some experts refer to both syndromes as "TTP-HUS."[138] From a practical perspective, the therapeutic approach to management of both disorders is similar,[138] although their pathogenesis may differ.

The incidence of both TTP and HUS is increased in pregnant patients.[2,139] Though it has been argued that because both TTP and HUS primarily affect women of child-bearing age, the association of these disorders with pregnancy may be coincidental. However, the fact that in some series 10 to 20% of all cases of TTP or HUS have occurred in pregnant patients suggests that this association is genuine.[84,135,140]

## 3. ADAMTS13 in Thrombotic Microangiopathies during Pregnancy.

Studies have found that a congenital or acquired deficiency of a specific protease that cleaves VWF is responsible, at least in part, for the development of TTP.[141,142] This protease, ADAMTS13, is a member of the ADAMTS family of metalloproteases.[143] Levels of ADAMTS13 are markedly decreased in most patients with TTP. In patients with sporadic TTP, this deficiency results from antibodies against the protease,[141,142] whereas patients with inherited forms of TTP harbor mutations in the ADAMTS13 gene.[143] Although an ADAMTS13 level of less than 5% is believed to be specific for TTP,[144,145] mild to moderately decreased levels of ADAMTS13 occur in several settings,[146] including pregnancy.[147] Detailed analyses have demonstrated that the level of ADAMTS13 drops progressively from the 12th to 16th week of normal pregnancy to the puerperium and that nulliparas have somewhat lower levels of ADAMTS13 than parous women. The mean level

of ADAMTS13 measured at term was 52% (range 22–89%) but increased after delivery.[148]

A severe deficiency of ADAMTS13 with resultant impaired cleavage of ultralarge VWF multimers (ULVWF) leads to increased levels of circulating ULVWF in the plasma of patients with TTP.[149,150] These ULVWF multimers have an increased avidity for platelets and cause agglutination of platelets in the microvasculature, particularly at areas of high shear stress where ULVWF may undergo conformational changes that normally make them more susceptible to cleavage.[151] Whether the increased incidence of TTP in pregnant patients reflects the pregnancy-related decrease in the plasma concentrations of ADAMTS13 is uncertain; whereas these levels do not decrease to the extent seen in patients with established TTP, pregnancy is also characterized by increased levels of VWF;[152] indeed, a reciprocal relationship between levels of ADAMTS13 and VWF has been reported.[153] Patients with HUS have not been found to have decreased levels of ADAMTS13, suggesting a different pathogenesis than TTP.[135,151] Of interest, one study reported that ADAMTS13 levels in patients with HELLP are significantly lower than healthy pregnant patients,[154] although these individuals did not have ADAMTS13 levels below 5% of normal.

## 4. Time of Onset of TTP and HUS during Pregnancy.

The most common time of onset of HUS in adults is in the postpartum period. In one series, the mean time from delivery to the onset of symptoms was 26.6 days,[155] a finding that helps to distinguish HUS from other causes of pregnancy-associated MAHA and thrombocytopenia, such as preeclampsia and HELLP. On the other hand, TTP appears to have a variable time of onset. In the classic series of Weiner, 40 of 45 cases of TTP developed antepartum with a mean gestational age at the onset of symptoms of 23.5 weeks.[155] In contrast, other reports have documented the frequent onset of TTP-HUS in the third trimester. A review of case series of TTP-HUS from 1964 to 2002 demonstrated that the time of greatest risk for development of these disorders was at term or postpartum, although many of these series did not attempt to distinguish TTP from HUS.[3] Several case reports appearing in the early 2000s confirm the fact that thrombotic microangiopathic syndromes consistent with TTP may develop as early as weeks 22 to 28 of pregnancy in a significant proportion of patients,[140,156–159] and thus TTP must be included in the differential diagnosis of pregnancy-associated thrombocytopenia developing within this time frame.

## 5. Management of TTP and HUS.

Unlike preeclampsia and HELLP, termination of pregnancy with delivery of the fetus does not lead to resolution of TTP or HUS. However, TTP responds equally well to plasma-based therapy in pregnant and nonpregnant patients.[9,84] Whereas the mortality of TTP exceeds 90% without treatment, survival exceeds 80%

in patients who receive plasma exchange.[138,160,161] A prospective trial reported in 1991 demonstrated that plasma exchange is superior to plasma infusion in the treatment of patients with TTP,[162] and thus in the absence of a known genetic deficiency of ADAMTS13, in which plasma infusion is sufficient and appropriate, plasma exchange is the treatment of choice for this disorder. Neurological improvement generally occurs rapidly in these patients,[130] although the platelet count may take 5 days or more to improve and weeks to return to normal values. Daily plasma exchange should be continued until symptoms have resolved and a normal LDH and platelet count have been maintained for at least 2 days.[138,160] Corticosteroids are of little benefit on their own,[161,163] and retrospective studies do not provide compelling evidence that they improve the response to plasma exchange.[161] Though their anti-inflammatory and immunosuppressive effects may make them attractive adjunctive therapy for acquired TTP associated with antibodies against ADAMTS13, the use of corticosteroids in pregnancy is associated with unique toxicities such as pregnancy-induced hypertension, and they should be used judiciously in this setting.

Plasma exchange is recommended for patients with HUS as well, particularly in light of the difficulty distinguishing HUS from TTP. Responses to plasma exchange or infusion occur,[2] though their frequency is uncertain and the prognosis of pregnancy-associated HUS is poor, with a reported mortality rate of 25% and chronic renal insufficiency in 50% of survivors.[164] Maternal TTP and HUS are both associated with poor fetal outcomes.

**6. Recurrence of TTP and HUS in Subsequent Pregnancies.** Complete maternal care should involve counseling of patients who have experienced a pregnancy complicated by TTP or HUS concerning the risk of developing the disorder in a subsequent pregnancy, and one review has provided estimates of this risk.[165] These authors found that the incidence of recurrent TTP-HUS during a subsequent pregnancy in patients whose initial episode of TTP-HUS had been idiopathic, and not pregnancy-related, ranged from 43 to 91%. In patients whose prior episode of TTP-HUS had been pregnancy associated, the incidence of recurrent disease in a subsequent pregnancy ranged from 18 to 56%. There were no maternal deaths from recurrent disease in this Oklahoma-TTP registry, whereas the maternal death rate from recurrent disease in other published series ranged from 10 to 13%.[165]

### E. Other Causes of Pregnancy-Associated Thrombocytopenia

Several other causes of pregnancy-associated thrombocytopenia, though rare, should be noted (see Table 51-1).[84]

**1. Acute Fatty Liver of Pregnancy.** Acute fatty liver of pregnancy (AFLP) is a rare disorder with an incidence of 1 in every 7000 to 16,000 pregnancies.[166] This disorder generally occurs in the third trimester, when patients present with nausea, vomiting, right upper quadrant pain, anorexia and malaise, jaundice, and cholestatic liver dysfunction.[167] Primiparas and women with twin gestations are more commonly affected.[168,169] Decreased levels of antithrombin and fibrinogen accompanied by laboratory evidence of DIC are present in most patients. Diabetes insipidus and hypoglycemia may be present in more than 50% or more patients.[2] The thrombocytopenia in patients with AFLP is characteristically mild, with an average nadir platelet count in one series of $88 \times 10^9$/L, although the platelet count may occasionally fall to levels as low as $20 \times 10^9$/L.[166] However, maternal bleeding is common, most likely reflecting thrombocytopenia accompanied by a severe coagulopathy occurring secondary to diminished hepatic synthesis of coagulation proteins and a systemic consumptive process due to acquired antithrombin deficiency.[2,9] Data suggest that some cases of AFLP and possibly HELLP may result from fetal mitochondrial fatty acid oxidation disorders (FAOD),[170] in which a FAOD in the fetus induces disease in the heterozygous mother. Of these, the most common is a long chain 3-hydroxyacyl-CoA dehydrogenase deficiency.[171] Although an association of these rare fetal disorders with maternal disease has been established,[171] others have argued that they are likely to account for only a small minority of AFLP and HELLP cases.[172] Therapy for AFLP involves intense supportive care and management of the coagulopathy, followed by emergent delivery of the fetus.[166,167,173] Antithrombin concentrates have not been shown to be of benefit, although experience is limited.[174] In rare cases, maternal hepatic transplantation has been employed, with variable success.[166]

**2. Type IIB von Willebrand Disease.** Patients with type IIB von Willebrand disease express a mutant VWF molecule that causes agglutination and enhanced clearance of circulating platelets. Most of the responsible mutations reside in the A1 domain, resulting in a gain-of-function that leads to increased binding of the mutant VWF to platelet glycoprotein (GP) Ibα.[175] Due to the increased levels of VWF that occur during pregnancy, this process leads to enhanced platelet agglutination and progressive thrombocytopenia throughout pregnancy. Platelet counts in patients with type IIB von Willebrand disease may fall to levels as low as 20 to $30 \times 10^9$/L at term, but they improve after delivery.[176] DDAVP (Chapter 67) is contraindicated in this disorder, as it leads to enhanced release of endogenous, mutant VWF, which can worsen the thrombocytopenia.[175] Patients who require treatment usually receive VWF-rich factor VIII concentrates, which in some cases have been reported to stabilize the thrombocytopenia.[177] Platelet concentrates have also been used, but with variable success, as the transfused platelets are also agglutinated by the mutant VWF protein.[175]

**3. HIV Infection.** A number of other processes that cause thrombocytopenia in nonpregnant individuals may also be first diagnosed during pregnancy and should be considered in the differential diagnosis of thrombocytopenia. Human immunodeficiency virus (HIV) infection may induce thrombocytopenia (as discussed in detail in Chapter 47) by enhancing peripheral platelet destruction through the effects of circulating immune complexes containing antibodies reactive with amino acids 49 to 66 of GPIIIa, a peptide that mimics an epitope of the HIV nef protein.[178,179] In addition, HIV infects megakaryocytes, leading to impaired thrombopoiesis.[180] Patients with HIV infection also develop TTP with increased frequency, and the occurrence of HIV-related TTP in pregnant patients has been reported.[159]

**4. Immune-Mediated Thrombocytopenias.** As discussed earlier, antiphospholipid antibodies are associated with early-onset HELLP[98] and preeclampsia,[43,58] as well as recurrent fetal loss.[58] Approximately 10% of patients with ITP have coexisting antiphospholipid antibodies and constitute a subgroup of ITP patients that may be at increased risk for thrombotic events.[99,181] Thrombocytopenia may complicate up to 25% of cases of systemic lupus erythematosus (SLE), and more than half of patients with SLE may experience exacerbation of their disease during pregnancy.[182] Drug-induced thrombocytopenia (see Chapters 48 and 49) must be considered in any thrombocytopenic patient exposed to a potential offending medication such as quinine, heparin, or trimethoprim-sulfamethoxazole, among others.[183]

**5. Congenital Thrombocytopenias.** Though the majority of congenital thrombocytopenias are diagnosed during childhood (see Chapter 54), some disorders are relatively asymptomatic and come to medical attention for the first time when thrombocytopenia is noted during pregnancy.[184] Perhaps the most common of these are the congenital macrothrombocytopenias characterized by mutations in the *MYH9* gene.[185] Patients with the mildest manifestations of these disorders may have isolated macrothrombocytopenia without other clinical findings and are classified as having the May–Hegglin anomaly (see Chapter 54).

**6. Primary Hematologic Disorders.** Primary hematologic disorders such as leukemias or myelodysplastic syndromes, though uncommon in women of child-bearing age, may occasionally present with isolated thrombocytopenia. Likewise, though chronic liver disease is uncommon in this population, patients with these disorders may develop portal hypertension leading to splenic sequestration and thrombocytopenia. Finally, although uncommon in the United States,

deficiencies of nutrients such as vitamin B12 and folic acid may lead to thrombocytopenia or pancytopenia in pregnant patients.

## III. Summary and Conclusions

Thrombocytopenia in the pregnant patient may be the result of many causes. The most common of these is incidental thrombocytopenia of pregnancy, followed by the hypertensive disorders and the HELLP syndrome. ITP is less common, whereas TTP and HUS are quite rare.

In some cases, it may be difficult, if not impossible, to discern the cause of thrombocytopenia in a pregnant woman due to the broad overlap in signs and symptoms among the various disorders. Nevertheless, it is important to make every attempt to diagnose these disorders as accurately as possible, because management strategies vary substantially for disorders that appear clinically similar. For example, HELLP and TTP may share thrombocytopenia, MAHA, renal dysfunction, and, in some cases, neurologic symptoms, although the optimal management of HELLP revolves around delivery of the fetus, whereas that for TTP is focused on plasma exchange. Hence, elevations of AST in a patient with these clinical findings may be particularly useful, tipping the diagnostic balance toward HELLP.

Other than incidental thrombocytopenia of pregnancy, thrombocytopenia in pregnancy is best managed by a team of physicians that includes high-risk obstetricians, hematologists, anesthesiologists, pediatricians, and transfusion medicine specialists. Inclusion of individuals with expertise in these areas will ensure that the patient receives the complex care required to secure an optimal pregnancy outcome.

## References

1. McCrae, K. R., Bussel, J. B., Mannucci, P. M., et al. (2001). Platelets: An update on diagnosis and management of thrombocytopenic disorders. *Hematology (Am Soc Hematol Educ Program)*, 282–305.
2. McCrae, K. R., & Cines, D. B. (1997). Thrombotic microangiopathy during pregnancy. *Sem Hematol, 34*, 48–158.
3. McMinn, J. R., & George, J. N. (2001). Evaluation of women with clinically suspected thrombotic thrombocytopenic purpura-hemolytic uremic syndrome during pregnancy. *J Clin Apheresis, 16*, 202–209.
4. Crowther, M. A., Burrows, R. F., Ginsberg, J., et al. (1996). Thrombocytopenia in pregnancy: Diagnosis, pathogenesis and managment. *Blood Rev, 10*, 8–18.
5. Kelton, J. G. (2002). Idiopathic thrombocytopenic purpura complicating pregnancy. *Blood Rev, 16*, 43–46.
6. Provan, D., Newland, A., Norfolk, D., et al. (2003). Guidelines for the investigation and managment of idiopathic

thrombocytopenic purpura in adults, children and in pregnancy. *Br J Haematol, 120,* 574–596.

7. Shehata, N., Burrows, R. F., & Kelton, J. G. (1999). Gestational thrombocytopenia. *Clin Obstet Gynecol, 42,* 327–334.

8. Burrows, R. F., & Kelton, J. G. (1988). Incidentally detected thrombocytopenia in healthy mothers and their infants. *N Engl J Med, 319,* 142–145.

9. McCrae, K. R., Samuels, P., & Schreiber, A. D. (1992). Pregnancy-associated thrombocytopenia: Pathogenesis and management. *Blood, 80,* 2697–2714.

10. Burrows, R. F., & Kelton, J. G. (1990). Thrombocytopenia at delivery: A prospective survey of 6715 deliveries. *Am J Obstet Gynecol, 162,* 731–734.

11. Matthews, J. H., Benjamin, S., Gill, D. S., et al. (1990). Pregnancy-associated thrombocytopenia: Definition, incidence and natural history. *Acta Haemat, 84,* 24–29.

12. Boehlen, F., Hohlfeld, H., Extermann, P., et al. (2000). Platelet count at term pregnancy: A reappraisal of the threshold. *Obstet Gynecol, 95,* 29–33.

13. Sainio, S., Kekomäki, R., Riikonon, S., et al. (2000). Maternal thrombocytopenia at term: A population-based study. *Acta Obstet Gynecol Scand, 79,* 744–749.

14. Burrows, R. F., & Kelton, J. G. (1993). Fetal thrombocytopenia and its relation to maternal thrombocytopenia. *N Engl J Med, 329,* 1463–1466.

15. Letsky, E. A., & Greaves, M. (1996). Guidelines on the investigation and management of thrombocytopenia in pregnancy and neonatal alloimmune thrombocytopenia. *Br J Haematol, 95,* 21–36.

16. Ruggeri, M., Schiavotto, C., Castaman, G., et al. (1997). Gestational thrombocytopenia: A prospective study. *Haematologica, 82,* 341–342.

17. Cines, D. B., & Blanchette, V. S. (2002). Immune thrombocytopenic purpura. *N Engl J Med, 346,* 13–995.

18. Provan, D., & Newland, A. (2003). Idiopathic thrombocytopenic purpura in adults. *J Ped Hematol Oncol, 25* (suppl 1), S34–S38.

19. Harrington, W. J., Minnich, V., Hollingsworth, J. W., et al. (1951). Demonstration of a thrombocytopenic factor in the blood of patients with thrombocytopenic purpura. *J Lab Clin Med, 38,* 1–8.

20. Cines, D. B., & McMillan, R. (2005). Management of adult idiopathic thrombocytopenic purpura. *Ann Rev Med, 56,* 425–442.

21. Chang, M., Nakagawa, P. A., Williams, S. A., et al. (2003). Immune thrombocytopenic purpura (ITP) plasma and purified ITP monoclonal autoantibodies inhibit megakaryocytopoiesis *in vitro. Blood, 102,* 887–895.

22. McMillan, R., Wang, L., Tomer, A., et al. (2004). Suppression of *in vitro* megakaryocyte production by antiplatelet autoantibodies from adult patients with chronic ITP. *Blood, 103,* 1364–1369.

23. Gill, K. K., & Kelton, J. G. (2000). Management of idiopathic thrombocytopenic purpura in pregnancy. *Sem Hematol, 37,* 275–283.

24. Karim, R., & Sacher, R. A. (2004). Thrombocytopenia in pregnancy. *Curr Hematol Rep, 3,* 128–133.

25. British Committee for Standards in Haematology General Haematology Task Force. (2003). Guidelines for the investigation and management of idiopathic thrombocytopenic purpura in adults, children and in pregnancy. *Br J Haematol, 120,* 574–596.

26. Webert, K. E., Mittal, R., Siguoin, C., et al. (2003). A retrospective 11-year analysis of obstetric patients with idiopathic thrombobocytopenic purpura. *Blood, 102,* 4306–4311.

27. Samuels, P., Main, E. K., Tomaski, A., et al. (1987). Abnormalities in platelet antiglobulin tests in preeclamptic mothers and their neonates. *Am J Obstet Gynecol, 107,* 109–113.

28. Fujimura, K., Harada, Y., Fujimoto, T., et al. (2002). Nationwide study of idiopathic thrombocytopenic purpura in pregnant women and the clinical influence on neonates. *Int J Haematol, 75,* 426–433.

29. George, J. N., Woolf, S. H., Raskob, G. E., et al. (1996). Idiopathic thrombocytopenic purpura: A practice guideline developed by explicit methods for the American Society of Hematology. *Blood, 88,* 3–40.

30. Moeller-Bertram, T., Kuczkowski, K. M., Benumof, J. L. (2004). Uneventful epidural labor analgesia in a parturient with immune thrombocytopenic purpura and platelet count of 26,000/mm$^3$ which was unknown preoperatively. *J Clin Anesthesiol, 16,* 51–53.

31. Michel, M., Novoa, M. V., & Bussel, J. B. (2003). Intravenous anti-D as a treatment for immune thrombocytopenic purpura (ITP) during pregnancy. *Br J Haematol, 123,* 142–146.

32. Kahn, M. J., & McCrae, K. R. (2004). Splenectomy in immune thrombocytopenic purpura: Recent controversies and long-term outcomes. *Curr Hematol Rep, 3,* 317–323.

33. Gottlieb, P., Axelsson, O., Bakos, O., et al. (1999).Splenectomy during pregnancy: An option in the treatment of autoimmune thrombocytopenic purpura. *Br J Obstet Gynaecol, 106,* 373–375.

34. Stasi, R., Pagano, A., Stipa, E., et al. (2001). Rituximab chimeric anti-CD20 monoclonal antibody treatment for adults with chronic idiopathic thrombocytopenic purpura. *Blood, 98,* 952–957.

35. Kimby, E., Sverrisdotter, E., & Elinder, G. (2004). Safety of rituximab therapy during the first trimester of pregnancy: A case history. *Eur J Haematol, 72,* 292–295.

36. Burrows, R. F., & Kelton, J. G. (1993). Pregnancy in patients with idiopathic thrombocytopenic purpura: Assessing the risks for the infant at delivery. *Obstet Gynecol Surv, 48,* 781–788.

37. Kelton, J. G. (2002). Idiopathic thrombocytopenic purpura complicating pregnancy. *Blood Rev, 16,* 43–46.

38. Godelieve, C., Christiaens, M. L., Nieuwenhuis, H. K., et al. (1997). Comparison of platelet counts in first and second newborns of mothers with immune thrombocytopenic purpura. *Obstet Gynecol, 90,* 546–552.

39. Payne, S. D., Resnik, R., Moore, T. R., et al. (1997). Maternal characteristics and risk of severe neonatal thrombocytopenia and intracranial hemorrhage in pregnancies complicated by autoimmune thrombocytopenia. *Am J Obstet Gynecol, 177,* 149–155.

40. Peleg, D., & Hunter, S. K. (1999). Perinatal management of women with immune thrombocytopenic purpura: Survey of United States perinatologists. *Am J Obstet Gynecol, 180,* 645–650.

41. Zhang, J., Meikle, S., & Trumble, A. (2003). Severe maternal morbidity associated with hypertensive disorders in pregnancy in the United States. *Hypertens Preg,* 203–212.

42. Lain, K. Y., & Roberts, J. M. (2002). Contemporary concepts of the pathogenesis and management of preeclampsia. *J Am Med Assoc, 287,* 3183–3186.

43. Duckitt, K., & Harrington, D. (2005). Risk factors for preeclampsia at antenatal booking: Systemic review of controlled studies. *Br Med J, 330,* 595–601.

44. Working Group Report on High Blood Pressure in Pregnancy (NIH Publication No. 00-3029). (2000). Bethesda, MD: National Institutes of Health, National Heart, Lung and Blood Institute.

45. Pridjian, G., & Puschett, J. B. (2002). Preeclampsia. Part 1: Clinical and pathophysiological considerations. *Obstet Gynecol Surv, 57,* 598–618.

46. American College of Obstet Gynecol. (2002). ACOG practice bulletin: Diagnosis and management of preeclampsia in pregnancy. *Obstet Gynecol, 99,* 159–167.

47. Mahmoudi, N., Graves, S. W., & Solomon, C. G. (1999). Eclampsia: A 13-year experience at a United States tertiary care center. *J Womens Health Gender Based Med, 8,* 495–500.

48. Esplin, M. S., Fausett, M. B., Fraser, A., et al. (2001). Paternal and maternal components of the predisposition to preeclampsia. *N Engl J Med, 344,* 867–872.

49. Mills, J. L., Klebanoff, M. A., Graubard, B. I., et al. (1991). Barrier contraceptive methods and preeclampsia. *J Am Med Assoc, 265,* 70–73.

50. Li, D. K., & Wi, S. (2000). Changing paternity and the risk of preeclampsia/eclampsia in the subsequent pregnancy. *Am J Epidemiol, 151,* 57–62.

51. Skjærven, R., Wilcox, A. J., & Lie, R. T. (2002). The interval between pregnancies and the risk of preeclampsia. *N Engl J Med, 346,* 33–38.

52. Robillard, P. Y., Dekker, G. A., & Hulsey, T. C. (1999). Revisiting the epidemiological standard of preeclampsia: Primigravidity of primipaternity. *Eur J Obstet Gynecol Repro Biol, 84,* 37–41.

53. Roberts, J. M., & Cooper, D. W. (2001). Pathogenesis and genetics of preeclampsia. *Lancet, 357,* 53–56.

54. Pridjian, G., & Puschett, J. B. (2002). Preeclampsia. Part 2: Experimental and genetic considerations. *Obstet Gynecol Surv, 57,* 619–640.

55. GOPEC consortium. (2005). Disentangling fetal and maternal susceptibility for pre-eclampsia: A British multicenter candidate gene study. *Am J Human Genet, 77,* 127–131.

56. Saftlas, A. F., Beydoun, H., & Triche, E. (2005). Immuno-genetic determinants of preeclampsia and related pregnancy disorders: A systematic review. *Obstet Gynecol, 106,* 162–172.

57. Alsulyman, O. M., Castro, M. A., Zuckerman, E., et al. (1996). Preeclampsia and liver infarction in early pregnancy associated with the antiphospholipid syndrome. *Obstet Gynecol, 88,* 644–646.

58. Wilson, W. A., Gharavi, A. E., Koike, T., et al. (1999). International consensus statement on preliminary classification criteria for definite antiphospholipid syndrome. *Arth Rheum, 42,* 1309–1311.

59. Shehata, H. A., Nelson-Piercy, C., & Khamashta, M. A. (2001). Management of pregnancy in the antiphospholipid syndrome. *Rheum Dis Clin NA, 27,* 643–649.

60. Dekker, G. A., deVries, J. I., Doelitzsch, P. M., et al. (1995). Underlying disorders associated with severe, early onset preeclampsia. *Am J Obstet Gynecol, 173,* 1042–1048.

61. Dizon-Townson, D. S., Nelson, L. M., Easton, K., et al. (1996). The factor V Leiden mutation may predispose women to severe preeclampsia. *Am J Obstet Gynecol, 175,* 902–905.

62. Grandone, E., Margaglione, M., Colaizzo, D., et al. (1997). Factor V Leiden, C > T MTHFR polymorphism and genetic susceptibility to preeclampsia. *Thromb Haemost, 77,* 1052–1054.

63. Kupferminc, M. J., Eldor, A., Steinman, M., et al. (1999). Increased frequency of genetic thrombophilia in women with complications of pregnancy. *N Engl J Med, 1999,* 340–349.

64. D'Elia, A. V., Driul, L., Giacomello, R., et al. (2002). Frequency of factor V, prothrombin and methylenetetrahydrofolate reductase gene variants in preeclampsia. *Gynecol Obstet Invest, 53,* 84–87.

65. Anonymous. Is there an increased maternal-infant prevalence of factor V Leiden in association with severe preeclampsia? *Br J Obstet Gynaecol, 109,* 191–196.

66. Livingston, J. C., Barton, J. R., Park, V., et al. (2001). Maternal and fetal inherited thrombophilias are not related to the development of severe preeclampsia. *Am J Obstet Gynecol, 185,* 153–157.

67. Alfirevic, Z., Roberts, D., & Martlew, V. (2002). How strong is the association between maternal thrombophilia and adverse pregnancy outcome? A systematic review. *Eur J Obstet Gynecol Repro Biol, 101,* 6–14.

68. Stanley-Christian, H., Ghidini, A., Sacher, R., et al. (2005). Fetal genotype for specific thrombophilias is not associated with severe preeclampsia. *J Soc Gynecol Invest, 12,* 198–201.

69. Witsenburg, C. P. J., Rosendall, F. R., Middeldorp, J. M., et al. (2005). Factor VIII levels and the risk of preeclampsia, HELLP syndrome, pregnancy related hypertension and severe intrauterine growth retardation. *Thromb Res, 115,* 382–392.

70. Goldman-Wohl, D., & Yagel, S. (2002). Regulation of trophoblast invasion: From normal implantation to preeclampsia. *Mol Cell Endocrinol, 187,* 233–238.

71. Brosens, I. O. (1988). The utero-placental vessels at term — the distribution and extent of physiological changes. *Troph Res, 3,* 61–67.

72. Khong, T. Y., De Wolf, F., Robertson, W. B., et al. (1987). Inadequate maternal vascular response to placentation in pregnancies complicated by preeclampsia and by small for gestational age infants. *Am J Obstet Gynecol, 157,* 360–363.

73. Zhou, Y., Damsky, C. H., Chiu, K., et al. (1993). Preeclampsia is associated with abnormal expression of adhesion molecules by invasive cytotrophoblasts. *J Clin Invest, 91,* 950–960.

74. Zhou, Y., Damsky, C. H., & Fisher, S. J. (1997). Preeclampsia is associated with failure of human cytotrophoblasts to mimic a vascular adhesion phenotype. *J Clin Invest, 99,* 2152–2164.

75. Zhou, Y., Fisher, S. J., Janatpour, M., et al. (1997). Human cytotrophoblasts adopt a vascular phenotype as they differentiate. A strategy for successful endovascular invasion? *J Clin Invest, 99,* 2139–2151.

76. Graham, C. H., & McCrae, K. R. (1996). Expression of gelatinase and plasminogen activator activity by trophoblast cells isolated from the placentae of normal and preeclamptic women. *Am J Obstet Gynecol, 175,* 555–562.

77. Tsatsiris, V., Goffin, F., Munuat, C., et al. (2003). Overexpression of the soluble vascular endothelial growth factor receptor in preeclamptic patients: Pathophysiological consequences. *J Clin Endo Metab, 88,* 5555–5563.

78. Levine, R. J., Maynard, S. E., Qian, C., et al. (2004). Circulating angiogenic factors and the risk of preeclampsia. *N Engl J Med, 350,* 672–683.

79. Soleymanlou, N., Jurisica, I., Nevo, O., et al. (2005). Molecular evidence of placental hypoxia in preeclampsia. *J Clin Endocrinol Metab, 90,* 4299–4308.

80. Taylor, R. N., Varma, M., Teng, N. N. H., et al. (1990). Women with preeclampsia have higher plasma endothelin levels than women with normal pregnancies. *J Clin Endocrinol Metab, 71,* 1675–1677.

81. Nova, A., Sibai, B. M., Barton, J. R., et al. (1991). Maternal plasma level of endothelin is increased in preeclampsia. *Am J Obstet Gynecol, 165,* 724–727.

82. Dekker, G. A., Kraayenbrink, A. A., Zeeman, G. G., et al. (1991). Increased plasma levels of the novel vasoconstrictor peptide endothelin in severe preeclampsia. *Eur J Obstet Gynecol Repro Biol, 40,* 215–220.

83. Myatt, L., & Cui, X. (2004). Oxidative stress in the placenta. *Histochem Cell Biol, 122,* 369–382.

84. McCrae, K. R. (2003). Thrombocytopenia in pregnancy: Differential diagnosis, pathogenesis and management. *Blood Rev, 17,* 7–14.

85. deBoer, K., ten Cate, J. W., Sturk, A., et al. (1989). Enhanced thrombin generation in normal and hypertensive pregnancy. *Am J Obstet Gynecol, 160,* 95–100.

86. Brenner, B., Zwang, E., Bronshtein, M., et al. (1989). Von Willebrand factor multimer patterns in pregnancy-induced hypertension. *Thromb Haemost, 62,* 715–717.

87. Thorp, J. M., White, G. C., Moake, J. L., et al. (1990). Von Willebrand factor multimeric levels and patterns in patients with severe preeclampsia. *Obstet Gynecol, 75,* 163–167.

88. Lockwood, C. J., & Peters, J. H. (1990). Increased plasma levels of ED1[+] cellular fibronectin precede the clinical signs of preeclampsia. *Am J Obstet Gynecol, 162,* 358–362.

89. Redman, C. W. G., Beilin, L. J., Denson, K. W. E., et al. (1977). Factor VIII consumption in pre-eclampsia. *Lancet, ii,* 1249–1252.

90. Curtin, W. M., & Weinstein, L. (1999). A review of the HELLP syndrome. *J Perinatol, 19,* 138–143.

91. O'Brien, J. M., & Barton, J. R. (2005). Controversies with the diagnosis and management of HELLP syndrome. *Clin Obstet Gynecol, 48,* 460–477.

92. Barton, J. R., & Sibai, B. M. (2004). Diagnosis and management of hemolysis, elevated liver enzymes and low platelets syndrome. *Clin Perinatol, 31,* 807–833.

93. Sibai, B. M. (1990). The HELLP syndrome (hemolysis, elevated liver enzymes, and low platelets): Much ado about nothing? *Am J Obstet Gynecol, 162,* 311–316.

94. Martin, J. N., Blake, P. G., Lowry, S. L., et al. (1990). Pregnancy complicated by preeclampsia with the syndrome of hemolysis, elevated liver enzymes, and low platelet counts: How rapid is postpartum recovery. *Obstet Gynecol, 76,* 737–741.

95. Rath, W., Faridi, A., & Dudenhausen, J. W. (2000). HELLP syndrome. *J Perinatal Med, 28,* 249–260.

96. McMahon, L. P., & Smith, J. (1997). The HELLP syndrome at 16 weeks gestation: Possible association with the antiphospholipid syndrome. *Aust NZJ Obstet Gynaecol, 37,* 313–314.

97. Haram, K., Trovik, J., Sanset, P. M., et al. (2003). Severe syndrome of hemolysis, elevated liver enzymes and low platelets (HELLP) in the 18th week of pregnancy associated with the antiphospholipid-antibody syndrome. *Acta Obstet Gynecol Scand, 82,* 679–680.

98. Li Thi Thoung, D., Tieulie, N., Costedoat, N., et al. (2005). The HELLP syndrome in the antiphospholipid syndrome: Retrospective study of 16 cases in 15 women. *Ann Rheum Dis, 64,* 273–278.

99. Nagayama, K., Izumi, N., Miyasaka, Y., et al. (1997). Hemolysis, elevated liver enzymes, and low platelets syndrome associated with primary anti-phospholipid antibody syndrome. *Intern Med, 36,* 661–667.

100. Magann, E. F., & Martin, J. R. (1999). Twelve steps to optimal management of HELLP syndrome. *Clin Obstet Gynecol, 42,* 532–550.

101. Sibai, B. M. (2004). Diagnosis, controversies and management of the syndrome of hemolysis, elevated liver enzymes, and low platelet count. *Obstet Gynecol, 103,* 981–991.

102. Murphy, M. A., & Ayazifar, M. (2005). Permanent visual defects secondary to the HELLP syndrome. *J Neuro-Opthalmol, 25,* 122–127.

103. Rinehart, B. K., Terrone, D. A., May, W. L., et al. (2001). Change in platelet count predicts eventual maternal outcome with syndrome of hemolysis, elevated liver enzymes and low platelet count. *J Mat Fet Med, 10,* 28–34.

104. Martin, J. N., Blake, P. G., Perry, K. G., et al. (1991). The natural history of HELLP syndrome: Patterns of disease progression and regression. *Am J Obstet Gynecol, 164,* 1500–1513.

105. Faridi, A., Heyl, W., & Rath, W. (2000). Preliminary results of the International HELLP-Multicenter-Study. *Int J Gynaecol Obstet, 69,* 279–280.

106. Sibai, B. M., Taslimi, M. M., El-Nazer, A., et al. (1986). Maternal-perinatal outome associated with the syndrome of

hemolysis, elevated liver enzymes, and low platelets in severe preeclampsia-eclampsia. *Am J Obstet Gynecol, 155,* 501–509.

107. Schwartz, M. L., & Brenner, W. E. (1978). The obfuscation of eclampsia by thrombotic thrombocytopenic purpura. *Am J Obstet Gynecol, 131,* 18–10.

108. Coppage, K. H., & Sibai, B. M. (2005). Treatment of hypertensive complications of pregnancy. *Current Pharm Des, 11,* 749–757.

109. Visser, W., & Wallenburg, H. C. S. (1995). Temporizing management of severe pre-eclampsia with and without the HELLP syndrome. *Br J Obstet Gynaecol, 102,* 111–117.

110. Hogg, B., Hauth, J. C., Caritis, S. N., et al. (1999). Safety of labor epidural anesthesia for women with severe hypertensive disease: National Institute of Child Health and Human Development Maternal-Fetal Medicine Units Network. *Am J Obstet Gynecol, 181,* 1096–1101.

111. Barilleaux, P. S., Martin, J. N., Klauser, C. K., et al. (2005). Postpartum intravenous dexamethasone for severely preeclamptic patients without hemolysis, elevated liver enzymes, low platelets (HELLP) syndrome: A randomized trial. *Obstet Gynecol, 105,* 843–848.

112. van Runnard Heimel, P. J., Franx, A., Schobben, A. F. A. M., et al. (2004). Corticosteroids, pregnancy, and HELLP syndrome: A review. *Obstet Gynecol Surv, 60,* 57–70.

113. Rose, C. H., Thigpen, B. D., Bofill, J. A., et al. (2004). Obstetric implications of antepartum corticosteroid therapy for HELLP syndrome. *Obstet Gynecol, 104,* 1011–1014.

114. O'Brien, J. M., Milligan, D. A., & Barton, J. R. (2000). Impact of high-dose corticosteroid therapy for patients with HELLP (hemolysis, elevated liver enzymes, and low platelet count) syndrome. *Am J Obstet Gynecol, 183,* 304–309.

115. Martin, J. N., Perry, K. G., Blake, P. G., et al. (1997). Better maternal outcomes are achieved with dexamethasone therapy for postpartum HELLP (hemolysis, elevated liver enzymes, and thrombocytopenia) syndrome. *Am J Obstet Gynecol, 177,* 1011–1017.

116. Martin, J. N., Jr., Files, J. C., Blake, P. G., et al. (1990). Plasma exchange for preeclampsia. I. Postpartum use for persistently severe preeclampsia-eclampsia with HELLP syndrome. *Am J Obstet Gynecol, 162,* 126–137.

117. Martin, J. N., Jr., Files, J. C., Blake, P. G., et al. (1995). Postpartum plasma exchange for atypical preeclampsia-eclampsia as HELLP (hemolysis, elevated liver enzymes and low platelets syndrome). *Am J Obstet Gynecol, 172,* 1107–1112.

118. Eser, B., Guven, M., Unal, A., et al. (2005). The role of plasma exchange in HELLP syndrome. *Clin Applied Thromb/ Hemost, 11,* 211–217.

119. CLASP Collaborative Group. (1994). CLASP: A randomized trial of low-dose aspirin for the prevention and treatment of pre-eclampsia among 9364 pregnant women. *Lancet, 343,* 619–629.

120. Roberts, J. M., Pearson, G., Cutler, J., et al. (2003). Summary of the NHLBI working group on research on hypertension during pregnancy. *Hypertension, 41,* 437–445.

121. Chappell, L. C., Seed, P. T., Briley, A. L., et al. (1999). Effect of antioxidants on the occurrence of pre-eclampsia in women at increased risk: A randomized trial. *Lancet, 354,* 810–816.

122. Rumbold, A., & Crowther, C. A. (2005). Vitamin E supplementation in pregnancy. *Cochrane Database Sys Rev,* April 18.

123. Miller, D. P., Kaye, J. A., Ziyadeh, N., et al. (2004). Incidence of thrombotic thrombocytopenic purpura/hemolytic uremic syndrome. *Epidemiol, 15,* 208–215.

124. Torok, T. J., Holman, R. C., & Chorba, T. L. (1995). Increasing mortality from thrombotic thrombocytopenic purpura in the United States — analysis of national mortality data, 1968–1961. *Am J Hematol, 5084,* 90.

125. Ridolfi, R. L., & Bell, W. R. (1981). Thrombotic thrombocytopenia purpura: Report of 25 cases and a review of the literature. *Medicine, 60,* 413–428.

126. Vesely, S. K., George, J. N., Lämmle, B., et al. (2003). ADAMTS13 activity in thrombotic thrombocytopenic purpura-hemolytic uremic syndrome: Relation to presenting features and clinical outcomes in a prospective cohort of 142 patients. *Blood, 102,* 60–68.

127. Nicol, K. K., Shelton, B. J., Knovich, M. S., et al. (2003). Overweight individuals are at increased risk for thrombotic thrombocytopenic purpura. *Am J Hematol, 74,* 170–174.

128. Amarosi, E. L., & Ultmann, J. E. (1966). Thrombotic thrombocytopenic purpura: Report of 16 cases and review of the literature. *Medicine, 45,* 139–159.

129. Nabhan, C., & Kwaan, H. C. (2003). Current concepts in the diagnosis and management of thrombotic thrombocytopenic purpura. *Hematol/Oncol Clin NA, 17,* 177–199.

130. Thompson, C. E., Damon, L. E., Ries, C. A., et al. (1992). Thrombotic microangiopathies in the 1980s: Clinical features, response to treatment, and the impact of the human immunodeficiency virus epidemic. *Blood, 80,* 1890–1895.

131. Cohen, J. D., Brechter, M. E., Bandarenko, N. (1998). Cellular source of serum lactate dehydrogenase elevation in patients with thrombotic thrombocytopenic purpura. *J Clin Apheresis, 13,* 16–19.

132. Berns, J. S., Kaplan, B. S., Mackow, R. C., et al. (1992). Inherited hemolytic uremic syndrome in adults. *Am J Kidney Dis, XIX,* 331–334.

133. Proulx, F., Siefdman, E. G., & Karpman, D. (2001). Pathogenesis of shiga toxin-associated hemolytic uremic syndrome. *Ped Res, 50,* 163–171.

134. Taylor, C. M. (2001). Hemolytic uremic syndrome and complement factor H deficiency: Clinical aspects. *Sem Thromb Hemost, 27,* 185–190.

135. McCrae, K. R., Sadler, J. E., & Cines, D. B. (2005). Thrombotic thrombocytopenic purpura and the hemolytic uremic syndrome. In R. Hoffman, E. J. Benz, Jr., S. J. Shattil, et al. (Eds.), *Hematology: Basic principles and practice* (pp. 2287–2304). Philadelphia: Elsevier, Churchill, Livingstone.

136. Neild, G. (1987). The haemolytic uraemic syndrome: A review. *Quart J Med, 241,* 367–376.

137. Schieppati, A., Ruggenenti, P., Plata Cornejo, R., et al. (1992). Renal function at hospital admission as a prognostic factor in adult hemolytic uremic syndrome. *J Am Soc Nephrol, 2,* 1640–1644.

138. George, J. N. (2000). How I treat patients with thrombotic thrombocytopenic purpura-hemolytic uremic syndrome. *Blood, 96,* 1223–1229.

139. Esplin, M. S., & Branch, D. W. (1999). Diagnosis and management of thrombotic microangiopathies during pregnancy. *Clin Obstet Gynecol, 42,* 360–368.

140. Castella, M., Pujol, M., Massague, I., et al. (2004). Thrombotic thrombocytopenic purpura and pregancy: A review of ten cases. *Vox Sang, 87,* 287–290.

141. Furlan, M., Robles, R., Galbusera, M., et al. (1998). von Willebrand factor-cleaving protease in thrombotic thrombocytopenic purpura and the hemolytic-uremic syndrome. *N Engl J Med, 339,* 1578–1584.

142. Tsai, H.-M., & Lian, E. C. Y. (1998). Antibodies to von Willebrand factor-cleaving protease in acute thrombotic thrombocytopenic purpura. *N Engl J Med, 339,* 1585–1594.

143. Levy, G. G., Nichols, W. C., Lian, E. C., et al. (2001). Mutations in a member of the ADAMTS gene family cause thrombotic thrombocytopenic purpura. *Nature, 413,* 488–494.

144. Bianchi, V., Robles, R., Alberio, L., et al. (2002). von Willebrand factor-cleaving protease (ADAMTS13) in thrombocytopenic disorders: A severely deficient activity is specific for thrombotic thrombocytopenic purpura. *Blood, 100,* 710–713.

145. Tsai, H.-M. (2003). Is severe deficiency of von Willebrand factor cleaving protease (ADAMTS-13) specific for thrombotic thrombocytopenic purpura? Yes. *J Thromb Haemost, 1,* 625–631.

146. Mannucci, P. M., Vanoli, M., Forza, I., et al. (2003). von Willebrand factor cleaving protease (ADAMTS-13) in 123 patients with connective tissue diseases (systemic lupus erythematosus and systemic sclerosis). *Haematologica, 88,* 914–918.

147. Mannucci, P. M., Canciani, T., Forza, I., et al. (2001). Changes in health and disease of the metalloprotease that cleaves von Willebrand factor. *Blood, 98,* 2730–2735.

148. Sanchez-Luceros, A., Farias, C. E., Amaral, M. M., et al. (2004). von Willebrand factor-cleaving protease (ADAMTS13) activity in normal non-pregnant women, pregnant and post-delivery women. *Thromb Haemost, 92,* 1320–1326.

149. Moake, J. L., Byrnes, J. J., Troll, J. H., et al. (1994). Abnormal VIII: von Willebrand factor patterns in the plasma of patients with the hemolytic-uremic syndrome. *Blood, 64,* 592–598.

150. Moake, J. L. (2002). Thrombotic microangiopathies. *N Engl J Med, 347,* 589–600.

151. Tsai, H.-M. (2005). Current concepts in thrombotic thrombocytopenic purpura. *Ann Rev Med,* e-publication.

152. George, J. N. (2003). The association of pregnancy with thrombotic thrombocytopenic purpura-hemolytic uremic syndrome. *Curr Opin Hematol, 10,* 339–344.

153. Reiter, R. A., Varadi, K., Turecek, P. L., et al. (2005). Changes in ADAMTS13 (von-Willebrand-factor-cleaving protease) activity after induced release of von Willebrand factor during acute systemic inflammation. *Thromb Haemost, 93,* 554–558.

154. Lattuada, A., Rossi, E., Calzarossa, C., et al. (2003). Mild to moderate reduction of a von Willebrand factor cleaving protease (ADAMTS-13) in pregnant women with HELLP microangiopathic syndrome. *Haematologica, 88,* 1029–1034.

155. Weiner, C. P. (1987). Thrombotic microangiopathy in pregnancy and the postpartum period. *Sem Hematol, 24,* 119–129.

156. Proia, A., Paesano, F., Torcia, L., et al. (2002). Thrombotic thrombocytopenic purpura and pregnancy: A case report and a review of the literature. *Ann Haematol, 81,* 210–214.

157. Ducloy-Bouthors, A. S., Caron, C., Subtil, D., et al. (2003). Thrombotic thrombocytopenic purpura: Medical and biological monitoring of six pregnancies. *Eur J Obstet Gynecol Repro Biol, 111,* 146–152.

158. Clark, K. (2004). TTP and me. *The Practising Midwife, 7,* 13–17.

159. Sherer, D. M., Sanmugarajah, J., Dalloul, M., et al. (2005). Thrombotic thrombocytopenic purpura in a patient with acquired immunodeficiency syndrome at 28 weeks gestation. *Am J Perinatol, 22,* 223–225.

160. Allford, S. L., Hunt, B. J., Rose, P., et al. (2003). Haemostasis and Thrombosis Task Force of the British Committee for Standards in Haematology: Guidelines on the diagnosis and management of the thrombotic microangiopathic haemolytic anaemias. *Br J Haematol, 120,* 556–573.

161. Kwaan, H. C., & Soff, G. A. (1997). Management of thrombotic thrombocytopenic purpura and hemolytic uremic syndrome. *Sem Hemato, 34,* 159–166.

162. Rock, G. A., Shumak, K. H., Buskard, N. A., et al. (1991). Comparison of plasma exchange with plasma infusion in the treatment of thrombotic thrombocytopenic purpura. *N Engl J Med, 325,* 393–397.

163. Ruggenenti, P., & Remuzzi, G. (1996). The pathophysiology and management of thrombotic thrombocytopenic purpura. *Eur J Haematol, 56,* 191–207.

164. Remuzzi, G. (1995). The hemolytic uremic syndrome. *Kidney Int, 47,* 2–19.

165. Vesely, S. K., McMinn, J. R., Terrell, D. R., et al. (2004). Pregnancy outcomes after recovery from thrombotic thrombocytopenic purpura-hemolytic uremic syndrome. *Transfusion, 44,* 1149–1158.

166. Fesenmeir, M. F., Coppage, K. H., Lambers, D. S., et al. (2005). Acute fatty liver of pregnancy in three tertiary care centers. *Am J Obstet Gynecol, 192,* 1416–1419.

167. Knox, T. A., & Olans, L. B. (1996). Liver disease in pregnancy. *N Engl J Med, 335,* 569–576.

168. Burroughs, A. K., Seong, N. H., Dojcinov, D. M., et al. (1982). Idiopathic acute fatty liver of pregnancy in 12 patients. *Quart J Med, 51,* 481–497.

169. Usta, I. M., Barton, J. R., Amon, E. A., et al. (1994). Acute fatty liver of pregnancy: An experience in the diagnosis and managment of fourteen cases. *Am J Obstet Gynecol, 171,* 1342–1347.

170. Ibdah, J. A., Yang, Z., & Bennett, M. J. (2000). Liver disease in pregnancy and fetal fatty acid oxidation defects. *Molec Genet Metab, 71,* 182–187.

171. Ibdah, J. A., Bennett, M. J., Rinaldo, P., et al. (1999). A fetal fatty-acid oxidation disorder as a cause of liver disease in pregnant women. *N Engl J Med, 340,* 1723–1731.

172. Holub, M., Bodamer, O. A., Item, C., et al. (2005). Lack of correlation between fatty acid oxidation disorders and haemolysis, elevated liver enzymes, low platelets (HELLP) syndrome? *Acta Paeds Scand, 94,* 48–52.

173. Tan, A. C., van Krieken, J. H., Peters, W. H., et al. (2002). Acute fatty liver in pregnancy. *Netherlands J Med, 60,* 370–373.

174. Castro, M. A., Goodwin, T. M., Shaw, K. J., et al. (1996). Disseminated intravascular coagulation and antithrombin III depression in acute fatty liver of pregnancy. *Am J Obstet Gynecol, 174,* 211–216.

175. Hepner, D. L., & Tsen, L. C. (2004). Severe thrombocytopenia, type 2B von Willebrand disease and pregnancy. *Anesthesiol, 101,* 1465–1467.

176. Casonato, A., Sartori, M. T., Bertomoro, A., et al. (1991). Pregnancy-induced worsening of thrombocytopenia in a patient with type IIB von Willebrand's disease. *Blood Coag Fibrinol, 2,* 33–40.

177. Icko, M., Sakurama, S., Sagawa, A., et al. (1990). Effect of a factor VIII concentrate on type IIB von Willebrand's disease-associated thrombocytopenia presenting during pregnancy in identical twin mothers. *Am J Hematol, 35,* 26–31.

178. Nardi, M., & Karpatkin, S. (2000). Antiidiotype antibody against platelet anti-GPIIIa contributes to the regulation of thrombocytopenia in HIV-1-ITP patients. *J Exp Med, 191,* 2093–2100.

179. Li, Z., Nardi, M. A., & Karpatkin, S. (2005). Role of molecular mimicry to HIV-1 peptides in HIV-1 related immunologic thrombocytopenia. *Blood, 106,* 572–576.

180. Chelucci, C., Federico, M., Guerriero, R., et al. (1999). Productive human immunodeficiency virus-1 infection of purified megakaryocytic progenitors/precursors and maturing megakaryocytes. *Blood, 91,* 1225–1234.

181. Diz-Kucukkaya, R., Hacehanefioglu, A., Yenerel, M., et al. (2001). Antiphospholipid antibodies and antiphospholipid syndrome in patients presenting with immune thrombocytopenic purpura: A prospective cohort study. *Blood, 98,* 1760–1764.

182. Chakravarty, E. F., Colon, I., Langen, E. S., et al. (2005). Factors that predict prematurity and preeclampsia in pregnancies that are complicated by systemic lupus erythematosus. *Am J Obstet Gynecol, 192,* 1897–1904.

183. Aster, R. H. (2005). Drug-induced immune cytopenias. *Toxicology, 209,* 149–153.

184. Balduini, C. L., Iolascon, A., & Savoia, A. (2002). Inherited thrombocytopenias. *Haematologica, 87,* 860–880.

185. Seri, M., Cusano, R., Gangarossa, S., et al. (2002). Mutations in MYH9 result in the May-Hegglin anomaly, and Fechtner and Sebastian syndromes: The May-Hegglin/Fechtner Syndrome Consortium. *Nat Genet, 26,* 103–105.

186. Sibai, B. M., Ramadan, M. K., Chari, R. S., et al. (1995). Pregnancies complicated by HELLP syndrome (hemolysis, elevated liver enzymes, and low platelets): Subsequent pregnancy outcome and long term prognosis. *Am J Obstet Gynecol, 172,* 125–129.

187. Sullivan, C. A., Magann, E. F., Perry, K. G., et al. (1994). The recurrence risk of the syndrome of hemolysis, elevated liver enzymes and low platelets (HELLP) in subsequent gestations. *Am J Obstet Gynecol, 171,* 940–943.

188. van Pampus, M. G., Wolf, H., Mayruu, G., et al. (2001). Long-term follow-up in patients with a history of (H)ELLP syndrome. *Hypertens Preg, 20,* 15–23.

189. Chames, M. C., Haddad, B., Barton, J. R., et al. (2003). Subsequent pregnancy outcome in women with a history of HELLP syndrome at <28 weeks of gestation. *Am J Obstet Gynecol, 188,* 1504–1508.

# CHAPTER 52

# Thrombocytopenia in the Newborn

**Irene A. G. Roberts and Neil A. Murray**

*Departments of Paediatric Haematology and Neonatal Medicine, Imperial College School of Medicine,
Hammersmith Hospital, London, United Kingdom*

## I. Introduction

All neonatal pediatricians are familiar with the problem of neonatal thrombocytopenia and recognize that it occurs in a significant number of their patients, particularly those who are sicker and more preterm. All also recognize that a long list of conditions is associated with neonatal thrombocytopenia and that establishing a precise definition for the cause of every episode is both time and resource consuming and impractical. In addition, experience shows that the majority of neonatal thrombocytopenias resolve during the course of standard neonatal care, without any apparent short- or long-term clinical sequelae. Therefore, at one end of the spectrum, neonatal thrombocytopenia can be considered as a physiological response to adverse neonatal conditions that requires little or no consideration other than to treat the underlying conditions.

By contrast, all neonatal pediatricians will also be familiar with the sick patient with profound thrombocytopenia in whom the cause of thrombocytopenia is uncertain, for whom specific management remains poorly defined, but for whom the potential clinical impact of severe hemorrhage is devastating both for short- and long-term outcome. Of equal importance is the neonate with profound thrombocytopenia who is otherwise well and for whom any clinical sequelae would be the sole result of the thrombocytopenia. At this end of the spectrum, careful evaluation and management of every episode of neonatal thrombocytopenia is essential to reduce the risk of significant bleeding. However, as outlined here, this represents a difficult task for most busy neonatal nurseries.

Effective assessment and management of neonates with significant thrombocytopenia therefore depends on neonatal pediatricians and hematologists knowing the common conditions and situations that lead to thrombocytopenia, the most useful investigations, and the most appropriate treatment, including an understanding of contemporary neonatal platelet transfusion practice. In addition, it is important to be able to recognize patterns of benign neonatal thrombo-cytopenia, in which a "wait and see" policy remains the best clinical approach.

This chapter summarizes current knowledge about the incidence, causes, and mechanisms of thrombocytopenia in the newborn. It outlines the presentations and clinical impact of the common forms of neonatal thrombocytopenia, paying particular attention to those conditions that give rise to severe thrombocytopenia, but also discussing rare forms of thrombocytopenia presenting in the neonatal period. Finally, the role of platelet transfusion and future possibilities for prevention and treatment of neonatal thrombocytopenia are discussed as a practical guide for those involved in the care of thrombocytopenic neonates. Chapter 53 specifically discusses fetal and neonatal alloimmune thrombocytopenia, Chapter 54 specifically discusses inherited thrombocytopenia, and Chapter 22 discusses platelet function in the newborn.

## II. Fetal Megakaryocytopoiesis and Platelet Production

Platelets first appear in the fetal circulation at 5 to 6 weeks postconceptual age.[1] By the end of the first trimester of pregnancy, the mean fetal platelet count is already above $150 \times 10^9$/L.[2] During the second trimester, the mean fetal platelet count rises to between 175 to $250 \times 10^9$/L and by this stage the great majority of all healthy fetuses have a platelet count above $150 \times 10^9$/L.[3-5] Although smaller studies suggest a slow linear rise in platelet count with advancing gestation,[6] the largest reported study shows no further significant increase in fetal platelet count through the second and third trimesters to term.[7] Thus, these independent studies indicate that a platelet count $<150 \times 10^9$/L defines thrombocytopenia in any neonate regardless of gestational age.

Although the healthy fetus and neonate can generate and maintain a circulating platelet count equivalent to that of adults from early in gestation, studies show that there are important differences in the process and regulation of fetal and neonatal megakaryocytopoiesis and platelet production that may predispose the sick fetus or neonate to develop thrombocytopenia. First, despite their normal platelet count, the numbers of reticulated (young) platelets during fetal life are markedly raised, suggesting that even in the healthy fetus platelet production may be set close to maximal capacity, perhaps to keep pace with the constantly growing fetus.[8,9] Second, megakaryocytes isolated from fetal liver, bone marrow or peripheral blood are significantly smaller than those found in adult bone marrow.[10,11] Third, fewer mature megakaryocytes with a lower ploidy distribution are seen in the fetus,[10–12] and mature fetal and neonatal megakaryocytes appear to have reduced proplatelet formation.[13]

The numbers of megakaryocytes and of megakaryocyte progenitors (burst-forming units-megakaryocyte [BFU-MK] and colony-forming units-megakaryocyte [CFU-MK]) are high early in fetal life and fall toward term.[14,15] Similarly, healthy preterm babies have higher numbers of megakaryocyte progenitors than healthy term babies.[16,17] Fetal megakaryocytes and their progenitors respond normally to thrombopoietin (TPO)[18–22] and may even be more sensitive to TPO than adult bone marrow-derived progenitors.[22] However, a number of studies have suggested that the ability of the fetus and neonate to produce large amounts of TPO in response to thrombocytopenia is reduced.[8,9,20,23,24] For example, the levels of TPO in preterm neonates with thrombocytopenia in association with reduced platelet production are modest compared to those in children undergoing chemotherapy with similar degrees of thrombocytopenia.[20,23] Together, these data suggest that the fetus and neonate may have limited potential to increase platelet production over baseline and that this becomes particularly evident at times of increased platelet demand.

## III. Incidence of Neonatal Thrombocytopenia

### A. Normal Platelet Count at Birth

The well-known large population studies of Burrows and Kelton,[25–27] reinforced by a more recent report,[28] show that more than 98% of term neonates born to mothers with normal platelet counts have platelets above $150 \times 10^9$/L at birth. Furthermore, studies of both healthy term and preterm neonates demonstrate platelet counts in the normal adult range throughout the neonatal period.[29,30]

Care must be taken to distinguish the rare cases of pseudothrombocytopenia due to platelet clumping, which has been documented in a preterm neonate[31] and was previously reported in a term neonate[32] (see Chapter 55). In these cases, there are no bleeding manifestations, despite an apparently extremely low peripheral blood platelet count, and large numbers of platelet clumps are visible on the blood smear.

Interestingly, in the case reported by Chiurazzi and colleagues the pseudothrombocytopenia was found to be due to transplacental passage of an EDTA-dependent antibody, which resolved within a month of birth.[32] By contrast, in the preterm baby reported by Christensen and colleagues, the thrombocytopenia was EDTA-independent and the infant still had severe pseudothrombocytopenia at the time of reporting (17 weeks of age).[31]

### B. Incidence of Thrombocytopenia at Birth

Thrombocytopenia is the most common hematological abnormality detected in the fetus or neonate. A platelet count of $<150 \times 10^9$/L is present in 1 to 5% of newborns, the incidence depending on the population studied.[7,25,28,33] The incidence of severe thrombocytopenia at birth (platelets $<50 \times 10^9$/L) is 0.1 to 0.5% overall.[27,28,33–35] In an unselected population of 1350 neonates born in one center over a period of 1 year, Burrows and Kelton[25] reported an overall incidence of thrombocytopenia in cord blood samples of 4.1%. A slightly higher incidence of thrombocytopenia (4.75%) was recorded by Hohlfeld et al.[7] in a study of 5194 fetal samples in a higher risk population (the majority of fetuses were screened for suspected congenital infection). In a population-based study confined to term neonates born to native Finnish women in Helsinki in a single year,[28] 2% of newborns (89/4,489) were found to have a cord blood platelet count of $<150 \times 10^9$/L, 11 of whom (0.24%) had severe thrombocytopenia (platelets $<50 \times 10^9$/L). This lower overall incidence is supported by the findings of de Moerloose et al.[33] who found an incidence of thrombocytopenia of 0.9% in 8388 consecutively delivered newborns in two university hospitals in Switzerland. Data from other studies are more difficult to interpret because they include only neonates with platelet counts $<100 \times 10^9$/L[24] or, because of methodological difficulties, have underrepresented neonates of low birth weight,[35] the high-risk population for neonatal thrombocytopenia.

### C. Thrombocytopenia in NICU Patients

A number of studies have shown that the incidence of thrombocytopenia in sick neonates is much higher than documented in population studies of largely healthy newborns assessed at birth. Thrombocytopenia develops in between 22% and 35% of all admissions to neonatal intensive care units (NICUs).[36–38] This incidence is relatively constant in different geographical populations.[39] Furthermore, in those neonates who go on to require intensive care, the overall incidence of thrombocytopenia can be as high as 50%,[37,38] with the sicker and more preterm neonates developing the most thrombocytopenic episodes.[40] Twenty percent of episodes of thrombocytopenia in NICU patients are severe (platelets $<50 \times 10^9/L$),[36] placing these patients at increased risk of hemorrhage. In tandem with treatment of the underlying cause of their thrombocytopenia, these neonates often receive multiple platelet transfusions. The fact that neither the optimal platelet transfusion regimen nor the clinical impact of thrombocytopenia in such neonates is well defined is another reason why neonatal thrombocytopenia remains an important clinical problem.

## IV. Natural History of Neonatal Thrombocytopenias

Perhaps the most important reason for understanding fetal and neonatal thrombocytopenia is to enable clinicians involved to avoid, or rapidly ameliorate, clinical situations in which thrombocytopenia could lead to significant hemorrhage and its sequelae. In the fetal and neonatal setting, it is also imperative to achieve this without subjecting patients to unnecessary investigations or treatment. Although a large and varied number of conditions can result in fetal and neonatal thrombocytopenia,[41] there is only a limited number of patterns of presentation. In general these are related either to the time of onset or to the severity of thrombocytopenia. In addition, for each pattern of presentation, a relatively small number of conditions accounts for the majority of cases.

The natural history of fetal and neonatal thrombocytopenia is characterized by three main patterns of presentation: (a) fetal thrombocytopenia, (b) early-onset neonatal thrombocytopenia (present at birth or developing by 72 hours of age), and (c) late-onset neonatal thrombocytopenia (developing after 72 hours of age). Table 52-1 shows the

**Table 52-1: Classification of Fetal and Neonatal Thrombocytopenias**

| | Condition |
|---|---|
| **Fetal** | **Alloimmune** |
| | **Congenital infection** (e.g., CMV, toxoplasma, rubella, HIV) |
| | **Aneuploidy** (e.g., trisomies 18, 13, 21, or triploidy) |
| | **Autoimmune** (e.g., maternal ITP, SLE) |
| | Severe Rhesus disease |
| | Congenital/inherited (e.g., Wiskott–Aldrich syndrome) |
| **Early-onset neonatal (<72 hours of age)** | **Placental insufficiency** (e.g., PET, IUGR, diabetes) |
| | **Perinatal asphyxia and whole body cooling** |
| | **Perinatal infection** (e.g., E. coli, GBS, Haemophilus influenzae) |
| | **DIC** |
| | **Alloimmune** |
| | **Autoimmune** (e.g., maternal ITP or SLE, neonatal lupus) |
| | Congenital infection (e.g., CMV, toxoplasma, rubella, HIV, enteroviruses, dengue) |
| | Trisomy 21-associated TMD or AMKL |
| | Thrombosis (e.g., HIT, TTP, renal vein thrombosis) |
| | Bone marrow replacement (e.g., congenital leukemia, osteopetrosis, HLH) |
| | Kasabach–Merritt syndrome |
| | Metabolic disease (e.g., proprionic and methylmalonic acidemia, Gaucher disease, subcutaneous fat necrosis of the newborn) |
| | Congenital/inherited (e.g., TAR, CAMT, X-linked macrothrombocytopenia) |
| **Late-onset neonatal (>72 hours of age)** | **Late-onset sepsis** |
| | **NEC** |
| | Congenital infection (e.g., CMV, toxoplasma, rubella, HIV, enteroviruses, dengue) |
| | Autoimmune |
| | Kasabach–Merritt syndrome |
| | Metabolic disease (e.g., proprionic and methylmalonic acidemia) |
| | Congenital/inherited (e.g., TAR, CAMT) |

The most frequently occurring conditions are in **bold**.

Abbreviations: AMKL, acute megakaryoblastic leukemia; CAMT, congenital amegakaryocytic thrombocytopenia; CMV, cytomegalovirus; GBS, group B streptococcus; HIT, heparin-induced thrombocytopenia; HLH, hemophagocytic lymphohistiocytosis; ITP, immune thrombocytopenic purpura; IUGR, intrauterine growth restriction; NEC, necrotizing enterocolitis; PET, preeclampsia; SLE, systemic lupus erythematosus; TAR, thrombocytopenia with absent radii; TMD, transient myeloproliferative disorder; TTP, thrombotic thrombocytopenic purpura.

**Table 52-2: Congenital and Inherited Thrombocytopenias That May Present in the Fetus or Neonate**

| | |
|---|---|
| Thrombocytopenia (with abnormal platelet function) | Bernard–Soulier syndrome<br>Wiskott–Aldrich syndrome<br>X-linked thrombocytopenia<br>Chediak–Higashi syndrome<br>Quebec platelet disorder<br>Some giant platelet syndromes (e.g., Montreal syndrome) |
| Thrombocytopenia (without markedly abnormal platelet function) | Fanconi anemia<br>TAR syndrome<br>Amegakaryocytic thrombocytopenia<br>ATRUS<br>Autosomal dominant thrombocytopenia<br>FPS/AML<br>Giant platelet syndromes (e.g., May–Hegglin anomaly, Sebastian syndrome, Fechtner syndrome) |

Abbreviations: ATRUS, amegakaryocytic thrombocytopenia with radio-ulnar synostosis; FPS/AML, familial platelet syndrome with predisposition to acute myelogenous leukemia; TAR, thrombocytopenia with absent radii.

most common causes of fetal and neonatal thrombocytopenia by age at presentation. Table 52-2 provides a comprehensive list of causes of congenital and inherited thrombocytopenias that may present in the fetus or neonate.

## A. Fetal Thrombocytopenia

With developments in fetal medicine, the diagnosis of fetal thrombocytopenia is increasing. Fetal blood sampling and measurement of hematology indices are now commonly undertaken in the assessment of fetuses with suspected congenital infections or aneuploidy. Many such fetuses are thrombocytopenic.[7] Some, but not all (see Chapter 53), authorities recommend that at least one fetal platelet count is essential in suspected cases of alloimmune thrombocytopenia in order to guide management and possible feto-maternal therapy.[42,43] Indeed, fetal platelet counts and platelet antigen genotyping/phenotyping are undertaken as primary diagnostic investigations in fetuses with intracranial hemorrhage (ICH), specifically to look for evidence of this condition. However, the practice of assessing fetal platelet counts in cases of maternal autoimmune thrombocytopenia is becoming less common as accumulating evidence suggests that fetal or neonatal hemorrhage in this situation is uncommon (see Chapter 51).[44,45] Finally, fetal thrombocytopenia may also be discovered in the assessment of fetuses at risk

of inherited thrombocytopenias such as Wiskott–Aldrich syndrome (see Chapter 54).[3]

## B. Early-Onset Neonatal Thrombocytopenia (Less Than 72 Hours of Age)

In patients admitted to NICUs, 75% of all episodes of neonatal thrombocytopenia are either present at birth or develop by 72 hours of life.[36,38] Commonly cited causes of neonatal thrombocytopenia (e.g., immunological disorders, congenital and perinatal infections, and disseminated intravascular coagulation [DIC]) account for only a minority of these early-onset cases (see Section V).[19,38,46] Our group's work shows that the vast majority of the remaining patients are preterm neonates born following pregnancies complicated by placental insufficiency or fetal hypoxia, such as maternal preeclampsia (PET), fetal intrauterine growth restriction (IUGR), and, less commonly, maternal diabetes mellitus.[19,38] Classically, such neonates have either a low normal platelet count at birth ($150–200 \times 10^9$/L) or borderline thrombocytopenia ($100–150 \times 10^9$/L). Their platelet count then falls slowly to reach a platelet nadir at day 4 to 5 of life before recovering to $>150 \times 10^9$/L by 7 to 10 days.[20,36] The diagnosis is usually straightforward because the thrombocytopenia is accompanied by other characteristic hematological features seen on routine blood smear, including increased numbers of circulating normoblasts, neutropenia, and red cell changes of hyposplenism.[47]

Thrombocytopenia associated with placental insufficiency is the commonest pattern of early-onset thrombocytopenia seen in neonates. In the absence of conditions causing platelet consumption (e.g., infection), precipitous falls in platelet count are uncommon, the platelet nadir rarely falls below $50 \times 10^9$/L, and platelet recovery is spontaneous.[47] Thus, this pattern of early-onset thrombocytopenia in neonates is predictable, evolves slowly, and in the majority of cases is a benign phenomenon as far as platelet count and hemostasis are concerned. As such it requires no specific investigation, and in the vast majority of cases it requires no specific treatment.

By contrast, early neonatal thrombocytopenias that do not conform to this pattern are likely to be a marker of significant pathology that may require specific management. Platelet counts of less than $50 \times 10^9$/L at birth or by 72 hours of age are uncommon.[28,36] When severe cases of placental insufficiency are excluded, the most likely causes are either

- conditions that also present as fetal thrombocytopenia, such as allo- and autoimmune thrombocytopenia, congenital infections (particularly with cytomegalovirus [CMV]), aneuploidy (particularly trisomies 18, 13, and 21 or triploidy), and congenital/inherited thrombocytopenias; or

- conditions not usually predicted by fetal screening, such as perinatal asphyxia, perinatal infections (e.g., group B Streptococcus, Escherichia coli, and Haemophilus influenzae), and congenital/ inherited thrombocytopenias (e.g., congenital amegakaryocytic thrombocytopenia [CAMT] and thrombocytopenia with absent radii [TAR]).

Early-onset neonatal thrombocytopenias that persist for more than 7 to 10 days are also unusual and warrant further investigation. As most other forms of thrombocytopenia will have resolved by this time, the likely causes of prolonged thrombocytopenia are immune thrombocytopenias, congenital infections, and congenital/inherited thrombocytopenias. Rare causes of congenital and inherited thrombocytopenia also have to be considered; these are listed in Table 52-2.

## C. Late-Onset Neonatal Thrombocytopenia (More Than 72 Hours of Age)

Late-onset thrombocytopenia in NICU patients is mostly caused by late-onset sepsis or necrotizing enterocolitis (NEC).[37,48] This type of thrombocytopenia has a distinctly different pattern to that seen in early-onset thrombocytopenia associated with placental insufficiency. Isolated thrombocytopenia can be the first sign of sepsis or NEC, but more commonly thrombocytopenia occurs in association with the early signs of these conditions. The degree of thrombocytopenia progresses rapidly with a platelet nadir reached within 24 to 48 hours. Thrombocytopenia is often severe (platelet count $<50 \times 10^9$/L), prolonged,[48,49,50] and prompts treatment by platelet transfusion. However, there are no data to show that platelet transfusion improves outcome in this situation and indeed some data suggest poorer outcomes in sick neonates that receive multiple platelet transfusions.[51] In our NICU, this pattern of late-onset thrombocytopenia occurs in approximately 6% of all admissions, and in affected preterm neonates sepsis or NEC it is the cause of the thrombocytopenia in almost 90% of cases.[48]

## V. Conditions Leading to Clinically Significant Neonatal Thrombocytopenia

Although the patterns of neonatal thrombocytopenia often point to the likely cause and thereby obviate the need for a battery of investigations, in a small number of conditions the precise diagnosis of the cause of thrombocytopenia or its specific therapy is relevant to clinical practice.

## A. Immune Thrombocytopenias

A well term neonate presenting with clinical signs of severe thrombocytopenia (e.g., hemorrhage or purpura) is uncom-

mon. However, it represents a neonatal emergency to avoid the potential for severe hemorrhage.[52] In practice, the usual cause is neonatal alloimmune thrombocytopenia (NAIT),[28,53] although a small number of cases will be due to maternal autoimmune disease.[44]

**1. Neonatal Alloimmune Thrombocytopenia (NAIT) (see also Chapter 53).** This condition is discussed in detail in Chapter 53, but because of its clinical importance the most clinically relevant information is summarized here. NAIT is caused by maternal sensitization to fetal platelet antigens inherited from the father. In contrast to alloimmune anemia, NAIT is observed in the first pregnancy in 50% of cases.[54] Maternal antiplatelet antibodies are detectable in 1 : 350 pregnancies, and NAIT occurs in approximately 1 : 1000 live births, although approximately 25% of cases may be clinically silent.[54–56] In Caucasians, antibodies are most commonly directed against HPA-1a (80%), HPA-5b (10% to 15%) and occasionally anti-HPA-3a, anti-HPA-1b and anti-HPA15.[54–57] In Asian populations, antibodies are more commonly directed against HPA-4.[54] The development of antibodies against HPA-1a in HPA-1a-negative women is strongly associated with HLA DRB3 0101 (odds ratio 140).[56] In some cases of NAIT the causative antibody is directed at low frequency human platelet antigens, such as CD36 (glycoprotein IV), or cannot be identified despite a convincing clinical presentation.[57–59]

NAIT usually presents in otherwise well term neonates with unexplained bruising and purpura. The platelet count is usually $<30 \times 10^9$/L.[53] Up to 20% of affected infants will suffer serious bleeding including ICH.[53] This occurs *in utero* in 25 to 50% of cases and is associated with long-term neurodevelopmental sequelae in 20% of survivors.[54] The laboratory diagnosis of NAIT is made by demonstrating antiplatelet antibodies (usually HPA-1a) in maternal serum which are directed against paternal antigens and by platelet genotyping usually by polymerase chain reaction (PCR) (see Chapter 53).[33,54,60]

*Management.* Treatment of mildly affected babies (platelet count $>50 \times 10^9$/L and no evidence of ICH or other significant hemorrhage) is not required, although the platelet count should be monitored for the first 5 days after delivery as it usually falls over this time.[54,60,61] There is also no evidence that treatment of moderately affected infants (platelet count $30–50 \times 10^9$/L) is necessary unless they are bleeding.[61] By contrast, severely affected infants with NAIT (platelet count $<30 \times 10^9$/L and/or ICH or other major bleeding) should be promptly transfused with HPA-compatible platelets[54,60,61] or washed maternal platelets.[61] The results of serological investigations are not necessary before commencement of treatment. If severe thrombocytopenia or hemorrhage persist despite HPA-compatible platelets, intravenous IgG (IVIG) is often useful in ameliorating the thrombocytopenia until spontaneous recovery occurs 1 to 6 weeks after birth.[41,54] The management of women with previously affected neonates during subsequent pregnancies is a complex issue that is reviewed in Chapter 53.

**2. Neonatal Autoimmune Thrombocytopenia (see also Chapter 51).** Maternal platelet autoantibodies complicate a number of conditions (principally immune thrombocytopenic purpura and systemic lupus erythematosus) and occur in 1 to 2 : 1000 pregnancies.[62–64] Although transplacental passage of maternal autoantibodies can cause neonatal thrombocytopenia, this is much less of a clinical problem than NAIT. It is now recognized that neonatal thrombocytopenia only occurs in approximately 10% of neonates whose mothers have autoantibodies, and the incidence of ICH is 1% or less.[45] In recognition of these data, most authorities no longer recommend fetal blood sampling or cesarean delivery in mothers with autoimmune thrombocytopenia, irrespective of their platelet count during pregnancy.[62,65] However, maternal disease severity or platelet count during pregnancy and the occurrence of severe thrombocytopenia in a previous neonate are the most useful indicators of the likelihood of significant fetal and neonatal thrombocytopenia complicating the current pregnancy.[44] Interestingly, a retrospective study that included mainly mothers who were thrombocytopenic during their pregnancy showed a higher incidence of affected babies: 25% of infants had thrombocytopenia, 9% had a platelet count of $<50 \times 10^9$/L, 15% received treatment for bleeding in association with thrombocytopenia, and two fetuses died, one with extensive hemorrhage.[66]

*Management.* All neonates of mothers with autoimmune disease should have their platelet count determined by cord blood sampling. In neonates with normal platelet counts ($>150 \times 10^9$/L), no further action is necessary. In those with thrombocytopenia, a platelet count should be repeated after 2 to 3 days as platelet counts are commonly at their lowest at this time before rising spontaneously by day 7 in most cases.[67] However, as in NAIT, thrombocytopenia may persist in a small number of cases for several weeks.[62,67] Neonates with severe thrombocytopenia (platelet count $<30 \times 10^9$/L) usually respond promptly to treatment with IVIG[68] (2 gm/kg over 2–5 days).

### B. Late-Onset Sepsis and NEC

Late-onset sepsis and NEC remain two of the most serious complications in preterm neonates. Sixteen to 25% of all preterm neonates develop late-onset sepsis (i.e., sepsis developing at more than 72 hours of age).[69–71] The incidence of NEC varies between NICUs but occurs in up to 10% of very low birth weight (birth weight <1500 gm) neonates managed

in tertiary centers.[72] It is well known that thrombocytopenia frequently complicates these conditions,[37] but until recently their overwhelming contribution to severe neonatal thrombocytopenia was poorly defined.[46,48,73–75]

Our group reviewed the causes of severe and prolonged thrombocytopenia (platelet count $<50 \times 10^9$/L) occurring in our NICU patients.[48] We found that in preterm neonates late-onset sepsis was by far the most common clinical condition occurring in association with severe thrombocytopenia. Thirty-four of 44 (77%) preterm neonates had evidence of sepsis (31 had organisms isolated from blood cultures and three had culture-negative sepsis) (Fig. 52-1). Five neonates developed severe thrombocytopenia during NEC (two of these neonates also had organisms isolated from blood cultures). Of the remaining five neonates, four had severe IUGR and developed early-onset severe thrombocytopenia and one had AIT. Thus, in our unit, 39/44 (89%) of all episodes of severe thrombocytopenia in preterm neonates occurred during late-onset sepsis or NEC (Fig. 52-1).

We also evaluated the natural history of the thrombocytopenia in this cohort of 44 preterm neonates.[48] Thrombocytopenia developed at a median postnatal age of 8 days (range 1–37 days) and in over 75% of cases the platelet count dropped precipitously to below $50 \times 10^9$/L within 48 hours. Severe thrombocytopenia was prolonged, the platelet count taking a median of 8 days (range 2–50 days) to rise again consistently above $50 \times 10^9$/L. Evidence of DIC was seen in fewer than 10% of neonates, confirming previous studies indicating that DIC is not a frequent occurrence in this setting.[36,37] Significant hemorrhage (e.g., IVH grades 3–4) was common, occurring in 7/44 (16%) preterm neonates, although the initial bleeding episode had almost invariably occurred prior to the development of severe thrombocytopenia. Five of the 44 preterm neonates with severe thrombocytopenia died prior to discharge: all 5 had a platelet nadir $<30 \times 10^9$/L and all had received platelet transfusions. These data confirm both the association of severe thrombocytopenia in preterm babies with poor outcome,[73–75] and also the apparent lack of benefit conveyed by platelet transfusion in this clinical situation,[48,51] as confirmed by others.[76] More important, these data also suggest more work needs to be done to identify the mechanisms of thrombocytopenia in this setting and to modify management accordingly if outcome is to be improved.

There is now evidence that impaired platelet production is an important component of the marked thrombocytopenia seen in association with neonatal sepsis and NEC. First, during sepsis and NEC, thrombocytopenia often persists long after these conditions have been controlled, suggesting that ongoing platelet consumption is unlikely to be the reason for the continued low platelet count. Second, a number of groups have now reported raised TPO levels in septic neonates with thrombocytopenia, again suggesting a degree of impairment of megakaryocytopoiesis and platelet

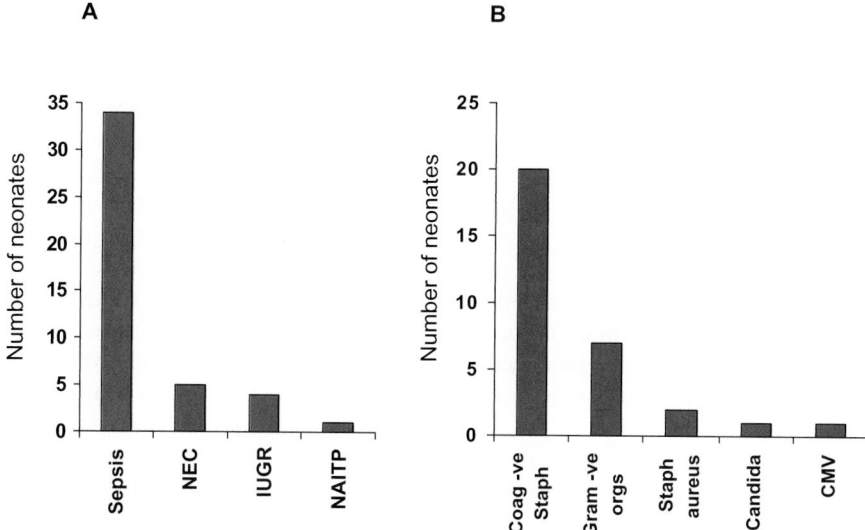

**Figure 52-1.** A. Causes and B. isolated organisms in 44 preterm neonates developing severe thrombocytopenia. Abbreviations: coag -ve staph, coagulase negative staphylococcus; CMV, cytomegalovirus; Gram -ve orgs, Gram negative organisms; IUGR, intrauterine growth restriction; NAITP, neonatal alloimmune thrombocytopenia; NEC, necrotizing enterocolitis; Staph aureus, staphylococcus aureus.

production.[23,77] Our group has also shown lower numbers of circulating megakaryocyte progenitor cells in preterm neonates with sepsis-associated thrombocytopenia when compared to septic controls who maintained normal platelet counts.[78] Together, these finding suggest that platelet production is compromised in preterm neonates who develop prolonged thrombocytopenia during sepsis or NEC.

### C. Congenital Infections

Published reports suggest that a large number of congenitally acquired viral infections can cause fetal and neonatal thrombocytopenia. Although the most commonly identified viral causes of neonatal thrombocytopenia are CMV and rubella,[79–85] cases due to parvovirus and enteroviruses (Coxsackie A and B and echovirus) can also cause severe, acute thrombocytopenia.[86–89] In addition, adenoviruses,[90] mumps,[91] HIV[84,92–94] and dengue[95–96] have all been reported in small series by a number of authors. Therefore, the presence of thrombocytopenia *per se* does not point to the diagnosis of any specific infection. In most cases, thrombocytopenia will be present in combination with other clinical features suggestive of congenital infections, such as intracranial calcification, hepatosplenomegaly, jaundice, or "viral" lymphocytes on the blood smear. However, severe thrombocytopenia (platelet count $<50 \times 10^9$/L) which is present in the first days of life and persists for more than the first week is a common feature in congenital infections[97,98] and may then suggest the diagnosis if other more common clinical features are absent or minimal.

Primary or secondary maternal CMV infection during pregnancy results in congenital infection in approximately 0.5 to 1% of all newborns.[79–81] Only 10 to 15% of such infants have symptomatic disease,[82] but up to 75% of these will have significant thrombocytopenia (platelet count $<100 \times 10^9$/L).[80,83,97] This suggests that up to 1 : 1000 newborns will have significant neonatal thrombocytopenia purely as a result of congenital CMV infection. The thrombocytopenia is often severe[84] and can sometimes persist for several months. The mechanism of thrombocytopenia is poorly defined, but recent reports suggest that CMV directly infects and inhibits megakaryocytes and their precursors, resulting in impaired platelet production.[99]

Congenital toxoplasmosis affects 1 : 2000 to 1 : 3000 newborns, depending on geographical location and dietary practices.[100–102] Screening programs and increased public awareness have reduced the incidence in some countries, such as France,[103] which previously suffered a high prevalence. The incidence of thrombocytopenia in affected neonates is approximately 40%.[98] Congenital rubella is now very rare in countries with an active immunization program,[104,105] but persistent thrombocytopenia is a prominent feature of neonates with congenital rubella syndrome.[85] The mechanism of thrombocytopenia in these conditions is also likely to be direct infection of megakaryocytes and their precursors, although there is as yet no direct evidence of this.

### D. Perinatal Asphyxia

Perinatal asphyxia — characterized by fetal distress, fetal and neonatal acidemia, low Apgar scores at birth, and neonatal encephalopathy — is a common precipitant of neonatal thrombocytopenia. Reviews of neonates with this degree of asphyxia suggest that up to 30% will develop thrombocytopenia.[106–108] In this situation, thrombocytopenia is often severe and prolonged and principally appears to be precipitated by DIC.[109]

## E. Perinatal Infection

Perinatal infection is commonly related to prolonged rupture of membranes and usually presents as early-onset neonatal infection (present by 72 hours). Perinatal infection occurs in 1 : 1000 to 1 : 2000 of all newborns[110] and in up to 2% of very low birth weight neonates.[111] Reflecting the precipitating perinatal factors, the commonest causative organisms are group B Streptococcus, Escherichia coli, and Haemophilus influenzae.[110–112] Infections with these organisms often cause serious neonatal morbidity and thrombocytopenia develops in approximately 50% of cases.[113] In contrast to late-onset neonatal sepsis, DIC appears to be an important mechanism of thrombocytopenia, probably because of the severity of the sepsis syndrome induced by the causative organisms and the fact that infection often begins prior to delivery and therefore before effective treatment can be instituted. The blood smear morphology is usually characteristic and reveals prominent left shifted neutrophils with or without toxic granulation. Worsening thrombocytopenia and neutropenia are poor prognostic signs.

## F. Chromosomal Abnormalities, Including Trisomy 21 (Down Syndrome)

Aneuploidy is increasingly a fetal diagnosis in developed countries. Hohlfeld et al.[7] found thrombocytopenia in association with trisomies 18 (86%), 13 (31%), and 21 (6%), triploidy (75%), and Turner syndrome (31%) at fetal diagnosis. However, thrombocytopenia was rarely severe, as only one of 43 fetuses with aneuploidy complicated by thrombocytopenia had a platelet count $<50 \times 10^9$/L. Thrombocytopenia is also sometimes seen with partial trisomies, such as isochromosome 18q.[114] For trisomies 13 and 18 the mechanism of thrombocytopenia is unknown, although the frequent association of the thrombocytopenia with neonatal polycythemia, neutropenia,[115] and IUGR suggests that it is likely to be due to reduced platelet production and that the pathogenesis may be similar to that seen in placental insufficiency.[47]

By contrast, there have been several developments in our understanding of the thrombocytopenia seen in association with trisomy 21 (Down syndrome) (reviewed in refs.[116–118]). Approximately 10% of neonates with Down syndrome develop a clonal preleukemic disorder, transient myeloproliferative disorder (TMD), which is characterized by an increased percentage of myeloblasts, abnormal megakaryocytes, and variable thrombocytopenia. In most cases, this resolves spontaneously, but approximately 30% of neonates with TMD subsequently develop acute megakaryocytic leukemia (AMKL), sometimes within the neonatal period.[116–118] Evidence has shown that in both TMD and AMKL, there are acquired somatic mutations in the key megakaryocytic transcription factor GATA-1 that lead to the transcription of a short GATA-1 mRNA (GATA-1s) and protein.[119–123] Infants who progress from TMD to AMKL have the same GATA-1 mutation at both stages. Because this mutation disappears as the TMD resolves or the AMKL enters remission, GATA-1s can be viewed as a key initiating step of the leukemia.[123] Studies using mouse models suggest that GATA-1s exerts its leukemogenic effect by altering the regulation of fetal megakaryocyte progenitor differentiation and proliferation.[124]

## G. Congenital/Inherited Thrombocytopenia (see also Chapter 54)

A number of congenital/inherited disorders may present with thrombocytopenia in the fetus or in the neonatal period (Tables 52-1 and 52-2). Unlike most of the disorders described earlier, these disorders are all rare. In most cases, the thrombocytopenia is due to reduced platelet production secondary to abnormal hematopoietic stem cell development, and in most neonates, but not all, there are associated congenital anomalies that are useful in guiding investigations and establishing the diagnosis. Congenital/inherited thrombocytopenias are commonly classified into two groups depending on whether they have or do not have an associated abnormality of platelet function (Table 52-2).

**1. Thrombocytopenia with Abnormal Platelet Function.** The best known examples of those disorders in which there is a thrombocytopathy and may present in the neonatal period are Bernard–Soulier syndrome (BSS) and Wiskott–Aldrich syndrome (and its variants, including X-linked thrombocytopenia) (reviewed in ref.[125] and Chapters 54 and 57). Chediak-Higashi syndrome usually presents later in infancy.

*a. Bernard–Soulier Syndrome.* BSS (discussed in detail in Chapter 57) is an autosomal recessive disorder characterized by mild to moderate thrombocytopenia, giant platelets, and a prolonged bleeding time. BSS is due to qualitative or quantitative defects in the glycoprotein (GP) Ib-IX-V complex as a result of mutations in one of the four genes encoding the complex.[126] Most of the mutations identified in BSS patients are in the GPIbα gene, and some are also found in the GPIbβ gene and GPIX genes, but none to date have been found in the GPV gene.[127] Bleeding is not usually severe in the neonatal period but may occur.[126] Treatment by platelet transfusion is effective but should be reserved for life-threatening hemorrhage because transfused patients may form allo-antibodies against GPIb, GPIX, or GPV.[128] In the offspring of women with BSS, alloantibodies against GPIb-IX-V are a rare cause of fetal/neonatal thrombocytopenia and may cause severe, even fatal, fetal ICH.[129]

*b. Wiskott–Aldrich Syndrome (see also Chapters 54 and 57).* Wiskott–Aldrich syndrome and X-linked thrombocytopenia form a spectrum of disorders resulting from mutations in the Wiskott–Aldrich syndrome protein (WASP) gene at band Xp11–12.[130] More than 100 different mutations have been identified, and evidence suggests that genotype and phenotype are closely linked.[131] Classical Wiskott–Aldrich syndrome is characterized by microthrombocytopenia, eczema, recurrent bacterial and viral infections, and a propensity to develop autoimmune disorders in a male infant.[132] Wiskott–Aldrich syndrome usually presents during the first year of life with bleeding problems that may be severe and include gastrointestinal hemorrhage and purpura; most cases do not present in the neonatal period unless there is a known family history.[132,133] The pathogenesis of hemorrhage in Wiskott-Aldrich syndrome is incompletely understood and appears to be multifactorial, reflecting the degree of thrombocytopenia, but also reduced platelet survival, reduced platelet turnover, and abnormal platelet function.[132] X-linked thrombocytopenia is characterized by an isolated thrombocytopenia, which is usually milder;[130,132] the WASP mutations are usually, but not always, missense mutations clustered in the EVH1 domain rather than frameshift, nonsense, and splice site mutations, which are more often found in severe disease.[131]

*c. Chediak–Higashi Syndrome (See Also Chapters 15 and 57).* Chediak–Higashi syndrome is an autosomal recessive disease caused by mutations in the lysosomal trafficking regulator (LYST) gene.[134] Chediak–Higashi syndrome is characterized by partial oculocutaneous albinism, predisposition to pyogenic infections, abnormal large granules in many cell types, platelet dysfunction, and, in the later stages, thrombocytopenia.[135] The bleeding problems of Chediak–Higashi syndrome usually present in infants rather than newborn babies and mainly reflect the associated platelet storage pool defect, as thrombocytopenia is not a feature until the accelerated phase of the disease.[135]

**2. Thrombocytopenia without Marked Thrombocytopathy.** There have been many advances in our understanding of the molecular basis of a number of these inherited disorders of platelet production. The best known examples of congenital/inherited platelet disorders in which there are no significant platelet function abnormalities and that may present in the neonate and occasionally the fetus are Fanconi anemia, TAR syndrome, amegakaryocytic thrombocytope-

nia with radio-ulnar synostosis (ATRUS), and congenital amegakaryocytic thrombocytopenia (Table 52-2). All of these disorders are rare. The other disorders in this category tend to present in older children unless there is a family history or they are identified as an incidental finding, such as familial platelet syndrome with predisposition to acute myelogenous leukemia (FPS/AML),[136] X-linked thrombocytopenia with dyserythropoiesis,[137] autosomal dominant thrombocytopenia linked to chromosome 10,[138] and giant platelet syndromes, such as May–Hegglin anomaly.[139]

*a. Fanconi Anemia.* The haematological manifestations of Fanconi anemia do not usually appear until approximately 7 years of age.[140,141] However, thrombocytopenia has been reported in neonates,[142–144] and this diagnosis should always be considered in unexplained neonatal thrombocytopenia, particularly if there are typical dysmorphic features such as malformations of the skin, thumb, face, or eyes, or if there is parental consanguinity. The diepoxybutane test is nearly always diagnostic,[145] and the bone marrow shows reduced cellularity, dyshematopoiesis, and reduced numbers of megakaryocytes. Treatment is rarely necessary in the neonatal period. Curative treatment by stem cell transplantation is successful in the majority of children once severe bone marrow failure has developed.[146] The molecular basis of Fanconi anemia is heterogeneous. Current evidence supports the existence of 12 Fanconi anemia genes, 10 of which have been characterized (FANCA, C, D2, E, F, G, J, L, M, and BRCA2).[147,148,148b,148c,148d] Mutations in these 10 genes account for approximately 99% of cases.[147,148] The Fanconi anemia proteins form a complex that is known to be essential for maintenance of chromosomal stability, although precisely how they do this is unknown.[147–149]

*b. TAR Syndrome (see also Chapter 54).* TAR syndrome is characterized by bilateral absence of the radii (Fig. 54-2). Both thumbs are present, which may be useful in distinguishing this syndrome from Fanconi anemia.[150] The majority of babies with TAR syndrome develop thrombocytopenia within the first week of life, and 95% do so within the first 4 months of life.[150] The platelet count is usually $<50 \times 10^9$/L, and the white cell count is elevated in more than 90% of patients, sometimes exceeding $100 \times 10^9$/L and mimicking congenital leukemia.[151] A review of 34 patients with TAR found a high prevalence of associated abnormalities including cow's milk intolerance (47%), lower limb anomalies (47%), renal anomalies (23%), and cardiac anomalies (15%).[152] Although historical data show an infant mortality rate of 30%, those infants that survive the first year of life generally do well because the platelet count spontaneously improves and is usually maintained at low normal levels thereafter. In the Greenhalgh study, only 2 of the 34 cases died prematurely, one from cardiac disease and the other from ICH.[152] Inheritance appears to be autosomal recessive, but neither the molecular basis nor the pathogenesis of TAR syndrome have yet been identified. HOX A10, A11, and D11, interesting candidate genes in view of their role in radio-ulnar aplasia in mice,[153] have now been ruled out as the cause of TAR,[154] although HOXA11 mutations are associated with a similar but distinct inherited disorder, amegakaryocytic thrombocytopenia with radio-ulnar synostosis (see below).[155] Megakaryocytes and megakaryocyte progenitors are reduced in TAR.[156,157] However, TPO levels are increased, excluding a TPO production defect.[156] Expression of the TPO receptor, c-mpl, appears to be normal,[156] and no mutations or rearrangements of the c-mpl gene have been found by Southern blotting or sequence analysis of the coding region.[157] Nevertheless, megakaryocyte colony growth *in vitro* in response to TPO is suboptimal, and there is a profound defect in megakaryocyte precursor differentiation, suggesting a defect in the TPO signaling pathway.[156–158]

*c. Amegakaryocytic Thrombocytopenia with Radio-Ulnar Synostosis (ATRUS) (see also Chapter 54).* Patients with ATRUS present at birth with severe thrombocytopenia, absent bone marrow megakaryocytes, and characteristic skeletal abnormalities.[159] In addition to the radio-ulnar synostosis, affected infants may have clinodactyly and shallow acetabulae.[159] Two kindreds show that ATRUS is caused by mutations in the HOXA11 gene,[155,159] thus distinguishing it from TAR.

*d. Congenital Amegakaryocytic Thrombocytopenia (see also Chapter 54).* Congenital amegakaryocytic thrombocytopenia is an autosomal recessive bone marrow failure syndrome that presents with isolated thrombocytopenia.[160] It nearly always presents in the neonatal period since the platelet count is usually $<20 \times 10^9/L$ at birth and most affected babies therefore have petechiae or other evidence of bleeding. Physical anomalies are present in approximately 50% of children.[160] Approximately 50% of patients subsequently develop aplastic anemia, usually in the first few years of life, and there are several reports of leukemia or myelodysplasia developing later in childhood.[160] Treatment in the neonatal period is with platelet transfusion for bleeding episodes. Beyond the neonatal period, stem cell transplantation is curative for those patients with clinically severe disease or aplasia.[161]

Congenital amegakaryocytic thrombocytopenia has been shown to be caused by mutations in the c-mpl gene in the majority of patients.[162–165] Both frameshift and nonsense mutations predicted to result in complete loss of c-Mpl function and missense mutations in the extracellular domain of the receptor have been described.[163,164] As a result, megakaryocytes and their progenitors are reduced[166] and plasma TPO levels are high.[163,164,167] CD34-positive hematopoietic progenitor cells cultured *in vitro* in the presence of TPO fail to differentiate, consistent with a failure of normal signal transduction through the mutated receptor.[163]

*e. X-linked Macrothrombocytopenia due to GATA-1 Mutation (see also Chapter 54).* Several families with a familial macrothrombocytopenia have been described with or without associated anemia in which there is a mutation in the FOG-1 binding site of GATA-1.[137,168,169] This form of thrombocytopenia may present with severe thrombocytopenia at birth and be associated with profound bleeding.[169]

*f. Autosomal Dominant Thrombocytopenias (see also Chapter 54).* Autosomal dominant thrombocytopenias represent a heterogeneous group of disorders characterized by lifelong mild to moderate thrombocytopenia with normal platelet function.[170,171] Clinical features depend on the degree of thrombocytopenia but usually consist of a moderate propensity to easy bruising and minor bleeding. Large numbers of affected individuals have been identified within families and one study demonstrates markedly delayed megakaryocyte nuclear and cytoplasmic differentiation in affected individuals in one such family.[170] Linkage analysis studies within this family suggested a defective gene on the short arm of chromosome 10,[170] which has now been putatively identified as FLJ14813.[138,172] In view of the clinical features, this gene is likely to be involved in megakaryocyte endomitosis and terminal differentiation, although other families with dominant thrombocytopenia are likely to have separate gene defects controlling other aspects of megakaryocytopoiesis.[171]

*g. Giant Platelet Syndromes (see Chapters 54 and 57 for more detail).* A number of rare giant platelet syndromes have been described, which may present with thrombocytopenia in the neonatal period or in the fetus, although many cases are not diagnosed until adulthood.[173] The best known of these syndromes is the May–Hegglin anomaly, which is characterized by the diagnostic triad of thrombocytopenia, giant platelets, and leukocyte Dohle-like inclusion bodies (Fig. 54-3). The platelet count varies from $<20 \times 10^9/L$ up to normal, but is usually 40 to $80 \times 10^9/L$. Thrombocytopenia may develop *in utero*, and May-Hegglin anomaly is a rare cause of fetal or neonatal ICH.[174,175] The genetic basis of May–Hegglin anomaly mutations in the MY H9 gene on chromosome 22q, which encodes nonmuscle myosin heavy chain A.[176] Interestingly, mutations in MY H9 also occur in two other giant platelet syndromes, Fechtner syndrome and Sebastian syndrome, confirming earlier suspicions based on linkage analysis and clinical similarities that these are closely related disorders.[176,177]

### H. Placental Insufficiency

Neonates with IUGR are now recognized to have a number of distinctive hematological abnormalities that are present at birth.[47] These include neonatal thrombocytopenia, neutropenia, increased erythropoiesis (high numbers of circulating nucleated red cells with or without associated polycythemia), and evidence on the blood smear of hyposplenism (spherocytes, target cells, and Howell–Jolly bodies).[47] The underlying cause of the hematological abnormalities appears to be chronic fetal hypoxia, because the same pattern of abnormalities occurs in a number of forms of placental insufficiency: both maternal disorders (including PET, hypertension, and diabetes mellitus) and fetal disorders manifest as "idiopathic" IUGR. Erythropoietin levels are increased in affected fetuses and neonates.[178–180] In addition, the severity of the hematological abnormalities correlates both with serum erythropoietin levels and with the severity of placental insufficiency.[178]

We and others have shown that megakaryocytopoiesis is severely impaired at birth in such neonates (as shown by a marked reduction in circulating megakaryocytes and their precursor and progenitor cells) and that this is likely to be the principal reason for the neonatal thrombocytopenia, as there is no evidence of increased platelet destruction/consumption.[19,23,38] These patients also exhibit significantly raised plasma TPO levels at the height of their thrombocytopenia,[20] a finding associated with hyporegenerative thrombocytopenias.[181–185] As both the megakaryocyte progenitor abnormalities and the raised TPO levels resolve in tandem with platelet recovery, these findings suggest that these neonates have underlying impaired platelet production originating in fetal life. The precise hematopoietic mechanism responsible for these abnormalities remains uncertain, although a number of lines of evidence point to disruption of the commitment of fetal multipotent hematopoietic progenitor cells to megakaryocytopoiesis, probably mediated by an abnormal fetal hematopoietic growth factor (HGF) environment.[47]

### I. Miscellaneous Rare Causes of Congenital Thrombocytopenia Presenting in the Neonatal Period

**1. Metabolic Disorders.** Thrombocytopenia is a common presenting feature in a number of inborn errors of metabolism, including propionic, methylmalonic and isovaleric academia,[186,187] and Gaucher disease.[188] Reports also show that thrombocytopenia is a complication of induced hypothermia used to improve outcome in neonates with hypoxic ischemic encephalopathy.[189,190]

**2. Hemangiomas.** Kasabach–Merritt syndrome typically presents in the neonatal period with profound thrombocytopenia together with microangiopathic anemia, DIC, and an enlarging vascular lesion.[191,192] In the majority of cases, the diagnosis is straightforward because the hemangiomas are cutaneous,[193] but in around 20% of cases, there

is visceral involvement without any cutaneous signs and the diagnosis may be delayed.[191,193] Thrombocytopenia in Kasabach–Merritt syndrome occurs principally due to trapping of platelets on the endothelium of the vascular tumor,[194,195] but this can be exacerbated by the concurrent development of a systemic coagulopathy.[196] Treatment with steroids followed by interferon or vincristine leads to resolution of the thrombocytopenia and the hemangiomas in more than 50% of cases, but the reported mortality in most series is 20 to 30%.[191,192,197]

**3. Thrombotic Disorders.** A number of thrombotic disorders well known in adults and older children have been reported in neonates. Thrombotic thrombocytopenic purpura (TTP) (see Chapter 50) due to an inherited deficiency of the von Willebrand factor cleaving protease ADAMTS13 may present in the neonatal period with thrombocytopenia, hyperbilirubinemia, and anemia.[198,199] In the absence of a family history, diagnosis may be delayed because these are all common clinical features in sick neonates. Data have shown that ADAMTS13 activity in neonates is usually within the normal range for adults; in the small proportion of neonates who had slightly reduced activity, this normalized within 3 days; so diagnosis of significant ADAMTS13 deficiency in the first week of life is possible.[200] Hemolytic-uremic syndrome (HUS) (see Chapter 50) has also been reported in a neonate, possibly triggered by Bordetella pertussis infection.[201] Heparin-induced thrombocytopenia (HIT) (see Chapter 48) has also been reported in a number of term and preterm neonates, often with associated arterial thrombosis.[202–204] Spadone et al. identified 14 neonates with HIT, 11 of whom had aortic thrombosis, out of 930 neonates who had received heparin; these data suggest that HIT may develop in 1.5% of those exposed to heparin during their NICU stay.[202] Neonatal thrombocytopenia may also occur as a secondary event after thrombosis of a major vessel. Renal vein thrombosis in particular is associated with a high incidence of thrombocytopenia and should be considered in any neonate with thrombocytopenia in association with renal failure.[205] The majority of neonates with renal vein thrombosis also have thrombosis at other sites.[205]

## VI. Clinical Impact of Neonatal Thrombocytopenia

Whereas a number of studies highlight the association between thrombocytopenia and poor outcome in neonates,[36–38,40,206] the direct contribution of thrombocytopenia to the clinical course of these neonates is difficult to assess. The majority of studies suggest that neonatal thrombocytopenia in general is a risk factor for significant hemorrhage (particularly ICH),[36,37,61,206–210] for mortality,[26–28,211,212] and for adverse neurodevelopmental outcome.[206,213] NAIT is a specific diagnosis for which there is clear evidence of the role played by thrombocytopenia (see Chapter 53). In this setting the otherwise well fetus or neonate who develops an ICH is likely to suffer direct neurodevelopmental consequences, although even here altered platelet-vascular interactions may contribute to the bleeding secondary to thrombocytopenia.[53,214] However, it is far from clear whether neonatal thrombocytopenia in general directly contributes to adverse outcome or is simply a marker of the severity of concurrent neonatal complications (e.g., birth asphyxia or severe sepsis), which themselves carry a poor prognosis.[212,215]

One way to determine the clinical impact of thrombocytopenia in neonates would be to prevent its development by the use of platelet transfusion in a high-risk population to see if the overall outcome improves. In the only randomized trial designed to address this question, Andrew et al.[216] found no benefit (i.e., reduction in significant hemorrhage) in preterm neonates in whom moderate thrombocytopenia (platelets $50–150 \times 10^9$/L) was prevented by platelet transfusion. However, appropriate lower limit platelet counts for all groups of preterm neonates were not defined, as all neonates were transfused if their platelet count fell to $<50 \times 10^9$/L because of the potential for hemorrhage. This study[216] is therefore unable to answer questions about the clinical impact of severe thrombocytopenia and the level of thrombocytopenia associated with a significant risk of bleeding. Because few babies develop hemorrhage (especially ICH) outside the first few days of life and because the majority of episodes of severe thrombocytopenia are late-onset (see Section V.B), it is difficult to be certain about the contribution of neonatal thrombocytopenia to overall outcome. Although potentially difficult to accomplish without putting babies at risk of hemorrhage, there therefore remains a pressing need for trials to define the safe lower limit for platelet counts in sick neonates and thereby which neonates will benefit from specific therapy for their thrombocytopenia.

# VII. Principles of Management of Neonatal Thrombocytopenia

As outlined earlier, studies from several groups demonstrate that thrombocytopenia occurs in 22 to 35% of all NICU admissions.[36–38] Severe thrombocytopenia (platelet count $<50 \times 10^9$/L) occurs during 20% of these episodes,[36] such that severe thrombocytopenia develops in approximately 1 in 20 of all NICU admissions.[48] Furthermore, because this degree of thrombocytopenia commonly persists for many days, severe thrombocytopenia presents a common management problem to every neonatal pediatrician. Identifying the common patterns of presentation of neonatal thrombocytopenia and its principal causes is clearly important. However, logical management of individual neonates also requires an understanding of the main mechanism(s) underlying their

thrombocytopenia and an evaluation of the likely role of platelet transfusion for each baby.

## A. Defining Important Mechanisms of Neonatal Thrombocytopenia

Until recently, the mechanism underlying many neonatal thrombocytopenias remained unknown.[38] Consequently, many previous classifications of neonatal thrombocytopenia based primarily on mechanism proved of little practical help to neonatal pediatricians because of an overemphasis on rare conditions with a known mechanism of thrombocytopenia (e.g., impaired platelet production in TAR syndrome). These classifications have also done little to dispel the widely held belief that many common neonatal conditions (e.g., sepsis) cause thrombocytopenia because of platelet consumption principally due to DIC, despite the fact that there is little evidence to support this.[36,37] An alternative practical approach to the classification of neonatal thrombocytopenia based on the age at presentation and focused on the most common causes has therefore been proposed by us (Table 52-1) and by Sola and colleagues[217] as being a more useful scheme for neonatal pediatricians caring for sick neonates. Nevertheless, a more precise definition of the fundamental mechanisms underlying the common forms of neonatal thrombocytopenia should lead to progress in developing more rational and innovative management and therapy of these neonates.

**1. The Role of Impaired Platelet Production.** As discussed earlier, studies show important differences in fetal and neonatal megakaryocytopoiesis and platelet production that seem likely to predispose the sick fetus or newborn baby to develop thrombocytopenia.[10–13,20,23] Fetal megakaryocytes have reduced proplatelet formation,[10–13] the fetus and neonate have a reduced ability to produce large amounts of TPO at times of increased platelet demand,[20,23] and circulating megakaryocytes and their progenitor cells are markedly reduced at birth in sick preterm neonates destined to become thrombocytopenic early in the neonatal period.[38] There is also emerging evidence of impaired platelet production contributing to the marked thrombocytopenia seen in association with neonatal sepsis and NEC.[23,77,78,218] These data suggest that two of the main aims in the management of neonatal thrombocytopenias should be (a) optimal platelet transfusion therapy (see Section VIII.B) and (b) development of the therapeutic potential of HGFs to stimulate platelet production (see Section X).

**2. The Role of Platelet Consumption and Sequestration.** Whereas impaired platelet production is the major mechanism in the most common form of early-onset neonatal thrombocytopenia, immunologically mediated platelet consumption occurring in allo- and autoimmune thrombocytopenia remains the most important cause of severe thrombocytopenia at birth. The most common consumptive thrombocytopenias in the first few days of life are immune mediated. Reports suggest that 15 to 20% of all neonatal thrombocytopenias present at birth result from transplacental passage of maternal platelet antibodies.[28,34] The most common consumptive thrombocytopenias after the first 72 hours of life are DIC and thrombocytopenia associated with thrombosis. The role of DIC in neonatal thrombocytopenia has undoubtedly been overstated in the past; it is the principal cause of thrombocytopenia in some sick newborns (e.g., following perinatal asphyxia) but probably only contributes to 10 to 15% of neonatal thrombocytopenias as a whole.[36,37] Thrombocytopenia is a presenting feature in approximately one third of cases of thrombosis in the neonatal period;[219] however, thrombosis sufficient to produce clinical symptoms and signs is rare, and most cases are associated with intravascular catheters.[219] Finally, there is evidence that a degree of splenic sequestration of platelets occurs in sick neonates.[220] Thus, although a number of conditions cause platelet consumption and sequestration, together they appear to be the major mechanisms of thrombocytopenia in well under 50% of neonatal thrombocytopenias. Once the cause of the thrombocytopenia has been established, specific optimal treatment strategies are now well defined (e.g., see Sections V.A.1 and V.A.2).

**3. Combined Mechanisms.** Although the major mechanisms involved in some neonatal thrombocytopenias are now more clearly defined, many neonates develop thrombocytopenia as a result of multiple concurrent mechanisms. A preterm neonate from a mother with PET who develops early bacterial sepsis or a neonate with IUGR who develops NEC may become thrombocytopenic as a result of both impaired platelet production (following PET or IUGR) and increased platelet consumption (during sepsis or NEC). Indeed, it seems clear that the majority of neonates who develop thrombocytopenia do so because their adverse fetal environment causes impaired megakaryocytopoiesis or thrombocytopenia at birth, with a predisposition for the thrombocytopenia to worsen when the neonate is exposed to concurrent neonatal platelet consumptive "stress." Defining the precise contribution of each mechanism is a difficult task that can make the logical management of an individual thrombocytopenic neonate problematic. However, a consistent approach to investigation and treatment will normally ensure optimal management and can often aid in the definition of the mechanism of the thrombocytopenia.

### B. The Role of Platelet Transfusion in Neonatal Thrombocytopenia

Despite the prevalence of neonatal thrombocytopenia and its relevance to short-term and long-term clinical outcome, there is no clear evidence-based practice to guide clinicians in the appropriate use of platelet transfusion therapy in neonates. This is reflected in a fairly marked divergence of practice between different institutions and countries.[48,221–223]

In the absence of evidence, a number of expert and consensus-based guidelines for platelet transfusion therapy in neonates have been devised.[61,224–228] These guidelines all set out relatively conservative transfusion thresholds in an attempt to minimize the risk of hemorrhage, with most suggesting platelet transfusion at platelet counts below $30 \times 10^9$/L in stable neonates, or when counts fall below $50 \times 10^9$/L in very sick or preterm neonates.

These guidelines are consistent with previous studies identifying thrombocytopenia as a risk factor for both neonatal mortality and long-term morbidity.[36–38,206] More important, they highlight the fact that serious systemic illness, coupled with profound thrombocytopenia, predicts poor outcome in preterm neonates, without indicating which of these factors are causal. This suggests that overall neonatal outcome might be improved if the clinical combination of serious systemic illness and profound thrombocytopenia can be prevented or rapidly controlled, for example, by combining platelet transfusion with therapy aimed both at ameliorating the underlying conditions precipitating severe thrombocytopenia (principally sepsis and NEC) and stimulating platelet production. Interleukin (IL)-11, which stimulates platelet production and is also known to convey survival benefit during sepsis[229–232] and bowel injury[233] in animals, and has recently been shown to reduce sepsis during chemotherapy for hematological malignant disease in humans,[233] is one such possibility (see Section X.C). Once again, only well-designed clinical trials can show whether this approach will improve outcome in such patients.

## VIII. Specific Approach to the Neonate with Significant Thrombocytopenia

The aim in the management of neonatal thrombocytopenia must be to prevent, or minimize, periods of significant thrombocytopenia in order to reduce the risk of significant hemorrhage. To achieve this goal, we suggest the following approach.

### A. Identify the Likely Clinical Cause for All Episodes of Neonatal Thrombocytopenia

This would seem to be an obvious part of good neonatal practice. However, the fact that thrombocytopenia is so common in NICU patients, and usually self-limiting, leads to a situation in which thrombocytopenia often remains unconsidered as a clinical problem until severe thrombocytopenia develops. When this stage is reached, the opportunity for planned investigation and management is often lost and the patient may then simply receive a platelet transfusion, whether or not this is clinically indicated. Perhaps more important, diagnosis of potentially important underlying conditions may be missed, including conditions with implications for future children such as NAIT or inherited thrombocytopenia. Identification of the cause of the thrombocytopenia will also predict its likely pattern and enable appropriate monitoring of the platelet count, which is particularly important in conditions in which precipitous falls in platelet count are common (e.g., sepsis). This allows the planned use of prophylactic platelet transfusion. In addition, deviations from the expected pattern of thrombocytopenia alert the clinician to possible alternative reasons for thrombocytopenia that may require more specific investigations or management (e.g., severe immune-mediated thrombocytopenia).

## B. Administer Platelet Transfusions Appropriately

Platelet transfusions confer one of the highest risks of transfusion-related infections and reactions of all blood products[235-238] (see Chapter 69). Therefore they should be administered to thrombocytopenic neonates when the degree of thrombocytopenia alone or in combination with other neonatal complications results in an unacceptable risk of hemorrhage. As discussed earlier, this is often difficult to assess because the risk of hemorrhage is closely related to the gestational and postnatal age of the neonate, the cause of the thrombocytopenia, and the severity of concurrent conditions. In practice, the decision to administer platelets is governed by two main questions, for which the answers may differ (Table 52-3): When is it appropriate to administer platelets to nonbleeding patients (prophylactic transfusions)? When is it appropriate to administer platelets to actively bleeding patients?

### 1. Prophylactic Platelet Transfusions

*a. General NICU Patients.* The only prospective randomized trial addressing this point[216] clearly shows that maintaining a normal platelet count (platelet count $>150 \times 10^9$/L) by prophylactic platelet transfusion in preterm neonates undergoing standard NICU care during the first week of life confers no clinically apparent hemostatic benefit compared to controls with moderate thrombocytopenia (platelet count $50-150 \times 10^9$/L). As term neonates are certainly at no greater risk of hemorrhage than preterm neonates, these findings firmly suggest that there is no place for prophylactic platelet transfusions in nonbleeding NICU patients with platelet counts of $50 \times 10^9$/L and above.

*b. Stable NICU Patients with Severe Thrombocytopenia (Platelet Count $<50 \times 10^9$/L).* This represents a more difficult problem. Many neonates developing severe thrombocytopenia are sick or unstable as a result of the conditions causing their thrombocytopenia, whereas others (e.g., neonates recovering from sepsis) may have a platelet count $<50 \times 10^9$/L for several days but can be otherwise clinically stable. Transfusing all such patients with platelets until a sustained count $>50 \times 10^9$/L is achieved is one option to manage this clinical problem. However, clinical experience, as reflected in reported platelet transfusion practice in neo-

### Table 52-3: Guidelines for Platelet Transfusion in the Neonate

| Platelet count ($\times 10^9$/L) | Nonbleeding Neonate | Bleeding Neonate | NAITP (proven or suspected) |
|---|---|---|---|
| <30 | Consider transfusion in all patients | Transfuse | Transfuse (with HPA compatible platelets) |
| 30–49 | Do not transfuse if clinically stable<br>Consider transfusion if<br>• <1000 gm birth weight and <1 week of age<br>• clinically unstable (e.g., fluctuating BP)<br>• previous major bleeding tendency (e.g., grade 3–4 IVH)<br>• current minor bleeding (e.g., petechiae, puncture site oozing)<br>• concurrent coagulopathy<br>• requires surgery or exchange transfusion | Transfuse | Transfuse (with HPA compatible platelets) |
| 50–99 | Do not transfuse | Transfuse | If bleeding, transfuse (with HPA compatible platelets) |
| >99 | Do not transfuse | Do not transfuse | Do not transfuse |

Abbreviations: BP, blood pressure; HPA, human platelet antigen; IVH, intraventricular hemorrhage; NAITP, neonatal alloimmune thrombocytopenia.

nates,[61,221–223,239] suggests that not all such neonates receive, or require, prophylactic platelet transfusions. In practice, only approximately 60% of preterm neonates with severe thrombocytopenia receive a platelet transfusion, whereas the remaining 40% do not, with no apparent adverse sequelae.[48] As outlined earlier, we now recommend that stable NICU patients with no evidence of clinical bleeding should not be considered for prophylactic platelet transfusions until the platelet count has fallen below $30 \times 10^9$/L (Table 52-3).

*c. Unstable NICU Patients with Severe Thrombocytopenia (Platelet Count $<50 \times 10^9$/L).* In view of the available evidence, we recommend that all unstable NICU patients with severe thrombocytopenia be given platelet transfusions (Table 52-3). Precise definitions of clinical instability related to this population are difficult but in this instance must reflect principally the risk of bleeding.

We currently use the following as guidelines:

- any extremely low birth weight infant (<1000 g) in the first week of life
- any infant undergoing intensive care (e.g., mechanical ventilation) who has fluctuating vital signs, particularly blood pressure and peripheral perfusion
- previous clinical evidence of a significant bleeding tendency (e.g., grade 3 to 4 IVH)
- current evidence of minor bleeding tendency (e.g., blood-stained gastric residuals, petechiae, bruising, excessive oozing from puncture sites)
- laboratory evidence of coagulopathy
- severe thrombocytopenia prior to procedures that are liable to reduce platelet count (e.g., surgery, exchange transfusion)

**2. Platelet Transfusions in Actively Bleeding Neonates.** All neonates who have major active hemorrhage (e.g., pulmonary hemorrhage, hematuria, or rapidly evolving IVH) should be given platelet transfusions at platelet counts below $100 \times 10^9$/L (Table 52-3).

### C. Monitor Response to Platelet Transfusions

For optimum practice, a 1-hour posttransfusion platelet count should be performed following all platelet transfusions in NICU patients. This provides useful information that guides further therapy. First, it provides an assessment of the effectiveness of the transfusion (by documenting the platelet increment). Second, it provides information regarding the timing of the next platelet count and necessity for further platelet transfusions. Lastly, it aids in the definition of the major underlying mechanism of thrombocytopenia. A good platelet increment suggests a hyporegenerative thrombocytopenia with platelet underproduction and predicts a good response to platelet transfusion therapy, whereas a poor platelet increment suggests major platelet consumption that is unlikely to respond to platelet transfusion but may require other specific therapies.

### D. Regularly Review the Clinical Cause of Thrombocytopenia

The clinical cause of thrombocytopenia in an individual neonate may change, as evidenced by the pattern of thrombocytopenia, investigation results, or response to treatment. Only by defining and regularly reviewing the cause of thrombocytopenia in each patient can optimal management of thrombocytopenia be achieved.

## IX. TPO in the Fetus and Neonate

TPO is the major regulator of platelet production in adults[240] (see Chapter 66), and there is increasing evidence that it plays the same role in the fetus and neonate.[20] However, data suggest that the fetus and neonate may have a reduced ability to produce TPO, which limits their capacity to upregulate platelet production at times of increased demand,[20,23,24] such as during periods of rapid platelet consumption in the initial phase of thrombocytopenia in neonatal sepsis.

### A. Fetal and Neonatal TPO Production

TPO mRNA has been demonstrated in many fetal and neonatal tissues[241,242] and, as in adults, the liver appears to be the major site of fetal and neonatal TPO production. A study assessing fetuses following elective termination and preterm neonates following early neonatal death (gestational age 17–36 weeks) found that the level of TPO mRNA expression per microgram total RNA and per gram of tissue studied was similar in most tissues examined.[242] However, as liver hepatocytes produce TPO,[243] the large mass of this organ in the fetus and neonate results in the liver being by far the most important site of TPO production, calculated to provide more than 80% of total body production.[242]

### B. Fetal and Neonatal TPO Function

The function of TPO in the fetus and neonate appears to be the same as in older age groups. Megakaryocyte progenitor cells (BFU-MK and CFU-MK) and more mature megakaryocyte precursors derived from fetal, umbilical cord, and neonatal peripheral blood and from neonatal bone marrow all show sensitivity to TPO.[12,37,179–182] During *in vitro* culture, these cells proliferate markedly in response to TPO[19,21–23] and in suspension cultures large numbers of megakaryo-cytes are generated.[19] The study by Sola et al.[244] that dem-onstrates a marked rise in platelet count in newborn rhesus monkeys given pegylated recombinant human megakaryo-cyte growth and development factor (PEG-rHuMGDF), a truncated pegylated derivative of the full length TPO mol-ecule that retains full biological activity, confirms these actions in an appropriate animal model. There are no pub-lished trials of the effects of TPO therapy *in vivo* in human neonates.

### C. Fetal and Neonatal TPO Regulation

There are few data on which to base models of TPO regula-tion in the fetus and neonate. Circulating TPO levels have been measured in healthy term and preterm neonates by a number of groups, including our own.[19,20,23,77,241,245,246] These studies show that TPO is detectable in the circulation of the vast majority of newborns. Preterm neonates appear to have marginally higher TPO levels compared to term neonates,[246] although there is a wide range in both groups.[19,20,241,245] However, it is unusual for a healthy newborn of any gesta-tional age to have a TPO level at birth >250 pg/mL, and this can be considered to be the upper limit of normal in the newborn. As has been found in adults, there is no correlation between circulating TPO level and platelet count in healthy neonates with normal platelet counts.[77,241,245]

### D. Fetal and Neonatal TPO Levels During Thrombocytopenia

TPO levels have now been measured during an increasing number of thrombocytopenic states in neonates: early-onset thrombocytopenia in preterm neonates associated with maternal PET or IUGR,[20,42,247] neonatal intensive care,[23] neonatal sepsis,[77] and in NAIT.[248]

**1. Early-Onset Thrombocytopenia.** Our group mea-sured TPO levels and circulating megakaryocyte progenitor cells serially over the first 12 days of life in a group of preterm neonates with early-onset thrombocytopenia and compared them to controls with normal platelet counts.[20] The neonates with early-onset thrombocytopenia had both significantly lower numbers of megakaryocyte progenitors and raised TPO levels compared to controls. Peak TPO levels (mean 425 pg/mL) were measured on days 4 and 5, when platelet counts and progenitor numbers were at their lowest. TPO levels and progenitor numbers returned to control values by day 12, in conjunction with platelet recov-

ery. These findings are consistent with the current concept of TPO regulation in older age groups, as the TPO level rose significantly only when the platelet count and circulating megakaryocyte progenitors were both low (i.e., presumed reduced total body c-mpl mass). We also noted that the peak TPO levels observed in the thrombocytopenic neonates were relatively modest,[20] suggesting that preterm neonates may have a reduced TPO production capacity compared to adults.

**2. Neonatal Intensive Care.** Sola et al.[23] measured TPO levels and assessed megakaryocytopoiesis in bone marrow aspirates in a group of thrombocytopenic neonates admitted to an NICU. In just over half of the neonates studied, there was no obvious cause for the thrombocytope-nia, but TPO levels in this group were marginally raised (median approximately 300 pg/mL). In the remaining neo-nates, two had evidence of DIC and normal TPO levels (<200 pg/mL), whereas the rest had either sepsis or alloim-mune thrombocytopenia and demonstrated a wide range of TPO levels (normal to >1000 pg/mL). In over half of the cases in which bone marrow aspirates were obtained, Sola et al.[23] reported reduced megakaryocyte mass, and the majority of these neonates had raised TPO levels, thereby supporting the contention that TPO levels in the neonate reflect total body c-mpl mass as in adults. In agreement with our findings,[20] Sola et al.[23] also noted that neonates with a hyporegenerative thrombocytopenia (as evidenced by decreased bone marrow megakaryocyte mass) appeared to have lower TPO levels than would be expected in adults with a similar mechanism of thrombocytopenia.

**3. Neonatal Sepsis.** Colarizi et al.[77] measured TPO levels in neonates (gestational age 24 to 42 weeks) with proven (blood or cerebrospinal fluid culture positive) and suspected neonatal sepsis with and without thrombocytope-nia. They found a trend toward the highest TPO levels in the neonates with proven sepsis, and in the septic neonates as a whole they found an inverse correlation between TPO level and platelet numbers. Following recovery from sepsis, TPO levels fell, and there was no longer any correlation between TPO levels and platelet count. Neonates with sepsis (proven plus suspected) who did not develop thrombocyto-penia had low TPO levels, with no significant difference between TPO levels during sepsis and upon recovery. The authors commented that significantly increased TPO levels are frequently found in neonates with sepsis but that this may be more closely related to accompanying thrombocy-topenia, as septic neonates without thrombocytopenia showed no significant difference between TPO levels during sepsis and upon recovery. These findings[77] suggest that preterm neonates who develop sepsis-associated thrombo-cytopenia may have reduced platelet production as reflected by their raised TPO levels, although no confirmatory assessment of megakaryocytopoiesis was carried out in this study. Although high TPO levels (>1000 pg/mL) have been recorded in some thrombocytopenic neonates (with or without sepsis), this is unusual, and most such neonates have TPO levels well below 1000 pg/mL, even those with profound thrombocytopenia.[20,23] There are several possible explanations for this: Sick neonates, particularly if preterm, may be unable to upregulate TPO production in response to an increased demand for platelets; their constitutive production of TPO may be relatively low; the rate of TPO clearance may be increased; the total c-mpl mass may be less compromised due to the transient nature and less severe degree of impairment of megakaryocytopoiesis (e.g., compared to bone marrow failure in adults); or there may be a reduced requirement for TPO due to increased sensitivity of progenitors. Given what we know about the frequency of thrombocytopenia secondary to impaired megakaryocytopoiesis in the neonate, it seems possible that sick thrombocytopenic neonates have little excess TPO production above that required for their normal level of platelet production and have an impaired ability to increase available TPO during thrombocytopenic episodes. This situation would make them potential candidates for treatment with recombinant human TPO or other thrombopoietic growth factors.

**4. NAIT.** Porcelijn et al.[248] have measured TPO levels in fetuses and neonates with NAIT. They found no difference in TPO levels in subjects with NAIT compared to healthy controls. However, supporting previous findings, TPO levels in neonatal subjects (NAIT and controls) were higher than in healthy adult controls.

## X. Future Therapeutic Options for Neonatal Thrombocytopenia

The evidence that impaired platelet production underlies many neonatal thrombocytopenias strongly suggests that appropriately targeted thrombopoietic growth factor therapy may be therapeutically useful in the management of these conditions. Three main options have been investigated: recombinant human (rh) TPO (rhTPO), TPO mimetics,[249-253] and rhIL-11.

### A. rhTPO (see also Chapter 66)

There are no clinical studies in human neonates. However, a study employing an appropriate neonatal animal model showed that rhTPO produced a rise in platelet count beginning 6 to 7 days after starting therapy,[244] a lag time during which many neonatal thrombocytopenias will resolve. In fact, this significant lag time is a major problem for the use

of all HGF therapy for neonatal thrombocytopenia, unless prophylactic HGF therapy is undertaken, which may be hard to justify for the treatment of thrombocytopenia in the absence of bleeding. In addition, clinical trials of one form of rhTPO (PEG-rHuMGDF) have been associated with the development of neutralizing anti-TPO antibodies causing thrombocytopenia in treated subjects.[254] This has led to the cancellation of further development of this product. Although development of the full-length recombinant protein is continuing, no form of rhTPO is currently available for therapeutic use.

### B. TPO Mimetic Peptides (see also Chapter 66)

Peptide and nonpeptide TPO mimetics have been developed by several groups.[249-253] These bear no structural resemblance to TPO but act as competitive agonists for the TPO receptor. They have varying potency compared to natural TPO, but some appear to be at least equally as potent as the natural protein.[250] Their therapeutic potential lies in the fact that they appear to be nonimmunogenic.[255-256] Indeed, if antimimetic antibodies did develop in treated individuals, they should only lead to neutralization of the mimetic rather than to thrombocytopenia, as has been the case with PEG-rHuMGDF. Our group has demonstrated that the TPO mimetic (GW395058) exhibits highly potent agonist effects on neonatal megakaryocyte progenitor and precursor cells.[257] However, the future role of TPO-mimetic peptides in the management of neonatal thrombocytopenia is also likely to be limited, because there is no suggestion from current *in vitro* studies that TPO-mimetic peptides would be able to increase the platelet count more quickly than rhTPO.

### C. rhIL-11 (see also Chapter 66)

Given the limitations of rhTPO and TPO mimetics, rhIL-11 may represent the best immediate hope for HGF therapy that could significantly improve the treatment of neonatal thrombocytopenia. IL-11 stimulates platelet production from megakaryocytes and ameliorates chemotherapy-induced thrombocytopenia in adults and children.[233,258-260] Our work shows that IL-11 stimulates megakaryocyte progenitor and precursor cells from term and preterm neonates and also significantly enhances TPO-induced proliferation and differentiation of these cells.[261] IL-11 is also involved in the cytokine response to sepsis and NEC by preterm neonates[262] — the high-risk population for severe and prolonged neonatal thrombocytopenia. Although there are no published reports of neonates receiving IL-11, these findings strongly suggest that IL-11 would stimulate platelet production in these subjects.

Intriguingly, IL-11 therapy also conveys survival benefit during sepsis (at least in animals)[229-232] and has been reported to reduce the incidence of sepsis in adults during chemotherapy for hematological malignancies.[234] IL-11 also protects against bowel injury following a range of insults[233] and promotes bowel growth.[263] In addition, it has been shown to ameliorate a rat model of NEC[264] and this effect appeared to be greatest in those animals who received IL-11 by oral administration. We have shown that circulating IL-11 levels are raised in neonates with sepsis and NEC, but IL-11 levels in these clinical conditions do not correlate with platelet count.[262] IL-11 is known to downregulate the proinflammatory response,[233,265] and our findings suggest that this is likely to be its primary role in neonates with sepsis and NEC, in view of the lack of correlation with platelet count. However, as is the situation with a number of other HGFs, it appears that sick neonates may not be able to endogenously produce large amounts of IL-11.[261] As severe neonatal thrombocytopenia is highly correlated with late-onset sepsis and NEC, these findings[261,264] suggest that rhIL-11 therapy during these conditions may ameliorate thrombocytopenia while also potentially benefiting the underlying conditions.

In adults, rhIL-11 has been associated with moderate toxicities, mostly related to fluid retention[258,259] (see Chapter 66); but in the small number of children treated so far, side effects appear to be less of a problem.[260] Many neonates who develop severe thrombocytopenia in association with late-onset sepsis and NEC require intensive care, and in our unit this combination is associated with a 10 to 15% mortality. Therefore, a potentially beneficial therapy with some moderate toxicities may well be acceptable in these circumstances. In addition, the finding that at least some of the benefits of IL-11 can be seen following oral administration[264] raises the possibility that prophylactic therapy could be given to high-risk neonates.

## XI. Summary

Neonatal thrombocytopenia remains a common clinical problem. Fortunately most episodes are mild or moderate and resolve spontaneously without apparent clinical sequelae. Evidence shows that the majority of these episodes are early-onset thrombocytopenias due to the impaired fetal megakaryocytopoiesis now known to be associated with common feto-maternal conditions resulting in placental insufficiency/fetal hypoxia (e.g., maternal PET).

However, severe thrombocytopenia occurs in approximately 1 : 20 NICU admissions and when present often lasts for many days, presenting an ongoing management problem. The causes of severe thrombocytopenia depend on the age at presentation (Table 52-1). At birth, the most common causes of severe thrombocytopenia are immune thrombocytopenias, congenital infections, perinatal infections, and asphyxia (the latter two often complicated by DIC). By

contrast, severe thrombocytopenia presenting after the first few days of life is associated with late-onset sepsis and NEC in 90% of cases. DIC does not commonly appear to be responsible for the thrombocytopenia in these conditions.

Although generally stable neonates appear to tolerate relatively low platelet counts without a significant risk of hemorrhage, ill or clinically unstable neonates developing profound thrombocytopenia often have a poor outcome. The only available therapy in these situations is platelet transfusion, which raises the difficult clinical question of the platelet count threshold that justifies transfusion at the cost of an increased risk of transmissible diseases and transfusion reactions. Despite the many guidelines for platelet transfusion therapy in the newborn that have been proposed, there remains a pressing need for well-designed and conducted trials to define the safe lower limit for platelet counts in sick neonates and to define which neonates will benefit from specific therapy of their thrombocytopenia.

In tandem with these considerations, preventing or ameliorating the clinical conditions that lead to the combination of overwhelming illness and profound thrombocytopenia remains a major goal. The increasing demonstration of impaired megakaryocytopoiesis and platelet production as a major contributor to neonatal thrombocytopenias is an important advance both in terms of mechanistic accuracy and the potential for innovative therapies. The use of HGFs to stimulate neonatal platelet production, and particularly the potential of IL-11, remains an exciting prospect for the future. Finally, ensuring accurate diagnosis and determining effective fetal and neonatal therapy for NAIT, the neonatal thrombocytopenia currently identifiable as directly causing most mortality and morbidity, should remain the goals for all fetal medicine specialists, hematologists, and neonatal pediatricians.

## References

1. Hann, I. M. (1991). Development of blood in the fetus. In I. M. Hann, B. E. S. Gibson, & E. Letsky (Eds.), *Fetal and neonatal haematology* (pp. 1–28). London: Bailliere Tindall.

2. Pahal, G., Jauniaux, E., Kinnon, C., et al. (2000). Normal development of human fetal hematopoiesis between eight and seventeen weeks' gestation. *Am J Obstet Gynecol, 183,* 1029–1034.

3. Holmberg, L., Gustavii, B., & Jonsson, A. (1983). A prenatal study of fetal platelet count and size with application to the fetus at risk of Wiskott–Aldrich syndrome. *J Pediatr, 102,* 773–781.

4. Forestier, F., Daffos, F., & Galacteros, F. (1986). Haematological values of 163 normal fetuses between 18 and 30 weeks of gestation. *Pediatr Res, 20,* 342–346.

5. Forestier, F., Daffos, F., Catherine, N., et al. (1991). Developmental hematopoiesis in normal human fetal blood. *Blood, 77,* 2360–2363.

6. Van den Hof, M. C., & Nicolaides, K. H. (1990). Platelet count in normal, small, and anemic fetuses. *Am J Obstet Gynecol, 162,* 735–739.

7. Hohlfeld, P., Forestier, F., Kaplan, C., et al. (1994). Fetal thrombocytopenia: A retrospective survey of 5194 fetal blood samplings. *Blood, 84,* 1851–1856.

8. Saxonhouse, M. A., Sola, M. C., Pastos, K. M., et al. (2004). Reticulated platelet percentages in term and preterm neonates. *J Pediatr Hematol Oncol, 26,* 797–802.

9. Jilma-Stohlawetz, P., Homoncik, M., Jilma, B., et al. (2001). High levels of reticulated platelets and thrombopoietin characterize fetal thrombopoiesis. *Br J Haematol, 112,* 466–468.

10. Hegyi, E., Nakazawa, M., Debili, N., et al. (1991). Developmental changes in human megakaryocyte ploidy. *Exp Hematol, 19,* 87–94.

11. Allen Graeve, J. L., & de Alarcon, P. A. (1989). Megakaryocytopoiesis in the human fetus. *Arch Dis Child, 64,* 481–484.

12. de Alarcon, P. A., & Graeve, J. L. (1996). Analysis of megakaryocyte ploidy in fetal bone marrow biopsies using a new adaptation of the Feulgen technique to measure DNA content and estimate megakaryocyte ploidy from biopsy specimens. *Pediatr Res, 39,* 166–170.

13. Miyazaki, R., Ogata, H., Iguchi, T., et al. (2000). Comparative analyses of megakaryocytes derived from cord blood and bone marrow. *Br J Haematol, 108,* 602–609.

14. Saxonhouse, M. A., Rimsza, L. M., Christensen, R. D., et al. (2003). Effects of anoxia on megakaryocyte progenitors derived from cord blood CD34pos cells. *Eur J Haematol, 71,* 359–365.

15. Muench, M. O., & Barcena, A. (2004). Megakaryocyte growth and development factor is a potent growth factor for primitive hematopoietic progenitors in the human fetus. *Pediatr Res, 55,* 1050–1056.

16. Murray, N. A., & Roberts, I. A. G. (1995). Circulating megakaryocytes and their progenitors (BFU-MK and CFU-MK) in term and pre-term neonates *Br J Haematol, 89,* 41–46.

17. Saxenhouse, M. A., Christensen, R. D., Walker, D. M., et al. (2004). The concentration of circulating megakaryocyte progenitors in preterm neonates is a function of post-conceptional age. *Early Hum Dev, 78,* 119–124.

18. Campagnoli, C., Fisk, N., Overton, T., et al. (2000). Circulating hematopoietic progenitor cells in first trimester fetal blood. *Blood, 95,* 1967–1972.

19. Murray, N. A., Watts, T. L., & Roberts, I. A. G. (1998). Endogenous thrombopoietin levels and effect of recombinant human thrombopoietin on megakaryocyte precursors in term and preterm babies. *Pediatr Res, 43,* 148–151.

20. Watts, T. L., Murray, N. A., & Roberts, I. A. G. (1999). Thrombopoietin has a primary role in the regulation of platelet production in preterm babies. *Pediatr Res, 46,* 28–32.

21. Nishihira, H., Toyoda, Y., Miyazaki, H., et al. (1996). Growth of macroscopic human megakaryocyte colonies from cord blood with recombinant human thrombopoietin (c-mpl ligand) and the effects of gestational age on the frequency of colonies. *Br J Haematol, 92,* 23–28.

22. Sola, M. C., Du, Y., Hutson, A. D., et al. (2000). Dose-response relationship of megakaryocyte progenitors from the bone marrow of thrombocytopenic and non-thrombocytopenic neonates to recombinant thrombopoietin. *Br J Haematol, 110,* 449–453.

23. Sola, M. C., Calhoun, D. A., Hutson, A. D., et al. (1999). Plasma thrombopoietin concentrations in thrombocytopenic and non-thrombocytopenic patients in a neonatal intensive care unit. *Br J Haematol, 104,* 90–92.

24. Albert, T. S. E., Meng, G., Simms, P., et al. (2000). Thrombopoietin in the thrombocytopenic term and preterm newborn. *Pediatrics, 105,* 1286–1291.

25. Burrows, R. F., & Kelton, J. G. (1988). Incidentally detected thrombocytopenia in healthy mothers and their infants. *N Engl J Med, 319,* 142–145.

26. Burrows, R. F., & Kelton, J. G. (1990). Thrombocytopenia at delivery: A prospective survey of 6715 deliveries. *Am J Obstet Gynecol, 162,* 731–734.

27. Burrows, R. F., & Kelton, J. G. (1993). Fetal thrombocytopenia and its relation to maternal thrombocytopenia. *N Engl J Med, 329,* 1463–1466.

28. Sainio, S., Jarvenpaa, A.-S., Renlund, M., et al. (2000). Thrombocytopenia in term infants: A population-based study. *Obstet Gynecol, 95,* 441–446.

29. Sasanakul, W., Singalavanija, S., Hathirat, P., et al. (1993). Hemogram in normal newborn babies with special reference to platelet. *Southeast Asian J Trop Med Public Health, 24* (suppl 1), 237–240.

30. McIntosh, N., Kempson, C., & Tyler, R. M. (1988). Blood counts in extremely low birth weight infants. *Arch Dis Child, 63,* 74–76.

31. Christensen, R., Sola, M. C., Rimsza, L. M., et al. (2004). Pseudothrombocytopenia in a preterm neonate. *Pediatrics, 114,* 273–275.

32. Chiurazzi, F., Villa, M. R., & Rotoli, B. (1999). Transplacental transmission of EDTA-dependent pseudothrombocytopenia. *Haematologica, 84,* 664–665.

33. de Moerloose, P., Boehlen, F., Extermann, P., et al. (1998). Neonatal thrombocytopenia: Incidence and characterization of maternal antiplatelet antibodies by MAIPA assay. *Br J Haematol, 100,* 735–740.

34. Uhrynowska, M., Niznikowska-Marks, M., & Zupanska, B. (2000). Neonatal and maternal thrombocytopenia: Incidence and immune background. *Eur J Haematol, 64,* 42–46.

35. Dreyfus, M., Kaplan, C., Verdy, E., et al. (1997). Frequency of immune thrombocytopenia in newborns: A prospective study. Immune Thrombocytopenia Working Group. *Blood, 89,* 4402–4406.

36. Castle, V., Andrew, M., Kelton, J., et al. (1986). Frequency and mechanism of neonatal thrombocytopenia. *J Pediatr, 108,* 749–755.

37. Mehta, P., Rohitkumar, V., Neumann, L., et al. (1980). Thrombocytopenia in the high risk infant. *J Pediatr, 97,* 791–794.

38. Murray, N. A., & Roberts, I. A. G. (1996). Circulating megakaryocytes and their progenitors in early thrombocytopenia in preterm neonates. *Pediatr Res, 40,* 112–119.

39. Aman, I., Hassan, K. A., & Ahmad, T. M. (2004). The study of thrombocytopenia in sick neonates. *J Coll Physicians Surg Pak, 14,* 282–285.

40. Kenton, A. B., O'Donovan, D., Cass, D. L., et al. (2005). Severe thrombocytopenia predicts outcome in neonates with necrotizing enterocolitis. *J Perinatol, 25,* 14–20.

41. Tunstall-Pedoe, O., & Roberts, I. A. G. (in press). Neonatal thrombocytopenia. *Adv. Pediatrics, 22.*

42. Forestier, F., & Hohlfeld, P. (1998). Management of fetal and neonatal alloimmune thrombocytopenia. *Biol Neonate, 74,* 395–401.

43. Murphy, M. F., & Williamson, L. M. (2000). Antenatal screening for fetomaternal alloimmune thrombocytopenia: An evaluation using the criterion of the UK national screening committee. *Br J Haematol, 111,* 726–732.

44. Valat, A. S., Caulier, M. T., Devos, P., et al. (1998). Relationships between severe neonatal thrombocytopenia and maternal characteristics in pregnancies associated with autoimmune thrombocytopenia. *Br J Haematol, 103,* 397–401.

45. Bussel, J. B. (1997). Immune thrombocytopenia in pregnancy: Autoimmune and alloimmune. *J Reprod Immunol, 37,* 35–61.

46. Beiner, M. E., Simchen, M. J., Sivan, E., et al. (2003). Risk factors for neonatal thrombocytopenia in preterm infants. *Am J Perinatol, 20,* 49–54.

47. Watts, T. L., & Roberts, I. A. G. (1999). Haematological abnormalities in the growth-restricted infant. *Semin Neonatol, 4,* 41–54.

48. Murray, N. A., Howarth, L. J., McCloy, M. P., et al. (2002). Platelet transfusion in the management of severe thrombocytopenia in neonatal intensive care unit patients. *Transfus Med, 12,* 35–41.

49. Zipursky, A., Palko, J., Milner, R., et al. (1976). The hematology of bacterial infections in premature infants. *Pediatrics, 57,* 839–853.

50. McPherson, R. J., & Juul, S. (2005). Patterns of thrombocytosis and thrombocytopenia in hospitalized neonates. *J Perinatol, 25,* 166–172.

51. Kenton, A. B., Hegemier, S., Smith, E. O., et al. (2005). Platelet transfusions in infants with necrotizing enterocolitis do not lower mortality but may increase morbidity. *J Perinatol, 25,* 173–177.

52. Mueller-Eckhardt, C., Kiefel, V., Grubert, A., et al. (1989). 348 cases of suspected neonatal alloimmune thrombocytopenia. *Lancet, 1*(8634), 363–366.

53. Bussel, J. B., Zacharoulis, S., Kramer, K., et al. (2005). Clinical and diagnostic comparison of neonatal alloimmune thrombocytopenia to non-immune cases of thrombocytopenia. *Pediatr Blood Cancer, 45,* 176–183.

54. Kaplan, C. (2002). Alloimmune thrombocytopenia of the fetus and the newborn. *Blood Rev, 16,* 69–72.

55. Davoren, A., Curtis, B. R., Aster, R. H., et al. (2004). Human platelet antigen-specific alloantibodies implicated in 1162 cases of neonatal alloimmune thrombocytopenia. *Transfusion, 44,* 1220–1225.

56. Williamson, L. M., Hackett, G., Rennie, J., et al. (1998). The natural history of fetomaternal alloimmunization to the platelet-specific antigen HPA-1a (PlA1, Zwa) as determined by antenatal screening. *Blood, 92,* 2280–2287.

57. Mandelbaum, M., Koren, D., Eichelberger, B., et al. (2005). Frequencies of maternal platelet alloantibodies and autoantibodies in suspected fetal/neonatal isoimmune thrombocytopenia, with emphasis on human platelet antigen-15 alloimmunization. *Vox Sang, 89,* 39–43.

58. Curtis, B. R., Ali, S., Glazier, A. M., et al. (2002). Isoimmunization against CD36 (glycoprotein IV): Description of four cases of neonatal isoimmune thrombocytopenia and brief review of the literature. *Transfusion, 42,* 1173–1179.

59. Kroll, H., Yates, J., & Santoso, S. (2005). Immunization against a low-frequency human platelet alloantigen in fetal alloimmune thrombocytopenia is not a single event: Characterization by the combined use of refernce DNA and novel allele-specific cell lines expressing recombinant antigens. *Transfusion, 45,* 353–358.

60. Blanchette, V. S., Johnson, J., & Rand, M. (2000). The management of alloimmune neonatal thrombocytopenia. *Baillieres Clin Haematol, 13,* 365–390.

61. Gibson, B. E., Todd, A., Roberts, I., et al. (2004). Transfusion guidelines for neonates and older children. *Br J Haematol, 124,* 433–453.

62. Gill, K. K., & Kelton, J. G. (2000). Management of idiopathic thrombocytopenic purpura in pregnancy. *Semin Hematol, 37,* 275–289.

63. George, D., & Bussel, J. B. (1995). Neonatal thrombocytopenia. *Semin Thromb Hemost, 21,* 276–293.

64. Kelton, J. G. (2002). Idiopathic thrombocytopenic purpura complicating pregnancy. *Blood Rev, 16,* 43–46.

65. Payne, S. D., Resnik, R., Moore, T. R., et al. (1997). Maternal characteristics and risk of severe neonatal thrombocytopenia and intracranial hemorrhage in pregnancies complicated by autoimmune thrombocytopenia. *Am J Obstet Gynecol, 177,* 149–155.

66. Webert, K. E., Mittal, R., Sigouin, C., et al. (2003). A retrospective 11-year analysis of obstetric patients with idiopathic thrombocytopenic purpura. *Blood, 102,* 4306–4311.

67. Burrows, R. F., & Kelton, J. G. (1990). Low fetal risks in pregnancies associated with idiopathic thrombocytopenic purpura. *Am J Obstet Gynecol, 163,* 1147–1150.

68. Ballin, A., Andrew, M., Ling, E., et al. (1988). High dose intravenous gammaglobulin therapy for neonatal idiopathic autoimmune thrombocytopenia. *J Pediatr, 112,* 789–792.

69. Fanaroff, A. A., Korones, S. B., Wright, L. L., et al. (1998). Incidence, presenting features, risk factors and significance of late onset septicemia in very low birth weight infants. The National Institute of Child Health and Human Development Neonatal Research Network. *Pediatr Infect Dis J, 17,* 593–598.

70. Isaacs, D., Barfield, C., Clothier, T., et al. (1996). Late-onset infections of infants in neonatal units. *J Paediatr Child Health, 32,* 158–161.

71. Stoll, B. J., Gordon, T., Korones, S. B., et al. (1996). Late-onset sepsis in very low birth weight neonates: A report from the National Institute of Child Health and Human Development Neonatal Research Network. *J Pediatr, 129,* 63–71.

72. Uauy, R. D., Fanaroff, A. A., Korones, S. B., et al. (1991). Necrotizing enterocolitis in very low birth weight infants: Biodemographic and clinical correlates. National Institute of Child Health and Human Development Neonatal Research Network. *J Pediatr, 119,* 630–638.

73. Guida, J. D., Kunig, A. M., Leef, K. H., et al. (2003). Platelet count and sepsis in very low birth weight neonates: Is there an organism-specific response? *Pediatrics, 111* (6 Pt, 1), 141–145.

74. Ragazzi, S., Pierro, A., Peters, M., et al. (2003). Early full blood count and severity of disease in neonates with necrotizing enterocolitis. *Pediatr Surg Int, 19,* 376–379.

75. Kenton, A. B., O'Donovan, D., Cass, D. L., et al. (2005). Severe thrombocytopenia predicts outcome in neonates with necrotizing enterocolitis. *J Perinatol, 25,* 14–20.

76. Kenton, A. B., Hegemier, S., Smith, E. O., et al. (2005). Platelet transfusions in infants with necrotizing enterocolitis do not lower mortality but may increase morbidity. *J Perinatol, 25,* 173–177.

77. Colarizi, P., Fiorucci, P., Caradonna, A., et al. (1999). Circulating thrombopoietin levels in neonates with infection. *Acta Paediatr, 88,* 332–337.

78. Murray, N. A., Watts, T. L., & Roberts, I. A. G. (1999). Inhibition of megakaryocytopoiesis in "late," sepsis-associated thrombocytopenia in preterm babies. *Blood, 94* (suppl 1), 450a.

79. Casteels, A., Naessens, A., Gordts, F., et al. (1999). Neonatal screening for congenital cytomegalovirus infections. *J Perinat Med, 27,* 116–121.

80. Boppana, S. B., Fowler, K. B., Britt, W. J., et al. (1999). Symptomatic congenital cytomegalovirus infection in infants born to mothers with preexisting immunity to cytomegalovirus. *Pediatrics, 104,* 55–60.

81. Barbi, M., Binda, S., Primache, V., et al. (1998). Congenital cytomegalovirus infection in a northern Italian region: NEOCMV Group. *Eur J Epidemiol, 14,* 791–796.

82. Brown, H. L., & Abernathy, M. P. (1998). Cytomegalovirus infection. *Semin Perinatol, 22,* 260–266.

83. Liesnard, C., Donner, C., Brancart, F., et al. (2000). Prenatal diagnosis of congenital cytomegalovirus infection: Prospective study of 237 pregnancies at risk. *Obstet Gynecol, 95,* 881–888.

84. Tighe, P., Rimsza, L. M., Christensen, R. D., et al. (2005). Severe thrombocytopenia in a neonate with congenital HIV infection. *J Pediatr, 146,* 408–413.

85. Janner, D. (1991). Growth retardation, congenital heart disease and thrombocytopenia in a newborn infant. *Pediatr Infect Dis J, 10,* 874–877.

86. Abzug, M. J., Levin, M. J., & Rotbart, H. A. (1993). Profile of enterovirus disease in the first two weeks of life. *Pediatr Infect Dis J, 12,* 820–824.

87. Abzug, M. J. (2001). Prognosis for neonates with enterovirus hepatitis and coagulopathy. *Pediatr Infect Dis J, 20,* 758–763.

88. Bryant, P. A., Tingay, D., Dargaville, P. A., et al. (2004). Neonatal coxsackie B virus infection — a treatable disease. *Eur J Pediatr, 163,* 223–228.

89. Chen, C. A., Tsao, P. N., Chou, H. C., et al. (2003). Severe echovirus 30 infection in twin neonates. *J Formosa Med Assoc, 102,* 59–61.

90. Abzug, M. J., & Levin, M. J. (1991). Neonatal adenovirus infection: Four patients and review of the literature. *Pediatrics, 87,* 890–896.

91. Lacour, M., Maherzi, M., Vienny, H., et al. (1993). Thrombocytopenia in a case of neonatal mumps infection: Evidence for further clinical presentations. *Eur J Pediatr, 152,* 739–741.

92. Tovo, P. A., de-Martino, M., Gabiano, C., et al. (1992). Prognostic factors and survival in children with perinatal HIV-1 infection: The Italian Register for HIV Infections in Children. *Lancet, 339,* 1249–1253.

93. Labrune, P., Blanche, S., Catherine, N., et al. (1989). Human immunedeficiency virus-associated thrombocytopenia in infants. *Acta Paed Scand, 78,* 811–814.

94. Roux, W., Pieper, C., & Cotton, M. (2001). Thrombocytopenia as a marker for HIV exposure in the neonate. *J Trop Pediatr, 47,* 208–210.

95. Sirinavin, S., Nuntnarumit, P., Supapannachart, S., et al. (2004). Vertical dengue infection: Case reports and review. *Pediatr Infect Dis J, 23,* 1042–1047.

96. Petdachai, W., Sila'on, J., Nimmannitya, S., et al. (2004). Neonatal dengue infection: Reprt of dengue fever in a 1-day-old infant. *Southeast Asian J Trop Med Public Health, 35,* 403–407.

97. Whitley, R. J., Cloud, G., Gruber, W., et al. (1997). Ganciclovir treatment of symptomatic congenital cytomegalovirus infection: Results of a phase II study: National Institute of Allergy and Infectious Diseases Collaborative Antiviral Study Group. *J Infect Dis, 175,* 1080–1086.

98. McAuley, J., Boyer, K. M., Patel, D., et al. (1994). Early and longitudinal evaluations of treated infants and children and untreated historical patients with congenital toxoplasmosis: The Chicago Collaborative Treatment Trial. *Clin Infect Dis, 18,* 38–72.

99. Crapnell, K., Zanjani, E. D., Chaudhuri, A., et al. (2000). *In vitro* infection of megakaryocytes and their precursors by human cytomegalovirus. *Blood, 95,* 487–493.

100. Paul, M., Petersen, E., Pawlowski, Z. S., et al. (2000). Neonatal screening for congenital toxoplasmosis in the Poznan region of Poland by analysis of Toxoplasma gondii-specific IgM antibodies eluted from filter paper blood spots. *Pediatr Infect Dis J, 19,* 30–36.

101. Walpole, I. R., Hodgen, N., & Bower, C. (1991). Congenital toxoplasmosis: A large survey in western Australia. *Med J Aust, 154,* 720–724.

102. Swisher, C. N., Boyer, K., & McLeod, R. (1994). Congenital toxoplasmosis: The Toxoplasmosis Study Group. *Semin Pediatr Neurol, 1,* 4–25.

103. Thulliez, P. (1992). Screening programme for congenital toxoplasmosis in France. *Scand J Infect Dis Suppl, 84,* 43–45.

104. Sullivan, E. M., Burgess, M. A., & Forrest, J. M. (1999). The epidemiology of rubella and congenital rubella in Australia, 1992 to 1997. *Commun Dis Intell, 23,* 209–214.

105. Tookey, P. A., & Peckham, C. S. (1999). Surveillance of congenital rubella in Great Britain, 1971–96. *BMJ, 318,* 769–770.

106. Debillon, T., Daoud, P., Durand, P., et al. (2003). Whole-body cooling after perinatal asphyxia: A pilot study in term neonates. *Dev Med Child Neurol, 45,* 17–23.

107. Gunn, A. J., Gluckman, P. D., & Gunn, T. R. (1998). Selective head cooling in newborn infants after perinatal asphyxia: A safety study. *Pediatrics, 102,* 885–892.

108. Carter, B. S., McNabb, F., & Merenstein, G. B. (1998). Prospective validation of a scoring system for predicting neonatal morbidity after acute perinatal asphyxia. *J Pediatr, 132,* 619–623.

109. Suzuki, S., & Morishita, S. (1998). Hypercoagulability and DIC in high-risk infants. *Semin Thromb Hemost, 24,* 463–466.

110. Leibovitz, E., Flidel-Rimon, O., Juster-Reicher, A., et al. (1997). Sepsis at a neonatal intensive care unit: A four-year retrospective study (1989–1992). *Isr J Med Sci, 33,* 734–738.

111. Stoll, B. J., Gordon, T., Korones, S. B., et al. (1996). Early-onset sepsis in very low birth weight neonates: A report from the National Institute of Child Health and Human Development Neonatal Research Network. *J Pediatr, 129,* 72–80.

112. Hershckowitz, S., Elisha, M. B., Fleisher-Sheffer, V., et al. (2004). A cluster of early neonatal sepsis and pneumonia caused by nontypable Haemophilus influenzae. *Pediatr Infect Dis J, 23,* 1061–1062.

113. Miura, E., Procianoy, R. S., Bittar, C., et al. (2001). A randomised, double-masked, placebo-controlled trail of recombinant granulocyte colony-stimulating factor administration to preterm infants with the clinical signs of early-onset sepsis. *Pediatrics, 107,* 30–35.

114. Sahoo, T., Naeem, R., Pham, K., et al. (2005). A patient with isochromosome 18q, radial-thumb aplasia, thrombocytopenia, and an unbalanced 10;18 chromosome translocation. *Am J Med Genet, 133A,* 93–98.

115. Kivivuori, S. M., Rajantie, J., & Siimes, M. A. (1996). Peripheral blood cell counts in infants with Down's syndrome. *Clin Genet, 49,* 15–19.

116. Gurbuxani, S., Vyas, P., & Crispino, J. D. (2004). Recent insights into the mechanisms of myeloid leukemogenesis in Down syndrome. *Blood, 103,* 399–406.

117. Hitzler, J. K., & Zipursky, A. (2005). Origins of leukaemia in children with Down syndrome. *Nat Rev Cancer, 5,* 11–20.

118. Vyas, P., & Roberts, I. A. G. (in press). Transient myeloproliferative disorder and acute megakaryocytic leukaemia in Down syndrome: Insights into leukaemogenesis.

119. Wechsler, J., Greene, M., McDevitt, M. A., et al. (2002). Acquired mutations in GATA1 in the megakaryoblastic leukemia of Down syndrome. *Nat Genet, 32,* 148–152.

120. Hitzler, J. K., Cheung, J., Li, Y., et al. (2003). GATA1 mutations in transient leukemia and acute megakaryoblastic leukemia of Down syndrome. *Blood, 101,* 4301–4304.

121. Rainis, L., et al. (2003). Mutations in exon 2 of GATA1 are early events in megakaryocytic malignancies associated withtrisomy 21. *Blood, 102,* 981–986.

122. Xu, G., et al. (2003). Frequent mutations in the *GATA-1* gene in the transient myeloproliferative disorder of Down syndrome. *Blood, 102,* 2960–2968.

123. Ahmed, M., Sternberg, A., Hall, G., et al. (2004). Natural history of GATA1 mutations in Down syndrome. *Blood, 103,* 2480–2489.

124. Li, Z., Godinho, F. J., Klusmann, J. H., et al. (2005). Developmental stage-selective effect of somatically mutated leukemogenic transcription factor GATA1. *Nat Genet, 37,* 613–619.

125. Balduini, C. L., & Savoia, A. (2004). Inherited thrombocytopenias: Molecular mechanisms. *Semin Thromb Hemost, 30,* 513–523.

126. Lopez, J. A., Andrews, R. K., Afshar-Kharghan, V., et al. (1998). Bernard-Soulier syndrome. *Blood, 91,* 4397–4418.

127. Kunishima, S., Kamiya, T., & Saito, H. (2002). Genetic abnormalities of Bernard-Soulier syndrome. *Int J Haematol, 76,* 319–327.

128. Simsek, S., Admiraal, L. G., Modderman, P. W., et al. (1994). Identification of a homozygous single base pair deletion in the gene coding for the human platelet glycoprotein Ib alpha causing Bernard-Soulier syndrome. *Thromb Haemost, 72,* 444–449.

129. Fujimori, K., Ohto, H., Honda, S., et al. (1999). Antepartum diagnosis of fetal intracranial hemorrhage due to maternal Bernard-Soulier syndrome. *Obstet Gynecol, 94,* 817–819.

130. Villa, A., Notarangelo, L., Macchi, P., et al. (1995). X-linked thrombocytopenia and Wiskott-Aldrich syndrome are allelic diseases with mutations in the WASP gene. *Nat Genet, 9,* 414–417.

131. Lutskiy, M. I., Rosen, F. S., & Remold-O'Donnell, E. (2005). Genotype-proteotype linkage in the Wiskott-Aldrich syndrome. *J Immunol, 175,* 1329–1336.

132. Ochs, H. (1998). The Wiskott-Aldrich syndrome. *Semin Hematol, 35,* 332–345.

133. Dupuis-Girod, S., Medioni, J., Haddad, E., et al. (2003). Autoimmunity in Wiskott-Aldrich syndrome: Risk factors, clinical features, and outcome in a single-center cohort of 55 patients. *Pediatrics, 111* (5 Pt, 1), e622–627.

134. Nagle, D. L., Karim, A. M., & Woolf, E. A. (1996). Identification and mutation analysis of the complete gene for Chediak-Higashi syndrome. *Nat Genet, 14,* 307–311.

135. Introne, W., Boissy, R. E., & Gahl, W. A. (1999). Clinical, molecular and cell biological aspects of Chediak-Higashi syndrome. *Mol Genet Metab, 68,* 283–303.

136. Song, W. J., Sullivan, M. G., Legare, R. D. et al. (1999). Haploinsufficiency of CBFA2 causes familial thrombocytopenia with propensity to develop acute myelogenous leukaemia. *Nat Genet, 23,* 166–175.

137. Nichols, K. E., Crispino, J. D., Poncz, M., et al. (2000). Familial dyserythropoietic anaemia and thrombocytopenia due to an inherited mutation in GATA1. *Nat Genet, 24,* 266–270.

138. Drachman, J. G., Jarvik, G. P., & Mehaffey, M. G. (2000). Autosomal dominant thrombocytopenia: Incomplete mega-

karyocyte differentiation and linkage to human chromosome 10. *Blood, 96,* 118–125.

139. Heath, K. E., Campos-Barros, A., Toren, A. et al. (2001). Nonmuscle myosin heavy chainIIa mutations define a spectrum of autosomal dominant macrothrombocytopenias: May-Hegglin anomaly, and Fechtner, Sebastian, Epstein, and Alport-like syndromes. *Am J Hum Genet, 69,* 1033–1045.

140. Butturini, A., Gale, R. P., Verlander, P. C., et al. (1994). Hematologic abnormalities in Fanconi anemia: An international Fanconi anemia registry study. *Blood, 84,* 1650–1655.

141. Faivre, L., Guardiola, P., Lewis, C., et al. (2000). Association of complementation group and mutation type with clinical outcome in Fanconi anemia. *Blood, 96,* 4064–4070.

142. Gershanik, J. J., Morgan, S. K., & Akers, R. (1972). Fanconi's anemia in a neonate. *Acta Paediatr Scand, 61,* 623–625.

143. Gibson, B. E. S. (1991). Inherited disorders. In I. M. Hann, B. E. S. Gibson, & E. Letsky (Eds.), *Fetal and neonatal haematology* (pp. 219–275). London: Bailliere Tindall.

144. Landmann, E., Bluetters-Sawatzki, R., Schindler, D., et al. (2004). Fanconi anemia in a neonate with pancytopenia. *J Pediatr, 145,* 125–127.

145. Auerbach, A. D., Rogatko, A., & Schroeder-Kurth, T. M. (1989). International Fanconi anemia registry: Relation of clinical symptoms to diepoxybutane sensitivity. *Blood, 73,* 391–396.

146. Guardiola, P., Pasquini, R., Dokal, I., et al. (2000). Outcome of 69 allogeneic stem cell transplantations for Fanconi anemia using HLA-matched unrelated donors: A study on behalf of the European Group for Blood and Marrow Transplantation. *Blood, 15,* 422–429.

147. Tischkowitch, M., & Dokal, I. (2004). Fanconi anaemia and leukaemia: Clinical and molecular aspects. *Br J Haematol, 126,* 176–91.

148. Bagby, G. C., Jr. (2003). Genetic basis of Fanconi anemia. *Curr Opin Hematol, 10,* 68–76.

148b. Meetei, A. R., Medhurst, A. L., Ling, C., et al. (2005). A human ortholog of archaeal DNA repair protein Hef is defective in Fanconi anemia complementation group M. *Nat Genet, 37,* 958–963.

148c. Levran, O., Attwooll, C., Henry, R. T., et al. (2005). The BRCA1-interacting helicase BRIP1 is deficient in Fanconi anemia. *Nat Genet, 37,* 931–933.

148d. Levitus, M., Waisfisz, Q., Godthelp, B. C., et al. (2005). The DNA helicase BRIP1 is defective in Fanconi anemia complementation group J. *Nat Genet, 37,* 934–935.

149. Venkitaraman, A. R. (2004). Tracing the network connecting BRCA and Fanconi anaemia proteins. *Nat Rev Cancer, 4,* 266–276.

150. Hedberg, V. A., & Lipton, J. M. (1988). Thrombocytopenia with absent radii: A review of 100 cases. *Am J Pediatr Hematol Oncol, 10,* 51–64.

151. Hall, J. G. (1987). Thrombocytopenia and absent radius (TAR) syndrome. *J Med Genet, 24,* 79–83.

152. Greenhalgh, K. L., Howell, R. T., Bottani, A., et al. (2002). Thrombocytopenia-absent radius syndrome: A clinical genetic study. *J Med Genet, 39,* 876–881.

153. Davis, A. P., Witte, D. P., Hsieh-Li, H. M., et al. (1995). Absence of radius and ulna in mice lacking hoxa-11 and hoxd-11. *Nature, 375,* 791–795.

154. Fleischman, R., Letestu, R., Mi, X., et al. (2002). HoxA-10 is the predominant Hox gene expressed in human megakaryocytic cells but is normal in the thrombocytopenia absent radius syndrome. *Br J Haematol, 116,* 367–375.

155. Thompson, A. A., Woodruff, K., Feig, S. A., et al. (2001). Congenital thrombocytopenia and radio-ulnar synostosis: A new familial syndrome. *Br J Haematol, 113,* 866–870.

156. Ballmaier, M., Schulze, H., Strauss, G., et al. (1997). Thrombopoietin in patients with congenital thrombocytopenia and absent radii: Elevated serum levels, normal receptor expression, but defective reactivity to thrombopoietin. *Blood, 90,* 612–619.

157. al-Jefri, A. H., Dror, Y., Bussel, J. B., et al. (2000). Thrombocytopenia with absent radii: Frequency of marrow megakaryocyte progenitors, proliferative characteristics, and megakaryocyte growth and development factor responsiveness. *Pediatr Hematol Oncol, 17,* 299–306.

158. Letestu, R., Vitrat, N., Masse, A., et al. (2000). Existence of a differentiation blockage at the stage of a megakaryocyte precursor in the thrombocytopenia and absent radii (TAR) syndrome. *Blood, 95,* 1633–1641.

159. Thompson, A. A., & Nguyen, L. T. (2000). Amegakaryocytic thrombocytopenia and radio-ulnar synostosis are associated with *HOXA11* mutation. *Nature Genetics, 26,* 397–398.

160. Freedman, M. H., & Doyle, J. J. (1999). Inherited bone marrow failure syndromes. In J. S. Lilleyman, I. M. Hann, V. S. Blanchette (Eds.), *Pediatric hematology* (pp. 23–49). London: Churchill Livingstone.

161. Lackner, A., Basu, O., Bierings, M., et al. (2000). Haematopoietic stem cell transplantation for amegakaryocytic thrombocytopenia. *Br J Haematol, 109,* 773–775.

162. Ihara, K., Ishii, E., Eguchi, M., et al. (1999). Identification of mutations in the c-mpl gene in congenital amegakaryocytic thrombocytopenia. *Proc Natl Acad Sci USA, 96,* 3132–3136.

163. van den Oudenrijn, S., Bruin, M., Folman, C. C., et al. (2000). Mutations in the thrombopoietin receptor, Mpl, in children with congenital amegakaryocytic thrombocytopenia. *Br J Haematol, 110,* 441–448.

164. Ballmaier, M., Germeshausen, M., Schulze, H., et al. (2001). c-mpl mutations are the cause of congenital amegakaryocytic thrombocytopenia. *Blood, 97,* 139–146.

165. Tonelli, R., Scardovi, A. L., Pession, A., et al. (2000). Compound heterozygosity for two different amino-acid substitutions in the thrombopoietin receptor (c-mpl gene) in congenital amegakaryocytic thrombocytopenia (CAMT). *Hum Genet, 107,* 225–233.

166. Muraoka, K., Ishii, E., Tsuji, K., et al. (1997). Defective response to thrombopoietin and impaired expression of c-mpl mRNA of bone marrow cells in congenital amegakaryocytic thrombocytopenia. *Br J Haematol, 96,* 287–292.

167. Mukai, H. Y., Kojima, H., Todokoro, K., et al. (1996). Serum thrombopoietin (TPO) levels in patients with amegakaryocytic thrombocytopenia are much higher than those with

immune thrombocytopenic purpura. *Thromb Haemost, 76,* 675–678.

168. Freson, K., Devriendt, K., Matthijs, G., et al. (2001). Platelet characteristics in patients with X-linked macrothrombocytopenia because of a novel GATA1 mutation. *Blood, 98,* 85–92.

169. Mehaffey, M. G., Newton, A. L., Gandhi, M. J., et al. (2001). X-linked thrombocytopenia caused by a novel mutation of GATA-1. *Blood, 98,* 2681–2688.

170. Drachman, J. G., Jarvik, G. P., & Mehaffey, M. G. (2000). Autosomal dominant thrombocytopenia: Incomplete megakaryocyte differentiation and linkage to human chromosome 10. *Blood, 96,* 118–125.

171. Iolascon, A., Perrotta, S., Amendola, G., et al. (1999). Familial dominant thrombocytopenia: Clinical, biologic, and molecular studies. *Pediatr Res, 46,* 548–552.

172. Gandhi, M. H., Cummings, C. L., & Drachman, J. G. (2003). FLJ14813 missense mutation: A candidate for autosomal dominant thombocytopenia on chromosome 10. *Hum Hered, 55,* 66–70.

173. Noris, P., Spedini, P., Belletti, S., et al. (1998). Thrombocytopenia, giant platelets, and leukocyte inclusion bodies (May-Hegglin anomaly): Clinical and laboratory findings. *Am J Med, 104,* 355–360.

174. Takashima, T., Maeda, H., Koyanagi, T., et al. (1992). Prenatal diagnosis and obstetrical management of May-Hegglin anomaly: A case report. *Fetal Diagn Ther, 7,* 186–189.

175. Urato, A. C., & Repke, J. T. (1998). May-Hegglin anomaly: A case of vaginal delivery when both mother and fetus are affected. *Am J Obstet Gynecol, 179,* 260–261.

176. Kelley, M. J., Jawian, W., Ortel, T. L., et al. (2000). Mutation of MY H9, encoding non-muscle myosin heavy chain A, in May-Hegglin anomaly. *Nat Genet, 26,* 106–108.

177. Toren, A., Rozenfeld-Granot, G., Rocca, B., et al. (2000). Autosomal-dominant giant platelet syndromes: A hint of the same genetic defect as in Fechtner syndrome owing to a similar genetic linkage to chromosome 22q11–13. *Blood, 96,* 3447–3451.

178. Murray, N. A., Watts, T. L., & Roberts, I. A. G. (1999). Early hematological problems in preterm babies usually have their origin in fetal rather than post-natal life. *Blood, 94* (suppl 1) (2), 82b.

179. Teramo, K. A., Widness, J. A., Clemons, G. K., et al. (1987). Amniotic fluid erythropoietin correlates with umbilical plasma erythropoietin in normal and abnormal pregnancy. *Obstet Gynecol, 69,* 710–716.

180. Salvesen, D. R., Brudenell, J. M., Snijders, R. J., et al. (1993). Fetal plasma erythropoietin in pregnancies complicated by maternal diabetes mellitus. *Am J Obstet Gynecol, 168,* 88–94.

181. Gonen, C., Haznedaroglu, I. C., Aksu, S., et al. (2005). Endogenous thrombopoietin levels during the clinical management of acute myeloid leukaemia. *Platelets, 16,* 31–37.

182. Emmons, R. V., Reid, D. M., Cohen, R. L., et al. (1996). Human thrombopoietin levels are high when thrombocytopenia is due to megakaryocyte deficiency and low when due to increased platelet destruction. *Blood, 87,* 4068–4071.

183. Tahara, T., Usuki, K., Sato, H., et al. (1996). A sensitive sandwich ELISA for measuring thrombopoietin in human serum: Serum thrombopoietin levels in healthy volunteers and in patients with haemopoietic disorders. *Br J Haematol, 93,* 783–788.

184. Hou, M., Andersson, P. O., Stockelberg, D., et al. (1998). Plasma thrombopoietin levels in thrombocytopenic states: Implication for regulatory role of bone marrow megakaryocytes. *Br J Haematol, 101,* 420–424.

185. Nagata, Y., Shozaki, Y., Nagahisa, H., et al. (1997). Serum thrombopoietin level is not regulated by transcription but by the total counts of both megakaryocytes and platelets during thrombocytopenia and thrombocytosis. *Thromb Haemost, 77,* 808–814.

186. Burlina, A. B., Bonafe, L., & Zacchello, F. (1999). Clinical and biochemical approach to the neonate with a suspected inborn error of amino acid and organic acid metabolism. *Semin Perinatol, 23,* 162–173.

187. Gilbert-Barness, E., & Barness, L. A. (1999). Isovaleric acidemia with promyelocytic myeloproliferative syndrome. *Pediatr Dev Pathol, 2,* 286–291.

188. Roth, P., Sklower Brooks, S., Potaznik, D., et al. (2005). Neonatal Gaucher disease presenting as persistent thrombocytopenia. *J Perinatol, 25,* 356–358.

189. Debillon, T., Daoud, P., Durand, P., et al. (2003). Whole-body cooling after perinatal asphyxia: A pilot study in term neonates. *Dev Med Child Neurol, 45,* 17–23.

190. Eicher, D. J., Wagner, C. L., Katikaneni, L. P., et al. (2005). Moderate hypothermia in neonatal encephalopathy: Safety outcomes. *Pediatr Neurol, 32,* 18–24.

191. Hall, G. W. (2001). Kasabach-Merritt syndrome: Pathogenesis and management. *Br J Haematol, 112,* 851–862.

192. Haisley-Royster, C., Enjolas, O., Frieden, I. J., et al. (2002). Kasabach-Merritt phenomenon: A retrospective study of treatment with vincristine. *J Pediatr Hematol Oncol, 24,* 459–462.

193. Enjolras, O., Mulliken, J. B., Wassef, M., et al. (2000). Residual lesions after Kasabach-Merritt phenomenon in 41 patients. *J Am Acad Dermatol, 42,* 225–235.

194. Pampin, C., Devillers, A., Treguier, C., et al. (2000). Intratumoral consumption of indium-111-labeled platelets in a child with splenic hemangioma and thrombocytopenia. *J Pediatr Hematol Oncol, 22,* 256–258.

195. Seo, S. K., Suh, J. C., Na, G. Y., et al. (1999). Kasabach-Merritt syndrome: Identification of platelet trapping in a tufted angioma by immunohistochemistry technique using monoclonal antibody to CD61. *Pediatr Dermatol, 16,* 392–394.

196. Hosono, S., Ohno, T., Kimoto, H., et al. (1999). Successful transcutaneous arterial embolization of a giant hemangioma associated with high-output cardiac failure and Kasabach-Merritt syndrome in a neonate: A case report. *J Perinat Med, 27,* 399–403.

197. el-Dessouky, M., Azmy, A. F., Raine, P. A., et al. (1988). Kasabach-Merritt syndrome. *J Ped Surg, 23,* 109–111.

198. Jubinsky, P. T., Moraille, R., & Tsai, H. M. (2003). Thrombotic thrombocytopenic purpura in a newborn. *J Perinatol, 23,* 85–87.

199. Schiff, D. E., Roberts, W. D., Willert, J., et al. (2004). Thrombocytopenia and severe hyperbilirubinemia in the neonatal period secondary to congenital thrombotic thrombocytopenic purpura and ADAMTS13 deficiency. *J Pediatr Hematol Oncol, 26,* 535–538.

200. Schmugge, M., Dunn, M. S., Amankwah, K. S., et al. (2004). The activity of the von Willebrand factor cleaving protease ADAMTS-13 in newborn infants. *J Thromb Haemost, 2,* 228–233.

201. Berner, R., Krause, M. F., Gordjani, N., et al. (2002). Hemolytic uremic syndrome due to an altered factor H triggered by neonatal pertussis. *Pediatr Nephrol, 17,* 190–192.

202. Spadone, D., Clark, F., James, E., et al. (1992). Heparin-induced thrombocytopenia in the newborn. *J Vasc Surg, 15,* 306–311.

203. Ranze, O., Ranze, P., Magnani, H. N., et al. (1999). Heparin-induced thrombocytopenia in paediatric patients — a review of the literature and a new case treated with danaparoid sodium. *Eur J Pediatr, 158,* S130–133.

204. Frost, J., Mureebe, L., Russo, P., et al. (2005). *Pediatr Crit Care Med, 6,* 216–219.

205. Marks, S. D., Massocote, M. P., Steele, B. T., et al. (2005). Neonatal renal venous thrombosis: Clinical outcomes and prevalence of prothrombotic disorders. *J Pediatr, 146,* 811–816.

206. Andrew, M., Castle, V., Saigal, S., et al. (1987). Clinical impact of neonatal thrombocytopenia. *J Pediatr, 110,* 457–464.

207. Amato, M., Fauchere, J. C., & Herman, U., Jr. (1988). Coagulation abnormalities in low birth weight infants with peri-intraventricular hemorrhage. *Neuropediatrics, 19,* 154–157.

208. Van De Bor, M., Briet, E., Van Bel, F., et al. (1986). Hemostasis and periventricular-intraventricular hemorrhage of the newborn. *Am J Dis Child, 140,* 1131–1134.

209. Setzer, E. S., Webb, I. B., Wassenaar, J. W., et al. (1982). Platelet dysfunction and coagulopathy in intraventricular hemorrhage in the premature infant. *J Pediatr, 100,* 599–605.

210. Jhawar, B. S., Ranger, A., Steven, D., et al. (2003). Risk factors for intracranial hemorrhage among full-term infants: A case-control study. *Neurosurgery, 52,* 581–590.

211. Garcia, M. G., Duenas, E., Sola, M., et al. (2001). Epidemiologic and outcome studies of patients who received platelet transfusions in the neonatal intensive care unit. *J Perinatol, 21,* 415–420.

212. Hall, R. W., Kronberg, S. S., Barton, B. A., et al. (2005). Morphine, hypotension, and adverse outcomes among preterm neonates: Who's to blame? Secondary results from the NEOPAIN trial. *Pediatrics, 115,* 1351–1359.

213. Castro Conde, J. R., Martinez, E. D., Rodriguez, R. C., et al. (2005). CNS siderosis and dandy-walker variant after neonatal alloimmune thrombocytopenia. *Pediatr Neurol, 32,* 346–349.

214. Abel, M., Bona, M., Zawodniak, L., et al. (2003). Cervical spinal cord hemorrhage secondary to neonatal allimmune thrombocytopenia. *J Pediatr Hematol Oncol, 25,* 340–342.

215. Khan, D. J., Richardson, D. K., & Billett, H. H. (2003). Inter-NICU variation in rates and management of thrombocytopenia among very low birth-weight infants. *J Perinatol, 23,* 312–316.

216. Andrew, M., Vegh, P., Caco, V. C., et al. (1993). A randomized, controlled trial of platelet transfusions in thrombocytopenic premature infants. *J Pediatr, 123,* 285–291.

217. Sola, M. C., Del Vecchio, A., & Rimsza, L. M. (2000). Evaluation and treatment of thrombocytopenia in the neonatal intensive care unit. *Clin Perinatol, 27,* 655–679.

218. Watts, T. L., Roberts, I. A. G., & Murray, N. A. (1999). Thrombopoietin response to thrombocytopenia in preterm babies. *Early Hum Dev, 54,* 66–67.

219. Schmidt, B., & Andrew, M. (1995). Neonatal thrombosis: Report of a prospective Canadian and international registry. *Pediatrics, 96,* 939–943.

220. Castle, V., Coates, G., Kelton, J. G., et al. (1987). [111]In-oxine platelet survivals in thrombocytopenic infants. *Blood, 70,* 652–656.

221. Kahn, D. J., Richardson, D. K., & Billett, H. H. (2003). Inter-NICU variation in rates and management of thrombocytopenia among very low birth-weight infants. *J Perinatol, 23,* 312–316.

222. Del Vecchio, A., Sola, M. C., Theriaque, D. W., et al. (2001). Platelet transfusions in the neonatal intensive care unit: Factors predicting which patients will require multiple transfusions. *Transfusion, 41,* 803–808.

223. Garcia, M. G., Duenas, E., Sola, M. C., et al. (2001). Epidemiologic and outcome studies of patients who received platelet transfusions in the neonatal intensive care unit. *J Perinatol, 21,* 415–420.

224. Chakravorty, S., Murray, N., & Roberts, I. (2005). Neonatal thrombocytopenia. *Early Hum Dev, 81,* 35–41.

225. Roberts, I. A. G., & Murray, N. A. (1999). Management of thrombocytopenia in neonates. *Br J Haematol, 105,* 864–870.

226. Sola, M. C. (2004). Evaluation and treatment of severe and prolonged thrombocytopenia in neonates. *Clin Perinatol, 31,* 1–14.

227. Murray, N. A., & Roberts, I. A. G. (2004). Neonatal transfusion practice. *Arch Dis Child Fetal Neonatal Ed, 89,* F101–F107.

228. Calhoun, D. A., Christensen, R. D., Edstrom, C. S., et al. (2000). Consistent approaches to procedures and practices in neonatal hematology. *Clin Perinatol, 27,* 733–753.

229. Barton, B. E., Shortall, J., & Jackson, J. V. (1996). Interleukins 6 and 11 protect mice from mortality in a staphylococcal enterotoxin-induced toxic shock model. *Infect Immun, 64,* 714–718.

230. Chang, M., Williams, A., Ishizawa, L., et al. (1996). Endogenous interleukin-11 (IL-11) expression is increased and prophylactic use of exogenous IL-11 enhances platelet recovery and improves survival during thrombocytopenia associated with experimental group B sepsis in neonatal rats. *Blood Cells Mol Dis, 22,* 57–67.

231. Opal, S. M., Jhung, J. W., Keith, J. C. Jr., et al. (1998). Recombinant human interleukin-11 in experimental pseudo-

monas aeruginosa sepsis in immunocompromised animals. *J Infect Dis, 178,* 1205–1208.

232. Opal, S. M., Jhung, J. W., Keith, J. C., Jr., et al. (1999). Additive effects of human recombinant interleukin-11 and granulocyte colony-stimulating factor in experimental gram-negative sepsis. *Blood, 93,* 3467–3472.

233. Schwertschlag, U. S., Trepicchio, W. L., Dykstra, K. H., et al. (1999). Hemopoietic, immunomodulatory and epithelial effects of interleukin-11. *Leukemia, 13,* 1307–1315.

234. Ellis, M., Zwaan, F., Hedstrom, U., et al. (2003). Recombinant human interleukin 11 and bacterial infection in patients with [correction of] haematological malignant disease undergoing chemotherapy: A double-blind placebo-controlled randomised trial. *Lancet, 361,* 275–280.

235. Burns, K. H., & Werch, J. B. (2004). Bacterial contamination of platelet units: A case report and literature survey with review of upcoming American Association of Blood Banks requirements. *Arch Pathol Lab Med, 128,* 279–281.

236. Larsson, L. G., Welsh, V. J., & Ladd, D. J. (2000). Acute intravascular hemolysis secondary to out-of-group platelet transfusion. *Transfusion, 40,* 902–906.

237. Kluter, H., Bubel, S., Kirchner, H., et al. (1999). Febrile and allergic transfusion reactions after the transfusion of white cell-poor platelet preparations. *Transfusion, 39,* 1179–1184.

238. McCullough, J. (2000). Current issues with platelet transfusion in patients with cancer. *Semin Hematol, 37,* 3–10.

239. Strauss, R. G., Levy, G. J., Sotelo-Avila, C., et al. (1993). National survey of neonatal transfusion practices: II. Blood component therapy. *Pediatrics, 91,* 530–536.

240. Kaushansky, K. (1998). Thrombopoietin. *N Engl J Med, 339,* 746–754.

241. Sola, M. C., Juul, S. E., Meng, Y. G., et al. (1999). Thrombopoietin (TPO) in the fetus and neonate: TPO concentrations in preterm and term neonates, and organ distribution of TPO receptor (c-mpl) during human fetal development. *Early Hum Dev, 53,* 239–250.

242. Wolber, E.-M., Bame, C., Fahnenstich, H., et al. (1999). Expression of the thrombopoietin gene in human fetal and neonatal tissues. *Blood, 94,* 97–105.

243. Nomura, S., Ogami, K., Kawamura, K., et al. (1997). Cellular localisation of thrombopoietin mRNA in the liver by in situ hybridization. *Exp Hematol, 25,* 565–572.

244. Sola, M. C., Christensen, R. D., Hutson, A. D., et al. (2000). Pharmacokinetics, pharmacodynamics, and safety of administering pegylated recombinant megakaryocyte growth and development factor to newborn rhesus monkeys. *Pediatr Res, 47,* 208–214.

245. Walka, M. M., Sonntag, J., Dudenhausen, J. W., et al. (1999). Thrombopoietin concentration in umbilical cord blood of healthy term newborns is higher than in adult controls. *Biol Neonate, 75,* 54–58.

246. Paul, D. A., Leef, K. H., Taylor, S., et al. (2002). Thrombopoietin in preterm infants: Gestational age-dependent response. *J Pediatr Hematol Oncol, 24,* 304–309.

247. Tsao, P. N., Teng, R. J., Chou, H. C., et al. (2002). The thrombopoietin level in the cord blood in premature infants born to mothers with pregnancy-induced hypertension. *Biol Neonate, 82,* 217–221.

248. Porcelijn, L., Folman, C. C., de Haas, M., et al. (2002). Fetal and neonatal thrombopoietin levels in alloimmune thrombocytopenia. *Pediatr Res, 52,* 105–108.

249. Inagaki, K., Oda, T., Naka, Y., et al. (2004). Induction of megakaryocytopoiesis and thrombocytopoiesis by JTZ-132, a novel small molecule with thrombopoietin mimetic activities. *Blood, 104,* 58–64.

250. Cwirla, S. E., Balasubramanian, P., Duffin, D. J., et al. (1997). Peptide agonist of the thrombopoietin receptor as potent as the natural cytokine. *Science, 13,* 1696–1699.

251. Kimura, T., Kaburaki, H., Miyamoto, S., et al. (1997). Discovery of a novel thrombopoietin mimic agonist peptide. *J Biochem (Tokyo), 122,* 1046–1051.

252. Kimura, T., Kaburaki, H., Tsujino, T., et al. (1998). A nonpeptide compound which can mimic the effect of thrombopoietin via c-Mpl. *FEBS Lett, 29,* 250–254.

253. Dower, W. J., Cwirla, S. E., Balasubramanian, P., et al. (1998). Peptide agonists of the thrombopoietin receptor. *Stem Cells, 16* (suppl 2), 21–29.

254. Li, J., Yang, C., Xia, Y., et al. (2001). Thrombocytopenia caused by the development of antibodies to thrombopoietin. *Blood, 98,* 3241–3248.

255. de Serres, M., Ellis, B., Dillberger, J. E., et al. (1999). Immunogenicity of thrombopoietin mimetic peptide GW395058 in BALB/c mice and New Zealand white rabbits: Evaluation of the potential for thrombopoietin neutralizing antibody production in man. *Stem Cells, 17,* 203–209.

256. de Serres, M., Yeager, R. L., Dillberger, J. E., et al. (1999). Pharmacokinetics and hematological effects of the PEGylated thrombopoietin peptide mimetic GW395058 in rats and monkeys after intravenous or subcutaneous administration. *Stem Cells, 17,* 16–26.

257. Howarth, L. J., McCloy, M. P., Murray, N. A., et al. (2000). Thrombopoietin mimetic peptide (GW 395058) is a potent stimulant of human megakaryocyte colony production. *Blood, 96* (suppl 1), 676a.

258. Tepler, I., Elias, L., Smith, J. W., et al. (1996). A randomized placebo-controlled trial of recombinant human interleukin-11 in cancer patients with severe thrombocytopenia due to chemotherapy. *Blood, 87,* 3607–3614.

259. Isaacs, C., Robert, N. J., Bailey, F. A., et al. (1997). Randomized placebo-controlled study of recombinant human interleukin-11 to prevent chemotherapy-induced thrombocytopenia in patients with breast cancer receiving dose-intensive cyclophosphamide and doxorubicin. *J Clin Oncol, 15,* 3368–3377.

260. Cairo, M. S., Davenport, V., Bessmertny, O., et al. (2005). Phase I/II dose escalation study of recombinant human interleukin-11 following ifosfamide, carboplatin and etoposide in children, adolescents and young adults with solid tumours or lymphoma: A clinical, haematological and biological study. *Br J Haematol, 128,* 49–58.

261. McCloy, M. P., Howarth, L. J., Watts, T. L., et al. (2000). The role of IL-11 in neonatal thrombocytopenia. *Blood, 96* (suppl 1), 564a.

262. McCloy, M. P., Roberts, I. A., Howarth, L. J., et al. (2002). Interleukin-11 levels in healthy and thrombocytopenic neonates. *Pediatr Res, 51,* 756–760.

263. Fiore, N. F., Ledniczky, G., Liu, Q., et al. (1998). Comparison of interleukin-11 and epidermal growth factor on residual small intestine after massive small bowel resection. *J Pediatr Surg, 33,* 24–29.

264. Claud, E., Jackson, M., Jilling, T., et al. (2000). Interleukin-11 diminishes intestinal injury in a rat model of necrotizing enterocolitis. *Pediatr Res, 47,* 163A.

265. Peterson, R. L., Wang, L., Albert, L., et al. (1998). Molecular effects of recombinant human interleukin-11 in the HLA-B27 rat model of inflammatory bowel disease. *Lab Invest, 78,* 1503–1512.

# CHAPTER 53

# Alloimmune Thrombocytopenia

## Cecile Kaplan[1] and John Freedman[2]

[1]Platelet Immunology Department, I.N.T.S., Paris, France
[2]Toronto Platelet Immunobiology Group, St. Michael's Hospital, Toronto, Ontario, Canada

## I. Introduction

Alloimmune thrombocytopenias are disorders in which the platelet life span is shortened by alloantibodies elicited during the recipient's immune response against platelets from a genetically different individual. Two major clinical conditions are recognized: fetal and neonatal alloimmune thrombocytopenia and posttransfusion purpura. Since the first description of these conditions in the 1950s,[1,2] significant progress has been made in the understanding of the underlying mechanisms, the laboratory diagnosis, and the management of patients with alloimmune thrombocytopenia, but some questions remain unresolved. The first part of this chapter addresses fetal and neonatal alloimmune thrombocytopenia, followed by reviews of posttransfusion purpura and passive alloimmune thrombocytopenia (Table 53-1).

## II. Fetal and Neonatal Alloimmune Thrombocytopenia

Fetal and neonatal alloimmune thrombocytopenia (FNAIT), resulting from the maternal immune response, affects the fetus, an "innocent bystander." Fetal platelet destruction is mediated following transplacental passage of specific antiplatelet maternal alloantibodies.[1] Hence, the syndrome may be regarded as the platelet counterpart of hemolytic disease of the fetus and newborn (HDFN),[3] but, in contrast to HDFN, FNAIT frequently occurs during first pregnancies.[4] As most cases are first diagnosed at birth, the term *neonatal alloimmune thrombocytopenia (NAIT)* has been widely used. However, because it has been shown that the fetus may be affected, we prefer the term *FNAIT*. An alternative term, feto-maternal alloimmune thrombocytopenia, has also been employed.[5]

FNAIT is the most common cause of severe fetal thrombocytopenia[6] and of neonatal thrombocytopenia in maternity wards.[7] The most feared complication of FNAIT is the occurrence of intracranial hemorrhage (ICH), leading to death or neurological sequelae,[4,8] although with advances in diagnosis and management of pregnancies at risk, the prognosis of the affected infants has improved.

### A. Definition of Thrombocytopenia

The mean fetal platelet count is above $150 \times 10^9$/L by the end of the first trimester of pregnancy[9] and ranges from $241 \pm 45 \times 10^9$/L at 18 to 23 weeks of gestation to $265 \pm 59 \times 10^9$/L at 30 to 35 weeks of gestation.[10] Thrombocytopenia has therefore been defined in the fetus and the neonate as a platelet count less than $150 \times 10^9$/L, irrespective of the gestational age.

### B. Incidence of Fetal and Neonatal Alloimmune Thrombocytopenia

Neonatal thrombocytopenia, due to different causes (see Chapter 52), is not a rare event in intensive care units, being seen in one third of infants.[11] In maternity wards, a prospective study indicated that 0.9% of the neonates were thrombocytopenic and the thrombocytopenia was of immune origin in one third of cases.[7] The incidence of FNAIT has been estimated by large prospective studies to be about 1:800 to 1:1500 births in Caucasians.[12,13] However, because most cases have no clinical evidence of bleeding, they would be overlooked in the absence of routine screening to detect thrombocytopenia in the newborn or immunization in pregnant women.[7,14]

The rate of recurrence among subsequent platelet antigen-positive siblings is close to 100%, with a similar or more severe thrombocytopenia than in the previously affected infant.[15]

### C. Pathophysiology

The permissiveness of the maternal immune system toward the fetus as an allogeneic "graft" is governed by complex

**Table 53-1: Alloimmune Thrombocytopenias: Main Characteristics**

|  | FNAIT | PTP | Passive |
|---|---|---|---|
| Incidence | 1 : 800 to 1 : 1500 live births | ? | ? |
| Pathophysiology | Maternal immunization | Allo and auto(?) immunization | Passive transmission |
| Thrombocytopenia | Moderate to severe | Severe | Severe |
| Laboratory diagnosis | Maternal specific anti-HPA-alloantibody<br>Offending antigen present in the fetus/neonate | Anti-HPA alloantibody and autoantibody (?)<br>Offending antigen in the blood component | Alloantibody from the blood donor |
| First-line management | Transfusion of washed, compatible platelets | IVIG | ? |
| Prevention | Antenatal management for subsequent pregnancies | Transfusion with compatible blood products | Detection of immunized female donors? |

immunologic interactions that ultimately allow successful pregnancy in most cases. Contributing to these processes are fetal and maternal factors that define immunogenicity and immunoreactivity in the bidirectional maternal-fetal relationship. A breakdown of maternal immune permissiveness (rejection of the fetal allograft) can have devastating consequences for the fetus and result in pregnancy loss. Some fetal elements do not become immunogenic to the mother or vulnerable to maternal immune responses until pregnancy progresses. This may occur concomitantly with developmentally regulated changes in fetal antigen expression or may be due to a breakdown of barriers that normally separate the fetal and maternal circulations. Certain maternal immune responses elicited by fetal antigens, such as RhD antigens, do not become clinically significant until subsequent pregnancies, whereas others, such as FNAIT due to human platelet antigens (HPA) (Table 53-2), cause disease in the first pregnancy.

Although platelets express HLA class I and ABH blood group antigens on their surface,[16,17] FNAIT is mainly due to the transplacental passage of maternal alloantibodies directed against platelet-specific alloantigens. The pathogenic relevance of HLA antibodies in FNAIT remains controversial, because the HLA antigens are less expressed on platelets than on other cell types. Nonetheless, during investigation of FNAIT, some mothers are found to have anti-HLA antibodies, either alone or together with platelet-specific antibodies. When anti-HLA is found alone in cases of FNAIT, consideration must be given to the potential lack of sensitivity or inability of the platelet-specific antibody detection technique to recognize platelet-specific HPA antibody. Several prospective studies have failed to document a relationship between maternal HLA immunization and neonatal thrombocytopenia, although the relationship has been implied in individual cases[18] and has been suggested to be

**Table 53-2: Human Platelet Antigens (HPA)**

| System | Antigen | Old Nomenclature | Glycoprotein |
|---|---|---|---|
| HPA-1 | HPA-1a | Zw$^a$, Pl$^{A1}$ | GPIIIa |
|  | HPA-1b | Zw$^b$, Pl$^{A2}$ |  |
| HPA-2 | HPA-2a | Ko$^b$ | GPIbα |
|  | HPA-2b | Ko$^a$, Sib$^a$ |  |
| HPA-3 | HPA-3a | Bak$^a$, Lek$^a$ | GPIIb |
|  | HPA-3b | Bak$^b$ |  |
| HPA-4 | HPA-4a | Yuk$^b$, Pen$^a$ | GPIIIa |
|  | HPA-4b | Yuk$^a$, Pen$^b$ |  |
| HPA-5 | HPA-5a | Br$^b$, Zav$^b$ | GPIa |
|  | HPA-5b | Br$^a$, Zav$^a$, Hc$^a$ |  |
|  | HPA-6bw | Ca$^a$, Tu$^a$ | GPIIIa |
|  | HPA-7bw | Mo$^a$ | GPIIIa |
|  | HPA-8bw | Sr$^a$ | GPIIIa |
|  | HPA-9bw | Max$^a$ | GPIIb |
|  | HPA-10bw | La$^a$ | GPIIIa |
|  | HPA-11bw | Gro$^a$ | GPIIIa |
|  | HPA-12bw | ly$^a$ | GPIbβ |
|  | HPA-13bw | Sit$^a$ | GPIa |
|  | HPA-14bw | Oe$^a$ | GPIIIa |
| HPA-15 | HPA-15a | Gov$^b$ | CD109 |
|  | HPA-15b | Gov$^a$ |  |
|  | HPA-16bw | Duv$^a$ | GPIIIa |

a side effect of allogeneic leukocyte immunization in unexplained recurrent aborters.[19] Although blood group A and B antigens are expressed on platelets, there is no consensus on their possible role in FNAIT.

The transplacental transfer of maternal specific antiplatelet-IgG can occur from 14 weeks of pregnancy onwards and the fetal platelet alloantigens are fully expressed as early as 18 weeks of pregnancy.[20,21] Although maternal sensitization to platelet alloantigens has been detected as early as 6 weeks in a primigravida woman,[22] it more commonly occurs around 16 to 20 weeks of gestation,[12,22] and ICH has been documented before 20 weeks of gestation.[23]

It has been demonstrated that the MHC class I-related neonatal Fc receptor (FcRn) plays an important role in maternofetal IgG antibody transfer.[24] FcRn prevents IgG degradation during transcytosis (the vesicular transport of macromolecules from one side of a cell to the other) by binding to the CH2 and CH3 regions of IgG antibodies. IgG unbound to FcRn will be degraded. This mechanism is also hypothesized for maintenance of IgG homeostasis by FcRn.[25,26] It is currently unknown whether the different HPA alloantibodies are equally transferred to the fetus by FcRn.

Transplacentally transferred maternal alloantibody binds to "paternal" antigen on the fetal platelets via the Fab portion of the immunoglobulin, then bridges the opsonized platelet to macrophages via the antibody's Fc portion and the Fc receptor on the macrophage surface. The platelet is subsequently cleared in the reticuloendothelial system, mainly in the spleen.[27] The interaction of the Fc portion of platelet-bound antibody with different Fc receptors on the macrophage may either initiate (FcRIIIA) or prevent (FcRIIb) macrophage phagocytosis,[28,29] thus controlling platelet clearance.

The mechanisms leading to the fetal and neonatal thrombocytopenia include not only the increased sensitized-platelet destruction by the fetal reticuloendothelial system,[30] but probably also an impaired platelet production. Megakaryocytes express similar membrane glycoprotein antigens as do platelets, and the potential interference by the maternal alloantibodies with fetal platelet production[31] and with reduced megakaryocytic differentiation compared to adults[32] should be considered in the pathophysiology of FNAIT.

To date, 24 platelet-specific alloantigens have been described, 12 with a bi-allelic polymorphism. The human platelet antigen (HPA) nomenclature was adopted in 1990 to replace the personalized nomenclature that previously existed.[33] The molecular basis is known in 22 of these 24 antigens, and a single nucleotide polymorphism in the gene encoding the membrane protein has been found in 21 out of 22.[34] Frequencies of platelet antigens vary among different populations (see also Chapter 14).[35–39] In Caucasians, HPA-1a is by far the most common antigen implicated in FNAIT,[4] followed at much lower frequency by HPA-5b,[40] then HPA-3.[41] In contrast, in Asians, FNAIT is essentially linked with HPA-4 and HPA-5b.[42] FNAIT has been reported involving rare or private antigens, which have been discovered by performing a crossmatch analysis between the maternal serum and the paternal platelets in antigen capture assays.[34,43–47] Three studies have shown that these low-frequency antigens are not restricted to single families; they

must therefore not be ignored in the screening of obvious cases of FNAIT with a negative initial laboratory investigation.[48–50] Relevance of the anti-HPA-15a and -15b alloantibodies in FNAIT requires further evaluation.[51] Isoantibodies occurring in mothers deficient for platelet CD36 (3–5% of Asian or African ancestry[42]) or deficient for the platelet glycoprotein (GP) IIb-IIIa complex (as observed in Glanzmann thrombasthenia [see Chapter 57]) also may be involved in FNAIT.[52,53]

Initiation of the humoral immune response to a foreign antigen is a complex biological association of many cell types and their secreted products (reviewed in refs.[54,55]). This response occurs when an antigen, such as a cell surface glycoprotein, first interacts with an antigen presenting cell (APC). APCs are MHC class II-positive macrophages or dendritic cells and, sometimes, B cells. After internalization by the APC, the antigen is processed by proteolytic degradation into smaller antigenic fragments. Antigen processing within the APC is critical for generating peptide determinants, which can be loaded and bound within the antigen-binding grooves of either MHC class I or II molecules. The antigenic peptides are then transported to the cell membrane of the APC, where they are reexpressed in conjunction with MHC-encoded class II molecules for presentation to antigen-specific T helper (Th) lymphocytes. When the MHC-peptide complexes are recognized with sufficient affinity by T cell receptors (TcR) on a CD4 Th cell, antigen-specific signal 1 is met, initiating coordinated molecular events in both the APC and T-cell that culminate as signal 2 (costimulation) and full T-cell activation (e.g., CD40 upregulation of B7 molecules on the APC's surface for interaction with CD28 molecules on the T-cell). These T-cell/APC events can stimulate antigen-primed B cells to differentiate into plasma cells and secrete antigen-specific IgG antibodies. Thus, a defect in, or abnormal stimulation of, antigen-specific Th cells can significantly alter the course of an immune response. This underscores the importance of Th cells in determining whether antibodies will be generated against a foreign antigen and, once generated, regulating the response by stimulating events such as antibody affinity maturation, regulatory T cells, or secretion of soluble cytokines. The immune response against HPA-1a in an HPA-1b mother occurs because the mother is not tolerant to the missing HPA-1a antigen. Therefore, as fetal platelets "leak" into the mother's circulation during pregnancy, they are processed by the mother's APC and the processed HPA-1a-derived peptide is presented to the mother's circulating T cells. Support for this notion comes from murine models of leukoreduced platelet alloimmunization, in which it was shown that platelets generate anti-MHC alloantibodies via indirect allorecognition, where recipient CD4 Th cells recognize donor platelet-derived alloantigens on recipient APCs.[56] The T cells, in turn, become activated and help in the production of anti-HPA-1a alloantibodies.

A genetic background for maternal alloimmunization has been investigated, as not all women with incompatible fetuses will develop antiplatelet alloantibodies. Fetal and neonatal alloimmune thrombocytopenia induced by HPA-1a is characterized by generation of alloantibodies by a mother who is homozygous for the HPA-1b alloantigen and who is commonly HLA-DR3*0101 or DQB1*0201.[57,58] The positive predictive value for developing alloimmunization when these phenotypes are present is, however, only 35%, and their usefulness for screening is therefore limited.[13] With regard to immunization against HPA-1b, no association with known HLA class II molecules has been found.[59] The immune response to HPA-5b antigen has also been shown to be HLA associated,[60] but studies of maternal immunization due to other HPA specificities are too small to allow conclusions in this regard.

Whereas FNAIT is generally considered to be antibody mediated, the linkage with HLA class II restriction suggests a role for T cells.[61] In studies examining T cell-mediated responses that regulate humoral reactions in alloimmune responses, it has been found that the polymorphic residue 33 of integrin β3 (GPIIIa, on which the HPA-1 alleles are found) is responsible for both a B-cell and a T-cell response.[62] Leu at position 33 (HPA-1a) controls the epitopes recognized by alloantibodies generated in individuals homozygous for Pro at position 33 (HPA-1b). Certain alloresponses can be viewed as entirely foreign because the polymorphism generates a functional anchor residue in a peptide that would otherwise not bind to MHC and a unidirectional alloantibody response occurs as the lack of binding of the Pro 33 peptide would preclude T-cell help.[62] This could explain the observation that alloantibodies to the Pro 33 allele (HPA-1b) are very rare.

Mutagenesis experiments have shown that the polymorphism accounting for the HPA-1a and -1b phenotypes is sufficient for inducing alloantibody recognition.[63] The same polymorphism as is seen for the HPA-1a and -1b allotypes (Leu and Pro, respectively, at amino acid 33 of the β chain of platelet integrin αIIbβ3 [GPIIb-IIIa]) may control both the B-cell epitope and constitute the MHC-bound peptide. Using defined peptides for T-cell stimulation, T-cell lineages specifically responsive to the HPA polymorphism responsible for HPA-1a alloimmunization in FNAIT have been identified.[64] Interestingly, the clonotypes constituted more of the circulating repertoire close to the time of delivery (when stimulation of maternal T-cells by fetal platelets might be expected to be higher) and decreased postpartum. The relevant T-cells may be providing the HLA-DR restricted help for generation of antiplatelet antibodies associated with FNAIT. A better understanding of the interactions between these cell-mediated mechanisms may be important for developing future antigen-specific immunotherapies to treat FNAIT.

As observed in prospective studies,[12,13,65] it is important to recognize that not all potentially significant antibodies observed in pregnancy will cause FNAIT and it is likely that different antibodies affect the fetal platelet and its destruction to differing degrees. *In vitro* experiments have identified two categories of anti-HPA-1a antibodies: those for which the binding requires only an intact amino-acid terminus and those for which binding depends on other structural requirements within the entire glycoprotein.[66] Moreover, studies with murine monoclonal antibodies have shown that recognition of the HPA-1a, HPA-3a, and HPA-9bw epitopes is not uniform in FNAIT,[50,67,68] and further study of this heterogenous behavior and its relevance to the clinical condition would be of interest.

### D. Clinical Description

**1. Fetal Thrombocytopenia.** As shown in a retrospective survey of 5194 fetal blood samplings,[6] alloimmune thrombocytopenia accounts for the most severe thrombocytopenia in the fetus, and no spontaneous correction of the fetal thrombocytopenia has been observed as gestation progresses.[69] When thrombocytopenia discovered incidentally by fetal blood sampling performed for another reason is more severe than usually observed in other causes, such as infections or chromosomal abnormalities, the diagnosis of FNAIT should be suspected.

During pregnancy, FNAIT should be suspected in a variety of circumstances. It could be among the causes for recurrent miscarriages, especially when these occur late in pregnancy. Unfortunately, FNAIT is more often discovered when its deleterious complication of ICH has occurred. The ICH may be diagnosed by sonography after the mother has observed a decrease of fetal activity or when a change in fetal heart rate alerts the clinician; ultrasonography and magnetic resonance imaging reveals ventriculomegaly, fetal hydrocephalus, or porencephalic cysts. Retrospective studies indicate that 80% of ICH occur *in utero* and 40% are diagnosed before 30 weeks gestation.[70] Fetal ICH has been mainly observed when the HPA-1a alloantigen is implicated, but FNAIT linked with HPA-3a is also severe and ICH has been reported with low-frequency platelet antigens. ICH appears to occur less frequently with FNAIT linked to HPA-5b.[40,70] When ICH is suspected, investigations for FNAIT are warranted to afford better management for the current and subsequent pregnancies. ICH can be associated with poor outcome; deaths have been documented in 10% of affected infants and neurological sequelae in 20%.[4,70] Unexplained fetal anemia or hydrops fetalis have also been reported in this setting.[71,72]

**2. Neonatal Thrombocytopenia.** Because there is no screening program for FNAIT, it is usually discovered incidentally when a full-term neonate born to a first time pregnant healthy mother has petechiae, purpura or, less frequently, overt visceral bleeding at birth or a few hours

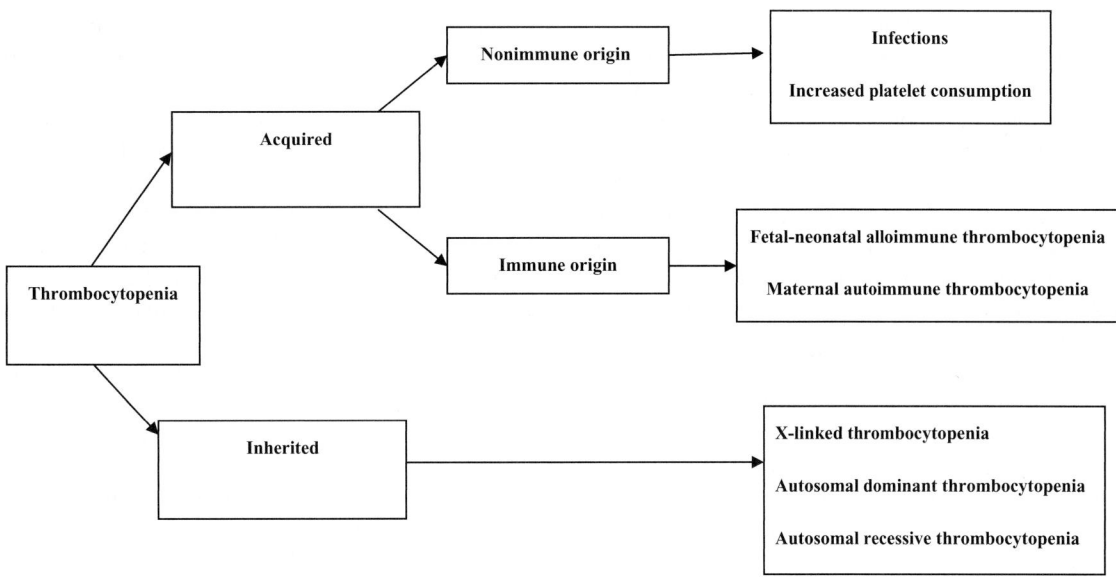

**Figure 53-1.** Major causes of thrombocytopenia in the fetus and newborn.

afterwards. ICH may be present at birth or can occur as long as the newborn is thrombocytopenic. The diagnosis of FNAIT is suspected when other causes of thrombocytopenia (Fig. 53-1) are excluded[73] (see Chapter 52). Careful examination of the neonate can provide important information. In FNAIT, the newborn has no evidence of infection, hepatosplenomegaly, disseminated intravascular coagulation, skeletal anomalies (e.g., the thrombocytopenia with absent radii [TAR] syndrome [see Chapter 54]), or dysmorphic features associated with chromosomal disorders (e.g., trisomy 13.18.21, Di George syndrome). Nonetheless, associations with other disorders do occur, especially with maternal autoimmune thrombocytopenia.[7] Cases of FNAIT have been reported with HDFN, infections, or chromosomal abnormalities, and alloimmune thrombocytopenia should be suspected when thrombocytopenia is more severe or prolonged than anticipated for the diagnosis.

The risk of life-threatening hemorrhage necessitates prompt diagnosis and effective therapy. On the other hand, the thrombocytopenia may be asymptomatic and unnoticed unless a routine platelet count is performed. Any unexpected or unexplained neonatal thrombocytopenia or severe early onset thrombocytopenia in both preterm and term babies should raise the possibility of FNAIT and guide investigations accordingly (see also Chapter 52).[73]

## E. Laboratory Diagnosis

The platelet count is low in the fetus or neonate, but normal in the mother. The thrombocytopenia is usually isolated, and anemia, if present, reflects bleeding. In one survey,[74] the mean platelet count at birth in 110 neonates with FNAIT was $26 \times 10^9$/L, and a literature review indicated that in 69% of cases the platelet count at birth was less than $50 \times 10^9$/L, especially when linked with HPA-1a alloimmunization.[70] The identification of the likely cause of the thrombocytopenia enables appropriate management of the index case and of future pregnancies. In severely affected infants with a presumptive diagnosis of FNAIT, because of the risk of ICH, it is not necessary to wait for the laboratory diagnosis before therapy, especially when there are difficulties in establishing the immune origin of thrombocytopenia.

The aim of the laboratory testing is to confirm the alloimmune origin of the thrombocytopenia by detecting circulating maternal alloantibody directed against a paternal HPA present in the infant (and which is absent in the mother). Detection of the causative alloantibody is usually achieved by incubating the paternal platelets with the maternal serum and analyzing the antigen-antibody complex by a global technique such as flow cytometry. Antibody identification is done with an antigen capture assay such as the monoclonal antibody-specific immobilization of platelet antigens (MAIPA)[75] or modified antigen capture ELISA (MACE) techniques,[76] with a panel of typed group O donor platelets and the paternal platelets; this ensures the detection of alloantibody to private antigens and may be helpful in case of paternity exclusion. These techniques using murine monoclonal antibodies (MoAbs) allow identification of specific antibodies or a mixture of antibodies directed against HPA located on the platelet glycoproteins, as well as HLA antibodies. Due to the drawbacks of the techniques, such as steric hindrance between the human alloantibody and the MoAbs,[67] careful choice of the MoAbs is essential, and it is best to employ at least two MoAbs for each glycoprotein, although these may be difficult to obtain routinely. Publications and workshops have shown that some alloantibodies may be difficult to detect due to modification of the antigen during storage[68] or to variability of antigen expression.[51] Moreover, in some cases, "private" antigens are involved and reference platelets are not readily available. It has therefore been suggested that recombinant cell lines may be useful for diagnosis.[49] Platelet phenotyping may be performed with the same techniques, although, due to shortage of serological reagents and advances in molecular biology,[34] platelet antigen genotyping is now usually done in experienced laboratories; international workshops on platelet immunology have shown the accuracy of the different techniques.[77,78] Although rarely encountered, we have, during laboratory investigation for FNAIT, observed discrepancies between phenotyping and genotyping in a mother who was a Glanzmann thrombasthenia carrier.[79] This could lead to incorrect assignment of platelet type and emphasizes the need to resolve discrepant reactions. FNAIT has also been reported in pregnancies involving *in vitro* fertilization[80] and attention must be paid in this situation to the determination of the offending antigen.

The diagnosis of FNAIT is straightforward when the maternal serum reacts positively with the father's platelets in the absence of maternal autoantibodies and when the identified maternal alloantibody is directed against a paternal HPA antigen expressed by the fetus or neonate but which the mother lacks. However, specific situations are encountered when, despite parental antigenic incompatibility, the maternal alloantibody is not detectable or is detectable only with a specific technique or when rare or private antigens are tested. In those situations, in order to ascertain the diagnosis, it could be necessary to obtain new samples from the family in the postnatal period, usually up to 2 months, and to combine different techniques. It is important to establish the diagnosis in order to provide best management for the next pregnancy.

To enable appropriate counseling for subsequent pregnancies, the father's genotype should be determined wherever possible. If he is homozygous and particularly if the mother still has the relevant antiplatelet antibodies, the subsequent fetus is at risk. In the case of paternal heterozygosity for the offending antigen, or when paternity is uncertain, fetal genotyping can be done on chorionic villi or amniotic cells.[81] A noninvasive procedure, such as fetal platelet genotyping from maternal serum, as performed for RhD status,[82] would be very advantageous, but there are no large-scale studies reported on the use of this approach.

## F. Management

**1. Postnatal Management.** Severely affected infants with bleeding or with a platelet count less than $30 \times 10^9$/L

during the first 24 hours of life should be promptly transfused with platelets, which will not be destroyed by the maternal alloantibody. Therefore the mother is the best donor. However, the maternal platelets must be washed to eliminate the antibody and irradiated to prevent graft versus host disease. As it could be difficult to rapidly organize a maternal platelet donation, alternatives such as provision of homozygous HPA-1b and -5a frozen-thawed platelet concentrates (for the most commonly implicated alloantibodies), which could be stored up to 3 years, have been proposed and this approach has been developed in France.[83] The National Blood Service in England has also developed a registry of such donors,[84] with good effectiveness.[85] The development and maintenance of such a registry is, however, a huge task.

Due to logistic difficulties in obtaining phenotyped platelets in emergency situations, transfusion of random platelet concentrates combined with intravenous immunoglobulin (IVIG; 0.8 gm/kg body weight ×2 days) has been proposed. Although the results are variable, this approach may overcome the emergency situation.[70,86] Despite some limited evidence of success, IVIG (with or without steroids) is likely not indicated as sole therapy, because the response is delayed up to 18 hours after injection, during which time the infant is at risk of bleeding.[87]

If the baby has no bleeding and has a platelet count above $30 \times 10^9$/L, no therapy is usually required. In that situation, as for after transfusion, it is important to monitor the platelet count to ensure that it does not, with time, fall to a level requiring therapy. The clinical outcome in the absence of bleeding is generally favorable and in most cases a normal platelet count is obtained within a week of life.

Ultrasound examination or magnetic resonance imaging is recommended to exclude ICH. When ICH is present, the mortality rate has been estimated in retrospective studies to be 10% of cases and neurological sequelae occur in 20% of cases,[4,8,70] although prospective studies report a lower rate of ICH.

**2. Antenatal Management of High-Risk Pregnancies.** The rationale for antenatal therapy is the high rate of recurrence for the subsequent incompatible fetuses, with usually a more severe condition. A search of the literature undertaken for the estimation of ICH in a subsequent untreated pregnancy found that 48% of infants with ICH had previous siblings with thrombocytopenia but without ICH.[88] When, however, there was a history of ICH in a previous sibling, the risk of ICH was 72% (79% when intrauterine fetal deaths were included in the analysis).[88]

The first attempts to avoid ICH were *in utero* platelet transfusions in the preterm period to reach a safe fetal platelet count (above $50 \times 10^9$/L), allowing vaginal delivery, or cesarean section when the safe threshold was not obtained.[69] However, this management could not prevent the deleterious consequences of severe thrombocytopenia during pregnancy. Therefore, different protocols have been developed, either maternal therapy with IVIG with or without corticosteroids, direct fetal injection of IVIG, or repeated *in utero* platelet transfusion.[89-92]

*a. Rationale for the Use of IVIG.* The mechanism of action of IVIG in ameliorating immune thrombocytopenia remains incompletely understood.[93] Even in autoimmune thrombocytopenia (ITP), there are at least 10 competing theories that could account for its action. The leading hypotheses, supported by experimental evidence, are (a) reticuloendothelial Fc receptor blockade,[94,95] (b) antiidiotypic antibodies,[96,97] and (c) inhibitory FcγRIIB,[98-100] but it is evident that these theories do not fully explain the mechanism of action.[93,101] They are likely not mutually exclusive. Of additional potential relevance to FNAIT in particular is the observation that administration of IVIG to the mother may be more effective than when given directly to the fetus.[102] IVIG can induce immune tolerance[93,101] and may decrease maternal antibody generation. IVIG may also be able to enhance maternal platelet alloantibody degradation; in a rodent model of ITP, high concentrations of plasma IVIG could "exhaust" FcRn and accelerate degradation of unbound IgG.[103] Furthermore, exhausted FcRn may reduce placental transportation of the maternal antibody.[26]

In contrast to its clear efficacy in autoimmune thrombocytopenias, IVIG is generally less effective with alloantibody mediated disease. Why this is so is unknown, but, interestingly, it has been shown that anti-GPIb antibodies induce thrombocytopenia in an Fc-independent manner, whereas anti-GPIIb-IIIa antibodies elicit thrombocytopenia by Fc-dependent clearance of antibody-coated platelets.[104] The HPA phenotypes reflect differences in amino acid sequences on the major platelet membrane glycoproteins. Whereas the major target antigen in FNAIT is platelet GPIIb-IIIa (at least 10 HPAs have been located on this integrin), GPIbα may also be a target (e.g., HPA-2), as may GPIa (HPA-5 and -13) (Table 53-2). The relationship of these observations to the pathophysiology of fetal and neonatal alloimmune thrombocytopenia, or to the effectiveness of IVIG in these disorders, has not as yet been investigated.

*b. The Fetal Status.* The main difficulties in antenatal management are the evaluation of the fetal status, identification of those fetuses requiring therapy, and the likely response to therapy. At present, there is no maternal parameter predictive of the fetal status. Discrepant results have been reported concerning the significance of the maternal alloantibody titration during pregnancy, and it is evident that some antibodies do not cause fetal thrombocytopenia.[12,13,22,69,105] To identify fetuses at risk of ICH, the effect of anti-HPA-1a antibodies on vascular endothelial cells has been studied; in a model incubating human umbilical cord endothelial cells with platelet alloantibodies, the alloantibodies involved were not predictive of endothelial cell activation or damage.[106] Currently, the only way to assess the fetal status is to perform fetal blood sampling (FBS), but in a thrombocytopenic fetus the risk of serious adverse events from this technique *per se* is significant, up to 10% in some reports.[107]

*c. Toward a Consensus on Antenatal Management.* An international forum concerning prenatal management in this setting has highlighted the absence of standardization.[108] The consensus is that pregnant alloimmunized women should be followed in referral centers where they may be offered antenatal therapy and can receive information concerning the risks of such pregnancies. There is a general trend to minimize invasive procedures and the decision of whether to perform FBS may be influenced by additional factors such as the past history of a sibling, especially the occurrence of ICH.[109] In the absence of randomized controlled trials on the efficacy of different management protocols, the general tendency currently is to provide maternal therapy with IVIG, with or without corticosteroids, as the first line of management.[91,92] The dose and timing of such therapy depends on the past history of any siblings and on the fetal platelet counts when performed. New strategies take into account the estimated risk of the current pregnancy — that is, high risk when there has been a sibling with ICH during pregnancy, standard risk with a thrombocytopenic sibling who does not have ICH. Repeated *in utero* transfusions are only used as salvage therapy when maternal therapy has failed.[91,92] *In utero* transfusion may, however, expose the infant to the risks of the FBS and to the side effects of the platelet transfusion, such as prolonged bradycardia. Two studies of the long-term follow-up of children managed with maternal IVIG therapy during the fetal stage did not show deleterious consequences (e.g., in neurological development or in immunologic profiles).[110,111] Proposed regimens for antenatal management should refer to current guidelines or published protocols.

In general, vaginal delivery is preferred when the fetal platelet count is above $50 \times 10^9$/L. Delivery by cesarean section is proposed when the fetal status is unknown at the end of therapy, or when there is treatment failure with fetal platelet count below $50 \times 10^9$/L. In this situation, compatible platelets are prepared to be available for transfusion immediately after birth if necessary. After birth, close monitoring of the newborn is necessary and silent ICH must be searched for when there is no response to therapy and platelet count is low at birth.

In our practice, for a first pregnancy in a woman discovered to be potentially at risk of alloimmunization (e.g., positive testing in a sister of a woman who gave birth to an infant with FNAIT or at-risk phenotype discovered during prospective studies), the decision for antenatal management relies on the detection of alloantibodies.

### G. Antenatal Screening

Most cases of FNAIT are discovered after birth and, therefore, in the absence of an antenatal screening program, the first affected infant will be at risk of severe thrombocytopenia with its possible deleterious consequences. The reduction of neonatal death and disability is a public health issue and the impact of a systematic detection program for FNAIT has been examined. Until methods are found to predict which immunized women will have thrombocytopenic infants, maternal screening has a low sensitivity. It has been suggested that it is more cost-effective to screen neonates than primiparous women and that such a program is not more expensive than other programs in newborns.[12,112] Evaluating only thrombocytopenic newborns at birth does increase sensitivity of screening programs, but it does not allow for intervention in the index pregnancy.[12] Introduction of a screening program does, however, raise several questions that require addressing, including the absence of parameters indicative of severe disease, the availability of a uniform policy on investigations, and the optimal approach to antenatal therapy.[113,114]

### H. Conclusion

Although real progress has been achieved in the more accurate diagnosis and management of this disorder, FNAIT is still underdiagnosed and methods for prevention of the maternal immunization do not exist. Further research must focus on the mechanisms of the maternal sensitization; the identification of sensitive, reliable, and noninvasive predictive parameters for severity of the fetal disease and response to therapy and for the occurrence of ICH and disability; and the determination of optimal antenatal therapy. Such answers can only be found by collaborative and large-scale studies. A novel murine model of FNAIT[115] may help to explain the pathophysiology and management of the disorder in humans.

## III. Posttransfusion Purpura

Posttransfusion purpura (PTP), first described in the early 1960s,[2,116,117] is a rare but serious complication of blood transfusion. It is characterized by the sudden onset of thrombocytopenia with purpura in the 7 to 10 days following a blood transfusion. The syndrome typically occurs in middle-aged women with a past history of pregnancy. More than 250 cases have been reported, but the true incidence of PTP

is unknown, and the disorder is probably underdiagnosed or misdiagnosed much of the time. The reported incidence seems less than would be expected from the prevalence of the platelet-associated antigen frequencies, and the reason for this is unclear. In the United Kingdom's adverse event reporting system, Serious Hazards of Transfusion (SHOT), there have been fewer case reports since 1998 following the implementation of the leukoreduced blood components; 10 cases were reported annually until 1998–1999, but only one case for the 2002–2003 period.[118]

Patients with PTP are "sensitized" by either prior pregnancy or blood product transfusion. When reexposed to the offending platelet antigen, typically by a transfusion, there is an anamnestic rise in antibody titre, which leads to immune clearance of the transfused platelets. There is, as well, concomitant immune destruction of the endogenous (autologous) antigen-negative platelets.

### A. Clinical Presentation

Although predominantly female patients are affected, some male patients have also been reported; in a retrospective European survey, 99 of 104 patients with PTP were females.[119] Typically, a middle-aged woman with prior pregnancies exhibits widespread purpura 7 to 10 days after being transfused. Fever, chills, and bronchospasm may accompany the disorder. In severe cases, visceral organ or intracranial hemorrhages have been described, sometimes with a fatal outcome. Transfusions of whole blood, packed red blood cells, platelet concentrates, and even plasma infusion have been implicated in the development of PTP. The time interval between the sensitization event (i.e., pregnancy or prior blood transfusion) and the onset of PTP ranges from weeks to years. The duration of PTP and its outcome is influenced by therapy. Although there are reports of spontaneous recovery after 2 weeks, the severity of thrombocytopenia and the significant impact of the bleeding diathesis mandate therapy to correct the problem.

Most patients with PTP have had surgery, and a diagnosis to be considered in that setting is that of heparin-induced thrombocytopenia (HIT) (see Chapter 48).[120] In PTP, the platelet count is usually very low, below $10 \times 10^9$/L,[119] whereas in HIT the platelet count is usually higher. There are, however, reports of misdiagnoses of PTP as HIT, resulting in delayed diagnosis and therapy with deleterious consequences.[121,122] A case of PTP following liver transplantation has also been reported.[123]

### B. Pathophysiology

The pathophysiology of PTP, in which the patient's own platelets are destroyed concurrently with the transfused platelets, is still incompletely understood. An anamnestic response seems to be the precondition for the onset of PTP, the initial sensitizing event having been either a pregnancy or a prior transfusion. The recent transfusion event precipitating PTP stimulates a platelet-specific immune response. As in FNAIT, involvement of a genetic background has been suspected. The same genetic background has been reported for anti-HPA-1a PTP and anti-HPA-1a FNAIT; however, when the past history of female patients with PTP has been evaluated, despite their immunization, there is no report of these patients giving birth to a thrombocytopenic infant.

Although an alloantibody to a platelet antigen is usually found in patients with PTP, the mechanism of destruction of the patient's own antigen-negative platelets is unexplained. Several different hypotheses have been proposed:

- The innocent bystander mechanism, in which the immune complexes generated by the interaction of the alloantibody and the fragments of the transfused platelets bind to the autologous platelets, causing their destruction[124]
- An autoimmune response, in which the patient produces autoantibodies as well as alloantibodies[125]
- Conversion of the antigen-negative platelets to antigen-positive platelets by adsorption of soluble antigens present in the plasma of transfused stored blood
- Production of alloantibodies with pseudo-specificity

### C. Laboratory Diagnosis

The confirmation of a diagnosis of PTP relies on the identification of the alloantibody and absence of the offending antigen on the patient's own platelets. In the majority of cases, the HPA-1a antigen is implicated. However, anti-HPA antibodies of other specificities have been identified, as have mixtures of antibodies and even anti-CD36 isoantibodies.[126] The antibodies are usually identified by an antigen capture assay MAIPA[75] or MACE technique,[76] and patient genotyping is usually performed.

### D. Therapy and Prognosis

Because of the small number of patients with PTP, there are no clinical trials that have addressed management. Therapy is therefore empiric, based on case reports, one case series, and literature review. Corticosteroids do not shorten the period of thrombocytopenia. Infusion of IVIG, with or without corticosteroids, is considered to be first-line therapy for PTP, resulting in improved platelet counts within a few days;[127] a dose of 1 gm/kg for two days is reasonable, in keeping with dosages recommended for other immuno-

hematologic disorders, including ITP. Because transfusion with platelets lacking the offending antigen is usually ineffective and may aggravate the condition, their use is usually contraindicated.[119,128] Nonetheless, there are reports of successful response to antigen-negative platelet transfusion.[129] Removal of antibody by plasma exchange may also induce a response in PTP.

PTP may recur with subsequent transfusions, but its recurrence is highly variable. It is recommended that PTP patients be informed of the potential risks of future transfusion. Further transfusion of cellular products (especially platelets) should be avoided, but if transfusion is necessary, the patients should ideally be transfused with HPA-compatible blood products or receive autologous transfusion. Whereas frozen-thawed red cells may be considered if red cell transfusion is required, relapse of PTP has been observed in those circumstances.[130]

## IV. Passive Alloimmune Thrombocytopenia

Passive alloimmune thrombocytopenia in a transfusion recipient has been reported to result from the transmission of antiplatelet alloantibodies in a blood product. Packed red blood cells, as well as plasma infusions, have been implicated. In this condition, the patient develops a transient, often severe, acute thrombocytopenia, which may be associated with significant bleeding. The onset of the thrombocytopenia is within hours of the transfusion or infusion — a timing that is quite different from that seen in PTP. Laboratory investigations reveal that the donor has a platelet-specific alloantibody.[131–133]

As most antiplatelet alloimmunizations occur during pregnancy, the presence or absence of alloantibodies has been evaluated in female blood donors.[134] Alloantibodies were found in 2.5% of HPA-homozygous female blood donors who have been pregnant; however, in this study,[134] none of the transfusion recipients developed passive immune thrombocytopenia. Nonetheless, further studies are needed to evaluate the risk of transfusions from such donations.

## V. Unresolved Questions and Future Directions

For many years, innate immunity has been considered as a separate entity from the adaptive immune response and has been regarded to be of secondary importance in the hierarchy of immune functions. However, evidence is accumulating that the innate immune system may be the ultimate controller of adaptive responses; for example, cells of the innate immune system can significantly influence the type of adaptive immune response by controlling differentiation of naïve T-helper (Th0) lymphocytes into effector cells of a particular type

(Th1 or Th2). Experiments have suggested that transfused platelets first interact with and stimulate innate immune responses such as nitric oxide production, which eventually culminates in the production of IgG antidonor antibodies by the adaptive immune system.[135,136] Evidence suggests that platelets themselves may bridge the innate and adaptive immune systems by expressing immunostimulatory molecules such as CD40 ligand (CD154)[137,138] and help stimulate adaptive immunity. Natural killer (NK) cells also play a critical role in bridging innate and adaptive immunity through the modulation of cytokine networks but, despite progress in understanding the role of NK cells in immunoregulation, little is known about their function in IgG alloantibody production. Data suggest that antiplatelet adaptive immune responses are ultimately stimulated and regulated by signals from NK cells of the innate immune system.[139]

Much has been learned about the mechanisms involved in alloimmunization, but much remains to be elucidated. Whereas previous work has focused on the humoral aspects of antibody production, more recently there has been investigation on the role of the cellular immune system. Although the focus has been on platelet transfusion-induced anti-HLA alloimmunization, investigators are now turning to alloimmunization to the platelet-specific antigens, which will be important in understanding the mechanisms of FNAIT and development of immunospecific therapies for this disorder.

Further studies in FNAIT should address improvements in detection and management in order to reduce the mortality and morbidity of this condition. In alloimmune thrombocytopenia resulting from pregnancy, transfusion, or transplantation, there is a need for study of the similarities and differences in platelet autoantibodies and alloantibodies. For example, in FNAIT, ICH is more frequent and more severe than in ITP, although the degree of thrombocytopenia may be similar; why this is so when the alloantibodies are often, as in ITP, directed to platelet membrane GPIIb-IIIa deserves investigation. Other issues requiring further clarification include the role of anti-HLA antibodies in FNAIT, the role of molecular genotyping in confirming the diagnosis, particularly in cases without detectable antibodies, and, importantly, the appropriate management of FNAIT during pregnancy. The interesting syndrome of PTP, despite many years of recognition, remains incompletely elucidated pathogenetically and optimal treatment options need to be confirmed. Although high-dose IVIG is generally effective, some patients do relapse, and whether anti-D is effective in this syndrome (as it is in ITP) remains to be shown.

# References

1. Harrington, W. J., Sprague, C. C., & Minnich, V. et al. (1953). Immunologic mechanisms in idiopathic neonatal thrombocytopenic purpura. *Ann Intern Med, 38,* 433–469.
2. Shulman, N. R., Aster, R. H., Leitner, A., et al. (1960). A new syndrome of post-transfusion immunologic purpura. *J Clin Invest, 39,* 1028–1029.
3. Levine, P., Burnham, L., Katzin, E. M., et al. (1941). The role of iso-immunization in the pathogenesis of erythroblastosis fetalis. *Am J Obstet Gynecol, 42,* 925–927.
4. Mueller-Eckhardt, C., Kiefel, V., Grubert, A., et al. (1989). 348 cases of suspected neonatal alloimmune thrombocytopenia. *Lancet, i,* 363–366.
5. Kaplan, C., Forestier, F., Daffos, F., et al. (1996). Management of fetal and neonatal alloimmune thrombocytopenia. *Transfus Med Rev, 10,* 233–240.
6. Hohlfeld, P., Forestier, F., Kaplan, C., et al. (1994). Fetal thrombocytopenia: A retrospective survey of 5,194 fetal blood samplings. *Blood, 84,* 1851–1856.
7. Dreyfus, M., Kaplan, C., Verdy, E., et al. (1997). Frequency of immune thrombocytopenia in newborns: A prospective study. *Blood, 89,* 4402–4406.
8. Kaplan, C., Daffos, F., Forestier, F., et al. (1991). Current trends in neonatal alloimmune thrombocytopenia diagnosis and therapy. In C. Kaplan-Gouet, N. Schlegel, C. Salmon, & J. Mac Gregor (Eds.), *Platelet immunology: Fundamental and clinical aspects* (Vol. 206, pp. 267–278). Paris: John Libbey-Eurotex/ed.INSERM.
9. Pahal, G. S., Jauniaux, E., Kinnon, C., et al. (2000). Normal development of human fetal hematopoiesis between eight and seventeen weeks gestation. *Am J Obstet Gynecol, 183,* 1029–1034.
10. Forestier, F., Daffos, F., Galacteros, F., et al. (1986). Hematological values of 163 normal fetuses between 18 and 30 weeks of gestation. *Pediatr Res, 20,* 342–346.
11. Castle, V., Andrew, M., Kelton, J., et al. (1986). Frequency and mechanism of neonatal thrombocytopenia. *J Pediatr, 108,* 749–755.
12. Durand-Zaleski, I., Schlegel, N., Blum-Boisgard, C., et al. (1996). Screening primiparous women and newborns for fetal/neonatal alloimmune thrombocytopenia: A prospective comparison of effectiveness and costs. *Am J Perinatol, 13,* 423–431.
13. Williamson, L. M., Hackett, G., Rennie, J., et al. (1998). The natural history of fetomaternal alloimmunization to the platelet-specific antigen HPA-1a (PLA1, Zwa) as determined by antenatal screening. *Blood, 92,* 2280–2287.
14. Davoren, A., McParland, P., Barnes, C. A., et al. (2002). Neonatal alloimmune thrombocytopenia in the Irish population: A discrepancy between observed and expected cases. *J Clin Pathol, 55,* 289–292.
15. Bussel, J. B. (1988). Neonatal alloimmune thrombocytopenia (NAIT): A prospective case accumulation study [abstract]. *Pediatr Res, 23,* 337a.
16. Curtis, B. R., Edwards, J. T., Hessner, M. J., et al. (2000). Blood group A and B antigens are strongly expressed on platelets of some individuals. *Blood, 96,* 1574–1581.
17. Cooling, L. L. W., Kelly, K., Barton, J., et al. (2005). Determinants of ABH expression on human blood platelets. *Blood, 105,* 3356–3364.

18. Taaning, E. (2000). HLA antibodies and fetomaternal allo-immune thrombocytopenia: Myth or meaningful? *Transfus Med Rev, 14,* 275–280.

19. Tanaka, T., Umesaki, N., Nishio, J., et al. (2000). Neonatal thrombocytopenia induced by maternal anti-HLA antibodies: A potential side effect of allogenic leukocyte immunization for unexplained recurrent aborters. *J Reprod Immun, 46,* 51–57.

20. Kaplan, C., Patereau, C., Reznikoff-Etievant, M., F., et al. (1985). Antenatal PLA1 typing and detection of GP IIb-IIIa complex. *Br J Haematol, 60,* 586–588.

21. Gruel, Y., Boizard, B., Daffos, F., et al. (1986). Determination of platelet antigens and glycoproteins in the human fetus. *Blood, 68,* 488–492.

22. Jaegtvik, S., Husebekk, A., Aune, B., et al. (2000). Neonatal alloimmune thrombocytopenia due to anti HPA-1a antibodies: The level of maternal antibodies predicts the severity of thrombocytopenia in the newborn. *BJOG, 107,* 691–694.

23. Giovangrandi, Y., Daffos, F., Kaplan, C., et al. (1990). Very early intracranial haemorrhage in alloimmune thrombocytopenia. *Lancet, 336,* 310.

24. Firan, M., Bawdon, R., Radu, C., et al. (2001). The MHC class I-related receptor, FcRn, plays an essential role in the maternofetal transfer of gamma-globulin in humans. *Int Immunol, 13,* 993–1002.

25. Ghetie, V., & Ward, E. S. (1997). FcRn: The MHC class I-related receptor that is more than an IgG transporter. *Immunol Today, 18,* 592–598.

26. Yoshida, M., Claypool, S. M., Wagner, J. S., et al. (2004). Human neonatal Fc receptor mediates transport of IgG into luminal secretions for delivery of antigens to mucosal dendritic cells. *Immunity, 20,* 769–783.

27. Horsewood, P., & Kelton, J. G. (1993). Macrophage-mediated cell destruction. In G. Garratty (Ed.), *Immunology of transfusion medicine* (pp. 435–464). New York: Marcel Dekker.

28. Raretch, J. V., & Lanier, L. L. (2000). Immune inhibitory receptors. *Science, 290,* 84–89.

29. Lin, S. Y., & Kinet, J. P. (2001). Immunology: Giving inhibitory receptors a boost. *Science, 291,* 445–446.

30. Wiener, E., Abeyakoon, O., Benchetrit, G., et al. (2003). Anti-HPA-1a-mediated platelet phagocytosis by monocytes *in vitro* and its inhibition by Fc gamma receptor (FcγR) reactive reagents. *Eur J Haematol, 70,* 67–74.

31. Warwick, R. M., Vaughan, J., Murray, N., et al. (1994). *In vitro* culture of colony forming unit-megakaryocyte (CFU-MK) in fetal alloimmune thrombocytopenia. *Br J Haematol, 88,* 874–877.

32. Miyazaki, R., Ogata, H., Iguchi, T., et al. (2000). Comparative analyses of megakaryocytes derived from cord blood and bone marrow. *Br J Haematol, 108,* 602–609.

33. von dem Borne, A. E. G. K. R., Kaplan, C., & Minchinton, R. (1995). Nomenclature of human platelet alloantigens. *Blood, 85,* 1409–1410.

34. Metcalfe, P., Watkins, N. A., Ouwehand, W. H., et al. (2003). Nomenclature of human platelet antigens. *Vox Sang, 85,* 240–245.

35. Simsek, S., Faber, N. M., Bleeker, P. M., et al. (1993). Determination of human platelet antigen frequencies in the Dutch population by immunophenotyping and DNA (allele-specific restriction enzyme) analysis. *Blood, 81,* 835–840.

36. Liu, T. C., Shih, M. C., Lin, C. L., et al. (2002). Gene frequencies of the HPA-1 to HPA-8w platelet antigen alleles in Taiwanese, Indonesian, and Thai. *Ann Hematol, 81,* 244–248.

37. Shih, M. C., Liu, T. C., Ling Lin, I., et al. (2003). Gene frequencies of the HPA-1 to HPA-13, Oe and Gov platelet antigen alleles in Taiwanese, Indonesian, Filipino and Thai populations. *Internat J Molec Med, 12,* 609–614.

38. Halle, L., Bach, K. H., Martageix, C., et al. (2004). Eleven human platelet systems studied in the Vietnamese and Ma'ohis Polynesian populations. *Tissue Antigens, 63,* 34–40.

39. Halle, L., Bigot, A., Mulen-Imandy, G., et al. (2005). HPA polymorphism in sub-Saharan African populations: Beninese, Cameroonians, Congolese, and Pygmies. *Tissue Antigens, 65,* 295–298.

40. Kaplan, C., Morel-Kopp, M. C., Kroll, H., et al. (1991). HPA-5b (Bra) neonatal alloimmune thrombocytopenia — Clinical and immunological analysis of 39 cases. *Br J Haematol, 78,* 425–429.

41. Glade-Bender, J., McFarland, J. G., Kaplan, C., et al. (2001). Anti-HPA-3a induces severe neonatal alloimmune thrombocytopenia. *J Pediatr, 138,* 862–867.

42. Ohto, H., Miura, S., Ariga, H., et al. (2004). The natural history of maternal immunization against foetal platelet alloantigens. *Transfus Med, 14,* 399–408.

43. Peyruchaud, O., Bourre, F., Morel-Kopp, M. C., et al. (1997). HPA-10w$^b$ (La$^a$): Genetic determination of a new platelet-specific alloantigen on glycoprotein IIIa and its expression in cos-7 cells. *Blood, 89,* 2422–2428.

44. Kroll, H., Kiefel, V., Santoso, S., et al. (1990). Sra, a private platelet antigen on glycoprotein-IIIa associated with neonatal alloimmune thrombocytopenia. *Blood, 76,* 2296–2302.

45. Berry, J. E., Murphy, C. M., Smith, G. A., et al. (2000). Detection of Gov system antibodies by MAIPA reveals an immunogenicity similar to the HPA-5 alloantigens. *Br J Haematol, 110,* 735–742.

46. Jallu, V., Meunier, M., Brement, M., & Kaplan, C. (2002). A new platelet polymorphism Duva$^+$, localized within the RGD binding domain of glycoprotein IIIa, is associated with neonatal thrombocytopenia. *Blood, 99,* 4449–4456.

47. Santoso, S., Kiefel, V., Richter, I. G., et al. (2002). A functional platelet fibrinogen receptor with a deletion in the cysteine-rich repeat region of the beta(3) integrin: The Oe(a) alloantigen in neonatal alloimmune thrombocytopenia. *Blood, 99,* 1205–1214.

48. Davoren, A., Curtis, B. R., Aster, R. H., et al. (2004). Human platelet antigen-specific alloantibodies implicated in 1162 cases of neonatal alloimmune thrombocytopenia. *Transfusion, 44,* 1220–1225.

49. Kroll, H., Yates, J., & Santoso, S. (2005). Immunization against a low-frequency human platelet alloantigen in fetal alloimmune thrombocytopenia is not a single event: Characterization by the combined use of reference DNA and novel

allele-specific cell lines expressing recombinant antigens. *Transfusion, 45,* 353–358.

50. Kaplan, C., Porcelijn, L., Vanlieferinghen, P. H., et al. (2005). Anti-HPA-9bw (Max[a]) feto-maternal alloimmunization, a clinically severe neonatal thrombocytopenia: Difficulties in diagnosis and therapy, report on 8 families. *Transfusion, 45,* 1799–1803.

51. Ertel, K., Al-Tawil, M., Santoso, S., et al. (2005). Relevance of the HPA-15 (Gov) polymorphism on CD109 in alloimmune thrombocytopenic syndromes. *Transfusion, 45,* 366–373.

52. Curtis, B. R., Ali, S., Glazier, A. M., et al. (2002). Isoimmunization against CD36 (glycoprotein IV): Description of four cases of neonatal isoimmune thrombocytopenia and brief review of the literature. *Transfusion, 42,* 1173–1179.

53. Léticée, N., Kaplan, C., & Lémery, D. (2005). Pregnancy in mother with Glanzmann's thrombasthenia and isoantibody against GPIIb–IIIa: Is there a foetal risk? *Eur J Obstet Gynecol Reproduc Biol, 121,* 139–142.

54. Semple, J. W., & Freedman, J. (2002). Recipient antigen-processing pathways of allogeneic platelet antigens: Essential mediators of immunity. *Transfusion, 42,* 958–961.

55. Semple, J. W., Bang, K. W. A., Speck, E. R., et al. (2000). Recipient mechanisms of alloimmunity: Antigen processing mediates IgG immunity against donor platelets. In C. Smit Sibinga & H. G. Klein (Eds.), *Molecular biology in blood transfusion* (pp. 135–145). Dordrecht, Netherlands: Kluwer Academic.

56. Semple, J. W., & Freedman, J. (1999). The basic immunology of platelet-induced alloimmunisation. In T. S. Kickler & J. H. Herman (Eds.), *Current Issues in Platelet Transfusion therapy and platelet alloimmunity* (pp. 77–101). Bethesda, MD: AABB Press.

57. Valentin, N., Vergracht, A., Bignon, J. D., et al. (1990). HLA DRw52a is involved in alloimmunization against PL-A1 antigen. *Hum Immunol, 27,* 73–79.

58. L'Abbé, D., Tremblay, L., Filion, M., et al. (1992). Alloimmunization to platelet antigen HPA-1a (Pl[A1]) is strongly associated with both HLA-DR3*0101 and HLA-DQB1*0201. *Hum Immunol, 34,* 107–114.

59. Kuijpers, R. W., von dem Borne, A. E., Kiefel, V., et al. (1992). Leucine33-proline33 substitution in human platelet glycoprotein IIIa determines HLA-DRw52a (Dw24) association of the immune response against HPA-1a (Zwa/PLA1) and HPA-1b (Zwb/PLA2). *Hum Immunol, 34,* 253–356.

60. Semana, G., Zazoun, T., Alizadeh, M., et al. (1996). Genetic susceptibility and anti-human platelet antigen 5b alloimmunization: Role of HLA class II and TAP genes. *Hum Immunol, 46,* 114–119.

61. Sayeh, E., Aslam, R., Speck, E. R., et al. (2004). Immune responsiveness against allogeneic platelet transfusions is determined by the recipient's major histocompatibility complex class II phenotype. *Transfusion, 44,* 1572–1578.

62. Wu, S., Maslanka, K., & Gorski, J. (1997). An integrin polymorphism that defines reactivity with alloantibodies generates an anchor for MHC class II peptide binding: A model for unidirectional alloimmune responses. *J Immunol, 158,* 3221–3226.

63. Goldberger, A., Kolodziej, M., Poncz, M., et al. (1991). Effect of single amino acid substitutions on the formation of PL alloantigenetic and Bakepitopes. *Blood, 78,* 681–687.

64. Maslanka, K., Yassai, M., & Gorski, J. (1996). Molecular identification of T cells that respond in a primary bulk culture to a peptide derived from a platelet glycoprotein implicated in neonatal alloimmune thrombocytopenia. *J Clin Invest, 98,* 1802–1808.

65. Panzer, S., Auerbach, L., Cechova, E., et al. (1995). Maternal alloimmunization against fetal platelet antigens: A prospective study. *Br J Haematol, 90,* 655–660.

66. Valentin, N., Visentin, G. P., & Newman, P. J. (1995). Involvement of the cystein-rich domain of glycoprotein IIIa in the expression of the human platelet alloantigen, Pl[A1]: Evidence for heterogeneity in the humoral response. *Blood, 85,* 3028–3033.

67. Morel-Kopp, M. C., Daviet, L., McGregor, J., et al. (1996). Drawbacks of the MAIPA technique in characterising human antiplatelet antibodies. *Blood Coagul Fibrinolysis, 7,* 144–146.

68. Harrison, C. R., Curtis, B. R., McFarland, J. G., et al. (2003). Severe neonatal alloimmune thrombocytopenia caused by antibodies to human platelet antigen 3a (Baka) detectable only in whole platelet assays. *Transfusion, 43,* 1398–1402.

69. Kaplan, C., Daffos, F., Forestier, F., et al. (1988). Management of alloimmune thrombocytopenia: Antenatal diagnosis and in utero transfusion of maternal platelets. *Blood, 72,* 340–343.

70. Spencer, J. A., & Burrows, R. F. (2001). Feto-maternal alloimmune thrombocytopenia: A literature review and statistical analysis. *Aust N Z J Obstet Gynaecol, 41,* 45–55.

71. Murphy, M. F., Waters, A. H., Doughty, H. A., et al. (1994). Antenatal management of fetomaternal alloimmune thrombocytopenia — report of 15 affected pregnancies. *Transfus Med, 4,* 281–292.

72. Stanworth, S. J., Hackett, G. A., & Williamson, L. M. (2001). Fetomaternal alloimmune thrombocytopenia presenting antenatally as hydrops fetalis. *Prenat Diagn, 21,* 423–424.

73. Chakravorty, S., Murray, N., & Roberts, I. (2005). Neonatal thrombocytopenia. *Early Hum Dev, 81,* 35–41.

74. Bussel, J. B., Zacharoulis, S., Kramer, K., et al. (2005). Clinical and diagnostic comparison of neonatal alloimmune thrombocytopenia to non-immune cases of thrombocytopenia. *Pediatr Blood Cancer, 45,* 176–183.

75. Kiefel, V., Santoso, S., Weisheit, M., et al. (1987). Monoclonal antibody-specific immobilization of platelet antigens (MAIPA): A new tool for the identification of platelet-reactive antibodies. *Blood, 70,* 1722–1726.

76. McMillan, R. (1990). Antigen-specific assays in immune thrombocytopenia. *Transfus Med Rev, 4,* 136–143.

77. Panzer, S., & Kaplan, C. (1998). On behalf of the platelet and granulocyte workshop ISBT: Report on the 1997 international society of blood transfusion workshop for genotyping of platelet alloantigens. *Transfus Med, 8,* 125–128.

78. Goldman, M., Trudel, E., & Richard, L. (2003). Report on the Eleventh International Society of Blood Transfusion Platelet Genotyping and Serology Workshop. *Vox Sang, 85,* 149–155.

79. Jallu, V., Bianchi, F., & Kaplan, C. (2005). Fetal-neonatal alloimmune thrombocytopenia and unexpected Glanzmann thrombasthenia carrier: Report of two cases. *Transfusion, 45,* 550–553.

80. Curtis, B. R., Bussel, J. B., Manco-Johnson, M. J., et al. (2005). Fetal and neonatal alloimmune thrombocytopenia in pregnancies involving *in vitro* fertilization: A report of four cases. *Am J Obstet Gynecol, 192,* 543–547.

81. Hurd, C., & Lucas, G. (2004). Human platelet antigen genotyping by PCR-SSP in neonatal/fetal alloimmune thrombocytopenia. In N. J. Goulden & C. G. Steward (Eds.), *Pediatric hematology: Methods and protocols* (Vol. 91, pp. 71–78). Totowa, NJ: Humana Press.

82. Gautier, E., Benachi, A., Giovangrandi, Y., et al. (2005). Fetal RhD genotyping by maternal serum analysis: A two-year experience. *Am J Obstet Gynecol, 192,* 666–669.

83. Lee, K., Beaujean, F., & Bierling, P. (2002). Treatment of severe fetomaternal alloimmune thrombocytopenia with compatible frozen-thawed platelet concentrates. *Br J Haematol, 117,* 482–483.

84. Ranasinghe, E., Walton, J. D., Hurd, C. M., et al. (2001). Provision of platelet support for fetuses and neonates affected by severe fetomaternal alloimmune thrombocytopenia. *Br J Haematol, 113,* 40–42.

85. Allen, D. L., Samol, J., Benjamin, S., et al. (2004). Survey of the use and clinical effectiveness of HPA-1a/5b-negative platelet concentrates in proven or suspected platelet alloimmunization. *Transfus Med, 14,* 409–417.

86. Win, N. (1996). Provision of Random-Donor Platelets (HPA-1a Positive) in neonatal alloimmune thrombocytopenia due to anti HPA-1a alloantibodies. *Vox Sang, 71,* 130–131.

87. Massey, G. V., McWilliams, N. B., Mueller, D. G., et al. (1987). Intravenous immunoglobulin in treatment of neonatal isoimmune thrombocytopenia. *J Pediatr, 111,* 133–135.

88. Radder, C. M., Brand, A., & Kanhai, H. H. (2001). A less invasive treatment strategy to prevent intracranial hemorrhage in fetal and neonatal alloimmune thrombocytopenia. *Am J Obstet Gynecol, 185,* 683–688.

89. Bussel, J. B. (2001). Alloimmune thrombocytopenia in the fetus and newborn. *Semin Thromb Hemost, 27,* 245–252.

90. Kaplan, C. (2002). Alloimmune thrombocytopenia of the fetus and the newborn. *Blood Rev, 16,* 69–72.

91. Birchall, J. E., Murphy, M. F., Kaplan, C., et al. (2003). On behalf of the European fetomaternal alloimmune thrombocytopenia study group: European collaborative study of the antenatal management of feto-maternal alloimmune thrombocytopenia. *Br J Haematol, 122,* 275–288.

92. Rayment, R., Brunskill, S. J., Stanworth, S., et al. (2005). Antenatal interventions for fetomaternal alloimmune thrombocytopenia. *Cochrane Database Syst Rev, 1,* CD004226.

93. Blanchette, V. S., Semple, J. W., & Freedman, J. (1996). Intravenous immunoglobulin and Rh immunoglobulin as immunomodulators of autoimmunity to blood elements. In L. E. Silberstein (Ed.), *Autoimmune disorders of blood* (pp. 35–77). Bethesda: AABB Press.

94. Salama, A., Mueller-Eckhardt, C., & Kiefel, V. (1983). Effect of intravenous immunoglobulin in immune thrombocytopenia: Competitive inhibition of reticuloendothelial system function by sequestration of autologous red blood cells? *Lancet, 2,* 193–195.

95. Crow, A., Song, S., Semple, J. W., et al. (2001). IVIg inhibits reticuloendothelial system function and ameliorates murine passive-immune thrombocytopenia independent of anti-idiotype reactivity. *Br J Haematol, 115,* 679–686.

96. Berchtold, P., Dale, G. L., Tani, P., et al. (1989). Inhibition of autoantibody binding to platelet glycoprotein IIb-IIIa by anti-idiotypic antibodies in intravenous gammaglobulin. *Blood, 74,* 2414–2417.

97. Rossi, F., & Kazatchkine, M. D. (1989). Anti idiotypes against autoantibodies in pooled normal human polyspecific Ig *J Immunol, 143,* 4104–4109.

98. McKenzie, S. E., Taylor, S. M., Malladi, P., et al. (1999). The role of the human Fc receptor Fc gamma RIIA in the immune clearance of platelets: A transgenic mouse model. *J Immunol, 162,* 4311–4318.

99. Samuelsson, A., Towers, T. L., Ravetch, J. V. (2001). Anti-inflammatory activity of IVIG mediated through the inhibitory Fc receptor. *Science, 291,* 484–486.

100. Ravetch, J. V., & Bolland, S. (2001). IgG Fc receptors. *Annu Rev Immunol, 19,* 275–290.

101. Lazarus, A. H., Freedman, J., & Semple, J. W. (1998). Intravenous immunoglobulin and anti-D in idiopathic thrombocytopenic purpura (ITP): Mechanisms of action. *Transfus Sci, 19,* 289–294.

102. Bowman, J., Harman, C., Mentigolou, S., et al. (1992). Intravenous fetal transfusion of immunoglobulin for alloimmune thrombocytopenia. *Lancet, 340,* 1034–1035.

103. Hansen, R. J., & Balthasar, J. P. (2002). Effects of intravenous immunoglobulin on platelet count and antiplatelet antibody disposition in a rat model of immune thrombocytopenia. *Blood, 100,* 2087–2093.

104. Nieswandt, B., Bergmeier, W., Rackebrandt, K., et al. (2000). Identification of critical anti-gen-specific mechanisms in the development of immune thrombocytopenic purpura in mice. *Blood, 96,* 2520–2527.

105. Kurz, M., Stockelle, E., Eichelberger, B., et al. (1999). IgG titer, subclass, and light-chain phenotype of pregnancy-induced HPA-5b antibodies that cause or do not cause neonatal alloimmune thrombocytopenia. *Transfusion, 39,* 379–382.

106. Radder, C. M., Beekhuizen, H., Kanhai, H. H., et al. (2004). Effect of maternal anti-HPA-1a antibodies and polyclonal IVIG on the activation status of vascular endothelial cells. *Clin Exp Immunol, 137,* 216–222.

107. Paidas, M. J., Berkowitz, R. L., Lynch, L., et al. (1995). Alloimmune thrombocytopenia: Fetal and neonatal losses related to cordocentesis. *Am J Obstet Gynecol, 172,* 475–479.

108. Engelfriet, C. P., Reesink, H. W., Kroll, H., et al. (2003). International forum: Prenatal management of alloimmune thrombocytopenia of the fetus. *Vox Sang, 84,* 142–149.

109. Radder, C. M., Brand, A., & Kanhai, H. H. (2003). Will it ever be possible to balance the risk of intracranial haemorrhage in fetal or neonatal alloimmune thrombocytopenia against the risk of treatment strategies to prevent it? *Vox Sang, 84,* 318–325.

110. Radder, C. M., de Hann, M. J., Brand, A., et al. (2004). Follow up of children after antenatal treatment for alloimmune thrombocytopenia. *Early Hum Dev, 80,* 65–76.

111. Radder, C. M., Roelen, D. L., van de Meers-Prins, E. M., et al. (2004). The immunologic profile of infants born after maternal immunoglobulin treatment and intrauterine platelet transfusions for fetal/neonatal alloimmune thrombocytopenia. *Am J Obstet Gynecol, 191,* 815–820.

112. Doughty, H. A., Murphy, M. F., Metcalfe, P., et al. (1995). Antenatal screening for fetal alloimmune thrombocytopenia: The results of a pilot study. *Br J Haematol, 90,* 321–325.

113. Murphy, M. F., & Williamson, L. M. (2000). Antenatal screening for fetomaternal alloimmune thrombocytopenia: An evaluation using the criteria of the UK national screening committee. *Br J Haematol, 111,* 726–732.

114. Murphy, M. F., Williamson, L. M., & Urbaniak, S. J. (2002). Antenatal screening for fetomaternal alloimmune thrombocytopenia: Should we be doing it? *Vox Sang, 83* (suppl. 1), 409–416.

115. Ni, H., Chen P., Spring C. M., et al. (2006). A novel murine model of fetal and neonatal alloimmune thrombocytopenia: response to intravenous IgG therapy. *Blood, 107,* 2976–2983.

116. Zucker, M. B., Ley, A. B., Borrelli, J., et al. (1959). Thrombocytopenia with a circulating platelet agglutinin, platelet lysin, and a clot retraction inhibitor. *Blood, 14,* 148–161.

117. van Loghem, J. J., Dorfmeijer, H., van Hart, M., et al. (1959). Serological and genetical studies on a platelet antigen (Zw). *Vox Sang, 4,* 161–169.

118. Serious Hazards of Transfusion. (2004). Annual report 2003: Post-transfusion Purpura, 65–66.

119. Mueller-Eckhardt, C., Kroll, H., Kiefel, V., et al. (1991). Post-transfusion purpura. In C. Kaplan-Gouet, N. Schlegel, C. Salmon, & J. McGregor (Eds.), *Clinical and fundamental aspects* (Vol. 206, pp. 249–255). Paris: John Libbey.

120. Warkentin, T. E. (2005). New approaches to the diagnosis of heparin-induced thrombocytopenia. *Chest, 127,* 35S–45S.

121. Lubenow, N., Eichler, P., Albrecht, D., et al. (2000). Very low platelet counts in post-transfusion purpura falsely diagnosed as heparin-induced thrombocytopenia: Report of four cases and review of literature. *Thromb Res, 100,* 115–125.

122. Araujo, F., Sa, J. J., Araujo, V., et al. (2000). Post-transfusion purpura vs. heparin-induced thrombocytopenia: Differential diagnosis in clinical practice. *Transfus Med, 10,* 323–324.

123. Wernet, D., Sessler, M., Dette, S., et al. (2003). Post-transfusion purpura following liver transplantation. *Vox Sang, 85,* 117–118.

124. Shulman, N. R., Aster, R. H., Leithner, A., et al. (1961). Immunoreactions involving platelets. V. Post-transfusion purpura due to a complement-fixing antibody against a genetically controlled platelet antigen: A proposed mechanism for thrombocytopenia and its relevance in "autoimmunity." *J Clin Invest, 40,* 1597–1620.

125. Watkins, N. A., Smethurst, P. A., Allen, D., et al. (2002). Platelet αIIbβ3 recombinant autoantibodies from the B-cell repertoire of a post-transfusion purpura patient. *Br J Haematol, 116,* 677–685.

126. McFarland, J. G. (2003). Detection and identification of platelet antibodies in clinical disorders. *Transfus Apher Sci, 28,* 297–305.

127. Mueller-Eckhardt, C., & Kiefel, V. (1988). High-dose IgG for post-transfusion purpura — revisited. *Blut, 57,* 163–167.

128. McFarland, J. G. (2001). Posttransfusion purpura. In M. A. Popovsky (Ed.), *Transfusion reactions* (pp. 187–212). Bethesda, MD: AABB Press.

129. Loren, A. W., & Abrams, C. S. (2004). Efficacy of HPA-1a(Pl$^{A1}$)-negative platelets in a patient with post-transfusion purpura. *Am J Hematol, 76,* 258–262.

130. Godeau, B., Fromont, P., Bettaieb, A., et al. (1991). Relapse of posttransfusion purpura after transfusion with frozen-thawed red cells. *Transfusion, 31,* 189–190.

131. Ballem, P. J., Buskard, N. A., Decary, F., et al. (1987). Post-transfusion purpura secondary to passive transfer of anti-P1A1 by blood transfusion. *Br J Haematol, 66,* 113–114.

132. Brunner-Bolliger, S., Kiefel, V., Horber, F. F., et al. (1997). Antibody studies in a patient with acute thrombocytopenia following infusion of plasma containing anti-(PlA1). *Am J Hematol, 56,* 119–121.

133. Warkentin, T. E., Smith, J. W., Hayward, C. P. M., et al. (1992). Thrombocytopenia caused by passive transfusion of anti-glycoprotein Ia/IIa alloantibody (anti-HPA-5b). *Blood, 79,* 2480–2484.

134. Boehlen, F., Bulla, O., Michel, M., et al. (2003). HPA-genotyping and antiplatelet antibodies in female blood donors. *Hematol J, 4,* 441–444.

135. Bang, A., Speck, E. R., Blanchette, V. S., et al. (1996). Recipient humoral immunity against leukoreduced allogeneic platelets is suppressed by aminoguanidine, a selective inhibitor of inducible nitric oxide synthase. *Blood, 88,* 2959–2966.

136. Bang, K. W. A., Speck, E. R., Blanchette, V. S., et al. (2000). Unique processing pathways within recipient antigen-presenting cells determine IgG immunity against donor platelet MHC antigens. *Blood, 95,* 1735–1742.

137. Henn, V., Slupsky, J. R., Grafe, M., et al. (1998). CD40 ligand on activated platelets triggers inflammatory reactions of endothelial cells. *Nature, 391,* 591–594.

138. Büchner, K., Henn, V., Gräfe, M., et al. (2003). CD40 ligand is selectively expressed on CD4$^+$ T cells and platelets: Implications for CD40-CD40L signaling in atherosclerosis. *J Pathol, 201,* 288–295.

139. Sayeh, E., Sterling, K., Speck, E., et al. (2004). IgG antiplatelet immunity is dependent on an early innate natural killer cell-derived interferon-γ response that is regulated by CD8+ T cells. *Blood, 103,* 2705–2709.

# CHAPTER 54

# Inherited Thrombocytopenias

**Michele P. Lambert and Mortimer Poncz**

*Department of Pediatrics, University of Pennsylvania School of Medicine, Philadelphia, Pennsylvania*

## I. Introduction

Thrombocytopenia is a fairly common hematological problem. However, rather than an inherited thrombocytopenia, a patient at any age is far more likely to have an acquired form of thrombocytopenia as a result of autoimmune disease, increased platelet consumption, bone marrow suppression, or bone marrow failure. For example, the incidence of acquired, immune thrombocytopenia purpura (ITP) is between 4 and 5 per 10,000 children per year[1,2] (see Chapter 46). Less than 5% of patients presenting with a chief complaint of thrombocytopenia have an inherited disorder. Two important clues to inherited thrombocytopenias are its chronicity/duration of symptoms and other associated symptoms or signs.

Symptoms of thrombocytopenia include superficial bruises, petechiae, and mucosal bleeding, particularly epistaxes, gastrointestinal hemorrhages, and bleeding with tooth brushing[3,4] (see Chapter 45). Intracranial hemorrhage and hemarthroses occur rarely. When an inherited thrombocytopenia is particularly severe or if the disorder is also associated with a qualitative defect in the platelets, the diagnosis may be made during the perinatal period (see Chapter 52) or in late infancy as the child begins to be mobile. Disorders that are milder may not be recognized until there is some hematologic stress (e.g., surgery, trauma, or childbirth). Some patients may have no or insignificant bleeding and may only be identified on a "routine" blood count or while being evaluated for another complication of their inherited condition. This chapter on the inherited thrombocytopenias has been organized to present the disorders based on platelet size (Table 54-1). Most modern electronic blood cell counters can accurately determine the size of platelets, and this beginning point in the diagnostic evaluation allows one to sort these disorders into three main categories, micro-, normo-, and macrothrombocytopenias.

## II. The Microthrombocytic Thrombocytopenias

### A. Wiskott–Aldrich Syndrome (WAS) and X-Linked Thrombocytopenia (XLT)

WAS and XLT are closely related, X-linked disorders characterized by mild to severe thrombocytopenia with significantly reduced platelet volumes (3.5 to 5 fL).[5] The combination of microthrombocytes with significant thrombocytopenia results in a substantial decrease in the total platelet mass or thrombocrit. Consequently, the risk of hemorrhage in these patients can be high. WAS patients also may have eczema and immunodeficiency, particularly an inability to make antipolysaccharide antibodies. This results in a susceptibility to infections, which continue to be a significant cause of mortality in this disorder.[6] Additionally, the immunodeficiency has been associated with autoimmune disease and malignancies of the lymphoreticular system (lymphomas). Not infrequently, part of the autoimmunity in these patients is ITP, which only worsens their degree of thrombocytopenia and makes therapeutic intervention problematic.

WAS is a rare disorder, occurring in 1 : 250,000 people of European descent. WAS and XLT result from a mutation in the WAS gene, located on the short arm of the X chromosome (Xp11.22).[7] The resulting protein, WAS protein (WASP), interacts with multiple other proteins and plays a role in signal transduction from cell surface receptors to the actin cytoskeleton through Rho family GTPases.[8,9] Mutations resulting in essentially no protein expression generally give the full WAS phenotype that includes eczema and immunodeficiency and is associated with a significantly shortened life span of affected individuals. However, single amino acid substitutions, particularly in exons 1 to 3, most commonly result in XLT with isolated thrombocytopenia.[8,9] The mechanism for the microthrombocytosis is complex as

**Table 54-1: Inherited Thrombocytopenias Classified by Platelet Size**

| Inherited Condition | Gene (Location) | Inheritance | Key Features |
|---|---|---|---|
| *Microthrombocytic* | | | |
| Wiskott–Aldrich Syndrome | WAS (Xp11) | X-linked | Thrombocytopenia, eczema, severe immunodeficiency, small platelets |
| X-linked thrombocytopenia | WAS (Xp11-exon2) | X-linked | Small platelets, thrombocytopenia, mild immunodeficiency |
| *Normothrombocytic* | | | |
| Familial platelet disorder with predisposition to AML | CBFA2 (21q22) | AD | Thrombocytopenia, myelodysplasia or even AML, platelet dysfunction |
| GATA-1 mutation of X-linked thrombocytopenia with thalassemia | GATA-1 or FOG-1 (Xp11.23) | X-linked | Thrombocytopenia with variable anemia |
| Paris–Trousseau/Jacobsen syndrome | Ets-1 or Fli-1 (11q23.3–24.2) | AR | Thrombocytopenia with large granules |
| Congenital amegakaryocytic thrombocytopenia | c-Mpl (1p34) | AR | Hypomegakaryocytic thrombocytopenia with eventual development of bone marrow failure |
| Amegakaryocytic thrombocytopenia with radio-ulnar synostosis | HOXA11 (7p15) | AD | Severe thrombocytopenia that improves with age, skeletal abnormalities (radio-ulnar synostosis, clinodactyly, syndactyly, hip dysplasia), hearing loss |
| Thrombocytopenia with absent radii | Unknown | AR | Thrombocytopenia that improves with age, limb anomalies (but normal thumbs) |
| Autosomal dominant thrombocytopenia with linkage to chromosome 10 | FLJ14813 (10p11–12) | AD | Mild to moderate thrombocytopenia with mild bleeding symptoms |
| *Macrothrombocytopenic* | | | |
| Bernard–Soulier syndrome | GPIbα (17), GPIbβ (22), GPIX (3) | AR | Platelet dysfunction with large platelets |
| Velocardiofacial syndrome | 22q11 | AD | Cardiac anomalies, cleft palate, hypocalcemia, thymic aplasia and typical facies. BSS-like thrombocytopenia +/− autoimmune |
| Benign mediterranean macrothrombocytopenia | GPIb/IX/V | AD | For some patients may represent heterozygous BSS mutations |
| Platelet-type von Willebrand disease | GPIbα (17) | AD | Decreased high molecular weight VWF multimers with thrombocytopenia because of increased platelet affinity for VWF |
| MYH9-related disease | NMMHC IIA (22q11.2) | AD | Large platelets, leukocyte inclusions, may have sensorineural hearing loss, cataracts, glomerulonephritis, or renal failure |
| Gray platelet syndrome | Unknown | AD | Large, pale platelets with absence of α granules |
| Montreal platelet syndrome | Unknown | AD | Spontaneous *in vitro* platelet agglutination with severe thrombocytopenia |
| Macrothrombocytopenia with platelet expression of glycophorin A | Unknown | AD | Mild bleeding with large platelets expressing surface glycophorin A |
| Macrothrombocytopenia with platelet β1 tubulin Q43P polymorphism | *TUBB1* 20q13.32 | AD | Spherocytic platelets and decreased cardiovascular disease in males |

Abbreviations: AD, autosomal dominant; AML, acute myelogenous leukemia; AR, autosomal recessive; BSS, Bernard-Soulier syndrome.; GP, glycoprotein; VWF, von Willebrand factor.

normal megakaryocytes development is seen *in vitro* using hematopoietic progenitors from WAS patients.[10] The reticuloendothelial system seems to be involved, as splenectomy often results in an increase in both platelet count and size.[5,11] Additionally, normal to increased numbers of megakaryocytes on the bone marrow biopsy of patients with WAS/XLT suggests increased peripheral destruction.[12–15] A proposed mechanism is increased destruction as a result of abnormal organization of the platelet cytoskeleton.[16]

The combination of immunodeficiency and thrombocytopenia in a male child should immediately raise a concern for WAS although, rarely, skewed X-inactivation can result in a female with WAS.[17] Presentation is generally within the first year of life, either because of bruising and bleeding or recurrent bacterial infections. Some neonates or very young infants may present with hematochezia, milk allergy, and thrombocytopenia. Life-threatening intracranial or gastrointestinal bleeding may occur, and approximately one quarter of patients die of bleeding complications.[18] In very young infants, platelet size may be near normal because of splenic immaturity. Clinical manifestations of the immunodeficiency may range from mild to severe and include susceptibility to infection (particularly with *Streptococcus pneumoniae*), eczema, autoimmune phenomena (including vasculitis, arthritis, inflammatory bowel disease, and ITP), and an approximately 10% cumulative risk of malignancy (generally lymphoma).[18,19]

Patients suspected of this disorder may be diagnosed by sequencing of the WAS gene or by flow cytometry of lymphocytes demonstrating absent or diminished expression of WASP.[20] Treatment is largely supportive: platelet transfusion or splenectomy (especially in the face of ITP) and monthly intravenous immunoglobulin infusions, as well as vigilance in monitoring for infection.[10,21] Treatment of eczema is also important as skin breakdown can provide a site of entry for bacteria.[21] Hematopoietic stem cell transplantation has also been used and is the only curative option to date,[22–24] although studies are ongoing using lentiviruses for directed gene therapy.[25] Median life expectancy has increased from less than 5 years (in the 1970s) to 10 to 20 years.[26,27]

## III. The Normothrombocytic Thrombocytopenias

There are multiple inherited thrombocytopenias that result in platelets that are normal in size.[3,4] These disorders are often associated with other features (such as skeletal abnormalities or a predisposition to malignancy), and the thrombocytopenia is often the result of disordered megakaryocytopoiesis or platelet release. A number of the disorders involve either mutations of hematopoietic transcription

factors or cytokines/cytokine receptors involved in normal megakaryocytopoiesis.

### A. Familial Platelet Disorder with Predisposition to Acute Myelogenous Leukemia (FPD/AML)

FPD/AML is a rare, autosomal dominant disorder characterized by mild thrombocytopenia with disproportionate bleeding often due to an aspirin-like platelet dysfunction and a predisposition to malignancy.[28] Platelet counts are generally between 80 to $150 \times 10^9$/L. Approximately half of these patients eventually develop AML (Fig. 54-1); one third develop solid tumors.[29]

This disorder has been shown in several pedigrees to be due to mutations in CBFA2 (chromosome 21q22.1–22.2), which is a part of the transcription regulation complex (CBFA2/CBFb) that is expressed in neural tissues, gonads, and thymus,[30] in hemangioblast stem cells,[31,32] and in multiple adult lineages, including the myeloid lineage and megakaryocyte-erythroid lineage.[33,34] Haploinsufficiency of this gene results in inappropriate levels of expression of downstream genes involved in megakaryocytopoiesis.[32] Bone marrow from patients with FPD/AML has been shown *in vitro* to have decreased numbers and size of megakaryocyte colonies. Platelet dysfunction is thought to be due to a decreased expression of platelet-specific genes that are regulated by (CBFA2/CBFb), perhaps related to decreased surface levels of c-mpl expression.[35] Additionally, CBFA2 (also called AML1) has been implicated in the development of myelogenous leukemia, particularly when part of a translocation (8;21 or 12;21).[36]

### B. GATA-1 Mutation of X-Linked Thrombocytopenia and Thalassemia

For many years, there had been descriptions of X-linked disorders characterized not only by thrombocytopenia but also by variable degrees of anemia or dyserythropoiesis.[37,38] Recently, the underlying defect in a number of these patients has been determined and appears to involve the hematopoietic-specific transcription factor GATA-1.[39–42] The GATA-1 gene encodes for a two zinc-finger protein that is an essential megakaryocyte- and erythroid-specific transcription factor.[43,44] Amino acid substitutions in or close to the N-terminal zinc-finger have been associated in these cases with either loss of DNA binding or loss of binding of GATA-1 to another hematopoietic transcription factor FOG-1 (Friend of GATA-1)[45,46] that is also pivotal for both megakaryocyte and erythrocyte differentiation.

These patients have defective megakaryocyte maturation and large misshapen megakaryocytes (Fig. 54-1) that proliferate abnormally in culture.[39] Platelet formation is

**A**                      **B**                      **C**

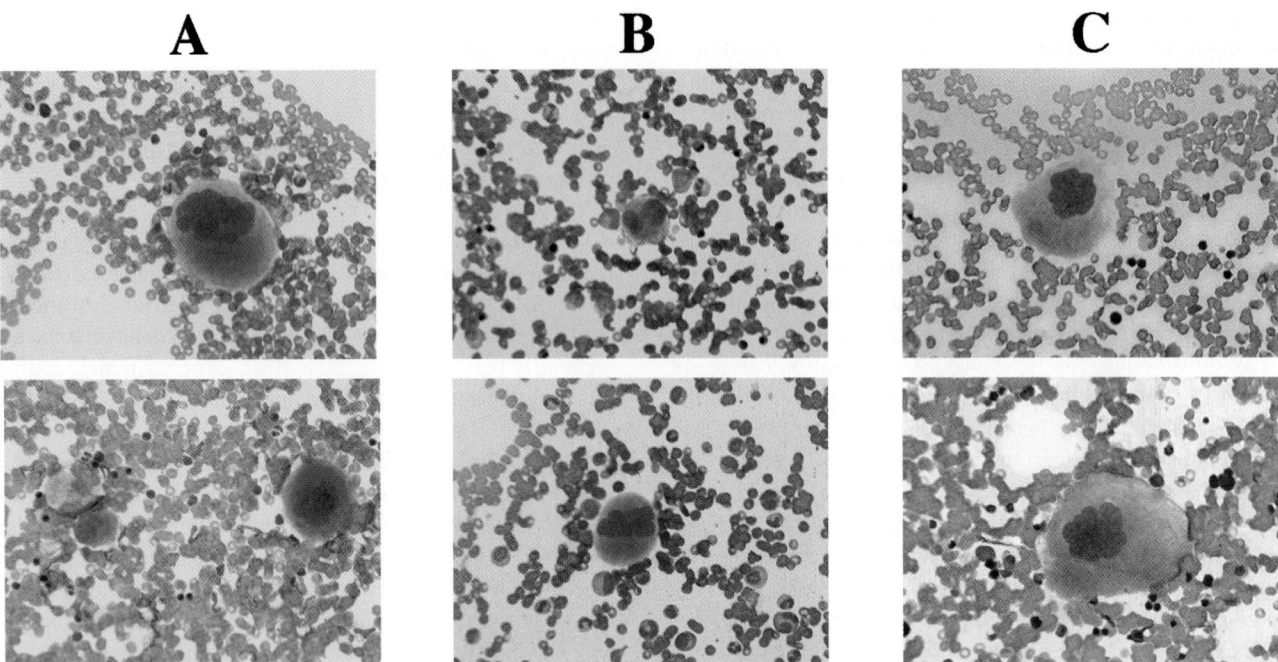

**Figure 54-1.**   Light microscopic analysis of megakaryocytes from patients with disorders of CBFA2 or GATA-1. Sample from A. normal marrow, B. a patient in remission from acute myelogenous leukemia with FPD-AML from a previously described family with a C→T nonsense mutation in Exon 5 of CBFA2,[32] C. an infant with a GATA-1$^{V205M}$ that was also previously described.[39] In the wild-type marrow, megakaryocytes were larger and approximately 5 to 10 times more frequent than morphologically recognizable megakaryocytes in the patients with the CBFA2 and the GATA-1 mutations. The patients with the FPD-AML had the smallest megakaryocytes. The GATA-1 megakaryocytes had nuclei that tended to be disproportionally small for the cytoplasmic size and more frequently set off to the side. Hematoxylin and eosin stain, original magnification ×500.

markedly curtailed in some of these patients, and the platelet count can be extremely low.[39] Normal platelet granular and membrane content is present, but in variable amounts and often somewhat disordered fashion. Additionally, anemia may be present with microcytosis due to a concurrent thalassemia. Some patients are transfusion dependent[37,39] and a few have undergone bone marrow transplantation.[39] Some of the patients also have an intrinsic platelet function defect.[47] Because this disorder can have variable thrombocytopenia and variable anemia, clinical diagnosis can be difficult. Patients have been misdiagnosed as having WAS/XLT and others have carried a misdiagnosis of thalassemia.[37]

### C. Paris–Trousseau Thrombocytopenia and the Jacobsen Syndrome

Paris–Trousseau thrombocytopenia and the Jacobsen syndrome are related rare disorders that result in thrombocytopenia with large granules that are the result of α granular fusion.[48,49] These disorders result from chromosomal deletions within breakpoints 11q23.3–24.2.[50] In Paris-Trousseau thrombocytopenia, the platelet defect is the predominant

syndrome. Paris-Trousseau thrombocytopenia requires only one defective chromosome as Fli-1 appears to undergo monoallelic expression.[51] In Jacobsen syndrome, there is a larger deletion, resulting in a more severe phenotype that also includes mental retardation, trigonocephaly, facial dysmorphism, and cardiac anomalies. These patients may also present with pancytopenia.

The underlying molecular defect responsible for the changes in platelets appears to be deletion of one or both of the Ets family members that map to chromosome 11, Fli-1, and Ets-1. Fli-1 has been shown to be critical for hematopoietic stem cell differentiation,[52] lymphoid differentiation,[53,54] and megakaryocyte maturation.[55,56] Fli-1 may be critical for final commitment of the megakaryocyte-erythroid progenitor cell down the megakaryocyte lineage.[57,58] Bone marrow from Fli-1 knockout embryos show either no recognizable megakaryocytes[59] or a dramatic decrease in megakaryocyte numbers,[60] and those that are present are not normal, demonstrating small size with a round nucleus. These micromegakaryocytes lyse at the end of maturation. There is significantly less expression of megakaryocyte-specific late genes including glycoprotein (GP) IX, though early megakaryocyte-specific genes such as GPIIb and c-mpl appear to be expressed normally.[60]

## D. Congenital Amegakaryocytic Thrombocytopenia (CAMT)

CAMT often presents with severe thrombocytopenia recognized very early in life because of the significant petechiae, bruising, and mucosal bleeding.[61] Classically, the platelet count is $<10 \times 10^9$/L at diagnosis. A large percentage of patients (5/24 in one series) present with intracranial hemorrhage. CAMT is inherited in an autosomal recessive manner, and therefore the parents have normal platelet counts. There is therefore seldom a family history for thrombocytopenia. If the children affected with CAMT survive long enough, they can develop pancytopenia during the first or second decade of life.[62]

On bone marrow examination, megakaryocytes are markedly decreased to absent.[63] CAMT often results from mutations in the thrombopoietin (TPO) receptor (c-mpl).[64] As a result, there is initially little to no differentiation of stem cells to megakaryocytes. However, TPO is also required for stem cell renewal, and the inability of hematopoietic cells to signal through c-mpl results in a gradual decline and eventual loss of stem cells resulting in a general bone marrow failure.

The differential diagnosis in infants who present with thrombocytopenia within the first few days of life includes neonatal alloimmune thrombocytopenia (NAIT) and maternal ITP (in which case the mother usually has thrombocytopenia) (see Chapters 51 and 52). In NAIT and maternal ITP, because the thrombocytopenia is antibody mediated, platelet counts gradually recover within the first few months of life. Treatment for CAMT in the first few months is supportive with platelet transfusion. Platelet survival should be normal initially. Once the diagnosis is definitively established, the only curative treatment for the thrombocytopenia and the pending pancytopenia is hematopoietic stem cell transplantation.[65,66]

## E. Amegakaryocytic Thrombocytopenia with Radio-Ulnar Synostosis (ATRUS)

ATRUS is an autosomal dominant disorder characterized by amegakaryocytic thrombocytopenia, skeletal anomalies (including radio-ulnar synostosis, hip dysplasia, clinodactyly, and syndactyly), and sensorineural hearing loss.[67] This disorder often presents initially with severe thrombocytopenia that tends to improve with time. The cause appears to be (in the majority of patients) a mutation in the Hox 11a gene,[68] a part of the homeobox group of regulatory proteins important in hematopoietic differentiation and proliferation. Studying animals in which the Hox 11a gene was selectively inactivated has revealed Hox 11a's role in skeletal development of the forearm,[69] but its role in hematopoiesis has not yet been fully clarified. However, bone marrow from Hox

11a-overexpressing mice show increased, immature megakaryocyte colonies,[70] and so lack of Hox 11a might result in decreased megakaryocytopoiesis. Treatment for ATRUS requires significant patience, because the majority of affected children will spontaneously recover their platelet counts sometime within the first years of life.

## F. Thrombocytopenia with Absent Radii (TAR)

TAR syndrome is usually inherited in an autosomal recessive manner.[71] There is a low incidence of siblings with TAR, suggesting that there is an increase in fetal demise.[72] Thrombocytopenia develops in the third trimester or early after birth and is symptomatic by 4 months of age in approximately 90% of patients. The thrombocytopenia is often initially severe ($<30 \times 10^9$/L), but it slowly improves with time so that platelet counts are nearing the normal range by 1 to 2 years of age.[73] There is a high incidence of serious bleeding in early infancy, especially gastrointestinal bleeding and more rarely intracranial hemorrhage.[74]

TAR patients are often easily identified. Besides having spontaneous petechiae and purpura, they often have obvious bilateral radial anomalies (Fig. 54-2). Abnormalities of the ulna, humerus, and tibia may also occur. However, abnormalities do not affect the fingers or hands. Other associated features in this disorder include a high incidence of milk protein allergy that may present as diarrhea and failure to thrive.[74] Additionally, there is a report of a 30% incidence of recurrent gastroenteritis that may complicate their thrombocytopenia-induced mucosal bleeding.[74] Other TAR patients have been reported with malformations of the genitourinary system including duplex ureter, mild renal pelvis dilatation and horseshoe kidneys, and cardiac anomalies with variants of atrioventriculoseptal defects.[75–78]

The molecular defect in TAR syndrome remains unknown. Serum levels of TPO are elevated in patients with TAR syndrome,[79,80] and bone marrow cultures from these patients have reduced megakaryocyte numbers,[81,82] whereas colony-forming units-megakaryocytes do not grow in response to TPO.[83] These findings together suggest a defect in the TPO/c-mpl signaling pathway downstream of the Mpl receptor, although patients with TAR have normal expression of mpl on megakaryocytes.[80] The platelets that form in TAR patients appear to be normal, although there has been at least one report of an associated storage pool defect.[84]

## G. Autosomal Dominant Thrombocytopenia with Linkage to Human Chromosome 10

This is a form of inherited mild to moderate thrombocytopenia with normal-sized platelets.[85] Clinically, these patients have mild bleeding symptoms that are generally restricted

**A**                                                    **B**

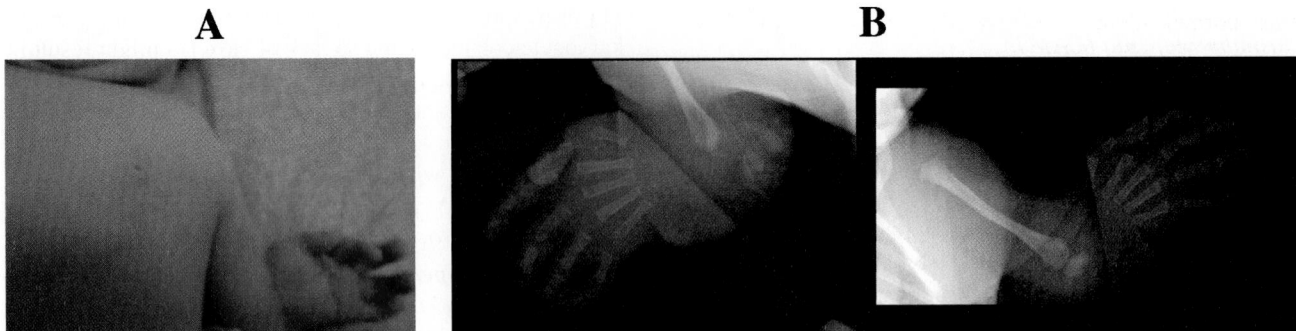

**Figure 54-2.** An infant with TAR syndrome. A. Photographic demonstration of diffuse petechiae on the chest wall at a platelet count of $35 \times 10^9$/L and a truncated left forearm with spared hand. B. X-rays of the arms of the same patient showing absence of radius and ulnar on the right and absence of the radius on the left.

to easy bruising, petechiae, and bleeding under hemostatic stress. Bone marrow examination reveals the presence of megakaryocytes that are small with hypolobulated nuclei, and growth assays reveal an increase in megakaryocyte progenitor cells.[86] This disorder has been shown by linkage analysis to be located on chromosome 10p11–12.[85] In one kindred, a missense mutation was reported in the gene FLJ14813, which is predicted to encode a kinase.[87]

## IV. The Macrothrombocytic Thrombocytopenias

Macrothrombocytosis can theoretically compensate for a decrease in platelet number so that the thrombocrit may remain constant and the patient does not have an associated bleeding diathesis. One possible mechanism that may underlie a number of these disorders is a defect in the megakaryocyte cytoskeleton, which prevents normal platelet release from megakaryocytic proplatelet extensions.[88] Because the TPO/mpl axis is intact in these patients, it would explain why the thrombocrit is normal in these patients. However, cytoskeletal defects can also affect platelet function, and this group of inherited thrombocytopenic disorders includes disorders that have associated platelet dysfunction in which the patients have a bleeding tendency in spite of a normal thrombocrit. Many of the macrothrombocytic thrombocytopenias discussed later have an associated qualitative defect often resulting in a mild to moderate bleeding diathesis.

A technical issue seen with macrothrombocytes is that electronic blood cell counters may underestimate the actual platelet count.[89] The recorded mean platelet volume may therefore be artificially low as well. Manual platelet counts are often needed to define the actual platelet count in many of these disorders.

### A. Bernard–Soulier Syndrome (BSS) (see also Chapter 57)

BSS was first described in 1948 by Jean Bernard and Jean-Pierre Soulier in a case report of a patient with a macro-thrombocytopenia and bleeding out of proportion to what was expected for the platelet count.[90] BSS is an autosomal recessive macrothrombocytopenia involving dysfunction in the von Willebrand factor (VWF) receptor, also known as the GPIb-IX-V complex.[91] Mutations of GPIbα, located on chromosome 17, are most commonly seen in BSS,[92] but mutations affecting GPIbβ, located on chromosome 22, and GPIX, on chromosome 3, also lead to BSS.[93–96] In murine studies, loss of GPV does not affect surface expression of the receptor, but appears to have a mild prothrombotic effect.[97,98] Most of the mutations involve either missense or nonsense mutations. Deletions occur rarely. Compound heterozygotes have been described.[99,100] The frequency of homozygous BSS has been reported to be approximately $1:10^6$,[101] so that approximately 1:500 individuals is heterozygous for BSS. Heterozygosity for a BSS mutation is associated with two other clinical states, the DiGeorge syndrome and benign Mediterranean macrothrombocytopenia, both of which will be discussed here.

The platelet GPIb-IX-V complex plays an important role in platelet activation by VWF[102] (see Chapter 7). Loss of one of the components of the complex, other than GPV, results in loss of expression of the components of the complex on the platelet surface.[103] Platelets then cannot bind well to VWF (although activated integrin αIIbβ3 also binds VWF[104]), as evidenced by lack of agglutination with ristocetin *in vitro*.[105] Additionally, the GPIbα subunit is also connected to the cytoskeleton via its C-terminal cytoplasmic tail to a network of actin filaments.[106] This may be part of the mechanism by which absence of this complex on the cell surface results in macrocytic platelets.[107] Patients with BSS have normal numbers of megakaryocytes in the bone marrow.[108] Peripheral destruction by the spleen of the abnormal platelets may play a role in the thrombocytopenia,[109] and the maintenance of a constant thrombocrit may also contribute to the mild thrombocytopenia.

Patients with BSS generally present in early childhood with easy bruisability, petechiae, epistaxes, and gastrointestinal bleeds.[101] In older females, menorrhagia can be problematic.[110,111] Joint bleeds and intracranial hemorrhages are rare but can occur.[91,112] The most severe bleeding occurs under hematologic stress (surgery, trauma, tooth extraction, delivery). It is important to note that even in patients with homozygous disease, the severity of the bleeding diathesis can vary from patient to patient (even within the same family), and symptom severity for any individual may lessen as the patient ages.

Platelet counts are generally mildly decreased (80–$100 \times 10^9$/L) but can vary considerably and even be normal.[91] Examination of the peripheral smear will invariably show marked macrocytosis of the platelets, even in patients with a normal or near normal platelet count. One third of platelets may be as large as half a red blood cell, and a few may even be as large as a lymphocyte.[112–114] Bleeding time is prolonged,[101] and platelet function studies reveal a defect in ristocetin-induced agglutination.[115] Patients heterozygous for a BSS mutation are generally asymptomatic (although there are patients with symptomatic heterozygous mutations[116]), but they may have large platelets and variable thrombocytopenia.[117]

Care of patients affected with BSS begins with good prophylaxis including good dental hygiene, education on how to locally control nosebleeds, and frequent follow-up for menstruating women. Platelet transfusion is the only consistently effective treatment for severe bleeding[101,114] but is reserved for situations with uncontrolled or life-threatening bleeds to avoid unnecessary exposure and possible loss of responsiveness to platelet transfusion via immunization. However, many BSS patients will require transfusion at some point in their lives. DDAVP therapy has been used effectively in BSS[118,119] (see Chapter 67). Activated Factor VIIa infusions (NovoSeven) have also been used[120,121] (see Chapter 68). Menorrhagia is responsive to hormonal regulation in most cases,[111] although there may occasionally be the need for supplementation with one of the preceding therapies.

## B. DiGeorge Syndrome (Velocardiofacial Syndrome (VCFS))

Hemizygous deletion in chromosome 22q11 results in a constellation of disorders now grouped together under the velo-cardiofacial syndrome.[123,124] This disorder is characterized by typical facies, cardiac anomalies, cleft palate, thymic

aplasia, neonatal hypocalcemia, learning disabilities, immune dysfunction that results in both immunodeficiency and autoimmune disorders, and thrombocytopenia. The thrombocytopenia in VCFS results from one of several mechanisms. More than half of the deletions of chromosome 22q11 result in loss of the GPIbβ gene and may result in a heterozygous BSS-like disorder.[125] Patients generally have only hemizygous loss of the allele, and therefore a significant bleeding disorder is unlikely. At least one patient has been described who had deletion of the gene on the VCFS chromosome and a point mutation in the promoter of the other chromosome.[126] However, VCFS is complicated by the occurrence of two other relatively common causes of thrombocytopenia: an ITP-like autoimmune thrombocytopenia and leukemic transformation.[127–129] The autoimmune thrombocytopenia is not necessarily associated with macro-thrombocytes, but it can exacerbate the benign and mild macrothrombocytopenia of the hemizygous loss of one functional GPIbβ gene and result in a bleeding diathesis. Usually the macrothrombocytopenia alone does not need treatment, whereas the other two forms require specific therapy.

## C. Benign Mediterranean Macrothrombocytopenia

This disorder was originally named Mediterranean macro-thrombocytopenia after a study by Behrens in 1975 that examined 145 "healthy" patients from both Italy and the Balkans in which the Italian patients were noted to have macrothrombocytopenia not seen in the Northern Europeans.[130] It once included those patients recently classified as being heterozygous for dysfunctional mutations in the GPIb-IX-V complex, now called heterozygous BSS. However, there are still a percentage of patients without defined molecular defects that present with mild thrombocytopenia with a variable percentage of large platelets.[131] Other names to describe this disorder include genetic thrombocytopenia with autosomal dominant transmission and chronic isolated macrothrombocytopenia with autosomal dominant transmission. This disorder is characterized by normal platelet survival, mild or no bleeding symptoms, and essentially normal megakaryocytes on bone marrow examination. However, one study showed that, at least in some patients, megakaryocyte colonies evidence abnormal apoptosis as well as abnormally increased proliferation.[132]

## D. Platelet-Type von Willebrand Disease

Described initially in 1982 by Weiss and colleagues, platelet-type von Willebrand disease (VWD), also called pseudo-VWD, is caused by specific mutations in the GPIbα subunit of the GPIb-IX-V complex (Gly$^{233Val}$, Gly$^{233Ser}$ or

Met[239Val], and a 27 bp deletion) that results in gain of function and increased affinity of this receptor for VWF without the necessity for shear stress or receptor or ligand activation.[133,134] As a result, high molecular weight multimers of VWF spontaneously bind to circulating platelets leading to intermittent thrombocytopenia. Perhaps because of rapid platelet turnover, the remaining platelets tend to be variably macrocytic. Platelets with VWF already bound to their surface are less likely to bind to sites of injury, resulting in a mild to moderate bleeding diathesis, although this disorder can also be also associated with prothrombotic episodes.[135,136] Platelet-type VWD is inherited in an autosomal dominant fashion.

Clinically, patients present with frequent and severe nosebleeds, menorrhagia, or excessive bleeding following tooth extraction, tonsillectomy, or other surgical procedures. Patients are unlikely to respond to infusion of FVIII : VWF and may even demonstrate a further drop in platelet count. There are reports of thrombosis in patients with the conceptually related disorder type 2B VWD who have been treated with FVIII:VWF.[137,138] Treatment of serious bleeding in platelet-type VWD generally requires platelet infusions.[139–141]

Laboratory studies in platelet-type VWD are similar to type 2B VWD, in which there is an increased affinity of VWF for GPIbα due to specific mutations of the A1 domain of VWF,[142] resulting in enhanced platelet aggregation in response to ristocetin, intermittent thrombocytopenia, and mildly decreased amounts of VWF with a disproportional decrease in high molecular weight multimers.[114,140] Spontaneous platelet aggregation may be seen. Differentiating platelet-type VWD from type 2B VWD may be difficult, but washing the platelets and adding normal plasma should result in normal platelet aggregation in type 2B VWD but will not change the defect in platelet-type VWD. Conversely, adding normal plasma to platelets from patients with platelet-type VWD results in spontaneous platelet agglutination but not in type 2B VWD.

### E. MYH9-Related Diseases (May-Hegglin Anomaly, Sebastian Syndrome, Fechtner Syndrome, and Epstein Syndrome) (see also Chapter 57)

What had once been separate, but overlapping rare syndromes, May–Hegglin anomaly, Sebastian syndrome, Fechtner syndrome, and Epstein syndrome are now grouped together because they have all been shown to involve mutations in the MYH9 gene that encodes for the nonmuscle myosin heavy chain IIA (myosin IIA).[143–145] Clinically, patients may present at birth with macrothrombocytopenia and spindle Döhle leukocyte inclusions (Fig. 54-3). In childhood or adult life, patients may develop sensorineural hearing loss, cataracts, and glomerulonephritis that may progress to severe renal failure.[146] In May-Hegglin anomaly or Sebastian

syndrome, patients classically have macrothrombocytopenia with leukocyte inclusions (the difference being only in subtle ultrastructural changes in the leukocyte inclusions). Patients with the other associated features were classified as Fechtner syndrome or Epstein syndrome based on the presence or absence of leukocyte inclusions. Thrombocytopenia is usually noted incidentally or on routine screening as patients with MHY9-related disease and does not usually cause a symptomatic bleeding diathesis, although a few patients may have severe bleeding.[147] Reports suggest there they may be a correlation between the particular mutation and phenotype: mutations of the C-terminal region appear to result in predominately hematologic manifestations, whereas mutations of the head ATPase domain are more commonly associated with nephropathy or hearing loss.[148]

On laboratory evaluation, patients present with variable thrombocytopenia with platelet counts ranging from 10 to $150 \times 10^9$/L.[147] However, macrothrombocytes are always present, but to a variable degree (Fig. 54-3). Platelets larger than a red cell can vary from 3% to 45% of all platelets.[143] Audiometric evaluation identified hearing loss in approximately 75% of patients identified with MHY9-related disease[143] and, interestingly, approximately 80% of patients initially diagnosed with May–Hegglin anomaly or Sebastian syndrome develop multiorgan involvement.[147] The same study also revealed microscopic hematuria and proteinuria even in patients initially identified as having only hematologic abnormalities.[147] Variable phenotypic expression in this disorder is seen in those with the same mutation and even within the same family.[114,149] Platelet survival is normal although bone marrow examination shows increased numbers of megakaryocytes.[150] In vitro platelet aggregation studies are usually normal, although shape change as a result of activation may be abnormal.[150,151]

### F. Gray Platelet Syndrome (see also Chapters 15 and 57)

Although there are more than 50 families identified with this disorder, the genetic defect has not yet been identified. The inheritance seems to be autosomal dominant in some families and autosomal recessive in other families, and sporadic mutations have also been described.[114,152] Gray platelet syndrome derives its name from the appearance of platelets on May–Grünwald–Giemsa staining.[153,154] Because the proteins that are normally within α granules are not packaged appropriately (resulting in empty granules), platelets appear gray and may be difficult to identify on blood smear. These proteins instead are redirected to the lumen of the demarcation membrane and are then secreted into the extracellular bone marrow space, resulting in mild bone marrow reticular fibrosis.[155] The resulting disruption of the normal marrow architecture is not usually enough to cause anemia. Because

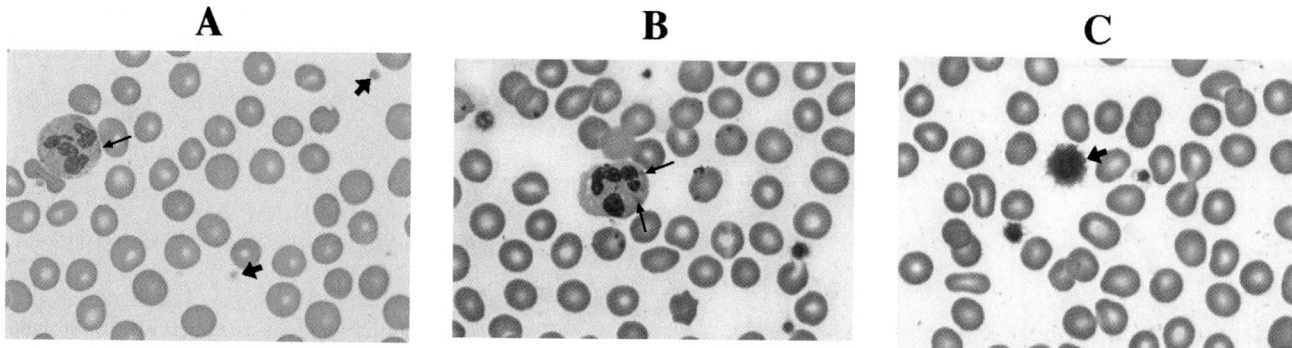

**Figure 54-3.** Peripheral blood smears from a patient with May–Hegglin anomaly. A. Normal blood smear with a leukocyte present (thin arrow) and platelets (thick arrows). B and C. Blood smears from a patient with MHY9 syndrome and a platelet count of $24 \times 10^9$/L and an MPV of 20.6 µm³. Thin arrows indicate blue spindle Döhle bodies in the leukocyte, the thick arrow points to a platelet larger than the adjacent red cell. Hematoxylin and eosin stain, original magnification ×500.

the α granules are empty, platelets do not release hemostatic proteins following platelet activation, resulting in a mild, but variable bleeding diathesis.[156] The cause of thrombocytopenia is unclear and may be partially related to peripheral destruction/shortened life span or defects in the cytoskeleton resulting in decreased platelet production.

Clinically, patients present with a prolonged bleeding time and bleeding that ranges from mild to severe (even with a normal platelet count). Thrombocytopenia may be mild to moderate. Macrothrombocytopenia is a typical feature, but some platelets are small or normal sized. Platelet aggregation studies may show an impaired response to thrombin. There is at least one case report of associated splenomegaly and extramedullary hematopoiesis.[157] Another patient has been reported with associated pulmonary fibrosis.[158]

### G. Montreal Platelet Syndrome

Montreal platelet syndrome is a very rare autosomal dominant disorder (described in two families).[114,159] Patients have severe thrombocytopenia (5 to $40 \times 10^9$/L) and platelet macrocytosis (median platelet diameter approximately 3 µm). The distinguishing feature is spontaneous *in vitro* platelet agglutination. The bleeding time is prolonged, and patients have excessive bruising and episodes of hemorrhage. A partial defect of calpain (calcium-activated neutral protease) has been described,[160] but the underlying genetic defect is unknown. This enzyme is involved in cleavage of cytoskeletal proteins.

### H. Macrothrombocytopenia with Platelet Expression of Glycophorin A

Hereditary macrothrombocytopenia with platelet expression of glycophorin A is an autosomal dominant disorder

described in one pedigree with a mild bleeding tendency.[161] Platelet counts were between 50 and $120 \times 10^9$/L, and platelet diameters were enlarged. Platelet morphology was otherwise normal with no defect in aggregation. The only known platelet abnormality is the surface presence of glycophorin A, which is normally present on the erythroid lineage, on multiple cell lines capable of differentiating into megakaryocytes, and is postulated to be present on immature megakaryocytes in normal bone marrow.[162] Whether this abnormality underlies the pathogenesis of this disorder or is secondary to abnormal megakaryocytopoiesis is unknown at present.

### I. Macrothrombocytopenia with Platelet β1 Tubulin Q43P Polymorphism

In a single report, a polymorphism in β1 tubulin (Q43P) has been described that results in spherocytic platelets.[163] Although this mutation occurs in 10.6% of the general population, it was found in 24.2% of 33 unrelated patients with congenital macrothrombocytopenia. This β1 tubulin polymorphism results in a single amino acid change and is associated with a decrease in platelet aggregation to thrombin. The authors[163] discovered that this β1 tubulin polymorphism is found infrequently in adult male patients hospitalized with cardiovascular disease. Further studies will need to define the full clinical implications of this mutation and its importance in patients with otherwise undefined macrothrombocytopenia.

### J. Other Macrothrombocytopenias

Some patients who present with macrothrombocytopenia have no currently identifiable explanation for their disorder. In the future, the additional discovery of polymorphisms (such as the emerging β1 tubulin story[163]) that may affect platelet formation, or that may enhance platelet turnover, could account for cases of macrothrombocytopenia. Characterization of the cause of macrothrombocytopenia may be helpful not only for understanding the patient's bleeding risk but also for the possible other-organ effects associated with the mutation.

# References

1. Zeller, B., et al. (2005). NOPHO ITP. Childhood idiopathic thrombocytopenic purpura in the Nordic countries: Epidemiology and predictors of chronic disease. *Acta Paediatr, 94,* 178–184.

2. Sutor, A., Harms, A., & Kaufmehl, K. (2001). Acute immune thrombocytopenia (ITP) in childhood: Retrospective and prospective survey in Germany. *Semin Thromb Hemost, 27,* 253–267.

3. Drachman, J. (2004). Inherited thrombocytopenia: When a low platelet count does not mean ITP. *Blood, 103,* 390–398.

4. Geddis, A. E., & Kaushansky, K. (2004). Inherited thrombocytopenias: Toward a molecular understanding of disorders of platelet production. *Curr Opin Pediatr, 16,* 15–22.

5. Corash, L., Shafer, B., & Blaese, R. (1985). Platelet-associated immunoglobulin, platelet size, and the effect of splenectomy in the Wiskott-Aldrich syndrom. *Blood, 65,* 1439–1443.

6. Imai, K., et al. (2004). Clinical course of patients with WASP gene mutations. *Blood, 103,* 456–464.

7. Ochs, H. (2001). The Wiskott-Aldrich syndrome. *Clin Rev Allergy Immunol, 20,* 61–86.

8. Derry, J., et al. (1995). WASP gene mutations in Wiskott-Aldrich syndrome and X-linked thrombocytopenia. *Hum Mol Genet, 4,* 1127–1135.

9. Snapper, S., & Rosen, F. (2003). A family of WASPs. *N Engl J Med, 348,* 350–351.

10. Thrasher, A., & Kinnon, C. (2000). The Wiskott-Aldrich Syndrome. *Clin Exp Immunol, 120,* 2–9.

11. Haddad, E., et al. (1999). The trhombocytopenia of Wiskott Aldrich syndrome is not related to a defect in proplatelet formation. *Blood, 94,* 509–518.

12. Litzman, J., et al. (1996). Intravenous immunoglobulin, splenectomy, and antibiotic prophylaxis in Wiskott-Aldrich syndrome. *Arch Dis Child, 75,* 436–439.

13. Sullivan, K. (1999). Recent advances in our understanding of Wiskott-Aldrich syndrome. *Curr Opin Hematol, 6,* 8–14.

14. Mullen, C., Anderson, K., & Blaese, R. (1993). Splenectomy and/or bone marrow transplantation in the management of the Wiskott-Aldrich syndrome: Long-term follow-up of 62 cases. *Blood, 82,* 2961–2966.

15. Lum, L., et al. (1980). Splenectomy in the management of the thrombocytopenia of the Wiskott-Aldrich syndrome. *N Engl J Med, 302,* 892–896.

16. Snapper, S., & Rosen, F. (1999). The Wiskott-Aldrich syndrome protein (WASP): Roles in signaling and cytoskeletal organization. *Annu Rev Immunol, 17,* 905–929.

17. Parolini, O., et al. (1998). X-linked Wiskott-Aldrich syndrome in a girl. *N Engl J Med, 338,* 291–295.

18. Orange, J., et al. (2004). The Wiskott-Aldrich syndrome. *Cell Mol Life Sci, 61,* 2361–2385.

19. Shcherbina, A., et al. (2003). The incidence of lymphomas in a subgroup of Wiskott-Aldrich syndrome patients. *Br J Haematol, 121,* 529–530.

20. Ariga, T., et al. (2004). Confirming or excluding the diagnosis of Wiskott-Aldrich syndrome in children with thrombocytopenia of an unknown etiology. *J Pediatr Hematol Oncol, 26,* 435–440.

21. Srivastava, A., et al. (1996). Management of Wiskott-Aldrich syndrome. *Indian J Pediatr, 63,* 709–712.

22. Parkman, R., et al. (1978). Complete correction of the WiskottAldrich syndrome by allogeneic bone-marrow transplantation. *N Engl J Med, 298,* 921–927.

23. Rimm, I., & Rappeport, J. (1990). Bone marrow transplantation for the Wiskott-Aldrich syndrome: Long-term follow-up. *Transplantation, 50,* 617–620.

24. Ozsahin, H., et al. (1996). Bone marrow transplantation in 26 patients with Wiskott-Aldrich syndrome from a single center. *J Pediatr, 129,* 238–244.

25. Wada, T., et al. (2002). Retrovirus-mediated WASP gene transfer corrects Wiskott Aldrich syndrome T-cell dysfunction. *Hum Gene Ther, 13,* 1039–1046.

26. Perry, G. R., et al. (1980). The Wiskott Aldrich syndrome in the United States and Canada (1892–1979). *J Pediatr, 97,* 72–78.

27. Sullivan, K., et al. (1994). A multiinstitutional survey of the Wiskott-Aldrich syndrome. *J Pediatr, 125,* 876–885.

28. Dowton, S., et al. (1985). Studies of a familial platelet disorder. *Blood, 65,* 557.

29. Gerrard, J., et al. (1991). Inherited platelet-storage pool deficiency associated with a high incidence of acute myeloid leukaemia. *Br J Haematol, 79,* 246–255.

30. Simeone, A., Daga, A., & Calabi, F. (1995). Expression of runt in the mouse embryo. *Dev Dyn, 203,* 61–70.

31. Ho, C., et al. (1996). Linkage of a familial platelet disorder with a propensity to develop myeloid malignancies to human chromosome 21q22.1–22.2. *Blood, 87,* 5218–5224.

32. Song, W., et al. (1999). Haploinsufficiency of CBFA2 causes familial thrombocytopenia with propensity to develop acute myelogenous leukaemia. *Nat Genet, 23,* 166–175.

33. Baron, M. (2001). Molecular regulation of embryonic hematopoiesis and vascular development: A novel pathway. *J Hematother Stem Cell Res, 10,* 587–594.

34. Basecke, J., et al. (2002). Transcription of AML1 in hematopoietic subfractions of normal adults. *Ann Hematol, 81,* 254–257.

35. Heller, P. G., et al. (2005). Low Mpl receptor expression in a pedigree with familial platelet disorder with predisposition to acute myelogenous leukemia and a novel AML1 mutation. *Blood, 105,* 4664–4670.

36. Miyoshi, H., et al. (1991). t(8;21) breakpoints on chromosome 21 in acute myeloid leukemia are clustered within a

limited region of a single gene, AML1. *Proc Natl Acad Sci,* *88,* 10431–10434.

37. Thompson, A., Wood, W., & S. G. (1977). X-linked syndrome of platelet dysfunction, thrombocytopenia, and imbalanced globin chain synthesis with hemolysis. *Blood, 50,* 303–316.

38. Raskind, W., et al. (2000). Mapping of a syndrome of X-linked thrombocytopenia with Thalassemia to band Xp11–12: Further evidence of genetic heterogeneity of X-linked thrombocytopenia. *Blood, 95,* 2262–2268.

39. Nichols, K. E., et al. (2000). Familial dyserythropoietic anaemia and thrombocytopenia due to an inherited mutation in GATA1. *Nat Genet, 24,* 266–270.

40. Mehaffey, M., et al. (2001). X-linked thrombocytopenia caused by a novel mutation of GATA-1. *Blood, 98,* 2681–2688.

41. Freson, K., et al. (2001). Platelet characteristics in patients with X-linked macrothrombocytopenia because of a novel GATA1 mutation. *Blood, 98,* 85–92.

42. Yu, C., et al. (2002). X-linked thrombocytopenia with thalassemia from a mutation in the amino finger of GATA-1 affecting DNA binding rather than FOG-1 interaction. *Blood, 100,* 2040–2045.

43. Weiss, M. J., & Orkin, S. H. (1995). GATA transcription factors: Key regulators of hematopoiesis. *Exp Hematol, 23,* 99–107.

44. Vyas, P., et al. (1999). Consequences of GATA-1 deficiency in megakaryocytes and platelets. *Blood, 93,* 2867–2875.

45. Chang, A., et al. (2002). GATA-factor dependence of the multitype zinc-finger protein FOG-1 for its essential role in megakaryopoiesis. *Proc Natl Acad Sci USA, 99,* 9237–9342.

46. Tsang, A., et al. (1997). FOG, a multitype zinc finger protein, acts as a cofactor for transcription factor GATA-1 in erythroid and megakaryocytic differentiation. *Cell Mol Life Sci, 90,* 109–119.

47. Hughan, S. C., et al. (2005). Selective impairment of platelet activation to collagen in the absence of GATA1. *Blood, 105,* 4369–4376.

48. Favier, R., et al. (1993). A novel genetic thrombocytopenia (Paris-Trousseau) associated with platelet inclusions, dysmegakaryopoiesis and chromosome deletion AT 11q23. *CR Acad Sci III, 316,* 698–701.

49. Breton-Gorius, J., et al. (1995). A new congenital dysmegakaryopoietic thrombocytopenia (Paris-Trousseau) associated with giant platelet alpha-granules and chromosome 11 deletion at 11q23. *Blood, 85,* 1805–1814.

50. Favier, R., et al. (2003). Paris-Trousseau syndrome: Clinical, hematological, molecular data of ten new cases. *Thromb Haemost, 90,* 893–897.

51. Raslova, H., et al. (2004). FLI1 monoallelic expression combined with its hemizygous loss underlies Paris-Trousseau/Jacobsen thrombopenia. *J Clin Invest, 114,* 77–84.

52. Truong, A. H., & Ben-David, Y. (2000). The role of Fli-1 in normal cell function and malignant transformation. *Oncogene, 19,* 6482–6489.

53. Masuya, M., et al. (2005). Dysregulation of granulocyte, erythrocyte, and NK cell lineages in Fli-1 gene-targeted mice. *Blood, 105,* 95–102.

54. Anderson, M. K., et al. (1999). Precise developmental regulation of Ets family transcription factors during specification and commitment to the T cell lineage. *Development, 126,* 3131–3148.

55. Jackers, P., et al. (2004). Ets-dependent regulation of target gene expression during megakaryopoiesis. *J Biol Chem, 279,* 52183–52190.

56. Athanasiou, M., et al. (1996). Increased expression of the ETS-related transcription factor FLI-1/ERGB correlates with and can induce the megakaryocytic phenotype. *Cell Growth Differ, 7,* 1525–1534.

57. Wang, X., et al. (2002). Control of megakaryocyte-specific gene expression by GATA-1 and FOG-1: Role of Ets transcription factors. *Embo J, 21,* 5225–5234.

58. Hu, W., et al. (2005). Identification of nuclear import and export signals within Fli-1: Roles of the nuclear import signals in Fli-1-dependent activation of megakaryocyte-specific promoters. *Mol Cell Biol, 25,* 3087–3108.

59. Kawada, H., et al. (2001). Defective megakaryopoiesis and abnormal erythroid development in Fli-1 gene-targeted mice. *Int J Hematol, 73,* 463–468.

60. Hart, A., et al. (2000). Fli-1 is required for murine vascular and megakaryocytic development and is hemizygously deleted in patients with thrombocytopenia. *Immunity, 13,* 167–177.

61. Cines, D. B., et al. (2004). Congenital and acquired thrombocytopenia. *Hematology (Am Soc Hematol Educ Program),* 390–406.

62. Ballmaier, M., et al. (2003). Thrombopoietin is essential for the maintenance of normal hematopoiesis in humans: Development of aplastic anemia in patients with congenital amegakaryocytic thrombocytopenia. *Ann NY Acad Sci, 996,* 17–25.

63. Muraoka, K., et al. (1997). Defective response to thrombopoietin and impaired expression of c-mpl mRNA of bone marrow cells in congenital amegakaryocytic thrombocytopenia. *Br J Haematol, 96,* 287–292.

64. Ballmaier, M., et al. (2001). c-mpl mutations are the cause of congenital amegakaryocytic thrombocytopenia. *Blood, 97,* 139–146.

65. Yesilipek, et al. (2000). Peripheral stem cell transplantation in a child with amegakaryocytic thrombocytopenia. *Bone Marrow Transplant, 26,* 571–572.

66. Muraoka, K., et al. (2005). Successful bone marrow transplantation in a patient with c-mpl-mutated congenital amegakaryocytic thrombocytopenia from a carrier donor. *Pediatr Transplant, 9,* 101–103.

67. Thompson, A. A., et al. (2001). Congenital thrombocytopenia and radio-ulnar synostosis: A new familial syndrome. *Br J Haematol, 113,* 866–870.

68. Thompson, A. A., & Nguyen, L. T. (2000). Amegakaryocytic thrombocytopenia and radio-ulnar synostosis are associated with HOXA11 mutation. *Nat Genet, 26,* 397–398.

69. Davis, A. P., et al. (1995). Absence of radius and ulna in mice lacking hoxa-11 and hoxd-11. *Nature, 375,* 791–795.

70. Thorsteinsdottir, U., et al. (1997). Overexpression of HOXA10 in murine hematopoietic cells perturbs both

myeloid and lymphoid differentiation and leads to acute myeloid leukemia. *Mol Cell Biol, 17,* 495–505.

71. Gounder, D. S., et al. (1989). Clinical manifestations of the thrombocytopenia and absent radii (TAR) syndrome. *Aust NZJ Med, 19,* 479–482.

72. Hall, J. G., et al. (1969). Thrombocytopenia with absent radius (TAR). *Medicine (Baltimore), 48,* 411–439.

73. Hedberg, V. A., & Lipton, J. M. (1988). Thrombocytopenia with absent radii: A review of 100 cases. *Am J Pediatr Hematol Oncol, 10,* 51–64.

74. Greenhalgh, K. L., et al. (2002). Thrombocytopenia-absent radius syndrome: A clinical genetic study. *J Med Genet, 39,* 876–881.

75. Bradshaw, A., Donnelly, L. F., & Foreman, J. W. (2000). Thrombocytopenia and absent radii (TAR) syndrome associated with horseshoe kidney. *Pediatr Nephrol, 14,* 29–31.

76. van Haeringen, A., et al. (1989). Intermittent thrombocytopenia and absent radii: Report of a patient with additional unusual manifestations. *Am J Med Genet, 34,* 202–206.

77. Leclerc, J., & Toth, J. (1982). Thrombocytopenia with absent radii. *Can Med Assoc J, 126,* 506–508.

78. Anyane-Yeboa, K., et al. (1985). Tetraphocomelia in the syndrome of thrombocytopenia with absent radii (TAR syndrome). *Am J Med Genet, 20,* 571–576.

79. Sekine, I., et al. (1998). Thrombocytopenia with absent radii syndrome: Studies on serum thrombopoietin levels and megakaryopoiesis in vitro. *J Pediatr Hematol Oncol, 20,* 74–78.

80. Ballmaier, M., et al. (1997). Thrombopoietin in patients with congenital thrombocytopenia and absent radii: Elevated serum levels, normal receptor expression, but defective reactivity to thrombopoietin. *Blood, 90,* 612–619.

81. Linch, D. C., Stewart, J. W., & West, C. (1982). Blood and bone marrow cultures in a case of thrombocytopenia with absent radii. *Clin Lab Haematol, 4,* 313–317.

82. al-Jefri, A. H., et al. (2000). Thrombocytopenia with absent radii: Frequency of marrow megakaryocyte progenitors, proliferative characteristics, and megakaryocyte growth and development factor responsiveness. *Pediatr Hematol Oncol, 17,* 299–306.

83. Letestu, R., et al. (2000). Existence of a differentiation blockage at the stage of a megakaryocyte precursor in the thrombocytopenia and absent radii (TAR) syndrome. *Blood, 95,* 1633–1641.

84. Zahavi, J., Gale, R., & Kakkar, V. V. (1981). Storage pool disease of platelets in an infant with thrombocytopenic absent radii (TAR) syndrome simulating Fanconi's anaemia. *Haemostasis, 10,* 121–133.

85. Savoia, A., et al. (1999). An autosomal dominant thrombocytopenia gene maps to chromosomal region 10p. *Am J Hum Genet, 65,* 1401–1405.

86. Drachman, J. G., Jarvik, G. R., & Mehaffey, M. G. (2000). Autosomal dominant thrombocytopenia: Incomplete megakaryocyte differentiation and linkage to human chromosome 10. *Blood, 96,* 118–125.

87. Gandhi, M. J., Cummings, C. L., & Drachman, J. G. (2003). FLJ14813 missense mutation: A candidate for autosomal dominant thrombocytopenia on human chromosome 10. *Hum Hered, 55,* 66–70.

88. Sabri, S., et al. (2004). Differential regulation of actin stress fiber assembly and proplatelet formation by alpha2beta1 integrin and GPVI in human megakaryocytes. *Blood, 104,* 3117–3125.

89. Oliveira, R. A., et al. (2003). Is automated platelet counting still a problem in thrombocytopenic blood? *Sao Paulo Med J, 121,* 19–23.

90. Bernard, J., & Soulier, J.-P. (1948). Sur une nouvelle variete de dystrophie thrombocytaire-hemoragipare congenitale. *Sem Hop Paris, 24,* 3217–3223.

91. Lopez, J. A., et al. (1998). Bernard-Soulier syndrome. *Blood, 91,* 4397–4418.

92. Kunishima, S., Kamiya, T., & Saito, H. (2002). Genetic abnormalities of Bernard-Soulier syndrome. *Int J Hematol, 76,* 319–327.

93. Hillmann, A., et al. (2002). A novel hemizygous Bernard-Soulier Syndrome (BSS) mutation in the amino terminal domain of glycoprotein (GP)Ibbeta–platelet characterization and transfection studies. *Thromb Haemost, 88,* 1026–1032.

94. Watanabe, R., et al. (2003). Bernard-soulier syndrome with a homozygous 13 base pair deletion in the signal peptide-coding region of the platelet glycoprotein Ib(beta) gene. *Blood Coagul Fibrinolysis, 14,* 387–394.

95. Sachs, U. J., et al. (2003). Bernard-Soulier syndrome due to the homozygous Asn-45Ser mutation in GPIX: An unexpected, frequent finding in Germany. *Br J Haematol, 123,* 127–131.

96. Lanza, F., et al. (2002). A Leu7Pro mutation in the signal peptide of platelet glycoprotein (GP)IX in a case of Bernard-Soulier syndrome abolishes surface expression of the GPIb-V-IX complex. *Br J Haematol, 118,* 260–266.

97. Ni, H., et al. (2001). Increased thrombogenesis and embolus formation in mice lacking glycoprotein V. *Blood, 98,* 368–373.

98. Kahn, M. L., et al. (1999). Glycoprotein V-deficient platelets have undiminished thrombin responsiveness and do not exhibit a Bernard-Soulier phenotype. *Blood, 94,* 4112–4121.

99. Drouin, J., Carson, N. L., & Laneuville, O. (2005). Compound heterozygosity for a novel nine-nucleotide deletion and the Asn45Ser missense mutation in the glycoprotein IX gene in a patient with Bernard-Soulier syndrome. *Am J Hematol, 78,* 41–48.

100. Gonzalez-Manchon, C., et al. (2001). Compound heterozygosity of the GPIbalpha gene associated with Bernard-Soulier syndrome. *Thromb Haemost, 86,* 1385–1391.

101. Bernard, J. (1983). History of congenital hemorrhagic thrombocytopathic dystrophy. *Blood Cells, 9,* 179–193.

102. Canobbio, I., Balduini, C., & Torti, M. (2004). Signalling through the platelet glycoprotein Ib-V-IX complex. *Cell Signal, 16,* 1329–1344.

103. Kenny, D., et al. (1999). The critical interaction of glycoprotein (GP) IBbeta with GPIX-a genetic cause of Bernard-Soulier syndrome. *Blood, 93,* 2968–2975.

104. Plow, E. F., et al. (1985). Related binding mechanisms for fibrinogen, fibronectin, von Willebrand factor, and thrombospondin on thrombin-stimulated human platelets. *Blood, 66,* 724–727.

105. Nichols, W. L., et al. (1989). Bernard-Soulier syndrome: Whole blood diagnostic assays of platelets. *Mayo Clin Proc, 64,* 522–530.

106. Jackson, S. P., Nesbitt, W. S., & Kulkarni, S. (2003). Signaling events underlying thrombus formation. *J Thromb Haemost, 1,* 1602–1612.

107. Ware, J., Russell, S., & Ruggeri, Z. M. (2000). Generation and rescue of a murine model of platelet dysfunction: The Bernard-Soulier syndrome. *Proc Natl Acad Sci USA, 97,* 2803–2828.

108. Hourdille, P., et al. (1990). Studies on the megakaryocytes of a patient with the Bernard-Soulier syndrome. *Br J Haematol, 76,* 521–530.

109. Tomer, A., et al. (1994). Bernard-Soulier syndrome: Quantitative characterization of megakaryocytes and platelets by flow cytometric and platelet kinetic measurements. *Eur J Haematol, 52,* 193–200.

110. Khalil, A., et al. (1998). Bernard-Soulier syndrome in pregnancy: Case report and review of the literature. *Clin Lab Haematol, 20,* 125–128.

111. Sharma, J. B., Buckshee, K., & Sharma, S. (1991). Puberty menorrhagia due to Bernard Soulier syndrome and its successful treatment by "Ovral" hormonal tablets. *Aust NZJ Obstet Gynaecol, 31,* 369–370.

112. Mitsui, T., et al. (1998). Severe bleeding tendency in a patient with Bernard-Soulier syndrome associated with a homozygous single base pair deletion in the gene coding for the human platelet glycoprotein Ibalpha. *J Pediatr Hematol Oncol, 20,* 246–251.

113. McGill, M., et al. (1984). Morphometric analysis of platelets in Bernard-Soulier syndrome: Size and configuration in patients and carriers. *Thromb Haemost, 52,* 37–41.

114. Balduini, C. L., Iolascon, A., & Savoia, A. (2002). Inherited thrombocytopenias: From genes to therapy. *Haematologica, 87,* 860–880.

115. Jenkins, C. S., et al. (1976). Platelet membrane glycoproteins implicated in ristocetin-induced aggregation: Studies of the proteins on platelets from patients with Bernard-Soulier syndrome and von Willebrand's disease. *J Clin Invest, 57,* 112–124.

116. Kunishima, S., et al. (2001). Novel heterozygous missense mutation in the platelet glycoprotein Ib beta gene associated with isolated giant platelet disorder. *Am J Hematol, 68,* 249–255.

117. Savoia, A., et al. (2001). Autosomal dominant macrothrombocytopenia in Italy is most frequently a type of heterozygous Bernard-Soulier syndrome. *Blood, 97,* 1330–1335.

118. Kemahli, S., et al. (1994). DDAVP shortens bleeding time in Bernard-Soulier syndrome. *Thromb Haemost, 71,* 675.

119. Greinacher, A., et al. (1993). Evidence that DDAVP transiently improves hemostasis in Bernard-Soulier syndrome independent of von Willebrand-factor. *Ann Hematol, 67,* 149–150.

120. Tonda, R., et al. (2004). Hemostatic effect of activated recombinant factor VIIa in Bernard-Soulier syndrome: Studies in an in vitro model. *Transfusion, 44,* 1790–1791.

121. Almeida, A. M., et al. (2003). The use of recombinant factor VIIa in children with inherited platelet function disorders. *Br J Haematol, 121,* 477–481.

122. Monroe, D. M., et al. (2000). The factor VII-platelet interplay: Effectiveness of recombinant factor VIIa in the treatment of bleeding in severe thrombocytopathia. *Semin Thromb Hemost, 26,* 373–377.

123. Shprintzen, R. J., et al. (1981). The velo-cardio-facial syndrome: A clinical and genetic analysis. *Pediatrics, 67,* 167–172.

124. Stevens, C. A., Carey, J. C., & Shigeoka, A. O. (1990). Di George anomaly and velocardiofacial syndrome. *Pediatrics, 85,* 526–530.

125. Van Geet, C., et al. (1998). Velocardiofacial syndrome patients with a heterozygous chromosome 22q11 deletion have giant platelets. *Pediatr Res, 44,* 607–611.

126. Ludlow, L. B., et al. (1996). Identification of a mutation in a GATA binding site of the platelet glycoprotein Ibbeta promoter resulting in the Bernard-Soulier syndrome. *J Biol Chem, 271,* 22076–22080.

127. Levy, A., et al. (1997). Idiopathic thrombocytopenic purpura in two mothers of children with DiGeorge sequence: A new component manifestation of deletion 22q11? *Am J Med Genet, 69,* 356–359.

128. DePiero, A. D., et al. (1997). Recurrent immune cytopenias in two patients with DiGeorge/velocardiofacial syndrome. *J Pediatr, 131,* 484–486.

129. Kratz, C. P., et al. (2003). Evans syndrome in a patient with chromosome 22q11.2 deletion syndrome: A case report. *Pediatr Hematol Oncol, 20,* 167–172.

130. Behrens, W. E. (1975). Mediterranean macrothrombocytopenia. *Blood, 46,* 199–208.

131. Fabris, F., et al. (1997). Chronic isolated macrothrombocytopenia with autosomal dominant transmission: A morphological and qualitative platelet disorder. *Eur J Haematol, 58,* 40–45.

132. Fabris, F., et al. (2002). Autosomal dominant macrothrombocytopenia with ineffective thrombopoiesis. *Haematologica, 87,* ELT27.

133. Weiss, H. J., et al. (1982). Pseudo-von Willebrand's disease: An intrinsic platelet defect with aggregation by unmodified human factor VIII/von Willebrand factor and enhanced adsorption of its high-molecular-weight multimers. *N Engl J Med, 306,* 326–333.

134. Russell, S. D., & Roth, G. J. (1993). Pseudo-von Willebrand disease: A mutation in the platelet glycoprotein Ib alpha gene associated with a hyperactive surface receptor. *Blood, 81,* 1787–1791.

135. Sadler, J. E. (2005). New concepts in von Willebrand disease. *Annu Rev Med, 56,* 173–179.

136. Pareti, F. I., et al. (1990). Spontaneous platelet aggregation during pregnancy in a patient with von Willebrand disease type IIB can be blocked by monoclonal antibodies to both platelet glycoproteins Ib and IIb/IIIa. *Br J Haematol, 75,* 86–91.

137. Rodeghiero, F., & Castaman, G. (2005). Treatment of von Willebrand disease. *Semin Hematol, 42,* 29–35.

138. Holmberg, L., et al. (1983). Platelet aggregation induced by 1-desamino-8-D-*arginine* vasopressin (DDAVP) in Type IIB von Willebrand's disease. *N Engl J Med, 309,* 816–821.

139. Takahashi, H. (1985). Replacement therapy in platelet-type von Willebrand disease. *Am J Hematol, 18,* 351–362.

140. Favaloro, E. J., et al. (2004). von Willebrand disease: Laboratory aspects of diagnosis and treatment. *Haemophilia, 10,* 164–168.

141. Mathew, P., et al. (2003). Type 2B vWD: The varied clinical manifestations in two kindreds. *Haemophilia, 9,* 137–144.

142. Keeney, S., & Cumming, A. M. (2001). The molecular biology of von Willebrand disease. *Clin Lab Haematol, 23,* 209–230.

143. Seri, M., et al. (2003). MYH9-related disease: May-Hegglin anomaly, Sebastian syndrome, Fechtner syndrome, and Epstein syndrome are not distinct entities but represent a variable expression of a single illness. *Medicine* (Baltimore), *82,* 203–215.

144. Heath, K. E., et al. (2001). Nonmuscle myosin heavy chain IIA mutations define a spectrum of autosomal dominant macrothrombocytopenias: May-Hegglin anomaly and Fechtner, Sebastian, Epstein, and Alport-like syndromes. *Am J Hum Genet, 69,* 1033–1045.

145. Seri, M., et al. (2000). Mutations in MYH9 result in the May-Hegglin anomaly, and Fechtner and Sebastian syndromes: The May-Hegglin/Fechtner Syndrome Consortium. *Nat Genet, 26,* 103–105.

146. Ghiggeri, G. M., et al. (2003). Genetics, clinical and pathological features of glomerulonephritis associated with mutations of nonmuscle myosin IIA (Fechtner syndrome). *Am J Kidney Dis, 41,* 95–104.

147. Di Pumpo, M., et al. (2002). Defective expression of GPIb/IX/V complex in platelets from patients with May-Hegglin anomaly and Sebastian syndrome. *Haematologica, 87,* 943–947.

148. Dong, F., et al. (2005). Genotype-phenotype correlation in MYH9-related thrombocytopenia. *Br J Haematol, 130,* 620–627.

149. Marigo, V., et al. (2004). Correlation between the clinical phenotype of MYH9-related disease and tissue distribution of class II nonmuscle myosin heavy chains. *Genomics, 83,* 1125–1133.

150. Hamilton, R. W., et al. (1980). Platelet function, ultrastructure, and survival in the May-Hegglin anomaly. *Am J Clin Pathol, 74,* 663–668.

151. Wei, Q., & Adelstein, R. S. (2000). Conditional expression of a truncated fragment of nonmuscle myosin II-A alters cell shape but not cytokinesis in HeLa cells. *Mol Biol Cell, 11,* 3617–3627.

152. Falik-Zaccai, T. C., et al. (2001). A new genetic isolate of gray platelet syndrome (GPS): Clinical, cellular, and hematologic characteristics. *Mol Genet Metab, 74,* 303–313.

153. Raccuglia, G. (1971). Gray platelet syndrome: A variety of qualitative platelet disorder. *Am J Med, 51,* 818–828.

154. White, J. G. (1979). Ultrastructural studies of the gray platelet syndrome. *Am J Pathol, 95,* 445–462.

155. Drouin, A., et al. (2001). Newly recognized cellular abnormalities in the gray platelet syndrome. *Blood, 98,* 1382–1391.

156. Smith, M. P., Cramer, E. M., & Savidge, G. F. (1997). Megakaryocytes and platelets in alpha-granule disorders. *Baillieres Clin Haematol, 10,* 125–148.

157. Jantunen, E., et al. (1994). Gray platelet syndrome with splenomegaly and signs of extramedullary hematopoiesis: A case report with review of the literature. *Am J Hematol, 46,* 218–224.

158. Facon, T., et al. (1990). Simultaneous occurrence of grey platelet syndrome and idiopathic pulmonary fibrosis: A role for abnormal megakaryocytes in the pathogenesis of pulmonary fibrosis? *Br J Haematol, 74,* 542–543.

159. Milton, J. G., et al. (1984). Spontaneous platelet aggregation in a hereditary giant platelet syndrome (MPS). *Am J Pathol, 114,* 336–345.

160. Okita, J. R., et al. (1989). Montreal platelet syndrome: A defect in calcium-activated neutral proteinase (calpain). *Blood, 74,* 715–721.

161. Gilman, A. L., et al. (1995). A novel hereditary macrothrombocytopenia. *J Pediatr Hematol Oncol, 17,* 296–305.

162. Greenberg, S. M., et al. (1988). Characterization of a new megakaryocytic cell line: The Dami cell. *Blood, 72,* 1968–1977.

163. Freson, K., et al. (2005). The TUBB1 Q43P functional polymorphism reduces the risk of cardiovascular disease in men by modulating platelet function and structure. *Blood, 106,* 2356–2362.

# CHAPTER 55

# Pseudothrombocytopenia

## Nicola Bizzaro

*Laboratorio di Patologia Clinica, Ospedale Civile, Tolmezzo, Italy*

## I. Introduction

Pseudothrombocytopenia is a relatively uncommon phenomenon in which the number of platelets reported by automated cell counters is much lower than the real number of platelets circulating *in vivo*. Awareness of this phenomenon is important because pseudothrombocytopenia may lead to the erroneous diagnosis of thrombocytopenia, with resultant unnecessary and costly additional laboratory testing, and inappropriate treatment.

Pseudothrombocytopenia is the result of platelet clumping *in vitro*, which may be induced either by (a) antibody-mediated agglutination, the most important causes of which are ethylene-diamine-tetra-acetic acid (EDTA)-dependent agglutination[1] and platelet satellitism,[2] or (b) aggregation secondary to platelet activation resulting from improper blood sampling techniques or delayed mixing with anticoagulant in the test tubes (i.e., preanalytical errors).[3]

## II. EDTA-Dependent Pseudothrombocytopenia

The frequency of EDTA-dependent pseudothrombocytopenia ranges from 0.07 to 0.11% of all blood counts.[3–9] The phenomenon is not confined to humans as it has also been observed in horses[10] and cats.[11]

EDTA-dependent pseudothrombocytopenia (Fig. 55-1) was first described in 1969 by Gowland et al.[12] The phenomenon has since been widely studied, and it is now appreciated that the mechanism leading to the formation of platelet clumps is complex and influenced by immunological (antiplatelet autoantibodies), chemical (anticoagulants), and physical (temperature) factors that together make the phenomenon possible only *in vitro*.

Pseudothrombocytopenia is caused by the presence of antiplatelet autoantibodies, which recognize platelet antigens that are modified or exposed by the combined action of EDTA and low temperature on platelet membrane glyco-proteins.[1,13–15] All the major antibody classes can be involved in the platelet clumping phenomenon, but IgG antibodies are much more frequent than IgM, and IgA are rarely involved. Most of the autoantibodies behave as cold agglutinins, showing maximum activity between 4°C and 20°C.[16] However, in 20% of the cases, agglutinins that are reactive both at room temperature (approximately 22°C) and 37°C (wide heat range agglutinins) are involved;[4,17,18] when this event occurs, IgM antibodies are almost always present.[4,13] The antibodies are directed against ubiquitous antigens[19] as the phenomenon is easily reproduced by incubating patient plasma or serum, with EDTA-anticoagulated blood from normal subjects.[4,20] The target structure, the platelet membrane glycoprotein (GP) IIb-IIIa complex (integrin αIIbβ3), was first identified by Pegels et al.,[1] who observed that the EDTA-dependent antibodies did not react with platelets from patients with Glanzmann thrombasthenia (that lack the GPIIb-IIIa complex). Subsequently, it was shown that the autoantibody specificity is directed against the GPIIb subunit.[21,22] The epitope is normally cryptic within the GPIIb-IIIa complex, but it becomes accessible to the cold antibody after dissociation of the glycoprotein complex due to the chelating effect of EDTA on calcium ions,[23–26] associated with the alterations in protein conformation induced by low temperature.[16,21]

The etiological importance of EDTA was further confirmed by the demonstration that the platelet binding of several monoclonal antibodies directed against epitopes on the dissociated GPIIb-IIIa complex was enhanced in its presence.[27,28]

Interestingly, Casonato et al.[29] showed that pseudothrombocytopenia was almost completely abolished after the addition of the RGD peptide, the recognition sequence of cytoadhesive proteins, suggesting that the agglutinating antibodies might recognize the cytoadhesive receptor on platelet GPIIb-IIIa. Even though GPIIb is the most frequently involved antigen, antibodies directed against a 78-kD glycoprotein on the platelet membrane[30] and other antigens including phospholipids have been described.[31]

**999**

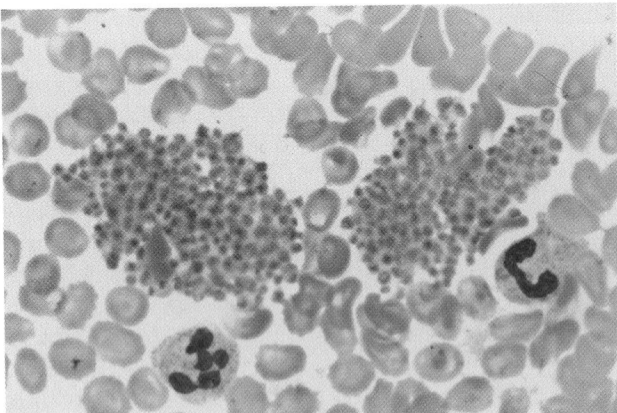

**Figure 55-1.** Peripheral blood film showing two large platelet clumps in EDTA-anticoagulated blood (May–Grünwald–Giemsa [MGG] 1000×).

## III. Citrate-Dependent Pseudothrombocytopenia

Although the antibodies are generally active only in the presence of EDTA, in about 10 to 20% of cases, platelet clumping also occurs in citrate-anticoagulated blood,[4,9,14,32,33] which also has a chelating effect on calcium ions, albeit to a lesser extent. These antibodies are almost always IgM and are probably the same antibodies responsible for EDTA-dependent clumping at 37°C.[4]

## IV. Other Anticoagulants

To avoid EDTA-induced platelet clumping, alternative chelating and nonchelating anticoagulants have been proposed, such as sodium fluoride,[34] acid-citrate-dextrose (ACD),[6] sodium heparin,[7] citrate-theophylline-adenosine-dipyridamole (CTAD),[35] heparin-β-hydroxy-ethyl-theophylline,[36] EGTA,[13] hirudin, and phenylalanine-proline-arginine chloromethyl-ketone (PPACK).[37] However, in addition to the fact that many of these anticoagulants may induce pseudothrombocytopenia, they are all unsuitable for multiparametric hematological analysis and cannot be used in routine clinical practice. Citrate-pyridoxalphosphate-Tris (CPT), a more recently developed anticoagulant,[38] has been demonstrated to be an effective alternative to EDTA for the prevention of pseudothrombocytopenia and, at the same time, for obtaining accurate routine hematology data,[38,39] but it is not yet widely used.

It has also been suggested that adding aminoglycosides to the EDTA-containing test tubes might prevent the formation of platelet clumps.[40] However, the mechanism of action of the antibiotics is not clear, and the methodology is more complicated than that recommended and normally

followed for resolving cases of pseudothrombocytopenia (see Section VII).

## V. Clinical Aspects

Pseudothrombocytopenia has been documented in patients affected by a wide variety of disorders, including autoimmune disease, chronic inflammatory disease, viral and bacterial infections, metabolic syndromes, and neoplastic diseases, as well as after allogeneic stem cell transplantation and in healthy subjects. These highly dissimilar clinical situations confirm that pseudothrombocytopenia is not caused by or associated with a specific disease or the use of specific drugs, although occasional cases have been described in which pseudothrombocytopenia has appeared after the administration of valproic acid,[41,42] olanzapine,[43] or levofloxacin.[44] Pseudothrombocytopenia does not pose a hemorrhagic or thrombotic risk to the patient. The only clinical implication resides in its lack of recognition, which potentially could cause inappropriate platelet transfusion, unnecessary additional testing, and delays in diagnostic or therapeutic procedures.

Indeed, reports describe patients with pseudothrombocytopenia who have been hospitalized and subjected to unnecessary transfusions of platelet concentrates;[35,45–49] others have undergone bone marrow aspiration and bone biopsy,[50–52] whereas others still have been administered long-term cortisone therapy or have even been splenectomized.[3,13,45] Delayed recognition of pseudothrombocytopenia may also determine a state of unjustified alarm in the patient, as well as precautionary behavior for fear of hemorrhage due to thrombocytopenia.

In some healthy subjects, the presence of pseudothrombocytopenia could be documented for more than 20 years.[4] This further indicates that the finding of pseudothrombocytopenia autoantibodies in an apparently healthy subject does not mean a higher risk for the future development of a particular disease or an overt autoimmune condition.[53] However, in at least half of the patients, it was determined that pseudothrombocytopenia was not present in previous years but had appeared suddenly with no apparent cause. After pseudothrombocytopenia appears, it usually remains indefinitely,[4,46] but in a few cases it is transient and disappears after some time.[4,54–57] Obviously, a subject with pseudothrombocytopenia does not require clinical or laboratory monitoring for this completely benign condition.

A particular potential issue arises when subjects with pseudothrombocytopenia require surgical intervention under hypothermia, as occurs in heart surgery. However, we and others have reported that surgery can be performed without complications in patients who undergo cardiac surgery for coronary artery bypass grafting and valve replacement even though body temperature was lowered to 28 to 30°C.[58–62]

Another interesting aspect is that related to the possible implications of transfusion of blood obtained from a donor with pseudothrombocytopenia. In our experience (unpublished observations of two cases) and in that of Sweeney et al.,[63] no problems arose in the recipients, not even the appearance of pseudothrombocytopenia, indicating that the blood of healthy subjects with pseudothrombocytopenia can be used in transfusion therapy.

It is noteworthy that on rare occasions pseudothrombocytopenia can be associated with chronic autoimmune thrombocytopenia,[4,47,64] in which case pseudothrombocytopenia artifactually further lowers the platelet count. Familial occurrence has never been documented, but a case of transient congenital pseudothrombocytopenia due to transplacental transmission of IgG antibodies from a mother to child has been described.[65]

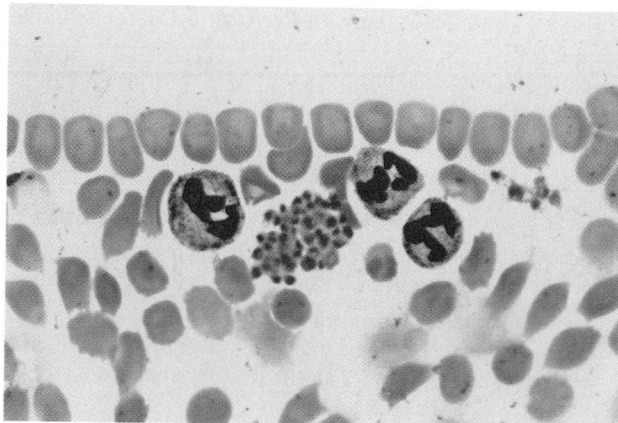

**Figure 55-2.** A very small platelet agglutinate the size of a neutrophil. When such clumps are present, automatic blood counters count them as leukocytes, leading to a pattern of pseudothrombocytopenia and pseudoleukocytosis (MGG, 1000×).

## VI. Pseudothrombocytopenia Agglutinins as Natural Antibodies

In view of their lack of association with specific diseases and their presence also in healthy subjects, pseudothrombocytopenia antibodies are believed to be of no pathological significance. The most likely hypothesis is that they are naturally occurring autoantibodies, physiologically delegated to the clearance of aged or damaged platelets.[14,66] Indeed, several studies since the early 1990s have described natural autoantibodies in the sera of healthy subjects,[67–69] and in normal individuals it was shown that at least 20% of all the immunoglobulins correspond to polyreactive natural autoantibodies able to recognize self-structures.[53,70] Moreover, EDTA-dependent autoantibodies directed against platelet cryptoantigens are detectable in the blood of many patients without causing pseudothrombocytopenia.[14] These natural autoantibodies are directed against self-cryptoantigens — that is, antigens normally not present on the platelet membrane surface that are exposed following an alteration in the cell membrane that may be the result of an aging process, or, as in the case of pseudothrombocytopenia, the effects of EDTA. Moreover, studies on the lymphocyte immunophenotype in healthy subjects with pseudothrombocytopenia have shown a normal pattern, both in percentage and absolute numbers, for all the B, T, and NK markers,[4] thus demonstrating that natural agglutinins are not the product of a clonal expansion.

It remains unclear why some subjects have these antibodies or why these subjects have high enough quantities to produce the pseudothrombocytopenia phenomenon. It is possible that an unrecognized viral infection could act as an immune stimulus against self-cryptoantigens, with a cross-reactive mechanism.

## VII. Clinical Laboratory Procedures

Pseudothrombocytopenia is typically manifested by the presence of a low platelet count produced by electronic cell counters, which are unable to differentiate platelet agglutinates from individual cells, leading to spuriously low platelet counts. In most cases, the platelet clumps are readily detectable due to the presence of specific warning flags and typical scattergram patterns produced by all modern blood counters. However, false negatives (as well as false positives) are frequent.[71–76] For example, when the clumps are small and their size is equal to or slightly larger than a leukocyte (Fig. 55-2), no instrumental signal of pseudothrombocytopenia may be present, and the clumps will be counted by automated instruments as white cells; in these cases a false increase in white cells (pseudoleukocytosis) may be recorded.[7,37,50,77] Therefore, identification of pseudothrombocytopenia requires an increased index of suspicion, and because inspection and interpretation of scattergrams is operator dependent, it may vary according to operator experience and type of autoanalyzer.[71] When suspicion is generated, pseudothrombocytopenia can be rapidly confirmed by the microscopic finding of platelet agglutinates on blood smears.

Although the diagnosis of pseudothrombocytopenia on the basis of instrumental findings is on the whole reasonably straightforward, the problem of obtaining an accurate platelet count in these patients is more complex. The most reliable and rapid method for obtaining accurate platelet counts in pseudothrombocytopenia patients is to collect and examine EDTA-blood at 37°C. In those cases that still show clumping at 37°C, and in which reactive IgM antibodies are usually present even in citrated samples, blood should be collected by finger stick into ammonium oxalate solution,

**Table 55-1: How to Perform a Correct Platelet Count in the Presence of Pseudothrombocytopenia**

After microscopic examination of the blood smear has confirmed pseudothrombocytopenia by evidencing platelet clumps, perform the following steps:

1. Exclude improper blood sampling techniques, such as traumatic venipuncture, low anticoagulant concentration in the test tube, or delayed mixing.

2. Collect blood in a prewarmed tube at 37°C and keep it at 37°C in a thermostat until processing.

3. Process the sample on an automated blood counter immediately after removal from the thermostat, using preferably the Stat procedure. Be sure that the sample does not cool; rewarming of a cooled sample is inappropriate for an accurate platelet count.

4. If pseudothrombocytopenia is still present in the sample processed at 37°C (10–20% of cases), this means that IgM antibodies, which are also reactive in citrate, are present. The use of citrated-anticoagulated blood is therefore useless. Collect capillary blood by finger stick using the Unopette Reagent System or similar device, and perform a manual platelet count at the light microscope in a Burker or Neubauer chamber.

5. Mark the patient's chart, and instruct him or her that every future blood count must be performed at 37°C or by finger stick, as appropriate.

and the platelets should be counted at the light microscope in a Burker or Neubauer chamber, which at present constitutes the gold standard for differentiating pseudothrombocytopenia from true thrombocytopenia (Table 55-1).

Sequential studies show that the agglutination is time dependent, in most cases taking place within a few minutes and becoming more pronounced upon standing at room temperature. Furthermore, the degree of the clumping, and subsequently the platelet count, may also vary in intensity in counts performed over time in the same patient. We have noticed that when examinations are conducted during the summer months, when temperatures may reach 32 to 35°C and higher, the phenomenon may attenuate and not be detected in the autoanalyzers. An increase in room temperature, therefore, could be one of the reasons why the phenomenon has been observed to vary over time.

Pseudothrombocytopenia has been observed together with erythrocyte agglutination[78,79] and with leukocyte agglutination;[80] moreover, isolated EDTA-dependent leukocyte clumping has also been reported.[81–85] No platelet function abnormalities have been described in association with EDTA-dependent pseudothrombocytopenia, nor have platelet activation and release of mediators of aggregating activity been documented.[86] That the phenomenon is completely independent of physiological platelet aggregation is also documented by the finding that platelet clumping can be induced in blood from afibrinogenemic patients by sera from patients with pseudothrombocytopenia.[29]

## VIII. Pseudothrombocytopenia due to Platelet Satellitism

Like platelet clumping, platelet satellitism is also an immunological phenomenon caused by the presence of natural antibodies, in which the platelets arrange themselves selec-

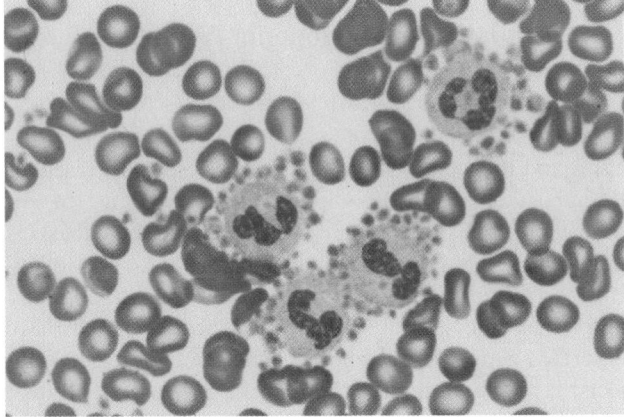

**Figure 55-3.** Peripheral blood film of EDTA-anticoagulated blood kept at room temperature, showing PMNs with platelet satellitism (MGG, 1000×).

tively around polymorphonuclear neutrophils (PMN)[2,87,88] (Fig. 55-3). Platelet satellitism also occurs *in vitro* in EDTA-anticoagulated blood, but differs from platelet clumping, at least in the cases described to date, because the antibodies are consistently class IgG, inactive at 37°C, and do not react when citrate is employed as the anticoagulant. In this very characteristic situation, the antibody recognizes a common antigenic structure on both platelets and PMN and, therefore, forms a bridge between the two cells.[89] The target antigens have been identified as the GPIIb-IIIa complex on the platelet surface and the Fcγ, receptor III (FcγRIII) on neutrophils.[89] The selectivity of the platelet satellitism phenomenon for PMN could be due to the specific characteristics of FcγRIII on the PMN, which express only the phosphatidyl inositol-linked form of FcγRIII (FcγRIIIb), whereas natural killer cells and macrophages express the transmembrane form (FcγRIIIa).[90]

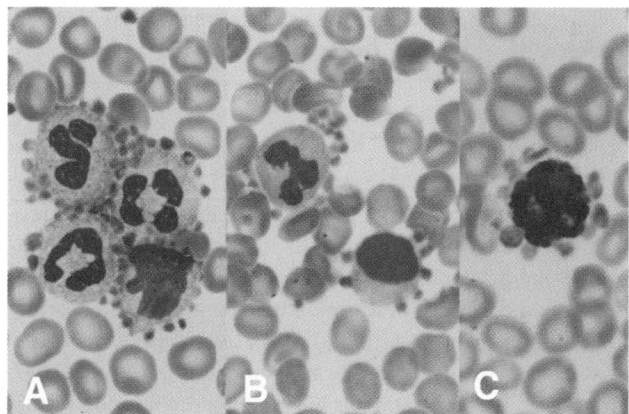

**Figure 55-4.** In very rare cases, platelet satellitism can also be observed around isolated monocytes (A), lymphocytes (B), or basophils (C) (MGG, 1000×).

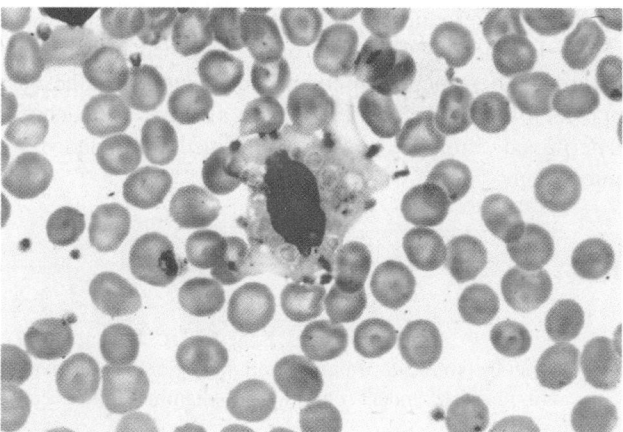

**Figure 55-5.** A monocyte showing intense platelet phagocytosis. (MGG, 1000×).

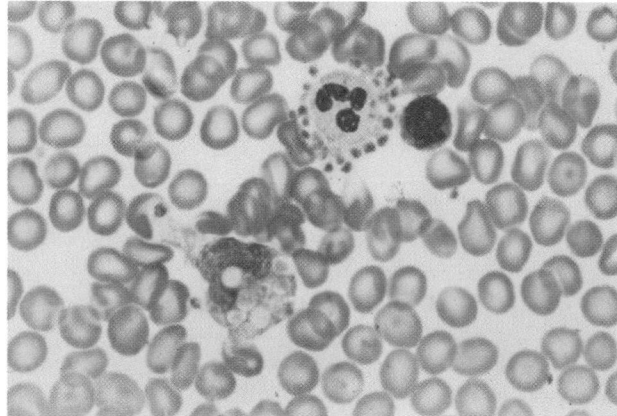

**Figure 55-6.** A PMN exhibiting platelet satellitism, a normal lymphocyte, and a monocyte containing phagocytosed platelets. Platelet phagocytosis by monocytes is also a distinctive feature of platelet satellitism (MGG, 1000×).

Even though other leukocytes are generally not involved, the sporadic observation of platelet satellitism involving lymphocytes,[91] monocytes,[92] and basophils[93] has been reported (Fig. 55-4). The reason for this occasional involvement of these other cells is unknown. Interestingly, over the years we have observed that even if monocytes apparently are not surrounded by platelets like the PMN, they are instead involved in a pronounced phagocytosing activity that is so striking that this morphological picture may be considered another distinctive sign of the phenomenon (Figs. 55-5 and 55-6). The most probable hypothesis is that while the antibody bond to platelets and PMN is $F_{ab}$ dependent, the antibody-covered platelets are actively phagocytosed by monocytes through an $F_c$-dependent mechanism.

The reason why some individuals have agglutinating antibodies that cause platelet clumping and others have antibodies that cause platelet satellitism is unknown. One possible explanation is that there are two different subsets of

natural autoantibodies directed against different epitopes on the platelet GPIIb-IIIa complex. The frequency of platelet satellitism is much lower than that of EDTA-dependent platelet clumping (approximately 1 : 30,000 blood counts),[20] but the real incidence of the phenomenon may be underestimated. Indeed, the number of platelets that are involved in platelet satellitism and thus escape automated counting is much lower and often not enough to result in pseudothrombocytopenia. In these cases, the platelet satellitism can be detected, other than by morphology, only by the characteristic cell distribution in the scattergrams provided by the analyzers.[20]

Even though both platelet clumping and platelet satellitism require EDTA and room temperature to occur, the two phenomena have never been observed together on the same blood smear, even if neutrophils surrounded by platelets and adjacent to small platelet clumps are occasionally found in platelet satellitism samples (Fig. 55-7). Moreover, in the case of a 78-year-old woman with hypertension who had platelet satellitism for more than 8 years, we observed a pattern change, with the sudden appearance of platelet clumps and the disappearance of platelet satellitism. Interestingly, Christopoulos and Mattock described a case of platelet satellitism in EDTA-anticoagulated blood, which instead formed platelet clumps in citrated blood.[94]

## IX. Pseudothrombocytopenia due to Therapeutic Monoclonal Antibodies

A new iatrogenic cause of pseudothrombocytopenia has been described in relation to the therapeutic use of drugs specifically designed to alter platelet function.[95,96] Thus, pseudothrombocytopenia was observed after the administration of abciximab, a human-mouse chimeric $F_{ab}$ fragment

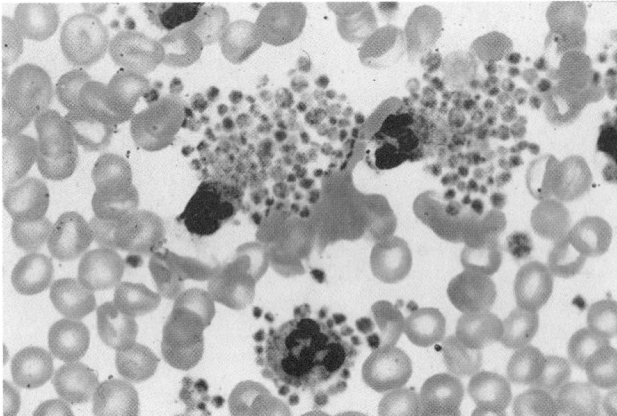

**Figure 55-7.** A typical platelet satellitism pattern is exhibited by the PMN in the lower half of the micrograph. Two other PMNs, in contact with platelet clumps, have a plume-like appearance (MGG, 1000×).

that binds to the $\beta_3$ subunit of the platelet GPIIb-IIIa complex[95] and is used to prevent thromboembolic complications in patients undergoing percutaneous coronary interventions. This phenomenon usually occurs in EDTA-anticoagulated blood[72,96–98] and in rare cases also in citrate-anticoagulated blood.[99]

It has been reported that pseudothrombocytopenia occurs in approximately 2% of abciximab-treated patients and it accounts for more than one third of abciximab-treated patients presenting with a low platelet count.[100] However, adopting the criteria of a platelet count of less than $150 \times 10^9$/L or a decrease from baseline of greater than 40% to define thrombocytopenia, Schell et al. have found a much higher incidence: Of 66 subjects treated with abciximab, 26 (39%) developed thrombocytopenia and 18 (27%) had pseudothrombocytopenia.[101]

This drug-induced pseudothrombocytopenia might have the same pathogenic mechanism as that caused by natural autoantibodies; the binding of abciximab $F_{ab}$ fragments to the GPIIb-IIIa receptor could alter the conformation of the molecule[102] and, in concert with the anticoagulant-induced changes, enhance the access of natural agglutinin autoantibodies to the target antigen.[100] An alternative hypothesis, which addresses the relatively high frequency of pseudothrombocytopenia as a cause of low platelet counts in patients receiving abciximab, is that the chimeric antibody itself might act as an agglutinating antibody after EDTA-induced conformational changes in the GPIIb-IIIa complex have occurred. Irrespective of the mechanism, the distinction between pseudothrombocytopenia and true thrombocytopenia in abciximab-treated patients is critical; if thrombocytopenia is present, the drug must be discontinued and platelet concentrates may be transfused, whereas if pseudothrombocytopenia is detected, the risk of hemorrhage is not increased, antithrombotic and antiplatelet therapy can be continued, and invasive procedures can be performed.

## X. Conclusions

Pseudothrombocytopenia is an uncommon and often unrecognized phenomenon that results in errors in the interpretation of platelet counts with consequent inappropriate clinical decisions or useless, and sometimes dangerous, therapeutic interventions. The widespread use of automatic instruments for the analysis and counting of blood cells in clinical laboratories on the one hand has enabled the recognition of this phenomenon, but on the other hand it has decreased the frequency of microscopic observation of blood films, which is the only procedure that can demonstrate the presence of platelet clumps with certainty. Although almost all modern blood analyzers are equipped with alarms to signal the presence of platelet clumps, these alarms are nonspecific and relatively inaccurate. In view of the practical implications of failure to recognize pseudothrombocytopenia, the best approach to the diagnosis is to have all cases of machine-determined thrombocytopenia confirmed or excluded by microscopy.

## References

1. Pegels, J. G., Bruynes, E. C. E., Engelfriet, C. P., et al. (1982). Pseudothrombocytopenia: An immunologic study on platelet antibodies dependent on ethylene diamine tetra-acetate. *Blood, 59,* 157–161.
2. Field, E. J., & McLeod, I. (1963). Platelet adherence to polymorphs. *Br Med J, 2,* 388.
3. Payne, B. A., & Pierre, R. V. (1984). Pseudothrombocytopenia: A laboratory artifact with potentially serious consequences. *Mayo Clin Proc, 59,* 123–125.
4. Bizzaro, N. (1995). EDTA-dependent pseudothrombocytopenia: A clinical and epidemiological study of 112 cases, with 10-year follow-up. *Am J Hematol, 50,* 103–109.
5. Mant, M. J., Doery, J. C. G., Gauldie, J., et al. (1975). Pseudothrombocytopenia due to platelet aggregation and degranulation in blood collected in EDTA. *Scand J Haematol, 15,* 161–170.
6. Manthorpe, R., Kofod, B., Wiik, A., et al. (1981). Pseudothrombocytopenia. In vitro studies on the underlying mechanism. *Scand J Haematol, 26,* 385–392.
7. Savage, R. A. (1984). Pseudoleukocytosis due to EDTA-induced platelet clumping. *Am J Clin Pathol, 81,* 317–322.
8. Vicari, A., Banfi, G., & Bonini, A. (1988). EDTA-dependent pseudothrombocytopaenia: A 12-month epidemiological study. *Scand J Clin Lab Invest, 48,* 537–452.
9. Garcia Suarez, J., Calero, M. A., Ricard, M. P., et al. (1992). EDTA-dependent pseudothrombocytopenia in ambulatory patients: Clinical characteristics and role of new automated cell-counting in its detection. *Am J Hematol, 39,* 146–147.

10. Hinchcliff, K. W., Kociba, G. J., & Mitten, L. A. (1993). Diagnosis of EDTA-dependent pseudothrombocytopenia in horse. *J Am Vet Med Assoc, 203,* 1715–1716.

11. Norman, E. J., Barron, R. C., Nash, A. S., et al. (2001). Evaluation of a citrate-based anticoagulant with platelet inhibitory activity for feline blood cell counts. *Vet Clin Pathol, 30,* 124–132.

12. Gowland, E., Kay, H. E. M., Spillman, J. C., et al. (1969). Agglutination of platelets by a serum factor in the presence of EDTA. *J Clin Pathol, 22,* 460–464.

13. Onder, O., Weinstein, A., & Hoyer, L. W. (1980). Pseudo-thrombocytopenia caused by platelet agglutinins that are reactive in blood anticoagulated with chelating agents. *Blood, 56,* 177–182.

14. von dem Borne, A. E. G. Kr., van der Lelie, J., Vos, J. J. E., et al. (1986). Antibodies against crypt-antigens of platelets: Characterization and significance for the serologist. *Curr Stud Hematol Blood Transfus, 52,* 33–46.

15. Veenhoven, W. A., van der Schans, G. S., Huiges, W., et al. (1979). Pseudothrombocytopenia due to agglutinins. *Am J Clin Pathol, 72,* 1005–1008.

16. Pidard, D., Didry, D., Kunicki, T. J., et al. (1986). Temperature-dependent effects of EDTA on the membrane glycoprotein IIb/IIIa complex and platelet aggregability. *Blood, 67,* 604–611.

17. Shreiner, D. P., & Bell, W. R. (1973). Pseudothrombocytopenia: Manifestation of a new type of platelet agglutinin. *Blood, 42,* 541–549.

18. Chang, Y. W. (1981). A case of thrombocytopenia. *Lab Med, 12,* 508–510.

19. Bizzaro, N. (1999). Pseudothrombocytopenia and platelet antigens. *Clin Appl Thromb Hemost, 5,* 141.

20. Bizzaro, N. (1991). Platelet satellitosis to polymorphonuclears: Cytochemical, immunological, and ultrastructural characterization of eight cases. *Am J Hematol, 36,* 235–242.

21. van Vliet, H. H. D., Kappers-Klunne, M. C., & Abels, J. (1986). Pseudothrombocytopenia: A cold autoantibody against platelet glycoprotein IIb. *Br J Haematol, 62,* 501–511.

22. Fiorin, F., Steffan, A., Pradella, P., et al. (1998). IgG platelet antibodies in EDTA-dependent pseudothrombocytopenia bind to platelet membrane glycoprotein IIb. *Am J Clin Pathol, 110,* 178–183.

23. Howard, L., Shulman, S., Sadanandan, S., et al. (1982). Crossed immunoelectrophoresis of human platelet membranes: The major antigen consists of a complex of glycoproteins GPIIb and GPIIIa, held together by Ca++ and missing in Glanzmann's thrombasthenia. *J Biol Chem, 257,* 8331–8336.

24. Kunicki, Th. J., Pidard, D., Rosa, J. Ph., et al. (1981). The formation of Ca++-dependent complexes of platelet membrane glycoproteins IIb and IIIa in solution as determined by crossed immunoelectrophoresis. *Blood, 58,* 268–278.

25. Ginsberg, M. H., Lightsey, A., Kunicki, Th. J., et al. (1986). Divalent cation regulation of the surface orientation of platelet membrane glycoprotein IIb. *J Clin Invest, 78,* 1103–1111.

26. Fitzgerald, L. A., & Phillips, D. R. (1985). Calcium regulation of the platelet membrane glycoprotein IIb-IIIa complex. *J Biol Chem, 260,* 11366–11374.

27. Khasperova, S. G., Vlasik, T. N., Byzova, T. V., et al. (1993). Detection of an epitope specific for the dissociated form of glycoprotein IIIa of platelet membrane glycoprotein IIb-IIIa complex and its expression on the surface of adherent platelets. *Br J Haematol, 85,* 332–340.

28. Kouns, W. C., Wall, D. C., White, M. M., et al. (1990). A conformation-dependent epitope of human glycoprotein IIIa. *J Biol Chem, 265,* 20594–20601.

29. Casonato, A., Bertomoro, A., Pontara, E., et al. (1994). EDTA dependent pseudothrombocytopenia caused by antibodies against the cytoadhesive receptor of platelet gpIIB-IIIA. *J Clin Pathol, 47,* 625–630.

30. De Caterina, M., Fratellanza, G., Grimaldi, E., et al. (1993). Evidence of a cold immunoglobulin M autoantibody against 78-kD platelet glycoprotein in a case of EDTA-dependent pseudothrombocytopenia. *Am J Clin Pathol, 99,* 163–167.

31. Bizzaro, N., & Brandalise, M. (1995). EDTA-dependent pseudothrombocytopenia: Association of anti-platelet and anti-phospholipid antibodies. *Am J Clin Pathol, 103,* 103–107.

32. Cunningham, V. L., & Brandt, J. T. (1992). Spurious thrombocytopenia due to EDTA-independent cold-reactive agglutinins. *Am J Clin Pathol, 97,* 359–362.

33. Kuijpers, R. W. A. M., Ouwehand, W. H., Peelen, W., et al. (1992). Thrombocytopenia due to platelet glycoprotein IIb/IIIa-reactive autoantibodies non-reactive with platelets from EDTA blood. *Vox Sang, 63,* 119–121.

34. Kabutomori, O., Koh, T., Amino, N., et al. (1995). Correct platelet count in EDTA-dependent pseudothrombocytopenia. *Eur J Haematol, 55,* 67–68.

35. Lombarts, A. J. P. F., Zijlstra, J. J., Peters, R. H. M., et al. (1999). Accurate counting in an insidious case of pseudothrombocytopenia. *Clin Chem Lab Med, 37,* 1063–1066.

36. Onhuma, O., Shirata, Y., & Miyazawa, K. (1988). Use of theophylline in the investigation of pseudothrombocytopenia induced by edetic acid (EDTA-2 K). *J Clin Pathol, 41,* 915–917.

37. Schrezenmeier, H., Muller, H., Gunsilius, E., et al. (1995). Anticoagulant-induced pseudothrombocytopenia and pseudoleukocytosis. *Thromb Haemost, 73,* 506–513.

38. Lippi, U., Schinella, M., Nicoli, M., et al. (1990). EDTA-induced platelet aggregation can be avoided by a new anticoagulant also suitable for automated complete blood count. *Haematologica, 75,* 38–41.

39. Lippi, G., Guidi, G., & Nicoli, M. (1996). Platelet count in EDTA-dependent pseudothrombocytopenia. *Eur J Haematol, 56,* 112–113.

40. Sakurai, S., Shiojima, I., Tanigawa, T., et al. (1997). Aminoglycosides prevent and dissociate the aggregation of platelets in patients with EDTA-dependent pseudothrombocytopenia. *Br J Haematol, 99,* 817–823.

41. Nagata, M., Hara, T., Mizuno, Y., et al. (1989). Valproic acid and pseudothrombocytopenia. *Jpn J Clin Hematol, 30,* 1675.

42. Yoshikawa, H. (2003). EDTA-dependent pseudothrombocytopenia induced by valproic acid. *Neurology, 61,* 579–580.

43. Tu, C. H., & Yang, S. (2002). Olanzapine-induced EDTA-dependent pseudothrombocytopenia. *Psychosomatics, 43,* 421–423.

44. Kinoshita, Y., Yamane, T., Kamimoto, A., et al. (2004). A case of pseudothrombocytopenia during antibiotic administration. *Rinsho Byori, 52,* 120–123.

45. Nilsson, T., & Norberg, B. (1986). Thrombocytopenia and pseudothrombocytopenia: A clinical and laboratory problem. *Scand J Haematol, 37,* 341–346.

46. Berkman, N., Michaeli, Y., Or, R., et al. (1991). EDTA-dependent pseudothrombocytopenia: A clinical study of 18 patients and review of the literature. *Am J Hematol, 36,* 195–201.

47. Edelmann, B., & Kickler, T. (1993). Sequential measurement of anti-platelet antibodies in a patient who developed EDTA-dependent pseudothrombocytopenia. *Am J Clin Pathol, 99,* 87–89.

48. Lau, L. G., Chng, W. J., & Liu, T. C. (2004). Unnecessary transfusions due to pseudothrombocytopenia. *Transfusion, 44,* 801.

49. Christensen, R. D., Sola, M. C., Rimsza, L. M., et al. (2004). Pseudothrombocytopenia in a preterm neonate. *Pediatrics, 114,* 273–275.

50. Solanki, D. L., & Blackburn, B. C. (1983). Spurious leukocytosis and thrombocytopenia. A dual phenomenon caused by clumping of platelets *in vitro. JAMA, 250,* 2514–2515.

51. Mancini, S., d'Onofrio, G., Zini, G., et al. (1986). Pseudothrombocytopenia and pseudoleukocytosis: Study of 17 cases of EDTA-dependent platelet agglutination. *Rec Prog Med, 77,* 573–579.

52. Carrillo-Esper, R., & Contreras-Dominguez, V. (2004). Pseudothrombocytopenia induced by ethylendiaminetetraacetic acid in burn patients. *Cir Cir, 72,* 335–338.

53. Shoenfeld, Y., & Isenberg, D. (1993). Natural autoantibodies: Their physiological role and regulatory significance. Boca Raton, FL: CRC Press.

54. Mori, M., Kudo, H., Yoshitake, S., et al. (2000). Transient EDTA-dependent pseudothrombocytopenia in a patient with sepsis. *Intensive Care Med, 26,* 218–220.

55. Kjeldsberg, C. R., & Hershgold, E. J. (1974). Spurious thrombocytopenia. *JAMA, 227,* 628–630.

56. Hsieh, A. T., Chao, T. Y., & Chen, Y. C. (2003). Pseudothrombocytopenia associated with infectious mononucleosis. *Arch Pathol Lab Med, 127,* 17–18.

57. Gillis, S., Eisenberg, M. E., Shapira, M. Y., et al. (2003). Pseudothrombocytopenia after allogeneic non-myeloablative stem cell transplantation. *Isr Med Ass J, 5,* 671–673.

58. Wilkes, N. J., Smith, N. A., & Mallett, S. V. (2000). Anticoagulant-induced pseudothrombocytopenia in a patient presenting for coronary artery bypass grafting. *Br J Anaesth, 84,* 640–642.

59. Bizzaro, N. (1999). Platelet cold agglutinins and cardiac surgery hypothermia. *Am J Hematol, 60,* 80.

60. Dalamangas, L. C., & Slaughter, T. F. (1998). Ethylenediaminetetraacetic acid-dependent pseudothrombocytopenia in a cardiac surgical patient. *Anesth Analg, 86,* 1210–1211.

61. Mert, M., Ozkara, A., & Arat-Ozkan, A. (2004). Very-low platelet counts are not always a contra-indication for cardiac surgery: A case report. *Haema, 7,* 509–511.

62. Satoh, M., Hirose, Y., Gamo, M., et al. (2003). Sudden onset of EDTA-dependent pseudothrombocytopenia in a patient scheduled for open heart surgery. *Masui, 52,* 402–405.

63. Sweeney, J. D., Holme, S., Heaton, W. A. L., et al. (1995). Pseudothrombocytopenia in plateletpheresis donors. *Transfusion, 35,* 46–49.

64. Forscher, C. A., Sussman, I. I., Friedman, E. W., et al. (1985). Pseudothrombocytopenia masking true thrombocytopenia. *Am J Hematol, 18,* 313–317.

65. Chiurazzi, F., Villa, M. R., & Rotoli, B. (1999). Transplacental transmission of EDTA-dependent pseudothrombocytopenia. *Haematologica, 84,* 664.

66. Kunicki, Th. J., & Newman, P. J. (1992). The molecular immunology of human platelet proteins. *Blood, 80,* 1386–1404.

67. Tchernia, G., Morel-Kopp, M. C., Yvart, J., et al. (1993). Neonatal thrombocytopenia and hidden maternal autoimmunity. *Br J Haematol, 84,* 457–463.

68. Morel-Kopp, M. C., Tchernia, G., & Lambert, T. (1992). Autoantibodies directed against the platelet GP complex Ib/IX in normal individuals. *Blood, 80,* 49a.

69. Yadin, O., Sarov, B., Naggan, L., et al. (1989). Natural autoantibodies in the serum of healthy women — a five year follow up. *Clin Exp Immunol, 75,* 402–406.

70. Dighiero, G., Lymberi, P., Guilbert, B., et al. (1986). Natural autoantibodies constitute a substantial part of normal circulating immunoglobulin. *Ann NY Acad Sci, 475,* 135–145.

71. Bartels, P. C. M., Schoorl, M., & Lombarts, A. J. P. F. (1997). Screening for EDTA-dependent deviations in platelet counts and abnormalities in platelet distribution histograms in pseudothrombocytopenia. *Scand J Clin Invest, 57,* 629–636.

72. Stiegler, H. M., Fischer, Y., Steiner, S., et al. (2000). Sudden onset of EDTA-dependent pseudothrombocytopenia after therapy with the glycoprotein IIb-IIIa antagonist c7E3 Fab. *Ann Hematol, 79,* 161–164.

73. Cohen, A. M., Cycowitz, Z., Mittelmann, M., et al. (2000). The incidence of pseudothrombocytopenia in automatic blood analyzers. *Haematologia, 30,* 117–121.

74. Bowen, K. L., Procopio, N., Wystepek, E., et al. (1998). Platelet clumps, nucleated red cells, and leukocyte counts: A comparison between the Abbott CELL-DYN 4000 and Coulter STKS. *Lab Hematol, 4,* 7–16.

75. Lombarts, A. J., De Kieviet, W., Franck, P. F., et al. (1992). Recognition and prevention of two cases of erroneous haemocytometry counts due to platelet and white blood cell aggregation: The use of acid citrate dextrose as an auxiliary anticoagulant. *Eur J Clin Chem Clin Biochem, 30,* 429–432.

76. Stiegler, H. M., Fischer, Y., & Steiner, S. (1999). Thrombocytopenia and glycoprotein IIb-IIIa receptor antagonists. *Lancet, 353,* 1185.

77. Lombarts, A. J. P. F., & De Kieviet, W. (1988). Recognition and prevention of pseudothrombocytopenia and concomitant pseudoleukocytosis. *Am J Clin Pathol, 89,* 634–639.

78. Pujol, M., Ma Ribera, J., Jimenez, C., et al. (1998). Essential monoclonal gammopathy with an IgM paraprotein that is a cryoglobulin with cold agglutinin and EDTA-dependent platelet antibody properties. *Br J Haematol, 100,* 603–604.

79. Bizzaro, N., & Fiorin, F. (1999). Coexistence of erythrocyte agglutination and EDTA-dependent platelet clumping in a patient with thymoma and plasmocytoma. *Arch Pathol Lab Med, 123,* 159–162.

80. Moraglio, D., Banfi, G., & Arnelli, A. (1994). Association of pseudothrombocytopenia and pseudoleukopenia: Evidence for different pathogenic mechanisms. *Scand J Clin Lab Invest, 54,* 257–265.

81. Bizzaro, N. (1993). Granulocyte aggregation is edetic acid and temperature dependent. *Arch Pathol Lab Med, 117,* 528–530.

82. Hillyer, C. D., Knopf, A. N., & Berkman, E. M. (1990). EDTA-dependent leukoagglutination. *Am J Clin Pathol, 94,* 458–461.

83. Hoffmann, J. J. M. L. (2001). EDTA-induced pseudoneutropenia resolved with kanamycin. *Clin Lab Haem, 23,* 193–196.

84. Lesesve, J. F., Haristoy, X., & Lecompte, T. (2002). EDTA-dependent leukoagglutination. *Clin Lab Haem, 24,* 67–69.

85. Wenburg, J. J., & Go, R. S. (2003). EDTA-dependent lymphocyte clumping. *Haematologica, 88,* EIM09.

86. Ryo, R., Sugano, W., Goto, M., et al. (1994). Platelet release reaction during EDTA-induced platelet agglutination by anti-glycoprotein IIb-IIIa complex monoclonal antibody. *Thromb Res, 74,* 265–272.

87. Greipp, P. R., & Gralnick, H. R. (1976). Platelet leukocyte adherence phenomena associated with thrombocytopenia. *Blood, 47,* 513–521.

88. Zeigler, Z. (1974). *In vitro* granulocyte-platelet rosette formation mediated by an IgG immunoglobulin. *Haemostasis, 3,* 282–287.

89. Bizzaro, N., Goldschmeding, R., & von dem Borne, A. E. G. Kr. (1995). Platelet satellitism is Fc gamma RIII (CD16) receptor mediated. *Am J Clin Pathol, 103,* 740–744.

90. Ravetch, J. V., & Perussia, B. (1989). Alternative membrane forms of FcRIII (CD16) on human natural killer cells and neutrophils. *J Exp Med, 170,* 481–485.

91. Español, I., Muñiz-Diaz, E., & Domingo-Claros, A. (2000). Platelet satellitism to granulated lymphocytes. *Haematologica, 85,* 1322.

92. Cohen, A. M., Lewinski, U. H., Klein, B., et al. (1980). Satellitism of platelets to monocytes. *Acta Haematol, 64,* 61–64.

93. Liso, V., & Bonomo, L. (1982). Platelet satellitism to basophils in a patient with chronic myelocytic leukaemia. *Blut, 45,* 347–350.

94. Christopoulos, C., & Mattock, C. (1991). Platelet satellitism and α granule proteins. *J Clin Pathol, 44,* 788–789.

95. Michelson, A. D. (2000). *GPIIb-IIIa antagonists. Hematology 2000.* San Francisco: American Society of Hematology Education Program Book.

96. Christopoulos, C. G., & Machin, S. J. (1994). A new type of pseudothrombocytopenia: EDTA-mediated aggregation of platelets bearing Fab fragments of a chimaeric antibody. *Br J Haematol, 87,* 650–652.

97. Moll, S., Poepping, I., Hauck, S., et al. (1999). Pseudothrombocytopenia after abciximab (ReoPro) treatment. *Circulation, 100,* 1460.

98. Holmes, M. B., Kabbani, S., Watkins, M. W., et al. (2000). Abciximab-associated pseudothrombocytopenia. *Circulation, 101,* 938–939.

99. Wool, R. L., Coleman, T. A., & Hamill, R. L. (2002). Abciximab-associated pseudothrombocytopenia. *Am J Med, 113,* 697–698.

100. Sane, D. C., Damaraju, L. V., Topol, E. J., et al. (2000). Occurrence and clinical significance of pseudothrombocytopenia during Abciximab therapy. *J Am Coll Cardiol, 36,* 75–83.

101. Schell, D. A., Ganti, A. K., Levitt, R., et al. (2002). Thrombocytopenia associated with c7E3 Fab (abciximab). *Ann Hematol, 81,* 76–79.

102. Gawaz, M., Ruf, A., Neumann, F. J., et al. (1998). Effect of glycoprotein IIb/IIIa receptor antagonism on platelet membrane glycoproteins after coronary stent placement. *Thromb Haemostas, 80,* 994–1001.

# CHAPTER 56

# Thrombocytosis and Essential Thrombocythemia

## Ayalew Tefferi

*Department of Internal Medicine, Mayo Clinic, Rochester, Minnesota*

## I. Introduction

The normal platelet count, based on modern automated cell counters, is higher in females than in males, but the 99th percentile value in both sexes as well as across different ethnic backgrounds is less than $400 \times 10^9$/L.[1-5] Therefore, it is reasonable to define thrombocytosis as a platelet count of above $400 \times 10^9$/L. In an individual patient, however, a biologically relevant increase in platelet count might occur without exceeding the population reference range, and this possibility has to be taken into consideration when evaluating a clinical occurrence that is characteristic of a myeloproliferative disorder (MPD) such as essential thrombocythemia (ET) or polycythemia vera (PV).[6,7]

Thrombocytosis can be either congenital or acquired. Congenital thrombocytosis is very rare, and its pathogenesis may involve gain-of-function mutations in either thrombopoietin (TPO)[8-11] or its receptor (Mpl).[12] Acquired thrombocytosis may represent either one of the inherent manifestations of a clonal myeloid process (primary thrombocythemia, or PT) or a secondary process (reactive thrombocytosis, or RT) related to a variety of clinical conditions including infection, tissue damage, malignancy, chronic inflammation, hemolysis, drugs, iron deficiency anemia, and splenectomy. The increase in platelet count associated with PT is attributed to either a growth factor-independent or growth factor-hypersensitive proliferation of megakaryocytes (autonomous platelet production).[13,14] In contrast, RT is often considered as a cytokine-mediated process.[15-19] The distinction between PT and RT is clinically relevant because the former but not the latter is considered to be associated with an increased risk of thrombohemorrhagic complications.[20-23]

## II. Congenital Thrombocytosis

Familial forms of thrombocytosis have been described and mutations of the TPO gene have been identified in some but not all studied families.[8-11,24] Transmission is usually auto-somal dominant, and the described mutation involves the 5′-untranslated regions of the TPO mRNA (a donor splice site) resulting in efficient ligand production.[9] Serum TPO concentrations are accordingly elevated in only those patients with hereditary thrombocytosis associated with a TPO gene mutation. Similarly, TPO production was shown to be increased in cell lines transfected with the mutant gene.[11] Activating *C-Mpl* gene mutations have also been described in affected but not unaffected family members in familial thrombocytosis.[12] In this instance, plasma TPO levels are expected to be normal, but information supporting such contention is currently not available. Interestingly, certain *C-Mpl* mutations have been shown to promote TPO-hypersensitivity.[25] Regardless, gene mutations involving either TPO[26] or Mpl[12,13,27] have yet to be identified in patients with PT.

## III. Reactive Thrombocytosis

### A. Prevalence and Relevance

RT is much more frequent than PT in both children[28-30] and adults[31,32] (Table 56-1). This is true even when only patients with extreme thrombocytosis (platelet count $>1000 \times 10^9$/L) are considered.[33,34] In routine clinical practice, RT accounts for more than 85% of cases with thrombocytosis.[20] The distinction between RT and PT is important because, among other reasons, thromboembolic complications occur significantly more often in patients with PT and high-risk patients with PT often require cytoreductive chemotherapy.[20] In contrast, the prothrombotic potential of RT may be too low to justify specific platelet-directed therapy[29,33-36] even in the context of surgical procedures.[35,37] Instead, the degree of thrombocytosis in RT might indicate either favorable recovery[23] or a grim prognosis[38] from the underlying disease process that is independent of a vascular event. Furthermore, even if one was to entertain a small risk of thrombosis associated with RT,[39] there is no evidence to

## Table 56-1: Causes of Thrombocytosis (Platelet Count of $500 \times 10^9$/L or above) in Unselected Cohorts of Consecutive Patients (Approximate Percentages)

| Condition | Adults (n = 777)[31] | Platelet Count of $\geq1000 \times 10^9$/L (n = 280)[33] | Children (n = 663)[28] |
|---|---|---|---|
| Infection | 22% | 31% | 31% |
| Rebound thrombocytosis | 19% | 3% | 15% |
| Tissue damage (surgery, etc.) | 18% | 14% | 15% |
| Chronic inflammation | 13% | 9% | 4% |
| Malignancy | 6% | 14% | 2% |
| Renal disorders | 5% | NS | 4% |
| Hemolytic anemia | 4% | NS | 19% |
| Post-splenectomy | 2% | 19% | 1% |
| Blood loss | NS | 6% | NS |
| Primary thrombocythemia | 3% | 14% | 0% |

NS, not specified.

support the utilization of cytoreductive therapy in such cases. In contrast, the association of PT with vasomotor symptoms, thrombosis, and bleeding is well established and some, but not all, patients with PT require cytoreductive therapy to prevent catastrophic thrombohemorrhagic complications.[40–43]

### B. Pathogenesis

Table 56-1 lists the most frequent causes of RT. RT that is associated with infection, tissue damage, chronic inflammation, and malignancy is considered an epiphenomenon of a systemic acute phase reaction that is accompanied by an excess of thrombopoietic growth factors including interleukin (IL)-6, TPO, IL-1, IL-4, and tumor necrosis factor-α.[15,17,19,44–52] Among these cytokines, IL-6 is considered the most important, and its mechanism of action may include the induction of TPO production, primarily from the liver.[44,48,49,53] The pleiotropic activity of IL-6 has also been implicated in other nonspecific inflammatory reactions including plasmacytosis, polyclonal hypergammaglobulinemia, leukemoid reaction, and paraneoplastic fever.[50,54] These inflammatory responses, including RT, have been shown to regress with either anti-IL-6 monoclonal antibody treatment[50] or removal of an IL-6 secreting tumor.[51] Conversely, the administration of human recombinant IL-6 in

patients with metastatic cancer has been associated with induction of thrombocytosis and other acute phase reactions.[55] However, cases of thrombocytosis associated with high TPO levels must be distinguished from those that represent a congenital TPO mutation.[30,56] In addition to TPO, IL-1β is another cytokine that has frequently been implicated as a mediator of both acute phase reaction and RT.[57]

Hyposplenism-associated RT may reflect platelet redistribution in the peripheral blood, as well as altered metabolism of thrombopoietic cytokines.[47,58] The incidence of postsplenectomy thrombocytosis depends on the surgical indications and may approach 50% in large unselected series.[59] Postsplenectomy thrombocytosis reaches its peak within a few days of surgery, and the platelet count returns to normal in the majority of the cases within a few months.[60] However, postsplenectomy thrombocytosis may persist for several years after surgery, and in some cases the phenomenon has been ascribed to persistent anemia after the surgical procedure.[61] In addition to surgical asplenia, persistent thrombocytosis has been observed in functional (celiac sprue,[62] amyloidosis[63]) as well as congenital asplenia.[64] Regardless of cause, hyposplenic thrombocytosis, in the absence of a chronic myeloproliferative disorder, is seldom associated with an increased risk of thrombosis.[65–68]

A cytokine-mediated process is also implicated in rebound thrombocytosis following drug- or alcohol-related myelosuppression.[69–71] In this instance, the thrombocytosis that follows thrombocytopenia may be preceded by excess TPO as a feedback response. A similar mechanism might explain the cyclic thrombocythemia that is occasionally seen in patients with myeloproliferative diseases.[72] On the other hand, the mechanism of RT in iron deficiency anemia is poorly understood and may not involve thrombopoietic cytokines.[73,74] In any event, the thrombocytosis subsides with supplemental iron therapy.[74]

## IV. Primary Thrombocythemia and the Myeloproliferative Disorders

PT is one aspect of clonal (neoplastic) myeloproliferation that is seen with chronic myeloid disorders (CMD).[75] The CMD are operationally classified into myelodysplastic syndrome (MDS) and myeloproliferative disorders (MPD) depending on the presence or absence, respectively, of trilineage morphological dysplasia (Table 56-2).[76] The MPD are further classified into classic MPD and atypical MPD. Included in the classic MPD category are chronic myeloid leukemia (CML), essential thrombocythemia (ET), polycythemia vera (PV), and myelofibrosis with myeloid metaplasia (MMM).[77] Atypical MPD include the chronic myelomonocytic leukemias, chronic neutrophilic leukemia, chronic eosinophilic leukemia, hypereosinophilic syndrome, systemic mastocytosis, and others listed in Table 56-2. Both

**Table 56-2: A Semimolecular Classification of Chronic Myeloid Disorders**

1) Myelodysplastic syndrome
2) Myeloproliferative disorders
   a. Classic myeloproliferative disorders
      i. Molecularly-defined
         1. Chronic myeloid leukemia (*Bcr/Abl*⁺)
      ii. Clinicopathologically assigned
         (*Bcr/Abl*⁻ *and frequently associated with JAK2*$^{V617F}$
         *mutation*)
         1. Essential thrombocythemia
         2. Polycythemia vera
         3. Myelofibrosis with myeloid metaplasia
   b. Atypical myeloproliferative disorders
      i. Molecularly defined
         1. PDGFRA-rearranged eosinophilic/mast cell
          disorders (e.g., *FIP1L1-PDGFRA*)
         2. *PDGFRB*-rearranged eosinophilic disorders (e.g.,
          *TEL/ETV6-PDGFRB*)
         3. Systemic mastocytosis associated with *c-kit*
          mutation (e.g., *c-kit*$^{D816V}$)
         4. *8p11* myeloproliferative syndrome (e.g.,
          *ZNF198/FIM/RAMP-FGFR1*)
      ii. Clinicopathologically assigned
         1. Chronic neutrophilic leukemia
         2. Chronic eosinophilic leukemia, molecularly not
          defined
         3. Hypereosinophilic syndrome
         4. Chronic basophilic leukemia
         5. Chronic myelomonocytic leukemia
         6. Juvenile myelomonocytic leukemia (associated with
          recurrent mutations of RAS signaling pathway
          molecules including *PTPN11* and *NF1*)
         7. Systemic mastocytosis, molecularly not defined
         8. Unclassified myeloproliferative disorder

Reproduced with permission.[76]

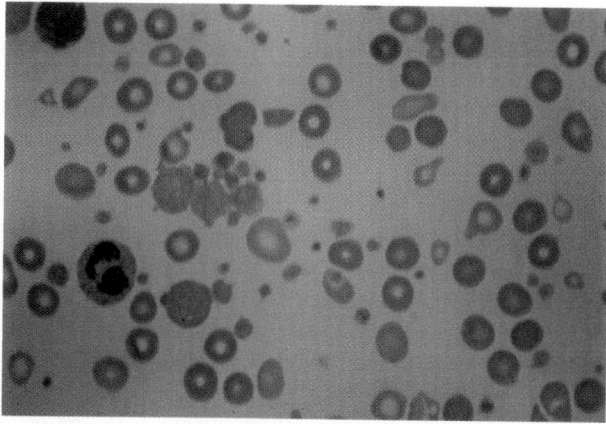

**Figure 56-1.** A peripheral blood smear from a patient with a chronic myeloid disorder showing an increased number of platelets, giant platelets, and marked red blood cell anisocytosis and poikilocytosis.

ated with certain cytogenetic abnormalities including trisomy 8,[97] deletion of the long arm of chromosome 5 (5q-syndrome),[98,99] and abnormalities of chromosome 3,[100,101] as well as the presence of ringed sideroblasts.[96,102] The underlying pathogenetic connection between thrombocytosis and these cytogenetic abnormalities remains to be determined.

## V. Distinguishing Reactive Thrombocytosis from Primary Thrombocythemia

Patient history and physical findings are the most important elements in differentiating PT from RT. The presence of a comorbid condition including acute or subacute infection, a connective tissue disorder, vasculitis, hemolysis, active bleeding, and the immediate postsurgical period or a history of splenectomy favors the diagnosis of RT over PT. The absence of these comorbid conditions associated with a previously documented persistent increase in platelet count suggests PT. Similarly, a historical account of vasomotor symptoms and the presence of positive physical findings, such as acral erythema or splenomegaly, are both highly suggestive of PT. Thrombocytosis that accompanies a thrombohemorrhagic event should always raise the possibility of PT.

Complete blood count, white blood cell differential count, red blood cell indices, and the peripheral blood smear are complementary to the clinical information and provide important clues for the differential diagnosis between RT and PT (Fig. 56-1). An associated increase in either hematocrit value or leukocyte count, for example, suggests the presence of a CMD such as PV or CML. Red blood cell macrocytosis may accompany MDS, whereas microcytosis suggests the presence of iron deficiency anemia as a possible cause of RT. A leukoerythroblastic blood smear suggests

classic and atypical MPD are further divided into molecularly defined and clinicopathologically assigned subcategories (Table 56-2).[76] Most categories of MPD, including ET, represent a clonal stem cell process that involves both the myeloid and lymphoid lineage.[78–85] However, the primary clonogenic mutation has been identified only for CML[86] and a few other atypical MPDs.[87] In this regard, the association of an activating JAK2 tyrosine kinase mutation (JAK2$^{V617F}$) with ET and a spectrum of other MPD promises an additional level of understanding about the molecular pathogenesis of these disorders.[88–94]

Clonal thrombocytosis is an integral feature of ET, and it also occurs in approximately 50% of patients with either PV or MMM.[20,95] Similarly, an increased platelet count is observed in as many as 35% of patients with CML.[95] The incidence of thrombocytosis is much less in both MDS and atypical MPD.[96] In MDS, thrombocytosis has been associ-

MMM, whereas the presence of Howell–Jolly bodies supports a diagnosis of RT from either surgical or functional hyposplenism. On the other hand, the degree of thrombocytosis is a poor discriminator of PT from RT, regardless of how high the platelet count might be.[33,103] Furthermore, some patients with ET may have a clinically relevant increase in platelet count of less than $500 \times 10^9$/L.[6,104]

The diagnostic value of platelet indices (mean volume, size distribution width) is undermined by excess overlap in the measured values between RT and PT.[105–107] Similarly, although platelet function tests are often abnormal in PT (prolonged bleeding time,[108] defects in epinephrine-, collagen-, and ADP-induced platelet aggregation,[109,110] decreased ATP secretion,[111] altered thromboxane generation,[112] increased spontaneous whole blood platelet aggregation[113]), these assays require substantial expertise and are not widely used.

In routine clinical practice, the most cost-effective initial step in the diagnostic workup of thrombocytosis is to determine its duration. Every effort should be made to obtain previous records. In the absence of an obvious explanation, persistent thrombocytosis is highly suggestive of PT. The initial laboratory tests should include the measurement of serum ferritin concentration. A normal serum ferritin level excludes the possibility of iron deficiency anemia-associated RT. However, a low level does not exclude the possibility of PT. The possibility of functional hyposplenism is addressed by examination of the peripheral blood smear for Howell–Jolly bodies. The measurement of C-reactive protein (CRP) is helpful in attending to the possibility of an occult inflammatory or malignant process.[114] Similarly, levels of other acute phase reactants including erythrocyte sedimentation rate,[115] plasma fibrinogen,[116] and plasma IL-6 levels[114] have been shown to be increased during RT. In contrast, plasma TPO levels are not helpful in distinguishing PT from RT.[117–119]

In rare instances, a bone marrow examination might be necessary to exclude the possibility of RT. Although bone marrow histology appears normal in RT, it displays a spectrum of abnormalities in PT depending on the specific underlying CMD. These include increased numbers of megakaryocytes and other myeloid cells, abnormality in cellular morphology, megakaryocyte clusters (Fig. 56-2), and reticulin fibrosis.[95,120,121] In addition to morphological analysis, bone marrow examination allows the performance of cytogenetic studies and special immunohistochemical stains. For example, the detection of a clonal cytogenetic abnormality is diagnostic of PT. However, although some causes of PT are always associated with an abnormal cytogenetic lesion (e.g., CML), less than 5% of patients with ET have detectable cytogenetic abnormalities.[122]

Several studies have reported on the association of an activating *JAK2* mutation with both ET and other MPD.[88–92,94,123,124] The newly identified somatic point muta-

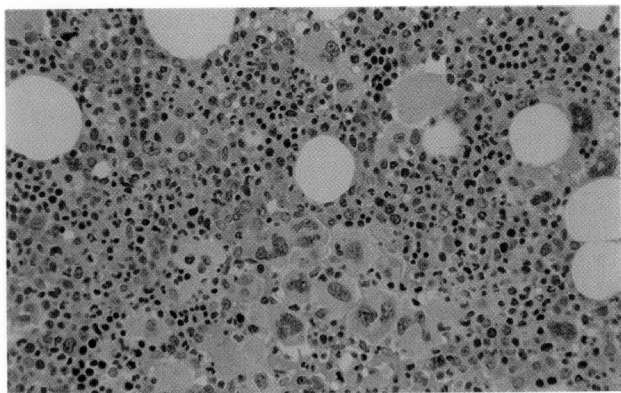

**Figure 56-2.** A hematoxylin-eosin-stained bone marrow biopsy showing megakaryocyte clusters in essential thrombocythemia.

tion is a G-C to T-A transversion, at nucleotide 1849 of exon 12, resulting in the substitution of valine by phenylalanine at codon 617 (JAK2[V617F]). To date, JAK2[V617F] has not been reported in either normal controls[88–91] or patients with secondary erythrocytosis,[89,90,94] which offers an opportunity to incorporate mutation screening for JAK2[V617F] into current diagnostic algorithms for thrombocytosis (Fig. 56-3).[42,124] However, peripheral blood mutation screening cannot currently substitute for bone marrow histology, because JAK2[V617F] is not always detected in patients with PT and is absent in almost half of the patients with ET.[90,91,94,126] Finally, in a research setting, endogenous (without the addition of growth factors) megakaryocyte or erythroid colony growth may be demonstrated in PT but not in RT.[127]

## VI. Essential Thrombocythemia

### A. Introduction and Distinction from Other Causes of Primary Thrombocythemia

ET is currently defined as a persistent thrombocythemic state that is neither reactive nor associated with an otherwise defined CMD including PV, MMM, CML, MDS (Table 56-2). In this regard, it is to be noted that CML,[128,129] MDS,[102,130] and the cellular phase of MMM[95] can all mimic ET in their presentation. Therefore, before making a working diagnosis of ET, in the context of PT, one must exclude CML not only by conventional cytogenetics but also by fluorescence *in situ* hybridization (FISH) test for *BCR/ABL*.[131] Similarly, bone marrow histology should be carefully scrutinized for the presence of both trilineage dysplasia that would suggest MDS and intense marrow cellularity accompanied by atypical megakaryocytic hyperplasia that would suggest cellular phase MMM. The latter and not ET is often accompanied by elevated levels of serum lactate dehydrogenase level, increased peripheral blood CD34 cell count, and a leukoerythroblastic peripheral blood smear.[132,133]

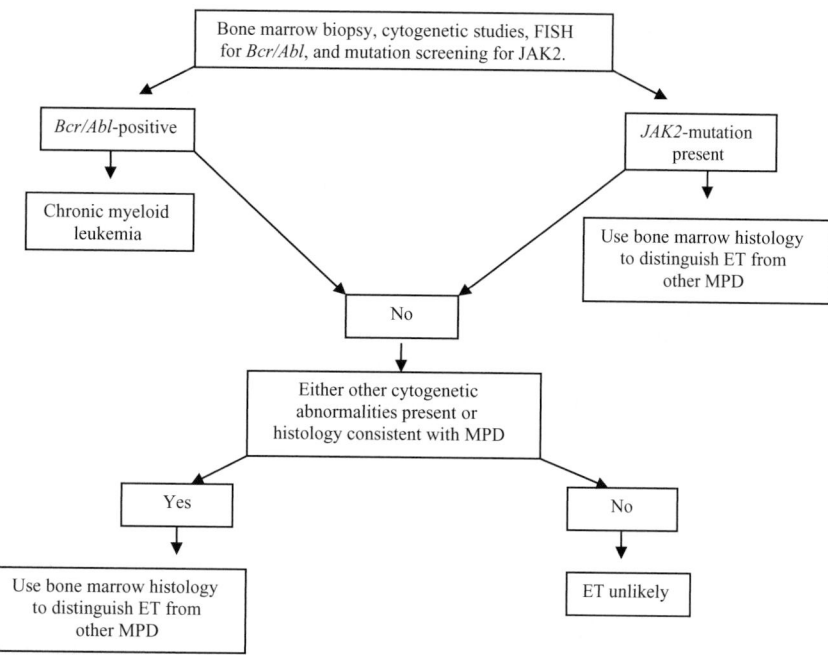

**Figure 56-3.** A diagnostic algorithm for thrombocythemia that is clinically not consistent with reactive thrombocytosis. Abbreviations: ET, essential thrombocythemia; MPD, myeloproliferative disorder. Reproduced with permission from A. Tefferi & D. G. Gilliland, *Mayo Clin Proc, 80,* 2005, 947 (modified).

## B. Epidemiology

Among the classic MPD, ET carries the best prognosis and is therefore the most prevalent (point prevalence rate of above 10/100,000) with annual incidence figures that range from 0.2 to 2.5/100,000.[134–139] Large cohort studies suggest a median age at diagnosis of 48 to 61 years.[140–147] ET is extremely rare in children[148] in whom almost all cases of thrombocytosis are reactive.[149]

## C. Molecular Pathogenesis

The first evidence of monoclonality in ET was suggested in the early 1980s by clonality studies that utilized glucose-6-phosphate dehydrogenase isoenzyme analysis.[85,150] These early observations were subsequently confirmed by X-linked DNA analysis.[151–155] However, such studies have also suggested the existence of both "monoclonal" and "polyclonal" ET based on X chromosome inactivation patterns derived from granulocyte and T lymphocytes.[156,157] Unlike CML, the primary molecular abnormality in ET has not been identified. In this regard, cytogenetic studies have been of limited value because they occur in less than 5% of patients at diagnosis and are nonspecific.[122,158,159] Both structural and numerical abnormalities involving many individual chromosomes, including trisomies 9 and 8, and long-arm deletions of chromosomes 5, 7, 13, 17, and 20 have been associated with ET.[122]

An acquired point mutation involving an autoinhibitory domain of the *JAK2* kinase (*JAK2*[V617F]) has been described in approximately 23 to 57% of patients with ET, but its pathogenetic relevance, especially in view of its frequent occurrence in PV and MMM as well as its infrequent occurrence in atypical MPD, remains to be clarified.[88–94] Homozygosity for the mutant allele is rare in ET.[89–91] JAK2 belongs to the Janus family of cytoplasmic protein tyrosine kinases, and its structure contains two homologous kinase-like domains, JH1 and JH2.[160–162] However, only one (JH1) is functionally intact, whereas JH2 is a pseudo-kinase domain that interacts with JH1 to negatively regulate its kinase activity.[163] Several studies have reported on the association of an activating mutation of JAK2 with both classic (including ET) and atypical MPDs.[88–94] The newly identified somatic point mutation is a G-C to T-A transversion, at nucleotide 1849 of exon 12, resulting in the substitution of valine by phenylalanine at codon 617 (JAK2[V617F]). The JAK2[V617F] occurs within the JH2 domain and interferes with its autoinhibitory function.[164–167] *In vitro,* JAK2[V617F] is associated with constitutive phosphorylation of JAK2 and its downstream effectors, as well as induction of erythropoietin (EPO) hypersensitivity.[89,91,92] *In vivo,* murine bone marrow transduced with a retrovirus containing JAK2[V617F] induced erythrocytosis in the transplanted mice.[89] Taken together, these observations suggest a pathogenetic relevance for the particular mutation in MPD. However, it is currently not clear if JAK2[V617F] represents either a disease-causing mutation or a modifier subclone.

## D. Myeloid Colony Growth and Cytokine Response

Endogenous (i.e., growth factor-independent) *in vitro* erythroid or megakaryocyte colony formation is not seen in either normal subjects or reactive myeloproliferation.[168] EPO-independent erythroid proliferation is primarily seen in PV[169-171] but is also seen in a proportion of patients with ET.[172-175] Furthermore, endogenous colony growth is also manifested by granulocyte[176] and megakaryocyte[177] progenitors. Such growth factor-independence has not been attributed to mutations in the ligand receptor[13,178] or receptor-associated signal transducer molecules.[179] It is well known that erythroid cells in PV are hypersensitive to a variety of cytokines including insulin-like growth factor (IGF-1),[180] stem cell factor,[181] granulocyte-monocyte colony-stimulating factor (GM-CSF),[182] IL-3,[183,184] and TPO.[185] However, a similar growth factor hypersensitivity to IL-3[186] or TPO[14] has also been suggested in ET. The genes for the receptors of both EPO[178,187,188] and TPO[13,26] have been examined in patients with MPD and found to be intact.

Circulating TPO concentration is primarily regulated by megakaryocyte/platelet binding and catabolism[189] and as a result is inversely correlated with platelet and megakaryocyte mass (see Chapter 66). However, in patients with ET,[117] PV,[190] and MMM,[191] serum TPO levels are usually normal or elevated despite an increased megakaryocyte mass. This has been attributed to the markedly decreased megakaryocyte/platelet expression of Mpl in PV and other related MPD.[192-195] Although the specific trait may be used to complement morphological diagnosis in PV and ET, its pathogenetic relevance remains unclear.[196,197]

## E. Pathogenesis of Vascular Events

The pathogenesis of vascular events in ET is not well understood. However, laboratory observations currently suggest better understanding of the bleeding diathesis in affected patients as compared to their hypercoagulable state. In general, bleeding manifestations in ET are often mucocutaneous and exacerbated by the use of aspirin.[21,146,198] The bleeding diathesis in ET is currently believed to involve altered von Willebrand factor (VWF) function in the presence of extreme thrombocytosis (platelet count >1000 × 10$^9$/L).[199-201] Acquired von Willebrand syndrome (AVWS) in ET and other MPD is characterized by the loss of large VWF multimers, which results in a functionally more relevant defect that may not be apparent when measuring VWF antigen and factor VIII coagulant levels alone.[202,203] On the other hand, the assays used to assess VWF function, including collagen binding activity (VWF: CBA) and ristocetin cofactor activity (VWF: RCoA), show that VWF function declines with increasing platelet counts.[204-206] The mechanism of AVWS in ET is currently believed to involve a platelet count-dependent increased proteolysis of high molecular weight VWF by the ADAMTS13 cleaving protease.[202,207-210]

Other potential causes of excessive bleeding in ET include a spectrum of qualitative platelet defects (see also Chapter 58). Acquired storage pool deficiency is a characteristic feature of ET and related MPD, and it is believed to involve abnormal *in vivo* platelet activation with resultant granule release.[211] Evidence for platelet activation in MPD comes from demonstration of increased levels of plasma and urinary arachidonate metabolites (thromboxane [TX] B$_2$), α granule proteins (platelet-derived growth factor, β-thromboglobulin, platelet factor 4), and membrane markers of platelet activation.[212-215] The latter are usually assessed by flow cytometry using monoclonal antibodies to P-selectin, thrombospondin, and the activated fibrinogen receptor, GPIIb-IIIa (see Chapter 30).[216] Other platelet abnormalities in ET and related MPD include decreased adrenergic receptor expression, impaired response to epinephrine, and decreased platelet membrane GPIb and GPIIb-IIIa receptor expression.[216-219]

The pathogenesis of thrombosis in ET remains enigmatic, for the most part. Disease-specific defects in arachidonic acid metabolism have been described and might result in abnormal TXA$_2$ generation.[220-222] Although endogenous TXA$_2$ has not been clinically proven to predict thrombotic complications, the demonstration, in a randomized setting, of the value of low-dose aspirin in reducing thrombosis risk in PV lends support for the clinical significance of spontaneous TXA$_2$ generation in the pathogenesis of the thrombotic phenotype observed in PV and ET.[223] The potential role of granulocytes and monocytes in MPD-associated thrombophilia has been entertained, and such a mechanism, if it existed, would be consistent with the well-established antithrombotic effect of myelosuppressive therapy.[224,225] Furthermore, affected patients display alterations in several neutrophil activation parameters (CD11b, leukocyte alkaline phosphatase, cellular elastase, plasma elastase, and myeloperoxidase), markers of both endothelial damage (thrombomodulin, VWF antigen) and thrombophilic state (thrombin-antithrombin complex, prothrombin fragment 1+2, D-dimer), and the presence of circulating platelet-leukocyte aggregates.[226,227]

Finally, vasomotor symptoms in ET, including headaches, paresthesia, and erythromelalgia, are believed to be linked to small vessel-based abnormal platelet-endothelial interactions.[228] Histopathological studies in erythromelalgia have revealed platelet-rich arteriolar microthrombi with endothelial inflammation and intimal proliferation accompanied by increased platelet consumption that is coupled with abundant VWF deposition.[228-230]

## F. Clinical Features

The majority of patients with ET are asymptomatic at diagnosis.[143] Symptoms and signs of disease at presentation as well as during the clinical course of the disease are generally classified as being life threatening or not. Non-life-threatening events occur in 30 to 50% of patients and include vasomotor symptoms (headache, visual symptoms, lightheadedness, atypical chest pain, acral dysesthesia, erythromelalgia)[142,143,231] and an increased risk of first-trimester miscarriages.[232–234] Erythromelalgia is a vasomotor symptom that is defined as a burning dysesthesia associated with a red discoloration of the hands or feet (Fig. 56-4).[228] The underling cause may involve an arteriolar inflammation secondary to an abnormal platelet-endothelium interaction that may lead to digital ischemic changes.[228] Erythromelalgia is effectively treated with low-dose aspirin (40–325 mg/day), and a prompt response to such treatment favors a diagnosis

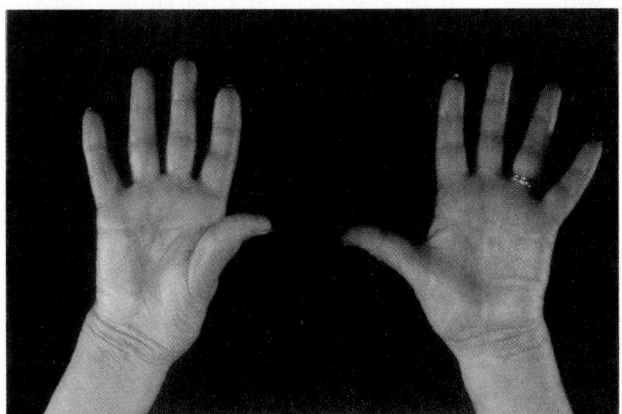

**Figure 56-4.** Erythromelalgia of the hands in a patient with essential thrombocythemia.

of a vasomotor symptom in general over that of a thrombotic vascular episode.[235]

Life-threatening complications of ET include thrombosis, bleeding, and disease transformation into either acute myeloid leukemia (AML) or MMM.[40,41,236] Tables 56-3 and 56-4 list the incidences of both thrombotic and hemorrhagic events in ET.[237] In general, major thrombotic events are more frequent than major bleeding episodes, and arterial events predominate over venous events. Abdominal large vessel and cerebral sinus thrombosis are potentially catastrophic events that occasionally occur in patients with ET.[238–240] With regard to disease transformation, a large retrospective study of 435 patients with ET revealed a 15-year cumulative risk of clonal evolution into either AML or MMM of 2% and 4%, respectively.[236] The results from most other studies are consistent with this observation.[40,41]

## G. Diagnosis

The first step in approaching a patient with thrombocytosis is to entertain the possibility of RT (see Section V). In this regard, it is worth reiterating the value of mutation screening for JAK2$^{V617F}$ in distinguishing ET from RT but not from other MPD (Fig. 56-3).[88–94] In addition, rare cases of genetically defined ET (e.g., activating mutation of the *c-Mpl* gene) have been described[12] and must be kept in mind while evaluating a patient with either a lifelong history of thrombocytosis or a family history of the same.[241] The second step in evaluating a patient with thrombocytosis is to confirm the diagnosis of ET with a bone marrow examination and exclude the possibility of other myeloid disorders. Although detailed analysis of megakaryocyte morphology might assist in distinguishing CML (dwarf megakaryocytes and not too many clusters) from ET (giant megakaryocytes with cluster

**Table 56-3: Thrombotic and Hemorrhagic Events in Essential Thrombocythemia Reported at Diagnosis**

| Reference | n | Platelets × 10$^9$/L (median/mean) | Asymptomatic (%) | Major Thrombosis (%) | Major Arterial Thrombosis$^a$ (%) | Major Venous Thrombosis* (%) | MVD (%) | Total Bleeds (%) (Major) |
|---|---|---|---|---|---|---|---|---|
| 144 | 94 | 1200 | 67 | 22 | 81 | 19 | 43 | 37 (3) |
| 142 | 147 | 1150 | 36 | 18 | 83 | 17 | 34 | 18 (4) |
| 145 | 100 | 1135 | 34 | 11 | 91 | 9 | 30 | 9 (3) |
| 146 | 103 | 1200 | 73 | 23 | 87 | 12 | 33 | 4 (2) |
| 143 | 148 | 898 | 57 | 25 | NA | NA | 29 | 6 (NA) |
| 134 | 96 | 1102 | 52 | 14 | 85 | 15 | 23 | 9 (5) |

Abbreviations: MVD, microvascular disturbances; NA, not available.
$^a$Percentage of total major thrombotic events.
Modified with permission.[237]

**Table 56-4: Thrombotic and Hemorrhagic Events in Essential Thrombocythemia Reported at Follow-up**

| Reference | n | Major Thrombosis (%) | Major Arterial Thrombosis (%)* | Major Venous Thrombosis (%)* | MVD (%) | Total Bleeds (%) (Major) | Deaths from Hemorrhage (%) | Deaths from Thrombosis (%) |
|---|---|---|---|---|---|---|---|---|
| 144 | 94 | 17 | 62 | 38 | 17 | 14 (3) | 0 | 0 |
| 142 | 147 | 14 | 86 | 14 | 4 | NA (1) | 0 | 25 |
| 145 | 100 | 20 | 71 | 29 | NA | NA (1) | 0 | 100 (one pt IAVT) |
| 146 | 103 | 11 | 91 | 9 | 33 | 9 (6) | 0 | 27 |
| 143 | 148 | 22 | 94 | 6 | 28 | 11 (4) | 0 | 13 |
| 134 | 96 | 17 | 69 | 31 | 17 | 14 (7) | 3 | 17 |

Abbreviations: IAVT, intraabdominal venous thrombosis; MVD, microvascular disturbances; NA, not available.
*Percentage of total major thrombotic events.
Modified with permission.[237]

**Table 56-5: Risk Stratification in Essential Thrombocythemia**

| | |
|---|---|
| **Low risk** | Age below 60 years, **and** No history of thrombosis, **and** Platelet count below $1500 \times 10^9$/L, **and** Absence of cardiovascular risk factors (smoking, hypertension, hyperlipidemia, diabetes) |
| **Indeterminate risk** | Neither low risk nor high risk |
| **High risk** | Age 60 years or older **or** A positive history of thrombosis |

formation), cytogenetic studies and FISH for *BCR/ABL* should accompany bone marrow examination to rule out the possibility of CML.[128] Mild reticulin fibrosis (grades 1 or 2) is detected in approximately 14% of patients with ET at diagnosis and does not portend an unusual outcome.[231] Clonal cytogenetic lesions in ET are detected in fewer than 5% of cases and are nonspecific.[122]

## H. Prognosis

ET is associated with a near-normal life expectancy and, therefore, treatment is never instituted for the purpose of improving survival.[236,242] Instead, specific therapy is sought to either alleviate microvascular disturbances (e.g., headache, lightheadedness, acral paresthesia, erythromelalgia, atypical chest pain) or prevent thrombohemorrhagic complications. Treatment to prevent vascular events is dictated by the presence or absence of defined risk factors for thrombosis (Table 56-5).[43] Accordingly, high-risk disease is defined by the presence of either age ≥60 years or history of throm-

bosis.[141,143,144,243] In the absence of these two adverse features, patients are assigned to either a low-risk or indeterminate-risk disease category based on an unsubstantiated concern that extreme thrombocytosis (platelet count $\geq 1500 \times 10^9$/L) as well as the presence of cardiovascular risk factors might be detrimental in terms of thrombohemorrhagic complications (Table 56-5).[41,141,145,243] In the latter regard, hypercholesterolemia, smoking, and hypertension have all been shown to influence the risk of thrombosis by different studies.[243–246] However, other large retrospective studies could not confirm these associations.[247–250]

Several studies have explored the contribution of hereditary and acquired causes of thrombophilia to the occurrence of thrombotic events in MPD, and the findings have so far been inconsistent. For example, two prospective studies found no difference in the allele frequencies of factor V Leiden, prothrombin G20210A, and methylene tetrahydrofolate reductase mutations among ET patients with and without thrombotic complications,[251,252] whereas another retrospective study suggested an increase in the prevalence of the factor V Leiden mutation in patients with a history of venous thrombotic events.[253] Similarly, although several studies have demonstrated elevated levels of homocysteine among patients with MPD,[254–256] the clinical relevance of this observation to arterial thrombosis is suggested by one[256] but not other studies.[254,255] An increased prevalence of antiphospholipid antibodies in patients with ET has also been described, but its clinical relevance remains to be carefully evaluated.[257,258]

## I. Treatment

Before considering any form of specific therapy for the patient with ET, one must define the goal of therapy, as well

**Table 56-6: Treatment Algorithm in Essential Thrombocythemia**

| Risk Category | Age <60 years | Age ≥60 years | Women of Childbearing Age |
|---|---|---|---|
| Low risk | Low-dose aspirin[a] | Not applicable | Low-dose aspirin[a] |
| Indeterminate risk[b] | Low-dose aspirin[a] | Not applicable | Low-dose aspirin[a] |
| High risk | Hydroxyurea **and** Low-dose aspirin | Hydroxyurea **and** Low-dose aspirin | Interferon alfa **and** Low-dose aspirin |

[a]In the absence of a contraindication, including acquired von Willebrand syndrome — that is, a ristocetin cofactor activity of less than 50%.
[b]The decision to use cytoreductive agents in indeterminate-risk patients should be made on an individual basis (see text for details).

as produce the evidence that supports such an action. If the goal is to alleviate microvascular symptoms such as headaches or erythromelalgia, then the use of low-dose aspirin (40–100 mg/day) is appropriate after excluding the possibility of clinically significant AVWS in patients with extreme thrombocytosis.[237,259] If such non-life-threatening symptoms are not effectively controlled by aspirin, then it is reasonable to institute a therapeutic trial with a cytoreductive agent to determine if increased platelet count can be blamed for the symptoms. If so, the goal of such therapy would be to decrease the platelet count to a target that provides relief for the patient. In asymptomatic cases of ET-associated AVWS, prophylactic cytoreduction is advised only in the presence of a clinically relevant reduction in VWF function (e.g., ristocetin cofactor activity less than 20%).[237]

As discussed in Section H, cytoreductive therapy in ET is never instituted to either prolong life or prevent clonal evolution into AML.[236] The usual current indication for cytoreductive therapy in ET is to prevent thrombohemorrhagic events. Accordingly, patients with low-risk disease are usually not offered cytoreductive therapy because their risk of thrombosis might not be significantly different than that of the control population (Tables 56-5 and 56-6).[41,141–143,145,260–263] On the other hand, the use of low-dose aspirin is encouraged in such patients based on its antithrombotic value in PV.[264] Similarly, the low-risk pregnant patient should not receive any cytoreductive agent, and the use of aspirin is optional and may not influence outcome of pregnancy.[265]

The risk of thrombosis in high-risk disease (Table 56-5) is substantial enough to consider cytoreductive therapy.[145] In this regard, the antithrombotic value of cytoreductive therapy in high-risk ET has been addressed by two randomized treatment trials.[261,266] In the first study, treatment with hydroxyurea was compared to observation alone, and the risk of thrombosis was significantly less in the treated group (3.6% versus 24%).[261] In the second study, hydroxyurea was compared to anagrelide, both in combination with aspirin therapy.[266] After a median follow-up of 39 months, the composite risk of both thrombosis and bleeding was once again

favorably affected by hydroxyurea treatment (36 versus 55 events in the anagrelide arm). In the case of hydroxyurea intolerance,[267] interferon alfa is a reasonable alternative and is the drug of choice during pregnancy (Table 56-6).[268,269] When both hydroxyurea and interferon alfa are not tolerated, other drugs including anagrelide and pipobroman might be considered (Table 56-7). Once cytoreductive therapy is initiated, the therapeutic goal in terms of platelet count, based on anecdotal evidence of optimal thrombosis control, is $<400 \times 10^9/L$.[104,270] Although antiplatelet therapy has not yet prospectively been shown to reduce the incidence of thrombosis in ET, I recommend its use in all high-risk patients along with cytoreductive therapy (Table 56-6).[249,250,271,272]

The use of cytoreductive therapy in indeterminate-risk patients with ET is controversial. In general, I recommend the use of low-dose aspirin in the absence of the clinically relevant AVWS (ristocetin cofactor activity of less than 20%) that may occur in a minority of patients with extreme thrombocytosis.[273] In addition, it is reasonable to consider cytoreductive treatment in patients with extreme thrombocytosis (platelet count over $1500 \times 10^9/L$) that is associated with a bleeding diathesis or aspirin-resistant microvascular symptoms. The target platelet count in this instance is the level that results in symptom relief or correction of the bleeding diathesis.

Physicians in practice are often concerned about the possibility of leukemia arising from the use of hydroxyurea, which is the current choice of initial cytoreductive therapy in ET. This represents an unsubstantiated fear that unfortunately led to the use of alternative drugs that might have been detrimental to patients. First, it should be remembered that the occurrence of disease transformation in ET is very low and estimated at 2% for leukemic transformation and 4% for fibrotic transformation at 15 years.[236] Second, none of the large retrospective[236,274] and prospective controlled studies[266] have ever shown an association between hydroxyurea use and the development of AML in ET. On the other hand, new information suggests that both anagrelide and interferon alfa may increase the risk of transformation into MMM.[266,275]

**Table 56-7: Clinical Properties of Platelet-Lowering Agents**

| Drug | Hydroxyurea | Anagrelide | Interferon alfa | Phosphorus-32 | Pipobroman |
|---|---|---|---|---|---|
| **Class** | Myelosuppressive | Unknown | Myelosuppressive | Myelosuppressive | Myelosuppressive |
| **Mechanism of action** | Antimetabolite | Unknown | Biologic agent | Radionuclide | Alkylating agent |
| **Pharmacology** | Half-life ≈5 hours, renal excretion | Half-life ≈1.5 hours, renal excretion | Kidney is main site of metabolism | Half-life ≈14 days | Insufficient information |
| **Starting dose** | 500 mg PO BID | 0.5 mg PO TID | 5 million units SC TIW | 2.3 mCi/m2 IV | 1 mg/kg/day PO |
| **Onset of action** | ≈3 to 5 days | ≅6 to 10 days | 1 to 3 weeks | 4 to 8 weeks | ≈16 days |
| **Frequent side effects** | Leukopenia, oral ulcers, anemia, hyperpigmentation, nail discoloration (Fig. 56-5), xerodermia | Headache, palpitations, diarrhea, fluid retention, anemia | Flulike syndrome, fatigue, anorexia, weight loss, lack of ambition, alopecia | Transient mild cytopenia | Nausea, abdominal pain, diarrhea |
| **Infrequent side effects** | Leg ulcers, alopecia, skin atrophy | Arrhythmias | Confusion, depression, autoimmune thyroiditis, myalgia, arthritis | Prolonged pancytopenia in elderly patients | Leukopenia, thrombocytopenia, hemolysis |
| **Rare side effects** | Fever, cystitis platelet oscillations | Cardiomyopathy | Pruritis, hyperlipidemia, transaminasemia | Leukemogenic | |
| **Cost[a]** | Annual = $1714, for 500 mg TID dose | Annual = $8500, for 0.5 mg QID dose | Annual = $10,500, for 3 million units 5 days/ week | Approximately $1025 for 4 mCi | Not available in the United States |

References: Hydroxyurea,[72,283–290] anagrelide,[270,291–297] interferon alfa,[269,298–301] phosphorus-32,[302–304] pipobroman.[287,305–308]

BID, twice a day; IV, intravenous; PO, oral; QID, four times a day; SC, subcutaneous; TID, three times a day; TIW, three times a week.

[a]Current cost to patient in the United States.

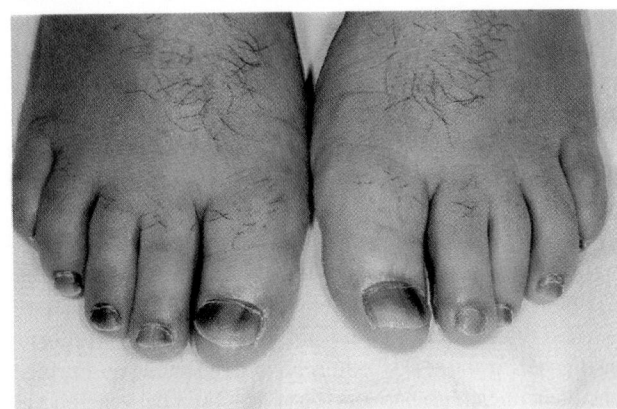

**Figure 56-5.** Hydroxyurea-induced nail discoloration.

### J. Management of ET-Associated Thrombosis and Hemorrhage

ET-associated acute thrombosis should be managed with both systemic anticoagulation and concomitant cytoreductive therapy.[225] In addition, although the use of aspirin in combination with oral anticoagulant therapy is discouraged in most instances, it is not unreasonable to consider such combination therapy in individual cases when indicated. Similarly, although there is no controlled evidence that supports the use of platelet apheresis in any situation, I currently recommend platelet apheresis for the acute management of hemorrhage or thrombosis that is accompanied by a platelet count >1000 × 10$^9$/L along with the prompt institution of cytoreductive therapy.[276] Other current indications for platelet apheresis include ET-associated AVWS associated with major hemorrhage and as a prophylactic measure before major surgery.[202,276–279] With regard to symptomatic ET-associated AVWS, there is usually no need for the application of therapeutic approaches that are used in the management of congenital von Willebrand disease. Furthermore, the increased rate of VWF proteolysis seen in this scenario would render treatment with either desmopressin or VWF-containing therapeutic products of limited value.[202,205,208,277,280–282]

## VII. Conclusions

In 2005, two major events occurred that are pertinent to ET. In one, a new activating mutation of the JAK2 tyrosine kinase (JAK2$^{V617F}$) was identified as a molecule of interest, and its clinical relevance is being defined.[124,126] The second event involves the completion and publication of the final results from the UK MRC PT1 trial that demonstrated the superiority of hydroxyurea over anagrelide in the treatment of patients with high-risk ET.[266] For now, the practical value of mutation screening is limited to disease diagnosis, whereas the results of the MRC PT1 study support the use of hydroxyurea as the initial treatment of choice for patients who are at increased risk of thrombosis. However, much remains unknown both in the science and practice of ET. JAK2$^{V617F}$ is also found in other MPDs, and it is unlikely that it represents a disease-causing mutation in ET. However, preliminary data from the Mayo Clinic indicates that ET patients carrying the JAK2$^{V617F}$ mutation are more likely to display a higher hematocrit level at diagnosis as well as transform to MMM. Thus, the specific mutation might be responsible for the PV phenotype that is shared by patients with ET. Studies are ongoing to molecularly characterize JAK2$^{V617F}$–negative ET patients, as well as identify additional mutations in those who carry the JAK2$^{V617F}$ mutant allele. With regard to treatment, based on current information, it might be more productive as well as cost-effective to divert resources into laboratory investigations of molecular pathogenesis rather than to pursue additional large-scale phase III clinical trials.

# References

1. Ruocco, L., Del Corso, L., Romanelli, A. M., et al. (2001). New hematological indices in the healthy elderly. *Minerva Med, 92,* 69–73.
2. Brummitt, D. R., & Barker, H. F. (2000). The determination of a reference range for new platelet parameters produced by the Bayer ADVIA120 full blood count analyser. *Clin Lab Haematol, 22,* 103–107.
3. Lozano, M., Narvaez, J., Faundez, A., et al. (1998). Platelet count and mean platelet volume in the Spanish population. *Med Clin (Barc), 110,* 774–777.
4. Gevao, S. M, Pabs-Garnon, E., & Williams, A. C. (1996). Platelet counts in healthy adult Sierra Leoneans. *West Afr J Med, 15,* 163–164.
5. Ross, D. W., Ayscue, L. H., Watson, J., et al. (1998). Stability of hematologic parameters in healthy subjects: Intraindividual versus interindividual variation. *Am J Clin Pathol, 90,* 262–267.
6. Lengfelder, E., Hochhaus, A., Kronawitter, U., et al. (1998). Should a platelet limit of $600 \times 10(9)/L$ be used as a diagnostic criterion in essential thrombocythaemia? An analysis of the natural course including early stages. *Br J Haematol, 100,* 15–23.
7. Sacchi, S., Vinci, G., Gugliotta, L., et al. (2000). Diagnosis of essential thrombocythemia at platelet counts between 400 and $600 \times 10(9)/L$. Gruppo Italiano Malattie Mieloproliferative Croniche (GIMMC). *Haematologica, 85,* 492–495.
8. Kondo, T., Okabe, M., Sanada, M., et al. (1998). Familial essential thrombocythemia associated with one-base deletion in the 5′-untranslated region of the thrombopoietin gene. *Blood, 92,* 1091–1096.
9. Wiestner, A., Schlemper, R. J., Vandermaas, A. P. C., et al. (1998). An activating splice donor mutation in the thrombopoietin gene causes hereditary thrombocythaemia. *Nat Genet, 18,* 49–52.
10. Wiestner, A., Padosch, S. A., Ghilardi, N., et al. (2000). Hereditary thrombocythaemia is a genetically heterogeneous disorder: Exclusion of TPO and MPL in two families with hereditary thrombocythaemia. *Br J Haematol, 110,* 104–109.
11. Ghilardi, N., Wiestner, A., Kikuchi, M., et al. (1999). Hereditary thrombocythaemia in a Japanese family is caused by a novel point mutation in the thrombopoietin gene. *Br J Haematol, 107,* 310–316.
12. Ding, J., Komatsu, H., Wakita, A., et al. (2004). Familial essential thrombocythemia associated with a dominant-positive activating mutation of the c-MPL gene, which encodes for the receptor for thrombopoietin. *Blood, 103,* 4198–4200.
13. Taksin, A. L., Couedic, J. P. L., Dusanter-Fourt, I., et al. (1999). Autonomous megakaryocyte growth in essential thrombocythemia and idiopathic myelofibrosis is not related to a c-mpl mutation or to an autocrine stimulation by Mpl-L. *Blood, 93,* 125–139.
14. Axelrad, A. A., Eskinazi, D., Correa, P. N., et al. (2000). Hypersensitivity of circulating progenitor cells to megakaryocyte growth and development factor (PEG-rHu MGDF) in essential thrombocythemia. *Blood, 96,* 3310–3321.
15. Dan, K., Gomi, S., Inokuchi, K., et al. (1995). Effects of interleukin-1 and tumor necrosis factor on megakaryocytopoiesis: Mechanism of reactive thrombocytosis. *Acta Haematol, 93,* 67–72.
16. Hollen, C. W., Henthorn, J., Koziol, J. A., et al. (1991). Elevated serum interleukin-6 levels in patients with reactive thrombocytosis. *Br J Haematol, 79,* 286–290.
17. Haznedaroglu, I. C., Ertenli, I., Ozcebe, O. I., et al. (1996). Megakaryocyte-related interleukins in reactive thrombocytosis versus autonomous thrombocythemia. *Acta Haematol, 95,* 107–111.
18. Ertenli, I., Kiraz, S., Ozturk, M. A., et al. (2003). Pathologic thrombopoiesis of rheumatoid arthritis. *Rheumatol Int, 23,* 49–60.
19. Wolber, E. M., Fandrey, J., Frackowski, U., et al. (2001). Hepatic thrombopoietin mRNA is increased in acute inflammation. *Thromb Haemost, 86,* 1421–1424.
20. Griesshammer, M., Bangerter, M., Sauer, T., et al. (1999). Aetiology and clinical significance of thrombocytosis: Analysis of 732 patients with an elevated platelet count. *J Intern Med, 245,* 295–300.
21. Randi, M. L., Stocco, F., Rossi, C., et al. (1991). Thrombosis and hemorrhage in thrombocytosis: Evaluation of a large cohort of patients (357 cases). *J Med, 22,* 213–223.
22. Buss, D. H., Stuart, J. J., & Lipscomb, G. E. (1985). The incidence of thrombotic and hemorrhagic disorders in association with extreme thrombocytosis: An analysis of 129 cases. *Am J Hematol, 20,* 365–372.

23. Valade, N., Decailliot, F., Rebufat, Y., et al. (2005). Thrombocytosis after trauma: Incidence, aetiology, and clinical significance. *Br J Anaesth, 94,* 18–23.

24. Stuhrmann, M., Bashawri, L., Ahmed, M. A., et al. (2001). Familial thrombocytosis as a recessive, possibly X-linked trait in an Arab family. *Br J Haematol, 112,* 616–620.

25. Onishi, M., Mui, A. L., Morikawa, Y., et al. (1996). Identification of an oncogenic form of the thrombopoietin receptor MPL using retrovirus-mediated gene transfer. *Blood, 88,* 1399–1406.

26. Harrison, C. N., Gale, R. E., Wiestner, A. C., et al. (1998). The activating splice mutation in intron 3 of the thrombopoietin gene is not found in patients with non-familial essential thrombocythaemia. *Br J Haematol, 102,* 1341–1343.

27. Kiladjian, J. J., Elkassar, N., Hetet, G., et al. (1997). Study of the thrombopoietin receptor in essential thrombocythemia. *Leukemia, 11,* 1821–1826.

28. Yohannan, M. D., Higgy, K. E., al-Mashhadani, S. A., et al. (1994). Thrombocytosis. Etiologic analysis of 663 patients. *Clin Pediatr (Phila), 33,* 340–343.

29. Chen, H. L., Chiou, S. S., Sheen, J. M., et al. (1999). Thrombocytosis in children at one medical center of southern Taiwan. *Acta Paediatr Taiwan, 40,* 309–313.

30. Dame, C., & Sutor, A. H. (2005). Primary and secondary thrombocytosis in childhood. *Br J Haematol, 129,* 165–177.

31. Santhosh-Kumar, C. R., Yohannan, M. D., Higgy, K. E., et al. (1991). Thrombocytosis in adults: Analysis of 777 patients. *J Intern Med, 229,* 493–495.

32. Robbins, G., & Barnard, D. L. (1983). Thrombocytosis and microthrombocytosis: A clinical evaluation of 372 cases. *Acta Haematol, 70,* 175–182.

33. Buss, D. H., Cashell, A. W., O'Connor, M. L., et al. (1994). Occurrence, etiology, and clinical significance of extreme thrombocytosis: A study of 280 cases. *Am J Med, 96,* 247–253.

34. Chuncharunee, S., Archararit, N., Ungkanont, A., et al. (2000). Etiology and incidence of thrombotic and hemorrhagic disorders in Thai patients with extreme thrombocytosis. *J Med Assoc Thai, 83* (Suppl 1), S95–S100.

35. Coon, W. W., Penner, J., Clagett, P., et al. (1978). Deep venous thrombosis and postsplenectomy thrombocytosis. *Arch Surg, 113,* 429–431.

36. Saathoff, A. D., Elkins, S. L., Chapman, S. W., et al. (2005). Thrombocytosis during antifungal therapy of candidemia. *Ann Pharmacother, 39,* 1238–1243.

37. Meekes, I., van der Staak, F., & van Oostrom, C. (1995). Results of splenectomy performed on a group of 91 children. *Eur J Pediatr Surg, 5,* 19–22.

38. Shimada, H., Oohira, G., Okazumi, S., et al. (2004). Thrombocytosis associated with poor prognosis in patients with esophageal carcinoma. *J Am Coll Surg, 198,* 737–741.

39. Keung, Y. K., & Owen, J. (2004). Iron deficiency and thrombosis: Literature review. *Clin Appl Thromb Hemost, 10,* 387–391.

40. Harrison, C. N. (2005). Essential thrombocythaemia: Challenges and evidence-based management. *Br J Haematol, 130,* 153–165.

41. Barbui, T., Barosi, G., Grossi, A., et al. (2004). Practice guidelines for the therapy of essential thrombocythemia: A statement from the Italian Society of Hematology, the Italian Society of Experimental Hematology and the Italian Group for Bone Marrow Transplantation. *Haematologica, 89,* 215–232.

42. Tefferi, A. (2003). Polycythemia vera: A comprehensive review and clinical recommendations. *Mayo Clin Proc, 78,* 174–194.

43. Tefferi, A., & Murphy, S. (2001). Current opinion in essential thrombocythemia: Pathogenesis, diagnosis, and management. *Blood Rev, 15,* 121–131.

44. Ishiguro, A., Ishikita, T., Shimbo, T., et al. (1998). Elevation of serum thrombopoietin precedes thrombocytosis in Kawasaki disease. *Thromb Haemost, 79,* 1096–1100.

45. Heits, F., Stahl, M., Ludwig, D., et al. (1999). Elevated serum thrombopoietin and interleukin-6 concentrations in thrombocytosis associated with inflammatory bowel disease. *J Interferon Cytokine Res, 19,* 757–760.

46. Ertenli, I., Haznedaroglu, I. C., Kiraz, S., et al. (1996). Cytokines affecting megakaryocytopoiesis in rheumatoid arthritis with thrombocytosis. *Rheumatol Int, 16,* 5–8.

47. Chuncharunee, S., Archararit, N., Hathirat, P., et al. (1997). Levels of serum interleukin-6 and tumor necrosis factor in postsplenectomized thalassemic patients. *J Med Assoc Thai, 80* (Suppl 1), S86–S91.

48. Estrov, Z., Talpaz, M., Mavligit, G., et al. (1995). Elevated plasma thrombopoietic activity in patients with metastatic cancer-related thrombocytosis. *Am J Med, 98,* 551–558.

49. Wolber, E. M., & Jelkmann, W. (2000). Interleukin-6 increases thrombopoietin production in human hepatoma cells HepG2 and Hep3B. *J Interferon Cytokine Res, 20,* 499–506.

50. Blay, J. Y., Rossi, J. F., Wijdenes, J., et al. (1997). Role of interleukin-6 in the paraneoplastic inflammatory syndrome associated with renal-cell carcinoma. *Int J Cancer, 72,* 424–430.

51. Takagi, M., Egawa, T., Motomura, T., et al. (1997). Interleukin-6 secreting phaeochromocytoma associated with clinical markers of inflammation. *Clin Endocrinol (Oxf), 46,* 507–509.

52. Hwang, S. J., Luo, J. C., Li, C. P., et al. (2004). Thrombocytosis: A paraneoplastic syndrome in patients with hepatocellular carcinoma. *World J Gastroenterol, 10,* 2472–2477.

53. Ishiguro, A., Suzuki, Y., Mito, M., et al. (2002). Elevation of serum thrombopoietin precedes thrombocytosis in acute infections. *Br J Haematol, 116,* 612–618.

54. Barton, B. E. (1996). The biological effects of interleukin 6. *Med Res Rev, 16,* 87–109.

55. Stouthard, J. M., Goey, H., de Vries E. G., et al. (1996). Recombinant human interleukin 6 in metastatic renal cell cancer: A phase II trial. *Br J Cancer, 73,* 789–793.

56. Hankins, J., Naidu, P., Rieman, M., et al. (2004). Thrombocytosis in an infant with high thrombopoietin concentrations. *J Pediatr Hematol Oncol, 26,* 142–145.

57. Chuen, C. K., Li, K., Yang, M., et al. (2004). Interleukin-1beta up-regulates the expression of thrombopoietin and

transcription factors c-Jun, c-Fos, GATA-1, and NF-E2 in megakaryocytic cells. *J Lab Clin Med, 143,* 75–88.

58. Ichikawa, N., Kitano, K., Shimodaira, S., et al. (1998). Changes in serum thrombopoietin levels after splenectomy. *Acta Haematol, 100,* 137–141.

59. Traetow, W. D., Fabri, P. J., & Carey, L. C. (1980). Changing indications for splenectomy: 30 years' experience. *Arch Surg, 115,* 447–451.

60. Breslow, A., Kaufman, R. M., & Lawsky, A. R. (1968). The effect of surgery on the concentration of circulating mega-karyocytes and platelets. *Blood, 32,* 393–401.

61. Hirsh, J., & Dacie, J. V. (1966). Persistent post-splenectomy thrombocytosis and thrombo-embolism: A consequence of continuing anaemia. *Br J Haematol, 12,* 44–53.

62. Croese, J., Harris, O., & Bain, B. (1999). Coeliac disease: Haematological features, and delay in diagnosis. *Med J Aust, 2,* 335–338.

63. Gertz, M. A., Kyle, R. A., & Greipp, P. R. (1983). Hyposplen-ism in primary systemic amyloidosis. *Ann Intern Med, 98,* 475–477.

64. Chanet, V., Tournilhac, O., Dieu-Bellamy, V., et al. (2000). Isolated spleen agenesis: A rare cause of thrombocytosis mimicking essential thrombocythemia. *Haematologica, 85,* 1211–1213.

65. Gordon, D. H., Schaffner, D., Bennett, J. M., et al. (1998). Postsplenectomy thrombocytosis: Its association with mes-enteric, portal, and/or renal vein thrombosis in patients with myeloproliferative disorders. *Arch Surg, 113,* 713–715.

66. Visudhiphan, S., Ketsa-Ard, K., Piankijagum, A., et al. (1985). Blood coagulation and platelet profiles in persistent post-splenectomy thrombocytosis: The relationship to throm-boembolism. *Biomed Pharmacother, 39,* 264–271.

67. Dawson, A. A., Bennett, B., Jones, P. F., et al. (1981). Throm-botic risks of staging laparotomy with splenectomy in Hodg-kin's disease. *Br J Surg, 68,* 842–845.

68. Boxer, M. A., Braun, J., & Ellman, L. (1978). Thromboem-bolic risk of postsplenectomy thrombocytosis. *Arch Surg, 113,* 808–809.

69. Schmitt, M., Gleiter, C. H., Nichol, J. L., et al. (1999). Haematological abnormalities in early abstinent alcoholics are closely associated with alterations in thrombopoietin and erythropoietin serum profiles. *Thromb Haemost, 82,* 1422–1427.

70. Radley, J. M., Hodgson, G. S., Thean, L. E., et al. (1980). Increased megakaryocytes in the spleen during rebound thrombocytosis following 5-fluorouracil. *Exp Hematol, 8,* 1129–1138.

71. Haselager, E. M., & Vreeken, J. (1977). Rebound thrombo-cytosis after alcohol abuse: A possible factor in the pathogenesis of thromboembolic disease. *Lancet, 1,* 774–775.

72. Tefferi, A., Elliott, M. A., Kao, P. C., et al. (2000). Hydroxy-urea-induced marked oscillations of platelet counts in patients with polycythemia vera. *Blood, 96,* 1582–1584.

73. Akan, H., Guven, N., Aydogdu, I., et al. (2000). Thrombo-poietic cytokines in patients with iron deficiency anemia with or without thrombocytosis. *Acta Haematol, 103,* 152–156.

74. Deray, G., Lejonc, J. L., & Galacteros, F. (1984). Thrombo-cytosis as a feature of iron-deficiency therapeutic correction. *Arch Intern Med, 144,* 414–415.

75. Vardiman, J. W., Harris, N. L., & Brunning, R. D. (2002). The World Health Organization (WHO) classification of the myeloid neoplasms. *Blood, 100,* 2292–2302.

76. Tefferi, A., & Gilliland, D. G. (in press). Classification of myeloproliferative disorders: From Dameshek towards a semi-molecular system. *Best Hematology Practice & Research in Clinical Haematology.*

77. Dameshek, W. (1951). Some speculations on the myelopro-liferative syndromes. *Blood, 6,* 372–375.

78. Fialkow, P. J., Gartler, S. M., & Yoshida, A. (1967). Clonal origin of chronic myelocytic leukemia in man. *Proc Natl Acad Sci USA, 58,* 1468–1471.

79. Barr, R. D., & Fialkow, P. J. (1973). Clonal origin of chronic myelocytic leukemia. *N Engl J Med, 289,* 307–309.

80. Adamson, J. W., Fialkow, P. J., Murphy, S., et al. (1976). Polycythemia vera: Stem-cell and probable clonal origin of the disease. *N Engl J Med, 295,* 913–916.

81. Fialkow, P. J., Jacobson, R. J., & Papayannopoulou, T. (1977). Chronic myelocytic leukemia: Clonal origin in a stem cell common to the granulocyte, erythrocyte, platelet and mono-cyte/macrophage. *Am J Med, 63,* 125–130.

82. Fialkow, P. J., Denman, A. M., Jacobson, R. J., et al. (1978). Chronic myelocytic leukemia: Origin of some lymphocytes from leukemic stem cells. *J Clin Invest, 62,* 815–823.

83. Jacobson, R. J., Salo, A., & Fialkow, P. J. (1978). Agnogenic myeloid metaplasia: A clonal proliferation of hematopoietic stem cells with secondary myelofibrosis. *Blood, 51,* 189–194.

84. Martin, P. J., Najfeld, V., Hansen, J. A., et al. (1980). Involve-ment of the B-lymphoid system in chronic myelogenous leu-kaemia. *Nature, 287,* 49–50.

85. Fialkow, P. J., Faguet, G. B., Jacobson, R. J., et al. (1981). Evidence that essential thrombocythemia is a clonal disorder with origin in a multipotent stem cell. *Blood, 58,* 916–919.

86. Nowell, P. C., & Hungerford, D. A. (1960). A minute chro-mosome in human chronic granulocytic leukemia. *J Natl Cancer Inst, 25,* 85.

87. Cools, J., DeAngelo, D. J., Gotlib, J., et al. (2003). A tyrosine kinase created by fusion of the PDGFRA and FIP1L1 genes as a therapeutic target of imatinib in idiopathic hypereosino-philic syndrome. *N Engl J Med, 348,* 1201–1214.

88. Baxter, E. J., Scott, L. M., Campbell, P. J., et al. (2005). Acquired mutation of the tyrosine kinase JAK2 in human myeloproliferative disorders. *Lancet, 365,* 1054–1061.

89. James, C., Ugo, V., Le Couedic, J. P., et al. (2005). A unique clonal JAK2 mutation leading to constitutive signalling causes polycythemia vera. *Nature, 434,* 1144–1148.

90. Kralovics, R., Passamonti, F., Buser, A. S., et al. (2005). A gain of function mutation in Jak2 is frequently found in patients with myeloproliferative disorders. *N Engl J Med, 352,* 1779–1790.

91. Levine, R. L., Wadleigh, M., Cools, J., et al. (2005). Activat-ing mutation in the tyrosine kinase JAK2 in polycythemia vera, essential thrombocythemia, and myeloid metaplasia with myelofibrosis. *Cancer Cell, 7,* 387–397.

92. Zhao, R., Xing, S., Li, Z., et al. (2005). Identification of an acquired JAK2 mutation in polycythemia vera. *J Biol Chem, 280,* 22788–22792.

93. Steensma, D. P., Dewald, G. W., Lasho, T. L., et al. (2005). The JAK2 V617F activating tyrosine kinase mutation is an infrequent event in both "atypical" myeloproliferative disorders and the myelodysplastic syndrome. *Blood, 106,* 1207–1209.

94. Jones, A. V., Kreil, S., Zoi, K., et al. (2005). Widespread occurrence of the JAK2 V617F mutation in chronic myeloproliferative disorders. *Blood, 106,* 2162–2168.

95. Thiele, J., Kvasnicka, H. M., Diehl, V., et al. (1999). Clinicopathological diagnosis and differential criteria of thrombocythemias in various myeloproliferative disorders by histopathology, histochemistry and immunostaining from bone marrow biopsies. *Leukemia & Lymphoma, 33,* 207–218.

96. Cabello, A. I., Collado, R., Ruiz, M. A., et al. (2005). A retrospective analysis of myelodysplastic syndromes with thrombocytosis: Reclassification of the cases by WHO proposals. *Leuk Res, 29,* 365–370.

97. Patel, K., & Kelsey, P. (1997). Primary acquired sideroblastic anemia, thrombocytosis, and trisomy 8. *Ann Hematol, 74,* 199–201.

98. Brusamolino, E., Orlandi, E., Morra, E., et al. (1988). Hematologic and clinical features of patients with chromosome 5 monosomy or deletion (5q). *Med Pediatr Oncol, 16,* 88–94.

99. Tefferi, A., Mathew, P., & Noel, P. (1994). The 5q-syndrome: A scientific and clinical update. *Leuk Lymphoma, 14,* 375–378.

100. Jenkins, R. B., Tefferi, A., Solberg, L. A., Jr., et al. (1989). Acute leukemia with abnormal thrombopoiesis and inversions of chromosome 3. *Cancer Genet Cytogenet, 39,* 167–179.

101. Jotterand Bellomo, M., Parlierm, V., Muhlematter, D., et al. (1992). Three new cases of chromosome 3 rearrangement in bands q21 and q26 with abnormal thrombopoiesis bring further evidence to the existence of a 3q21q26 syndrome. *Cancer Genet Cytogenet, 59,* 138–160.

102. Gupta, R., Abdalla, S. H., & Bain, B. J. (1994). Thrombocytosis with sideroblastic erythropoiesis: A mixed myeloproliferative myelodysplastic syndrome. *Leuk Lymphoma, 34,* 615–619.

103. Schilling, R. F. (1980). Platelet millionaires. *Lancet, 2,* 372–373.

104. Regev, A., Stark, P., Blickstein, D., et al. (1997). Thrombotic complications in essential thrombocythemia with relatively low platelet counts. *Am J Hematol, 56,* 168–172.

105. Small, B. M., & Bettigole, R. E. (1981). Diagnosis of myeloproliferative disease by analysis of the platelet volume distribution. *Am J Clin Pathol, 76,* 685–691.

106. Sehayek, E., Ben-Yosef, N., Modan, M., et al. (1988). Platelet parameters and aggregation in essential and reactive thrombocytosis. *Am J Clin Pathol, 90,* 431–436.

107. Osselaer, J. C., Jamart, J., & Scheiff, J. M. (1997). Platelet distribution width for differential diagnosis of thrombocytosis. *Clin Chem, 43,* 1072–1076.

108. Murphy, S., Davis, J. L., Walsh, P. N., et al. (1978). Template bleeding time and clinical hemorrhage in myeloproliferative disease. *Arch Intern Med, 138,* 1251–1253.

109. Boneu, B., Nouvel, C., Sie, P., et al. (1980). Platelets in myeloproliferative disorders. I. A comparative evaluation with certain platelet function tests. *Scandinavian J Haematology, 25,* 214–220.

110. Waddell, C. C., Brown, J. A., & Repinecz, Y. A. (1981). Abnormal platelet function in myeloproliferative disorders. *Arch Pathol Lab Med, 105,* 432–435.

111. Lofvenberg, E., & Nilsson, T. K. (1989). Qualitative platelet defects in chronic myeloproliferative disorders: Evidence for reduced ATP secretion. *Eur J Haematol, 43,* 435–440.

112. Zahavi, J., Zahavi, M., Firsteter, E., et al. (1991). An abnormal pattern of multiple platelet function abnormalities and increased thromboxane generation in patients with primary thrombocytosis and thrombotic complications. *Eur J Haematol, 47,* 326–332.

113. Balduini, C. L., Bertolino, G., Noris, P., et al. (1991). Platelet aggregation in platelet-rich plasma and whole blood in 120 patients with myeloproliferative disorders. *Am J Clin Pathol, 95,* 82–86.

114. Tefferi, A., Ho, T. C., Ahmann, G. J., et al. (1994). Plasma interleukin-6 and C-reactive protein levels in reactive versus clonal thrombocytosis. *Am J Med, 97,* 374–378.

115. Espanol, I., Hernandez, A., Cortes, M., et al. (1999). Patients with thrombocytosis have normal or slightly elevated thrombopoietin levels. *Haematologica, 84,* 312–316.

116. Messinezy, M., Westwood, N., Sawyer, B., et al. (1994). Primary thrombocythaemia: A composite approach to diagnosis. *Clin Lab Haematol, 16,* 139–148.

117. Wang, J. C., Chen, C., Novetsky, A. D., et al. (1998). Blood thrombopoietin levels in clonal thrombocytosis and reactive thrombocytosis. *Am J Med, 104,* 451–455.

118. Uppenkamp, M., Makarova, E., Petrasch, S., et al. (1998). Thrombopoietin serum concentration in patients with reactive and myeloproliferative thrombocytosis. *Ann Hematol, 77,* 217–223.

119. Hou, M., Carneskog, J., Mellqvist, U. H., et al. (1998). Impact of endogenous thrombopoietin levels on the differential diagnosis of essential thrombocythaemia and reactive thrombocytosis. *Eur J Haematol, 61,* 119–122.

120. Buss, D. H., O'Connor, M. L., Woodruff, R. D., et al. (1991). Bone marrow and peripheral blood findings in patients with extreme thrombocytosis: A report of 63 cases. *Arch Pathol Lab Med, 115,* 475–480.

121. Annaloro, C., Lambertenghi, Deliliers, G., Oriani, A., et al. (1999). Prognostic significance of bone marrow biopsy in essential thrombocythemia. *Haematologica, 84,* 17–21.

122. Steensma, D. P., & Tefferi, A. (2002). Cytogenetic and molecular genetic aspects of essential thrombocythemia. *Acta Haematol, 108,* 55–65.

123. Tefferi, A., & Gilliland, D. G. (in press). JAK2 in myeloproliferative disorders is not just another kinase. *Cell Cycle.*

124. Tefferi, A., & Gilliland, D. G. (2005). The JAK2 V617F tyrosine kinase mutation in myeloproliferative disorders: Status report and immediate implications for disease classification and diagnosis. *Mayo Clin Proc, 80,* 947–958.

125. Reference deleted in proof.
126. Tefferi, A., & Gilliland, D. G. (2005). JAK2 in myeloproliferative disorders is not just another kinase. *Cell Cycle, 4.*
127. Rolovic, Z., Basara, N., Gotic, M., et al. (1995). The determination of spontaneous megakaryocyte colony formation is an unequivocal test for discrimination between essential thrombocythaemia and reactive thrombocytosis. *Br J Haematol, 90,* 326–331.
128. Stoll, D. B., Peterson, P., Exten, R., et al. (1988). Clinical presentation and natural history of patients with essential thrombocythemia and the Philadelphia chromosome. *Am J Hematol, 27,* 77–83.
129. Michiels, J. J., Berneman, Z., Schroyens, W., et al. (2004). Philadelphia (Ph) chromosome-positive thrombocythemia without features of chronic myeloid leukemia in peripheral blood: Natural history and diagnostic differentiation from Ph-negative essential thrombocythemia. *Ann Hematol, ••.*
130. Koike, T., Uesugi, Y., Toba, K., et al. (1995). 5q-syndrome presenting as essential thrombocythemia: Myelodysplastic syndrome or chronic myeloproliferative disorders? *Leukemia, 9,* 517–518.
131. Tefferi, A., Dewald, G. W., Litzow, M. L., et al. (2005). Chronic myeloid leukemia: Current application of cytogenetics and molecular testing for diagnosis and treatment. *Mayo Clin Proc, 80,* 390–402.
132. Tefferi, A., & Elliott, M. A. (2004). Schistocytes on the peripheral blood smear. *Mayo Clin Proc, 79,* 809.
133. Arora, B., Sirhan, S., Hoyer, J. D., et al. (2005). Peripheral blood CD34 count in myelofibrosis with myeloid metaplasia: A prospective evaluation of prognostic value in 94 patients. *Br J Haematol, 128,* 42–48.
134. Jensen, M. K., de Nully Brown, P., Nielsen, O. J., et al. (2000). Incidence, clinical features and outcome of essential thrombocythaemia in a well defined geographical area. *Eur J Haematol, 65,* 132–139.
135. Ridell, B., Carneskog, J., Wedel, H., et al. (2000). Incidence of chronic myeloproliferative disorders in the city of Goteborg, Sweden 1983–1992. *Eur J Haematol, 65,* 267–271.
136. McNally, R. J., Rowland, D., Roman, E., et al. (1997). Age and sex distributions of hematological malignancies in the U.K. *Hematol Oncol, 15,* 173–189.
137. Chaiter, Y., Brenner, B., Aghai, E., et al. (1992). High incidence of myeloproliferative disorders in Ashkenazi Jews in northern Israel. *Leuk Lymphoma, 7,* 251–255.
138. Heudes, D., Carli, P. M., Bailly, F., et al. (1989). Myeloproliferative disorders in the department of Cote d'Or between 1980 and 1986. *Nouv Rev Fr Hematol, 31,* 375–378.
139. Mesa, R. A., Silverstein, M. N., Jacobsen, S. J., et al. (1999). Population-based incidence and survival figures in essential thrombocythemia and agnogenic myeloid metaplasia: An Olmsted County study, 1976–1995. *Am J Hematol, 61,* 10–15.
140. Jantunen, R., Juvonen, E., Ikkala, E., et al. (2001). The predictive value of vascular risk factors and gender for the development of thrombotic complications in essential thrombocythemia. *Ann Hematol, 80,* 74–78.
141. Bazzan, M., Tamponi, G., Schinco, P., et al. (1999). Thrombosis-free survival and life expectancy in 187 consecutive patients with essential thrombocythemia. *Ann Hematol, 78,* 539–543.
142. Fenaux, P., Simon, M., Caulier, M. T., et al. (1990). Clinical course of essential thrombocythemia in 147 cases. *Cancer, 66,* 549–556.
143. Besses, C., Cervantes, F., Pereira, A., et al. (1999). Major vascular complications in essential thrombocythemia: A study of the predictive factors in a series of 148 patients. *Leukemia, 13,* 150–154.
144. Bellucci, S., Janvier, M., Tobelem, G., et al. (1986). Essential thrombocythemias: Clinical evolutionary and biological data. *Cancer, 58,* 2440–2447.
145. Cortelazzo, S., Viero, P., Finazzi, G., et al. (1990). Incidence and risk factors for thrombotic complications in a historical cohort of 100 patients with essential thrombocythemia. *J Clin Oncol, 8,* 556–562.
146. Colombi, M., Radaelli, F., Zocchi, L., et al. (1991). Thrombotic and hemorrhagic complications in essential thrombocythemia: A retrospective study of 103 patients. *Cancer, 67,* 2926–2930.
147. Chistolini, A., Mazzucconi, M. G., Ferrari, A., et al. (1990). Essential thrombocythemia: A retrospective study on the clinical course of 100 patients. *Haematologica, 75,* 537–540.
148. Hasle, H. (2000). Incidence of essential thrombocythaemia in children. *Br J Haematol, 110,* 751.
149. Matsubara, K., Fukaya, T., Nigami, H., et al. (2004). Age-dependent changes in the incidence and etiology of childhood thrombocytosis. *Acta Haematol, 111,* 132–137.
150. Raskind, W. H., Jacobson, R., Murphy, S., et al. (1985). Evidence for the involvement of B lymphoid cells in polycythemia vera and essential thrombocythemia. *J Clin Invest, 75,* 1388–1390.
151. Anger, B., Janssen, J. W., Schrezenmeier, H., et al. (1990). Clonal analysis of chronic myeloproliferative disorders using X-linked DNA polymorphisms. *Leukemia, 4,* 258–261.
152. Tsukamoto, N., Morita, K., Maehara, T., et al. (1994). Clonality in chronic myeloproliferative disorders defined by X-chromosome linked probes: Demonstration of heterogeneity in lineage involvement. *Br J Haematol, 86,* 253–258.
153. el Kassar, N., Hetet, G., Li, Y., et al. (1995). Clonal analysis of haemopoietic cells in essential thrombocythaemia. *Br J Haematol, 90,* 131–137.
154. Elkassar, N., Hetet, G., Briere, J., et al. (1997). Clonality analysis of hematopoiesis in essential thrombocythemia — advantages of studying T lymphocytes and platelets. *Blood, 89,* 128–134.
155. Shih, L. Y., Lin, T. L., Dunn, P., et al. (2001). Clonality analysis using X-chromosome inactivation patterns by HUMARA-PCR assay in female controls and patients with idiopathic thrombocytosis in Taiwan. *Exp Hematol, 29,* 202–208.
156. Harrison, C. N., Gale, R. E., Machin, S. J., et al. (1999). A large proportion of patients with a diagnosis of essential thrombocythemia do not have a clonal disorder and may be at lower risk of thrombotic complications. *Blood, 93,* 417–424.

157. Chiusolo, P., La Barbera, E. O., Laurenti, L., et al. (2001). Clonal hemopoiesis and risk of thrombosis in young female patients with essential thrombocythemia. *Exp Hematol, 29,* 670–676.

158. Bacher, U., Haferlach, T., Kern, W., et al. (2005). Conventional cytogenetics of myeloproliferative diseases other than CML contribute valid information. *Ann Hematol,* ••.

159. Sessarego, M., Defferrari, R., Dejana, A. M., et al. (1989). Cytogenetic analysis in essential thrombocythemia at diagnosis and at transformation: A 12-year study. *Cancer Genet Cytogenet, 43,* 57–65.

160. Rane, S. G., & Reddy, E. P. (2000). Janus kinases: Components of multiple signaling pathways. *Oncogene, 19,* 5662–5679.

161. Yeh, T. C., & Pellegrini, S. (1999). The Janus kinase family of protein tyrosine kinases and their role in signaling. *Cell Mol Life Sci, 55,* 1523–1534.

162. Rane, S. G., & Reddy, E. P. (2002). JAKs, STATs and Src kinases in hematopoiesis. *Oncogene, 21,* 3334–3358.

163. Saharinen, P., Vihinen, M., & Silvennoinen, O. (2003). Autoinhibition of Jak2 tyrosine kinase is dependent on specific regions in its pseudokinase domain. *Mol Biol Cell, 14,* 1448–1459.

164. Lindauer, K., Loerting, T., Liedl, K. R., et al. (2001). Prediction of the structure of human Janus kinase 2 (JAK2) comprising the two carboxy-terminal domains reveals a mechanism for autoregulation. *Protein Eng, 14,* 27–37.

165. Saharinen, P., Takaluoma, K., & Silvennoinen, O. (2000). Regulation of the Jak2 tyrosine kinase by its pseudokinase domain. *Mol Cell Biol, 20,* 3387–3395.

166. Saharinen, P., & Silvennoinen, O. (2002). The pseudokinase domain is required for suppression of basal activity of Jak2 and Jak3 tyrosine kinases and for cytokine-inducible activation of signal transduction. *J Biol Chem, 277,* 47954–47963.

167. Feener, E. P., Rosario, F., Dunn, S. L., et al. (2004). Tyrosine phosphorylation of Jak2 in the JH2 domain inhibits cytokine signaling. *Mol Cell Biol, 24,* 4968–4978.

168. Reid, C. D. (1987). The significance of endogenous erythroid colonies (EEC) in haematological disorders. *Blood Rev, 1,* 133–140.

169. Prchal, J. F, & Axelrad, A. A. (1974). Letter: Bone-marrow responses in polycythemia vera. *N Engl J Med, 290,* 1382.

170. Fisher, M. J., Prchal, J. F., Prchal, J. T., et al. (1994). Antierythropoietin (EPO) receptor monoclonal antibodies distinguish EPO-dependent and EPO-independent erythroid progenitors in polycythemia vera. *Blood, 84,* 1982–1991.

171. Juvonen, E., Partanen, S., Ikkala, E., et al. (1988). Megakaryocytic colony formation in polycythaemia vera and secondary erythrocytosis. *Br J Haematol, 69,* 441–444.

172. Juvonen, E., Ikkala, E., Oksanen, K., et al. (1993). Megakaryocyte and erythroid colony formation in essential thrombocythaemia and reactive thrombocytosis: Diagnostic value and correlation to complications. *Br J Haematol, 83,* 192–197.

173. Juvonen, E., Partanen, S., & Ruutu, T. (1987). Colony formation by megakaryocytic progenitors in essential thrombocythaemia. *Br J Haematol, 66,* 161–164.

174. Battegay, E. J., Thomssen, C., Nissen, C., et al. (1989). Endogenous megakaryocyte colonies from peripheral blood in precursor cell cultures of patients with myeloproliferative disorders. *Eur J Haematol, 42,* 321–326.

175. Florensa, L., Besses, C., Woessner, S., et al. (1995). Endogenous megakaryocyte and erythroid colony formation from blood in essential thrombocythaemia. *Leukemia, 9,* 271–273.

176. Siitonen, T., Zheng, A., Savolainen, E. R., et al. (1996). Spontaneous granulocyte-macrophage colony growth by peripheral blood mononuclear cells in myeloproliferative disorders. *Leuk Res, 20,* 187–195.

177. Li, Y., Hetet, G., Maurer, A. M., et al. (1994). Spontaneous megakaryocyte colony formation in myeloproliferative disorders is not neutralizable by antibodies against IL3, IL6 and GM-CSF. *Br J Haematol, 87,* 471–476.

178. Hess, G., Rose, P., Gamm, H., et al. (1994). Molecular analysis of the erythropoietin receptor system in patients with polycythaemia vera. *Br J Haematol, 88,* 794–802.

179. Asimakopoulos, F. A., Hinshelwood, S., Gilbert, J. G. R., et al. (1997). The gene encoding hematopoietic cell phosphatase (Shp-1) is structurally and transcriptionally intact in polycythemia vera. *Oncogene, 14,* 1215–1222.

180. Correa, P. N., Eskinazi, D., & Axelrad, A. A. (1994). Circulating erythroid progenitors in polycythemia vera are hypersensitive to insulin-like growth factor-1 *in vitro:* Studies in an improved serum-free medium. *Blood, 83,* 99–112.

181. Dai, C. H., Krantz, S. B., Green, W. F., et al. (1994). Polycythaemia vera. III. Burst-forming units-erythroid (BFU-E) response to stem cell factor and c-kit receptor expression. *Br J Haematol, 86,* 12–21.

182. Dai, C. H., Krantz, S. B., Dessypris, E. N., et al. (1992). Polycythemia vera. II. Hypersensitivity of bone marrow erythroid, granulocyte-macrophage, and megakaryocyte progenitor cells to interleukin-3 and granulocyte-macrophage colony-stimulating factor. *Blood, 80,* 891–899.

183. Dai, C. H., Krantz, S. B., Means, R. T. Jr., et al. (1991). Polycythemia vera blood burst-forming units-erythroid are hypersensitive to interleukin-3. *J Clin Invest, 87,* 391–396.

184. Montagna, C., Massaro, P., Morali, F., et al. (1994). *In vitro* sensitivity of human erythroid progenitors to hemopoietic growth factors: Studies on primary and secondary polycythemia. *Haematologica, 79,* 311–318.

185. Martin, J. M., Gandhi, K., Jackson, W. R., et al. (1996). Hypersensitivity of polycythemia vera megakaryocytic progenitors to thrombopoietin. *Blood, 88,* 363a.

186. Kobayashi, S., Teramura, M., Hoshino, S., et al. (1993). Circulating megakaryocyte progenitors in myeloproliferative disorders are hypersensitive to interleukin-3. *Br J Haematol, 83,* 539–544.

187. Lecouedic, J. P., Mitjavila, M. T., Villeval, J. L., et al. (1996). Missense mutation of the erythropoietin receptor is a rare event in human erythroid malignancies. *Blood, 87,* 1502–1511.

188. Mittelman, M., Gardyn, J., Carmel, M., et al. (1996). Analysis of the erythropoietin receptor gene in patients with myeloproliferative and myelodysplastic syndromes. *Leuk Res, 20,* 459–466.

189. Fielder, P. J., Hass, P., Nagel, M., et al. (1997). Human platelets as a model for the binding and degradation of thrombopoietin. *Blood, 89*, 2782–2788.

190. Cerutti, A., Custodi, P., Duranti, M., et al. (1997). Thrombopoietin levels in patients with primary and reactive thrombocytosis. *Br J Haematol, 99*, 281–284.

191. Wang, J. C., Chen, C., Lou, L. H., et al. (1997). Blood thrombopoietin, IL-6 and IL-11 levels in patients with agnogenic myeloid metaplasia. *Leukemia, 11*, 1827–1832.

192. Horikawa, Y., Matsumura, I., Hashimoto, K., et al. (1997). Markedly reduced expression of platelet C-Mpl receptor in essential thrombocythemia. *Blood, 90*, 4031–4038.

193. Moliterno, A. R., Hankins, W. D., & Spivak, J. L. (1998). Impaired expression of the thrombopoietin receptor by platelets from patients with polycythemia vera. *N Engl J Med, 338*, 572–580.

194. Harrison, C. N., Gale, R. E., Pezella, F., et al. (1999). Platelet c-mpl expression is dysregulated in patients with essential thrombocythaemia but this is not of diagnostic value. *Br J Haematol, 107*, 139–147.

195. Yoon, S. Y., Li, C. Y., & Tefferi, A. (2000). Megakaryocyte c-Mpl expression in chronic myeloproliferative disorders and the myelodysplastic syndrome: Immunoperoxidase staining patterns and clinical correlates. *Eur J Haematol, 65*, 170–174.

196. Tefferi, A., Yoon, S. Y., & Li, C. Y. (2000). Immunohistochemical staining for megakaryocyte c-mpl may complement morphologic distinction between polycythemia vera and secondary erythrocytosis. *Blood, 96*, 771–772.

197. Mesa, R. A., al. e. (2002). Diagnostic and prognostic value of bone marrow angiogenesis and megakaryocyte c-Mpl expression in essential thrombocythemia. *Blood, 99*, 4131–4137.

198. Kessler, C. M., Klein, H. G., & Havlik, R. J. (1982). Uncontrolled thrombocytosis in chronic myeloproliferative disorders. *Br J Haematol, 50*, 157–167.

199. Budde, U., Scharf, R. E., Franke, P., et al. (1993). Elevated platelet count as a cause of abnormal von Willebrand factor multimer distribution in plasma. *Blood, 82*, 1749–1757.

200. Budde, U., & van Genderen, P. J. (1997). Acquired von Willebrand disease in patients with high platelet counts. *Semin Thromb Hemost, 23*, 425–431.

201. Sato, K. (1988). Plasma von Willebrand factor abnormalities in patients with essential thrombocythemia. *Keio J Med, 37*, 54–71.

202. Budde, U., Schaefer, G., Mueller, N., et al. (1984). Acquired von Willebrand's disease in the myeloproliferative syndrome. *Blood, 64*, 981–985.

203. Michiels, J. J., Budde, U., van der Planken, M., et al. (2001). Acquired von Willebrand syndromes: Clinical features, aetiology, pathophysiology, classification and management. *Bailliere's Best Practice in Clinical Haematology, 14*, 401–436.

204. van Genderen, P. J., Budde, U., Michiels, J. J., et al. (1996). The reduction of large von Willebrand factor multimers in plasma in essential thrombocythaemia is related to the platelet count. *Br J Haematol, 93*, 962–965.

205. van Genderen, P. J., Prins, F. J., Lucas, I. S., et al. (1997). Decreased half-life time of plasma von Willebrand factor collagen binding activity in essential thrombocythaemia: Normalization after cytoreduction of the increased platelet count. *Br J Haematol, 99*, 832–836.

206. Favaloro, E. J. (2000). Collagen binding assay for von Willebrand factor (VWF:CBA): Detection of von Willebrands disease (VWD), and discrimination of VWD subtypes, depends on collagen source. *Thromb Haemost, 83*, 127–135.

207. Budde, U., Dent, J. A., Berkowitz, S. D., et al. (1986). Subunit composition of plasma von Willebrand factor in patients with the myeloproliferative syndrome. *Blood, 68*, 1213–1217.

208. Lopez-Fernandez, M. F., Lopez-Berges, C., Martin, R., et al. (1987). Abnormal structure of von Willebrand factor in myeloproliferative syndrome is associated to either thrombotic or bleeding diathesis. *Thromb Haemost, 58*, 753–757.

209. Tsai, H. M. (1996). Physiologic cleavage of von Willebrand factor by a plasma protease is dependent on its conformation and requires calcium ion. *Blood, 87*, 4235–4244.

210. Levy, G. G., Nichols, W. C., Lian, E. C., et al. (2001). Mutations in a member of the ADAMTS gene family cause thrombotic thrombocytopenic purpura. *Nature, 413*, 488–494.

211. Wehmeier, A., Fricke, S., Scharf, R. E., et al. (1990). A prospective study of haemostatic parameters in relation to the clinical course of myeloproliferative disorders. *Eur J Haematol, 45*, 191–197.

212. Wehmeier, A., Scharf, R. E., Fricke, S., et al. (1989). Bleeding and thrombosis in chronic myeloproliferative disorders: Relation of platelet disorders to clinical aspects of the disease. *Haemostasis, 19*, 251–259.

213. Gersuk, G. M., Carmel, R., & Pattengale, P. K. (1989). Platelet-derived growth factor concentrations in platelet-poor plasma and urine from patients with myeloproliferative disorders. *Blood, 74*, 2330–2334.

214. Burstein, S. A., Malpass, T. W., Yee, E., et al. (1984). Platelet factor-4 excretion in myeloproliferative disease: Implications for the aetiology of myelofibrosis. *Br J Haematol, 57*, 383–392.

215. Wehmeier, A., Tschope, D., Esser, J., et al. (1991). Circulating activated platelets in myeloproliferative disorders. *Thromb Res, 61*, 271–278.

216. Jensen, M. K., de Nully, Brown, P., Lund, B. V., et al. (2000). Increased platelet activation and abnormal membrane glycoprotein content and redistribution in myeloproliferative disorders. *Br J Haematol, 110*, 116–124.

217. Kaywin, P., McDonough, M., Insel, P. A., et al. (1978). Platelet function in essential thrombocythemia: Decreased epinephrine responsiveness associated with a deficiency of platelet alpha-adrenergic receptors. *N Engl J Med, 299*, 505–509.

218. Mazzucato, M., De Marco, L., De Angelis, V., et al. (1989). Platelet membrane abnormalities in myeloproliferative disorders: Decrease in glycoproteins Ib and IIb/IIIa complex is associated with deficient receptor function. *Br J Haematol, 73*, 369–374.

219. Le Blanc, K., Lindahl, T., Rosendahl, K., et al. (1998). Impaired platelet binding of fibrinogen due to a lower number

of GPIIB/IIIA receptors in polycythemia vera. *Thromb Res, 91,* 287–295.

220. Schafer, A. I. (1982). Deficiency of platelet lipoxygenase activity in myeloproliferative disorders. *N Engl J Med, 306,* 381–386.

221. Landolfi, R., Ciabattoni, G., Patrignani, P., et al. (1992). Increased thromboxane biosynthesis in patients with polycythemia vera: Evidence for aspirin-suppressible platelet activation *in vivo. Blood, 80,* 1965–1971.

222. Rocca, B., Ciabattoni, G., Tartaglione, R., et al. (1995). Increased thromboxane biosynthesis in essential thrombocythemia. *Thromb Haemost, 74,* 1225–1230.

223. Landolfi, R., Marchioli, R., Kutti, J., et al. (2004). Efficacy and safety of low-dose aspirin in polycythemia vera. *N Engl J Med, 350,* 114–124.

224. Berk, P. D., Goldberg, J. D., Silverstein, M. N., et al. (1981). Increased incidence of acute leukemia in polycythemia vera associated with chlorambucil therapy. *N Engl J Med, 304,* 441–447.

225. Cortelazzo, S., Finazzi, G., Ruggeri, M., et al. (1995). Hydroxyurea for patients with essential thrombocythemia and a high risk of thrombosis. *N Engl J Med, 332,* 1132–1136.

226. Jensen, M. K., de Nully Brown, P., Lund, B. V., et al. (2001). Increased circulating platelet-leukocyte aggregates in myeloproliferative disorders is correlated to previous thrombosis, platelet activation and platelet count. *Eur J Haematol, 66,* 143–151.

227. Falanga, A., Marchetti, M., Evangelista, V., et al. (2000). Polymorphonuclear leukocyte activation and hemostasis in patients with essential thrombocythemia and polycythemia vera. *Blood, 96,* 4261–4266.

228. Michiels, J. J., Abels, J., Steketee, J., et al. (1985). Erythromelalgia caused by platelet-mediated arteriolar inflammation and thrombosis in thrombocythemia. *Ann Intern Med, 102,* 466–471.

229. van Genderen, P. J., Lucas, I. S., van Strik, R., et al. (1996). Erythromelalgia in essential thrombocythemia is characterized by platelet activation and endothelial cell damage but not by thrombin generation. *Thromb Haemost, 76,* 333–338.

230. van Genderen, P. J., Michiels, J. J., van Strik, R., et al. (1995). Platelet consumption in thrombocythemia complicated by erythromelalgia: Reversal by aspirin. *Thromb Haemost, 73,* 210–214.

231. Tefferi, A., Fonseca, R., Pereira, D. L., et al. (2001). A long-term retrospective study of young women with essential thrombocythemia. *Mayo Clin Proc, 76,* 22–28.

232. Wright, C. A., & Tefferi, A. (2001). A single institutional experience with 43 pregnancies in essential thrombocythemia. *Eur J Haematol, 66,* 152–159.

233. Elliott, M. A., & Tefferi, A. (2003). Thrombocythaemia and pregnancy. *Best Pract Res Clin Haematol, 16,* 227–242.

234. Harrison, C. (2005). Pregnancy and its management in the Philadelphia negative myeloproliferative diseases. *Br J Haematol, 129,* 293–306.

235. Michiels, J. J., van Genderen, P. J., Lindemans, J., et al. (1996). Erythromelalgic, thrombotic and hemorrhagic manifestations in 50 cases of thrombocythemia. *Leuk Lymphoma, 1,* 47–56.

236. Passamonti, F., Rumi, E., Pungolino, E., et al. (2004). Life expectancy and prognostic factors for survival in patients with polycythemia vera and essential thrombocythemia. *Am J Med, 117,* 755–761.

237. Elliott, M. A., & Tefferi, A. (2005). Thrombosis and haemorrhage in polycythaemia vera and essential thrombocythaemia. *Br J Haematol, 128,* 275–290.

238. Anger, B. R., Seifried, E., Scheppach, J., et al. (1989). Budd-Chiari syndrome and thrombosis of other abdominal vessels in the chronic myeloproliferative diseases. *Klinische Wochenschrift, 67,* 818–825.

239. Bazzan, M., Tamponi, G., Schinco, P., et al. (1989). Thrombosis-free survival and life expectancy in 187 consecutive patients with essential thrombocythemia. *Ann Hematol, 78,* 539–543.

240. Lengfelder, E., Hochhaus, A., Kronawitter, U., et al. (1998). Should a platelet limit of $600 \times 10(9)/L$ be used as a diagnostic criterion in essential thrombocythaemia? An analysis of the natural course including early stages. *Br J Haematol, 100,* 15–23.

241. Florensa, L., Besses, C., Zamora, L., et al. (2004). Endogenous erythroid and megakaryocytic circulating progenitors, HUMARA clonality assay, and PRV-1 expression are useful tools for diagnosis of polycythemia vera and essential thrombocythemia. *Blood, 103,* 2427–2428.

242. Rozman, C., Giralt, M., Feliu, E., et al. (1991). Life expectancy of patients with chronic nonleukemic myeloproliferative disorders. *Cancer, 67,* 2658–2663.

243. Watson, K. V., & Key, N. (1993). Vascular complications of essential thrombocythaemia: A link to cardiovascular risk factors. *Br J Haematol, 83,* 198–203.

244. Besses, C., Cervantes, F., Pereira, A., et al. (1999). Major vascular complications in essential thrombocythemia: A study of the predictive factors in a series of 148 patients. *Leukemia, 13,* 150–154.

245. Jantunen, R., Juvonen, E., Ikkala, E., et al. (2001). The predictive value of vascular risk factors and gender for the development of thrombotic complications in essential thrombocythemia. *Ann Hematol, 80,* 74–78.

246. Shih, L. Y., Lin, T. L., Lai, C. L., et al. (2002). Predictive values of X-chromosome inactivation patterns and clinicohematologic parameters for vascular complications in female patients with essential thrombocythemia. *Blood, 100,* 1596–1601.

247. Cortelazzo, S., Viero, P., Finazzi, G., et al. (1990). Incidence and risk factors for thrombotic complications in a historical cohort of 100 patients with essential thrombocythemia. *J Clin Oncol, 8,* 556–562.

248. Fenaux, P., Simon, M., Caulier, M. T., et al. (1990). Clinical course of essential thrombocythemia in 147 cases. *Cancer, 66,* 549–556.

249. van Genderen, P. J., Mulder, P. G., Waleboer, M., et al. (1997). Prevention and treatment of thrombotic complications in essential thrombocythaemia: Efficacy and safety of aspirin. *Br J Haematol, 97,* 179–184.

250. Jensen, M. K., de Nully Brown, P., Nielsen, O. J., et al. (2000). Incidence, clinical features and outcome of essential

thrombocythaemia in a well defined geographical area. *Eur J Haematol, 65,* 132–139.

251. Afshar-Kharghan, V. L. A., Gray, L., Padilla, A., et al. (2001). Hemostatic gene polymorphisms and the prevalence of thrombohemorrhagic complications in polycythemia vera and essential thrombocythemia. *Blood, 98,* 471a.

252. Dicato, M. A. S. B., Berchem, G. J., et al. (1999). V Leiden mutations, prothromin and methylene-tetrahydrofolate reductase are not risk factors for thromboembolic disease in essential thrombocythemia. *Blood, 94.*

253. Ruggeri, M., Gisslinger, H., Tosetto, A., et al. (2002). Factor V Leiden mutation carriership and venous thromboembolism in polycythemia vera and essential thrombocythemia. *Am J Hematol, 71,* 1–6.

254. Faurschou, M., Nielsen, O. J., Jensen, M. K., et al. (2000). High prevalence of hyperhomocysteinemia due to marginal deficiency of cobalamin or folate in chronic myeloproliferative disorders. *Am J Hematol, 65,* 136–140.

255. Gisslinger, H., Rodeghiero, F., Ruggeri, M., et al. (1999). Homocysteine levels in polycythaemia vera and essential thrombocythaemia. *Br J Haematol, 105,* 551–555.

256. Amitrano, L., Guardascione, M. A., Ames, P. R., et al. (2003). Thrombophilic genotypes, natural anticoagulants, and plasma homocysteine in myeloproliferative disorders: Relationship with splanchnic vein thrombosis and arterial disease. *Am J Hematol, 72,* 75–81.

257. Harrison, C. N., Donohoe, S., Carr, P., et al. (2002). Patients with essential thrombocythaemia have an increased prevalence of antiphospholipid antibodies which may be associated with thrombosis. *Thromb Haemost, 87,* 802–807.

258. Jensen, M. K., de Nully Brown, P., Thorsen, S., et al. (2002). Frequent occurrence of anticardiolipin antibodies, Factor V Leiden mutation, and perturbed endothelial function in chronic myeloproliferative disorders. *Am J Hematol, 69,* 185–191.

259. McCarthy, L., Eichelberger, L., Skipworth, E., et al. (2002). Erythromelalgia due to essential thrombocythemia. *Transfusion, 42,* 1245.

260. Ruggeri, M., Finazzi, G., Tosetto, A., et al. (1998). No treatment for low-risk thrombocythaemia: Results from a prospective study. *Br J Haematol, 103,* 772–777.

261. Cortelazzo, S., Finazzi, G., Ruggeri, M., et al. (1995). Hydroxyurea for patients with essential thrombocythemia and a high risk of thrombosis. *N Engl J Med, 332,* 1132–1136.

262. Ruggeri, M., Finazzi, G., Tosetto, A., et al. (1998). No treatment for low-risk thrombocythaemia: Results from a prospective study. *Br J Haematol, 103,* 772–777.

263. Tefferi, A., Solberg, L. A., & Silverstein, M. N. (2000). A clinical update in polycythemia vera and essential thrombocythemia. *Am J Med, 109,* 141–149.

264. Landolfi, R., Marchioli, R., Kutti, J., et al. Efficacy and safety of low-dose aspirin in polycythemia vera. *N Engl J Med, 350,* 114–124.

265. Beressi, A. H., Tefferi, A., Silverstein, M. N., et al. (1995). Outcome analysis of 34 pregnancies in women with essential thrombocythemia. *Arch Intern Med, 155,* 1217–1222.

266. Green, A. R., Campbell, P., Buck, G., et al. (2004). The Medical Research Council PT1 trial in essential thrombocythemia. *Blood, 104,* Abstract #6.

267. Demircay, Z., Comert, A., & Adiguzel, C. (2002). Leg ulcers and hydroxyurea: Report of three cases with essential thrombocythemia. *Int J Dermatol, 41,* 872–874.

268. Alvarado, Y., Cortes, J., Verstovsek, S., et al. (2003). Pilot study of pegylated interferon-alpha 2b in patients with essential thrombocythemia. *Cancer Chemother Pharmacol, 51,* 81–86.

269. Elliott, M. A., & Tefferi, A. (1997). Interferon-alpha therapy in polycythemia vera and essential thrombocythemia. *Sem Thromb Hemostas, 23,* 463.

270. Storen, E. C., & Tefferi, A. (2001). Long-term use of anagrelide in young patients with essential thrombocythemia. *Blood, 97,* 863–866.

271. Willoughby, S., & Pearson, T. C. (1998). The use of aspirin in polycythaemia vera and primary thrombocythaemia. *Blood Rev, 12,* 12–22.

272. Randi, M. L., Rossi, C., Fabris, F., et al. (1999). Aspirin seems as effective as myelosuppressive agents in the prevention of rethrombosis in essential thrombocythemia. *Clin Appl Thromb Hemost, 5,* 131–135.

273. Anonymous. (1994). Antiplatelet trialist's collaboration collaborative overview of randomised trials of antiplatelet therapy-1. *Br Med J, 308,* 81–106.

274. Finazzi, G., Caruso, V., Marchioli, R., et al. (2004). Acute leukemia in polycythemia vera. An analysis of 1638 patients enrolled in a prospective observational study. *Blood,* ••.

275. Gugliotta, L., Bulgarelli, A., Tieghi, A., et al. (2004). Bone marrow biopsy and aspirate evaluation in 90 patients with essential thrombocythemia treated with PEG interferon alpha-2b. Preliminary results. *Blood, 104,* Abstract #1523.

276. Adami, R. (1993). Therapeutic thrombocytapheresis: A review of 132 patients. *Artificial Organs, 16* (Suppl 5), 183–184.

277. van Genderen, P. J., Leenknegt, H., & Michiels, J. J. (1997). The paradox of bleeding and thrombosis in thrombocythemia: Is von Willebrand factor the link? *Semin Thromb Hemost, 23,* 385–389.

278. Grima, K. M. (2000). Therapeutic apheresis in hematological and oncological diseases. *J Clinical Apheresis, 15,* 28–52.

279. Greist, A. (2002). The role of blood component removal in essential and reactive thrombocytosis. *Therapeutic Apheresis, 6,* 36–44.

280. van Genderen, P. J., van Vliet, H. H., Prins, F. J., et al. (1997). Excessive prolongation of the bleeding time by aspirin in essential thrombocythemia is related to a decrease of large von Willebrand factor multimers in plasma. *Ann Hematol, 75,* 215–220.

281. Bond, L., & Bevan, D. (1988). Myocardial infarction in a patient with hemophilia treated with DDAVP. *N Engl J Med, 318,* 121.

282. Byrnes, J. J., Larcada, A., & Moake, J. L. (1988). Thrombosis following desmopressin for uremic bleeding. *Am J Hematol, 28,* 63–65.

283. Kennedy, B. J., Smith, L. R., & Goltz, R. W. (1975). Skin changes secondary to hydroxyurea therapy. *Arch Dermatol, 111,* 183–187.

284. Yarbro, J. W. (1992). Mechanism of action of hydroxyurea. *Semin Oncol, 19,* 1–10.

285. Nguyen, T. V., & Margolis, D. J. (1993). Hydroxyurea and lower leg ulcers. *Cutis, 52,* 217–219.

286. Lossos, I. S., & Matzner, Y. (1995). Hydroxyurea-induced fever: Case report and review of the literature. *Ann Pharmacother, 29,* 132–133.

287. Najean, Y., & Rain, J. D. (1997). Treatment of polycythemia vera — the use of hydroxyurea and pipobroman in 292 patients under the age of 65 years. *Blood, 90,* 3370–3377.

288. Daoud, M. S., et al. (1997). Hydroxyurea dermopathy: A unique lichenoid eruption complicating long-term therapy with hydroxyurea. *J Am Acad Dermatol, 36,* 178–182.

289. Best, P. J., et al. (1998). Hydroxyurea-induced leg ulceration in 14 patients. *Ann Intern Med, 128,* 29–32.

290. Stevens, M. R. (1999). Hydroxyurea: An overview. *J Biol Regul Homeost Agents, 13,* 172–175.

291. Silverstein, M. N., Petitt, R. M., Solberg, L. A. Jr., et al. (1988). Anagrelide: A new drug for treating thrombocytosis. *N Engl J Med, 318,* 1292–1294.

292. Anonymous. (1992). Anagrelide, a therapy for thrombocythemic states: Experience in 577 patients. Anagrelide Study Group. *Am J Med, 92,* 69–76.

293. Mazur. (1992). Analysis of the mechanism of anagrelide-induced thrombocytopenia in humans. *Blood, 29,* 1931–1937.

294. Spencer, C. M., & Brogden, R. N. (1994). Anagrelide: A review of its pharmacodynamic and pharmacokinetic properties, and therapeutic potential in the treatment of thrombocythaemia. *Drugs, 47,* 809–822.

295. Solberg, L. A., Tefferi, A., Oles, K. J., et al. (1997). The effects of anagrelide on human megakaryocytopoiesis. *Br J Haematol, 99,* 174–180.

296. Tefferi, A., Silverstein, M. N., Petitt, R. M., et al. (1997). Anagrelide as a new platelet-lowering agent in essential thrombocythemia: Mechanism of action, efficacy, toxicity, current indications. *Semin Thromb Hemost, 23,* 379.

297. Jurgens, D. J., Moreno-Aspitia, A., & Tefferi, A. (2004). Anagrelide-associated cardiomyopathy in polycythemia vera and essential thrombocythemia. *Haematologica, 89,* 1394–1395.

298. Quesada, J. R. (1986). Clinical toxicity of interferons in cancer patients: A review. *J Clin Oncol, 4,* 234–243.

299. Gugliotta. (1989). *In vivo* and *in vitro* inhibitory effect of alpha-interferon on megakaryocyte colony growth in essential thrombocythemia. *Br J Haematol, 71,* 177–181.

300. Lengfelder, E., Berger, U., & Hehlmann, R. (2000). Interferon alpha in the treatment of polycythemia vera [In Process Citation]. *Ann Hematol, 79,* 103–109.

301. Gilbert, H. S. (1998). Long term treatment of myeloproliferative disease with interferon-alpha-2b — feasibility and efficacy. *Cancer, 83,* 1205–1213.

302. Arthur, K. (1967). Radioactive phosphorus in the treatment of polycythaemia: A review of ten years' experience. *Clin Radiol, 18,* 287–291.

303. Roberts, B. E., & Smith, A. H. (1967). Use of radioactive phophorus in haematology. *Blood Rev, 11,* 146–153.

304. Wagner, S., Waxman, J., & Sikora, K. (1989). The treatment of essential thrombocythaemia with radioactive phosphorus. *Clin Radiol, 40,* 190–192.

305. Najman, A., Stachowiak, J., Parlier, Y., et al. (1982). Pipobroman therapy of polycythemia vera. *Blood, 59,* 890–894.

306. Anonymous. (1967). Evaluation of two antineoplastic agents: Pipobroman (Vercyte) and thioguanine. *JAMA, 200,* 619–620.

307. Passamonti, F., Brusamolino, E., Lazzarino, M., et al. (2000). Efficacy of pipobroman in the treatment of polycythemia vera: Long-term results in 163 patients. *Haematologica, 85,* 1011–1018.

308. Mazzucconi, M. G., Francesconi, M., Chistolini, A., et al. (1983). Pipobroman therapy of essential thrombocythemia. *Scand J Haematol, 37,* 306–309.

CHAPTER 57

# Inherited Disorders of Platelet Function

**Alan T. Nurden and Paquita Nurden**

*Centre de Référence des Pathologies Plaquettaires, Hôpital Cardiologique, Pessac, France*

## I. Introduction

Genetic defects of platelets give rise to bleeding syndromes of varying severity. Deficiencies of glycoprotein (GP) mediators of adhesion and aggregation, especially the Bernard-Soulier syndrome (BSS, a disorder of the GP Ib-IX-V complex) and Glanzmann thrombasthenia (GT, a disorder of the integrin αIIbβ3) are discussed in detail in this chapter. Recent studies have highlighted defects of primary receptors for stimuli, including a pathology of the P2Y$_{12}$ adenosine diphosphate (ADP) receptor. These lead to agonist-specific deficiencies in the platelet aggregation response. Likewise, abnormalities of signaling pathways that send messages to targets within the platelet give rise to functional defects only when that pathway is used. Defects of secretion from dense bodies and α-granules, of adenosine triphosphate (ATP) production, and of the generation of procoagulant activity are also encountered. Some disorders are exclusive to megakaryocytes and platelets; in others (e.g., the Chediak–Higashi, Hermansky–Pudlak, and Wiskott–Aldrich syndromes), the molecular lesion extends to other cell types. Molecular defects responsible for familial thrombocytopenias (Chapter 54) are also sometimes associated with reduced platelet function.

Inherited platelet disorders are rare diseases, yet studies on them have been instrumental in the elucidation of molecular pathways essential for platelet function. Bleeding is mostly mucocutaneous in nature and is usually life threatening only when associated with trauma or surgical intervention. Treatment is based on transfusion of platelets (Chapter 69), although substitute therapies including the use of desmopressin (DDAVP, Chapter 67) and recombinant factor VIIa (Chapter 68) are becoming more common. Bone marrow transplantation is used in some diseases, whereas gene therapy (Chapter 71) may be on the horizon.

This chapter on inherited abnormalities begins by describing disorders of platelet adhesion to the vessel wall, proceeds to abnormalities of the signaling mechanisms through which agonists induce platelet activation, continues with defects of secretion, discusses in some detail defective platelet aggregation in thrombasthenia, and concludes by mentioning a disorder of platelet procoagulant activity. Table 57-1 summarizes the principal diagnostic characteristics of the diseases to be discussed. The final section of this chapter discusses treatment.

## II. Defects of Platelet Adhesion

### A. Bernard–Soulier Syndrome

BSS is characterized by autosomal recessive inheritance, a prolonged bleeding time, thrombocytopenia, decreased platelet survival in most patients, and giant platelets.[1,2] Platelet counts range from as low as $20 \times 10^9$/L to near normal. The disease is characterized by quantitative or qualitative defects within the GPIb-IX-V complex of platelets. Historically, the original description in the 1970s of the then-called "GP I" deficiency in BSS was a major step in showing that platelet membrane glycoproteins mediate platelet function.[3] BSS is a rare disease and may be confused in early life with ITP; on occasion this has erroneously led to splenectomy. Platelets on peripheral blood smears can be up to 20 μm in diameter, and giant forms may represent up to 80% of the total platelet population. Electron microscopy often highlights cytoplasmic vacuoles and zones enriched in membrane complexes in the giant platelets (Fig. 57-1), and these abnormalities extend to megakaryocytes where a dystrophied demarcation membrane system is a common feature.[4] According to one report, mature megakaryocytes in BSS show an increased ploidy.[5] The abnormality in BSS is restricted to the megakaryocyte lineage.

**1. Platelet Function Abnormalities.** BSS platelets aggregate and secrete in response to ADP, epinephrine, arachidonic acid and collagen, but they fail to agglutinate when platelet-rich plasma (PRP) is stirred with ristocetin or botrocetin.[1] The fact that von Willebrand factor (VWF), normally

1029

PLATELETS

**Table 57-1: Essential Characteristics of the Principal Inherited Disorders of Platelets**

| Disorder | Bleeding Syndrome | Platelet Count ×10$^9$/L | Platelet Characteristics | Associations | Platelet Functional Abnormality | Inheritance | Structural Defects and Gene Affected | Refs |
|---|---|---|---|---|---|---|---|---|
| Bernard–Soulier syndrome | Moderate to severe | 20–100 | Giant platelets | DiGeorge/velocardial facial syndrome[a] | Abnormal adhesion, no agglutination with ristocetin or botrocetin, retarded response to thrombin | Autosomal recessive (1 report of autosomal dominant) | GPIb-IX GPIbα (chr 17) GPIbβ (chr 22) GPIX (chr 3) | 1–5, 10, 12, 15–32, 34–39, 42 |
| May–Hegglin anomaly | Mild | 30–100 | Large size | Neutrophil inclusions | No consistent defects | Autosomal dominant | Non-muscle myosin heavy chain IIA MYH9 | 1, 42–46 |
| Fechtner syndrome (FS), Sebastian syndrome (SS) | Mild | 30–100 | Large size, large vacuoles | Hereditary nephritis and hearing loss (FS) Leukocyte inclusions (FS, SS) | No consistent defects | Autosomal dominant | Non-muscle myosin heavy chain IIA MYH9 | 1, 42–46 |
| Epstein syndrome | Mild | 5–100 | Large size, large vacuoles | Hereditary nephritis and hearing loss | Impaired response to collagen (not consistent) | Autosomal dominant | Nonmuscle myosin heavy chain IIA MYH9 | 1, 4, 42–46 |
| Montreal platelet syndrome | Moderate to severe | 5–40 | Large size | Not known | Spontaneous agglutination, decreased response to thrombin | Autosomal dominant | Unknown, reduced calpain activity | 1, 42 |
| Platelet-type von Willebrand disease | Moderate | Normal or decreased | Platelet size heterogeneity | Absence of high mol wt VWF multimers | Increased sensitivity to agglutination by ristocetin | Autosomal dominant | GPIbα | 1, 2, 33, 40 |
| α2β1 collagen receptor | Mild | Normal | Normal | Modification in receptor density according to haplotype | Decreased response to collagen | ? | Probably α2 (chr 5) | 1, 50, 67 |
| GPVI collagen receptor | Mild | Normal | Normal | Linked to FcR-γ Receptor density depends on haplotype | Decreased response to collagen | ? | Probably GPVI (chr 19) Absence can be secondary to proteolytic cleavage | 1, 52, 54, 67 |
| P2Y$_{12}$ ADP receptor | Mild | Normal | Normal | None known, but receptor found in brain | Small and unstable aggregates with ADP | Autosomal recessive | P2Y$_{12}$ receptor (chr 3) | 1, 57–63, 65, 67 |
| TPα, thromboxane (TX) A$_2$ receptor | Mild | Normal | Normal | Unknown | Absence of response to TXA$_2$ and analogues, Diminished response to collagen | Autosomal recessive | TPα | 66, 67 |

| | | | | | | | |
|---|---|---|---|---|---|---|---|
| Intracellular signaling mechanisms[b] | Mild | Normal or decreased | None | Not reported | Variable aggregation and secretion defects to multiple agonists | Not reported for most families | Phospholipase C-β2, Gαq protein among others | 60, 67–70, 74, 155 |
| Cyclooxygenase[c] | Moderate | Decreased on occasion | Normal | None known | No aggregation with arachidonic acid, reduced response to collagen and ADP | Probably autosomal recessive | Cyclooxygenase enzyme | 67, 155 |
| Gray platelet syndrome | Mild to moderate | 30–100 | Empty α–granules | Myelofibrosis | Abnormal but variable; can be decreased with thrombin, epinephrine and/or collagen | Recessive or dominant | Unknown, but prevents storage of α-granule proteins | 78–80, 155 |
| Quebec syndrome | Moderate to severe | ~100 | Abnormal content of α–granules | None known | Absent aggregation with epinephrine | Autosomal dominant | Multimerib Increased urokinase-type activator in α–granules, degraded proteins | 78, 84, 85, 155 |
| Wiskott–Aldrich syndrome | Mild to severe | 10–100 | Small size, fewer granules | Usually eczema, infections, immunodeficiency | Decreased aggregation and secretion | X-linked | WAS protein (a signaling molecule) | 89–92 |
| Hermansky–Pudlak (HP) and Chediak–Higashi (CH) syndromes[d] | Mild to moderate | Normal | Reduced number of abnormal dense granules, giant granules (CH) | Oculocutaneous albinism Ceroid-lipofuscinosis (HP), infections (CH) | Decreased aggregation and secretion especially with collagen | Autosomal recessive | Proteins involved in vesicle formation and trafficking (see text) | 92–97 |
| Glanzmann thrombasthenia | Moderate to severe | Normal | None | Possibly increased bone thickening and low fertility (β3) | Absent aggregation with all physiological agonists, clot retraction often defective | Autosomal recessive | αIIb and β3 (chr 17q21-23) | 1, 99, 101, 102–115, 117–129, 133, 137–148, 154, 156–159 |
| Scott syndrome | Moderate to severe | Normal | Normal | Defects extend to other blood cells | Low expression of procoagulant activity and microparticle release | Autosomal recessive | ATP-binding cassette transporter A1 (chr 9) | 1, 151–153 |

[a]Condition associated with loss of GPIbβ gene in chromosome 22q11 deletion.

[b]Two examples only are given.

[c]Other enzyme deficiencies have been reported (e.g., thromboxane synthetase, lipoxygenase, glycogen synthetase).

[d]Other types of storage pool disease affecting dense bodies are described in the text.

Familial thrombocytopenias are discussed in Chapter 54.

Abbreviations: ADP, adenosine diphosphate; CH, Chediak–Higashi; GP, glycoprotein; chr, chromosome; FS, Fechtner syndrome f; HP, Hermansky–Pudlak; SS, Sebastian syndrome; TXA₂, thromboxane A₂; VWF, von Willebrand factor.

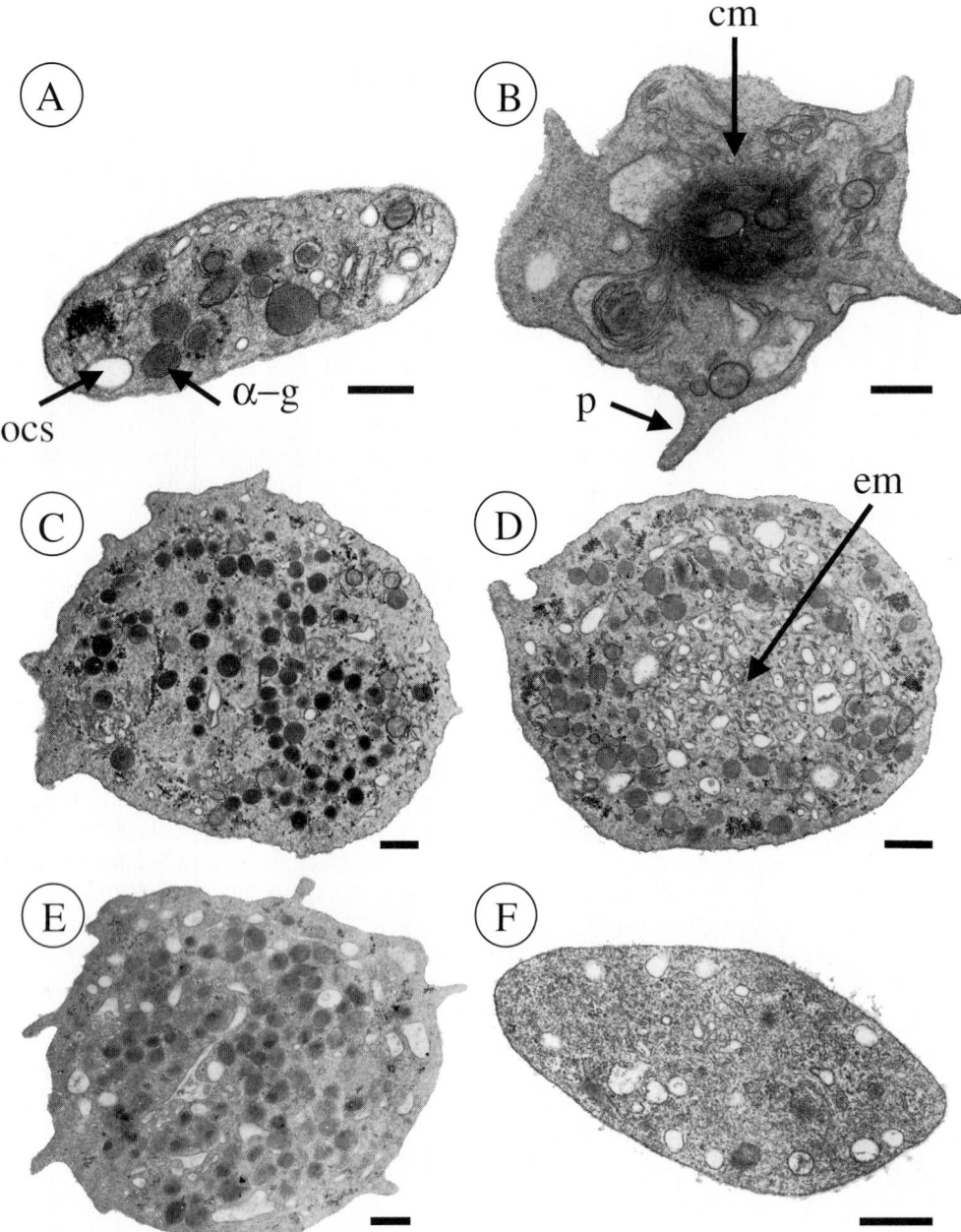

**Figure 57-1.**   Platelet morphology in the giant platelet syndromes as visualized by electron microscopy. A. A normal discoid platelet showing α-granules (α-g) and channels of the open canalicular system (ocs). B. A normal platelet seen after stimulation for 3 minutes with a high dose (0.5 U) α-thrombin. Note the rounded shape, the dark central mass (cm) of contractile protein, the absence of granules, and the long pseudopods (p). C–E. Giant platelets from patients with Bernard–Soulier syndrome (C), May–Hegglin anomaly (D), and Epstein syndrome (E). Note the rounded shapes, the abundance of granules distributed throughout the cytoplasm except for zones rich in entangled membranes (em). F. A platelet from a patient with gray platelet syndrome, which, although enlarged, has retained a discoid appearance. Note the absence of α-granules. Images were obtained on ultrathin sections by standard transmission electron microscopy. Bars = 0.5 μm. Technical details for the electron microscopy are given in ref. 4.

present in plasma, cannot bind to GPIbα accounts for this lack of agglutination and the decreased adherence to sub-endothelium that characterizes this disorder. The platelet adhesion defect in BSS occurs at all shear rates, although its effect is most apparent at the high shear rates of the microcirculation. In normal primary haemostasis, this step represents an initial transient attachment of platelets that roll and whose firm attachment requires the involvement of other receptors such as α2β1, GPVI, αIIbβ3[6] (see Chapter 18 for further details). Because the GPIb-VWF axis fails to function, not only adhesion but also thrombus formation is defective, and BSS is a relatively severe bleeding disorder.

GPIbα contains two thrombin-binding sites located within the N-terminal domain, and these bind to exosite I and exosite II of two distinct α-thrombin molecules, respectively.[7,8] BSS platelets bind decreased amounts of α-thrombin and aggregate poorly following a prolonged lag phase. The residual response of BSS platelets to thrombin is mediated by the moderate-affinity receptors of the protease-activated receptor (PAR) family (PAR-1 and PAR-4)[9,10] (Chapter 9). GPIbα clustering may help present thrombin to the PAR receptors and play a role in accelerating the speed and the intensity of platelet activation under normal conditions. Diminished platelet prothrombin consumption suggestive of reduced thrombin generation was reported in early studies of BSS; this may be attributed to a defective binding of factor XI due to the absence of GPIb[11] and to a decrease in GPIb-fibrin-dependent thrombin generation.[12] The functional consequences of newly reported functions of GPIbα, which can act as a receptor for P-selectin and bind thrombospondin-1, high-molecular-weight kininogen, and the β2 integrin Mac-1, remain to be elucidated (see Chapter 7).

**2. Molecular Defects.** The products of four separate genes (Ibα, Ibβ, IX, and V) assemble within the maturing megakaryocyte in the bone marrow to form the GPIb-IX-V complex (Chapter 7). The genes for GPIbα and GPIbβ map to chromosomes 17p12 and 22q11.2, respectively, whereas those encoding GPIX and GPV are on chromosome 3. Each subunit is a member of the leucine-rich motif superfamily, suggesting a common ancestral gene. Coding sequences are contained within a single exon with the exception of the gene for GPIbβ, which contains a short intron beginning 10 bases after the start codon.[13] Flow cytometry and immunoblotting can demonstrate GPIb-IX-V deficiencies and should precede screening of subunit genes by molecular biology procedures. Flow cytometry (Chapter 30) also allows an evaluation of platelet size, whereas double labeling permits the dual assessment of GPIb-IX-V and integrin αIIbβ3.[14] Obligate heterozygotes for the classic form of the disease have platelets with intermediate amounts of GPIb-IX-V, and whereas platelet size heterogeneity is common, an abundance of giant forms is rare in carriers.[1] Although mild bleeding has been reported, the majority of heterozygotes do not experience hemorrhage.

*a. Defects of the GPIbα Gene.* Probably because of the larger size of the GPIbα gene, defects of this gene are the most frequent genetic cause of BSS. Mutations affect both the N-terminal flanking sequence, eight tandem leucine-rich repeats, and most of the remaining extracellular domain including the heavily glysosylated macroglycopeptide region[1,2,15] (Fig. 57-2). Common are insertions or deletions followed by frameshifts leading to novel peptide sequences and a premature stop. Such a patient was studied by Simsek et al.,[16] who amplified genomic DNA and observed a homo-

zygous deletion of $T^{317}$ in codon 76 of the GPIbα gene. Loss of this nucleotide caused a frameshift and a premature stop codon after 19 altered amino acids. Nonsense mutations are also common (Fig. 57-2). An example is a patient with GPIbα stopping at $Ser^{444}$ and for whom a normally glycosylated but truncated form lacking the transmembrane domain was found in the plasma.[17] A missense mutation leading to a $Cys^{209} \rightarrow Ser$ substitution was thought to prevent complex formation by affecting the folding of GPIbα.[18] In fact, transfection studies showed that coexpression of GPIbα, GPIbβ, and GPIX is required for efficient trafficking across the endoplasmic reticulum and the Golgi apparatus.[19] GPV is the only one of the four polypeptides that can be expressed alone in significant amounts on the surface of transfected cells; even so, its presence results in a more efficient expression of the rest of the complex.[20]

*b. Defects of the GPIbβ and GPIX Genes.* The first report of a GPIbβ gene defect was not a classic case, for although possessing giant platelets (and decreased ristocetin-induced platelet aggregation), the patient suffered from a developmental disorder, the DiGeorge/velo-cardial facial syndrome.[21] A hemizygous microdeletion at 22q11.2 is characteristic of the DiGeorge syndrome, and this deletion includes the GPIbβ gene. For this patient, the second allele had a $C \rightarrow G$ mutation at −133 of the Ibβ-gene and within a binding site for the GATA-1 transcription factor.[22] This was the first description of BSS linked to a promoter defect. A series of f mutations in the coding region of the GPIbβ gene are shown in Fig. 57-2; of these, four are associated with the DiGeorge syndrome and affect but one chromosome. Missense mutations in the GPIbβ gene can give rise to BSS; an example from our work is a homozygous $Asn^{64} \rightarrow Thr$ substitution in the single leucine-rich motif flanking domain.[23] The abnormal GPIbβ failed to react with GPIX, with neither being expressed on the surface of CHO cells cotransfected with wild-type GPIbα and GPIX genes. Typically, the abnormal GPIbβ was retained in the endoplasmic reticulum as was GPIX, but for this patient some poorly glycosylated GPIbα did aberrantly make it to the giant platelet surface. Studies on patients with GPIbβ defects show just how the surface expression of GPIbα and GPIX is dependent on GPIbβ.[24] Of novel interest is a homozygous 13 base pair deletion in the signal peptide-coding region of the GPIbβ gene in a patient with giant platelets and giant α-granules.[25]

The first reported defect in the GPIX gene involved double heterozygosity arising from two missense mutations: (a) an $A \rightarrow G$ transition in codon 21 resulting in an $Asp \rightarrow Gly$ conversion and (b) an $A \rightarrow G$ change in codon 45 that converted $Asn \rightarrow Ser$.[26] These mutations alter conserved residues in or flanking the single leucine-rich motif of GPIX. The $Asn^{45} \rightarrow Ser$ substitution was subsequently shown to be the most common defect giving rise to BSS, particularly in families of Northern European origin.[27]

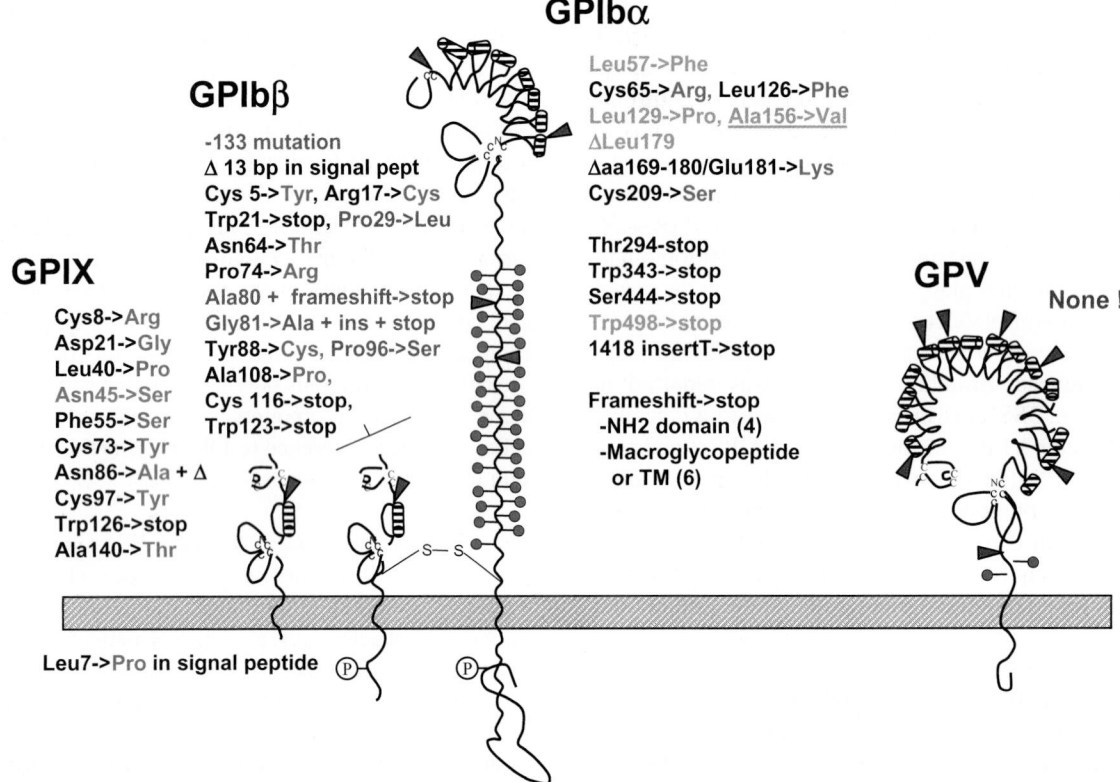

**Figure 57-2.** Genetic defects known to cause Bernard–Soulier syndrome (BSS). Abnormalities are most frequently found in the gene coding GPIbα, but they also occur in those coding GPIbβ and GPIX. So far, no defects have been detected in the gene for GPV in BSS patients. Mutations giving rise to platelets with significant or even normal amounts of nonfunctional GPIb-IX-V complexes (variant forms) are in brown. A common defect responsible for BSS in North Europe is in green. Mutations in the gene coding GPIbβ and associated with the chromosome 22 breakage found in the DiGeorge/velo-cardial facial syndrome are in mauve. Red circles: O-linked oligosaccharide chains. Blue triangles: N-glycosylation sites. This figure was modified from a composition originally provided by François Lanza (U311 INSERM, Strasbourg).

Expression studies with this mutant strongly suggested that GPIX interacts directly with GPIbβ and is necessary for the stability of the GPIb-IX complex.[19] Other mutations within the leucine-rich repeat and adjoining parts of the extracellular domain have since been described, as has a nonsense mutation in codon 126 resulting in a truncated protein without the transmembrane domain, thus preventing insertion of the subunit into the membrane.[15] A Leu[7] → Pro mutation in the signal peptide of GPIX was shown to be the cause of one of the first-reported cases of BSS.[28]

*c. Variant-Type Bernard–Soulier Syndrome.* In variant BSS, qualitative defects lead to a nonfunctional GPIb-IX-V complex that is at least partially expressed in platelets that remain giant. In the so-called Bolzano variant, the binding of thrombin to the patient's platelets was normal, yet the ability of GPIbα to bind VWF had been lost, confirming that these functions are independent. A homozygous Ala[156] → Val substitution in a late leucine-rich repeat of GPIbα (Fig. 57-2) is responsible for this unique phenotype.[29]

A second variant possessed a unique autosomal dominant form of the disease and a Leu[57] → Phe substitution within the first leucine-rich repeat.[30] The patient possessed enlarged platelets, but ristocetin-induced platelet agglutination was only slightly decreased suggesting residual GPIb function. GPIbα was indeed present, but it was more susceptible to proteolysis. In variant Nancy I, a 3-bp deletion resulted in the loss of Leu[179] in a highly conserved region of another late leucine-rich repeat of GPIbα.[31] This curtailed the expression of GPIb-IX complexes on the surface of transfected CHO cells. Significantly, CHO cells transfected with the mutated GPIbα (and wild-type GPIbβ and GPIX) failed to attach to (or roll on) a VWF substratum. For two other patients, platelets show about 40% residual VWF binding and normal surface GPV, confirming that this gene product can be processed independently.[32] Here, the first leucine within the then-termed fifth leucine repeat sequence of GPIbα was substituted by a Pro. As implied from the crystal structure, the β-strands of the leucine-rich repeats of GPIbα form a concave face that, when unmasked, participates directly in the binding of the VWF A1 domain.[7,33] Therefore, these mutations may interfere directly or indirectly with this interaction. The Karlstrad variant results from a Trp[498] → Stop mutation, leading to a truncated GPIbα.[34] In contrast to other truncated forms of GPIbα (discussed earlier), this variant contains part of the transmembrane domain as well as the disulphide linking the Ibα and β subunits. It appears that the cytoplasmic domain is not essential for GPIb-IX expression but may be required for GPIb function.

*d. Transgenic Mouse Models of BSS.* Mice deficient in GPIbα present a mild thrombocytopenia, giant platelets, and a bleeding diathesis typical of BSS.[35] Mature megakaryocytes in the GPIbα "null" mice possess marked morphological abnormalities and abnormal membrane complexes could be seen in immature cells of this lineage.[36] These changes were reversed after rescue with the human GPIbα transgene, proving that GPIb-IX complexes are essential for normal thrombocytopoiesis and membrane development. Later work showed that megakaryocyte proliferation and ploidy were regulated by signaling through the cytoplasmic tail of GPIbα.[37] Genetically modified mice whose platelets lack GPV expressed GP Ib-IX normally and failed to show any of the platelet and megakaryocyte functional and morphological abnormalities that characterize the BSS phenotype.[38] This may explain why defects in the GPV gene have yet to be described in patients with BSS (Fig. 57-2). In contrast, genetic deletion of mouse GPIbβ produced a BSS phenotype with an unexpected side effect, increased α-granule size.[39] Although this has been observed in rare patients with BSS,[25] such a finding may also depend on other genetic factors in the mice (or the individual).

### B. Platelet-type von Willebrand Disease

Platelet-type or "pseudo" von Willebrand disease (VWD) is an autosomal dominant disease in which GPIbα has a gain-of-function mutation.[1] This leads to spontaneous binding of plasma VWF and a hypersensitivity to ristocetin. GPIbα mutations giving rise to this disorder are Gly233Val or Ser and Met239Val.[1,2,33] In terms of crystallographic data, these substitutions change the conformation of the regulator "thumb" incorporating a disulfide-linked double loop region and allowing high affinity binding of VWF through a switch of the β-hairpin.[7,33] This also leads to a low "off-rate" and a consequence is that large VWF multimers are cleared from the plasma; the outcome is a mild thrombocytopenia and a clinical condition that resembles type 2B VWD. Platelet adhesion is affected because the active site of GPIbα is blocked. A similar phenotype is brought about by loss of residues 421 to 429 in the macroglycopeptide domain, a finding that emphasizes how long-range allosteric interactions can influence GPIbα function.[40] Although platelets are not truly giant, abnormal size heterogeneity has been reported.[1] This implies interference with an unrecognized function of GPIb during megakaryocytopoiesis.

## III. Other Giant Platelet Syndromes

BSS must be differentiated from other congenital thrombocytopenias and diseases associated with giant platelets. Yet, curiously, a heterozygous Ala[156] → Val mutation affecting GPIbα is a common cause of an autosomal dominant form of Mediterranean macrothrombocytopenia.[41] The same but homozygous mutation was previously shown to give rise to the Bolzano variant of BSS in a family with apparently autosomal recessive inheritance (see Section II.A.2.c). A group of autosomal dominant disorders is characterized by large platelets, thrombocytopenia, and various combinations of Döhle-like inclusion bodies (Fig. 54-C) in leukocytes (May–Hegglin anomaly and Sebastian syndrome), sensorineural hearing loss, cataracts, and progressive nephritis (Fechtner and Epstein syndromes).[1,42] Originally thought to represent distinct diseases, mutations in the MYH9 gene encoding the heavy chain of nonmuscle myosin heavy chain IIA (NMMHC-IIA) expressed in platelets and leukocytes have been reported in each of the above-named syndromes.[43,44] Thus, they more likely represent different subtypes of the same syndrome. Interestingly, rod mutations associated with MYH9-related diseases disrupt filament assembly and function of NMMHC-IIA.[45] Aberrant antisense gene regulation of the fibulin-1 gene (encoding fibulin-1, an extracellular matrix protein) has been suggested to be a disease modifier.[46] In the related Alport syndrome, glomerulonephritis and deafness are associated with mutations in collagen IV, principally involving the α3, α4, and α5 genes depending on the mode of inheritance, but without giant platelets.[47] Actinomyosin complexes in the submembranous cytoskeleton are directly attached to the GPIb-IX complex through filamin (Chapters 4 and 7) and may drive the redistribution of developing membrane systems in maturing megakaryocytes (Chapter 2). Giant platelet syndromes may arise when this function is interfered with. Alternatively, defects in GPIb receptor signaling could lead to specific losses in transcription factor activation and gene transcription in marrow cells. Gene expression profiling may be required to understand the molecular etiology underlying the development of giant platelet syndromes and to predict if, for example, a patient with an MYH9 mutation will go on to develop deafness, cataracts, or renal disease.

The range of genetic abnormalities now known to give rise to inherited thrombocytopenias is large, and these are reviewed in detail in Chapter 54. Nevertheless, it should be emphasized that some of these abnormalities also give rise to defects in platelet function, such as the abnormal collagen-induced platelet aggregation in patients with mutations localized to the GATA-1 gene,[48] and the impaired receptor-mediated activation of integrin αIIbβ3 in patients with defects of the AML1 (RUNX1, CBFA2) gene.[49] Finally, in the Montreal platelet syndrome, another example of macrothrombocytopenia, spontaneous agglutination, and a decreased response to thrombin are primary features of the enlarged platelets.[1,42]

## IV. Inherited Disorders of Agonist Receptors and Signaling Pathways

### A. Collagen Receptors

Integrin α2β1 (GPIa-IIa) is a platelet surface collagen receptor that is shared with a wide variety of cell types (see Chapter 6). In the mid-1980s, a patient was described with a mild, lifelong bleeding disorder and a selective absence of aggregation and adhesion to collagen *in vitro*.[50] Platelets of this patient lacked α2β1 on biochemical analysis; however, the molecular basis of this deficiency has never been reported. This is emphasized, for care must be taken in such patients to rule out changes in receptor density linked to natural polymorphisms of the α2 gene (see Chapter 14). Low α2β1 levels can result in less efficient primary platelet adhesion and in increased bleeding in patients already at risk, as for example in type I VWD.[51]

Patients with platelets deficient in GPVI and lacking a collagen-induced aggregation response have also been described.[52] GPVI is a cloned member of the immunoglobulin superfamily of cell membrane receptors[53] (Chapter 6). Platelet-collagen interaction occurs through a multistep process (Chapter 18). Platelets attach through both α2β1 and GPVI, where activation primarily depends on GPVI-mediated signaling by a mechanism requiring Fc receptor γ chain and induced *in vitro* by convulxin A, a protein isolated from snake venom.[53] Interestingly, the GPVI gene also encodes a series of haplotypes that can influence receptor density and bleeding risk.[51] These polymorphisms need to be ruled out when assessing a patient, as do acquired antibodies, for these can lead to metalloprotease-induced cleavage of the receptor.[54] Finally, studies with transgenic mice show that loss of either integrin α2 or GPVI function in platelets does not lead to major bleeding,[55,56] suggesting that in humans α2β1 or GPVI loss is associated with a mild bleeding risk.

### B. ADP Receptors

Platelets possess three classes of receptor for adenine nucleotides (Chapter 10): P2X₁, a ligand-gated cation channel, is activated by ATP; P2Y₁ mediates ADP-induced Ca²⁺-mobilization and shape change; P2Y₁₂ is responsible for macro-scopic platelet aggregation.[57] P2Y₁₂ is linked to adenylate cyclase through Gᵢ and was only recently cloned.[58] Both P2Y₁ and P2Y₁₂ belong to the seven transmembrane domain family of G protein-linked receptors. Two groups have reported patients with a hereditary disease linked to a much decreased and reversible platelet aggregation to ADP despite a normal shape change and Ca²⁺ mobilization.[59–61] A reduced binding of radiolabeled 2-MeS-ADP (a stable analogue of ADP), together with the inability of ADP to lower the cyclic adenosine monophosphate (cAMP) content of prostaglandin (PG) E₁-stimulated platelets, suggested a specific receptor defect. Although platelets from these patients respond in part to ADP, integrin αIIbβ3 shows a limited activation, and the rapidly dissociated aggregates are composed of loosely bound platelets attached by few contact points.[59] The central role of ADP in mediating thrombus formation was reflected by the formation of smaller and unstable aggregates when blood from such patients was perfused over collagen.[62]

Analysis of polymerase chain reaction (PCR) products from the P2Y₁₂ coding region of genomic DNA, or following the isolation of platelet mRNA, has led to the identification of mutations in patients with P2Y₁₂-deficient platelets.[58,60,61] Featured are single base-pair or short deletions with a frameshift and the introduction of a stop codon that leads to the synthesis of a truncated and nonfunctional protein. Missense mutations have also been described that do not interfere with ADP binding but prevent P2Y₁₂ signaling.[61] These patients are of particular interest, for their platelets show identical functional changes to normal platelets treated with the antiplatelet drugs, ticlopidine and clopidogrel (Chapter 61), thereby helping to prove that P2Y₁₂ is the molecular target for the drugs. So far, no human pathology of the P2Y₁ receptor has been reported. However, studies on P2Y₁ receptor-null mice have shown a severe aggregation defect with ADP and an absence of shape change.[63] The description of a patient with an abnormal ADP-induced platelet response and a dominant negative disease linked to the P2X₁ gene and leading to the deletion of a single amino acid residue requires further explanation.[64] Interestingly, a partial deficiency of the surface pool of the adenylate cyclase-linked P2Y₁₂ receptor has been reported to give a secretory defect in platelets.[65]

### C. Other Agonist Receptors

A defective platelet aggregation to thromboxane (TX) A₂ has been reported in a Japanese family for whom a Arg⁶⁰ → Leu mutation has been identified in the TXA₂ receptor.[66] This receptor is present in two isoforms in platelets that differ only in their carboxyl-terminal tails and in their capacity to activate adenylate cyclase (Chapters 6 and 31). Interestingly, the mutated α-form impairs adenylate cyclase stimulation. Congenital defects of the α₂-adrenergic receptor associated with a decreased platelet response to epinephrine have been described, as have defects of receptors for platelet activating factor (PAF) and for serotonin. All of these receptors belong to the seven transmembrane domain receptor family of which other major examples in human platelets are the thrombin receptors, PAR-1 and PAR-4[9] (Chapter 9). So far, human pathologies of the PAR receptors have not been described, perhaps because their absence is incompatible with fetal development.

### D. Signaling Pathways

Platelet pathologies may also concern the signal transduction pathways into which surface receptors are locked. These represent examples of intracellular defects (Table 57-1) and are observed in patients with normal granule contents. These patients usually have mild bleeding disorders and defects of platelet aggregation that affect individual stimuli more than others. Featured in one review[67] are patients with an impaired G protein activation, an altered phosphatidylinositol metabolism, a reduced protein phosphorylation, or a defective calcium mobilization. One patient had a lineage-specific deficiency of the phospholipase C-$\beta$2 isoform, thereby highlighting its role in G protein-mediated platelet activation.[68] Another patient had a specific decrease in platelet membrane $G_{\alpha q}$ and platelets that responded less well to several agonists, including a decreased activation of $\alpha$IIb$\beta$3.[69] A series of other patients were shown to have abnormal aggregation and secretion in response to several agonists, and receptor-mediated calcium mobilization or plekstrin phosphorylation were deficient in most of them.[70] These selected examples of reports of abnormalities in signaling pathways may be just the beginning of what may eventually prove to be a long list of rare congenital disorders of platelets. However, signaling pathways in platelets are complex, and defects in any one protein may be compensated for in part by other pathways. The key to identifying these disorders may be via the simultaneous assessment of a wide range of gene products using proteomics or microarray-based technology (Chapter 5). Studies on mice suggest that phosphoinositide 3-kinases, the serine/threonine kinase, Akt, and the $G_{\alpha 13}$ subunit are at the forefront of a long list of proteins to examine.[71–73] A report of decreased $G_{\alpha i1}$ in a patient with a decreased platelet response to weak agonists is surprising, because $G_{\alpha i1}$ has not been recognized to have a role in platelet aggregation.[74]

It should also be emphasized that many of the activation pathways may be regulated by naturally occurring antagonists themselves reacting through metabolic pathways in platelets. Examples are $PGI_2$ and nitric oxide (Chapter 13), whereas the vasodilator-stimulated phosphoprotein (VASP) protein, that is phosphorylated by both cAMP- and cyclic guanosine 5′-monophosphate (cGMP)-dependent mecha-

nisms, regulates adhesion in mice in its phosphorylated form.[75] The possible upregulation of such pathways is little studied in man but could also give rise to bleeding syndromes.

### E. Enzyme Deficiencies

Patients with congenital deficiencies of cyclooxygenase, PGH synthetase-1, thromboxane synthetase, lipoxygenase, glycogen-6 synthetase, and of ATP metabolism have all been reported (Table 57-1) and lead to platelet function abnormalities, often resembling those seen in storage pool disease or after aspirin ingestion.[67]

## V. Defects of Secretion (Storage Pool Disease)

Defects of secretion, or storage pool disease (see also Chapter 15), are a heterogeneous collection of inherited disorders that include some well-characterized examples of intracellular defects of platelets but in which the affected gene encodes a protein whose function extends to several cell types.

### A. α-Granule Disorders

$\alpha$-Granules (Chapters 3 and 15) are the storage site for proteins that are either synthesized in megakaryocytes (e.g., platelet factor 4 [PF4], $\beta$-thromboglobulin [$\beta$TG], VWF, platelet-derived growth factor [PDGF]) or endocytosed from plasma (e.g., fibrinogen, albumin, immunoglobulin). As well as containing proteins that will promote platelet aggregation, $\alpha$-granules also store proteins that regulate cell proliferation and angiogenesis depending on the local environment[76] (Chapter 41). The organelle membranes contain a variety of glycoproteins, some of which are common to the plasma membrane. Others, such as P-selectin (Chapter 12), are exclusive to granule membranes in unactivated cells, although they may be shared by $\alpha$- and dense granules. An inherited deficiency of P-selectin has yet to be described. Another potentially active membrane component of $\alpha$-granules is the triggering receptor expressed on myeloid cells (TREM) family member, TLT-1 (TREM-like transcript-1).[77] Isolated deficiencies of $\alpha$-granule-stored proteins can occur associated with inherited deficiencies of the corresponding plasma proteins (e.g., factor V deficiency, fibrinogen in GT or afibrinogenemia, VWF in VWD). Only disorders unique to platelet $\alpha$-granules will be described here.

**1. Gray Platelet Syndrome.** A mild bleeding disorder with autosomal recessive inheritance in most cases, gray platelet syndrome (GPS, see also Chapter 15) is characterized by the platelet's inability to store α-granule proteins, and this leads to the platelets appearing gray on stained blood smears.[78] A feature of most GPS patients is the early onset of myelofibrosis, a finding attributed to the inability of megakaryocytes to store newly synthesized growth factors and cytokines such as PDGF-AB, PDGF-C, and TGF-β1. There is a tendency for secretion-dependent platelet aggregation to be abnormal in GPS. Whereas thrombin-induced platelet aggregation appears particularly affected in some patients, in others it is the response to collagen that is lacking. This heterogeneity may be due to the fact that, in some patients, the GPVI collagen receptor is lacking from the platelets, probably as a result of metalloprotease-induced cleavage.[79] Immunoelectron microscopy has shown that residual α-granules in GPS platelets are small and almost empty, and many vacuoles are observed.[80] Residual α-granule proteins can be detected in the surface-connected canalicular system (SCCS). In the absence of the α-granules, P-selectin is widely distributed in the SCCS. The aberrant distribution of P-selectin in the megakaryocytes can result in the capture of neutrophils (emperipolesis), and proteases from the latter can contribute to the marrow fibrosis just as in GATA-1 (low) mice or in idiopathic myelofibrosis.[81] The basic molecular defect(s) in GPS is unknown, but it appears to involve packaging or storage of the α-granule contents.[78] SNARE proteins (so-called after their initial description as receptors for attachment proteins) constitute a superfamily whose members direct membrane fusion events involved in vesicle trafficking and exocytosis. Several SNARE proteins are present in platelets (Chapter 15) and they, along with small molecular mass GTP-binding proteins, are candidates susceptible to be modified in GPS.[82] Interestingly, mutations in VPS33B, encoding a regulator of SNARE-dependent membrane fusion, have been described in the arthrogryposis-renal dysfunction-cholestasis (ARC) syndrome in which platelet dysfunction and low granule content are associated with a multisystem disorder featuring renal tubular and other dysfunctions.[83]

**2. Quebec Platelet Disorder.** The Quebec platelet disorder is an autosomal dominant bleeding disorder that, as its name suggests, was first described in two French-Canadian families. Originally thought to involve factor V, platelets are also severely deficient in multimerin, a high molecular mass protein that is stored as a complex with factor V in α-granules.[78] The situation became clearer when it was shown that platelets of these patients show protease-related degradation of many α-granule proteins (including P-selectin) even though α-granule ultrastructure is preserved.[78,84] Thrombocytopenia is sometimes observed. The platelet aggregation deficiency is, surprisingly, most striking with epinephrine, and the reason for this is unknown. Inter-estingly, the bleeding syndrome responds to fibrinolytic inhibitors rather than platelet transfusions. This observation led to the discovery that platelets in the Quebec platelet disorder possess unusually large amounts of urokinase-type plasminogen activator.[84,85] The molecular basis for this remains to be elucidated, although localized plasmin generation is thought to account for the protein degradation.

### B. Dense (δ) Granule Disorders

Dense (δ) granules are the storage sites for serotonin and nucleotides such as ADP and ATP (Chapter 15). Storage pool disease affecting dense granules may be quite common and should always be considered when a patient presents with a deficiency of secretion-dependent aggregation, although a prolonged bleeding time may also be associated with a normal platelet aggregation in some cases.[86] The granule deficiency may be severe or partial, and in occasional patients it may also extend to α-granules (αδ-storage pool deficiency) in which inheritance may be autosomal and dominant.[87] Secretion-dependent platelet aggregation can be more severely affected than in GPS, and platelet adhesion may also be reduced (perhaps explaining the effect on bleeding time). The molecular basis for the defect(s) is unknown and may concern granule formation or the packaging of their contents. Inherited diseases in which platelet deficiencies of dense granules are associated with abnormalities of other cells and which lead to a clearly defined phenotype will now be described.

**1. Wiskott–Aldrich Syndrome (WAS).** Wiskott–Aldrich syndrome (WAS, see also Chapter 54) is an X-linked recessive disease in which thrombocytopenia and small platelets are associated with eczema, recurrent infections, and an increased risk for autoimmunity and malignancy.[88,89] A milder form without the immune and other problems is known as hereditary X-linked thrombocytopenia (see also Chapter 54). WAS platelets show a decreased aggregation response to ADP, epinephrine, and collagen and have a reduced dense granule number. The disease is not exclusive to platelets, and T lymphocytes among other blood cells also show defective function. The gene responsible for WAS has been cloned and a total of 12 exons encode a 502 amino acid protein termed WASP. A large number of genetic defects have been found in WAS that result either in the decreased expression or absence of WASP.[89] There is a strong correlation between genotype and phenotype. Thus, mutations in exons 1 and 2 are more likely to give rise to hereditary X-linked thrombocytopenia, whereas a more severe phenotype is associated with mutations causing the absence of WASP. WASP is involved in signal transduction, possessing tyrosine phosphorylation sites and adapter protein function. It binds through proline-rich motifs to SH3-containing proteins such as Fyn, Lck, phospholipase C-γ, and Grb2. WASP is particularly known to regulate actin filament assembly and stimulates actin nucleation by the so-called Arp2/3 complex[90] (see also Chapter 4). In some respects, WAS could be classified as a pathology of the cytoskeleton. Interestingly, multiple patients with revertant T-cell mosaicism due to DNA polymerase slippage and progressive clinical improvement have been reported in a WAS family.[91]

### 2. Hermansky–Pudlak, Chediak–Higashi, and Griscelli Syndromes.

Hermansky–Pudlak, Chediak–Higashi, and Griscelli syndromes are distinctive autosomal recessive diseases in which a bleeding diathesis resulting from a platelet dense granule deficiency is accompanied by deficient pigmentation of the skin and hair (affecting melanosomes) and defective lysosomes.[92] In the Hermansky–Pudlak syndrome (HPS), oculocutaneous albinism is a common characteristic, and some patients also develop granulomatous colitis or pulmonary fibrosis. It is common on the Caribbean island of Puerto Rico. Originally, a single HPS gene was detected, localized to chromosome 19, and predicted to encode a 79 kDa protein with two membrane-spanning domains. A 16-base duplication in exon 15 of this gene resulting in a frameshift was then shown to largely account for HPS in Puerto Rico. However, as time went by, more molecular defects were described in HPS. In HPS2, it is the beta3A subunit of the AP-3 adaptor complex that is abnormal and causes an increased routing of lysosomal membrane proteins, such as LAMP-3 (CD63), to the plasma membrane.[93] In severe cases of HPS2, neutropenia is an additional feature. Now, at least five further subtypes of HPS (HPS3-7) are known and the affected gene products are all involved in vesicle trafficking from the transgolgi apparatus (Table 15-3). These proteins along with HPS1 normally form complexes (BLOC-1, -2, and -3) all of which are involved in granule biogenesis[94,95] (Chapter 15).

In the Chediak–Higashi syndrome (CHS), the bleeding disorder is accompanied by severe immunologic defects and progressive neurological dysfunction if the patient survives to adulthood.[92,96,97] Morbidity is often linked to repeated bacterial infections or to an "accelerated phase" lymphoproliferation into major body organs. An ultrastructural characteristic of CHS is the giant inclusion bodies in a variety of granule-containing cells including platelets. The CHS gene is located on chromosome 13. A series of frameshift and nonsense mutations have been described that are distributed along the length of the 13.5-kb CHS cDNA, and many of these result in a truncated or absent CHS protein.[92,96,97] Very rare patients show a milder phenotype associated with missense mutations that encode a protein with partial function. In the normal state, the CHS gene encodes a large protein (LYST) with structural domains that include a series of hydrophobic helixes and hydrophobic

repeat motifs (for example, HEAT, BEACH, and WD40 domains or repeat motifs). These suggest a function in membrane contact interactions and organelle protein trafficking.[97] With regard to platelet function, the scarcity of dense bodies in HPS and CHS means that the secretable pool of ADP, an important cofactor in aggregation induced by low doses of most stimuli, is not available.

Finally, in the Griscelli syndrome, in which patients with partial albinism of the skin and hair have life-threatening defects of cytotoxic T lymphocytes, an associated dense granule defect (largely seen in mouse models of the disease) is due to mutations which abrogate expression of Rab27a, a small GTPase known to regulate vesicle fusion.[98] Defects in the myosin Va gene have also been described in some patients. As illustrated in the previously cited works, mouse models of storage pool disease have contributed in an important way to our understanding of HPS, CHS, and the Griscelli syndrome.

## VI. Glanzmann Thrombasthenia

As in BSS, the nature of bleeding in GT is clearly defined. Purpura, epistaxes, gingival hemorrhage, and menorrhagia are nearly constant features.[1,99] Gastrointestinal bleeding is also a problem for some patients, but hemarthroses and deep visceral hematomas are rare. Platelets from patients with GT show quantitative or qualitative defects of the integrin αIIbβ3. The virtual absence of platelet aggregation to all agonists is the hallmark of this disease. Platelet aggregation fails to occur because in normal hemostasis αIIbβ3 on activated platelets binds the adhesive proteins, which form the protein bridges that link platelets within the aggregate (Chapter 8). The principal adhesive proteins in this regard are fibrinogen and, under conditions of high flow, VWF. However, others, including fibronectin, vitronectin, and even CD40 ligand, most certainly play a role[100] (Chapter 18). Ristocetin-induced binding of VWF to GPIbα on GT platelets is normal, although platelet "agglutination" induced by ristocetin and VWF *in vitro* may occur in cycles in a process that may mimic transient GPIb-dependent platelet adherence to VWF in subendothelium. Whereas GT platelets bind to exposed subendothelial tissue, platelet spreading on the vessel wall surface, a process that involves αIIbβ3, fails to occur.[101] The net result of the αIIbβ3 deficiency is that thrombus formation is severely affected.

Ligand-bound αIIbβ3 mediates "outside-in" signaling (Chapter 17), a phenomenon that may influence thrombus stability under normal conditions.[102] Thus, when GT platelets are incubated with thrombin, a decreased tyrosine phosphorylation of several high-molecular-mass cytosolic proteins is observed[103] — phosphorylations that in normal platelets depend on adhesive protein binding to αIIbβ3 or aggregation. The αIIbβ3 integrin also provides another link between platelets and the coagulation system. Prothrombin can bind directly to αIIbβ3, an interaction that cannot occur on platelets from patients with classic GT.[104] Most reports confirm that the ability of GT platelets to generate thrombin is decreased, and they produce fewer procoagulant microparticles when stimulated.[1] Yet GT platelets appear able to attach to fibrin independently of activated αIIbβ3 under flow, suggesting the presence of an alternative platelet receptor for fibrin and an explanation for the observed benefit of recombinant factor VIIa in the cessation of bleeding in these patients[105] (Chapter 68).

## A. Molecular Defects

Much has been learned from crystallography and computer-based modeling about the structure of αIIbβ3 and its activation mechanisms.[106] That a defined tertiary structure is required for receptor activity is demonstrated by patients with variant forms of GT with qualitative defects of αIIbβ3 unable to bind fibrinogen. Nevertheless, the majority of patients have a severely decreased surface expression of the integrin on their platelets, and bleeding is a manifestation of this deficiency.[1,107] The gene location of a large series of mutations and other defects giving rise to GT is shown in Fig. 57-3. Large deletions are rare, whereas splice defects, nonsense mutations, small deletions and inversions, and point mutations are abundant. The gene encoding αIIb spans 17 kb and is composed of 30 exons; the gene encoding β3 is much larger but only has 15 exons. Both genes colocalize to 17q 21–23 (see Chapter 8).

After routine platelet aggregation, screening for GT involves analyzing platelet surface glycoproteins and flow cytometry has become the method of choice[14] (Chapter 30), particularly as the analysis can be performed on small blood samples thereby permitting diagnosis in children. Accompanying Western blot procedures will allow the detection of residual amounts of the αIIb or β3 subunits, and the screening for αvβ3 in platelets or activated lymphocytes is useful. In terms of molecular biology, we and others chose a PCR-SSCP ("single chain conformation polymorphism") procedure using isolated DNA and screening each exon (+ splice sites) for mutations, followed by the sequencing of amplified products in the event of migration changes.[108] But as automated procedures have evolved, the direct sequencing of PCR amplified products encoding individual exons of both subunits, or the sequencing of cDNA after RNA extraction from platelets, is now the method of choice. In general, if αIIbβ3 complex formation cannot occur or is incorrect, both mutated and normal (but noncomplexed) polypeptides are retained in the cytoplasm and are degraded intracellularly.

**1. Defects in the αIIb Gene (Fig. 57-3A).** In a pioneering study, Israeli Arab kindred were shown to possess a 13-bp deletion encompassing the splice acceptor site of exon 4 and resulting in alternative splicing to a downstream nucleotide AG receptor, producing a 6-aa deletion in the αIIb protein.[109] No surface αIIbβ3 was observed and the deleted sequence was shown to be critical for posttranslational processing of this subunit. Defects leading to type I disease (for a description of early nomenclature in GT, see ref. 99) often involve splice site mutations accompanied by frameshifts and the production of a truncated protein. This is so in the French gypsy population.[110] As shown on Figure 57-3A, nonsense mutations giving rise to stop codons are another frequent cause of the disease and again mostly lead to a severe platelet depletion of αIIbβ3.[107] Several missense mutations within or near the N-terminal Ca$^{2+}$-binding domains of αIIb have been reported. Site-directed mutation and expression of recombinant glycoproteins in heterologous cell lines show how these amino acid substitutions allow complex formation but inhibit the transport of the complexes from the endoplasmic reticulum to the Golgi apparatus or their export to the cell surface.[111,112] Occasionally, sufficient mutated αIIbβ3 is processed for there to be greater than 5% surface expression on platelets, thus accounting for the previously described type II GT. Mutation analysis of αIIbGlu$^{374}$ has shown that αIIbβ3 surface expression occurs in CHO cells when this amino acid is replaced by other negatively charged or polar amino acids, but that surface expression does not occur when a positively charged amino acid is substituted.[113] Thus, with missense mutations, the GT phenotype depends not only on the position of the mutated residue but also on the nature of the substituted amino acid. The early interaction between the mutated protein and molecular chaperones such as BiP may determine its fate.[114] Other mutations may render mRNA unstable and prevent protein synthesis in adequate amounts.[115] In the absence of consanguinity, compound heterozygotes resulting from two defects within the αIIb gene are a common cause of GT.[107] Some of the described mutations affect extracellular regions other than the αIIb N-terminal β-propeller;[106,116] these can include the calf-1 and calf-2 domains, both of which have contact sites with β3; such mutations, although they allow complex formation, interfere with transport from the endoplasmic reticulum.[117]

**2. Defects in the β3 Gene.** As shown in Fig. 57-3B, mutations are found across the β3 gene. A recurring mutation in Iraqi Jews is an 11 base deletion within exon 13 (originally exon 12) that results in a frame-shift and protein termination just before the transmembrane domain.[109] This defect prevents normal membrane insertion of β3 and expression of both αIIbβ3 and αvβ3 (discussed later). Several other splice site mutations and DNA deletions and inversions resulting in premature termination of β3 have since been reported.[107] Jin et al.[118] described a splice mutation associated with two different base changes, which acted

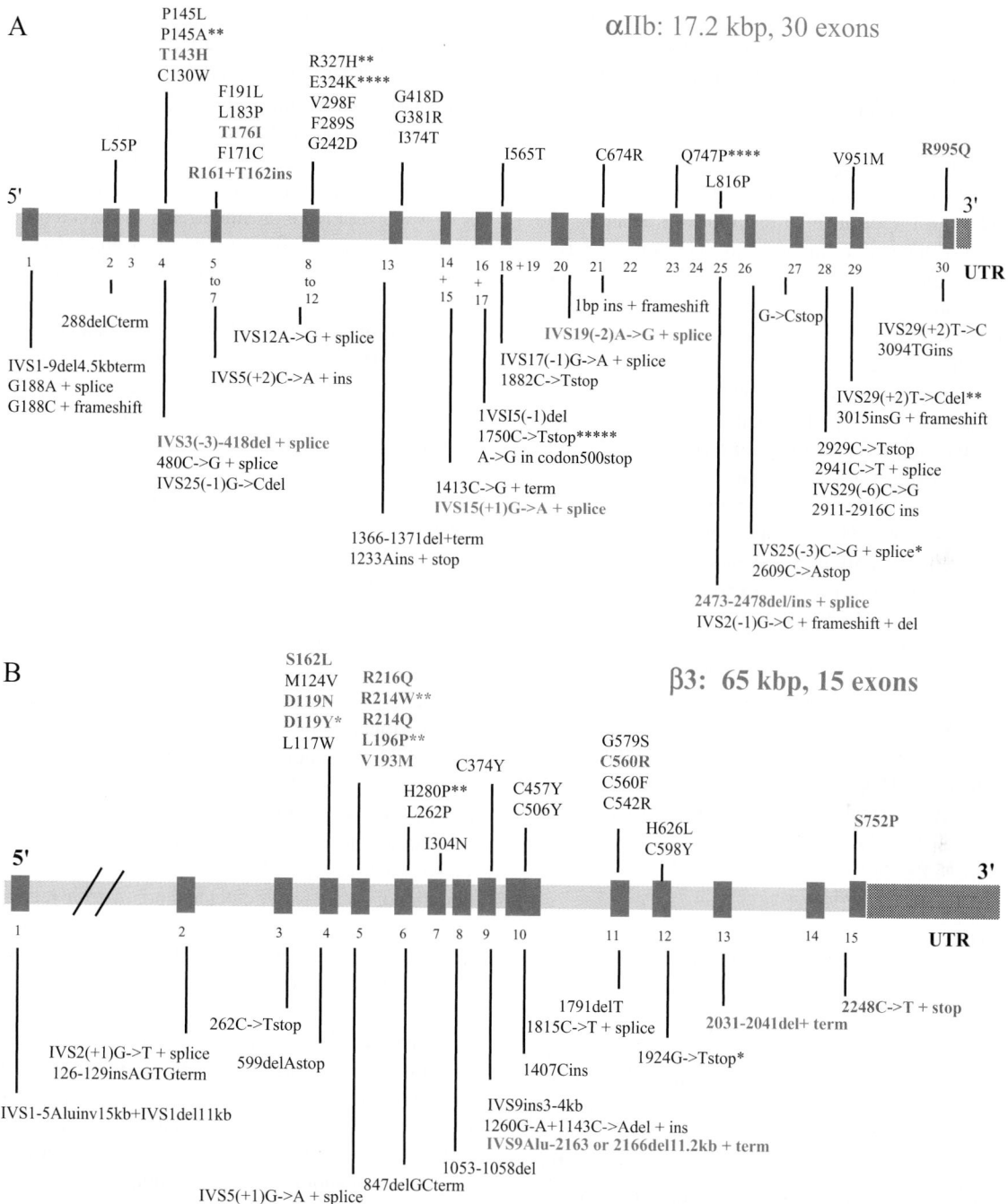

**Figure 57-3.** Genetic defects known to cause Glanzmann thrombasthenia; A: abnormalities in the αIIb (GPIIb) gene. B. Abnormalities in the β3 (GPIIIa) gene. The exons are shown by red boxes. (For the αIIb gene, several small exons are grouped together to allow the whole gene to be shown.) Missense mutations are shown as amino acid substitutions above the gene and identified by the codon number. The amino acid substitutions are designated according to the single letter code. Leader sequences are not included and thus, on occasion, the numbering may differ from that in the original reports. Variants with qualitative defects of at least a partially expressed integrin are shown in blue. Mutations that are common within ethnic groups are in green. Nucleotide changes involved in nonsense mutations, frameshifts, insertions (ins), and deletions (del) that often are followed by stop codons are shown below the gene. The nucleotide numbering is as described by French.[107] Nucleotide sequences within acceptor sites are given a number (−1, −2 . . .) from the exon border, and those within donor splice sites are designated (+1, +2 . . .). Also to be considered for the list of αIIb mutations are 15 novel candidate mutations reported in an Italian study,[158] whereas to the β3 mutations may be added 11 novel mutations reported following screening in India.[159] The number of asterisks indicates the number of unrelated cases that have been reported. Abbreviations: IVS, intervening sequence; UTR, untranslated region.

in concert and resulted in deletion of exon 9 and a 5-bp insertion that restored the reading frame. The result was an abnormal β3 unable to pair with αIIb. Although the deletion was homozygous, consanguinity was denied in the family. DNA microsatellite dinucleotide polymorphism analysis suggested the presence of two copies of the maternal chromosome. This study shows the complexity that can be at the basis of GT. The importance of disulfides in assuring the conformation of β3 is well known, and mutations affecting cysteines have an important influence on the structure of this subunit and its capacity to complex with αIIb. In an example from our own work, a homozygous $Cys^{347} \rightarrow Arg$ mutation in the β3 gene led to classic GT and a much decreased αIIbβ3 expression in platelets.[119] Studies on $Cys^{598}$, mutated in a GT type II patient, helped define a role for the cysteine-rich EGF domains in αIIbβ3 activation.[120] In another patient, a $Cys^{506} \rightarrow Met$ missense mutation resulted in an unpaired cysteine and the appearance of a disulfide-linked β3 dimer.[121] Of much interest in terms of their influence on phenotype, some missense mutations in β3 have a different impact on the expression and function between αIIbβ3 and αvβ3.[122]

**3. Variant Glanzmann Thrombasthenia.** In all but one of the reported GT variants, αIIbβ3 is qualitatively defective in its fibrinogen-binding function despite significant and sometimes normal platelet surface expression. In this way, this category of patient differs from the type I and type II GT. The variants are a heterogeneous group of patients, and bleeding symptoms range from severe to mild. Studies on these patients have been most useful in determining how integrins function.

*a. MIDAS-domain Mutations in β3.* The first report of variant GT described an $Asp^{119} \rightarrow Tyr$ substitution in β3, a mutation that helped to identify an RGD-binding site on β3.[123] Three unrelated patients were then reported for whom αIIb and β3 subunits dissociate much more readily from each other on $Ca^{2+}$ chelation. Studies on these patients revealed that the codon for $Arg^{214}$ of the β3 gene is a mutational hotspot.[1] Mutations at $Arg^{214}$ may influence three potential metal binding sites on β3 including the so-called MIDAS (metal-ion-dependent adhesion site) which resembles that present in the I domain (inserted domain) of VWF and certain integrin α-subunits and also a ligand-induced metal binding site (LIMBS).[106] Substitutions in the β3 I domain not only prevent the expression of the ligand-binding pocket in response to platelet stimulation but also may make the complex unstable.

*b. Cytoplasmic Domain Mutations.* Studies on patients with cytoplasmic domain mutations have made an important contribution not only to the understanding of how signaling from primary receptors transforms the αIIbβ3 integrin from its "bent" inactivated form to the extended confirmation necessary to bind fibrinogen but also to unraveling the mechanisms of outside-in signaling from activated and occupied integrin.[102,106] The story began with a patient from Argentina whose platelets possess half the normal αIIbβ3 content, and which fail to bind fibrinogen or the activation-dependent monoclonal antibody PAC-1 when stimulated by ADP.[124] Nevertheless, αIIbβ3 isolated from his platelets bound normally to an RGDS affinity column. A heterozygous $Ser^{752} \rightarrow Pro$ point mutation was detected in the cytoplasmic domain of β3. Studies on transfected CHO cells confirmed a role for signaling through the β3 cytoplasmic domain in the expression of the extracellular ligand-binding conformation of αIIbβ3. The transfected cells also showed reduced αIIbβ3-mediated cell spreading on immobilized fibrinogen.[123,125] Similar findings were reported for another variant by Wang et al.[126] Flow cytometry confirmed that fibrinogen binding to platelets occurred in the presence of monoclonal antibodies that directly activate αIIbβ3, yet physiologic stimuli were unable to reproduce this effect. A heterozygous mutation within exon 13 of the maternal allele of the β3 gene gave a stop codon at $Arg^{774}$ and a truncated protein lacking most of the cytoplasmic domain. Significantly, in both of these patients, and the variant to follow, only the affected allele was expressed.

A young Italian man with a thrombasthenia-like syndrome showed a platelet aggregation response that was not null but much reduced in velocity and intensity. Total platelet αIIbβ3 was about 50% of normal, but the surface pool was decreased to 18%. The residual complexes were functional, but their density appeared at the threshold of that needed for platelet aggregation. A heterozygous $Arg^{995} \rightarrow Gln$ substitution in the GFFKR sequence of the cytoplasmic domain of αIIb was detected and in transfection experiments led to an altered partitioning of the residual receptor.[127] Previously, alanine substitutions within this sequence had been shown to favor the activated state of the integrin,[128] although others claimed that complex assembly is impaired when the GFFKR region is mutated.[129] Awaiting discovery, perhaps, are abnormalities in the cytoplasmic proteins (e.g., β3-endonexin, calreticulin, calcium, and integrin-binding protein) now thought to regulate the activation state of αIIbβ3 or even in cytoskeletal proteins such as talin that associate with and activate the integrin and possibly help maintain its configuration[130] (Chapters 8 and 17).

Also to be noted is a novel form of integrin dysfunction that involves β1, β2, and β3 integrins and that associates the GT and leukocyte adhesion deficiency-1 syndromes.[131] An unknown intracellular defect essential for integrin activation or clustering is thought to be at the origin of this disease.

*c. Upregulated αIIbβ3.* A pioneering observation came from Shattil's group who showed that a stable CHO cell transfectant in which wild-type αIIb was coexpressed with β3 containing a $Thr^{562} \rightarrow Asn$ substitution (a substitution within the cysteine-rich EGF domains) spontaneously bound fibrinogen (and monoclonal antibody PAC-1) and the cells agglutinated when stirred with fibrinogen.[132] That gain-of-function mutations in the extracellular domains of αIIbβ3 might exist in pathology was confirmed when a patient was described with platelets that spontaneously bound PAC-1 and fibrinogen.[133] This patient has a homozygous Cys[560] → Arg mutation in β3 and platelets that express about 20% of the normal amount of surface αIIbβ3. This situation recalls platelet-type von Willebrand disease in which normal VWF multimers spontaneously bind to a mutated GPIbα subunit (Section II.B). Not least, these results elegantly show how the MIDAS-like domain is influenced by long-range constraints within the three-dimensional structure of "unactivated" αIIbβ3. Presumably, the patient has a bleeding syndrome because all available αIIbβ3 molecules are occupied monovalently with fibrinogen, thus preventing platelet-to-platelet cross-linking.

**4. Platelet Fibrinogen Transport, Fibrin, and Clot Retraction.** Fibrinogen is stored within the α-granules of platelets and is acquired by αIIbβ3-mediated uptake.[134] Platelets from thrombasthenic patients who lack αIIbβ3 are unable to take up fibrinogen, so the α-granule pool of fibrinogen is absent. Surprisingly, patients whose platelets have a low residual surface expression of functional αIIbβ3 may have near normal α-granule fibrinogen.[99] This is evidence that αIIbβ3 complexes recycle between the surface and the α-granules. Our demonstration of fine channels containing αIIbβ3 leading from the surface to the storage pool compartment of normal platelets suggests guided transport to and from the granules.[135] Clathrin-mediated and other more classic pathways of endocytosis may also be involved.[136] Interestingly, for certain variants, α-granule fibrinogen is present despite the inability of the surface αIIbβ3 to bind fibrinogen upon platelet activation.[1,137] For example, the Ser[752] → Pro mutation in the cytoplasmic domain of β3 clearly allows uptake of fibrinogen. This suggests that uptake occurs by low-affinity activation-independent binding to αIIbβ3. A Leu[262] → Pro substitution within β3 resulted in another variant form of GT.[138] Platelets of this patient possess reduced but clearly detectable amounts of αIIbβ3; they support clot retraction but do not bind fibrinogen when stimulated (and they do not aggregate). The retained ability of platelets to mediate clot retraction implies that the abnormal complexes were able to interact with fibrin even if they did not bind soluble fibrinogen. In fact, the presence of an αIIbβ3-independent receptor for fibrin on GT platelets[105] probably means that platelets of all patients may bind to this protein but that residual αIIbβ3 is needed for retraction.

**5. Vitronectin Receptor.** The β3 subunit in integrin nomenclature is also a component of the vitronectin receptor (VnR or αvβ3). Expression of the αIIb gene (and therefore of αIIbβ3) is normally restricted to cells of the megakaryocytic lineage. In contrast, αvβ3 is found in many cell types,

including endothelial cells, chondrocytes, fibroblasts, monocytes, and activated B lymphocytes, where it acts as a promiscuous receptor for adhesive protein ligands. Yet αvβ3 has but a minor presence in platelets, about 50 copies per platelet being found at the surface compared to the 50,000 or more copies of αIIbβ3 per platelet. In fact, much of the αvβ3 appears confined to vesicular bodies that appear to differ from α-granules.[139] In GT, αvβ3 is absent from platelets and probably from other cells of most patients when the genetic lesion affects the β3 gene; in contrast, αvβ3 is present in about twice the usual amount in platelets of patients with αIIb gene defects.[1] Despite this, patients with β3 gene defects do not have a distinctive phenotype. Although αvβ3 has been implicated in angiogenesis in mouse models (see Section VI.C), no evidence for abnormal vessel development has been forthcoming in GT patients. Interestingly, in contrast to β3 deficient mice, upregulation of α2β1 compensates for lack of αvβ3 in osteoclasts of Iraqi-Jewish GT patients with a β3 defect.[140]

### B. Genetic Counseling

Only in ethnic groups in whom consanguinity is high can procedures permitting the rapid screening of given mutations such as PCR-single strand conformation polymorphism (PCR-SSCP) or allele-specific restriction enzyme analysis be used.[1] This is so for the Iraqi-Jewish and Arab groups in Israel and the French gypsy population.[107,109,110,141] Genetic counseling can be given with the following reservations: (a) when screening is followed for a single mutation, the presence of a second defect would go undetected; (b) individuals with the same genetic lesion may differ widely in the frequency and severity of bleeding. Peretz et al. have used restriction digest analysis of PCR-amplified fragments from DNA isolated from blood or urine to screen subjects for the Iraqi-Jewish and Arab mutations.[142] An alternative source for prenatal diagnosis is DNA extracted from chorionic villi. Prenatal diagnosis of GT has also been achieved using the polymorphic markers BRAC1 and THRA1 on chromosome 17.[143] As is also the case for BSS, for patients in previously unstudied families, there is no alternative but to sequence and identify the mutation prior to screening potential parents. Platelet membrane GP density shows too much variability in a normal population for carrier determination by flow cytometry alone.

### C. Transgenic Murine Models of GT

Results obtained for αIIb "null" mice show that they mimic the classic GT phenotype.[144] Although β3 integrin-deficient mice also mimic GT, results suggest accelerated reepithelialization during wound healing and enhanced TGF-β1 signaling.[145] The implication is that αvβ3 can suppress TGF-β1 signaling. Whether this will apply to GT patients with β3 gene defects will require further study. Previously, others have shown accelerated angiogenesis, placental defects, and increased fetal mortality, as well as an altered bone remodeling by dysfunctional osteoclasts in β3$^{-/-}$ mice, suggesting new clinical areas for study in GT.[146,147] A defective clot retraction was also observed in a transgenic mouse model in which the tyrosines in the cytoplasmic tyrosine motif of β3 had been replaced by phenylalanines.[148] Interestingly, platelet aggregation induced by low and medium doses of thrombin as well as ADP was more readily reversible in this model, while mice homozygous for the double tyrosine substitution showed an excessive tendency to rebleed at sites of injury. This suggests a subtle GT phenotype for human subjects possessing mutations at one or the other of these residues perhaps affecting clot retraction more than the primary phase of platelet aggregation. One other possibility is that integrin avidity changes and clustering do not occur in the presence of β3 subunit cytoplasmic domain deletions or mutations.

## VII. Scott Syndrome

Scott syndrome is a rare inherited disorder of $Ca^{2+}$-induced phospholipid scrambling and assembly of the prothombinase complex on blood cells including platelets.[149] When activated, Scott platelets are unable to translocate phosphatidylserine from the inner to the outer phospholipid leaflet of the membrane bilayer, with the result that factors Va and Xa fail to bind, leading to the incapacity of the platelets to transform prothrombin to thrombin. This lack of thrombin generation is sufficient to induce a bleeding diathesis. Physiologic stimuli that induce this translocation include a thrombin and collagen mixture and complement C5b-9. Microvesiculation and the diffusion of the procoagulant activity in the circulation accompany the phosphatidylserine expression; both phenomena can be readily measured by flow cytometry using fluoroscein isothiocyanate-conjugated annexin V. The defect in the Scott syndrome has been variously described to concern the capacitative $Ca^{2+}$ entry into cells and the subsequent regulated activation of the scramblase enzyme thought to be responsible for phosphatidylserine mobilization.[150–152] A novel heterozygous missense mutation in the ATP-binding cassette transporter A1 (ABCA1) has been said to be associated with the phenotype of a Scott syndrome patient and to give rise to impaired trafficking of ABCA1.[153] A putative second mutation in a transacting regulatory gene was speculated on.

## VIII. Therapy

### A. General Principles

Inherited platelet function disorders are often mild bleeding disorders. As Michelson[154] discussed, the general principles

of treatment of platelet function defects are avoidance of major body contact activities, avoidance of drugs that interfere with platelet function (e.g., aspirin and other nonsteroidal antiinflammatory drugs) and other components of the hemostatic system, avoidance of intramuscular injections, good dental care (to minimize gum bleeding and avoid tooth extraction), local hemostatic measures, and oral iron (to prevent or treat iron deficiency secondary to blood loss); oral contraceptive pills may be necessary for the control of menorrhagia, immunizations can be given subcutaneously with direct pressure using ice, and consideration should be given to the wearing of a medical bracelet or necklace with the words "platelet function defect" inscribed or carrying an information card to that effect.

### B. Antifibrinolytic Agents

Oral aminocaproic acid is often useful for oral bleeding.[154] In the Quebec syndrome, the bleeding responds to antifibrinolytic agents rather than to platelet transfusions, because platelets in this disorder have unusually large amounts of urokinase-type plasminogen activator that is released upon platelet activation and whose presence in excess of the natural inhibitors of this protease accounts for the *in situ* protein degradation in the α-granules.[84]

### C. DDAVP

DDAVP (desmopressin), either intravenously or by nasal spray (Stimate), is often helpful for patients with a platelet function defect (see Chapter 67 for details). There should be a major response by 30 minutes with a maximum response at 90 minutes. Plasma levels last at least 8 to 10 hours. DDAVP can be repeated every 12 hours, but tolerance may occur after 72 hours.[154] The beneficial effect of desmopressin in platelet function disorders may be related to an increased release of factor VIII and VWF from endothelial Weibel–Palade bodies or a poorly characterized direct effect on platelets (Chapter 67).

### D. Activated Factor VII

NovoSeven (intravenous activated factor VII [VIIa]) is effective in many platelet function disorders (see Chapter 68 for details). The beneficial effect of factor VIIa may be via a tissue factor-dependent mechanism or a tissue factor-independent, platelet-dependent mechanism.[154]

### E. Platelet Transfusion

If necessary, platelet transfusions (preferably HLA-matched) can be given for major bleeding episodes. The platelet transfusion will temporarily correct the platelet function defect. Platelets should be available for transfusion during any surgery.[154] There is evidence that leukocyte depletion of platelet and red cell concentrates by filtration reduces the likelihood of alloimmunization as well as of febrile transfusion reactions (Chapter 69). A major problem with platelet transfusion in BSS and GT is alloimmunization, with the patients forming antibodies against the glycoproteins missing from their own platelets. Some of these antibodies may recognize epitopes on active sites of the glycoproteins involved in adhesion and aggregation, thus rendering transfused platelets refractory as well as leading to their accelerated destruction.[154] Therefore, platelet transfusions in BSS and GT should be reserved for life-threatening emergencies. Platelet concentrates are the treatment of choice in platelet-type VWD, because the defect is in the platelets. Both desmopressin and cryoprecipitate should be avoided in platelet-type VWD because the resultant increase in circulating large VWF multimers may result in worsening thrombocytopenia.[154]

### F. Bone Marrow Transplantation

Bone marrow transplantation is used in children with severe diseases such as Chediak–Higashi syndrome,[155] Wiskott–Aldrich syndrome,[156] and, occasionally, GT.[157] The risks of bone marrow transplantation confine its use to cases in which severe bleeding cannot be adequately controlled by more conservative measures.

### G. Gene Therapy

Diseases such as BSS and GT are good candidates for gene therapy, and progress is being made in this regard (Chapter 71).

### H. Genetic Counseling

See Section VI.B.

# References

1. Nurden, A. T., & George, J. N. (2005). Inherited abnormalities of the platelet membrane: Glanzmann thrombasthenia, Bernard-Soulier syndrome, and other disorders. In R. W. Colman, V. J. Marder, A. W. Clowes, J. N. George, & S. Z. Goldhaber (Eds.), *Hemostasis and thrombosis, basic principles and clinical practice,* 5th ed. Philadelphia: Lippincott, Williams & Wilkins.
2. Lopez, J. A., Andrews, R. K., Afshar-Kharghan, V., et al. (1998). Bernard-Soulier syndrome. *Blood, 91,* 4397–4418.
3. Nurden, A. T., & Caen, J. P. (1975). Specific roles for platelet surface glycoproteins in platelet function. *Nature, 255,* 720–722.
4. Nurden, P., & Nurden, A. T. (1996). Giant platelets, megakaryocytes and the expression of glycoprotein Ib-IX complexes. *C R Acad Sci Paris, 319,* 717–726.
5. Tomer, A., Scharf, R. E., McMillan, R., et al. (1994). Bernard-Soulier syndrome: Quantitative characterization of megakaryocytes and platelets by flow cytometric and platelet kinetic measurements. *Eur J Haematol, 52,* 193–200.
6. Cruz, M. A., Chen, J., Whitelock, J. L., et al. (2005). The platelet glycoprotein Ib-von Willebrand factor interaction activates the collagen receptor $\alpha2\beta1$ to bind collagen: Activation-dependent conformational change of the $\alpha2$-I domain. *Blood, 105,* 1986–1991.
7. Uff, S., Clemetson, J. M., Harrison, T., et al. (2002). Crystal structure of the platelet glycoprotein Ib$\alpha$ N-terminal domain reveals an unmasking mechanism for receptor activation. *J Biol Chem, 277,* 35657–35663.
8. Celikel, R., McClintock, R. A., Roberts, J. R., et al. (2003). Modulation of $\alpha$-thrombin function by distinct interactions with platelet glycoprotein Ib$\alpha$. *Science, 301,* 218–221.
9. Kahn, M. L., Zheng, Y-W., Huang, W., et al. (1998). A dual thrombin receptor system for platelet activation. *Nature, 394,* 690–694.
10. McNichol, A., Sutherland, M., Zou, R., et al. (1996). Defective thrombin-induced calcium changes and aggregation of Bernard-Soulier platelets are not associated with deficient moderate-affinity receptors. *Arterioscler Thromb Vasc Biol, 16,* 628–632.
11. Baglia, F. A., Shrimpton, C. N., Emsley, J., et al. (2004). Factor XI interacts with the leucine-rich repeats of glycoprotein Ib$\alpha$ on the activated platelet. *J Biol Chem, 279,* 49323–49329.
12. Beguin, S., Keularts, I., Al Dieri, R., et al. (2004). Fibrin polymerization is crucial for thrombin generation in platelet-rich plasma in a VWF-GPIb-dependent process, defective in Bernard-Soulier syndrome. *J Thromb Haemost, 2,* 170–176.
13. Yagi, M., Edelhoff, S., Disteche, C. M., et al. (1994). Structural characterization and chromosomal localization of the gene encoding human platelet glycoprotein Ib$\beta$. *J Biol Chem, 269,* 17424–17427.
14. Michelson, A. D. (1996). Flow cytometry: A clinical test of platelet function. *Blood, 87,* 4925–4936.
15. Lanza, F. (2003). Hemorrhagipous thrombocytic dystrophy. *Orphanet Encyclopedia,* www.orpha.net.
16. Simsek, S., Admiraal, L. G., Modderman, P. W., et al. (1994). Identification of a homozygous single base pair deletion in the gene coding for the human platelet glycoprotein Ib$\alpha$ causing Bernard-Soulier syndrome. *Thromb Haemost, 72,* 444–449.
17. Kunishima, S., Miura, H., Fukutani, H., et al. (1994). Bernard-Soulier syndrome Kagoshima: Ser444 $\rightarrow$ stop mutation of glycoprotein (GP) Ib$\alpha$ resulting in circulating truncated GP Ib$\alpha$ and surface expression of GPIb$\beta$ and GPIX. *Blood, 84,* 3356–3362.
18. Simsek, S., Noris, P., Lozano, M., et al. (1994). Cys209Ser mutation in the platelet membrane glycoprotein Ib$\alpha$ gene is

associated with Bernard-Soulier syndrome. *Br J Haematol*, *88*, 839–844.

19. Sae-Tung, G., Dong, J. F., & Lopez, J. A. (1996). Biosynthetic defect in platelet glycoprotein IX mutants associated with Bernard-Soulier syndrome. *Blood*, *87*, 1361–1367.

20. Li, C. Q., Dong, J-F., Lanza, F., et al. (1995). Expression of platelet glycoprotein (GP) V in heterologous cells and evidence for its association with GP Ibα in forming a GP Ib-IX-V complex on the cell surface. *J Biol Chem, 270,* 16302–16307.

21. Budarf, M. L., Konkle, B. A., Ludlow, L. B., et al. (1995). Identification of a patient with a Bernard-Soulier syndrome and a deletion in the DiGeorge/velo-cardial facial chromosomal region in 22q11.2. *Hum Mol Genet, 4,* 763–766.

22. Ludlow, L. B., Schick, B. P., Budarf, M. L., et al. (1996). Identification of a mutation in a GATA binding site of the platelet glycoprotein Ibβ promoter resulting in the Bernard-Soulier syndrome. *J Biol Chem, 271,* 22076–22080.

23. Strassel, C., Pasquet, J-M., Alessi, M-C., et al. (2003). A novel missense mutation shows that GPIbβ has a dual role in controlling the processing and stability of the platelet GPIb-IX adhesion receptor. *Biochemistry, 42,* 4452–4462.

24. Moran, N., Morateck, P. A., Deering, A., et al. (2000). The surface expression of glycoprotein Ibα is dependent on glycoprotein Ibβ: Evidence from a novel mutation causing Bernard-Soulier syndrome. *Blood, 96,* 532–569.

25. Watanabe, R., Ishibashi, T., Saitoh, Y., et al. (2003). Bernard-Soulier syndrome with a homozygous 13 base pair deletion in the signal peptide-coding region of the platelet glycoprotein Ibβ gene. *Blood Coagul Fibrinolysis, 14,* 1–8.

26. Wright, S. D., Michaelides, K., Johnson, D. J. D., et al. (1993). Double heterozygosity for mutations in the platelet glycoprotein IX gene in three siblings with Bernard-Soulier syndrome. *Blood, 81,* 2339–2347.

27. Sachs, U. J. H., Kroll, H., Matzdorff, A. C., et al. (2003). Bernard-Soulier syndrome due to the homozygous Asn-45Ser mutation in GPIX: An unexpected, frequent finding in Germany. *Br J Haematol, 123,* 127–131.

28. Lanza, F., De La Salle, C., Baas, M-J., et al. (2002). A Leu7Pro mutation in the signal peptide of platelet glycoprotein (GP)IX in a case of Bernard-Soulier syndrome abolishes surface expression of the GPIb-V-IX complex. *Br J Haematol, 118,* 260–266.

29. Ware, J., Russell, S. R., Marchese, P., et al. (1993). Point mutation in a leucine-rich repeat of platelet glycoprotein Ibα resulting in the Bernard-Soulier syndrome. *J Clin Invest, 92,* 1213–1220.

30. Miller, J. L., Lyle, V. A., & Cunningham, D. (1992). Mutation of leucine-57 to phenylalanine in a platelet glycoprotein Ibα leucine tandem repeat occurring in patients with an autosomal dominant variant of Bernard-Soulier disease. *Blood, 79,* 439–446.

31. Ulsemer, P., Lanza, F., Baas, M. J., et al. (2000). Role of the leucine-rich domain of platelet GP Ibα in correct post-translational processing — the Nancy I Bernard-Soulier mutation expressed on CHO cells. *Thromb Haemost, 84,* 104–111.

32. Li, C., Martin, S. E., & Roth, G. J. (1995). The genetic defect in two well-studied cases of Bernard-Soulier syndrome: A point mutation in the fifth leucine-rich repeat of platelet glycoprotein Ibα. *Blood, 86,* 3805–3814.

33. Dumas, J. J., Kumar, R., McDonagh, T., et al. (2004). Crystal structure of the wild-type von Willebrand factor A1-glycoprotein Ibα complex reveals conformation differences with a complex bearing von Willebrand disease mutations. *J Biol Chem, 279,* 23327–23334.

34. Holmberg, L., Karpman, D., Nilsson, I., et al. (1997). Bernard-Soulier syndrome Karlstrad: Trp 498 → Stop mutation resulting in a truncated glycoprotein Ibα that contains part of the transmembranous domain. *Br J Haematol, 98,* 57–63.

35. Ware, J., Russell, S., & Ruggeri, Z. M. (2000). Generation and rescue of a murine model of platelet dysfunction: The Bernard-Soulier syndrome. *Proc Natl Acad Sci USA, 97,* 2803–2808.

36. Poujol, C., Ware, J., Nieswandt, B., et al. (2002). Absence of GPIbα is responsible for aberrant membrane development during megakaryocyte maturation: Ultrastructural study using a transgenic model. *Exp Hematol, 30,* 352–360.

37. Kanaji, T., Russell, S., Cunningham, J., et al. (2004). Megakaryocyte proliferation and ploidy regulated by the cytoplasmic tail of GPIbα. *Blood, 104,* 3161–3168.

38. Kahn, M. L., Diacovo, T. G., Bainton, D. F., et al. (1999). Glycoprotein V-deficient platelets have undiminished thrombin responsiveness and do not exhibit a Bernard-Soulier phenotype. *Blood, 94,* 4112–4121.

39. Kato, K., Martinez, C., Russell, S., et al. (2004). Genetic deletion of mouse platelet glycoprotein Ibβ produces a Bernard-Soulier phenotype with increased α-granule size. *Blood, 104,* 2339–2344.

40. Othman, M., Notley, C., Lavender, F. L., et al. (2005). Identification and functional characterization of a novel 27-bp deletion in the macroglycopeptide-coding region of the GPIbα gene resulting in platelet-type von Willebrand disease. *Blood, 105,* 4330–4336.

41. Savoia, A., Balduini, C. L., Savino, M., et al. (2001). Autosomal dominant macrothrombocytopenia in Italy is most frequently a type of heterozygous Bernard-Soulier syndrome. *Blood, 97,* 1330–1335.

42. Balduini, C., Cattaneo, M., Fabris, F., et al. (2003). Italian Gruppo di Studio delle Piastrine. Inherited thrombocytopenias: A proposed diagnostic algorithm from the Italian Gruppo di Studio delle Piastrine. *Haematologica, 88,* 582–592.

43. The May-Hegglin/Fechtner syndrome consortium. (2000). Mutations in *MYH9* result in the May-Hegglin anomaly, and Fechtner and Sebastian syndromes. *Nature Genetics, 26,* 103–105.

44. Kelley, M. J., Jawien, W., Ortel, T. L., et al. (2000). Mutation of MYH9, encoding non-muscle myosin heavy chain A, in May-Hegglin anomaly. *Nature Genetics, 26,* 106–108.

45. Frannke, J. D., Dong, F., Rickoll, W. L., et al. (2005). Rod mutations associated with MYH9-related disorders disrupt non-muscle myosin-IIA assembly. *Blood, 105,* 161–169.

46. Toren, A., Rozenfeld-Granot, G., Heath, K. E., et al. (2003). MYH9 spectrum of autosomal-dominant giant platelet syndromes: Unexpected association with fibulin-1 variant-D inactivation. *Am J Hematol, 74,* 254–262.

47. Lemmink, H. H., Schroder, C. H., Monnens, L. A., et al. (1997). The clinical spectrum of type IV collagen mutations. *Hum Mutat, 9,* 477–499.

48. Hughan, S. C., Senis, Y., Best, D., et al. (2005). Selective impairment of platelet activation, to collagen in the absence of GATA1. *Blood, 105,* 4369–4376.

49. Sun, L., Mao, G., & Rao, A. K. (2004). Association of CBFA2 mutation with decreased platelet PKC-θ and impaired receptor-mediated activation of GPIIb-IIIa and plekstrin phosphorylation: Proteins regulated by CBFA2 play a role in GPIIb-IIIa activation. *Blood, 103,* 948–954.

50. Nieuwenhuis, H. K., Sakariassen, K. S., Houdijk, W. P. M., et al. (1986). Deficiency of platelet membrane glycoprotein Ia associated with a decreased platelet adhesion to subendothelium: A defect in platelet spreading. *Blood, 68,* 692–695.

51. Kunicki, T. J., Federici, A. B., Salomon, D. R., et al. (2004). An association of candidate gene haplotypes and bleeding severity in von Willebrand disease (VWD) type 1 pedigrees. *Blood, 104,* 2539–2567.

52. Moroi, M., Jung, S. M., Okuma, M., et al. (1989). A patient with platelets deficient in glycoprotein VI that lack both collagen-induced aggregation and adhesion. *J Clin Invest, 84,* 1440–1445.

53. Jandrot-Perrus, M., Susfield, S., Lagrue, A-H., et al. (2000). Cloning, characterization, and functional studies of human and mouse glycoprotein VI: A platelet-specific collagen receptor from the immunoglobulin superfamily. *Blood, 96,* 1798–1807.

54. Boylan, B., Chen, H., Rathore, V., et al. (2004). Anti-GPVI-associated ITP: An acquired platelet disorder caused by auto-antibody-mediated clearance of the GPVI/FcRγ-chain complex from the human platelet surface. *Blood, 104,* 1350–1355.

55. Niesswandt, B., Brakebusch, C., Bergmeier, W., et al. (2001). Glycoprotein VI but not α2β1 integrin is essential for platelet interaction with collagen. *EMBO J, 20,* 2120–2130.

56. Massberg, S., Gawaz, M., Gruner, S., et al. (2003). A crucial role of glycoprotein VI for platelet recruitment to the injured arterial wall *in vivo. J Exp Med, 197,* 41–49.

57. Hechler, B., Cattaneo, M., & Gachet, C. (2005). The P2 receptors in platelet function. *Semin Thromb Hemost, 31,* 150–161.

58. Hollopeter, G., Jantzen, H-M., Vincent, D., et al. (2001). Molecular identification of the platelet receptor targeted by antithrombotic drugs. *Nature, 409,* 202–207.

59. Nurden, P., Savi, P., Heilmann, E., et al. (1995). An inherited bleeding disorder linked to a defective interaction between ADP and its receptor on platelets. *J Clin Invest, 95,* 1612–1622.

60. Cattaneo, M. (2003). Inherited platelet-based bleeding disorders. *J Thromb Haemost, 1,* 1628–1636.

61. Cattaneo, M., Zighetti, M. L., Lombardi, R., et al. (2003). Molecular bases of defective signal transduction in the platelet P2Y12 receptor of a patient with congenital bleeding. *Proc Natl Acad Sci USA, 100,* 1978–1983.

62. Remijn, J. A., Wu, Y. P., Jeninga, E. H., et al. (2002). Role of ADP receptor P2Y(12) in platelet adhesion and thrombus formation in flowing blood. *Arterioscler Thromb Vasc Biol, 22,* 686–691.

63. Fabre, J. E., Nguyen, M., Latour, A., et al. (1999). Decreased platelet aggregation, increased bleeding time and resistance to thromboembolism in P2Y1-deficient mice. *Nat Med, 5,* 1199–1202.

64. Oury, C., Toth-Zsamboki, E., Van Geet, G., et al. (2000). A natural dominant negative P2X1 receptor due to deletion of a single amino acid residue. *J Biol Chem, 275,* 22611–22614.

65. Cattaneo, M., Lombardi, R., Zighetti, M. L., et al. (1997). Deficiency of (33P)2MeS-ADP binding sites on platelets with secretion defect, normal granule stores and normal thromboxane A2 production. Evidence that ADP potentiates platelet secretion independently of the formation of large platelet aggregates and thromboxane A2 production. *Thromb Haemost, 77,* 986–990.

66. Hirata, T., Ushikubi, F., Kakizuka, A., et al. (1996). Two thromboxane A2 receptor isoforms in human platelets. Opposite coupling to adenylate cyclase with different sensitivity to Arg[60] to Leu mutation. *J Clin Invest, 97,* 949–956.

67. Rao, A. K. (2003). Inherited defects in platelet signaling mechanisms. *J Thromb Haemost, 1,* 671–681.

68. Mao, G. F., Vaidyula, V. R., Kunapuli, S. P., et al. (2002). Lineage-specific defect in gene expression in human platelet phospholipase C-β2 deficiency. *Blood, 99,* 905–911.

69. Gabbeta, J., Yang, X., Kowalska, M. A., et al. (1997). Platelet signal transduction defect with Gα subunit dysfunction and diminished Gαq in a patient with abnormal platelet responses. *Proc Natl Acad Sci USA, 94,* 8750–8755.

70. Yang, X., Sun, L., Gabbeta, J., et al. (1996). Human platelet signaling defect characterized by impaired production of 1,4,5 inositoltriphosphate and phosphatidic acid, and diminished plekstrin phosphorylation. Evidence for defective phospholipase C activation. *Blood, 88,* 1676–1683.

71. Watanabe, N., Nakajima, H., Suzuki, H., et al. (2003). Functional phenotype of phosphoinositide 3-kinase p85α-null platelets characterized by an impaired response to GPVI stimulation. *Blood, 102,* 541–548.

72. Woulfe, D., Jiang, H., Morgans, A., et al. (2004). Defects in secretion, aggregation and thrombus formation in platelets from mice lacking Akt2. *J Clin Invest, 113,* 441–450.

73. Moers, A., Nieswandt, B., Massberg, S., et al. (2003). G13 is an essential mediator of platelet activation in hemostasis and thrombosis. *Nat Med, 9,* 1418–1422.

74. Patel, Y. M., Patel, K., Rahman, S., et al. (2003). Evidence for a role for Gαi in mediating weak agonist-induced platelet aggregation in human platelets: Reduced Gαi1 expression and defective Gi signaling in the platelets of a patient with a chronic bleeding disorder. *Blood, 101,* 4828–4835.

75. Massberg, S., Gruner, S., Konrad, I., et al. (2004). Enhanced in vivo platelet adhesion in vasodilator-stimulated phosphoprotein (VASP)-deficient mice. *Blood, 103,* 136–142.

76. Anitua, E., Andia, I., Ardanza, B., et al. (2004). Autologous platelets as a source of proteins for healing and tissue regeneration. *Thromb Haemost, 91,* 4–15.

77. Washington, A. V., Schubert, R. L., Quigley, L., et al. (2004). A TREM family member, TLT-1, is found exclusively in the α-granules of megakaryocytes and platelets. *Blood, 104,* 1042–1047.

78. Hayward, C. P. M. (1997). Inherited disorders of platelet α-granules. *Platelets, 8,* 197–209.

79. Nurden, P., Jandrot-Perrus, M., Combrié, R., et al. (2004). Severe deficiency of glycoprotein VI in a patient with gray platelet syndrome. *Blood, 104,* 107–114.

80. Gebrane-Younes, J., Cramer, E. M., Orcel, L., et al. (1993). Gray platelet syndrome: Dissociation between abnormal sorting in megakaryocyte α-granules and normal sorting in Weibel-Palade bodies of endothelial cells. *J Clin Invest, 92,* 3023–3028.

81. Centurione, L., Di Baldassarre, A., Zingariello, M., et al. (2004). Increased and pathologioc emperipolesis of neutrophils within megakaryocytes associated with marrow fibrosis in GATA-1$^{low}$ mice. *Blood, 104,* 3573–3580.

82. Lemons, P. P., Chen, D., & Whiteheart, S. W. (2000). Molecular mechanisms of platelet exocytosis: Requirements for α-granule release. *Biochem Biophys Res Commun, 267,* 875–880.

83. Gissen, P., Johnson, C. A., Morgan, N. V., et al. (2004). Mutations in VPS33B, encoding a regulator of SNARE-dependent membrane fusion, cause arthrogryposis-renal dysfunction-cholestasis (ARC) syndrome. *Nature Genetics, 36,* 400–404.

84. Kahr, W. H., Zheng, S., Sheth, P. M., et al. (2001). Platelets from patients with the Quebec platelet disorder contain and secrete abnormal amounts of urokinase-type plasminogen activator. *Blood, 98,* 257–265.

85. Sheth, P. M., Kahr, W. H., Haq, M. A., et al. (2003). Intracellular activation of the fibrinolytic cascade in the Quebec platelet disorder. *Thromb Haemost, 90,* 293–298.

86. Nieuwenhuis, H. K., Akkerman, J. W., & Sixma, J. J. (1987). Patients with a prolonged bleeding time and normal aggregation tests may have storage pool deficiency: Studies on 106 patients. *Blood, 70,* 620–623.

87. Weiss, H. J., Lages, B., Vicic, W., et al. (1993). Heterogeneous abnormalities of platelet dense granule ultrastructure in 20 patients with congenital storage pool deficiency. *Br J Haematol, 83,* 282–295.

88. Burns, S., Cory, G. O., Vainchenker, W., et al. (2004). Mechanism of WASp-mediated hematologic and immunologic disease. *Blood, 104,* 3454–3462.

89. Imai, K., Morio, T., Jin, Y., et al. (2004). Clinical course of patients with WASP gene mutations. *Blood, 103,* 456–464.

90. Higgs, J. N., & Pollard, T. D. (2000). Activation by Cdc42 and PIP$_2$ of Wiskott-Aldrich syndrome protein (WASP) stimulates actin nucleation by Arp2/3 complex. *J Cell Biol, 150,* 1311–1320.

91. Wada, T., Schurman, S. H., Jagadeesh, J., et al. (2004). Multiple patients with revertant mosaicism in a single Wiskott-Aldrich syndrome family. *Blood, 104,* 1270–1272.

92. Gunay-Aygun, M., Huizing, M., & Gahl, W. A. (2004). Molecular defects that effect platelet dense granules. *Semin Thromb Haemost, 30,* 537–547.

93. Huizing, M., Scher, C. D., Strovel, E., et al. (2002). Nonsense mutations in ADTB3A cause complete deficiency of the β3A subunit of adaptor complex-3 and severe Hermansky-Pudlak syndrome type 2. *Pediatric Res, 51,* 150–158.

94. Nazarian, R., Falcon-Perez, J. M., & Dell'Angelica, E. C. (2003). Biogenesis of lysosome-related organelles complex 3 (BLOC-3): A complex containing the Hermansky-Pudlak syndrome (HPS) proteins HPS1 and HPS4. *Proc Natl Acad Sci USA, 100,* 8770–8775.

95. Li, W., Zhang, Q., Oiso, N., et al. (2003). Hermansky-Pudlak syndrome type 7 (HPS-7) results from mutant dysbindin, a member of the biogenesis of lysosome-related organelles complex 1 (BLOC-1). *Nat Genet, 35,* 84–89.

96. Karim, M. A., Suzuki, K., Fukai, K., et al. (2002). Apparent genotype-phenotype correlation in childhood, adolescent, and adult Chediak-Higashi syndrome. *Am J Med Genet, 108,* 16–22.

97. Ward, D. M., Shiflett, S. L., & Kaplan, J. (2002). Chediak-Higashi syndrome: A clinical and molecular view of a rare lysosomal storage disorder. *Curr Mol Med, 2,* 469–477.

98. Ménasche, G., Feldmann, J., Houdusse, A., et al. (2003). Biochemical and functional characterization of Rab27a mutations occurring in Griscelli syndrome patients. *Blood, 101,* 2736–2742.

99. George, J. N., Caen, J-P., & Nurden, A. T. (1990). Glanzmann's thrombasthenia: The spectrum of clinical disease. *Blood, 75,* 1383–1395.

100. Wagner, D. D., & Burger, P. C. (2003). Platelets in inflammation and thrombosis. *Arterioscler Thromb Vasc Biol, 23,* 2131–2137.

101. Patel, D., Vaananen, H., Jirouskova, M., et al. (2003). Dynamics of GPIIb/IIIa-mediated platelet-platelet interactions in platelet adhesion/thrombus formation on collagen *in vitro* as revealed by videomicroscopy. *Blood, 101,* 929–936.

102. Shattil, S. J., Kashiwagi, H., & Pampori, N. (1998). Integrin signaling: The platelet paradigm. *Blood, 91,* 2645–2657.

103. Rosa, J-P., Artçanuthurry, V., Grelac, F., et al. (1997). Reassessment of protein tyrosine phosphorylation in thrombasthenic platelets: Evidence that phosphorylation of cortactin and a 64 kDa protein is dependent on thrombin activation and integrin αIIbβ3. *Thromb Haemost, 89,* 4385–4392.

104. Byzova, T. V., & Plow, E. F. (1997). Networking in the hemostatic system: Integrin αIIbβ3 binds prothrombin and influences its activation. *J Biol Chem, 272,* 27183–27188.

105. Lisman, T., Moschatsis, S., Adelmeijer, J., et al. (2003). Recombinant factor VIIa enhances deposition of platelets with congenital or acquired αIIbβ3 deficiency to endothelial cell matrix and collagen under conditions of flow via tissue factor-independent thrombin generation. *Blood, 101,* 1864–1870.

106. Xiao, T., Takagi, J., Wang, J-H., et al. (2004). Structural basis for allostery in integrins and binding to fibrinogen-mimetic therapeutics. *Nature, 432,* 59–67.

107. French, D. L. (1998). The molecular genetics of Glanzmann's thrombasthenia. *Platelets, 9,* 5–20.

108. Bray, P. (1994). Inherited diseases of platelet glycoproteins: Considerations for rapid molecular diagnosis. *Thromb Haemostas, 72,* 492–502.

109. Newman, P. J., Seligsohn, U., Lyman, S., et al. (1991). The molecular genetic basis of Glanzmann thrombasthenia in the Iraqi-Jewish and Arab populations in Israel. *Proc Natl Acad Sci USA, 88,* 3160–3164.

110. Schlegel, N., Gayet, O., Morel-Kopp, M. C., et al. (1995). The molecular genetic basis of Glanzmann's thrombasthenia

in a gypsy population in France: Identification of a new mutation on the αIIb gene. *Blood, 86,* 977–982.

111. Gidwitz, S., Temple, B., & White, G. C., 2nd (2004). Mutations in and near the second calcium-binding domain of integrin αIIb affect the structure and function of integrin αIIbβ3. *Biochem J, 379,* 449–459.

112. Mitchell, W. B., Li, J. H., Singh, F., et al. (2003). Two novel mutations in the αIIb calcium-binding domains identify hydrophobic regions essential for αIIbβ3 biogenesis. *Blood, 101,* 2268–2276.

113. Milet-Marsal, S., Breillat, C., Peyruchaud, O., et al. (2002). Analysis of the amino acid requirement for a normal αIIbβ3 maturation at αIIbGlu324 commonly mutated in Glanzmann thrombasthenia. *Thromb Haemost, 88,* 655–662.

114. Arias-Salgado, E. G., Butta, N., Gonzalez-Manchon, C., et al. (2001). Competition between normal (674C) and mutant (674)GPIIb subunits: The role of the molecular chaperone BiP in the processing of GPIIb-IIIa complexes. *Blood, 97,* 2640–2647.

115. Gonzalez-Manchon, C., Arias-Salgado, E. G., Butta, N., et al. (2003). A novel homozygous splice junction mutation in GPIIb associated with alternative splicing, nonsense-mediated decay of GPIIb-mRNA, and type II Glanzmann's thrombasthenia. *J Thromb Haemost, 1,* 1071–1078.

116. Springer, T. A. (1997). Folding of the N-terminal, ligand-binding region of integrin α-subunits into a β-propeller domain. *Proc Natl Acad Sci USA, 94,* 65–72.

117. Rosenberg, N., Yatuv, R., Sobolev, V., et al. (2003). Major mutations in calf-1 and calf-2 domains of glycoprotein IIb in patients with Glanzmann thrombasthenia enable GPIIb/IIIa complex formation, but impair its transport from the endoplasmic reticulum to the Golgi apparatus. *Blood, 101,* 4808–4815.

118. Jin, Y., Dietz, H. C., Montgomery, R. A., et al. (1996). Glanzmann thrombasthenia: Cooperation between sequence variants in *Cis* during splice selection. *J Clin Invest, 98,* 1745–1754.

119. Milet-Marsal, S., Breillat, C., Peyruchaud, O., et al. (2002). Two different β3 cysteine substitutions alter αIIbβ3 maturation and result in Glanzmann thrombasthenia. *Thromb Haemost, 88,* 104–110.

120. Chen, P., Melchior, C., Brons, N. H. C., et al. (2001). Probing conformational changes in the I-like domain and the cysteine-rich repeat of human β3 integrins following disulfide bond disruption by cysteine mutations: Identification of cysteine598 involved in αIIbβ3 activation. *J Biol Chem, 276,* 38268–38365.

121. Nair, S., Li, J., Mitchell, W. B., et al. (2002). Two new β3 mutations in Indian patients with Glanzmann's thrombasthenia: Localization of mutations affecting cysteine residues in integrin β3. *Thromb Haemost, 88,* 503–509.

122. Tadokoroa, S., Tomiyama, Y., Honda, S., et al. (2002). Missense mutations in the β3 subunit have a different impact on the expression and function between αIIbβ3 and αvβ3. *Blood, 99,* 931–938.

123. Loftus, J. C., O'Toole, T. E., Plow, E. F., et al. (1990). A β3 integrin mutation abolishes ligand binding and alters divalent cation-dependent conformation. *Science, 249,* 915–918.

124. Chen, Y., Djaffar, I., Pidard, D., et al. (1992). Ser$^{752}$ → Pro mutation in the cytoplasmic domain of integrin β3 subunit and defective activation of platelet integrin αIIbβ3 (glycoprotein IIb-IIIa) in a variant of Glanzmann's thrombasthenia. *Proc Natl Acad Sci USA, 89,* 10169–10173.

125. Chen, Y-P., O'Toole, T. E., Ylänne, J., et al. (1994). A point mutation in the integrin β3 cytoplasmic domain (S$^{752}$ → P) impairs bidirectional signaling through, αIIbβ3 (platelet glycoprotein IIb-IIIa). *Blood, 84,* 1857–1865.

126. Wang, R., Shattil, S. J., Ambruso, D. R., et al. (1997). Truncation of the cytoplasmic domain of β3 in a variant form of Glanzmann thrombasthenia abrogates signaling through the integrin αIIbβ3 complex. *J Clin Invest, 100,* 2393–2403.

127. Peyruchaud, O., Nurden, A. T., Milet, S., et al. (1998). R to Q aminoacid substitution in the GFFKR sequence of the cytoplasmic domain of the integrin αIIb subunit in a patient with a Glanzmann's thrombasthenia-like syndrome. *Blood, 92,* 4178–4187.

128. Hughes, P. E., Diaz-Gonzalez, F., Leong, L., et al. (1996). Breaking the integrin hinge: A defined structural constraint regulates integrin signaling. *J Biol Chem, 271,* 6571–6574.

129. Low, E., Qi, W., Vilaire, G., et al. (1996). Effect of cytoplasmic domain mutations on the agonist-stimulated ligand-binding activity of the platelet integrin αIIbβ3. *J Biol Chem, 271,* 30233–30241.

130. Shattil, S. J., & Newman, P. J. (2004). Integrins: Dynamic scaffolds for adhesion and signaling in platelets. *Blood, 104,* 1606–1615.

131. McDowall, A., Inwald, D., Leitinger, B., et al. (2003). A novel form of integrin dysfunction involving β1, β2, and β3 integrins. *J Clin Invest, 111,* 51–60.

132. Kashiwagi, H., Tomiyama, Y., Tadokoro, S., et al. (1999). A mutation in the extracellular cysteine-rich repeat region of the β3 subunit activates integrins αIIbβ3 and αvβ3. *Blood, 93,* 2559–2568.

133. Ruiz, C., Liu, C-Y., Sun, Q-H., et al. (2001). A point mutation in the cysteine-rich domain of glycoprotein (GP) IIIa results in the expression of a GP IIb-IIIa (αIIbβ3) integrin receptor locked in a high affinity state and a Glanzmann thrombasthenia-like phenotype. *Blood, 98,* 2432–2441.

134. Handagama, P. J., Scarborough, R. A., Shuman, M. A., et al. (1993). Endocytosis of fibrinogen into megakaryocyte and platelet α-granules is mediated by αIIbβ3 (glycoprotein IIb-IIIa). *Blood, 82,* 135–138.

135. Nurden, P., Poujol, C., Durrieu-Jais, C., et al. (1999). Labeling of the internal pool of GP IIb-IIIa in platelets by c7E3 Fab fragments (abciximab, ReoPro): Flow and endocytic mechanisms contribute to the transport. *Blood, 93,* 1622–1633.

136. Pelkman, L., Fava, E., Grabner, H., et al. (2005). Genome-wide analysis of human kinases in clathrin- and caveolae/raft-mediated endocytosis. *Nature, 436,* 78–86.

137. Nurden, P., Poujol, C., Winckler, J., et al. (2002). A Ser752 → Pro substitution in the cytoplasmic domain of β3 in a Glanzmann thrombasthenia variant fails to prevent interactions between the αIIbβ3 integrin and the platelet granule pool of fibrinogen. *Br J Haematol, 118,* 1143–1151.

138. Ward, C. M., Kestin, A. S., & Newman, P. J. (2000). A Leu262Pro mutation in the integrin β3 subunit resulted in an

$\alpha_{IIb}\beta_3$ complex that binds fibrin but not fibrinogen. *Blood, 96,* 161–169.

139. Poujol, C., Nurden, A. T., & Nurden, P. (1997). Ultrastructural analysis of the distribution of the vitronectin receptor ($\alpha v\beta_3$) in human platelets and megakaryocytes reveals an intracellular pool and labelling of the $\alpha$-granule membrane. *Br J Haematol, 96,* 823–835.

140. Horton, M. A., Massey, H. M., Rosenberg, N., et al. (2003). Upregulation of osteoclast $\alpha 2\beta 1$ integrin compensates for lack of $\alpha v\beta 3$ vitronectin receptor in Iraqi-Jewish-type Glanzmann thrombasthenia. *Br J Haematol, 122,* 950–957.

141. Ruan, J., Peyruchaud, O., Nurden, P., et al. (1998). Family screening for a Glanzmann's thrombasthenia mutation using PCR-SSCP. *Platelets, 9,* 129–136.

142. Peretz, H., Seligsohn, U., Zwang, E., et al. (1991). Detection of the Glanzmann's thrombasthenia mutations in Arab and Iraqi-Jewish patients by polymerase chain reaction and restriction analysis of blood and urine samples. *Thromb Haemost, 616,* 500–504.

143. French, D., Coller, B. S., Usher, S., et al. (1998). Prenatal diagnosis of Glanzmann thrombasthenia using the polymorphic markers BRCA1 and THRAI on chromosome 17. *Br J Haematol, 102,* 582–587.

144. Tronik-Le Roux, D., Roullot, V., Poujol, C., et al. (2000). Thrombasthenic mice generated by replacement of the integrin $\alpha$IIb gene: Demonstration that transcriptional activation at this megakaryocytic locus precedes lineage commitment. *Blood, 96,* 1399–1408.

145. Reynolds, L. E., Conti, F. J., Lucas, M., et al. (2005). Accelerated re-epithelialisation in $\beta_3$-integrin-deficient mice is associated with enhanced TGF-$\beta_1$ signaling. *Nature Med, 11,* 167–174.

146. McHugh, K. P., Hodivala-Dilke, K., Zheng, M-H., et al. (2000). Mice lacking $\beta_3$-integrins are osteosclerotic because of dysfunctional osteoclasts. *J Clin Invest, 105,* 433–440.

147. Reynolds, A. R., Reynolds, L. E., Nagel, T. E., et al. (2004). Elevated Flk1 (vascular endothelial growth factor receptor 2) signaling mediates enhanced angiogenesis in $\beta_3$-integrin-deficient mice. *Cancer Res, 64,* 8643–8650.

148. Law, D. A., DeGuzman, F. R., Heiser, P., et al. (1999). Integrin cytoplasmic tyrosine motif is required for outside-in signalling and platelet function. *Nature, 401,* 808–811.

149. Zwaal, R. F. A., Comfurius, P., & Schroit, A. J. (1997). Scott syndrome, a bleeding disorder caused by defective scrambling of membrane phospholipids. *Biochim Biophys Acta, 1636,* 119–128.

150. Dachary-Prigent, J., Pasquet, J. M., Fressinaud, E., et al. (1997). Aminophospholipid exposure, microvesiculation and abnormal protein tyrosine phosphorylation in the platelets of a patient with Scott syndrome: A study using physiologic agonists and local anaesthetics. *Br J Haematol, 99,* 959–967.

151. Zhou, Q., Sims, P. J., Wiedmer, T. (1998). Expression of proteins controlling transbilayer movement of plasma membrane phospholipids in the B lymphocytes from a patient with Scott syndrome. *Blood, 92,* 1707–1712.

152. Martinez, M. C., Martin, S., Toti, F., et al. (1999). Significance of capacitative $Ca^{2+}$ entry in the regulation of phosphatidylserine expression at the surface of stimulated cells. *Biochemistry, 38,* 10092–10098.

153. Albrecht, C., McVey, J. H., Elliott, J. I., et al. (2005). A novel missense mutation in ABCA1 results in altered protein trafficking and reduced phosphatidylserine translocation in a patient with Scott syndrome. *Blood, 106,* 542–549.

154. Michelson, A. D. (2006). Platelet function disorders. In I. Hann, R. Arceci, & O. Smith (Eds.), *Pediatric hematology,* (3rd ed.) Oxford: Blackwell.

155. Haddad, E., Le Deist, F., Blanche, S., et al. (1995). Treatment of Chediak-Higashi syndrome by allogenic bone marrow transplantation: Report of 10 cases. *Blood, 85,* 3328–3333.

156. Filipovich, A. H., Stone, J. V., Tomany, S. C., et al. (2001). Impact of donor type on outcome of bone marrow transplantation for Wiskott-Aldrich syndrome: Collaborative study of the International Bone Marrow Transplant Registry and the National Marrow Donor Program. *Blood, 97,* 1598–1603.

157. Bellucci, S., Damaj, G., Boval, B., et al. (2000). Bone marrow tranplantation in severe Glanzmann's thrombasthenia with antiplatelet alloimmunization. *Bone Marrow Transplant, 25,* 327–330.

158. D'Andrea, G., Colaizzo, D., Vecchione, G., et al. (2002). Glanzmann's thrombasthenia: Identification of 19 new mutations in 30 patients. *Thromb Haemost, 87,* 1034–1042.

159. Nair, S., Ghosh, K., Shetty, S., et al. (2005). Mutations in GPIIIa molecule as a cause for Glanzmann thrombasthenia in Indian patients. *J Thromb Haemost, 3,* 482–488.

# CHAPTER 58

# Acquired Disorders of Platelet Function

## A. Koneti Rao

*Hematology Division and Sol Sherry Thrombosis Research Center,
Temple University School of Medicine, Philadelphia, Pennsylvania*

## I. Introduction

Acquired disorders of diverse etiologies are associated with altered platelet function and a bleeding diathesis. In most of these disorders, the specific biochemical and pathophysiological aberrations leading to platelet dysfunction are poorly understood. In several disorders, abnormalities have been described in multiple aspects of platelet function, including adhesion, aggregation, and secretion, and in the platelet contribution to blood coagulation reactions. Although it would be preferable to consider the acquired platelet function disorders based on the nature of the specific biochemical defect, the presence of multiple and poorly characterized platelet abnormalities even in a single disease state makes such a classification difficult. In this chapter, the acquired qualitative platelet disorders are therefore described according to the disease states in which they are recognized (Table 58-1). The platelet abnormalities in these diseases arise because of a variety of factors. In some, such as the myeloproliferative disorders, there is production of intrinsically abnormal platelets by the bone marrow. In others, the dysfunction results from the interaction of platelets with exogenous factors such as pharmacologic agents, artificial surfaces (cardiopulmonary bypass, discussed in Chapter 59), compounds that accumulate in plasma due to impaired renal function, and antibodies.

In most acquired disorders of platelet dysfunction, the bleeding is mucocutaneous and there is a wide and often unpredictable spectrum. In most situations, the bleeding symptoms are mild to moderate, but the potential for severe life-threatening bleeding exists. The usual laboratory tests that lead to the identification of the platelet dysfunction are the bleeding time (see Chapter 25) and/or *in vitro* platelet aggregation (see Chapter 26). However, the bleeding time is variably prolonged and often normal even in individuals with impaired platelet aggregation responses. For example, aspirin ingestion impairs platelet aggregation or secretion responses to standard agonists (e.g., adenosine diphosphate [ADP], epinephrine, arachidonic acid, and low doses of collagen) in almost all individuals, but it prolongs the bleeding time in only approximately half of the subjects.[1,2] The correlation between the abnormalities observed in platelet aggregation studies and clinical bleeding (or prolongation of the bleeding time) remains poor. Despite these shortcomings, the bleeding time and *in vitro* platelet aggregation and secretion studies have been the main tools used to diagnose abnormalities in platelet function. Newer methods (e.g., the platelet function analyzer (PFA)-100 [see Chapter 28], the Impact cone and plate(let) analyzer [see Chapter 29], the VerifyNow rapid platelet function analyzer [see Chapter 27], and flow cytometry [see Chapter 30]) are being increasingly utilized for the assessment of acquired platelet function disorders.

## II. Uremia

### A. Clinical Features

Bleeding has been a frequent and sometimes fatal complication of uremia in the past, but its incidence has declined substantially because of better treatment of associated clinical features, including anemia. Nevertheless, in uremic patients undergoing surgery or invasive procedures, excessive bleeding remains a concern. Bleeding in uremia is generally mucocutaneous but rarely may be intracranial or pericardial. The gastrointestinal tract is a major site of hemorrhage.

### B. Platelet Function Defects

The pathogenesis of the hemostatic defect in uremia remains unclear, but platelet dysfunction and impaired platelet–vessel wall interaction are considered the major causes of the bleeding tendency. Thrombocytopenia occurs in 16 to 55% of patients with uremia.[3] However, coagulation abnormalities are not a major factor in the uremic hemostatic defect. Patients on dialysis have an increased number of circulating

## Table 58-1: Disorders Associated with Acquired Defects in Platelet Function

Uremia
Myeloproliferative disorders
  Essential thrombocythemia
  Polycythemia vera
  Chronic myelogenous leukemia
  Agnogenic myeloid metaplasia
Acute leukemias and myelodysplastic syndromes
Dysproteinemias
Cardiopulmonary bypass
Acquired von Willebrand disease
Acquired storage pool deficiency
Antiplatelet antibodies
Liver disease
Drugs and other agents

reticulated platelets, indicating accelerated platelet turnover.[4] The bleeding time is prolonged in uremia and appears to correlate with the severity of renal failure and clinical bleeding.[5] The prolongation of the bleeding time is also independently related to the anemia and is corrected by transfusion of washed packed red blood cells or treatment with erythropoietin.[6,7]

Multiple platelet abnormalities have been recognized in uremia. However, many more patients have abnormalities in in vitro platelet function tests than have clinically significant bleeding. Adhesion of platelets to the subendothelium is an initial step in hemostasis and involves the interaction of von Willebrand factor (VWF), platelets, and the subendothelium (see Chapter 18). Adhesion of platelets from uremic whole blood perfused over de-endothelialized rabbit aorta has been reported to be impaired; platelet adhesion was decreased when perfusions were carried out with mixtures containing either washed uremic platelets and normal plasma or normal platelets and uremic plasma.[8] This may be related to components in uremic plasma directly inhibiting the platelet interaction with artery segments. Most studies have reported normal or increased VWF antigen levels and activity (ristocetin cofactor) in uremic plasma; however, in some, VWF activity was consistently lower than the antigen level.[3] The multimeric structure of VWF has been reported to be normal in some studies and abnormal in others, with a decrease in the largest multimers.[3] The number of glycoprotein (GP) Ib and integrin $\alpha IIb\beta3$ (GPIIb-IIIa) sites per platelet has been reported to be normal in uremia.[9] However, in one study, platelet GPIb levels were reported to be decreased.[10] Uremic platelets have also been shown to have a functional defect affecting the interaction of VWF with integrin $\alpha IIb\beta3$.[11,12]

There have been many studies of platelet aggregation responses in uremia, with sometimes conflicting results.[13] In

general, there is no correlation between the severity or cause of the renal disease and the platelet aggregation defects. Moreover, in vitro platelet aggregation defects in uremia do not correlate with clinical bleeding. In response to ADP, impaired primary aggregation and subsequent rapid disaggregation, and an increase in the ADP requirement for secondary aggregation, have been described. Reduced aggregation in response to epinephrine, collagen, and arachidonic acid has also been reported. Thrombin- and ristocetin-induced platelet aggregation have more often been shown to be normal than reduced. Although a deficiency of fibrinogen receptors (integrin $\alpha IIb\beta3$) on platelets has been considered in uremia, quantitatively the platelet integrin $\alpha IIb\beta3$ has been normal.[10,12] However, receptor-mediated activation of $\alpha IIb\beta3$ and fibrinogen binding are impaired in uremia.[12] Elevated levels of fibrin(ogen) degradation products have been reported in uremia and the platelet defect has been attributed to an inhibitory effect of these fragments.[14,15]

Studies of dense and $\alpha$-granule secretion in uremia have yielded divergent results.[16] The findings have ranged from storage pool deficiency with membrane ATPase deficiency to partial storage pool deficiency, to normal nucleotide content and normal numbers of platelet dense bodies. In some instances, impaired platelet dense granule secretion has been attributed to prior platelet activation by membrane dialyzers. With regard to $\alpha$-granule constituents, plasma levels of platelet factor 4 and $\beta$-thromboglobulin are elevated in uremia and increase further with hemodialysis.[17–19] Accumulation of these platelet proteins in the circulation may result from reduced renal clearance and/or from platelet activation by membrane dialyzers. The increase in plasma $\beta$-thromboglobulin in uremia correlated with the raised serum creatinine in some studies but not others.[18–20]

Several studies have shown consistent reductions in products of arachidonic acid metabolism (prostaglandin [PG] $G_2$/$PGH_2$, thromboxane [TX] $A_2$) in uremic platelets and have implicated this in the observed defective aggregation and secretion.[21–24] The reduction in $TXA_2$ has been attributed to a dysfunctional cyclooxygenase.[23] However, because $TXA_2$ production has been noted to be normal on platelet activation with arachidonic acid, a defect proximal to cyclooxygenase (e.g., in the liberation of arachidonic acid from phospholipids by phospholipase $A_2$ activation) could also explain the impaired $TXA_2$ production. Another study reported increased release of arachidonic acid from phospholipids associated with increased production of lipoxygenase products known to inhibit platelets.[25] Enhanced prostacyclin ($PGI_2$)-like activity has been observed in uremic plasma.[22] The imbalance between reduced platelet $TXA_2$ production and enhanced endothelial prostacyclin ($PGI_2$) formation has been postulated to lead to the bleeding diathesis of uremia.[22]

Other platelet defects described in uremia include increased levels of intracellular cyclic adenosine monophosphate (cAMP) and adenylyl cyclase, which can inhibit platelet responses.[26] Uremic platelets have been shown to have a defect in agonist-induced $Ca^{2+}$ mobilization, which correlated with prolongation of the bleeding time.[27] The ability of platelets to retract a clot *in vitro* is diminished in uremia.[28] Defective cytoskeletal assembly and impaired agonist-induced tyrosine phosphorylation have been reported in uremic platelets.[29] Platelets contribute in a major way to specific steps in blood coagulation (see Chapter 19), a property referred to as platelet procoagulant activity. This aspect of platelets has been assessed by tests, such as platelet factor 3 activity, and is reduced in uremia.[30,31] Overall, platelet dysfunction in uremia is complex and involves alterations in multiple aspects of platelet function in hemostasis, including adhesion, aggregation, secretion, signal transduction mechanisms, and platelet procoagulant activity.

The hemostatic defect in uremia is linked to accumulation of dialyzable and nondialyzable molecules in the uremic plasma. Uremic plasma inhibits platelet aggregation and platelet factor 3 activation in normal platelets,[31,32] and guanidinosuccinic acid (GSA), a compound that accumulates in uremic plasma, consistently inhibits both responses in normal platelets.[32] Some observations provide insights into the mechanisms by which GSA induces platelet dysfunction[33,34] and implicate a role of nitric oxide (NO) in platelet function. NO is a potent inhibitor of platelet adhesion to endothelium and of platelet–platelet interactions, which are related to increased levels of cellular cyclic guanosine 5′-monophosphate (cGMP) (see Chapter 13). In human subjects, inhalation of NO prolongs the bleeding time and inhibits *ex vivo* platelet aggregation, P-selectin expression, and fibrinogen binding.[35,36] Platelets generate more NO in uremic patients than in healthy subjects,[33,34] and levels of NO are higher in plasma and exhaled air.[33,34,37,38] Exposure of human endothelial cells to uremic plasma leads to enhanced NO production.[34] Moreover, GSA, but not some of the other guanidines, induces relaxation of the isolated rat aortas related to formation of NO.[33] GSA-induced inhibition of platelet responses is blocked by an NO inhibitor,[33] indicating that NO mediates the effect of GSA on platelet function. Together, these and other studies provide support for the role of GSA as one of the mediators of the uremic platelet dysfunction. Further studies are needed to characterize the other molecules that accumulate in uremia and contribute to the hemostatic defects.

### C. Therapy

Aggressive dialysis is an important component of the overall management of the hemostatic defect. Intensive dialysis corrects the bleeding diathesis in some patients but is only partially effective in others.[13] Both hemo- and peritoneal dialysis are effective, with the former being superior.

Elevation of the hematocrit with packed red blood cells or recombinant erythropoietin shortens the bleeding time, improves platelet adhesion, and corrects mild bleeding in uremic patients.[6,7,39] A simplistic explanation for this is that the repletion of red blood cells leads to a mechanical displacement of platelets toward the vessel wall and facilitates the platelet–vessel wall interaction. Therapy with recombinant erythropoietin may be associated with an increased risk of thrombosis and worsening of hypertension.[7] Other treatments for uremic patients with significant bleeding include platelet transfusion, desmopressin, cryoprecipitate, and conjugated estrogens. The role of platelet transfusions in uremia is somewhat limited because of the availability of other modalities, the transfusion-related risks, and the possibility that the transfused platelets acquire the uremic defects *in vivo*. Intravenous infusion of desmopressin (1-desamino-8-D arginine vasopressin [DDAVP]) (see Chapter 67) induces the release of VWF from endothelial cells and shortens the bleeding time in 75% of uremic patients.[40,41] The effect of DDAVP correlates with a rise in circulating VWF, including its largest multimeric forms, which mediate platelet adhesion. DDAVP is commonly administered intravenously but is also effective if given subcutaneously or intranasally. The effect of DDAVP on the bleeding time occurs within 1 hour and lasts up to 4 hours. The intranasal route requires higher dosages and has variable absorption. DDAVP does not appear to affect the quantity or quality of the patient's platelets.[40] The side effects of DDAVP include reduction of arterial blood pressure, facial flushing, water retention, hyponatremia, and, rarely, cerebral thrombosis.

Cryoprecipitate transfusions in uremic patients shorten the bleeding time and improve clinical bleeding.[42] This effect occurs within 1 to 24 hours and may last up to 36 hours. However, some patients fail to correct the bleeding time or stop bleeding.[21] Cryoprecipitate administration has the same risks inherent in all blood product transfusions.

Conjugated estrogens shorten the bleeding time and reduce clinical bleeding in uremic patients.[43,44] In these studies, a single intravenous infusion of 0.6 mg/kg of Emopremarn (Ayerst, New York) shortened the bleeding time within 6 hours, but this effect disappeared within 72 hours.[44] Repeated daily infusions for 5 days shortened the bleeding time by 50%, an effect that lasted 14 days.[43] Oral conjugated estrogens (premarin 10–50 mg/day) have also corrected the bleeding time to normal.[43] Low-dose transdermal estrogen (estradiol) administration has been reported to reduce clinical gastrointestinal bleeding and shorten the bleeding time.[45] Estrogen therapy may be particularly useful in uremic patients who need longer lasting hemostatic control, such as those with gastrointestinal telangiectasias, intracranial bleeding, or undergoing major surgery.[41,44] The side effects of estrogens include hot flashes, fluid retention, elevation of blood pressure, and abnormal liver function tests.[43,44] The mechanism by which estrogens shorten the bleeding time remains unknown. In uremic rats, conjugated estrogens lower the elevated plasma NO metabolites and limit vascular endothelial overexpression of NO-forming enzymes.[46]

Activated factor VII may also prove to have a role in the treatment of uremic bleeding (see Chapter 68).

## III. Myeloproliferative Diseases

Myeloproliferative disorders (MPDs), including essential thrombocythemia (ET), polycythemia vera (PV), agnogenic myeloid metaplasia (AMM), and chronic myelogenous leukemia (CML), are associated with a bleeding tendency, thromboembolic complications, and qualitative platelet defects[47–50] (see Chapter 56). The observed platelet defects and clinical manifestations may vary in these entities. The platelet abnormalities, which are noted even in asymptomatic individuals, likely result from an abnormal clone of stem cells. However, some of the alterations may be secondary to enhanced platelet activation *in vivo*. The clinical significance of the abnormal platelet responses on *in vitro* testing is often unclear, with widely divergent findings in different studies. The occurrence of bleeding and thrombotic events in the same patient creates further complexity in the interpretation of the laboratory findings. The platelet abnormalities likely contribute to the excessive morbidity and mortality of these disorders, but the precise mechanisms are poorly understood.

### A. Clinical Features

Although both bleeding and thrombotic complications occur in MPD patients, there is greater resulting morbidity from the latter.[47,48] CML patients have less frequent bleeding and thrombosis compared to those with other MPDs. Bleeding appears more prevalent in AMM, whereas patients with other MPDs are more prone to thrombosis. Bleeding is predominantly mucocutaneous, with particular involvement of the gastrointestinal and genitourinary tracts. Deep hematomas, hemarthroses, and retroperitoneal hemorrhages are distinctly unusual. The risk of spontaneous hemorrhage may be increased when platelet counts are in excess of $2000 \times 10^9/L$.[51] Aspirin ingestion is an important contributing factor to the overall hemorrhage in MPDs.[47,50,52]

Venous thrombotic events (deep vein thrombosis of the extremities and pulmonary embolism) occur in MPDs and thrombosis may be at otherwise unusual sites, such as the splenic, hepatic, and mesenteric vessels and cerebral venous sinuses. Arterial events involving the peripheral, coronary, and cerebral vessels have also been documented.[47–50] Clinical manifestations may be associated with large-vessel arterial thrombosis[53] as well as with microcirculatory arterial events that result in erythromelalgia and neurologic symp-

toms. Erythromelalgia occurs predominantly in ET and is characterized by intense burning or throbbing pain in the extremities, predominantly the feet, and warmth and mottled erythema.[54] It may progress to digital ischemia and necrosis. The arterial pulses in the extremities are generally normal in these patients. On biopsy, erythromelalgia is characterized by fibromuscular intimal proliferation, endothelial swelling, and thrombotic occlusions.[54] Patients with ET and erythromelalgia have decreased platelet survival compared to patients with asymptomatic ET or reactive thrombocytosis.[54] Aspirin provides symptomatic relief of erythromelalgia[47,49,54] and reverses the shortened platelet survival.[55] Neurologic symptoms are frequently noted in MPD patients and range from nonspecific headaches and dizziness to focal neurologic events, such as transient ischemic events, seizures, and monocular blindness.[48,49,56] The transient neurologic events are also highly responsive to aspirin therapy.[56]

Risk factors for thrombosis identified in patients with ET include increasing age, a prior thrombotic event, inadequate control of thrombocytosis, the presence of risk factors for atherosclerotic heart disease, and spontaneous *in vitro* megakaryocyte colony formation.[49,50,52,57–62] The degree of thrombocytosis[52,57,58,63] and the abnormal *in vitro* platelet responses[51,57,58] do not correlate with risk of thrombosis in MPD patients. In contrast, the degree of erythrocytosis in PV correlates with an increased risk of thrombosis.[64] In patients with thrombocytosis, there is a correlation of thrombosis with increased platelet turnover, reflected by measurement of reticulated platelets.[65]

MPDs constitute a frequent underlying disease in patients presenting with the Budd–Chiari syndrome and portal vein thrombosis.[47,66] This association is supported by the finding of a high frequency of endogenous erythropoietin-independent erythroid colony growth in patients presenting with thrombosis in these areas.[66] Hepatic vein and/or portal vein thrombosis may be an initial presentation of MPDs, especially PV. ET patients have been reported to have an increased risk of recurrent spontaneous abortions, fetal growth retardation, premature deliveries, and abruptio placentae.[49]

The life expectancy of ET patients is normal despite the hemorrhagic and thrombotic events.[67] The incidence of thrombosis in ET is approximately 6.6 episodes per 100 patient-years of observation.[52] In different studies, 20 to 50% of ET patients had thrombotic events.[48] In CML, the prognosis is primarily determined by progression to blast crisis. In the chronic phase, the incidence of thrombotic and hemorrhagic complications is low;[48] bleeding increases in the accelerated phase, largely due to thrombocytopenia. Survival in AMM is determined by disease progression with increasing spleen size, decreasing hemoglobin, and bone marrow failure. Bleeding is common in the advanced thrombocytopenic stage. Patients with PV have a prolonged survival, and morbidity and mortality are influenced by vascular events. In two large trials of PV patients, thrombotic events occurred in 34 and 41% of patients, respectively.[68,69]

## B. Platelet Morphology and Function in MPD

Several studies have examined platelet function and morphology in MPD patients.[47–50,70] Under light microscope, heterogeneity in platelet size and morphology may be observed. This is reflected by alterations in mean platelet volume and in its distribution, as detected by electronic platelet counters.[71,72] Under the electron microscope, the dense granules, α-granules, and mitochondria are reduced, with alterations in the open canalicular and dense tubular systems.[73] Depletion of dense granule contents of ADP, ATP, and serotonin has also been documented.[74–77] The decreased platelet buoyant density described in MPD patients may reflect these alterations in the dense granules.[78]

A minority of patients with MPD (approximately 17%) have a prolonged bleeding time, and it appears to be more often prolonged in AMM than in other MPDs.[47,79] The bleeding time does not correlate with an increased risk of bleeding symptoms.[47,79,80] Platelet aggregation responses are also highly variable in MPD patients and often vary in the same patient over time. Decreased platelet responses are more common,[81] although some patients demonstrate enhanced responses to agonists or spontaneous aggregation.[82,83] In one analysis,[47] responses to ADP, collagen, and epinephrine were decreased in 39, 37, and 57% of patients, respectively. The impairment in aggregation is more commonly encountered in response to epinephrine than with other agonists; however, a diminished response to epinephrine is not pathognomonic of a MPD because it may be encountered in otherwise normal subjects or as a familial defect. The impaired epinephrine responsiveness in MPD has been attributed to a decrease in the number of platelet $\alpha_2$-adrenergic receptors, which has been observed in some studies[84–86] but not others,[74] and has depended on the ligand used: $^3$H-dihydroergocryptine[84,86] or $^3$H-yohimbine.[74] In addition to aggregation, epinephrine induces several other responses, including the inhibition of adenylyl cyclase and cAMP levels. Even in patients with impaired aggregation response, epinephrine-induced inhibition of an adenylyl cyclase has been normal,[87,88] reflecting the differential receptor requirements for the two responses. Diminished epinephrine-induced $TXA_2$ production, dense granule secretion, and $Ca^{2+}$ mobilization have also been reported in patients with impaired epinephrine-induced aggregation response.[74,88] The finding of decreased dense granule contents in some of these patients suggests that the impaired functional responses may not necessarily be due to a decrease in the surface receptors alone.

Alterations in membrane glycoproteins have been described in platelets from MPD patients. Decreases in

platelet integrin αIIbβ3 and GPIb have been documented[89–93]; in some studies,[90,94,95] there is decreased platelet fibrinogen binding, indicating a functional concomitant decrease in αIIbβ3. Moreover, receptor-mediated signal transduction-dependent activation of αIIbβ3 may be impaired, in addition to a quantitative decrease in the surface αIIbβ3.[93] One patient has been described[96] with deficiency of GPIa–IIa and markedly abnormal response to collagen. Platelet GPIV (CD36), a membrane GP related to platelet–collagen and platelet–thrombospondin interactions, has been noted to be increased in ET[97–99]; increased thrombospondin binding to MPD platelets has been noted in some[97] but not other studies.[98,100] Interestingly, platelet receptors for the Fc portion of IgG are increased in MPD platelets.[101]

Free arachidonic acid is mobilized from phospholipids upon platelet activation, and it is metabolized via two well-recognized pathways: (1) the cyclooxygenase pathway leading to $TXA_2$ production and (2) the 12-lipoxygenase pathway leading to formation of 12-hydroperoxyeicosatetraenoic acid (12-HPETE) and 12-hydroxyeicosatetraenoic acid (12-HETE). Reduced platelet formation of lipoxygenase products has been reported in MPD patients.[102–104] In one study,[104] the synthesis of $TXA_2$ was enhanced in the lipoxygenase-deficient platelets. Another study demonstrated substantial production of 12-HETE following aspirin treatment of MPD platelets to block the cyclooxygenase pathway.[105] Agonist-induced platelet $TXA_2$ production in MPD platelets has generally been normal[74,104,106] and there is evidence for an enhanced $TXA_2$ production *in vivo*,[105,107,108] suggesting ongoing platelet activation. Consistent with the platelet abnormality, a lipoxygenase deficiency has been described in leukocytes from MPD patients.[109] Although a selective lipoxygenase deficiency with intact or enhanced cyclooxygenase may intuitively suggest enhanced platelet function, patients with the lipoxygenase deficiency have had bleeding rather than thrombotic symptoms.[104] The role of the lipoxygenase products 12-HETE and 12-HPETE in platelets is not fully known. In addition, platelets produce other products (lipoxins) related to the action of 12-lipoxygenase that have diverse biologic effects. In one study,[110] deficiency in lipoxin synthesis was observed in the blast crisis of CML, with improvement during conversion to a second chronic phase.

Defects in platelet signaling mechanisms reported in MPD platelets include agonist-induced $Ca^{2+}$ mobilization and $Ca^{2+}$ fluxes across platelet membranes,[111] signaling through the $TXA_2$ receptor,[112] and protein phosphorylation due to deficiency of cGMP-dependent protein kinase.[113] In normal platelets, $PGD_2$ increases cAMP levels due to stimulation of adenylyl cyclase; this results in inhibition of platelet responses. Cooper et al.[114] reported a decrease in $PGD_2$-induced activation of adenylyl cyclase associated with a 50% reduction in $PGD_2$ receptors on platelets; responses to $PGE_2$ and $PGI_2$ were normal. In another study,[115] the cAMP response to all three of these inhibitory prostaglandins was blunted, suggesting impaired responses via multiple receptors. These findings indicate a defect in the normal platelet inhibitory mechanisms in MPD. Platelets from patients with PV and idiopathic myelofibrosis, but not ET or CML, have been shown to have reduced expression of the thrombopoietin receptor (Mpl) and reduced thrombopoietin-induced tyrosine phosphorylation of proteins.[116] Interestingly, another study showed enhanced constitutive phosphorylation in CML of Crkl, a protein that is phosphorylated on platelet activation.[117]

Patients with MPDs may have abnormalities in plasma VWF. A decrease in plasma VWF, particularly in the large VWF multimers, has been described; these changes have been inversely related to the platelet counts and have improved following cytoreduction.[118] The changes in plasma VWF may contribute to the hemostatic defect in MPD but have also been observed in patients with reactive thrombocytosis.[118]

Several studies have documented mutations in Janus kinase 2 (JAK2), a protein tyrosine kinase, in patients with MPD.[119] The impact of these alterations on platelet production and function remains to be delineated.

### C. Therapy

The two major pharmacologic approaches are cytoreduction and platelet inhibition. The treatment of ET is discussed in detail in Chapter 56.

## IV. Acute Leukemias and Myelodysplastic Syndromes

Thrombocytopenia is the major cause of bleeding in acute leukemias and myelodysplastic syndromes. However, bleeding complications may occur even in patients with normal platelet counts as a result of platelet dysfunction. Acquired platelet defects associated with clinical bleeding are more commonly found in acute myelogenous leukemia but have been reported in acute lymphoblastic and myelomonoblastic leukemias, hairy cell leukemia, and MDS.[120a–120g] The reported abnormalities in these patients include impaired platelet aggregation responses to ADP, epinephrine, and collagen and diminished nucleotide secretion, serotonin uptake and release, $TXA_2$ production, and platelet PDGF and β-thromboglobulin levels. Impaired platelet procoagulant activities have been described in both acute leukemias and MDS. In patients with hairy cell leukemia, the platelet dysfunction appears to improve postsplenectomy.[121–123] Acquired forms of von Willebrand disease (VWD) and a Bernard–Soulier-like platelet defect have been described in hairy cell leukemia and juvenile MDS, respectively.[124,125]

In acute leukemias and MDS, platelets may be large and morphologically abnormal, with a balloon-like appearance. Ultrastructural studies have shown decreased microtubules, reduced number and abnormal sizes of dense granules, and excessive membranous systems[120a–120d]; the megakaryocytes have shown dysplasia.[120a,120b] Bleeding in the acute leukemias and MDS usually responds to transfusion of platelets.

## V. Dysproteinemias

A hemostatic defect may occur in patients with dysproteinemias, and this is related to multiple mechanisms, including platelet dysfunction, specific coagulation abnormalities, hyperviscosity, and amyloid deposition in the vessel wall.[126] Qualitative platelet defects occur in 33% of patients with IgA myeloma or Waldenstrom macroglobulinemia, 15% of patients with IgG multiple myeloma, and infrequently in patients with monoclonal gammopathy of undetermined significance. Several coagulation abnormalities have been reported in patients with dysproteinemias, including factor X deficiency related to amyloidosis,[127] a circulating heparin-like anticoagulant,[128] and impaired fibrin polymerization.[129] An acquired form of VWD has been described in some patients.[130–134] Platelets possess Fc receptors (see Chapter 6) and it has been proposed that binding of the paraprotein to the platelet membrane interferes with normal platelet membrane functions. In some cases, the paraprotein bound specifically to platelet GPIIIa or interfered with the VWF–GPIb interaction.[130–133]

### A. Therapy

The platelet dysfunction appears to correlate with serum paraprotein levels. Thus, acute bleeding episodes can be managed with plasmapheresis to lower the paraprotein levels, and chronic bleeding may be controlled by chemotherapy aimed at reducing the concentration of the abnormal protein. DDAVP may be transiently effective in patients with acquired VWD[131,133] (see Chapter 67).

## VI. Acquired von Willebrand Disease

Most patients with the acquired VWD (AVWD) are older than age 40 years without previous manifestations or a family history of a bleeding diathesis. The associated disorders in these patients are diverse,[135,136] with half of them having an underlying lymphoproliferative disorder or plasma cell proliferative disorder. AVWD has been reported in chronic lymphocytic leukemia, hairy cell leukemia, acute myeloid and lymphoblastic leukemias, and non-Hodgkin's lymphoma. In patients with myeloproliferative diseases (CML, ET, or PV) and in reactive thrombocytosis, there is an impressive correlation between the abnormalities in plasma VWF and the elevated platelet counts.[118,137,138] AVWD has been observed in patients with autoimmune disorders including systemic lupus erythematosus, scleroderma and mixed connective tissue disease, hypothyroidism, and antiphospholipid antibody syndrome,[135,136] and following administration of ciprofloxacin,[139] valproic acid,[140] griseofulvin,[141] and the plasma expander hydroxyethyl starch.[142,143] AVWD has been reported in patients with solid tumors, most notably Wilms' tumor, but case reports have associated it with others as well (adrenocortical carcinoma, lung carcinoma, and gastric carcinoma).[135,144] Patients with aortic stenosis and congenital valvular heart disease may have excess bleeding (particularly gastrointestinal) and several reports have documented AVWD in such patients.[145,146] Vincentelli et al.[146] studied 50 consecutive patients with aortic stenosis and found skin or mucosal bleeding in 21% of patients with severe aortic stenosis. Abnormal platelet function documented by PFA-100, decreased VWF collagen-binding activity, and loss of the largest multimers or a combination of these occurred in 67 to 92% of patients with severe aortic stenosis. These abnormalities were corrected following surgery.

AVWD arises by diverse mechanisms. Some patients (especially those with dysproteinemias or lymphoproliferative and myeloproliferative diseases) have antibodies directed against functional domains of VWF[135,147]; the patient's plasma inhibits the ristocetin cofactor activity in normal plasma. In some patients, enhanced clearance of VWF has been mediated by antibodies that are not targeted to a functional VWF region. Increased shear stress-induced proteolysis of VWF is another mechanism leading to AVWD, such as in patients with severe aortic stenosis[146,148] and congenital cardiac diseases.[145] In some patients, cellular adsorption and removal of VWF by malignant or other cells (platelets in MPD) is the mechanism.[135] Nonimmunologic binding and precipitation of VWF has been noted following administration of hydroxyethyl starch.[142] Lastly, decreased VWF synthesis has been invoked in patients with hypothyroidism[149] and those receiving valproate.[140]

Laboratory findings for AVWD have included various combinations of a prolonged bleeding time, decreased plasma levels of VWF (VWF antigen and ristocetin cofactor activity) and factor VIII coagulant, and a selective reduction in the large VWF multimers in plasma.

### A. Therapy

The goals of treatment in patients with AVWD are (1) to quickly raise plasma VWF levels to treat or prevent bleeding and (2) to address the underlying associated conditions. The first goal is achieved by administration of DDAVP (see Chapter 67)[118,135,136,147] or factor VIII concentrates that contain VWF. Cryoprecipitate is also effective, but it is associated with the risk of transfusion-transmitted diseases. Recombinant factor VIIa (see Chapter 68) has also been used to control bleeding in a patient with AVWD.[150] Several reports have found intravenous immunoglobulin to be effective, including in patients with plasma cell dyscrasias.[135,136,147,151,152] Other modalities utilized include plasma exchange[153] and extracorporeal immunoadsorption.[154]

Treatment of the underlying disorder is an important aspect of management of patients with AVWD. In patients with MPDs and elevated platelet counts, cytoreduction is effective in reversing the plasma VWF abnormalities[118,155,156] as well as the decreased half-life.[118,157] Remissions have occurred spontaneously and after therapy of the underlying disease, such as lymphoproliferative disease, Wilms' tumor, multiple myeloma, hypothyroidism, CML, or cardiac valve abnormalities.[135,146]

## VII. Acquired Storage Pool Disease

Inherited deficiency of the platelet dense granule pool of ATP and ADP (storage pool deficiency [SPD]) is associated with impaired platelet function and a bleeding diathesis (see Chapter 57). Several patients have been reported in whom the dense granule SPD appears to be acquired as a result of *in vivo* activation and release of platelet dense granule contents or due to production of abnormal platelets by the bone marrow. Acquired SPD may therefore occur in diverse clinical states. The presence of antiplatelet antibodies has been associated with acquired SPD.[158,159] SPD has also been reported in patients with disseminated intravascular coagulation, hemolytic–uremic syndrome, renal transplant rejection, multiple congenital cavernous hemangioma, myeloproliferative diseases, chronic granulocytic leukemia, hairy cell leukemia, and acute nonlymphocytic leukemia.[159a–159h] Depletion of platelet granule contents has also been reported in patients with severe valvular disease and those with Dacron aortic grafts, in patients undergoing cardiopulmonary bypass, and in platelet concentrates stored for transfusion.[160–163]

## VIII. Antiplatelet Antibodies and Platelet Dysfunction

Antibodies induce several effects in platelets, including accelerated destruction, cell lysis, aggregation, secretion of granule contents, and expression of platelet factor-3 activity. These interactions may lead to impaired function as a consequence of platelet activation and binding of antibodies to platelet surface glycoproteins. The overall impact may be thrombocytopenia and/or impaired platelet function, as noted in a wide range of autoimmune disorders, including immune thrombocytopenic purpura (ITP), collagen vascular diseases, and AIDS. In addition, these antibodies inhibit megakaryocytopoiesis and proplatelet formation.[164,165] Patients with ITP have decreased platelet survival, with increased platelet-associated antibodies demonstrable in most patients. It has been generally considered that because of the increased rate of thrombopoiesis, these patients have young platelets in circulation with an enhanced functional ability. Bleeding times have been reported to be disproportionately shortened in relation to the platelet counts in patients with ITP compared to those with regenerative thrombocytopenia.[166] However, ITP patients may have impaired platelet function and disproportionately long bleeding times. In one study, defective *in vitro* platelet aggregation was observed in 9 of 11 patients with chronic ITP and normal platelet counts.[167] Serum globulin fractions isolated from some of the patients inhibited the aggregation responses of normal platelets to collagen and ADP. Impaired platelet aggregation responses to agonists (ADP, epinephrine, and collagen) with and without prolonged bleeding times have been reported in ITP and in disorders such as systemic lupus erythematosus and Graves' disease, which are associated with increased platelet-associated antibodies.[167a–167i]

Antibody-induced platelet dysfunction arises by multiple mechanisms. In many patients, the antibodies are directed against specific platelet surface membrane glycoproteins (GPIb,[168,169] αIIbβ3,[170,171] GPIa-IIa,[172,173] GPIV,[174] and GPVI[175,176]) or glycosphingolipids.[177,178] In some patients, the antibody specifically blocked platelet aggregation induced by collagen alone through interaction with GPVI,[179] GPIV,[174] or GPIa/IIa.[172,173] In one report, the anti-GPVI antibody induced a selective clearance of the GPVI/FcRr complex from the platelet surface,[176] a novel mechanism. Antibodies against GPIb-IX and integrin αIIbβ3 have been detected in 5 to 29% and 10 to 75% of ITP patients, respectively.[168,170,180] These antibodies in effect induced acquired forms of the Bernard–Soulier syndrome and Glanzmann thrombasthenia, respectively. One study noted that the severity of thrombocytopenia was not related to the glycoprotein specificity of the autoantibody.[181] Most of the platelet antibodies in ITP have been IgG, with IgM and IgA occurring less frequently.[182] Other functional defects in ITP platelets include impaired platelet arachidonate metabolism, as reflected by increased lipoxygenase activity, and decreased cyclooxygenase products (TXA$_2$ and hydroxyheptadecatrienoic acid).[183] Another defect is the antibody-induced deficiency of platelet granule contents in some patients.[158,159,184] Overall, antiplatelet antibodies not only induce thrombocytopenia but also may impair platelet function.

## IX. Liver Disease

The hemostatic defect in chronic liver disease results from multiple mechanisms, including deficiencies of blood coagulation factors and their inhibitors, disseminated intravascular coagulation, decreased fibrinolytic inhibitors and increased fibrinolysis, thrombocytopenia, and qualitative platelet defects. The overall contribution of platelet dysfunction to the hemostatic defect in severe liver disease is unknown. Platelet abnormalities noted in patients with chronic liver disease of various etiologies include prolongation of the bleeding time, diminished platelet survival, and reduced platelet aggregation, platelet adhesiveness, and platelet procoagulant activities.[185–190] In some studies, platelet aggregation responses were normal.[191] Platelet TXA$_2$ production has been reported to be decreased, associated with a defect in the release of free arachidonic acid from phospholipids and in phosphatidic acid formation.[189,190,192] The platelet dysfunction in severe liver disease has been attributed to elevated plasma fibrin degradation products,[185,187] abnormal high-density lipoproteins,[193] and ethanol.[194,195] Therapy for the hemostatic defect in severe liver disease involves administration of products to correct the coagulopathy. Platelet transfusions are indicated in patients with serious bleeding and severe thrombocytopenia. Intravenous and subcutaneous DDAVP shortens the bleeding time of patients with cirrhosis[196–198] and may be a possible treatment in the management of some patients (see Chapter 67).

## X. Miscellaneous Conditions with Impaired Platelet Function

Platelet function abnormalities have been described in several other disease states, such as severe vitamin B$_{12}$ deficiency,[199] eosinophilia,[200–202] atopia,[203] asthma and hay fever,[203,204] and Bartter syndrome.[205] The clinical importance of these observations is uncertain.

## XI. Drugs That Inhibit Platelet Function

A large number of drugs affect platelet function (Table 58-2). For several drugs, the antiplatelet effects have been established largely *in vitro*, and the relevance of such findings to the drug levels achieved in clinical practice is not well established. Information regarding the impact on hemostasis is unavailable for many drugs, even when they have been shown to alter platelet responses *ex vivo*. Moreover, the impact of concomitant administration of multiple drugs, each with a mild effect on platelet function, is unknown, even though this is a clinically relevant issue.

**Table 58-2: Drugs That Affect Platelet Function**$^a$

**Cyclooxygenase inhibitors**
Aspirin
Nonsteroidal anti-inflammatory drugs (indomethacin, phenylbutazone, ibuprofen, sulfinpyrazone, sulindac, meclofenamic acid, naproxen, diflunisal, piroxicam, tolmetin, zomepirac)

**ADP receptor antagonists**
Ticlopidine, clopidogrel

**GPIIb-IIIa antagonists**
Abciximab, tirofiban, eptifibatide

**Drugs that increase platelet cyclic AMP or cyclic GMP**
Prostaglandin $I_2$ and analogues
Phosphodiesterase inhibitors (dipyridamole, cilostazol, caffeine, theophylline, aminophylline)
Nitric oxide and nitric oxide donors

**Antimicrobials**
Penicillins
Cephalosporins
Nitrofurantoin
Hydroxychloroquine
Miconazole

**Cardiovascular drugs**
β-Adrenergic blockers (propranolol)
Vasodilators (nitroprusside, nitroglycerin)
Diuretics (furosemide)
Calcium channel blockers
Quinidine
Angiotensin converting enzyme inhibitors

**Anticoagulants**
Heparin

**Thrombolytic agents**
Streptokinase, tissue plasminogen activator, urokinase

**Psychotropics and anesthetics**
Tricyclic antidepressants (imipramine, amitryptyline, nortriptyline)
Phenothiazines (chlorpromazine, promethazine, trifluoperazine)
Selective serotonin reuptake inhibitors (fluoxetine, paroxetine, etc.)
General anesthesia (halothane)

**Chemotherapeutic agents**
Mithramycin
BCNU
Daunorubicin

**Miscellaneous agents**
Dextrans and hydroxyethyl starch
Lipid-lowering agents (clofibrate, halofenate)
Epsilon-aminocaproic acid
Antihistamines
Ethanol
Vitamin E
Radiographic contrast agents
Food items and food supplements (omega-3 fatty acids, vitamin E, onions, garlic, ginger, cumin, turmeric clove, black tree fungus, Ginko)

$^a$Modified from Rao.[70]

## A. Aspirin

Aspirin is an important cause of platelet dysfunction in clinical practice because of its widespread use. Aspirin irreversibly acetylates and inactivates the platelet cyclooxygenase, leading to the inhibition of endoperoxide (PGG$_2$ and PGH$_2$) and TXA$_2$ synthesis.[206,207] Much of this effect of orally administered aspirin on platelet cyclooxygenase occurs in the presystemic (portal) circulation.[208] As little as a single 80-mg tablet is adequate to block platelet TXA$_2$ production.[209,210] The impressive efficacy of aspirin in the primary and secondary prevention of arterial vascular events is well established[211,212] (see Chapter 60). The efficacy of aspirin as an antithrombotic agent and its effects on impairing primary hemostasis are both related to the inhibition of platelet TXA$_2$ production, although aspirin induces other effects including acetylation of plasma proteins (prothrombin and fibrinogen).[212] In otherwise normal individuals, the overall result of aspirin ingestion is a mild impairment of hemostasis. Aspirin prolongs the bleeding time in many, but not all, normal individuals; effect on this test is influenced by the direction of the incision on the forearm and the application of venostasis during the procedure[213–215] (see Chapter 25). Its ingestion results in inhibition of platelet aggregation and secretion upon stimulation with ADP, epinephrine, low concentrations of collagen and thrombin, and, especially, arachidonic acid. The closure times measured by the PFA-100 are often prolonged using the cartridge coated with collagen and epinephrine but not the one coated with collagen and ADP.[216] The risk of bleeding with aspirin is dose related. In a meta-analysis of 192,036 patients from 31 clinical trials, the risk of hemorrhage was lowest with an aspirin dose of <100 mg compared to higher doses.[217] The overall risk of a major bleed with aspirin is <1% per year.[212] Evidence for aspirin's adverse effect on hemostasis, particularly the risk of major extracranial and intracranial hemorrhages, derives from large aspirin trials. Several observations point to the clinical significance of the aspirin-induced platelet defect. Ingestion of aspirin during pregnancy has been reported to result in excessive bleeding at delivery in both the neonate and the mother.[218–220] In the Physicians' Health Study, 22,071 healthy male physicians were randomized to receive placebo or aspirin (325 mg every other day); over a 5-year follow-up there was a significantly higher incidence of easy bruising, mucosal bleeding (including gastrointestinal bleeding), and blood transfusion requirement.[221] The rate of hemorrhagic stroke was higher with aspirin (0.04% per year) compared to placebo (0.02% per year, $p = 0.06$).[221] Preoperative ingestion of aspirin increases blood loss, transfusions, and repeated operations in patients undergoing cardiac surgery.[222,223] However, one large study[224] on the early use of aspirin after coronary artery bypass grafting (CABG) showed that patients who received aspirin during the first 48 hours had a 60% lower mortality compared to those who did not, and the rates of other nonfatal ischemic events were also lower. Interestingly, the frequency of bleeding complications and reoperations was significantly lower among aspirin-treated patients. This study may, indeed, question the generally held reluctance to use aspirin in the immediate CABG postoperative period.

A major adverse effect of aspirin is gastrointestinal (GI) bleeding.[212,221,225–229] Aspirin-induced GI toxicity, but not its antithrombotic effect, is dose related in the range of 30 to 1300 mg daily.[212,225] However, even at low doses (30–50 mg daily) aspirin can cause serious GI bleeding.[226,227] The overall relative risk of GI complications with the use of nonsteroidal antiinflammatory drugs (NSAIDs) (including aspirin) has been estimated to be between 3.0 and 5.0 compared to nonusers.[229]

Individuals with underlying hemostatic defects, such as VWD, hemophilia, or mild platelet function defects, and those on oral anticoagulant therapy may develop considerable postoperative or even spontaneous bleeding with concomitant aspirin therapy. In these patients, aspirin ingestion often leads to a striking prolongation of the bleeding time and increase in bleeding manifestations, and it should be avoided. In otherwise normal subjects who are on aspirin and need to undergo elective surgery, aspirin should be discontinued 7 to 10 days prior to the procedure. This is often not feasible in patients with arterial diseases. In patients undergoing CABG, one set of guidelines recommends starting 325 mg of aspirin 6 hours after the surgery.[230] Based on the previously mentioned study on the beneficial effects of early administration of aspirin,[224] there is a need to reevaluate the current practice regarding use of antiplatelet agents in the immediate post-CABG period.[231] If excessive perioperative hemorrhage is encountered, it is generally responsive to platelet transfusions. Moreover, DDAVP infusion shortens the aspirin-induced prolongation in bleeding time.[232]

## B. Nonsteroidal Anti-inflammatory Drugs

Platelet function is also inhibited by several NSAIDs, which inhibit cyclooxygenase and may prolong the bleeding time. These drugs include indomethacin, ibuprofen, sulfinpyrazone, meclofenamic acid, phenylbutazone, sulindac, naproxen, mefanamic acid, piroxicam, tolmetin, and diflunisal.[233–242] The inhibition of cyclooxygenase by these agents is generally short-lived and reversible. For example, the effect of indomethacin is not detectable after 6 hours.[235] In contrast, the effect of piroxicam may last for several days, related to its long half-life.[243] Exposure to NSAIDs other than aspirin is also recognized to increase the risk of upper GI bleeding and/or perforation. However, there is variability in this effect with different NSAIDs.[228,229] A meta-analysis[228] found the relative risk to be lowest for ibuprofen and diclofenac, intermediate for aspirin, indomethacin, naproxen and sulindac, and higher for azapropazone, tolmetin, ketoprofen, and piroxicam. Ibuprofen has been administered to hemophiliacs without a major increase in bleeding,[244] although enhanced bleeding was noted in those receiving ibuprofen and zidovudine.[245] These findings may be relevant to the choice of drug for use in patients with impaired hemostasis. Concomitant administration of ibuprofen and aspirin antagonizes the irreversible cyclooxygenase inhibition of aspirin and may impact its antithrombotic efficacy.[246] Like aspirin, most of these NSAIDs inhibit both forms of cyclooxygenase (cyclooxygenase-1 and -2).[247] The selective cyclooxygenase-2 inhibitors, celexocib, rofecoxib, and valdecoxib, do not have an antiplatelet activity. Acetaminophen does not inhibit or impair platelet function at levels attained *in vivo,* although in some patients it may inhibit *in vitro* platelet responses as a weak cyclooxygenase inhibitor.[248]

## C. ADP Receptor Antagonists

Thienopyridine derivatives, ticlopidine and clopidogrel, inhibit platelet function by inhibiting the binding of ADP to the $P2Y_{12}$ receptor (see Chapter 61 for details).[212,249,250] In patients treated with a thienopyridine, platelet aggregation responses to several agonists, including ADP, collagen, epinephrine, and thrombin, are inhibited to various extents depending on agonist concentrations. Both ticlopidine and clopidogrel mediate their effects through active metabolites generated *in vivo* by hepatic transformation. With repeated twice-daily dosing, plasma ticlopidine levels increase approximately three-fold over 2 or 3 weeks because of drug accumulation. Likewise, the elimination half-life is up to 96 hours after 14 days of repeated dosing. Clopidogrel differs from ticlopidine in its pharmacokinetics. It is rapidly metabolized and the main carboxylic derivative has a plasma elimination half-life of approximately 8 hours. On repeated daily dosing (50–100 mg) in healthy volunteers, ADP-induced platelet aggregation is inhibited from the second day and reaches steady state after 4 to 7 days.[212] After a single dose of 400 mg clopidogrel, platelet aggregation is inhibited at 2 hours and this persists up to 48 hours. The platelet inhibition persists for 7 to 10 days after therapy is discontinued. Ticlopidine and clopidogrel prolong the bleeding time by 1.5- to 2.0-fold over control, but a more striking prolongation has been noted.[250–252] The closure times as measured by the PFA are insensitive to therapy with these agents.[216] These drugs need to be discontinued for 7 to 10 days prior to elective surgery. Both ticlopidine and clopidogrel have been associated with thrombotic thrombocytopenic purpura[253–255] (see Chapter 50).

### D. GPIIb-IIIa Antagonists

GPIIb-IIIa antagonists are a class of compounds that inhibit fibrinogen binding and platelet aggregation on activation (see Chapter 62 for details).[212] The clinically available compounds are the humanized chimeric monoclonal antibody against the GPIIb-IIIa receptor c7E3 (abciximab), the synthetic cyclic heptapeptide containing the KGD sequence (eptifibatide), and the nonpeptide RGD mimetic (tirofiban). These intravenous agents are used in acute coronary syndromes and in the context of percutaneous coronary interventions.[212] All of these agents are potent inhibitors of aggregation (both primary and secondary) in response to all platelet agonists, and they all prolong bleeding time.[212] Not unexpectedly, bleeding is a potential complication with these agents, and it is responsive to platelet transfusion. Thrombocytopenia is another potential feature with all of the GPIIb-IIIa antagonists and it is immune-mediated[212,256] (see Chapter 49).

### E. Drugs That Increase Platelet Cyclic AMP or Cyclic GMP

Several agents elevate intracellular cAMP levels by activating adenylyl cyclase (e.g., $PGI_2$, $PGE_1$, and $PGD_2$) or inhibiting the cyclic nucleotide phosphodiesterases, and this results in inhibition of platelet responses. Intravenous infusion of prostacyclin ($PGI_2$) and its stable analogues prolong bleeding time.[257,258] Dipyridamole (see Chapter 63), a weak phosphodiesterase inhibitor, appears not to inhibit aggregation responses to collagen, epinephrine, and ADP at usual doses but has a synergistic effect with aspirin in prolonging the shortened platelet survival in thromboembolic disorders.[259] Cilostazol (see Chapter 64) is a phosphodiesterase-3 inhibitor that suppresses platelet aggregation and is currently used for therapy of peripheral arterial disease.[260,261] It does not

prolong bleeding time.[252] Other weak phosphodiesterase inhibitors, such as caffeine, theophylline, and aminophylline, inhibit ADP-induced platelet aggregation *in vitro*, but this may not have clinical significance.[262]

Sildenafil (Viagra), a selective inhibitor of phosphodiesterase type 5 that increases platelet cyclic GMP levels, was found not to have a direct effect on platelet function but potentiated the *in vitro* antiaggregatory effect of sodium nitroprusside.[263] Another study found that sildenafil inhibits ADP-induced activation of integrin $\alpha IIb\beta 3$.[264] Some reports have associated sildenafil with epistaxis, hemorrhoidal, and variceal bleeding. NO and NO donors inhibit platelet activation and responses by elevating cyclic cGMP levels[35,36,265–267] (see Chapter 13). Elevated cGMP inhibits platelet cAMP phosphodiesterase, thereby elevating cAMP. Administration of NO is associated with a prolonged bleeding time.[35,265]

### F. Antimicrobial Agents

β-Lactam antibiotics, including penicillins and cephalosporins, inhibit platelet aggregation responses, and some induce a bleeding diathesis when given in high doses. These include carbenicillin, penicillin G, ticarcillin, ampicillin, nafcillin, cloxacillin, mezlocillin, oxacillin, and piperacillin.[268–278] The effects on platelet function appear to be dose dependent, taking approximately 2 or 3 days to manifest and 3 to 10 days to abate after discontinuation of the drug.[268,273,274,277,278] Penicillins lacking the α-carboxy group (mezlocillin, piperacillin, and apalcillin) appear to adversely affect platelet function less often than carboxypenicillins or moxalactam.[277] Cephalosporins may also impair platelet function.[277,279,280] Moxalactam has been reported to induce platelet dysfunction associated with prolonged bleeding times and clinical hemorrhage.[277,280] However, other third-generation cephalosporins appear to show little effect on platelet function.[277] Some β-lactam antibiotics (e.g., moxalactam, cefamandole, cefoperazone, and cefotetan) may induce hypoprothrombinemia and thereby contribute to hemorrhage.[277,280] These compounds contain an *N*-methylthiotetrazole side chain, which inhibits vitamin K epoxide reductase, thus interfering with the carboxylation of the vitamin K-dependent coagulation factors.

The following generalizations can be made regarding the impact of β-lactam antibiotics on platelets. First, the effects on platelet function and hemostasis appear to be dose dependent and time dependent, with effects becoming discernible over several days.[268,273,274,277,278] Second, the patients at particular risk of bleeding appear to be those with concurrent illnesses, including sepsis, malnourishment, thrombocytopenia, and malignancy; intensive care units are the typical setting. Third, the platelet inhibitory effect of β-lactam antibiotics is influenced by plasma albumin. Both the platelet binding of the antibiotic and the impact on platelet responses are inversely related to albumin concentration.[281] Lastly, with some of the β-lactam antibiotics, the bleeding is related to a concomitant inhibition of synthesis of vitamin K-dependent coagulation factors.[277,279,280]

A number of mechanisms have been invoked to explain β-lactam antibiotic-induced platelet inhibition. These drugs inhibit platelet aggregation and secretion as well as platelet adherence to subendothelial structures and collagen-coated surfaces.[272] Many studies have evaluated the effects on platelets only *in vitro* and often at concentrations higher than those attained *in vivo*. Short-term *in vitro* exposure of platelets to penicillin has been reported to result in impaired aggregation responses and decreased binding of agonists to specific receptors (ADP, VWF, α₂-adrenergic) on platelets.[282,283] Although the latter effect of penicillins was found to be rapidly reversible,[282,283] the platelet inhibitory effect of β-lactam antibiotics appears to persist for several days after discontinuation, indicating that the sustained effect is related to other mechanisms. Other reported effects of β-lactam antibiotics include inhibition of intracellular signaling events, calcium mobilization,[284] and TXA$_2$ synthesis[283,284] and alterations in activation-induced changes in membrane integrin αIIbβ3 and GPIb-IX.[285]

The context in which the bleeding events are encountered in patients on antibiotics precludes clear definition of the precise role played by the antimicrobials because of the simultaneous presence of thrombocytopenia, disseminated intravascular coagulation, infection, and/or vitamin K deficiency. Discontinuation of a specifically indicated antibiotic may not be an option or necessary. Supportive measures using blood products and other interventions (correcting metabolic abnormalities and vitamin K deficiency) are indicated in the overall management.

In addition to the β-lactam antibiotics, other antimicrobials shown to inhibit platelet function include nitrofurantoin,[286] hydroxychloroquine,[287] and miconazole.[288]

## G. Cardiovascular Drugs

*In vitro*, propranolol inhibits the aggregation responses of normal platelets to ADP and epinephrine and signaling events only at relatively high concentrations.[289–291] At lower concentrations, propranolol inhibits secondary aggregation and secretion in patients whose platelets are susceptible to a low threshold concentration of agonists, but not uniformly in all patients taking the drug.[289,292] At conventional doses, propranolol inhibits and normalizes the increased sensitivity to agonists of platelets from patients with angina pectoris.[293] The mechanism by which propranolol alters platelet behavior is probably unrelated to its β-adrenergic blocking effects and may be due to an effect on platelet membranes.[289,292] Propranolol inhibits the release of free arachidonic acid

from phospholipids and TXA$_2$ production during platelet activation.[294,295] Overall, propranolol is a weak inhibitor of platelet function and does not influence the bleeding time. Other β-blockers (pindolol, atenolol, metoprolol, and carvedilol) have also been reported to inhibit platelet responses when added *in vitro*.[290,296] Other cardiovascular drugs known to inhibit platelet responses include furosemide,[297] nitroprusside,[298–301] nitroglycerin and isosorbide dinitrate,[298,302] and quinidine.[303,304] Although calcium channel blockers (nifedipine, verapamil, and diltiazem) inhibit platelet responses, this is seen *in vitro* at drug concentrations higher than those achieved during therapy.[305–309] These agents inhibit platelet α₂-adrenergic receptors.[307,308]

Some angiotensin converting enzyme (ACE) inhibitors (captopril,[310,311] perindopril,[312] and fosinopril[313]), but not others (quinapril,[312] enalapril,[314] and lisinopril[315]), have been reported to inhibit platelet aggregation in human studies.[316] Some ACE inhibitors (captopril[311,314] and fosinopril[314]) decreased TXA$_2$ production. Interestingly, reduced circulating levels of plasma fibrinogen, factor XII, factor XI, tissue plasminogen activator, and plasminogen activator inhibitor-I have been observed with the use of ACE inhibitors.[316]

## H. Anticoagulants and Thrombolytic Agents

Heparin induces a number of effects in platelets including those leading to severe thrombocytopenia (see Chapter 48). Heparin binds to platelets[317] and *in vitro* enhances aggregation responses to platelet agonists.[318] A bolus injection of heparin into normal subjects prolongs the bleeding time.[319] Heparin inhibits VWF-dependent platelet responses *in vitro* and *ex vivo*.[320] The impact of these disparate effects on hemostatic mechanisms during heparin therapy is complex and unclear. Protamine sulfate, widely used to neutralize heparin, also inhibits thrombin-induced platelet responses *in vitro*.[321] The major complication following the administration of thrombolytic agents (streptokinase, urokinase, and recombinant tissue plasminogen activator [rt-PA]) is hemorrhage, which arises from multiple mechanisms, including the effect of plasmin on the plasma coagulation system and platelets and the dissolution of blood clots providing hemostasis at the site of vascular breach or abnormality. The bleeding time in patients receiving rt-PA may be prolonged.[322] *In vitro*, plasmin induces several effects in platelets: initial platelet activation followed by inhibition, cleavage of membrane glycoproteins, inhibition of TXA$_2$ production, and disaggregation of platelet clumps.[323–327] Plasmin-induced platelet aggregation has been attributed to cleavage of protease-activated receptor-4.[328] Urinary metabolites of TXA$_2$ are increased during thrombolytic therapy with rt-PA or streptokinase, indicating *in vivo* activation of platelets.[329,330]

### I. Psychotropic Drugs and Anesthetics

This heterogeneous group consists of drugs that are generally considered to stabilize biological membranes by increasing their fluidity. Tricyclic antidepressants (imipramine, amitriptyline, and nortriptyline) inhibit platelet aggregation and secretion responses to ADP, epinephrine, and collagen.[331,332] Selective serotonin reuptake inhibitors (SSRIs; e.g., fluoxetine and paroxetine) are widely used antidepressants that cause an acquired but reversible serotonin deficiency in platelet dense granules and this may be related to the apparent benefit of SSRIs in coronary artery disease.[332a] Fluoxetine has been associated with prolongation of bleeding time, impaired aggregation, and excessive bruising.[333,334] Paroxetine prolongs the PFA-100 closure time and decreases platelet surface expression of CD63 in response to thrombin.[335] Phenothiazines (chlorpromazine, promethazine, and trifluoperazine) inhibit platelet responses to stimulation *in vitro* and perhaps *in vivo*.[336–340] Some studies indicate that trifluoperazine may impair platelet responses by inhibiting calmodulin-dependent processes in platelets.[339] Halothane has been reported to increase bleeding times during general anesthesia[341,342] but probably without much effect on surgical hemostasis. Thiopental appears to inhibit platelet responses in cardiac surgery patients.[343]

### J. Miscellaneous Drugs

Dextrans are branched polysaccharides of 40 or 70 kDa that have been used as antithrombotic agents. They prolong the bleeding time after infusion and inhibit platelet factor-3 availability, glass bead retention, aggregation, and secretion.[344–349] The prolongation of the bleeding time and inhibition of platelet function are most pronounced 4 to 8 hours after transfusion.[347] The mechanisms by which dextrans induce platelet dysfunction are unclear but include adsorption to platelets and a decrease in plasma factor VIII–VWF.[348,350] However, it is unclear if their use, even in the perioperative state, leads to excessive hemorrhage.[351] In one multicenter study of surgical patients, prophylaxis with dextran-70 was associated with less bleeding compared to low-dose heparin.[352] Hydroxyethyl starch, another volume expander, may enhance bleeding and prolong bleeding time,[353–355] as well as induce AVWD.[142,143]

Other drugs that impair platelet function include lipid-lowering agents (clofibrate and halofenate)[356,357]; chemotherapeutic agents such as daunomycin, mithramycin, and 1,3-*bis* (2-chloroethyl)-1-nitrosurea[358–363]; and antihistamines.[364,365] The fibrinolytic inhibitor epsilon aminocaproic acid (EACA), which has been used in patients with subarachnoid hemorrhage, can prolong the bleeding time at the generally administered dose.[366]

Ethanol inhibits platelet responses *in vitro*.[367–369] Ethanol alone does not prolong the bleeding time; however, acute ingestion of 50 g of ethanol potentiates the aspirin-induced prolongation of the bleeding time even in otherwise normal subjects.[370] Other agents reported to inhibit platelet function include radiographic contrast agents[371–373] (although some contrast agents activate platelets[374]); vitamin E[375,376]; fish oil with omega-3 fatty acids[377]; and food items and their extracts, such as onions,[378–380] garlic,[380,381] ginger,[382] cumin,[383] turmeric,[383,384] clove,[385] flavanol-rich cocoa beverage,[386] Chinese food, and black tree fungus.[387,388] Inhibition of eicosanoid synthesis appears to be the mechanism of action for several of the spices, although they have other effects, such as lowering of plasma lipids. For most of these agents, there is little evidence that they impair clinical hemostasis.

Herbal medicines, such as Ginkgo (*Gingko biloba*) and garlic (*Allium sativum*), inhibit platelet function and have been associated with increased bleeding, particularly when administered concomitantly with other medicines (warfarin and aspirin).[389] With the increasing use of herbal medicines and food supplements, their role and interaction with pharmaceutical drugs needs to be considered in the evaluation of patients with unexplained bleeding.

## Acknowledgments

This work was supported by grant HL56724 from the National Heart, Lung, and Blood Institute. The excellent secretarial assistance of Ms. Denise Tierney is gratefully acknowledged.

## References

1. Mielke, C. H., Kaneshiro, M. M., Maher, I. A., et al. (1969). The standardized normal Ivy bleeding time and its prolongation by aspirin. *Blood, 34,* 204–215.
2. Mielke, C. H. (1983). Influence of aspirin on platelets and the bleeding time. *Am J Med, 74,* 72–78.
3. Boccardo, P., Remuzzi, G., & Galbusera, M. (2004). Platelet dysfunction in renal failure. *Semin Thromb Hemost, 30,* 579–589.
4. Himmelfarb, J., Holbrook, D., McMonagle, E., et al. (1997). Increased reticulated platelets in dialysis patients. *Kidney Int, 51,* 834–839.
5. Steiner, R. W., Coggins, C., & Carvalho, A. C. (1979). Bleeding time in uremia: A useful test to assess clinical bleeding. *Am J Hematol, 7,* 107–117.
6. Livio, M., Gotti, E., Marchesi, D., et al. (1982). Uraemic bleeding: Role of anemia and beneficial effect of red cell transfusions. *Lancet, 2,* 1013–1015.
7. Moia, M., Mannucci, P. M., & Vizzotto, L. (1987). Improvement in the haemostatic defect of uraemia after treatment with recombinant human erythropoietin. *Lancet, 8570,* 1227–1229.

8. Castillo, R., Lozano, T., Escolar, G., et al. (1986). Defective platelet adhesion on vessel subendothelium in uremic patients. *Blood,68,* 337–342.

9. Gralnick, H. P., McKeown, L. P., Williams, S. B., et al. (1988). Plasma and platelet von Willebrand's factor defects in uremia. *Am J Med,85,* 806–810.

10. Salvati, F., & Liani, M. (2001). Role of platelet surface receptor abnormalities in the bleeding and thrombotic diathesis of uremic patients on hemodialysis and peritoneal dialysis. *Int J Artif Organs, 24,* 131–135.

11. Escolar, G., Cases, A., Bastida, E., et al. (1990). Uremic platelets have a functional defect affecting the interaction of von Willebrand factor with glycoprotein IIb-IIIa. *Blood, 73,* 1336–1340.

12. Benigni, A., Boccardo, P., Galbusera, M., et al. (1993). Reversible activation defect of the platelet glycoprotein IIb-IIIa complex in patients with uremia. *Am J Kidney Dis, 22,* 668–676.

13. Joist, J. H., & George, J. N. (2001). Hemostatic abnormalities in liver and renal disease. In R. W. Colman, J. Hirsh, V. J. Marder, et al. (Eds.), *Hemosasis and thrombosis: Basic principles and clinical practice* (4th ed., pp. 955–973). Philadelphia: Lippincott.

14. Sreedhara, R., Itagaki, I., & Hakim, R. M. (1996). Uremic patients have decreased shear-induced platelet aggregation mediated by decreased availability of glycoprotein IIb-IIIa receptors. *Am J Kidney Dis, 27,* 355–364.

15. Kozek-Langenecker, S. A., Masaki, T., Mohammad, H., et al. (1999). Fibrinogen fragments and platelet dysfunction in uremia. *Kidney Int, 56,* 299–305.

16. Rao, A. K., & Carvalho, A. (1994). Acquired qualitative platelet defects. In R. W. Colman, J. Hirsh, V. J. Marder, et al. (Eds.), *Hemostasis and thrombosis: Basic principles and clinical practice* (3rd ed., pp. 685–704). Philadelphia: Lippincott.

17. Knudsen, F., Nielsen, A. H., & Kristensen, S. D. (1985). The effect of dialyser membrane material on intradialytic changes in platelet count, platelet aggregation, circulating platelet aggregates, and antithrombin III. *Scand J Urol Nephrol, 19,* 227–232.

18. Guzzo, J., Niewiarowski, S., Musial, J., et al. (1980). Secreted platelet proteins with anti-heparin and mitogenic activities in chronic renal failure. *J Lab Clin Med, 16,* 102–112.

19. Green, D., Santhanam, S., Krumlovsky, F. A., et al. (1979). β-Thromboglobulin in patients with chronic renal failure: Effect of hemodialysis. *Thromb Haemost, 41,* 416.

20. Kubisz, P., Parizek, M., Seghier, F., et al. (1985). Relationship between platelet aggregation and plasma β-thromboglobulin levels in arteriovascular and renal disease. *Atherosclerosis, 55,* 363–368.

21. DiMinno, G., Martinez, J., McKean, M. L., et al. (1985). Platelet dysfunction in uremia. Multifaceted defect partially corrected by dialysis. *Am J Med, 79,* 552–559.

22. Remuzzi, G. D., Marchesi, M., Livio, A. E., et al. (1978). Altered platelet and vascular prostaglandin-generation in patients with chronic renal failure and prolonged bleeding times. *Thromb Res, 13,* 1007–1115.

23. Remuzzi, G., Benigni, A., Dodesini, P., et al. (1983). Reduced thromboxane formation in uremia: Evidence for functional cyclooxygenase-defect. *J Clin Invest, 71,* 762–768.

24. Rao, A. K. (1986). Uraemic platelets. *Lancet,1,* 913–914.

25. Vecino, A. M., Teruel, J. L., Navarro, J. L., et al. (2002). Phospholipase A2 activity in platelets of patients with uremia. *Platelets, 13,* 415–418.

26. Lubbecke, F., Sandlebaum, J., Schutterle, G., et al. (1988). Adenylate cyclase and α-2 adrenoreceptors in thrombocytes of chronic uremic patients. *Blood Purif, 6,* 269–275.

27. Ware, J. A., Clark, B. A., Smith, M., et al. (1989). Abnormalities of cytoplasmic $Ca^{2+}$ in platelets from patients with uremia. *Blood, 73,* 172–176.

28. Castaldi, P. A., Rozenberg, M. C., & Stewart, J. H. (1966). The bleeding disorder of uraemia. *Lancet, 2,* 66–69.

29. Diaz-Ricart, M., Estebanell, E., Cases, A., et al. (2000). Abnormal platelet cytoskeletal assembly in hemodialyzed patients results in deficient tyrosine phosphorylation signaling. *Kidney Int, 57,* 1905–1914.

30. Horowitz, H. I., Cohen, B. D., Martinez, P., et al. (1967). Defective ADP-induced platelet factor 3 activation in uremia. *Blood, 30,* 331–340.

31. Rabiner, S. F., & Hrodek, O. (1968). Platelet factor 3 in normal subjects and patients with renal failure. *J Clin Invest, 47,* 901–912.

32. Horowitz, H. I., Stein, I. M., Cohen, B. D., et al. (1970). Further studies on the platelet inhibitory effect of guanidinosuccinic acid: Its role in uremic bleeding. *Am J Med, 49,* 336–345.

33. Noris, M., & Remuzzi, G. (1999). Uremic bleeding: Closing the circle after 30 years of controversies? *Blood, 94,* 2569–2574.

34. Noris, M., Benigni, A., Boccardo, P., et al. (1993). Enhanced nitric oxide synthesis in uremia: Implications for platelet dysfunction and dialysis hypotension. *Kidney Int, 44,* 445–450.

35. Hogman, M., Frostell, C., Arnberg, H., et al. (1993). Bleeding time prolongation and NO inhalation. *Lancet, 341,* 1664–1665.

36. Gries, A., Bode, C., Peter, K., et al. (1998). Inhaled nitric oxide inhibits human platelet aggregation, P-selectin expression, and fibrinogen binding *in vitro* and *in vivo*. *Circulation, 97,* 1481–1487.

37. Matsumoto, A., Hirata, Y., Kakoki, M., et al. (1999). Increased excretion of nitric oxide in exhaled air of patients with chronic renal failure. *Clin Sci (Colch), 96,* 67–74.

38. Madore, F., Prud'homme, L., Austin, J. S., et al. (1997). Impact of nitric oxide on blood pressure in hemodialysis patients. *Am J Kidney Dis,30,* 665–671.

39. Fernandez, F., Goudrable, C., Sie, P., et al. (1985). Low haematocrit and prolonged bleeding time in uraemic patients: Effect of red cell transfusions. *Br J Haematol, 59,* 139–148.

40. Mannucci, P. M., Remuzzi, G., Pusinera, F., et al. (1983). Desamino-8-D-arginine vasopressin shortens the bleeding time in uremia. *N Engl J Med, 308,* 8–12.

41. Mannucci, P. M. (1998). Hemostatic drugs. *N Engl J Med, 339,* 245–253.

42. Janson, P. A., Jubelirer, J. J., Weinstein, M. J., et al. (1980). Treatment of the bleeding tendency in uremia with cryoprecipitate. *N Engl J Med, 303*, 1318–1322.

43. Liu, Y., Kosfeld, R. E., & Marcum, S. G. (1984). Treatment of uraemic bleeding with conjugated oestrogen. *Lancet, 2*, 887–890.

44. Livio, M., Mannucci, P. M., Vigano, G., et al. (1986). Conjugated estrogens for the management of bleeding associated with renal failure. *N Engl J Med, 315*, 731–735.

45. Sloand, J. A., & Schiff, M. J. (1995). Beneficial effect of low-dose transdermal estrogen on bleeding time and clinical bleeding in uremia. *Am J Kidney Dis, 26*, 22–26.

46. Noris, M., Todeschini, M., Zappella, S., et al. (2000). 17β-Estradiol corrects hemostasis in uremic rats by limiting vascular expression of nitric oxide synthases. *Am J Physiol Renal Physiol, 279*, F626–F635.

47. Schafer, A. I. (1984). Bleeding and thrombosis in myeloproliferative disorders. *Blood, 64*, 1–12.

48. Wehmeier, A., Sudhoff, T., & Meierkord, F. (1997). Relation of platelet abnormalities to thrombosis and hemorrhage in chronic myeloproliferative disorders. *Semin Thromb Hemost, 23*, 391–402.

49. Ravandi-Kashani, F., & Schafer, A. I. (1997). Microvascular disturbances, thrombosis, and bleeding in thrombocythemia: Current concepts and perspectives. *Semin Thromb Hemost, 23*, 479–488.

50. Landolfi, R., Marchioli, R., & Patrono, C. (1997). Mechanisms of bleeding and thrombosis in myeloproliferative disorders. *Thromb Haemost, 78*, 617–621.

51. Fenaux, P., Simon, M., Caulier, M. T., et al. (1990). Clinical course of essential thrombocythemia in 147 cases. *Cancer, 66*, 549–556.

52. Cortelazzo, S., Viero, P., Finazzi, G., et al. (1990). Incidence and risk factors for thrombotic complications in a historical cohort of 100 patients with essential thrombocythemia. *J Clin Oncol, 8*, 556–562.

53. Johnson, M., Gernsheimer, T., & Johansen, K. (1995). Essential thrombocytosis: Underemphasized cause of large-vessel thrombosis. *J Vasc Surg, 22*, 443–449.

54. van Genderen, P. J., & Michiels, J. J. (1997). Erythromelalgia: A pathognomonic microvascular thrombotic complication in essential thrombocythemia and polycythemia vera. *Semin Thromb Hemost, 23*, 357–363.

55. van Genderen, P. J., Michiels, J. J., van Strik, R., et al. (1995). Platelet consumption in thrombocythemia complicated by erythromelalgia: Reversal by aspirin. *Thromb Haemost, 73*, 210–214.

56. Koudstaal, P. J., & Koudstaal, A. (1997). Neurologic and visual symptoms in essential thrombocythemia: Efficacy of low-dose aspirin. *Semin Thromb Hemost, 23*, 365–370.

57. Tefferi, A., & Hoagland, H. C. (1994). Issues in the diagnosis and management of essential thrombocythemia. *Mayo Clinic Proc, 69*, 651–655.

58. Colombi, M., Radaelli, F., Zocchi, L., et al. (1991). Thrombotic and hemorrhagic complications in essential thrombocythemia. A retrospective study of 103 patients. *Cancer, 67*, 2926–2930.

59. Lahuerta-Palacios, J. J., Bornstein, R., Fernandez-Debora, F. J., et al. (1988). Controlled and uncontrolled thrombocytosis. Its clinical role in essential thrombocythemia. *Cancer, 61*, 1207–1212.

60. Juvonen, E., Ikkala, E., Oksanen, K., et al. (1993). Megakaryocyte and erythroid colony formation in essential thrombocythaemia and reactive thrombocytosis: Diagnostic value and correlation to complications. *Br J Haematol, 83*, 192–197.

61. Watson, K. V., & Key, N. (1993). Vascular complications of essential thrombocythaemia: A link to cardiovascular risk factors. *Br J Haematol, 83*, 198–203.

62. Besses, C., Cervantes, F., Pereira, A., et al. (1999). Major vascular complications in essential thrombocythemia: A study of the predictive factors in a series of 148 patients. *Leukemia, 13*, 150–154.

63. Wehmeier, A., Daum, I., Jamin, H., et al. (1991). Incidence and clinical risk factors for bleeding and thrombotic complications in myeloproliferative disorders. A retrospective analysis of 260 patients. *Ann Hematol, 63*, 101–106.

64. Pearson, T. C. (1997). Hemorheologic considerations in the pathogenesis of vascular occlusive events in polycythemia vera. *Semin Thromb Hemost, 23*, 433–439.

65. Rinder, H. M., Schuster, J. E., Rinder, C. S., et al. (1998). Correlation of thrombosis with increased platelet turnover in thrombocytosis. *Blood, 91*, 1288–1294.

66. De Stefano, V., Teofili, L., Leone, G., et al. (1997). Spontaneous erythroid colony formation as the clue to an underlying myeloproliferative disorder in patients with Budd–Chiari syndrome or portal vein thrombosis. *Semin Thromb Hemost, 23*, 411–418.

67. Rozman, C., Giralt, M., Feliu, E., et al. (1991). Life expectancy of patients with chronic nonleukemic myeloproliferative disorders. *Cancer, 67*, 2658–2663.

68. Berk, P. D., Wasswerman, L. R., Fruchtman, S. M., et al. (1995). Treatment of polycythemia vera: A summary of clinical trials conducted by the Polycythemia Vera Study Group. In L. R. Wasserman, P. D. Berk, & N. I. Berlin (Eds.), *Polycythemia vera and the myeloproliferative disorders* (pp. 166–194). Philadelphia: Saunders.

69. Gruppo Italiano Studio Policitemia. (1995). Polycythemia vera: The natural history of 1213 patients followed for 20 years. *Ann Intern Med, 123*, 654–656.

70. Rao, A. K. (2001). Acquired qualitative platelet defects. In R. W. Colman, J. Hirsh, V. J. Marder, et al. (Eds.), *Hemostasis and thrombosis: Basic principles and clinical practice* (4th ed., pp. 905–920). Philadelphia: Lippincott.

71. Small, B. M., & Bettigole, R. E. (1981). Diagnosis of myeloproliferative disease by analysis of the platelet volume distribution. *Am J Clin Pathol, 76*, 685–691.

72. Van der Lelie, J., & Von dem Borne, A. K. (1986). Platelet volume analysis for differential diagnosis of thrombocytosis. *J Clin Pathol, 39*, 129–133.

73. Maldonado, J. E., Pintado, T., & Pierre, R. V. (1974). Dysplastic platelet and circulating megakaryocytes in chronic myeloproliferative diseases. I. The platelets: Ultrastructure and peroxidase reaction. *Blood, 43*, 797–809.

74. Swart, S. S., Pearson, D., Wood, J. K., et al. (1984). Functional significance of the platelet α2-adrenoceptor: Studies

in patients with myeloproliferative disorders. *Thromb Res,* *33,* 531–541.

75. Rendu, F., Lebret, M., Nurden, A., et al. (1979). Detection of an acquired platelet storage pool disease in three patients with a myeloproliferative disorder. *Thromb Haemostas, 42,* 794–796.

76. Caranobe, C., Sie, P., Nouvel, C., et al. (1980). Platelets in myeloproliferative disorders. II. Serotonin uptake and storage: Correlation with mepacrine labelled dense bodies and with platelet density. *Scand J Haematol, 25,* 289–295.

77. Malpass, T. W., Savage, B., Hanson, S. R., et al. (1984). Correlation between prolonged bleeding time and depletion of platelet dense granule ADP in patients with myelodysplastic and myeloproliferative disorders. *J Lab Clin Med, 103,* 894–904.

78. Holme, S., & Murphy, S. (1984). Studies of the platelet density abnormality in myeloproliferative disease. *J Lab Clin Med, 103,* 373–383.

79. Murphy, S., Davis, J. L., Walsh, P. N., et al. (1978). Template bleeding time and clinical hemorrhage in myeloproliferative disease. *Arch Intern Med, 138,* 1251–1253.

80. Boneu, B., Nouvel, C., Sie, P., et al. (1980). Platelets in myeloproliferative disorders. I. A comparative evaluation with certain platelet function tests. *Scand J Haematol, 25,* 214–220.

81. Balduini, C. L., Bertolino, G., Noris, P., et al. (1991). Platelet aggregation in platelet-rich plasma and whole blood in 120 patients with myeloproliferative disorders. *Am J Clin Pathol, 95,* 82–86.

82. Waddell, C. C., Brown, J. A., & Repinecz, Y. A. (1981). Abnormal platelet function in myeloproliferative disorders. *Arch Path Lab Med, 105,* 432–435.

83. Wu, K. K. (1978). Platelet hyperaggregability and thrombosis in patients with thrombocythemia. *Ann Intern Med, 88,* 7–11.

84. Kaywin, P., McDonough, M., Insel, P. A., et al. (1978). Platelet function in essential thrombocythemia: Decreased ephinephrine responsiveness associated to α adrenergic receptors. *N Engl J Med, 299,* 505–509.

85. Pfeifer, M. A., Ward, K., Malpass, T., et al. (1984). Variations in circulating catecholamines fail to alter human platelet α-2-adrenergic receptor number or affinity for [3H]yohimbine or [3H]dihydroergocryptine. *J Clin Invest, 74,* 1063–1072.

86. Swart, S. S., Wood, J. K., & Barnett, D. B. (1985). Differential labeling of platelet α 2 adrenoceptors by 3H dihydroergocryptine and 3H yohimbine in patients with myeloproliferative disorders. *Thromb Res, 40,* 623–629.

87. Swart, S. S., Maguire, M., Wood, J. K., et al. (1985). α 2-adrenoceptor coupling to adenylate cyclase in adrenaline insensitive human platelets. *Eur J Pharmacol, 116,* 113–119.

88. Ushikubi, F., Okuma, M., Ishibashi, T., et al. (1990). Deficient elevation of the cytoplasmic calcium ion concentration by epinephrine in epinephrine-insensitive platelets of patients with myeloproliferative disorders. *Am J Hematol, 33,* 96–100.

89. Gugliotta, L., Pickering, C., Greaves, M., et al. (1983). Abnormality of platelet membrane glycoproteins in essential thrombocythemia. *Thromb Haemostas, 50,* 216.

90. Mazzucato, M., De Marco, L., De Angelis, V., et al. (1989). Platelet membrane abnormalities in myeloproliferative disorders: Decrease in glycoproteins Ib and IIb/IIIa complex is associated with deficient receptor function. *Br J Haematol, 73,* 369–374.

91. Eche, N., Sie, P., Caranobe, C., et al. (1981). Platelets in myeloproliferative disorders. III: Glycoprotein profile in relation to platelet function and platelet density. *Scand J Haematol, 26,* 123–129.

92. Clezardin, P., McGregor, J. L., Dechavanne, M., et al. (1985). Platelet membrane glycoprotein abnormalities in patients with myeloproliferative disorders and secondary thrombocytosis. *Br J Haematol, 60,* 331–344.

93. Kaplan, R., Gabbeta, J., Sun, L., et al. (2000). Combined defect in membrane expression and activation of platelet GPIIb-IIIa complex without primary sequence abnormalities in myeloproliferative disease. *Br J Haematol, 111,* 954–964.

94. Landolfi, R. (1998). Bleeding and thrombosis in myeloproliferative disorders. *Curr Opin Hematol, 5,* 327–331.

95. Mistry, R., Cahill, M., Chapman, C., et al. (1991). 125I-fibrinogen binding to platelets in myeloproliferative disease. *Thromb Haemost, 66,* 329–333.

96. Handa, M., Watanabe, K., Kawai, Y., et al. (1995). Platelet unresponsiveness to collagen: Involvement of glycoprotein Ia–IIa (α 2b 1 integrin) deficiency associated with a myeloproliferative disorder. *Thromb Haemost, 73,* 521–528.

97. Legrand, C., Bellucci, S., Disdier, M., et al. (1991). Platelet thrombospondin and glycoprotein IV abnormalities in patients with essential thrombocythemia: Effect of α-interferon treatment. *Am J Hematol, 38,* 307–313.

98. Thibert, V., Bellucci, S., Cristofari, M., et al. (1995). Increased platelet CD36 constitutes a common marker in myeloproliferative disorders. *Br J Haematol, 91,* 618–624.

99. Bolin, R. B., Okumura, T., & Jamieson, G. A. (1977). Changes in the distribution of platelet membrane glycoproteins in patients with myeloproliferative disorders. *Am J Hematol, 3,* 63–71.

100. Wehmeier, A., Tschope, D., Esser, J., et al. (1991). Circulating activated platelets in myeloproliferative disorders. *Thromb Res, 61,* 271–278.

101. Moore, A., & Nachman, R. L. (1981). Platelet Fc receptor: Increased expression in myeloproliferative disease. *J Clin Invest, 67,* 1064–1071.

102. Okuma, M., Hirata, T., Ushikubi, F., et al. (1996). Molecular characterization of a dominantly inherited bleeding disorder with impaired platelet responses to thromboxane A2. *Pol J Pharmacol, 48,* 77–82.

103. Russel, N. H., Salmon, J., Keenan, J. P., et al. (1981). Platelet adenine nucleotides and arachidonic acid metabolism in the myeloproliferative disorders. *Thromb Res, 22,* 389–397.

104. Schafer, A. I. (1982). Deficiency of platelet lipoxygenase activity in myeloproliferative disorders. *N Engl J Med, 306,* 381–386.

105. van Genderen, P. J., Michiels, J. J., & Zijlstra, F. J. (1994). Lipoxygenase deficiency in primary thrombocythemia is not a true deficiency. *Thromb Haemost, 71,* 803–804.

106. Smith, I. L., & Martin, T. J. (1982). Platelet thromboxane synthesis and release reactions in myeloproliferative disorders. *Haemostasis, 11,* 119–127.

107. Zahavi, J., Zahavi, M., Firsteter, E., et al. (1991). An abnormal pattern of multiple platelet function abnormalities and increased thromboxane generation in patients with primary thrombocytosis and thrombotic complications. *Eur J Haematol, 47,* 326–332.

108. Rocca, B., Ciabattoni, G., Tartaglione, R., et al. (1995). Increased thromboxane biosynthesis in essential thrombocythemia. *Thromb Haemost, 74,* 1225–1230.

109. Takayama, H., Okuma, M., Kanaji, K., et al. (1983). Altered arachidonate metabolism by leukocytes and platelets in myeloproliferative disorders. *Prostaglandins Leukot Med, 12,* 261–272.

110. Stenke, L., Edenius, C., Samuelsson, J., et al. (1991). Deficient lipoxin synthesis: A novel platelet dysfunction in myeloproliferative disorders with special reference to blastic crisis of chronic myelogenous leukemia. *Blood, 78,* 2989–2995.

111. Fujimoto, T., Fujimura, K., & Kuramoto, A. (1989). Abnormal $Ca^{2+}$ homeostasis in platelets with myeloproliferative disorders: Low levels of $Ca^{2+}$ influx and efflux across the plasma membrane and increased $Ca^{2+}$ accumulation into the dense tubular system. *Thromb Res, 53,* 99–108.

112. Ushikubi, F., Ishibashi, T., Narumiya, S., et al. (1992). Analysis of the defective signal transduction mechanism through the platelet thromboxane $A_2$ receptor in a patient with polycythemia vera. *Thromb Hemostas, 67,* 144–146.

113. Eigenthaler, M., Ullrich, H., Geiger, J., et al. (1993). Defective nitrovasodilator-stimulated protein phosphorylation and calcium regulation in cGMP-dependent protein kinase-deficient human platelets of chronic myelocytic leukemia. *J Biol Chem, 268,* 13526–13531.

114. Cooper, B., Schafer, A. I., Puchalsky, D., et al. (1978). Platelet resistance to protaglandin D2 in patients with myeloproliferative disorders. *Blood, 52,* 618–626.

115. Cortelazzo, S., Galli, M., Castagna, D., et al. (1988). Increased response to arachidonic acid and U-46619 and resistance to inhibitory prostaglandins in patients with chronic myeloproliferative disorders. *Thromb Haemost, 59,* 73–76.

116. Moliterno, A. R., Siebel, K. E., Sun, A. Y., et al. (1998). A novel thrombopoietin signaling defect in polycythemia vera platelets. *Stem Cells, 16,* 185–192.

117. Best, D., Pasquet, S., Littlewood, T. J., et al. (2001). Platelet activation via the collagen receptor GPVI is not altered in platelets from chronic myeloid leukaemia patients despite the presence of the constitutively phosphorylated adapter protein CrkL. *Br J Haematol, 112,* 609–615.

118. Budde, U., & van Genderen, P. J. (1997). Acquired von Willebrand disease in patients with high platelet counts. *Semin Thromb Hemost, 23,* 425–431.

119. Tefferi, A., & Gilliland, D. G. (2005). The JAK2$^{V617F}$ tyrosine kinase mutation in myeloproliferative disorders: Status report and immediate implications for disease classification and diagnosis. *Mayo Clinic Proc, 80,* 947–958.

120a. Cowan, D. H., & Graham, R. C., Jr. (1975). Structural–functional relationships in platelets in acute leukemia and related disorders. *Series Haematologica, 8,* 68.

120b. Cowan, D. H., Graham, R. C., Jr., & Baunach, D. (1975). The platelet defect in leukemia, platelet ultrastructure, adenine nucleotide metabolism and the release reaction. *J Clin Invest, 56,* 188.

120c. Stuart, J. J., & Lewis, J. C. (1982). Platelet aggregation and electron microscopic studies of platelets in preleukemia. *Arch Path Lab Med, 106,* 458.

120d. Levine, P. H., & Katayama, I. (1975). The platelet in leukemic reticuloendotheliosis. Functional and morphological evidence of a qualitative disorder. *Cancer, 36,* 1353.

120e. Nouvel, C., Caronobe, C., Sie, P., et al. (1978). Platelet volume, density and 5HT organelles (Mepacrine test) in acute leukemia. *Scand J Haematol, 21,* 421.

120f. Russell, N. H., Keenan, J. P., & Bellingham, A. J. (1981). Platelet adenine nucleotides and arachidonic acid metabolism in myeloproliferative diseases. *Thromb Res, 22,* 389.

120g. Maldonado, J. E., & Pierre, R. V. (1975). The platelets in preleukemia and myelomonocytic leukemia. Ultrastructural cytochemistry and cytogenetics. *Mayo Clinic Proc, 50,* 573.

121. Sweet, D. L., & Golomb, H. M. (1979). Correction of platelet defect after splenectomy in hairy cell leukemia. *J Am Med Assoc, 241,* 1684.

122. Rosove, M. H., Naeim, F., Harwig, S., et al. (1980). Severe platelet dysfunction in hairy cell leukemia with improvement after splenectomy. *Blood, 55,* 903–906.

123. Zuzel, M., Cawley, J. C., Paton, R. C., et al. (1979). Platelet function in hairy-cell leukaemia. *J Clin Pathol, 32,* 814–821.

124. Roussi, J. H., Houbouyan, L. L., Alterscu, R., et al. (1980). Acquired von Willebrand's syndrome associated with hairy cell leukemia. *Br J Haematol, 46,* 503–506.

125. Berndt, M. C., Kabral, A., Grimsley, P., et al. (1988). An acquired Bernard–Soulier-like platelet defect associated with juvenile myelodysplastic syndrome. *Br J Haematol, 68,* 97–101.

126. Robert, F., Mignucci, M., McCurdy, S. A., et al. (1993). Hemostatic abnormalities associated with monoclonal gammopathies. *Am J Med Sci, 306,* 359–366.

127. Furie, B., Greene, E., & Furie, B. C. (1977). Syndrome of acquired factor X deficiency and systemic amyloidosis *in vivo* studies of the metabolic fate of factor X. *N Engl J Med, 297,* 81–85.

128. Palmer, R. N., Rick, M. E., Rick, P. D., et al. (1984). Circulating heparan sulfate anticoagulant in a patient with a fatal bleeding disorder. *N Engl J Med, 310,* 1696–1699.

129. Lackner, H. (1973). Hemostatic abnormalities associated with dysproteinemias. *Semin Hematol, 10,* 125–133.

130. DiMinno, G., Coraggio, F., Cerbone, A. M., et al. (1986). A myeloma paraprotein with specificity for platelet glycoprotein IIIa in a patient with a fatal bleeding disorder. *J Clin Invest, 77,* 157–164.

131. Mohri, H., Noguchi, T., Kodoma, F., et al. (1987). Acquired von Willebrand disease due to inhibitor of human myeloma

protein specific for von Willebrand factor. *Am J Pathol, 97*, 663–668.

132. Vigliano, E. M., & Horowitz, H. I. (1967). Bleeding syndrome in a patient with IgA myeloma: Interaction of protein and connective tissue. *Blood, 29*, 823–836.

133. Bovill, E. G., Ershler, W. B., Golden, E. A., et al. (1986). A human myeloma produced monoclonal protein directed against the active subpopulation of von Willebrand factor. *Am J Clin Pathol, 85*, 115–123.

134. Bovill, E. G., Ershler, W. B., Golden, E. A., et al. (1986). DDAVP in acquired von Willebrand syndrome associated with multiple myeloma. *Am J Hematol, 122*, 421–423.

135. Kumar, S., Pruthi, R. K., & Nichols, W. L. (2002). Acquired von Willebrand disease. *Mayo Clinic Proc, 77*, 181–187.

136. Federici, A. B., Rand, J. H., Bucciarelli, P., et al. (2000). Acquired von Willebrand syndrome: Data from an international registry. *Thromb Haemost, 84*, 345–349.

137. Budde, U., Scharf, R. E., Franke, P., et al. (1993). Elevated platelet count as a cause of abnormal von Willebrand factor multimer distribution in plasma. *Blood, 82*, 1749–1757.

138. van Genderen, P. J., Budde, U., Michiels, J. J., et al. (1996). The reduction of large von Willebrand factor multimers in plasma in essential thrombocythaemia is related to the platelet count. *Br J Haematol, 93*, 962–965.

139. Castaman, G., Lattuada, A., Mannucci, P. M., et al. (1995). Characterization of two cases of acquired transitory von Willebrand syndrome with ciprofloxacin: Evidence for heightened proteolysis of von Willebrand factor. *Am J Hematol, 49*, 83–86.

140. Kreuz, W., Linde, R., Funk, M., et al. (1990). Induction of von Willebrand disease type I by valproic acid. *Lancet, 335*, 1350–1351.

141. Conrad, M. E., & Latour, L. F. (1992). Acquired von Willebrand's disease, IgE polyclonal gammopathy and griseofulvin therapy. *Am J Hematol, 41*, 143.

142. Lazarchick, J., & Conroy, J. M. (1995). The effect of 6% hydroxyethyl starch and desmopressin infusion on von Willebrand factor: Ristocetin cofactor activity. *Ann Clin Lab Sci, 25*, 306–309.

143. Dalrymple-Hay, M., Aitchison, R., Collins, P., et al. (1992). Hydroxyethyl starch induced acquired von Willebrand's disease. *Clin Lab Haematol, 14*, 209–211.

144. Jonge Poerink-Stockschlader, A. B., Dekker, I., Risseeuw-Appel, I. M., et al. (1996). Acquired Von Willebrand disease in children with a Wilms' tumor. *Med Pediatr Oncol, 26*, 238–243.

145. Gill, J. C., Wilson, A. D., Endres-Brooks, J., et al. (1986). Loss of the largest von Willebrand factor multimers from plasma of patients with congenital cardiac defects. *Blood, 67*, 758–761.

146. Vincentelli, A., Susen, S., Le Tourneau, T., et al. (2003). Acquired von Willebrand syndrome in aortic stenosis. *N Engl J Med, 349*, 343–349.

147. Mohri, H., Motomura, S., Kanamori, H., et al. (1998). Clinical significance of inhibitors in acquired von Willebrand syndrome. *Blood, 91*, 3623–3629.

148. Sadler, J. E. (2003). Aortic stenosis, von Willebrand factor, and bleeding. *N Engl J Med, 349*, 323–325.

149. Levesque, H., Borg, J. Y., Cailleux, N., et al. (1993). Acquired von Willebrand's syndrome associated with decrease of plasminogen activator and its inhibitor during hypothyroidism. *Eur J Med, 2*, 287–288.

150. Friederich, P. W., Wever, P. C., Briet, E., et al. (2001). Successful treatment with recombinant factor VIIa of therapy-resistant severe bleeding in a patient with acquired von Willebrand disease. *Am J Hematol, 66*, 292–294.

151. Federici, A. B., Stabile, F., Castaman, G., et al. (1998). Treatment of acquired von Willebrand syndrome in patients with monoclonal gammopathy of uncertain significance: Comparison of three different therapeutic approaches. *Blood, 92*, 2707–2711.

152. Agarwal, N., Klix, M. M., & Burns, C. P. (2004). Successful management with intravenous immunoglobulins of acquired von Willebrand disease associated with monoclonal gammopathy of undetermined significance. *Ann Intern Med, 141*, 83–84.

153. Bovill, E. G., Ershler, W. B., Golden, E. A., et al. (1986). A human myeloma-produced monoclonal protein directed against the active subpopulation of von Willebrand factor. *Am J Clin Pathol, 85*, 115–123.

154. Uehlinger J, Button G. R., McCarthy J, et al. (1991). Immunoadsorption for coagulation factor inhibitors. *Transfusion, 31*, 265–269.

155. Budde, U., Schaefer, G., Mueller, N., et al. (1984). Acquired von Willebrand's disease in the myeloproliferative syndrome. *Blood, 64*, 981–985.

156. Budde, U., Dent, J. A., Berkowitz, S. D., et al. (1986). Subunit composition of plasma von Willebrand factor in patients with myeloproliferative syndrome. *Blood, 68*, 1213–1217.

157. van Genderen, P. J., Prins, F. J., Lucas, I. S., et al. (1997). Decreased half-life time of plasma von Willebrand factor collagen binding activity in essential thrombocythaemia: Normalization after cytoreduction of the increased platelet count. *Br J Haematol, 99*, 832–836.

158. Zahavi, J., & Marder, V. J. (1974). Acquired storage pool disease of platelets associated with circulating anti-platelet antibodies. *Am J Med, 56*, 883–889.

159. Weiss, H. J., Rosove, M. H., Lages, B. A., et al. (1980). Acquired storage pool deficiency with increased platelet-associated IgG. Report of five cases. *Am J Med, 69*, 711–717.

159a. Rendu, F., Lebret, M., Nurden, A., et al. (1979). Detection of an acquired platelet storage pool disease in three patients with a myeloproliferative disorder. *Thromb Haemost, 42*, 794.

159b. Russel, N. H., Salmon, J., Keenan, J. P., et al. (1981). Platelet adenine nucleotides and arachidonic acid metabolism in the myeloproliferative disorders. *Thromb Res, 22*, 389.

159c. Cowan, D. H., Graham, R. C., Jr., & Baunach, D. (1975). The platelet defect in leukemia, platelet ultrastructure, adenine nucleotide metabolism and the release reaction. *J Clin Invest, 56*, 188.

159d. Pareti, F. I., Gugliotta, L., Mannucci, L., et al. (1982). Biochemical and metabolic aspects of platelet dysfunction in myeloproliferative disorders. *Thromb Haemost, 47*, 84.

159e. Pareti, F. I., Capitanio, A., & Mannucci, P. M. (1976). Acquired storage pool disease in platelets during disseminated intravascular coagulation. *Blood, 48,* 511.

159f. Pareti, F. I., Capitanio, A., Mannucci, C., et al. (1980). Acquired dysfunction due to circulation of "exhausted" platelets. *Am J Med, 69,* 235.

159g. Khurana, M. S., Lian, E. C. Y., & Harkness, D. R. (1980). Storage pool disease of platelets associated with multiple congenital cavernous hemangiomas. *J Am Med Assoc, 244,* 169.

159h. Nenci, G. G., Gresele, P., Agnetti, F., et al. (1981). Intrinsically defective or exhausted platelets in hairy cell leukemia? *Thromb Haemost, 46,* 572.

160. Harker, L. A., Malpass, T. W., Branson, H. E., et al. (1980). Mechanism of abnormal bleeding in patients undergoing cardiopulmonary bypass: Acquired transient platelet dysfunction associated with selective α granule release. *Blood, 55,* 824–834.

161. Beurling-Harbury, C., & Galvan, C. A. (1978). Acquired decreased in platelet secretory ADP associated with increased postoperative bleeding in post-cardiopulmonary bypass patients and in patients with severe valvular heart disease. *Blood, 52,* 13–23.

162. Savage, B., Malpass, T. W., Stratton, J. R., et al. (1983). Platelet adenine nucleotide levels in patients with Dacron vascular prostheses. *Thromb Res, 32,* 365–372.

163. Rao, A. K., Niewiarowski, S., & Murphy, S. (1981). Acquired granular pool defect in stored platelets. *Blood, 57,* 203–208.

164. Chang, M., Nakagawa, P. A., Williams, S. A., et al. (2003). Immune thrombocytopenic purpura (ITP) plasma and purified ITP monoclonal autoantibodies inhibit megakaryocytopoiesis *in vitro. Blood, 102,* 887–895.

165. Takahashi, R., Sekine, N., & Nakatake, T. (1999). Influence of monoclonal antiplatelet glycoprotein antibodies on *in vitro* human megakaryocyte colony formation and proplatelet formation. *Blood, 93,* 1951–1958.

166. Harker, L. A., & Slichter, S. J. (1972). The bleeding time as a screening test for evaluation of platelet function. *N Engl J Med, 287,* 155–159.

167. Clancy, R., Jenkins, E., & Firkin, B. (1972). Qualitative platelet abnormalities in idiopathic thrombocytopenic purpura. *N Engl J Med, 286,* 622–626.

167a. Zahavi, J., & Marder, V. J. (1974). Acquired storage pool disease of platelets associated with circulating anti-platelet antibodies. *Am J Med, 56,* 883.

167b. Weiss, H. J., Rosove, M. H., Lages, B. A., et al. (1980). Acquired storage pool deficiency with increased platelet-associated IgG. Report of five cases. *Am J Med, 69,* 711.

167c. Karpatkin, S., & Lackner, H. L. (1975). Association of antiplatelet antibody with functional platelet disorders: Autoimmune thrombocytopenic purpura, systemic lupus erythematosus and thrombopathia. *Am J Med, 59,* 599–604.

167d. Lackner, H. L., & Karpatkin, S. (1975). On the "easy bruising" syndrome with normal platelet count. A study of 75 patients. *Ann Intern Med, 83,* 190.

167e. Heyns, A., Fraser, J., & Retief, F. P. (1978). Platelet aggregation in chronic idiopathic thrombocytopenic purpura. *J Clin Pathol, 31,* 1239.

167f. Stuart, M. J., Kelton, J. G., & Allen, J. B. (1981). Abnormal platelet function and arachidonate metabolism in chronic idiopathic thrombocytopenic purpura. *Blood, 58,* 326.

167g. Regan, M. G., Lackner, H. L., & Karpatkin, S. (1974). Platelet function and coagulation profile in lupus erythematosus. *Ann Intern Med, 81,* 462.

167h. Dorsch, C. A., & Meyerhoff, J. (1982). Mechanisms of abnormal platelet aggregation in systemic lupus erythematosus. *Arthritis Rheum, 25,* 966.

167i. Kurata, Y., Nishioeda, Y., Tusubakio, T., et al. (1980). Thrombocytopenia in Graves' disease: Effect of T3 on platelet kinetics. *Acta Haematol, 63,* 185.

168. Woods, V. L., Jr., Kurata, Y., Montgomery, R. R., et al. (1984). Autoantibodies against platelet glycoprotein Ib in patients with chronic immune thrombocytopenic purpura. *Blood, 64,* 156–160.

169. Szatkowski, N. S., Kunicki, T. J., & Aster, R. H. (1986). Identification of glycoprotein Ib as a target for autoantibody in idiopathic (autoimmune) thrombocytopenic purpura. *Blood, 67,* 310–315.

170. Woods, V. L., Jr., Oh, E. H., Mason, D., et al. (1984). Autoantibodies against the platelet glycoprotein IIb/IIIa complex in patients with chronic ITP. *Blood, 63,* 368–375.

171. Berchtold, P., McMillan, R., Tani, P., et al. (1989). Autoantibodies against platelet membrane glycoproteins in children with acute and chronic immune thrombocytopenic purpura. *Blood, 74,* 1600–1602.

172. Deckmyn, H., Zhang, J., Van Houtte, E., et al. (1994). Production and nucleotide sequence of an inhibitory human IgM autoantibody directed against platelet glycoprotein Ia/IIa. *Blood, 84,* 1968–1974.

173. Dromigny, A., Triadou, P., Lesavre, P., et al. (1996). Lack of platelet response to collagen associated with autoantibodies against glycoprotein (GP) Ia/IIa and Ib/IX leading to the discovery of SLE. *Hematol Cell Ther, 38,* 355–357.

174. Rao, A. K., Kowalska, M. A., Karczewski, J., et al. (1989). Impaired platelet response to collagen and human antibody against an 88-kilodalton platelet membrane glycoprotein. *Thromb Haemost, 62,* 506.

175. Sugiyama, T., Okuma, M., Ushikubi, F., et al. (1987). A novel platelet aggregating factor found in a patient with defective collagen-induced platelet aggregation and autoimmune thrombocytopenia. *Blood, 69,* 1712–1720.

176. Boylan, B., Chen, H., Rathore, V., et al. (2004). Anti-GPVI-associated ITP: An acquired platelet disorder caused by autoantibody-mediated clearance of the GPVI/FcRgamma-chain complex from the human platelet surface. *Blood, 104,* 1350–1355.

177. van Vliet, H. H., Kappers-Klunne, M. C., van der Hel, J. W., et al. (1987). Antibodies against glycosphingolipids in sera of patients with idiopathic thrombocytopenic purpura. *Br J Haematol, 67,* 103–108.

178. Koerner, T. A., Weinfeld, H. M., Bullard, L. S., et al. (1989). Antibodies against platelet glycosphingolipids: Detection in serum by quantitative HPTLC–autoradiography and asso-

ciation with autoimmune and alloimmune processes. *Blood, 74,* 274–284.

179. Sugiyama, T., Okuma, M., Ushikubi, F., et al. (1987). A novel platelet aggregating factor found in a patient with defective collagen-induced platelet aggregation and autoimmune thrombocytopenia. *Blood, 69,* 1712–1720.

180. McMillan, R., Tani, P., Millard, F., et al. (1987). Platelet-associated and plasma anti-glycoprotein autoantibodies in chronic ITP. *Blood, 70,* 1040–1045.

181. Kiefel, V., Santoso, S., Kaufmann, E., et al. (1991). Auto-antibodies against platelet glycoprotein Ib/IX: A frequent finding in autoimmune thrombocytopenic purpura. *Br J Haematol, 79,* 256–262.

182. Kiefel, V., Freitag, E., Kroll, H., et al. (1996). Platelet auto-antibodies (IgG, IgM, IgA) against glycoproteins IIb/IIIa and Ib/IX in patients with thrombocytopenia. *Ann Hematol, 72,* 280–285.

183. Stuart, M. J., Kelton, J. G., & Allen, J. B. (1981). Abnormal platelet function and arachidonate metabolism in chronic idiopathic thrombocytopenic purpura. *Blood, 58,* 326–329.

184. Meyerhoff, J., & Dorsch, C. A. (1981). Decreased platelet serotonin levels in systemic lupus erythematosus. *Arthritis Rheum, 24,* 1495–1500.

185. Thomas, D. P., Ream, V. J., & Stuart, R. K. (1967). Platelet aggregation in patients with Laennec's cirrhosis of the liver. *N Engl J Med, 276,* 1344–1348.

186. Thomas, D. P. (1967). Abnormalities of platelet aggregation in patients with alcoholic cirrhosis. *Ann N Y Acad Sci, 201,* 243.

187. Ballard, H. S., & Marcus, A. J. (1976). Platelet aggregation in portal cirrhosis. *Arch Intern Med, 136,* 316–319.

188. Ingeberg, S., Jacobsen, P., Fisher, E., et al. (1985). Platelet aggregation and release of ATP in patients with hepatic cirrhosis. *Scand J Gastroenterol, 20,* 285–288.

189. Laffi, G., Cominelli, F., Ruggiero, M., et al. (1987). Molecular mechanism underlying platelet responsiveness in liver cirrhosis. *Fed Eur Biochem Soc, 220,* 217–219.

190. Hillbom, M., Muuronen, A., & Neiman, J. (1987). Liver disease and platelet function in alcoholics. *Br Med J (Clin Res Ed), 295,* 581.

191. Stein, S. F., & Harker, L. A. (1982). Kinetic and functional studies of platelets, fibrinogen and plasminogen in patients with hepatic cirrhosis. *J Lab Clin Med, 99,* 217–230.

192. Laffi, G., Cominelli, F., Ruggiero, M., et al. (1988). Altered platelet function in cirrhosis of the liver: Impairment of inositol lipid and arachidonic acid metabolism in response to agonists. *Hepatology, 8,* 1620–1626.

193. Desai, K., Mistry, P., Bagget, C., et al. (1989). Inhibition of platelet aggregation by abnormal high density lipoprotein particles in plasma from patients with hepatic cirrhosis. *Lancet, 1,* 693–695.

194. Cowan, D. H. (1980). Effect of alcoholism on hemostasis. *Semin Hematol, 17,* 137–147.

195. Hillbom, M., & Neiman, J. (1988). Platelet thromboxane formation capacity after ethanol withdrawal in chronic alcoholics. *Haemostasis, 18,* 170–178.

196. Mannucci, P. M., Vicente, V., Vianello, L., et al. (1986). Controlled trial of desmopressin in liver cirrhosis and other conditions associated with a prolonged bleeding time. *Blood, 67,* 1148–1153.

197. Lopez, P., Otaso, J. C., Alvarez, D., et al. (1997). Hemostatic and hemodynamic effects of vasopressin analogue DDAVP in patients with cirrhosis. *Acta Gastroenterol Latinoam, 27,* 59–62.

198. Cattaneo, M., Tenconi, P. M., Alberca, I., et al. (1990). Subcutaneous desmopressin (DDAVP) shortens the prolonged bleeding time in patients with liver cirrhosis. *Thromb Haemost, 64,* 358–360.

199. Ingeberg, S., & Stoffersen, E. (1979). Platelet dysfunction in patients with vitamin B12 deficiency. *Acta Haematol, 61,* 75–79.

200. Laosombat, V., Wongchanchailert, M., Sattayasevana, B., et al. (2001). Acquired platelet dysfunction with eosinophilia in children in the south of Thailand. *Platelets, 12,* 5–14.

201. Lim, S. H., Tan, C. E., Agasthian, T., et al. (1989). Acquired platelet dysfunction with eosinophilia: Review of seven adult cases. *J Clin Pathol, 42,* 950–952.

202. Poon, M. C., Ng, S. C., & Coppes, M. J. (1995). Acquired platelet dysfunction with eosinophilia in white children. *J Pediatr, 126,* 959–961.

203. Solinger, A., Bernstein, I. L., & Glueck, H. I. (1973). The effect of epinephrine on platelet aggregation in normal and atopic patients. *J Allergy Clin Immunol, 51,* 29–34.

204. Szczeklik, A., Milner, P. C., Birch, J., et al. (1986). Prolonged bleeding time, reduced platelet aggregation, altered PAF-acether sensitivity and increased platelet mass are a trait of asthma and hay fever. *Thromb Haemost, 56,* 283–287.

205. Stoff, J. S., Stomerman, M., Steer, M., et al. (1980). A defect in platelet aggregation in Bartter's syndrome. *Am J Med, 68,* 171–180.

206. Smith, J. B., & Willis, A. L. (1971). Aspirin selectively inhibits prostaglandin production in human platelets. *Nat New Biol, 231,* 235–237.

207. Roth, G. J., & Majerus, P. W. (1975). The mechanism of the effect of aspirin on human platelets: I. Acetylation of a particular fraction protein. *J Clin Invest, 56,* 624–632.

208. Pedersen, A. K., & Fitzgerald, G. A. (1984). Dose-related kinetics of aspirin. Presystemic acetylation of platelet cyclooxygenase. *N Engl J Med, 311,* 1206–1211.

209. Patrigrani, P., Filabozzi, P., & Patrono, C. (1982). Selective cumulative inhibition of platelet thromboxane production by low-dose aspirin in healthy subjects. *J Clin Invest, 69,* 1366–1372.

210. Weksler, B. B., Pett, S. B., Alonso, D., et al. (1983). Differential inhibition of aspirin of vascular prostaglandin synthesis in atherosclerotic patients. *N Engl J Med, 308,* 800–805.

211. Antiplatelet Trialists' Collaboration. (2002). Collaborative meta-analysis of randomised trials of antiplatelet therapy for prevention of death, myocardial infarction, and stroke in high risk patients. *Br Med J, 324,* 71–86.

212. Patrono, C., Coller, B., Fitzgerald, G. A., et al. (2004). Platelet-active drugs: The relationships among dose, effectiveness, and side effects: The Seventh ACCP Conference

on Antithrombotic and Thrombolytic Therapy. *Chest, 126,* 234S–264S.

213. O'Grady, J., & Moncada, S. (1978). Aspirin: A paradoxical effect on bleeding time. *Lancet, 2,* 780.

214. Rajah, S. M., Penny, S. M., & Kester, R. (1978). Aspirin and bleeding time. *Lancet, 2,* 1104.

215. Mielke, C. H. (1982). Aspirin prolongation of the template bleeding time: Influence of venostasis and direction of incision. *Blood, 60,* 1139–1142.

216. Hayward, C. P., Harrison, P., Cattaneo, M., et al. (2006). The closure times in the evaluation of platelet disorders and function: A report from the Working Group on the PFA-100, Platelet Physiology Subcommittee of the Scientific and Standardization Committee of the International Society on Thrombosis and Haemostasis. *J Thromb Haemost, 4,* 312–319.

217. Serebruany, V. L., Steinhubl, S. R., Berger, P. B., et al. (2005). Analysis of risk of bleeding complications after different doses of aspirin in 192, 036 patients enrolled in 31 randomized controlled trials. *Am J Cardiol, 95,* 1218–1222.

218. Rumack, C. M., Guggenheim, M. A., Rumack, B. H., et al. (1981). Neonatal intracranial hemorrhage and maternal use of aspirin. *Obstet Gynecol, 58,* 52S–56S.

219. Bleyer, W. A., & Breckenridge, R. T. (1970). Studies on the detection of adverse drug reactions in the newborn: II. The effects of prenatal aspirin on newborn hemostasis. *J Am Med Assoc, 213,* 2049–2053.

220. Stuart, M. J., Gross, S. J., Elrad, H., et al. (1982). Effects of acetylsalicylic-acid ingestion on maternal and neonatal hemostasis. *N Engl J Med, 307,* 909–912.

221. Steering Committee of the Physicians' Health Study Research Group. (1989). Final report on the aspirin component of the ongoing Physicians' Health Study. *N Engl J Med, 321,* 129–135.

222. Merritt, J. C., & Bhatt, D. L. (2002). The efficacy and safety of perioperative antiplatelet therapy. *J Thromb Thrombolysis, 13,* 97–103.

223. Sethi, G. K., Copeland, J. G., Goldman, S., et al. (1990). Implications of preoperative administration of aspirin in patients undergoing coronary artery bypass grafting. Department of Veterans Affairs Cooperative Study on Antiplatelet Therapy. *J Am Coll Cardiol, 15,* 15–20.

224. Mangano, D. T. (2002). Aspirin and mortality from coronary bypass surgery. *N Engl J Med, 347,* 1309–1317.

225. Roderick, P. J., Wilkes, H. C., & Meade, T. W. (1993). The gastrointestinal toxicity of aspirin: An overview of randomised controlled trials. *Br J Clin Pharmacol, 35,* 219–226.

226. Diener, H. C., Cunha, L., Forbes, C., et al. (1996). European Stroke Prevention Study. 2. Dipyridamole and acetylsalicylic acid in the secondary prevention of stroke. *J Neurol Sci, 143,* 1–13.

227. The Dutch TIA Trial Study Group. (1991). A comparison of two doses of aspirin (30 mg vs 823 mg a day) in patients after a transient ischemic attack or minor ischemic stroke. *N Engl J Med, 325,* 1261–1266.

228. Garcia Rodriguez, L. A., Cattaruzzi, C., Troncon, M. G., et al. (1998). Risk of hospitalization for upper gastrointestinal tract bleeding associated with ketorolac, other nonsteroidal anti-inflammatory drugs, calcium antagonists, and other antihypertensive drugs. *Arch Intern Med, 158,* 33–39.

229. Henry, D., Lim, L. L., Garcia Rodriguez, L. A., et al. (1996). Variability in risk of gastrointestinal complications with individual non-steroidal anti-inflammatory drugs: Results of a collaborative meta-analysis. *Br Med J, 312,* 1563–1566.

230. Stein, P. D., Dalen, J. E., Goldman, S., et al. (2001). Antithrombotic therapy in patients with saphenous vein and internal mammary artery bypass grafts. *Chest, 119,* 278S–282S.

231. Topol, E. J. (2002). Aspirin with bypass surgery — From taboo to new standard of care. *N Engl J Med, 347,* 1359–1360.

232. Mannucci, P. M. (1997). Desmopressin (DDAVP) in the treatment of bleeding disorders: The first 20 years. *Blood, 90,* 2515–2521.

233. Simon, L. S., & Mills, J. A. (1980). Drug therapy: Nonsteroidal antiinflammatory drugs (first of two parts). *N Engl J Med, 302,* 1179–1185.

234. Zucker, M. B., & Peterson, J. (1970). Effect of acetylsalicylic acid, other nonsteroidal anti-inflammatory agents and dipyridamole on human blood platelets. *J Lab Clin Med, 76,* 66–75.

235. O'Brien, J. R., Finch, W., & Clark, E. (1970). A comparison of an effect of different anti-inflammatory drugs on human platelets. *J Clin Pathol, 23,* 522–525.

236. O'Brien, J. R. (1968). Effect of anti-inflammatory agents on platelets. *Lancet, 1,* 894–895.

237. Ali, M., & McDonald, J. W. D. (1978). Reversible and irreversible inhibition of platelet cyclooxygenase and serotonin release by nonsteroidal anti-inflammatory drugs. *Thromb Res, 13,* 1057.

238. Nishizawa, E. E., & Wynalda, D. J. (1981). Inhibitory effect of ibuprofen (Motrin) on platelet function. *Thromb Res, 21,* 347–356.

239. Green, D., Given, K. M., Ts'ao, C. H., et al. (1977). The effect of a new non-steroidal anti-inflammatory agent, sulindac, on platelet function. *Thromb Res, 10,* 283–289.

240. Ali, M., & McDonald, J. W. D. (1977). Effects of sulfinpyrazone on platelet prostaglandin synthesis and platelet release of serotonin. *J Lab Clin Med, 89,* 868–875.

241. Cerskus, A. L., Ali, M., & McDonald, J. W. D. (1980). Thromboxane $B_2$ and 6-keto-prostaglandin $F_1$ synthesis during infusion of collagen and arachidonic acid in rabbits: Inhibition by aspirin and sulfinpyrazone. *Thromb Res, 18,* 693–705.

242. Simon, L. S., & Mills, J. A. (1980). Nonsteroidal antiinflammatory drugs (second of two parts). *N Engl J Med, 302,* 1237–1243.

243. McQueen, E. G., & Facoory, B. (1986). Nonsteroidal anti-inflammatory drugs and platelet function. *N Z Med J, 99,* 358–360.

244. Thomas, P., Hepburn, B., Kim, H. C., et al. (1982). Nonsteroidal anti-inflammatory drugs in the treatment of hemophilic arthropathy. *Am J Hematol, 12,* 131–137.

245. Ragni, M. V., Miller, B. J., Whalen, R., et al. (1992). Bleeding tendency, platelet function, and pharmacokinetics of ibuprofen and zidovudine in HIV(+) hemophilic men. *Am J Hematol, 40,* 176–182.

246. Catella-Lawson, F., Reilly, M. P., Kapoor, S. C., et al. (2001). Cyclooxygenase inhibitors and the antiplatelet effects of aspirin. *N Engl J Med, 345,* 1809–1817.

247. Feldman, M., & McMahon, A. T. (2000). Do cyclooxygenase-2 inhibitors provide benefits similar to those of traditional nonsteroidal anti-inflammatory drugs, with less gastrointestinal toxicity? *Ann Intern Med, 132,* 134–143.

248. Lages, B., & Weiss, H. J. (1989). Inhibition of human platelet function *in vitro* and *ex vivo* by acetaminophen. *Thromb Res, 53,* 603–613.

249. Sharis, P. J., Cannon, C. P., & Loscalzo, J. (1998). The antiplatelet effects of ticlopidine and clopidogrel. *Ann Intern Med, 129,* 394–405.

250. Coukell, A. J., & Markham, A. (1997). Clopidogrel. *Drugs, 54,* 745–750.

251. Mills, D. C., Puri, R., Hu, C. J., et al. (1992). Clopidogrel inhibits the binding of ADP analogues to the receptor mediating inhibition of platelet adenylate cyclase. *Arterioscler Thromb, 12,* 430–436.

252. Wilhite, D. B., Comerota, A. J., Schmieder, F. A., et al. (2003). Managing PAD with multiple platelet inhibitors: The effect of combination therapy on bleeding time. *J Vasc Surg, 38,* 710–713.

253. Bennett, C. L., Weinberg, P. D., Rozenberg-Ben-Dror, K., et al. (1998). Thrombotic thrombocytopenic purpura associated with ticlopidine. A review of 60 cases. *Ann Intern Med, 128,* 541–544.

254. Bennett, C., Kiss, J., Weinberg, P., et al. (1998). Thrombotic thrombocytopenic purpura after stenting and ticlopidine. *Lancet, 352,* 1036–1037.

255. Bennett, C. L., Connors, J. M., Carwile, J. M., et al. (2000). Thrombotic thrombocytopenic purpura associated with clopidogrel. *N Engl J Med, 342,* 1773–1777.

256. Aster, R. H., Curtis, B. R., & Bougie, D. W. (2004). Thrombocytopenia resulting from sensitivity to GPIIb-IIIa inhibitors. *Semin Thromb Hemost, 30,* 569–577.

257. Szczeklik, A., Gryglewski, R. J., Nizankowski, R., et al. (1978). Circulatory and anti-platelet effects of intravenous prostacyclin in healthy men. *Pharmacol Res Commun, 10,* 545–556.

258. Fitzgerald, G. A., Friedman, L. A., Miyamori, I., et al. (1979). A double blind placebo controlled crossover study of prostacylin in man. *Life Sci, 25,* 665–672.

259. Fitzgerald, G. A. (1987). Dipyridamole. *N Engl J Med, 316,* 1247–1257.

260. Kim, J. S., Lee, K. S., Kim, Y. I., et al. (2004). A randomized crossover comparative study of aspirin, cilostazol and clopidogrel in normal controls: Analysis with quantitative bleeding time and platelet aggregation test. *J Clin Neurosci, 11,* 600–602.

261. Dawson, D. L., Cutler, B. S., Meissner, M. H., et al. (1998). Cilostazol has beneficial effects in treatment of intermittent claudication: Results from a multicenter, randomized, prospective, double-blind trial. *Circulation, 98,* 678–686.

262. Ardlie, N. G., Glew, G., Schultz, B. G., et al. (1968). Inhibition and reversal of platelet aggregation by methylxanthines. *Thrombos Diath Haemorrh, 18,* 670.

263. Wallis, R. M., Corbin, J. D., Francis, S. H., et al. (1999). Tissue distribution of phosphodiesterase families and the effects of sildenafil on tissue cyclic nucleotides, platelet function, and the contractile responses of trabeculae carneae and aortic rings *in vitro. Am J Cardiol, 83,* 3C–12C.

264. Halcox, J. P., Nour, K. R., Zalos, G., et al. (2002). The effect of sildenafil on human vascular function, platelet activation, and myocardial ischemia. *J Am Coll Cardiol, 40,* 1232–1240.

265. Cheung, P. Y., Salas, E., Schulz, R., et al. (1997). Nitric oxide and platelet function: Implications for neonatology. *Semin Perinatol, 21,* 409–417.

266. Freedman, J. E., Loscalzo, J., Barnard, M. R., et al. (1997). Nitric oxide released from activated platelets inhibits platelet recruitment. *J Clin Invest, 100,* 350–356.

267. Samama, C. M., Diaby, M., Fellahi, J. L., et al. (1995). Inhibition of platelet aggregation by inhaled nitric oxide in patients with acute respiratory distress syndrome. *Anesthesiology, 83,* 56–65.

268. Brown, C. H., III, Natelson, E. A., Bradshaw, M. W., et al. (1974). The hemostatic defect produced by carbenicillin. *N Engl J Med, 291,* 265–270.

269. Haburchak, D. R., Head, D. R., & Everett, E. D. (1977). Postoperative hemorrhage associated with carbenicillin administration — Report of two cases and review of the literature. *Am J Surg, 134,* 630–634.

270. Brown, C. H., III, Bradshaw, M. W., Natelson, E. A., et al. (1976). Defective platelet function following the administration of penicillin compounds. *Blood, 47,* 949–956.

271. Andrassy, K., Ritz, E., Hasper, B., et al. (1976). Penicillin-induced coagulation disorder. *Lancet, 2,* 1039–1041.

272. Cazenave, J. P., Guccione, M. A., Packham, M. A., et al. (1977). Effects of cephalothin and penicillin G on platelet function *in vitro. Br J Haematol, 35,* 135–152.

273. Fass, R. J., Copelan, E. A., Brandt, J. T., et al. (1987). Platelet mediated bleeding caused by broad spectrum penicillins. *J Infect Dis, 155,* 1242–1248.

274. Brown, C. H., III, Natelson, E. A., & Bradshaw, M. W. (1975). A study of the effects of ticarcillin on blood coagulation and platelet function. *Antimicrob Agents Chemother, 7,* 652–657.

275. Pillgram-Larsen, J., Wisloff, F., Jorgensen, J. J., et al. (1985). Effect of high-dose ampicillin and cloxacillin on bleeding time and bleeding in open-heart surgery. *Scand J Thorac Cardiovasc Surg, 19,* 45–48.

276. Alexander, D. P., Russo, M. E., Fohrman, D. E., et al. (1983). Nafcillin induced platelet dysfunction and bleeding. *Antimicrob Agents Chemother, 23,* 59–62.

277. Sattler, F. R., Weitekamp, M. R., & Ballard, J. O. (1986). Potential for bleeding with the new β-lactam antibiotics. *Ann Intern Med, 105,* 924–931.

278. Johnson, G. J. (1993). Platelets, penicillins, and purpura: What does it all mean? *J Lab Clin Med, 121,* 531–533.

279. Natelson, E. A., Brown, C. H., III, Bradshaw, M. W., et al. (1976). Influence of cephalosporin antibiotics on blood coagulation and platelet function. *Antimicrob Agents Chemother, 9,* 91–93.

280. Weitekamp, M. R., & Aber, R. C. (1983). Prolonged bleeding times and bleeding diathesis associated with moxalactam administration. *J Am Med Assoc, 249,* 69–71.

281. Sloand, E. M., Klein, H. G., Pastakia, K. B., et al. (1992). Effect of albumin on the inhibition of platelet aggregation by β-lactam antibiotics. *Blood, 79,* 2022–2027.

282. Shattil, S. J., Bennett, J. S., McDonaugh, M., et al. (1980). Carbenicillin and penicillin G inhibit platelet function *in vitro* by impairing the interaction of agonists with the platelet surface. *J Clin Invest, 65,* 329–337.

283. Burroughs, S. F., & Johnson, G. J. (1990). β-Lactam antibiotic-induced platelet dysfunction: Evidence for irreversible inhibition of platelet activation *in vitro* and *in vivo* after prolonged exposure to penicillin. *Blood, 75,* 1473–1480.

284. Burroughs, S. F., & Johnson, G. J. (1993). β-Lactam antibiotics inhibit agonist-stimulated platelet calcium influx. *Thromb Haemost, 69,* 503–508.

285. Pastakia, K. B., Terle, D., & Produoz, K. N. (1993). Penicillin-induced dysfunction of platelet membrane glycoproteins. *J Lab Clin Med, 121,* 546–554.

286. Rossi, E. C., & Levin, N. W. (1973). Inhibition of primary ADP-induced platelet aggregation in normal subjects after administration of nitrofurtoin (Furadantin). *J Clin Invest, 52,* 2457–2467.

287. Cummins, D., Faint, R., Yardumian, D. A., et al. (1990). The *in-vitro* and *ex-vivo* effects of chloroquine sulphate on platelet function: Implications for malaria prophylaxis in patients with impaired haemostasis. *J Trop Med Hyg, 93,* 112–115.

288. Ishikawa, S., Manabe, S., & Wada, O. (1986). Miconazole inhibition of platelet aggregation by inhibiting cyclooxygenase. *Biochem Pharmacol, 35,* 1787–1792.

289. Weksler, B. B., Gillik, M., & Pink, J. (1977). Effect of propranolol on platelet function. *Blood, 49,* 185–196.

290. Petrikova, M., Jancinova, V., Nosal, R., et al. (2002). Antiplatelet activity of carvedilol in comparison to propranolol. *Platelets, 13,* 479–485.

291. Dash, D., & Rao, K. (1995). Effect of propranolol on platelet signal transduction. *Biochem J, 309(Pt 1),* 99–104.

292. Leon, R., Tiarks, C. Y., & Pechet, L. (1978). Some observations on the *in vivo* effect of propranolol on platelet aggregation and release. *Am J Hematol, 5,* 117–121.

293. Frishman, W. H., Weksler, B., Christodoulo, J. B., et al. (1974). Reversal of abnormal platelet aggregability and change in exercise tolerance in patients with angina pectoris. *Circulation, 50,* 887–896.

294. Vanderhoek, J. Y., & Feinstein, M. B. (1979). Local anesthetics, chlorpromazine and propranolol inhibit stimulus-activation of phospholipase A$_2$ in human platelets. *Mol Pharmacol, 16,* 171–180.

295. Mehta, J., & Mehta, P. (1982). Effects of propranolol therapy on platelet release and prostaglandin generation in patients with coronary heart disease. *Circulation, 66,* 1294–1299.

296. Srivastava, K. C. (1987). Influence of some β blockers (pindolol, atenolol, timolol and metoprolol) on aggregation and arachidonic acid metabolism in human platelets. *Prostaglandins Leukot Med, 29,* 79–84.

297. Ingerman, C. M., Smith, J. B., & Silver, M. J. (1976). Inhibition of the platelet release reaction and platelet prostaglandin synthesis by furosemide. *Thromb Res, 8,* 417–419.

298. Pfister, B., & Imhof, P. (1979). Influence of vasodilators used in the therapy of heart failure on platelet aggregation. *Agents Actions, 9,* 217–219.

299. Saxon, A., & Kattlove, H. (1976). Platelet inhibition by sodium nitroprusside, a smooth muscle inhibitor. *Blood, 47,* 957–961.

300. Mehta, J., & Mehta, P. (1979). Platelet function in heart disease. VI. Enhanced platelet aggregate formation activity in congestive heart failure. Inhibition by sodium nitroprusside. *Circulation, 60,* 497–503.

301. Hines, R., & Barash, P. G. (1989). Infusion of sodium nitroprusside induces platelet dysfunction *in vitro. Anesthesiology, 70,* 611–615.

302. Schafer, A. J., Alexander, R. W., & Handin, R. I. (1980). Inhibition of platelet function by organic nitrate vasodilators. *Blood, 55,* 649–654.

303. Lawson, D., Mehta, J., Mehta, P., et al. (1986). Cumulative effects of quinidine and aspirin on bleeding time and platelet α2-adrenoreceptors: Potential mechanism of bleeding diathesis in patients receiving this combination. *J Lab Clin Med, 108,* 581–586.

304. Motulsky, H. J., Maisel, A. S., Snavely, M. D., et al. (1984). Quinidine is a competitive antagonist at α1- and α2-adrenergic receptors. *Circ Res, 55,* 376–381.

305. Ring, M. E., Corrigan, J. J., Jr., & Fenster, P. E. (1986). Effects of oral diltiazem on platelet function: Alone and in combination with "low dose" aspirin. *Thromb Res, 44,* 391–400.

306. Ware, J. A., Johnson, P. C., Smith, M., et al. (1986). Inhibition of human platelet aggregation and cytoplasmic calcium response by calcium antagonists: Studies with aequorin and quin2. *Circ Res, 59,* 39–42.

307. Barnathan, E. S., Addonizio, V. P., & Shattil, S. J. (1982). Interaction of verapamil with human platelet α-adrenergic receptors. *Am J Physiol, 242,* H19–H23.

308. Johnson, G. J., Leis, L. A., & Francis, G. S. (1986). Disparate effects of the calcium-channel blockers, nifedipine and verapamil, on α2-adrenergic receptors and thromboxane A$_2$-induced aggregation of human platelets. *Circulation, 73,* 847–854.

309. Glusa, E., Bevan, J., & Heptinstall, S. (1989). Verapamil is a potent inhibitor of 5-HT-induced platelet aggregation. *Thromb Res, 55,* 239–245.

310. Ambrosioni, E., & Borghi, C. (1989). Potential use of ACE inhibitors after acute myocardial infarction. *J Cardiovasc Pharmacol, 14,* S92–S94.

311. James, I. M., Dickenson, E. J., Burgoyne, W., et al. (1988). Treatment of hypertension with captopril: Preservation of regional blood flow and reduced platelet aggregation. *J Hum Hypertens, 2,* 21–25.

312. Okrucka, A., Pechan, J., & Kratochvilova, H. (1998). Effects of the angiotensin-converted enzyme inhibitor perindopril on endothelial and platelet function in essential hypertension. *Platelets, 9,* 395–396.

313. Keidar, S., Oiknine, J., Leiba, A., et al. (1996). Fosinopril reduces ADP-induced platelet aggregation in hypertensive patients. *J Cardiovasc Pharmacol, 27,* 183–186.

314. Moser, L., Callahan, K. S., Cheung, A. K., et al. (1997). ACE inhibitor effects on platelet function in stages I–II hypertension. *J Cardiovasc Pharmacol, 30,* 461–467.

315. Zannad, F., Bray-Desboscs, L., el Ghawi, R., et al. (1993). Effects of lisinopril and hydrochlorothiazide on platelet function and blood rheology in essential hypertension: A randomly allocated double-blind study. *J Hypertens, 11,* 559–564.

316. Jagroop, I. A., Papadaksi, J. A., & Mikhailidis, D. P. (1998). Effects of the angiotensin-converting enzyme inhibitor perindopril on endotheial and platelet function in essential hypertension. *Platelets, 9,* 395–396.

317. Horne, M. K. D., & Chao, E. S. (1989). Heparin binding to resting and activated platelets. *Blood, 74,* 238–243.

318. Salzman, E. W., Rosenberg, R. D., Smith, M. H., et al. (1980). Effect of heparin and heparin fractions on platelet aggregation. *J Clin Invest, 65,* 64–73.

319. Heiden, D., Mielke, C. H., Jr., & Rodvien, R. (1977). Impairment by heparin of primary haemostasis and platelet [14C]5-hydroxytryptamine release. *Br J Haematol, 36,* 427–436.

320. Sobel, M., McNeill, P. M., Carlson, P. L., et al. (1991). Heparin inhibition of von Willebrand factor-dependent platelet function *in vitro* and *in vivo*. *J Clin Invest, 87,* 1787–1793.

321. Lindblad, B., Wakefield, T. W., Whitehouse, W. M., et al. (1988). The effect of protamine sulfate on platelet function. *Scand J Thorac Cardiovasc Surg, 22,* 55–58.

322. Gimple, L. W., Gold, H. K., Leinbach, R. C., et al. (1989). Correlation between template bleeding times and spontaneous bleeding during treatment of acute myocardial infarcation with recombinant tissue-type plasminogen activator. *Circulation, 80,* 581–588.

323. Coller, B. S. (1990). Platelets and thrombolytic therapy. *N Engl J Med, 322,* 33–42.

324. Schafer, A. I., & Adelman, B. (1985). Plasmin inhibition of platelet function and of arachidonic acid metabolism. *J Clin Invest, 75,* 456–461.

325. Schafer, A. I., Maas, A. K., Ware, J. A., et al. (1986). Platelet protein phosphorylation, elevation of cytosolic calcium, and inositol phospholipid breakdown in platelet activation induced by plasmin. *J Clin Invest, 78,* 73–79.

326. Penny, W. F., & Ware, J. A. (1992). Platelet activation and subsequent inhibition by plasmin and recombinant tissue-type plasminogen activator. *Blood, 79,* 91–98.

327. Loscalzo, J., & Vaughan, D. E. (1987). Tissue plasminogen activator promotes platelet disaggregation in plasma. *J Clin Invest, 79,* 1749–1755.

328. Quinton, T. M., Kim, S., Derian, C. K., et al. (2004). Plasmin-mediated activation of platelets occurs by cleavage of protease-activated receptor 4. *J Biol Chem, 279,* 18434–18439.

329. Fitzgerald, D. J., Catella, F., Roy, L., et al. (1988). Marked platelet activation *in vivo* after intravenous streptokinase in patients with acute myocardial infarction. *Circulation, 77,* 142–150.

330. Kerins, D. M., Roy, L., Fitzgerald, G. A., et al. (1989). Platelet and vascular function during coronary thrombolysis with tissue-type plasminogen activator. *Circulation, 80,* 1718–1725.

331. Mills, D. C. B., & Roberts, G. C. K. (1967). Membrane active drugs and the aggregation of human blood platelets. *Nature, 213,* 35.

332. Mills, D. C. B., Robb, I. A., & Roberts, G. C. K. (1968). The release of nucleotides, 5-hydroxytryptamine and enzymes from human blood platelets during aggregation. *J Physiol, 195,* 715–729.

332a. Maurer-Spurej, E. (2005). Serotonin reuptake inhibitors and cardiovascular diseases: A platelet connection. *Cell Mol Life Sci, 62,* 159–170.

333. Humphries, J. E., Wheby, M. S., & VandenBerg, S. R. (1990). Fluoxetine and the bleeding time. *Arch Pathol Lab Med, 114,* 727–728.

334. Alderman, C. P., Moritz, C. K., & Ben-Tovim, D. I. (1992). Abnormal platelet aggregation associated with fluoxetine therapy. *Ann Pharmacother, 26,* 1517–1519.

335. Hergovich, N., Aigner, M., Eichler, H. G., et al. (2000). Paroxetine decreases platelet serotonin storage and platelet function in human beings. *Clin Pharmacol Ther, 68,* 435–442.

336. Rysanek, R., Svehla, C., Spankova, H., et al. (1966). The effect of tricyclic antidepresive drugs on adrenaline and adenosine diphosphate induced platelet aggregation. *J Pharmacol Exp Ther, 18,* 616.

337. O'Brien, J. R. (1961). The adhesiveness of native platelets and its prevention. *J Clin Pathol, 14,* 140.

338. Jain, M. F., Eskow, E., Kuchibhotla, J., et al. (1978). Correlation of inhibition of platelet aggregation by phenothiazines and local anesthetics with their effects on a phospholipid bilayer. *Thromb Res, 13,* 1067–1075.

339. White, G. C., & Raynor, S. T. (1980). The effects of trifluoperazine, an inhibitor of calmodulin on platelet function. *Thromb Res, 18,* 279–284.

340. Warlow, C., Ogston, D., & Douglas, A. S. (1976). Platelet function after administration of chloropromazine to human subjects. *Haemostasis, 5,* 21–26.

341. Dalsgaard-Nielsen, J., Risbo, A., Simmelkjaer, P., et al. (1981). Impaired platelet aggregation and increased bleeding time during general anesthesia with halothane. *Br J Anaesth, 53,* 1039–1042.

342. Corbin, F., Blaise, G., & Sauve, R. (1998). Differential effect of halothane and forskolin on platelet cytosolic $Ca^{2+}$ mobilization and aggregation. *Anesthesiology, 89,* 401–410.

343. Parolari, A., Guarnieri, D., Alamanni, F., et al. (1999). Platelet function and anesthetics in cardiac surgery: An *in vitro* and *ex vivo* study. *Anesth Analg, 89,* 26–31.

344. Langdell, R. D., Adelson, E. A., & Furth, F. W. (1958). Dextran and prolonged bleeding time. *J Am Med Assoc, 166,* 346.

345. Ewald, R. A., Eichelberger, J. W., Young, A. A., et al. (1965). The effect of dextran on platelet factor 3 activity: *In vitro* and *in vivo* studies. *Transfusion, 5,* 109.

346. Bygdeman, S., & Eliasson, R. (1967). Effect of dextrans on platelet adhesiveness and aggregation. *Scand J Clin Lab Invest, 20,* 17–23.

347. Weiss, H. J. (1967). The effect of clinical dextran on platelet aggregation, adhesion and ADP release in man: *In vivo* and *in vitro* studies. *J Lab Clin Med, 69,* 37–46.

348. Evans, R. J., & Gordon, J. D. (1974). Mechanisms of the antithrombotic action of dextran. *N Engl J Med, 290,* 748.

349. Mishler, J. M. (1984). Synthetic plasma volume expanders — Their pharmacology, safety and clinical efficacy. *Clin Haematol, 13,* 75–92.

350. Aberg, M., Hedner, U., & Bergentz, S. E. (1978). Effect of dextran 70 on factor VIII and platelet function in von Willebrand's disease. *Thromb Res, 12,* 629–634.

351. Kelton, J. G., & Hirsch, J. (1980). Bleeding associated with antithrombotic therapy. *Semin Hematol, 17,* 259–291.

352. Gruber, U. F., Saldeen, T., Brokop, T., et al. (1980). Incidences of fatal postoperative pulmonary embolism after prophylaxis with dextran 70 and low-dose heparin: An international multicentre study. *Br Med J, 280,* 69–72.

353. Roberts, J. S., & Bratton, S. L. (1998). Colloid volume expanders. Problems, pitfalls and possibilities. *Drugs, 55,* 621–630.

354. Cope, J. T., Banks, D., Mauney, M. C., et al. (1997). Intraoperative hetastarch infusion impairs hemostasis after cardiac operations. *Ann Thorac Surg, 63,* 78–82.

355. Ruttmann, T. G., James, M. F., & Aronson, I. (1998). *In vivo* investigation into the effects of haemodilution with hydroxyethyl starch (200/0.5) and normal saline on coagulation. *Br J Anaesth, 80,* 612–616.

356. Favis, G. R., & Colman, R. W. (1978). The action of halofenate on platelet shape change and prostaglandin synthesis. *J Lab Clin Med, 92,* 45–52.

357. Colman, R. W., Bennett, J. S., Sheridan, J. S., et al. (1976). Halofenate: A potent inhibitor of normal and hypersensitive platelets. *J Lab Clin Med, 88,* 282–291.

358. Klener, P., Kubisz, P., & Suranova, J. (1977). Influence of cytotoxic drugs on platelet functions and coagulation *in vitro*. *Thromb Haemostas, 37,* 53–61.

359. Pogliani, E. M., Fantasia, R., Lambertenghi-Deliliers, G., et al. (1981). Daunorubicin and platelet function. *Thromb Haemost, 45,* 38–42.

360. Kubisz, P., Klener, P., & Gronberg, S. (1980). Influence of mithramycin on some platelet functions *in vitro*. *Acta Haematol, 63,* 101–106.

361. Ahr, D. J., Scialla, S. J., & Kimball, D. B., Jr. (1978). Acquired platelet dysfunction following mithramycin therapy. *Cancer, 41,* 448–454.

362. McKenna, R., Ahmad, T., Ts'ao, C. H., et al. (1983). Glutathione reductase deficiency and platelet dysfunction induced by 1, 3-bis (2-chloroethyl)-1-nitrosourea. *J Lab Clin Med, 102,* 102–115.

363. Karolak, L., Chandra, A., Khan, W., et al. (1993). High-dose chemotherapy-induced platelet defect: Inhibition of platelet signal transduction pathways. *Mol Pharmacol, 43,* 37–44.

364. Thomson, C., Forbes, C. D., & Prentice, C. R. M. (1973). A comparison of the effects of antihistamines on platelet function. *Thromb Diath Haemorrh, 30,* 547–556.

365. Herrmann, R. G., & Frank, J. D. (1966). Effect of adenosine derivatives and antihistamines on platelet aggregation. *Proc Soc Exp Biol Med, 123,* 654–660.

366. Green, D., Ts'ao, C. H., Cerullo, L., et al. (1985). Clinical and laboratory investigation of the effects of E-aminocaproic acid on hemostasis. *J Lab Clin Med, 105,* 321–327.

367. Rubin, R., & Rand, M. L. (1994). Alcohol and platelet function. *Alcohol Clin Exp Res, 18,* 105–110.

368. Rand, M. L., Packham, M. A., Kinlough-Rathbone, R. L., et al. (1988). Effects of ethanol on pathways of platelet aggregation *in vitro*. *Thromb Haemost, 59,* 383–387.

369. Mikhailidis, D. P., Barradas, M. A., & Jeremy, J. Y. (1990). The effect of ethanol on platelet function and vascular prostanoids. *Alcohol, 7,* 171–180.

370. Deykin, D., Janson, P., & McMahon, L. (1982). Ethanol potentiation of aspirin-induced prolongation of the bleeding time. *N Engl J Med, 306,* 852–854.

371. Rao, A. K., Rao, V. M., Willis, J., et al. (1985). Inhibition of platelet function by contrast media. Iopamidol and Hexabrix are less inhibitory then Conray-60. *Radiology, 156,* 311–313.

372. Li, X., & Gabriel, D. A. (1997). Differences between contrast media in the inhibition of platelet activation by specific platelet agonists. *Acad Radiol, 4,* 108–114.

373. Parvez, Z., Moncada, R., Fareed, J., et al. (1984). Antiplatelet action of intravascular contrast media. Implications in diagnostic procedures. *Invest Radiol, 19,* 208–211.

374. Grabowski, E. F., Jang, I. K., Gold, H., et al. (1996). Variability of platelet degranulation by different contrast media. *Acad Radiol, 3*(Suppl. 3), S485–S487.

375. Jandak, J., Steiner, M., & Richardson, P. D. (1989). α-Tocopherol, an effective inhibitor of platelet adhesion. *Blood, 73,* 141–149.

376. Steiner, M., & Anastasi, J. (1976). Vitamin E. An inhibitor of the platelet release reaction. *J Clin Invest, 57,* 732–737.

377. Goodnight, S. H., Jr. (1988). Effects of dietary fish oil and omega-3 fatty acids on platelets and blood vessels. *Semin Thromb Hemost, 14,* 285–289.

378. Srivastava, K. C. (1986). Onion exerts antiaggregatory effects by altering arachidonic acid metabolism in platelets. *Prostaglandins Leukotr Med, 24,* 43–50.

379. Phillips, C., & Poyser, N. L. (1978). Inhibition of platelet aggregation by onion extracts. *Lancet, 1,* 1051–1052.

380. Makheja, A. N., Vanderhock, J. Y., & Bailey, J. M. (1979). Inhibition of platelet aggregation and thromboxane synthesis by onion and garlic. *Lancet, 1,* 781.

381. Bordia, A., Verma, S. K., & Srivastava, K. C. (1998). Effect of garlic (*Allium sativum*) on blood lipids, blood sugar, fibrinogen and fibrinolytic activity in patients with coronary artery disease. *Prostaglandins Leukotr Essent Fatty Acids, 58,* 257–263.

382. Bordia, A., Verma, S. K., & Srivastava, K. C. (1997). Effect of ginger (*Zingiber officinale* Rosc.) and fenugreek (*Trigo-*

*nella foenumgraecum* L.) on blood lipids, blood sugar and platelet aggregation in patients with coronary artery disease. *Prostaglandins Leukotr Essent Fatty Acids, 56,* 379–384.

383.  Srivastava, K. C. (1989). Extracts from two frequently consumed spices — cumin (*Cuminum cyminum*) and turmeric (*Curcuma longa*) — inhibit platelet aggregation and alter eicosanoid biosynthesis in human blood platelets. *Prostaglandins Leukotr Essent Fatty Acids, 37,* 57–64.

384.  Srivastava, K. C., Bordia, A., & Verma, S. K. (1995). Curcumin, a major component of food spice turmeric (*Curcuma longa*), inhibits aggregation and alters eicosanoid metabolism in human blood platelets. *Prostaglandins Leukotr Essent Fatty Acids, 52,* 223–227.

385.  Srivastava, K. C. (1993). Antiplatelet principles from a food spice clove (*Syzygium aromaticum* L). *Prostaglandins Leukotr Essent Fatty Acids, 48,* 363–372.

386.  Pearson, D. A., Paglieroni, T. G., Rein, D., et al. (2002). The effects of flavanol-rich cocoa and aspirin on *ex vivo* platelet function. *Thromb Res, 106,* 191–197.

387.  Dorso, C. R., Levin, R. I., Eldor, A., et al. (1980). Chinese food and platelets. *N Engl J Med, 303,* 756–757.

388.  Hammerschmidt, D. E. (1980). Szechwan purpura. *N Engl J Med, 302,* 1191–1193.

389.  Fugh-Berman, A. (2000). Herb–drug interactions. *Lancet, 355,* 134–138.

# CHAPTER 59

# Cardiopulmonary Bypass

**Brian Richard Smith, Henry M. Rinder, and Christine S. Rinder**

*Department of Laboratory Medicine, Yale University School of Medicine, New Haven, Connecticut*

## I. Introduction

Cardiopulmonary bypass (CPB), first used successfully in the 1950s, remains the keystone on which much of modern cardiac surgery rests. On a worldwide basis, approximately 2000 CPB procedures are performed every 24 hours.[1] In addition to its primary role in cardiac surgery, CPB has also found a place in the management of severe respiratory failure, particularly in neonates, and in emergency hemodynamic support following massive pulmonary embolism, trauma, environmental hypothermia, and cardiac arrest.[2,3] Although the procedure has clearly benefited millions of patients, it is not without complications. These include primarily pulmonary dysfunction, renal dysfunction, myocardial ischemia, neurocognitive defects,[4] and hemostatic abnormalities.[5-10] Although this chapter focuses on hemostatic abnormalities, particularly platelet effects, it is important to recognize that the normal physiologic interaction of the hemostatic and inflammatory systems[11,12] implies a possible role for platelet pathobiology in almost all of the clinical abnormalities seen with CPB.

## II. Clinical Aspects of CPB Relevant to Platelet Pathophysiology

The CPB circuit, which includes an oxygenator, a heating/cooling unit, a circulatory pump, and associated tubing, represents a highly thrombogenic biomaterial stimulus. Hence, it is necessary for all CPB patients to undergo systemic anticoagulation, nearly always with high-dose unfractionated heparin. Although there is significant variability between centers, and randomized, controlled trials are lacking, typical protocols initially bolus the patient with 300 U heparin per kilogram and then continue to administer heparin boluses as needed to maintain an activated clotting time (ACT) of more than 400 seconds during normothermic CPB and more than 480 seconds during hypothermic CPB. There are also a variety of alternate approaches to monitor-

ing anticoagulation during the procedure.[13] The extracorporeal circuit is generally initially primed with 3 U heparin per milliliter, and monitoring with the ACT is performed every 30 minutes. Even with this level of heparinization, studies have shown significant fibrin formation during CPB as measured by fibrinopeptide A levels,[14] albeit the observed levels remain lower than those following other surgical procedures carried out without heparinization of the patient. Attempts to reduce the thrombogenicity of the CPB system, including the use of heparin-bonded tubing and other biomaterial modifications, have had only modest impact, allowing the reduction of total heparin dosing in some studies.[15-17] These biomaterial alterations appear to have a greater effect on reducing the systemic inflammatory response syndrome (SIRS) produced by CPB than they do on the thrombogenicity of the procedure, although platelet loss to the circuit may indeed be less with nontraditional biomaterials.[18] Recognized clinical hematologic complications of heparin administration include heparin-induced thrombocytopenia with or without thrombosis (see Chapter 48).

At the end of CPB, heparin reversal is carried out by administration of protamine. Multiple regimens have been suggested for clinical use. The simplest regimens administer 1.0 to 1.3 mg protamine for each 100 U heparin given to the patient. More complex approaches involve various ways of estimating heparin levels, including dose–response curves.[19] Complications of protamine administration are traditionally divided into three major categories: transient hypotension secondary to rapid administration; classic anaphylactic reactions mediated via IgE antibodies and "anaphylactoid" reactions, either immediate or delayed, not mediated by IgE; and catastrophic pulmonary vasoconstriction.[20] Transient hypotension is thought by many to be related to protamine-induced histamine release. True IgE-mediated anaphylaxis is seen predominantly in diabetics receiving NPH insulin (which contains protamine) and only very rarely in CPB patients. Anaphylactoid reactions, the most common type of protamine complication in CPB, are believed to be secondary to excessive complement activation induced not by protamine alone but by heparin–protamine complexes mediated through a complement component C1 activation route (discussed later). Protamine can also inhibit the inactivation of anaphylatoxin and kinin mediators, perhaps exacerbating the problem of simultaneous complement activation.[21] Finally, a large amount of animal model work has attempted to address the potentially fatal and idiosyncratic type III protamine reaction (summarized in Patla et al.[22]). The weight of evidence suggests that thromboxane $A_2$ ($TXA_2$) release is at the core of this complication in that $TXA_2$ receptor and thromboxane synthetase blockers abrogate the complication in animal studies, and clinical epidemiologic data suggest that this complication is more unusual in aspirinized patients.[23] Although it is tempting to postulate that circulating platelets might be the crucial source of $TXA_2$, animal data have shown production of the clinical effect and $TXA_2$ in isolated lung preparations perfused with an acellular medium, suggesting that pulmonary intravascular macrophages are the major culprit.

CPB patients sometimes exhibit a clinical phenomenon known as "heparin rebound," defined as the prolongation of coagulation times, often in the setting of clinical bleeding, that follows prior complete heparin neutralization. This may occur as soon as 1 hour after neutralization and may last for up to 6 hours. The phenomenon is the result of the reappearance of unneutralized heparin in the circulation, probably due to a combination of more rapid clearance of protamine compared to heparin and the release of heparin from endothelium or other tissue sources.[24]

In addition to the biomaterials used in the circuit and the nature of the anticoagulation, other mechanical aspects of the CPB procedure affect hemostasis. Bubble oxygenators generally result in greater activation of inflammatory and hemostatic cells and proteins, but these have been almost exclusively replaced by membrane oxygenators in modern clinical practice. Controversy exists over the benefits of pulsatile versus nonpulsatile pump approaches,[25] with some studies demonstrating evidence for reduced endothelial injury with pulsatile bypass. In addition, the best fluid with which to initially "prime" the CPB pump is uncertain. For example, although most centers avoid the use of hetastarch because of concern about its possible adverse effects on coagulation, some continue to find it a useful adjustment.[26]

CPB is carried out under conditions of hemodilution with isotonic nonplasma fluids, although the specifics are quite variable from institution to institution. Typically in an adult, this results in a reduction of the hematocrit from approximately 40% to approximately 25% and similar reductions in circulating plasma proteins, including coagulation factors, complement components, and members of the contact system. Fibrinogen is also selectively lost to the foreign biomaterial surfaces. However, none of these soluble coagulation protein decreases approach the levels usually associated with a clinical bleeding diathesis. In addition, systemic

hypothermia is used during the procedure. Some centers reduce the patient and circuit temperature in a relatively rapid, scheduled fashion to the 26 or 27°C range, although in recent years the trend in clinical practice has been to allow patients to slowly "drift" toward 30°C throughout the procedure. Rewarming at the end of bypass is generally carried out as an active process. Both leukocyte and platelet physiology can be significantly altered by these changes in temperature.[27] Indeed, there has been renewed interest in the specific physiologic effects of hypothermia since several studies have shown clinical benefit from the use of hypothermia in patients who are comatose following cardiac arrest as well as in subsets of patients with stroke, head trauma, and myocardial infarction (summarized in Gazmuri and Gopalakrishnan[28]). In experimental models of cardiac surgery, hypothermia increases hepatic interleukin-10 (IL-10) while decreasing tumor necrosis factor-α (TNF-α), perhaps in part related to increased suppression of cytokine signaling (SOCS-3) activity and increased signal transducer and activator of transcription (STAT-3) activity.[29] In contrast to the potentially beneficial effects on reducing overall inflammation, hypothermia may have adverse affects on platelet activation and function and may be clinically associated with increased bleeding.[27,30]

Although again quite variable between centers, up to 35% of patients undergoing CPB also receive allogeneic red cell transfusions — on average, two or three transfusions per patient.[31] There are no clear guidelines regarding the appropriate transfusion "triggers" for cardiac surgery patients; therefore, there have been calls for an appropriate multicenter randomized trial to establish evidence-based guidelines.[32] Clinical parameters associated with transfusion of more than 2 units of packed red blood cells include increased age, increased preoperative creatinine, low body surface area, preoperative hematocrit, nonelective surgery, hypothermic bypass, and prolonged duration of bypass.[33] Transfusion of platelets, fresh frozen plasma, cryoprecipitate, or surgical reexploration for bleeding are more likely to occur in patients with lower body surface area, those undergoing repeat surgery or nonelective surgery, and those with greater time on bypass. Interestingly, elevated preoperative prothrombin time (PT), international normalized ratio, and preoperative heparin were not more likely to result in increased transfusion requirements in one study.[33]

In addition to allogeneic transfusion, autologous transfusion via intraoperative cell-saving (blood scavenging) mechanisms and perioperative autologous transfusion of previously stored blood are common concomitants of cardiac surgical procedures.[34,35] Since blood scavenger products are washed of most plasma constituents and filtered, they are regarded as a relatively neutral aspect of the procedure with regard to coagulation and inflammation,[36] even though the initially shed blood contains increased amounts of cytokines and other bioactive molecules.[37] With the universal recognition of the risks of transfusion, advances are being made in moving toward more "bloodless" surgical approaches.[38]

Although the focus of this chapter is on CPB, the trend in recent years toward the use of alternative "off-pump" coronary artery bypass grafting (OPCAB) in appropriately selected patients raises the question of how this procedure may compare to CPB-dependent cardiac surgery in overall platelet pathophysiology.[39] It is difficult to compare the procedures for a number of reasons: Randomized trials in which relevant hemostatic parameters were measured are limited; comparisons in the literature are frequently between CPB performed without additional anti-inflammatory techniques (e.g., aprotinin) and pharmacologic use of antiplatelet and other agents is not always similar in the two groups; differences relate not just to lack of the heart–lung machine but also to the lack of high-dose heparinization, hypothermia, and hemodilution in the OPCAB patient; and patients who require conversion from off-pump to on-pump surgery may be eliminated from hemostatic analysis. Of note, OPCAB adds new pathophysiologies that may affect hemostasis. Hemodynamic compromise due to manipulation of the heart and immobilization of the operative area is greater than in CPB procedures. Intraluminal shunting, passive distal coronary perfusion or active distal perfusion using an in-line pump is often employed. Hence, it is not surprising that there is no consensus on how these procedures differ with respect to many aspects of inflammatory and hemostatic physiology. However, most data suggest that OPCAB results in less fibrinolytic and complement activation and less leukocyte activation but not necessarily less platelet activation (measured by platelet surface P-selectin expression) or reproducible improvement in platelet functional alterations in the immediate peri- and postoperative period.[40–42] Platelet counts are likely initially better preserved with OPCAB. Overall, results of clinical trials have shown generally similar outcomes between the procedures, with one trial suggesting less myocardial damage but also lower 3-month graft patency rates with OPCAB.[43,44]

## III. Inflammation and CPB

### A. The Whole Body Inflammatory Response of CPB

The sudden and widespread exposure of normal blood protein and cellular constituents to solid and gaseous foreign surfaces during CPB results in extensive alterations to all the involved elements. This reaction to CPB has been termed a "whole body inflammatory response" and the clinical consequences have been termed systemic inflammatory response syndrome (SIRS).[45] In addition to the initiation of the process by CPB artificial devices, many other factors contribute, including ischemia–reperfusion injury, operative

trauma and manipulation, and even endotoxin release from the gut.[46] SIRS involves activation of the contact,[47] complement,[48] coagulation, cytokine, and fibrinolytic systems. The complement system is activated in two phases, initially via the alternate pathway secondary to biomaterial interactions and later via the classical pathway secondary to protamine–heparin complexes.[49,50] Cellular elements are also activated, probably both directly and indirectly, including endothelial cells, neutrophils, and monocytes. Although more details regarding cellular activation and its relationship to platelet–leukocyte adhesion and alterations in platelet physiology are presented later, it is worth recognizing at the onset that this cellular inflammatory response results in the release of a large number of active mediators that can influence platelet behavior. These include prostacyclin (prostaglandin [PG] $I_2$) and other prostaglandins and leukotrienes[51]; endothelin-1[52]; products of neutrophil granules, including elastase, acid hydrolases, and collagenases[53]; and cytokines such as monocyte chemoattractant protein-1 (MCP-1), TNF-$\alpha$, and IL-1, -6, -8, and -10.[54–57] Studies are not always congruent with respect to the exact time course and degree of elevation of these and other mediators during CPB. This is not surprising given the significant differences between centers in surgical and anesthetic technique, the site of blood drawing, the complicating factors of transfusion and hemodilution, and the myriad of vasoactive and other drugs that may be administered to these patients. In contrast to the generalized activation of phagocytic cells, whose functional life span may be increased by the evolving cytokine environment following CPB surgery,[58] lymphocytes tend to be selectively lost and have decreased function over the course of CPB, resulting in a relative lymphocytic immunosuppressed state.[12]

In considering the effects of *in vivo* CPB on both inflammatory and hemostatic physiology, it must also be recognized that CPB is not undertaken for its own sake but is always part of other therapeutic procedures.[59] Surgery undertaken in the absence of CPB is associated with significant alterations in coagulation factors.[60] Indeed, as noted previously, the reintroduction of off-pump coronary artery surgery has gradually made it clear that pathophysiologic events such as myocardial ischemia and mechanical trauma can mimic some of the same effects on protein and cellular activation that are currently attributed predominantly to CPB, albeit with different quantitative alterations in these systems.[61–70]

### B. Platelets and the Inflammatory Response of CPB

It has become clear in recent years that the traditional conceptual separation of the hemostatic and inflammatory systems is arbitrary because there are numerous pathways through which these two systems constantly interact. In addition to soluble mediator links such as those of the contact and complement pathways, cellular interactions are also important. In particular, platelets adhere to neutrophils, monocytes, and lymphocytes, as well as to endothelial cells, and this adhesive process results in functional changes in both participating cell types (see Chapters 12 and 39). In addition, this adhesive interaction facilitates targeting of cells to areas of thrombosis and/or inflammation.[71–82] Platelet microparticles (see Chapter 20) may be a crucial mechanism involved in neutrophil–neutrophil homotypic adhesion, independent of the L-selectin/P-selectin glycoprotein ligand-1 (PSGL-1) interaction.[83,84] Finally, other platelet–leukocyte interactions that may not require adhesion, such as metabolic cooperation and mediator release,[85–93] are important aspects of the regulation of both physiologic and pathophysiologic processes, and platelet–endothelial interactions similarly influence the inflammatory response as well as the hemostatic response to injury (see Chapters 13 and 39).[72]

How do these various interactions of the endothelial cell–leukocyte–platelet triad influence CPB-related pathophysiology? It is well established that CPB induces platelet–leukocyte adhesion,[94,95] a phenomenon that is also observed in other disease states, including acute coronary syndromes,[75,96–99] platelet collection and storage,[100] sepsis,[101] hemodialysis,[102,103] peripheral vascular disease,[104] and thrombotic thrombocytopenic purpura.[105,106] Platelet–monocyte adhesion can occur through a number of receptor–ligand pairs (see Chapter 12), including P-selectin–PSGL-1,[107–111] glycoprotein (GP) Ibα–CD11b/CD18,[112] GPIV–thrombospondin GPIV, and GPIIb-IIIa–fibrinogen CD11b/CD18. Engagement of P-selectin by monocytes results in functional alterations in monocytes, including tissue factor upregulation,[113–115] TNF-α release,[116] promotion of monocyte differentiation,[117] and the production of other cytokines and chemokines[118–120] — all phenomena associated with SIRS. Platelet–neutrophil adhesion occurs predominantly through the interaction of P-selectin with its counterligand PSGL-1 but also via platelet intercellular adhesion molecule-2 (ICAM-2) and GPIIb-IIIa-bound fibrinogen to neutrophils CD11a/CD18 and CD11b/CD18 (see Chapter 12). This interaction has a variety of functional consequences, including enhanced neutrophil locomotion,[121] enhanced superoxide production,[122] phosphorylation of a number of neutrophil proteins,[123] and cytokine production[118] — again, all potential participants in ischemia–reperfusion injury.

Platelets also directly release inflammatory mediators upon activation (see Chapter 39), independent of their effects on leukocytes, and thus contribute to SIRS. These include prominently CD40 ligand (CD40L)[124] and IL-1β,[125] among many others.[126] Monocyte recruitment to endothelial sites is similarly promoted through platelet-derived RANTES (regulated upon activation, normal T cell expressed and presumably secreted) and platelet factor 4.[127]

All these observations suggest that clinical improvement in cardiac surgery outcomes may follow an improved under-

standing of the interactions of platelets and leukocytes, such that what may be initially thought of as an action to benefit hemostasis may in fact abrogate (or perhaps worsen) the inflammatory response associated with CPB. Recent work, for example, has shown that there is a correlation between platelet activation and cytokine release during CPB.[128] It may therefore be prudent to routinely measure important parameters of both the inflammatory and the hemostatic systems whenever one is considering a therapeutic intervention to affect one of the two systems.

## IV. Clinical Hemostatic Alterations during and after CPB

### A. Bleeding during and after CPB

In some centers, more than 35% of patients undergoing cardiac surgery with CPB bleed more than 1 L in the 24-hour postoperative period.[129] Although preoperative laboratory testing is of controversial utility for predicting bleeding during and after CPB, several clinical parameters placing patients at increased risk can be identified: female gender, preoperative aspirin or clopidogrel use, higher concentrations of heparin used during the procedure, increased duration of CPB, and hypothermia.[129,130]

In addition to quantitative and qualitative abnormalities of platelets, discussed in detail later, there are multiple abnormalities of coagulation that may contribute to the bleeding diathesis. Despite high-dose heparinization, the coagulation system is still activated on bypass and, indeed, thrombin is generated in all patients. The contact system is activated,[47] as is the tissue factor pathway — the latter both locally at the level of the pericardium and, to a more limited extent, systemically.[131–134] Data suggest that tissue factor-bearing leukocytes and leukocyte microparticles appear both locally at the wound and systemically. Systemic appearance may occur in two phases. The first phase is immediately concomitant with surgery and instigated by platelet–leukocyte conjugate formation and cell salvage from the pericardial area. The latter phase, a monocyte phenomenon dependent on new tissue factor synthesis secondary to other monocyte-activating stimuli (cytokines and complement), begins on approximately day 1 after surgery.[135]

Fibrinolytic activity increases both during and for some time after bypass[136] as a consequence of tissue plasminogen activator (TPA) release from the endothelium.[137] Multiple causes of this TPA release can be postulated, and all of them may well contribute: contact system activation with stimulation of TPA release by kallikrein and bradykinin; thrombin, catecholamine, leukotriene, angiotensin II, and hypoxic stimulation of the endothelium; and heparinization.[138,139] In addition to release of TPA by the vascular endothelium, blood deposited in the pericardial cavity during surgery activates both the coagulation and the fibrinolytic systems.[131,140]

There is evidence that fibrinolysis contributes in a major way to the bleeding diathesis observed with CPB. Correlative studies show a relationship between blood levels of fibrinolytic products and blood loss during CPB.[141,142] Furthermore, fibrinolytic inhibitors, including ε-amino caproic acid (EACA) and tranexamic acid, have shown efficacy in reducing postoperative bleeding.[143,144] Aprotonin has similarly been shown in several randomized, controlled trials to significantly reduce bleeding, although it is likely that this broad protease inhibitor also works through additional, predominantly anti-inflammatory, mechanisms.[145–148] Indeed, evidence suggests that EACA, long considered more fibrinolytic specific than broad serine protease inhibitors such as aprotonin, results in abrogation of inflammatory parameters such as cytokine production[149,150] in addition to its hemostatic effects. Aprotonin has received the highest clinical acceptance as an adjunctive agent in minimizing CPB-induced hemostatic complications and in improving overall outcomes. Although there remains concern that such complete blockade of fibrinolysis and other protease systems could result in increased late thrombotic complications of CPB, the majority of studies do not support this concern.[151]

An often overlooked potential player in the pathogenesis of post-CPB hemostatic abnormalities is the endothelium. Several lines of evidence point to significant endothelial activation during bypass. Animal models directly demonstrate this activation.[152,153] In humans, the fact that there is release of TPA from endothelium as outlined previously, in addition to data suggesting the release of endothelial-associated adhesion moieties,[153] confirms that a similar process occurs in clinical bypass. CPB results in increased endothelial expression of platelet activating factor, nitric oxide, and arachidonic acid metabolites in addition to the adhesion molecule and other alterations.[154] Some data indicate the possibility of induction of apoptosis in the endothelium during cardiac surgery.[155] Whether this endothelial activation is in part related to nonpulsatile flow and other mechanical factors, or whether it is a secondary consequence of changes in soluble factors such as complement and cytokines, is uncertain.

## B. Embolism during CPB

As noted previously, one of the most pervasive adverse consequences of CPB is the development of neurologic complications, including frank stroke (with an estimated incidence of 1.5–5.2% of patients), postoperative delirium (10–30% of patients), short-term cognitive changes (33–88% of patients), and long-term neurocognitive dysfunction (possibly almost 50% of patients).[2,156] Although the cause of all these events is likely multifactorial, evidence suggests that a significant portion of the risk may be attributable to embolic phenomena.[157–160] The majority of frank strokes are detected early postoperatively, and further evidence suggests that in addition to the major risk factor of postoperative atrial fibrillation,[161,162] aortic manipulation, particularly application and/or removal of the aortic cross-clamp,[163–165] may dislodge atheromatous, highly prothrombotic material from the ascending aorta.

Three types of emboli are associated with CPB: blood-borne, foreign material, and gaseous.[166] The blood-derived microemboli consist of both aggregates of various types of autologous cells and autologous cells complexed with lipids, plasma proteins, and even extracellular matrix material.[167] The cellular aggregates include platelet–platelet aggregates, which have been causally associated by some authors with neurologic events,[168] leukocyte–leukocyte aggregates,[169,170] and platelet–leukocyte aggregates.[94] Although there is general agreement based on animal model work and clinical observations that these cellular emboli, combined with other particulate and air-bubble emboli, are associated with adverse neurologic outcomes, the specific pathophysiologic pathways involved are not well-defined. However, it is interesting to postulate a significant role for platelet activation in this pathophysiology based on animal models of stroke.[171,172]

## C. Thrombosis during and after CPB

Following coronary artery bypass grafting, occlusion of the aortocoronary saphenous vein grafts is an important aspect of both early and late morbidity and mortality associated with the procedure: 1 month after surgery, occlusion rates are 8 to 18%, and at 1 year they are 16 to 26%.[173] Perioperative myocardial infarction is an additional risk. It is difficult to determine which aspects of this local thrombotic diathesis may be attributable to events that occur in direct relationship to CPB, as opposed to the myriad other surgical and medical variables in these patients. Nevertheless, the fact that in other relevant clinical settings such as coronary angioplasty, platelet activation,[174] leukocyte activation,[175,176] and endothelial activation[176,177] seem to play a major role in restenosis suggests that events that commence during CPB may well influence at least short-term and possibly long-term success rates for coronary revascularization. For example, data suggest that increased levels of soluble P-selectin, which occur as a part of CPB pathophysiology,[178] promote a procoagulant state in animal models,[179] possibly because they induce production of tissue factor-expressing leukocyte microparticles.[180] In nonsurgical settings, increased soluble P-selectin has been shown to be associated with adverse cardiovascular outcomes.[181]

Platelet-specific physiology is implicated in postoperative thrombotic events by several additional lines of evidence. The PL$^{A2}$ GPIIb-IIIa (integrin $\alpha$IIb$\beta$3) polymorphism (see Chapter 14) is associated with increased myocardial damage following CPB and is also associated with increased neurocognitive deficits.[182,183] One of the most important clinical observations concerning the appropriate management of cardiac surgery patients in recent years has been the demonstration that rapid reinstitution of aspirin therapy following surgery (within 48 hours) is both safe and critical to reduce ischemic morbidity and mortality.[184] Indeed, overall mortality in this study[184] was 3.2% without aspirin and 1.3% with it, and there was a 48% or greater reduction in multiple ischemic complications (myocardial infarction, renal impairment, bowel ischemia, and stroke) when aspirin was rapidly restarted. Although the mechanism of action may be in part the anti-inflammatory activity of salicylates, this is nonetheless a cogent argument for the critical importance of platelet activity in this setting.

## V. Quantitative Platelet Abnormalities during CPB

Within 5 minutes of initiation, CPB induces a rapid decrease in total platelet count that exceeds the decrease that would be explicable on the basis of hemodilution alone. However, the degree of thrombocytopenia is seldom sufficient to result in a bleeding diathesis solely on this basis. The platelet nadir is generally reached approximately 30 minutes into the procedure, but thrombocytopenia may persist for many days postoperatively.[141,185–187] This persistent thrombocytopenia is both poorly understood and relatively poorly studied. Acute decreases in platelet count, however, are predominantly (although not exclusively) the result of platelet adhesion to the biomaterials of the CPB circuit. The platelet loss is proportional to the flow rate and surface area of the circuit.[188,189] Platelets appear to adhere before their activation but likely adhere to an even greater extent following activation — blocking activation of the platelets through pharmacologic means decreases overall adhesion.

The underlying mechanisms of platelet adhesion to the foreign biomaterials are partially understood. Plasma proteins are adsorbed onto the surface of foreign biomaterials very rapidly (within seconds) and form a protein layer that is approximately 200 Å thick.[190–195] This adsorption, the detailed molecular basis for which is not well characterized, is selective. Fibrinogen is generally adsorbed in the highest concentration; once adsorbed, its conformational structure appears to be altered.[192,196,197] Adsorption differs between biomaterials, with hydrophobic surfaces adsorbing more fibrinogen than hydrophilic surfaces. The location and conformation of deposited fibrinogen may also relate to shear forces imparted to the blood as it circulates over the artificial surface.[198] In addition to fibrinogen, albumin, gamma globulin, fibronectin, lipoproteins, thrombospondin, factor XII, high-molecular-weight kininogen, prekallikrein, plasminogen, and von Willebrand factor (VWF)/factor VIII complex are also adsorbed onto artificial surfaces. These proteins are in a dynamic equilibrium with those in the plasma.[199] For example, over time, fibrinogen is partially replaced with high-molecular-weight kininogen, a process referred to as the "Vroman effect." The presence of these adsorbed blood proteins has a profound effect on the biocompatibility of a specific material and its interaction with blood components. As a rule, gamma globulin and fibrinogen promote platelet and leukocyte adhesion and possibly activation, whereas albumin tends to neutralize these effects.[200] Thus, the relative capacity of an artificial surface to affect blood components may change significantly based on the adsorbed protein composition of its blood-exposed surfaces. In addition, it has become increasingly clear that just as inflammation and coagulation are intrinsically interrelated in normal physiology and pathophysiology, leukocyte adhesiveness on a biomaterial surface along with complement activation profoundly affect platelet deposition and activation. Hence, all aspects of material "biocompatibility" are interwoven from the mechanistic perspective.[201]

Platelets usually attach to a biomaterial and spread during the initial 15- to 90-minute period of blood–biomaterial contact.[202,203] During this time, activated platelets generally appear in the circulating blood, presumably reflecting both platelets activated by surface contact that did not achieve permanent attachment and those in the circulation that are secondarily activated by soluble products of the attached platelets. Platelet attachment may be accompanied by extensive spreading and degranulation. In other circumstances, attachment may occur with relatively modest effects on shape and granule content. After the platelet attachment phase, most biomaterials become resistant to further platelet attachment,[204,205] a phenomenon sometimes referred to as "passivation." However, some materials (e.g., hydrogel poly [vinyl alcohol]) result in continuous activation of platelets that fail to adhere to the surface of the biomaterial; in these circumstances, passivation fails[206] and the material may be less suitable for intravascular use. Other materials have been described that show very little platelet activation,[207] at least under specific temperature and flow conditions. Heparin coating of biomaterials, utilized in some CPB circuits, may result in less platelet adhesiveness.[208] Some evidence suggests that the degree and quality of platelet adhesion to artificial surfaces relate directly to the conformation of fibrinogen adsorbed to the surface and to the presence or absence of other plasma protein constituents.[209] It is reasonable to speculate that the relative composition of the protein layer in terms of other extracellular matrix molecules and coagulation factors will affect platelet attachment and activation.

In various clinical circumstances, fragmentation of platelets may also occur. In CPB, the generation of platelet microparticles (see Chapter 20) has been observed in blood and pericardial fluid.[210–212] Shear forces, detachment of platelets from nonpassivating biomaterials, platelet activation *per se,* and complement activation may all play a significant role in the generation of these potentially procoagulant entities during CPB, although the predominant mechanism remains undefined.[213] Morphologic evidence suggests that some nonpassivating biomaterials may permit transient attachment of platelets followed by partial release of the platelet and the consequent generation of microparticles on that basis, both on the biomaterial surface and released into the blood.

There are additional, relatively common, mechanisms responsible for thrombocytopenia in some patients undergoing CPB, including disseminated intravascular coagulation and primary fibrinolysis, heparin-induced thrombocytopenia (see Chapter 48), other drug-induced thrombocytopenias (e.g., antiplatelet agents) (see Chapter 49), and thrombocytopenia associated with an underlying disorder such as cyanotic congenital heart disease.[214]

## VI. Qualitative Platelet Abnormalities during CPB

### A. Platelet Hypofunction

Although the decrease in total platelet count attributable to CPB is unlikely to contribute significantly to perioperative bleeding, alteration in platelet function has been cited by numerous investigators since at least the 1970s as a significant contributor to the clinically relevant hemostatic defect of CPB.[215–218] The bleeding time is markedly prolonged during CPB (out of proportion to the degree of thrombocytopenia). Although the bleeding time normalizes 2 to 24 hours following completion of the procedure in most patients, some patients may demonstrate prolongation up to 72 hours. Although this abnormal bleeding time is usually considered to be mainly the result of platelet dysfunction, it is possible, as noted previously, that endothelial dysfunction may also contribute to this laboratory finding. By contrast, it seems less likely that quantitative abnormalities in VWF contribute significantly to the prolonged bleeding time since VWF levels even at their nadir generally do not reach those thought to prolong the bleeding time.[219] Whether qualitative abnormalities in VWF could be a contributor is uncertain.[7,8]

Standard platelet aggregation studies performed on the blood of patients undergoing CPB show a reduced response to *in vitro* stimulation with most agonists and also show refractoriness to repeated stimulation with the same agonist.[8,9,220] Clot strength is also reduced during CPB.[221] GPIb-IX, GPIIb-IIIa, catecholamine, and thrombin receptors on the surface of circulating platelets are modulated in

CPB, although the extent of these alterations is generally mild and they seem unlikely, by themselves, to result in a major clinical effect, with the possible exception of patients with cyanotic congenital heart disease.[222–226] The observed platelet dysfunction has been etiologically attributed to one or more of the following: mechanical trauma and shear stress, decreased available thrombin and ADP, contact activation, complement activation, fibrinolytic moieties such as plasmin and fibrin-split products, hypothermia, simultaneous administration of pump priming solutions and other medications, and biomaterial interactions.[7,8] There are *in vitro* model systems that support any or all of these as viable possibilities, and hypothermia in particular has been studied both *in vitro* and directly *in vivo.*[227] Many of these proposed mechanisms may operate through the production of partially activated "spent" platelets that can continue to circulate but that are in fact hypofunctional. The as-yet-unactivated platelets might remain intrinsically functionally normal, and the results of bulk platelet functional assays would then reflect the relative proportions of these different subsets of platelets. Against this hypothesis is evidence that in the setting of CPB, the platelets showing the greatest defect in aggregability are not the platelets showing the greatest degree of activation, at least as measured by P-selectin (CD62P) expression.[228] The situation may be further complicated by the fact that platelet age (time since release from the bone marrow) also affects platelet function, with the youngest platelets demonstrating the most hemostatic potential.[228] Unpublished data from our laboratory suggest that 36 hours or more is required for a significant marrow response to the thrombocytopenia of CPB, leaving a higher proportion of "older" platelets in the circulation and hence a higher proportion of unactivated but functionally less sufficient platelets. Whether the magnitude of any of these changes is sufficient to explain the observed *in vivo* and *in vitro* platelet qualitative lesions remains controversial.

In recent years, the potential role of both heparin and protamine in the genesis of the CPB-induced platelet lesion has received increased attention (summarized in Griffin et al.[229]). The molecular basis for the adverse influence of these pharmacologic agents on platelets remains speculative.[229] However, the data suggest that both these agents can impair platelet function,[230] especially under conditions of arterial-level shear, and that the optimal heparin : protamine ratio for reversal of heparin's anti-factor II and anti-factor X effects may not be the same as the optimal heparin : protamine ratio for reversal of heparin's platelet effects and minimization of the adverse platelet effects of protamine. Such data raise an additional caveat to the significance of standard platelet functional analyses in that these studies can be difficult to interpret in view of the high levels of heparin used during CPB (and hence the low levels of endogenous thrombin activity). These considerations, combined with results from whole blood studies and the analysis of platelets derived from bleeding time wounds, have led some investigators to suggest that the majority of platelets are functionally quite well preserved in CPB but that much of the apparent platelet dysfunction is in fact a result of an *in vivo* lack of appropriate platelet agonist activity.[8]

## B. Platelet Activation

Regardless of the controversy surrounding the intrinsic functional capacity of the majority of circulating platelets during CPB, there are numerous studies to support the notion that many platelets are activated by CPB,[186,226,231–233] resulting in release of granule contents and surface membrane alterations. This activation is accompanied by the production of platelet microparticles.[210] Since these circulating activated platelets may now express the adhesive ligands P-selectin and conformationally altered GPIIb-IIIa, they may be gradually lost to exposed subendothelium, fibrinogen-coated biomaterial surfaces, resident macrophages, activated endothelium, and circulating monocytes and neutrophils.[7] In addition, they will represent partially "spent" platelets that are hypofunctional relative to unactivated cells. Hence, a number of investigators have postulated that if one can prevent platelet activation on CPB, one might be able to improve both the quantitative and the qualitative platelet lesion. $PGE_1$ and $PGI_2$ were shown to reduce platelet loss in simulated CPB *in vitro* model systems, but clinical trials did not demonstrate a clear benefit of this approach and the vasodilatory effects of $PGI_2$ and its congeners are particularly problematic.[234,235] Other investigators suggested the use of disintegrins and other platelet blockers as potential therapeutic agents.[236] Although concern must be raised about the possibility that platelet functional blockade would result in increased rather than decreased bleeding, this has not been the empiric result. Even when surgery is performed after the administration of potent platelet antagonists such as abciximab, bleeding is generally no worse than usual.[237]

Another approach to decreasing platelet activation during CPB would be to attempt to block the initiating events rather than to block the function of the platelets. This approach has the attraction of leaving the unactivated platelets fully intact to provide adequate postoperative hemostasis. This raises the question of what might be the major factor involved in platelet activation in this setting. As noted previously, there are multiple possible pathways that may contribute. Increased fibrinolysis resulting in plasmin production can activate platelets.[238] Hypothermia may have the same effect,[227] as may heparin and protamine.[229,230] Direct contact with the foreign biomaterials activates platelets.[7] The contact system also produces important mediators that participate in platelet activation.[47,190] Although all these pathways may be important, activation of complement by CPB was one of the first pathophysiologic results of the procedure noted to have

an association with some clinical complications of CPB.[48,239] Combined with the fact that terminal complement components are known potent activators of platelets and can also result in platelet microparticle formation (which is prominently seen on CPB), these observations suggested that a more detailed analysis of the role of complement in CPB-induced platelet activation might be necessary.[7,240]

*In vitro* simulated cardiopulmonary bypass models, which duplicate many of the abnormalities seen in CPB patients (but which do not result in fibrinolysis since no endothelium is included), can be used to study the effects of complement blockade on platelet activation as well as on a variety of other protein and cellular phenomena. In this system, as in clinical bypass, complement is activated by the circuit, initially predominantly through the alternate pathway. If complement is blocked at the level of component C5, thus abrogating the production of both the anaphylatoxin C5a and the terminal membrane attack complex (MAC) C5b-9, there is a resultant decrease in the number of P-selectin-positive platelets as well as in the numbers of platelet–neutrophil and platelet–monocyte conjugates.[240] Neutrophil CD11b/CD18 upregulation (reflecting neutrophil activation) is similarly abrogated by C5 blockade. In addition, these alterations in the expression of the adhesion moieties P-selectin and CD11b on platelets and leukocytes, respectively, are also accompanied by concomitant changes in platelet and leukocyte adhesion to the CPB circuit; in the presence of the C5 blocking antibody, fewer platelets and leukocytes are "lost" to the circuit. If the complement cascade is blocked at the level of C8, abrogating C5b-9 production but permissive of C5a generation, platelet activation defined by P-selectin expression is abrogated but CD11b/CD18 upregulation on leukocytes is unaffected,[241] suggesting, in keeping with other *in vitro* systems, that MAC formation is an important initiator of platelet activation in this setting. With C8 blockade, as with C5 blockade, fewer platelets adhere to the CPB circuit compared to control. Interestingly, even though leukocyte adhesion molecule expression is unaffected by C8 blockade, the number of leukocytes that adhere to the circuit is reduced under these conditions, perhaps suggesting that some of the leukocyte adhesion to the CPB circuit is mediated by initial activated platelet "tethering" of leukocytes to the artificial biomaterial. Finally, at the level of C3, inhibiting C3a as well as C5a and C5b-9 generation also blocks platelet activation, as expected. This inhibition may result in more effective blockade of monocyte activation compared to that seen with blockade at the level of C5.[242] Other investigators have studied the effects of complement blockade using soluble complement receptor-1 and have demonstrated similar decreases in leukocyte activation.[243]

These data suggested that, at least in terms of the inflammatory response associated with CPB and possibly in terms of the bleeding diathesis, a clinical trial of complement blockade in bypass may have utility. In a small (35 patient) phase I clinical trial of *in vivo* complement component C5 blockade during cardiac surgery with CPB, chest tube blood drainage was shown to be decreased compared to control, although decreases in platelet activation measured by P-selectin expression did not reach statistical significance.[244] Leukocyte upregulation of CD11b/CD18 was effectively reduced by this manipulation, and preliminary trial data also showed a decrease in myocardial ischemia (measured by creatinine kinase-myocardial band [CK-MB] integration over several time points) and a decrease in neurocognitive decline. A subsequent phase II trial with a larger number of patients failed to show clear efficacy of the complement C5 blocking strategy in terms of the primary clinical end points initially chosen for the trial,[245] although there was a reduction in both hemolysis and bleeding as assessed by chest tube drainage in the patients receiving C5 blockade.[246] Nevertheless, since questions have been raised concerning the adequacy of the end points chosen, and since a post-hoc analysis revealed a greater than 60% reduction in mortality and/or myocardial infarction (defined as CK-MB ≥100 ng/ml) on postoperative days 4 and 30, a phase III trial has been initiated.

## VII. Diagnostic and Therapeutic Issues in CPB Hemostasis

As noted previously, management of heparinization during the CPB procedure is generally carried out by monitoring the ACT, although a number of alternative point-of-care approaches are also available.[13] For patients who cannot receive heparin, direct thrombin inhibitors such as hirudin or argatroban can be used. Although these agents are often more effective than heparin in reducing thrombin production and activity, the lack of a means for rapid reversal has limited their utility.

Presurgical coagulation management of the cardiac surgery patient remains an area of controversy in this era of routine use of antiplatelet therapy for patients with coronary artery, peripheral vascular, and/or cerebrovascular disease. For patients on clopidogrel, receiving the drug within 5 days preceding surgery increases the risk of major CPB-related bleeding by 50%,[247] and some experts recommend waiting at least 7 days before elective surgery. In the case of patients on aspirin, data suggest that patients who stop aspirin as recently as 3 days before surgery do not have an increased transfusion requirement.[248]

Excessive bleeding after cardiac surgery is reported in up to 20% of patients, with 2 to 6% requiring reexploration of the mediastinum.[249,250] Fifty percent of these cases of bleeding are due to an identifiable anatomic source. As discussed previously, bleeding in the postoperative (and perioperative) CPB patient is a multifactorial clinical issue,

although most authorities continue to believe that the platelet lesion predominates.[251] Unfortunately, no single test or combination of tests are clearly able to pinpoint the dominant lesion in any one patient, nor are there reproducible clinical criteria to accomplish this purpose.[252] Although there remains great potential for both laboratory-based and point-of-care approaches to better define the need for specific interventions in a bleeding post-CPB patient, and some data suggest the possibility that new instrumentation will be useful[13,253–256] (see Chapters 23 and 27–29), acute management of the bleeding postsurgical patient remains a clinical art.

For the bleeding patient in the immediate postoperative period, administration of additional protamine is often the initial intervention (generally up to a maximum dose of 1 mg per 100 U heparin previously administered), followed by empiric platelet transfusion if the former fails. If bleeding persists, many clinicians will repeat the platelet transfusion if the platelet count remains $<100 \times 10^9$/L or administer fresh frozen plasma if the PT or partial thromboplastin time is >1.5 that of control. Depending on the clinical scenario, surgical exploration may be necessary if these interventions are ineffective.

Truly refractory and intractable bleeding after CPB is rare; when it occurs, however, it may be devastating and its origin, even in retrospect, often remains obscure. Experience has been reported with the use of recombinant activated factor VII (factor VIIa; see Chapter 68) in this setting.[257] In 20 literature cases, 14 patients had reversal of bleeding after a single dose (mean, 57 μg/kg) and the remainder after multiple doses (mean, 3.4 doses). Of note, however, 2 patients experienced thromboembolic complications, one of which was fatal. Hence, although factor VIIa may be effective, it is not without risk of significant complications.

Because of the difficulty in management once bleeding begins, most effort is directed at prophylactic approaches. The close interrelationship of hemostasis with inflammation in the setting of cardiac surgery makes it incumbent to evaluate the efficacy of such hemostatic interventions in terms of their effect on inflammation as well as on bleeding and thrombosis. Indeed, several investigators have proposed that antiplatelet therapy could result in amelioration of the systemic inflammatory response, reduce ischemia–reperfusion injury, and accomplish these goals without a concomitant increase in hemorrhage. Studies suggest that perioperative administration of the GPIIb-IIIa antagonists abciximab, tirofiban, or eptifibatide does not increase bleeding but does lead to improved preservation of platelet counts and perhaps a reduction in measured inflammatory parameters,[258–260] a phenomenon referred to as "platelet anesthesia" by Edmunds' group.[261] *In vitro* data obtained from a simulated CPB model suggest that GPIIb-IIIa antagonists abrogate P-selectin expression on the platelets, block platelet–leukocyte conjugate formation, and reduce leukocyte activation measured by elastase release.[262] Similarly, experimental work with nitric oxide, nitroprusside, or iloprost to improve hemostasis is active, with some interesting preclinical and pilot findings.[263–265] All of these direct antiplatelet interventions, however, remain in the experimental stage and are not part of routine care at most centers.

Therapeutically, the most successful agent in prophylactically alleviating the bleeding diathesis associated with CPB is the general serine protease inhibitor aprotinin.[145–151] Although the primary mechanism of action of aprotinin is probably inhibition of plasmin, some investigators have suggested that aprotinin may have an additional direct beneficial effect on platelet function, whereas others have failed to find such an association.[266,267] Aprotinin also has significant anti-inflammatory effects, and there is evidence of a myocardial protective effect. As discussed previously, other fibrinolytic inhibitors have also shown efficacy, and in some centers EACA is used routinely for patients undergoing their first procedure, with aprotinin used for patients undergoing repeat CPB cardiac surgery ("redo" surgery) and for higher risk patients. Although tempting from a theoretical perspective, attempts to minimize heparin dose by infusing antithrombin III have met with only modest success[268] and this strategy is not routinely employed.

There have been attempts to improve the platelet–vascular lesion. Desmopressin (see Chapter 67) has failed to show significant clinical efficacy in the prophylactic setting, although hope remains that further laboratory testing might be able to identify individual patients who would benefit from such intervention.[269] Leukofiltration strategies,[270] low-dose statin administration,[271] and the use of nitric oxide donors, antioxidants, and phosphodiesterase inhibitors all have their advocates for further experimental work. Note that routine corticosteroid administration is utilized in some centers (for its anti-inflammatory properties), although its overall benefit remains controversial.

Overall, hemostatic management, particularly management of the platelet abnormalities associated with CPB, remains as much medical art as clear application of underlying pathophysiologic principles. The hope remains, however, that further elucidation of underlying pathophysiologies will continue to provide improved outcomes for these cardiac surgical patients.

# References

1. Lillehei, C. W. (2000). Historical development of cardiopulmonary bypass. In G. P. Gravlee, R. F. Davis, M. Kurusz, et al. (Eds.), *Cardiopulmonary bypass: Principles and practice* (2nd ed.). Philadelphia: Lippincott Williams & Wilkins.
2. Kurusz, M. (2004). Cardiopulmonary bypass: Past, present, and future. *ASAIO J, 50,* 33–36.
3. Chughati, T. S., Gilardino, M. S., Fleiszer, D. M., et al. (2002). An expanding role for cardiopulmonary bypass in trauma. *Can J Surg, 45,* 95–103.
4. Newman, M. F., Kirchner, J. L., Phillips-Bute, B., et al. (2001). Longitudinal assessment of neurocognitive function after coronary artery bypass surgery. *N Engl J Med, 344,* 395–402.
5. Rinder, C. S. (2000). Hematologic effects of cardiopulmonary bypass. In G. P. Gravlee, R. F. Davis, M. Kurusz, et al. (Eds.), *Cardiopulmonary bypass: Principles and practice.* Philadelphia: Lippincott Williams & Wilkins.
6. Rinder, C. S., & Rinder, H. M. (2000). Preservation of platelet function after cardiac surgery and apheresis. In J. Seghatchian, E. L. Snyder, & P. Krailadsiri (Eds.), *Platelet therapy: Current status and future trends.* New York: Elsevier.
7. Smith, B. (1998). Interaction of blood and artificial surfaces. In A. I. Schafer & J. Loscalzo (Eds.), *Thrombosis and hemorrhage* (2nd ed., pp. 925–941). Baltimore: Williams & Wilkins.
8. Khuri, S. F., Michelson, A. D., & Valeri, C. R. (1998). Effects of cardiopulmonary bypass on hemostasis. In A. I. Schafer & J. Loscalzo (Eds.), *Thrombosis and hemorrhage* (2nd ed., pp. 1091–1117). Baltimore: Williams & Wilkins.
9. Edmunds, L. H., Jr. (2001). Hemostatic problems in surgical patients. In R. W. Colman, J. Hirsh, V. J. Marder, et al. (Eds.), *Hemostasis and thrombosis: Basic principles and clinical practice* (4th ed., pp. 1031–1043). Philadelphia: Lippincott Williams & Wilkins.
10. Rinder, C. S., Rinder, H. M., & Smith, B. R. (2003). Neutrophil activation and CD11b upregulation during cardiopulmonary bypass is associated with greater postoperative renal injury. *Ann Thorac Surg, 75,* 899–905.
11. Rinder, C. S. (2000). Platelets and their interactions. In B. D. Spiess (Ed.), *The relationship between coagulation, inflammation and endothelium — A pyramid towards outcome.* Philadelphia: Lippincott Williams & Wilkins.
12. Rinder, C. S., Mathew, J. P., Rinder, H. M., et al. (1997). Immunosuppression and immune stimulation during cardiopulmonary bypass: Effects of aging and gender. *J Lab Clin Med, 129,* 592–602.
13. Shore-Lesserson, L. (2003). Monitoring anticoagulation and hemostasis in cardiac surgery. *Anesthesiol Clin N Am, 21,* 511–526.
14. Gravlee, G. P., Haddon, W. S., Rothberger, H. K., et al. (1990). Heparin dosing and monitoring for cardiopulmonary bypass. A comparison of techniques with measurement of subclinical plasma coagulation. *J Thorac Cardiovasc Surg, 99,* 518–527.
15. Belboul, A., Akbar, O., Lofgren, C., et al. (2000). Improved blood cellular biocompatibility with heparin coated circuits during cardiopulmonary bypass. *J Cardiovasc Surg (Torino), 41,* 357–362.
16. Videm, V., Mollnes, T. E., Bergh, K., et al. (1999). Heparin-coated cardiopulmonary bypass equipment: II. Mechanisms for reduced complement activation *in vivo. J Thorac Cardiovasc Surg, 117,* 803–809.

17. Rubens, F. D., Labow, R. S., Lavallee, G. R., et al. (1999). Hematologic evaluation of cardiopulmonary bypass circuits prepared with a novel block copolymer. *Ann Thorac Surg, 67,* 689–698.

18. Ikuta, T., Fujii, H., Shibata, T., et al. (2004). A new poly-2-methoxyethylacrylate-coated cardiopulmonary bypass circuit possesses superior platelet preservation and inflammatory suppression efficacy. *Ann Thorac Surg, 77,* 1678–1683.

19. Bull, B. S., Huse, W. M., Brauer, F. S., et al. (1975). Heparin therapy during extracorporeal circulation. *J Thorac Surg, 69,* 685–689.

20. Horrow, J. C. (1988). Protamine allergy. *J Cardiothor Anesth, 2,* 225–252.

21. Tan, F., Jackman, H., Skidgel, R. A., et al. (1989). Protamine inhibits plasma carboxypeptidase N, the inactivator of anaphylatoxins and kinins. *Anesthesiology, 70,* 267–275.

22. Patla, V., Comunale, M. E., & Lowenstein, E. (2000). Heparin neutralization. In G. P. Gravlee, R. F. Davis, M. Kurusz, et al. (Eds.), *Cardiopulmonary bypass: Principles and practice* (2nd ed.). Philadelphia: Lippincott Williams & Wilkins.

23. Comunale, M. E., Haering, J. M., Robertson, L. K., et al. (1997). Aspirin prevents protamine-induced pulmonary hypertension. *Anesth Analg, 84,* S66.

24. Kesteven, P. J., Ahmed, A., Aps, C., et al. (1986). Protamine sulphate and heparin rebound following open-heart surgery. *J Cardiovasc Surg (Torino), 27,* 600–603.

25. Undar, A., Rosenberg, G., & Myers, J. L. (2005). Major factors in the controversy of pulsatile versus nonpulsatile flow during acute and chronic cardiac support. *ASAIO J, 51,* 173–175.

26. Canver, C. C., & Nichols, R. D. (2000). Use of intraoperative hetastarch priming during coronary bypass. *Chest, 118,* 1616–1620.

27. Michelson, A. D., MacGregor, H., Barnard, M. R., et al. (1994). Reversible inhibition of human platelet activation by hypothermia *in vivo* and *in vitro. Thromb Haemost, 71,* 633–640.

28. Gazmuri, R. J., & Gopalakrishnan, P. (2003). Hypothermia: Cooling down inflammation. *Crit Care Med, 31,* 2811–2812.

29. Qing, M., Nimmesgern, A., Heinrich, P. C., et al. (2003). Intrahepatic synthesis of tumor necrosis factor-alpha related to cardiac surgery is inhibited by interleukin-10 via the Janus kinase (Jak)/signal transducers and activator of transcription (STAT) pathway. *Crit Care Med, 31,* 2769–2775.

30. Faraday, N., & Rosenfeld, B. A. (1998). *In vitro* hypothermia enhances platelet GPIIb–IIIa activation and P-selectin expression. *Anesthesiology, 88,* 1579–1585.

31. Goodnough, L. T., Despotis, G. J., Hogue, C. W., Jr., et al. (1995). On the need for improved transfusion indicators in cardiac surgery. *Ann Thorac Surg, 60,* 473–480.

32. Shuhaiber, J. H. (2005). Randomized prospective trial for blood transfusion during adult cardiopulmonary bypass surgery. *J Thorac Cardiovasc Surg, 129,* 1200–1201.

33. Parr, K. G., Patel, M. A., Dekker, R., et al. (2003). Multivariate predictors of blood product use in cardiac surgery. *J Cardiothorac Vasc Anesth, 17,* 176–181.

34. Laub, G. W., Dharan, M., Riebman, J. B., et al. (1993). The impact of intraoperative autotransfusion on cardiac surgery. A prospective randomized double-blind study. *Chest, 104,* 686–689.

35. Desmond, M. J., Thomas, M. J., Gillon, J., et al. (1996). Consensus conference on autologous transfusion. Perioperative red cell salvage. *Transfusion (Paris), 36,* 644–651.

36. Tylman, M., Bengtson, J. P., & Bengtsson, A. (2003). Activation of the complement system by different autologous transfusion devices: An *in vitro* study. *Transfusion (Paris), 43,* 395–399.

37. Arnestad, J. P., Bengtsson, A., Bengtson, J. P., et al. (1994). Formation of cytokines by retransfusion of shed whole blood. *Br J Anaesth, 72,* 422–425.

38. Shander, A., & Rijhwani, T. S. (2005). Clinical outcomes in cardiac surgery: Conventional surgery versus bloodless surgery. *Anesthesiol Clin North Am, 23,* 327–345.

39. Verma, S., Fedak, P. W., Weisel, R. D., et al. (2004). Off-pump coronary artery bypass surgery: Fundamentals for the clinical cardiologist. *Circulation, 109,* 1206–1211.

40. Ngaage, D. L. (2003). Off-pump coronary artery bypass grafting: The myth, the logic and the science. *Eur J Cardiothorac Surg, 24,* 557–570.

41. Biglioli, P., Cannata, A., Alamanni, F., et al. (2003). Biological effects of off-pump vs. on-pump coronary artery surgery: Focus on inflammation, hemostasis and oxidative stress. *Eur J Cardiothorac Surg, 24,* 260–269.

42. Englberger, L., Immer, F. F., Eckstein, F. S., et al. (2004). Off-pump coronary artery bypass operation does not increase procoagulant and fibrinolytic activity: Preliminary results. *Ann Thorac Surg, 77,* 1560–1566.

43. Legare, J. F., Buth, K. J., King, S., et al. (2004). Coronary bypass surgery performed off pump does not result in lower in-hospital morbidity than coronary artery bypass grafting performed on pump. *Circulation, 109,* 887–892.

44. Khan, N. E., De Souza, A., Mister, R., et al. (2004). A randomized comparison of off-pump and on-pump multivessel coronary-artery bypass surgery. *N Engl J Med, 350,* 21–28.

45. Levy, J. H., & Tanaka, K. A. (2003). Inflammatory response to cardiopulmonary bypass. *Ann Thorac Surg, 75,* S715–S720.

46. Pintar, T., & Collard, C. D. (2003). The systemic inflammatory response to cardiopulmonary bypass. *Anesthesiol Clin North Am, 21,* 453–464.

47. Wachtfogel, Y. T., Harpel, P. C., Edmunds, L. H., Jr., et al. (1989). Formation of C1s–C1-inhibitor, kallikrein–C1-inhibitor, and plasmin–alpha 2-plasmin-inhibitor complexes during cardiopulmonary bypass. *Blood, 73,* 468–471.

48. Chenoweth, D. E., Cooper, S. W., Hugli, T. E., et al. (1981). Complement activation during cardiopulmonary bypass: Evidence for generation of C3a and C5a anaphylatoxins. *N Engl J Med, 304,* 497–503.

49. Kirklin, J. K., Chenoweth, D. E., Naftel, D. C., et al. (1986). Effects of protamine administration after cardiopulmonary bypass on complement, blood elements, and the hemodynamic state. *Ann Thorac Surg, 41,* 193–199.

50. Cavarocchi, N. C., Schaff, H. V., Orszulak, T. A., et al. (1985). Evidence for complement activation by protamine–

heparin interaction after cardiopulmonary bypass. *Surgery, 98*, 525–531.

51. Faymonville, M. E., Deby-Dupont, G., Larbuisson, R., et al. (1986). Prostaglandin E2, prostacyclin, and thromboxane changes during nonpulsatile cardiopulmonary bypass in humans. *J Thorac Cardiovasc Surg, 91*, 858–866.

52. Hashimoto, K., Miyamoto, H., Suzuki, K., et al. (1992). Evidence of organ damage after cardiopulmonary bypass. The role of elastase and vasoactive mediators. *J Thorac Cardiovasc Surg, 104*, 666–673.

53. Wachtfogel, Y. T., Kucich, U., Greenplate, J., et al. (1987). Human neutrophil degranulation during extracorporeal circulation. *Blood, 69*, 324–330.

54. Haeffner-Cavaillon, N., Roussellier, N., Ponzio, O., et al. (1989). Induction of interleukin-1 production in patients undergoing cardiopulmonary bypass. *J Thorac Cardiovasc Surg, 98*, 1100–1106.

55. Steinberg, J. B., Kapelanski, D. P., Olson, J. D., et al. (1993). Cytokine and complement levels in patients undergoing cardiopulmonary bypass. *J Thorac Cardiovasc Surg, 106*, 1008–1016.

56. Frering, B., Philip, I., Dehoux, M., et al. (1994). Circulating cytokines in patients undergoing normothermic cardiopulmonary bypass. *J Thorac Cardiovasc Surg, 108*, 636–641.

57. Kawahito, K., Kawakami, M., Fujiwara, T., et al. (1995). Interleukin-8 and monocyte chemotactic activating factor responses to cardiopulmonary bypass. *J Thorac Cardiovasc Surg, 110*, 99–102.

58. Chello, M., Mastroroberto, P., Quirino, A., et al. (2002). Inhibition of neutrophil apoptosis after coronary bypass operation with cardiopulmonary bypass. *Ann Thorac Surg, 73*, 123–130.

59. Murray, M. J., & Cook, D. J. (2000). Noncardiovascular applications of cardiopulmonary bypass. In G. P. Gravlee, R. F. Davis, M. Kurusz, et al. (Eds.), *Cardiopulmonary bypass: Principles and practice* (2nd ed.). Philadelphia: Lippincott Williams & Wilkins.

60. Seyfer, A. E., Seaber, A. V., Dombrose, F. A., et al. (1981). Coagulation changes in elective surgery and trauma. *Ann Surg, 193*, 210–213.

61. Czerny, M., Baumer, H., Kilo, J., et al. (2000). Inflammatory response and myocardial injury following coronary artery bypass grafting with or without cardiopulmonary bypass. *Eur J Cardiothorac Surg, 17*, 737–742.

62. Kshettry, V. R., Flavin, T. F., Emery, R. W., et al. (2000). Does multivessel, off-pump coronary artery bypass reduce postoperative morbidity? *Ann Thorac Surg, 69*, 1725–1730.

63. Ascione, R., Lloyd, C. T., Underwood, M. J., et al. (2000). Inflammatory response after coronary revascularization with or without cardiopulmonary bypass. *Ann Thorac Surg, 69*, 1198–1204.

64. Matata, B. M., Sosnowski, A. W., & Galinanes, M. (2000). Off-pump bypass graft operation significantly reduces oxidative stress and inflammation. *Ann Thorac Surg, 69*, 785–791.

65. Taggart, D. P., Browne, S. M., Halligan, P. W., et al. (1999). Is cardiopulmonary bypass still the cause of cognitive dysfunction after cardiac operations? *J Thorac Cardiovasc Surg, 118*, 414–420.

66. Wan, S., Izzat, M. B., Lee, T. W., et al. (1999). Avoiding cardiopulmonary bypass in multivessel CABG reduces cytokine response and myocardial injury. *Ann Thorac Surg, 68*, 52–57.

67. Gu, Y. J., Mariani, M. A., Boonstra, P. W., et al. (1999). Complement activation in coronary artery bypass grafting patients without cardiopulmonary bypass: The role of tissue injury by surgical incision. *Chest, 116*, 892–898.

68. Struber, M., Cremer, J. T., Gohrbandt, B., et al. (1999). Human cytokine responses to coronary artery bypass grafting with and without cardiopulmonary bypass. *Ann Thorac Surg, 68*, 1330–1335.

69. Liebold, A., Keyl, C., & Birnbaum, D. E. (1999). The heart produces but the lungs consume proinflammatory cytokines following cardiopulmonary bypass. *Eur J Cardiothorac Surg, 15*, 340–345.

70. Diegeler, A., Tarnok, A., Rauch, T., et al. (1998). Changes of leukocyte subsets in coronary artery bypass surgery: Cardiopulmonary bypass versus "off-pump" techniques. *Thorac Cardiovasc Surg, 46*, 327–332.

71. Laurenzi, I. J., & Diamond, S. L. (1999). Monte Carlo simulation of the heterotypic aggregation kinetics of platelets and neutrophils. *Biophys J, 77*, 1733–1746.

72. Wagner, D. D., & Burger, P. C. (2003). Platelets in inflammation and thrombosis. *Arterioscler Thromb Vasc Biol, 23*, 2131–2137.

73. Schmidtke, D. W., & Diamond, S. L. (2000). Direct observation of membrane tethers formed during neutrophil attachment to platelets or P-selectin under physiological flow. *J Cell Biol, 149*, 719–730.

74. Palabrica, T., Lobb, R., Furie, B. C., et al. (1992). Leukocyte accumulation promoting fibrin deposition is mediated *in vivo* by P-selectin on adherent platelets. *Nature, 359*, 848–851.

75. Lefer, A. M., Campbell, B., Scalia, R., et al. (1998). Synergism between platelets and neutrophils in provoking cardiac dysfunction after ischemia and reperfusion: Role of selectins. *Circulation, 98*, 1322–1328.

76. Kirchhofer, D., Riederer, M. A., & Baumgartner, H. R. (1997). Specific accumulation of circulating monocytes and polymorphonuclear leukocytes on platelet thrombi in a vascular injury model. *Blood, 89*, 1270–1278.

77. Hagberg, I. A., Roald, H. E., & Lyberg, T. (1998). Adhesion of leukocytes to growing arterial thrombi. *Thromb Haemost, 80*, 852–858.

78. Diacovo, T. G., Roth, S. J., Buccola, J. M., et al. (1996). Neutrophil rolling, arrest, and transmigration across activated, surface-adherent platelets via sequential action of P-selectin and the beta 2-integrin CD11b/CD18. *Blood, 88*, 146–157.

79. Kuijper, P. H., Gallardo Torres, H. I., Lammers, J. W., et al. (1997). Platelet and fibrin deposition at the damaged vessel wall: Cooperative substrates for neutrophil adhesion under flow conditions. *Blood, 89*, 166–175.

80. Ostrovsky, L., King, A. J., Bond, S., et al. (1998). A juxtacrine mechanism for neutrophil adhesion on platelets involves

platelet-activating factor and a selectin-dependent activation process. *Blood, 91,* 3028–3036.

81. Rinder, H. M., Tracey, J. L., Rinder, C. S., et al. (1994). Neutrophil but not monocyte activation inhibits P-selectin-mediated platelet adhesion. *Thromb Haemost, 72,* 750–756.

82. Lorant, D. E., McEver, R. P., McIntyre, T. M., et al. (1995). Activation of polymorphonuclear leukocytes reduces their adhesion to P-selectin and causes redistribution of ligands for P-selectin on their surfaces. *J Clin Invest,96,* 171–182.

83. Forlow, S. B., McEver, R. P., & Nollert, M. U. (2000). Leukocyte–leukocyte interactions mediated by platelet microparticles under flow. *Blood, 95,* 1317–1323.

84. Konstantopoulos, K., Neelamegham, S., Burns, A. R., et al. (1998). Venous levels of shear support neutrophil–platelet adhesion and neutrophil aggregation in blood via P-selectin and beta2-integrin. *Circulation, 98,* 873–882.

85. Brandt, E., Petersen, F., Ludwig, A., et al. (2000). The beta-thromboglobulins and platelet factor 4: Blood platelet-derived CXC chemokines with divergent roles in early neutrophil regulation. *J Leukocyte Biol, 67,* 471–478.

86. Sambrano, G. R., Huang, W., Faruqi, T., et al. (2000). Cathepsin G activates protease-activated receptor-4 in human platelets. *J Biol Chem, 275,* 6819–6823.

87. Hirayama, A., Noronha-Dutra, A. A., Gordge, M. P., et al. (1999). S-nitrosothiols are stored by platelets and released during platelet–neutrophil interactions. *Nitric Oxide, 3,* 95–104.

88. Gronert, K., Clish, C. B., Romano, M., et al. (1999). Transcellular regulation of eicosanoid biosynthesis. *Methods Mol Biol, 120,* 119–144.

89. Marcus, A. (1990). Thrombosis and inflammation as multicellular processes: Pathophysiologic significance of transcellular metabolism. *Blood, 76,* 1903–1907.

90. Piccardoni, P., Evangelista, V., Piccoli, A., et al. (1996). Thrombin-activated human platelets release two NAP-2 variants that stimulate polymorphonuclear leukocytes. *Thromb Haemost, 76,* 780–785.

91. Valles, J., Santos, M. T., Marcus, A. J., et al. (1993). Down-regulation of human platelet reactivity by neutrophils. Participation of lipoxygenase derivatives and adhesive proteins. *J Clin Invest, 92,* 1357–1365.

92. Kowalska, M. A., Ratajczak, M. Z., Majka, M., et al. (2000). Stromal cell-derived factor-1 and macrophage-derived chemokine: two chemokines that activate platelets. *Blood, 96,* 50–57.

93. Flad, H. D., Harter, L., Petersen, F., et al. (1997). Regulation of neutrophil activation by proteolytic processing of platelet-derived alpha-chemokines. *Adv Exp Med Biol, 421,* 223–230.

94. Rinder, C., Bonan, J., Rinder, J., et al. (1992). Cardiopulmonary bypass induces leukocyte–platelet adhesion. *Blood, 79,* 1201–1205.

95. Zahler, S., Massoudy, P., Hartl, H., et al. (1999). Acute cardiac inflammatory responses to postischemic reperfusion during cardiopulmonary bypass. *Cardiovasc Res, 41,* 722–730.

96. Xiao, Z., Theroux, P., & Frojmovic, M. (1999). Modulation of platelet–neutrophil interaction with pharmacological inhi-bition of fibrinogen binding to platelet GPIIb/IIIa receptor. *Thromb Haemost, 81,* 281–285.

97. Kogaki, S., Sawa, Y., Sano, T., et al. (1999). Selectin on activated platelets enhances neutrophil endothelial adherence in myocardial reperfusion injury. *Cardiovasc Res, 43,* 968–973.

98. Chauvet, P., Bienvenu, J. G., Theoret, J. F., et al. (1999). Inhibition of platelet–neutrophil interactions by Fucoidan reduces adhesion and vasoconstriction after acute arterial injury by angioplasty in pigs. *J Cardiovasc Pharmacol, 34,* 597–603.

99. Ott, I., Neumann, F. J., Gawaz, M., et al. (1996). Increased neutrophil–platelet adhesion in patients with unstable angina. *Circulation, 94,* 1239–1246.

100. Gutensohn, K., Alisch, A., Krueger, W., et al. (2000). Extracorporeal platelet pheresis induces the interaction of activated platelets with white blood cells. *Vox Sang, 78,* 101–105.

101. Kirschenbaum, L. A., Aziz, M., Astiz, M. E., et al. (2000). Influence of rheologic changes and platelet–neutrophil interactions on cell filtration in sepsis. *Am J Respir Crit Care Med, 161,* 1602–1607.

102. Gawaz, M. P., Mujais, S. K., Schmidt, B., et al. (1999). Platelet–leukocyte aggregates during hemodialysis: Effect of membrane type. *Artif Organs, 23,* 29–36.

103. Minamino, T., Kitakaze, M., Asanuma, H., et al. (1998). Endogenous adenosine inhibits P-selectin-dependent formation of coronary thromboemboli during hypoperfusion in dogs. *J Clin Invest, 101,* 1643–1653.

104. Esposito, C. J., Popescu, W. M., Rinder, H. M., et al. (2003). Increased leukocyte–platelet adhesion in patients with graft occlusion after peripheral vascular surgery. *Thromb Haemost, 90,* 1128–1134.

105. Valant, P. A., Jy, W., Horstman, L. L., et al. (1998). Thrombotic thrombocytopenic purpura plasma enhances platelet–leucocyte interaction *in vitro. Br J Haematol, 100,* 24–32.

106. Smith, B. R., & Rinder, H. M. (1997). Interactions of platelets and endothelial cells with erythrocytes and leukocytes in thrombotic thrombocytopenic purpura. *Semin Hematol, 34,* 90–97.

107. Rinder, H. M., Bonan, J. L., Rinder, C. S., et al. (1991). Activated and unactivated platelet adhesion to monocytes and neutrophils. *Blood, 78,* 1760–1769.

108. Rinder, H. M., Bonan, J. L., Rinder, C. S., et al. (1991). Dynamics of leukocyte–platelet adhesion in whole blood. *Blood, 78,* 1730–1737.

109. Rinder, C. S., Student, L. A., Bonan, J. L., et al. (1993). Aspirin does not inhibit adenosine diphosphate-induced platelet alpha-granule release. *Blood, 82,* 505–512.

110. Larsen, E., Celi, A., Gilbert, G. E., et al. (1989). PADGEM protein: A receptor that mediates the interaction of activated platelets with neutrophils and monocytes. *Cell, 59,* 305–312.

111. Moore, K. L., Eaton, S. F., Lyons, D. E., et al. (1994). The P-selectin glycoprotein ligand from human neutrophils displays sialylated, fucosylated, O-linked poly-N-acetyllactos-amine. *J Biol Chem, 269,* 23318–23327.

112. Simon, D. I., Chen, Z., Xu, H., et al. (2000). Platelet glycoprotein ibalpha is a counterreceptor for the leukocyte integrin Mac-1 (CD11b/CD18). *J Exp Med, 192,* 193–204.

113. Celi, A., Pellegrini, G., Lorenzet, R., et al. (1994). P-selectin induces the expression of tissue factor on monocytes. *Proc Natl Acad Sci USA, 91,* 8767–8771.

114. Amirkhosravi, A., Alexander, M., May, K., et al. (1996). The importance of platelets in the expression of monocyte tissue factor antigen measured by a new whole blood flow cytometric assay. *Thromb Haemost, 75,* 87–95.

115. Osterud, B. (1998). Tissue factor expression by monocytes: Regulation and pathophysiological roles. *Blood Coagul Fibrinolysis,9* (Suppl 1), S9–S14.

116. Koike, J., Nagata, K., Kudo, S., et al. (2000). Density-dependent induction of TNF-alpha release from human monocytes by immobilized P-selectin. *FEBS Lett, 477,* 84–88.

117. Ammon, C., Kreutz, M., Rehli, M., et al. (1998). Platelets induce monocyte differentiation in serum-free coculture. *J Leukocyte Biol, 63,* 469–476.

118. Neumann, F. J., Marx, N., Gawaz, M., et al. (1997). Induction of cytokine expression in leukocytes by binding of thrombin-stimulated platelets. *Circulation, 95,* 2387–2394.

119. Weyrich, A. S., McIntyre, T. M., McEver, R. P., et al. (1995). Monocyte tethering by P-selectin regulates monocyte chemotactic protein-1 and tumor necrosis factor-alpha secretion. Signal integration and NF-kappa B translocation. *J Clin Invest, 95,* 2297–2303.

120. Weyrich, A. S., Elstad, M. R., McEver, R. P., et al. (1996). Activated platelets signal chemokine synthesis by human monocytes. *J Clin Invest, 97,* 1525–1534.

121. Bengtsson, T., Fryden, A., Zalavary, S., et al. (1999). Platelets enhance neutrophil locomotion: Evidence for a role of P-selectin. *Scand J Clin Lab Invest, 59,* 439–449.

122. Nagata, K., Tsuji, T., Todoroki, N., et al. (1993). Activated platelets induce superoxide anion release by monocytes and neutrophils through P-selectin (CD62). *J Immunol, 151,* 3267–3273.

123. Evangelista, V., Manarini, S., Sideri, R., et al. (1999). Platelet/polymorphonuclear leukocyte interaction: P-selectin triggers protein-tyrosine phosphorylation-dependent CD11b/CD18 adhesion: Role of PSGL-1 as a signaling molecule. *Blood, 93,* 876–885.

124. Henn, V., Slupsky, J. R., Grafe, M., et al. (1998). CD40 ligand on activated platelets triggers an inflammatory reaction of endothelial cells. *Nature, 391,* 591–594.

125. Lindemann, S., Tolley, N. D., Dixon, D. A., et al. (2001). Activated platelets mediate inflammatory signaling by regulated interleukin 1beta synthesis. *J Cell Biol, 154,* 485–490.

126. Ross, R. (1985). Platelets, platelet-derived growth factor, growth control, and their interactions with the vascular wall. *J Cardiovasc Pharmacol, 7*(Suppl 3), S186–S190.

127. von Hundelshausen, P., Weber, K. S., Huo, Y., et al. (2001). RANTES deposition by platelets triggers monocyte arrest on inflamed and atherosclerotic endothelium. *Circulation, 103,* 1772–1777.

128. Ferroni, P., Speziale, G., Ruvolo, G., et al. (1998). Platelet activation and cytokine production during hypothermic cardiopulmonary bypass — A possible correlation? *Thromb Haemost, 80,* 58–64.

129. Despotis, G. J., Filos, K. S., Zoys, T. N., et al. (1996). Factors associated with excessive postoperative blood loss and hemostatic transfusion requirements: A multivariate analysis in cardiac surgical patients. *Anesth Analg, 82,* 13–21.

130. Paparella, D., Brister, S. J., & Buchanan, M. R. (2004). Coagulation disorders of cardiopulmonary bypass: A review. *Intensive Care Med, 30,* 1873–1881.

131. Chung, J. H., Gikakis, N., Rao, A. K., et al. (1996). Pericardial blood activates the extrinsic coagulation pathway during clinical cardiopulmonary bypass. *Circulation, 93,* 2014–2018.

132. Barstad, R. M., Ovrum, E., Ringdal, M. A., et al. (1996). Induction of monocyte tissue factor procoagulant activity during coronary artery bypass surgery is reduced with heparin-coated extracorporeal circuit. *Br J Haematol, 94,* 517–525.

133. Ernofsson, M., Thelin, S., & Siegbahn, A. (1997). Monocyte tissue factor expression, cell activation, and thrombin formation during cardiopulmonary bypass: A clinical study. *J Thorac Cardiovasc Surg, 113,* 576–584.

134. Boisclair, M. D., Lane, D. A., Philippou, H., et al. (1993). Mechanisms of thrombin generation during surgery and cardiopulmonary bypass. *Blood, 82,* 3350–3357.

135. Shibamiya, A., Tabuchi, N., Chung, J., et al. (2004). Formation of tissue factor-bearing leukocytes during and after cardiopulmonary bypass. *Thromb Haemost, 92,* 124–131.

136. Kongsgaard, U. E., Smith-Erichsen, N., Geiran, O., et al. (1989). Changes in the coagulation and fibrinolytic systems during and after cardiopulmonary bypass surgery. *Thorac Cardiovasc Surg, 37,* 158–162.

137. Stibbe, J., Kluft, C., Brommer, E. J., et al. (1984). Enhanced fibrinolytic activity during cardiopulmonary bypass in open-heart surgery in man is caused by extrinsic (tissue-type) plasminogen activator. *Eur J Clin Invest, 14,* 375–382.

138. Khuri, S. F., Valeri, C. R., Loscalzo, J., et al. (1995). Heparin causes platelet dysfunction and induces fibrinolysis before cardiopulmonary bypass. *Ann Thorac Surg, 60,* 1008–1014.

139. Upchurch, G. R., Valeri, C. R., Khuri, S. F., et al. (1996). Effect of heparin on fibrinolytic activity and platelet function *in vivo. Am J Physiol, 271,* H528–H534.

140. Tabuchi, N., de Haan, J., Boonstra, P. W., et al. (1993). Activation of fibrinolysis in the pericardial cavity during cardiopulmonary bypass. *J Thorac Cardiovasc Surg, 106,* 828–833.

141. Khuri, S. F., Wolfe, J. A., Josa, M., et al. (1992). Hematologic changes during and after cardiopulmonary bypass and their relationship to the bleeding time and nonsurgical blood loss. *J Thorac Cardiovasc Surg, 104,* 94–107.

142. Ray, M. J., Marsh, N. A., & Hawson, G. A. (1994). Relationship of fibrinolysis and platelet function to bleeding after cardiopulmonary bypass. *Blood Coagul Fibrinolysis, 5,* 679–685.

143. Arom, K. V., & Emery, R. W. (1994). Decreased postoperative drainage with addition of epsilon-aminocaproic acid

before cardiopulmonary bypass. *Ann Thorac Surg, 57,* 1108–1103.

144. Karski, J. M., Teasdale, S. J., Norman, P., et al. (1995). Prevention of bleeding after cardiopulmonary bypass with high-dose tranexamic acid. Double-blind, randomized clinical trial. *J Thorac Cardiovasc Surg, 110,* 835–842.

145. Mastroroberto, P., Chello, M., Zofrea, S., et al. (1995). Suppressed fibrinolysis after administration of low-dose aprotinin: Reduced level of plasmin–alpha2-plasmin inhibitor complexes and postoperative blood loss. *Eur J Cardiothorac Surg, 9,* 143–145.

146. Kawasuji, M., Ueyama, K., Sakakibara, N., et al. (1993). Effect of low-dose aprotinin on coagulation and fibrinolysis in cardiopulmonary bypass. *Ann Thorac Surg, 55,* 1205–1209.

147. Asimakopoulos, G., Kohn, A., Stefanou, D. C., et al. (2000). Leukocyte integrin expression in patients undergoing cardiopulmonary bypass. *Ann Thorac Surg, 69,* 1192–1197.

148. Landis, R. C., Haskard, D. O., & Taylor, K. M. (2001). New antiinflammatory and platelet-preserving effects of aprotinin. *Ann Thorac Surg, 72,* S1808–S1813.

149. Greilich, P. E., Brouse, C. F., Rinder, C. S., et al. (2004). Effects of epsilon-aminocaproic acid and aprotinin on leukocyte–platelet adhesion in patients undergoing cardiac surgery. *Anesthesiology, 100,* 225–233.

150. Greilich, P. E., Brouse, C. F., Whitten, C. W., et al. (2003). Antifibrinolytic therapy during cardiopulmonary bypass reduces proinflammatory cytokine levels: A randomized, double-blind, placebo-controlled study of epsilon-aminocaproic acid and aprotinin. *J Thorac Cardiovasc Surg, 126,* 1498–1503.

151. Alderman, E. L., Levy, J. H., Rich, J. B., et al. (1998). Analyses of coronary graft patency after aprotinin use: Results from the International Multicenter Aprotinin Graft Patency Experience (IMAGE) trial. *J Thorac Cardiovasc Surg, 116,* 716–730.

152. Boyle, E. M., Jr., Pohlman, T. H., Johnson, M. C., et al. (1997). Endothelial cell injury in cardiovascular surgery: The systemic inflammatory response. *Ann Thorac Surg, 63,* 277–284.

153. Dreyer, W. J., Burns, A. R., Phillips, S. C., et al. (1998). Intercellular adhesion molecule-1 regulation in the canine lung after cardiopulmonary bypass. *J Thorac Cardiovasc Surg, 115,* 689–698.

154. Paparella, D., Yau, T. M., & Young, E. (2002). Cardiopulmonary bypass induced inflammation: Pathophysiology and treatment. An update. *Eur J Cardiothorac Surg, 21,* 232–244.

155. Valen, G. (2003). The basic biology of apoptosis and its implications for cardiac function and viability. *Ann Thorac Surg, 75,* S656–S660.

156. Selnes, O. A., & McKhann, G. M. (2001). Coronary-artery bypass surgery and the brain. *N Engl J Med, 344,* 451–452.

157. Likosky, D. S., Marrin, C. A., Caplan, L. R., et al. (2003). Determination of etiologic mechanisms of strokes secondary to coronary artery bypass graft surgery. *Stroke, 34,* 2830–2834.

158. Stump, D. A., Rogers, A. T., Hammon, J. W., et al. (1996). Cerebral emboli and cognitive outcome after cardiac surgery. *J Cardiothorac Vasc Anesth, 10,* 113–119.

159. Hill, J. D., Aguilar, M. J., Baranco, A., et al. (1969). Neuropathological manifestations of cardiac surgery. *Ann Thorac Surg, 7,* 409–419.

160. Sylivris, S., Levi, C., Matalanis, G., et al. (1998). Pattern and significance of cerebral microemboli during coronary artery bypass grafting. *Ann Thorac Surg, 66,* 1674–1678.

161. Fontes, M. L., Mathew, J. P., Rinder, H. M., et al. (2005). Atrial fibrillation after cardiac surgery/cardiopulmonary bypass is associated with monocyte activation. *Anesth Analg, 101,* 17–23.

162. Mathew, J. P., Fontes, M. L., Tudor, I. C., et al. (2004). A multicenter risk index for atrial fibrillation after cardiac surgery. *J Am Med Assoc, 291,* 1720–1729.

163. Barbut, D., Hinton, R. B., Szatrowski, T. P., et al. (1994). Cerebral emboli detected during bypass surgery are associated with clamp removal. *Stroke, 25,* 2398–2402.

164. Kapetanakis, E. I., Stamou, S. C., Dullum, M. K., et al. (2004). The impact of aortic manipulation on neurologic outcomes after coronary artery bypass surgery: A risk-adjusted study. *Ann Thorac Surg, 78,* 1564–1571.

165. Mackensen, G. B., Ti, L. K., Phillips-Bute, B. G., et al. (2003). Cerebral embolization during cardiac surgery: Impact of aortic atheroma burden. *Br J Anaesth, 91,* 656–661.

166. Butler, B. D., & Kurusz, M. (2000). Embolic events. In G. P. Gravlee, R. F. Davis, M. Kurusz, et al. (Eds.), *Cardiopulmonary bypass: Principles and practice* (2nd ed.). Philadelphia: Lippincott Williams & Wilkins.

167. Solis, R. T., Kennedy, P. S., Beall, A. C., Jr., et al. (1975). Cardiopulmonary bypass. Microembolization and platelet aggregation. *Circulation, 52,* 103–108.

168. Preston, F. E., Martin, J. F., Stewart, R. M., et al. (1979). Thrombocytosis, circulating platelet aggregates, and neurological dysfunction. *Br Med J, 2,* 1561–1563.

169. Hicks, R. E., Dutton, R. C., Ries, C. A., et al. (1973). Production and fate of platelet aggregate emboli during venovenous perfusion. *Surg Forum, 24,* 250–252.

170. Ratliff, N. B., Young, W. G., Jr., Hackel, D. B., et al. (1973). Pulmonary injury secondary to extracorporeal circulation. An ultrastructural study. *J Thorac Cardiovasc Surg, 65,* 425–432.

171. Choudhri, T. F., Hoh, B. L., Zerwes, H. G., et al. (1998). Reduced microvascular thrombosis and improved outcome in acute murine stroke by inhibiting GP IIb/IIIa receptor-mediated platelet aggregation. *J Clin Invest, 102,* 1301–1310.

172. Abumiya, T., Fitridge, R., Mazur, C., et al. (2000). Integrin alpha(IIb)beta(3) inhibitor preserves microvascular patency in experimental acute focal cerebral ischemia. *Stroke, 31,* 1402–1410.

173. Fuster, V., & Chesebro, J. H. (1986). Role of platelets and platelet inhibitors in aortocoronary artery vein-graft disease. *Circulation, 73,* 227–232.

174. Chandrasekar, B., & Tanguay, J. F. (2000). Platelets and restenosis. *J Am Coll Cardiol, 35,* 555–562.

175. Inoue, T., Hoshi, K., Yaguchi, I., et al. (1999). Serum levels of circulating adhesion molecules after coronary angioplasty. *Cardiology, 91,* 236–242.

176. Serrano, C. V., Jr., Ramires, J. A., Venturinelli, M., et al. (1997). Coronary angioplasty results in leukocyte and platelet activation with adhesion molecule expression. Evidence of inflammatory responses in coronary angioplasty. *J Am Coll Cardiol, 29,* 1276–1283.

177. Ishiwata, S., Tukada, T., Nakanishi, S., et al. (1997). Postangioplasty restenosis: Platelet activation and the coagulation–fibrinolysis system as possible factors in the pathogenesis of restenosis. *Am Heart J, 133,* 387–392.

178. Hayashi, Y., Sawa, Y., Nishimura, M., et al. (2000). P-selectin participates in cardiopulmonary bypass-induced inflammatory response in association with nitric oxide and peroxynitrite production. *J Thorac Cardiovasc Surg, 120,* 558–565.

179. Hartwell, D. W., Mayadas, T. N., Berger, G., et al. (1998). Role of P-selectin cytoplasmic domain in granular targeting *in vivo* and in early inflammatory responses. *J Cell Biol, 143,* 1129–1141.

180. Andre, P., Hartwell, D., Hrachovinova, I., et al. (2000). Pro-coagulant state resulting from high levels of soluble P-selectin in blood. *Proc Natl Acad Sci USA, 97,* 13835–13840.

181. Ridker, P. M., Buring, J. E., & Rifai, N. (2001). Soluble P-selectin and the risk of future cardiovascular events. *Circulation, 103,* 491–495.

182. Rinder, C. S., Mathew, J. P., Rinder, H. M., et al. (2002). Platelet PlA2 polymorphism and platelet activation are associated with increased troponin I release after cardiopulmonary bypass. *Anesthesiology, 97,* 1118–1122.

183. Mathew, J. P., Rinder, C. S., Howe, J. G., et al. (2001). Platelet GP IIIa PlA2 polymorphism enhances risk of neurocognitive decline after cardiopulmonary bypass. Multicenter Study of Perioperative Ischemia (McSPI) Research Group. *Ann Thorac Surg, 71,* 663–666.

184. Mangano, D. T. (2002). Aspirin and mortality from coronary bypass surgery. *N Engl J Med, 347,* 1309–1317.

185. Colman, R. W. (1990). Platelet and neutrophil activation in cardiopulmonary bypass. *Ann Thorac Surg, 49,* 32–34.

186. Harker, L. A., Malpass, T. W., Branson, H. E., et al. (1980). Mechanism of abnormal bleeding in patients undergoing cardiopulmonary bypass: Acquired transient platelet dysfunction associated with selective alpha-granule release. *Blood, 56,* 824–834.

187. Gelb, A. B., Roth, R. I., Levin, J., et al. (1996). Changes in blood coagulation during and following cardiopulmonary bypass: Lack of correlation with clinical bleeding. *Am J Clin Pathol, 106,* 87–99.

188. Addonizio, V. P., Jr., Colman, R. W., & Edmunds, L. H., Jr. (1978). The effect of blood flow rate and circuit surface area on platelet loss during extracorporeal circulation. *Trans Am Soc Artif Intern Organs, 24,* 650–655.

189. Edmunds, L. H., Jr., Ellison, N., Colman, R. W., et al. (1982). Platelet function during cardiac operation: Comparison of membrane and bubble oxygenators. *J Thorac Cardiovasc Surg, 83,* 805–812.

190. Edmunds, L. H., Jr. (1995). Why cardiopulmonary bypass makes patients sick: Strategies to control the blood–synthetic surface interface. *Adv Card Surg, 6,* 131–167.

191. Baier, R. E., & Dutton, R. C. (1969). Initial events in interactions of blood with a foreign surface. *J Biomed Mater Res, 3,* 191–206.

192. Uniyal, S., & Brash, J. L. (1982). Patterns of adsorption of proteins from human plasma onto foreign surfaces. *Thromb Haemost, 47,* 285–290.

193. Horbett, T. A. (1993). Principles underlying the role of adsorbed plasma proteins in blood interactions with foreign materials. *Cardiovasc Pathol, 2,* 137S–148S.

194. Ziats, N. P., Pankowsky, D. A., Tierney, B. P., et al. (1990). Adsorption of Hageman factor (factor XII) and other human plasma proteins to biomedical polymers. *J Lab Clin Med, 116,* 687–696.

195. George, J. N. (1972). Direct assessment of platelet adhesion to glass: A study of the forces of interaction and the effects of plasma and serum factors, platelet function, and modification of the glass surface. *Blood, 40,* 862–874.

196. Brash, J. L., Scott, C. F., ten Hove, P., et al. (1988). Mechanism of transient adsorption of fibrinogen from plasma to solid surfaces: Role of the contact and fibrinolytic systems. *Blood, 71,* 932–939.

197. Lindon, J. N., McManama, G., Kushner, L., et al. (1986). Does the conformation of adsorbed fibrinogen dictate platelet interactions with artificial surfaces? *Blood, 68,* 355–362.

198. Ryu, G. H., Kim, J., Ruggeri, Z. M., et al. (1995). Effect of shear stress on fibrinogen adsorption and its conformational change. *ASAIO J, 41,* M384–M388.

199. Courtney, J. M., & Forbes, C. D. (1994). Thrombosis on foreign surfaces. *Br Med Bull, 50,* 966–981.

200. Engbers, G. H., & Feijen, J. (1991). Current techniques to improve the blood compatibility of biomaterial surfaces. *Int J Artif Organs, 14,* 199–215.

201. Gorbet, M. B., & Sefton, M. V. (2004). Biomaterial-associated thrombosis: Roles of coagulation factors, complement, platelets and leukocytes. *Biomaterials, 25,* 5681–5703.

202. Merhi, Y., Bernier, J., Marois, Y., et al. (1995). Acute thrombogenicity of arterial prostheses exposed to reduced blood flow in dogs: Effects of heparin, aspirin, and prostacyclin. *J Cardiovasc Pharmacol, 26,* 1–5.

203. Schoephoerster, R. T., Oynes, F., Nunez, G., et al. (1993). Effects of local geometry and fluid dynamics on regional platelet deposition on artificial surfaces. *Arterioscler Thromb, 13,* 1806–1813.

204. Edmunds, L. H., Jr. (1993). Blood–surface interactions during cardiopulmonary bypass. *J Card Surg, 8,* 404–410.

205. Zucker, M. B., & Vroman, L. (1969). Platelet adhesion induced by fibrinogen adsorbed onto glass. *Proc Soc Exp Biol Med, 131,* 318–320.

206. Haycox, C. L., & Ratner, B. D. (1993). *In vitro* platelet interactions in whole human blood exposed to biomaterial surfaces: Insights on blood compatibility. *J Biomed Mater Res, 27,* 1181–1193.

207. Shiba, E., Lindon, J. N., Kushner, L., et al. (1991). Antibody-detectable changes in fibrinogen adsorption affecting platelet

activation on polymer surfaces. *Am J Physiol, 260,* C965–C974.

208. Hietala, E. M., Maasilta, P., Juuti, H., et al. (2004). Platelet deposition on stainless steel, spiral, and braided polylactide stents. A comparative study. *Thromb Haemost, 92,* 1394–1401.

209. Sheppard, J. I., McClung, W. G., & Feuerstein, I. A. (1994). Adherent platelet morphology on adsorbed fibrinogen: Effects of protein incubation time and albumin addition. *J Biomed Mater Res, 28,* 1175–1186.

210. George, J. N., Pickett, E. B., Saucerman, S., et al. (1986). Platelet surface glycoproteins. Studies on resting and activated platelets and platelet membrane microparticles in normal subjects, and observations in patients during adult respiratory distress syndrome and cardiac surgery. *J Clin Invest, 78,* 340–348.

211. Abrams, C. S., Ellison, N., Budzynski, A. Z., et al. (1990). Direct detection of activated platelets and platelet-derived microparticles in humans. *Blood, 75,* 128–138.

212. Nieuwland, R., Berckmans, R. J., Rotteveel-Eijkman, R. C., et al. (1997). Cell-derived microparticles generated in patients during cardiopulmonary bypass are highly procoagulant. *Circulation, 96,* 3534–3541.

213. van den Goor, J. M., van den Brink, A., Nieuwland, R., et al. (2003). Generation of platelet-derived microparticles in patients undergoing cardiac surgery is not affected by complement activation. *J Thorac Cardiovasc Surg, 126,* 1101–1106.

214. Ekert, H., Gilchrist, G. S., Stanton, R., et al. (1970). Hemostasis in cyanotic congenital heart disease. *J Pediatr, 76,* 221–230.

215. Beurling-Harbury, C., & Galvan, C. A. (1978). Acquired decrease in platelet secretory ADP associated with increased postoperative bleeding in post-cardiopulmonary bypass patients and in patients with severe valvular heart disease. *Blood, 52,* 13–23.

216. Friedenberg, W. R., Myers, W. O., Plotka, E. D., et al. (1978). Platelet dysfunction associated with cardiopulmonary bypass. *Ann Thorac Surg, 25,* 298–305.

217. McKenna, R., Bachmann, F., Whittaker, B., et al. (1975). The hemostatic mechanism after open-heart surgery: II. Frequency of abnormal platelet functions during and after extracorporeal circulation. *J Thorac Cardiovasc Surg, 70,* 298–308.

218. Woodman, R. C., & Harker, L. A. (1990). Bleeding complications associated with cardiopulmonary bypass. *Blood, 76,* 1680–1697.

219. Jones, D. K., Luddington, R., Higenbottam, T. W., et al. (1988). Changes in factor VIII proteins after cardiopulmonary bypass in man suggest endothelial damage. *Thromb Haemost, 60,* 199–204.

220. Mammen, E. F., Koets, M. H., Washington, B. C., et al. (1985). Hemostasis changes during cardiopulmonary bypass surgery. *Semin Thromb Hemost, 11,* 281–292.

221. Greilich, P. E., Carr, M. E., Jr., Carr, S. L., et al. (1995). Reductions in platelet force development by cardiopulmonary bypass are associated with hemorrhage. *Anesth Analg, 80,* 459–465.

222. Wachtfogel, Y. T., Musial, J., Jenkin, B., et al. (1985). Loss of platelet alpha 2-adrenergic receptors during simulated extracorporeal circulation: Prevention with prostaglandin E1. *J Lab Clin Med, 105,* 601–607.

223. Wenger, R. K., Lukasiewicz, H., Mikuta, B. S., et al. (1989). Loss of platelet fibrinogen receptors during clinical cardiopulmonary bypass. *J Thorac Cardiovasc Surg, 97,* 235–239.

224. Ferraris, V. A., Ferraris, S. P., Singh, A., et al. (1998). The platelet thrombin receptor and postoperative bleeding. *Ann Thorac Surg, 65,* 352–358.

225. Rinder, C. S., Mathew, J. P., Rinder, H. M., et al. (1991). Modulation of platelet surface adhesion receptors during cardiopulmonary bypass. *Anesthesiology, 75,* 563–570.

226. Rinder, C. S., Gaal, D., Student, L. A., et al. (1994). Platelet–leukocyte activation and modulation of adhesion receptors in pediatric patients with congenital heart disease undergoing cardiopulmonary bypass. *J Thorac Cardiovasc Surg, 107,* 280–288.

227. Valeri, C. R., Khabbaz, K., Khuri, S. F., et al. (1992). Effect of skin temperature on platelet function in patients undergoing extracorporeal bypass. *J Thorac Cardiovasc Surg, 104,* 108–116.

228. Rinder, H. M., Tracey, J. B., Recht, M., et al. (1998). Differences in platelet alpha-granule release between normals and immune thrombocytopenic patients and between young and old platelets. *Thromb Haemost, 80,* 457–462.

229. Griffin, M. J., Rinder, H. M., Smith, B. R., et al. (2001). The effects of heparin, protamine, and heparin/protamine reversal on platelet function under conditions of arterial shear stress. *Anesth Analg, 93,* 20–27.

230. Sobel, M., McNeill, P. M., Carlson, P. L., et al. (1991). Heparin inhibition of von Willebrand factor-dependent platelet function *in vitro* and *in vivo. J Clin Invest, 87,* 1787–1793.

231. Mathew, J. P., Rinder, C. S., Tracey, J. B., et al. (1995). Acadesine inhibits neutrophil CD11b up-regulation *in vitro* and during *in vivo* cardiopulmonary bypass. *J Thorac Cardiovasc Surg, 109,* 448–456.

232. Harris, S. N., Rinder, C. S., Rinder, H. M., et al. (1995). Nitroprusside inhibition of platelet function is transient and reversible by catecholamine priming. *Anesthesiology, 83,* 1145–1152.

233. Addonizio, V. P., Jr., Strauss, J. F., 3rd, Colman, R. W., et al. (1979). Effects of prostaglandin E1 on platelet loss during *in vivo* and *in vitro* extracorporeal circulation with a bubble oxygenator. *J Thorac Cardiovasc Surg, 77,* 119–126.

234. Addonizio, V. P., Jr., Fisher, C. A., Jenkin, B. K., et al. (1985). Iloprost (ZK36374), a stable analogue of prostacyclin, preserves platelets during simulated extracorporeal circulation. *J Thorac Cardiovasc Surg, 89,* 926–933.

235. Fish, K. J., Sarnquist, F. H., van Steennis, C., et al. (1986). A prospective, randomized study of the effects of prostacyclin on platelets and blood loss during coronary bypass operations. *J Thorac Cardiovasc Surg, 91,* 436–442.

236. Musial, J., Niewiarowski, S., Rucinski, B., et al. (1990). Inhibition of platelet adhesion to surfaces of extracorporeal

circuits by disintegrins. RGD-containing peptides from viper venoms. *Circulation, 82,* 261–273.

237. Lincoff, A. M., LeNarz, L. A., Despotis, G. J., et al. (2000). Abciximab and bleeding during coronary surgery: Results from the EPILOG and EPISTENT trials. Improved long-term outcome with abciximab GP IIb/IIIa blockade. Evaluation of platelet IIb/IIIa inhibition in STENTing. *Ann Thorac Surg, 70,* 516–526.

238. Niewiarowski, S., Senyi, A. F., & Gillies, P. (1973). Plasmin-induced platelet aggregation and platelet release reaction. Effects on hemostasis. *J Clin Invest, 52,* 1647–1659.

239. Kirklin, J. K., Westaby, S., Blackstone, E. H., et al. (1983). Complement and the damaging effects of cardiopulmonary bypass. *J Thorac Cardiovasc Surg, 86,* 845–857.

240. Rinder, C. S., Rinder, H. M., Smith, B. R., et al. (1995). Blockade of C5a and C5b-9 generation inhibits leukocyte and platelet activation during extracorporeal circulation. *J Clin Invest, 96,* 1564–1572.

241. Rinder, C. S., Rinder, H. M., Smith, M. J., et al. (1999). Selective blockade of membrane attack complex formation during simulated extracorporeal circulation inhibits platelet but not leukocyte activation. *J Thorac Cardiovasc Surg, 118,* 460–466.

242. Rinder, C. S., Rinder, H. M., Johnson, K., et al. (1999). Role of C3 cleavage in monocyte activation during extracorporeal circulation. *Circulation, 100,* 553–558.

243. Larsson, R., Elgue, G., Larsson, A., et al. (1997). Inhibition of complement activation by soluble recombinant CR1 under conditions resembling those in a cardiopulmonary circuit: Reduced up-regulation of CD11b and complete abrogation of binding of PMNs to the biomaterial surface. *Immunopharmacology, 38,* 119–127.

244. Fitch, J. C., Rollins, S., Matis, L., et al. (1999). Pharmacology and biological efficacy of a recombinant, humanized, single-chain antibody C5 complement inhibitor in patients undergoing coronary artery bypass graft surgery with cardiopulmonary bypass. *Circulation, 100,* 2499–2506.

245. Shernan, S. K., Fitch, J. C., Nussmeier, N. A., et al. (2004). Impact of pexelizumab, an anti-C5 complement antibody, on total mortality and adverse cardiovascular outcomes in cardiac surgical patients undergoing cardiopulmonary bypass. *Ann Thorac Surg, 77,* 942–950.

246. Chen, J. C., Rollins, S. A., Shernan, S. K., et al. (2005). Pharmacologic C5-complement suppression reduces blood loss during on-pump cardiac surgery. *J Card Surg, 20,* 35–41.

247. Yusuf, S., Zhao, F., Mehta, S. R., et al. (2001). Effects of clopidogrel in addition to aspirin in patients with acute coronary syndromes without ST-segment elevation. *N Engl J Med, 345,* 494–502.

248. Weightman, W. M., Gibbs, N. M., Weidmann, C. R., et al. (2002). The effect of preoperative aspirin-free interval on red blood cell transfusion requirements in cardiac surgical patients. *J Cardiothorac Vasc Anesth, 16,* 54–58.

249. Maslow, A., & Schwartz, C. (2004). Cardiopulmonary bypass-associated coagulopathies and prophylactic therapy. *Int Anesthesiol Clin, 42,* 103–133.

250. Green, J. A., & Spiess, B. D. (2003). Current status of anti-fibrinolytics in cardiopulmonary bypass and elective deep hypothermic circulatory arrest. *Anesthesiol Clin North Am, 21,* 527–551.

251. Despotis, G. J., & Goodnough, L. T. (2000). Management approaches to platelet-related microvascular bleeding in cardiothoracic surgery. *Ann Thorac Surg, 70,* S20–S32.

252. Despotis, G. J., Gravlee, G., Filos, K., et al. (1999). Anticoagulation monitoring during cardiac surgery: A review of current and emerging techniques. *Anesthesiology, 91,* 1122–1151.

253. Forestier, F., Coiffic, A., Mouton, C., et al. (2002). Platelet function point-of-care tests in post-bypass cardiac surgery: Are they relevant? *Br J Anaesth, 89,* 715–721.

254. Faraday, N., Guallar, E., Sera, V. A., et al. (2002). Utility of whole blood hemostatometry using the clot signature analyzer for assessment of hemostasis in cardiac surgery. *Anesthesiology, 96,* 1115–1122.

255. Slaughter, T. F., Sreeram, G., Sharma, A. D., et al. (2001). Reversible shear-mediated platelet dysfunction during cardiac surgery as assessed by the PFA-100 platelet function analyzer. *Blood Coagul Fibrinolysis, 12,* 85–93.

256. Chen, L., Bracey, A. W., Radovancevic, R., et al. (2004). Clopidogrel and bleeding in patients undergoing elective coronary artery bypass grafting. *J Thorac Cardiovasc Surg, 128,* 425–431.

257. DiDomenico, R. J., Massad, M. G., Kpodonu, J., et al. (2005). Use of recombinant activated factor VII for bleeding following operations requiring cardiopulmonary bypass. *Chest, 127,* 1828–1835.

258. Vahl, C. F., Kayhan, N., Thomas, G., et al. (2001). Myocardial revascularization after pretreatment with the GPIIb/IIIa receptor blocker abciximab. *Z Kardiol, 90,* 889–897.

259. Bizzarri, F., Scolletta, S., Tucci, E., et al. (2001). Perioperative use of tirofiban hydrochloride (Aggrastat) does not increase surgical bleeding after emergency or urgent coronary artery bypass grafting. *J Thorac Cardiovasc Surg, 122,* 1181–1185.

260. Dyke, C. M., Bhatia, D., Lorenz, T. J., et al. (2000). Immediate coronary artery bypass surgery after platelet inhibition with eptifibatide: Results from PURSUIT. Platelet Glycoprotein IIb/IIIa in Unstable Angina: Receptor Suppression Using Integrelin Therapy. *Ann Thorac Surg, 70,* 866–871.

261. Hiramatsu, Y., Gikakis, N., Anderson, H. L., 3rd, et al. (1997). Tirofiban provides "platelet anesthesia" during cardiopulmonary bypass in baboons. *J Thorac Cardiovasc Surg, 113,* 182–193.

262. Straub, A., Wendel, H. P., Azevedo, R., et al. (2005). The GP IIb/IIIa inhibitor abciximab (ReoPro) decreases activation and interaction of platelets and leukocytes during *in vitro* cardiopulmonary bypass simulation. *Eur J Cardiothorac Surg, 27,* 617–621.

263. Suzuki, Y., Malekan, R., Hanson, C. W., 3rd, et al. (1999). Platelet anesthesia with nitric oxide with or without eptifibatide during cardiopulmonary bypass in baboons. *J Thorac Cardiovasc Surg, 117,* 987–993.

264. Massoudy, P., Zahler, S., Barankay, A., et al. (1999). Sodium nitroprusside during coronary artery bypass grafting: Evidence for an antiinflammatory action. *Ann Thorac Surg, 67,* 1059–1064.

265. Chung, A., Wildhirt, S. M., Wang, S., et al. (2005). Combined administration of nitric oxide gas and iloprost during cardiopulmonary bypass reduces platelet dysfunction: A pilot clinical study. *J Thorac Cardiovasc Surg, 129,* 782–790.

266. Basora, M., Gomar, C., Escolar, G., et al. (1999). Platelet function during cardiac surgery and cardiopulmonary bypass with low-dose aprotinin. *J Cardiothorac Vasc Anesth, 13,* 382–387.

267. Shore-Lesserson, L. (1999). Aprotinin has direct platelet protective properties: Fact or fiction? *J Cardiothorac Vasc Anesth, 13,* 379–381.

268. Rossi, M., Martinelli, L., Storti, S., et al. (1999). The role of antithrombin III in the perioperative management of the patient with unstable angina. *Ann Thorac Surg, 68,* 2231–2236.

269. Despotis, G. J., Levine, V., Saleem, R., et al. (1999). Use of point-of-care test in identification of patients who can benefit from desmopressin during cardiac surgery: A randomised controlled trial. *Lancet, 354,* 106–110.

270. Chen, Y. F., Tsai, W. C., Lin, C. C., et al. (2004). Effect of leukocyte depletion on endothelial cell activation and transendothelial migration of leukocytes during cardiopulmonary bypass. *Ann Thorac Surg, 78,* 634–642.

271. Chello, M., Mastroroberto, P., Patti, G., et al. (2003). Simvastatin attenuates leucocyte–endothelial interactions after coronary revascularisation with cardiopulmonary bypass. *Heart, 89,* 538–543.

# Pharmacology: Antiplatelet Therapy

PART FIVE

# Pharmacology: Antiplatelet Therapy

# CHAPTER 60

# Aspirin

## Eric H. Awtry[1] and Joseph Loscalzo[2]

[1]*Evans Department of Medicine, Boston University School of Medicine, Boston, Massachusetts*
[2]*Department of Medicine, Brigham and Women's Hospital, Harvard Medical School, Boston, Massachusetts*

## I. Introduction

Acetylsalicylic acid, or aspirin, is a synthetic compound with antipyretic, analgesic, anti-inflammatory, and antiplatelet properties. Since its introduction into the commercial market in June 1899, it has become one of the most versatile pharmaceutical agents known, with demonstrated efficacy in a myriad of clinical settings. The central role of platelets in the pathogenesis and pathophysiology of cardiovascular disease (see Chapters 35 and 36), combined with the current understanding of the antiplatelet effects of aspirin, has placed this agent at the cornerstone of treatment for cardiovascular disorders.

## II. Historical Perspectives

Salicylates are naturally occurring compounds that are found in the bark and leaves of several plants, including willow, poplar, birch, and beech trees, wintergreen, coffee, licorice, and several shrubs and grasses. Evidence suggests that these heterocyclic compounds serve a role in innate immunity in the plant species.[1] In various forms, salicylates have been used therapeutically for more than 2400 years. It was Hippocrates who first recorded the beneficial analgesic effects of chewing willow leaves and Galen who first recognized their antipyretic and anti-inflammatory properties.[2] For the next two millennia, natural sources of salicylates were used to treat a variety of maladies.

In 1763, Reverend Edward Stone, in a letter to the Earl of Macclesfield, reported his success in using willow bark to treat 50 patients who were suffering from febrile or inflammatory diseases,[3] thus providing scientific evidence to support its use. Throughout the following century, various synthetic salicylates were derived, and although they produced potent analgesic and antipyretic effects, their use was often marred by severe gastric toxicity. Acetylsalicylic acid, after being synthesized by several chemists in the 1850s and 1860s, was rediscovered by Felix Hoffman in

1897 and marketed to the public by the Bayer Company in 1899.

The beneficial antithrombotic effects of aspirin were not recognized until the mid-20th century when Gibson reported success in using aspirin to treat a small group of patients with vascular diseases,[4] and Craven reported that aspirin protected patients against acute myocardial infarction (AMI) and major stroke.[5,6] During the past half century, the mechanism of aspirin's antiplatelet effect has been elucidated, and multiple clinical trials have demonstrated its efficacy when used as acute treatment for, and primary and secondary prophylaxis against, a variety of cardiovascular disorders. This chapter discusses the pharmacology and mechanism of action of aspirin and reviews its clinical utility.

## III. Chemical Structure

Salicylic acid (*o*-hydroxybenzoic acid) (Fig. 60-1) is extremely irritating to the gastric mucosa and cannot safely be taken orally. This parent compound has been modified through esterification to produce various derivatives that are better tolerated and can be given enterally. Substitution of the carboxyl group of this compound with acetic acid results in the production of acetylsalicylic acid, or aspirin (Fig. 60-1).

## IV. Mechanism of Action

The pharmacological effects of aspirin are mediated primarily through its interference with prostaglandin biosynthesis (Fig. 60-2).[7] This synthetic pathway begins with the generation of arachidonic acid from membrane phospholipids by the enzyme phospholipase A. Arachidonic acid then provides the substrate upon which a series of enzymatically catalyzed reactions occur. The enzyme 12-lipoxygenase converts arachidonic acid to the unstable intermediates 12-hydroperoxyeicosatetraenoic acid and 12-

hydroxyeicosatetraenoic acid, which are further metabolized by 5-lipoxygenase, resulting in the formation of various leukotrienes. Conversely, metabolism of arachidonic acid by the enzyme prostaglandin (PG) H synthase results in the formation of prostaglandins. The cyclooxygenase activity of PGH synthase converts arachidonic acid to $PGG_2$, which is then metabolized by the peroxidase activity of this enzyme to $PGH_2$. The modification of $PGH_2$ by specific synthases results in the production of $PGD_2$, $PGE_2$, $PGF_{2\alpha}$, $PGI_2$ (prostacyclin), and thromboxane $A_2$ ($TXA_2$).[8] These substances are important mediators of a variety of cellular functions (Fig. 60-2).

PGH synthase, also referred to as cyclooxygenase (COX), exists in two forms, termed COX-1 and COX-2.[9] COX-1 is a constitutive isoform expressed on the endoplasmic reticular membrane of all cells, including gastric, kidney, and vascular cells, as well as in platelets,[10] and it results in the formation of a variety of prostaglandins that mediate normal cellular functions. For example, maintenance of renal blood flow, gastric mucosal protection, and platelet activation and aggregation are all affected by COX-1-catalyzed prostaglandin synthesis (Fig. 60-2).[8] COX-2 is expressed constitutively in microvascular endothelial cells, in which it is a source of

$PGI_2$ under normal conditions[11]; this COX-2-derived $PGI_2$ may have an atheroprotective effect in laboratory animals.[12] Low levels of COX-2 mRNA and protein have been identified in human platelets[13]; however, COX-2 is not routinely expressed in most normally functioning cells. Rather, it is induced in the inflammatory state by growth factors, cytokines, and other inflammatory stimuli and results in the synthesis of prostaglandins that mediate the inflammatory response.[14,15]

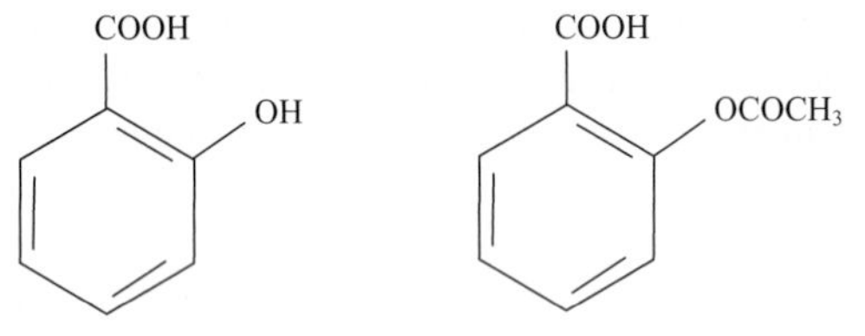

Aspirin's inhibitory effect on prostaglandin biosynthesis is mediated by its covalent modification of COX through the irreversible acetylation of a specific serine moiety.[16,17] Aspirin-induced acetylation of serine 529 of COX-1 results in a conformational change in the active site of the enzyme, thus preventing it from binding arachidonic acid. Conversely, aspirin-induced acetylation of serine 516 of COX-2 alters the function of the enzyme so that it metabolizes arachidonic acid to 15-$R$-hydroxyeicosatetraenoic acid.[18] Although aspirin irreversibly inhibits both COX-1 and COX-2, its inhibitory effect on COX-1 is approximately 170-fold greater.[19] In the presence of aspirin, both COX-1 and COX-2 become incapable of converting arachidonic acid to $PGH_2$, a necessary precursor of prostaglandin biosynthesis. The inhibition of COX-2 likely accounts for the beneficial anti-inflammatory effects of aspirin, whereas inhibition of COX-1 accounts for its antithrombotic effect. Inhibition of COX-1 also interferes with production of the homeostatic prostanoids, thereby potentially resulting in serious toxicities, including renal failure, gastric mucosal ulceration, and impaired hemostasis. Aspirin does not inhibit the activity of lipoxygenase; thus, it does not affect the generation of leukotrienes.

Physiologically, aspirin is a relatively weak inhibitor of platelets because it blocks only thromboxane-dependent

**Figure 60-1.** The chemical structure of salicylic acid and aspirin.

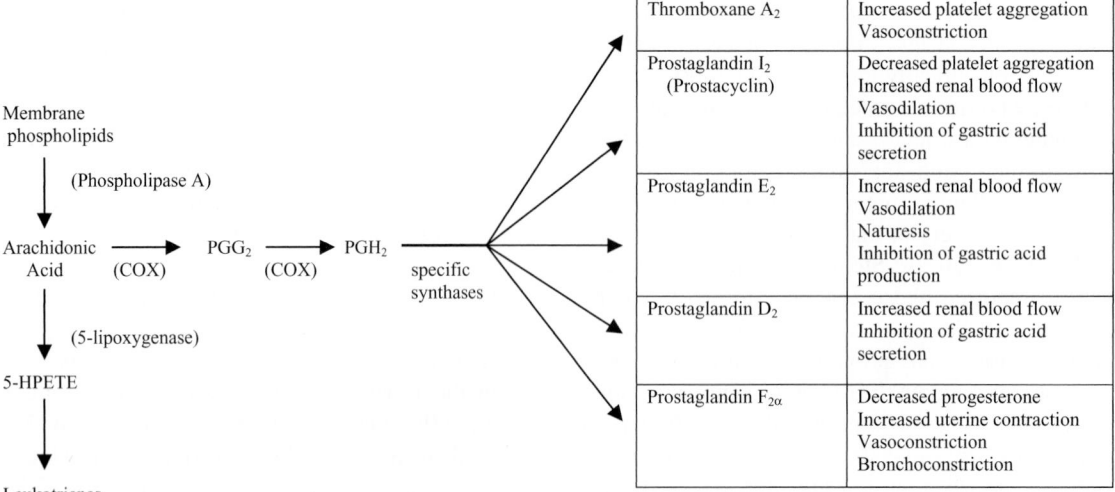

**Figure 60-2.** Biosynthetic pathway of prostaglandins and their physiological effects. COX, cyclooxygenase; HPETE, hydroperoxyeicosatetraenoic acid; PG, prostaglandin.

platelet activation and aggregation. Several other stimuli, including shear force, increased plasma catecholamines, thrombin, and ADP, can activate platelets despite aspirin therapy and likely contribute to thrombosis in acute coronary syndromes.[20] Other agents have been developed that inhibit platelet activation and aggregation through blockade of these various stimuli and of the glycoprotein (GP) IIb-IIIa receptor, the final common pathway of platelet aggregation. These agents are discussed in Chapters 61, 62, and 65.

Selective inhibitors of the COX isomers have been developed and owe their specificity to a single amino acid substitution at the catalytic site of the isoenzymes.[21,22] COX-2-selective agents appear to be effective anti-inflammatory agents but have a lower risk of gastric side effects than does aspirin.[23–26] These agents may have a beneficial effect on endothelial function;[27] however, they do not inhibit platelet $TXA_2$ production. There is an increasing body of evidence that demonstrates an increased risk of cardiovascular events in patients treated with COX-2-selective agents. In a subgroup analysis of the Physician's Health Study, the regular use of nonsteroidal anti-inflammatory agents (NSAIDs) was associated with an increased risk of AMI and inhibited the beneficial effects of aspirin on the primary prevention of cardiovascular events.[28] In prospective randomized studies aimed at evaluating the potential beneficial effect of the COX-2 inhibitors celecoxib or rofecoxib on the incidence of colon cancer,[29,30] as well as in population-based case-control studies,[31,32] the use of these agents has been associated with an increased risk of thrombotic cardiovascular events. The mechanisms responsible for the adverse cardiovascular effects of COX-2 inhibition have not been completely elucidated but may involve destabilization of atherosclerotic plaque, a reduction in the production of prostacyclin, a shift in the metabolism of arachidonic acid favoring production of proinflammatory mediators, and increased generation of reactive oxygen species resulting in increased oxidative stress in the vasculature.[7] Owing to these considerations, COX-2 inhibitors should not be administered routinely to patients with cardiovascular disease, and they should be avoided in patients who are at high risk of thrombotic events.

## V. Mechanisms of Benefit in Cardiovascular Disease

The beneficial cardiovascular effects of aspirin are mediated primarily by its antithrombotic action. Platelets aggregate when exposed to collagen, thrombin, ADP, and various other stimuli. These agents also induce an increase in the platelet production of $TXA_2$, which then amplifies the aggregation response and stimulates vasoconstriction.[33,34] These actions favor thrombus formation but are balanced in the normal state by vasodilation and the inhibition of platelet aggregation induced by vascular endothelial cell production of prostacyclin ($PGI_2$) and nitric oxide (NO) (see Chapter 13). The inhibition of $TXA_2$ and $PGI_2$ by aspirin results in competing effects: Decreased platelet $TXA_2$ production limits thrombosis, whereas decreased endothelial cell $PGI_2$ production favors thrombosis. The duration of aspirin's effect, however, differs between the two cell types. Platelets are anuclear and therefore lack the synthetic machinery to generate new COX; consequently, aspirin-induced COX inhibition lasts for the lifetime of the platelet. Conversely, vascular endothelial cells retain their capacity to generate new COX and recover normal function soon after exposure to aspirin.[35] This differential aspirin effect likely accounts, in part, for the observation that the potentially prothrombotic effects of (low-dose) aspirin appear to be clinically unimportant and the antithrombotic effects are dominant.[36]

Mechanisms other than $TXA_2$ inhibition may also contribute to the antithrombotic effect of aspirin. Neutrophils inhibit platelet activation through a process that appears to be mediated by an NO/cyclic guanosine 5′-monophosphate (cGMP)-dependent process.[37] Aspirin may facilitate this antithrombotic effect by inhibition of endothelial cell prostacyclin synthesis, an effect that would increase NO production.[38] Additionally, at high doses, aspirin decreases thrombin formation[39] and enhances fibrinolysis[40]; however, the significance of these effects at the doses of aspirin used clinically is not clear.

The antithrombotic effects of aspirin are unlikely to be the sole mechanism whereby it exerts clinical benefit in the treatment of cardiovascular disease. Aspirin may affect atherosclerogenesis by making low-density lipoprotein (LDL) less susceptible to oxidative modification[41] and, thereby, less likely to be taken up by macrophages and incorporated into the atherosclerotic plaque. The exact mechanism by which this effect occurs is not clear but likely involves aspirin's antioxidant function. When aspirin is exposed to hydroxyl radicals, the 2,3- and 2,5-dihydroxybenzoate derivatives form[42] and serve as markers of oxidative stress *in vivo*.[43] It is possible that this scavenging effect of aspirin neutralizes oxygen radicals, thus limiting oxidation of LDL. Additionally, PGH synthase-catalyzed reactions generate free radicals[44] that are capable of initiating lipid peroxidation. Aspirin's inhibition of prostaglandin synthesis may further limit oxidative stress. In the atherogenic process, the lysine residues of LDL are modified, allowing for the further oxidation of LDL. Aspirin acetylates the ε-amino lysine residues of human proteins including LDL,[45,46] an effect that limits the higher order oxidation of these functionalities. A similar antioxidant effect on fibrinogen may also be of importance, resulting in reduced fibrin formation[45,47] and enhanced fibrinolysis.[40] Lastly, aspirin improves endothelial function in patients with atherosclerosis, an effect that appears to be mediated through the inhibition of a COX-dependent, endothelium-derived vasoconstrictor,[48] and possibly through modulation of the cytokine system.[49] It is likely that aspirin imparts its clinical benefit in patients with coronary artery disease (CAD) through a combination of these effects.[50]

## VI. Pharmacodynamics

After oral ingestion, aspirin is rapidly absorbed by the mucosa of the stomach and upper intestine. It is detectable in plasma within 20 minutes of ingestion, reaches peak plasma level within 30 to 40 minutes, and produces measurable platelet inhibition within 1 hour.[36] Aspirin has a bioavailability of approximately 40 to 50%; however, the inhibition of $TXA_2$-dependent platelet aggregation and prolongation of the bleeding time are largely independent of the systemic bioavailability because aspirin-induced COX inhibition occurs in the portal circulation.

Enteric coating causes delayed release of aspirin into the small bowel, where it is hydrolyzed in the alkaline environment, resulting in lower bioavailability compared to regular aspirin.[51] Additionally, enteric coating significantly delays aspirin's absorption[52] and its therapeutic effect as measured by prolongation of the bleeding time[53]; it may take 4 or 5 days for coated aspirin to produce its maximal antiplatelet effect.[54] However, chewing a coated aspirin results in similarly rapid inhibition of platelet aggregation as that produced by noncoated tablets,[55,56] and during chronic aspirin use, coated and uncoated formulations appear to be equally effective at inhibiting platelet aggregation and platelet $TXA_2$ production.[57] This is clearly the case at higher doses (>300 mg) when the platelet inhibitory effect of enteric-coated aspirin is equivalent to that of uncoated formulations.[51,58] However, at lower doses (<100 mg) the efficacy of enteric-coated tablets has not been clearly established. Thus, if coated preparations are prescribed, larger doses may be required to obtain the desired antiplatelet effect.

The plasma half-life of aspirin is approximately 20 minutes; however, the effective half-life is much longer. Aspirin-induced inhibition of COX is irreversible due to the inability of platelets to generate new enzyme. Thus, the effect of a single dose of aspirin lasts for the lifetime of the platelet — approximately 10 days. Following a single inhibitory dose of aspirin, platelet COX activity recovers by approximately 10% per day as a reflection of platelet turnover.[59] However, hemostasis may be normal in the presence of as little as 20% of normally functioning platelets.[60,61] Thus, the majority of patients will have recovery of a normal bleeding time within 3 or 4 days after ingesting a single dose of aspirin.[55] Enteric coating of aspirin may prolong its effects and require 2 to 4 weeks before complete normalization of platelet function.[54]

The platelet inhibitory efficacy of various doses of aspirin has been extensively investigated and most studies demonstrate equal platelet inhibition by a wide range of aspirin dosages.[54,62] In patients without prior aspirin exposure, a single dose of 100 mg completely abolishes platelet $TXA_2$ production.[63,64] Single doses of less than 100 mg have a lesser acute effect, may require more than 24 hours to attain their maximal effect, and require repeated daily doses to produce maximal COX inhibition.[61,63,65] Low-dose aspirin as well as coated formulations may have differential inhibitory effects on platelets and vascular endothelial cells. These preparations may preferentially inhibit platelet COX while leaving endothelial COX relatively unaffected.[64,66,67] This would have the effect of inhibiting $TXA_2$-induced platelet aggregation and vasoconstriction while leaving the production of endothelial prostacyclin intact and thereby stimulating vasodilation. Nonetheless, the clinical importance of preserved prostacyclin production is unproven, and the very low doses of aspirin (40 mg daily) necessary to achieve this effect may, in fact, be inadequate to produce sufficient platelet inhibition even when used chronically.[56] Additionally, although 325- and 81-mg dosages of aspirin equally inhibit ADP- and epinephrine-induced platelet aggregation, the higher dose is more effective at inhibiting collagen-induced aggregation[68] and may therefore be more effective during acute coronary syndromes when plaque rupture results in collagen exposure. Daily doses of aspirin that are higher than 325 mg have not been shown to have incremental benefit over the lower dosing regimens, but they appear to increase the risk of gastric toxicity.[36,69]

### A. Aspirin Resistance

Variability in individual responsiveness to aspirin has been noted, and a small subset of patients appear to be relatively resistant to its antiplatelet effects as assessed by various *ex vivo* measures of platelet function.[70,71] Between 5 and 25% of patients with stable cardiovascular disease who receive chronic aspirin therapy have incomplete inhibition of platelet aggregation.[72,73] Furthermore, almost one third of patients who initially respond to aspirin may develop aspirin resistance while receiving chronic therapy.[72] These patients may respond favorably to escalating doses of aspirin; however, as many as 8% of patients may still demonstrate aspirin resistance at daily doses as high as 1300 mg. This phenomenon of aspirin resistance appears to be clinically important because aspirin-treated patients with laboratory evidence of inadequate platelet inhibition have a higher incidence of AMI, stroke, and cardiovascular death than do aspirin-sensitive individuals.[73,74]

The mechanism underlying aspirin resistance is not well understood but is likely multifactorial. Compliance with therapy may underlie some of the variability in aspirin responsiveness,[75] as may pharmacological interference by other NSAIDs.[76] Additionally, the increased sensitivity of aspirin-resistant platelets to collagen[71] and the limited efficacy of aspirin in the inhibition of platelet COX-2[13] may allow for platelet activation even in the face of adequate aspirin dosing. It is also feasible that resistance to aspirin-induced inhibition of platelet aggregation and the occurrence of cardiovascular events despite treatment with adequate doses of aspirin may have a molecular and/or genetic basis. Genetic polymorphisms of cyclooxygenase and of platelet membrane glycoproteins have been described, some of which appear to be associated with decreased aspirin sensitivity as measured by its effect on platelet function.[77,78] Nonetheless, the clinical importance of these genetic polymorphisms is yet to be determined. "Aspirin resistance" is also discussed in Chapter 23.

## VII. Aspirin in the Treatment of Acute Cardiovascular Disease

### A. Myocardial Infarction

**1. Aspirin Therapy Alone.** Aspirin can be expected to have a beneficial effect on the treatment of AMI given what is now known about the central role of platelets in this condition (see Chapter 35). However, early trials in this regard were either too small to demonstrate a mortality benefit or yielded conflicting results. The Second International Study of Infarct Survival (ISIS-2)[79] first definitively established the integral role of aspirin in this setting (Table 60-1). In this study, 17,187 patients presenting within 24 hours of an AMI were randomized to receive streptokinase (1.5 MU), aspirin (162.5 mg daily for 1 month), both, or neither. The 5-week vascular mortality rate among patients randomized to double-placebo was 13.2%. Aspirin decreased this rate to 9.4% (25% risk reduction), an effect that occurred irrespective of concurrent heparin use. In addition, aspirin decreased the rate of nonfatal reinfarction and of nonfatal stroke by 50%, and it was not associated with an increase in major bleeding or hemorrhagic stroke. Among more than 19,000 patients in the Antiplatelet Trialists' Collaboration meta-analysis who received aspirin for the treatment of AMI, the risk of vascular events (fatal or nonfatal stroke or AMI) was decreased from 14.2 to 10.4% corresponding to 38 events prevented per 1000 persons treated with aspirin for 1 month after their acute event ($p < 0.001$).[80] Overall, there was no significant difference in the magnitude of benefit obtained over a wide range of aspirin doses; low-dose aspirin (81 mg) appears to be as effective as higher doses at decreasing in-hospital mortality. Nonetheless, daily doses <75 mg are less well studied and may have a smaller beneficial effect.

**2. Aspirin as an Adjunct to Thrombolysis.** In 1981, DeWood and colleagues[81] demonstrated the presence of thrombus in the infarct-related arteries of patients suffering from an AMI, thus ushering in the era of thrombolytic therapy. In the past two decades, multiple studies have documented the beneficial effects of thrombolytic therapy in decreasing mortality, improving ventricular function, and improving the functional status of patients with AMI.[79,82,83] It has also become apparent that the magnitude of the benefit of thrombolytic therapy is dependent in part on the concurrent use of antiplatelet agents. In the ISIS-2 trial, the administration of streptokinase alone resulted in a 25% reduction in vascular deaths (9.2 vs. 12%).[79] As noted previously, the effect of aspirin alone was equivalent; furthermore, its effect was additive to that of thrombolytic therapy, with the combined regimen of streptokinase plus aspirin resulting in a mortality of 8% — corresponding to a 42% mortality reduction. Additionally, the use of streptokinase alone resulted in an excess of recurrent infarctions in the weeks following thrombolysis. This presumably occurred as a result of thrombin-induced platelet activation and was completely prevented by the concurrent use of aspirin.[79] A meta-analysis of 32 thrombolytic studies with angiographic follow-up revealed a significantly lower angiographic reocclusion rate and clinical recurrent ischemic event rate in patients treated with aspirin versus those on no aspirin therapy (11 vs. 25% and 25 vs. 41%, respectively), and the effect was similar in trials of streptokinase or tissue plasminogen activator (tPA).[84]

The long-term use of aspirin (81 mg daily) decreases the incidence of recurrent AMI ($p = 0.0045$) after successful treatment with thrombolytic therapy.[85] At the 10-year follow-up of ISIS-2, the mortality benefit of aspirin therapy persisted and corresponded to 26 fewer deaths per 1000 patients treated with 1 month of aspirin therapy.[86] The decrease in mortality seen with aspirin therapy likely relates in large part to the prevention of reocclusion of the infarct-related artery; however, it may also involve aspirin-associated preservation of left ventricular function.[87,88]

**3. Comparison with Other Agents.** When compared to aspirin as adjunctive therapy to thrombolysis with tPA in the setting of an AMI, heparin therapy is associated with a greater short-term (7–24 hours) patency rate (82% with heparin and 52% with aspirin), whereas aspirin therapy (80 mg daily) is associated with a lower 1-week reocclusion rate (95 vs. 88%).[89] When added to aspirin in patients receiving thrombolysis, unfractionated heparin is associated with a slightly lower reinfarction rate but does not decrease the 35-day mortality.[82] Similarly, low-molecular-weight heparin, when administered daily for 1 month after thrombolysis, may offer greater protection against angina and reinfarction than aspirin therapy alone.[90] However, combined therapy with aspirin and heparin in this setting is associated with an excess of major bleeding and cerebral hemorrhage. In the Antithrombotics in the

Prevention of Reocclusion in Coronary Thrombolysis (APRICOT) study,[91] patients who had been successfully treated with thrombolytic therapy for an AMI were randomized to receive warfarin, aspirin (325 mg daily), or placebo and were reassessed with angiography 3 months later. The rate of reocclusion of the infarct-related artery was not different among the three regimens. However, there was a trend toward lower reinfarction and lower revascularization rates in patients treated with aspirin compared to those treated with warfarin.

The addition of clopidogrel to aspirin therapy appears to impart further benefit (see Chapter 61). In the CLARITY-TIMI 28 trial,[92] 3491 patients who presented within 12 hours of an ST elevation AMI were treated with aspirin and thrombolytic therapy and randomized to treatment with clopidogrel (300 mg loading dose followed by 75 mg daily) or placebo. All patients underwent angiography within 192 hours of enrollment. Treatment with clopidogrel was associated with a 36% reduction in the combined end point of an occluded infarct-related artery at angiography, death, or recurrent AMI before angiography (15 vs. 21.7%, $p < 0.001$). The rates of major bleeding were not increased by the addition of clopidogrel to the standard regimen.

### B. Unstable Angina and Non-ST-Elevation AMI

Unstable angina and non-ST-elevation AMI represent part of the spectrum of acute coronary syndromes and pathophysiologically are characterized by platelet activation and intracoronary thrombosis.[93,94] Patients who present with these syndromes have a high rate of recurrent ischemic events.[95-97] Aspirin has been evaluated extensively in this setting, and several studies have demonstrated that treatment with aspirin at various dosages (75–1300 mg daily) and for varying durations of time (5 days to 2 years) reduces the risk of subsequent nonfatal AMI and cardiac death (Table 60-1).[96-100] During the acute phase of unstable angina, treatment with aspirin (325 mg twice daily) for 6 days has been shown to decrease the incidence of AMI to 3% versus 12% with placebo ($p = 0.01$).[100] In the Veterans Administration (VA) Cooperative Study,[96] 1266 men with unstable angina were randomized to treatment with placebo or 325 mg of buffered aspirin daily for 12 weeks. Aspirin therapy resulted in a significant 51% decrease in the incidence of nonfatal AMI and in the combined end point of death or AMI. There was also a 51% decrease in mortality, although this was not statistically significant. In the Canadian multicenter trial,[99] patients with unstable angina who were treated with aspirin (200 mg four times daily) for an average of 18 months had an 8.6% incidence of cardiac death or nonfatal AMI versus 17.0% for patients not receiving aspirin, again corresponding to a 51% risk reduction. Additionally, in this study there was a 71% decrease in mortality in the aspirin-treated group (3.0 vs. 11.7%, $p = 0.004$).

The benefits of low-dose aspirin (75 mg daily) for the treatment of patients with unstable angina or non-ST-elevation AMI have been evaluated by the Research Group on Instability in Coronary Artery Disease in Southeast

### Table 60-1: Benefit of Aspirin for the Acute Treatment of Cardiovascular Disease

| Setting | Reference | Aspirin Dose (mg) | Duration of Treatment | Total Mortality (RR) | Death or Nonfatal AMI (RR) | Death or Nonfatal CVA (RR) |
|---|---|---|---|---|---|---|
| AMI | ISIS-2[79] | 160 qd | 4 weeks | 0.77 ($2p < 0.00001$) | | |
| Unstable angina | VA Cooperative Study[96] | 325 qd | 12 weeks | 0.49 ($p = 0.054$) | 0.49 ($p = 0.0005$) | |
| | Canadian Multicenter Trial[99] | 325 qid | 24 months | 0.29 ($p = 0.004$) | 0.49 ($p = 0.008$) | |
| | Théroux et al.[100] | 325 bid | 6 days | — | 0.28 ($p = 0.01$) | |
| | RISC[97,98,101a] | 75 qd | 5 days | 0 (NS) | 0.43 ($p = 0.033$) | |
| | | | 6 months | 0.53 (NS) | 0.47 ($p < 0.0001$) | |
| | | | 12 months | 0.62 (NS) | 0.51 ($p = 0.0001$) | |
| Acute ischemic | IST[119] | 300 qd | 14 days | 0.96 (NS) | | 0.93 ($2p < 0.05$) |
| Stroke | CAST[118] | 160 qd | 4 weeks | 0.86 ($2p = 0.04$) | | 0.88 ($2p = 0.03$) |

[a]Included patients with unstable angina and non-Q wave infarctions. Results were similar in both groups.
AMI, acute myocardial infarction; CAST, Chinese Acute Stroke Trial Collaboration Group; CVA, cerebrovascular accident; ISIS-2, Second International Study of Infarct Survival; IST, International Stroke Trial Collaborative Group; NS, not significant; RISC, Research Group on Instability in Coronary Artery Disease in Southeast Sweden; RR: relative risk; VA, Veteran's Administration.

Sweden (RISC Group). In a series of articles, this group has shown that low-dose aspirin therapy decreases the combined risk of AMI or death after 5 days (2.5 vs. 5.8%, $p = 0.033$), 6 months (8.9 vs. 19.0%, $p < 0.0001$), and 1 year of treatment (11.0 vs. 21.4%, $p = 0.0001$).[97,98] Additionally, after 3 months of therapy, the risk of developing severe angina necessitating coronary angiography is decreased when aspirin is given during the acute event (10.7 vs. 18.0%, $p = 0.004$), an effect that persists after at least 1 year of therapy.[97] In a group of unstable angina and non-ST-elevation AMI patients considered to be at high risk by virtue of ischemia documented on a treadmill test at the time of hospital discharge, low-dose aspirin decreased the subsequent risk of AMI or death irrespective of whether the ischemia was symptomatic or "silent."[101]

In the Antiplatelet Trialists' Collaboration overview of randomized trials of antiplatelet (mainly aspirin) therapy, the subset of approximately 5000 patients with unstable angina had a 5.3% absolute reduction in the risk of stroke, AMI, or vascular death when treated with aspirin (8.0 vs. 13.3%, $p = 0.001$).[69] This outcome corresponded to 50 fewer vascular events per 1000 patients treated with 6 months of aspirin therapy.

**1. Comparison with Other Agents.** The Clopidogrel in Unstable Angina to Prevent Recurrent Events in Patients with Acute Coronary Syndromes without ST-Segment Elevation (CURE) trial randomized 12,596 patients with unstable angina or non-ST-elevation AMI to treatment with clopidogrel (300-mg loading dose, then 75 mg/day) and aspirin (75–325 mg/day) versus aspirin alone.[102] Combination therapy significantly reduced the incidence of stroke, AMI, or vascular death (9.3 vs. 11.4%, $p < 0.001$), and the benefit was seen regardless of the dose of aspirin given.[103] The rate of in-hospital refractory ischemia, heart failure, or need for revascularization was also significantly reduced by the addition of clopidogrel to aspirin therapy. However, this combined antiplatelet therapy was associated with an increased risk of major bleeding (3.7 vs. 2.7%, $p = 0.001$), especially at higher aspirin doses.[104]

There are conflicting data regarding the use of heparin for the treatment of unstable angina. In the RISC trial, heparin therapy alone resulted in no significant benefit with regard to the risk of AMI or death, although the group treated with aspirin and heparin had the lowest event rate.[98] A significant delay in instituting heparin (average delay, 33 hours) may account for this inefficacy of heparin. Other studies have demonstrated a significant benefit of heparin for the treatment of unstable angina, with intravenous heparin resulting in a decrease in the incidence of refractory angina[100] and AMI.[100,105] In still other studies, heparin has produced even greater benefit than that of aspirin alone.[100,105,106] Several small studies have suggested that when combined with aspirin therapy for the treatment of

unstable angina, heparin appears to offer no significant further reduction in AMI but is associated with an increased bleeding risk.[100] However, a meta-analysis of trials of combination therapy revealed that the addition of heparin to aspirin therapy for unstable angina and non-ST-elevation AMI results in a nonsignificant 33% decrease in the risk of AMI or death.[107] This benefit of combined therapy occurred without a significant increase in bleeding complications. Low-molecular-weight heparin has been reported to be at least as effective as unfractionated heparin in this setting,[108] although an extensive review found no significant difference between these two agents with regard to their reduction in the risk of AMI or death after unstable coronary syndromes.[109] Furthermore, the use of long-term low-molecular-weight heparin is no more effective at preventing vascular events than aspirin alone.[110–112] When heparin is used for the treatment of unstable angina, its discontinuation is associated with anginal reactivation; this is prevented by concomitant aspirin therapy.[113]

In the Oral Anticoagulant Strategies for Ischemic Syndromes (OASIS) pilot study of oral anticoagulation in patients with unstable angina or non-ST-elevation AMI,[114] the addition of low-dose warfarin (mean international normalized ratio [INR] 1.5) to standard treatment including aspirin was of no benefit. At higher doses (mean INR 2.3), coumadin appeared to decrease the incidence of adverse cardiovascular outcomes; however, significant differences in the event rates of the placebo groups have raised questions about the validity of this finding.

Compared to a regimen of aspirin and intravenous heparin, treatment with eptifibatide (a peptide GPIIb-IIIa antagonist) and heparin decreases the frequency and duration of ischemic episodes in patients with unstable angina.[115] Studies comparing the efficacy of the GPIIb-IIIa antagonist tirofiban with that of heparin as adjunctive treatment to aspirin in patients with unstable angina and non-ST-elevation AMI have yielded conflicting results. Some have demonstrated a reduction in death, AMI, or refractory ischemia at 48 hours in the tirofiban plus aspirin group,[116] whereas others have noted an increase in mortality with this regimen compared with a regimen of aspirin plus intravenous heparin.[117] Triple therapy with tirofiban, aspirin, and heparin, however, is associated with a 32% reduction in the combined end point of death, AMI, or refractory ischemia at 7 days (12.9 vs. 17.9%, $p = 0.004$), and this benefit persists for at least 6 months after treatment.[117] Nonetheless, the benefit of GPIIb-IIIa antagonists in this setting appears to be limited to patients who undergo percutaneous coronary interventions (PCIs). A meta-analysis of the randomized, placebo-controlled trials of GPIIb-IIIa antagonists demonstrated a significant reduction in the risk of death or nonfatal AMI in patients with acute coronary syndromes who underwent PCIs but no benefit in patients who did not undergo revascularization.[103]

## C. Acute Cerebrovascular Accident

The role of aspirin in the treatment of acute ischemic stroke has been evaluated in two large trials. The Chinese Acute Stroke Trial[118] enrolled 21,106 patients within 48 hours of the onset of an acute ischemic stroke and randomized them to treatment with aspirin (160 mg daily) or placebo. After a maximal treatment period of 4 weeks, aspirin was associated with a significant reduction in mortality (3.3 vs. 3.9%, $2p = 0.04$), significantly fewer recurrent ischemic strokes (1.6 vs. 2.1%, $2p = 0.01$), but slightly more hemorrhagic strokes (1.1 vs. 0.9%, $2p > 0.1$). The combined end point of in-hospital death or recurrent stroke was reduced by 12% with aspirin therapy (5.3 vs. 5.9%, $2p = 0.03$). In the International Stroke Trial,[119] 19,435 patients were enrolled and randomized to aspirin (300 mg daily), subcutaneous unfractionated heparin (5000 or 12,500 IU twice daily), both, or neither. After a maximum of 14 days of therapy, aspirin resulted in a nonsignificant decrease in mortality (9.0 vs. 9.4%) but a significant decrease in recurrent ischemic strokes (2.8 vs. 3.9%, $2p < 0.00001$). After 6 months of follow-up, there was a nonsignificant trend toward a lower rate of mortality or physical dependence in the aspirin-treated group. Combined analysis of these trials demonstrated a highly significant reduction in recurrent stroke, a borderline significant reduction in mortality, and a net benefit of approximately nine fewer deaths or nonfatal strokes per 1000 patients treated with 2 to 4 weeks of aspirin therapy.[118] These benefits occurred irrespective of the patients' age, gender, level of consciousness, blood pressure at presentation, the presence of atrial fibrillation, CT scan findings, or concomitant heparin use.[120] Antiplatelet therapy in this setting was associated with an absolute excess of 1.9 hemorrhagic strokes per 1000 patients treated, but this was offset by an absolute reduction of 6.9 ischemic strokes per 1000.[80]

**1. Comparison with Other Agents.** In the International Stroke Trial, heparin therapy did not significantly reduce mortality or the combined end point of mortality or physical dependence.[119] Heparin did reduce the incidence of recurrent ischemic stroke (2.9 vs. 3.8%, $2p < 0.01$); however, this was offset by an increase in hemorrhagic strokes (1.2 vs. 0.4%, $2p < 0.00001$). Higher dose heparin was associated with significantly more extracranial bleeding as well as hemorrhagic strokes compared with the lower dose regimen. Also, the addition of heparin to aspirin offered no further benefit over treatment with aspirin alone.[119] Similarly, when started within 30 hours of the onset of an ischemic stroke, low-molecular-weight heparin (dalteparin 100 IU twice daily) offers no benefit over aspirin alone with regard to recurrent ischemic stroke or functional outcome.[121]

Several studies have demonstrated a beneficial effect of thrombolysis with tPA in the treatment of acute ischemic stroke.[122,123] These agents appear to improve neurological recovery, although without an improvement in mortality, in patients who present within 3 hours of a large cerebral infarction. However, when aspirin is used as adjunctive therapy to thrombolysis with streptokinase, it is associated with a significantly higher risk of early death.[124,125]

## VIII. Aspirin for Secondary Prevention of Cardiovascular Events

### A. Myocardial Infarction/Ischemia

Six large, randomized, placebo-controlled trials have been performed evaluating the role of aspirin therapy as long-term secondary prophylaxis against vascular events in patients following an AMI.[126–131] In the Aspirin Myocardial Infarction Study,[126] in which 4524 patients were randomized to aspirin (1000 mg daily) or placebo and followed for a minimum of 3 years, aspirin therapy resulted in no difference in total mortality, a trend toward a lower incidence of AMI, and a higher rate of gastrointestinal (GI) irritation. In all five other studies, aspirin therapy produced at least a trend toward decreased mortality.[127–131]

An overview of these studies and 190 others that evaluated the effects of long-term antiplatelet therapy (predominantly aspirin) was performed by the Antiplatelet Trialists' Collaboration.[80] This meta-analysis included 135,640 patients who were treated with antiplatelet agents for a variety of indications, at a wide range of dosing regimens, and for varying durations of time. Among almost 19,000 patients who had a prior history of an AMI, aspirin given as secondary prophylaxis decreased the risk of vascular events (fatal or nonfatal stroke or AMI) from 17 to 13.5% ($p < 0.0001$). This outcome corresponds to approximately 36 events prevented per 1000 patients treated with aspirin for 2 years. Additionally, antiplatelet therapy led to significant reductions in the risk of vascular events among patients with a history of unstable angina (8 vs. 13.3%), stable angina (9.9 vs. 14.1%), or percutaneous revascularization (2.7 vs. 5.5%). These effects occurred irrespective of patient age or gender, and they were unaffected by the presence of hypertension. Although this study found no differential effect in diabetic patients, other studies suggest that the absolute benefit of aspirin therapy may be greater in diabetic patients than in their nondiabetic counterparts.[132]

The results of these studies clearly support the use of aspirin for secondary prevention in patients with established CAD. The beneficial effects of aspirin are observed at a wide range of dosages (75–1550 mg daily). Available data suggest that lower doses (75–150 mg/day) are equally effective as higher dose regimens but are associated with less toxicity. Thus, it seems reasonable to use low-dose regimens (75–160 mg) when prescribing aspirin for the secondary prophylaxis of ischemic heart disease.

**1. Comparison with Other Agents.** In the Clopidogrel versus Aspirin in Patients at Risk for Ischemic Events (CAPRIE) study, 19,185 patients with symptomatic peripheral arterial disease or with a recent stroke or AMI were randomized to treatment with clopidogrel (75 mg daily) versus aspirin (325 mg daily).[133] After an average follow-up of almost 2 years, clopidogrel reduced the combined incidence of ischemic stroke, AMI, or vascular death by 8.7% (5.32 vs. 5.83%, $p = 0.043$) compared to treatment with aspirin. However, in the subgroup of patients with a prior AMI, the beneficial effects of clopidogrel were not significant. When added to aspirin therapy for patients with a prior AMI, clopidogrel produced synergistic inhibitory effects on *ex vivo* platelet activation and aggregation, suggesting that combination therapy may be superior to either agent alone.[134]

The Coumadin Aspirin Reinfarction Study (CARS) compared the effect of aspirin therapy alone (160 mg daily) with that of low-dose aspirin (80 mg daily) in addition to low-dose warfarin (1 or 3 mg daily).[135] After a mean follow-up of 14 months, there was no clinical benefit of warfarin beyond that of aspirin; however, at the 3-mg dose, warfarin was associated with a significant increase in spontaneous major bleeding. A similar lack of benefit of warfarin was found in the CHAMP study, comparing aspirin monotherapy (162 mg/day) with a regimen of aspirin (81 mg/day) and low-dose warfarin (goal INR, 1.5–2.5).[136] Nonetheless, in a study of 3630 patients treated with aspirin (160 mg/day), warfarin (goal INR, 2.8–4.2), or both (aspirin 81 mg/day; goal INR, 2.0–2.5) after AMI, warfarin therapy offered a significant benefit.[137] In this study, the risk of death, nonfatal AMI, or stroke was 20% in the aspirin group, 16.7% in the warfarin group ($p = 0.003$), and 15% in the group receiving combination therapy ($p = 0.001$); however, there was a significant increase in major nonfatal bleeding in the warfarin group (0.62 vs. 0.17% with aspirin, $p < 0.001$). These data appear to confirm the results of a prior meta-analysis suggesting that when used in combination with aspirin, higher dose warfarin (INR > 2), but not low-dose warfarin (INR < 2), decreases the combined risk of death, AMI, or stroke, albeit at a higher rate of hemorrhagic complications.[138]

**2. Effect of Prior Aspirin Use on Ischemic Presentation.** In patients presenting with symptoms of acute coronary ischemia, the Thrombolysis in Myocardial Infarction (TIMI)-7 investigators found that those who were taking long-term aspirin therapy before hospitalization were less likely to be proven to have suffered an AMI (5 vs. 14%, $p = 0.004$) and correspondingly more likely to have had unstable angina.[139] This occurred despite the fact that patients taking aspirin were more likely to be older and have risk factors for CAD. In addition, among patients who suffer an AMI, those who have been taking aspirin prior to their hospitalization may have smaller infarcts and less ST-elevation infarctions than their non-aspirin-using counterparts.[140]

Nonetheless, despite this apparent protective effect of aspirin in patients presenting with unstable coronary syndromes, prior aspirin use is associated with a higher risk of death or AMI in this setting.[141] The reason for this finding is not clear but may relate to aspirin resistance (see Section VI.A) or an increased thrombus burden in the patients already taking aspirin. Alternatively, prior aspirin use may be a marker for patients who have more severe CAD.

### B. Cerebral Infarction/Ischemia

Early studies evaluating the utility of aspirin for the secondary prevention of stroke in patients with a prior history of stroke or transient ischemic attack (TIA) yielded conflicting results, with several studies failing to demonstrate a significant beneficial effect.[142–144] Subsequently, a meta-analysis of 10 of these studies including 4192 patients in placebo-controlled trials of aspirin therapy revealed a highly significant 13% relative reduction in the risk of vascular events ($p = 0.0011$).[145] Among the 195 studies reviewed by the Antiplatelet Trialists' Collaboration were 21 trials of antiplatelet therapy in patients with a previous stroke or TIA.[80] In the majority of these studies, aspirin (50–1500 mg daily) was the antiplatelet agent used, although dipyridamole, sulphinpyrazone, and ticlopidine were also used, either alone or in combination with aspirin. These 21 studies included more than 18,000 patients treated with antiplatelet agents for a mean of 29 months, and among these patients there was a highly significant 23% reduction in nonfatal stroke ($p < 0.0001$), a 26% decrease in nonfatal AMI ($p = 0.0009$), and a 12% reduction in all-cause mortality ($p = 0.002$). These results corresponded to 36 fewer vascular events (fatal and nonfatal stroke and AMI) per 1000 patients treated with antiplatelet agents ($p < 0.0001$). A previous analysis demonstrated that the benefits of aspirin therapy were similar for patients who presented with a completed stroke (23% risk reduction) and for those who presented with a TIA (22% risk reduction).[69]

Carotid endarterectomy benefits certain patients with carotid disease; however, there is a significant risk of perioperative cerebrovascular events.[146] In a small study of patients who had recently undergone carotid endarterectomy, aspirin (50–100 mg daily) produced no significant reduction in the risk of vascular events.[147] In the Aspirin and Carotid Endarterectomy Trial,[148] 2849 patients scheduled for endarterectomy were randomized to receive 81, 325, 650, or 1200 mg of aspirin daily, starting preoperatively. After 3 months of follow-up, the risk of stroke, AMI, and death was significantly lower for patients taking 81 or 325 mg daily versus higher doses (4.2 vs. 10.0%, $p = 0.0002$).

**1. Comparison with Other Agents.** In the Antiplatelet Trialists' Collaboration analysis, several of the included studies compared the effect of aspirin to that of dipyrida-mole, sulphinpyrazone, ticlopidine, and clopidogrel either alone or in combination with aspirin. Although the total number of patients treated with these other agents was small, there was no significant difference in outcome when com-pared with aspirin, and there were no benefits beyond that of aspirin alone; however, the results of the CURE trial were not included in this analysis.[102] Individual trials comparing ticlopidine with aspirin as secondary stroke prevention have not found ticlopidine to be significantly more effective; however, it is associated with an increased risk of side effects including severe neutropenia.[149–151] In the CAPRIE trial, clopidogrel (75 mg daily) was more effective than aspirin in the secondary prevention of vascular events (stroke, AMI, and vascular death), with less side effects than ticlopidine. However, in the subgroup of patients with stroke, the benefi-cial effect of clopidogrel was not significantly different from that of aspirin.[133] When combined with aspirin in patients with symptomatic carotid stenosis[152] and in those under-going carotid endarterectomy,[153] clopidogrel significantly reduced asymptomatic microembolization as assessed by transcranial Doppler ultrasound, suggesting a beneficial effect of combination therapy.[152] However, in a trial of high-risk patients after stroke or TIA, the addition of aspirin (75 mg daily) to clopidogrel (75 mg daily) did not significantly reduce the risk of major vascular events (15.7% with com-bination therapy vs. 16.7% with clopidogrel alone) but increased the risk of life-threatening bleeding (2.6 vs. 1.3%, $p < 0.001$).[154]

The Second European Stroke Prevention Study (ESPS-2)[155] randomized 6602 patients with a prior stroke or TIA to treatment with aspirin (50 mg daily), dipyridamole (400 mg daily), both, or neither. After 2 years of treatment, single-drug therapy with aspirin or dipyridamole resulted in a highly significant reduction in the risk of stroke (18% reduction with aspirin and 16% reduction with dipyrida-mole) and in the combined risk of stroke or death (13% reduction with aspirin and 15% reduction with dipyridam-ole). Furthermore, combined therapy with aspirin and dipyridamole was the most effective regimen, producing a 37% reduction in stroke risk and a 24% reduction in the combined end point. A meta-analysis of trials of aspirin with or without dipyridamole in the secondary prevention of stroke demonstrated a 13% relative risk reduction in vas-cular events with aspirin therapy alone ($p = 0.0011$) and a 30% relative risk reduction with combination therapy ($p < 0.00001$).[145] Additionally, compared to aspirin alone, com-bination therapy produced an additional 15% reduction in the risk of vascular events ($p = 0.012$); however, this result was largely dependent on the outcome of the ESPS-2 trial and was not supported by findings from other individual trials. Controversy regarding the additive benefits of combi-nation antiplatelet therapy for patients after a TIA or stroke will hopefully be resolved when results of ongoing trials (the European/Australian Stroke Prevention in Reversible Ischemia Trial [ESPRIT][156] and the Prevention Regimen for Effectively Avoiding Second Strokes [ProFESS] trial) become available.

Aspirin (1300 mg daily) was compared with warfarin (goal INR 2.0–3.0) in the Warfarin–Aspirin Symptomatic Intracranial Disease trial.[157] This trial was stopped prema-turely due to an increased risk of death (4.3 vs. 9.3%, $p = 0.02$), major bleeding (3.2 vs. 8.3%, $p = 0.01$), and AMI or sudden death (2.9 vs. 7.3%, $p = 0.02$) in the warfarin-treated group. Lower dose warfarin (goal INR, 1.4–2.8) was com-pared to aspirin (325 mg daily) in the Warfarin–Aspirin Recurrent Stroke study and demonstrated no difference in the rate of recurrent stroke or death or in the rate of major bleeding.[158] Furthermore, in the Coumadin Aspirin Rein-farction Study, the incidence of ischemic stroke was lower in patients receiving 160 mg aspirin than in the group receiving combination therapy with 80 mg aspirin and 1 mg warfarin (0.6 vs. 1.1%, $p = 0.53$).[159] These data suggest no benefit of warfarin over aspirin for secondary prevention of is-chemic stroke.

**2. Aspirin Dose for Secondary Prevention of Cere-brovascular Disease.** The optimal dose of aspirin required to produce beneficial reductions in vascular events in patients with a prior stroke or TIA has been a widely debated topic.[160,161] The Antiplatelet Trialists' Collaboration found no significant benefit of high-dose aspirin (500–1500 mg daily) over medium-dose (160–325 mg daily) or low-dose (75–150 mg daily) regimens.[80] In the UK-TIA Trial, there was no difference in efficacy between a 300- and 1200-mg daily aspirin dose in preventing serious vascular events or death.[162] In the Swedish Aspirin Low-Dose Trial (SALT), 75 mg of aspirin daily resulted in a significant reduction in the risk of stroke or death in patients with ischemic cerebrovascular events.[163] Furthermore, in the Dutch TIA Trial,[164] 30 mg of aspirin daily was equally effective in the prevention of vas-cular events as a 238-mg dose. These lower dose regimens were also associated with significantly less adverse effects. In a meta-analysis of placebo-controlled trials, the relative risk reductions with regard to vascular event rates were similar for high, medium, and low doses of aspirin.[145] Thus, as for the secondary prevention of coronary events, it appears appropriate to use lower dose aspirin for the secondary prevention of cerebrovascular disease.

# IX. Aspirin for Primary Prevention of Cardiovascular Disease

## A. Myocardial Infarction/Ischemia

Following the success of the trials of aspirin for the treat-ment of acute myocardial and cerebral ischemia and for the secondary prevention of recurrent cardiovascular events, several studies were developed to investigate the utility of aspirin for primary prevention in patients at risk for cardiovascular disease. The initial Antiplatelet Trialists' Collaboration review of 28,000 low-risk patients treated with aspirin for an average of 5 years demonstrated a significant reduction in the risk of nonfatal AMI ($2p < 0.0005$) but an increase in hemorrhagic strokes ($2p < 0.05$).[69] Two large randomized trials evaluated the utility of prophylactic aspirin therapy in male physicians (Table 60-2).[165,166] The participants had no prior history of stroke or AMI and had a low incidence of angina or TIA; thus, they represented a low-risk patient population. In the British Physician's Study,[165] 5139 participants were randomized to treatment with 500 mg of aspirin daily and followed for an average of 6 years. Although the total mortality was 10% lower in the treatment group, this difference was not significant. There was no reduction in the incidence of nonfatal AMI or stroke, although there was an increase in disabling strokes in the treatment group. The Physician's Health Study[166] was a placebo-controlled trial that randomized 22,071 men aged 40 to 84 years. After an average follow-up of 5 years, the study was stopped prematurely due to a highly significant 44% reduction (0.26 vs. 0.44% per year, $p < 0.00001$) in the risk of first AMI in the aspirin-treated group (325 mg every other day). This effect was only apparent among patients older than the age of 50 years. Furthermore, in the active treatment group there was no reduction in cardiovascular mortality, a nonsignificant increase in the risk of hemorrhagic stroke (RR, 2.14), and a

significant increase in bleeding complications including some that required transfusion. An overview of both these trials demonstrated a significant one-third reduction in the risk of nonfatal AMI ($p < 0.0001$).[167] However, there was no reduction in the risk of stroke or vascular mortality, and the risk of hemorrhagic stroke was slightly increased.

Several primary prevention studies enrolled patients who were at a higher risk of developing cardiac events (Table 60-2). As part of the Hypertension Optimal Treatment (HOT) study,[168] 18,790 hypertensive patients aged 50 to 80 years were randomized to treatment with aspirin (75 mg daily) or placebo. Aspirin therapy reduced major cardiovascular events by 15% ($p = 0.03$) and the risk of AMI by 36% ($p = 0.002$). There was no effect on stroke, and there was a significant increase in nonfatal major bleeding. The Primary Prevention Project[169] enrolled 4494 patients considered to be at high risk for cardiovascular events due to their risk factor profile and randomized them to treatment with aspirin (100 mg daily), vitamin E (300 mg daily), both, or neither. After an average follow-up of 3.6 years, aspirin-treated patients had a significantly lower risk of cardiovascular death (RR, 0.56; $p = 0.042$) and total cardiovascular events (RR, 0.77; $p = 0.14$), although the risk of severe bleeding was increased (1.1 vs. 0.3%, $p < 0.0008$). The Thrombosis Prevention Trial[170] enrolled 5499 men aged 45 to 69 years who had no prior history of stroke or AMI but were considered to be at high risk for cardiovascular events due to their risk factor

**Table 60-2: Benefit of Aspirin for the Primary Prevention of Cardiovascular Events**

| Setting | Reference | Aspirin Dose (mg) | Duration of Follow-Up (Years) | Total Mortality (RR) | Nonfatal AMI (RR) | Stroke (RR) |
|---|---|---|---|---|---|---|
| Low-risk men | Physicians' Health Study[166] | 325 qod | 5 | 0.96 (NS) | 0.59 ($p < 0.00001$) | 1.22 (NS) |
| | Peto et al.[165] | 500 qd | 6 | 0.89 (NS) | 0.97 (NS) | 1.15 (NS) |
| Low-risk women | Women's Health Study[175] | 100 qod | 10 | 0.95 (NS) | 1.01 (NS) | 0.83 ($p = 0.04$) |
| | Manson et al.[173c] | Varied | 5.4 | 0.86 (NS) | 0.68 ($p = 0.005$) | 0.99 (NS) |
| Multiple risk factors | Thrombosis Prevention Trial[170] | 75 qd | 6.4 | 1.06 (NS) | 0.68 (CI, 0.52–0.88)[a] | 0.98 (NS) |
| | HOT[168] | 75 qd | 3.8 | 0.93 (NS) | 0.64 ($p = 0.002$)[b] | 0.88 (NS) |
| | Primary Prevention Project[169] | 100 qd | 3.6 | 0.81 (NS) | 0.69 (NS) | 0.67 (NS) |
| Chronic stable angina | SAPAT[172] | 75 qd | 4.2 | 0.78 (NS) | 0.61 ($p = 0.006$) | 0.75 (NS) |
| Atrial fibrillation | SPAF[178] | 325 qd | 1.3 | 0.80 (NS) | 0.6 (NS)[b] | 0.58 ($p = 0.02$)[d] |

[a]$p$ value not available.
[b]Data is for total MI.
[c]Relative risks for subgroup taking up to six aspirin per week *vs.* nonaspirin group.
[d]Includes stroke plus systemic embolism.
AMI, acute myocardial infarction; HOT, Hypertension Optimal Treatment randomized trial; NS, not significant; RR, relative risk for aspirin versus nonaspirin groups; SAPAT, Swedish Angina Pectoris Aspirin Trial; SPAF, Stroke Prevention in Atrial Fibrillation trial.

profile. These subjects were randomized to receive warfarin (mean INR approximately 1.5), aspirin (75 mg daily), both, or neither. After more than 6 years of follow-up, aspirin therapy was associated with a 20% reduction in the combined end point of AMI and coronary death ($p = 0.04$). This was due almost entirely to a 32% reduction in nonfatal events. Conversely, warfarin therapy resulted in a 21% reduction in the combined end point ($p = 0.02$), primarily as a result of a 39% decrease in fatal events. Combined treatment with warfarin and aspirin produced the greatest risk reduction (34% decrease in combined end point, $p = 0.006$) but was associated with an increase in hemorrhagic and fatal strokes. Subsequent analysis of this study suggests that aspirin has a significantly greater beneficial effect in patients with lower (<133 mmHg) rather than higher (>145 mmHg) systolic blood pressure (RR, 0.59 vs. 1.08, respectively; $p = 0.0001$).[171]

Patients who have chronic stable angina represent an even higher risk group of patients. In the Swedish Angina Pectoris Aspirin Trial (SAPAT),[172] 2035 patients with stable angina were treated with sotalol for symptom control and randomized to treatment with aspirin (75 mg daily). After an average follow-up of 50 months, aspirin reduced the risk of first AMI and sudden death by 34% ($p = 0.003$) while producing a nonsignificant increase in major bleeding. Similarly, in patients with chronic stable angina who were enrolled in the Physicians' Health Study,[166] aspirin therapy resulted in an 87% decrease in the incidence of first AMI.

Several studies have addressed the use of aspirin for the primary prevention of vascular events in women. In a prospective cohort study of 87,678 U.S. registered nurses aged 35 to 65 years,[173] the use of one to six aspirin per week was associated with a lower risk of a first AMI (RR, 0.68; $p = 0.02$). Similar to the primary prevention trials in men, this effect was limited to subjects older than 50 years of age. There was no significant reduction in the risk of stroke, cardiovascular death, or important vascular events in this study. A post hoc analysis of the HOT study suggests a differential effect of prophylactic aspirin based on gender.[174] In women, treatment with 75 mg of aspirin daily had no effect on AMI prevention, whereas in men there was a highly significant 42% risk reduction ($p = 0.001$). The results of the Women's Health Study have recently been reported.[175] In this study, 39,876 female health professionals older than 45 years of age were randomized to treatment with low-dose aspirin (100 mg every other day) versus placebo and followed for an average of 10 years. Overall, aspirin use was associated with a nonsignificant 9% reduction in the risk of a first major cardiovascular event (fatal or nonfatal stroke or AMI), although a significant reduction was noted among women 65 years of age or older (RR, 0.74; $p = 0.008$). When individual end points were evaluated, aspirin had no significant effect on the risk of AMI or death from cardiovascular causes.

Taken together, these studies offer little support for the routine use of aspirin therapy in the primary prevention of coronary events, especially in patients who are at a low risk of events. For the high-risk patient, the benefits of prophylactic aspirin therapy may outweigh the well-documented risks of bleeding, including cerebral hemorrhage, but need to be considered on an individual basis. Whether combination antiplatelet therapy (clopidogrel and aspirin) is a more effective primary prevention regimen for high-risk patients will be addressed by the CHARISMA trial.[176] Patients with stable angina do appear to benefit from prophylactic aspirin therapy; however, one could argue that such therapy represents secondary prevention because these patients have established CAD.

### B. Cerebral Infarction/Ischemia

There are conflicting data regarding the role of aspirin in the primary prevention of cerebrovascular disease. In the Antiplatelet Trialists' Collaboration review, not only was there no beneficial effect of aspirin in low-risk patients but also there was a significant increase in hemorrhagic strokes and a trend toward an excess of other strokes.[69] In the two primary prevention trials in male physicians,[165,166] aspirin therapy decreased the risk of TIA (15.9 vs. 27.5%, $p < 0.05$) but did not reduce the incidence of stroke. Furthermore, there was an increase in the incidence of disabling stroke (19 vs. 7.4%, $p < 0.05$). In the aforementioned Women's Health Study,[175] aspirin effected a 17% reduction in the risk of stroke ($p = 0.04$) due primarily to a 24% reduction in the risk of ischemic stroke ($p = 0.009$). Aspirin had no significant effect on the risk of AMI or death from cardiovascular causes, and it was associated with a slight increase in the risk of hemorrhagic stroke (RR, 1.24; $p = 0.31$) and a significant increase in the risk of major bleeding (RR, 1.4; $p = 0.02$). The authors of this study performed a meta-analysis of primary prevention trials attempting to identify gender discrepancies and found that aspirin was associated with a significant 19% reduction in the risk of ischemic stroke in women, without a reduction in the risk of AMI. In contrast, in men aspirin was associated with a significant 32% reduction in the risk of AMI and a nonsignificant increase in the risk of stroke. The reasons for these differences remain uncertain. In patients who have carotid bruits and >50% carotid artery stenosis, aspirin does not appear to prevent subsequent cerebrovascular events.[177]

Several studies have evaluated the role of aspirin for stroke prevention in patients with atrial fibrillation. In the original Stroke Prevention in Atrial Fibrillation (SPAF) study, aspirin therapy for a mean of 1.3 years decreased the stroke risk by 42% (3.6 vs. 6.3%, $p = 0.02$).[178] In later trials and a subsequent meta-analysis, warfarin was shown to be more effective than aspirin at reducing atrial fibrillation-associated cerebrovascular events.[179–182] When added to low-dose warfarin (mean INR, 1.3), aspirin is insufficient for stroke prevention compared to treatment with adjusted-dose warfarin alone (mean INR, 2.4).[183] In younger patients (<75 years) who have no high-risk markers for stroke (history of hypertension, diabetes, AMI, heart failure, or prior embolic event), the risk of atrial fibrillation-associated stroke is low and aspirin therapy (325 mg daily) may be equally effective as coumadin in stroke prevention, with less risk of major bleeding.[179] Nonetheless, the vast majority of patients with atrial fibrillation receive greater protection with warfarin than with aspirin therapy.

## X. Aspirin Therapy after Revascularization

### A. *Percutaneous Revascularization*

Percutaneous transluminal coronary angioplasty and intracoronary stent placement are associated with platelet activation as a result of vascular endothelial trauma (see Chapter 35). Platelet aggregation at the site of this vascular injury results in a 3.5 to 8.6% incidence of acute or subacute thrombosis following these procedures, resulting in a significant risk of periprocedural AMI.[184–186] Studies of antiplatelet therapy with aspirin and dipyridamole have routinely demonstrated a significant decrease in the risk of periprocedural AMI following angioplasty,[187,188] although the combination is not significantly superior to treatment with aspirin alone.[189] Conversely, the addition of ticlopidine to aspirin produces synergistic inhibitory effects on platelet activation and aggregation[190] and results in a lower incidence of stent thrombosis compared with aspirin treatment alone.[191] Clopidogrel (300 mg loading dose, then 75 mg daily for 1 month) has been demonstrated to be equally effective as ticlopidine when added to aspirin (325 mg daily) for the prevention of stent thrombosis and reduction of adverse cardiovascular events after intracoronary stent placement.[192,193] Furthermore, the safety/tolerability profile of clopidogrel is superior to that of ticlopidine, making it the preferred agent in this setting (see Chapter 61).[194] Among patients enrolled in the CURE trial were 2658 patients with non-ST-elevation acute coronary syndromes who were treated with aspirin and randomized to treatment with clopidogrel (300 mg loading dose, then 75 mg/day) versus placebo prior to undergoing PCI.[195] The addition of clopidogrel was associated with a 30% reduction in the risk of cardiovascular death, AMI, or target vessel revascularization within 30 days of the initial procedure ($p = 0.03$), and the benefit was maintained during long-term clopidogrel administration (average duration of therapy, 8 months). These results were largely driven by a significant reduction in AMI without a reduction in death. Similar results were obtained in the Clopidogrel for the Reduction of Events during Observation (CREDO) trial, which demonstrated that the addition of clopidogrel to aspirin for 12 months after elective PCI results in a sustained reduction in adverse ischemic events.[196]

The addition of warfarin to aspirin does not further reduce the rate of subacute coronary thrombosis, is associated with a significant increase in bleeding complications,[197] and is an inferior regimen compared to the combination of aspirin (250 mg daily) and ticlopidine (500 mg daily) for the prevention of cardiac events following stent placement.[191,198] GPIIb-IIIa antagonists have been studied extensively as an adjunct to aspirin (325 mg daily) and heparin in the setting of percutaneous revascularization with balloon angioplasty or intracoronary stents (see Chapter 62). These agents have routinely demonstrated benefit beyond that of aspirin and heparin in reducing the risk of ischemic complications (death, AMI, repeat revascularization, or refractory angina) in patients undergoing revascularization electively or in the setting of unstable angina or an evolving AMI.[199–202]

The minimal dose of aspirin necessary to maintain benefit in the setting of percutaneous revascularization has not been studied. However, given the efficacy of 325 mg of aspirin, and the 4 to 6 weeks required for the artery and stent to reendothelialize after the intervention, it seems reasonable to administer full-dose aspirin (325 mg) at the time of intervention and for 4 to 6 weeks following and then decrease to low-dose aspirin (75 or 81 mg) thereafter. Whether full-dose aspirin should be continued for a longer period of time in patients with drug-eluting stents remains to be determined.

### B. *Surgical Revascularization*

Coronary artery bypass graft surgery using venous or arterial conduits is an effective method of reestablishing adequate blood flow to diseased coronary arteries. However, the long-term success of the surgery depends on the patency of the grafts (see Chapter 59). Procedural complications, vascular injury, endothelial disruption, and surgery-associated hypercoagulable state all predispose to thrombosis at the anastomotic site and contribute to the reported 5 to 15% rate of graft occlusion in the first postoperative month.[203–205] Among approximately 4000 patients with coronary artery grafts included in a review of antiplatelet therapy, platelet inhibition significantly reduced the incidence of confirmed graft occluson.[206] Preoperative aspirin administration may be associated with decreased in-hospital mortality[207]; however, it appears to offer no significant benefit in early graft patency compared to postoperative therapy[208] and may result in a higher bleeding risk. When started within 1 hour after surgery, aspirin (324 mg daily) reduces the incidence of angiographically proven 1-week graft occlusion by 74% (1.6 vs. 6.2%, $p = 0.004$), and continued treatment reduces the 1-year incidence of graft occlusion by 50% (5.8 vs.

11.2%, $p = 0.01$).[203] Furthermore, when started within 48 hours after bypass surgery, aspirin is associated with a significant reduction in in-hospital mortality (1.3 vs. 4.0%, $p < 0.001$), as well as significant reductions in the incidence of AMI, stroke, renal failure, and bowel infarction.[209] Long-term aspirin therapy does not appear to have an effect on graft patency beyond the first year of treatment[210]; however, these patients require continued aspirin therapy for secondary prevention of coronary events. Compared to treatment with aspirin in a group of patients with prior coronary artery bypass procedures, clopidogrel significantly reduced the incidence of vascular death, AMI, stroke, or rehospitalization (15.9 vs. 22.3%, $p = 0.001$).[211] Aspirin may play a role in addition to myocardial protection and maintenance of graft patency in the bypass patient. Intraoperative data suggest that much of the increase in systemic thromboxane levels seen after cardiopulmonary bypass reflects increased production in the lungs, presumably by ischemic pulmonary tissue[212] or as a result of platelet activation on the bypass membrane. In a small, prospective study of patients undergoing coronary bypass surgery, those patients who received aspirin until the day of the operation had a significantly shorter duration of postoperative mechanical ventilation compared to patients who had not received aspirin for at least 1 week preoperatively.[213] Aspirin may therefore play an additional role in the prevention of pulmonary injury after bypass procedures.

## XI. Aspirin in Other Clinical Settings

Aspirin has been studied in a variety of other settings, albeit to a lesser extent than for coronary and cerebrovascular diseases.

### A. Other Vascular Disorders

There are limited data to support the use of aspirin for the control of claudication in patients with peripheral vascular disease (see Chapter 37). However, the chronic use of aspirin in apparently healthy men may reduce their future need for peripheral arterial surgery.[214] Following peripheral arterial bypass surgery, aspirin appears to be superior for the prevention of nonvenous graft occlusion, whereas warfarin is superior in preventing venous graft thrombosis.[215] Low-dose aspirin (100 mg daily) appears as effective as high-dose aspirin (1000 mg daily) at maintaining long-term vascular graft patency.[216] Although aspirin does not appear to protect against vascular events in patients with asymptomatic carotid bruits,[177] it may attenuate the progression of intima-media thickness in diabetic patients with early atherosclerotic carotid artery disease.[217] In this regard, high-dose aspirin (900 mg daily) may be more effective at slowing carotid plaque progression than low-dose aspirin (50 mg

daily) ($p = 0.011$), and it may lead to plaque regression.[218] Among the 9214 patients with peripheral arterial disease in the Antiplatelet Trialists' Collaboration study, there was a 23% reduction in serious vascular events in patients treated with antiplatelet therapy ($p = 0.004$).[80] The results were similar among patients with symptomatic claudication and those who had undergone surgical or percutaneous revascularization. Of the patients enrolled in the CAPRIE trial, more than 6000 had a primary diagnosis of peripheral arterial disease.[133] Compared with aspirin therapy, clopidogrel resulted in a 24% reduction in the risk of ischemic stroke, AMI, or ischemic death ($p = 0.003$). This benefit was significantly greater than that seen in patients enrolled in the study for a primary diagnosis of AMI or stroke, suggesting that clopidogrel may have a preferential effect on peripheral vascular disease.

Several small studies and a meta-analysis of 53 trials of antiplatelet therapy (mainly aspirin) versus control following general or orthopedic surgery and in high-risk, immobilized medical patients reported significant reductions in the risk of deep venous thrombosis and pulmonary embolism associated with aspirin use.[219–221] Additionally, in the Pulmonary Embolism Prevention (PEP) trial of 17,444 patients undergoing hip arthroplasty, 160 mg of aspirin started preoperatively and continued for 35 days following surgery reduced the risk of symptomatic deep vein thrombosis by 29% ($p = 0.03$) and pulmonary embolism by 43% ($p = 0.002$).[222] Nonetheless, methodological problems were present in many of these studies, and recent data suggest that aspirin is inferior to treatment with low-molecular-weight heparin or pneumatic compression devices for the prevention of venous thromboembolism.[223–225] Current evidence-based guidelines therefore argue against the use of aspirin alone as prophylactic therapy for venous thromboembolic disease.[226]

Initial trials investigating the use of aspirin in conjunction with oral anticoagulation after mechanical valve replacement demonstrated a reduction in thromboembolic events with aspirin therapy[227]; however, there was excess morbidity as a result of hemorrhagic complications.[228] A meta-analysis of 10 randomized trials revealed that compared with anticoagulation alone, the addition of an antiplatelet agent (aspirin or dipyridamole) significantly reduced the risk of thromboembolic events (OR, 0.41; $p < 0.001$) and total mortality (OR, 0.49; $p < 0.001$) but increased the risk of major bleeding (OR, 1.5; $p = 0.033$).[229] The benefit was similar with aspirin or dipyridamole, although the bleeding risk was reduced when lower doses of aspirin (<100 mg daily) were used.

In patients with Kawasaki disease, aspirin is initially given in high doses (80–100 mg/kg daily during the acute phase for up to 14 days) as an anti-inflammatory agent and then in lower doses (3–5 mg/kg daily for 7 weeks or longer) as an antiplatelet agent to prevent coronary aneurysm thrombosis and subsequent AMI (the major cause of death in

Kawasaki disease).[230] Although no randomized controlled trials have been performed, data are available to support the concept that aspirin can reduce the coronary involvement in Kawasaki disease.[231,232]

## B. Preeclampsia

Preeclampsia is a pregnancy-related complication characterized by hypertension and proteinuria. The pathophysiology of preeclampsia is thought to involve the activation of platelets,[233] inadequate production of prostacyclin, and excessive production of $TXA_2$.[234] It was therefore hypothesized that aspirin may decrease the risk of preeclampsia by inhibiting platelet function and decreasing thromboxane production. Early small trials suggested significant benefit in terms of decreasing the risk of proteinuria and hypertension; however, larger randomized studies have not confirmed the benefits of aspirin in decreasing preeclampsia in either nulliparous women[235] or women who are otherwise at high risk for preeclampsia.[236] Some investigators have suggested that inadequate dosing or insufficient duration of aspirin administration may have led to the negative results.[237,238] A review of 42 randomized trials of antiplatelet therapy (mainly low-dose aspirin) in 32,000 pregnant women demonstrated mild benefits, with a 15% reduction in the risk of preeclampsia, an 8% reduction in the risk of preterm delivery, and a 14% reduction in fetal demise.[239] It therefore appears that aspirin may play a role in the prophylaxis of preeclampsia; however, it is not clear which women obtain the most benefit, what dose of aspirin is most effective, or how early in pregnancy aspirin therapy should be started.

## C. Essential Thrombocythemia

The role of aspirin in the treatment of essential thrombocythemia is discussed in Chapter 56.

## D. Prevention of Colonic Neoplasia

Several lines of investigation support a role for aspirin and other NSAIDs in the chemoprevention of colonic malignancies. In a rat model of familial adenomatous polyposis, aspirin appears to decrease the rate of tumor formation.[240] Several human studies also demonstrate a beneficial effect of these agents on the development of colorectal cancer[241] and adenomas.[242,243] Few data exist regarding the specific aspirin dose and duration of treatment required to reduce cancer risk. However, in the Nurse's Health Study, the maximum protective effect was observed in patients who took aspirin at least four times per week and for at least a decade.[244] A report from the U.K. General Practice Research Database[245] on more than 943,000 people suggests that a

beneficial effect may be seen after as few as 6 months of aspirin use but is not seen with lower (75–150 mg) aspirin doses. Additionally, the benefit is lost within 1 year after stopping aspirin therapy. However, there are conflicting data on this point. A retrospective analysis of the Physician's Health Study failed to demonstrate an association between aspirin use and the incidence of colorectal malignancy.[246] A similar lack of a chemoprevention effect was reported by the Women's Health Study of approximately 40,000 women randomized to aspirin (100 mg every other day) versus placebo and followed for 10 years.[247] This study found no reduction in the risk of total, breast, colorectal, or other site-specific cancers in the aspirin-treated group, although a reduction in lung cancer risk could not be excluded. It is unclear if the lack of findings relate to the low dose of aspirin used in these studies; however, the increased risk of bleeding associated with the use of higher dose aspirin must be considered before committing an otherwise healthy individual to long-term therapy.

The mechanism by which aspirin and other NSAIDs may prevent colorectal cancer has not been completely elucidated but is unlikely to be the result of earlier tumor detection because of NSAID-induced GI bleeding. Evidence suggests that the putative mechanism may also involve NSAID-induced cell cycle arrest,[248] increased apoptosis,[249] inhibition of angiogenesis,[250] and/or attenuation of epidermal growth factor-induced cellular proliferation.[251] Additionally, COX-2 appears to be overexpressed in colonic cancer cells[252] and may, in part, mediate the malignant process. Williams and colleagues[253] proposed a hypothesis whereby a genetic abnormality produces an overexpression of COX-2 in intestinal cells resulting in changes in cellular adhesion characteristics and a resistance to apoptosis, thereby predisposing to the development of malignant cellular growth. Inhibition of COX-2 and its downstream effects may therefore afford a specific mechanism of tumor suppression.[254]

## XII. Adverse Effects

Aspirin-induced inhibition of prostaglandin synthesis mediates the beneficial effects of this agent. However, the attendant alterations in protective homeostatic prostaglandins result in potentially serious side effects, including gastric mucosal injury and renal insufficiency. Furthermore, aspirin's platelet-inhibitory effects predispose to both minor bleeding and major hemorrhage.

### A. Gastrointestinal Toxicity

In almost all clinical trials, aspirin therapy is associated with a greater incidence of GI side effects compared to the corresponding control agent. This occurs as a result of COX inhibition and the subsequent reduction in $PGE_2$, $PGI_2$ and

PGD$_2$, resulting in the loss of cytoprotective effects on the gastric mucosa (Fig. 60-2). Among patients treated with aspirin, 5.2 to 40% report heartburn, indigestion, nausea, or vomiting versus 0.7 to 34% of controls.[98,99,126,144,166] Gastric ulcers are reported in 0.8 to 2.6% of patients receiving aspirin versus 0 to 1.2% of patients receiving placebo.[143,165,166] Major bleeding (hematemesis or melena requiring transfusion) occurs in <1% of patients in both groups.[96,119,166,170,172] Episodes of minor bleeding including epistaxis and melena not requiring transfusion, as well as bruising and hematuria, are significantly more common in patients treated with aspirin than among placebo-treated controls.[155,168,170] The incidence of GI toxicity increases across a wide range of aspirin doses (30–1300 mg daily), with a significantly higher rate of GI side effects occurring in patients treated with high-dose aspirin than in their counterparts receiving low-dose regimens.[255] Nonetheless, even at low doses (50–75 mg daily), aspirin may increase the risk of GI bleeding,[155,163,168] frequently resulting in the discontinuation of therapy.[97,98]

Enteric coating of aspirin tablets may result in less gastric toxicity as evidenced by less endoscopically documented gastric mucosal erosions[256] and less GI blood loss.[257] Nonetheless, enteric-coated aspirin is still associated with significantly greater gastric toxicity compared to placebo[256] and results in an equivalent risk of major upper GI bleeding to that of uncoated aspirin.[258] Sustained release and topical aspirin preparations appear to semiselectively inhibit platelet TXA$_2$ production while producing minimal effects on vasodilatory and gastroprotective prostaglandins.[67,259] These formulations may therefore result in less gastric toxicity; however, their clinical benefit in the treatment of cardiovascular diseases is less well studied. Attempts to decrease gastric injury by administering aspirin along with the synthetic PGE$_2$ analogue misoprostol have produced encouraging results. Combined aspirin and misoprostol therapy decreases the risk of gastric erosion, ulceration, and hemorrhage in dogs.[260,261] In humans given very high-dose aspirin (3900 mg daily), misoprostol (200 mg daily) decreases endoscopic evidence of gastric and duodenal mucosal injury ($p < 0.006$).[262]

Patients who develop GI bleeding on aspirin therapy present a particular problem because most require long-term antiplatelet therapy for treatment of cardiovascular disease. The addition of a proton pump inhibitor to low-dose aspirin (100 mg daily) after healing of gastric ulcers has been shown to be particularly effective in this setting (1.6% recurrence over 12 months with lansoprazole treatment vs. 14.8% recurrence with placebo).[263] Clopidogrel has been suggested as an alternative to aspirin based on a lower risk of GI bleeding associated with this agent in large randomized trials.[133] However, a recent report suggests that clopidogrel is still associated with a significant risk of recurrent bleeding.[264] In this study, 320 patients on aspirin and with ulcer-related bleeding were randomized to treatment with clopidogrel (75 mg daily) versus aspirin (80 mg daily) and the proton pump inhibitor esomeprazole. After 1 year of follow-up, recurrent ulcer bleeding occurred in 8.6% of the clopidogrel-treated patients versus 0.7% of the aspirin/esomeprazole group ($p = 0.001$). Thus, the combination of low-dose aspirin and a proton pump inhibitor appears to be the best alternative.

### B. Renal Toxicity

Nonaspirin NSAIDs may inhibit renal vasodilatory prostaglandins resulting in the development of renal insufficiency, adversely affecting hypertension control.[251,265] Conversely, when used in low to moderate doses (75–325 mg daily) for the treatment and prevention of cardiovascular disease, aspirin only weakly inhibits renal prostaglandin synthesis and produces no significant effect on renal function,[266] even in diabetic patients with microalbuminuria.[267] However, when used in higher doses (1500 mg daily) in patients with heart failure, aspirin can reduce renal sodium excretion significantly.[268] Furthermore, in elderly patients with hypoalbuminemia, even low-dose aspirin (75 mg daily) may result in significant changes in renal function.[269] In a review, short-term aspirin administration to otherwise healthy adults had no effect on renal function; however, in individuals with chronic renal insufficiency, glomerulonephritis, and cirrhosis, and in children with heart failure, aspirin use was associated with an increased risk of reversible acute renal failure.[270]

### C. Hemorrhagic Stroke

Although aspirin decreases the overall risk of stroke in various settings, it may also increase the specific risk of hemorrhagic stroke. The incidence of cerebral hemorrhage in most individual studies is too low for a significant treatment effect of aspirin to be detected. Nonetheless, a trend toward an increased risk of hemorrhagic stroke has been noted when aspirin is used for the treatment of AMI[79] or acute ischemic stroke.[118,119,144] Additionally, aspirin may increase the risk of cerebral bleeding when it is used for the primary[165,166,175] or secondary[163] prevention of cardiovascular events. The risk of cerebral bleeding appears to increase with higher doses of aspirin and when aspirin is used in conjunction with warfarin therapy.[170] In an analysis of 16 randomized, placebo-controlled trials with a combined total of 55,462 patients, treatment with aspirin resulted in a significant increase in hemorrhagic strokes (RR, 1.84; $p < 0.001$), corresponding to an increase of 12 hemorrhagic strokes per 10,000 patients treated with aspirin.[271] Similarly, the Antiplatelet Trialists' Collaboration reported a 22% increase in the risk of hemorrhagic stroke in patients on antiplatelet therapy.[80] However, in both these studies the risk was outweighed by a significant reduction in ischemic strokes, total strokes, and AMI in aspirin-treated patients.

### D. Interaction with ACE Inhibitors

Initial reports suggested that aspirin may attenuate the mortality benefit of angiotensin-converting enzyme (ACE) inhibitors in the treatment of patients with congestive heart failure[272–274]; however, several subsequent meta-analyses have failed to demonstrate a detrimental effect of aspirin in this setting.[275,276] Thus, it appears reasonable to continue aspirin in CAD patients taking ACE inhibitors but to withdraw aspirin from patients without CAD who require treatment with ACE inhibitors.

## XIII. Epidemiology of Aspirin Use

During the past several decades, the use of aspirin for the treatment of cardiovascular disease has increased dramatically.[70,277,278] Nonetheless, as recently as 1996, aspirin was used in only 26.2% of outpatients with CAD and without contraindications to its use.[279] Despite the strength of supporting data, aspirin remains an underutilized therapy not only for the secondary prevention of recurrent cardiovascular events[280,281] but also for the treatment of acute coronary syndromes.[282,283] Importantly, patients who suffer an AMI and are not discharged on aspirin therapy have a high risk of subsequent vascular events and 1-year mortality approximately twice that of their aspirin-treated counterparts.[284] Younger patients and male patients are more likely to receive aspirin than their older or female counterparts.[279] Although in the past cardiovascular specialists were more likely to prescribe aspirin than were primary care physicians,[279] recent prescribing trends are similar across the two specialties.[285] Furthermore, one-third of primary care physicians and cardiologists recommend aspirin at a dose of 325 mg daily, despite the suggestion of published guidelines to use lower doses.[285]

The lack of aspirin use in the elderly population is particularly troublesome. Although aspirin use in Medicare patients with AMI is relatively high, there are regional differences throughout the United States and, on average, only 85% of these patients receive aspirin within 24 hours of presentation or at the time of hospital discharge.[283] Data from the Global Registry of Acute Coronary Events (GRACE) demonstrated that 96% of patients <45 years old who presented with an acute coronary syndrome received aspirin therapy; this decreased to 88% in the >85-year-old group. Furthermore, <20% of elderly patients with a history of AMI are receiving aspirin at the time of their admission to a nursing home.[286] Although complicating factors and comorbidities likely contribute to this lack of aspirin

prescription, absolute contraindications to aspirin use are rarely present.[284]

## XIV. Cost

Aspirin is exceedingly inexpensive, with hospital costs of approximately $0.01 per dose (81 or 325 mg). Other antiplatelet agents are significantly more expensive; the hospital cost of either clopidogrel (75 mg) or the combination dipyridamole/aspirin (200 mg/25 mg) is approximately $2 or $3 per dose. A formal cost-effectiveness analysis of antiplatelet therapy for the secondary prevention of stroke suggests that the combination of aspirin and dipyridamole may be more effective and less costly than aspirin therapy alone, whereas clopidogrel is more effective but more costly.[287] Similar analyses suggest that compared to aspirin alone for the treatment of acute coronary syndromes, treatment with aspirin plus clopidogrel for 1 year provides an increase in quality-adjusted life expectancy at a cost that is in a traditionally acceptable range.[288,289]

## XV. Conclusions/Recommendations

After several decades of intensive study, the bulk of the data clearly support the use of aspirin in the treatment of AMI, unstable angina, TIAs, and acute ischemic stroke. Additionally, aspirin therapy is clearly beneficial for the secondary prevention of cardiovascular events in patients with established coronary or cerebrovascular disease. The initial dose of aspirin in the acute setting should be high enough to inhibit completely platelet $TXA_2$ production (160–325 mg) and should be followed by the lowest daily dose that has proven efficacy (≥75 mg) (Table 60-3).

### Table 60-3: Recommendations for Aspirin Use

| Indication | Recommended Dose |
| --- | --- |
| Treatment of AMI | 160–325 mg acutely followed by 75–160 mg daily |
| Treatment of acute ischemic stroke | 160–325 mg acutely followed by 75–160 mg daily |
| Treatment of unstable angina | 160–325 mg acutely followed by 75–160 mg daily |
| Secondary prevention following AMI, stroke, or TIA and in patients with chronic stable angina | 75–160 mg daily |
| Primary prevention | No indication for routine use |

AMI, acute myocardial infarction; TIA, transient ischemic attack.

Although there may be benefit of aspirin therapy in the primary prevention of cardiovascular events in some subsets of patients, the well-documented increase in bleeding risk and trend toward increased hemorrhagic stroke risk must be considered before committing a patient to empiric therapy. However, there is clearly a gradient of risk. For patients who are at a low risk for cardiovascular events by virtue of their lack of prior cardiovascular disease and absence of significant cardiac risk factors, the prophylactic use of aspirin would result in exposure to an increased bleeding risk with little expected cardiovascular benefit. Patients with multiple risk factors have a higher likelihood of subsequent cardiovascular events and the risk benefit profile for aspirin therapy may be more favorable in this setting. Prophylactic aspirin therapy may be considered in older patients (>50 years) who are considered to be at the highest risk providing they do not have uncontrolled hypertension.

With the advent of more potent antiplatelet agents, including clopidogrel (see Chapter 61) and the GPIIb-IIIa antagonists (see Chapter 62), the role of aspirin has evolved from single-agent antiplatelet therapy for cardiovascular disease to a component of a more effective antiplatelet regimen. Additionally, as safer aspirin formulations or alternatives are devised and hemorrhagic risks are minimized, the improved risk benefit profile may support the use of aspirin for the primary prevention of cardiovascular disease in lower risk patients.

# References

1. Dangl, J. L., & Jones, J. D. (2001). Plant pathogens and integrated defense responses to infection. *Nature, 411*, 826–833.
2. Gross, M., & Greenberg, L. A. (1948). *The salicylates: A critical biographical review.* New Haven, CT: Hillhouse.
3. Stone, E. (1763). An account of the success of the bark of the willow tree in the cure of agues. *Philo Trans R Soc London, 53*, 195–200.
4. Gibson, P. C. (1949). Aspirin in the treatment of vascular disease. *Lancet, 2*, 1172–1174.
4. Craven, L. L. (1950). Acetylsalicylic acid, possible prevention of coronary thrombosis. *Ann West Med Surg, 4*, 95.
6. Craven, L. L. (1956). Prevention of coronary and cerebral thrombosis. *Miss Valley Med J, 78*, 213–215.
7. Antman, E. M., DeMets, D., & Loscalzo, J. (2005). Cyclooxygenase inhibition and cardiovascular risk. *Circulation, 112*, 759–770.
8. Smith, W. L. (1992). Prostanoid biosynthesis and the mechanism of action. *Am J Physiol, 263*, F118–F191.
9. Williams, C. S., & DuBois, R. N. (1996). Prostaglandin endoperoxide synthase: Why two isoforms? *Am J Physiol, 270*, G393–G400.
10. Morita, I., Schindler, M., Regier, M. K., et al. (1995). Different intracellular locations for prostaglandin endoperoxide H synthase-1 and -2. *J Biol Chem, 270*, 10902–10908.
11. McAdam, B. F., Catella-Lawson, F., Mardini, I. A., et al. (1999). Systemic biosynthesis of prostacyclin by cyclooxygenase (COX)-2: The human pharmacology of a selective inhibitor of COX-2. *Proc Natl Acad Sci USA, 96*, 272–277.
12. Egan, K. M., Lawson, J. A., Fries, S., et al. (2004). COX-2-derived prostacyclin confers atheroprotection on female mice. *Science, 306*, 1954–1957.
13. Weber, A. A., Zimmermann, K. C., Meyer-Kirchrath, J., et al. (1999). Cyclooxygenase-2 in human platelets as a possible factor in aspirin resistance. *Lancet, 353*, 900.
14. Xie, W. L., Chipman, J. G., Robertson, D. L., et al. (1991). Expression of a mitogen-responsive gene encoding prostaglandin synthase is regulated by mRNA splicing. *Proc Natl Acad Sci USA, 88*, 2692–2696.
15. Kujubu, D. A., Fletcher, B. S., Varnum, B. C., et al. (1991). TIS10, a phobol ester tumor promoter-inducible mRNA from Swiss 3T3 cells, encodes a novel prostaglandin synthase/cyclooxygenase homologue. *J Biol Chem, 266*, 12866–12872.
16. Roth, G. J., & Majerus, P. W. (1975). The mechanism of the effect of aspirin on human platelets: I. Acetylation of a particulate fraction protein. *J Clin Invest, 56*, 624–632.
17. Loll, P. J., Picot, D., & Garavito, R. M. (1995). The structural basis of aspirin activity inferred from the crystal structure of inactivated prostaglandin H2 synthase. *Nature Struct Biol, 2*, 637–643.
18. Smith, W. L., & DeWitt, D. L. (1995). Biochemistry of prostaglandin endoperoxidase H synthase-1 and -2 and their differential susceptibility to nonsteroidal antiinflammatory drugs. *Semin Nephrol, 15*, 179–194.
19. Vane, J. R., Bakhle, Y. S., & Botting, R. M. (1998). Cyclooxygenases 1 and 2. *Annu Rev Pharmacol Toxicol, 38*, 97–120.
20. Folts, J. D., Schafer, A. L., Loscalzo, J., et al. (1999). A perspective on the potential problems with aspirin as an antithrombotic agent: A comparison of studies in an animal model with clinical trials. *J Am Coll Cardiol, 33*, 295–303.
21. Gierse, J. K., McDonald, J. J., Hauser, S. D., et al. (1996). A single amino acid difference between cyclooxygenase-1 (COX-1) and -2 (COX-2) reverses the selectivity of COX-2 specific inhibitors. *J Biol Chem, 271*, 15810–15814.
22. Hawkey, C. J. (1999). COX-2 inhibitors. *Lancet, 353*, 307–314.
23. Meade, E. A., Smith, W. L., & DeWitt, D. L. (1993). Differential inhibition of prostaglandin endoperoxide synthase (cyclooxygenase) isozymes by aspirin and other nonsteroidal anti-inflammatory drugs. *J Biol Chem, 268*, 6610–6614.
24. DeWitt, D. L., Meade, E. A., & Smith, W. L. (1993). PGH synthase isoenzyme selectivity: The potential for safer nonsteroidal anti-inflammatory drugs. *Am J Med, 95*(Suppl. 2A), 40S–44S.
25. Lanza, F. L. (1993). Gastrointestinal toxicity of newer nonsteroidal anti-inflammatory drugs. *Am J Gastroenterol, 88*, 1318–1323.
26. Roth, S. H., Tindall, E. A., Jain, A. K., et al. (1993). A controlled study comparing the effects of nabumetone, ibuprofen, and ibuprofen plus misoprostol on the upper

gastrointestinal tract mucosa. *Arch Intern Med, 153,* 2565–2571.

27. Chenevard, R., Hurlimann, D., Bechir, M., et al. (2003). Selective COX-2 inhibition improves endothelial function in coronary artery disease. *Circulation, 107,* 405–409.

28. Kurth, T., Glynn, R. J., Walker, A. M., et al. (2003). Inhibition of clinical benefits of aspirin on first myocardial infarction by nonsteroidal antiinflammatory drugs. *Circulation, 108,* 1191–1195.

29. Bresalier, R. S., Sandler, R. S., Quan, H., et al. (2005). Cardiovascular events associated with rofecoxib in a colorectal adenoma chemoprevention trial. *N Engl J Med, 352,* 1092–1102.

30. Solomon, S. D., McMurray, J. J., Pfeffer, M. A., et al. (2005). Cardiovascular risk associated with celcoxib in a clinical trial for colorectal adenoma prevention. *N Engl J Med, 352,* 1071–1080.

31. Hippisley-Cox, J., & Coupland, C. (2005). Risk of myocardial infarction in patients taking cyclooxygenase-2 inhibitors or conventional non-steroidal anti-inflammatory drugs: Population based nested case control analysis. *Br Med J, 330,* 1366.

32. Johnson, S. P., Larsson, H., Tarone, R. E., et al. (2005). Risk of hospitalization for myocardial infarction among users of rofecoxib, celecoxib, and other NSAIDs: A population-based case-controlled study. *Arch Intern Med, 165,* 978–984.

33. Hamberg, M., Svensson, J., & Samuelsson, B. (1975). Thromboxanes: A new group of biologically active compounds derived from prostaglandin endoperoxides. *Proc Natl Acad Sci USA, 72,* 2294–2298.

34. Fitzgerald, G. A. (1991). Mechanisms of platelet activation: Thromboxane A2 as an amplifying signal for other agonists. *Am J Cardiol, 68,* 11B–15B.

34. Jaffe, E. A., & Weksler, B. B. (1979). Recovery of endothelial cell prostacyclin production after inhibition by low doses of aspirin. *J Clin Invest, 63,* 532–535.

36. Patrono, C., Coller, B., Fitzgerald, G. A., et al. (2004). Platelet-active drugs: The relationships among dose, effectiveness, and side effects: The seventh ACCP conference on antithrombotic and thrombolytic therapy. *Chest, 126,* 234–264.

37. Lopez-Farre, A., Caramelo, C., Esteban, A., et al. (1995). Effects of aspirin on platelet–neutrophil interactions. Role of nitric oxide and endothelin-1. *Circulation, 91,* 2080–2088.

38. Bolz, S. S., & Pohl, U. (1997). Indomethacin enhances endothelial NO release — Evidence for a role of PGI₂ in the autocrine control of calcium-dependent autocoid production. *Cardiovasc Res, 36,* 437–444.

39. Szczeklik, A., Krzanowski, M., Gora, P., et al. (1992). Antiplatelet drugs and generation of thrombin in clotting blood. *Blood, 80,* 2006–2011.

40. Bjornsson, T. D., Schneider, D. E., & Berger, H., Jr. (1989). Aspirin acetylates fibrinogen and enhances fibrinolysis. Fibrinolytic effect is independent of changes in plasminogen activator levels. *J Pharmacol Exp Ther, 250,* 154–161.

41. Steer, K. A., Wallace, T. M., Bolton, C. H., et al. (1997). Aspirin protects low density lipoprotein from oxidative modification. *Heart, 77,* 333–337.

42. Grootveld, M., & Halliwell, B. (1986). Aromatic hydroxylation as a potential measure of hydroxyl-radical formation *in vivo. Biochem J, 237,* 499–504.

43. Ghiselli, A., Laurenti, O., De Mattia, G., et al. (1992). Salicylate hydroxylation as an early marker of *in vivo* oxidative stress in diabetic patients. *Free Radical Biol Med, 13,* 621–626.

44. Kureja, R. C., Knotas, H. A., Hess, M. L., et al. (1986). PGH synthase and lipoxygenase generate superoxide in the presence of NADH or NADPH. *Circ Res, 59,* 612–619.

45. Pinckard, R. N., Hawkins, D., & Farr, R. S. (1968). *In vitro* acetylation of plasma proteins, enzymes and DNA by aspirin. *Nature, 219,* 68–69.

46. Maziere, C., Goldstein, S., Moreau, M., et al. (1987). Aspirin induces alterations in low-density lipoprotein and decreases its catabolism by cultured human fibroblasts. *FEBS Lett, 218,* 243–246.

47. Upchurch, G. R., Jr., Ramdev, N., Walsh, M. T., et al. (1998). Prothrombotic consequences of the oxidation of fibrinogen and their inhibition by aspirin. *J Thromb Thrombol, 5,* 9–14.

48. Husain, S., Andrews, N. P., Mulcahy, D., et al. (1998). Aspirin improves endothelial dysfunction in atherosclerosis. *Circulation, 97,* 716–720.

49. Kharbanda, R. K., Walton, B., Allen, M., et al. (2002). Prevention of inflammation-induced endothelial dysfunction — A novel vasculo-protective action of aspirin. *Circulation, 105,* 2600–2604.

50. Ridker, P. M., Cushman, M., Stampfer, M. J., et al. (1997). Inflammation, aspirin, and the risks of cardiovascular disease in apparently healthy men. *N Engl J Med, 336,* 973–979.

51. Hawthorne, A. B., Mahida, Y. R., Cole, A. T., et al. (1991). Aspirin-induced gastric mucosal damage: Prevention by enteric-coating and relation to prostaglandin synthesis. *Br J Clin Pharm, 32,* 77–83.

52. Latini, R., Cerletti, C., de Gaetano, G., et al. (1986). Comparative bioavailability of aspirin from buffered, enteric-coated and plain preparations. *Intern J Clin Pharm Ther Toxicol, 24,* 313–318.

53. Gantt, A. J., & Gantt, S. (1998). Comparison of enteric-coated aspirin and uncoated aspirin effect on bleeding time. *Cath Cardiovasc Diag, 45,* 396–399.

54. Vanags, D., Rodgers, S. E., Lloyd, J. V., et al. (1990). The antiplatelet effect of low dose enteric-coated aspirin in man: A time course of onset and recovery. *Thromb Res, 59,* 995–1005.

55. Jimenez, A. H., Stubbs, M. E., Tofler, G. H., et al. (1992). Rapidity and duration of platelet suppression by enteric-coated aspirin in healthy young men. *Am J Cardiol, 69,* 258–262.

56. Feldman, M., & Cryer, B. (1999). Aspirin absorption rates and platelet inhibition times with 325-mg buffered aspirin tablets (chewed or swallowed intact) and with buffered aspirin solution. *Am J Cardiol, 84,* 404–409.

57. Bode-Boger, S. M., Boger, R. H., Schubert, M., et al. (1998). Effects of very low dose and enteric-coated acetylsalicylic acid on prostacyclin and thromboxane formation and on

bleeding time in healthy subjects. *Eur J Clin Pharmacol, 54,* 707–714.

58. Stampfer, M. J., Jakubowski, J. A., Deykin, D., et al. (1986). Effect of alternate-day regular and enteric-coated aspirin on platelet aggregation, bleeding time, and thromboxane A2 levels in bleeding-time blood. *Am J Med, 81,* 400–404.

59. Burch, J. W., Stanford, N., & Majerus, P. W. (1979). Inhibition of platelet prostaglandin synthase by oral aspirin. *J Clin Invest, 61,* 314–319.

60. Bradlow, B. A., & Chetty, N. (1982). Dosage frequency for suppression of platelet function by low dose aspirin therapy. *Thromb Res, 27,* 99–110.

61. Patrono, C., Ciabattoni, G., Patrignani, P., et al. (1985). Clinical pharmacology of platelet cyclooxygenase inhibition. *Circulation, 72,* 177–184.

62. May, J. A., Heptinstall, S., Cole, A. T., et al. (1997). Platelet responses to several agonists and combination of agonists in whole blood: A placebo controlled comparison of the effects of once daily dose of plain aspirin 300 mg, plain aspirin 75 mg and enteric coated aspirin 300 mg, in man. *Thromb Res, 88,* 183–192.

63. Patrignani, P., Filabozzi, P., & Patrono, C. (1982). Selective cumulative inhibition of platelet thromboxane production by low-dose aspirin in healthy subjects. *J Clin Invest, 69,* 1366–1372.

64. Weksler, B. B., Pett, S. B., Alonso, D., et al. (1983). Differential inhibition by aspirin of vascular and platelet prostaglandin synthesis in atherosclerotic patients. *N Engl J Med, 308,* 800–805.

65. Tohgi, H., Konno, S., Tamura, K., et al. (1992). Effects of low-to-high doses of aspirin on platelet aggregability and metabolites of thromboxane A2 and prostacyclin. *Stroke, 23,* 1400–1403.

66. Fitzgerald, G. A., Oates, J. A., Hawiger, J., et al. (1983). Endogenous biosynthesis of prostacyclin and thromboxane and platelet function during chronic administration of aspirin in man. *J Clin Invest, 71,* 676–688.

67. Clarke, R. J., Mayo, G., Price, P., et al. (1991). Suppression of thromboxane A2 but not systemic prostacyclin by controlled-release aspirin. *N Engl J Med, 325,* 1137–1141.

68. Feng, D., McKenna, C., Murillo, J., et al. (1997). Effect of aspirin dosage and enteric coating on platelet reactivity. *Am J Cardiol, 80,* 189–193.

69. Antiplatelet Trialists' Collaboration. (1994). Collaborative overview of randomised trials of antiplatelet therapy — I: Prevention of death, myocardial infarction, and stroke by prolonged antiplatelet therapy in various categories of patients. *Br Med J, 308,* 81–106.

70. Helgason, C. M., Tortorice, K. L., Winkler, S. R., et al. (1993). Aspirin response and failure in cerebral infarction. *Stroke, 24,* 345–350.

71. Kawasaki, T., Ozeki, Y., Igawa, T., et al. (2000). Increased platelet sensitivity to collagen in individuals resistant to low-dose aspirin. *Stroke, 31,* 591–595.

72. Helgason, C. M., Bolin, K. M., Hoff, J. A., et al. (1994). Development of aspirin resistance in persons with previous ischemic stroke. *Stroke, 25,* 2331–2336.

73. Gum, P. A., Kotte-Marchant, K., Welsh, P. A., et al. (2003). A prospective, blinded determination of the natural history of aspirin resistance among stable patients with cardiovascular disease. *J Am Coll Cardiol, 41,* 961–965.

74. Eikelboom, J. W., Hirsh, J., Wietz, J. I., et al. (2002). Aspirin-resistant thromboxane biosynthesis and the risk of myocardial infarction, stroke, or cardiovascular death in patients at high risk for cardiovascular events. *Circulation, 105,* 1650–1655.

75. Schwartz, K. A., Schwartz, D. E., Ghosheh, K., et al. (2005). Compliance as a critical consideration in patients who appear to be resistant to aspirin after healing of myocardial infarction. *Am J Cardiol, 95,* 973–975.

76. Catella-Lawson, F., Reilly, M. P., & Kapoor, S. C. (2001). Cyclooxygenase inhibitors and the antiplatelet effects of aspirin. *N Engl J Med, 345,* 1809–1817.

77. Cambria-Kiely, J. A., & Gandhi, P. J. (2002). Aspirin resistance and genetic polymorphisms. *J Thromb Thrombol, 14,* 51–58.

78. Macchi, L., Christiaens, L., Brabant, S., et al. (2003). Resistance *in vitro* to low-dose aspirin is associated with platelet PlA1 (GPIIIa) polymorphism but not with C807T (GP)Ia/IIa) and C-5T Kozak (GP Ibalpha) polymorphisms. *J Am Coll Cardiol, 42,* 1115–1119.

79. ISIS-2 (Second International Study of Infarct Survival) Collaborative Group. (1988). Randomised trial of intravenous streptokinase, oral aspirin, both, or neither among 17,187 cases of suspected acute myocardial infarction: ISIS-2. *Lancet, 2,* 349–360.

80. Antiplatelet Trialists' Collaboration (2002). Collaborative meta-analysis of randomised trials of antiplatelet therapy for prevention of death, myocardial infarction, and stroke in high risk patients. *Br Med J, 324,* 71–86.

81. DeWood, M. A., Spores, J., Notske, R., et al. (1980). Prevalence of total coronary occlusion during the early hours of transmural myocardial infarction. *N Engl J Med, 303,* 897–902.

82. ISIS-3 (Third International Study of Infarct Survival) Collaborative Group (1992). ISIS-3: A randomised comparison of streptokinase vs tissue plasminogen activator vs anistreplase and of aspirin plus heparin vs aspirin alone among 41,299 cases of suspected acute myocardial infarction. *Lancet, 339,* 753–770.

83. Fibrinolytic Therapy Trialists' (FTT) Collaborative Group (1994). Indications for fibrinolytic therapy in suspected acute myocardial infarction: Collaborative overview of early mortality and major morbidity results from all randomised trials of more than 1000 patients. *Lancet, 343,* 311–322.

84. Roux, S., Christeller, S., & Lüdin, E. (1992). Effects of aspirin on coronary reocclusion and recurrent ischemia after thrombolysis: A meta-analysis. *J Am Coll Cardiol, 19,* 671–677.

85. Yasue, H., Ogawa, H., Tanaka, H., et al. (1999). Effects of aspirin and trapidil on cardiovascular events after acute myocardial infarction. Japanese Antiplatelets Myocardial Infarction Study (JAMIS) investigators. *Am J Cardiol, 83,* 1308–1313.

86. Baigent, C., Collins, R., Appleby, P., et al. (1998). ISIS-2: 10 year survival among patients with suspected acute myocar-

dial infarction in randomised comparison of intravenous streptokinase, oral aspirin, both, or neither. *Br Med J, 316,* 1337–1343.

87. Verheugt, F. W., van der Laarse, A., Funke-Kupper, A. J., et al. (1990). Effects of early intervention with low-dose aspirin (100 mg) on infarct size, reinfarction and mortality in anterior wall acute myocardial infarction. *Am J Cardiol, 66,* 267–270.

88. Verheugt, F. W., Kupper, A. J., Galema, T. W., et al. (1988). Low dose aspirin after early thrombolysis in anterior wall acute myocardial infarction. *Am J Cardiol, 61,* 904–906.

89. Hsia, J., Hamilton, W. P., Kleiman, N. S., et al. (1990). A comparison between heparin and low-dose aspirin as adjunctive therapy with tissue plasminogen activator for acute myocardial infarction. *N Engl J Med, 323,* 1433–1437.

90. Glick, A., Kornowski, R., Michowich, Y., et al. (1996). Reduction of reinfarction and angina with use of low-molecular-weight heparin therapy after streptokinase (and heparin) in acute myocardial infarction. *Am J Cardiol, 77,* 1145–1148.

91. Meijer, A., Verheugt, F. W. A., Werter, C. J. P. J., et al. (1993). Aspirin versus coumadin in the prevention of reocclusion and recurrent ischemia after successful thrombolysis: A prospective study. Results of the APRICOT study. *Circulation, 87,* 1524–1530.

92. Sabatine, M. S., Cannon, C. P., Gibson, C. M., et al. (2005). Addition of clopidogrel to aspirin and fibrinolytic therapy for myocardial infarction with ST-segment elevation. *N Engl J Med, 352,* 1179–1189.

93. Fuster, V., Badimon, L., Badimon, J., et al. (1992). The pathogenesis of coronary artery disease and the acute coronary syndromes [Part 1]. *N Engl J Med, 326,* 242–250.

94. Kirshenbaum, L. A., & Singal, P. K. (1993). Increase in endogenous antioxidant enzymes protects hearts against reperfusion injury. *Am J Physiol, 265*(2 Pt 2), H484–H493.

95. Gazes, P. C., Mobley, E. M., Jr., Farris, H. M., Jr., et al. (1973). Preinfarction (unstable) angina — A prospective study — Ten year follow-up. *Circulation, 48,* 331–337.

96. Lewis, H. D., Davis, J. W., Archibald, D. G., et al. (1983). Protective effects of aspirin against acute myocardial infarction and death in men with unstable angina. *N Engl J Med, 309,* 396–403.

97. Wallentin, L. C., & the Research Group on Instability in Coronary Artery Disease in Southeast Sweden. (1991). Aspirin (75 mg/day) after an episode of unstable coronary artery disease: Long-term effects on the risk for myocardial infarction, occurrence of severe angina and the need for revascularization. *J Am Coll Cardiol, 18,* 1587–1593.

98. The RISC Group (1990). Risk of myocardial infarction and death during treatment with low dose aspirin and intravenous heparin in men with unstable coronary artery disease. *Lancet, 336,* 827–830.

99. Cairns, J. A., Gent, M., Singer, J., et al. (1985). Aspirin, sulfinpyrazone, or both in unstable angina: Results of a Canadian multicenter trial. *N Engl J Med, 313,* 1369–1375.

100. Théroux, P., Ouimet, H., McCans, J., et al. (1988). Aspirin, heparin, or both to treat acute unstable angina. *N Engl J Med, 319,* 1105–1111.

101. Nyman, I., Larsson, H., Wallentin, L., & the Research Group on Instability in Coronary Artery Disease in Southeast Sweden. (1992). Prevention of serious cardiac events by low-dose aspirin in patients with silent myocardial ischaemia. *Lancet, 340,* 497–501.

102. Yusef, S., Zhao, F., Mehta, S. R., et al. (2001). Effects of clopidogrel in addition to aspirin in patients with acute coronary syndromes without ST segment elevation. *N Engl J Med, 345,* 494–502.

103. Boersma, E., Harrington, R. A., Moliterno, D. J., et al. (2002). Platelet glycoprotein IIb/IIIa inhibitors in acute coronary syndromes: A meta-analysis of all major randomized clinical trials. *Lancet, 359,* 189–198.

104. Peters, R. J., Mehta, S. R., Fox, K. A., et al. (2003). Effects of aspirin dose when used alone or in combination with clopidogrel in patients with acute coronary syndromes: Observations from the Clopidogrel in Unstable Angina to Prevent Recurrent Events (CURE) study. *Circulation, 108,* 1682–1687.

105. Théroux, P., Waters, D., Qui, S., et al. (1993). Aspirin versus heparin to prevent myocardial infarction during the acute phase of unstable angina. *Circulation, 881,* 2045–2048.

106. Serneri, G. G. N., Gensini, G. F., Poggesi, L., et al. (1990). Effect of heparin, aspirin, or alteplase in reduction of myocardial ischaemia in refractory unstable angina. *Lancet, 335,* 615–618.

107. Oler, A., Whooley, M. A., Oler, J., et al. (1996). Adding heparin to aspirin reduces the incidence of myocardial infarction and death in patients with unstable angina: A meta-analysis. *J Am Med Assoc, 272,* 811–815.

108. Cohen, M., Demers, C., Gurfinkel, E. P., et al. (1997). A comparison of low-molecular weight heparin with unfractionated heparin for unstable coronary artery disease. *N Engl J Med, 337,* 447–452.

109. Eikelboom, J. W., Anand, S. S., Malmberg, K., et al. (2000). Unfractionated heparin and low-molecular-weight heparin in acute coronary syndrome without ST elevation: A meta-analysis. *Lancet, 355,* 1926–1928.

110. FRISC Study Group (1996). Low-molecular-weight heparin during instability in coronary artery disease. *Lancet, 347,* 561–568.

111. FRISC II Investigators (1999). Long-term low-molecular-mass heparin in unstable coronary-artery disease: FRISC II prospective randomised multicentre study. *Lancet, 354,* 701–707.

112. Antman, E. M., McCabe, C. H., Gurfinkel, E. P., et al. (1999). Enoxaparin prevents death and cardiac ischemic events in unstable angina/non-Q-wave myocardial infarction. Results of the thrombolysis in myocardial infarction (TIMI) 11B trial. *Circulation, 100,* 1593–1601.

113. Théroux, P., Water, D., Lam, J., et al. (1992). Reactivation of unstable angina after the discontinuation of heparin. *N Engl J Med, 327,* 141–145.

114. Anand, S. S., Yusuf, S., Pogue, J., et al. (1998). Long-term oral anticoagulant therapy in patients with unstable angina or suspected non-Q-wave myocardial infarction: Organization to Assess Strategies for Ischemic Syndromes (OASIS) pilot study results. *Circulation, 98,* 1064–1070.

115. Schulman, S. P., Goldschmidt-Clermont, P. J., Topol, E., et al. (1996). Myocardial ischemia/coronary artery vasoconstriction/thrombosis/myocaridal infarction: Effects of Integrelin, a platelet glycoprotein IIb/IIIa receptor antagonist, in unstable angina: A randomized multicenter trial. *Circulation, 94,* 2083–2089.

116. The PRISM Investigators (1998). A comparison of aspirin plus tirofiban with aspirin plus heparin for unstable angina. The Platelet Receptor Inhibition in Ischemic Syndrome Management (PRISM) study investigators. *N Engl J Med, 338,* 1498–1505.

117. The PRISM-PLUS Investigators (1998). Inhibition of the platelet glycoprotein IIb/IIIa receptor with tirofiban in unstable angina and non-Q-wave myocardial infarction. The Platelet Receptor Inhibition in Ischemic Syndrome Management in Patients Limited by Unstable Signs and Symptoms (PRISM-PLUS) study investigators. *N Engl J Med, 338,* 1488–1497.

118. Chinese Acute Stroke Trial Collaborative Group (1997). CAST: Randomised placebo-controlled trial of early aspirin use in 20,000 patients with acute ischaemic stroke. *Lancet, 349,* 1641–1649.

119. International Stroke Trial Collaborative Group (1997). The International Stroke Trial (IST): A randomised trial of aspirin, subcutaneous heparin, both, or neither among 19,435 patients with acute ischaemic stroke. *Lancet, 349,* 1569–1581.

120. Chen, Z. M., Sandercodk, P., Pan, H. C., et al. (2000). Indications for early aspirin use in acute ischemic stroke: A combined analysis of 40,000 randomized patients from the Chinese Acute Stroke Trial and the International Stroke Trial. On behalf of the CAST and IST collaborative groups. *Stroke, 31,* 1240–1249.

121. Berge, E., Abdeinoor, M., Nakstad, P. H., et al. (2000). Low molecular-weight heparin versus aspirin in patients with acute ischaemic stroke and atrial fibrillation: A double-blind randomised study. HAEST study group. Heparin in Acute Embolic Stroke Trial. *Lancet, 355,* 1205–1210.

122. Hacke, W., Kaste, M., Fieschi, C., et al. (1995). Intravenous thrombolysis with recombinant tissue plasminogen activator for acute hemispheric stroke. The European Cooperative Acute Stroke Study (ECASS). *J Am Med Assoc, 274,* 1017–1025.

123. Mori, E., Yoneda, Y., Tabuchi, M., et al. (1992). Intravenous recombinant tissue plasminogen activator in acute carotid artery territory stroke. *Neurology, 42,* 976–982.

124. Multicentre Acute Stroke Trial-Italy (MAST-I) group (1995). Randomised controlled trial of streptokinase, aspirin, and combination of both in treatment of acute ischaemic stroke. *Lancet, 346,* 1509–1514.

125. Ciccone, A., Motto, C., Aritzu, E., et al. (2000). Negative interaction of aspirin and streptokinase in acute ischemic stroke: Further analysis of the Multicenter Acute Stroke Trial-Italy. *Cerebrovasc Dis, 10,* 61–64.

126. The Aspirin Myocardial Infarction Study Research Group. (1980). The aspirin myocardial infarction study: Final results. *Circulation, 62*(Suppl. V), V79–V84.

127. The Coronary Drug Project Research Group (••). Aspirin in coronary heart disease. *J Chronic Dis, 29,* 625–642.

128. Elwood, P. C., Cochrane, A. L., Burr, M. L., et al. (1974). A randomized controlled trial of acetylsalicylic acid in the secondary prevention of mortality from myocardial infarction. *Br Med J, 1,* 436–440.

129. The Persantine-Aspirin Reinfarction Study Research Group (1980). Persantine and aspirin in coronary heart disease. *Circulation, 62,* 449–461.

130. Elwood, P. C., & Sweetnam, P. M. (1979). Aspirin and secondary mortality after myocardial infarction. *Lancet, 2,* 1313–1315.

131. Breddin, K., Loew, D., Lechner, K., et al. on behalf of the German–Austrian Study Group. (1980). The German–Austrian aspirin trial: A comparison of acetylsalicylic acid, placebo, and phenprocoumon in secondary prevention of myocardial infarction. *Circulation, 62*(Suppl. V), V63–V72.

132. Harpaz, D., Gottlieb, S., Graff, E., et al. (1998). Effects of aspirin treatment on survival in non-insulin-dependent diabetic patients with coronary artery disease. Israeli Bezafibrate Infarction Prevention Study Group. *Am J Med, 105,* 494–495.

133. CAPRIE Steering Committee (1996). A randomised, blinded, trial of clopidogrel versus aspirin in patients at risk if ischaemic events (CAPRIE). *Lancet, 348,* 1329–1339.

134. Moshfegh, K., Redondo, M., Julmy, F., et al. (2000). Antiplatelet effects of clopidogrel compared with aspirin after myocardial infarction: Enhanced inhibitory effects of combination therapy. *J Am Coll Cardiol, 36,* 699–705.

135. Coumadin Aspirin Reinfarction Study (CARS) investigators (1997). Randomised double-blind trial of fixed low-dose warfarin with aspirin after myocardial infarction. *Lancet, 350,* 389–396.

136. Fiore, L. D., Ezekowitz, M. D., Brophy, M. T., et al. (2002). Department of Veterans Affairs cooperative studies program clinical trial comparing combined warfarin and aspirin with aspirin alone in survivors of acute myocardial infarction. *Circulation, 105,* 557–563.

137. Hurlen, M., Abdelnoor, M., Smith, P., et al. (2002). Warfarin, aspirin, or both after myocardial infarction. *N Engl J Med, 347,* 969–974.

138. Anand, S. S., & Yusuf, S. (1999). Oral anticoagulant therapy in patients with coronary artery disease: A meta analysis. *J Am Med Assoc, 282,* 2058–2067.

139. Borzak, S., Cannon, C. P., Kraft, P. L., et al. (1998). Effects of prior aspirin and anti-ischemic therapy on outcome of patients with unstable angina. TIMI 7 Investigators. Thrombin Inhibition in Myocardial Ischemia. *Am J Cardiol, 81,* 678–681.

140. Garcia-Dorado, D., Theroux, P., Tornos, P., et al. (1995). Previous aspirin use may attenuate the severity of the manifestation of acute ischemic syndromes. *Circulation, 92,* 1743–1748.

141. Alexander, J. H., Harrington, R. A., Tuttle, R. H., et al. (1999). Prior aspirin use predicts worse outcomes in patients with non-ST-elevation acute coronary syndromes. *Am J Cardiol, 83,* 1147–1151.

142. Sorensen, P. S., Pedersen, H., Marquardsen, J., et al. (1983). Acetylsalicylic acid in the prevention of stroke in patients with reversible cerebral ischemic attacks. A Danish cooperative study. *Stroke,14,* 15–21.

143. The Swedish Cooperative Study Group (1986). High-dose acetylsalicylic acid after cerebral infarction. A Swedish cooperative study. *Stroke, 18,* 325–334.

144. UK-TIA Study Group (1988). United Kingdom transient ischaemic attack (UK-TIA) aspirin trial: Interim results. *Br Med J, 296,* 316–320.

145. Tijssen, J. G. P. (1998). Low-dose and high-dose acetylsalicylic acid, with and without dipyridamole. A review of clinical trial results. *Neurology, 51*(Suppl. 3), S15–S16.

146. Rothwell, P. M., Slattery, J., & Warlow, C. P. (1996). A systematic comparison of the risks of stroke and death due to carotid endarterectomy for asymptomatic stenosis. *Stroke, 27,* 266–269.

147. Boysen, G., Sorensen, P. S., Juhler, M., et al. (1988). Danish very-low-dose aspirin after carotid endarterectomy trial. *Stroke, 19,* 1211–1215.

148. Taylor, D. W., Barnett, H. J. M., Haynes, R. B., et al. (1999). Low-dose and high-dose acetylsalicylic acid for patients undergoing carotid endarterectomy: A randomised controlled trial. *Lancet, 353,* 2179–2184.

149. Gent, M., Blakely, J. A., & Easton, J. D. (1989). The Canadian American Ticlopidine Study (CATS) in thromboembolic stroke. *Lancet, 1,* 1215–1220.

150. Hass, W. K., Easton, D., Adams, H. P., et al. for the Ticlopidine Aspirin Stroke Study Group. (1989). A randomized trial comparing ticlopidine hydrochloride with aspirin for the prevention of stroke in high-risk patients. *N Engl J Med, 321,* 501–507.

151. Gorelick, P. B., Richardson, D., Kelly, M., et al. (2004). Aspirin and ticlopidine for prevention of recurrent stroke in black patients: A randomized trial. *J Am Med Assoc, 289,* 2947–2957.

152. Markus, H. S., Droste, D. W., Kaps, M., et al. (2005). Dual antiplatelet therapy with clopidogrel and aspirin in symptomatic carotid stenosis evaluated using Doppler embolic signal detection: The CARESS Trial. *Circulation, 111,* 2233–2240.

153. Payne, D. A., Jones, C. I., Hayes, P. D., et al. (2004). Beneficial effects of clopidogrel combined with aspirin in reducing cerebral emboli in patients undergoing carotid endarterectomy. *Circulation, 109,* 1476–1481.

154. Diener, H. C., Bogousslavsky, J., Brass, L. M., et al. (2004). Aspirin and clopidogrel compared with clopidogrel alone after recent ischaemic stroke or transient ischaemic attack in high-risk patients (MATCH): Randomised, double-blind, placebo-controlled trial. *Lancet, 364,* 331–337.

155. Diener, H. C., Cunha, L., Forbes, C., et al. (1996). European Stroke Prevention Study 2. Dipyridamole and acetylsalicylic acid in the secondary prevention of stroke. *J Neurol Sci, 143,* 1–13.

156. De Schryver, E. L. (2000). Design of ESPRIT: An international randomized trial for secondary prevention after non-disabling cerebral ischaemia of arterial origin. European/Australian Stroke Prevention in Reversible Ischaemia Trial (ESPRIT) group. *Cerebrovasc Dis, 10,* 147–150.

157. Chimowitz, M. I., Lynn, M. J., Howlett-Smith, H., et al. (2005). Comparison of warfarin and aspirin for symptomatic intracranial arterial stenosis. *N Engl J Med, 352,* 1305–1316.

158. Mohr, J. P., Thompson, J. L. P., Lazar, R. M., et al. (2001). A comparison of warfarin and aspirin for the prevention of recurrent ischemic stroke. *N Engl J Med, 345,* 1444–1451.

159. O'Connor, C. M., Gattis, W. A., Hellkamp, A. S., et al. (2001). Comparison of two aspirin doses on ischemic stroke in post-myocardial infarction patients in the warfarin (Coumadin) Aspirin Reinfarction Study (CARS). *Am J Cardiol, 88,* 541–546.

160. Dyken, M. L., Barnett, H. J. M., Easton, D., et al. (1992). Low-dose aspirin and stroke: "It ain't necessarily so." *Stroke, 23,* 1395–1399.

161. Patrono, C., & Roth, G. J. (1996). Aspirin in ischemic cerebrovascular disease: How strong is the case for a different dosing regimen? *Stroke, 27,* 756–760.

162. Farrell, B., Goodwin, J., Richards. S., et al. (1991). The United Kingdom transient ischemia attack (UK-TIA) aspirin trial: Final results. *J Neurol Neurosurg Psych, 54,* 1044–1054.

163. The SALT Collaborative Group (1991). Swedish aspirin low-dose trial (SALT) of 75 mg aspirin as secondary prophylaxis after cerebrovascular ischaemic events. *Lancet, 338,* 1345–1349.

164. The Dutch TIA Trial Study Group (1991). A comparison of two doses of aspirin (30 mg vs 283 mg a day) in patients after a transient ischemic attack or minor stroke. *N Engl J Med, 325,* 1261–1266.

165. Peto, R., Gray, R., Collins, R., et al. (1988). Randomised trial of prophylactic daily aspirin in British male doctors. *Br Med J, 926,* 313–316.

166. Steering Committee of the Physicians' Health Study Research Group (1989). Final report on the aspirin component of the ongoing Physicians' Health Study. *N Engl J Med, 321,* 129–135.

167. Hennekens, C. H., Peto, R., Hutchison, G. B., et al. (1988). An overview of the British and American aspirin studies. *N Engl J Med, 318,* 923–924.

168. Hansson, L., Zanchetti, A., Carruthers, S. G., et al. (1998). Effects of intensive blood-pressure lowering and low-dose aspirin in patients with hypertension: Principal results of the Hypertension Optimal Treatment (HOT) randomised trial. *Lancet, 351,* 1755–1762.

169. Collaborative Group of the Primary Prevention Project (2001). Low-dose aspirin and vitamin E in people at cardiovascular risk: A randomised trial in general practice. *Lancet, 357,* 89–95.

170. The Medical Research Council's General Practice Research Framework (1998). Thrombosis prevention trial: Randomised trial of low-intensity oral anticoagulation with warfarin and low-dose aspirin in the primary prevention of ischaemic heart disease in men at increased risk. *Lancet, 351,* 233–241.

171. Meade, T. W., & Brennan, P. J. (2000). Determination of who may derive most benefit from aspirin in primary prevention:

Subgroup results from a randomised controlled trial. *Br Med J, 321,* 13–17.

172. Juul-Möller, S., Edvardson, N., Jahnmatz, B., et al. for the Swedish Angina Pectoris Aspirin Trial (SAPAT) group (1992). Double-blind trial of aspirin in primary prevention of myocardial infarction in patients with stable chronic angina pectoris. *Lancet, 340,* 1421–1425.

173. Manson, J. E., Stampfer, M. J., Colditz, G. A., et al. (1991). A prospective study of aspirin use and primary prevention of cardiovascular disease in women. *J Am Med Assoc, 266,* 521–527.

174. Kjeldsen, S. E., Killoch, R. E., Leonetti, G., et al. (2000). Influence of gender and age on preventing cardiovascular disease by antihypertensive treatment and acetylsalicylic acid. The HOT study. Hypertension Optimal Treatment. *J Hypertens, 18,* 629–642.

175. Ridker, P. M., Cook, N. R., Lee, I. M., et al. (2005). A randomized trial of low-dose aspirin in the primary prevention of cardiovascular disease in women. *N Engl J Med, 352,* 1293–1304.

176. Bhatt, D. L., & Topol, E. J. (2004). Clopidogrel added to aspirin versus aspirin alone in secondary prevention and high-risk primary prevention: Rationale and design of the Clopidogrel for High Atherothrombotic Risk and Ischemic Stabilization, Management and Avoidance (CHARISMA) trial. *Am Heart J, 148,* 263–268.

177. Côté, R., Battista, R. N., Abrahamowicz, M., et al. and the Asymptomatic Cervical Bruit Study Group. (1995). Lack of effect of aspirin in asymptomatic patients with carotid bruits and substantial carotid narrowing. *Ann Intern Med, 123,* 649–655.

178. Stroke Prevention in Atrial Fibrillation Investigators (1991). Stroke Prevention in Atrial Fibrillation Study. Final results. *Circulation, 84,* 527–539.

179. Stroke Prevention in Atrial Fibrillation Investigators (1994). Warfarin versus aspirin for prevention of thromboembolism in atrial fibrillation: Stroke Prevention in Atrial Fibrillation II Study. *Lancet, 343,* 687–691.

180. Petersen, P., Boysen, G., Godtfredsen, J., et al. (1989). Placebo-controlled, randomised trial of warfarin and aspirin for prevention of thromboembolic complications in chronic atrial fibrillation. The Copenhagen AFASAK study. *Lancet, 1,* 175–179.

181. Singer, D. E., Hughes, R. A., Gress, D. R., et al. (1992). The effect of aspirin on the risk of stroke in patients with non-rheumatic atrial fibrillation. *Am Heart J, 124,* 1567.

182. van Walraven, C., Hart, R. G., Singer, D. E., et al. (2002). Oral anticoagulants versus aspirin in nonvalvular atrial fibrillation: An individual patient meta-analysis. *J Am Med Assoc, 288,* 2441–2448.

183. Stroke Prevention in Atrial Fibrillation Investigators. (1996). Adjusted-dose warfarin versus low-intensity, fixed-dose warfarin plus aspirin for high-risk patients with atrial fibrillation: Stroke Prevention in Atrial Fibrillation III randomised clinical trial. *Lancet, 348,* 633–638.

184. Baim, D. S., & Carrozza, J. P. (1997). Stent thrombosis: Closing in on the best preventive treatment [editorial]. *Circulation, 95,* 1098–1100.

185. de Feyter, P. J., van den Brand, M., Laarman, G. J., et al. (1991). Acute coronary artery occlusion during and after percutaneous transluminal coronary angioplasty: Frequency, prediction, clinical course management and follow-up. *Circulation, 83,* 927–936.

186. Serruys, P. W., de Jaegere, P., Kiemeneij, F., et al. (1994). A comparison of balloon-expandable-stent implantation with balloon angioplasty in patients with coronary disease. *N Engl J Med, 331,* 489–495.

187. Bourassa, M. G., Schwartz, L., Lesperance, J., et al. (1990). Prevention of acute complications after percutaneous transluminal coronary angioplasty. *Thromb Res, 12,* 51–58.

188. Schwartz, L., Bourassa, M. G., Lesperance, J., et al. (1988). Aspirin and dipyridamole in the prevention of restenosis after percutaneous transluminal coronary angioplasty. *N Engl J Med, 318,* 1714–1719.

189. Lembo, N. J., Black, A. J., Roubin, G. S., et al. (1990). Effect of pretreatment with aspirin versus aspirin plus dipyridamole on frequency and type of acute complications of percutaneous transluminal coronary angioplasty. *Am J Cardiol, 65,* 422–426.

190. Rupprecht, H. J., Darius, H., Borkowski, U., et al. (1998). Comparison of antiplatelet effects of aspirin, ticlopidine, or their combination after stent placement. *Circulation, 97,* 1046–1052.

191. Leon, M. B., Baim, D. S., Popma, J. J., et al. (1998). A clinical trial comparing three antithrombotic-drug regimens after coronary-artery stenting. Stent Anticoagulation Restenosis Study Investigators. *N Engl J Med, 339,* 1665–1671.

192. Moussa, I., Oetgen, M., Roubin, G., et al. (1999). Effectiveness of clopidogrel and aspirin versus ticlopidine and aspirin in preventing stent thrombosis after coronary stent implantation. *Circulation, 99,* 2364–2366.

193. Muller, C., Buttner, H. J., Petersen, J., et al. (2000). A randomized comparison of clopidogrel and aspirin versus ticlopidine and aspirin after the placement of coronary-artery stents. *Circulation, 101,* 590–593.

194. Bertrand, M. E., Rupprecht, H. J., Urban, P., et al. (2000). Double-blind study of the safety of clopidogrel with and without a loading dose in combination with aspirin compared with ticlopidine in combination with aspirin after coronary stenting. *Circulation, 102,* 624–629.

195. Mehta, S. R., Yusuf, S., Peters, R. J. G., et al. (2001). Effects of pretreatment with clopidogrel and aspirin followed by long-term therapy in patients undergoing percutaneous intervention: The PCI-CURE study. *Lancet, 358,* 527–533.

196. Steinhubl, S. R., Berger, P. B., Mann, J. T., et al. (2002). Early and sustained dual oral antiplatelet therapy following percutaneous coronary intervention. *J Am Med Assoc, 288,* 2411–2420.

197. Machraoui, A., Germing, A., von Dryander, S., et al. (1999). Comparison of the efficacy and safety of aspirin alone with commadin plus aspirin after provisional coronary stenting: Final and follow-up results of a randomized study. *Am Heart J, 138,* 663–669.

198. Urban, P., Macaya, C., Rupprecht, H. J., et al. (1998). Randomized evaluation of anticoagulation versus antiplatelet

therapy after coronary stent implantation in high-risk patients: The multicenter aspirin and ticlopidine trial after intracoronary stenting (MATTIS). *Circulation, 98,* 2126–2132.

199. The EPIC Investigators (1994). Use of a monoclonal antibody directed against the platelet glycoprotein IIb/IIIa receptor in high-risk coronary angioplasty. *N Engl J Med, 330,* 956–961.

200. The EPILOG Investigators (1997). Platelet glycoprotein IIb/IIIa receptor blockade and low-dose heparin during percutaneous coronary revascularization. *N Engl J Med, 336,* 1689–1696.

201. The RESTORE Investigators (••). Effects of platelet glycoprotein IIb/IIIa blockade with tirofiban on adverse cardiac events in patients with unstable angina or acute myocardial infarction undergoing coronary angioplasty. *Circulation, 96,* 1445–1453.

202. The EPISTENT Investigators (1998). Randomised placebo-controlled and balloon-angioplasty-controlled trial to assess safety of coronary stenting with use of platelet glycoprotein-IIb/IIIa blockade. The EPISTENT Investigators. Evaluation of Platelet IIb/IIIa Inhibitor for Stenting. *Lancet, 352,* 87–92.

203. Gavaghan, T. P., Gebski, V., & Baron, D. W. (1991). Immediate postoperative aspirin improves vein graft patency early and late after coronary artery bypass graft surgery. A placebo-controlled, randomized study. *Circulation, 83,* 1526–1533.

204. Goldman, S., Copeland, J., Morwitz, T., et al. (1988). Improvement in early saphenous vein graft patency after coronary artery bypass surgery with antiplatelet therapy: Results of a Veterans Administration Cooperative Study. *Circulation, 77,* 1324–1332.

205. Fuster, V., & Chesebro, J. H. (1986). Role of platelets and platelet inhibitors in aortocoronary artery vein-graft disease. *Circulation, 73,* 227–232.

206. Antiplatelet Trialists' Collaboration (1994). Collaborative overview of randomised trials of antiplatelet therapy — II: Maintenance of vascular graft or arterial patency by antiplatelet therapy. *Br Med J, 308,* 159–168.

207. Dacey, L. J., Munoz, J. J., Johnson, E. R., et al. (2000). Effect of preoperative aspirin use on mortality in coronary artery bypass grafting patients. *Ann Thorac Surg, 70,* 1986–1990.

208. Goldman, S., Copeland, J., Moritz, T., et al. (1991). Starting aspirin therapy after operation. Effects on early graft patency. Department of Veterans Affairs Cooperative Study Group. *Circulation, 84,* 520–526.

209. Mangano, D. T. (2002). Aspirin and mortality from coronary bypass surgery. *N Engl J Med, 347,* 1309–1617.

210. Goldman, S., Copeland, J., Morwitz, T., et al. (1994). Long-term graft patency (3 years) after coronary artery surgery. Effects of aspirin: Results of a VA Cooperative study. *Circulation, 89,* 1138–1143.

211. Bhatt, D. L., Chew, D. P., Hirsch, A. T., et al. (2001). Superiority of clopidogrel versus aspirin in patients with prior cardiac surgery. *Circulation, 103,* 363–368.

212. Erez, E., Erman, A., Snir, E., et al. (1998). Thromboxane production in human lung during cardiopulmonary bypass: Beneficial effect of aspirin? *Ann Thorac Surg, 65,* 101–106.

213. Gerrah, R., Elami, A., Stamler, A., et al. (2005). Preoperative aspirin administration improves oxygenation in patients undergoing coronary artery bypass grafting. *Chest, 127,* 1622–1626.

214. Goldhaber, S. Z., Manson, J. E., Stampfer, M. J., et al. (1992). Low-dose aspirin and subsequent peripheral arterial surgery in the Physicians' Health Study. *Lancet, 340,* 43–45.

215. The Dutch Bypass Oral Anticoagulants Investigators (2000). Efficacy of oral anticoagulants compared with aspirin after infrainguinal bypass surgery: A randomised trial. *Lancet, 355,* 346–351.

216. Minar, E., Ahmadi, A., Koppensteiner, R., et al. (1995). Comparison of effects of high-dose and low-dose aspirin on restenosis after femoropopliteal percutaneous transluminal angioplasty. *Circulation, 91,* 2167–2173.

217. Kodama, M., Yamasaki, Y., Sakamoto, K., et al. (2000). Antiplatelet drugs attenuate progression of carotid intima-media thickness in subjects with type 2 diabetes. *Thromb Res, 97,* 239–245.

218. Ranke, C., Hecker, H., Creutzig, A., et al. (1993). Dose-dependent effect of aspirin on carotid atherosclerosis. *Circulation, 87,* 1873–1879.

219. Lotke, P. A., Palevsky, H., Keenan, A. M., et al. (1996). Aspirin and warfarin for thromboembolic disease after total joint arthroplasty. *Clin Orthop,* 251–258.

220. Powers, P. J., Gent, M., Jay, R. M., et al. (1989). A randomized trial of less intense postoperative warfarin or aspirin therapy in the prevention of venous thromboembolism after surgery for fractured hip. *Arch Intern Med, 149,* 771–774.

221. Antiplatelet Trialists' Collaboration (1994). Collaborative overview of randomized trials of antiplatelet therapy: III. Reduction in venous thrombosis and pulmonary embolism by antiplatelet prophylaxis among surgical and medical patients. *Br Med J, 308,* 235–246.

222. Pulmonary Embolism Prevention (PEP) trial (2000). Prevention of pulmonary embolism and deep vein thrombosis with low dose aspirin. *Lancet, 355,* 1295–1302.

223. Westrich, G. H., & Sculco, T. P. (1996). Prophylaxis against deep venous thrombosis after total knee arthroplasty: Pneumatic plantar compression and aspirin compared with aspirin alone. *J Bone Joint Surg Br, 78,* 826–834.

224. Graor, R. A., Stewart, J. H., & Lotke, P. A. (1992). RD heparin (ardeparin sodium) vs aspirin to prevent deep venous thrombosis after hip or knee replacement surgery [abstract]. *Chest, 102,* 118S.

225. Gent, H., Hirsh, J., & Ginsberg, J. S. (1996). Low-molecular-weight heparinoid organan is more effective that aspirin in the prevention of venous thromboembolism after surgery for hop fracture. *Circulation, 93,* 80–84.

226. Geerts, W. H., Pineo, G. F., Heit, J. A., et al. (2005). Prevention of venous thromboembolism: The Seventh ACCP Conference on Antithrombotic and Thrombolytic Therapy. *Chest, 126*(Suppl. 3), 338S–400S.

227. Altman, R., Rouvie, J., Gurfinkel, E., et al. (1996). Comparison of high-dose with low-dose aspirin in patients with

mechanical heart valve replacement treated with oral anti-coagulant. *Circulation, 94,* 2113–2116.

228. Laffort, P., Roudaut, R., Roques, X., et al. (2000). Early and long-term (one-year) effects of the association of aspirin and oral anticoagulant on thrombi and morbidity after replacement of the mitral valve with the St. Jude medical prosthesis: A clinical and transesophageal echocardiographic study. *J Am Coll Cardiol, 35,* 739–746.

229. Massel, D., & Little, S. H. (2001). Risks and benefits of adding anti-platelet therapy to warfarin among patients with prosthetic heart valves: A meta-analysis. *J Am Coll Cardiol, 37,* 569–578.

230. Monagle, P., Chan, A., Massicotte, P., et al. (2004). Antithrombotic therapy in children: The Seventh ACCP Conference on Antithrombotic and Thrombolytic Therapy. *Chest, 126*(Suppl. 3), 645S–687S.

231. Koren, G., Rose, V., & Lavi, S. (1985). Probable efficacy of high-dose salicylates in reducing coronary involvement in Kawasaki disease. *J Am Med Assoc, 254,* 767–769.

232. Daniels, S. R., Specker, B., & Capannari, T. E. (1987). Correlates of coronary artery aneurysm formation in patients with Kawasaki disease. *Am J Dis Child, 141,* 205–207.

233. Janes, S., Kyle, P., Redman, C., et al. (1995). Flow cytometric detection of activated platelets in pregnant women prior to the development of pre-eclampsia. *Thromb Haemost, 74,* 1059–1063.

234. Bussolino, F., Benedetto, C., Massobrio, M., et al. (1980). Maternal vascular prostacyclin activity in pre-eclampsia. *Lancet, 2,* 702.

235. Subtil, D., Goeusse, P., Puech, F., et al. (2003). Aspirin (100 mg) used for prevention of pre-eclampsia in nulliparous woman: The Essai Regional Aspirine Mere-Enfant (ERASME) Collaborative Group. *Br J Obstet Gynaecol, 110,* 475–484.

236. Caritis, S., Sibi, B., Hauth, J., et al. (1998). Low-dose aspirin to prevent preeclampsia in woman at high risk. *N Engl J Med, 338,* 701–705.

237. Dumont, A., Flahault, A., Beaufils, M., et al. (1999). Effect of aspirin in pregnant women is dependent on increase in bleeding time. *Am J Obstet Gynecol, 180*(1 Pt 1), 135–140.

238. Heyborne, K. D. (2000). Preeclampsia prevention: Lessons from the low-dose aspirin therapy trials. *Am J Obstet Gynecol, 183,* 523–528.

239. Duley, L., Henderson-Smart, D., Knight, M., et al. (2001). Antiplatelet drugs for prevention of pre-eclampsia and its consequences: Systemic review. *Br Med J, 322,* 329–333.

240. Mahmoud, N. N., Dannenberg, A. J., Mestre, J., et al. (1998). Aspirin prevents tumors in a murine model of familial adenomatous polyposis. *Surgery, 124,* 225–231.

241. Rosenberg, L., Louik, C., & Shapiro, S. (1998). Nonsteroidal antiinflammatory drug use and reduced risk of large bowel carcinoma. *Cancer, 82,* 2326–2333.

242. Sandler, R. S., Galanko, J. C., Murray, S. C., et al. (1998). Aspirin and nonsteroidal anti-inflammatory agents and risk for colorectal adenomas. *Gastroenterology, 114,* 441–447.

243. Martinez, M. E., McPherson, R. S., Levin, B., et al. (1997). A case-control study of dietary intake and other lifestyle risk factors for hyperplastic polyps. *Gastroenterology, 113,* 423–429.

244. Giovannucci, E., Egan, K. M., Hunter, D. J., et al. (1995). Aspirin and the risk of colorectal cancer in women. *N Engl J Med, 333,* 609–614.

245. Garcia-Rodriguez, L. A., & Huerta-Alvarez, C. (2001). Reduced risk of colorectal cancer among long-term users of aspirin and nonaspirin nonsteroidal antiinflammatory drugs. *Epidemiology, 12,* 88–93.

246. Stumer, T., Glynn, R. J., Lee, I. M., et al. (1998). Aspirin use and colorectal cancer: Post-trial follow-up data from the Physicians' Health Study. *Ann Intern Med, 128,* 713–720.

247. Cook, N. R., Lee, I. M., Gaziano, J. M., et al. (2005). Low-dose aspirin in the primary prevention of cancer. *J Am Med Assoc, 294,* 47–55.

248. Law, B. K., Waltner-Law, M. E., Entingh, A. J., et al. (2000). Salicylate-induced growth arrest is associated with inhibition of p70s6k and down-regulation of c-myc, cyclin D1, cyclin A, and proliferating cell nuclear antigen. *J Biol Chem, 275,* 38261–38267.

249. Qiao, L., Hanif, R., Sphicas, E., et al. (1998). Effect of aspirin on induction of apoptosis in HT-29 human colon adenocarcinoma cells. *Biochem Pharm, 55,* 53–64.

250. Jones, M. K., Wang, H., Peskar, B. M., et al. (1999). Inhibition of angiogenesis by nonsteroidal anti-inflammatory drugs: Insight into mechanisms and implications for cancer growth and ulcer healing. *Nat Med, 5,* 1418–1423.

251. Patrono, C., & Dunn, M. J. (1987). The clinical significance of inhibition of renal prostaglandin synthesis. *Kidney Int, 32,* 1–12.

252. Sano, H., Kawahito, Y., Wilder, R. L., et al. (1995). Expression of cyclooxygenase-1 and -2 in human colorectal cancer. *Cancer Res, 55,* 3785–3789.

253. Williams, C. S., Smalley, W., & DuBois, R. N. (1997). Aspirin use and potential mechanisms for colorectal cancer prevention. *J Clin Invest, 100,* 1326–1329.

254. Sheng, H., Shao, J., Kirkland, S. C., et al. (1997). Inhibition of human colon cancer cell growth by selective inhibition of cyclooxygenase-2. *J Clin Invest, 99,* 2254–2259.

255. Serebruany, V. L., Steinhubl, S. R., Berger, P. B., et al. (2005). Analysis of risk of bleeding complications after different doses of aspirin in 192,036 patients enrolled in 31 randomized controlled trials. *Am J Cardiol, 95,* 1218–1222.

256. Jaszewski, R. (1990). Frequency of gastroduodenal lesions in asymptomatic patients on chronic aspirin or nonsteroidal antiinflammatory drug therapy. *J Clin Gastroent, 12,* 10–13.

257. Savon, J. J., Allen, M. L., DiMarino, A. J., Jr., et al. (1995). Gastrointestinal blood loss with low dose (325 mg) plain and enteric-coated aspirin administration. *Am J Gastroenterol, 90,* 581–585.

258. Kelly, J. P., Kaufman, D. W., Jurgelon, J. M., et al. (1996). Risk of aspirin-associated major upper-gastrointestinal bleeding with enteric-coated or buffered product. *Lancet, 348,* 1413–1416.

259. Keimowitz, R. M., Pulvermacher, G., Mayo, G., et al. (1993). Aspirin and platelets: Transdermal modification of platelet function: A dermal aspirin preparation selectively inhibits platelet cyclooxygenase and preserves prostacyclin biosynthesis. *Circulation, 88,* 556–561.

260. Johnson, S. A., Leib, M. S., Forrester, S. D., et al. (1995). The effect of misoprostol on aspirin-induced gastroduodenal lesions in dogs. *J Vet Intern Med, 9,* 32–38.

261. Bowersox, T. S., Lipowitz, A. J., Hardy, R. M., et al. (1996). The use of a synthetic prostaglandin E1 analog as a gastric protectant against aspirin-induced hemorrhage in the dog. *J Am Anim Hosp Assoc, 32,* 401–407.

262. Lanza, F. L., Kochman, R. L., Geis, G. S., et al. (1991). A double-blind, placebo-controlled, 6-day evaluation of two doses of misoprostol in gastroduodenal mucosal protection against damage from aspirin and effect on bowel habits. *Am J Gastroenterol, 86,* 1743–1748.

263. Lai, K. C., Lam, S. K., Chu, K. M., et al. (2002). Lansoprazole for the prevention of recurrences of ulcer complications from long-term low-dose aspirin. *N Engl J Med, 346,* 2033–2038.

264. Chan, F. K. L., Ching, J. Y. L., Hung, L. C. T., et al. (2005). Clopidogrel versus aspirin and esomeprazole to prevent recurrent ulcer bleeding. *N Engl J Med, 352,* 238–244.

265. Sandler, D. P., Burr, F. R., & Weinberg, C. R. (1991). Nonsteroidal anti-inflammatory drugs and the risk for chronic renal disease. *Ann Intern Med, 115,* 165–172.

266. Mene, P., Pugliese, F., & Patrono, C. (1995). The effects of nonsteroidal anti-inflammatory drugs on human hypertensive vascular disease. *Semin Nephrol, 15,* 244–252.

267. Hansen, H. P., Gaede, P. H., Jensen, B. R., et al. (2000). Lack of impact of low-dose acetylsalicylic acid on kidney function in type 1 diabetic patients with microalbuminuria. *Diabetes Care, 23,* 1742–1745.

268. Riegger, G. A., Kahles, H. W., Elsner, D., et al. (1991). Effects of acetylsalicylic acid on renal function in patients with chronic heart failure. *Am J Med, 90,* 571–575.

269. Caspi, D., Lubart, E., Graff, E., et al. (2000). The effect of mini-dose aspirin on renal function and uric acid handling in elderly patients. *Arthritis Rheum, 43,* 103–108.

270. D'Agati, V. (1996). Does aspirin cause acute or chronic renal failure in experimental animals and in humans? *Am J Kidney Dis, 28*(Suppl. 1), S24–S29.

271. He, J., Whelton, P. K., Vu, B., et al. (1998). Aspirin and risk of hemorrhagic stroke: A meta-analysis of randomized controlled trials. *J Am Med Assoc, 280,* 1930–1935.

272. Hall, D., Zeitler, H., & Rudolph, W. (1992). Counteraction of the vasodilator effects of enalapril by aspirin in severe heart failure. *J Am Coll Cardiol, 20,* 1549–1555.

273. Al-Khadra, A. S., Salem, D. N., Rand, W. M., et al. (1998). Antiplatelet agents and survival: A cohort analysis from the Studies of Left Ventricular Dysfunction (SOLVD) trial. *J Am Coll Cardiol, 31,* 419–425.

274. Nguyen, K. N., Aursnes, I., & Kjekshus, J. (1997). Interaction between enalapril and aspirin on mortality after acute myocardial infarction: Subgroup analysis of the Cooperative New Scandinavian Enalapril Survival Study II. *Am J Cardiol, 79,* 115–119.

275. Lantini, R., Tognoni, G., Maggioni, A. P., et al. (2000). Clinical effects of early angiotensin-converting enzyme inhibitor treatment for acute myocardial infarction are similar in the presence and absence of aspirin: Systematic overview of individual data from 96,712 randomized patients. Angiotensin-converting Enzyme Inhibitor Myocardial Infarction Collaborative Group. *J Am Coll Cardiol, 35,* 1801–1807.

276. Teo, K. K., Yusuf, S., Pfeffer, M., et al. (2002). Effects of long-term treatment with angiotensin-converting-enzyme inhibitors in the presence of aspirin: A systemic review. *Lancet, 360,* 1037–1043.

277. Rogers, W. J., Canto, J. G., Lambrew, C. T., et al. (2000). Temporal trends in the treatment of over 1.5 million patients with myocardial infarction in the US from 1990 through 1999: The National Registry of Myocardial Infarction 1, 2 and 3. *J Am Coll Cardiol, 36,* 2056–2063.

278. Jackson, E. A., Sivasubramian, R., Spencer, F. A., et al. (2002). Changes over time in the use of aspirin in patients hospitalized with acute myocardial infarction (1975 to 1997): A population-based perspective. *Am Heart J, 144,* 259–268.

279. Stafford, R. S. (2000). Aspirin use is low among United States outpatients with coronary artery disease. *Circulation, 101,* 1097–1101.

280. Shahar, E., Folsom, A. R., Romm, F. J., et al. (1996). Patterns of aspirin use in middle-aged adults: The Atherosclerosis Risk in Communities (ARIC) study. *Am Heart J, 131,* 915–922.

281. Bowker, T. J., Clayton, T. C., Ingham, J., et al. (1996). A British Cardiac Society survey of the potential for the secondary prevention of coronary disease: ASPIRE (Action on Secondary Prevention through Intervention to Reduce Events). *Heart, 75,* 334–342.

282. Krumholtz, H. M., Philbin, D. M., Jr., Wang, Y., et al. (1998). Trends in the quality of care for Medicare beneficiaries admitted to the hospital with unstable angina. *J Am Coll Cardiol, 31,* 957–963.

283. Jencks, S. F., Huff, E. D., & Cuerdon, T. (2003). Change in the quality of care delivered to Medicare beneficiaries, 1998–1999 to 2000–2001. *J Am Med Assoc, 289,* 305–312.

284. Frilling, B., Schiele, R., Gitt, A. K., et al. (2004). Too little aspirin for secondary prevention after acute myocardial infarction in patients at high risk for cardiovascular events: Results of the MITRA study. *Am Heart J, 148,* 306–311.

285. Mosca, L., Linfante, A. H., Benjamin, E. J., et al. (2005). National Study of Physician Awareness and Adherence to Cardiovascular Disease Prevention Guidelines. *Circulation, 111,* 499–510.

286. Aronow, W. S. (1998). Underutilization of aspirin in older patients with prior myocardial infarction at the time of admission to a nursing home. *J Am Geriatr Soc, 46,* 615–616.

287. Sarasin, F. P., Gaspoz, J. M., & Bounameaux, H. (2000). Cost-effectiveness of new antiplatelet regimens used as secondary prevention of stroke and transient ischemic attack. *Arch Intern Med, 160,* 2773–2778.

288. Weintraub, W. S., Mahoney, E. M., Lamy, A., et al. (2005). Long-term cost-effectiveness of clopidogrel given for up to one year in patients with acute coronary syndromes without ST-segment elevation. *J Am Coll Cardiol, 45,* 838–845.

289. Schleinitz, M. D., & Heidenreich, P. A. (2005). A cost-effectiveness analysis of combination antiplatelet therapy for high-risk acute coronary syndromes: Clopidogrel plus aspirin versus aspirin alone. *Ann Intern Med, 142,* 251–259.

# CHAPTER 61

# ADP Receptor Antagonists

## Marco Cattaneo

*Unità di Ematologia e Trombosi, Ospedale San Paolo, Università di Milano, Milan, Italy*

## I. Introduction

Adenosine-5′-diphosphate (ADP) plays a key role in platelet function (see Chapter 10). Although ADP is a weak platelet agonist, when it is secreted from the platelet dense granules where it is stored, ADP amplifies the platelet responses induced by other platelet agonists. The amplifying effect of ADP on platelet aggregation accounts for the critical role played by ADP in hemostasis, as documented by the observation that patients lacking releasable ADP in granule stores or with congenital abnormalities of platelet ADP receptors have a bleeding diathesis. In addition, ADP plays a key role in the pathogenesis of arterial thrombosis because pharmacologic inhibition of ADP-induced platelet aggregation decreases the risk of arterial thrombosis. The pharmaceutical compounds that antagonize platelet ADP receptors, both those already marketed and those under development, are reviewed in this chapter.

## II. The Platelet Purinergic Receptors for ADP

Purinergic receptors are classified as P1, which are receptors for adenosine, and P2, which interact with purine and pyrimidine nucleotides.[1] The P2 receptors are subdivided into two main groups: G protein-linked or "metabotropic," designated P2Y, and ligand-gated ion channels or "ionotropic," termed P2X. P2Y receptors are seven-membrane-spanning proteins with a molecular mass of 41 to 53 kDa after glycosylation; their carboxyl terminal domain is on the cytoplasmic side, whereas the amino terminal domain is exposed to the extracellular environment. Common mechanisms of signal transduction are shared by most seven-membrane-spanning receptors, including activation of phospholipase C and/or regulation of adenylyl cyclase activity. P2X receptors are ligand-gated ion channels that mediate rapid changes in the membrane permeability of monovalent and divalent cations, including $Na^+$, $K^+$, and $Ca^{2+}$. P2X

receptors range from 379 to 595 amino acids and have two transmembrane domains separated by a large extracellular region. Unlike the P2Y receptors, the amino and carboxyl terminal domains are both on the cytoplasmic side of the plasma membrane. Human platelets express at least three distinct receptors stimulated by adenosine nucleotides: $P2Y_1$ and $P2Y_{12}$, which interact with ADP, and $P2X_1$, which interacts with ATP.[2]

The transduction of the ADP signal involves both a transient rise in free cytoplasmic calcium, mediated by the $G_q$-coupled $P2Y_1$ receptor, and inhibition of adenylyl cyclase, which is mediated by the $G_i$-coupled $P2Y_{12}$ receptor. Concomitant activation of both the $G_q$ and $G_i$ pathways by ADP is necessary to elicit normal platelet aggregation. Activation of the $G_q$ pathway through $P2Y_1$ leads to platelet shape change and rapidly reversible aggregation, whereas activation of the $G_i$ pathway through $P2Y_{12}$ elicits a slowly progressive and sustained platelet aggregation not preceded by shape change. In addition to its role in ADP-induced platelet aggregation, $P2Y_{12}$ mediates (1) the potentiation of platelet secretion by ADP, which is independent of the formation of large aggregates and thromboxane $A_2$ synthesis, and (2) the stabilization of thrombin-induced platelet aggregates.[2]

The platelet P2 receptors ($P2Y_1$, $P2Y_{12}$, and $P2X_1$) and their roles in platelet function and thrombosis are discussed in more detail in Chapter 10.

$P2Y_{12}$ has a more selective tissue distribution than $P2Y_1$, making it an attractive molecular target for therapeutic intervention. $P2Y_{12}$ is the target of efficacious antithrombotic agents such as ticlopidine and clopidogrel, which are already used in clinical practice, and of other compounds that are currently under evaluation in clinical trials.

The results of studies of experimental thrombosis in animals suggest that antagonists of $P2Y_1$ could also prove to be highly potent antithrombotic agents. However, no specific $P2Y_1$ antagonist has been evaluated in clinical studies.

# III. P2Y$_{12}$ Antagonists

## A. Thienopyridines

### 1. Ticlopidine and Clopidogrel

*a. Pharmacokinetics.* Ticlopidine (5-[2-chlorobenzyl]-4,5,6,7-tetrahydrothieno-[3,2-c] pyridine hydrochloride) and clopidogrel (methyl [+]-[S]-a-[2-chlorophenyl]-6,7-dihydrothieno [3,2-c] pyridin-5 [4H]-acetate hydrogen sulfate) are structurally related compounds belonging to the thienopyridine family of ADP receptor antagonists. Clopidogrel differs structurally from ticlopidine by the addition of a carboxymethyl side group (Fig. 61-1). Clopidogrel is a chiral drug with S-configuration; the R-enantiomer (SR-25989) is devoid of antiplatelet activity.

Ticlopidine and clopidogrel are rapidly absorbed and metabolized after oral administration. After a single 250-mg dose of ticlopidine, the mean area under the curve (AUC; 0–12 hours) was 1.11 μg ml$^{-1}$ hour in young patients and 2.04 μg ml$^{-1}$ hour in elderly patients; the mean plasma half-lives in young and elderly subjects were 7.9 and 12.6 hours, respectively.[3] Steady-state plasma drug concentrations were attained after 14 days of dosing with ticlopidine. After repeated dosing, AUC values in elderly patients were two or three times those in young patients, and the plasma half-lives averaged 4.0 days for young patients and 3.8 days for elderly patients. The longer half-life and higher AUC values after multiple dosing probably reflect an increase in bioavailability of ticlopidine or saturation of metabolism.[3] The bioavailability of ticlopidine is increased by food and decreased by antacids.[4]

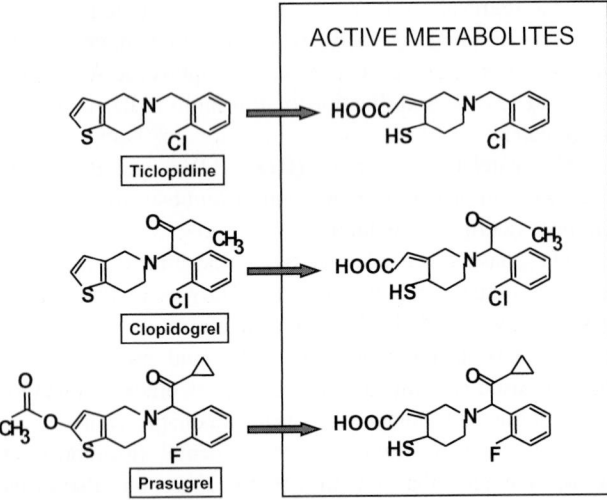

**Figure 61-1.** Chemical structures of three thienopyridyl compounds — ticlopidine, clopidogrel, and prasugrel — and their active metabolites.

The major circulating metabolite of clopidogrel is the inactive carboxylic acid derivative SR26334. The mean $C_{max}$ values for SR26334 following single doses of 50, 75, 100, and 150 mg clopidogrel to healthy male volunteers were 1.6, 2.9, 3.1, and 4.9 mg/L, demonstrating a dose-proportional increase of $C_{max}$ in this range of clopidogrel doses. Median $T_{(max)}$ (0.8–1.0 hours) and mean plasma half-life (7.2–7.6 hours) were not significantly different between doses.[5] Following repeated dosing of healthy volunteers with clopidogrel 75 mg for 14 days or 12 weeks, mean $C_{trough}$ values for SR26334 at steady state were approximately 0.1 mg/L, indicating that steady-state values are reproducible and that the esterasic biotransformation of clopidogrel into its carboxylic acid metabolite remains constant over several weeks of treatment.[5] Net absorption of clopidogrel is not significantly modified by food or by prior antacid ingestion.[6] Mean cumulative urinary excretion of clopidogrel over 120 hours represents 41% of the dose after a single-dose administration and 46% after administration at steady state, whereas the cumulative fecal recovery over 120 hours ranges from 35 to 57% after a single-dose administration and from 39 to 59% after administration at steady state. Mean total excretion is approximately 90% of the dose both after single dose and for the steady state.[7]

*b. Pharmacodynamics.* Ticlopidine and clopidogrel are prodrugs that are inactive *in vitro* and need to be metabolized *in vivo* by the hepatic cytochrome P450 1A enzymatic pathway to active metabolites to exert their inhibitory effect on platelet function.[8–11] One study reported that clopidogrel inhibits platelet aggregation *in vitro*,[12] but this effect was later shown to be artifactual.[13] Both drugs irreversibly and selectively inhibit ADP-induced platelet aggregation (30–60% inhibition) and ADP-induced adenylyl cyclase downregulation[14–18] and cause a dose-dependent reduction of the platelet binding sites for ADP or its poorly hydrolysable analog 2-methylthio-ADP (2MeS-ADP).[19–22] The administration of high-dose clopidogrel to rats reduced the number of platelet binding sites for 2-MeS-ADP by approximately 70% of the total but did not inhibit ADP-induced platelet shape change, transient platelet aggregation, or Ca$^{2+}$ mobilization, whereas it completely abolished the ability of ADP to inhibit platelet adenylyl cyclase.[2] This pharmacological effect of clopidogrel reproduces the platelet function abnormalities that are observed in patients who are congenitally deficient in P2Y$_{12}$ and also in P2Y$_{12}$ knockout mice (see Chapter 10),[2] demonstrating that the drug selectively antagonizes the platelet P2Y$_{12}$ receptor for ADP without affecting the other platelet ADP receptor, P2Y$_1$. The ability of thienopyridines to inhibit platelet aggregation induced by several platelet agonists (e.g., thromboxane A$_2$ analogs, collagen, and low concentrations of thrombin) is accounted for by suppression of the amplifying effect on platelet aggregation of ADP released from platelet dense granules.[15,23] Ticlopidine treatment renders thrombin-induced platelet aggregates more susceptible to deaggregation[24] and inhibits shear-induced platelet aggregation.[25] A similar inhibitory effect on shear-induced platelet aggregation was later shown after treatment with clopidogrel.[26] Both drugs cause a two- or threefold prolongation of the bleeding time.[8]

The need for metabolism of thienopyridines to active metabolites accounts for the delay until their antiplatelet effects are observed. The antiplatelet effect of ticlopidine is manifest 2 or 3 days after oral administration of 500 mg daily, reaches its plateau after 4 days of treatment, has a half-life of approximately 5 days, and persists for approximately 10 days. This time course corresponds to the life span of a circulating platelet.[8] Doses of 250 mg ticlopidine daily induce comparable inhibition of platelet aggregation, whereas 100 mg daily does not inhibit platelet function even after 14 days of administration.[27] The pharmacological effect of ticlopidine is similar in both sexes[27] and is not significantly affected by renal function.[28]

A maximum degree of inhibition of platelet aggregation similar to that obtained with ticlopidine is observed approximately 5 days after treatment of healthy volunteers with clopidogrel 75 mg.[8] The delayed onset of action of the antiplatelet effect can be reduced to approximately 2 to 5 hours by a loading dose of 300 to 600 mg both in healthy volunteers and in patients with atherothrombotic disease.[29–33] The activity of clopidogrel on the inhibition of platelet aggregation is maintained with long-term treatment[34,35] and is independent of age and the presence of atherosclerosis[36] or cirrhosis.[37]

The active metabolites of ticlopidine and clopidogrel have been identified (Fig. 61-1). They are generated from their prodrugs through a two-step metabolism in the liver. First, the aromatic thiophene ring is oxidized to generate 2-hydroxy thiophene and isomerized to the 2-oxo form through a cytochrome P450-dependent pathway. Second, the 2-oxo-thiophene ring is hydrolyzed. Both metabolites have a carboxylic acid and a thiol group as a result of 2-oxo-thiophene ring opening. This is consistent with the structure of the active metabolite of another thienopyridine, prasugrel (Fig. 61-1), confirming that the mode of action and the target molecule of these drugs are identical (see below). The irreversible modification of the ADP $P2Y_{12}$ receptor site that is responsible for the biological activity of the metabolites of the thienopyridines could be explained by the formation of a disulfide bridge between the reactive thiol group of the active metabolite and a cysteine residue of the $P2Y_{12}$ receptor.[38–40]

The active metabolite of clopidogrel was identified in 2000 by Savi et al.[38] It belongs to a family of eight stereoisomers with the following primary chemical structure: 2-[1-[1-(2-chlorophenyl)-2-methoxy-2-oxoethyl]-4-sulfanyl-3-piperidinylidene]acetic acid. The observation that only one isomer (of S configuration at C7 and Z con-

figuration at C3–C16 double bond) retained biological activity underlies the critical importance of associated absolute configuration.[41] The active metabolite is a highly unstable compound, probably due to a very reactive thiol function, that exhibits *in vitro* all the biological activities of clopidogrel observed *ex vivo*: irreversible inhibition of the binding of radiolabeled 2MeS-ADP to washed human platelets ($IC_{50}$, 0.53 μM), selective inhibition of ADP-induced platelet aggregation ($IC_{50}$, 1.8 μM), and downregulation of ADP-induced adenylyl cyclase.

UR-4501, the active metabolite of ticlopidine, was identified by Yoneda et al. in 2004.[39] Generated from 2-oxo-ticlopidine, it displayed characteristics of irreversible inhibition of platelet aggregation that were consistent with those observed *ex vivo* after oral administration of ticlopidine.[39]

*c. Other Pharmacological Effects.* Ticlopidine and clopidogrel have additional pharmacological effects that may contribute to their antithrombotic activities, including a reduction in the level of circulating fibrinogen,[42–44] which may be associated with hemorheological modifications (decreases in whole blood and plasma viscosity),[45] reduction of erythrocyte aggregation,[46] stimulation of nitric oxide production,[47] inhibition of platelet-dependent tissue factor expression on endothelial cells,[48] and inhibition of fibronectin synthesis by endothelial cells in culture.[49]

*d. Side Effects.* The most common side effects associated with ticlopidine treatment are diarrhea, nausea, and vomiting, which can occur in up to 50% of treated patients. Bone marrow suppression, primarily involving granulopoiesis, is the most serious, potentially fatal toxic effect reported with ticlopidine, which can occur in approximately 1% of treated patients. Bone marrow aplasia or isolated neutropenia, which can be severe (<450 neutrophils/μL), develops in the first 3 months of treatment in most instances, and it may reverse upon drug discontinuation. Therefore, full blood counts must be obtained every 10 to 15 days during the first 3 months of treatment so that treatment can be discontinued as early as possible if signs of cytopenia become manifest.[50,51]

Thrombotic thrombocytopenic purpura (TTP; see Chapter 50) is another very harmful complication of ticlopidine treatment.[52,53] In a series of 60 cases of ticlopidine-associated TTP, ticlopidine had been prescribed for less than 1 month in 80% of the patients; mortality rates were higher among patients who were not treated with plasmapheresis than among those who underwent plasmapheresis (50 vs. 24%, $p < 0.05$).[52] The prevalence of TTP in a cohort of ticlopidine-treated patients following coronary stenting was 0.02%, 20-fold higher than that estimated in the general population (0.0004%).[54] In the same cohort, the mortality rate for this complication exceeded 20%. Limiting ticlopidine therapy to 2 weeks after stenting did not prevent the development of TTP; therefore, rapid diagnosis and treatment are critical for improved survival. As in most patients with idiopathic acute TTP, autoantibodies to ADAMTS13, the von Willebrand factor-cleaving metalloproteinase, are usually found in patients who develop ticlopidine-associated TTP.[55] Microvascular end cell (MVEC) apoptosis related to altered endothelial cell matrix–MVEC interactions may be a key part of the pathology of ticlopidine-linked TTP.[56]

Ticlopidine has also been associated with cholestatic jaundice,[57] elevated levels of liver enzymes,[58] skin rash,[59] colitis,[60] and arthritis.[61]

In general, clopidogrel is a better tolerated and safer drug than ticlopidine. However, fatal cases of aplastic anemia and TTP or hemolytic–uremic syndrome (HUS) have also been reported during clopidogrel treatment (see Chapter 50).[62–67] Bone marrow toxicity associated with clopidogrel therapy may be fatal and usually occurs weeks to months following therapy initiation.[68–70] Clopidogrel-associated TTP/HUS often occurs within 2 weeks of drug initiation, occasionally relapses, and has a high mortality if not treated promptly.[63,65–67]

As for ticlopidine, gastrointestinal side effects are common during clopidogrel treatment. However, in the Clopidogrel versus Aspirin in Patients at Risk of Ischaemic Events (CAPRIE) trial, clinically severe gastrointestinal hemorrhage was more common in aspirin-treated patients than in clopidogrel-treated patients (0.71 vs. 0.49%, $p < 0.05$).[71] These findings are compatible with the observation that in patients without gastroduodenal disease, clopidogrel (75 mg daily), but not aspirin (325 mg daily), does not induce any gastroscopically evident erosions during short-term treatment,[72] and they led the American College of Cardiology/American Heart Association to recommend the use of clopidogrel for hospitalized patients with acute coronary syndrome who are unable to take aspirin because of major gastrointestinal intolerance.[73] However, this recommendation was not supported by the findings of a study that showed that among patients with a history of aspirin-induced ulcer bleeding whose ulcers had healed before they received the study treatment, aspirin plus esomeprazole was superior to clopidogrel in the prevention of recurrent ulcer bleeding.[74,75]

In the CAPRIE trial, clinically severe rash was more frequent with clopidogrel than with aspirin (0.26 vs. 0.10%). A possible link between autoimmune acquired hemophilia and clopidogrel has been postulated.[76] In contrast to ticlopidine, no increase in liver enzymes is observed in clopidogrel-treated patients.[77]

Treatment with clopidogrel, alone or in combination with aspirin, within 4 or 5 days of coronary bypass surgery is associated with increased blood loss, reoperation for bleeding, increased transfusion requirements, and prolonged intensive care unit and hospital length of stay.[78–82] In these patients, the intraoperative use of aprotinin signifi-

cantly reduces postoperative bleeding and transfusion requirements.[83,84]

The overall incidence of bleeding complications during ticlopidine or clopidogrel treatment in patients not undergoing surgery is low but could be slightly increased by the combination of aspirin or anticoagulants.[85] In the case of severe hemorrhage that requires medical intervention, infusion of desmopressin, which shortens the prolonged bleeding times of patients treated with ticlopidine and other congenital or acquired defects of primary hemostasis,[25,86] could be useful (see Chapter 67).

*e. Drug Interactions.* No interaction was observed between clopidogrel and several drugs, including phenobarbital, antacids, estrogen, digoxin, theophylline, β-blockers, calcium antagonists, angiotensin-converting enzyme inhibitors, lipid-lowering agents (including some statins), antivitamin K compounds, N-acetyl L-cysteine, or in patients with hepatic insufficiency.[8,87–89] Atorvastatin was shown to reduce the activity of clopidogrel through competitive inhibition of the cytochrome P450 3A4 isoenzyme.[90] However, this issue is controversial.

Several studies showed that the combined administration of clopidogrel or ticlopidine with aspirin produced a much higher inhibition of platelet aggregation and platelet thrombus formation in perfusion chambers than either drug alone.[91–100] Similarly, ticlopidine has been shown to enhance the inhibitory effects of a glycoprotein (GP) IIb-IIIa antagonist.[101–103] This synergism between thienopyridines and aspirin is explained by the fact that they inhibit the two main physiologic pathways of amplification of platelet activation: the ADP pathway (thienopyridines) and the arachidonic acid/thromboxane $A_2$ pathway (aspirin). These observations set the stage for the combined use of thienopyridines and aspirin for the treatment of patients at high risk of vascular occlusion, such as in acute coronary syndromes.

*f. Experimental Thrombosis.* Several studies demonstrated that both ticlopidine and clopidogrel inhibit arterial and venous thrombus formation in different models of experimental thrombosis in several animal species.[8] Clopidogrel inhibited thrombus formation under conditions in which thrombin production played an important pathogenic role,[104–108] suggesting that the ADP interaction with platelet $P2Y_{12}$ plays a role in the generation of thrombin. Indeed, the important role of $P2Y_{12}$ in promoting platelet procoagulant activity was shown in several studies (see Chapter 10).[48,109–111] Thienopyridines usually proved more effective than aspirin; however, their antithrombotic effects were potentiated by the addition of aspirin in several experimental models.[112–114]

*g. Clinical Trials.* Both ticlopidine and clopidogrel proved effective antithrombotic agents in several randomized clinical trials. The largest trial involving a thienopyridine, the CAPRIE trial, enrolled 19,185 patients at risk of ischemic events due to previous acute myocardial infarction (MI), ischemic stroke, or peripheral artery disease.[71] The trial showed an 8.7% relative risk reduction of the major end points (MI, ischemic stroke, and vascular death) within a mean follow-up of 1.9 years for patients treated with clopidogrel (75 mg daily) compared to patients treated with aspirin (325 mg daily). The benefit associated with clopidogrel therapy seemed to be confined to patients with previous peripheral artery disease. Secondary analyses of the trial showed that clopidogrel demonstrated an amplified clinical benefit versus aspirin in patients with prior cardiac surgery[115] and in those at high risk of atherothrombotic events, such as those with a previous history of symptomatic atherothrombotic disease or with major risk factors such as diabetes mellitus or hypercholesterolemia.[116,117] However, the 95% confidence intervals (Cis) of the relative risk reduction for high-risk patients with preexistent atherosclerotic disease overlap substantially with those for the entire CAPRIE population, thus failing to prove convincingly that the effect of clopidogrel is amplified in high-risk patients.[118]

Cerebrovascular Disease. The Canadian American Ticlopidine Study — a randomized, double-blind, placebo-controlled trial with a mean clinical follow-up of 24 months — showed that ticlopidine (250 mg b.i.d.), compared to placebo, significantly reduced the incidence of stroke, MI, or vascular death in men and women who had suffered a recent thromboembolic stroke (relative risk reduction [RRR] 23.3%).[119] In the Ticlopidine Aspirin Stroke Study (TASS), ticlopidine proved more effective than aspirin (650 mg b.i.d.) in reducing the risk of stroke and death in patients with recent transient or mild persistent focal cerebral or retinal ischemia, although the risk of side effects was greater for ticlopidine than for aspirin.[120] The benefit was manifest after the first year of treatment and persisted for the 6 years of follow-up. A subanalysis of the same study showed that ticlopidine is more effective than aspirin in reducing the risk of stroke in a subgroup of patients with a recent minor completed stroke as the qualifying ischemic event.[121] In contrast to the results of TASS, the African American Antiplatelet Stroke Prevention Study found no statistically significant difference between ticlopidine and aspirin (650 mg daily) in the prevention of recurrent stroke, MI, and vascular death in black patients with previous noncardioembolic stroke.[122] Considering the higher incidence of side effects (and the greater cost) associated with the use of ticlopidine, aspirin should therefore be regarded as a better prophylaxis for stroke, at least in black patients.

The MATCH trial showed that adding aspirin to clopidogrel (both given at 75 mg daily) in high-risk patients with recent ischemic stroke or transient ischemic attack (TIA) and at least one additional risk factor was associated with a nonsignificant difference in reducing major vascular events (RRR 6.4%; 95% CI −4.6 to 16.3) but doubled the relative risk of life-threatening hemorrhages from 1.3% (clopidogrel) to 2.6% (aspirin plus clopidogrel).[123] Predefined subgroup analyses did not identify patient groups in whom the combination therapy was advantageous compared to clopidogrel alone. In contrast, the combination of aspirin and clopidogrel was more effective than aspirin alone in reducing asymptomatic embolization, detected by transcranial Doppler ultrasound, in patients with a recently symptomatic carotid stenosis of ≥50%.[124] In this study, the number of strokes was smaller in the combination therapy group (zero vs. four), although the number of patients (110) was too small to reach a statistically significant reduction in clinical end points.

A systematic review of randomized trials showed that the thienopyridines are modestly but significantly more effective than aspirin in preventing serious vascular events in patients at high risk (specifically in TIA/ischemic stroke patients), but there is uncertainty regarding the importance of the additional benefit.[125]

Coronary Artery Disease. In a randomized, unblinded trial not controlled with placebo, ticlopidine (250 b.i.d.) plus conventional therapy was compared to conventional therapy alone in the treatment of 652 patients with unstable angina.[126] The study showed that ticlopidine significantly reduced (RRR 46.3%) the incidence of vascular death and nonfatal MI. The role of ticlopidine in the secondary prevention of myocardial infarction was studied in a randomized, double-blind, multicenter trial, which allocated 1470 patients with acute MI treated with thrombolysis to treatment with either aspirin (160 mg/day) or ticlopidine (500 mg/day).[127] No difference was found between the ticlopidine and aspirin groups in the rate of the primary combined end point of death, recurrent MI, stroke, and angina.

The Clopidogrel in Unstable Angina to Prevent Recurrent Events (CURE) trial demonstrated the sustained, incremental benefit of clopidogrel in addition to standard therapy (including aspirin) in patients with unstable angina and MI without ST segment elevation.[128] After a maximum duration of treatment and follow-up of 1 year (mean, 9 months), clopidogrel therapy reduced the combined risk of stroke, MI, or cardiovascular death by 20% compared with placebo ($p < 0.001$). The benefit was independent of the dose of aspirin that was chosen by the physicians, although the incidence of bleeding was lowest with low-dose aspirin (75–100 mg daily).[129] In an analysis that examined the onset and duration of the therapeutic effect, the benefit of the clopidogrel loading dose was observed as early as 24 hours after randomization and continued throughout the entire follow-up period.[130] In addition, Budaj et al.[131] showed that the benefit of clopidogrel demonstrated in the CURE trial was consistent in low-, intermediate-, and high-risk patients with acute coronary syndromes (as strati-fied by TIMI risk score), thus supporting its use in all patients with documented non-ST-elevation acute coronary syndromes.

Clopidogrel was studied in the setting of acute MI with ST segment elevation.[132] When given to patients who pre-sented within 12 hours after the onset of MI at a loading dose of 300 mg, followed by 75 mg daily, in addition to a fibrinolytic agent, aspirin, and, when appropriate, heparin, clopidogrel reduced the primary efficacy end point, which was defined as a composite of an occluded infarct-related artery (defined by a TIMI flow grade of 0 or 1) on angiog-raphy (performed 48–192 hours after the start of study medication) or death or recurrent MI before angiography (RRR 36%; 95% CI 24–47%; $p < 0.001$). By 30 days, clopi-dogrel therapy reduced the odds of the composite end point of death from cardiovascular causes, recurrent MI, or recur-rent ischemia leading to the need for urgent revasculariza-tion by 20% (95% CI 14.1–11.6%; $p = 0.03$). The rates of major bleeding and intracranial hemorrhage were similar between the two groups.[132]

The Clopidogrel and Metoprolol in Myocardial Infarc-tion Trial randomly allocated 45,852 patients with suspected MI to clopidogrel 75 mg daily ($n = 22,961$) or matching placebo ($n = 22,891$) in addition to aspirin 162 mg daily. Treatment was continued until discharge or up to 4 weeks in the hospital (mean, 15 days in survivors). Clopidogrel treatment was associated with a statistically significant reduction in the incidence of the two coprimary end points: (1) death, reinfarction, or stroke: 9% (3–14, $p = 0.002$); and (2) death from any cause: 7% (1–13, $p = 0.03$). There was no significant increase in major bleeding complications (0.58 vs. 0.55%, $p = 0.59$).[133]

Coronary Artery Revascularization. The greater effi-cacy of combined antiplatelet therapy with aspirin plus ticlopidine compared to conventional anticoagulant therapy after coronary artery stenting was proven in several ran-domized, double-blind, placebo-controlled trials published in the late 1990s. Clopidogrel has progressively replaced ticlopidine because of its better safety profile and apparent equivalent efficacy. However, reports have raised concern regarding the possibility of excess long-term mortality in stented patients given clopidogrel rather than ticlopidine.

The Intracoronary Stenting and Antithrombotic Regimen study showed that treatment with ticlopidine plus aspirin, compared with oral anticoagulants plus aspirin, dramati-cally decreased the incidence of death from cardiac causes or the occurrence of MI, aortocoronary bypass surgery, or repeat angioplasty 30 days after the procedure (relative risk 0.25; 95% CI 0.06–0.77).[134] However, the incidence of reste-nosis at 6 months was not different between the two treat-ment groups.[135] Hemorrhagic complications occurred in the anticoagulation group only. Similar results were obtained in

the Full Anticoagulation versus Aspirin and Ticlopidine after Stent Implantation[136] and Multicenter Aspirin and Ticlopidine Trial after Intracoronary Stenting[137] studies, which enrolled high-risk patients. Finally, the Stent Antico-agulation Restenosis study compared three antithrombotic regimens: aspirin alone (557 patients), aspirin and warfarin (550 patients), and aspirin and ticlopidine (546 patients).[138] All clinical events reflecting stent thrombosis were included in the prespecified primary end point: death, revasculariza-tion of the target lesion, angiographically evident thrombo-sis, or MI within 30 days. Treatment with aspirin plus ticlopidine resulted in a much lower rate of stent thrombosis compared to the other treatments (0.5 vs. 3.6% with aspirin alone and 2.6% with aspirin plus warfarin; $p < 0.001$ for the comparison of all three groups). Hemorrhagic complica-tions occurred in 10 patients (1.8%) who received aspirin alone, 34 (6.2%) who received aspirin and warfarin, and 30 (5.5%) who received aspirin and ticlopidine ($p < 0.001$ for all three groups). However, in this trial the combined aspirin–ticlopidine therapy, despite its superiority in reduc-ing stent thrombosis (which was still evident after 4 years of follow-up), again did not reduce the incidence of restenosis.

PCI-CURE, a prespecified substudy of the CURE trial, assessed the benefit of a preprocedural clopidogrel loading dose (300 mg) and long-term continuation of clopidogrel (75 mg) in patients who underwent percutaneous coronary intervention (PCI) during the study.[139] Patients who were randomized to clopidogrel experienced a 31% RRR for the composite of cardiovascular death or MI and an RRR of 30% for the composite of cardiovascular death, MI, or urgent target vessel revascularization within 30 days of PCI.[139] In addition, the Clopidogrel for the Reduction of Events during Observation trial showed that dual antiplatelet therapy should be continued beyond the usual 30 days because after 1 year of treatment, patients on dual therapy experienced a 27% relative risk reduction in death, MI, and stroke compared to patients who were assigned to aspirin alone after the first 30 days of treatment with clopidogrel and aspirin.[140]

The Percutaneous Coronary Intervention-Clopidogrel as Adjunctive Reperfusion Therapy (PCI-CLARITY) study was a prospectively planned analysis of the 1863 patients undergoing PCI after mandated angiography in CLARITY-TIMI 28, a randomized, double-blind, placebo-controlled trial of clopidogrel in patients receiving fibrinolytics for ST-elevation MI.[141] Patients received aspirin and were random-ized to either clopidogrel (300 mg loading dose, then 75 mg once daily) or placebo initiated with fibrinolysis and given until coronary angiography, which was performed 2 to 8 days after initiation of the study drug. The study showed that clopidogrel pretreatment significantly reduced the inci-dence of cardiovascular death or ischemic complications both before and after PCI and without a significant increase in major or minor bleeding.[141] In the Antiplatelet Therapy for Reduction of Myocardial Damage during Angioplasty-2 study, pretreatment with a 600-mg loading dose of clopidogrel 4 to 8 hours before the procedure, compared to the conventional 300-mg loading dose, significantly reduced periprocedural MI in patients undergoing PCI without increasing the incidence of side effects.[142]

The combination of aspirin and clopidogrel is so effective at reducing major adverse cardiovascular events in patients undergoing coronary artery revascularization that the addition of abciximab did not further reduce their incidence in low- to intermediate-risk patients[143] and in diabetics.[144]

Several studies have compared aspirin plus ticlopidine therapy with aspirin plus clopidrogrel after coronary stenting. A meta-analysis of most of these studies showed that clopidogrel, in addition to being better tolerated, was at least as efficacious as ticlopidine in reducing vascular events.[145] Therefore, it was concluded that clopidogrel plus aspirin should replace ticlopidine plus aspirin as the standard antiplatelet regimen after stent deployment.[145] This conclusion was challenged by the results of two studies: Long-term follow-up after stent implantation in patients receiving the traditional 2- to 4-week course of combined antiplatelet therapy revealed increased rates of mortality[146,147] and thrombotic stent occlusion[147] in patients given clopidogrel as opposed to ticlopidine. These alarming observations require confirmation from further studies.

**Peripheral Artery Disease.** The Swedish Ticlopidine Multicenter Study — a double-blind, placebo-controlled trial — showed that ticlopidine reduces the risk of MI, stroke, or TIA[148] and the requirement for leg vascular surgery[149] in patients with intermittent claudication. Ticlopidine has also been reported to improve the maximum walking distance in these patients[150] and the long-term patency of femoropopliteal or femorotibial saphenous vein bypass grafts in patients with peripheral artery disease.[151]

As discussed previously, in the 19,185-patient CAPRIE trial, the benefit associated with clopidogrel over aspirin therapy seemed to be limited to patients with previous peripheral artery disease.[71]

*h. Cost-Effectiveness.* An analysis based on the results of the CAPRIE study suggested that the cost-effectiveness of clopidogrel in these patients at risk of future vascular events is at best unattractive.[152] However, two subsequent analyses supported the use of clopidogrel as a clinically efficient and cost-effective option for secondary prevention of atherothrombotic disease, particularly in high-risk patients.[153,154] There is no controversy among the published studies regarding the favorable cost-effectiveness of clopidogrel in patients with acute coronary syndromes treated with medical treatment alone or in combination with coronary revascularization.[155–161]

*i. Resistance to Thienopyridines.* In the past few years, the issue of resistance to antiplatelet agents has been emphasized in the medical literature,[162] although its definition is still uncertain. The degree of inhibition of platelet function observed after administration of clopidogrel (or ticlopidine) varies widely among patients. However, there is no consensus on a cutoff value of platelet function inhibition for the definition of "clopidogrel resistance." In addition, there is no consensus on the type of screening laboratory test that should be used to measure individual response to the drug. The extent of the platelet aggregation response to ADP *in vitro* has been used in the majority of studies. However, the results of *in vitro* platelet aggregation are influenced by many preanalytical and analytical variables, and even when all of them are well standardized, the accuracy and reproducibility of the technique are very poor.[162] In addition, although ADP is the most appropriate aggregating agent in the context of studying the response to the $P2Y_{12}$ antagonist clopidogrel, it must be remembered that platelets also express a second ADP receptor, $P2Y_1$, which causes the initial wave of ADP-induced platelet aggregation. Since the extent of residual, $P2Y_1$-dependent platelet aggregation induced by ADP varies widely among patients with congenital $P2Y_{12}$ deficiency or normal subjects in whom $P2Y_{12}$ function has been completely blocked *in vitro* by saturating concentrations of specific antagonists, ADP-induced platelet aggregation may not be the most suitable test to measure the individual response to clopidogrel. A better and more specific test would be measurement of the extent of ADP-induced inhibition of adenylyl cyclase, which is uniquely mediated by $P2Y_{12}$. This could be accomplished by measuring the inhibition by ADP of prostaglandin-induced platelet cyclic adenosine monophosphate increase or phosphorylation of vasodilator-stimulated phosphoprotein,[163] which has proved to be a reliable test to measure the variability of response to clopidogrel.[164–167]

In published studies, a variable proportion (up to 50%) of patients were either clopidogrel nonresponders or low responders, depending on the method of assessment and the definition of nonresponsiveness, even when a 300-mg loading dose was used.[164–169] A loading dose of 600 mg reduced the frequency of nonresponsiveness but did not eliminate it.[170]

Interindividual differences in the extent of metabolism of the prodrug to the active metabolite are the most plausible mechanism for the observed variability in platelet inhibition by clopidogrel. In fact, the extent of inhibition of ADP-induced platelet aggregation by clopidogrel correlated well with the metabolic activity of the hepatic cytochrome P450, which activates the prodrug to its active metabolite.[168] Differences in drug absorption[171] and the presence of the PL[A2] allele of GPIIIa[172] are probably involved, whereas known polymorphisms of $P2Y_{12}$ do not seem to play additional roles in modulating the individual response.[166,173] Interference with clopidogrel metabolism by other drugs that are frequently given to patients with atherosclerosis, such as atorvastatin[90,174] or other statins,[175,176] may result in an increased number of patients who are resistant to clopidogrel, although this is still a controversial issue.[33,177–182]

Preliminary, small studies demonstrated an association between insufficient platelet function inhibition by clopidogrel and increased incidence of vascular events.[164,183] However, further studies are needed to clarify better the clinical relevance of clopidogrel resistance. "Clopidogrel resistance" is also discussed in Chapter 23.

## 2. Prasugrel

*a. Pharmacokinetics and Pharmacodynamics.* Prasugrel, previously known as CS-747 (2-acetoxy-5-[α-cyclopropylcarbonyl-2-fluorobenzyl]-4,5,6,7-tetrahydrothieno [3,2-c]pyridine), is a thienopyridyl prodrug that is metabolized *in vivo* to the active platelet-inhibitory metabolite, R-99224/R-138727, with specific $P2Y_{12}$ receptor antagonistic activity (Fig. 61–1).[184–186]

When administered orally to experimental animals, prasugrel caused dose-related inhibition of platelet aggregation, with an approximately 10- and 100-fold higher potency than those of the other thienopyridyl prodrugs, clopidogrel and ticlopidine, respectively. In addition, prasugrel has a more rapid onset of action than clopidogrel. A time course study of the antiaggregatory effects after administration of prasugrel (1–10 mg/kg, PO) and clopidogrel (10–100 mg/kg, PO) showed that more than 80% inhibition was observed in prasugrel (10 mg/kg)-treated rats 30 minutes after the dosing, whereas at the same time point clopidogrel exhibited minimal inhibition even at the highest dose used, suggesting earlier onset of action of prasugrel compared to clopidogrel.[186] The duration of the antiaggregatory effects of prasugrel is comparable to the life span of circulating platelets in the rat, suggesting that, like other thienopyridyl compounds, prasugrel irreversibly inhibits platelet $P2Y_{12}$ receptors.

The rank order of the antithrombotic and antihemostatic potencies (evaluated using a rat arteriovenous shunt thrombosis model and the rat tail transection bleeding time) among prasugrel, clopidogrel, and ticlopidine agents was the same as the rank order for ADP antiaggregatory potencies.[186] Prasugrel also proved to be a potent antithrombotic agent in other models of experimental thrombosis.[186] Further studies on experimental animals suggested that the combination of prasugrel with aspirin may provide a greater antithrombotic effect than either agent alone without a concomitant, significant increase in the bleeding time.[186]

*b. Clinical Pharmacology.* A single oral administration of escalating doses of prasugrel to healthy male volunteers dose-dependently inhibited platelet aggregation induced by 5 μM ADP. Although 2.5- and 10-mg doses of prasugrel caused a modest inhibition, the higher doses of 30 and 75 mg

achieved approximately 70% inhibition of platelet aggregation. The onset of action of prasugrel was rapid, with the maximal level of inhibition for each dose being achieved within a few hours after dosing.[186]

In a multiple-dose study, healthy volunteers were administered 2.5 or 10 mg prasugrel orally at 24-hour intervals for 10 days. Inhibition of 5 μM ADP-induced platelet aggregation accumulated over the days of dosing, reaching a steady state by days 2 to 4, at which time 2.5 mg provided approximately 40% inhibition, whereas 10 mg resulted in approximately 60% inhibition, which is equal to or greater than that expected after the administration of 75 mg clopidogrel.[186] In a crossover study, it was demonstrated that a 60-mg loading dose of prasugrel provided rapid and high-grade inhibition of ADP-induced platelet aggregation even in those subjects who responded poorly to a standard loading dose of clopidogrel.[187] It has been demonstrated that the increased *in vivo* potency of prasugrel compared to clopidogrel reflects more efficient conversion of the prodrug to the active metabolite.[188]

In a phase II, randomized, dose-ranging, double-blind safety trial (JUMBO-TIMI 26), three doses of prasugrel were compared to clopidogrel in 904 patients undergoing elective or urgent PCI.[189] Clopidogrel was administered at standard doses (300-mg loading dose followed by 75 mg daily), whereas prasugrel was given at low dose (40-mg loading dose followed by 7.5 mg daily), intermediate dose (60-mg loading dose followed by 10 mg daily), and high dose (60-mg loading dose followed by 15 mg daily). The incidence of a clinically significant (TIMI major plus minor) bleeding event, which was the primary end point of the trial, was not significantly different between patients treated with prasugrel and those treated with clopidogrel (1.7 vs. 1.2%; HR, 1.42; 95% CI, 0.40–5.08). However, there was more minimal bleedings in the high-dose prasugrel group (3.6%) compared to the low-dose (2.0%) and intermediate-dose (1.5%) groups and the clopidogrel group (2.4%). In prasugrel-treated patients, there were numerically lower incidences of the primary efficacy end point (30-day major adverse cardiac events) and of the secondary end points of MI, recurrent ischemia, and clinical target vessel thrombosis.

Prasugrel is currently undergoing phase III evaluation (TRITON TIMI-38 trial) in patients with acute coronary syndromes undergoing PCI. This study will determine whether the more rapid and potent platelet inhibition achievable with prasugrel provides superior benefit over the approved dose of clopidogrel in a safe and tolerable fashion.

## B. Direct $P2Y_{12}$ Antagonists

### 1. Cangrelor

*a. Pharmacokinetics and Pharmacodynamics.* Cangrelor (N6-[2-methylthioethyl]-2-[3,3,3-trifluoropropylthio]-5′-

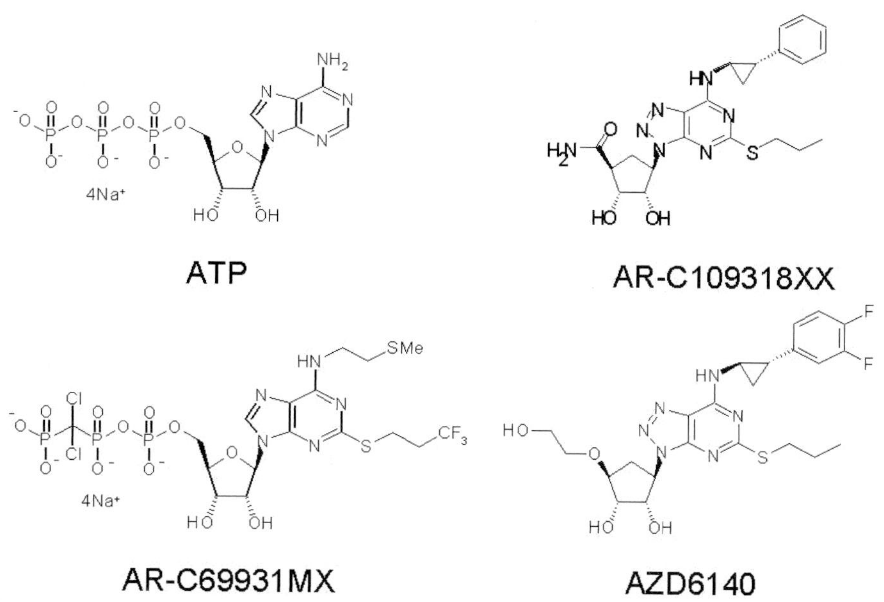

**Figure 61-2.** Chemical structures of ATP and its analogs with high affinity for the P2Y₁₂ receptor: AR-C69931MX (cangrelor), AR-C109318XX, and AZD6140.

adenylic acid, monoanhydride with dichloromethylenebis [phosphonic acid]), previously known as ARC69931[MX], belongs to a family of analogs of ATP that are relatively resistant to breakdown by ectonucleotidases and display high affinity for the P2Y₁₂ receptor (Fig. 61-2).[190] They are characterized by the presence of substituents in the 2-position of the adenine ring that increase the affinity properties of the molecule and of β,γ-methylene substitutions in the triphosphate part of the molecule, which increase the metabolic stability.[190] Cangrelor is a potent inhibitor of ADP-induced aggregation of human washed platelets (pIC₅₀ of 9.4 with 30 μM ADP). Initial studies showed that cangrelor displays no significant affinity for other P2 receptors at concentrations >30 μM.[189] However, further studies showed that it is a potent (IC₅₀, 4 nM), noncompetitive antagonist of P2Y₁₃, which is not expressed on the platelet membrane.[191,192]

*b. Experimental Thrombosis.* In a model of cyclic flow reductions in the femoral artery of anesthetized male beagle dogs, cangrelor showed a better separation between antithrombotic effect and prolongation of the tongue bleeding time compared to clopidogrel. Both drugs behaved much better than the GPIIb-IIIa antagonist orbofiban.[190] Cangrelor also displayed good antithrombotic effects in other models of experimental thrombosis, including a carotid artery thrombosis model in beagles[193] and thrombosis of the rabbit mesenteric arteries induced by vessel wall puncture[194] and, given in combination with thrombolytic therapy, in a canine coronary electronic injury thrombosis model.[195]

*c. Clinical Pharmacology.* Intravenous infusion of cangrelor was well tolerated in healthy volunteers, resulting in dose-dependent inhibition of ADP-induced platelet aggregation at doses up to 4 μg/kg/min[190] and, at the highest dose, in a 3.2- or 2.9-fold increase in the bleeding time in men and women. The short half-life (mean, 2.6 minutes) resulted in a rapid reversal of both the platelet inhibitory effect and the effect on the bleeding time within 20 minutes after cessation of the infusion.

An open multicenter ascending-dose study of 39 acute coronary syndrome patients[196] showed dose-dependent and predictable plasma levels of cangrelor, resulting in complete or near complete inhibition of 3 μM ADP-induced platelet aggregation in all patients at doses of 2 μg/kg/min or higher. Bleeding times were increased three- to fivefold at the 2 μg/kg/min dose and approximately sevenfold at the 4 μg/kg/min dose. Although trivial bleeding was common in these patients (22/39), no major or minor bleeding was recorded as defined by the TIMI criteria.

A double-blind, placebo-controlled study of cangrelor as adjunctive therapy to aspirin and either heparin or low-molecular-weight heparin in patients with non-Q-wave MI showed that minor bleeding was slightly increased from 26% in the placebo-treated group to 38% in the cangrelor-treated group.[197]

*d. Comparison of Cangrelor with Thienopyridines.* In a study that directly compared the effects of clopidogrel and cangrelor administration in patients with ischemic heart disease, Storey et al.[198] showed that cangrelor infusion at 2 and 4 μg/ml/min resulted in near complete inhibition of platelet aggregation measured at 4 minutes after the addition of 10 μM ADP, whereas 4 to 7 days of clopidogrel treatment resulted in only approximately 60% inhibition. Addition of cangrelor *in vitro* to blood from the clopidogrel-treated patients resulted in near complete inhibition of $P2Y_{12}$-dependent platelet function.[199]

## 2. AZD6140

*a. Pharmacokinetics and Pharmacodynamics.* AZD6140, (1S,2S,3R,5S)-3-{[(1R,2S)-2-(3,4-difluorophenyl)cyclopropyl]amino}-5-(propylthio)-3H-[1,2,3,]-triazolo[4,5-d]pyrimidin-3-yl]-5-(2-hydroxyethoxy)cyclopentane-1,2-diol, belongs to the same family as cangrelor of stable ATP analogs with high affinity for $P2Y_{12}$ (Fig. 61-2). Efforts to identify compounds for oral administration led to the discovery of AR-C109318XX, the first selective and stable, nonphosphate, competitive $P2Y_{12}$ antagonist. Further refinement to increase the oral bioavailability resulted in the discovery of the selective $P2Y_{12}$ competitive antagonist AZD6140, which is currently being developed as an oral antiplatelet agent. AR-C109318XX and AZD6140 both belong to the new chemical class cyclopentyl-triazolo-pyrimidines (Fig. 61-2).[190]

AZD6140 is a less potent antagonist than cangrelor, with a $pIC_{50}$ of 7.9 for inhibition of 30 µM ADP-induced aggregation of human washed platelets. AZD6140 displays no significant affinity for other P2 receptors at concentrations >3 µM.[190]

*b. Experimental Thrombosis.* In a model of cyclic flow reductions in the femoral artery of anesthetized male beagles, AZC6140 displayed good separation between the antithrombotic effect and the prolongation of the tongue bleeding time, which was intermediate between that of cangrelor and clopidogrel.[190] Both drugs behaved much better than the GPIIb-IIIa antagonist orbofiban.[190]

*c. Clinical Pharmacology.* In a randomized, double-blind, parallel-group study, 200 stable atherosclerotic outpatients, on treatment with aspirin 75 to 100 mg once daily, received AZD6140 (50 mg b.i.d., 100 mg b.i.d., 200 mg b.i.d., or 400 mg once daily) or clopidogrel 75 mg once daily for 28 days. The study showed that AZD6140 at doses higher than 50 mg b.i.d. more effectively inhibited platelet aggregation and with less variability than clopidogrel. In addition, the inhibition of platelet aggregation by AZD6140 was very rapid (2 hours postdose, 96 ± 6.1% for 400 mg once daily) compared to that of clopidogrel. Only one major, nonfatal hemorrhage occurred in a patient treated with AZD6140 400 mg once daily.[200]

## IV. $P2Y_1$ Antagonists

The demonstration that $P2Y_1$ knockout mice display mild prolongation of the bleeding time, absence of ADP-induced platelet shape change and aggregation, and resistance to

experimental thrombosis[201,202] elicited a search for specific $P2Y_1$ receptor antagonists that could be used as potential antithrombotic drugs in human disease. The infusion of wild-type mice with MRS2179, the first $P2Y_1$ antagonist tested *in vivo* in models of experimental thrombosis, reproduced the results obtained in $P2Y_1$ knockout mice.[203,204] MRS2179 also reduced thrombus formation in mouse mesenteric arteries that had been injured by ferric chloride.[205] Studies of a much more potent and stable $P2Y_1$ antagonist, MRS2500,[206] showed that it strongly inhibited both systemic thromboembolism after the intravenous injection of collagen and epinephrine and thrombus formation in laser-injured mesenteric arteries.[207]

Based on studies of $P2Y_1$ knockout mice treated with clopidogrel,[201] it is possible that a combination of P2 receptor antagonists could improve antithrombotic strategies.

## V. Conclusions

Several lines of evidence indicate that the interaction of ADP with its platelet receptors plays a very important role in thrombogenesis. The thienopyridine ticlopidine was the first specific antagonist of the platelet $P2Y_{12}$ ADP receptor to be tested in randomized clinical trials for the prevention of arterial thrombotic events. Ticlopidine reduces the incidence of vascular events in patients with a history of cerebrovascular or coronary artery diseases, but it also has important drawbacks: a relatively high incidence of toxic effects, which may be fatal in some cases; delayed onset of action; and high interindividual response variability. A second thienopyridine, clopidogrel, has superseded ticlopidine, particularly in the management of patients undergoing coronary stenting, because it is also an efficacious antithrombotic drug and is less toxic than ticlopidine. However, clopidogrel is not completely free from drawbacks: Severe toxic effects, although they occur much less frequently than with ticlopidine, may still complicate treatment with clopidogrel; the onset of pharmacologic action is accelerated by the use of large loading doses, but it still may not be optimal; and the high interpatient response variability remains an important issue. These drawbacks justify the continuing search for agents that can further improve the clinical outcome of patients with atherosclerosis through greater efficacy and/or safety. A new thienopyridyl compound, prasugrel, which is characterized by higher potency and faster onset of action compared to clopidogrel, is currently under clinical evaluation. Two direct and reversible $P2Y_{12}$ antagonists, cangrelor and AZD6140, have very rapid onset and reversal of platelet inhibition, which make them attractive alternatives to thienopyridines, especially when rapid inhibition of platelet aggregation or its quick reversal are required. Along with new $P2Y_{12}$ antagonists, inhibitors of the other platelet receptor for ADP, $P2Y_1$, are under development and may prove to be effective antithrombotic agents.

# References

1. Ralevic, V., & Burnstock, G. (1998). Receptors for purines and pyrimidines. *Pharmacol Rev, 50,* 413–492.

2. Cattaneo, M., & Gachet, C. (1999). ADP receptors and clinical bleeding disorders. *Arterioscler Thromb Vasc Biol, 19,* 2281–2285.

3. Shah, J., Teitelbaum, P., Molony, B., et al. (1991). Single and multiple dose pharmacokinetics of ticlopidine in young and elderly subjects. *Br J Clin Pharmacol, 32,* 761–764.

4. Shah, J., Fratis, A., Ellis, D., et al. (1990). Effect of food and antacid on absorption of orally administered ticlopidine hydrochloride. *J Clin Pharmacol, 30,* 733–736.

5. Caplain, H., Donat, F., Gaud, C., et al. (1999). Pharmacokinetics of clopidogrel. *Semin Thromb Hemost, 25*(Suppl. 2), 25–28.

6. McEwen, J., Strauch, G., Perles, P., et al. (1999). Clopidogrel bioavailability: Absence of influence of food or antacids. *Semin Thromb Hemost, 25*(Suppl. 2), 47–50.

7. Lins, R., Broekhuysen, J., Necciari, J., et al. (1999). Pharmacokinetic profile of 14C-labeled clopidogrel. *Semin Thromb Hemost, 25*(Suppl. 2), 29–33.

8. Savi, P., & Herbert, J. M. (2005). Clopidogrel and ticlopidine: P2Y12 adenosine diphosphate-receptor antagonists for the prevention of atherothrombosis. *Semin Thromb Hemost, 31,* 174–183.

9. Savi, P., Combalbert, J., Gaich, C., et al. (1994). The antiaggregating activity of clopidogrel is due to a metabolic activation by the hepatic cytochrome P450-1A. *Thromb Haemost, 72,* 313–317.

10. Dalvie, D. K., & O'Connell, T. N. (2004). Characterization of novel dihydrothienopyridinium and thienopyridinium metabolites of ticlopidine *in vitro:* Role of peroxidases, cytochromes p450, and monoamine oxidases. *Drug Metab Dispos, 32,* 49–57.

11. Ha-Duong, N. T., Dijols, S., Macherey, A. C., et al. (2001). Ticlopidine as a selective mechanism-based inhibitor of human cytochrome P450 2C19. *Biochemistry, 40,* 12112–12122.

12. Weber, A. A., Reimann, S., & Schror, K. (1999). Specific inhibition of ADP-induced platelet aggregation by clopidogrel *in vitro. Br J Pharmacol, 126,* 415–420.

13. Herbert, J. M., & Savi, P. (1999). Non-specific inhibition of ADP-induced platelet antiaggregation by clopidogrel *in vitro. Thromb Haemost, 82,* 156–157.

14. Gachet, C., Cazenave, J. P., Ohlmann, P., et al. (1990). The thienopyridine ticlopidine selectively prevents the inhibitory effects of ADP but not of adrenaline on cAMP levels raised by stimulation of the adenylate cyclase of human platelets by PGE1. *Biochem Pharmacol, 40,* 2683–2687.

15. Cattaneo, M., Akkawat, B., Lecchi, A., et al. (1991). Ticlopidine selectively inhibits human platelet responses to adenosine diphosphate. *Thromb Haemost, 66,* 694–699.

16. Gachet, C., Savi, P., Ohlmann, P., et al. (1992). ADP receptor induced activation of guanine nucleotide binding proteins in rat platelet membranes — An effect selectively blocked by the thienopyridine clopidogrel. *Thromb Haemost, 68,* 79–83.

17. Gachet, C., Stierle, A., Cazenave, J. P., et al. (1990). The thienopyridine PCR 4099 selectively inhibits ADP-induced platelet aggregation and fibrinogen binding without modifying the membrane glycoprotein IIb-IIIa complex in rat and in man. *Biochem Pharmacol, 40,* 229–238.

18. Defreyn, G., Gachet, C., Savi, P., et al. (1991). Ticlopidine and clopidogrel (SR 25990C) selectively neutralize ADP inhibition of PGE1-activated platelet adenylate cyclase in rats and rabbits. *Thromb Haemost, 65,* 186–190.

19. Lips, N. P., Sixma, J. J., & Schiphorst, M. E. (1980). The effect of ticlopidine administration to humans on the binding of adenosine diphosphate to blood platelets. *Thromb Res, 17,* 19–27.

20. Mills, D. C., Puri, R., Hu, C. J., et al. (1992). Clopidogrel inhibits the binding of ADP analogues to the receptor mediating inhibition of platelet adenylate cyclase. *Arterioscler Thromb, 12,* 430–436.

21. Savi, P., Laplace, M. C., Maffrand, J. P., et al. (1994). Binding of [3H]-2-methylthio ADP to rat platelets — Effect of clopidogrel and ticlopidine. *J Pharmacol Exp Ther, 269,* 772–777.

22. Gachet, C., Cattaneo, M., Ohlmann, P., et al. (1995). Purinoceptors on blood platelets: Further pharmacological and clinical evidence to suggest the presence of two ADP receptors. *Br J Haematol, 91,* 434–444.

23. Feliste, R., Delebassee, D., Simon, M. F., et al. (1987). Broad spectrum anti-platelet activity of ticlopidine and PCR 4099 involves the suppression of the effects of released ADP. *Thromb Res, 48,* 403–415.

24. Cattaneo, M., Akkawat, B., Kinlough-Rathbone, R. L., et al. (1994). Ticlopidine facilitates the deaggregation of human platelets aggregated by thrombin. *Thromb Haemost, 71,* 91–94.

25. Cattaneo, M., Lombardi, R., Bettega, D., et al. (1993). Shear-induced platelet aggregation is potentiated by desmopressin and inhibited by ticlopidine. *Arterioscler Thromb, 13,* 393–397.

26. Orford, J. L., Kinlay, S., Adams, M. R., et al. (2002). Clopidogrel inhibits shear-induced platelet function. *Platelets, 13,* 187–189.

27. Kuzniar, J., Splawinska, B., Malinga, K., et al. (1996). Pharmacodynamics of ticlopidine: Relation between dose and time of administration to platelet inhibition. *Int J Clin Pharmacol Ther, 34,* 357–361.

28. Buur, T., Larsson, R., Berglund, U., et al. (1997). Pharmacokinetics and effect of ticlopidine on platelet aggregation in subjects with normal and impaired renal function. *J Clin Pharmacol, 37,* 108–115.

29. Thebault, J. J., Kieffer, G., & Cariou, R. (1999). Single-dose pharmacodynamics of clopidogrel. *Semin Thromb Hemost, 25*(Suppl. 2), 3–8.

30. Savcic, M., Hauert, J., Bachmann, F., et al. (1999). Clopidogrel loading dose regimens: Kinetic profile of pharmacody-

namic response in healthy subjects. *Semin Thromb Hemost,* *25*(Suppl. 2), 15–19.

31. Gawaz, M., Seyfarth, M., Muller, I., et al. (2001). Comparison of effects of clopidogrel versus ticlopidine on platelet function in patients undergoing coronary stent placement. *Am J Cardiol, 87,* 332–336, A9.

32. Helft, G., Osende, J. I., Worthley, S. G., et al. (2000). Acute antithrombotic effect of a front-loaded regimen of clopidogrel in patients with atherosclerosis on aspirin. *Arterioscler Thromb Vasc Biol, 20,* 2316–2321.

33. Hochholzer, W., Trenk, D., Frundi, D., et al. (2005). Time dependence of platelet inhibition after a 600-mg loading dose of clopidogrel in a large, unselected cohort of candidates for percutaneous coronary intervention. *Circulation, 111,* 2560–2564.

34. Caplain, H., & Cariou, R. (1999). Long-term activity of clopidogrel: A three-month appraisal in healthy volunteers. *Semin Thromb Hemost, 25*(Suppl. 2), 21–24.

35. Thebault, J. J., Kieffer, G., Lowe, G. D., et al. (1999). Repeated-dose pharmacodynamics of clopidogrel in healthy subjects. *Semin Thromb Hemost, 25*(Suppl. 2), 9–14.

36. Denninger, M. H., Necciari, J., Serre-Lacroix, E., et al. (1999). Clopidogrel antiplatelet activity is independent of age and presence of atherosclerosis. *Semin Thromb Hemost, 25*(Suppl. 2), 41–45.

37. Slugg, P. H., Much, D. R., Smith, W. B., et al. (2000). Cirrhosis does not affect the pharmacokinetics and pharmacodynamics of clopidogrel. *J Clin Pharmacol, 40,* 396–401.

38. Savi, P., Pereillo, J. M., Uzabiaga, M. F., et al. (2000). Identification and biological activity of the active metabolite of clopidogrel. *Thromb Haemost, 84,* 891–896.

39. Yoneda, K., Iwamura, R., Kishi, H., et al. (2004). Identification of the active metabolite of ticlopidine from rat *in vitro* metabolites. *Br J Pharmacol, 142,* 551–571.

40. Ding, Z., Kim, S., Dorsam, R. T., et al. (2003). Inactivation of the human P2Y12 receptor by thiol reagents requires interaction with both extracellular cysteine residues, Cys17 and Cys270. *Blood, 101,* 3908–3914.

41. Pereillo, J. M., Maftouh, M., Andrieu, A., et al. (2002). Structure and stereochemistry of the active metabolite of clopidogrel. *Drug Metab Dispos, 30,* 1288–1295.

42. Palareti, G., Poggi, M., Torricelli, P., et al. (1988). Long-term effects of ticlopidine on fibrinogen and haemorheology in patients with peripheral arterial disease. *Thromb Res, 52,* 621–629.

43. Kroft, L. J., de Maat, M. P., & Brommer, E. J. (1993). The effect of ticlopidine upon plasma fibrinogen levels in patients undergoing suprapubic prostatectomy. *Thromb Res, 70,* 349–354.

44. de Maat, M. P., Arnold, A. E., van Buuren, S., et al. (1996). Modulation of plasma fibrinogen levels by ticlopidine in healthy volunteers and patients with stable angina pectoris. *Thromb Haemost, 76,* 166–170.

45. Mazoyer, E., Ripoll, L., Boisseau, M. R., et al. (1994). How does ticlopidine treatment lower plasma fibrinogen? *Thromb Res, 75,* 361–370.

46. Hayakawa, M., & Kuzuya, F. (1991). Effects of ticlopidine on erythrocyte aggregation in thrombotic disorders. *Angiology, 42,* 747–753.

47. de Lorgeril, M., Bordet, J. C., Salen, P., et al. (1998). Ticlopidine increases nitric oxide generation in heart-transplant recipients: A possible novel property of ticlopidine. *J Cardiovasc Pharmacol, 32,* 225–230.

48. Savi, P., Bernat, A., Dumas, A., et al. (1994). Effect of aspirin and clopidogrel on platelet-dependent tissue factor expression in endothelial cells. *Thromb Res, 73,* 117–124.

49. Piovella, F., Ricetti, M. M., Almasio, P., et al. (1984). The effect of ticlopidine on human endothelial cells in culture. *Thromb Res, 33,* 323–332.

50. Paradiso-Hardy, F. L., Angelo, C. M., Lanctot, K. L., et al. (2000). Hematologic dyscrasia associated with ticlopidine therapy: Evidence for causality. *Can Med Assoc J, 163,* 1441–1448.

51. Symeonidis, A., Kouraklis-Symeonidis, A., Seimeni, U., et al. (2002). Ticlopidine-induced aplastic anemia: Two new case reports, review, and meta-analysis of 55 additional cases. *Am J Hematol, 71,* 24–32.

52. Bennett, C. L., Weinberg, P. D., Rozenberg-Ben-Dror, K., et al. (1998). Thrombotic thrombocytopenic purpura associated with ticlopidine. A review of 60 cases. *Ann Intern Med, 128,* 541–544.

53. Chen, D. K., Kim, J. S., & Sutton, D. M. (1999). Thrombotic thrombocytopenic purpura associated with ticlopidine use: A report of 3 cases and review of the literature. *Arch Intern Med, 159,* 311–314.

54. Steinhubl, S. R., Tan, W. A., Foody, J. M., et al. (1999). Incidence and clinical course of thrombotic thrombocytopenic purpura due to ticlopidine following coronary stenting. EPISTENT Investigators. Evaluation of Platelet IIb/IIIa Inhibitor for Stenting. *J Am Med Assoc, 281,* 806–810.

55. Tsai, H. M., Rice, L., Sarode, R., et al. (2000). Antibody inhibitors to von Willebrand factor metalloproteinase and increased binding of von Willebrand factor to platelets in ticlopidine-associated thrombotic thrombocytopenic purpura. *Ann Intern Med, 132,* 794–799.

56. Mauro, M., Zlatopolskiy, A., Raife, T. J., et al. (2004). Thienopyridine-linked thrombotic microangiopathy: Association with endothelial cell apoptosis and activation of MAP kinase signalling cascades. *Br J Haematol, 124,* 200–210.

57. Skurnik, Y. D., Tcherniak, A., Edlan, K., et al. (2003). Ticlopidine-induced cholestatic hepatitis. *Ann Pharmacother, 37,* 371–375.

58. Martinez Perez-Balsa, A., De Arce, A., Castiella, A., et al. (1998). Hepatotoxicity due to ticlopidine. *Ann Pharmacother, 32,* 1250–1251.

59. McTavish, D., Faulds, D., & Goa, K. L. (1990). Ticlopidine. An updated review of its pharmacology and therapeutic use in platelet-dependent disorders. *Drugs, 40,* 238–259.

60. Berrebi, D., Sautet, A., Flejou, J. F., et al. (1998). Ticlopidine induced colitis: A histopathological study including apoptosis. *J Clin Pathol, 51,* 280–283.

61. Dakik, H. A., Salti, I., Haidar, R., et al. (2002). Drug points: Ticlopidine associated with acute arthritis. *Br Med J, 324,* 27.

62. Bennett, C. L., Connors, J. M., Carwile, J. M., et al. (2000). Thrombotic thrombocytopenic purpura associated with clopidogrel. *N Engl J Med, 342,* 1773–1777.

63. Andersohn, F., Hagmann, F. G., & Garbe, E. (2004). Thrombotic thrombocytopenic purpura/haemolytic uraemic syndrome associated with clopidogrel: Report of two new cases. *Heart, 90,* e57.

64. Hankey, G. J. (2000). Clopidogrel and thrombotic thrombocytopenic purpura. *Lancet, 356,* 269–270.

65. Manor, S. M., Guillory, G. S., & Jain, S. P. (2004). Clopidogrel-induced thrombotic thrombocytopenic purpura–hemolytic uremic syndrome after coronary artery stenting. *Pharmacotherapy, 24,* 664–667.

66. Paradiso-Hardy, F. L., Papastergiou, J., Lanctot, K. L., et al. (2002). Thrombotic thrombocytopenic purpura associated with clopidogrel: Further evaluation. *Can J Cardiol, 18,* 771–773.

67. Zakarija, A., Bandarenko, N., Pandey, D. K., et al. (2004). Clopidogrel-associated TTP: An update of pharmacovigilance efforts conducted by independent researchers, pharmaceutical suppliers, and the Food and Drug Administration. *Stroke, 35,* 533–537.

68. McCarthy, M. W., & Kockler, D. R. (2003). Clopidogrel-associated leukopenia. *Ann Pharmacother, 37,* 216–219.

69. Meyer, B., Staudinger, T., & Lechner, K. (2001). Clopidogrel and aplastic anaemia. *Lancet, 357,* 1446–1447.

70. Trivier, J. M., Caron, J., Mahieu, M., et al. (2001). Fatal aplastic anaemia associated with clopidogrel. *Lancet, 357,* 446.

71. CAPRIE Steering Committee. (1996). A randomised, blinded, trial of clopidogrel versus aspirin in patients at risk of ischaemic events (CAPRIE). *Lancet, 348,* 1329–1339.

72. Fork, F. T., Lafolie, P., Toth, E., et al. (2000). Gastroduodenal tolerance of 75 mg clopidogrel versus 325 mg aspirin in healthy volunteers. A gastroscopic study. *Scand J Gastroenterol, 35,* 464–469.

73. Braunwald, E., Antman, E. M., Beasley, J. W., et al.; American College of Cardiology/American Heart Association Task Force on Practice Guidelines (Committee on the Management of Patients with Unstable Angina). (2002). ACC/AHA guideline update for the management of patients with unstable angina and non-ST-segment elevation myocardial infarction — 2002: Summary article: A report of the American College of Cardiology/American Heart Association Task Force on Practice Guidelines (Committee on the Management of Patients with Unstable Angina). *Circulation, 106,* 1893–1900.

74. Chan, F. K., Ching, J. Y., Hung, L. C., et al. (2005). Clopidogrel versus aspirin and esomeprazole to prevent recurrent ulcer bleeding. *N Engl J Med, 352,* 238–244.

75. Doggrell, S. A. (2005). Aspirin and esomeprazole are superior to clopidogrel in preventing recurrent ulcer bleeding. *Expert Opin Pharmacother, 6,* 1253–1256.

76. Haj, M., Dasani, H., Kundu, S., et al. (2004). Acquired haemophilia A may be associated with clopidogrel. *Br Med J, 329,* 323.

77. Pierce, C. H., Houle, J. M., Dickinson, J. P., et al. (1999). Clopidogrel and drug metabolism: Absence of effect on hepatic enzymes in healthy volunteers. *Semin Thromb Hemost, 25*(Suppl. 2), 35–39.

78. Ascione, R., Ghosh, A., Rogers, C. A., et al. (2005). In-hospital patients exposed to clopidogrel before coronary artery bypass graft surgery: A word of caution. *Ann Thorac Surg, 79,* 1210–1216.

79. Chen, L., Bracey, A. W., Radovancevic, R., et al. (2004). Clopidogrel and bleeding in patients undergoing elective coronary artery bypass grafting. *J Thorac Cardiovasc Surg, 128,* 425–431.

80. Chu, M. W., Wilson, S. R., Novick, R. J., et al. (2004). Does clopidogrel increase blood loss following coronary artery bypass surgery? *Ann Thorac Surg, 78,* 1536–1541.

81. Hongo, R. H., Ley, J., Dick, S. E., et al. (2002). The effect of clopidogrel in combination with aspirin when given before coronary artery bypass grafting. *J Am Coll Cardiol, 40,* 231–237.

82. Kapetanakis, E. I., Medlam, D. A., Boyce, S. W., et al. (2005). Clopidogrel administration prior to coronary artery bypass grafting surgery: The cardiologist's panacea or the surgeon's headache? *Eur Heart J, 26,* 576–583.

83. Lindvall, G., Sartipy, U., & van der Linden, J. (2005). Aprotinin reduces bleeding and blood product use in patients treated with clopidogrel before coronary artery bypass grafting. *Ann Thorac Surg, 80,* 922–927.

84. van der Linden, J., Lindvall, G., & Sartipy, U. (2005). Aprotinin decreases postoperative bleeding and number of transfusions in patients on clopidogrel undergoing coronary artery bypass graft surgery: A double-blind, placebo-controlled, randomized clinical trial. *Circulation, 112*(9 Suppl.), I276–I280.

85. Buresly, K., Eisenberg, M. J., Zhang, X., et al. (2005). Bleeding complications associated with combinations of aspirin, thienopyridine derivatives, and warfarin in elderly patients following acute myocardial infarction. *Arch Intern Med, 165,* 784–789.

86. Mannucci, P. M., Vicente, V., Vianello, L., et al. (1986). Controlled trial of desmopressin in liver cirrhosis and other conditions associated with a prolonged bleeding time. *Blood, 67,* 1148–1153.

87. Caplain, H., Thebault, J. J., & Necciari, J. (1999). Clopidogrel does not affect the pharmacokinetics of theophylline. *Semin Thromb Hemost, 25*(Suppl. 2), 65–68.

88. Forbes, C. D., Lowe, G. D., MacLaren, M., et al. (1999). Clopidogrel compatibility with concomitant cardiac co-medications: A study of its interactions with a beta-blocker and a calcium uptake antagonist. *Semin Thromb Hemost, 25*(Suppl. 2), 55–60.

89. Lale, A., Herbert, J. M., & Savi, P. (2003). The antiaggregating activity of clopidogrel is not affected by N-acetyl L-cysteine. *Thromb Haemost, 90,* 839–843.

90. Lau, W. C., Waskell, L. A., Watkins, P. B., et al. (2003). Atorvastatin reduces the ability of clopidogrel to inhibit

platelet aggregation: A new drug–drug interaction. *Circulation, 107,* 32–37.

91. Bossavy, J. P., Thalamas, C., Sagnard, L., et al. (1998). A double-blind randomized comparison of combined aspirin and ticlopidine therapy versus aspirin or ticlopidine alone on experimental arterial thrombogenesis in humans. *Blood, 92,* 1518–1525.

92. Cadroy, Y., Bossavy, J. P., Thalamas, C., et al. (2000). Early potent antithrombotic effect with combined aspirin and a loading dose of clopidogrel on experimental arterial thrombogenesis in humans. *Circulation, 101,* 2823–2828.

93. Grau, A. J., Reiners, S., Lichy, C., et al. (2003). Platelet function under aspirin, clopidogrel, and both after ischemic stroke: A case-crossover study. *Stroke, 34,* 849–854.

94. Jagroop, I. A., Matsagas, M. I., Geroulakos, G., et al. (2004). The effect of clopidogrel, aspirin and both antiplatelet drugs on platelet function in patients with peripheral arterial disease. *Platelets, 15,* 117–125.

95. Moshfegh, K., Redondo, M., Julmy, F., et al. (2000). Antiplatelet effects of clopidogrel compared with aspirin after myocardial infarction: Enhanced inhibitory effects of combination therapy. *J Am Coll Cardiol, 36,* 699–705.

96. Rupprecht, H. J., Darius, H., Borkowski, U., et al. (1998). Comparison of antiplatelet effects of aspirin, ticlopidine, or their combination after stent implantation. *Circulation, 97,* 1046–1052.

97. Serebruany, V. L., Malinin, A. I., Jerome, S. D., et al. (2003). Effects of clopidogrel and aspirin combination versus aspirin alone on platelet aggregation and major receptor expression in patients with heart failure: The Plavix Use for Treatment of Congestive Heart Failure (PLUTO-CHF) trial. *Am Heart J, 146,* 713–720.

98. Splawinska, B., Kuzniar, J., Malinga, K., et al. (1996). The efficacy and potency of antiplatelet activity of ticlopidine is increased by aspirin. *Int J Clin Pharmacol Ther, 34,* 352–356.

99. Thebault, J. J., Blatrix, C. E., Blanchard, J. F., et al. (1977). The interactions of ticlopidine and aspirin in normal subjects. *J Int Med Res, 5,* 405–411.

100. van de Loo, A., Nauck, M., Noory, E., et al. (1998). Enhancement of platelet inhibition of ticlopidine plus aspirin vs aspirin alone given prior to elective PTCA. *Eur Heart J, 19,* 96–102.

101. Umemura, K., Kondo, K., Ikeda, Y., et al. (1997). Enhancement by ticlopidine of the inhibitory effect on *in vitro* platelet aggregation of the glycoprotein IIb/IIIa inhibitor tirofiban. *Thromb Haemost, 78,* 1381–1384.

102. Fredrickson, B. J., Turner, N. A., Kleiman, N. S., et al. (2000). Effects of abciximab, ticlopidine, and combined abciximab/ticlopidine therapy on platelet and leukocyte function in patients undergoing coronary angioplasty. *Circulation, 101,* 1122–1129.

103. Dalby, M., Montalescot, G., Bal dit Sollier, C., et al. (2004). Eptifibatide provides additional platelet inhibition in non-ST-elevation myocardial infarction patients already treated with aspirin and clopidogrel. Results of the Platelet Activity Extinction in Non-Q-Wave Myocardial Infarction with Aspirin, Clopidogrel, and Eptifibatide (PEACE) study. *J Am Coll Cardiol, 43,* 162–168.

104. Freund, M., Mantz, F., Nicolini, P., et al. (1993). Experimental thrombosis on a collagen coated artrioarterial shunt in rats: A pharmacological model to study antithrombotic agents inhibiting thrombin formation and platelet deposition. *Thromb Haemost, 69,* 515–521.

105. Leger, P., Magues, J. P., Freund, M., et al. (1999). Comparison of the antithrombotic effects of heparin, hirudin and clopidogrel according to the nature of the thrombogenic surface in an arterioarterial shunt in rats. *Thromb Haemost, 82,* 1203–1204.

106. Bernat, A., Vallee, E., Maffrand, J. B., et al. (1988). The role of platelets and ADP in experimental thrombosis induced by venous stasis in the rat. *Thromb Res, 52,* 65–70.

107. Herbert, J. M., Bernat, A., & Maffrand, J. B. (1992). Importance of platelets in experimental venous thrombosis in the rat. *Blood, 80,* 2281–2286.

108. Maffrand, J. P., Bernat, A., Delebassee, D., et al. (1988). ADP plays a key role in thrombogenesis in rats. *Thromb Haemost, 59,* 225–230.

109. Hérault, J. P., Dol, F., Gaich, C., et al. (1999). Effect of clopidogrel on thrombin generation in platelet-rich plasma in the rat. *Thromb Haemost, 81,* 957–960.

110. Léon, C., Ravanat, C., Freund, M., et al. (2003). Differential invovement of the P2Y1 and P2Y12 receptors in platelet procoagulant activity. *Arterioscler Thromb Vasc Biol, 23,* 1941–1947.

111. Léon, C., Alex, M., Klocke, A., et al. (2004). Platelet ADP receptors contribute to the initiation of intravascular coagulation. *Blood, 103,* 594–600.

112. Harker, L. A., Marzec, U. M., Kelly, A. B., et al. (1998). Clopidogrel inhibition of stent, graft, and vascular thrombogenesis with antithrombotic enhancement by aspirin in non-human primates. *Circulation, 98,* 2461–2469.

113. Herbert, J. M., Bernat, A., Samama, M., et al. (1996). The antiaggregating and antithrombotic activity of ticlopidine is potentiated by aspirin in the rat. *Thromb Haemost, 76,* 94–98.

114. Herbert, J. M., Dol, F., Bernat, A., et al. (1998). The antiaggregating and antithrombotic activity of clopidogrel is potentiated by aspirin in several experimental models in the rabbit. *Thromb Haemost, 80,* 512–518.

115. Bhatt, D. L., Chew, D. P., Hirsch, A. T., et al. (2001). Superiority of clopidogrel versus aspirin in patients with prior cardiac surgery. *Circulation, 103,* 363–368.

116. Hirsh, J., & Bhatt, D. L. (2004). Comparative benefits of clopidogrel and aspirin in high-risk patient populations: Lessons from the CAPRIE and CURE studies. *Arch Intern Med, 164,* 2106–2110.

117. Ringleb, P. A., Bhatt, D. L., Hirsch, A. T., et al.; Clopidogrel Versus Aspirin in Patients at Risk of Ischemic Events Investigators. (2004). Benefit of clopidogrel over aspirin is amplified in patients with a history of ischemic events. *Stroke, 35,* 528–532.

118. Hankey, G. J. (2005). Is clopidogrel the antiplatelet drug of choice for high-risk patients with stroke/TIA? No. *J Thromb Haemost, 3,* 1137–1140.

119. Gent, M., Blakely, J. A., Easton, J. D., et al. (1989). The Canadian American Ticlopidine Study (CATS) in thromboembolic stroke. *Lancet, 1,* 1215–1220.

120. Hass, W. K., Easton, J. D., Adams, H. P., Jr., et al. (1989). A randomized trial comparing ticlopidine hydrochloride with aspirin for the prevention of stroke in high-risk patients. Ticlopidine Aspirin Stroke Study Group. *N Engl J Med, 321,* 501–507.

121. Harbison, J. W. (1992). Ticlopidine versus aspirin for the prevention of recurrent stroke. Analysis of patients with minor stroke from the Ticlopidine Aspirin Stroke Study. *Stroke, 23,* 1723–1727.

122. Gorelick, P. B., Richardson, D., Kelly, M., et al.; African American Antiplatelet Stroke Prevention Study Investigators. (2003). Aspirin and ticlopidine for prevention of recurrent stroke in black patients: A randomized trial. *J Am Med Assoc, 289,* 2947–2957.

123. Diener, H. C., Bogousslavsky, J., Brass, L. M., et al.; MATCH Investigators. (2004). Aspirin and clopidogrel compared with clopidogrel alone after recent ischaemic stroke or transient ischaemic attack in high-risk patients (MATCH): Randomised, double-blind, placebo-controlled trial. *Lancet, 364,* 331–337.

124. Markus, H. S., Droste, D. W., Kaps, M., et al. (2005). Dual antiplatelet therapy with clopidogrel and aspirin in symptomatic carotid stenosis evaluated using Doppler embolic signal detection: The Clopidogrel and Aspirin for Reduction of Emboli in Symptomatic Carotid Stenosis (CARESS) trial. *Circulation, 111,* 2233–2240.

125. Hankey, G. J., Sudlow, C. L., & Dunbabin, D. W. (2000). Thienopyridine derivatives (ticlopidine, clopidogrel) versus aspirin for preventing stroke and other serious vascular events in high vascular risk patients. *Cochrane Database Syst Rev,* CD001246.

126. Balsano, F., Cocchieri, S., Libretti, G., et al. (1989). Ticlopidine in the treatment of intermittent claudication: A 21-month double-blind trial. *J Lab Clin Med, 114,* 84–91.

127. Scrutinio, D., Cimminiello, C., Marubini, E., et al. (2001). Ticlopidine versus aspirin after myocardial infarction (STAMI) trial. *J Am Coll Cardiol, 37,* 1259–1265.

128. Yusuf, S., Zhao, F., Mehta, S. R., et al.; Clopidogrel in Unstable Angina to Prevent Recurrent Events Trial Investigators. (2001). Effects of clopidogrel in addition to aspirin in patients with acute coronary syndromes without ST-segment elevation. *N Engl J Med, 345,* 494–502.

129. Peters, R. J., Mehta, S. R., Fox, K. A., et al.; Clopidogrel in Unstable Angina to Prevent Recurrent Events (CURE) Trial Investigators. (2003). Effects of aspirin dose when used alone or in combination with clopidogrel in patients with acute coronary syndromes: Observations from the Clopidogrel in Unstable Angina to Prevent Recurrent Events (CURE) study. *Circulation, 108,* 1682–1687.

130. Yusuf, S., Mehta, S. R., Zhao, F., et al.; Clopidogrel in Unstable Angina to Prevent Recurrent Events Trial Investigators. (2003). Early and late effects of clopidogrel in patients with acute coronary syndromes. *Circulation, 107,* 966–972.

131. Budaj, A., Yusuf, S., Mehta, S. R., et al.; Clopidogrel in Unstable Angina to Prevent Recurrent Events (CURE) Trial Investigators. (2002). Benefit of clopidogrel in patients with acute coronary syndromes without ST-segment elevation in various risk groups. *Circulation, 106,* 1622–1626.

132. Sabatine, M. S., Cannon, C. P., Gibson, C. M., et al.; CLARITY-TIMI 28 Investigators. (2005). Addition of clopidogrel to aspirin and fibrinolytic therapy for myocardial infarction with ST-segment elevation. *N Engl J Med, 352,* 1179–1189.

133. COMMIT (Clopidogrel and Metoprolol in Myocardial Infarction Trial) collaborative group. (2005). Addition of clopidogrel to aspirin in 45,852 patients with acute myocardial infarction: Randomised placebo-controlled trial. *Lancet, 366,* 1607–1621.

134. Schomig, A., Neumann, F. J., Kastrati, A., et al. (1996). A randomized comparison of antiplatelet and anticoagulant therapy after the placement of coronary-artery stents. *N Engl J Med, 334,* 1084–1089.

135. Kastrati, A., Schuhlen, H., Hausleiter, J., et al. (1997). Restenosis after coronary stent placement and randomization to a 4-week combined antiplatelet or anticoagulant therapy: Six-month angiographic follow-up of the Intracoronary Stenting and Antithrombotic Regimen (ISAR) trial. *Circulation, 96,* 462–467.

136. Bertrand, M. E., Legrand, V., Boland, J., et al. (1998). Randomized multicenter comparison of conventional anticoagulation versus antiplatelet therapy in unplanned and elective coronary stenting. The Full Anticoagulation versus Aspirin and Ticlopidine (FANTASTIC) study. *Circulation, 98,* 1597–1603.

137. Urban, P., Macaya, C., Rupprecht, H. J., et al. (1998). Randomized evaluation of anticoagulation versus antiplatelet therapy after coronary stent implantation in high-risk patients: The Multicenter Aspirin and Ticlopidine Trial after Intracoronary Stenting (MATTIS). *Circulation, 98,* 2126–2132.

138. Leon, M. B., Baim, D. S., Popma, J. J., et al. (1998). A clinical trial comparing three antithrombotic-drug regimens after coronary-artery stenting. Stent Anticoagulation Restenosis Study Investigators. *N Engl J Med, 339,* 1665–1671.

139. Mehta, S. R., Yusuf, S., Peters, R. J., et al.; Clopidogrel in Unstable Angina to Prevent Recurrent Events (CURE) Trial Investigators. (2001). Effects of pretreatment with clopidogrel and aspirin followed by long-term therapy in patients undergoing percutaneous coronary intervention: The PCI-CURE study. *Lancet, 358,* 527–533.

140. Steinhubl, S. R., Berger, P. B., Mann, J. T., 3rd, et al.; CREDO Investigators. (2002). Clopidogrel for the Reduction of Events during Observation. Early and sustained dual oral antiplatelet therapy following percutaneous coronary intervention: A randomized controlled trial. *J Am Med Assoc, 288,* 2411–2420.

141. Sabatine, M. S., Cannon, C. P., Gibson, C. M., et al.; Clopidogrel as Adjunctive Reperfusion Therapy (CLARITY)–Thrombolysis in Myocardial Infarction (TIMI) 28 Investigators. (2005). Effect of clopidogrel pretreatment before percutaneous coronary intervention in patients with

ST-elevation myocardial infarction treated with fibrinolytics: The PCI-CLARITY study. *JAMA, 294,* 1224–1232.

142. Patti, G., Colonna, G., Pasceri, V., et al. (2005). Randomized trial of high loading dose of clopidogrel for reduction of periprocedural myocardial infarction in patients undergoing coronary intervention: Results from the ARMYDA-2 (Antiplatelet Therapy for Reduction of Myocardial Damage during Angioplasty) study. *Circulation, 111,* 2099–2106.

143. Kastrati, A., Mehilli, J., Schuhlen, H., et al.; Intracoronary Stenting and Antithrombotic Regimen–Rapid Early Action for Coronary Treatment Study Investigators. (2004). A clinical trial of abciximab in elective percutaneous coronary intervention after pretreatment with clopidogrel. *N Engl J Med, 350,* 232–238.

144. Mehilli, J., Kastrati, A., Schuhlen, H., et al.; Intracoronary Stenting and Antithrombotic Regimen: Is Abciximab a Superior Way to Eliminate Elevated Thrombotic Risk in Diabetics (ISAR-SWEET) Study Investigators. (2004). Randomized clinical trial of abciximab in diabetic patients undergoing elective percutaneous coronary interventions after treatment with a high loading dose of clopidogrel. *Circulation, 110,* 3627–3635.

145. Bhatt, D. L., Bertrand, M. E., Berger, P. B., et al. (2002). Meta-analysis of randomized and registry comparisons of ticlopidine with clopidogrel after stenting. *J Am Coll Cardiol, 39,* 9–14.

146. Mueller, C., Roskamm, H., Neumann, F. J., et al. (2003). A randomized comparison of clopidogrel and aspirin versus ticlopidine and aspirin after the placement of coronary artery stents. *J Am Coll Cardiol, 41,* 969–973.

147. Wolak, A., Amit, G., Cafri, C., et al. (2005). Increased long term rates of stent thrombosis and mortality in patients given clopidogrel as compared to ticlopidine following coronary stent implantation. *Int J Cardiol, 103,* 293–297.

148. Janzon, L., Bergqvist, D., Boberg, J., et al. (1990). Prevention of myocardial infarction and stroke in patients with intermittent claudication; Effects of ticlopidine. Results from STIMS, the Swedish Ticlopidine Multicentre Study. *J Intern Med, 227,* 301–308. [Erratum in *J Intern Med, 228,* 659, 1990.]

149. Bergqvist, D., Almgren, B., & Dickinson, J. P. (1995). Reduction of requirement for leg vascular surgery during long-term treatment of claudicant patients with ticlopidine: Results from the Swedish Ticlopidine Multicentre Study (STIMS). *Eur J Vasc Endovasc Surg, 10,* 69–76.

150. Balsano, F., Rizzon, P., Violi, F., et al. (1990). Antiplatelet treatment with ticlopidine in unstable angina. A controlled multicenter clinical trial. The Studio della Ticlopidina nell'Angina Instabile Group. *Circulation, 82,* 17–26.

151. Becquemin, J. P. (1997). Effect of ticlopidine on the long-term patency of saphenous-vein bypass grafts in the legs. Etude de la Ticlopidine apres Pontage Femoro-Poplite and the Association Universitaire de Recherche en Chirurgie. *N Engl J Med, 337,* 1726–1731.

152. Gaspoz, J. M., Coxson, P. G., Goldman, P. A., et al. (2002). Cost effectiveness of aspirin, clopidogrel, or both for secondary prevention of coronary heart disease. *N Engl J Med, 346,* 1800–1806.

153. Durand-Zaleski, I., & Bertrand, M. (2004). The value of clopidogrel versus aspirin in reducing atherothrombotic events: The CAPRIE study. *Pharmacoeconomics, 22*(Suppl. 4), 19–27.

154. Karnon, J., Brennan, A., Pandor, A., et al. (2005). Modelling the long term cost effectiveness of clopidogrel for the secondary prevention of occlusive vascular events in the UK. *Curr Med Res Opin, 21,* 101–112.

155. Beinart, S. C., Kolm, P., Veledar, E., et al. (2005). Long-term cost effectiveness of early and sustained dual oral antiplatelet therapy with clopidogrel given for up to one year after percutaneous coronary intervention: Results from the Clopidogrel for the Reduction of Events during Observation (CREDO) trial. *J Am Coll Cardiol, 46,* 761–769.

156. Cowper, P. A., Udayakumar, K., Sketch, M. H., Jr., et al. (2005). Economic effects of prolonged clopidogrel therapy after percutaneous coronary intervention. *J Am Coll Cardiol, 45,* 369–376.

157. Karnon, J., Bakhai, A., Brennan, A., et al. (2006). A cost-utility analysis of clopidogrel in patients with non-ST-segment-elevation acute coronary syndromes in the UK. *Int J Cardiol, 109,* 307–316.

158. Lamy, A., Jonsson, B., Weintraub, W. S., et al.; The CURE Economic Group. (2004). The cost-effectiveness of the use of clopidogrel in acute coronary syndromes in five countries based upon the CURE study. *Eur J Cardiovasc Prev Rehab, 11,* 460–465.

159. Lindgren, P., Jonsson, B., & Yusuf, S. (2004). Cost-effectiveness of clopidogrel in acute coronary syndromes in Sweden: A long-term model based on the CURE trial. *J Intern Med, 255,* 562–570.

160. Schleinitz, M. D., & Heidenreich, P. A. (2005). A cost-effectiveness analysis of combination antiplatelet therapy for high-risk acute coronary syndromes: Clopidogrel plus aspirin versus aspirin alone. *Ann Intern Med, 142,* 251–259.

161. Weintraub, W. S., Mahoney, E. M., Lamy, A., et al.; CURE Study Investigators. (2005). Long-term cost-effectiveness of clopidogrel given for up to one year in patients with acute coronary syndromes without ST-segment elevation. *J Am Coll Cardiol, 45,* 838–845.

162. Cattaneo, M. (2004). Aspirin and clopidogrel. Efficacy, safety and the issue of drug resistance. *Arterioscler Thromb Vasc Biol, 24,* 1980–1987.

163. Schwarz, U. R., Geiger, J., Walter, U., et al. (1999). Flow cytometry analysis of intracellular VASP phosphorylation for the assessment of activating and inhibitory signal transduction pathways in human platelets — Definition and detection of ticlopidine/clopidogrel effects. *Thromb Haemost, 82,* 1145–1152.

164. Barragan, P., Bouvier, J. L., Roquebert, P. O., et al. (2003). Resistance to thienopyridines: Clinical detection of coronary stent thrombosis by monitoring of vasodilator-stimulated phosphoprotein phosphorylation. *Cathet Cardiovasc Interv, 59,* 295–302.

165. Aleil, B., Ravanat, C., Cazenave, J. P., et al. (2005). Flow cytometric analysis of intraplatelet VASP phosphorylation for the detection of clopidogrel resistance in patients with

ischemic cardiovascular diseases. *J Thromb Haemost, 3,* 85–92.

166. Grossmann, R., Sokolova, O., Schnurr, A., et al. (2004). Variable extent of clopidogrel responsiveness in patients after coronary stenting. *Thromb Haemost, 92,* 1201–1206.

167. Geiger, J., Teichmann, L., Grossmann, R., et al. (2005). Monitoring of clopidogrel action: Comparison of methods. *Clin Chem, 51,* 957–965.

168. Lau, W. C., Gurbel, P. A., Watkins, P. B., et al. (2004). Contribution of hepatic cytochrome P450 3A4 metabolic activity to the phenomenon of clopidogrel resistance. *Circulation, 109,* 166–171.

169. Serebruany, V. L., Steinhubl, S. R., Berger, P. B., et al. (2005). Variability in platelet responsiveness to clopidogrel among 544 individuals. *J Am Coll Cardiol, 45,* 246–251.

170. Gurbel, P. A., Bliden, K. P., Hayes, K. M., et al. (2005). The relation of dosing to clopidogrel responsiveness and the incidence of high post-treatment platelet aggregation in patients undergoing coronary stenting. *J Am Coll Cardiol, 45,* 1392–1396.

171. Taubert, D., Kastrati, A., Harlfinger, S., et al. (2004). Pharmacokinetics of clopidogrel after administration of a high loading dose. *Thromb Haemost, 92,* 311–316.

172. Angiolillo, D. J., Fernandez-Ortiz, A., Bernardo, E., et al. (2004). PLA polymorphism and platelet reactivity following clopidogrel loading dose in patients undergoing coronary stent implantation. *Blood Coagul Fibrinolysis, 15,* 89–93.

173. von Beckerath, N., von Beckerath, O., Koch, W., et al. (2005). P2Y12 gene H2 haplotype is not associated with increased adenosine diphosphate-induced platelet aggregation after initiation of clopidogrel therapy with a high loading dose. *Blood Coagul Fibrinolysis, 16,* 199–204.

174. Clarke, T. A., & Waskell, L. A. (2003). The metabolism of clopidogrel is catalyzed by human cytochrome P450 3A and is inhibited by atorvastatin. *Drug Metab Dispos, 31,* 53–59.

175. Neubauer, H., Gunesdogan, B., Hanefeld, C., et al. (2003). Lipophilic statins interfere with the inhibitory effects of clopidogrel on platelet function — A flow cytometry study. *Eur Heart J, 24,* 1744–1749.

176. Mach, F., Senouf, D., Fontana, P., et al. (2005). Not all statins interfere with clopidogrel during antiplatelet therapy. *Eur J Clin Invest, 35,* 476–481.

177. Mitsios, J. V., Papathanasiou, A. I., Rodis, F. I., et al. (2004). Atorvastatin does not affect the antiplatelet potency of clopidogrel when it is administered concomitantly for 5 weeks in patients with acute coronary syndromes. *Circulation, 109,* 1335–1338.

178. Serebruany, V. L., Midei, M. G., Malinin, A. I., et al. (2004). Absence of interaction between atorvastatin or other statins and clopidogrel: Results from the interaction study. *Arch Intern Med, 164,* 2051–2057.

179. Vinholt, P., Poulsen, T. S., Korsholm, L., et al. (2005). The antiplatelet effect of clopidogrel is not attenuated by statin treatment in stable patients with ischemic heart disease. *Thromb Haemost, 94,* 438–443.

180. Muller, I., Besta, F., Schulz, C., et al. (2003). Effects of statins on platelet inhibition by a high loading dose of clopidogrel. *Circulation, 108,* 2195–2197.

181. Gorchakova, O., von Beckerath, N., Gawaz, M., et al. (2004). Antiplatelet effects of a 600 mg loading dose of clopidogrel are not attenuated in patients receiving atorvastatin or simvastatin for at least 4 weeks prior to coronary artery stenting. *Eur Heart J, 25,* 1898–1902.

182. Poulsen, T. S., Vinholt, P., Mickley, H., et al. (2005). Existence of a clinically relevant interaction between clopidogrel and HMG-CoA reductase inhibitors? Re-evaluating the evidence. *Basic Clin Pharmacol Toxicol, 96,* 103–110.

183. Matetzky, S., Shenkman, B., Guetta, V., et al. (2004). Clopidogrel resistance is associated with increased risk of recurrent atherothrombotic events in patients with acute myocardial infarction. *Circulation, 109,* 3171–3175.

184. Sugidachi, A., Asai, F., Ogawa, T., et al. (2000). The in vivo pharmacological profile of CS-747, a novel antiplatelet agent with platelet ADP receptor antagonist properties. *Br J Pharmacol, 129,* 1439–1446.

185. Sugidachi, A., Asai, F., Yoneda, K., et al. (2001). Antiplatelet action of R-99224, an active metabolite of a novel thienopyridine-type G(i)-linked P2T antagonist, CS-747. *Br J Pharmacol, 132,* 47–54.

186. Niitsu, Y., Jakubowski, J. A., Sugidachi, A., et al. (2005). Pharmacology of CS-747 (Prasugrel, LY640315), a novel, potent antiplatelet agent with in vivo P2Y12 receptor antagonist activity. *Semin Thromb Hemost, 31,* 184–194.

187. Brandt, J. T., Payne, C. D., Weerakkody, G., et al. (2005). Superior responder rate for inhibition of platelet aggregation with a 60 mg loading dose of prasugrel (CS-747, LY640315) compared with a 300 mg loading dose of clopidogrel. *J Am Coll Cardiol, 45*(Suppl.1), 87A.

188. Payne, C. D., Brandt, J. T., Weerakkody, G. J., et al. (2005). Superior inhibition of platelet aggregation following a loading dose of CS-747 (prasugrel, LY640315) versus clopidogrel: Correlation with the pharmacokinetics of active metabolite generation. *J Thromb Haemost, 3*(Suppl. 1), P0952.

189. Wiviott, S. D., Antman, E. M., Winters, K. J., et al.; JUMBO-TIMI 26 Investigators. (2005). Randomized comparison of prasugrel (CS-747, LY640315), a novel thienopyridine P2Y12 antagonist, with clopidogrel in percutaneous coronary intervention: Results of the Joint Utilization of Medications to Block Platelets Optimally (JUMBO)-TIMI 26 trial. *Circulation, 111,* 3366–3373.

190. van Giezen, J. J. J., & Humphries, R. G. (2005). Preclinical and clinical studies with selective reversible direct P2Y12 antagonists. *Semin Thromb Hemost, 31,* 195–204.

191. Marteau, F., Le Poul, E., Communi, D., et al. (2003). Pharmacological characterization of the human P2Y13 receptor. *Mol Pharmacol, 64,* 104–112.

192. Fumagalli, M., Trincavelli, L., Lecca, D., et al. (2004). Cloning, pharmacological characterization and distribution of the rat G-protein-coupled P2Y(13) receptor. *Biochem Pharmacol, 68,* 113–124.

193. Huang, J., Driscoll, E. M., Gonzales, M. L., et al. (2000). Prevention of arterial thrombosis by intravenously adminis-

tered platelet P2T receptor antagonist AR-C69931MX in a canine model. *J Pharmcol Exp Ther, 295,* 492–499.

194. van Gestel, M. A., Heemskerk, J. W., Slaaf, D. W., et al. (2003). *In vivo* blockade of platelet ADP receptor P2Y12 reduces embolus and thrombus formation but not thrombus stability. *Arterioscler Thromb Vasc Biol, 23,* 518–523.

195. Wang, K., Zhou, X., Zhiou, Z., et al. (2003). Blockade of the platelet P2Y12 receptor by AR-C69931MX sustains coronary artery recanalozation and improves the myocardial tissue perfusion in a canine thrombosis model. *Arterioscler Thromb Vasc Biol, 23,* 357–362.

196. Storey, R. F., Oldroyd, K. G., & Wilcox, R. G. (2001). Open multicentre study of the P2T receptor antagonist AR-C69931MX assessing safety, tolerability and activity in patients with acute coronary syndromes. *Thromb Haemost, 85,* 401–407.

197. Jacobsson, F., Swahn, E., Wallentin, L., et al. (2002). Safety profile and tolerability of intravenous ARC69931MX, a new antiplatelet drug, in unstable angina pectoris and non-Q-wave myocardial infarction. *Clin Ther, 24,* 752–765.

198. Storey, R. F., Wilcox, R. G., & Heptinstall, S. (2002). Comparison of the pharmacodynamic effects of the platelet ADP receptor antagonists clopidogrel and AR-C69931MX in patients with ischaemic heart disease. *Platelets, 13,* 407–413.

199. Beham, M., Fox, S., Sanderson, H., et al. (2002). Effect of clopidogrel on procoagulant activity in acute coronary syndromes: Evidence for incomplete P2Y12 receptor blockade. *Circulation, 106,* II-149–II-150.

200. Husted, S., Emanuelsson, H., Heptinstall, S., et al. (2005). Greater and less variable inhibition of platelet aggregation (IPA) with AZD6140, the first oral reversible ADP receptor antagonist, compared with clopidogrel in patients with atherosclerosis. *Eur Heart J, 26,* 643. [abstracts supplement]

201. Fabre, J. E., Nguyen, M., Latour, A., et al. (1999). Decreased platelet aggregation, increased bleeding time and resistance to thromboembolism in P2Y1-deficient mice. *Nat Med, 5,* 1199–1202.

202. Léon, C., Hechler, B., Freund, M., et al. (1999). Defective platelet aggregation and increased resistance to thrombosis in purinergic P2Y(1) receptor-null mice. *J Clin Invest, 104,* 1731–1737.

203. Baurand, A., Raboisson, P., Freund, M., et al. (2001). Inhibition of platelet function by administration of MRS2179, a P2Y1 receptor antagonist. *Eur J Pharmacol, 412,* 213–221.

204. Baurand, A., & Gachet, C. (2003). The P2Y(1) receptor as a target for new antithrombotic drugs: A review of the P2Y(1) antagonist MRS-2179. *Cardiovasc Drug Rev, 21,* 67–76.

205. Lenain, N., Freund, M., Leon, C., et al. (2003). Inhibition of localized thrombosis in P2Y1-deficient mice and rodents treated with MRS2179, a P2Y1 receptor antagonist. *J Thromb Haemost, 1,* 1144–1149.

206. Cattaneo, M., Lecchi, A., Ohno, M., et al. (2004). Anti-aggregatory activity in human platelets of potent antagonists of the P2Y(1) receptor. *Biochem Pharmacol, 68,* 1995–2002.

207. Hechler, B., Nonne, C., Roh, E. J., et al. (2006). MRS2500, a potent, selective and stable antagonist of the P2Y₁ receptor, with strong antithrombotic activity in mice. *J Pharmacol Exp Ther, 316,* 556–563.

# CHAPTER 62

# αIIbβ3 (GPIIb-IIIa) Antagonists

**Ramtin Agah, Edward F. Plow, and Eric J. Topol**

*Departments of Cardiovascular Medicine and Molecular Cardiology, Cleveland Clinic Foundation, Cleveland, Ohio*

## I. Introduction

The discovery of coronary thrombosis as the underlying cause of myocardial infarction (MI) was first reported by Parkinson and Bedford in 1928.[1] However, it was not for another 50 years that the broad acceptance of this notion led to the treatment of patients with acute MI and unstable angina by directly targeting the thrombotic process. With better understanding of the molecular mechanisms of thrombosis, therapeutic targets to inhibit this process have emerged, including thrombin inhibitors, fibrinolytics, and platelet antagonists (Fig. 62-1). Among the latter class of agents, aspirin was the first to show significant clinical benefit in treatment of patients with acute MI (see Chapter 60). A remarkable 19% reduction in mortality was observed with aspirin in the International Study of Infarct Survival-2 trial.[2] This significant benefit from aspirin, a relatively weak antiplatelet agent, provided a clear impetus to target platelets with more potent and specific drugs to treat patients with coronary syndromes. Advancement in the basic understanding of platelet aggregation provided the field with such targets. Although many agonists are able to activate platelets and induce their aggregation, the final common pathway underlying platelet aggregation is binding of an adhesive protein set, primarily fibrinogen or von Willebrand factor (VWF), to the platelet surface.[3–5] The demonstration that integrin αIIbβ3 (glycoprotein [GP] IIb-IIIa) served as the fibrinogen receptor on the platelet surface and blockade of this interaction inhibited platelet aggregation[6] defined this specific membrane protein as a target of antithrombotic therapy. However, complete abrogation of platelet aggregation, although attractive in the treatment of the thrombotic state, raised major safety concerns. Would such blockade lead to major, and ultimately unacceptable, bleeding complications? The reports that natural mutations in αIIbβ3 in patients with Glanzmann thrombasthenia[7–9] (see Chapter 57) lead to only sporadic and minor bleeding tendencies and rarely to major internal bleeding provided some comfort, and efforts proceeded throughout the pharmaceutical and biotechnology sectors to develop potent, specific, and safe αIIbβ3 antagonists. These efforts led to development of the first clinically acceptable antagonist of αIIbβ3, abciximab (ReoPro), for clinical use.[10] After extensive animal studies,[11–14] including in primates,[15] to establish efficacy and safety, the clinical benefit of abciximab was first tested in preventing thrombotic complications in patients undergoing percutaneous coronary intervention (PCI) in the Evaluation of c7E3 for Prevention of Ischemic Complications (EPIC) trial.[16] In 1994, the results of the EPIC trial demonstrated the efficacy of αIIbβ3 blockade in reducing thrombotic complications. Within the next 5 years, more than 40,000 patients were randomized in studies, which demonstrated the clinical efficacy, and ultimately U.S. Food and Drug Administration (FDA) approval, of three intravenous αIIbβ3 antagonists in the reduction of thrombotic complications in patients undergoing PCI or in the medical management of patients with acute coronary syndromes (ACS).

At the beginning of the 21st century, the future of αIIbβ3 antagonists appeared to be particularly bright. An expanding utility of the approved intravenous αIIbβ3 antagonists to broader indications and patient populations was anticipated. Furthermore, looming on the horizon was the promise of oral αIIbβ3 antagonists, which were under active development throughout the pharmaceutical industry and offered the potential for extended suppression of receptor function to prevent thrombotic events in at-risk patients. Today, however, the use of intravenous agents has settled into a relatively narrow niche, and all of the oral αIIbβ3 antagonist programs have been abandoned. This chapter reviews the clinical trials that supported the efficacy of the intravenous agents, led to the apparent demise of the oral αIIbβ3 antagonists, and now define the utility of the αIIbβ3 antagonists.

## II. αIIbβ3 Signaling Mechanisms

The detailed signaling mechanisms of integrin αIIbβ3 are discussed in Chapters 8 and 17; however, a brief overview

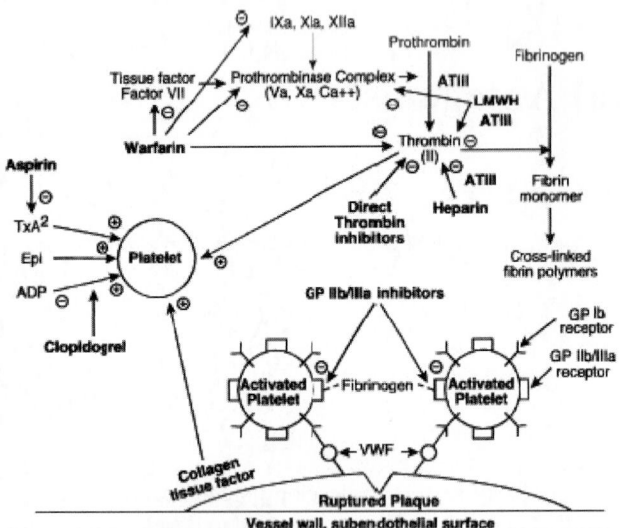

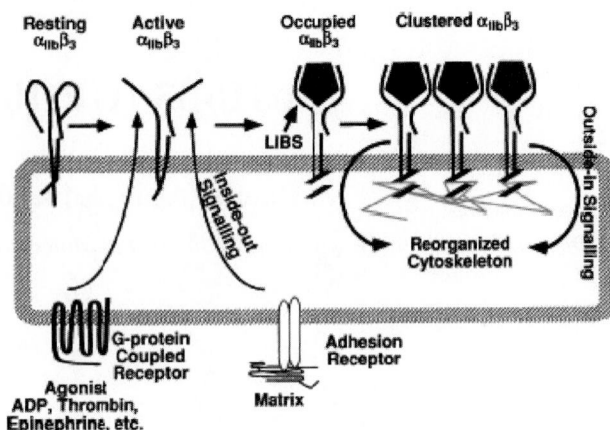

**Figure 62-2.** The conformational states of the integrin $\alpha_{IIb}\beta_3$ (GPIIb-IIIa) receptor. LIBS, ligand-induced binding site. Adapted with permission from Topol et al. (1999). *Lancet, 353,* 227–231.

**Figure 62-1.** Interaction between antiplatelet and antithrombotic agents in vascular thrombosis. ADP, adenosine diphosphate; ATIII, antithrombin III; Epi, epinephrine; GP, glycoprotein; LMWH, low-molecular-weight heparin; TxA2, thromboxane A2; VWF, von Willebrand factor. Adapted with permission from Alexander et al. (1997). *Curr Opin Cardiol, 12,* 427–437.

is useful here. The receptor in its native state on nonactivated platelets exhibits very low affinity for fibrinogen or other plasma ligands, a desirable feature that precludes spontaneous platelet aggregation under normal hemostatic conditions. With vascular injury, there is exposure of subendothelium matrix, which contains platelet adhesive ligands such as collagen and surface-bound VWF (Fig. 62-1). With binding of these ligands to their respective receptors on the platelet surface, notably GPVI and GPIb-IX-V, platelets become activated (see Chapters 6, 7, 16, and 18). Other activating stimuli include soluble agonists such as adenosine diphosphate (ADP), thrombin, and epinephrine and VWF under high shear conditions (simulating a physiologically stenosed artery), which can all be generated rapidly within the microenvironment. With platelet activation, $\alpha_{IIb}\beta_3$ undergoes a conformational metamorphosis, which depends on the transmission of *inside-out signals* from the cytoplasmic domain of $\alpha_{IIb}\beta_3$ within the platelets to the extracellular domain of the receptor (see Chapter 8). This transformation enhances the affinity of $\alpha_{IIb}\beta_3$ for its main ligands, fibrinogen and VWF, leading to receptor occupancy (Fig. 62-2). Divalent fibrinogen and multivalent VWF can act as bridging molecules to bind platelets together to form aggregates and can induce additional responses within the platelets (see Chapter 16) that rapidly lead to formation of a stable thrombus (Fig. 62-1).[17]

The signaling that occurs upon fibrinogen or VWF binding to $\alpha_{IIb}\beta_3$, referred to as *outside-in signaling* (Fig. 62-2), involves transmission of information to the cytoplas-

mic domain of $\alpha_{IIb}\beta_3$, which in turn initiates a cascade of intracellular events (see Chapter 17). Consequently, platelets secrete their dense and $\alpha$-granules, activate other $\alpha_{IIb}\beta_3$ receptors, promote further platelet aggregation, and retract clots. In addition, platelets recruit and activate other cells (e.g., platelets become able to bind and activate leukocytes [see Chapter 12] and stimulate smooth muscle cell migration and proliferation through their secretion of platelet-derived growth factor and other growth factors). Furthermore, there is induction of neoepitopes on $\alpha_{IIb}\beta_3$, collectively called ligand-induced binding sites (LIBS),[18,19] which are expressed only by the occupied receptor (Fig. 62-2). Different ligands and antagonists can induce different LIBS,[20] and there are reports that naturally occurring LIBS antibodies can induce thrombocytopenia in some patients administered $\alpha_{IIb}\beta_3$ antagonists (see Chapter 49).[21]

In addition to their inhibition of platelet aggregation, $\alpha_{IIb}\beta_3$ antagonists have a specific anticoagulant action, as evidenced by prolongation of the activated clotting time,[22] inhibition of thrombin generation,[23] and inhibition of platelet procoagulant activity.[24] The mechanisms of this anticoagulant effect of $\alpha_{IIb}\beta_3$ antagonists include inhibition of prothrombin binding to $\alpha_{IIb}\beta_3$ with less resultant thrombin generation and inhibition of procoagulant platelet-derived microparticle formation.[23]

## III. $\alpha_{IIb}\beta_3$ Ligand Binding Sites: Targets for Inhibition

Throughout the years, it has been determined that the binding of the multiple large adhesive ligands to $\alpha_{IIb}\beta_3$ can be effectively blocked by small peptides or certain monoclonal antibodies to the receptor (Fig. 62-3).[25–28] These antagonists also block the functional response of platelets, their aggregation, elicited by the binding of the adhesive ligands. Hence, these antagonists demonstrated the potential of αIIbβ3 blockade as an antithrombotic strategy. Two of the peptides that received early attention are referred to as the RGD and the γ-chain peptides (reviewed in Gartner and Bennett[29]). RGD, Arg-Gly-Asp, is a sequence found in fibrinogen as well as many of the other ligands of αIIbβ3, including VWF, vitronectin, and fibronectin. Peptides containing RGD block the interactions of multiple ligands with

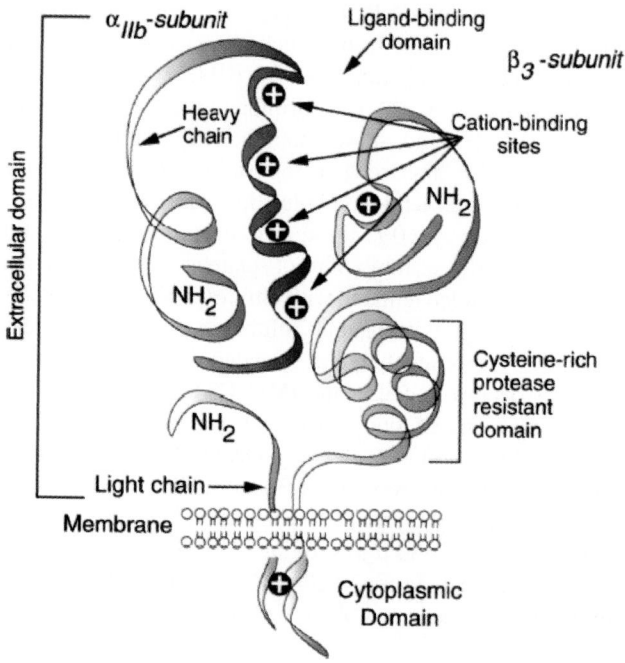

**Figure 62-3.** The heterodimeric structure of the integrin αIIbβ3 (GPIIb-IIIa) receptor. Adapted with permission from Topol et al. (1999). *Lancet, 353,* 227–231.

αIIbβ3, bind directly to the receptor, and can fully inhibit platelet aggregation.[28–31] Several other integrin family members also recognize RGD sequences.[32] The minimum sequence of the γ-chain peptide is KQAGDV, the sequence at the carboxyl terminus of the γ-chain of fibrinogen.[26,33,34] This sequence is unique to fibrinogen, but the γ-chain peptide blocks the binding of not only fibrinogen but also the other previously mentioned ligands of the receptor.[35] RGD and the γ-chain peptides inhibit the binding of each other to αIIbβ3.[36] The sites in αIIbβ3 that bind RGD and γ-chain peptides are not necessarily identical and may be allosterically linked.[37,38] High-affinity mimetics of these two recognition peptides have been developed as antagonists of αIIbβ3 and platelet aggregation and have led to the development of drugs used clinically (see Section IV). Also, the function-blocking properties of a monoclonal antibody to αIIbβ3 were exploited in the development of a therapeutic (see Section IV). Crystal structures of representative agonists bound to the extracellular domain of αIIbβ3 have been published. Multiple contact sites in the extracellular domain of the receptor are involved in engagement of these antagonists.[39–41] These sites reside in both the αIIb and β3 subunits. Details of the structure of αIIbβ3 are discussed in Chapter 8.

## IV. The αIIbβ3 (GPIIb-IIIa) Antagonists

By 1998, three αIIbβ3 (GPIIb-IIIa) antagonists were licensed by the FDA for clinical use (Table 62-1), and these remain the only three approved drugs of this class. These three agents are a humanized antibody, abciximab; a nonpeptide, tirofiban, based on RGD; and a peptide, eptifibatide, based on a snake venom sequence and presumably a mimetic of the γ-chain peptide. It is relevant to discuss the

**Table 62-1: Receptor Binding Properties of FDA-Approved αIIbβ3 Antagonists**

| Antagonist | Structure | Receptor Affinity ($K_d$ nM) | Receptor Selectivity | % Inhibition of Aggregation 24 Hours Postinfusion | FDA-Approved Indications | Cost[a] ($) |
|---|---|---|---|---|---|---|
| Abciximab (ReoPro) | Chimeric monoclonal antibody | 5 | αIIbβ3 αVβ3 αMβ2 | Approximately 50 | PCI | 1317 |
| Eptifibatide (Integrilin) | Cyclic heptapeptide | 120 | αIIbβ3 | Approximately 0 | PCI Acute coronary syndrome | 535 |
| Tirofiban (Aggrastat) | Nonpeptide derivative | 15 | αIIbβ3 | Approximately 0 | Acute coronary syndrome | 358 |

[a]Based on treatment of a 75 kg patient with bolus plus 12-hour infusion of abciximab; double-bolus plus infusion of eptifibatide for 18 hours; and bolus plus infusion of tirofiban for 18 hours.

basic structure and function of these agents because they shed light on their clinical effects and potential differences.

## A. Abciximab

Abciximab (ReoPro) is a murine human chimeric Fab fragment that was derived from the murine monoclonal antibody 7E3.[6,10,42] Its binding site resides primarily in the midsegment of the β3 subunit,[43] and specific contact residues within this region have been identified.[44] Since the β3 subunit is also present in integrin αvβ3, abciximab also binds to this vitronectin receptor with high affinity. Cross-reaction of abciximab with the active conformer of the leukocyte integrin αMβ2 (CDllb/CD18, Mac-1) has also been demonstrated.[45,46] In terms of pharmacodynamics, multiple animal studies and subsequent human trials have established that a bolus dose of 0.25 μg/kg followed by an infusion of 0.125 μg/kg/min for 12 hours provides efficacious therapy.[11-14] Approximately two thirds of the bolus dose binds to platelets in the circulation within minutes, with >80% αIIbβ3 blockade resulting in >80% inhibition of 5 to 20 μM ADP-induced platelet aggregation. The antibody has a short plasma half-life, with the free plasma abciximab level decreasing rapidly after the infusion (initial half-life, 30 minutes), and only 4% of the injected dose is present as free antibody in plasma after 2 hours.[47] However, the platelet-bound antibody persists and can exchange from one platelet to another, thereby redistributing to new platelets entering the circulation. This exchange may account for the persistence of platelet-associated abciximab in the circulation for more than 14 days beyond the initial infusion. The platelet aggregation function, however, returns toward baseline within 12 to 26 hours after discontinuation of the abciximab infusion.

## B. Tirofiban

Tirofiban (Aggrastat) is a nonpeptide derivative. Structure–activity studies with an RGD-containing disintegrin, echistatin, as a starting point[48] resulted in identification of a compound with the ability to inhibit aggregation of human gel-filtered platelets and to bind αIIbβ3 with high affinity ($K_d$ 9 nM). Furthermore, this compound was selective for αIIbβ3 and showed no reactivity with αIIbβ3. Dose-ranging studies for platelet inhibition in patients undergoing angioplasty revealed that a 10 μg/kg bolus and 0.15 μg/kg/min infusion of tirofiban resulted in 96% inhibition of platelet aggregation (in response to 5 μM ADP) within minutes of infusion.[49,50] This dosing regimen was subsequently used in the clinical trial Randomized Efficacy Study of Tirofiban for Outcomes and Restenosis (RESTORE). The plasma half-life of tirofiban is 1.6 hours, and rapid dissociation of this molecule from platelets decreases inhibition of platelet aggregation to less than 30% by 4 hours postinfusion.

## C. Eptifibatide

Barbourin is a disintegrin that contains a KGD rather than an RGD sequence. This single lysine-for-arginine substitution gives this molecule high specificity in binding αIIbβ3. A KGD-containing cyclic heptapeptide, eptifibatide (Integrilin), was developed from structure–activity analyses of this sequence. Eptifibatide binds to αIIbβ3 with a dissociation constant of 120 nM. Like tirofiban, eptifibatide is selective for αIIbβ3 and shows no reactivity with αvβ3. In IMPACT Hi/Lo, a phase II dosing trial of patients undergoing coronary intervention, a bolus dose of 135 μg/kg eptifibatide followed by a 0.75 μg/kg/min infusion produced and maintained >80% inhibition of platelet aggregation (in whole blood anticoagulated with citrate). Four hours postinfusion, the degree of platelet inhibition decreased to less than 50%. Subsequent studies revealed that platelet inhibition was overestimated in the IMPACT Hi/Lo studies due to the low concentration of free calcium in citrated blood. At physiological blood calcium concentrations, the platelet inhibition achieved was in the 30 to 50% range. A subsequent dosing study, using heparinized blood, resulted in a bolus regimen of 180 μg/kg followed by an infusion of 2.0 μg/kg/min. This dosing regimen achieved >80% platelet inhibition in noncitrated blood.

## D. Summary

All three agents (abciximab, tirofiban, and eptifibatide) achieve effective inhibition of platelet aggregation by binding to αIIbβ3. However, the exact binding sites of each agent in the receptor may be different. A potential consequence of their different binding sites may be the induction of different functional responses in platelets. For example, eptifibatide and tirofiban both induce the LIBS D3 epitope upon binding the receptor but abciximab does not, and antecedent therapy with eptifibatide or tirofiban does not alter the pharmacodynamic properties of abciximab.[51] As a consequence, the spectrum of inhibition of platelet activity by these blocking agents may be different. In addition, as noted previously, these agents have different specificities for other integrins, with abciximab cross-reacting with integrins αvβ3 and αMβ2. As a result, some of the clinical effects of these agents may go beyond their binding to αIIbβ3. Thus, the clinical efficacy of these drugs in platelet inhibition and decreasing clinical end points may be different due to both inter- and intrareceptor binding differences.

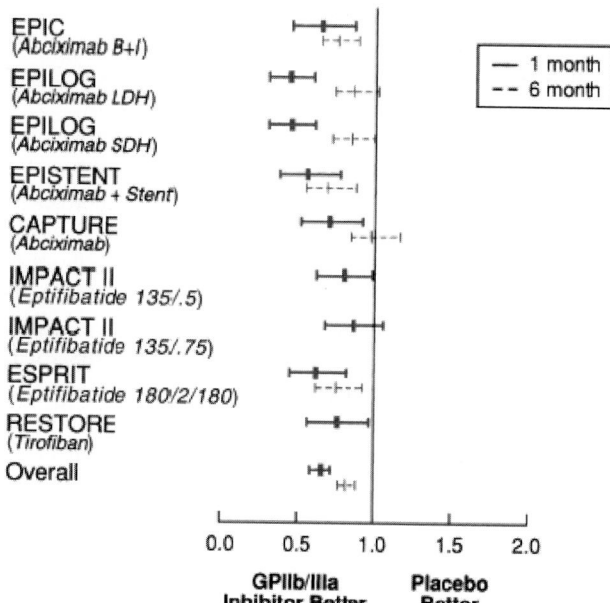

**Figure 62-4.** Comparison of the 1- and 6-month composite end point (death, MI, and revascularization) event rates for the GPIIb-IIIa antagonist interventional trials. B, bolus; I, infusion; LDH, low-dose heparin; SDH, standard-dose heparin. See text for details.

## V. αIIbβ3 Blockade during Coronary Interventions

The EPIC trial[16] using abciximab in patients undergoing high-risk coronary intervention was the first application of αIIbβ3 inhibition in a clinical setting. The success of this trial was followed by multiple trials in which the utility of the parenteral αIIbβ3 antagonists was demonstrated in a variety of clinical settings, including elective PCI and unstable angina (Fig. 62-4). These trials and their outcomes are discussed next.

### A. EPIC (Evaluation of c7E3 for Prevention of Ischemic Complications)

Patients undergoing high-risk PCI were envisaged to obtain the maximum benefit from GPIIb-IIIa blockade by avoidance of periprocedural ischemic events. The criteria for entry into the trial included acute or recent MI, unstable angina, or moderately complex or complex target lesion angiographic morphology (modified American College of Cardiology/American Heart Association criteria B1, B2, or C) in association with advanced age, female gender, or diabetes mellitus. Exclusion criteria included age older than 80 years, a known bleeding diathesis, major surgery within 6 weeks, or stroke within 2 years. The study enrolled 2099

patients between November 1991 and November 1992. Patients were randomized in a double-blinded fashion to one of three arms: placebo, a 0.25 mg/kg bolus of abciximab, or a 0.25 mg/kg bolus of abciximab followed by a 10 μg/min infusion for 12 hours. All patients received aspirin and heparin, which was started during the procedure and continued for at least 12 hours thereafter.

The prespecified study end points were any of the following events in the first 30 days after randomization: death, MI, or recurrent ischemia (as judged by repeat PCI or coronary arterial bypass grafting [CABG], insertion of an intra-aortic balloon pump, or bailout stenting secondary to recurrent or refractory ischemia). A graded effect of abciximab was found among the treatment groups. Patients receiving the bolus of abciximab derived a 12% relative reduction of the composite 30-day end point (12.8 vs. 11.3%, $p = 0.43$), and patients receiving the bolus plus infusion of abciximab derived a pronounced 35% relative reduction (12.8 vs. 8.3%, $p = 0.008$). This treatment effect was homogeneous among all prespecified subgroups, including unstable angina, high-risk anatomical features, age, and sex. Of note, both major and minor bleeding were more common in the groups receiving abciximab compared to those receiving placebo.

Subsequent follow-up studies at 6 months and 3 years revealed that the clinical efficacy of abciximab was maintained.[52,53] At 6 months, there was a 23% relative reduction in the composite end points in the patients receiving the bolus plus infusion of abciximab compared to those receiving placebo (35.1 vs. 27.0%, $p < 0.004$). At 3 years, there was a 13% relative reduction in composite events (47.2 vs. 41.1%, $p = 0.009$). Of note, the 6-month analysis revealed a 26% relative reduction in target vessel revascularization after successful PCI in patients receiving abciximab (22.3 vs. 16.5%, $p = 0.007$). This finding suggested that abciximab reduced the incidence of clinical restenosis. The success of the EPIC study led to FDA approval of abciximab as the first αIIbβ3 antagonist for coronary intervention.

### B. EPILOG (Evaluation of PTCA to Improve Long-Term Outcome with Abciximab GPIIb-IIIa Blockade)

The major criticism of the outcome of the EPIC study was reflected in the accompanying editorial upon its publication:[54]

> Any antithrombotic benefit (of abciximab) may be accompanied by a corresponding risk of hemorrhage, as demonstrated by the twofold increase in risk of both major and minor bleeding in patients receiving abciximab compared to placebo. Furthermore, the usefulness of these agents may be limited only to patients undergoing high-risk intervention, where the incidence of thrombotic complications involving PCI is high.

The EPILOG study was designed to address these two criticisms.[55] First, through a pilot study, PROLOG,[56] the hypothesis was tested that the increased incidence of bleeding in EPIC could be ameliorated through a more conservative use of heparin during the procedure, with no postprocedure heparin therapy and with early sheath removal. Second, the study aimed to show that the benefits of abciximab could be extended to all patients undergoing coronary intervention, regardless of the risk of ischemic complications. Accordingly, EPILOG was designed to exclude patients with acute MI and ongoing ischemia as demonstrated by dynamic ECG changes within 24 hours of enrollment. Once enrolled, the patients were randomized to three different treatment arms: heparin only ("standard" heparin treatment was weight-adjusted at 100 units/kg with additional boluses as needed to target an activated clotting time [ACT] of >300 seconds during PCI), standard heparin plus bolus and infusion of abciximab (same dose as EPIC), and low-dose heparin (defined as 75 units/kg of heparin with additional boluses as needed to target an ACT of >200 seconds during PCI) plus bolus and infusion of abciximab. The heparin therapy was not continued after the procedure, and sheaths were removed once the ACT was <175 seconds (in most cases within 6 hours of the procedure).

The primary end points of EPILOG were similar to those of EPIC and included a composite of death, nonfatal MI, or recurrent ischemia at 30 days. However, the definition for recurrent ischemia was limited to the need for urgent repeat PCI or CABG and excluded the need for bailout stenting and insertion of an intra-aortic balloon pump, as previously specified in EPIC. Since only a very small subset of patients (<0.7%) in EPIC required intra-aortic balloon pump and/or stenting, comparison between the two trials was still relevant but the definition for the composite end point in EPILOG was simplified.

The planned enrollment of EPILOG was for 4800 patients, but the trial was terminated on the recommendation of the independent Data and Safety Monitoring Board after enrollment of 2792 patients when an unexpectedly strong clinical benefit was observed at the first interim analysis. There was a 54% relative risk reduction of the composite end point with the standard heparin plus abciximab compared to heparin alone (11.7 vs. 5.4%, $p < 0.0001$) and a 56% relative reduction of events with the low-dose heparin and abciximab compared to heparin (11.7 vs. 5.2%, $p < 0.0001$). Furthermore, there was no statistically significant difference in bleeding risk (major and minor) in patients receiving low-dose heparin plus abciximab versus standard heparin (6.1 vs. 6.8%). Hence, this study was able to address the two perceived limitations raised after the EPIC trial. First, the benefit of $\alpha IIb\beta 3$ receptor blockade during PCI is not limited to patients with a high-risk clinical profile. Second, the increased incidence of bleeding seen

in the abciximab arm of EPIC is not an inherent limitation of the drug but can be overcome by more judicious use of heparin during the procedure and aggressive sheath management.

The secondary 6-month analysis revealed that the suppression of ischemic events was maintained with 31% relative risk reduction in both abciximab arms compared to heparin only.[57] However, the decreased rate of revascularization seen at 6 months with the use of abciximab in EPIC was not observed in EPILOG, suggesting that there was no effect on "clinical restenosis" with the use of abciximab in this study. The reason for this disparate effect was not evident at the time; however, among potential explanations were the increased use of coronary stenting (13% in EPILOG compared to <1% in EPIC), differences in the study patient population, and the longer duration of heparin.

## C. EPISTENT (Evaluation of Platelet Inhibition in Stenting)

During the EPILOG study, the practice of interventional cardiology was changing from use of stents only for suboptimal results and threatened vessel closure to the routine use of stents for improved clinical outcome, specifically for reducing the incidence of restenosis. Thus, routine stenting was replacing the practice of bailout stenting. This rapid transformation was supported by two studies, Benistent (520 patients) and the Stent Restenosis Study (410 patients).[58,59] These two studies, although relatively small in size, demonstrated that the routine use of stents decreases the rate of both clinical and angiographic restenosis. Since only a small proportion of patients in EPIC and EPILOG (<1 and 13%, respectively) received a coronary stent, the benefits of $\alpha IIb\beta 3$ blockade, if any, with the advent of routine stenting became an open question. Although the data available at the time suggested that these two treatment modalities were complimentary, with abciximab reducing periprocedural events (the rates of MI and death) and stents offering more long-term benefits in reducing the need for repeat revascularization, this hypothesis needed to be tested. Hence, the EPISTENT trial was initiated to address this clinical question.[60] The study was designed with three different treatment arms. Patients were randomized to stenting with adjunctive therapy with abciximab, balloon angioplasty with adjunctive therapy with abciximab, or stenting alone. Randomization to placebo or abciximab was blinded in the stented patients. Patients randomized to abciximab received the low-dose weight-adjusted heparin regimen according to the EPILOG dosing, whereas those randomized to placebo received the standard-dose heparin regimen (100 units/kg bolus, target ACT >300 seconds). The inclusion criteria were broader than either EPIC or EPILOG in order to reflect the "real-world" broad spectrum of patients and make the results more clinically applicable. Specifically, the exclusion criteria included acute MI, prior stenting in the target lesion, any intervention in the past 3 months, and planned rotational arterectomy. With the exception of acute MI, the other exclusions were in place due to the use of a stent as a study variable. The end points of EPISTENT were defined to reflect the anticipated dichotomy of benefits between abciximab and stents. Specifically, two principal end points were defined for assessment at 30 days and at 6 months: death or MI as a combined end point and repeat revascularization of the target vessel.

The study was completed with enrollment of 2399 patients. At 30 days, the incidence of the composite end point of death or MI was 10.8% in the group that received a stent and placebo, compared to 5.3% in the group that received a stent and abciximab ($p < 0.001$) and 6.9% in the group assigned to balloon angioplasty and abciximab ($p = 0.007$). The rate for repeat revascularization of the target vessel at 6 months was 10.6% in the stent plus placebo group, compared to 8.7% in the stent plus abciximab group ($p = 0.22$) and 15.4% in the angioplasty plus abciximab group ($p = 0.005$).[60] Hence, EPISTENT showed that periprocedural benefits derived from abciximab were independent of coronary stenting (specifically in terms of death or MI) and that coronary stenting offered the best protection from clinical restenosis. As such, the investigators concluded that for coronary revascularization, abciximab and stent implantation confer complimentary long-term clinical benefits. With regard to any role of abciximab in preventing clinical restenosis in conjunction with coronary stenting, there was a 22% relative risk reduction in favor of abciximab, a difference that was not statistically significant. However, a subset analysis of diabetic versus nondiabetic patients revealed that diabetic patients did receive a significant reduction in both clinical and angiographic restenosis with abciximab in conjunction with coronary stenting compared with stenting alone (8.1 vs. 16.6%, $p = 0.02$ for clinical restenosis).

Together, EPIC, EPILOG, and EPISTENT randomized more than 7000 patients in placebo-controlled trials, and these trials consistently documented the efficacy of abciximab in reducing periprocedural ischemic complications of death, MI, and need for repeat revascularization. These findings were consistent in a diverse group of patients, and the benefits were sustained for at least 3 years of follow-up. A long-term benefit from abciximab in reducing the risk of clinical restenosis in all patients was suggested by the EPIC trial; however, in EPISTENT, with routine stenting, this benefit appeared limited to diabetic patients. The clinical success of these abciximab studies resulted in other trials of small-molecule inhibitors of αIIbβ3 in the setting of PCI.

### D. IMPACT II (Integrilin to Minimize Platelet Aggregation and Coronary Thrombosis II)

IMPACT was a multicenter, double-blinded randomized trial assessing the benefit of two regimens of eptifibatide infusion in reducing PCI events compared to placebo.[2] The dosing regimen was based on a smaller phase II trial, IMPACT Hi/Lo.[61] Patients were randomized to one of three regimens: a bolus of 135 μg/kg of eptifibatide with an infusion of 0.5 μg/kg/min, a bolus of 135 μg/kg eptifibatide with an infusion of 0.75 μg/kg/min, or a bolus and infusion of placebo. The infusion was maintained for 24 hours in all three arms. The duration of the infusion was chosen to achieve approximately the same duration of platelet inhibition as obtained with the 12-hour abciximab infusion. The clinical end point was a composite of death, MI, or repeat revascularization or coronary stent implantation for abrupt vessel closure at 30 days. The study was based on intention-to-treat analysis, with secondary analysis based on the treatment received. Analysis of the composite end points at 24 hours and 6 months was also prespecified in the study protocol.

IMPACT II was the largest αIIbβ3 trial to that date with enrollment of 4010 patients. The primary composite end point occurred in 11.4% of patients in the placebo group compared to 9.2% in the 135/0.5 eptifibatide group and 9.9% in the 135/0.75 eptifibatide group ($p = 0.062$ and $p = 0.22$, respectively). By treatment-received analysis, the 135/0.5 group achieved statistically significant benefit in outcome (11.6 vs. 9.1%, $p = 0.035$). The secondary analysis also revealed a statistically significant event rate at 24 hours between placebo groups (9.3%) versus either the 135/0.5 group (6.6%, $p = 0.006$) or the 135/0.75 group (6.9%, $p = 0.014$). However, at 6 months there was no statistically significant difference between the treatment arms and placebo for the major composite end point. In terms of safety, there was no difference among any of the groups with respect to stroke or the rate of major or minor bleeding.

The results of this trial were somewhat disappointing for two reasons. First, although there was evidence of clinical benefit from utilization of the drug in lowering clinical end points, the relative benefit derived from eptifibatide appeared less robust than seen with abciximab at 30 days (13–19 vs. 35%). Furthermore, there was no persistence of this benefit at 6 months, as previously observed with abciximab. Second, there was no dose–response relationship between the treatment regimen (135/0.5 vs. 135/0.75) and the clinical benefit, as one may have expected. Both of these points were subsequently addressed through careful study of the pharmacodynamic effects of eptifibatide. First, while the IMPACT II study was still ongoing, it was found that the initial dosing regimen to establish 80% αIIbβ3 blockade in IMPACT Hi/Lo based on blood samples collected into sodium citrate lowered the free calcium concentration and increased the apparent sensitivity of αIIbβ3 to blockade by eptifibatide. At physiological blood levels of calcium, eptifibatide only inhibited platelet aggregation by 30 to 50%. Second, because 60 to 70% of all the 30-day events occurred within the first 6 hours of treatment, the infusion had little effect, and all the benefits derived would arise from the bolus dose only. Hence, no dose–response should be expected from the two treatment arms using the same bolus dose with suboptimal different infusion rates. These findings resulted in a redosing phase II study, Platelet Aggregation and Receptor Occupancy with Integrilin a Dynamic Evaluation, that established a new dosing regimen for eptifibatide using noncitrated blood to measure platelet aggregation as an end point. This new dosing regimen was used in the subsequent Enhanced Suppression of the Platelet IIb-IIIa Receptor with Integrilin Therapy (ESPRIT) trial to study eptifibatide in the setting of PCI.

## E. ESPRIT (Enhanced Suppression of the Platelet IIb-IIIa Receptor with Integrilin Therapy)

This trial was the second large-scale trial using eptifibatide in patients undergoing PCI.[62] There were several novel aspects of this trial. First, the trial used a higher dosing regimen of eptifibatide, designed to achieve 90% inhibition of platelet aggregation in >90% of patients. This dosing regimen was sought by administration of a 180 μg/kg bolus followed by a 2.0 μg/kg/min infusion and a second bolus of 180 μg/kg 10 minutes after the first bolus. It was hoped that with higher dosing of eptifibatide, more complete platelet inhibition would be achieved, thereby improving the modest clinical benefit seen in IMPACT II. Another aspect of this trial was the targeting of patients receiving coronary stents, similar to EPISTENT. With rapid improvement of interventional techniques and stent technology, the notion that the αIIbβ3 blockers should be utilized only as a bailout strategy when there was a poor outcome was being advocated, without clear data for or against such an approach. ESPRIT was designed to address this uncertainty by allowing both a bailout strategy and an open-label α_{IIb}β_3 antagonist therapy for periprocedural ischemic complications of PCI (abrupt closure, no reflow, coronary thrombosis, or other similar complications). In ESPRIT, 2604 patients were randomized in a double-blinded fashion to either the 180/2/180 dosing regimen of eptifibatide or placebo. All patients received aspirin, a thienopyridine (clopidogrel or ticlopidine), and weight-adjusted heparin at 60 units/kg. The study primary end point was the composite of death, MI, urgent target vessel revascularization, and thrombotic bailout with αIIbβ3 antagonist therapy within 48 hours after randomization. The prespecified secondary end point was the composite of death, MI, and urgent target vessel revascularization within 30 days after randomization. The primary end point allowed assessment of a bailout strategy with use of eptifibatide, and

the secondary end point allowed the indirect comparison between the clinical efficacy of eptifibatide in this trial to prior abciximab trials with similar end points. The major exclusion criteria were MI within 24 hours and refractory ischemia requiring urgent PCI.

The trial was terminated early due to efficacy associated with eptifibatide. The primary end point was reduced from 10.5% in the placebo group to 6.6% in the treatment group, a relative risk reduction of 37% ($p = 0.0015$). The secondary end point was reduced from 10.5% in the placebo group to 6.8% in the treatment arm, a relative risk reduction of 35% ($p = 0.0034$). Although infrequent (<1.5%), there was a statistically significant increased incidence of major bleeding in the treatment group despite the use of a conservative heparin regimen of 60 units/kg (compared to 75 units/kg in EPILOG and EPISTENT). The results of the trial supported both of its major objectives. First, it showed that routine use of an αIIbβ3 inhibitor is superior to a conservative strategy of reserving treatment as a bailout in patients with ischemic complications during PCI. Second, it showed that eptifibatide, with the new dosing regimen, can achieve significant reduction of clinical ischemic end points of similar magnitude to that obtained in previous trials using abciximab. However, as a point of caution, this study was not designed to demonstrate clinical equivalence and/or superiority of eptifibatide to abciximab.

## F. RESTORE (Randomized Efficacy Study of Tirofoban for Outcomes and Restenosis)

RESTORE was the first large-scale randomized trial of tirofiban in patients undergoing PCI.[63] It was targeted toward patients with ACS (unstable angina or MI) who were scheduled to undergo PCI within 72 hours of presentation. Patients were randomized to either a treatment arm of a bolus of 10 μg/kg tirofiban followed by an infusion of 0.15 μg/kg/min for 36 hours or bolus and infusion of placebo. The end points of the study were death from any cause, MI, repeat PCI due to recurrent ischemia, CABG due to PCI failure or recurrent ischemia, and insertion of a stent to abort acute vessel closure. The primary end point was the composite of these events at 30 days. The study also prespecified assessment of the same events at 2, 7, and 180 days. A total of 2139 patients were randomized to tirofiban or placebo. The primary composite end point at 30 days was reduced from 12.2% in the placebo group to 10.3% in the tirofiban group, a 16% relative reduction ($p = 0.16$). However, 2 days after angioplasty, the tirofiban group had a 38% ($p < 0.05$) relative reduction of events and at 7 days there was a 27% ($p = 0.022$) relative reduction of events. The incidence of major bleeding was not statistically different between the tirofiban and the control groups. Based on these results, the RESTORE trial showed that tirofiban produced a short-term improvement in clinical outcome. However, this benefit did not persist at 6 months, as previously shown for the abciximab trials (EPIC, EPILOG, and EPISTENT) and the subsequent eptifibatide trial (ESPRIT). There were two methodological differences between the end point definition in these trials versus the RESTORE trial. First, the incidence of non-Q wave MI was based solely on sampling for creatine kinase-myocardial band (CK-MB) at the end of the 36-hour infusion or with recurrent ischemia postintervention, and it was not based on routine blood sampling within the first 24 hours as in the abciximab trials. This difference could lead to an underestimation of the event rate in RESTORE and, hence, an underestimation of the subsequent benefit of the drug and its overall clinical efficacy. The consequence of this methodological difference is apparent in the frequency of MI in the control group in RESTORE versus that in EPILOG and EPIC (5.7 vs. 8.7 and 8.6%, respectively). Second, the need for revascularization in RESTORE was based on a broader definition of the presence of ischemia rather than the urgent need for repeat revascularization as in the other trials of abciximab and IMPACT II. Post hoc analysis of the data using this more strict definition for urgent revascularization changed the results of the composite event rate in the RESTORE trial to 10.5% in the placebo versus 8.0% in the tirofiban group, a relative risk reduction of 24% ($p = 0.052$). Nevertheless, a 6-month angiographic restenosis subgroup analysis did not reveal any long-term additional benefit in reducing the rate of restenosis (angiographic) in patients treated with tirofiban, in contrast to the abciximab trials.

The results of RESTORE, although broadening the spectrum of αIIbβ3 antagonists that can achieve reduction of clinical ischemic end points during PCI, left unanswered the question as to whether all these agents are clinically equivalent. Despite differences in reported reduction of relative risk with these agents, the trials were of sufficient difference in design, patient population, and definition of clinical end points that it was not possible to meaningfully compare these inhibitors. Ultimately, this uncertainty gave impetus to the first head-to-head comparison of these agents, abciximab and tirofiban, in a later trial, Do Tirofiban and ReoPro Give Similar Efficacy Trial (TARGET).

### G. TARGET (Do Tirofoban and ReoPro Give Similar Efficacy Trial)

This was the first study in which a head-to-head comparison of the efficacy of two αIIbβ3 antagonists was conducted in patients undergoing coronary stenting.[64] The trial was designed to show the noninferiority of tirofiban to abciximab in such a setting. The dose of abciximab was similar to that used in previous PCI trials (for 12 hours) and the dose of tirofiban was identical to the dosing regimen of RESTORE (for 18–24 hours). The patients were randomized to either tirofiban or abciximab in a double-blind fashion. The primary end point was a composite of death,

MI, and target vessel revascularization within 30 days of the procedure. A total of 2398 patients were randomized to tirofiban and 2411 patients to abciximab.

The primary composite end point occurred more frequently among the tirofiban group than among the abciximab group (7.6 vs. 6.0%), demonstrating the superiority of abciximab over tirofiban ($p = 0.038$). This superiority of abciximab to tirofiban was evident in all prespecified subgroups except for patients who underwent coronary stenting for indications other than ACS. There was no significant difference in the rates of major bleeding between the two groups.

Despite the low event rates in both treatment arms, this relatively large trial was able to establish that in patients undergoing PCI, abciximab is superior to tirofiban in reducing ischemic event rates. Among potential explanations for superiority of abciximab in such a setting is the difference in the binding sites for the two drugs on αIIbβ3 and the ability of abciximab to bind to additional integrin receptors (see Section IV.A and Table 62-1). Furthermore, the pharmacokinetics of these two agents may also alter their therapeutic efficacy because different "dose-finding" regimens were used to establish their respective therapeutic platelet inhibition doses. The results of TARGET demonstrating the clinical superiority of abciximab over tirofiban should not be extrapolated to all small molecules, including eptifibatide. Overall, the results of TARGET ushered in the era in which we have moved from testing the benefits of αIIbβ3 inhibition in reducing ischemic events in patients undergoing PCI to tailoring the differences between these agents to specific patient populations to achieve maximal clinical benefit.

## VI. αIIbβ3 Blockade in the Medical Management of Acute Coronary Syndromes

### A. CAPTURE (C7E3 Anti-Platelet Therapy in Unstable Refractory Angina)

This trial assessed the benefit of αIIbβ3 blockade with abciximab in patients who were scheduled to undergo PCI to treat angina that was refractory to standard medical treatment.[65] The study design required pretreatment with abciximab for 18 to 24 hours between the time of the first angiogram and the actual angioplasty. The abciximab infusion was then continued for 1 hour after angioplasty. The major clinical end point was the composite of death, MI, and repeat revascularization within 30 days. Since all patients were pretreated with abciximab for at least 18 hours prior to PCI, this was the first study to allow assessment of the benefit of αIIbβ3 blockade in patients with medically refractory angina independent of PCI.

The study enrolled 1265 patients and was terminated early due to the significant treatment benefit with abciximab. The primary end point occurred in 15.9% of patients who received placebo compared to 11.3% of patients who received abciximab, for a relative risk reduction of 29% ($p = 0.012$). Furthermore, there was a benefit in pretreatment with abciximab in reducing the incidence of MI prior to PCI — 2.1 versus 0.6% ($p < 0.03$). However, unlike the earlier studies with abciximab, benefit was not sustained at 6 months. One explanation for the lack of efficacy at 6 months in CAPTURE as contrasted with EPIC, EPILOG, and EPISTENT was the differences in dosing regimens. In the latter trials, abciximab was continued for 12 hours after initiation of PCI rather than termination of the drug 1 hour after PCI as in CAPTURE. The ability of abciximab to decrease the incidence of MI before PCI raised the possibility that up-front treatment with drug could even further increase the treatment benefit seen in EPIC, EPILOG, and EPISTENT. This was addressed in the Global Use of Strategies to Open Occluded Coronary Arteries-IV (GUSTO-IV) trial, in which patients were pretreated with abciximab prior to PCI (see Section VI.D). The results of CAPTURE opened a new arena for the use of αIIbβ3 antagonists in the medical management of patients with ACS (unstable angina and non-Q wave MI) outside the catheterization laboratory.

### B. PURSUIT (Platelet IIb-IIIa in Unstable Angina: Receptor Suppression Using Integrilin Therapy)

This trial was the largest single study of αIIbβ3 antagonists, enrolling a total of 10,498 patients. This trial was designed to assess the treatment benefits of eptifibatide in the medical management of ACS in patients who were already receiving aspirin and heparin. The inclusion criteria included presentation with ischemic chest pain lasting at least 10 minutes and either ECG changes suggestive of acute ischemia or elevation of enzyme markers (CK-MB above the upper limit of normal for the enrolling hospital or total CK elevation two times above the upper limit of normal if MB isoform was not available). The patients were randomized either to bolus and infusion of placebo or to a 180 µg/kg bolus of eptifibatide followed by 2.0 µg/kg/min infusion. (A third arm with 1.3 µg/kg/min infusion of eptifibatide was dropped after the safety of the higher infusion regimen was established at interim analysis.) The major end point of the trial was the composite of death and MI at 30 days. Of note, the definition of MI using the serum marker was more liberal than in CAPTURE and previous PCI trials. The primary end point occurred in 15.7% of the placebo group and 14.2% of the treatment arm, for a relative risk reduction of 10% ($p = 0.04$). All the benefit was achieved by 96 hours after the index event and persisted through the 30-day end point. The 6-month analysis of outcome revealed that the early benefit was maintained.[66]

Most of the benefit in PURSUIT was derived from a decrease in the incidence of non-Q wave MI. Furthermore,

although not prespecified in the study design, analysis of the outcomes in patients treated with eptifibatide demonstrated that the subgroup of patients with the most pronounced benefit were those who subsequently underwent PCI — 31% relative risk reduction ($p = 0.01$) versus 7% for patients who were only treated medically ($p = 0.23$).[66] Although achieving its major clinical end point, the treatment benefit from the PURSUIT study was modest and limited to non-Q wave MI (with a liberal definition of MI) patients, especially in those who subsequently underwent PCI. However, the study was important for achieving this benefit in a "practice-based" design, with no protocol-specified strategy for catheterization and revascularization. Furthermore, it validated the findings of CAPTURE that treatment with a αIIbβ3 antagonist can reduce ischemic events in patients with ACS.

### C. PRISM (Platelet Receptor Inhibition in Ischemic Syndrome Management) and PRISM-Plus (Platelet Receptor Inhibition in Ischemic Syndrome Management in Patients Limited by Unstable Angina)

These two studies examined the efficacy of tirofiban in patients with ACS.[67,68] Both studies enrolled patients with ischemic chest pain in addition to ECG changes or CK-MB markers. The PRISM study had a more liberal inclusion criterion by including patients with a history of coronary artery disease (as documented by angiogram, previous revascularization, or positive stress test) in lieu of ECG or enzyme markers. In both studies, patients were randomized to tirofiban for 48 hours versus heparin for 48 hours. The PRISM-Plus study also had a third arm of tirofiban in addition to heparin. The dose of tirofiban in both studies was the same — 0.6 µg/kg/min for 30 minute bolus followed by a 0.15 µg/kg/min infusion. The primary end point was a composite of death, MI, and refractory ischemia. PRISM used a 48 hour time point to assess drug efficacy, and PRISM-Plus used a 7 day time point. The definition of refractory ischemia was based on strict prespecified guidelines to include patients with ongoing evidence of ischemia despite "full" medical therapy. The departure from the 30 day time point was advocated by the investigators because they expected the benefit of treatment to be present early during the acute phase of the coronary syndrome. Furthermore, the study design did not prespecify patient management beyond the 48 hour initial medical treatment (during which time revascularization was actively discouraged) and any treatment benefit may have been confounded at later time points by the benefits of αIIbβ3 blockade with routine PCI. However, the same end points were collected and reported for the 30 day time point in both studies.

The PRISM study enrolled 3232 patients. It achieved its primary end point by demonstrating a 32% relative risk reduction at 48 hours with tirofiban treatment (5.6 vs. 3.8%,

$p = 0.01$). However, this early benefit was not sustained at 30 days, with an event rate of 17.1 versus 15.9% (placebo vs. tirofiban, $p = 0.34$). Of note, the rate of PCI during the 30 day time point was 22% in the cohort.

The PRISM-Plus trial enrolled 1915 patients. It was stopped prematurely after the first interim analysis due to an increased incidence of death at 7 days in the patient arm receiving tirofiban only. This was surprising, especially because the results from PRISM showed benefit in a similar patient population with tirofiban alone. Without any clear explanation, the investigators attributed this to chance, with a small number of events analyzed in the first interim analysis (16 vs. 4 events). However, there was a treatment benefit in patients receiving tirofiban with heparin — a 28% relative risk reduction in the heparin–tirofiban treatment group at 7 days (17.9 vs. 12.9%, $p = 0.004$). This benefit was maintained at 30 days (22.3 vs. 18.5%, $p = 0.02$) and at 6 months (32.1 vs. 27.7%, $p = 0.02$). Again, similar to the results of the PURSUIT trial, the 31% of patients who subsequently underwent PCI after the initial medical management with tirofiban derived the most benefit in the study in terms of reduction of death and/or MI at the 30 day time point — 42 versus 23% for patients who were only treated medically.[69] The 6 month analysis of outcome showed persistence of this early benefit, again especially in patients who underwent PCI.

### D. GUSTO-IV (Global Use of Strategies to Open Occluded Coronary Arteries-IV)

GUSTO-IV tested the hypothesis that abciximab can have a clinical benefit when used solely in medical management of ACS.[70] The premise of the trial was based on the results of CAPTURE demonstrating abciximab reduced ischemic events in patients with refractory angina awaiting PCI and on the success of other αIIbβ3 antagonists in this setting, as demonstrated in PURSUIT, PRISM, and PRISM-Plus. Patients with ACS (excluding patients with ST-segment elevation MI) were enrolled if they met the following enrollment criteria: one or more episodes of angina lasting at least 5 minutes in the preceding 24 hours and either ECG changes suggestive of acute ischemia or elevation of cardiac troponin T or I at the time of evaluation. The primary end point was a composite of death and MI at 30 days. Of note, the definition of MI (CK-MB elevation three times the upper limit of normal) was stricter than that in PURSUIT, PRISM, or PRISM-Plus. The study enrolled 7800 patients and randomized them to three arms: a bolus and infusion of abciximab for 24 hours followed by a 24-hour infusion of placebo, a bolus and infusion of abciximab for 48 hours, or a bolus and infusion of placebo for 48 hours. The dosing of abciximab was 0.25 mg/kg bolus and 0.125 µg/kg/min infusion. Angiography and/or revascularization was discouraged during the infusion period and 12 hours postinfusion. The primary end point occurred in 8.0% of patients in the placebo group,

8.2% of patients with the 24-hour infusion of abciximab, and 9.1% of patients with the 48-hour infusion of abciximab. There was no statistical difference in reduction of events in either abciximab arm compared to placebo.

The lack of any efficacy, specifically in reduction of MI, of either abciximab regimen compared to placebo was in stark contrast to previous trials using small molecules (tirofiban and eptifibatide) in this setting. There was no clear explanation for this. Among possible explanations ruled out by the investigators was the definition of MI. Using a more liberal definition for MI similar to PRISM-Plus (MB twice the upper limit of normal) or PURSUIT (any MB elevation above the upper limit of normal), the outcome of the trial remained the same.

The lack of clinical benefits in GUSTO-IV has to be attributed to one of two general explanations: (1) a unique aspect of this trial that, in contrast to prior trials, diminished any beneficial effect of abciximab in reducing ischemic events in patients with ACS or (2) an agent-specific phenomenon unique to abciximab (i.e., a lack of therapeutic effect of abciximab when used in medical management of patients with ACS). Possible trial-specific explanations include the characteristics of the patient population enrolled (i.e., enrollment criteria using troponin in contrast to CK-MB in prior trials). Other unique aspects in GUSTO-IV were the longer infusion duration of abciximab than in previous trials (48 vs. 12 hours in all previous PC1 trials and 18–24 hours in CAPTURE) and completion of infusion for at least 12 hours prior to any planned intervention (98% of the patients). This longer infusion of abciximab may have an untoward effect. Furthermore, among all trials using αIIbβ3 antagonists in initial medical management of patients with ACS, patients in GUSTO-IV had the longest wait period before an invasive strategy could be employed (60 hours), and an invasive strategy was adopted 12 hours or more after the drug infusion had been completed. Since both PURSUIT and PRISM-Plus derived the largest benefit in patients who underwent PCI early in the trial while on the study drug, this wait period may have diminished the drug benefit in GUSTO-IV. Although the exact basis for lack of clinical benefit in GUSTO-IV remains uncertain, the practical consequence was a more restricted use of abciximab to patients undergoing PCI as opposed to its anticipated broader use in the medical management of patients with ACS.

## VII. Utilization of αIIbβ3 Antagonists in ST-Elevation MI

The current treatment of acute MI remains timely reperfusion therapy either with a fibrinolytic agent or by mechanical intervention. The strategy of adding αIIbβ3 antagonists to augment reperfusion therapy appears attractive. The most recent trials of αIIbβ3 antagonists focus on this clinical question, with most trials utilizing abciximab and only a few small trials using tirofiban or eptifibatide.[71–77] Broadly, we can break down these trials into those using abciximab to augment primary PTCA/stenting versus trials using abciximab along with reduced-dose fibrinolytic agents to augment pharmacologic reperfusion.

### A. Mechanical Reperfusion

The first trial to directly address the role of abciximab in primary PCI was the ReoPro and Primary PTCA Organization and Randomized Trial (RAPPORT), a relatively small trial of 483 patients.[75] The findings from this trial confirmed the benefits of abciximab in primary PCI by demonstrating a 45% reduction in the composite 30-day end point of death, reinfarction, and/or urgent target vessel revascularization (TVR) (11.2 vs. 5.8%, $p = 0.03$). These benefits were extended to primary stenting in two subsequent trials, ISAR-2 and ACE.[74,77] Both of these trials demonstrated a similar 50% reduction in the same composite end point as RAPPORT (10.5 vs. 5.0%, $p = 0.038$ in ISAR-2 and 10 vs. 4.5%, $p = 0.05$ in ACE in favor of abciximab). A third study, ADMIRAL, which used the novel strategy of starting abciximab immediately after the patient had been stratified to primary PCI, demonstrated a 60% reduction in the same composite end point at 30 days (14.6 vs. 6.0%, $p = 0.01$), with persistence of this benefit at 6 months.[76] Of note, 70% of the patients in ADMIRAL who were stratified to receive abciximab had it administered outside the catheterization laboratory, and 100% had it administered prior to sheath insertion.

The findings from the previously mentioned four trials contrast with the results from the large primary PCI trial, Controlled Abciximab and Device Investigation to Lower Late Angiographic Complications (CADILLAC), a trial of 2082 patients undergoing primary angioplasty or stenting with adjunctive therapy with abciximab.[78,79] The clinical end point was a composite of death, reinfarction, TVR, or stroke. The addition of stroke to the composite end point, a rare event, did not preclude the comparison of the findings in this trial to the those of other primary PCI trials of abciximab. Similar to RAPPORT, at 30 days there was a significant reduction in events with primary PCI with adjunctive abciximab (8.3 vs. 4.8%, $p < 0.02$), but with primary stenting there was no benefit with adjunctive abciximab (5.7 vs. 4.4%, $p = 0.2$). Of note, the absolute event rate in the stent only group in this study was significantly lower than those of other comparable trials (5.7% in CADILLAC vs. 10–14.6% in ISAR-2, ACE, and ADMIRAL). A potential explanation for this divergence and lack of benefit from abciximab with stenting in CADILLAC may lie in the early administration of thienopyridines (prior to catheterization) as part of the study protocol, a point further discussed later.

A meta-analysis of all trials discussed previously (as well as three smaller studies) showed that abciximab in the

setting of primary PCI/stenting is associated with a statistically significant absolute reduction in mortality at both 30 days and 6 months (1% at 30 days and 1.8% at 6 months; $p = 0.047$ and $p = 0.01$, respectively).[80]

### B. Pharmacological Reperfusion

The role of abciximab in this setting has been investigated in two major trials, ASSENT III and GUSTO-V, enrolling more than 22,000 patients combined. Both trials compared full-dose fibrinolytics (reteplase or tenecteplase) versus half-dose fibrinolytics combined with full-dose abciximab.[72,73] The results have been uniform across both trials: Abciximab use is associated with decreased reinfarction rate at a cost of increased major bleeding events, with no overall difference in mortality or the composite end point of death, reinfarction, or urgent TVR. A meta-analysis combining the results of these two trials, along with those of a much smaller TIMI-23 trial of 483 patients, revealed an absolute 1.2% reduction in 30-day reinfarction rate (2.1 vs. 3.3%, $p < 0.001$) and an absolute 2.1% increase in major bleeding events (5.2 vs. 3.1%, $p < 0.001$), with no change in 30-day mortality rates (5.2 vs. 5.5%, $p = 0.61$) with utilization of abciximab in this setting.[80–82]

In summary, there is little role for "adjunctive" abciximab therapy in the setting of pharmacological reperfusion therapy. However, because primary PCI has become the reperfusion therapy of choice for ST-elevation MI, the role of adjunctive αIIbβ3 antagonists initiated prior to presentation to the catheterization laboratory needs to be addressed. The results of ongoing trials, including FINESSE and CARESS, should shed light on this possibility.

## VIII. Benefits of αIIbβ3 Antagonists in the Setting of Other Adjunctive Pharmacotherapy

Several trials have shown that pretreatment with clopidogrel prior to PCI may attenuate the benefit derived from abciximab.[83,84] Initially, a subgroup analysis of the CURE trial (a study designed to address the role of clopidogrel in ACS; see Chapter 61) revealed that patients pretreated with clopidogrel for at least 48 hours prior to PCI had a significant reduction in the combined end point of death/MI/urgent revascularization at 30 days (6.4 vs. 4.5%, $p = 0.03$); this was in the setting of decreased utilization of αIIbβ3 antagonist in the clopidogrel group (20.9 vs. 26.6%, $p = 0.001$).[85] However, the overall low rate of αIIbβ3 antagonist use in this trial precluded any analysis of additional and/or additive benefit of αIIbβ3 antagonism during PCI with clopidogrel pretreatment. Two subsequent studies have directly addressed whether routine αIIbβ3 antagonism is associated with any benefit in nonemergent PCI in patients adequately pretreated with clopidogrel. In ISAR-REACT, patients pretreated with 600 mg of clopidogrel for at least 2 hours prior to PCI did not demonstrate any additional benefit with routine use of abciximab in decreasing the composite clinical end point (death, MI, or urgent TVR) at 30 days (4.0 vs. 4.0%, $p = 0.82$).[83] The ISAR-SWEET trial extended the same findings to diabetic patients.[85] These findings must be considered in the context of the population enrolled — relatively low-risk patients without clinical evidence of ACS. As such, the strategy of substituting clopidogrel pretreatment for an $\alpha_{IIb}\beta_3$ antagonist during PCI in patients with ACS and/or acute MI remains untested.

Bivalirudin, a direct thrombin inhibitor, is being developed as an anticoagulant with the potential to replace unfractionated heparin. The clinical equivalence (noninferiority) of bivalirudin versus provisional use of αIIbβ3 antagonists with unfractionated heparin was tested in the REPLACE-II trial, in which 6010 patients were randomized between the two treatment arms.[86] The provisional use of αIIbβ3 antagonists took place in approximately 6.2% of the patients. There was no difference in the composite end point of MI, urgent TVR, or death at 30 days in the bivalirudin versus the αIIbβ3 antagonist plus heparin arm (7.6 vs. 7.1%, $p = 0.4$); however, there was a significant decrease in the rate of major bleeding with bivalirudin use (2.4 vs. 4.1%, $p < 0.001$). Of note, patients with acute MI or ACS requiring αIIbβ3 antagonist prior to catheterization were excluded from the study and ACS patients represented 21% of the overall participants. Furthermore, 85% of the patients in the trial were pretreated with a thienopyridine prior to the procedure (the majority of them for more than 2 hours prior to the procedure). Based on the study results, the investigators concluded that the strategy of bivalirudin with provisional αIIbβ3 antagonist is an attractive alternative to the use of unfractionated heparin with routine use of αIIbβ3 antagonist in patients with low to moderate risk characteristics. Six- and 12-month follow-up data from REPLACE-2 have continued to support the clinical equivalence of these two clinical strategies.[87]

# IX. Side Effects

The incidence of hemorrhage with αIIbβ3 antagonist therapy is discussed in Sections V and VI. In at least one meta-analysis, the incidence of severe bleeding appears more significant with abciximab compared to small-molecule αIIbβ3 antagonists.[88] For patients who develop life-threatening bleeding, the antiplatelet effect of abciximab may be reversed by discontinuation of drug infusion and by platelet transfusion (the latter being more effective after the approximately 10–30 minutes required for clearance of the drug from plasma). After transfusion, abciximab redistributes from the patient's platelets to the transfused platelets, thereby reducing the mean level of receptor block-

ade. Platelet transfusions are rarely necessary with the rapidly reversible agents eptifibatide and tirofiban and might not be effective during the approximately 2 hours required for clearance of these agents from plasma. αIIbβ3 antagonist-induced thrombocytopenia and pseudothrombocytopenia are discussed in Chapters 49 and 55, respectively.

# X. Oral αIIbβ3 Antagonists

Collectively, the consistent efficacy of intravenous αIIbβ3 antagonists in decreasing clinical events in patients undergoing PCI and in patients with ACS (Fig. 62-4) led to anticipation of orally active αIIbβ3 antagonists. From a clinical standpoint, two obvious benefits of oral αIIbβ3 agents were envisioned. First, in patients with ACS, platelets can continue to be activated for extended periods after the initiating event, long after the intravenous αIIbβ3 antagonists have been cleared. Continuing administration of oral agents could potentially provide protection for the extended period of vulnerability. Second, if oral agents could provide the same clinical efficacy as intravenous agents, they could become the agents of choice during initial treatment for ACS and PCI because of their ease of administration, and they could then be continued for long-term use beyond the acute hospitalization. However, the results of the first five major trials that were completed phase III studies with oral αIIbβ3 antagonists were extremely disappointing. Four different agents were used in these trials — sibrafiban, xemilofiban, orbofiban, and lotrafiban. Each of these agents is a prodrug that releases an active metabolite that inhibits αIIbβ3 function.

The first of these trials, Sibrafiban Versus Aspirin to Yield Maximum Protection from Ischemic Heart Events Post-Acute Coronary Syndromes (SYMPHONY), used sibrafiban in patients after an episode of an ACS.[89] The study had three arms: a placebo, a low-dose regimen designed to achieve 25% platelet inhibition, and a high-dose regimen designed to achieve 50% inhibition. The primary end point was the composite of death, nonfatal MI, and severe recurrent ischemia at 90 days. A total of 9233 patients were randomized between these treatment groups. The primary end point at 90 days did not differ significantly between the two treatment groups and the placebo group. However, there was a higher incidence of major bleeding in the sibrafiban-treated patients.

Evaluation of Oral Xemilofiban in Controlling Thrombotic Events (EXCITE) was the second large trial of an oral GPIIb-IIIa antagonist.[90] This trial was designed to address the long-term benefits of xemilofiban in patients undergoing PCI. The study drug was initiated 30 to 90 minutes before PCI and continued for 6 months. Similar to the SYMPHONY trial, there was a low-dose (20 mg before PCI followed by maintenance dose of 10 mg) and a high-dose (20 mg before PCI with maintenance dose of 20 mg) treatment group. The major composite end point was death, nonfatal MI, and urgent revascularization at 6 months. A total of 7232 patients were randomly assigned among the three arms. There was no statistically significant difference between the treatment groups and the placebo group in the incidence of the major composite end point at 6 months (13.5, 13.9, and 12.7% for placebo, low-dose xemilofiban group, and high-dose xemilofiban group, respectively). However, there was a disturbing trend toward a higher incidence of death in both xemilofiban-treated groups (0.3, 0.8, and 0.6% for placebo, low-dose group, and high-dose group, respectively). Furthermore, there was a consistent incidence of moderate to major bleeding with treatment with xemilofiban compared to placebo (1.8, 5.1, and 7.1% for placebo, low-dose xemilofiban, and high-dose xemilofiban, respectively).

The Orbofiban in Patients with Unstable Coronary Syndromes–Thrombolysis in Myocardial Infarction 16 (OPUS-TIMI 16) trial, similar in design to the other two oral agent trials, used orbofiban in patients with ACS, with a placebo arm, a low-dose arm (50 mg initially for 30 days followed by 30 mg thereafter for 1 year), and a high-dose arm (50 mg for the duration of the planned 1 year).[91] The primary end point was a composite of death, MI, recurrent ischemia, and urgent revascularization (similar to EXCITE and SYMPHONY). The trial was terminated prematurely because of an unexpected increase in mortality in the 50/30 orbofiban group (2.4 vs. 1.4% in placebo). Further subgroup analysis (not prespecified) revealed that the majority of increased deaths in the 10-month follow-up were the result of new thrombotic events. Furthermore, among all subgroups analyzed, only one treatment group derived benefit in term of decreased incidence of death with orbofiban treatment — patients who underwent PCI. Similar to the previous two trials, the patients receiving the oral agents had a higher incidence of bleeding episodes.

The last two trials of oral agents, SYMPHONY II and BRAVO (Blockade of the IIb-IIIa Receptor to Avoid Vascular Occlusion), enrolled patients with ACS. Both trials were terminated prematurely at interim analysis due to an increased incidence of death at 30 days in the treatment groups.

Several explanations have been put forward to account for the disappointing results of these five trials of first-generation oral αIIbβ3 antagonists. First, because of differences in pharmacodynamics, dosing with the oral agents is more susceptible to fluctuations in drug levels compared to intravenous agents and hence may provide subtherapeutic platelet inhibition. For example, the targeted platelet inhibition in the first SYMPHONY trial ranged between 20 and 50%. However, based on the findings of the intravenous agents, the maximum clinical benefits of αIIbβ3 blockade are attained with platelet inhibition greater than 80%. Interindividual variations can enhance these fluctuations even

more (e.g., because of renal function, weight, or baseline platelet number). Another possible explanation is that the overall patient population represented a lower risk group, as indicated by the fewer composite end point events in the placebo group in these trials compared to placebo groups of previous trials of intravenous agents (7.9% in EXCITE vs. 11.7% in EPILOG and 12.8% in EPISTENT). Hence, any treatment benefit with αIIbβ3 inhibition in patients at risk for events might be lessened by inclusion of a more heterogeneous patient population with less risk of clinical events. This was corroborated by the results of the OPUS-TIMI 16 trial,[91] in which the subgroup of patients undergoing PCI did derive benefit with orbofiban treatment. Another explanation is that long-term blockade of αIIbβ3 might have deleterious effects that are not seen with the short-term treatment regimens of antagonists. One rationale for this explanation is that long-term receptor blockade could lead to expression of LIBS epitopes (Fig. 62-2), thereby eliciting an immune response. Another possibility is that long-term outside-in signaling induced by chronic receptor blockade could lead to a paradoxical activation of $\alpha_{IIb}\beta_3$ with a prothrombotic consequence. The prothrombotic effect of long-term treatment is supported by the results from OPUS-TIMI 16, in which most deaths in the orbofiban-treated group were due to new thrombotic events.[91]

A meta-analysis of four trials of oral αIIbβ3 antagonists found a highly significant excess mortality associated with these agents.[92] Based on these data, broad concerns were raised that oral αIIbβ3 antagonists were neither efficacious nor safe. The extensive development of oral agents, concluding with the development of several second-generation oral αIIbβ3 antagonists with different pharmacodynamic profiles, was abandoned throughout the pharmaceutical industry. Currently, no oral αIIbβ3 antagonist is in phase II or phase III clinical trials. Whether more incisive and closer laboratory monitoring of the oral αIIbβ3 antagonists could have influenced the outcome of these trials (e.g., identified patients at risk) is uncertain. Flow cytometry, VerifyNow, the platelet function analyzer (PFA)-100, and the cone and plate(let) analyzer represent examples of such approaches, and these assays are discussed in Chapters 30, 27, 28, and 29, respectively.

## XI. Emerging Indications: Cerebrovascular Disease

As percutaneous interventions for cerebrovascular arterial disease gain wider acceptance, the potential for αIIbβ3 inhibition in this setting represents an interesting target. There are case reports of benefits of αIIbβ3 antagonists as a bailout strategy in cerebrovascular interventions complicated by acute thrombosis.[93–95] However, in percutaneous carotid stenting, a rapidly expanding alternative to surgical carotid endarterectomy, the utilization of abciximab has been tested in several small studies as a means of decreasing ischemic/embolic complications of the procedure.[96,97] So far, these limited studies have not shown any beneficial effect for αIIbβ3 antagonists in this setting.

In the setting of nonhemorrhagic stroke, early utilization of αIIbβ3 antagonists may have a significant benefit in re-establishing cerebral perfusion. Furthermore, these agents may have less potential for causing intracerebral hemorrhage than has been encountered with fibrinolytic therapy in this setting. A dose-escalation study using abciximab within 3 hours of ischemic stroke showed its safety and a trend toward improved functional outcome.[98] A multicenter phase II randomized trial, the Abciximab Emergent Stroke Treatment Trial (AbeSTT), assessed the benefit and safety of abciximab in patients with ischemic stroke who were not candidates for fibrinolytic agents.[99] Four hundred patients were randomized to placebo versus standard-dose abciximab (0.25 mg/kg bolus followed by 12-hour infusion at a rate of 0.125 μg/kg/min). There was a trend toward increased intracranial hemorrhage in patients treated with abciximab (3.6 vs. 1%, $p = 0.09$). However, at 3-month follow-up, the percentage of responders (improvement in baseline NIHSS score) was higher in the abciximab group compared to placebo (37 vs. 27%, $p = 0.04$). Unfortunately, the prospective, blinded phase III controlled study of abciximab in early ischemic stroke (ABeSTT-II), at a dose applied for cardiac indications, was terminated because of significant safety (hemorrhagic) concerns.[100]

## XII. αIIbβ3 Antagonists: Current Status and Future Prospects

Integrin αIIbβ3 (GPIIb-IIIa) antagonists have been used in patients for more than a decade. Although concerns for safety dictated the cautious and careful evaluation of these drugs, major and life-threatening bleeding complications have not emerged as a major limitation. However, the anticipated broad use of this class of drugs has not come to fruition. This status has evolved from the lack of a striking benefit of αIIbβ3 antagonism in other than the highest risk patients with coronary disease and from the emergence of safe, effective, and low-cost alternative therapies, especially clopidogrel (see Chapter 61). Hence, the intravenous αIIbβ3 antagonists are now being used for relatively restricted indications, namely the prevention of thrombosis in the setting of PCI and acute MI. Although narrower than originally anticipated, these indications still result in substantial use of the approved αIIbβ3 antagonists. Their use varies geographically and by practice within individual hospitals. It is estimated that 45% of PCI patients in the northeastern United States currently receive an αIIbβ3 antagonist. Ongoing clinical trials to address the most efficacious strategy for the use of αIIbβ3 antagonists as adjunctive agents and for new indications will ultimately establish the breadth of their usage.

## References

1. Parkinson, J., & Bedford, C. (1928). Cardiac infarction and coronary thrombosis. *Lancet, 14,* 195–239.
2. IMPACT-II Investigators. (1997). Randomised placebo-controlled trial of effect of eptifibatide on complications of percutaneous coronary intervention: IMPACT-II. Integrilin to Minimise Platelet Aggregation and Coronary Thrombosis — II. *Lancet, 349*(9063), 1422–1428.
3. Phillips, D. R., et al. (1988). The platelet membrane glycoprotein IIb-IIIa complex. *Blood, 71,* 831–843.
4. Coller, B. S. (1997). GPIIb/IIIa antagonists: Pathophysiologic and therapeutic insights from studies of c7E3 Fab. *Thromb Haemost, 78,* 730–735.
5. Marguerie, G. A., & Plow, E. F. (1983). The fibrinogen-dependent pathway of platelet aggregation. *Ann N Y Acad Sci, 408,* 556–566.
6. Coller, B. S., et al. (1983). A murine monoclonal antibody that completely blocks the binding of fibrinogen to platelets produces a thrombasthenic-like state in normal platelets and binds to glycoproteins IIb and/or IIIa. *J Clin Invest, 72,* 325–338.
7. Phillips, D. R., & Agin, P. P. (1977). Platelet membrane defects in Glanzmann's thrombasthenia. Evidence for decreased amounts of two major glycoproteins. *J Clin Invest, 60,* 535–545.
8. Caen, J. P., et al. (1996). Congenital bleeding disorders with long bleeding time and normal platelet count: I. Glanzmann's thrombasthenia. *Am J Med, 41,* 2–26.
9. Nurden, A. T., & Caen, J. P. (1974). An abnormal platelet glycoprotein pattern in three cases of Glanzmann's thrombasthenia. *Br J Haematol, 28,* 253–260.
10. Jordan, R. E., et al. (1996). Preclinical development of c7E3 Fab, a mouse/human chimeric monoclonal antibody fragment that inhibits platelet function by blockade of GPIIb-IIIa receptors with observations on the immunogenicity of c7E3 Fab in humans. In H. Ma (Ed.), *Adhesion receptors as therapeutic targets* (pp. 281–305). Boca Raton, FL: CRC Press.
11. Anderson, H. V., et al. (1994). Cyclic flow variations after coronary angioplasty in humans: Clinical and angiographic characteristics and elimination with 7E3 monoclonal antiplatelet antibody. *J Am Coll Cardiol, 23,* 1031–1037.
12. Anderson, H. V., et al. (1992). Intravenous administration of monoclonal antibody to the platelet GP IIb/IIIa receptor to treat abrupt closure during coronary angioplasty. *Am J Cardiol, 69,* 1373–1376.
13. Coller, B. S., et al. (1986). Antithrombotic effect of a monoclonal antibody to the platelet glycoprotein IIb/IIIa receptor in an experimental animal model. *Blood, 68,* 783–786.
14. Gold, H. K., et al. (1988). Rapid and sustained coronary artery recanalization with combined bolus injection of recombinant tissue-type plasminogen activator and

monoclonal antiplatelet GPIIb/IIIa antibody in a canine preparation. *Circulation, 77,* 670–677.

15. Coller, B. S., et al. (1989). Abolition of *in vivo* platelet thrombus formation in primates with monoclonal antibodies to the platelet GPIIb/IIIa receptor. Correlation with bleeding time, platelet aggregation, and blockade of GPIIb/IIIa receptors. *Circulation, 80,* 1766–1774.

16. EPIC Investigators. (1994). Use of a monoclonal antibody directed against the platelet glycoprotein IIb/IIIa receptor in high-risk coronary angioplasty. The EPIC Investigation. *N Engl J Med, 330,* 956–961.

17. Peerschke, E. I., et al. (1980). Correlation between fibrinogen binding to human platelets and platelet aggregability. *Blood, 55,* 841–847.

18. Frelinger, A. L., 3rd, et al. (1990). Selective inhibition of integrin function by antibodies specific for ligand-occupied receptor conformers. *J Biol Chem, 265,* 6346–6352.

19. Frelinger, A. L., 3rd, et al. (1991). Monoclonal antibodies to ligand-occupied conformers of integrin alpha IIb beta 3 (glycoprotein IIb-IIIa) alter receptor affinity, specificity, and function. *J Biol Chem, 266,* 17106–17111.

20. Jennings, L. K., & White, M. M. (1998). Expression of ligand-induced binding sites on glycoprotein IIb/IIIa complexes and the effect of various inhibitors. *Am Heart J, 135*(5 Pt 2), S179–S183.

21. Bednar, B., et al. (1999). Fibrinogen receptor antagonist-induced thrombocytopenia in chimpanzee and rhesus monkey associated with preexisting drug-dependent antibodies to platelet glycoprotein IIb/IIIa. *Blood, 94,* 587–599.

22. Ammar, T., Scudder, L. E., & Coller, B. S. (1997). *In vitro* effects of the platelet glycoprotein IIb/IIIa receptor antagonist c7E3 Fab on the activated clotting time. *Circulation, 95,* 614–617.

23. Reverter, J. C., et al. (1996). Inhibition of platelet-mediated, tissue factor-induced thrombin generation by the mouse/human chimeric 7E3 antibody. Potential implications for the effect of c7E3 Fab treatment on acute thrombosis and "clinical restenosis." *J Clin Invest, 98,* 863–874.

24. Furman, M. I., et al. (2000). GPIIb-IIIa antagonist-induced reduction in platelet surface factor V/Va binding and phosphatidylserine expression in whole blood. *Thromb Haemost, 84,* 492–498.

25. Plow, E. F., & Marguerie, G. (1982). Inhibition of fibrinogen binding to human platelets by the tetrapeptide glycyl-L-prolyl-L-arginyl-L-proline. *Proc Natl Acad Sci USA, 79,* 3711–3715.

26. Kloczewiak, M., Timmons, S., & Hawiger, J. (1982). Localization of a site interacting with human platelet receptor on carboxy-terminal segment of human fibrinogen gamma chain. *Biochem Biophys Res Commun, 107,* 181–187.

27. Ginsberg, M., et al. (1985). Inhibition of fibronectin binding to platelets by proteolytic fragments and synthetic peptides which support fibroblast adhesion. *J Biol Chem, 260,* 3931–3936.

28. Plow, E. F., et al. (1985). The effect of Arg-Gly-Asp-containing peptides on fibrinogen and von Willebrand factor binding to platelets. *Proc Natl Acad Sci USA, 82,* 8057–8061.

29. Gartner, T. K., & Bennett, J. S. (1985). The tetrapeptide analogue of the cell attachment site of fibronectin inhibits platelet aggregation and fibrinogen binding to activated platelets. *J Biol Chem, 260,* 11891–11894.

30. Plow, E. F., Marguerie, G., & Ginsberg, M. (1987). Fibrinogen, fibrinogen receptors, and the peptides that inhibit these interactions. *Biochem Pharmacol, 36,* 4035–4040.

31. D'Souza, S. E., et al. (1988). Chemical cross-linking of arginyl-glycyl-aspartic acid peptides to an adhesion receptor on platelets. *J Biol Chem, 263,* 3943–3951.

32. Ruoslahti, E. (1996). RGD and other recognition sequences for integrins. *Annu Rev Cell Dev Biol, 12,* 697–715.

33. Kloczewiak, M., Timmons, S., & Hawiger, J. (1983). Recognition site for the platelet receptor is present on the 15-residue carboxy-terminal fragment of the gamma chain of human fibrinogen and is not involved in the fibrin polymerization reaction. *Thromb Res, 29,* 249–255.

34. Tranqui, L., et al. (1989). Differential structural requirements for fibrinogen binding to platelets and to endothelial cells. *J Cell Biol, 108,* 2519–2527.

35. Timmons, S., Kloczewiak, M., & Hawiger, J. (1984). ADP-dependent common receptor mechanism for binding of von Willebrand factor and fibrinogen to human platelets. *Proc Natl Acad Sci USA, 81,* 4935–4939.

36. Lam, S. C., et al. (1987). Evidence that arginyl-glycyl-aspartate peptides and fibrinogen gamma chain peptides share a common binding site on platelets. *J Biol Chem, 262,* 947–950.

37. Hu, D. D., et al. (1999). A new model of dual interacting ligand binding sites on integrin alphaIIbbeta3. *J Biol Chem, 274,* 4633–4639.

38. Cierniewski, C. S., et al. (1999). Peptide ligands can bind to distinct sites in integrin alphaIIbbeta3 and elicit different functional responses. *J Biol Chem, 274,* 16923–16932.

39. Plow, E. F., D'Souza, S. E., & Ginsberg, M. H. (1992). Ligand binding to GPIIb-IIIa: A status report. *Semin Thromb Hemost, 18,* 324–332.

40. Plow, E. F., et al. (2000). Ligand binding to integrins. *J Biol Chem, 275,* 21785–21788.

41. Xiao, T., et al. (2004). Structural basis for allostery in integrins and binding to fibrinogen-mimetic therapeutics. *Nature, 432*(7013), 59–67.

42. Knight, D. M., et al. (1995). The immunogenicity of the 7E3 murine monoclonal Fab antibody fragment variable region is dramatically reduced in humans by substitution of human for murine constant regions. *Mol Immunol, 32,* 1271–1281.

43. Puzon-McLaughlin, W., Kamata, T., & Takada, Y. (2000). Multiple discontinuous ligand-mimetic antibody binding sites define a ligand binding pocket in integrin alpha(IIb)beta(3). *J Biol Chem, 275,* 7795–7802.

44. Artoni, A., et al. (2004). Integrin beta3 regions controlling binding of murine mAb 7E3: Implications for the mechanism of integrin alphaIIbbeta3 activation. *Proc Natl Acad Sci USA, 101,* 13114–13120.

45. Plescia, J., et al. (1998). Molecular identification of the cross-reacting epitope on alphaM beta2 integrin I domain recognized by anti-alphaIIb beta3 monoclonal antibody 7E3 and

its involvement in leukocyte adherence. *J Biol Chem, 273,* 20372–20377.

46. Simon, D. I., et al. (1997). 7E3 monoclonal antibody directed against the platelet glycoprotein IIb/IIIa cross-reacts with the leukocyte integrin Mac-1 and blocks adhesion to fibrinogen and ICAM-1. *Arterioscler Thromb Vasc Biol, 17,* 528–535.

47. Scarborough, R. M., Kleiman, N. S., & Phillips, D. R. (1999). Platelet glycoprotein IIb/IIIa antagonists. What are the relevant issues concerning their pharmacology and clinical use? *Circulation, 100,* 437–444.

48. Egbertson, M. S., et al. (1994). Non-peptide fibrinogen receptor antagonists: 2. Optimization of a tyrosine template as a mimic for Arg-Gly-Asp. *J Med Chem, 37,* 2537–2551.

49. Kereiakes, D. J., et al. (1996). Randomized, double-blind, placebo-controlled dose-ranging study of tirofiban (MK-383) platelet IIb/IIIa blockade in high risk patients undergoing coronary angioplasty. *J Am Coll Cardiol, 27,* 536–542.

50. Lynch, J. J., Jr., et al. (1995). Nonpeptide glycoprotein IIb/IIIa inhibitors: 5. Antithrombotic effects of MK-0383. *J Pharmacol Exp Ther, 272,* 20–32.

51. Nakada, M. T., et al. (2002). Abciximab pharmacodynamics are unaffected by antecedent therapy with other GPIIb/IIIa antagonists in non-human primates. *J Thromb Thromboly, 14,* 15–24.

52. Topol, E. J., et al. (1997). Long-term protection from myocardial ischemic events in a randomized trial of brief integrin beta3 blockade with percutaneous coronary intervention. EPIC Investigator Group. Evaluation of Platelet IIb/IIIa Inhibition for Prevention of Ischemic Complication. *J Am Med Assoc, 278,* 479–484.

53. Topol, E. J., et al. (1994). Randomised trial of coronary intervention with antibody against platelet IIb/IIIa integrin for reduction of clinical restenosis: Results at six months. The EPIC Investigators. *Lancet, 343*(8902), 881–886.

54. Harker, L. A. (1994). Platelets and vascular thrombosis. *N Engl J Med, 330,* 1006–1007.

55. The EPILOG Investigators. (1997). Platelet glycoprotein IIb/IIIa receptor blockade and low-dose heparin during percutaneous coronary revascularization. The EPILOG Investigators. *N Engl J Med, 336,* 1689–1696.

56. Lincoff, A. M., et al. (1997). Standard versus low-dose weight-adjusted heparin in patients treated with the platelet glycoprotein IIb/IIIa receptor antibody fragment abciximab (c7E3 Fab) during percutaneous coronary revascularization. PROLOG Investigators. *Am J Cardiol, 79,* 286–291.

57. Lincoff, A. M., et al. (1999). Sustained suppression of ischemic complications of coronary intervention by platelet GP IIb/IIIa blockade with abciximab: One-year outcome in the EPILOG trial. Evaluation in PTCA to Improve Long-term Outcome with abciximab GP IIb/IIIa blockade. *Circulation, 99,* 1951–1958.

58. Serruys, P. W., et al. (1994). A comparison of balloon-expandable-stent implantation with balloon angioplasty in patients with coronary artery disease. Benestent Study Group. *N Engl J Med, 331,* 489–495.

59. Fischman, D. L., et al. (1994). A randomized comparison of coronary-stent placement and balloon angioplasty in the treatment of coronary artery disease. Stent Restenosis Study Investigators. *N Engl J Med, 331,* 496–501.

60. Lincoff, A. M., et al. (1999). Complementary clinical benefits of coronary-artery stenting and blockade of platelet glycoprotein IIb/IIIa receptors. Evaluation of Platelet IIb/IIIa Inhibition in Stenting Investigators. *N Engl J Med, 341,* 319–327.

61. Harrington, R. A., et al. (1995). Immediate and reversible platelet inhibition after intravenous administration of a peptide glycoprotein IIb/IIIa inhibitor during percutaneous coronary intervention. *Am J Cardiol, 76,* 1222–1227.

62. ESPRIT Investigators. (2000). Novel dosing regimen of eptifibatide in planned coronary stent implantation (ESPRIT): A randomised, placebo-controlled trial. *Lancet, 356*(9247), 2037–2044.

63. RESTORE Investigators. (1997). Effects of platelet glycoprotein IIb/IIIa blockade with tirofiban on adverse cardiac events in patients with unstable angina or acute myocardial infarction undergoing coronary angioplasty. The RESTORE Investigators. Randomized Efficacy Study of Tirofiban for Outcomes and Restenosis. *Circulation, 96,* 1445–1453.

64. Topol, E. J., et al. (2001). Comparison of two platelet glycoprotein IIb/IIIa inhibitors, tirofiban and abciximab, for the prevention of ischemic events with percutaneous coronary revascularization. *N Engl J Med, 344,* 1888–1894.

65. CAPTURE Study Investigators. (1997). Randomised placebo-controlled trial of abciximab before and during coronary intervention in refractory unstable angina: The CAPTURE Study. *Lancet, 349*(9063), 1429–1435.

66. Harrington, R. A., et al. (1998). Maintenance of clinical benefit at six months in patients treated with the platelet glycoprotein IIb/IIIa inhibitor eptifibatide versus placebo during an acute ischemic coronary event. *Circulation,* I-359.

67. PRISM Study Investigators. (1998). A comparison of aspirin plus tirofiban with aspirin plus heparin for unstable angina. Platelet Receptor Inhibition in Ischemic Syndrome Management (PRISM) Study Investigators. *N Engl J Med, 338,* 1498–1505.

68. PRISM-PLUS Study Investigators. (1998). Inhibition of the platelet glycoprotein IIb/IIIa receptor with tirofiban in unstable angina and non-Q-wave myocardial infarction. Platelet Receptor Inhibition in Ischemic Syndrome Management in Patients Limited by Unstable Signs and Symptoms (*PRISM-PLUS*) Study Investigators. *N Engl J Med, 338,* 1488–1497.

69. Barr, E., et al. (1998). Benefit of tirofiban + heparin therapy in unstable angina/non-Q-wave myocardial infarction patients is observed regardless of interventional treatment strategy. *Circulation,* I-504.

70. Simoons, M. L. (2001). Effect of glycoprotein IIb/IIIa receptor blocker abciximab on outcome in patients with acute coronary syndromes without early coronary revascularisation: The GUSTO IV-ACS randomised trial. *Lancet, 357*(9272), 1915–1924.

71. Giugliano, R. P., et al. (2003). Combination reperfusion therapy with eptifibatide and reduced-dose tenecteplase for ST-elevation myocardial infarction: Results of the Integrilin and Tenecteplase in Acute Myocardial Infarction (INTEGRITI) Phase II Angiographic Trial. *J Am Coll Cardiol, 41,* 1251–1260.

72. Topol, E. J. (2001). Reperfusion therapy for acute myocardial infarction with fibrinolytic therapy or combination reduced fibrinolytic therapy and platelet glycoprotein IIb/IIIa inhibition: The GUSTO V randomised trial. *Lancet, 357*(9272), 1905–1914.

73. ASSENT-3 Investigators. (2001). Efficacy and safety of tenecteplase in combination with enoxaparin, abciximab, or unfractionated heparin: The ASSENT-3 randomised trial in acute myocardial infarction. *Lancet, 358*(9282), 605–613.

74. Neumann, F. J., et al. (2000). Effect of glycoprotein IIb/IIIa receptor blockade with abciximab on clinical and angiographic restenosis rate after the placement of coronary stents following acute myocardial infarction. *J Am Coll Cardiol, 35,* 915–921.

75. Brener, S. J., et al. (1998). Randomized, placebo-controlled trial of platelet glycoprotein IIb/IIIa blockade with primary angioplasty for acute myocardial infarction. ReoPro and Primary PTCA Organization and Randomized Trial (RAPPORT) Investigators. *Circulation, 98,* 734–741.

76. Montalescot, G., et al. (2001). Platelet glycoprotein IIb/IIIa inhibition with coronary stenting for acute myocardial infarction. *N Engl J Med, 344,* 1895–1903.

77. Antoniucci, D., et al. (2003). A randomized trial comparing primary infarct artery stenting with or without abciximab in acute myocardial infarction. *J Am Coll Cardiol, 42,* 1879–1885.

78. Stone, G. W., et al. (2002). Comparison of angioplasty with stenting, with or without abciximab, in acute myocardial infarction. *N Engl J Med, 346,* 957–966.

79. Tcheng, J. E., et al. (2003). Benefits and risks of abciximab use in primary angioplasty for acute myocardial infarction: The Controlled Abciximab and Device Investigation to Lower Late Angioplasty Complications (CADILLAC) trial. *Circulation, 108,* 1316–1323.

80. De Luca, G., et al. (2005). Abciximab as adjunctive therapy to reperfusion in acute ST-segment elevation myocardial infarction: A meta-analysis of randomized trials. *J Am Med Assoc, 293,* 1759–1765.

81. Antman, E. M., et al. (2000). Combination reperfusion therapy with abciximab and reduced dose reteplase: Results from TIMI 14. The Thrombolysis in Myocardial Infarction (TIMI) 14 Investigators. *Eur Heart J, 21,* 1944–1953.

82. Antman, E. M., et al. (2002). Enoxaparin as adjunctive antithrombin therapy for ST-elevation myocardial infarction: Results of the ENTIRE-Thrombolysis in Myocardial Infarction (TIMI) 23 Trial. *Circulation, 105,* 1642–1649.

83. Mehilli, J., et al. (2004). Randomized clinical trial of abciximab in diabetic patients undergoing elective percutaneous coronary interventions after treatment with a high loading dose of clopidogrel. *Circulation, 110,* 3627–3635.

84. Kastrati, A., et al. (2004). A clinical trial of abciximab in elective percutaneous coronary intervention after pretreatment with clopidogrel. *N Engl J Med, 350,* 232–238.

85. Mehta, S. R., et al. (2001). Effects of pretreatment with clopidogrel and aspirin followed by long-term therapy in patients undergoing percutaneous coronary intervention: The PCI-CURE study. *Lancet, 358*(9281), 527–533.

86. Lincoff, A. M., et al. (2003). Bivalirudin and provisional glycoprotein IIb/IIIa blockade compared with heparin and planned glycoprotein IIb/IIIa blockade during percutaneous coronary intervention: REPLACE-2 randomized trial. *J Am Med Assoc, 289,* 853–863.

87. Lincoff, A. M., et al. (2004). Long-term efficacy of bivalirudin and provisional glycoprotein IIb/IIIa blockade vs heparin and planned glycoprotein IIb/IIIa blockade during percutaneous coronary revascularization: REPLACE-2 randomized trial. *J Am Med Assoc, 292,* 696–703.

88. Brown, D. L., Fann, C. S., & Chang, C. J. (2001). Meta-analysis of effectiveness and safety of abciximab versus eptifibatide or tirofiban in percutaneous coronary intervention. *Am J Cardiol, 87,* 537–541.

89. SYMPHONY Investigators. (2000). Comparison of sibrafiban with aspirin for prevention of cardiovascular events after acute coronary syndromes: A randomised trial. The SYMPHONY Investigators. Sibrafiban Versus Aspirin to Yield Maximum Protection from Ischemic Heart Events Post-acute Coronary Syndromes. *Lancet, 355*(9201), 337–345.

90. O'Neill, W. W., et al. (2000). Long-term treatment with a platelet glycoprotein-receptor antagonist after percutaneous coronary revascularization. EXCITE Trial Investigators. Evaluation of Oral Xemilofiban in Controlling Thrombotic Events. *N Engl J Med, 342,* 1316–1324.

91. Cannon, C. P., et al. (2000). Oral glycoprotein IIb/IIIa inhibition with orbofiban in patients with unstable coronary syndromes (OPUS-TIMI 16) trial. *Circulation, 102,* 149–156.

92. Chew, D. P., et al. (2001). Increased mortality with oral platelet glycoprotein IIb/IIIa antagonists: A meta-analysis of phase III multicenter randomized trials. *Circulation, 103,* 201–206.

93. Qureshi, A. I., et al. (2000). Abciximab as an adjunct to high-risk carotid or vertebrobasilar angioplasty: Preliminary experience. *Neurosurgery, 46,* 1316–1325.

94. Schneiderman, J., et al. (2000). Abciximab in carotid stenting for postsurgical carotid restenosis: Intermediate results. *J Endovasc Ther, 7,* 263–272.

95. Tong, F. C., et al. (2000). Abciximab rescue in acute carotid stent thrombosis. *Am J Neuroradiol, 21,* 1750–1752.

96. Hofmann, R., et al. (2002). Abciximab bolus injection does not reduce cerebral ischemic complications of elective carotid artery stenting: A randomized study. *Stroke, 33,* 725–727.

97. Chan, A. W., et al. (••). Comparison of the safety and efficacy of emboli prevention devices versus platelet glycoprotein IIb/IIIa inhibition during carotid stenting. *Am J Cardiol, 95,* 791–795.

98. Abciximab in Ischemic Stroke Investigators. (2000). Abciximab in acute ischemic stroke: A randomized, double-blind, placebo-controlled, dose-escalation study. The Abciximab in Ischemic Stroke Investigators. *Stroke, 31,* 601–609.

99. AbESTT Investigators. (2005). Emergency administration of abciximab for treatment of patients with acute ischemic stroke: Results of a randomized phase 2 trial. AbESTT Investigators. *Stroke, 36,* 880–890.

100. Anonymous (2005, October 28). Patient enrollment permanently discontinued in investigational clinical trial of ReoPro for acute ischemic stroke. Available at www.centocor.com/news/news_103105.jsp.

# CHAPTER 63

# Dipyridamole

**Wolfgang G. Eisert**

*Department of Biophysics, University of Hannover, Hannover, and Boehringer-Ingelheim GmbH, Ingelheim, Germany*

## I. Introduction

Dipyridamole had been in clinical use for some time with variable success before it was investigated for the prevention of stroke. In the European Stroke Prevention Study 2 (ESPS-2), a trial involving 6602 patients with transient ischemic attacks or stroke, treatment with dipyridamole alone was as effective as low-dose aspirin in the reduction of stroke risk, and combination therapy with dipyridamole and aspirin was more than twice as effective as aspirin alone.[1,2] Long-term administration of stronger inhibitors of platelet function in a preventive setting, such as oral fibrinogen receptor antagonists[3] or the combination of aspirin and clopidogrel (despite its success in acute/interventional settings), failed to show the expected clinical improvement over conventional aspirin therapy in long-term prevention, particularly of stroke,[4] or had to be terminated prematurely due to excessive bleeding, such as in the prevention of graft failure.[5]

The major benefit of strong inhibition of platelet function seems to be the prevention of acute thrombosis following the rupture of an unstable atherosclerotic plaque or intravascular interventions (Fig. 63-1). The antithrombotic benefit of dipyridamole particularly when combined with low-dose aspirin suggests that in addition to its antiplatelet effects, other beneficial mechanisms may play an important role without increasing the risk of bleeding.[6] The conclusions of the ESPS-2 trial — a well-designed, large-scale study — as well as a meta-analysis[7] contradict previous reports indicating that dipyridamole may be only a weak antithrombotic agent that adds nothing to the effect of aspirin.[8–10] The difference in results from older clinical studies can be attributed to a change in formulation of dipyridamole tested for the first time, in ESPS-2. Compensating for elevated gastric pH ensured reliable absorption and significant sustained plasma levels, now shown to be critical for several antithrombotic and anti-inflammatory mechanisms. This is in contrast to the early understanding of platelet inhibition as the only mode of action of dipyridamole. Reduced gastric absorption in animals may be the reason why many earlier animal studies suffered from inadequate dosing or experi-

mental design, and may therefore have shown negative results. Coadministration of high doses of aspirin and its negative effect on endothelial cells seem to reduce the benefit of dipyridamole *in vivo*.[11] With the development of *in vitro* model systems that better reflect the physiologic interactions of platelets and other blood cells with the vessel wall, dipyridamole was shown to be more effective in studies involving whole blood, endothelium, and/or smooth muscle cells, whereas traditional aggregation studies in platelet-rich plasma (PRP) show marginal effects. Other studies investigated the effect of dipyridamole on endothelial cells, smooth muscle cells, as well as platelet–monocyte interactions and revealed not only its antithrombotic but also its anti-inflammatory, antioxidative, and neuroprotective[12] effects, as well as its indirect fibrinolytic activity. These findings have been confirmed *in vivo* by studies showing increased ischemic tolerance in animals[13] as well as in humans,[14] a reduction of reperfusion injury,[15,16] and an increase in perfusion of ischemic tissue in patients.[17] Dipyridamole results in a decreased density of platelet thrombin receptors[18] and may correct for insufficient response to inhibition by aspirin.[19]

Dipyridamole was originally synthesized almost half a century ago. As is the case with many well-established compounds, only in the recent past have we begun to better understand its different modes of action and its capacity to modulate the dynamic processes that control not only hemostasis by several pathways, including thrombin formation, but also innate inflammation, modulation of receptor expression in precursor cells, matrix degradation, oxidative damage, smooth muscle cell proliferation and subsequently the development of atherosclerosis, as well as destabilization of vessel wall integrity. Dipyridamole was introduced into clinical medicine in the early 1960s as a coronary vasodilator. Selective dilatation of the cardiac vessel was thought to be of clinical value when treating patients with symptoms of heart disease. At approximately the same time, it was shown to inhibit the uptake of adenosine,[20] which in turn was found to inhibit platelet aggregation[21] and to exhibit other positive properties in cardiovascular regulation.[22]

**Need for acute
antiplatelet treatment**

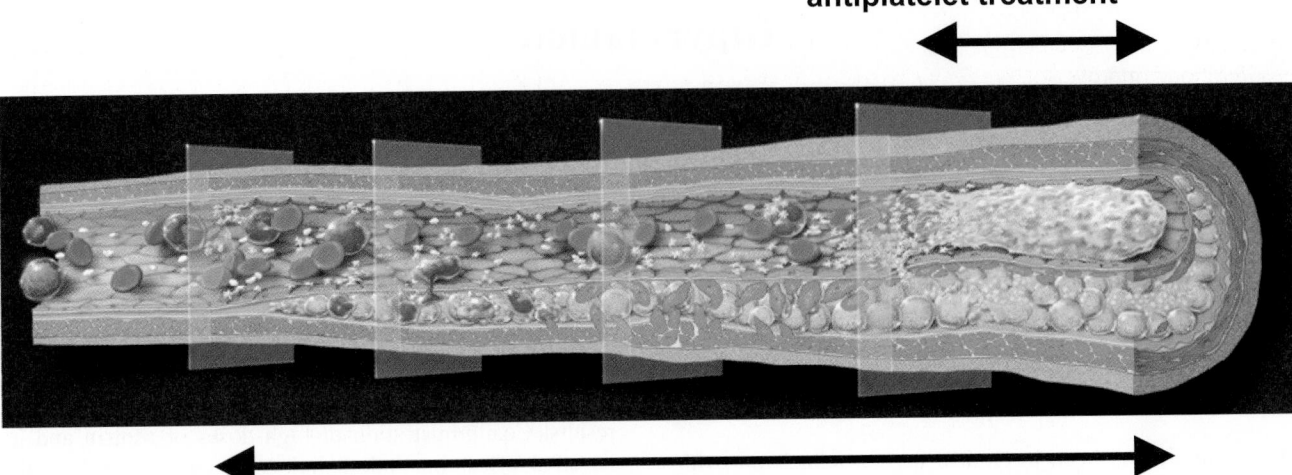

**Need for chronic antithrombotic, anti-inflammatory and vessel wall
treatment
(Pleiotropic effects of dipyridamole add to inhibition of platelets)**

**Figure 63-1.**    Treatment of atherothrombosis is aimed at different targets during the course of vascular disease development. Although substantial inhibition of platelets is favorable during acute thrombosis over limited time intervals (weeks to months), chronic prevention of thromboembolism and atherosclerosis may require differentiated therapeutic approaches directed toward endothelial and vessel wall dysfunction. Perfusion of microvessels seems to be of particular importance. Dipyridamole's pleiotropic antithrombotic, antioxidant, and anti-inflammatory effects add to platelet inhibition and may intervene in disease progression at earlier stages of vascular disease and thrombosis.

Recently, it was demonstrated that only in the presence of dipyridamole could significant interstitial concentrations of adenosine be achieved.[23] Elevation of interstitial adenosine has been associated with improved ischemic tolerance, such as in preconditioning. When tested for its vasodilatory properties in the arterial circulation of the brain in a rabbit model,[24] dipyridamole was found to reduce the formation of the thrombus that generally formed as a result of the mechanical manipulation during the test. The antithrombotic effect of dipyridamole was thus discovered *in vivo* and not by *in vitro* platelet function testing.

These investigations led to the use of dipyridamole as an antithrombotic agent for stroke prevention,[25] maintaining the patency of coronary bypass,[26] and valve replacement,[27] as well as for treatment prior to coronary angioplasty.[28,29] Although the 6-month rate of restenosis after successful coronary angioplasty was not changed by ticlopidine or dipyridamole/aspirin combination treatment,[30] treatment with dipyridamole combined with aspirin markedly reduced the incidence of transmural myocardial infarction[29] and decreased the likelihood of severe restenosis.[31] The inhibition of smooth muscle cell proliferation independent of other effects has been shown, supporting dipyridamole's benefit on remodeling, particularly in dialysis shunt maintenance. In angioplasty of peripheral arteries, dipyridamole in combination with low-dose aspirin (50 mg/day) was found to be as effective as treatment with anticoagulants in secondary prophylaxis.[32]

Despite the popularity of dipyridamole in clinical practice, investigators were unable to demonstrate a significant antiplatelet effect in early *in vitro* studies, except at high doses. At the same time, stroke prevention trials involving both dipyridamole and aspirin seemed to indicate that combination therapy showed a trend but no significant benefit over aspirin.[33,34] These studies compared combination dipyridamole/aspirin therapy with aspirin alone but not dipyridamole alone; thus, the specific efficacy of dipyridamole was not known. In other cases, end points such as the rate of restenosis were chosen when detailed analysis of the severity of stenosis might have shown benefit with combined treatment.[20] Also, the dipyridamole dose used in the AICLA study[35] and the American-Canadian Co-Operative Study in Cerebral Ischemia[36] was relatively low, which, in addition to the bioavailability problems with the instant-release preparations used at this time, most likely failed to reach pharmacologically meaningful plasma levels. This might be even more relevant if the drug exerts antithrombotic effects via the vessel wall. Finally, the studies may have included too few patients to draw well-powered conclusions.

The issue of the efficacy of dipyridamole alone and combined with aspirin in stroke prevention appears to have been answered by the ESPS-2 trial.[1,2] Although the first ESPS trial also showed a marked risk reduction with combined therapy compared to placebo, it did not test dipyridamole alone.[37] The literature suggests that dipyridamole shows an array of actions that not only inhibit the aggregation of platelets indirectly but also improve the antithrombotic properties of the vascular wall, control smooth muscle cell proliferation, improve tolerance of tissue to hypoxia, and increase perfusion under chronic ischemic conditions. Thus, dipyridamole shows antithrombotic, anti-inflammatory, and, potentially, antiatherosclerotic benefit.

## II. Absorption and Metabolism

After oral ingestion, dipyridamole (Fig. 63-2) requires low pH for absorption. At approximately pH 4, the absorption

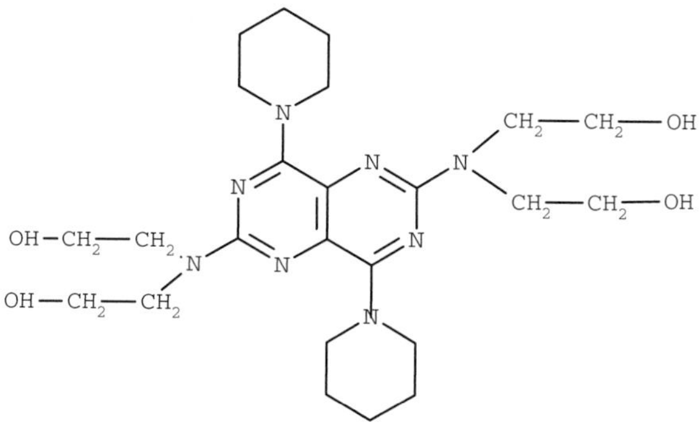

**Figure 63-2.** Dipyridamole (2,6-bis(diathanolamino)-4,8-dipiperidino-pyrimido 5,4-*d* pyrimidin).

decreases to less than 50% of that at pH 2.[38] Anacidic stomach conditions are seen in approximately 30% of elderly patients. This implies that in previous clinical studies or studies employing instant-release and nonmodified preparations, a significant percentage of patients in the dipyridamole group may not have achieved relevant plasma levels and may therefore not have received the full benefit of the dipyridamole treatment. Also, the increasing popularity of gastric acid suppressant treatment further amplifies the problem of the old instant-release preparation. This has been taken into consideration in the previously completed clinical trial (ESPS-2) as well as in a large ongoing trial (PRoFESS) using a new sustained-release preparation with a tartaric acid core around which dipyridamole is layered (Fig. 63-3). Regardless of the ambient pH, the acid core ensures a local pH of approximately 2 and thereby ensures complete absorption of dipyridamole.

The local acid release also explains why, in rare cases with very fast transit of ingested matter, dipyridamole and tartaric acid might be released in the upper bowel, which might result in irritation by the tartaric acid. Absolute absorption of dipyridamole is approximately 70%. The sustained-release preparation now used in the Aggrenox preparation allows twice-daily dosing with trough levels on average not lower than 1.4 μM. Distribution and metabolism have been described,[39] revealing that dipyridamole is eliminated by conjugation with glucuronic acid to form mainly a monoglucuronide and only small amounts of diglucuronide.

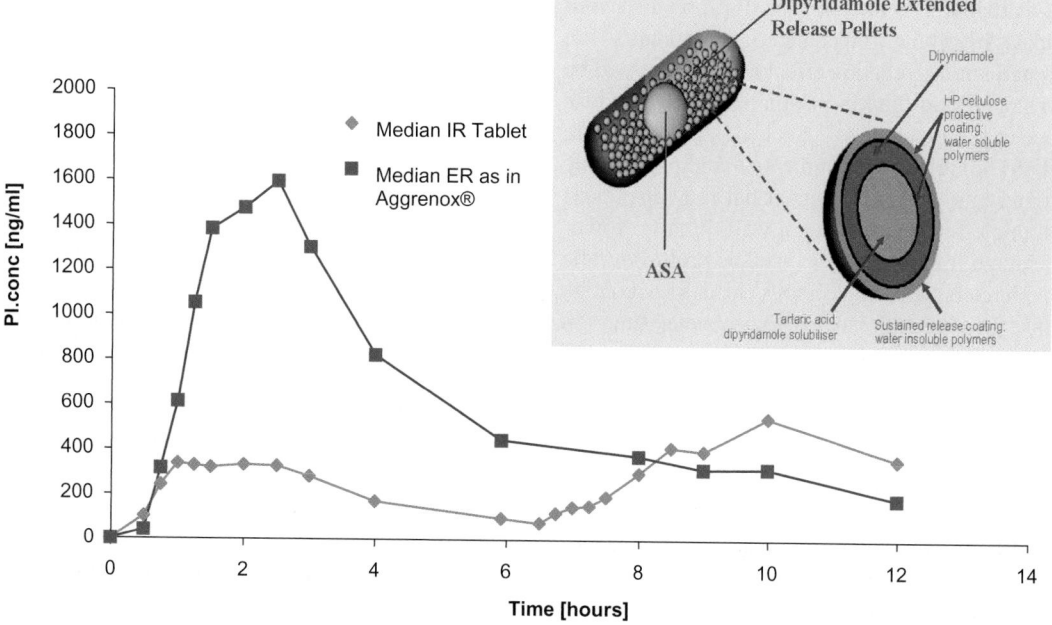

**Figure 63-3.** Bioavailability of dipyridamole at different gastric pH. Dipyridamole is absorbed from the stomach only at low pH. Volunteers were treated with gastric acid suppressant to lock the stomach pH at 4.2. Instant-release (IR) dipyridamole tablets 2 × 100 mg given every 6 hours were compared to the modified-release preparation as used in Aggrenox, which utilizes a tartaric acid core to compensate for elevated ambient pH (see insert). ER, extended release.

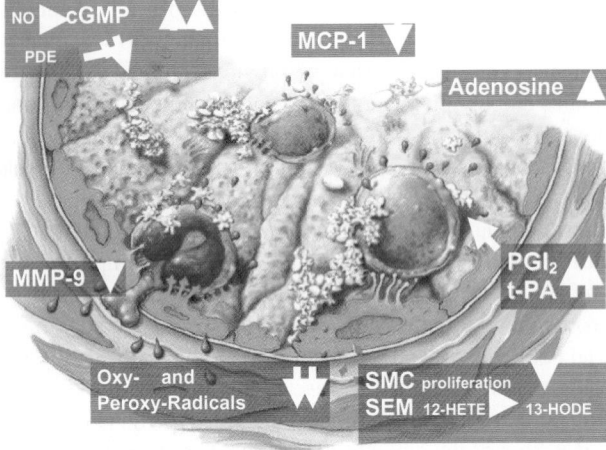

**Figure 63-4.** Dipyridamole's mode of action in the early phase of vessel wall injury and thrombus formation, as well as atherosclerosis and vessel wall thickening. cGMP, cyclic guanosine 5'-monophosphate; 12-HETE, 12-hydroxyeicosatetraenoic acid; 13-HODE, 13-hydroxyoctadecadienoic acid; MCP-1, monocyte chemotactic protein-1; MMP-9, matrix metalloproteinase 9; NO, nitric oxide; PDE, phosphodiesterase; PGI$_2$, prostaglandin I$_2$ (prostacyclin); SEM, subendothelial matrix; SMC, smooth muscle cell; t-PA, tissue plasminogen activator.

## III. Mechanisms of Action

A number of mechanisms of action have been reported for dipyridamole, of which the ones more recently discovered are depicted in Fig. 63-4. The lack of effectiveness observed with dipyridamole in early platelet aggregometry studies may be attributed to two reasons. First, as discussed later, in studies involving impedance aggregometry, an adenosine-related antiaggregation effect is more likely to be observed in whole blood than in plasma, probably because of the presence of red blood cells.[34] Second, studies in more sophisticated physiologic models indicate that dipyridamole may not be purely an antiplatelet agent but may also exert an antithrombotic effect via the vessel wall. Finally, downregulation of innate inflammation as well as prothrombotic receptors on platelets will not directly lead to a change in the response in conventional *ex vivo* platelet function testing.

### A. Increase of Adenosine

The inhibition of the adenosine transporter and thereby an increase in the local extracellular concentration of adenosine has traditionally been viewed as one of the major modes of action of dipyridamole. Adenosine stimulates adenylate cyclase via adenosine receptors localized on the cell membrane.[40] Adenylate cyclase activation results in an increase in intraplatelet cyclic adenosine monophosphate (cAMP), a

very potent inhibitor of platelet activation (see Chapter 13).[21,22]

*In vitro,* in human PRP, active adenosine concentrations are in the range of 1 to 40 µM.[40] Adenosine is released from the cytoplasm into the extracellular space as a breakdown product of adenosine triphosphate (ATP), for example, during ischemia. Another source of adenosine is mechanically stressed cells (e.g., by shear forces [turbulent flow]) or cells with plasma membrane alterations due to osmotic disturbances. Released adenine nucleotides are rapidly reduced to adenosine by nucleases. In circulating blood, the largest amounts of adenosine are thought to be located in red blood cells. Upon increased shear stress, as observed in the vicinity of flow-limiting vascular stenosis, adenine nucleotides will be released.

Free adenosine is rapidly removed from plasma by a specific adenosine carrier (half-life of <10 seconds).[41] Inhibition of adenosine uptake by dipyridamole is concentration dependent and reaches 90% at 1 µM dipyridamole in whole blood[41] or in blood from volunteers pretreated with dipyridamole (200 mg b.i.d. for 3 days). At the resulting dipyridamole plasma level of 0.5 to 3 µg/mL (1–6 µM), adenosine uptake in whole blood *ex vivo* was reduced by 92%.[42] The increased concentration of adenosine in the presence of dipyridamole inhibits platelet aggregation in whole blood *in vitro* and potentiates the antiaggregatory effect of adenosine *in vitro*.

With the development of impedance aggregometry in the late 1970s, dipyridamole was shown to inhibit platelet aggregation in whole blood but not plasma, and the difference was attributed to prevention of the uptake of adenosine by red blood cells.[43] Adenosine deaminase, an enzyme that facilitates the conversion of adenosine to inosine, or the adenosine antagonist MTA can markedly attenuate the efficacy of dipyridamole.[44] Later it was found that inhibition of adenosine catabolism by dipyridamole can also explain the observed increase in adenosine.[45] The interaction of dipyridamole and adenosine can therefore only be observed in whole blood and not in PRP because only in the presence of red blood cells can a sufficient amount of adenosine be recruited.[44]

Measurements of total plasma adenosine in healthy volunteers following oral dipyridamole treatment (100 mg four times daily) revealed an increase in adenosine plasma level by 0.13 µM or 60%.[46] Adenosine concentrations >0.1 µM have been shown to substantially influence whole blood aggregation *ex vivo*.[47] However, the increases in plasma adenosine by dipyridamole treatment[48] may well underestimate local adenosine concentration at the site of a significant stenosis. Investigations in humans showed that a desirable increase in the interstitial adenosine concentration, even during infusion of adenosine, can only be achieved in the presence of dipyridamole.[49] Also, in flow-induced vasodilatation in the forearm, the dipyridamole-induced increase in adenosine seems to be more prominent than the flow-induced release of nitric oxide (NO) and subsequent vasodilatation.[50] However, the effect may be more in favor of the NO-driven cyclic guanosine 5′-monophosphate (cGMP) elevation in much smaller vessel lumen, as discussed in the following section.

### B. Inhibition of Phosphodiesterases and Potentiation of the NO System

An increase in intracellular cAMP or cGMP inhibits platelet aggregation (see Chapters 13 and 17). The concentration of intraplatelet cGMP is regulated by the activity of the cytosolic guanylate cyclase generating cGMP and by cGMP phosphodiesterase (PDE) inactivating this cyclic nucleotide. The physiological stimulator of guanylate cyclase is endothelium-derived relaxing factor (EDRF), which has been identified as NO. EDRF is released from endothelial cells by chemical (e.g., acetylcholine) or physical (e.g., shear stress) stimuli, and then it diffuses through the plasma membrane and activates guanylate cyclase, thus increasing cGMP.

EDRF can be pharmacologically mimicked by NO donors such as glyceryltrinitrate or the sydnonimine SIN-1. Dipyridamole inhibits cGMP PDE in different tissues, such as isolated human platelets (IC$_{50}$, 1.6 μM),[51] rabbit aorta (IC$_{50}$, 7 μM), or rod cells (IC$_{50}$, 0.4 μM).[52] Comparing 11 different PDE enzymes, selectivity of dipyridamole for cGMP PDE-5 could be demonstrated.[51] However, inhibition of PDE-6, -10, and -11 is also found in the same order of magnitude.

At a concentration of 1 to 30 μM, dipyridamole potentiates the effect of synthetic NO on inhibition of the aggregation of human or rabbit[53] platelets *in vitro*. Moreover, EDRF released from rabbit aorta showed increased activity in inhibiting platelets if these platelets were isolated from volunteers after 5 days of dipyridamole treatment (200 mg sustained-release b.i.d.) compared to control platelets.

Inhibition of platelet aggregation paralleled intraplatelet levels of cGMP.[54] Dipyridamole potentiated the platelet effect of not only NO[55–58] but also a combination of NO and prostaglandin (PG) I$_2$ (prostacyclin).[56–59] Also, vasodilator-stimulated phosphoprotein (VASP) Ser 239 phosphorylation was shown to be a sensitive parameter for detecting defective NO/cGMP signaling and endothelial dysfunction.[60] It was shown that dipyridamole treatment *in vitro* and in volunteers increased phosphorylated VASP, indicating that dipyridamole amplifies the antiaggregatory action of PG and NO at physiological concentrations.[61]

Treatment with lipid-lowering 3-hydroxy-3-methylglutaryl coenzyme-A reductase inhibitors, or statins, elevates endothelial NO synthase (eNOS), thereby increasing NO.[62,63] Treatment with statins, leading to an increase in NO and subsequently intracellular cGMP, combined with inhibition of cGMP-dependent PDE-5 by dipyridamole showed addi-

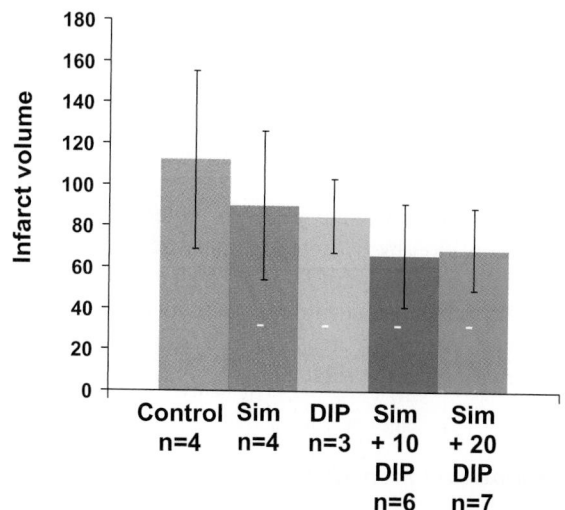

**Figure 63-5.** Additivity by dual activation of the NO system: Elevation of NO production following statin treatment and reduced deactivation of cGMP-dependent phosphodiesterase show additivity in reducing ischemia in experimental stroke. DIP, dipyridamole; Sim, simvastatin. Modified from Liao et al.[64]

tive reduction of experimental stroke volume *in vivo* (Fig. 63-5).[64]

Dipyridamole at higher concentrations was also shown to inhibit cAMP PDE, thereby increasing intracellular cAMP.[65] However, sufficient inhibition is only achieved at concentrations generally not reached by oral therapy. It is therefore not surprising that dipyridamole did not exhibit significant inhibition of platelet aggregation in early studies. Such studies required the use of PRP and therefore only revealed direct inhibitory effects on platelet function.[66]

### C. Stimulation of PGI$_2$ Production

Dipyridamole stimulates prostacyclin production in Adenosia indirectly by increasing intracellular levels of cAMP.[67–69] PGI$_2$ is a very potent endogenous inhibitor of platelet aggregation (as discussed in Chapter 13). PGI$_2$ interacts with platelets by specific receptors and induces the generation of cAMP. Prostacyclin is generated by a cyclooxygenase-dependent pathway in a variety of cells, including endothelial cells, but not in platelets. *In vitro*, 1 and 10 μM dipyridamole enhances by 13 and 49%, respectively, the release of prostacyclin from rat aortic tissue, with and without endothelium.[70] This observation was confirmed in rabbit vascular tissue.[71] Human veins in the presence of arachidonic acid also responded to dipyridamole (5 and 10 μM) by an increase in PGI$_2$ release (79–142%).[72] A similar effect was observed in veins from individuals pretreated with oral dipyridamole (4 × 100 mg/day for 2 days) *ex vivo*; PGI$_2$ production increased by 150%.[72]

Employing an *ex vivo* bioassay system, Neri Serneri et al.[67] proved dipyridamole's action on plasma PGI$_2$ in the *in vivo* situation by treating healthy volunteers with either a short infusion of dipyridamole (resulting in a 1.5–2 µg/mL plasma level of dipyridamole) or oral dipyridamole (375 mg/day for 7 days), resulting in elevations of PGI$_2$ plasma levels by 137 and 40%, respectively.

Studies by Marnett et al.[73] offer an explanation for this effect of dipyridamole: It seems to protect PGI$_2$ synthase from inactivation by hydroperoxy fatty acids and thus can enhance the production of PGI$_2$ by this enzyme. Extrapolation of the reported concentration–activity relationship indicates that a twofold increase in PGI$_2$ generation could be achieved at 1 µM dipyridamole.

### D. Antioxidant Properties

Dipyridamole has antioxidant properties.[74] Besides reducing potential damage by reactive oxidative species (ROS), these antioxidant properties may contribute to dipyridamole's antithrombotic effect.[75–77] Antioxidants have been shown to protect low-density lipoproteins (LDL) from oxidation and reduce the progression of atherosclerosis in experimental animals.[78,79] In a preliminary report of the Harvard Physicians' Health Study, treatment with β-carotene reduced the combined end point of myocardial infarction, revascularization, stroke, or coronary death.[80] Organic peroxy free radicals have been described as the ultimate agent in oxygen toxicity.[81] It has been suggested that inhibition of lipid peroxidation is responsible for the observed dipyridamole-induced increase in 6-keto-PGF-1α.[82]

In addition to elevating extracellular adenosine, which reduces superoxide anion generation by human neutrophils,[83] dipyridamole was shown to scavenge both oxygen radicals[84] and hydroxyl radicals,[85] as well as to inhibit lipid peroxidation[84] and oxidative modification of LDL.[86] Dipyridamole has been reported to be a more effective antioxidant than probucol, ascorbic acid, and α-tocopherol.[87] Fluorescence-quenching studies showed dipyridamole to be preferentially located at the polar micellar interface.[88] In studies of mitochondria, dipyridamole was found to be located in the lipid bilayer and not interacting with membrane proteins.[89,90] These investigators also noted that dipyridamole inhibited the decrease in membrane fluidity in a concentration-dependent manner. This observation may explain the reported dipyridamole-induced improvement in the deformability of red blood cells[91] and inhibition of deoxygenation-induced Na, K, and Ca fluxes contributing to the dehydration of sickle red blood cells.[92] The inhibition of free radical formation by dipyridamole has been found to inhibit fibrinogenesis in experimental liver fibrosis[93] and to suppress oxygen radicals and proteinuria in experimental animals with aminonucleoside nephropathy.[94,95] Inhibition of lipid peroxidation has also been observed in human nonneoplastic lung tissue.[96] At therapeutically relevant concentrations, dipyridamole not only reduced intracellular basal ROS generation from endo-

thelial cells and attenuated *t*-butylhydroperoxide-induced oxidative stress but also suppressed platelet-soluble CD40 ligand release without a reduction of NO, suggesting that redox-dependent properties of dipyridamole have a direct effect on vascular cells.[97] Direct scavenging of oxy as well as lipophilic peroxy radicals by dipyridamole also seems to explain its direct neuroprotective effect.[12] *In vivo* inhibition of reperfusion injury by dipyridamole has been described in animal experiments[14] and organ transplantation,[15] as well as experimentally in volunteers.[16]

### E. Anti-inflammatory Properties

Activation of monocytes by chemical stimuli such as lipopolysaccharide or smoking, or by activated platelets, leads to the production of cytokines in monocytes and resultant inflammation of the vessel wall and/or other tissues.[98,99] Certain genes have been found to be dysregulated in vascular disease and may contribute to endothelial/vessel wall dysfunction. Thus, the enhanced release of proinflammatory cytokines may lead to persistent innate inflammation and an atherosclerotic phenotype.[100] Dipyridamole has been shown to modulate cytokine release and to selectively inhibit the expression of the gene encoding for the adhesive and prothrombotic/proatherosclerotic cytokine, monocyte chemotactic protein-1 (MCP-1) (Fig. 63-6A), whereas other cytokines such as tissue necrosis factor-α were not affected.[101] This study also showed that dipyridamole inhibits the release of matrix metalloproteinase-9 (Fig. 63-6B), suggesting that dipyridamole may stabilize the blood–brain barrier as well as unstable plaques, and it may even control vessel wall integrity and the risk of spontaneous rupture. Dipyridamole also attenuated the release of sCD40L from activated platelets.[97]

### F. Release of Tissue Plasminogen Activator

With decreasing vessel diameter, the surface area presented by the endothelial lining becomes exponentially more important for the control of local thrombus formation and the lysis of emboli. Fibrin is cleaved by plasmin and the endogenous activator of plasminogen is tissue plasminogen activator (t-PA), synthesized and released from endothelial cells (see Chapter 21). *Ex vivo*, dipyridamole increases the release of t-PA from brain microvascular endothelial cells in a dose-dependent manner.[102]

### G. Reduction of Procoagulant Activities on the Surface of Injured Vessels and Improved Tissue Perfusion In Vivo

Studies in physiologic models show dipyridamole's indirect antiplatelet effect but also show its antithrombotic effect

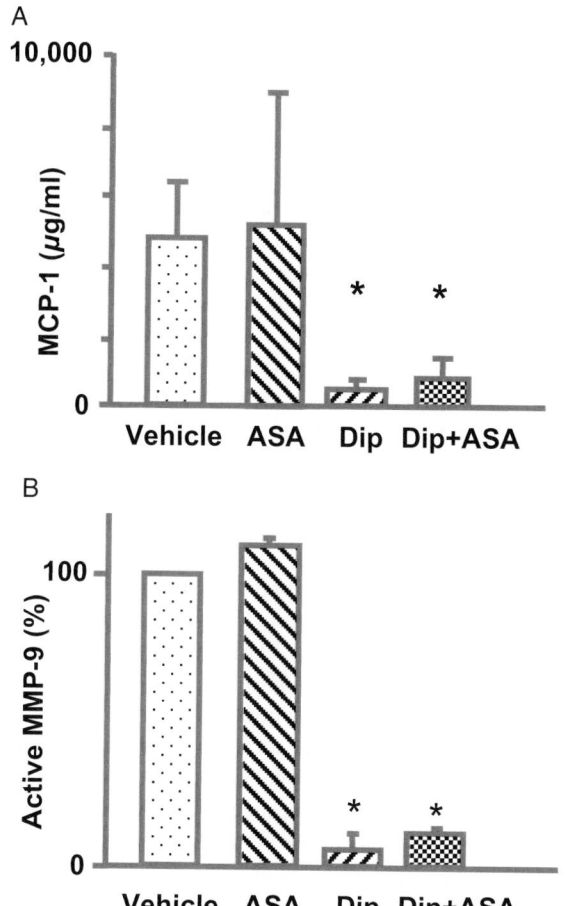

**Figure 63-6.** (A) Thrombin-stimulated platelets increased monocyte chemotactic protein-1 (MCP-1) expression by monocytes. Dipyridamole (Dip), but not aspirin (ASA), blocked the synthesis of MCP-1. *$p < 0.05$ vs. vehicle or ASA. (B) Percentage of active matrix metalloproteinase 9 (MMP-9) generated in platelet–monocyte samples. *$p < 0.05$ vs. vehicle or ASA. Modified with permission from Weyrich and Zimmerman.[98]

through control of prothrombotic activities of the vessel wall.[103] Thrombogenic aortic subendothelium was used to study the reduction of adherent platelets by dipyridamole and aspirin alone and in combination.[104] Subendothelial matrix produced by endothelial cells was used to study the formation of mural thrombus *in vitro*.[105] The development of a method of seeding a subendothelial matrix (SEM) with bovine cells[106,107] enabled investigators to combine adhesion of platelets and formation of microaggregates on vessel wall structures with the local regulatory influence of endothelial cells. Computer imaging was employed to generate detailed information on the size distribution of microthrombi forming on SEM in this physiological model of mural thrombus formation.[108,109] In contrast to other *in vitro* systems,[110] this system facilitated the assessment of the inhibition of platelet–vessel wall interaction by dipyridamole and aspirin in whole blood *ex vivo* by employing computer-controlled microscopy, allowing formation of larger thrombi

to be differentiated from platelet adhesion and very small thrombi. Dipyridamole was tested in this setting in a larger clinical pharmacological study.[111] This *ex vivo* study showed the inhibition of the formation of larger thrombi on subendothelial matrix alone and the additive inhibitory effect of low-dose aspirin.

These results confirmed earlier observations by Eldor et al.,[112] who used radiolabeled platelets to study the effect of dipyridamole on endothelial cells *in vitro*. These investigators seeded bovine endothelial cells in various densities on extracellular matrix and then incubated the cultures with PRP. The interaction of the plasma with the intima was evaluated by phase microscopy, in addition to testing for thromboxane and prostacyclin. Platelet aggregation, but not adhesion, was inhibited by the presence of endothelial cells, indicating that the endothelium plays a role in regulating thrombotic mechanisms. Dipyridamole markedly inhibited platelet activation by the extracellular matrix when added to citrated whole blood before the preparation of PRP. However, pretreatment of the endothelial cells with aspirin abolished production of prostacyclin and was associated with platelet aggregation.

The effect of dipyridamole on the endothelium was further evaluated using subendothelial matrix covered with first-passage human umbilical vein endothelial cells (HUVEC).[113] Whole blood was flowed over the subendothelial matrix, and thrombus formation was measured adjacent to a monolayer of HUVEC by automated quantitative microscopy. The presence of untreated endothelial cells led to a significant reduction of mean thrombus size formed on the subendothelial matrix (Fig. 63-7). Pretreatment of HUVEC with dipyridamole led to further suppression of thrombus formation in a dose-dependent fashion.[114]

Weber et al.[115] investigated the thrombogenicity of injured vascular walls in rabbits pretreated with dipyridamole or salicylate. Subendothelial basement membrane was exposed by air injury in the carotid artery. Thrombogenicity was decreased by half in rabbits treated with dipyridamole and increased twofold in rabbits treated with salicylate. Levels of cAMP were correlated with both dipyridamole levels and synthesis of 13-hydroxyoctadecadienoic acid (13-HODE). The authors concluded that there is a relationship between cAMP levels and 13-HODE synthesis that influences vessel wall thrombogenicity. A similar study by other investigators showed that the effect of reducing subendothelial thrombogenicity by dipyridamole was more pronounced at high shear rates.[116]

Early clinical studies showed that treatment with dipyridamole and aspirin reduced the uptake of radiolabeled platelets to Dacron or polytetrafluoroethylene (PTFE) grafts.[117] In patients with atherosclerotic lesions, dipyridamole and aspirin individually showed a trend toward reduction of platelet accretion and a significant reduction was shown when administered together.[118] In rabbits, the effects of dipyridamole and heparin on fibrin deposition after balloon

**Mean area
of all platelets
and thrombi**

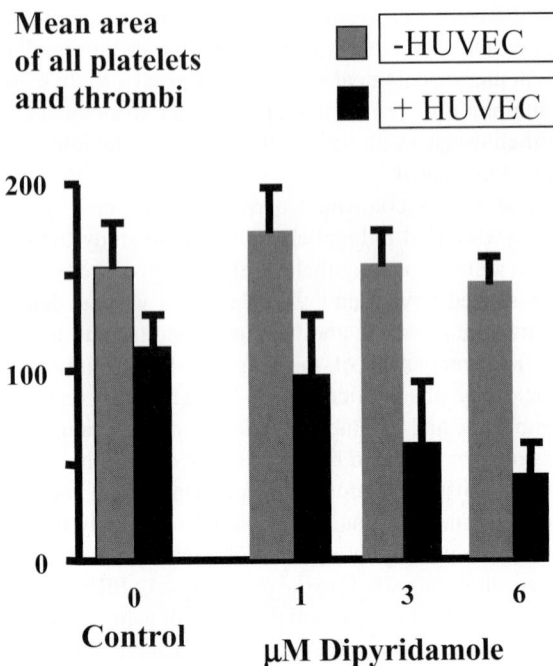

**Figure 63-7.** Microthrombus formation *ex vivo* controlled by pretreatment of human umbilical vein endothelial cells (HUVECs) with dipyridamole. Untreated whole blood flowing over subendothelial matrix results in microthrombus formation. HUVECs reduce thrombus size. This reduction is amplified by dipyridamole pretreatment of HUVECs in a dose-dependent manner. Vertical axis:Mean area of all microaggregates (thrombi) and platelets was quantified by computer-aided microscopy (arbitrary units ± SEM, *n* = 5).

angioplasty were investigated using fibrinogen labeled with [123]iodine. The accretion of fibrinogen was recorded more than 3.5 hours after balloon injury. Different dose regimens of unfractionated heparin (100 IU/kg bolus treatment prior to and 1 hour after balloon injury and 100 IU/kg prior to injury followed by 25 IU/kg/hour infusion) showed differing reductions of accreted fibrinogen (72 vs. 49%, respectively) compared to the untreated control (100%). However, dipyridamole resulted in far greater reduction in accretion of fibrinogen than was the case with heparin.[119,120] These results confirmed other observations.[103,121–123]

Similar observations have been published in animal models of hepatic sinusoidal fibrosis. Here, the suppression of fibrogenesis by dipyridamole has been reported. The capacity of dipyridamole to stabilize cellular membranes may play a role.[89] It is known that the prothrombinase complex needs to settle on negatively charged phospholipids of the cellular membrane (see Chapter 19). Acceleration of thrombin formation as one of the consequences of loss of asymmetry of cellular membranes has been investigated for many years.[124–128] The strong link between alteration of cell membranes, particularly the formation of oxidatively

damaged phospholipids, and accelerated thrombin formation has been described.[129] Vitamin E was found to reduce the increase of thrombin formation.[130] Platelet aggregation induced by oxygen-derived free radicals was inhibited by dipyridamole, indicating its unique ability to disrupt the highly dynamic surface-dependent acceleration of thrombin formation.[131] Dipyridamole's strong ability to block oxy as well as peroxy radicals may prevent membrane damage and its prothrombotic consequences.

After 6 months of oral dipyridamole (200 mg slow-release b.i.d.) in patients with chronic stable angina, tissue perfusion in the poststenotic area was significantly increased.[132] Increased utilization of collaterals following chronically elevated adenosine plasma levels has been suggested as the mode of action for improved tissue perfusion, as well as the anti-ischemic properties of dipyridamole.[133–135] Dipyridamole was shown to reduce pulmonary vascular resistance *in situ*.[136] Dipyridamole has also been reported to mimic the effect of ischemic preconditioning[137] and enhance the infarct size-limiting effect of preconditioning when given prior to the procedure.[138] This improved microcirculation after dipyridamole therapy seems to correlate well with three other studies showing improvement in heart failure patients and reduced complications in recanalization procedures, particularly of small-caliber vessels of the heart.[14] Intracoronary dipyridamole reduced the incidence of adverse cardiovascular events in the first 48 hours after balloon angioplasty of small coronary arteries.[139] Patients with ischemic cardiomyopathy showed improved collateral score, myocardial thallium uptake, and left ventricular systolic performance. All of these parameters were further improved by exercise training.[140] Improved tissue perfusion may also account for clinical observations that dipyridamole significantly retards the evolution of glomerulonephritis[141] and proteinuria,[142,143] as well as a reduction of albuminuria in diabetics without nephropathy.[144]

### H. Inhibition of Smooth Muscle Cell Proliferation

Ingerman-Wojenski and Silver[145] demonstrated that dipyridamole, in addition to its inhibitory effect on platelet adherence to damaged subendothelium, inhibits smooth muscle cell proliferation. These investigators induced injury in anesthetized rabbits by putting a forceps over the central artery in the ear for 30 minutes. In vessels injured two or more times, smooth muscle cells migrated into the intima and proliferated for up to 3 weeks. Pretreatment with dipyridamole limited proliferation to a small area, whereas aspirin did not. In another study, dipyridamole, but not other antithrombotic drugs, was shown to decrease platelet-derived growth factor levels and inhibit vascular smooth muscle cell proliferation *in vitro* and *in vivo*.[146,147] Based on local delivery in experimental animals, dipyridamole has

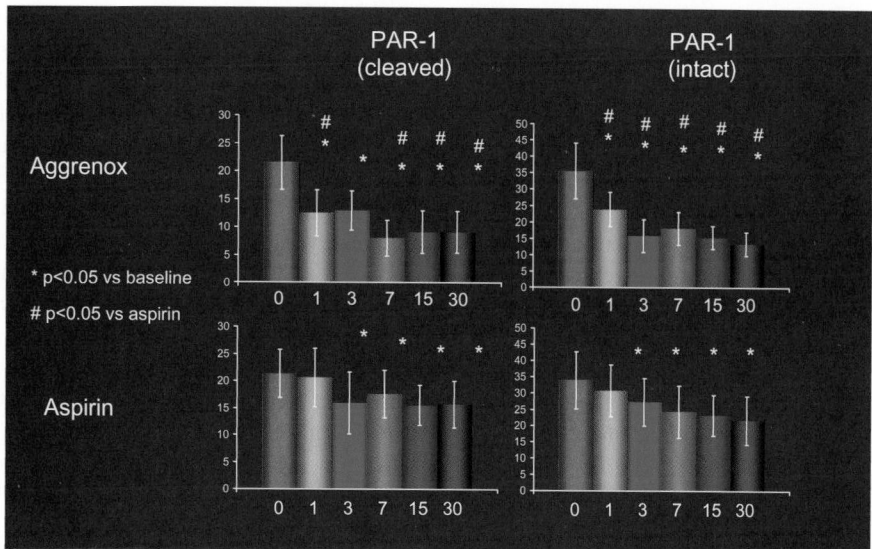

**Figure 63-8.** Density of thrombin protease-activated receptor (PAR)-1 on platelets following treatment with Aggrenox or aspirin. Thrombin receptors are reduced in second and subsequent generations of platelets during chronic therapy with dipyridamole and to a much lesser degree with aspirin therapy. Horizontal axis indicates the days after initiation of Aggrenox or aspirin therapy.

been suggested as treatment of restenosis after angioplasty.[148] Dipyridamole is apparently not cytotoxic but results in a reversible block of the cell cycle during cell proliferation. Dipyridamole has also been used to prevent smooth muscle cell proliferation and subsequent PTFE graft occlusion in dialysis patients.[149] A large multicenter clinical trial is being conducted to verify this effect in a large patient cohort using the currently available sustained-modified-release preparation to ensure sufficient plasma level. In coronary bypass patients, dipyridamole reduced the adherence of neutrophils to the endothelium and the generation of superoxide anions.[150]

### I. Modulation of Platelet Surface Receptors

The density of receptors on the surface of platelets is of importance in activation or inhibition of platelet function (see Chapter 6). Blockade of prothrombotic receptors, such as the ADP, thrombin (protease-activated receptor [PAR]), or thromboxane receptors, results in significant inhibition of platelet activation. During several weeks of therapy with modified-release dipyridamole in combination with low-dose aspirin, compared to treatment with low-dose aspirin alone, the density of PAR-1 thrombin receptors on platelets was studied. Both groups showed an initial decrease in the density of intact and cleaved thrombin receptors during the first 3 days. The group treated with the combination, however, showed significant further decline in thrombin receptor density that was not observed in the aspirin-treated group. Given a half-life of platelets in humans of approximately 3.5 days, any changes observed beyond day 7 must

reflect properties of platelets segregated during the treatment period and therefore may reflect the long-term influence of treatment on the megakaryocytic precursor cells. The decline at day 7 and thereafter until the last day of observation (day 30) reflects changes in subsequent generations of platelets (Fig. 63-8). Intact thrombia receptor as well as the cleaved thrombin receptor densities were found to be significantly reduced with dipyridamole treatment.[18]

## IV. Side Effects

Headache at the onset of therapy with rapid development of tolerance[151] is the most common adverse effect of dipyridamole. However, headache was reported in the ESPS-2 trial at a similar frequency in the treatment arms not containing dipyridamole. Studies of volunteers showed that infusion of adenosine at different rates dilated the middle cerebral artery without impairment of cerebral perfusion; however, this did not result in headache.[152] This led to the conclusion that the increased utilization of NO by rapid blockade of PDE-5 is the most likely cause of headache, and tolerance development should be understood by analogy to the well-known tolerance to nitrate therapy.

Other less frequently reported side effects of dipyridamole include dizziness, hypotension, nausea, headache, blood pressure lability, flushing, vomiting, diarrhea, abdominal pain, and rash.

In the ESPS-2 study of 6602 patients, bleeding at any site was almost doubled in the two aspirin arms, which correlated well with the elevated risk of bleeding in all other aspirin trials but was surprisingly indistinguishable from placebo in the dipyridamole-treated patients. Dipyridamole did not increase the bleeding risk of aspirin when combined with low-dose aspirin,[1] which has been demonstrated to have the lowest risk of bleeding among all antiplatelet therapies.[6,153]

Diener et al.[154] have examined the cardiac and found no safety of dipyridamole, corss of angina or MI in cardiac setting. Rapid intravenous infusion of dipyridamole over 3 minutes for the purpose of overriding the vasoconstrictor autoregulation for blood pressure control by quickly elevating circulating adenosine levels involves a very different set of physiologic responses and is in sharp contrast to chronic oral therapy. Oral dipyridamole has been observed to be safe in patients with ischemic heart disease;[155] in fact, these patients can be expected to benefit from dipyridamole treatment.

# V. Conclusions

The antithrombotic benefit from dipyridamole is most evident in combination with low-dose aspirin.[1,156] Dipyridamole's antithrombotic efficacy was originally discovered in an *in vivo* laboratory experiment. Conventional platelet aggregometry studies did not show a convincing direct platelet inhibitory effect. Only by utilizing modern experimental approaches to study interactions of the blood and various elements of the vessel wall was dipyridamole's antithrombotic effect evident. However, it also became clear that sufficient plasma levels of a reversible inhibitor or stimulant are essential to translate these findings into clinical benefit. Therefore, support for the findings of ESPS-2[1] do not only come from the different antithrombotic mechanisms summarized in Figure 63-4 but also from the development of a formulation that allows optimal absorption of dipyridamole. Such formulation had not been previously available, which still today leads to much confusion about the benefit of dipyridamole.

Most of the attention in antithrombotic therapy has been focused on the development of thromboembolic complications in larger arteries; the importance of the microvasculature has often been overlooked (Fig. 63-1). With an increasing health care burden caused by the high prevalence of obesity resulting in early and sustained vascular damage manifested in early onset diabetes, as well as cerebrovascular and cardiovascular disease, more refined antithrombotic therapy will be of ever-greater importance. Conventional direct platelet inhibition has already been shown to be of lesser value in diabetic cohorts.[157]

The data obtained from many laboratories support the concept that dipyridamole is more than an antiplatelet agent; it is also an antithrombotic compound exerting therapeutic effects at least in part through the vessel wall. The concept of treating the vessel wall is not new, but no other compound has shown a similar array of antithrombotic, antiinflammatory, antiproliferative, thrombolytic, and antioxidative properties as dipyridamole. Finally, dipyridamole has

most recently again showed its clinical benefit in the ESPRIT trial, including cardiac benfits, particularly when given as sustaine release preparation and in combination with very low dose of ASA (158, 159), with 83% of the patients receiving modified release Dipyridomole as in the unvailable product (Aggrenox®) ESPRIT showed very similar results as ESPS-2 highlighting the importance of proper drug formulation.

Finally, dipyridamole has an excellent safety profile compared to other antithrombotic and vascular treatments, as demonstrated by approximately 50 years of widespread clinical use.

This should allow dipyridamole to be considered an antithrombotic compound of a different categomy as compared to conventional platelet inhibitions.

# References

1. Diener, H. C., Cunha, L., Forbes, C., et al. (1996). European Stroke Prevention Study: 2. Dipyridamole and acetylsalicylic acid in the secondary prevention of stroke. *J Neurol Sci, 143,* 1–13.
2. Diener, H. C. (1998). Dipyridamole trials in stroke prevention. *Neurology, 51,* 17–19.
3. Topol, E., et al. (2003). Randomized, double-blind, placebo-controlled, international trial of the oral IIb/IIIa antagonist lotrafiban in coronary and cerebrovascular disease. *Circulation, 108,* 399–406.
4. Diener, H. C., et al. (2004). Aspirin and clopidogrel compared with clopidogrel alone after recent ischaemic stroke or transient ischaemic attack in high-risk patients (MATCH): Randomized, double-blind, placebo-controlled trial. *Lancet, 364,* 331–337.
5. Kaufman, J., et al. (2003). Randomized controlled trial of clopidogrel plus aspirin to prevent hemodialysis access graft thrombosis. *J Am Soc Nephrol, 14,* 2313–2321.
6. Serebruany, V., et al. (2004). Risk of bleeding complications with antiplatelet agents: Meta-analysis of 338,191 patients enrolled in 50 randomized controlled trials. *Am J Hematol, 75,* 40–47.
7. Leonardi-Bee, J., et al. (2005). Dipyridamole for preventing recurrent ischemic stroke and other vascular events. *Stroke, 36,* 162–168.
8. Fitzgerald, G. A. (1987). Drug therapy: Dipyridamole. *N Engl J Med, 316,* 1247–1257.
9. Gibbs, C. R., & Lip, G. Y. (1998). Do we still need dipyridamole? *Br J Clin Pharmacol, 145,* 323–328.
10. Stein, B., Fuster, V., Israel, D. H., et al. (1989). Platelet inhibitor agents in cardiovascular disease: An update. *J Am Coll Cardiol, 14,* 813–836.
11. Green, D., & Miller, V. (1993). The role of dipyridamole in the therapy of vascular disease. *Geriatrics, 48,* 46–59.
12. Blake, A. (2004). Dipyridamole is neuroprotective for cultured rat embryonic cortical neurons. *Biochem Biophys Res Commun, 314,* 501–504.
13. Figueredo, V. M., et al. (1999). Chronic dipyridamole therapy produces sustained protection against cardiac ischemia–

reperfusion injury. *Am J Physiol, 277 (Heart Circ Physiol, 46)*, H2091–H2097.

14. Kitakaze, M., Minamino, T., Node, K., et al. (1998). Elevation of plasma adenosine levels may attenuate the severity of chronic heart failure. *Cardiovasc Drug Ther, 12*, 307–309.

15. Taniguchi, M., et al. (2004). Dipyridamole protects the liver against warm ischemia and reperfusion injury. *J Am Coll Surg, 198*, 758–769.

16. Riksen, N. P., et al. (2005). Oral therapy with dipyridamole limits ischemia–reperfusion injury in humans. *Clin Pharmacol Ther, 78*, 52–59.

17. Picano, E., & Jagathesan, R. (2000). The beneficial effects of long-term chronic oral dipyridamole therapy on coronary vasodilator reserve in patients with chronic stable angina: The PISA (Persantine in Stable Angina)–PET study. *Circulation, 102*, 707.

18. Serebruany, V., et al. (2004). Magnitude and time course of platelet inhibition with Aggrenox® and aspirin in patients after ischemic stroke: The Aggrenox Versus Aspirin Therapy Evaluation (AGATE) trial. *Eur J Pharmacol, 499*, 315–324.

19. Serebruany, V., et al. (2006). Dipyridamole decreases protease-activated receptor and annexin-V binding on platelets of poststroke patients with aspirin nonresponsiveness. *Cerebrovasc Dis, 21*, 98–105.

20. Bunag, R. D., Douglas, C. R., Imai, S., et al. (1964). Influence of a pyrimidopyrimidine derivative on deamination of adenosine by blood. *Circ Res, 15*, 83–88.

21. Born, G. V. R., & Cross, M. J. (1963). Inhibition of the aggregation of blood platelets by substances related to adenosine diphosphate. *J Physiol, 166*, 29P–30P.

22. Ohisalo, J. J. (1987). Regulatory functions of adenosine. *Med Biol, 65*, 181–191.

23. Gamboa, A., et al. (2003). Blockade of nucleoside transport is required for delivery of intraarterial adenosine into the interstitium. *Circulation, 108*, 2631–2635.

24. Elkeles, R. S., Hampton, J. R., Honour, A. J., et al. (1968). Effect of a pyrimido–pyrimidine compound on platelet behaviour *in vitro* and *in vivo*. *Lancet, 2*, 751–754.

25. Olsson, J. E., Brechter, C., Backlund, H., et al. (1980). Anticoagulant vs anti-platelet therapy as prophylactic against cerebral infarction in transient ischemic attacks. *Stroke, 11*, 4–9.

26. Chesebro, J. H., Fuster, V., Elveback, L. R., et al. (1984). Effect of dipyridamole and aspirin on late vein-graft patency after coronary bypass operations. *N Engl J Med, 310*, 209–214.

27. Sullivan, J. M., Harken, D. E., & Gorlin, R. (1971). Pharmacologic control of thromboembolic complications of cardiac-valve replacement. *N Engl J Med, 284*, 1391–1394.

28. Barnathan, E. S., Schwartz, J. S., Taylor, L., et al. (1987). Aspirin and dipyridamole in the prevention of acute thrombosis complicating coronary angioplasty. *Circulation, 76*, 125–134.

29. Schwartz, L., Bourassa, G., Lespérance, J., et al. (1988). Aspirin and dipyridamole in the prevention of restenosis after percutaneous transluminal coronary angioplasty. *N Engl J Med, 318*, 1714–1719.

30. White, C. W., Knudson, M., Schmidt, D., et al. (1987). Neither ticlopidine nor aspirin–dipyridamole prevents restenosis post PTCA: Results from a randomized, placebo-controlled multicenter trial. *Circulation, 76*(Suppl.), IV-213.

31. Schwartz, L., Lespérance Bourassa, M. G., et al. (1990). The role of antiplatelet agents in modifying the extent of restenosis following percutaneous transluminal coronary angioplasty. *Am Heart J, 119*, 232–236.

32. Mahler, F., & Do-Dai-Do. (1992). Comparison of the combination of aspirin (50 mg)/dipyridamole (400 mg) with anticoagulants in secondary prevention following PTCA. Swiss Society for Internal Medicine, 60th Ann Mtg, Geneve, 12–13 Jun 1992. *Schweiz Med Wochenschr, 122*(Suppl.), 6.

33. Antiplatelet Trialists' Collaboration. (1994). Collaborative overview of randomised trials of antiplatelet therapy — I: Prevention of death, myocardial infarction, and stroke by prolonged antiplatelet therapy in various categories of patients. *Br Med J, 308*, 81–106.

34. Forrester, J. S., Merz Bairey, C. N., Bush, T. L., et al. (1996). Task Force 4. Efficacy of risk factor management. *J Am Coll Cardiol, 27*, 964–1047.

35. Bousser, M. G., Eschwege, E., Haguenau, M., et al. (1983). "AICLA" controlled trial of aspirin and dipyridamole in the secondary prevention of athero-thrombotic cerebral ischemia. *Stroke, 14*, 5–14.

36. American-Canadian Co-Operative Study Group. (1985). Persantine Aspirin Trial in cerebral ischemia. Part II: Endpoint results. *Stroke, 16*, 406–415.

37. ESPS Study Group. (1990). European Stroke Prevention Study. *Stroke, 21*, 1122–1130.

38. Derendorf, H., et al. (2005). Dipyridamole bioavailability in subjects with reduced gastric acidity. *J Clin Pharmacol, 45*, 845–850.

39. Mellinger, T. J., & Bohorfoush, J. G. (1965). Pathways and tissue distribution of dipyridamole (Persantine). *Arch Int Pharmacodyn Ther, 156*, 380–388.

40. Haslam, R. J., & Rosson, G. M. (1975). Effects of adenosine on levels of adenosine cyclic 3′, 5′-monophosphate in human blood platelets in relation to adenosine incorporation and platelet aggregation. *Mol Pharmacol, 11*, 528–544.

41. Klabunde, R. E. (1993). Dipyridamole inhibition of adenosine metabolism in human blood. *Eur J Pharmacol, 93*, 21–26.

42. Dresse, A., Chevolet, C., Delapierre, D., et al. (1982). Pharmacokinetics of oral dipyridamole (Persantine) and its effect on platelet adenosine uptake in man. *Eur J Clin Pharmacol, 23*, 229–234.

43. Gresele, P., Zoja, C., Deckmyn, H., et al. (1983). Dipyridamole inhibits platelet aggregation in whole blood. *Thromb Haemost, 50*, 852–856.

44. Gresele, P., Arnout, J., Deckmyn, H., et al. (1986). Mechanism of the antiplatelet action of dipyridamole in whole blood: Modulation of adenosine concentration and activity. *Thromb Haemost, 55*, 12–18.

45. Ferrandon, P., Barcelo, B., Perche, J. C., et al. (1994). Effects of dipyridamole, soluflazine and related molecules on

adenosine uptake and metabolism by isolated human red blood cells. *Fundam Clin Pharmacol, 8,* 446–452.

46. German, D. C., Kredich, N. M., & Bjornsson, T. D. (1989). Oral dipyridamole increases plasma adenosine levels in human beings. *Clin Pharmacol Ther, 45,* 80–84.

47. Edlund, A., Siden, A., & Sollevi, A. (1987). Evidence for an anti-aggregatory effect of adenosine at physiological concentrations and for its role in the action of dipyridamole. *Thromb Res, 45,* 183–190.

48. Sollevi, A., Link, H., & Fredholm, B. B. (1983). Dipyridamole treatment doubles the plasma adenosine in patients with minor stroke. *Thromb Haemost, 50,* 20.

49. Gamboa, A., et al. (2003). Blockade of nucleoside transport is required for delivery of intraarterial adenosine into the interstitium. *Circulation, 108,* 2631–2635.

50. Gamboa, A., et al. (2005). Role of adenosine and nitric oxide on the mechanisms of action of dipyridamole. *Stroke, 36,* 2170–2175.

51. Ahn, H. S., Crim, W., Romano, M., et al. (1989). Effects of selective inhibitors on cyclic nucleotide phosphodiesterase of rabbit aorta. *Biochem Pharmacol, 38,* 3331–3339.

52. Gillespie, P. G., & Beavo, J. A. (1989). Inhibition and stimulation of photoreceptor phosphodiesterases by dipyridamole and M&B 22,948. *Mol Pharmacol, 36,* 773–781.

53. Bult, H., Fret, H. R. L., Jordaens, F. H., et al. (1991). Dipyridamole potentiates the anti-aggregating and vasodilator activity of nitric oxide. *Eur J Pharmacol, 199,* 1–8.

54. Akaishi, Y., Fukao, M., Matsuno, K., et al. (1989). Dipyridamole potentiates EDRF-mediated inhibition of platelet aggregation: The role of dipyridamole as a cyclic GMP phosphodiesterase inhibitor. *Cardiovasc Drug Ther, 3*(Suppl. 2), 625.

55. Sakuma, I., Akaishi, Y., Fukao, M., et al. (1990). Dipyridamole potentiates the anti-aggregating effect of endothelium-derived relaxing factor. *Thromb Res, 12*(Suppl.), 87–90.

56. Rand, J. H., Glanville, R. W., Wu, X. X., et al. (1997). The significance of subendothelial von Willebrand factor. *Thromb Haemost, 78,* 434–438.

57. Bult, H., Fret, H. R., Jordaens, F. H., et al. (1991). Dipyridamole potentiates platelet inhibition by nitric oxide. *Thromb Haemost, 66,* 343–349.

58. Vane, J. R., & Meade, T. W. (1997). Second European Stroke Prevention Study (ESPS 2): Clinical and pharmacological implications. *J Neurol Sci, 145,* 123–125.

59. Potel, G., Maulaz, B., Paboeuf, C., et al. (1989). Potentiation of acenocoumarol after cutaneous application of a semi-synthetic heparinoid. *Therapie, 44,* 67–68.

60. Oelze, M., Mollnau, H., Hoffmann, N., et al. (2000). Vasodilator-stimulated phosphoprotein serine 239 phosphorylation as a sensitive monitor of defective nitric oxide/cGMP signaling and endothelial dysfunction. *Circ Res, 87,* 999–1005.

61. Utz, A., Aktas, B., Honig-Liedl, P., et al. (2001). Dipyridamole effects at physiological plasma concentrations on platelets *in vitro* and *in vivo. Naunyn-Schmiedebergs Arch Pharmacol, 363,* 3.

62. Takemoto, M., & Liao, J. K. (2001). Pleiotropic effects of 3-hydroxy-3-methylglutaryl coenzyme A reductase inhibitors. *Arteriosclr Thromb Vasc Biol, 21,* 1712.

63. Rikitake, Y., & Liao, J. K. (2005). Rho GTPases, statins, and nitric oxide. *Circ Res, 97,* 1232.

64. Liao, J. K. (2006). *Stroke,* in press.

65. Kadatz, R., & Diederen, W. (1967). The influence of coronary-active drugs on circulation and oxygen tension of myocardium in experimental coronary insufficiency. *Arztl Forsch, 21,* 51–60.

66. Niewiarowski, S., Lukasiewicz, H., Nath, N., et al. (1975). Inhibition of human platelet aggregation by dipyridamole and two related compounds and its modification by acid glycoproteins of human plasma. *J Lab Clin Med, 86,* 64–76.

67. Neri Serneri, G., Masotti, G., Poggesi, L., et al. (1981). Enhanced prostacyclin production by dipyridamole in man. *Eur J Pharmacol, 21,* 9–15.

68. Blass, K. E., Block, H. U., Förster, W., et al. (1980). Dipyridamole: A potent stimulator of prostacyclin ($PGI_2$) biosynthesis. *Br J Pharmacol, 68,* 71–73.

69. Mehta, J., & Mehta, P. (1982). Dipyridamole and aspirin in relation to platelet aggregation and vessel wall prostaglandin generation. *J Cardiovasc Pharmacol, 4,* 688–693.

70. Velde, V. J. S. van de, Bult, H., Weisenberger, H., et al. (1982). Dipyridamole stimulates prostacyclin production in isolated rat aortic tissue. *Arch Int Pharmacodyn Ther, 256,* 327–328.

71. Neri Serneri, G. G., Masotti, G., Poggesi, L., et al. (1981). Enhanced prostacyclin production by dipyridamole in man. *Eur J Clin Pharmacol, 21,* 9–15.

72. Costantini, V., Talpacci, A., Bastiano, M. L., et al. (1990). Increased prostacyclin production from human veins by dipyridamole: An *in vitro* and *ex vivo* study. *Biomed Biochim Acta, 49,* 263–271.

73. Marnett, L. J., Siedlik, P. H., Ochs, R. C., et al. (1984). Mechanism of the stimulation of prostaglandin H synthase and prostacyclin synthase by the antithrombotic and anti-metastatic agent, nafazatrom. *Mol Pharmacol, 26,* 328–335.

74. Iuliano, L., Pedersen, J. Z., Rotilio, G., et al. (1995). A potent chain-breaking antioxidant activity of the cardiovascular drug dipyridamole. *Free Radic Biol Med, 18,* 239–247.

75. Parthasarathy, S., Steinberg, D., & Witztum, J. L. (1992). The role of oxidized low-density lipoproteins in the pathogenesis of atherosclerosis. *Annu Rev Med, 43,* 219–225.

76. Morisaki, N., Stitts, J. M., Bartels-Tomei, L., et al. (1982). Dipyridamole: An antioxidant that promotes the proliferation of aorta smooth muscle cells. *Artery, 11,* 88–107.

77. Esterbauer, H., Gebicki, J., Puhl, H., et al. (1992). The role of lipid peroxidation and antioxidants in oxidative modification of LDL. *Free Radic Biol Med, 13,* 341–390.

78. Sparrow, C. P., Doebber, T. W., Olszewski, J., et al. (1992). Low density lipoprotein is protected from oxidation and the progression of atherosclerosis is slowed in cholesterol-fed rabbits by the antioxidant *N,N*-diphenyl-phenylenediamine. *J Clin Invest, 89,* 1885–1891.

79. Kuzuya, M., & Kuzuya, F. (1993). Probucol as an antioxidant and antiatherogenic drug. *Free Radic Biol Med, 14,* 67–77.

80. Steinberg, D. (1992). Antioxidants in the prevention of human atherosclerosis. Summary of the proceedings of a National Heart, Lung, and Blood Institute Workshop: Sep-

tember 5–6, 1991, Bethesda, Maryland. *Circulation, 85,* 2337–2344.

81. Willson, R. L. (1985). Organic peroxy free radicals as ultimate agents in oxygen toxicity. In H. Seis (Ed.), *Oxidative stress* (pp. 41–72). London: Academic Press.

82. Cruz, J. P. de la, Ortega, G., & Sanchez de la Cuesta, F. (1994). Differential effects of the pyrimido-pyrimidine derivatives, dipyridamole and mopidamol, on platelet and vascular cyclooxygenase activity. *Biochem Pharmacol, 47,* 209–215.

83. Cronstein, B. N., Kramer, S. B., Weissmann, G., et al. (1983). Adenosine: A physiological modulator of superoxide anion generation by human neutrophils. *J Exp Med, 158,* 1160–1177.

84. Iuliano, L., Violi, F., Ghiselli, A., et al. (1989). Dipyridamole inhibits lipid peroxidation and scavenges oxygen radicals. *Lipids, 24,* 430–433.

85. Iuliano, L., Pratico, D., Ghiselli, A., et al. (1992). Reaction of dipyridamole with the hydroxyl radical. *Lipids, 27,* 349–353.

86. Selly, M. L., Czeti, A. L., McGuiness, J. A., et al. (1994). Dipyridamole inhibits the oxidative modification of low density lipoprotein. *Atherosclerosis, 111,* 91–97.

87. Iuliano, L., Colavita, A. R., Camastra, C., et al. (1996). Protection of low density lipoprotein oxidation at chemical and cellular level by the antioxidant drug dipyridamole. *Br J Pharmacol, 119,* 1438–1446.

88. Tabak, M., & Borisevitch, I. E. (1992). Interaction of dipyridamole with micelles of lysophosphatidylcholine and with bovine serum albumin: Fluorescence studies. *Biochim Biophys Acta, 1116,* 241–249.

89. Nepomuceno, M. F., Alonso, A., Pereira-Da-Silva, L., et al. (1997). Inhibitory effect of dipyridamole and its derivates on lipid peroxidation in mitochondria. *Free Radic Biol Med, 23,* 1046–1054.

90. Nepumuceno, M. F., de Oliveira Mamede, M. E., Vaz de Macedo, D., et al. (1999). Antioxidant effect of dipyridamole and its derivative RA-25 in mitochondria: Correlation of activity and location in the membrane. *Biochim Biophys Acta, 1418,* 285–294.

91. Bozzo, J., Hernandez, M. R., & Ordinas, A. (1995). Reduced red cell deformability associated with blood flow and platelet activation: Improved by dipyridamole alone or combined with aspirin. *Cardiovasc Res, 30,* 725–730.

92. Joiner, C. H., Claussen, W., Yasin, Z., et al. (1997). Dipyridamole inhibits *in vitro* deoxygenation-induced cation fluxes in sickle red blood cells at membrane concentrations achievable *in vivo*. *Blood, 90*(Suppl.), 125A.

93. Wanless, I. R., Belgiorno, J., & Huet, P. M. (1996). Hepatitic sinusoidal fibrosis induced by cholesterol and stilbestrol in the rabbit: 1. Morphology and inhibition of fibrogenesis by dipyridamole. *Hepatology, 24,* 855–864.

94. Nakamura, K., Kojima, K., Shirai, M., et al. (1998). Dipyridamole and dilazep suppress oxygen radicals in puromycin aminonucleoside nephrosis rats. *Eur J Clin Invest, 28,* 877–883.

95. Nagase, M., Kumagi, H., & Honda, N. (1984). Suppression of proteinuria by dipyridamole in rats with aminonucleoside nephropathy. *Renal Physiol, 7,* 218–226.

96. De la Cruz, J. P., Olveira, C., Gonzales-Correa, J. A., et al. (1996). Inhibition of ferrous-induced lipid peroxydation by dipyridamole, RA-642 and mopidamol in human lung tissue. *Gen Pharmacol, 27,* 855–859.

97. Chakrabarti, S., et al. (2005). The effect of dipyridamole on vascular cell-derived reactive oxygen species. *J Pharmacol Exp Ther, 315*(2), 494–500.

98. Weyrich, A. S., & Zimmerman, G. A. (2004). Platelets: Signaling cells in the immune continuum. *Trends Immunol, 25,* 489–495.

99. Lindemann, S. W., Yost, C. C., Denis, M. M., et al. (2004). Neutrophils alter the inflammatory milieu by signal-dependent translation of constitutive messenger RNAs. *Proc Natl Acad Sci USA, 101,* 7076–7081.

100. Lindemann, S. W., Weyrich, A. S., & Zimmerman, G. A. (2005). Signaling to translational control pathways: Diversity in gene regulation in inflammatory and vascular cells. *Trends Cardiovasc Med, 15,* 1–9.

101. Weyrich, A. S., Denis, M. M., Kuhlmann-Eyre, J. R., et al. (2005). Dipyridamole selectively inhibits inflammatory gene expression in platelet–monocyte aggregates. *Circulation, 111,* 633–642.

102. Kim, J. A., Tran, N. D., Zhou, W., et al. (2005). Dipyridamole enhances tissue plasminogen activator release by brain capillary endothelial cells. *Thromb Res, 115,* 435–438.

103. Eisert, W. (2001). How to get from antiplatelet to antithrombotic treatment. *Am J Ther, 8,* 443–449.

104. Lauri, D., Zanetti, A., Dejama, E., et al. (1986). Effects of dipyridamole and low-dose aspirin therapy on platelet adhesion to vascular subendothelium. *Am J Cardiol, 58,* 1261–1264.

105. Vlodavsky, I., Fuks, Z., Eisert, W. G., et al. (1986). Platelet interaction with subendothelial extracellular matrix: Effects of platelet inhibitor drugs. In A. Heyns (Ed.), *Proceedings of the Conference on Radionuclide Labeled Cellular Blood Elements.* South Africa: Bloomfontain.

106. Gospodarowicz, D., Vlodavsky, I., & Savion, N. (1980). The extracellular matrix and the control of proliferation of vascular endothelial and vascular smooth muscle cells. *J Supramol Struct, 13,* 339–372.

107. Vlodavsky, I., Eldor, A., Hyam, E., et al. (1982). Platelet interaction with the extracellular matrix produced by cultured endothelial cells: A model to study the thrombogenicity of isolated subendothelial basal lamina. *Thromb Res, 28,* 179–191.

108. Müller, T. H., Rühr, K., Callisen, H., et al. (1987). Modulation of antithrombotic effects of cultured human endothelial cells by inhibitors of cyclooxygenase or phosphodiesterase. *Thromb Haemost, 58,* 155. [Abstract]

109. Eisert, W. G., & Müller, T. H. (1987). Dipyridamole and aspirin show more than additive effect in acute thrombosis model *in vivo*. *Blood, 70*(Suppl.), 370a.

110. Kirchmaier, C. M., Altorjay, J., Bellinger, O., et al. (1991). Dipyridamole inhibits platelet aggregation in the presence of endothelial cells. *Vasa, 20*(Suppl.), 318–321.

111. Muller, T. H., Su, C. A., Weisenberger, H., et al. (1990). Dipyridamole alone or combined with low-dose acetylsalicylic acid inhibits platelet aggregation in human whole blood *ex vivo*. *Br J Clin Pharmacol, 30,* 179–186.

112. Eldor, A., Vlodavsky, I., Fuks, Z., et al. (1986). Different effects of aspirin, dipyridamole and UD-CG 115 on platelet activation in a model of vascular injury: Studies with extracellular matrix covered with endothelial cells. *Thromb Haemost, 56,* 333–339.

113. Eisert, W. G., & Muller, T. H. (1990). Dipyridamole — Evaluation of an established antithrombotic drug in view of modern concepts of blood cell–vessel wall interactions. *Thromb Res, 12*(Suppl.), 65–72.

114. Eisert, W. G. (2001). Near-field amplification of antithrombotic effects of dipyridamole through vessel wall cells. *Neurology, 57*(Suppl. 2), S20–S23.

115. Weber, E., Haas, T. A., Muller, T. H., et al. (1990). Relationship between vessel wall 13-HODE synthesis and vessel wall thrombogenicity following injury: Influence of salicylate and dipyridamole treatment. *Thromb Res, 57,* 383–392.

116. Aznar-Salatti, J., Bastida, E., Escolar, G., et al. (1991). Dipyridamole induces changes in the thrombogenic properties of extracellular matrix generated by endothelial cells in culture. *Thromb Res, 64,* 341–353.

117. Goldman, M., Hall, C., Dykes, J., et al. (1983). Does [111]indium-platelet deposition predict patency in prosthetic arterial grafts? *Br J Surg, 70,* 635–638.

118. Sinzinger, H., O'Grady, J., & Fitscha, P. (1988). Platelet deposition on human atherosclerotic lesions is decreased by low-dose aspirin in combination with dipyridamole. *J Int Med Res, 16,* 39–43.

119. Lorenz, M., Van Ryn, J., Merk, H., et al. (1994). Portable gamma spectrometry system for simultaneous monitoring of radiotracers *in vivo* using CdTE and CdZnTe radiation detectors probes. *Nuclear Instruments Methods Phys Res A, 353,* 448–452.

120. Van Ryn, J., Lorenz, M., Merk, H., et al. (1993). The continuous accumulation of $^{99m}$Tc-platelet and $^{123}$I-fibrin after balloon injury in the carotid artery of rabbits. *Thromb Haemost, 69,* 569.

121. Plate, G., Stanson, A. W., Hollier, L. H., et al. (1989). Effect of platelet inhibitors on platelet and fibrin deposition following transluminal angioplasty of the atherosclerotic rabbit aorta. *Eur J Vasc Surg, 3,* 127–133.

122. Hasday, J. D., & Sitrin, R. G. (1987). Dipyridamole stimulates urokinase production and suppresses procoagulant activity of rabbit alveolar macrophages: A possible mechanism of antithrombotic action. *Blood, 69,* 660–667.

123. Iomhair, M. M., & Lavelle, S. M. (1996). Effect of aspirin–dipyridamole and heparin and their combination on venous thrombosis in hypercoagulable or thrombotic animals. *Thromb Res, 83*(6), 479–483.

124. Gerads, I., Govers-Riemslag, J. W. P., Tans, G., et al. (1990). Prothrombin activation on membranes with anionic lipids containing phosphate, sulfate and/or carboxyl groups. *Biochemistry, 29,* 7967–7974.

125. Govers-Riemslag, J. W. P., Janssen, M. P., Waal, R. F. A., et al. (1992). Effect of membrane fluidity and fatty acid composition on the prothrombin-converting activity of phospholipid vesicles. *Biochemistry, 31,* 10000–10008.

126. Zwaal, R. F. A., & Schroit, A. J. (1997). Pathophysiologic implications of membrane phospholipid asymmetry in blood cells. *Blood, 89,* 1121–1132.

127. Mann, K. G., Nesheim, M. E., Church, W. R., et al. (1990). Surface-dependent reactions of the vitamin K-dependent enzyme complexes. *Blood, 76,* 1–16.

128. Koppaka, V., Wang, J., Banerjee, M., et al. (1996). Soluble phospholipids enhance factor $X_a$-catalyzed prothrombin activation in solution. *Biochemistry, 35,* 7482–7491.

129. Weinstein, E. A., Li, H., Lawson, J. A., et al. (2000). Prothrombinase acceleration by oxidatively damaged phospholipids. *J Biol Chem, 275,* 22925–22930.

130. Rota, S., McWilliam, N. A., Baglin, T. P., et al. (1998). Atherogenic lipoproteins support assembly of the prothrombinase complex and thrombin generation: Modulation by oxidation and vitamin E. *Blood, 91,* 505–515.

131. De la Cruz, J. P., García, P. J., & Sánchez de la Cuesta, F. (1992). Dipyridamole inhibits platelet aggregation induced by oxygen-derived free radicals. *Thromb Res, 66,* 277–285.

132. Picano, E., & Jagathesan, R. (2000). The beneficial effects of long-term chronic oral dipyridamole therapy on coronary vasodilator reserve in patients with chronic stable angina: The PISA (Persantine in Stable Angina)–PET study. *Circulation, 102,* 707.

133. Picano, E., & Michelassi, C. (1997). Chronic oral dipyridamole as a "novel" antianginal drug: The collateral hypothesis. *Cardiovasc Res, 33,* 666–670.

134. Picano, E. (1989). Dipyridamole-echocardiography test: Historical background and physiologic basis. *Eur Heart J, 10,* 365–376.

135. Picano, E., & Abbracchio, M. P. (2000). Adenosine, the imperfect endogenous anti-ischemic cardio-neuroprotector. *Brain Res Bull, 52*(3), 75–82.

136. Clarke, W. R., Uezono, S., Chambers, A., et al. (1994). The type III phosphodiesterase inhibitor milrinone and type V PDE inhibitor dipyridamole individually and synergistically reduce elevated pulmonary vascular resistance. *Pulmonary Pharmacol, 7,* 81–89.

137. Mosca, S. M., Gelpi, R. J., & Cingolani, H. E. (1994). Adenosine and dipyridamole mimic the effects of ischemic preconditioning. *J Mol Cell Cardiol, 26,* 1403–1409.

138. Suzuki, K., Miura, T., Miki, T., et al. (1998). Infarct-size limitation by preconditioning is enhanced by dipyridamole administered before but not after preconditioning: Evidence for the role of interstitial adenosine level during preconditioning as a primary determinant of cardioprotection. *J Cardiovasc Pharmacol, 31,* 1–9.

139. Heidland, U. E., Heintzen, M. P., Michel, C. J., et al. (2000). Adjunctive intracoronary dipyridamole in the interventional treatment of small coronary arteries: A prospectively randomized trial. *Am Heart J, 139,* 1039–1045.

140. Belardinelli, R., Belardinelle, L., & Shryock, J. C. (2001). Effects of dipyridamole on coronary collateralization and myocardial perfusion in patients with ischemic cardiomyopathy. *Eur Heart J, 22,* 1203–1213.

141. Camara, S., de la Cruz, J. P., Frutos, M. A., et al. (1991). Effects of dipyridamole on the short term evolution of glomerulonephritis. *Nephron, 58,* 13–16.

142. Ueda, N., Kawaguchi, S., Niiomi, Y., et al. (1986). Effect of dipyridamole treatment on proteinuria in pediatric renal disease. *Nephron, 44,* 174–179.

143. Zäuner, I., Böhler, J., Braun, N., et al. (1994). Effect of aspirin and dipyridamole on proteinuria in idiopahtic membrane proliferative glomerulonephritis: A multicentre prospective clinical trial. *Nephrol Dial Transplant, 9,* 619–622.

144. Aizawa, T., Suzuki, S., Asawa, T., et al. (1990). Dipyridamole reduces urinary albumin excretion in diabetic patients with normo- or microalbuminuria. *Clin Nephrol, 33,* 130–135.

145. Ingerman-Wojenski, C. M., & Silver, M. J. (1988). Model system to study interaction of platelets with damaged arterial wall: II. Inhibition of smooth muscle cell proliferation by dipyridamole and AH-P719. *Exp Mol Pathol, 48,* 116–134.

146. Takehara, K., Igarashi, A., & Ishibashi, Y. (1990). Dipyridamole specifically decreases platelet derived growth factor release from platelets. *Thromb Res, 12*(Suppl.), 73–79.

147. Takehara, K., Grotendorst, G. R., Silver, R., et al. (1987). Dipyridamole decreases platelet-derived growth factor levels in human serum. *Arteriosclerosis, 7,* 152–155.

148. Singh, J. P., Rothfuss, K. J., Wiernicki, T. R., et al. (1994). Dipyridamole directly inhibits vascular smooth muscle cell proliferation *in vitro* and *in vivo:* Implications in the treatment of restenosis after angioplasty. *J Am Coll Cardiol, 23,* 665–671.

149. Himmelfarb, J., & Couper, L. (1997). Dipyridamole inhibits PDGF- and bFGF-induced vascular smooth muscle cell proliferation. *Kidney Int, 52,* 1671–1677.

150. Chello, M., Mastroroberto, P., Malta Emanuelle Cirillo, F., et al. (1999). Inhibition by dipyridamole of neutrophil adhesion to vascular endothelium during coronary bypass surgery. *Ann Thorac Surg, 67,* 1277–1282.

151. Theis, J. G. W., Diechsel, G., & Marshall, S. (1999). Rapid development of tolerance to dipyridamole-associated headaches. *Br J Clin Pharmacol, 48,* 750–755.

152. Birk, S., Petersen, K. A., Kruuse, C., et al. (2005). The effect of circulating adenosine on cerebral haemodynamics and headache generation in healthy subjects. *Cephalalgia, 25,* 369–377.

153. Serebruany, V. L., Steinhubl, S. R., Berger, P. B., et al. (2005). Analysis of risk of bleeding complications after different doses of aspirin in 192,036 patients enrolled in 31 randomized controlled trials. *Am J Cardiol, 95,* 1218–1222.

154. Diener, H.-C., Darius, H., Bertrand-Hardy, J. M., et al. (2001). Cardiac safety in the European Stroke Prevention Study 2 (ESPS2). *Int J Clin Pract, 55,* 162–163.

155. Humphreys, D. M., Street, J., Schumacher, H., et al. (2002). Dipyridamole may be used safely in patients with ischemic heart disease. *Int J Clin Pract, 56,* 121–127.

156. Albers, G. W., Amarenco, P., Easton, J. D., et al. (2004). Antithrombotic and thrombolytic therapy for ischemic stroke. *Chest, 126,* 483S–512S.

157. Angiolillo, D. J., Fernandez-Ortiz, A., Bernardo, E., et al. (2005). Platelet function profiles in patients with type 2 diabetes and coronary artery disease on combined aspirin and clopidogrel treatment. *Diabetes, 54,* 2430–2435.

158. The ESPRIT Study groups (2006). Aspirin plus dipyridamole versers Aspirin alone after cercloral ischemia of arterial origin (ESPRIT): randomized, controlled trial. *Lancet, 367,* 1665–1673.

159. Norving, B. (2006). Dipyridamole with Aspirin for secondary stroke prevention. *Lancet, 367,* 1638–1639.

# CHAPTER 64

# Cilostazol

**Yasuo Ikeda,[1] Toshiki Sudo,[2] and Yukio Kimura[2]**

[1]*Division of Hematology, Keio University School of Medicine, Tokyo, Japan*
[2]*Otsuka Pharmaceutical Company, Ltd., Tokushima, Japan*

## I. Introduction

Platelet adhesion to blood vessel walls and aggregation are crucial physiological events in thrombosis and hemostasis. Excessive platelet accumulation at sites of atherosclerotic plaque rupture causes occlusion of blood vessels and leads to ischemia. In addition, platelet aggregates release substances acting on vascular tissues such as the platelet-derived growth factor (PDGF) that induce intimal hyperplasia. These phenomena are responsible for cardiovascular ischemic diseases such as acute coronary syndromes (see Chapter 35), peripheral arterial diseases (see Chapter 37), and stroke (see Chapter 36). In order to treat ischemic diseases resulting from platelet aggregation, many kinds of antiplatelet drugs are widely used in clinical situations.[1]

Cilostazol (Pletal) (Fig. 64-1) is an oral selective cyclic nucleotide phosphodiesterase 3 (PDE3) inhibitor with antiplatelet, vasodilatory, and antimitogenic effects.[2–5] Cilostazol has been clinically used for treatment of chronic peripheral arterial occlusion and stroke in 13 countries, including Japan, the United States, and the United Kingdom. Cilostazol was first approved in Japan in 1988 and has subsequently been approved in 12 other countries. In the United States, cilostazol has been clinically investigated since 1993 in patients with intermittent claudication, and it was approved by the Food and Drug Administration (FDA) in 1999 for this indication. This chapter reviews the pharmacology and clinical utility of cilostazol.

## II. Mechanism of Action

It is well established that cyclic adenosine monophosphate (cAMP)-elevating agents, such as adenosine, prostaglandin I$_2$ (PGI$_2$), PGE$_1$, and PDE inhibitors, and cyclic guanosine 5′-monophosphate (cGMP)-elevating agents, such as nitric oxide (NO) donors, are able to inhibit platelet functions. Elevation of cAMP/cGMP concentrations can be accom-

plished either directly by stimulation of adenylate/guanylate cyclases or indirectly by inhibition of PDEs.[6] It has been considered that increased cAMP/cGMP in platelets activates cAMP-dependent protein kinase (PKA) and cGMP-dependent protein kinase (PKG) and thus regulates platelet activation and aggregation responses by phosphorylating intracellular protein substrates, such as the IP3 receptor, phospholipase Cβ, glycoprotein (GP) Ibβ, Gα$_{13}$, RapIb, actin binding protein, vasodilator-stimulated phosphoprotein (VASP), and PDE3.[7] In most cells, including platelets, the intracellular effects of cAMP are primarily mediated by PKA and the protein substrates. However, the detailed signal transduction pathways remain unclear.

The intracellular levels of cAMP and cGMP are regulated by synthesizing systems, which include adenylate cyclase and guanylate cyclase, and hydrolytic systems, which include PDEs. In mammalian tissues, 11 isozymes of PDE (PDE1 through PDE11) have been identified according to their primary structures, substrate affinities, and inhibitor sensitivities (Table 64-1).[8,9] Most cell types express one or more PDE isozymes, which are expressed in tissue- and cell-specific distribution patterns, each regulating intracellular cAMP and/or cGMP levels in different cellular compartments and in different manners. The PDE activity of platelets has been reported to be mainly due to PDE3 and PDE5, with a minor activity for PDE2.[10,11] PDE2 and PDE3 preferentially hydrolyze cAMP, whereas PDE5 specifically hydrolyzes cGMP. It has been reported that the inhibition of PDE3 is important for the suppression of adenosine diphosphate (ADP)-induced platelet aggregation, whereas that of PDE5 is linked to a reduction of serotonin release.[12]

Cilostazol selectively inhibits PDE3 isozyme by a cAMP-competitive mechanism (Table 64-2).[5] Cilostazol potently inhibits the activity of PDE3A, a cardiovascular subtype of PDE3 (IC$_{50}$ 0.20 μM), and increases intracellular cAMP concentrations and activates PKA in human platelets. Tyr751, Thr844, Asp950, Phe972, and Gln975 in the catalytic domain of PDE3A are key residues for the binding of cilostazol.[13] Inhibitory effects of PDE3 inhibitors on PDE3A activity are highly correlated with their inhibition of platelet aggregation induced by thrombin, ADP, or collagen.[5] The pharmacological effect of cilostazol is therefore considered to be due to elevation of intracellular cAMP levels by inhibition of PDE3A activity in platelets.

Cilostazol induces the phosphorylation of VASP in platelets, mediated by PKA activation, despite the weak

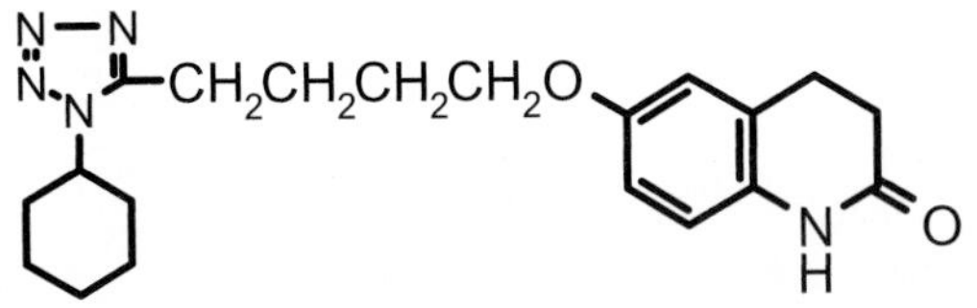

Figure 64-1. Chemical structure of cilostazol: 6-[4-(1-cyclohexyl-1H-tetrazol-5-yl)butoxy]-3,4-dihydro-2(1H)-quinolinone.

stimulatory effect on cAMP accumulation in comparison with forskolin, $PGE_1$, or $PGI_2$.[5,14] VASP is a 46 to 50 kDa protein, which was found concentrated along highly dynamic filamentous membrane structures, in focal adhesions and cell–cell contacts.[7,15] VASP modulates actin polymerization and actin filament bundling and integrin activation. VASP is phosphorylated in human platelets at Ser157, Ser239, and Thr278 with different affinities by both PKA and PKG. VASP phosphorylation downregulates its interaction with actin filaments.[16] VASP phosphorylation has been shown to closely correlate with the inhibition of fibrinogen binding to integrin $\alpha IIb\beta 3$ and the inhibition of platelet aggregation and adhesion.[17,18] The VASP phosphorylation assay is also useful for quantifying the antiplatelet effect of clopidogrel.[19] NO donors induce VASP phosphorylation with a marked increase in cGMP levels, without influencing cAMP. The fact that VASP phosphorylation in response to various plate-

## Table 64-1: PDE Isozyme Family

| Isozyme | Affinity, $K_m$ (µM) | | Subtype | Major Tissue Distribution |
| --- | --- | --- | --- | --- |
| | cAMP | cGMP | | |
| PDE1 (Ca$^{2+}$/CaM-dependent) | 1–12 | 1–3 | 1A, 1B, 1C | Brain, heart, SMC, lung |
| PDE2 (cGMP-stimulated) | 30 | 15 | 2A | Adrenal, heart, brain, kidney, liver, platelet |
| PDE3 (cGMP-inhibited) | 0.1-0.4 | 0.03-0.3 | 3A, 3B | Platelet, heart, SMC, adipocyte, liver, β cell |
| PDE4 (cAMP-specific) | 1–3 | >300 | 4A, 4B, 4C, 4D | Brain, leukocyte, testis, EC, SMC, heart |
| PDE5 (cGMP-specific) | >100 | 0.5-5 | 5A | Lung, platelet, SMC |
| PDE6 (photoreceptor cGMP-specific) | >100 | 5-20 | α, β, γ | Retina |
| PDE7 (cAMP-specific, rolipram-insensitive) | 0.2 | — | 7A, 7B | Skeletal muscle, T cell |
| PDE8 (cAMP-specific, IBMX-insensitive) | 0.055 | 124 | 8A, 8B | Testis, thyroid, liver, kidney, ovary, brain |
| PDE9 (cGMP-specific, IBMX-insensitive) | 230 | 0.07-0.17 | 9A | Small intestine, kidney, liver, lung, brain, heart |
| PDE10 (cAMP/cGMP-specific) | 0.05 | 3 | 10A | Brain, thyroid, testis |
| PDE11 (cAMP/cGMP-specific) | 3.3 | 5.7 | 11A | Prostate, testis, pituitary, skeletal muscle, liver |

CaM, calmodulin; cAMP, cyclic adenosine monophosphate; cGMP, cyclic guanosine monophosphate; EC, endothelial cells; IBMX, isobutyl methyl xanthine; PDE, phosphodiesterase; SMC, smooth muscle cells.

## Table 64-2: Inhibitory Effects of Cilostazol on PDE Isozymes

| | $IC_{50}$ (µM) | | | | | | |
| --- | --- | --- | --- | --- | --- | --- | --- |
| | PDE-1 | PDE-2 | PDE-3A | PDE-3B | PDE-4 | PDE-5 | PDE-7 |
| Cilostazol | >100 | 45.2 | 0.20 | 0.38 | 88.0 | 4.4 | 21.4 |
| Specific inhibitor | Vinpocetine, 23.2 | EHNA, 9.2 | Cilostamide, 0.027 | Cilostamide, 0.075 | Rolipram, 0.45 | Dipyridamole, 0.26 | |

let inhibitors is known to be mediated by either PKA or PKG[7] suggests that a small increase in cAMP due to PDE3 inhibition is sufficient to activate PKA/PKG. The activity of PDE3 may be restricted to distinct compartments from PDE2 and PDE5 within platelets.

The inhibition of platelet function by cilostazol appears to be due to decreased intracellular $Ca^{2+}$ levels resulting from increased cAMP levels.[5,20] However, cilostazol also inhibits calcium ionophore-induced platelet aggregation, so the detailed mechanism of action remains unclear. It is reported that the inhibition of agonist-evoked elevations in intracellular $Ca^{2+}$ concentration by cAMP/cGMP-elevating agents might have two components — a PKA/PKG-dependent mechanism for the inhibition of $Ca^{2+}$ release from intracellular stores and a PKA/PKG-independent mechanism for the direct inhibition of $Ca^{2+}$ entry mediated by protein tyrosine phosphatases.[21] These results suggest that the inhibitory effect of cilostazol on platelet aggregation might be mediated by PKA/PKG-dependent mechanisms, such as VASP phosphorylation and the inhibition of $Ca^{2+}$ release, and PKA/PKG-independent mechanisms.

## III. Antiplatelet Effects

Cilostazol inhibits platelet aggregation induced by various agonists in humans, mice, rabbits, and dogs.[3] $IC_{50}$ values of cilostazol range from 3.6 to 15.0 $\mu M$ for human platelet aggregation induced by various agonists in platelet-rich plasma and whole blood (Table 64-3).[3,22] Unlike aspirin, which inhibits only secondary platelet aggregation, cilostazol, with its different mode of action, inhibits not only primary but also secondary platelet aggregation induced by ADP or epinephrine. It is known that platelet aggregation is also induced by the high shear stress generated by blood flow. This aggregation appears to play an important role in the pathogenesis of arterial thrombosis.[23] Cilostazol inhibits shear stress-induced platelet aggregation (SIPA), with an $IC_{50}$ of 15.0 $\mu M$, whereas aspirin has no effect.[24] Cilostazol inhibits the collagen- or thrombin-induced binding to platelets of PAC-1 (a monoclonal antibody specific for the activated GPIIb-IIIa complex [integrin $\alpha IIb\beta 3$]).[20] Therefore, it is suggested that cilostazol inhibits platelet aggregation by suppressing GPIIbIIIa activation (Fig. 64-2). Cilostazol

### Table 64-3: Inhibitory Effects of Cilostazol on Platelet Aggregation in Human PRP

| | $IC_{50}$ ($\mu M$) | | | | | | | |
| --- | --- | --- | --- | --- | --- | --- | --- | --- |
| | ADP | Collagen | Thrombin | Arachidonic Acid | Epinephrine | U46619[a] | A23187[b] | SIPA[c] |
| Cilostazol | 12.8 | 3.9 | 4.9 | 3.6 | 11.5 | 15.0 | 14.7 | 15.0 |
| Aspirin | >1000 | 148 | >1000 | 98.6 | 83.0 | | | NE |

[a]Thromboxane $A_2$ analogue.
[b]Calcium ionophore.
[c]Shear stress-induced platelet aggregation. NE, no effect.

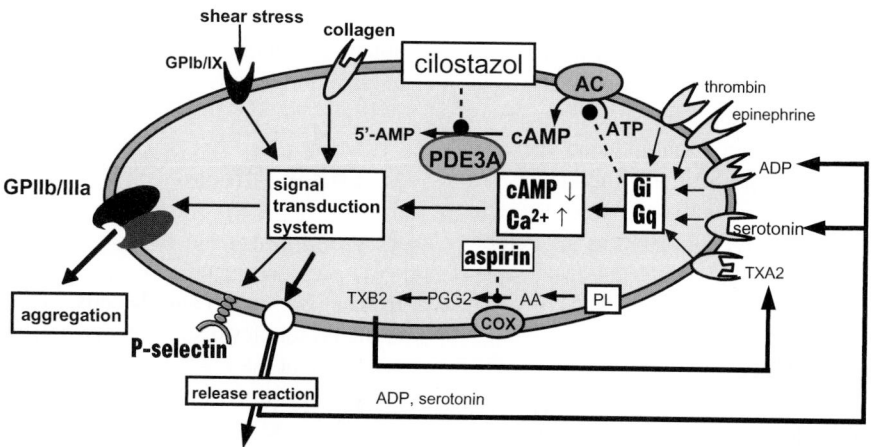

**Figure 64-2.** Inhibitory effects of cilostazol on agonist-induced platelet aggregation and release reaction. Cilostazol inhibits cyclic adenosine monophosphate (cAMP) hydrolysis via PDE3A inhibition and increases cAMP concentrations. As a result, cAMP inhibits platelet aggregation and the release reaction (including PDGF, P-selectin, and $TXB_2$ production). AA, arachidonic acid; AC, adenylate cyclase; ADP, adenosine diphosphate; ATP, adenosine triphosphate; COX, cyclooxygenase; PDE, phosphodiesterase; PG, prostaglandin; PL, phospholipid; TX, thromboxane.

(3 mg P.O. [oral administration]) inhibits *ex vivo* ADP- and collagen-induced platelet aggregation in dogs.[3]

Cilostazol inhibits not only platelet aggregation but also other aspects of platelet activation, such as thromboxane $B_2$ ($TXB_2$) production, PDGF release, expression and release of P-selectin (CD62P), platelet–leukocyte interaction, and microparticle generation.[20,25–28] Effects of cilostazol on platelets are reversible, depending on the blood concentration of cilostazol. In contrast, the effects of aspirin (see Chapter 60) and clopidogrel (see Chapter 61) on platelets are irreversible. Cilostazol might attenuate the responsiveness of platelets to various agonists.

The antiaggregation effect of cilostazol is enhanced in the presence of $PGE_1$ and adenosine, which activate adenylate cyclase.[5,29] This antiplatelet effect is also enhanced in the presence of cultured vascular endothelial cells.[25] This latter phenomenon is the result of a synergistic effect between the activation of cAMP synthesis by $PGI_2$ generated from vascular endothelial cells and the inhibition of cAMP hydrolysis by cilostazol.

These antiplatelet effects of cilostazol have been demonstrated in patients with thrombotic diseases. In patients with cerebral infarction or arteriosclerosis obliterans, cilostazol (100–200 mg/day) inhibited ADP-, collagen-, arachidonic acid-, and epinephrine-induced *ex vivo* platelet aggregation.[30,31] In patients with cerebral arteriosclerosis, cilostazol (100 mg/day for 2 weeks) decreased plasma $TXB_2$ levels.[32] Unlike aspirin, cilostazol does not decrease endothelial cell-derived $PGI_2$ levels. In patients with cerebral thrombosis or cerebral arteriosclerosis, cilostazol (100 mg/day for 4 weeks) significantly decreased plasma β-thromboglobulin and platelet factor 4 levels.[33] In patients with type II diabetes, cilostazol (150 mg/day for 4 weeks) decreased the levels of platelet activation markers such as serum-soluble vascular cell adhesion molecule-1, intercellular adhesion molecule-1, and P-selectin; platelet-derived microparticles; platelet surface markers P-selectin and CD63; and platelet membrane-expressed annexin V.[34] Using flow cytometry analysis of whole blood obtained from the coronary sinus of patients undergoing coronary stenting, the stent-induced increase in platelet surface P-selectin expression and the increase in neutrophil Mac-1 (CD11b) expression were suppressed in the cilostazol group (200 mg/day) compared to the ticlopidine group (200 mg/day).[35]

## IV. Antithrombotic Effects

By virtue of the antiplatelet effects described previously, cilostazol has been shown to have antithrombotic effects *in vivo*. Cilostazol inhibited ADP- and collagen-induced pulmonary thrombotic embolism and reduced mortality in mice at doses of 10 and 3 mg/kg P.O., respectively, with more potent effects than aspirin and pentoxyfylline.[3] In the laser-induced thrombosis model with mouse cremaster artery (see Chapter 34), cilostazol (10 mg/kg intravenously [IV]) inhibited platelet accumulation at the site of vascular injury site.[36,37] Analysis of the kinetics of individual platelets at injury sites using intravital microscopy demonstrates that cAMP directs the rate at which platelets attach to and detach from thrombi. This study demonstrated that cAMP in circulating platelets controls attachment to and detachment from sites of arteriolar injury.

In rabbits with cerebral infarction induced by injection of arachidonic acid into the unilateral internal carotid artery, cilostazol (1 mg/kg IV) reduced the area of infarction identified by perfusion with India ink by 55%.[38] In canine models of peripheral circulatory insufficiency in a hind leg caused by injection of lauric acid into the peripheral end of the ligated femoral artery, cilostazol (10 mg/kg/day P.O.) inhibited the progression of ischemic ulcers and the decrease in skin temperature in obstructed hindlimbs.[39] In dogs, cilostazol (100 mg/kg/day P.O.) inhibited thrombotic occlusion in an artificial vessel transplanted as a replacement for the femoral artery.[40] The efficacy of cilostazol in preventing abrupt reocclusion after percutaneous coronary intervention (PCI) was examined in dogs.[41,42] In this model, Saitoh et al. examined the efficacies of antiplatelet drugs in inhibiting abrupt platelet thrombus reformation after tissue plasminogen activator dissolved thrombin-induced thrombi in a coronary artery, in which high shear stress is also important. Reocclusion occurred in six of the seven animals given aspirin (35 mg/kg IV), a result not significantly different from that of the control group. Similarly, beraprost (12 μg/kg IV) failed to prevent reocclusion in five of seven animals. However, cilostazol (1.8 mg/kg IV) prevented reocclusion in six of seven animals. These antithrombotic effects of cilostazol clearly reflect its antiplatelet properties. It is suggested that cilostazol might have synergistic effect with $PGI_2$ generated from endothelial cells, as described previously. It is therefore possible that cilostazol may have a more potent inhibitory effect on platelet function *in vivo* than *in vitro* and *ex vivo*.

## V. Other Effects

### A. *Vasodilation*

In vascular smooth muscle cells, PDE1, 3, 4, and 5 are found at the protein level and PDE1A, 1B, 1C, 3A, 4B, 4D, and 5A are found at the mRNA level.[43–45] The vasodilating effect of cilostazol, similar to its effect on platelet aggregation, is due to an increase in intracellular cAMP levels caused by inhibition of PDE-3A activity in vascular smooth muscle cells.[46] Cilostazol induces the relaxation of rabbit mesenteric artery strips precontracted by KCl. Cilostazol dilates precontracted guinea pig cerebral basilar arteries.[47] The response to cilostazol is independent of the endothelium and of the NO–cGMP pathway in cerebral arteries. In anesthetized dogs,

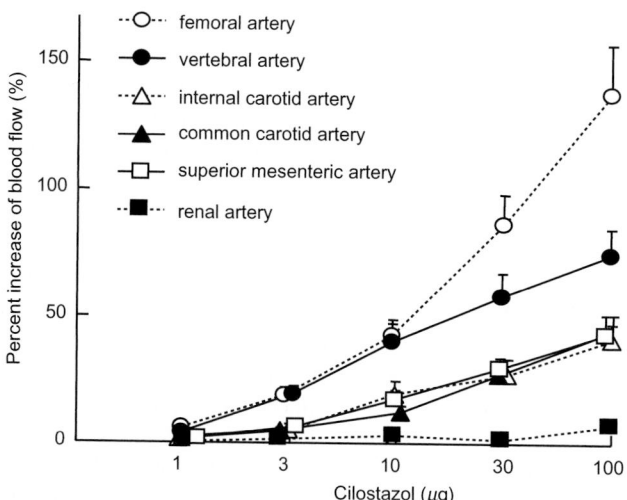

**Figure 64-3.** Vasodilating effects of cilostazol by intraarterial administration in anesthetized dogs. Data are mean ± SE (*n* = 5).

intraarterially administered cilostazol dose dependently increased the blood flow in the vertebral artery, the internal carotid artery, and the common carotid artery (Fig. 64-3).[48] This vasodilatory effect of cilostazol was particularly pronounced in the vertebral artery and the femoral artery, but it was minimal in the renal artery. Intravenously administered cilostazol also significantly increased cerebral and femoral arterial blood flow in dogs.

In patients with chronic arterial occlusion in the lower extremities, blood flow to the lower extremities significantly increased after administration of cilostazol 150 to 200 mg/day for 2 weeks,[49,50] and skin blood flow was also increased by cilostazol 200 mg/day for 6 weeks.[51] In a double-blind, randomized, crossover study, cilostazol (200 mg) or placebo was administered orally to 12 healthy participants.[52] Mean flow velocity in the middle cerebral arteries (MCA) was measured with transcranial Doppler (TCD), and the diameters of the superficial temporal and radial arteries were measured by ultrasonography. Velocity in the MCA decreased 21.5 ± 5.7% after cilostazol and 5.5 ± 12.2% after placebo (*p* = 0.02 vs. placebo), without any change in global or regional cerebral blood flow. The superficial temporal artery diameter increased 17.6 ± 12.3% (*p* < 0.001 vs. baseline) and radial artery diameter increased 12.6 ± 8.6% (*p* < 0.001 vs. baseline). This study suggested that cilostazol dilates the MCA without affecting cerebral blood flow or blood pressure.

## B. Inhibition of Vascular Smooth Muscle Cell Proliferation

In a study that assessed [3H] thymidine uptake and cell counting, cilostazol inhibited proliferation of rat aortic smooth muscle cells in culture.[53] Cilostazol potently inhib-

ited [3H] thymidine uptake by smooth muscle cells stimulated by various growth factors, such as PDGF and insulin, and exhibited no growth factor specificity. PDE3 inhibitors, including cilostazol, also inhibit smooth muscle cell proliferation and were found to inhibit intimal hyperplasia in rat and mouse models of vascular injury.[54-57] Their inhibitory effects on intimal hyperplasia are considered dependent on their inhibition of cell proliferation by increasing cAMP levels in vascular smooth muscle cells. Clinically, cilostazol also significantly decreased the incidence of intimal hyperplasia and restenosis after PCI, directional coronary atherectomy, and stent implantation, as described in Section VI.

## C. Cytoprotective Effects

Cilostazol has cytoprotective effects on cultured endothelial cell (EC) dysfunction. Through mediation of the stimulation of hepatocyte growth factor production, cilostazol prevents EC death induced by high glucose or hypoxia.[56,58] Cilostazol significantly attenuates the dose-dependent increment of monocyte chemoattractant protein-1 production by tumor necrosis factor (TNF)-α.[59] Cilostazol also prevents EC death induced by lipopolysaccharide (LPS).[60] Cilostazol reduces the increases in TNF-α production, Bax protein expression, and cytochrome c release induced by LPS, and it reverses the decrease of Bcl-2 protein. Cilostazol suppresses remnant lipoprotein particle-induced apoptosis of EC.[61]

To investigate the effects of cilostazol on hemispheric ischemic lesions, the apparent diffusion coefficient (ADC) and T2 images by magnetic resonance imaging (MRI) techniques were compared with histology at the termination of and 24 hours after reperfusion following a 2-hour occlusion of rat MCA.[62] Cilostazol (30 mg/kg P.O. at 5 minutes and 4 hours after reperfusion) significantly suppressed the hemispheric lesion area and volumes when detected by ADC, T2 images, and histology. Cilostazol significantly reduced the increased cerebral water content at the ischemic hemisphere. The neurological deteriorations were much improved in the cilostazol-treated group. These investigators[62] suggested that posttreatment with cilostazol exerts a potent protective effect against cerebral infarct size by reducing the cytotoxic edema.

## D. Effect on Lipid Metabolism

By improving lipid metabolism, cilostazol decreases plasma triglycerides and remnant lipoprotein cholesterol, and it increases high-density lipoprotein cholesterol levels in patients with peripheral arterial diseases and those with type 2 diabetes mellitus.[63-66] Cilostazol increases apolipoprotein A₁ and decreases apolipoprotein B levels without affecting the low-density lipoprotein cholesterol level.

The lipid metabolism-improving effect of cilostazol is considered to be due to enhanced lipoprotein lipase activity[67] and is particularly pronounced in patients with hypertriglyceridemia.

# VI. Clinical Results

By virtue of the pharmacological effects described previously, cilostazol has been shown to be effective in various clinical disorders of thrombosis and circulatory insufficiency.

## A. Prevention of Recurrence of Cerebral Infarction

To evaluate the efficacy of cilostazol in preventing recurrent cerebral infarction, a multicenter, double-blind, placebo-controlled trial — the Cilostazol Stroke Prevention Study (CSPS) — was performed from 1992 to 1996 and included 1095 patients.[68,69] Patients were randomized to treatment with cilostazol (100 mg twice daily) or placebo for at least 1 year and up to 5 years (mean treatment period, 632 days). The primary outcome of recurrence of cerebral infarction (fatal and nonfatal) is shown in Fig. 64-4. Cerebral infarction recurred in 30 of 526 patients (5.7%) receiving cilostazol and in 57 of 526 patients (10.8%) receiving placebo. This difference was statistically significant ($p = 0.0149$), with a relative risk reduction of 41.7%. The annual recurrence rate was 3.37% in the cilostazol group and 5.78% in the placebo group.

In the CSPS trial, there were four major hemorrhagic events (three cerebral hemorrhages and one subarachnoid hemorrhage) in the cilostazol group and seven (cerebral hemorrhages) in the control group. Adverse events of bleeding tendency excluding the previously mentioned major

hemorrhagic events were reported in 2.8% (15/526) of the cilostazol-treated patients and 2.1% (11/526) of the controls, a nonsignificant difference. These outcomes concerning bleeding events are noteworthy because clinical studies on other antiplatelet drugs have shown a significant difference in the rate of hemorrhagic events between the placebo and the antiplatelet drug-treated groups. In a small study,[70,71] cilostazol (200 mg/day), aspirin (330 mg/day), and ticlopidine (300 mg/day) were administered to 10 healthy men for 3 days at the respective doses found to have an antiplatelet effect. Pre- and postdrug administration, bleeding time, and volumes were measured with a quantitative bleeding time test apparatus. A small incision (1 mm deep and 1 cm long) was made on the forearm by the Simplate and was covered by a small cup. Physiological saline was continuously perfused into the cup. The amount of hemoglobin from the incision was continually and quantitatively measured to calculate bleeding time and volume. Bleeding time changed as follows (predrug to postdrug): $359.0 \pm 95.8$ to $646.0 \pm 248.0$ seconds for aspirin, $323.3 \pm 99.9$ to $528.7 \pm 180.2$ seconds for ticlopidine, and $313.0 \pm 112.5$ to $343.3 \pm 154.0$ seconds for cilostazol. Bleeding volumes before and after administration of aspirin, ticlopidine, and cilostazol were $14.5 \pm 4.9$ and $30.2 \pm 18.8$ μL, $12.5 \pm 5.0$ and $19.2 \pm 7.2$ μL, and $12.4 \pm 5.2$ and $13.4 \pm 6.8$ μL, respectively. Both bleeding time and volume were significantly changed after administration of aspirin and ticlopidine. By contrast, cilostazol at a dose found to show the same antiplatelet effect as aspirin and ticlopidine in the same study did not significantly change bleeding time or volume.

The effect of cilostazol on the progression of intracranial arterial stenosis (IAS) was investigated in 135 patients with acute symptomatic stenosis in the M1 segment of MCA or the basilar artery randomized to either cilostazol (200 mg/day) or placebo for 6 months.[72] Aspirin (100 mg/day) was given to all patients. IAS was assessed by magnetic resonance angiogram and TCD at the time of recruitment and 6 months later. In the cilostazol group, 3 (6.7%) of 45 symptomatic IASs progressed and 11 (24.4%) regressed. In the placebo group, 15 (28.8%) of symptomatic IASs progressed and 8 (15.4%) regressed. Progression of symptomatic IASs in the cilostazol group was significantly lower than that in the placebo group ($p = 0.008$).

## B. Intermittent Claudication in Chronic Arterial Occlusion

Using the primary end point of maximum walking distance on a treadmill, the efficacy of cilostazol for the treatment of intermittent claudication due to lower limb blood flow insufficiency has been evaluated in the United States since 1993.[73,74] In a clinical trial for this purpose, the absolute claudication distance (ACD) was measured in 239 patients

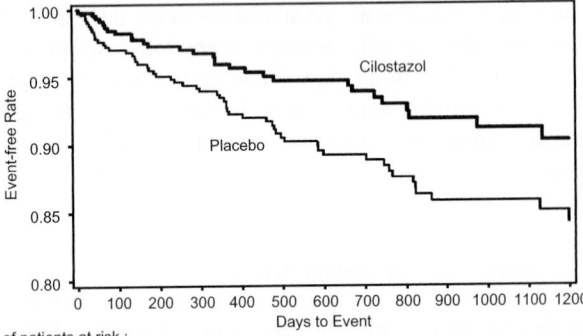

**Figure 64-4.** Kaplan–Meier plot for the primary outcome (recurrence of cerebral infarction) according to assigned treatment in the Cilostazol Stroke Prevention Study (CSPS). The numbers of patients at risk are shown at the bottom.

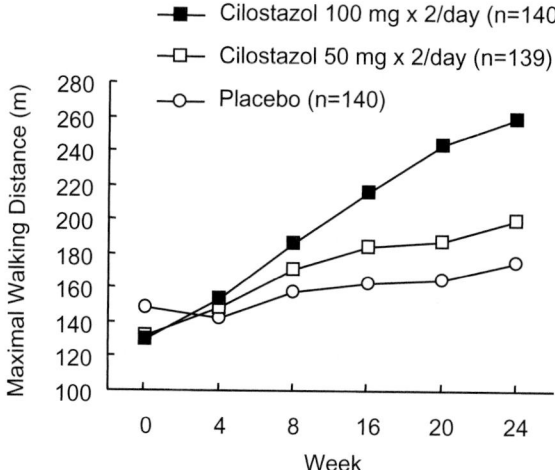

**Figure 64-5.** Effect of cilostazol on maximal walking distance in patients with intermittent claudication. Data are from Beebe et al.[75]

with chronic peripheral arterial occlusion who received cilostazol (200 mg/day) or placebo for 12 weeks.[74] ACD increased by 47.0% in the cilostazol group (119 patients) and by 12.9% in the placebo group (120 patients) ($p < 0.001$). This result appears to be associated with cilostazol's vasodilatory effect on the femoral artery. Based on these results and those of seven other large-scale clinical trials (Fig. 64-5),[73–76] oral cilostazol was granted approval by the FDA in January 1999 for the treatment of various symptoms in patients with intermittent claudication.

A meta-analysis of the results from these eight randomized, placebo-controlled trials has been performed.[77] The meta-analysis examined the effect of cilostazol on pain-free and maximal walking distance, quality-of-life (QOL) measures, and adverse effects in 2702 patients with stable, moderate to severe claudication. Treatment ranged from 12 to 24 weeks. Cilostazol therapy increased maximal and pain-free walking distances by 50 and 67%, respectively. In subgroup analysis, cilostazol increased pain-free and maximal distances similarly in men and women, in older and younger patients, and in patients with and without diabetes. QOL assessments revealed enhanced scores for physical well-being. Cilostazol-treated patients reported a higher incidence of headache, bowel complaints, and palpitations than patients given placebos. This meta-analysis demonstrated that cilostazol significantly increases walking distances and QOL measures in patients with claudication without major adverse effects.

## C. Restenosis after PCI and Stent Implantation

Restenosis after PCI and stenting appears to be the result, at least in part, of proliferation of vascular tissues.[78] In most clinical trials, aspirin and ticlopidine have not shown a clear benefit in preventing this type of restenosis. However, cilostazol has been demonstrated in many studies to prevent restenosis through its inhibitory effect on vascular smooth muscle cell proliferation. Kunishima et al.[79] studied the effect of cilostazol and aspirin on restenosis approximately 5 months after stent implantation. The minimal lumen diameter (MLD) at follow-up was $2.34 \pm 0.74$ mm in the cilostazol group (200 mg/day, $n = 28$ patients with 35 lesions) and $1.89 \pm 1.08$ mm in the aspirin group (81 mg/day, $n = 37$ patients with 41 lesions), revealing significant dilation in the cilostazol group. The restenosis rate was 8.6% in the cilostazol group compared to 26.8% in the aspirin group. This study suggests that administration of cilostazol alone after the implantation of an intracoronary Palmaz–Schatz stent is useful for the prevention of restenosis.

Tsuchikane et al.[80] studied the effect of cilostazol and aspirin on restenosis 3 months after PCI. Late loss, which was calculated as the difference between post-PCI MLD and MLD at follow-up angiography, was $0.15 \pm 0.45$ mm in 123 patients treated with cilostazol (200 mg/day) and $0.45 \pm 0.52$ mm in 129 patients treated with aspirin (250 mg/day) ($p < 0.0001$). The angiographic restenosis rate, defined as follow-up diameter stenosis exceeding 50%, was significantly lower in the cilostazol group (17.9 vs. 39.5%, $p < 0.0001$), which demonstrated efficacy of cilostazol in preventing restenosis.

In a study conducted by Ochiai et al.,[81] aspirin (81 mg) was administered to all patients who received primary Palmaz–Schatz stenting within 12 hours after acute myocardial infarction, and the patients were randomized to receive cilostazol (200 mg/day for 6 months) or ticlopidine (200 mg/day for 1 month) to prevent subacute stent thrombosis. Clinical and angiographic outcomes at 6 months were analyzed. Late loss was $0.49 \pm 0.40$ mm in the cilostazol group and $0.88 \pm 0.52$ mm in the ticlopidine group. Restenosis rates were 0 and 20%, respectively ($p = 0.05$), demonstrating the inhibitory effect of cilostazol on restenosis.

In another study of patients who were implanted with a Palmaz–Schatz stent, 56 patients received cilostazol 200 mg/day and 58 patients received ticlopidine 200 mg/day (the standard dose in Japan) for 6 months.[82] Late loss in the cilostazol and ticlopidine groups was $0.58 \pm 0.52$ and $1.09 \pm 0.65$ mm, respectively ($p < 0.0001$). The restenosis rate was significantly lower in the cilostazol group (16 vs. 33%, $p = 0.0044$). Analysis of outcome at 6 months (total 1-year follow-up) showed that the target vessel revascularization rate at 1 year was 23% in the cilostazol group and 42% in the ticlopidine group ($p = 0.03$), also suggesting that cilostazol is effective in the prevention of restenosis after stenting. Other studies have also reported this inhibitory effect of cilostazol on restenosis after coronary intervention.[83–85]

The Cilostazol for Restenosis (CREST) trial is in progress to evaluate more definitively the ability of cilostazol to prevent restenosis following uncomplicated stent implantation for *de novo* coronary artery stenosis.[86,87] In this randomized, double-blind, multicenter study, 700 patients will receive clopidogrel, aspirin, and either cilostazol or placebo after successful intracoronary stent implantation. The primary end point is MLD of the first lesion stented after 6 months; secondary end points include MLD in all lesions, mean percentage diameter stenosis, target lesion revascularization, and major angiographic end points.

### D. Adverse Effects

Side effects are infrequent with cilostazol, but they include headache, palpitations, and diarrhea. Cilostazol is contraindicated in patients with congestive heart failure. The lack of cilostazol-induced hemorrhagic side effect in the 1095-patient CSPS study is discussed in Section VI.A.

## References

1. Jackson, S. P., & Schoenwaelder, S. M. (2003). Antiplatelet therapy: In search of the "magic bullet." *Nat Rev Drug Discov, 2,* 775–789.
2. Nishi, T., Tabusa, F., Tanaka, T., et al. (1983). Studies on 2-oxoquinoline derivatives as blood platelet aggregation inhibitors: II. 6-[3-(1-cyclohexyl-5-tetrazolyl)propoxy]-1, 2-dihydro-2-oxoquinoline and related compounds. *Chem Pharm Bull, 31,* 1151–1157.
3. Kimura, Y., Tani, T., Kanbe, T., et al. (1985). Effect of cilostazol on platelet aggregation and experimental thrombosis. *Arzneimittelforschung, 35,* 1144–1149.
4. Nishi, T., Kimura, Y., & Nakagawa, K. (2000). Research and development of cilostazol: An antiplatelet agent. *Yakugaku Zasshi, 120,* 1247–1260.
5. Sudo, T., Tachibana, K., Toga, K., et al. (2000). Potent effects of novel anti-platelet aggregatory cilostamide analogues on recombinant cyclic nucleotide phosphodiesterase isozyme activity. *Biochem Pharmacol, 59,* 347–356.
6. Beavo, J. A., & Brunton, L. L. (2002). Cyclic nucleotide research — Still expanding after half a century. *Nat Rev Mol Cell Biol, 3,* 710–718.
7. Schwarz, U. R., Walter, U., & Eigenthaler, M. (2001). Taming platelets with cyclic nucleotides. *Biochem Pharmacol, 62,* 1153–1161.
8. Soderling, S. H., & Beavo, J. A. (2000). Regulation of cAMP and cGMP signaling: New phosphodiesterases and new functions. *Curr Opin Cell Biol, 12,* 174–179.
9. Maurice, D. H., Palmer, D., Tilley, D. G., et al. (2003). Cyclic nucleotide phosphodiesterase activity, expression, and targeting in cells of the cardiovascular system. *Mol Pharmacol, 64,* 533–546.
10. Hidaka, H., & Asano, T. (1976). Human blood platelet 3′: 5′-cyclic nucleotide phosphodiesterase. Isolation of low-Km and high-Km phosphodiesterase. *Biochim Biophys Acta, 429,* 485–497.
11. Tani, T., Sakurai, K., Kimura, Y., et al. (1992). Pharmacological manipulation of tissue cyclic AMP by inhibitors. Effects of phosphodiesterase inhibitors on the functions of platelets and vascular endothelial cells. *Adv Second Messenger Phosphoprotein Res, 25,* 215–227.
12. Ashida, S., & Sakuma, K. (1992). Demonstration of functional compartments of cyclic AMP in rat platelets by the use of phosphodiesterase inhibitors. *Adv Second Messenger Phosphoprotein Res, 25,* 229–239.
13. Zhang, W., Ke, H., & Colman, R. W. (2002). Identification of interaction sites of cyclic nucleotide phosphodiesterase type 3A with milrinone and cilostazol using molecular modeling and site-directed mutagenesis. *Mol Pharmacol, 62,* 514–520.
14. Sudo, T., Ito, H., & Kimura, Y. (2003). Phosphorylation of the vasodilator-stimulated phosphoprotein (VASP) by the anti-platelet drug, cilostazol, in platelets. *Platelets, 14,* 381–390.
15. Reinhard, M., Jarchau, T., & Walter, U. (2001). Actin-based motility: Stop and go with Ena/VASP proteins. *Trends Biochem Sci, 26,* 243–249.
16. Harbeck, B., Huttelmaier, S., Schluter, K., et al. (2000). Phosphorylation of the vasodilator-stimulated phosphoprotein regulates its interaction with actin. *J Biol Chem, 275,* 30817–30825.
17. Horstrup, K., Jablonka, B., Hönig-Liedl, P., et al. (1994). Phosphorylation of focal adhesion vasodilator-stimulated phosphoprotein at Ser157 in intact human platelets correlates with fibrinogen receptor inhibition. *Eur J Biochem, 225,* 21–27.
18. Massberg, S., Gruner, S., Konrad, I., et al. (2004). Enhanced *in vivo* platelet adhesion in vasodilator-stimulated phosphoprotein (VASP)-deficient mice. *Blood, 103,* 136–142.
19. Geiger, J., Teichmann, L., Grossmann, R., et al. (2005). Monitoring of clopidogrel action: Comparison of methods. *Clin Chem, 51,* 957–965.
20. Ito, H., Miyakoda, G., & Mori, T. (2004). Cilostazol inhibits platelet–leukocyte interaction by suppression of platelet activation. *Platelets, 15,* 293–301.
21. Rosado, J. A., Porras, T., Conde, M., et al. (2001). Cyclic nucleotides modulate store-mediated calcium entry through the activation of protein-tyrosine phosphatases and altered actin polymerization in human platelets. *J Biol Chem, 276,* 15666–15675.
22. Sudo, T., Ito, H., Ozeki, Y., et al. (2001). Estimation of anti-platelet drugs on human platelet aggregation with a novel whole blood aggregometer by a screen filtration pressure method. *Br J Pharmacol, 133,* 1396–1404.
23. Ikeda, Y., Handa, M., Kawano, K., et al. (1991). The role of von Willebrand factor and fibrinogen in platelet aggregation under varying shear stress. *J Clin Invest, 87,* 1234–1240.
24. Minami, N., Suzuki, Y., Yamamoto, M., et al. (1997). Inhibition of shear stress-induced platelet aggregation by cilostazol, a specific inhibitor of cGMP-inhibited phosphodiesterase, *in vitro* and *ex vivo*. *Life Sci, 61,* PL383–PL389.
25. Igawa, T., Tani, T., Chijiwa, T., et al. (1990). Potentiation of anti-platelet aggregating activity of cilostazol with vascular endothelial cells. *Thromb Res, 57,* 617–623.

26. Inoue, T., Sohma, R., & Morooka, S. (1999). Cilostazol inhibits the expression of activation-dependent membrane surface glycoprotein on the surface of platelets stimulated *in vitro*. *Thromb Res, 93*, 137–143.

27. Kariyazono, H., Nakamura, K., Shinkawa, T., et al. (2001). Inhibition of platelet aggregation and the release of P-selectin from platelets by cilostazol. *Thromb Res, 101*, 445–453.

28. Yamazaki, M., Uchiyama, S., Xiong, Y., et al. (2005). Effect of remnant-like particle on shear-induced platelet activation and its inhibition by antiplatelet agents. *Thromb Res, 115*, 211–218.

29. Sun, B., Le, S. N., Lin, S., et al. (2002). New mechanism of action for cilostazol: Interplay between adenosine and cilostazol in inhibiting platelet activation. *J Cardiovasc Pharmacol, 40*, 577–585.

30. Yasunaga, K., & Mase, K. (1985). Antiaggregatory effect of oral cilostazol and recovery of platelet aggregability in patients with cerebrovascular disease. *Arzneimittelforschung, 35*, 1189–1192.

31. Ikeda, Y., Kikuchi, M., Murakami, H., et al. (1987). Comparison of the inhibitory effects of cilostazol, acetylsalicylic acid and ticlopidine on platelet functions *ex vivo:* Randomized, double-blind cross-over study. *Arzneimittelforschung, 37*, 563–566.

32. Nagakawa, Y., Konuki, Y., Orimo, H., et al. (1986). The effect of cilostazol (OPC-13013) on arachidonic acid metabolism. *Jpn Pharmacol Ther, 14*, 6319–6324.

33. Uehara, S., & Hirayama, A. (1989). Effects of cilostazol on platelet function. *Arzneimittelforschung, 39*, 1531–1534.

34. Nomura, S., Shouzu, A., Omoto, S., et al. (1998). Effect of cilostazol on soluble adhesion molecules and platelet-derived microparticles in patients with diabetes. *Thromb Haemost, 80*, 388–392.

35. Inoue, T., Uchida, T., Sakuma, M., et al. (2004). Cilostazol inhibits leukocyte integrin Mac-1, leading to a potential reduction in restenosis after coronary stent implantation. *J Am Coll Cardiol, 44*, 1408–1414.

36. Falati, S., Gross, P., Merrill-Skoloff, G., et al. (2002). Real-time *in vivo* imaging of platelets, tissue factor and fibrin during arterial thrombus formation in the mouse. *Nat Med, 8*, 1175–1180.

37. Sim, D. S., Merrill-Skoloff, G., Furie, B. C., et al. (2004). Initial accumulation of platelets during arterial thrombus formation *in vivo* is inhibited by elevation of basal cAMP levels. *Blood, 103*, 2127–2134.

38. Watanabe, K., Nakase, H., & Kimura, Y. (1986). Effect of cilostazol on experimental cerebral infarction in rabbits. *Arzneimittelforschung, 36*, 1022–1024.

39. Kawamura, K., Fujita, S., Tani, T., et al. (1985). Effect of cilostazol, a new antithrombotic drug, on an experimental model of peripheral circulation insufficiency. *Arzneimittelforschung, 35*, 1154–1156.

40. Yasuda, K., Tanabe, T., Hashimoto, M., et al. (1985). Effect of cilostazol, a new antithrombotic drug, on small arterial replacement. *Thromb Haemost, 54*, 211.

41. Saitoh, S., Saitoh, T., Otake, A., et al. (1993). Cilostazol, a novel cyclic AMP phosphodiesterase inhibitor, prevents reocclusion after coronary arterial thrombolysis with recombinant tissue-type plasminogen activator. *Arterioscler Thromb, 13*, 563–570.

42. Saitoh, T., Saitoh, S., Yaoita, H., et al. (1993). Effects of antiplatelet agents to prevent immediate reocclusion after thrombolysis. *J Med Pharm Sci, 29*, 89–93.

43. Souness, J. E., Maslen, C., Webber, S., et al. (1995). Suppression of eosinophil function by RP 73401, a potent and selective inhibitor of cyclic AMP-specific phosphodiesterase: Comparison with rolipram. *Br J Pharmacol, 115*, 39–46.

44. Degerman, E., Belfrage, P., & Manganiello, V. C. (1997). Structure, localization, and regulation of cGMP-inhibited phosphodiesterase (PDE3). *J Biol Chem, 272*, 6823–6826.

45. Rybalkin, S. D., Bornfeldt, K. E., Sonnenburg, W. K., et al. (1997). Calmodulin-stimulated cyclic nucleotide phosphodiesterase (PDE1C) is induced in human arterial smooth muscle cells of the synthetic, proliferative phenotype. *J Clin Invest, 100*, 2611–2621.

46. Tanaka, T., Ishikawa, T., Hagiwara, M., et al. (1988). Effect of cilostazol, a selective cAMP phosphodiesterase inhibitor, on the contraction of vascular smooth muscle. *Pharmacology, 36*, 313–320.

47. Birk, S., Edvinsson, L., Olesen, J., et al. (2004). Analysis of the effects of phosphodiesterase type 3 and 4 inhibitors in cerebral arteries. *Eur J Pharmacol, 489*, 93–100.

48. Kawamura, K., Watanabe, K., & Kimura, Y. (1985). Effect of cilostazol, a new antithrombotic drug, on cerebral circulation. *Arzneimittelforschung, 35*, 1149–1154.

49. Yasuda, K., Sakuma, M., & Tanabe, T. (1985). Hemodynamic effect of cilostazol on increasing peripheral blood flow in arteriosclerosis obliterans. *Arzneimittelforschung, 35*, 1198–1200.

50. Kamiya, T., & Sakaguchi, S. (1985). Hemodynamic effects of antithrombotic drug cilostazol in chronic arterial occlusion in the extremities. *Arzneimittelforschung, 35*, 1201–1203.

51. Ohashi, S., Iwatani, M., Hyakuna, Y., et al. (1985). Thermographic evaluation of the hemodynamic effect of the antithrombotic drug cilostazol in peripheral arterial occlusion. *Arzneimittelforschung, 35*, 1203–1208.

52. Birk, S., Kruuse, C., Petersen, K. A., et al. (2004). The phosphodiesterase 3 inhibitor cilostazol dilates large cerebral arteries in humans without affecting regional cerebral blood flow. *J Cereb Blood Flow Metab, 24*, 1352–1358.

53. Takahashi, S., Oida, K., Fujiwara, R., et al. (1992). Effect of cilostazol, a cyclic AMP phosphodiesterase inhibitor, on the proliferation of rat aortic smooth muscle cells in culture. *J Cardiovasc Pharmacol, 20*, 900–906.

54. Ishizaka, N., Taguchi, J., Kimura, Y., et al. (1999). Effects of a single local administration of cilostazol on neointimal formation in balloon-injured rat carotid artery. *Atherosclerosis, 142*, 41–46.

55. Inoue, Y., Toga, K., Sudo, T., et al. (2000). Suppression of arterial intimal hyperplasia by cilostamide, a cyclic nucleotide phosphodiesterase 3 inhibitor, in a rat balloon double-injury model. *Br J Pharmacol, 130*, 231–241.

56. Aoki, M., Morishita, R., Hayashi, S., et al. (2001). Inhibition of neointimal formation after balloon injury by cilostazol, accompanied by improvement of endothelial dysfunction and

induction of hepatocyte growth factor in rat diabetes model. *Diabetologia, 44,* 1034–1042.

57. Kim, M. J., Park, K. G., Lee, K. M., et al. (2005). Cilostazol inhibits vascular smooth muscle cell growth by downregulation of the transcription factor E2F. *Hypertension, 45,* 552–556.

58. Morishita, R., Higaki, J., Hayashi, S. I., et al. (1997). Role of hepatocyte growth factor in endothelial regulation: Prevention of high D-glucose-induced endothelial cell death by prostaglandins and phosphodiesterase type 3 inhibitor. *Diabetologia, 40,* 1053–1061.

59. Nishio, Y., Kashiwagi, A., Takahara, N., et al. (1997). Cilostazol, a cAMP phosphodiesterase inhibitor, attenuates the production of monocyte chemoattractant protein-1 in response to tumor necrosis factor-alpha in vascular endothelial cells. *Horm Metab Res, 29,* 491–495.

60. Kim, K. Y., Shin, H. K., Choi, J. M., et al. (2002). Inhibition of lipopolysaccharide-induced apoptosis by cilostazol in human umbilical vein endothelial cells. *J Pharmacol Exp Ther, 300,* 709–715.

61. Shin, H. K., Kim, Y. K., Kim, K. Y., et al. (2004). Remnant lipoprotein particles induce apoptosis in endothelial cells by NAD(P)H oxidase-mediated production of superoxide and cytokines via lectin-like oxidized low-density lipoprotein receptor-1 activation: Prevention by cilostazol. *Circulation, 109,* 1022–1028.

62. Lee, J. H., Lee, Y. K., Ishikawa, M., et al. (2003). Cilostazol reduces brain lesion induced by focal cerebral ischemia in rats — An MRI study. *Brain Res, 994,* 91–98.

63. Elam, M. B., Heckman, J., Crouse, J. R., et al. (1998). Effect of the novel antiplatelet agent cilostazol on plasma lipoproteins in patients with intermittent claudication. *Arterioscler Thromb Vasc Biol, 18,* 1942–1947.

64. Iwasaki, K., Mochizuki, K., Iwasaki, M., et al. (2002). Cilostazol, a potent phosphodiesterase type III inhibitor, selectively increases antiatherogenic high-density lipoprotein subclass LpA-I and improves postprandial lipemia in patients with type 2 diabetes mellitus. *Metabolism, 51,* 1348–1354.

65. Wang, T., Elam, M. B., Forbes, W. P., et al. (2003). Reduction of remnant lipoprotein cholesterol concentrations by cilostazol in patients with intermittent claudication. *Atherosclerosis, 171,* 337–342.

66. Nakamura, N., Hamazaki, T., Johkaji, H., et al. (2003). Effects of cilostazol on serum lipid concentrations and plasma fatty acid composition in type 2 diabetic patients with peripheral vascular disease. *Clin Exp Med, 2,* 180–184.

67. Tani, T., Uehara, K., Sudo, T., et al. (2000). Cilostazol, a selective type III phosphodiesterase inhibitor, decreases triglyceride and increases HDL cholesterol levels by increasing lipoprotein lipase activity in rats. *Atherosclerosis, 152,* 299–305.

68. Gotoh, F., Tohgi, H., Hirai, S., et al. (2000). Cilostazol Stroke Prevention Study: A placebo-controlled double-blind trial for secondary prevention of cerebral infarction. *J Stroke Cerebrovasc Dis, 9,* 147–157.

69. Gotoh, F., Ohashi, Y., and the Cilostazol Stroke Prevention Study Group. (2000). Design and organization of the Cilostazol Stroke Prevention Study. *J Stroke Cerebrovasc Dis, 9,* 36–44.

70. Tamai, Y., Takami, R., Nakahata, R., et al. (1999). Comparison of the effects of acetylsalicylic acid, ticlopidine and cilostazol on primary hemostasis using a quantitative bleeding time test apparatus. *Haemostasis, 29,* 269–276.

71. Kim, J. S., Lee, K. S., Kim, Y. I., et al. (2004). A randomized crossover comparative study of aspirin, cilostazol and clopidogrel in normal controls: Analysis with quantitative bleeding time and platelet aggregation test. *J Clin Neurosci, 11,* 600–602.

72. Kwon, S. U., Cho, Y. J., Koo, J. S., et al. (2005). Cilostazol prevents the progression of the symptomatic intracranial arterial stenosis: The multicenter double-blind placebo-controlled trial of cilostazol in symptomatic intracranial arterial stenosis. *Stroke, 36,* 782–786.

73. Dawson, D. L., Cutler, B. S., Meissner, M. H., et al. (1998). Cilostazol has beneficial effects in treatment of intermittent claudication. Results from a multicenter, randomized, prospective, double-blind trial. *Circulation, 98,* 678–686.

74. Money, S., Herd, A., Isaacsohn, J. L., et al. (1998). Effect of cilostazol on walking distances in patients with intermittent claudication caused by peripheral vascular disease. *J Vasc Surg, 27,* 267–275.

75. Beebe, H. G., Dawson, D. L., Cutler, B. S., et al. (1999). A new pharmacological treatment for intermittent claudication: Results of a randomized, multicenter trial. *Arch Intern Med, 159,* 2041–2050.

76. Dawson, D. L., DeMaioribus, C. A., Hagino, R. T., et al. (1999). The effect of withdrawal of drugs treating intermittent claudication. *Am J Surg, 178,* 141–146.

77. Thompson, P. D., Zimet, R., Forbes, W. P., et al. (2002). Meta-analysis of results from eight randomized, placebo-controlled trials on the effect of cilostazol on patients with intermittent claudication. *Am J Cardiol, 90,* 1314–1319.

78. Lefkovits, J., & Topol, E. J. (1997). Pharmacological approaches for the prevention of restenosis after percutaneous coronary intervention. *Prog Cardiovasc Dis, 40,* 141–158.

79. Kunishima, T., Musha, H., Eto, F., et al. (1997). A randomized trial of aspirin versus cilostazol therapy after successful coronary stent implantation. *Clin Ther, 19,* 1058–1066.

80. Tsuchikane, E., Fukuhara, A., Kobayashi, T., et al. (1999). Impact of cilostazol on restenosis after percutaneous coronary balloon angioplasty. *Circulation, 100,* 21–26.

81. Ochiai, M., Eto, K., Takeshita, S., et al. (1999). Impact of cilostazol on clinical and angiographic outcome after primary stenting for acute myocardial infarction. *Am J Cardiol, 84,* 1074–1076.

82. Kozuma, K., Hara, M., Yamasaki, M., et al. (2001). Effects of cilostazol on late lumen loss and repeat revascularization after Palmaz–Schatz coronary stent implantation. *Am Heart J, 141,* 124–130.

83. Tsuchikane, E., Katoh, O., Sumitsuji, S., et al. (1998). Impact of cilostazol on intimal proliferation after directional coronary atherectomy. *Am Heart J, 135,* 495–502.

84. Sekiya, M., Funada, J., Watanabe, K., et al. (1998). Effects of probucol and cilostazol alone and in combination on frequency of poststenting restenosis. *Am J Cardiol, 82,* 144–147.

85. Tsuchikane, E., Kobayashi, T., & Awata, N. (2000). The potential of cilostazol in interventional cardiology. *Curr Interv Cardiol Rep, 2,* 143–148.

86. Douglas, J. S., Weintraub, W. S., & Holmes, D. (2003). Rationale and design of the randomized, multicenter, Cilostazol for Restenosis (CREST) trial. *Clin Cardiol, 26,* 451–454.

87. Weintraub, W. S., Foster, J., Culler, S. D., et al. (2004). Methods for the economic and quality of life supplement to the Cilostazol for Restenosis (CREST) trial. *J Invasive Cardiol, 16,* 257–259.

# CHAPTER 65

# Experimental Antiplatelet Therapy

**Anthony A. Bavry, Deepak L. Bhatt, and Eric J. Topol**

*Department of Cardiovascular Medicine, Cleveland Clinic Foundation, Cleveland, Ohio*

## I. Introduction

The platelet plays a central role in the pathophysiology of atherosclerosis through its effects on inflammation.[1,2] In addition to modulating inflammation, the platelet is directly involved in thrombosis and subsequent acute vascular events (including acute coronary syndromes, ischemic strokes, and symptomatic peripheral arterial disease).[3] Together, this spectrum of disease is termed atherothrombosis, and the goal of antiplatelet therapy is to reduce the development and clinical expression of atherothrombotic disease.

Significant improvements in cardiovascular morbidity and mortality have occurred through the use of antiplatelet agents, of which aspirin is the prototypical agent (see Chapter 60). Despite high use of aspirin in individuals hospitalized for acute coronary syndromes, its use remains less than optimal in outpatients.[4,5] Although aspirin is effective at reducing cardiovascular outcomes, it is limited by an increased risk for major bleeding including a modest increase in hemorrhagic stroke.[6] Resistance to aspirin is also increasingly recognized as an important clinical entity.[7,8] This was shown in a prospective study of 325 patients with stable coronary disease who received 325 mg of aspirin daily for at least 7 days.[8] This study reported that 5 to 9% of individuals were aspirin resistant, whereas an additional 23% were classified as semiresponders. This finding may help to explain why prior aspirin use predicts worse outcomes in individuals with non-ST-elevation acute coronary syndromes.[9] Using data from the Platelet IIb/IIIa in Unstable Angina: Receptor Suppression Using Integrilin Therapy (PURSUIT) trial, the incidence for death at 30 days for individuals presenting with a non-ST-elevation acute coronary syndrome was 4.0% for previous aspirin users versus 2.9% ($p = 0.003$) for nonusers. Thus, although the first goal of the medical community is to increase use of proven existing therapies such as aspirin, a second goal is to design more effective antiplatelet agents with improved safety profiles.

This chapter briefly reviews general platelet biology. Current antiplatelet therapies are discussed briefly, highlighting their mechanism of action, key landmark trials or systematic reviews that support their clinical use, and their limitations. This chapter then explores experimental antiplatelet therapies. Agents that directly modulate platelet enzymes or receptors are discussed first, followed by agents that indirectly affect platelet function.

## II. Platelet Biology

Platelets are anucleate 1- to 3-μm cells formed in the bone marrow from megakaryocytes (see Chapter 2). Platelets circulate in an inactive state predominantly in a layer of blood that is closest to the endothelium. Platelets remain in an inactive state unless they are exposed to agonists or foreign substances such as subendothelial components. The life span of an inactive platelet is approximately 7 to 10 days. The critical processes of platelet-mediated thrombosis/hemostasis are adhesion, secretion, activation, and aggregation, although there is considerable overlap in these processes.[9,10]

Thrombosis is initiated when a vulnerable plaque ruptures (see Chapter 35).[3] This exposes platelets to substances such as collagen and von Willebrand factor (VWF) in the subendothelium of the atherosclerotic plaque. Subendothelial collagen binds to glycoprotein (GP) Ia-IIa and GPVI receptors on the platelet surface under both stasis and high shear conditions (see Chapter 18), whereas VWF binds to the platelet GPIb-IX-V complex under high shear conditions (see Chapter 7).[11–15] Interaction with either the GPIb-IX-V complex or the GPIa-IIa complex initiates the process of adhesion, which causes a conformational shape change in the platelet.[12]

This process is followed by platelet secretion and activation (see Chapters 15 and 16). Platelets may have been activated prior to adhesion by exposure to circulating mediators, thus highlighting the extensive overlap in the processes of thrombosis/hemostasis. Activated platelets secrete a variety of substances involved in inflammation and hemostasis (see

Chapters 15 and 39). Substances released from dense granules include adenosine diphosphate (ADP) and serotonin, whereas substances released from α granules include fibrinogen, VWF, and inflammatory mediators such as interleukin-1β, platelet factor-4, and platelet-derived growth factor.[16–18] CD40 ligand and thromboxane A$_2$ (TXA$_2$) (see Chapter 31) are also released from activated platelets.[19] During activation, actin–myosin complexes shorten, which causes retraction and stabilization of the early clot (see Chapter 4).

This process is intimately involved with and followed by platelet aggregation (see Chapters 8 and 26).[17,20] With platelet activation, GPIIb-IIIa (integrin αIIbβ3) receptors expressed on the surface of the platelet change their conformation to allow their binding to fibrinogen, VWF, and fibronectin. This cross-links platelets together, forming an interconnected layer of platelets as part of the thrombus.

In addition to platelets' role in thrombosis, they also play an integral role in inflammation (see Chapter 39).[2] The inflammatory mediators secreted by activated platelets promote the development of atherosclerosis by attracting leukocytes to platelets and the endothelium, further activating more platelets and enhancing the inflammatory process. Platelet P-selectin is instrumental in this regard by providing a site for leukocyte tethering and rolling (see Chapter 12).[18] Areas of vascular atherosclerosis further activate platelets through altered flow of blood (increased shear stress activates platelets) and impaired nitric oxide (NO) production (see Chapter 35). Impaired NO inhibits platelet guanylate cyclase and decreases cyclic GMP levels (see Chapter 13). Reduced NO also increases platelet intracellular calcium content, thereby increasing surface P-selectin expression and converting GPIIb-IIIa receptors to their activated state. Thus, the overall result of impaired NO production is to shift platelets into a more activated state (see Chapter 13).[17,20]

The end result of platelet membrane phospholipid metabolism is the production of TXA$_2$ and other prostaglandins/prostanoids (see Chapter 31).[10,17] This process starts by the conversion of cell membrane phospholipids into arachidonic acid by the phospholipase A enzyme. Arachidonic acid then undergoes metabolism by cyclooxygenase (COX) to form prostaglandin (PG) G$_2$ and finally PGH$_2$. Two COX isoenzymes exist (each 72 kDa), termed COX-1 and COX-2. The former is constitutively expressed in most cells, particularly platelets and endothelial cells. This enzyme is responsible for the homeostatic production of prostaglandins that are involved in important functions such as maintaining renal blood flow and supporting the gastric mucosa. On the other hand, the COX-2 isoenzyme is induced by inflammatory stimuli and accordingly helps to regulate inflammation. COX-2 is found in a wide variety of cells, including monocytes and macrophages, but little if any is present in resting platelets. PGH$_2$ is acted upon by specific enzymes that form specific prostaglandins. Two principal and opposing mediators in thrombosis are TXA$_2$ and PGI$_2$ (prostacyclin) (see Chapters 13 and 31). TXA$_2$ acts to increase platelet aggregation and cause vasoconstriction, whereas prostacyclin produces opposite effects. TXA$_2$ production is increased in smooth muscle cells of atherosclerotic areas.[21]

## III.  Current Antiplatelet Therapies

### A.  Cyclooxygenase Inhibitors

Acetylsalicylic acid, commonly referred to as aspirin, is the cornerstone of antiplatelet therapy (Table 65-1) (as discussed in detail in Chapter 60). It is also the oldest and best studied agent in this class of medicines. Aspirin irreversibly acetylates platelet COX enzyme for the life of the platelet.[10] Specifically, acetylation occurs at serine position 529 of COX-1 and Ser516 of COX-2. Irreversible inhibition of COX decreases the production of downstream prostanoids, including TXA$_2$, which thereby inhibits platelet aggregation and decreases vasoconstriction. This helps to prevent or minimize intravascular thrombus formation upon plaque rupture.

Although the plasma half-life of aspirin is only 15 to 20 minutes, COX is inhibited for the life of the platelet because, lacking a nucleus, the platelet cannot synthesize significant quantities of new COX. This makes daily dosing of aspirin sufficient. Aspirin produces a measurable decrease in platelet aggregation within 60 minutes of ingestion. Low-dose aspirin (75–150 mg per day) appears to be equally effective as higher doses in reducing adverse clinical events.[22] Since aspirin is approximately 100- to 200-fold more potent at inhibiting COX-1 than COX-2, in order for aspirin to have a potent anti-inflammatory effect, markedly higher doses would be required.

Aspirin is a relatively weak antiplatelet agent because high concentrations of collagen and thrombin are able to activate platelets through alternative pathways. For example, aspirin is not able to inhibit ADP-mediated α granule release.[23] Also, aspirin does not inhibit adhesion and activation that have been mediated by high shear stress, and it does not inhibit the binding of fibrinogen.[24,25] Work has also elucidated that diabetic patients exhibit a relatively impaired antiplatelet effect in response to aspirin.[26] Despite aspirin's limitations, a large body of data supports its clinical efficacy in preventing cardiovascular events among high-risk individuals.

The Anti-Thrombotic Trialist Collaboration published a large meta-analysis that documented the beneficial effects of aspirin in more than 100,000 individuals with acute or previous symptomatic vascular disease.[22] This meta-analysis found that aspirin significantly reduced mortality

**Table 65-1: Current Antiplatelet Therapies**

| Mechanism of Action | Drugs | Clinical Use | Limitations |
|---|---|---|---|
| Cyclooxygenase inhibition | Acetylsalicylic acid (aspirin) | Survival benefit in secondary prevention of vascular disease.<br>Primary prevention of myocardial infarction and ischemic stroke in high-risk individuals. | Irreversible effect.<br>Increased major bleeding.<br>Resistance/relative lack of effect in certain individuals. |
| P2Y$_{12}$ ADP receptor inhibition | Ticlopidine (Ticlid)<br>Clopidogrel (Plavix) | Reduction in composite cardiac events in conservatively treated non-ST-elevation acute coronary syndromes.<br>Reduction in stent thrombosis when used adjunctively with aspirin for coronary revascularization (i.e., intracoronary stents). | Irreversible effect.<br>Increased major bleeding when used in close proximity to significant surgical procedures.<br>Increased bleeding when used with higher dose aspirin.<br>Resistance/relative lack of effect in certain individuals. |
| GPIIb-IIIa inhibition | Abciximab (ReoPro)<br>Eptifibatide (Integrilin)<br>Tirofiban (Aggrastat) | Survival benefit in high-risk acute coronary syndrome patients managed by percutaneous coronary intervention. | Increased major bleeding/thrombocytopenia.<br>Use limited to short duration (i.e., generally less than 24 hours).<br>Equivocal efficacy in conservatively treated acute coronary syndromes. |
| Phosphodiesterase inhibition | Dipyridamole (Persantine)<br>Dipyridamole + aspirin (Aggrenox)<br>Cilostazol (Pletal) | Aggrenox used as an alternative to aspirin in secondary prevention of ischemic stroke.<br>Cilostazol rarely used except for intermittent claudication and during percutaneous coronary intervention in Asian countries. Some evidence for decreased restenosis. | Aggrenox has not conclusively been shown to be more effective than aspirin and is limited by significant headaches.<br>Higher rate of stent thrombosis with cilostazol compared to P2Y$_{12}$ ADP receptor antagonists. |

by 15%, myocardial infarction by 35%, and stroke by 31% compared to placebo ($p < 0.0001$ for each end point). Multiple clinical trials have also examined the role of low-dose aspirin in primary prevention. In aggregate, these studies showed that aspirin reduces myocardial infarction in people at high risk.[27,28]

## B. ADP Receptor Antagonists

ADP released from dense granules of activated platelets and from damaged endothelium binds to platelet ADP receptors, thus stimulating more ADP release.[29,30] This acts to further enhance platelet activation and recruit more platelets. Aspirin-resistant patients have increased platelet sensitivity to ADP, so these individuals may especially benefit from ADP receptor antagonism.[31]

There are two known ADP receptors — P2Y$_1$ and P2Y$_{12}$ (see Chapter 10). The P2Y receptors are G protein-coupled receptors. Binding to the P2Y$_1$ receptor results in mobilization of intracellular calcium and causes a major change in

the platelet and a mild, transient inhibition of platelet aggregation. In contrast to the weak effect on inhibiting platelet function by P2Y$_1$, binding to the P2Y$_{12}$ receptor results in more potent and sustained inhibition of platelet aggregation. The effect of inhibition of the P2Y$_{12}$ receptor is to inhibit cAMP formation, which keeps GPIIb-IIIa receptors in an inactive state.

The two commercially available ADP receptor antagonists are thienopyridines — clopidogrel and ticlopidine (Table 65-1) (as discussed in detail in Chapter 61).[33] Thienopyridines are metabolized by hepatic cytochrome P450-1A to an active metabolite that irreversibly binds to the P2Y$_{12}$ ADP receptor. Thienopyridines have a long half-life and so they are dosed on a daily basis. Unlike aspirin, ADP receptor antagonists have a slower onset of action, although this can be partially overcome by giving a higher clopidogrel loading dose of 300 mg, 600 mg, or perhaps even higher in the future.[34] The degree of platelet activation when clopidogrel therapy is initiated may also help determine the subsequent effect of the drug.[35–37] Patients with unstable angina at the time of percutaneous coronary intervention (PCI) have been shown to have less inhibition of platelet aggregation compared to more stable patients undergoing coronary revascularization. Since thienopyridines irreversibly block the P2Y$_{12}$ ADP receptor, some authors have suggested that major surgery should be delayed for approximately 5 to 7 days after the last dose since there is an increased risk for major bleeding during this time period.[38]

Additionally, as many as 30% of individuals treated with clopidogrel have been found to be resistant, as measured by standard platelet aggregation studies.[39,40] Potential mechanisms for clopidogrel resistance include genetic polymorphisms of the P2Y$_{12}$ receptor and the hepatic enzyme CYP3A that metabolizes thienopyridines into the active metabolite. Polymorphism of the P2Y$_{12}$ receptor is thought to increase the number of ADP receptors on the surface of the platelet. This knowledge could help to identify individuals who would require a higher dose of a thienopyridine or alternatively a different antiplatelet regimen.

Two landmark trials that established the effectiveness of clopidogrel in patients with coronary disease were the Clopidogrel in Unstable Angina to Prevent Recurrent Events (CURE) and the Clopidogrel in Unstable Angina to Prevent Recurrent Events (CREDO) trials.[41,42] The CURE trial randomized more than 12,000 non-ST-elevation acute coronary syndrome patients within 24 hours of onset of chest pain to clopidogrel (mean 9 months) versus placebo. This was largely a conservatively treated population because fewer than 50% of the patients underwent early angiography and fewer than 10% of patients were treated with a GPIIb-IIIa antagonist. Cardiac outcomes were significantly reduced at 30 days (relative risk [RR] 0.79; 95% confidence interval [CI], 0.67–0.92) and at 1 year by the regimen of clopidogrel plus aspirin versus aspirin alone (RR, 0.8; 95% CI, 0.72–0.9; $p < 0.001$). Important safety data also come from the CURE trial. Major bleeding was increased when clopidogrel was stopped within 5 days of coronary artery bypass grafting (CABG) (9.6% in clopidogrel vs. 6.3% in placebo; RR 1.53; $p = 0.06$), although patients who had clopidogrel discontinued more than 5 days before CABG did not have an increase in major postoperative bleeding.[39] Additionally, patients on clopidogrel who were treated with "high-dose" aspirin (325 mg per day) suffered more bleeding than individuals on lower doses.

The CREDO trial enrolled more than 2000 patients who were scheduled to undergo elective PCI.[42] A loading dose of clopidogrel did not significantly reduce 28-day events, although when the analysis was restricted to individuals who received clopidogrel 6 or more hours prior to the procedure, there was a significant reduction in 28-day events (RR 0.61; 95% CI, 0.37–0.98; $p = 0.051$). At 1 year, there was a significant reduction in all-cause mortality, myocardial infarction, or stroke from the long-term use of clopidogrel (RR 0.73; 95% CI, 0.56–0.96). Major bleeding that occurred from the use of clopidogrel was usually in the setting of CABG.

## C. GPIIb-IIIa Antagonists

Platelet activation results in a conformational change in the platelet GPIIb-IIIa receptor (see Chapter 8). Once conformational change occurs, the GPIIb-IIIa receptor is able to block the RGD binding site for fibrinogen with high affinity.[17,43,44] Fibrinogen is bivalent, so two platelets are able to link together, thus initiating the process of aggregation. GPIIb-IIIa antagonists have little direct effect on platelet activation.

GPIIb-IIIa antagonists reversibly bind at the GPIIb-IIIa receptor and prevent fibrinogen/platelet cross-linking from occurring. The commercially available GPIIb-IIIa antagonists are abciximab (ReoPro), eptifibatide (Integrilin), and tirofiban (Aggrastat) (Table 65-1) (as discussed in detail in Chapter 62).[17,43,44] Abciximab is a humanized monoclonal antibody fragment directed against the receptor, whereas the other agents are small molecules. Continuous infusion of the small-molecule agents is needed to produce a sustained inhibition of platelet function, whereas the effect of abciximab lasts more than 24 hours.

GPIIb-IIIa antagonists are associated with a reduction in mortality among patients undergoing PCI. In this setting, the use of GPIIb-IIIa antagonists reduced mortality by approximately 30% at 30 days (RR 0.69; 95% CI 0.53–0.90), 20% at 6 months (RR 0.79; 95% CI 0.64–0.97), and 20% during longer follow-up (RR 0.79; 95% CI 0.66–0.94).[45–47] The Do Tirofiban and ReoPro Give Similar Efficacy (TARGET) trial was designed to assess the noninferiority of tirofiban compared to abciximab when used during PCI.[48] The primary end point of death, myocardial infarction, or urgent target vessel revascularization occurred in 7.6% of the tirofiban group and 6.0% of the abciximab group (hazard ratio [HR], 1.26; 95% CI 1.01–1.57; $p = 0.038$). Data on the use of GPIIb-IIIa antagonists outside of coronary intervention are less robust. An analysis of more than 31,000 acute coronary syndrome patients who were largely conservatively treated revealed a 9% reduction in the composite outcome of death and myocardial infarction at 30 days (OR 0.91; 95% CI 0.84–0.98; $p = 0.015$), although this approach may confer a more pronounced survival advantage to diabetics.[49,50]

## D. Phosphodiesterase Inhibitors

This class is represented by dipyridamole (Persantine) and the combination agent dipyridamole and aspirin (Aggrenox) (Table 65-1) (as discussed in detail in Chapter 63). Phosphodiesterase inhibitors interfere with the breakdown of intraplatelet cyclic nucleotides and increase intracellular cGMP and cAMP levels. An increase in cGMP potentiates the action of NO, whereas an increase in cAMP enhances prostacyclin production.[17] Cilostazol (Pletal) is a related agent that is a selective inhibitor of the phosphodiesterase-3 isoenzyme (as discussed in detail in Chapter 64). Cilostazol inhibits ADP-induced platelet aggregation, although it does not decrease the production of prostacyclin.[17]

Dipyridamole plus aspirin may have slightly better efficacy than aspirin in patients with cerebrovascular disease,[51] although improved efficacy has not been convincingly shown. This agent may be considered a potential alternative agent in the secondary prevention of ischemic stroke and transient ischemic attacks.[51,52] Unfortunately, dipyridamole plus aspirin is limited by a high proportion of headaches. Cilostazol is effective in the treatment of intermittent claudication (see Chapter 37). Cilostazol may have a modest effect in reducing restenosis following PCI, although this agent is not as effective as P2Y$_{12}$ ADP receptor antagonists in preventing stent thrombosis.[53,54]

## IV. Experimental Antiplatelet Therapies

### A. Direct Antiplatelet Agents

**1. Selective P2Y$_{12}$ ADP Receptor Antagonists.** Currently available P2Y$_{12}$ ADP receptor antagonists (clopidogrel and ticlopidine) are effective in decreasing urgent target vessel revascularization and stent thrombosis, although they are limited by variability in response and slow onset of action (especially during states of increased platelet activation) (see Chapter 61).[34] Novel P2Y$_{12}$ ADP receptor antagonists have shown promise in overcoming some of these limitations. Three new agents currently being investigated are briefly discussed in Table 65-2, with a more detailed description of these in Chapter 61.

A 60-mg loading dose of the P2Y$_{12}$ antagonist prasugrel (CS-747, LY640315) resulted in greater inhibition of platelet aggregation than a 300-mg loading dose of clopidogrel, whereas chronic dosing with prasugrel resulted in more sustained inhibition of platelet function. The use of prasugrel versus clopidogrel during PCI has been reported in the Joint Utilization of Medications to Block Platelets Optimally (JUMBO)-Thrombolysis in Myocardial Infarction (TIMI) 26 trial, a double-blind phase II trial designed to address the safety of prasugrel in patients undergoing elective or urgent coronary revascularization. Prasugrel resulted in major and minor bleeding in 1.7% of patients compared to 1.2% in the clopidogrel group (HR 1.42; 95% CI 0.40–5.08), whereas major adverse cardiac outcomes occurred in 7.2 and 9.4%, respectively (HR 0.76; 95% CI 0.46–1.24). The TRITON-TIMI 38 trial randomized 13,000 patients with an acute coronary syndrome and planned PCI to prasugrel or clopidogrel for 1 year.[55–57]

AZD-6140 is an oral P2Y$_{12}$ receptor antagonist. Because this agent is not a thienopyridine, hepatic metabolism is not required for its activity. Another distinct advantage of this agent is that it reversibly blocks the P2Y$_{12}$ receptor; its

clinical effect should therefore be minimal by 24 hours. A phase II trial is ongoing with this agent.[58]

Cangrelor (AR-C69931MX) is an ATP analogue that has the advantage of being given parenterally.[59–61] This agent has a molecular weight of 800 Da and has a plasma half-life of only 5 to 9 minutes. This short plasma half-life and reversible antagonism of the P2Y$_{12}$ receptor translates into a rapid return to normal platelet function. Studies have confirmed that platelet function returns to normal within 20 minutes. This represents a major advance in P2Y$_{12}$ ADP receptor antagonists used in the treatment of acute coronary syndromes because clopidogrel and ticlopidine are often withheld in acute coronary syndromes until coronary anatomy has been defined and there is no longer considered to be a need for urgent CABG.

**2. Oral GPIIb-IIIa Antagonists.** Since intravenous GPIIb-IIIa antagonist have been effective at reducing cardiac outcomes in high-risk patients undergoing PCI, the logical extension was to design oral GPIIb-IIIa antagonists that could be used chronically in patients with vascular disease. Five trials of oral GPIIb-IIIa antagonists have been studied in more than 45,000 patients (Table 65-2) (see also Chapter 62). Most of these trials were in patients with acute coronary syndromes. In aggregate, these studies revealed a 35% increase in mortality (OR, 1.35; $p < 0.001$) and a twofold increase in bleeding from the use of these agents.[62,63]

This paradoxical finding of increased ischemic events from the use of oral GPIIb-IIIa antagonists was unexpected. Although the mechanism for this finding is unknown, several factors from our knowledge of intravenous GPIIb-IIIa antagonists help to explain it. First, there is a loss of antiplatelet effect after prolonged infusion with some GPIIb-IIIa antagonists.[64] More than 80% platelet inhibition is required for GPIIb-IIIa antagonists to be effective in the setting of PCI. With a 36-hour infusion of abciximab, loss of inhibition of platelet function becomes apparent by 12 hours and becomes increasingly pronounced for the remainder of the infusion.[65] This has been termed "platelet escape." Second, GPIIb-IIIa antagonists may become partial agonists at subthreshold levels. At high levels of platelet inhibition, GPIIb-IIIa antagonists reduce the inflammatory mediator interleukin-1β, decrease platelet–leukocyte interactions, and inhibit thrombin formation.[66] In contrast, at low levels of platelet inhibition, GPIIb-IIIa antagonists may function as partial agonists by increasing the expression of platelet inflammatory markers. An *ex vivo* analysis from the Oral Glycoprotein IIb/IIIa Inhibition with Orbofiban in Patients with Unstable Coronary Syndromes Thrombolysis in Myocardial Infarction (OPUS-TIMI 16) trial showed that patients treated with the oral GPIIb-IIIa antagonist orbofiban had increased platelet surface expression of P-selectin and CD63, which help mediate platelet–leukocyte interactions.[67,68] Third, low levels of GPIIb-IIIa antagonists have

## Table 65-2: Experimental Antiplatelet Agents That Directly Modulate Platelet Function

| Mechanism of Action | Agents | *In Vitro, ex Vivo, in Vivo,* and Clinical Studies |
|---|---|---|
| Selective P2Y$_{12}$ receptor inhibition | Prasugrel, cangrelor, and AZD-6140 | Prasugrel completed phase II clinical trial (JUMBO-TIMI 26), phase III clinical trial (TRITON-TIMI 38) ongoing. Cangrelor, phase II clinical trial completed. AZD-6140, phase II clinical trial ongoing. |
| Oral GPIIb-IIIa inhibition | Orofiban, lotrafiban, sibrafiban, and zemilofiban | ↑ Mortality in clinical trials. ↑ Bleeding in clinical trials. |
| Phosphoinositide 3-kinase p110β inhibition | TGX-221 | ↓ Platelet aggregation during periods of sustained high shear flow. Prevents occlusive arterial thrombus formation in several animal models, without changing bleeding times or hemodynamic parameters. |
| Thromboxane A$_2$ synthase inhibition | S-18886, U-3405 | S-18886 ↓ platelet aggregation and regresses advanced atherosclerosis in a rabbit model; may also cause plaque stabilization. U-3405 ↓ thrombosis in baboon model. |
| Thromboxane A$_2$ synthase and receptor inhibition | Ridogrel, terbogrel, and picotamide | ↓ Platelet aggregation and ↓ Thromboxanes. No improvement in TIMI-2 or -3 flow in setting of acute myocardial infarction with ridogrel. No improvement with terbogrel in 6 minute walk in patients with primary pulmonary hypertension; limited by significant leg pain. Mortality reduction in diabetics with peripheral arterial disease with picotamide. |
| GPIb–VWF inhibition | GPIb antibody — 6B4 GPIb antagonists — crotolin and mamushigin VWF antibody — AjvW-2 VWF antagonist — aurin-tricarboxylic acid VWF peptide — VCL | All agents ↓ ristocetin-induced platelet aggregation. GPIb antibody limited by severe and irreversible thrombocytopenia. Crotolin ↓ thrombus in baboon model. Mamushigin is limited by increased mouse tail bleeding times and potential for platelet activation (↓ and ↑ platelet aggregation at high shear stress and low shear stress, respectively). AjvW-2 ↓ arterial/venous thrombus in hamster model; no prolongation of bleeding times. Aurin tricarboxylic acid ↓ thrombus in rabbit model. VCL ↓ thrombus in primates and guinea pig models. |
| PAR inhibition | PAR-1 and -4 antibody and pepducins | Rapid and transient ↓ platelet aggregation by PAR-1 antibody, slower and sustained ↓ platelet aggregation by PAR-4 antibody. Blocking both PAR-1 and -4 produces synergistic ↓ platelet aggregation. Anti-PAR-4 pepducin ↓ platelet aggregation, although ↑ tail bleeding times. |
| Leptin receptor inhibition | Studied in leptin-deficient mice | This agent ↓ thrombosis in response to arterial injury. |
| Arp 2/3 complex inhibition | Arp 2 antibody | "Freezes" platelets in rounded early stage of activation. |
| Gas6 protein | Gas6 protein antibody | ↓ Platelet aggregation and thrombosis in mouse model. ↓ Arterial and venous thrombosis in response to vascular injury. Unaltered bleeding and coagulation parameters; no thrombocytopenia. |
| Miscellaneous | Gingerol | ↓ Platelet aggregation and prevents degranulation. |

GP, glycoprotein; PAR, protease activated receptor; TIMI, thrombolysis in myocardial infarction; VWF, von Willebrand factor.

been reported to promote shedding of CD40 ligand, which may produce prothrombotic and proinflammatory effects.[69] Because of the previously discussed results, further development of oral GPIIb-IIIa antagonists seems highly unlikely.

### 3. Phosphoinositide 3-Kinase p110β Inhibitors.
Since intravenous GPIIb-IIIa antagonists have a narrow window of therapeutic efficacy and are limited by bleeding complications, current research is focused on modulating GPIIb-IIIa receptors only during certain flow states so that clinical efficacy may be retained or enhanced while improving safety. Lipid kinases are important in cellular signaling pathways by producing phosphoinositides.[70–72] The β isoform of phosphoinositide 3-kinase (PI3K) p110 is primarily involved in platelet function and is regulated by tyrosine kinases. The p110α isoform is involved in oncogenesis, whereas the p110δ isoform is involved in immune function and the p110γ isoform is involved in inflammation. These latter isoforms are regulated by G protein-coupled receptors.[70–72]

Experiments have shown that PI3K p110β, but not the other isoforms, is important in modulating intracellular calcium flux and GPIIb-IIIa adhesive bond stability.[73] PI3K p110β is predominantly required for sustained platelet aggregation under high shear conditions. A specific cell-permeable inhibitor of PI3K p110β, termed TGX-221, has been developed and tested *in vitro* and *in vivo* (Table 65-2). TGX-221 effectively inhibited platelet aggregation under high shear conditions across a range of platelet agonists. A 2 mg/kg intravenous bolus was also effective in preventing occlusive arterial thrombus formation within 5 minutes of administration in several animal models. A dose of TGX-221 as low as 0.5 mg/kg still showed antithrombotic activity.[73] There was no change in heart rate or blood pressure in the studied animals, and there was no change in bleeding times with TGX-221 in doses as high as 20 mg/kg or with coadministration with heparin.

### 4. TXA₂ Synthase Inhibitors and TXA₂ Receptor Antagonists.
Aspirin is limited by increased major bleeding, which is in large part secondary to gastric and duodenal ulcers that form as a result of the reduction in protective prostacyclins. In addition to the harm that is associated with the use of this agent, there is also concern that aspirin is relatively ineffective in certain populations, such as diabetics.[26] This led to attempts to specifically block the production of $TXA_2$ while leaving the formation of beneficial prostanoids intact. Initial attempts at blocking only the $TXA_2$ synthase enzyme led to the discovery that the upstream accumulation of prostaglandin endoperoxides was able to exert an agonist effect on the thromboxane receptor and induce platelet aggregation.[74] Agents that block the thromboxane receptor, while leaving the thromboxane synthase

enzyme intact, reduce thrombosis in experimental animal models.[75–77]

S-18886 is a $TXA_2$ receptor antagonist that is reigniting interest in this class of antiplatelet agents (Table 65-2). This oral compound exerts dose-dependent antiplatelet effects with collagen and ADP stimulation that is similar to clopidogrel, although superior to aspirin, under both high and low shear conditions.[78] In a related experiment, 6 months of treatment with S-18886 caused regression and stabilization of advanced atherosclerotic plaques in the rabbit aorta.[79] Such a compound may extend the benefit of an antiplatelet agent beyond antithrombotic effects to include regression and stabilization of atherosclerosis.

Experimental studies of a combined agent that produces both $TXA_2$ synthase inhibition and thromboxane receptor blockade have also shown promise and have been performed clinically in humans (Table 65-2). *In vitro* studies with this agent revealed that it decreased thromboxane production (≥90% for 48 hours) and inhibited platelet aggregation (>90% for 18 hours).[74]

Ridogrel is a combined $TXA_2$ synthase inhibitor and thromboxane receptor antagonist that has been tested in the Ridogrel Versus Aspirin Patency Trial (RAPT).[80] In the RAPT trial, 907 individuals with acute myocardial infarction were randomized to receive streptokinase and aspirin or streptokinase and ridogrel. The primary end point of TIMI 2 or 3, flow in the infarct-related artery between 7 and 14 days after admission, was achieved in 75.5% of the aspirin group and 72.2% of the ridogrel group ($p$ = not significant [NS]). The composite end point of reinfarction, recurrent angina, and stroke was 32% lower in the ridogrel group ($p < 0.025$), whereas mortality occurred in 6.4% of the ridogrel group and 7.1% of the aspirin group ($p$ = NS). There was no increase in bleeding complications from the use of ridogrel, although hemorrhagic strokes were very infrequent (one in each group).

Another trial examined a similar agent, terbogrel, in the treatment of primary pulmonary hypertension.[81] Pulmonary hypertension patients are characterized by excess $TXA_2$ and decreased production of prostacyclin. Accordingly, a combined $TXA_2$ synthase inhibitor and thromboxane receptor blocker made theoretical sense as a way to reorient the balance between $TXA_2$ and prostacyclin. Unfortunately, only 52 out of a planned 135 patients completed the trial due to severe leg pain in the terbogrel group. Although terbogrel was effective at reducing levels of $TXB_2$ (the stable metabolite of $TXA_2$), there was no difference in a 6-minute walk or in hemodynamics from the use of the agent.

The Drug Evaluation in Atherosclerotic Vascular Disease in Diabetics (DAVID) study evaluated the combined $TXA_2$ synthase and receptor inhibitor picotamide in 1209 diabetics with peripheral arterial disease.[82] Patients received picotamide 600 mg twice daily or aspirin 320 mg every other day for 2 years. The cumulative mortality was 3.0% in the picotamide group and 5.5% in the aspirin group (RR 0.55; 95% CI 0.31–0.98). Also, the incidence of severe bleeding that resulted in hospitalization was 0.2% in the picotamide group and 1.2% in the aspirin group, with significantly less dyspepsia in the picotamide group versus the aspirin group (10.9 vs.18.3%; $p < 0.0001$). Although these results need to be replicated, picotamide holds promise in certain high-risk individuals for potentially being more effective yet having fewer bleeding side effects.

**5. GPIb–VWF Inhibitors.** Although GPIIb-IIIa antagonists are effective pharmacologically at inhibiting platelet aggregation and clinically at preventing adverse cardiac events during PCI, they are not able to prevent platelet adhesion and activation because both of these processes occur earlier than aggregation. Platelet GPIb receptors bind at areas of damaged vasculature by interacting with VWF (and thrombin) in the subendothelial matrix (see Chapter 7). Blockade of the GPIb receptor or VWF is therefore a potential mechanism to block platelet adhesion and activation upstream of aggregation (Table 65-2).[1,17]

A monoclonal antibody, 6B4-Fab, directed against GPIb has been tested in a baboon thrombosis model. In this experiment, 6B4-Fab was given to baboons at a dose of 0.6 mg/kg. This dose significantly reduced thrombus formation in an injured femoral artery,[83,84] without prolonging the bleeding time. Although this antibody was effective in preventing platelet aggregation and reducing thrombosis, it also caused immediate and irreversible thrombocytopenia. In primates, this antibody reduced the platelet count from $383 \times 10^9/L$ to less than $40 \times 10^9/L$ within 5 minutes. The GPIb antibody markedly decreases proplatelet formation after 24 hours of incubation.[85,86] Electron micrographs revealed that antibodies against GPIb result in adhesion of platelets to the surface of the megakaryocyte, whereas circulating platelets were found clumped together in an inactive state.

Another approach to blockade of the GPIb receptor is to use proteins derived from snake venoms. Crotalin is a 30-kDa protein derived from *Crotalus atrox* snake venom that is effective at blocking GPIb receptors.[87] Experiments have confirmed the potential clinical efficacy of this protein. *In vitro* experiments revealed that crotalin dose dependently inhibited ristocetin-induced platelet agglutination. This interaction was reversible because crotalin–GPIb binding was blocked by administering a monoclonal antibody against the GPIb receptor. Unfortunately, bleeding was increased with this agent. *In vivo* experiments showed that when crotalin was intravenously administered to mice at a dose of 100 to 300 μg/kg, there was a dose-dependent prolongation in tail bleeding times. The duration of effect of crotalin at 300 μg/kg in prolonging tail bleeding time was 4 hours.

Another GPIb-binding protein, termed mamushigin, is derived from the venom of *Agkistrodon halys blomhoffii*. Mamushigin is able to differentially inhibit platelet aggrega-

tion according to shear stress.[88] This differential effect based on shear stress is unlike those of other snake venom-derived GPIb-binding proteins. Mamushigin binds at (or closely to) the VWF binding site of the GPIb receptor. *In vitro* studies indicate that at low shear stress, mamushigin stimulates platelet aggregation.[88] This effect was reversed by the presence of GPIb receptor blockers but not GPIIb-IIIa receptor blockers. In contrast, at high shear stress, mamushigin prevented platelet aggregation in a dose-dependent manner. This property could make mamushigin effective as an antiplatelet agent in individuals with stenotic lesions that cause high shear stress, although the potential for stimulating platelet aggregation at low shear stress is concerning.

Another approach to blocking the GPIb complex is to interfere with the receptor–ligand interaction at the level of VWF.[89–91] A monoclonal antibody directed against the GPIb binding site on VWF, referred to as AJvW-2, has been tested in various models. AJvW-2 reduces platelet agglutination by ristocetin and botrocetin. This antibody not only prevented arterial thrombus formation in hamsters but also prevented venous thrombus formation. This is interesting given that the GPIb–VWF interaction is mediated under high shear conditions. AJvW-2 was also shown to prevent platelet aggregation and thrombus formation in the guinea pig, but unlike proteins derived from snake venom, it did not prolong bleeding time. A dog experiment characterized the percentage of VWF occupancy needed to achieve efficacy while minimizing the risk for bleeding. AJvW-2 maximally inhibited platelet aggregation at a dose of 0.18 mg/kg. This corresponded to 53% plasma VWF occupancy. The dose required to prevent thrombus formation was only 0.06 mg/kg. Increased bleeding was observed at a dose of 1.8 mg/kg, or 95% occupancy of plasma VWF. Since this dose was 30-fold greater than the dose needed to prevent thrombus formation, increased bleeding might be minimal with the clinical use of this agent.

The interaction between GPIb and VWF can also be blocked by aurin tricarboxylic acid.[92,93] Aurin tricarboxylic acid is a polyphenolic, polycarboxylated compound that binds to the GPIb binding site on VWF. In rabbit experiments in which the aorta was denuded by passage of an inflated balloon catheter and platelets were labeled with [111]indium oxide, aurin tricarboxylic acid was able to inhibit ristocetin- and thrombin-induced platelet aggregation in a dose-dependent manner. The compound also inhibited shear-induced platelet aggregation. In an experimentally stenosed and denuded rabbit carotid artery, platelet deposition was more effectively reduced by aurin tricarboxylic acid than a thromboxane synthase/thromboxane receptor blocker, a platelet activating factor receptor antagonist, a GPIIb-IIIa antagonist, or an antibody against tissue factor.[92]

Lastly, a novel approach that uses a peptide, VCL, derived from the VWF binding domain may be able to block the

GPIb–VWF interaction.[94,95] Such an approach does not rely on directly antagonizing the GPIb receptor or VWF but, rather, saturates the GPIb receptor so that platelet adhesion and activation are effectively prevented. The VWF domain that interacts with the GPIb receptor involves amino acids valine-449 to lysine-728. VCL is a recombinant peptide that is composed of amino acids leucine-504 to lysine-728 and represents 10% of the overall VWF domain. The administration of VCL to primates reduced platelet aggregation (induced by botrocetin) and also prevented thrombus formation in a stenosed coronary artery denuded of endothelium. Similarly, in guinea pigs VCL was shown to reduce platelet aggregation (induced by botrocetin and ristocetin) and to reduce thrombus formation in nitrogen laser-damaged mesenteric arteries. Importantly, platelet counts remained stable and bleeding times were unaffected in this experiment.

### 6. Protease-Activated Receptor Antagonists.

Thrombin is recognized as the most potent platelet activator.[96] Accordingly, agents that modulate thrombin–platelet interactions may serve as effective antiplatelet agents (Table 65-2). Thrombin interacts with the G protein-coupled protease-activated receptor (PAR), of which several subtypes have been identified (PAR-1, PAR-3, and PAR-4) (see Chapter 9).[97] The protease enzyme on the surface of PAR is activated through proteolytic cleavage of an N-terminal exodomain. This exposes a new N-terminal peptide that serves as a tethered peptide ligand. Although thrombin also interacts with the GPIb receptor, its main effects are mediated through the PAR system. Blocking PAR-1 by antibody inhibits platelet aggregation at low thrombin concentrations (1 nM) but not at high thrombin concentrations (30 n$M$). Human platelets carry PAR-1 and PAR-4, although only PAR-1 exists on the surface of smooth muscle cells. In addition to decreasing platelet aggregation by low-dose thrombin, blocking PAR-1 may also be useful as a way to prevent vascular restenosis.[97,98] Blocking the high-affinity PAR-1 by an antagonist (RWJ-58259 at 10 mg/kg) has been shown to only modestly reduce thrombus formation in a guinea pig shunt model. However, RWJ-58259 had a more profound effect on decreasing the amount of restenosis following balloon angioplasty in rats, presumably through inhibition of PAR-1 on smooth muscle cells. This antagonist reduced neointimal thickness from 77 μm in control rats to 45 μm in actively treated rats. This property could make a PAR-1 antagonist a potentially useful agent in limiting restenosis following PCI.

Blocking PAR-4 by antibody has little effect on inhibiting platelet aggregation, although it does prevent the release of endostatin.[99] Endostatin (see Chapter 41) is an endogenous inhibitor of angiogenesis that is stored in platelets and released by thrombin stimulation, although not by ADP. The most potent inhibition of platelet aggregation (even at high-dose thrombin) is accomplished by blocking both PAR-1 and PAR-4 simultaneously.[100,101]

### 7. Pepducins.

Pepducins are novel cell-penetrating peptides that are able to modulate thrombin/PAR-induced intracellular signaling while bypassing the extracellular PAR receptors.[102] Pepducins are formed by attaching a palmitate lipid to peptides that have been based on the third intracellular loop of the PAR. An anti-PAR-4 pepducin has been shown to inhibit platelet aggregation, although with prolonged tail bleeding times. This system may afford another mechanism for inhibiting receptor function outside of the standard receptor antagonists and antibodies.

### 8. Leptin Receptor Antagonists.

Leptin levels are increased in obesity, and the finding of leptin receptors on platelets identified another potential site for modulation of platelet function. In murine experiments, the thrombotic response to arterial injury was decreased in leptin-deficient mice.[103] Leptin was also shown to enhance platelet aggregation through its effects on the leptin receptor. A leptin receptor antagonist might be useful in the care of obese individuals (Table 65-2).

### 9. Arp 2/3 Complex Inhibitors.

Activated platelets undergo a shape change from discoid to round, followed by the formation of filopodia, lamellipodia, stress fibers, and contractile rings (see Chapter 4). This shape change is brought about by polymerization of actin monomers and is regulated by the Arp 2/3 complex. Recombinant Arp2 was used to create an antibody that could be used to study the effects of blocking this complex (Table 65-2).[104] Pretreatment with this antibody, followed by activation by thrombin receptor activating peptide or by spreading platelets on glass, froze them in the early rounded stage of activation. The later stage that is characterized by filopodia, lamellipodia, stress fibers, and contractile rings did not occur. Using morphometric analysis, treatment with this antibody increased the percentage of platelets in the rounded stage from 3 to 63%. This effect was reversed by the addition of recombinant Arp2 protein. This antibody represents a novel approach in the modulation of platelet function by inhibiting the shape change that is necessary for stable platelet aggregation to occur while keeping other critical platelet functions intact.

### 10. Growth Arrest-Specific Gene 6 Product.

The product of growth arrest-specific gene 6 (Gas6) is a member of the vitamin K-dependent protein family and is similar to protein S.[105,106] This protein interacts with the Axl/Sky subfamily of receptor tyrosine kinases. Unlike protein S, Gas6 protein lacks a loop that is necessary for anticoagulation activity. To study the effects of this protein *in vivo,* a genetic knockout mouse that lacked the Gas6 gene was created

(Table 65-2). The Gas6 knockout mouse had normal coagulation parameters (i.e., prothrombin and activated partial thromboplastin time) and did not have an increase in either spontaneous bleeding or induced bleeding from tail clipping. Additionally, this mouse had normal platelet counts and normal number and morphology of megakaryocytes. Platelets from the Gas6 knockout mouse had impaired degranulation, as shown by measuring ATP release in response to various agonists. ATP secretion was significantly reduced after stimulation by low-dose ADP, collagen, or U46619 (a stable thromboxane $A_2$ analogue) or from stimulation by high-dose ADP or thrombin. In contrast, ATP secretion was essentially normal after stimulation by high-dose collagen or U46619. Gas6 knockout platelets failed to aggregate in response to lose-dose ADP, collagen, or U46619, although high doses of the agonists were able to inhibit platelet aggregation. Additionally, thrombin caused platelet aggregation at all doses tested, although the platelet aggregates were abnormal. In contrast to platelet aggregates from wild-type mice, Gas6 knockout mice had loosely packed platelet aggregates. Gas6 knockout mice also had reduced levels of fibrinogen on their surface, in contrast to wild-type mice, which may have contributed to the finding of loosely packed platelet aggregates.[105,106]

Thrombosis models confirmed that Gas6 knockout mice were protected against arterial and venous thrombosis. Gas6 knockout mice had 85% smaller ($p < 0.001$) venous thrombi as a result of stasis that had been induced by ligation of the inferior vena cava. These mice also had 60% smaller ($p < 0.05$) arterial thrombi as a result of photochemical denudation of the carotid artery. Fatal thromboembolism induced by injection of collagen or epinephrine was also significantly reduced in Gas6 knockout mice because as 80% of wild-type mice died within 1 to 3 minutes compared to 20% of Gas6 knockout mice ($p < 0.03$).

Recombinant Gas6 protein infused into mice did not stimulate platelet aggregation by itself, although it amplified the response to known platelet agonists. Also, Gas6 antibody produced similar effects that were observed in the Gas6 knockout mice, namely inhibition of platelet aggregation and reduction of thrombosis. In summary, Gas6 can be viewed as a platelet-response amplifier that has a significant role in thrombosis. Inactivation of Gas6 through antibody or antagonist holds promise as a mechanism for preventing thrombus formation while retaining critical platelet functions.

**11. Gingerols.** Gingerol is the active component of ginger and has been shown to inhibit platelet aggregation and prevent ADP release by dose-dependently inhibiting COX-1 activity (Table 65-2).[107] In fact, gingerol was more potent than aspirin at inhibiting platelet aggregation, and it also showed antiulcer activity. As mentioned previously, peptic ulcer disease is a significant limitation of aspirin.

### B. Indirect Antiplatelet Agents

**1. ATPase/CD39 Inhibitors.** Adenosine triphosphate diphosphohydrolase (ATPase) or CD39 is an enzyme on the endothelial surface that functions as an important regulator of thrombosis and inflammation (see Chapter 13). Adenosine triphosphate (ATP) and ADP are potent stimuli for platelet activation and leukocyte recruitment. ATP and ADP are degraded by CD39 into inactive adenosine monophosphate (AMP). This system directs thrombosis to sites of vascular injury by suppressing thrombus formation in areas of normal vasculature.

In a murine stroke experiment, recombinant soluble CD39 reduced platelet aggregation *ex vivo* and decreased cerebral infarct size when given as late as 3 hours after transient middle cerebral artery occlusion (Table 65-3).[108,109] Aspirin did not reduce the infarct size, although it did increase the risk for hemorrhagic stroke. Additionally, CD39 knockout mice were shown to have unaltered bleeding times and hematological profiles compared to normal mice. These mice had increased cerebral infarct size and reduced postischemic perfusion compared to controls. Soluble CD39 was able to restore cerebral perfusion and prevent cerebral injury.

A related compound is modeled after the hematophagous arthropod apyrase, which functions as an ATPase enzyme.[110] Using recombinant technology, this enzyme has been redesigned to have 100-fold increased ATPase activity. This apyrase is distinct from soluble CD39 because there is no shared sequence homology between the two enzymes. *In vitro* studies have shown that this apyrase (at concentrations of 0.25–0.5 $\mu M$) inhibits platelet aggregation induced by 10 $\mu M$ ADP and also disperses existing platelet aggregates.

**2. CD40/CD40 Ligand System.** The CD40/CD40 ligand system plays an important role in inflammation (see Chapter 39).[19,111] This transmembrane protein is a member of the tumor necrosis superfamily. More than 90% of CD40 ligand is shed as a result of platelet activation. This ligand is a key mediator in inflammation since it increases the production of chemokines, E-selectin, VCAM, and ICAM, resulting in leukocyte recruitment to sites of inflammation. CD40 ligand is elevated in patients with acute coronary syndromes and predicts individuals with a poor prognosis.[19] Such patients may especially derive benefit from early invasive therapy. Identifying therapy that may specifically decrease the shedding of CD40 ligand or block the ligand–receptor interaction may represent a novel approach to antiplatelet therapy. Of note, however, antibodies against CD40 ligand have been associated with rare thromboembolic events.

**3. Antioxidation/NO Donors.** The general level of oxidative stress that surrounds platelets helps to modulate

**Table 65-3: Experimental Antiplatelet Agents That Indirectly Modulate Platelet Function**

| Mechanism of Action | Agents | *In Vitro, ex Vivo, in Vivo,* and Clinical Studies |
|---|---|---|
| ATPase/CD39 inhibition | Recombinant ATPase<br>Recombinant soluble CD39 | ↓ Platelet aggregation and disperses existing platelet aggregates.<br>↓ Platelet aggregation and ↓ cerebral infarct size in mouse model; unchanged bleeding times. |
| NO donors | Amifostine, NCX-4016, and LA-810 | ↓ Platelet aggregation and thromboxane production; ↑ NO production with amifostine.<br>↓ Restenosis and atherosclerotic area in mouse/rat carotid injury model with NCX-4016.<br>↓ Thrombosis in pig model without hypotensive or tachycardic effects with LA-810. |
| RANTES inhibition | Met-RANTES antagonist | ↓ Macrophage infiltration in atherosclerotic areas and ↓ neointimal formation with Met-RANTES antagonist. |
| Collagen inhibition | CTRP-1 | ↓ Platelet aggregation and thrombosis in animal model. |

ATPase, adenosine triphosphate diphosphohydrolase; NO, nitric oxide; RANTES, regulated upon activation, normal T cell expressed and presumably secreted.

thrombosis (Table 65-3). Amifostine, an organic thiophosphate prodrug, is a reactive oxygen species scavenger that was originally developed by the Army Institute of Research to protect against the harmful effects of radiation in the event of nuclear fallout.[112] Reactive oxygen species cause irreversible damage to DNA and activate platelets. Compounds with antioxidation properties might be able to prevent or minimize platelet activation that would otherwise occur as a result of pro-oxidative effects.

Experiments with amifostine have shown that this compound is capable of preventing ADP-, collagen-, and platelet activating factor-mediated platelet aggregation.[112] This effect was most pronounced with ADP-induced platelet aggregation. Amifostine also prevented $TXB_2$ production by the same mediators, although the most pronounced effect was seen with ADP-stimulated platelets. Additionally, amifostine stimulated endogenous platelet NO production. For example, the production of NO from the stimulation of control platelets by 1 μM ADP was $51.0 \pm 6.6$ pmol/$10^6$, whereas platelets treated with 5 μM amifostine and stimulated by the same agonist produced $85.3 \pm 11.7$ pmol/$10^6$ of NO. NO has marked platelet inhibitory and antithrombotic effects (see Chapter 13).

NO-releasing aspirin (NCX-4016) is an experimental aspirin that has the capability of donating NO.[113–116] The pharmacological and clinical effects of NO-releasing aspirin have been explored in mice and rats. In a carotid artery injury model, NO-releasing aspirin, but not regular aspirin, increased plasma NO levels and decreased experimental restenosis by reducing vascular smooth muscle cell production. In hypercholesterolemic mice, chronic treatment (12 weeks) with NO-releasing aspirin reduced

the cumulative atherosclerotic aortic lesion area by 40% ($p < 0.001$). In addition, there was a reduction in plasma low-density lipoprotein oxidation and oxidative stress in these mice compared to mice treated with regular aspirin.

LA810 is an *S*-nitrosothiol compound that functions as a pure NO donor.[117] An *ex vivo* porcine model perfused pigs with LA810 and then, after they were anesthetized, connected their carotid artery to a Badimon perfusion chamber, which was then connected to an internal jugular vein. The Badimon perfusion chamber exposed the circulating blood to a separate porcine aorta that had been denuded of endothelium. Immunohistochemistry was performed on the exposed aorta in the perfusion chamber and the amount of thrombus formation quantitated. This experiment found that LA810 decreased thrombus formation independent of the degree of vascular damage or shear stress. Additionally, LA810 did not produce any hypotensive or tachycardic effects as a similar compound, GSNO, had been known to do.

**4. RANTES.** RANTES (Regulated upon Activation, Normal T Cell Expressed and Presumably Secreted) is a chemokine secreted by platelets that have been activated predominantly during flow conditions.[118–121] This chemokine interacts with P-selectin in mediating monocyte/macrophage infiltration into atherosclerotic lesions. Separate experiments have shown that P-selectin-deficient mice are protected against the development of atherosclerosis and neointima formation (Table 65-3). An experiment on apolipoprotein E-deficient mice that involved wire-induced damage of carotid arteries was performed to study the effects of blocking RANTES by the peptide antagonist Met-RANTES on neointima formation and macrophage infiltration. Mice treated with daily intraperitoneal injections of Met-RANTES for 28 days had a 40% reduction in macrophage infiltration in areas of atherosclerosis. Additionally, these mice had a reduction in neointima formation ($p < 0.05$). Mice that were absent P-selectin did not show this reduction in neointima formation and macrophage infiltration.

**5. Collagen Blockers.** CTRP-1 (also known as zsig37 and C1QTNF1) is a tumor necrosis factor-related protein.[122] This is a novel protein that has been shown in animal experiments to modulate thrombosis through high-affinity blockade of collagen that is exposed at sites of vascular injury (Table 65-3). *In vitro* experiments with this compound revealed a dose-dependent reduction in platelet aggregation. *Ex vivo* experiments were performed using an arteriovenous fistula model in baboons. This showed that blood that had been circulated though a collagen-coated graft pretreated with CTRP-1 protein exhibited a reduction in the accumulation of radiolabeled platelets. In a related experiment, CTRP-1 (at a dose of 1 mg/kg) prevented thrombus formation in rabbits fed a high-cholesterol diet after balloon injury to the iliac artery followed by experimental stenosis. A protein derived from the snake venom of *Echis multisquamatus*, EMS16, has also been shown to be a potent and selective inhibitor of the GPIa-IIa collagen receptor and is currently being evaluated.[123]

## V. Summary

Platelets play a central and active role in the development of atherosclerosis and the clinical expression of thrombotic events. Current antiplatelet therapy, as represented by aspirin, $P2Y_{12}$ receptor antagonists, and GPIIb-IIIa antagonists, has improved survival and reduced adverse cardiac events across a range of patients with vascular disease. Despite their clinical effectiveness, these agents are limited by increased major and minor bleeding, nonbleeding-related side effects, relative lack of efficacy, and resistance. This has provided the motivation to design more effective and safer antiplatelet medications. This chapter reviewed the antiplatelet agents that are under study. Some of these agents act directly, whereas others indirectly modulate platelet function. Agents such as the oral GPIIb-IIIa antagonists have been disappointing, although they have enhanced our knowledge of subthreshold inhibition of GPIIb-IIIa antagonists. Experimental antiplatelet agents are in various stages of development, with some having completed phase II evaluation. In the future, the treatment of acute vascular events, as well as chronic therapy for secondary and primary prevention of vascular disease, will likely look different from our current landscape of treatment, with the use of safer and more effective agents.

## References

1. Bhatt, D. L., & Topol, E. J. (2003). Scientific and therapeutic advances in antiplatelet therapy. *Nat Rev Drug Discov, 2,* 15–28.
2. Libby, P. (2002). Inflammation in atherosclerosis. *Nature, 420,* 868–874.
3. Topol, E. J., & Van de Werf, F. J. (2002). Acute myocardial infarction: Early diagnosis and management. In E. J. Topol (Ed.), *Textbook of cardiovascular medicine* (2nd ed., pp. 385–419). Philadelphia: Lippincott Williams & Wilkins.
4. Califf, R. M., DeLong, E. R., Ostbye, T., et al. (2002). Underuse of aspirin in a referral population with documented coronary artery disease. *Am J Cardiol, 89,* 653–661.
5. Stafford, R. S. (2000). Aspirin use is low among United States outpatients with coronary artery disease. *Circulation, 101,* 1097–1101.
6. He, J., Whelton, P. K., Vu, B., et al. (1998). Aspirin and risk of hemorrhagic stroke: A meta-analysis of randomized controlled trials. *J Am Med Assoc, 280,* 1930–1935.
7. Bhatt, D. L. (2004). Aspirin resistance: More than just a laboratory curiosity. *J Am Coll Cardiol, 43,* 1127–1129.
8. Gum, P. A., Kottke-Marchant, K., Poggio, E. D., et al. (2001). Profile and prevalence of aspirin resistance in patients with cardiovascular disease. *Am J Cardiol, 88,* 230–235.
9. Alexander, J. H., Harrington, R. A., Tuttle, R. H., et al. (1999). Prior aspirin use predicts worse outcomes in patients with non-ST-elevation acute coronary syndromes. PURSUIT Investigators. Platelet IIb/IIIa in Unstable Angina: Receptor Suppression Using Integrilin Therapy. *Am J Cardiol, 83,* 1147–1151.
10. Awtry, E. H., & Loscalzo, J. (2000). Aspirin. *Circulation, 101,* 1206–1218.
11. Tcheng, J. E., & Campbell, M. E. (2003). Platelet inhibition strategies in percutaneous coronary intervention: Competition or coopetition? *J Am Coll Cardiol, 42,* 1196–1198.
12. Kulkarni, S., Dopheide, S. M., Yap, C. L., et al. (2000). A revised model of platelet aggregation. *J Clin Invest, 105,* 783–791.
13. Weiss, H. J. (1995). Flow-related platelet deposition on subendothelium. *Thromb Haemost, 74,* 117–122.
14. Fressinaud, E., Sakariassen, K. S., Rothschild, C., et al. (1992). Shear rate-dependent impairment of thrombus growth on collagen in nonanticoagulated blood from patients with von Willebrand disease and hemophilia A. *Blood, 80,* 988–994.
15. Saelman, E. U., Nieuwenhuis, H. K., Hese, K. M., et al. (1994). Platelet adhesion to collagen types I through VIII under conditions of stasis and flow is mediated by GPIa/IIa (alpha 2 beta 1-integrin). *Blood, 83,* 1244–1250.
16. Messmore, H. L., Jr., Jeske, W. P., Wehrmacher, W., et al. (2005). Antiplatelet agents: Current drugs and future trends. *Hematol Oncol Clin North Am, 19,* 87–117.

17. Jneid, H., & Bhatt, D. L. (2003). Advances in antiplatelet therapy. *Expert Opin Emerg Drugs, 8,* 349–363.
18. Furie, B., Furie, B. C., & Flaumenhaft, R. (2001). A journey with platelet P-selectin: The molecular basis of granule secretion, signalling and cell adhesion. *Thromb Haemost, 86,* 214–221.
19. Heeschen, C., Dimmeler, S., Hamm, C. W., et al. (2003). Soluble CD40 ligand in acute coronary syndromes. *N Engl J Med, 348,* 1104–1111.
20. Patrono, C., Coller, B., Dalen, J. E., et al. (2001). Platelet-active drugs: The relationships among dose, effectiveness, and side effects. *Chest, 119,* 39S–63S.
21. Fitzgerald, D. J., Roy, L., Catella, F., et al. (1986). Platelet activation in unstable coronary disease. *N Engl J Med, 315,* 983–989.
22. Antithrombotic Trialists' Collaboration. (2002). Collaborative meta-analysis of randomised trials of antiplatelet therapy for prevention of death, myocardial infarction, and stroke in high risk patients. *Br Med J, 324,* 71–86.
23. Rinder, C. S., Student, L. A., Bonan, J. L., et al. (1993). Aspirin does not inhibit adenosine diphosphate-induced platelet alpha-granule release. *Blood, 82,* 505–512.
24. Maalej, N., & Folts, J. D. (1996). Increased shear stress overcomes the antithrombotic platelet inhibitory effect of aspirin in stenosed dog coronary arteries. *Circulation, 93,* 1201–1205.
25. Chronos, N. A., Wilson, D. J., Janes, S. L., et al. (1994). Aspirin does not affect the flow cytometric detection of fibrinogen binding to, or release of alpha-granules or lysosomes from, human platelets. *Clin Sci (London), 87,* 575–580.
26. Watala, C., Golanski, J., Pluta, J., et al. (2004). Reduced sensitivity of platelets from type 2 diabetic patients to acetylsalicylic acid (aspirin) — Its relation to metabolic control. *Thromb Res, 113,* 101–113.
27. Eidelman, R. S., Hebert, P. R., Weisman, S. M., et al. (2003). An update on aspirin in the primary prevention of cardiovascular disease. *Arch Intern Med, 163,* 2006–2010.
28. Ridker, P. M., Cook, N. R., Lee, I. M., et al. (2005). A randomized trial of low-dose aspirin in the primary prevention of cardiovascular disease in women. *N Engl J Med, 352,* 1293–1304.
29. Kunapuli, S. P., Dorsam, R. T., Kim, S., et al. (2003). Platelet purinergic receptors. *Curr Opin Pharmacol, 3,* 175–180.
30. Geiger, J., Brich, J., Honig-Liedl, P., et al. (1999). Specific impairment of human platelet P2Y(AC) ADP receptor-mediated signaling by the antiplatelet drug clopidogrel. *Arterioscler Thromb Vasc Biol, 19,* 2007–2011.
31. Macchi, L., Christiaens, L., Brabant, S., et al. (2002). Resistance to aspirin *in vitro* is associated with increased platelet sensitivity to adenosine diphosphate. *Thromb Res, 107,* 45–49.
32. Jin, J., Daniel, J. L., & Kunapuli, S. P. (1998). Molecular basis for ADP-induced platelet activation: II. The P2Y1 receptor mediates ADP-induced intracellular calcium mobilization and shape change in platelets. *J Biol Chem, 273,* 2030–2034.
33. Bhatt, D. L., Bertrand, M. E., Berger, P. B., et al. (2002). Meta-analysis of randomized and registry comparisons of ticlopidine with clopidogrel after stenting. *J Am Coll Cardiol, 39,* 9–14.
34. Muller, I., Seyfarth, M., Rudiger, S., et al. (2001). Effect of a high loading dose of clopidogrel on platelet function in patients undergoing coronary stent placement. *Heart, 85,* 92–93.
35. Soffer, D., Moussa, I., Harjai, K. J., et al. (2003). Impact of angina class on inhibition of platelet aggregation following clopidogrel loading in patients undergoing coronary intervention: Do we need more aggressive dosing regimens in unstable angina? *Catheter Cardiovasc Interv, 59,* 21–25.
36. Gurbel, P. A., & Bliden, K. P. (2003). Durability of platelet inhibition by clopidogrel. *Am J Cardiol, 91,* 1123–1125.
37. Muller, I., Besta, F., Schulz, C., et al. (2003). Prevalence of clopidogrel non-responders among patients with stable angina pectoris scheduled for elective coronary stent placement. *Thromb Haemost, 89,* 783–787.
38. Hongo, R. H., Ley, J., Dick, S. E., et al. (2002). The effect of clopidogrel in combination with aspirin when given before coronary artery bypass grafting. *J Am Coll Cardiol, 40,* 231–237.
39. Wiviott, S. D., & Antman, E. M. (2004). Clopidogrel resistance: A new chapter in a fast-moving story. *Circulation, 109,* 3064–3067.
40. Nguyen, T. A., Diodati, J. G., & Pharand, C. (2005). Resistance to clopidogrel: A review of the evidence. *J Am Coll Cardiol, 45,* 1157–1164.
41. Yusuf, S., Zhao, F., Mehta, S. R., et al. (2001). Effects of clopidogrel in addition to aspirin in patients with acute coronary syndromes without ST-segment elevation. *N Engl J Med, 345,* 494–502.
42. Steinhubl, S. R., Berger, P. B., Mann, J. T., 3rd, et al. (2002). Early and sustained dual oral antiplatelet therapy following percutaneous coronary intervention: A randomized controlled trial. *J Am Med Assoc, 288,* 2411–2420.
43. Coller, B. S. (2003). Glycoprotein IIb/IIIa antagonists. Development of abciximab and pharmacology of abciximab, tirofiban, and eptifibatide. In A. M. Lincoff (Ed.), *Platelet glycoprotein IIb/IIIa inhibitors in cardiovascular disease* (2nd ed., pp. 73–101). Totowa, NJ: Humana Press.
44. Scarborough, R. M., Kleiman, N. S., & Phillips, D. R. (1999). Platelet glycoprotein IIb/IIIa antagonists. What are the relevant issues concerning their pharmacology and clinical use? *Circulation, 100,* 437–444.
45. Karvouni, E., Katritsis, D. G., & Ioannidis, J. P. (2003). Intravenous glycoprotein IIb/IIIa receptor antagonists reduce mortality after percutaneous coronary interventions. *J Am Coll Cardiol, 41,* 26–32.
46. Topol, E. J., Lincoff, A. M., Kereiakes, D. J., et al. (2002). Multi-year follow-up of abciximab therapy in three randomized, placebo-controlled trials of percutaneous coronary revascularization. *Am J Med, 113,* 1–6.
47. Bhatt, D. L., Marso, S. P., Lincoff, A. M., et al. (2000). Abciximab reduces mortality in diabetics following percutaneous coronary intervention. *J Am Coll Cardiol, 35,* 922–928.

48. Topol, E. J., Moliterno, D. J., Herrmann, H. C., et al. (2001). Comparison of two platelet glycoprotein IIb/IIIa inhibitors, tirofiban and abciximab, for the prevention of ischemic events with percutaneous coronary revascularization. *N Engl J Med, 344*, 1888–1894.

49. Boersma, E., Harrington, R. A., Moliterno, D. J., et al. (2002). Platelet glycoprotein IIb/IIIa inhibitors in acute coronary syndromes: A meta-analysis of all major randomised clinical trials. *Lancet, 359*, 189–198.

50. Roffi, M., Chew, D. P., Mukherjee, D., et al. (2001). Platelet glycoprotein IIb/IIIa inhibitors reduce mortality in diabetic patients with non-ST-segment-elevation acute coronary syndromes. *Circulation, 104*, 2767–2771.

51. Diener, H. C., Cunha, L., Forbes, C., et al. (1996). European Stroke Prevention Study. 2. Dipyridamole and acetylsalicylic acid in the secondary prevention of stroke. *J Neurol Sci, 143*, 1–13.

52. Moonis, M., & Fisher, M. (2003). Antiplatelet treatment for secondary prevention of acute ischemic stroke and transient ischemic attacks: Mechanisms, choices and possible emerging patterns of use. *Expert Rev Cardiovasc Ther, 1*, 611–615.

53. Park, S. J., Shim, W. H., Ho, D. S., et al. (2003). A paclitaxel-eluting stent for the prevention of coronary restenosis. *N Engl J Med, 348*, 1537–1545.

54. Sekiguchi, M., Hoshizaki, H., Adachi, H., et al. (2004). Effects of antiplatelet agents on subacute thrombosis and restenosis after successful coronary stenting: A randomized comparison of ticlopidine and cilostazol. *Circ J, 68*, 610–614.

55. Wiviott, S. D., Antman, E. M., Winters, K. J., et al. (2005). Randomized comparison of prasugrel (CS-747, LY640315), a novel thienopyridine P2Y12 antagonist, with clopidogrel in percutaneous coronary intervention: Results of the Joint Utilization of Medications to Block Platelets Optimally (JUMBO)-TIMI 26 trial. *Circulation, 111*, 3366–3373.

56. Brandt, J. T., Payne, C. D., Weerakkody, G., et al. (2005). Superior responder rate for inhibition of platelet aggregation with a 60 mg loading dose of prasugrel (CS-747, LY640315) compared with a 300 mg loading dose of clopidogrel. *J Am Coll Cardiol, 45*(Suppl. A), 87A.

57. Asai, F., Jakubowski, J. A., Hirota, T., et al. (2005). A comparison of prasugrel (CS-747, LY640315) with clopidogrel on platelet function in healthy male volunteers. *J Am Coll Cardiol, 45*(Suppl. A), 87A.

58. Kelly, R. V., & Steinhubl, S. (2005). Changing roles of anticoagulant and antiplatelet treatment during percutaneous coronary intervention. *Heart, 91*(Suppl. III), III16–III19.

59. Storey, R. F., Oldroyd, K. G., & Wilcox, R. G. (2001). Open multicentre study of the P2T receptor antagonist AR-C69931MX assessing safety, tolerability and activity in patients with acute coronary syndromes. *Thromb Haemost, 85*, 401–407.

60. Storey, R. F., Judge, H. M., Wilcox, R. G., et al. (2002). Inhibition of ADP-induced P-selectin expression and platelet-leukocyte conjugate formation by clopidogrel and the P2Y12 receptor antagonist AR-C69931MX but not aspirin. *Thromb Haemost, 88*, 488–494.

61. Bhatt, D. L. (2005). *Cangrelor, a novel intravenous ADP platelet antagonist.* Available at www.tctmd.com/csportal/appmanager/tctmd/main?_nfpb=true&_pageLabel=TCTMDContent&hdCon=757408. Accessed October 12, 2005.

62. Chew, D. P., Bhatt, D. L., Sapp, S., et al. (2001). Increased mortality with oral platelet glycoprotein IIb/IIIa antagonists: A meta-analysis of phase III multicenter randomized trials. *Circulation, 103*, 201–206.

63. Chew, D. P., Bhatt, D. L., & Topol, E. J. (2001). Oral glycoprotein IIb/IIIa inhibitors: Why don't they work? *Am J Cardiovasc Drugs, 1*, 421–428.

64. Quinn, M. J., Plow, E. F., & Topol, E. J. (2002). Platelet glycoprotein IIb/IIIa inhibitors: Recognition of a two-edged sword? *Circulation, 106*, 379–385.

65. Quinn, M. J., Murphy, R. T., Dooley, M., et al. (2001). Occupancy of the internal and external pools of glycoprotein IIb/IIIa following abciximab bolus and infusion. *J Pharmacol Exp Ther, 297*, 496–500.

66. Neumann, F. J., Zohlnhofer, D., Fakhoury, L., et al. (1999). Effect of glycoprotein IIb/IIIa receptor blockade on platelet–leukocyte interaction and surface expression of the leukocyte integrin Mac-1 in acute myocardial infarction. *J Am Coll Cardiol, 34*, 1420–1426.

67. Cox, D., Smith, R., Quinn, M., et al. (2000). Evidence of platelet activation during treatment with a GPIIb/IIIa antagonist in patients presenting with acute coronary syndromes. *J Am Coll Cardiol, 36*, 1514–1519.

68. Li, N., Hu, H., Lindqvist, M., et al. (2000). Platelet-leukocyte cross talk in whole blood. *Arterioscler Thromb Vasc Biol, 20*, 2702–2708.

69. Nannizzi-Alaimo, N., Alves, V. L., Prasad, S., et al. (2001). GP IIb-IIIa antagonists demonstrate a dose-dependent inhibition and potentiation of soluble CD40L (CD154) release during platelet stimulation. *Circulation, 104*, II-318.

70. Okkenhaug, K., Bilancio, A., Farjot, G., et al. (2002). Impaired B and T cell antigen receptor signaling in p110delta PI 3-kinase mutant mice. *Science, 297*, 1031–1034.

71. Sasaki, T., Irie-Sasaki, J., Jones, R. G., et al. (2000). Function of PI3Kgamma in thymocyte development, T cell activation, and neutrophil migration. *Science, 287*, 1040–1046.

72. Stein, R. C., & Waterfield, M. D. (2000). PI3-kinase inhibition: A target for drug development? *Mol Med Today, 6*, 347–357.

73. Jackson, S. P., Schoenwaelder, S. M., Goncalves, I., et al. (2005). PI 3-kinase p110beta: A new target for antithrombotic therapy. *Nat Med, 11*, 507–514.

74. De Clerck, F., Beetens, J., Van de Water, A., et al. (1989). R 68 070: Thromboxane A2 synthetase inhibition and thromboxane A2/prostaglandin endoperoxide receptor blockade combined in one molecule — II. Pharmacological effects *in vivo* and *ex vivo*. *Thromb Haemost, 61*, 43–49.

75. Takiguchi, Y., Wada, K., & Nakashima, M. (1992). Comparison of the inhibitory effects of the TXA2 receptor antagonist, vapiprost, and other antiplatelet drugs on arterial thrombosis in rats: Possible role of TXA2. *Thromb Haemost, 68*, 460–463.

76. Kotze, H. F., Lamprecht, S., Badenhorst, P. N., et al. (1993). *In vivo* inhibition of acute platelet-dependent thrombosis in

a baboon model by Bay U3405, a thromboxane A2-receptor antagonist. *Thromb Haemost, 70,* 672–675.

77. White, B. P., Sullivan, A. T., & Lumley, P. (1994). Prevention of intra-coronary thrombosis in the anaesthetised dog: The importance of thromboxane A2 and thrombin. *Thromb Haemost, 71,* 366–374.

78. Osende, J. I., Shimbo, D., Fuster, V., et al. (2004). Antithrombotic effects of S 18886, a novel orally active thromboxane A2 receptor antagonist. *J Thromb Haemost, 2,* 492–498.

79. Viles-Gonzalez, J. F., Fuster, V., Corti, R., et al. (2005). Atherosclerosis regression and TP receptor inhibition: Effect of S18886 on plaque size and composition — A magnetic resonance imaging study. *Eur Heart J, 6,* 1557–1561.

80. Ridogrel Versus Aspirin Patency Trial (RAPT) Investigators. (1994). Randomized trial of ridogrel, a combined thromboxane A2 synthase inhibitor and thromboxane A2/prostaglandin endoperoxide receptor antagonist, versus aspirin as adjunct to thrombolysis in patients with acute myocardial infarction. The Ridogrel Versus Aspirin Patency Trial (RAPT). *Circulation, 89,* 588–595.

81. Langleben, D., Christman, B. W., Barst, R. J., et al. (2002). Effects of the thromboxane synthetase inhibitor and receptor antagonist terbogrel in patients with primary pulmonary hypertension. *Am Heart J, 143,* E4.

82. Neri Serneri, G. G., Coccheri, S., Marubini, E., et al. (2004). Picotamide, a combined inhibitor of thromboxane A2 synthase and receptor, reduces 2-year mortality in diabetics with peripheral arterial disease: The DAVID study. *Eur Heart J, 25,* 1845–1852.

83. Cadroy, Y., Hanson, S. R., Kelly, A. B., et al. (1994). Relative antithrombotic effects of monoclonal antibodies targeting different platelet glycoprotein–adhesive molecule interactions in nonhuman primates. *Blood, 83,* 3218–3224.

84. Wu, D., Meiring, M., Kotze, H. F., et al. (2002). Inhibition of platelet glycoprotein Ib, glycoprotein IIb/IIIa, or both by monoclonal antibodies prevents arterial thrombosis in baboons. *Arterioscler Thromb Vasc Biol, 22,* 323–328.

85. Alimardani, G., Guichard, J., Fichelson, S., et al. (2002). Pathogenic effects of anti-glycoprotein Ib antibodies on megakaryocytes and platelets. *Thromb Haemost, 88,* 1039–1046.

86. Takahashi, R., Sekine, N., & Nakatake, T. (1999). Influence of monoclonal antiplatelet glycoprotein antibodies on *in vitro* human megakaryocyte colony formation and proplatelet formation. *Blood, 93,* 1951–1958.

87. Chang, M. C., Lin, H. K., Peng, H. C., et al. (1998). Antithrombotic effect of crotalin, a platelet membrane glycoprotein Ib antagonist from venom of *Crotalus atrox. Blood, 91,* 1582–1589.

88. Sakurai, Y., Fujimura, Y., Kokubo, T., et al. (1998). The cDNA cloning and molecular characterization of a snake venom platelet glycoprotein Ib-binding protein, mamushigin, from *Agkistrodon halys blomhoffii* venom. *Thromb Haemost, 79,* 1199–1207.

89. Yamamoto, H., Nagano, M., Kitoh, M., et al. (1995). Monoclonal antibody against von Willebrand factor inhibits thrombus formation without prolongation of bleeding time. *Blood, 86*(Suppl.), 352, 84a.

90. Kageyama, S., Yamamoto, H., Nakazawa, H., et al. (2001). Anti-human vWF monoclonal antibody, AJvW-2 Fab, inhibits repetitive coronary artery thrombosis without bleeding time prolongation in dogs. *Thromb Res, 101,* 395–404.

91. Yamamoto, H., Vreys, I., Stassen, J. M., et al. (1998). Antagonism of vWF inhibits both injury induced arterial and venous thrombosis in the hamster. *Thromb Haemost, 79,* 202–210.

92. Golino, P., Ragni, M., Cirillo, P., et al. (1995). Aurintricarboxylic acid reduces platelet deposition in stenosed and endothelially injured rabbit carotid arteries more effectively than other antiplatelet interventions. *Thromb Haemost, 74,* 974–979.

93. Owens, M. R., & Holme, S. (1996). Aurin tricarboxylic acid inhibits adhesion of platelets to subendothelium. *Thromb Res, 81,* 177–185.

94. Azzam, K., Garfinkel, L. I., Bal dit Sollier, C., et al. (1995). Antithrombotic effect of a recombinant von Willebrand factor, VCL, on nitrogen laser-induced thrombus formation in guinea pig mesenteric arteries. *Thromb Haemost, 73,* 318–323.

95. McGhie, A. I., McNatt, J., Ezov, N., et al. (1994). Abolition of cyclic flow variations in stenosed, endothelium-injured coronary arteries in nonhuman primates with a peptide fragment (VCL) derived from human plasma von Willebrand factor–glycoprotein Ib binding domain. *Circulation, 90,* 2976–2981.

96. Davey, M. G., & Luscher, E. F. (1967). Actions of thrombin and other coagulant and proteolytic enzymes on blood platelets. *Nature, 216,* 857–858.

97. Kahn, M. L., Nakanishi-Matsui, M., Shapiro, M. J., et al. (1999). Protease-activated receptors 1 and 4 mediate activation of human platelets by thrombin. *J Clin Invest, 103,* 879–887.

98. Andrade-Gordon, P., Derian, C. K., Maryanoff, B. E., et al. (2001). Administration of a potent antagonist of protease-activated receptor-1 (PAR-1) attenuates vascular restenosis following balloon angioplasty in rats. *J Pharmacol Exp Ther, 298,* 34–42.

99. Ma, L., Hollenberg, M. D., & Wallace, J. L. (2001). Thrombin-induced platelet endostatin release is blocked by a proteinase activated receptor-4 (PAR4) antagonist. *Br J Pharmacol, 134,* 701–704.

100. Covic, L., Gresser, A. L., & Kuliopulos, A. (2000). Biphasic kinetics of activation and signaling for PAR1 and PAR4 thrombin receptors in platelets. *Biochemistry, 39,* 5458–5467.

101. Covic, L., Singh, C., Smith, H., et al. (2002). Role of the PAR4 thrombin receptor in stabilizing platelet–platelet aggregates as revealed by a patient with Hermansky–Pudlak syndrome. *Thromb Haemost, 87,* 722–727.

102. Covic, L., Misra, M., Badar, J., et al. (2002). Pepducin-based intervention of thrombin-receptor signaling and systemic platelet activation. *Nat Med, 8,* 1161–1165.

103. Konstantinides, S., Schafer, K., Koschnick, S., et al. (2001). Leptin-dependent platelet aggregation and arterial

thrombosis suggests a mechanism for atherothrombotic disease in obesity. *J Clin Invest, 108,* 1533–1540.

104. Li, Z., Kim, E. S., & Bearer, E. L. (2002). Arp2/3 complex is required for actin polymerization during platelet shape change. *Blood, 99,* 4466–4474.

105. Angelillo-Scherrer, A., de Frutos, P., Aparicio, C., et al. (2001). Deficiency or inhibition of Gas6 causes platelet dysfunction and protects mice against thrombosis. *Nat Med, 7,* 215–221.

106. Evenas, P., Garcia de Frutos, P., Nicolaes, G. A., et al. (2000). The second laminin G-type domain of protein S is indispensable for expression of full cofactor activity in activated protein C-catalysed inactivation of factor Va and factor VIIIa. *Thromb Haemost, 84,* 271–277.

107. Nurtjahja-Tjendraputra, E., Ammit, A. J., Roufogalis, B. D., et al. (2003). Effective antiplatelet and COX-1 enzyme inhibitors from pungent constituents of ginger. *Thromb Res, 111,* 259–265.

108. Pinsky, D. J., Broekman, M. J., Peschon, J. J., et al. (2002). Elucidation of the thromboregulatory role of CD39/ectoapyrase in the ischemic brain. *J Clin Invest, 109,* 1031–1040.

109. Enjyoji, K., Sevigny, J., Lin, Y., et al. (1999). Targeted disruption of CD39/ATP diphosphohydrolase results in disordered hemostasis and thromboregulation. *Nat Med, 5,* 1010–1017.

110. Dai, J., Liu, J., Deng, Y., et al. (2004). Structure and protein design of a human platelet function inhibitor. *Cell, 116,* 649–659.

111. Henn, V., Slupsky, J. R., Grafe, M., et al. (1998). CD40 ligand on activated platelets triggers an inflammatory reaction of endothelial cells. *Nature, 391,* 591–594.

112. Porta, C., Maiolo, A., Tua, A., et al. (2000). Amifostine, a reactive oxygen species scavenger with radiation- and chemoprotective properties, inhibits *in vitro* platelet activation induced by ADP, collagen or PAF. *Haematologica, 85,* 820–825.

113. Napoli, C., Aldini, G., Wallace, J. L., et al. (2002). Efficacy and age-related effects of nitric oxide-releasing aspirin on experimental restenosis. *Proc Natl Acad Sci USA, 99,* 1689–1694.

114. Napoli, C., Ackah, E., De Nigris, F., et al. (2002). Chronic treatment with nitric oxide-releasing aspirin reduces plasma low-density lipoprotein oxidation and oxidative stress, arterial oxidation-specific epitopes, and atherogenesis in hypercholesterolemic mice. *Proc Natl Acad Sci USA, 99,* 12467–12470.

115. Napoli, C., Cirino, G., Del Soldato, P., et al. (2001). Effects of nitric oxide-releasing aspirin versus aspirin on restenosis in hypercholesterolemic mice. *Proc Natl Acad Sci USA, 98,* 2860–2864.

116. Di Napoli, M., & Papa, F. (2003). NCX-4016 NicOx. *Curr Opin Invest Drugs, 4,* 1126–1139.

117. Vilahur, G., Baldellou, M. I., Segales, E., et al. (2004). Inhibition of thrombosis by a novel platelet selective S-nitrosothiol compound without hemodynamic side effects. *Cardiovasc Res, 61,* 806–816.

118. Kumar, A., Hoover, J. L., Simmons, C. A., et al. (1997). Remodeling and neointimal formation in the carotid artery of normal and P-selectin-deficient mice. *Circulation, 96,* 4333–4342.

119. Collins, R. G., Velji, R., Guevara, N. V., et al. (2000). P-selectin or intercellular adhesion molecule (ICAM)-1 deficiency substantially protects against atherosclerosis in apolipoprotein E-deficient mice. *J Exp Med, 191,* 189–194.

120. Schober, A., Manka, D., von Hundelshausen, P., et al. (2002). Deposition of platelet RANTES triggering monocyte recruitment requires P-selectin and is involved in neointima formation after arterial injury. *Circulation, 106,* 1523–1529.

121. von Hundelshausen, P., Weber, K. S., Huo, Y., et al. (2001). RANTES deposition by platelets triggers monocyte arrest on inflamed and atherosclerotic endothelium. *Circulation, 103,* 1772–1777.

122. Specht, Susan. (2006). ZymoGenetics reports new findings on anti-thrombotic activities of CTRP 7; Novel protein prevents platelet thrombosis without causing bleeding. Paper presented at the International Society on Thrombosis and Haemostasis Congress 2003. Available at www.zymogenetics.com/ir/newsItem.php?id=432558. Accessed July 19, 2006.

123. Horii, K., Okuda, D., Morita, T., et al. (2003). Structural characterization of EMS16, an antagonist of collagen receptor (GPIa/IIa) from the venom of *Echis multisquamatus. Biochemistry, 42,* 12497–12502.

# PART SIX

# Pharmacology: Therapy to Increase Platelet Numbers and/or Function

# CHAPTER 66

# Platelet Growth Factors

## David J. Kuter

*Hematology Unit, Massachusetts General Hospital, Harvard Medical School, Boston, Massachusetts*

## I. Introduction

During the past 50 years, a number of substances that increase platelet production have been identified, including recombinant granulocyte-macrophage colony-stimulating factor (GM-CSF), stem cell factor (c-kit ligand or steel factor), interleukin (IL)-1, IL-3, IL-6, and IL-11, and thrombopoietin (TPO).[1–11] Except for TPO and IL-11, these substances have either been ineffective or associated with excessive toxicity when administered to patients. This chapter focuses on the physiologically relevant platelet growth factor TPO and its derivatives that are under development but not yet approved for clinical use. Attention is also given to recombinant IL-11, which has been approved for clinical use but is associated with a number of significant side effects and is therefore not widely used.[12,13]

The name "thrombopoietin" was first proposed in 1958 for the growth factor that was presumed to exist for regulation of platelet production, analogous to the role erythropoietin (EPO) serves in the regulation of red cell production.[14–16] After more than 30 years of research efforts, five separate groups using several different approaches finally purified TPO in 1994. Two of the research groups purified the molecule directly from the plasma of thrombocytopenic mice[6] or sheep[5] using bioassays to detect the stimulation of megakaryocyte growth. The other groups purified TPO from thrombocytopenic animal plasma by affinity purification methods that used the previously described, presumed TPO receptor, c-mpl.[2,4] Finally, one group subjected a BaF3 cell line expressing the presumed TPO receptor, c-mpl, to mutagenesis and selected for clones exhibiting exogenous factor-independent autocrine growth.[3]

Although all of the molecules initially discovered had virtually the same amino acid sequence, they were called by different names — TPO,[6] megakaryocyte growth and development factor (MGDF),[2] c-mpl ligand,[4] or megapoietin.[5] MGDF was the name given to the 25- and 31-kDa amino-terminal fragments of full-length TPO purified from thrombocytopenic canine plasma that stimulated an increase in the size and number of megakaryocyte colonies *in vitro.*[2,17] Megapoietin was the name given the 31-kDa amino-terminal TPO fragment purified from thrombocytopenic sheep plasma based on its ability to stimulate an increase in the number and ploidy of megakaryocytes in culture.[5,18]

The name c-mpl ligand derived from the fact that prior to its identification as the TPO receptor, v-mpl was found to be a truncated hematopoietic cytokine receptor encoded by a murine retroviral oncogene that caused a myeloproliferative leukemia (hence the name "mpl") in mice.[19] When the full-length cellular homologue, c-mpl, was cloned in 1992,[20] it was found to be present in megakaryocytes and platelets and this led to the presumption, now known to be correct, that it was the TPO receptor. This led many to try to find the ligand ("c-mpl ligand") that bound to the c-mpl receptor. It is now known that the c-mpl ligand is TPO and that c-mpl is the TPO receptor. The term c-mpl ligand is often used interchangeably with TPO, but it best describes a family of ligands that bind to c-mpl, including endogenous TPO, recombinant human TPO (rhuTPO), pegylated recombinant human MGDF (PEG-rHuMGDF), promegapoietin (a fusion protein of IL-3 and TPO), as well as TPO peptide and non-peptide mimetics.

## II. TPO Structure

TPO is synthesized in the liver as a 353-amino acid precursor protein with a molecular weight of 36 kDa.[2,4,21] Following removal of the 21-amino acid signal peptide, the remaining 332 amino acids undergo glycosylation to produce a glycoprotein with a molecular weight of 95 kDa on sodium dodecyl sulfate polyacrylamide gel electrophoresis[22] and 57.5 kDa by mass spectrometry (Fig. 66-1).[23] The glycoprotein is then released into the circulation with no apparent intracellular storage in the liver.

TPO is a member of the four-helix-bundle cytokine superfamily and has several unusual properties. First, it is much larger than most other regulators of blood cell

1211

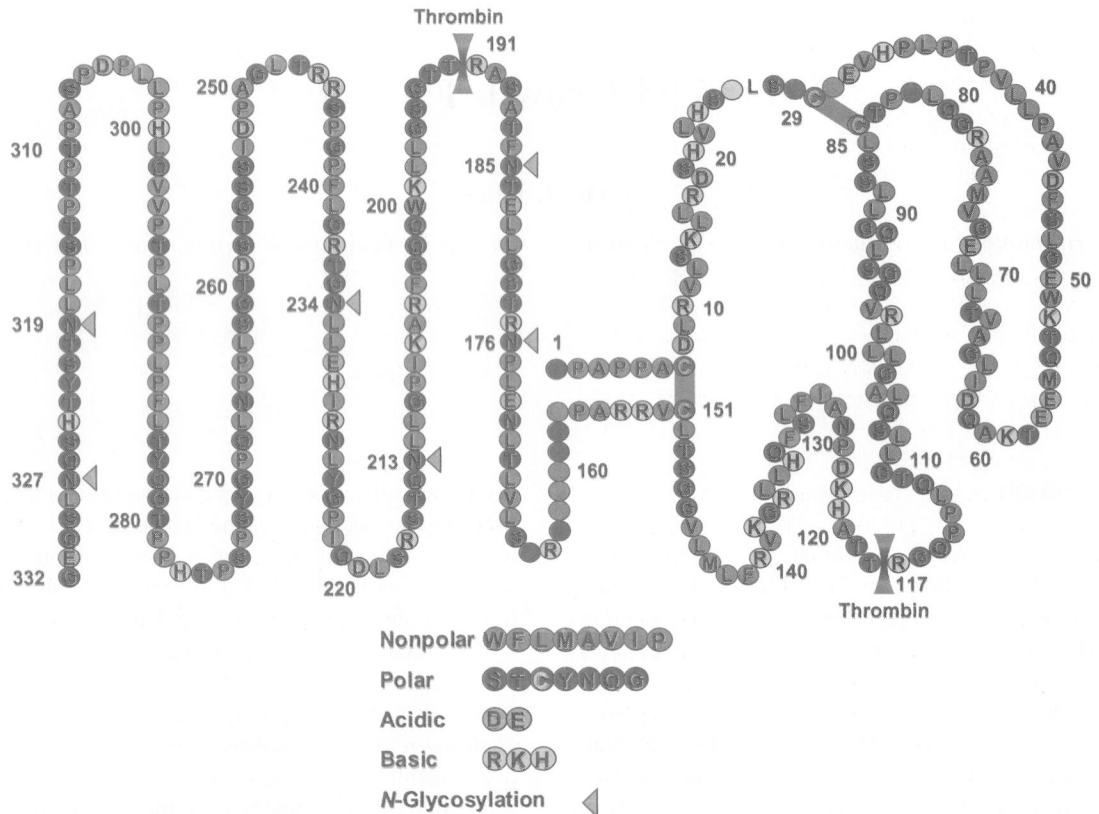

**Figure 66-1.** The primary structure of human TPO. Courtesy of Dr. T. Kato, Pharmaceutical Research Laboratory, Kirin Brewery Company, Ltd., Takasaki, Gunma, Japan.

production, such as granulocyte colony-stimulating factor (G-CSF) and EPO. Second, it has two distinct domains, an EPO-like domain (residues 1–153) and a carbohydrate-rich domain (residues 154–332), separated by a dibasic site of potential proteolytic cleavage (Arg153–Arg154). It is unclear whether this cleavage site is biologically relevant.[24]

### A. The "EPO-Like" Domain of TPO

The first 153 amino acids of the mature protein are 23% identical with human EPO[25] and probably 50% similar if conservative amino acid substitutions are considered. This region is also highly conserved between species. In addition, there is a 39-amino acid domain within the first 153 amino acids of TPO that shares significant homology with neurotrophins.[26] Their shared amphipathic region is highly conserved and in the neurotrophins is involved in receptor binding. One of the neurotrophins, brain-derived neurotrophic factor, is 36% identical, and if conservative substitutions are taken into account it is 62% similar. EPO also shares similar homology with the neurotrophins.

The EPO-like region also contains four cysteine residues — Cys7, Cys151, Cys29, and Cys85[27] — just like those in EPO, which are highly conserved among different species. Replacement of Cys7 or Cys151 abrogates all activity, whereas disruption of Cys29 or Cys85 reduces activity.[28] This region also contains four α-helices, just as in EPO. All of the TPO receptor binding activity is located in this region. Despite these similarities with EPO, TPO does not bind the EPO receptor and EPO does not bind the TPO receptor.

Structure–function analysis of the EPO-like domain of TPO suggests a number of important amino acids. Three amino acids in helix A (Arg10, Lys14, and Arg17) and four in helix D (Gln132, His133, Lys138, and Phe141) are crucial for activity.[29] Arg10 appears to be of particular interest; mutation of Arg10 to Ala results in normal receptor binding but no biological activity, suggesting that this mutation prevents receptor dimerization.[29] In a second study,[30] mutation of Arg10, Pro42, Glu50, and Lys138 to Ala completely abolished activity, whereas mutation of Lys14 modestly reduced activity. These areas are highly conserved between species. These findings have been confirmed in a third study in which mutations at the mostly positively charged residues (Asp8, Lys14, Lys52, Lys59, Lys136, Lys138, and Arg140) decreased receptor binding activity.[31] These residues are mostly in helix A and helix D as well as in a loop region between helix A and helix B.

The areas of TPO that bind the TPO receptor are probably analogous to those of other members of this cytokine superfamily, such as human growth hormone and EPO. Modeling studies of TPO suggested two regions that may be important for binding and causing homodimerization of the TPO receptor.[27] It has been proposed that Lys138 of helix D and Pro42 and Glu50 of loop AB may constitute one binding region, whereas Arg10 and Lys14 of helix A may constitute a second binding region that participates in the dimerization of TPO receptors.[30] This suggestion has been partly confirmed by monoclonal antibody studies. Two monoclonal antibodies that block TPO binding to its receptor have determinants in helix D.[31] These results again suggest that TPO has two different receptor binding sites that help dimerize its receptor.

The crystal structure of the receptor-binding domain (residues 1–163) has been determined and confirms some of the structure–function relationships described previously.[32] TPO has an antiparallel four-helix bundle fold and can interact with a soluble c-mpl construct with 1 : 2 stoichiometry with one high-affinity ($3.3 \times 10^9$ M$^{-1}$) and one low-affinity ($1.1 \times 10^9$ M$^{-1}$) binding site. There are 26 residues in this region that are identical with EPO. Of these 26 invariant, shared residues, 13 (L12, P42, W51, A60, L69, L70, L86, L104, L107, F128, L135, G138, and K138) are buried in the hydrophobic core and establish the basic structure. There are an additional 24 residues that determine the specificity of TPO for c-mpl (as determined by their constancy across different species and their lack of conservation across species for EPO). These 24 residues define two different clusters that account for the high-affinity binding site (13 residues: V44, D45, F46, S47, L48, E50, A126, L129, Q132, H133, R140, F141, and L144) and the low-affinity binding site (11 residues: K14, R17, G91, Q92, S94, G95, L99, L101, G102, A103, and Q105) (Fig. 66-2).

### B. The "Carbohydrate-Rich" Domain of TPO

Amino acids 154 to 332 comprise a unique sequence that contains six N-linked glycosylation sites and is rich in serine, threonine, and proline residues. It is less well conserved among different species: Murine and human TPO are 84% identical in the EPO-like domain but only 62% identical in the carbohydrate-rich domain. Although murine TPO has 335 amino acids and the rat has 305 amino acids, the 30-amino acid difference is completely within the carbohydrate-rich domain.

Structure–function studies have demonstrated that although the first 153 amino acids of the c-mpl ligand are all that are required for its thrombopoietic effect *in vitro*,[2,4] this truncated molecule has a markedly decreased circulatory half-life compared to the 20- to 40-hour half-life of the native protein.[33] Paradoxically, the truncated molecule has

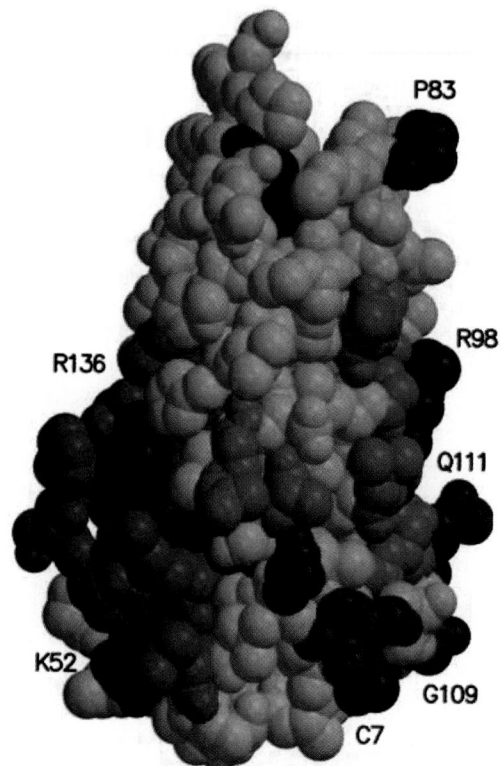

**Figure 66-2.** The partial crystal structure of TPO (residues 1–163). Residues that are common and invariant for both TPO and EPO are depicted in black. The 13 residues that comprise the high-affinity site are in blue and the 11 residues that compose the low-affinity site are in green. Courtesy of Dr. T. Kato, Pharmaceutical Research Laboratory, Kirin Brewery Company, Ltd., Takasaki, Gunma, Japan.

a specific activity *in vitro* that is 20-fold greater than the full-length molecule.[34] Presumably, the glycosylated second half of the molecule confers stability and prolongs the circulatory half-life. Similar carbohydrate sequences regulate the stability of EPO.[35] In addition, this part of the molecule assists in the secretion of the intact molecule from hepatocytes by serving as a molecular chaperone or guide in protein folding; truncated muteins of this portion of the molecule have diminished secretion.[36]

### C. Importance of TPO Glycosylation

TPO undergoes significant posttranslational modification by glycosylation, especially in residues 154 to 332. Failure to undergo glycosylation reduces hepatic secretion[36] and produces a molecule with a greatly reduced half-life.[34] All 6 sites of *N*-glycosylation are found in residues 154 to 332. The *N*-glycans are of the complex type, with the core-fucosylated disialylated biantennary and trisialylated triantennary structures predominating. $N_{185}FT$ and $N_{234}GT$

appear to be 100% occupied, whereas $N_{176}RT$ and $N_{213}QT$ are partially occupied and the status of $N_{319}TS$ and $N_{327}LS$ is uncertain.[23] In addition, there are 3 $O$-glycosylation sites (S1, T37, and T110) in the EPO-like domain and at least 5 (T158, T159, S163, T192, and T244) and possibly 16 others in the carbohydrate-rich domain. The $O$-glycans are of the mucin type, with the monosialylated and disialylated Gal-GalNAc-S/T structures predominating. This has resulted in the proposal that the carbohydrate-rich domain of TPO can be further divided into two subdomains on the basis of sequence homology among the cloned sequences and glycosylation: an N-glycan domain (154–246) and an $O$-glycan domain (247–332).[23] The previous analyses were obtained using recombinant human TPO expressed in Chinese hamster ovary (CHO) cells; the glycosylation pattern of endogenous TPO is different[22] but has not been studied in detail.

## III. The TPO Gene and Its Regulation

### A. TPO Gene Structure

There is a single copy of the gene for TPO on human chromosome 3q27–28.[21,25,37] The gene spans approximately 7 kb with seven exons, the first two of which are noncoding. The third exon contains part of the 5′-untranslated sequence and part of the signal peptide. The EPO-like region is coded for by exons 4 to 7 and the entire carbohydrate-rich domain is encoded by exon 7. Comparison with the EPO gene shows conservation of the boundaries of the coding exons except for the addition of the carbohydrate domain sequence in the final exon of the TPO gene.

Although human studies are limited, it is unclear whether polymorphisms in the TPO open reading frame exist. TPO cloned from a Chinese human fetal liver had four base substitutions (sites 497, 595, 767, and 795 bp) that differed from the GenBank database and led to a change of three amino acids in the predicted protein.[38] However, more than 20 different TPO cDNAs have been sequenced in various myeloproliferative disorders[39] and normal individuals[22] with no variations from the wild-type TPO coding sequence found.

### B. Regulatory Elements of the TPO Gene

Hepatic TPO synthesis does not appear to be significantly regulated in normal physiology but is constitutively produced in the liver. The 5′-flanking region of the TPO gene contains several possible regulatory regions, including SP-1, AP-2, and NF-κB. Downstream of exon 1 are several GATA and Ets elements but no TATA box or CAAT motif. The gene is located in a region of chromosome 3 in which there are many other genes containing iron-responsive elements, but no such elements have been detected in the TPO promoter.

### C. TPO mRNA Species

In addition to the mRNA (TPO-1) encoding the functional TPO protein, two other mRNA sequences encoding different TPO proteins are present due to alternative splicing.[21,25,40] One variant (TPO-2) has a deletion of 12 bp (residues 112–115 [LPPQ] of the mature protein) at the junction of exons 6 and 7 but maintains the normal reading frame. In transfected cells, this protein is secreted 500-fold less than TPO-1 and the secreted protein is probably inactive.[25] The second variant (TPO-3) is produced by a splice site within exon 7 causing a 116-bp deletion and a frameshift that produces a protein predicted to contain 286 amino acids, the first 159 of which are identical to the native TPO (including the signal peptide). This protein is synthesized in transfected cells but is poorly secreted and may not have any biological effect.[40] Additional TPO variants (TPO-4, TPO-5, and TPO-6) have been identified in some cancer cell lines but are probably not secreted or functional.[41]

Further analysis of the expression of the human TPO gene has documented several TPO-1 mRNA transcripts. Ten percent of all hepatic TPO-1 transcripts originate from a promoter in exon 1 (P1) and 90% from the dominant promoter site in exon 2 (P2). In addition, a splice variant in which exon 2 (containing much of the 5′-untranslated region) is skipped (P1ΔE2) is made 2% of the time.[42] Although low in abundance, this last transcript has much greater translational ability than does either of the other mRNAs. As discussed later, the TPO 5′-untranslated region contains eight translation initiation sites, only one of which yields an intact protein; TPO mRNA transcripts with fewer "nonproductive" translation initiation sites are more efficiently translated.[42]

## IV. Functions of TPO

### A. Cellular Effects of TPO

The TPO receptor is found in a wide range of hematopoietic cells, including mature megakaryocytes and platelets, immature cells of all lineages, CD34+ cells, and pluripotential stem cells. Binding of TPO to its megakaryocyte receptor produces the following major effects: prevention of megakaryocyte apoptosis[43]; increased megakaryocyte number, size, and ploidy[5]; increased rate of megakaryocyte maturation; and internalization of the receptor–ligand complex.[44] Multiple signal transduction pathways, some involving JAK, STAT, MAP kinase as well as other intracellular mediators, mediate the first three of these effects. Addition of TPO to CD34$^+$ cells can result in the majority of cells becoming megakaryocytes and then shedding platelets.[45] The shedding of platelets from megakaryocytes does not require, and actually may be inhibited by, the presence of TPO.[46]

The mechanism by which TPO binding initiates signal transduction by c-mpl is not fully established. Based on EPO,[47] it was initially proposed that one site on TPO bound the first receptor followed by a second site on the same TPO molecule binding a second receptor, resulting in the dimerized receptor becoming active. However, studies have shown that the cytokine receptor homology (CRH) domains of the EPO receptor preform an inactive dimeric receptor in which the intracellular regions are sufficiently distant from one another to prevent phosphorylation and activation of JAK2.[48,49] Subsequent binding of EPO to the preformed dimeric receptor changes the structure of the dimeric receptor and initiates signal transduction.

The TPO receptor c-mpl contains two CRH regions. Biochemical and crystallographic data show that TPO binds only to the distal CRH of c-mpl and not to the proximal CRH and thereby initiates signal transduction. In the absence of the distal CRH (or its replacement by the proximal CRH), c-mpl becomes active, suggesting that the distal CRH functions as an inhibitor of c-mpl until relieved by TPO binding. It is not known whether inactive, preformed c-mpl dimers exist on the surface of target cells, as shown for the EPO receptor.[27,29,32]

The platelet and possibly the megakaryocyte TPO receptor also serve as the major means of TPO catabolism.[5,18,44,50,51] Although not enumerated on the megakaryocyte, each platelet contains $56 \pm 17$ TPO (c-mpl) receptors with a $K_D$ of $163 \pm 31$ pM.[44] Upon binding the TPO receptor, the TPO receptor/TPO complex is rapidly internalized, the TPO is degraded, and the receptor is not recycled to the surface. It has been estimated that the normal total amount of circulating platelet receptors is able to clear more than 95% of the daily hepatic production of TPO.[44] Two intracytoplasmic motifs in c-mpl are important for this receptor internalization.[52]

In addition to its effect on megakaryocyte growth and maturation, TPO also stimulates early megakaryocytopoiesis by increasing the number of megakaryocyte colony-forming cells (Meg-CFC) (see Chapter 2). TPO can also stimulate early precursors of all other lineages, as well as prevent the apoptosis of pluripotential stem cells.[53] Approximately 70% of CD34$^+$ cells express the TPO receptor c-mpl.[54] In general, TPO affects the growth of early progenitors of all lineages but affects the late maturation only of megakaryocytes. Hence, TPO stimulates the production of only platelets, not red or white blood cells.

Studies have explored the effect of TPO on nonhematopoietic tissues and have illustrated a novel inverse relationship in neurons between TPO and c-mpl expression versus EPO and the EPO receptor (EPO-R).[55] During development of the normal rat brain, EPO and EPO-R mRNA and protein decline, whereas TPO and c-mpl mRNA and protein increase. With hypoxia of neurons in culture, TPO/c-mpl decrease and EPO/EPO-R increase. In addition, treating neurons with TPO increased programmed cell death, an effect that was reversed completely by the addition of EPO. These effects were mediated by the countervailing effects of EPO on the phosphatidylinositide 3-kinase (PI3K)-Akt/protein kinase B pathway (increased survival) versus the effect of TPO on the Ras-ERK1/2 pathway (decreased survival).

### B. Effects of the In Vivo Administration of TPO

Several important principles of the platelet response to TPO were demonstrated in baboons treated daily with a recombinant form of TPO (Fig. 66-3).[56,57] During the first 4 days of administration, bone marrow megakaryocyte ploidy rose to a maximum but there was no change in the platelet count. On day 5, the platelet count began to rise and did so at a dose-dependent rate. With continued administration of TPO,

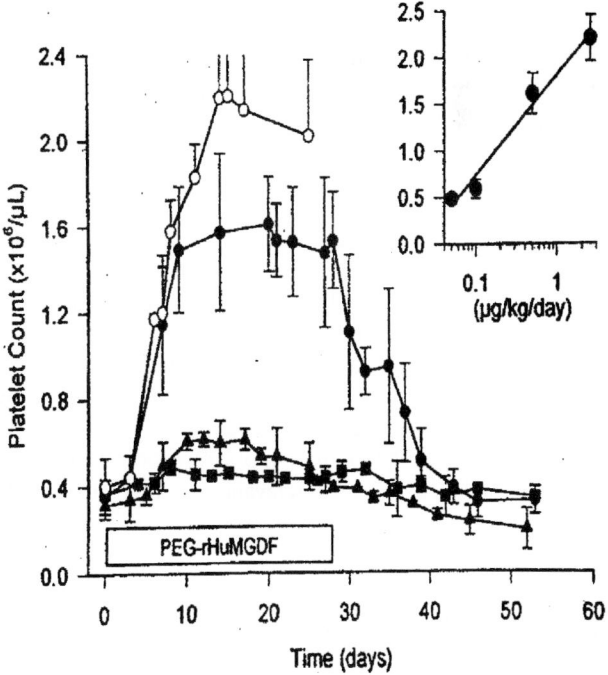

**Figure 66-3.** The effect of recombinant TPO on the platelet count *in vivo*. Baboons were injected daily for 28 days with PEG-rHuMGDF at doses of 0.05 (■, $n = 3$), 0.10 (▲, $n = 3$), 0.50 (●, $n = 4$), or 2.50 (○, $n = 6$) μg/kg/day. The inset shows the log-linear relationship between the administered dose and the fold-increase in platelet counts. There was no effect on the white cell or red cell count. Reprinted with permission from Harker et al.[57]

a dose-dependent plateau platelet count was reached on days 8 to 12. There was an initial log-linear relationship between the TPO dose and the plateau platelet count, with a maximum sixfold increase in the rate of platelet production. At higher doses of TPO, the peak platelet count declined, probably due to the inhibitory effect of high doses of TPO on platelet shedding from megakaryocytes.[46] Upon stopping the growth factor, the platelet count returned to its baseline over 10 days without a rebound thrombocytopenia. A similar time course and platelet response have been demonstrated in humans.[58–60]

In both animals and humans, TPO administration also stimulates increases in bone marrow and peripheral blood Meg-CFC, as well as erythroid and multipotential precursor cells. However, there is no effect on the white or red blood cell count.

In addition to increasing the number of megakaryocytes and platelets, TPO also affects the function of platelets. When TPO binds to its platelet receptor, it induces phosphorylation of the c-mpl receptor and a number of other molecules in several different signal transduction pathways[61–63] but does not directly cause platelet activation. However, TPO treatment reduces by 50% the threshold for activation by other platelet agonists, such as ADP and collagen.[56,57] This effect may be mediated by TPO-dependent activation of PI3K, which in turn phosphorylates Thr306 and Ser473 of platelet protein kinase Bα, an important anti-apoptotic protein.[64] However, TPO does not prevent apoptosis of platelets during *ex vivo* storage.[65–67] It is unclear if this is a clinically relevant effect since other hematopoietic growth factors also have the same effect and have not been associated with thrombosis.

The finding that TPO increases apoptosis in neurons has been only preliminarily explored *in vivo*.[55] In a model of mild to moderate hypoxia/ischemic brain damage in juvenile rats, treatment with TPO increased programmed cell death in the distressed neurons, whereas EPO was protective. Of interest is the additional finding that c-mpl expression was increased in TPO-treated hypoxic mice, suggesting that the normal "protective" physiological response to hypoxia (decreased TPO and c-mpl expression in neurons) was breached. The clinical relevance of this unexpected observation is unclear.

## V. Physiology of TPO

### A. The Liver is the Primary Site of TPO Production

TPO mRNA is expressed in low abundance, primarily in liver parenchymal cells, and even smaller amounts are made in the kidney.[68] Although probably only hepatic TPO production contributes to the circulating amounts of this cytokine, TPO transcripts have also been detected in murine skeletal muscle, heart, brain, testes, spleen,[69] and bone marrow.[70,71]

Hematologically relevant protein expression has been documented only in hepatocytes and possibly bone marrow cells,[71,72] with the liver accounting for more than 50% of total body production.[73] The hematopoietic importance of the nonhepatic sites of TPO production is unclear, but human clinical studies suggest little significant role. For example, in patients with hepatic failure undergoing liver transplantation, the low platelet counts and undetectable TPO levels pretransplant become normal posttransplant.[74,75]

The physiological expression of TPO does not appear to be regulated by any known stimuli at either a translational or a transcriptional level. TPO production appears to be constitutive, and the circulating levels are directly determined primarily by the circulating platelet mass.[18,76,77] A large number of hematopoietic growth factors, cytokines, and small molecules have been tested and none has any effect on TPO mRNA or protein production in hepatic cell lines or primary hepatic cell cultures. Bone marrow stromal cell production of TPO is increased slightly (1.5- to 2-fold) by exposure to CD40 ligand, platelet-derived growth factor, and fibroblast growth factor-2 and is inhibited by thrombospondin, platelet factor-4, and transforming growth factor-β.[78–80] The physiological relevance of these effects has not been demonstrated.

### B. Normal Physiology: Evidence from TPO and c-mpl Knockout Mice

TPO is the major physiologically relevant regulator of platelet production and acts to "amplify" the basal production rate of megakaryocytes and platelets. When TPO or its receptor are "knocked out" by homologous recombination in mice,[81–83] the megakaryocyte and platelet mass are reduced to approximately 10% of normal but the animals are healthy and do not spontaneously bleed (Fig. 66-4). The neutrophil and erythrocyte counts are normal. In animals in which only one of the TPO genes has been deleted, the platelet count is reduced to approximately 65% of normal. Such TPO-deficient mice can modestly increase their platelet count if treated with other thrombopoietic growth factors, such as IL-6, IL-11, or stem cell factor.[84]

What determines the TPO-independent rate of platelet production is unclear. "Double-knockout" mice that lack the genes for c-mpl and one other growth factor (IL-3, IL-11, IL-6, or leukemia inhibitory factor [LIF]) have been created to investigate this observation. These double-knockout mice had no additional defect in platelets or their precursors, indicating that IL-3, IL-11, IL-6, or LIF alone are not responsible for the basal platelet production seen in the absence of TPO signaling.[85,86]

Chemokines may play a prominent role in platelet production.[87] Stromal cell-derived factor-1 (SDF-1) and fibroblast growth factor-4 (FGF-4) can normalize the platelet count in TPO-deficient mice. These two chemokines appear

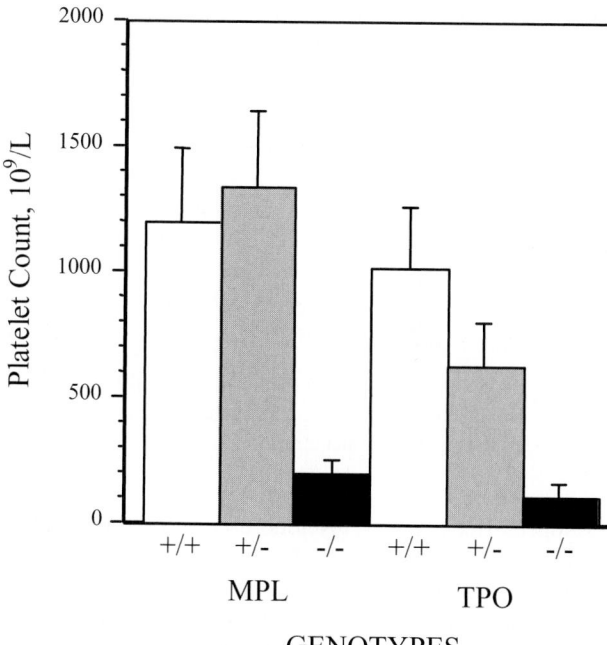

**Figure 66-4.** Mice deficient in TPO or its receptor (MPL) are thrombocytopenic. Platelet counts (± SD) in normal (+/+) mice were compared with those in TPO or MPL homozygous (−/−) or heterozygous (+/−) deficient mice. Adapted from Gurney et al.[82] and de Sauvage et al.[83]

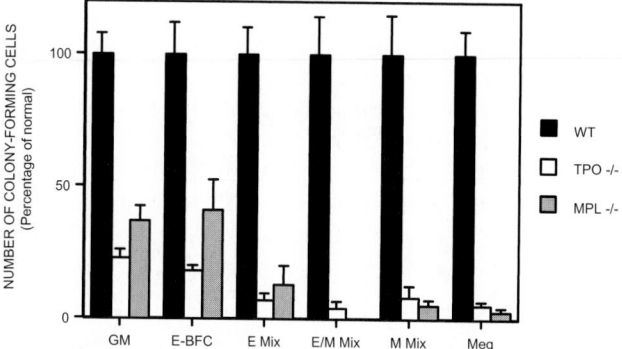

**Figure 66-5.** Bone marrow progenitor cells of all lineages are decreased in mice deficient in TPO (TPO−/−) or its receptor (MPL−/−) compared to normal wild-type (WT) mice. Granulocyte-macrophage colony-forming cells (GM), erythroid burst-forming cells (E-BFC), megakaryocyte colony-forming cells (Meg), as well as mixed colonies with erythroid component (E Mix), erythroid and megakaryocytic components (E/M Mix), or megakaryocytic with nonerythroid components (M Mix), were enumerated (± SD). Adapted from Carver-Moore et al.[84]

vital for the chemotaxis of megakaryocyte progenitors to the "vascular niche" wherein other cell–cell interactions probably promote survival, maturation, and platelet release across the endothelial sinusoidal barrier. Disruption of this vascular niche and inhibition of chemokine activity reduce platelet production and cause thrombocytopenia. These chemokines may therefore be an alternative means to increase platelet production, independent of TPO.

TPO affects bone marrow precursor cells of all lineages (Fig. 66-5). In animals deficient in TPO or c-mpl, the megakaryocyte precursor cells (Meg-CFC) are reduced by 90 to 95%, as expected. However, the myeloid and erythroid precursor cells are also reduced by 60 to 80%.[81,84] Presumably, the normal neutrophil and erythrocyte counts in these animals are maintained by the intact feedback mechanisms mediated by G-CSF and EPO. These findings again support the concept that TPO is vital for the proliferation of all early hematopoietic cells but affects late maturation of only megakaryocytes.

Although the production of red blood cells is regulated by a cytochrome system that senses changes in the hematocrit and alters the rate of transcription of the EPO gene, there is no such "sensor" of the platelet mass.[5,76,88–92] Rather, TPO production is constitutive and the circulating levels are directly determined by the circulating platelet mass (Fig. 66-6). TPO mRNA is produced at the same rate in normal and thrombocytopenic individuals.[76,77] No drug or clinical

condition has been shown to significantly increase hepatic TPO production. Platelets and megakaryocytes contain high-affinity TPO (c-mpl) receptors that bind and clear TPO from the circulation and thereby directly determine the circulating TPO level. When platelet production is decreased, clearance of TPO is reduced and levels rise.

In support of this model is the finding that transfusion of platelets into thrombocytopenic animals or humans has resulted in a decrease in plasma TPO levels,[5,18,88,90,93] and similar results have been observed when normal platelets are transfused into c-mpl-deficient mice.[51] Upon binding of TPO to its receptor on platelets, and probably megakaryocytes, the ligand–receptor complex is internalized.[5,18,88] Once internalized, the TPO is degraded,[51,94,95] with no recycling of the TPO receptor.[44] These findings indicate that TPO is constitutively synthesized in the liver and removed from circulation by binding to the c-mpl receptor on platelets and possibly bone marrow megakaryocytes.

This type of feedback system is not unusual in hematology. Indeed, both monocyte (M)-CSF and G-CSF are regulated in normal physiology, primarily by the amount of circulating monocytes and neutrophils, respectively. Only for EPO is there a true sensor of the circulating blood cell mass that in turn alters production of this hematopoietic growth factor.

### C. Normal Physiology: Effects of TPO Overexpression

Overexpression of TPO has been studied in three separate models. Mice in which TPO was overexpressed following transplantation of bone marrow cells transfected with MGDF (the first 163 amino acids of TPO) had a more rapid

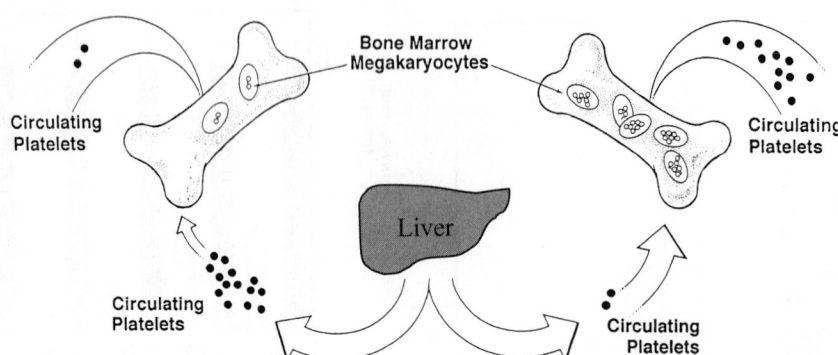

**Figure 66-6.** The physiological regulation of TPO levels. The constitutive hepatic production of TPO (center) is cleared by avid TPO receptors on platelets, resulting in normal levels when the platelet production is normal (left) and elevated levels when the platelet production is reduced (right). The bone marrow megakaryocytes are stimulated as the circulating TPO concentration rises.

platelet reconstitution than did control mice.[96] Platelet counts increased four- to eightfold and remained elevated. These mice demonstrated increased numbers of bone marrow and splenic megakaryocytes but ultimately developed marrow fibrosis, extramedullary hematopoiesis, hepatosplenomegaly, osteosclerosis, and anemia.[96]

In a second model, murine bone marrow cells were transfected to a much greater extent with TPO cDNA and transplanted into mice.[97] Whereas levels of circulating TPO were barely detectable in the preceding experiment, TPO levels were persistently highly elevated in these mice during the entire course of the experiment. Early after transplantation, platelet and white blood cell counts increased, whereas the hematocrits decreased. Megakaryocytes and granulocytes and their respective progenitor cells were markedly increased in the spleen, but erythroblasts and their precursor cells were decreased in the marrow. Later, progenitor cells in the spleen decreased, extramedullary hematopoiesis was observed, marrow and spleen developed marked fibrosis, and the bones developed osteosclerosis. The mice had reduced survival and some developed leukemia.[97]

In a third model, normal mice and syngeneic mice with variable degrees of immune dysfunction (nude, SCID, and NOD-SCID) were infected with adenovectors carrying the human TPO cDNA.[98] Platelet counts increased in all mice but were much higher in the SCID (T and B cell defect and minimal antibody response) and NOD-SCID (T and B cell defect, minimal antibody response, and mononuclear phagocytes diminished in number and function) mice compared with the BALB/c (control) and nude mice (T cell defect and impaired antibody production), and the increase was proportional to the concentration of circulating TPO. The control mice subsequently developed antibodies to the human TPO that cross-reacted with murine TPO and the mice became thrombocytopenic. The SCID and NOD-SCID mice continued to express high levels of TPO and did not develop antibodies. They both developed hypercellular bone marrows; however, only the SCID mice

developed osteosclerosis, myelofibrosis, and extramedullary hematopoiesis.[98]

These results suggest that chronic overexpression of low amounts of TPO in normal animals leads to thrombocytosis, marrow fibrosis, and osteosclerosis that mimics the human disorder of agnogenic myelofibrosis with myeloid metaplasia. With higher chronic expression, such changes may evolve into more significant fibrosis and possibly leukemic transformation. However, the fibrosis is not entirely mediated by TPO or the increased megakaryocyte mass but requires concomitant functional monocytes and/or macrophages.

## VI. Pathophysiology of TPO

### A. TPO Mutations

Inherited thrombocythemia is an uncommon disorder that clinically affects a few families just like the more common, sporadic cases of essential thrombocythemia (see Chapter 56). Analysis of one of these families[99] has identified a single point mutation in the splice donor site of intron 3 of the TPO gene that produces a new TPO mRNA with a normal protein coding region but with a shortened 5′-untranslated region. This transcript is more efficiently translated than normal TPO transcripts and leads to more TPO protein synthesis, higher plasma TPO levels, and chronically elevated platelet counts. Similar mutations have been described in other families[100,101] but do not appear to be involved in the more common, sporadic cases of essential thrombocythemia.[102]

The nature of this translational regulation has been elucidated.[42] The TPO 5′-untranslated region is unusual in that it has eight AUG codons from which translation may be initiated. Only the eighth AUG results in a successful protein product, whereas the other AUG sites compete for ribosomal binding sites and markedly suppress effective TPO protein production. Elimination of some of the upstream AUG sites by mutation, deletion, or alternative splicing produces transcripts with fewer AUG sites, less competition for ribosome binding, and increased effective production of TPO protein. The families in which there is overproduction of TPO are characterized by the loss of some of these AUG sites.[99–101] Since TPO is such a highly potent hematopoietic growth factor, it has been suggested that the presence of these extra AUG sites serves to guarantee against overproduction of TPO.

A number of hematopoietic disorders associated with thrombocythemia or abnormal megakaryocyte formation have been associated with defects involving chromosome 3q,[103] and some myeloid leukemias associated with thrombocytosis have a characteristic rearrangement of chromosome 3q21 and 3q26.[37] Since the TPO gene is located on chromosome 3q27–28,[21,25,37] it has been suggested that the TPO gene might be mediating these effects. However, closer analysis of these chromosome regions in these patients has not demonstrated involvement of the TPO gene,[37,104] and blood TPO levels have been normal. These results suggest that other genes close to the TPO gene may be responsible for other aspects of megakaryocyte differentiation and growth.

Although some patients with familial thrombocytopenia have been thought to have mutations of the TPO gene that decrease its expression, most such thrombocytopenic patients have had mutations in c-mpl[105–108] that reduce its function. Indeed, mutations in c-mpl, not TPO, have been a much more common finding to explain familial thrombocythemia and thrombocytopenia (see Chapter 54).[109–113]

### B. Circulating TPO Levels in Hematological Disorders

Measurement of the circulating level of TPO has not been proven to be clinically useful. Published normal values have varied widely with the assays used but generally average 64 ± 41 pg/ml (range, 27–188; $n = 40$) for the most widely used enzyme-linked immunosorbent assay.[39,114] Although this area remains controversial, several conclusions may be reached:

- TPO levels are increased 10- to 20-fold over normal in bone marrow failure states such as aplastic anemia or following myeloablative chemotherapy.[115,116] In these conditions, there is an inverse relationship between the platelet count and the concentration of TPO.

- TPO levels are only slightly elevated above normal in immune thrombocytopenic purpura,[115,116] at platelet counts as low as those seen in the bone marrow failure states. One explanation for these relatively normal levels may be that total TPO clearance is normal due to the increased mass of bone marrow megakaryocytes or, more likely, to the increased flux of platelets through the circulation.[90]

- TPO levels are normal or slightly elevated in essential thrombocythemia.[39,116,117] The platelet TPO receptor is reduced 10-fold in these patients but with a normal binding affinity. This produces a 4-fold decrease in platelet-dependent TPO clearance but, given the 2.7-fold increase in platelet count, results in a total TPO clearance that is normal.[39]

- Since TPO is produced primarily in the liver, TPO levels appear to be inappropriately low in patients with liver failure.[75] After liver transplantation, TPO levels and the platelet counts become normal.[74]

- TPO levels may allow the distinction between states of increased (normal levels) and decreased platelet production (elevated levels) and possibly reduce the need for bone marrow examination. This concept has not been clinically validated.

## VII. Therapeutic Thrombopoietins

### A. First-Generation Thrombopoietins

Soon after the discovery of TPO, two recombinant proteins were developed and have been subjected to extensive clinical investigation (Fig. 66-7). One is produced in CHO cells and consists of the full-length, glycosylated, native human amino acid sequence (recombinant thrombopoietin, rhuTPO) that has a circulatory half-life of 20 to 40 hours. The other is produced in *Escherichia coli* and is a nonglycosylated, truncated molecule composed of the first 163 amino acids of the native TPO molecule. This part of the native molecule is 50% similar to EPO, contains the entire receptor-binding domain, but has a very short circulatory half-life and little biological activity *in vivo* due to the absence of the carbohydrate-rich portion of the native molecule. The addition of a 20-kDa polyethylene glycol moiety to the amino terminus replaces the carbohydrate domain and serves to stabilize the molecule in the circulation. This molecule is called pegylated recombinant megakaryocyte growth and development factor (PEG-rHuMGDF). It has a half-life of 30 to 40 hours.

In addition to rhuTPO and PEG-rHuMGDF, a number of fusion proteins of TPO with other hematopoietic growth factors have been created. For example, promegapoietin[118] is a molecule in which the TPO receptor-binding region is coupled to the hematopoietic growth factor IL-3. This molecule can bind to and activate both the TPO and IL-3 receptors, but clinical development of this molecule had been discontinued due to antibody formation.

### B. Second-Generation Thrombopoietins

After the development of rhuTPO and PEG-rHuMGDF, a number of other molecules that bind and activate the TPO

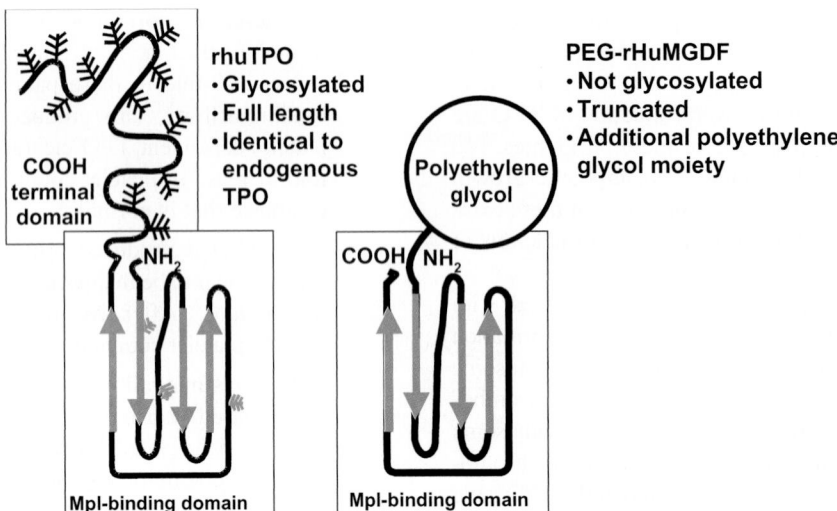

**Figure 66-7.** Structures of the recombinant thrombopoietins rhuTPO and PEG-rHuMGDF. Purple indicates areas of α helix, whereas orange-and-black "feathers" denote glycosylation sites. Courtesy of Amgen, Inc., Thousand Oaks, CA.

### Table 66-1: Platelet Growth Factors

mpl Ligands[a]

   Endogenous TPO

   rhuTPO

   PEG-rHuMGDF

   TPO/IL-3 fusion protein

   TPO peptide mimetics

   TPO nonpeptide mimetics

   TPO-Ig fusion proteins

     *AMG 531*

   Minibodies *vs.* c-mpl

Non-mpl Ligands[b]

   IL-11

   Chemokines

     *SDF-1*

     *FGF-4*

[a]Defined as those that bind c-mpl and increase its activity.

[b]Defined as those that increase platelet production but not through activation of c-mpl.

FGF-4, fibroblast growth factor-4; IL, interleukin; PEG-rHuMGDF, pegylated recombinant human megakaryocyte growth and development factor; rhuTPO, recombinant human TPO; SDF-1, stromal cell-derived factor-1; TPO, thrombopoietin.

receptor, c-mpl, were developed (Table 66-1) in an attempt to eliminate the immunogenicity found with PEG-rHuMGDF (see Section X) or improve efficacy. Despite intense clinical investigation, none of the first- or second-generation thrombopoietins has been approved for clinical use.

Great interest has been focused on the development of TPO peptide[119–121] and nonpeptide, small molecule[122] mimetics. These mimetics were designed to bind to the TPO receptor but have no sequence homology with endogenous TPO, reducing the likelihood that any antibody that might arise would cross-react with endogenous TPO. They have been identified by screening peptide and small molecule libraries for the ability to activate the TPO receptor. Many lead compounds have been identified and further modified to increase their efficacy and stability.

Early preclinical work with one of these TPO peptide mimetics[119,120] demonstrated that to be effective, a TPO peptide mimetic needed to have a long half-life and possess a dimeric structure in order to activate the TPO receptor. Unmodified monomeric peptides poorly activated the TPO receptor and were rapidly cleared from the circulation.

Although pegylated dimeric TPO peptide mimetics have been developed, a more common approach has been to create TPO mimetics by inserting small peptides containing the receptor-binding region of TPO into the IgG Fc region or into the CDR domain of $(Fab)_2$. This provides a dimeric structure with a prolonged half-life. One of these, AMG 531 (formerly AMP 2), consists of four identical peptides (with no sequence homology with TPO) that avidly bind c-mpl and are inserted into a dimerized immunoglobulin Fc domain (Fig. 66-8). AMG 531 has a molecular weight of 60 kDa and a $t_{1/2}$ in humans of more than 100 hours.[123] It is cleared and recycled by endothelial FcRn receptors. *In vitro,* AMG 531 binds the TPO receptor and competes with TPO, activates the JAK2/STAT5 pathway, stimulates the growth of TPO-dependent cell lines, promotes the growth of Meg-CFC, and increases the ploidy and maturation of megakaryocytes.[124] When given to healthy human volunteers, it produces a dose-dependent increase in platelet count with no adverse effects.[123] AMG 531 is in clinical trials and has been shown to increase the platelet count in some patients with immune thrombocytopenic purpura

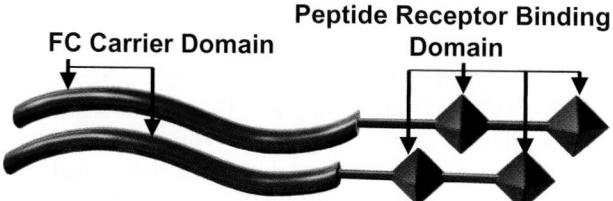

**Figure 66-8.** Structure of AMG 531. This "peptibody" is composed of a dimerized immunoglobulin Fc domain into which is inserted four identical peptides that bind c-mpl. None of the peptides has sequence homology with TPO. This 60-kDa protein is made in *E. coli* and is recycled by endothelial FcRn receptors to give a >100 hour half-life. Courtesy of Amgen, Inc., Thousand Oaks, CA.

(ITP) (see below).[125,126] To date, no antibody formation has been identified in animals or humans, despite repetitive subcutaneous administration.

Another approach has been to develop small nonpeptide TPO mimetics that are potentially orally available. This approach has been surprisingly effective in identifying substances that bind to the TPO receptor, in contrast to the unrewarding search for EPO small molecule mimetics. By using cell lines expressing the TPO receptor c-mpl, many small molecules have been identified that stimulate STAT5 phosphorylation. Families of hydrazinonaphthalene, azonaphthalene, semicarbazone, and naphtho[1,2-d]imidazole TPO mimetics have been described that possess low molecular weights (<600 Da) and EC$_{50}$ values of 1 to 20 nM.[122,127–130] Preclinical studies have demonstrated a stimulation of platelet production identical with TPO. One of these, SB-497115, has a molecular weight of 546 Da (Fig. 66-9), stimulates the growth of TPO-dependent cell lines, promotes the growth of human Meg-CFC and megakaryocytes in culture, and demonstrates marked species specificity in that it activates human and chimpanzee TPO receptors but not those of any other species.[131–133] SB-497115 increases the platelet count in healthy humans[134] and is in clinical trials.

Many of these TPO small molecule mimetics possess several distinct attributes. One attribute is that they bind the TPO receptor at a site distant from the binding site for TPO and appear to induce signal transduction by a mechanism different from recombinant TPO. It is not clear whether, like the TPO peptide mimetics, they need to dimerize the TPO receptor. By not competing with TPO for binding, they may be active in clinical settings in which TPO is ineffective.

A second attribute of these TPO small molecule mimetics is that they are highly species specific.[133] For example, nonpeptide TPO mimetics developed for the human TPO receptor bind to human and chimpanzee TPO receptors but do not bind the murine or cynomolgus monkey TPO receptor.[133] This has been shown to be due to a single amino acid difference in the transmembrane domain of the TPO recep-

**Figure 66-9.** Structure of SB-497115. This 546 D molecule binds to c-mpl at a site different from TPO and is a potent stimulator of platelet production in healthy subjects.[134] Courtesy of J. Jenkins, GlaxoSmithKline, Inc., Philadelphia, PA.

tor. Human and chimpanzee TPO receptors have a histidine at residue 499, whereas all of the other species have a leucine. A current model suggests that binding of nonpeptide TPO mimetics to His499 and Thr496 in the transmembrane region either induces dimerization of the TPO receptor or directly activates the signal transduction mechanism.[133] This species specificity has frustrated preclinical efforts to demonstrate biological effects *in vivo,* and therefore most proof of effect is based on stimulation of human cell lines or growth of Meg-CFC from human CD34$^+$ cells.[127–129]

Other important aspects of these nonpeptide TPO mimetics are that they are orally available and, given their structure, not antigenic. Furthermore, these nonpeptide mimetics do not activate platelets. Although the recombinant thrombopoietins and the TPO peptide mimetics do not directly activate platelets, they do alter the threshold for other agonists such as ADP. This potentiating effect is not observed with the nonpeptide TPO mimetics, probably due to the aforementioned difference in their mechanism of action.

In addition to the TPO mimetics, monoclonal antibodies binding to c-mpl have been developed.[135,136] The intact IgG monoclonal antibodies have relatively weak affinity for c-mpl but can be converted into "minibodies" composed of the V$_H$ and V$_L$ regions that have enhanced binding affinity.

These minibodies appear not to be antigenic. Unlike TPO, some can bind and activate the dysfunctional mutant c-mpl that causes some cases of congenital amegakaryocytic thrombocytopenia (see Chapter 54).

## VIII. Clinical Uses of Therapeutic Thrombopoietins: Animal Studies

The first-generation recombinant thrombopoietins (rhuTPO and PEG-rHuMGDF) have undergone extensive testing in animal models of chemotherapy, radiation, bone marrow transplantation, and HIV infection. Except for AMG 531, the second-generation thrombopoietins have been less extensively tested in animal models. This is largely due to the aforementioned finding that most of the TPO small molecule mimetics are so species specific that they work only in chimps and humans, a fact that has severely limited their preclinical development.

### A. Chemotherapy

When mice were treated with a combination of carboplatin and irradiation, they developed a life-threatening thrombocytopenia that resulted in the death of 95% of the animals. When the animals were treated with MGDF (the first 163 amino acids of TPO without pegylation or glycosylation) daily after the cytotoxic therapy, mortality was reduced to 15% and the animals experienced a reduction in the depth and duration of the thrombocytopenia. In addition, the severity and duration of the leukopenia and anemia were reduced.[33] In most similar animal models, the administration of rhuTPO also reduced the depth and duration of thrombocytopenia and demonstrated multilineage recovery,[53,137] as suggested by *in vitro* data showing that TPO stimulates the growth of progenitor cells of most lineages.

### B. Radiotherapy

Nonhuman primates were treated with 700 cGy $^{60}$Co gamma, total body irradiation, and then daily with PEG-rHuMGDF, MGDF, G-CSF, a combination of PEG-rHuMGDF and G-CSF, or placebo. Treatment with MGDF, PEG-rHuMGDF, or PEG-rHuMGDF plus G-CSF significantly decreased the duration of thrombocytopenia (platelet count $<20 \times 10^9$/L for 0.25, 0, or 0.5 days, respectively) and the severity of the platelet nadir (28, 43, and $30 \times 10^9$/L, respectively) compared to those of the controls (12.2 days' duration; nadir, $4 \times 10^9$/L) and elicited an earlier platelet recovery. Neutrophil regeneration was augmented in all cytokine protocols, and

the combination of PEG-rMGDF and r-metHuG-CSF (filgrastim) further decreased the duration of neutropenia compared with r-metHuG-CSF alone.[138]

The appropriate scheduling of TPO administration has been studied in animal models of radiotherapy with or without chemotherapy. When mice were given a high dose (950 rad) of total body irradiation (TBI), 83% of them died. However, if mice were given recombinant murine TPO (rMuTPO) 2 hours before the irradiation, only 25% died. If mice were given 500 rad of TBI plus a dose of carboplatin, the administration of rMuTPO from 2 hours before to 4 hours, but not 24 hours, after TBI markedly increased the number of platelets, red cells, and white cells 10 days later.[139–141] Given the antiapoptotic effect of TPO on pluripotential stem cells,[53,142] these studies suggest that there may be a narrow window of time during which administration of TPO may reverse the apoptotic effects on stem cells of irradiation and possibly chemotherapy.

### C. Bone Marrow Transplantation

In several animal models, recombinant TPO was administered after bone marrow infusion with no significant effect on the recovery of any lineage. These unexpected results led to further investigations in mice in which the donor animals were treated with recombinant TPO and the TPO-stimulated marrow was then transplanted into recipient animals. This produced a reduction in the duration of thrombocytopenia and earlier recovery of erythrocytes as well.[143]

### D. Ex Vivo *Expansion of Bone Marrow Stem Cells*

Yagi and colleagues[144] demonstrated that administration of TPO alone can sustain *ex vivo* expansion of hematopoietic stem cells for 4 or 5 months in long-term bone marrow cultures (LTBMCs) from mice. In this study, the continuous presence of TPO resulted in the generation of long- and short-term colony-forming cells (CFCs) and maintained the relative amount of high-proliferative-potential CFCs. Most important, competitive repopulation studies using LTBMCs found that the TPO-treated LTBMC cells were as effective as fresh marrow. Subsequent data obtained by this research group suggest that the expanded population of stem cells, when transplanted into recipient mice, is adequate for long-term repopulation.

### E. HIV Infection

The thrombocytopenia of HIV-infected primates could be rapidly reversed by the infusion of PEG-rHuMGDF.[145]

# IX. Clinical Uses of Therapeutic Thrombopoietins: Human Studies

## A. Chemotherapy for Solid Tumors

Approximately 8% of the U.S. platelet supply is used in supporting patients undergoing chemotherapy for solid tumors.[146] In studies of the administration of recombinant thrombopoietins to humans following chemotherapy, platelet recovery was enhanced but, unlike the animal models, there was no change in the recovery of white or red blood cells. In one study, 53 lung cancer patients were treated with carboplatin and paclitaxel.[147] Thirty-eight patients received doses of PEG-rHuMGDF ranging from 0.03 to 5 μg/kg daily after chemotherapy and 15 patients received placebo. Those patients who received PEG-rHuMGDF had a significantly higher platelet nadir than those who received placebo (188 × 10$^9$/L vs. 111 × 10$^9$/L), and their platelet counts returned to baseline in 14 days compared to more than 21 days in the placebo-treated group. In a second study, 41 cancer patients were treated with carboplatin and cyclophosphamide followed by the administration of either PEG-rHuMGDF or placebo.[148] There was no effect on the platelet nadir, but in patients who received PEG-rHuMGDF platelet count returned to baseline sooner than in patients who received placebo (17 vs. 22 days, $p = 0.014$).

The relevance of these studies is the demonstration that TPO has activity in the chemotherapy setting. These studies were not intended to demonstrate a decrease in platelet transfusions. Indeed, few of the patients in these two studies received transfusions since the dose intensity of the chemotherapy was not high enough. However, using a dose-intense chemotherapy regimen to treat ovarian cancer patients, Vadhan-Raj and colleagues demonstrated that rhuTPO elevated nadir platelet counts, shortened the duration of thrombocytopenia, and reduced platelet transfusions from 75% to 25% (Fig. 66-10).[149–151]

An important principle has emerged from several dose-finding studies with rhuTPO and PEG-rHuMGDF in the chemotherapy setting: A single dose of recombinant TPO is as effective as multiple daily doses given after chemotherapy. The former regimen ameliorates the rebound thrombocytosis that is routinely seen with the latter regimen.

## B. Chemotherapy for Acute Leukemia

Approximately 16% of the total U.S. platelet supply is used in the treatment of acute leukemia.[146] Unfortunately, the administration of TPO following chemotherapy for acute leukemia has provided disappointing results. Several trials found no effect on platelet recovery to 20 × 10$^9$/L, platelet transfusions, or remission rate when PEG-rHuMGDF was administered after standard induction regimens or during

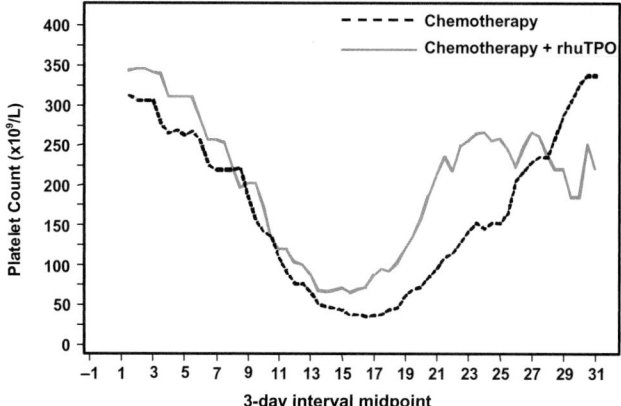

**Figure 66-10.** Recombinant human TPO (rhuTPO) reduces the platelet nadir in patients undergoing dose-intense chemotherapy. Patients undergoing chemotherapy for gynecological malignancy were treated with rhuTPO or placebo on days 2, 4, 6, and 8 after chemotherapy and the platelet count response followed.[149] Unpublished data courtesy of Pharmacia Corp., Peapack, NJ.

consolidation.[152,153] The reason for this failure is not entirely clear. It may relate to the absence of target marrow progenitors upon which to act, high endogenous TPO levels, or an inappropriate administration scheme. However, attempts to increase the TPO dose and alter the dosing scheme have not been successful.

## C. Stem Cell Transplantation

Although comprising a small number of patients, stem cell transplantation patients, especially those patients who have failed or delayed engraftment of platelets, consume a disproportionately large amount (7%) of the national supply of platelets.[146] In multiple studies with PEG-rHuMGDF or rhuTPO, there has been no effect on the time to platelet transfusion independence, platelet count >20 × 10$^9$/L, or platelet transfusions.[154–157] All that was seen was a dose-dependent rebound in the posttransplant platelet count. Furthermore, in patients who failed to engraft after 30 days, administration of rhuTPO showed no significant benefit.[158] These results parallel those seen in the murine model discussed previously,[143] in which it was found that TPO was more effective in treating the stem cell donor than in treating the stem cell recipient.

The ultimate role of TPO in stem cell transplantation may be to stimulate the number of progenitor cells in the marrow or peripheral stem cell population prior to harvesting. TPO appears to be effective in mobilizing peripheral blood progenitor cells. When combined with chemotherapy and filgrastim in oncology patients, PEG-rHuMGDF produced a 250-fold increase in the number of circulating Meg-CFC, a 190-fold increase in GM-CFC, a 65-fold increase in

E-BFC, and a 24-fold increase in CD34+ cells compared to patients receiving only chemotherapy and filgrastim.[148] Compared to chemotherapy mobilization with G-CSF alone, rhuTPO plus G-CSF required fewer aphereses to attain the target CD34+ cell count (one vs. three, $p < 0.001$) and gave a higher number of CD34+ cells ($4.1 \times 10^6$ vs. $0.8 \times 10^6$/kg, $p < 0.0003$). Upon stem cell transplantation, those patients who received cells mobilized with G-CSF plus TPO had a slightly earlier recovery of absolute neutrophil count (8 vs. 9 days, $p < 0.001$) and platelets (9 vs. 10 days, $p < 0.07$). These patients also received fewer red cell (three vs. four, $p < 0.02$) and platelet (four vs. five, $p < 0.02$) transfusions than did those patients who received cells mobilized only with G-CSF.[159]

### D. Acute Treatment of Thrombocytopenia

Despite much early enthusiasm for using TPO in the acute treatment of thrombocytopenia,[89] TPO will not replace platelet transfusions in this situation. TPO does not hasten megakaryocyte fragmentation into platelets but acts on earlier progenitors and takes 5 days to begin to increase the platelet count.

### E. Other Hematological Disorders with Thrombocytopenia

Thrombocytopenia is a common problem in many chronic hematological conditions ranging from aplastic anemia to myelodysplasia and drug-induced thrombocytopenia. Although only a few human studies have been reported, it is anticipated that if adequate amounts of responsive precursor cells are present, TPO may be helpful in these patients.

**1. HIV-Associated Thrombocytopenia.** Harker et al.[160] administered PEG-rHuMGDF to six patients with thrombocytopenia due to HIV infection; two patients received placebo injections. With an average pretreatment platelet count of $46 \times 10^9$/L, all six PEG-rHuMGDF-treated patients normalized their platelet counts to a median of $456 \times 10^9$/L; the placebo-injected patients had no change in platelet count.[160]

**2. Immune Thrombocytopenic Purpura.** Since TPO levels in ITP patients are normal[115,116,161–164] and since some ITP patients might not be producing platelets at a maximal rate,[165–171] TPO may be of potential benefit in this setting.

One patient with a form of cyclic thrombocytopenia with platelet autoantibodies has been treated twice weekly for more than 7 years with subcutaneous PEG-rHuMGDF.[172] Without treatment the platelet count was $5 \times 10^9$/L, and with

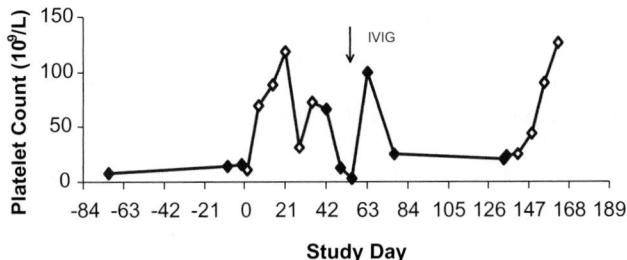

**Figure 66-11.** Platelet count response to AMG 531. A woman with ITP for 4 years and a chronic platelet count of 15 to 30 × $10^9$/L was given six doses of AMG 531 on the days indicated by the open symbols. Her platelet count increased. Upon discontinuation of therapy, her platelet count decreased to <5 × $10^9$/L and she was given intravenous immunoglobulin (IVIG) rescue therapy. She returned to her baseline platelet count. Starting on day 140, the patient was given titrated doses of AMG 531 weekly (again indicated by the open symbols) to maintain a normal platelet count and was able to discontinue her long-standing corticosteroids. She remained on weekly AMG 531 therapy 1 year later with a normal platelet count. Solid symbols indicate the platelet count on days when no AMG 531 was given.

PEG-rHuMGDF the platelet count cycled between $50 \times 10^9$ and $100 \times 10^9$/L. Four Japanese patients with refractory ITP have been treated with PEG-rHuMGDF and three demonstrated a marked improvement in platelet count.

In a phase I trial, 24 patients with chronic ITP were treated in groups of 4 with six different single doses of the novel TPO mimetic AMG 531. At concentrations ≥3 µg/kg, 8/12 patients (67%) achieved a platelet response (defined as doubling from baseline and an increase above $50 \times 10^9$/L).[125] Five of the 8 responders (63%) achieved platelet counts >150 × $10^9$/L. A subsequent phase II randomized, placebo-controlled, double-blind study of weekly injections of AMG 531 versus placebo for 6 weeks showed that 75% of the 16 AMG 531 patients versus 25% of the 4 placebo-treated patients had their platelet count double from baseline and increase above $50 \times 10^9$/L.[126] An illustrative platelet count response to AMG 531 is shown in Fig. 66-11. There are several ongoing studies with AMG 531, as well as with SB-497115, in ITP patients.

**3. Thrombocytopenia with Absent Radii Syndrome.** The thrombocytopenia in patients with the thrombocytopenia with absent radii syndrome is due to a failure of cell signaling events downstream from the TPO receptor (see Chapter 54), making it unlikely that exogenous TPO will be useful.[173]

**4. Thrombocytopenia in Liver Disease.** The thrombocytopenia in chronic liver disease may be responsive to TPO. Although it has long been assumed that the thrombocytopenia in liver disease is due to sequestration of platelets in the spleen,[174] the fact that most TPO is produced in the liver suggests that insufficient TPO production might be an additional cause. Measurements of TPO levels in this patient group confirm that the levels are inappropriately low[75] and that TPO levels and the platelet count return to normal after liver transplantation, despite no significant change in spleen size.[74]

**5. Myelodysplastic Syndromes and Aplastic Anemia.** Bone marrow from only some patients with myelodysplastic syndromes and aplastic anemia can be stimulated in vitro to form megakaryocytes, but this suggests that these patients might benefit from TPO administration.[175,176] In a pilot study of 20 thrombocytopenic Japanese patients with myelodysplastic syndrome, daily intravenous administration of PEG-rHuMGDF for 1 week increased the median platelet count from $15.7 \pm 11.0 \times 10^9/L$ at baseline to $30.8 \pm 20.5 \times 10^9/L$ by week 5.[177] Additionally, some patients experienced multilineage responses. A similar protocol in 16 Japanese aplastic anemia patients showed a minor effect on the platelet count ($10.8 \pm 6.3 \times 10^9/L$ at baseline, rising to $19.5 \pm 15.4 \times 10^9/L$ by week 5).

### F. Surgical Disorders with Thrombocytopenia

Patients undergoing cardiovascular operations, liver transplantation, and other major surgery consume approximately 40% of all platelets transfused in the United States,[146] and this significant need for platelet transfusions might be ameliorated by TPO administration. No clinical studies have been undertaken, but in one animal model of cardiac surgery precisely timed administration of PEG-rHuMGDF prior to bypass reduced bleeding, improved platelet function, and reduced thrombocytopenia.[178] Given the approximately 5-day lead time for TPO to start to increase platelet production, these animal studies suggest that similar precise timing of TPO prior to cardiac surgery or other major surgery may be beneficial. However, there remains the concern, discussed later, that administration of TPO to patients with active vascular disease might exacerbate their disease.

### G. Transfusion Medicine

**1. *Ex Vivo* Expansion of Peripheral Blood or Cord Blood Progenitor Cells.** TPO in combination with other hematopoietic growth factors, such as flt-3 ligand and IL-3, may be used to expand cord blood or peripheral blood progenitors *ex vivo*. A combination of flt-3 ligand and TPO can expand cord blood progenitors several hundred thousand-fold over 25 weeks in culture.[179] Whether these expanded progenitor populations will provide clinical benefit after transplantation is unclear. The murine studies

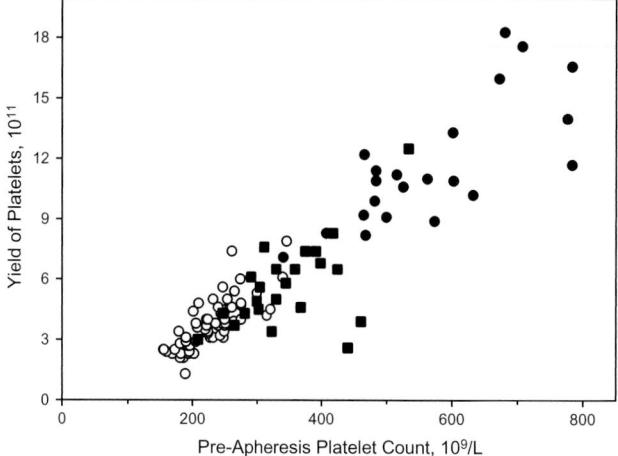

**Figure 66-12.** Recombinant TPO increases the platelet number and platelet apheresis yield. Healthy apheresis donors were treated with 1 (■) or 3 μg/kg (●) PEG-rHuMGDF or placebo (○) on day 1 and subjected to a standardized apheresis on day 15.[58] The apheresis yield is compared with the platelet count on day 15.

previously described suggest that this approach might be possible.[144]

**2. *Ex Vivo* Platelet Production.** Platelets can be produced *in vitro* under the stimulation of TPO. Although probably not an economical source of platelets, CD34+ cells can be grown such that most of the cells become megakaryocytes and in turn shed platelets. These shed platelets have normal ultrastructure and function compared with their *in vivo* counterparts.[45]

**3. Increasing Platelet Apheresis Yields.** TPO stimulates platelet production in normal apheresis donors and increases the apheresis yield (Fig. 66-12).[58,180] Treatment of apheresis donors with a single dose of PEG-rHuMGDF on day 1 produces a dose-dependent increase in the platelet count that peaks after 12 to 15 days. Compared with placebo-treated donors who had platelet counts of $225 \times 10^9/L$, donors treated with 1 or 3 μg/kg of PEG-rHuMGDF had median platelet counts of $336 \times 10^9$ and $599 \times 10^9/L$, respectively. There was a direct relationship between the platelet count and the yield at apheresis, with the placebo-treated donors providing $3.7 \times 10^{11}$ platelets and the 1 and 3 μg/kg donors providing $5.6 \times 10^{11}$ and $11 \times 10^{11}$ platelets, respectively. The donors had no adverse effects. The PEG-rHuMGDF apheresis product had normal platelet aggregation responses *in vitro* and when transfused into thrombocytopenic recipients gave a dose-dependent increase in platelet count. The absolute platelet increments 1 to 4 hours after transfusion were $19 \times 10^9$, $41 \times 10^9$, and $82 \times 10^9/L$ for the platelets obtained from donors treated with placebo or 1 or 3 μg/kg PEG-rHuMGDF, respectively. There was a

statistically significant increase in the corrected count increment for both PEG-rHuMGDF platelet products ($14 \times 10^9$/L
for the combined 1 and 3 µg/kg groups) versus the placebo
product ($10 \times 10^9$/L).

In a similar study, rhuTPO was given to cancer patients
before undergoing chemotherapy. Large amounts of platelets were harvested by apheresis, cryopreserved, and successfully transfused back into the donors when they
subsequently developed chemotherapy-induced thrombocytopenia. This form of autologous platelet donation may
prove to be an important method to support platelet-
refractory, alloimmunized patients undergoing dose-intense
chemotherapy.[181]

# X.  Adverse Effects of Thrombopoietins

Except for two problems (antibody formation and bone
marrow fibrosis), the therapeutic thrombopoietins have been
remarkably free of significant side effects in clinical studies.
There have been no acute phase reaction responses with the
thrombopoietins, as there were with the interleukins.[12,182]
This lack of adverse events probably indicates the safety of
the thrombopoietins, but it may also reflect the carefully
selected patient populations studied. What follows is a
review of the potential adverse events of TPO administration, as well as an analysis of the two situations (antibody
formation and bone marrow fibrosis) in which problems
have been identified.

## A. Thrombocytosis and Thrombosis

In none of the closely followed animal or human studies
with the thrombopoietins has there been any evidence
for increased thrombotic events, despite platelet counts as
high as $3000 \times 10^9$/L. However, all patients in trials have
been screened to exclude a history of active or prior
thrombotic or cardiovascular disease. In a more representative general medical population with cardiovascular
disease, several prothrombotic attributes of TPO deserve
attention.

First, recombinant thrombopoietins are very potent
growth factors and can markedly elevate the platelet count
in a short period of time. The deposition of platelets in an
extravascular shunt in baboons was directly related to the
platelet count after PEG-rHuMGDF administration.[56,57]
Since the extravascular shunt mimics an ulcerated atheroma
in humans, these results show that, except for its ability to
elevate the platelet count, PEG-rHuMGDF does not synergize with or exacerbate platelet deposition. Nonetheless,
increasing the platelet count in individuals with active
arterial thrombotic disease may exacerbate the vascular
disease.

Second, as mentioned previously, TPO can decrease by
50% the threshold for platelet activation by various agonists
(e.g., ADP and collagen).[56,57]

Finally, TPO stimulates the production of young platelets
("platelet tide") and these peak in the circulation 4 or 5 days
after administration of PEG-rHuMGDF to normal baboons
or humans.[56,57] These younger platelets have a lower threshold for agonists and are more active in *in vitro* platelet
aggregation experiments, but these effects have not resulted
in increased clinical thrombosis in either normal animals or
humans.

## B. Tumor Growth

Since many hematopoietic malignancies express the TPO
receptor c-mpl,[183] caution must be exercised in using TPO
in this patient population. However, in the acute leukemia
studies discussed previously, administration of PEG-
rHuMGDF did not result in any acceleration of leukemic
growth or increase the relapse rate. There is little likelihood
that TPO will stimulate nonhematopoietic tumor growth
since c-mpl has not been detected on nonhematological solid
tumors.[184]

## C. Bone Marrow Fibrosis

One problem with recombinant TPO administration in
some patients has been increased bone marrow fibrosis. In
preclinical studies in animals, long-term administration
of PEG-rHuMGDF produced reversible bone marrow
fibrosis.[185] Indeed, overexpression of TPO by transplantation of c-mpl-transfected murine bone marrow cells[97] or by
c-mpl retroviral infection in mice[98] creates extensive
bone marrow fibrosis, osteosclerosis, and extramedullary
hematopoiesis, like the human disease agnogenic myelofibrosis with myeloid metaplasia.[96,186,187] In humans given
rhuTPO and PEG-rHuMGDF, there has been little clinical
evidence for bone marrow fibrosis. However, given the
brevity of most exposures to recombinant TPO and the lack
of bone marrow analysis, this problem has not been fully
explored.

In the only study in which bone marrow analysis was
performed, increased reticulin was seen in most patients
treated with rhuTPO. Serial analyses of bone marrow and
peripheral blood were conducted in 9 patients who received
rhuTPO after induction therapy for acute myeloid leukemia
(AML) and in 8 patients undergoing the same AML induction treatment but without rhuTPO treatment.[188] Eight of the
9 TPO-treated and 5 of the 8 control patients had increased
bone marrow cellularity; 8 of 9 treated and 2 of 6 untreated
patients had increased bone marrow reticulin staining. A
semiquantitative measurement of the number of bone

marrow megakaryocytes showed that treated patients had 19.5 megakaryocytes per high power field (MHPF) versus 3.7 MHPF for the AML controls and 2.95 MHPF for patients without any bone marrow disease. All of these morphological findings resolved within 42 days of the last dose of rhuTPO.

### D. Antibody Formation

A more important complication of TPO treatment has emerged. Administration of multiple doses of one recombinant TPO, PEG-rHuMGDF, to some cancer patients and healthy volunteers was associated with an abrogation of its pharmacologic effect as a result of the development of neutralizing antibodies.[22,189–191] These antibodies neutralized both the recombinant and the endogenous TPO, resulting in thrombocytopenia. Thrombocytopenia occurred in 4 of 665 (0.6%) cancer/stem cell transplantation/leukemia patients given multiple doses, 2 of 210 (1.2%) healthy volunteers who received two doses, and 11 of 124 (8.9%) healthy volunteers given three doses of PEG-rHuMGDF.[22,190] No subject developed neutralizing antibodies or thrombocytopenia after a single injection. Evaluation of these thrombocytopenic subjects showed that the thrombocytopenia was due to the formation of an IgG antibody to PEG-rHuMGDF that cross-reacted with endogenous TPO and neutralized its biologic activity.[22,190,191] Because endogenous TPO is produced in a constitutive fashion by the liver, megakaryocyte number and ploidy decrease and thrombocytopenia ensues (Fig. 66-6). In 3 patients, thrombocytopenia was also associated with anemia and neutropenia, suggesting an additional effect on a stem cell population.[22,191] Because of this unexpected effect, PEG-rHuMGDF was withdrawn from clinical trials in the United States in September 1998.[192]

The development of neutralizing antibodies in patients treated with intravenous rhuTPO has not been reported, although one nonneutralizing antibody was found after subcutaneous injection of rhuTPO.[149,151]

A possible explanation for the immunogenicity of PEG-rHuMGDF may simply be that this molecule is truncated, nonglycosylated, and pegylated, in contrast to the full-length, glycosylated, more "native" rhuTPO molecule. However, PEG-rHuMGDF has usually been administered subcutaneously, whereas full-length native rhuTPO has been injected intravenously. Because TPO is a potent mobilizer of dendritic cells, injection of any form of TPO subcutaneously might enhance its immunogenicity. Support for this latter hypothesis comes from experiments in which PEG-ratMGDF was injected into rats once monthly for 3 months by either a subcutaneous or an intravenous route. Most animals treated subcutaneously developed neutralizing antibodies and thrombocytopenia, whereas those treated intravenously did not.[92]

No antibody formation has been observed in the relatively small number of patients treated with AMG 531 and the nonpeptide TPO mimetics such as SB-497115. For AMG 531, the major adverse effect has been mild headache, a symptom reported with many hematopoietic growth factors.[125,126]

### E. Central Nervous System Injury

The finding in rats that TPO administration may increase apoptosis in ischemic neurons[55] has not been clinically tested or observed in human trials. All trials have excluded individuals with active thrombotic or cardiovascular disease.

## XI. Non-TPO Platelet Growth Factors

### A. Interleukin-11

Although interleukins stimulate thrombopoiesis, their action on platelets is not their principal physiologic function. Gene-targeting studies have shown that the primary physiologic function of IL-11 is to maintain female fertility; IL-11 is not essential for hematopoiesis either in normal physiology or in response to hematopoietic stress.[193,194] Furthermore, the pleiotropic effect of interleukins often results in unwanted or unacceptable toxic effects, including hyperbilirubinemia, rapid induction of anemia, fever, fatigue, chills, hypotension, and headache.[195–199] Although administration of IL-11 reduces the need for platelet transfusions by approximately one third in patients with severe chemotherapy-induced thrombocytopenia, it is associated with peripheral edema, dyspnea, conjunctival redness, and a low incidence of atrial arrhythmias and syncope.[13,182] Thus, despite the ability of interleukins to ameliorate thrombocytopenia in a subset of patients treated with conventional chemotherapy, the moderate toxicity encountered with interleukin treatment may interfere with its therapeutic effect and potential use as a thrombopoietic agent.

Oprelvekin (Neumega) is a recombinant IL-11 that is approved for the prevention of severe thrombocytopenia and also for the reduction of the need for platelet transfusions following myelosuppressive chemotherapy in adult patients with nonmyeloid malignancies who are at high risk of severe thrombocytopenia. Its efficacy was demonstrated in patients who had experienced severe thrombocytopenia following the previous chemotherapy cycle. Oprelvekin is not indicated following myeloablative chemotherapy. The safety and effectiveness of oprelvekin have not been established in pediatric patients.

Oprelvekin is contraindicated in patients with a history of hypersensitivity to oprelvekin or any component of the product. It is associated with significant toxicities that may limit its use, including the following:

- Allergic or hypersensitivity reactions, including anaphylaxis
- Fluid retention that can result in peripheral edema, dyspnea on exertion, pulmonary edema, capillary leak syndrome, and atrial arrhythmias
- Cardiovascular events, especially exacerbation of congestive heart failure due to fluid retention
- Atrial fibrillation or atrial flutter, which occurred in 15% of patients
- Sudden death
- Dilutional anemia; usual decreases in hematocrit of 10 to 15% due to increased plasma volume beginning within 3 to 5 days of starting treatment

Oprelvekin is administered to adults as a daily subcutaneous dose of 50 μg/kg starting 6 to 24 hours after the administration of chemotherapy. There is no established dose for children. All patients need to be warned of and monitored for the potential side effects discussed previously, especially fluid retention, dilutional anemia, and arrhythmias. Dosing is continued until the postnadir counts exceed $50 \times 10^9$/L and should not exceed a total of 21 days. Oprelvekin should be discontinued at least 2 days before the next chemotherapy cycle. Oprelvekin is available as vials containing 5 mg of lyophilized protein and is reconstituted with 1 mL of sterile water before injection.

### B. Chemokines

The chemokines SDF-1 and FGF-4 have demonstrated thrombopoietic activity in animals (see Section V.B).[87] As currently formulated, these molecules have too brief a circulatory half-life to be clinically useful. Development of longer acting congeners is awaited with interest since they work by a mechanism independent of TPO and would be expected to synergize with TPO.

# References

1. Kimura, H., Ishibashi, T., Shikama, Y., et al. (1990). Interleukin-1 beta (IL-1 beta) induces thrombocytosis in mice: Possible implication of IL-6. *Blood, 76,* 2493–2500.
2. Bartley, T. D., Bogenberger, J., Hunt, P., et al. (1994). Identification and cloning of a megakaryocyte growth and development factor that is a ligand for the cytokine receptor Mpl. *Cell, 77,* 1117–1124.
3. Lok, S., Kaushansky, K., Holly, R. D., et al. (1994). Cloning and expression of murine thrombopoietin cDNA and stimulation of platelet production *in vivo. Nature, 369,* 565–568.
4. de Sauvage, F. J., Hass, P. E., Spencer, S. D., et al. (1994). Stimulation of megakaryocytopoiesis and thrombopoiesis by the c-Mpl ligand. *Nature, 369,* 533–538.
5. Kuter, D. J., Beeler, D. L., & Rosenberg, R. D. (1994). The purification of megapoietin: A physiological regulator of megakaryocyte growth and platelet production. *Proc Natl Acad Sci USA, 91,* 11104–11108.
6. Kato, T., Ogami, K., Shimada, Y., et al. (1995). Purification and characterization of thrombopoietin. *J Biochem, 118,* 229–236.
7. Metcalf, D., Burgess, A. W., Johnson, G. R., et al. (1986). *In vitro* actions on hemopoietic cells of recombinant murine GM-CSF purified after production in *Escherichia coli*: Comparison with purified native GM-CSF. *J Cell Physiol, 128,* 421–431.
8. Debili, N., Masse, J. M., Katz, A., et al. (1993). Effects of the recombinant hematopoietic growth factors interleukin-3, interleukin-6, stem cell factor, and leukemia inhibitory factor on the megakaryocytic differentiation of CD34+ cells. *Blood, 82,* 84–95.
9. Leonard, J. P., Quinto, C. M., Kozitza, M. K., et al. (1994). Recombinant human interleukin-11 stimulates multilineage hematopoietic recovery in mice after a myelosuppressive regimen of sublethal irradiation and carboplatin. *Blood, 83,* 1499–1506.
10. Metcalf, D., Begley, C. G., Williamson, D. J., et al. (1987). Hemopoietic responses in mice injected with purified recombinant murine GM-CSF. *Exp Hematol, 15,* 1–9.
11. Metcalf, D., Begley, C. G., Johnson, G. R., et al. (1986). Effects of purified bacterially synthesized murine multi-CSF (IL-3) on hematopoiesis in normal adult mice. *Blood, 68,* 46–57.
12. Tepler, I., Elias, L., Smith, J. W., 2nd, et al. (1996). A randomized placebo-controlled trial of recombinant human interleukin-11 in cancer patients with severe thrombocytopenia due to chemotherapy. *Blood, 87,* 3607–3614.
13. Vredenburgh, J. J., Hussein, A., Fisher, D., et al. (1998). A randomized trial of recombinant human interleukin-11 following autologous bone marrow transplantation with peripheral blood progenitor cell support in patients with breast cancer. *Biol Blood Marrow Transplant, 4,* 134–141.
14. Kelemen, E., Cserhati, I., & Tanos, B. (1958). Demonstration and some properties of human thrombopoietin in thrombocythaemic sera. *Acta Haematol, 20,* 350–355.
15. Kelemen, E. (1970). Thrombopoietin. *Br Med J, 2,* 733–734.
16. Kelemen, E. (1995). Specific thrombopoietin cloned and sequenced — With personal retrospect and clinical prospects. *Leukemia, 9,* 1–2.
17. Hunt, P., Li, Y. S., Nichol, J. L., et al. (1995). Purification and biologic characterization of plasma-derived megakaryocyte growth and development factor. *Blood, 86,* 540–547.
18. Kuter, D. J., & Rosenberg, R. D. (1994). Appearance of a megakaryocyte growth-promoting activity, megapoietin, during acute thrombocytopenia in the rabbit. *Blood, 84,* 1464–1472.
19. Souyri, M., Vigon, I., Penciolelli, J. F., et al. (1990). A putative truncated cytokine receptor gene transduced by the

myeloproliferative leukemia virus immortalizes hematopoietic progenitors. *Cell, 63,* 1137–1147.

20. Vigon, I., Mornon, J. P., Cocault, L., et al. (1992). Molecular cloning and characterization of MPL, the human homolog of the v-mpl oncogene: Identification of a member of the hematopoietic growth factor receptor superfamily. *Proc Natl Acad Sci USA, 89,* 5640–5644.

21. Foster, D. C., Sprecher, C. A., Grant, F. J., et al. (1994). Human thrombopoietin: Gene structure, cDNA sequence, expression, and chromosomal localization. *Proc Natl Acad Sci USA, 91,* 13023–13027.

22. Li, J., Yang, C., Xia, Y., et al. (2001). Thrombocytopenia caused by the development of antibodies to thrombopoietin. *Blood, 98,* 3241–3248.

23. Hoffman, R. C., Andersen, H., Walker, K., et al. (1996). Peptide, disulfide, and glycosylation mapping of recombinant human thrombopoietin from Ser1 to Arg246. *Biochemistry, 35,* 14849–14861.

24. Kato, T., Oda, A., Inagaki, Y., et al. (1997). Thrombin cleaves recombinant human thrombopoietin: One of the proteolytic events that generates truncated forms of thrombopoietin. *Proc Natl Acad Sci USA, 94,* 4669–4674.

25. Gurney, A. L., Kuang, W. J., Xie, M. H., et al. (1995). Genomic structure, chromosomal localization, and conserved alternative splice forms of thrombopoietin. *Blood, 85,* 981–988.

26. Li, B., & Dai, W. (1995). Thrombopoietin and neurotrophins share a common domain. *Blood, 86,* 1643–1644.

27. Deane, C. M., Kroemer, R. T., & Richards, W. G. (1997). A structural model of the human thrombopoietin receptor complex. *J Mol Graph Model, 15,* 170–178, 185–188.

28. Wada, T., Nagata, Y., Nagahisa, H., et al. (1995). Characterization of the truncated thrombopoietin variants. *Biochem Biophys Res Commun, 213,* 1091–1098.

29. Jagerschmidt, A., Fleury, V., Anger-Leroy, M., et al. (1998). Human thrombopoietin structure–function relationships: Identification of functionally important residues. *Biochem J, 333,* 729–734.

30. Park, H., Park, S. S., Jin, E. H., et al. (1998). Identification of functionally important residues of human thrombopoietin. *J Biol Chem, 273,* 256–261.

31. Pearce, K. H., Jr., Potts, B. J., Presta, L. G., et al. (1997). Mutational analysis of thrombopoietin for identification of receptor and neutralizing antibody sites. *J Biol Chem, 272,* 20595–20602.

32. Feese, M. D., Tamada, T., Kato, Y., et al. (2004). Structure of the receptor-binding domain of human thrombopoietin determined by complexation with a neutralizing antibody fragment. *Proc Natl Acad Sci USA, 101,* 1816–1821.

33. Hokom, M. M., Lacey, D., Kinstler, O. B., et al. (1995). Pegylated megakaryocyte growth and development factor abrogates the lethal thrombocytopenia associated with carboplatin and irradiation in mice. *Blood, 86,* 4486–4492.

34. Foster, D., & Hunt, P. (1997). The biological significance of truncated and full-length forms of Mpl ligand. In D. J. Kuter, P. Hunt, W. Sheridan, et al. (Eds.), *Thrombopoiesis and thrombopoietins: Molecular, cellular, preclinical, and clinical biology* (pp. 203–214). Totowa, NJ: Humana Press.

35. Spivak, J. L., & Hogans, B. B. (1989). The *in vivo* metabolism of recombinant human erythropoietin in the rat. *Blood, 73,* 90–99.

36. Foster, D., & Lok, S. (1996). Biological roles for the second domain of thrombopoietin. *Stem Cells, 14*(Suppl. 1), 102–107.

37. Schnittger, S., de Sauvage, F. J., Le Paslier, D., et al. (1996). Refined chromosomal localization of the human thrombopoietin gene to 3q27–q28 and exclusion as the responsible gene for thrombocytosis in patients with rearrangements of 3q21 and 3q26. *Leukemia, 10,* 1891–1896.

38. Fan, Y., Gao, T., Ye, Y., et al. (1997). Cloning and sequencing of human thrombopoietin cDNA. *Chin J Biotechnol, 13,* 219–223.

39. Li, J., Xia, Y., & Kuter, D. J. (2000). The platelet thrombopoietin receptor number and function are markedly decreased in patients with essential thrombocythaemia. *Br J Haematol, 111,* 943–953.

40. Chang, M. S., McNinch, J., Basu, R., et al. (1995). Cloning and characterization of the human megakaryocyte growth and development factor (MGDF) gene. *J Biol Chem, 270,* 511–514.

41. Sasaki, Y., Takahashi, T., Miyazaki, H., et al. (1999). Production of thrombopoietin by human carcinomas and its novel isoforms. *Blood, 94,* 1952–1960.

42. Ghilardi, N., Wiestner, A., & Skoda, R. C. (1998). Thrombopoietin production is inhibited by a translational mechanism. *Blood, 92,* 4023–4030.

43. Zauli, G., Vitale, M., Falcieri, E., et al. (1997). *In vitro* senescence and apoptotic cell death of human megakaryocytes. *Blood, 90,* 2234–2243.

44. Li, J., Xia, Y., & Kuter, D. J. (1999). Interaction of thrombopoietin with the platelet c-mpl receptor in plasma: Binding, internalization, stability and pharmacokinetics. *Br J Haematol, 106,* 345–356.

45. Choi, E. S., Nichol, J. L., Hokom, M. M., et al. (1995). Platelets generated *in vitro* from proplatelet-displaying human megakaryocytes are functional. *Blood, 85,* 402–413.

46. Choi, E. S., Hokom, M. M., Chen, J. L., et al. (1996). The role of megakaryocyte growth and development factor in terminal stages of thrombopoiesis. *Br J Haematol, 95,* 227–233.

47. Matthews, D. J., Topping, R. S., Cass, R. T., et al. (1996). A sequential dimerization mechanism for erythropoietin receptor activation. *Proc Natl Acad Sci USA, 93,* 9471–9476.

48. Remy, I., Wilson, I. A., & Michnick, S. W. (1999). Erythropoietin receptor activation by a ligand-induced conformation change. *Science, 283,* 990–993.

49. Livnah, O., Stura, E. A., Middleton, S. A., et al. (1999). Crystallographic evidence for preformed dimers of erythropoietin receptor before ligand activation. *Science, 283,* 987–990.

50. Broudy, V. C., Lin, N. L., Sabath, D. F., et al. (1997). Human platelets display high-affinity receptors for thrombopoietin. *Blood, 89,* 1896–1904.

51. Fielder, P. J., Gurney, A. L., Stefanich, E., et al. (1996). Regulation of thrombopoietin levels by c-mpl-mediated binding to platelets. *Blood, 87,* 2154–2161.

52. Dahlen, D. D., Broudy, V. C., & Drachman, J. G. (2003). Internalization of the thrombopoietin receptor is regulated by 2 cytoplasmic motifs. *Blood, 102,* 102–108.

53. Kaushansky, K., Lin, N., Grossmann, A., et al. (1996). Thrombopoietin expands erythroid, granulocyte-macrophage, and megakaryocytic progenitor cells in normal and myelosuppressed mice. *Exp Hematol, 24,* 265–269.

54. Solar, G. P., Kerr, W. G., Zeigler, F. C., et al. (1998). Role of c-mpl in early hematopoiesis. *Blood, 92,* 4–10.

55. Ehrenreich, H., Hasselblatt, M., Knerlich, F., et al. (2005). A hematopoietic growth factor, thrombopoietin, has a proapoptotic role in the brain. *Proc Natl Acad Sci USA, 102,* 862–867.

56. Harker, L. A., Hunt, P., Marzec, U. M., et al. (1996). Regulation of platelet production and function by megakaryocyte growth and development factor in nonhuman primates. *Blood, 87,* 1833–1844.

57. Harker, L. A., Marzec, U. M., Hunt, P., et al. (1996). Dose–response effects of pegylated human megakaryocyte growth and development factor on platelet production and function in nonhuman primates. *Blood, 88,* 511–521.

58. Kuter, D. J., Goodnough, L. T., Romo, J., et al. (2001). Thrombopoietin therapy increases platelet yields in healthy platelet donors. *Blood, 98,* 1339–1345.

59. Basser, R. L., Rasko, J. E., Clarke, K., et al. (1996). Thrombopoietic effects of pegylated recombinant human megakaryocyte growth and development factor (PEG-rHuMGDF) in patients with advanced cancer. *Lancet, 348,* 1279–1281.

60. Tomita, D., Petrarca, M., Paine, T., et al. (1997). Effect of a single dose of pegylated human recombinant megakaryocyte growth and development factor (PEG-rHuMGDF) on platelet counts: Implications for platelet apheresis. *Transfusion, 37,* 2S.

61. Chen, J., Herceg-Harjacek, L., Groopman, J. E., et al. (1995). Regulation of platelet activation *in vitro* by the c-Mpl ligand, thrombopoietin. *Blood, 86,* 4054–4062.

62. Kubota, Y., Arai, T., Tanaka, T., et al. (1996). Thrombopoietin modulates platelet activation *in vitro* through protein-tyrosine phosphorylation. *Stem Cells, 14,* 439–444.

63. Montrucchio, G., Brizzi, M. F., Calosso, G., et al. (1996). Effects of recombinant human megakaryocyte growth and development factor on platelet activation. *Blood, 87,* 2762–2768.

64. Kroner, C., Eybrechts, K., & Akkerman, J. W. (2000). Dual regulation of platelet protein kinase B. *J Biol Chem, 275,* 27790–27798.

65. Snyder, E., Perrotta, P., Rinder, H., et al. (1999). Effect of recombinant human megakaryocyte growth and development factor coupled with polyethylene glycol on the platelet storage lesion. *Transfusion, 39,* 258–264.

66. Xia, Y., Li, J., Bertino, A., et al. (2000). Thrombopoietin and the TPO receptor during platelet storage. *Transfusion, 40,* 976–987.

67. Bertino, A. M., Qi, X. Q., Li, J., et al. (2003). Apoptotic markers are increased in platelets stored at 37 degrees C. *Transfusion, 43,* 857–866.

68. Nomura, S., Ogami, K., Kawamura, K., et al. (1997). Cellular localization of thrombopoietin mRNA in the liver by *in situ* hybridization. *Exp Hematol, 25,* 565–572.

69. Lok, S., & Foster, D. C. (1994). The structure, biology and potential therapeutic applications of recombinant thrombopoietin. *Stem Cells, 12,* 586–598.

70. McCarty, J. M., Sprugel, K. H., Fox, N. E., et al. (1995). Murine thrombopoietin mRNA levels are modulated by platelet count. *Blood, 86,* 3668–3675.

71. Guerriero, A., Worford, L., Holland, H. K., et al. (1997). Thrombopoietin is synthesized by bone marrow stromal cells. *Blood, 90,* 3444–3455.

72. Sungaran, R., Markovic, B., & Chong, B. H. (1997). Localization and regulation of thrombopoietin mRNa expression in human kidney, liver, bone marrow, and spleen using *in situ* hybridization. *Blood, 89,* 101–107.

73. Quin, S., Fu, F., Li, W., et al. (1998). Primary role of the liver in thrombopoietin production shown by tissue-specific knockout. *Blood, 92,* 2189–2191.

74. Peck-Radosavljevic, M., Wichlas, M., Zacherl, J., et al. (2000). Thrombopoietin induces rapid resolution of thrombocytopenia after orthotopic liver transplantation through increased platelet production. *Blood, 95,* 795–801.

75. Peck-Radosavljevic, M., Zacherl, J., Meng, Y. G., et al. (1997). Is inadequate thrombopoietin production a major cause of thrombocytopenia in cirrhosis of the liver? *J Hepatol, 27,* 127–131.

76. Yang, C., Li, Y. C., & Kuter, D. J. (1999). The physiological response of thrombopoietin (c-Mpl ligand) to thrombocytopenia in the rat. *Br J Haematol, 105,* 478–485.

77. Stoffel, R., Wiestner, A., & Skoda, R. C. (1996). Thrombopoietin in thrombocytopenic mice: Evidence against regulation at the mRNA level and for a direct regulatory role of platelets. *Blood, 87,* 567–573.

78. Sungaran, R., Chisholm, O. T., Markovic, B., et al. (2000). The role of platelet alpha-granular proteins in the regulation of thrombopoietin messenger RNA expression in human bone marrow stromal cells. *Blood, 95,* 3094–3101.

79. Solanilla, A., Dechanet, J., El Andaloussi, A., et al. (2000). CD40-ligand stimulates myelopoiesis by regulating flt3-ligand and thrombopoietin production in bone marrow stromal cells. *Blood, 95,* 3758–3764.

80. Sakamaki, S., Hirayama, Y., Matsunaga, T., et al. (1999). Transforming growth factor-beta1 (TGF-beta1) induces thrombopoietin from bone marrow stromal cells, which stimulates the expression of TGF-beta receptor on megakaryocytes and, in turn, renders them susceptible to suppression by TGF-beta itself with high specificity. *Blood, 94,* 1961–1970.

81. Alexander, W. S., Roberts, A. W., Nicola, N. A., et al. (1996). Deficiencies in progenitor cells of multiple hematopoietic lineages and defective megakaryocytopoiesis in mice lacking the thrombopoietic receptor c-Mpl. *Blood, 87,* 2162–2170.

82. Gurney, A. L., Carver-Moore, K., de Sauvage, F. J., et al. (1994). Thrombocytopenia in c-mpl-deficient mice. *Science, 265,* 1445–1447.

83. de Sauvage, F. J., Carver-Moore, K., Luoh, S. M., et al. (1996). Physiological regulation of early and late stages of

megakaryocytopoiesis by thrombopoietin. *J Exp Med, 183,* 651–656.

84. Carver-Moore, K., Broxmeyer, H. E., Luoh, S. M., et al. (1996). Low levels of erythroid and myeloid progenitors in thrombopoietin- and c-mpl-deficient mice. *Blood, 88,* 803–808.

85. Gainsford, T., Roberts, A. W., Kimura, S., et al. (1998). Cytokine production and function in c-mpl-deficient mice: No physiologic role for interleukin-3 in residual megakaryocyte and platelet production. *Blood, 91,* 2745–2752.

86. Gainsford, T., Nandurkar, H., Metcalf, D., et al. (2000). The residual megakaryocyte and platelet production in c-mpl-deficient mice is not dependent on the actions of interleukin-6, interleukin-11, or leukemia inhibitory factor. *Blood, 95,* 528–534.

87. Avecilla, S. T., Hattori, K., Heissig, B., et al. (2004). Chemokine-mediated interaction of hematopoietic progenitors with the bone marrow vascular niche is required for thrombopoiesis. *Nat Med, 10,* 64–71.

88. Kuter, D. J., & Rosenberg, R. D. (1995). The reciprocal relationship of thrombopoietin (c-Mpl ligand) to changes in the platelet mass during busulfan-induced thrombocytopenia in the rabbit. *Blood, 85,* 2720–2730.

89. Kuter, D. J. (1996). Thrombopoietin: Biology, clinical applications, role in the donor setting. *J Clin Apheresis, 11,* 149–159.

90. Kuter, D. J. (1996). The physiology of platelet production. *Stem Cells, 14,* 88–101.

91. Kuter, D. J. (1996). Thrombopoietin: Biology and clinical applications. *Oncologist, 1,* 98–106.

92. Kuter, D. J., & Begley, C. G. (2002). Recombinant human thrombopoietin: Basic biology and evaluation of clinical studies. *Blood, 100,* 3457–3469.

93. Scheding, S., Bergmann, M., Shimosaka, A., et al. (2002). Human plasma thrombopoietin levels are regulated by binding to platelet thrombopoietin receptors *in vivo. Transfusion, 42,* 321–327.

94. Li, J., Xia, Y., & Kuter, D. (1999). Interaction of thrombopoietin with the platelet c-mpl receptor in plasma: Binding, internalization, stability and pharmacodynamics. *Br J Haematol, 106,* 345–356.

95. Fielder, P. J., Hass, P., Nagel, M., et al. (1997). Human platelets as a model for the binding and degradation of thrombopoietin. *Blood, 89,* 2782–2788.

96. Yan, X. Q., Lacey, D., Fletcher, F., et al. (1995). Chronic exposure to retroviral vector encoded MGDF (mpl-ligand) induces lineage-specific growth and differentiation of megakaryocytes in mice. *Blood, 86,* 4025–4033.

97. Villeval, J. L., Cohen-Solal, K., Tulliez, M., et al. (1997). High thrombopoietin production by hematopoietic cells induces a fatal myeloproliferative syndrome in mice. *Blood, 90,* 4369–4383.

98. Frey, B. M., Rafii, S., Teterson, M., et al. (1998). Adenovector-mediated expression of human thrombopoietin cDNA in immune-compromised mice: Insights into the pathophysiology of osteomyelofibrosis. *J Immunol, 160,* 691–699.

99. Wiestner, A., Schlemper, R. J., van der Maas, A. P., et al. (1998). An activating splice donor mutation in the thrombopoietin gene causes hereditary thrombocythaemia. *Nature Genet, 18,* 49–52.

100. Kondo, T., Okabe, M., Sanada, M., et al. (1998). Familial essential thrombocythemia associated with one-base deletion in the 5′-untranslated region of the thrombopoietin gene. *Blood, 92,* 1091–1096.

101. Ghilardi, N., Wiestner, A., Kikuchi, M., et al. (1999). Hereditary thrombocythaemia in a Japanese family is caused by a novel point mutation in the thrombopoietin gene. *Br J Haematol, 107,* 310–316.

102. Allen, A. J., Gale, R. E., Harrison, C. N., et al. (2001). Lack of pathogenic mutations in the 5′-untranslated region of the thrombopoietin gene in patients with non-familial essential thrombocythaemia. *Eur J Haematol, 67,* 232–237.

103. Pinto, M. R., King, M. A., Goss, G. D., et al. (1985). Acute megakaryoblastic leukaemia with 3q inversion and elevated thrombopoietin (TSF): An autocrine role for TSF? *Br J Haematol, 61,* 687–694.

104. Bouscary, D., Fontenay-Roupie, M., Chretien, S., et al. (1995). Thrombopoietin is not responsible for the thrombocytosis observed in patients with acute myeloid leukemias and the 3q21q26 syndrome. *Br J Haematol, 91,* 425–427.

105. Ballmaier, M., Germeshausen, M., Schulze, H., et al. (2001). c-mpl mutations are the cause of congenital amegakaryocytic thrombocytopenia. *Blood, 97,* 139–146.

106. van den Oudenrijn, S., Bruin, M., Folman, C. C., et al. (2000). Mutations in the thrombopoietin receptor, Mpl, in children with congenital amegakaryocytic thrombocytopenia. *Br J Haematol, 110,* 441–448.

107. Wiestner, A., Padosch, S. A., Ghilardi, N., et al. (2000). Hereditary thrombocythaemia is a genetically heterogeneous disorder: Exclusion of TPO and MPL in two families with hereditary thrombocythaemia. *Br J Haematol, 110,* 104–109.

108. Ihara, K., Ishii, E., Eguchi, M., et al. (1999). Identification of mutations in the c-mpl gene in congenital amegakaryocytic thrombocytopenia. *Proc Natl Acad Sci USA, 96,* 3132–3136.

109. Heller, P. G., Glembotsky, A. C., Gandhi, M. J., et al. (2005). Low Mpl receptor expression in a pedigree with familial platelet disorder with predisposition to acute myelogenous leukemia and a novel AML1 mutation. *Blood, 105,* 4664–4670.

110. Abe, M., Suzuki, K., Inagaki, O., et al. (2002). A novel MPL point mutation resulting in thrombopoietin-independent activation. *Leukemia, 16,* 1500–1506.

111. Ding, J., Komatsu, H., Wakita, A., et al. (2004). Familial essential thrombocythemia associated with a dominant-positive activating mutation of the c-MPL gene, which encodes for the receptor for thrombopoietin. *Blood, 103,* 4198–4200.

112. Moliterno, A. R., Williams, D. M., Gutierrez-Alamillo, L. I., et al. (2004). Mpl Baltimore: A thrombopoietin receptor polymorphism associated with thrombocytosis. *Proc Natl Acad Sci USA, 101,* 11444–11447.

113. Tonelli, R., Strippoli, P., Grossi, A., et al. (2000). Hereditary thrombocytopenia due to reduced platelet production — Report on two families and mutational screening of the

thrombopoietin receptor gene (c-mpl). *Thromb Haemost, 83,* 931–936.

114. Vonderheide, R. H., Thadhani, R., & Kuter, D. J. (1998). Association of thrombocytopenia with the use of intra-aortic balloon pumps. *Am J Med, 105,* 27–32.

115. Emmons, R. V., Reid, D. M., Cohen, R. L., et al. (1996). Human thrombopoietin levels are high when thrombocytopenia is due to megakaryocyte deficiency and low when due to increased platelet destruction. *Blood, 87,* 4068–4071.

116. Nichol, J. L. (1998). Thrombopoietin levels after chemotherapy and in naturally occurring human diseases. *Curr Opin Hematol, 5,* 203–208.

117. Horikawa, Y., Matsumura, I., Hashimoto, K., et al. (1997). Markedly reduced expression of platelet c-mpl receptor in essential thrombocythemia. *Blood, 90,* 4031–4038.

118. Giri, J. G., Smith, W. G., Kahn, L. E., et al. (1997). Promegapoietin, a chimeric growth factor for megakaryocyte and platelet restoration. *Blood, 90,* 580a.

119. Cwirla, S. E., Balasubramanian, P., Duffin, D. J., et al. (1997). Peptide agonist of the thrombopoietin receptor as potent as the natural cytokine. *Science, 276,* 1696–1699.

120. de Serres, M., Ellis, B., Dillberger, J. E., et al. (1999). Immunogenicity of thrombopoietin mimetic peptide GW395058 in BALB/c mice and New Zealand white rabbits: Evaluation of the potential for thrombopoietin neutralizing antibody production in man. *Stem Cells, 17,* 203–209.

121. Case, B. C., Hauck, M. L., Yeager, R. L., et al. (2000). The pharmacokinetics and pharmacodynamics of GW395058, a peptide agonist of the thrombopoietin receptor, in the dog, a large-animal model of chemotherapy-induced thrombocytopenia. *Stem Cells, 18,* 360–365.

122. Erickson-Miller, C. L., Delorme, E., Tian, S. S., et al. (2000). Discovery and characterization of a selective, non-peptidyl thrombopoietin receptor agonist. *Blood, 96,* 675a.

123. Wang, B., Nichol, J. L., & Sullivan, J. T. (2004). Pharmacodynamics and pharmacokinetics of AMG 531, a novel thrombopoietin receptor ligand. *Clin Pharmacol Ther, 76,* 628–638.

124. Broudy, V. C., & Lin, N. L. (2004). AMG531 stimulates megakaryopoiesis *in vitro* by binding to Mpl. *Cytokine, 25,* 52–60.

125. Bussel, J. B., George, J. N., Kuter, D. J., et al. (2003). An open-label, dose-finding study evaluating the safety and platelet response of a novel thrombopoietic protein (AMG 531) in thrombocytopenic adult patients with immune thrombocytopenic purpura (ITP). *Blood, 102,* 234b.

126. Kuter, D. J., Bussel, J., Aledort, L., et al. (2004). A phase 2 placebo controlled study evaluating the platelet response and safety of weekly dosing with a novel thrombopoietic protein (AMG 531) in thrombocytopenic adult patients with immune thrombocytopenic purpura. *Blood, 104,* 148a.

127. Duffy, K. J., Darcy, M. G., Delorme, E., et al. (2001). Hydrazinonaphthalene and azonaphthalene thrombopoietin mimics are nonpeptidyl promoters of megakaryocytopoiesis. *J Med Chem, 44,* 3730–3745.

128. Duffy, K. J., Shaw, A. N., Delorme, E., et al. (2002). Identification of a pharmacophore for thrombopoietic activity of small, non-peptidyl molecules: 1. Discovery and optimization of salicylaldehyde thiosemicarbazone thrombopoietin mimics. *J Med Chem, 45,* 3573–3575.

129. Duffy, K. J., Price, A. T., Delorme, E., et al. (2002). Identification of a pharmacophore for thrombopoietic activity of small, non-peptidyl molecules: 2. Rational design of naphtho[1,2-d]imidazole thrombopoietin mimics. *J Med Chem, 45,* 3576–3578.

130. Erickson-Miller, C. L., DeLorme, E., Tian, S. S., et al. (2005). Discovery and characterization of a selective, non-peptidyl thrombopoietin receptor agonist. *Exp Hematol, 33,* 85–93.

131. Luengo, J. I., Duffy, K. J., Shaw, A. N., et al. (2004). Discovery of SB-497115, a small-molecule thrombopoietin (TPO) receptor agonist for the treatment of thrombocytopenia. *Blood, 104,* 795a.

132. Erickson-Miller, C., Delorme, E., Giampa, L., et al. (2004). Biological activity and selectivity for Tpo receptor of the orally bioavailable, small molecule Tpo receptor agonist, SB-497115. *Blood, 104,* 796a.

133. Erickson-Miller, C., Delorme, E., Iskander, M., et al. (2004). Species specificity and receptor domain interaction of a small molecule TPO receptor agonist. *Blood, 104,* 795a.

134. Jenkins, J., Nicholl, R., Williams, D., et al. (2004). An oral, non-peptide, small molecule thrombopoietin receptor agonist increases platelet counts in healthy subjects. *Blood, 104,* 797a.

135. Deng, B., Banu, N., Malloy, B., et al. (1998). An agonist murine monoclonal antibody to the human c-Mpl receptor stimulates megakaryocytopoiesis. *Blood, 92,* 1981–1988.

136. Orita, T., Tsunoda, H., Yabuta, N., et al. (2005). A novel therapeutic approach for thrombocytopenia by minibody agonist of the thrombopoietin receptor. *Blood, 105,* 562–566.

137. Ulich, T. R., del Castillo, J., Yin, S., et al. (1995). Megakaryocyte growth and development factor ameliorates carboplatin-induced thrombocytopenia in mice. *Blood, 86,* 971–976.

138. Farese, A. M., Hunt, P., Grab, L. B., et al. (1996). Combined administration of recombinant human megakaryocyte growth and development factor and granulocyte colony-stimulating factor enhances multilineage hematopoietic reconstitution in nonhuman primates after radiation-induced marrow aplasia. *J Clin Invest, 97,* 2145–2151.

139. Stefanich, E. G., Carlson-Zermeno, C. C., McEvoy, K., et al. (2001). Dose schedule of recombinant murine thrombopoietin prior to myelosuppressive and myeloablative therapy in mice. *Cancer Chemother Pharmacol, 47,* 70–77.

140. Neelis, K. J., Visser, T. P., Dimjati, W., et al. (1998). A single dose of thrombopoietin shortly after myelosuppressive total body irradiation prevents pancytopenia in mice by promoting short-term multilineage spleen-repopulating cells at the transient expense of bone marrow-repopulating cells. *Blood, 92,* 1586–1597.

141. Mouthon, M. A., Van der Meeren, A., Gaugler, M. H., et al. (1999). Thrombopoietin promotes hematopoietic recovery and survival after high-dose whole body irradiation. *Int J Radiat Oncol Biol Phys, 43,* 867–875.

142. Kaushansky, K. (1997). Thrombopoietin: More than a lineage-specific megakaryocyte growth factor. *Stem Cells, 15,* 97–103.

143. Fibbe, W. E., Heemskerk, D. P., Laterveer, L., et al. (1995). Accelerated reconstitution of platelets and erythrocytes after syngeneic transplantation of bone marrow cells derived from thrombopoietin pretreated donor mice. *Blood, 86,* 3308–3313.

144. Yagi, M., Ritchie, K. A., Sitnicka, E., et al. (1999). Sustained *ex vivo* expansion of hematopoietic stem cells mediated by thrombopoietin. *Proc Natl Acad Sci USA, 96,* 8126–8131.

145. Harker, L. A., Marzec, U. M., Novembre, F., et al. (1998). Treatment of thrombocytopenia in chimpanzees infected with human immunodeficiency virus by pegylated recombinant human megakaryocyte growth and development factor. *Blood, 91,* 4427–4433.

146. Kuter, D. J. (1998). The use of PEG-rhuMGDF in platelet apheresis. *Stem Cells, 16,* 231–242.

147. Fanucchi, M., Glaspy, J., Crawford, J., et al. (1997). Effects of polyethylene glycol-conjugated recombinant human megakaryocyte growth and development factor on platelet counts after chemotherapy for lung cancer. *N Engl J Med, 336,* 404–409.

148. Basser, R. L., Rasko, J. E., Clarke, K., et al. (1997). Randomized, blinded, placebo-controlled phase I trial of pegylated recombinant human megakaryocyte growth and development factor with filgrastim after dose-intensive chemotherapy in patients with advanced cancer. *Blood, 89,* 3118–3128.

149. Vadhan-Raj, S., Verschraegen, C. F., Bueso-Ramos, C., et al. (2000). Recombinant human thrombopoietin attenuates carboplatin-induced severe thrombocytopenia and the need for platelet transfusions in patients with gynecologic cancer. *Ann Intern Med, 132,* 364–368.

150. Vadhan-Raj, S. (1998). Recombinant human thrombopoietin: Clinical experience and *in vivo* biology. *Semin Hematol, 35,* 261–268.

151. Vadhan-Raj, S., Murray, L. J., Bueso-Ramos, C., et al. (1997). Stimulation of megakaryocyte and platelet production by a single dose of recombinant human thrombopoietin in patients with cancer. *Ann Int Med, 126,* 673–681.

152. Schiffer, C. A., Miller, K., Larson, R. A., et al. (2000). A double-blind, placebo-controlled trial of pegylated recombinant human megakaryocyte growth and development factor as an adjunct to induction and consolidation therapy for patients with acute myeloid leukemia. *Blood, 95,* 2530–2535.

153. Archimbaud, E., Ottmann, O. G., Yin, J. A., et al. (1999). A randomized, double-blind, placebo-controlled study with pegylated recombinant human megakaryocyte growth and development factor (PEG-rHuMGDF) as an adjunct to chemotherapy for adults with de novo acute myeloid leukemia. *Blood, 94,* 3694–3701.

154. Beveridge, R., Schuste, R. M., Waller, E., et al. (1997). Randomized, double-blind, placebo-controlled trial of pegylated recombinant megakaryocyte growth and development factor (PEG-rHuMGDF) in breast cancer patients following autologous bone marrow transplantation. *Blood, 90,* 580a.

155. Bolwell, B., Vredenburgh, J., Overmoyer, B., et al. (1997). Safety and biological effect of pegylated recombinant megakaryocyte growth and development factor (PEG-rHuMGDF) in breast cancer patients following autologous peripheral blood progenitor cell transplantation (PBPC). *Blood, 90,* 171a.

156. Glaspy, J., Vredenburgh, J., Demetri, G. D., et al. (1997). Effects of PEGylated recombinant human megakaryocyte growth and development factor (PEG-rHuMGDF) before high dose chemotherapy (HDC) with peripheral blood progenitor cell (PBPC) support. *Blood, 90,* 580a.

157. Schuster, M. W., Beveridge, R., Frei-Lahr, D., et al. (2002). The effects of pegylated recombinant human megakaryocyte growth and development factor (PEG-rHuMGDF) on platelet recovery in breast cancer patients undergoing autologous bone marrow transplantation. *Exp Hematol, 30,* 1044–1050.

158. Nash, R. A., Kurzrock, R., DiPersio, J., et al. (2000). A phase I trial of recombinant human thrombopoietin in patients with delayed platelet recovery after hematopoietic stem cell transplantation. *Biol Blood Marrow Transplant, 6,* 25–34.

159. Somlo, G., Sniecinski, I., ter Veer, A., et al. (1999). Recombinant human thrombopoietin in combination with granulocyte colony-stimulating factor enhances mobilization of peripheral blood progenitor cells, increases peripheral blood platelet concentration, and accelerates hematopoietic recovery following high-dose chemotherapy. *Blood, 93,* 2798–2806.

160. Harker, L. A., Carter, R. A., Marzec, U. M., et al. (1998). Correction of thrombocytopenia and ineffective platelet production in patients infected with human immunodeficiency virus (HIV) by PEG-rHuMGDF therapy. *Blood, 92,* 707a.

161. Usuki, K., Tahara, T., Iki, S., et al. (1996). Serum thrombopoietin level in various hematological diseases. *Stem Cells, 14,* 558–565.

162. Kosugi, S., Kurata, Y., Tomiyama, Y., et al. (1996). Circulating thrombopoietin level in chronic immune thrombocytopenic purpura. *Br J Haematol, 93,* 704–706.

163. Chang, M., Suen, Y., Meng, G., et al. (1996). Differential mechanisms in the regulation of endogenous levels of thrombopoietin and interleukin-11 during thrombocytopenia: Insight into the regulation of platelet production. *Blood, 88,* 3354–3362.

164. Marsh, J. C., Gibson, F. M., Prue, R. L., et al. (1996). Serum thrombopoietin levels in patients with aplastic anaemia. *Br J Haematol, 95,* 605–610.

165. Ballem, P. J., Belzberg, A., Devine, D. V., et al. (1992). Kinetic studies of the mechanism of thrombocytopenia in patients with human immunodeficiency virus infection. *N Engl J Med, 327,* 1779–1784.

166. Ballem, P. J., Segal, G. M., Stratton, J. R., et al. (1987). Mechanisms of thrombocytopenia in chronic autoimmune thrombocytopenic purpura. Evidence of both impaired platelet production and increased platelet clearance. *J Clin Invest, 80,* 33–40.

167. Heyns, A. D., Lotter, M. G., Badenhorst, P. N., et al. (1982). Kinetics and sites of destruction of 111Indium-oxine-labeled

platelets in idiopathic thrombocytopenic purpura: A quantitative study. *Am J Hematol, 12,* 167–177.

168. Heyns, A. P., Badenhorst, P. N., Lotter, M. G., et al. (1986). Platelet turnover and kinetics in immune thrombocytopenic purpura: Results with autologous 111In-labeled platelets and homologous 51Cr-labeled platelets differ. *Blood, 67,* 86–92.

169. Chang, M., Nakagawa, P. A., Williams, S. A., et al. (2003). Immune thrombocytopenic purpura (ITP) plasma and purified ITP monoclonal autoantibodies inhibit megakaryocytopoiesis *in vitro. Blood, 102,* 887–895.

170. Houwerzijl, E. J., Blom, N. R., van der Want, J. J., et al. (2004). Ultrastructural study shows morphologic features of apoptosis and para-apoptosis in megakaryocytes from patients with idiopathic thrombocytopenic purpura. *Blood, 103,* 500–506.

171. McMillan, R., Wang, L., Tomer, A., et al. (2004). Suppression of *in vitro* megakaryocyte production by antiplatelet autoantibodies from adult patients with chronic ITP. *Blood, 103,* 1364–1369.

172. Rice, L., Nichol, J. L., McMillan, R., et al. (2001). Cyclic immune thrombocytopenia responsive to thrombopoietic growth factor therapy. *Am J Hematol, 68,* 210–214.

173. Ballmaier, M., Schulze, H., Strauss, G., et al. (1997). Thrombopoietin in patients with congenital thrombocytopenia and absent radii: Elevated serum levels, normal receptor expression, but defective reactivity to thrombopoietin. *Blood, 90,* 612–619.

174. Aster, R. H. (1966). Pooling of platelets in the spleen: Role in the pathogenesis of "hypersplenic" thrombocytopenia. *J Clin Invest, 45,* 645–657.

175. Adams, J. A., Liu Yin, J. A., Brereton, M. L., et al. (1997). The *in vitro* effect of pegylated recombinant human megakaryocyte growth and development factor (PEG rHuMGDF) on megakaryopoiesis in normal subjects and patients with myelodysplasia and acute myeloid leukaemia. *Br J Haematol, 99,* 139–146.

176. Fontenay-Roupie, M., Dupont, J. M., Picard, F., et al. (1998). Analysis of megakaryocyte growth and development factor (thrombopoietin) effects on blast cell and megakaryocyte growth in myelodysplasia. *Leuk Res, 22,* 527–535.

177. Komatsu, N., Okamoto, T., Yoshida, T., et al. (2000). Pegylated recombinant human megakaryocyte growth and development factor (PEG-rHuMGDF) increased platelet counts (plt) in patients with aplastic anemia (AA) and myelodysplastic syndrome (MDS). *Blood, 96,* 296a.

178. Nakamura, M., Toombs, C. F., Duarte, I. G., et al. (1998). Recombinant human megakaryocyte growth and development factor attenuates postbypass thrombocytopenia. *Ann Thorac Surg, 66,* 1216–1223.

179. Piacibello, W., Sanavio, F., Garetto, L., et al. (1997). Extensive amplification and self-renewal of human primitive hematopoietic stem cells from cord blood. *Blood, 89,* 2644–2653.

180. Goodnough, L. T., Kuter, D. J., McCullough, J., et al. (2001). Prophylactic platelet transfusions from healthy apheresis platelet donors undergoing treatment with thrombopoietin. *Blood, 98,* 1346–1351.

181. Vadhan-Raj, S., Kavanagh, J. J., Freedman, R. S., et al. (2002). Safety and efficacy of transfusions of autologous cryopreserved platelets derived from recombinant human thrombopoietin to support chemotherapy-associated severe thrombocytopenia: A randomised cross-over study. *Lancet, 359,* 2145–2152.

182. Gordon, M. S., McCaskill-Stevens, W. J., Battiato, L. A., et al. (1996). A phase I trial of recombinant human interleukin-11 (neumega rhIL-11 growth factor) in women with breast cancer receiving chemotherapy. *Blood, 87,* 3615–3624.

183. Vigon, I., Dreyfus, F., Melle, J., et al. (1993). Expression of the c-mpl proto-oncogene in human hematologic malignancies. *Blood, 82,* 877–883.

184. Columbyova, L., Loda, M., & Scadden, D. T. (1995). Thrombopoietin receptor expression in human cancer cell lines and primary tissues. *Cancer Res, 55,* 3509–3512.

185. Ulich, T. R., del Castillo, J., Senaldi, G., et al. (1996). Systemic hematologic effects of PEG-rHuMGDF-induced megakaryocyte hyperplasia in mice. *Blood, 87,* 5006–5015.

186. Abina, M. A., Tulliez, M., Lacout, C., et al. (1998). Major effects of TPO delivered by a single injection of a recombinant adenovirus on prevention of septicemia and anemia associated with myelosuppression in mice: Risk of sustained expression inducing myelofibrosis due to immunosuppression. *Gene Therapy, 5,* 497–506.

187. Yan, X. Q., Lacey, D., Hill, D., et al. (1996). A model of myelofibrosis and osteosclerosis in mice induced by overexpressing thrombopoietin (mpl ligand): Reversal of disease by bone marrow transplantation. *Blood, 88,* 402–409.

188. Douglas, V. K., Tallman, M. S., Cripe, L. D., et al. (2002). Thrombopoietin administered during induction chemotherapy to patients with acute myeloid leukemia induces transient morphologic changes that may resemble chronic myeloproliferative disorders. *Am J Clin Pathol, 117,* 844–850.

189. Crawford, J., Glaspy, J., Belani, C., et al. (1998). A randomized, placebo-controlled, blinded, dose scheduling trial of pegylated recombinant human megakaryocyte growth and development factor (PEG-HUMGDF) with filgrastim support in non-small cell lung cancer (NSCLC) patients treated with paclitaxel and carboplatin during multiple cycles of chemotherapy. *Proc ASCO, 17,* 73a.

190. Yang, C., Xia, Y., Li, J., et al. (1999). The appearance of anti-thrombopoietin antibody and circulating thrombopoietin-IgG complexes in a patient developing thrombocytopenia after the injection of PEG-rHuMGDF. *Blood, 94,* 681a.

191. Basser, R. L., O'Flaherty, E., Green, M., et al. (2002). Development of pancytopenia with neutralizing antibodies to thrombopoietin after multicycle chemotherapy supported by megakaryocyte growth and development factor. *Blood, 99,* 2599–2602.

192. F-D-C- Reports. (1998). In brief: Amgen Megagen. *Pink Sheet, 60,* 27.

193. Nandurkar, H. H., Robb, L., Tarlinton, D., et al. (1997). Adult mice with targeted mutation of the interleukin-11 receptor (IL11Ra) display normal hematopoiesis. *Blood, 90,* 2148–2159.

194. Robb, L., Li, R., Hartley, L., et al. (1998). Infertility in female mice lacking the receptor for interleukin 11 is due to a defective uterine response to implantation. *Nat Med, 4,* 303–308.

195. Smith, J. W. D., Longo, D. L., Alvord, W. G., et al. (1993). The effects of treatment with interleukin-1 alpha on platelet recovery after high-dose carboplatin. *N Engl J Med, 328,* 756–761.

196. Vadhan-Raj, S., Kudelka, A. P., Garrison, L., et al. (1994). Effects of interleukin-1 alpha on carboplatin-induced thrombocytopenia in patients with recurrent ovarian cancer. *J Clin Oncol, 12,* 707–714.

197. Gordon, M. S., Nemunaitis, J., Hoffman, R., et al. (1995). A phase I trial of recombinant human interleukin-6 in patients with myelodysplastic syndromes and thrombocytopenia. *Blood, 85,* 3066–3076.

198. Lazarus, H. M., Winton, E. F., Williams, S. F., et al. (1995). Phase I multicenter trial of interleukin 6 therapy after autologous bone marrow transplantation in advanced breast cancer. *Bone Marrow Transplant, 15,* 935–942.

199. Nieken, J., Mulder, N. H., Buter, J., et al. (1995). Recombinant human interleukin-6 induces a rapid and reversible anemia in cancer patients. *Blood, 86,* 900–905.

# CHAPTER 67

# Desmopressin (DDAVP)

## Pier Mannuccio Mannucci[1] and Marco Cattaneo[2]

[1]IRCCS Maggiore Hospital, Mangiagalli and Regina Elena Foundation, University of Milan, Milan, Italy
[2]Ospedale San Paolo, University of Milan, Milan, Italy

## I. Introduction

Desmopressin (1-deamino-8-D-arginine vasopressin; DDAVP) is a synthetic analogue of the antidiuretic hormone vasopressin. Like the natural antidiuretic hormone, desmopressin increases the plasma levels of factor VIII and von Willebrand factor (VWF) with the advantage, compared to vasopressin, that it produces no vasoconstriction, no increase in blood pressure, and no contraction of the uterus or gastrointestinal tract so that it is well tolerated when administered to humans.[1,2]

In 1977, desmopressin was used for the first time in patients with mild hemophilia A and von Willebrand disease (VWD) for the prevention and treatment of bleeding, first during dental extractions and then during major surgical procedures.[3] Surgery was performed without blood products, demonstrating that autologous factor VIII and VWF increased in patient plasma by desmopressin infusion could effectively replace allogeneic factors infused with blood products.[3] After the original clinical study performed in Italy, desmopressin was used in many other countries and the World Health Organization included it in its list of essential drugs. A drug that could be used in the prophylaxis and treatment of the two most common congenital bleeding disorders without the need of blood products was very attractive in view of its safety (no risk of transmission of blood-borne viral diseases) and cost (desmopressin is relatively inexpensive compared to blood products).

The clinical indications for desmopressin subsequently expanded beyond hemophilia and VWD. The compound was used in bleeding disorders not involving a deficiency or dysfunction of factor VIII or VWF, including congenital and acquired defects of platelet function and such frequent acquired abnormalities of hemostasis as those associated with chronic kidney and liver disease. Desmopressin has also been used prophylactically in patients undergoing surgical operations characterized by large blood loss and transfusion requirements. Some of these indications of desmopressin have been subsequently strengthened by the

experience accumulated, whereas others have not been supported by rigorous clinical trials or have been superseded by the advent of more efficacious treatments. This chapter reviews the spectrum of indications of desmopressin in the prevention and treatment of bleeding, with emphasis on its use in patients with congenital or acquired defects of primary hemostasis.

## II. Desmopressin in the Management of Hemophilia A and von Willebrand Disease

### A. Congenital Hemophilia A and von Willebrand Disease

Well-established indications of desmopressin are prophylaxis and treatment of spontaneous, posttraumatic, and postsurgical bleeding in patients with mild hemophilia A and type 1 VWD. An intravenous dose of 0.3 µg/kg should be used in patients with plasma levels of at least 10 U/mL of the deficient protein. Factor VIII and VWF increase on average to approximately three to five times resting levels, with little intrapatient variability in responses.[4,5] Patients with severe deficiencies do not respond at all. The biological half-lives of endogenous factor VIII and VWF released by desmopressin infusion are similar to those of plasma-derived proteins.[5] Release into plasma of tissue plasminogen activator (tPA) is another short-lived effect of desmopressin.[6,7] Plasminogen activator does generate plasmin *in vivo*, but most of the plasmin is quickly complexed to $\alpha_2$-antiplasmin and does not produce fibrin(ogen)olysis in circulating blood.[8] Accordingly, it is usually unnecessary to inhibit fibrinolysis with antifibrinolytic amino acids when desmopressin is used for clinical purposes.

In mild hemophilia A, the efficacy of desmopressin usually correlates with the postinfusion plasma levels of factor VIII.[3,9–11] Therefore, therapeutic indications are defined by the nature of the bleeding episode, the baseline factor VIII levels, and the levels that must be attained and

**Table 67-1: Summary of Recommended Treatments for Different Types of von Willebrand Disease**

| Type | Treatment of Choice | Alternative Therapy |
|------|---------------------|---------------------|
| 1 | Desmopressin | Factor VIII–VWF concentrates |
| 2A | Factor VIII–VWF concentrates | Desmopressin |
| 2B | Factor VIII–VWF concentrates | None |
| 2M | Factor VIII–VWF concentrates | Desmopressin |
| 2N | Factor VIII–VWF concentrates | Desmopressin |
| 3 | | |
|    In patients without alloantibodies | Factor VIII–VWF concentrates | Platelet concentrates |
|    In patients with alloantibodies | Recombinant activated factor VII | Recombinant factor VIII |

VWF, von Willebrand factor.

maintained for hemostasis. Clinical failures of desmopressin can usually be explained by the attainment of factor VIII levels in plasma that are insufficient to control bleeding.[3,9–11] For example, a major surgical procedure in a patient with factor VIII levels of 10 U/dL may not be successfully managed with desmopressin because the expected posttreatment levels of 30 to 50 U/dL are not high enough for normal hemostasis. On the other hand, these levels should be sufficient for the patient to have such minor procedures as dental extractions.

Most patients with VWD type 1, who have a functionally normal VWF, respond to desmopressin with relative increases in factor VIII and VWF that are usually larger than those seen in hemophiliacs[12] and with a normalization of their prolonged bleeding times.[3,10,11,13] Hence, desmopressin should be the first choice for treatment for these patients, but it appears to be used less often than it should be. This may be related to the results of a prospective study on the response of factor VIII and VWF measurements to desmopressin showing that only 27% of VWD type 1 patients could be classified as responders.[14] However, the criteria adopted in this study for biological responsiveness were strict and demanding,[14] and lower responses than those adopted may be sufficient for hemostasis. The defects of patients with VWD type 3 and of those with dysfunctional molecules (VWD type 2) are usually not corrected by desmopressin, with some exceptions.[13,14] General guidelines for the use of desmopressin or plasma products in the different subtypes of VWD are given in Table 67-1. However, there is some variability in the response to desmopressin between patients with VWD type 1.[14] The reasons for these different behaviors are not clear, and a test dose is the only way to differentiate responders from nonresponders. Table 67-2 shows the schedule of desmopressin administration and blood sampling recommended to evaluate the degree of laboratory response to a test dose. On the basis of the results

**Table 67-2: Schedule of the Test Dose of Desmopressin Given to Assess Responsiveness in Patients with von Willebrand Disease (and Mild Hemophilia)[a]**

| | |
|---|---|
| Step 1 | Infuse 0.3 μg/kg of desmopressin in 100 mL saline over 30 minutes in newly diagnosed patients or in those who must undergo an elective treatment |
| Step 2 | Obtain citrated blood samples at 60 minutes after starting desmopressin (to assess the postinfusion peak) and at 4 hours (to assess the rate of factor clearance) |
| Step 3 | Measure factor VIII coagulant activity and ristocetin cofactor activity |

[a]If the subcutaneous or intranasal routes are preferred for desmopressin administration, the same schedule should be followed.

obtained, one can predict whether the attained factor levels and the duration of their persistence in plasma are of such a degree that the successful management of a given clinical situation can be predicted (Table 67-3).

Patients treated repeatedly with desmopressin may become less responsive, perhaps because stores are exhausted.[12] The average factor VIII and VWF responses obtained when desmopressin is repeated three or four times at 24-hour intervals are approximately 30% lower than those obtained after the first dose.[12] In general, treatment with desmopressin can be usefully repeated two to four times, but it is preferable to monitor factor VIII and/or VWF responses and tailor repeated treatments on the basis of the results obtained. In rare situations in which factor VIII and VWF levels must be maintained above baseline for a prolonged period of time, it may become necessary to use plasma-derived or recombinant factors or to supplement desmopressin with them.

**Table 67-3: Target Levels of Factor VIII and von Willebrand Factor Recommended in Patients with Mild Hemophilia A and von Willebrand Disease for Different Clinical Situations**[a]

| Clinical Situation | Target |
|---|---|
| Major surgery | Peak factor levels[a] of 100% and trough daily levels of at least 50% until healing is complete (usually 5–10 days) |
| Minor surgery | Peak factor levels of 60% and trough daily levels of at least 30% until healing is complete (usually 2–4 days) |
| Dental extractions | Peak factor levels of 60% (single dose) |
| Spontaneous bleeding episodes | Peak factor levels higher than 50% until bleeding stops (usually 2–4 days) |
| Delivery and puerperium | Peak factor levels higher than 80% and trough levels of at least 30%, usually for 3 or 4 days |

[a]In patients with von Willebrand disease, the same target levels of ristocetin cofactor are recommended.

Desmopressin is also available in formulations for subcutaneous injections (at a dose of 0.3 µg/kg) and nasal inhalation (at fixed doses of 300 µg in adults and 150 µg in children). These formulations are at least as efficacious as intravenous desmopressin and perhaps more clinically practical when they are self-administered at home to prevent or treat such minor bleeding episodes as excessive bleeding at menstruation in women with VWD.[15–17]

### B. Acquired Hemophilia and von Willebrand Syndrome

Desmopressin is of some clinical usefulness for treatment of patients with acquired von Willebrand syndrome (AVWS), which occurs in association with a variety of underlying diseases, particularly in lymphoproliferative, cardiovascular, and myeloproliferative disorders. Management of bleeding episodes in these patients is based on desmopressin, factor VIII/VWF concentrates, and high-dose intravenous immunoglobulins, but no single drug is effective for all.[18] In a large series of 186 cases of AVWS treated with desmopressin to control bleeding episodes, treatment was effective in 38 cases (20%). The variability of the clinical response to desmopressin was related to the variably short half-life of endogenous factor VIII and VWF released by the compound.[19] In some patients with acquired hemophilia due to anti-factor VIII autoantibodies, desmopressin elicits a rise in plasma factor VIII levels that may be relatively long

lasting and sufficient to attain hemostasis. It is difficult to predict the degree and duration of the response, but patients with low titer inhibitor and residual factor VIII are more likely to respond. A test dose should be given to each candidate patient to establish the potential usefulness of desmopressin (Table 67-2).

## III. Desmopressin as a General Hemostatic Agent

### A. Congenital and Acquired Bleeding Diathesis Not Due to Defects of Factor VIII and von Willebrand Factor

**1. Congenital Disorders of Primary Hemostasis.** Desmopressin shortens or normalizes the bleeding time and/or the PFA-100 platelet function test in most patients with congenital defects of platelet function (Table 67-4). There is usually a good response in patients with defects of the release reaction and in those with isolated and unexplained prolongations of the bleeding time. Most patients with storage pool deficiency respond to desmopressin but a few do not, particularly those with severe deficiencies of platelet δ-granules content. Most patients with Glanzmann thrombasthenia are not responsive. The demonstration that the increased levels of VWF induced by desmopressin could substitute for fibrinogen at the glycoprotein (GP) IIb/IIIa level in patients with afibrinogenemia, thereby correcting their prolonged bleeding times,[23] was not confirmed by a subsequent study.[35] The documented efficacy in patients with Bernard–Soulier syndrome,[28,34,37,40,44] who lack GPIb-IX-V, the platelet receptor for VWF that is essential for platelet adhesion to the vessel wall at high shear forces, supports the demonstration that the drug can shorten the prolonged bleeding time through a mechanism(s) that is independent of released VWF. Whether the effect on a laboratory test such as the bleeding time corresponds to a hemostatic effect is not established. Although there are anecdotal reports of desmopressin successfully stopping or preventing bleeding in these patients (Table 67-4), the clinical efficacy of desmopressin is not formally proven.

**2. Acquired and Drug-Induced Disorders of Hemostasis.** The hemostatic defect in uremia is characterized by a prolonged bleeding time, a laboratory abnormality that correlates strongly with the hemorrhagic symptoms of these patients, mainly epistaxis and bleeding from the gastrointestinal tract (see Chapter 58). Even though there is no definite evidence that the derangement of primary hemostasis in uremia is due to the deficiency or abnormality of factor VIII and/or VWF, the effectiveness of infusion of cryopre-

**Table 67-4: Reports on the Hemostatic Effects of Desmopressin in Patients with Congenital Disorders of Platelet Function**

| Condition | Effects on the Prolonged Bleeding Time or PFA-100 (No. of Studies [Refs.]) | | Case Reports of Clinical Efficacy (No. of Studies [Refs.])[a] | |
| --- | --- | --- | --- | --- |
| | Shortening/Normalization | No Effects | Prevention/Control of Bleeding | No Effects/Worsening |
| δ-Storage pool deficiency | 8[20,22,25,29,33,36,42,43] | 3[22,36,42] | 1[25] | — |
| Hemansky–Pudlak syndrome | 2[31,32] | 2[32,49] | 1[46] | 1[46] |
| Gray platelet syndrome | 1[24] | 1[21] | — | — |
| Other secretion defects | 7[20,25,33,42,43,45,47] | — | 2[25,42] | — |
| Glanzmann thrombasthenia | 1[33] | 2[22,25] | — | — |
| Bernard–Soulier syndrome | 6[28,33,34,37,40,44] | — | — | — |
| Defect of $P2Y_{12}$ receptors | 1[41] | — | 1[41] | — |
| MYH9-related disorders | 1[33] | — | 1[50] | — |
| Defective calcium ionophore-induced platelet aggregation | 1[48] | — | — | — |
| Easy bruisability (aspirin intolerance) | 1[38] | — | — | — |
| Unexplained prolonged bleeding time | 5[20,22,30,36,144] | 1[36] | 5[30,39,144,145,146] | — |
| Afibrinogenemia | 1[23] | 1[35] | — | — |

[a]Some studies reported high variability of responses among patients with the same abnormality of primary hemostasis.

cipitate[51] suggested that the effects of desmopressin could also be tested in these patients.[52] After the intravenous infusion of 0.3 μg/kg desmopressin, the prolonged bleeding time became normal in approximately 75% of uremic patients.[53] Similar results were subsequently obtained after administering desmopressin subcutaneously[54] or intranasally.[55,56] It is important to note, however, that closely spaced infusions of desmopressin may fail to consistently shorten the bleeding time.[57] Uncontrolled clinical studies in uremic patients have suggested that desmopressin can be successfully used to prevent bleeding before invasive procedures (biopsies and major surgery) and to stop spontaneous bleeding,[58–63] with only one exception.[64] Conjugated estrogens are a long-acting alternative to desmopressin in uremic patients because they shorten the bleeding time with a more sustained effect lasting 10 to 15 days.[65] The two products can be given together, exploiting the different timings of their maximal effects. Currently, most patients with chronic renal insufficiency are regularly treated with erythropoietin, which leads to the sustained improvement not only of anemia but also of the hemostatic defect[66] so that compounds such as desmopressin and conjugated estrogens are now rarely needed.

The bleeding time is prolonged in some patients with liver cirrhosis with mild or moderate thrombocytopenia.

Factor VIII and VWF are in the high normal range or even higher in these patients, yet both intravenous and subcutaneous desmopressin shortens the bleeding time of cirrhotic patients.[22,33,67–69] Hence, desmopressin is a possible prophylactic treatment for cirrhotics who have a prolonged bleeding time and need invasive diagnostic procedures. In contrast, it is not useful in the management of acute variceal bleeding.[70]

The efficacy of desmopressin to shorten the prolonged bleeding time in patients with thrombocytopenia varies according to the baseline platelet count.[22,71,72] A very good response is usually obtained in patients with platelet counts higher than $50 \times 10^9$/L, whereas a variable degree of shortening can be observed for platelet counts lower than $50 \times 10^9$/L.

Desmopressin counteracts the effects on hemostasis measurements of some antithrombotic drugs. It shortens the prolonged bleeding time of individuals taking such widely used antiplatelet agents as aspirin and ticlopidine,[20,22,73] the prolonged bleeding time and activated partial thromboplastin time of patients receiving heparin,[74] and the prolonged bleeding time of rabbits treated with aspirin and streptokinase[75] or hirudin[76] (without corresponding human data). Desmopressin also accelerates normalization of the *in vitro* platelet dysfunction in patients receiving GPIIb-IIIa antago-

**Table 67-5: Reports on the Hemostatic Effects of Desmopressin in Patients with Acquired Defects of Primary Hemostasis and Other Conditions at Risk of Bleeding Not Related to Abnormalities of the Hemostatic System**[a]

| Condition | Effects on the Prolonged Bleeding Time or PFA-100 (No. of Studies [Refs.]) | | Case Reports of Clinical Efficacy (No. of Studies [Refs.]) | |
| --- | --- | --- | --- | --- |
| | Shortening/Normalization | No Effects | Prevention/Control of Bleeding | No Effects/Worsening |
| Uremia | 9[52–60] | — | 6[58–63] | 1[64] |
| Liver cirrhosis | 5[22,33,67–69] | — | — | 1[70] |
| Thrombocytopenia[b] | 3[22,71,72] | 2[22,71] | 1[72] | — |
| Drug-induced | | | | |
|   Aspirin | 3[20,22,77] | 1[25] | 2[78,79] | — |
|   Ticlopidine | 2[22,73] | — | — | — |
|   Clopidogrel | — | — | 1[149] | — |
|   Heparin | 1[74] | — | — | — |
|   Dextran | 1[147] | — | — | 1[114] |
|   GPIIb–IIIa antagonists | 1[77] | — | — | — |
| Sickle cell trait | — | — | 2[82,83] | — |
| Ehlers–Danlos syndrome | 1[148] | — | 2[80,148] | — |
| Hereditary telangiectasia | — | — | 1[81] | — |

[a]Conditions associated with acquired VWD are not included in this table. Desmopressin is of proven clinical efficacy in patients with acquired VWD.
[b]Good responses are usually obtained in patients with platelet counts higher than $50 \times 10^9/L$.

nists.[77] A few small studies have shown that desmopressin controls postsurgical bleeding complications in patients treated with aspirin.[78,79] Finally, desmopressin has been reported to be beneficial for bleeding episodes in patients with congenital or acquired bleeding diathesis not due to abnormalities of the hemostatic system, such as Ehlers–Danlos syndrome,[80] Rendu–Osler telangiectasia,[81] and sickle cell trait.[82,83]

In summary, in chronic renal disease desmopressin is still indicated only for those patients with renal failure not treated or unresponsive to erythropoietin. Desmopressin is a possible treatment for patients with liver cirrhosis and prolonged bleeding time who need invasive diagnostic procedures such as liver biopsies. Notwithstanding the shortened bleeding times and improved PFA-100, there is little clinical evidence that desmopressin prevents or stops bleeding complications that develop in association with the use of antithrombotic agents. The compound may provide an opportunity to control drug-induced bleeding without stopping treatment and perhaps avoiding recurrence or progression of thrombosis. Data obtained from a few nonrandomized studies indicate that desmopressin can be a useful alternative to blood products during or after surgery or delivery, ensuring satisfactory hemostasis in patients with congenital or acquired defects of primary hemostasis (Tables 67-4 and 67-5).

## B. Desmopressin as a Blood-Saving Agent

The broadening indications of desmopressin, since the first use in hemophilia and VWD in 1977, led several investigators to evaluate whether the compound was beneficial during surgical operations in which blood loss is large and for which multiple blood transfusions are needed.

Open-heart surgery with extracorporeal circulation is the epitome of operations that warrant the adoption of blood-saving measures. In addition to such techniques as presurgical removal of autologous blood for postsurgical retransfusion, returning all oxygenator and tubing contents to the patient, and autotransfusion of the mediastinal shed blood, prophylaxis with pharmacological agents might help reduce blood transfusion further. Since 1986, desmopressin has been evaluated for this purpose. In the first controlled randomized study carried out in patients undergoing complex cardiac operations associated with large blood losses, results were impressive.[84] Given at the time of chest closure, desmopressin reduced dramatically perioperative and early (12 hours) postoperative blood loss and transfusion requirements by approximately 40%.[84] On the other hand, in two subsequent large studies of patients undergoing less complex operations with less blood loss, there were no significant differences between desmopressin- and placebo-treated patients in terms of total blood loss or transfusion requirements.[85,86] Other studies, mainly in patients undergoing coronary artery bypass grafting and uncomplicated valve replacement, gave conflicting results.[87–105]

The conflicting results of desmopressin in open-heart surgery might be due to the fact that most studies were of small size and had insufficient statistical power to detect differences in blood loss. A systemic review of 17 randomized, double-blind, placebo-controlled trials, which included 1171 patients undergoing open-heart surgery, attempted to overcome this pitfall.[106] Overall, desmopressin reduced postoperative blood loss by 9%, which was statistically significant but of little clinical impact. A subgroup analysis showed that although desmopressin had no blood-saving effect when the 24-hour postoperative total blood loss in placebo-treated patients was in the lower or middle third of distribution (687–1108 mL), the compound reduced blood losses by 34% in those trials in which the 24-hour blood loss in the placebo-treated patients was in the upper third of distribution (>1108 mL),[106] suggesting that the drug may be clinically efficacious only in patients at risk of excessive postoperative bleeding. These data were substantially confirmed in a later prospective, double-blind, placebo-controlled trial, in which the effects of desmopressin were compared to those of placebo in patients identified as being at high risk of postoperative bleeding with the point-of-care test hemoSTATUS.[104] The trial showed that the average 24-hour blood loss in desmopressin-treated, high-risk patients was 39% lower than that in placebo-treated, high-risk patients (624 vs. 1028 mL, $p = 0.0004$) and was comparable to that of low-risk, untreated patients (656 mL). The study also showed that transfusion requirements were significantly reduced by desmopressin (total donor exposures of 1.6 in desmopressin-treated, high-risk patients vs. 5.2 in placebo-treated, high-risk patients [$p = 0.0001$] and 1.6 in untreated, low-risk patients).

Two subsequent systemic reviews confirmed the mild clinical effect of desmopressin in reducing peri- and postoperative blood loss[107,108] but did not repeat the subgroup analysis of the effects of the drug as a function of the severity of blood losses in the placebo-treated patients. In these systemic reviews, no demonstrable effect of the drug was shown on transfusion requirements, need for rethoracotomy, or mortality.[107,108]

Desmopressin is not the only blood-saving agent that can be used in cardiac surgery. The synthetic antifibrinolytic amino acid ε-aminocaproic acid (EACA), tranexamic acid, and the broad-spectrum protease inhibitor aprotinin have also been used. A few direct comparison studies and two systemic reviews[107,108] have shown that the order of efficacy of these hemostatic agents (greatest to least) is aprotinin, tranexamic acid, EACA, and desmopressin. On the other hand, the order of drug cost is also the same. Cost-effectiveness analysis is necessary to help clinicians make a choice that currently would be directed to aprotinin, but with formidable costs.

The efficacy of desmopressin has also been evaluated in noncardiac surgical operations characterized by large blood loss. In 1987, Kobrinsky et al.[109] showed that when administered to hemostatically normal children before spinal fusion for idiopathic scoliosis, desmopressin reduced their average operative blood loss by approximately one third. However, these favorable results could not be confirmed in subsequent studies.[110–112] Contrasting results were also reported by other small trials in total hip or knee arthroplasty.[113–115] Preoperative desmopressin failed to reduce blood loss in patients undergoing debridement and grafting of burn wounds, a procedure in which extreme blood loss is a frequent occurrence,[116] and in patients undergoing elective aortic operations.[117,118] It must be considered, however, that most of these trials had insufficient statistical power to detect true differences in postsurgical blood loss.

In summary, the efficacy of desmopressin as a blood-saving agent in cardiac and noncardiac surgical operations appears of doubtful clinical significance.

## IV. Side Effects

Side effects of desmopressin include mild facial flushing, transient headache, and 10 to 20% increases in heart rate and diastolic blood pressure.[1] Because of its potent antidiuretic effects and the large doses needed for hemostatic efficacy (15 times higher than for treatment of diabetes insipidus), there is a risk of water retention, which can lead to severe hyponatremia.[119,120] Because several cases of water intoxication have been reported, fluid restriction, avoidance of hyponatremic solutions, and monitoring of serum electrolytes and body weight for 24 hours after treatment are warranted, especially when desmopressin is used in children younger than the age of 2 years.

Desmopressin has been used with caution in the first two trimesters of pregnancy in women with bleeding disorders because there is concern that the compound may cause placental insufficiency due to arterial vasoconstriction and increase the risk of miscarriage due to an oxytocic effect and of maternal and/or neonatal hyponatremia.[121,122] However, the compound is practically devoid of these biological activities, so its potential for vasoconstriction and uterine contraction is negligible. Evidence of its safety during pregnancy in women with diabetes insipidus is now available.[123] Desmopressin was also used in 32 pregnant women with low factor VIII levels (27 obligatory carriers of hemophilia A and 5 with VWD type 1) in order to improve hemostasis at the time of such invasive procedures such as chorionic villus sampling or amniocentesis.[124] In all patients there was no abnormal bleeding and 20 pregnancies went successfully to term with the delivery of healthy newborns. In the remaining 12 women, male fetuses found to be affected by hemophilia on genotyping were aborted under coverage with additional doses of desmopressin.[124] There were no side effects in the treated women other than mild facial flushing and headache and no significant increase in body weight. Hence, desmopressin can be used during the first and second trimesters of pregnancy and is safe during invasive procedures that increase the risk of miscarriage *per se*.

Several case reports indicate that desmopressin increases the risk of arterial thrombosis, especially stroke and myocardial infarction. A systemic review of 31 clinical trials of patients who had undergone cardiac surgery, vascular surgery, orthopedic surgery, and other high-risk surgical interventions showed that the risk of thrombosis in desmopressin-treated and placebo-treated patients was 3.4 and 2.7%, respectively; this difference was not statistically significant.[125] In contrast, a systematic review of 16 trials evaluating the efficacy of desmopressin to reduce blood loss after cardiac surgery showed that desmopressin treatment was associated with an almost 2.4-fold increased risk of perioperative myocardial infarction compared with placebo (2.39; 95% confidence interval, 1.02–5.6).[108] Due to the increased risk for arterial thrombosis, the use of desmopressin should be considered cautiously in the elderly and in patients with clinical overt atherosclerosis.

## V. Mechanisms of Action of Desmopressin

### A. Mechanisms of Desmopressin-Induced Increase in Plasma Factor VIII and VWF

Because factor VIII, VWF, and tPA are rapidly and transiently raised after desmopressin administration, it is most likely that desmopressin causes them to be released from storage sites. The storage site for VWF is probably the Weibel–Palade bodies of the endothelial cells.[1] This hypothesis is supported by the observation that in rats injections of desmopressin elicit biological responses that are clearly related to the activation of endothelial cells, including surface expression of P-selectin and subsequent margination of leukocytes.[126] In normal individuals, desmopressin infusion produces important changes in VWF present in vascular endothelial cells. There is a reduction in the amount of this protein and a change in its localization, which results in a tendency for the protein to move abluminally toward the cellular basement membrane.[127] Notwithstanding these early data focusing on the endothelial cell as the most likely source of VWF, addition of desmopressin to cultured human umbilical vein endothelial cells (HUVEC) *in vitro* does not release VWF.[128] The apparent paradox was solved by the demonstration that the lack of direct effect of desmopressin on HUVEC is attributable to the fact that these cells do not express the V2 receptor (V2R).[129] When desmopressin was added to cultured HUVEC that had been transfected to

express V2R, or to lung microvascular endothelial cells (which express endogenous V2R), the compound did elicit the release of VWF, which was mediated by an increase in intracellular cAMP.[129] The interactions between released factor VIII and concomitantly released VWF and tPA are not well established. The observation that patients with VWD type 3 treated with desmopressin fail to release not only VWF (which is not synthesized in these patients) but also factor VIII and tPA (which are normally synthesized and stored in different tissues) supports the hypothesis that these effects are regulated by a single mechanism, which is probably defective in type 3 VWD.[130]

The site of cellular storage and release of factor VIII is less well established than that of VWF.[131] Desmopressin did elicit the expected VWF rise but did not elicit a factor VIII rise in dogs with hemophilia A after hepatocyte-driven neonatal gene therapy. This observation suggests that the increase in factor VIII induced by desmopressin in normal dogs (and perhaps in humans) is due to its release from cells other than the hepatocyte, such as endothelial cells, in which factor VIII is colocalized and complexed with VWF.[132] That cellular colocalization of factor VIII and VWF is required for the rise in factor VIII after desmopressin is also demonstrated by the observation that following liver transplantation, patients with hemophilia A infused with desmopressin showed the expected VWF rise but no change in plasma factor VIII.[133] Because factor VIII is synthesized only in the transplanted liver, this observation supports the hypothesis that colocalization of factor VIII and VWF in extrahepatic cells is necessary for *in vivo* release of factor VIII after desmopressin.[133]

### B. Mechanisms of Desmopressin Efficacy in Patients with Normal Factor VIII and VWF

**1. Released VWF: A Biologically Plausible but Unproven Mediator.** A puzzling question is how desmopressin is efficacious in patients who have normal or even high levels of factor VIII and VWF. It is biologically plausible that the favorable effects of the compound may be mediated by increased platelet adhesion to the vessel wall[134] due not only to the rise in plasma VWF but also to the abluminal secretion of the protein toward the subendothelium[127] and by the fresh appearance in plasma of ultralarge VWF multimers.[14] These are hemostatically very effective because they support platelet adhesion to the vascular subendothelium to a higher degree than other VWF multimers and induce directly platelet aggregation under conditions of high shear. It has been shown that the infusion of desmopressin improves the formation of platelet aggregates that form at the high shear rates that can be found in the microcirculation. This effect of desmopressin can be observed not only in patients with VWD type 1[135,136] but also in those with normal VWF, whose impairment of platelet aggregation at high shear is due to congenital or drug-induced abnormalities of the secretory mechanisms or of the interaction of released ADP with its platelet receptors. The improvement of platelet aggregation at high shear after desmopressin administration to these patients correlated with the shortening of the bleeding time and the increase in the plasma levels of VWF with ultralarge multimers, suggesting that these changes in VWF may indeed be responsible, at least in part, for the observed effects of desmopressin on primary hemostasis.

### 2. Mechanisms Independent of Released VWF: Proven but Uncharacterized.

Although it is biologically plausible that the potentiation of platelet function is mediated by desmopressin-induced release of VWF, there is no direct evidence that it is responsible for the effects of the drug observed *in vivo*. In contrast, clear evidence exists that other, unknown mechanisms are operating *in vivo*. Desmopressin infusion in patients with VWD type 3 further shortened their prolonged bleeding times, which had been partially corrected by the administration of cryoprecipitate.[137] Since type 3 VWD patients lack VWF in tissue stores even after replacement therapy with cryoprecipitate, the beneficial effect of desmopressin on bleeding times was not associated with an increase in plasma VWF level, nor with the appearance of ultralarge VWF multimers. These results unequivocally indicated that the drug can affect primary hemostasis independently of released VWF. Subsequent studies in rabbits, which do not respond to desmopressin infusion with an increase in the plasma levels of factor VIII and VWF, gave further support to the demonstration that the drug can affect primary hemostasis independently of released VWF. In rabbits whose bleeding times had been prolonged by the combined treatment with aspirin and the thrombolytic agent streptokinase, desmopressin infusion shortened the prolonged bleeding times without increasing the plasma levels of VWF.[75] Another, more indirect, demonstration of a VWF-independent mechanism derives from the observation that desmopressin shortens the prolonged bleeding times of patients with Bernard–Soulier syndrome (Table 67-4), who lack GPIb, the platelet receptor for VWF that is essential for platelet adhesion and activation at high shear. In addition, it has been shown that the enhancement of platelet interaction with the subendothelium that was observed after desmopressin infusion in some studies could not be mimicked by the addition of VWF *in vitro* in amounts designed to match the postinfusion concentrations of the protein attained *in vivo*.[137,138] Effects of desmopressin on the platelet count and on agonist-induced platelet aggregation have been ruled out by many studies.[1,43,139] Several putative mechanisms or mediators other than VWF have been proposed. For example, the compound has been shown to induce the adhesion of erythrocytes to the endothelium[140]; decrease the endothelial production of 13-hydroxyoctadeca-

dienoic acid, a derivative of linoleic acid that powerfully inhibits platelet adhesion to the vessel wall[141]; and increase the expression of tissue factor by cultured endothelial cells.[142] The role of these mechanisms is uncertain and the search for additional or alternative mechanisms of action has been unfruitful.

## VI. Conclusions

Desmopressin is efficacious in mild hemophilia and VWD type 1 and usually permits the avoidance of factor VIII/VWF concentrates, with significant reductions in costs. The benefits of this nontransfusional hemostatic agent are not limited to cost savings. Desmopressin may be needed to meet religious requests, such as the avoidance of blood products in Jehovah's Witnesses. Desmopressin is likely to have spared many patients from infection with the human immunodeficiency virus type 1 (HIV). In Italy, where desmopressin was used earlier and more extensively than in other areas of the world, the prevalence of HIV infection in patients with mild hemophilia A (2.1%) is much lower than in the United States.[143]

In patients with congenital defects of platelet function, with the hemostatic abnormalities associated with chronic liver disease and with those induced by the therapeutic use of antiplatelet and anticoagulant agents, desmopressin has been used to prevent or stop bleeding. However, no well-designed clinical trial truly shows the hemostatic efficacy of the compound in these conditions. The widespread use of erythropoietin and the resulting sustained correction of the hemostatic defect make the use of desmopressin unnecessary in the majority of patients with chronic renal insufficiency.

Antifibrinolytic amino acids and aprotinin should be preferred to desmopressin in reducing blood loss and transfusion requirements during cardiac surgery with extracorporeal circulation. The use of desmopressin in surgical operations other than cardiac surgery is not warranted.

## References

1. Mannucci, P. M. (1990). Desmopressin: A nontransfusional agent. *Annu Rev Med, 41,* 55–64.
2. Mannucci, P. M. (1997). Desmopressin (DDAVP) in the treatment of bleeding disorders: The first 20 years. *Blood, 90,* 2515–2521.
3. Mannucci, P. M., Ruggeri, Z. M., Pareti, F. I., et al. (1977). 1-Deamino-8-D-arginine vasopressin: A new pharmacological approach to the management of haemophilia and von Willebrand's diseases. *Lancet, 1,* 869–872.
4. Mannucci, P. M. (1986). Desmopressin (DDAVP) for treatment of disorders of hemostasis. *Prog Hemost Thromb, 8,* 19–45.

5. Mannucci, P. M., Canciani, M. T., Rota, L., et al. (1981). Response of factor VIII/von Willebrand factor to DDAVP in healthy subjects and patients with hemophilia A and von Willebrand's disease. *Br J Haematol, 47,* 283–293.

6. Cash, J. D., Gader, A. M. A., & de Costa, J. (1974). The release of plasminogen activator and factor VIII by LVP, AVP and DDAVP, ATIII, and OT in man. *Br J Haematol, 27,* 363–364.

7. Mannucci, P. M., Aberg, M., Nilsson, I. M., et al. (1975). Mechanism of plasminogen activator and factor VIII increase after vasoactive drugs. *Br J Haematol, 30,* 81–93.

8. Levi, M., de Boer, J. P., Roem, D., et al. (1992). Plasminogen activation *in vivo* upon intravenous infusion of DDAVP. Quantitative assessment of plasmin-alpha2-antiplasmin complex with a novel monoclonal antibody based radio-immunoassay. *Thromb Haemost, 67,* 111–116.

9. Warrier, L., & Lusher, J. M. (1983). DDAVP: A useful alternative to blood components in moderate hemophilia A and von Willebrand's disease. *J Pediatr, 102,* 228–233.

10. Mariani, G., Ciavarella, N., Mazzucconi, M. G., et al. (1984). Evaluation of the effectiveness of DDAVP in surgery and bleeding episodes in hemophilia and von Willebrand's disease. A study of 43 patients. *Clin Lab Hematol, 6,* 229–238.

11. de la Fuente, B., Kasper, C. K., Rickles, F. R., et al. (1985). Response of patients with mild and moderate hemophilia A and von Willebrand disease to treatment with desmopressin. *Ann Intern Med, 103,* 6–14.

12. Mannucci, P. M., Bettega, D., & Cattaneo, M. (1992). Patterns of development of tachyphylaxis in patients with haemophilia and von Willebrand disease after repeated doses of desmopressin (DDAVP). *Br J Haematol, 82,* 87–93.

13. Ruggeri, Z. M., Mannucci, P. M., Lombardi, R., et al. (1982). Multimeric composition of factor VIII/von Willebrand factor following administration of DDAVP: Implications for pathophysiology and therapy of von Willebrand's disease subtypes. *Blood, 58,* 1272–1278.

14. Federici, A. B., Mazurier, C., Berntorp, E., et al. (2004). Biologic response to desmopressin in patients with severe type 1 and type 2 von Willebrand disease: Results of a multicenter European study. *Blood, 103,* 2032–2038.

15. Rose, E. H., & Aledort, L. M. (1993). Nasal spray desmopressin (DDAVP) for mild hemophilia A and von Willebrand disease. *Ann Intern Med, 114,* 563–568.

16. Lethagen, S., Ragnarson, U., & Tennvall, G. (1993). Self-treatment with desmopressin intranasal spray in patients with bleeding disorders: Effect on bleeding symptoms and socioeconomic factors. *Ann Hematol, 66,* 257–260.

17. Rodeghiero, F., Castaman, G., & Mannucci, P. M. (1996). Prospective multicenter study of subcutaneous concentrated desmopressin for home treatment of patients with von Willebrand disease and mild or moderate hemophilia A. *Thromb Haemost, 76,* 692–696.

18. Federici, A. B., Rand, J. H., Bucciarelli, P., et al. (2000). Acquired von Willebrand syndrome: Data from an international registry. *Thromb Haemost, 84,* 345–349.

19. Federici, A. B., Stabile, F., Castaman, G., et al. (1998). Treatment of acquired von Willebrand syndrome in patients with monoclonal gammopathy of uncertain significance: Comparison of three different therapeutic approaches. *Blood, 92,* 2707–2711.

20. Kobrinsky, N. L., Israels, E. D., Gerrard, J. M., et al. (1984). Shortening of bleeding time by 1-deamino-8-D-arginine vasopressin in various bleeding disorders. *Lancet, 1,* 1145–1148.

21. Kohler, M., Hellstern, P., Morgenstern, E., et al. (1985). Gray platelet syndrome: Selective alpha-granule deficiency and thrombocytopenia due to increased platelet turnover. *Blut, 50,* 331–340.

22. Mannucci, P. M., Vicente, V., Vianello, L., et al. (1986). Controlled trial of desmopressin in liver cirrhosis and other conditions associated with a prolonged bleeding time. *Blood, 67,* 1148–1153.

23. De Marco, L., Girolami, A., Zimmerman, T. S., et al. (1986). von Willebrand factor interaction with the glycoprotein IIb/IIIa complex. Its role in platelet function as demonstrated in patients with congenital afibrinogenemia. *J Clin Invest, 77,* 1272–1277.

24. Pfueller, S. L., Howard, M. A., White, J. G., et al. (1987). Shortening of bleeding time by 1-deamino-8-arginine vaso-pressin (DDAVP) in the absence of platelet von Willebrand factor in Gray platelet syndrome. *Thromb Haemost, 58,* 1060–1063.

25. Schulman, S., Johnsson, H., Egberg, N., et al. (1987). DDAVP-induced correction of prolonged bleeding time in patients with congenital platelet function defects. *Thromb Res, 45,* 165–174.

26. Kentro, T. B., Lottenberg, R., & Kitchens, C. S. (1987). Clinical efficacy of desmopressin acetate for hemostatic control in patients with primary platelet disorders undergoing surgery. *Am J Hematol, 24,* 215–219.

27. Sieber, P. R., Belis, J. A., Jarowenko, M. V., et al. (1988). Desmopressin control of surgical hemorrhage secondary to prolonged bleeding time. *J Urol, 139,* 1066–1067.

28. Cuthbert, R. J., Watson, H. H., Handa, S. I., et al. (1988). DDAVP shortens the bleeding time in Bernard–Soulier syndrome. *Thromb Res, 49,* 649–650.

29. Nieuwenhuis, H. K., & Sixma, J. J. (1988). 1-Desamino-8-D-arginine vasopressin (desmopressin) shortens the bleeding time in storage pool deficiency. *Ann Intern Med, 108,* 65–67.

30. Kim, H. C., Salva, K., Fallot, P. L., et al. (1988). Patients with prolonged bleeding time of undefined etiology, and their response to desmopressin. *Thromb Haemost, 59,* 221–224.

31. Wijermans, P. W., & van Dorp, D. B. (1989). Hermansky–Pudlak syndrome: Correction of bleeding time by 1-desamino-8D-arginine vasopressin. *Am J Hematol, 30,* 154–157.

32. Van Dorp, D. B., Wijermans, P. W., Meire, F., et al. (1990). The Hermansky–Pudlak syndrome. Variable reaction to 1-desamino-8D-arginine vasopressin for correction of the bleeding time. *Ophthalmic Paediatr Genet, 11,* 237–244.

33. Di Michele, D. M., & Hathaway, W. E. (1990). Use of DDAVP in inherited and acquired platelet dysfunctions. *Am J Hematol, 33,* 39–45.

34. Waldenstrom, E., Holmberg, L., Axelsson, U., et al. (1991). Bernard–Soulier syndrome in two Swedish families: Effect of DDAVP on bleeding time. *Eur J Haematol, 46,* 182–187.

35. Castaman, G., & Rodeghiero, F. (1992). Failure of DDAVP to shorten the prolonged bleeding time of two patients with congenital afibrinogenemia. *Thromb Res, 68,* 309–315.

36. Castaman, G., & Rodeghiero, F. (1993). Consistency of responses to separate desmopressin infusions in patients with storage pool disease and isolated prolonged bleeding time. *Thromb Res, 69,* 407–412.

37. Greinacher, A., Potzsch, B., Kiefel, V., et al. (1993). Evidence that DDAVP transiently improves hemostasis in Bernard–Soulier syndrome independent of von Willebrand-factor. *Ann Hematol, 67,* 149–150.

38. Lekas, M. D., & Crowley, J. P. (1993). Easy bruisability, aspirin intolerance, and response to DDAVP. *Laryngoscope, 103,* 156–159.

39. Prinsley, P., Wood, M., & Lee, C. A. (1993). Adenotonsillectomy in patients with inherited bleeding disorders. *Clin Otolaryngol, 18,* 206–208.

40. Kemahli, S., Canatan, D., Uysal, Z., et al. (1994). DDAVP shortens the bleeding time in Bernard–Soulier syndrome. *Thromb Haemost, 71,* 675.

41. Cattaneo, M., Zighetti, M. L., Lombardi, R., et al. (1994). Role of ADP in platelet aggregation at high shear: Studies in a patient with congenital defect of platelet responses to ADP. *Br J Haematol, 88,* 826–829.

42. Rao, A. K., Ghosh, S., Sun, L., et al. (1995). Mechanisms of platelet dysfunction and response to DDAVP in patients with congenital platelet function defects. A double-blind placebo-controlled trial. *Thromb Haemost, 74,* 1071–1078.

43. Cattaneo, M., Pareti, F. I., Zighetti, M., et al. (1995). Platelet aggregation at high shear is impaired in patients with congenital defects of platelet secretion and is corrected by DDAVP: Correlation with the bleeding time. *J Lab Clin Med, 125,* 540–547.

44. Noris, P., Arbustini, E., Spedini, P., et al. (1998). A new variant of Bernard–Soulier syndrome characterized by dysfunctional glycoprotein (GP) Ib and severely reduced amounts of GPIX and GPV. *Br J Haematol, 103,* 1004–1013.

45. Cattaneo, M., Lecchi, A., Agati, B., et al. (1999). Evaluation of platelet function with the PFA-100 system in patients with congenital defects of platelet secretion. *Thromb Res, 96,* 213–217.

46. Zatik, J., Poka, R., Borsos, A., et al. (2002). Variable response of Hermansky–Pudlak syndrome to prophylactic administration of 1-desamino-8D-arginine in subsequent pregnancies. *Eur J Obstet Gynecol Reprod Biol, 104,* 165–166.

47. Fuse, I., Higuchi, W., Mito, M., et al. (2003). DDAVP normalized the bleeding time in patients with congenital platelet TxA2 receptor abnormality. *Transfusion, 43,* 563–567.

48. Fuse, I., Higuchi, W., & Aizawa, Y. (2003). 1-deamino-8-D-arginine vasopressin (DDAVP) normalized the bleeding time in patients with platelet disorder characterized by defective calcium ionophore-induced platelet aggregation. *Br J Haematol, 122,* 870–871.

49. Cordova, A., Barrios, N. J., Ortiz, I., et al. (2005). Poor response to desmopressin acetate (DDAVP) in children with Hermansky–Pudlak syndrome. *Pediatr Blood Cancer, 44,* 51–54.

50. Sehbai, A. S., Abraham, J., & Brown, V. K. (2005). Perioperative management of a patient with May–Hegglin anomaly requiring craniotomy. *Am J Hematol, 79,* 303–308.

51. Janson, P. A., Jubelirer, S. J., Weinstein, M. S., et al. (1980). Treatment of bleeding tendency in uremia with cryoprecipitate. *N Engl J Med, 303,* 1318–1321.

52. Watson, A. J., & Keogh, J. A. (1984). 1-Deamino-8-D-arginine vasopressin as a therapy for the bleeding diathesis of renal failure. *Am J Nephrol, 4,* 49–51.

53. Mannucci, P. M., Remuzzi, G., Pusineri, F., et al. (1983). Deamino-8-D-arginine vasopressin shortens the bleeding time in uremia. *N Engl J Med, 308,* 8–12.

54. Viganò, G. L., Mannucci, P. M., Lattuada, A., et al. (1989). Subcutaneous desmopressin (DDAVP) shortens the bleeding time in uremia. *Am J Hematol, 31,* 32–35.

55. Shapiro, M. D., & Kelleher, S. P. (1984). Intranasal deamino-8-D-arginine vasopressin shortens the bleeding time in uremia. *Am J Nephrol, 4,* 260–261.

56. Rydzewski, A., Rowinski, M., & Mysliwiec, M. (1986). Shortening of bleeding time after intranasal administration of 1-deamino-8-D-arginine vasopressin to patients with chronic uremia. *Folia Haematol Int Mag Klin Morphol Blutforsch, 113,* 823–830.

57. Canavese, C., Salomone, M., Pacitti, A., et al. (1985). Reduced response of uraemic bleeding time to repeated doses of desmopressin. *Lancet, 1,* 867–868.

58. Gotti, E., Mecca, G., Valentino, C., et al. (1984). Renal biopsy in patients with acute renal failure and prolonged bleeding time. *Lancet, 2,* 978–979.

59. Watson, A. J., & Keogh, J. A. (1982). Effect of 1-deamino-8-D-arginine vasopressin on the prolonged bleeding time in chronic renal failure. *Nephron, 32,* 49–52.

60. Gotti, E., Mecca, G., Valentino, C., et al. (1985). Renal biopsy in patients with acute renal failure and prolonged bleeding time: A preliminary report. *Am J Kidney Dis, 6,* 397–399.

61. Buckley, D. J., Barrett, A. P., Koutts, J., et al. (1986). Control of bleeding in severely uremic patients undergoing oral surgery. *Oral Surg Oral Med Oral Pathol, 61,* 546–549.

62. Juhl, A., & Jorgensen, F. (1987). DDAVP and life-threatening diffuse gastric bleeding in uraemia. Case report. *Acta Chir Scand, 153,* 75–77.

63. Donovan, K. L., Moore, R. H., Mulkerrin, E., et al. (1992). An audit of appropriate tests in renal biopsy coagulation screens. *Am J Kidney Dis, 19,* 335–338.

64. Sateriale, M., Cronan, J. J., & Savadler, L. D. (1991). A 5-year experience with 307 CT-guided renal biopsies: Results and complications. *J Vasc Interv Radiol, 2,* 401–407.

65. Livio, M., Mannucci, P. M., Viganò, G., et al. (1986). Conjugated estrogens for the management of bleeding associated with renal failure. *N Engl J Med, 315,* 731–735.

66. Moia, M., Mannucci, P. M., Vizzotto, L., et al. (1987). Improvement in the hemostatic defect of uremia after

treatment with recombinant human erythropoietin. *Lancet, 2*, 1227–1229.

67. Burroughs, A. K., Matthews, K., Qadiri, M., et al. (1985). Desmopressin and bleeding time in patients with cirrhosis. *Br J Med, 291*, 1377–1381.

68. Lopez, P., Otaso, J. C., Alvarez, D., et al. (1997). Hemostatic and hemodynamic effects of vasopressin analogue DDAVP in patients with cirrhosis. *Acta Gastroenterol Latinoam, 27*, 59–62.

69. Cattaneo, M., Tenconi, P. M., Alberca, I., et al. (1990). Subcutaneous desmopressin (DDAVP) shortens the prolonged bleeding time in patients with liver cirrhosis. *Thromb Haemost, 64*, 358–360.

70. de Franchis, F., Arcidiacono, P. G., Carpinelli, P. G., et al. (1993). Randomized controlled trial of desmopressin plus terlipressin and terlipressin alone for the treatment of acute variceal hemorrhage in cirrhotic patients: A multicenter, double blind study. *Hepatology, 18*, 1102–1107.

71. Parker, R. I., Grewal, R. P., McKeown, L. P., et al. (1992). Effect of platelet count on the DDAVP-induced shortening of the bleeding time in thrombocytopenic Gaucher's patients. *Am J Pediatr Hematol Oncol, 14*, 39–43.

72. Castaman, G., Bona, E. D., Schiavotto, C., et al. (1997). Pilot study on the safety and efficacy of desmopressin for the treatment or prevention of bleeding in patients with hematologic malignancies. *Haematologica, 82*, 584–587.

73. Cattaneo, M., Lombardi, R., Bettega, D., et al. (1993). Shear-induced platelet aggregation is potentiated by desmopressin and inhibited by ticlopidine. *Arterioscler Thromb, 13*, 393–397.

74. Schulman, S., & Johnsson, H. (1991). Heparin, DDAVP and the bleeding time. *Thromb Hemost, 65*, 242–244.

75. Johnstone, M. T., Andrews, T., Ware, J. A., et al. (1990). Bleeding time prolongation with streptokinase and its reduction with 1-deamino-8-D-arginine vasopressin. *Circulation, 82*, 2142–2151.

76. Bove, C. M., Casey, B., & Marder, V. J. (1996). DDAVP reduces bleeding during continued hirudin administration in the rabbit. *Thromb Haemost, 75*, 471–475.

77. Reiter, R. A., Mayr, F., Blazicek, H., et al. (2003). Desmopressin antagonizes the *in vitro* platelet dysfunction induced by GPIIb/IIIa inhibitors and aspirin. *Blood, 102*, 4594–4599.

78. Chard, R. B., Kam, C. A., Nunn, G. R., et al. (1990). Use of desmopressin in the management of aspirin-related and intractable haemorrhage after cardiopulmonary bypass. *Aust N Z J Surg, 60*, 125–128.

79. Kam, P. C. (1994). Use of desmopressin (DDAVP) in controlling aspirin-induced coagulopathy after cardiac surgery. *Heart Lung, 23*, 333–336.

80. Stine, K. C., & Becton, D. L. (1997). DDAVP therapy controls bleeding in Ehlers–Danlos syndrome. *J Pediatr Hematol Oncol, 19*, 156–158.

81. Quitt, M., Froom, P., Veisler, A., et al. (1990). The effect of desmopressin on massive gastrointestinal bleeding in hereditary telangiectasia unresponsive to treatment with cryoprecipitate. *Arch Intern Med, 150*, 1744–1746.

82. Baldree, L. A., Ault, B. H., Chesney, C. M., et al. (1990). Intravenous desmopressin acetate in children with sickle trait and persistent macroscopic hematuria. *Pediatrics, 86*, 238–243.

83. Moudgil, A., & Kamil, E. S. (1996). Protracted, gross hematuria in sickle cell trait: Response to multiple doses of 1-desamino-8-D-arginine vasopressin. *Pediatr Nephrol, 10*, 210–212.

84. Salzman, E. W., Weinstein, M. J., Weintraub, R. M., et al. (1986). Treatment with desmopressin acetate to reduce blood loss after cardiac surgery. A double-blind randomized trial. *N Engl J Med, 314*, 1402–1406.

85. Rocha, E., Llorens, R., Paramo, J. A., et al. (1988). Does desmopressin acetate reduce blood loss after surgery in patients on cardiopulmonary bypass? *Circulation, 77*, 1319–1323.

86. Hackmann, T., Gascoyne, R. D., Naiman, S. C., et al. (1989). A trial of desmopressin (1-desamino-8-D-arginine vasopressin) to reduce blood loss in uncomplicated cardiac surgery. *N Engl J Med, 321*, 1437–1443.

87. Seear, M. D., Wadsworth, L. D., Rogers, P. C., et al. (1989). The effect of desmopressin acetate (DDAVP) on postoperative blood loss after cardiac operations in children. *J Thorac Cardiovasc Surg, 98*, 217–219.

88. Lazenby, W. D., Russo, I., Zadeh, B. J., et al. (1990). Treatment with desmopressin acetate in routine coronary artery bypass surgery to improve postoperative hemostasis. *Circulation, 82*, IV413–IV419.

89. Andersson, T. L., Solem, J. O., Tengborn, L., et al. (1990). Effects of desmopressin acetate on platelet aggregation, von Willebrand factor, and blood loss after cardiac surgery with extracorporeal circulation. *Circulation, 81*, 872–878.

90. Hedderich, G. S., Petsikas, D. J., Cooper, B. A., et al. (1990). Desmopressin acetate in uncomplicated coronary artery bypass surgery: A prospective randomized clinical trial. *Can J Surg, 33*, 33–36.

91. Reich, D. L., Hammerschlag, B. C., Rand, J. H., et al. (1991). Desmopressin acetate is a mild vasodilator that does not reduce blood loss in uncomplicated cardiac surgical procedures. *J Cardiothorac Vasc Anesth, 5*, 142–145.

92. Horrow, J. C., van Riper, D. F., Strong, M. D., et al. (1991). Hemostatic effects of tranexamic acid and desmopressin during cardiac surgery. *Circulation, 84*, 2063–2070.

93. Gratz, I., Koehler, J., Olsen, D., et al. (1992). The effect of desmopressin acetate on postoperative hemorrhage in patients receiving aspirin therapy before coronary artery bypass operations. *J Thorac Cardiovasc Surg, 104*, 1417–1422.

94. de Prost, D., Barbier-Boehm, G., Hazebroucq, J., et al. (1992). Desmopressin has no beneficial effect on excessive postoperative bleeding or blood product requirements associated with cardiopulmonary bypass. *Thromb Haemost, 68*, 106–110.

95. Ansell, J., Klassen, V., Lew, R., et al. (1992). Does desmopressin acetate prophylaxis reduce blood loss after valvular heart operations? A randomized, double-blind study. *J Thorac Cardiovasc Surg, 104*, 117–123.

96. Mongan, P. D., & Hosking, M. P. (1992). The role of desmopressin acetate in patients undergoing coronary artery bypass surgery. A controlled clinical trial with thromboelastographic risk stratification. *Anesthesiology, 77,* 38–46.

97. Marquez, J., Koehler, S., Strelec, S. R., et al. (1992). Repeated dose administration of desmopressin acetate in uncomplicated cardiac surgery: A prospective, blinded, randomized study. *J Cardiothorac Vasc Anesth, 6,* 674–676.

98. Dilthey, G., Dietrich, W., Spannagl, M., et al. (1993). Influence of desmopressin acetate on homologous blood requirements in cardiac surgical patients pretreated with aspirin. *J Cardiothorac Vasc Anesth, 7,* 425–430.

99. Reynolds, L. M., Nicolson, S. C., Jobes, D. R., et al. (1993). Desmopressin does not decrease bleeding after cardiac operation in young children. *J Thorac Cardiovasc Surg, 106,* 954–958.

100. Rocha, E., Hidalgo, F., Llorens, R., et al. (1994). Randomized study of aprotinin and DDAVP to reduce postoperative bleeding after cardiopulmonary bypass surgery. *Circulation, 90,* 921–927.

101. Sheridan, D. P., Card, R. T., Pinilla, J. C., et al. (1994). Use of desmopressin acetate to reduce blood transfusion requirements during cardiac surgery in patients with acetylsalicylic-acid-induced platelet dysfunction. *Can J Surg, 37,* 33–36.

102. Temeck, B. K., Bachenheimer, L. C., Katz, N. M., et al. (1994). Desmopressin acetate in cardiac surgery: A double-blind, randomized study. *South Med J, 87,* 611–615.

103. Casas, J. I., Zuazu-Jausoro, I., Mateo, J., et al. (1995). Aprotinin versus desmopressin for patients undergoing operations with cardiopulmonary bypass. A double-blind placebo-controlled study. *J Thorac Cardiovasc Surg, 110,* 1107–1117.

104. Despotis, G. J., Levine, V., Saleem, R., et al. (1999). Use of point-of-care test in identification of patients who can benefit from desmopressin during cardiac surgery: A randomised controlled trial. *Lancet, 354,* 106–110.

105. Oliver, W. C., Jr., Santrach, P. J., Danielson, G. K., et al. (2000). Desmopressin does not reduce bleeding and transfusion requirements in congenital heart operations. *Ann Thorac Surg, 70,* 1923–1930.

106. Cattaneo, M., Harris, A. S., Stromberg, U., et al. (1995). The effect of desmopressin on reducing blood loss in cardiac surgery — A meta-analysis of double-blind, placebo-controlled trials. *Thromb Haemost, 74,* 1064–1070.

107. Fremes, S. E., Wong, B. I., Lee, E., et al. (1994). Metaanalysis of prophylactic drug treatment in the prevention of postoperative bleeding. *Ann Thorac Surg, 58,* 1580–1588.

108. Levi, M., Cromheecke, M. E., de Jonge, E., et al. (1999). Pharmacological strategies to decrease excessive blood loss in cardiac surgery: A meta-analysis of clinically relevant endpoints. *Lancet, 354,* 1940–1947.

109. Kobrinsky, N. L., Letts, R. M., Patel, L. R., et al. (1987). 1-Desamino-8-D-arginine vasopressin (desmopressin) decreases operative blood loss in patients having Harrington rod spinal fusion surgery. A randomized, double-blinded, controlled trial. *Ann Intern Med, 107,* 446–450.

110. Guay, J., Rainberg, C., Poitras, B., et al. (1992). A trial of desmopressin to reduce blood loss in patients undergoing spinal fusion for idiopathic scoliosis. *Anesth Analg, 75,* 405–410.

111. Theroux, M. C., Corddry, D. H., Tietz, A. E., et al. (1997). A study of desmopressin and blood loss during spinal fusion for neuromuscular scoliosis: A randomized, controlled, double-blinded study. *Anesthesiology, 87,* 260–267.

112. Alanay, A., Acaroglu, E., Ozdemir, O., et al. (1999). Effects of deamino-8-D-arginin vasopressin on blood loss and coagulation factors in scoliosis surgery. A double-blind randomized clinical trial. *Spine, 24,* 877–882.

113. Flordal, P. A., Ljungstrom, K. G., Ekman, B., et al. (1992). Effects of desmopressin on blood loss in hip arthroplasty. Controlled study in 50 patients. *Acta Orthop Scand, 63,* 381–385.

114. Schott, U., Sollen, C., Axelsson, K., et al. (1995). Desmopressin acetate does not reduce blood loss during total hip replacement in patients receiving dextran. *Acta Anaesthesiol Scand, 39,* 592–598.

115. Karnezis, T. A., Stulberg, S. D., Wixson, R. L., et al. (1994). The hemostatic effects of desmopressin on patients who had total joint arthroplasty. A double-blind randomized trial. *J Bone Joint Surg Am, 76,* 1545–1450.

116. Haith, L. R., Patton, M. L., Goldman, W. T., et al. (1993). Diminishing blood loss after operation for burns. *Surg Gynaecol Obstet, 176,* 119–123.

117. Lethagen, S., Rugarn, P., & Bergqvist, D. (1991). Blood loss and safety with desmopressin or placebo during aorto-iliac graft surgery. *Eur J Vasc Surg, 5,* 173–178.

118. Clagett, G. P., Valentine, R. J., Myers, S. I., et al. (1995). Does desmopressin improve hemostasis and reduce blood loss from aortic surgery? A randomized, double-blind study. *J Vasc Surg, 22,* 223–229.

119. Smith, T. J., Gill, J. C., Ambruso, D. R., et al. (1989). Hyponatremia and seizures in young children given DDAVP. *Am J Hematol, 31,* 199–202.

120. Bertholini, D. M., & Butler, C. S. (2000). Severe hyponatraemia secondary to desmopressin therapy in von Willebrand's disease. *Anaesth Intens Care, 28,* 199–201.

121. Cohen, A. J., Kessler, C. M., & Ewenstein, B. M. (2001). Management of von Willebrand disease: A survey on current clinical practice from the haemophilia centres of North America. *Haemophilia, 7,* 235–241.

122. Kujovich, J. L. (2005). Von Willebrand disease and pregnancy. *J Thromb Haemost, 3,* 246–253.

123. Ray, J. G. (1998). DDAVP use during pregnancy: An analysis of its safety for mother and child. *Obstet Gynecol Surv, 53,* 450–455.

124. Mannucci, P. M. (2005). Use of desmopressin (DDAVP) during early pregnancy in factor VIII-deficient women. *Blood, 105,* 3382.

125. Mannucci, P. M., Carlsson, S., & Harris, A. S. (1994). Desmopressin, surgery and thrombosis. *Thromb Haemost, 71,* 154–155.

126. Kanwar, S., Woodman, R. C., Poon, M. C., et al. (1995). Desmopressin induces endothelial P-selectin expression and leukocyte rolling in postcapillary venules. *Blood, 86,* 2760–2766.

127. Takeuchi, M., Naguza, H., & Kanedu, T. (1988). DDAVP and epinephrine induce changes in the localization of von Willebrand factor antigen in endothelial cells of human oral mucosa. *Blood, 72,* 850–854.

128. Booyse, E. M., Osikowicz, G., & Fedr, S. (1981). Effects of various agents on ristocetin–Willebrand factor activity in long-term cultures of von Willebrand and normal human umbilical vein endothelial cells. *Thromb Haemost, 46,* 668.

129. Kaufmann, J. E., Oksche, A., Wollheim, C. B., et al. (2000). Vasopressin-induced von Willebrand factor secretion from endothelial cells involves V2 receptors and cAMP. *J Clin Invest, 106,* 107–116.

130. Cattaneo, M., Simoni, L., Gringeri, A., et al. (1994). Patients with severe von Willebrand disease are insensitive to the releasing effect of DDAVP: Evidence that the DDAVP-induced increase in plasma factor VIII is not secondary to the increase in plasma von Willebrand factor. *Br J Haematol, 86,* 333–337.

131. Kaufmann, J. E., & Vischer, U. M. (2003). Cellular mechanisms of the hemostatic effects of desmopressin (DDAVP). *J Thromb Haemostas, 1,* 682–689.

132. Xu, L., Nichols, T. C., Sarkar, R., et al. (2005). Absence of a desmopressin response after therapeutic expression of factor VIII in hemophilia A dogs with liver-directed neonatal gene therapy. *Proc Natl Acad Sci USA, 102,* 6080–6085.

133. Lamont, P. A., & Ragni, M. V. (2005). Lack of desmopressin (DDAVP) response in men with hemophilia A following liver transplantation. *J Thromb Haemost, 3,* 2259–2263.

134. Sakariassen, K. S., Cattaneo, M., van den Berg, A., et al. (1984). DDAVP enhances platelet adherence and platelet aggregate growth on human artery subendothelium. *Blood, 64,* 229–236.

135. Pareti, F. I., Cattaneo, M., Carpinelli, L., et al. (1996). Evaluation of the abnormal platelet function in von Willebrand disease by the blood filtration test. *Thromb Haemost, 76,* 460–468.

136. Cattaneo, M., Federici, A. B., Lecchi, A., et al. (1999). Evaluation of the PFA-100 system in the diagnosis and therapeutic monitoring of patients with von Willebrand disease. *Thromb Haemost, 82,* 35–39.

137. Cattaneo, M., Moia, M., Delle Valle, P., et al. (1989). DDAVP shortens the prolonged bleeding times of patients with severe von Willebrand disease treated with cryoprecipitate. Evidence for a mechanism of action independent of released von Willebrand factor. *Blood, 74,* 1972–1975.

138. Escolar, G., Cases, A., Monteagudo, J., et al. (1989). Uremic plasma after infusion of desmopressin (DDAVP) improves the interaction of normal platelets with vessel subendothelium. *J Lab Clin Med, 114,* 36–42.

139. Lethagen, S., & Nilssonm, I. M. (1992). DDAVP-induced enhancement of platelet retention: Its dependence on platelet-von Willebrand factor and the platelet receptor GP IIb/IIIa. *Eur J Haematol, 49,* 7–13.

140. Tsai, J.-M., Sussman, I. I., Nagel, R. L., et al. (1990). Desmopressin induces adhesion of normal human erythrocytes to the endothelial surface of a perfused microvascular preparation. *Blood 75,* 261–265.

141. Setty, B. N., Dampier, C. D., & Stuart, M. J. (1992). 1-Deamino-8-D-arginine vasopressin decreases the production of 13-hydroxyoctadecadienoic acid by endothelial cells. *Thromb Res, 67,* 545–558.

142. Galvez, A., Gomez-Ortiz, G., Diaz-Ricart, M., et al. (1997). Desmopressin (DDAVP) enhances platelet adhesion to the extracellular matrix of cultured human endothelial cells through increased expression of tissue factor. *Thromb Haemost, 77,* 975–980.

143. Mannucci, P. M., & Ghirardini, A. (1997). Desmopressin: Twenty years after. *Thromb Haemost, 78,* 958.

144. Edlund, M., Blomback, M., & Fried, G. (2002). Desmopressin in the treatment of menorrhagia in women with no common coagulation factor deficiency but with prolonged bleeding time. *Blood Coagul Fibrinolysis, 13,* 225–231.

145. Kentro, T. B., Lottenberg, R., & Kitchens, C. S. (1987). Clinical efficacy of desmopressin acetate for hemostatic control in patients with primary platelet disorders undergoing surgery. *Am J Hematol, 24,* 215–219.

146. Sieber, P. R., Belis, J. A., Jarowenko, M. V., et al. (1988). Desmopressin control of surgical hemorrhage secondary to prolonged bleeding time. *J Urol, 139,* 1066–1067.

147. Flordal, P. A., Ljungstrom, K. G., & Svensson, J. (1989). Desmopressin reverses effects of dextran on von Willebrand factor. *Thromb Hemost, 61,* 541.

148. Yasui, H., Adachi, Y., Minami T, et al. (2003). Combination therapy of DDAVP and conjugated estrogens for a recurrent large subcutaneous hematoma in Ehlers–Danlos syndrome. *Am J Hematol, 72,* 71–72.

149. Nacul, F. E., de Moraes, E., Penido, C., et al. (2004). Massive nasal bleeding and hemodynamic instability associated with clopidogrel. *Pharm World Sci, 26,* 6–7.

# CHAPTER 68

# Factor VIIa

## Man-Chiu Poon

*Departments of Medicine, Pediatrics and Oncology, University of Calgary and Calgary Health Region, Calgary, Alberta, Canada*

## I. Introduction

Bleeding in patients with platelet disorders can often be treated by conservative means, including local compression, gelatin sponge, local hemostatic agents (e.g., topical thrombin or fibrin glue), hormonal manipulation (e.g., birth control pills), and antifibrinolytics (e.g., ε-aminocaproic acid or tranexamic acid). In some patients, desmopressin acetate (DDAVP) may be effective (see Chapter 67). When these measures fail, or in the case of severe bleeding or surgery, transfusion of platelet concentrates is required. Repeated transfusion of platelets, however, may result in the development of allergic reactions as well as the development of alloantibodies that may render future platelet transfusion ineffective (see Chapter 69). Blood products also carry a risk of transmission of blood-borne infection.[1,2] It is therefore desirable to have available a safe and effective alternative agent to platelet transfusions. This is especially important for patients who are already refractory to platelet transfusion and for patients who live in areas where platelets are not readily available. Recombinant activated factor VII (rFVIIa) has been used extensively in patients with inhibitors to factor VIII (FVIII) or factor IX (FIX) (congenital or acquired hemophilia A and B) with good efficacy and a good safety record.[3–8] In the past several years, investigators have begun to explore the use of rFVIIa as an alternative to platelet transfusion. This chapter focuses on the experience with rFVIIa for the treatment of bleeding in patients with quantitative and qualitative platelet defects and the possible mechanisms of action of high-dose rFVIIa in these platelet disorders.

## II. Pharmacology

Factor VII (FVII), a vitamin K-dependent coagulation protein, is a single-chain glycoprotein (approximately 50 kDa) that is synthesized in the liver.[9] FVII circulates as a zymogen composed of 406 amino acids at a plasma concentration of approximately 10 nM.[9] Activated factor VII (FVIIa), a disulfide-linked two-chain procoagulant enzyme active only when complexed to tissue factor (TF) (Fig. 68-1), is formed by cleavage of the zymogen at Arg152–Ile153 and circulates in a low concentration (approximately 100 pM).[9] Hemostasis is initiated by the interaction between TF and FVIIa. TF is normally not exposed to flowing blood, but it is found in various cells in the deeper layer of the blood vessel wall. Thus, in normal physiological circumstances, clotting is initiated at the site of tissue injury, and FVIIa in the circulation is otherwise proteolytically inert. In this TF-dependent pathway, normal thrombin generation also requires the presence of factors IX, VIII, X, and V and prothrombin. Other cells that express TF include some tumor cells, monocytes, and neutrophils that have been stimulated by bacterial endotoxin and certain other inflammatory mediators.[10–13] The significance to normal hemostasis of free plasma TF, reported by Giesen et al.,[12] remains to be clarified.

rFVIIa is produced by biotechnology[14] and is supplied lyophilized in vials for intravenous injection after reconstitution. Baby hamster kidney (BHK) cells cultured in medium containing fetal calf serum are transfected with the human FVII gene. The rFVII produced undergoes autoactivation to rFVIIa during the purification procedure, which includes detergent treatment, ion-exchange chromatography, and mouse monoclonal antibody immunoaffinity chromatography. Because human proteins or their derivatives are absent in the culturing, processing, purification, and formulation procedures, there is virtually no risk of transmission of infectious agents of human origin.

The pharmacokinetics of rFVIIa have been studied in patients with hemophilia and FVII deficiency. In adult hemophilia patients with or without inhibitors, after a bolus intravenous rFVIIa infusion at doses of 17.5, 35, and 70 μg/kg, median *in vivo* plasma recovery in the bleeding and nonbleeding state was approximately 46 and approxi-

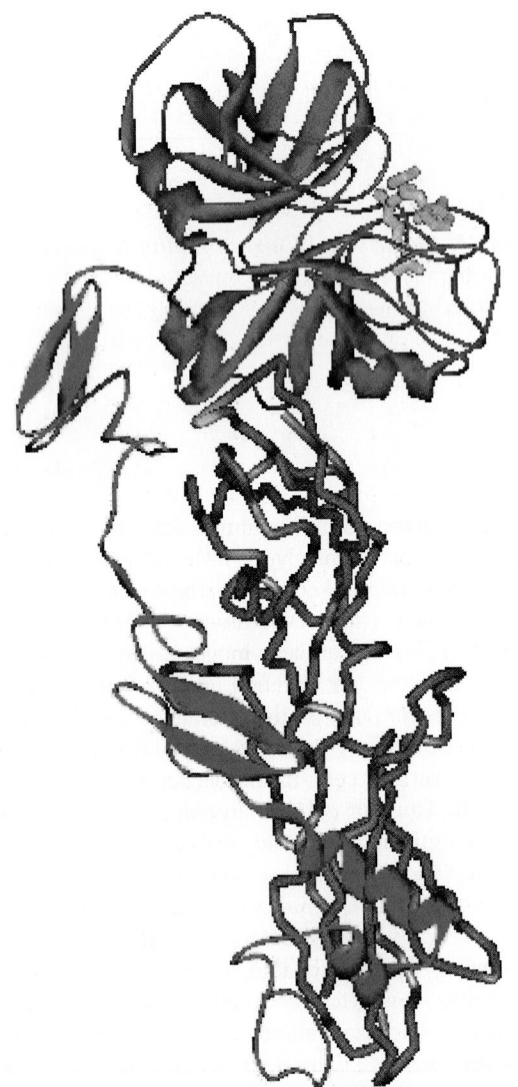

**Figure 68-1.** Molecular structural model of activated factor VII (FVIIa)/tissue factor complex. Shown are the light chain (green) and heavy chain (red) of FVIIa complexed with tissue factor (gray). The catalytic site of FVIIa is occupied by an active site inhibitor (blue). In the light chain (green), the two sets of sheets represent the two EGF-like domains, and the curled ribbons represent helices in the Gla domain. (Figure by Dr. Egon Persson, Department of Haemostasis Biochemistry, Novo Nordisk A/S, Malov, Denmark; generated using WebLab ViewerPro 3.2 and publicly available structural data [PDB code 1dan].)

mately 44%, respectively, and the median half disappearance time ($t_{1/2}$) was approximately 2.3 and approximately 2.9 hours, respectively.[15] The $t_{1/2}$ in patients with FVII deficiency was similar (mean approximately 3 hours, nonbleeding state).[16] In one study, the clearance of rFVIIa was increased in hemophiliac children younger than 15 years of age, with a $t_{1/2}$ of 1.3 or 1.4 hours, determined within 4 hours following infusion of rFVIIa at 114 to 196 μg/kg.[17] Another study[18] showed a $t_{1/2}$ in children similar to that in adults (2.6 vs. 3.1 hours) when the blood sampling following infusion of rFVIIa at 90 to 180 μg/kg was extended to 12 hours. Also, the $t_{1/2}$ in both children and adults was shortened by approximately 1 hour if the calculation was based on blood sample data from within 4 hours of infusion.

The potential for thrombogenicity of rFVIIa has been examined in animal studies. In a rabbit stasis model,[19] infusion of 100 to 1000 μg/kg rFVIIa did not result in significant clot formation over 10 minutes of stasis, although significant clotting occurred over 30 minutes of stasis. Significant fibrinogen and platelet changes were not observed 3 hours after infusion. In another animal study, Turacek et al.[20] found rFVIIa to have low thrombogenic potential, which was enhanced by the addition of soluble TF.

## III. rFVIIa in the Treatment of Thrombocytopenia

Animal studies suggest that high dose rFVIIa could shorten the total hemostasis time[21] and nail cuticle bleeding time[22] in rabbits with thrombocytopenia induced, respectively, by platelet antiserum alone or in combination with gamma radiation. Kristensen and colleagues[23] subsequently showed that in some patients with thrombocytopenia, the bleeding time could be modestly shortened (by at least 2 minutes) after administration of rFVIIa at a dose of either 50 or 100 μg/kg. In general, the response rate increased with higher platelet count but was similar with both rFVIIa doses. The difficulty in interpreting the bleeding time results in this study is that the bleeding time method was not the same in all patients; venostasis was performed in one group of patients but not in another. These investigators also treated nine bleeding episodes (three epistaxes, four neck incisions, one epistaxis plus neck incision, and one uterine bleeding) in eight patients with platelet counts between 5 and $33 \times 10^9$/L. Bleeding stopped promptly after rFVIIa in six episodes (platelet counts, $5–33 \times 10^9$/L) and slowed in two (platelet counts, 13 and $19 \times 10^9$/L). Cessation of bleeding was not necessarily accompanied by a shortening or normalization of the bleeding time. No conclusion could be drawn as to which rFVIIa dose (50 or 100 μg/kg) was more effective. Since then, a number of case reports on the use of high-dose rFVIIa in bleeding thrombocytopenic patients have been published.[24] Successful treatment was documented in bleeding episodes in patients with immune thrombocytopenic purpura (ITP)[25–29]; thrombocytopenia induced by chemotherapy, stem cell transplantation, or leukemia[25,30–36]; thrombocytopenia related to hemolysis, elevated liver enzymes, low platelet count (HELLP) syndrome[37–40]; and the Wiskott–Aldrich syndrome.[41] Most of the patients had platelet counts

$<20 \times 10^9$/L. Treatment dosage varied, but most were in the 90-µg/kg range. Failure has been reported for three episodes of hemorrhagic cystitis (with or without gastrointestinal [GI] or pulmonary bleeding) associated with thrombocytopenia following bone marrow transplantation using high-dose rFVIIa (90–270 µg/kg) together with platelet transfusion to maintain a platelet count $>50 \times 10^9$/L,[42] and also for uterine bleeding in a patient with thrombocytopenia (platelet count, $4$–$13 \times 10^9$/L) during induction chemotherapy for acute leukemia.[43] rFVIIa was also used successfully to cover emergency splenectomy and abdominal surgery in three thrombocytopenic patients (two with ITP and one with concurrent platelet transfusion).[44,45] In summary, clinical data on rFVIIa treatment for bleeding and for surgical prophylaxis in patients with thrombocytopenia are scant and are restricted to individual case reports, with the inherent shortcoming of reporting bias in which negative outcomes tend not to be published. There is an urgent need for clinical trials to better assess the clinical efficacy, safety, and optimal treatment regimen of rFVIIa in thrombocytopenic patients.

## IV. rFVIIa in the Treatment of Platelet Function Disorders

Tengborn and Petruson[46] were the first to report the effectiveness of rFVIIa in the treatment of a platelet function disorder in a 2-year-old male with Glanzmann thrombasthenia (GT) and severe epistaxis. Since then, other patients with congenital or acquired platelet function disorders have been treated with rFVIIa for acute bleeding and for surgical prophylaxis. These include patients with GT,[47–61] Bernard–Soulier syndrome,[52,57] platelet storage pool defect,[62,63] platelet-type (pseudo-) von Willebrand disease,[64] and acquired platelet disorders.[65–67]

### A. Congenital Platelet Function Disorders

The majority of reports on the use of rFVIIa in patients with platelet function disorders have been in GT (see Chapter 57), a disorder characterized by a quantitative or qualitative defect in the platelet membrane glycoprotein (GP) IIb-IIIa (integrin αIIbβ3). The rarity of this disorder and the low number of bleeding episodes requiring systemic hemostatic therapy make a randomized clinical trial comparing the efficacy and safety of rFVIIa to platelet transfusion difficult. An international survey[56] included 59 GT patients from 49 centers in 15 countries treated for 108 bleeding episodes and 34 surgical/invasive procedures. Some previously published cases[46,47,49,50,52,68,69] were included in this survey, and patient characteristics and outcomes for the different types of procedures and bleeding episodes are shown in Table 68-1. There are limitations to this survey. First, reporting was retrospective, so reporting bias may exist. Second, treatment regimens were heterogeneous in terms of dosage, dosage intervals, and the number of doses used before switching to alternative therapy (hence considered treatment failure). Despite these limitations, valuable information was obtained. rFVIIa was particularly effective as prophylaxis for a wide variety of surgical and other procedures (94% of evaluable episodes not receiving platelet transfusions concurrently). The overall success rate for bleeding episodes (64%) based on intention-to-treat analysis was lower than the 96% (23 of 24 episodes) obtained in a Canadian pilot study[47] on four GT children, perhaps because of the heterogeneity of treatment regimen and a lack of uniformity in definition of when to declare failure. Ten episodes were declared a failure after only one or two doses, so the treatment may be insufficient. The success rate of treating severe bleeding episodes (defined as intracranial bleeding, bleeding that resulted in a decline in hemoglobin level by ≥20 g/L within 4 days, bleeding that resulted from severe trauma, or bleeding that compressed a vital organ) with bolus rFVIIa administration at a minimum of 80 µg/kg at 2.5-hour intervals or less for at least three doses was significantly higher than those by other modalities, including continuous infusion (77% [24/31] vs. 48% [19/40], $p = 0.010$). Patients given maintenance doses had significantly fewer recurrences compared to those not given any. As indicated in Table 68-1, GI bleeding appeared to be particularly difficult to stop by rFVIIa. Possible reasons for this include the ineffectiveness of continuous infusion (used in four episodes), inadequate number of doses used (two episodes using one and three doses, respectively), and the severe nature of the bleeding lesion (four with angiodysplasia and one Mallory–Weiss tear). The overall success rate in this international survey[56] was nonetheless higher than that in a UK study[52] reporting excellent/good results in only 12 of 25 bleeding episodes (48%) in five GT children. The poor results in the latter study were likely related to delayed treatment because good/excellent responses were observed in 10 of 14 episodes (71%) treated within 12 hours of bleeding onset but in only 2 of 11 (18%) episodes treated after 12 hours, confirming the importance of early treatment.

In summary, the following observations can be made:

1. In patients with congenital platelet function disorders and a normal coagulation cascade, rFVIIa appears to be a safe and effective alternative to platelet transfusion for treating bleeding episodes and for surgical prophylaxis with the following indications (in descending order of importance): refractoriness to platelet transfusion, the presence of alloantibodies to HLA or platelet antigens, avoidance of alloimmunization; and

## Table 68-1: rFVIIa Treatment of Patients with Glanzmann Thrombasthenia[a]

**Patient characteristics (N = 59)**

| Male/Female | Age in Years (range) | Age Group, ≤15 years / >15 years | Type 1 Severity[b] | Platelet Antibodies and/or Platelet Refractoriness[c] |
|---|---|---|---|---|
| 24/35 | 22 (1–70) | 27/32 | 39 of 49 known | 35 |

**rFVIIa treatment outcome**

| | n | Success[d] (% of total[e]/% of evaluable) | Recurrence[f] | Failure[g] | Not Evaluable[h] |
|---|---|---|---|---|---|
| Procedures (all) | 34 | 29 (85/94) | — | 2 | 3 |
| Major | 9 | 6 (67/86) | — | 1 | 2 |
| Minor, dental extraction | 9 | 9 (100/100) | — | 0 | 0 |
| Minor, other | 16 | 14 (85/93) | — | 1 | 1 |
| Bleeding (all) | 108 | 69 (64/67) | 8 | 26 | 5 |
| GI | 17 | 8 (47/47) | 1 | 8 | 6 |
| Nose | 45 | 28 (62/65) | 4 | 11 | 2 |
| Mouth | 29 | 21 (72/72) | 2 | 6 | 0 |
| Other | 17 | 12 (71/86) | 1 | 1 | 3 |

[a]Data from Poon et al.[56]
[b]Type 1: <5% platelet membrane GPIIb-IIIa.
[c]Antiplatelet antibodies: 29 (anti-GPIIb-IIIa, 16; anti-HLA, 8; both, 5). Refractoriness to platelets: 23 (with antiplatelet antibodies, 17; without antiplatelet antibodies, 6).
[d]Bleeding stopped in ≤48 hours after the initiation of rFVIIa treatment and without a recurrence.
[e]Based on total number (intention to treat).
[f]Bleeding stopped in ≤48 hours but with recurrence of bleeding in ≤48 hours following cessation of initial bleeding.
[g]Bleeding stopped >48 hours after the initiation of rFVIIa treatment and/or if another treatment (other than antifibrinolytic drugs) was needed.
[h]Platelet transfusion given concurrently with rFVIIa.

avoidance of exposure to residual risk of blood-borne infections.

2. As for treatment of any bleeding disorder, early treatment is important, as indicated by the analysis of the UK study.[52]

3. Although many bleeds can be stopped with one to three doses of rFVIIa, some bleeds need more doses. In clinical practice, the exact timing of changes in therapy will depend on the available alternative therapeutic options and the relative efficacy, risk, and cost of each option.

4. The experience with continuous infusion of rFVIIa is limited, but it suggests that although continuous infusion appears to be effective in preventing bleeding (as in surgical prophylaxis), it may not be effective in stopping bleeding. In the international survey,[56] continuous infusion was successful in each of six invasive procedures but the total amount of rFVIIa used was not reduced relative to that for bolus administration. Continuous infusion failed to stop bleeding in six of seven evaluable bleeding episodes in which this mode of administration was used.

5. For bolus injections, a dose of approximately 90 μg/kg body weight every 2 hours until bleeding stops should be considered. Maintenance doses should be considered for severe bleeding episodes, bleeding episodes related to trauma, and GI bleeds.

6. The use of antifibrinolytic agents with rFVIIa appears to be safe, although their contribution to hemostasis together with rFVIIa is not clear because the success rate for those patients who were given antifibrinolytics was not significantly different from that of patients who were not given antifibrinolytics. Antifibrinolytics should still be considered, especially for mucosal bleeds or surgery involving mucosal sites.

7. A high proportion (58/80, 83%) of the successfully treated bleeds stopped within 6 hours after the first rFVIIa injection, suggesting that rFVIIa could be used as first-line therapy while waiting for adequate apheresis platelet concentrates. On the contrary, high-dose HLA-compatible platelet transfusions, with or without antibody removal therapy,[70,71] should not be delayed in case of life-threatening bleeding or when rFVIIa therapy fails.

In 2004, the European Medical Evaluation Agency (EMEA) approved the use of rFVIIa for European Union patients with GT with platelet antibodies and past and/or present history of platelet refractoriness. As required by the EMEA, an international postmarketing pharmacovigilance study (www.glanzmann-reg.org) has been launched to assess the efficacy and safety of rFVIIa and other systemic hemostatic agents (including platelet transfusion) on a larger number of GT patients with or without platelet antibodies or platelet refractoriness.[72]

### B. Acquired Platelet Function Disorders

rFVIIa has been used effectively in a limited number of patients with acquired platelet functional disorders. At least three patients have been treated for acute bleeding — a 76-year-old man with myelodysplastic syndrome and persistent small intestinal bleeding after laparotomy,[73] a 12-year-old girl with uremia and pulmonary bleeding complicating cytomegalovirus pneumonitis following a renal transplant,[65] and a 69-year-old woman with essential thrombocythemia and uncontrolled gynecological surgical bleeding.[67] The first two patients were refractory to desmopressin, and all three were refractory to platelet transfusions. All bleeding responded promptly to one dose of rFVIIa at 40 to 90 μg/kg, but maintenance doses were given to the third patient. In a 9-year-old boy with myelodysplastic syndrome associated with storage pool disorder and mild thrombocytopenia (platelet count, $75 \times 10^9$/L), the extraction of six teeth was successful with two doses of rFVIIa, one pre- and one postoperatively.[41]

## V. Adverse Events

When used in patients with platelet disorders, rFVIIa appears to be safe. In the international survey,[56] one patient with GT developed clots in the right renal pelvis and ureter after gynecological surgery covered with rFVIIa by continuous infusion (25 μg/kg/hour tapered gradually to 12 μg/kg/hour) for 4 days and antifibrinolytics.[49] This patient likely had trauma to the kidney during surgery, with bleeding and clotting in the renal pelvis and ureter that did not lyse. A 72-year-old woman with GT developed deep vein thrombosis and pulmonary embolism after bowel resection surgery covered with rFVIIa.[50] She received a bolus of 90 μg/kg rFVIIa followed by high-dose continuous infusion at 30 μg/kg/hour for 16 days. The thrombotic event occurred 6 days after rFVIIa was discontinued. Among patients with other platelet disorders receiving rFVIIa, thrombotic events in one patient with Bernard–Soulier syndrome and one with

uremic platelet dysfunction have been reported to the U.S. Food and Drug Administration (FDA) MedWatch Pharmacovigilance Program.[74] Thus, the prevalence of thrombotic events in patients with platelet disorders receiving rFVIIa appears to be low, but experience is limited.

rFVIIa has been used more extensively in hemophilia with inhibitors, and thrombotic complications are rare except in unusual circumstances.[8,74] In hemophilia patients with inhibitors, prothrombin time (PT) and activated partial thromboplastin time (aPTT) were shortened following infusion of rFVIIa.[3-5,7] Generally, there were no significant changes in the mean level of antithrombin, fibrinogen, and platelet counts over 48 hours after surgery covered with rFVIIa,[7] although the coagulation activation marker prothrombin $F_{1+2}$ fragment may increase without clinical sequalae.[16,75] A review by Abshire and Kenet[8] suggested that since 1996 more than 700,000 standard doses of rFVIIa (each equivalent to 90 μg/kg × 40 kg) have been administered to patients (number not specified) with hemophilia and inhibitors or acquired hemophilia. There were 20 thrombotic events (seven acute myocardial infarctions, six cerebral vascular thromboses, and seven deep venous thromboembolisms) reported spontaneously to the manufacturer or in clinical trials. Most of these patients apparently had comorbid or predisposing factors, such as diabetes mellitus, coronary artery disease, atherosclerosis, hypertension, obesity, advanced age, or indwelling catheter. There was also one case of bowel gangrene and four cases of disseminated intravascular coagulation (DIC)[8] associated with other DIC risk factors, including septic arthritis, hemicolectomy, hypovolemic cardiovascular collapse from GI bleeding,[4] and surgery for massive abscess in the thigh and *Salmonella septicemia*.[76]

rFVIIa has also been used on an investigational (off-label) basis in patients without hemophilia, thrombocytopenia, or platelet function disorders for diverse bleeding situations, such as GI, intracranial, and postoperative bleeds; bleeding after trauma or after bone marrow/stem cell or organ transplantation; and von Willebrand disease or factor XI deficiency. Thirty-eight thrombotic events (including cerebrovascular thrombosis, myocardial infarction, deep vein thrombosis/pulmonary embolism, and DIC) related to these off-label treatments were reported spontaneously to the FDA MedWatch Pharmacovigilance Program or as published case reports during the period from April 1999 to June 2002.[74,77] The contribution of patient factors relative to rFVIIa in these adverse events was not clear, and the true incidence of rFVIIa-induced thrombosis is unknown. In a clinical trial of 309 patients with intracranial hemorrhage, venous thrombotic events in the patients receiving rFVIIa (5 of 303, 2%) were similar to those in patients receiving placebo (2 of 96, 2%), but arterial thrombosis was significantly higher in the rFVIIa group (16 of 303 [5%] vs. 0 of 96 [0%]).[78] In summary, caution should be exercised when using rFVIIa in patients with underlying conditions that may predispose them to arterial or venous thrombosis or DIC and in patients of advanced age.

## VI. Mechanisms of Action

### A. High-Dose rFVIIa in Hemophilia: TF-Dependent and TF-Independent Models

The mechanisms by which high-dose rFVIIa enhances thrombin generation have been studied in patients with hemophilia. van't Veer et al.[79] showed that at physiologic ratios of FVII (10 nM) to FVIIa (100 pM), FVII competes with FVIIa for TF binding, resulting in downregulation of thrombin generation. High-dose rFVIIa could overcome the FVII inhibition, and in hemophilia patients FVIIa at 10 nM (attained by therapeutic rFVIIa doses) normalized the thrombin generation profile to that observed in the presence of FVIII and normal concentrations of FVII (10 nM) and FVIIa (100 pM).[79,80] An alternative mechanism was based on the observation that FVIIa can bind to activated platelet weakly ($K_d$ approximately 100 nM)[81] and that at high concentrations FVIIa on the platelet surface can activate FX and mediate thrombin generation sufficiently to effect hemostasis in the absence of FVIII or FIX and independent of TF. It is likely that both the TF-dependent and TF-independent mechanisms contribute to thrombin generation in hemophilia patients treated with high-dose rFVIIa. Butenas et al.[82] showed that in a synthetic mixture corresponding to hemophilia B and "acquired hemophilia B" blood produced in vitro by an anti-FIX antibody, the delay in thrombin generation in the presence of 5 pM TF could be normalized in the presence of 10 nM rFVIIa and 6 to $8 \times 10^8$ activated platelets. TF, rFVIIa, and activated platelets are all required, as thrombin generation was abolished by omitting TF, and was substantially decreased in the absence of either rFVIIa or activated platelets in the mixture. Enhanced thrombin generation by rFVIIa in hemophilia patients also improves fibrin clot structure[83] and decreases fibrinolysis through activation of the thrombin-activatable fibrinolytic inhibitor,[84] further contributing to hemostasis.

### B. High-Dose rFVIIa in Thrombocytopenia

Kjalke et al.[85] used their cell-based in vitro model to study the effect of high-dose rFVIIa on thrombocytopenia using unactivated platelets and TF-bearing monocytes in the presence of physiologic concentrations of clotting factors and natural inhibitors. In this model, decreasing platelet concentration resulted in a dose-dependent decrease in the rate of thrombin generation, prolongation of time to maximal

thrombin generation, and a lower peak level of thrombin formed, together with a decreased rate of platelet activation as monitored by surface expression of CD62P (P-selectin). A high concentration of rFVIIa at 50 to 100 nM (at low platelet concentration) increased the initial rate of thrombin generation (Fig. 68-2A, insert) and shortened the lag phase of platelet activation (Fig. 68-2B) without influencing the time to maximal thrombin generation or the peak level of thrombin (Fig. 68-2A). Kjalke et al.[85] therefore proposed that high-dose rFVIIa may ensure hemostasis in thrombocytopenic patients by increasing the initial thrombin generation, resulting in faster platelet activation (and thereby compensating for the lower number of platelets). Improved thrombin generation represented by enhanced fibrin deposition was also observed in an ex vivo model of Galán et al.,[86] perfusing thrombocytopenic blood through an annular chamber containing damaged vascular segments with high-concentration rFVIIa, even though there was no improvement in platelet deposition. Lisman et al.[87] reported that high-concentration rFVIIa (1.2 µg/mL) in the presence of factors X (10 µg/ml) and II (20 ng/mL) significantly increased adhesion of platelets (but not thrombus height) to type III collagen- or fibrinogen-coated slides perfused with washed red cells and platelets at all platelet concentrations from 10 to $200 \times 10^9$/L. This platelet adhesion was prevented by the omission of rFVIIa or the addition of the thrombin inhibitor hirudin, confirming that it was the result of thrombin generation mediated by high-dose rFVIIa. Platelet activation was enhanced during platelet adhesion as evident by calcium flux and enhanced expression of negatively charged procoagulant phospholipid, as shown by annexin A5 binding. These investigators[87] also demonstrated binding of fluoroscein isothiocyanate (FITC)-labeled rFVIIa to the deposited platelet surface. They suggested that enhanced generation of a procoagulant phospholipid surface would further facilitate thrombin and fibrin generation mediated by the bound rFVIIa in a TF-independent manner.

### C. High-Dose FVIIa in Platelet Function Defects

The mechanism of action of high-dose factor VIIa in platelet function disorders has been more extensively studied in GT with impaired thrombin generation.[88] The data suggest a TF-independent mechanism of high-dose rFVIIa-mediated thrombin generation that results in platelet activation, adhesion, and aggregation (Fig. 68-3).

**1. Adhesion of Glanzmann Thrombasthenia Platelets Mediated by High-Dose rFVIIa.** In an in vitro perfusion model,[89] in which washed red cells and platelets deficient in αIIbβ3 (from GT patients or normal platelets treated with αIIbβ3 antagonists) were perfused over an extracellular

A

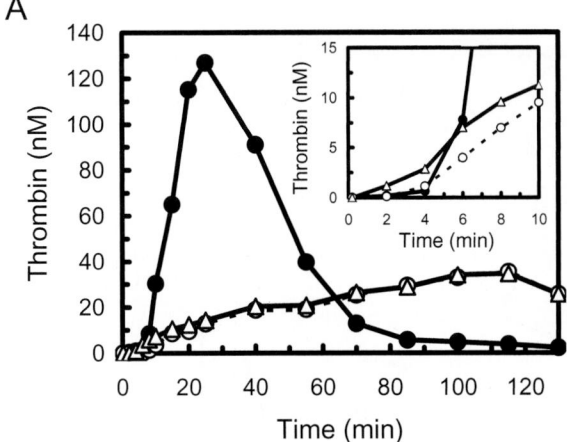

B

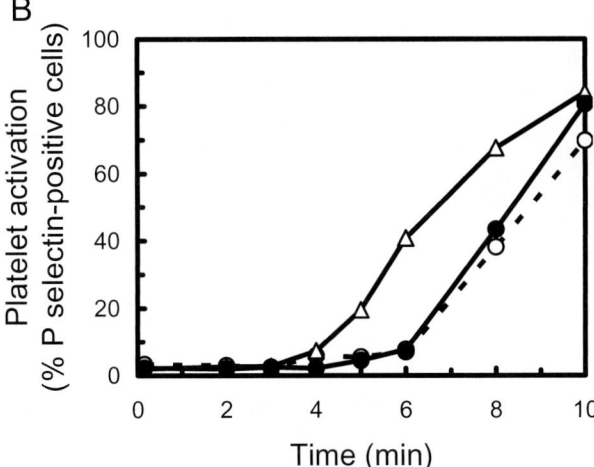

**Figure 68-2.** The effect of high-dose rFVIIa on thrombin generation and platelet activation. Unactivated platelets at various concentrations were mixed with factors V, VIII, IX, X, and XI, prothrombin, antithrombin, tissue factor pathway inhibitor, calcium, and various concentrations of rFVIIa and added to tissue factor-expressing monocytes. Aliquots were removed and analyzed for thrombin generation by amidolytic activity (A) and for platelet activation, measured as P-selectin-positive cells by flow cytometry (B). A depicts thrombin generation over a 120-minute assay, with the first 10-minute assay shown in the insert, whereas B depicts platelet activation over a 10-minute assay. Solid circles, normal platelet count with rFVIIa at 0.2 nM; open circles, low platelet count with rFVIIa at 0.2 nM; open triangles, low platelet count with rFVIIa at 50 nM. Adapted from Poon.[24] (Original figure by Dr. Marianne Kjalke, Department of Hemostasis Biology, Novo Nordisk A/S, Maaloev, Denmark.)

matrix of stimulated human umbilical vein endothelial cells or type III collagen, adhesion of these defective platelets to the matrix was significantly increased by a high concentration of rFVIIa (1.2 µg/mL) in the presence of factors X (10 µg/mL) and II (20 ng/mL). This improvement in adhesion required the participation of the von Willebrand factor–

GPIb interaction because it could be blocked by anti-GPIb and anti-von Willebrand factor antibodies. Thrombin generation on the activated platelet surface, mediated by bound rFVIIa, was responsible for the improvement in adhesion. This is because adhesion was blocked by hirudin and by annexin V, capable of covering the phospholipid coagulant surface on the activated platelets, and FITC-labeled rFVIIa was demonstrated on the activated platelet surface. This TF-independent mechanism was further confirmed by the finding that improvement of platelet adhesion was not affected by anti-TF sufficient to abolish FVIIa-mediated FXa generation on extracellular matrix in a static experiment and to prolong the PT from 12 seconds to >200 seconds.

**2. Aggregation of Glanzmann Thrombasthenia Platelets Mediated by High-Dose FVIIa.** GT platelets are deficient in integrin αIIbβ3, the primary binding site for fibrinogen in aggregation of normal activated platelets for primary hemostatic plug formation (see Chapters 8 and 57). However, in the *ex vivo* model of GT blood perfusing through an annular chamber containing damaged vascular segments, a high concentration of rFVIIa improved thrombin generation, increased fibrin deposition, and partially restored platelet aggregation.[86] A number of studies have also demonstrated that fibrin, particularly polymeric fibrin, can, independently of fibrinogen, mediate agglutination of GT platelets or platelets with αIIbβ3 inhibited.[90–92] Lisman et al.[93] further showed that high-dose rFVIIa could mediate fibrin-mediated agglutination of GT platelets via TF-independent thrombin generation. In their experimental model, when washed GT platelets were activated with collagen or thrombin receptor activating peptide SFLLRN, full aggregation did not occur until a high concentration of rFVIIa (1.2 µg) was present together with FX (10 µg/mL), FII (20 ng/mL), and fibrinogen (0.5 mg/mL). The platelet aggregation phase was not abolished by anti-TF but was partially dependent on platelet activation, formation of thromboxane $A_2$, secretion of ADP, activation of protease activated receptor-1, as well as binding of thrombin to GPIb,[93] which has been shown in normal platelets to mediate binding of fibrin to an unidentified platelet receptor.[94] Thrombin generation and fibrin formation are both required in this rFVIIa-mediated platelet aggregation, as the time course of the aggregation phase correlated with the generation of FII, $F_{1+2}$ as well as fibrinopeptide, and aggregation could be abolished by the addition of hirudin or omission of any of the clotting factors (FX, FII, or fibrinogen). Electron microscopy of the αIIbβ3-deficient platelet aggregates showed platelet packing and immunogold-identifiable fibrin(ogen) present at some platelet–platelet contact sites, albeit less so than that in similarly prepared normal platelet aggregates. Fibrin appeared to participate actively to mediate platelet aggregation partly

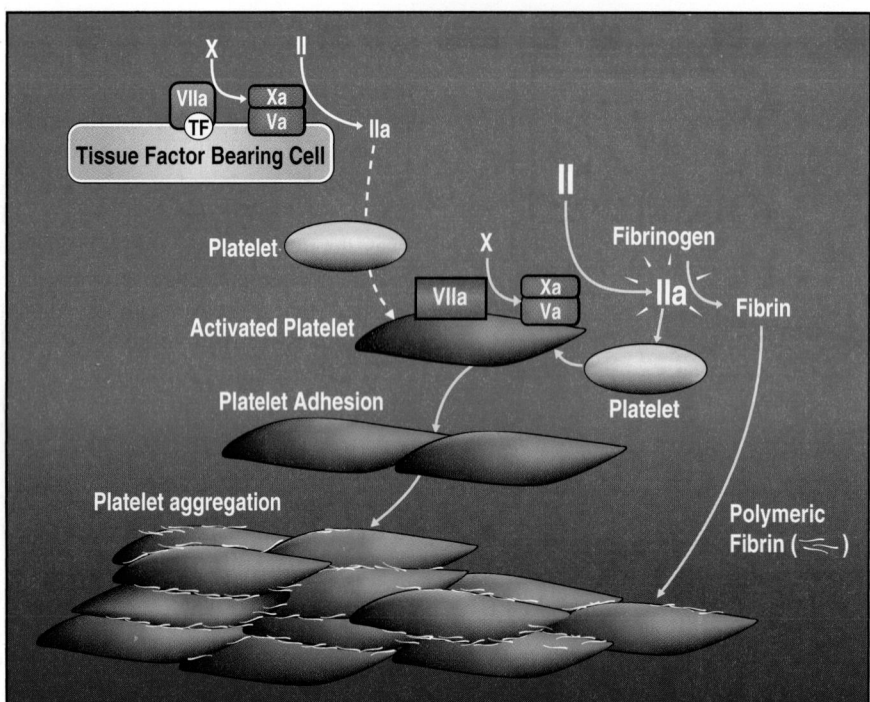

**Figure 68-3.** Schematic tissue factor-independent, platelet-dependent model of primary hemostatic plug formation in Glanzmann thrombasthenia (GT) platelets deficient in GPIIb-IIIa (integrin αIIbβ3). FVIIa-tissue factor (TF) complex on TF-bearing cells at the site of vascular injury activates FX to FXa. FXa-FVa on the TF-bearing cells initiates generation of a small amount of thrombin (FIIa) that is insufficient to provide fibrin formation but sufficient to activate the GT platelets, causing degranulation and release of FV. FVIIa binds to activated platelets weakly, and at high concentration (attained by high-dose rFVIIa therapy) it can directly activate FX to FXa to mediate generation of a high concentration of thrombin (thrombin burst). The augmented thrombin generation results in an increased number of activated platelets deposited (adhesion) to the wound site and increased available platelet procoagulant surface to facilitate more thrombin generation and more platelet activation. The augmented thrombin generated also converts fibrinogen to fibrin. GT platelets lack the fibrinogen receptor (integrin αIIbβ3) and these platelets therefore cannot utilize fibrinogen for aggregation. However, binding of fibrin/polymeric fibrin to an unidentified platelet surface receptor can mediate aggregation of the GT platelets at the wound site (albeit less potently than fibrinogen-mediated aggregation of normal platelets), resulting in primary hemostatic plug formation.

in a receptor-mediated manner, as opposed to being passively trapped during platelet aggregation, since aggregation was less efficient if viable platelets were replaced with fixed platelets.

## VII. Conclusions

Data suggest that rFVIIa is an attractive alternative to platelet transfusion for the treatment of patients with platelet disorders. However, there is a major need for clinical studies, particularly clinical trials to formally assess efficacy, safety, and optimal treatment regimens in patients with thrombocytopenia and in patients with congenital and acquired platelet function disorders. The data need to be stratified according to mild/moderate and severe bleeding episodes, as well as minor and major surgical procedures. Finally, the mechanisms of action of rFVIIa need to be clarified in the different types of platelet function disorders (e.g., GT,

Bernard–Soulier syndrome, platelet-type von Willebrand disease, and other congenital and acquired platelet disorders) given that they have different pathophysiology, and the efficacy, safety, and optimal treatment regimen for each may not necessarily be the same.

## References

1. Palavecino, E., Jacobs, M., & Yomtovian, R. (2004). Bacterial contamination of blood products. *Curr Hematol Rep, 3,* 450–455.
2. Dodd, R. Y., Notari, E. P., & Stramer, S. L. (2002). Current prevalence and incidence of infectious disease markers and estimated window-period risk in the American Red Cross blood donor population. *Transfusion, 42,* 975–979.
3. Hedner, U., Glazer, S., Pingel, K., et al. (1988). Successful use of recombinant factor VIIa in patient with severe haemophilia A during synovectomy. *Lancet, 2,* 1193.

4. Hay, C. R., Negrier, C., & Ludlam, C. A. (1997). The treatment of bleeding in acquired haemophilia with recombinant factor VIIa: A multicentre study. *Thromb Haemost, 78,* 1463–1467.

5. Lusher, J. M., Roberts, H. R., Davignon, G., et al. (1998). A randomized, double-blind comparison of two dosage levels of recombinant factor VIIa in the treatment of joint, muscle and mucocutaneous haemorrhages in persons with haemophilia A and B, with and without inhibitors. rFVIIa Study Group. *Haemophilia, 4,* 790–798.

6. Key, N. S., Aledort, L. M., Beardsley, D., et al. (1998). Home treatment of mild to moderate bleeding episodes using recombinant factor VIIa (Novoseven) in haemophiliacs with inhibitors. *Thromb Haemost, 80,* 912–918.

7. Shapiro, A. D., Gilchrist, G. S., Hoots, W. K., et al. (1998). Prospective, randomised trial of two doses of rFVIIa (NovoSeven) in haemophilia patients with inhibitors undergoing surgery. *Thromb Haemost, 80,* 773–778.

8. Abshire, T., & Kenet, G. (2004). Recombinant factor VIIa: Review of efficacy, dosing regimens and safety in patients with congenital and acquired factor VIII or IX inhibitors. *J Thromb Haemost, 2,* 899–909.

9. Morrissey, J. H. (2001). Tissue factor and factor VII initiation of coagulation. In R. W. Colman, J. Hirsh, V. J. Marder, et al. (Eds.), *Hemostasis and Thrombosis. Basic Principles and Clinical Practice* (pp. 89–101). Philadelphia: Lippincott Williams & Wilkins.

10. Geczy, C. L. (1994). Cellular mechanisms for the activation of blood coagulation. *Int Rev Cytol, 152,* 49–108.

11. Camerer, E., Lolsto, A. B., & Prydz, H. (1996). Cell biology of tissue factor, the principal initiator of blood coagulation. *Thromb Res, 81,* 1–41.

12. Giesen, P. L., Rauch, U., Bohrmann, B., et al. (1999). Blood-borne tissue factor: Another view of thrombosis. *Proc Natl Acad Sci USA, 96,* 2311–2315.

13. Todoroki, H., Nakamura, S., Higure, A., et al. (2000). Neutrophils express tissue factor in a monkey model of sepsis. *Surgery, 127,* 209–216.

14. Jurlander, B., Thim, L., Klausen, N. K., et al. (2001). Recombinant activated factor VII (rFVIIa): Characterization, manufacturing, and clinical development. *Semin Thromb Hemost, 27,* 373–384.

15. Lindley, C. M., Sawyer, W. T., Macik, B. G., et al. (1994). Pharmacokinetics and pharmacodynamics of recombinant factor VIIa. *Clin Pharmacol Ther, 55,* 638–648.

16. Berrettini, M., Mariani, G., Schiavoni, M., et al. (2001). Pharmacokinetic evaluation of recombinant, activated factor VII in patients with inherited factor VII deficiency. *Haematologica, 86,* 640–645.

17. Hedner, U., Kristensen, H., Berntorp, E., et al. (1998). Pharmacokinetics of rFVIIa in children [Abstract]. *Haemophilia, 4,* 244.

18. Villar, A., Aronis, S., Morfini, M., et al. (2004). Pharmacokinetics of activated recombinant coagulation factor VII (NovoSeven) in children versus adults with haemophilia A. *Haemophilia, 10,* 352–359.

19. Diness, V., Bregengaard, C., Erhardtsen, E., et al. (1992). Recombinant human factor VIIa (rFVIIa) in a rabbit stasis model. *Thromb Res, 67,* 233–241.

20. Turecek, P. L., Richter, G., Muchitsch, E. M., et al. (1997). Thrombogenicity of recombinant factor VIIa and recombinant soluble tissue factor in an *in vivo* rabbit model [Abstract]. *Thromb Haemost Suppl, 78*(suppl.), 222.

21. Hedner, U., Bergqvist, D., Ljungberg, J., et al. (1985). Haemostatic effect of factor VIIa in thrombocytopenic rabbits [Abstract]. *Blood, 86*(Suppl. 1), 1043.

22. Tranholm, M., Rojkjaer, R., Pyke, C., et al. (2003). Recombinant factor VIIa reduces bleeding in severely thrombocytopenic rabbits. *Thromb Res, 109,* 217–223.

23. Kristensen, J., Killander, A., Hippe, E., et al. (1996). Clinical experience with recombinant factor VIIa in patients with thrombocytopenia. *Haemostasis, 26*(Suppl. 1), 159–164.

24. Poon, M. C. (2003). Management of thrombocytopenic bleeding: Is there a role for recombinant coagulation factor VIIa? *Curr Hematol Rep, 2,* 139–147.

25. Gerotziafas, G. T., Zervas, C., Gavrielidis, G., et al. (2002). Effective hemostasis with rFVIIa treatment in two patients with severe thrombocytopenia and life-threatening hemorrhage. *Am J Hematol, 69,* 219–222.

26. Waddington, D. P., McAuley, F. T., Hanley, J. P., et al. (2002). The use of recombinant factor VIIa in a Jehovah's Witness with auto-immune thrombocytopenia and post-splenectomy haemorrhage. *Br J Haematol, 119,* 286–288.

27. Culic, S. (2003). Recombinant factor VIIa for refractive haemorrhage in autoimmune idiopathic thrombocytopenic purpura. *Br J Haematol, 120,* 909–910.

28. Virchis, A., Hughes, C., & Berney, S. (2004). Severe gastrointestinal haemorrhage responding to recombinant factor VIIa in a Jehovah's Witness with refractory immune thrombocytopenia. *Hematol J, 5,* 281–282.

29. Busani, S., Marietta, M., Pasetto, A., et al. (2005). Use of recombinant factor VIIa in a thrombocytopenic patient with spontaneous intracerebral haemorrhage. *Thromb Haemost, 93,* 381–382.

30. Vidarsson, B., & Onundarson, P. T. (2000). Recombinant factor VIIa for bleeding in refractory thrombocytopenia. *Thromb Haemost, 83,* 634–635.

31. Hoffman, R., Eliakim, R., Zuckerman, T., et al. (2003). Successful use of recombinant activated factor VII in controlling upper gastrointestinal bleeding in a patient with relapsed acute myeloid leukemia. *J Thromb Haemost, 1,* 606–608.

32. Culligan, D. J., Salamat, A., Tait, J., et al. (2003). Use of recombinant factor VIIa in life-threatening bleeding following autologous peripheral blood stem cell transplantation complicated by platelet refractoriness. *Bone Marrow Transplant, 31,* 1183–1184.

33. De Fabritis, P., Dentamaro, T., Picardi, A., et al. (2004). Recombinant factor VIIa for the management of severe hemorrhages in patients with hematologic malignancies. *Haematologica, 89,* 243–245.

34. Zulfikar, B., & Kayran, S. M. (2004). Successful treatment of massive gastrointestinal hemorrhage in acute biphenotypic leukemia with recombinant factor VIIa (NovoSeven). *Blood Coagul Fibrinolysis, 15,* 261–263.

35. Osborne, W., Bhandari, S., Tait, R. C., et al. (2004). Immediate haemostasis with recombinant factor VIIa for haemor-

rhage following Hickman line insertion in acute myeloid leukaemia. *Clin Lab Haematol, 26,* 229–231.

36. Kurekci, A. E., Atay, A. A., Okutan, V., et al. (2005). Recombinant activated factor VII for severe gastrointestinal bleeding after chemotherapy in an infant with acute megakaryoblastic leukemia. *Blood Coagul Fibrinolysis, 16,* 145–147.

37. Sokolic, V., Bukovic, D., Fures, R., et al. (2002). Recombinant factor VIIa (rFVIIa) is effective at massive bleeding after caesarean section — A case report. *Coll Antropol, 26*(Suppl.), 155–157.

38. Zupancic-Salek, S., Sokolic V., Viskovic, T., et al. (2002). Successful use of recombinant factor VIIa for massive bleeding after caesarean section due to HELLP syndrome. *Acta Haematol, 108,* 162–163.

39. Dart, B. W., Cockerham, W. T., Torres, C., et al. (2004). A novel use of recombinant factor VIIa in HELLP syndrome associated with spontaneous hepatic rupture and abdominal compartment syndrome. *J Trauma, 57,* 171–174.

40. Merchant, S. H., Mathew, P., Vanderjagt, T. J., et al. (2004). Recombinant factor VIIa in management of spontaneous subcapsular liver hematoma associated with pregnancy. *Obstet Gynecol, 103,* 1055–1058.

41. Thomas, A. E., & Plews, D. E. (2001). Use of recombinant factor VIIa in platelet disorders — A single centre experience [Abstract]. *Blood Coagul Fibrinolysis, 12,* A14.

42. Blatt, J., Gold, S. H., Wiley, J. M., et al. (2001). Off-label use of recombinant factor VIIa in patients following bone marrow transplantation. *Bone Marrow Transplant, 28,* 405–407.

43. Phelan, J. T., Broder, J., & Kouides, P. A. (2004). Near-fatal uterine hemorrhage during induction chemotherapy for acute myeloid leukemia: A case report of bilateral uterine artery embolization. *Am J Hematol, 77,* 151–155.

44. Minniti, C., & Weinthal, J. (2001). Use of recombinant activated factor VII (rFVIIa) in two children with idiopathic thrombocytopenic purpura (ITP) [Abstract]. *Blood, 98*(suppl.), 62b.

45. Conesa, V., Navarro-Ruiz, A., Borras-Blasco, J., et al. (2005). Recombinant factor VIIa is an effective therapy for abdominal surgery and severe thrombocytopenia: A case report. *Int J Hematol, 81,* 75–76.

46. Tengborn, L., & Petruson, B. (1996). A patient with Glanzmann thrombasthenia and epistaxis successfully treated with recombinant factor VIIa [Letter]. *Thromb Haemost, 75,* 981–982.

47. Poon, M. C., Demers, C., Jobin, F., et al. (1999). Recombinant factor VIIa is effective for bleeding and surgery in patients with Glanzmann thrombasthenia. *Blood, 94,* 3951–3953.

48. Chuansumrit, A., Sangkapreecha, C., & Hathirat, P. (1999). Successful epistaxis control in a patient with Glanzmann thrombasthenia by increased bolus injection dose of recombinant factor VIIa. *Thromb Haemost, 82,* 1778.

49. Robinson, K. L., Savoia, H., & Street, A. M. (2000). Thrombotic complications in two patients receiving NovoSeven [Abstract]. *Haemophilia, 6,* 349.

50. d'Oiron, R., Menart, C., Trzeciak, M. C., et al. (2000). Use of recombinant factor VIIa in 3 patients with inherited type I Glanzmann's thrombasthenia undergoing invasive procedures. *Thromb Haemost, 83,* 644–647.

51. van Buuren, H. R., & Wielenga, J. J. (2002). Successful surgery using recombinant factor VIIa for recurrent, idiopathic nonulcer duodenal bleeding in a patient with Glanzmann's thrombasthenia. *Dig Dis Sci, 47,* 2134–2136.

52. Almeida, A. M., Khair, K., Hann, I., et al. (2003). The use of recombinant factor VIIa in children with inherited platelet function disorders. *Br J Haematol, 121,* 477–481.

53. Chuansumrit, A., Suwannuraks, M., Sri-Udomporn, N., et al. (2003). Recombinant activated factor VII combined with local measures in preventing bleeding from invasive dental procedures in patients with Glanzmann thrombasthenia. *Blood Coagul Fibrinolysis, 14,* 187–190.

54. Caglar, K., Cetinkaya, A., Aytac, S., et al. (2003). Use of recombinant factor VIIa for bleeding in children with Glanzmann thrombasthenia. *Pediatr Hematol Oncol, 20,* 435–438.

55. Bell, J. A., & Savidge, G. F. (2003). Glanzmann's thrombasthenia proposed optimal management during surgery and delivery. *Clin Appl Thromb Hemost, 9,* 167–170.

56. Poon, M. C., d'Oiron, R., von Depka, M., et al. (2004). Prophylactic and therapeutic recombinant factor VIIa administration to patients with Glanzmann's thrombasthenia: Results of an international survey. *J Thromb Haemost, 2,* 1096–1103.

57. Kaleelrahman, M., Minford, A., & Parapia, L. A. (2004). Use of recombinant factor VIIa in inherited platelet disorders. *Br J Haematol, 125,* 95–96.

58. Kale, A., Bayhan, G., Yalinkaya, A., et al. (2004). The use of recombinant factor VIIa in a primigravida with Glanzmann's thrombasthenia during delivery. *J Perinat Med, 32,* 456–458.

59. Uzunlar, H. I., Eroglu, A., Senel, A. C., et al. (2004). A patient with Glanzmann's thrombasthenia for emergent abdominal surgery. *Anesth Analg, 99,* 1258–1260.

60. Coppola, A., Tufano, A., Cimino, E., et al. (2004). Recombinant factor VIIa in a patient with Glanzmann's thrombasthenia undergoing gynecological surgery: Open issues in light of successful treatment. *Thromb Haemost, 92,* 1450–1452.

61. Yilmaz, B. T., Alioglu, B., Ozyurek, E., et al. (2005). Successful use of recombinant factor VIIa (NovoSeven) during cardiac surgery in a pediatric patient with Glanzmann thrombasthenia. *Pediatr Cardiol, 26,* 843–845.

62. Pozo Pozo, A. I., Jimenez-Yuste, V., Villar, A., et al. (2002). Successful thyroidectomy in a patient with Hermansky–Pudlak syndrome treated with recombinant activated factor VII and platelet concentrates. *Blood Coagul Fibrinolysis, 13,* 551–553.

63. Langendonck, L., & Appel, I. M. (2005). Modification of biological parameters after treatment with recombinant factor VIIa in a patient with thrombocytopathy due to storage pool disease. *Pediatr Blood Cancer, 44,* 676–678.

64. Fressinaud, E., Sigaud-Fiks, M., Le Boterff, C., et al. (1998). Use of recombinant factor VIIa (NovoSeven®) for dental extraction in a patient affected by platelet-type (pseudo-) von Willebrand disease [Abstract]. *Haemophilia, 4,* 299.

65. Révész, T., Arets, B., Bierings, M., et al. (1998). Recombinant factor VIIa in severe uremic bleeding [Letter]. *Thromb Haemost, 80,* 353.

66. Plews, D. E., & Thomas, A. E. (1999). Novel uses of recombinant factor VIIa [Abstract]. *Blood, 94*(Suppl. 1), 83b.

67. Cervera, J. S., Mena-Duran, A. V., & Piqueras, C. S. (2005). The use of recombinant factor VIIa in a patient with essential thrombocythaemia with uncontrolled surgical bleeding. *Thromb Haemost, 93,* 383–384.

68. Musso, R., Cultrera, D., Russo, M., et al. (1999). Recombinant activated factor VII as haemostatic agent in Glanzmann's thrombasthenia [Abstract]. *Thromb Haemost, 82*(Suppl.), 621.

69. Devecioglu, O., Unuvar, A., Anak, S., et al. (2003). Pyelolithotomy in a patient with Glanzmann thrombasthenia and antiglycoprotein IIb/IIIa antibodies: The shortest possible duration of treatment with recombinant activated factor VII and platelet transfusions. *Turk J Pediatr, 45,* 64–66.

70. Ito, K., Yoshida, H., Hatoyama, H., et al. (1991). Antibody removal therapy used successfully at delivery of a pregnant patient with Glanzmann's thrombasthenia and multiple antiplatelet antibodies. *Vox Sang, 61,* 40–46.

71. Martin, I., Kriaa, F., Proulle, V., et al. (2002). Protein A sepharose immunoadsorption can restore the efficacy of platelet concentrates in patients with Glanzmann's thrombasthenia and anti-glycoprotein IIb-IIIa antibodies. *Br J Haematol, 119,* 991–997.

72. Poon, M.-C., Zotz, R., Di Minno, R., et al. (2006). Glanzmann's thrombasthenia treatment: A prospective observational registry on the use of recombinant human activated factor VII (rFVIIa) and other hemostatic agents. *Semin Hematol, 43*(Suppl. 1), S35–S36.

73. Meijer, K., Sieders, E., Slooff, M. J., et al. (1998). Effective treatment of severe bleeding due to acquired thrombocytopathia by single dose administration of activated recombinant factor VII [Letter]. *Thromb Haemost, 80,* 204–205.

74. Aledort, L. M. (2004). Comparative thrombotic event incidence after infusion of recombinant factor VIIa versus factor VIII inhibitor bypass activity. *J Thromb Haemost, 2,* 1700–1708.

75. Baudo, F., Redaelli, R., Caimi, T. M., et al. (2000). The continuous infusion of recombinant activated factor VIIa (rFVIIa) in patients with factor VIII inhibitors activates the coagulation and fibrinolytic systems without clinical complications. *Thromb Res, 99,* 21–24.

76. Stein, S. F., Duncan, A., Cutler, D., et al. (1990). Disseminated intravascular coagulation (DIC) in a hemophiliac treated with recombinant factor [Abstract]. *Blood, 76*(Suppl. 1), 438a.

77. Sallah, S., Isaksen, M., Seremetis, S., et al. (2005). Comparative thrombotic event incidence after infusion of recombinant factor VIIa vs. factor VIII inhibitor bypass activity — A rebuttal. *J Thromb Haemost, 3,* 820–822.

78. Mayer, S. A., Brun, N. C., Begtrup, K., et al. (2005). Recombinant activated factor VII for acute intracerebral hemorrhage. *N Engl J Med, 352,* 777–785.

79. van't Veer, C., Golden, N. J., & Mann, K. G. (2000). Inhibition of thrombin generation by the zymogen factor VII: Implications for the treatment of hemophilia A by factor VIIa. *Blood, 95,* 1330–1335.

80. Butenas, S., Brummel, K. E., Branda, R. F., et al. (2002). Mechanism of factor VIIa-dependent coagulation in hemophilia blood. *Blood, 99,* 923–930.

81. Monroe, D. M., Hoffman, M., Oliver, J. A., et al. (1997). Platelet activity of high-dose factor VIIa is independent of tissue factor. *Br J Haematol, 99,* 542–547.

82. Butenas, S., Brummel, K. E., Bouchard, B. A., et al. (2003). How factor VIIa works in hemophilia. *J Thromb Haemost, 1,* 1158–1160.

83. He, S., Blomback, M., Jacobsson, E. G., et al. (2003). The role of recombinant factor VIIa (FVIIa) in fibrin structure in the absence of FVIII/FIX. *J Thromb Haemost, 1,* 1215–1219.

84. Lisman, T., Mosnier, L. O., Lambert, T., et al. (2002). Inhibition of fibrinolysis by recombinant factor VIIa in plasma from patients with severe hemophilia A. *Blood, 99,* 175–179.

85. Kjalke, M., Ezban, M., Monroe, D. M., et al. (2001). High-dose factor VIIa increases initial thrombin generation and mediates faster platelet activation in thrombocytopenia-like conditions in a cell-based model system. *Br J Haematol, 114,* 114–120.

86. Galán, A. M., Tonda, R., Pino, M., et al. (2003). Increased local procoagulant action: A mechanism contributing to the favorable hemostatic effect of recombinant FVIIa in PLT disorders. *Transfusion, 43,* 885–892.

87. Lisman, T., Adelmeijer, J., Cauwenberghs, S., et al. (2005). Recombinant factor VIIa enhances platelet adhesion and activation under flow conditions at normal and reduced platelet count. *J Thromb Haemost, 3,* 742–751.

88. Reverter, J. C., Beguin, S., Kessels, H., et al. (1996). Inhibition of platelet-mediated, tissue factor-induced thrombin generation by the mouse/human chimeric 7E3 antibody. Potential implications for the effect of c7E3 Fab treatment on acute thrombosis and "clinical restenosis." *J Clin Invest, 98,* 863–874.

89. Lisman, T., Moschatsis, S., Adelmeijer, J., et al. (2003). Recombinant factor VIIa enhances deposition of platelets with congenital or acquired alpha IIb beta 3 deficiency to endothelial cell matrix and collagen under conditions of flow via tissue factor-independent thrombin generation. *Blood, 101,* 1864–1870.

90. Niewiarowski, S., Levy-Toledano, S., & Caen, J. P. (1981). Platelet interaction with polymerizing fibrin in Glanzmann's thrombasthenia. *Thromb Res, 23,* 457–463.

91. McGregor, L., Hanss, M., Sayegh, A., et al. (1989). Aggregation to thrombin and collagen of platelets from a Glanzmann thrombasthenic patient lacking glycoproteins IIb and IIIa. *Thromb Haemost, 62,* 962–967.

92. Osdoit, S., & Rosa, J.-P. (2001). Polymeric fibrin interacts with platelets independently from integrin $\alpha_{IIb}\beta_3$ [Abstract]. *Blood, 98,* 518a.

93. Lisman, T., Adelmeijer, J., Heijnen, H. F., et al. (2004). Recombinant factor VIIa restores aggregation of alphaIIbbeta3-deficient platelets via tissue factor-independent fibrin generation. *Blood, 103,* 1720–1727.

94. Soslau, G., Class, R., Morgan, D. A., et al. (2001). Unique pathway of thrombin-induced platelet aggregation mediated by glycoprotein Ib. *J Biol Chem, 276,* 21173–21183.

# Platelet Transfusion Medicine

PART SEVEN

# Platelet Transfusion Medicine

# CHAPTER 69

# Platelet Storage and Transfusion

Peter L. Perrotta[1] and Edward L. Snyder[2]

[1]*Department of Pathology, West Virginia University, Morgantown, West Virginia*
[2]*Department of Laboratory Medicine, Yale University School of Medicine, New Haven, Connecticut*

## I. Introduction

Platelets maintain normal hemostasis through their elaborate responses to vascular injury. Accordingly, patients with low numbers of circulating platelets or functionally hyperactive platelets are at increased risk of spontaneous bleeding or hemorrhage following traumatic injuries or during surgical procedures. Thrombocytopenic bleeding was a major cause of death in patients with acute leukemia until platelet concentrates (PC) became widely available in the early 1970s.[1] Before then, the only source of viable platelets was freshly drawn whole blood.[2] Routine platelet transfusion therapy was made possible in large part by the development of gas-permeable plastic containers that facilitate the collection, separation, and storage of platelets from whole blood. Today, PC are used extensively to support patients who receive thrombocytopenia-inducing, intensive therapies for hematologic malignancies and solid tumors.[3,4]

Transfusion services seek to maintain an adequate supply of PC that are both safe and efficacious. Unfortunately, the constrained shelf-life of PC (5–7 days) makes it impractical for blood banks to maintain large reserves of products. The safety of platelet transfusions, at least in terms of viral transmission, has improved with advances in donor selection and testing. Methods for inactivating contaminating bacteria and monitoring bacterial growth are being implemented in many countries, and this will further increase PC safety. Current techniques used to collect and store platelets limit the detrimental effects of storage on platelets. However, a correlation of improved platelet processing with enhanced efficacy remains elusive. Despite vast clinical experience with platelet transfusion therapy, transfusion practices vary widely and evidence-based guidelines are often not followed.[5]

## II. Platelet Preparation

Platelets used in transfusion therapy are prepared by centrifuging whole blood obtained from a single blood donation or by apheresis using automated cell separators.

The processes used to separate platelets from whole blood have evolved during the past several decades to maximize the yield of platelets while limiting the number of red and white blood cells. The isolation of platelets from other blood cells also allows platelets to be stored under optimal conditions, which are different than the conditions used to store other blood components (e.g., red blood cells and plasma).

### A. Methods

**1. Platelet Concentrate Preparation from Whole Blood by Platelet-Rich Plasma and Buffy Coat Methods.** Blood is drawn through wide-bore, siliconized needles to minimize the activation of platelets and clotting proteins. Removed blood is immediately mixed with citrate anticoagulant. Before separation by differential separation, whole blood must be left undisturbed at room temperature for a short period because platelets are activated during blood collection. Approximately 450 mL of whole blood is withdrawn during allogeneic donation, which is then separated to yield a unit each of red cells, platelets, and plasma. Platelets prepared in this manner are often referred to as "random donor" platelets (RDP) because the human leukocyte antigen (HLA) type of the donor is unknown (i.e., of random HLA type). Each RDP contains on the order of $5.5 \times 10^{10}$ platelets suspended in approximately 50 to 60 mL of the donor's plasma. This volume of suspending plasma is required to maintain platelet viability during storage.

The two major methods used to isolate platelets from whole blood are the platelet-rich plasma (PRP) and the buffy coat (BC) methods (Fig. 69-1).[6] Most PCs in the United States are prepared by the PRP method, whereas the BC method is preferred in Europe. Anticoagulated blood is separated based on differential sedimentation, a process that is accelerated by centrifugation. The sedimentation rate is most heavily influenced by physical properties of cells (specific gravity, size, and deformability) and the

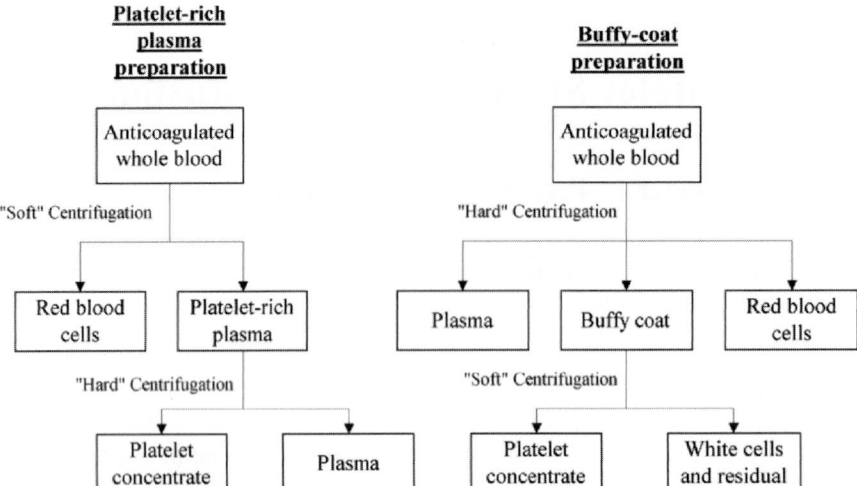

**Figure 69-1.** Preparation of platelet concentrates from anticoagulated whole blood by the platelet-rich plasma and buffy coat techniques.

viscosity of the medium. Separation times and speeds are optimized to maximize platelet yields within shorter time periods.

In the PRP technique, whole blood first undergoes a low g force ("soft") spin, which separates red cells from PRP (Fig. 69-1). Low-speed centrifugation results in a supernatant (PRP) that contains the majority of suspended platelets, 30 to 50% of the original white cells, and some red cells. The PRP is transferred to a satellite bag and centrifuged at a higher g force ("hard" spin). The concentrated supernatant platelet-poor plasma is removed, and the resulting platelet pellet is gently resuspended in 50 to 60 mL residual plasma.[7] Leaving the product undisturbed for 1 hour prior to resuspension minimizes platelet aggregation and damage.[8] The PRP method yields approximately 5.0 to 7.5 × 10^{10} platelets, or 60 to 75% of the platelets found in the whole blood unit before separation. These products typically contain many (10^8–10^9) white cells. Before transfusion, multiple (e.g., four to six) PRP units are combined to create a random-donor platelet "pool." PRP pools are more convenient to transfuse than single PRP units. However, they must be used within 4 hours of preparation because of the risks of bacterial contamination during the pooling process. During neonatal and pediatric transfusion, it may be necessary to limit plasma volume.[9] Plasma can be removed by centrifuging a single PC or a platelet pool at 2000 g for 10 minutes before transfusion. The spun platelets must "rest" undisturbed for up to 1 hour before they are resuspended.[10] As might be expected, up to 50% of the original number of platelets are lost during this additional centrifugation.

When platelets are manufactured by the BC method, whole blood is first centrifuged at high force (3000 g for 7–10 minutes) to create a buffy coat layer where platelets and leukocytes reside (Fig. 69-1). Higher speed centrifugation separates cells somewhat differently than slow-speed centrifugation. During high-speed centrifugation, white cells initially sediment with red cells; platelets remain in the supernatant plasma. Next, red cells are packed closely together and rapidly fall to the bag bottom. This process forces plasma and white cells upwards to the plasma interface. The platelets eventually accumulate on this interface. The settling of platelets on the red cell interface may explain the lesser degree of platelet activation when platelets are prepared by the BC compared to the PRP method.[11] The resultant buffy coat, consisting of platelets and white cells, is removed along with small portions of the lower plasma layer and the upper red cell layer. The BC is then centrifuged at low g force to separate the platelets from leukocytes and red cells. Top and bottom bag systems facilitate separation by allowing the platelets in the buffy coat to remain relatively undisturbed in the primary separation bag (with the plasma and red cells removed from the top and bottom ports, respectively).[12] In practice, four to six BCs are pooled, diluted in plasma, and centrifuged at low speed.[13] The resulting platelet-rich supernatant is then transferred to a larger volume storage bag. Alternatively, crystalloid platelet additive solutions are used instead of plasma to resuspend pooled BCs.[14,15] This effectively limits problems associated with plasma incompatibility in platelet recipients.

**2. Single-Donor Platelet Concentrates Prepared by Apheresis.** Platelets collected by cell separators are generally referred to as "platelets, apheresis" or "single-donor platelets (SDP) by apheresis." Blood removed through a catheter in one arm vein is passed through an instrument in which platelets are separated from other cellular components by differential centrifugation.[16] Instruments are automated in that separation parameters are determined by the instrument's electronics based on input parameters, including the donor's weight, platelet count, and hematocrit. PRP is separated from whole blood in the centrifuge; the residual red cells and plasma are returned through a catheter in the donor's other arm. Most apheresis systems directly collect platelets as PRP, whereas other systems yield a concentrated platelet pellet that must be gently resuspended. Approximately 4 or 5 L of donor blood is processed during the 1.5- to 2-hour collection. Other systems have been developed that can collect not only platelets but also red cells and plasma during a single donation.[17] Although platelet apheresis is well tolerated by most donors, adverse reactions related to citrate toxicity (hypocalcemia) and hemodynamic instability can occur.[18-20] Interestingly, there is evidence of platelet activation in donors several days after platelets are collected by apheresis.[21] Moreover, repeated platelet donations by apheresis result in the transient appearance of platelet dysfunction in the donor as assessed under conditions of shear in the platelet function analyzer 100 (PFA-100).[22,23] Regular plateletpheresis donors may also develop small decreases in their platelet count.[24]

United States standards require that an apheresis platelet product contain at least $3 \times 10^{11}$ platelets suspended in approximately 200 mL of the donor's plasma. Thus, one apheresis SDP approximates six RDP, but because of the variability in platelet collection, the range is four to eight RDP. Because of increased machine efficiency, many blood centers collect two, or up to three, SDP doses ($6-9 \times 10^{11}$ platelets) from a single apheresis procedure. Plateletpheresis products are stored at 20 to 24°C for up to 5 days, identical to platelets prepared by other methods. Apheresis-derived platelets, like BC- and PRP-prepared PC, contain few red cells and red cell cross-matching is therefore not necessary. Transfusing apheresis-derived platelets that are ABO incompatible typically does not produce measurable hemolysis in adults.[25] Some centers titer group O platelets to avoid the uncommon situation in which the platelet donor carries an exceptionally high-titer anti-A that could cause hemolysis if infused to a group A recipient.[26,27] Several percent of group O single-donor platelets may have high-titer anti-A.[28] This situation is more dangerous when group O apheresis-derived platelets are given to group A infants and small children.[29] Most instruments collect concentrated platelets that contain few white cells, typically less than $5 \times 10^6$. Technologic advances have allowed even further degrees of leukoreduction, to white cell levels below $1 \times 10^6$, and these SDP are considered leukocyte reduced by most standards.[30,31]

## B. Quality Control

Institutions that prepare PC should implement a quality control (QC) program that can detect problems in the collection, processing, or storage of such products.[32,33] Guidelines for monitoring the quality of PC vary slightly across different countries and are specific for the preparation technique.[34,35] Most standards include measurements of platelet numbers, pH, and, when products are labeled leukoreduced, white blood cell number (Table 69-1). Platelet and white cell numbers reflect the efficiency of the particular platelet collection and/or leukocyte reduction process utilized. Unfortunately, platelet and white cell counting is limited by the variability of the techniques used to perform these tests. The concentrations of platelets prepared from PRP (150,000–450,000/μL per unit) are less than those collected by apheresis (>1,000,000/μL per unit). Processes that increase the yield of platelets per unit are preferred in order to limit the degree of recipient donor exposure.

As a QC tool, PC pH is measured at the end of the storage period because a pH below 6.2 or above 7.6 correlates with decreased *in vivo* efficacy.[36] Component volumes for whole blood-derived PCs are largely based on the ability of residual plasma to maintain product pH within an acceptable range during storage. However, pH measurements alone may not have the ability to detect a problem in the platelet manufacturing process.[37] Although the volume of apheresis products is not specified by the U.S. Food and Drug Administration (FDA) standards, it is typically determined based on the size of the product. Temperature control charts verify that the platelets were stored between 20 and 24°C.

**1. Platelet Counting in Platelet Concentrates.** The original methods employed to enumerate platelets used small-volume chamber hemacytometers and phase contrast microscopy. These techniques required removal of red cells by lysis or sedimentation before counting and were time-consuming, labor-intensive, and highly variable (coefficients of variation >10%). Currently, automated hematology analyzers are commonly used by blood banks to determine platelet counts in PCs. The results are rapid and, in general, reproducible over time. Most analyzers determine cell numbers based on the principle of impedance: Particles passing through a small orifice in which an electric field is maintained cause a change in electrical resistance (see Chapter 24). The change in resistance is detected and the magnitude of the signal is proportional to particle size. Other instruments utilize differences in cell conductance and/or laser light scatter to better discriminate cell types. Although the precision and accuracy of most instruments used to count platelet numbers in PCs are acceptable, there are several limitations. First, the instruments are designed for whole blood samples — not PRP — and thus are cali-

**Table 69-1: Standards for Platelet Concentrate Quality Assurance**

| | United States[34] | | Europe[35] | |
|---|---|---|---|---|
| | Whole blood derived | Apheresis derived | Whole blood derived | Apheresis derived |
| Platelets ($\times 10^{10}$) | $\geq 5.5$ | $\geq 30$ | Varies[a] | Varies[a] |
| Volume (mL) | Not specified[b] | Not specified[b] | Not specified[c] | Not specified[c] |
| White cells ($\times 10^6$) to label leukoreduced | $<0.83^d$ | $<5.0^d$ | $<1.0^b$ | $<1.0^b$ |
| pH | $>6.2$ | $>6.2$ | 6.4–7.4 | 6.4–7.4 |

[a]Platelet number within limits that comply with validated preparation and preservation conditions.
[b]Standard met if 90% of units tested fall within indicated values.
[c]Volume must maintain product within specified pH for duration of storage.
[d]In 95% of units tested.

brated using whole blood controls. Second, analytic errors can be introduced when PC samples are diluted before analysis. Third, platelet aggregates can form during product sampling, which results in falsely low platelet counts.

Other techniques developed to quantify platelets utilize fluorescent dyes and monoclonal antibodies conjugated to fluorochromes. These techniques typically require a flow cytometer or other instrument to detect fluorescence at an appropriate wavelength.[38] Flow cytometric methods are not commonly used in platelet quality monitoring due to the increased cost and requirements for specialized equipment.[39] Microvolumetric fluorimetry has also been applied to platelet counting. Cellular samples contained within a volumetric capillary are stained with appropriate antibodies and fluorescent dyes and then scanned by a moving laser beam.[40] Although this technique has been used to count platelets, it is better suited for white cell enumeration and has also been used to detect bacteria in PCs.[41]

**2. Residual White Cell Counting in Platelet Concentrates.** Centers that produce leukoreduced RDP should use validated QC procedures to ensure that a sufficient number of white cells have been removed from products.[42] QC is also required to document leukoreduction of platelets collected by apheresis.[43] Automated hematology analyzers are inappropriate for determining white cell numbers in leukoreduced platelet components because of their lack of sensitivity.[44] Most residual white cell counts are therefore performed by manual chamber techniques using larger volume chambers such as the 50 μl Nageotte-type hemacytometer.[45] Chamber white cell counts have been validated in leukoreduced platelets.[46] However, the lower limits of detection and variability of this technique may not be optimal for determining the extremely low white cell numbers found in properly prepared PC.[47]

Therefore, alternative techniques and devices have been developed that are more precise and accurate at the low white cell counts found in leukoreduced PCs. These include flow cytometry, microvolume fluorimetry, and quantitative polymerase chain reaction (PCR). Flow cytometry has been used to count residual white cells in PCs.[48] Most methods use dyes that stain nuclear compounds (propidium iodide) and/or broad leukocyte antibodies (e.g., CD45) conjugated to a fluorochrome. A commercial kit has been developed for counting white cells in PCs (LEUKO-Count). Commercial systems based on microvolumetric fluorimetry have been developed to enumerate white cells in leukocyte-reduced blood components (IMAGN2000).[49] A nucleic acid dye is used to stain white cells, after which the sample is placed in a volumetric capillary and scanned using the instrument's laser and detectors.[50] These systems are highly automated and may be more reproducible than manual techniques in determining extremely low white cell numbers.[51]

## III. Platelet Storage and Storage Injury

PCs undergo alterations during collection, processing, and storage that adversely affect their structure and function.[52] These changes, commonly referred to as the platelet storage defect (PSD) or platelet storage lesion (PSL), are important because they are associated with decreased posttransfusion *in vivo* survival.[53–57] The mechanisms responsible for PSD have been better defined during the past decade (Table 69-2).[58] For example, centrifugation may damage platelets by exposing them to conditions of high shear stress. Shear stresses may cause the release of both cytosolic lactate dehydrogenase (LDH) and platelet granule contents.[59,60] Residual leukocytes and platelets remain metabolically active during

**Table 69-2: Factors That Influence Development of the Platelet Storage Defect**

| Collection and Separation Techniques | Storage Conditions and Processing |
|---|---|
| Anticoagulant/preservative solution | Storage temperature |
| Type of anticoagulant | Storage duration |
| Ratio of anticoagulant to whole blood | Type of agitation |
|  | Volume of suspending plasma |
| Blood drawing flow rate | Composition of storage container (permeability) |
| Time between whole blood collection and separation | Leukodepletion technique |
| Processing temperature | Irradiation (gamma and UVB) |
| Centrifugation forces (time and g force) | Cryopreservation |

**Table 69-3: *In Vitro* Tests of Platelet Concentrate Quality**

Platelet structure
  Cellular content (platelet count)
  Visual inspection for swirling phenomena
  Platelet morphology by microscopy
  Platelet size distribution by automated counters
Functional tests
  Platelet aggregation, spontaneous and to agonists
  Hypotonic shock response
  Extent of shape change
  Thrombin-stimulated ATP release
Metabolic status
  Supernatant pH, $pO_2$, $pCO_2$, $HCO_3$
  Glucose consumption
  Lactate production
Platelet activation
  P-selectin (CD62P) surface expression
  Soluble P-selectin in supernatant
  Platelet factor 4 and β-thromboglobulin
  Annexin V binding
  Lactate dehydrogenase in supernatant
  Platelet microparticle formation

storage and continue to consume nutrients and produce potentially harmful metabolic products. Cellular debris and proteolytic enzymes are also found in the surrounding plasma. Interactions between stored platelets and suspending plasma may activate clotting factors and, hence, the coagulation system.

Stored PCs can be evaluated using a wide variety of laboratory assays,[61] including platelet morphology by microscopy, pH, LDH, platelet activation markers, osmotic recovery, platelet aggregation, and extent of shape change (Table 69-3). Simple evaluations include visually inspecting PCs before they are transfused. Normal discoid platelets, when exposed to a light source and gently rotated or squeezed, refract that light and produce a "swirling" phenomenon that can be identified by trained personnel.[62] The lack of swirling may correlate with a less than predicted posttransfusion platelet increment.[63,64] In clinical practice, however, only platelet number, concentrate volume, supernatant pH, and white cell number are routinely measured in PCs (see Section II.B). These relatively simple tests may reflect only a small subset of changes that occur during platelet storage.

One of the difficulties in correlating *in vitro* tests of platelet function with *in vivo* survival is that some of the changes that occur during platelet storage may be reversible after transfusion.[65] Thus, many investigators believe that *in vivo* autologous recovery of radiolabeled platelets is the gold standard test of platelet viability and must be conducted when storage conditions and/or platelet substitutes are evaluated.[66] The survival curves of modified platelets should ideally approximate those curves obtained when unaltered autologous platelets are infused.[67] Standardized methods utilize radiolabeling of platelets with [51]chromium or

[111]indium.[68,69] Other methods for labeling platelets with the nonradioactive compounds biotin and streptavidin have been developed.[70] However, humans may develop circulating antibodies directed against biotin.[71] Polymorphisms in the noncoding region of the human mitochondrial genome have been explored to discriminate between donor and recipient platelets.[72]

### A. Platelet Concentrate Storage Conditions

**1. Platelet Storage Temperature and Platelet Cold Injury.** PCs are stored at 20 to 24°C because their function and viability are severely compromised when stored at colder refrigerated temperatures.[73] Moreover, posttransfusion viability is also markedly improved when PCs are stored at room temperature.[74] These observations were confirmed by later clinical studies that showed platelets stored for 72 hours at room temperature produced higher platelet increments and greater reductions in bleeding times when transfused than platelets stored at refrigerated temperatures.[75] Alternatively, platelets stored at physiologic temperatures (closer to 37°C) have reduced viability compared to platelets stored at 22°C.[74] This observation may be related to the normal high metabolic rate of platelets with rapid adenosine triphosphate (ATP) turnover[76] — a rate that can be decreased by maintaining platelets at lower than physiologic temperatures.[77]

It has long been recognized that platelets stored at 4°C develop a spherelike morphology, which is evidence of irreversible physical damage.[78] Even when platelets are briefly exposed to temperatures less than 20°C for 24 hours, as may occur during transport of platelets from a supplier to a hospital, there are observable differences in platelet morphology. These shape changes may involve actin assembly.[79,80] Furthermore, refrigerated platelets are rapidly cleared from the circulation upon transfusion. Evidence suggests that this clearance is mediated by the integrin αMβ2 (Mac-1) on Kupfer cells in the liver that recognize clustered glycoprotein (GP) Ib receptors on chilled platelets,[81] resulting in rapid platelet phagocytosis. The circulation of functional cooled platelets in mice can be prolonged through enzymatic galactosylation of chilled platelets. This effectively blocks a lectin that recognizes exposed β-$N$-acetylglucosamine residues of N-linked glycans on GPIbα.[82] Experimental studies in humans are in progress.

**2. Agitation of Platelet Concentrates.** PCs are stored with continuous gentle agitation because most measures of *in vitro* platelet function and structure deteriorate more rapidly when platelets are stored without agitation.[83] In addition, agitation appears to prolong platelet survival following transfusion.[84] Horizontal agitation on a flat-bed agitator is generally preferred over circular rolling "Ferris wheel"-type agitation.[85,86] Agitation is thought to enhance the transport of gases such as $O_2$ through the storage bag. However, surface interactions between the agitated platelets and the plastic storage bag can also result in shear stresses capable of activating platelets. Most evidence supports the standard that platelets be routinely stored with agitation.[87]

### B. Platelet Storage Containers

One of the most significant advances in platelet transfusion therapy was the development of plastic, gas-permeable storage containers. These storage bags are specifically designed to allow adequate gas (e.g., $O_2$ and $CO_2$) exchange. The earliest storage bags composed of polyvinyl chloride (PVC) and a 2-diethylhexyl phthalate (DEHP) plasticizer did not allow platelet storage beyond 3 days. Aerobic metabolism was not maintained, which resulted in lactic acid production and a rapidly falling pH. These changes adversely affected *in vivo* platelet recovery and survival.[88,89] The next generation of storage containers, composed of PVC and non-DEHP plasticizers such as butyryl-tri-hexyl citrate, was more gas permeable, allowing the storage of platelets for 5 days at 20 to 24°C, while maintaining acceptable degrees of *in vitro* function and *in vivo* survival.[90–92]

### C. Metabolic Changes during Platelet Storage

The metabolic processes of human platelets continue after they are removed from the body and stored as a PC.[93] Moreover, the white cells found in PCs also maintain activity. Thus, the pH of whole blood begins to decline soon after collection at a rate dependent on the buffering capacities of these cells and the suspending solution. Decrements in pH are relatively small within the first 14 to 24 hours after whole blood collection because of the buffering capacity of plasma and red cells. As in cold injury, platelets exposed to a pH <6.3 exhibit morphologic changes and have diminished *in vivo* survival upon transfusion. Most of the pH-related effects of stored platelets appear permanent, although more modest damage may be partially reversible.[94] Platelets exposed to excessively alkaline conditions (pH > 7.4) may become activated and form small platelet aggregates.[36] Platelets are typically separated from whole blood within 24 hours of phlebotomy. The amount of plasma required in a PC is selected to ensure adequate buffering capacity to maintain the pH of the PC > 6.0 during the storage period. In general, 35 mL of plasma is sufficient for a single RDP unit, but 50 to 70 mL is typically provided in practice. Platelets stored at room temperatures have lower metabolic activity than platelets that circulate *in vivo* at body temperature.[57] Hence, the metabolic requirements of platelets stored at 22°C are less than those of platelets in a 37°C environment. Finally, the detrimental effects on platelet metabolism during storage appear at least partially reversible by "rescuing" the platelets with fresh plasma.[95]

### D. Anticoagulants

The anticoagulant-preservative solutions into which whole blood is drawn contain either of two formulations of citrate–phosphate–dextrose (CPD or CP2D) or CPD with adenine (CPDA-1). The cardiotoxic agent EDTA, used in the early days of platelet transfusion, is inappropriate because EDTA-anticoagulated platelets are rapidly removed from the circulation. The concentration of citrate in CPD plasma is usually 20 to 22 mM. Apheresis platelets are typically collected into solutions containing citric acid, trisodium citrate, and dextrose (ACD-A). Dextrose provides a source of energy and phosphate serves as a buffer. Adenine, which is added to blood collection bags to improve red cell survival during storage by increasing cellular ATP levels, does not appear to enhance platelet survival during storage. By maintaining pH levels and concentrations of calcium ions lower than in the physiologic state, platelets are less likely to become activated, release their intracellular contents, and undergo irreversible aggregation. The amount of citrate found in PCs does not usually produce hypocalcemia or systemic anticoagulation in recipients of PCs because it is rapidly metabolized to bicarbonate in the liver. However, patients with end-stage liver disease are more prone to citrate toxicity, which can cause transient hemodynamic depression.[96]

## E. Platelet Storage Solutions

PCs are typically stored suspended in autologous plasma to better maintain pH and cell viability. However, a number of platelet storage solutions (PSS) have been developed that, like plasma, maintain platelet structure and function.[97] A major reason for developing PSSs is that plasma not used to suspend platelets could be used for other purposes.[98] This was particularly true in the 1970s and 1980s, when large amounts of plasma were needed to manufacture factor VIII concentrates for patients with hemophilia. In addition, patients transfused with platelets stored in PSSs may be less likely to experience certain adverse reactions (e.g., allergic reactions) associated with plasma exposure. Crystalloid solutions that do not contain glucose or bicarbonate do not maintain PC pH, and platelets stored in such solutions therefore exhibit reduced *in vivo* recovery after transfusion.[99,100] Furthermore, at least some glucose is required in the Krebs cycle for oxidative processes and in glycolytic reactions that result in lactic acid production.

Most PSS are buffered salt solutions that contain various additives (e.g., gluconate and acetate), designed to reduce oxygen consumption, glucose utilization, and lactate production, and also to limit *in vitro* platelet activation.[101] Acetate also participates in platelet metabolism through the tricarboxylic acid (citrate) cycle and is oxidized through the respiratory chain.[102] The use of acetate in PSS is thought to reduce production of lactate and hence to partially replace glucose as a substrate for energy production.[103] Removing plasma may decrease the incidence of common febrile and allergic reactions, but posttransfusion increments may be lower when platelets are resuspended in additive solutions alone.[104] Other investigators, however, have been unable to demonstrate a difference in platelet recovery when PSS are used to suspend platelets.[105]

PSS have been employed to prepare buffy coat platelets,[106] although they can also be used to produce platelets by the PRP technique. Various inhibitors of platelet and coagulation factor activation, such as theophylline, prostaglandin (PG) $E_1$, and aprotinin, have been added experimentally to PSS in attempts to further preserve platelet function.[107,108] $PGE_1$ stimulates adenyl cyclase, whereas theophylline inhibits platelet phosphodiesterase; the net effect of both additives is increased cyclic adenosine monophosphate (cAMP) availability. cAMP at least partially inhibits calcium release from the dense tubular system membrane.[109]

The addition of aprotinin, a broad-spectrum serine protease inhibitor, and a thrombin inhibitor (e.g., hirudin) to $PGE_1$/ theophylline PSS has also been studied in an attempt to further improve *in vitro* platelet preservation.[110] Other PSSs containing apyrase, aprotinin, and ascorbic acid have been evaluated in buffy coat-prepared PCs to reduce platelet activation during storage.[111]

It does not appear that plasma can be entirely replaced by storage solutions because plasma provides additional buffering capacity that better maintains platelet membrane integrity — approximately 30% plasma is required for this purpose.[112] PSS are not commonly used today, but they remain under development. If PSS are developed that effectively extend the viability of stored platelets past 5 days while maintaining *in vivo* viability, they will need to be implemented along with a system that limits or detects bacterial proliferation (see Section VI.B). For example, PSS are being used to suspend and store platelets that undergo photochemical treatment to inactivate infectious pathogens (e.g., viruses and bacteria) and leukocytes.[113,114]

## F. Platelet Activation during Storage

Platelets become activated during the preparation process and during prolonged storage.[59,115] Platelet activation can be measured by flow cytometric analysis of the platelet surface expression of the α-granule membrane protein P-selectin (CD62P)[116] (see Chapter 30). Alternatively, soluble P-selectin can be quantified in supernatant plasma.[117] P-selectin surface expression is one of the most commonly applied measures of platelet activation in PCs, and efforts have been made to standardize these measurements.[118] Other proteins found within platelet granules, such as β-thromboglobulin, platelet factor-4, and serotonin, have also been examined in stored platelets.[59] Annexin V binding, which is used as a marker of phosphatidylserine exposure on the platelet surface (see Chapters 19 and 30), has been examined as an alternative marker of platelet activation within the context of platelet preparation and storage.[115] In general, annexin V binding increases during 5 days of platelet storage at room temperature.[119] However, the degree of change in annexin V binding may be related in part to the method used to prepare platelets (e.g., PRP vs. apheresis-collected platelets).[120]

In general, increased platelet activation correlates with other adverse changes that occur during platelet storage. However, there is little evidence that the degree of platelet activation associated with platelet preparation and storage impairs the ability of transfused platelets to produce acceptable posttransfusion recovery or to arrest bleeding. Studies have been performed to determine if activated, P-selectin-positive platelets are preferentially removed from the circulation in proportion to P-selectin surface expression. However, these studies are limited by the difficulty in controlling other factors (e.g., supernatant pH) that affect the survival of transfused platelets in the circulation.[121]

Michelson et al.[122] demonstrated in a nonhuman primate model of platelet transfusion that transfused degranulated platelets rapidly lose surface P-selectin to the plasma pool but continue to circulate and function *in vivo*. This study demonstrated that platelet surface P-selectin molecules, rather than degranulated platelets, are rapidly cleared. These results were subsequently independently confirmed by Berger et al.,[123] who found that the platelets of both wild-type and P-selectin knockout mice had identical life spans. When platelets were isolated, activated with thrombin, and reinjected into mice, the rate of platelet clearance was unchanged. The infused thrombin-activated platelets rapidly lost their surface P-selectin in circulation, and this loss was accompanied by the simultaneous appearance of a 100-kDa P-selectin fragment in the plasma.[123] Storage of platelets at 4°C caused a significant reduction in their life span *in vivo,* but again no significant differences were observed between the two genotypes. Thus, the results of Berger et al. confirm that P-selectin does not mediate platelet clearance.

Furthermore, in a thrombocytopenic rabbit kidney injury model, Krishnamurti et al.[124] reported that thrombin-activated human platelets lose platelet surface P-selectin in the (reticuloendothelial system-inhibited) rabbit circulation, survive in the circulation just as long as fresh human platelets, and, most important, are just as effective as fresh human platelets at decreasing blood loss. Taken together, these studies[122–124] strongly suggest that the measurement of platelet surface P-selectin in platelet concentrates stored in the blood bank should not be used as a predictor of platelet survival or function *in vivo*. However, platelet surface P-selectin could still be a useful measure of QC during processing, storage, and manipulation (filtration and washing). This is because, in contrast to the situation *in vivo*,[122–124] the activation-dependent increase in platelet surface P-selectin is not reversible over time under standard blood banking conditions.[125]

## G. Platelet-Derived Microparticle Formation in Platelet Concentrates

Platelet-derived microparticles (PMP) (also known as platelet-derived microvesicles), small particles derived from the membranes of intact platelets, are present in PC.[126] PMP are strongly procoagulant and there is evidence that they retain many of the biologic properties of intact platelets (see Chapter 20). Various mechanisms may contribute to PMP formation in PC, including direct mechanical injury and exposure to stresses during component preparation. In addition, PMP are formed as a result of the inevitable degrees of platelet activation that occur during platelet processing and storage — activation that partially depends on interactions between platelets and the plastic storage container.[127] PMP formation is most conveniently quantified by flow cytometric methods based on light-scattering properties and surface expression of GPIb-IX or GPIIb-IIIa (integrin $\alpha$IIb$\beta$3)[128] (see Chapters 20 and 30). Differences in PMP formation observed when platelets are prepared using different component preparation techniques (e.g., apheresis versus whole blood-derived PC) appear to be related to separation forces and anticoagulant concentrations.[129]

Thus, patients transfused with PC prepared using modern techniques are also exposed to significant numbers of PMP. However, it remains unclear to what degree these procoagulant PMP contribute to the hemostatic effectiveness of transfused platelets. In addition, fragments of white cells and platelets appear capable of provoking primary alloimmunization to HLA antigens in a transfusion recipient.[130,131] These fragments are not entirely removed by the currently used leukoreduction filters and thus may be responsible for the development of platelet alloimmunization in patients who receive leukoreduced platelet products.

## H. Apoptotic Activity of Stored Platelets

Apoptosis, or programmed cell death, plays an important role in many biological processes. During the past several years, it has been recognized that anucleate platelets undergo apoptoticlike changes in response to chemical and physical stimulation.[132] Platelets contain enzymes that are central to apoptotic execution,[133] such as caspase-3, and key elements of the mitochondrial death pathway including cytochrome c, Apaf-1, and death regulators of the Bcl-2 family.[134–137] Since platelets are anucleate, it is hypothesized that this apoptotic machinery originally resided in, and was programmed by, nucleated megakaryocytes from which platelets are derived. *In vitro* experiments demonstrate that platelet apoptosis can be induced by calcium ionophores, other platelet agonists, and following storage at room temperature under standard blood bank conditions. Platelet caspase activity is enhanced during storage at 37°C; however, inhibition of caspase activity does not improve platelet viability at 37°C despite clear decreases in caspase activity.[138] Evidence of platelet apoptosis also exists at the mitochondrial level in stored and experimentally stressed platelets.[139,140] Understanding the role of apoptotic mechanisms in platelet survival could help to better understand the platelet storage lesion.

# IV. Postcollection Processing

## A. Leukocyte Reduction of Platelet Components

In many countries, platelets, like red blood cells, are commonly leukoreduced. The United Kingdom implemented universal leukodepletion in 1999, largely based on the theoretical benefit of reducing the risk of variant Creutzfeld–Jacob disease. The clearest benefits of leukoreduction include decreasing the risks of febrile nonhemolytic transfusion reactions, cytomegalovirus (CMV) transmission,[141] and HLA alloimmunization.[141,142] Platelets can be leukoreduced soon after collection and before storage (prestorage leukoreduction) or immediately before transfusion (poststorage leukoreduction). Prestorage leukoreduction can be performed when platelets are prepared by either the PRP method or the BC method. For the latter technique, BC concentrates are pooled and filtered through a single filter, either on the day of processing or after overnight storage.[143] Process leukoreduction refers to the ability to collect platelets, by apheresis, with very little white cell contamination.[144] Platelets collected by apheresis are considered prestorage leukoreduced. There is increasing evidence that leukoreduction should be performed before storage to further reduce risks of febrile reactions that are at least partially attributed to cytokines produced by leukocytes during storage.[145–147]

Cotton wool was the first material used to remove white blood cells from donated whole blood.[148] Currently, the three most commonly used leukoreduction filters are composed of negatively charged polyester, positively charged polyester, and noncharged polyurethane. Some polyester filters are constructed as a nonwoven mesh, whereas polyurethane filters form multilayer sponge networks. White blood cells are removed from blood by filters through several mechanisms, including simple sieving/mechanical retention based on cell size, direct adhesion of white cells to fibers, and indirect adhesion through platelets.[149,150] Complex interactions between cells and plasma proteins with the artificial surface affect the ability of filters constructed of such materials to remove white cells.[151] Earlier developed leukoreduction filters removed not only leukocytes but also platelets, which adhered to the polyester fiber surface. Fibers were later modified, for example, by coating with polymers such as polyhydroxyethyl methacrylate/polystyrene, to minimize the loss of platelets during blood filtration.[152] Two filters remain in widespread use — one to remove white cells from red cell concentrates and another to remove white cells from PC.

## B. Gamma and Ultraviolet Irradiation of Platelets

Residual white blood cells found in PCs can cause transfusion-associated graft-versus-host disease (TA-GVHD) in susceptible patients.[153] This rare, albeit usually fatal, reaction is prevented by gamma-irradiating platelets before transfusion. Ionizing radiation inactivates residual T cells, largely by damaging nuclear DNA. Platelets that are stored for 1 to 5 days and then irradiated with 5000 cGy (1 rad = 1 cGy) maintain normal *in vitro* measures of platelet structure and function.[154] More important clinically, irradiated platelets (5000 cGy) produce the expected platelet increments and appear hemostatically effective when transfused to thrombocytopenic patients.[155] When irradiation is performed, the FDA requires that a dose of 2500 cGy be delivered to the midplane of the irradiation canister and that a minimum dose of 1500 cGy be delivered to any other point in the canister. This irradiation dose effectively inactivates T cells found in PC[156] and is well tolerated by both whole blood- and apheresis-derived products in terms of *in vivo* platelet recovery or platelet survival.[157–159] Gamma-irradiation of platelets does not destroy bacteria and cannot be used to prevent bacterial proliferation in platelet components.[160]

The exposure of platelet transfusion recipients to foreign major histocompatibility complex (MHC) antigens is a major cause of platelet alloimmunization. This in turn can lead to a platelet refractory state in which patients do not respond to platelet transfusions. The Trial to Prevent Alloimmunization to Platelets (TRAP) showed that ultraviolet B (UVB) irradiation at 1480 mJ/cm² was equivalent to leukofiltration for decreasing the incidence of platelet refractoriness in patients with acute myelogenous leukemia.[141] Other investigators have also suggested that platelet alloimmunization is limited by UVB irradiation.[161] Animal experimentation has shown that UVB-irradiated leukocytes induce a state of humoral immune tolerance in which recipients cannot respond to foreign MHC antigens.[162,163] The lymphocytotoxic effects of UVB occur without affecting the *in vitro* ability of platelets to aggregate in response to common agonists.[164] *In vitro* studies have shown that medium-wavelength (280–320 nm) UVB light inactivates leukocytes found in PCs, but the necessary dose is related to the type and size of the plastic container in which platelets are stored. When exposed to 3000 J/m² UVB, PCs stored for up to 5 days demonstrate no adverse effects on pH or aggregation responses.[165] Higher doses of UVB irradiation (100,000 J/m²) cause changes in platelet structure and affect the expression of various platelet membrane proteins including GPIb.[166] However, UVB radiation at lower doses does not appear to diminish the clinical efficacy of platelet transfusion in patients with hematologic malignancies.[167]

## C. Photochemical Inactivation

Photochemical treatment of blood bank products is being developed for the primary purpose of inactivating/

decontaminating viruses and bacteria.[168,169] An important feature of any photochemical treatment developed for platelet therapy is its ability to leave platelet function intact so that posttransfusion recovery and survival are not impaired.[170] Platelet pathogen inactivation methods under investigation include riboflavin plus UVA irradiation,[171,172] amotosalen-HCL (S-59) plus UVA irradiation,[173–176] L-carnitine, and gamma-irradiation.[177] This technology is discussed in Chapter 70.

### D. Volume-Reduced Platelet Concentrates

The volume of PC prepared by the PRP technique ranges from 40 to 60 mL. This volume is required to maintain PC pH > 6.0 during 5 days of storage, although smaller (35–40 mL) volumes of donor plasma may be adequate.[178] In certain clinical situations, it may be necessary to reduce the volume of PC even further. For example, ABO-incompatible plasma can potentially harm neonates and small infants.[179] Removal of platelet-specific antibodies (e.g., anti-HPA-1a) from platelets obtained from a mother who has delivered a child with neonatal alloimmune thrombocytopenia may be necessary when the mother's platelets are needed for transfusion because there are no other readily available sources of HPA-1a-negative platelets (see Chapter 53). Finally, patients with circulatory overload may not tolerate excessive volume. In all these cases, donor plasma can be removed by centrifugation. Recentrifugation of stored PC at 1500 g for 7 minutes, 2000 g for 10 minutes, or 5000 g for 6 minutes, followed by a rest period of 1 hour and, finally, resuspension in 10 mL plasma, results in platelet loss between 5 and 20%.[10] Mean platelet loss is <15% when PC are centrifuged at lower forces over longer periods (580 g for 20 minutes).[9] In vitro testing and in vivo survival of volume-reduced platelets is considered to be clinically acceptable.[180]

## V. Platelet Transfusion Therapy

The earliest reports of the potential benefits of platelet transfusions utilized freshly drawn whole blood as a source of viable platelets.[181] Bleeding times were reduced and hemorrhage stopped when thrombocytopenic patients were transfused with whole blood. Fresh whole blood, however, was neither a convenient nor an optimal source of platelets, and thrombocytopenic bleeding remained a major cause of death in patients with acute leukemia.[1] PC became widely available in the late 1960s and early 1970s with the development of plastic collection/storage containers that facilitated the separation of platelets from whole blood. These new sources of concentrated and viable platelets improved the outcomes of patients with hematologic malignancies and solid tumors who received intensive chemotherapy. Hematology/oncology

patients are the major recipients of PCs, although significant numbers of PCs are utilized by trauma, general surgery, cardiothoracic surgery, and solid-organ transplant services.[182]

Platelet transfusions are primarily used to treat or prevent bleeding in patients with thrombocytopenia or platelet function defects. The majority of hospitals (>70%) transfuse platelets primarily as a prophylactic precaution.[183] The effectiveness of platelet transfusions in bleeding thrombocytopenic patients, however, has been well established through an accumulation of clinical experience, not through controlled trials.[184] It is generally agreed that increasing platelet counts to 40 to $50 \times 10^9$/L stops major bleeding. The benefit of prophylactic platelet transfusions in preventing hemorrhage in thrombocytopenic patients with bone marrow failure is more controversial. Importantly, the cause of thrombocytopenia should be established before initiating platelet transfusions because platelets are often ineffective in some thrombocytopenic conditions, such as immune thrombocytopenic purpura (ITP) (see Chapter 46). In other conditions, such as thrombotic thrombocytopenic purpura (see Chapter 50) and heparin-induced thrombocytopenia (see Chapter 48), platelet transfusions could be harmful. Patients with congenital or acquired platelet function defects (see Chapters 57 and 58) will usually have normal numbers of circulating platelets. However, since these platelets have decreased hemostatic capabilities, platelet transfusions can control or arrest bleeding in many circumstances.

### A. General Considerations in Platelet Transfusion Therapy

Most currently used PC are prepared by the PRP or BC techniques.[6] Each RDP prepared by either method typically contains more than $5.5 \times 10^{10}$ platelets. When transfused, one RDP unit is expected to increase a recipient's platelet count by 5 to $10 \times 10^9$/L in the absence of conditions associated with decreased platelet survival (e.g., fever, sepsis, and splenomegaly). Historically, RDPs have been administered in "pools" of 6 to 10 individual units. Some hospitals have decreased pool sizes to 4 or 5 RDP units because processes used to separate platelets from whole blood have become more efficient; an RDP often contains >8 to $10 \times 10^{10}$ platelets per concentrate.[185] Single-donor platelets collected by apheresis are highly concentrated and contain more than $3 \times 10^{11}$ platelets, equivalent to 4 to 8 average RDP units.

Testing requirements of platelet donors are the same as for any blood donor and include ABO and Rh typing, as well as screens for transfusion-transmitted diseases such as human immunodeficiency virus (HIV)-1, HIV-2, hepatitis B, hepatitis C, syphilis, human T cell lymphotropic virus (HTLV)-1, and HTLV-2. PC can transmit CMV, and a portion of donors are screened for CMV IgG antibodies. Products drawn from CMV seronegative donors are labeled

"CMV negative." PC that are leukoreduced under carefully controlled and monitored conditions are considered "CMV safe." CMV safe units are increasingly being used as an alternative to CMV seronegative products to avoid CMV transmission during transfusion.[186] In particular, leukoreduced apheresis platelet products do not appear to be a risk factor for transfusion-transmitted CMV infection after stem cell transplantation.[187]

Many hospitals transfuse ABO-matched platelets whenever possible because data suggest that ABO-mismatched platelet transfusions are associated with decreased platelet survival.[188,189] Transfusing ABO-unmatched platelets may also lead to increased levels of circulating immune complexes, the effects of which are unclear.[190] Most PC contain few red cells and red cell cross-matching is unnecessary. However, there are sufficient red cells in random donor PC to cause RhD alloimmunization in Rh-negative individuals, including those patients with nonhematologic disease[191] and infants.[192] If Rh-negative PCs are not available for an Rh-negative patient, Rh-positive platelets can be transfused followed by administration of Rh immune globulin (RhIG) within 72 hours of transfusion. An intravenous preparation is available to prevent Rh alloimmunization, which is a safe and effective alternative to intramuscular RhIG preparations for thrombocytopenic patients.[193] Apheresis-derived platelets do not appear to contain enough red cells to cause Rh alloimmunization in pediatric oncology patients[194] and thus, D immunoprophylaxis is generally unnecessary in this situation. Similarly, the risk of D alloimmunization is low in patients with hematologic disorders who are transfused with D-incompatible PC prepared by the BC technique.[195]

The practical aspects of transfusing platelets are similar to those used when other cellular products are transfused. Both the transfusion component and the intended recipient must be accurately identified before starting the transfusion. Vital signs should be taken before the transfusion and then soon thereafter or if there is any evidence of a transfusion reaction. Common signs of a reaction include temperature elevations and changes in blood pressure. Patients may develop symptoms such as chills, pruritis, rashes, and shortness of breath. Platelets, like red cells and fresh-frozen plasma (FFP), must be transfused through an infusion set that contains a microaggregate filter (e.g., approximately 170 μm) that removes fibrin clots and larger debris. Routine platelet transfusions must be completed within 4 hours, although most require less than 2 hours.

## B. Prophylactic Platelet Transfusion

Decisions to transfuse any blood product, including platelets, should not be made based purely on transfusion "triggers."[196] The overall status of the patient (e.g., disease, medications, and coagulation status) must be considered when determining the need for platelet transfusions. Historically, many physicians have transfused platelets to maintain platelet counts higher than $20 \times 10^9$/L, believing that this level was required to prevent spontaneous bleeding.[197] However, serious bleeding may not occur until platelet counts are $<5 \times 10^9$/L in the absence of other conditions that impair hemostasis.[198] The earliest efforts to determine thresholds for prophylactic platelet transfusion were complicated by the widespread use of aspirin as an antipyretic agent, before its detrimental effects on platelets were established.

There are three major difficulties encountered when attempting to delineate an optimal prophylactic platelet level: (1) serious hemorrhage is rare, even at very low platelet numbers; (2) minor clinical bleeding is difficult to quantify[199]; and (3) platelet counts are less accurately determined at the very low platelet numbers found in severely thrombocytopenic patients.[200] Measurement of stool red cell loss by $^{51}$Cr red cell labeling to estimate spontaneous bleeding in patients with aplastic anemia revealed that stool blood loss was not significantly elevated until platelet counts fell below $5 \times 10^9$/L.[201] A small clinical trial reported at the same time suggested that major bleeding, mortality, and the use of red blood cell transfusions were not different in patients who received prophylactic platelet transfusions ($<20 \times 10^9$/L) versus those who received platelet transfusions only when bleeding (other than from the skin or mucous membranes) developed.[202] The most widely studied patients are those with acute leukemias, although similar studies of patients with solid tumors suggest that serious bleeding is uncommon until platelet counts are $<10 \times 10^9$/L.[203]

One of the earlier larger trials designed to study the safety of lowering platelet transfusion thresholds prospectively followed 102 consecutive patients with acute leukemia.[204] Patients with platelet counts $<6 \times 10^9$/L received prophylactic transfusions, whereas those with counts $>20 \times 10^9$/L were transfused only for major bleeding or before significant invasive procedures. Other thresholds were 6 to $11 \times 10^9$/L for patients with fever or minor bleeding and 11 to $20 \times 10^9$/L for patients with coagulation disorders and minor procedures. Thirty-one major bleeding episodes occurred on 1.9% of the study days when platelet counts were $\leq 10 \times 10^9$/L and on only 0.07% of study days when counts were between 10 and $20 \times 10^9$/L. The investigators concluded that a prophylactic level of $5 \times 10^9$/L was safe in the absence of fever or bleeding. However, several of the major bleeds occurred in the 6 to $10 \times 10^9$/L group of patients who were not receiving prophylactic platelet transfusions.

Prospective randomized platelet transfusion trials have compared the bleeding risks and platelet transfusion needs of groups of thrombocytopenic patients who received platelets at either the $10 \times 10^9$/L or $20 \times 10^9$/L thresholds (Table 69-4).[205–208] These studies suggest that there are no

**Table 69-4: Summary of Platelet Transfusion Trigger Trials**

| Study | No. of Patients | Type | Design | Platelet Count Trigger ($\times 10^9$/L) | Findings |
|---|---|---|---|---|---|
| Gil-Fernandez et al., 1996[205] | 190 | Bone marrow transplant | Nonrandomized | 10 vs. 20 | No difference in bleeding Fewer platelet transfusions in 10K group |
| Heckman et al., 1997[206] | 78 | Acute leukemia | Randomized | 10 vs. 20 | No difference in bleeding |
| Rebulla et al., 1997[207] | 255 | Newly diagnosed acute myeloid leukemia | Randomized, multi-institution | 10 vs. 20 | No difference in major bleeding No difference in red cell transfusions |
| Wandt et al., 1998[208] | 105 | Acute myeloid leukemia | Prospective comparison | 10 vs. 20 | No difference in bleeding Fewer platelet transfusions in 10K group |

differences in hemorrhagic morbidity and mortality rates when the lower platelet transfusion trigger values are used. By applying lower prophylactic platelet transfusion thresholds, recipients are exposed to less donor blood, which in turn will decrease the risk of transfusion-transmitted disease and other complications associated with blood transfusion. Restrictive platelet transfusion practices have also been applied to patients with severe aplastic anemia who require long-term platelet support.[209]

There is no clear consensus on clinical indications for prophylactic platelet transfusions, largely because available objective data are insufficient to develop evidence-based recommendations.[210–212] Guidelines have been developed over the years by a number of different groups,[213–215] but there is wide variation in the thresholds selected for prophylactic platelet transfusions.[216] Overall, thresholds of $10 \times 10^9$/L are probably as safe as higher levels (e.g., $20 \times 10^9$/L) for preventing hemorrhage.[217]

If other risk factors for bleeding exist, such as high fever, sepsis, hyperleukocytosis, or other hemostatic abnormalities, higher thresholds for prophylactic transfusion are indicated, although specific thresholds have not been defined for these patients.[218,219] Platelet levels higher than $20 \times 10^9$/L are indicated before invasive procedures. Platelet counts are generally increased to at least $50 \times 10^9$/L before procedures such as lumbar puncture, indwelling catheter insertion, liver biopsy, thoracentesis, or transbronchial biopsy.[211] However, lower platelet counts may also be acceptable in these circumstances,[220] particularly for adults with acute leukemia who undergo lumbar puncture.[221]

Children with acute lymphoblastic leukemia and platelet counts $>10 \times 10^9$/L may tolerate lumbar puncture without serious complication.[222] Higher platelet levels ($100 \times 10^9$/L) are generally recommended for procedures involving the central nervous system. Platelet transfusions are not considered necessary prior to bone marrow aspiration or biopsy if adequate surface pressure is applied to the site after the procedure. Finally, profound anemia can alter hemostasis and thus should be avoided in patients with thrombocytopenia or platelet function defects[223] because red blood cells enhance the movement of platelets across parallel streamlines in flowing blood. The number of collisions between platelets and a vessel wall is directly related to the number of red cells (hematocrit), as demonstrated by a 50-fold increase in platelet deposition when blood is compared to PRP.[224]

### C. Efficacy of Platelet Transfusion

In clinical practice, it is difficult to evaluate the efficacy of platelet transfusions for several reasons. First, severe bleeding due to thrombocytopenia alone is rare. Second, hemorrhagic death in thrombocytopenia is even more uncommon in the absence of vascular damage or a coagulopathic state. Finally, estimating the extent of bleeding is difficult.[225] Tests used to evaluate the effectiveness of platelet transfusions are classified into four general categories: (1) *in vitro* platelet function and biochemistry; (2) *in vivo* platelet survival in the circulation; (3) clinical hemostatic efficacy; and (4) clinical tests of hemostatic efficacy, such as monitoring of epistaxis, hematuria, and petechiae (measures that are imprecise and difficult to reproduce).

The bleeding time is not helpful in determining the effectiveness of platelet transfusions (see Chapter 25). The bleeding time becomes prolonged in a linear fashion as the platelet count falls below $100 \times 10^9$/L,[226] and it is prolonged beyond measure (>30 minutes) when platelet counts fall below $10 \times 10^9$/L. In addition, the bleeding time is difficult to reproduce, has a wide normal range, is affected by a variety of drugs, is affected by variations in technique, and is prolonged in anemia (see Chapter 25).[227] Although stool blood loss has been used to evaluate hemostasis in thrombocytopenic patients, this technique is not widely available.[73] The PFA-100 and other automated platelet function testing devices are under investigation as possible measures to determine the effectiveness of platelet transfusions (see Chapter 23).

The response to prophylactic platelet transfusions is often evaluated by determining the corrected-count increment (CCI). This is performed by measuring a platelet count 10 to 60 minutes posttransfusion and then calculating the CCI according to the following formula:

$$CCI = \frac{Post(/\mu L) - Pre(/\mu L)}{\# platelets \times 10^{-11}} \times BSA(m^2)$$

where Post is the posttransfusion platelet count/$\mu$L drawn 1 hour after completing the transfusion, Pre is the pretransfusion platelet count/$\mu$L, # platelets is the number of platelets transfused (1 RDP $\approx 0.5 \times 10^{11}$; 1 apheresis platelet $\approx 3.0 \times 10^{11}$ platelets), and BSA is body surface area in square meters. In general, patients with a low CCI (<5000) have a less than expected response to platelet transfusion. Patients with a low 1-hour posttransfusion platelet increment may have become alloimmunized to platelet transfusions.[228] These alloantibodies most commonly develop from previous transfusions or pregnancies (see Section VI.C). Autoantibodies encountered in ITP can also hasten platelet removal. Nonimmune conditions, such as fever, sepsis, and disseminated intravascular coagulation (DIC), may also dramatically reduce the expected increment. Overall, the CCI is considered a relatively crude estimate of platelet survival.[229]

Animal models have been used to study the efficacy of platelet components (see Chapter 33).[124,230,231] These models can be used to detect changes in the survival of platelets processed or stored under different conditions in a much more controlled setting than the human clinical setting. They have also been applied to studies of platelet substitutes and lyophilized platelets (see Chapter 70). Rabbit models have been used to monitor the *in vivo* viability of human platelet concentrates. Survival of human platelets in rabbits has been studied by flow cytometry using antibodies specific for GPIX.[231,232] Ethyl palmitate must be administered to animals before transfusion to inhibit uptake of transfused human platelets by the reticuloendothelial system. These models could prove useful in comparing the viability of different platelet preparations that cannot be easily tested in humans.

### D. Platelet Dosing

Doses of transfused platelets are typically in the range of $3 \times 10^{11}$ platelets, which is approximately one single-donor apheresis product or six pooled random-donor PC. Optimal adult doses of platelets have not been clearly defined, and doses are typically based on a number of factors unrelated to efficacy, such as cost and availability.[233,234] Unlike children, an adult patient's body weight or body surface area is not usually considered when determining the dose of platelets to administer.[235] Of importance in this regard is the observation that, in general, the higher a patient's posttransfusion platelet count, the longer it will be until the patient's transfusion trigger level is reached.[236] Normal platelet survival is approximately 9 days. However, patients undergoing induction chemotherapy for leukemia often require platelet transfusions at least every 3 days.[236] Moreover, many patients require daily platelet transfusions during periods of severe bone marrow hypoplasia.

Some practitioners advocate providing larger doses of platelets (e.g., 10–12 RDP) every 2 or 3 days.[237] Although this practice may not reduce the risk of spontaneous bleeding, larger platelet doses may result in higher CCI and thus longer intervals between each transfusion.[238,239] Others have suggested that smaller doses of platelets (e.g., 3 or 4 RDP) can reduce the total number of platelets required during a patient's thrombocytopenic period.[240] Although decreasing pool size provides a more economical dose, this practice may result in an increased number of individual transfusions, which can actually increase overall costs.[241] In summary, the number of platelets provided per transfusion will remain largely based on local preferences until optimal platelet doses are better defined.

### E. Platelet Transfusions in Other Settings

**1. Immune Thrombocytopenia.** Platelet transfusions are rarely indicated in patients with autoimmune thrombocytopenias such as ITP. Platelet transfusions are used only when these patients develop major bleeding, not as a prophylactic measure (see Chapter 46). Similarly, platelet transfusions are often ineffective in patients with posttransfusion purpura (see Chapter 53), even when donor platelets are negative for the implicated platelet antigen (e.g., anti-HPA-1a). Selected donor platelets, however, play an important role in treating newborns with neonatal alloimmune thrombocytopenia.[242] The majority of these cases are attributed to fetomaternal incompatibility for the HPA-1a platelet-specific antigen. If infants are severely affected, they must be transfused with compatible HPA-1a-negative platelets collected from the mother or a phenotyped donor (see Chapter 53).

**2. DIC and Massive Transfusion.** PC are often transfused in acute DIC when patients are actively bleeding and severely thrombocytopenic ($<50 \times 10^9$/L). Platelet transfusions do not, however, appear useful in preventing bleeding in chronic DIC. Massive blood transfusion, in which more than 1.5 times a patient's blood volume is replaced with blood products and crystalloid, can cause significant thrombocytopenia. In this situation, platelets are being consumed and there is a compounding dilution effect caused by transfusing products such as red cells and FFP that do not contain viable platelets. Platelet transfusions should be considered during massive transfusion when counts fall below $50 \times 10^9$/L. Fibrinogen and coagulation factors are similarly diluted during massive transfusions and, if not replaced, will further diminish hemostatic capabilities. Replacement of these factors requires FFP because although PC contain human plasma, there is marked degradation of coagulation factors — especially factors V and VIII — during storage of PC at room temperature.[243,244] Platelet transfusions are frequently used during and following liver transplantation. These patients, like those with DIC, have complex hemostatic abnormalities, including coagulation factor defects and hyperfibrinolysis, that can further contribute to thrombocytopenic bleeding.

**3. Bone Marrow Transplantation.** Patients who undergo autologous and, especially, allogeneic bone marrow transplantation have prolonged periods of thrombocytopenia that require platelet transfusion support. Following autologous peripheral blood stem cell transplants, most patients recover platelet counts to levels above 20 to $50 \times 10^9$/L within 14 days of receiving stem cells. However, platelet recovery time may be twice as long when using allogeneic bone marrow, likely due to decreased collection of CD34$^+$ stem cells by this technique.[245] There are several risk factors for prolonged thrombocytopenia ($<20 \times 10^9$/L) in this patient population, but it is difficult to identify all patients at risk.[246] For example, autologous transplant patients who receive numerous cycles of high-dose chemotherapy are predisposed to poorer CD34$^+$ cell mobilization. This leads to lower CD34$^+$ cell yields during harvesting and subsequently delayed platelet recovery following stem cell infusion.[247] In contrast, patients who receive higher numbers of stem cells (e.g., $\geq 5 \times 10^6$ CD34$^+$ cells/kg body weight) have shorter periods of severe thrombocytopenia.[248]

**4. Cardiac Surgery.** PC are commonly provided during or soon after cardiac surgery in patients undergoing cardiopulmonary bypass (CPB) in an effort to limit postoperative bleeding. The hemostatic defects associated with CPB are complex but are generally related to anticoagulation, temperature changes, length of time on bypass, and preoperative aspirin use (see Chapter 59). Effects of CPB on platelets include platelet activation due to exposure to foreign biomaterial surfaces, platelet fragmentation, and impaired aggregation responses to most agonists.[249] There is usually some degree of platelet damage and thrombocytopenia during bypass surgery, and platelet function defects may persist for days after the procedure. However, the treatment of post-CPB bleeding is usually not based on laboratory monitoring and is largely empiric and institution dependent.[250] Although clinical trials are limited, several studies have been unable to demonstrate benefits of routine platelet administration during open-heart surgery or trauma, as determined by decreased red cell transfusions, chest tube bleeding, or microvascular bleeding.[251,252] Thus, many surgeons reserve platelet transfusions for patients who are actually bleeding despite adequate surgical hemostasis. Patients who have received aspirin or GPIIb-IIIa antagonists[253] immediately before cardiac surgery are at additional risk for bleeding. In the latter situation, fibrinogen supplementation with FFP and/or cryoprecipitate either alone or in combination with platelet transfusion may help to restore the hemostatic defect associated with the administration of GPIIb-IIIa antagonists.[254] Like CPB, extracorporeal membrane oxygenation in neonatal patients can result in severe hemostatic defects that require frequent platelet transfusions.[255]

**5. Inherited and Acquired Platelet Function Defects.** Patients with inherited (see Chapter 57) or acquired (see Chapter 58) platelet function defects do not usually require prophylactic platelet transfusions. However, platelet transfusions are often used to treat bleeding episodes or are administered prior to surgery. Following platelet transfusion, patients with Glanzmann thrombasthenia (see Chapter 57) may develop GPIIb-IIIa-specific antibodies that can compromise survival and, thus, the efficacy of further platelet transfusions.[256] Desmopressin acetate[257] and recombinant activated factor VII (FVIIa),[258,259] which shorten the bleeding time of many patients with congenital platelet function disorders, are potential alternatives or adjuncts to platelet transfusion therapy (see Chapters 67 and 68). The efficacy of VIIa in decreasing bleeding has been demonstrated using a rabbit model of severe thrombocytopenia.[260] Patients with acquired platelet defects caused by the myriad of drugs with antiplatelet activity (see Chapter 58) should have these medications discontinued whenever possible prior to invasive procedures. Finally, platelet transfusions can increase platelet counts to safer levels without apparent adverse effects in patients with abciximab-induced thrombocytopenia (see Chapter 49).[261] Platelet transfusions are not recommended in the setting of heparin-induced thrombocytopenia because they can precipitate acute thrombosis (see Chapter 48).

## F. Autologous Platelet Donation

Although autologous red cell donation is a standard practice, autologous platelet donations are used only rarely. Liquid stored platelets are not practical for these purposes because of their limited 5-day shelf-life. Thus, patients are unable to "bank" enough of their own platelets prior to prolonged periods of thrombocytopenia. However, platelets collected by apheresis can be placed in a dimethyl sulfoxide (DMSO) cryoprotectant and stored frozen at −80°C.[262] They are then thawed, washed to remove DMSO, and resuspended in autologous plasma or other solution before transfusion. The techniques are not difficult to perform, but most blood banks have not developed procedures for preparing and storing such products. In addition, frozen and thawed platelets undergo a number of structural and metabolic changes that adversely affect their *in vivo* survival.[263] Collection of autologous platelets can be considered in patients receiving high-dose chemotherapy, especially those who are alloimmunized to platelet transfusion.[264] However, most oncology patients cannot donate enough platelets to support their needs through an entire course of chemotherapy. Autologous platelet donation has also been used to decrease the number of allogeneic platelet and red blood cell exposures required during cardiac surgery (see Chapter 59).[265,266] In this scenario, platelets are collected by apheresis shortly before surgery and are then reinfused after surgery.[267] Lastly, autologous platelets have been studied to aid wound healing and tissue regeneration.[268] In one such approach, a platelet "gel" is produced by treating platelets with thrombin.[269]

## VI. Adverse Reactions to Platelet Transfusion

Platelet transfusion therapy is not without risk (Table 69-5). Common, but non-life-threatening, complications of platelet transfusions include febrile and allergic transfusion reac-

### Table 69-5: Adverse Consequences of Platelet Transfusion Therapy

Febrile transfusion reactions

Allergic reactions (urticarial and anaphylactic)

Septic reactions (bacterial)

Viral transmission (hepatitis C, hepatitis B, HIV-1, HIV-2, HTLV-1, HTLV-2, CMV)

Parasitic infection (*Babesia microti, Trypanosoma cruzi, Plasmodium sp.*)

Transfusion-associated graft-versus-host disease

Transfusion-related acute lung injury

Platelet refractory state (alloimmunization)

Transfusion-associated immunosuppression

Hypotensive reactions associated with ACE inhibition

Post-transfusion purpura

ACE, angiotensin-converting enzyme; CMV, cytomegalovirus; HIV, human immunodeficiency virus; HTLV; human T cell lymphotropic virus.

tions.[270] Although PC are capable of transmitting viral disease, improved donor testing has markedly decreased this risk. For example, the estimated risk of HIV-1 transmission from a platelet transfusion is between 1 in 1 million and 1 in 3 million.[271,272] Similarly, the risk of contracting hepatitis C from donated blood has been reduced to approximately 1 in 1 million to 1 in 2 million exposures since nucleotide testing for this virus was implemented. Transfused PC are now far more likely to be contaminated by bacteria than by known viruses. Patients who receive such bacterially contaminated products can develop serious septic reactions. Because PC contain viable lymphocytes, platelet transfusion can also cause transfusion-associated graft-versus-host disease in susceptible patients, a rare but usually fatal reaction that is prevented by irradiating PC before transfusion (see Section IV.B).[273] Transfusion-related acute lung injury (TRALI), an often unrecognized cause of noncardiogenic pulmonary edema, can follow platelet, plasma, or red cell transfusion.[274–276] The infusion of bioactive lipids found in stored platelets has been implicated in the pathogenesis of TRALI. Of particular concern for frequently transfused patients is the development of antibodies directed against platelet membrane components. These antibodies can result in a "platelet refractory state" in which recipients rapidly remove transfused platelets through immune-mediated mechanisms.

### A. Febrile Reactions to Platelet Transfusion

Febrile transfusion reactions (FTR) are commonly encountered with platelet transfusions. They usually occur during the transfusion, but they may develop minutes to several hours after the transfusion is completed. The frequency of FTR is estimated to be between 4 and 30% of platelet transfusions.[277] FTR typically present as fever (>1°C increase) and shaking chills, and they may be accompanied by nausea, vomiting, dyspnea, and hypotension. The severity of symptoms appears to be related to the number of white cells found in the product and/or the rate of transfusion. Severe rigors promptly resolve when meperidine is administered.[278] Although patients are often medicated with an antipyretic such as acetaminophen before platelet transfusions to minimize FTR, it is unclear if this practice actually minimizes these reactions.[279] Premedication with intravenous corticosteroids in patients who have had repeated severe FTR may be useful; however, to be maximally effective, these medications must be administered several hours before transfusion. Antihistamines do not appear useful in preventing or treating FTR.

The mechanism of FTR was originally reported to involve interactions between cytotoxic antibodies in the platelet recipient and HLA and/or leukocyte-specific antigens on donor white cells found in the product.[280] The formation of white cell antigen–antibody complexes results in complement binding and subsequent release of endogenous pyrogens such as tumor necrosis factor-α (TNF-α), interleukin (IL)-1, and IL-6. It is increasingly clear that the direct activity of various biological response modifiers including cytokines plays a role in FTR.[281] The incidence of FTR may be directly related to the duration of platelet storage.[282] For example, there is a progressive increase in the relative risk of developing FTR when BC platelets are stored more than 5 days.[283] In addition, transfusing fresher (3-day-old or less) pooled RDP may decrease the incidence of FTR.[284]

These observations may be related to the continued elaboration of biologically active cytokines by residual white cells, which include activated monocytes.[285] The levels of these cytokines found in stored PC, particularly IL-1β and IL-6, strongly correlate with the frequency of FTR.[145] Development of FTR may also be related to preparation technique. For example, one prospective study designed to compare single-donor apheresis platelets and PC prepared by the BC and PRP techniques noted significantly more FTR in patients who received PRP-prepared PC.[286] However, there was no difference in platelet survival between the different preparations based on CCI measured 1 to 6 and 18 to 24 hours posttransfusion. It has also been suggested that the level of cytokines (e.g., IL-1β and TNF-α) in the plasma of stored PC may be influenced by specific cytokine gene polymorphisms.[287]

It is generally accepted that leukoreduction of PC can help minimize FTR. The degree of leukoreduction is limited by the available techniques. Third-generation filters eliminate >99.9% of white cells in a unit of red cells obtained from a whole blood donation, leaving $<5 \times 10^6$ residual leukocytes. Prestorage leukoreduction, in which white cells are removed soon after blood collection, may even further reduce the likelihood of FTR.[147,288] Again, this benefit appears to correlate with the reduction of cytokines in the plasma portion of PC during storage.[289]

Leukoreduction filters not only remove white cells but also have a varying ability to directly remove biological response modifiers, such as the chemokines IL-8 and RANTES (regulated upon activation, normal T cell expressed and presumably secreted) and the anaphylatoxins C3a and C5a.[290,291] However, the additional benefit of direct removal of biological response modifiers by leukoreduction filters in reducing the incidence of FTR is unclear. If a patient continues to have severe febrile reactions despite leukoreduction, washed platelets that have had most of the suspending plasma removed can be considered.[292]

### B. Bacterial Contamination of Platelet Concentrates

Bacterial contamination of stored PC occurs because platelets are stored at room temperature.[293,294] The shelf-life of PC was decreased from 7 to 5 days in the mid-1980s, largely

**Table 69-6: Contaminating Organisms Found in Platelet Concentrates**[a]

| | |
|---|---|
| Bacillus species | Propionibacterium acnes |
| B. cereus | Pseudomonas aeruginosa |
| B. subtilis | Salmonella species |
| Candida albicans | Staphylococcus species |
| Clostridium perfringens | S. aureus |
| Corynebacterium species | S. epidermidis |
| Enterobacter cloacae | Serratia marcescens |
| Escherichia coli | Streptococcus species |
| Klebsiella oxytoca | S. pyogenes |
| Leishmaniasis | S. viridans |
| Morganella morganii | |

[a]Data from references 293–311 and 366–368.

based on observations that many cases of septic transfusion reactions occurred in patients who received PCs stored for more than 5 days.[293,295] Moreover, *in vitro* studies performed by inoculating PC with bacteria suggest that a significant number of PC experience the most rapid rate of bacterial growth on days 6 and 7 of storage.[296] The most commonly implicated organisms in platelet septic transfusion reactions are gram-positive bacteria (*Staphylococcus* sp.), although gram-negative organisms are also isolated (Table 69-6).[297,298] Platelets, and other blood products, can be contaminated by bacteria if a donor is bacteremic during blood collection or if the arm is improperly cleansed before venipuncture.[299] Blood donors must be in good health on the day of donation. However, asymptomatic donors who may have had gastroenteritis of short duration several days before donation are potentially infectious. These donors may have a prolonged period of asymptomatic bacteremia that allows transmission of organisms.[300]

Studies indicate that as many as 1:1000 to 1:3000 PC may be contaminated by bacteria, depending on the method used to detect organisms.[301,302] However, the incidence of transfusion-related sepsis following platelet transfusion is much less — recognized and culture confirmed in at least 1:100,000 recipients.[303] These cases can be fatal.[304] Standards have been adopted in the United States that require blood collection facilities and transfusion services to limit and detect bacterial contamination in all platelet components. No detection method has been universally adopted and, regardless of the method, no bacterial screening system is likely to be 100% sensitive to bacterial pathogens.[305] Different methods are used, depending on the type of platelet donation (e.g., whole blood-derived vs. apheresis). Most blood centers directly culture apheresis PC[306] for bacteria and release the unit after the culture has incubated between 12 and 36 hours.[307] The most commonly cultured bacteria

**Table 69-7: Conditions Associated with a Platelet Refractory State**

| Non-Immune Related | Immune Related |
|---|---|
| Splenomegaly | Alloantibodies to class I HLA antigens |
| Disseminated intravascular coagulation | Alloantibodies to platelet-specific antigens |
| Drugs (antibiotics, amphotericin B) | Autoantibodies to platelet-specific antigens |
| Sepsis, fever, viremia | Circulating immune complexes |
| Graft-versus-host disease | |
| Veno-occlusive disease of the liver | |

are gram-positive organisms such as *Staphylococcus epidermidis*, *S. aureus*, and *Streptococcus pneumoniae;* these are often skin commensals. Skin disinfection by povidone–iodine and isopropyl alcohol helps to minimize venipuncture-related contamination of platelet concentrates by skin flora.[308] The presence of gram-negative organisms such as *Escherichia coli* usually signifies occult bacteremia in the blood donor. Screening of whole blood-derived platelets for bacteria is typically conducted by individual hospital transfusion services. Simpler point-of-use techniques based on lowering of pH and glucose levels in bacterially contaminated platelet units are often used by hospital blood banks.[309] However, these tests are insensitive, providing unacceptable false-negative results.[310] Detection of bacterial contamination in platelet concentrates using real-time reverse transcriptase PCR remains at the investigational level.[311]

## C. Platelet Refractory State

Early in the history of platelet transfusion therapy it was recognized that patients who receive long-term platelet support may develop a refractory state in which transfused platelets undergo accelerated destruction.[312] Refractoriness to platelet transfusion is often related to clinical conditions that hasten platelet removal through either immune or non-immune mechanisms (Table 69-7).[313–316] In the former situations, foreign donor HLA appears to evoke an immune response in the recipient that leads to the rapid removal of transfused platelets.[317] Presumably, donor platelets become coated by HLA-specific, or in some cases platelet-specific, antibodies. The antibody-coated platelets are then removed by the reticuloendothelial system.[318] In addition, platelet activation and subsequent aggregation may play a role in immune-mediated platelet destruction.[319] Platelet alloimmu-

nization rates are higher in patients with prior pregnancies or transfusion. Alternatively, alloimmunization rates may be lower in thrombocytopenic patients who are receiving high-dose immunosuppressive chemotherapy.[320–322] Overall, nonimmune platelet destruction caused by conditions such as splenomegaly, infection, and antibiotic treatment with amphotericin B is more frequent than that related to immune removal.[323–325] The effects of amphotericin B on platelet survival can be lessened by transfusing platelets 2 hours after completing the amphotericin infusion.[326]

Platelets express ABH, Lewis, P, and I blood system antigens, as well as class I HLA-A, -B, and -C and platelet-specific antigens (HPA). The ABH, class I HLA, and HPA antigens are most relevant to allogeneic platelet survival.[327] However, the presence of recipient antibodies directed against these antigens may or may not be associated with a bleeding risk. Platelet survival is clearly impaired when recipients with higher titer anti-A or anti-B IgG antibodies are transfused with platelets carrying one of these antigens.[328] Although this problem is avoided by using ABO-identical platelets, it must be recognized that this option may not always be available during blood shortages.[329]

Alloimmunization against HLA-A and HLA-B class I antigens is the most common cause of an immune-mediated platelet refractory state; mismatches of HLA-C antigens are less important.[330] Even before these immune processes were well understood, a small trial conducted in the 1970s suggested that platelet refractoriness rates could be lowered by removing white blood cells from blood products through a cotton filter.[331] However, these findings were not reproduced in a randomized trial of leukocyte-reduced blood components.[332] Patients at particular risk for HLA alloantibody formation include those previously transfused with non-leukocyte-reduced blood components.[333] Primary HLA alloimmunization can be reduced by removing antigen-presenting cells (leukocytes) from PC before transfusion and UVB irradiating transfused platelets to inactivate antigen-presenting cells.[141,334–338] Strategies used to reduce HLA alloimmunization are especially effective in patients with acute myeloid leukemia.[141] It is more difficult to identify significant differences in immunization rates using leukoreduced blood products when broader groups of patients are studied.[338] Evidence suggests that transfusing filtered apheresis platelets reduces the rate of platelet refractoriness.[339]

The management of platelet-refractory patients varies across institutions and is often related to the availability of specialized testing and products.[340] In nonbleeding patients, prophylactic platelet transfusions are often withheld. However, patients with significant bleeding, other risk factors for bleeding, or those undergoing an invasive procedure may require transfusion. Platelet transfusion strategies in these cases include increasing the dose by providing larger numbers of platelets, providing HLA-selected platelets, and providing cross-match-compatible platelets. Ideally, anti-HLA antibodies should be identified in the patient before HLA-specific products are obtained.[341] Products are collected by apheresis and are selected based on the HLA-A and HLA-B types of the donor and intended recipient. HLA-C antigens are weakly expressed by platelets and do not seem to affect platelet survival.[342] Because the HLA system is extremely polymorphic, a large number of HLA-typed donors is needed to support refractory patients.[343] The use of HLA-selected or "matched" platelets in patients who are not alloimmunized with the goal of preventing such immunization is not warranted.[344,345]

Platelet cross-matching is performed by reacting recipient serum with potential donor platelets that are fixed to a solid support. The compatibility is determined based on the presence or absence of reactivity.[346] An incompatible platelet cross-match predicts a poor response in more than 90% of transfusions, whereas a compatible cross-match is 50% predictive of a successful transfusion.[347] Large collection facilities often combine the two approaches: Platelet units are selected for cross-matching based on the class I HLA types of the donor and recipient. More unusual and difficult to manage is the patient who has developed both HLA and platelet-specific antibodies.[348] Other approaches that have been explored to support HLA-alloimmunized patients are based on reducing the expression of class I HLA antigens on platelets.[349] Although they are effective treatments for many autoimmune thrombocytopenias, corticosteroids, chemotherapy, and splenectomy are not useful in most patients who have become alloimmunized to platelet transfusions.

Platelet autoantibodies also occur in conditions such as ITP (see Chapter 46) in which patients produce antibodies directed against common platelet antigens. These individuals usually respond very poorly to platelet transfusions because donor platelets are rapidly removed from the circulation. Interestingly, patients with ITP and extremely low platelet counts infrequently experience serious bleeding and hence do not usually require prophylactic platelet transfusions, despite extremely low platelet counts. Intravenous immunoglobulin, which is often used as a treatment for ITP (see Chapter 46), may improve the response to transfused platelets. Patients with ITP who are bleeding and responding poorly to standard platelet transfusions may benefit from continuous 24-hour infusions of both intravenous immunoglobulin and platelets.[350] Continuous platelet infusions, or "platelet drips," are typically performed by infusing three pooled RDPs or one-half of an apheresis-derived PC every 4 hours. They can be infused through electromechanical pumps because these devices do not appear to harm platelets.[351]

### D. Hypotensive Reactions during Platelet Transfusions

A number of serious hypotensive episodes have been described in patients on angiotensin-converting enzyme (ACE) inhibitors who received platelet transfusions through negatively charged bedside leukocyte reduction filters.[352] These reactions began soon after the infusion was initiated and subsided rapidly after the transfusion was terminated. The mechanisms responsible for these reactions appear to involve bradykinin, which is a vasodilator that can be generated when blood contacts a negatively charged artificial surface. Bradykinin is then capable of activating the kallikrein–kinin cascade. Bradykinin and its active metabolites are rapidly degraded to inactive compounds through reactions that require ACE. Therefore, patients on ACE inhibitor medications are less able to degrade these vasodilatory substances and, presumably, are at risk for hypotension caused by decreased systemic vascular resistance. *In vitro* studies have shown that supernatant bradykinin levels markedly increase when PC are passed through certain negatively charged filters.[353] In addition, patients who receive PC through a negatively charged filter experience significant increases in bradykinin levels during the first 5 minutes of transfusion.[354] Bradykinin levels were particularly high in two platelet recipients who were receiving ACE inhibitor therapy at the time of transfusion. However, other mechanisms may be involved because similar hypotensive reactions have occurred in patients who were not receiving ACE inhibitors.[355] A metabolic abnormality has been identified in several blood recipients who have developed such reactions.[356] This defect affects the degradation of des-Arg9-bradykinin, a metabolically active metabolite of bradykinin that is primarily inactivated by ACE and aminopeptidase P. Accordingly, the half-life of des-Arg9-bradykinin was significantly longer in patients on ACE inhibitors who experienced severe hypotensive reactions than in control patients. Based on these studies, it appears that both patient and product-related factors contribute to the pathogenesis of severe hypotensive reactions to platelet transfusion.

## VII. Thrombopoietic Growth Factors in Platelet Transfusion Therapy

One of the primary reasons for developing hematopoietic growth factors is to limit the exposure of patients to allogeneic blood components. For example, the use of recombinant erythropoietin has dramatically decreased the need for red cell transfusions in a variety of patients. Efforts continue to produce similar agents that would reduce the need for platelet transfusion therapy (see Chapter 66).[357] Other potential applications of thrombopoietic growth factors include stimulating platelet apheresis donors, which would increase platelet yields during a single collection; increasing stem cell harvest yields, which in turn would decrease the period of severe thrombocytopenia that occurs following stem cell transplantation; and expanding progenitor cells *ex vivo*.[358] In the latter situation, thrombopoietic growth factors are combined with other growth factors to produce megakaryocytic progenitors *ex vivo* that can then be administered to the autologous donor.[359] IL-11, which directly stimulates the proliferation of hematopoietic stem cells and megakaryocyte progenitors and hastens megakaryocyte maturation, was approved by the FDA in 1997 for preventing severe thrombocytopenia in patients receiving myelosuppressive chemotherapy (see Chapter 66).

Thrombopoietin is an endogenous protein whose effects include increasing the number, size, and ploidy of megakaryocytes in culture (see Chapter 66). Its receptor, c-mpl, was discovered during studies of a viral oncogene v-mpl in the myeloproliferative (mpl) leukemia virus. A recombinant human, truncated form of the mpl ligand has been developed and evaluated for human use. This form is covalently linked to polyethylene glycol (pegylated megakaryocyte growth and development factor [PEG-MGDF]) in order to extend the half-life of the molecule. This agent stimulates the production of platelets that are not activated, as determined by surface P-selectin expression and surface binding of annexin V.[360] In addition, direct exposure of stored PC to PEG-MGDF does not appear to hasten development of the platelet storage defect, as assessed by a battery of *in vitro* tests of platelet structure and function.[361]

Thrombopoietin has been shown to stimulate the production of platelets in thrombocytopenic oncology patients following chemotherapy[362,363] (see Chapter 66). The platelets produced in response to this growth factor demonstrate normal responses to platelet agonists and exhibit normal ATP release *in vivo*.[364] PEG-MGDF has been administered to normal donors, in which case a single 2 μg/kg dose produces a doubling of the platelet count within 9 to 14 days.[364] In the same study, single 3 μg/kg doses produced platelet counts of approximately $600 \times 10^9$/L, showing that MGDF could be used to increase the yield of platelets from volunteer platelet apheresis donors. Platelets collected from these "stimulated" donors are viable and improve platelet counts when transfused to thrombocytopenic recipients. However, several normal platelet donors have developed neutralizing antibodies directed against endogenous thrombopoietin.[365] These antibodies have produced severe thrombocytopenia ($<10 \times 10^9$/L), and subsequently, trials in normal volunteer donors were suspended. Alternatives to PEG-MGDF that remain under development (see Chapter 66) could have a significant impact on platelet transfusion therapy by reducing platelet transfusion needs of thrombocytopenic patients and by increasing platelet yields in volunteer donors.

## VIII. Conclusions

Platelet transfusion therapy has benefited large numbers of patients who are thrombocytopenic or have platelet function defects. In most cases, these transfusions occur without serious complications, as a direct result of efforts that have improved the safety and efficacy of liquid PC. Undoubtedly, platelet transfusions will become even safer with further advances in donor testing and the development of techniques that detect or eliminate bacteria and other pathogens that may contaminate PC. The efficacy of platelet transfusion therapy has been advanced to a large extent during the past 35 years by optimizing the conditions under which platelets are stored. Storage conditions have been selected to minimize the changes in platelet structure and function that occur during room temperature storage. These changes, which adversely affect platelet viability and hemostatic function, could be further reduced by developing other technologies such as platelet additive solutions that better preserve platelets (see Chapter 70). A large number of *in vitro* laboratory tests have been used to compare different methods for PC preparation and storage. These tests correlate with tests of *in vivo* hemostatic function to varying degrees, but no single test provides a complete measure of platelet quality. Innovative, yet practical, *in vitro* methods are needed that more accurately predict and assess platelet function and viability *in vivo*. Better markers of the *in vivo* hemostatic efficacy of platelet therapy could serve as alternatives to posttransfusion increments and bleeding times. These tests could help to better define both platelet transfusion thresholds and optimal platelet transfusion doses. Although alternatives to liquid-stored platelets (e.g., cryopreserved platelets, liposomes, and thrombopoietic growth factors) are under active development (see Chapters 66 and 70), it appears that whole blood-derived and apheresis PC will remain in widespread clinical use for the foreseeable future.

## References

1. Hersh, E. M., Bodey, G. P., Boyd, A. N., et al. (1965). Causes of death in acute leukaemia: A ten year study of 414 patients from 1954–1963. *JAMA, 193,* 99–103.
2. Minor, A. H., & Burnett, L. (1952). Method for separating and concentrating platelets from human blood. *Blood, 7,* 693–699.
3. Wallace, E. L., Churchill, W. H., Surgenor, D. M., et al. (1998). Collection and transfusion of blood and blood components in the United States, 1994. *Transfusion, 38,* 625–636.
4. Meehan, K. R., Matias, C. O., Rathore, S. S., et al. (2000). Platelet transfusions: Utilization and associated costs in a tertiary care hospital. *Am J Hematol, 64,* 251–256.
5. Schiffer, C. A., Anderson, K. C., Bennett, C. L., et al. (2001). Platelet transfusion for patients with cancer: Clinical practice guidelines of the American Society of Clinical Oncology. *J Clin Oncol, 19,* 1519–1538.
6. Murphy, S., Heaton, W. A., & Rebulla, P. (1996). Platelet production in the Old World — And the New. *Transfusion, 36,* 751–754.

7. Slichter, S. J., & Harker, L. A. (1976). Preparation and storage of platelet concentrates. *Transfusion, 16,* 8–12.

8. Mourad, N. (1968). Studies on release of certain enzymes from certain enzymes from human platelets. *Transfusion, 8,* 363–367.

9. Moroff, G., Friedman, A., Robkin-Kline, L., et al. (1984). Reduction of the volume of stored platelet concentrates for use in neonatal patients. *Transfusion, 24,* 144–146.

10. Simon, T. L., & Sierra, E. R. (1984). Concentration of platelet units into small volumes. *Transfusion, 24,* 173–175.

11. Fijnheer, R., Pietersz, R. N., de Korte, D., et al. (1990). Platelet activation during preparation of platelet concentrates: A comparison of the platelet-rich plasma and the buffy coat methods. *Transfusion, 30,* 634–638.

12. Hogman, C. F., Eriksson, L., Hedlund, K., et al. (1988). The bottom and top system: A new technique for blood component preparation and storage. *Vox Sang, 55,* 211–217.

13. Pietersz, R. N., Loos, J. A., & Reesink, H. W. (1985). Platelet concentrates stored in plasma for 72 hours at 22 degrees C prepared from buffycoats of citrate-phosphate-dextrose blood collected in a quadruple-bag saline-adenine-glucose-mannitol system. *Vox Sang, 49,* 81–85.

14. Bertolini, F., Rebulla, P., Porretti, L., et al. (1992). Platelet quality after 15-day storage of platelet concentrates prepared from buffy coats and stored in a glucose-free crystalloid medium. *Transfusion, 32,* 9–16.

15. Eriksson, L., Shanwell, A., Gulliksson, H., et al. (1993). Platelet concentrates in an additive solution prepared from pooled buffy coats. *In vivo* studies. *Vox Sang, 64,* 133–138.

16. McLeod, B. C., Price, T. H., & Drew, M. J. (1997). *Apheresis: Principles and practice of apheresis.* Bethesda, MD: AABB Press.

17. Elfath, M. D., Whitley, P., Jacobson, M. S., et al. (2000). Evaluation of an automated system for the collection of packed RBCs, platelets, and plasma. *Transfusion, 40,* 1214–1222.

18. McLeod, B. C., Price, T. H., Owen, H., et al. (1998). Frequency of immediate adverse effects associated with apheresis donation. *Transfusion, 38,* 938–943.

19. Despotis, G. J., Goodnough, L. T., Dynis, M., et al. (1999). Adverse events in platelet apheresis donors: A multivariate analysis in a hospital-based program. *Vox Sang, 77,* 24–32.

20. Bolan, C. D., Greer, S. E., Cecco, S. A., et al. (2001). Comprehensive analysis of citrate effects during plateletpheresis in normal donors. *Transfusion, 41,* 1165–1171.

21. Gutensohn, K., Bartsch, N., & Kuehnl, P. (1997). Flow cytometric analysis of platelet membrane antigens during and after continuous-flow plateletpheresis. *Transfusion, 37,* 809–815.

22. Harrison, P., Segal, H., Furtado, C., et al. (2004). High incidence of defective high-shear platelet function among platelet donors. *Transfusion, 44,* 764–770.

23. Jilma-Stohlawetz, P., Hergovich, N., Homoncik, M., et al. (2001). Impaired platelet function among platelet donors. *Thromb Haemost, 86,* 880–886.

24. Lazarus, E. F., Browning, J., Norman, J., et al. (2001). Sustained decreases in platelet count associated with multiple, regular plateletpheresis donations. *Transfusion, 41,* 756–761.

25. Mair, B., & Benson, K. (1998). Evaluation of changes in hemoglobin levels associated with ABO-incompatible plasma in apheresis platelets. *Transfusion, 38,* 51–55.

26. McManigal, S., & Sims, K. L. (1999). Intravascular hemolysis secondary to ABO incompatible platelet products. An underrecognized transfusion reaction. *Am J Clin Pathol, 111,* 202–206.

27. Larsson, L. G., Welsh, V. J., & Ladd, D. J. (2000). Acute intravascular hemolysis secondary to out-of-group platelet transfusion. *Transfusion, 40,* 902–906.

28. Josephson, C. D., Mullis, N. C., Van Demark, C., et al. (2004). Significant numbers of apheresis-derived group O platelet units have "high-titer" anti-A/A,B: Implications for transfusion policy. *Transfusion, 44,* 805–808.

29. Angiolillo, A., & Luban, N. L. (2004). Hemolysis following an out-of-group platelet transfusion in an 8-month-old with Langerhans cell histiocytosis. *J Pediatr Hematol Oncol, 26,* 267–269.

30. Zingsem, J., Glaser, A., Weisbach, V., et al. (1998). Evaluation of a platelet apheresis technique for the preparation of leukocyte-reduced platelet concentrates. *Vox Sang, 74,* 189–192.

31. Moog, R., & Muller, N. (1999). White cell reduction during plateletpheresis: A comparison of three blood cell separators. *Transfusion, 39,* 572–577.

32. Seghatchian, J., Krailadsiri, P., & McCall, M. (2001). Statistical process monitoring of WBC-reduced blood components assessed by two types of software. *Transfusion, 41,* 102–105.

33. Perrotta, P. L., Ozcan, C., Whitbread, J. A., et al. (2002). Applying a novel statistical process control model to platelet quality monitoring. *Transfusion, 42,* 1059–1066.

34. American Association of Blood Banks. (2002). *Technical manual* (14th ed.). Bethesda, MD.

35. Council of Europe. (2004). *Guide to the preparation, use and quality assurance of blood components* (10th. ed.). Strasbourg: Council of Europe Publishing.

36. Murphy, S., Rebulla, P., Bertolini, F., et al. (1994). *In vitro* assessment of the quality of stored platelet concentrates. The BEST (Biomedical Excellence for Safer Transfusion) Task Force of the International Society of Blood Transfusion. *Transfus Med Rev, 8,* 29–36.

37. Tudisco, C., Jett, B. W., Byrne, K., et al. (2005). The value of pH as a quality control indicator for apheresis platelets. *Transfusion, 45,* 773–778.

38. Dickerhoff, R., and Von Ruecker, A. (1995). Enumeration of platelets by multiparameter flow cytometry using platelet-specific antibodies and fluorescent reference particles. *Clin Lab Haematol, 17,* 163–172.

39. Dijkstra-Tiekstra, M. J., van der Meer, P. F., Pietersz, R. N., et al. (2004). Multicenter evaluation of two flow cytometric methods for counting low levels of white blood cells. *Transfusion, 44,* 1319–1324.

40. Dietz, L. J., Dubrow, R. S., Manian, B. S., et al. (1996). Volumetric capillary cytometry: A new method for absolute cell enumeration. *Cytometry, 23,* 177–186.

41. Brecher, M. E., Wong, E. C., Chen, S. E., et al. (2000). Antibiotic-labeled probes and microvolume fluorimetry for the rapid detection of bacterial contamination in platelet components: A preliminary report. *Transfusion, 40,* 411–413.

42. Dumont, L. J., Dzik, W. H., Rebulla, P., et al. (1996). Practical guidelines for process validation and process control of white cell-reduced blood components: Report of the Biomedical Excellence for Safer Transfusion (BEST) Working Party of the International Society of Blood Transfusion (ISBT). *Transfusion, 36,* 11–20.

43. Moog, R., & Muller, N. (1999). Cost effectiveness of quality assurance in plateletpheresis. *Transfus Sci, 21,* 141–145.

44. Dzik, S. (1997). Principles of counting low numbers of leukocytes in leuko reduced blood components. *Transfus Med Rev, 11,* 44–55.

45. Lutz, P., & Dzik, W. H. (1993). Large-volume hemocytometer chamber for accurate counting of white cells (WBCs) in WBC-reduced platelets: Validation and application for quality control of WBC-reduced platelets prepared by apheresis and filtration. *Transfusion, 33,* 409–412.

46. Moroff, G., Eich, J., & Dabay, M. (1994). Validation of use of the Nageotte hemocytometer to count low levels of white cells in white cell-reduced platelet components. *Transfusion, 34,* 35–38.

47. Hanseler, E., Fehr, J., & Keller, H. (1996). Estimation of the lower limits of manual and automated platelet counting. *Am J Clin Pathol, 105,* 782–787.

48. Dzik, W. H., Ragosta, A., & Cusack, W. F. (1990). Flow-cytometric method for counting very low numbers of leukocytes in platelet products. *Vox Sang, 59,* 153–159.

49. Krailadsiri, P., Seghatchian, J., Rigsby, P., et al. (2002). A national quality assessment scheme for counting residual leucocytes in unfixed leucodepleted products: The effect of standardisation and 48 hour storage. *Transfus Apheresis Sci 26,* 73–81.

50. Adams, M. R., Johnson, D. K., Busch, M. P., et al. (1997). Automatic volumetric capillary cytometry for counting white cells in white cell-reduced plateletpheresis components. *Transfusion, 37,* 29–37.

51. Dzik, S., Moroff, G., & Dumont, L. (2000). A multicenter study evaluating three methods for counting residual WBCs in WBC-reduced blood components: Nageotte hemocytometry, flow cytometry, and microfluorometry. *Transfusion, 40,* 513–520.

52. Klinger, M. H. (1996). The storage lesion of platelets: Ultrastructural and functional aspects. *Ann Hematol, 73,* 103–112.

53. Hogge, D. E., Thompson, B. W., & Schiffer, C. A. (1986). Platelet storage for 7 days in second-generation blood bags. *Transfusion, 26,* 131–135.

54. Schiffer, C. A., Lee, E. J., Ness, P. M., et al. (1986). Clinical evaluation of platelet concentrates stored for one to five days. *Blood, 67,* 1591–1594.

55. Chernoff, A., & Snyder, E. L. (1992). The cellular and molecular basis of the platelet storage lesion: A symposium summary. *Transfusion, 32,* 386–390.

56. Leach, M. F., & AuBuchon, J. P. (1993). Effect of storage time on clinical efficacy of single-donor platelet units. *Transfusion, 33,* 661–664.

57. Holme, S., & Heaton, A. (1995). *In vitro* platelet ageing at 22 degrees C is reduced compared to *in vivo* ageing at 37 degrees C. *Br J Haematol, 91,* 212–218.

58. Rebulla, P. (1998). *In vitro* and *in vivo* properties of various types of platelets. *Vox Sang, 74,* 217–222.

59. Snyder, E. L., Hezzey, A., Katz, A. J., et al. (1981). Occurrence of the release reaction during preparation and storage of platelet concentrates. *Vox Sang, 41,* 172–177.

60. Triulzi, D. J., Kickler, T. S., & Braine, H. G. (1992). Detection and significance of alpha granule membrane protein 140 expression on platelets collected by apheresis. *Transfusion, 32,* 529–533.

61. Seghatchian, J., & Krailadsiri, P. (1997). The platelet storage lesion. *Transfus Med Rev, 11,* 130–144.

62. Brecher, M. E., & Hay, S. N. (2004). Transfusion medicine illustrated. Platelet swirling. *Transfusion, 44,* 627.

63. Bertolini, F., & Murphy, S. (1996). A multicenter inspection of the swirling phenomenon in platelet concentrates prepared in routine practice. Biomedical Excellence for Safer Transfusion (BEST) Working Party of the International Society of Blood Transfusion. *Transfusion, 36,* 128–132.

64. Bertolini, F., Agazzi, A., Peccatori, F., et al. (2000). The absence of swirling in platelet concentrates is highly predictive of poor posttransfusion platelet count increments and increased risk of a transfusion reaction. *Transfusion, 40,* 121–122.

65. Owens, M., Holme, S., Heaton, A., et al. (1992). Posttransfusion recovery of function of 5-day stored platelet concentrates. *Br J Haematol, 80,* 539–544.

66. Murphy, S. (2004). Radiolabeling of PLTs to assess viability: A proposal for a standard. *Transfusion, 44,* 131–133.

67. Heyns, A. D., Lotter, M. G., Badenhorst, P. N., et al. (1980). Kinetics, distribution and sites of destruction of 111indium-labelled human platelets. *Br J Haematol, 44,* 269–280.

68. Snyder, E. L., Moroff, G., Simon, T., et al. (1986). Recommended methods for conducting radiolabeled platelet survival studies. *Transfusion, 26,* 37–42.

69. Holme, S., Heaton, A., & Roodt, J. (1993). Concurrent label method with 111In and 51Cr allows accurate evaluation of platelet viability of stored platelet concentrates. *Br J Haematol, 84,* 717–723.

70. Heilmann, E., Friese, P., Anderson, S., et al. (1993). Biotinylated platelets: A new approach to the measurement of platelet life span. *Br J Haematol, 85,* 729–735.

71. Dale, G. L., Gaddy, P., & Pikul, F. J. (1994). Antibodies against biotinylated proteins are present in normal human serum. *J Lab Clin Med, 123,* 365–371.

72. Garritsen, H. S., Hoerning, A., Hellenkamp, F., et al. (2001). Polymorphisms in the non-coding region of the human mitochondrial genome in unrelated plateletapheresis donors. *Br J Haematol, 112,* 995–1003.

73. Slichter, S. J., & Harker, L. A. (1976). Preparation and storage of platelet concentrates: II. Storage variables influencing platelet viability and function. *Br J Haematol, 34,* 403–419.

74. Murphy, S., & Gardner, F. H. (1969). Effect of storage temperature on maintenance of platelet viability — Deleterious effect of refrigerated storage. *N Engl J Med, 280,* 1094–1098.

75. Filip, D. J., & Aster, R. H. (1978). Relative hemostatic effectiveness of human platelets stored at 4 degrees and 22 degrees C. *J Lab Clin Med, 91,* 618–624.

76. Holmsen, H., Kaplan, K. L., & Dangelmaier, C. A. (1982). Differential energy requirements for platelet responses. A simultaneous study of aggregation, three secretory processes, arachidonate liberation, phosphatidylinositol breakdown and phosphatidate production. *Biochem J, 208,* 9–18.

77. Badlou, B. A., Ijseldijk, M. J., Smid, W. M., et al. (2005). Prolonged platelet preservation by transient metabolic suppression. *Transfusion, 45,* 214–222.

78. White, J. G., & Krivit, W. (1967). An ultrastructural basis for the shape changes induced in platelets by chilling. *Blood, 30,* 625–635.

79. Winokur, R., & Hartwig, J. H. (1995). Mechanism of shape change in chilled human platelets. *Blood, 85,* 1796–1804.

80. Hoffmeister, K. M., Falet, H., Toker, A., et al. (2001). Mechanisms of cold-induced platelet actin assembly. *J Biol Chem, 27,* 27.

81. Hoffmeister, K. M., Felbinger, T. W., Falet, H., et al. (2003). The clearance mechanism of chilled blood platelets. *Cell, 112,* 87–97.

82. Hoffmeister, K. M., Josefsson, E. C., Isaac, N. A., et al. (2003). Glycosylation restores survival of chilled blood platelets. *Science, 301,* 1531–1534.

83. Vainer, H., Prost-Dvojakovic, R. J., Jaisson, F., et al. (1976). Studies on human platelets stored at 20–22 degrees C without agitation. *Acta Haematol, 56,* 160–173.

84. Kunicki, T. J., Tuccelli, M., Becker, G. A., et al. (1975). A study of variables affecting the quality of platelets stored at "room temperature." *Transfusion, 15,* 414–421.

85. Holme, S., Vaidja, K., & Murphy, S. (1978). Platelet storage at 22 degrees C: Effect of type of agitation on morphology, viability, and function *in vitro*. *Blood, 52,* 425–435.

86. Snyder, E. L., Pope, C., Ferri, P. M., et al. (1986). The effect of mode of agitation and type of plastic bag on storage characteristics and *in vivo* kinetics of platelet concentrates. *Transfusion, 26,* 125–130.

87. Hunter, S., Nixon, J., & Murphy, S. (2001). The effect of the interruption of agitation on platelet quality during storage for transfusion. *Transfusion, 41,* 809–814.

88. Murphy, S., & Gardner, F. H. (1975). Platelet storage at 22 degrees C: Role of gas transport across plastic containers in maintenance of viability. *Blood, 46,* 209–218.

89. Kilkson, H., Holme, S., & Murphy, S. (1984). Platelet metabolism during storage of platelet concentrates at 22 degrees C. *Blood, 64,* 406–414.

90. Murphy, S., Kahn, R. A., Holme, S., et al. (1982). Improved storage of platelets for transfusion in a new container. *Blood, 60,* 194–200.

91. Snyder, E. L., Ezekowitz, M., Aster, R., et al. (1985). Extended storage of platelets in a new plastic container: II. *In vivo* response to infusion of platelets stored for 5 days. *Transfusion, 25,* 209–214.

92. Snyder, E. L., Aster, R. H., Heaton, A., et al. (1992). Five-day storage of platelets in a non-diethylhexyl phthalate-plasticized container. *Transfusion, 32,* 736–741.

93. Cohen, P., & Wittels, B. (1970). Energy substrate metabolism in fresh and stored human platelets. *J Clin Invest, 49,* 119–127.

94. Rudderow, D., & Soslau, G. (1998). Permanent lesions of stored platelets correlate to pH and cell count while reversible lesions do not. *Proc Soc Exp Biol Med, 217,* 219–227.

95. Rinder, H. M., Snyder, E. L., Tracey, J. B., et al. (2003). Reversibility of severe metabolic stress in stored platelets after *in vitro* plasma rescue or *in vivo* transfusion: Restoration of secretory function and maintenance of platelet survival. *Transfusion, 43,* 1230–1237.

96. Marquez, J., Martin, D., Virji, M. A., et al. (1986). Cardiovascular depression secondary to ionic hypocalcemia during hepatic transplantation in humans. *Anesthesiology, 65,* 457–461.

97. Murphy, S. (1999). The efficacy of synthetic media in the storage of human platelets for transfusion. *Transfus Med Rev, 13,* 153–163.

98. Rock, G., Swenson, S. D., & Adams, G. A. (1985). Platelet storage in a plasma-free medium. *Transfusion, 25,* 551–556.

99. Murphy, S., Kagen, L., Holme, S., et al. (1991). Platelet storage in synthetic media lacking glucose and bicarbonate. *Transfusion, 31,* 16–20.

100. Gulliksson, H. (2003). Defining the optimal storage conditions for the long-term storage of platelets. *Transfus Med Rev, 17,* 209–215.

101. Holme, S., Bode, A., Heaton, W. A., et al. (1992). Improved maintenance of platelet *in vivo* viability during storage when using a synthetic medium with inhibitors. *J Lab Clin Med, 119,* 144–150.

102. Murphy, S. (1995). The oxidation of exogenously added organic anions by platelets facilitates maintenance of pH during their storage for transfusion at 22 degrees C. *Blood, 85,* 1929–1935.

103. Shimizu, T., & Murphy, S. (1993). Roles of acetate and phosphate in the successful storage of platelet concentrates prepared with an acetate-containing additive solution. *Transfusion, 33,* 304–310.

104. de Wildt-Eggen, J., Nauta, S., Schrijver, J. G., et al. (2000). Reactions and platelet increments after transfusion of platelet concentrates in plasma or an additive solution: A prospective, randomized study. *Transfusion, 40,* 398–403.

105. Adams, G. A., Swenson, S. D., & Rock, G. (1986). Survival and recovery of human platelets stored for five days in a non-plasma medium. *Blood, 67,* 672–675.

106. Gulliksson, H., Eriksson, L., Hogman, C. F., et al. (1995). Buffy-coat-derived platelet concentrates prepared from half-strength citrate CPD and CPD whole-blood units. Comparison between three additive solutions: *In vitro* studies. *Vox Sang, 68,* 152–159.

107. Menitove, J. E., Frenzke, M., & Aster, R. H. (1986). Use of PGE1 for preparation of platelet concentrates. *Transfusion, 26,* 346–350.

108. Bode, A. P., Holme, S., Heaton, W. A., et al. (1991). Extended storage of platelets in an artificial medium with the platelet activation inhibitors prostaglandin E1 and theophylline. *Vox Sang, 60,* 105–112.

109. Tertyshnikova, S., & Fein, A. (1998). Inhibition of inositol 1,4,5-trisphosphate-induced $Ca^{2+}$ release by cAMP-dependent protein kinase in a living cell. *Proc Natl Acad Sci USA, 95,* 1613–1617.

110. Bode, A. P., & Miller, D. T. (1989). The use of thrombin inhibitors and aprotinin in the preservation of platelets stored for transfusion. *J Lab Clin Med, 113,* 753–758.

111. Mrowiec, Z. R., Oleksowicz, L., Zuckerman, D., et al. (1996). Buffy coat platelets stored in apyrase, aprotinin, and ascorbic acid in a suspended bag: Combined strategies for reducing platelet activation during storage. *Transfusion, 36,* 5–10.

112. Klinger, M. H., Josch, M., & Kluter, H. (1996). Platelets stored in a glucose-free additive solution or in autologous plasma — An ultrastructural and morphometric evaluation. *Vox Sang, 71,* 13–20.

113. Knutson, F., Alfonso, R., Dupuis, K., et al. (2000). Photochemical inactivation of bacteria and HIV in buffy-coat-derived platelet concentrates under conditions that preserve *in vitro* platelet function. *Vox Sang, 78,* 209–216.

114. van Rhenen, D. J., Vermeij, J., Mayaudon, V., et al. (2000). Functional characteristics of S-59 photochemically treated platelet concentrates derived from buffy coats. *Vox Sang, 79,* 206–214.

115. Metcalfe, P., Williamson, L. M., Reutelingsperger, C. P., et al. (1997). Activation during preparation of therapeutic platelets affects deterioration during storage: A comparative flow cytometric study of different production methods. *Br J Haematol, 98,* 86–95.

116. Divers, S. G., Kannan, K., Stewart, R. M., et al. (1995). Quantitation of CD62, soluble CD62, and lysosome-associated membrane proteins 1 and 2 for evaluation of the quality of stored platelet concentrates. *Transfusion, 35,* 292–297.

117. Kostelijk, E. H., Fijnheer, R., Nieuwenhuis, H. K., et al. (1996). Soluble P-selectin as parameter for platelet activation during storage. *Thromb Haemost, 76,* 1086–1089.

118. Dumont, L. J., VandenBroeke, T., & Ault, K. A. (1999). Platelet surface P-selectin measurements in platelet preparations: An international collaborative study. Biomedical Excellence for Safer Transfusion (BEST) Working Party of the International Society of Blood Transfusion (ISBT). *Transfus Med Rev, 13,* 31–42.

119. Shapira, S., Friedman, Z., Shapiro, H., et al. (2000). The effect of storage on the expression of platelet membrane phosphatidylserine and the subsequent impact on the coagulant function of stored platelets. *Transfusion, 40,* 1257–1263.

120. Gutensohn, K., Alisch, A., Geidel, K., et al. (1999). Annexin V and platelet antigen expression is not altered during storage of platelet concentrates obtained with the AMICUS cell separator. *Transfus Sci, 20,* 113–120.

121. Rinder, H. M., Murphy, M., Mitchell, J. G., et al. (1991). Progressive platelet activation with storage: Evidence for shortened survival of activated platelets after transfusion. *Transfusion, 31,* 409–414.

122. Michelson, A. D., Barnard, M. R., Hechtman, H. B., et al. (1996). *In vivo* tracking of platelets: Circulating degranulated platelets rapidly lose surface P-selectin but continue to circulate and function. *Proc Natl Acad Sci USA, 93,* 11877–11882.

123. Berger, G., Hartwell, D. W., & Wagner, D. D. (1998). P-selectin and platelet clearance. *Blood, 92,* 4446–4452.

124. Krishnamurti, C., Maglasang, P., & Rothwell, S. W. (1999). Reduction of blood loss by infusion of human platelets in a rabbit kidney injury model. *Transfusion, 39,* 967–974.

125. Holme, S., Sweeney, J. D., Sawyer, S., et al. (1997). The expression of p-selectin during collection, processing, and storage of platelet concentrates: Relationship to loss of *in vivo* viability. *Transfusion, 37,* 12–17.

126. Wolf, P. (1967). The nature and significance of platelet products in human plasma. *Br J Haematol, 13,* 269–288.

127. Bode, A. P., Orton, S. M., Frye, M. J., et al. (1991). Vesiculation of platelets during *in vitro* aging. *Blood, 77,* 887–895.

128. Bode, A. P., & Hickerson, D. H. (2000). Characterization and quantitation by flow cytometry of membranous microparticles formed during activation of platelet suspensions with ionophore or thrombin. *Platelets, 11,* 259–271.

129. Sloand, E. M., Yu, M., & Klein, H. G. (1996). Comparison of random-donor platelet concentrates prepared from whole blood units and platelets prepared from single-donor apheresis collections. *Transfusion, 36,* 955–959.

130. Blajchman, M. A., Bardossy, L., Carmen, R. A., et al. (1992). An animal model of allogeneic donor platelet refractoriness: The effect of the time of leukodepletion. *Blood, 79,* 1371–1375.

131. Bordin, J. O., Bardossy, L., & Blajchman, M. A. (1993). Experimental animal model of refractoriness to donor platelets: The effect of plasma removal and the extent of white cell reduction on allogeneic alloimmunization. *Transfusion, 33,* 798–801.

132. Leytin, V., Allen, D. J., Mykhaylov, S., et al. (2004). Pathologic high shear stress induces apoptosis events in human platelets. *Biochem Biophys Res Commun, 320,* 303–310.

133. Li, J., Xia, Y., Bertino, A. M., et al. (2000). The mechanism of apoptosis in human platelets during storage. *Transfusion, 40,* 1320–1329.

134. Vanags, D. M., Orrenius, S., & Aguilar-Santelises, M. (1997). Alterations in Bcl-2/Bax protein levels in platelets form part of an ionomycin-induced process that resembles apoptosis. *Br J Haematol, 99,* 824–831.

135. Shcherbina, A., & Remold-O'Donnell, E. (1999). Role of caspase in a subset of human platelet activation responses. *Blood, 93,* 4222–4231.

136. Wolf, B. B., Goldstein, J. C., Stennicke, H. R., et al. (1999). Calpain functions in a caspase-independent manner to promote apoptosis-like events during platelet activation. *Blood, 94,* 1683–1692.

137. Plenchette, S., Moutet, M., Benguella, M., et al. (2001). Early increase in DcR2 expression and late activation of caspases in the platelet storage lesion. *Leukemia, 15,* 1572–1581.

138. Bertino, A. M., Qi, X. Q., Li, J., et al. (2003). Apoptotic markers are increased in platelets stored at 37 degrees C. *Transfusion, 43,* 857–866.

139. Perrotta, P. L., Perrotta, C. L., & Snyder, E. L. (2003). Apoptotic activity in stored human platelets. *Transfusion, 43,* 526–535.

140. Verhoeven, A. J., Verhaar, R., Gouwerok, E. G., et al. (2005). The mitochondrial membrane potential in human platelets: A sensitive parameter for platelet quality. *Transfusion, 45,* 82–89.

141. The Trial to Reduce Alloimmunization to Platelets Study Group. (1997). Leukocyte reduction and ultraviolet B irradiation of platelets to prevent alloimmunization and refractoriness to platelet transfusions. *N Engl J Med, 337,* 1861–1869.

142. Seftel, M. D., Growe, G. H., Petraszko, T., et al. (2004). Universal prestorage leukoreduction in Canada decreases platelet alloimmunization and refractoriness. *Blood, 103,* 333–339.

143. Pietersz, R. N., van der Meer, P. F., Steneker, I., et al. (1999). Preparation of leukodepleted platelet concentrates from pooled buffy coats: Prestorage filtration with Autostop BC. *Vox Sang, 76,* 231–236.

144. Burgstaler, E. A., Pineda, A. A., & Wollan, P. (1997). Platelet-apheresis: Comparison of processing times, platelet yields, and white blood cell content with several commonly used systems. *J Clin Apheresis, 12,* 170–178.

145. Heddle, N. M., Klama, L., Singer, J., et al. (1994). The role of the plasma from platelet concentrates in transfusion reactions. *N Engl J Med, 331,* 625–628.

146. Ogawa, Y., Wakana, M., Tanaka, K., et al. (1998). Clinical evaluation of transfusion of prestorage-leukoreduced apheresis platelets. *Vox Sang, 75,* 103–109.

147. Heddle, N. M., Klama, L., Meyer, R., et al. (1999). A randomized controlled trial comparing plasma removal with white cell reduction to prevent reactions to platelets. *Transfusion, 39,* 231–238.

148. Diepenhorst, P., Sprokholt, R., & Prins, H. K. (1972). Removal of leukocytes from whole blood and erythrocyte suspensions by filtration through cotton wool: I. Filtration technique. *Vox Sang, 23,* 308–320.

149. Dzik, S. (1993). Leukodepletion blood filters: Filter design and mechanisms of leukocyte removal. *Transfus Med Rev, 7,* 65–77.

150. Bruil, A., Beugeling, T., Feijen, J., et al. (1995). The mechanisms of leukocyte removal by filtration. *Transfus Med Rev, 9,* 145–166.

151. Courtney, J. M., Lamba, N. M., Sundaram, S., et al. (1994). Biomaterials for blood-contacting applications. *Biomaterials, 15,* 737–744.

152. Kickler, T. S., Bell, W., Drew, H., et al. (1989). Depletion of white cells from platelet concentrates with a new adsorption filter. *Transfusion, 29,* 411–414.

153. Moroff, G., & Luban, N. L. (1997). The irradiation of blood and blood components to prevent graft-versus-host disease: Technical issues and guidelines. *Transfus Med Rev, 11,* 15–26.

154. Moroff, G., George, V. M., Siegl, A. M., et al. (1986). The influence of irradiation on stored platelets. *Transfusion, 26,* 453–456.

155. Button, L. N., DeWolf, W. C., Newburger, P. E., et al. (1981). The effects of irradiation on blood components. *Transfusion, 21,* 419–426.

156. Luban, N. L., Drothler, D., Moroff, G., et al. (2000). Irradiation of platelet components: Inhibition of lymphocyte proliferation assessed by limiting-dilution analysis. *Transfusion, 40,* 348–352.

157. Read, E. J., Kodis, C., Carter, C. S., et al. (1988). Viability of platelets following storage in the irradiated state. A pair-controlled study. *Transfusion, 28,* 446–450.

158. Rock, G., Adams, G. A., & Labow, R. S. (1988). The effects of irradiation on platelet function. *Transfusion, 28,* 451–455.

159. Sweeney, J. D., Holme, S., & Moroff, G. (1994). Storage of apheresis platelets after gamma radiation. *Transfusion, 34,* 779–783.

160. Huston, B. M., Brecher, M. E., & Bandarenko, N. (1998). Lack of efficacy for conventional gamma irradiation of platelet concentrates to abrogate bacterial growth. *Am J Clin Pathol, 109,* 743–747.

161. Pamphilon, D. H. (1999). The rationale and use of platelet concentrates irradiated with ultraviolet-B light. *Transfus Med Rev, 13,* 323–333.

162. Lindahl-Kiessling, K., & Safwenberg, J. (1971). Inability of UV-irradiated lymphocytes to stimulate allogeneic cells in mixed lymphocyte culture. *Int Arch Allergy Appl Immunol, 41,* 670–678.

163. Kao, K. J. (1996). Induction of humoral immune tolerance to major histocompatibility complex antigens by transfusions of UVB-irradiated leukocytes. *Blood, 88,* 4375–4382.

164. Kahn, R. A., Duffy, B. F., & Rodey, G. G. (1985). Ultraviolet irradiation of platelet concentrate abrogates lymphocyte activation without affecting platelet function *in vitro. Transfusion, 25,* 547–550.

165. Pamphilon, D. H., Corbin, S. A., Saunders, J., et al. (1989). Applications of ultraviolet light in the preparation of platelet concentrates. *Transfusion, 29,* 379–383.

166. Snyder, E. L., Beardsley, D. S., Smith, B. R., et al. (1991). Storage of platelet concentrates after high-dose ultraviolet B irradiation. *Transfusion, 31,* 491–496.

167. Blundell, E. L., Pamphilon, D. H., Fraser, I. D., et al. (1996). A prospective, randomized study of the use of platelet concentrates irradiated with ultraviolet-B light in patients with hematologic malignancy. *Transfusion, 36,* 296–302.

168. Corash, L. (1999). Inactivation of viruses, bacteria, protozoa, and leukocytes in platelet concentrates: Current research perspectives. *Transfus Med Rev, 13,* 18–30.

169. Corash, L. (2000). Inactivation of viruses, bacteria, protozoa and leukocytes in platelet and red cell concentrates. *Vox Sang, 78,* 205–210.

170. van Rhenen, D., Gulliksson, H., Cazenave, J. P., et al. (2003). Transfusion of pooled buffy coat platelet components prepared with photochemical pathogen inactivation treatment: The euroSPRITE trial. *Blood, 101,* 2426–2433.

171. Corbin, F., 3rd. (2002). Pathogen inactivation of blood components: Current status and introduction of an approach using riboflavin as a photosensitizer. *Int J Hematol, 76*(Suppl. 2), 253–257.

172. Ruane, P. H., Edrich, R., Gampp, D., et al. (2004). Photochemical inactivation of selected viruses and bacteria in platelet concentrates using riboflavin and light. *Transfusion, 44,* 877–885.

173. Janetzko, K., Lin, L., Eichler, H., et al. (2004). Implementation of the INTERCEPT Blood System for Platelets into routine blood bank manufacturing procedures: Evaluation of apheresis platelets. *Vox Sang, 86,* 239–245.

174. Lin, L., Dikeman, R., Molini, B., et al. (2004). Photochemical treatment of platelet concentrates with amotosalen and long-wavelength ultraviolet light inactivates a broad spectrum of pathogenic bacteria. *Transfusion, 44,* 1496–1504.

175. McCullough, J., Vesole, D. H., Benjamin, R. J., et al. (2004). Therapeutic efficacy and safety of platelets treated with a photochemical process for pathogen inactivation: The SPRINT Trial. *Blood, 104,* 1534–1541.

176. Lin, L., Hanson, C. V., Alter, H. J., et al. (2005). Inactivation of viruses in platelet concentrates by photochemical treatment with amotosalen and long-wavelength ultraviolet light. *Transfusion, 45,* 580–590.

177. Blajchman, M. A., Goldman, M., & Baeza, F. (2004). Improving the bacteriological safety of platelet transfusions. *Transfus Med Rev, 18,* 11–24.

178. Holme, S., Heaton, W. A., & Moroff, G. (1994). Evaluation of platelet concentrates stored for 5 days with reduced plasma volume. *Transfusion, 34,* 39–43.

179. Blanchette, V. S., Kuhne, T., Hume, H., et al. (1995). Platelet transfusion therapy in newborn infants. *Transfus Med Rev, 9,* 215–230.

180. Schoenfeld, H., Muhm, M., Doepfmer, U. R., et al. (2005). The functional integrity of platelets in volume-reduced platelet concentrates. *Anesth Analg, 100,* 78–81.

181. Duke, W. W. (1910). The relation of blood platelets to hemorrhagic disease. Description of a method for determining the bleeding time and coagulation time and report of three cases of hemorrhagic disease relieved by transfusion. *J Am Med Assoc, 55,* 1185–1192.

182. Triulzi, D. J., & Griffith, B. P. (1998). Blood usage in lung transplantation. *Transfusion, 38,* 12–15.

183. Pisciotto, P. T., Benson, K., Hume, H., et al. (1995). Prophylactic versus therapeutic platelet transfusion practices in hematology and/or oncology patients. *Transfusion, 35,* 498–502.

184. Hunt, B. J. (1998). Indications for therapeutic platelet transfusions. *Blood Rev, 12,* 227–233.

185. Kelley, D. L., Fegan, R. L., Ng, A. T., et al. (1997). High-yield platelet concentrates attainable by continuous quality improvement reduce platelet transfusion cost and donor exposure. *Transfusion, 37,* 482–486.

186. Ronghe, M. D., Foot, A. B., Cornish, J. M., et al. (2002). The impact of transfusion of leucodepleted platelet concentrates on cytomegalovirus disease after allogeneic stem cell transplantation. *Br J Haematol, 118,* 1124–1127.

187. Nichols, W. G., Price, T. H., Gooley, T., et al. (2003). Transfusion-transmitted cytomegalovirus infection after receipt of leukoreduced blood products. *Blood, 101,* 4195–4200.

188. Heal, J. M., Rowe, J. M., McMican, A., et al. (1993). The role of ABO matching in platelet transfusion. *Eur J Haematol, 50,* 110–117.

189. Heal, J. M., Kenmotsu, N., Rowe, J. M., et al. (1994). A possible survival advantage in adults with acute leukemia receiving ABO-identical platelet transfusions. *Am J Hematol, 45,* 189–190.

190. Heal, J. M., Masel, D., Rowe, J. M., et al. (1993). Circulating immune complexes involving the ABO system after platelet transfusion. *Br J Haematol, 85,* 566–572.

191. Atoyebi, W., Mundy, N., Croxton, T., et al. (2000). Is it necessary to administer anti-D to prevent RhD immunization after the transfusion of RhD-positive platelet concentrates? *Br J Haematol, 111,* 980–983.

192. Haspel, R. L., Walsh, L., & Sloan, S. R. (2004). Platelet transfusion in an infant leading to formation of anti-D: Implications for immunoprophylaxis. *Transfusion, 44,* 747–749.

193. Anderson, B., Shad, A. T., Gootenberg, J. E., et al. (1999). Successful prevention of post-transfusion Rh alloimmunization by intravenous Rho (D) immune globulin (WinRho SD). *Am J Hematol, 60,* 245–247.

194. Molnar, R., Johnson, R., Sweat, L. T., et al. (2002). Absence of D alloimmunization in D-pediatric oncology patients receiving D-incompatible single-donor platelets. *Transfusion, 42,* 177–182.

195. Cid, J., Ortin, X., Elies, E., et al. (2002). Absence of anti-D alloimmunization in hematologic patients after D-incompatible platelet transfusions. *Transfusion, 42,* 173–176.

196. Clark, P., & Mintz, P. D. (2001). Transfusion triggers for blood components. *Curr Opin Hematol, 8,* 387–391.

197. Gaydos, L. A., Freireich, E. J., & Mantel, N. (1962). The quantitative relation between platelet count and hemorrhage in patients with acute leukemia. *N Engl J Med, 13,* 283–290.

198. Beutler, E. (1993). Platelet transfusions: The 20,000/microL trigger. *Blood, 81,* 1411–1413.

199. Heddle, N. M., Cook, R. J., Webert, K. E., et al. (2003). Methodologic issues in the use of bleeding as an outcome in transfusion medicine studies. *Transfusion, 43,* 742–752.

200. Segal, H. C., Briggs, C., Kunka, S., et al. (2005). Accuracy of platelet counting haematology analysers in severe thrombocytopenia and potential impact on platelet transfusion. *Br J Haematol, 128,* 520–525.

201. Slichter, S. J., & Harker, L. A. (1978). Thrombocytopenia: Mechanisms and management of defects in platelet production. *Clin Haematol, 7,* 523–539.

202. Solomon, J., Bofenkamp, T., Fahey, J. L., et al. (1978). Platelet prophylaxis in acute non-lymphoblastic leukaemia. *Lancet, 1,* 267.

203. Belt, R. J., Leite, C., Haas, C. D., et al. (1978). Incidence of hemorrhagic complications in patients with cancer. *J Am Med Assoc, 239,* 2571–2574.

204. Gmur, J., Burger, J., Schanz, U., et al. (1991). Safety of stringent prophylactic platelet transfusion policy for patients with acute leukaemia. *Lancet, 338,* 1223–1226.

205. Gil-Fernandez, J. J., Alegre, A., Fernandez-Villalta, M. J., et al. (1996). Clinical results of a stringent policy on prophylactic platelet transfusion: Non-randomized comparative analysis in 190 bone marrow transplant patients from a single institution. *Bone Marrow Transplant, 18,* 931–935.

206. Heckman, K. D., Weiner, G. J., Davis, C. S., et al. (1997). Randomized study of prophylactic platelet transfusion threshold during induction therapy for adult acute leukemia: 10,000/microL versus 20,000/microL. *J Clin Oncol, 15,* 1143–1149.

207. Rebulla, P., Finazzi, G., Marangoni, F., et al. (1997). The threshold for prophylactic platelet transfusions in adults with acute myeloid leukemia. Gruppo Italiano Malattie Ematologiche Maligne dell'Adulto. *N Engl J Med, 337,* 1870–1875.

208. Wandt, H., Frank, M., Ehninger, G., et al. (1998). Safety and cost effectiveness of a $10 \times 10(9)$/L trigger for prophylactic platelet transfusions compared with the traditional $20 \times 10(9)$/L trigger: A prospective comparative trial in 105 patients with acute myeloid leukemia. *Blood, 91,* 3601–3606.

209. Sagmeister, M., Oec, L., & Gmur, J. (1999). A restrictive platelet transfusion policy allowing long-term support of outpatients with severe aplastic anemia. *Blood, 93,* 3124–3126.

210. (1987). Consensus conference. Platelet transfusion therapy. *J Am Med Assoc, 257,* 1777–1780.

211. Murphy, M. F., Brozovic, B., Murphy, W., et al. (1992). Guidelines for platelet transfusions. British Committee for Standards in Haematology, Working Party of the Blood Transfusion Task Force. *Transfus Med, 2,* 311–318.

212. Norfolk, D. R., Ancliffe, P. J., Contreras, M., et al. (1998). Consensus Conference on Platelet Transfusion, Royal College of Physicians of Edinburgh, 27–28 November 1997. Synopsis of background papers. *Br J Haematol, 101,* 609–617.

213. American Society of Anesthesiologists Task Force on Blood Component Therapy. (1996). Practice guidelines for blood component therapy: A report by the American Society of Anesthesiologists Task Force on Blood Component Therapy. *Anesthesiology, 84,* 732–747.

214. British Committee for Standards in Haematology, Blood Transfusion Task Force (2003). Guidelines for the use of platelet transfusions. *Br J Haematol, 122,* 10–23.

215. Rebulla, P. (2001). Platelet transfusion trigger in difficult patients. *Transfus Clin Biol, 8,* 249–254.

216. Murphy, M. F., Murphy, W., Wheatley, K., et al. (1998). Survey of the use of platelet transfusions in centres participating in MRC leukaemia trials. *Br J Haematol, 102,* 875–876.

217. Lawrence, J. B., Yomtovian, R. A., Hammons, T., et al. (2001). Lowering the prophylactic platelet transfusion threshold: A prospective analysis. *Leuk Lymphoma, 41,* 67–76.

218. Elting, L. S., Martin, C. G., Kurtin, D. J., et al. (2002). The Bleeding Risk Index: A clinical prediction rule to guide the prophylactic use of platelet transfusions in patients with lymphoma or solid tumors. *Cancer, 94,* 3252–3262.

219. Avvisati, G., Tirindelli, M. C., & Annibali, O. (2003). Thrombocytopenia and hemorrhagic risk in cancer patients. *Crit Rev Oncol Hematol, 48,* S13–S16.

220. Waldman, S. D., Feldstein, G. S., Waldman, H. J., et al. (1987). Caudal administration of morphine sulfate in anticoagulated and thrombocytopenic patients. *Anesth Analg, 66,* 267–268.

221. Vavricka, S. R., Walter, R. B., Irani, S., et al. (2003). Safety of lumbar puncture for adults with acute leukemia and restrictive prophylactic platelet transfusion. *Ann Hematol, 82,* 570–573.

222. Howard, S. C., Gajjar, A., Ribeiro, R. C., et al. (2000). Safety of lumbar puncture for children with acute lymphoblastic leukemia and thrombocytopenia. *J Am Med Assoc, 284,* 2222–2224.

223. Ho, C. H. (1998). The hemostatic effect of packed red cell transfusion in patients with anemia. *Transfusion, 38,* 1011–1014.

224. Turitto, V. T., & Baumgartner, H. R. (1975). Platelet interaction with subendothelium in a perfusion system: Physical role of red blood cells. *Microvasc Res, 9,* 335–344.

225. Cook, R. J., Heddle, N. M., Rebulla, P., et al. (2004). Methods for the analysis of bleeding outcomes in randomized trials of PLT transfusion triggers. *Transfusion, 44,* 1135–1142.

226. Harker, L. A., & Slichter, S. J. (1972). The bleeding time as a screening test for evaluation of platelet function. *N Engl J Med, 287,* 155–159.

227. Livio, M., Gotti, E., Marchesi, D., et al. (1982). Uraemic bleeding: Role of anaemia and beneficial effect of red cell transfusions. *Lancet, 2,* 1013–1015.

228. Daly, P. A., Schiffer, C. A., Aisner, J., et al. (1980). Platelet transfusion therapy. One-hour posttransfusion increments are valuable in predicting the need for HLA-matched preparations. *J Am Med Assoc, 243,* 435–438.

229. Davis, K. B., Slichter, S. J., & Corash, L. (1999). Corrected count increment and percent platelet recovery as measures of posttransfusion platelet response: Problems and a solution. *Transfusion, 39,* 586–592.

230. Blajchman, M. A., & Lee, D. H. (1997). The thrombocytopenic rabbit bleeding time model to evaluate the *in vivo* hemostatic efficacy of platelets and platelet substitutes. *Transfus Med Rev, 11,* 95–105.

231. Rothwell, S. W., Maglasang, P., & Krishnamurti, C. (1998). Survival of fresh human platelets in a rabbit model as traced by flow cytometry. *Transfusion, 38,* 550–556.

232. Leytin, V., Allen, D. J., Mody, M., et al. (2002). A rabbit model for monitoring *in vivo* viability of human platelet concentrates using flow cytometry. *Transfusion, 42,* 711–718.

233. Rinder, H. M., Arbini, A. A., & Snyder, E. L. (1999). Optimal dosing and triggers for prophylactic use of platelet transfusions. *Curr Opin Hematol, 6,* 437–441.

234. Heal, J. M., & Blumberg, N. (2004). Optimizing platelet transfusion therapy. *Blood Rev, 18,* 149–165.

235. Askari, S., Weik, P. R., & Crosson, J. (2002). Calculated platelet dose: Is it useful in clinical practice? *J Clin Apheresis, 17,* 103–105.

236. Hanson, S. R., & Slichter, S. J. (1985). Platelet kinetics in patients with bone marrow hypoplasia: Evidence for a fixed platelet requirement. *Blood, 66,* 1105–1109.

237. Sensebe, L., Giraudeau, B., Bardiaux, L., et al. (2005). The efficiency of transfusing high doses of platelets in hematologic patients with thrombocytopenia: Results of a prospective, randomized, open, blinded end point (PROBE) study. *Blood, 105,* 862–864.

238. Norol, F., Bierling, P., Roudot-Thoraval, F., et al. (1998). Platelet transfusion: A dose–response study. *Blood, 92,* 1448–1453.

239. Klumpp, T. R., Herman, J. H., Gaughan, J. P., et al. (1999). Clinical consequences of alterations in platelet transfusion dose: A prospective, randomized, double-blind trial. *Transfusion, 39,* 674–681.

240. Hersh, J. K., Hom, E. G., & Brecher, M. E. (1998). Mathematical modeling of platelet survival with implications for optimal transfusion practice in the chronically platelet transfusion-dependent patient. *Transfusion, 38,* 637–644.

241. Ackerman, S. J., Klumpp, T. R., Guzman, G. I., et al. (2000). Economic consequences of alterations in platelet transfusion dose: Analysis of a prospective, randomized, double-blind trial. *Transfusion, 40,* 1457–1462.

242. Blanchette, V. S., Johnson, J., & Rand, M. (2000). The management of alloimmune neonatal thrombocytopenia. *Baillieres Best Pract Res Clin Haematol, 13,* 365–390.

243. Simon, T. L., & Henderson, R. (1979). Coagulation factor activity in platelet concentrates. *Transfusion, 19,* 186–189.

244. Ciavarella, D., Lavallo, E., & Reiss, R. F. (1986). Coagulation factor activity in platelet concentrates stored up to 7 days: An *in vitro* and *in vivo* study. *Clin Lab Haematol, 8,* 233–242.

245. Schmitz, N., Linch, D. C., Dreger, P., et al. (1996). Randomised trial of filgrastim-mobilised peripheral blood progenitor cell transplantation versus autologous bone-marrow transplantation in lymphoma patients. *Lancet, 347,* 353–357.

246. Blay, J. Y., Le Cesne, A., Mermet, C., et al. (1998). A risk model for thrombocytopenia requiring platelet transfusion after cytotoxic chemotherapy. *Blood, 92,* 405–410.

247. Moskowitz, C. H., Glassman, J. R., Wuest, D., et al. (1998). Factors affecting mobilization of peripheral blood progenitor cells in patients with lymphoma. *Clin Cancer Res, 4,* 311–316.

248. Meldgaard Knudsen, L., Jensen, L., Jarlbaek, L., et al. (1999). Subsets of CD34$^+$ hematopoietic progenitors and platelet recovery after high dose chemotherapy and peripheral blood stem cell transplantation. *Haematologica, 84,* 517–524.

249. Rinder, C. S., & Rinder, H. M. (2000). Preservation of platelet function after cardiac surgery and apheresis In J. Seghatchian, E. L. Snyder, & P. Krailadsiri (Eds.), *Platelet therapy* (pp. 170–198). Amsterdam: Elsevier.

250. Stover, E. P., Siegel, L. C., Parks, R., et al. (1998). Variability in transfusion practice for coronary artery bypass surgery persists despite national consensus guidelines: A 24-institution study. Institutions of the Multicenter Study of Perioperative Ischemia Research Group. *Anesthesiology, 88,* 327–333.

251. Simon, T. L., Akl, B. F., & Murphy, W. (1984). Controlled trial of routine administration of platelet concentrates in cardiopulmonary bypass surgery. *Ann Thorac Surg, 37,* 359–364.

252. Reed, R. L., 2nd, Ciavarella, D., Heimbach, D. M., et al. (1986). Prophylactic platelet administration during massive transfusion. A prospective, randomized, double-blind clinical study. *Ann Surg, 203,* 40–48.

253. Lee, L. Y., DeBois, W., Krieger, K. H., et al. (2002). The effects of platelet inhibitors on blood use in cardiac surgery. *Perfusion, 17,* 33–37.

254. Li, Y. F., Spencer, F. A., & Becker, R. C. (2002). Comparative efficacy of fibrinogen and platelet supplementation on the *in vitro* reversibility of competitive glycoprotein IIb/IIIa receptor-directed platelet inhibition. *Am Heart J, 143,* 725–732.

255. Zavadil, D. P., Stammers, A. H., Willett, L. D., et al. (1998). Hematological abnormalities in neonatal patients treated with extracorporeal membrane oxygenation (ECMO). *J Extra Corpor Technol, 30,* 83–90.

256. George, J. N., Caen, J. P., & Nurden, A. T. (1990). Glanzmann's thrombasthenia: The spectrum of clinical disease. *Blood, 75,* 1383–1395.

257. Rao, A. K., Ghosh, S., Sun, L., et al. (1995). Mechanisms of platelet dysfunction and response to DDAVP in patients with congenital platelet function defects. A double-blind placebo-controlled trial. *Thromb Haemost, 74,* 1071–1078.

258. Almeida, A. M., Khair, K., Hann, I., et al. (2003). The use of recombinant factor VIIa in children with inherited platelet function disorders. *Br J Haematol, 121,* 477–481.

259. Poon, M. C., D'Oiron, R., Von Depka, M., et al. (2004). Prophylactic and therapeutic recombinant factor VIIa administration to patients with Glanzmann's thrombasthenia: Results of an international survey. *J Thromb Haemost, 2,* 1096–1103.

260. Tranholm, M., Rojkjaer, R., Pyke, C., et al. (2003). Recombinant factor VIIa reduces bleeding in severely thrombocytopenic rabbits. *Thromb Res, 109,* 217–223.

261. Berkowitz, S. D., Harrington, R. A., Rund, M. M., et al. (1997). Acute profound thrombocytopenia after C7E3 Fab (abciximab) therapy. *Circulation, 95,* 809–813.

262. Bock, M., Schleuning, M., Heim, M. U., et al. (1995). Cryopreservation of human platelets with dimethyl sulfoxide: Changes in biochemistry and cell function. *Transfusion, 35,* 921–924.

263. Funke, I., Wiesneth, M., Koerner, K., et al. (1995). Autologous platelet transfusion in alloimmunized patients with acute leukemia. *Ann Hematol, 71,* 169–173.

264. Torretta, L., Perotti, C., Pedrazzoli, P., et al. (1998). Autologous platelet collection and storage to support thrombocytopenia in patients undergoing high-dose chemotherapy and circulating progenitor cell transplantation for high-risk breast cancer. *Vox Sang, 75,* 224–229.

265. Rubens, F. D., Fergusson, D., Wells, P. S., et al. (1998). Platelet-rich plasmapheresis in cardiac surgery: A meta-analysis of the effect on transfusion requirements. *J Thorac Cardiovasc Surg, 116,* 641–647.

266. Masuda, H., Kobayashi, A., Moriyama, Y., et al. (1999). Effectiveness of preoperatively obtained autologous platelet concentrates in open heart surgery. *Thorac Cardiovasc Surg, 47,* 298–301.

267. Christenson, J. T., Reuse, J., Badel, P., et al. (1996). Platelet-pheresis before redo CABG diminishes excessive blood transfusion. *Ann Thorac Surg, 62,* 1373–1378.

268. Anitua, E., Andia, I., Ardanza, B., et al. (2004). Autologous platelets as a source of proteins for healing and tissue regeneration. *Thromb Haemost, 91,* 4–15.

269. Mazzucco, L., Medici, D., Serra, M., et al. (2004). The use of autologous platelet gel to treat difficult-to-heal wounds: A pilot study. *Transfusion, 44,* 1013–1018.

270. Perrotta, P. L., & Snyder, E. L. (2001). Non-infectious complications of transfusion therapy. *Blood Rev, 15,* 69–83.

271. Dodd, R. Y., Notari, E. P. t., & Stramer, S. L. (2002). Current prevalence and incidence of infectious disease markers and estimated window-period risk in the American Red Cross blood donor population. *Transfusion, 42,* 975–979.

272. Busch, M. P., Glynn, S. A., Stramer, S. L., et al. (2005). A new strategy for estimating risks of transfusion-transmitted viral infections based on rates of detection of recently infected donors. *Transfusion, 45,* 254–264.

273. Luban, N. L. (2001). Prevention of transfusion-associated graft-versus-host disease by inactivation of T cells in platelet components. *Semin Hematol, 38,* 34–45.

274. Medeiros, B. C., Kogel, K. E., & Kane, M. A. (2003). Transfusion-related acute lung injury (TRALI) following platelet transfusion in a patient receiving high-dose interleukin-2 for treatment of metastatic renal cell carcinoma. *Transfus Apheresis Sci, 29,* 25–27.

275. Ramanathan, R. K., Triulzi, D. J., & Logan, T. F. (1997). Transfusion-related acute lung injury following random donor platelet transfusion: A report of two cases. *Vox Sang, 73,* 43–45.

276. Lucas, G., Rogers, S., Evans, R., et al. (2000). Transfusion-related acute lung injury associated with interdonor incompatibility for the neutrophil-specific antigen HNA-1a. *Vox Sang, 79,* 112–115.

277. Heddle, N. M., Klama, L. N., Griffith, L., et al. (1993). A prospective study to identify the risk factors associated with acute reactions to platelet and red cell transfusions. *Transfusion, 33,* 794–797.

278. Winqvist, I. (1991). Meperidine (pethidine) to control shaking chills and fever associated with non-hemolytic transfusion reactions. *Eur J Haematol, 47,* 154–155.

279. Patterson, B. J., Freedman, J., Blanchette, V., et al. (2000). Effect of premedication guidelines and leukoreduction on the rate of febrile nonhaemolytic platelet transfusion reactions. *Transfus Med, 10,* 199–206.

280. Payne, R. (1957). The association of febrile transfusion reactions with leukoagglutanins. *Vox Sang, 2,* 233.

281. Perrotta, P. L., Feldman, D., & Snyder, E. L. (2000). Biological response modifiers in platelet transfusion therapy. In J. Seghatchian, E. L. Snyder, & P. Krailadsiri (Eds.), *Platelet therapy* (pp. 227–262). Amsterdam: Elsevier.

282. Sarkodee-Adoo, C. B., Kendall, J. M., Sridhara, R., et al. (1998). The relationship between the duration of platelet storage and the development of transfusion reactions. *Transfusion, 38,* 229–235.

283. Riccardi, D., Raspollini, E., Rebulla, P., et al. (1997). Relationship of the time of storage and transfusion reactions to platelet concentrates from buffy coats. *Transfusion, 37,* 528–530.

284. Kelley, D. L., Mangini, J., Lopez-Plaza, I., et al. (2000). The utility of < or = 3-day-old whole-blood platelets in reducing the incidence of febrile nonhemolytic transfusion reactions. *Transfusion, 40,* 439–442.

285. Hartwig, D., Hartel, C., Hennig, H., et al. (2002). Evidence for *de novo* synthesis of cytokines and chemokines in platelet concentrates. *Vox Sang, 82,* 182–190.

286. Anderson, N. A., Gray, S., Copplestone, J. A., et al. (1997). A prospective randomized study of three types of platelet concentrates in patients with haematological malignancy: Corrected platelet count increments and frequency of non-haemolytic febrile transfusion reactions. *Transfus Med, 7,* 33–39.

287. Addas-Carvalho, M., Origa, A. F., & Saad, S. T. (2004). Interleukin 1 beta and tumor necrosis factor levels in stored platelet concentrates and the association with gene polymorphisms. *Transfusion, 44,* 996–1003.

288. Federowicz, I., Barrett, B. B., Andersen, J. W., et al. (1996). Characterization of reactions after transfusion of cellular blood components that are white cell reduced before storage. *Transfusion, 36,* 21–28.

289. Aye, M. T., Palmer, D. S., Giulivi, A., et al. (1995). Effect of filtration of platelet concentrates on the accumulation of cytokines and platelet release factors during storage. *Transfusion, 35,* 117–124.

290. Snyder, E. L., Mechanic, S., Baril, L., et al. (1996). Removal of soluble biologic response modifiers (complement and chemokines) by a bedside white cell-reduction filter. *Transfusion, 36,* 707–713.

291. Geiger, T. L., Perrotta, P. L., Davenport, R., et al. (1997). Removal of anaphylatoxins C3a and C5a and chemokines interleukin 8 and RANTES by polyester white cell-reduction and plasma filters. *Transfusion, 37,* 1156–1162.

292. Vo, T. D., Cowles, J., Heal, J. M., et al. (2001). Platelet washing to prevent recurrent febrile reactions to leucocyte-reduced transfusions. *Transfus Med, 11,* 45–47.

293. Anderson, K. C., Lew, M. A., Gorgone, B. C., et al. (1986). Transfusion-related sepsis after prolonged platelet storage. *Am J Med, 81,* 405–411.

294. Blajchman, M. A. (1998). Bacterial contamination and proliferation during the storage of cellular blood products. *Vox Sang, 74,* 155–159.

295. Braine, H. G., Kickler, T. S., Charache, P., et al. (1986). Bacterial sepsis secondary to platelet transfusion: An adverse effect of extended storage at room temperature. *Transfusion, 26,* 391–393.

296. Heal, J. M., Singal, S., Sardisco, E., et al. (1986). Bacterial proliferation in platelet concentrates. *Transfusion, 26,* 388–390.

297. Brecher, M. E., Means, N., Jere, C. S., et al. (2001). Evaluation of an automated culture system for detecting bacterial contamination of platelets: An analysis with 15 contaminating organisms. *Transfusion, 41,* 477–482.

298. Brecher, M. E., Heath, D. G., Hay, S. N., et al. (2002). Evaluation of a new generation of culture bottle using an automated bacterial culture system for detecting nine common contaminating organisms found in platelet components. *Transfusion, 42,* 774–779.

299. Puckett, A., Davison, G., Entwistle, C. C., et al. (1992). Post transfusion septicaemia 1980–1989: Importance of donor arm cleansing. *J Clin Pathol, 45,* 155–157.

300. Heal, J. M., Jones, M. E., Forey, J., et al. (1987). Fatal Salmonella septicemia after platelet transfusion. *Transfusion, 27,* 2–5.

301. Engelfriet, C. P., Reesink, H. W., Blajchman, M. A., et al. (2000). Bacterial contamination of blood components. *Vox Sang, 78,* 59–67.

302. Dodd, R. Y. (2003). Bacterial contamination and transfusion safety: Experience in the United States. *Transfus Clin Biol, 10,* 6–9.

303. Kuehnert, M. J., Roth, V. R., Haley, N. R., et al. (2001). Transfusion-transmitted bacterial infection in the United States, 1998 through 2000. *Transfusion, 41,* 1493–1499.

304. (2005). Fatal bacterial infections associated with platelet transfusions — United States, 2004. *Morbidity Mortality Weekly Rep, 54,* 168–170.

305. te Boekhorst, P. A., Beckers, E. A., Vos, M. C., et al. (2005). Clinical significance of bacteriologic screening in platelet concentrates. *Transfusion, 45,* 514–519.

306. Liu, H. W., Yuen, K. Y., Cheng, T. S., et al. (1999). Reduction of platelet transfusion-associated sepsis by short-term bacterial culture. *Vox Sang, 77,* 1–5.

307. AuBuchon, J. P., Cooper, L. K., Leach, M. F., et al. (2002). Experience with universal bacterial culturing to detect contamination of apheresis platelet units in a hospital transfusion service. *Transfusion, 42,* 855–861.

308. Lee, C. K., Ho, P. L., Chan, N. K., et al. (2002). Impact of donor arm skin disinfection on the bacterial contamination rate of platelet concentrates. *Vox Sang, 83,* 204–208.

309. Burstain, J. M., Brecher, M. E., Workman, K., et al. (1997). Rapid identification of bacterially contaminated platelets using reagent strips: Glucose and pH analysis as markers of bacterial metabolism. *Transfusion, 37,* 255–258.

310. Chambers, L. A., Long, K. F., Wissel, M. E., et al. (2004). Multisite trial of a pH paper and a pH value of less than 7.0 to screen whole-blood-derived platelets: Implications for bacterial detection. *Transfusion, 44,* 1261–1263.

311. Dreier, J., Stormer, M., & Kleesiek, K. (2004). Two novel real-time reverse transcriptase PCR assays for rapid detection of bacterial contamination in platelet concentrates. *J Clin Microbiol, 42,* 4759–4764.

312. Howard, J. E., & Perkins, H. A. (1978). The natural history of alloimmunization to platelets. *Transfusion, 18,* 496–503.

313. Klingemann, H. G., Self, S., Banaji, M., et al. (1987). Refractoriness to random donor platelet transfusions in patients with aplastic anaemia: A multivariate analysis of data from 264 cases. *Br J Haematol, 66,* 115–121.

314. Bishop, J. F., McGrath, K., Wolf, M. M., et al. (1988). Clinical factors influencing the efficacy of pooled platelet transfusions. *Blood, 71,* 383–387.

315. Bishop, J. F., Matthews, J. P., McGrath, K., et al. (1991). Factors influencing 20-hour increments after platelet transfusion. *Transfusion, 31,* 392–396.

316. Ishida, A., Handa, M., Wakui, M., et al. (1998). Clinical factors influencing posttransfusion platelet increment in patients undergoing hematopoietic progenitor cell transplantation — A prospective analysis. *Transfusion, 38,* 839–847.

317. Yankee, R. A., Grumet, F. C., & Rogentine, G. N. (1969). Platelet transfusion. The selection of compatible platelet donors for refractory patients by lymphocyte HL-A typing. *N Engl J Med, 281,* 1208–1212.

318. Green, D., Tiro, A., Basiliere, J., et al. (1976). Cytotoxic antibody complicating platelet support in acute leukemia. Response to chemotherapy. *J Am Med Assoc, 236,* 1044–1046.

319. Brandt, J. T., Julius, C. J., Osborne, J. M., et al. (1996). The mechanism of platelet aggregation induced by HLA-related antibodies. *Thromb Haemost, 76,* 774–779.

320. Slichter, S. J., Deeg, H. J., & Kennedy, M. S. (1987). Prevention of platelet alloimmunization in dogs with systemic cyclosporine and by UV-irradiation or cyclosporine-loading of donor platelets. *Blood, 69,* 414–418.

321. Klumpp, T. R., Herman, J. H., Innis, S., et al. (1996). Factors associated with response to platelet transfusion following hematopoietic stem cell transplantation. *Bone Marrow Transplant, 17,* 1035–1041.

322. Toor, A. A., Choo, S. Y., & Little, J. A. (2000). Bleeding risk and platelet transfusion refractoriness in patients with acute myelogenous leukemia who undergo autologous stem cell transplantation. *Bone Marrow Transplant, 26,* 315–320.

323. Doughty, H. A., Murphy, M. F., Metcalfe, P., et al. (1994). Relative importance of immune and non-immune causes of platelet refractoriness. *Vox Sang, 66,* 200–205.

324. Bock, M., Muggenthaler, K. H., Schmidt, U., et al. (1996). Influence of antibiotics on posttransfusion platelet increment. *Transfusion, 36,* 952–954.

325. Levin, M. D., de Veld, J. C., van der Holt, B., et al. (2003). Immune and nonimmune causes of low recovery from leukodepleted platelet transfusions: A prospective study. *Ann Hematol, 82,* 357–362.

326. Hussein, M. A., Fletcher, R., Long, T. J., et al. (1998). Transfusing platelets 2 h after the completion of amphotericin-B decreases its detrimental effect on transfused platelet recovery and survival. *Transfus Med, 8,* 43–47.

327. Cooling, L. L., Kelly, K., Barton, J., et al. (2005). Determinants of ABH expression on human blood platelets. *Blood, 105,* 3356–3364.

328. Lee, E. J., & Schiffer, C. A. (1989). ABO compatibility can influence the results of platelet transfusion. Results of a randomized trial. *Transfusion, 29,* 384–389.

329. Heal, J. M., Blumberg, N., & Masel, D. (1987). An evaluation of crossmatching, HLA, and ABO matching for platelet transfusions to refractory patients. *Blood, 70,* 23–30.

330. Saito, S., Ota, S., Seshimo, H., et al. (2002). Platelet transfusion refractoriness caused by a mismatch in HLA-C antigens. *Transfusion, 42,* 302–308.

331. Eernisse, J. G., & Brand, A. (1981). Prevention of platelet refractoriness due to HLA antibodies by administration of leukocyte-poor blood components. *Exp Hematol, 9,* 77–83.

332. Schiffer, C. A., Dutcher, J. P., Aisner, J., et al. (1983). A randomized trial of leukocyte-depleted platelet transfusion to modify alloimmunization in patients with leukemia. *Blood, 62,* 815–820.

333. Novotny, V. M., van Doorn, R., Witvliet, M. D., et al. (1995). Occurrence of allogeneic HLA and non-HLA antibodies after transfusion of prestorage filtered platelets and red blood cells: A prospective study. *Blood, 85,* 1736–1741.

334. Andreu, G., Dewailly, J., Leberre, C., et al. (1988). Prevention of HLA immunization with leukocyte-poor packed red cells and platelet concentrates obtained by filtration. *Blood, 72,* 964–969.

335. Sniecinski, I., O'Donnell, M. R., Nowicki, B., et al. (1988). Prevention of refractoriness and HLA-alloimmunization using filtered blood products. *Blood, 71,* 1402–1407.

336. Oksanen, K., Kekomaki, R., Ruutu, T., et al. (1991). Prevention of alloimmunization in patients with acute leukemia by use of white cell-reduced blood components — A randomized trial. *Transfusion, 31,* 588–594.

337. van Marwijk Kooy, M., van Prooijen, H. C., Moes, M., et al. (1991). Use of leukocyte-depleted platelet concentrates for the prevention of refractoriness and primary HLA allo-immunization: A prospective, randomized trial. *Blood, 77,* 201–205.

338. Williamson, L. M., Wimperis, J. Z., Williamson, P., et al. (1994). Bedside filtration of blood products in the prevention of HLA alloimmunization — A prospective randomized study. Alloimmunisation Study Group. *Blood, 83,* 3028–3035.

339. Slichter, S. J., Davis, K., Enright, H., et al. (2005). Factors affecting post-transfusion platelet increments, platelet refractoriness, and platelet transfusion intervals in thrombocytopenic patients. *Blood, 105,* 4106–4114.

340. McFarland, J. G. (1996). Alloimmunization and platelet transfusion. *Semin Hematol, 33,* 315–328.

341. Levin, M. D., Kappers-Klunne, M., Sintnicolaas, K., et al. (2004). The value of alloantibody detection in predicting response to HLA-matched platelet transfusions. *Br J Haematol, 124,* 244–250.

342. Mueller-Eckhardt, G., Hauck, M., Kayser, W., et al. (1980). HLA-C antigens on platelets. *Tissue Antigens, 16,* 91–94.

343. Bolgiano, D. C., Larson, E. B., & Slichter, S. J. (1989). A model to determine required pool size for HLA-typed community donor apheresis programs. *Transfusion, 29,* 306–310.

344. Messerschmidt, G. L., Makuch, R., Appelbaum, F., et al. (1988). A prospective randomized trial of HLA-matched versus mismatched single-donor platelet transfusions in cancer patients. *Cancer, 62,* 795–801.

345. Schonewille, H., Haak, H. L., & van Zijl, A. M. (1999). Alloimmunization after blood transfusion in patients with hematologic and oncologic diseases. *Transfusion, 39,* 763–771.

346. (2003). Detection of platelet-reactive antibodies in patients who are refractory to platelet transfusions, and the selection of compatible donors. *Vox Sang, 84,* 73–88.

347. Friedberg, R. C., Donnelly, S. F., & Mintz, P. D. (1994). Independent roles for platelet crossmatching and HLA in the selection of platelets for alloimmunized patients. *Transfusion, 34,* 215–220.

348. Kekomaki, S., Volin, L., Koistinen, P., et al. (1998). Successful treatment of platelet transfusion refractoriness: The use of platelet transfusions matched for both human leucocyte antigens (HLA) and human platelet alloantigens (HPA) in alloimmunized patients with leukaemia. *Eur J Haematol, 60,* 112–118.

349. Novotny, V. M., Doxiadis, I. I., & Brand, A. (1999). The reduction of HLA class I expression on platelets: A potential approach in the management of HLA-alloimmunized refractory patients. *Transfus Med Rev, 13,* 95–105.

350. Chandramouli, N. B., & Rodgers, G. M. (2000). Prolonged immunoglobulin and platelet infusion for treatment of immune thrombocytopenia. *Am J Hematol, 65,* 85–86.

351. Snyder, E. L., Rinder, H. M., & Napychank, P. (1990). *In vitro* and *in vivo* evaluation of platelet transfusions administered through an electromechanical infusion pump. *Am J Clin Pathol, 94,* 77–80.

352. Mair, B., & Leparc, G. F. (1998). Hypotensive reactions associated with platelet transfusions and angiotensin-converting enzyme inhibitors. *Vox Sang, 74,* 27–30.

353. Takahashi, T. A., Abe, H., Hosoda, M., et al. (1995). Bradykinin generation during filtration of platelet concentrates with a white cell-reduction filter. *Transfusion, 35,* 967.

354. Shiba, M., Tadokoro, K., Sawanobori, M., et al. (1997). Activation of the contact system by filtration of platelet concentrates with a negatively charged white cell-removal filter and measurement of venous blood bradykinin level in patients who received filtered platelets. *Transfusion, 37,* 457–462.

355. Belloni, M., Alghisi, A., Bettini, C., et al. (1998). Hypotensive reactions associated with white cell-reduced apheresis platelet concentrates in patients not receiving ACE inhibitors. *Transfusion, 38,* 412–415.

356. Cyr, M., Hume, H. A., Champagne, M., et al. (1999). Anomaly of the des-Arg9-bradykinin metabolism associated with severe hypotensive reactions during blood transfusions: A preliminary study. *Transfusion, 39,* 1084–1088.

357. Webb, I. J., & Anderson, K. C. (1999). Risks, costs, and alternatives to platelet transfusions. *Leuk Lymphoma, 34,* 71–84.

358. Kuter, D. J. (1998). Thrombopoietins and thrombopoiesis: A clinical perspective. *Vox Sang, 74,* 75–85.

359. Bertolini, F., Battaglia, M., Pedrazzoli, P., et al. (1997). Megakaryocytic progenitors can be generated *ex vivo* and safely administered to autologous peripheral blood progenitor cell transplant recipients. *Blood, 89,* 2679–2688.

360. Harker, L. A., Marzec, U. M., Novembre, F., et al. (1998). Treatment of thrombocytopenia in chimpanzees infected with human immunodeficiency virus by pegylated recombinant human megakaryocyte growth and development factor. *Blood, 91,* 4427–4433.

361. Snyder, E., Perrotta, P., Rinder, H., et al. (1999). Effect of recombinant human megakaryocyte growth and development factor coupled with polyethylene glycol on the platelet storage lesion. *Transfusion, 39,* 258–264.

362. Basser, R. L., Rasko, J. E., Clarke, K., et al. (1996). Thrombopoietic effects of pegylated recombinant human megakaryocyte growth and development factor (PEG-rHuMGDF) in patients with advanced cancer. *Lancet, 348,* 1279–1281.

363. Fanucchi, M., Glaspy, J., Crawford, J., et al. (1997). Effects of polyethylene glycol-conjugated recombinant human megakaryocyte growth and development factor on platelet

counts after chemotherapy for lung cancer. *N Engl J Med, 336,* 404–409.

364. O'Malley, C. J., Rasko, J. E., Basser, R. L., et al. (1996). Administration of pegylated recombinant human mega-karyocyte growth and development factor to humans stimulates the production of functional platelets that show no evidence of *in vivo* activation. *Blood, 88,* 3288–3298.

365. Kuter, D. J., Cebon, J., Harker, L. A., et al. (1999). Platelet growth factors: Potential impact on transfusion medicine. *Transfusion, 39,* 321–332.

366. Jafari, M., Forsberg, J., Gilcher, R. O., et al. (2002). Salmonella sepsis caused by a platelet transfusion from a donor with a pet snake. *N Engl J Med, 347,* 1075–1078.

367. Mathur, P., & Samantaray, J. C. (2004). The first probable case of platelet transfusion-transmitted visceral leishmaniasis. *Transfus Med, 14,* 319–321.

368. Golubic-Cepulic, B., Budimir, A., Plecko, V., et al. (2004). *Morganella morganii* causing fatal sepsis in a platelet recipient and also isolated from a donor's stool. *Transfus Med, 14,* 237–240.

# CHAPTER 70

# Platelet Substitutes and Novel Methods of Platelet Preservation

## David H. Lee[1] and Morris A. Blajchman[2]

[1]Departments of Medicine, Pathology and Molecular Medicine, Queen's University, Kingston, Ontario, Canada
[2]Departments of Medicine and Pathology, McMaster University, Hamilton, Ontario, Canada

## I. Introduction

Despite advances in the safety, processing, and storage of conventional 22°C liquid-stored allogeneic platelet concentrates (see Chapter 69), there are still significant limitations to standard platelet concentrates (Table 70-1). These limitations have led to efforts to minimize the exposure of recipients to allogeneic blood products and to develop safe and effective platelet products and substitutes with longer shelf-lives. This chapter reviews advances in the development of new platelet products, synthetic platelet substitutes, new storage methods, and novel pathogen inactivation technology for platelet concentrates.

What are the desirable properties of platelet products and substitutes? These agents should function hemostatically to offset the hemostatic defect associated with thrombocytopenia or platelet function defects, without causing pathologic thrombosis or consumptive coagulopathy. They should be immunologically silent. They should not be immunogenic, nor should they evoke an inflammatory response. Any additives or components of the infused product should be non-toxic. A platelet substitute should not transmit infection or cause reticuloendothelial blockade. Ideally, novel platelet products and platelet substitutes should have a long duration of action to allow long dosing intervals. They should have simple storage requirements, a long shelf-life, and should be easy to administer (Table 70-2). The novel platelet products and substitutes and their current developmental status are listed in Table 70-3.

## II. Novel Platelet Products

### A. Frozen Platelets

**1. Platelets Cryopreserved in 5 or 6% Dimethyl Sulfoxide.** The transfusion of cryopreserved platelets into thrombocytopenic patients was first reported half a century ago,[1,2] and by the 1970s the methodology had evolved sufficiently to permit its integration into the transfusion program of some centers. However, cryopreservation is cumbersome to use and expensive. As a result, freezing has generally been reserved for storing autologous platelets collected from alloimmunized patients with acute leukemia in remission,[3] although a few studies have evaluated its use in cardiac surgery.[4,5]

Freezing platelets in 5 or 6% dimethyl sulfoxide (DMSO) currently represents the standard for long-term platelet cryopreservation.[3,6–9] Platelets can be stored at −80°C in 6% DMSO for up to 2 years without significant loss of recovery, whereas platelets frozen in 5% DMSO can be stored for 3 years at −150°C.[8] Polyolefin or polyvinylchloride (PVC) bags are used, but storage at −135°C is associated with a high incidence of breakage of PVC bags in comparison with −80°C storage.[10] Formal controlled-rate freezing does not appear to be essential since placement in the gas phase of liquid nitrogen with attention to consistent location,[3] or placement in a −80°C mechanical freezer, produces satisfactory results.[8,11,12] Thawing is performed by immersion in a water bath.

A single dilution centrifugation wash step removes 95% of the DMSO prior to transfusion. *In vitro* platelet recovery postthaw is approximately 75%, and after transfusion into normal volunteers, *in vivo* platelet recovery at 1 hour is approximately 33% of the total number of frozen platelets. In practical terms, it takes twice as many cryopreserved platelets to obtain a platelet count increment that is equivalent to that observed with conventional liquid-stored platelet concentrates. The *in vivo* survival of cryopreserved platelets has been reported to be similar to that of liquid-preserved platelet concentrates in some studies[9,13] but shortened in others.[4]

Cryopreserved platelets acquire a variety of morphologic and functional defects *in vitro*. Reductions in platelet

**Table 70-1: Limitations and Adverse Effects of Conventional Platelet Concentrates**

Short shelf-life

Inadequate supply to meet demands

Alloimmunization and refractoriness

Bacterial, viral, protozoan infections

Febrile nonhemolytic transfusion reactions

Immunosuppression

Graft-versus-host disease

Transfusion-associated lung injury

**Table 70-2: Desirable Properties of a Platelet Product or Substitute**

Hemostatic efficacy without pathologic thrombogenicity

No immunogenicity

No reticuloendothelial blockade

No immune suppression

Nontoxic

Sterile

Long duration of action

Long shelf-life

Simple storage requirements

Easy to administer

**Table 70-3: Novel Platelet Products and Platelet Substitutes**

| Novel Platelet Product or Substitute | Developmental Status[a] |
|---|---|
| *Platelet Products* | |
| Frozen platelets (5–6% DMSO) | Advanced clinical studies |
| Frozen platelets (ThromboSol/2% DMSO) | Early clinical studies |
| Cold-stored liquid platelets | Preclinical |
| Amotosalen/UVA-treated platelets | Advanced clinical trials; approved for use clinically in some European countries |
| Riboflavin/UVA-treated platelets | Early clinical studies |
| Infusible platelet membranes | Early clinical studies |
| Lyophilized platelets | Preclinical |
| *Platelet Substitutes* | |
| Ligand-coated red blood cells | Preclinical |
| Ligand-coated albumin microspheres | Preclinical |
| Liposome-based agents | Preclinical |

[a]For some preparations, further active development has not been pursued. DMSO, dimethyl sulfoxide; UVA, ultraviolet-A light.

adhesion,[14] stimulus response coupling,[15] aggregation,[16,17] and release[9,17] and increased numbers of discoid forms[18] occur with freezing in DMSO. Despite these *in vitro* abnormalities, cryopreserved platelets are hemostatically effective *in vivo*.[3,9] In one study, cardiac surgery patients who were randomized to receive cryopreserved platelets were found to have less blood loss and required fewer transfusions of red cells, platelets, and plasma than those randomized to receive standard liquid-stored platelets, suggesting superior hemostatic function for cryopreserved over liquid-stored platelets.[4]

What is the explanation for the hemostatic effect of frozen platelets? Clues come from the observation that cryopreserved platelets generate more surface procoagulant activity (as measured by increased factor V binding) and thromboxane $A_2$ generation than liquid-stored platelets.[4]

Freezing also produces a subpopulation of glycoprotein (GP) Ib-reduced platelets that are rapidly cleared from the circulation in baboons, leaving a GPIb-normal subpopulation that has a longer *in vivo* survival.[19,20] It has been hypothesized that the hemostatic efficacy of cryopreserved platelets may be attributed to the rapid hemostatic effect of the GPIb-reduced subpopulation secondary to increased binding of factor V and P-selectin expression.[20]

For most clinical circumstances, DMSO-cryopreserved platelets do not offer advantages over conventional liquid-stored platelet concentrates and are more laborious and expensive to prepare. Moreover, most centers lack the additional facilities and personnel needed to prepare, store, and process frozen platelets. For these reasons, the use of cryopreserved platelets is not widespread.

**2. Platelets Cryopreserved in ThromboSol and 2% DMSO.** The inclusion of a cryoprotectant cocktail consisting of amiloride, adenosine, and sodium nitroprusside (ThromboSol) together with 2% DMSO was formulated to stimulate second messenger pathways that may inhibit cold-induced platelet activation. Specifically, amiloride inhibits the sodium/hydrogen ion pump; adenosine inhibits intracellular cyclic adenosine monophosphate; and sodium nitroprusside releases nitric oxide, which increases intracellular cyclic guanosine monophosphate. Although the components of ThromboSol are approved for clinical use and the concentration of DMSO is reduced, the protocols used in human studies published to date have included a postthaw washing step to remove the various cryoprotectants.

Platelets cryopreserved in ThromboSol/2% DMSO have been reported to have equivalent or superior postthaw *in vitro* platelet recovery and function in comparison with platelets frozen in 6% DMSO.[21–23] After transfusion of thawed autologous platelets into normal volunteers, the recovery of platelets cryopreserved in ThromboSol/2% DMSO was superior to that of platelets frozen in 6% DMSO, whereas platelet survival was similar for both.[24]

Preservation of *in vitro* platelet characteristics appears to be improved further when platelets harvested for ThromboSol/2% DMSO cryopreservation follows treatment of the autologous donor with recombinant human thrombopoietin (rhTPO).[25] A subsequent randomized trial in patients with chemotherapy-associated thrombocytopenia evaluated thawed cryopreserved platelets that had been collected after rhTPO treatment in comparison to conventional allogeneic platelet concentrates.[26] The corrected count increment following transfusion of thawed autologous cryopreserved platelets was found to be similar to that observed following transfusion of conventional allogeneic platelet concentrates.[26]

**3. Cryopreservation of Platelets with Trehalose.** Trehalose is a glucose disaccharide that can be found within the cells of a variety of organisms that are capable of surviving dehydration as well as other environmental stresses. Trehalose plus phosphate has been examined as a strategy for platelet cryopreservation.[27] Platelets preloaded with trehalose and frozen in the presence of phosphate demonstrated partial preservation of *in vitro* platelet characteristics and function upon thawing. (Further details on trehalose are in Section II.E.2.)

#### B. Cold Liquid-Stored Platelets

Because of the risk of bacterial growth, the storage of platelets suspended in plasma at room temperature is limited to 5 days in the absence of additional bacterial detection or

pathogen reduction technology. Lowering the storage temperature of platelets to 4°C inhibits bacterial growth but results in an unacceptably short platelet survival following transfusion.[28–30] Chilling also induces a variety of *in vitro* changes, many of which are associated with platelet activation. However, chilled platelets are functional *in vitro* and *in vivo* and may be more hemostatically effective than room temperature-stored platelets, despite a short life span in the circulation.[31] Hence, several groups have attempted to devise a means of preventing the deleterious effects of chilling on platelets to allow cold storage.

An important step was the identification of the mechanism by which chilled mouse platelets are removed from the circulation.[32,33] Chilling causes the clustering of GPIbα on the platelet plasma membrane. Recognition of clustered GPIbα by complement type 3 (CR3) receptors (also known as integrin αMβ2, CD11b/CD18, and Mac-1) on hepatic macrophages results in their phagocytosis. Furthermore, this recognition is lectin mediated and is dependent on exposed β-*N*-acetylglucosamine residues on GPIbα. Concealing these critical residues by enzymatic galactosylation has been shown to block the recognition by CR3 receptors of chilled murine platelets and restore their survival, without impairing hemostatic function. Galactosylation can be achieved by the simple addition of UDP-galactose to murine platelets, without the need for exogenous galactosyltransferase.[34] Whether this approach can be successfully translated into the human clinical setting remains to be determined.

A number of other approaches to liquid cold storage of platelets have been reported. A combination of amiloride, adenosine, and sodium nitroprusside (ThromboSol) has been evaluated as an additive to permit the cold storage of liquid-suspended platelets, in addition to its potential use as a cryoprotectant (see Section II.A.2). Platelets stored at 4°C in ThromboSol had a loss of discoid shape but greater retention of *in vitro* function with less evidence of cold-induced activation.[22,35,36] Again, the clinical hemostatic efficacy of such human platelets has yet to be demonstrated.

Another strategy to prevent the deleterious effects of cold temperature on cold liquid-stored platelets has been to inhibit cytoskeleton actin assembly associated with cold-induced shape change. Platelets cooled to 4°C in the presence of cytochalasin B and a cytoplasmic $Ca^{2+}$ chelator remain discoid and functional[37]; however, they were shown to be cleared from the circulation of mice and baboons as rapidly as chilled untreated platelets.[32,38]

Antifreeze glycoproteins (AFGP) have been shown to inhibit cold-induced platelet shape change and activation,[39] but the large amount of AFGP that would be required for clinical use and the need to wash the platelets are impediments to further development. Immunogenicity related to the use of AFGP is also a concern.

## C. Photochemically Treated Platelets

The current strategy for minimizing the risk of transfusion-associated viral infection is to exclude high-risk donors and to test donors for infectious agents. This reactive approach is most effective for preventing the transmission of well-known pathogens for which screening tests exist but are at risk of failure when new and emerging pathogens (for which there is no test) infect donors. Pathogen inactivation by photochemical treatment is a proactive line of defense in which most known and unknown pathogens are genetically and transcriptionally inactivated before collected blood products are stored.[40] Two leading pathogen inactivation technologies for platelet concentrates are described here.

**1. Amotosalen and Ultraviolet-A.** Photochemical treatment of liquid-stored platelets with amotosalen hydrochloride followed by illumination with long-wavelength ultraviolet-A radiation (UVA, 320–400 nm) is a pathogen and leukocyte inactivation technology that has reached advanced stages of clinical development and is approved for use in Europe.

Psoralens such as amotosalen are tricyclic compounds that interact with helical DNA and RNA in a two-step reaction. The psoralen molecule first intercalates within a double-helical region of the nucleic acid. When exposed to irradiation with UVA, psoralens become covalently bound to pyrimidine bases, forming monovalent adducts and divalent adducts that crosslink strands of nucleic acid.[41] Bacteria, viruses, protozoa, and leukocytes are thus inactivated because the psoralen-modified template cannot be transcribed or replicated.[42–44] This strategy is employed by the INTERCEPT Blood System of pathogen reduction for buffy coat or apheresis platelets. After treatment of platelets, the concentration of residual amotosalen and its photoproducts in the platelet product is reduced using a compound adsorption device that contains a binding resin to which the residual psoralen is bound.

Platelets treated with amotosalen and UVA light continue to function adequately in vitro,[42] although small treatment-associated differences in the stored platelet product have been reported.[45] Cytokine production during platelet storage and transfusion-associated graft-versus-host disease in an animal model have been reported to be reduced by photochemical treatment with amotosalen/UVA.[43,46,47]

The results of three randomized double-blind phase III clinical trials comparing amotosalen/UVA-treated platelets to untreated platelets have been reported.[48–50] Photochemical treatment of platelets was associated with lower platelet count increments and corrected count increments, but this was not observed across all analyses. In the largest of the studies,[49] a greater number of platelets and platelet transfusions were utilized by the group receiving amotosalen/UVA-treated platelets, in keeping with the observation that such

treatment is associated with a slight reduction in the recovery and survival of [111]In-labeled platelets.[51] Despite the effects on platelet recovery and kinetics, amotosalen/UVA-treated platelets appear to be both hemostatically effective and safe. No differences in bleeding or adverse effects were detected between groups across the studies.[48–50]

**2. Riboflavin and UV Light.** Riboflavin (vitamin B$_2$) covalently binds to nucleic acids upon exposure to UV or visible light, resulting in nucleic acid fragmentation and impaired transcription, replication, and translation.[52–54] This strategy is employed by the Mirasol PRT system of pathogen reduction. Viruses, bacteria, and leukocytes are effectively inactivated with this technology, with adequate retention of platelet quiescence and function ex vivo except at high illumination energies.[53,55–57] The level of toxicologic concern for riboflavin is low, and no washing step is included in the processing of the platelets. Riboflavin/UV treatment of platelets is associated with an increase in glucose consumption and a reduction in pH in the stored platelet product.[58–60] Recovery and survival of treated autologous platelets in normal subjects was reduced in comparison with control platelets,[60] but platelet viability appears to be adequate to justify further clinical studies.

## D. Platelet-Derived Microparticles and Infusible Platelet Membranes

When platelets are activated by strong agonists or mechanically disrupted, platelet-derived microparticles are generated[61] (see Chapter 20). These are particles of platelet origin that are heterogeneous in size and composition but are commonly membranous vesicles that bear platelet membrane glycoproteins. They spontaneously form during platelet storage and can be found in platelet concentrates,[62–64] fresh-frozen plasma, and cryoprecipitate.[65] Microparticles have been shown to have procoagulant activity,[66–68] adhere to vascular subendothelium, and enhance platelet adhesion.[69]

Since platelet-derived microparticles possess many similar hemostatic properties to intact platelets, they provide a rational strategy for the development of a platelet substitute. However, the lack of efficacy and considerable toxicity of the earliest preparations of platelet microparticles[70] led to a long hiatus in this line of research. A report published in 1987 provided preclinical evidence of hemostatic efficacy without significant morbidity in thrombocytopenic rabbits.[71] This concept was further developed in the 1990s into a preparation known as infusible platelet membrane (IPM; Cyplex), which is a lyophilized human platelet membrane microvesicle preparation derived from outdated platelet concentrate units.[72]

Although IPM lack many platelet structural and functional components, such as integrin αIIbβ3 (GPIIb-IIIa), α-granule factor V, and serotonin, others such as GPIb-IX-V functionality, procoagulant activity, and platelet recruitment are at least partially retained.[73,74] IPM shortened the prolonged bleeding time in thrombocytopenic rabbits for at least 6 hours without demonstrating excessive thrombogenicity.[72] Early clinical studies were conducted in the 1990s,[75,76] but detailed clinical experience has not been reported in full form in the literature.

### E. Lyophilized Platelets

Lyophilization is a platelet preservation strategy that is approximately half a century old. Early reports of the apparent effectiveness of rehydrated lyophilized platelets in a small number of thrombocytopenic patients[77] were followed by a series of studies that failed to confirm the hemostatic efficacy of reconstituted lyophilized platelets.[78–80] Thereafter, lyophilized platelets suffered a long absence from the literature. However, two contemporary lyophilized platelet preparations have emerged with more encouraging preclinical results — lyophilized paraformaldehyde-fixed platelets and lyophilized trehalose-loaded platelets.

**1. Lyophilized Paraformaldehyde-Fixed Platelets.** Read and colleagues developed a platelet lyophilization procedure that, upon reconstitution, permits the preservation of structural and several functional properties with a 50 to 60% yield of platelets from whole blood.[81,82] Washed platelets are fixed with paraformaldehyde (1.8% for human platelets or 0.68% for canine) and then frozen and lyophilized.

When reconstituted, lyophilized paraformaldehyde-fixed platelets resemble fresh platelets morphologically, with mild to moderate reductions in GPIb-IX, GPIV, and integrin αIIbβ3.[81,82] Rehydrated lyophilized human platelets do not appear to aggregate, but there is partial agglutination in response to ristocetin.[83] The capacity to express P-selectin and generate thromboxane is reduced[84] or absent.[83] However, they appear to accumulate more procoagulant factor V than fresh platelets,[83,85] are able to bind fibrinogen, and can be incorporated into clots that appear to be more susceptible to fibrinolysis than clots containing fresh platelets.[86] Furthermore, rehydrated lyophilized platelets have been shown to adhere to damaged arterial vessel segments in an *ex vivo* perfusion model.[81,84]

Thrombin stimulation of rehydrated lyophilized platelets is associated with phosphorylation of several intracellular proteins with depletion of free platelet adenosine diphosphate (ADP) and adenosine triphosphate (ATP).[87] This suggests that these platelets retain some metabolic activity and appear to be capable of some stimulus-response signaling rather than simply functioning as procoagulant platforms or passive structural participants during platelet aggregation.

Rehydrated lyophilized platelets transfused into animals circulate for less than 15 minutes,[85,88] likely due to phagocytic clearance by splenic macrophages.[88] Despite the short survival of rehydrated lyophilized platelets, they appear to be hemostatically effective. In dogs, rehydrated lyophilized platelets adhere to exposed subendothelium and are incorporated into thrombi, without producing a consumptive coagulopathy.[81] In thrombocytopenic rabbits, reconstituted lyophilized human platelets have been shown to shorten the microvascular ear bleeding time, but this effect was found to be inferior to an equivalent dose of fresh human platelets.[89]

**2. Lyophilized Trehalose-Loaded Platelets.** Trehalose is a disaccharide that is found in high concentrations in organisms that are capable of surviving extreme dehydration and, like other small carbohydrates, is an effective stabilizer of cells undergoing freezing, drying, and freeze-drying. Although the precise mechanisms by which trehalose stabilizes cells remain unclear, trehalose preserves the integrity of membranes during cooling and freezing and appears to prevent the aggregation of lipid microdomains that normally occurs under such conditions. Dessication tolerance may relate to the ability of trehalose to replace the water shell of macromolecules and its high glass transition temperature. Detailed descriptions of trehalose and its stabilizing properties can be found elsewhere.[90–92]

Intracellular trehalose is believed to be required for intact cells to be stabilized. External trehalose is spontaneously internalized by human platelets at 37°C, likely by fluid phase endocytosis.[93] When trehalose-loaded platelets are lyophilized and then rehydrated, *in vitro* recovery of intact platelets is approximately 85%. Rehydrated trehalose-loaded lyophilized platelets were found to have aggregation responses and biophysical membrane properties that were almost identical to those of fresh platelets.[93] To date, *in vivo* studies have not been reported. It remains to be seen whether the short circulation times of rehydrated paraformaldehyde-fixed lyophilized platelets are also a property of rehydrated trehalose-loaded lyophilized platelets. As a potential means of long-term platelet preservation, trehalose is appealing because, as a food additive, it is generally regarded as safe.[94]

### F. Platelets Produced In Vitro

Megakaryocyte progenitors isolated from normal human blood by leukapheresis can generate platelet-like particles *in vitro* that are structurally identical to platelets and are capable of aggregation.[95] Functional platelets have also been produced from murine embryonic stem cells when cocultured with stromal cells.[96] However, it remains to be determined whether this approach to platelet replacement therapy is feasible.

## III. Platelet Substitutes

The hemostatic agents that are designed to function as platelet substitutes seek to reproduce one or more aspects of normal platelet hemostatic function. Many of the current agents are designed to interact with platelets to augment the aggregation or recruitment of platelets, using a biodegradable carrier platform (red blood cells, albumin microspheres, or liposomes) to which proteins capable of interacting with platelets are attached (Table 70-4). Others are intended to provide a procoagulant phospholipid surface to enhance localized fibrin deposition as a means of ameliorating the impaired hemostasis associated with thrombocytopenia. This section reviews these agents.

### A. Red Blood Cells with Surface-Bound Fibrinogen or RGD Peptides

The observation that platelets can agglutinate or coaggregate with fixed cells or inert particles coated with fibrinogen[110–112] provided an interesting approach to designing a putative platelet substitute.

The use of autologous red blood cells (RBCs) with fibrinogen covalently bound to the surface was evaluated as a hemostatic agent in thrombocytopenia by Agam and Livne.[97] Fibrinogen was crosslinked to RBCs by incubation with formaldehyde or transglutaminase, yielding a fibrinogen density of 58 to 1400 molecules per RBC. Such cells enhanced the aggregation of stimulated platelets in a fibrino-

gen-dependent manner. RBC osmotic fragility and other parameters were unaltered by the treatment process. Infusion of these modified RBCs into thrombocytopenic rats shortened the tail bleeding time with a duration of action that was longer than that of fresh rat platelets. Moreover, this effect was observed using RBC preparations bearing as few as 58 fibrinogen molecules per cell.[97]

A slightly different approach was used by Coller et al.,[104] who attached RGD-containing peptides to the RBC surface. In a prior study, peptides of varying lengths bearing a carboxyl-terminal RGD sequence were bound via the amino terminus to polyacrylonitrile beads, and the optimum peptide length was determined for selective interaction with activated, but not resting, platelets.[113] Subsequently, the RGD peptides of appropriate length were covalently coupled to the surface of RBCs via glycophorin A, and the term *thromboerythrocytes* was coined to describe this putative platelet substitute.[104]

Thromboerythrocytes were shown to enhance the aggregation of ADP-stimulated platelets and became incorporated into the aggregates, an effect that was dependent on $\alpha IIb\beta 3$ receptors on activated platelets.[104] Despite the *in vitro* effectiveness of thromboerythrocytes, this agent did not appear to be hemostatically effective in thrombocytopenic primates.[114]

### B. Ligand-Coated Albumin Microspheres

The observations that either inert beads or cells bearing fibrinogen on the surface could enhance platelet aggregation[110–112] also provided a rational basis for the evaluation of ligand-coated albumin microspheres as a putative platelet substitute. Albumin can be polymerized by various methods to form solid or air-filled microspheres in the nanometer to micrometer size range. A variety of ligands capable of binding to platelet receptors can thus be covalently bound to the surface of these albumin microspheres to create a different class of putative platelet substitute. To date, several different preparations have undergone *in vitro* and limited animal testing.

Synthocytes[98] are fibrinogen-coated microspheres similar to air-filled albumin microspheres or microcapsules that are used as an ultrasound contrast agent.[115] They are manufactured by spray-drying a 10% solution of human albumin to form albumin microspheres. Human fibrinogen is then immobilized to the surface under specific ionic conditions and pH, yielding a fibrinogen content that represents less than 2% of the total protein of the product. The median diameter of the microspheres is 3.5 to 4.5 $\mu m$, with fewer than 2% having a diameter in excess of 6 $\mu m$.[98] Synthocytes coaggregate with platelets and have been reported to enhance the responsiveness of platelets to agonists such as ADP.[116]

### Table 70-4: Platelet Substitutes Based on Ligand- or Receptor-Coated Biodegradable Carrier Platforms

| Ligand or receptor | Platform (References) | | |
| --- | --- | --- | --- |
| | Erythrocytes | Albumin Microspheres | Liposomes |
| Fibrinogen | 97 | 98–100 | 101, 102 |
| HHLGGAKQAGDV | — | 103 | 102 |
| RGD peptide | 104 | — | — |
| rGPIbα | — | 105 | 106–108 |
| Combinations of platelet membrane glycoproteins | — | — | 101, 109 |

Perfusion chamber experiments suggest that Synthocytes enhance the adhesion of platelet-containing aggregates to activated endothelial cells in thrombocytopenic blood and become incorporated into the clot.[98] In a thrombocytopenic rabbit model, a single bolus infusion of 1.5 or $0.75 \times 10^9$ Synthocytes/kg shortened the prolonged microvascular ear bleeding time and decreased blood loss from a standard abdominal wall surgical wound. This hemostatic effect was still measurable 3 hours postinfusion, but it was no longer detectable 8 hours postinfusion. Biopsies of the ear bleeding time wounds also revealed Synthocytes surrounded by platelets and fibrin. No thrombogenicity was observed using a rabbit jugular vein thrombosis model, and no cardiopulmonary toxicity was seen in single-dose toxicity studies.[98]

Similar results have been reported with Thrombospheres, a fibrinogen-coated albumin microsphere preparation with a mean diameter of 1.2 μm, manufactured by a different process.[99] Thrombospheres also shorten the microvascular ear bleeding time in thrombocytopenic rabbits, but the hemostatic effect persists for at least 72 hours after a single bolus injection. The mechanism of action is poorly understood because the hemostatic effect of Thrombospheres is detectable even after they have disappeared from circulating blood. In normal rabbits, Thrombospheres did not shorten platelet survival and were not found to be thrombogenic.[117]

Another group used *N*-succinimidyl 3-(2-pyridyldithio) propionate (SPDP) to introduce disulfide linkages to conjugate different proteins and peptides to microspheres of recombinant human albumin.[100,103,105] This preparation of fibrinogen–microspheres was reported to bind to platelets adherent to a collagen surface under flow conditions, and it enhanced the recruitment of platelets to the platelet-immobilized surface.[100]

Like the RGDS and RGDF sequences of the fibrinogen α-chain, the fibrinogen γ-chain C-terminal dodecapeptide (HHLGGAKQAGDV) also binds to integrin αIIbβ3 (see Chapter 8). HHLGGAKQAGDV-coated albumin microspheres with a mean diameter of 260 nm had no effect on platelet aggregation but enhanced the recruitment of platelets to a collagen surface under flow conditions. *In vivo,* HHLGGAKQAGDV–microspheres shortened the prolonged tail bleeding time of thrombocytopenic rats, with 20 microspheres being hemostatically equivalent to one platelet.[103]

A recombinant form of the von Willebrand factor (VWF) binding domain of GPIbα (rGPIbα) has also been conjugated to microspheres of recombinant human albumin using SPDP.[105] Approximately 2500 copies of the rGPIbα are estimated to be bound to the surface of each 240-nm microsphere. These rGPIbα–microspheres agglutinate in response to ristocetin in the presence of VWF and enhance the ristocetin-induced agglutination of platelets.[105] No *in vivo* studies have been reported for rGPIbα–microspheres.

## C. Hemostatic Agents Based on Synthetic Phospholipid Vesicles/Liposomes

**1. Liposomes Carrying Platelet Membrane Glycoproteins.** In the 1980s, the incorporation of platelet membrane glycoproteins into lipid vesicles was originally used to study the structure and function of membrane glycoproteins.[118–121] These studies demonstrated that platelet–platelet interactions, such as aggregation and ristocetin-induced agglutination, could be emulated using reconstituted liposomes.

Subsequently, a liposome-based agent called the plateletsome was evaluated as a platelet substitute.[109] Plateletsomes measuring 10 to 200 nm were produced by incorporating a deoxycholate extract of platelet membrane containing GPIb, integrin αIIbβ3, and GPIV into unilamellar lipid vesicles consisting of sphingomyelin, phosphatidylcholine, monosialylganglioside, or egg phosphatide. Although plateletsomes did not enhance *in vitro* platelet aggregation, there was evidence of hemostatic efficacy *in vivo*. Infusion of plateletsomes decreased blood loss from tail incisions in thrombocytopenic rats and rats with platelet storage pool disease. No evidence of consumptive coagulopathy was observed after infusion of plateletsomes, nor was there evidence of pathologic thrombosis on postmortem examination.[109]

A liposomal preparation that incorporates rGPIbα onto liposomes measuring 200 to 300 nm (rGPIbα–liposomes) has been evaluated *in vitro* as a potential platelet substitute and drug delivery system.[106–108] The rGPIbα copy number was 200 to 2000 per liposome. In the presence of VWF, rGPIbα–liposomes agglutinate when ristocetin is added and coaggregate with platelets.[106] Studies using similar preparations suggest that rGPIbα–liposomes adhere to and roll on immobilized VWF under flow conditions. The extent of these interactions was dependent on shear, VWF density, rGPIbα density, and liposomal membrane flexibility.[107,108] Studies evaluating the hemostatic efficacy of rGPIbα–liposomes *in vivo* have not been reported.

Liposomes carrying a combination of recombinant GPIa-IIa and rGPIbα, fibrinogen, or the fibrinogen γ-chain C-terminal dodecapeptide (HHLGGAKQAGDV) enhance the aggregation of platelets and bind to activated adherent platelets.[101,102] Interestingly, the binding of HHLGGAKQAGDV–liposomes to platelets is associated with the release of liposome contents, a property that can be further manipulated to vary the extent of liposome release and internalization by platelets.[102]

**2. Phospholipid Vesicle Infusion.** The administration of phospholipid extracts from human brain[122] and soybeans[123] to thrombocytopenic patients nearly 50 years ago represents some of the earliest attempts to improvise a platelet substitute based on the recognition that platelet procoagulant activity contributes to hemostasis. Today, this general approach is no less valid.

Synthetic phospholipid vesicles have been found to promote coagulation and cause the deposition of fibrin on exposed aortic subendothelium perfused with thrombocytopenic blood *ex vivo*.[124,125] The procoagulant potential in this model was found to be dependent on the phospholipid composition of vesicles.

**3. Factor Xa and Phospholipid Vesicles.** The combination of factor Xa and phosphatidylcholine–phosphatidylserine vesicles (PCPS) was first proposed as a procoagulant strategy to correct hemophilic bleeding.[126,127] When infused into fawn-hooded rats with platelet storage pool disease, factor Xa/PCPS had a modest hemostatic effect,[114] but concerns about toxicity in animals have impeded further development of factor Xa/PCPS.

### D. Recombinant Factor VIIa

The use of recombinant activated factor VII (rFVIIa) to decrease bleeding in patients with qualitative platelet disorders or thrombocytopenia is described in detail in Chapter 68. In general, except for some anecdotal reports, rFVIIa does not appear to ameliorate bleeding in thrombocytopenia.[128]

## IV. Challenges in the Development of Novel Platelet Products and Substitutes

### A. What Are the Minimal Requirements for Hemostatic Efficacy in Thrombocytopenia?

Platelets support hemostasis by adhering, aggregating, releasing, providing a procoagulant surface membrane, shedding microparticles, and effecting clot retraction. Yet the *in vitro* and *in vivo* properties of the novel platelet products and substitutes suggest that the full repertoire of platelet function is not essential to achieve partial hemostatic compensation or correction in thrombocytopenia. Our understanding of the integration of these hemostatic responses is inadequate to decide which of these components of platelet function can suffice or can be dispensed with when considering a putative hemostatic product. Although minimally modified platelet products such as photochemically treated platelets still possess the entire repertoire of hemostatic responses, platelet substitutes have generally been designed to exploit a single hemostatic mechanism. However, the preclinical evidence indicates that such diverse "single-pronged" agents ameliorate the hemostatic

defect associated with thrombocytopenia. By providing a reductionist strategy to reconstituting various components of normal hemostasis, the study of novel single-pronged agents may advance the current understanding of platelet function and hemostasis.

### B. The Challenge of Evaluation

As platelet products and substitutes become less similar to conventional platelet concentrates, *in vitro* tests of platelet function will become less useful in their evaluation. Although this is self-evident for platelet substitutes that bear little resemblance to platelets, it also applies to platelet-derived products (e.g., frozen platelets) that decrease thrombocytopenic bleeding *in vivo* but do not perform well on standard *in vitro* measures of platelet function. New and novel methods of *in vitro* evaluation that are suitable for the type of product tested will need to be introduced.

The challenge of preclinical testing using animal models can be characterized as one of establishing the validity, reliability, and measurement of thrombocytopenic bleeding (see Chapter 33). Thrombocytopenic bleeding has been measured by determining the RBC concentration in thoracic duct lymph[79,129] or, more commonly, by determining the bleeding time or blood loss from standardized incisions or injuries.[89,130] The assumptions when using such animal models are that the measured end point reflects general hemostatic function and that the model simulates the human condition. Defining optimal animal models and understanding their limitations thus remains a challenge.

As new platelet products and substitutes enter clinical trials, researchers and regulators are faced with the difficult task of the clinical evaluation of hemostatic efficacy. Similar to the challenge of animal models, the principal challenge in the clinical evaluation of platelet products and substitutes is the measurement of thrombocytopenic bleeding.[131,132] Major bleeding events occur infrequently in thrombocytopenic patients, and using such end points will require large trials. Minor bleeding is difficult to quantify in a standard and reproducible manner, and valid surrogate measures have been difficult to identify. Such issues will need to be resolved to allow for the proper clinical evaluation of novel platelet products and substitutes.

### C. Toxicity

The principal toxicity concerns of novel platelet products and substitutes involve pathologic thrombosis, coagulopathy, immunogenicity, immune suppression, and infection. However, the potential toxicity profile for these agents may be broader, particularly for the platelet substitutes that bear little resemblance to agents that clinicians are accustomed to using. Chemical additives that are not removed prior to infusion need to be evaluated for specific toxicity, as well as for mutagenicity and carcinogenicity.

# References

1. Klein, E., Toch, R., Farber, S., et al. (1956). Hemostasis in thrombocytopenic bleeding following infusion of stored, frozen platelets. *Blood, 11*, 693–999.

2. Tullis, J. L., Surgenor, D. M., & Baudanza, P. (1959). Preserved platelets: Their preparation, storage and clinical use. *Blood, 14*, 456–475.

3. Schiffer, C. A., Aisner, J., & Wiernik, P. H. (1978). Frozen autologous platelet transfusion for patients with leukemia. *N Engl J Med, 299*, 7–12.

4. Khuri, S. F., Healey, N., MacGregor, H., et al. (1999). Comparison of the effects of transfusions of cryopreserved and liquid-preserved platelets on hemostasis and blood loss after cardiopulmonary bypass. *J Thorac Cardiovasc Surg, 117*, 172–184.

5. Yokomuro, M., Ebine, K., Shiroma, K., et al. (1999). Safety and efficacy of autologous platelet transfusion in cardiac surgery: Comparison of cryopreservation, blood collection on the day before surgery, and blood collection during surgery. *Cryobiology, 38*, 236–242.

6. Valeri, C. R., Feingold, H., & Marchionni, L. D. (1974). A simple method for freezing human platelets using 6 percent dimethylsulfoxide and storage at −80 degrees C. *Blood, 43*, 131–136.

7. Daly, P. A., Schiffer, C. A., Aisner, J., et al. (1979). Successful transfusion of platelets cryopreserved for more than 3 years. *Blood, 54*, 1023–1027.

8. Valeri, C. R. (1981). The current state of platelet and granulocyte cryopreservation. *Crit Rev Clin Lab Sci, 14*, 21–74.

9. Melaragno, A. J., Carciero, R., Feingold, H., et al. (1985). Cryopreservation of human platelets using 6% dimethyl sulfoxide and storage at −80 degrees C. Effects of 2 years of frozen storage at −80 degrees C and transportation in dry ice. *Vox Sang, 49*, 245–258.

10. Valeri, C. R., Srey, R., Lane, J. P., et al. (2003). Effect of WBC reduction and storage temperature on PLTs frozen with 6 percent DMSO for as long as 3 years. *Transfusion, 43*, 1162–1167.

11. Melaragno, A. J., Abdu, W. A., Katchis, R. J., et al. (1982). Cryopreservation of platelets isolated with the IBM 2997 blood cell separator: A rapid and simplified approach. *Vox Sang, 43*, 321–326.

12. Angelini, A., Dragani, A., Berardi, A., et al. (1992). Evaluation of four different methods for platelet freezing. *In vitro* and *in vivo* studies. *Vox Sang, 62*, 146–151.

13. Vecchione, J. J., Melaragno, A. J., Hollander, A., et al. (1982). Circulation and function of human platelets isolated from units of CPDA-1, CPDA-2, and CPDA-3 anticoagulated blood and frozen with DMSO. *Transfusion, 22*, 206–209.

14. Owens, M., Cimino, C., & Donnelly, J. (1991). Cryopreserved platelets have decreased adhesive capacity. *Transfusion, 31*, 160–163.

15. Dullemond-Westland, A. C., van Prooijen, H. C., Riemens, M. I., et al. (1987). Cryopreservation disturbs stimulus-response coupling in a platelet subpopulation. *Br J Haematol, 67*, 325–333.

16. Shepherd, K. M., Sage, R. E., Barber, S., et al. (1984). Platelet cryopreservation: 1. *In vitro* aggregation studies. *Cryobiology, 21*, 39–43.

17. Spector, J. I., Skrabut, E. M., & Valeri, C. R. (1977). Oxygen consumption, platelet aggregation and release reactions in platelets freeze-preserved with dimethylsulfoxide. *Transfusion, 17*, 99–109.

18. Spector, J. I., Flor, W. J., & Valeri, C. R. (1979). Ultrastructural alterations and phagocytic function of cryopreserved platelets. *Transfusion, 19*, 307–312.

19. Owens, M., Werner, E., Holme, S., et al. (1994). Membrane glycoproteins in cryopreserved platelets. *Vox Sang, 67*, 28–31.

20. Barnard, M. R., MacGregor, H., Ragno, G., et al. (1999). Fresh, liquid-preserved, and cryopreserved platelets: Adhesive surface receptors and membrane procoagulant activity. *Transfusion, 39*, 880–888.

21. Currie, L. M., Livesey, S. A., Harper, J. R., et al. (1998). Cryopreservation of single-donor platelets with a reduced dimethyl sulfoxide concentration by the addition of second-messenger effectors: Enhanced retention of *in vitro* functional activity. *Transfusion, 38*, 160–167.

22. Lozano, M., Escolar, G., Mazzara, R., et al. (2000). Effects of the addition of second-messenger effectors to platelet concentrates separated from whole-blood donations and stored at 4 degrees C or −80 degrees C. *Transfusion, 40*, 527–534.

23. Lozano, M. L., Rivera, J., Corral, J., et al. (1999). Platelet cryopreservation using a reduced dimethyl sulfoxide concentration and second-messenger effectors as cryopreserving solution. *Cryobiology, 39*, 1–12.

24. Currie, L. M., Lichtiger, B., Livesey, S. A., et al. (1999). Enhanced circulatory parameters of human platelets cryopreserved with second-messenger effectors: An *in vivo* study of 16 volunteer platelet donors. *Br J Haematol, 105*, 826–831.

25. Vadhan-Raj, S., Currie, L. M., Bueso-Ramos, C., et al. (1999). Enhanced retention of *in vitro* functional activity of platelets from recombinant human thrombopoietin-treated patients following long-term cryopreservation with a platelet-preserving solution (ThromboSol) and 2% DMSO. *Br J Haematol, 104*, 403–411.

26. Vadhan-Raj, S., Kavanagh, J. J., Freedman, R. S., et al. (2002). Safety and efficacy of transfusions of autologous cryopreserved platelets derived from recombinant human thrombopoietin to support chemotherapy-associated severe thrombocytopenia: A randomised cross-over study. *Lancet, 359*, 2145–2152.

27. Nie, Y., de Pablo, J. J., & Palecek, S. P. (2005). Platelet cryopreservation using a trehalose and phosphate formulation. *Biotechnol Bioeng, 92*, 79–90.

28. Murphy, S., & Gardner, F. H. (1969). Effect of storage temperature on maintenance of platelet viability — Deleterious effect of refrigerated storage. *N Engl J Med, 280*, 1094–1098.

29. Slichter, S. J., & Harker, L. A. (1976). Preparation and storage of platelet concentrates: II. Storage variables influencing platelet viability and function. *Br J Haematol, 34*, 403–419.

30. Filip, D. J., & Aster, R. H. (1978). Relative hemostatic effectiveness of human platelets stored at 4 degrees and 22 degrees C. *J Lab Clin Med, 91*, 618–624.

31. Kaufman, R. M. (2005). Uncommon cold: Could 4 degrees C storage improve platelet function? *Transfusion, 45*, 1407–1412.

32. Hoffmeister, K. M., Felbinger, T. W., Falet, H., et al. (2003). The clearance mechanism of chilled blood platelets. *Cell, 112*, 87–97.

33. Josefsson, E. C., Gebhard, H. H., Stossel, T. P., et al. (2005). The macrophage alphaMbeta2 integrin alphaM lectin domain mediates the phagocytosis of chilled platelets. *J Biol Chem, 280*, 18025–18032.

34. Hoffmeister, K. M., Josefsson, E. C., Isaac, N. A., et al. (2003). Glycosylation restores survival of chilled blood platelets. *Science, 301*, 1531–1534.

35. Connor, J., Currie, L. M., Allan, H., et al. (1996). Recovery of *in vitro* functional activity of platelet concentrates stored at 4 degrees C and treated with second-messenger effectors. *Transfusion, 36*, 691–698.

36. Rivera, J., Lozano, M. L., Corral, J., et al. (1999). Quality assessment of platelet concentrates supplemented with second-messenger effectors. *Transfusion, 39*, 135–143.

37. Winokur, R., & Hartwig, J. H. (1995). Mechanism of shape change in chilled human platelets. *Blood, 85*, 1796–1804.

38. Valeri, C. R., Ragno, G., Marks, P. W., et al. (2004). Effect of thrombopoietin alone and a combination of cytochalasin B and ethylene glycol bis(beta-aminoethyl ether) *N,N'*-tetraacetic acid-AM on the survival and function of autologous baboon platelets stored at 4 degrees C for as long as 5 days. *Transfusion, 44*, 865–870.

39. Tablin, F., Oliver, A. E., Walker, N. J., et al. (1996). Membrane phase transition of intact human platelets: Correlation with cold-induced activation. *J Cell Physiol, 168*, 305–313.

40. Allain, J. P., Bianco, C., Blajchman, M. A., et al. (2005). Protecting the blood supply from emerging pathogens: The role of pathogen inactivation. *Transfus Med Rev, 19*, 110–126.

41. Cimino, G. D., Gamper, H. B., Isaacs, S. T., et al. (1985). Psoralens as photoactive probes of nucleic acid structure and function: Organic chemistry, photochemistry, and biochemistry. *Annu Rev Biochem, 54*, 1151–1193.

42. Lin, L., Cook, D. N., Wiesehahn, G. P., et al. (1997). Photochemical inactivation of viruses and bacteria in platelet concentrates by use of a novel psoralen and long-wavelength ultraviolet light. *Transfusion, 37*, 423–435.

43. Grass, J. A., Hei, D. J., Metchette, K., et al. (1998). Inactivation of leukocytes in platelet concentrates by photochemical treatment with psoralen plus UVA. *Blood, 91*, 2180–2188.

44. Van Voorhis, W. C., Barrett, L. K., Eastman, R. T., et al. (2003). *Trypanosoma cruzi* inactivation in human platelet concentrates and plasma by a psoralen (amotosalen HCl) and long-wavelength UV. *Antimicrob Agents Chemother, 47*, 475–479.

45. Picker, S. M., Speer, R., & Gathof, B. S. (2004). Functional characteristics of buffy-coat PLTs photochemically treated with amotosalen-HCl for pathogen inactivation. *Transfusion, 44*, 320–329.

46. Hei, D. J., Grass, J., Lin, L., et al. (1999). Elimination of cytokine production in stored platelet concentrate aliquots by photochemical treatment with psoralen plus ultraviolet A light. *Transfusion, 39*, 239–248.

47. Grass, J. A., Wafa, T., Reames, A., et al. (1999). Prevention of transfusion-associated graft-versus-host disease by photochemical treatment. *Blood, 93*, 3140–3147.

48. van Rhenen, D., Gulliksson, H., Cazenave, J. P., et al. (2003). Transfusion of pooled buffy coat platelet components prepared with photochemical pathogen inactivation treatment: The euroSPRITE trial. *Blood, 101*, 2426–2433.

49. McCullough, J., Vesole, D. H., Benjamin, R. J., et al. (2004). Therapeutic efficacy and safety of platelets treated with a photochemical process for pathogen inactivation: The SPRINT trial. *Blood, 104*, 1534–1541.

50. Janetzko, K., Cazenave, J. P., Kluter, H., et al. (2005). Therapeutic efficacy and safety of photochemically treated apheresis platelets processed with an optimized integrated set. *Transfusion, 45*, 1443–1452.

51. Snyder, E., Raife, T., Lin, L., et al. (2004). Recovery and life span of 111indium-radiolabeled platelets treated with pathogen inactivation with amotosalen HCl (S-59) and ultraviolet A light. *Transfusion, 44*, 1732–1740.

52. Hoffmann, M. E., & Meneghini, R. (1979). DNA strand breaks in mammalian cells exposed to light in the presence of riboflavin and tryptophan. *Photochem Photobiol, 29*, 299–303.

53. Kumar, V., Lockerbie, O., Keil, S. D., et al. (2004). Riboflavin and UV-light based pathogen reduction: Extent and consequence of DNA damage at the molecular level. *Photochem Photobiol, 80*, 15–21.

54. Tsugita, A., Okada, Y., & Uehara, K. (1965). Photosensitized inactivation of ribonucleic acids in the presence of riboflavin. *Biochim Biophys Acta, 103*, 360–363.

55. Goodrich, R. P. (2000). The use of riboflavin for the inactivation of pathogens in blood products. *Vox Sang, 78*(Suppl. 2), 211–215.

56. Ruane, P. H., Edrich, R., Gampp, D., et al. (2004). Photochemical inactivation of selected viruses and bacteria in platelet concentrates using riboflavin and light. *Transfusion, 44*, 877–885.

57. Perez-Pujol, S., Tonda, R., Lozano, M., et al. (2005). Effects of a new pathogen-reduction technology (Mirasol PRT) on functional aspects of platelet concentrates. *Transfusion, 45*, 911–919.

58. Li, J., de Korte, D., Woolum, M. D., et al. (2004). Pathogen reduction of buffy coat platelet concentrates using riboflavin and light: Comparisons with pathogen-reduction technology-treated apheresis platelet products. *Vox Sang, 87*, 82–90.

59. Li, J., Lockerbie, O., de Korte, D., et al. (2005). Evaluation of platelet mitochondria integrity after treatment with Mirasol pathogen reduction technology. *Transfusion, 45,* 920–926.

60. AuBuchon, J. P., Herschel, L., Roger, J., et al. (2005). Efficacy of apheresis platelets treated with riboflavin and ultraviolet light for pathogen reduction. *Transfusion, 45,* 1335–1341.

61. Owens, M. R. (1994). The role of platelet microparticles in hemostasis. *Transfus Med Rev, 8,* 37–44.

62. Solberg, C., Holme, S., & Little, C. (1986). Morphological changes associated with pH changes during storage of platelet concentrates in first-generation 3-day container. *Vox Sang, 50,* 71–77.

63. George, J. N., Pickett, E. B., & Heinz, R. (1988). Platelet membrane glycoprotein changes during the preparation and storage of platelet concentrates. *Transfusion, 28,* 123–126.

64. Bode, A. P., Orton, S. M., Frye, M. J., et al. (1991). Vesiculation of platelets during *in vitro* aging. *Blood, 77,* 887–895.

65. George, J. N., Pickett, E. B., & Heinz, R. (1986). Platelet membrane microparticles in blood bank fresh frozen plasma and cryoprecipitate. *Blood, 68,* 307–309.

66. Wolf, P. (1967). The nature and significance of platelet products in human plasma. *Br J Haematol, 13,* 269–288.

67. Solberg, C., Osterud, B., & Little, C. (1987). Platelet storage lesion: Formation of platelet fragments with platelet factor 3 activity. *Thromb Res, 48,* 559–565.

68. Bode, A. P., & Miller, D. T. (1986). Analysis of platelet factor 3 in platelet concentrates stored for transfusion. *Vox Sang, 51,* 299–305.

69. Owens, M. R., Holme, S., & Cardinali, S. (1992). Platelet microvesicles adhere to subendothelium and promote adhesion of platelets. *Thromb Res, 66,* 247–258.

70. Hjort, P. F., Perman, V., & Cronkite, E. P. (1959). Fresh, disintegrated platelets in radiation thrombocytopenia: Correction of prothrombin consumption without correction of bleeding. *Proc Soc Exp Biol Med, 102,* 31–35.

71. McGill, M., Fugman, D. A., Vittorio, N., et al. (1987). Platelet membrane vesicles reduced microvascular bleeding times in thrombocytopenic rabbits. *J Lab Clin Med, 109,* 127–133.

72. Chao, F. C., Kim, B. K., Houranieh, A. M., et al. (1996). Infusible platelet membrane microvesicles: A potential transfusion substitute for platelets. *Transfusion, 36,* 536–542.

73. Graham, S. S., Gonchoroff, N. J., & Miller, J. L. (2001). Infusible platelet membranes retain partial functionality of the platelet GPIb/IX/V receptor complex. *Am J Clin Pathol, 115,* 144–147.

74. Galan, A. M., Bozzo, J., Hernandez, M. R., et al. (2000). Infusible platelet membranes improve hemostasis in thrombocytopenic blood: Experimental studies under flow conditions. *Transfusion, 40,* 1074–1080.

75. Goodnough, L. T., Kolodziej, M., & Ehlenbeck, C. (1995). A phase I study of safety and efficacy for infusible platelet membrane in patients [Abstract]. *Blood, 86*(Suppl. 1), 610.

76. Scigliano, E., Enright, H., Telen, M., et al. (1997). Infusible platelet membrane for control of bleeding in thrombocytopenic patients [Abstract]. *Blood, 90*(Suppl. 1), 267.

77. Arnold, P., Djerassi, I., Farber, S., et al. (1956). The preparation and clinical administration of lyophilized platelet material to children with acute leukemia and aplastic anemia. *J Pediatr, 49,* 517–522.

78. Fliedner, T. M., Sorensen, D. K., Bond, V. P., et al. (1958). Comparative effectiveness of fresh and lyophilized platelets in controlling irradiation hemorrhage in the rat. *Proc Soc Exp Biol Med, 99,* 731–733.

79. Jackson, D. P., Sorensen, D. K., Cronkite, E. P., et al. (1959). Effectiveness of transfusions of fresh and lyophilized platelets in controlling bleeding due to thrombocytopenia. *J Clin Invest, 38,* 1689–1697.

80. Firkin, B. G., Arimura, G., & Harrington, W. J. (1960). A method for evaluating the hemostatic effect of various agents in thrombocytopenic rats and mice. *Blood, 15,* 388–394.

81. Read, M. S., Reddick, R. L., Bode, A. P., et al. (1995). Preservation of hemostatic and structural properties of rehydrated lyophilized platelets: Potential for long-term storage of dried platelets for transfusion. *Proc Natl Acad Sci USA, 92,* 397–401.

82. Bode, A. P. (1995). Preclinical testing of lyophilized platelets as a product for transfusion medicine. *Transfus Sci, 16,* 183–185.

83. Valeri, C. R., MacGregor, H., Barnard, M. R., et al. (2004). *In vitro* testing of fresh and lyophilized reconstituted human and baboon platelets. *Transfusion, 44,* 1505–1512.

84. Bode, A. P., Read, M. S., & Reddick, R. L. (1999). Activation and adherence of lyophilized human platelets on canine vessel strips in the baumgartner perfusion chamber. *J Lab Clin Med, 133,* 200–211.

85. Valeri, C. R., MacGregor, H., Barnard, M. R., et al. (2005). Survival of baboon biotin-X-N-hydroxysuccinimide and (111)In-oxine-labelled autologous fresh and lyophilized reconstituted platelets. *Vox Sang, 88,* 122–129.

86. Fischer, T. H., Merricks, E. P., Bode, A. P., et al. (2002). Thrombus formation with rehydrated, lyophilized platelets. *Hematology, 7,* 359–369.

87. Fischer, T. H., Merricks, E. P., Russell, K. E., et al. (2000). Intracellular function in rehydrated lyophilized platelets. *Br J Haematol, 111,* 167–174.

88. Fischer, T. H., Merricks, E., Bellinger, D. A., et al. (2001). Splenic clearance mechanisms of rehydrated, lyophilized platelets. *Artif Cells Blood Substit Immobil Biotechnol, 29,* 439–451.

89. Blajchman, M. A., & Lee, D. H. (1997). The thrombocytopenic rabbit bleeding time model to evaluate the *in vivo* hemostatic efficacy of platelets and platelet substitutes. *Transfus Med Rev, 11,* 95–105.

90. Crowe, J. H., Carpenter, J. F., & Crowe, L. M. (1998). The role of vitrification in anhydrobiosis. *Annu Rev Physiol, 60,* 73–103.

91. Crowe, J. H., Crowe, L. M., Oliver, A. E., et al. (2001). The trehalose myth revisited: Introduction to a symposium on stabilization of cells in the dry state. *Cryobiology, 43,* 89–105.

92. Crowe, J. H., Tablin, F., Wolkers, W. F., et al. (2003). Stabilization of membranes in human platelets freeze-dried with trehalose. *Chem Phys Lipids, 122,* 41–52.

93. Wolkers, W. F., Walker, N. J., Tablin, F., et al. (2001). Human platelets loaded with trehalose survive freeze-drying. *Cryobiology, 42,* 79–87.

94. U.S. Food and Drug Administration, Department of Health and Human Services. (2000, October 5). Agency response letter — GRAS notice No. GRN 000045. Rockville, MD: Author. Accessed September 18, 2005. Available at www.cfsan.fda.gov/~rdb/opa-g045.html.

95. Choi, E. S., Nichol, J. L., Hokom, M. M., et al. (1995). Platelets generated *in vitro* from proplatelet-displaying human megakaryocytes are functional. *Blood, 85,* 402–413.

96. Fujimoto, T. T., Kohata, S., Suzuki, H., et al. (2003). Production of functional platelets by differentiated embryonic stem (ES) cells *in vitro*. *Blood, 102,* 4044–4051.

97. Agam, G., & Livne, A. A. (1992). Erythrocytes with covalently bound fibrinogen as a cellular replacement for the treatment of thrombocytopenia. *Eur J Clin Invest, 22,* 105–112.

98. Levi, M., Friederich, P. W., Middleton, S., et al. (1999). Fibrinogen-coated albumin microcapsules reduce bleeding in severely thrombocytopenic rabbits. *Nat Med, 5,* 107–111.

99. Yen, R. C. K., Ho, T. W. C., & Blajchman, M. A. (1995). A new hemostatic agent: Thrombospheres shorten bleeding time in thrombocytopenic rabbits [Abstract]. *Thromb Haemost, 73,* 986.

100. Takeoka, S., Teramura, Y., Okamura, Y., et al. (2001). Fibrinogen-conjugated albumin polymers and their interaction with platelets under flow conditions. *Biomacromolecules, 2,* 1192–1197.

101. Nishiya, T., Kainoh, M., Murata, M., et al. (2001). Platelet interactions with liposomes carrying recombinant platelet membrane glycoproteins or fibrinogen: Approach to platelet substitutes. *Artif Cells Blood Substit Immobil Biotechnol, 29,* 453–464.

102. Nishiya, T., & Toma, C. (2004). Interaction of platelets with liposomes containing dodecapeptide sequence from fibrinogen. *Thromb Haemost, 91,* 1158–1167.

103. Okamura, Y., Takeoka, S., Teramura, Y., et al. (2005). Hemostatic effects of fibrinogen gamma-chain dodecapeptide-conjugated polymerized albumin particles *in vitro* and *in vivo*. *Transfusion, 45,* 1221–1228.

104. Coller, B. S., Springer, K. T., Beer, J. H., et al. (1992). Thromboerythrocytes. *In vitro* studies of a potential autologous, semi-artificial alternative to platelet transfusions. *J Clin Invest, 89,* 546–555.

105. Takeoka, S., Teramura, Y., Ohkawa, H., et al. (2000). Conjugation of von Willebrand factor-binding domain of platelet glycoprotein Ib alpha to size-controlled albumin microspheres. *Biomacromolecules, 1,* 290–295.

106. Kitaguchi, T., Murata, M., Iijima, K., et al. (1999). Characterization of liposomes carrying von Willebrand factor-binding domain of platelet glycoprotein Ibalpha: A potential substitute for platelet transfusion. *Biochem Biophys Res Commun, 261,* 784–789.

107. Nishiya, T., Murata, M., Handa, M., et al. (2000). Targeting of liposomes carrying recombinant fragments of platelet membrane glycoprotein Ibalpha to immobilized von Wille-

brand factor under flow conditions. *Biochem Biophys Res Commun, 270,* 755–760.

108. Takeoka, S., Teramura, Y., Okamura, Y., et al. (2002). Rolling properties of rGPIbalpha-conjugated phospholipid vesicles with different membrane flexibilities on vWf surface under flow conditions. *Biochem Biophys Res Commun, 296,* 765–770.

109. Rybak, M. E., & Renzulli, L. A. (1993). A liposome based platelet substitute, the plateletsome, with hemostatic efficacy. *Biomater Artif Cells Immobilization Biotechnol, 21,* 101–118.

110. Coller, B. S. (1980). Interaction of normal, thrombasthenic, and Bernard–Soulier platelets with immobilized fibrinogen: Defective platelet–fibrinogen interaction in thrombasthenia. *Blood, 55,* 169–178.

111. Agam, G., & Livne, A. (1983). Passive participation of fixed platelets in aggregation facilitated by covalently bound fibrinogen. *Blood, 61,* 186–191.

112. Agam, G., & Livne, A. (1984). Platelet–platelet recognition during aggregation: Distinct mechanisms determined by the release reaction. *Thromb Haemost, 51,* 145–149.

113. Beer, J. H., Springer, K. T., & Coller, B. S. (1992). Immobilized arg-gly-asp (RGD) peptides of varying lengths as structural probes of the platelet glycoprotein IIb/IIIa receptor. *Blood, 79,* 117–128.

114. Alving, B. M., Reid, T. J., Fratantoni, J. C., et al. (1997). Frozen platelets and platelet substitutes in transfusion medicine. *Transfusion, 37,* 866–876.

115. Perkins, A. C., Frier, M., Hindle, A. J., et al. (1997). Human biodistribution of an ultrasound contrast agent (quantison) by radiolabelling and gamma scintigraphy. *Br J Radiol, 70,* 603–611.

116. Davies, A. R., Judge, H. M., May, J. A., et al. (2002). Interactions of platelets with Synthocytes, a novel platelet substitute. *Platelets, 13,* 197–205.

117. Yen, R. K. C., Ho, T. W. C., & Blajchman, M. A. (1995). A novel approach to correcting the bleeding associated with thrombocytopenia [Abstract]. *Transfusion, 35,* 41S.

118. Sie, P., Gillois, M., Boneu, B., et al. (1980). Reconstitution of liposomes bearing platelet receptors for human von Willebrand factor. *Biochem Biophys Res Commun, 97,* 133–138.

119. Baldassare, J. J., Kahn, R. A., Knipp, M. A., et al. (1985). Reconstruction of platelet proteins into phospholipid vesicles. Functional proteoliposomes. *J Clin Invest, 75,* 35–39.

120. Parise, L. V., & Phillips, D. R. (1985). Platelet membrane glycoprotein IIb-IIIa complex incorporated into phospholipid vesicles. Preparation and morphology. *J Biol Chem, 260,* 1750–1756.

121. Rybak, M. E. (1986). Glycoproteins IIb and IIIa and platelet thrombospondin in a liposome model of platelet aggregation. *Thromb Haemost, 55,* 240–245.

122. Hayhoe, F. G., & Whitby, L. (1955). The management of acute leukaemia in adults. *Br J Haematol, 1,* 1–19.

123. Schulman, I., Currimbhoy, Z., Smith, C. H., et al. (1958). Phosphatides as platelet substitutes in blood coagulation. *Ann N Y Acad Sci, 75,* 195–202.

124. Galan, A. M., Hernandez, M. R., Bozzo, J., et al. (1998). Preparations of synthetic phospholipids promote procoagulant activity on damaged vessels: Studies under flow conditions. *Transfusion, 38,* 1004–1010.

125. Galan, A. M., Casals, E., Estelrich, J., et al. (2002). Possible hemostatic effect of synthetic liposomes in experimental studies under flow conditions. *Haematologica, 87,* 615–623.

126. Giles, A. R., Mann, K. G., & Nesheim, M. E. (1988). A combination of factor Xa and phosphatidylcholine–phosphatidylserine vesicles bypasses factor VIII *in vivo. Br J Haematol, 69,* 491–497.

127. Ni, H. Y., & Giles, A. R. (1992). Normalization of the haemostatic plugs of dogs with haemophilia A (factor VIII deficiency) following the infusion of a combination of factor Xa and phosphatidylcholine/phosphatidylserine vesicles. *Thromb Haemost, 67,* 264–271.

128. Tranholm, M., Røjkjaer, R., Pyke, C., et al. (2003). Recombinant factor VIIa reduces bleeding in severely thrombocytopenic rabbits. *Thromb Res, 109,* 217–223.

129. Woods, M. C., Gamble, F. N., Furth, J., et al. (1953). Control of the postirradiation hemorrhagic state by platelet transfusions. *Blood, 8,* 545–553.

130. Krishnamurti, C., Maglasang, P., & Rothwell, S. W. (1999). Reduction of blood loss by infusion of human platelets in a rabbit kidney injury model. *Transfusion, 39,* 967–974.

131. Vostal, J. G., Reid, T. J., & Mondoro, T. H. (2000). Summary of a workshop on *in vivo* efficacy of transfused platelet components and platelet substitutes. *Transfusion, 40,* 742–750.

132. Heddle, N. M., Cook, R. J., Webert, K. E., et al.; Biomedical Excellence for Safer Transfusion Working Party of the International Society for Blood Transfusion. (2003). Methodologic issues in the use of bleeding as an outcome in transfusion medicine studies. *Transfusion, 43,* 742–752.

# PART EIGHT

# Gene Therapy for Platelet Disorders

# Chapter 71

# Gene Therapy for Platelet Disorders

## David A. Wilcox[1,2] and Gilbert C. White, II[2]

[1]*Department of Pediatrics, Children's Hospital of Wisconsin, Milwaukee, Wisconsin*
[2]*BloodCenter of Wisconsin, Blood Research Institute, Medical College of Wisconsin, Milwaukee, Wisconsin*

## I. Introduction

The current treatment of inherited platelet disorders such as Glanzmann thrombasthenia and Bernard–Soulier syndrome (see Chapter 57) consists of platelet transfusions, with supplemental use of antifibrinolytic agents, local hemostatic agents, and possibly recombinant factor VIIa (see Chapter 68) or desmopressin (see Chapter 67). Although these remedies may be highly effective and even life saving, they can be very expensive and use of platelet transfusions is often associated with the development of an immune response to proteins present on the surface of the transfused platelets but deficient on the surface of the patient's platelets: glycoprotein (GP) IIb-IIIa (integrin $\alpha$IIb$\beta$3) in Glanzmann thrombasthenia and GPIb-IX-V in Bernard–Soulier syndrome. Subsequent to this alloimmunization, platelet transfusions are of limited benefit. For such patients, there may be no reproducibly effective treatment. Developments in gene transfer technology suggest that this could be a useful approach for patients with Glanzmann thrombasthenia, Bernard–Soulier syndrome, or other platelet defects. In this chapter, we review the current state of gene therapy as a potential treatment for inherited platelet bleeding disorders.

There have been a series of crucial advances within the past 20 years that have increased the likelihood for genetic transfer to emerge as a viable approach for treating individuals with inherited platelet defects. The application of reverse-transcriptase polymerase chain reaction techniques to platelet RNA in 1988 facilitated the characterization of molecular defects responsible for several inherited bleeding disorders.[1,2] Advances in the procurement of hematopoietic stem cell populations for bone marrow transplantation also led to important implications for gene transfer approaches. Although the quest for unique markers for stem cells continues, the observation that CD34[+]/CD38[-] peripheral blood and bone marrow cells are enriched in long-term repopulating cells enables strategies to pursue these cells for gene transfer. The identification in 1994 of the long-sought thrombopoietin as the ligand for c-mpl was a remarkable event[3] (see Chapter 66). This and the identification of other cytokines and growth factors responsible for megakaryocyte differentiation, including interleukin (IL)-11, IL-3, GATA1 and friend of GATA (FOG), nuclear factor-erythroid 2 (NF-E2), megakaryocyte endoreduplication inducing factor (MERIF), and others (see Chapters 2 and 66), have increased our understanding of megakaryocytopoiesis and permitted growth and differentiation of megakaryocytes *in vitro*. Use of the c-mpl ligand in tissue culture resulted in production of sufficient numbers of megakaryocytes to detect transgene expression and examine the function of corrected cells.[4] Finally, there have been significant developments in gene transfer technology, including better and safer vectors, more efficient vector production resulting in higher titers, the identification of cell surface receptors for viruses, and the use of promoters to direct tissue-specific transgene expression.

A number of inherited disorders affecting platelets may be amenable to gene therapy strategies (Table 71-1).[5-21] Theoretically, a disorder that can be corrected with bone marrow transplantation has the potential to also be corrected by gene therapy. Bone marrow transplantation has been successfully used to treat genetic diseases affecting hematopoietic cells.[22] To date, there have been seven cases of Glanzmann thrombasthenia reported to have been corrected with bone marrow transplantation.[23-27] However, the risks of graft rejection, graft-versus-host disease, and radiation-related toxicities reduce the overall potential benefits of using an allogeneic transplant to correct a patient's bleeding disorder.

**Table 71-1: Inherited Disorders of Platelet Function**

| Inherited Disorder | Defect | Function Disrupted | References |
|---|---|---|---|
| G-protein disorder | $G_{\alpha q}$, $G_{\alpha i1}$ | Activation | 5–7 |
| ADP receptor defect | $P2Y_{12}$ | Activation | 8 |
| Bernard–Soulier syndrome | Glycoproteins Ib–IX | Adhesion | 9 |
| Collagen receptor deficiency | Glycoproteins Ia–IIa | Adhesion | 10 |
| Glanzmann thrombasthenia | Glycoproteins IIb–IIIa | Aggregation | 11 |
| Gray platelet syndrome | $\alpha$-Granules | $\alpha$-Granule formation/storage | 12 |
| Quebec platelet disorder | $\alpha$-Granules | $\alpha$-Granule storage | 13 |
| Scott syndrome | Phosphatidylserine translocation | Coagulation | 14 |
| May–Hegglin anomaly | MYH9 | Cytoskeleton/platelet formation | 15–17 |
| Fechtner syndrome | | | |
| Sebastian platelet syndrome | | | |
| Epstein syndrome | | | |
| Wiskott–Aldrich syndrome | WAS protein | Cytoskeleton | 18 |
| Chediak–Higashi syndrome | CHS protein | Dense body formation/storage | 19 |
| Hermansky–Pudlak syndrome | HPS1, HPS3-7, AP-3 | Dense body formation/storage | 20 |
| Thromboxane deficiency | Thromboxane $A_2$ | Signal transduction | 21 |

## II. Approach to Gene Transfer for Platelet Disorders

The goal of gene therapy for platelet disorders is to achieve lifelong correction of the phenotypic defect. Treating monogenic disorders of other tissues may be accomplished by targeting the affected cells in the tissue, but for terminally differentiated bone marrow cells such as platelets, long-term correction would most likely be carried out by targeting a precursor cell that has the ability to both self-replicate and differentiate. This must be a self-replicating stem cell to ensure persistence of the gene, and the cell must be able to differentiate normally down a megakaryocytic and platelet pathway to generate the corrected phenotype. One of the problems with gene therapy of hematopoietic disorders is the current inability to precisely define such self-replicating, pluripotent cells. Such cells are present in $CD34^+/CD38^-$ cell preparations, but not all $CD34^+/CD38^-$ cells are true self-replicating, pluripotent hematopoietic progenitor cells.

If targeting pluripotent progenitor cells is one goal of gene therapy for platelet disorders, than another aim should be to direct transgene expression with gene promoters of megakaryocyte-specific proteins since premature expression of some proteins in progenitor cells might affect the function of those cells and alter their ability to self-replicate or differentiate. For example, the expression of GPIIb-IIIa in progenitor cells might affect their adhesive properties in a way that could perturb their function or generate a signal that might alter function. The possibility of this event occurring indicates that transgene expression of platelet-specific proteins should be controlled, ideally in a manner that is similar to the way the normal gene is regulated. Although the promoters for genes such as GPIIb have been identified and partially characterized, a better understanding of these regulatory elements is needed to ensure proper gene regulation.

Based on the preceding principles, a potential strategy for gene therapy of platelet disorders is illustrated in Fig. 71-1. Hematopoietic progenitor cells in the $CD34^+/CD38^-$ fraction of cells obtained by peripheral blood leukopheresis from the patient are placed in culture under conditions to maintain the cells in an undifferentiated state. While in culture, the corrective gene under control of a platelet-specific promoter/enhancer element (e.g., GPIIb promoter-driven gene) is introduced into the cells. The modified progenitor cells are then reintroduced into the patient by peripheral blood injection and they home to the bone marrow, where they set up residence as a self-renewing population of cells. The modified progenitor cells may also be maintained in culture and induced to differentiate *in vitro* to examine gene expression, transduction efficiency, and other properties.

### A. Hematopoietic Stem Cells as Targets

Stem cells of hematopoietic origin have the ability to generate cells of all hematopoietic and lymphoid lineages (plu-

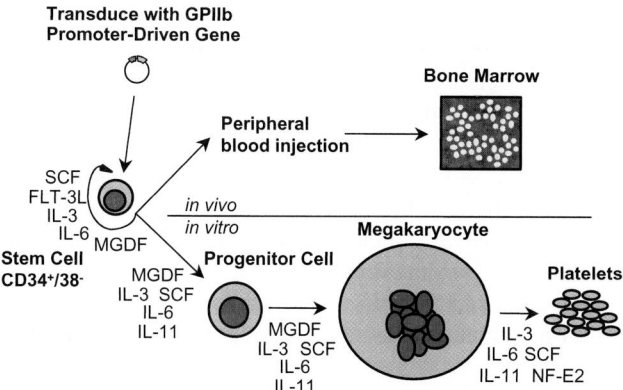

**Figure 71-1.** Gene therapy strategy for platelet disorders. Hematopoietic progenitor cells contained in the CD34⁺/CD38⁻ fraction of cells obtained by peripheral bleed leukopheresis from the patient are placed in culture under conditions to maintain the cells in an undifferentiated state. While in culture, the corrective gene under control of a platelet-specific promoter/enhancer element (e.g., GPIIb promoter-driven gene) is introduced into the cells. The modified progenitor cells are then reintroduced into the patient by peripheral blood injection, and they home to the bone marrow, where they set up residence as a self-renewing population of cells. The modified progenitor cells may also be maintained in culture and induced to differentiate *in vitro* to examine gene expression, transduction efficiency, and other properties. FLT-3L, flt-3 ligand; IL, interleukin; MGDF, megakaryocyte growth and development factor; NF-E2, nuclear factor-erythroid 2; SCF, stem cell factor.

ripotent) in a manner that results in the production of cells for an individual's lifetime.[28] The use of hematopoietic stem cells as targets for gene therapy is facilitated by techniques developed for bone marrow transplantation to safely remove these cells from the body, maintain the cells in tissue culture, and then reinfuse the cells into the patient.[29–31] The progeny of transduced cells then have the potential to express a stably incorporated transgene as treatment to correct the disorder.

For a number of reasons, hematopoietic cells have proved to be an elusive target for gene therapy strategies.[30] First, hematopoietic stem cells divide infrequently and therefore are not transduced efficiently with vectors that require cell division for stable incorporation of proviral DNA into the genome. One strategy is focused on mobilizing cells from a quiescent to an activated state by preconditioning the cells with cytokines. Using this approach, Heim and Dunbar have been able to detect persistent *in vivo* gene transfer efficiency in 10% of hematopoietic cells from primate transplant recipients.[32] Another method to efficiently transduce quiescent stem cells is to use a vector that does not require the cell to divide (e.g., lentivirus).[33] Second, receptors for viral vectors may not be present on stem cells. As a result, strategies are in progress to pseudotype viruses with coat proteins that are recognized on the surface of stem cells.[34]

Embryonic and other pluripotent stem cells, umbilical cord blood, bone marrow, and bone marrow cells mobilized by growth factors into the peripheral blood can each serve as a source of hematopoietic stem cells.[30] Although embryonic stem cells and cord blood are a rich source of hematopoietic pluripotent cells, neither is likely to be a source of target cells for current trials in gene therapy of platelet disorders. Bone marrow has been successfully modified with lentivirus vectors, leading to correction of the murine model for Glanzmann thrombasthenia.[35] However, bone marrow collection may be too invasive for patients with hemorrhagic platelet defects. There is abundant evidence that cells with pluripotent capacity are present in peripheral blood and can be obtained using apheresis. Apheresis of peripheral blood containing cytokine-mobilized stem cells has been successfully performed in thrombasthenic individuals and appears to be a safe and efficient means to obtain stem cells from individuals at risk of bleeding as a result of their platelet disorder.[36]

Clinical studies with human cancer patients demonstrated that the number of stem cells obtained from peripheral blood can be increased substantially by the use of growth factors.[37] Stem cell factor (SCF) and granulocyte colony-stimulating factor (G-CSF) are typically used to increase the number of long-term repopulating cells harvested from these individuals.[38,39] SCF and/or G-CSF are routinely administered for 5 days prior to apheresis. The mobilized population contains cells that are positive for the CD34 antigen,[40] and these cells have been shown to engraft and recapitulate the stem cell compartment of lethally irradiated marrow recipients.[41] When commercial kits and antibodies became readily available for selection of CD34⁺ cells from bone marrow,[42,43] this technology was quickly adapted to protocols for targeting gene transfer into hematopoietic cells.[44] Typically, one CD34⁺ cell is obtained from 10⁴ or 10⁵ nucleated marrow cells. Thus, enrichment for the CD34⁺ population ultimately allows conservation of viral supernatant. CD34 and CD38 are expressed in a reciprocal manner, with the CD34⁺/CD38⁻ cell thought to be an activated subset of progenitor cells that result from cytokine-induced mobilization of quiescent CD34⁻/CD38⁺ marrow.[38] Efforts to further define the more primitive cell types have resulted in identification of subfractions of murine bone marrow following further selection of CD34⁺/CD38⁻ cells with additional surface markers (cKit⁺, Sca-1⁺, Lin⁻, Thy-1^low).[45,46] Cells of this distinction have been used to repopulate the murine stem cell compartment, although identification of a true stem cell becomes complicated since cells from the CD34⁻ fraction selected with identical subclass markers also possess pluripotent ability.[45,46] Goodell et al.[47] demonstrated the presence of stem cells in a small subset of the CD34⁻ cell fraction that stain differentially with Hoechst 33342 dye (termed side population cells) in mice, nonhuman primates, and humans. This CD34⁻ population can convert to CD34⁺

cells following long-term culture, indicating the presence of a more primitive and uncommitted cell type.[48] Since none of these fractions represents a pure population of stem cells, the search for a distinct marker that is specific for stem cells continues.[49]

It is important to maintain CD34[+] cells obtained by any of the proposed techniques as pluripotent cells during the *in vitro* culture and transduction with the gene transfer vector. Differentiation of the cells during this time would result in loss of their ability to self-replicate and their pluripotent capacity. There are numerous recipes for culture media and combinations of growth factors to induce proliferation and cycling of progenitor cells during the transduction period.[30] Generally, these include IL-6, FLT-3 ligand, SCF, and pegylated-recombinant human megakaryocyte growth and development factor (MGDF) (Fig. 71-1). Variations in growth factors used to culture cells for gene transduction protocols are consistent with the choice of viral vector used to transduce the cells. For example, protocols employing oncoretrovirus vectors use high concentrations of growth factors because the target cell must divide for stable integration of the transgene, whereas strategies using other vectors (lentivirus and adeno-associated virus) do not require cell division for stable transgene integration and therefore typically use lower concentrations of growth factors.[30]

Current protocols for gene transfer usually consist of 48 hours of prestimulation of CD34[+] cells in media containing cytokines followed by 18 to 24 hours of transduction of cells immobilized to tissue culture plates coated with a fragment of fibronectin (CH296, Retronectin).[50–52] This method seems to work as well as previous protocols that incubated target cells in plates containing a monolayer of virus-producing fibroblasts. Use of the fibronectin fragment prevents contamination of the human hematopoietic cells with cells that produce virus, and it is therefore preferred for protocols requiring transplantation of transduced cells.[53]

### B. Preparation of the Recipient's Marrow

For some bone marrow disorders, such as Fanconi anemia, it is important to create a population of normal hematopoietic progenitor cells to correct the pancytopenia, but it is also important to remove the abnormal hematopoietic cells that may predispose to malignant myeloid transformation. With most platelet disorders, the presence of the abnormal platelets does not confer any long-term problems other than the bleeding disorder. Thus, the abnormal marrow elements need not necessarily be removed. Using Glanzmann thrombasthenia as an example, hemostasis was restored in GPIIb-deficient dogs producing 20 to 30% normal circulating peripheral blood platelets derived from a nonmyeloablative "mini" transplantation of bone marrow obtained from DLA-matched littermates that were heterozygous for Glanzmann

thrombasthenia.[54] This outcome is consistent with previous work indicating that the thrombasthenic defect may be obviated at moderate transduction efficiencies since measurable aggregation was detected *in vitro* with a mixture of only 10% normal human platelets and 90% thrombasthenic platelets.[36] It has yet to be demonstrated whether 10% corrected platelets will be sufficient to support normal platelet aggregation *in vivo,* although this may be tested in the near future since ongoing trials in large animal models have attained transduction efficiencies of 10 to 15% with marker genes in hematopoietic cells for more than 6 months after transplant.[52]

There is evidence that the level of transgene expression, as well as transduction efficiency, is essential for maintaining hemostasis. Although it has been known for quite some time that genetic carriers for thrombasthenia typically express only 50% of normal levels of GPIIb-IIIa on the platelet surface, this is sufficient for normal platelet aggregation and bleeding times. Mascelli et al.[55] showed that platelet aggregation was diminished when more than 80% of GPIIb-IIIa surface receptors were occupied with an antibody (abciximab) that recognizes GPIIIa (integrin β3) in complex with GPIIb (integrin αIIb) or integrin αv. This result is consistent with observations of Fang et al.[35] demonstrating that bleeding times could be partially corrected in GPIIIa-deficient mice (the murine model for Glanzmann thrombasthenia) when platelets expressed GPIIb-IIIa at a moderate receptor density (7–11% of normal levels) following transplantation of bone marrow transduced with a lentivirus transfer vector. Since previous studies showed that retrovirus transduction of human megakaryocytes from two Glanzmann thrombasthenia patients restored GPIIb-IIIa to the cell surface at 34% of normal receptor levels, it is likely that genetic therapy for humans with Glanzmann thrombasthenia may also lead to improved bleeding times.[36]

Complete myeloablation of the recipient prior to autologous transplant could be performed to create a niche for the transduced marrow and allow the highest possible transduction efficiencies as a result of total replenishment of the stem cell pool with genetically altered marrow. However, since platelet disorders are often not lethal in nature and only the presence of a fraction of normal functioning platelets is required to induce measurable aggregation of human platelets *in vitro* (10%)[36] and canine platelets *in vivo* (20–30%),[54] it may be practical to adopt nonmyeloablative mini-transplantation techniques to correct platelet disorders with a chimeric graft of genetically altered marrow, thereby reducing the risk of side effects of the transplantation protocols (e.g., graft-versus–host disease, graft rejection, and radiation-induced toxicity).[56–58] Current mini-protocols include a short posttransplant delivery of immunosuppressive drugs, which has been shown to result in the creation of a mixed hematopoietic chimerism, possibly by initiating a tolerance for allogeneic marrow in recipients. Results from studies using allogeneic mini-transplant indicate that these techniques may be useful for autologous transplant of genetically modified marrow, with the aim of creating conditions for engraftment and eventual immune tolerance for transduced cells. This may be necessary because even though the marrow is autologously derived, there is a potential for an immune response since the cells would be genetically altered in a way that progeny megakaryocyte and platelets express a new receptor on the platelet surface (e.g., GPIIb-IIIa in Glanzmann thrombasthenia). From the standpoint of avoiding an immune response, the best candidates for gene therapy of thrombasthenia are likely to be individuals that express missense mutations in the extracellular domain or defects within the cytoplasmic domains of GPIIb-IIIa (see Chapter 57). The immune system of these individuals may be more likely to tolerate an alteration of integrin structure than appearance of an entirely new molecule on the surface of the platelets as would occur in patients with type I Glanzmann thrombasthenia.[59] However, the occurrence of platelet alloantigens from single amino acid changes demonstrates that minor differences do not ensure tolerance of a molecule, especially in the case of GPIIb-IIIa.[60]

## C. Analysis of the Correction of the Phenotype for Platelet Disorders

The development of techniques to culture megakaryocytes has greatly facilitated the ex viro analysis of phenotypic correction of platelet defects.[3] Although adequate numbers of platelets can be difficult to obtain *in vitro* for analysis of

function,[4] it is possible to activate megakaryocytes in culture with physiological agonists and study the megakaryocyte responses as representative of platelet responses.[36,61,62] Cellular adhesion and aggregation, binding of soluble fibrinogen and fibrinogen-mimetics, retraction of a fibrin clot, and the integrity of signaling pathways can all be examined. Studies using cells from thrombasthenic patients have shown that transduced cultured megakaryocytes may be used to demonstrate correction of defective phenotypes. In addition, pre- and posttransduction testing of animals subjected to gene therapy protocols can be accomplished with standard techniques for mucosal, cuticle, and tail bleeding time analysis.[35,54,63,64]

## D. Vectors

Table 71-2 lists some of the viral vectors currently being used for gene transfer studies and their advantages and disadvantages. In 1999, three groups demonstrated efficient transgene expression in megakaryocytes following transduction of cultured primary hematopoietic cells with vectors derived from adenovirus,[61] Sindbis virus,[62,65] and retrovirus.[66] Sindbis and adenovirus are suitable vectors for *in vitro* studies because they can generate high-level protein expression. However, these vectors have limited applications for gene therapy strategies since they have significant side effects and do not incorporate genetic material into the cellular genome, resulting in a limited life span of transgene expression (Table 71-2). Retroviral vectors are commonly used for gene therapy protocols since the genetic material

## Table 71-2: Viral Vectors Used for Transgene Expression in Megakaryocytes

| Virus | Advantages | Disadvantages |
|---|---|---|
| Adenovirus | High titer<br>High gene expression<br>Can infect nondividing cells<br>Accepts very large cassettes (40 kb) | Immunogenic<br>Does not integrate into genome |
| Adeno-associated virus | Can infect nondividing cells<br>Relatively safe in humans | Accepts small cassettes (4 kb)<br>Low transduction efficiency in hematopoietic cells |
| Alphavirus (Sindbis) | Can infect nondividing cells<br>High titer<br>High transduction efficiency<br>High gene expression | Toxic to cells<br>Does not integrate into genome |
| Lentivirus | Stably incorporated into genome<br>Can infect nondividing cells | New to field<br>Safety uncertain in humans |
| Retrovirus | Stably incorporated into genome<br>Relatively safe in humans<br>High titer<br>Accepts large cassettes (8 kb) | Infects only dividing cells |

can be stably integrated into the host cell's genome with the potential to produce sustained transgene expression in lineages derived from transduced cells.[67] Applications using retroviral vectors have shown potential usefulness for human gene therapy of platelet disorders. Functional correction of the phenotype for Glanzmann thrombasthenia was demonstrated following retroviral transduction and differentiation of CD34[+] cells to megakaryocytes *in vitro* from two patients with defects in the GPIIIa gene.[36] Retrovirus-based vectors are likely to continue to be improved and utilized for platelet gene therapy protocols since these constructs mediated the first successful human gene therapy trials in individuals with a hematological disorder (X-linked severe combined immunodeficiency [X-SCIDS]).[44] Since retrovirus incorporates randomly into the host genome, use of this construct places the patient at risk for insertional mutagenesis. For example, 3 of 10 patients treated with gene therapy for X-SCIDS developed leukemia as a result of mutagenesis of the patient's genome.[68] In two of the patients, the vector inserted into and activated a T-cell protooncogene (LMO2).[69] X-SCID may be particularly susceptible to leukemogenesis since correction of the gamma (c) gene defect confers a survival advantage to the transduced cells; this, plus alterations in genes such as LMO-2 that control cell division, may result in uncontrolled growth of cells that have a survival advantage.[69,70] Another limitation of oncoretrovirus vectors is that they only integrate into cycling cells, and a majority of normal hematopoietic stem cells cycle very slowly (30 days/cycle on average). Since few stem cells are transduced with this system, it is not surprising that the first success within gene therapy clinical trials was for a disorder in which the transgene product provided a growth advantage for transduced hematopoietic stem cells.[71] Lentiviruses (e.g., HIV-1 and EIAV) are unique retroviruses that in the future may become more useful for gene therapy of platelet disorders because they can transduce nondividing cells.[72,73] For example, the bleeding phenotype of the murine model for Glanzmann thrombasthenia was corrected following genetic transfer with a lentivirus vector into the bone marrow of GPIIIa-deficient animals.[35] As an alternative, improvements are also being reported in the efficiency of transduction of hematopoietic cells using nonviral techniques (electroporation, liposomes, and artificial chromosomes), which makes it likely that there will be many choices in the future for vehicles to transfer genetic materials leading to transgene expression in megakaryocytes.[31]

## III. Advances in Retrovirus-Mediated Transduction for Platelets

To obtain sustained transgene expression in humans, the ideal vector for gene therapy of platelet disorders should allow megakaryocyte-specific transgene expression, which would preferentially be maintained at levels equal to those of the normal endogenous gene for the lifetime of the recipient. Improvements in retrovirus vectors to achieve this goal are outlined here.

### A. Promoters and Locus Control Regions for Megakaryocyte-Specific Transgene Expression

For genetic correction of platelet disorders, high-level, megakaryocyte-specific transgene expression should be obtained from integration of a single copy of proviral DNA. Transcription from the retroviral long terminal repeat (LTR) is dependent on the viral promoter–enhancer elements located in the LTR. Retroviral LTR enhancers often deregulate lineage-specific promoters. Therefore, self-inactivating (SIN) vectors have been utilized for megakaryocyte-specific gene expression because these vectors delete the three viral enhancers and allow gene transcription to be promoted from inserted tissue-specific promoters.[35,36,74,75]

There are several megakaryocyte-specific gene promoters that could potentially direct transgene transcription, including members of the GPIb-IX-V complex,[76] integrin αIIb (GPIIb),[77] GPVI,[78] c-mpl,[79] and platelet factor 4 (PF4).[80] These promoters bind GATA-1, Ets (Fli-1), and FOG-1 factors that induce transcription in early and midstages of megakaryocytopoiesis.[81,82] For example, PF4 is expressed at high levels during megakaryocytopoiesis and stored in platelet α-granules.[80] Its gene promoter[83] and distal regulatory regions[84] are well characterized and have been shown to be useful for controlling megakaryocyte-specific transgene expression. Nguyen et al.[85] demonstrated that the PF4 gene promoter sequence can be modified with an "on/off" switch using the tetracycline/doxycycline (Tet)–off system for conditional overexpression of a transgene in murine megakaryocytes and platelets.

A locus control region may be incorporated into genetic transfer vectors used for platelet gene therapy because the region enhances gene expression, but it does not have classic enhancer activity since it does not function equally well in either orientation.[86] These elements may help stabilize transgene expression derived from proviral constructs in much the same manner as locus control regions have been utilized for β-globin constructs for erythrocyte-specific expression to correct thalassemia.[87] Provirus randomly integrates into the genome and therefore has been observed to suffer from chromatin rearrangements that occur during stem cell differentiation, leading to silencing of transgene expression in progeny lymphocytes and myocytes *in vivo*.[88] Genes that are actively transcribed are located in "open" chromatin containing highly acetylated nucleosomes that are more accessible to *trans*-acting factors. In contrast, "closed" chromatin generally has deacetylated histones that make the DNA less accessible to binding factors, and genes in these regions are therefore not expressed.[89] Advances in the use of locus control elements for improving gene expression indicate that these sequences may be keeping chromatin open following hematopoietic stem cell transduction, as evidenced by work with globin gene expression in erythroid progeny of transduced hematopoietic stem cells to treat sickle cell disease and thalassemia.[87,89,90]

The human GPIIb gene promoter was utilized in retroviral constructs to direct transgene expression in megakaryocytes of animals and humans since this promoter was previously observed to drive "species-independent," tissue-specific transgene expression in murine megakaryocytes.[91] Expression vectors for megakaryocyte-targeted transgene expression have also been constructed from human GPIIb promoter sequences that produce detectable levels of GPIIb-IIIa on the cell surface of megakaryocytes of individuals with Glanzmann thrombasthenia.[36] This promoter binds GATA and Ets factors, which induce a high level of gene transcription that is restricted to developing megakaryocytes due to a repressor element localized to the immediate 5 upstream region of the GPIIb gene. A fragment (900 base pairs) of the human GPIIb gene's promoter sequence (1.2 kb) is sufficient to direct transgene expression in megakaryocytes and platelets.[66,92] Results of *in vitro* studies demonstrate that the human GPIIb promoter-controlled retrovirus SIN vector can also target transgene synthesis preferentially in megakaryocytes in humans and mice with Glanzmann thrombasthenia.[35,36] The GPIIb promoter fragment directed transgene expression that resulted in 34% of normal receptor density levels for GPIIIa on human thrombasthenic megakaryocytes, leading to appreciable retraction of a fibrin clot and binding of megakaryocytes to fibrinogen mimetics in *in vitro* assays. Individuals who are heterozygous for thrombasthenia are disease-free with 50% of normal levels of GPIIb-IIIa on the platelet surface,[36] demonstrating that normal receptor density is not required for adequate platelet function. This suggests that the GPIIb promoter-driven construct could potentially produce therapeutic benefit for correction the thrombasthenic phenotype *in vivo*. A 3′ distal regulatory element of the integrin $\alpha_{IIb}$ gene was discovered that is highly conserved between humans and mice. This element has been shown to enhance transgene expression 1000-fold, leading to mRNA levels that were approximately 30% of native levels within transfected human megakaryocyte cell lines.[93] Future studies will likely investigate whether this distal regulatory element could increase the level of transgene expression within megakaryocytes modified with a genetic transfer vector under the control of the GPIIb gene promoter.

### B. Prevention of Silencing of the Transgene

Despite early optimism concerning the transfer and expression of a variety of gene sequences in hematopoietic cells, progress in obtaining stable and long-term expression of a transgene in progeny of human hematopoietic stem cells has been slow, with only limited success. Surprisingly, inclusion of the appropriate regulatory elements may not be sufficient to obtain therapeutic levels during gene therapy.[87] Retrovirus-transduced genes are often silenced in hematopoietic stem cells.[67,88] It has been demonstrated that retrovirus and lentivirus sequences are silenced in mice because these vectors have elements that may be recognized by specific factors, leading to methylation and decreased gene transcription in mammalian stem cells.[94] The silencers tend to be located in viral LTRs. Therefore, utilization of SIN vectors, which delete viral enhancer sequences in the LTRs and rely on endogenous promoters for transgene expression, has resulted in a reduction but not obliteration of gene silencing. To further reduce transgene silencing, insulator elements have been incorporated to protect transgenes from "bystander effects" as a result of silencing of nearby viral elements. Chromatin insulators are protein-binding DNA elements that lack intrinsic promoter/enhancer activity but shelter genes from the transcriptional influence of surrounding chromatin. Silencing of transgene expression has not been examined in megakaryocytes. It is likely that some form of chromatin insulator would be helpful for sustained transgene expression in megakaryocytes, similar to transgene expression in erythrocyte progenitors.[87,89]

### C. Tolerance for the Transgene Product

Development of tolerance for proviral-derived surface antigen is likely to be critical for sustained expression of the transgene. There may be differences in the mechanisms and requirements for induction of immune tolerance to a single neo-antigen (as in a gene therapy setting) compared to the allogeneic transplant where donor-versus-host and host-versus-donor phenomena coexist.[95] Although those differences are not well characterized, Rosenzweig et al.[56] observed that a neo-antigen (murine CD24) could be tolerated on the surface of monkey hematopoietic cells, as evidenced by sustained expression in three out of four animals following autologous transplantation of transduced marrow into animals conditioned with nonmyeloablative irradiation. The murine CD24 antigen was detected 6 months posttransplant on the surface of 5 to 10% of circulating blood cells.[56] Since the megakaryocyte-specific promoters are not normally activated in self-renewing hematopoietic stem cells, transgene products under the promoter's control are presented to the immune system at the stages of megakaryocyte differentiation, and therefore recognition and rejection of genetically modified, "parental" stem cells by the immune system should not occur.

### D. Viral Envelope Proteins

The choice of envelope protein (pseudotype) displayed on the virus vehicle can have a significant effect on the efficiency of gene transfer. Initial gene therapy protocols utilized murine retrovirus vectors packaged in a murine amphotropic envelope to transduce hematopoietic stem cells because the envelope's cellular phosphate receptor is present on murine, primate, and human cells.[96] The density of this receptor on the cell surface correlates with the transduction efficiency of the target cells. Therefore, viral vectors are currently being pseudotyped with alternative envelope proteins that recognize a high number of receptors on the human CD34$^+$/CD38$^-$ target cells since these cells normally express a low level of receptor for the amphotropic envelope protein. Significant improvement in transduction efficiencies has been observed with envelope proteins derived from gibbon ape leukemia virus[97] and feline endogenous retrovirus (RD114).[34,98] Envelope protein from the vesicular stomatitis virus (VSV-G) (which transduces cells with comparable efficiency to amphotropic envelope)[99] has found widespread acceptance due to its versatility, through direct membrane fusion by recognizing glycolipid on the surface of virtually any cell from any species, and as a result of its ability to be concentrated by centrifugation.[100]

## IV. State-of-the-Art Animal Models for Disorders Affecting Platelets

A few animal models have been established to examine the molecular basis of inherited platelet bleeding disorders.[101] These models also appear to be ideally suited to test all aspects of gene therapy, such as the efficacy and persistence of gene transfer and expression, as well as safety.

Hermansky–Pudlak syndrome (HPS; see Chapters 15 and 57) is a disorder affecting platelet granule storage pools that has been attributed to at least 15 separate genetic defects in murine models.[102] Defects in three organelles (lysosomes, melanosomes, and platelet dense bodies) result in lysosome abnormalities, albinism, and prolonged bleeding in HPS.[20] The organelles are defective as a result of abnormalities in proteins that play a role in vesicle formation, storage, and trafficking (e.g., HPS1, components of the AP-3 complex, and HPS3).[20,103,104]

A murine model for Bernard–Soulier syndrome has been developed that presents as mild thrombocytopenia, circulating giant platelets, and a bleeding phenotype associated with molecular genetic defects in the GPIbα subunit of the GPIb-IX-V complex.[105] Since the diseased phenotype was corrected when the human GPIbα transgene was knocked into these mice,[105] this model should prove useful for investigations to evaluate vectors that are being developed for genetic correction of Bernard–Soulier syndrome in humans.[75]

One murine model for Glanzmann thrombasthenia was developed when Tronik-Le Roux et al.[106] knocked in the herpes virus thymidine kinase gene into the GPIIb locus

resulting in the knockout of the GPIIb gene. Another model was produced when Hodivala-Dilke et al.[107] knocked out the gene encoding GPIIIa. These mice exhibit a condition that is essentially identical to the Glanzmann thrombasthenia phenotype in humans: Defective platelet function leads to prolonged bleeding times compared to those of normal and heterozygous animals. Platelets isolated from these mice are unable to retract a fibrin clot, whereas normal and heterozygous mice retract a clot appreciably.[107] GPIIIa-deficient mice also display abnormalities in placental development,[107] osteosclerosis,[108] and increased tumor hypervascularization[109] and growth,[110] thus highlighting a vital role for integrin αvβ3 (αv-GPIIIa complex) in these processes.[111]

Glanzmann thrombasthenia resulting from a naturally occurring defect in GPIIb in a Great Pyrenees dog was reported by Boudreaux et al.[112] The dog exhibited classical symptoms of thrombasthenia, including excessive gingival bleeding upon shedding of teeth. Western blot analysis of normal and thrombasthenic canine $^{125}$I-labeled platelet surface proteins showed a total absence of GPIIb and GPIIIa.[112] Analysis of the cDNA sequence from the affected dog identified a molecular genetic defect in GPIIb DNA consisting of an insertion of 13 nucleotides in exon 13 of GPIIb.[113] Boudreaux et al.[114] also detected a point mutation in Otterhounds within the third calcium-binding domain of GPIIb associated with type I Glanzmann thrombasthenia. These are the first reported large animals with well-characterized molecular defects resulting in thrombasthenia. The colony is currently being utilized as a large animal model for genetic correction of this disorder.[115,116]

### A. Results of Gene Therapy for GPIIIa-Deficient Mice

A retrovirus vector encoding human GPIIIa under the transcriptional control of the human GPIIb promoter was transduced into bone marrow cells from GPIIIa-null mice to determine the feasibility of correcting thrombasthenia due to defects in GPIIIa.[35] In that study, murine GPIIb associated with human GPIIIa to form a functional hybrid complex on the surface of GPIIIa-transduced murine megakaryocytes and derivative circulating blood platelets. This result was most likely due to the high degree of amino acid sequence homology between human GPIIb and GPIIIa and murine GPIIb and GPIIIa (78 and 92%, respectively).[117,118] Integrin function (inside-out and outside-in signaling), as well as receptor-mediated storage of fibrinogen, was restored in transduced platelets. In addition, GPIIIa-transduced platelets mediated aggregation similar to platelets from normal mice, whereas untransduced thrombasthenic platelets were unable to aggregate. This resulted in reduced bleeding times for these animals, thus demonstrating correction of the Glanzmann thrombasthenia phenotype.[35]

### B. Xeno-Animal Systems to Examine the Basis and Correction of Platelet Disorders

Most platelet disorders have not been redefined in an animal model. This has not inhibited the development of strategies to examine sustained correction of hematopoietic disorders *in vivo* since immune-deficient animal models such as the nonobese diabetic severe combined immunodeficient (NOD/ SCID) mice allow xenografting of human hematopoietic cells for study of the efficiency and effects of various transduction conditions of these cells.[119–121] NOD/SCID mice provide an ideal *in vivo* condition for transplanted human hematopoietic stem cells, which are supported in the microenvironment of the bone marrow. To show efficacy for this concept, we induced the synthesis of human factor VIII in human platelets produced in NOD/SCID mice. Factor VIII colocalization with von Willebrand factor was observed in the platelet α-granules 3 weeks following transplantation of transduced human CD34⁺ cells.[122] Since platelets do not normally synthesize factor VIII, this result suggests that peripheral blood CD34⁺ cells from patients with other known defects in platelet proteins could be transduced, transplanted, and examined for functional correction of those disorders. Further support for use of platelets as vehicles to deliver therapeutic agents to the site of a vascular injury was provided in recent studies showing factor VIII-deficient mice expressing human factor VIII as potential treatment for hemophilia A (123, 124). In addition, other proteins with therapeutic value could potentially be synthesized and stored in platelets to correct disorders accessible to platelets (e.g., cancer and heart disease).

## V. Conclusions

Encouraging results have been reported from studies of gene transfer in large animals. Using improvements in vectors and transduction conditions, investigators have been consistently obtaining 1 to 10% gene marking of all hematopoietic and lymphoid lineages after retrovirus-mediated transduction of monkey and canine hematopoietic stem cells.[56,97] The concept of correcting platelet and other disorders of the hematopoietic system has been slowly evolving during the past 15 years. Studies for developing methods for gene therapy of platelet disorders have begun at an ideal time since transduction efficiencies have increased to levels that could result in therapeutic benefit for treated patients.

The development of a lineage-targeted therapy for well-characterized, rare disorders, in addition to its intrinsic benefit, has the potential to serve as a model for the correction of a variety of inherited hematological diseases. Although the benefits of this approach are of clinical relevance, there are basic scientific principles that address both the cell and molecular biology of platelet integrins, as well as the potential for lineage-specific gene expression as a scientifically practical and achievable outcome, that can be extrapolated to other platelet and nonplatelet diseases.

## References

1. Newman, P. J., Gorski, J., White, G. C., 2nd, et al. (1988). Enzymatic amplification of platelet-specific messenger RNA using the polymerase chain reaction. *J Clin Invest, 82,* 739–743.
2. Peretz, H., Seligsohn, U., Zwang, E., et al. (1991). Detection of the Glanzmann's thrombasthenia mutations in Arab and Iraqi–Jewish patients by polymerase chain reaction and restriction analysis of blood or urine samples. *Thromb Haemost, 66,* 500–504.
3. Bartley, T. D., Bogenberger, J., Hunt, P., et al. (1994). Identification and cloning of a megakaryocyte growth and development factor that is a ligand for the cytokine receptor Mpl. *Cell, 77,* 1117–1124.
4. Choi, E. S., Nichol, J. L., Hokom, M. M., et al. (1995). Platelets generated *in vitro* from proplatelet-displaying human megakaryocytes are functional. *Blood, 85,* 402–413.
5. Freson, K., Hoylaerts, M. F., Jaeken, J., et al. (2001). Genetic variation of the extra-large stimulatory G protein alpha-subunit leads to Gs hyperfunction in platelets and is a risk factor for bleeding. *Thromb Haemost, 86,* 733–738.
6. Gabbeta, J., Vaidyula, V. R., Dhanasekaran, D. N., et al. (2002). Human platelet Galphaq deficiency is associated with decreased Galphaq gene expression in platelets but not neutrophils. *Thromb Haemost, 87,* 129–133.
7. Patel, Y. M., Patel, K., Rahman, S., et al. (2003). Evidence for a role for Galphai1 in mediating weak agonist-induced platelet aggregation in human platelets: Reduced Galphai1 expression and defective Gi signaling in the platelets of a patient with a chronic bleeding disorder. *Blood, 101,* 4828–4835.
8. Cattaneo, M., Zighetti, M. L., Lombardi, R., et al. (2003). Molecular bases of defective signal transduction in the platelet P2Y12 receptor of a patient with congenital bleeding. *Proc Natl Acad Sci USA, 100,* 1978–1983.
9. Lopez, J. A., Andrews, R. K., Afshar-Kharghan, V., et al. (1998). Bernard–Soulier syndrome. *Blood, 91,* 4397–4418.
10. Nurden, A. T. (2005). Qualitative disorders of platelets and megakaryocytes. *J Thromb Haemost, 3,* 1773–1782.
11. Wilcox, D. A., Paddock, C. M., Lyman, S., et al. (1995). Glanzmann thrombasthenia resulting from a single amino acid substitution between the second and third calcium-binding domains of GPIIb. Role of the GPIIb amino terminus in integrin subunit association. *J Clin Invest, 95,* 1553–1560.
12. Lages, B., Sussman, I. I., Levine, S. P., et al. (1997). Platelet alpha granule deficiency associated with decreased P-selectin and selective impairment of thrombin-induced activation in a new patient with gray platelet syndrome (alpha-storage pool deficiency). *J Lab Clin Med, 129,* 364–375.
13. Hayward, C. P., Cramer, E. M., Kane, W. H., et al. (1997). Studies of a second family with the Quebec platelet disorder:

Evidence that the degradation of the alpha-granule membrane and its soluble contents are not secondary to a defect in targeting proteins to alpha-granules. *Blood, 89,* 1243–1253.

14. Ahmad, S. S., Rawala-Sheikh, R., Ashby, B., et al. (1989). Platelet receptor-mediated factor X activation by factor IXa. High-affinity factor IXa receptors induced by factor VIII are deficient on platelets in Scott syndrome. *J Clin Invest, 84,* 824–828.

15. Seri, M., Cusano, R., Gangarossa, S., et al. (2000). Mutations in MYH9 result in the May–Hegglin anomaly, and Fechtner and Sebastian syndromes. The May–Hegglin/Fechtner Syndrome Consortium. *Nat Genet, 26,* 103–105.

16. Heath, K. E., Campos-Barros, A., Toren, A., et al. (2001). Nonmuscle myosin heavy chain IIA mutations define a spectrum of autosomal dominant macrothrombocytopenias: May–Hegglin anomaly and Fechtner, Sebastian, Epstein, and Alport-like syndromes. *Am J Hum Genet, 69,* 1033–1045.

17. Balduini, C. L., Cattaneo, M., Fabris, F., et al. (2003). Inherited thrombocytopenias: A proposed diagnostic algorithm from the Italian Gruppo di Studio delle Piastrine. *Haematologica, 88,* 582–592.

18. Oda, A., & Ochs, H. D. (2000). Wiskott–Aldrich syndrome protein and platelets. *Immunol Rev, 178,* 111–117.

19. Ward, D. M., Shiflett, S. L., & Kaplan, J. (2002). Chediak–Higashi syndrome: A clinical and molecular view of a rare lysosomal storage disorder. *Curr Mol Med, 2,* 469–477.

20. Huizing, M., Anikster, Y., & Gahl, W. A. (2001). Hermansky–Pudlak syndrome and Chediak–Higashi syndrome: Disorders of vesicle formation and trafficking. *Thromb Haemost, 86,* 233–245.

21. Hirata, T., Kakizuka, A., Ushikubi, F., et al. (1994). Arg60 to Leu mutation of the human thromboxane A2 receptor in a dominantly inherited bleeding disorder. *J Clin Invest, 94,* 1662–1667.

22. Parkman, R. (1986). The application of bone marrow transplantation to the treatment of genetic diseases. *Science, 232,* 1373–1378.

23. Bellucci, S., Devergie, A., Gluckman, E., et al. (1985). Complete correction of Glanzmann's thrombasthenia by allogeneic bone-marrow transplantation. *Br J Haematol, 59,* 635–641.

24. Johnson, A., Goodall, A. H., Downie, C. J., et al. (1994). Bone marrow transplantation for Glanzmann's thrombasthenia. *Bone Marrow Transplant, 14,* 147–150.

25. McColl, M. D., & Gibson, B. E. (1997). Sibling allogeneic bone marrow transplantation in a patient with type I Glanzmann's thrombasthenia. *Br J Haematol, 99,* 58–60.

26. Bellucci, S., Damaj, G., Boval, B., et al. (2000). Bone marrow transplantation in severe Glanzmann's thrombasthenia with antiplatelet alloimmunization. *Bone Marrow Transplant, 25,* 327–330.

27. Flood, V. H., Johnson, F. L., Boshkov, L. K., et al. (2005). Sustained engraftment post bone marrow transplant despite anti-platelet antibodies in Glanzmann thrombasthenia. *Pediatr Blood Cancer, 45,* 971–975.

28. Orlic, D., & Bodine, D. M. (1994). What defines a pluripotent hematopoietic stem cell (PHSC): Will the real PHSC please stand up! *Blood, 84,* 3991–3994.

29. Beutler, E. (1999). Gene therapy. *Biol Blood Marrow Transplant, 5,* 273–276.

30. Larochelle, A., & Dunbar, C. E. (2004). Genetic manipulation of hematopoietic stem cells. *Semin Hematol, 41,* 257–271.

31. Nathwani, A. C., Davidoff, A. M., & Linch, D. C. (2005). A review of gene therapy for haematological disorders. *Br J Haematol, 128,* 3–17.

32. Heim, D. A., & Dunbar, C. E. (2000). Hematopoietic stem cell gene therapy: Towards clinically significant gene transfer efficiency. *Immunol Rev, 178,* 29–38.

33. Naldini, L., Blomer, U., Gallay, P., et al. (1996). *In vivo* gene delivery and stable transduction of nondividing cells by a lentiviral vector. *Science, 272,* 263–267.

34. Gatlin, J., Melkus, M. W., Padgett, A., et al. (2001). Engraftment of nod/scid mice with human CD34(+) cells transduced by concentrated oncoretroviral vector particles pseudotyped with the feline endogenous retrovirus (rd114) envelope protein. *J Virol, 75,* 9995–9999.

35. Fang, J., Hodivala-Dilke, K., Johnson, B. D., et al. (2005). Therapeutic expression of the platelet-specific integrin, $\alpha$IIb$\beta$3, in a murine model for Glanzmann thrombasthenia. *Blood, 106,* 2671–2679.

36. Wilcox, D. A., Olsen, J. C., Ishizawa, L., et al. (2000). Megakaryocyte-targeted synthesis of the integrin beta(3)-subunit results in the phenotypic correction of Glanzmann thrombasthenia. *Blood, 95,* 3645–3651.

37. Urbano-Ispizua, A., Rozman, C., Martinez, C., et al. (1997). Rapid engraftment without significant graft-versus-host disease after allogeneic transplantation of CD34+ selected cells from peripheral blood. *Blood, 89,* 3967–3973.

38. Dunbar, C. E., Seidel, N. E., Doren, S., et al. (1996). Improved retroviral gene transfer into murine and Rhesus peripheral blood or bone marrow repopulating cells primed *in vivo* with stem cell factor and granulocyte colony-stimulating factor. *Proc Natl Acad Sci USA, 93,* 11871–11876.

39. Bodine, D. M., Seidel, N. E., & Orlic, D. (1996). Bone marrow collected 14 days after *in vivo* administration of granulocyte colony-stimulating factor and stem cell factor to mice has 10-fold more repopulating ability than untreated bone marrow. *Blood, 88,* 89–97.

40. Strauss, L. C., Trischmann, T. M., Rowley, S. D., et al. (1991). Selection of normal human hematopoietic stem cells for bone marrow transplantation using immunomagnetic microspheres and CD34 antibody. *Am J Pediatr Hematol Oncol, 13,* 217–221.

41. Brugger, W., Heimfeld, S., Berenson, R. J., et al. (1995). Reconstitution of hematopoiesis after high-dose chemotherapy by autologous progenitor cells generated *ex vivo*. *N Engl J Med, 333,* 283–287.

42. Rowley, S. D., Loken, M., Radich, J., et al. (1998). Isolation of CD34+ cells from blood stem cell components using the Baxter Isolex system. *Bone Marrow Transplant, 21,* 1253–1262.

43. Laurenti, L., Sora, F., Piccirillo, N., et al. (2001). Immune reconstitution after autologous selected peripheral blood progenitor cell transplantation: Comparison of two CD34⁺ cell-selection systems. *Transfusion, 41*, 783–789.

44. Cavazzana-Calvo, M., Hacein-Bey, S., de Saint Basile, G., et al. (2000). Gene therapy of human severe combined immunodeficiency (SCID)-X1 disease. *Science, 288*, 669–672.

45. Osawa, M., Hanada, K., Hamada, H., et al. (1996). Long-term lymphohematopoietic reconstitution by a single CD34-low/negative hematopoietic stem cell. *Science, 273*, 242–245.

46. Morel, F., Galy, A., Chen, B., et al. (1998). Equal distribution of competitive long-term repopulating stem cells in the CD34⁺ and CD34⁻ fractions of Thy-1lowLin-/lowSca-1+ bone marrow cells. *Exp Hematol, 26*, 440–448.

47. Goodell, M. A., Rosenzweig, M., Kim, H., et al. (1997). Dye efflux studies suggest that hematopoietic stem cells expressing low or undetectable levels of CD34 antigen exist in multiple species. *Nat Med, 3*, 1337–1345.

48. Tajima, F., Deguchi, T., Laver, J. H., et al. (2001). Reciprocal expression of CD38 and CD34 by adult murine hematopoietic stem cells. *Blood, 97*, 2618–2624.

49. Goodell, M. A. (1999). Introduction: Focus on hematology. CD34(+) or CD34(-): Does it really matter? *Blood, 94*, 2545–2547.

50. Moritz, T., Dutt, P., Xiao, X., et al. (1996). Fibronectin improves transduction of reconstituting hematopoietic stem cells by retroviral vectors: Evidence of direct viral binding to chymotryptic carboxy-terminal fragments. *Blood, 88*, 855–862.

51. Hanenberg, H., Xiao, X. L., Dilloo, D., et al. (1996). Colocalization of retrovirus and target cells on specific fibronectin fragments increases genetic transduction of mammalian cells. *Nat Med, 2*, 876–882.

52. Horn, P. A., Keyser, K. A., Peterson, L. J., et al. (2004). Efficient lentiviral gene transfer to canine repopulating cells using an overnight transduction protocol. *Blood, 103*, 3710–3716.

53. Kiem, H. P., Andrews, R. G., Morris, J., et al. (1998). Improved gene transfer into baboon marrow repopulating cells using recombinant human fibronectin fragment CH-296 in combination with interleukin-6, stem cell factor, FLT-3 ligand, and megakaryocyte growth and development factor. *Blood, 92*, 1878–1886.

54. Niemeyer, G. P., Boudreaux, M. K., Goodman-Martin, S. A., et al. (2003). Correction of a large animal model of type I Glanzmann's thrombasthenia by nonmyeloablative bone marrow transplantation. *Exp Hematol, 31*, 1357–1362.

55. Mascelli, M. A., Lance, E. T., Damaraju, L., et al. (1998). Pharmacodynamic profile of short-term abciximab treatment demonstrates prolonged platelet inhibition with gradual recovery from GP IIb/IIIa receptor blockade. *Circulation, 97*, 1680–1688.

56. Rosenzweig, M., MacVittie, T. J., Harper, D., et al. (1999). Efficient and durable gene marking of hematopoietic progenitor cells in nonhuman primates after nonablative conditioning. *Blood, 94*, 2271–2286.

57. Storb, R., Yu, C., Wagner, J. L., et al. (1997). Stable mixed hematopoietic chimerism in DLA-identical littermate dogs given sublethal total body irradiation before and pharmacological immunosuppression after marrow transplantation. *Blood, 89*, 3048–3054.

58. Zaucha, J. A., Yu, C., Lothrop, C. D., Jr., et al. (2001). Severe canine hereditary hemolytic anemia treated by nonmyeloablative marrow transplantation. *Biol Blood Marrow Transplant, 7*, 14–24.

59. Wang, R., Shattil, S. J., Ambruso, D. R., et al. (1997). Truncation of the cytoplasmic domain of beta3 in a variant form of Glanzmann thrombasthenia abrogates signaling through the integrin alpha(IIb)beta3 complex. *J Clin Invest, 100*, 2393–2403.

60. Valentin, N., & Newman, P. J. (1994). Human platelet alloantigens. *Curr Opin Hematol, 1*, 381–387.

61. Faraday, N., Rade, J. J., Johns, D. C., et al. (1999). *Ex vivo* cultured megakaryocytes express functional glycoprotein IIb-IIIa receptors and are capable of adenovirus-mediated transgene expression. *Blood, 94*, 4084–4092.

62. Shiraga, M., Ritchie, A., Aidoudi, S., et al. (1999). Primary megakaryocytes reveal a role for transcription factor NF-E2 in integrin alpha IIb beta 3 signaling. *J Cell Biol, 147*, 1419–1430.

63. Jergens, A. E., Turrentine, M. A., Kraus, K. H., et al. (1987). Buccal mucosa bleeding times of healthy dogs and of dogs in various pathologic states, including thrombocytopenia, uremia, and von Willebrand's disease. *Am J Vet Res, 48*, 1337–1342.

64. Connelly, S., Mount, J., Mauser, A., et al. (1996). Complete short-term correction of canine hemophilia A by *in vivo* gene therapy. *Blood, 88*, 3846–3853.

65. Polo, J. M., Belli, B. A., Driver, D. A., et al. (1999). Stable alphavirus packaging cell lines for Sindbis virus and Semliki Forest virus-derived vectors. *Proc Natl Acad Sci USA, 96*, 4598–4603.

66. Wilcox, D. A., Olsen, J. C., Ishizawa, L., et al. (1999). Integrin alphaIIb promoter-targeted expression of gene products in megakaryocytes derived from retrovirus-transduced human hematopoietic cells. *Proc Natl Acad Sci USA, 96*, 9654–9659.

67. Verma, I. M., & Somia, N. (1997). Gene therapy — Promises, problems and prospects. *Nature, 389*, 239–242.

68. Kaiser, J. (2005). Gene therapy. Panel urges limits on X-SCID trials. *Science, 307*, 1544–1545.

69. Hacein-Bey-Abina, S., von Kalle, C., Schmidt, M., et al. (2003). A serious adverse event after successful gene therapy for X-linked severe combined immunodeficiency. *N Engl J Med, 348*, 255–256.

70. Kaiser, J. (2003). Gene therapy. Seeking the cause of induced leukemias in X-SCID trial. *Science, 299*, 495.

71. Bradford, G. B., Williams, B., Rossi, R., et al. (1997). Quiescence, cycling, and turnover in the primitive hematopoietic stem cell compartment. *Exp Hematol, 25*, 445–453.

72. Naldini, L., Blomer, U., Gallay, P., et al. (1996). *In vivo* gene delivery and stable transduction of nondividing cells by a lentiviral vector. *Science, 272*, 263–267.

73. Olsen, J. C. (1998). Gene transfer vectors derived from equine infectious anemia virus. *Gene Therapy, 5,* 1481–1487.

74. Salmon, P., Kindler, V., Ducrey, O., et al. (2000). High-level transgene expression in human hematopoietic progenitors and differentiated blood lineages after transduction with improved lentiviral vectors. *Blood, 96,* 3392–3398.

75. Shi, Q., Wilcox, D. A., Morateck, P. A., et al. (2004). Targeting platelet GPIbalpha transgene expression to human megakaryocytes and forming a complete complex with endogenous GPIbbeta and GPIX. *J Thromb Haemost, 2,* 1989–1997.

76. Roth, G. J., Yagi, M., & Bastian, L. S. (1996). The platelet glycoprotein Ib-V-IX system: Regulation of gene expression. *Stem Cells, 14,* 188–193.

77. Uzan, G., Prenant, M., Prandini, M. H., et al. (1991). Tissue-specific expression of the platelet GPIIb gene. *J Biol Chem, 266,* 8932–8939.

78. Holmes, M. L., Bartle, N., Eisbacher, M., et al. (2002). Cloning and analysis of the thrombopoietin-induced megakaryocyte-specific glycoprotein VI promoter and its regulation by GATA-1, Fli-1, and Sp1. *J Biol Chem, 277,* 48333–48341.

79. Kaushansky, K., & Drachman, J. G. (2002). The molecular and cellular biology of thrombopoietin: The primary regulator of platelet production. *Oncogene, 21,* 3359–3367.

80. Doi, T., Greenberg, S. M., & Rosenberg, R. D. (1987). Structure of the rat platelet factor 4 gene: A marker for megakaryocyte differentiation. *Mol Cell Biol, 7,* 898–904.

81. Romeo, P. H., Prandini, M. H., Joulin, V., et al. (1990). Megakaryocytic and erythrocytic lineages share specific transcription factors. *Nature, 344,* 447–449.

82. Wang, X., Crispino, J. D., Letting, D. L., et al. (2002). Control of megakaryocyte-specific gene expression by GATA-1 and FOG-1: Role of Ets transcription factors. *EMBO J, 21,* 5225–5234.

83. Ravid, K., Beeler, D. L., Rabin, M. S., et al. (1991). Selective targeting of gene products with the megakaryocyte platelet factor 4 promoter. *Proc Natl Acad Sci USA, 88,* 1521–1525.

84. Zhang, C., Thornton, M. A., Kowalska, M. A., et al. (2001). Localization of distal regulatory domains in the megakaryocyte-specific platelet basic protein/platelet factor 4 gene locus. *Blood, 98,* 610–617.

85. Nguyen, H. G., Yu, G., Makitalo, M., et al. (2005). Conditional overexpression of transgenes in megakaryocytes and platelets *in vivo. Blood, 106,* 1559–1564.

86. Tanimoto, K., Liu, Q., Bungert, J., et al. (1999). Effects of altered gene order or orientation of the locus control region on human beta-globin gene expression in mice. *Nature, 398,* 344–348.

87. Ellis, J., & Pannell, D. (2001). The beta-globin locus control region versus gene therapy vectors: A struggle for expression. *Clin Genet, 59,* 17–24.

88. Klug, C. A., Cheshier, S., & Weissman, I. L. (2000). Inactivation of a GFP retrovirus occurs at multiple levels in long-term repopulating stem cells and their differentiated progeny. *Blood, 96,* 894–901.

89. Bell, A. C., West, A. G., & Felsenfeld, G. (2001). Insulators and boundaries: Versatile regulatory elements in the eukaryotic genome. *Science, 291,* 447–450.

90. Emery, D. W., & Stamatoyannopoulos, G. (1999). Stem cell gene therapy for the beta-chain hemoglobinopathies. Problems and progress. *Ann N Y Acad Sci, 872,* 94–108.

91. Tropel, P., Roullot, V., Vernet, M., et al. (1997). A 2.7-kb portion of the 5′ flanking region of the murine glycoprotein alphaIIb gene is transcriptionally active in primitive hematopoietic progenitor cells. *Blood, 90,* 2995–3004.

92. Tronik-Le Roux, D., Roullot, V., Schweitzer, A., et al. (1995). Suppression of erythro-megakaryocytopoiesis and the induction of reversible thrombocytopenia in mice transgenic for the thymidine kinase gene targeted by the platelet glycoprotein alpha IIb promoter. *J Exp Med, 181,* 2141–2151.

93. Thornton, M. A., Zhang, C., Kowalska, M. A., et al. (2002). Identification of distal regulatory regions in the human alpha IIb gene locus necessary for consistent, high-level megakaryocyte expression. *Blood, 100,* 3588–3596.

94. Challita, P. M., & Kohn, D. B. (1994). Lack of expression from a retroviral vector after transduction of murine hematopoietic stem cells is associated with methylation *in vivo. Proc Natl Acad Sci USA, 91,* 2567–2571.

95. Salama, A. D., Remuzzi, G., Harmon, W. E., et al. (2001). Challenges to achieving clinical transplantation tolerance. *J Clin Invest, 108,* 943–948.

96. Muul, L. M., Tuschong, L. M., Soenen, S. L., et al. (2003). Persistence and expression of the adenosine deaminase gene for 12 years and immune reaction to gene transfer components: Long-term results of the first clinical gene therapy trial. *Blood, 101,* 2563–2569.

97. Kiem, H. P., McSweeney, P. A., Bruno, B., et al. (1999). Improved gene transfer into canine hematopoietic repopulating cells using CD34-enriched marrow cells in combination with a gibbon ape leukemia virus-pseudotype retroviral vector. *Gene Therapy, 6,* 966–972.

98. Goerner, M., Horn, P. A., Peterson, L., et al. (2001). Sustained multilineage gene persistence and expression in dogs transplanted with CD34(+) marrow cells transduced by RD114-pseudotype oncoretrovirus vectors. *Blood, 98,* 2065–2070.

99. von Laer, D., Corovic, A., Vogt, B., et al. (2000). Amphotropic and VSV-G-pseudotyped retroviral vectors transduce human hematopoietic progenitor cells with similar efficiency. *Bone Marrow Transplant, 25,* S75–S79.

100. Guenechea, G., Gan, O. I., Inamitsu, T., et al. (2000). Transduction of human $CD34^+$ $CD38^-$ bone marrow and cord blood-derived SCID-repopulating cells with third-generation lentiviral vectors. *Mol Ther, 1,* 566–573.

101. Ware, J. (2004). Dysfunctional platelet membrane receptors: From humans to mice. *Thromb Haemost, 92,* 478–485.

102. Swank, R. T., Novak, E. K., McGarry, M. P., et al. (2000). Abnormal vesicular trafficking in mouse models of Hermansky–Pudlak syndrome. *Pigment Cell Res, 13,* 59–67.

103. Gwynn, B., Ciciotte, S. L., Hunter, S. J., et al. (2000). Defects in the cappuccino (cno) gene on mouse chromosome 5 and human 4p cause Hermansky–Pudlak syndrome by an AP-3-independent mechanism. *Blood, 96,* 4227–4235.

104. Anikster, Y., Huizing, M., White, J., et al. (2001). Mutation of a new gene causes a unique form of Hermansky–Pudlak syndrome in a genetic isolate of central Puerto Rico. *Nat Genet, 28,* 376–380.

105. Ware, J., Russell, S., & Ruggeri, Z. M. (2000). Generation and rescue of a murine model of platelet dysfunction: The Bernard–Soulier syndrome. *Proc Natl Acad Sci USA, 97,* 2803–2808.

106. Tronik-Le Roux, D., Roullot, V., Poujol, C., et al. (2000). Thrombasthenic mice generated by replacement of the integrin alpha(IIb) gene: Demonstration that transcriptional activation of this megakaryocytic locus precedes lineage commitment. *Blood, 96,* 1399–1408.

107. Hodivala-Dilke, K. M., McHugh, K. P., Tsakiris, D. A., et al. (1999). Beta3-integrin-deficient mice are a model for Glanzmann thrombasthenia showing placental defects and reduced survival. *J Clin Invest, 103,* 229–238.

108. McHugh, K. P., Hodivala-Dilke, K., Zheng, M. H., et al. (2000). Mice lacking beta3 integrins are osteosclerotic because of dysfunctional osteoclasts. *J Clin Invest, 105,* 433–440.

109. Reynolds, L. E., Wyder, L., Lively, J. C., et al. (2002). Enhanced pathological angiogenesis in mice lacking beta3 integrin or beta3 and beta5 integrins. *Nat Med, 8,* 27–34.

110. Taverna, D., Moher, H., Crowley, D., et al. (2004). Increased primary tumor growth in mice null for beta3- or beta3/beta5-integrins or selectins. *Proc Natl Acad Sci USA, 101,* 763–768.

111. Hynes, R. O. (2002). Integrins: Bidirectional, allosteric signaling machines. *Cell, 110,* 673–687.

112. Boudreaux, M. K., Kvam, K., Dillon, A. R., et al. (1996). Type I Glanzmann's thrombasthenia in a Great Pyrenees dog. *Vet Pathol, 33,* 503–511.

113. Lipscomb, D. L., Bourne, C., & Boudreaux, M. K. (2000). Two genetic defects in alphaIIb are associated with type I Glanzmann's thrombasthenia in a Great Pyrenees dog: A 14-base insertion in exon 13 and a splicing defect of intron 13. *Vet Pathol, 37,* 581–588.

114. Boudreaux, M. K., & Catalfamo, J. L. (2001). Molecular and genetic basis for thrombasthenic thrombopathia in Otterhounds. *Am J Vet Res, 62,* 1797–1804.

115. Wilcox, D. A. (2002). Targeting transgene expression in canine megakaryocytes derived from lentivirus-transduced G-CSF mobilized CD34$^+$ peripheral blood cells. *Blood, 100,* 1713a.

116. Boudreaux, M. K., & Lipscomb, D. L. (2001). Clinical, biochemical, and molecular aspects of Glanzmann's thrombasthenia in humans and dogs. *Vet Pathol, 38,* 249–260.

117. Poncz, M., & Newman, P. J. (1990). Analysis of rodent platelet glycoprotein IIb: Evidence for evolutionarily conserved domains and alternative proteolytic processing. *Blood, 75,* 1282–1289.

118. Cieutat, A. M., Rosa, J. P., Letourneur, F., et al. (1993). A comparative analysis of cDNA-derived sequences for rat and mouse beta 3 integrins (GPIIIA) with their human counterpart. *Biochem Biophys Res Commun, 193,* 771–778.

119. Hogan, C. J., Shpall, E. J., McNiece, I., et al. (1997). Multilineage engraftment in NOD/LtSz-SCID/SCID mice from mobilized human CD34$^+$ peripheral blood progenitor cells. *Biol Blood Marrow Transplant, 3,* 236–246.

120. Barquinero, J., Segovia, J. C., Ramirez, M., et al. (2000). Efficient transduction of human hematopoietic repopulating cells generating stable engraftment of transgene-expressing cells in NOD/SCID mice. *Blood, 95,* 3085–3093.

121. Schiedlmeier, B., Kuhlcke, K., Eckert, H. G., et al. (2000). Quantitative assessment of retroviral transfer of the human multidrug resistance 1 gene to human mobilized peripheral blood progenitor cells engrafted in nonobese diabetic/severe combined immunodeficient mice. *Blood, 95,* 1237–1248.

122. Wilcox, D. A., Shi, Q., Nurden, P., et al. (2003). Induction of megakaryocytes to synthesize and store a releasable pool of human factor VIII. *J Thromb Haemost, 1,* 2477–2489.

123. Yarovoi, H., Kufrin, D., Eslin, D. E., et al. (2003). Factor VIII ectopically expressed in platelets: Efficacy in hemophilia A treatment. *Blood, 102,* 4006–4013.

124. Shi, Q., Wilcox, D. A., Fahs, S. A., et al. (2006). Factor VIII ectopically targeted to platelets is therapeutic in hemophilia A with high-titer inhibitory antibodies. *J Clin Invest, 116,* 1974–1982.

# Index